ENGINEERING MATHEMATICS

with APPLICATIONS

Useful for

- BE/B Tech and other Engineering courses

Dr. SUDHIR KUMAR PUNDIR

M.Sc., M.Phil, NET (JRF), Ph.D.
Head
Department of Mathematics
S.D. (P.G.) College
Muzaffarnagar (U.P.)

CBS

CBS Publishers & Distributors Pvt Ltd

New Delhi • Bengaluru • Chennai • Kochi • Kolkata • Mumbai
Bhopal • Bhubaneswar • Hyderabad • Jharkhand • Nagpur • Patna
• Pune • Uttarakhand • Dhaka (Bangladesh) • Kathmandu (Nepal)

ENGINEERING MATHEMATICS

with APPLICATIONS

ISBN: 978-93-89688-87-0

Copyright © Author and Publisher

First Edition: 2020

Published by Satish Kumar Jain and produced by Varun Jain for

CBS Publishers & Distributors Pvt Ltd

4819/XI Prahlad Street, 24 Ansari Road, Daryaganj, New Delhi 110 002, India.
Ph: 23289259, 23266861, 23266867   Website: www.cbspd.com
Fax: 011-23243014   e-mail: delhi@cbspd.com; cbspubs@airtelmail.in.

Corporate Office: 204 FIE, Industrial Area, Patparganj, Delhi 110 092

Ph: 4934 4934   Fax: 4934 4935   e-mail: publishing@cbspd.com;
publicity@cbspd.com

Branches

- **Bengaluru:** Seema House 2975, 17th Cross, K.R. Road,
  Banasankari 2nd Stage, Bengaluru 560 070, Karnataka
  Ph: +91-80-26771678/79   Fax: +91-80-26771680   e-mail: bangalore@cbspd.com
- **Chennai:** 7, Subbaraya Street, Shenoy Nagar, Chennai 600 030, Tamil Nadu
  Ph: +91-44-26680620, 26681266   Fax: +91-44-42032115   e-mail: chennai@cbspd.com
- **Kochi:** 68/1534, 35, 36, Power House Road, Opp KSEB Power House,
  Ernakulam 682 018, Kochi, Kerala
  Ph: +91-484-4059061-65   Fax: +91-484-4059065   e-mail: kochi@cbspd.com
- **Kolkata:** 6/B, Ground Floor, Rameswar Shaw Road, Kolkata-700 014, West Bengal
  Ph: +91-33-22891126, 22891127, 22891128   e-mail: kolkata@cbspd.com
- **Mumbai:** 83-C, Dr E Moses Road, Worli, Mumbai-400018, Maharashtra
  Ph: +91-22-24902340/41   Fax: +91-22-24902342   e-mail: mumbai@cbspd.com

Representatives

• Bhopal	0-8319310552	• Bhubaneswar	0-9911037372	• Hyderabad	0-9885175004
• Jharkhand	0-9811541605	• Nagpur	0-9421945513	• Patna	0-9334159340
• Pune	0-9623451994	• Uttarakhand	0-9716462459	• Dhaka (Bangladesh)	01912-003485
• Kathmandu (Nepal)	977-9181742655				

Printed at: Glorious Printers, Daryaganj, Delhi

Preface

The book entitled 'ENGINEERING MATHEMATICS with Applications' is meant for the students of B.E., B.Tech. and all other engineering courses. The book covers almost the entire syllabi of various technical universities in India.

In this book special and conscious efforts have been made to keep the writing style simple. It is a collection and compilation work from various sources and has been endeavoured to include as much as information as could be possible. Different concepts have been explained with the help of detailed theory and examples. There is a plenty of scope in the form of exercise for the readers to try and solve the problem on his own.

I express my gratitude to the authors and publishers of various books, I consulted during the preparation of the book.

I wish sincerely thank **Sh. S.K. Jain** and **Sh. Varun Jain**, Managing directors, CBS publishers and Distributors, New Delhi for their encouragement and help in bringing out this publication in a present nice form.

My special thank to Sh. Y.N. Arjuna, Senior director, publishing, editorial and publicity and Smt. Ritu Chawla, publishing head, CBS publishers and distributors, New Delhi whose encouragement and unstinted support enabled me to complete my book. Sh. Sunil Dutt and Sh. Suresh Sharma, CBS publishers and distributors deserve special mention for their kind help and support. Mr. Peeyush Goel, M/s Dreamshapers also deserve special mention for nice typesetting.

I must also record my appreciation due to my wife **Dr. Rimple**, daughter **Rijuta** and son **Shrish** for their understanding and love during the long period that I have taken to complete this book. I am indebted to my colleagues and research students who generously shared their views on the need of a comprehensive book on ENGINEERING MATHEMATICS.

Above all I am thankful to 'The Almighty God' without whose grace nothing is possible for anyone .

Further suggestions and comments for improvement of the book will be thankfully received and duly incorporated in the next edition.

DR. SUDHIR KUMAR PUNDIR
email : skpundir05@yahoo.co.in

Contents

Ch. 11   COMPLEX NUMBERS   1005-1132

⭘   APPENDIX   1133-1140

⭘   INDEX   1141-1144

CHAPTER
Algebra
1

A set of mn numbers either real or complex arranged in the form of a reactangular array in which there are m rows and n columns, is called a matrix of order $m \times n$ which is denoted by $[a_{ij}]_{m \times n}$ where $i = 1, 2, 3, ..., m$ represents the number of rows and $j = 1, 2, 3, ..., n$ represents the number of columns and thus a matrix of order $m \times n$ is usually written as

$$[a_{ij}]_{m \times n} = \begin{bmatrix} a_{11} & a_{12} & \cdots & a_{1n} \\ a_{21} & a_{22} & \cdots & a_{2n} \\ \vdots & \vdots & \vdots & \vdots \\ a_{m1} & a_{m2} & \cdots & a_{mn} \end{bmatrix}_{m \times n}$$

☞ REMARK

- Sometimes, a matrix is a rectangular array of numbers enclosed in double straight lines shown as '‖ ‖' or enclosed in parenthesis '( )'.

1.2 TYPE OF MATRICES

(i) **Null matrix (or zero matrix):** A matrix of order $m \times n$ is called a *null matrix* if it contains all mn elements zero. It is denoted by O and is usually written as

$$O = \begin{bmatrix} 0 & 0 & \cdots & 0 \\ 0 & 0 & \cdots & 0 \\ \vdots & \vdots & \vdots & \vdots \\ 0 & 0 & \cdots & 0 \end{bmatrix}_{m \times n}$$

(ii) **Row matrix:** A matrix having only one row and n columns is called a *row matrix* of order $1 \times n$.

For example : The matrix $A = \begin{bmatrix} a_{11} & a_{12} & a_{13} & \cdots & a_{1n} \end{bmatrix}_{1 \times n}$ is a row matrix.

(iii) **Column matrix :** A matrix having m rows and only one column is called a *column matrix* of order $m \times 1$.

For example : The matrix $A = \begin{bmatrix} a_{11} \\ a_{21} \\ a_{31} \\ \vdots \\ a_{m1} \end{bmatrix}_{m \times 1}$ is a column matrix.

(iv) **Horizontal matrix :** A matrix having more columns than the number of its rows, is called *Horizontal matrix*.

For example : The matrix $A = \begin{bmatrix} a_{11} & a_{12} & a_{13} \\ a_{21} & a_{22} & a_{23} \end{bmatrix}_{2 \times 3}$ is a horizontal matrix.

(v) **Vertical matrix :** A matrix having more number of rows than its columns, *is called vertical matrix*.

For exmaple : The matrix $A = \begin{bmatrix} a_{11} & a_{12} \\ a_{21} & a_{22} \\ a_{31} & a_{32} \end{bmatrix}_{3 \times 2}$ is a vertical matrix.

☞ REMARK

- Row matrix is also a horizontal matrix and column matrix is also a vertical matrix.

(vi) **Square matrix :** A matrix having a number of rows equal to number of columns, is called *square matrix*.

For example : The matrix $A = \begin{bmatrix} a_{11} & a_{12} & a_{13} \\ a_{21} & a_{22} & a_{23} \\ a_{31} & a_{32} & a_{33} \end{bmatrix}_{3\times 3}$ is a square matrix.

Here, the matrix A has 3 rows and 3 columns, so it is a square matrix. Also the elements a_{11}, a_{22}, a_{33} are placed in the diagonal, so these elements are known as *diagonal elements*.

(vii) Diagonal matrix : A matrix of order $n \times n$ is called a *diagonal matrix* if it contains all its off diagonal elements equal to zero. Suppose $A = [a_{ij}]_{n\times n}$ and if $a_{ij} = 0$ for all $i \neq j$, then A is a diagonal matrix. Diagonal matrix of order $n \times n$ is usually written as

$$\text{Diag } [a_{11} \quad a_{22} \quad a_{33} \quad \dots \quad a_{nn}]$$

For example :

The matrix $A = \begin{bmatrix} 1 & 0 & 0 \\ 0 & 2 & 0 \\ 0 & 0 & 3 \end{bmatrix}_{3\times 3}$ = Diag [1 2 3] is a diagonal matrix.

(viii) Scalar matrix : A diagonal matrix whose diagonal elements are all equal but not equal to 1 is called a *scalar matrix*.

For example : The matrix $A = \begin{bmatrix} k & 0 & 0 \\ 0 & k & 0 \\ 0 & 0 & k \end{bmatrix}$, $k \neq 1$ is a scalar matrix.

(ix) Unit matrix : A square matrix of order $n \times n$ having all off-diagonal elements equal to zero and each of the diagonal elements equal to 1, is called a *unit matix*. It is usually denoted by I_n and is written as

$$I_n = \begin{bmatrix} 1 & 0 & \dots & 0 \\ 0 & 1 & \dots & 0 \\ 0 & 0 & \dots & 0 \\ \vdots & \vdots & \vdots & \vdots \\ 0 & 0 & \dots & 1 \end{bmatrix}_{n\times n}$$

☞ REMARK

- Unit matrix can also be denoted by I.

(x) Triangular matrix : A matrix in which the elements lying above or below principal diagonal are all zero, is called a *triangular matrix*. There are two kinds of triangular matrix.

(a) Upper triangular matrix : A matrix of order $n \times n$ is called an *upper triangular matrix* if it contains all its elements below the diagonal elements equal to zero.
Suppose $A = [a_{ij}]_{n\times n}$ and if $a_{ij} = 0$ for all $i > j$, then A is an upper triangular matrix.

For example : The matrix $A = \begin{bmatrix} 2 & 3 & 4 \\ 0 & 1 & 5 \\ 0 & 0 & 3 \end{bmatrix}_{3\times 3}$ is an upper triangular matrix of order 3×3.

(b) Lower triangular matrix : A matrix of order $n \times n$ is called a *lower triangular matrix* if it contains all its elements above the diagonal elements equal to zero.

Suppose $A = [a_{ij}]_{n\times n}$ and if $a_{ij} = 0$ for all $i < j$, then A is called lower triangular matrix.

For example : The matrix $A = \begin{bmatrix} 1 & 0 & 0 \\ 3 & 4 & 0 \\ 5 & 6 & 7 \end{bmatrix}_{3\times 3}$ is a lower triangular matrix of order 3×3.

1.3 OPERATION ON MATRICES

1.3.1 ADDITION OF MATRICES

Suppose A and B are two matrices of same order, then the addition of these two matrices is obtained by adding corresponding elements of A and B. It is denoted by $A + B$. If the order of A and B is $m \times n$, then the order of $A+B$ will be $m \times n$.

Suppose $\quad\quad A = [a_{ij}]_{m\times n}$ and $B = [b_{ij}]_{m\times n}$ then $A + B = [a_{ij} + b_{ij}]_{m\times n}$

For example: If $\quad A = \begin{bmatrix} 1 & 2 & 3 \\ 5 & 1 & 4 \\ 7 & 8 & 9 \end{bmatrix}$ and $B = \begin{bmatrix} 1 & 3 & 5 \\ 5 & 0 & 1 \\ 3 & 2 & 12 \end{bmatrix}$

then
$$A + B = \begin{bmatrix} 1 & 2 & 3 \\ 5 & 1 & 4 \\ 7 & 8 & 9 \end{bmatrix} + \begin{bmatrix} 1 & 3 & 5 \\ 5 & 0 & 1 \\ 3 & 2 & 12 \end{bmatrix} = \begin{bmatrix} 1+1 & 2+3 & 3+5 \\ 5+5 & 1+0 & 4+1 \\ 7+3 & 8+2 & 9+12 \end{bmatrix} = \begin{bmatrix} 2 & 5 & 8 \\ 10 & 1 & 5 \\ 10 & 10 & 21 \end{bmatrix}$$

☞ REMARK
- If the orders of the matrices are different, then they are not conformable for addition.

1.3.2 SUBSTRACTION OF MATRICES

Suppose A and B are two matrices of same order, then the substraction of A and B, i.e., $A–B$ is obtained by substracting each element of B from the corresponding element of A. If A and B are of order $m \times n$, then the order of $A – B$ will be of order $m \times n$.

Let $\qquad A = [a_{ij}]_{m \times n}$ and $B = [b_{ij}]_{m \times n}$

then $\qquad A - B = [a_{ij} - b_{ij}]_{m \times n}$

For example: If $\quad A = \begin{bmatrix} 1 & 2 & 3 \\ 3 & 4 & 5 \\ 5 & 6 & 7 \end{bmatrix}$ and $B = \begin{bmatrix} 0 & 5 & 2 \\ 3 & -2 & 2 \\ 5 & 7 & 8 \end{bmatrix}$

then
$$A - B = \begin{bmatrix} 1 & 2 & 3 \\ 3 & 4 & 5 \\ 5 & 6 & 7 \end{bmatrix} - \begin{bmatrix} 0 & 5 & 2 \\ 3 & -2 & 2 \\ 5 & 7 & 8 \end{bmatrix} = \begin{bmatrix} 1-0 & 2-5 & 3-2 \\ 3-3 & 4-(-2) & 5-2 \\ 5-5 & 6-7 & 7-8 \end{bmatrix} = \begin{bmatrix} 1 & -3 & 1 \\ 0 & 6 & 3 \\ 0 & -1 & -1 \end{bmatrix}$$

☞ REMARK
- If the order of matrices are different, then they are not conformable for subtraction.

1.3.3 MULTIPLICATION OF A MATRIX BY A SCALAR

Suppose A is a matrix of order $m \times n$ and k is a scalar, then the multiplication of A by k, i.e. kA is obtained by multiplying each element of A by k.

Let $\qquad A = [a_{ij}]_{m \times n} \; \forall \; 1 \le i \le m$ and $i \le j \le m$

For example : If $A = \begin{bmatrix} 1 & 2 & 3 \\ 4 & 5 & 6 \\ 7 & 8 & 9 \end{bmatrix}$ and $k = 3$, then $3A = 3\begin{bmatrix} 1 & 2 & 3 \\ 4 & 5 & 6 \\ 7 & 8 & 9 \end{bmatrix} = \begin{bmatrix} 3\times1 & 3\times2 & 3\times3 \\ 3\times4 & 3\times5 & 3\times6 \\ 3\times7 & 3\times8 & 3\times9 \end{bmatrix} = \begin{bmatrix} 3 & 6 & 9 \\ 12 & 15 & 18 \\ 21 & 24 & 27 \end{bmatrix}$

1.3.4 EQUALITY OF MATRICES

Two matrices are said to be equal if both have same order and having same corresponding elements.

For example : The matrices $A = \begin{bmatrix} 1 & 2 \\ -3 & 4 \end{bmatrix}$ and $B = \begin{bmatrix} x & y \\ z & 4 \end{bmatrix}$ are said to be equal if $x = 1, y = 2$ and $z = -3$.

1.3.5 PROPERTIES OF MATRIX ADDITION

(i) Commutative Law: *If A and B are two matrices of same order $m \times n$, then $A + B = B + A$*

Proof. Let $A = [a_{ij}]_{m \times n}$ and $B = [b_{ij}]_{m \times n}$ where $1 \le i \le m$ and $1 \le j \le n$. Then
$$A+B = [a_{ij}]_{m \times n} + [b_{ij}]_{m \times n}$$
$$= [a_{ij} + b_{ij}]_{m \times n} \quad \text{(By definition of addition)}$$
$$= [b_{ij} + a_{ij}]_{m \times n} \quad \text{(Real numbers are always commutative.)}$$
$$= [b_{ij}]_{m \times n} + [a_{ij}]_{m \times n} = B + A$$

Hence, $\qquad A + B = B + A$

(ii) Associative Law: *If A, B and C are three matrices of same order $m \times n$, then $(A + B) + C = A + (B + C)$*

Proof. Let $A = [a_{ij}]_{m \times n}$, $B = [b_{ij}]_{m \times n}$ and $C = [c_{ij}]_{m \times n}$ where $1 \le i \le m$ and $1 \le j \le n$. Then
$$(A+B)+C = ([a_{ij}]_{m \times n} + [b_{ij}]_{m \times n}) + [c_{ij}]_{m \times n}$$
$$= [a_{ij} + b_{ij}]_{m \times n} + [c_{ij}]_{m \times n}$$
$$= [(a_{ij} + b_{ij}) + (c_{ij})]_{m \times n} \quad \text{(Numbers are always associative.)}$$
$$= [a_{ij}]_{m \times n} + ([b_{ij} + c_{ij}]_{m \times n}) = [a_{ij}]_{m \times n} + ([b_{ij}]_{m \times n} + [c_{ij}]_{m \times n})$$
$$= A + (B + C)$$

Hence, $\qquad (A+B) + C = A + (B+C)$

(iii) Additive identity: *If A is a matrix of order m × n and O is a null matrix of the same order m ×n, then A + O = A = O + A*

Proof. Let $A = [a_{ij}]_{m \times n}$ and $O = [0]_{m \times n}$, then

$$A + O = [a_{ij}]_{m \times n} + [0]_{m \times n} = [a_{ij} + 0]_{m \times n} = [a_{ij}]_{m \times n} = A$$

Also $\qquad\qquad\qquad O + A = [0]_{m \times n} + [a_{ij}]_{m \times n} = [0 + a_{ij}]_{m \times n} = [a_{ij}]_{m \times n} = A$

Hence $\qquad\qquad\qquad A + O = A = O + A$

Therefore, the null matrix O is treated as an additive identity.

(iv) Additive Inverse: *If A is a matrix of order m×n and –A is the negative of A, so its order is also m×n, then –A + A = O (Null matrix)*

Hence, –A is the additive inverse of A.

(v) Cancellation Law: *If A, B and C are three matrices of order m × n then*

(i) $A + B = A + C \Rightarrow B = C$ (Left cancellation law) $\qquad$ (ii) $B + A = C + A \Rightarrow B = C$ (Right cancellation law)

Proof.

(i) It is given that $\qquad\qquad A + B = A + C$ $\qquad\qquad\qquad\qquad\qquad\qquad\qquad\qquad$ ...(1)

Adding – A to the left of both sides, we get

$$-A + (A + B) = -A + (A + C)$$

$\Rightarrow \qquad\qquad (-A + A) + B = (-A + A) + C \qquad\qquad\qquad$ (By associative law)

$\Rightarrow \qquad\qquad\qquad O + B = O + C \qquad\qquad\qquad\qquad\qquad$ (By additive inverse)

$\Rightarrow \qquad\qquad\qquad\qquad B = C \qquad\qquad\qquad\qquad\qquad\qquad$ (By additive inverse)

Similarly, we can prove that if $\qquad B + A = C + A$, then $B = C$.

1.3.6 PROPERTIES OF MULTIPLICATION OF MATRIX BY A SCALAR

Prop. (i) $\qquad$ If A and B are two matrices of order $m \times n$ and k is any scalar, then $\quad k(A + B) = kA + kB$

Prop. (ii) $\qquad$ If A is a matrix of order $m \times n$ and a, b are two scalars, then $(a + b)A = aA + bA$

Prop. (iii) $\qquad$ If A is a matrix of order $m \times n$ and a, b are two scalars, then $a(bA) = (ab)A$

Prop. (iv) $\qquad$ If A is a matrix of order $m \times n$ and k is any scalar, then $(-k)A = -(kA) = k(-A)$

Solved Examples

EXAMPLE 1. *Find the number of rows and columns in the following matrices :*

(i) [1 2 3 4] $\qquad$ (ii) $\begin{bmatrix} -1 \\ 3 \\ 0 \\ 4 \end{bmatrix}$

(iii) $\begin{bmatrix} -1 & 0 & 3 & 4 \\ 5 & 6 & 7 & 8 \end{bmatrix}$

SOLUTION. (i) In the matrix [1 2 3 4], there is one row and four columns.

(ii) In the matrix $\begin{bmatrix} -1 \\ 3 \\ 0 \\ 4 \end{bmatrix}$ there are four rows and one column.

(iii) In the given matrix $\begin{bmatrix} -1 & 0 & 3 & 4 \\ 5 & 6 & 7 & 8 \end{bmatrix}$.

There are 2 rows and 4 columns.

EXAMPLE 2. If $A = \begin{bmatrix} 1 & 0 \\ 2 & -1 \end{bmatrix}$, $B = \begin{bmatrix} 3 & 7 \\ 4 & 8 \end{bmatrix}$, $C = \begin{bmatrix} -1 & 1 \\ 0 & 0 \end{bmatrix}$ *then find :*

(i) 7 A; $\qquad$ (ii) –3B $\qquad$ (iii) 2C

(iv) A–5B $\qquad$ (v) 4A+3C.

SOLUTION. (i) $\qquad 7A = 7\begin{bmatrix} 1 & 0 \\ 2 & -1 \end{bmatrix} = \begin{bmatrix} 7 \times 1 & 7 \times 0 \\ 7 \times 2 & 7 \times -1 \end{bmatrix}$

$\qquad\qquad\qquad = \begin{bmatrix} 7 & 0 \\ 14 & -7 \end{bmatrix}$

(ii) $-3B = -3\begin{bmatrix} 3 & 7 \\ 4 & 8 \end{bmatrix} = \begin{bmatrix} -3 \times 3 & -3 \times 7 \\ -3 \times 4 & -3 \times 8 \end{bmatrix}$

$\qquad\qquad\qquad = \begin{bmatrix} -9 & -21 \\ -12 & -24 \end{bmatrix}$

(iii) $2C = 2\begin{bmatrix} -1 & 1 \\ 0 & 0 \end{bmatrix} = \begin{bmatrix} 2 \times -1 & 2 \times 1 \\ 2 \times 0 & 2 \times 0 \end{bmatrix}$

$\qquad\qquad\qquad = \begin{bmatrix} -2 & 2 \\ 0 & 0 \end{bmatrix}$

(iv) $A - 5B = \begin{bmatrix} 1 & 0 \\ 2 & -1 \end{bmatrix} - 5\begin{bmatrix} 3 & 7 \\ 4 & 8 \end{bmatrix}$

$\qquad\qquad = \begin{bmatrix} 1 & 0 \\ 2 & -1 \end{bmatrix} - \begin{bmatrix} 15 & 35 \\ 20 & 40 \end{bmatrix}$

$\qquad\qquad = \begin{bmatrix} 1-15 & 0-35 \\ 2-20 & -1-40 \end{bmatrix}$

$\qquad\qquad = \begin{bmatrix} -14 & -35 \\ -18 & -41 \end{bmatrix}$

(v) $4A + 3C = 4\begin{bmatrix} 1 & 0 \\ 2 & -1 \end{bmatrix} + 3\begin{bmatrix} -1 & 1 \\ 0 & 0 \end{bmatrix}$

$$= \begin{bmatrix} 4 & 0 \\ 8 & -4 \end{bmatrix} + \begin{bmatrix} -3 & 3 \\ 0 & 0 \end{bmatrix}$$

$$= \begin{bmatrix} 4-3 & 0+3 \\ 8+0 & -4+0 \end{bmatrix} = \begin{bmatrix} 1 & 3 \\ 8 & -4 \end{bmatrix}$$

EXAMPLE 3. *Find the additive inverse of the matrix*

$$A = \begin{bmatrix} 2 & -3 & -1 & -1 \\ -3 & 1 & -2 & 2 \\ 1 & -2 & -8 & 7 \end{bmatrix}$$

SOLUTION. The additive inverse of $A = -A$

$$\therefore \quad -A = - \begin{bmatrix} 2 & -3 & -1 & -1 \\ -3 & 1 & -2 & 2 \\ 1 & -2 & -8 & 7 \\ -2 & 3 & 1 & 1 \end{bmatrix}$$

$$= \begin{bmatrix} 3 & -1 & 2 & -2 \\ -1 & 2 & 8 & -7 \end{bmatrix}$$

1.4 MULTIPLICATION OF MATRICES

Let A and B be two matrices of order $m \times n$ and $n \times p$ respectively. Then a matrix C of order $m \times p$ is obtained by multiplying each row of A to each column of B.

Suppose $A = [a_{ij}]_{m \times n}$, $B = [b_{jk}]_{n \times p}$, then $C = [c_{ik}]_{m \times p}$ is known as the multiplication of A and B where

$$C_{ik} = \sum_{j=1}^{n} a_{ij} b_{jk}$$

and hence, we can write $\qquad C = AB$

WORKING PROCEDURE

First we check whether the matrices are conformable for multiplication or not. For this we check that if the number of columns of first matrix is equal to the number of rows of the second matrix, then the matrices can be multiplied. Multiplication is operated by the rule (row × Column). In this rule, we first put the first row of the first matrix next to the first column of the second matrix and the corresponding elements are now multiplied and then summed up which gives the first element of the first row of the product matrix. This process runs till the first row of the first matrix is operated to all columns of the second matrix. After that the first process is applied to the second, third, etc. rows of the first matrix.

For example : If $A = \begin{bmatrix} 2 & 1 & 5 \\ 6 & 2 & 3 \end{bmatrix}_{2 \times 3}$ and $B = \begin{bmatrix} 3 & 4 \\ 5 & 6 \\ 7 & 8 \end{bmatrix}_{3 \times 2}$, then

$$AB = \begin{bmatrix} 2 & 1 & 5 \\ 6 & 2 & 3 \end{bmatrix} \begin{bmatrix} 3 & 4 \\ 5 & 6 \\ 7 & 8 \end{bmatrix} = \begin{bmatrix} 2 \times 3 + 1 \times 5 + 5 \times 7 & 2 \times 4 + 1 \times 6 + 5 \times 8 \\ 6 \times 3 + 2 \times 5 + 3 \times 7 & 6 \times 4 + 2 \times 6 + 3 \times 8 \end{bmatrix}$$

$$= \begin{bmatrix} 6+5+35 & 8+6+40 \\ 18+10+21 & 24+12+24 \end{bmatrix} = \begin{bmatrix} 49 & 54 \\ 49 & 60 \end{bmatrix}$$

REMARKS

- If the number of columns of the matrix A is equal to the number of rows of matrix B, then A and B are conformable for the multiplication AB but not for BA.
- Square matrices are always conformable for multiplication both ways.

1.4.1 PROPERTIES OF MATRIX MULTIPLICATION

(i) Associative law : *If A, B and C are the matrices of order $m \times n$, $n \times p$ and $p \times q$, then $(AB)C = A(BC)$.* (JNTU 2002, 14)

Proof. Let $A = [a_{ij}]_{m \times n}$, $B = [b_{jk}]_{n \times p}$ and $C = [c_{kl}]_{p \times q}$, then

$$AB = [a_{ij}]_{m \times n} \cdot B = [b_{jk}]_{n \times p} = [x_{ik}]_{m \times p} \text{ where } x_{ik} = \sum_{j=1}^{n} a_{ij} b_{jk} \qquad \dots(1)$$

$$\therefore \qquad (AB)C = [x_{ik}]_{m \times p} \cdot [c_{kl}]_{p \times q} = [u_{il}]_{m \times q} \text{ where } u_{il} = \sum_{k=1}^{p} x_{ik} c_{kl} \qquad \dots(2)$$

Now from (1) and (2), we get $\qquad u_{il} = \sum_{k=1}^{p} \left(\sum_{j=1}^{n} a_{ij} b_{jk} \right) c_{kl} = \sum_{j=1}^{n} a_{ij} \left(\sum_{k=1}^{p} b_{jk} c_{kl} \right) \qquad \dots(3)$

Now $\qquad BC = [b_{jk}]_{m \times p} \cdot [c_{kl}]_{p \times q} = [v_{jl}]_{m \times q}$

where

$$v_{jl} = \sum_{k=1}^{p} b_{jk} c_{kl} \qquad \ldots(4)$$

From equations (3) and (4), we get

$$u_{il} = \sum_{j=1}^{n} a_{ij} v_{jl} \qquad \ldots(5)$$

Equation (2) implies that u_{il} is the (i, l)th element in $(AB)C$ and the equation (5) implies that u_{il} is (i, l)th element in $A(BC)$. Therefore by the equality of two matrices, we obtain $(AB)C = A(BC)$.

(ii) **Distributive law :** *If A, B and C are the matrices of order* $m \times n$, $n \times p$ *and* $n \times p$ *respectively, then*

$$A(B+C) = AB + AC$$

Proof. Let $A = [a_{ij}]_{m \times n}$, $B = [b_{jk}]_{n \times p}$ and $C = [c_{jk}]_{n \times p}$, then

$$B+C = [b_{jk}]_{n \times p} + [c_{jk}]_{n \times p} = [b_{jk} + c_{jk}]_{n \times p}$$

$$\therefore \qquad A(B+C) = [a_{ij}]_{m \times n} + [b_{jk} + c_{jk}]_{n \times p}$$

$$= [x_{ik}]_{m \times p} \qquad \ldots(1)$$

where

$$x_{ik} = \sum_{j=1}^{n} a_{ij}(b_{jk} + c_{jk}) = \sum_{j=1}^{n} (a_{ij} b_{jk} + a_{ij} c_{jk}) = \sum_{j=1}^{n} a_{ij} b_{jk} + \sum_{j=1}^{n} a_{ij} c_{jk}$$

$$= (i, k)^{\text{th}} \text{ element of } AB + (i, k) \text{ the element of } AC$$

$$= (i, k)^{\text{th}} \text{ element of } (AB + BC)$$

But equation (1) implies that x_{ik} is the (i, k) the element of $A(B+C)$, therefore, by the definition of equality of two matrices, we must have

$$A(B + C) = AB + AC.$$

📧 **REMARK**

• Matrix multiplication is not commutative in general.

Solved Examples

EXAMPLE 1. *If* $A = \begin{bmatrix} 2 & 3 & 4 \\ 1 & 2 & 3 \\ -1 & 1 & 2 \end{bmatrix}$, $B = \begin{bmatrix} 1 & 3 & 0 \\ -1 & 2 & 1 \\ 0 & 0 & 2 \end{bmatrix}$ *then find AB and BA and show that AB ≠ BA.*

SOLUTION. Since A and B are square matrices of order 3×3 so that multiplication of AB and BA is possible.

Now,

$$AB = \begin{bmatrix} 2 & 3 & 4 \\ 3 & 2 & 3 \\ -1 & 1 & 2 \end{bmatrix} \begin{bmatrix} 1 & 3 & 0 \\ -1 & 2 & 1 \\ 1 & 0 & 2 \end{bmatrix}$$

$$= \begin{bmatrix} 2 \times 1 + 3 \times (-1) + 4 \times 0 & 2 \times 3 + 3 \times 2 + 4 \times 0 \\ 1 \times 1 + 2 \times (-1) + 3 \times 0 & 1 \times 3 + 2 \times 2 + 3 \times 0 \\ -1 \times 1 + 1 \times (-1) + 2 \times 0 & -1 \times 3 + 1 \times 2 + 2 \times 0 \end{bmatrix}$$

$$\begin{matrix} 2 \times 0 + 3 \times 1 + 4 \times 2 \\ 1 \times 0 + 2 \times 1 + 3 \times 2 \\ -1 \times 0 + 1 \times 1 + 2 \times 2 \end{matrix}$$

$$= \begin{bmatrix} 2 - 3 + 0 & 6 + 6 + 0 & 0 + 3 + 8 \\ 1 - 2 + 0 & 3 + 4 + 0 & 0 + 2 + 6 \\ -1 - 1 + 0 & -3 + 2 + 0 & 0 + 1 + 4 \end{bmatrix}$$

$$= \begin{bmatrix} -1 & 12 & 11 \\ -1 & 7 & 8 \\ -2 & -1 & 5 \end{bmatrix}_{3 \times 3}$$

and

$$BA = \begin{bmatrix} 1 & 3 & 0 \\ -1 & 2 & 1 \\ 0 & 0 & 2 \end{bmatrix} \begin{bmatrix} 2 & 3 & 4 \\ 1 & 2 & 3 \\ -1 & 1 & 2 \end{bmatrix}$$

$$= \begin{bmatrix} 1 \times 2 + 3 \times 1 + 0 \times (-1) & 1 \times 3 + 3 \times 2 + 0 \times 1 & 1 \times 4 + 3 \times 3 + 0 \times 2 \\ -1 \times 2 + 2 \times 1 + 1 \times (-1) & -1 \times 3 + 2 \times 2 + 1 \times 1 & -1 \times 4 + 2 \times 3 + 1 \times 2 \\ 0 \times 2 + 0 \times 1 + 2 \times (-1) & 0 \times 3 + 0 \times 2 + 2 \times 1 & 0 \times 4 + 0 \times 3 + 2 \times 2 \end{bmatrix}$$

$$= \begin{bmatrix} 2 + 3 - 0 & 3 + 6 + 0 & 4 + 9 + 0 \\ -2 + 2 - 1 & -3 + 4 + 1 & -4 + 6 + 2 \\ 0 + 0 - 2 & 0 + 0 + 2 & 0 + 0 + 4 \end{bmatrix}$$

$$= \begin{bmatrix} 5 & 9 & 13 \\ -1 & 2 & 4 \\ -2 & 2 & 4 \end{bmatrix}_{3 \times 3}$$

Here AB and BA have same order but different corresponding elements. Hence $AB \neq BA$.

1.5 DETERMINANT OF A SQUARE MATRIX

Let A be a square matrix. Then the determinant which is formed by the elements of matrix A is usually denoted by $|A|$.

For example : If $A = \begin{bmatrix} a_{11} & a_{12} & a_{13} \\ a_{21} & a_{22} & a_{23} \\ a_{31} & a_{32} & a_{33} \end{bmatrix}$, then its determinant is $|A| = \begin{vmatrix} a_{11} & a_{12} & a_{13} \\ a_{21} & a_{22} & a_{23} \\ a_{31} & a_{32} & a_{33} \end{vmatrix}$

📧 **REMARK**

• The determinant of a matrix is reduced to a number.

1.5.1 PROPERTIES OF DETERMINANTS

(1) The value of a determinant is zero if all the elements of a row or column are zero.

(2) The value of a determinant remain unchanged when rows are changed into corresponding columns.

(3) If any two rows or columns of a determinant are interchanged, the sign of the determinant is changed.

(4) If any two rows or columns of a determinant are identical, then the value of the determinant is zero.

(5) If every element of same columns or row is the sum of two terms, then determinant is equal to the sum of two determinant are containing only the first term and other the second term only in place of each sum.

(6) If each element of a row (or column) is multiplied by a constant k, then the value of new determinant will be k times the value of original determinant.

(7) If each element of a row (or column) of a determinant multiplied by a constant k and then added to the corresponding elements of some other row (or column) then the value of the determinant remain same.

(8) If the elements of the determinant are the polynomial in a variable x and if by putting $x = a$, the determinant vanishes. Then $(x - a)$ will be a factor of determinant.

1.5.2 EVOLUTION OF A DETERMINANT BY SARRUS DIAGRAM

$$\begin{vmatrix} a_{11} & a_{12} & a_{13} \\ a_{21} & a_{22} & a_{23} \\ a_{31} & a_{32} & a_{33} \end{vmatrix} = a_{11}(a_{22}a_{33} - a_{32}a_{23}) - a_{12}(a_{21}a_{33} - a_{31}a_{23}) + a_{13}(a_{21}a_{32} - a_{31}a_{22})$$

$$= a_{11}a_{22}a_{33} + a_{12}a_{31}a_{23} + a_{13}a_{21}a_{32} - (a_{11}a_{32}a_{23} + a_{12}a_{21}a_{33} + a_{13}a_{31}a_{22})$$

WORKING PROCEDURE

• Write the columns of the determinant and again write the first and second columns on the right side and draw the lines as shown in the following figure :

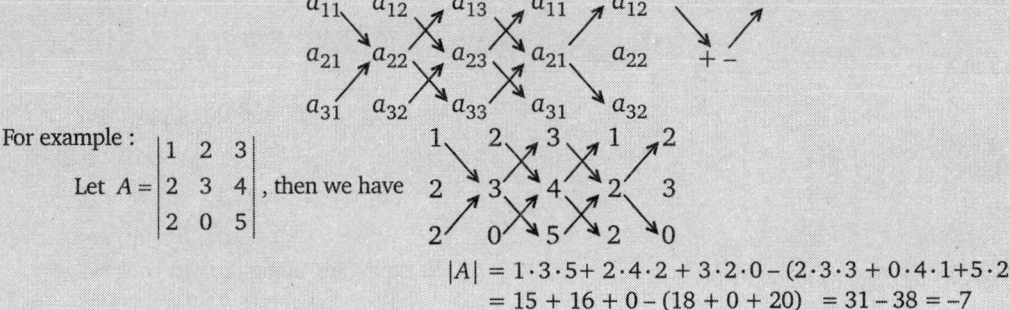

$$|A| = 1 \cdot 3 \cdot 5 + 2 \cdot 4 \cdot 2 + 3 \cdot 2 \cdot 0 - (2 \cdot 3 \cdot 3 + 0 \cdot 4 \cdot 1 + 5 \cdot 2 \cdot 2)$$
$$= 15 + 16 + 0 - (18 + 0 + 20) = 31 - 38 = -7$$

Solved Examples

EXAMPLE 1. Evaluate the following determinant

$$\begin{vmatrix} 3 & -2 \\ 4 & 5 \end{vmatrix}$$

SOLUTION. We have $|A| = \begin{vmatrix} 3 & -2 \\ 4 & 5 \end{vmatrix}$

$$= 3 \times 5 - 4 \times (-2) = 15 + 8 = 23$$

EXAMPLE 2. Find the value of the determinant of the matrix

$$A = \begin{bmatrix} 1 & 2 & 3 \\ 2 & 3 & 1 \\ 3 & 1 & 2 \end{bmatrix}$$

SOLUTION. We have $A = \begin{bmatrix} 1 & 2 & 3 \\ 2 & 3 & 1 \\ 3 & 1 & 2 \end{bmatrix}$

On expanding the determinant along the first row, we get

$$= 1 \begin{vmatrix} 3 & 1 \\ 1 & 2 \end{vmatrix} - 2 \begin{vmatrix} 2 & 1 \\ 3 & 2 \end{vmatrix} + 3 \begin{vmatrix} 2 & 3 \\ 3 & 1 \end{vmatrix}$$

$$= 1.(6 - 1) - 2.(4 - 3) + 3.(2 - 9) = -18$$

EXAMPLE 3. Show that

$$\begin{vmatrix} 1 & x & y \\ 0 & \cos x & \sin y \\ 0 & \sin x & \cos y \end{vmatrix} = \cos(x + y)$$

SOLUTION. We have $\begin{vmatrix} 1 & x & y \\ 0 & \cos x & \sin y \\ 0 & \sin x & \cos y \end{vmatrix}$

On expanding the determinant along the first column, we get

$$= 1 \begin{vmatrix} \cos x & \sin y \\ \sin x & \cos y \end{vmatrix} - 0 \begin{vmatrix} x & y \\ \sin x & \sin y \end{vmatrix} + 0 \begin{vmatrix} x & y \\ \cos x & \cos y \end{vmatrix}$$

$$= \cos x \cos y - \sin x \sin y = \cos (x+y)$$

EXAMPLE 4. Show that $\begin{vmatrix} 1 & 1 & 1 \\ 1 & 1+x & 1 \\ 1 & 1 & 1+y \end{vmatrix} = xy$

SOLUTION. We have L.H.S. $= \begin{vmatrix} 1 & 1 & 1 \\ 1 & 1+x & 1 \\ 1 & 1 & 1+y \end{vmatrix}$

Applying $C_2 - C_1$ and $C_3 - C_1$ in the given determinant, we get

$$= \begin{vmatrix} 1 & 0 & 0 \\ 1 & x & 0 \\ 1 & 0 & y \end{vmatrix}$$

On expanding the determinant along the first row, we get

$$= 1\begin{vmatrix} x & 0 \\ 0 & y \end{vmatrix} - 0\begin{vmatrix} 1 & 0 \\ 1 & y \end{vmatrix} - 0\begin{vmatrix} 1 & x \\ 1 & 0 \end{vmatrix}$$

$$= xy = \text{R.H.S.}$$

EXAMPLE 5. *Without expanding, show that*

$$\begin{vmatrix} b-c & c-a & a-b \\ c-a & a-b & b-c \\ a-b & b-c & c-a \end{vmatrix} = 0$$

SOLUTION. We have

$$\begin{vmatrix} b-c & c-a & a-b \\ c-a & a-b & b-c \\ a-b & b-c & c-a \end{vmatrix} = \begin{vmatrix} 0 & c-a & a-b \\ 0 & a-b & b-c \\ 0 & b-c & c-a \end{vmatrix}$$

(Operating $C_1 \to C_1 + C_2 + C_3$, we get)

$$= 0$$

EXAMPLE 6. *Without expanding, show that*

$$\begin{vmatrix} b^2c^2 & bc & b+c \\ c^2a^2 & ca & c+a \\ a^2b^2 & ab & a+b \end{vmatrix} = 0$$

SOLUTION. Consider

$$\begin{vmatrix} b^2c^2 & bc & b+c \\ c^2a^2 & ca & c+a \\ a^2b^2 & ab & a+b \end{vmatrix} = \frac{abc}{abc}\begin{vmatrix} b^2c^2 & bc & b+c \\ c^2a^2 & ca & c+a \\ a^2b^2 & ab & a+b \end{vmatrix}$$

(Multiply R_1 by a, R_2 by b and R_3 by c)

$$= \frac{1}{abc}\begin{vmatrix} ab^2c^2 & abc & ab+ca \\ bc^2a^2 & abc & bc+ab \\ ca^2b^2 & abc & ca+bc \end{vmatrix}$$

(Take abc out from C_1 and C_2)

$$= \frac{abc.abc}{abc}\begin{vmatrix} bc & 1 & ab+ca \\ ca & 1 & bc+ab \\ ab & 1 & ca+bc \end{vmatrix}$$

$$= abc\begin{vmatrix} bc & 1 & ab+bc+ca \\ ca & 1 & bc+ab+ca \\ ab & 1 & ab+ca+bc \end{vmatrix}$$

(Operate $C_3 \to C_3 + C_1$)

$$= abc(ab+bc+ca)\begin{vmatrix} bc & 1 & 1 \\ ca & 1 & 1 \\ ab & 1 & 1 \end{vmatrix}$$

$$= abc(ab+bc+ca) \times 0$$

$$= 0$$

EXAMPLE 7. *If a, b, c are in A.P, prove that*

$$\begin{vmatrix} x+1 & x+2 & x+a \\ x+2 & x+3 & x+b \\ x+3 & x+4 & x+c \end{vmatrix} = 0$$

SOLUTION. Given a, b, c are in A.P. therefore,

$$a+c = 2b$$
$$\Rightarrow \quad a+c-2b = 0$$

Operating $R_1 \to R_1 + R_3 - 2R_2$, we get

$$\begin{vmatrix} x+1 & x+2 & x+a \\ x+2 & x+3 & x+b \\ x+3 & x+4 & x+c \end{vmatrix} = \begin{vmatrix} 0 & 0 & a+c-2b \\ x+2 & x+3 & x+b \\ x+3 & x+4 & x+c \end{vmatrix}$$

$$= \begin{vmatrix} 0 & 0 & 0 \\ x+2 & x+3 & x+b \\ x+3 & x+4 & x+c \end{vmatrix} = 0$$

EXAMPLE 8. *Prove that*

$$\begin{vmatrix} a & b & c \\ a^2 & b^2 & c^2 \\ a^3 & b^3 & c^3 \end{vmatrix} = abc\begin{vmatrix} 1 & 1 & 1 \\ a & b & c \\ a^2 & b^2 & c^2 \end{vmatrix}$$

$$= abc(a-b)(b-c)(c-a)$$

SOLUTION. We have

$$|A| = \begin{vmatrix} a & b & c \\ a^2 & b^2 & c^2 \\ a^3 & b^3 & c^3 \end{vmatrix} = = abc\begin{vmatrix} 1 & 1 & 1 \\ a & b & c \\ a^2 & b^2 & c^2 \end{vmatrix}$$

Now again, $|A| = abc\begin{vmatrix} 1 & 1 & 1 \\ a & b & c \\ a^2 & b^2 & c^2 \end{vmatrix}$

Applying $C_2 - C_1$ and $C_3 - C_1$, we get

$$= abc\begin{vmatrix} 1 & 0 & 0 \\ a & b-a & c-a \\ a^2 & b^2-a^2 & c^2-a^2 \end{vmatrix}$$

On expanding along the first row, we get

$$= abc\begin{vmatrix} b-a & c-a \\ b^2-a^2 & c^2-a^2 \end{vmatrix}$$

$$= abc[(b-a)(c^2-a^2)-(b^2-a^2)(c-a)]$$
$$= abc[(b-a)(c-a)\{(c+a)-(b+a)\}]$$
$$= abc(b-a)(c-a)(c+a-b-a)$$
$$= abc\,(a-b)\,(b-c)\,(c-a)$$

EXAMPLE 9. *Prove that*

$$\begin{vmatrix} a+b+2c & a & b \\ c & b+c+2a & b \\ c & a & c+a+2b \end{vmatrix} = 2(a+b+c)^3$$

SOLUTION. Let $|A| = \begin{vmatrix} a+b+2c & a & b \\ c & b+c+2a & b \\ c & a & c+a+2b \end{vmatrix}$

Applying C_2 and C_3 in C_1, we get

$$= \begin{vmatrix} 2(a+b+c) & a & b \\ 2(a+b+c) & b+c+2a & b \\ 2(a+b+c) & a & c+a+2b \end{vmatrix}$$

$$= 2(a+b+c)\begin{vmatrix} 1 & a & b \\ 1 & b+c+2a & b \\ 1 & a & c+a+2b \end{vmatrix}$$

Applying $(R_2 - R_1)$ and $(R_3 - R_1)$, we get

$$= 2(a+b+c)\begin{vmatrix} 1 & a & b \\ 0 & b+c+a & 0 \\ 0 & 0 & c+a+b \end{vmatrix}$$

On expanding determinant along the first column, we get

$$= 2(a+b+c)\begin{vmatrix} b+c+a & 0 \\ 0 & a+b+c \end{vmatrix}$$

$$= 2(a+b+c)(a+b+c)^2 = 2(a+b+c)^3$$

EXAMPLE 10. *Prove that*

$$\begin{vmatrix} 1+a & 1 & 1 \\ 1 & 1+b & 1 \\ 1 & 1 & 1+c \end{vmatrix} = abc\left(1 + \frac{1}{a} + \frac{1}{b} + \frac{1}{c}\right)$$

SOLUTION. Operating $C_1 \to C_1 - C_3$ and $C_2 \to C_2 - C_3$, we get

$$\begin{vmatrix} 1+a & 1 & 1 \\ 1 & 1+b & 1 \\ 1 & 1 & 1+c \end{vmatrix} = \begin{vmatrix} a & 0 & 1 \\ 0 & b & 1 \\ -c & -c & 1+c \end{vmatrix}$$

$$= a[b.(1+c) - (-c).1] + 1[0.(-c) - (-c)b]$$

$$= a(b + bc + c) + bc = abc + bc + ca + ab$$

$$= abc\left(1 + \frac{1}{a} + \frac{1}{b} + \frac{1}{c}\right)$$

EXAMPLE 11. *Prove that*

$$\begin{vmatrix} a-b-c & 2a & 2a \\ 2b & b-c-a & 2b \\ 2c & 2c & c-a-b \end{vmatrix} = (a+b+c)^3$$

SOLUTION. Operating $R_1 \to R_1 + R_2 + R_3$, we get

$$\begin{vmatrix} a-b-c & 2a & 2a \\ 2b & b-c-a & 2b \\ 2c & 2c & c-a-b \end{vmatrix}$$

$$= \begin{vmatrix} a+b+c & a+b+c & a+b+c \\ 2b & b-c-a & 2b \\ 2c & 2c & c-a-b \end{vmatrix}$$

(Taking $(a+b+c)$ out from R_1)

$$= (a+b+c)\begin{vmatrix} 1 & 1 & 1 \\ 2b & b-c-a & 2b \\ 2c & 2c & c-a-b \end{vmatrix}$$

(Operate $C_2 \to C_2 - C_1$ and $C_3 \to C_3 - C_1$)

$$= (a+b+c)\begin{vmatrix} 1 & 0 & 0 \\ 2b & -b-c-a & 0 \\ 2c & 0 & -a-b-c \end{vmatrix}$$

(expand by R_1)

$$= (a+b+c).1(-a-b-c)(-a-b-c)$$

$$= (a+b+c)^3$$

EXAMPLE 12. *Show that*

$$\begin{vmatrix} a & b & c \\ a-b & b-c & c-a \\ b+c & c+a & a+b \end{vmatrix} = a^3 + b^3 + c^3 - 3abc$$

SOLUTION. Operating $R_2 \to R_2 - R_1$ and $R_3 \to R_3 + R_1$, we get

$$= \begin{vmatrix} a & b & c \\ -b & -c & -a \\ a+b+c & a+b+c & a+b+c \end{vmatrix}$$

[Take $(a+b+c)$ out from R_3 and (-1) from R_2)

$$= -(a+b+c).\begin{vmatrix} a & b & c \\ b & c & a \\ 1 & 1 & 1 \end{vmatrix} \qquad \text{(Expand by } R_3)$$

$$= -(a+b+c).[1.(ab - c^2) - 1(a^2 - bc) + 1.(ca - b^2)]$$

$$= -(a+b+c).(ab + bc + ca - a^2 - b^2 - c^2)$$

$$= (a+b+c).(a^2 + b^2 + c^2 - ab - bc - ca)$$

$$= a^3 + b^3 + c^3 - 3abc$$

EXAMPLE 13. *Prove that*

$$\begin{vmatrix} a^2+1 & ab & ac \\ ab & b^2+1 & bc \\ ac & bc & c^2+1 \end{vmatrix} = 1 + a^2 + b^2 + c^2$$

SOLUTION. We have $|A| = \begin{vmatrix} a^2+1 & ab & ac \\ ab & b^2+1 & bc \\ ac & bc & c^2+1 \end{vmatrix}$

Now multiply the column 1st, 2nd and 3rd by a, b and c respectively, we get

$$|A| = \frac{1}{abc}\begin{vmatrix} a(a^2+1) & ab^2 & ac^2 \\ a^2b & b(b^2+1) & bc^2 \\ a^2c & b^2c & c(c^2+1) \end{vmatrix}$$

To take a, b, c common from 1st, 2nd and 3rd rows respectively, we get

$$= \frac{abc}{abc}\begin{vmatrix} a^2+1 & b^2 & c^2 \\ a^2 & (b^2+1) & c^2 \\ a^2 & b^2 & c^2+1 \end{vmatrix}$$

Now applying $C_1 \to C_1 + C_2 + C_3$, we get

$$= \begin{vmatrix} a^2+b^2+c^2+1 & b^2 & c^2 \\ a^2+b^2+c^2+1 & b^2+1 & c^2 \\ a^2+b^2+c^2+1 & b^2 & c^2+1 \end{vmatrix}$$

$$= (a^2+b^2+c^2+1)\begin{vmatrix} 1 & b^2 & c^2 \\ 1 & b^2+1 & c^2 \\ 1 & b^2 & c^2+1 \end{vmatrix}$$

Now apply $R_2 \to R_2 - R_1$ and $R_3 \to R_3 - R_1$ we get

$$= (a^2+b^2+c^2)\begin{vmatrix} 1 & 0 \\ 0 & 1 \end{vmatrix}$$

$$= a^2 + b^2 + c^2 + 1$$

EXAMPLE 14. *Prove that* $\begin{vmatrix} b+c & c+a & a+b \\ q+r & r+p & p+q \\ y+z & z+x & x+y \end{vmatrix} = 2\begin{vmatrix} a & b & c \\ p & q & r \\ x & y & z \end{vmatrix}$

SOLUTION. We have

L.H.S. $= \begin{vmatrix} b+c & c+a & a+b \\ q+r & r+p & p+q \\ y+z & z+x & x+y \end{vmatrix}$

Applying $C_1 \to C_1 + C_2 - C_3$, we get

$$= \begin{vmatrix} 2c & c+a & a+b \\ 2r & r+p & p+q \\ 2z & z+x & x+y \end{vmatrix} = 2\begin{vmatrix} c & c+a & a+b \\ r & r+p & p+q \\ z & z+x & x+y \end{vmatrix}$$

Now applying $C_2 \to C_2 - C_1$, we get

$$= 2\begin{vmatrix} c & a & a+b \\ r & p & p+q \\ z & x & x+y \end{vmatrix}$$

Applying $C_3 \to C_3 - C_2$, we get

$$= 2\begin{vmatrix} c & a & b \\ r & p & q \\ z & x & y \end{vmatrix} = 2\begin{vmatrix} a & b & c \\ p & q & r \\ x & y & z \end{vmatrix}$$

(By interchanging the columns)

= R.H.S.

EXAMPLE 15. *If x, y, z are all different and* $\begin{vmatrix} x & x^2 & 1+x^3 \\ y & y^2 & 1+y^3 \\ z & z^2 & 1+z^3 \end{vmatrix} = 0$

Show that $xyz = -1$ (ANDHRA–1999, 2015, ASSAM–1999)

SOLUTION. Given

$$\begin{vmatrix} x & x^2 & 1+x^3 \\ y & y^2 & 1+y^3 \\ z & z^2 & 1+z^3 \end{vmatrix} = 0$$

$$\Rightarrow \begin{vmatrix} x & x^2 & 1 \\ y & y^2 & 1 \\ z & z^2 & 1 \end{vmatrix} + \begin{vmatrix} x & x^2 & x^3 \\ y & y^2 & y^3 \\ z & z^2 & z^3 \end{vmatrix} = 0$$

[Take x, y, z out from R_1, R_2 and R_3 respectively from the second determinant]

$$\Rightarrow \begin{vmatrix} 1 & x & x^2 \\ 1 & y & y^2 \\ 1 & z & z^2 \end{vmatrix} + xyz\begin{vmatrix} 1 & x & x^2 \\ 1 & y & y^2 \\ 1 & z & z^2 \end{vmatrix} = 0$$

$$\Rightarrow \begin{vmatrix} 1 & x & x^2 \\ 1 & y & y^2 \\ 1 & z & z^2 \end{vmatrix}(1+xyz) = 0$$

$$\Rightarrow (x-y)(y-z)(z-x).(1+xyz) = 0$$

$$\Rightarrow (1+xyz) = 0$$

(Because x, y, z are all distinct, so
$x-y \neq 0, y-z \neq 0, z-x \neq 0$)

$$\Rightarrow xyz = -1$$

EXAMPLE 16. *Show that*

$$\begin{vmatrix} (b+c)^2 & a^2 & bc \\ (c+a)^2 & b^2 & ca \\ (a+b)^2 & c^2 & ab \end{vmatrix}$$

$$= (a^2+b^2+c^2)(a+b+c)(b-c)(c-a)(a-b)$$

SOLUTION. Let $\Delta = \begin{vmatrix} (b+c)^2 & a^2 & bc \\ (c+a)^2 & b^2 & ca \\ (a+b)^2 & c^2 & ab \end{vmatrix}$

Applying $C_1 \to C_1 - 2C_3$, we get

$$= \begin{vmatrix} b^2+c^2+a^2 & a^2 & bc \\ c^2+a^2+b^2 & b^2 & ca \\ a^2+b^2+c^2 & c^2 & ab \end{vmatrix}$$

Operating $C_1 \to C_1 + C_2$, we get

$$= (a^2+b^2+c^2)\begin{vmatrix} 1 & a^2 & bc \\ 1 & b^2 & ca \\ 1 & c^2 & ab \end{vmatrix}$$

Operating $R_2 \to R_2 - R_1$ and $R_3 \to R_3 - R_2$

$$= (a^2+b^2+c^2)\begin{vmatrix} 1 & a^2 & bc \\ 0 & b^2-a^2 & (ca-bc) \\ 0 & c^2-a^2 & (ab-bc) \end{vmatrix}$$

$$= (a^2+b^2+c^2)(b-a)(c-a)\begin{vmatrix} 1 & a^2 & bc \\ 0 & b+a & -c \\ 0 & c+a & -b \end{vmatrix}$$

$R_3 \to R_3 - R_2$, we get

$$= (a^2+b^2+c^2)(b-a)(c-a)\begin{vmatrix} 1 & a^2 & bc \\ 0 & b+a & -c \\ 0 & c-b & c-b \end{vmatrix}$$

$$= (a^2+b^2+c^2)(b-a)(c-a)(c-b)\begin{vmatrix} 1 & a^2 & bc \\ 0 & b+a & -c \\ 0 & 1 & 1 \end{vmatrix}$$

Expanding along first column, we get
$$\Delta = (a^2+b^2+c^2)(b-a)(c-a)(c-b)(a+b+c)$$

EXAMPLE 17. *Show that*

$$\begin{vmatrix} a+b & b+c & c+a \\ b+c & c+a & a+b \\ c+a & a+b & b+c \end{vmatrix} = 2\begin{vmatrix} a & b & c \\ b & c & a \\ c & a & b \end{vmatrix}$$

SOLUTION. Let $\Delta = \begin{vmatrix} a+b & b+c & c+a \\ b+c & c+a & a+b \\ c+a & a+b & b+c \end{vmatrix}$

Applying $C_1 \to C_1 + C_2 + C_3$, we get

$$= \begin{vmatrix} 2(a+b+c) & b+c & c+a \\ 2(a+b+c) & c+a & a+b \\ 2(a+b+c) & a+b & b+c \end{vmatrix}$$

$$= 2\begin{vmatrix} (a+b+c) & -a & -b \\ (a+b+c) & -b & -c \\ (a+b+c) & -c & -a \end{vmatrix}$$

Applying $C_2 \to C_2 - C_1, C_3 \to C_3 - C_1$, we get

$$= 2(-1)(-1)\begin{vmatrix} a+b+c & a & b \\ a+b+c & b & c \\ a+b+c & c & a \end{vmatrix}$$

Applying $C_1 \to C_1 - C_2 - C_3$, we get

$$= 2\begin{vmatrix} c & a & b \\ a & b & c \\ b & c & a \end{vmatrix}$$

Applying $C_1 \to C_2$, we get

$$= -2\begin{vmatrix} a & c & b \\ b & a & c \\ c & b & a \end{vmatrix} = 2\begin{vmatrix} a & b & c \\ b & c & a \\ c & b & a \end{vmatrix}$$

(On using $C_2 \to C_3$)

EXAMPLE 18. *If a, b, c (all positive) are the pth, qth and rth terms respectively of a geometric progression, show that*

$$\begin{vmatrix} \log a & p & 1 \\ \log b & q & 1 \\ \log c & r & 1 \end{vmatrix} = 0$$

SOLUTION. Consider the terms of G.P. which are

$$A, AR, AR^2, \dots.$$

$$a = T_p = AR^{p-1}$$
$$b = T_q = AR^{q-1}$$
$$c = T_r = AR^{r-1}$$

Consider $\begin{vmatrix} \log a & p & 1 \\ \log b & q & 1 \\ \log c & r & 1 \end{vmatrix} = \begin{vmatrix} \log AR^{p-1} & p & 1 \\ \log AR^{q-1} & q & 1 \\ \log AR^{r-1} & r & 1 \end{vmatrix}$

$$= \begin{vmatrix} \log A + (p-1)\log R & p & 1 \\ \log A + (q-1)\log R & q & 1 \\ \log A + (r-1)\log R & r & 1 \end{vmatrix}$$

$$= \begin{vmatrix} \log A & p & 1 \\ \log A & q & 1 \\ \log A & r & 1 \end{vmatrix} + \begin{vmatrix} (p-1)\log R & p & 1 \\ (q-1)\log R & q & 1 \\ (r-1)\log R & r & 1 \end{vmatrix}$$

$$= \log A \times 0 + \log R\begin{vmatrix} p & p & 1 \\ q & q & 1 \\ r & r & 1 \end{vmatrix}$$

$$= 0 + \log R \times 0 = 0$$

EXAMPLE 19. *Show that*

$$\begin{vmatrix} (y+z)^2 & xy & zx \\ xy & (x+z)^2 & yz \\ xz & yz & (x+y)^2 \end{vmatrix} = 2xyz(x+y+z)^2$$

SOLUTION. Let $\Delta = \begin{vmatrix} (y+z)^2 & xy & zx \\ xy & (x+z)^2 & yz \\ xz & yz & (x+y)^2 \end{vmatrix}$

Operating $R_1 \to xR_1, R_2 \to yR_2, R_3 \to zR_3$, we get

$$\Delta = \frac{1}{xyz}\begin{vmatrix} x(y+z)^2 & x^2 y & x^2 z \\ xy^2 & y(x+z)^2 & y^2 z \\ xz^2 & yz^2 & z(x+y)^2 \end{vmatrix}$$

Taking x, y, z common from C_1, C_2, C_3 respectively, we get

$$\Delta = \frac{xyz}{xyz}\begin{vmatrix} (y+z)^2 & x^2 & x^2 \\ y^2 & (x+z)^2 & y^2 \\ z^2 & z^2 & (x+y)^2 \end{vmatrix}$$

$$= \begin{vmatrix} (y+z)^2 & x^2 & x^2 \\ y^2 & (x+z)^2 & y^2 \\ z^2 & z^2 & (x+y)^2 \end{vmatrix}$$

$$= \begin{vmatrix} (y+z)^2 - x^2 & 0 & x^2 \\ 0 & (z+x)^2 - y^2 & y^2 \\ z^2 - (x+y)^2 & z^2 - (x+y)^2 & (x+y)^2 \end{vmatrix}$$

Operating $C_1 \to C_1 - C_3, C_2 \to C_2 - C_3$, we get

$$= \begin{vmatrix} (y+z+x)(y+z-x) & 0 & x^2 \\ 0 & (z+x+y)(z+x-y) & y^2 \\ (z+x+y)(z-x-y) & (z+x+y)(z-x-y) & (x+y)^2 \end{vmatrix}$$

Taking $(x+y+z)$ common from C_1 and C_2 each, we get

$$\Delta = (x+y+z)^2\begin{vmatrix} y+z-x & 0 & x^2 \\ 0 & z+x-y & y^2 \\ z-x-y & z-x-y & (x+y)^2 \end{vmatrix}$$

Operating $R_3 \to R_3 - R_1 - R_2$, we get

$$\Delta = (x+y+z)^2\begin{vmatrix} y+z-x & 0 & x^2 \\ 0 & z+x-y & y^2 \\ -2y & -2x & 2xy \end{vmatrix}$$

Operating $C_1 \to C_1 + \dfrac{C_3}{x}, C_2 \to C_2 + \dfrac{C_3}{y}$, we get

$$= (x+y+z)^2\begin{vmatrix} y+z & x^2/y & x^2 \\ y^2/x & z+x & y^2 \\ 0 & 0 & 2xy \end{vmatrix}$$

Expanding along R_1, we get

$$\Delta = (x+y+z)^2 . 2xy\begin{vmatrix} y+z & x^2/y \\ y^2/x & z+x^2 \end{vmatrix}$$

$$= (x+y+z)^2 . 2xy[(y+z)(z+x) - xy]$$

$$= (x+y+z)^2 . 2xy(yz + z^2 + zx)$$

$$= 2xyz(x+y+z)^2$$

Exercise-1.1

Evaluate the following determinants. (1 to 7)

1. $\begin{vmatrix} 1 & 8 \\ 2 & \\ 4 & 2 \end{vmatrix}$

2. $\begin{vmatrix} -2 & 3 \\ 4 & -9 \end{vmatrix}$

3. $\begin{vmatrix} \cos\theta & -\sin\theta \\ \sin\theta & \cos\theta \end{vmatrix}$

4. $\begin{vmatrix} x^2-x+1 & x-1 \\ x+1 & x-1 \end{vmatrix}$

5. $\begin{vmatrix} 1 & 0 & 6 \\ 3 & 4 & 15 \\ 5 & 6 & 21 \end{vmatrix}$

6. $\begin{vmatrix} 23 & 12 & 11 \\ 36 & 10 & 26 \\ 63 & 26 & 37 \end{vmatrix}$

7. $\begin{vmatrix} 3 & 1 & -4 \\ 3 & 2 & 5 \\ 1 & 1 & 3 \end{vmatrix}$

Write the minor and co-factor of each element of the following determinants and also evaluate the determinants in each case. (8 to 11):

8. $\begin{vmatrix} 5 & -10 \\ 0 & 3 \end{vmatrix}$

9. $\begin{vmatrix} 1 & 3 & -2 \\ 4 & -5 & 6 \\ 3 & 5 & 2 \end{vmatrix}$

10. $\begin{vmatrix} 1 & 0 & 0 \\ 0 & 1 & 0 \\ 0 & 0 & 1 \end{vmatrix}$

11. $\begin{vmatrix} 1 & 0 & 4 \\ 3 & 5 & -1 \\ 0 & 1 & 2 \end{vmatrix}$

12. Evaluate $\begin{vmatrix} x+1 & x+2 & x+4 \\ x+5 & x+6 & x+8 \\ x+7 & x+10 & x+14 \end{vmatrix}$

13. Evaluate $\begin{vmatrix} 1 & a & bc \\ 1 & b & ca \\ 1 & c & ab \end{vmatrix}$

14. Evaluate $\begin{vmatrix} x+\lambda & x & x \\ x & x+\lambda & x \\ x & x & x+\lambda \end{vmatrix}$

15. Evaluate $\begin{vmatrix} b+c & a & a \\ b & c+a & b \\ c & c & a+b \end{vmatrix}$

16. Prove that $\begin{vmatrix} 1 & x & x^2 \\ 1 & y & y^2 \\ 1 & z & z^2 \end{vmatrix} = (x-y)(y-z)(z-x)$

17. Prove that $\begin{vmatrix} -a^2 & ab & ac \\ ba & -b^2 & bc \\ ac & bc & -c^2 \end{vmatrix} = 4a^2b^2c^2$

18. Prove that $\begin{vmatrix} x & x^2 & yz \\ y & y^2 & zx \\ z & z^2 & xy \end{vmatrix} = (x-y)(y-z)(z-x)(xy+yz+zx)$

19. Using properties of determinants, prove that
$$\begin{vmatrix} y+z & x & y \\ z+x & z & x \\ x+y & y & z \end{vmatrix} = (x+y+z)(x-z)^2$$

20. Using properties of determinants, prove that
$$\begin{vmatrix} a-b-c & 2a & 2a \\ 2b & b-c-a & 2b \\ 2c & 2c & c-a-b \end{vmatrix} = (a+b+c)^3$$

21. Solve the following determinant
$$\begin{vmatrix} x-2 & 2x-3 & 3x-4 \\ x-4 & 2x-9 & 3x-16 \\ x-8 & 2x-27 & 3x-64 \end{vmatrix} = 0$$

22. Prove that using properties of determinants
$$\begin{vmatrix} 1+a^2-b^2 & 2ab & -2b \\ 2ab & 1-a^2+b^2 & 2a \\ 2b & -2a & 1-a^2-b^2 \end{vmatrix} = (1+a^2+b^2)^3$$

23. Prove that
$$\begin{vmatrix} x & x^2 & 1+px^3 \\ y & y^2 & 1+py^3 \\ z & z^2 & 1+pz^3 \end{vmatrix} = (1+pxyz)(x-y)(y-z)(z-x)$$
(ANDHRA 1999, ASSAM 1999)

24. Prove that using properties of determinants
$$\begin{vmatrix} 3a & -a+b & -a+c \\ -b+a & 3b & -b+c \\ -c+a & -c+b & 3c \end{vmatrix} = 3(a+b+c)(ab+bc+ca)$$

25. Prove that
$$\begin{vmatrix} \sin\alpha & \cos\alpha & \cos(\alpha+\delta) \\ \sin\beta & \cos\beta & \cos(\beta+\delta) \\ \sin\gamma & \cos\gamma & \cos(\gamma+\delta) \end{vmatrix} = 0$$

Hints to Selected Problems

12. Applying $R_3 \to R_3 - R_1$ and $R_2 \to R_2 - R_1$
$C_3 \to C_3 - C_1$ and $C_2 \to C_2 - C_1$
We get value of let $= -24$

13. Applying $R_2 \to R_2 - R_1$ and $R_3 \to R_3 - R_1$
and after expansion, we get the required result.

14. Applying $R_1 \to R_1 + R_2 + R_3$
$C_2 \to C_2 - C_1 . C_3 \to C_3 - C_1$

15. Applying $R_1 \to R_1 + R_2 - R_3$ and expanding.

16. Applying $R_1 \to R_2 - R_1, R_3 \to R_3 - R_1$.

17. Taking a, b, c common from first, second and third column and after than applying $R_2 \to R_2 + R_1$ and $R_3 \to R_3 + R_1$.

18. Multiplying first, second and third rows of the determinant by x, y, z respectively, and thus Applying $C_2 \to C_2 - C_1, C_3 \to C_3 - C_1$

19. Applying $R_1 \to R_1 + R_2 + R_3$

Then $C_1 \to C_1 - C_2 - C_3$.

20. Applying $R_1 \to R_1 + R_2 + R_3$,

$C_2 \to C_2 - C_1$

$C_3 \to C_3 - C_1$

22. Applying $C_1 \to C_1 - C_3$,

$R_3 \to R_3 - R_1$

24. $C_1 \to C_1 + C_2 + C_3$

$R_2 \to R_2 - R_1, R_3 \to R_3 - R_1$

Answers

1. –31   **2.** 6   **3.** 1   **4.** $x^3 - x^2 + 2$ **5.** –18   **6.** 0   **7.** 49

8. $M_{11} = 3, M_{12} = 0, M_{21} = -10, M_{22} = 5, A_{11} = 3, A_{12} = 0, A_{21} = 10, A_{22} = 5, 15$

9. $M_{11} = -40, M_{12} = -10, M_{13} = 35, M_{21} = 16, M_{22} = 8, M_{23} = -4, M_{31} = 8, M_{32} = 14, M_{33} = -17$

$A_{11} = -40, A_{12} = 10, A_{13} = 35, A_{21} = -16, A_{22} = 8, A_{23} = 4, A_{31} = 8, A_{32} = -14, A_{33} = -17; -80$

10. $M_{11} = 1, M_{12} = 0, M_{13} = 0, M_{21} = 0, M_{22} = 1, M_{23} = 0, M_{31} = 0, M_{32} = 0, M_{33} = 1$

$A_{11} = 1, A_{12} = 0, A_{13} = 0, A_{21} = 0, A_{22} = 1, A_{23} = 0, A_{31} = 0, A_{32} = 0, A_{33} = 1; 1$

11. $M_{11} = 11, M_{12} = 6, M_{13} = 3, M_{21} = -4, M_{22} = 2, M_{23} = 1, M_{31} = -20, M_{32} = -13, M_{33} = 5$

$A_{11} = 11, A_{12} = -6, A_{13} = 3, A_{21} = 4, A_{22} = 2, A_{23} = -1, A_{31} = 20, A_{32} = 13, A_{33} = 5; 23$

12. –24   **13.** $(a-b)(b-c)(c-a)$ **14.** $\lambda^2(3x+\lambda)$ **15.** $4abc$ **21.** $x = 4$

1.6 MINOR AND COFACTORS

1.6.1 MINOR

In determinant, $$\Delta = \begin{vmatrix} a_{11} & a_{12} & a_{13} \\ a_{21} & a_{22} & a_{23} \\ a_{31} & a_{32} & a_{33} \end{vmatrix} \qquad \text{...(1)}$$

If we leave the row and column passing through the element a_{ij} then we obtained the second order determinant, which is called the minor of the element a_{ij}. It is denoted by M_{ij}. Therefore, in a determinant of order 3, we may get 9 minors corresponding to the 9 elements of the determinant.

For example, in determinant (1)

$$\text{Minor of } a_{21} = \begin{vmatrix} a_{12} & a_{13} \\ a_{32} & a_{33} \end{vmatrix} = M_{21}$$

and

$$\text{Minor of } a_{32} = \begin{vmatrix} a_{11} & a_{13} \\ a_{21} & a_{23} \end{vmatrix} = M_{32}$$

If we expand the determinant along the first row, then

$$\Delta = (-1)^{1+1} a_{11}M_{11} + (-1)^{1+2} a_{12}M_{12} + (-1)^{1+3} a_{13}M_{13} = a_{11}M_{11} - a_{12}M_{12} + a_{13}M_{13}$$

Similarly, along second column, we can write

$$\Delta = -a_{12}M_{12} + a_{22}M_{22} - a_{32}M_{32}$$

1.6.2 COFACTOR

If we multiply the minor M_{ij} by $(-1)^{i+j}$. Then resulting value is called cofactor of the element a_{ij}. If A_{ij} is the cofactor of a_{ij} then we write

$$\text{Cofactor of } a_{ij} = A_{ij} = (-1)^{i+j} M_{ij}$$

$$\text{Cofactor of } a_{21} = A_{21} = (-1)^{2+1} M_{21} = -\begin{vmatrix} a_{12} & a_{13} \\ a_{32} & a_{33} \end{vmatrix}$$

$$\text{Cofactor of } a_{32} = A_{32} = (-1)^{3+2} M_{32} = -\begin{vmatrix} a_{11} & a_{13} \\ a_{21} & a_{23} \end{vmatrix}$$

Hence, cofactor of $a_{ij} = (-1)^{i+j}$ determinant obtained by leaving row and column passing through that element. Therefore, we can write

$$\Delta = a_{11}A_{11} + a_{12}A_{12} + a_{13}A_{13}$$

$$\Delta = a_{21}A_{21} + a_{22}A_{22} + a_{23}A_{23}$$

$$\Delta = a_{31}A_{31} + a_{32}A_{32} + a_{33}A_{33}$$

and

$$a_{11}A_{21} + a_{12}A_{22} + a_{13}A_{23} = 0$$

$$a_{11}A_{31} + a_{12}A_{32} + a_{13}A_{33} = 0$$

1.7 SINGULAR AND NON-SINGULAR MATRIX

A matrix whose determinant value is zero, is said to be singular matrix.
If the matrix is not singular then it is said to be non-singular.

For example : If $A = \begin{bmatrix} 2 & 3 \\ 6 & 9 \end{bmatrix}$, then its determinant value $|A| = \begin{vmatrix} 2 & 3 \\ 6 & 9 \end{vmatrix} = 2 \times 9 - 3 \times 6 = 18 - 18 = 0$

Thus the matrix A is singular.

1.8 TRANSPOSE OF A MATRIX

Definition 1. *Consider a matrix $A = [a_{ij}]_{m \times n}$. Then a matrix which is obtained by interchanging the rows and columns of A is called the transpose of A. It is denoted by A' or A^T.*

That is , if $A = [a_{ij}]_{m \times n}$, then $A' = [a_{ji}]_{n \times m}$.

For example : If $A = \begin{bmatrix} 2 & 3 & 5 \\ 1 & 6 & 7 \end{bmatrix}_{2 \times 3}$, then its transpose is $A' = \begin{bmatrix} 2 & 3 & 5 \\ 1 & 6 & 7 \end{bmatrix}' = \begin{bmatrix} 2 & 1 \\ 3 & 6 \\ 5 & 7 \end{bmatrix}_{3 \times 2}$

Definition 2. *Transpose of the matrix of cofactors is called adjoint of the given matrix.*

REMARKS
- Transpose of row matrix is a column matrix and transpose of a column matrix is a row matrix.
- If a matrix is square then its transpose will be a square matrix of same order.

1.8.1 PROPERTIES OF TRANSPOSE OF A MATRIX

THEOREM 1. *If A' and B' are the transpose of the matrix A and B respectively, then :*

(i) $(A')' = A$

(ii) $(A+B)' = A' + B'$, *(here A and B must be of same order).*

(iii) $(kA)' = kA'$, *here k is any scalar.*

(iv) $(AB)' = B'A'$, *here AB and B'A' are conformable for multiplication.*

Proof.

(i) Let $A = [a_{ij}]_{m \times n}$, then $A' = [a_{ji}]_{n \times m}$

since (i, j)th element in $(A')' = (j, i)$th element in $A' = (i, j)$th element in A

Thus by the definition of equality of matrices, we must have $(A')' = A$.

(ii) Let $A = [a_{ij}]_{m \times n}$, $B = [b_{ij}]_{m \times n}$. So, $A' = [a_{ji}]_{n \times m}$ *and* $B' = [b_{ji}]_{n \times m}$, then

(i, j)th element in $(A+B)' = (j, i)$th element in $(A+B)$

$= (j, i)$th element in $A + (j, i)$th element in B

$= (i, j)$th element in $A' + (i, j)$th element in B'

$= (i, j)$th element in $(A'+B')$

Thus by the definition of equality of matrices, we get

$(A+B)' = A' + B'$

(iii) Let $A = [a_{ij}]_{m \times n}$ so that $A' = [a_{ji}]_{n \times m}$ and k be a scalar, then

(i, j)th element in $(kA)' = (j, i)$th element in $(kA) = (i, j)$th element in kA'

Thus by the defintion of equality of matrices, we get

$(kA)' = kA'$

(iv) Let $A = [a_{ij}]_{m \times n}$ and $B = [b_{ij}]_{n \times p}$ then AB is conformable for multiplication and having the order $m \times p$. Therefore, the order of $(AB)'$ is $p \times m$. Since the orders of A' and B' are respectively $n \times m$ and $p \times n$ so $B'A'$ is conformable for multiplication and having the order $p \times m$.

Now (k, i)th element in $(AB)' = (i, k)$th element in $AB = \sum_{j=1}^{n} a_{ij} b_{jk}$

(By the definition of multiplication of matrices)

But (k, i)th element in $B'A' = \sum_{j=1}^{n} b_{jk} b_{ij} = \sum_{j=1}^{n} a_{ij} b_{jk} = (i, k)$th elemetn in AB

∴ (k, i)th element in $(AB)' = (k, i)$th element in $B'A'$

Thus by the definition of equality of matrices, we must have $(AB)' = B'A'$

1.9 SYMMETRIC AND SKEW-SYMMETRIC MATRIX

Definition: *A matrix A is said to be a symmetric matrix if A' = A, that is, the transpose of a matrix is equal to the matrix itself.*

For exmaple : If $A = \begin{bmatrix} 1 & 2 & 3 \\ 2 & 4 & 5 \\ 3 & 5 & 6 \end{bmatrix}$, then $A' = \begin{bmatrix} 1 & 2 & 3 \\ 2 & 4 & 5 \\ 3 & 5 & 6 \end{bmatrix}$ so that $A' = A$

Hence, *A* is symmetric.

Definition: *A matrix 'A' is said to be a skew-symmetric matrix if A' = –A.*

For exmaple : If $A = \begin{bmatrix} 0 & 2 & 3 \\ -2 & 0 & 4 \\ -3 & -4 & 0 \end{bmatrix}$, then $A' = \begin{bmatrix} 0 & -2 & -3 \\ 2 & 0 & -4 \\ 3 & 4 & 0 \end{bmatrix} = \begin{bmatrix} 0 & 2 & 3 \\ -2 & 0 & 4 \\ 3 & -4 & 0 \end{bmatrix} = -A$

Hence *A* is skew-symmetric matrix.

REMARK
- The diagonal elements of a skew-symmetric matrix are all zero.

1.9.1 PROPERTIES OF SYMMETRIC AND SKEW-SYMMETRIC MATRIX

(i) *If A is a symmetric (skew-symmetric) matrix, then kA is symmetric (skew-symmetric) matrix, where k is any scalar.*

Proof. Let *A* be a symmetric matrix, then $A' = A$, since we have

$$(kA)' = kA' = kA \qquad (\because A' = A)$$
$$\Rightarrow \qquad (kA)' = (kA)$$
$\Rightarrow kA$ is symmetric matrix.

Also, if *A* is a skew-symmetric matrix, then $A' = -A.$

Since we have
$$(kA)' = kA' = k(-A) \qquad (\because A' = -A)$$
$$= -(kA)$$
$$\therefore \qquad (kA') = -(kA)$$
$\Rightarrow kA$ is skew-symmetric matrix.

(ii) *If A and B are symmetric (skew-symmetric) matrices then A+ B is symmetric (skew- symmetric) matrix.*

Proof. Let *A* and *B* be symmetric matrices, then $A' = A$ and $B' = B$.

Since we have
$$(A+B)' = A' + B'$$
$$\Rightarrow \qquad (A+B)' = (A+B) \qquad (\because A' = A, B' = A)$$
$\Rightarrow A+B$ is symmetric.

Similarly, if *A* and *B* are skew-symmetric matrices, then $A' = -A$ and $B' = -B$.

But
$$(A+B)' = A' + B'$$
$$\Rightarrow \qquad (A+B)' = -A - B \qquad (\because A' = -A, B' = -B)$$
$$\Rightarrow \qquad (A+B)' = -(A+B)$$
$\Rightarrow A+B$ is skew-symmetric matrix.

(iii) *If A is any matrix, then AA' and A'A both are symmetric matrices.*

Proof. If
$$(AA')' = (A')'A' \qquad [\because (AB)' = B'A']$$
$$= AA'$$
$\Rightarrow AA'$ is symmetric.

Also
$$(A'A)' = A'(A')' \qquad [\because (AB)' = B'A']$$
$$= A'A \qquad [\because (A')' = A]$$
$\Rightarrow AA'$ is symmetric.

(iv) *If A is any square matrix, then A + A' is symmetric and A – A' is skew-symmetric.*

Proof. We have
$$(A+A')' = A' + (A')' \qquad [\because (A+B)' = A' + B']$$
$$= A' + A \qquad [\because (A')' = A]$$
$$= A + A' \qquad \text{[Matrix addition is commutative.]}$$
$\Rightarrow A + A'$ is symmetric.

Similarly,
$$(A-A')' = A' - (A')'$$
$$= A' - A = -(A - A')$$
$\Rightarrow A - A'$ is skew-symmetric.

(v) *All positive integral powers of a symmetric matrix are symmetric.*

Proof. Let *A* be a symmetric matrix and *m* be any positive integer, then

$$A^m = A . A . A . \ldots . A \ (m \text{ times})$$

$$\Rightarrow \qquad (A^m)' = [A \, . \, A \, . \, A \, . \, ... \, A \text{ (}m\text{ times)}]'$$
$$= A' \, . \, A' \, . \, A' \, ... \, A' \text{ (}m\text{ times)} \quad (\because A' = A)$$
$$= A \, . \, A \, . \, A \, . \, ... \, A \text{ (}m\text{ times)} = A^m$$

$\Rightarrow \quad A^m$ is symmetric matrix.

1.10 COMPLEX MATRIX

A matrix A is said to be *complex matrix* if it contains some of its elements equal to a complex number.

1.10.1 CONJUGATE OF A COMPLEX MATRIX

Let A be a complex matrix, then a matrix which is obtained by replacing all the complex elements of A by their complex conjugate number, is called conjugate of a matrix. It is denoted by A.

For example: If $A = \begin{bmatrix} 1+2i & 3i & 6 \\ 7 & 2+4i & 1+i \end{bmatrix}$, then $\bar{A} = \begin{bmatrix} 1-2i & -3i & 6 \\ 7 & 2-4i & 1-i \end{bmatrix}$

1.10.2 TRANSPOSE CONJUGATE OF A MATRIX

The transpose of the conjugate matrix is called the transposed conjugate of that matrix. It is denoted by A^θ.

Therefore, we must have $(A^\theta) = (\bar{A})'$.

1.10.3 PROPERTIES OF TRANSPOSE CONJUGATE OF MATRIX

(1) $(A^\theta)^\theta = A$

(2) $(A + B)^\theta = A^\theta + B^\theta$, A, B being of same order

(3) $(kA)^\theta = \bar{k}A^\theta$, k being a complex number

(4) $(AB)^\theta = B^\theta A^\theta$, A, B being conformable for multiplication.

1.10.4 HERMITIAN AND SKEW-HERMITIAN MATRICES

(1) Hermitian matrix : A matrix A is said to be Hermitian if $A^\theta = A$.

(2) Skew-Hermitian matrix : A matrix A is said to be skew-Hermitian if $A^\theta = -A$.

1.10.5 ORTHOGONAL AND UNITARY MATRICES

(i) Orthogonal matrix : A matrix A is said to be orthogonal if $A'A = I$, where I is a unit matrix of order same as of order A.

(ii) Unitary Matrix : A square matrix A is said to be unitary if $A^\theta A = I$.

For example : If $A = \begin{bmatrix} \cos\theta & \sin\theta \\ -\sin\theta & \cos\theta \end{bmatrix}$, then $A' = \begin{bmatrix} \cos\theta & -\sin\theta \\ \sin\theta & \cos\theta \end{bmatrix}$

$$\Rightarrow \qquad A'A = \begin{bmatrix} \cos\theta & -\sin\theta \\ \sin\theta & \cos\theta \end{bmatrix}\begin{bmatrix} \cos\theta & \sin\theta \\ -\sin\theta & \cos\theta \end{bmatrix}$$
$$= \begin{bmatrix} \cos^2\theta + \sin^2\theta & 0 \\ 0 & \sin^2\theta + \cos^2\theta \end{bmatrix} = \begin{bmatrix} 1 & 0 \\ 0 & 1 \end{bmatrix} = I$$

Hence, A is orthogonal matrix.

For example : If $A = \begin{bmatrix} 0 & -i \\ i & 0 \end{bmatrix}$, then $\bar{A} = \begin{bmatrix} 0 & i \\ -i & 0 \end{bmatrix}$

Now $\qquad\qquad A^\theta = (\bar{A})' = \begin{bmatrix} 0 & i \\ i & 0 \end{bmatrix}$

Then $\qquad\qquad A^\theta A = \begin{bmatrix} 0 & -i \\ i & 0 \end{bmatrix}\begin{bmatrix} 0 & -i \\ i & 0 \end{bmatrix} = \begin{bmatrix} 1 & 0 \\ 0 & 1 \end{bmatrix} \qquad (\because i^2 = -1)$
$$= I$$

$\Rightarrow \quad A$ is unitary.

1.10.6 PROPERTIES OF HERMITIAN AND SKEW-HERMITIAN MATRICES

(1) If A is Hermitian (skew-Hermitian) matrix, then iA is Hermitian (skew-Hermitian) matrix.

(2) If A, B are Hermitian or skew-Hermitian, then $A + B$ is Hermitian or skew-Hermitian.

(3) If A is Hermitian or skew-Hermitian, then $\bar{A}$ is Hermitian or skew-Hermitian.

Solved Examples

EXAMPLE 1. *If A be any square matrix, prove that $A + A^\theta$, AA^θ, $A^\theta A$ are all Hermitian and $A - A^\theta$ is skew-Hermitian.*

SOLUTION. (i) $(A+A^\theta)^\theta = A^\theta + (A^\theta)^\theta$

$\qquad\qquad\qquad\qquad [\because (A+B)^\theta = A^\theta + B^\theta]$

$\qquad\qquad = A^\theta + A \qquad\qquad [\because (A^\theta)^\theta = A]$

$\qquad\qquad = A + A^\theta$

$\Rightarrow A + A^\theta$ is Hermitian.

(ii) $(AA^\theta)^\theta = (A^\theta)^\theta A^\theta \qquad [\because (AB)^\theta = B^\theta A^\theta]$

$\qquad\qquad = AA^\theta \qquad\qquad [\because (A^\theta)^\theta = A]$

$\Rightarrow AA^\theta$ is Hermitian.

(iii) $(A^\theta A)^\theta = A^\theta (A^\theta)^\theta = A^\theta A$

$\Rightarrow A^\theta A$ is Hermitian.

(iv) $(A-A^\theta) = A^\theta - (A^\theta)^\theta$

$\qquad\qquad = A^\theta - A = -(A - A^\theta)$

$\Rightarrow A - A^\theta$ is skew-Hermitian.

EXAMPLE 2. *Prove that the matrix* $\begin{bmatrix} \dfrac{1+i}{2} & \dfrac{-1+i}{2} \\ \dfrac{1+i}{2} & \dfrac{1-i}{2} \end{bmatrix}$ *is unitary.*

SOLUTION. Let us suppose $A = \begin{bmatrix} \dfrac{1+i}{2} & \dfrac{-1+i}{2} \\ \dfrac{1+i}{2} & \dfrac{1-i}{2} \end{bmatrix}$

$\therefore \quad A' = \begin{bmatrix} \dfrac{1+i}{2} & \dfrac{1+i}{2} \\ \dfrac{-1+i}{2} & \dfrac{1-i}{2} \end{bmatrix}$

$A^\theta = \overline{A'} = \begin{bmatrix} \dfrac{1-i}{2} & \dfrac{1-i}{2} \\ \dfrac{-1-i}{2} & \dfrac{1+i}{2} \end{bmatrix}$

Now

$A^\theta A = \begin{bmatrix} \dfrac{1-i}{2} & \dfrac{1-i}{2} \\ \dfrac{-1-i}{2} & \dfrac{1-i}{2} \end{bmatrix}\begin{bmatrix} \dfrac{1+i}{2} & \dfrac{-1+i}{2} \\ \dfrac{1+i}{2} & \dfrac{1-i}{2} \end{bmatrix}$

$= \begin{bmatrix} \dfrac{1}{4}(1-i^2)+\dfrac{1}{4}(1-i^2) & -\dfrac{1}{4}(1-i)^2+\dfrac{1}{4}(1-i)^2 \\ -\dfrac{1}{4}(1+i)^2+\dfrac{1}{4}(1+i)^2 & \dfrac{1}{4}(1-i^2)+\dfrac{1}{4}(1-i)^2 \end{bmatrix}$

$= \begin{bmatrix} \dfrac{1}{4}(1+1)+\dfrac{1}{4}(1+1) & 0 \\ 0 & \dfrac{1}{4}(1+1)+\dfrac{1}{4}(1+1) \end{bmatrix}$

$= \begin{bmatrix} 1 & 0 \\ 0 & 1 \end{bmatrix} = I$

Hence A is unitary.

Exercise-1.2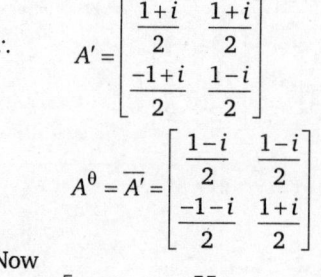

1. If $A = \begin{bmatrix} 1 & 0 \\ 2 & -1 \end{bmatrix}$, $B = \begin{bmatrix} 3 & 7 \\ 4 & 8 \end{bmatrix}$, $C = \begin{bmatrix} -1 & 1 \\ 0 & 0 \end{bmatrix}$,

then prove that

$A + (B+C) = (A+B) + C$.

2. If $A = \begin{bmatrix} 2 & 3 \\ 4 & -5 \end{bmatrix}$, $B = \begin{bmatrix} 8 & 9 \\ 6 & 7 \end{bmatrix}$, then find

(i) $2A + 3B$ \qquad (ii) $5A - 3B$

3. Find $A + B$ and show that $A+B = B+A$, when

(i) $A = \begin{bmatrix} 7 & 8 \\ 9 & 2 \\ 3 & 4 \end{bmatrix}$, $B = \begin{bmatrix} 2 & 3 \\ 4 & 5 \\ 6 & 7 \end{bmatrix}$

(ii) $A = \begin{bmatrix} 1 & 2 & -3 \\ 4 & 1 & 5 \\ -3 & -2 & 2 \end{bmatrix}$, $B = \begin{bmatrix} 3 & -1 & 2 \\ 4 & 2 & 5 \\ 2 & 0 & 3 \end{bmatrix}$

4. If $A = \begin{bmatrix} \cos\alpha & -\sin\alpha \\ \sin\alpha & \cos\alpha \end{bmatrix}$, $B = \begin{bmatrix} \cos\beta & -\sin\beta \\ \sin\beta & \cos\beta \end{bmatrix}$,

then show that $AB = BA$.

5. If $A = \begin{bmatrix} 1 & 1 & -1 \\ 2 & 0 & 3 \\ 3 & -1 & 2 \end{bmatrix}$, $B = \begin{bmatrix} 1 & 3 \\ 0 & 2 \\ -1 & 4 \end{bmatrix}$ and

$C = \begin{bmatrix} 1 & 2 & 3 & -4 \\ 2 & 0 & -2 & 1 \end{bmatrix}$, then find $A(BC)$,

$(AB)C$ and show that $A(BC) = (AB)C$.

6. (i) If $A = \begin{bmatrix} 1 & 2 \\ 3 & 4 \end{bmatrix}$, $B = \begin{bmatrix} 2 & 1 \\ 4 & 2 \end{bmatrix}$, $C = \begin{bmatrix} 5 & 1 \\ 7 & 4 \end{bmatrix}$,

then verify that :

$A(B+C) = AB + BC$.

(ii) For the matrices :

$A = \begin{bmatrix} 3 & 2 \\ 1 & 5 \end{bmatrix}$, $B = \begin{bmatrix} 2 & -4 \\ -3 & 1 \end{bmatrix}$, $C = \begin{bmatrix} 1 & 2 \\ 3 & -1 \end{bmatrix}$

Verify $(A+B)C = AC + BC$.

7. If $\begin{bmatrix} 4 & 1 & 2 \\ 0 & 5 & 3 \end{bmatrix}\begin{bmatrix} 3 & 4 & 5 \\ -1 & 0 & -2 \\ 3 & 4 & 7 \end{bmatrix} = \begin{bmatrix} 8x+3y & 6z & 32 \\ 4 & 12 & 26x-5y \end{bmatrix}$

find the values of x, y, z.

8. For what values of a, b, c, d, e the following matrices are same ?

$A = \begin{bmatrix} a & 1 & 2 \\ c & b & 3 \\ 1 & -1 & 0 \end{bmatrix}$, $B = \begin{bmatrix} 0 & e & 2 \\ 7 & 9 & d \\ e & -1 & a \end{bmatrix}$

9. If $A = \begin{bmatrix} 2 & 3 \\ 1 & -4 \end{bmatrix}$, $B = \begin{bmatrix} 1 & 0 \\ 0 & 1 \end{bmatrix}$, then find

(i) $10B$ \qquad\qquad (ii) $4A$

(iii) $A+B$ \qquad\qquad (iv) $A-B$

(v) $2A-3B$ \qquad\qquad (vi) $5A + 3B$.

10. Find the product of

(i) $A = \begin{bmatrix} 1 & 2 \\ 3 & 4 \end{bmatrix}$, $B = \begin{bmatrix} 1 & 7 \\ 2 & 3 \end{bmatrix}$

(ii) $A = \begin{bmatrix} 1 & 2 & 1 \\ 4 & 0 & 2 \end{bmatrix}$, $B = \begin{bmatrix} 3 & -4 \\ 1 & 5 \\ -2 & 2 \end{bmatrix}$

(iii) $A = \begin{bmatrix} x \\ y \\ z \end{bmatrix}$, $B = \begin{bmatrix} x & y & z \end{bmatrix}$

11. If $A = \begin{bmatrix} 1 & 0 & 0 \\ 0 & 1 & 0 \\ 0 & 0 & 1 \end{bmatrix}$, then prove that $A^3 = A$.

12. Find the number of rows and columns in the following matrices :

(i) $\begin{bmatrix} 2 & 6 & 7 & 8 \\ 2 & 5 & 11 & 6 \end{bmatrix}$

(ii) $\begin{bmatrix} 1 & 1 & 1 & 1 \end{bmatrix}$

(iii) $\begin{bmatrix} 2 & -1 & 3 \\ 0 & 3 & 4 \\ 2 & 3 & 7 \\ 2 & 5 & 11 \end{bmatrix}$

(iv) $\begin{bmatrix} x \\ y \\ z \end{bmatrix}$

13. Find the order of the following matrices :

(i) $\begin{bmatrix} 1 & 1 & 1 & 1 \end{bmatrix}$

(ii) $\begin{bmatrix} 1 & a & b & 0 \\ 0 & c & d & 1 \\ 1 & a & b & 0 \end{bmatrix}$

(iii) $\begin{bmatrix} 2 \\ 3 \\ 4 \\ 5 \end{bmatrix}$

(iv) $\begin{bmatrix} 2 & 3 \\ 3 & 4 \\ 5 & 6 \\ 7 & 8 \end{bmatrix}$

14. Are the following matrices conformable for addition ?

(i) $\begin{bmatrix} 1 & 2 & 3 \\ 4 & 5 & 6 \\ 7 & 8 & 9 \end{bmatrix}, \begin{bmatrix} 6 & 4 \\ 7 & 4 \\ 7 & 3 \end{bmatrix}$

(ii) $\begin{bmatrix} 5 & 6 \\ 7 & 8 \end{bmatrix}, \begin{bmatrix} 9 & 10 \\ 11 & 12 \end{bmatrix}$

(iii) $\begin{bmatrix} 3 & 2 & -1 \\ 0 & 4 & 0 \end{bmatrix}, \begin{bmatrix} 2 & 1 \\ 5 & 2 \\ 7 & 8 \end{bmatrix}$

(iv) $\begin{bmatrix} x \\ y \\ z \end{bmatrix}, \begin{bmatrix} a \\ b \\ c \end{bmatrix}$

15. Find the additive inverse of matrix :

$A = \begin{bmatrix} 1 & 2 & -7 & 5 \\ 0 & 5 & 0 & 8 \\ 0 & 0 & 0 & -8 \end{bmatrix}$

16. Are the following matrices conformable for the product AB ?

(i) $A = \begin{bmatrix} 5 & 7 \\ 8 & 9 \end{bmatrix}$, $B = \begin{bmatrix} 3 & 5 & 7 \\ 1 & 0 & 1 \end{bmatrix}$

(ii) $A = \begin{bmatrix} -1 & 2 \\ 3 & 4 \end{bmatrix}$, $B = \begin{bmatrix} 3 \\ 4 \end{bmatrix}$

(iii) $A = \begin{bmatrix} 2 & 1 & 0 \\ 3 & 2 & 1 \\ 1 & 0 & 1 \end{bmatrix}$, $B = \begin{bmatrix} 1 & 2 & 3 & 4 \\ 2 & 0 & 1 & 2 \\ 3 & 1 & 0 & 5 \end{bmatrix}$

(iv) $A = \begin{bmatrix} 1 & 1 & 1 & 1 \end{bmatrix}$, $B = \begin{bmatrix} 1 \\ 1 \\ 1 \\ 1 \end{bmatrix}$

17. Find the values of x and y if :

$\begin{bmatrix} x & 3 \\ 5 & x-y \end{bmatrix} = \begin{bmatrix} 2 & 3 \\ 5 & 1 \end{bmatrix}$

18. Find the transpose of the matrix :

$A = \begin{bmatrix} 1 & 3 & 5 \\ -2 & 5 & 6 \\ 7 & 0 & 3 \end{bmatrix}$.

19. If $A = \begin{bmatrix} 2 & 1 & 4 \\ 5 & -3 & 7 \end{bmatrix}$, $B = \begin{bmatrix} 1 & 2 & 3 \\ 2 & -4 & 6 \end{bmatrix}$

Find $2A - 3B$ and $B - \dfrac{A}{2}$. Also write down the unit matrix of order 3.

20. Prove that the matrix $\dfrac{1}{3}\begin{bmatrix} 1 & 2 & 2 \\ 2 & 1 & -2 \\ -2 & 2 & 1 \end{bmatrix}$ is orthogonal.

21. If $A = \begin{bmatrix} 1 & 2 & 1 \\ a & 0 & 4 \\ 1 & 1 & 1 \end{bmatrix}$ and $adj.\ adj(A) = A$, find a.   (MTU–2012)

22. If $A = \begin{bmatrix} 0 & 1+2i \\ -1+2i & 0 \end{bmatrix}$ obtain the matrix $(I-N)\ (I+N)^{-1}$ and show that it is unitary.   (GBTU–2011)

23. Show that the matrix $A = \begin{bmatrix} 2 & 3-4i \\ 3+4i & 2 \end{bmatrix}$ is Hermitian and iA is skew Hermitian.   (UKTU–2012)

24. Show that the matrix $A = \begin{bmatrix} i & 0 & 0 \\ 0 & 0 & i \\ 0 & i & 0 \end{bmatrix}$ is skew Hermition.   (MTU–2012)

25. Express the Hermitian matrix $A = \begin{bmatrix} 1 & -i & 1+i \\ i & 0 & 2-3i \\ 1-i & 2+3i & 2 \end{bmatrix}$ and $P+iQ$ where P is real symmetric and Q is real skew-symmetric matrix.   (GBTU–2012, 14)

Answers

2. (i) $\begin{bmatrix} 28 & 33 \\ 26 & 11 \end{bmatrix}$ (ii) $\begin{bmatrix} -14 & -12 \\ 2 & -46 \end{bmatrix}$ **3.** (i) $\begin{bmatrix} 9 & 11 \\ 13 & 7 \\ 9 & 11 \end{bmatrix}$ (ii) $\begin{bmatrix} 4 & 1 & -1 \\ 8 & 3 & 10 \\ -1 & -2 & 5 \end{bmatrix}$ **7.** $x=1, y=3, z=4$ **8.** $a=0, b=9, c=7, d=3, e=1$

9. (i) $\begin{bmatrix} 10 & 0 \\ 0 & 10 \end{bmatrix}$ (ii) $\begin{bmatrix} 8 & 12 \\ 4 & -16 \end{bmatrix}$ (iii) $\begin{bmatrix} 3 & 3 \\ 1 & -3 \end{bmatrix}$ (iv) $\begin{bmatrix} 1 & 3 \\ 1 & -5 \end{bmatrix}$ (v) $\begin{bmatrix} 1 & 6 \\ 2 & -11 \end{bmatrix}$ (vi) $\begin{bmatrix} 13 & 15 \\ 5 & -17 \end{bmatrix}$ **10.** (i) $\begin{bmatrix} 5 & 13 \\ 11 & 33 \end{bmatrix}$ (ii) $\begin{bmatrix} 3 & 8 \\ 8 & -12 \end{bmatrix}$ (iii) $\begin{bmatrix} x^2 & xy & xz \\ yx & y^2 & yz \\ \delta x & zy & z^2 \end{bmatrix}$

12. (i) Rows = 2, Columns = 4, (ii)Row = 1, Columns = 4 (iii) Rows =4, Columns = 3 (ii)Rows = 3, Column = 1.

13. (i) 1×4 (ii) 3×4 (iii) 4×1 (iv) 4×2 **14.** (i) No (ii) Yes (iii) No (iv) Yes

15. $\begin{bmatrix} -1 & -2 & 7 & -5 \\ 0 & -5 & 0 & -8 \\ 0 & 0 & 0 & 8 \end{bmatrix}$ **16.** (i) Yes (ii) Yes (iii) Yes (iv) Yes

17. $x = 2, y = 1$ **18.** $\begin{bmatrix} 1 & -2 & 7 \\ 3 & 5 & 0 \\ 5 & 6 & 3 \end{bmatrix}$ **19.** $\begin{bmatrix} 1 & -4 & -1 \\ 4 & 6 & -4 \end{bmatrix}$, $\begin{bmatrix} 0 & \frac{3}{2} & 1 \\ -\frac{1}{2} & -\frac{5}{3} & -\frac{5}{2} \end{bmatrix}$, $\begin{bmatrix} 1 & 0 & 0 \\ 0 & 1 & 0 \\ 0 & 0 & 1 \end{bmatrix}$ **21.** 3 **22.** $\frac{1}{6}\begin{bmatrix} -4 & -2-4i \\ 2-4i & -4 \end{bmatrix}$

1.11 SUBMATRIX OF A MATRIX

Let A be a matrix of order $m \times n$, then a matrix obtained from A by removing some rows and columns, is called a submatrix of the matrix A.

For Example :

(i) Consider a matrix $A = \begin{bmatrix} 1 & 2 & 3 & 6 \\ 5 & 7 & 9 & 9 \\ 4 & 5 & 6 & 12 \end{bmatrix}$ of order 3×4, then a matrix $\begin{bmatrix} 1 & 2 & 3 \\ 5 & 7 & 9 \end{bmatrix}$ is a submatrix of A, which is obtained from A

by removing third row and fourth column.

(ii) Consider a matrix $A = \begin{bmatrix} 1 & 5 & 2 \\ 0 & 1 & 3 \\ 0 & 0 & 1 \end{bmatrix}$ of order 3×3, then a matrix $\begin{bmatrix} 1 & 5 \\ 0 & 1 \\ 0 & 0 \end{bmatrix}$ is submatrix of A, which is obtained from A by

removing third column.

✎ REMARK
- If the given matrix A is a square matrix, then a square submatrix of A is known as principal submatrix.

1.11.1 MINORS OF A MATRIX

Let A be a matrix of order $m \times n$, then the determinant of every square submatrix of A is called a minor of A. If the order of the determinant of square submatrix of A is $r \times r$, then it is denoted as r-minor of A.

For example : (i) Consider a matrix $A = \begin{bmatrix} 1 & 2 & 3 \\ 2 & 4 & 6 \end{bmatrix}$. Then all the 2-minors of A are $\begin{vmatrix} 1 & 2 \\ 2 & 4 \end{vmatrix}, \begin{vmatrix} 2 & 3 \\ 4 & 6 \end{vmatrix}, \begin{vmatrix} 1 & 3 \\ 2 & 6 \end{vmatrix}$

(ii) Consider a matrix $A = \begin{bmatrix} 1 & 2 & 3 \\ 5 & 7 & 9 \\ 4 & 5 & 6 \\ 6 & 9 & 12 \end{bmatrix}$.

Then all the 3-minors of A are $\begin{vmatrix} 1 & 2 & 3 \\ 5 & 7 & 9 \\ 4 & 5 & 6 \end{vmatrix}, \begin{vmatrix} 1 & 2 & 3 \\ 5 & 7 & 9 \\ 6 & 9 & 12 \end{vmatrix}, \begin{vmatrix} 1 & 2 & 3 \\ 4 & 5 & 6 \\ 6 & 9 & 12 \end{vmatrix}, \begin{vmatrix} 5 & 7 & 9 \\ 4 & 5 & 6 \\ 6 & 9 & 12 \end{vmatrix}$ (PTU–2005, 06)

1.12 RANK OF A MATRIX

Let A be a matrix of order $m \times n$, then a non-negative integer r is said to the rank of matrix A if it possesses the following two properties :

(i) There exists at least one r-minor of A which is not equal to zero.

(ii) Every s-minor of A for all $s > r$ is zero.

We denote the rank of A by $\rho(A)$.

In other words, the rank of a matrix is the order of any highest order of a non-zero minor of the matrix.

✎ REMARKS
- If the order of a matrix A is $m \times n$, then $\rho(A) \le$ min. $\{m, n\}$
- A is a null matrix iff $\rho(A) = 0$.
- If A is any non-zero matrix, then $\rho(A) \ge 1$.
- $\rho(A) \ge r$, if there exists a non-zero r-minor of A.
- For any square matrix A of order n, $\rho(A) = n$ iff A is non-singular.
- For any square matrix A of order n, $\rho(A) < n$ iff A is singular.
- $r(A) \le r$ if every s-minor of A is zero, where $s > r$.
- Every $(r+1)$- rowed minor of A can be expressed as a linear combination of its r-rowed minors, therefore if every r-minor of A is zero, then its every $(r+1)$-minor is also zero.

1.12.1 Echelon Form of a Matrix

A matrix A is said to be in Echelon form if :
(i) every row of A has all its entries 0 (zero) which occurs below every row having a non-zero entry and
(ii) the number of zeros before the first non-zero entry in a row is less than the number of such zeros in the next row.

REMARK

- The rank of a matrix is equal to the number of non-zero rows in Echelon form of that matrix.
  For example: Consider a matrix

$$A = \begin{bmatrix} 0 & 2 & 3 & 5 \\ 0 & 0 & 3 & 2 \\ 0 & 0 & 0 & 0 \end{bmatrix}$$

Clearly, A is in Echelon form which has 2 non-zero rows, hence the rank of A is 2.

THEOREM 1. *The rank of the transpose of a matrix is equal to the rank of that matrix.*

PROOF. Let A be a marix, then A' is its transpose and let $\rho(A) = r$, then there exists an r-rowed minor of A which is not equal to zero and all s-rowed minors of A are zero, where $s > r$. Let $|B|$ be a r-rowed minor of A such that $|B| \neq 0$. Since A' is the transpose of A, then $|B'|$ is the r-rowed minor of A' but $|B'| = |B| \neq 0$, therefore $\rho(A') \geq r$. Suppose there is an s-minor $|C|$ of A' such that $|C| \neq 0$, where $s > r$, then $|C'|$ will be an s-minor of A such that $|C'| = |C| \neq 0$, therefore $\rho(A) > r$ which is a contradiction, hence $\rho(A') = r$.

Solved Examples

EXAMPLE 1. *Find the rank of the following matrices :*

(i) $\begin{bmatrix} 3 & 0 & 0 \end{bmatrix}$ (ii) $\begin{bmatrix} 1 & 2 & 3 \\ 2 & 4 & 5 \end{bmatrix}$

(iii) $\begin{bmatrix} 1 & 2 & 3 \\ 3 & 4 & 5 \\ 4 & 5 & 6 \end{bmatrix}$ (iv) $\begin{bmatrix} 1 & 5 & 2 & 4 \\ 0 & 1 & 3 & 1 \\ 0 & 0 & 1 & 3 \end{bmatrix}$

SOLUTION. (i) Let $A = \begin{bmatrix} 3 & 0 & 0 \end{bmatrix}$, then A is the non-zero rowed matrix, thereofore $\rho(A) \geq 1$. Also A is a matrix of order of 1×3, then $\rho(A) \leq 1$, hence $\rho(A) = 1$.

(ii) Let $A = \begin{bmatrix} 1 & 2 & 3 \\ 2 & 4 & 5 \end{bmatrix}$

The order of A is 2×3, then $\rho(A) \leq 2$.

Also there is a 2-minor $\begin{vmatrix} 2 & 3 \\ 4 & 5 \end{vmatrix}$ of A which is not equal to zero, then $\rho(A) \geq 2$, hence $\rho(A) = 2$.

(iii) Let $A = \begin{bmatrix} 1 & 2 & 3 \\ 3 & 4 & 5 \\ 4 & 5 & 6 \end{bmatrix}$. The order of A is 3×3, then $\rho(A) \leq 3$.

Now,

$$|A| = \begin{vmatrix} 1 & 2 & 3 \\ 3 & 4 & 5 \\ 4 & 5 & 6 \end{vmatrix}$$

$$= 1(24 - 25) - 2(18 - 20) + 3(15 - 16) = 0$$

∴ The only 3-minor $|A|$ of A is zero, thus $\rho(A) <$

3. Further, there is a 2-minor $\begin{vmatrix} 1 & 2 \\ 3 & 4 \end{vmatrix}$ of A which is not equal to zero, hence $\rho(A) = 2$.

(iv) Let $A = \begin{bmatrix} 1 & 5 & 2 & 4 \\ 0 & 1 & 3 & 1 \\ 0 & 0 & 1 & 3 \end{bmatrix}$

The order of A is 3×4, then $\rho(A) \leq 3$.

Now there is a 3-minor $\begin{vmatrix} 1 & 5 & 2 \\ 0 & 1 & 3 \\ 0 & 0 & 1 \end{vmatrix}$ of A which is not equal to zero, then $\rho(A) \geq 3$.

Hence $\rho(A) = 3$.

EXAMPLE 2. *Let A and B be two square matrices of order n. If $\rho(A) = \rho(B) = n$, then prove that $\rho(AB) = n$ and conversely.*

SOLUTION. Suppose $\rho(A) = \rho(B) = n$, then both A and B are non-singular.

∴ $|A| \neq 0$ and $|B| \neq 0$

⟹ $|AB| = |A||B| \neq 0$

Since the order of AB is n and $|AB| \neq 0$, therefore
$$\rho(AB) = n.$$

Conversely, suppose that
$$\rho(AB) = n$$

∴ $|AB| \neq 0$ ⟹ $|A||B| \neq 0$

⟹ $|A| \neq 0$ and $|B| \neq 0$

⟹ $\rho(A) = n, \rho(B) = n.$

EXAMPLE 3. *Prove that every skew-symmetric matrix of odd order has rank less than its order.*

SOLUTION. Let A be a skew-symmetric matrix of order n, where n is an odd natural number, then
$$A' = -A.$$

Now $A' = -A$ ⟹ $|A'| = |-A|$

⟹ $|A| = |(-1)A|$ [∵ $|A'| = |A|$]

⟹ $|A| = (-1)^n |A|$

⟹ $|A| = -|A|$ [∵ n is odd]

$$\Rightarrow \qquad 2|A| = 0$$
$$\Rightarrow \qquad |A| = 0$$
$$\Rightarrow \qquad \rho(A) \neq n$$

But $\rho(A) \leq n$, hence $\rho(A) < n$.

EXAMPLE 4. *If A be a non-zero column and B is a non-zero row matrix, then show that $\rho(AB)=1$.*

SOLUTION. Let $A = \begin{bmatrix} a_{11} \\ a_{21} \\ \vdots \\ a_{m1} \end{bmatrix}$ and $B = \begin{bmatrix} b_{11} & b_{12} & \cdots & b_{1n} \end{bmatrix}$

be two non-zero column and row matrices respectively.

Then we have $AB = \begin{bmatrix} a_{11} \\ a_{21} \\ \vdots \\ a_{m1} \end{bmatrix} \begin{bmatrix} b_{11} & b_{12} & \cdots & b_{1n} \end{bmatrix}$

$$= \begin{bmatrix} a_{11}b_{11} & a_{11}b_{12} & \cdots & a_{11}b_{1n} \\ a_{21}b_{11} & a_{21}b_{12} & \cdots & a_{21}b_{1n} \\ \vdots & \vdots & \vdots & \vdots \\ a_{m1}b_{11} & a_{m1}b_{12} & \cdots & a_{m1}b_{1n} \end{bmatrix}$$

Clearly, AB is a matrix of order $m \times n$ and AB is a non-zero matrix since A and B are non-zero matrices, then
$$\rho(AB) \geq 1 \qquad \qquad ..(1)$$

also every 2-minor of AB vanishes, then
$$\rho(AB) \leq 1 \qquad \qquad ...(2)$$
From (1) and (2) we have
$$\rho(AB) = 1.$$

EXAMPLE 5. *If A is n-rowed square matrix of rank n −1, then show that adj A is non-zero matrix.*

SOLUTION. Since the rank of A is $n–1$, i.e., $\rho(A) = n–1$, then there exists a non-zero $(n–1)$ minor of A, therefore there exists at least one element of *adj A* which is non-zero, hence *adj. A* is a non-zero matrix.

EXAMPLE 6. *Let A be a square matrix of order n. Show that r(adj. A) is n or 0 in accordance with r(A) is n or less than n–1.*

SOLUTION. Suppose $\rho(A) = n$, then $|A| \neq 0$
But we know that
$$|adj. A| = |A|^{n-1}$$
$$\Rightarrow \qquad |adj. A| \neq 0 \qquad [\because |A| \neq 0]$$
$$\therefore \qquad \rho(adj. A) = n,$$
since the order of *adj. A* = n.

Next, when $\rho(A) < n –1$, then $|A| = 0$ and every r-minor of A is zero, where $r \geq n–1$, therefore every element of *adj. A* is zero so that *adj. A* is a null matrix, hence
$$\rho(adj. A) = 0.$$

Exercise-1.3

1. Find the rank of the following matrices :

(i) $\begin{bmatrix} 0 & 0 \\ 0 & 0 \end{bmatrix}$

(ii) $\begin{bmatrix} 5 & 10 \\ 3 & 6 \end{bmatrix}$

(iii) $\begin{bmatrix} 1 & -3 & 4 & 7 \\ 9 & 1 & 2 & 0 \end{bmatrix}$

(iv) $\begin{bmatrix} 1 & 2 & 3 \\ 2 & 1 & 0 \\ 0 & 1 & 2 \end{bmatrix}$

(v) $\begin{bmatrix} 1 & 2 & -7 & 5 \\ 0 & 5 & 0 & 8 \\ 0 & 0 & 0 & -3 \end{bmatrix}$

(vi) $\begin{bmatrix} 0 & 1 & 2 & 1 \\ 1 & 2 & 3 & 2 \\ 3 & 1 & 1 & 3 \end{bmatrix}$

(vii) $\begin{bmatrix} 1 & 2 & 3 & 4 \\ 2 & 4 & 6 & 8 \\ 3 & 6 & 9 & 12 \end{bmatrix}$

(viii) $\begin{bmatrix} 1 & 5 & 4 & 6 \\ 2 & 7 & 5 & 9 \\ 3 & 9 & 6 & 12 \end{bmatrix}$

(ix) $\begin{bmatrix} 1 & x & x^2 \\ 1 & y & y^2 \\ 1 & z & z^2 \end{bmatrix}$

(x) $\begin{bmatrix} 1 & 1 & 1 & 1 \\ 1 & 1 & 1 & 1 \\ 1 & 1 & 1 & 1 \\ 1 & 1 & 1 & 1 \end{bmatrix}$

(xi) $\begin{bmatrix} 1 & 0 & 0 & 0 \\ 0 & 1 & 0 & 0 \\ 0 & 0 & 1 & 0 \\ 0 & 0 & 0 & 1 \end{bmatrix}$

2. If $A = \begin{bmatrix} 0 & 1 & 0 & 0 \\ 0 & 0 & 1 & 0 \\ 0 & 0 & 0 & 1 \\ 0 & 0 & 0 & 0 \end{bmatrix}$,

find $\rho(A)$, $\rho(A^2)$, $\rho(A^3)$ and $\rho(A^4)$.

3. Show that the rank of a matrix does not alter on affixing any number of additional rows or columns of zeros.

4. Show that the rank of a matrix is greater than or equal to the rank of its every submatrix.

5. Find the rank of A, B, $A+B$ and AB, where

$$A = \begin{bmatrix} 1 & 1 & -1 \\ 2 & -3 & 4 \\ 3 & -2 & 3 \end{bmatrix} \text{ and } B = \begin{bmatrix} -1 & -2 & -1 \\ 6 & 12 & 6 \\ 5 & 10 & 5 \end{bmatrix}$$

Answers

1. (i) 0   (ii) 1   (iii) 2   (iv) 2   (v) 3   (vi) 3   (vii) 1   (viii) 2   (ix) Rank =3

if $x \neq y \neq z$; Rank =2 if only two of x, y, z are different; Rank = 1 if $x = y = z$. (x) 1   (xi) 4

2. $\rho(A) = 3$, $\rho(A^2) = 2$, $\rho(A^3) = 1$, $\rho(A^4) = 0$ **5.** $\rho(A) = 2$, $\rho(B) = 1$, $\rho(A+B) = 2$, $\rho(AB) = 0$

1.13 ELEMENTARY TRANSFORMATIONS (OR E-TRANSFORMATIONS) OF A MATRIX

Consider the matrices (UPTU 2007)

$$A = \begin{bmatrix} 1 & 2 & -3 \\ 3 & 0 & 1 \end{bmatrix}, B = \begin{bmatrix} 3 & 0 & 1 \\ 1 & 2 & -3 \end{bmatrix}, C = \begin{bmatrix} -3 & 2 & 1 \\ 1 & 0 & 3 \end{bmatrix}, D = \begin{bmatrix} 4 & 8 & -12 \\ 3 & 0 & 1 \end{bmatrix}$$

$$E = \begin{bmatrix} -3 & 2 & 7 \\ 1 & 0 & 21 \end{bmatrix}, F = \begin{bmatrix} 1 & 2 & -3 \\ 6 & 6 & -8 \end{bmatrix} \text{ and } G = \begin{bmatrix} 7 & 0 & 1 \\ -11 & 0 & -3 \end{bmatrix}$$

From above matrices, we observe that :

(1) B can be obtained from A by interchanging the first and second row.

(2) C can be obtained from A by interchanging the first and third column.

(3) D can be obtained from A by multiplying each element of the first row by 4.

(4) E can be obtained from C by multiplying each element of third column by 7.

(5) F can be obtained from A by adding to the elements of second row, 3 times the corresponding elements of the first row.

(6) G can be obtained from B by adding to the elements of first column, 4 times the corresponding elements of the third column.

Such transformations as performed above are known as elementary transformations (or E-operations or E-transformations).

Elementary transformations on rows are known as elementary row transformations whereas the transformations on columns are known as elementary column transformations.

Thus we may define E-transformations as follows:

Definition: *An elementary transformation (or E-transformation) is an operation of any one of the following types*:

(i) *The interchange of any two rows (or columns).*

(ii) *The multiplication of any row (or column) by any non-zero number.*

(iii) *The addition of non-zero scalar multiple of any row (or column) to another row (or column).*

1.13.1 NOTATIONS FOR E-TRANSFORMATIONS

E-transformations can be denoted by the following notations:

(i) The transformation of interchanging i^{th} and j^{th} row of a matrix is denoted by $R_i \leftrightarrow R_j$.

(ii) The transformation of interchanging i^{th} and j^{th} column is denoted by $C_i \leftrightarrow C_j$.

(iii) The transformation of multiplication of ith row of a matrix by non-zero scalar k is denoted by $R_i \rightarrow kR_i$.

(iv) The transformation of multiplication of jth column by a non-zero scalar k is denoted by $C_j \rightarrow kC_j$.

(v) The transformation of addition of a non-zero scalar k multiple of jth row to another i^{th} row of a matrix is denoted by $R_i \rightarrow R_i + kR_j \ (i \neq j)$.

(vi) The transformation of addition of a non-zero scalar k multiple of jth column to another i^{th} column of a matrix is denoted by $C_i \rightarrow C_i + kC_j \ (i \neq j)$.

1.13.2 ELEMENTARY MATRICES

A matrix which is obtained from a unit (identity) matrix by a single E-transformation is known as an elementary matrix.

For example : Consider the matrices

$$\begin{bmatrix} 0 & 0 & 1 \\ 0 & 1 & 0 \\ 1 & 0 & 0 \end{bmatrix}, \begin{bmatrix} 1 & 0 & 0 \\ 0 & 1 & 0 \\ 3 & 0 & 1 \end{bmatrix}, \begin{bmatrix} 1 & 0 & 0 \\ 0 & 3 & 0 \\ 0 & 0 & 1 \end{bmatrix}$$

Clearly, these matrices are elementary matrices because these are obtained from a unit matrix I_3 (the identity matrix of order 3×3) by performing the E-transformations $C_1 \rightarrow C_3, R_3 \rightarrow R_3 + 3R_1$ and $R_2 \rightarrow 3R_2$.

The elementary matrices of different types can be denoted by the following notations:

(i) The elementary matrix obtained by interchanging i^{th} and j^{th} rows (or columns) of a unit matrix is denoted by E_{ij}.

(ii) The elementary matrix obtained by multiplying i^{th} row (or column) of a unit matrix by a non-zero scalar k is denoted by $E_i(k)$.

(iii) The elementary matrix obtained by adding a non-zero scalar k multiple of jth row (or column) to i^{th} row (or column) of a unit matrix is denoted by $E_{ij}(k)$.

Obviously, $\qquad |E_{ij}| = -1, \ |E_i(k)| = k \neq 0 \text{ and } |E_{ij}(k)| = 1$

Hence we can say that all the elementary matrices are non-singular and hence they possess their inverse.

THEOREM 1. *Every E-row (column) transformation of a matrix can be obtained by pre-multiplication (post-multiplication) with the corresponding elementary matrix.*

PROOF. Let A be a matrix of order $m \times n$, then we can write $A = I_m A I_n$

So that any elementary row transformation can be obtained by subjecting the pre-factor I_m and the same elementary column transformation can be obtained by subjecting the post-factor I_n.

In order to prove this result we shall first prove that any E-row transformation of a product AB can be obtained by subjecting the pre-factor A to the same E-row transformation and any E-column transformation of a product AB can be obtained by subjecting the post-factor B to the same E-column transformation, where B is a matrix of order $n \times p$.

Let $A = [a_{ij}]_{m \times n}$ and $B = [b_{ij}]_{n \times p}$, then the product AB is conformable.

We can write A and B as follows :

$$A = \begin{bmatrix} R_1 \\ R_2 \\ \vdots \\ R_m \end{bmatrix} \text{ and } B = \begin{bmatrix} C_1 & C_2 & \cdots & C_p \end{bmatrix}$$

where $R_1, R_2, ..., R_m$ are row vectors of A and $C_1, C_2, ..., C_p$ are column vectors of B.

Now
$$AB = \begin{bmatrix} R_1C_1 & R_1C_2 & \cdots & R_1C_p \\ R_2C_1 & R_2C_2 & \cdots & R_2C_p \\ \vdots & \vdots & & \vdots \\ R_mC_1 & R_mC_2 & \cdots & R_mC_p \end{bmatrix}_{m \times p}$$

If σ be any E-row transformation, then
$$(\sigma A)B = \sigma(AB)$$

[**Note:** If σ denotes the operation $R_1 \leftrightarrow R_2$, then

$$\sigma A = \begin{bmatrix} R_2 \\ R_1 \\ \vdots \\ R_m \end{bmatrix}$$

$$\therefore \quad (\sigma A)B = \begin{bmatrix} R_2 \\ R_1 \\ \vdots \\ R_m \end{bmatrix} \begin{bmatrix} C_1 & C_2 & \cdots & C_p \end{bmatrix} = \begin{bmatrix} R_2C_1 & R_2C_2 & \cdots & R_2C_p \\ R_1C_1 & R_1C_2 & \cdots & R_1C_p \\ \vdots & & & \\ R_mC_1 & R_mC_2 & \cdots & R_mC_p \end{bmatrix}$$

Clearly $\quad\quad (\sigma A)B = \sigma(AB)$

Similarly, if the columns $C_1, C_2, ..., C_p$ of B be subjected to any E-column transformation, the columns of AB are also subjected to the same E-column transformation.

i.e., $\quad\quad (\sigma A)B = \sigma(AB)$ where σ denotes $C_1 \leftrightarrow C_2$.

Now we move to main theorem, if A is a matrix of order $m \times n$, then we can write
$$A = I_m A \text{ where } I_m \text{ is a unit matrix of order } n \times m.$$

If σ be any E-row transformation, then $\sigma(I_m A) = (\sigma I_m)A = EA$

where E is the elementary matrix corresponding to the same row transformation σ.

Similarly, we can also write $A = AI_m$ and if σ dentoes the E-column transformation, then $\sigma(AI_m) = A(\sigma I_m) = AE_1$

where E_1 is the elementary matrix corresponding to the same column transformation.

1.13.3 Invariance of Rank Under E-Transformations

Elementary transformation (E-transformation) do not change the rank of matrix. i.e.
(i) Interchanging the rows (or columns) does not change the rank.
(ii) Multiplication of the elements of a row by a non-zero number does not change the rank.
(iii) Addition of any row to the product of any number k and other row does not change the rank.

Remarks

- The rank of a matrix does not change by a series of E-transformation.
- The rank of a matrix does not change by a column-transformation.

1.13.4 Normal Form

If a matrix is reduced to the form $\begin{pmatrix} I_r & O \\ O & O \end{pmatrix}$. Then this form is called normal form of the given matrix.

THEOREM 1. *Every matrix of order $m \times n$ of rank r can be reduced to the form $\begin{pmatrix} I_r & O \\ O & O \end{pmatrix}$ by a finite number of E-transformations, where I_r is the unit matrix of order $r \times r$.*

PROOF. Let $A = [a_{ij}]_{m \times n}$ be a matrix of order $m \times n$ and of rank r. If A is a zero matrix, then its rank is zero and thus A can be written

as $\begin{pmatrix} I_r & O \\ O & O \end{pmatrix}$.

Let us suppose A is a non-zero matrix. It means that it has at least one of its elements non-zero. Let this non-zero element be $a_{ij} = k \neq 0$. Let B be a matrix which is obtained from A by E-transformations $R_1 \leftrightarrow R_i$ and and $C_1 \leftrightarrow C_j$ and whose leading element is k.

Again using the E-transformation $R_1 \to \dfrac{1}{k} R_1$ on B we get a matrix C whose leading element becomes 1. Let this matrix C be

$$C = \begin{bmatrix} 1 & c_{12} & c_{13} & \cdots & c_{1n} \\ c_{21} & c_{22} & c_{23} & \cdots & c_{2n} \\ c_{31} & c_{32} & c_{33} & \cdots & c_{3n} \\ \cdots & \cdots & \cdots & \cdots & \cdots \\ c_{m1} & c_{m2} & c_{m3} & \cdots & c_{mn} \end{bmatrix}_{m \times n}$$

Now subtracting first column after multiplying by suitable number from remaining columns of C and subtracting first row after multiplying by suitable number from remaining rows of C, we obtain a matrix D whose elements of the first row and first column are zero except the leading element. Let D be given as

$$D = \begin{bmatrix} 1 & 0 & 0 & \cdots & 0 \\ 0 & & & & \\ 0 & & A_1 & & \\ \vdots & & & & \\ 0 & & & & \end{bmatrix}_{m \times n}$$

where A_1 is a matrix of order $(m-1) \times (n-1)$.

If this matrix A_1 is non-zero matrix, then we shall apply above process on A_1. Since we know that E-transformation will not effect the first row and first column of D, so that we shall apply E-transformations on D and there is no need to take A_1 separately. Continuing this process finitely we obtain a matrix M such that

$$M = \begin{pmatrix} I_k & O \\ O & O \end{pmatrix}$$

This implies that matrix M has a rank k. But M is obtained from A by a finite number of E-transformations and we know that E-tansformations do not change the rank, therefore k must be equal to r.

Hence the matrix A of order $m \times n$ of rank r can be reduced to the form $\begin{pmatrix} I_r & O \\ O & O \end{pmatrix}$ by a finite number of E-transformations.

☞ REMARKS

- The form $\begin{pmatrix} I_r & O \\ O & O \end{pmatrix}$ of A is also called first canonical form.
- The rank of a matrix of order $m \times n$ is r if and only if it can be reduced to the form $\begin{bmatrix} I_r & O \\ O & O \end{bmatrix}$ by a finite chain of E-transformations.
- If A is a matrix of order $m \times n$ of rank r, then there exists two non-singular matrices P and Q such that $PAQ = \begin{bmatrix} I_r & O \\ O & O \end{bmatrix}$

1.13.5 EQUIVALENCE OF MATRICES

Let A be a matrix of order $m \times n$. If a matrix B of order $m \times n$ is obtained from A by a finite chain of elementary transformation on A, then A is said to be equivalent to B. We write symbolically as $A \sim B$ which is read as 'A is equivalent to B'.

THEOREM 1. *The relation '$\sim$' in the set of all $m \times n$ matrices is an equivalenvr relation.*

PROOF. We shall prove that the relation '$\sim$' is   (i) reflexive   (ii) symmetric   (iii) transitive.

(i) **Reflexivity:** Let A be an $m \times n$ matrix, then A can be obtained from A itself by the elementary transformation $R_i \to R_i$, for all $i = 1, 2, \ldots, m$

$$\therefore \qquad\qquad A \sim A$$

(ii) **Symmetry:** Let A and B be any two $m \times n$ matrices such that $A \sim B$.

Now $A \sim B \Rightarrow B$ can be obtained from A by a finite chain of elementary transformations on A

$\Rightarrow A$ can also be obtained from B by a finite chain of elementary transformations on B.

$\Rightarrow \qquad\qquad\qquad B \sim A \qquad\qquad\qquad$ *i.e.*, If $A \sim B$, then $\qquad\qquad B \sim A$.

(iii) **Transitivity:** Let A, B and C be any three $m \times n$ matrices such that $A \sim B$ and $B \sim C$.

$A \sim B \Rightarrow B$ can be obtained from A by a finite chain of elementary transformations on A.

$B \sim C \Rightarrow C$ can be obtained from B by a finite chain of elementary transformations on B.

On combining these two statements we can say that C can also be obtained from A by a finite chain of elementary transformations on A.

$$\therefore \quad A \sim C$$
$$i.e., \quad A \sim B, B \sim$$
$$C \Rightarrow A \sim C$$

Hence the relation '$\sim$' is an equivalence relation.

1.13.6 ROW AND COLUMN EQUIVALENCE OF MATRICES

(i) Row equivalence of matrix : A matrix A is said to be row equivalent to a matrix B, if B can be obtained from A by a finite chain of elementary row transformations on A and we write $A \overset{R}{\sim} B$.

(ii) Column equivalence of matrix : A matrix A is said to be column equivalent to a matrix B, if B can be obtained from A by a finite chain of elementary column transformations on A, we write $A \overset{C}{\sim} B$.

SOME RESULTS

1. **(Employment of only row transformations) :** Let A be a matrix of order $m \times n$ of rank r, then there exists a non-singular matrix P such that $PA = \begin{bmatrix} G \\ O \end{bmatrix}$ where G is a matrix of order $r \times n$ of rank r and O is a null matrix of order $(m-r) \times n$.

2. **(Employment of only column transformations):** Let A be a matrix of order $m \times n$ of rank r, then there exists a non-singular matrix Q such that $AQ = [H \quad O]$ where H is a matrix of order $m \times n$ and O is a null matrix of order $m \times (n - r)$.

3. **(Rank of product of matrices):** The rank of a product of two matrices cannot exceed the rank of either matrix, *i.e.*, if A and B be two matrices conformable for the product AB, then $\rho(AB) \leq \rho(A)$, $\rho(AB) \leq \rho(B)$, *i.e.*, $\rho(AB) \leq \min \{\rho(A), \rho(B)\}$.

4. Every non-singular matrix is row equivalent to a unit matrix.

REMARKS

- If A is a non-singular matrix of order $n \times n$, there exist 8 elementary matrices $E_1, E_2, ..., E_s$ such that
$$E_s E_{s-1} ... E_2 E_1 A = I_n$$

- Every non-singular matrix A is expressible as the product of elementary matrices.

- The rank of a matrix does not change by pre-multiplication or postmultiplication with a non-singular matrix.

Solved Examples

EXAMPLE 1. *Show that the matrices* $\begin{bmatrix} 1 & 2 & 3 \\ 2 & 4 & 6 \end{bmatrix}$ *and* $\begin{bmatrix} 0 & 3 & 2 \\ 0 & 6 & 4 \end{bmatrix}$ *are equivalent.*

SOLUTION. Let $A = \begin{bmatrix} 1 & 2 & 3 \\ 2 & 4 & 6 \end{bmatrix}$ and $B = \begin{bmatrix} 0 & 3 & 2 \\ 0 & 6 & 4 \end{bmatrix}$

Applying $R_2 \to R_2 - 2R_1$ on A, we get
$$A \sim \begin{bmatrix} 1 & 2 & 3 \\ 0 & 0 & 0 \end{bmatrix}$$

Again applying $C_1 \to C_1 - \frac{1}{2}C_2$, we get
$$A \sim \begin{bmatrix} 0 & 2 & 3 \\ 0 & 0 & 0 \end{bmatrix}$$

Again applying $R_2 \to R_2 + 2R_1$, we get
$$A \sim \begin{bmatrix} 0 & 2 & 3 \\ 0 & 4 & 6 \end{bmatrix}$$

Again applying $C_2 \leftrightarrow C_3$, we get
$$A \sim \begin{bmatrix} 0 & 3 & 2 \\ 0 & 6 & 4 \end{bmatrix} = B$$

$$A \sim B$$

Thus B can be obtained from A by a finite number of elementary transformations on A. Hence A and B are equivalent.

EXAMPLE 2. *If A and B be two equivalent matrices, then show that* $\rho(A) = \rho(B)$.

SOLUTION. Since A and B are equivalent, therefore B can be obtained from A by a finite chain of elementary transformations on A and elementary transformations do not change the rank of the matrices, hence $\rho(A) = \rho(B)$.

EXAMPLE 3. *Show that if two matrices A and B have the same size and the same rank they are equivalent.*

SOLUTION. Let A and B be two matrices of order $m \times n$ and $\rho(A) = \rho(B) = r$. Then we have
$$A \sim \begin{bmatrix} I_r & O \\ O & O \end{bmatrix} \text{ and } B \sim \begin{bmatrix} I_r & O \\ O & O \end{bmatrix}$$

Since '$\sim$' is symmetric, then
$$B \sim \begin{bmatrix} I_r & O \\ O & O \end{bmatrix} \Leftrightarrow \begin{bmatrix} I_r & O \\ O & O \end{bmatrix} \sim B$$

Again '~' is transitive, then

$$A \sim \begin{bmatrix} I_r & O \\ O & O \end{bmatrix} \text{ and } \begin{bmatrix} I_r & O \\ O & O \end{bmatrix} \sim B$$

$$\Rightarrow \quad A \sim B$$

Hence A and B are equivalent.

EXAMPLE 4. *Use E-transformations to reduce the following matrices to triangular form and hence find their rank.*

$$(i) \begin{bmatrix} 5 & 3 & 14 & 4 \\ 0 & 1 & 2 & 1 \\ 1 & -1 & 2 & 0 \end{bmatrix} \quad (ii) \begin{bmatrix} 8 & 1 & 3 & 6 \\ 0 & 3 & 2 & 2 \\ -8 & -1 & -3 & 4 \end{bmatrix}$$

(KURUKSHETRA-2005)

SOLUTION. (i) Let $A = \begin{bmatrix} 5 & 3 & 14 & 4 \\ 0 & 1 & 2 & 1 \\ 1 & -1 & 2 & 0 \end{bmatrix}$

Applying $R_1 \leftrightarrow R_3$

$$A \sim \begin{bmatrix} 1 & -1 & 2 & 0 \\ 0 & 1 & 2 & 1 \\ 5 & 3 & 14 & 4 \end{bmatrix}$$

Again applying $R_3 \to R_3 - 5R_1$

$$A \sim \begin{bmatrix} 1 & -1 & 2 & 0 \\ 0 & 1 & 2 & 1 \\ 0 & 8 & 4 & 4 \end{bmatrix}$$

Again applying $R_3 \to R_3 - 8R_2$

$$A \sim \begin{bmatrix} 1 & -1 & 2 & 0 \\ 0 & 1 & 2 & 1 \\ 0 & 0 & -12 & -4 \end{bmatrix}$$

The last equivalent matrix is in Echelon form (or triangular form) which has three non-zero rows. Hence $\rho(A) = 3$.

(ii) Let $A = \begin{bmatrix} 8 & 1 & 3 & 6 \\ 0 & 3 & 2 & 2 \\ -8 & -1 & -3 & 4 \end{bmatrix}$

Applying $C_1 \to \dfrac{1}{8} C_1$

$$A \sim \begin{bmatrix} 1 & 1 & 3 & 6 \\ 0 & 3 & 2 & 2 \\ -1 & -1 & -3 & 4 \end{bmatrix}$$

Again applying $R_3 \to R_3 + R_1$

$$A \sim \begin{bmatrix} 1 & 1 & 3 & 6 \\ 0 & 3 & 2 & 2 \\ 0 & 0 & 0 & 10 \end{bmatrix}$$

The last equivalent matrix is in Echelon form (or triangular form) which has three non-zero rows, hence $\rho(A) = 3$.

EXAMPLE 5. *Is the matrix* $\begin{bmatrix} 1 & 2 & 1 \\ -1 & 0 & 2 \\ 2 & 1 & -3 \end{bmatrix}$ *equivalent to* I_3 ?

SOLUTION. Let $A = \begin{bmatrix} 1 & 2 & 1 \\ -1 & 0 & 2 \\ 2 & 1 & -3 \end{bmatrix}$

Then $|A| = 1(0 - 2) - 2(3 - 4) + 1(-1 - 0)$
$$= -2 + 2 + 1 = 1 \neq 0$$

Therefore, A is a non-singular matrix of order 3×3, so it is row equivalent to a unit matrix. Hence $A \sim I_3$.

EXAMPLE 6. *If A and B are two matrices of the same type, then*
$$\rho(A+B) \leq \rho(A) + \rho(B)$$

SOLUTION. Let A and B be two matrices of order $m \times n$ and let $\rho(A) = r_1$ and $\rho(B) = r_2$.

Now $\rho(A) = r_1$

$\Rightarrow A$ contains r_1 linearly independent rows.

and $\rho(B) = r_2$

$\Rightarrow B$ contains r_2 linearly independent rows.

Therefore $A + B$ will contain atmost $r_1 + r_2$ linearly independent rows

$\therefore \quad \rho(A+B) \leq r_1 + r_2$

Hence $\rho(A+B) \leq \rho(A) + \rho(B)$.

EXAMPLE 7. *If A be any non-singular matrix and B a matrix such that AB exists, then show that*
$$\rho(AB) = \rho(B).$$

SOLUTION. Since A is a non-singular matrix, then

$$B = A^{-1}(AB)$$

We know that

$$\rho(AB) \leq \rho(B) \qquad \qquad \dots(1)$$

Now $\rho(B) = \rho(A^{-1}(AB)) \leq \rho(AB)$

or $\rho(B) \leq \rho(AB) \qquad \qquad \dots(2)$

From (1) and (2), we have

$$\rho(AB) = \rho(B).$$

EXAMPLE 8. *If A is a square matrix of order $n \times n$ such that $A^2 = A$, then show that*
$$\rho(A) + \rho(I_n - A) = n.$$

SOLUTION. We have $A^2 = A$

$\Rightarrow \quad A - A^2 = 0$

$\Rightarrow \quad AI_n - A^2 = 0 \qquad \qquad [\because AI_n = A]$

$\Rightarrow \quad A(I_n - A) = 0$

$\therefore \quad P(A(I_n - A)) = 0$

Also we know that

$$\rho(A(I_n - A)) \geq \rho(A) + \rho(I_n - A) - n$$

$\Rightarrow \quad \rho(A) + \rho(I_n - A) - n \leq 0$

$\therefore \quad \rho(A) + \rho(I_n - A) \leq n \qquad \qquad \dots(1)$

Again we know that

$$\rho(A + I_n - A) = \rho(A) + \rho(I_n - A)$$

$$\Rightarrow \qquad \rho(I_n) \leq \rho(A) + \rho(I_n - A)$$

$$\therefore \qquad \rho(A) + \rho(I_n - A) \geq n \qquad \qquad ...(2)$$

$$[\because \rho(I_n) = n]$$

From (1) and (2), we get

$$\rho(A) + \rho(I_n - A) = n$$

EXAMPLE 9. *If A is a square matrix of order n × n and* $\rho(A) = n - 1$, *show that* $\rho(adj. A) = 1$.

SOLUTION. Since A is an $n \times n$ matrix and $\rho(A) = n - 1$,

then we have $\qquad |A| = 0$

But we know that $A(adj. A) = |A| I_n$

$$\Rightarrow \qquad A(adj. A) = O$$

$$\therefore \qquad \rho(A \, adj. A) = 0$$

Also $\qquad \rho(A \, adj. A) \geq \rho(A) + \rho(adj. A) - n$

$$\Rightarrow \qquad \rho(A) + \rho(adj. A) - n \leq O$$

$$\Rightarrow \qquad \rho(A) + \rho(adj. A) \leq n$$

$$\Rightarrow \qquad \rho(adj. A) \leq n - \rho(A)$$

$$\Rightarrow \qquad \rho(adj. A) \leq n - (n - 1)$$

$$\therefore \qquad \rho(adj. A) \leq 1 \qquad \qquad ...(1)$$

Since $\rho(A) = n - 1$, then there exists at least one minor of order $n - 1$ of A not equal to zero, therefore there exists at least one element of *adj. A* which is non-zero, it follows that

$$\rho(adj. A) > 0 \qquad \qquad ...(2)$$

From (1) and (2), we get

$$\rho(adj. A) = 1.$$

EXAMPLE 10. *Determine the rank of the following matrices :*

(i) $\begin{bmatrix} 2 & -1 & 3 & 4 \\ 0 & 3 & 4 & 1 \\ 2 & 3 & 7 & 5 \\ 2 & 5 & 11 & 6 \end{bmatrix}$ (ii) $\begin{bmatrix} -2 & -1 & -3 & -1 \\ 1 & 2 & 3 & -1 \\ 1 & 0 & 1 & 1 \\ 0 & 1 & 1 & -1 \end{bmatrix}$

SOLUTION. (i) Let $\qquad A = \begin{bmatrix} 2 & -1 & 3 & 4 \\ 0 & 3 & 4 & 1 \\ 2 & 3 & 7 & 5 \\ 2 & 5 & 11 & 6 \end{bmatrix}$

Applying $R_3 \to R_3 - R_1$, $R_4 \to R_4 - R_1$

$$A \sim \begin{bmatrix} 2 & -1 & 3 & 4 \\ 0 & 3 & 4 & 1 \\ 0 & 4 & 4 & 1 \\ 0 & 6 & 8 & 2 \end{bmatrix}$$

Again applying $R_3 \to R_3 - \dfrac{4}{3} R_2$, $R_4 \to R_4 - 2R_2$

$$A \sim \begin{bmatrix} 2 & -1 & 3 & 4 \\ 0 & 3 & 4 & 1 \\ 0 & 0 & -4/3 & -1/3 \\ 0 & 0 & 0 & 0 \end{bmatrix}$$

Again applying $R_3 \to 3R_3$

$$A \sim \begin{bmatrix} 2 & -1 & 3 & 4 \\ 0 & 3 & 4 & 1 \\ 0 & 0 & -4 & -1 \\ 0 & 0 & 0 & 0 \end{bmatrix}$$

The last equivalent matrix is in Echelon form which has 3 non-zero rows, hence $\quad \rho(A) = 3$.

(ii) Let $\qquad A = \begin{bmatrix} -2 & -1 & -3 & -1 \\ 1 & 2 & 3 & -1 \\ 1 & 0 & 1 & 1 \\ 0 & 1 & 1 & -1 \end{bmatrix}$

Applying $R_1 \to R_2$

$$A \sim \begin{bmatrix} 1 & 2 & 3 & -1 \\ -2 & -1 & -3 & -1 \\ 1 & 0 & 1 & 1 \\ 0 & 1 & 1 & -1 \end{bmatrix}$$

Again applying $R_2 \to R_2 + 2R_1$, $R_3 \to R_3 - R_1$

$$A \sim \begin{bmatrix} 1 & 2 & 3 & -1 \\ 0 & 3 & 3 & -3 \\ 0 & -2 & -2 & 2 \\ 0 & 1 & 1 & -1 \end{bmatrix}$$

Again applying $R_2 \to R_2 + R_3$

$$A \sim \begin{bmatrix} 1 & 2 & 3 & -1 \\ 0 & 1 & 1 & -1 \\ 0 & -2 & -2 & 2 \\ 0 & 1 & 1 & -1 \end{bmatrix}$$

Again applying $R_3 \to R_3 + 2R_2$, $R_4 \to R_4 - R_2$

$$A \sim \begin{bmatrix} 1 & 2 & 3 & -1 \\ 0 & 1 & 1 & -1 \\ 0 & 0 & 0 & 0 \\ 0 & 0 & 0 & 0 \end{bmatrix}$$

The last equivalent matrix is in Echelon form which has 2 non-zero rows, hence $\rho(A) = 2$.

EXAMPLE 11. *Reduce the matrix* $A = \begin{bmatrix} 1 & -1 & 2 & -3 \\ 4 & 1 & 0 & 2 \\ 0 & 3 & 0 & 4 \\ 0 & 1 & 0 & 2 \end{bmatrix}$ *to the normal form* $\begin{bmatrix} I_r & O \\ O & O \end{bmatrix}$ *and hence determine its rank.*

SOLUTION. We have $\qquad A = \begin{bmatrix} 1 & -1 & 2 & -3 \\ 4 & 1 & 0 & 2 \\ 0 & 3 & 0 & 4 \\ 0 & 1 & 0 & 2 \end{bmatrix}$

Applying $R_2 \to R_2 - 4R_1$

$$A \sim \begin{bmatrix} 1 & -1 & 2 & -3 \\ 0 & 5 & -8 & 14 \\ 0 & 3 & 0 & 4 \\ 0 & 1 & 0 & 2 \end{bmatrix}$$

Applying $C_2 \to C_2 + C_1$, $C_3 \to C_3 - 2C_1$, $C_4 \to C_4 + 3C_1$

$$A \sim \begin{bmatrix} 1 & 0 & 0 & 0 \\ 0 & 5 & -8 & 14 \\ 0 & 3 & 0 & 4 \\ 0 & 1 & 0 & 2 \end{bmatrix}$$

Applying $R_2 \leftrightarrow R_4$

$$A \sim \begin{bmatrix} 1 & 0 & 0 & 0 \\ 0 & 1 & 0 & 2 \\ 0 & 3 & 0 & 4 \\ 0 & 5 & -8 & 14 \end{bmatrix}$$

Applying $R_3 \to R_3 - 3R_2$, $R_4 \to R_4 - 5R_2$

$$A \sim \begin{bmatrix} 1 & 0 & 0 & 0 \\ 0 & 1 & 0 & 2 \\ 0 & 0 & 0 & -2 \\ 0 & 0 & -8 & 4 \end{bmatrix}$$

Applying $C_4 \to C_4 - 2C_2$

$$A \sim \begin{bmatrix} 1 & 0 & 0 & 0 \\ 0 & 1 & 0 & 0 \\ 0 & 0 & 0 & -2 \\ 0 & 0 & -8 & 4 \end{bmatrix}$$

Applying $C_3 \leftrightarrow C_4$

$$A \sim \begin{bmatrix} 1 & 0 & 0 & 0 \\ 0 & 1 & 0 & 0 \\ 0 & 0 & -2 & 0 \\ 0 & 0 & 4 & -8 \end{bmatrix}$$

Applying $R_4 \to R_4 + 2R_3$

$$A \sim \begin{bmatrix} 1 & 0 & 0 & 0 \\ 0 & 1 & 0 & 0 \\ 0 & 0 & -2 & 0 \\ 0 & 0 & 0 & -8 \end{bmatrix}$$

Applying $R_3 \to \dfrac{1}{2}R_3, R_4 \to -\dfrac{1}{8}R_4$

$$A \sim \begin{bmatrix} 1 & 0 & 0 & 0 \\ 0 & 1 & 0 & 0 \\ 0 & 0 & 1 & 0 \\ 0 & 0 & 0 & 1 \end{bmatrix}$$

$\therefore \qquad\qquad A \sim I_4$

Hence $\qquad\qquad \rho(A) = 4$

EXAMPLE 12. *Find two non-singular matrices P and Q such that PAQ is in the normal form where*

$$A = \begin{bmatrix} 1 & 1 & 1 \\ 1 & -1 & -1 \\ 3 & 1 & 1 \end{bmatrix}$$

Also find the rank of the matrix A.

(KURUKSHETRA 2005)

SOLUTION. We write $\qquad A = I_3 A I_3$

or $\begin{bmatrix} 1 & 1 & 1 \\ 1 & -1 & -1 \\ 3 & 1 & 1 \end{bmatrix} = \begin{bmatrix} 1 & 0 & 0 \\ 0 & 1 & 0 \\ 0 & 0 & 1 \end{bmatrix} A \begin{bmatrix} 1 & 0 & 0 \\ 0 & 1 & 0 \\ 0 & 0 & 1 \end{bmatrix}$ ...(1)

In order to find P and Q such that $PAQ = \begin{bmatrix} I & O \\ O & O \end{bmatrix}$ we shall reduce the matrix on LHS of (1) by using elementary transformations, while in doing so we shall apply elementary row transformation to pre-factor of A and elementary-column transformation to post-factor of A on RHS of (1).

Now applying $R_2 \to R_2 - R_1$, $R_3 \to R_3 - 3R_1$

$$\begin{bmatrix} 1 & 1 & 1 \\ 1 & -2 & -2 \\ 0 & -2 & -2 \end{bmatrix} = \begin{bmatrix} 1 & 0 & 0 \\ -1 & 1 & 0 \\ -3 & 0 & 1 \end{bmatrix} A \begin{bmatrix} 1 & 0 & 0 \\ 0 & 1 & 0 \\ 0 & 0 & 1 \end{bmatrix}$$

Applying $C_2 \to C_2 - C_1$, $C_3 \to C_3 - C_1$

$$\begin{bmatrix} 1 & 0 & 0 \\ 0 & -2 & -2 \\ 0 & -2 & -2 \end{bmatrix} = \begin{bmatrix} 1 & 0 & 0 \\ -1 & 1 & 0 \\ -3 & 0 & 1 \end{bmatrix} A \begin{bmatrix} 1 & -1 & -1 \\ 0 & 1 & 0 \\ 0 & 0 & 1 \end{bmatrix}$$

Applying $R_2 \to \left(-\dfrac{1}{2}\right)R_2$

$$\begin{bmatrix} 1 & 0 & 0 \\ 0 & 1 & 1 \\ 0 & -2 & -2 \end{bmatrix} = \begin{bmatrix} 1 & 0 & 0 \\ \frac{1}{2} & -\frac{1}{2} & 0 \\ -3 & 0 & 1 \end{bmatrix} A \begin{bmatrix} 1 & -1 & -1 \\ 0 & 1 & 0 \\ 0 & 0 & 1 \end{bmatrix}$$

Applying $R_3 \to R_3 + 2R_2$

$$\begin{bmatrix} 1 & 0 & 0 \\ 0 & 1 & 1 \\ 0 & 0 & 0 \end{bmatrix} = \begin{bmatrix} 1 & 0 & 0 \\ \frac{1}{2} & -\frac{1}{2} & 0 \\ -2 & -1 & 1 \end{bmatrix} A \begin{bmatrix} 1 & -1 & -1 \\ 0 & 1 & 0 \\ 0 & 0 & 1 \end{bmatrix}$$

Applying $C_2 \to C_3 - C_2$

$$\begin{bmatrix} 1 & 0 & 0 \\ 0 & 1 & 0 \\ 0 & 0 & 0 \end{bmatrix} = \begin{bmatrix} 1 & 0 & 0 \\ 1/2 & -1/2 & 0 \\ -2 & -1 & 1 \end{bmatrix} A \begin{bmatrix} 1 & -1 & 0 \\ 0 & 1 & -1 \\ 0 & 0 & 1 \end{bmatrix}$$

or $\begin{bmatrix} I_2 & O \\ O & O \end{bmatrix} = PAQ$

where $P = \begin{bmatrix} 1 & 0 & 0 \\ 1/2 & -1/2 & 0 \\ -2 & -1 & 1 \end{bmatrix}$ and $Q = \begin{bmatrix} 1 & -1 & 0 \\ 0 & 1 & -1 \\ 0 & 0 & 1 \end{bmatrix}$

$$A \sim \begin{bmatrix} I_2 & O \\ O & O \end{bmatrix}$$

Hence $\rho(A) = 2$.

Exercise-1.4

1. Are the followings pairs of matrices equivalent ?

(i) $\begin{bmatrix} 4 & 0 & 2 \\ 3 & 1 & 0 \\ 5 & 2 & 0 \end{bmatrix}, \begin{bmatrix} 3 & 9 & 0 & 2 \\ 7 & -2 & 0 & 1 \\ 8 & 1 & 1 & 5 \end{bmatrix}$

(ii) $\begin{bmatrix} 2 & -1 & 3 & 4 \\ 0 & 3 & 4 & 1 \\ 2 & 3 & 7 & 5 \\ 2 & 5 & 11 & 5 \end{bmatrix}, \begin{bmatrix} 1 & 0 & -5 & 6 \\ 3 & -2 & 1 & 2 \\ 5 & -2 & -9 & 14 \\ 4 & -2 & -4 & 8 \end{bmatrix}$

Determine the rank of the following matrices:

2. $\begin{bmatrix} 1 & 1 & 1 \\ 2 & 2 & 2 \\ 3 & 3 & 3 \end{bmatrix}$

3. $\begin{bmatrix} 2 & 1 & 3 \\ 4 & 7 & 13 \\ 4 & -3 & -1 \end{bmatrix}$

4. $\begin{bmatrix} 4 & 5 & 6 \\ 5 & 6 & 7 \\ 7 & 8 & 9 \end{bmatrix}$

5. $\begin{bmatrix} 1 & 2 & 3 \\ 2 & 3 & 4 \\ 3 & 5 & 7 \end{bmatrix}$

6. $\begin{bmatrix} 2 & 3 & 7 \\ 3 & -2 & 4 \\ 1 & -3 & -1 \end{bmatrix}$

7. $\begin{bmatrix} 3 & -1 & 2 \\ -6 & 2 & -4 \\ -3 & 1 & -2 \end{bmatrix}$

8. $\begin{bmatrix} 1 & 2 & 3 & 1 \\ 2 & 4 & 6 & 2 \\ 1 & 2 & 3 & 2 \end{bmatrix}$

9. $\begin{bmatrix} 1 & 3 & 4 & 3 \\ 3 & 9 & 12 & 9 \\ 1 & 3 & 4 & 1 \end{bmatrix}$

10. $\begin{bmatrix} 1 & 2 & -1 & 4 \\ 2 & 4 & 3 & 5 \\ -1 & -2 & 6 & -7 \end{bmatrix}$

11. $\begin{bmatrix} 1 & 2 & -4 & 5 \\ 2 & -1 & 3 & 6 \\ 8 & 1 & 9 & 7 \end{bmatrix}$

12. $\begin{bmatrix} 1 & -1 & 3 & 6 \\ 1 & 3 & -3 & -4 \\ 5 & 3 & 3 & 11 \end{bmatrix}$

13. $\begin{bmatrix} 1 & 2 & 3 & 0 \\ 2 & 4 & 3 & 2 \\ 3 & 2 & 1 & 3 \\ 6 & 8 & 7 & 5 \end{bmatrix}$

14. $\begin{bmatrix} 2 & 3 & -1 & -1 \\ 1 & -1 & -2 & -4 \\ 3 & 1 & 3 & -2 \\ 6 & 3 & 0 & -7 \end{bmatrix}$ (UPTU-2005)

15. $\begin{bmatrix} 1 & 2 & 1 & 2 \\ 1 & 3 & 2 & 2 \\ 2 & 4 & 3 & 4 \\ 3 & 7 & 4 & 6 \end{bmatrix}$

16. $\begin{bmatrix} 3 & -2 & 0 & -1 \\ 0 & 2 & 2 & 1 \\ 1 & -2 & -3 & 2 \\ 0 & 1 & 2 & 1 \end{bmatrix}$

17. $\begin{bmatrix} 0 & 1 & -3 & -1 \\ 1 & 0 & 1 & 1 \\ 3 & 1 & 0 & 2 \\ 1 & 1 & -2 & 0 \end{bmatrix}$

18. $\begin{bmatrix} 1 & 2 & -1 & 3 \\ 4 & 1 & 2 & 1 \\ 3 & -1 & 1 & 2 \\ 1 & 2 & 0 & 1 \end{bmatrix}$

19. $\begin{bmatrix} 1 & 0 & 2 & 1 \\ 0 & 1 & -2 & 1 \\ 1 & -1 & 4 & 0 \\ -2 & 2 & 8 & 0 \end{bmatrix}$

20. $\begin{bmatrix} 8 & 0 & 0 & 1 \\ 1 & 0 & 8 & 1 \\ 0 & 0 & 1 & 8 \\ 0 & 1 & 1 & 8 \end{bmatrix}$

21. $\begin{bmatrix} 6 & 1 & 3 & 8 \\ 4 & 2 & 6 & -1 \\ 10 & 3 & 9 & 7 \\ 16 & 4 & 12 & 15 \end{bmatrix}$

22. Reduce the matrix to normal form and find its rank.

$\begin{bmatrix} 0 & 1 & -3 & -1 \\ 1 & 0 & 1 & 1 \\ 3 & 1 & 0 & 2 \\ 1 & 1 & -2 & 0 \end{bmatrix}$ (UPTU-2007)

23. Find the rank of the matrix after reducing it to normal form :

$A = \begin{bmatrix} 1 & 2 & 3 \\ 2 & 3 & 4 \\ 3 & 5 & 7 \end{bmatrix}$

24. Reduce the matrix to normal form and find its rank.

$A = \begin{bmatrix} 9 & 7 & 3 & 6 \\ 5 & -1 & 4 & 1 \\ 6 & 8 & 2 & 4 \end{bmatrix}$

25. Use elementary row or column transformations to find the rank of the matrix

$\begin{bmatrix} 1 & 1 & 2 & 3 \\ 1 & 3 & 0 & 3 \\ 1 & -2 & -3 & -3 \\ 1 & 1 & 2 & 3 \end{bmatrix}$

26. Find the rank of A, B, $A+B$, AB and BA where

$A = \begin{bmatrix} 1 & 1 & -1 \\ 2 & -3 & 4 \\ 3 & -2 & 3 \end{bmatrix}, B = \begin{bmatrix} -1 & -2 & -1 \\ 6 & 12 & 6 \\ 5 & 10 & 5 \end{bmatrix}$

27. Find two non-singular matrices P and Q such that PAQ is in the normal form where $A = \begin{bmatrix} 1 & -1 & 2 & -1 \\ 4 & 2 & -1 & 2 \\ 2 & 2 & -2 & 0 \end{bmatrix}$

Also find the rank of the matrix A.

28. Show that if A and B are equivalent matrices, then there exists non-singular matrices P and Q such that $B = PAQ$.

29. Show that the rank of a matrix is not altered if a column of a matrix is multiplied by a non-zero scalar.

30. Find matrices P and Q such that $P \begin{bmatrix} 2 & 2 & -6 \\ -1 & 2 & 2 \end{bmatrix} Q$ is in the normal form.

31. Transform $\begin{bmatrix} 1 & 3 & 3 \\ 2 & 4 & 10 \\ 3 & 8 & 4 \end{bmatrix}$ into a unit matrix by using elementary transformation. (UPTU-2011)

32. Reduce the following matrix into normal form and find its rank

$$\begin{bmatrix} 1 & 2 & -1 & 4 \\ 2 & 4 & 3 & 4 \\ 1 & 2 & 3 & 4 \\ -1 & -2 & 6 & 7 \end{bmatrix}$$ (GBTU-2010)

33. Find the rank of the following matrices

(i) $$\begin{bmatrix} 1 & -3 & 1 & 2 \\ 0 & 1 & 2 & 3 \\ 3 & 4 & 1 & -2 \end{bmatrix}$$ (GBTU-2010)

(ii) $$\begin{bmatrix} 0 & 1 & 2 & -2 \\ 4 & 0 & 2 & 6 \\ 2 & 1 & 3 & 1 \end{bmatrix}$$ (UPTU-2007)

(iii) $$\begin{bmatrix} 5 & 3 & 14 & 4 \\ 0 & 1 & 2 & 1 \\ 1 & -1 & 2 & 0 \end{bmatrix}$$ (GBTU-2011)

(iv) $$\begin{bmatrix} 1 & 2 & 1 & 0 \\ -2 & 4 & 3 & 0 \\ 1 & 0 & 2 & 8 \end{bmatrix}$$ (MTU-2011)

(v) $$\begin{bmatrix} 0 & 1 & 2 & -1 \\ 1 & 0 & 1 & 1 \\ 3 & 1 & 0 & 2 \\ 1 & 1 & -2 & 0 \end{bmatrix}$$ (UKTU-2011)

34. (i) Find all values of μ for which rank of the matrix is equal to 3. (UPTU-2009)

$$\begin{bmatrix} \mu & -1 & 0 & 0 \\ 0 & \mu & -1 & 0 \\ 0 & 0 & \mu & -1 \\ -6 & 11 & -6 & 1 \end{bmatrix}$$

(ii) Find the value of k for which the matrix

$$\begin{bmatrix} 3 & k & k \\ k & 3 & k \\ k & k & 3 \end{bmatrix}$$ is of rank 1. (MTU-2012)

Answers

1. (i) Not equivalent   (ii) Not equivalent
2. 1   **3.** 2   **4.** 2   **5.** 2   **6.** 1   **7.** 2   **8.** 2   **9.** 2   **10.** 3   **11.** 3
12. 3   **13.** 3   **14.** 3   **15.** 4   **16.** 2   **17.** 3   **18.** 3   **19.** 4   **20.** 2   **21.** 3
22. 2   **23.** 3   **24.** 3
25. $\rho(A) = 2, \rho(B) = 1, \rho(A+B) = 2, \rho(AB) = 0, \rho(BA) = 1$ **26.** $\rho(A) = 2.29, R = \begin{bmatrix} 1 & 1 \\ 1/2 & 0 \end{bmatrix}, S = \begin{bmatrix} 1 & 4 & 8 \\ 0 & 0 & 1 \\ 0 & 1 & 3 \end{bmatrix}$

31. $\begin{bmatrix} 1 & 0 & 0 \\ 0 & 1 & 0 \\ 0 & 0 & 1 \end{bmatrix}$ **32.** 3   **33.** (i) 3 (ii) 2 (iii) 3 (iv) 3 (v) 3 **34.** (i) μ = 1, 2, 3 (ii) k = 3

1.14 INVERSE OF A MATRIX

Let A be a non-singular matrix of order $n \times n$. Then it is said to be invertible if there exists a non-singular square matrix of order $n \times n$ such that $AB = I_n = BA$ where I_n is the unit matrix of order $n \times n$.

The matrix B is the inverse of A, we write $B = A^{-1}$.

THEOREM 1. *The inverse of a matrix, if it exists, is unique.*

PROOF. Let A be a non-singular matrix of order $n \times n$ and if possible, let B and C be its inverses, then we have

$$AB = I_n = BA \qquad \ldots(1)$$
and $$AC = I_n = CA \qquad \ldots(2)$$
From (1) and (2) we get, $$AB = AC$$
$\Rightarrow \qquad B(AB) = B(AC)$
$\Rightarrow \qquad (BA)B = (BA)C$ (By associative law)
$\Rightarrow \qquad I_nB = I_nC$ [Using (1)]
$\Rightarrow \qquad B = C.$

THEOREM 2. *A square matrix is invertible if and only if it is non-singular.*

PROOF. Let A be a square matrix of order $n \times n$ and suppose that A is invertible, then there exists a matrix B of order n such that

$$AB = I_n = BA$$
$\Rightarrow \qquad |AB| = |I_n|$

$$\Rightarrow \qquad |A||B| = 1$$

$$\Rightarrow \qquad |A| \neq 0$$

Thus, A is non-singular.

Conversely, suppose that A is non-singular matrix, then we have

$$A(adj.\ A) = |A|\ I_n = (adj.\ A)A$$

$$\Rightarrow \qquad A\left(\frac{adj.\ A}{|A|}\right) = I_n = \left(\frac{adj.\ A}{|A|}\right)A \qquad\qquad \left[\because |A| \neq 0 \Rightarrow \frac{1}{|A|}\ \text{exists}\right]$$

$$\Rightarrow \qquad AB = I_n = BA, \text{ if } \quad B = \frac{adj.\ A}{|A|}$$

Thus, A is invertible.

THEOREM 3. *If A is an invertible matrix, then $(A^{-1})^{-1} = A$.*

PROOF. Since A is invertible, then we have $\qquad AA^{-1} = I = AA^{-1}$

$$\Rightarrow \quad A \text{ is the inverse of } A^{-1}$$

$$\therefore \qquad\qquad (A^{-1})^{-1} = A$$

THEOREM 4. **(Reversal law):** *If A and B are invertible matrices of the same order, then AB is invertible and $(AB)^{-1} = B^{-1}A^{-1}$.* (DELHI–2012)

PROOF. Since A and B are invertible, therefore we have

$$|A| \neq 0,\ |B| \neq 0$$

$$\Rightarrow \qquad |AB| = |A|\ |B|$$

$$\Rightarrow \qquad |AB| \neq 0$$

$\therefore \quad AB$ is invertible.

Now	$(AB)(B^{-1}A^{-1}) = A(BB^{-1})A^{-1}$	[By associative law]
$\Rightarrow$	$(AB)(B^{-1}A^{-1}) = A(I)A^{-1}$	$[\because BB^{-1} = I]$
$\Rightarrow$	$(AB)(B^{-1}A^{-1}) = (AI)A^{-1}$	[By associative law]
$\Rightarrow$	$(AB)(B^{-1}A^{-1}) = AA^{-1}$	$[\because AI = A]$
$\Rightarrow$	$(AB)(B^{-1}A^{-1}) = I$	$[\because AA^{-1} = I]$
Also,	$(B^{-1}A^{-1})(AB) = B^{-1}(A^{-1}A)B$	[By associative law]
$\Rightarrow$	$(B^{-1}A^{-1})(AB) = B^{-1}(I)B$	$[\because A^{-1}A = I]$
$\Rightarrow$	$(B^{-1}A^{-1})(AB) = B^{-1}(IB)$	[By associative law]
$\Rightarrow$	$(B^{-1}A^{-1})(AB) = B^{-1}B$	$[\because IB = B]$
$\Rightarrow$	$(B^{-1}A^{-1})(AB) = I$	$[\because B^{-1}B = I]$
$\therefore$	$(AB)(B^{-1}A^{-1}) = I = (B^{-1}A^{-1})(AB)$	
$\Rightarrow$	$(AB)^{-1} = B^{-1}A^{-1}$	

REMARK

- If A, B and C are three invertible matrices of the same order, then $(ABC)^{-1} = C^{-1}B^{-1}A^{-1}$.

THEOREM 5. *If A is an invertible square matrix, then A' is also invertible and $(A')^{-1} = (A^{-1})'$, where A' is the transpose of A.*

PROOF. Since A is an invertible matrix, then we have $|A| \neq 0$

Now $\qquad\qquad |A| = |A'| \quad \Rightarrow \quad |A'| \neq 0 \Rightarrow A'$ is invertible.

Also $\qquad\qquad AA^{-1} = I = A^{-1}A$

$$\Rightarrow \qquad (AA^{-1})' = (I)' = (A^{-1}A)'$$

$$\Rightarrow \qquad (A^{-1})'\ A' = I = A'\ (A^{-1})' \qquad\qquad \text{(By reversal rule of transpose)}$$

$\Rightarrow (A^{-1})'$ is the inverse of $A' \quad \Rightarrow \quad (A')^{-1} = (A^{-1})'$

THEOREM 6.   *The inverse of an invertible symmetric matrix is a symmetric matrix.*

PROOF.   Let A be an invertible symmetric matrix, then

$$|A| \neq 0 \text{ and } A' = A$$

Now by above theorem.

$$(A')^{-1} = (A^{-1})'$$

$\Rightarrow$ $(A^{-1})' = A^{-1}$ $[\because A' = A]$

$\Rightarrow$ A^{-1} is a symmetric matrix.

THEOREM 7.   *If A is an invertible matrix, then (adj. A)' = adj (A')*

PROOF.   Since A is an invertible matrix, then $|A| \neq 0$

Now $|A'| = |A| \Rightarrow |A'| \neq 0$

$\Rightarrow$ A' is invertible $\Rightarrow$ $(A')^{-1}$ exists.

We have $A(adj. A) = |A|I$

$\Rightarrow$ $(A \, adj. A) = (|A|I)' = |A|I' = |A|I$

$\Rightarrow$ $(adj. A') A' = |A|I$ ...(1)

Also $(adj. A)' A' = |A'|I$

$\Rightarrow$ $(adj. A)'A' = |A|I$ $[\because |A'| = |A|]$ ... (2)

From (1) and (2), we get, $(adj. A)' A' = (adj. A')A'$

$\Rightarrow$ $(adj. A)' A' (A')^{-1} = (adj. A') A' (A')^{-1}$

$\Rightarrow$ $(adj. A)' I = (adj. A')I$

$\Rightarrow$ $(adj. A)' = adj. (A)$

THEOREM 8.   *The adjoint of a symmetric matrix is also a symmetric matrix,*

PROOF.   Let A be a symmetric matrix of order $n \times n$, then $A' = A$

Now by above theorem, $(adj. A)' = adj.(A')$

$\Rightarrow$ $(adj. A)' = adj. A$ $[\because A' = A]$

$\Rightarrow$ $adj. A$ is a symmetric matrix.

THEOREM 9.   *If A is a non-singular matrix, then* $|A^{-1}| = |A|^{-1}.$

PROOF.   Since A is a non-singular matrix, then

$$|A| \neq 0 \qquad \Rightarrow A^{-1} \text{ exists.}$$

Also $AA^{-1} = |I| = A^{-1}A \Rightarrow |AA^{-1}| = |I| = 1$

$\Rightarrow$ $|A| \, |A^{-1}| = 1$ $\Rightarrow$ $|A^{-1}| = \dfrac{1}{|A|} = |A|^{-1}$

THEOREM 10. *If A and B are non-singular matrices of the same order, then*

$$adj. (AB) = (adj. B) (adj. A)$$

PROOF.   Since A and B are non-singular matrices of the same order, then AB exists.

Also $|A| \neq 0, |B| \neq 0 \Rightarrow |AB| = |A||B| \neq 0$

$\Rightarrow$ $(AB)^{-1}$ exists.

Now we have $A(adj. A) = |A| I$ ...(1)

and $B(adj. B) = |B| I$ ...(2)

Also $AB (adj. B) = |AB| I$ ...(3)

We have $(AB) (adj. B \, adj. A) = A(adj. B) \, adj. A$ (By associative law)

$\Rightarrow$ $(AB) (adj. B \, adj. A) = A(|B| I) \, adj. A$ [Using (1)]

$\Rightarrow$ $(AB) (adj. B \, adj. A) = |B|(AI) \, adj. A$

$\Rightarrow$ $(AB) (adj. B \, adj. A) = |B|(A \, adj. A)$

$\Rightarrow$ $(AB) (adj. B \, adj. A) = |B||A| I$

$$\therefore \qquad (AB)\ (adj.\ B\ adj.\ A) = |AB|\ I \qquad\qquad ...(4)$$

From (3) and (4), we get

$$(AB)\ (adj.\ AB) = (AB)\ (adj.\ B\ adj.\ A)$$

$$\Rightarrow \qquad (AB)^{-1}\ (AB)(adj.\ AB) = (AB)^{-1}\ (AB)\ (adj.\ B\ adj.\ A) \qquad [\because (AB)^{-1}\ \text{exists.}]$$

$$\Rightarrow \qquad I(adj.\ AB) = I(adj.\ B\ adj.\ A)$$

$$\Rightarrow \qquad adj.\ (AB) = (adj.\ B)\ (adj.\ A)$$

THEOREM 11. (Cancellation laws) : *Let A, B and C be three square matrices of the same order. If A is a non-singular matrix, then*

(i) $AB = AC \Rightarrow B = C$ *[Left cancellation law]*   (ii) $BA = CA \Rightarrow B = C$ *[Right cancellation law]*

PROOF.   Since A is a non-singular matrix, then $|A| \neq 0 \Rightarrow A^{-1}$ exists.

(i)   We have $\qquad\qquad\qquad AB = AC$

$$\Rightarrow \qquad\qquad A^{-1}(AB) = A^{-1}(AC) \qquad\qquad\qquad [\because A^{-1}\ \text{exists.}]$$

$$\Rightarrow \qquad\qquad (A^{-1}A)B = (A^{-1}A)C \qquad\qquad\qquad [\text{By associative law}]$$

$$\Rightarrow \qquad\qquad IB = IC \qquad\qquad\qquad [\because A^{-1}A = I]$$

$$\Rightarrow \qquad\qquad B = C \qquad\qquad\qquad [\because IB = B, IC = C]$$

(ii)   We have $\qquad\qquad\qquad BA = CA$

$$\Rightarrow \qquad\qquad (BA)\ A^{-1} = (CA)\ A^{-1} \qquad\qquad\qquad [\because A^{-1}\ \text{exists.}]$$

$$\Rightarrow \qquad\qquad B\ (AA^{-1}) = C(AA^{-1}) \qquad\qquad\qquad [\text{By associative law}]$$

$$\Rightarrow \qquad\qquad BI = CI \qquad\qquad\qquad [\because AA^{-1} = I]$$

$$\Rightarrow \qquad\qquad B = C \qquad\qquad\qquad [\because BI = B, CI = C]$$

THEOREM 12. *If the product of two non-null square matrices is a null matrix, then both of them must be singular.*

PROOF.   Let A and B be two non-null matrices of the same order $n \times n$ such that

$$AB = O \qquad\qquad ...(1)$$

where O is a null matrix of order $n \times n$.

Let, if possible B be a non-singular matrix, then B^{-1} exists.

From (1) we have $\qquad\qquad\qquad AB = O$

$$\Rightarrow \qquad\qquad (AB)B^{-1} = OB^{-1}$$

$$\Rightarrow \qquad\qquad A(BB^{-1}) = O \qquad\qquad\qquad [\text{By associative law and } OB^{-1} = O]$$

$$\Rightarrow \qquad\qquad AI_n = O \qquad\qquad\qquad [\because BB^{-1} = I_n]$$

$$\Rightarrow \qquad\qquad A = O \qquad\qquad\qquad [\because AI_n = A]$$

which is a contradiction because A is a non-null matrix.

Therefore, B is a singular matrix.

Similarly, we can prove that A is a singular matrix.

1.14.1 Inverse of a Matrix by Elementary Transformations

Let A be a non-singular matrix of order $n \times n$, then there exists a finite number of elementary matrices $E_1, E_2, ..., E_3, ..., E_s$ such that

$$E_s E_{s-1} ...E_2 E_1 A = I_n$$

$$\Rightarrow \qquad\qquad E_s E_{s-1} = E_2 E_1 AA^{-1} = I_n A^{-1} \qquad\qquad\qquad [\because |A| \neq 0 \Rightarrow A^{-1}\ \text{exists.}]$$

$$\Rightarrow \qquad\qquad (E_s E_{s-1} ... E_2 E_1)\ (AA^{-1}) = I_n A^{-1} \qquad\qquad\qquad [\text{By associative law}]$$

$$\Rightarrow \qquad\qquad (E_s E_{s-1} ... E_2 E_1)\ I_n = A^{-1} \qquad\qquad\qquad [\because AA^{-1} = I_n, A^{-1} = A^{-1}]$$

Hence, $\qquad\qquad\qquad A^{-1} = (E_s E_{s-1} ... E_2 E_1)\ I_n \qquad\qquad ...(1)$

We know that every non-singular matrix of order $n \times n$ can be reduced to the unit matrix I_n by a finite chain of elementary row-transformations only and each elementary row-transformation of a matrix is equivalent to pre-multiplication by the corresponding elementary matrix.

From (1) it follows that if a non-singular matrix A of order $n \times n$ is reduced to the unit matrix I_n by a finite chain of elementary row-transformations only, then the same chain of elementary row-transformations applied to the unit matrix I_n gives the inverse of A.

WORKING PROCEDURE

Let A be a non-singular matrix of order $n \times n$, then we follow the following steps :

Step 1. Write $\qquad A = I_n A$ ...(1)

Step 2. Apply elementary row-transformations on A on L.H.S. of (1) and reduce it to I_n and apply corresponding elementary row-transformations on the pre-factor I_n on R.H.S. of (1) till we obtain $I_n = BA$.

Step 3. Finally, we write $A^{-1} = B$.

Solved Examples

EXAMPLE 1. *By using elementary row-transformations find the inverse of the following matrices:*

$$\text{(i)} \begin{bmatrix} 1 & 2 \\ 3 & 7 \end{bmatrix} \qquad \text{(ii)} \begin{bmatrix} 1 & 2 \\ 2 & -1 \end{bmatrix}$$

SOLUTION. (i) We write $A = I_2 A$

or $\begin{bmatrix} 1 & 2 \\ 3 & 7 \end{bmatrix} = \begin{bmatrix} 1 & 0 \\ 0 & 1 \end{bmatrix} A$

Applying $R_2 \to R_2 - 3R_1$, we get

$$\begin{bmatrix} 1 & 2 \\ 0 & 1 \end{bmatrix} = \begin{bmatrix} 1 & 0 \\ -3 & 1 \end{bmatrix} A$$

Again applying $R_1 \to R_1 - 2R_2$, we get

$$\begin{bmatrix} 1 & 0 \\ 0 & 1 \end{bmatrix} = \begin{bmatrix} 7 & -2 \\ -3 & 1 \end{bmatrix} A$$

$\Rightarrow \qquad I_2 = BA$

$\Rightarrow \qquad A^{-1} = B = \begin{bmatrix} 7 & -2 \\ -3 & 1 \end{bmatrix}.$

(ii) We write $A = I_2 A$

or $\begin{bmatrix} 1 & 2 \\ 2 & -1 \end{bmatrix} = \begin{bmatrix} 1 & 0 \\ 0 & 1 \end{bmatrix} A$

Applying $R_2 \to R_2 - 2R_1$, we get

$$\begin{bmatrix} 1 & 2 \\ 0 & -5 \end{bmatrix} = \begin{bmatrix} 1 & 0 \\ -2 & 1 \end{bmatrix} A$$

Applying $R_2 \to -\dfrac{1}{5} R_2$, we get

$$\begin{bmatrix} 1 & 2 \\ 0 & 1 \end{bmatrix} = \begin{bmatrix} 1 & 0 \\ 2/5 & -1/5 \end{bmatrix} A$$

Applying $R_1 \to R_1 - 2R_2$, we get

$$\begin{bmatrix} 1 & 0 \\ 0 & 1 \end{bmatrix} = \begin{bmatrix} 1/5 & 2/5 \\ 2/5 & -1/5 \end{bmatrix} A$$

$\Rightarrow \qquad I_2 = BA$

$\Rightarrow \qquad A^{-1} = B = \begin{bmatrix} 1/5 & 2/5 \\ 2/5 & -1/5 \end{bmatrix}$

EXAMPLE 2. *Using elementary transformations, find the inverse of the following matrix :*

$$A = \begin{bmatrix} 1 & 2 & 3 \\ 2 & 5 & 7 \\ -2 & -4 & -5 \end{bmatrix}$$

SOLUTION. We write $\qquad A_3 = I_3 A$

or $\begin{bmatrix} 1 & 2 & 3 \\ 2 & 5 & 7 \\ -2 & -4 & -5 \end{bmatrix} = \begin{bmatrix} 1 & 0 & 0 \\ 0 & 1 & 0 \\ 0 & 0 & 1 \end{bmatrix} A$

Applying $R_2 \to R_2 - 2R_1$, $R_3 \to R_3 + 2R_1$, we get

$$\begin{bmatrix} 1 & 2 & 3 \\ 0 & 1 & 1 \\ 0 & 0 & 1 \end{bmatrix} = \begin{bmatrix} 1 & 0 & 0 \\ -2 & 1 & 0 \\ 2 & 0 & 1 \end{bmatrix} A$$

Applying $R_1 \to R_1 - 2R_2$, we get

$$\begin{bmatrix} 1 & 0 & 1 \\ 0 & 1 & 1 \\ 0 & 0 & 1 \end{bmatrix} = \begin{bmatrix} 5 & -2 & 0 \\ -2 & 1 & 0 \\ 2 & 0 & 1 \end{bmatrix} A$$

Applying $R_1 \to R_1 - R_3$, $R_2 \to R_2 - R_3$, we get

$$\begin{bmatrix} 1 & 0 & 0 \\ 0 & 1 & 0 \\ 0 & 0 & 1 \end{bmatrix} = \begin{bmatrix} 3 & -2 & -1 \\ -4 & 1 & -1 \\ 2 & 0 & 1 \end{bmatrix} A$$

$\Rightarrow \qquad I_3 = BA$

$\Rightarrow \qquad A^{-1} = B = \begin{bmatrix} 3 & -2 & -1 \\ -4 & 1 & -1 \\ 2 & 0 & 1 \end{bmatrix}$

EXAMPLE 3. *Find the inverse of the matrix*

$$A = \begin{bmatrix} 0 & 1 & 2 & 2 \\ 1 & 1 & 2 & 3 \\ 2 & 2 & 2 & 3 \\ 2 & 3 & 3 & 3 \end{bmatrix}$$

by using elementary transformations.

SOLUTION. We write $A = I_4 A$

or $\begin{bmatrix} 0 & 1 & 2 & 2 \\ 1 & 1 & 2 & 3 \\ 2 & 2 & 2 & 3 \\ 2 & 3 & 3 & 3 \end{bmatrix} = \begin{bmatrix} 1 & 0 & 0 & 0 \\ 0 & 1 & 0 & 0 \\ 0 & 0 & 1 & 0 \\ 0 & 0 & 0 & 1 \end{bmatrix} A$

Applying $R_2 \leftrightarrow R_1$, we get

$$\begin{bmatrix} 1 & 1 & 2 & 3 \\ 0 & 1 & 2 & 2 \\ 2 & 2 & 2 & 3 \\ 2 & 3 & 3 & 3 \end{bmatrix} = \begin{bmatrix} 0 & 1 & 0 & 0 \\ 1 & 0 & 0 & 0 \\ 0 & 0 & 1 & 0 \\ 0 & 0 & 0 & 1 \end{bmatrix} A$$

Applying $R_3 \to R_3 - 2R_1, R_4 \to R_4 - 2R_1$, we get

$$\begin{bmatrix} 1 & 1 & 2 & 3 \\ 0 & 1 & 2 & 2 \\ 0 & 0 & -2 & -3 \\ 0 & 1 & -1 & -3 \end{bmatrix} = \begin{bmatrix} 0 & 1 & 0 & 0 \\ 1 & 0 & 0 & 0 \\ 0 & -2 & 1 & 0 \\ 0 & -2 & 0 & 1 \end{bmatrix} A$$

Applying $R_1 \to R_1 - R_2, R_4 \to R_4 - R_2$, we get

$$\begin{bmatrix} 1 & 0 & 0 & 1 \\ 0 & 1 & 2 & 2 \\ 0 & 0 & -2 & -3 \\ 0 & 0 & -3 & -5 \end{bmatrix} = \begin{bmatrix} -1 & 1 & 0 & 0 \\ 1 & 0 & 0 & 0 \\ 0 & -2 & 1 & 0 \\ -1 & -2 & 0 & 1 \end{bmatrix} A$$

Applying $R_3 \to -\dfrac{1}{2} R_3$, we get

$$\begin{bmatrix} 1 & 0 & 0 & 1 \\ 0 & 1 & 2 & 2 \\ 0 & 0 & 1 & 3/2 \\ 0 & 0 & -3 & -5 \end{bmatrix} = \begin{bmatrix} -1 & 1 & 0 & 0 \\ 1 & 0 & 0 & 0 \\ 0 & 1 & -1/2 & 0 \\ -1 & -2 & 0 & 1 \end{bmatrix} A$$

Applying $R_2 \to R_2 - 2R_3, R_4 \to R_4 + 3R_3$, we get

$$\begin{bmatrix} 1 & 0 & 0 & 1 \\ 0 & 1 & 0 & -1 \\ 0 & 0 & 1 & 3/2 \\ 0 & 0 & 0 & -1/2 \end{bmatrix} = \begin{bmatrix} -1 & 1 & 0 & 0 \\ 1 & -2 & 1 & 0 \\ 0 & 1 & -1/2 & 0 \\ -1 & 1 & -3/2 & 1 \end{bmatrix} A$$

Applying $R_4 \to -2R_4$, we get

$$\begin{bmatrix} 1 & 0 & 0 & 1 \\ 0 & 1 & 0 & -1 \\ 0 & 0 & 1 & 3/2 \\ 0 & 0 & 0 & 1 \end{bmatrix} = \begin{bmatrix} -1 & 1 & 0 & 0 \\ 1 & -2 & 1 & 0 \\ 0 & 1 & -1/2 & 0 \\ 2 & -2 & 3 & -2 \end{bmatrix} A$$

Applying $R_1 \to R_1 - R_4, R_2 \to R_2 + R_4, R_3 \to R_3 - \dfrac{3}{2} R_4$,

we get

$$\begin{bmatrix} 1 & 0 & 0 & 0 \\ 0 & 1 & 0 & 0 \\ 0 & 0 & 1 & 0 \\ 0 & 0 & 0 & 1 \end{bmatrix} = \begin{bmatrix} -3 & 3 & -3 & 2 \\ 3 & -4 & 4 & -2 \\ -3 & 4 & -5 & 3 \\ 2 & -2 & 3 & -2 \end{bmatrix} A$$

$$\Rightarrow \qquad I_4 = BA$$

$$\Rightarrow \qquad A^{-1} = B = \begin{bmatrix} -3 & 3 & -3 & 2 \\ 3 & -4 & 4 & -2 \\ -3 & 4 & -5 & 3 \\ 2 & -2 & 3 & -2 \end{bmatrix}$$

 Exercise-1.5

Using elementary row-transformations, find the inverse of each of the following matrices, if it exists :

1. $\begin{bmatrix} 5 & 2 \\ 2 & 1 \end{bmatrix}$
2. $\begin{bmatrix} 2 & 3 \\ 0 & 1 \end{bmatrix}$

3. $\begin{bmatrix} 1 & 6 \\ -3 & 5 \end{bmatrix}$
4. $\begin{bmatrix} 1 & 2 & 3 \\ 2 & 4 & 5 \\ 3 & 5 & 6 \end{bmatrix}$

5. $\begin{bmatrix} 1 & 2 & -1 \\ -1 & 1 & 2 \\ 2 & -1 & 1 \end{bmatrix}$
6. $\begin{bmatrix} 1 & -1 & 0 \\ 1 & -3 & 9 \\ 8 & 9 & 2 \end{bmatrix}$

7. $\begin{bmatrix} 0 & 1 & 2 \\ 1 & 2 & 3 \\ 3 & 1 & 1 \end{bmatrix}$
8. $\begin{bmatrix} 2 & 3 & 1 \\ 2 & 4 & 1 \\ 3 & 7 & 2 \end{bmatrix}$

9. $\begin{bmatrix} 1 & -3 & 2 \\ 2 & 0 & 0 \\ 1 & 4 & 1 \end{bmatrix}$
10. $\begin{bmatrix} 2 & -1 & 3 \\ 1 & 2 & 4 \\ 3 & 1 & 1 \end{bmatrix}$

11. $\begin{bmatrix} 1 & 1 & 1 \\ 2 & 2 & 3 \\ 2 & 4 & 9 \end{bmatrix}$
12. $\begin{bmatrix} 2 & 0 & -1 \\ 5 & 1 & 0 \\ 0 & 1 & 3 \end{bmatrix}$

13. $\begin{bmatrix} 1 & 2 & 0 \\ 2 & 3 & -1 \\ 1 & -1 & 3 \end{bmatrix}$
14. $\begin{bmatrix} 1 & 1 & 2 \\ 3 & 1 & 1 \\ 2 & 3 & 1 \end{bmatrix}$

15. $\begin{bmatrix} 1 & 2 & 1 \\ 3 & 2 & 3 \\ 1 & 1 & 2 \end{bmatrix}$
16. $\begin{bmatrix} 3 & 0 & -1 \\ 2 & 3 & 0 \\ 0 & 4 & 1 \end{bmatrix}$

17. $\begin{bmatrix} -1 & -3 & 3 & -1 \\ 1 & 1 & -1 & 0 \\ 2 & -5 & 2 & -3 \\ -1 & 1 & 0 & 1 \end{bmatrix}$
18. $\begin{bmatrix} 1 & 1 & 2 & 0 \\ 0 & 1 & 1 & -1 \\ 2 & 1 & 2 & 1 \\ 3 & -2 & 1 & 6 \end{bmatrix}$

19. $\begin{bmatrix} 3 & -3 & 4 \\ 2 & -3 & 4 \\ 0 & -1 & 1 \end{bmatrix}$ (GBTU-2010)

20. $\begin{bmatrix} 1 & 3 & 3 \\ 1 & 4 & 3 \\ 1 & 3 & 4 \end{bmatrix}$ (UPTU-2008)
21. $\begin{bmatrix} \dfrac{1}{3} & \dfrac{1}{5} & \dfrac{1}{7} \\ \dfrac{1}{5} & \dfrac{1}{7} & \dfrac{1}{11} \\ -\dfrac{1}{7} & \dfrac{1}{11} & \dfrac{1}{13} \end{bmatrix}$ (UPTU-2008)

Answers

1. $\begin{bmatrix} 1 & -2 \\ -2 & 5 \end{bmatrix}$ **2.** $\begin{bmatrix} 1 & -3 \\ 0 & 2 \end{bmatrix}$ **3.** $\dfrac{1}{23}\begin{bmatrix} 5 & -6 \\ 3 & 1 \end{bmatrix}$ **4.** $\begin{bmatrix} 1 & -3 & 2 \\ -3 & 3 & -1 \\ 2 & -1 & 0 \end{bmatrix}$ **5.** $\dfrac{1}{14}\begin{bmatrix} 3 & -1 & 5 \\ 5 & 3 & -1 \\ -1 & 5 & 3 \end{bmatrix}$ **6.** $\dfrac{1}{157}\begin{bmatrix} 87 & -2 & 9 \\ -70 & -2 & 9 \\ -33 & 17 & 2 \end{bmatrix}$ **7.** $\dfrac{1}{2}\begin{bmatrix} 1 & -1 & 1 \\ -8 & 6 & -2 \\ 5 & -3 & 1 \end{bmatrix}$

8. $\begin{bmatrix} 1 & 1 & -1 \\ -1 & 1 & 0 \\ 2 & -5 & 2 \end{bmatrix}$ **9.** $\dfrac{1}{22}\begin{bmatrix} 0 & 11 & 0 \\ -2 & -1 & 4 \\ 8 & -7 & 6 \end{bmatrix}$ **10.** $\dfrac{1}{30}\begin{bmatrix} 2 & -4 & 10 \\ -11 & 7 & 5 \\ 5 & 5 & -5 \end{bmatrix}$ **11.** $\dfrac{1}{3}\begin{bmatrix} -6 & 5 & -1 \\ 15 & -8 & 1 \\ -6 & 3 & 0 \end{bmatrix}$ **12.** $\begin{bmatrix} 3 & -1 & 1 \\ -15 & 6 & -5 \\ 5 & -2 & 2 \end{bmatrix}$ **13.** $\dfrac{1}{6}\begin{bmatrix} -8 & 6 & 2 \\ 7 & -3 & -1 \\ 5 & -3 & 1 \end{bmatrix}$

14. $\dfrac{1}{11}\begin{bmatrix} -2 & 5 & -1 \\ -1 & -3 & 5 \\ 7 & -1 & -2 \end{bmatrix}$ **15.** $\dfrac{1}{4}\begin{bmatrix} -1 & 3 & -4 \\ 3 & -1 & 0 \\ -1 & -1 & 4 \end{bmatrix}$ **16.** $\begin{bmatrix} 3 & -4 & -3 \\ -2 & 3 & 2 \\ 8 & -12 & 9 \end{bmatrix}$ **17.** $\begin{bmatrix} 0 & 2 & 1 & 3 \\ 1 & 1 & -1 & -2 \\ 1 & 2 & 0 & 1 \\ -1 & 1 & 2 & 6 \end{bmatrix}$ **18.** $\begin{bmatrix} 2 & -1 & 1 & -1 \\ -5 & -3 & 1 & 1 \\ 2 & 3 & -1 & 0 \\ -3 & -1 & 0 & 1 \end{bmatrix}$

19. $\begin{bmatrix} 1 & -1 & 0 \\ -2 & 3 & -4 \\ -2 & 3 & -3 \end{bmatrix}$ **20.** $\begin{bmatrix} 7 & -3 & -3 \\ -1 & 1 & 0 \\ -1 & 0 & 1 \end{bmatrix}$ **21.** $\dfrac{1}{2238}\begin{bmatrix} 55125 & -48510 & -45045 \\ -48510 & 105875 & -35035 \\ -45045 & -35035 & 154154 \end{bmatrix}$

1.15 SYSTEM OF LINEAR EQUATIONS

In this section we shall study the nature of solutions of a system of linear equations with the help of the theory of matrices discussed in previous chapters. Before going into details of solutions of linear equations we shall try to understand the concepts of linearly dependent and independent set of vectors.

1.15.1 Vectors and their Dependence and Independence

An ordered set of n numbers $(x_1, x_2, x_3, ..., x_n)$ is known as a vector of order n.

The n numbers $x_1, x_2, x_3, ..., x_n$ are called the components of the vector. We denote this vector by a single letter X. Conveniently, we may write the components of a vector X in the form of a row or in the form of a column.

Therefore, we may write $\qquad X = [x_1, x_2, x_3, ..., x_n]$

which is known as a n-dimensional row vector or it may be written as $X = \begin{bmatrix} x_1 \\ x_2 \\ x_3 \\ \vdots \\ x_n \end{bmatrix}$ which is known as an n-dimensional column vector.

If we consider an $m \times n$ matrix, then it contains m-row vectors and n-column vectors, each row vector consists of the components of an n-vector and each column vector consists of the components of an m-vector.

For example: Consider a matrix of order 3×4 given by

$$A = \begin{bmatrix} 1 & -1 & 4 & 5 \\ 2 & 3 & 0 & -7 \\ 3 & 2 & 2 & 6 \end{bmatrix}$$

Then, $\qquad R_1 = [1 \ -1 \ 4 \ 5], R_2 = [2 \ 3 \ 0 \ -7], R_3 = [3 \ 2 \ 2 \ 6]$

$$C_1 = \begin{bmatrix} 1 \\ 2 \\ 3 \end{bmatrix}, C_2 = \begin{bmatrix} -1 \\ 3 \\ 2 \end{bmatrix}, C_3 = \begin{bmatrix} 4 \\ 0 \\ 2 \end{bmatrix}, C_4 = \begin{bmatrix} 5 \\ -7 \\ 6 \end{bmatrix}$$

Thus A can be written as $\qquad A = \begin{bmatrix} R_1 \\ R_2 \\ R_3 \end{bmatrix}$ or $\qquad A = \begin{bmatrix} C_1 & C_2 & C_3 & C_4 \end{bmatrix}$

Definiton : *If all the components of a vector are zero, then it is called a null vector or a zero vector. It is usually denoted by capital letter O.*

For example : The vectors $\begin{bmatrix} 0 & 0 & 0 & 0 \end{bmatrix}$ and $\begin{bmatrix} 0 \\ 0 \\ 0 \\ 0 \end{bmatrix}$ are both null vectors.

1.15.2 SUM OF TWO VECTORS

Let $X = (x_1, x_2, x_3, ..., x_n)$ and $Y = (y_1, y_2, y_3, ..., y_n)$ be two vectors, then $X + Y$ is obtained by adding their corresponding components. Thus

$$X + Y = (x_1 + y_1, x_2 + y_2, ..., x_n + y_n)$$

REMARK

- If two vectors are of different dimensions then they cannot be added up.

1.15.3 MULTIPLICATION OF A VECTOR BY A SCALAR

Let $X = (x_1, x_2, ..., x_n)$ be an n-vector and λ be a scalar, then λX can be obtained on multiplication of each component of X by λ. Thus

$$\lambda X = (\lambda x_1, \lambda x_2, ..., \lambda x_n)$$

1.15.4 LINEAR DEPENDENCE AND INDEPENDENCE OF VECTORS

Let $x_1, x_2, x_3, ..., x_m$ be m vectors. Then they are said to be linearly independent if

$$\lambda_1 x_1 + \lambda_2 x_2 + ... + \lambda_m x_m = 0 \implies \qquad \lambda_1 = \lambda_2 = \lambda_3 = ... = \lambda_m = 0$$

If atleast one of $\lambda_1, \lambda_2, ..., \lambda_m$ is non-zero, then the vectors $x_1, x_2, ..., x_m$ are called linearly dependent.

1.15.5 LINEAR COMBINATION OF VECTORS

A vector X is said to be a linear combination of the vectors $x_1, x_2, ..., x_m$ if there exists scalars $\lambda_1, \lambda_2, ..., \lambda_m$ such that

$$X = \lambda_1 x_1 + \lambda_2 x_2 + ... + \lambda_m x_m$$

Suppose that the vectors $x_1, x_2, ..., x_m$ are linearly dependent, then in the equation.

$$\lambda_1 x_1 + \lambda_2 x_2 + ... + \lambda_m x_m = O \qquad \qquad ... (1)$$

there is at least one of $\lambda_1, \lambda_2, ... \lambda_m$ is non-zero, let it be λ_r, then equation (1) can be written as

$$\lambda_r X_r = -\lambda_1 x_1 - \lambda_2 x_2 - ... \lambda_{r-1} x_{r-1} - \lambda_{r+1} x_{r+1} - ... - \lambda_m x_m$$

$$\implies \qquad X_r = \left(-\frac{\lambda_1}{\lambda_r}\right) x_1 + \left(-\frac{\lambda_2}{\lambda_r}\right) x_2 + ... + \left(-\frac{\lambda_{r-1}}{\lambda_r}\right) x_{r-1} + \left(-\frac{\lambda_{r+1}}{\lambda_r}\right) x_{r+1} + ... + \left(-\frac{\lambda_m}{\lambda_r}\right) x_m$$

$$\implies \qquad X_r = k_1 x_1 + k_2 x_2 + ... + k_{r-1} x_{r-1} + k_{r+1} x_{r+1} + ... + k_m x_m$$

It follows that X_r is a linear combination of vectors $x_1, x_2, .., x_{r-1}, x_{r+1}, ..., x_m$.

Hence if a set of vectors is linearly dependent, then at least one member of the set can be expressed as a linear combination of the remaining vectors.

1.15.6 LINEAR DEPENDENCE OF THE ROWS AND COLUMNS OF A SQUARE MATRIX

Consider a square matrix of order 3×3, namely $A = \begin{bmatrix} a_{11} & a_{12} & a_{13} \\ a_{21} & a_{22} & a_{23} \\ a_{31} & a_{32} & a_{33} \end{bmatrix}$ or $A = \begin{bmatrix} C_1 & C_2 & C_3 \end{bmatrix}$

where $\qquad C_1 = \begin{bmatrix} a_{11} \\ a_{21} \\ a_{31} \end{bmatrix}, C_2 = \begin{bmatrix} a_{12} \\ a_{22} \\ a_{32} \end{bmatrix}$ and $C_3 = \begin{bmatrix} a_{13} \\ a_{23} \\ a_{33} \end{bmatrix}$

The columns C_1, C_2, C_3 are linearly dependent if there exist scalars k_1, k_2, k_3 not all zero such that

$$k_1 C_1 + k_2 C_2 + k_3 C_3 = O$$

$$\Rightarrow \qquad k_1 \begin{bmatrix} a_{11} \\ a_{21} \\ a_{31} \end{bmatrix} + k_2 \begin{bmatrix} a_{12} \\ a_{22} \\ a_{32} \end{bmatrix} + k_3 \begin{bmatrix} a_{13} \\ a_{23} \\ a_{33} \end{bmatrix} = \begin{bmatrix} 0 \\ 0 \\ 0 \end{bmatrix}$$

$$\therefore \qquad \begin{aligned} k_1 a_{11} + k_2 a_{12} + k_3 a_{13} &= 0 \\ k_1 a_{21} + k_2 a_{22} + k_3 a_{23} &= 0 \\ k_1 a_{31} + k_2 a_{32} + k_3 a_{33} &= 0 \end{aligned}$$

i.e., if $\qquad\qquad\qquad\qquad |A| = 0$

Hence, the columns of A are linearly dependent if $|A| = 0$. Since $|A'| = |A|$, if $|A| = 0$, then $|A'| = 0$. Now if $\qquad |A'| = 0$, then the columns of A' are linearly dependent but the columns of A' are the rows of A. Hence if $|A| = 0$, then both the rows and columns of A are linearly dependent. It follows that if $|A| \neq 0$, then its rows and columns are linearly independent and *vice-versa*.

1.15.7 LINEAR DEPENDENCE AND INDEPENDENCE OF ANY MATRIX

Consider a matrix of order $m \times n$, given by

$$A = \begin{bmatrix} a_{11} & a_{12} & \cdots & a_{1n} \\ a_{21} & a_{22} & \cdots & a_{2n} \\ \vdots & \vdots & & \vdots \\ a_{m1} & a_{m2} & \cdots & a_{mn} \end{bmatrix}$$

Let the rank of A be r, then there exists at least one r-minor of A which is non-zero. If A_r be a square submatrix of order $r \times r$ such that $|A_r| \neq 0$, then r rows and columns of A_r are linearly independent, it follows that the matrix A has r rows and columns which are linearly independent. As the rank of A is r so that no set of $(r+1)$ rows and columns of A can be linearly independent. Hence the rank of a matrix A is defined to be the maximum number of linearly independent rows and columns of A.

Since on interchanging rows, the rank of A does not change so without loss of generality we may suppose that the first r rows of A are linearly independent. Let $x_1, x_2, x_3, \dots, x_r$ denote the r independent vectors and let x_t be one of the remaining $(m-r)$ vectors, then the vectors $x_1, x_2, x_3, \dots, x_r, x_t$ are linearly dependent, therefore there exists scalars $\lambda_1, \lambda_2, \lambda_3, \dots \lambda_r, \lambda_t$, not all zero such that

$$\lambda_1 x_1 + \lambda_2 x_2 + \dots + \lambda_r x_r + \lambda_t x_t = O$$

Since $x_1, x_2, \dots, x_r$ are linearly dependent, so we take $\lambda_t \neq 0$, thus

$$x_t = \left(-\frac{\lambda_1}{\lambda_t}\right) x_1 + \left(-\frac{\lambda_2}{\lambda_t}\right) x_2 + \dots + \left(-\frac{\lambda_r}{\lambda_t}\right) x_r$$

It follows that x_t is a linear combination of $x_1, x_2, \dots, x_r$.

Hence if the rank of a matrix of order $m \times n$ is r, then it has a set of r linearly independent rows (or columns) and $(m-r)$ linearly dependent rows (or columns).

1.16 HOMOGENEOUS LINEAR EQUATIONS

Let us consider a system of linear homogeneous equations as follows

$$\left.\begin{aligned} a_{11}x_1 + a_{12}x_2 + \dots + a_{1n}x_n &= 0 \\ a_{21}x_1 + a_{22}x_2 + \dots + a_{2n}x_n &= 0 \\ \dots\dots\dots\dots\dots\dots\dots\dots\dots \\ a_{m1}x_1 + a_{m2}x_2 + \dots + a_{mn}x_n &= 0 \end{aligned}\right\} \qquad \dots(1)$$

These equations are m equations in n unknowns. Any set of numbers $x_1, x_2, \dots, x_n$ that satisfies all the equations (1) is called a solution of (1).

(I) TRIVIAL SOLUTION

The solution $x_1 = 0, x_2 = 0, \dots x_n = 0$ of the equations (1) is called *trivial* solution.

(II) NON-TRIVIAL SOLUTION

Any solutions other than trivial, if exists, is called a *non-trivial* solution of equation (1).

Let the coefficient matrix be $\qquad\qquad A = \begin{bmatrix} a_{11} & a_{12} & \cdots & a_{1n} \\ a_{21} & a_{22} & \cdots & a_{2n} \\ \vdots & \vdots & & \vdots \\ a_{m1} & a_{m2} & \cdots & a_{mn} \end{bmatrix}_{m \times n}$

and
$$X = \begin{bmatrix} x_1 \\ x_2 \\ x_3 \\ \vdots \\ x_n \end{bmatrix}_{n \times 1} , O = \begin{bmatrix} 0 \\ 0 \\ 0 \\ \vdots \\ 0 \end{bmatrix}_{m \times 1}$$

Then the system of equation (1) can also be written as
$$AX = O \qquad \qquad \qquad \ldots (2)$$

This equation (2) is called a *matrix equation*.

THEOREM 1. *If X_1 and X_2 are two non-trivial solutions of $AX = O$, then $k_1X_1 + k_2X_2$ is also a solution of $AX = O$, where k_1 and k_2 are any arbitrary numbers.*

PROOF. Here $AX = O$ and $AX_1 = O, AX_2 = O$ are given.

Now consider, $A(k_1X_1 + k_2X_2) = k_1(AX_1) + k_2(AX_2) = k_1(O) + k_2(O) = O$

Hence $k_1X_1 + k_2X_2$ is the solution of $AX = O$.

THEOREM 2. *If the rank of A is r, then the number of linearly independent solutions of the equation $AX = O$ which is a system of m homogeneous linear equations in n unknowns is $(n - r)$.*

PROOF. Since the equation is $AX = O \qquad \qquad \qquad \ldots (1)$

where
$$A = \begin{bmatrix} a_{11} & a_{12} & \cdots & a_{1n} \\ a_{21} & a_{22} & \cdots & a_{2n} \\ \vdots & \vdots & & \vdots \\ a_{m1} & a_{m2} & \cdots & a_{mn} \end{bmatrix}_{m \times n} \quad \text{and} \quad X = \begin{bmatrix} x_1 \\ x_2 \\ x_3 \\ \vdots \\ x_n \end{bmatrix}_{n \times 1}, O = \begin{bmatrix} 0 \\ 0 \\ 0 \\ \vdots \\ 0 \end{bmatrix}_{m \times 1}$$

Since the rank of $A = r$, so A has r linearly independent columns. Suppose the matrix A can be written as
$$A = [c_1 c_2 \ldots c_r \ldots c_n]_{1 \times n}$$

where $c_1, c_2, \ldots c_r, \ldots c_n$ are column vectors of the matrix A. Each $c_1, c_2, \ldots c_n$ has m vectors. Thus the equation (1) can be written as
$$x_1 c_1 + x_2 c_2 + \ldots + x_r c_r + \ldots + x_n c_n = 0 \qquad \qquad \ldots (2)$$

But each $c_{r+1}, c_{r+2}, \ldots c_n$ is a linear combination of $c_1, c_2, \ldots c_r$. Then
$$\left. \begin{aligned} c_{r+1} &= p_{11}c_1 + p_{12}c_2 + \ldots + P_{1r}c_r \\ c_{r+2} &= p_{21}c_1 + p_{22}c_2 + \ldots + P_{2r}c_r \\ &\cdots\cdots\cdots\cdots\cdots\cdots\cdots\cdots\cdots \\ C_n &= p_{k1}c_1 + p_{k2}c_2 + \ldots + P_{kr}c_r \end{aligned} \right\} \qquad \ldots (3)$$

where $k = (n - r)$

Now (3) can be written as
$$\left. \begin{aligned} p_{11}c_1 + p_{12}c_2 + \ldots + p_{1r}c_r - 1.c_{r+1} + 0.c_{r+2} + \ldots + 0.c_n &= 0 \\ p_{21}c_1 + p_{22}c_2 + \ldots + p_{2r}c_r + 0.c_{r+1} + 1.c_{r+2} + \ldots + 0.c_n &= 0 \\ \cdots\cdots\cdots\cdots\cdots\cdots\cdots\cdots\cdots\cdots\cdots\cdots\cdots\cdots\cdots\cdots \\ p_{k1}c_1 + p_{k2}c_2 + \ldots + P_{kr}c_r + 0.c_{r+1} - 0.c_{r+2} - \ldots + -1.c_n &= 0 \end{aligned} \right\} \qquad \ldots (4)$$

Thus equation (2) and (4) are same, so comparing we get
$$X_1 = \begin{bmatrix} p_{11} \\ p_{12} \\ \vdots \\ p_{1r} \\ -1 \\ 0 \\ \vdots \\ 0 \end{bmatrix}, X_2 = \begin{bmatrix} p_{21} \\ p_{22} \\ \vdots \\ p_{2r} \\ 0 \\ -1 \\ \vdots \\ 0 \end{bmatrix}, \ldots X_{n-r} = \begin{bmatrix} p_{k1} \\ p_{k2} \\ \vdots \\ p_{kr} \\ 0 \\ 0 \\ \vdots \\ -1 \end{bmatrix}$$

where $\qquad\qquad\qquad\qquad\qquad k = (n - r)$.

Hence, we obtained $(n - r)$ solutions of the equation $AX = O$. Next we have to show that X_1, X_2, X_{n-r} are linearly independent.

For this let us have $\qquad l_1X_1 + l_2X_2 + ... + l_{n-r}X_{n-r} = O$ $\qquad\qquad$ ... (5)

Now comparing the $(r+1)^{th}, (r+2)^{th}, n^{th}$ components on both sides of (5), we get

$$l_1 = 0 = l_2 = ... = l_{n-r}$$

Hence $X_1, X_2, X_3, ..., X_{n-r}$ are linearly independent. Finally we shall have that every solution of the equation $AX = O$ is a linear combination of $X_1, X_2, ..., X_{n-r}$.

Suppose X is any solution of $AX = O$ with components $x_1, x_2, ... x_n$. Then

$$X + x_{r+1}X_1 + x_{r+2}X_2 + ... + x_nX_{n-r} \qquad\qquad ... (6)$$

is also a solution of $AX = O$.

Obviously, let $(n - r)$ components of the vector (6) be all equal to zero. Let $z_1, z_2, ... z_r$ be the first r components of (6). Then $(z_1, z_2, ..., z_r, 0, 0, ... 0)$ is a solution of $AX = O$. Therefore from (2), we get

$$z_1c_1 + z_2c_2 + ... + z_rc_r = O$$

This implies $z_1 = 0 = z_2 = ... = z_r$ because $c_1, c_2, ... c_r$ are linearly independent, and hence (6) comes out to be zero, then

$$X = -x_{r+1}X_1 - x_{r+2}X_2 - ... - x_nX_{n-r}$$

This shows that every solution of $AX = O$ is a linear combination of $X_1, X_2, ... X_{n-r}$.

1.16.1 NATURE OF THE SOLUTION OF THE EQUATION Ax = 0

Since $AX = O$ is a matrix equation of a system of m homogeneous linear equations in n unknowns and A is a coefficient matrix of order $m \times n$. Let the rank of A be r. Then obviously r cannot be greater than n. So that either r is n or r is less than n. Therefore these are some cases.

Case 1. If $r = n$, then the equation $AX = O$, will have no linearly independent solution. So in this case only trivial solution will exist.

Case 2. If $r < n$, then there will be $(n - r)$ linearly independent solution of $AX = O$ and thus in this case we shall have infinite solutions.

Case 3. Suppose the number of equations is less than number of unknowns, i.e., $m < n$ and since $r \leq m$, then obviously $r < n$. Thus in this case a non-zero solution will exist. Therefore, the equation $AX = O$ will have infinite solutions.

WORKING PROCEDURE

In order to determine the solutions of the equation **$AX = O$**, we proceed to the following steps :

Step 1. Reduce the matrix A to Echelon form by applying E-row transformations only. The Echelon form gives the rank of A.

Step 2. Let A be matrix of order $m \times n$ and let $\rho(A) = r$. If $r = n$, then $AX = O$ will have zero solution only. If $r < n$, then we will assign $n - r$ arbitrarily chosen values to $n - r$ unknowns.

Step 3. Let B be the Echelon form of A, then the equation $AX = O$ is equivalent to the equation $BX = O$. Reduce $BX = O$ to a system of equations and choose $n - r$ unknowns in this system of equations for assigning arbitrary values like $c_1, c_2..., c_{n-r}$.

Step 4. By back substitution of $(n-r)$ unknowns to the system of equations reduced from $BX = O$, we finally obtain the solutions. In case of $r < n$, we get infinite solutions.

Solved Examples

EXAMPLE 1. *Find the non-trivial solutions of the equations*:
$$x + y - 6z = 0$$
$$-3x + y + 2z = 0$$
$$x - y + 2z = 0$$

SOLUTION. The given system of equations can be written as
$$AX = O \qquad\qquad ...(1)$$
where $\quad A = \begin{bmatrix} 1 & 1 & -6 \\ -3 & 1 & 2 \\ 1 & -1 & 2 \end{bmatrix}, X = \begin{bmatrix} x \\ y \\ z \end{bmatrix}$ and $O = \begin{bmatrix} 0 \\ 0 \\ 0 \end{bmatrix}$

Reducing the matrix A into Echelon form, we have
Appling $R_2 \to R_2 + 3R_1, R_3 \to R_3 - R_1$, we get

$$A \sim \begin{bmatrix} 1 & 1 & -6 \\ 0 & 4 & -16 \\ 0 & -2 & 8 \end{bmatrix}$$

Again applying $R_2 \to \dfrac{1}{4}R_2$, we get

$$A \sim \begin{bmatrix} 1 & 1 & -6 \\ 0 & 1 & -4 \\ 0 & -2 & 8 \end{bmatrix}$$

Again applying $R_3 \to R_3 + 2R_2$ we get

$$A \sim \begin{bmatrix} 1 & 1 & -6 \\ 0 & 1 & -4 \\ 0 & 0 & 0 \end{bmatrix}$$

The last equivalent matrix in Echelon form with two non-zero rows, therefore $\rho(A) = 2$

Thus the given system of equations is equivalent to

$$\begin{bmatrix} 1 & 1 & -6 \\ 0 & 1 & -4 \\ 0 & 0 & 0 \end{bmatrix} \begin{bmatrix} x \\ y \\ z \end{bmatrix} = \begin{bmatrix} 0 \\ 0 \\ 0 \end{bmatrix}$$

$\Rightarrow \qquad x + y - 6z = 0 \qquad \qquad \text{... (2)}$

$\qquad \qquad y - 4z = 0 \qquad \qquad \text{... (3)}$

Let us put $z = c$ in (3), we get

$$y = 4c$$

Now putting $y = 4c$ and $z = c$ in (2), we get

$$x = 2c$$

Hence, the non-trival solutions of the given system of equatons are $x = 2c, y = 4c, z = c$, where c is a non-zero arbitrary number.

EXAMPLE 2. *Show that the only real value of λ for which the following equations have non-zero solutions is 6:*

$x + 2y + 3z = \lambda x,\ 3x + y + 2z = \lambda y,\ 2x + 3y + z = \lambda z$

SOLUTION. The given system of equations can be rewritten as

$(1 - \lambda)x + 2y + 3z = 0 \qquad \text{...(1)}$

$3x + (1 - \lambda)y + 2z = 0 \qquad \text{...(2)}$

$2x + 3y + (1 - \lambda)z = 0 \qquad \text{...(3)}$

This system of equations can be written as

$$AX = O \qquad \text{...(4)}$$

where

$$A = \begin{bmatrix} 1-\lambda & 2 & 3 \\ 3 & 1-\lambda & 2 \\ 2 & 3 & 1-\lambda \end{bmatrix}, X = \begin{bmatrix} x \\ y \\ z \end{bmatrix} \text{and} O = \begin{bmatrix} 0 \\ 0 \\ 0 \end{bmatrix}$$

For non-zero solutions, we must have $|A| = 0$

i.e., $\qquad \begin{vmatrix} 1-\lambda & 2 & 3 \\ 3 & 1-\lambda & 2 \\ 2 & 3 & 1-\lambda \end{vmatrix} = 0$

$\Rightarrow (1-\lambda)(1-\lambda)(1-\lambda) + 8 + 27 - 6(1-\lambda)$
$\qquad \qquad - 6(1-\lambda) - 6(1-\lambda) = 0$

$\Rightarrow \qquad 1 - \lambda^3 - 3\lambda + 3\lambda^2 + 35 - 18(1-\lambda) = 0$

$\Rightarrow \qquad -\lambda^3 + 3\lambda^2 + 15\lambda + 18 = 0$

$\Rightarrow \qquad \lambda^3 - 3\lambda^2 - 15\lambda - 18 = 0$

$\Rightarrow \qquad (\lambda - 6)(\lambda^2 + 3\lambda + 3) = 0$

Since $\lambda^2 + 3\lambda + 3 = 0$ given imaginary roots, therefore the only real value of λ for which the system of equations is to have a non-zero solution is 6.

EXAMPLE 3. *Find all the solutions of the following system of linear homogeneous equations.*

$x - 2y + z - w = 0$
$x + y - 2z + 3w = 0$
$4x + y - 5z + 8w = 0$
$5x - 7y + 2z - w = 0$

SOLUTION. The coefficient matrix is given by

$$A = \begin{bmatrix} 1 & -2 & 1 & -1 \\ 1 & 1 & -2 & 3 \\ 4 & 1 & -5 & 8 \\ 5 & -7 & 2 & -1 \end{bmatrix}$$

Change this matrix into Echelon form as follows:

Performing $\qquad R_2 \to R_2 - R_1, R_3 \to R_3 - 4R_1$ and $R_4 \to R_4 - 5R_1$

$$\sim \begin{bmatrix} 1 & -2 & 1 & -1 \\ 0 & 3 & -3 & 4 \\ 0 & 9 & -9 & 12 \\ 0 & 3 & -3 & 4 \end{bmatrix}$$

Performing $R_2 \to \dfrac{1}{3} R_2$

$$\sim \begin{bmatrix} 1 & -2 & 1 & -1 \\ 0 & 1 & -1 & \dfrac{4}{3} \\ 0 & 9 & -9 & 12 \\ 0 & 3 & -3 & 4 \end{bmatrix}$$

Performing $R_3 \to R_3 - 9R_2, R_4 \to R_4 - 3R_2$

$$\sim \begin{bmatrix} 1 & -2 & 1 & -1 \\ 0 & 1 & -1 & \dfrac{4}{3} \\ 0 & 0 & 0 & 0 \\ 0 & 0 & 0 & 0 \end{bmatrix}$$

This is an Echelon form having two non-zero rows. Hence rank of $A = 2$.

Therefore the given system of equation is equivalent to

$$\begin{bmatrix} 1 & -2 & 1 & -1 \\ 0 & 1 & -1 & \dfrac{4}{3} \\ 0 & 0 & 0 & 0 \\ 0 & 0 & 0 & 0 \end{bmatrix} \begin{bmatrix} x \\ y \\ z \\ w \end{bmatrix} = 0$$

or $\qquad \qquad x - 2y + z - w = 0 \qquad \text{... (1)}$

$\qquad \qquad y - z + \dfrac{4}{3}w = 0 \qquad \text{... (2)}$

Let $\qquad \qquad z = c_1, w = c_2$

From (2) $\qquad \qquad y = c_1 - \dfrac{4}{3}c_2$

and from (1) $\qquad \qquad x = c_1 - \dfrac{5}{3}c_2$

Hence, solution is

$$x = c_1 - \dfrac{5}{3}c_2, y = c_1 - \dfrac{4}{3}c_2, z = c_1, w = c_2$$

where c_1 and c_2 are arbitrary numbers.

Exercise-1.6

Find the solution of the following system of linear homogeneous equations:

1.
$$x + 2y + 3z = 0$$
$$3x + 4y + 4z = 0$$
$$7x + 10y + 12z = 0$$

2.
$$x + y - 3z + 2w = 0$$
$$2x - y + 2z - 3w = 0$$
$$3x - 2y + z - 4w = 0$$
$$-4x + y - 3z + w = 0$$

3.
$$x + y + z = 0$$
$$2x + 5y + 7z = 0$$
$$2x - 5y + 3z = 0$$

4.
$$3x + 4y - z - 6w = 0$$
$$2x + 3y + 2z - 3w = 0$$
$$2x + y + 4z - 9w = 0$$
$$x + 3y + 13z + 3w = 0$$

5.
$$2x - 3y + z = 0$$
$$x + 2y - 3z = 0$$
$$4x - y - 2z = 0$$

6.
$$x + 2y + 3z = 0$$
$$2x + 3y + 4z = 0$$
$$7x + 13y + 19z = 0$$

7.
$$x + 3y - 2z = 0$$
$$2x - y + 4z = 0$$
$$x - 11y + 14z = 0$$

8.
$$2x - 2y + 5z + 3w = 0$$
$$4x - y + z + w = 0$$
$$3x - 2y + 3z + 4w = 0$$
$$x - 3y + 7z + 6w = 0$$

 Hints to Selected Problems

1. Performing $R_2 \to R_2 - 3R_1, R_3 \to R_3 - 7R_1$, we get

$$A = \begin{bmatrix} 1 & 2 & 3 \\ 0 & -2 & -5 \\ 0 & -4 & -9 \end{bmatrix} \Rightarrow |A| = 10$$

$\Rightarrow$ Rank of A is 3 which is equal to the number of unknown. Therefore, the only solution is $x = y = z = 0$.

3. Do same as (1).

 Answers

1. $x = 0 = y = z$ **2.** $x = 0 = y = z = w$ **3.** $x = 0 = y = z$ **4.** $x = 11c_1 + 6c_2, y = -8c_1 - 3c_2, z = c_1, w = c_2$

5. $x = 0 = y = z$ **6.** $x = c, y = -2c, z = c$ **7.** $x = -\dfrac{10}{7}c, y = \dfrac{8}{7}c, z = c$ **8.** $x = \dfrac{5}{9}c, y = 4c, z = \dfrac{7}{9}c, w = c$

1.17 NON-HOMOGENEOUS EQUATIONS

Let us consider a system of equations which are non-homogeneous as follows:

$$\left. \begin{array}{l} a_{11}x_1 + a_{12}x_2 + \ldots + a_{1n}x_n = b_1 \\ a_{21}x_1 + a_{22}x_2 + \ldots + a_{2n}x_n = b_2 \\ \cdots\cdots\cdots\cdots\cdots\cdots\cdots\cdots\cdots\cdots\cdots\cdots \\ a_{m1}x_1 + a_{m2}x_2 + \ldots + a_{mn}x_n = b_m \end{array} \right\} \qquad \ldots (1)$$

These are m equations in n unknowns. Let

$$A = \begin{bmatrix} a_{11} & a_{12} & \cdots & a_{1n} \\ a_{21} & a_{22} & \cdots & a_{2n} \\ \vdots & \vdots & \vdots & \vdots \\ a_{m1} & a_{m2} & \cdots & a_{mn} \end{bmatrix}_{m \times n}$$

$$X = \begin{bmatrix} x_1 \\ x_2 \\ \vdots \\ x_n \end{bmatrix}_{n \times 1}, B = \begin{bmatrix} b_1 \\ b_1 \\ \vdots \\ b_m \end{bmatrix}_{m \times 1}$$

Then the system of equations (1) can also be written as

$$AX = B \qquad \ldots (2)$$

This equation is called a matrix equation. If $x_1, x_2, \ldots x_n$ simultaneously satisfy the equation (2), then $(x_1, x_2, \ldots x_n)$ is called the solution of (2).

1.17.1 CONSISTENCY AND INCONSISTENCY

When there exists one or more than one solution of the equation $AX = B$, then the equations are said to be consistent otherwise they are said to be inconsistent.

1.17.2 Augmented matrix

The matrix of the type

$$[A \mid B] = \begin{bmatrix} a_{11} & a_{12} & \cdots & a_{1n} & b_1 \\ a_{21} & a_{22} & \cdots & a_{2n} & b_2 \\ \vdots & \vdots & \cdots & \cdots & \cdots \\ a_{m1} & a_{m2} & \cdots & a_{mn} & b_m \end{bmatrix}$$

is called the augmented matrix of the equations.

1.17.3 Condition for Consistency

THEOREM **(Rouche's Theorem).** *The equation $AX = B$ is consistent if and only if the rank of A and the rank of the augmented matrix* $[A \mid B]$ *are same.*

(NIT(KURUKSHETRA)–2008, 18)

PROOF. Since the equation is $\qquad AX = B$

The matrix A can be written as $\qquad A = [c_1, c_2, \ldots c_n]$ ... (1)

where $c_1, c_2, \ldots c_n$ are column vectors. Then the equation (1) can be written as

$$[c_1, c_2, \ldots, c_n] \cdot \begin{bmatrix} x_1 \\ x_2 \\ \vdots \\ x_n \end{bmatrix} = B$$

or $\qquad x_1 c_1 + x_2 c_2 + \ldots + x_n c_n = B$ ... (2)

Suppose the rank of A is r, then A has r linearly independent columns. Let these columns be $c_1, c_2, \ldots c_r$ and these $c_1, c_2, \ldots c_r$ are linearly independent and remaining $(n - r)$ columns are linear combination of $c_1, c_2, \ldots c_r$.

Necessary condition. Suppose the equations are consistent, there must exist $k_1, k_2, \ldots, k_n$ such that

$$k_1 c_1 + k_2 c_2 + \ldots + k_n c_n = B \qquad \ldots (3)$$

But $c_{r+1}, c_{r+2}, \ldots c_n$ is a linear combination of $c_1, c_2, \ldots c_r$ then from (2) it is obvious that B is also a linear combination of $c_1, c_2, \ldots c_r$ and thus $[A \mid B]$ has the rank r. Hence, the rank of A is same as the rank of $[A \mid B]$.

Sufficient condition. Suppose rank A = rank $[A \mid B] = r$. This implies that $[A \mid B]$ has r linearly independent columns. But $c_1, c_2, \ldots c_r$ of $[A \mid B]$ are already linearly independent.

Thus B can be expressed as $\qquad B = k_1 c_1 + k_2 c_2 + \ldots + k_r c_r$; where $k_1, k_2, \ldots k_r$ are scalars. ... (4)

Now, equation (4) becomes $\qquad B = k_1 c_1 + k_2 c_2 + \ldots + k_r c_r + 0 . c_{r+1} + \ldots + 0 . c_n$ ... (5)

Comparing (2) and (5), we get $x_1 = k_1, x_2 = k_2, \ldots x_r = k_r, x_{r+1} = 0, \ldots = x_n = 0$ and these values of $x_1, x_2, \ldots x_n$ are the solution of $AX = B$. Hence, the equations are consistent.

☞ Remarks

- The n equations in n unknowns have a unique solution.
- If rank of A < rank of $[A \mid B]$, then there is no solution.
- If $r = n$, then there will be a unique solution.
- If $r < n$, then $(n - r)$ variables can be assigned arbitrary values. Thus there will be infinite solutions and $(n - r + 1)$ solutions will be linearly independent.
- If $m < n$ and $r \leq m \leq n$, then equations will have infinite solutions.

WORKING PROCEDURE

In order to determine the solutions of the equation $AX = B$, we proceed the following steps:

Step 1. Reduce the augmented matrix $[A \mid B]$ to Echelon form by applying E-row transformations only. The Echelon form gives the rank of A and augmented matrix $[A \mid B]$.

Step 2. (i) If the rank of A is not equal to the rank of $[A \mid B]$, then the system of equations has no solution, *i.e.*, equations are inconsistent.

(ii) If the rank of A is equal to the rank of $[A \mid B]$, then the equations are consistent and they will have unique solution if

rank of A = rank of $[A \mid B]$ = number of unknowns

and then will have infinite solutions if

rank of A = rank of $[A \mid B]$ = number of unknowns

Step 3. Let $[A' \mid B']$ be the reduced Echelon form of $[A \mid B]$. Now reduce the equation $A'X = B'$ to a system of equations, after solving these equations we get the required solution.

Solved Examples

EXAMPLE 1. *Show that the equations*
$$x + 2y - z = 3, 3x - y + 2z = 1,$$
$$2x - 2y + 3z = 2, x - y + z = -1$$
are consistent and solve them.

(BHILAI-2005; VIT−2017; MADRAS-2002)

SOLUTION. The given equations can be written as:

$$\begin{bmatrix} 1 & 2 & -1 \\ 3 & -1 & 2 \\ 2 & -2 & 3 \\ 1 & -1 & 1 \end{bmatrix} \begin{bmatrix} x \\ y \\ z \end{bmatrix} = \begin{bmatrix} 3 \\ 1 \\ 2 \\ -1 \end{bmatrix}, i.e., AX = B$$

Therefore, augmented matrix is

$$[A|B] = \begin{bmatrix} 1 & 2 & -1 & \vdots & 3 \\ 3 & -1 & 2 & \vdots & 1 \\ 2 & -2 & 3 & \vdots & 2 \\ 1 & -1 & 1 & \vdots & -1 \end{bmatrix}$$

Performing $R_2 \rightarrow R_2 - 3R_1, R_3 \rightarrow R_3 - 2R_1, R_4 \rightarrow R_4 - R_1$
we get

$$[A|B] = \begin{bmatrix} 1 & 2 & -1 & \vdots & 3 \\ 0 & -7 & 5 & \vdots & -8 \\ 0 & -6 & 5 & \vdots & -4 \\ 0 & -3 & 2 & \vdots & -4 \end{bmatrix}$$

Performing $R_2 \rightarrow R_2 - R_3$

$$\sim \begin{bmatrix} 1 & 2 & -1 & \vdots & 3 \\ 0 & -1 & 0 & \vdots & -4 \\ 0 & -6 & 5 & \vdots & -4 \\ 0 & -3 & 2 & \vdots & -4 \end{bmatrix}$$

Performing $R_3 \rightarrow R_3 - 6R_2, R_4 \rightarrow R_4 - 3R_2$

$$\sim \begin{bmatrix} 1 & 2 & -1 & \vdots & 3 \\ 0 & -1 & 0 & \vdots & -4 \\ 0 & 0 & 5 & \vdots & 20 \\ 0 & 0 & 2 & \vdots & 8 \end{bmatrix}$$

Performing $R_3 \rightarrow \dfrac{1}{5}R_3, R_4 \rightarrow \dfrac{1}{2}R_4$

$$\sim \begin{bmatrix} 1 & 2 & -1 & \vdots & 3 \\ 0 & -1 & 0 & \vdots & -4 \\ 0 & 0 & 1 & \vdots & 4 \\ 0 & 0 & 1 & \vdots & 4 \end{bmatrix}$$

Performing $R_4 \rightarrow R_4 - R_3$

$$\sim \begin{bmatrix} 1 & 2 & -1 & \vdots & 3 \\ 0 & -1 & 0 & \vdots & -4 \\ 0 & 0 & 1 & \vdots & 4 \\ 0 & 0 & 0 & \vdots & 0 \end{bmatrix}$$

This is an Echelon form and having three non-zero rows. Thus rank A = rank of $[A|B]$ = 3. Therefore the equations are consistent

and $$\begin{bmatrix} 1 & 2 & -1 \\ 0 & -1 & 0 \\ 0 & 0 & 1 \\ 0 & 0 & 0 \end{bmatrix} \begin{bmatrix} x \\ y \\ z \end{bmatrix} = \begin{bmatrix} 3 \\ -4 \\ 4 \\ 0 \end{bmatrix}$$

$$\therefore \qquad x + 2y - z = 3, -y = -4, z = 4$$
Hence, the solution is $x = -1, \ y = 4, z = 4$.

EXAMPLE 2. *Investigate for what values of* λ, μ *the simulataneous equations*
$$x + y + z = 6, x + 2y + 3z = 10, x + 2y + \lambda z = \mu$$
have (i) no solution (ii) a unique solution (iii) infinite solution.

(UKTU-2011, UPTU-2006, 14, MUMBAI-2007, PTU-2005, 07, ROHTAK-2004)

SOLUTION. The given equations can be written as

$$\begin{bmatrix} 1 & 1 & 1 \\ 1 & 2 & 3 \\ 1 & 2 & \lambda \end{bmatrix} \begin{bmatrix} x \\ y \\ z \end{bmatrix} = \begin{bmatrix} 6 \\ 10 \\ \mu \end{bmatrix}$$

i.e., $\qquad AX = B$

Therefore, augmented matrix is

$$[A|B] = \begin{bmatrix} 1 & 1 & 1 & \vdots & 6 \\ 1 & 2 & 3 & \vdots & 10 \\ 1 & 2 & \lambda & \vdots & \mu \end{bmatrix}$$

Performing $R_2 \rightarrow R_2 - R_1, R_3 \rightarrow R_3 - R_1$, we get

$$\sim \begin{bmatrix} 1 & 1 & 1 & \vdots & 6 \\ 0 & 1 & 2 & \vdots & 4 \\ 0 & 1 & \lambda - 1 & \vdots & \mu - 6 \end{bmatrix}$$

Performing $R_3 \rightarrow R_3 - R_2$

$$\sim \begin{bmatrix} 1 & 1 & 1 & \vdots & 6 \\ 0 & 1 & 2 & \vdots & 4 \\ 0 & 0 & \lambda - 3 & \vdots & \mu - 10 \end{bmatrix}$$

If $\lambda \neq 3$, then rank A = rank $[A|B]$ = 3. Thus in this case a unique solution exists. If $\lambda = 3$ and $\mu_0 \neq 10$, then rank $A = 2$, rank $[A|B]$ is 3. Thus rank $A \neq$ rank $[A|B]$. Hence, in this case equations are inconsistent.

If $\lambda = 3$ and $\mu = 10$, then rank A = rank $[A|B]$ = 2. Thus in this case infinite solutions exist.

EXAMPLE 3. *For what values of* η *the equations* $x+y+z=1$, $x+2y+4z = \eta$, $x+4y+10z = \eta^2$ *have a solution? Solve them completely in each case.*

(GBTU-2011, BHOPAL-2000, MUMBAI-2008, VTU-2006)

SOLUTION. The given system of equations can be written as
$$AX=B \qquad \qquad ... (1)$$

where $\quad A = \begin{bmatrix} 1 & 1 & 1 \\ 1 & 2 & 4 \\ 1 & 4 & 10 \end{bmatrix}, X = \begin{bmatrix} x \\ y \\ z \end{bmatrix}, B = \begin{bmatrix} 1 \\ \eta \\ \eta^2 \end{bmatrix}$

Augmented matrix $[A|B]$ is given by

$$\begin{bmatrix} 1 & 1 & 1 & \vdots & 1 \\ 1 & 2 & 4 & \vdots & \eta \\ 1 & 4 & 10 & \vdots & \eta^2 \end{bmatrix}$$

Applying $R_2 \rightarrow R_2 - R_1, R_3 \rightarrow R_3 - R_1$, we get

$$\sim \begin{bmatrix} 1 & 1 & 1 & \vdots & 1 \\ 0 & 1 & 3 & \vdots & \eta - 1 \\ 0 & 3 & 9 & \vdots & \eta^2 - 1 \end{bmatrix}$$

Applying $R_3 \to R_3 - 3R_2$, we get

$$\sim \begin{bmatrix} 1 & 1 & 1 & : & 1 \\ 0 & 1 & 3 & : & \eta-1 \\ 0 & 0 & 0 & : & \eta^2-3\eta+2 \end{bmatrix}$$

This last equivalent matrix is in Echelon form. The given system of equations will have the solutions if

rank of A = rank of $[A|B]$

For Echelon form, the rank of A is 2 and the augmented matrix $[A|B]$ will have rank 2 if

$$\eta^2 - 3\eta + 2 = 0$$

i.e., if $(\eta-2)(\eta-1) = 0$

i.e., if $\eta = 1, 2$

The last equivalent matrix gives the system of equations as follows:

$$\begin{bmatrix} 1 & 1 & 1 \\ 0 & 1 & 3 \\ 0 & 0 & 0 \end{bmatrix}\begin{bmatrix} x \\ y \\ z \end{bmatrix} = \begin{bmatrix} 1 \\ \eta-1 \\ \eta^2-3\eta+2 \end{bmatrix}$$

$$\Rightarrow \quad \left.\begin{array}{r} x+y+z = 1 \\ y+3z = \eta-1 \end{array}\right\} \qquad \dots(2)$$

Since rank of A = rank of $[A|B]$ if $\eta = 1$ and $\eta = 2$
Now we have two cases:

Case I: When $\eta = 1$

From (2), we have

$$\left.\begin{array}{r} x+y+z = 1 \\ y+3z = 0 \end{array}\right\} \qquad \dots(3)$$

Since rank of A = rank of $[A|B] = 2$ and number of unknowns is 3, therefore we will have $3 - 2 = 1$ unknown to be assigned.

Let us assign z to be c_1, therefore put $z = c_1$ in $y + 3z = 0$, we get $y = -3c_1$.

Again putting $y = -3c_1$ and $z = c_1$ in $x + y + z = 1$, we get $x = 1 + 2c_1$

Thus, in this case the solutions are

$$x = 1 + 2c_1, y = -3c_1, z = c_1$$

where c_1 is an arbitrary number.

Case II : When $\eta = 2$

From (2), we have

$$\left.\begin{array}{r} x+y+z = 1 \\ y+3z = 1 \end{array}\right\} \qquad \dots(4)$$

Let us assign z to be c_2, therefore, putting $z = c_2$ in $y + 3z = 1$, we get $y = 1 - 3c_2$.

Again, putting $z = c_2, y = 1 - 3c_2$ in $x + y + z = 1$, we get $x = 2c_2$.

Thus, in this case the solutions are

$$x = 2c_2, y = 1 - 3c_2, z = c_2$$

where c_2 is an arbitrary number.

Exercise-1.7

1. Use matrix method to solve the equations
   $$2x - y + 3z = 9, x + y + z = 6, x - y + z = 2.$$

2. Show that the equations $x - 3y - 8z + 10 = 0$, $3x + y - 4z = 0$, $2x + 5y + 6z - 13 = 0$ are consistent and solve them.

3. Examine if the system of equations
   $$x + y + 4z = 6, 3x + 2y - 2z = 9, 5x + y + 2z = 13$$
   is consistent. Find also the solution if it exists.

4. For what values of λ will the following equations fail to have a unique solution
   $$3x - y + \lambda z = 1, 2x + y + z = 2, x + 2y - \lambda z = -1$$
   Will the equations have any solution for these values of λ?

5. Solve $2x + 3y + z = 9, x + 2y + 3z = 6, 3x + y + 2z = 8.$
   Solve the following equations by matrix method:

6. $5x + 3y + 7z = 4, 3x + 26y - 2z = 9, 7x + 2y + 10z = 5.$
   (JNTU-2005, PTU-2005, 12, BHOPAL-2008)

7. $5x - 6y + 4z = 15, 7x + 4y - 3z = 19, 2x + y + 6z = 46.$

8. $x - y + 2z = 4, 3x + y + 4z = 6, x + y + z = 1.$

9. $x + y + z = 6, x + 2y + 3z = 4, 3x + y - 4z = 0.$
   (UPTU-2007)

10. $2x - y + 3z = 8, -x + 2y + z = 4, 3x + y - 4z = 0.$

11. Show that the following equations are inconsistent
    $$2x - y + z = 4, 3x - y + z = 6, 4x - y + 2z = 7, -x + y - z = 9.$$

12. Show that the equations are inconsistent
    $$x - 4y + 7z = 14, 3x + 8y - 2z = 13, 7x - 8y + 26z = 5.$$

13. Prove that the following system of equations have a unique solution
    $$5x + 3y + 14z = 4, y + 2z = 1, x - y + 2z = 0.$$

14. Solve the following equations by matrix mehod:
    $$x + y + z = 9, 2x + 5y + 7z = 52, 2x + y - z = 0.$$

15. Using matrix method, show that the equations are consistant and hence find the solutions.
    $$3x + 3y + 2z = 1, x + 2y = 4, 10y + 3z = -2,$$
    $2x - 3y - z = 5$ are consistant and hence find the solutions.
    (UKTU-2010, GBTU-2010, NAGARJUNA-2008, 14)

16. Show that the equations $2x + 6y + 11 = 0$, $6x + 20y - 6z + 3 = 0$ and $6y - 18z + 1 = 0$ are not consistent.
    (UKTU-2011, RAJASTHAN-2005, 13)

17. Show that the system of equations $2x - 3y + 7z = 5$; $3x + y - 3z = 13$ and $2x + 19y - 47z = 32$ is not consistent.
    (GBTU-2010)

18. For what value of λ, the system of equations
    $$2x - 2y + z = \lambda x, 2x - 3y + 2z = \lambda y, -x + 2y + 0z = \lambda z$$
    posses a non-trivial solution. Obtain its general solution
    (MTU-2011)

19. Find the value of k so that the equation $x + y + 3z = 0$, $4x + 3y + kz = 0$ and $2x + y + 2z = 0$ have a non-trivial solution.
    (UPTU-2008)

20. Show that the system of equations $3x + 4y + 5z = a$, $4x + 5y + 6z = b$, $5x + 6y + 7z = c$ does not have a solution unless $a + c = 2b$.
    (UPTU-2008, MTU-2009, RAIPUR-2004, 14, NAGPUR-2001)

Hints to Selected Problems

1. Consider the augmented matrix and perform the following opeartions sequentially

$$R_1 \leftrightarrow R_2,\ R_2 \rightarrow R_2 - 2R_1,\ R_3 \rightarrow R_3 - R_1,\ R_3 \rightarrow R_3 - \frac{2}{3}R_2.$$

2. Here, Rank $(A) = 2$, which is less than the number of unknowns, therefore, given system of equations have infinite number of solutions.

3. The rank of augmented matrix is equal to the rank of (A).

Therefore, the given system of equation is consistent.

4. The coefficient matrix A is non-singular if $\lambda \neq -\frac{7}{2}$. Thus the given system of equations have a unique solution if $\lambda \neq \frac{7}{2}$.

11. The rank of augmented matrix = 4

Rank of A is 3.

Hence the given system of equations is inconsistent.

 **Answers**

1. $x = 1, y = 2, z = 3$ **2.** $x = 2c - 1, y = 3 - 2c, z = c$ **3.** Consistent; $x = 2, y = 2, z = \frac{1}{2}$

4. $\lambda \neq -\frac{7}{2}$ solution is unique; $\lambda = -\frac{7}{2}$, no solution. **5.** $x = \frac{35}{18}, y = \frac{29}{18}, z = \frac{5}{18}$ **6.** $x = \frac{7}{11}, y = \frac{3}{11}, z = 0$

7. $x = 3, y = 4, z = 6$ **8.** $x = \frac{5}{2} - \frac{3}{2}c, y = -\frac{3}{2} + \frac{1}{2}c, z = c$ **9.** $x = c - 2, y = 8 - 2c, z = c$ **10.** $x = 2, y = 2, z = 2$ **14.** $x = 1, y = 3, z = 4$

15. $2, 1, -4$ **18.** $\lambda = 1, x = 2k_1 - k_2, y = k_1, z = k_2, \lambda = -3, x = -k', y' = -2k, z = k$ **19.** 8

1.18 GAUSS ELIMINATION METHOD

In this method, the variables from the system of linear equations are eliminated successively and the system of equations is therefore reduced to an upper triangular system from which the variable are determined by back substitution. This method is described as follows: Let us consider a system of linear equation

$$AX = B \qquad \qquad \ldots(1)$$

Assuming det $A \neq 0$. Equation (1) has the following form:

$$\left.\begin{aligned}
a_{11}x_1 + a_{12}x_2 + \ldots + a_{1n}x_n &= b_1 \\
a_{21}x_1 + a_{22}x_2 + \ldots + a_{2n}x_n &= b_2 \\
\ldots \quad \ldots \quad \ldots \quad \ldots \quad \ldots \quad \ldots \\
\ldots \quad \ldots \quad \ldots \quad \ldots \quad \ldots \quad \ldots \\
a_{n1}x_1 + a_{n2}x_2 + \ldots + a_{nn}x_n &= b_n
\end{aligned}\right\} \qquad \ldots(2)$$

Assuming $a_{11} \neq 0$ and divide the first equation by a_{11} and then we subtract this equation multiplied by $a_{21}, a_{31}, \ldots, a_{n1}$ from second, third ... nth equation of (2), we get

$$\left.\begin{aligned}
x_1 + a'_{12}x_2 + \ldots + a'_{1n}x_n &= b'_1 \\
a'_{22}x_2 + \ldots + a'_{2n}x_n &= b'_2 \\
\ldots \quad \ldots \quad \ldots \quad \ldots \quad \ldots \\
\ldots \quad \ldots \quad \ldots \quad \ldots \quad \ldots \\
a'_{n2}x_2 + \ldots + a'_{n2}x_n &= b'_n
\end{aligned}\right\} \qquad \ldots(3)$$

Next, we divide second equation of (3) by a'_{22} (assuming $a'_{22} \neq 0$) and subtract this equation multiplied by $a'_{32}, a'_{42}, \ldots, a'_{n2}$ from third, fourth ... nth equation of (3), we get

$$\left.\begin{aligned}
x_1 + a'_{12}x_2 + \ldots + a'_{1n}x_n &= b'_1 \\
x_2 + a''_{23}x_3 + \ldots + a''_{2n}x_n &= b''_2 \\
a''_{33}x_3 + \ldots + a''_{3n}x_n &= b''_3 \\
\ldots \quad \ldots \quad \ldots \quad \ldots \quad \ldots \\
a''_{3n}x_3 + \ldots + a''_{nn}x_n &= b''_n
\end{aligned}\right\} \qquad \ldots(4)$$

Continuing in this way, we get a system of equation as follows:

$$\left.\begin{aligned}
x_1 + c_{12}x_2 + c_{13}x_3 + \ldots + c_{1n}x_n &= d_1 \\
x_2 + c_{23}x_3 + \ldots + c_{2n}x_n &= d_2 \\
\vdots \\
\vdots \\
c_{nn}x_n &= d_n
\end{aligned}\right\} \qquad \ldots(5)$$

This is a form of upper triangular system. From back substitution we can find the solution of the system of given equations.

REMARKS

- The coefficient a_{11}, a'_{22} and a''_{33} are called pivots.
- This method will fail if any one of the pivots a_{11}, a'_{22} and a''_{33} becomes zero. In such cases, we rewrite the equations in a different order so that the pivots are non-zero.
- From each of the procedure, the largest coefficient of x is chosen as pivot element.
- This method proposes a systematic astrology for reducing the system of equations to the upper triangular form using the forward elimination approach and then for obtaining values of unknowns using back substitution process.

WORKING PROCEDURE

Let us consider these equations

$$\left.\begin{array}{l} a_{11}x_1 + a_{12}x_2 + a_{13}x_3 = b_1 \\ a_{21}x_1 + a_{22}x_2 + a_{23}x_3 = b_2 \\ a_{31}x_1 + a_{32}x_2 + a_{33}x_3 = b_3 \end{array}\right\} \qquad \text{...(6)}$$

Step 1. First, eliminate x_1 from second and third equations. Assuming $a_{11} \neq 0$, now dividing first equation by a_{11} and then subtract from second and third after multiplied by a_{21} and respectively, we get

$$\left.\begin{array}{l} x_1 + a'_{12}x_2 + a'_{13}x_3 = b'_1 \\ a'_{22}x_2 + a'_{23}x_3 = b'_2 \\ a'_{32}x_2 + a'_{33}x_3 = b'_3 \end{array}\right\} \qquad \text{...(7)}$$

where $a'_{12} = \dfrac{a_{12}}{a_{11}}$, $a'_{13} = \dfrac{a_{13}}{a_{11}}$, $a'_{22} = a_{22} - a_{21}a'_{12}$, $a'_{23} = a_{23} - a_{21}a'_{13}$

$a'_{32} = a_{32} - a_{31}a'_{12}$, $a'_{33} = a_{33} - a_{31}a'_{13}$, $b'_1 = \dfrac{b_1}{a_{11}}$, $b'_2 = b_2 - a_{21}b'_1$, $b'_3 = b_3 - a_{31}b'_1$

Step 2. Now eliminating x_2 from third equation in (7).

Again assuming $a'_{22} \neq 0$. Dividing second equation in (7) by $a_{22}{}'$ and then subtract from third equation after multiplied by $a_{32}{}'$ we get

$$\left.\begin{array}{l} x_1 + a'_{12}x_2 + a'_{13}x_3 = b'_1 \\ x_2 + a''_{23}x_3 = b'_2 \\ a''_{33}x_3 = b'_3 \end{array}\right\} \qquad \text{...(8)}$$

where $a''_{23} = \dfrac{a'_{23}}{a'_{22}}$, $a'_{33} = a'_{33} - a'_{32}a''_{23}$, $b''_2 = \dfrac{b'_2}{a'_{22}}$, $b''_3 = b_3{}' - a'_{32}b''_2$.

Step 3. Evaluating x_1, x_2 and x_3 from (8) by back substitution.

Solved Examples

EXAMPLE 1. *Solve the following equations by Gauss's elimination method*

$$6x + 3y + 2z = 6$$
$$6x + 4y + 3z = 0$$
$$20x + 15y + 12z = 0.$$

SOLUTION. Here pivot element is 6. Now Divide first equation by 6, we get

$$x + \frac{1}{2}y + \frac{1}{3}z = 1 \qquad \text{...(1)}$$

Now eliminating x from second and third equation with the help of (1). Subtract (1) multiplied by 6 and 20 from second and third equation, respectively we get

$$y + z = -6 \qquad \text{...(2)}$$

$$5y + \frac{16}{3}z = -20 \qquad \text{...(3)}$$

Now eliminating y from (3) with the help of (2), we get

$$\left(\frac{16}{3} - 5\right)z = -20 + 30$$

$$\frac{1}{3}z = 10 \Rightarrow z = 30$$

Substitute the value of z into (2), we get

$$y = -6 - 30 = -36$$

and again substitute the values of y and z into (1), we get

$$x + \frac{1}{2}(-36) + \frac{1}{3}(30) = 1$$

$$x - 18 + 10 = 1 \Rightarrow x = 9$$

Hence, the solution of the equations are

$$x = 9, y = -36, z = 30.$$

EXAMPLE 2. *Solve the following system by Gauss's elimination method :*

$$2x + y + z = 10$$
$$3x + 2y + 3z = 18$$
$$x + 4y + 9z = 4$$

SOLUTION. We have $\quad 2x + y + z = 10 \qquad \text{...(1)}$
$$3x + 2y + 3z = 18 \qquad \text{...(2)}$$
$$x + 4y + 9z = 4 \qquad \text{...(3)}$$

Divide (1) and 2 and subtract after multiplied by 3 from (2) then subtract from (3), we get

$$x + \frac{1}{2}y + \frac{1}{2}z = 5 \qquad \text{...(4)}$$

$$\frac{1}{2}y + \frac{3}{2}z = 3 \qquad \text{...(5)}$$

$$\frac{7}{2}y + \frac{17}{2}z = 11 \qquad \qquad \dots(6)$$

Now divide (5) by $\frac{1}{2}$ and then subtract after multiplied

by $\frac{7}{2}$ from (6) we get,

$$x + \frac{1}{2}y + \frac{1}{2}z = 5 \qquad \dots(7)$$
$$y + 3z = 6 \qquad \dots(8)$$
$$-2z = -10 \qquad \dots(9)$$

From back substitution in (9), (8) and (7) we get

$$z = 5$$
and $$y + 3z = 6$$
$$y + 3(5) = 6$$
$$y = 6 - 15 \quad \Rightarrow y = -9$$
$$y = -9$$
and $$x + \frac{1}{2}(-9) + \frac{1}{2}(5) = 5$$

$$x = 5 + \frac{9}{2} - \frac{5}{2}$$

Hence, the solution is $x = 7, y = -9, z = 5$.

EXAMPLE 3. *Solve by Gauss's elimination method*
$$x + 2y + z = 3$$
$$2x + 3y + 3z = 10$$
$$3x - y + 2z = 13$$

SOLUTION. Given equation are
$$x + 2y + z = 3 \qquad \dots(1)$$
$$2x + 3y + 3z = 10 \qquad \dots(2)$$
$$3x - y + 2z = 13 \qquad \dots(3)$$

Here pivot element of (1) is 1. Now eliminating x from

(2) and (3) by subtracting (1) after multiplied by 2 and 3 respectively from (2) and (3), we get
$$x + 2y + z = 3 \qquad \dots(4)$$
$$-y + z = 4 \qquad \dots(5)$$
$$-7y - z = 4 \qquad \dots(6)$$

Now eliminating y from (6) with the help of (5) by subtracting (5) after multiplied by -7 from (6), we get
$$x + 2y + z = 3 \qquad \dots(7)$$
$$-y + z = 4 \qquad \dots(8)$$
$$6z = 32 \qquad \dots(9)$$

By back substitution from (7), (8) and (9), we get

From (9) $$z = \frac{32}{6} = \frac{16}{3}$$

From (8) $$-y = 4 - z$$

$$-y = 4 - \frac{32}{6} = -\frac{8}{6}$$

$$\therefore \qquad y = \frac{8}{6} = \frac{4}{3}$$

From (7)

$$x + 2y + z = 3$$

$$x + 2\left(\frac{8}{6}\right) + \frac{32}{6} = 3$$

$$x = 3 - \frac{32}{6} - \frac{16}{6}$$

$$= \frac{18 - 48}{6} = -\frac{30}{6}$$

$$x = -5$$

Hence, the solution is $x = -5, y = \frac{4}{3}, z = \frac{16}{3}$.

Exercise-1.8

1. Solve the following equations by Gauss's elimination method :

(i) $x_1 + x_2 + 2x_3 = 4$
$3x_1 + x_2 - 3x_3 = -4$
$2x_1 - 3x_2 - 5x_3 = -5$

(ii) $2x_1 + x_2 + 4x_3 = 12$
$8x_1 - 3x_2 + 2x_3 = 20$
$4x_1 + 11x_2 - x_3 = 33$

(iii) $x_1 + x_2 + x_3 = 10$
$2x_1 + x_2 + 2x_3 = 17$
$3x_1 + 2x_2 + x_3 = 17$

(iv) $2x + 3y - z = 5$
$4x + 4y - 3z = 3$
$2x - 3y + 2z = 2$

(v) $2x + y + z = 10$
$x + 2y + 3z = 18$
$x + 4y + 9z = 16$

(vi) $2x_1 + 4x_2 + x_3 = 2$
$3x_1 + 2x_2 - 2x_3 = -2$
$x_1 - x_2 + x_3 = 6$

Answers

1. (i) $x_1 = 1, x_2 = -1, x_3 = 2$

(ii) $x_1 = 3, x_2 = 2, x_3 = 1$ (iii) $x_1 = 2, x_2 = 3, x_3 = 5$

(iv) $x = 1, y = 2, z = 3$

(v) $x = 7, y = -9, z = 5$ (vi) $x_1 = 2, x_2 = -1, x_3 = 3$

1.19 EIGENVALUE AND EIGENVECTORS OF A MATRIX

A polynomial in indeterminate λ of the form $f(\lambda) = A_0 + A_1\lambda + A_2\lambda^2 + \dots + A_n\lambda^n$ where $A_0, A_1, A_2, \dots, A_n$ are all square matrices of the same order, is called a matrix polynomial of degree n if $A_n \neq O$ (null matrix).

From above definition it is clear that every square matrix can be expressed as a matrix polynomial of zero degree. If A is a square matrix, then we can write

$$A = \lambda° A$$

☞ REMARK

• Two matrix polynomials are said to be equal if and only if the coefficients of like powers of λ are the same.

1.19.1 THE CHARACTERISTIC EQUATION OF A MATRIX

(UPTU 2008)

Let A be a square matrix of order $n \times n$ and let

$$A = \begin{bmatrix} a_{11} & a_{12} & \cdots & a_{1n} \\ a_{21} & a_{22} & \cdots & a_{2n} \\ \vdots & \vdots & & \vdots \\ a_{n1} & a_{n2} & \cdots & a_{nn} \end{bmatrix}$$

If λ is indeterminate, then the matrix $A - \lambda I$ is called the characteristic matrix of A, where I is the unit matrix of order $n \times n$.

The determinant $|A - \lambda I| = \begin{bmatrix} a_{11} - \lambda & a_{12} & \cdots & a_{1n} \\ a_{21} & a_{22} - \lambda & \cdots & a_{2n} \\ \vdots & \vdots & & \vdots \\ a_{n1} & a_{n2} & \cdots & a_{nn} - \lambda \end{bmatrix}$ is an ordinary polynomial in λ which is called the characteristic polynomial of

A and the equation

$$|A - \lambda I| = 0$$

i.e.,

$$\begin{vmatrix} a_{11} - \lambda & a_{12} & \cdots & a_{1n} \\ a_{21} & a_{22} - \lambda & \cdots & a_{2n} \\ \vdots & \vdots & & \vdots \\ a_{n1} & a_{n2} & \cdots & a_{nn} - \lambda \end{vmatrix} = 0$$

is known as the characteristic equation of A. The roots of the equation $|A - \lambda I| = 0$ are called characteristic roots or latent roots or eigenvalues of A. The set of all eigenvalues of a matrix A is called spectrum of A.

1.19.2 CHARACTERISTIC VECTORS OR EIGENVECTORS OF A MATRIX

Let $A = [a_{ij}]$ be a matrix of order $n \times n$ and let $X = \begin{bmatrix} x_1 \\ x_2 \\ \vdots \\ x_n \end{bmatrix}$ be a column vector. Consider a vector equation

$$AX = \lambda X \, ; \text{ where } \lambda \text{ is a scalar.} \qquad \qquad \ldots(1)$$

It is evident that $X = O$ satisfies the equation (1) for every value of λ, thus $X = O$ is a solution of (1). A value of λ for which a non-zero vector, *i.e.*, $X \neq O$ satisfies (1) is called an eigenvalue of the matrix A and the non-zero vector X is called an eigenvector of A corresponding to that eigenvalue λ.

Now equation (1) can be written as $\qquad AX = \lambda IX$

or $\qquad\qquad\qquad\qquad\qquad (A - \lambda I)X = O \qquad\qquad\qquad\qquad\qquad \ldots(2)$

where I is the unit matrix of order $n \times n$. Equation (2) represents a matrix equation of a system of n homogeneous equations. The necessary and sufficient condition for the equation (2) to possess a non-zero solution, *i.e.*, $X \neq O$ is that $|A - \lambda I| = 0$, which is a characteristic equation of matrix A.

REMARKS

- The eigenvector is also known as proper vector.
- If X is an eigenvector of a matrix corresponding to eigenvalue λ, then for any non-zero scalar kX is also an eigenvector of A corresponding to the same eigenvalue λ.
- Corresponding to an eigenvalue of a matrix A, there will be different eigenvectors of A.
- For a given eigenvector of a matrix A there corresponds one and only one eigenvalue of A.

1.19.3 RELATION BETWEEN EIGENVALUES AND EIGENVECTORS

(1) λ is an eigenvalue of a matrix A if and only if there exists a non-zero vector X such that $AX = \lambda X$.

(2) If X is an eigenvector of a matrix A corresponding to an eigenvalue of A, then kX is also an eigenvector of A corresponding to the same eigenvalue A, where k is any non-zero number.

(4) If X is a non-zero eigenvector of a matrix A, then X cannot correspond to more than one eigenvalue of A.

(4) If X_1 and X_2 be non-zero eigenvectors of a matrix A corresponding to an eigenvalue λ of A, then $k_1 X_1 + k_2 X_2$ is also an eigenvector of A corresponding to eigenvalue λ, where k_1 and k_2 are non-zero numbers.

(5) Let A be an $n \times n$ matrix. Then the distinct eigenvectors corresponding to distinct eigenvalues of A are linearly independent.

1.19.4 EIGENVALUE OF SPECIAL TYPE OF MATRICES

1. The eigenvalues of a Hermitian matrix are real.　　　　　　　　　　　　　　　(UKTU – 2011)
2. The eigenvalues of a real symmetric matrix are all real.
3. The eigenvalues of a skew-Hermitian matrix are either purely imaginary or zero.
4. The eigenvalues of a real skew-symmetric matrix are either purely imaginary or zero.
5. The eigenvalues of a unitary matrix are of unit modulus.　　　　　　　　　　(UKTU–2010, 12)
6. The eigenvalues of an orthogonal matrix are of unit modulus.
7. If λ is a non-zero eigenvalue of a matrix A, then $\dfrac{1}{\lambda}$ is an eigenvalue of A^{-1}.
8. Let A be an $n \times n$ matrix. Then zero is an eigenvalue of A iff A is singular.
9. If $\lambda_1, \lambda_2, ..., \lambda_n$ are the eigenvalues of A, then $k\lambda_1, k\lambda_2, ..., k\lambda_n$ are eigenvalues of kA, where k is any number.
10. If X be a non-zero eigenvector of an $n \times n$ matrix A, then for each positive integer n, X is an eigenvector of A^n corresponding to the eigenvalue λ^n.
11. Similar matrices have the same eigenvalues.
12. Let A and B be two matrices of order $n \times n$. Let $X \neq O$ be an eigenvector of A and B corresponding to the eigenvalues λ_1 and λ_2 respectively, then X is an eigenvector of AB corresponding to the eigenvalue $\lambda_1\lambda_2$ of AB.

WORKING PROCEDURE

To find the eigenvalue and eigenvectors

Let A be an $n \times n$ matrix, then it will have n eigenvalues. In order to find the eigenvalues and eigenvectors of A, we use the following steps :

Step 1.　Find the roots of the characteristic equation $|A - \lambda I| = 0$, the roots of λ give the eigenvalues of A.

Step 2.　Let $X = \begin{bmatrix} x_1 \\ x_2 \\ \vdots \\ x_n \end{bmatrix} \neq O$ be an eigenvector of A corresponding to an eigenvalue λ_1 (say).

Then X can be determined from the equation $(A - \lambda_1 I)\,X = O$

which is a system of n homogeneous equations in $x_1, x_2, ..., x_n$. If the rank of $(A - \lambda_1 I)$ is r, then the number of linearly independent solutions is $n - r$.

Solved Examples

EXAMPLE 1. *Determine the eigenvalues of the matrix* :
$$A = \begin{bmatrix} 1 & 2 & 3 \\ 0 & -4 & 2 \\ 0 & 0 & 7 \end{bmatrix}$$

SOLUTION. The characteristic equation of A is given by
$$|A - \lambda I| = 0$$

i.e.,　$\begin{vmatrix} 1-\lambda & 2 & 3 \\ 0 & -4-\lambda & 2 \\ 0 & 0 & 7-\lambda \end{vmatrix} = 0$

i.e.,　$(1-\lambda)(-2-\lambda)(7-\lambda) = 0$

The roots of this characteristic equation are given by $\lambda = 1, -4, 7$.
These are the required eigenvalues of A.

REMARK

• It is clear that the given matrix A is an upper triangular matrix so that the principal diagonal elements $1, -4, 7$ will be the eigenvalues of A.

EXAMPLE 2. *Determine the eigenvalues and eigenvectors of the matrix*
$$A = \begin{bmatrix} 8 & -6 & 2 \\ -6 & 7 & -4 \\ 2 & -4 & 3 \end{bmatrix}.$$　　(GBTU-2011)

SOLUTION. The characteristic equation of A is given by

$|A - \lambda I| = 0$

or　$\begin{vmatrix} 8-\lambda & -6 & 2 \\ -6 & 7-\lambda & -4 \\ 2 & -4 & 3-\lambda \end{vmatrix} = 0$

or $(8-\lambda)((7-\lambda)(3-\lambda)-16)$
$\qquad + 6(-18+6\lambda+8)+2(24-14+2\lambda) = 0$

or　$\lambda^3 - 18\lambda^2 + 45\lambda = 0$

or　$\lambda(\lambda - 3)(\lambda - 15) = 0$

The roots of this equation are $\lambda = 0, 3, 15$.
Thus, the eigenvalues of A are
$$\lambda_1 = 0, \lambda_2 = 3, \lambda_3 = 15.$$
Eigenvector corresponding to $\lambda_1 = 0$:

Let $X_1 = \begin{bmatrix} x_1 \\ x_2 \\ x_3 \end{bmatrix} \neq O$ be an eigenvector corresponding to

the eigenvalue $\lambda_1 = 0$, then we have
$$AX_1 = \lambda_1 X1.$$
or $\qquad AX_1 = 0X_1$
or $\qquad (A - 0I)\, X_1 = O$

or $\begin{bmatrix} 8 & -6 & 2 \\ -6 & 7 & -4 \\ 2 & -4 & 3 \end{bmatrix} \begin{bmatrix} x_1 \\ x_2 \\ x_3 \end{bmatrix} = \begin{bmatrix} 0 \\ 0 \\ 0 \end{bmatrix}$ ...(1)

The non-zero solution of (1) will give X_1.
Reducing the coefficient matrix of (1) in Echeleon form by applying elementary row transformations.
Applying $R_1 \leftrightarrow R_3$, we get

$$\begin{bmatrix} 2 & -4 & 3 \\ -6 & 7 & -4 \\ 8 & -6 & 2 \end{bmatrix} \begin{bmatrix} x_1 \\ x_2 \\ x_3 \end{bmatrix} = \begin{bmatrix} 0 \\ 0 \\ 0 \end{bmatrix}$$

Applying $R_2 \to R_2 + 3R_1, R_3 \to R_3 - 4R_1$, we get

$$\begin{bmatrix} 2 & -4 & 3 \\ 0 & -5 & 5 \\ 0 & 10 & -10 \end{bmatrix} \begin{bmatrix} x_1 \\ x_2 \\ x_3 \end{bmatrix} = \begin{bmatrix} 0 \\ 0 \\ 0 \end{bmatrix}$$

Applying $R_3 \to R_3 + 2R_2$, we get

$$\begin{bmatrix} 2 & -4 & 3 \\ 0 & -5 & 5 \\ 0 & 0 & 0 \end{bmatrix} \begin{bmatrix} x_1 \\ x_2 \\ x_3 \end{bmatrix} = \begin{bmatrix} 0 \\ 0 \\ 0 \end{bmatrix}$$ ...(2)

Clearly $\rho(A - 0.I) = 2$, therefore the system of equations (2) will have $3 - 2 = 1$ (unknowns – rank) linearly independent solution.
From (2), we have
$$2x_1 - 4x_2 + 3x_3 = 0$$
$$- 5x_2 + 5x_3 = 0$$
Clearly, $x_1 = \dfrac{1}{2}, x_2 = 1$ and $x_3 = 1$ satisfy the above equations.
Hence, the eigenvector corresponding to eigenvalue $\lambda_1 = 0$ is
$$X_1 = \begin{bmatrix} 1/2 \\ 1 \\ 1 \end{bmatrix}$$

Eigenvector corresponding to $\lambda_2 = 3$:

Let $X_2 = \begin{bmatrix} x_1 \\ x_2 \\ x_3 \end{bmatrix} \neq O$ be an eigenvector of A corresponding

to $\lambda_2 = 3$, then we have
$$AX_2 = \lambda_2 X_2$$
or $\qquad (A - \lambda_2 I)X_2 = O$
or $\qquad (A - 3I)X_2 = O$

or $\begin{bmatrix} 8-3 & -6 & 2 \\ -6 & 7-3 & -4 \\ 2 & -4 & 3-3 \end{bmatrix} \begin{bmatrix} x_1 \\ x_2 \\ x_3 \end{bmatrix} = O$

or $\begin{bmatrix} 5 & -6 & 2 \\ -6 & 4 & -4 \\ 2 & -4 & 0 \end{bmatrix} \begin{bmatrix} x_1 \\ x_2 \\ x_3 \end{bmatrix} = O$ ...(3)

The non-zero solution of (3) will give X_2.
Applying $R_1 \to R_1 + R_2$, we get

$$\begin{bmatrix} -1 & -2 & -2 \\ -6 & 4 & -4 \\ 2 & -4 & 0 \end{bmatrix} \begin{bmatrix} x_1 \\ x_2 \\ x_3 \end{bmatrix} = \begin{bmatrix} 0 \\ 0 \\ 0 \end{bmatrix}$$

Applying $R_2 \to R_2 - 6R_1, R_3 \to R_3 + 2R_1$, we get

$$\begin{bmatrix} -1 & -2 & -2 \\ 0 & 16 & 8 \\ 0 & -8 & -4 \end{bmatrix} \begin{bmatrix} x_1 \\ x_2 \\ x_3 \end{bmatrix} = \begin{bmatrix} 0 \\ 0 \\ 0 \end{bmatrix}$$

Applying $R_2 \to \dfrac{1}{8}R_2$, we get

$$\begin{bmatrix} -1 & -2 & -2 \\ 0 & 2 & 1 \\ 0 & -8 & -4 \end{bmatrix} \begin{bmatrix} x_1 \\ x_2 \\ x_3 \end{bmatrix} = \begin{bmatrix} 0 \\ 0 \\ 0 \end{bmatrix}$$

Again applying $R_3 \to R_3 + 4R_2$, we get

$$\begin{bmatrix} -1 & -2 & -2 \\ 0 & 2 & 1 \\ 0 & 0 & 0 \end{bmatrix} \begin{bmatrix} x_1 \\ x_2 \\ x_3 \end{bmatrix} = \begin{bmatrix} 0 \\ 0 \\ 0 \end{bmatrix}$$ ...(4)

Clearly $\rho(A - 3I) = 2$, therefore the system of equations (3) will have $3 - 2 = 1$ linearly independent solution.
From (4), we have
$$- x_1 - 2x_2 - 2x_3 = 0$$
$$2x_2 + x_3 = 0$$
Clearly, $x_1 = -2, x_2 = -1$ and $x_3 = 2$ satisfy the above equations.
Hence, the eigenvector corresponding to eigenvalue $\lambda_2 = 3$ is
$$X_2 = \begin{bmatrix} -2 \\ -1 \\ 2 \end{bmatrix}$$

Eigenvector corresponding to $\lambda_3 = 15$:

Let $X_3 = \begin{bmatrix} x_1 \\ x_2 \\ x_3 \end{bmatrix} \neq O$ be an eigenvector of A corresponding

to $\lambda_3 = 15$, then we have
$$AX_3 = \lambda_3 X_3$$
or $\qquad (A - \lambda_3 I)X_3 = O$
or $\qquad (A - 15I)X_3 = O$

or $\begin{bmatrix} 8-15 & -6 & 2 \\ -6 & 7-15 & -4 \\ 2 & -4 & 3-15 \end{bmatrix} \begin{bmatrix} x_1 \\ x_2 \\ x_3 \end{bmatrix} = O$

or $\begin{bmatrix} -7 & -6 & 2 \\ -6 & -8 & -4 \\ 2 & -4 & -12 \end{bmatrix}\begin{bmatrix} x_1 \\ x_2 \\ x_3 \end{bmatrix} = O$ ...(5)

The non-zero solution of (5) will give X_3.

Applying $R_1 \leftrightarrow R_3$, we get

$\begin{bmatrix} 2 & -4 & -12 \\ -6 & -8 & -4 \\ -7 & -6 & 2 \end{bmatrix}\begin{bmatrix} x_1 \\ x_2 \\ x_3 \end{bmatrix} = \begin{bmatrix} 0 \\ 0 \\ 0 \end{bmatrix}$

Applying $R_1 \to \frac{1}{2}R_1$, we get

$\begin{bmatrix} 1 & -2 & -6 \\ -6 & -8 & -4 \\ -7 & -6 & 2 \end{bmatrix}\begin{bmatrix} x_1 \\ x_2 \\ x_3 \end{bmatrix} = \begin{bmatrix} 0 \\ 0 \\ 0 \end{bmatrix}$

Applying $R_2 \to R_2 + 6R_1, R_3 \to R_3 + 7R_1$, we get

$\begin{bmatrix} 1 & -2 & -6 \\ 0 & -20 & -40 \\ 0 & -20 & -40 \end{bmatrix}\begin{bmatrix} x_1 \\ x_2 \\ x_3 \end{bmatrix} = \begin{bmatrix} 0 \\ 0 \\ 0 \end{bmatrix}$

Applying $R_3 \to R_3 - R_2$, we get

$\begin{bmatrix} 1 & -2 & -6 \\ 0 & -20 & -40 \\ 0 & 0 & 0 \end{bmatrix}\begin{bmatrix} x_1 \\ x_2 \\ x_3 \end{bmatrix} = \begin{bmatrix} 0 \\ 0 \\ 0 \end{bmatrix}$ ...(6)

Clearly $\rho(A - 15I) = 2$, therefore the system of equations (5) will have $3 - 2 = 1$ linearly independent solution.

From (6), we have
$$x_1 - 2x_2 - 6x_3 = 0$$
$$-20x_2 - 40x_3 = 0$$

Clearly, $x_1 = 2, x_2 = -2$ and $x_3 = 1$ satisfy the above equations.

Hence, the eigenvector corresponding to eigenvalue $\lambda_3 = 15$ is

$$X_3 = \begin{bmatrix} 2 \\ -2 \\ 1 \end{bmatrix}.$$

EXAMPLE 3. *Find the eigenvalues and eigenvectors of the matrix*

$$A = \begin{bmatrix} 2 & 0 & 1 & -3 \\ 0 & 2 & 10 & 4 \\ 0 & 0 & 2 & 0 \\ 0 & 0 & 0 & 3 \end{bmatrix}.$$

SOLUTION. The characteristic equation of A is given by
$$|A - \lambda I| = 0$$

or $A = \begin{vmatrix} 2-\lambda & 0 & 1 & -3 \\ 0 & 2-\lambda & 10 & 4 \\ 0 & 0 & 2-\lambda & 0 \\ 0 & 0 & 0 & 3-\lambda \end{vmatrix} = 0$

or $(\lambda - 2)^2 (\lambda - 3) = 0$

The roots of characteristic equations are 2, 2, 2, 3.

Thus, the eigenvalues of A are $\lambda_1 = 2, \lambda_2 = 2, \lambda_3 = 2, \lambda_4 = 3$.

Eigenvector corresponding to $\lambda_1 = \lambda_2 = \lambda_3 = 2$:

Let $X = \begin{bmatrix} x_1 \\ x_2 \\ x_3 \\ x_4 \end{bmatrix} \neq O$ be an eigenvector of A corresponding to the eigenvalue 2, then

we have $AX = 2X.$
or $(A - 2I)X = O$

or $\begin{bmatrix} 2-2 & 0 & 1 & -3 \\ 0 & 2-2 & 10 & 4 \\ 0 & 0 & 2-2 & 0 \\ 0 & 0 & 0 & 3-2 \end{bmatrix}\begin{bmatrix} x_1 \\ x_2 \\ x_3 \\ x_4 \end{bmatrix} = \begin{bmatrix} 0 \\ 0 \\ 0 \\ 0 \end{bmatrix}$

or $\begin{bmatrix} 0 & 0 & 1 & -3 \\ 0 & 0 & 10 & 4 \\ 0 & 0 & 0 & 0 \\ 0 & 0 & 0 & 1 \end{bmatrix}\begin{bmatrix} x_1 \\ x_2 \\ x_3 \\ x_4 \end{bmatrix} = \begin{bmatrix} 0 \\ 0 \\ 0 \\ 0 \end{bmatrix}$ ...(1)

Applying $R_3 \leftrightarrow R_4$, we get

$\begin{bmatrix} 0 & 0 & 1 & -3 \\ 0 & 0 & 10 & 4 \\ 0 & 0 & 0 & 1 \\ 0 & 0 & 0 & 0 \end{bmatrix}\begin{bmatrix} x_1 \\ x_2 \\ x_3 \\ x_4 \end{bmatrix} = \begin{bmatrix} 0 \\ 0 \\ 0 \\ 0 \end{bmatrix}$

Applying $R_2 \to R_2 - 10R_1$, we get

$\begin{bmatrix} 0 & 0 & 1 & -3 \\ 0 & 0 & 0 & 34 \\ 0 & 0 & 0 & 1 \\ 0 & 0 & 0 & 0 \end{bmatrix}\begin{bmatrix} x_1 \\ x_2 \\ x_3 \\ x_4 \end{bmatrix} = \begin{bmatrix} 0 \\ 0 \\ 0 \\ 0 \end{bmatrix}$

Applying $R_2 \to \frac{1}{34}R_2$, we get

$\begin{bmatrix} 0 & 0 & 1 & -3 \\ 0 & 0 & 0 & 1 \\ 0 & 0 & 0 & 1 \\ 0 & 0 & 0 & 0 \end{bmatrix}\begin{bmatrix} x_1 \\ x_2 \\ x_3 \\ x_4 \end{bmatrix} = \begin{bmatrix} 0 \\ 0 \\ 0 \\ 0 \end{bmatrix}$

Applying $R_2 \to R_3 - R_2$, we get

$\begin{bmatrix} 0 & 0 & 1 & -3 \\ 0 & 0 & 0 & 1 \\ 0 & 0 & 0 & 0 \\ 0 & 0 & 0 & 0 \end{bmatrix}\begin{bmatrix} x_1 \\ x_2 \\ x_3 \\ x_4 \end{bmatrix} = \begin{bmatrix} 0 \\ 0 \\ 0 \\ 0 \end{bmatrix}$ ...(2)

From (2), it is clear that $\rho(A - 2I) = 2$, therefore the equation (1) will have $4 - 2 = 2$ linearly independent solution.

Equation (2) reduces to
$$\left.\begin{matrix} x_3 - 3x_4 = 0 \\ x_4 = 0 \end{matrix}\right\} \Rightarrow x_4 = 0, x_3 = 0$$

Let us take $x_1 = c_1, x_2 = c_2$, then we have

$$X = \begin{bmatrix} x_1 \\ x_2 \\ x_3 \\ x_4 \end{bmatrix} = \begin{bmatrix} c_1 \\ c_2 \\ 0 \\ 0 \end{bmatrix} = c_1\begin{bmatrix} 1 \\ 0 \\ 0 \\ 0 \end{bmatrix} + c_2\begin{bmatrix} 0 \\ 1 \\ 0 \\ 0 \end{bmatrix}$$

Hence, the eigenvector corresponding to eigenvalue 2 are

$$X_1 = \begin{bmatrix} 1 \\ 0 \\ 0 \\ 0 \end{bmatrix}, X_2 = \begin{bmatrix} 0 \\ 1 \\ 0 \\ 0 \end{bmatrix}$$

Eigenvector corresponding to the eigenvalue $\lambda_4 = 3$:

Let $X = \begin{bmatrix} x_1 \\ x_2 \\ x_3 \\ x_4 \end{bmatrix} \neq O$ *be an eigenvector*

corresponding to the eigenvalue 3, then we have

$$AX = 3X$$

or $(A - 3I)X = O$

or $\begin{bmatrix} 2-3 & 0 & 1 & -3 \\ 0 & 2-3 & 10 & 4 \\ 0 & 0 & 2-3 & 0 \\ 0 & 0 & 0 & 3-3 \end{bmatrix} \begin{bmatrix} x_1 \\ x_2 \\ x_3 \\ x_4 \end{bmatrix} = \begin{bmatrix} 0 \\ 0 \\ 0 \\ 0 \end{bmatrix}$

or $\begin{bmatrix} -1 & 0 & 1 & -3 \\ 0 & -1 & 10 & 4 \\ 0 & 0 & -1 & 0 \\ 0 & 0 & 0 & 0 \end{bmatrix} \begin{bmatrix} x_1 \\ x_2 \\ x_3 \\ x_4 \end{bmatrix} = \begin{bmatrix} 0 \\ 0 \\ 0 \\ 0 \end{bmatrix}$ ...(3)

From (3), it is clear that $\rho(A - 3I) = 3$, therefore the equation (3) will have $4 - 3 = 1$ linearly independent solution.
Equation (3) reduces to
$$-x_1 + x_3 - 3x_4 = 0$$
$$-x_2 + 10x_3 + 4x_4 = 0$$
$$-x_3 = 0$$
Let us take $x_4 = c$, then from above equaions, we get
$$x_1 = -3c, x_2 = 4c, x_3 = 0, x_4 = c$$
Therefore,
$$\begin{bmatrix} x_1 \\ x_2 \\ x_3 \\ x_4 \end{bmatrix} = \begin{bmatrix} -3c \\ 4c \\ 0 \\ c \end{bmatrix} = c \begin{bmatrix} -3 \\ 4 \\ 0 \\ 1 \end{bmatrix}$$
Hence, the eigenvector corresponding to eigenvalue 3 is
$$X = \begin{bmatrix} -3 \\ 4 \\ 0 \\ 1 \end{bmatrix}.$$

📨 REMARK

- Let A be an $n \times n$ matrix with real entries. If λ is a complex eigenvalue of A with associated eigenvector X, then $\bar{\lambda}$ is also an eigenvalue of A with associated eigenvector $\bar{X}$.

Exercise-1.9

1. Prove that a square matrix A and its transpose A' have the same set of eigenvalues.

2. Let A be an $n \times n$ matrix and let $g(x)$ be any polynomial. If λ is an eigenvalue of A, then prove that $g(\lambda)$ is an eigenvalue of $g(A)$.

3. Show that the eigenvalues of a triangular matrix are just the diagonal elements of the matrix.

4. Let $A = $ dig. $(\lambda_1, \lambda_2,..., \lambda_n)$ be a diagonal matrix. Prove that each λ_i $(i = 1, 2, 3,..., n)$ is an eigenvalue of A.

5. Let A be an 3×3 matrix. If $\lambda_1, \lambda_2, \lambda_3$ are the eigenvalues of A, then find the eigenvalues of the matrix $(I + aA)^{-1} (1 + bA)$, where a, b are scalars such that $a\lambda_i \neq -1$ for $i = 1, 2, 3$.

6. Let A and B be two $n \times n$ matrices. Let X be an eigenvector of A and B both. Show that X is also an eigenvector of $aA + bB$, where a, b are scalars.

7. Prove that the eigenvectors of a real symmetric matrix corresponding to two distinct eigenvalues are orthogonal.

8. Prove that the eigenvectors of a Hermitian matrix corresponding to two distinct eigenvalues are orthogonal.

9. (i) If λ is an eigenvalue of a matrix A, then show that $k + \lambda$ is an eigenvalue of $A + kI$.
   (ii) If the matrix A has characteristic roots $\lambda_1, \lambda_2,..., \lambda_n$ show that the matrix A^2 has such roots as $\lambda_1^2, \lambda_2^2,..., \lambda_n^2$.

10. (i) Find the eigenvalues of a matrix $\begin{bmatrix} 1 & 4 \\ 2 & 3 \end{bmatrix}$.
   (ii) Find the eigenvalues of the matrix $A = \begin{bmatrix} a & h & g \\ 0 & b & f \\ 0 & 0 & c \end{bmatrix}$.

11. Find the eigenvalues and eigenvectors of the following matrices :
   (i) $\begin{bmatrix} 2 & -4 \\ -1 & -1 \end{bmatrix}$
   (ii) $\begin{bmatrix} -1 & 0 \\ 0 & 1 \end{bmatrix}$
   (iii) $\begin{bmatrix} 1 & 1 \\ -2 & 4 \end{bmatrix}$
   (iv) $\begin{bmatrix} 10 & -18 \\ 6 & -11 \end{bmatrix}$

12. Find the eigenvalues and eigenvectors of the following matrices :
   (i) $\begin{bmatrix} 0 & 1 & 0 \\ 0 & 0 & 1 \\ 1 & -3 & 3 \end{bmatrix}$
   (ii) $\begin{bmatrix} 5 & 8 & 16 \\ 4 & 1 & 8 \\ -4 & -4 & -11 \end{bmatrix}$
   (iii) $\begin{bmatrix} 1 & -1 & -1 \\ -1 & 1 & -1 \\ -1 & -1 & 1 \end{bmatrix}$
   (iv) $\begin{bmatrix} 1 & 2 & 2 \\ 1 & 2 & -1 \\ -1 & 1 & 4 \end{bmatrix}$
   (v) $\begin{bmatrix} 6 & -2 & 2 \\ -2 & 3 & -1 \\ 2 & -1 & 3 \end{bmatrix}$
   (vi) $\begin{bmatrix} -2 & 2 & -3 \\ 2 & 1 & -6 \\ -1 & -2 & 0 \end{bmatrix}$

(UKTU-2011, GBTU-2010)

(vii) $\begin{bmatrix} 1 & 2 & 3 \\ 0 & 2 & 3 \\ 0 & 0 & 2 \end{bmatrix}$ (viii) $\begin{bmatrix} 1 & 1 & 0 \\ 0 & 2 & 2 \\ 0 & 0 & 3 \end{bmatrix}$

$A = \begin{bmatrix} 1 & 1 & 0 & 0 \\ 0 & 2 & 0 & 0 \\ 0 & 0 & 1 & 1 \\ 0 & 0 & -2 & 4 \end{bmatrix}$

(ix) $\begin{bmatrix} 3 & 1 & 1 \\ 2 & 4 & 2 \\ 1 & 1 & 3 \end{bmatrix}$ (x) $\begin{bmatrix} 2 & 1 & 0 \\ 0 & 2 & 1 \\ 0 & 0 & 2 \end{bmatrix}$

14. Find all the characteristic roots and the corresponding characteristic vectors of the matrix

13. Find the eigenvalues and eigenvectors of the matrix

$A = \begin{bmatrix} 2 & 1 & -1 \\ 0 & 3 & -2 \\ 2 & 4 & -3 \end{bmatrix}$

Answers

5. $\dfrac{1+b\lambda_1}{1+a\lambda_1}, \dfrac{1+b\lambda_2}{1+a\lambda_2}, \dfrac{1+b\lambda_3}{1+a\lambda_3}$ **10.** (i) $-1, 5$ (ii) a, b, c **11.** (i) $\lambda_1 = -2, X_1 = \begin{bmatrix} 1 \\ 1 \end{bmatrix}; \lambda_2 = 3, X_2 = \begin{bmatrix} -4 \\ 1 \end{bmatrix}$ (ii) $\lambda_1 = 1, X_1 = \begin{bmatrix} 0 \\ 1 \end{bmatrix}; \lambda_2 = -1, X_2 = \begin{bmatrix} 1 \\ 0 \end{bmatrix}$

(iii) $\lambda_1 = 2, X_1 = \begin{bmatrix} 1 \\ 1 \end{bmatrix}; \lambda_2 = 3, X_2 = \begin{bmatrix} 1 \\ 2 \end{bmatrix}$ (iv) $\lambda_1 = -2, X_1 = \begin{bmatrix} 3 \\ 2 \end{bmatrix}; \lambda_2 = 1, X_2 = \begin{bmatrix} 2 \\ 1 \end{bmatrix}$ **12.** (i) $\lambda_1 = \lambda_2 = \lambda_3 = 1, X = \begin{bmatrix} 1 \\ 1 \\ 1 \end{bmatrix}$

(ii) $\lambda_1 = 1, X_1 = \begin{bmatrix} -2 \\ -1 \\ 1 \end{bmatrix}; \lambda_2 = -3, X_2 = \begin{bmatrix} -1 \\ 1 \\ 0 \end{bmatrix}; \lambda_3 = -3, X_3 = \begin{bmatrix} -2 \\ 0 \\ 1 \end{bmatrix}$ (iii) $\lambda_1 = -1, X_1 = \begin{bmatrix} 1 \\ 1 \\ 1 \end{bmatrix}; \lambda_2 = 2, X_2 = \begin{bmatrix} -1 \\ 1 \\ 0 \end{bmatrix}; \lambda_3 = 2, X_3 = \begin{bmatrix} -1 \\ 0 \\ 1 \end{bmatrix}$

(iv) $\lambda_1 = 1, X_1 = \begin{bmatrix} 2 \\ -1 \\ 1 \end{bmatrix}; \lambda_2 = 3, X_2 = \begin{bmatrix} 1 \\ 1 \\ 0 \end{bmatrix}; \lambda_3 = 3, X_3 = \begin{bmatrix} 1 \\ 0 \\ 1 \end{bmatrix}$ (v) $\lambda_1 = 2, X_1 = \begin{bmatrix} -1 \\ 0 \\ 2 \end{bmatrix}; \lambda_2 = 2, X_2 = \begin{bmatrix} 1 \\ 2 \\ 0 \end{bmatrix}; \lambda_3 = 8, X_3 = \begin{bmatrix} 2 \\ -1 \\ 1 \end{bmatrix}$

(vi) $\lambda_1 = -3, X_1 = \begin{bmatrix} -2 \\ 1 \\ 0 \end{bmatrix}; \lambda_2 = -3, X_2 = \begin{bmatrix} 3 \\ 0 \\ 1 \end{bmatrix}; \lambda_3 = 5, X_3 = \begin{bmatrix} 1 \\ 2 \\ 1 \end{bmatrix}$ (vii) $\lambda_1 = \lambda_2 = 1, X = \begin{bmatrix} 1 \\ 0 \\ 0 \end{bmatrix}; \lambda_3 = 2, X_1 = \begin{bmatrix} 2 \\ 1 \\ 0 \end{bmatrix}$

(viii) $\lambda_1 = 1, X_1 = \begin{bmatrix} 1 \\ 0 \\ 0 \end{bmatrix}; \lambda_2 = 2, X_2 = \begin{bmatrix} 2 \\ 1 \\ 0 \end{bmatrix}; \lambda_3 = 3, X_3 = \begin{bmatrix} 1 \\ 2 \\ 1 \end{bmatrix}$ (ix) $\lambda_1 = 2, X_1 = \begin{bmatrix} -1 \\ 1 \\ 0 \end{bmatrix}; \lambda_2 = 2, X_2 = \begin{bmatrix} -1 \\ 0 \\ 1 \end{bmatrix}; \lambda_3 = 6, X_3 = \begin{bmatrix} 1 \\ 2 \\ 1 \end{bmatrix}$

(x) $\lambda_1 = \lambda_2 = \lambda_3 = 2, X = \begin{bmatrix} 1 \\ 0 \\ 0 \end{bmatrix}$ **13.** $\lambda_1 = 1, X_1 = \begin{bmatrix} 1 \\ 0 \\ 0 \\ 0 \end{bmatrix}; \lambda_2 = 2, X_2 = \begin{bmatrix} 1 \\ 1 \\ 0 \\ 0 \end{bmatrix}; \lambda_3 = 2, X_3 = \begin{bmatrix} 0 \\ 0 \\ 1 \\ 1 \end{bmatrix}; \lambda_4 = 3, X_4 = \begin{bmatrix} 0 \\ 0 \\ 0 \\ 1 \end{bmatrix}$

1.19.5 THE CAYLEY-HAMILTON THEOREM

THEOREM 1. *Every square matrix satisfies its characteristic equation.*   (UPTU-2006, 11, 12)

or let A be a square matrix of order n and the characteristic equation of A is

$$|A - \lambda I| = (-1^n) [\lambda^n + a_1 \lambda^{n-1} + a_2 \lambda^{n-2} + \dots + a_{n-1}\lambda + a_n] = 0$$

then its matrix equation $X^n + a_1 X^{n-1} + a_2 X^{n-2} + \dots + a_{n-1}X + a_n I = O$ *is satisfied by the matrix* $X = A$

i.e. $$A^n + a_1 A^{n-1} + a_2 A^{n-2} + \dots + a_{n-1}A + a_n I = O$$

where I is a unit matrix of order n and O is null matrix of order n.

PROOF.   Since A and I are two square matrices of order n and λ is any characteristic root of A, then the matrix $(A - \lambda I)$ is also a square matrix of order n whose elements are at most of degree one in λ. Therefore Adj. $(A - \lambda I)$ will have its elements a polynomials in λ of degree $n-1$ or less and thus Adj. $(A - \lambda I)$ can be expressed as a matrix polynomial in λ as follows :

$$\text{Adj. } (A - \lambda I) = B_0 \lambda^{n-1} + B_1 \lambda^{n-2} + \dots + B_{n-2}\lambda + B_{n-1} \qquad \dots(1)$$

where $B_0, B_1, \dots, B_{n-1}$ are the square matrices of order n.

Since we know that $\quad A(\text{Adj. } A) = |A|I_n$

$\therefore \quad (A - \lambda I) \text{ Adj. } (A - \lambda I) = |A - \lambda I|I$

or $\quad (A - \lambda I) \text{ Adj. } (A - \lambda I) = (-1^n)(\lambda^n + a_1\lambda^{n-1} + a_2\lambda^{n-2} + \dots + a_{n-1}\lambda + a_n)I \quad \dots(2)$

Multiplying both sides of (1) by $(A - \lambda I)$, we get

$\quad (A - \lambda I) \text{ Adj.} (A - \lambda I) = (A - \lambda I)(B_0\lambda^{n-1} + B_1\lambda^{n-2} + \dots + B_{n-2}\lambda + B_{n-1}) \quad \dots(3)$

From (2) and (3), we get

$\quad (A - \lambda I)(B_0\lambda^{n-1} + B_1\lambda^{n-2} + \dots + B_{n-2}\lambda + B_{n-1}) = (-1^n)(\lambda^n + a_1\lambda^{n-1} + a_2\lambda^{n-2} + \dots + a_{n-1}\lambda + a_n)I$

Now comparing the coefficients of like powers of λ, we get

$$\left.\begin{array}{l} -IB_0 = (-1)^n I \\[4pt] AB_0 - IB_1 = (-1)^n a_1 I \\[4pt] AB_1 - IB_2 = (-1)^n a_2 I \\[4pt] \dots\dots\dots\dots\dots\dots\dots \\[4pt] AB_{n-2} - IB_{n-3} = (-1)^n a_{n-1}I \\[4pt] AB_{n-1} = (-1)^n a_n I \end{array}\right\} \quad \dots(4)$$

Premultiplying first, second, third, etc. equations of (4) by A^n, A^{n-1}, A^{n-2}, etc. respectively and then adding, we get

$\quad -A^n B_0 + A^n B_0 - A^{n-1}B_1 + A^{n-1}B_1 + \dots = (-1)^n(A^n + a_1 A^{n-1} + \dots + a_n I)$

or $\quad\quad\quad 0 = (-1)^n(A^n + a_1 A^{n-1} + \dots + a_n I)$

Hence $\quad\quad A^n + a_1 A^{n-1} + \dots + a_n I = O$

SOME RESULTS

1. If A be a non-singular matrix of order $n \times n$ and its characteristic polynomial is

$|A - \lambda I| = (-1)^n(\lambda^n + a_1\lambda^{n-1} + \dots + a_{n-1}\lambda + a_n)$

then $\det(A) = (-1)^n a_n$.

2. If $\lambda_1, \lambda_2, \dots, \lambda_n$ are eigenvalues of a square matrix of order $n \times n$, then $\det(A) = \lambda_1\lambda_2\lambda_3 \dots \lambda_n$.

3. Let A be an $n \times n$ matrix with characteristic polynomial

$f(t) = (-1)^n(t^n + a_1 t^{n-1} + \dots + a_{n-1}t + a_n)$

Then A is invertible iff $a_n \neq 0$ and its inverse is

$A^{-1} = \left(\dfrac{-1}{a_n}\right)(A^{n-1} + a_1 A^{n-2} + \dots + a_{n-2}A + a_{n-1}I)$

4. If $\lambda_1, \lambda_2, \dots, \lambda_n$ are the eigenvalue of a matrix A of order $n \times n$,

then $\text{Tr}(A) = \text{Trace of } A = \sum\limits_{i=1}^{n} \lambda_i$

5. If the characteristic equation of a matrix A of order $n \times n$ is

$|A - \lambda I| = (-1)^n(\lambda^n + a_1\lambda^{n-1} + a_2\lambda^{n-2} + \dots + a_{n-1}\lambda + a_n) = 0$

then $\text{Tr}(A) = (-1)^n a_1$

6. Let A be a matrix of order $n \times n$. If m be a positive integer such that $m \geq n$, then A^m is linearly expressible in terms of those of lower order of A.

Solved Examples

EXAMPLE 1. *Find the characteristic equation of the matrix*

$$A = \begin{bmatrix} 2 & -1 & 1 \\ -1 & 2 & -1 \\ 1 & -1 & 2 \end{bmatrix} \text{ and verify that it is satisfied}$$

by A and hence find A^{-1}.

(UPTU-2006, GBTU-2012, UKTU-2011, MADRAS-2006)

SOLUTION. The characteristic equation of A is given by

$$|A - \lambda I| = 0$$

or $\quad \begin{vmatrix} 2-\lambda & -1 & 1 \\ -1 & 2-\lambda & -1 \\ 1 & -1 & 2-\lambda \end{vmatrix} = 0$

or $(2-\lambda)\{(2-\lambda)(2-\lambda)-1\}$

$\quad +1(-2+\lambda+2)+1(1-2+\lambda)=0$

or $\quad\quad -\lambda^3 + 6\lambda^2 - 9\lambda + 4 = 0$

or $\quad\quad \lambda^3 - 6\lambda^2 + 9\lambda - 4 = 0$

Next we have to show that $A^3 - 6A^2 + 9A - 4I = O$

Now $A^2 = A.A$

$$= \begin{bmatrix} 2 & -1 & 1 \\ -1 & 2 & -1 \\ 1 & -1 & 2 \end{bmatrix}\begin{bmatrix} 2 & -1 & 1 \\ -1 & 2 & -1 \\ 1 & -1 & 2 \end{bmatrix} = \begin{bmatrix} 6 & -5 & 5 \\ -5 & 6 & -5 \\ 5 & -5 & 6 \end{bmatrix}$$

and $\quad A^3 = A^2 A$

$$= \begin{bmatrix} 6 & -5 & 5 \\ -5 & 6 & -5 \\ 5 & -5 & 6 \end{bmatrix}\begin{bmatrix} 2 & -1 & 1 \\ -1 & 2 & -1 \\ 1 & -1 & 2 \end{bmatrix} = \begin{bmatrix} 22 & -21 & 21 \\ -21 & 22 & -21 \\ 21 & -21 & 22 \end{bmatrix}$$

Now $A^3 - 6A^2 + 9A - 5I$

$$= \begin{bmatrix} 22 & -21 & 21 \\ -21 & 22 & -21 \\ 21 & -21 & 22 \end{bmatrix} - 6 \begin{bmatrix} 6 & -5 & 5 \\ -5 & 6 & -5 \\ 5 & -5 & 6 \end{bmatrix}$$

$$+ 9 \begin{bmatrix} 2 & -1 & 1 \\ -1 & 2 & -1 \\ 1 & -1 & 2 \end{bmatrix} - 4 \begin{bmatrix} 1 & 0 & 0 \\ 0 & 1 & 0 \\ 0 & 0 & 1 \end{bmatrix}$$

$$= \begin{bmatrix} 22-36+18-4 & -21+30-9-0 & 21-30+9-0 \\ -21+30-9-0 & 22-36+18-4 & -21+30-9-0 \\ 21-30+9-0 & -21+30-9-0 & 22-36+18-4 \end{bmatrix}$$

$$= \begin{bmatrix} 0 & 0 & 0 \\ 0 & 0 & 0 \\ 0 & 0 & 0 \end{bmatrix} = O$$

Hence, $A^3 - 6A^2 + 9A - 4I = O$ ...(1)

Since $|A| = 2(4-1) + 1(-2+1) + 1(1-2)$

$$= 6 - 1 - 1 = 4 \neq 0 \Rightarrow A^{-1} \text{ exist.}$$

Premultiplying (1) by A^{-1}, we get

$$A^2 - 6A + 9I - 4A^{-1} = O$$

$$\Rightarrow \quad A^{-1} = +\frac{1}{4}[A^2 - 6A + 9I]$$

$$\Rightarrow \quad A^{-1} = \frac{1}{4} \left\{ \begin{bmatrix} 6 & -5 & 5 \\ -5 & 6 & -5 \\ 5 & -5 & 6 \end{bmatrix} - 6 \begin{bmatrix} 2 & -1 & 1 \\ -1 & 2 & -1 \\ 1 & -1 & 2 \end{bmatrix} \right.$$

$$\left. + 9 \begin{bmatrix} 1 & 0 & 0 \\ 0 & 1 & 0 \\ 0 & 0 & 1 \end{bmatrix} \right\}$$

$$= \frac{1}{4} \begin{bmatrix} 6-12+9 & -5+6+0 & 5-6+0 \\ -5+6+0 & 6-12+9 & -5+6+0 \\ 5-6+0 & -5+6+0 & 6-12+9 \end{bmatrix}$$

$$\therefore \quad A^{-1} = \frac{1}{4} \begin{bmatrix} 3 & 1 & -1 \\ 1 & 3 & 1 \\ -1 & 1 & 3 \end{bmatrix}.$$

EXAMPLE 2. *Show that the matrix* $A = \begin{bmatrix} 0 & c & -b \\ -c & 0 & a \\ b & -a & 0 \end{bmatrix}$ *satisfies*

Cayley-Hamilton Theorem.

SOLUTION. The characteristic equation of A is given by

$$|A - \lambda I| = 0$$

or $\begin{vmatrix} -\lambda & c & -b \\ -c & -\lambda & a \\ b & -a & -\lambda \end{vmatrix} = 0$

or $-\lambda(\lambda^2 + a^2) - c(c\lambda - ab) - b(ca + b\lambda) = 0$

or $\qquad\qquad -\lambda^3 - \lambda(a^2 + b^2 + c^2) = 0$

or $\qquad\qquad \lambda^3 + \lambda(a^2 + b^2 + c^2) = 0$

We have to show that

$$A^3 + A(a^2 + b^2 + c^2) = O$$

Now $\qquad\qquad A^2 = A.A$

$$= \begin{bmatrix} 0 & c & -b \\ -c & 0 & a \\ b & -a & 0 \end{bmatrix} \begin{bmatrix} 0 & c & -b \\ -c & 0 & a \\ b & -a & 0 \end{bmatrix}$$

$$= \begin{bmatrix} -(c^2+b^2) & ab & ac \\ ab & -(c^2+a^2) & bc \\ ac & bc & -(a^2+b^2) \end{bmatrix}$$

and $A^3 = A^2.A$

$$= \begin{bmatrix} -(c^2+b^2) & ab & ac \\ ab & -(c^2+a^2) & bc \\ ac & bc & -(a^2+b^2) \end{bmatrix} \begin{bmatrix} 0 & c & -b \\ -c & 0 & a \\ b & -a & 0 \end{bmatrix}$$

$$= \begin{bmatrix} 0 & -c^3-b^2c-a^2c & bc^2+b^2+a^2b \\ c^3+a^2c+b^2c & 0 & -ab^2-ac^2-a^3 \\ -bc^2-b^3-a^2b & ac^2+ab^2+a^3 & 0 \end{bmatrix}$$

$$= \begin{bmatrix} 0 & -c(a^2+b^2+c^2) & b(a^2+b^2+c^2) \\ c(a^2+b^2+c^2) & 0 & -a(a^2+b^2+c^2) \\ -b(a^2+b^2+c^2) & a(a^2+b^2+c^2) & 0 \end{bmatrix}$$

$$A^3 = -(a^2+b^2+c^2) \begin{bmatrix} 0 & c & -b \\ -c & 0 & a \\ b & -a & 0 \end{bmatrix} = -(a^2+b^2+c^2)A$$

Hence, $\qquad A^3 + (a^2 + b^2 + c^2)A = O$

EXAMPLE 3. *Verify Cayley-Hamilton theorem for the matrix*

$$A = \begin{bmatrix} 1 & 1 & 0 & 0 \\ 0 & 2 & 0 & 0 \\ 0 & 0 & -1 & 1 \\ 0 & 0 & -2 & 4 \end{bmatrix}.$$

SOLUTION. The characteristic equation of the matrix A is given by

$$|A - \lambda I| = 0$$

or $\begin{vmatrix} 1-\lambda & 1 & 0 & 0 \\ 0 & 2-\lambda & 0 & 0 \\ 0 & 0 & 1-\lambda & 1 \\ 0 & 0 & -2 & 4-\lambda \end{vmatrix} = 0$

or $(1-\lambda) \begin{vmatrix} 2-\lambda & 0 & 0 \\ 0 & 1-\lambda & 1 \\ 0 & -2 & 4-\lambda \end{vmatrix} = 0$

[Expanding along first column]

or $(1-\lambda)\left[(2-\lambda)\{(1-\lambda)(4-\lambda)+2\}\right] = 0$

or $(1-\lambda)\left[(2-\lambda)(1-\lambda)(4-\lambda) + 2(2-\lambda)\right] = 0$

or $(1-\lambda)(2-\lambda)\left(\lambda^2 - 5\lambda + 6\right) = 0$

or $\qquad (1-\lambda)(2-\lambda)(2-\lambda)(3-\lambda)=0$

or $\qquad (\lambda-1)(\lambda-2)^2(\lambda-3)=0$

We have to show that $(A-I)(A-2I)^2(A-3I)=O$

Now

$$A-I=\begin{bmatrix}1&1&0&0\\0&2&0&0\\0&0&1&1\\0&0&-2&4\end{bmatrix}-\begin{bmatrix}1&0&0&0\\0&1&0&0\\0&0&1&0\\0&0&0&1\end{bmatrix}=\begin{bmatrix}0&1&0&0\\0&1&0&0\\0&0&0&1\\0&0&-2&3\end{bmatrix}$$

and

$$A-2I=\begin{bmatrix}1&1&0&0\\0&2&0&0\\0&0&1&1\\0&0&-2&4\end{bmatrix}-2\begin{bmatrix}1&0&0&0\\0&1&0&0\\0&0&1&0\\0&0&0&1\end{bmatrix}=\begin{bmatrix}-1&1&0&0\\0&0&0&0\\0&0&-1&1\\0&0&-2&2\end{bmatrix}$$

$$\therefore\ (A-2I)^2=\begin{bmatrix}-1&1&0&0\\0&0&0&0\\0&0&-1&1\\0&0&-2&2\end{bmatrix}\begin{bmatrix}-1&1&0&0\\0&0&0&0\\0&0&-1&1\\0&0&-2&2\end{bmatrix}$$

$$=\begin{bmatrix}1&-1&0&0\\0&0&0&0\\0&0&-1&1\\0&0&-2&2\end{bmatrix}$$

$$A-3I=\begin{bmatrix}1&1&0&0\\0&2&0&0\\0&0&1&1\\0&0&-2&4\end{bmatrix}-3\begin{bmatrix}1&0&0&0\\0&1&0&0\\0&0&1&0\\0&0&0&1\end{bmatrix}$$

$$=\begin{bmatrix}-2&1&0&0\\0&-1&0&0\\0&0&-2&1\\0&0&-2&1\end{bmatrix}$$

Now $(A-I)(A-2I)^2(A-3I)$

$$=\begin{bmatrix}0&1&0&0\\0&1&0&0\\0&0&0&1\\0&0&-2&3\end{bmatrix}\begin{bmatrix}1&-1&0&0\\0&0&0&0\\0&0&-1&1\\0&0&-2&2\end{bmatrix}\begin{bmatrix}-2&1&0&0\\0&-1&0&0\\0&0&-2&1\\0&0&-2&1\end{bmatrix}$$

$$=\begin{bmatrix}0&1&0&0\\0&1&0&0\\0&0&0&1\\0&0&-2&3\end{bmatrix}\begin{bmatrix}-2&2&0&0\\0&0&0&0\\0&0&0&0\\0&0&0&0\end{bmatrix}=\begin{bmatrix}0&0&0&0\\0&0&0&0\\0&0&0&0\\0&0&0&0\end{bmatrix}=O$$

$\therefore\qquad (A-I)(A-2I)^2(A-3I)=O$

Hence, the Cayley-Hamilton theorem is verified.

EXAMPLE 4. *Use Cayley-Hamilton theorem to express* $2A^5-3A^4+A^2-4I$ *as a linear polynomial in* **A**, *where :*

$$A=\begin{bmatrix}3&1\\-1&2\end{bmatrix}.$$

SOLUTION. The characteristic equation of A is given by

$$|A-\lambda I|=0$$

or $\qquad \begin{vmatrix}3-\lambda&1\\-1&2-\lambda\end{vmatrix}=0$

or $\qquad (3-\lambda)(2-\lambda)+1=0$

or $\qquad \lambda^2-5\lambda+7=0$

By Cayley-Hamilton theorem, we have

$A^2-5A+7I=O \qquad\qquad …(1)$

$\Rightarrow \qquad A^2=5A-7I \qquad\qquad …(2)$

Now $\qquad\qquad A^3=A^2.A$

$\qquad\qquad =(5A-7I)A=5A^2-7A$

$\therefore\qquad A^3=5A^2-7A \qquad\qquad …(3)$

Again, $\qquad A^4=A^3.A=(5A^2-7A)A$

$\qquad\qquad =5A^3-7A^2$

$\Rightarrow\qquad A^4=5(5A^2-7A)-7(5A-7I)$

$\qquad\qquad$ [Using (2) and (3)]

$\Rightarrow\qquad A^4=25A^2-35A-35A+49I$

$\Rightarrow\qquad A^4=25(5A-7I)-70A+49I$

$\qquad\qquad$ [Using (2)]

$\Rightarrow\qquad A^4=125A-175I-70A+49I$

$\Rightarrow\qquad A^4=55A-126I \qquad\qquad …(4)$

Also $\qquad A^5=A^4.A=(55A-126I)A$

$\qquad\qquad A^5=55A^2-126A$

$\Rightarrow\qquad A^5=55(5A-7I)-126A \qquad$ [Using (2)]

$\therefore\qquad A^5=149A-385I \qquad\qquad …(5)$

Now $\qquad\qquad 2A^5-3A^4+A^2-4I$

$\qquad =2(149A-385I)-3(55A-126I)$

$\qquad\qquad\qquad +5A-7I-4I$

$\qquad\qquad$ [Using (2), (4) and (5)]

$\qquad =298A-770I-165A+378I+5A-11I$

$\qquad =138A-403I$

$\therefore\quad 2A^5-3A^4+A^2-4I=138A-403I$

which is a linear polynomial in A.

Exercise-1.10

1. Verify Cayley-Hamilton theorem for the matrix $A=\begin{bmatrix}1&1\\8&1\end{bmatrix}$ and use it to find A^{-1}.

2. Use Cayley-Hamilton theorem to find the inverse of the matrix $A=\begin{bmatrix}2&1\\5&3\end{bmatrix}$.

3. Verify Cayley-Hamilton theorem for the matrix $A=\begin{bmatrix}0&0&0\\3&1&0\\-2&1&4\end{bmatrix}$ and hence find A^{-1}.

4. Verify Cayley-Hamilton theorem for the following matrix:

$A=\begin{bmatrix}2&0\\0&1\end{bmatrix}.$

5. Show that the matrix $A = \begin{bmatrix} 1 & 2 \\ 1 & 1 \end{bmatrix}$ satisfies Cayley-Hamilton theorem.

6. State the Cayley-Hamilton theorem and verify it for the matrix

$$A = \begin{bmatrix} 1 & 0 & -2 \\ 0 & 0 & 0 \\ -2 & 0 & 4 \end{bmatrix}$$

7. Verify Cayley-Hamilton theorem for the matrix $A = \begin{bmatrix} 1 & 4 \\ 2 & 3 \end{bmatrix}$ and hence obtain A^{-1}.

8. Verify Cayley-Hamilton theorem for the matrix

$$A = \begin{bmatrix} 1 & 2 & 1 \\ 0 & 1 & -1 \\ 3 & -1 & 1 \end{bmatrix}$$

and hence find A^{-1}.

9. Verify that the matrix $A = \begin{bmatrix} 1 & 2 & 1 \\ -1 & 0 & 3 \\ 2 & -1 & 1 \end{bmatrix}$ satisfies its

characteristic equation.

10. Show that the matrix $A = \begin{bmatrix} 2 & 2 & 1 \\ 1 & 3 & 1 \\ 1 & 2 & 2 \end{bmatrix}$ satisfies Cayley-Hamilton theorem.

11. Verify Cayley-Hamilton theorem for the matrix

$$A = \begin{bmatrix} 1 & \sqrt{2} & 0 \\ \sqrt{2} & -1 & 0 \\ 0 & 0 & 1 \end{bmatrix}$$

and hence find A^{-1}.

12. Verify Cayley-Hamilton theorem for the matrix

$$A = \begin{bmatrix} 1 & 0 & 2 \\ 0 & -1 & 1 \\ 0 & 1 & 0 \end{bmatrix}$$

and hence find A^{-1}.

13. Verify Cayley-Hamilton theorem for the matrix

$$A = \begin{bmatrix} 1 & 3 & 7 \\ 4 & 2 & 3 \\ 0 & 2 & 1 \end{bmatrix}$$

and hence find A^{-1}.

14. Verify Cayley-Hamilton theorem for the matrix

$$A = \begin{bmatrix} 1 & 2 & 3 \\ 3 & -2 & 1 \\ 4 & 2 & 1 \end{bmatrix}$$

and hence find A^{-1}.

15. Verify Cayley-Hamilton theorem for the matrix

$$A = \begin{bmatrix} 1 & 1 & 3 \\ 5 & 2 & 6 \\ -2 & -1 & -3 \end{bmatrix}$$

16. Verify Cayley-Hamilton theorem for the matrix

$$A = \begin{bmatrix} 3 & 2 & 4 \\ 4 & 3 & 2 \\ 2 & 4 & 3 \end{bmatrix}$$

17. Verify Cayley-Hamilton theorem for the matrix

$$A = \begin{bmatrix} 2 & 3 & -2 \\ 0 & 5 & 4 \\ 1 & 0 & 1 \end{bmatrix}$$

18. If $A = \begin{bmatrix} 1 & 2 \\ -1 & 3 \end{bmatrix}$ express $A^6 - 4A^5 + 8A^4 - 12A^3 + 14A^2$ as a linear polynomial in A.

19. Find the characteristic equation of the matrix $A = \begin{bmatrix} 2 & 1 & 1 \\ 0 & 1 & 0 \\ 1 & 1 & 2 \end{bmatrix}$

and hence compute A^{-1}. Also find the value of

$A^8 - 5A^7 + 7A^6 - 3A^5 + A^4 - 5A^3 + 8A^2 - 2A + I$ (UKTU 2010)

20. If $A = \begin{bmatrix} 1 & 0 & 0 \\ 1 & 0 & 1 \\ 0 & 1 & 0 \end{bmatrix}$, show that for every integer $n \geqslant 3$

$A^n = A^{n-2} + A^2 - I$ (UPTU-2009, MUMBAI-2006)

21. Verify Cayleg-Hamilton theorem for the following matrices

(i) $\begin{bmatrix} 2 & 2 & 1 \\ 0 & 1 & -1 \\ 3 & -1 & 1 \end{bmatrix}$ (UPTU-2007)   (ii) $\begin{bmatrix} 1 & 0 & -4 \\ 0 & 5 & 4 \\ -4 & 4 & 3 \end{bmatrix}$ (GBTU-2010)

(iii) $\begin{bmatrix} 3 & 0 & 1 \\ 0 & 2 & 0 \\ 0 & 0 & 1 \end{bmatrix}$ (UPTU-2008)   (iv) $\begin{bmatrix} 7 & 2 & -2 \\ -6 & -1 & 2 \\ 6 & 2 & -1 \end{bmatrix}$ (UKTU-2012)

(v) $\begin{bmatrix} 2 & -1 & 1 \\ -1 & 2 & -1 \\ 1 & -1 & 2 \end{bmatrix}$ (SVTV-2008, ANNA-2009, MADRAS-2006)

(vi) $\begin{bmatrix} 3 & 2 & 4 \\ 4 & 3 & 2 \\ 2 & 4 & 3 \end{bmatrix}$ (PTU-2006)

(vii) $\begin{bmatrix} 1 & 3 & 7 \\ 4 & 2 & 3 \\ 1 & 2 & 1 \end{bmatrix}$ (ANNA-2005, BHOPAL-2008, KERALA 2005)

(viii) $\begin{bmatrix} 1 & 2 & 3 \\ 2 & 4 & 5 \\ 3 & 5 & 6 \end{bmatrix}$ (UPTU-2007)

Answers

1. $A^{-1} = -\dfrac{1}{7}\begin{bmatrix} 1 & -1 \\ -8 & 1 \end{bmatrix}$

2. $A^{-1} = \begin{bmatrix} 3 & -1 \\ -5 & 2 \end{bmatrix}$

3. $A^{-1} = \dfrac{1}{5}\begin{bmatrix} 4 & 1 & -1 \\ -12 & 2 & 3 \\ 5 & 0 & 0 \end{bmatrix}$

7. $A^{-1} = -\dfrac{1}{3}\begin{bmatrix} 3 & -4 \\ -2 & 1 \end{bmatrix}$

8. $A^{-1} = \dfrac{1}{9}\begin{bmatrix} 0 & 3 & 3 \\ 3 & 2 & -1 \\ 3 & -7 & -1 \end{bmatrix}$

11. $A^{-1} = -\dfrac{1}{3}\begin{bmatrix} -1 & -\sqrt{2} & 0 \\ -\sqrt{2} & 1 & 0 \\ 0 & 0 & -3 \end{bmatrix}$

13. $A^{-1} = \dfrac{1}{10}\begin{bmatrix} -4 & 11 & -5 \\ -4 & 1 & 25 \\ 8 & -2 & -10 \end{bmatrix}$

14. $A^{-1} = \dfrac{1}{36}\begin{bmatrix} -4 & 4 & 8 \\ 1 & -11 & 0 \\ 14 & 6 & -8 \end{bmatrix}$

18. $-4A + 5I$

19. $A^3 - 5A^2 + 7A - 3I = 0; \ A^{-1} = \dfrac{1}{3}\begin{bmatrix} 2 & -1 & -1 \\ 0 & 3 & 0 \\ -1 & -1 & 2 \end{bmatrix}, \begin{bmatrix} 8 & 5 & 5 \\ 0 & 3 & 0 \\ 5 & 5 & 8 \end{bmatrix}$

1.20 DIAGONALIZATION OF A MATRIX

Let A be a square matrix of order n. Then A is said to be diagonalizable iff it is similar to a diagonal matrix.

Therefore, A is diagonalizable if there exists a non-singular matrix P such that $P^{-1}AP = D$ where D is a diagonal matrix and the matrix P is said to transform A to diagonal form.

THEOREM 1. *An $n \times n$ matrix is diagonalizable if and only if it possesses n linearly independent eigenvectors.*

PROOF. Let A be $n \times n$ matrix. Suppose that A is diagonalizable, then it is similar to a diagonal matrix. Let $D = \text{diag}(\lambda_1, \lambda_2, ..., \lambda_n)$ be that diagonal matrix, therefore there exists a non-singular matrix P (say) such that

$$P^{-1}AP = D \ \Rightarrow \ AP = PD \qquad \qquad ...(1)$$

The eigenvalues of D are $\lambda_1, \lambda_2, ..., \lambda_n$ and A is similar to D, therefore $\lambda_1, \lambda_2, ..., \lambda_n$ are the only eigenvalues of A.

Suppose that $X_1, X_2, ..., X_n$ are column vectors of P *i.e.,* $P = [X_1, X_2, ..., X_n]$

Since P is invertible so that $X_1, X_2, ..., X_n$ are n linearly independent vectors.

Now from (1), we get

$$A[X_1, X_2, ..., X_n] = [X_1, X_2, ..., X_n]\text{dia.}[\lambda_1, \lambda_2, ..., \lambda_n]$$

or $\qquad \qquad [AX_1, AX_2, ..., AX_n] = [\lambda_1 X_1, \lambda_2 X_2, ..., \lambda_n X_n]$

$\Rightarrow \qquad \qquad AX_1 = \lambda_1 X_1, AX_2 = \lambda_2 X_2, ..., AX_n = \lambda_n X_n$

Therefore, $X_1, X_2, ..., X_n$ are the eigenvectors of A corresponding to the eigenvalues of A. Also $X_1, X_2, ..., X_n$ are linearly independent, hence A has n linearly independent eigenvectors.

Conversely. Suppose that A has n linearly independent eigenvectors $X_1, X_2, ..., X_n$ then there are scalars $\lambda_1, \lambda_2, ..., \lambda_n$ (not necessarily distinct) such that

$$AX_1 = \lambda_1 X_1, AX_2 = \lambda_2 X_2, ..., AX_n = \lambda_n X_n$$

Let $P = [X_1, X_2, ..., X_n]$. Since $X_1, X_2, ..., X_n$ are linearly independent eigenvectors, therefore P is invertible.

Let $\qquad \qquad \qquad \qquad \qquad D = \text{dia.}(\lambda_1, \lambda_2, ..., \lambda_n)$. Then

$$AP = A[X_1, X_2, ..., X_n] = [AX_1, AX_2, ..., AX_n]$$
$$= [\lambda_1 X_1, \lambda_2 X_2, ..., \lambda_n X_n] \ [X_1, X_2, ..., X_n]\text{dia.}[\lambda_1, \lambda_2, ..., \lambda_n]$$

$\Rightarrow \qquad \qquad AP = PD \ \Rightarrow \ P^{-1}AP = D \qquad \qquad [\because P^{-1} \text{ exists.}]$

$\Rightarrow A$ is similar to a diagonal matrix.

$\Rightarrow A$ is diagonalizable.

COROLLARY. *If the eigenvalues of an $n \times n$ matrix are all distinct then it is necessarily diagonalizable.*

PROOF. Let A be an $n \times n$ matrix. Since the eigenvectors of A corresponding to the distinct eigenvalues are linearly independent, therefore A has n linearly independent eigenvectors, hence by above theorem, A is diagonalizable.

REMARK

- In view of above theorem, if A is diagonalizable and P diagonalizes A, then

$$P^{-1}AP = \begin{bmatrix} \lambda_1 & 0 & 0 & \cdots & 0 \\ 0 & \lambda_2 & 0 & \cdots & 0 \\ 0 & 0 & \lambda_3 & \cdots & 0 \\ \vdots & & & & \\ 0 & 0 & 0 & \cdots & \lambda_n \end{bmatrix}$$

if and only if the *jth* column of P is an eigenvector of A corresponding to the eigenvalue λ_i of A for $i = 1, 2, 3, \ldots n$.

Solved Examples

EXAMPLE 1. *Consider the matrix* $A = \begin{bmatrix} 3 & 2 & 0 \\ 2 & 0 & 0 \\ 1 & 0 & 2 \end{bmatrix}$

Find an invertible matrix P such that $P^{-1}AP$ is a diagonal matrix. Also find the diagonal matrix.

SOLUTION. The characteristic equation of A is given by

$$|A - \lambda I| = 0$$

or

$$\begin{vmatrix} 3-\lambda & 2 & 0 \\ 2 & 0-\lambda & 0 \\ 1 & 0 & 2-\lambda \end{vmatrix} = 0$$

or $(3-\lambda)\{-\lambda(2-\lambda)\} - 2\{2(2-\lambda)\} = 0$

or $(2-\lambda)(4-\lambda)(\lambda+1) = 0$

∴ $-1, 2, 4$ are eigenvalues of A.

Since A has distinct eigenvalues so that A is diagonalizable, therefore there exists a non-singular matrix P such that $P^{-1}AP$ is diagonalizable.

Now to find P we shall find eigenvectors of A.

Eigenvector corresponding to $\lambda_1 = -1$

Let $X_1 = \begin{bmatrix} x_1 \\ x_2 \\ x_3 \end{bmatrix} \neq O$ be an eigenvector corresponding to

eigenvalue $\lambda_1 = -1$, then

or $\qquad AX_1 = (-1)X_1$

or $(A + I)X_1 = O$

or $\begin{bmatrix} 3-(-1) & 2 & 0 \\ 2 & 0-(-1) & 0 \\ 1 & 0 & 2-(-1) \end{bmatrix}\begin{bmatrix} x_1 \\ x_2 \\ x_3 \end{bmatrix} = \begin{bmatrix} 0 \\ 0 \\ 0 \end{bmatrix}$

or $\begin{bmatrix} 4 & 2 & 0 \\ 2 & 1 & 0 \\ 1 & 0 & 3 \end{bmatrix}\begin{bmatrix} x_1 \\ x_2 \\ x_3 \end{bmatrix} = \begin{bmatrix} 0 \\ 0 \\ 0 \end{bmatrix}$ ...(1)

Applying $R_1 \leftrightarrow R_3$, we get

$$\begin{bmatrix} 1 & 0 & 3 \\ 2 & 1 & 0 \\ 4 & 2 & 0 \end{bmatrix}\begin{bmatrix} x_1 \\ x_2 \\ x_3 \end{bmatrix} = \begin{bmatrix} 0 \\ 0 \\ 0 \end{bmatrix}$$

Applying $R_2 \to R_2 - 2R_1, R_3 \to R_3 - 4R_1$, we get

$$\begin{bmatrix} 1 & 0 & 3 \\ 0 & 1 & -6 \\ 0 & 2 & -12 \end{bmatrix}\begin{bmatrix} x_1 \\ x_2 \\ x_3 \end{bmatrix} = \begin{bmatrix} 0 \\ 0 \\ 0 \end{bmatrix}$$

Applying $R_3 \to R_3 - 2R_2$, we get

$$\begin{bmatrix} 1 & 0 & 3 \\ 0 & 1 & -6 \\ 0 & 0 & 0 \end{bmatrix}\begin{bmatrix} x_1 \\ x_2 \\ x_3 \end{bmatrix} = \begin{bmatrix} 0 \\ 0 \\ 0 \end{bmatrix}$$ ...(2)

Clearly $\rho(A + I) = 2$, therefore the equation (1) will have $3 - 2 = 1$ linearly independent solution.

From (2), we have $x_1 + 3x_3 = 0$

$$x_2 - 6x_3 = 0$$

Let us put $x_3 = c$ (an arbitrary constant), then

$$x_2 = 6c, x_1 = -3c$$

Now $\begin{bmatrix} x_1 \\ x_2 \\ x_3 \end{bmatrix} = \begin{bmatrix} -3c \\ 6c \\ c \end{bmatrix} = c\begin{bmatrix} -3 \\ 6 \\ 1 \end{bmatrix}$

∴ $X_1 = \begin{bmatrix} -3 \\ 6 \\ 1 \end{bmatrix}$

Eigenvector corresponding to $\lambda_2 = 2$:

Let $X_2 = \begin{bmatrix} x_1 \\ x_2 \\ x_3 \end{bmatrix} \neq O$ be an eigenvector corresponding to

eigenvalue $\lambda_2 = 2$, then

or $\qquad AX_2 = 2X_2$

or $(A - 2I)X_2 = O$

or $\begin{bmatrix} 3-2 & 2 & 0 \\ 2 & 0-2 & 0 \\ 1 & 0 & 2-2 \end{bmatrix}\begin{bmatrix} x_1 \\ x_2 \\ x_3 \end{bmatrix} = \begin{bmatrix} 0 \\ 0 \\ 0 \end{bmatrix}$

or $\begin{bmatrix} 1 & 2 & 0 \\ 2 & -2 & 0 \\ 1 & 0 & 0 \end{bmatrix}\begin{bmatrix} x_1 \\ x_2 \\ x_3 \end{bmatrix} = \begin{bmatrix} 0 \\ 0 \\ 0 \end{bmatrix}$ ...(3)

Applying $R_2 \to R_2 - 2R_1, R_3 \to R_3 - R_1$, we get

$$\begin{bmatrix} 1 & 2 & 0 \\ 0 & -6 & 0 \\ 0 & -2 & 0 \end{bmatrix}\begin{bmatrix} x_1 \\ x_2 \\ x_3 \end{bmatrix} = \begin{bmatrix} 0 \\ 0 \\ 0 \end{bmatrix}$$

Applying $R_3 \leftrightarrow R_2$ we get

$$\begin{bmatrix} 1 & 2 & 0 \\ 0 & -2 & 0 \\ 0 & -6 & 0 \end{bmatrix}\begin{bmatrix} x_1 \\ x_2 \\ x_3 \end{bmatrix} = \begin{bmatrix} 0 \\ 0 \\ 0 \end{bmatrix}$$

Applying $R_3 \to R_3 - 3R_2$, we get

$$\begin{bmatrix} 1 & 2 & 0 \\ 0 & -2 & 0 \\ 0 & 0 & 0 \end{bmatrix}\begin{bmatrix} x_1 \\ x_2 \\ x_3 \end{bmatrix} = \begin{bmatrix} 0 \\ 0 \\ 0 \end{bmatrix} \qquad \dots(4)$$

Clearly $\rho(A - 2I) = 2$, therefore the equation (3) will have $3 - 2 = 1$ linearly independent solution.

From (4), we have

$$x_1 + 2x_2 = 0$$
$$-2x_2 = 0$$

Let us put $x_3 = c$, we get $x_2 = 0$, $x_1 = 0$

Now

$$\begin{bmatrix} x_1 \\ x_2 \\ x_3 \end{bmatrix} = \begin{bmatrix} 0 \\ 0 \\ c \end{bmatrix} = c\begin{bmatrix} 0 \\ 0 \\ 1 \end{bmatrix}$$

$$\therefore \qquad X_2 = \begin{bmatrix} 0 \\ 0 \\ 1 \end{bmatrix}$$

Eigenvector corresponding to $\lambda_3 = 4$:

Let $X_3 = \begin{bmatrix} x_1 \\ x_2 \\ x_3 \end{bmatrix} \neq O$ be an eigenvector

corresponding to eigenvalue $\lambda_3 = 4$, then

or $\qquad\qquad AX_3 = 4X_3$

or $\qquad\qquad (A - 4I)X_3 = O$

or $\begin{bmatrix} 3-4 & 2 & 0 \\ 2 & 0-4 & 0 \\ 1 & 0 & 2-4 \end{bmatrix}\begin{bmatrix} x_1 \\ x_2 \\ x_3 \end{bmatrix} = \begin{bmatrix} 0 \\ 0 \\ 0 \end{bmatrix}$

or $\begin{bmatrix} -1 & 2 & 0 \\ 2 & -4 & 0 \\ 1 & 0 & -2 \end{bmatrix}\begin{bmatrix} x_1 \\ x_2 \\ x_3 \end{bmatrix} = \begin{bmatrix} 0 \\ 0 \\ 0 \end{bmatrix} \qquad \dots(5)$

Applying $R_2 \to R_2 + 2R_1$, $R_3 \to R_3 + R_1$, we get

$$\begin{bmatrix} -1 & 2 & 0 \\ 0 & 0 & 0 \\ 0 & 2 & -2 \end{bmatrix}\begin{bmatrix} x_1 \\ x_2 \\ x_3 \end{bmatrix} = \begin{bmatrix} 0 \\ 0 \\ 0 \end{bmatrix}$$

Applying $R_2 \leftrightarrow R_3$, we get

$$\begin{bmatrix} -1 & 2 & 0 \\ 0 & 2 & -2 \\ 0 & 0 & 0 \end{bmatrix}\begin{bmatrix} x_1 \\ x_2 \\ x_3 \end{bmatrix} = \begin{bmatrix} 0 \\ 0 \\ 0 \end{bmatrix} \qquad \dots(6)$$

Clearly $\rho(A - 4I) = 2$, therefore the equation (5) will have $3 - 2 = 1$ linearly independent solution.

From (6), we have $\quad -x_1 + 2x_2 = 0$
$$2x_2 - 2x_3 = 0$$

Let us put $x_2 = c$, we get $x_1 = 2c$, $x_3 = c$

Now

$$\begin{bmatrix} x_1 \\ x_2 \\ x_3 \end{bmatrix} = \begin{bmatrix} 2c \\ c \\ c \end{bmatrix} = c\begin{bmatrix} 2 \\ 1 \\ 1 \end{bmatrix}$$

$$\therefore \qquad X_3 = \begin{bmatrix} 2 \\ 1 \\ 1 \end{bmatrix}$$

Since $\qquad P = [X_1, X_2, X_3] = \begin{bmatrix} -3 & 0 & 2 \\ 6 & 0 & 1 \\ 1 & 1 & 1 \end{bmatrix}$

and $\qquad P^{-1}AP = \text{dia.}(-1, 2, 4) = \begin{bmatrix} -1 & 0 & 0 \\ 0 & 2 & 0 \\ 0 & 0 & 4 \end{bmatrix} = D$

EXAMPLE 2. *Show that the matrix* $A = \begin{bmatrix} 8 & -8 & -2 \\ 4 & -3 & -2 \\ 3 & -4 & 1 \end{bmatrix}$ *is diagonalizable.*

SOLUTION. If the matrix A has distinct eigenvalues, then it is essentially diagonalizable, so we shall find its eigenvalues.

The characteristic equation of A is given by

$$|A - \lambda I| = 0$$

or $\qquad \begin{vmatrix} 8-\lambda & -8 & -2 \\ 4 & -3-\lambda & -2 \\ 3 & -4 & 1-\lambda \end{vmatrix} = 0$

Applying $R_1 \to R_1 - (R_2 + R_3)$, we get

$$\begin{vmatrix} 1-\lambda & -1+\lambda & -1+\lambda \\ 4 & -3-\lambda & -2 \\ 3 & -4 & 1-\lambda \end{vmatrix} = 0$$

or $\qquad (1+\lambda)\begin{vmatrix} 1 & -1 & -1 \\ 4 & -3-\lambda & -2 \\ 3 & -4 & 1-\lambda \end{vmatrix} = 0$

Applying $C_2 \to C_2 + C_1$, $C_3 \to C_3 + C_1$, we get

or $\qquad (1-\lambda)\begin{vmatrix} 1 & 0 & 0 \\ 4 & 1-\lambda & 2 \\ 3 & -1 & 4-\lambda \end{vmatrix} = 0$

or $\qquad (1-\lambda)\{(1-\lambda)(4-\lambda)+2\} = 0$

or $\qquad (1-\lambda)(\lambda^2 -5\lambda + 6) = 0$

or $\qquad (1-\lambda)(2-\lambda)(3-\lambda) = 0$

$\therefore$ $\quad A$ has 1, 2, 3 eigenvalues which are all distinct, hence A is diagonalizable.

✎ REMARK

- If some eigenvalues of a matrix are the same, then it need not be diagonalizable.

1.20.1 ALGEBRAIC AND GEOMETRIC MULTIPLICITY OF AN EIGENVALUE

(i) **Algebraic multiplicity :** Let A be an $n \times n$ matrix and let

$$|A - \lambda I_n| = (\lambda - \lambda_1)^{n_1}(\lambda - \lambda_2)^{n_2}(\lambda - \lambda_3)^{n_3}\dots(\lambda - \lambda_k)^{n_k}$$

where $n_1 + n_2 + n_3 + \dots + n_k = n$. Then the numbers $n_1, n_2, n_3, \dots, n_k$ are the algebraic multiplicities of the eigenvalues $\lambda_1, \lambda_2, \lambda_3, \dots, \lambda_n$ respectively.

(ii) **Eigenspace of a matrix A corresponding to eigenvalue λ:**

Let A be an $n \times n$ matrix, let λ be an eigenvalue of A.

For eigenvalue λ, we find eigenvectors $X = \begin{bmatrix} x_1 \\ x_2 \\ x_3 \\ \vdots \\ x_n \end{bmatrix}$ by solving the linear system $(A - \lambda I)X = O$

The set of all vectors X satisfying $AX = \lambda X$ is called the eigenspace of A corresponding to eigenvalue λ, which is denoted by E_λ.

(iii) **Geometric multiplicity :**

Let A be an $n \times n$ matrix and λ be one of its eigenvalues, then the dimension of eigenspace E_λ of A corresponding to λ is called the geometric multiplicity of λ.

For eigenvalue λ, we find eigenvectors $X = \begin{bmatrix} x_1 \\ x_2 \\ x_3 \\ \vdots \\ x_n \end{bmatrix}$ by solving the linear system $(A - \lambda I)X = O$

The set of all vectors X satisfying $AX = \lambda X$ is called the eigenspace of A corresponding to eigenvalue λ, which is denoted by E_k.

REMARKS

- Dimension of E_λ = Dimension of the null space $(A - \lambda I)$.
- Geometric multiplicity of $\lambda = n - \rho(A - \lambda I)$.
- The geometric mulitplicity of an eigenvalue cannot exceed its algebraic multiplicity.

THEOREM 1.   *The necessary and sufficient conditions for a square matrix to be similar to a diagonal matrix is that the geometric multiplicity of each of its eigenvalues coincides with the algebraic multiplicity.*

PROOF.   Let A be an $n \times n$ matrix.

Necessary Condition : Suppose that A is similar to a diagonal matrix $D = $ diag. $(\lambda_1, \lambda_2, ..., \lambda_n)$, then there exists a non-singular matrix P such that

$$P^{-1}AP = D = \text{diag. } (\lambda_1, \lambda_2, ..., \lambda_n) \qquad ...(1)$$

$\therefore$ $\lambda_1, \lambda_2, ..., \lambda_n$ are the eigenvalues of A not necessarily distinct.

Let t be an eigenvalue of A of algebraic multiplicity p, then we have

$$\lambda_1 = t, \ \lambda_2 = t , ..., \lambda_p = t \qquad ...(2)$$

If $\rho(A - tI) = m$, then the system of equations

$$(A - tI)X = O$$

will have $n - m$ linearly independent solutions and therefore $n - m$ will be the geometric multiplicity of t.

So, we have to prove that $p = n - m$.

Since the rank of a matrix does not change on premultiplication and postmultiplication by a non-singular matrix.

$\therefore$

$$\rho\left(A - tI\right) = \rho\left[P^{-1}\left(A - tI\right)P\right] = \rho\left[P^{-1}AP - tI\right] = \rho\left[D - tI\right]$$

$$= \rho\left(\text{diag.}\left[\lambda_1 - t, \lambda_2 - t, ..., \lambda_p - t, \lambda_{p+1} - t, \lambda_n - t\right]\right) \qquad \text{[Using (2)]}$$

$$= \rho\left(\text{diag.}\left[0, 0, ..., 0, \lambda_{p+1} - t, \lambda_n - t\right]\right)$$

$$\rho\left(A - tI\right) = n - p$$

$\Rightarrow$ $\qquad\qquad m = n - p$ $\qquad\qquad\qquad\qquad\qquad\qquad\qquad\qquad \because \rho(A - tI) = m$

$\Rightarrow$ $\qquad\qquad p = m - n$

Sufficient Condition : Suppose that the geometric multiplicity of each eigenvalue of A coincides with its algebraic multiplicity. Then we have to show that A is similar to a diagonal matrix, *i.e.*, A is diagonalizable.

Suppose that A has $\lambda_1, \lambda_2, ... \lambda_k$ distinct eigenvalues of multiplicity $n_1, n_2, ... n_k$ respectively, then we have

$$n_1 + n_2 + ... + n_k = n$$

Let $\qquad\qquad X_{11}, X_{12}, ..., X_{1n_1}, X_{21}, X_{22}, ..., X_{2n_2}; .....; X_{k1}, X_{k2}, ..., X_{kn_k}\Big\}$

be linearly independent sets of eigenvectors corresponding to the eigenvalues $\lambda_1, \lambda_2, ... \lambda_k$ respectively.

Now we prove that the n vectors given by (3) are linearly independent.

Let

$$(a_{11}X_{11} + a_{12}X_{12} + \ldots + a_{1n_1}X_{1n_1}) + (a_{21}X_{21} + a_{22}X_{22} + \ldots + a_{2n_2}X_{2n_2})$$

$$+ \ldots + (a_{k1}X_{k1} + a_{k2}X_{k2} + \ldots + a_{kn_K}X_{kn_K}) = O \qquad \ldots(4)$$

or

$$X_1 + X_2 + \ldots + X_k = O \qquad \ldots(5)$$

where

$$\left. \begin{array}{l} X_1 = a_{11}X_{11} + a_{12}X_{12} + \ldots + a_{1n_1}X_{1n_1} \\ X_2 = a_{21}X_{21} + a_{22}X_{22} + \ldots + a_{2n_2}X_{2n_2} \\ \ldots\ldots\ldots\ldots\ldots\ldots\ldots\ldots\ldots\ldots\ldots\ldots\ldots\ldots \\ X_k = a_{k1}X_{k1} + a_{k2}X_{k2} + \ldots + a_{kn_k}X_{kn_k} \end{array} \right\} \qquad \ldots(6)$$

From (6), we see that X_1 is a linear combination of $X_{11}, X_{12}, \ldots, X_{1n_1}$ which are eigenvectors of A corresponding to the eigenvalue λ_1. If $X_1 \neq O$ then X_1 is also the eigenvector of A corresponding to eigenvalue λ_1, similarly for other eigenvectors X_2, $X_3, \ldots, X_k$.

Suppose $X_i \neq O$, then from (3), we see that a system of eigenvectors of A corresponding to distinct eigenvalues of A is linearly dependent which is not possible, hence $\quad X_i = O \; \forall \, i = 1, 2, 3, \ldots, k$

$$\Rightarrow \qquad a_{i1}X_{i1} + a_{i2}X_{i2} + \ldots + a_{in_1}X_{in_1} = O \qquad \text{[Using (6)]}$$

But $X_{i1}, X_{i2}, \ldots, X_{in_1}$ is a set of linearly independent vectors, therefore

$$a_{i1} = a_{i2} = \ldots = a_{in_1} = 0 \qquad \forall \, i = 1, 2, 3, \ldots, k$$

It follows that the n vectors given by (3) are linearly independent. Therefore A has n linearly independent eigenvectors, hence A is diagonalizable and hence A is similar to a diagonal matrix.

Solved Examples

EXAMPLE 1. *Show that the following matrices are not similar to diagonal matrices:*

(i) $\begin{bmatrix} 2 & 3 & 4 \\ 0 & 2 & -1 \\ 0 & 0 & 1 \end{bmatrix}$ **(ii)** $\begin{bmatrix} 2 & -1 & 1 \\ 2 & 2 & -1 \\ 1 & 2 & -1 \end{bmatrix}$

SOLUTION. (i) Let $\qquad A = \begin{bmatrix} 2 & 3 & 4 \\ 0 & 2 & -1 \\ 0 & 0 & 1 \end{bmatrix}$

Clearly, A is an upper triangular matrix therefore its diagonal elements 2, 2, 1 are the eigenvalues of A.

The algebraic multiplicity of eigenvalue 2 is 2.

Next, we find the geometric multiplicity of eigenvalue 2.

Let $X = \begin{bmatrix} x_1 \\ x_2 \\ x_3 \end{bmatrix} \neq O$ be an eigenvector corresponding

to eigenvalue 2, then

or $\qquad\qquad\qquad AX_1 = 2X_1$

or $\qquad\qquad\qquad (A - 2I)X_1 = O$

or $\quad \begin{bmatrix} 2-2 & 3 & 4 \\ 0 & 2-2 & -1 \\ 0 & 0 & 1-2 \end{bmatrix} \begin{bmatrix} x_1 \\ x_2 \\ x_3 \end{bmatrix} = \begin{bmatrix} 0 \\ 0 \\ 0 \end{bmatrix}$

or $\quad \begin{bmatrix} 0 & 3 & 4 \\ 0 & 0 & -1 \\ 0 & 0 & -1 \end{bmatrix} \begin{bmatrix} x_1 \\ x_2 \\ x_3 \end{bmatrix} = \begin{bmatrix} 0 \\ 0 \\ 0 \end{bmatrix}$

Applying $R_3 \to R_3 - R_2$, we get

$\quad \begin{bmatrix} 0 & 3 & 4 \\ 0 & 0 & -1 \\ 0 & 0 & 0 \end{bmatrix} \begin{bmatrix} x_1 \\ x_2 \\ x_3 \end{bmatrix} = \begin{bmatrix} 0 \\ 0 \\ 0 \end{bmatrix}$

Clearly $\rho(A - 2I) = 2$, therefore the geometric

multiplicity of 2 is $3 - 2 = 1$, which is not equal to the algebraic multiplicity of eigenvalue 2. Hence, A is not similar to the diagonal matrix.

(ii) Let $\qquad A = \begin{bmatrix} 2 & -1 & 1 \\ 2 & 2 & -1 \\ 1 & 2 & -1 \end{bmatrix}$

The characteristic equation of A is given by

$$|A - \lambda I| = 0$$

or $\quad \begin{vmatrix} 2-\lambda & -1 & 1 \\ 2 & 2-\lambda & -1 \\ 1 & 2 & -1-\lambda \end{vmatrix} = 0$

Applying $C_3 \to C_3 + C_2$, we get

$$\begin{vmatrix} 2-\lambda & -1 & 0 \\ 2 & 2-\lambda & 1-\lambda \\ 1 & 2 & 1-\lambda \end{vmatrix} = 0$$

or

$$(1-\lambda) \begin{vmatrix} 2-\lambda & -1 & 0 \\ 2 & 2-\lambda & 1 \\ 1 & 2 & 1 \end{vmatrix} = 0$$

Again applying $R_2 \to R_2 - R_3$, we get

or $\qquad (1-\lambda) \begin{vmatrix} 2-\lambda & -1 & 0 \\ 1 & -\lambda & 0 \\ 1 & 2 & 1 \end{vmatrix} = 0$

or $\qquad (1-\lambda)\{-\lambda(2-\lambda) + 1\} = 0$

or $\qquad (1-\lambda)(\lambda^2 - 2\lambda + 1) = 0$

or $\qquad (1-\lambda)^3 = 0$

$\therefore \quad$ 1, 1, 1 are the eigenvalues of A, thus the algebraic multiplicity of 1 is 3.

Next, we find the geometric multiplicity of eigenvalue 1.

Let $X = \begin{bmatrix} x_1 \\ x_2 \\ x_3 \end{bmatrix} \neq O$ be an eigenvector A corresponding to

eigenvalue 1, then

or $\qquad\qquad AX = (1)X$

or $\qquad\qquad (A - I)X = O$

or $\quad \begin{bmatrix} 2-1 & -1 & 1 \\ 2 & 2-1 & -1 \\ 1 & 2 & -1-1 \end{bmatrix} \begin{bmatrix} x_1 \\ x_2 \\ x_3 \end{bmatrix} = \begin{bmatrix} 0 \\ 0 \\ 0 \end{bmatrix}$

or $\quad \begin{bmatrix} 1 & -1 & 1 \\ 2 & 1 & -1 \\ 1 & 2 & -2 \end{bmatrix} \begin{bmatrix} x_1 \\ x_2 \\ x_3 \end{bmatrix} = \begin{bmatrix} 0 \\ 0 \\ 0 \end{bmatrix}$

Applying $R_2 \rightarrow R_2 - 2R_1, R_3 \rightarrow R_3 - R_1$, we get

$$\begin{bmatrix} 1 & -1 & 1 \\ 0 & 3 & -3 \\ 0 & 3 & -3 \end{bmatrix} \begin{bmatrix} x_1 \\ x_2 \\ x_3 \end{bmatrix} = \begin{bmatrix} 0 \\ 0 \\ 0 \end{bmatrix}$$

Again applying $R_3 \rightarrow R_3 - R_2$, we get

$$\begin{bmatrix} 1 & -1 & 1 \\ 0 & 3 & -3 \\ 0 & 0 & 0 \end{bmatrix} \begin{bmatrix} x_1 \\ x_2 \\ x_3 \end{bmatrix} = \begin{bmatrix} 0 \\ 0 \\ 0 \end{bmatrix}$$

Clearly $\rho(A - I) = 2$, therefore the geometric multiplicity of 2 is $3 - 2 = 1$, which is not equal to the algebraic multiplicity of eigenvalue 1. Hence, A is not similar to the diagonal matrix.

Exercise-1.11

1. Prove that the matrix $A = \begin{bmatrix} 0 & 1 \\ -1 & 0 \end{bmatrix}$ is not diagonalized over R the set of all real numbers, however A is diagonalizable over C the set of all complex numbers. Find an invertible matrix P over C such that $P^{-1}AP$ is a diagonal matrix.

2. Show that the matrix $A = \begin{bmatrix} 1 & -6 & -4 \\ 0 & 4 & 2 \\ 0 & -6 & -3 \end{bmatrix}$ is similar to a diagonal matrix. Also find the transforming matrix and diagonal matrix.

3. Transform the matrix $\begin{bmatrix} 8 & -12 & 5 \\ 15 & -25 & 11 \\ 24 & -42 & 19 \end{bmatrix}$ into diagonal form.

4. Show that each of the following matrices is similar to a diagonal matrix. Also in each case find the diagonal form D and a diagonalizing matrix P :

(i) $\begin{bmatrix} 8 & -6 & 2 \\ -6 & 7 & -4 \\ 2 & -4 & 3 \end{bmatrix}$ (ii) $\begin{bmatrix} 6 & -2 & 2 \\ -2 & 3 & -1 \\ 2 & -1 & 3 \end{bmatrix}$

(iii) $\begin{bmatrix} 17 & 18 & -6 \\ -18 & 19 & -6 \\ -9 & 9 & 2 \end{bmatrix}$ (iv) $\begin{bmatrix} 4 & 2 & -2 \\ -5 & 3 & 2 \\ -2 & 4 & 1 \end{bmatrix}$

5. Find the non-singualar matrix P such that $P^{-1}AP$ is a diagonal matrix, where $A = \begin{bmatrix} 1 & -3 & 3 \\ 3 & -5 & 3 \\ 6 & -6 & 6 \end{bmatrix}$.

6. Let $A = \begin{bmatrix} -9 & 4 & 4 \\ -8 & 3 & 4 \\ -16 & 8 & 7 \end{bmatrix}$. Find an invertible matrix P such that

$P^{-1}AP$ is a diagonal matrix.

7. Test the matrix $A = \begin{bmatrix} 3 & 1 & 0 \\ 0 & 3 & 0 \\ 0 & 0 & 4 \end{bmatrix}$ for diagonalizability.

8. Show that the following matrices are not similar to diagonal matrices :

(i) $\begin{bmatrix} 2 & 1 & 0 \\ 0 & 2 & 1 \\ 0 & 0 & 2 \end{bmatrix}$ (ii) $\begin{bmatrix} 3 & 10 & 5 \\ -2 & -3 & -4 \\ 3 & 5 & 7 \end{bmatrix}$

9. Reduce the matrix $A = \begin{bmatrix} -1 & 2 & -2 \\ 1 & 2 & 1 \\ -1 & -1 & 0 \end{bmatrix}$ to its diagonal form.

(UPTU-2006, UKTU-2010, 11, VTU -2011, BHOPAL-2011)

10. The matrix $A = \begin{bmatrix} a & h \\ h & b \end{bmatrix}$ is transformed to the diagonal form

$D = T^{-1}AT$ where $T = \begin{bmatrix} \cos\theta & \sin\theta \\ -\sin\theta & \cos\theta \end{bmatrix}$. Find the value of θ which gives the diagonal transformation.

11. For the matrix $A = \begin{bmatrix} 4 & 1 & 0 \\ 1 & 4 & 1 \\ 0 & 1 & 4 \end{bmatrix}$, determine a matrix P such that

$P^{-1}AP$ is a diagonal matrix.   (MTU-2011)

12. Show that the matrix $A = \begin{bmatrix} 3 & 1 & -1 \\ -2 & 1 & 2 \\ 0 & 1 & 2 \end{bmatrix}$ is diagonalisable.

Hence, find P such that $P^{-1}AP$ is a diagonal matrix.

Answers

1. $P = \begin{bmatrix} -i & i \\ 1 & 1 \end{bmatrix}$ **2.** $P = \begin{bmatrix} 1 & 2 & 2 \\ -2 & -2 & 1 \\ 3 & 3 & -2 \end{bmatrix}, D = \begin{bmatrix} 1 & 0 & 0 \\ 0 & 1 & 0 \\ 0 & 0 & 0 \end{bmatrix}$ **4. (i)** $D = \begin{bmatrix} 0 & 0 & 0 \\ 0 & 3 & 0 \\ 0 & 0 & 15 \end{bmatrix}, P = \begin{bmatrix} 1 & 2 & 2 \\ 2 & 1 & -2 \\ 2 & -2 & 1 \end{bmatrix}$ **(ii)** $D = \begin{bmatrix} 2 & 0 & 0 \\ 0 & 2 & 0 \\ 0 & 0 & 8 \end{bmatrix}, P = \begin{bmatrix} -1 & 1 & 2 \\ 0 & 2 & -1 \\ 2 & 0 & 1 \end{bmatrix}$

(iii) $D = \begin{bmatrix} -2 & 0 & 0 \\ 0 & 1 & 0 \\ 0 & 0 & 1 \end{bmatrix}, P = \begin{bmatrix} 2 & 1 & -1 \\ 2 & 1 & 0 \\ 1 & 0 & 3 \end{bmatrix}$ **(iv)** $D = \begin{bmatrix} 1 & 0 & 0 \\ 0 & 2 & 0 \\ 0 & 0 & 5 \end{bmatrix}, P = \begin{bmatrix} 2 & 1 & 0 \\ 1 & 1 & 1 \\ 4 & 2 & 1 \end{bmatrix}$ **5.** $P = \begin{bmatrix} 1 & -1 & 1 \\ 1 & 0 & 1 \\ 0 & 1 & 2 \end{bmatrix}$ **6.** $P = \begin{bmatrix} 1 & 1 & 1 \\ 0 & 2 & 1 \\ 2 & 0 & 2 \end{bmatrix}$

7. Not diagonalizable **9.** $\begin{bmatrix} 1 & 0 & 0 \\ 0 & \sqrt{5} & 0 \\ 0 & 0 & -\sqrt{5} \end{bmatrix}$ **10.** $\theta = \frac{1}{2}\tan^{-1}\left(\frac{2h}{b-a}\right)$ **11.** $\begin{bmatrix} -1 & 1 & 1 \\ 0 & \sqrt{2} & -\sqrt{2} \\ 1 & 1 & 1 \end{bmatrix}$ **12.** $\begin{bmatrix} 1 & 1 & 0 \\ -1 & 0 & 1 \\ 1 & 1 & 1 \end{bmatrix}$

1.21 QUADRATIC FORM AND MATRICES

An expression of the form : $\sum\limits_{i=1}^{n} \sum\limits_{j=1}^{n} a_{ij}x_i x_j$ where $a_{ij} \in F$ (a field), is called a quadratic form in n variables $x_1, x_2, ..., x_n$ over a field F, which is denoted by $Q(x_1, x_2, ..., x_n)$ or by Q.

1.21.1 REAL QUADRATIC FORM

The quadratic form $Q = \sum\limits_{i=1}^{n} \sum\limits_{j=1}^{n} a_{ij}x_i x_j$ is called a real quadratic form if a_{ij} are all real numbers i.e., $a_{ij} \in R$ (the field of all real numbers).

For example :

(1) $ax^2 + 2hxy + by^2$ is a real quadratic form in two variables x and y.

(2) $x^2 + y^2 + z^2 + 2yz + 2zx + 2xy$ is a real quadratic form in three variables x, y and z.

(3) $x_1x_2 + x_2x_3 + x_3x_4$ is a real quadratic form in four variables x_1, x_2, x_3 and x_4.

THEOREM 1. *Every quadratic form in n variables $x_1, x_2, ..., x_n$ over a field F can be expressed in the form $X'BX$, where $X = [x_1, x_2, ..., x_n]'$ is a column vector and B is a symmetric matrix of order n over the field F.*

PROOF. Let
$$Q = \sum_{i=1}^{n} \sum_{j=1}^{n} a_{ij}x_i x_j \qquad ...(1)$$
be a quadratic form in n variables $x_1, x_2, ..., x_n$ over a field F.

Writing the equation (1) such that the terms $a_{ij}x_ix_j$ and $a_{ji}x_jx_i$ are taken together, we get

$$Q = a_{11}x_1^2 + (a_{12} + a_{21})x_1x_2 + (a_{13} + a_{31})x_1x_3 + ... + (a_{1n} + a_{n1})x_1x_n$$
$$+ a_{22}x_2^2 + (a_{23} + a_{32})x_2x_3 + ... + (a_{2n} + a_{n2})x_2x_n$$
$$+ a_{33}x_3^2 + (a_{34} + a_{43})x_3x_4 + ... + (a_{3n} + a_{3n})x_3x_n \qquad ...(2)$$
$$+ .... + a_{nn}x_n^2$$

Set $b_{ij} = \frac{1}{2}(a_{ij} + a_{ji})$, then $b_{ij} = b_{ji}$ and $b_{ij} + b_{ji} = a_{ij} + a_{ji}$, using these relations, equations (2) can be written as

$$Q = b_{11}x_1^2 + b_{12}x_1x_2 + ... + b_{1n}x_1x_n + b_{21}x_2x_1 + b_{22}x_2^2 + ...$$
$$... + b_{2n}x_2x_n + ... + b_{n1}x_nx_1 + b_{n2}x_nx_2 + ... + b_{nn}x_n^2$$

$$\therefore \quad Q = \sum_{i=1}^{n} \sum_{j=1}^{n} b_{ij}x_i x_j \qquad ...(3)$$

Let $B = [b_{ij}]$ be a matrix of order $n \times n$. Clearly B is a symmetric matrix.

Let $X = \begin{bmatrix} x_1 \\ x_2 \\ \vdots \\ x_n \end{bmatrix}$, then $X' = [x_1, x_2, ..., x_n]$

Therefore $X'BX$ is a matrix of order 1×1 i.e., $X'BX$ has a single element and this single element is $\sum\limits_{i=1}^{n} \sum\limits_{j=1}^{n} b_{ij}x_i x_j$. If we regard a matrix of order 1×1 equal to its single element, then we have

$$X'BX = \sum_{i=1}^{n} \sum_{j=1}^{n} b_{ij} x_i x_j$$

But we have $b_{ii} = a_{ii}$ and $b_{ij} = b_{ji}, = \frac{1}{2}(a_{ij} + a_{ji})$, then we have $\sum_{i=1}^{n} \sum_{j=1}^{n} a_{ij} x_i x_j = \sum_{i=1}^{n} \sum_{j=1}^{n} b_{ij} x_i x_j$

Hence, $$X'BX = \sum_{i=1}^{n} \sum_{j=1}^{n} a_{ij} x_i x_j.$$

1.21.2 Matrix of Quadratic Form $\sum_{i=1}^{n} \sum_{j=1}^{n} a_{ij} x_i x_j$

If $Q = \sum_{i=1}^{n} \sum_{j=1}^{n} a_{ij} x_i x_j$ is a quadratic form in n variables $x_1, x_2, ..., x_n$ over a field, then there exists unique symmetric matrix B such that

$$X'BX = \sum_{i=1}^{n} \sum_{j=1}^{n} a_{ij} x_i x_j$$

The symmetric matrix B is called the matrix of the quadratic form.

In order to find the matrix of the quadratic form $\sum_{i=1}^{n} \sum_{j=1}^{n} a_{ij} x_i x_j$ we shall adjust the coefficients, a_{ij} in such a way that its coefficients form a symmetric matrix.

1.21.2 Conversion of a Symmetric Matrix into Quadratic Form

Let $A = [a_{ij}]$ be an $n \times n$ symmetric matrix over a field F and let $X = \begin{bmatrix} x_1 \\ x_2 \\ \vdots \\ x_n \end{bmatrix}$, then the quadratic form of A is given by

$$Q = X'AX = [x_1, x_2, ..., x_n] A \begin{bmatrix} \\ \\ \\ \end{bmatrix} = \sum_{i=1}^{n} \sum_{j=1}^{n} a_{ij} x_i x_j$$

 Solved Examples

EXAMPLE 1. *Find the matrices of the following quadratic forms and verify that they can be written as matrix products X′AX.*

(i) $x_1^2 - 18x_1x_2 + 5x_2^2$

(ii) $x_1^2 + 2x_2^2 - 5x_3^2 - x_1x_2 + 4x_2x_3 - 3x_3x_1$

SOLUTION. Let $Q = x_1^2 - 18x_1x_2 + 5x_2^2$

The given quadratic form Q can be written as

$Q = x_1^2 + (-9 - 9)x_1x_2 + 5x_2^2$

$= x_1^2 + (-9)x_1x_2 + (-9)x_1x_2 + 5x_2^2$

Therefore, the matrix corresponding the given quadratic form is

$$A = \begin{bmatrix} 1 & -9 \\ -9 & 5 \end{bmatrix}$$

Let $X = \begin{bmatrix} x_1 \\ x_2 \end{bmatrix}$, then $X' = [x_1 \quad x_2]$

Now $X'AX = [x_1 \quad x_2] \begin{bmatrix} 1 & -9 \\ -9 & 5 \end{bmatrix} \begin{bmatrix} x_1 \\ x_2 \end{bmatrix}$

$= [x_1 \quad x_2] \begin{bmatrix} x_1 - 9x_2 \\ -9x_1 + 5x_2 \end{bmatrix}$

$= x_1(x_1 - 9x_2) + x_2(-9x_1 + 5x_2)$

$= x_1^2 - 9x_1x_2 - 9x_2x_1 + 5x_2^2$

$= x_1^2 - 18x_1x_2 + 5x_2^2$

(ii) Let $Q = x_1^2 + 2x_2^2 - 5x_3^2 - x_1x_2 + 4x_2x_3 - 3x_3x_1$

The given quadratic form Q can be written as

$Q = x_1^2 + \left(-\frac{1}{2} - \frac{1}{2}\right)x_1x_2 + \left(-\frac{3}{2} - \frac{3}{2}\right)x_1x_3$

$\quad + 2x_2^2 + (2 + 2)x_2x_3 + (-5)x_3^2$

$= x_1^2 + \left(-\frac{1}{2}\right)x_1x_2 + \left(-\frac{3}{2}\right)x_1x_3 + \left(-\frac{1}{2}\right)x_2x_1$

$\quad + 2x_2^2 + 2x_2x_3 + \left(-\frac{3}{2}\right)x_3x_1 + 2x_3x_2 + (-5)x_3^2$

Therefore, the matrix corresponding to the given quadratic form Q is

$$A = \begin{bmatrix} 1 & -\dfrac{1}{2} & -\dfrac{3}{2} \\ -\dfrac{1}{2} & 2 & 2 \\ -\dfrac{3}{2} & 2 & -5 \end{bmatrix}$$

Let $X = \begin{bmatrix} x_1 \\ x_2 \\ x_3 \end{bmatrix}$, then $X' = \begin{bmatrix} x_1 & x_2 & x_3 \end{bmatrix}$

Now $X'AX = \begin{bmatrix} x_1 & x_2 & x_3 \end{bmatrix} \begin{bmatrix} 1 & -\dfrac{1}{2} & -\dfrac{3}{2} \\ -\dfrac{1}{2} & 2 & 2 \\ -\dfrac{3}{2} & 2 & -5 \end{bmatrix} \begin{bmatrix} x_1 \\ x_2 \\ x_3 \end{bmatrix}$

$= \begin{bmatrix} x_1 & x_2 & x_3 \end{bmatrix} \begin{bmatrix} x_1 - \dfrac{1}{2}x_2 - \dfrac{3}{2}x_3 \\ -\dfrac{1}{2}x_1 + 2x_2 + 2x_3 \\ -\dfrac{3}{2}x_1 + 2x_2 - 5x_3 \end{bmatrix}$

$= x_1\left(x_1 - \dfrac{1}{2}x_2 - \dfrac{3}{2}x_3 \right) + x_2\left(-\dfrac{1}{2}x_1 + 2x_2 + 2x_3 \right)$
$\qquad + x_3\left(-\dfrac{3}{2}x_1 + 2x_2 - 5x_3 \right)$

$= x_1^2 - \dfrac{1}{2}x_1x_2 - \dfrac{3}{2}x_1x_3 - \dfrac{1}{2}x_2x_1 + 2x_2^2 + 2x_2x_3$
$\qquad - \dfrac{3}{2}x_3x_1 + 2x_3x_2 - 5x_3^2$

$= x_1^2 + 2x_2^2 - 5x_3^2 - x_1x_2 + 4x_2x_3 - 3x_3x_1.$

EXAMPLE 2. *Obtain the matrices corresponding to the following quadratic forms* :

 (i) $x^2 + 2y^2 + 3z^2 + 4xy + 5yz + 6zx$
 (ii) $ax^2 + by^2 + cz^2 + 2fyz + 2gzx + 2hxy$
 (iii) $a_{11}x_1^2 + a_{22}x_2^2 + a_{33}x_3^2 + 2a_{12}x_1x_2$
 $\qquad + 2a_{23}x_2x_3 + 2a_{31}x_3x_1$
 (iv) $x_1^2 - 2x_2^2 + 4x_3^3 - 4x_4^4 - 2x_1x_2 + 3x_1x_4$
 $\qquad\qquad\qquad + 4x_2x_3 - 5x_3x_4$
 (v) $x_1^2 - 2x_2x_3 - x_3x_4$
 (vi) $d_1x_1^2 + d_2x_2^2 + d_3x_3^2 + d_4x_4^2$
 (vii) $x_1x_2 + x_2x_3 + x_3x_1 + x_1x_4 + x_2x_4 + x_3x_4$

SOLUTION. (i) The given quadratic form can be written as

$x^2 + (2+2)xy + (3+3)xz + 2y^2 + \left(\dfrac{5}{2} + \dfrac{5}{2} \right)yz + 3z^2$

$= x^2 + 2xy + 3xz + 2yx + 2y^2$
$\qquad + \dfrac{5}{2}yz + 3zx + \dfrac{5}{2}zy + 3z^2$

Therefore, the matrix corresponding to the given

quadratic form is $\begin{bmatrix} 1 & 2 & 3 \\ 2 & 2 & \dfrac{5}{2} \\ 3 & \dfrac{5}{2} & 3 \end{bmatrix}$

(ii) The given quadratic form can be written as

$ax^2 + (h+h)xy + (g+g)xz + by^2 + (f+f)yz + cz^2$

$= ax^2 + hxy + gxz + hyx + by^2 + fyz + gzx + fzy + cz^2$

Therefore, the matrix corresponding to the given quadratic form is
$\begin{bmatrix} a & h & g \\ h & b & f \\ g & f & c \end{bmatrix}$

(iii) The given quadratic form can be written as

$a_{11}x_1^2 + (a_{12} + a_{12})x_1x_2 + (a_{31} + a_{31})x_1x_3$
$\qquad + a_{22}x_2^2 + (a_{23} + a_{23})x_2x_3 + a_{33}x_3^2$
$= a_{11}x_1^2 + a_{12}x_1x_2 + a_{31}x_3x_1 + a_{12}x_2x_1 + a_{22}x_2^2$
$\qquad + a_{23}x_2x_3 + a_{31}x_3x_1 + a_{23}x_2x_3 + a_{33}x_3^2$

Therefore, the matrix corresponding to the given quadratic form is

$\begin{bmatrix} a_{11} & a_{12} & a_{31} \\ a_{12} & a_{22} & a_{23} \\ a_{31} & a_{23} & a_{33} \end{bmatrix}$

(iv) The given quadratic form can be written as

$x_1^2 + (-1-1)x_1x_2 + (0+0)x_1x_3$
$\qquad + \left(\dfrac{3}{2} + \dfrac{3}{2} \right)x_1x_4 + (-2)x_2^2 + (2+2)x_2x_3$
$\qquad\qquad + (0+0)x_2x_4 + 4x_3^2$
$\qquad\qquad + \left(-\dfrac{5}{2} - \dfrac{5}{2} \right)x_3x_4 + (-4)x_4^2$

$= x_1^2 - x_1x_2 + 0x_1x_3 + \dfrac{3}{2}x_1x_2 + (-1)x_2x_1$
$\qquad + (-2)x_2^2 + 2x_2x_3 + 0.x_2x_3 + 0.x_3x_1$
$\qquad + 2x_3x_2 + 4x_3^2 + \left(-\dfrac{5}{2} \right)x_3x_4 + \dfrac{3}{2}x_4x_1$
$\qquad + 0.x_4x_2 + \left(-\dfrac{5}{2} \right)x_4x_3 + (-4)x_4^2$

Therefore, the matrix corresponding to the given quadratic form is

$\begin{bmatrix} 1 & -1 & 0 & 3/2 \\ -1 & -2 & 2 & 0 \\ 0 & 2 & 4 & -5/2 \\ \dfrac{3}{2} & 0 & -5/2 & -4 \end{bmatrix}$

(v) The given quadratic form can be written as

$x_1^2 + (0+0)x_1x_2 + (0+0)x_1x_3 + (0+0)x_1x_4$
$\qquad + 0.x_2^2 + (-1-1)x_2x_3 + (0+0)x_2x_4$
$\qquad + 0.x_3^2 + \left(-\dfrac{1}{2} - \dfrac{1}{2} \right)x_3x_4 + 0.x_4^2$

$= x_1^2 + 0.x_1x_2 + 0.x_1x_3 + 0.x_1x_4 + 0.x_2x_1$
$\qquad + 0.x_2^2 + (-1)x_2x_3 + 0.x_2x_4 + 0.x_3x_1$
$\qquad + (-1)x_3x_2 + 0.x_3^2 + \left(-\dfrac{1}{2} \right)x_3x_4 + 0.x_4x_1$
$\qquad + 0.x_4x_2 + \left(-\dfrac{1}{2} \right)x_4x_3 + 0.x_4^2$

Therefore, the matrix corresponding to the given quadratic form is
$\begin{bmatrix} 1 & 0 & 0 & 0 \\ 0 & 0 & -1 & 0 \\ 0 & -1 & 0 & -\dfrac{1}{2} \\ 0 & 0 & -\dfrac{1}{2} & 0 \end{bmatrix}$

(vi) The given quadratic form can be written as

$d_1x_1^2 + 0.x_1x_2 + 0.x_1x_3 + 0.x_1x_4$
$\qquad + 0.x_2x_1 + d_2x_2^2 + 0.x_2x_3 + 0.x_2x_4$
$\qquad + 0.x_3x_1 + 0.x_3x_2 + d_3x_3^2 + 0.x_3x_4$
$\qquad + 0.x_4x_1 + 0.x_4x_2 + 0.x_4x_3 + d_4x_4^2$

Therefore, the matrix corresponding to the given quadratic form is

$$\begin{bmatrix} d_1 & 0 & 0 & 0 \\ 0 & d_2 & 0 & 0 \\ 0 & 0 & d_3 & 0 \\ 0 & 0 & 0 & d_4 \end{bmatrix} = \text{diag.}\left(d_1, d_2, d_3, d_4\right)$$

(vii) The given quadratic form can be written as

$$0.x_1^2 + \left(\frac{1}{2} + \frac{1}{2}\right) x_1 x_2 + \left(\frac{1}{2} + \frac{1}{2}\right) x_1 x_4$$

$$+ 0.x_2^2 + \left(\frac{1}{2} + \frac{1}{2}\right) x_2 x_3 + \left(\frac{1}{2} + \frac{1}{2}\right) x_2 x_4$$

$$+ \left(\frac{1}{2} + \frac{1}{2}\right) x_3 x_1 + 0.x_3^2 + \left(\frac{1}{2} + \frac{1}{2}\right) x_3 x_4 + 0.x_4^2$$

$$= 0.x_1^2 + \frac{1}{2} x_1 x_2 + \frac{1}{2} x_1 x_3 + \frac{1}{2} x_1 x_4$$

$$+ \frac{1}{2} x_2 x_1 + 0.x_2^2 + \frac{1}{2} x_2 x_3 + \frac{1}{2} x_2 x_4$$

$$+ \frac{1}{2} x_3 x_1 + \frac{1}{2} x_3 x_2 + 0.x_3^2 + \frac{1}{2} x_3 x_4 + \frac{1}{2} x_4 x_1$$

$$+ \frac{1}{2} x_4 x_2 + \frac{1}{2} x_4 x_3 + 0.x_4^2$$

Therefore, the matrix corresponding to the given quadratic form is

$$\begin{bmatrix} 0 & \frac{1}{2} & \frac{1}{2} & \frac{1}{2} \\ \frac{1}{2} & 0 & \frac{1}{2} & \frac{1}{2} \\ \frac{1}{2} & \frac{1}{2} & 0 & \frac{1}{2} \\ \frac{1}{2} & \frac{1}{2} & \frac{1}{2} & 0 \end{bmatrix}$$

EXAMPLE 3. *Find the matrix of the following quadratic form* :

$$(x_1 - x_2 + x_3)^2$$

SOLUTION. The given quadratic form can be written as

$$(x_1 - x_2 + x_3)^2 = x_1^2 + x_2^2 + x_3^2$$

$$- 2x_1 x_2 + 2x_1 x_3 - 2x_2 x_3$$

$$= x_1^2 + (-1 -1)x_1 x_2 + (1 + 1)$$
$$x_1 x_3 + x_2^2 + (-1 -1)x_2 x_3 + x_3^2$$

$$= x_1^2 - x_1 x_2 + x_1 x_3 - x_2 x_1 + x_2^2$$

$$- x_2 x_3 + x_3 x_1 - x_3 x_2 + x_3^2$$

Therefore, the matrix corresponding to the given quadratic form is

$$\begin{bmatrix} 1 & -1 & 1 \\ -1 & 1 & -1 \\ 1 & -1 & 1 \end{bmatrix}$$

EXAMPLE 4. *Find the quadratic form of the real symmetric matrix*

$$A = \begin{bmatrix} 2 & 2 & 0 \\ 2 & 0 & -3 \\ 0 & -3 & -7 \end{bmatrix}$$

SOLUTION. Let $X = \begin{bmatrix} x_1 \\ x_2 \\ x_3 \end{bmatrix}$, then $X' = \begin{bmatrix} x_1 & x_2 & x_3 \end{bmatrix}$

Now $X'AX = \begin{bmatrix} x_1 & x_2 & x_3 \end{bmatrix} \begin{bmatrix} 2 & 2 & 0 \\ 2 & 0 & -3 \\ 0 & -3 & -7 \end{bmatrix} \begin{bmatrix} x_1 \\ x_2 \\ x_3 \end{bmatrix}$

$$= \begin{bmatrix} x_1 & x_2 & x_3 \end{bmatrix} \begin{bmatrix} 2x_1 + 2x_2 + 0.x_3 \\ 2x_1 + 0.x_2 - 3x_3 \\ 0.x_1 - 3x_2 - 7x_3 \end{bmatrix}$$

$$= x_1(2x_1 + 2x_2 + 0.x_3) + x_2(2x_1 + 0.x_2 - 3x_3)$$
$$+ x_3(0.x_1 - 3x_2 - 7x_3)$$

$$= 2x_1^2 + 2x_1 x_2 + 0.x_1 x_3 + 2x_2 x_1 + 0.x_2^2$$
$$- 3x_2 x_3 + 0x_3 x_1 - 3x_3 x_2 - 7x_3^2$$

$$= 2x_1^2 + 4x_1 x_2 - 6x_2 x_3 - 7x_3^2$$

$$\therefore \quad X'AX = 2x_1^2 + 4x_1 x_2 - 6x_2 x_3 - 7x_3^2$$

which is the required quadratic form.

Exercise-1.12

1. Find the matrices corresponding to the following quadratic forms as :

(i) $ax^2 + 2hxy + by^2$ (ii) $5x_1^2 - 2x_1 x_2 + x_2^2$

(iii) $x_1^2 + 5x_2^2 - 7x_3^2$ (iv) $4x_1 x_3 + 2x_2 x_3 + x_3^2$

(v) $(x_1 + x_2)^2 - x_3^2$

(vi) $x_1^2 - 2x_2^2 - 3x_3^2 + 4x_1 x_2 + 6x_1 x_3 - 8x_2 x_3$

(vii) $2x_1 x_2 + 6x_1 x_3 - 4x_2 x_3$

(viii) $5x_1^2 + 3x_2^2 + 2x_3^2 - x_1 x_2 + 8x_2 x_3$

(ix) $8x_1^2 + 7x_2^2 - 3x_3^2 - 6x_1 x_2 + 4x_1 x_3 - 2x_2 x_3$

(x) $4x_1 x_2 + 6x_1 x_3 - 8x_2 x_3$

(xi) $5x_1^2 - x_2^2 + 7x_3^2 + 5x_1 x_2 - 3x_1 x_3$

2. Compute the quadratic form $X'AX$, when

(i) $A = \begin{bmatrix} 5 & 1/3 \\ 1/3 & 1 \end{bmatrix}$ (ii) $A = \begin{bmatrix} 4 & 0 \\ 0 & 3 \end{bmatrix}$ (iii) $A = \begin{bmatrix} 3 & -2 \\ -2 & 7 \end{bmatrix}$

3. Write down the quadratic forms corresponding to the following matrices :

(i) $\begin{bmatrix} 2 & 1 & 5 \\ 1 & 3 & -2 \\ 5 & -2 & 4 \end{bmatrix}$ (ii) $\begin{bmatrix} 1 & 0 & 0 \\ 0 & 2 & 0 \\ 0 & 0 & 3 \end{bmatrix}$

(iii) $\begin{bmatrix} 1 & 2 & 3 \\ 2 & 2 & 5/2 \\ 3 & 5/2 & 3 \end{bmatrix}$ (iv) $\begin{bmatrix} 0 & 5 & -1 \\ 5 & 1 & 6 \\ -1 & 6 & 2 \end{bmatrix}$

(v) $\begin{bmatrix} 2 & 2 & 0 \\ 2 & 0 & -3 \\ 0 & -3 & -7 \end{bmatrix}$ (vi) $\begin{bmatrix} 1 & -1/2 & -3/2 \\ -1/2 & 2 & 2 \\ -3/2 & 2 & -5 \end{bmatrix}$

(vii) $\begin{bmatrix} 0 & 1 & 3 \\ 1 & 0 & -2 \\ 3 & -2 & 0 \end{bmatrix}$ (viii) $\begin{bmatrix} 0 & a & b & c \\ a & 0 & l & m \\ b & l & 0 & p \\ c & m & p & 0 \end{bmatrix}$

Answers

1.(i) $\begin{bmatrix} a & h \\ h & b \end{bmatrix}$ (ii) $\begin{bmatrix} 5 & -1 \\ -1 & 1 \end{bmatrix}$ (iii) $\begin{bmatrix} 1 & 0 & 0 \\ 0 & 5 & 0 \\ 0 & 0 & -7 \end{bmatrix}$ (iv) $\begin{bmatrix} 0 & 0 & 2 \\ 0 & 0 & 1 \\ 2 & 1 & 1 \end{bmatrix}$ (v) $\begin{bmatrix} 1 & 1 & 0 \\ 1 & 1 & 0 \\ 0 & 0 & -1 \end{bmatrix}$ (vi) $\begin{bmatrix} 1 & 2 & 3 \\ 2 & -2 & -4 \\ 3 & -4 & -3 \end{bmatrix}$ (vii) $\begin{bmatrix} 0 & 1 & 3 \\ 1 & 0 & -2 \\ 3 & -2 & 0 \end{bmatrix}$

(viii) $\begin{bmatrix} 5 & -1/2 & 0 \\ -1/2 & 3 & 4 \\ 0 & 4 & 2 \end{bmatrix}$ (ix) $\begin{bmatrix} 8 & -3 & 2 \\ -3 & 7 & -1 \\ 2 & -1 & -3 \end{bmatrix}$ (x) $\begin{bmatrix} 0 & 2 & 3 \\ 2 & 0 & -4 \\ 3 & -4 & 0 \end{bmatrix}$ (xi) $\begin{bmatrix} 5 & 5/2 & -3/2 \\ 5/2 & -1 & 0 \\ -3/2 & 0 & 7 \end{bmatrix}$ **2.** (i) $5x_1^2 + \dfrac{2}{3}x_1x_2 + x_2^2$

(ii) $4x_1^2 + 3x_2^2$ (iii) $3x_1^2 + 7x_2^2 - 4x_1x_2$ **3.** (i) $2x_1^2 + 3x_2^2 + + 4x_3^2 + 2x_1x_2 + 10x_1x_3 - 4x_2x_3$ (ii) $x_1^2 + 2x_2^2 + 3x_3^2$

(iii) $x_1^2 + 2x_2^2 + 3x_3^2 + 4x_1x_2 + 6x_1x_3 + 5x_2x_3$ (iv) $x_2^2 + 2x_3^2 + 10x_1x_2 - 2x_1x_3 + 12x_2x_3$ (v) $2x_1^2 - 7x_3^2 + 4x_1x_2 - 6x_2x_3$

(vi) $x_1^2 + 2x_2^2 - 5x_3^2 - x_1x_2 - 3x_1x_3 + 4x_2x_3$ (vii) $2x_1x_2 + 6x_1x_3 - 4x_2x_3$

(viii) $2ax_1x_2 + 2bx_1x_3 + 2cx_1x_4 + 2lx_2x_3 + 2mx_2x_4 + 2px_3x_4$

1.21.3 RANK OF A QUADRATIC FORM

Let $Q(X) = X'AX$ be a quadratic form over a field F. Then the rank of $Q(X)$ is the rank of the matrix A.

If the rank of $Q(X)$ is r, then there exists a non-singular matrix P which reduces $Q(X)$ to a sum of r square terms.

i.e.,
$$Q(X) = \lambda_1 y_1^2 + \lambda_2 y_2^2 + \ldots \lambda_r y_r^2$$

WORKING PROCEDURE

Let A be $n \times n$ real symmetric matrix. In order to find a non-singular matrix P such that $P'AP$ = diagonal matrix, we use the following steps:

Step 1.   Write $A = I_n A I_n$

Step 2.   Apply congruent row operations on pre-factor I_n of A on RHS and congruent column operations on post-factor I_n of A. Applying such operations simultaneously on A on LHS till A reduces to a diagonal matrix. When the matrix A reduces to a diagonal matrix, the post-factor I_n ultimately gives the non-singular matrix P such that $P'AP$ = diagonal matrix.

Solved Examples

EXAMPLE 1. *Determine a non-singular matrix P such that P′AP is a diagonal matrix, where*

$$A = \begin{bmatrix} 0 & 1 & 2 \\ 1 & 0 & 3 \\ 2 & 3 & 0 \end{bmatrix}$$

SOLUTION. We have $A = I_3 A I_3$

or $\begin{bmatrix} 0 & 1 & 2 \\ 1 & 0 & 3 \\ 2 & 3 & 0 \end{bmatrix} = \begin{bmatrix} 1 & 0 & 0 \\ 0 & 1 & 0 \\ 0 & 0 & 1 \end{bmatrix} A \begin{bmatrix} 1 & 0 & 0 \\ 0 & 1 & 0 \\ 0 & 0 & 1 \end{bmatrix}$

Applying $R_1 \to R_1 + R_2$, we get

$\begin{bmatrix} 1 & 1 & 5 \\ 1 & 0 & 3 \\ 2 & 3 & 0 \end{bmatrix} = \begin{bmatrix} 1 & 1 & 0 \\ 0 & 1 & 0 \\ 0 & 0 & 1 \end{bmatrix} A \begin{bmatrix} 1 & 0 & 0 \\ 0 & 1 & 0 \\ 0 & 0 & 1 \end{bmatrix}$

Applying $C_1 \to C_1 + C_2$, we get

$\begin{bmatrix} 2 & 1 & 5 \\ 1 & 0 & 3 \\ 5 & 3 & 0 \end{bmatrix} = \begin{bmatrix} 1 & 1 & 0 \\ 0 & 1 & 0 \\ 0 & 0 & 1 \end{bmatrix} A \begin{bmatrix} 1 & 0 & 0 \\ 1 & 1 & 0 \\ 0 & 0 & 1 \end{bmatrix}$

Applying $R_2 \to R_2 - \dfrac{1}{2}R_1, R_3 \to R_3 - \dfrac{5}{2}R_1$, we get

$\begin{bmatrix} 2 & 1 & 5 \\ 0 & -1/2 & 1/2 \\ 0 & 1/2 & -25/2 \end{bmatrix} = \begin{bmatrix} 1 & 1 & 0 \\ -1/2 & 1/2 & 0 \\ -5/2 & -5/2 & 1 \end{bmatrix} A \begin{bmatrix} 1 & 0 & 0 \\ 1 & 1 & 0 \\ 0 & 0 & 1 \end{bmatrix}$

Applying $C_2 \to C_2 - \dfrac{1}{2}C_1, C_3 \to C_3 - \dfrac{5}{2}C_1$, we get

$\begin{bmatrix} 2 & 1 & 5 \\ 0 & -1/2 & 1/2 \\ 0 & 1/2 & -25/2 \end{bmatrix} = \begin{bmatrix} 1 & 1 & 0 \\ -1/2 & 1/2 & 0 \\ -5/2 & -5/2 & 1 \end{bmatrix} A \begin{bmatrix} 1 & -1/2 & -5/2 \\ 1 & 1/2 & -5/2 \\ 0 & 0 & 1 \end{bmatrix}$

Applying $R_3 \to R_3 + R_2$, we get

$\begin{bmatrix} 2 & 0 & 0 \\ 0 & -1/2 & 1/2 \\ 0 & 0 & -12 \end{bmatrix} = \begin{bmatrix} 1 & 1 & 0 \\ -1/2 & 1/2 & 0 \\ -3 & -2 & 1 \end{bmatrix} A \begin{bmatrix} 1 & -1/2 & -5/2 \\ 1 & 1/2 & -5/2 \\ 0 & 0 & 1 \end{bmatrix}$

Applying $C_3 \to C_3 + C_2$, we get

$\begin{bmatrix} 2 & 0 & 0 \\ 0 & -1/2 & 0 \\ 0 & 0 & -12 \end{bmatrix} = \begin{bmatrix} 1 & 1 & 0 \\ -1/2 & 1/2 & 0 \\ -3 & -2 & 1 \end{bmatrix} A \begin{bmatrix} 1 & -1/2 & -3 \\ 1 & 1/2 & -2 \\ 0 & 0 & 1 \end{bmatrix}$

$\Rightarrow$ diag. $\left(2, -\dfrac{1}{2}, -12\right) = P'AP$

where $P = \begin{bmatrix} 1 & -1/2 & -3 \\ 1 & 1/2 & -2 \\ 0 & 0 & 1 \end{bmatrix}$

1.21.4 Normal (or Canonical) Form of a Real Quadratic Matrix

Let $X'AX$ be a real quadratic form in n variables over the real field, then there exists a real non-singular linear transformation $X = PY$, which reduces the given quadratic form to the form

$$Y'(P'AP)Y = y_1^2 + y_2^2 + \dots + y_p^2 - y_{p+1}^2 - y_{p+2}^2 - \dots - y_r^2$$

This new form is known as the normal (or canonical) form of $X'AX$.

☞ Remarks

- The number of positive terms in any two normal forms of a real quadratic form is the same.
- The number of negative terms in any two normal forms of a quadratic form is the same.
- The excess of the number of positive terms over the number of negative terms in any two normal forms of a real quadratic form is the same.

1.21.5 Signature and Index of a Real Quadratic Form

Let $X'AX$ be a real quadratic form of rank r in n variables over the field of reals and its normal form be

$$y_1^2 + y_2^2 + \dots + y_p^2 - y_{p+1}^2 - y_{p+2}^2 - \dots - y_r^2$$

Then the number p of positive terms in a normal form is called the index of $X'AX$. The excess of the number of positive terms over the number of negative terms in a normal form, *i.e.*, $p - (r - p) = 2p - r$ is called the signature *of* $X'AX$.

THEOREM 1.   *The signature of a real quadratic form is invariant for its all normal forms.*

PROOF.   Since the number of positive and negative terms in any two normal forms of a real quadratic form are the same so that their difference are the same, *i.e.*, the signature of given quadratic is the same for its all normal forms.

☞ Remark

- Two real quadratic forms in n variables over the field of reals are real equivalent if and only if they have the same rank and signature (or index).

1.21.6 Reduction of a Real Quadratic Form over the Field of Complex Numbers

THEOREM 1.   *Let A be an $n \times n$ real symmetric matrix of rank r, then there exists a non-singular matrix P whose elements may be complex numbers such that $P'AP = diag.(1, 1, 1, \dots 1, 0, 0, \dots 0)$ where 1 appears r times.*

PROOF.   Since A is a real symmetric matrix of rank r, then there exists a non-singular real matrix Q such that

$$Q'AQ = \text{diag. } (\lambda_1, \lambda_2, \dots, \lambda_r, 0, 0, \dots 0) = D \text{ (say)}$$

where D is a diagonal matrix which has exactly r non-zero elements $\lambda_1, \lambda_2, \dots, \lambda_r$ and $\lambda_1, \lambda_2, \dots, \lambda_r$ may be positive or negative or both.

Let

$$S = \text{diag.} \left(\frac{1}{\sqrt{\lambda_1}}, \frac{1}{\sqrt{\lambda_2}}, \dots, \frac{1}{\sqrt{\lambda_r}}, 1, 1, \dots, 1 \right)$$

be an $n \times n$ complex diagonal matrix, which is obviously, a complex non-singular diagonal matrix and $S' = S$.

Let us take $P = QS$, clearly P is a complex non-singular matrix.

Now
$$P'AP = (QS)'A(QS)$$
$$= S'(Q'AQ)S = S'DS \qquad [\because Q'AQ = D]$$
$$= SDS \qquad [\because S' = S]$$
$$= \text{diag.} \left(\frac{1}{\sqrt{\lambda_1}}, \frac{1}{\sqrt{\lambda_2}}, \dots, \frac{1}{\sqrt{\lambda_r}}, 1, 1, \dots, 1 \right) \text{diag.}(\lambda_1, \lambda_2, \dots, \lambda_r, 0, 0, \dots 0)S$$
$$= \text{diag} \left(\sqrt{\lambda_1}, \sqrt{\lambda_2}, \dots, \sqrt{\lambda_r}, 0, 0, \dots, 0 \right) S$$
$$= \text{diag} \left(\sqrt{\lambda_1}, \sqrt{\lambda_2}, \dots, \sqrt{\lambda_r}, 0, 0, \dots, 0 \right) \text{diag} \left(\frac{1}{\sqrt{\lambda_1}}, \frac{1}{\sqrt{\lambda_2}}, \dots, \frac{1}{\sqrt{\lambda_r}}, 1, 1, \dots, 1 \right)$$
$$= \text{diag}(1, 1, 1, \dots, 1, 0, 0, \dots, 0)$$
$$\therefore \qquad P'AP = \text{diag}(1, 1, 1, \dots, 1, 0, 0, \dots, 0) \text{ where 1 appears } r \text{ times.}$$

☞ Remarks

- Every real quadratic form $X'AX$ is complex equivalent to the form $z_1^2 + z_2^2 + \dots + z_r^2$ where r is the rank of A.
- Two real quadratic forms in n variables are complex equivalent if and only if they have the same rank.

1.21.7 ORTHOGONAL REDUCTION OF A REAL QUADRATIC FORM

THEOREM 1. *If $X'AX$ be a real quadratic form of rank r in n variables, then there exists a real orthogonal transformation $X = PY$ which transforms $X'AX$ to the form $\lambda_1 y_1^2 + \lambda_2 y_2^2 + ... \lambda_r y_r^2$ where $l_1, l_2, ..., l_r$ are the r non-zero eigenvalues of A and $n - r$ eigenvalues of A being equal to zero.*

PROOF. Since A is a real symmetric matrix of order n, then there exists a real orthogonal matrix P such that $P^{-1}AP = D$

where D is a diagonal matrix, whose diagonal elements are the eigenvalues of A. Again, since the rank of A is r, then the rank of $P^{-1}AP$ is also r, therefore D will have exactly r non-zero diagonal elements, hence A has exactly r non-zero eigenvalues and remaining $n - r$ eigenvalues of A are all zero.

So we can take $D = \text{diag.}(\lambda_1, \lambda_2, ..., \lambda_r, 0, 0, .., 0)$ where $\lambda_1, \lambda_2, ..., \lambda_r$ are the r non-zero eigenvalues of A

$$\therefore \qquad P^{-1}AP = \text{diag.}(\lambda_1, \lambda_2, ..., \lambda_r, 0, 0, ... 0)$$

Since P is an orthogonal matrix, then $P^{-1} = P'$

$$\therefore \qquad P^{-1}AP = P'AP = \text{diag.}(\lambda_1, \lambda_2, ..., \lambda_r, 0, 0, ... 0) = D$$

It follows that A is congruent to D.

Now, let us take a real orthogonal transformation $X = PY$ such that

$$X'AX = (PY)'A(PY) = Y'(P'AP)Y = Y'DY$$

$\lambda_1 y_1^2 + \lambda_2 y_2^2 + ... \lambda_r y_r^2$, if $Y = [y_1, y_2, ... y_r, ..., y_n]$

THEOREM 2. *Every real quadratic form $X'AX$ in n variables is real equivalent to the form*

$$y_1^2 + y_2^2 + ... + y_p^2 - y_{p+1}^2 - y_{p+2}^2 - ... - y_r^2$$

where r is the rank of A, and p is the number of positive eigenvalues of A.

PROOF. Since A is a real symmetric matrix, then there exists a real orthogonal matrix Q such that

$$Q^{-1}AQ = D$$

or $$Q'AQ = D \qquad\qquad\qquad [\therefore Q^{-1} = Q']$$

where D is a diagonal matrix whose diagonal elements are the eigenvalues of A.

Since the rank of A is r so that D is also of rank r, therefore D has exactly r non-zero elements it follows that A has exactly r non-zero eigenvalues and remaining $n - r$ eigenvalues of A are all zero.

If $\lambda_1, \lambda_2, ..., \lambda_r$ be non-zero eigenvalues of A, then we have

$$Q'AQ = D = \text{diag.}(\lambda_1, \lambda_2, ..., \lambda_r, 0, 0, ... 0)$$

Suppose out of r eigenvalues of A, $\lambda_1, \lambda_2, ..., \lambda_p$ are positive eigenvalues and $\lambda_{p+1}, \lambda_{p+2}, ..., \lambda_r$ are negative eigenvalues of A.

Let $$S = \text{diag.}\left(\frac{1}{\sqrt{\lambda_1}}, \frac{1}{\sqrt{\lambda_2}}, ..., \frac{1}{\sqrt{\lambda_p}}, \frac{1}{\sqrt{-\lambda_{p+1}}}, ..., \frac{1}{\sqrt{-\lambda_r}}, 0, ..., 0\right)$$

Then S is a non-singular diagonal matrix, and $S' = S$.

Let us take $P = QS$, clearly P is a real non-singular matrix, then we have

$$P'AP = (QS)'A(QS) = S'(Q'AQ)S$$
$$= S'DS \qquad\qquad [\because Q'AQ = D]$$
$$= SDS \qquad\qquad [\because S' = S]$$
$$= S\,\text{diag.}(\lambda_1, \lambda_2, ..., \lambda_r, 0, 0, ..., 0)S$$
$$= \text{diag.}(1, 1, ..., 1, -1, -1, ... -1, 0, 0, ..., 0)$$

where 1 appears p times and -1 appears $r - p$ times.

Consider a real non-singular linear transformation $X = PY$ which reduces $X'AX$ to the form.

$$Y'(P'AP)Y = y_1^2 + y_2^2 + ... + y_p^2 - y_{p+1}^2 - y_{p+2}^2 - ... - y_r^2$$

REMARKS

- Two real quadratic forms $X'AX$ and $Y'BY$ in the same number of variables are real equivalent if and only if A and B have the same number of positive and negative eigenvalues.
- If $X'AX$ is a real quadratic form, the number of non-zero eigenvalues of A is equal to the rank of $X'AX$ and the number of positive eigenvalues of A is equal to the index of $X'AX$.
- Two real quadratic forms $X'AX$ and $Y'BY$ are orthogonally equivalent if and only if A and B have the same eigenvalues and these occur with the same multiplicities.

Solved Examples

EXAMPLE 1. *Reduce each of the following quadratic forms in three variables to real canonical form and find its rank and signature. Also write in each case the linear transformation which reduces to normal form.*

(i) $x^2 + 2y^2 + 2z^2 - 2xy - 2yz + zx$

(ii) $x^2 - 2y^2 + 3z^2 - 4yz + 6zx$

(iii) $2x_1^2 + x_2^2 - 3x_3^2 - 8x_2x_3 - 4x_3x_1 + 12x_1x_2$

(iv) $6x_1^2 + 3x_2^2 + 14x_3^2 + 4x_2x_3 + 18x_3x_1 + 4x_1x_2$

SOLUTION. (i) The given quadratic form is
$$X'AX = x^2 + 2y^2 + 2z^2 - 2xy - 2yz + zx$$

$$\therefore \quad A = \begin{bmatrix} 1 & -1 & \dfrac{1}{2} \\ -1 & 2 & -1 \\ \dfrac{1}{2} & -1 & 2 \end{bmatrix}$$

We write
$$A = IAI$$

or $\begin{bmatrix} 1 & -1 & 1/2 \\ -1 & 2 & -1 \\ 1/2 & -1 & 2 \end{bmatrix} = \begin{bmatrix} 1 & 0 & 0 \\ 0 & 1 & 0 \\ 0 & 0 & 1 \end{bmatrix} A \begin{bmatrix} 1 & 0 & 0 \\ 0 & 1 & 0 \\ 0 & 0 & 1 \end{bmatrix}$

Applying $R_2 \to R_2 + R_1$, we get
$$\begin{bmatrix} 1 & 1 & 1/2 \\ 0 & 1 & -1/2 \\ 1/2 & 1 & 2 \end{bmatrix} = \begin{bmatrix} 1 & 0 & 0 \\ 1 & 1 & 0 \\ 0 & 0 & 1 \end{bmatrix} \begin{bmatrix} 1 & 0 & 0 \\ 0 & 1 & 0 \\ 0 & 0 & 1 \end{bmatrix}$$

Applying $C_2 \to C_2 + C_1$, we get
$$\begin{bmatrix} 1 & 0 & 1/2 \\ 0 & 1 & -1/2 \\ 1/2 & -1/2 & 2 \end{bmatrix} = \begin{bmatrix} 1 & 0 & 0 \\ 1 & 1 & 0 \\ 0 & 0 & 1 \end{bmatrix} A \begin{bmatrix} 1 & 1 & 0 \\ 0 & 1 & 0 \\ 0 & 0 & 1 \end{bmatrix}$$

Applying $R_3 \to R_3 - \dfrac{1}{2} R_1$, we get
$$\begin{bmatrix} 1 & 0 & 1/2 \\ 0 & 1 & -1/2 \\ 0 & -1/2 & 7/4 \end{bmatrix} = \begin{bmatrix} 1 & 0 & 0 \\ 1 & 1 & 0 \\ -1/2 & 0 & 1 \end{bmatrix} A \begin{bmatrix} 1 & 1 & 0 \\ 0 & 1 & 0 \\ 0 & 0 & 1 \end{bmatrix}$$

Applying $C_3 \to C_3 - \dfrac{1}{2} C_1$, we get
$$\begin{bmatrix} 1 & 0 & 0 \\ 0 & 1 & -1/2 \\ 0 & -1/2 & 7/4 \end{bmatrix} = \begin{bmatrix} 1 & 0 & 0 \\ 1 & 1 & 0 \\ -1/2 & 0 & 1 \end{bmatrix} A \begin{bmatrix} 1 & 1 & -1/2 \\ 0 & 1 & 0 \\ 0 & 0 & 1 \end{bmatrix}$$

Applying $R_3 \to R_3 + \dfrac{1}{2} R_2$, we get
$$\begin{bmatrix} 1 & 0 & 0 \\ 0 & 1 & -1/2 \\ 0 & 0 & 3/2 \end{bmatrix} = \begin{bmatrix} 1 & 0 & 0 \\ 1 & 1 & 0 \\ 0 & 1/2 & 1 \end{bmatrix} A \begin{bmatrix} 1 & 1 & -1/2 \\ 0 & 1 & 0 \\ 0 & 0 & 1 \end{bmatrix}$$

Applying $C_3 \to C_3 + \dfrac{1}{2} C_2$, we get
$$\begin{bmatrix} 1 & 0 & 0 \\ 0 & 1 & 0 \\ 0 & 0 & 3/2 \end{bmatrix} = \begin{bmatrix} 1 & 0 & 0 \\ 1 & 1 & 0 \\ 0 & 1/2 & 1 \end{bmatrix} A \begin{bmatrix} 1 & 1 & 0 \\ 0 & 1 & 1/2 \\ 0 & 0 & 1 \end{bmatrix}$$

Applying $R_3 \to \sqrt{\dfrac{2}{3}} R_3$, we get
$$\begin{bmatrix} 1 & 0 & 0 \\ 0 & 1 & 0 \\ 0 & 0 & \sqrt{3/2} \end{bmatrix} = \begin{bmatrix} 1 & 0 & 0 \\ 1 & 1 & 0 \\ 0 & 1/\sqrt{6} & \sqrt{2/3} \end{bmatrix} A \begin{bmatrix} 1 & 1 & 0 \\ 0 & 1 & 1/2 \\ 0 & 0 & 1 \end{bmatrix}$$

Applying $C_3 \to \sqrt{\dfrac{2}{3}} C_3$, we get
$$\begin{bmatrix} 1 & 0 & 0 \\ 0 & 1 & 0 \\ 0 & 0 & 1 \end{bmatrix} = \begin{bmatrix} 1 & 0 & 0 \\ 1 & 1 & 0 \\ 0 & 1/\sqrt{6} & \sqrt{2/3} \end{bmatrix} A \begin{bmatrix} 1 & 1 & 0 \\ 0 & 1 & 1/\sqrt{6} \\ 0 & 0 & \sqrt{2/3} \end{bmatrix}$$

$$\Rightarrow \quad D = P'AP$$

where $\quad P = \begin{bmatrix} 1 & 1 & 0 \\ 0 & 1 & 1/\sqrt{6} \\ 0 & 0 & \sqrt{2/3} \end{bmatrix}$

Now the real non-singular linear transformation is
$$X = PY$$

i.e., $\quad \begin{bmatrix} x \\ y \\ z \end{bmatrix} = \begin{bmatrix} 1 & 1 & 0 \\ 0 & 1 & 1/\sqrt{6} \\ 0 & 0 & \sqrt{2/3} \end{bmatrix} \begin{bmatrix} y_1 \\ y_2 \\ y_3 \end{bmatrix}$

i.e., $\quad x = y_1 + y_2$
$$y = y_2 + \frac{1}{\sqrt{6}} y_3$$
$$z = \sqrt{\frac{2}{3}} y_3$$

which reduces $X'AX$ to the normal form
$$Y'(P'AP)Y = y_1^2 + y_2^2 + y_3^2$$

The rank of $X'AX$ = number of non-zero terms in normal form = 3

The signature of $X'AX$ = the excess of positive terms over the negative terms
$$= 3 - 0 = 3$$

The index of $X'AX$ = the number of positive terms
$$= 3$$

(ii) The given quadratic form is
$$X'AX = x^2 - 2y^2 + 3z^2 - 4yz + 6zx$$
$$= \begin{bmatrix} x & y & z \end{bmatrix} \begin{bmatrix} 1 & 0 & 0 \\ 0 & -2 & -2 \\ 3 & -2 & 3 \end{bmatrix} \begin{bmatrix} x \\ y \\ z \end{bmatrix}$$

$$\therefore \quad A = \begin{bmatrix} 1 & 0 & 3 \\ 0 & -2 & -2 \\ 3 & -2 & 3 \end{bmatrix}$$

We write
$$A = IAI$$

or $\begin{bmatrix} 1 & 0 & 3 \\ 0 & -2 & -2 \\ 3 & -2 & 3 \end{bmatrix} = \begin{bmatrix} 1 & 0 & 0 \\ 0 & 1 & 0 \\ 0 & 0 & 1 \end{bmatrix} A \begin{bmatrix} 1 & 0 & 0 \\ 0 & 1 & 0 \\ 0 & 0 & 1 \end{bmatrix}$

Applying $R_3 \to R_3 - 3R_1$, we get
$$\begin{bmatrix} 1 & 0 & 3 \\ 0 & -2 & -2 \\ 3 & -2 & -6 \end{bmatrix} = \begin{bmatrix} 1 & 0 & 0 \\ 0 & 1 & 0 \\ -3 & 0 & 1 \end{bmatrix} A \begin{bmatrix} 1 & 0 & 0 \\ 0 & 1 & 0 \\ 0 & 0 & 1 \end{bmatrix}$$

Applying $C_3 \rightarrow C_3 - 3C_1$, we get

$$\begin{bmatrix} 1 & 0 & 0 \\ 0 & -2 & -2 \\ 0 & -2 & -6 \end{bmatrix} = \begin{bmatrix} 1 & 0 & 0 \\ 0 & 1 & 0 \\ -3 & 0 & 1 \end{bmatrix} A \begin{bmatrix} 1 & 0 & -3 \\ 0 & 1 & 0 \\ 0 & 0 & 1 \end{bmatrix}$$

Applying $R_3 \rightarrow R_3 - R_2$, we get

$$\begin{bmatrix} 1 & 0 & 0 \\ 0 & -2 & -2 \\ 0 & 0 & -4 \end{bmatrix} = \begin{bmatrix} 1 & 0 & 0 \\ 0 & 1 & 0 \\ -3 & -1 & 1 \end{bmatrix} A \begin{bmatrix} 1 & 0 & -3 \\ 0 & 1 & 0 \\ 0 & 0 & 1 \end{bmatrix}$$

Applying $C_3 \rightarrow C_3 - C_2$, we get

$$\begin{bmatrix} 1 & 0 & 0 \\ 0 & -2 & 0 \\ 0 & 0 & -4 \end{bmatrix} = \begin{bmatrix} 1 & 0 & 0 \\ 0 & 1 & 0 \\ -3 & -1 & 1 \end{bmatrix} A \begin{bmatrix} 1 & 0 & -3 \\ 0 & 1 & -1 \\ 0 & 0 & 1 \end{bmatrix}$$

Applying $R_2 \rightarrow \dfrac{1}{\sqrt{2}} R_2$, we get

$$\begin{bmatrix} 1 & 0 & 0 \\ 0 & -\sqrt{2} & 0 \\ 0 & 0 & -4 \end{bmatrix} = \begin{bmatrix} 1 & 0 & 0 \\ 0 & 1/\sqrt{2} & 0 \\ -3 & -1 & 1 \end{bmatrix} A \begin{bmatrix} 1 & 0 & -3 \\ 0 & 1 & -1 \\ 0 & 0 & 1 \end{bmatrix}$$

Applying $C_2 \rightarrow \dfrac{1}{\sqrt{2}} C_2$, we get

$$\begin{bmatrix} 1 & 0 & 0 \\ 0 & -1 & 0 \\ 0 & 0 & -4 \end{bmatrix} = \begin{bmatrix} 1 & 0 & 0 \\ 0 & 1/\sqrt{2} & 0 \\ -3 & -1 & 1 \end{bmatrix} A \begin{bmatrix} 1 & 0 & -3 \\ 0 & 1/\sqrt{2} & -1 \\ 0 & 0 & 1 \end{bmatrix}$$

Applying $R_3 \rightarrow \dfrac{1}{\sqrt{4}} R_3$, we get

$$\begin{bmatrix} 1 & 0 & 0 \\ 0 & -1 & 0 \\ 0 & 0 & -\sqrt{4} \end{bmatrix} = \begin{bmatrix} 1 & 0 & 0 \\ 0 & 1/\sqrt{2} & 0 \\ -3/\sqrt{4} & -1/\sqrt{4} & 1/\sqrt{4} \end{bmatrix} A \begin{bmatrix} 1 & 0 & -3 \\ 0 & 1/\sqrt{2} & -1 \\ 0 & 0 & 1 \end{bmatrix}$$

Applying $C_3 \rightarrow \dfrac{1}{\sqrt{4}} C_3$, we get

$$\begin{bmatrix} 1 & 0 & 0 \\ 0 & -1 & 0 \\ 0 & 0 & -1 \end{bmatrix} = \begin{bmatrix} 1 & 0 & 0 \\ 0 & 1/\sqrt{2} & 0 \\ -3/2 & -1/2 & 1/2 \end{bmatrix} A \begin{bmatrix} 1 & 0 & -3/2 \\ 0 & 1/\sqrt{2} & -1/2 \\ 0 & 0 & 1/2 \end{bmatrix}$$

$\Rightarrow \qquad D = P'AP$

where $\quad P = \begin{bmatrix} 1 & 0 & -3/2 \\ 0 & 1/\sqrt{2} & -1/2 \\ 0 & 0 & 1/2 \end{bmatrix}$

Thus the real non-singular linear transformation is
$$X = PY$$

i.e., $\quad \begin{bmatrix} x \\ y \\ z \end{bmatrix} = \begin{bmatrix} 1 & 0 & -3/2 \\ 0 & 1/\sqrt{2} & -1/2 \\ 0 & 0 & 1/2 \end{bmatrix} \begin{bmatrix} y_1 \\ y_2 \\ y_3 \end{bmatrix}$

i.e., $\qquad x = y_1 - \dfrac{3}{2} y_3$

$$y = \dfrac{1}{\sqrt{2}} y_2 - \dfrac{1}{2} y_3$$

$$z = \dfrac{1}{2} y_3$$

which reduces $X'AX$ to the normal form

$$Y'(P'AP)Y = \begin{bmatrix} y_1 & y_2 & y_3 \end{bmatrix} \begin{bmatrix} 1 & 0 & 0 \\ 0 & -1 & 0 \\ 0 & 0 & -1 \end{bmatrix} \begin{bmatrix} y_1 \\ y_2 \\ y_3 \end{bmatrix}.$$

$$= y_1^2 - y_2^2 - y_3^2$$

The rank of $X'AX$ = number of non-zero terms in normal form
$$= 3$$

The signature of $X'AX$ = the excess of positive terms over the negative terms
$$= 1 - 2 = -1$$

The index of $X'AX$ = the number of positive terms in the normal form
$$= 1$$

(iii) The given quadratic form is

$$X'AX = 2x_1^2 + x_2^2 - 3x_3^2 - 8x_2 x_3 - 4x_3 x_1 + 12 x_1 x_2$$

$$= \begin{bmatrix} x_1 & x_2 & x_3 \end{bmatrix} \begin{bmatrix} 2 & 6 & -2 \\ 6 & 1 & -4 \\ -2 & -4 & -3 \end{bmatrix} \begin{bmatrix} x_1 \\ x_2 \\ x_3 \end{bmatrix}$$

$$\therefore A = \begin{bmatrix} 2 & 6 & -2 \\ 6 & 1 & -4 \\ -2 & -4 & -3 \end{bmatrix}$$

We write
$$A = IAI$$

or $\begin{bmatrix} 2 & 6 & -2 \\ 6 & 1 & -4 \\ -2 & -4 & -3 \end{bmatrix} = \begin{bmatrix} 1 & 0 & 0 \\ 0 & 1 & 0 \\ 0 & 0 & 1 \end{bmatrix} A \begin{bmatrix} 1 & 0 & 0 \\ 0 & 1 & 0 \\ 0 & 0 & 1 \end{bmatrix}$

Applying $R_2 \rightarrow R_2 - 3R_1, C_2 \rightarrow C_2 - 3C_1,$
$R_3 \rightarrow R_3 + R_1, C_3 \rightarrow C_3 + C_1$ we get

$$\begin{bmatrix} 2 & 0 & 0 \\ 0 & -17 & 2 \\ 0 & 2 & -5 \end{bmatrix} = \begin{bmatrix} 1 & 0 & 0 \\ -3 & 1 & 0 \\ 1 & 0 & 1 \end{bmatrix} A \begin{bmatrix} 1 & -3 & 1 \\ 0 & 1 & 0 \\ 0 & 0 & 1 \end{bmatrix}$$

Applying $R_3 \rightarrow R_3 + \dfrac{2}{17} R_2, C_3 \rightarrow C_3 + \dfrac{2}{17} C_2$, we get

$$\begin{bmatrix} 2 & 0 & 0 \\ 0 & -17 & 0 \\ 0 & 0 & -81/17 \end{bmatrix} = \begin{bmatrix} 1 & 0 & 0 \\ -3 & 1 & 0 \\ 11/17 & 2/17 & 1 \end{bmatrix} A \begin{bmatrix} 1 & -3 & 11/17 \\ 0 & 1 & 2/17 \\ 0 & 0 & 1 \end{bmatrix}$$

Applying $R_1 \rightarrow \dfrac{1}{\sqrt{2}} R_1, C_1 \rightarrow \dfrac{1}{\sqrt{2}} C_1,$

$R_2 \rightarrow \dfrac{1}{\sqrt{17}} R_2, C_2 \rightarrow \dfrac{1}{\sqrt{17}} C_2, R_3 \rightarrow \sqrt{\dfrac{17}{81}} R_3,$

$C_3 \rightarrow \sqrt{\dfrac{17}{81}} C_3$, we get

$$\begin{bmatrix} 1 & 0 & 0 \\ 0 & -1 & 0 \\ 0 & 0 & -1 \end{bmatrix} = \begin{bmatrix} 1/\sqrt{2} & 0 & 0 \\ -3/\sqrt{17} & 1/\sqrt{17} & 0 \\ 11\sqrt{17}/9 & 2\sqrt{17}/9 & 1\sqrt{17}/9 \end{bmatrix} A \begin{bmatrix} 1/\sqrt{2} & -3/\sqrt{17} & 11\sqrt{17}/9 \\ 0 & 1/\sqrt{17} & 2\sqrt{17}/9 \\ 0 & 0 & 1\sqrt{17}/9 \end{bmatrix}$$

$\Rightarrow \qquad D = P'AP$

where $P = \begin{bmatrix} 1/\sqrt{2} & 3/\sqrt{17} & 11\sqrt{17}/9 \\ 0 & 1/\sqrt{17} & 2\sqrt{17}/9 \\ 0 & 0 & 1\sqrt{17}/9 \end{bmatrix}$

Thus the real non-singular linear transformation is

$X = PY$

i.e., $\begin{bmatrix} x_1 \\ x_2 \\ x_3 \end{bmatrix} = \begin{bmatrix} 1/\sqrt{2} & -3/\sqrt{17} & 11\sqrt{17}/9 \\ 0 & 1/\sqrt{17} & 2\sqrt{17}/9 \\ 0 & 0 & 1\sqrt{17}/9 \end{bmatrix} \begin{bmatrix} y_1 \\ y_2 \\ y_3 \end{bmatrix}$

i.e., $x_1 = \dfrac{1}{\sqrt{2}} y_1 - \dfrac{3}{\sqrt{17}} y_2 + \dfrac{11}{9}\sqrt{17} y_3$

$x_2 = \dfrac{1}{\sqrt{17}} y_2 + \dfrac{2\sqrt{17}}{9} y_3$

$x_3 = \dfrac{1\sqrt{17}}{9} y_3$

These transformations reduce the given quadratic form $X'AX$ to the normal form

$Y'(P'AP)Y = Y'DY$

$= \begin{bmatrix} y_1 & y_2 & y_3 \end{bmatrix} \begin{bmatrix} 1 & 0 & 0 \\ 0 & -1 & 0 \\ 0 & 0 & -1 \end{bmatrix} \begin{bmatrix} y_1 \\ y_2 \\ y_3 \end{bmatrix}$

$= y_1^2 - y_2^2 - y_3^2$

The rank of $X'AX$ = number of non-zero terms in normal form = 3

The signature of $X'AX$ = the excess of the number of positive terms over the negative terms in the normal form = $1 - 2 = -1$

The index of $X'AX$ = the number of positive terms in the normal form = 1

(iv) The given quadratic form is

$X'AX = 6x_1^2 + 3x_2^2 + 14x_3^2 + 4x_2x_3 + 18x_3x_1 + 4x_1x_2$

$= \begin{bmatrix} x_1 & x_2 & x_3 \end{bmatrix} \begin{bmatrix} 6 & 2 & 9 \\ 2 & 3 & 2 \\ 9 & 2 & 14 \end{bmatrix} \begin{bmatrix} x_1 \\ x_2 \\ x_3 \end{bmatrix}$

$\therefore \quad A = \begin{bmatrix} 6 & 2 & 9 \\ 2 & 3 & 2 \\ 9 & 2 & 14 \end{bmatrix}$

We write $\quad A = IAI$

or $\begin{bmatrix} 6 & 2 & 9 \\ 2 & 3 & 2 \\ 9 & 2 & 14 \end{bmatrix} = \begin{bmatrix} 1 & 0 & 0 \\ 0 & 1 & 0 \\ 0 & 0 & 1 \end{bmatrix} A \begin{bmatrix} 1 & 0 & 0 \\ 0 & 1 & 0 \\ 0 & 0 & 1 \end{bmatrix}$

Applying $R_2 \to R_2 - \dfrac{1}{3} R_1$, $C_2 \to C_2 - \dfrac{1}{2} C_1$ and

$R_3 \to R_3 - \dfrac{3}{2} R_1$, $C_3 \to C_3 - \dfrac{3}{2} C_1$, we get

$\begin{bmatrix} 6 & 0 & 0 \\ 0 & 7/3 & -1 \\ 0 & -1 & 1/2 \end{bmatrix} = \begin{bmatrix} 1 & 0 & 0 \\ -1/3 & 1 & 0 \\ -3/2 & 0 & 1 \end{bmatrix} A \begin{bmatrix} 1 & -1/3 & -3/2 \\ 0 & 1 & 3/7 \\ 0 & 0 & 1 \end{bmatrix}$

Applying $R_3 \to R_3 + \dfrac{3}{7} R_2$, $C_3 \to C_3 + \dfrac{3}{7} C_1$, we get

$\begin{bmatrix} 6 & 0 & 0 \\ 0 & 7/3 & 0 \\ 0 & -1 & 1/14 \end{bmatrix} = \begin{bmatrix} 1 & 0 & 0 \\ -1/3 & 1 & 0 \\ -23/14 & 3/7 & 1 \end{bmatrix} A \begin{bmatrix} 1 & -1/3 & -23/14 \\ 0 & 1 & 3/7 \\ 0 & 0 & 1 \end{bmatrix}$

Applying $R_1 \to \dfrac{1}{\sqrt{6}} R_1$, $C_1 \to \dfrac{2}{\sqrt{6}} C_1$,

$R_2 \to \sqrt{\dfrac{3}{7}} R_2$, $C_2 \to \sqrt{\dfrac{3}{7}} C_2$ and $R_3 \to \dfrac{1}{\sqrt{14}} R_3$,

$C_3 \to \dfrac{1}{\sqrt{14}} C_3$, we get

$\begin{bmatrix} 1 & 0 & 0 \\ 0 & 1 & 0 \\ 0 & 0 & 1 \end{bmatrix} = \begin{bmatrix} \dfrac{1}{\sqrt{6}} & 0 & 0 \\ -\dfrac{1}{3}\sqrt{\dfrac{3}{7}} & \sqrt{\dfrac{3}{7}} & 0 \\ -\dfrac{23}{14}\dfrac{1}{\sqrt{14}} & \dfrac{3}{7}\dfrac{1}{\sqrt{14}} & 1 \end{bmatrix}$

$A \begin{bmatrix} \dfrac{1}{\sqrt{6}} & -\dfrac{1}{3}\sqrt{\dfrac{3}{7}} & -\dfrac{23}{14}\dfrac{1}{\sqrt{14}} \\ 0 & \sqrt{\dfrac{3}{7}} & \dfrac{3}{7}\dfrac{1}{\sqrt{14}} \\ 0 & 0 & 1 \end{bmatrix}$

$\Rightarrow \quad D = P'AP$

where $P = \begin{bmatrix} \dfrac{1}{\sqrt{6}} & -\dfrac{1}{3}\sqrt{\dfrac{3}{7}} & -\dfrac{23}{14}\dfrac{1}{\sqrt{14}} \\ 0 & \sqrt{\dfrac{3}{7}} & \dfrac{3}{7}\dfrac{1}{\sqrt{14}} \\ 0 & 0 & 1 \end{bmatrix}$

Thus the real non-singular linear transformation is

$X = PY$

i.e., $\begin{bmatrix} x_1 \\ x_2 \\ x_3 \end{bmatrix} = \begin{bmatrix} \dfrac{1}{\sqrt{6}} & -\dfrac{1}{3}\sqrt{\dfrac{3}{7}} & -\dfrac{23}{14}\dfrac{1}{\sqrt{14}} \\ 0 & \sqrt{\dfrac{3}{7}} & \dfrac{3}{7}\dfrac{1}{\sqrt{14}} \\ 0 & 0 & 1 \end{bmatrix} \begin{bmatrix} y_1 \\ y_2 \\ y_3 \end{bmatrix}$

or $\quad x_1 = \dfrac{1}{\sqrt{6}} y_1 - \dfrac{1}{3}\sqrt{\dfrac{3}{7}} y_2 - \dfrac{23}{14}\dfrac{1}{\sqrt{14}} y_3$

$x_2 = \sqrt{\dfrac{3}{7}} y_2 + \dfrac{3}{7}\dfrac{1}{\sqrt{14}} y_3 , x_3 = y_3$

These transformations reduce the given quadratic form $X'AX$ to the normal form

$Y'(P'AP)Y = Y'DY = y_1^2 + y_2^2 + y_3^2$

The rank of $X'AX$ = number of non-zero terms in normal form = 3

The signature of $X'AX$

$= 2(\text{positive terms in normal form}) - \text{rank}$

$= 2(3) - 3 = 6 - 3 = 3$

The index of $X'AX$

$= \text{the number of positive terms in the normal form}$

$= 3$

EXAMPLE 2. *Find an orthogonal matrix P that will diagonalize the real matrix*

$$A = \begin{bmatrix} 0 & 1 & 1 \\ 1 & 0 & -1 \\ 1 & -1 & 0 \end{bmatrix}$$

Interpret the result in terms of quadratic form.

SOLUTION. The characteristic equation of A is given by
$$|A - \lambda I| = 0$$

or
$$\begin{bmatrix} -\lambda & 1 & 1 \\ 1 & -\lambda & -1 \\ 1 & -1 & -\lambda \end{bmatrix} = 0$$

Applying $C_1 \to C_1 + C_2$, we get
$$\begin{bmatrix} 1-\lambda & 1 & 1 \\ 1-\lambda & -\lambda & -1 \\ 0 & -1 & -\lambda \end{bmatrix} = 0$$

or
$$(1-\lambda)\begin{bmatrix} 1 & 1 & 1 \\ 1 & -\lambda & -1 \\ 0 & -1 & -\lambda \end{bmatrix} = 0$$

or $(1-\lambda)(\lambda^2 - 1 + \lambda - 1) = 0$

or $(1-\lambda)^2(2 + \lambda) = 0$

Therefore, the eigenvalues of A are 1, 1 and –2.

Eigenvector corresponding to the eigen value 1

Let $X = \begin{bmatrix} x_1 \\ x_2 \\ x_3 \end{bmatrix} \neq 0$ be an eigenvector corresponding to

the eigenvalue 1, then X
is a solution of the equation
$$(A - I)X = 0$$

or
$$\begin{bmatrix} -1 & 1 & 1 \\ 1 & -1 & -1 \\ 1 & -1 & -1 \end{bmatrix}\begin{bmatrix} x_1 \\ x_2 \\ x_3 \end{bmatrix} = \begin{bmatrix} 0 \\ 0 \\ 0 \end{bmatrix}$$

Solving these equation, we have
$$-x_1 + x_2 + x_3 = 0 \qquad \qquad ...(1)$$

The two orthogonal solutions, which satisfy (1) are
$$X_1 = \begin{bmatrix} 1 \\ 0 \\ 1 \end{bmatrix} \text{ and } X_2 = \begin{bmatrix} 1 \\ 2 \\ -1 \end{bmatrix}$$

Thus, the two mutually orthogonal eigenvectors of A are
$$X_1 = \begin{bmatrix} 1 \\ 0 \\ 1 \end{bmatrix} \text{ and } X_2 = \begin{bmatrix} 1 \\ 2 \\ -1 \end{bmatrix}$$

Eigenvector corresponding to the eigen value –2

Let $X = \begin{bmatrix} x_1 \\ x_2 \\ x_3 \end{bmatrix} \neq 0$ be an eigenvector corresponding to

the eigenvalue –2, then X is a solution of the equation
$(A + 2I)X = 0$

or
$$\begin{bmatrix} 2 & 1 & 1 \\ 1 & 2 & -1 \\ 1 & -1 & 2 \end{bmatrix}\begin{bmatrix} x_1 \\ x_2 \\ x_3 \end{bmatrix} = \begin{bmatrix} 0 \\ 0 \\ 0 \end{bmatrix}$$

Applying $R_1 \leftrightarrow R_2$, we get
$$\begin{bmatrix} 1 & 2 & -1 \\ 2 & 1 & 1 \\ 1 & -1 & 2 \end{bmatrix}\begin{bmatrix} x_1 \\ x_2 \\ x_3 \end{bmatrix} = \begin{bmatrix} 0 \\ 0 \\ 0 \end{bmatrix}$$

Applying $R_2 \to R_2 - 2R_1, R_3 \to R_3 - R_1$, we get
$$\begin{bmatrix} 1 & 2 & -1 \\ 0 & -3 & 3 \\ 0 & -3 & 3 \end{bmatrix}\begin{bmatrix} x_1 \\ x_2 \\ x_3 \end{bmatrix} = \begin{bmatrix} 0 \\ 0 \\ 0 \end{bmatrix}$$

Applying $R_3 \to R_3 - R_2$, we get
$$\begin{bmatrix} 1 & 2 & -1 \\ 0 & -3 & 3 \\ 0 & 0 & 0 \end{bmatrix}\begin{bmatrix} x_1 \\ x_2 \\ x_3 \end{bmatrix} = \begin{bmatrix} 0 \\ 0 \\ 0 \end{bmatrix}$$

Solving these equation, we get
$$x_1 + 2x_2 - x_3 = 0$$
$$-3x_2 + 3x_3 = 0$$

Clearly, $x_1 = -1, x_2 = 1, x_3 = 1$ satisfy these equations, therefore, the eigenvector is

$$X_3 = \begin{bmatrix} -1 \\ 1 \\ 1 \end{bmatrix}$$

Thus, the required matrix P is a matrix whose column vectors are unit vectors which are scalar multiplies of X_1, X_2 and X_3.

$$\therefore \qquad P = \begin{bmatrix} 1/\sqrt{2} & 1/\sqrt{6} & -1/\sqrt{3} \\ 0 & 2/\sqrt{16} & 1/\sqrt{3} \\ 1/\sqrt{2} & -1/\sqrt{6} & 1/\sqrt{3} \end{bmatrix}$$

Now, we have
$$P'AP = D = \text{diag. } (1, 1, -2)$$

The quadratic form of the given symmetric matrix is
$$X'AX = 2x_1x_2 + 2x_1x_3 - 2x_2x_3$$

Thus, the orthogonal linear transformation $X = PY$ reduces the quadratic form $X'AX$ to the diagonal form
$$X'AX = Y'(P'AP)Y = Y'DY = y_1^2 + y_2^2 - 2y_3^2$$

The rank of $X'AX$ = number of non-zero eigenvalues of $A = 3$

The signature of $X'AX$ = the excess of the number of positive eigenvalues over the number of negative eigenvalues of A $= 2 - 1 = 1$

The diagonal form $y_1^2 + y_2^2 - 2y_3^2$ can be reduced to the normal form $z_1^2 + z_2^2 - z_3^2$.

EXAMPLE 3. *Reduce the following quadratic form in to canonical form and find its rank and signature:*
$$x^2 + 4y^2 + 9z^2 + t^2 - 12yz + 6zx - 4xy - 2xt - 6zt$$

SOLUTION. Let $X = \begin{bmatrix} x \\ y \\ z \\ t \end{bmatrix}$, then the given quadratic form can

be written as
$$x^2 + 4y^2 + 9z^2 + t^2 - 12yz + 6zx - 4xy - 2xt - 6zt$$

$$= X' \begin{bmatrix} 1 & -2 & 3 & -1 \\ -2 & 4 & -6 & 0 \\ 3 & -6 & 9 & -3 \\ -1 & 0 & -3 & 1 \end{bmatrix} X = X'AX$$

$$\therefore \quad A = \begin{bmatrix} 1 & -2 & 3 & -1 \\ -2 & 4 & -6 & 0 \\ 3 & -6 & 9 & -3 \\ -1 & 0 & -3 & 1 \end{bmatrix}$$

We write

$$A = IAI$$

$$\begin{bmatrix} 1 & -2 & 3 & -1 \\ -2 & 4 & -6 & 0 \\ 3 & -6 & 9 & -3 \\ -1 & 0 & -3 & 1 \end{bmatrix} = \begin{bmatrix} 1 & 0 & 0 & 0 \\ 0 & 1 & 0 & 0 \\ 0 & 0 & 1 & 0 \\ 0 & 0 & 0 & 1 \end{bmatrix} A \begin{bmatrix} 1 & 0 & 0 & 0 \\ 0 & 1 & 0 & 0 \\ 0 & 0 & 1 & 0 \\ 0 & 0 & 0 & 1 \end{bmatrix}$$

Applying
$$R_2 \to R_2 + 2R_1, C_2 \to C_2 + 2C_1;$$
$$R_3 \to R_3 + 3R_1, C_3 \to C_3 - 3C_1;$$
$$R_4 \to R_4 + R_1, C_4 \to C_4 + C_1$$

We get

$$\begin{bmatrix} 1 & 0 & 0 & 0 \\ 0 & 0 & 0 & -2 \\ 0 & 0 & 0 & 0 \\ 0 & -2 & 0 & 0 \end{bmatrix} = \begin{bmatrix} 1 & 0 & 0 & 0 \\ 2 & 1 & 0 & 0 \\ -3 & 0 & 1 & 0 \\ 1 & 0 & 0 & 1 \end{bmatrix} A \begin{bmatrix} 1 & 2 & -3 & 1 \\ 0 & 1 & 0 & 0 \\ 0 & 0 & 1 & 0 \\ 0 & 0 & 0 & 1 \end{bmatrix}$$

Applying $R_2 \to R_2 + R_4, C_2 \to C_2 + C_4,$ we get

$$\begin{bmatrix} 1 & 0 & 0 & 0 \\ 0 & -4 & 0 & -2 \\ 0 & 0 & 0 & 0 \\ 0 & -2 & 0 & 0 \end{bmatrix} = \begin{bmatrix} 1 & 0 & 0 & 0 \\ 3 & 1 & 0 & 1 \\ -3 & 0 & 1 & 0 \\ 1 & 0 & 0 & 1 \end{bmatrix} A \begin{bmatrix} 1 & 3 & -3 & 1 \\ 0 & 1 & 0 & 0 \\ 0 & 0 & 1 & 0 \\ 0 & 1 & 0 & 1 \end{bmatrix}$$

Applying $R_4 \to R_4 - \frac{1}{2}R_2, C_4 \to C_4 - \frac{1}{2}C_2,$ we get

$$\begin{bmatrix} 1 & 0 & 0 & 0 \\ 0 & -4 & 0 & 0 \\ 0 & 0 & 0 & 0 \\ 0 & 0 & 0 & 1 \end{bmatrix} = \begin{bmatrix} 1 & 0 & 0 & 0 \\ 3 & 1 & 0 & 1 \\ -3 & 0 & 1 & 0 \\ -1/2 & -1/2 & 0 & 1/2 \end{bmatrix}$$

$$\times A \begin{bmatrix} 1 & 3 & -3 & -1/2 \\ 0 & 1 & 0 & -1/2 \\ 0 & 0 & 1 & 0 \\ 0 & 1 & 0 & 1/2 \end{bmatrix}$$

Applying $R_2 \to \frac{1}{2}R_2, C_2 \to \frac{1}{2}C_2,$ we get

$$\begin{bmatrix} 1 & 0 & 0 & 0 \\ 0 & -1 & 0 & 0 \\ 0 & 0 & 0 & 0 \\ 0 & 0 & 0 & 1 \end{bmatrix} = \begin{bmatrix} 1 & 0 & 0 & 0 \\ \frac{3}{2} & \frac{1}{2} & 0 & \frac{1}{2} \\ -3 & 0 & 1 & 0 \\ -\frac{1}{2} & -\frac{1}{2} & 0 & \frac{1}{2} \end{bmatrix} A \begin{bmatrix} 1 & \frac{3}{2} & -3 & -\frac{1}{2} \\ 0 & \frac{1}{2} & 0 & -\frac{1}{2} \\ 0 & 0 & 1 & 0 \\ 0 & \frac{1}{2} & 0 & \frac{1}{2} \end{bmatrix}$$

$$\Rightarrow \quad D = P'AP$$

where
$$P = \begin{bmatrix} 1 & 3/2 & -3 & -1/2 \\ 0 & 1/2 & 0 & 1/2 \\ 0 & 0 & 1 & 0 \\ 0 & 1/2 & 0 & 1/2 \end{bmatrix}$$

Thus, the real non-singular linear transformation $X = PY$ reduces the given quadratic form $X'AX$ to the normal form

$$Y'(P'AP)Y = Y'DY = y_1^2 - y_2^2 + y_4^2$$

The rank of $X'AX$ = the number of non-zero terms in the normal form

$$= 3$$

The signature of $X'AX$

$$= 2(\text{positive terms}) - \text{rank}$$
$$= 2(2) - 3 = 4 - 3 = 1$$

 **Exercise-1.13**

1. Find the rank of each of the following quadratic forms :
   (i) $x^2 - 12xy - 4y^2$ (ii) $3x^2 + 2xy + 3y^2$
   (iii) $x^2 - 2xy + y^2$
   (iv) $x_1^2 - 2x_1x_2 + 2x_2^2$
   (v) $4x_1^2 + x_2^2 - 8x_3^2 + 4x_1x_2 - 4x_1x_3 + 8x_2x_3$

2. Find a real non-singular linear transformation $X = PY$ which reduces the given real quadratic form $x^2 + 2y^2 + 3z^2 + 4xy + 4yz$ to real canonical form. Also find the rank and signature of the given quadratic form.

3. Reduce each of the following quadratic forms to real canonical form and find its rank and signature. Also write in each case the linear transformation which reduce the normal form.

   (i) $x_1^2 + 2x_2^2 - 7x_3^2 - 4x_1x_2 + 8x_1x_3$
   (ii) $(x_1 + x_2 + x_3)^2 + (x_2 + x_3)^2 + 4x_3^2$
   (iii) $x_1^2 + 2x_2^2 + 3x_3^2 + 2x_2x_3 - 2x_3x_1 + 2x_1x_2$
   (iv) $4x_1^2 + 9x_2^2 + 2x_3^2 + 8x_2x_3 - 6x_3x_1 + 6x_1x_2$
   (v) $3x^2 + 3y^2 + 3z^2 - 2yz + 2zx + 2xy$
   (vi) $2x^2 + 9y^2 + 2z^2 - 2yz + 2zx + 6xy$
   (vii) $x^2 - 4y^2 + 6z^2 + 2xy - 4xz + 2w^2 - 6zw$
   (viii) $x_1x_2 - 4x_1x_4 - 2x_2x_3 + 12x_3x_4$

4. Reduce the quadratic form $7x^2 - 8y^2 - 8z^2 - 2yz - 8zx + 8xy$ to canonical form by an orthogonal transformation and hence find the signature of the quadratic form.

Answers

1. (i) rank = 2   (ii) rank = 2   (iii) rank = 1   (iv) rank = 2   (v) rank = 3

2. $X = \begin{bmatrix} 1 & -1/\sqrt{2} & -2/\sqrt{5} \\ 0 & 1/\sqrt{2} & 1/\sqrt{5} \\ 0 & 0 & 1/\sqrt{5} \end{bmatrix} Y$, canonical form is $y_1^2 - y_2^2 + y_3^2$, rank = 3, signature = 1

3. (i) rank = 3, signature = 1   (ii) rank = 3, signature = 3   (iii) rank = 3, signature = 3   (iv) rank = 3, signature = 1
   (v) rank = 3, signature = 3   (vi) rank = 3, signature = 3   (vii) rank = 4, signature = 0   (viii) rank = 4, signature = 0

1.22 SOLUTIONS OF EQUATIONS

A function $f(x)$ in the form $f(x) = a_0 x^n + a_1 x^{n-1} + a_2 x^{n-2} + ... + a_{n-1}x + a_n, a_0 \neq 0$ of degree n is said to be a rational integral function of x if all the coefficients $a_0, a_1, a_2 ..., a_{n-1}, a_n$ are supposed to be rational.

Definition. *The equation $f(x) = 0$ is called the general form of rational integral equation of nth degree.*

Definition. *Any value of x for which the value of $f(x)$ comes out to be zero, is called a root of the equation $f(x) = 0$.*

Since when $f(x)$ is divided by the factor $(x - a)$ then $f(a)$ is obtained as a remainder. If this remainder $f(a)$ becomes zero, then a is a root of the function $f(x) = 0$. Therefore, we can say that if 'a' is a root of the equation $f(x) = 0$, then $f(a) = 0$.

1.23 NUMBER OF ROOTS OF ANY EQUATION

THEOREM 1. *Every equation of degree n has n roots and no more.*

PROOF. Let the equation of degree n be

$$f(x) = a_0 x^n + a_1 x^{n-1} + a_2 x^{n-2} + ... + a_{n-1}x + a_n = 0; \text{ provided } a_0 \neq 0. \quad ...(1)$$

The equation $f(x) = 0$ has the roots, real as well as imaginary. Therefore if α_1 is any root of the equation (1), then $f(x)$ can be written as

$$f(x) = (x - \alpha_1)(a_0 x^{n-1} + ...) \text{ or } \qquad f(x) = (x - \alpha_1)\phi_1(x) \quad ...(2)$$

where $\phi_1(x)$ is a function of x of degree $n - 1$, such that $\phi_1(\alpha_1) \neq 0$. Further let α_2 be a root of $\phi_1(x) = 0$, then $\phi_1(x)$ can be written as

$$\phi_1(x) = (x - \alpha_2)\phi_2(x)$$
$$\therefore \qquad f(x) = (x - \alpha_1)(x - \alpha_2)\phi_2(x) \quad ...(3)$$

Continuing this process upto n times, we obtain

$$f(x) = a_0(x - \alpha_1)(x - \alpha_2)...(x - \alpha_n) \quad ...(4)$$

From equation (4) it is clear that when x take the values from α_1 to α_n, $f(x)$ comes out to be zero.

Hence, the equation $f(x)$ has n roots. Moreover if x takes any value different from $\alpha_1, \alpha_2, ... \alpha_n$, $f(x)$ cannot be zero so that $f(x) = 0$ has exactly n roots.

1.24 RELATION BETWEEN THE ROOT AND COEFFICIENTS

Let the general equation of degree n be given by

$$a_0 x^n + a_1 x^{n-1} + a_2 x^{n-2} + ... + a_{n-1}x + a_n = 0 \quad ...(1)$$

where $a_0, a_1, a_2 ... a_n$ are the coefficients and $a_0 \neq 0$. Let $\alpha_1, \alpha_2, \alpha_3 ... \alpha_n$ be the roots of the equation (1). The equation (1) can be identically written as

$$a_0 x^n + a_1 x^{n-1} + a_2 x^{n-2} + ... + a_{n-1}x + a_n = a_0(x - \alpha_1)(x - \alpha_2)...(x - \alpha_n)$$

or $\quad a_0 x^n + a_1 x^{n-1} + a_2 x^{n-2} + ... + a_{n-1}x + a_n = a_0[x^n - (\sum \alpha_1)x^{n-1} + (\sum \alpha_1 \alpha_2)x^{n-2} + ... + (-1)^n \alpha_1 \alpha_2 ... \alpha_n]$

where $\qquad \sum \alpha_1 = \alpha_1 + \alpha_2 + ... + \alpha_n$

$$\sum \alpha_1 \alpha_2 = \alpha_1 \alpha_2 + \alpha_1 \alpha_3 + ... \text{ etc.}$$

Now equating the coefficients of like powers of x of both sides we get

$$\sum \alpha_1 = -\frac{a_1}{a_0}; \ \sum \alpha_1 \alpha_2 = \frac{a_2}{a_0}; \ \sum \alpha_1 \alpha_2 \alpha_3 = -\frac{a_3}{a_0}; ...; \ \alpha_1 \alpha_2 \alpha_3 ... \alpha_n = (-1)^n \frac{a_n}{a_0} \Bigg\} \quad ...(2)$$

Hence, the equation (2) gives the required relation between the roots and the coefficients of the equation.

REMARK

- If the equation is not complete *i.e.*, some of the terms are missing, then we should first make this equation complete by adding the missing terms with zero coefficients.

Solved Examples

EXAMPLE 1. *Find the condition that the cubic $x^3 - lx^2 + mx - n = 0$ should have its roots in*

(1) *arithmetic Progression*

(2) *geometric Progression*   (MADRAS–2000, 09)

SOLUTION. (1) Let the roots in A.P be $a - d, a$ and $a + d$.

The sum of the roots
$$= a - d + a + a + d = 3a = l$$
$$\Rightarrow \qquad a = \frac{l}{3} \quad ...(1)$$

Since a is the root of the given equation, therefore
$$a^3 - la^2 + ma - n = 0$$
$$\Rightarrow \left(\frac{l}{3}\right)^3 - l\left(\frac{l}{3}\right)^2 + m\left(\frac{l}{a}\right) - n = 0 \qquad \text{(Using (1))}$$
$$\Rightarrow \qquad 2l^3 - 9lm + 27n = 0,$$
which is the required condition.

(2) Let the roots in G.P be $\dfrac{a}{r}, a$ and ar

Then   product of the roots $= \dfrac{a}{r}.a.ar = n$

$\Rightarrow$ $\qquad\qquad a^3 = n$

Then form (1)

$$n - ln^{2/3} + mn^{1/3} - n = 0$$

$\Rightarrow$ $\qquad\qquad m = ln^{1/3} \Rightarrow m^3 = l^3 n ,$

which is the required condition.

EXAMPLE 2. *Solve the equation* $x^4 - 2x^3 + 4x^2 + 6x - 21 = 0$ *given that the sum of two of its roots is zero.*

(MADRAS–2003, COCHIN–2005)

SOLUTION. Let α, β, γ and δ be the roots of the given equation such that
$$\alpha + \beta = 0$$

Then $\qquad \alpha + \beta + \gamma + \delta = 2 \Rightarrow \gamma + \delta = 2$

Therefore, the quadratic factor corresponding to α, β of the form $x^2 - 0x + p$ and that corresponding to γ, δ is of the form $x^2 - 2x + q$. Therefore, we can write
$$x^4 - 2x^3 + 4x^2 + 6x - 21 = (x^2 + p)(x^2 - 2x + q)$$

Equating the coefficient of x^2 and x from both sides of (1) we get
$$p + q = 4 , -2p = 6$$
$\Rightarrow$ $\qquad\qquad p = -3, q = 7$

So, the given equation is equivalent to
$$(x^2 - 3)(x^2 - 2x + 7) = 0$$

Hence, the required roots are $x = \pm\sqrt{3} , 1 \pm i\sqrt{6}$.

Exercise-1.14

1. Solve the equation $x^4 - 2x^3 - 21x^2 + 22x + 40 = 0$ whose roots are in A.P.

2. Solve the equation $2x^4 - 15x^3 + 35x^2 - 30x + 8 = 0$ whose roots are in G.P.

3. Form the equation of fourth degree whose roots are $3 + i$ and $\sqrt{7}$.

(MADRAS–2000)

4. Show that $x^7 - 3x^4 + 2x^3 - 1 = 0$ has at least four imaginary roots.

(COCHIN–2006, 14)

5. Solve the equation $x^3 - 4x^2 - 20x + 48 = 0$, given that the roots α and β connected by the relation $\alpha + 2\beta = 0$.

(SVTU–2007)

6. Solve the equation $x^4 - 6x^3 + 13x^2 - 12x + 4 = 0$ given that it has two parts of equal roots.

(MADRAS–2003)

7. If O, A, B, C are the four points on a straight line such that the distances A, B, C from O are the roots of the equation $ax^3 + 3bx^2 + 3cx + d = 0$. If B is the middle point of AC, show that $a^2d - 3abc + 2b^3 = 0$.

(SVTU–2006)

8. Solve the equation $x^4 - 8x^3 + 21x^2 - 20x + 5 = 0$, given that the sum of two of the roots is equal to the sum of the other two.

9. Solve the equation $8x^3 - 14x^2 + 7x - 1 = 0$, roots being in G.P.

(OSMANIA–1999, 2007)

10. Solve the equation $x^3 - 12x^2 + 39x - 28 = 0$, roots being in A.P.

(MADRAS–2001, 2011)

Answers

1. $-4, -1, 2, 5$	**2.** $\dfrac{1}{2}, 1, 2, 4$	**3.** $x^4 - 6x^3 + 3x^2 + 42x - 70 = 0$	**5.** $-4, 2, 6$
6. $1, 1, 2, 2$	**8.** $\dfrac{1}{2}(3 \pm \sqrt{5}), \dfrac{1}{2}(5 + \sqrt{5})$	**9.** $1, \dfrac{1}{2}, \dfrac{1}{4}$	**10.** $1, 4, 7$

1.24.1 Important Results

1. In an equation with real coefficients, imaginary roots occur in pair, that is if $\alpha + i\beta$ is one of the root of the equation $f(x) = 0$, then $\alpha - i\beta$ will also be a root of that equation.

2. If the equation $f(x) = 0$ has a pair of complex (imaginary) roots $\alpha \pm i\beta$ then $(x - \alpha)^2 + \beta^2$ will be a factor of $f(x)$.

3. If $\alpha + \sqrt{\beta}$ is a root of the equation $f(x) = 0$, then $\alpha - \sqrt{\beta}$ will also be a root of $f(x) = 0$.

4. Every equation of odd degree with real coefficients has at least one real root with the sign opposite to that of its last term.

5. Every equation of even degree with negative last term has at least two real roots with contrary sign.

6. If the equation $f(x) = 0$ and $g(x) = 0$ have common roots and their common roots are the roots of $h(x) = 0$, then $h(x)$ will be H.C.F. (G.C.D.) of $f(x)$ and $g(x)$.

7. If the equation $f(x) = 0$ has two roots equal, then the equation $f(x) = 0$, and $f'(x) = 0$ must have a common root.

1.24.2 Horner's Synthetic Division

In order to find the quotient and the remainder when a polynomial
$$f(x) = a_0 x^n + a_1 x^{n-1} + a_2 x^{n-2} + \ldots + a_{n-1}x + a_n, (a_0 \neq 0) \qquad \ldots(1)$$

of degree n is divided by a linear factor $(x - \alpha)$, we use a method given by Horner, called *synthetic division*. This method is being discussed as follows:

$$
\begin{array}{c|cccccc}
\alpha & a_0 & a_1 & a_2 \ldots & a_{n-1} & a_n \\
       &   & \alpha a_0 & \alpha b_1 \ldots & \alpha b_{n-2} & \alpha b_{n-1} \\
\hline
       & a_0 & b_1 & b_2 & b_{n-1} & R
\end{array}
$$

Step (1) If the equation (1) is not complete, then first make it complete by adding missing terms with zero coefficient.

Step (2) In the first horizontal line (row) we should write the coefficients $a_0, a_1, a_2, ... a_{n-1}, a_n$ of the polynomial $f(x)$.

Step (3) Since we have to divide to polynomial $f(x)$ by $x - \alpha$, so we should write α to the left to the vertical line as shown above.

Step (4) In the third horizontal line (row) we should write a_0 and the first term of the second horizontal line (row) is obtained by multiplying a_0 to α and then add this term with a_1 we obtain b_1 which is the second term of the third row. Next, we multiply b_1 and α and obtained the second term of the second row now adding this αb_1 with a_2, we obtain third terms of the third row. Continue the process in the same way we obtain the last term in the third row which is in fact the remainder R while the second last term in the same is b_{n-1}.

✒ REMARK

- If the remainder R comes out be zero, then α will be root of the equation $f(x) = 0$.

1.25 TRANSFORMATION OF EQUATION

Sometimes there arises some difficulties to find the roots of a given equation. In that case a process of transformation of a given equation into another equation plays an important role for finding the roots of given equation.

In this section we shall discuss some important transformation.

(i) To transform an equation into another equation whose roots are the roots of the given equation with different sign.

Let the given equation be $\quad f(x) = a_0 x^n + a_1 x^{n-1} + a_2 x^{n-2} + ... + a_{n-1} x + a_n = 0$ ...(1)

and let $\alpha_1, \alpha_2, \alpha_3, ..., \alpha_n$ be the roots of the equation (1).

Now put $x = -y$ in (1), we get

$$f(-y) = a_0(-y)^n + a_1(-y)^{n-1} + a_2(-y)^{n-2} + ... + a_{n-1}(-y) + a_n = 0$$

or $\quad f(-y) = (-1)^n[a_0 y^n - a_1 y^{n-1} + a_2 y^{n-2} - ... + (-1)^{n-1} a_{n-1} y + (-1)^n a_n] = 0$ ...(2)

This is the transformed equation.

Now we shall have to show that the equations (2) has the roots $-\alpha_1, -\alpha_2, -\alpha_3, ..., -\alpha_n$.

Since $\alpha_1, \alpha_2, ..., \alpha_n$ are the roots of equation (1), then (1) can also be written as

$$a_0 x^n + a_1 x^{n-1} + a_2 x^{n-2} + ... + a_{n-1} x + a_n = a_0(x - \alpha_1)(x - \alpha_2)...(x - \alpha_n)$$

Now putting $x = -y$ in both sides, we get

$$a_0(-y)^n + a_1(-y)^{n-1} + a_2(-y)^{n-2} + ... + a_{n-1}(-y) + a_n = a_0(-y - \alpha_1)(-y - \alpha_2)...(-y - \alpha_n)$$

or $\quad (-1)^n[a_0 y^n - a_1 y^{n-1} + a_2 y^{n-2} - ... + (-1)^{n-1} a_{n-1} y + (-1)^n a_0] = a_0(-1)^n(y + \alpha_1)(y + \alpha_2)...(y + \alpha_n)$

Using (2) $\qquad\qquad\qquad\qquad f(-y) = a_0(-1)^n(y + \alpha_1)(y + \alpha_2)...(y + \alpha_n)$

Thus the roots of the equation $f(-y) = 0$ are given by

$$(y + \alpha_1)(y + \alpha_2)...(y + \alpha_n) = 0 \qquad \text{or} \qquad y = -\alpha_1, -\alpha_2, ... -\alpha_n$$

Hence, the roots of the transformed equation (2) are the roots of the given equation with different sign.

(ii) To transform an equation into another equation whose roots are equal to the roots of the given equation multiplied by a given constant number m.

Let the given equation be $\qquad f(x) = a_0 x^n + a_1 x^{n-1} + a_2 x^{n-2} + ... a_{n-1} x + a_n = 0$ ...(1)

and let $\alpha_1, \alpha_2, ..., \alpha_n$ be its roots, then (1) can be written as

$$a_0 x^n + a_1 x^{n-1} + a_2 x^{n-2} + ... + a_{n-1} x + a_n = a_0(x - \alpha_1)(x - \alpha_2)...(x - \alpha_n)$$ ...(2)

Putting $y = mx$ or $x = \dfrac{y}{m}$ in (1), we get

$$f\left(\frac{y}{m}\right) = a_0\left(\frac{y}{m}\right)^n + a_1\left(\frac{y}{m}\right)^{n-1} + a_2\left(\frac{y}{m}\right)^{n-2} + ... + a_{n-1}\left(\frac{y}{m}\right) + a_n = 0$$

or $\quad f\left(\dfrac{y}{m}\right) = \dfrac{1}{m^n}[a_0 y^n + m a_1 y^{n-1} + m^2 a_2 y^{n-2} + ... + m^{n-1} y a_{n-1} + m^n a_n] = 0$

or $\qquad\qquad a_0 y^n + m a_1 y^{n-1} + m^2 a_2 y^{n-2} + ... + m^{n-1} y a_{n-1} + m^n a_n = 0$ ...(3)

This is the transformed equation. Now we shall show that the transformed equation has the roots $m\alpha_1, m\alpha_2, ... m\alpha_n$. For this let us put $x = \dfrac{y}{m}$ in (2), we get

$$a_0\left(\frac{y}{m}\right)^n + a_1\left(\frac{y}{m}\right)^{n-1} + a_2\left(\frac{y}{m}\right)^{n-2} + .. + a_{n-1}\left(\frac{y}{m}\right) + a_n = a_0\left(\frac{y}{m} - \alpha_1\right)\left(\frac{y}{m} - \alpha_2\right)...\left(\frac{y}{m} - \alpha_n\right)$$

or $\quad \dfrac{1}{m^n}[a_0 y^n + m a_1 y^{n-1} + m^2 a_2 y^{n-2} + ... + m^{n-1} a_{n-1} y + m^n a_n] = a_0 \dfrac{1}{m^n}(y - m\alpha_1)(y - m\alpha_2)...(y - m\alpha_n)$

or $\quad a_0 y^n + m a_1 y^{n-1} + m^2 a_2 y^{n-2} + ... + m^{n-1} a_{n-1} y + m^n a_n = a_0(y - m\alpha_1)(y - m\alpha_2)...(y - m\alpha_n)$

This shows that the transformed equation (3) has the roots $m\alpha_1, m\alpha_2, ... m\alpha_n$.

(iii) To transform an equation into another equation whose roots are the reciprocals of the roots of the given equation.

Let the given equation be

$$f(x) = a_0 x^n + a_1 x^{n-1} + a_2 x^{n-2} + ... + a_{n-1} x + a_n = 0 \qquad ...(1)$$

And let $\alpha_1, \alpha_2 ... \alpha_n$ be its roots, then, we have

$$a_0 x^n + a_1 x^{n-1} + a_2 x^{n-2} + ... + a_{n-1} x + a_n = a_0(x - \alpha_1)(x - \alpha_2)...(x - \alpha_n) \qquad ...(2)$$

Putting $x = \dfrac{1}{y}$ in (1), we get

$$f\left(\dfrac{1}{y}\right) = a_0\left(\dfrac{1}{y}\right)^n + a_1\left(\dfrac{1}{y}\right)^{n-1} + a_2\left(\dfrac{1}{y}\right)^{n-2} + ... + a_{n-1}\left(\dfrac{1}{y}\right) + a_n = 0$$

or $\qquad f\left(\dfrac{1}{y}\right) = \dfrac{1}{y^n}[a_0 + a_1 y + a_2 y^2 + ... + a_{n-1} y^{n-1} + a_n y^n] = 0$

or $\qquad a_n y^n + a_{n-1} y^{n-1} + ... + a_1 y + a_0 = 0$

This is the transformed equation. Now we shall show that this equation (3) has the roots $\dfrac{1}{\alpha_1}, \dfrac{1}{\alpha_2}, ..., \dfrac{1}{\alpha_n}$.

Let us put $x = \dfrac{1}{y}$ in (2), we get

$$a_0\left(\dfrac{1}{y}\right)^n + a_1\left(\dfrac{1}{y}\right)^{n-1} + a_2\left(\dfrac{1}{y}\right)^{n-2} + ... + a_{n-1}\left(\dfrac{1}{y}\right) + a_n = a_0\left(\dfrac{1}{y} - \alpha_1\right)\left(\dfrac{1}{y} - \alpha_2\right)...\left(\dfrac{1}{y} - \alpha_n\right)$$

or $\qquad \dfrac{1}{y^n}[a_0 + a_1 y + a_2 y^2 + ... + a_{n-1} y^{n-1} + a_n y^n] = \dfrac{a_0}{y^n}(1 - \alpha_1 y)(1 - \alpha_2 y)...(1 - \alpha_n y)$

or $\qquad a_0 + a_1 y + a_2 y^2 + ... + a_{n-1} y^{n-1} + a_n y^n = a_0(1 - \alpha_1 y)(1 - \alpha_2 y)...(1 - \alpha_n y)$

This shows that the equation (3) has the roots $\dfrac{1}{\alpha_1}, \dfrac{1}{\alpha_2}, ..., \dfrac{1}{\alpha_n}$.

Reciprocal Equation: We know that an equation which remains unchanged when x is replaced by $\dfrac{1}{x}$ is called a reciprocal equation.

Let the given equation be

$$f(x) \equiv a_0 x^n + a_1 x^{n-1} + a_2 x^{n-2} + ... + a_{n-1} x + a_n = 0 \qquad ...(1)$$

Replace x by $\dfrac{1}{x}$, we obtain

$$f\left(\dfrac{1}{x}\right) \equiv a_0 + a_1 x + a_2 x^2 + ... + a_{n-1} x^{n-1} + a_n x^n = 0 \qquad ...(2)$$

The equation (2) is an equation whose roots are the reciprocal of the roots of the equation (1). If both equations are same, then by comparing the coefficients of like powers of x we obtain

$$\dfrac{a_0}{a_n} = \dfrac{a_1}{a_{n-1}} = \dfrac{a_2}{a_{n-2}} ... \dfrac{a_{n-1}}{a_1} = \dfrac{a_n}{a_0}$$

From first and last fraction, we get

$$\dfrac{a_0}{a_n} = \dfrac{a_n}{a_0} \qquad \text{or} \qquad a_n^2 = a_0^2 \qquad \text{or} \qquad a_n = \pm a_0$$

Therefore from this result we have $a_n = a_0, a_n = -a_0$ and thus there are two classes of the reciprocal equations.

(a)   If $a_n = a_0$, then $\dfrac{a_1}{a_{n-1}} = \dfrac{a_2}{a_{n-2}} = ... = 1$ or $a_1 = a_{n-1}, a_2 = a_{n-2}...$

     This is, the coefficients of the terms in the equation equidistant from the beginning and the end are equal and the equation is therefore called the first class.

(b)   If $a_n = -a_0$, then $\dfrac{a_1}{a_{n-1}} = \dfrac{a_2}{a_{n-2}} = ... = -1$ or $a_1 = -a_{n-1}, a_2 = -a_{n-2}...$

     That is, the coefficients of the terms in the equation equidistant from the beginning and the end are equal in magnitude and opposite in sign. Therefore the reciprocal is called second class. In this case if the degree of the equation is **2m** (even) then, $a_m = -a_m$ or $a_m = 0$. Thus we can say that if the equation of second class and of even degree, then the middle term of the equation will be absent.

Standard form of the reciprocal equation. Let $f(x) = 0$ be a reciprocal equation and if $f(x) = 0$ is of first class and of an odd degree, then one of the roots of this equation $f(x) = 0$ must be its own reciprocal so it has a root –1 and thus $f(x)$ is divisible by the factor $x + 1$. If $\phi(x)$ is the quotient, then $\phi(x) = 0$ will be a reciprocal equation of first class and of an even degree.

On the other hand if the equation $f(x) = 0$ is of second class and of an odd degree, then it will have the root +1, and therefore $f(x)$ is divisible by the factor $(x - 1)$. If $\phi(x)$ is the quotient, then $f(x) = (x - 1)\phi(x)$

Thus $\phi(x) = 0$ is a reciprocal equation of first class and of even degree. And if the equation $f(x) = 0$ is of the second class and of an even degree, then it will have two roots –1 and +1. Therefore $f(x)$ is divisible by $(x + 1)$ and $(x - 1)$ or divisible by $(x^2 - 1)$.

If $\phi(x)$ is the quotient, then $f(x) = (x^2 - 1)\phi(x)$.

From this equation it is obvious that $\phi(x) = 0$ will be a reciprocal equation of first class and of even degree. Hence from above discussion we can say that every reciprocal equation can be reduced to a reciprocal of first class and of even degree which is known as the standard form.

THEOREM 1. *Every reciprocal equation of the standard form can be reduced to an equation of degree half of the degree of the original equation.*

PROOF. Let the reciprocal equation of the standard form be given by

$$a_0 x^{2m} + a_1 x^{2m-1} + a_2 x^{2m-2} + ... + a_m x^m + ... + a_2 x^2 + a_1 x + a_0 = 0 \qquad ...(1)$$

Divide this equation by x^m, we get

$$a_0 x^m + a_1 x^{m-1} + a_2 x^{m-2} + ... + a_m + ... + a_2 \frac{1}{x^{m-2}} + \frac{a_1}{x^{m-1}} + \frac{a_0}{x^m} = 0$$

$$a_0\left(x^m + \frac{1}{x^m}\right) + a_1\left(x^{m-1} + \frac{1}{x^{m-1}}\right) + a_2\left(x^{m-2} + \frac{1}{x^{m-2}}\right) + ... + a_m = 0 \qquad ...(2)$$

Since, we know that $\quad x^{k+1} + \dfrac{1}{x^{k+1}} = \left(x^k + \dfrac{1}{x^k}\right)\left(x + \dfrac{1}{x}\right) - \left(x^{k-1} + \dfrac{1}{x^{k-1}}\right)$

Putting $x + \dfrac{1}{x} = y$ for $k = 1, 2, 3...$ successively, such that

for $k = 1$, $\quad x^2 + \dfrac{1}{x^2} = \left(x + \dfrac{1}{x}\right)\left(x + \dfrac{1}{x}\right) - (1 + 1) = y^2 - 2$

for $k = 2$, $\quad x^3 + \dfrac{1}{x^3} = \left(x^2 + \dfrac{1}{x^2}\right)\left(x + \dfrac{1}{x}\right) - \left(x + \dfrac{1}{x}\right) = (y^2 - 2)y - y = y^3 - 3y$

for $k = 3$, $\quad x^4 + \dfrac{1}{x^4} = \left(x^3 + \dfrac{1}{x^3}\right)\left(x + \dfrac{1}{x}\right) - \left(x^2 + \dfrac{1}{x^2}\right) = (y^3 - 3y)y - (y^2 - 2) = y^4 - 4y^2 + 2$

and so on, we obtain $\left(x^m + \dfrac{1}{x^m}\right)$ is a polynomial of degree 'm'. Hence, the equation (2) is obtained an equation of degree m

which is half of the degree of the equation (1).

SOME FACTS

1. A reciprocal equation of an odd degree having coefficients of terms equidistant from the beginning and end equal, has a root = –1.
2. A reciprocal equation of an odd degree having coefficients of terms equidistant from the beginning and end but in opposite sign, has root = 1.
3. A reciprocal equation of an even degree having coefficients of terms equidistant from the beginning and end equal but opposite in sign has two roots 1 and –1.

(iv) **To transform an equation into another equation whose roots are any powers of the roots of the given equation.**

Let the given equation be $\quad f(x) \equiv a_0 x^n + a_1 x^{n-1} + a_2 x^{n-2} + ... + a_{n-1} x + a_n = 0 \qquad ...(1)$

And let $\alpha_1, \alpha_2, ... \alpha_n$ be its roots, then we have

$$f(x) \equiv a_0 x^n + a_1 x^{n-1} + a_2 x^{n-2} + ... + a_{n-1} x + a_n = a_0(x - \alpha_1)(x - \alpha_2)...(x - \alpha_n) \qquad ...(2)$$

The equation (2) can be modified as follows

$$f(x) \equiv a_0(x - \alpha_1)\left(\frac{x^m - \alpha_1^m}{x^m - \alpha_1^m}\right)(x - \alpha_2)\left(\frac{x^m - \alpha_2^m}{x^m - \alpha_2^m}\right)...(x - \alpha_n)\left(\frac{x^m - \alpha_n^m}{x^m - \alpha_n}\right)$$

or $\quad a_0(x^m - \alpha_1^m)(x^m - \alpha_2^m)...(x^m - \alpha_n^m) = f(x)\left(\dfrac{x^m - \alpha_1^m}{x - \alpha_1}\right)\left(\dfrac{x^m - \alpha_2^m}{x - \alpha_2}\right)...\left(\dfrac{x^m - \alpha_n^m}{x - \alpha_n}\right)$

or $\qquad a_0(x^m - \alpha_1{}^m)(x^m - \alpha_2{}^m)...(x^m - \alpha_n{}^m) = f(x)[x^{m+1} + \alpha_1 x^{m-2} + ... + \alpha_1{}^{m-1}]$

$$[x^{m-1} + \alpha_2 x^{m-2} + ... + \alpha_2{}^{m-1}]...[x^{m-1} + \alpha_n x^{m-2} + ... + \alpha_n{}^{m-1}] \qquad ...(3)$$

Let us assume $\qquad \phi(x^m) = a_0(x^m - \alpha_1^m)(x^m - \alpha_2^m)...(x^m - \alpha_n^m) \qquad\qquad ...(4)$

It is obvious that for $x = \alpha_1, \alpha_2,...\alpha_n$ equation (3) gives the identity so that the equation (4) is the transformed equation whose roots are $\alpha_1{}^m, \alpha_2{}^m,...\alpha_n{}^m$. Hence, if we put $x^m = y$ in (4), we obtain (4) as follows :

$$\phi(y) = a_0(y - \alpha_1^m)(y - \alpha_2^m)...(y - \alpha_n^m)$$

(v) To transform an equation into another equation whose roots exceed the roots of the given equation by a constant h.

Let the given equation be $f(x) \equiv a_0 x^n + a_1 x^{n-1} + a_2 x^{n-2} + ... + a_{n-1}x + a_n = 0 \qquad ...(1)$

and let $\alpha_1, \alpha_2,...\alpha$ be its roots. We have

$$f(x) \equiv a_0 x^n + a_1 x^{n-1} + a_2 x^{n-2} + ... + a_{n-1}x + a_n = a_0(x - \alpha_1)(x - \alpha_2)...(x - \alpha_n) \qquad ...(2)$$

Putting $y = x + h$, *i.e.*, $x = y - h$ in (1), we get

$$f(y - h) = a_0(y - h)^n + a_1(y - h)^{n-1} + a_2(y - h)^{n-2} + ... + a_{n-1}(y - h) + a_n = 0$$

The equation can be written in descending powers of y as follows:

$$A_0 y^n + A_1 y^{n-1} + A_2 y^{n-2} + ... + A_{n-1}y + A_n = 0 \qquad ...(3)$$

where $A_0, A_1, A_2,..., A_n$ are coefficients and constants and whose values depend upon $a_0, a_1, a_2,..., a_n$

Now put $y = x + h$ in (3), we get

$$f(x) = A_0(x + h)^n + A_1(x + h)^{n-1} + ... + A_{n-1}(x + h) + A_n = 0$$

$$f(x) = (x + h)[A_0(x + h)^{n-1} + A_1(x + h)^{n-2} + ... + A_{n-1}] + A_n$$

The equation gives that if $f(x)$ is divided by $x + h$, then A_n is obtained as remainder and the quotient is

$$A_0(x + h)^{n-1} + A_1(x + h)^{n-2} + ... + A_{n-2}(x + h) + A_{n-1}$$

Similarly if this quotient is divided by $(x + h)$, then we obtain A_{n-1} as remainder.

Continuing this process until we get all the constant $A_n, A_{n-1},..., A_2, A_1$ and we also obtain $A_0 = a_0$

Hence, the transformed equation is

$$f(y - h) \equiv A_0 y^n + A_1 y^{n-1} + A_2 y^{n-2} + ... + A_{n-1}y + A_n = 0$$

Now we have to show that $\alpha_1 + h, \alpha_2 + h,..., \alpha_n + h$ are the roots of this transformed equation.

For this put $x = y - h$ in (2), we get

$$f(y - h) \equiv a_0(y - h - \alpha_1)(y - h - \alpha_2)...(y - h - \alpha_n) \equiv a_0(y - (\alpha_1 + h)(y - (\alpha_2 + h))...(y - (\alpha_n + h)).$$

Hence, $\alpha_1 + h, \alpha_2 + h,... \alpha_n + h$ are the roots of the transformed equation.

1.25.1 REMOVAL OF TERMS OF AN EQUATION

Let the given equation be $f(x) \equiv a_0 x^n + a_1 x^{n-1} + a_2 x^{n-2} + ... + a_{n-1}x + a_n = 0 \qquad ...(1)$

If we put $x = y + h$, we get

$$a_0(y + h)^n + a_1(y + h)^{n-1} + a_2(y + h)^{n-2} + ... + a_{n-1}(y + h) + a_n = 0$$

This equation can be written in the descending powers of y as follows:

$$a_0 y^n + (na_0 h + a_1)y^{n-1} + \left\{ \frac{n(n-1)}{2!}a_0 h^2 + (n-1)a_1 h y^{n-1} + a_2 \right\} + ... = 0$$

Now, we want to remove second term, then we shall equal to zero the coefficient of y^{n-1}, we get $na_0 h + a_1 = 0$ or $h = -\dfrac{a_1}{na_0}$

Hence, we decreased all the roots of the given equation by a constant $-\dfrac{a_1}{na_0}$, the second term of the given equation can be removed.

Similarly, if we want to remove third term, we put $\dfrac{n(n-1)}{2!}a_0 h^2 + (n-1)a_1 h + a_2 = 0$

Solve this equation we get two values of h and similarly we can remove any term of the given equation.

(i) To remove the second term of the equation $a_0 x^3 + 3a_1 x^2 + 3a_2 x + a_3 = 0$ and form the obtained equation with integral coefficients having leading coefficient unity.

Consider $\qquad f(x) \equiv a_0 x^3 + 3a_1 x^2 + 3a_2 x + a_3 = 0 \qquad\qquad ...(1)$

Let $\alpha_1, \alpha_2, \alpha_3$ be its roots.

Put $x = y + h$ in (1), we get $\qquad a_0(y + h)^3 + 3a_1(y + h)^2 + 3a_2(y + h) + a_3 = 0$

or $\qquad a_0(y^3 + h^3 + 3y^2 h + 3yh^2) + 3a_1(y^2 + h^2 + 2yh) + 3a_2(y + h) + a_3 = 0$

or $\quad a_0 y^3 + (3ha_0 + 3a_1)y^2 + (3h^2 a_0 + 6a_1 h + 3a_2)y + (a_0 h^3 + 3a_1 h^2 + 3a_2 h + a_3) = 0$ ...(2)

Now, we want to remove second term, then put $\quad 3ha_0 + 3a_1 = 0 \quad$ or $\quad h = -\dfrac{a_1}{a_0}$

Substitute the value of h in (2), we get

$$a_0 y^3 + \left(\frac{3a_1^2}{a_0} - \frac{6a_1^2}{a_0} + 3a_2 \right)y + \left(-\frac{a_1^3}{a_0^2} + \frac{3a_1^2}{a_0^2} - \frac{3a_1 a_2}{a_0} + a_3 \right) = 0$$

or $\quad a_0 y^3 + \dfrac{3(a_0 a_2 - a_1^2)}{a_0}y + \dfrac{(a_0^2 a_3 - 3a_0 a_1 a_2 + 2a_1^2)}{a_0^2} = 0 \quad$ or $\quad a_0 y^3 + \dfrac{3H}{a_0}y + \dfrac{G}{a_0^2} = 0$ ...(3)

where $H = a_0 a_2 - a_1^2$, $\qquad\qquad G = a_0^2 a_3 - 3a_0 a_1 a_2 + 2a_1^2$

Thus the equation (3) is a transformed equation. Further, make all the coefficients of (3) integers, so that (3) can be written as

$$a_0^3 y^3 + 3Ha_0 y + G = 0$$

Let us put $z = a_0 y$

$$z^3 + 3Hz + G = 0$$ ...(4)

This is the transformed equation with integral coefficient and having leading coefficient unity. Now the roots of (4) are obtained

by the transformation $\quad z = a_0 y = a_0(x - h) = a_0 \left(x + \dfrac{a_1}{a_0} \right) = a_0 x + a_1 \qquad \left[\because h = -\dfrac{a_1}{a_0} \right]$

Since $\alpha_1, \alpha_2, \alpha_3$ are the roots of equation (1), then the roots of (4) are $a_0 \alpha_1 + a_1, a_0 \alpha_2 + a_1, a_2 \alpha_3 + a_1$

Further since we know that

$$\alpha_1 + \alpha_2 + \alpha_3 = -\frac{3a_1}{a_0} \qquad \text{or} \qquad \frac{a_1}{a_0} = -\frac{\alpha_1 + \alpha_2 + \alpha_3}{3}$$

then $\quad a_0 \alpha_1 + a_1 = a_0 \left(\alpha_1 + \dfrac{a_1}{a_0} \right) = a_0 \left(\alpha_1 - \dfrac{\alpha_1 + \alpha_2 + \alpha_3}{3} \right) = \dfrac{a_0}{3}(2\alpha_1 - \alpha_2 - \alpha_3)$

Similarly $\quad a_0 \alpha_2 + a_1 = \dfrac{a_0}{3}(2\alpha_2 - \alpha_1 - \alpha_3)$ and $a_0 \alpha_3 + a_1 = \dfrac{a_0}{3}(2\alpha_3 - \alpha_1 - \alpha_2)$

Hence, the roots of (4) can also be taken as $\dfrac{a_0}{3}(2\alpha_1 - \alpha_2 - \alpha_3), \dfrac{a_0}{3}(2\alpha_2 - \alpha_1 - \alpha_3), \dfrac{a_0}{3}(2\alpha_3 - \alpha_1 - \alpha_2)$

Now if we put $z = a_0 x + a_1$ in (4), we get

$$(a_0 x + a_1)^3 + 3H(a_0 x + a_1) + G \equiv a_0^2 [a_0 x^3 + 3a_1 x^2 + 3a_2 x + a_3]$$

(ii) **To remove the second term in the equation $a_0 x^4 + 4a_1 x^3 + 6a_2 x^2 + 4a_3 x + a_4 = 0$ with the binomial coefficients and to form the equation with integral coefficients having leading coefficients unity.**

Since the equation is $\quad f(x) \equiv a_0 x^4 + 4a_1 x^3 + 6a_2 x^2 + 4a_3 x + a_4 = 0$ ...(1)

And let $\alpha_1, \alpha_2, \alpha_3, \alpha_4$ be its roots

Put $x = y - h$ in (1), we obtain

$$f(y - h) \equiv a_0(y - h)^4 + 4a_1(y - h)^3 + 6a_2(y - h)^2 + 4a_3(y - h) + a_4 = 0$$

or $\quad f(y - h) \equiv a_0 y^4 + 4(a_0 h + a_1)y^3 + 6(a_0 h^2 + 2a_1 h + a_2)y^2 + 4(a_0 h^3 + 3a_1 h^2 + 3a_2 h + a_3)$

$$+ (a_0 h^4 + 4a_1 h^3 + 6a_2 h^2 + 4a_3 h + a_4) = 0$$ ...(2)

Now we want to remove second term by putting $\quad 4(a_0 h + a_1) = 0 \quad$ or $\quad h = -\dfrac{a_1}{a_0}$

Substitute the value of h in (2), we obtain

$$a_0 y^4 + \frac{6H}{a_0}y^2 + \frac{4G}{a_0^2}y + \frac{(a_0^2 I - 3H^2)}{a_0^3} = 0$$ ...(3)

where $H = a_0 a_1 - a_1^2, G = a_0^2 a_3 - 3a_0 a_1 a_2 + 2a_1^3$ and $I = a_0 a_4 - 4a_1 a_3 + 3a_2^2$

Equation (3) can also be written as $a_0^4 y^4 + 6Ha_0^2 y^2 + 4Ga_0 y + (a_0^2 I - 3H^2) = 0$

Let us put $z = a_0 y$ we get $\quad z^4 + 6Hz^2 + 4Gz + (a_0^2 I - 3H^2) = 0$ ...(4)

This is transformed equation whose leading coefficients being unity and all other coefficients are integers. Since we have

$$z = a_0 y = a_0(x - h) = a_0 \left(x + \frac{a_1}{a_0} \right) \qquad\qquad \left[\because h = -\frac{a_1}{a_0} \right]$$

$\therefore \qquad z = a_0 x + a_1$

Thus the roots of the equation (4) are obtained by the transformation $z = a_0x + a_1$

Since $\alpha_1, \alpha_2, \alpha_3, \alpha_4$ are the roots of (1), then $a_0\alpha_1 + a_1, a_0\alpha_2 + a_2, a_0\alpha_3 + a_3$ and $a_0\alpha_4 + a_4$ are the roots of (4). Further since we know that

$$\alpha_1 + \alpha_2 + \alpha_3 + \alpha_4 = -\frac{3a_1}{a_0} \quad \text{or} \quad \frac{a_1}{a_0} = -\frac{\alpha_1 + \alpha_2 + \alpha_3 + \alpha_4}{4}$$

$\therefore \qquad a_0\alpha_1 + a_1 = a_0\left(\alpha_1 + \frac{a_1}{a_0}\right) = a_0\left(\alpha_1 - \frac{\alpha_1 + \alpha_2 + \alpha_3 + \alpha_4}{4}\right) = \frac{a_0}{4}(3\alpha_1 - \alpha_2 - \alpha_3 - \alpha_4)$

Similarly $\qquad a_0\alpha_2 + a_1 = \frac{a_0}{4}(3\alpha_2 - \alpha_1 - \alpha_3 - \alpha_4)$

$$a_0\alpha_3 + a_1 = \frac{a_0}{4}(3\alpha_3 - \alpha_1 - \alpha_2 - \alpha_4)$$

and $\qquad a_0\alpha_4 + a_1 = \frac{a_0}{4}(3\alpha_4 - \alpha_1 - \alpha_2 - \alpha_3)$

Hence, the roots of equation (4) can also be taken as $\frac{a_0}{4}(3\alpha_1 - \alpha_2 - \alpha_3 - \alpha_4), \frac{a_0}{4}(3\alpha_2 - \alpha_1 - \alpha_3 - \alpha_4),$

$\frac{a_0}{4}(3\alpha_3 - \alpha_1 - \alpha_2 - \alpha_4)$ and $\frac{a_0}{4}(3\alpha_4 - \alpha_1 - \alpha_2 - \alpha_3)$.

1.25.2 An Important Relation

In order to discuss the biquadratic equation, a function of its coefficients plays a key role. This function is taken as

$$J = a_0a_2a_4 + 2a_1a_2a_3 - a_0a_3^2 - a_1^2a_4 - a_2^3$$

which can also be written in the form of a determinant as follows :

$$J = \begin{vmatrix} a_0 & a_1 & a_2 \\ a_1 & a_2 & a_3 \\ a_2 & a_3 & a_4 \end{vmatrix}$$

Further, we have an important relation between H, G, I and J as follows :

$$G^2 + 4H^3 = a_0^2(HI - a_0J)$$

Verification :   L.H.S. $= G^2 + 4H^3 = (a_0^2a_3 - 3a_0a_1a_2 + 2a_1^3)^2 + 4(a_0a_2 - a_1^2)^3$

$= (a_0^4a_3^2 + 9a_0^2a_1^2a_2^2 + 4a_1^6 - 6a_0^3a_1a_2a_3 + 4a_0^2a_1^3a_3 - 12a_0a_1^4a_2) + 4(a_0^3a_2^3 - a_1^6 - 3a_0^2a_2^2a_1^2 + 3a_0a_2a_1^4)$

$= a_0^4a_3^2 - 3a_0^2a_1^2a_2^2 - 6a_0^3a_1a_2a_3 + 4a_0^2a_1^3a_3 + 4a_0^3a_2^3 = a_0^2(a_0^2a_3^2 - 3a_1^2a_2^2 - 6a_0a_1a_2a_3 + 4a_1^3a_3 + 4a_0a_2^3)$

R.H.S. $= a_0^2(HI - a_0J) = a_0^2[(a_0a_2 - a_1^2)(a_0a_4 - 4a_1a_3 + 3a_2^2) - a_0(a_0a_2a_4 + 2a_1a_2a_3 - a_0a_3^2 - a_1^2a_4 - a_2^3)]$

$= a_0^2[a_0^2a_2a_4 - 4a_0a_1a_2a_3 + 3a_0a_2^3 - a_0a_1^2a_4 + 4a_1^3a_3 - 3a_1^2a_2^2 - a_0^2a_2a_4 - 2a_0a_1a_2a_3 + a_0^2a_3^2 + a_0a_1^2a_4 + a_0a_2^3]$

$= a_0^2(a_0^2a_3^2 - 3a_1^2a_2^2 - 6a_0a_1a_2a_3 + 4a_1^3a_3 + 4a_0a_2^3)$

Hence,   L.H.S. = R.H.S.

1.25.3 General Method of Transformation

Let the given equation be $\qquad\qquad\qquad\qquad f(x) = 0 \qquad\qquad\qquad\qquad\qquad\qquad\qquad$ ...(1)

and suppose y is a root of transformed equation such that x and y are related by some relation

$$\phi(x, y) = 0 \qquad\qquad\qquad\qquad\qquad\qquad \text{...(2)}$$

Eliminating x between (1) and (2), we get the transformed equation.

(i) **To form the equation whose roots are $(\alpha - \beta)^2, (\beta - \gamma)^2, (\gamma - \alpha)^2$, where α, β, γ, are the roots of the given equation $a_0x^3 + 3a_1x^2 + 3a_2x + a_3 = 0$ and to discuss the nature of the roots of the given equation.**

Since the given equation is

$$a_0x^3 + 3a_1x^2 + 3a_2x + a_3 = 0 \qquad\qquad\qquad\qquad \text{...(1)}$$

First remove the second term of (1) by diminishing it's root by h, we obtain

$$y^3 + \frac{3H}{a_0^2}y + \frac{G}{a_0^3} = 0 \quad \text{or} \quad y^3 + 2y + r = 0 \quad \text{where} \quad q = \frac{3H}{a_0^2}, r = \frac{G}{a_0^3} \qquad\qquad \text{...(2)}$$

and also $\qquad h = -\frac{a_1}{a_0}, H = a_0a_2 - a_1^2, G = a_0^2a_3 - 3a_0a_1a_2 + 2a_1^3$

The roots of the transformed equation (2) are $\alpha - h, \beta - h, \gamma - h$ respectively.

For simplicity let us take $\alpha_1 = \alpha - h, \beta_1 = \beta - h$ and $\gamma_1 = \gamma - h$.

Now $(\alpha - \beta)^2 = (\alpha - h - \beta + h)^2 = [(\alpha - h) - (\beta - h)]^2 = (\alpha_1 - \beta_1)^2.$

Similarly $(\beta - \gamma)^2 = (\beta_1 - \gamma_1)^2$ and $(\gamma - \alpha)^2 = (\gamma_1 - \alpha_1)^2.$

Hence, the equation of the squared differences of (1) is same as that of the equation (2). Therefore if $\alpha_1, \beta_1, \gamma_1$ are the roots of (2), then we have to find the equation whose roots are $(\alpha_1 - \beta_1)^2, (\beta_1 - \gamma_1)^2, (\gamma_1 - \alpha_1)^2$. Let z be one of the roots of required equation.

$$\therefore \qquad z = (\alpha_1 - \beta_1)^2 = \alpha_1^2 + \beta_1^2 - 2\alpha_1\beta_1 = (\alpha_1 + \beta_1)^2 - 4\alpha_1\beta_1 = (-\gamma_1)^2 - \frac{4\alpha_1\beta_1\gamma_1}{\gamma_1} \qquad [\because \alpha_1 + \beta_1 + \gamma_1 = 0]$$

$$z = \gamma_1^2 + \frac{4r}{\gamma_1} \qquad \text{or} \qquad \gamma_1^3 + z\gamma_1 - 4r = 0 \qquad [\because \alpha_1\beta_1\gamma_1 = -r]$$

Since γ_1 is the root of equation (2) so put $\gamma_1 = y$.

$$\therefore \qquad y^3 + zy - 4r = 0 \qquad \qquad \qquad \text{...(3)}$$

Eliminating y between (2) and (3), we get

$$z^3 + 6qz^2 + 9q^2z + 27r^2 + 4q^3 = 0 \qquad \qquad \text{...(4)}$$

Putting $q = \frac{3H}{a_0^2}, r = \frac{G}{a_0^3}$, we get $z^3 + \frac{18H}{a_0^2}z^2 + \frac{81H^2}{a_0^4}z + \frac{27}{a_0^6}(G^2 + 4H^3) = 0$ ...(5)

Hence, $(\alpha - \beta)^2, (\beta - \gamma)^2, (\gamma - \alpha)^2$ are the roots of (5)

$$\therefore \qquad (\alpha - \beta)^2(\beta - \gamma)^2(\gamma - \alpha)^2 = -\frac{27}{a_0^6}(G^2 + 4H^3) \qquad \text{...(6)}$$

1.25.4 Nature of Roots of the Given Equation

Since the degree of the given equation is odd so it has at least one of the roots α, β, γ say α real, and we know that complex roots lie in pair. If β, γ are also real, then $(\alpha - \beta)^2, (\beta - \gamma)^2, (\gamma - \alpha)^2$ must be positive. If β, γ are supposed to be imaginary and if $\beta = a + ib$, then $\gamma = a - ib$ where a, b are real.

$$\therefore \qquad (\alpha - \beta)^2(\beta - \gamma)^2(\gamma - \alpha)^2 = (\alpha - a - ib)^2(2ib)^2(a - ib - \alpha) = -4b^2[(\alpha - a)^2 + b^2] < 0.$$

Thus $(\alpha - \beta)^2(\beta - \gamma)^2(\gamma - \alpha)^2$ is negative.

Therefore we can say that if $(\alpha - \beta)^2(\beta - \gamma)^2(\gamma - \alpha)^2$ is positive then the roots of the given equation will be real and if $(\alpha - \beta)^2(\beta - \gamma)^2(\gamma - \alpha)^2$ is negative, then two roots of the given equation will be imaginary. But we have

$$(\alpha - \beta)^2(\beta - \gamma)^2(\gamma - \alpha)^2 = -\frac{27}{a_0^6}(G^2 + 4H^3)$$

So that we can discuss the nature of the roots of the given equation as follows :

1. If $G^2 + 4H^3 > 0$ then two roots of the given equation will be imaginary.
2. If $G^2 + 4H^3 < 0$ then all the roots of the given equation will be real.
3. If $G^2 + 4H^3 = 0$ then two roots of the given equation will be equal.
4. If $G = 0, H = 0$ then all the three roots of the given equation will be equal.

On the other hand we can discuss this case as follows :

Since we know that $G^2 + 4H^3 = a_0^2(HI - a_0J)$

$\therefore$ If $G = 0, H = 0$, then $a_0^2(HI - a_0J) = 0$ or $a_0^2\Delta = 0$ or $\Delta = 0$

where Δ is the discriminant of the given equation. Hence, we can say that if the discriminant of the cubic is zero, then all the root of the equation will be equal.

Solved Examples

EXAMPLE 1. *Change the signs of the roots of the equation*

$$x^7 + 5x^5 - x^3 + x^2 + 7x + 3 = 0.$$

SOLUTION. First making the equation complete by adding missing terms with zero coefficients, we get

$$f(x) \equiv x^7 + 0.x^6 + 5x^5 + 0.x^4 - x^3 + x^2 + 7x + 3 = 0 \qquad \text{...(1)}$$

Put $x = -y$ in (1), we get

$$(-y)^7 + 0.(-y)^6 + 5(-y)^5 + 0.(-y)^4$$
$$- (-y)^3 + (-y)^2 + 7(-y) + 3 = 0$$

or $-y^7 + 0.y^6 - 5y^5 + 0.y^4 + y^3 + y^2 - 7y + 3 = 0$

$\Rightarrow \qquad y^7 + 5y^5 - y^3 - y^2 + 7y - 3 = 0$

This is the required equation whose roots are same to the roots of the given equation with contrary signs.

EXAMPLE 2. *Transform the equation*

$$72x^3 - 54x^2 + 45x - 7 = 0$$

into another equation with integral coefficients and having the leading coefficient unity.

SOLUTION. The given equation can be written as

$$x^3 - \frac{54}{72}x^2 + \frac{45}{72}x - \frac{7}{72} = 0$$

or $\qquad x^3 - \dfrac{3}{4}x^2 + \dfrac{5}{8}x - \dfrac{7}{72} = 0 \qquad \ldots(1)$

Put $y = xm$ or $x = \dfrac{y}{m}$ in (1), we get

$$\left(\dfrac{y}{m}\right)^3 - \dfrac{3}{4}\left(\dfrac{y}{m}\right)^2 + \dfrac{5}{8}\left(\dfrac{y}{m}\right) - \dfrac{7}{72} = 0$$

or $\qquad y^3 - \dfrac{3}{4}my^2 + \dfrac{5}{8}m^2 y - \dfrac{7}{72}m^3 = 0 \qquad \ldots(2)$

Now to remove fractional coefficients let us put $m = 12$ in (2), we get

$$y^3 - \dfrac{3}{4}(12)y^2 + \dfrac{5}{8}(12)^2 y - \dfrac{7}{72}(12)^3 = 0$$

or $\qquad y^3 - 9y^2 + 90y - 168 = 0$

This is the required equation.

EXAMPLE 3. *Form the equation whose roots are the reciprocals of the roots of the equation*

$$x^4 - 3x^3 + 7x^2 + 5x - 2 = 0$$

SOLUTION. The given equation is

$$x^4 - 3x^3 + 7x^2 + 5x - 2 = 0 \quad \ldots(1)$$

Putting $x = \dfrac{1}{y}$ in (1), we get

$$\left(\dfrac{1}{y}\right)^4 - 3\left(\dfrac{1}{y}\right)^3 + 7\left(\dfrac{1}{y}\right)^2 + 5\left(\dfrac{1}{y}\right) - 2 = 0$$

or $\qquad 1 - 3y + 7y^2 + 5y^3 - 2y^4 = 0$

or $\qquad 2y^4 - 5y^3 - 7y^2 + 3y - 1 = 0$

This is the required equation whose roots are the reciprocals of the roots of (1).

EXAMPLE 4. *Remove the fractional coefficients from the equation* $2x^3 - \dfrac{3}{2}x^2 - \dfrac{1}{8}x + \dfrac{3}{16} = 0$.

SOLUTION. The given equation is

$$2x^3 - \dfrac{3}{2}x^2 - \dfrac{1}{8}x + \dfrac{3}{16} = 0 \quad \ldots(1)$$

Putting $x = \dfrac{y}{m}$ in (1) we get

$$2\left(\dfrac{y}{m}\right)^3 - \dfrac{3}{2}\left(\dfrac{y}{m}\right)^2 - \dfrac{1}{8}\left(\dfrac{y}{m}\right) + \dfrac{3}{16} = 0$$

or $\qquad 2y^3 - \dfrac{3}{2}my^2 - \dfrac{1}{8}m^2 y + \dfrac{3}{16}m^3 = 0$

Let us put $m = 4$, we get

$$2y^3 - \dfrac{3}{2}(4)y^2 - \dfrac{1}{8}(4)^2 y + \dfrac{3}{16}(4)^3 = 0$$

or $\qquad 2y^3 - 6y^2 - 2y + 12 = 0$

or $\qquad y^3 - 3y^2 - y + 6 = 0$

This is the required equation.

EXAMPLE 5. *Solve the following reciprocal equation*

$$x^4 - 10x^3 + 26x^2 - 10x + 1 = 0.$$

SOLUTION. The given equation can be written as

$$x^4 + 1 - 10(x^3 + x) + 26x^2 = 0$$

Divide by x^2, we get

$$\left(x^2 + \dfrac{1}{x^2}\right) - 10\left(x + \dfrac{1}{x}\right) + 26 = 0 \qquad \ldots(1)$$

Let us put $x + \dfrac{1}{x} = y$ and $x^2 + \dfrac{1}{x^2} = y^2 - 2$ in (1), we get

$$y^2 - 2 - 10y + 26 = 0$$

or $\qquad y^2 - 10y + 24 = 0$

or $\qquad (y - 6)(y - 4) = 0$

or $\qquad y = 4, 6$

Since $x + \dfrac{1}{x} = y$ if $y = 4$, then $x + \dfrac{1}{x} = 4$

or $\qquad x^2 - 4x + 1 = 0$

or $\qquad x = \dfrac{4 \pm \sqrt{16 - 4}}{2} = \dfrac{4 \pm 2\sqrt{3}}{2} = 2 \pm \sqrt{3}$

If $\quad y = 6$, then $x + \dfrac{1}{x} = 6$

or $\qquad x^2 - 6x + 1 = 0$

or $\qquad x = \dfrac{6 \pm \sqrt{36 - 4}}{2} = \dfrac{6 \pm 4\sqrt{2}}{2} = 3 \pm 2\sqrt{2}$

Hence, the roots of the given equation are $2 \pm \sqrt{3}, 3 \pm 2\sqrt{2}$.

EXAMPLE 6. *Find the equation whose roots are the cubes of the roots of the equation*

$$x^4 - x^3 + 2x^2 + 3x + 1 = 0$$

SOLUTION. Since the equation is

$$x^4 - x^3 + 2x^2 + 3x + 1 = 0 \qquad \ldots(1)$$

This equation can be written as

$$(1 - x^3) + x(x^3 + 3) + 2x^2 = 0$$

Let $P = (1 - x^3), Q = x(x^3 + 3), R = 2x^2$

Then we have $\quad P + Q + R = 0$ or $P + Q = -R$

Cubing of both sides, we get

$$(P + Q)^3 = -R^3$$

or $\quad P^3 + Q^3 + 3PQ(P + Q) = -R^3$

or $\quad P^3 + Q^3 + 3PQ(-R) = -R^3 \qquad [\because P + Q = -R]$

or $\quad P^3 + Q^3 + R^3 - 3PQR = 0$

Now substitute the value of P, Q and R in (3), we get

$$(1 - x^3)^3 + x^3(x^3 + 3)^3 + (2x^2)^3$$
$$- 3(1 - x^3)x(x^3 + 3)(2x^2) = 0$$
$$(1 - x^3)^3 + x^3(x^3 + 3)^3 + 8(x^3)^2$$
$$- 6(1 - x^3)(x^3 + 3)x^3 = 0$$

Let us put $x^3 = y$, we get

$$(1 - y)^3 + y(y + 3)^3 + 8y^2 - 6(1 - y)(y + 3)y = 0$$
$$1 - y^3 - 3y + 3y^2 + y(y^3 + 27 + 9y^2 + 27y)$$
$$+ 8y^2 - 6y(y + 3 - y^2 - 3y) = 0$$
$$1 - y^3 - 3y + 3y^2 + y^4 + 27y + 9y^3$$
$$+ 27y^2 + 8y^2 - 6y^2 - 18y + 6y^3 + 18y^2 = 0$$

or $\quad y^4 + 14y^3 + 50y^2 + 6y + 1 = 0$

This is the required equation whose roots are the cube of the roots of the given equation.

EXAMPLE 7. *Find the equation whose roots are the roots of the equation* $x^5 - 4x^4 + 3x^2 - 4x + 6 = 0$ *diminished by 3.*

SOLUTION. First complete the given equation as follows :

$$f(x) \equiv x^5 - 4x^4 + 0x^3 + 3x^2 - 4x + 6 = 0 \qquad \ldots(1)$$

•Suppose the required equation is

$$A_0y^5 + A_1y^4 + A_2y^3 + A_3y^2 + A_4y + A_5 = 0 \quad ...(2)$$

where $A_0, A_1, A_2, ...A_5$ are the constants which can be determined as follows :

Use synthetic division method as follows :

3	1	−4	0	3	−4	6
		3	−3	−9	−18	−66
3	1	−1	−3	−6	−22	−60 = A_5
		3	6	9	9	
3	1	2	3	3	−13 = A_4	
		3	15	54		
3	1	5	18	57 = A_3		
		3	24			
3	1	8	42 = A_2			
		3				
	1	11 = A_1				
	1 = A_0					

$$\therefore \ A_0 = 1, A_1 = 11, A_2 = 42, A_3 = 57, A_4 = -13,$$
$$A_5 = -60$$

Thus the required equation is

$$y^5 + 11y^4 + 42y^3 + 57y^2 - 13y - 60 = 0.$$

EXAMPLE 8. *If* $\alpha,\ \beta,\ \gamma$ *are the roots of the cubic* $x^3 - px^2 + qx - r = 0$, *form the equation whose roots are* $\beta\gamma + \dfrac{1}{\alpha}, \gamma\alpha + \dfrac{1}{\beta}, \alpha\beta + \dfrac{1}{\gamma}$. (SVTU–2008)

SOLUTION. Since the given equation is

$$x^3 - px^2 + qx - r = 0 \quad ...(1)$$

and α, β, γ are its roots, then

$$\alpha + \beta + \gamma = p, \ \alpha\beta + \beta\gamma + \alpha\gamma = q, \alpha\beta\gamma = r$$

Let y be a root of the required equation. Then

$$y = \beta\gamma + \frac{1}{\alpha} = \frac{\alpha\beta\gamma + 1}{\alpha}$$

$$y = \frac{r+1}{\alpha} \quad \Rightarrow \quad y = \frac{r+1}{x} \qquad [\because x = \alpha]$$

$$\therefore \qquad x = \frac{r+1}{y}$$

Substitute this value of x in (1), we get

$$\left(\frac{r+1}{y}\right)^3 - p\left(\frac{r+1}{y}\right)^2 + q\left(\frac{r+1}{y}\right) - r = 0$$

or $\quad \dfrac{(r+1)^3}{y^3} - \dfrac{p(r+1)^2}{y^2} + \dfrac{q(r+1)}{y} - r = 0$

or $\ (r+1)^3 - p(r+1)^2 y + q(r+1)y^2 - ry^3 = 0$

or $\ ry^3 - q(r+1)y^2 + p(r+1)^2 y - (r+1)^3 = 0$

This is the required equation.

Exercise-1.15

1. Change the sign of the roots of the equation $x^5 - 4x^3 + 3x^2 + 8x - 9 = 0$.

2. Transform the equation $x^3 - 4x^2 + \dfrac{1}{4}x - \dfrac{1}{9} = 0$ into another equation with integral coefficients and having leading coefficient unity.

3. Transform the equation $3x^4 - 5x^3 + x^2 - x + 1 = 0$ into another equation with integral coefficients having leading coefficient unity.

4. Find the equation whose roots are twice the reciprocals of the roots of $x^4 + 3x^3 - 6x^2 + 2x - 4 = 0$

5. Remove the fractional coefficients from the equation
$$x^3 - \frac{5}{2}x^2 - \frac{7}{18}x + \frac{1}{108} = 0$$

6. Remove the fractional coefficients from the equation
$$x^4 - \frac{5}{6}x^3 - \frac{13}{12}x^2 + \frac{1}{300} = 0$$

7. Solve the following reciprocal equations:
   (i) $6x^6 - 25x^5 + 31x^4 - 31x^2 + 25x - 6 = 0$ (MADRAS–2003)
   (ii) $x^5 - 5x^4 + 9x^3 - 9x^2 + 5x - 1 = 0$
   (iii) $6x^5 - 41x^4 + 97x^3 - 97x^2 + 41x - 6 = 0$
   (COIMBATORE–2003, 09)

8. Reduce the equation $4x^4 - 85x^3 + 357x^2 - 340x + 64 = 0$ into a reciprocal equation.

9. Find the equation whose roots are the squares of the roots of the equation $x^4 + x^3 + 2x^2 + x + 1 = 0$

10. Find the equation whose roots are the cubes of the roots of the following equations :
    (i) $x^3 + ax^2 + bx + ab = 0$ (ii) $x^3 + 3x^2 + 2 = 0$

11. Remove the second term form the following equations :
    (i) $x^3 - 6x^2 + 10x - 3 = 0$ (ii) $x^4 + 8x^3 + x - 5 = 0$
    (iii) $x^5 + 5x^4 + 3x^3 + x^2 + x - 1 = 0$
    (iv) $x^4 + 20x^3 + 143x^2 + 430x + 462 = 0$
    (v) $x^6 - 12x^5 + 3x^2 - 17x + 300 = 0$

12. Find the equation each of whose roots is greater than unity then a root of the equation $x^3 - 5x^2 + 2x - 3 = 0$.

13. Transform the equation $x^3 - \dfrac{x}{4} - \dfrac{3}{4} = 0$ into an equation whose roots increased by $\dfrac{3}{2}$ the corresponding roots of the given equation.

14. Find the equation whose roots are the roots of $3x^3 - 2x^2 + x - 9 = 0$ each diminished by 5.

15. If α, β, γ are the roots of the equation $x^3 + xq + r = 0$ form the equation whose roots are :
    (i) $\alpha(\beta + \gamma), \beta(\gamma + \alpha), \gamma(\alpha + \beta)$ (ii) $\left(\dfrac{\beta}{\gamma} + \dfrac{\gamma}{\beta}\right), \left(\dfrac{\gamma}{\alpha} + \dfrac{\alpha}{\gamma}\right), \left(\dfrac{\alpha}{\beta} + \dfrac{\beta}{\alpha}\right)$
    (iii) $\left(\alpha - \dfrac{1}{2}\right), \left(\beta - \dfrac{1}{2}\right), \left(\gamma - \dfrac{1}{2}\right)$

16. If α, β, γ are the roots of the equation $x^3 + px^2 + qx + r = 0$ form the equation whose roots are $\alpha - \dfrac{1}{\beta\gamma}, \beta - \dfrac{1}{\gamma\alpha}, \gamma - \dfrac{1}{\alpha\beta}$.

17. Show that the same transformation removes both second and fourth terms of the equation $x^4 + 16x^3 + 83x^2 + 152x + 84 = 0$.

18. Find the condition that the second and third terms of the equation $a_0x^3 + 3a_1x^2 + 3a_2x + a_3 = 0$ are removed by the same transformation.

19. If α, β, γ are the roots of the equation $x^3 - 6x^2 + 11x - 6 = 0$ form the equation whose roots are $\beta^2 + \gamma^2, \gamma^2 + \alpha^2, \alpha^2 + \beta^2$.

20. If α, β, γ are the roots of the equation $2x^3 + x^2 + x + 1 = 0$ form the equation whose roots are

$$\frac{1}{\beta^2} + \frac{1}{\gamma^2} - \frac{1}{\alpha^2}; \frac{1}{\gamma^2} + \frac{1}{\alpha^2} - \frac{1}{\beta^2}; \frac{1}{\alpha^2} + \frac{1}{\beta^2} - \frac{1}{\gamma^2}.$$

21. If α, β, γ are the roots of the equation $x^3 + px^2 + qx + r = 0$ form the equation whose roots are :

(i) $\alpha + \dfrac{1}{\beta\gamma}, \beta + \dfrac{1}{\gamma\alpha}, \gamma + \dfrac{1}{\alpha\beta}$ (ii) $\dfrac{\alpha}{\beta+\gamma}, \dfrac{\beta}{\gamma+\alpha}, \dfrac{\gamma}{\alpha+\beta}$

22. If α, β, γ are the roots of the equation $x^3 + px^2 + qx + r = 0$ form the equation whose roots are $\alpha^2 + 2\beta\gamma, \beta^2 + 2\alpha\gamma, \gamma^2 + 2\alpha\beta$.

23. If α, β, γ are the roots of the equation $x^3 - px^2 + qx - r = 0$

form the equation whose roots are $\beta + \gamma - \alpha, \gamma + \alpha - \beta, \alpha + \beta - \gamma$

Also find the value of $(\beta + \gamma - \alpha)(\gamma + \alpha - \beta)(\alpha + \beta - \gamma)$.

24. If the roots of $x^3 + 3px^2 + 2qx + r = 0$ are in harmonic progression, show that $2q^3 = r(3pq - r)$.

25. Transform the equation $x^3 - 6x^2 + 5x + 8 = 0$ into another in which the second term is missing, find the equation of squared difference. (COCHIN–2005)

26. Form the equation, whose roots are the reciprocal of the roots of $2x^5 + 4x^3 - 13x^2 + 7x - 6 = 0$. (SVTU–2009)

27. Solve the equation
(i) $4x^4 - 20x^3 + 33x^2 - 20x + 4 = 0$ (MADRAS–2003, 13)
(ii) $6x^5 + x^4 - 43x^3 - 43x^2 + x + 6 = 0$ (SVTU–2006)

Hints to Selected Problems

1. Put $x = -y$ in the given equation.

2. Put $y = mx$, i.e. $x = \dfrac{y}{m}$ in the given equation

7. (i) The given equation can be written as
$6(x^6 - 1) - 25x(x^4 - 1) + 31x^2(x^2 - 1) = 0$

9. The given equation can be written as
$(x^4 + 2x^2 + 1) = -x(x^2 + 1)$
On squaring both sides, we get
$x^8 + 3x^6 + 4x^4 + 3x^2 + 1 = 0$

Now put $x^2 = y$.

17. Suppose the roots of the given equation, are diminished by h, put $y = x - h$ or $x = y + h$ in the given equation.

21. If the roots of the given equation are α, β, γ. Then
$$\alpha + \beta + \gamma = -p$$
$$\alpha\beta + \beta\gamma + \gamma\alpha = q$$
$$\alpha\beta\gamma = -r$$
If y be the root of the required equation, the $y = \alpha + \dfrac{1}{\beta\gamma}$.

Answers

1. $y^5 - 4y^3 - 3y^2 + 8y + 9 = 0$ 2. $y^3 - 24y^2 + 9y - 24 = 0$ 3. $y^4 - 5y^3 + 3y^2 - 9y + 27 = 0$ 4. $y^4 - y^3 + 6y^2 - 6y - 4 = 0$

5. $y^3 - 15y^2 - 14y^2 + 2 = 0$ 6. $y^4 - 25y^3 - 975y^2 + 2700 = 0$ 7.(i) $\pm 1, 2, \dfrac{1}{2}, \dfrac{5 \pm i\sqrt{11}}{6}$ (ii) $1, \dfrac{1}{2}(1 \pm i\sqrt{3}), \dfrac{1}{2}(3 \pm \sqrt{5})$ (iii) $1, \dfrac{1}{3}, 3, \dfrac{1}{2}, 2$

8. $16y^4 - 170y^3 + 357y^2 - 170y + 16 = 0$ 9. $y^4 + 3y^3 + 4y^2 + 3y + 1 = 0$

10. (i) $y^3 + a^3y^2 + b^3y + a^3b^3 = 0$ (ii) $y^3 + 33y^2 + 12y + 8 = 0$ 11.(i) $y^3 - 2y + 1 = 0$ (ii) $y^4 - 24y^2 + 65y - 55 = 0$

(iii) $y^5 - 7y^3 + 12y^2 - 7y + 2 = 0$ (iv) $y^4 - 7y^2 + 12 = 0$ (v) $y^6 - 60y^4 - 320y^3 - 717y^2 - 773y - 42 = 0$

12. $y^3 - 8y^2 + 19y - 15 = 0$ 13. $y^3 - \dfrac{9}{2}y^2 + \dfrac{13}{2}y - \dfrac{15}{4} = 0$ 14. $3y^3 + 43y^2 + 206y + 321 = 0$

15. (i) $y^3 - 2qy^2 + q^2y + r^2 = 0$ (ii) $r^3y^3 + 3r^2y^2 + (3r^2 + q^3)y + (r^2 + 2q^3) = 0$ (iii) $8y^3 + 12y^2 + (6 + 8q)y + (8r + 4q + 1) = 0$

16. $r^3y^3 + pr(1 + r)y^2 + q(1 + r)^2y + (1 + r)^3 = 0$ 18. $a_0a_2 - a_1^2 = 0$ 19. $y^3 - 28y^2 + 245y - 650 = 0$ 20. $z^3 + z^2 - 13z + 19 = 0$

21. (i) $r^2y^3 + pr(1 - r)y^2 + q(1 - r)^2y + (1 - r)^3 = 0$ (ii) $(pq - r)y^3 + (2pq - p^3 - 3r)y^2 + (pq - 3r)y - r = 0$

22. $y^3 - p^2y^2 + q(2p^2 - 3q)y - (4p^3r - 18pqr + 2q^3 + 27r^2) = 0$ 23. $y^3 - py^2 + (4q - p^2)y + (8r - 4pq + p^3) = 0$; $4pq - p^3 - 8r$

25. $y^3 - 28y^2 + 245y - 682 = 0$ 26. $6x^5 - 7x^4 - 13x^3 + 4x^2 - 2 = 0$ 27. (i) $2, 2, \dfrac{1}{2}, \dfrac{1}{2}$ (ii) $-1, -2, 3, -\dfrac{1}{2}, \dfrac{1}{3}$

1.26 DESCARTE'S RULE OF SIGNS

We know that an equation $f(x) = 0$ cannot have more positive roots than the number of changes of signs form positive to negative or from negative to positive terms of its first number and an equation $f(x) = 0$ cannot have more negative roots than the number of changes of sign $f(-x) = 0$.

We shall simply verify the above statement.

Let the signs of a polynomial be $+ + - + - + - -$.

The given polynomial has five changes of sings. Now we shall multiply the given polynomial by binomial $x - h$ corresponding to the positive root h. The sings of this binomial are $+ -$. We are concerned only with the signs and hence we multiply as below :

$$+ \ + \ - \ + \ - \ - \qquad \text{5 changes of sign}$$
$$\underline{+ \ -}$$
$$\overline{+ \ + \ - \ + \ - \ + \ - \ -}$$
$$- \ - \ + \ - \ - \ + \ +$$
$$\overline{+ \ \pm \ - \ + \ - \ + \ - \ \pm \ +}$$

The resulting polynomial has two ambiguous sings and we can write in four different ways as follows :

$$+ \ + \ - \ + \ - \ + \ - \ + \ + \qquad \text{6 changes of sign}$$
$$+ \ + \ - \ + \ - \ + \ - \ - \ + \qquad \text{6 changes of sign}$$
$$+ \ - \ - \ + \ - \ + \ - \ - \ + \qquad \text{6 changes of sign}$$
$$+ \ - \ - \ + \ - \ + \ - \ + \ + \qquad \text{6 changes of sign}$$

Thus we see that in all the four possible ways the resulting polynomial has six changes of signs, *i.e.*, one more than the number of changes of signs in the original polynomial.

Hence, we conclude that corresponding to the introduction of a positive root the resulting polynomial has one more change of sign. Now if $\phi(x)$ be the product of factors corresponding to negative and complex roots and α, β, γ... be the positive roots, then if $\phi(x)$ be multiplied by $(x - \alpha), (x - \beta), (x - \gamma)$... in succession, then each multiplication will introduce one more change of sign. Hence, the number of positive roots cannot exceed the number of changes of signs in $f(x) = 0$.

NEGATIVE ROOTS

We know that negative roots of $f(x) = 0$ are positive roots of $f(-x) = 0$ and as such the number of negative roots of $f(x) = 0$ cannot exceed the number of changes of signs of $f(-x) = 0$.

COMPLEX ROOTS

If $f(x) = 0$ be an equation of nth degree and if it be complete, then the number of changes of signs in $f(x)$, *i.e.*, positive roots and numbers of sign in $f(-x)$, *i.e.*, –ve roots is equal to n, the degree of the equation and as such we cannot draw any definite conclusion regarding the existence of imaginary roots. In case the equation be incomplete, then the number of changes of signs in $f(x)$, *i.e.*, positive roots and the number of changes of signs in $f(-x)$, *i.e.*, negative roots is less than the degree n of the equation. If a and b be the number of changes of signs in $f(x)$ and $f(-x)$ respectively, *i.e.*, greatest number of positive roots is a and that of negative roots is b, then $n - (a + b)$ is the least number of imaginary roots. For example, consider the equation

$$f(x) = x^7 - 3x^5 + 4x^4 + 2x^3 - 11 = 0 .$$

The above equation has three changes of sign and as such if cannot have more than three positive roots. Again $f(-x) = 0$ *i.e.*, $-x^7 + 3x^5 + 4x^4 - 2x^3 - 11 = 0$ has only two changes of sign and as such $f(x) = 0$ cannot gave more than two negative roots. Thus the max. number of real roots is $3 + 2$, *i.e.*, 5 and the degree of the equation being 7 we conclude that the equation must have at least two imaginary roots.

1.26.1 CHANGE OF SIGN

Let two real numbers a and b be substituted for x in the polynomial $f(x)$, $f(a)$ and $f(b)$ are found to be of opposite signs, then at least one or an odd number of real roots of the equation $f(x) = 0$ lie between a and b. In case $f(a)$ and $f(b)$ of the same sign, then either no real root or an even number of roots of $f(x) = 0$ lie between a and b.

Case I. $f(a)$ and $f(b)$ of opposite signs.

Let $y = f(x)$ be a continuous function of x, it should assume all values between $f(a)$ and $f(b)$

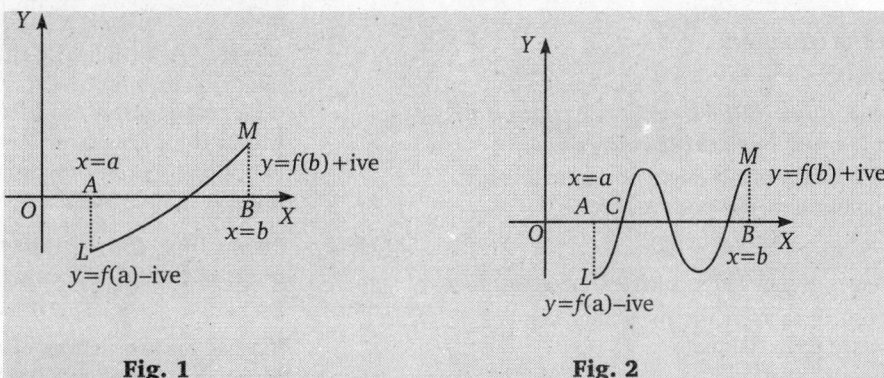

Fig. 1   **Fig. 2**

Now $f(a)$, $f(b)$ are of oppositsoe signs, *i.e.*, values of y corresponding to the values of x, *i.e.*, a and b are of opposite signs. From one side of x-axis to the other side of x-axis the curve $y = f(x)$ must cross the axis of x at least once as in fig.1 at C or an odd number of times as in fig. 2 at C, D and E at all such points where the curve crosses the axis of x, $y = 0$, *i.e.*, $f(x) = 0$ which means that $f(x)$ vanishes either at one or an odd number of times for values of x between a and b. Hence, at least one or an odd number of roots of $f(x) = 0$ lie between a and b.

Case II. **If** *f(a)* **and** *f(b)* **of same sign.**

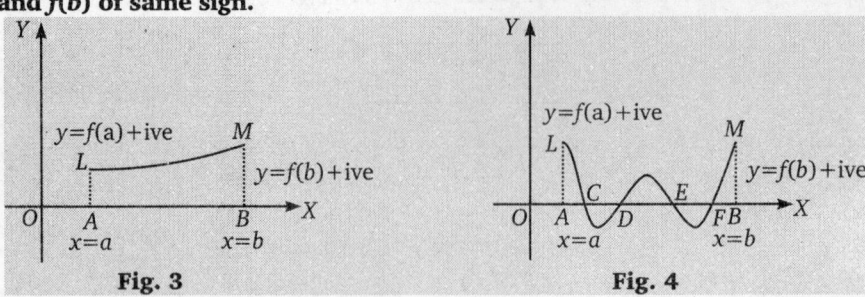

Fig. 3   **Fig. 4**

Here, $f(a)$ and $f(b)$ are of the same sign, the values of y corresponding to the values of x, *i.e.*, a and b are of the same sign which means that in passing from a point on one side of x-axis to the other point on the same side either the curve $y = f(x)$ will not cross the x-axis as in Fig.3 where $y = 0$, *i.e.*, $f(x) = 0$ or, it must cross an even number of times of Fig. 4. Hence we conclude that either $f(x)$ does not vanish for values of x between a and b. or, if it vanishes, it must vanish for even number of times.

Some Results

1. Every equation of an odd degree has at least one real root whose sign is opposite to that of its last term, the coefficients of the first term being positive.
2. Every equation of an even degree whose last terms is negative and the coefficient of the first term positive, has at least two real roots, one positive and other negative.
3. If an equation has only one change of sign, it must have one positive root and no more.
4. If all terms of an equation are positive and the equation involves no odd power of x, then all its root are complex.

📑 Remark

- If all the terms of an equation are positive and all involve odd powers of x, then 0 is the only real root.

🗂 Solved Examples

EXAMPLE 1. *Apply Descarte's rule of signs to discuss the nature of the roots of the equation*
$$x^4 + 15x^2 + 7x - 11 = 0.$$

SOLUTION. Let $f(x) = x^4 + 15x^2 + 7x - 11 = 0$. It has only one change of sign and hence it must have one positive root.

Now, $f(x) = x^4 + 15x^2 + 7x - 11 = 0$.

As above it must have one positive root, *i.e*, $f(x) = 0$ must have one negative root. Thus the equation has two real roots, one positive and one negative and hence, the other two roots must be imaginary.

EXAMPLE 2. *Show that the equation*
$$f(x) = x^5 + x^3 - 8x - 5 = 0$$

cannot have more than three real roots and prove that it must have three real roots.

SOLUTION. Let $f(x) = 0$ has only one changes of sign and hence, the equation must have one positive root.

Now, $f(-x) = -x^5 - x^3 + 8x - 5 = 0$

has two changes of sign and as such it can have at the most two positive roots or the maximum number of negative roots of $f(x) = 0$ is two.

Again

$$f(0) = \text{negative}, f(-1) = \text{positive}, f(-\infty) = \text{negative}$$

Since $f(0)$ and $f(-\infty)$ are of the same sign, so there lies either none or an even number of roots between 0 and $-\infty$ and now $f(0)$ and $f(-1)$ are of opposite signs and also $f(-1)$ and $f(-\infty)$ too are of opposite signs and

hence one negative root lies between 0 and -1 and the other between -1 and $-\infty$. Thus we conclude that the equation must have three real roots and hence two complex.

EXAMPLE 3. *Find the minimum number of imaginary roots which equation $f(x) = 2x^7 - x^4 + 4x^3 - 5 = 0$ must possess.*

SOLUTION. Here $f(x) = 0$ has three changes of signs and as such it can at the most have three positive roots.

Now, $f(-x) = -2x^7 - x^4 - 4x^3 - 5 = 0$

or $\qquad 2x^7 + x^4 + 4x^3 + 5 = 0$

Clearly, $f(-x) = 0$ has no changes of signs and as such it will have no positive root which means that $f(x) = 0$ has no negative roots. Hence, the given equation can at the most have three real, *i.e.*, positive roots and it being of 7th degree and hence, the minimum number of imaginary roots is $7 - 3$, *i.e*, 4.

EXAMPLE 4. *Prove that the equation $x^5 - x + 16 = 0$ has two pairs of complex roots.*

SOLUTION. Let $f(x) = x^5 - x + 16 = 0$

$f(x)$ has got two changes of signs and as such it cannot have more than two positive roots. Again $f(0) = $ positive and $f(\infty) = $ positive, since both $f(0)$ and $f(\infty)$ are of the same sign. Also we observe that for all values of x between 0 and ∞, $f(x)$ remain positive always, *i.e.*, the graph of the curve $y = f(x)$ never crosses the x-axis which in other words means that y or $f(x)$ never becomes negative or y or $f(x)$ is always positive for all values of x. Hence, the equation $f(x) = 0$ has no positive roots.

If $f(a)$ and $f(b)$ are of the same sign, then either no root or in general even number of roots of $f(x) = 0$ lie between a and b. In the later case the curve crosses the x-axis even times; but here we have shown $f(x)$ always remains positive and hence, we have established that the given equation has no positive root even though $f(x) = 0$ has two changes of signs.

Again $\qquad f(-x) = -x^5 + x + 16 = 0$

or $\qquad x^5 - x - 16 = 0$

$f(-x) = 0$ has only one change of sign and as such it must have one positive root or $f(x) = 0$ must have one negative root.

Thus in all the given equation has only one real root which is negative and it being of fifth degree we conclude that the remaining four roots must be imaginary. Again since imaginary roots occur in conjugate pairs we say that the

given equation has two pairs of complex roots.

EXAMPLE 5. *Show that the equation $f(x) = x^3 - qx + r = 0$ where q and r are essentially positive has one negative root and that the other two roots are either imaginary or both positive.*

SOLUTION. Clearly $f(x) = 0$ has got only two changes of signs and hence the number of positive roots cannot exceed two

Now, $\qquad f(-x) = -x^3 + qx + r = 0$

or $\qquad x^3 - qx - r = 0$

and it has only one change of sign; hence $f(x) = 0$ must have one negative root. Otherwise the equation must have one real root whose sign is opposite to that of the last terms and hence it should be negative.

Hence, we conclude that one root of $f(x) = 0$ is essentially negative and therefore the remaining two are either both positive or imaginary.

 **Exercise-1.16**

1. Show that the equation $x^7 - 3x^4 + 12x^2 + 5x - 4 = 0$ has at least two imaginary roots.

2. Show that the equation $2x^7 + 3x^4 + 3x + k = 0$ has at least four imaginary roots for all values of k (constant).

3. Prove if $q > r > 0$ the cubic $x^3 + 9x + r = 0$ has one negative and two imaginary roots.

4. Prove that the equation $x^9 - x^5 + x^4 + x + 1$ has one real root

which is negative and eight imaginary roots.

5. Find an equation whose roots are the squares of the roots $x^3 - x^2 + 8x - 6 = 0$ and hence deduce that the equation must have a pair of imaginary roots.

6. Find the equation whose roots are the squares of the roots of the equation $x^3 + 4x^2 + 9x + 10 = 0$ and hence, find the nature of the roots of the given equation.

Answers

5. $y^3 + 15y^2 + 52y - 36 = 0$ **6.** One real, two complex

1.26.2 MULTIPLE ROOTS

If an equation $f(x) = 0$ has exactly m roots equal to α, then $f(x)$ and its first $(m - 1)$ derivatives all vanish for $x = \alpha$ but the mth and all the following derivatives do not vanish and if $f(x)$ and its first $(m - 1)$ derivatives vanish for $x = \alpha$, then $f(x) = 0$ has m roots equal to α.

PROOF. Let $f(x) = (x - \alpha)^m \phi(x)$ $\qquad\qquad ...(1)$

where $\phi(x)$ does not vanish for $x = \alpha$, for if it vanishes, then it will contain the factor $(x - \alpha)$ and then $f(x)$ will have more than m equal roots. Differentiating both sides of (1), we get

$$f'(x) = m(x - \alpha)^{m-1}\phi(x) + (x - \alpha)^m \phi'(x) = (x - \alpha)^{m-1}[m\phi(x) + (x - \alpha)\phi'(x)]$$

$$= (x - \alpha)^{m-1}\psi_1(x) \qquad\qquad ...(2)$$

where $\psi_1(x)$ does not vanish for $x = \alpha$, because $\psi_1(\alpha) = m\phi(\alpha)$ and it can be zero only if $\phi(\alpha) = 0$ which is contrary to the supposition. From (2), we observe that a root which occurs exactly m times in $f(x) = 0$ occurs exactly $(m-1)$ times in $f'(x) = 0$. Hence we have the following :

A multiple root of order m of the equation $f(x) = 0$ is a multiple root of order $(m - 1)$ of the first derived equation $f'(x) = 0$ and hence $f(x)$ and $f'(x)$ have the common factor $(x - \alpha)^{m-1}$.

Again if we differentiate (2), we get $f''(x) = (x - \alpha)^{m-2}[(m - 1)\psi_1(x) + (x - \alpha)\psi_1'(x)] = (x - \alpha)^{m-2}\psi_2(x)$

where $\psi_2(x)$ does not vanish for $x = \alpha$, i.e., $f''(x) = 0$ has exactly $(m - 2)$ equal roots α. Similarly we can show $f'''(x)$ will have

the factor $(x - \alpha)^{m-3}$ and so on till $f \quad (x)$ will have the following :

Any root which occurs m times in $f(x) = 0$ occurs in degree of multiplicity diminishing by it in the first derived function.

Conversely Let $f(x) = f\{\alpha + (x - \alpha)\}$

Expanding the R.H.S by Taylor's Theorem, we get

$$f(x) = f(\alpha) + (x - \alpha)f'(\alpha) + \frac{(x - \alpha)^2}{2!}f''(a) + ... + \frac{(x - \alpha)^{m-1}}{(m-1)!}f^{m-1}(\alpha) + \frac{(x - \alpha)^m}{m!}f^m(\alpha) + ... + \frac{(x - \alpha)^n}{n!}f^n(\alpha)$$

Now $\qquad f(\alpha) = f'(\alpha) = f''(\alpha) = \ldots = f^{m-1}(\alpha) = 0$ (given)

$$\therefore \qquad f(x) = \frac{(x-\alpha)^m}{m!} f^m(\alpha) + \ldots + \frac{(x-\alpha)^n}{n!} f^m(\alpha)$$

Above shows that $(x - \alpha)^m$ is a factor of $f(x) = 0$ which therefore has m roots equal to α.

1.26.3 DETERMINATION OF MULTIPLE ROOTS

We have discussed above that multiple root of the order m is multiple root of order $(m - 1)$ of the equation $f'(x) = 0$. In order to find such roots we should find the H.C.F. of $f(x)$ and $f'(x)$. This H.C.F. will give the multiple root of $f(x)$ each repeated $(m - 1)$ times. Thus if $(x - 2)$ is the H.C.F. of $f(x)$ and $f'(x)$, then $f(x)$ contains $(x - 2)^2$, as a factor or 2 is a double root of $f(x) = 0$. If $(x - 2)^2(x - 1)^5$ is the H.C.F. of $f(x)$ and $f'(x)$, then $f(x) = 0$ has three roots equal to 2 and six roots equal to 1.

📝 **REMARK**
* In case $f(x)$ and $f'(x)$ have no common factor, then clearly $f(x)$ has no equal roots.

1.26.4 CONDITION FOR TWO OR THREE EQUAL ROOTS

If α be a double root of $f(x) = 0$, then $f(\alpha) = 0$ and $f'(\alpha) = 0$ and the required condition is obtained by eliminating α between $f(\alpha) = 0$ and $f'(\alpha) = 0$. Similarly if α is a triple root the required condition can be obtained by eliminating α between $f(\alpha) = 0$, $f'(\alpha) = 0$ and $f''(\alpha) = 0$.

 Solved Examples

EXAMPLE 1. *Solve the equation* $x^5 - 15x^3 + 10x^2 + 60x - 72 = 0$ *by testing for equal roots.*

SOLUTION. Let $f(x) = x^5 - 15x^3 + 10x^2 + 60x - 72 = 0$

$$f'(x) = x^4 - 9x^2 + 4x + 12 = 0$$

Let us find the H.C.F. of $f(x) = 0$ and $f'(x) = 0$

x	$x^5 - 15x^3 + 10x^2 + 60x - 72$	$x^4 - 9x^2 + 4x + 12$	x
	$x^5 - 9x^3 + 4x^2 + 12x$	$x^4 - x^3 - 8x^2 + 12x$	
	$-6x^3 + 6x^2 + 48x - 72$	$x^3 - x^2 - 8x + 12$	1
or	$x^3 - x^2 - 8x + 12$	$x^3 - x^2 - 8x + 12$	

Thus we find that H.C.F of $f(x) = 0$ and $f'(x) = 0$ is

$$x^3 - x^2 - 8x + 12 = 0$$

or $\quad (x - 2)(x^2 + x - 6) = 0$

or $\qquad (x - 2)^2(x + 3) = 0$

Giving $n = 2, 2, -3$. Now we know that H.C.F. gives a multiple root of $f(x) = 0$, $(m - 1)$ times. Hence, 2 is triple root and -3 is a double root of $f(x) = 0$.

EXAMPLE 2. *Solve the equation* $x^4 - 6x^3 + 12x^2 - 10x + 3 = 0$ *which has equal roots.*

SOLUTION. The H.C.F is $(x - 1)^2 = 0$ giving 1 as a triple root of $f(x) = 0$

Also $\alpha\beta\gamma\delta = 3$ or $1.1.1.\delta = 3$; $\therefore \delta = 3$

$\therefore$ Roots are 1, 1, 1, 3.

EXAMPLE 3. *Factorize the following :*

$$2x^5 - x^4 + 5x^2 - 4x + 3.$$

SOLUTION. The H.C.F. of $f(x)$ and $f'(x)$ is found to be $x^2 - x + 1$. Hence $(x^2 - x + 1)^2$ is a factor of $f(x)$. Let the other factor which will be linear is say $x + k$.

$\therefore f(x) = 2(x^2 - x + 1)^2(x + k)$ as the coefficient of x^5 is 2.
Comparing constant term on either side, we get

$$3 = 2k; \quad \therefore k = \frac{3}{2}.$$

$$\therefore f(x) = (x^2 - x + 1)^2(2x + 3)$$

EXAMPLE 4. *Show that the equation* $x^n - qx^{n-m} + r = 0$ *has two equal roots if* $\left\{\dfrac{q}{n}(n-m)\right\}^n = \left\{\dfrac{r}{m}(n-m)\right\}^m$.

SOLUTION. Let $\quad f(x) = x^n - qx^{n-m} + r = 0 \qquad \ldots(1)$

and $f'(x) = nx^{n-1} - q(n-m)x^{n-m-1} = 0 \qquad \ldots(2)$

The required condition is obtained by eliminating x between (1) and (2).

Form (2), $n = \dfrac{q(n-m)}{x^m}$ or $x^m = \dfrac{q}{n}(n-m) \qquad \ldots(3)$

and from (1) with the help of (3), we get

$$x^n = \left[1 - \frac{n}{n-m}\right] + r = 0$$

or $\qquad x^n = \dfrac{r}{m}(n-m) \qquad \ldots(4)$

or $(x^m)^n = (x^n)^m$ etc. (3) and (4).

EXAMPLE 5. *If the equation* $x^4 - 4p^3x + 1$ *has a pair of equal roots, find the value of p and solve the equation completely.*

SOLUTION. Let $f(x) = x^4 - 4p^3x + 1$

$\Rightarrow f'(x) = 4(x^3 - p^3) = 0$ giving $x = p$.

Eliminating x between $f(x)$ and $f'(x) = 0$, we get

$$p^4 - 4p^4 + 1 = 0$$

or $\quad 3p^4 = 1 \quad$ or $\quad p = 3^{-1/4} \qquad \ldots(1)$

Now let us find the H.C.F. of $f(x)$ and $f'(x)$.

$$
\begin{array}{r}
x \\
x^3 - p^3 \overline{\smash{\big)}\, x^4 - 4p^3x + 1} \\
\underline{x^4 - p^3x} \\
-3p^3x + 1
\end{array}
$$

or $\quad -3p^4x + p \qquad$ or $\quad -x + p$

$\therefore \quad 3p^4 = 1$, by (1)

or $-x+p\overline{)x^3-p^3}$ with $\dfrac{-x^2-px-p^2}{}$ above

$$\dfrac{x^3-p^3}{\times}$$

Thus the H.C.F. is $(x-p)$ and hence two equal roots are p and q, their sum being $2p$ and product p^2. Let the other roots be α and β.
Now sum of all roots $= 0$

$\therefore \quad \alpha+\beta+p+p = 0 \quad$ or $\quad \alpha+\beta = -2p$

and $\alpha\beta pp = $ product of all the roots $= 1$

$\therefore \quad \alpha\beta = \dfrac{1}{p^2} = 3p^2$ from (1), $\because 3p^4 = 1$

$\therefore \quad \alpha$ and β are the roots of the quadratic

$$t^2 + 2pt + 3p^2 = 0$$

or $\quad t = \dfrac{-2p \pm \sqrt{(4p^2-12p^2)}}{2} = p \pm ip\sqrt{2}$

Hence, the four roots are $p, p, -p \pm ip\sqrt{2}$
where $p = 3^{-(1/4)}$.

EXAMPLE 6. *Prove that if the coefficients of a given equation are all integers, an integral root is an exact divisor of the absolute term.*

SOLUTION. Let the equation be

$$a_0 x^n + a_1 x^{n-1} + a_2 x^{n-2} + ... + a_{n-1}x + a_n = 0$$

where all the a's are integers. If α be an integral root of above equation, then

$$a_0\alpha^n + a_1\alpha^{n-1} + a_2\alpha^{n-2} + ... + a_{n-1}\alpha + a_n = 0$$

or $\alpha(a_0\alpha^{n-1} + a_1\alpha^{n-2} + ... + a_{n-1}) = -a_n$

or $\dfrac{a_n}{\alpha} = -(a_0\alpha^{n-1} + a_1\alpha^{n-2} + ... + a_{n-1}) = $ an integer

Hence $\dfrac{a_n}{\alpha} = $ an integer . Therefore the integral root α divides a_n exactly.

EXAMPLE 7. *Find the values of a, for which the equation $ax^3 - 9x^2 + 12x - 5 = 0$ has two equal roots and solve the equation completely in one case.*

SOLUTION. Let $f(x) = ax^3 - 9x^2 + 12x - 5 = 0 \qquad ...(1)$

$\Rightarrow f'(x) = 3(ax^2 - 6x + 4) = 0$

or $\qquad ax^2 - 6x + 4 = 0 \qquad ...(2)$

Multiplying (2) by x and subtracting form (1), we get

$$3x^2 - 8x + 5 = 0$$

Solving (2) and (3) by method of cross-multiplication, we get

$$\dfrac{x^2}{-30+32} = \dfrac{x}{12-5a} = \dfrac{1}{18-8a}$$

Eliminating x, we get

$$(12-5a)^2 = 2(18-8a)$$

or $\quad 25a^2 - 104a + 108 = 0$

or $\quad (a-2)(25a-54) = 0 \quad \therefore \quad a = 2$ or $\dfrac{54}{25}$

Putting $a = 2$, we get

$$f(x) = 2x^3 - 9x^2 + 12x + 5 = 0$$

$\Rightarrow f'(x) = x^2 - 3x + 2 = 0$

or $\qquad (x-1)(x-2) = 0$

whose roots are 1 and 2. Either can be a double root of the given equation $f(x) = 0$. Since (2) does not divide the absolute term of the given equation, as such it cannot be its root. Hence, (1) is a double root of the given equation corresponding to the value 2 of a.

 **Exercise-1.17**

1. Factorize : $x^4 + 12x^3 + 32x^2 - 24x + 4$
2. Solve the equation :
   (i) $x^4 - 8x^3 + 24x^2 - 32x + 16 = 0$
   (ii) $x^6 - 6x^4 - 4x^3 + 9x^2 + 12x + 4 = 0$
3. Find the multiple root of :
   (i) $x^4 + 3x^3 - 7x^2 - 15x + 18 = 0$
   (ii) $x^4 + 7x^3 + 17x^2 + 17x + 6 = 0$
   (iii) $x^5 - x^4 + 2x^3 - 2x^2 + x - 1 = 0$

4. Show that the equation $x^n - nax + (n-1)b = 0$ will have a pair of equal roots of $a^n = b^{n-1}$.
5. Show that the equation
   (i) $x^n - 1 = 0$
   (ii) $1 + \dfrac{x}{1!} + \dfrac{x^2}{2!} + ... + \dfrac{x^n}{n!} = 0$
   cannot have equal roots.
6. If $x^5 + 5px^3 + 5qx^2 + r = 0$ has two equal roots prove that either of them is a root of the quadratic
   $$3qx^2 - 6p^2x - 4pq + r = 0$$

Answers

1. $(x^2 + 6x - 2)^2$
2. (i) All roots are equal to 2
   (ii) $(2, 2, -1, -1, -1, -1)$
3. (i) $-3, -3, 1, 2$
   (ii) $-1, -1, -2, -3$
   (iii) $\pm i, \pm i, 1$

1.26.5 MAXIMUM AND MINIMUM VALUES OF f(x)

We know from the definition of Maximum and Minimum that as x varies in the interval $(x-h, x+h)$ where h is small, then $f(x)$ is greater than both $f(x-h)$ and $f(x+h)$ if $f(x)$ be Maximum, i.e., $f(x-h) - f(x)$ and $f(x+h) - f(x)$ are of the same sign, i.e., negative. Similarly if $f(x)$ is Minimum, then $f(x)$ is less than both $f(x-h)$ and $f(x+h)$, i.e., $f(x-h) - f(x)$ and $f(x+h) - f(x)$ are of the same sign, i.e. positive.

Now by Taylor's Theorem

$$f(x+h) - f(x) = hf'(x) + \dfrac{h^2}{2!}f''(x) + \dfrac{h^3}{3!}f'''(x) + \dfrac{h^4}{4!}f^{iv}(x) + ...$$

$$f(x-h) - f(x) = -hf'(x) + \frac{h^2}{2!}f''(x) - \frac{h^3}{3!}f'''(x) + \frac{h^4}{4!}f^{iv}(x) - \dots$$

Now when h is made sufficiently small, the sign of the right hand side of each equation in (1) and therefore of the left hand side is ultimately dependent upon that of $hf'(x)$, that being the term of lowest degree in h. But both $f(x+h) - f(x)$ and $f(x-h) - f(x)$ are of opposite signs whereas they should have the same sign in $f(x)$ is either Max. or Min. It is therefore necessary that $f'(x)$ should vanish so that the lowest term of the right hand side of equation in (1) should depend upon an even power of h.

Hence $f'(x) = 0$ is the essential condition for the occurrence of Max. and Min. values of $f(x)$.

Let the roots of $f'(x) = 0$ be a, b, c etc.

Consider one of the roots of the above equation say b; then putting $x = b$ and $f'(x) = 0$ at $x = b$ in (1), we get

$$f(b+h) - f(b) = \frac{h^2}{2!}f''(b) + \frac{h^3}{3!}f'''(b) + \frac{h^4}{4!}f^{iv}(b) + \dots$$

$$f(b-h) - f(b) = \frac{h^2}{2!}f''(b) - \frac{h^3}{3!}f'''(b) + \frac{h^4}{4!}f^{iv}(b) + \dots$$

Now if $f''(b)$ is negative, then clearly $f(b+h)$ and $f(b-h)$ are both less than $f(b)$ so that $f(x)$ is maximum at $x = b$. In case $f''(b)$ is positive then both $f(b+h)$ and $f(b-h)$ are greater than $f(b)$, showing that $f(x)$ is minimum at $x = b$.

 Solved Examples

EXAMPLE 1. *Show that the maximum and minimum values of the biquadratic $ax^4 + 4bx^3 + 6cx^2 + 4dx + e$ are the roots of the equation*

$$a^2k^3 - 3(a^2I - 9H^2)k^2 + 3(aI^2 - 18H)k - \Delta = 0$$

where Δ is discriminant of the quadratic.

SOLUTION. We know that the discriminant Δ of the quadratic is $I^3 - 27J^2$,

where $I = ae - 4bd + 3c^2$,

and $J = ace + 2bcd - ad^2 - b^2e - c^3$.

Let $f(x) = ax^4 + 4bx^3 + 6cx^2 + 4dx + e$.

Now if the curve $y = f(x)$ be moved parallel to the axis of x through a distance k, where k is the maximum or minimum value of $f(x)$, then x-axis will become a tangent, *i.e.*, the two values of x will coincide. Now by moving the curve parallel to x-axis through a distance k, y becomes $y - k$. Then the equation $f(x) - k = 0$ will

have two roots equal, the condition for which is that its discriminant should be zero.

Now, $f(x) - k \equiv ax^4 + 4bx^3 + 6cx^2 + 4dx + (e - k) = 0$

Its discriminant is $I'^3 - 27J'^2$ where in the usual values of I and J, we have to put e equal to $e - k$ in order to get I' and J'.

$\therefore I' = a(e - k) - 4bd + 3c^2 = I - ak$

$J' = ac(e - k) + 2bcd - ad^2 - b^2(e - k) - c^3$

$\qquad = J - (ack - b^2k) = J - kH \qquad \therefore H = ac - b^2$

$\therefore I'^3 - 27J'^2 = 0$ gives

$\qquad I^3 - 3I^2ak + 3Ia^2k^3 - a^3k^3$

$\qquad - 27J^2 + 54JkH - 27k^2H^2 = 0$

Cancelling minus sign and putting $I^3 - 27J^2 = \Delta$, we get

$$a^3k^3 - 3(a^2I - 9H^2)k^2 + 3(aI^2 - 18HJ)k - \Delta = 0$$

The above equation gives the values of k.

1.26.6 LIMIT OF THE ROOTS OF AN EQUATION

1. **Definition 1.** *A number which is greater than all the positive roots of a given equation is called upper or superior limit of the positive roots of that equation.*
2. **Definition 2.** *A number which is less than all the positive roots of a given equation is called the inferior or lower limit of the positive roots of that equation.*
3. **Definition 3.** *A number which is greater than all the negative roots of the given equation is called an upper or superior limit of the negative roots of that equation. In other words, superior limit of negative roots of a given equation is that negative number below which (numerically) lie all the negative roots.*
4. **Definition 4.** *A number which is less than all the negative roots of the given equation is called the inferior or lower limit of negative roots of that equation. In other words, inferior limit of negative roots of a given equation is that negative number above which (numerically) lie all the negative roots.*

REMARK

- From the above definitions we can say that the superior limit of the real roots of a given equation is that number which is greater than all real roots of that equation and the inferior limit of the real roots in that number which is less than all the real roots of that equation. Thus superior limit of positive roots is the superior limit of the real roots and the inferior limit of the negative roots is the inferior limit of the real roots of a given equation.

SOME RESULTS

1. If the polynomial $a_0x^n + a_1x^{n-1} + a_2x^{n-2} + \dots + a_{n-1}x + a_n$ the value of $\frac{a_k}{a_0} + 1$ or any greater value be substituted for x where a_k is that one of the coefficient $a_1, a_2, \dots, a_n$ whose numerical value is greatest, irrespective of sign, the term containing the highest power of x will exceed the sum of all the terms which follows.

2. If in the polynomials $a_0x^n + a_1x^{n-1} + ... + a_{n-1}x + a_n$ the value $\dfrac{a_n}{a_n + a_k}$ or any smaller value be substituted for w where a_k is the greatest coefficient exclusive of a_n, the term a_n will numerically be greater than sum of all the others.

1.26.7 STURM'S METHOD OF FINDING THE EXACT NUMBER OF REAL ROOTS OF AN EQUATION

By Descarte's rule of signs we cannot get the exact number of real roots of a given equation $f(x) = 0$. Sturm's theorem gives us the exact number of real roots of a given equation.

1.26.8 STURM'S FUNCTIONS

Let $f(x)$ be any function of x of degree n and $f'(x)$ be its first derivative. Divide $f(x)$ by $f'(x)$ and let the remainder with sign changed be denoted by $f_2(x)$. Again divide $f'(x)$ by $f_2(x)$ and let the remainder with sign changed be denoted by $f_3(x)$.

Continue the process till you get the last remainder whose sign is also to be changed.

The above process is the same as that of finding the H.C.F. of $f(x)$ and $f'(x)$ with the modification that the sign of each remainder is to be changed before it becomes the divisor. Also we know that in process of finding the H.C.F. we can multiply and divide any remainder by any constant before using it as a divisor; but here in the above process we can only multiply or divide the remainder only by a positive constant before using it as a divisor or by a polynomial of x which is always positive real values of x say of the type $x^2 + 1$ or $x^4 + x^2 + 1$, etc.

The series of functions $f(x), f'(x), f_2(x), f_3(x), ... f_r(x)$ consisting of the given function, its derivative and remainder with their sign changed in process of finding the H.C.F. of $f(x)$ and $f'(x)$ are called Sturm's functions.

The functions $f_2(x), f_3(x), ... f_r(x)$ are called auxiliary functions.

(i) $f(x) = 0$ having equal roots.

In case $f(x) = 0$ has equal roots, then we know that $f(x)$ and $f'(x)$ will have some H.C.F. and hence the last of the Sturm's remainder will be a function of x.

(ii) $f(x) = 0$ having no equal roots.

In this case evidently the last of the Sturm's remainder will be numerical for if it is some function of x, then it would mean that $f(x)$ has got equal roots. In this case there will be $(n + 1)$ Sturm's function.

i.e., $$f(x),\ f'(x), f_2(x), f_3(x),\ ..., f_n(x)$$

1.26.9 STURM'S THEOREM

Case I. All roots unequal.

If $f(x)$ is a polynomial and a and b be any two real numbers, then the number of distinct real roots of the equation $f(x) = 0$ lying between a and b is exactly equal to the difference between the number of changes of sign when x is put equal to a and the number when x is put equal to b in the $(n+1)$ Sturm's functions $f(x), f'(x), f_2(x), f_3(x), ... f_n(x)$ consisting of the given function, its derivative and the $(n-1)$ remainders with their sign changed in the process of finding the H.C.F. of $f(x)$ and $f'(x)$.

From the definition of Sturm's functions we can establish the following relations between them :

$$\text{Dividend} = \text{Quotient} \times \text{Divisor} + \text{Remainder}$$

$$
\left.
\begin{aligned}
f(x) &= q_1 f'(x) - f_2(x),\\
f'(x) &= q_2 f_2(x) - f_3(x),\\
f_2(x) &= q_3 f_3(x) - f_4(x),\\
&\cdots\cdots\cdots\cdots\cdots\cdots\cdots\cdots\cdots\\
&\cdots\cdots\cdots\cdots\cdots\cdots\cdots\cdots\cdots\\
f_{r-1}(x) &= q_r f_r(x) - f_{r+1}(x)\\
&\cdots\cdots\cdots\cdots\cdots\cdots\cdots\cdots\cdots\\
f_{n-2}(x) &= q_{n-1} f_{n-1}(x) - f_n(x)
\end{aligned}
\right\} \qquad ...(A)
$$

From the relation (A) we have the following observations.

1. As $f(x) = 0$ has no equal roots, $f(x)$ and $f'(x)$ have no common factors and consequently $f_n(x)$ the last Strum's function is numerical having a definite sign + or –.

2. No two consecutive auxiliary functions vanish for the same value of x. If possible, suppose that when $x = \alpha$, both $f_2(x)$ and $f_3(x)$ vanish which shows that $f'(x)$ contains the factor $x - \alpha$ and from first of the relations (A), we find that $(x - \alpha)$ is also a factor of $f(x)$. Thus $(x - \alpha)$ is the H.C.F. of $f(x)$ and $f'(x)$ showing the existence of equal roots which is contrary to the hypothesis.

3. If any of the auxiliary functions vanishes, then the two adjacent functions, *i.e.*, one which precedes it and the one which follows it must have opposite signs. For example suppose that $f_3(x) = 0$, when $x = \alpha$, then from relation (A), $f_2(x) = -f_4(x)$ for $x = \alpha$ showing that $f_2(x)$ and $f_4(x)$ have opposite signs for $x = \alpha$.

4. The same reasoning applies if any of the q's vanishes. Since all the functions are polynomials in x, no one can changes sign as x increases continuously form a to b, when x assumes a value which causes that functions to vanish.

5. Now we shall show that when x in passing from a to b takes a values α such that $f_r(\alpha) = 0$, then no change of sign is gained or lost. Now since $f_r(\alpha) = 0$, we have from (3), $f_{r-1}(\alpha) = 0$ and $f_{r+1}(\alpha)$ must be of opposite signs. Now as $f_r(x)$ passes through zero, it changes sign either from + to – or from – to +. But $f_{r-1}(x)$, $f_{r+1}(x)$ are continuous at $x = \alpha$; so that each of them has an invariable sign near $x = \alpha$. Thus the three functions $f_{r-1}(x), f_r(x)$ and $f_{r+1}(x)$ will have only one changes of sign just before $x = \alpha$ and just after $x = \alpha$, *i.e.*, their signs can be either $+ +, -, - - +, + - -, - + +$.

Thus we find that whatever sign we may place between two unlike signs, we have only one change of sign. Hence, no change of sign in either lost or gained among Sturm's functions.

In case the value of x, *i.e.*, α be such that it causes more than one of the functions to vanish, then they cannot be consecutive, for in that case, by (2) the equation $f(x) = 0$ will have equal roots.

6. Now we shall show that when x in passing from a to b takes a values α which causes $f(x)$ to vanish, *i.e.*, α be a root of $f(x) = 0$, then a change of sign is lost. Now by Taylor's theorem,

$$f(a - h) = 0 - hf'(a) + \frac{h^2}{2!}f''(\alpha)...(\because f(\alpha) = 0)$$

$$f(a + h) = 0 + hf'(a) + \frac{h^2}{2!}f''(\alpha)...\because (f(\alpha) = 0)$$

Now let h be sufficiently small so that the sign of the L.H.S. is made to depend on the first term of R.H.S.

Therefore if $f'(\alpha)$ be positive then $f(\alpha - h)$ is negative and $f(\alpha + h)$ is positive, *i.e.*, in this case the signs of $f(x)$ and $f'(x)$ will – + just before $x = \alpha$ and + + just after $x = \alpha$. Thus one change of sign is lost. Again if $f'(x)$ be negative then $f(\alpha - h)$ is positive and $f(\alpha + h)$ is negative, *i.e.*, in this case the signs of $f(x)$ and $f'(x)$ will be + – just before $x = \alpha$ and – – just after $x = \alpha$. Here also one change of sign is lost. Thus we conclude that when x passes through a root α of the equation $f(x) = 0$, one change of sign is lost whether $f'(\alpha)$ be positive or negative. Hence, the number of variations lost as x goes from a real value a to a real value b is exactly equal to the number of real root; of the equation $f(x) = 0$ between a and b.

Case II. Equal roots.

If $f(x) = 0$ be an equation having equal roots and the Strum's functions be found as $f, f', f_2,...f_r$, the last of these being the H.C.F. of $f(x)$ and $f'(x)$ then the difference between the number of changes of signs when a and b are substituted in the Strum's functions is equal to the number of real roots of the equation $f(x) = 0$ which lie between a and b each multiple root being counted once only.

Let $f(x) = (x - \alpha)^p(x - \beta)^q(x - \gamma)(x - \delta)...$, then clearly $f(x)$ and $f'(x)$ will have an H.C.F $(x - \alpha)^{p-1}(x - \beta)^{q-1}$ which may be denoted by H. Thus H.C.F will be a factor of all Sturm's functions $f, f', f_2, f_3,...f_r$. Again let $\psi(x)$ stand for $(x - \alpha)(x - \beta)(x - \gamma)(x - \delta)$

then clearly $f(x) = H.\psi(x), f'(x) = H.\psi'(x), f_2(x) = H.\psi_2(x)...$ For any value of x, the number of changes of sign in the sequence of f's is the same as that in sequence of ψ's.

Now $\psi(x) = 0$ has all its roots unequal and all its roots are the same as those of $f(x) = 0$ only with the change that the multiple roots of $f(x) = 0$ occur in $\psi(x) = 0$ only once.

Now applying the reasoning of 1st case on $\psi(x) = 0$ we can say that difference between the number of changes of signs when x is put equal to a and b in the sequence of ψ's (hence of f's) represents exactly the number of real roots of the equation $\psi(x) = 0$ lying between a and b or the number of real roots of $f(x) = 0$ that lie between a and b but each multiple roots being counted only once.

TIME SAVING TRICKS

Trick 1. When $f(x) = 0$ has no repeated root, *i.e.*, the last Sturmain function be numerical, and we are concerne only with its sign. In order to get its sign we put $f_{n-1}(x) = 0$ and find the value of x; then we know that for this value of x, $f_{n-2}(x)$ and $f_n(x)$ must have the opposite sign. Thus if the value of x obtained from $f_{n-1}(x) = 0$ makes $f_{n-2}(x)$ positive, then $f_n(x)$ is negative and if it makes $f_{n-2}(x)$ negative, then $f_n(x)$ is positive. This device saves us the labour of actually calculating the value of $f_n(x)$.

Trick 2. If any of the Sturmain function say $f_2(x)$ has all roots imaginary, we may stop further calculation and we should see $f, f', f_2, f_3,...f_r$ functions only for Strum's theorem, because in this theorem the last of the function should be of invariable sign for all real values of x and we know from the properties of equation that if $f(x) = 0$ has all its roots imaginary, then $f(x)$ is always positive for all real values of x. The quadratic $ax^2 + bx + c = 0$ has its roots imaginary if $b^2 - 4ac < 0$.

Similarly the calculation of Sturmain function will stop at the stage when any of them become a perfect squares for it too will have an in variable sign for all real values of x.

Trick 3. In case any of the Sturm's functions $f_r(x)$ vanishes for $x = \alpha$; then for counting the number of changes of sign in the sequence

$f(\alpha), f'(\alpha), f_2(\alpha), f_3(\alpha), \ldots,$

📑 **REMARK**

- We may regard the sign of $f_r(\alpha)$ either positive of negative because the function that precedes it and the one which follows it have opposite signs.

WORKING PROCEDURE

Step 1. Find the Sturm's functions as explained.

Step 2. The last Sturm's function will be numerical in case $f(x) = 0$ has no equal roots. Otherwise it will be some function of x.

Step 3. The calculation of Sturm's functions should stop at the stage when any of them is either a perfect square or has all its roots imaginary.

Step 4. Make a table as explained below :

   (i) In the first column write down $x, f, f', f_2, f_3, f_4, \ldots$
   In the first row write down x and the various values that you may give to x.

   (ii) In the various other columns write down the signs of the Sturm's functions corresponding to the value of x written at the top that column.

   (iii) At the bottom of each column write down the number of changes of signs in that particular column.

   (iv) If we are to find only number of real roots, we put $x = \infty$ and $-\infty$ and find out the difference of the changes of signs corresponding to the two values.

   (v) If we want to find out the positive and negative roots, we put $x = \infty, 0$ and $-\infty$ and proceeding as above, we get the number of positive roots which will lie between 0 and ∞. Negative roots which lie between 0 and $-\infty$.

   (vi) If we want to find interval in which the roots lie, then for positive roots, we put $x = 1, 2, 3, \ldots$ and for negative roots, we put $x = -1, -2, -3 \ldots$

   (vii) If corresponding to the two values of x say 2 and 3 the changes of sign in Sturm's function be same, i.e., their differences be zero, then no root will lie between 2 and 3.

   (viii) We know that if $f(a)$ and $f(b)$ are of opposite signs, then at least one or in general odd number of roots of the equation $f(x) = 0$ lie between a and b. In case they be of the same sign, then either no root or an even number of roots of $f(x) = 0$ lie between a and b.

In case there be only one positive roots, then in order to find its location we need not put $x = 1, 2, 3 \ldots$ in all the Sturm's functions. We shall put only in $f(x)$ and in case of any two consecutive values of x the values of $f(x)$ are of opposite signs; then by the above theorem the root will lie between those two consecutive numbers. This will save us the labour of putting the consecutive numbers in all the series of Strum's functions.

🍎 Solved Examples

EXAMPLE 1. *Find the number and position of the real roots of the equation $x^6 - 2x^2 + 3x - 4 = 0$.*

SOLUTION. We have $f(x) = x^6 - 2x^2 + 3x - 4 = 0$
$$f'(x) = 6x^5 - 4x + 3$$
Multiplying $f(x)$ be 6 and then dividing by $f'(x)$

$$
\begin{array}{r}
x \\
6x^5 - 4x + 3 \overline{)6x^6 - 12x^2 + 18x - 24} \\
\underline{6x^5 - 4x^2 - 3x} \\
-8x^2 + 15x - 24
\end{array}
$$

Changing the sign of this remainder, we get
$$f_2(x) = 8x^2 - 15x + 24$$
Now since $b^2 - 4ac$, i.e., $(15)^2 - 4.8.24$ is nega- tive, we conclude that the roots of the Strum's functions $f_2(x) = 0$ are imaginary and hence we stop further calculations.

x	$-\infty$	0	∞	1	2	-1	-2
$f(x)$	+	–	+	–	+	–	+
$f'(x)$	–	+	+	+	+	+	–
$f_2(x)$	+	+	+	+	+	+	+
No. of changes of sign	2	1	0	1	0	1	2

Now, we shall discuss the nature of the roots

1. There are only two real roots (from column 2 and 4) as the difference of changes of signs when x is put $-\infty$ and ∞ in Sturm's functions $2 - 0 = 2$.

2. One of them is positive (from column 3 and 4) which lies between 1 and 2 (from column 5 and 6).

3. One of them is negative (from columns 2 and 3) which lies between –1 and –2 (from columns 7 and 8).

4. The equation being of sixth degree has only two real roots; therefore remaining four roots are imaginary.

☞ **REMARK**

• We have found above that the above equations has only one positive root and only one negative root. Also we know that if $f(a)$ and $f(b)$ be of opposite signs then at least one root and in general odd number of roots must lie between a and b.

Now $f(0) = -$ and $f(1) = -$

From above we could easily conclude that positive root does not lie between 0 and 1 as $f(0)$ and $f(1)$ are of the same sign. This would save us from calculating the signs of all the Sturm's functions corresponding to $x = 1$.

Again $f(1)$ = negative, $f(2)$ = positive. Since $f(1)$ and $f(2)$ are of opposite signs, hence the only positive roots of $f(x) = 0$ must lie between 1 and 2. This has saved us the labour of calculating the signs of all Sturm's functions corresponding to $x = 2$.

Again $f(0) = -$ and $f(-1) = -$.

Therefore arguing as above negative root does not lie between 0 and –1.

Again $f(-1)$ = negative and $f(-2)$ = positive

∴ Negative root lies between –1, –2.

Hence, the above chart could be easily put as :

x	$-\infty$	0	∞	1	2	–1	–2
$f(x)$	+	–	+	–	+	–	+
$f'(x)$	–	+	+				
$f_2(x)$	+	+	+				
No. of changes of sign	2	1	0				

EXAMPLE 2. *Find the number and position of real roots of the equation $x^3 - 7x + 7 = 0$.*

SOLUTION. We have $f(x) = x^3 - 7x + 7$, $f'(x) = 3x^2 - 7$

Let us find the H.C.F. of $f(x)$ and $f'(x)$ and change the sign of remainders to get Sturm's functions.

	$f(x)$	$f'(x)$	
x	$\dfrac{x^3 - 7x + 7}{3}$	$\dfrac{3x^2 - 7}{2}$	
	$\dfrac{3x^3 - 21x + 21}{3x^3 - 7x}$	$\dfrac{6x^2 - 14}{6x^2 - 9x}$	$3x$
	$\dfrac{-14x + 21}{\text{or } -2x + 3}$	$\dfrac{9x - 14}{2}$	
	$\therefore f_2(x) = 2x - 3$ after changing sign	$\dfrac{18x - 28}{18x - 27}$	9
		$\dfrac{-1}{\therefore f_2(x)}$ after changing sign	

∴ Sturm's functions are

$$x^3 - 7x + 7, \quad 3x^2 - 7, 2x - 3, 1$$

x	$-\infty$	0	∞	1	2	–1	–2	–3	–4
$f(x) = x^3 - 7x + 7$	–	+	+	+	+	+	+	+	–
$f'(x) = 3x^2 - 7$	+	–	+	–	+				
$f_2(x) = 2x - 3$	–	–	+	–	+				
$f_3(x) = 1$	+	+	+	+	+				
No. of changes of sign	3	2	0	2	0				

Above table shows that all the three roots are real out of which one is negative and two are positive which

clearly lie between 1 and 2. Also the negative root lies between –3 and –4. We have not tabulated the signs of all Sturm's functions for the only negative root because $f(0), f(-1), f(2), f(-3)$ are all of same signs and hence there arises no questions of a negative root lying between these numbers. Again $f(-3)$ = positive and $f(-4)$ = negative and hence a root lies between –3 and –4.

EXAMPLE 3. *Find the number of distinct real roots of the equation $x^3 - 3x + 1 = 0$ and locate them.*

SOLUTION. We have $f(x) = x^3 - 3x + 1$,

$f'(x) = 3(x^2 - 1)$ or $(x^2 - 1)$ on dividing by 3.

$f_2(x) = 2x - 1$ [After changing sign of the remainder]

$f_3(x) = 3$ [After changing the sign.]

x	$-\infty$	0	∞	1	2	–1	–2
$f(x)$	–	+	+	+	+	+	–
$f'(x)$	+	–	+	+	+	+ or –	+
				or –			
$f_2(x)$	–	–	+	+	+	–	–
$f_3(x)$	+	+	+	+	+	+	+
No. of changes of sign	3	2	0	1	0	2	3

As in example 1 the above table shows that the given equation has all its roots real and distinct out of which one is negative and the other two positive. The negative root lies in the interval (–2, –1) and one of the positive roots lies in the interval (1, 2) and the other in the interval (0, 1).

Since $f(0)$ = positive and $f(-1)$ = positive, *i.e.*, they are of the same sign, hence the root does not lie between 0 and –1 and as such we need not have completed wholly the column corresponding to $x = -1$.

Again $f(-1) = $ positive and $f(-2) = $ negative hence the only negative root lies between $(-1, -2)$ and we need not complete the column corresponding to $x = -2$.

EXAMPLE 4. *Apply Sturm's theorem to the analysis of the equation* $x^4 - 2x^3 - 3x^2 + 10x - 4 = 0$.

SOLUTION. We have $f(x) = x^4 - 2x^3 - 3x^2 + 10x - 4$

$f'(x) = 2x^3 - 3x^2 - 3x + 5$, after cancelling 2.

$f_2(x) = 9x^2 - 27x + 11$, after changing sign.

✍ REMARK

- If we actually proceed to calculate the value of $f_4(x)$, it will be −1433 (after changing the sign).

x	$-\infty$	0	∞	1	-1	-2	3
$f(x)$	+	−	+	+	−	−	+
$f'(x)$	−	+	+				
$f_2(x)$	+	−	+				
$f_3(x)$	+	−	−				
$f_4(x)$	−	−	−				
No. of changes of sign	3	2	1				

From above table we observe that there are only 3 − 1 = 2 real roots. One of them is positive and the other negative. Again $f(0) = $ negative and $f(1) = $ positive, *i.e.,* they are of opposite signs; so we conclude that the

$f_3(x) = -8x - 3$, after changing sign.

Now $f_4(x)$ will be numerical and we are concerned only with its sign.

Putting $f_3(x) = 0$, we get $x = -\dfrac{3}{8}$. This value of x makes $f_2(x) = 9 \cdot \dfrac{3}{64} + \dfrac{81}{8} + 11$, *i.e.,* positive and we know that when $f_3(x) = 0$, then $f_2(x)$ and $f_4(x)$ are of opposite signs. Since $f_2(x)$ is positive therefore $f_4(x)$ must be negative.

positive root lie between 0 and 1. Hence we need not complete the column corresponding to $x = 1$.

Again $f(0) = $ negative and $f(-1) = $ negative, *i.e.* they are of the same sign. Hence, either no root or an even no. of roots of $f(x) = 0$, lie between 0 and −1; but as there is only one negative root, as such it cannot lie between 0 and −1; therefore we have not completed the column corresponding to $x = -1$. Similarly for column corresponding to $x = -2$.

Again $f(-2) = $ negative and $f(-3) = $ positive.

As they are of opposite signs, therefore they only negative root lies between −2 and −3, and we need not complete the column corresponding to $x = -3$.

1.26.10 CONDITION FOR ALL THE ROOTS REAL AND DISTINCT

Case I. In case the equation has all its roots distinct, then the sequence of Sturm's function must in general consist of $(n + 1)$ functions, *i.e.,* the given function, its derivatives and $(n - 1)$ Sturm's remainders.

Case II. In case the equation has all its roots real, then the difference of the number of changes of signs when x is put ∞ in the sequence over the number when x is put $-\infty$ should be n. In other words it means that when x is put $+\infty$ no change of signs occurs, *i.e.,* the number of changes of signs be zero and when $x = -\infty$ in the $(n + 1)$ functions, it should be alternately positive and negative so that there be n changes of signs and the difference of these change of signs be $n - 0 = n$, the number of real and distinct roots.

The above conditions are satisfied if we say that the leading coefficients of all the Sturm's functions be positive (we always take the leading coefficients of a given equation to be positive).

🗔 Solved Examples

EXAMPLE. *Use Sturm's theorem to show that the equation* $z^3 + Hz + G = 0$ *has all the three roots real and distinct if and only if* $G^2 + 4H^3 < 0$.

SOLUTION. The leading coefficients of all the Sturm's functions is positive we get the condition as
$-H$ positive, $-(G^2 + 4H^3)$ positive

i.e., H negative and $G^2 + 4H^3$ negative

But $G^2 + 4H^3$ can be negative only if H is negative and hence the former condition is implied in the latter.

Hence $G^2 + 4H^3$ is negative *i.e.* < 0 if all the roots of the given cubic are real and distinct.

1.26.11 NATURE OF THE ROOTS OF BIQUADRATIC

By the help of Sturm's theorem, discuss the nature of the roots of the equation $ax^4 + 4bx^3 + 6cx^2 + 4dx + e = 0$

The above equation can be reduced to the form $f(z) = z^4 + 6Hz^2 + 4Gz + a^2I - 3H^2 = 0$

where $z = ax + b$ and the nature of the roots of the given equation and of the transformation equation is same the Sturm's functions are

$$f(z) = z^4 + 6Hz^2 + 4Gz + a^2I - 3H^2$$
$$f'(z) = z^3 + 3Hz + G$$
$$f_2(z) = -3Hz^2 - 3Gz - (a^2I - 3H^2)$$

$$f_3(z) = -z(2HI - 3aJ) - IG$$

$$f_4(z) = I^3 - 27J^2 = \Delta \text{, the discriminate.}$$

1. **All roots real and distinct.** The leading coefficients of Sturm's functions should be positive.

   Thus, H negative, $2HI - 3aJ$ negative, $I^3 - 27J^2$ positive.

2. **All roots imaginary.**

   $I^3 - 27J^2$ positive, and either H positive or $2HI - 3aJ$ positive.

   In this case corresponding to the $z = \infty$, number of changes of sign in the above sequence will be

   $$z = \infty + + - (+ \text{ or } -) + 2 \text{ changes}$$
   $$z = -\infty + - - (+ \text{ or } -) + 2 \text{ changes}$$

   The difference corresponding to the values ∞ and $-\infty$ of z is zero and hence the equation has no real roots. Thus all its roots are imaginary.

3. **Two roots real and two imaginary.**

   $I^3 - 27J^2$ negative : In this case we shall find that the difference of the changes of signs corresponding to $z = \infty$ and $-\infty$ in the above sequence is always 2. We may give H and $2HI - 3aJ$ any sign we like. Thus the equation has two real and two imaginary roots.

4. **Two equal roots.**

   In this case clearly $I^3 - 27J^2 = 0$ and then $f_3(z) = 0$ will give the H.C.F. of $f(z)$ and $f'(z)$ and thus proving the existence of two equal roots.

5. **Three equal roots or two pairs of equal roots.**

   In this case the H.C.F. should be of 2nd degree and also a perfect square or composed of two unequal factors according as three roots are equal to two pair of equal roots exist. Hence, $f_2(z)$ should vanish identically which will happen when either.

   (1) $I = 0$ and $J = 0$ or

   (2) $G = 0$ and $2HI - 3aJ = 0$

   when $I = 0$ and $J = 0$, we get $G^2 + 4H^3 = a^2(HI - aJ) = 0$ and $f_2(z) = 0$ becomes

   $$3Hz^2 + 3Gz - 3H^2 = 0$$

   and its discriminant is $9(G^2 + 4H^3)$ which is zero and hence its roots are equal, *i.e.*, it is a perfect square. Since the H.C.F is a perfect square, $f(z) = 0$, has three equal roots.

   When $G = 0$ and $2HI - 3aJ = 0$, then $f_2(z)$ is the H.C.F but is not a perfect square and hence the equation will have a pair of equal roots.

 REMARK

- We have $G^2 + 4h^3 = a^2HI - a^3J$. Putting $G = 0$ and $J = \dfrac{2HI}{3a}$, we get $4H^3 = a^2HI - a^3\dfrac{2HI}{3a}$ or $a^2I = 12H^2$ and this is same as the condition we had found.

6. **All roots equal.**

   If $I = 0$, $H = 0$ and $G = 0$ then $f_3(z)$ vanishes identically and the H.C.F. comes out to be $f'(z)$, *i.e.*, z^5 showing that all the roots are equal.

Exercise-1.18

1. Find the number and position of real roots of the following equations :

   (i) $x^4 + 15x^2 + 7x - 11 = 0$ (ii) $x^4 - 6x^3 + 5x^2 + 14x - 1 = 0$

   (iii) $x^4 - 4x^3 + 7x^2 - 6x - 4 = 0$

   (iv) $x^4 - 8x^3 + 25x^3 - 36x + 8 = 0$

   (v) $x^3 + x^2 - 2x - 1 = 0$ (vi) $x^4 - 7x^2 + 18x - 8 = 0$

2. Apply Sturm's theorem to prove that the equation $x^3 - 2x - 5 = 0$ has only one positive real root lying between 2 and 3.

3. Prove that the equation $x^5 - x + 16 = 0$ has two pair of complex roots.

4. Prove by Sturm's method that the equation $x^4 - 6x^3 + 13x^2 + 4 = 0$ has two pair of equal roots.

5. Varify by means of Sturm's remainders, the condition which must be fulfilled when the biquadratic $9x^4 + 4bx^3 + 6cx^2 + 4dx + e = 0$ is a perfect square and prove in that case $a^3 f(x) = \{(ax + b)^2 + 3H\}^2$.

6. Prove that when the biquadratic of the last example has a triple factor it may be expressed in the form

   $$a^3 f(x) = \{ax + b + \sqrt{(-H)}\}^3 \{ax + b - 3\sqrt{(-H)}\}$$

7. Find Sturm's function for the cubic $z^3 + 3Hz + G = 0$.

Answers

1. (i) One negative in $(-1, -2)$ and one positive in $(0, 1)$

   (ii) One negative in $(-1, -2)$, three positive out of which one is in $(0, 1)$ and two in $(2, 4)$

   (iii) One negative in $(0, -1)$ and one positive in $(2, 3)$ and two imaginary

(iv) Two imaginary, two real positive, one in (0, 1) and other two in (3, 4)

(v) All real, one positive in (1, 2) two negative in (0, 1) and other in (–1, –2)

(vi) Two imaginary, two real one positive in (0, 1) one negative in (3, 4)

4. 1, 1, 2, 2   **7.** $f(z) = z^3 + 3Hz + G$, $f_1(z) = z^2 + H$, $f_2(z) = -2Hz - G$, $f_3(z) = -(G^2 + 4H^3)$

1.27 CARDAN'S METHOD TO FIND THE ROOTS OF A CUBIC EQUATION

Let the general cubic equation be $a_0 x^3 + 3a_1 x^2 + 3a_2 x + a_3 = 0$ ...(1)

First reduce this equation (1) into an equation having no second degree term, i.e., $3a_1 x^2$. The equation (1) is reduced to the following equation

$$Z^3 + 3HZ + G = 0 \qquad \qquad ...(2)$$

where $H = a_0 a_2 - a_1^2$, $G = a_0^2 a_3 - 3a_0 a_1 a_2 + 2a_1^3$ and $Z = a_0 x + a_1$

Let as assume $z = u + v$.   ...(3)

Cubing both sides of (3), we get $z^3 = (u+v)^3 = u^3 + v^3 + 3uv(u+v) = u^3 + v^3 + 3uv(z)$

$$z^3 = u^3 + v^3 + 3uvz \qquad \text{or} \qquad z^3 - 3uvz - (u^3 + v^3) = 0 \qquad ...(4)$$

Comparing (2) and (4), we get

$$uv = -H, \ u^3 + v^3 = -G \quad \text{or} \quad u^3 v^3 = (-H)^3, \ u^3 + v^3 = -G$$

Hence u^3, v^3 are the roots of the quadratic equation given by

$$t^2 + Gt - H^3 = 0 \qquad \qquad ...(5)$$

Solving (5), we get

$$t = \frac{-G \pm \sqrt{G^2 + 4H^3}}{2}$$

$$\therefore \qquad u^3 = \frac{-G + \sqrt{G^2 + 4H^3}}{2} \qquad \qquad ...(6)$$

and

$$v^3 = \frac{-G - \sqrt{G^2 + 4H^3}}{2} \qquad \qquad ...(7)$$

From (3), we get

$$z = \left\{ -\frac{G}{2} + \frac{1}{2}\sqrt{G^2 + 4H^3} \right\}^{1/3} + \left\{ -\frac{G}{2} - \frac{1}{2}\sqrt{G^2 + 4H^3} \right\} \qquad ...(8)$$

From (6) and (7) it is obvious that each u and v will have three cube roots and hence from (8) z will have nine values. But the degree of the equation (2) in z is three so it must have three roots, i.e., three values of z. Since we have that $uv = -H$, therefore the cube roots are taken in pairs so that $uv = -H$. Hence we shall take the pair of cube roots as

$$u, v, u\omega, v\omega^2; u\omega^2, v\omega$$

where ω and ω^2 are the imaginary cube roots of unity. Therefore the roots of the equation (2) are $u + v$, $u\omega + v\omega^2$, $u\omega^2 + v\omega$

and hence we can find the roots of the equation (1) by the relation $z = a_0 x + a_1$ corresponding to $u + v$, $u\omega + v\omega^2$ and $u\omega^2 + v\omega$.

1.27.1 APPLICATION OF CARDAN'S METHOD

From equation (6) and (7), we have $u^3 = \dfrac{-G + \sqrt{G^2 + 4H^3}}{2}$ and $v^3 = \dfrac{-G - \sqrt{G^2 + 4H^3}}{2}$

Case I. If $G^2 + 4H^3 > 0$, i.e., the cubic equation (1) has a pair of imaginary roots, then R.H.S. of above two equations are real and hence by the some method we can extract the cube root of real quantities and consequently we get the values of z from equation (8).

Case II. If $G^2 + 4H^3 < 0$, i.e., if the roots of the cubic equation (1) are all real, then from above equations u^3 and v^3 are imaginary and there is no method in general for extracting the cube root of imaginary number so that Cardan's method fails to give the roots of the given cubic (1). However in this case we use De Moivre's theorem to find the cubic root of imaginary number.

Let us assume $-G = P$ and $G^2 + 4H^3 = -Q$ then from equation (8), we get

$$z = \left(\frac{P}{2} + \frac{i}{2}Q \right)^{1/3} + \left(\frac{P}{2} - \frac{i}{2}Q \right)^{1/3} \quad \text{where } i = \sqrt{-1}$$

Now put $\dfrac{P}{2} = r\cos\theta$, $\dfrac{Q}{2} = r\sin\theta$

$$\therefore \qquad r^2 = \frac{P^2 + Q^2}{4} = \frac{G^2 - G^2 - 4H^3}{4} = -H^3$$

and $\qquad \tan\theta = \dfrac{Q/2}{P/2} = \dfrac{Q}{P} \implies \qquad \tan\theta = \dfrac{-\sqrt{-(G^2+4H^3)}}{G}$

$\therefore \qquad z = (r\cos\theta + ir\sin\theta)^{1/3} + (r\cos\theta - ir\sin\theta)^{1/3}$

$$= r^{1/3}\left[\cos\left(\dfrac{2n\pi+\theta}{3}\right) + i\sin\left(\dfrac{2n\pi+\theta}{3}\right) + \cos\left(\dfrac{2n\pi+\theta}{3}\right) - i\sin\left(\dfrac{2n\pi+\theta}{3}\right)\right]$$

$$= 2r^{1/3}\cos\left(\dfrac{2n\pi+\theta}{3}\right) \text{ where } n = 0, 1, 2.$$

Hence, the roots of the equation (2) are

$$z_1 = 2r^{1/3}\cos\left(\dfrac{\theta}{3}\right) = 2(-H)^{1/2}\cos\left(\dfrac{\theta}{3}\right) ; \quad z_2 = 2r^{1/3}\cos\left(\dfrac{2n\pi+\theta}{3}\right) = 2(-H)^{1/2}\cos\left(\dfrac{2n\pi+\theta}{3}\right)$$

$$z_3 = 2r^{1/3}\cos\left(\dfrac{2\pi-\theta}{3}\right) = 2(-H)^{1/2}\cos\left(\dfrac{2\pi-\theta}{3}\right) \qquad \text{where } \theta = \tan^{-1}\left[\dfrac{-\sqrt{-(G^2+4H^3)}}{G}\right]$$

Consequently by the relation $z = a_0 x + a$, we can find all the roots of the given cubic.

1.27.2 Method by Expressing the Equation as Sum or Difference of two cubes

Let the given cubic equation be

$$a_0 x^3 + 3a_1 x^2 + 3a_2 x + a_3 = 0 \qquad \qquad \text{...(1)}$$

Let us suppose the equation (1) can be expressed as follows :

$$a_0 x^3 + 3a_1 x^2 + 3a_2 x + a_3 \equiv A(x-a)^3 + B(x-b)^3 \qquad \qquad \text{...(2)}$$

or $\qquad a_0 x^3 + 3a_1 x^2 + 3a_2 x + a_3 \equiv (A+B)x^3 - 3(Aa+Bb)x^2 + 3(Aa^2+Bb^2)x - (Aa^3+Bb^3)$.

Now equating the coefficients of like powers of x, we get

$$A + B = a_0 \qquad \qquad \text{...(3)}$$

$$Aa + Bb = -a_1 \qquad \qquad \text{...(4)}$$

$$Aa^2 + Bb^2 = a_2 \qquad \qquad \text{...(5)}$$

$$Aa^3 + Bb^3 = -a_3 \qquad \qquad \text{...(6)}$$

Multiplying (3), (4), (5) by a and subtracting respectively (4), (5) and (6), we get

$$B(a-b) = a_0 a + a_1 \qquad \qquad \text{...(7)}$$

$$-Bb(a-b) = a_1 a + a_2 \qquad \qquad \text{...(8)}$$

$$Bb^2(a-b) = a_2 a + a_3 \qquad \qquad \text{...(9)}$$

From (7), (8) and (9), we get $\qquad (a_1 a + a_2)^2 = (a_0 a + a_1)(a_2 a + a_3) \qquad \qquad \text{...(10)}$

Similarly multiply (3), (4), (5) by b and subtracting respectively (4), (5) and (6), we get

$$(a_1 b + a_2)^2 = (a_0 b + a_1)(a_2 b + a_3) \qquad \qquad \text{...(11)}$$

From equation (10) and (11) it is concluded that a and b are the roots of the equation

$$(a_1 x + a_2)^2 = (a_0 x + a_1)(a_2 x + a_3) \qquad \qquad \text{...(12)}$$

or $\qquad (a_0 a_2 - a_1^2)x^2 + (a_0 a_3 - a_1 a_2)x + (a_1 a_3 - a_2^2) = 0 \qquad \qquad \text{...(13)}$

This equation (13) is called the Hessian of the cubic.

From (13) we obtain the values of a and b and then we can find A and B from (3) and (4). Substitute these values of a, b, A and B in (2) we get the given equation as the sum or difference of two cubes as

$$A(x-a)^3 + B(x-b)^3 = 0 \quad \text{or} \quad \left(\dfrac{x\ a}{x\ b}\right) = \left(-\dfrac{B}{A}\right) \quad \text{or} \quad \left(\dfrac{x-a}{x-b}\right) = \left(-\dfrac{B}{A}\right)^{1/3} = -\left(\dfrac{B}{A}\right)^{1/3}k \qquad \qquad \text{...(14)}$$

where $k = 1, \omega, \omega^2$.

Therefore if a and b (provided $a \neq b$) are the roots of the equation,

$$(a_1 x + a_2)^2 = (a_0 x + a_1)(a_2 x + a_3)$$

Then the given equation (1) can be reduced to the form given by

$$A(x-a)^3 + B(x-b)^3 = 0$$

Now from (14) one values of x is obtained and then divide the given equation by so obtained factor we get a quadratic as quotient and the remaining roots of the given equation can be obtained from this quadratic quotient.

Further, if the obtained values of a and b are imaginary and so A and B, then we cannot find the cube roots of complex number. Hence, by above method we can find the roots of the given cubic if the roots of the quadratic equation (13) are real and distinct.

That is, if the condition $(a_0a_3 - a_1a_2)^2 - 4(a_1a_3 - a_2^2)(a_0a_2 - a_1^2) > 0$ holds then the equation (13) will have both real and distinct roots.

1.27.3 HESSIAN OF THE CUBIC EQUATION

The quadratic equation given by $(a_0a_2 - a_1^2)x^2 + (a_0a_3 - a_1a_2)x + (a_1a_3 - a_2^2) = 0$

is called the Hessian of the cubic equation $a_0x^3 + 3a_1x^2 + 3a_2x + a_3 = 0$.
This Hessian of cubic can also be obtained as follows.
First making the given cubic homogenous by introducing a new variable t as

$$f(x,t) \equiv a_0x^3 + 3a_1x^2t + 3a_2xt^2 + a_3t^3 = 0 \qquad \ldots(1)$$

Now differentiating (1) partially w.r.t. x and t and respectively, we get

$$\frac{\partial f}{\partial x} = 3a_0x^2 + 6a_1xt + 3a_2t^2 \qquad \ldots(2)$$

$$\frac{\partial f}{\partial t} = 3a_1x^2 + 6a_2xt + 3a_3t^2 \qquad \ldots(3)$$

Again differentiating (2) w.r.t. x and (3) w.r.t. t, we get

$$\frac{\partial^2 f}{\partial x^2} = 6a_0x + 6a_1t \qquad \ldots(4)$$

$$\frac{\partial^2 f}{\partial t^2} = 6a_2x + 6a_3t \qquad \ldots(5)$$

And differentiating (3) w.r.t. x, we get $\qquad \dfrac{\partial^2 f}{\partial x \partial t} = 6a_1x + 6a_2t \qquad \ldots(6)$

Now putting $t = 1$ and observed that the expression

$$\left(\frac{\partial^2 f}{\partial x \partial t}\right) \quad \left(\frac{\partial^2 f}{\partial x^2}\right)\left(\frac{\partial^2 f}{\partial t^2}\right) \qquad \ldots(7)$$

gives the Hessian of the cubic. Hence the Hessian of the cubic can also be obtained by (7).

☞ **REMARK**

- Hessian of the cubic can also be expressed in the form of a determinant as follows :

$$\begin{vmatrix} 1 & -x & x^2 \\ a_0 & a_1 & a_2 \\ a_1 & a_2 & a_3 \end{vmatrix} = 0$$

Solved Examples

EXAMPLE 1. *Solve the equation* $x^3 - 15x - 126 = 0$
by Cardan's method. (SVTU–2009)

SOLUTION. The given equation is
$$x^3 - 15x - 126 = 0 \qquad \ldots(1)$$
Let the solution of (1) be
$$x = u + v$$
Cubing (2), we get

$$x^3 = (u+v)^3 = u^3 + v^3 + 3uv(u+v)$$

or $\qquad x^3 = u^3 + v^3 + 3uv(x) \qquad [\because x = u + v]$

or $\quad x^3 - 3uvx - (u^3 + v^3) = 0$

The equations (1) and (3) are same so comparing the coefficients of like terms, we get
$$3uv = 15$$

or $\qquad uv = \dfrac{15}{3} \quad$ or $\quad u^3v^3 = 125$

and $\qquad u^3 + v^3 = 126$

Hence u^3, v^3 are the roots of the quadratic
$$t^2 - 126t + 125 = 0$$
$\therefore \quad (t-125)(t-1) = 0, \ t = 125, \ t = 1$
$\therefore \quad u^3 = 125, v^3 = 1 \ $ or $u = 5, \ v = 1$
Thus the roots of (1) are given by
$$u + v, \ u\omega + v\omega^2, \ u\omega^2 + v\omega$$

where $\quad \omega = -\dfrac{1}{2} + \dfrac{i\sqrt{3}}{2}$

$\therefore \qquad u + v = 5 + 1 = 6$

$u\omega + v\omega^2 = 5\omega + \omega^2 = 4\omega + \omega + \omega^2$

$\qquad\qquad = 4\omega - 1 \qquad (\because 1 + \omega + \omega^2 = 0)$

$\qquad\qquad = 4\left(-\dfrac{1}{2} + \dfrac{i\sqrt{3}}{2}\right) - 1 = -3 + i2\sqrt{3}$

and $\quad u\omega^2 + v\omega = 5\omega^2 + \omega = 4\omega^2 + \omega^2 + \omega = 4\omega^2 - 1$

$$= 4\left(-\frac{1}{2} + \frac{i\sqrt{3}}{2}\right)^2 - 1 = -3 - i2\sqrt{3}$$

Hence, roots are $6, -3 + 2i\sqrt{3}, -3 - 2i\sqrt{3}$.

EXAMPLE 2. *Show that the roots of the equation*

$$x^3 - 3x + 1 = 0 \text{ are } 2\cos\frac{2\pi}{9}, 2\cos\frac{8\pi}{9}, \cos\frac{14\pi}{9}.$$

SOLUTION. The given equation is

$$x^3 - 3x + 1 = 0 \qquad \text{...(1)}$$

Let $\qquad x = u + v \qquad \text{...(2)}$

Cubing (2) of both sides, we get

$$x^3 - 3uvx - (u^3 + v^3) = 0$$

Since (1) and (3) are same so we have

$$uv = 1, u^3 + v^3 = -1$$

or $\quad u^3v^3 = 1, u^3 + v^3 = -1$

$\therefore u^3, v^3$ are the roots of the equation $t^2 + t + 1 = 0$

$$\therefore \quad t = \frac{-1 \pm \sqrt{1-4}}{2} \Rightarrow t = \frac{-1 \pm i\sqrt{3}}{2}$$

$$\therefore \quad u^3 = -\frac{1}{2} + \frac{i}{2}\sqrt{3}, v^3 = -\frac{1}{2} - \frac{i}{2}\sqrt{3}$$

From (2) we get

$$x = \left(-\frac{1}{2} + \frac{i}{2}\sqrt{3}\right)^{1/3} + \left(-\frac{1}{2} - \frac{i}{2}\sqrt{3}\right)^{1/3} \quad \text{...(4)}$$

Change the complex number of R.H.S of (4) into polar form by putting

$$-\frac{1}{2} = r\cos\theta, \frac{\sqrt{3}}{2} = r\sin\theta$$

$$\therefore \quad r^2 = 1 \quad \Rightarrow \quad r = 1$$

and $\quad \tan\theta = -\sqrt{3} \Rightarrow \theta = \frac{2\pi}{3}$

$$\therefore \quad x = (r\cos\theta + ir\sin\theta)^{1/3} + (r\cos\theta - ir\sin\theta)^{1/3}$$

$$= r^{1/3}\left[\cos\frac{2n\pi + \theta}{3} + i\sin\frac{2n\pi + \theta}{3}\right.$$

$$\left. + \cos\frac{2n\pi + \theta}{3} - i\sin\frac{2n\pi + \theta}{3}\right]$$

$$= 2r^{1/3}\cos\frac{2n\pi + \theta}{3}, n = 0,1,2$$

Therefore, $x_1 = 2(1)^{1/3}\cos\frac{\theta}{3} = 2\cos\frac{2\pi}{9}$

$$x_2 = 2\cos\frac{2\pi + \theta}{3} = 2\cos\left(\frac{2\pi}{3} + \frac{2\pi}{9}\right)$$

$$= 2\cos\frac{8\pi}{9}$$

and $\quad x_3 = 2\cos\frac{4\pi + \theta}{3} = 2\cos\left(\frac{4\pi}{3} + \frac{2\pi}{9}\right)$

$$= 2\cos\frac{14\pi}{9}.$$

EXAMPLE 3. *Solve the cubic* $x^3 - 3a^2x - 2a^3\cos 3A = 0$ *by Cardan's method.*

SOLUTION. Since the given cubic is

$$x^3 - 3a^2x - 2a^3\cos 3A = 0 \qquad \text{...(1)}$$

Let the solution of (1) be

$$x = u + v \qquad \text{...(2)}$$

Cubing of both sides of (2), we get

$$x^3 - 3uvx - (u^3 + v^3) = 0 \qquad \text{...(3)}$$

The equation (1) and (3) are same so we have

$$uv = a^2, u^3 + v^3 = 2a^3\cos 3A$$

or $\quad u^3v^3 = a^6, u^3 + v^3 = 2a^3\cos 3A$

$\therefore u^3, v^3$ are the roots of the following equation

$$t^3 - 2a^3\cos 3A + a^6 = 0 \qquad \text{...(4)}$$

$$\therefore \quad t = \frac{2a^3\cos 3A \pm \sqrt{(4a^6\cos^2 3A - 4a^6)}}{2}$$

$$= \frac{2a^3\cos 3A \pm 2a^3 i\sin 3A}{2}$$

$$t = a^3(\cos 3A \pm i\sin 3A)$$

$$\therefore \quad u^3 = a^3(\cos 3A + i\sin 3A),$$

$$v^3 = a^3(\cos 3A - i\sin 3A)$$

or $\quad u = a(\cos 3A + i\sin 3A)^{1/3},$

$$v = a(\cos 3A - i\sin 3A)^{1/3}$$

Substitute these values of u and v in (2), we get

$$x = a(\cos 3A + i\sin 3A)^{1/3} + a(\cos 3A - i\sin 3A)^{1/3}$$

$$= a\left[\cos\left(\frac{2n\pi + 3A}{3}\right) + i\sin\left(\frac{2n\pi + 3A}{3}\right)\right.$$

$$\left. + \cos\left(\frac{2n\pi + 3A}{3}\right) - i\sin\left(\frac{2n\pi + 3A}{3}\right)\right]$$

$$= 2a\cos\left(\frac{2n\pi + 3A}{3}\right), \text{ where } n = 0, 1, 2$$

$$\therefore \quad x_1 = 2a\cos A \quad \text{when } n = 0$$

$$x_2 = 2a\cos\left(\frac{2\pi + 3A}{3}\right), \quad \text{when } n = 1$$

$$= 2a\cos\left(\frac{2\pi}{3} + 1\right)$$

and $\quad x_3 = 2a\cos\left(\frac{4\pi + 3A}{3}\right), \text{when } n = 2$

$$= 2a\cos\left(\frac{4\pi}{3} + A\right) = 2a\cos\left\{2\pi - \left(\frac{2\pi}{3} - A\right)\right\}$$

$$= 2a\cos\left(\frac{2\pi}{3} - A\right)$$

Hence, the solution of the given cubic is given by

$$2a\cos A, 2a\cos\left(\frac{2\pi}{3} \pm A\right).$$

EXAMPLE 4. *Solve the equation* $9x^3 - 30x^2 + 36x - 16 = 0$ *by expressing it as the sum or difference of two cubes.*

SOLUTION. The given cubic is $9x^3 - 30x^2 + 36x - 16 = 0$ ...(1)

Compare this equation with the equation

$$a_0x^3 + 3a_1x^2 + 3a_2x + a_3 = 0$$

we get $a_0 = 9, a_1 = -10, a_2 = 12, a_3 = -16$

Let the given cubic (1) can be expressed as

$$9x^3 - 30x^2 + 36x - 16$$

$$\equiv A(x - a)^3 + B(x - b)^3 = 0 \qquad \text{...(2)}$$

Then a, b are the roots of the following equation

$$(a_1 x + a_2)^2 = (a_0 x + a_1)(a_2 x + a_3)$$

or $(a_0 a_2 - a_1^2)x^2 + (a_0 a_3 - a_1 a_2)x + (a_1 a_3 - a_2^2) = 0$
$$\qquad \ldots(3)$$

Substitute the values of a_0, a_1, a_2, a_3 in (3), we get

$$(108 - 100)x^2 + (-144 + 120)x + (160 - 144) = 0$$

or $\qquad\qquad 8x^2 - 24x + 16 = 0$

or $\qquad\qquad x^2 - 3x + 2 = 0$

or $\qquad\qquad (x-1)(x-2) = 0$

or $\qquad\qquad x = 1, 2$

$\therefore \qquad\qquad a = 1, b = 2$

Now substitute these values of a and b in (2) and equation the coefficients of like powers of x of both sides, we get

$$9x^3 - 30x^2 + 36x - 16$$
$$\equiv A(x-1)^3 + B(x-2)^3 = 0 \qquad \ldots(4)$$

Taking the coefficients of x^3 and x^2, we get

$$A + B = 9$$
$$A + 2B = 10$$

Solving these two equations, we get $A = 8, B = 1$.
Substitute the values of A and B in (4), we get the given cubic as

$$8(x-1) + (x-2)^3 = 0 \text{ or } \left(\frac{x-1}{x-2}\right)^3 = -\frac{1}{8}$$

or $\qquad \dfrac{x-1}{x-2} = \left(-\dfrac{1}{8}\right)^{1/3}$

or $\qquad \dfrac{x-1}{x-2} = \left(-\dfrac{1}{2}\right)^{1/3}(1)^{1/3} = \left(-\dfrac{1}{2}\right)k$

where $k = 1, \omega, \omega^2$ and $\omega = \left(\dfrac{-1}{2} + \dfrac{i}{2}\sqrt{3}\right)$

or $\qquad 2(x-1) = -k(x-2)$

or $\qquad 2x - 2 = -kx + 2k$

or $\qquad x(2+k) = 2k + 2$

or $\qquad x = \dfrac{2k+2}{2+k} = \dfrac{2(k+1)}{(2+k)}$

when $k = 1$, $x_1 = \dfrac{2(1+1)}{2+1} = \dfrac{4}{3}$ when $k = \omega$

$$x_2 = \frac{2(\omega+1)}{(2+\omega)} = \frac{2(-\omega^2)}{(1-\omega^2)}$$
$$(1 + \omega + \omega^2 = 0 \text{ and } \omega^3 = 1)$$

$$= \frac{2(-\omega^3)}{(\omega - \omega^3)} = -\frac{2}{\omega - 1} = \frac{2}{1-\omega}$$

$$= \frac{2}{1 - \left(-\dfrac{1}{2} + \dfrac{i}{2}\sqrt{3}\right)} = \frac{2}{\dfrac{3}{2} - \dfrac{i}{2}\sqrt{3}}$$

$$= \frac{4}{3 - i\sqrt{3}} = \frac{4(3 + i\sqrt{3})}{9 + 3} = \frac{3 + i\sqrt{3}}{3}$$

when $k = \omega^2$

$$x_3 = \frac{2(\omega^2 + 1)}{(2 + \omega^2)} = \frac{2(\omega^3 + \omega)}{(2\omega + \omega^3)} = \frac{2(1+\omega)}{2\omega + 1}$$
$$(\because \omega^3 = 1)$$

$$= \frac{2\left(1 - \dfrac{1}{2} + \dfrac{i}{2}\sqrt{3}\right)}{2\left(-\dfrac{1}{2} + \dfrac{i}{2}\sqrt{3}\right) + 1}$$

$$\left[\because \omega = \left(-\dfrac{1}{2} + \dfrac{i}{2}\sqrt{3}\right)\right]$$

$$= \frac{1 + i\sqrt{3}}{i\sqrt{3}} = \frac{i\sqrt{3} - 3}{-3} = \frac{3 - i\sqrt{3}}{3}$$

Hence, the roots of the given cubic are $\dfrac{4}{3}, \dfrac{3 \pm i\sqrt{3}}{3}$.

 Exercise-1.19

Solve the following cubic equations by Cardan's method.

1. $x^3 + 6x^2 + 9x + 4 = 0$

2. $x^3 + 6x^2 - 12x + 32 = 0$

3. $x^3 - 21x - 344 = 0$

4. $x^3 - 12x^2 - 6x - 10 = 0$

5. $27x^3 + 54x^2 + 198x - 73 = 0$

6. $x^3 - 18x - 35 = 0$ $\qquad\qquad$ (OSMANIA–2003)

7. $9x^3 + 6x^2 - 1 = 0$ $\qquad\qquad$ (SVTU–2008)

8. $x^3 - 15x^2 - 357x + 5491 = 0$

9. $x^3 + 3x^2 - 27x + 104 = 0$

10. $x^3 - 6x - 9 = 0$

11. $2x^3 + 3x^2 + 3x + 1 = 0$

12. $8a^3 x^3 - 6ax + 2\sin 3A = 0$

13. $64x^3 - 144x^2 + 108x - 27 = 0$

14. $x^3 + 3ax^2 + 3(a^2 - bc)x + a^3 + b^3 + c^3 - 3abc = 0$

Solve the following equations by expressing them as the sum or difference of two cubes.

15. $x^3 - 3x^2 + 33x - 1 = 0$

16. $2x^3 + 3x^2 - 21x + 19 = 0$

17. $x^3 + 3x^2 - 27x + 104 = 0$

18. $152x^3 - 60x^2 - 606x - 485 = 0$

Hints to Selected Problems

1. $h = -\dfrac{a_0}{na_0} = -\dfrac{6}{3 \times 1} = -2$

The transformed equation is $z^3 - 3z + 2 = 0$
where $z = x + 2$

Assume that the solution of transformed equation is $z = u + v$

11. $h = -\dfrac{1}{2}$ Transformed equation is $2z^3 + \dfrac{3}{2}z = 0$

13. $h = \dfrac{3}{4}$ Transformed equation is $z^3 = 0$

14. $h = -a$ Transformed equation is $z^3 - 3bcz + b^3 + c^3 = 0$

18. The given equation can be expressed as

$$152x^3 - 60x^2 - 606x - 485 = A(x-a)^3 + B(x-b)^3$$

Then a, b are the roots of the following equation

$$(a_0 a_1 - a_1^2)x^2 + (a_0 a_3 - a_1 a_2)x + (a_1 x_3 - a_2^2) = 0$$

Answers

1. $-4, -1, -1$ **2.** $-8, (1 \pm i\sqrt{3})$ **3.** $8, (-4 \pm i3\sqrt{3})$

4. $4 + 3(2)^{1/3} + 3(4)^{1/3}, 4 + 3\omega(2)^{1/3} + 3\omega^2(4)^{1/3}, 4 + 3\omega^2(2)^{1/3} + 3\omega(4)^{1/3}$ where $\omega = \left(-\dfrac{1}{2} \pm \dfrac{i}{2}\sqrt{3}\right)$ **5.** $\dfrac{1}{3}, \left(-\dfrac{7}{6} \pm \dfrac{3\sqrt{3}}{2}i\right)$

6. $5, \left(-\dfrac{5}{2} \pm \dfrac{i\sqrt{3}}{2}\right)$ **7.** $\dfrac{1}{3}, \dfrac{-3 \pm i\sqrt{3}}{6}$ **8.** $-19, 17, 17$ **9.** $-8, \dfrac{1}{2}, (5 \pm i3\sqrt{3})$ **10.** $3, \left(-\dfrac{3}{2} \pm \dfrac{i\sqrt{3}}{2}\right)$ **11.** $-\dfrac{1}{2}, \left(-\dfrac{1}{2} \pm \dfrac{i\sqrt{3}}{2}\right)$

12. $\dfrac{1}{a}\sin A, \dfrac{1}{a}\sin\left(\dfrac{\pi}{3} - A\right), -\dfrac{1}{a}\sin\left(\dfrac{\pi}{3} + A\right)$ **13.** $\dfrac{3}{4}, \dfrac{3}{4}, \dfrac{3}{4}$ **14.** $-(a+b+c), -(a+b\omega+c\omega^2), -(a+b\omega^2+c\omega)$

15. $\dfrac{7 - 7k(2)^{1/3}(5)^{2/3} + 7k^2(2)^{2/3}(5)^{1/3}}{7}, k = 1, \omega, \omega^2$ **16.** $\dfrac{-1 - 3^{1/3}(5)^{2/3}\alpha - (3)^{2/3}(5)^{1/3}\alpha}{2}, \alpha = 1, \omega, \omega^2$ **17.** $-8, \dfrac{1}{2}(5 \pm i3\sqrt{3})$ **18.** $\dfrac{5}{2}, \dfrac{-40 \pm i7\sqrt{7}}{38}$

1.28 REDUCTION OF BIQUADRATIC EQUATION INTO EULER'S CUBIC AND REDUCING CUBIC

(i) Reduction into Euler's Cubic. Let the biquadratic equation be

$$a_0 x^4 + 4a_1 x^3 + 6a_2 x^2 + 4a_3 x + a_4 = 0 \qquad \ldots(1)$$

First we remove the second term, *i.e.*, $4a_1 x^3$ by diminishing each of its roots by a constant $h = -\dfrac{a_1}{4a_0}$, we get

$$z^4 + 6Hz^2 - 4Gz + (a_0^2 I - 3H^2) = 0 \qquad \ldots(2)$$

where $z = a_0 x + a_1$, $H = a_0 a_2 - a_1^2$, $G = a_0^2 a_3 - 3a_0 a_1 a_2 + 2a_1^3$ and $I = a_0 a_4 - 4a_1 a_3 + 3a_2^2$

Let the solution of (2) be $z = \sqrt{a} + \sqrt{b} + \sqrt{c}$. $\ldots(3)$

Squaring of both sides of (3), we get

$$z^2 = a + b + c + 2(\sqrt{a}\sqrt{b} + \sqrt{b}\sqrt{c} + \sqrt{c}\sqrt{a}) \quad \text{or} \quad z^2 - (a+b+c) = 2(\sqrt{a}\sqrt{b} + \sqrt{b}\sqrt{c} + \sqrt{c}\sqrt{a})$$

Again squaring of both sides, we get

$$[z^2 - (a+b+c)^2] = 4[ab + bc + ca + 2\sqrt{a}\sqrt{b}\sqrt{c} + (\sqrt{a} + \sqrt{b} + \sqrt{c})]$$

or $z^4 + (a+b+c)^2 - 2(a+b+c)z^2 = 4(ab+bc+ca) + 8\sqrt{a}\sqrt{b}\sqrt{c}z)$

or $z^4 - 2(a+b+c)z^2 - 8z\sqrt{a}\sqrt{b}\sqrt{c} + (a+b+c)^2 - 4(ab+bc+ca)^2 = 0$ $\ldots(4)$

The equation (2) and (4) are same, so comparing the coefficients of like powers of z, we get

$$(a+b+c) = -3H, \quad \sqrt{a}\sqrt{b}\sqrt{c} = -\dfrac{G}{2} \quad \text{and} \quad (a+b+c)^2 - 4(ab+bc+ca) = a_0^2 I - 3H^2$$

or $\sum a = -3H, \quad abc = \dfrac{G^2}{4} \quad \text{and} \quad (\sum a)^2 - 4(\sum ab) = a_0^2 I - 3H^2$

or $\sum ab = 3H^2 - \dfrac{a_0^2 I}{4}$

Therefore a, b, c are the roots of the equation

$$t^2 - (\sum a)t^2 + (\sum ab)t - abc = 0 \qquad \text{or} \qquad t^3 + 3Ht^2 + \left(3H^2 - \dfrac{a_0^2 I}{4}\right)t - \dfrac{G^2}{4} = 0 \qquad \ldots(5)$$

This equation is called Euler's cubic of biquadratic equation (1).

(ii) Reducing cubic. We have a relation $G^2 + 4H^3 = a_0^2(HI - a_0 J)$ where $J = a_0 a_2 a_3 + 2a_1 a_2 a_3 - a_0 a_3^2 - a_1^2 a_4 - a_2^3$

$\therefore$ $\dfrac{G^2}{4} + H^3 = \dfrac{a_0^2 HI}{4} - \dfrac{a_0^3 J}{4}$ or $-\dfrac{G^2}{4} = H^3 - \dfrac{a_0^2 HI}{4} + \dfrac{a_0^2 J}{4}$

Substitute the value $-\dfrac{G^2}{4}$ in (5), we get

$$t^3 + 3Ht^2 + \left(3H^2 - \frac{a_0^2 t}{4}\right)t + H^3 - \frac{a_0^2 HI}{4} + \frac{a_0^3 J}{4} = 0$$

or

$$t^3 + H^3 + 3H^2 t + 3Ht^2 - \frac{a_0^2 I}{4}t - \frac{a_0^2 HI}{4} + \frac{a_0^3 J}{4} = 0$$

or

$$(t+H)^3 - \frac{a_0^2 I}{4}(t+H) + \frac{a_0^3 J}{4} = 0$$

Let us put $t + H = a_0^2 \theta$, we get

$$a_0^6 \theta^3 - \frac{a_0^2 I}{4}(a_0^2 \theta) + \frac{a_0^3 J}{4} = 0 \quad \text{or} \quad 4a_0^3 \theta^3 - Ia_0\theta + J \qquad \qquad \text{...(6)}$$

This cubic equation of θ is called the reducing cubic of the biquadratic equation (1).

1.28.1 RELATION BETWEEN THE ROOTS OF BIQUADRATIC AND EULER'S CUBIC

The equation of the biquadratic and Euler's cubic are respectively given by

$$a_0 x^4 + 4a_1 x^3 + 6a_2 x^2 + 4a_3 x + a_4 = 0 \qquad \qquad \text{...(1)}$$

and

$$t^3 + 3Ht^2 + \left(3H^2 - \frac{a_0^2 I}{4}\right)t - \frac{G^2}{4} = 0 \qquad \qquad \text{...(2)}$$

Let α, β, γ, δ be the roots of the biquadratic and a, b, c the roots of the Euler's cubic.

Since we have taken $\qquad \qquad z = \sqrt{a} + \sqrt{b} + \sqrt{c}$ $\qquad \qquad$ ...(3)

and $\qquad \qquad z = a_0 x + a_1$ $\qquad \qquad$ ...(4)

It has been observed form equation (2) that z will have eight values because $\sqrt{a}, \sqrt{b}$ and $\sqrt{c}$ will have double signs. But we have

$$\sqrt{a}\sqrt{b}\sqrt{c} = -\frac{G}{2} \quad \text{therefore} \quad \sqrt{c} = \frac{G}{2\sqrt{a}\sqrt{b}}$$

$\therefore$ from (2), we get $\qquad \qquad z = \sqrt{a} + \sqrt{b} - \dfrac{G}{2\sqrt{a}\sqrt{b}}$ $\qquad \qquad$ ...(5)

Now from this equation it is observed that z will have four values. The signs of $\sqrt{a}, \sqrt{b}$ and $\sqrt{c}$ are taken in such a way that equation (4) should be satisfied. Therefore it G is negative, then $-\dfrac{G}{2}$ will be positive. Thus we should take the signs of $\sqrt{a}, \sqrt{b}$ and $\sqrt{c}$ so as to make the quantity $\sqrt{a}, \sqrt{b}, \sqrt{c}$ positive. Let z_1, z_2, z_3 and z_4 be the four values of z as the equation (5) indicates.

$\therefore$
$$\left. \begin{array}{l} z_1 = a_0 \alpha + a_1 = \sqrt{a} - \sqrt{b} - \sqrt{c} \\ z_2 = a_0 \beta + a_1 = -\sqrt{a} + \sqrt{b} - \sqrt{c} \\ z_3 = a_0 \gamma + a_1 = -\sqrt{a} - \sqrt{b} + \sqrt{c} \\ z_4 = a_0 \delta + a_1 = \sqrt{a} + \sqrt{b} + \sqrt{c} \end{array} \right\}$$
$\quad$ (Using (2) and (3) and $\sqrt{a}$ is taken positive.)
$\quad$ ($\sqrt{b}$ taken positive.)
$\quad$ ($\sqrt{c}$ is taken positive.)
$\quad$ (all are taken positive.)
$\qquad \qquad$ ...(A)

Adding first two equations and adding last two equations of above system of equations respectively, we get

$$a_0(\alpha + \beta) + 2a_1 = -2\sqrt{c} \quad \text{and} \quad a_0(\gamma + \delta) + 2a_1 = 2\sqrt{c}$$

Now substracting these equations, we get $a_0(\alpha + \beta - \gamma - \delta) = -4\sqrt{c}$

Squaring of both sides, we get $c = \dfrac{a_0^2}{16}(\alpha + \beta - \gamma - \delta)^2$

Similarly we get $\qquad a = \dfrac{a_0^2}{16}(\beta + \gamma - \alpha - \delta)^2 \quad \text{and} \quad = \dfrac{}{16}(\gamma + \alpha - \beta - \delta)$

Hence, we obtained the required relations as follows :

$$= \frac{}{16}(\beta + \gamma - \alpha - \delta) \;, b = \frac{a_0^2}{16}(\gamma + \alpha - \beta - \delta)^2, c = \frac{a_0^2}{16}(\alpha + \beta - \gamma - \delta)^2$$

It G is a positive, then $-\dfrac{G}{2}$ will be negative. Therefore we will take the signs of $\sqrt{a}, \sqrt{b}, \sqrt{c}$ either all negative or two positive and one negative. Thus we shall obtain

$$z_1 = a_0\alpha + a_1 = -\sqrt{a} + \sqrt{b} + \sqrt{c}$$
$$z_2 = a_0\beta + a_1 = \sqrt{a} - \sqrt{b} + \sqrt{c}$$
$$z_3 = a_0\gamma + a_1 = \sqrt{a} + \sqrt{b} - \sqrt{c}$$
$$z_4 = a_0\delta + a_1 = -\sqrt{a} - \sqrt{b} - \sqrt{c}$$

and ...(B)

Solving these equation's we obtain the relations as the same as obtained above.

1.28.2 Relation between the Roots of Biquadratic and the Reducing Cubic

The equations of biquadratic and reducing cubic are respectively given by

$$a_0 x^4 + 4a_1 x^3 + 6a_2 x^2 + 4a_3 x + a_4 = 0 \qquad \text{...(1)}$$

and

$$4a_0^3 \theta^3 - Ia_0\theta + J = 0 \qquad \text{...(2)}$$

Let α, β, γ, δ be the roots of (1) and θ_1, θ_2, θ_3 be the roots of (2) in the system of equations (A) of §24.26, we have

$$a_0(\alpha - \beta) = 2(\sqrt{a} - \sqrt{b}) \qquad \text{and} \qquad a_0(\gamma - \beta) = -2(\sqrt{a} + \sqrt{b})$$

$\therefore$
$$a_0^2(\alpha - \beta)(\gamma - \delta) = 4(a - b)$$

But we have
$$t + H = a_0^2\theta$$

$\therefore$
$$a + H = a_0^2\theta_1, \, b + H = a_0^2\theta_2, \, c + H = a_0^2\theta_3$$

$\therefore$
$$(a - b) = a_0^2(\theta_1 - \theta_2)$$

$$4(a - b) = 4a_0^2(\theta_1 - \theta_2) = -a_0^2(\alpha - \beta)(\gamma - \delta)$$

Similarly
$$4(b - c) = 4a_0^2(\theta_2 - \theta_3) = -a_0^2(\beta - \gamma)(\alpha - \delta) \qquad \text{...(A)}$$

$$4(c - a) = 4a_0^2(\theta_3 - \theta_1) = -a_0^2(\gamma - \alpha)(\beta - \delta)$$

Substracting first equation of (A) from third equation of (A), we get

$$4a_0^2(\theta_3 - \theta_1 - \theta_1 + \theta_2) = a_0^2[(\alpha - \beta)(\gamma - \delta) - (\gamma - \alpha)(\beta - \delta)]$$

$$4a_0^2(\theta_2 + \theta_3 - 2\theta_1) = a_0^2[(\alpha - \beta)(\gamma - \delta) - (\gamma - \alpha)(\beta - \delta)]$$

But we have
$$\theta_1 + \theta_2 + \theta_3 = 0, \, \theta_2 + \theta_3 = -\theta_1$$

$\therefore$
$$4a_0^2(-3\theta_1) = a_0^2[(\alpha - \beta)(\gamma - \delta) - (\gamma - \alpha)(\beta - \delta)]$$

or
$$12\theta_1 = (\gamma - \alpha)(\beta - \delta) - (\alpha - \beta)(\gamma - \delta)$$

Similarly
$$12\theta_2 = (\alpha - \beta)(\gamma - \delta) - (\beta - \gamma)(\alpha - \delta), \, 12\theta_3 = (\beta - \gamma)(\alpha - \delta) - (\gamma - \alpha)(\beta - \delta)$$

Hence, the required relations are

$$\theta_1 = \frac{1}{12}[(\gamma - \alpha)(\beta - \delta) - (\alpha - \beta)(\gamma - \delta)], \quad \theta_2 = \frac{1}{12}[(\alpha - \beta)(\gamma - \delta) - (\beta - \gamma)(\alpha - \delta)], \quad \theta_3 = \frac{1}{12}[(\beta - \gamma)(\alpha - \delta) - (\gamma - \alpha)(\beta - \delta)]$$

1.29 DESCARTE'S METHOD FOR FINDING THE ROOTS OF A BIQUADRATIC

Let the equation of a biquadratic be

$$a_0 x^4 + 4a_1 x^3 + 6a_2 x^2 + 4a_3 x + a_4 = 0 \qquad \text{...(1)}$$

First we remove the second term, i.e., $4a_1 x^3$ from (1) be diminishing each of root of (1) by a constant $h = -\dfrac{a_1}{na_0}$, we get

$$z^4 + 6Hz^2 + 4Gz + a_0^2 I - 3H^2 = 0 \qquad \text{...(2)}$$

where $H = a_0 a_2 - a_1^2$, $G = a_0^2 a_3 - 3a_0 a_1 a_2 + 2a_1^3$, $I = a_0 a_4 - 4a_1 a_3 + 3a_2^2$ and $z = a_0 x + a_1$.

Let us assume

$$z^4 + 6Hz^2 + 4Gz + a_0^2 I - 3H^2 \equiv (z^2 + kz + l)(z^2 - kz + m)$$

Now equating the coefficients of like powers of z, we get

$$l + m - k^2 = 6H, \, k(m - 1) = 4G, \, lm = a_0^2 I - 3H^2$$

Solving first two of these equations for l and m, we get

$$2l = k^2 + 6H - \frac{4G}{k} \text{ and } 2m = k^2 + 6H + \frac{4G}{k} \qquad \text{...(A)}$$

Substitute these values of l, m in the following equation $lm = a_0^2 I - 3H^2$, we get

$$\left(k^2 + 6H - \frac{4G}{k}\right)\left(k^2 + 6H + \frac{4G}{k}\right) = 4(a_0^2 I - 3H^2)$$

$$\Rightarrow \qquad (k^3 + 6Hk - 4G)(k^3 + 6Hk + 4G) = 4(a_0^2 I - 3H^2)k^2$$

$$k^6 + 12Hk^4 + 4k^2(12H^2 - a_0^2 I) - 16G^2 = 0 \qquad \qquad \qquad \dots(3)$$

This is cubic equation in k^2 so it will always have one positive real value of k^2 when k^2 is know, then the values of l and m are obtained from the equation (A). Thus the biquadratic (2) is obtained as the product of quadratics $(z^2 + kz + 1)$ and $(z^2 - kz + m)$.

Now solving these two quadratics

$$(z^2 + kz + l) \text{ and } (z^2 - kz + m) = 0$$

and finally from the transformation $z = a_0 x + a_1$ we obtain the solution of the given biquadratic (1) corresponding to the roots of the equations.

$$z^2 + kz + l = 0 \qquad \qquad \text{and} \qquad \qquad z^2 - kz + m = 0$$

1.30 FERRARI'S METHOD FOR FINDING THE ROOTS OF A BIQUADRATIC EQUATION

Let the equation of a biquadratic be $\quad x^4 + 2a_1 x^3 + a_2 x^2 + 2a_3 x + a_4 = 0 \qquad \dots(1)$

Now adding $(ax + b)^2$ to each side of (1), we get

$$x^4 + 2a_1 x^3 + a_2 x^2 + 2a_3 x + a_4 + (ax + b)^2 = (ax + b)^2$$

or $\qquad x^4 + 2a_1 x^3 + (a_2 + a^2)x^2 + 2(a_3 + ab)x + (a_4 + b)^2 = (ax + b)^2 \qquad \dots(2)$

In order to determine a and b make the left side of above equation of perfect square.

Suppose the perfect square of left side of (2) is $(x^2 + a_1 x + k)^2$, then

$$x^4 + 2a_1 x^3 + (a_2 + a^2)x^2 + 2(a_3 + ab)x + (a_4 + b)^2 \equiv (x^2 + a_1 x + k)^2 \qquad \dots(3)$$

Comparing the coefficients of like powers of x of (3), we get

$$a_1^2 + 2k = a_2 + a^2, \, a_1 k = a_3 + ab, \, k^2 = a_4 + b^2.$$

Eliminating a and b between these equations, we get

$$(2k + a_1^2 - a_2)(k^2 - a_4) = (a_1 k - a_3)^2 \quad \text{or} \quad 2k^3 - a_2 k^2 + 2(a_1 a_3 - a_4)k - a_1^2 a_4 + a_2 a_4 - a_3^2 = 0 \qquad \dots(4)$$

This is a cubic equation in k so it must have one real values of k. This real value is obtained by trial method. Once we obtained the value of k we thus obtain a and b and then put these values in (3) and using (2), we get

$$(x^2 + a_1 x + k^2) = (ax + b)^2 \qquad \text{or} \qquad x^2 + a_1 x + k = \pm(ax + b)$$

Thus the given biquadratic is obtained as the product of two quadratics.

$$\left.\begin{array}{l} x^2 + (a_1 - a)x + (k - b) = 0 \\ x^2 + (a_1 + a)x + (k + b) = 0 \end{array}\right\} \qquad \dots(5)$$

and

On solving these quadratics we finally obtained the solution of the given biquadratic equation.

Solved Examples

EXAMPLE 1. *Solve the equation $x^4 - 3x^2 - 42x - 40 = 0$ by Descarte's method.*

SOLUTION. The given equation is

$$x^4 - 3x^2 - 42x - 40 = 0 \qquad \dots(1)$$

Let us assume

$$x^4 - 3x^2 - 42x - 40 \equiv (x^2 + kx + l)(x^2 - kx + m) = 0 \qquad \dots(2)$$

Equating the coefficients of like powers of x, we get

$$l + m - k^2 = -3 \quad \text{or} \quad l + m = -3 + k^2 \qquad \dots(3)$$

and $\quad k(m - l) = -42 \quad \text{or} \quad m - l = -\dfrac{42}{k} \qquad \dots(4)$

and $\qquad lm = -40 \qquad \dots(5)$

Solving (3) and (4) we get

$$2m = -3 + k^2 - \frac{42}{k}$$

and $\qquad 2l = -3 + k^2 + \dfrac{42}{k}$

Substitute the values of l and m in (5), we get

$$\left(-3 + k^2 - \frac{42}{k}\right)\left(-3 + k^2 + \frac{42}{k}\right) = 4(-40)$$

or $\quad (k^3 - 3k - 42)(k^3 - 3k + 42) = -160k^2$

or $\quad (k^3 - 3k)^2 - (42)^2 = -160k^2$

or $\qquad k^6 - 6k^4 + 169k^2 - 1764 = 0$

Let $k^2 = t$, then we get

$$t^3 - 6t^2 + 169t - 1764 = 0$$

By trial method it is obvious that $t = 9$ satisfies above equation.

Hence $\qquad k^2 = 9 \text{ or } k = \pm 3$

Taking $k = 3$, in (3) and (4), we get

$$l + m = 6 \text{ and } m - l = -14$$

Solving these equation for l and m, we get $l = 10$, $m = -4$ therefore from (2) we obtain the given biquadratic as the product of two quadratics

$$(x^2 + 3x + 10)(x^2 - 3x - 4) = 0$$

Solving these quadratics respectively, we get the required solutions

$$x = 4, -1, \frac{-2 \pm i\sqrt{31}}{2}$$

EXAMPLE 2. *Solve the equation $x^4 + 8x^3 + 9x^2 - 8x - 10 = 0$ by Descrate's method.*

SOLUTION. The given equation is

$$x^4 + 8x^3 + 9x^2 - 8x - 10 = 0 \qquad \ldots(1)$$

First we remove the second term, *i.e.*, by diminishing each of its roots by a constant

$$h = -\frac{a_1}{na_0} = -\frac{8}{4.1} = -2$$

Using synthetic division method as follows :

```
-2 | 1    8     9    -8   -10
   |     -2   -12     6     4
   |_____
     1    6    -3    -2  | -6
         -2    -8    22
   |_____
     1    4   -11    20
         -2    -4
   |_____
     1    2  | -15
         -2
   |_____
     1  | 0
   |____
     1
```

Thus the transformed equation is

$$z^4 - 15z^2 + 20z - 6 = 0 \qquad \ldots(2)$$

where $z = x + 2$

let as assume

$$z^4 - 15z^2 + 20z - 6 \equiv (z^2 + kz + l)(z^2 - kz + m) = 0 \ldots(3)$$

Comparing the coefficients of like powers of z, we get

$$l + m - k^2 = -15$$

or

$$l + m = -15 + k^2 \qquad \ldots(4)$$

$$k(m - l) = 20$$

or

$$m - l = \frac{20}{k} \qquad \ldots(5)$$

and

$$lm = -6$$

Solving (4) and (5), we get

$$2l = k^2 - 15 - \frac{20}{k}$$

$$2m = k^2 - 15 + \frac{20}{k}$$

Substitute these values of l and m in (6), we get

$$\left(k^2 - 15 - \frac{20}{k}\right)\left(k^2 - 15 + \frac{20}{k}\right) = -24$$

or $(k^3 - 15k - 20)(k^3 - 15k + 20) = -24k^2$

or $\qquad (k^3 - 15k)^2 - 400 = -24k^2$

or $\quad k^6 - 30k^4 + 249k^2 - 400 = 0$

Let $k^2 = t$, then

$$t^3 - 30t^2 + 249t - 400 = 0 \qquad \ldots(7)$$

From (7) it is obvious that $t = 16$ satisfies the equation (7)

$\therefore \qquad k^2 = 16$ or $k = \pm 4$

Taking $k = 4$, in (4) and (5), we get

$$l + m = 1$$
$$m - l = 5$$

On solving these equations, we get

$$l = -2, m = 3$$

Substitute the values of l, m and k in (3), we get

$$z^4 - 15z^2 + 20z - 6 \equiv (z^2 + 4z - 2)(z^2 - 4z + 3) = 0$$

$\therefore \quad (z^2 + 4z - 2)(z^2 - 4z + 3) = 0$

and $\qquad z = 1, 3, -2 \pm \sqrt{6}$

But $\qquad z = x + 2 \quad \Rightarrow \quad x = z - 2$

Hence, the solution of the given biquadratic are

$$x = -1, 1, -4 \pm \sqrt{6}$$

EXAMPLE 3. *Solve the equation $x^4 - 2x^3 - 5x^2 + 10x - 3 = 0$ by Ferrari's method.*

SOLUTION. Since the equation is

$$x^4 - 2x^3 - 5x^2 + 10x - 3 = 0 \qquad \ldots(1)$$

Adding $(ax + b)^2$ of both sides, we get

$$x^4 - 2x^3 - 5x^2 + 10x - 3 + (ax + b)^2 = (ax + b)^2$$

or $x^4 - 2x^3 + (a^2 - 5)x^2 + 2(ab + 5)x + b^2 - 3 = (ax + b)^2$

$$\ldots(2)$$

Let as assume that L.H.S. of (2) must be a perfect square, let us suppose $(x^2 - a_1x + k^2)$ is a perfect square of L.H.S. of (2).

$$\therefore \quad x^4 - 2x^3 + (a^2 - 5)x^2$$
$$+ 2(ab + 5)x + b^2 - 3 \equiv (x^2 - x + k)^2 \qquad \ldots(3)$$

$$(\because a_1 = -1)$$

Equating the coefficients of like powers of x, we get

$$a^2 = 2k + 6, ab = -k - 5, b^2 = k^2 + 3$$

Now eliminating a and b between these three equations, we get

$$(2k + 6)(k^2 + 3) = (k + 5)^2$$

or $\qquad 2k^3 + 5k^2 - 4k - 7 = 0$

It is a cubic in k so it must have one real root, then by trial method, we get

$$k = -1$$

and hence $\qquad a^2 = 4, b^2 = 4, ab = -4$

or $\qquad a = 2, b = -2$

Substitute the values of k, a and b in (3) and (4), we get

$$(x^2 - x - 1)^2 = (2x - 2)^2$$

or $\qquad x^2 - x - 1 = \pm(2x - 2)$

or $\qquad x^2 - 3x + 1 = 0$

and $\qquad x^2 + x - 3 = 0$

Solving these quadratics, we get

$$x = \frac{3 \pm \sqrt{5}}{2}, \frac{-1 \pm \sqrt{13}}{2}$$

These are the solutions of the given biquadratic equation.

EXAMPLE 4. *Solve the equation $x^4 + 2x^3 - 7x^2 - 8x + 12 = 0$ by Ferrari's method.*

SOLUTION. Since the given biquadratic is

$$x^4 + 2x^3 - 7x^2 - 8x + 12 = 0 \qquad \ldots(1)$$

Adding $(ax + b)^2$ of both sides of (1), we get

$$x^4 + 2x^3 - 7x^2 - 8x + 12 + (ax + b)^2 = (ax + b)^2$$

or $x^4 + 2x^3 + (a^2 - 7)x^2 + (2ab - 8)x + b^2 + 12 = (ax + b)^2$

$$\ldots(2)$$

In order to determine a and b make the L.H.S. of (2) a perfect square. Let the perfect square be

$$(x^2 + a_1x + k)^2$$

$\therefore \quad x^4 + 2x^3 + (a^2 - 7)x^2 + (2ab - 8)x + b^2 + 12$
$$\equiv (x^2 + a_1x + k)^2$$

or $\qquad x^4 + 2x^3 + (a^2 - 7)x^2 + (2ab - 8)x$

$$+ b^2 + 12 \equiv (x^2 + x + k)^2 \qquad \ldots(3)$$

$(\because a_1 = 1 \text{ (from(1))})$

Equating the coefficients of like powers of x, we get

$$a^2 - 7 = 2k + 1, 2ab - 8 = 2k, b^2 + 12 = k^2$$

Eliminating a and b between above three equations, we get

$$(k+4)^2 = (2k+8)(k^2-12)$$

or $\quad k^3 + 16 + 8k = 2k^2 + 8k^2 - 24k - 96$

or $\quad 2k^3 + 7k^2 - 32k - 112 = 0$

This is a cubic in k so it must have one real root. By trial method, $k = -7/2$ satisfies above cubic.

Then $\qquad a^2 = 1, b^2 = \dfrac{1}{4}$

$\therefore \qquad\qquad a = 1, b = \dfrac{1}{2}$

Now substitute the values of k, a and b in (3) and using (2), we get

$$\left(x^2 + x - \frac{7}{2}\right)^2 = \left(x + \frac{1}{2}\right)^2$$

or $\qquad x^2 + x - \dfrac{7}{2} = \pm\left(x + \dfrac{1}{2}\right)$

or $\qquad x^2 - 4 = 0$ and $x^2 + 2x - 3 = 0$

Solving these quadratics, we get

$$x = -2, 2, \text{ and } x = 1, -3$$

Hence, the solution of given biquadratic are

$$x = -3, -2, 1, 2.$$

EXAMPLE 5. *Show that the equation*

$$x^4 + px^3 + qx^2 + rx + s = 0$$

may be solved a quadratic if $r^2 = p^2s$.

SOLUTION. Since the given equation is

$$x^4 + px^3 + qx^2 + rx + s = 0 \qquad\qquad …(1)$$

Let us assume

$$x^4 + px^3 + qx^2 + rx + s \equiv \left(x^2 + \frac{p}{2}x + l\right)^2 = 0$$

Comparing the coefficients of like power of x, we get

$$2l\left(q - \frac{p^2}{4}\right), pl = r, l^2 = s$$

Eliminating l between last two equations, we get

$$pl = r \qquad \Rightarrow \qquad p^2l^2 = r^2$$

or $\qquad p^2s = r^2 \qquad \because \qquad l^2 = s$

Exercise-1.20

Solve the following biquadratic equation by Descarte's method.

1. $x^4 - 6x^3 - 9x^2 + 66x - 22 = 0$

2. $x^4 - 8x^2 - 24x + 7 = 0 \qquad$ (UPTU–2001, 10)

3. $x^4 - 10x^2 - 20x - 16 = 0$

4. $x^4 + 2x^3 - 7x^2 - 8x + 12 = 0$

5. $x^4 - 8x^3 - 12x^2 + 60x + 63 = 0$

6. $x^4 - 3x^2 - 6x - 2 = 0$

7. $x^4 - 5x^2 - 6x - 5 = 0$

8. $x^4 - 12x - 5 = 0$

9. $4x^4 - 20x^3 + 33x^2 - 20x + 4 = 0$

10. $x^4 + 8x^3 + 9x^2 - 8x - 10 = 0$

11. $x^4 - 2x^2 + 8x - 3 = 0$

Solve the following biquadratic equation by Ferrari's method.

12. $x^4 - 8x^3 - 12x^2 + 60x + 63 = 0 \qquad$ (UPTU–2005)

13. $x^4 + 12x - 5 = 0$

14. $x^4 - 2x^3 - 5x^2 + 10x - 3 = 0$

15. $x^4 - 3x^2 - 42x - 40 = 0$

16. $x^4 + 9x^3 + 12x^2 - 80x - 192 = 0$

17. $x^4 - 2x^3 - 12x^2 + 10x + 3 = 0$

18. $x^4 - 10x^3 + 44x^2 - 104x + 96 = 0$

19. $x^4 + 4x^3 + 12x^2 - 8x + 95 = 0$

20. $x^4 + 3x^3 + x^2 - 2 = 0$

21. $2x^4 + 6x^3 - 3x^2 + 2 = 0$

22. $x^4 - 12x^3 + 41x^2 - 18x - 72 = 0 \qquad$ (SVTU–2007)

23. $x^4 - 10x^3 + 35x^2 - 50x + 24 = 0 \qquad$ (UPTU–2003, 08)

24. $x^4 + 2x^3 - 7x^2 - 8x + 12 = 0 \qquad$ (UPTU–2002)

Hints to Selected Problems

1. $h = \dfrac{3}{2}$ The transformed equation is

$$y^4 - \frac{90}{4}y^2 + \frac{96}{8}y + \frac{665}{16} = 0$$

$\Rightarrow \quad (2y)^4 - 90(2y)^2 + 96(2y) + 665 = 0$

Put $\qquad 2y = z$

5. $h = 2$

The transformed equation is

$$z^4 - 36z^2 - 52z + 87 = 0 \quad \text{where } z = x - 2$$

Let us write

$$z^4 - 36z^2 - 52z + 89 = (z^2 + kx + l)(z^2 - kz + m) = 0$$

11. Assume that $x^4 - 2x^2 + 8x - 3 = (x^2 + kx + l)(x^2 - kx + m) = 0$

13. Adding $(ax + b)^2$ of both the sides of the given equation.

Answers

1. $\pm\sqrt{11}, 3 \pm \sqrt{7}$ $\qquad$ **2.** $-2 \pm i\sqrt{3}, 2 \pm \sqrt{3}$ $\qquad$ **3.** $4, -2, -1 \pm i$ $\qquad$ **4.** $\pm 2, -3, 1$ $\qquad$ **5.** $-1, 3, 3 \pm \sqrt{30}$

6. $1 \pm \sqrt{2}, -1 \pm i$ $\qquad$ **7.** $\dfrac{-1 \pm \sqrt{3}}{2}, \dfrac{1 \pm \sqrt{21}}{2}$ $\qquad$ **8.** $-1 \pm 2i, 1 \pm \sqrt{2}$ $\qquad$ **9.** $2, 2, \dfrac{1}{2}, \dfrac{1}{2}$ $\qquad$ **10.** $\pm 1, -4 \pm \sqrt{6}$

11. $1 \pm i\sqrt{2}, -1 \pm \sqrt{2}$ $\qquad$ **12.** $-1, 3, 3 \pm \sqrt{30}$ $\qquad$ **13.** $-1 \pm \sqrt{2}, -1 \pm 2i$ $\qquad$ **14.** $\dfrac{3 \pm \sqrt{5}}{2}, \dfrac{-1 \pm \sqrt{13}}{2}$ $\qquad$ **15.** $4, -1, -\dfrac{1}{2}(3 \pm i\sqrt{31})$

16. $-4, -4, -4, 3$ **17.** $1, -3, 2 \pm \sqrt{5}$ **18.** $2, 4, 2 \pm i2\sqrt{2}$ **19.** $-3 \pm i\sqrt{10}, 1 \pm 2i$ **20.** $-1 \pm \sqrt{3}, \dfrac{-1 \pm i\sqrt{3}}{2}$

21. $-2 \pm \sqrt{2}, \dfrac{1}{2}(1 \pm i)$ **22.** $-1, 3, 4, 6$ **23.** $1, 2, 3, 4$ **24.** $-3, 1, 2, -2$

ARCHIVE

1. Solve $6x^5 - 41x^4 - 97x^3 - 97x^2 + 41x - 6 = 0$. (MADRAS–2000, 12)

2. Solve $6x^6 - 25x^5 + 31x^4 - 31x^2 + 25x - 6 = 0$. (MADRAS–2000)

3. Show that $\begin{vmatrix} A_1 & B_1 & C_1 \\ A_2 & B_2 & C_2 \\ A_3 & B_3 & C_3 \end{vmatrix} = \begin{vmatrix} a_1 & b_1 & c_1 \\ a_2 & b_2 & c_2 \\ a_3 & b_3 & c_3 \end{vmatrix}^2$ where A, B, C are the cofactors of a, b, c respectively in the determinants $(a_1 b_2 c_3)$. (RANCHI–2009)

4. Show that if $\lambda \neq 5$ the system of equations
$3x + 4y + 5z = a$, $4x + 5y + 6z = b$, $5x + 6y + 7z = c$
do not have a solution unless $a + c = 2b$. (RAIPUR–1998, 2014)

5. Find the rank of the following matrices:

(i) $\begin{bmatrix} 2 & 3 & -1 & -1 \\ 1 & -1 & -2 & -4 \\ 3 & 1 & 3 & -2 \\ 6 & 3 & 0 & -7 \end{bmatrix}$ (NITR–2005, 11; PTU–2007)

(ii) $\begin{bmatrix} 1 & 2 & 3 & 0 \\ 2 & 4 & 3 & 2 \\ 3 & 2 & 1 & 3 \\ 6 & 8 & 7 & 5 \end{bmatrix}$ (KOTTAYAM–2005)

6. Find the rank of the matrix $\begin{bmatrix} 2 & -2 & 0 & 6 \\ 4 & 2 & 0 & 2 \\ 1 & -1 & 0 & 3 \\ 1 & -2 & 1 & 2 \end{bmatrix}$ by reducing it into normal form. (NITK–2010)

7. Are the vectors $x_1 = (1, 2, 4)$, $x_2 = (0, 1, 2)$, $x_3 = (0, ,1, 2)$, $x_4 = (-3, 7, 2)$ linearly dependent? If so find the relation between them. (UPTU–2011)

8. For what value of k, the set of vectors $(k, 1, 1)$, $(0, 1, 1)$ and $(k, 0, k)$ is linearly independent. (PTU–2010, 12, 17)

9. For what value of k, the equations $x + y + z = 1$, $2x + y + 4z = k$, $4x + y + 10z = k^2$ have a solution? (NITU–2011; KUK–2005; PTU–2005)

10. Solve the system of equations $x_1 + 2x_3 - 2x_4 = 0$, $2x_1 - x_2 - x_4 = 0$, $x_1 + 2x_3 - x_4 = 0$, $4x_1 - x_2 + 3x_3 - x_4 = 0$. (NITK–2012)

11. Test whether the matrix $\begin{bmatrix} 3 & 1 & -1 \\ -2 & 1 & 2 \\ 0 & 1 & 2 \end{bmatrix}$ is diagonalizable or not. Find P such that $P'AP$ is a diagonal matrix. Then obtain the matrix $B = A^2 + 5A + 3I$. (PTU–2009, 12)

12. Prove that then matrix $A = \begin{bmatrix} -2/3 & 1/3 & 2/3 \\ 2/3 & 2/3 & 1/3 \\ 1/3 & -2/3 & 2/3 \end{bmatrix}$ is orthogonal. (PTU–2011)

13. Reduce the quadratic form $2xy + 2yz + 2zx$ to the Canonical form by an orthogonal reduction and state its nature. (NITK–2006; NITU–2011)

14. If $S = \begin{bmatrix} 1 & 1 & 1 \\ 1 & a^2 & a \\ 1 & a & a^2 \end{bmatrix}$ where $a = e^{2\pi i/3}$, show that $S^{-1} = \dfrac{1}{3}\bar{S}$. (NITK–2011)

15. Find the characteristic equation of the matrix $\begin{bmatrix} 2 & 1 & 1 \\ 0 & 1 & 0 \\ 1 & 1 & 2 \end{bmatrix}$ and hence find the matrix represented by
$A^8 - 5A^7 + 7A^6 - 3A^5 + A^4 - 5A^3 + 8A^2 - 2A + I$ (RAJASTHAN–2005; NIT(HAMIRPUR)–2011; NITK–2011)

16. If $A = \begin{bmatrix} 0 & 1+2i \\ -1+2i & 0 \end{bmatrix}$. Show that the matrix $(I - A)(I + A)^{-1}$ is unitary. (KUK–2008, 14)

Answers

1. $1, \dfrac{1}{3}, 3, \dfrac{1}{2}, 2$ **2.** $1, -1, 2, \dfrac{1}{2}, \dfrac{5 \pm i\sqrt{11}}{6}$ **5.** (i) 3   (ii) 3   **6.** 3   **7.** Yes, $9x_1 - 12x_2 + 5x_3 - 5x_4 = 0$

8. $k \neq 0$ **9.** 3   **10.** $x_1 = x_2 = x_3 = x_4 = 0$ **11.** $P = \begin{bmatrix} 1 & 1 & 0 \\ -1 & 0 & 1 \\ 1 & 1 & 1 \end{bmatrix}, B = \begin{bmatrix} 25 & 8 & -8 \\ -18 & 9 & 18 \\ -2 & 8 & 19 \end{bmatrix}$

13. The quadratic form in indefinite in nature, r = 3, index = 1   **15.** $A^3 - 5A^2 + 7A - 31 = 0, \begin{bmatrix} 8 & 5 & 5 \\ 0 & 3 & 0 \\ 5 & 5 & 8 \end{bmatrix}$

✳✳✳✳✳✳✳

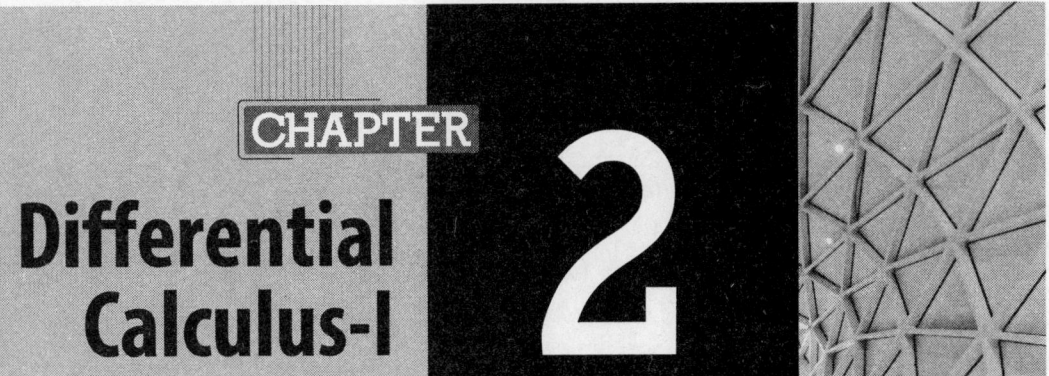

CHAPTER 2

Differential Calculus-I

2.1 SUCCESSIVE DIFFERENTIATION

Let $y = f(x)$ be a function, then the differential coefficient of $f(x)$ denoted by $f'(x)$ is defined as follows

$$f'(x) = \lim_{\delta x \to 0} \frac{f(x + \delta x) - f(x)}{\delta x} = \frac{dy}{dx}$$

If the limit exists (*i.e.*, limit is finite and unique), then $f'(x)$ is called *first differential coefficient of $f(x)$ with respect to x*. Similarly, if $f(x)$ is differentiable twice, it is denoted by $f''(x)$, if it is differentiable thrice, it is denoted by $f'''(x)$, *i.e.*,

$$f''(x) = \frac{d}{dx}\left(\frac{dy}{dx}\right) = \frac{d^2 y}{dx^2}$$

$$f'''(x) = \frac{d}{dx}\left(\frac{d^2 y}{dx^2}\right) = \frac{d^3 y}{dx^3}$$

If $y = f(x)$ be a function of x, then we adopt the following notations.

$$y_1 = f'(x) = \frac{dy}{dx} = Df(x) = \frac{d}{dx}(f(x))$$

$$y_2 = f''(x) = \frac{d^2 y}{dx^2} = D^2 f(x) = \frac{d^2}{dx^2}(f(x))$$

$$y_3 = f'''(x) = \frac{d^3 y}{dx^3} = D^3 f(x) = \frac{d^3}{dx^3}(f(x))$$

$$\cdots \qquad \cdots \qquad \cdots \qquad \cdots \ \cdots$$
$$\cdots \qquad \cdots \qquad \cdots \qquad \cdots \ \cdots$$

$$y_n = f^n(x) = \frac{d^n y}{dx^n} = D^n f(x) = \frac{d^n}{dx^n}(f(x))$$

This process of finding the differential coefficients of a function is called successive differentiation.

2.1.1 N^{TH} DIFFERENTIATION OF SOME STANDARD FUNCTIONS

(i) $y = f(x) = x^n$.

We have

$$y = f(x) = x^n$$
$$y_1 = f'(x) = nx^{n-1}$$
$$y_2 = f''(x) = n(n-1)x^{n-2}$$
$$y_3 = f'''(x) = n(n-1)(n-2)x^{n-3}$$
$$\cdots\cdots\cdots\cdots\cdots\cdots\cdots\cdots\cdots\cdots$$
$$\cdots\cdots\cdots\cdots\cdots\cdots\cdots\cdots\cdots\cdots$$
$$y_n = f^n(x) = n(n-1)(n-2)\ldots 3.2.1.x^0$$

$$\Rightarrow \qquad \frac{d^n}{dx^n}(x^n) = y_n = n!$$

(ii) $y = f(x) = x^m$.

We have $\qquad y_1 = f'(x) = mx^{m-1}, \qquad\qquad y_2 = f''(x) = m(m-1)x^{m-2},$

$$y_3 = f'''(x) = m(m-1)(m-2)x^{m-3}, \ldots, \ y_n = f^n(x) = m(m-1)(m-2)\ldots(m-n+1).x^{m-n}$$

$$= \left[\frac{m(m-1)(m-2)\ldots(m-n+1)(m-n)\ldots 3.2.1}{(m-n)(m-n-1)\ldots 3.2.1}\right]x^{m-n}$$

$$\Rightarrow \qquad y_n = \frac{d^n}{dx^n}(x^m) = \frac{m!}{(m-n)!}x^{m-n}$$

(iii) $y = f(x) = \dfrac{1}{(ax + b)}$.

We have $\quad y_1 = f'(x) = -\dfrac{a}{(ax+b)^2}$, $\quad y_2 = f''(x) = \dfrac{a^2 \cdot 2}{(ax+b)^3}$,

$$y_3 = f'''(x) = \dfrac{-a^3 \cdot 2.3}{(ax+b)^4} , \dots, y_n = f^n(x) = \dfrac{(-1)^n a^n \cdot 2.3.4 \dots n}{(ax+b)^{n+1}}$$

$\Rightarrow \qquad y_n = \dfrac{d^n}{dx^n}\left(\dfrac{1}{ax+b}\right) = \dfrac{(-1)^n \cdot a^n \cdot n!}{(ax+b)^{n+1}}$

(iv) $y = f(x) = \dfrac{1}{(ax + b)^m}$.

We have $\quad y_1 = f'(x) = -\dfrac{a.m}{(ax+b)^{m+1}}$, $\quad y_2 = f''(x) = \dfrac{a^2 . m(m+1)}{(ax+b)^{m+2}}$,

$$y_3 = f'''(x) = -\dfrac{a^3 . m(m+1)(m+2)}{(ax+b)^{m+3}} , \dots , y_n = f^n(x) = (-1)^n \dfrac{a^n . m(m+1)(m+2)\dots(m+n-1)}{(ax+b)^{m+n}}$$

$\Rightarrow \qquad y_n = \dfrac{d^n}{dx^n}\left(\dfrac{1}{(ax+b)^m}\right) = (-1)^n \dfrac{a^n .(m+n-1)!}{(m-1)!(ax+b)^{m+n}}$

(v) $y = f(x) = \sin(ax + b)$.

We have $\quad y_1 = f'(x) = a\cos(ax+b) = a\sin\left(\dfrac{\pi}{2}+ax+b\right)$, $y_2 = f''(x) = a^2\cos\left(\dfrac{\pi}{2}+ax+b\right) = a^2\sin\left(2.\dfrac{\pi}{2}+ax+b\right)$

$$y_3 = f'''(x) = a^3\cos\left(2.\dfrac{\pi}{2}+ax+b\right) = a^3\sin\left(3.\dfrac{\pi}{2}+ax+b\right)$$

$$\dots\dots\dots\dots\dots\dots\dots\dots\dots\dots\dots\dots\dots\dots$$
$$\dots\dots\dots\dots\dots\dots\dots\dots\dots\dots\dots\dots\dots\dots$$

$$y_n = f^n(x) = a^n\cos\left((n-1)\dfrac{\pi}{2}+ax+b\right) = a^n\sin\left(n.\dfrac{\pi}{2}+ax+b\right)$$

$\Rightarrow \qquad y_n = \dfrac{d^n}{dx^n}[\sin(ax+b)] = a^n\sin\left(\dfrac{n\pi}{2}+ax+b\right)$

(vi) $y = f(x) = \cos(ax + b)$.

We have $\quad y_1 = f'(x) = -a\sin(ax+b) = a\cos\left(\dfrac{\pi}{2}+ax+b\right)$, $\quad y_2 = f''(x) = -a^2\sin(\dfrac{\pi}{2}+ax+b) = a^2\cos\left(\dfrac{2\pi}{2}+ax+b\right)$

$$y_3 = f'''(x) = -a^3\sin(2.\dfrac{\pi}{2}+ax+b) = a^3\cos\left(3.\dfrac{\pi}{2}+ax+b\right)$$

$$\dots\dots\dots\dots\dots\dots\dots\dots\dots\dots\dots\dots\dots\dots$$
$$\dots\dots\dots\dots\dots\dots\dots\dots\dots\dots\dots\dots\dots\dots$$

$$y_n = f^n(x) = -a^n\sin\left((n-1)\dfrac{\pi}{2}+ax+b\right) = a^n\cos\left(\dfrac{n\pi}{2}+ax+b\right)$$

$\Rightarrow \qquad y_n = \dfrac{d^n}{dx^n}[\cos(ax+b)] = a^n\cos\left(\dfrac{n\pi}{2}+ax+b\right)$

(vii) $y = f(x) = e^{ax+b}$.

We have $\quad y_1 = f'(x) = a.e^{ax+b}$, $\quad y_2 = f''(x) = a^2.e^{ax+b}$

$$y_3 = f'''(x) = a^3.e^{ax+b}$$

$$\dots\dots\dots\dots\dots\dots\dots\dots\dots\dots\dots\dots\dots\dots$$

$$y_n = f^n(x) = a^n.e^{ax+b}$$

$\Rightarrow \qquad y_n = \dfrac{d^n}{dx^n}(e^{ax+b}) = a^n e^{ax+b}$

(viii) $y = f(x) = \log(ax + b)$.

We have $\quad y_1 = f'(x) = \dfrac{a}{ax+b}$

Now using result (iii), we get

$$y_n = f^n(x) = (-1)^{n-1}\dfrac{a^n(n-1)!}{(ax+b)^n}$$

$\Rightarrow \qquad y_n = \dfrac{d^n}{dx^n}[\log(ax+b)] = (-1)^{n-1}\dfrac{a^n(n-1)!}{(ax+b)^n}$

(ix) $y = f(x) = e^{ax} \sin(bx + c)$.

We have
$$y_1 = f'(x) = ae^{ax}.\sin(bx+c) + be^{ax}\cos(bx+c) = e^{ax}[a\sin(bx+c) + b\cos(bx+c)]$$

Put $a = r\cos\theta, b = r\sin\theta \Rightarrow r^2 = a^2 + b^2$

and $\tan\theta = b/a$ i.e., $\theta = \tan^{-1} b/a$

Therefore,
$$y_1 = f'(x) = r.e^{ax}\sin(bx+c+\theta)$$
$$= (a^2+b^2)^{1/2}.e^{ax}\sin\left(bx+c+\tan^{-1}\frac{b}{a}\right)$$

Similarly,
$$y_2 = f''(x) = (a^2+b^2)^{1/2}(a^2+b^2)^{1/2}.e^{ax}\sin(bx+c+\tan^{-1}b/a+\tan^{-1}b/a)$$
$$= (a^2+b^2)^{2/2}.e^{ax}\sin(bx+c+2\tan^{-1}b/a)$$
$$y_3 = f'''(x) = (a^2+b^2)^{3/2}.e^{ax}\sin(bx+c+3\tan^{-1}b/a)$$
$$\cdots\cdots\cdots\cdots\cdots\cdots\cdots\cdots\cdots\cdots\cdots\cdots\cdots\cdots\cdots\cdots\cdots\cdots$$
$$\cdots\cdots\cdots\cdots\cdots\cdots\cdots\cdots\cdots\cdots\cdots\cdots\cdots\cdots\cdots\cdots\cdots\cdots$$
$$y_n = f^n(x) = (a^2+b^2)^{n/2}.e^{ax}\sin(bx+c+n\tan^{-1}b/a)$$
$$\Rightarrow \quad y_n = \frac{d^n}{dx^n}[e^{ax}\sin(bx+c)] = (a^2+b^2)^{n/2}.e^{ax}\sin(bx+c+n\tan^{-1}b/a)$$

(x) $y = f(x) = e^{ax} \cos(bx + c)$.

We have
$$y_1 = f'(x) = ae^{ax}.\cos(bx+c) - be^{ax}\sin(bx+c) = e^{ax}[a\cos(bx+c) - b\sin(bx+c)]$$

Put $a = r\cos\theta, b = r\sin\theta \Rightarrow \theta = \tan^{-1}b/a$ and $r = (a^2+b^2)^{1/2}$

$\therefore$
$$y_1 = f'(x) = r.e^{ax}[\cos\theta\cos(bx+c) - \sin\theta\sin(bx+c)]$$
$$= re^{ax}\cos(bx+c+\theta) = (a^2+b^2)^{1/2}.e^{ax}\cos(bx+c+\tan^{-1}b/a)$$

Similarly,
$$y_2 = f''(x) = (a^2+b^2)^{2/2}.e^{ax}\cos(bx+c+2\tan^{-1}b/a)$$
$$y_3 = f'''(x) = (a^2+b^2)^{3/2}.e^{ax}\cos(bx+c+3\tan^{-1}b/a)$$
$$\cdots\cdots\cdots\cdots\cdots\cdots\cdots\cdots\cdots\cdots\cdots\cdots\cdots\cdots\cdots\cdots\cdots\cdots$$
$$\cdots\cdots\cdots\cdots\cdots\cdots\cdots\cdots\cdots\cdots\cdots\cdots\cdots\cdots\cdots\cdots\cdots\cdots$$
$$y_n = f^n(x) = (a^2+b^2)^{n/2}.e^{ax}\cos(bx+c+n\tan^{-1}b/a)$$
$$\Rightarrow \quad y_n = \frac{d^n}{dx^n}[e^{ax}\cos(bx+c)] = (a^2+b^2)^{n/2}.e^{ax}\cos(bx+c+n\tan^{-1}b/a)$$

 Solved Examples

EXAMPLE 1. *Find the nth differential coefficient of $\tan^{-1}\dfrac{x}{a}$*

SOLUTION. We have $\quad y = \tan^{-1}\dfrac{x}{a}$

$\Rightarrow \quad y_1 = \dfrac{a}{x^2+a^2} = \dfrac{a}{(x+ia)(x-ia)}$

Let us suppose

$$\frac{a}{(x+ia)(x-ia)} = \frac{A}{(x+ia)} + \frac{B}{(x-ia)}$$
$$\text{(Using partial fractions)}$$

$\Rightarrow \quad a = A(x-ia) + B(x+ia)$

To find the value of A, put $x = -ia$

We get $\quad A = -\dfrac{1}{2i}$

and for B, put $x = ia$, which gives $B = \dfrac{1}{2i}$ therefore, we have

$$y_1 = \frac{1}{2i}\left[\frac{1}{x-ia} - \frac{1}{x+ia}\right]$$
$$= \frac{1}{2i}[(x-ia)^{-1} - (x+ia)^{-1}]$$

Differentiating $(n-1)$ times, we get

$$y_n = \frac{1}{2i}[(-1)^{n-1}(n-1)!(x-ia)^{-n}$$
$$- (-1)^{n-1}(n-1)!(x+ia)^{-n}]$$
$$= \frac{(-1)^{n-1}(n-1)!}{2i}[(x-ia)^{-n} - (x+ia)^{-n}]$$

Put $x = r\cos\theta, a = r\sin\theta$, we have

$$y_n = \frac{(-1)^{n-1}(n-1)!}{2i}[r^{-n}(\cos\theta - i\sin\theta)^{-n}$$
$$- r^{-n}(\cos\theta + i\sin\theta)^{-n}]$$
$$= \frac{(-1)^{n-1}(n-1)!}{2i}r^{-n}[(\cos n\theta + i\sin n\theta)$$
$$- (\cos n\theta - i\sin n\theta)]$$
$$[\because \sin(-n\theta) = -\sin n\theta]$$
$$= \frac{(-1)^{n-1}(n-1)!}{2i}r^{-n}.2i\sin n\theta$$
$$= (-1)^{n-1}(n-1)!r^{-n}.\sin n\theta$$
$$= (-1)^{n-1}(n-1)!\left(\frac{a}{\sin\theta}\right)^{-n}\sin n\theta$$
$$\left[\text{since } r = \frac{a}{\sin\theta}\right]$$
$$= (-1)^{n-1}(n-1)!a^{-n}\sin^n\theta.\sin n\theta$$

EXAMPLE 2. *Find the nth differential coefficient of $\log(ax + x^2)$.*

SOLUTION. Let $y = \log(ax + x^2) = \log[x(a+x)]$
$$= \log x + \log(a+x)$$

Differentiating n times, we get

$$y_n = \frac{d^n}{dx^n}(\log x) + \frac{d^n}{dx^n}\log(a+x)$$
$$= \frac{(-1)^{n-1}(n-1)!.1^n}{x^n} + \frac{(-1)^{n-1}(n-1)!.1^n}{(x+a)^n}$$
$$= (-1)^{n-1}(n-1)!\left[\frac{1}{x^n} + \frac{1}{(x+a)^n}\right].$$

EXAMPLE 3. *Find the nth differential coefficients of*

    (i) $e^{ax} \sin bx \cos cx$ **(ii)** $e^{2x} \sin^3 x$

SOLUTION. (i) Let $y = e^{ax} \sin bx \cos cx$

$$= \frac{1}{2} e^{ax} [2 \sin bx \cos cx]$$

$$= \frac{1}{2} e^{ax} [\sin(bx + cx) + \sin(bx - cx)]$$

$$= \frac{1}{2} [e^{ax} \sin(b+c)x + e^{ax} \sin(b-c)x] \quad ...(1)$$

        Differentiating (1) n times, we get

$$\frac{d^n}{dx^n}[y] = y_n = \frac{1}{2}[\{a^2 + (b+c)^2\}^{n/2} e^{ax} \sin\{(b+c)x$$
$$+ n \tan^{-1}(b+c)/a\} + \{a^2 + (b-c)^2\}^{n/2}$$
$$e^{ax} \sin\{(b-c)x + n \tan^{-1}(b-c)/a\}]$$

(ii) Let $\quad y = e^{2x} \sin^3 x$.

    Now using the result

    $\sin 3x = 3 \sin x - 4 \sin^3 x$

    We have

    $4 \sin^3 x = 3 \sin x - \sin 3x$

$$\Rightarrow \sin^3 x = \frac{1}{4}(3 \sin x - \sin 3x)$$

    Therefore,

$$y = \frac{1}{4} e^{2x} [3 \sin x - \sin 3x]$$

$$= \frac{3}{4} e^{2x} \sin x - \frac{1}{4} e^{2x} \sin 3x.$$

    Now, differentiating n times, we get

$$y_n = \frac{3}{4} [(2^2 + 1^2)^{1/2}]^n e^{2x} \sin[x + n \tan^{-1} 1/2]$$

$$- \frac{1}{4} [(2^2 + 3^2)^{1/2}]^n e^{2x} \sin[3x + n \tan^{-1} 3/2].$$

EXAMPLE 4. *Find the nth differential coefficients of $\sin^5 x \cos^3 x$.*

SOLUTION. First we reduce $\sin^5 x \cos^3 x$ into a function consisting sine function of multiple of x.

Let $\quad\quad\quad z = \cos x + i \sin x$.

Then $\quad\quad z^{-1} = \cos x - i \sin x$

$\therefore \quad z + z^{-1} = 2\cos x$ and $z - z^{-1} = 2i\sin x$

Also, by De-Moivre's theorem, we have

$$z^m + z^{-m} = 2\cos mx$$

and $\quad z^m - z^{-m} = 2i\sin mx$

Now $(2i\sin x)^5 (2\cos x)^3 = (z - z^{-1})^5 + (z + z^{-1})^3$

$$\Rightarrow \quad 2^8 i \sin^5 x \cos^3 x = (z^8 - z^{-8}) - 2(z^6 - z^{-6})$$
$$- 2(z^4 - z^{-4}) + 6(z^2 - z^{-2})$$
$$= 2i \sin 8x - 4i \sin 6x$$
$$- 4i \sin 4x + 12i \sin 2x$$

$$\Rightarrow \quad \sin^5 x \cos^3 x = 2^{-7}[\sin 8x - 2\sin 6x$$
$$- 2\sin 4x + 6 \sin 2x].$$

Dfferentiating both sides n times w.r.t. x, we get

$$D^n(\sin^5 x \cos^3 x)$$

$$= 2^{-7}\left[8^n \sin\left(8x + \frac{n\pi}{2}\right) - 2.6^n \sin\left(6x + \frac{n\pi}{2}\right)\right.$$

$$\left. -2.4^n \sin\left(4x + \frac{n\pi}{2}\right) + 6.2^n \sin\left(2x + \frac{n\pi}{2}\right)\right]$$

2.1.2 USE OF PARTIAL FRACTIONS

To determine the nth derivative of any rational function, we have to split it into partial fractions.

Partial fractions for

(i) $\dfrac{f(x)}{(x-a)(x-b)(x-c)} = \dfrac{A}{(x-a)} + \dfrac{B}{(x-b)} + \dfrac{C}{(x-c)}$

(ii) $\dfrac{f(x)}{(x-a)^2(x-b)} = \dfrac{A}{(x-a)} + \dfrac{B}{(x-a)^2} + \dfrac{C}{(x-b)}$

(iii) $\dfrac{f(x)}{(x-a)^3(x-b)} = \dfrac{A}{(x-a)} + \dfrac{B}{(x-a)^2} + \dfrac{C}{(x-a)^3} + \dfrac{D}{(x-b)}$

(iv) $\dfrac{f(x)}{(x-a)(x-b)(px^2+qx+r)} = \dfrac{A}{(x-a)} + \dfrac{B}{(x-b)} + \dfrac{Cx+D}{(px^2+qx+r)}$

To find A, B, C, D etc., we put each linear factor of LCM equal to zero. The remaining constants are obtained by comparing coefficients of like powers on both sides.

REMARKS

- Forming partial fractions is converse process of taking LCM.
- To resolve a fraction into partial fractions, the degree of the numerator must be less than the degree of denominator.

Solved Examples

EXAMPLE 1. *Find the nth differential coefficients of*

    (i) $\dfrac{1}{1-5x+6x^2}$ **(ii)** $\dfrac{x^2}{[(x+2)(2x+3)]}$

                          (UKTU 2011)

SOLUTION. (i) Let $\quad y = \dfrac{1}{1-5x+6x^2} = \dfrac{1}{(3x-1)(2x-1)}$

$$= \dfrac{2}{2x-1} - \dfrac{3}{3x-1}$$

(By resolving into partial fractions)

$$= 2(2x-1)^{-1} - 3(3x-1)^{-1}$$

Differentaittng, n times, we get

$$y_n = 2(-1)^n n! 2^n (2x-1)^{-n-1}$$
$$-3(-1)^n n! 3^n (3x-1)^{-n-1}$$
$$= (-1)^n . n! [2^{n+1}(2x-1)^{-n-1}$$
$$- 3^{n+1}(3x-1)^{-n-1}]$$

(ii) Let $\quad y = \dfrac{x^2}{[(x+2)(2x+3)]}$

Since, the given fraction is not a proper one so, divide the Nr. by Dr., we observe that the quotient will be 1/2.

So let $\dfrac{x^2}{(x+2)(2x+3)} = \dfrac{1}{2} + \dfrac{A}{x+2} + \dfrac{B}{2x+3}$

which gives $\quad A = -4, B = 9/2$

Therefore,

$$y = \dfrac{1}{2} - \dfrac{4}{x+2} + \dfrac{9}{2(2x+3)}$$
$$= \dfrac{1}{2} - 4(x+2)^{-1} + \dfrac{9}{2}(2x+3)^{-1}$$

Differentiating n times, we get

$$y_n = -4(-1)^n n! (x+2)^{-n-1}$$
$$+ \dfrac{9}{2}(-1)^n . n! 2^n (2x+3)^{-n-1}$$
$$= (-1)^n n! \left[\dfrac{9.2^{n-1}}{(2x+3)^{n+1}} - \dfrac{4}{(x+2)^{n+1}} \right]$$

☞ REMARK

- If none of the standard formulae is applicable to find y_n in any problem, then find y_1, y_2, y_3 and generalise.

📂 More Solved Examples

EXAMPLE 1. *If* $y = \sqrt{x+a}$, *find* y_n.

SOLUTION. We have

$$y = \sqrt{x+a} = (x+a)^{1/2}$$
$$y = -(x+a)^{1/2}$$
$$y_2 = \left(\dfrac{1}{2}\right)\left(-\dfrac{1}{2}\right)(x+a)^{-3/2}$$
$$y_3 = \left(\dfrac{1}{2}\right)\left(-\dfrac{1}{2}\right)\left(-\dfrac{3}{2}\right)(x+a)^{-5/2}$$
$$= (-1)^2 \dfrac{1.3}{2^3}(x+a)^{-5/2}$$

....
....

$$y_n = (-1)^{n-1} \dfrac{1.3.5 \ldots \text{upto } (n-1) \text{ times}}{2^n}(x+a)^{-\frac{(2n-1)}{2}}$$

$$y_n = (-1)^{n-1} \dfrac{1.3 \ldots (2n-3)}{2^n}(x+a)^{-\left(\frac{2n-1}{2}\right)}$$

EXAMPLE 2. *If* $y = \tan^{-1}\left\{ \dfrac{\sqrt{(1+x^2)}-1}{x} \right\}$, *show that*

$$y_n = \dfrac{1}{2}(-1)^{n-1}(n-1)! \sin^n \theta \sin n\theta,$$

where $\theta = \cot^{-1}x$.

SOLUTION. We have $y = \tan^{-1}\left\{ \dfrac{\sqrt{(1+x^2)}-1}{x} \right\}$.

Put $x = \tan\phi$, then

$$y = \tan^{-1}\left\{ \dfrac{\sqrt{(1+\tan^2\phi)}-1}{\tan\phi} \right\} = \tan^{-1}\left[\dfrac{\sec\phi-1}{\tan\phi} \right]$$
$$= \tan^{-1}\left(\dfrac{1-\cos\phi}{\sin\phi} \right) = \tan^{-1}\left(\dfrac{2\sin^2(\phi/2)}{2\sin(\phi/2)\cos(\phi/2)} \right)$$
$$= \tan^{-1}\tan(\phi/2) = \phi/2 = \dfrac{1}{2}\tan^{-1}x$$

$$\Rightarrow \quad y_1 = \dfrac{1}{2(1+x^2)} = \dfrac{1}{2(x-i)(x+i)}$$
$$= \dfrac{1}{4i}\left(\dfrac{1}{x-i} - \dfrac{1}{x+i} \right)$$

Differentiating $(n-1)$ times, we get

$$y_n = \dfrac{(-1)^{n-1}(n-1)!}{4i}[(x-i)^{-n} - (x+i)^{-n}]$$

Now putting $x = r\cos\theta$, $1 = r\sin\theta$, we have

$$y_n = \dfrac{(-1)^{n-1}(n-1)!}{4i}\left[\begin{array}{c} r^{-n}(\cos\theta - i\sin\theta)^{-n} \\ -r^{-n}(\cos\theta + i\sin\theta)^{-n} \end{array} \right]$$
$$= \dfrac{(-1)^{n-1}(n-1)!}{4i}r^{-n}\left[\begin{array}{c} (\cos n\theta + i\sin n\theta) \\ -(\cos n\theta - i\sin n\theta) \end{array} \right]$$
$$= \dfrac{1}{2}(-1)^{n-1}(n-1)! r^{-n}\sin n\theta$$
$$= \dfrac{1}{2}(-1)^{n-1}(n-1)!\left(\dfrac{1}{\sin\theta} \right)^{-n}\sin n\theta \quad \left[\because r = \dfrac{1}{\sin\theta} \right]$$
$$= \dfrac{1}{2}(-1)^{n-1}(n-1)! \sin^n\theta \sin n\theta$$

where $\theta = \tan^{-1}\dfrac{1}{x} = \cot^{-1}x$.

EXAMPLE 3. *If* $y = \sin mx + \cos mx$, *prove that*

$$y_n = m^n [1 + (-1)^n \sin 2mx]^{1/2}. \quad \text{(MTU–2012)}$$

SOLUTION. We have

$$y_n = \dfrac{d^n}{dx^n}(\sin mx) + \dfrac{d^n}{dx^n}(\cos mx)$$
$$= m^n \sin\left(mx + n\dfrac{\pi}{2} \right) + m^n \cos\left(mx + n\dfrac{\pi}{2} \right)$$
$$= m^n\left[\left\{ \sin\left(mx + n\dfrac{\pi}{2} \right) + \cos\left(mx + n\dfrac{\pi}{2} \right) \right\}^2 \right]^{1/2}$$
$$= m^n\left[1 + 2\sin\left(mx + n\dfrac{\pi}{2} \right).\cos\left(mx + n\dfrac{\pi}{2} \right) \right]^{1/2}$$
$$= m^n[1 + \sin(2mx + n\pi)]^{1/2}$$
$$= m^n[1 \pm \sin 2mx]^{1/2}$$
$$= m^n[1 + (-1)^n \sin 2mx]^{1/2}.$$

EXAMPLE 4. *Find the* n^{th} *derivative of* $y = \cos^4 x$.

SOLUTION. Let $y = \cos^4 x = (\cos^2 x)^2$

$$= [(1/2)(1 + \cos 2x)]^2$$

$$= 1/4(1 + 2\cos 2x + \cos^2 2x)$$
$$= 1/4[1 + 2\cos 2x + (1/2)(1 + \cos 4x)]$$
$$= 1/4[3/2 + 2\cos 2x + 1/2\cos 4x]$$
$$= 3/8 + (1/2)\cos 2x + (1/8)\cos 4x.$$

Now,

$$y_n = 0 + \frac{1}{2}.2^n \cos\left(2x + \frac{1}{2}n\pi\right) + \frac{1}{8}.4^n \cos\left(4x + \frac{1}{2}n\pi\right)$$

$$= 2^{n-1}.\cos\left(2x + \frac{1}{2}n\pi\right) + 2^{2n-3}\cos\left(4x + \frac{1}{2}n\pi\right)$$

EXAMPLE 5. *If $y = x\log\dfrac{x-1}{x+1}$, show that*

$$y_n = (-1)^{n-2}(n-2)!\left[\frac{x-n}{(x-1)^n} - \frac{x+n}{(x+1)^n}\right]. \text{ (UPTU-2003,14)}$$

SOLUTION. Let $\quad y = x\log\dfrac{x-1}{x+1}$

$$\Rightarrow \quad y = x\log(x-1) - x\log(x+1) \qquad ...(1)$$

Differentiating (1) w.r.t. x we get

$$y_1 = \frac{x}{x-1} + \log(x-1) - \frac{x}{x+1} - \log(x+1)$$

$$= 1 + \frac{1}{x-1} + \log(x-1) - 1 + \frac{1}{x+1} - \log(x+1)$$

$$= \frac{1}{x-1} + \frac{1}{x+1} + \log(x-1) - \log(x+1) \quad ...(2)$$

Differentiating both sides of (2) w.r.t. x, $(n-1)$ times we get

$$y_n = \frac{(-1)^{n-1}(n-1)!}{(x-1)^n} + \frac{(-1)^{n-1}(n-1)!}{(x+1)^n}$$

$$+ \frac{(-1)^{n-2}(n-2)!}{(x-1)^{n-1}} - \frac{(-1)^{n-2}(n-2)!}{(x+1)^{n-1}}$$

$$= (-1)^{n-2}(n-2)!\left\{-\frac{(n-1) + x - 1}{(x-1)^n}\right\}$$

$$+ (-1)^{n-2}(n-2)!\left\{\frac{-(n-1) - (x+1)}{(x+1)^n}\right\}$$

$$\Rightarrow \quad y_n = (-1)^{n-2}(n-2)!\left\{\frac{x-n}{(x-1)^n} - \frac{x+n}{(x+1)^n}\right\}$$

EXAMPLE 6. *Find the nth derivative of $\dfrac{1}{x^2 + a^2}$.* (UKTU–2011, 17)

SOLUTION. Let $\quad y = \dfrac{1}{x^2 + a^2} = \dfrac{1}{(x + ia)(x - ia)}$

$$= \frac{1}{2ia}\left[\frac{1}{x - ia} - \frac{1}{x + ia}\right] \qquad ...(1)$$

Differentiating (1) n times w.r.t. x we get

$$y_n = \frac{1}{2ia}\left[\frac{(-1)^n n!}{(x - ia)^{n+1}} - \frac{(-1)^n n!}{(x + ia)^{n+1}}\right]$$

$$= \frac{(-1)^n.n!}{2ia}\left[\frac{1}{(x - ia)^{n+1}} - \frac{1}{(x + ia)^{n+1}}\right] \qquad ...(2)$$

Let $x = r\cos\theta$ and $a = r\sin\theta$ i.e., $\theta = \tan^{-1}\dfrac{a}{x}$ in (2), we get

$$y_n = \frac{(-1)^n.n!}{2iar^{n+1}}\left[\frac{1}{(\cos\theta - i\sin\theta)^{n+1}} - \frac{1}{(\cos\theta + i\sin\theta)^{n+1}}\right]$$

$$= \frac{(-1)^n.n!}{2iar^{n+1}}\left[\frac{1}{\cos(n+1)\theta - i\sin(n+1)\theta} - \frac{1}{\cos(n+1)\theta + i\sin(n+1)\theta}\right]$$

$$= \frac{(-1)^n.n!}{2iar^{n+1}}\left[\begin{array}{l}\{\cos(n+1)\theta + i\sin(n+1)\theta\} \\ -\{\cos(n+1)\theta - i\sin(n+1)\theta\}\end{array}\right]$$

$$= \frac{(-1)^n.n!}{2iar^{n+1}}[2i\sin(n+1)\theta]$$

$$= \frac{(-1)^n.n!\sin(n+1)\theta}{a\left(\dfrac{a^{n+1}}{\sin^{n+1}\theta}\right)} \qquad [\because a = \sin\theta]$$

$$= \frac{(-1)^n.n!\sin(n+1)\theta\sin^{n+1}\theta}{a^{n+2}},$$

where $\quad \theta = \tan^{-} —.$

![Exercise-2.1]

1. Find the nth derivatives of

(i) $\sin^3 x$ (ii) $\cos x \cos 2x \cos 3x$

(iii) $e^{ax}\cos^2 x \sin x$ (SVTU–2009) (iv) $\sin ax \cos bx$

(v) $\sin^2 x \sin 2x$

2. Find the nth derivatives of

(i) $\dfrac{x^4}{(x-1)(x-2)}$ (ii) $\dfrac{x}{1 + 3x + 2x^2}$

(iii) $\dfrac{1}{(x-2)(x-1)^3}$ (iv) $\dfrac{1}{x^2 - a^2}$

(v) $\dfrac{x^2}{(x-a)(x-b)}$ (vi) $\dfrac{17x^2 + 26x - 42}{6x^3 - 25x^2 - 29x + 20}$

3. Find the nth derivatives of

(i) $\tan^{-1}\left(\dfrac{1+x}{1-x}\right)$ (ii) $\tan^{-1}\left(\dfrac{2x}{1-x^2}\right)$ (UPTU–2002)

4. Show that the value of the nth differential coefficients of $\dfrac{x^3}{x^2 - 1}$ for $x = 0$, is zero if n is even and is $-n!$, if n is odd and greater than 1.

5. If $y = x(a^2 + x^2)^{-1}$, show that

$$y_n = (-1)^n n!a^{-n-1}\sin^{n+1}\theta\cos(n+1)\theta \text{ where } \theta = \tan^{-1}\left(\frac{a}{x}\right).$$

(MUMBAI–2007)

6. (i) If $x = a(t - \sin t)$ and $y = a(1 + \cos t)$, prove that

$$\frac{d^2y}{dx^2} = \frac{1}{4a}\text{cosec}^4\left(\frac{t}{2}\right). \qquad \text{(MADURAI–1990, 2004)}$$

(ii) If $x = a(\cos\theta + \theta\sin\theta), y = a(\sin\theta - \theta\cos\theta)$, find $\dfrac{d^2y}{dx^2}$.

(iii) If $y = \sin(\sin x)$, show that $\left(\dfrac{d^2y}{dx^2}\right) + \left(\dfrac{dy}{dx}\right)\tan x + y\cos^2 x = 0.$

(iv) If $y = A\sin mx + B\cos mx$, show that $\dfrac{d^2y}{dx^2} + m^2 y = 0.$

(v) If $y = e^{ax}\sin bx$, show that $\dfrac{d^2y}{dx^2} - 2a\dfrac{dy}{dx} + (a^2+b^2)y = 0$.

7. If $p^2 = a^2\cos^2\theta + b^2\sin^2\theta$, prove that $p + \dfrac{d^2p}{d\theta^2} = \dfrac{a^2 \cdot b^2}{p^3}$.

8. Prove that the value when $x = 0$ of $\dfrac{d^n}{dx^n}(\tan^{-1}x)$ is 0, $(n-1)!$ or $-(n-1)!$ according as n is of the form $2p$, $4p+1$ or $4p+3$ respectively.

Hints to Selected Problems

2. (i) Resolving into partial fractions, we get

$$y = (x^2 + 3x + 7) + \frac{15x - 14}{(x-1)(x-2)}.$$

Now differentiate successively.

(ii) $y = \dfrac{1}{(1+x)} - \dfrac{1}{(1+2x)}$. Now differentiate successively.

3. (i) $y = \tan^{-1}\dfrac{1+x}{1-x} \Rightarrow y_1 = \dfrac{1}{(1+x^2)}$

$$= \frac{1}{(x-i)(x+i)} = \frac{1}{2i}\left(\frac{1}{x-i} + \frac{1}{x+i}\right)$$

Then differentiate successively.

4. The given function can be written as $y = x + \dfrac{1}{2}\left(\dfrac{1}{x-1} + \dfrac{1}{x+1}\right)$.

5. The given function can be written as $y = \dfrac{1}{2}\left(\dfrac{1}{x-ai} + \dfrac{1}{x+ai}\right)$.

6. (i) Find $\dfrac{dx}{dt}$ and $\dfrac{dy}{dt}$ such that $\dfrac{dx}{dt} = a(1-\cos t)$ and $\dfrac{dy}{dt} = -a\sin t$.

Then $\dfrac{dy}{dx} = \dfrac{dy/dt}{dx/dt} = \dfrac{a\sin t}{a(1-\cos t)} = -\cot\ /2.$

Now differentiate successively.

8. Let $y = \tan^{-1}x$. Then $y_1 = \dfrac{1}{1+x^2}$

$$= \frac{1}{(x-i)(x+i)} = \frac{1}{2i}\left[\frac{1}{x-i} - \frac{1}{x+i}\right].$$

Now differentiate successively.

Answers

1. (i) $y_n = \dfrac{3}{4}\sin\left(x + \dfrac{n\pi}{2}\right) - \dfrac{1}{4} \cdot 3^n\sin\left(3x + \dfrac{n\pi}{2}\right)$

(ii) $y_n = \dfrac{1}{4}\left\{6^n\cos\left(6x + \dfrac{1}{2}n\pi\right) + 4^n\cos\left(4x + \dfrac{n\pi}{2}\right) + 2^n\cos\left(2x + \dfrac{n\pi}{2}\right)\right\}$

(iii) $y_n = \dfrac{1}{4}[(a^2+1)^{n/2}e^{ax}\sin\{x + n\tan^{-1}1/a\} + (a^2+9)^{n/2}e^{ax}\sin(3x + n\tan^{-1}3/a)]$

(iv) $y_n = \dfrac{1}{2}\left[(a+b)^n\sin\left\{(a+b)x + \dfrac{1}{2}n\pi\right\} + (a-b)^n\sin\left\{(a-b)x + \dfrac{1}{2}n\pi\right\}\right]$

(v) $y_n = 2^{n-1}\sin\left(2x + \dfrac{1}{2}n\pi\right) - 4^{n-1}\sin\left(4x + \dfrac{1}{2}n\pi\right)$

2. (i) $y_n = (-1)^n n![16(x-2)^{-n-1} - (x-1)^{-n-1}]$

(ii) $y_n = (-1)^n n!\left[\dfrac{1}{(x+1)^{n+1}} - \dfrac{2^n}{(2x+1)^{n+1}}\right]$

(iii) $y_n = (-1)^{n+1} \cdot n!\left[\dfrac{(n+2)(n+1)}{2(x-1)^{n+3}} + \dfrac{(n+1)}{(x-1)^{n+2}} + \dfrac{1}{(x-1)^{n+1}} - \dfrac{1}{(x-2)^{n+1}}\right]$

(iv) $y_n = \dfrac{1}{2a}(-1)^n \cdot n!\{(x-a)^{-n-1} - (x+a)^{-n-1}\}$

(v) $y_n = \dfrac{(-1)^n \cdot n!}{(a-b)}\left[\dfrac{a^2}{(x-a)^{n+1}} - \dfrac{b^2}{(x-b)^{n+1}}\right]$

(vi) $y_n = (-1)^n \cdot n!\left[\dfrac{2^n}{(2x-1)^{n+1}} - \dfrac{2 \cdot 3^n}{(3x+4)^{n+1}} + \dfrac{3}{(x-5)^{n+1}}\right]$

3. (i) $y_n = (-1)^{n-1} \cdot (n-1)!\sin^n\theta \sin n\theta$, where $\theta = \tan^{-1}\dfrac{1}{x}$

(ii) $y_n = 2(-1)^{n-1}(n-1)!\sin^n\theta \sin n\theta$, where $\theta = \tan^{-1}\dfrac{1}{x}$

6. (ii) $y_2 = \dfrac{1}{a} \cdot \dfrac{\sec^3\theta}{\theta}$

2.1.3 Leibnitz's Theorem

(UPTU–2008, 12, 14, 17, GBTU–2011, 16)

This theorem help us to find the nth differential coefficient of the product of two functions in terms of the successive derivatives of the functions.

Statement. *If u, v be two functions of x, having derivative of nth order, then*

$$D^n(uv) = u_n v + {}^nC_1 u_{n-1}v_1 + {}^nC_2 u_{n-2}v_2 + \ldots + {}^nC_r u_{n-r}v_r + \ldots + {}^nC_n uv_n$$

where suffixes of u and v denote differentiations w.r.t. x.

Step 1. Let $y = uv$

$\Rightarrow$ $\quad y_1 = u_1 v + uv_1$

and $\quad y_2 = u_2 v + u_1 v_1 + u_1 v_1 + uv_2 = u_2 v + 2u_1 v_1 + uv_2$

$\quad\quad = u_2 v + {}^2C_1 u_1 v_1 + {}^2C_2 uv_2.$

Thus the theorem is true for $n = 1, 2$.

Step 2. Let us assume that the theorem is true for a particular value of n say m, then we have

$$y_m = u_m v + {}^mC_1 u_{m-1}v_1 + {}^mC_2 u_{m-2}v_2 + \ldots + {}^mC_{r-1} u_{m-r+1}v_{r-1} + {}^mC_r u_{m-r}v_r + \ldots + {}^mC_m uv_m. \qquad \ldots(1)$$

Step 3.   Now, differentiating (1), we have

$$y_{m+1} = u_{m+1}v + u_m v_1 + {}^mC_1 u_m v_1 + {}^mC_1 u_{m-1}v_2 + {}^mC_2 u_{m-1}v_2 + {}^mC_2 u_{m-2}v_3 + \ldots + {}^mC_{r-1}u_{m-r+2}v_{r-1}$$

$$+ {}^mC_{r-1}u_{m-r+1}v_r + {}^mC_r u_{m-r+1}v_r + {}^mC_r u_{m-r}v_{r+1} + \ldots + {}^mC_m u_1 v_m + {}^mC_m uv_{m+1}.$$

$$= u_{m+1}.v + ({}^mC_1 + 1)u_m v_1 + ({}^mC_2 + {}^mC_1)u_{m-1}v_2 + \ldots + ({}^mC_r + {}^mC_{r-1})u_{m-r+1}v_r + \ldots + {}^mC_m uv_{m+1}.$$

Now using Pascal's law, given by

$$^mC_{r-1} + {}^mC_r = {}^{m+1}C_r$$

For $r = 1, 2, 3, \ldots$

We have $^mC_0 + {}^mC_1 = {}^{m+1}C_1 \Rightarrow 1 + {}^mC_1 = {}^{m+1}C_1$

$$^mC_1 + {}^mC_2 = {}^{m+1}C_2$$

....

and $^mC_m = 1 = {}^{m+1}C_{m+1}$

Therefore, $y_{m+1} = u_{m+1}.v + {}^{m+1}C_1 u_m v_1 + {}^{m+1}C_2 u_{m-1}v_2 + \ldots + {}^{m+1}C_r u_{m-r+1}v_r + \ldots + {}^{m+1}C_{m+1}uv_{m+1}$

$\Rightarrow$ If the theorem is true for $n = m$, then it is also true for the next higher value $n = m + 1$.

Then, by the principle of mathematical induction, we can say that theorem is true for any positive integer n.

Solved Examples

EXAMPLE 1.   *Find the nth derivative of $x^2 \sin x$.*   (UPTU–2009)

SOLUTION .   Let $u = \sin x$ and $v = x^2$.

Then, $u_n = \sin\left[x + \dfrac{n\pi}{2}\right]$

$$u_{n-1} = \sin\left(x + (n-1)\dfrac{\pi}{2}\right)$$

$$u_{n-2} = \sin\left[x + (n-2)\dfrac{\pi}{2}\right]$$

Also, $v_1 = 2x, v_2 = 2, v_3 = 0$

Now, by Leibnitz's theorem, we have

$$\dfrac{d^n}{dx^n}(uv) = u_n.v + {}^nC_1 u_{n-1}.v_1 + {}^nC_2 u_{n-2}.v_2$$

$$\dfrac{d^n}{dx^n}(x^2 \sin x) = \sin\left(x + \dfrac{n\pi}{2}\right)x^2 + {}^nC_1 \sin\left[x + (n-1)\dfrac{\pi}{2}\right]2x$$

$$+ {}^nC_2 \sin\left[x + (n-2)\dfrac{\pi}{2}\right]2$$

$$= x^2 \sin\left(x + \dfrac{n\pi}{2}\right) + 2nx \sin\left[x + (n-1)\dfrac{\pi}{2}\right]$$

$$+ n(n-1)\sin\left[x + (n-2)\dfrac{\pi}{2}\right]$$

EXAMPLE 2.   *If $y = a \cos(\log x) + b \sin(\log x)$, show that*
$x^2 y_2 + xy_1 + y = 0$
and $x^2 y_{n+2} + (2n+1)xy_{n+1} + (n^2 + 1)y_n = 0$.

                 (UPTU–2004, 2012, MADRAS–2000)

SOLUTION .   We have $y = a \cos(\log x) + b \sin(\log x)$ ...(1)

Differentiating (1) with respect to x, we have

$$y_1 = -\dfrac{a}{x}\sin(\log x) + \dfrac{b}{x}\cos(\log x)$$

$$xy_1 = -a\sin(\log x) + b\cos(\log x)$$

Again, differentiating w.r.t. x, we get

$$xy_2 + y_1 = -\dfrac{a}{x}\cos(\log x) - \dfrac{b}{x}\sin(\log x)$$

$$\Rightarrow \quad x^2 y_2 + xy_1 = -a\cos(\log x) - b\sin(\log x) = -y$$

$$\Rightarrow \quad x^2 y_2 + xy_1 + y = 0 \qquad \qquad \text{...(2)}$$

Now, differentiating (2) both sides n times by

Leibnitz's theorem, we get

$$D^n(x^2 y_2) + D^n(xy_1) + D^n(y) = 0$$

$$\Rightarrow \quad (D^n y_2)x^2 + {}^nC_1(D^{n-1}y_2)(Dx^2)$$

$$+ {}^nC_2(D^{n-2}y_2)(D^2 x^2) + (D^n y_1)x$$

$$+ {}^nC_1(D^{n-1}y_1)(Dx) + D^n y = 0$$

$$\Rightarrow x^2 y_{n+2} + 2nxy_{n+1} + \dfrac{n(n-1)}{2}2y_n + xy_{n+1} + ny_n + y_n = 0$$

$$\Rightarrow \quad x^2 y_{n+2} + (2n+1)xy_{n+1} + (n^2 + 1)y_n = 0$$

EXAMPLE 3.   *If $y = e^{a\sin^{-1}x}$, show that*
$(1 - x^2)y_{n+2} - (2n+1)xy_{n+1} - (n^2 + a^2)y_n = 0$.

           (MDU–1998, VTU–2003, KURUKSHETRA–1999)

SOLUTION .   We have

$$y = e^{a\sin^{-1}x} \Rightarrow y_1 = e^{a\sin^{-1}x}.\dfrac{a}{\sqrt{1 - x^2}}$$

$$y_1\sqrt{1 - x^2} = ae^{a\sin^{-1}x} = ay$$

$$\Rightarrow y_1^2(1 - x^2) = a^2 y^2 \qquad \qquad \text{...(1)}$$

Now differentiating (1) with respect to x, we get

$$2y_1 y_2(1 - x^2) + y_1^2(-2x) = 2a^2 yy_1$$

$$\Rightarrow 2y_1[y_2(1 - x^2) - xy_1 - a^2 y] = 0 \qquad [\because 2y_1 \neq 0]$$

$$\Rightarrow \quad [y_2(1 - x^2) - xy_1 - a^2 y] = 0 \qquad \text{...(2)}$$

Using Leibnitz's theorem, differentiating (2), n times, we get

$$D^n[y_2(1 - x^2)] - D^n(y_1 x) - a^2 D^n y = 0$$

$$\Rightarrow \left[y_{n+2}(1 - x^2) + ny_{n+1}(-2x) + \dfrac{n(n-1)}{2}y_n(-2)\right]$$

$$- [y_{n+1}x + ny_n] - a^2 y_n = 0$$

$$\Rightarrow (1 - x^2)y_{n+2} - (2n+1)xy_{n+1} - (n^2 + a^2)y_n = 0$$

EXAMPLE 4.   *If $y = \sin(m\sin^{-1}x)$, Prove that*
$(1 - x^2)y_2 - xy_1 + m^2 y = 0$
and $(1 - x^2)y_{n+2} - (2n+1)xy_{n+1} - (n^2 - m^2)y_n = 0$

           (UKTU–2012, UPTU–2007, 2009, GBTU–2011)

SOLUTION .   Let $y = \sin(m\sin^{-1}x)$ ...(1)

Differentiating w.r.t. x we get

$$y_1 = \cos(m\sin^{-1}x).\frac{m}{\sqrt{1-x^2}}$$

$$\Rightarrow \quad y_1\sqrt{1-x^2} = m\cos(m\sin^{-1}x)$$

$$\Rightarrow \quad y_1^2(1-x^2) = m^2\cos^2(m\sin^{-1}x)$$

$$= m^2[1-\sin^2(m\sin^{-1}x)]$$

$$= m^2(1-y^2) \qquad \ldots(2)$$

Again, differentiating both sides of (2) w.r.t. x we get

$$(1-x^2)2y_1y_2 - 2xy_1^2 = -2m^2yy_1$$

$$\Rightarrow \quad (1-x^2)y_2 - xy_1 = -m^2y$$

$$\Rightarrow \quad (1-x^2)y_2 - xy_1 + m^2y = 0 \qquad \ldots(3)$$

Finally, differentiating (3) n times, by Leibnitz's theorem, we get

$$\left[y_{n+2}(1-x^2) + {}^nC_1y_{n+1}(-2x) + {}^nC_2y_n(-2)\right]$$

$$- \left[y_{n+1}x + {}^nC_1y_n\right] + m^2y_n = 0$$

$$\Rightarrow \quad (1-x^2)y_{n+2} - 2nxy_{n+1}$$

$$- n(n+1)y_n - xy_{n+1} - ny_n + m^2y_n = 0$$

or $\quad (1-x^2)y_{n+2} - (2n+1)xy_{n+1} - (n^2-m^2)y_n = 0$

EXAMPLE 5. *If $y = (x^2-1)^n$, Prove that*

$$(x^2-1)y_{n+2} + 2xy_{n+1} - n(n+1)y_n = 0.$$

(VTU–2010, UPTU–2010)

Hence if $P_n = \dfrac{d^n}{dx^n}(x^2-1)^n$ show that

$$\frac{d}{dx}\left\{(1-x^2)\frac{dP_n}{dx}\right\} + n(n+1)P_n = 0$$

SOLUTION. We have $\quad y = (x^2-1)^n \qquad \ldots(1)$

Therefore $\quad y_1 = n(x^2-1)^{n-1}.2x$

or $\quad (x^2-1)y_1 = n(x^2-1)^n.2x$

$\Rightarrow \quad (x^2-1)y_1 = 2nxy. \qquad \ldots(2)$

Differentiating (2), $(n+1)$ times by Leibnitz's theorem, we get

$$D^{n+1}[y_1(x^2-1)] - 2nD^{n+1}(yx) = 0$$

or $\quad y_{n+2}(x^2-1) + (n+1)y_{n+1}.2x$

$$+ \frac{n(n+1)}{2}.y_n.2 - 2ny_{n+1}.x - 2n(n+1)y_n.1 = 0$$

or $\quad (x^2-1)y_{n+2} + 2xy_{n+1} - n(n+1)y_n = 0$

Hence, the first result from (2), we get

$$(x^2-1)D^2y_n + 2xDy_n - n(n+1)y_n = 0. \ldots(3)$$

Putting $y_n = \dfrac{d^n}{dx^n}(x^2-1)^n = P_n$; equation (3) becomes

$$(x^2-1)D^2P_n + 2xDP_n - n(n+1)P_n = 0$$

or $\quad -(1-x^2)D^2P_n + 2xD(P_n) - n(n+1)P_n = 0$

or $\quad -\frac{d}{dx}\left\{(1-x^2)DP_n\right\} - n(n+1)P_n = 0$

or $\quad \frac{d}{dx}\left\{(1-x^2)\frac{d}{dx}P_n\right\} + n(n+1)P_n = 0$

EXAMPLE 6. *If $\cos^{-1}\left(\dfrac{y}{b}\right) = \log\left(\dfrac{x}{m}\right)^m$, Show that*

$$x^2y_{n+2} + (2n+1)xy_{n+1} + (n^2+m^2)y_n = 0.$$

(UPTU–2006)

SOLUTION. We have

$$\cos^{-1}\left(\frac{y}{b}\right) = \log\left(\frac{x}{m}\right)^m$$

$$\Rightarrow \quad y = b\cos\left(m\log\left(\frac{x}{m}\right)\right)$$

$$\therefore \quad y_1 = -b\sin\left(m\log\left(\frac{x}{m}\right)\right).m\frac{1}{(x/m)}.\frac{1}{m}$$

$$\Rightarrow \quad xy_1 = -bm\sin\left(m\log\left(\frac{x}{m}\right)\right)$$

Again differentiating, we get

$$xy_2 + y_1 = -bm\cos\left\{m\log\left(\frac{}{m}\right)\right\}.m.\frac{1}{(x/m)}.\frac{1}{m}$$

$$\Rightarrow x^2y_2 + xy_1 = -m^2b\cos\left\{m\log\left(\frac{x}{m}\right)\right\} = -m^2y$$

$$\therefore \quad x^2y_2 + xy_1 + m^2y = 0$$

Differentiating both sides of the above equation, n times by Leibnitz's theorem, we get

$$[y_{n+2}.x^2 + {}^nC_1y_{n+1}(2x) + {}^nC_2y_n(2)]$$

$$+ [y_{n+1}(x) + {}^nC_1y_n(1)] + m^2y_n = 0$$

$$\Rightarrow x^2y_{n+2} + (2n+1)xy_{n+1} + (n^2+m^2)y_n = 0$$

EXAMPLE 7. *If $x = \tan(\log y)$, prove that*

$$(1+x^2)y_{n+1} + (2nx-1)y_n + n(n-1)y_{n-1} = 0.$$

(UPTU–2007)

SOLUTION. Let $\quad x = \tan(\log y)$

$$\Rightarrow \quad y = e^{\tan^{-1}x} \qquad \ldots(1)$$

$$\Rightarrow \quad y_1 = e^{\tan^{-1}x}.\frac{1}{(1+x^2)}$$

$$\therefore \quad (1+x^2)y_1 = y \qquad \ldots(2)$$

Differentiating (2) n times by Leibnitz's theorem, we get

$$y_{n+1}(1+x^2) + {}^nC_1y_n(2x) + {}^nC_2y_{n-1}(2) = y_n$$

$$\Rightarrow (1+x^2)y_{n+1} + (2nx-1)y_n + n(n-1)y_{n-1} = 0$$

EXAMPLE 8. *If $y = (1-x)^\alpha e^{-\alpha x}$, prove that*

$$(1-x)y_{n+1} - (n+\alpha x)y_n - n\alpha y_{n-1} = 0.$$

(UKTU–2011)

SOLUTION. We have

$$y = (1-x)^\alpha e^{-\alpha x} \qquad \ldots(1)$$

$$\Rightarrow y_1 = (1-x)^{-\alpha}(-\alpha e^{-\alpha x}) + e^{-\alpha x}(-\alpha)(1-x)^{-\alpha-1}(-1)$$

$$= e^{-\alpha x}(1-x)^{-\alpha}\left(-\alpha + \frac{\alpha}{1-x}\right)$$

$$\Rightarrow \quad y_1(1-x) = \alpha xy \qquad \ldots(2)$$

Differentiating (2) n times by Leibnitz's theorem, we get

$$y_{n+1}(1-x) + {}^nC_1y_n(-1) = \alpha[y_n(x) + {}^nC_1y_{n+1}(1)]$$

$$\therefore \quad (1-x)y_{n+1} + (-n-\alpha x)y_n - n\alpha y_{n-1} = 0.$$

$$\Rightarrow \quad (1-x)y_{n+1} - (n+\alpha x)y_n - n\alpha y_{n-1} = 0$$

EXAMPLE 9. *If* $y = \tan^{-1}\left(\dfrac{a+x}{a-x}\right)$, *prove that*

$$(a^2 + x^2)y_{n+2} + 2(n+1)xy_{n+1} + n(n+1)y_n = 0.$$ (GBTU–2012)

SOLUTION. We have $y = \tan^{-1}\left(\dfrac{a+x}{a-x}\right)$ ...(1)

Differentiating (1) w.r.t. x, we get

$$y_1 = \frac{a}{(a^2 + x^2)} \Rightarrow (a^2 + x^2)y_1 = a$$

Again differentiating w.r.t. x, we get

$$(a^2 + x^2)y_2 + 2xy_1 = 0 \qquad ...(2)$$

Now, differentiating (2), n times by Leibnitz's theorem, we get

$$\left[(x^2 + a^2)y_{n+2} + {}^nC_1(2x)y_{n+1} + {}^nC_2(2)y_n\right]$$
$$+ \left[2xy_{n+1} + {}^nC_1 . 2 . y_n\right] = 0$$
$$\Rightarrow (x^2 + a^2)y_{n+2} + 2nxy_{n+1}$$
$$+ n(n-1)y_n + 2xy_{n+1} + 2ny_n = 0$$
$$\Rightarrow (x^2 + a^2)y_{n+2} + 2x(n+1)y_{n+1} + [n(n-1) + 2n]y_n = 0$$
$$\Rightarrow (x^2 + a^2)y_{n+2} + 2x(n+1)y_{n+1} + n(n+1)y_n = 0$$

Exercise-2.2

1. Use Leibnitz's theorem, to find y_n in the following cases :

  (i) $x^3 e^{ax}$ (ii) $x^2 e^x$

  (iii) $x^3 \sin ax$ (iv) $x^3 \log x$

  (v) $x^2 e^x \cos x$ (vi) $e^x \log x$

  (vii) $x^n \log x$ (UPTU 2008) (viii) $x^2 \tan^{-1} x$

2. If $I_n = \dfrac{d^n}{dx^n}(x^n \log x)$, prove that $I_n = nI_{n-1} + (n-1)!$

and hence show that $I_n = n!\left(\log x + 1 + \dfrac{1}{2} + \dfrac{1}{3} + ... + \dfrac{1}{n}\right)$

            (MTU–2011, VTU–2001, MUMBAI–2008)

3. If $y = e^{\tan^{-1} x}$, prove that

$$(1 + x^2)y_{n+2} + [2(n+1)x - 1]y_{n+1} + n(n+1)y_n = 0.$$

4. If $y = (\sin^{-1} x)^2$, prove that $(1 - x^2)y_2 - xy_1 - 2 = 0$

and $(1 - x^2)y_{n+2} - x(2n+1)y_{n+1} - n^2 y_n = 0.$

                (UPTU–2005, 2009, 2010)

5. If $y = \dfrac{\sin^{-1} x}{\sqrt{(1 - x^2)}}$, prove that

$$(1 - x^2)y_{n+1} - (2n+1)xy_n - n^2 y_{n-1} = 0.$$

6. If $y = [\log\{x + \sqrt{(1 + x^2)}\}]^2$, prove that

$$(1 + x^2)y_{n+2} + (2n+1)xy_{n+1} + n^2 y_n = 0.$$

      (VTU–2007, BHILLAI–2005, GBTU(AG)–2010)

7. Differentiating n times the equation :

  (i) $(1 + x^2)\dfrac{d^2 y}{dx^2} - x\dfrac{dy}{dx} + a^2 y = 0.$

  (ii) $x^2 \dfrac{d^2 y}{dx^2} + x\dfrac{dy}{dx} + y = 0.$

8. If $y = [x + \sqrt{(1 + x^2)}]^m$, prove that

$$(1 + x^2)y_{n+2} + (2n+1)xy_{n+1} + (n^2 - m^2)y_n = 0.$$

          (VTU–2009, MADRAS–2000, 2004)

9. If $y^{1/m} + y^{-1/m} = 2x$, prove that

$$(x^2 - 1)y_{n+2} + (2n+1)xy_{n+1} + (n^2 - m^2)y_n = 0.$$

         (UPTU–2008, UKTU–2011, VTU–2008, SVTU–2007,

                  SRM-2006, 10, MUMBAI–2007)

10. If $y = \cos(\log x)$, prove that

$$x^2 y_{n+2} + (2n+1)xy_{n+1} + (n^2 + 1)y_n = 0.$$

11. If $x + y = 1$, prove that

$$\frac{d^n}{dx^n}(x^n y^n) = n!\left[y^n - \left({}^nC_1\right)^2 y^{n-1}x\right.$$
$$\left. + \left({}^nC_2\right)^2 y^{n-2}x^2 ... + (-1)^n x^n\right].$$

12. If $y = x\cos(\log x)$, prove that

$$x^2 y_{n+2} + (2n+1)xy_{n+1} + (n^2 - 2n + 2)y_n = 0. \qquad \text{(UPTU–2006)}$$

13. If $y = \left(\dfrac{1+x}{1-x}\right)^{1/2}$, prove that

$$(1 - x^2)y_n - [2(n-1)x + 1]y_{n-1} - (n-1)(n-2)y_{n-2} = 0. \text{ (MTU –2012)}$$

14. If $y = \dfrac{\sinh^{-1} x}{\sqrt{1 + x^2}}$, prove that

$$(1 + x^2)y_{n+2} + (2n+3)xy_{n+1} + (n+1)^2 y_n = 0. \qquad \text{(VTU–2010)}$$

15. If $x = \sin t, y = \cos pt$, prove that $(1 - x^2)y_2 - xy_1 + p^2 y = 0.$
Hence, show that

$$(1 - x^2)y_{n+2} - (2n+1)xy_{n+1} - (n^2 - p^2)y_n = 0.$$

               (VTU 2005, RAIPUR– 2005)

16. If $y = \sinh[m \log(x + \sqrt{x^2 + 1})]$, prove that

$$(x^2 + 1)y_{n+2} + (2n+1)xy_{n+1} + (n^2 - m^2)y_n = 0. \qquad \text{(VTU– 2010)}$$

17. If $\sin^{-1} y = 2\log(x + 1)$, prove that

$$(x+1)^2 y_{n+2} + (2n+1)(x+1)y_{n+1} + (x^2 + 4)y_n = 0 \quad \text{(VTU –2003)}$$

18. Prove the following

$$\frac{d^n}{dx^n}\left[\frac{\log x}{x}\right] = \frac{(-1)^n . n!}{x^{n+1}}\left(\log x - 1 - \frac{1}{2} - \frac{1}{3} - ... - \frac{1}{n}\right)$$

                        (VTU –2006)

Hints to Selected Problems

2. Since $I_n = \dfrac{d^n}{dx^n}(x^n \log x)$

$$\Rightarrow I_n = \frac{d^{n-1}}{dx^{n-1}}\left[\frac{d}{dx}(x^n \log x)\right] = nI_{n-1} + (n-1)!$$

Now replace $(n-1)$, $(n-2)$ in place of n successively.

3. $y = e^{\tan^{-1} x} \Rightarrow y_1 = e^{\tan^{-1} x} . \dfrac{1}{(1+x^2)} \Rightarrow y_1(1 + x^2) = y.$

Differentiating, we get
$(1 + x^2)y_2 + (2x - 1)y_1 = 0.$
Now apply Leibnitz theorem.

4. $y = (\sin^{-1} x)^2 \Rightarrow y_1 = 2\sin^{-1} x . \dfrac{1}{\sqrt{1 - x^2}}$

$$\Rightarrow y_1^2 = \frac{4(\sin^{-1} x)^2}{(1 - x^2)} \Rightarrow (1 - x^2)y_1^2 = 4(\sin^{-1} x)^2$$

$\Rightarrow (1-x^2)y_1^2 = 4y$. Again differentiating, we get

$(1-x^2)2y_1.y_2 - 2xy_1^2 = 4y_1 \Rightarrow (1-x^2)y_2 - xy_1 - 2 = 0$.

Now apply Leibnitz's theorem.

5. $y = \dfrac{\sin^{-1}x}{\sqrt{1-x^2}} \Rightarrow y_1 = \dfrac{1}{(1-x^2)} + \dfrac{x}{(1-x^2)}\dfrac{\sin^{-1}x}{\sqrt{1-x^2}}$

$\Rightarrow \quad (1-x^2)y_1 = 1 + xy$.

Now apply Leibnitz's theorem.

6. $y = [\log\{x+\sqrt{1+x^2}\}]^2 \Rightarrow \sqrt{1+x^2}\,y_1 = 2[\log\{x+\sqrt{1+x^2}\}]$

On squaring, we get

$(1+x^2)y_1 = 4[\log\{x+(1+x^2)\}]^2 \Rightarrow (1+x^2)y_1^2 = 4y$.

Again differentiating, we get

$(1+x^2)y_2 + xy_1 - 2 = 0$.

Now applying Leibnitz's theorem.

7. Apply directly Leibnitz's theorem.

8. $y = [x+\sqrt{1+x^2}]^m \Rightarrow y_1 = \dfrac{m[x+\sqrt{1+x^2}]^m}{\sqrt{1+x^2}} \Rightarrow \sqrt{1+x^2}\,y_1 = my$.

On squaring, we get $(1+x^2)y_1^2 = m^2 y^2$

Again differentiating, we get

$(1+x^2)y_2 + xy_1 - m^2 y = 0$.

Now apply Leibnitz theorem.

9. $y^{1/m} + y^{-1/m} = 2x \Rightarrow 2xy^{1/m} = y^{2/m} + 1$

Let $t = y^{1/m}$. Then $t^2 + 1 = 2xt$.

Solving for t, we get $t = x \pm \sqrt{x^2-1} \Rightarrow y = [x+\sqrt{x^2-1}]^m$.

On differentiating, we get $y_1 = \dfrac{m[x+\sqrt{x^2-1}]^m}{\sqrt{x^2-1}}$.

Differentiating w.r.t. x after squaring, we get

$(x^2-1)y_2 + xy_1 - m^2 y = 0$. Now apply Leibnitz theorem.

10. $y = \cos(\log x) \Rightarrow y_1 = -\sin(\log x).\dfrac{1}{x} \Rightarrow xy_1 = -\sin(\log x)$.

Again differentiating, we get

$xy_2 + y_1 = -\cos(\log x).\dfrac{1}{x} \Rightarrow x^2 y_2 + xy_1 + y = 0$.

Now apply Leibnitz's theorem.

Answers

1. (i) $e^{ax}a^{n-3}[a^3x^3 + 3na^2x^2 + 3n(n-1)ax + n(n-1)(n-2)]$ (ii) $e^x[x^2 + 2nx + n(n-1)]$

(iii) $a^{n-3}\left[a^3x^3\sin\left(ax+\dfrac{n\pi}{2}\right) + 3na^2x^2\sin\left(ax+(n-1)\dfrac{\pi}{2}\right) + 3n(n-1)ax\sin\left\{ax+(n-2)\dfrac{\pi}{2}\right\} + n(n-1)(n-2)\sin\left(ax+(n-3)\dfrac{\pi}{2}\right)\right]$

(iv) $\dfrac{(-1)^{n-1}n!}{x^{n-3}}\left[\dfrac{1}{n} - \dfrac{3}{n-1} + \dfrac{3}{n-2} - \dfrac{1}{n-3}\right]$

(v) $e^x\left[2^{n/2}x^2\cos\left(x+\dfrac{n\pi}{4}\right) + 2^{(n-1)/2}2nx\cos\left(x+(n-1)\dfrac{\pi}{4}\right) + 2^{(n-2)/2}n(n-1)\cos\left(x+(n-2)\dfrac{\pi}{4}\right)\right]$

(vi) $e^x[\log x + {}^nC_1 x^{-1} - {}^nC_2 x^{-2} + {}^nC_3 2! x^{-3} - ... + {}^nC_n(-1)^{n-1}(n-1)! x^{-n}]$ (vii) $y_{n+1} = \dfrac{n!}{x}$

(viii) $(-1)^{n-1}(n-3)![(n-1)(n-2)x^2\sin^n\phi\sin n\phi - 2n(n-1)\sin^{n-1}\phi\sin(n-1)\phi + n(n-1)\sin^{n-2}\phi\sin(n-2)\phi]$ where $\phi = \tan^{-1}\dfrac{1}{x}$

7. (i) $(1-x^2)y_{n+2} - (2n+1)xy_{n+1} - (n^2-a^2)y_n = 0$ (ii) $x^2 y_{n+2} + (2n+1)xy_{n+1} + (n^2+1)y_n = 0$.

2.1.4 DETERMINATION OF THE VALUE OF N^TH DERIVATIVE OF A FUNCTION AT x = 0

WORKING PROCEDURE

Step 1. Put the given function equal to y.

Step 2. Find $y_1 = \dfrac{dy}{dx}$. Then

   (i) Take L.C.M. (if required).

   (ii) Square both sides, if square roots are there.

   (iii) Try to get y in R.H.S. (if possible).

Step 3. Again differentiating both sides w.r.t. x and get an equation in y_2, y_1 and y.

Step 4. Differentiate both sides n times w.r.t. x by Leibnitz's theorem.

Step 5. Put $x = 0$ in equations of step 1, 2, 3, 4.

Step 6. Put $n = 1, 2, 3, 4, ...$ in last equation of step 5.

Step 7. Discuss the two cases, when n is even and when n is odd.

Solved Examples

EXAMPLE 1. *If $y = e^{a\cos^{-1}x}$, show that*

$$(1-x^2)y_{n+2} - (2n+1)xy_{n+1} - (n^2+a^2)y_n = 0$$

and hence calculate y_n at $x = 0$.

(GBTU–2010, MDU–1997)

SOLUTION. We have $\quad y = e^{a\cos^{-1}x}$ ...(1)

$\therefore \qquad y_1 = e^{a\cos^{-1}x}.\dfrac{-a}{\sqrt{1-x^2}} = -\dfrac{ya}{\sqrt{1-x^2}}$ ...(2)

$\Rightarrow y_1\sqrt{1-x^2} = -ya$

Now squaring both sides we get

$y_1^2(1-x^2) = y^2 a^2$

Differentiating w.r.t. x, we have

$(1-x^2)2y_1 y_2 - 2xy_1^2 = 2a^2 yy_1$

$\Rightarrow \quad (1-x^2)y_2 - xy_1 = a^2 y$ ...(3)

Now, using Leibnitz's theorem, differentiating (3), n times, we get

$(1-x^2)y_{n+2} - 2nxy_{n+1} - n(n-1)y_n - xy_{n+1} - ny_n = a^2 y_n$

$\Rightarrow (1-x^2)y_{n+2} - (2n+1)xy_{n+1} - (n^2+a^2)y_n = 0$ ...(4)

By putting $x = 0$ in (1), (2), (3) and (4), we get

$$y(0) = e^{a.\pi/2}$$
$$y_1(0) = -ae^{a.\pi/2}$$
$$y_2(0) = a^2 y(0) = a^2.e^{a.\pi/2}$$
$$\Rightarrow \quad y_{n+2}(0) = (n^2 + a^2)y_n(0) \qquad ...(5)$$

Put $n - 2$ for n in (5), we get
$$y_n(0) = [(n-2)^2 + a^2]y_{n-2}(0) \qquad ...(6)$$

Again put $n - 4$ for n in (5), we get
$$y_{n-2}(0) = [(n-4)^2 + a^2]y_{n-4}(0) \qquad ...(7)$$

From (6) and (7), we get
$$y_n(0) = [(n-2)^2 + a^2][(n-4)^2 + a^2]y_{n-4}(0) \quad ...(8)$$

Again put $n - 6$ for n in (5), we get
$$y_{n-4}(0) = [(n-6)^2 + a^2]y_{n-6}(0) \qquad ...(9)$$

From (8) and (9), we get
$$y_n(0) = [(n-2)^2 + a^2][(n-4)^2 + a^2]$$
$$[(n-6)^2 + a^2]y_{n-6}(0) \qquad ...(10)$$

Now there are following two cases :

Case I. When n is even.
$$y_n(0) = [(n-2)^2 + a^2][(n-4)^2 + a^2]$$
$$[(n-6)^2 + a^2]...[2^2 + a^2]a^2 e^{a\pi/2}$$

Case II. When n is odd.
$$y_n(0) = [(n-2)^2 + a^2][(n-4)^2 + a^2]$$
$$[(n-6)^2 + a^2]...[1^2 + a^2](-ae^{a\pi/2})$$

EXAMPLE 2. *If* $y = \tan^{-1} x$, *prove that*
$$(1 + x^2)y_{n+1} + 2nxy_n + n(n-1)y_{n-1} = 0.$$
Hence, determine the values of all the derivatives of y with respect to x when x = 0.

(MUMBAI–2008)

SOLUTION . We have $\qquad y = \tan^{-1} x.$ ...(1)
$$\therefore \qquad y_1 = \frac{1}{1 + x^2} \qquad ...(2)$$
$$\Rightarrow \qquad y_1(1 + x^2) = 1.$$

Differentiating, n times by Leibnitz's theorem, we have
$$y_{n+1}(1 + x^2) + ny_n.2x + \frac{n(n-1)}{2}y_{n-1}.2 = 0$$
$$\Rightarrow \quad (1 + x^2)y_{n+1} + 2nxy_n + n(n-1)y_{n-1} = 0 \quad ...(3)$$

Putting $x = 0$ in (1), (2) and (3), we get
$$y(0) = 0$$
$$y_1(0) = 1$$
$$\overline{\qquad\qquad\qquad\qquad\qquad\qquad}$$
$$y_{n+1}(0) = -n(n-1)y_{n-1}(0) \quad ...(4)$$

Put $n = 1$ in (4), we get
$$y_2(0) = 0.$$

Put $n - 1$ for n in (4), we get
$$y_n(0) = -(n-1)(n-2)y_{n-2}(0) \qquad ...(5)$$

Put $n - 3$ for n in (4), we get
$$y_{n-2}(0) = -(n-3)(n-4)y_{n-4}(0) \qquad ...(6)$$

From (5) and (6), we get
$$y_n(0) = (n-1)(n-2)(n-3)(n-4)y_{n-4}(0) \quad ...(7)$$

There arise following two cases :

Case I. When n is even.
$$y_n(0) = (-1)^{(n-2)/2}(n-1)(n-2)(n-3)(n-4)...4.2y_2(0)$$
$$= (-1)^{(n-2)/2}(n-1)(n-2)(n-3)(n-4)...3.2.0$$
$$= 0 \qquad\qquad\qquad [\because y_2(0) = 0]$$

Case II. When n is odd.
$$y_n(0) = (-1)^{(n-1)/2}(n-1)(n-2)(n-3)...3.2.1y_1(0)$$
$$= (-1)^{(n-1)/2}(n-1)! y_1(0)$$
$$= (-1)^{(n-1)/2}(n-1)! \qquad\qquad [\because y_1(0) = 1]$$

EXAMPLE 3. *If* $y = \sin(a \sin^{-1} x)$, *then, prove that*
$$(1 - x^2)y_2 - xy_1 + a^2 y = 0 \qquad\qquad and$$
$$(1 - x^2)y_{n+2} - (2n+1)xy_{n+1} + (a^2 - n^2)y_n = 0.$$
Hence, find $y_n(0)$.   (UPTU–2009, 2011)

SOLUTION . We have $y = \sin(a \sin^{-1} x)$ ...(1)

Differentiating (1) w.r.t. x we get
$$y_1 = \cos(a \sin^{-1} x).\frac{a}{\sqrt{1 - x^2}} = \frac{a}{\sqrt{1 - x^2}}\cos(a \sin^{-1} x)$$
$$\Rightarrow \quad (\sqrt{1 - x^2})y_1 = a \cos(a \sin^{-1} x)$$
$$\Rightarrow \quad (1 - x^2)y_1^2 = a^2 \cos^2(a \sin^{-1} x)$$
$$= a^2(1 - \sin^2(a \sin^{-1} x))$$
$$\Rightarrow \quad (1 - x^2)y_1^2 = a^2(1 - y^2) \qquad ...(2)$$

Differentiating (2) w.r.t. x, we get   (Using (1))
$$(1 - x^2)2y_1y_2 - 2xy_1^2 = a^2(-2yy_1)$$
$$\Rightarrow (1 - x^2)y_2 - xy_1 + a^2 y = 0 \qquad ...(3)$$

Now differentiating (3) n times by Leibnitz's theorem, we get
$$\left[(1 - x^2)y_{n+2} + {}^nC_1(-2x)y_{n+1} + {}^nC_2(-2)y_n\right]$$
$$- \left[xy_{n+1} + {}^nC_1(1)y_n\right] + a^2 y_n = 0$$
$$\Rightarrow (1 - x^2)y_{n+2} + n(-2x)y_{n+1} + \frac{n(n-1)}{2}(-2)y_n$$
$$- xy_{n+1} - n.1.y_n + a^2 y_n = 0$$
$$\Rightarrow (1 - x^2)y_{n+2} - (2n+1)xy_{n+1} + (a^2 - n^2 - n + n)y_n = 0$$
$$\Rightarrow (1 - x^2)y_{n+2} - (2n+1)xy_{n+1} + (a^2 - n^2)y_n = 0 \quad ...(4)$$

From (1), $y(0) = \sin(a \sin^{-1} 0) = 0$

From (2), $y_1(0) = \dfrac{a}{\sqrt{1 - 0}}\cos(a \sin^{-1} 0) = a \cos 0 = a$

From (3), $(1 - 0^2)y_2(0) - 0.y_1(0) + a^2 y(0) = 0$
$$\Rightarrow \qquad y_2(0) = 0$$

Form (4),
$$(1 - 0^2)y_{n+2}(0) - (2n+1).0 + (a^2 - n^2)y_n(0) = 0$$
$$\Rightarrow \qquad y_{n+2}(0) = (n^2 - a^2)y_n(0) \qquad ...(5)$$

Case I. If n is even.

Put $n = 2$ in equation (5), we get
$$y_4(0) = (2^2 - a^2)y_2(0) = 0$$

Put $n = 4$ in equation (5), we get
$$y_6(0) = (4^2 - a^2)y_4(0) = 0$$

Put $n = 6$ in equation (5), we get
$$y_8(0) = (6^2 - a^2)y_6(0) = 0$$
$$\Rightarrow \quad y_n(0) = 0, \text{ if } n \text{ is even}$$

Case II. If n is odd.

Put $n = 1$ in equation (5), we get
$$y_3(0) = (1^2 - a^2)y_1(0) = (1^2 - a^2).a$$

Put $n = 3$ in equation (5), we get
$$y_5(0) = (3^2 - a^2)y_3(0) = (1^2 - a^2)(3^2 - a^2).a$$

Put $n = 5$ in equation (5), we get
$$y_7(0) = (5^2 - a^2)y_5(0)$$
$$= (1^2 - a^2)(3^2 - a^2)(5^2 - a^2).a$$
$$\Rightarrow y_n(0) = (1^2 - a^2)(3^2 - a^2)(5^2 - a^2)...[(n-2)^2 - a^2]a$$
if n is odd and $n \neq 1$

Hence,
$$y_n(0) = \begin{cases} 0 & \text{if } n \text{ is even} \\ (1^2 - a^2)(3^2 - a^2)(5^2 - a^2) & \text{if } n \text{ is odd} \\ \quad ...[(n-2)^2 - a^2]a & \text{and } n \neq 1 \end{cases}$$

Exercise-2.3

1. If $y = \sin^{-1}x$, prove that $(1-x^2)y_{n+2} - (2n+1)xy_{n+1} - n^2y_n = 0$ and also find the value of $y_n(0)$. (SVTU–2009)

2. (i) If $y = [\log\{x + \sqrt{(1+x^2)}\}]^2$, find all the derivatives of y w.r.t. x when $x = 0$.

(ii) If $y = (\sinh^{-1}x)^2$, prove that
$(1+x^2)y_{n+2} + (2n+1)xy_{n+1} + n^2y_n = 0$ Hence, find $y_n(0)$.

3. (i) If $y = \sin(m\sin^{-1}x)$, then prove that
$(y_{n+2})_0 = (n^2 - m^2)(y_n)_0$ and find $y_n(0)$. (MUMBAI- 2008)

(ii) If $y = \cos(m\sin^{-1}x)$, find $y_n(0)$. (COCHIN- 2005, VTU-2009)

4. If $y = e^{a\sin^{-1}x}$, show that
$(1-x^2)y_{n+2} - x(2n+1)y_{n+1} - (n^2 + a^2)y_n = 0$
and hence, find the value of $y_n(0)$.

5. If $x = \sin\left(\dfrac{1}{a}\log y\right)$, find $(y_n)0$.

Hints to Selected Problems

1. $y = \sin^{-1}x \Rightarrow y_1 = \dfrac{1}{\sqrt{1-x^2}} \Rightarrow (1-x^2)y_1^2 = 1$

Differentiating, we get $(1-x^2)y_2 - xy_1 = 0$
Now apply Leibnitz's theorem and put $x = 0$.

2. (i) $y = [\log x + \sqrt{1+x^2}]^2$

$\Rightarrow y_1 = \dfrac{2}{\sqrt{1+x^2}}\log(x + \sqrt{1+x^2}) \Rightarrow (1+x^2)y_1^2 = 4y$

Again differentiating, we get
$(1 + x^2)y_2 + xy_1 - 2 = 0.$

Now apply Leibnitz's theorem to find y_n and then put $x = 0$.

(ii) $\sinh^{-1}x = \log[x + \sqrt{1+x^2}]$

3. (i) $y = \sin(m\sin^{-1}x)$

$\Rightarrow y_1 = \dfrac{m\cos(m\sin^{-1}x)}{\sqrt{1-x^2}} \Rightarrow y_1\sqrt{(1-x^2)} = m\cos(m\sin^{-1}x)$

Again differentianting, we get $(1-x^2)y_2 - xy_1 - m^2y = 0$
Now apply Leibnitz's theorem.

4. $y = e^{a\sin^{-1}x} \Rightarrow y_1 = \dfrac{ay}{\sqrt{1-x^2}} \Rightarrow (1-x^2)y_1^2 = a^2y^2$

Differentiating w.r.t. x, we get
$(1-x^2)y_2 - xy_1 - a^2y = 0$
Now apply Leibnitz's theorem.

Answers

1. When n is even, $\quad y_n(0) = 0$; $\quad$ When n is odd $\quad y_n(0) = 1^2.3^2.5^2...(n-2)^2$

2. (i),(ii) when n is even, $y_n(0) = (-1)^{n/2-1}.2.2^2.4^2...(n-2)^2$, when n is odd $y_n(0) = 0$

3. (i) When n is even, $\quad y_n(0) = 0$, When n is odd, $y_n(0) = [(n-2)^2 - m^2][(n-4)^2 - m^2][(n-6)^2 - m^2]...[(3^2 - m^2)(1^2 - m^2)]m$

(ii) When n is even, $\quad y_n(0) = -[(n-2)^2 - m^2][(n-4)^2 - m^2]...[(2^2 - m^2)m^2$; When n is odd, $\quad y_n(0) = 0$

4. When n is even, $\quad y_n(0) = [(n-2)^2 + a^2][(n-4)^2 + a^2]...(4^2 + a^2)(2^2 + a^2).a^2$

When n is odd, $\quad y_n(0) = [(n-2)^2 + a^2][(n-4)^2 + a^2]...(3^2 + a^2)(1^2 + a^2).a$

2.2 MEAN VALUE THEOREMS AND EXPANSION OF FUNCTIONS

2.2.1 ROLLE'S THEOREM

If a function f defined on [a,b] is such that it is
(i) continuous in [a,b], (ii) differentiable in]a,b[and (iii) f(a) = f(b),
then there exists at least one vlaue of x, say c,(a<c<b) such that f'(c) = 0

PROOF. Since, the function $f(x)$ is continuous on $[a, b]$

$\Rightarrow f(x)$ is bounded $\qquad\qquad$ [∵ Every continuous function is bounded.]

$\Rightarrow f(x)$ attains its bounds [∵ A function, which is continuous on a closed bounded interval $[a, b]$, then it attains its bound on $[a, b]$.]

Let M and m are the supremum and infimum of $f(x)$ respectively.

Now there are two possibilities:

(i) If $M=m$, then obviously $f(x)$ is a constant function, and therefore its derivative is zero, i.e., $f'(x) = 0 \;\forall x \in]a, b[.$

(ii) If $M \neq m$, then at least one of the numbers M and m must be different from the equal values $f(a)$ and $f(b)$.

Let us assume $\qquad\qquad\qquad M \neq f(a).$

Now, since, every continuous function on a closed interval attains its supremum, therefore, there exists a real number c in $[a,b]$ such that $f(c)=M$. Also since $f(a) \neq M \neq f(b)$.Therefore $c \neq a$ and $c \neq b$, this implies that $c \in]a,b[.$

Now, $f(c)$ is the supremum of f on $[a, b]$

$\therefore \qquad\qquad\qquad f(x) \leq f(c) \;\forall x \in [a, b]$ $\qquad\qquad\qquad\qquad$ ...(1)

[By definition of supremum]

In particular, $\qquad\qquad f(c–h) \leq f(c)\; h>0$

$\Rightarrow \qquad\qquad \dfrac{f(c-h) - f(c)}{-h} \geq 0$ $\qquad\qquad\qquad\qquad\qquad$ ...(2)

Since $f'(x)$ exists at each point of $]a, b[$, and hence, $f'(c)$ exists.

Therefore, from (2) $\qquad\qquad Lf'(c) \geq 0$ $\qquad\qquad\qquad\qquad\qquad$ ...(3)

Similarly, from (1) $f(c+h) \leq f(c) \, h > 0$

Then by the same arguments $Rf'(c) \leq 0.$ ...(4)

Since $f(x)$ is differentiable in $]a, b[$ $\Rightarrow f'(c)$ exist

$\Rightarrow$ $Lf'(c) = f'(c) = Rf'(c)$ ...(5)

Now from (3), (4) and (5) $f'(c) = 0.$

Similarly we can consider the case $M = f(a) \neq m.$

REMARKS

- Converse of Rolle's theorem is not true, *i.e.*, $f'(x)$ may vanish at a point $c \in]a, b[$ without $f(x)$ satisfying the three conditions of Rolle's theorem.
- There may be more than one point like c at which $f'(x)$ vanishes but Rolle's theorem ensures the existence of at least one such c.
- Rolle's theorem will not hold good if
  (a) $f(x)$ is discontinuous at some point in the interval $[a, b]$
  (b) $f'(x)$ does not exist at some point in the interval $]a, b[$
  (c) $f(a) \neq f(b)$
- The hypothesis of Rolle's theorem cannot be weakened.

  For example, if $f(x) = 1 - |x|, -1 \leq x \leq 1$, then $f(-1) = f(1) = 0$ and f is continuous on $[-1,1]$. Also if $f'(x)$ exist $\forall x \in]-1, 1[$ except at $x = 0$. Then, f satisfies all the condition of Rolle's theorem except that f is not differentiable at $x = 0$. For this f, there is no c in $]-1,1[$ for which $f'(c) = 0.$

GEOMETRICAL INTERPRETATION OF ROLLE'S THEOREM

Geometrically, Rolle's theorem means that if the curve $y = f(x)$ is continuous from $x = a$ to $x = b$, has a definite tangent at each point of $]a,b[$ and the ordinates at the extremities are equal, then there exists at least one point between a and b at which the tangent is parallel to x-axis.

ALGEBRAIC INTERPRETATION OF ROLLE'S THEOREM

Algebraically, Rolle's theorem means that if $f(x)$ is a polynomial function in x and $x = a$ and $x = b$ are two roots of the equation $f(x) = 0$, then, there is at least one root of the equation $f'(x) = 0$ which lies between a and b.

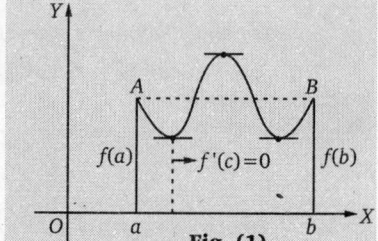

Fig. (1)

2.2.2 LAGRANGE'S MEAN VALUE THEOREM

Let f be a function defined on $[a, b]$ such that

   (i) f is continuous on $[a, b]$,   (ii) f is differentiable on $]a, b[$.

Then, there exists a real number $c \in]a,b[$ such that $\dfrac{f(b) - f(a)}{b - a} = f'(c)$

PROOF.   Let us define a function $F(x)$ such that

$$F(x) = f(x) + Ax \; \forall \, x \in [a,b]$$ ...(1)

where A is a constant to be suitably chosen such that $F(a) = F(b)$.
Now
   (i) Since, f is continuous on $[a,b]$ and Ax is continuous on $[a,b]$ therefore, F is continuous on $[a,b]$

$$[\because \text{ Sum of two continuous functions is again continuous.}]$$

   (ii) Similarly F is differentiable on (a, b)

   (iii) $F(a) = F(b) \Rightarrow \qquad -A = \dfrac{f(b) - f(a)}{b - a}$ ...(2)

Hence, we find that F satisfy all the conditions of Rolle's Theorem on $[a,b]$ and consequently, there exists a real number $c \in]a,b[$ such that $F'(c) = 0$, this gives

$$f'(c) + A = 0 \quad \Rightarrow \quad -A = f'(c).$$ ...(3)

Now, from (2) and (3), we have $\dfrac{f(b) - f(a)}{b - a} = f'(c)$

REMARKS

- If we take $b = a + h$ and c can be written as $a + \theta h$, where θ is some real number such that $0 < \theta < 1$. Lagrange's theorem then read as follows :

  " Let f be defined and continuous on $[a, a + h]$ and differentiable on $]a, a + h[$, then for some real number $\theta (0 < \theta < 1)$

$$\frac{f(a+h) - f(a)}{h} = f'(a + \theta h).$$

- The hypothesis of the Lagrange's mean value theorem cannot be weakened, as it is clear from the following examples :
  " Let f be the function defined on $[-1,2]$ by setting $f(x) = |x|$, $\forall x \in [-1,2]$.
  Here, f is continuous on $[-1,2]$ and differentiable at all points of $]-1, 2[$ except at $x=0$ (so that second condition is violated.)
  Now $\qquad f'(x) = \begin{cases} -1 & \text{if } x \in]-1,0[\\ 1 & \text{if } x \in]0,2[\end{cases}$; Also $\dfrac{f(2) - f(-1)}{2 - (-1)} \neq f'(x)$ for any x in $]-1, 2[$.

- Lagrange's mean value theorem is known as first mean value theorem.
- The result $f(b) - f(a) = (b-a)f'(c)$ is also known as the formula for finite increment.
- For $f(a) = f(b)$, the Lagrange's mean value theorem yields Rolle's theorem

GEOMETRICAL INTERPRETATION OF LAGRANGE'S MEAN VALUE THEOREM

If the curve $y = f(x)$ is continuous from $x=a$ and $x=b$ and has a tangent at each point on the curve between $x=a$ and $x=b$, then, geometrically, the first mean value theorem means that there is at least one point between $x=a$ and $x=b$ on the curve where the tangent to the curve parallel to the chord joining the points $(a, f(a))$ and $(b, f(b))$.

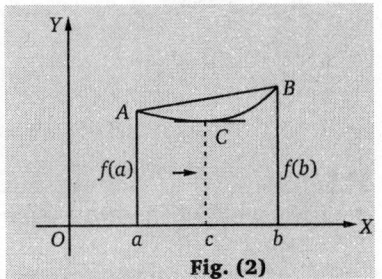

Fig. (2)

Let ACB be the graph of the function $y = f(x)$ then the co-ordinate of the points A and B are given by $(a, f(a))$ and $(b, f(b))$ respectively. If the chord AB makes an angle θ with the x-axis, then

$$\tan \theta = \frac{f(b) - f(a)}{b - a} = f'(c), \text{ where } a < c < b.$$

DEDUCTIONS FROM THE FIRST MEAN VALUE THEOREM

THEOREM 1. *If a function $f(x)$ satisfies the conditions of mean value theorem then*
$\qquad$ (i) $f'(x) = 0 \ \forall x \in]a, b[\Rightarrow f$ *is constant on $[a, b]$.*
$\qquad$ (ii) $f'(x) > 0 \ \forall x \in]a,b[\Rightarrow f$ *is strictly increasing on $[a,b]$.*
and (iii) $f'(x) < 0 \ \forall x \in]a,b[\Rightarrow f$ *is strictly decreasing on $[a,b]$.*

PROOF. $\quad$ (i) Let x_1, x_2 (where $x_1 > x_2$) be any two distinict points of $[a,b]$, then by Lagrange's mean value theorem,

$$\frac{f(x_2) - f(x_1)}{x_2 - x_1} = f'(c) = 0, \ x_1 < c < x_2 \qquad \qquad \ldots(1)$$

$\Rightarrow \qquad\qquad\qquad\qquad f(x_2) = f(x_1).$

$\Rightarrow \qquad$ function keeps the same value. Therefore $f(x)$ is constant on $[a,b]$.

$\qquad$ (ii) From (1), we have

$$\frac{f(x_2) - f(x_1)}{x_2 - x_1} = f'(c) \text{ for some } c \in]x_1, x_2[$$

But $\qquad\qquad\qquad\qquad f'(c) > 0 \qquad\qquad\qquad\qquad\qquad [\because f'(x) > 0 \ \forall x \in [a, b]]$

$\Rightarrow \qquad\qquad\qquad f(x_2) - f(x_1) > 0$

$\Rightarrow \qquad\qquad\qquad f(x_2) > f(x_1)$

Thus $\qquad\qquad\qquad x_2 > x_1 \Rightarrow f(x_2) > f(x_1) \ \forall x_1, x_2 \in [a,b]$

$\qquad$ Hence, f is strictly increasing on $[a,b]$.

$\qquad$ (iii) Same as (ii).

✍ REMARK

- For a strictly increasing function f, the derivative $f'(x)$ need not be strictly positive. For example, consider $f(x) = x^3$, $x \in]-1, 1[$. Here, $f(x)$ is strictly increasing but $f'(x) = 3x^2$, which is zero at $x=0 \in]-1, 1[$.

2.2.3 CAUCHY'S MEAN VALUE THEOREM

Let f and g be two functions defined on $[a,b]$ such that
$\qquad$ (i) f and g are continuous on $[a, b]$, $\quad$ (ii) f and g aredifferentiabfe on $]a, b[$,
and $\quad$ (iii) $g'(x) \neq 0$ for any point of $]a, b[$. $\qquad\qquad\qquad\qquad$ (PTU–2007, 11, VTU–2006)

Then, there exists a real number $c \in]a, b[$ such that $\dfrac{f(b) - f(a)}{g(b) - g(a)} = \dfrac{f'(c)}{g'(c)}$

PROOF. $\qquad$ Let us define a function $\qquad F(x) = f(x) + A.g(f) \qquad\qquad\qquad \ldots(1)$

$\qquad\qquad$ where A is a constant, to be suitably chosen such that
$\qquad\qquad\qquad\qquad\qquad\qquad F(a) = F(b) \qquad\qquad\qquad\qquad\qquad\qquad \ldots(2)$

Now, the function F is the sum of two continuous and differentiable functions. Therefore
      (i) F is continuous on $[a,b]$,
      (ii) F is differentiable on $]a,b[$,
and   (iii) $F(a)=F(b)$.

Then, by Rolle's theorem, there must exists a real number c between a and b such that
$$F'(c)=0$$
Here,
$$F'(x)=f'(x)+Ag'(x)$$
$$F'(c)=0 \qquad \Rightarrow f'(c)+Ag'(c)=0$$
$$\Rightarrow \qquad -A=\frac{f'(c)}{g'(c)} \qquad\qquad \text{...(3)}$$
Now
$$F(a)=F(b) \qquad \Rightarrow f(a)+Ag(a)=f(b)+Ag(b)$$
$$\Rightarrow \qquad -A=\frac{f(b)-f(a)}{g(b)-g(a)} \qquad\qquad \text{...(4)}$$

From (3) and (4), we have
$$\frac{f(b)-f(a)}{g(b)-g(a)}=\frac{f'(c)}{g'(c)}$$

REMARKS

- If we put $b=a+h$, then c can be written as $a+\theta h$, where $\theta \in R$ such that $0 < \theta < 1$, then Cauchy's mean value theorem can be restated as
  "If f and g are continuous on $[a, a+h]$ and are differentiable on $]a, a+h[$ and $g'(x) \neq 0$ for any $x \in]a, a+h[$ then, $\exists$ a $\theta \in R:0 < \theta < 1$ such that
  $$\frac{f(a+h)-f(a)}{g(a+h)-g(a)}=\frac{f'(a+\theta h)}{g'(a+\theta h)}.$$

- If we take $g(a)=g(b)$, then the function g would satisfy all the conditions of Rolle's theorem and consequently for some x in $]a,b[$, we would have $g'(x)=0$. In view of this we take $g(a) \neq g(b)$.

- In some cases, the Lagrange's mean value theorem is a particular case of Cauchy's mean value theorem (*e.g.*, take $g(x)=k$).

- Cauchy's mean value theorem cannot be deduced by applying Lagrange's mean value theorem to two functions f and g seperately and then dividing. It can be easily seen that the desired result cannot be obtained in this manner. In this way, we get $\frac{f(b)-f(a)}{g(b)-g(a)}=\frac{f'(c_1)}{g'(c_2)}$ where $a < c_1 < b$, and $a < c_2 < b$. But, it is not necessary that c_1 and c_2 are equal.
  Hence, Cauchy's means value theorem is not directly deduceable from the first one.

- The conditions in the theorem are sufficient one. The conclusion may still hold even when the function involved do not satisfy the condition on $[a,b]$.

GEOMETRICAL INTERPRETATION OF CAUCHY'S MEAN VALUE THEOREM

(1) Under suitable conditions, Cauchy's mean value theorem geometrically means that there is an ordinate $x=c$ between $x=a$ and $x=b$, such that the tangents at the points where $x=c$ cut the graphs of the function $f(x)$ and $\frac{f(b)-f(a)}{g(b)-g(a)}g(x)$ are mutually parallel.

(2) The ratio of the mean rates of increase of two functions in an interval is equal to the ratio of the actual rates of increase of the functions at some point within the interval.

Solved Examples

EXAMPLE 1. *Discuss the applicability of Rolle's theorem in the interval* $[-1,1]$ *to the function* $f(x)=|x|$.

SOLUTION. Here, we have $\qquad f(x)=|x|$
$$\Rightarrow \qquad \left.\begin{array}{l} f(-1)=1 \\ f(1)=1 \end{array}\right\}$$
and
$$\Rightarrow \qquad f(1)=f(-1)$$

Now, the function $f(x)$ is continuous throughout the closed interval $[-1,1]$ but $f(x)$ is not differentiable at $x=0 \in]-1,1[$. Hence, Rolle's theorem is not satisfied (due to the second condition).

EXAMPLE 2. *Verify Rolle's theorem the function* $f(x)=x^3-4x$ *on* $[-2, 2]$.

SOLUTION. The function $f(x)=x^3-4x$ is a polynomial and so it is continuous and differentiable at all $x \in R$. In particular it is continuous in the closed interval $[-2,2]$ and differentiable in the open interval $]-2,2[$. Also $f(-2)=0=f(2)$.

Thus, $f(x)$ satisfies all the three conditions of Rolle's theorem in $[-2,2]$. Therefore, there must exist at least one real number 'x' in the open interval $]-2,2[$ for which
$$f'(x)=0.$$
Now $\qquad f'(x)=0$ gives $3x^2-4=0$
or $\qquad x=\pm\dfrac{2}{\sqrt{3}}=\pm1.55$.

Both these values lie in the open interval $]-2, 2[$ and thus the conclusion of Rolle's theorem is verified.

EXAMPLE 3. *Discuss the applicability of Rolle's theorem to the function* $f(x) = \log\left[\dfrac{x^2 + ab}{(a+b)x}\right]$ *in the interval* $[a, b]$.

(VTU–2005)

SOLUTION. Here, we have

$$f(a) = \log\left[\frac{a^2 + ab}{(a+b)a}\right] = \log 1 = 0$$

and $f(b) = \log\left[\dfrac{b^2 + ab}{(a+b)b}\right] = \log 1 = 0$

Also, it can be easily seen that $f(x)$ is continuous on $[a,b]$ and differentiable on $]a,b[$.

Thus all the three conditions of Rolle's theorem are satisfied. Hence $f'(x) = 0$ for at least one value of x in $]a, b[$.

Now $\qquad f'(x) = 0 \Rightarrow \dfrac{2x}{x^2 + ab} - \dfrac{1}{x} = 0$

$\Rightarrow \qquad 2x^2 - (x^2 + ab) = 0$

$\Rightarrow \qquad x^2 = ab$ or $x = \sqrt{ab}$.

Obviously $\qquad \sqrt{ab} \in]a,b[$

[Being the geometric mean of a and b]

Hence, the Rolle's theorem is verified.

EXAMPLE 4. *Show that there is no real number p for which the equation* $x^3 - 3x + p = 0$ *has two distinct roots in* $]0,1[$.

SOLUTION. Let, if possible, there are two distinct roots a and b of the given equation in $]0, 1[$, such that $0 < a < b < 1$.

Now, let $\qquad f(x) = x^3 - 3x + p$

Obviously, $f(x)$ is continuous and differentiable for all values of x (Being a polynomial).

Also, we have $f(a) = f(b) = 0$

$\Rightarrow f$ satisfies all the conditions of Rolle's theorem in $[a,b]$ hence, $\exists$ a point $c \in]a,b[$ such that $f'(c) = 0$.

Now $\qquad\qquad f'(x) = 0 \Rightarrow \qquad 3x^2 - 3 = 0$

$\Rightarrow \qquad\qquad x = \pm 1$

which is a contradiction

$(\because a < c < b$ as $0 < a < b < 1)$

$\Rightarrow$ our assumption is wrong. Hence, there cannot be two distinct roots of $f(x) = 0$ in $]0, 1[$ for any value of p.

EXAMPLE 5. *If* $a + b + c = 0$, *then show that the quadratic equation* $3ax^2 + 2bx + c = 0$ *has at least one root in* $]0, 1[$.

SOLUTION. Let us define a function $f(x)$ such that

$$f(x) = ax^3 + bx^2 + cx + d.$$

Here we have $\qquad f(0) = d$

and $f(1) = a + b + c + d = d \qquad (\because a + b + c = 0)$

Obviously, $f(x)$ is continuous and differentiable in $]0, 1[$ (Being a polynomial).

Thus, $f(x)$ satisfies all the three conditions of Rolle's theorem in $[0, 1]$. Hence, there is at least one value of x in the open interval $]0, 1[$ where $f'(x) = 0$

i.e., $3ax^2 + 2bx + c = 0$ has at least one root in $]0, 1[$.

EXAMPLE 6. *Find 'c' of the mean value theorem, if* $f(x) = x(x-1) \cdot (x-2); a = 0, b = 1/2$. (GORAKHPUR–1999)

SOLUTION. Here, we have $f(a) = f(0) = 0$

and $\qquad f(b) = f\left(\dfrac{1}{2}\right) = \dfrac{3}{8}$

$\therefore \qquad \dfrac{f(b) - f(a)}{b - a} = \dfrac{\frac{3}{8} - 0}{\frac{1}{2} - 0} = \dfrac{3}{4}$

Now $\qquad\qquad f(x) = x^3 - 3x^2 + 2x$

$\therefore \qquad\qquad f'(x) = 3x^2 - 6x + 2$

$\Rightarrow \qquad\qquad f'(c) = 3c^2 - 6c + 2$

Putting all these values in the Lagrange's mean value theorem

$$\frac{f(b) - f(a)}{b - a} = f'(c), (a < c < b)$$

We get $\qquad \dfrac{3}{4} = 3c^2 - 6c + 2$ or $c = 1 \pm \dfrac{\sqrt{21}}{6}$

Hence, $c = \dfrac{1 - \sqrt{21}}{6}$ lies in the open interval $]0, \dfrac{1}{2}[$ therefore, it is the required value.

EXAMPLE 7. *If* $f(x) = \log x$, *find all numbers strictly between* e^2 *and* e^3 *such that*

$$f'(x) = \frac{f(e^3) - f(e^2)}{e^3 - e^2} \qquad \text{(BURDWAN–2003)}$$

SOLUTION. Obviously $f(x) = \log x$ is continuous in $[e^2, e^3]$ and differentiable in $]e^2, e^3[$.

Then by Lagrange's mean value theorem. There exist $c \in]e^2, e^3[$, such that

$$f'(c) = \frac{f(e^3) - f(e^2)}{e^3 - e^2}$$

$\Rightarrow \qquad\qquad \dfrac{1}{c} = \dfrac{3 - 2}{e^3 - e^2}$

$\therefore \qquad\qquad c = (e^3 - e^2).$

There exists only one value $c = (e^3 - e^2)$ in $]e^2, e^3[$.

EXAMPLE 8. *Use the function* $f(x) = x^{1/x}$, $x > 0$ *show that* $e^\pi > \pi^e$.

SOLUTION. Here $\qquad f(x) = x^{1/x}, x > 0$

$\therefore \qquad \log f(x) = \dfrac{1}{x} \log_e x$

Differentiating w.r.t. x, we get

$$\frac{1}{f(x)} f'(x) = \frac{1}{x} \cdot \frac{1}{x} - \frac{1}{x^2} \log_e x$$

$$f'(x) = \frac{x^{1/x}}{x^2}[1 - \log_e x]$$

For $\qquad x > e, f'(x) < 0 \quad [\because \log_e x > 1$ for $x > e]$

$\therefore \quad f(x)$ is a decreasing function of x for $x > e$.

Hence $\qquad \pi < e \Rightarrow f(\pi) < f(e) \Rightarrow \pi^{1/\pi} < e^{1/e}$

$\Rightarrow \left(\pi^{1/\pi}\right)^{e\pi} < \left(e^{1/e}\right)^{e\pi}$

$\Rightarrow \pi^e < e^\pi$

$\Rightarrow e^\pi > \pi^e$

EXAMPLE 9. *Show that* $\dfrac{x}{1+x} < \log(1+x) < x$, *for* $x > 0$.

(MUMBAI–2008)

SOLUTION. Let, $\qquad f(x) = \log(1+x) - \dfrac{x}{1+x}$

Obviously, $f(0) = 0$.

and $\qquad f'(x) = \dfrac{1}{1+x} - \dfrac{1 \cdot (1+x) - x \cdot 1}{(1+x)^2}$

$$= \frac{1}{1+x} - \frac{1}{(1+x)^2} = \frac{x}{(1+x)^2}$$

Here, we observe that $f'(x)>0$, for $x>0$.

$\Rightarrow$ $f(x)$ is monotonically increasing in the interval $[0,\infty[$. Therefore

$$f(x)>f(0), \qquad \text{for } x>0$$

$$\Rightarrow \left[\log(1+x)-\frac{x}{1+x}\right]>0, \qquad \text{for } x>0$$

$$\Rightarrow \qquad \log(1+x)>\frac{x}{1+x}, \qquad \text{for } x>0 \qquad \ldots(1)$$

Now let $\qquad F(x)=x-\log(1+x)$

Obviously $\quad F(0)=0$

Then $\qquad F'(x)=1-\dfrac{1}{1+x}=\dfrac{x}{1+x}$

Here, we observe that $F'(x)>0$, for $x>0$. Hence $F(x)$ is monotonically increasing in the interval $[0,\infty[$.

$\therefore \qquad\qquad F(x)>F(0), \qquad \text{for } x>0$

$\Rightarrow \quad [x-\log(1+x)]>0, \qquad \text{for } x>0$

$\Rightarrow \qquad\qquad x>\log(1+x), \quad \text{for } x>0 \qquad \ldots(2)$

Now from (1) and (2), we get

$$\frac{x}{1+x}<\log(1+x)<x, \text{ for } x>0$$

EXAMPLE 10. *Prove that* $(1+x)<e^x<1+xe^x,\ \forall\ x>0$.

SOLUTION. Let us consider the function $f(x)=e^x$ in $[0,x]$.

Obviously $f(x)$ is continuous as well as differentiable in $]0,x[$.

Then, by Lagrange's theorem $\exists\ c\in\ ]0,x[$, such that

$$f'(c)=\frac{f(x)-f(0)}{x-0}$$

or $\qquad e^c=\dfrac{e^x-1}{x} \qquad \ldots(1)$

$$0<c<x \ \Rightarrow\ e^0<e^c<e^x$$

$$(\because e^x \text{ is an increasing function.}) \ \ldots(2)$$

Now, from (1) and (2), we have

$$e^0<\frac{e^x-1}{x}<e^x,\forall x>0$$

$$\Rightarrow \quad 1<\frac{e^x-1}{x}<e^x \ \Rightarrow\ x<e^x-1<xe^x$$

$$\Rightarrow \qquad (1+x)<e^x<xe^x$$

EXAMPLE 11. *Verify Lagrange's mean value theorem for the function* $f(x)=\sin x$ *in* $\left[0,\dfrac{\pi}{2}\right]$.   (NAGPUR–2008)

SOLUTION. The function $f(x)=\sin x$ is continuous and differentiable on R. Hence it is continuous as well as differentiable in $[0,\pi/2]$. Then, by Lagrange's mean value theorem, there must exist at least one c in $]0,\pi/2[$ such that

$$\frac{f(\pi/2)-f(0)}{\pi/2-0}=f'(c) \qquad \ldots(1)$$

Here $\qquad\qquad f(0)=0, f(\pi/2)=1$

$$f'(x)=\cos x \ \Rightarrow\ f'(c)=\cos c$$

Put all these values in (1), we have

$$\frac{1-0}{\pi/2}=\cos c \Rightarrow \cos c=\frac{2}{\pi}$$

$$\Rightarrow c=\cos^{-1}\left(\frac{2}{\pi}\right)$$

Since, $0<2/\pi<1$, therefore the value of $c=\cos^{-1}\left(\dfrac{2}{\pi}\right)$

lies in $\left]0,\dfrac{\pi}{2}\right[$, which is the required value of c.

Hence, Lagrange's mean value theorem is satisfied.

EXAMPLE 12. *If* $f''(x)$ *exists for all points in* [a, b] *and* $\dfrac{f(c)-f(a)}{c-a}=\dfrac{f(b)-f(c)}{b-c}$ *where* a < c < b, *then, there is a number l such that*

$$a<l<b \text{ and } f''(l)=0.$$

SOLUTION. Since $f''(x)$ exist for all points in $[a, b]$,

$\Rightarrow$ $f'(x)$ is continuous in $[a, b]$

$\Rightarrow$ $f(x)$ is continuous in $[a, b]$.

Now, applying Lagrange's mean value theorem to $f(x)$ in $[a, c]$ and $[c, b]$ respectively, we get

$$\frac{f(c)-f(a)}{c-a}=f'(l_1), a<l_1<c \qquad \ldots(1)$$

and $\qquad \dfrac{f(b)-f(c)}{b-c}=f'(l_2),\ c<l_2<b \qquad \ldots(2)$

Then, from (1) and (2), we get

$$f'(l_1)=f'(l_2) \quad \left[\because \frac{f(c)-f(a)}{c-a}=\frac{f(b)-f(c)}{b-c}\right]$$

Now $f'(x)$ satisfies all the conditions of Rolle's theorem in the interval $[l_1, l_2]$.

Hence, $\quad f''(l)=0$ where $l\in\ ]l_1, l_2[$ and $l\in\ ]a, b[$.

EXAMPLE 13. *If in the Cauchy's mean value theorem, we write* $f(x)=e^x$ *and* $g(x)=e^{-x}$, *show that 'c' is the arithmetic mean between a and b.*   (MUMBAI-2008)

SOLUTION. Since, we have

$$f(x)=e^x \text{ and } g(x)=e^{-x}$$

$\therefore \quad \dfrac{f(b)-f(a)}{g(b)-g(a)}=\dfrac{e^b-e^a}{e^{-b}-e^{-a}}$

$$=-e^a e^b=-e^{a+b}$$

$\dfrac{f'(x)}{g'(x)}=\dfrac{e^x}{-e^{-x}}$

and

so that $\dfrac{f'(c)}{g'(c)}=\dfrac{e^c}{-e^{-c}}=-e^{2c}$

After putting all these values in Cauchy's mean value theorem, we get

$$-e^{a+b}=-e^{2c} \quad \Rightarrow \quad a+b=2c$$

$$\Rightarrow \quad c=\frac{a+b}{2}$$

Hence, c is the arithmetic mean between a and b.

EXAMPLE 14. *Verify Cauchy's mean value for* $f(x)=\sin x$ *and* $g(x)=\cos x$ *in* $\left[-\dfrac{\pi}{2},0\right]$.   (JNTU–2006)

SOLUTION. It can be easily seen that $f(x)$ and $g(x)$ both are continuous on $\left[-\dfrac{\pi}{2},0\right]$ and differentiable on $\left]-\dfrac{\pi}{2},0\right[$.

Also, $g'(x)=-\sin x\neq0$ for any point in the interval $\left]-\dfrac{\pi}{2},0\right[$.

Then, by Cauchy's mean value theorem, $\exists$ at least one

$c \in \left] -\dfrac{\pi}{2}, 0 \right[$ such that

$$\frac{f(0) - f\left(-\dfrac{\pi}{2}\right)}{g(0) - g\left(-\dfrac{\pi}{2}\right)} = \frac{f'(c)}{g'(c)}$$

Putting all the values and after simplification, we have

$$\cot c = -1 \Rightarrow c = -\pi/4.$$

Since $c = -\pi/4$ lies in $]-\pi/2, 0[$, hence, Cauchy mean value theorem is verified.

EXAMPLE 15. *Find 'c' of Cauchy's mean value theorem for the functions*

$$f(x) = \sqrt{x}, \ \phi(x) = \frac{1}{x} \text{ in } [a,b]$$

and show that it is the G.M. of a and b.

SOLUTION . We have

(i) $f(x)$ and $\phi(x)$ are continuous in the closed interval $[a,b]$.

(ii) $f'(x)$ and $\phi'(x)$ exist in the open interval (a,b).

(iii) $\phi'(x) = -1/2\, x^{-3/2} \neq 0$ for any x in $]a,b[$.

Therefore $f(x)$ and $\phi(x)$ satisfies all the conditions of Cauchy's mean value theorem. Hence there exist a point $c \in]a,b[$ such that

$$\frac{f(b) - f(a)}{\phi(b) - \phi(a)} = \frac{f'(c)}{\phi'(c)} \qquad \ldots(1)$$

Also here $\qquad f'(x) = \dfrac{1}{2}x^{-1/2}, \phi'(x) = -\dfrac{1}{2}x^{-3/2}$

From (1), we get

$$\frac{\sqrt{b} - \sqrt{a}}{(1/\sqrt{b}) - (1/\sqrt{a})} = \frac{1/2c^{-1/2}}{-1/2c^{-3/2}}$$

or $\qquad \dfrac{(\sqrt{b} - \sqrt{a})\sqrt{a}.\sqrt{b}}{\sqrt{a} - \sqrt{b}} = -\dfrac{c^{3/2}}{c^{1/2}}$

Hence, $\qquad\qquad c = \sqrt{ab}$.

2.2.4 Taylor's Theorem

Let $f(x)$ be a single valued function defined on $[a, a+h]$ such that

(i) all the derivative of $f(x)$ upto $(n-1)^{th}$ order are continuous in $[a, a+h]$, and

(ii) $f^n(x)$ exists in $(a, a+h)$, then there exists a real number $\theta, 0 < \theta < 1$, such that

$$f(a+h) = f(a) + hf'(a) + \frac{h^2}{2!}f''(a) + \ldots + \frac{h^{n-1}}{(n-1)!}f^{n-1}(a) + \frac{h^n(1-\theta)^{n-p}}{p(n-1)!}f^n(a+\theta h)$$

where p is a given positive integer.

PROOF. Since, f^n exists, all the derivative $f', f''\ldots f^{n-1}$ exist and continuous on $[a, a+h]$, consider a function f defined on $[a, a+h]$ such that

$$\phi(x) = f(x) + (a+h-x)f'(x) + \frac{(a+h-x)^2}{2!}f''(x) + \ldots + \frac{(a+h-x)^{n-1}}{(n-1)!}f^{n-1}(x) + A(a+h-x)^p \qquad \ldots(1)$$

where A is a constant to be determined such that $\phi(a+h) = \phi(a)$

Now $\qquad\qquad \phi(a) = f(a) + hf'(a) + \dfrac{h^2}{2!}f''(a) + \ldots + \dfrac{h^{n-1}}{(n-1)!}f^{n-1}(a) + Ah^p$

and $\qquad\qquad \phi(a) = \phi(a+h)$

$\Rightarrow \qquad\qquad \phi(a+h) = f(a) + hf'(a) + \dfrac{h^2}{2!}f''(a) + \ldots + \dfrac{h^{n-1}}{(n-1)!}f^{n-1}(a) + Ah^p \qquad \ldots(2)$

Now (i) $f, f', f'', \ldots, f^{n-1}$ being all continuous on $[a, a+h]$ the function ϕ is continuous on $[a, a+h]$,

(ii) Similarly the function ϕ is differentiable on $]a, a+h[$,

and (iii) $\phi(a+h) = \phi(a)$

Thus, the function ϕ satisfies all the conditions of Rolle's theorem and hence $\exists$ a real number $\theta(0 < \theta < 1)$ such that $\phi'(a+\theta h) = 0$.

Here $\qquad \phi'(x) = f'(x) + (-f'(x) + (a+h-x)f''(x))$

$$+ \frac{1}{2!}\left[-2(a+h-x)f''(x) + (a+h-x)^2 f'''(x) \right] + \ldots$$

$$+ \frac{1}{(n-1)!}\left[-(n-1)(a+h-x)^{n-2}f^{n-1}(x) + (a+h-x)^{n-1}f^n(x) \right] - Ap(a+h-x)^{p-1}$$

$$= \frac{(a+h-x)^{n-1}}{(n-1)!}f^n(x) - Ap(a+h-x)^{p-1} \quad \text{[Other terms canceled in pairs]}$$

$\therefore \qquad\qquad 0 = \phi'(a+\theta h) = \dfrac{h^{n-1}(1-\theta)^{n-1}}{(n-1)!}f^n(a+\theta h) - Aph^{p-1}(1-\theta)^{p-1}$

$\Rightarrow \qquad\qquad A = \dfrac{h^{n-p}(1-\theta)^{n-p}}{p(n-1)!}f^n(a+\theta h), h \neq 0, \theta \neq 1$

Now, putting the values of A in (2), we get

$$f(a+h) = f(a) + hf'(a) + \frac{h^2}{2!}f''(a) + \ldots + \frac{h^{n-1}}{(n-1)!}f^{n-1}(a) + \frac{h^n(1-\theta)^{n-p}}{p(n-1)!}f^n(a+\theta h)$$

FORMS OF REMAINDER AFTER n TERMS

(i) The term $R_n = \dfrac{h^n (1-\theta)^{n-1}}{(n-1)!} f^n (a+\theta h)$ which occur after n terms, is called the Taylor's remainder after n terms. The theorem with this form of remainder is called Taylor's theorem with Schlomilch and Roche form of remainder.

(ii) For $p=1$, we get $R_n = \dfrac{h^n (1-\theta)^{n-1}}{(n-1)!} f^n (a+\theta h)$

Then, R_n is called Cauchy's form of remainder.

(iii) For $p=n$, we get $R_n = \dfrac{h^n}{n!} f^n (a+\theta h)$

then, R_n is called Lagrange's form of remainder.

ANOTHER FORM OF TAYLOR'S THEOREM

Replacing h by $(x-a)$ in Taylor's theorem, we get

$$f(x) = f(a) + (x-a) f'(a) + \frac{(x-a)^2}{2!} f''(a) + \ldots + \frac{(x-a)^{n-1}}{(n-1)!} f^{n-1}(a) + \frac{(x-a)^n}{p(n-1)!} f^n (1-\theta)^{n-p}$$

The remainder, after n terms can be written as

$$R_n = \frac{(x-a)^n (1-\theta)^{n-1}}{p(n-1)!} f^n (c), a < c < x.$$

DEDUCTIONS

Putting $a=0$ in second form of Taylor's theorem, we get (Maclaurin's theorem)

$$f(x) = f(0) + x f'(0) + \frac{x^2}{2!} f''(0) + \ldots + \frac{x^{n-1}}{(n-1)!} f^{n-1}(0) + R_n \qquad \ldots(1)$$

(i) If $R_n = \dfrac{x^n (1-\theta)^{n-p}}{p(n-1)!} f^n (\theta x)$, then (1) is known as Maclaurin's theorem with Schlomilch and Roche's form of remainder.

(ii) For $p=1$, $R_n = \dfrac{x^n (1-\theta)^{n-p}}{p(n-1)!} f^n (\theta x)$ is called Cauchy's form of remainder.

(iii) For $p=n$, $R_n = \dfrac{x^n}{n!} f^n (\theta x)$, is called Lagrange's form of remainder.

TAYLOR'S SERIES

Let $f(x)$ possesess continuous derivatives of all orders in the interval $[a, a+h]$, then for every positive integral value of n, we have

$$f(a+h) = f(a) + h f'(a) + \frac{h^2}{2!} f''(a) + \ldots + \frac{h^{n-1}}{(n-1)!} f^{n-1}(a) + R_n$$

where, $R_n = \dfrac{h^n}{n!} f^n (a+\theta h), (0 < \theta < 1).$ ...(1)

Equation (1) can also be written as $S_n = f(a) + h f'(a) + \dfrac{h^2}{2!} f''(a) + \ldots + \dfrac{h^{n-1}}{(n-1)!} f^{n-1}(a)$

Then $f(a+h) = S_n + R_n.$

Let us suppose $R_n \to 0$ as $n \to \infty$, then $\lim\limits_{n \to \infty} S_n = f(a+h)$

i.e., the series $f(a) + h f'(a) + \dfrac{h^2}{2!} f''(a) + \ldots + \dfrac{h^{n-1}}{(n-1)!} f^{n-1}(a) + \ldots$ converges to $f(a+h)$.
Thus,

(i) If f possess a continuous derivative of every order in $[a, a+h]$.

(ii) The remainder after n terms $R_n \to 0$ as $n \to \infty$, then

$$f(a+h) = f(a) + h f'(a) + \frac{h^2}{2!} f''(a) + \ldots + \frac{h^n}{n!} f^n (a) + \ldots$$

This series is known as Taylor's series for the expansion of $f(a+h)$ as a power series in h.
Maclaurin's series. If we put $a=0$ and replace h by x in Taylor's series, we get

$$f(x) = f(0) + x f'(0) + \frac{x^2}{2!} f''(0) + \ldots + \frac{x^n}{n!} f^n (0) + \ldots$$

This Series is known as Maclaurin's series for the expansion of $f(x)$ as a power series in x.

REMARKS

- Maclaurin's series is a particular case of Taylor's series.
- Maclaurin's expansions of $f(x)$ fails if any of the functions $f(x), f'(x), f''(x) \ldots$ becomes infinite or discontinuous at any point of the interval $[0, x]$ or if R_n does not tends to zero as n tends to infinity.

2.2.5 Maclaurin's Theorem

Let $f(x)$ be a function of x which possesses continuous derivatives of all orders in the interval $[0, x]$ and can be expanded as an infinite series in x, then

$$f(x)= f(0)+ x f'(0)+\frac{x^2}{2!} f''(0)+...+\frac{x^n}{n!} f^n(0)+...$$

PROOF. Let us define

$$f(x) = A_0+A_1 x+A_2 x^2+A_3 x^3+... \qquad ...(1)$$

Let the expression (1) be differentiable term by term any number of times. Then by successive differentiation, we have

$$f'(x)= A_1+2A_2 x+3A_3 x^2+4A_4 x^3+...$$
$$f''(x)= 2.1.A_2+3.2.A_3 x+4.3.A_4 x^2+...$$
$$f'''(x)= 3.2.A_3+4.3.2.A_4 x+...$$

Putting $x=0$, we get

$$f(0)=A_0, f'(0)=A_1, f''(0)=2!A_2, f'''(0)=3!A_3....$$

$$\Rightarrow \qquad A_0= f(0), A_1=f'(0), A_2=\frac{f''(0)}{2!}, A_3=\frac{f'''(0)}{3!}...$$

Substitute all these values in (1), we get

$$f(x)= f(0)+ x f'(0)+\frac{x^2}{2!} f''(0)+...+\frac{x^n}{n!} f^n(0)+...$$

✒ Remarks

- The Maclaurin's theorem is a particular case of Taylor's Theorem, and can be obtained by replacing $a=0$ and $h=x$ in Taylor's theorem.
- If the function $f(x)$ is denoted by y, then the expansion may be written in the form

$$y= y(0)+ x y_1(0)+\frac{x^2}{2!} y_2(0)+...+\frac{x^n}{n!} y_n(0)+...$$

where $y(0), y_1(0), y_2(0),..., y_n(0)$ etc. denotes values of $y, y_1, y_2,...,y_n$ respectively for $x=0$.

2.2.6 Failure of Taylor's and Maclaurin's Theorem

(a) Taylor's theorem fails to expand $f(a+h)$ in an infinite power series in the following cases :
- If any of the function $f(x), f'(x), f''(x)...$ become infinite or does not exist for any value of x in the given interval.
- If R_n does not tends to zero as $n\to\infty$.

(b) Maclaurin's theorem fails to expand $f(x)$ in an infinite power series in the following cases :
- If any of the function $f(x), f'(x), f''(x)...$ becomes infinite or does not exist in interval $[0, x]$.
- If R_n does not tends to zero as $n\to\infty$.

✒ Remarks

- Before expanding a given function as an infinite Taylor's or Maclaurin's series, it is essential to examine the behaviour of R_n as $n\to\infty$, which is not simple in many cases. We, therefore, generally obtain the expansion by assuming the possibility of expanding it in an infinite series by assuming that $R_n\to 0$ as $n\to\infty$.

2.2.7 Power Series Expansion of some Standard Functions

WORKING PROCEDURE

To find the power series expansion we shall use the following procedure :
Step 1. Put the given function equal to $f(x)$.
Step 2. Differentiate $f(x)$, a number of times and obtain $f'(x), f''(x), f'''(x)...$ and so on.
Step 3. Put $x = 0$ and find $f(0), f'(0), f''(0)...$ and so on.
Step 4. Now substitute the values of $f(0), f'(0), f''(0), f'''(0),...$ in $f(x)= f(0)+ x f'(0)+\frac{x^2}{2!} f''(0)+...$

We shall now consider Maclaurin's series expansions of the function $e^x, \sin x, \cos x, (1+x)^m$ and $\log x$.

(i) Expansion of e^x.

Let $\qquad f(x)= e^x \ \forall x\in\mathbf{R}$
(COCHIN–2005)

Then $\qquad f^n(x)= e^x \ \forall x\in\mathbf{R}$

Thus, for each positive n, f^n is defined in the interval $[-h, h]$.

Writing, Lagrange's form of remainder, after n terms

$$R_n(x) = \frac{x^n}{n!} f^n(\theta x), \theta \in R, 0 < \theta < 1 = \frac{x^n}{n!} e^{\theta x}$$

Now, we shall show that $\lim_{n \to \infty} R_n(x) = 0$. Here, it is enough to show that $e^{\theta x}$ is bounded in $[-h, h]$ and $\lim_{n \to \infty} \frac{x^n}{n!} = 0$.

Since, $0 < \theta < 1$ and $x \in [-h, h]$, therefore $|\theta x| < h$ and consequently, $0 < e^{\theta x} < e^h$, hence $e^{\theta x}$ is bounded.

Now, let us write

$$a_n = \frac{x^n}{n!} \ \forall n \in N$$

Then

$$\frac{a_{n+1}}{a_n} = \frac{x}{n+1} \Rightarrow \lim_{n \to \infty} \frac{a_{n+1}}{a_n} = 0$$

$\Rightarrow \lim_{n \to \infty} a_n$ exists and equal to zero.

Now,

$$\lim_{n \to \infty} R_n(x) = e^{\theta x} \left[\lim \frac{x^n}{n!} \right] = 0$$

Hence, we find that the function $f(x)$ has a Maclaurin's series expansions for each $x \in [-h, h]$. This implies

$$f(x) = f(0) + x f'(0) + \frac{x^2}{2!} f''(0) + \dots + \frac{x^{n-1}}{(n-1)!} f^{n-1}(0) + \dots \quad \forall x \in R.$$

Substituting $f(x) = e^x, f'(x) = e^x, \dots, f^n(x) = e^x$, we have

$$e^x = 1 + x + \frac{x^2}{2!} + \frac{x^3}{3!} + \dots + \frac{x^{n-1}}{(n-1)!} + \dots \quad \forall x \in \mathbf{R}$$

(ii) Expansion of sin x.

Let

$$f(x) = \sin x, \ \forall x \in \mathbf{R}$$ (PTU–2005)

$\Rightarrow$

$$f^n(x) = \sin\left(x + \frac{n\pi}{2}\right), \quad \forall x \in \mathbf{R}$$

Writing, Lagrange's form of remainder after n terms, we have

$$R_n(x) = \frac{x^n}{n!} f^n(\theta x), \text{ where } 0 < \theta < 1 = \frac{x^n}{n!} \sin\left(\theta x + \frac{n\pi}{2}\right)$$

Now, for all $x \in \mathbf{R}$, $\left| R_n(x) \right| \le \left| \frac{x^n}{n!} \right|$ and $\lim_{n \to \infty} \frac{x^n}{n!} = 0$ [As in (i)]

Thus, we find that, the function $f(x)$ has a Maclaurin's series expansions for each x in $[-h, h]$. Hence, we have

$$f(x) = f(0) + x f'(0) + \frac{x^2}{2!} f''(0) + \dots + \frac{x^{n-1}}{(n-1)!} f^{n-1}(0) + \dots \quad \forall x \in \mathbf{R}.$$

Now, substituting $f(x) = \sin x$, $f^n(x) = \sin \frac{n\pi}{2}$, we have

$$\sin x = x - \frac{x^3}{3!} + \frac{x^5}{5!} - \dots \ \forall x \in \mathbf{R}.$$

(iii) Expansion of cos x.

Let $f(x) = \cos x, \ \forall x \in \mathbf{R}$, then $f^n(x) = \cos\left(x + \frac{n\pi}{2}\right)$

Thus, for eaeh n, f^n is defined in every interval $[-h, h]$.

Writing, Lagrange's remainder after n terms, we have

$$R_n(x) = \frac{x^n}{n!} f^n(\theta x) = \frac{x^n}{n!} \cos\left(\theta x + \frac{n\pi}{2}\right), \quad \text{where } 0 < \theta < 1$$

Now, for all $x \in \mathbf{R}$, $\left| R_n(x) \right| \le \left| \frac{x^n}{n!} \right|$ and $\lim_{n \to \infty} \frac{x^n}{n!} = 0$ [As in (i)]

Thus, we find that, the function f has a Maclaurin's series expansions for each $x \in [-h, h]$, which gives

$$f(x) = f(0) + x f'(0) + \frac{x^2}{2!} f''(0) + \dots + \frac{x^n}{n!} f^n(0) + \dots \quad \forall x \in \mathbf{R}.$$

Now, substituting $f(x) = \cos x \dots, f^n(0) = \cos \frac{n\pi}{2}$, we have $\cos x = 1 - \frac{x^2}{2!} + \frac{x^4}{4!} - \dots \ \forall x \in \mathbf{R}.$

(iv) Expansion of $(1+x)^m$.

Case (i). Let m is a positive integer, then letting

$$f(x) = (1+x)^m, \ \forall x \in \mathbf{R}$$

We find that for each $n \in \mathbf{N}, f^n(x)$ exists for all $x \in \mathbf{R}$, and whenever $n > m, f^n(x) = 0 \ \forall x \in \mathbf{R}.$

$\Rightarrow$

$$R_n(x) = 0, \text{ whenever } n > m$$

Hence, $\lim_{n \to \infty} R_n(x) = 0$ and for all $x \in R$, we have

$$f(x) = f(0) + x f'(0) + \dots + \frac{x^m}{m!} f^m(0), \quad (\because \text{ All other terms must vanish.})$$

Substituting the value of $f(x), f(0), ..., f^m(0)$, We have

$$(1+x)^m = 1 + mx + \frac{m(m-1)}{2!}x^2 + ... + x^m.$$

Case (ii). Let m not be a positive integer (may be a fraction or negative integer).
Here, we find that, if we write $\qquad f(x) = (1+x)^m$, whenever $x \neq -1$
then $\qquad\qquad\qquad f^n(x) = m(m-1)...(m-n+1)(1+x)^{m-n}$, whenever $x \neq -1$

Thus, for each positive integer n, f^n is defined in $[-h, h]$ for each h between 0 and 1.
Now, writing Cauchy's form of remainder after n terms, we have

$$= \frac{x^n(1-\theta)^{n-1}}{(n-1)!}f^n(\theta x), \text{where } 0 < \theta < 1$$

$$R_n(x) = \frac{x^n(1-\theta)^{n-1}}{(n-1)!}m(m-1)...(m-n+1)(1+\theta x)^{m-n}$$

$$= \frac{m(m+1)...(m+n+1)}{(n-1)!}x^n\left(\frac{1-\theta}{1+\theta x}\right)^{n-1}.(1+\theta x)^{m-1}$$

Now, we observe the following:

(a) $\lim\limits_{n\to\infty} \dfrac{m(m-1)...(m-n+1)}{(n-1)!}x^n = 0$

If we write $\qquad\qquad\qquad a_n = \dfrac{m(m+1)...(m-n+1)}{(n-1)!}x^n$

Then, we have $\qquad\qquad \dfrac{a_{n+1}}{a_n} = \dfrac{(m-n)x}{n} \Rightarrow \lim\limits_{n\to\infty} \dfrac{a_{n+1}}{a_n} = -x$

If follows that if $|x| < 1$, then $\lim\limits_{n\to\infty} a_n = 0$.

(b) $\lim\limits_{n\to\infty}\left(\dfrac{1-\theta}{1+\theta x}\right)^{n-1} = 0$

In fact, since $0 < \theta < 1$ and $-1 < x < 1$, therefore, $0 < \left[\dfrac{1-\theta}{1+\theta x}\right] < 1$ and hence $\lim\limits_{n\to\infty}\left[\dfrac{1-\theta}{1+\theta x}\right]^{n-1} = 0$

(c) If $m > 1$, then $\qquad\qquad (1+\theta x)^{m-1} < (1-|x|)^{m-1}$

For (a), (b) and (c), we find that for all x in $]-1, 1[$ $\lim\limits_{n\to\infty} R_n(x) = 0$

Thus, we find that for each h between 0 and 1, the function f has Maclaurin's series expansion for all $x \in [-h, h]$.
Hence, we have

$$f(x) = f(0) + xf'(0) + \frac{x^2}{2!}f''(0) + ... + \frac{x^{n-1}}{(n-1)!}f^{n-1}(0) + ... \quad \forall x \in]-1, 1[.$$

Substituting the values of $f(x), f'(0), ..., f^{n-1}(0)$, we have

$$(1+x)^m = 1 + mx + \frac{m(m-1)}{2!}x^2 + \frac{m(m-1)(m-2)}{3!}x^3 + ... + \frac{m(m-1)...(m-n+1)}{n!}x^n + ...\text{whenever} -1 < x < 1$$

(v) Expansion of $\log_e(1+x)$.
Let $\qquad\qquad\qquad\qquad f(x) = \log(1+x), \; -1 < x < 1.$

Then $\qquad\qquad\qquad\qquad f^n(x) = \dfrac{(-1)^{n-1}(n-1)!}{(1+x)^n}$, whenever $x > -1$.

Now, we shall consider the following cases :
Case (a) Let $0 \leq x \leq 1$. Writing Lagrange's form of remainder after n terms, we have

$$R_n = \frac{x^n}{n!}f^n(\theta x) = \frac{x^n}{n!}(-1)^{n-1}\frac{(n-1)!}{(1+\theta x)^n} = \frac{(-1)^{n-1}}{n}\cdot\left(\frac{x}{1+\theta x}\right)^n$$

Since, $0 \leq x \leq 1, 0 < \theta < 1$, therefore $\qquad 0 < \dfrac{x}{1+\theta x} < 1$

$\therefore \qquad\qquad\qquad\qquad |R_n| < \dfrac{1}{n}, \text{and} \dfrac{1}{n} \to 0 \text{ as } n \to \infty$

Therefore $\qquad\qquad\qquad \lim\limits_{n\to\infty} R_n = 0.$

Case (b)   Let $-1 < x < 0$. Since in this case $\left|\dfrac{x}{1+\theta x}\right|$ need not be less than unity, therefore, we may not be able to show easily that $R_n \to 0$ as $n \to \infty$ by considering Lagrange's remainder.

Now, writing Cauchy's form of remainder, we have

$$R_n = \frac{x^n}{(n-1)!}(1-\theta)^{n-1} f^n(\theta x) = (-1)^{n-1} x^n \left(\frac{1-\theta}{1+\theta x}\right)^{n-1} \cdot \frac{1}{1+\theta x} \text{ since } |x| < 1$$

therefore $\left|\dfrac{1-\theta}{1+\theta x}\right| < 1$, so that $\left|\left(\dfrac{1-\theta}{1+\theta x}\right)^{n-1}\right| < 1$ and $\left|\dfrac{1}{1+\theta x}\right| < \dfrac{1}{1-|x|}$

Thus $|R_n| < \dfrac{|x|^n}{1-|x|}$

This implies that $\lim\limits_{n \to \infty} R_n = 0.$, since $|x| < 1$. Thus we find that if $-1 \le x \le 1$, then $\lim\limits_{n \to \infty} R_n = 0$.

$$f(x) = f(0) + xf'(0) + \frac{x^2}{2!}f''(0) + \dots + \frac{x^{n-1}}{(n-1)!}f^{n-1}(0) + \dots \text{ whenever } -1 < x \le 1.$$

Substituting the values of $f(x), f(0), f'(0), \dots, f^{n-1}(0), \dots$, we get

$$\log(1+x) = x - \frac{x^2}{2} + \frac{x^3}{3} - \dots, \text{ whenever } -1 < x \le 1.$$

Solved Examples

EXAMPLE 1. *Expand $e^{a\sin^{-1}x}$ by Maclaurin's series and find the general term. Hence, show that*

$$e^\theta = 1 + \sin\theta + \frac{1}{2!}\sin^2\theta + \frac{2}{3!}\sin^3\theta + \dots$$

SOLUTION. Here $y = e^{a\sin^{-1}x}$ ...(1)

Then $y_1 = e^{a\sin^{-1}x} \cdot \dfrac{a}{\sqrt{1-x^2}} = \dfrac{ay}{\sqrt{1-x^2}}$ ...(2)

$\Rightarrow$ $\left(\sqrt{1-x^2}\right)y_1 = ay$

$\Rightarrow$ $\left(1-x^2\right)y_1^2 - a^2y^2 = 0$ ...(3)

Now, differentiating both the sides, we have

$\Rightarrow$ $\left(1-x^2\right)2y_1y_2 - 2xy_1^2 - 2a^2yy_1 = 0$

$2y_1\left[\left(1-x^2\right)y_2 - xy_1 - a^2y\right] = 0$ ...(4)

Since $2y_1 \ne 0$ hence $[(1-x^2)y_2 - xy_1 - a^2y] = 0$.

Now, differentiating n times by Leibnitz's theorem, we get

$\left(1-x^2\right)y_{n+2} + ny_{n+1}(-2x)$

$+\dfrac{n(n-1)}{2}y_n(-2) - y_{n+1}x - ny_n \cdot 1 - a^2y_n = 0$

$\Rightarrow \left(1-x^2\right)y_{n+2} - (2n+1)xy_{n+1} - \left(n^2+a^2\right)y_n = 0$ ...(5)

Now, we can easily find, (from (1) to (5)) the following values

$(y)_0 = 1, \quad (y_1)_0 = a, \quad (y_2)_0 = a^2$

$(y_{n+2})_0 = (n^2+a^2)(y_n)_0$ ...(6)

Replacing n by $(n-2)$ in (6), we get

$(y_n)_0 = [(n-2)^2 + a^2](y_{n-2})_0$

$= [(n-2)^2 + a^2][(n-4)^2 + a^2](y_{n-4})_0$

If n is odd, then

$(y_n)_0 = [(n-2)^2 + a^2][(n-4)^2 + a^2]\dots(3^2+a^2)(1^2+a^2)(y_1)_0$

$= [(n-2)^2 + a^2][(n-4)^2 + a^2]\dots[(3^2+a^2)(1^2+a^2)].a$

If n is even, then

$(y_n)_0 = [(n-2)^2 + a^2][(n-4)^2 + a^2]\dots(4^2+a^2)(2^2+a^2)(y_2)_0$

$= [(n-2)^2 + a^2][(n-4)^2 + a^2]\dots[(4^2+a^2)(2^2+a^2)].a^2$

Hence,

$$y_n(0) = \begin{cases} a\left(1^2+a^2\right)\left(3^2+a^2\right)\dots\left[(n-2)^2+a^2\right], & \text{if } n \text{ is odd} \\ a^2\left(2^2+a^2\right)\left(4^2+a^2\right)\dots\left[(n-2)^2+a^2\right], & \text{if } n \text{ is even} \end{cases}$$

Putting $n = 1, 2, 3, 4, \dots$ in (6), we get

$(y_3)_0 = (3^2+a^2)(1^2+a^2)a,$

$(y_6)_0 = (4^2+a^2)(2^2+a^2)a^2$ etc.

Now putting all these values in the Maclaurin's theorem

$$y = (y)_0 + x \cdot (y_1)_0 + \frac{x^2}{2!}(y_2)_0 + \dots + \frac{x^n}{n!}(y_n)_0 + \dots$$

We have

$$e^{a\sin^{-1}x} = 1 + ax + \frac{a^2}{2!}x^2 + \frac{a(1^2+a^2)}{3!}x^3 + \frac{a(2^2+a^2)}{4!}x^4 + \dots$$

The general term is $\dfrac{x^n}{n!}(y_n)_0$.

Now putting $x = \sin\theta$ and $a = 1$, in the above equation, we get $e^\theta = 1 + \sin\theta + \dfrac{1}{2!}\sin^2\theta + \dfrac{2}{3!}\sin^3\theta + \dots$

EXAMPLE 2. *Expand $\log\sin(x+h)$ in powers of h by Taylor's theorem.*

SOLUTION. Let $f(x) = \log\sin(x)$

$\Rightarrow$ $f(x+h) = \log\sin(x+h)$

Expanding $f(x+h)$ by Taylor's theorem in powers of h, we have

$$f(x+h) = f(x) + hf'(x) + \frac{h^2}{2!}f''(x) + \frac{h^3}{3!}f'''(x) + \dots \dots (1)$$

Now $f(x) = \log\sin x$

$\Rightarrow$ $f'(x) = \cot x$

         $f''(x) = -\text{cosec}^2 x$

$\Rightarrow$ $f'''(x) = 2\,\text{cosec}^2 x \cot x$ etc.

Substituting all these values in equation (1),

we get $\log\sin(x+h) = \log\sin x + h\cot x - \dfrac{h^2}{2!}\text{cosec}^2 x$

$$+ \frac{2h^3}{3!}\text{cosec}^2 x \cot x + \dots$$

Exercise-2.4

1. Discuss the applicability of, Rolle's theorem of the following functions :
   (a) $f(x)=2+(x-1)^{2/3}$ in the interval $[0,2]$
   (b) $f(x)=x^2$ in $2\le x\le 3$
   (c) $f(x)=\tan x$ in $0\le x\le \pi$
   (d) $f(x)=x^4-3x^2+4$ in the interval $[-4,4]$
   (e) $f(x)=1/(x^2+1)$ in the interval $[-3,3]$
   (f) $f(x)=e^x \sin x$ in the interval $[0,\pi]$ (JNTU–2003)
   (g) $f(x)=|x|$ in the interval $[-1,1]$
   (h) $f(x)=(x-2)\sqrt{x}$ in the interval $[0,2]$
   (i) $f(x)=(x-a)^m(x-b)^n, m,n\in \mathbf{Z}^+$ in the interval $[a,b]$.
       (VTU–2010, NAGARJUNA–2008)

2. Show that between any two roots of $e^x \cos x = 1$, there exists at least one root of $e^x \sin x - 1 = 0$.

3. Let $\dfrac{a_0}{n+1}+\dfrac{a_1}{n}+\dfrac{a_2}{n-1}+...+\dfrac{a_{n-1}}{2}+a_n = 0$.

   Show that there exists at least one real x between 0 and 1 such that $a_0 x^n + a_1 x^{n-1}+...+a_n = 0$.

4. Verify the Rolle's theorem for the following functions:
   (a) $f(x)=x^4-1$ on the interval $[-1,1]$
   (b) $f(x)=e^x(\sin x-\cos x)$ in $\left(\dfrac{\pi}{4},\dfrac{5\pi}{4}\right)$

5. If $f(x)=\begin{vmatrix} \sin x & \sin \alpha & \sin \beta \\ \cos x & \cos \alpha & \cos \beta \\ \tan x & \tan \alpha & \tan \beta \end{vmatrix}$ where $\quad 0< \alpha < \beta < \dfrac{\pi}{2}$. Show that $f'(l) = 0$, where $\alpha<l<\beta$.

6. A function $f(x)$ is continuous in the closed interval $[0,1]$ and differentiable in the open interval $]0,1[$ prove that $f'(x_1)=f(1)-f(0), 0<x_1<1$.

7. Show that the set of all x for which $\log(1+x)\le x$ is equal to $[0,\infty[$.

8. Compute the value of θ in the first mean value theorem $f(x+h)=f(x)+hf'(x+\theta h)$ if $f(x)=ax^2+bx+c$.

9. Show that $x^n-a=0$ has atmost one real positive root if n is a positive integer.

10. Show that the function f', if it exists in an interval, cannot have an ordinary or removable discontinuity in that interval.

11. Verify the Lagrange's mean value theorem for the following functions :
    (a) $f(x)=x^3$ in $[-1,1]$
    (b) $f(x)=\sin x$ in $[0,\pi/2]$ (NAGPUR–2008)
    (c) $f(x)=x^n$ in $[-1,1], n\in \mathbf{Z}^+$

(d) $f(x)=2x^2-7x+10, x\in[2,5]$

12. Find the value of c, of mean value theorem, when
    (a) $f(x)=\sqrt{x^2-4}$ in the interval $[2,4]$
    (b) $f(x)=2x^2+3x+4$ in the interval $[1,2]$
    (c) $f(x)=x(x-1)$ in the interval $[1,2]$

13. (a) If $f(x)=\sqrt{x}$ and $g(x)=1/\sqrt{x}$, then show by Cauchy's mean value theorem that c is the geometric mean between a and b.
    (b) If $f(x)=\dfrac{1}{x^2}$ and $g(x)=\dfrac{1}{x}$, then show that c is the harmonic mean between a and b.

14. If f'' exists and continuous on $[a,b]$ and differentiable on $]a,b[$, then prove that
    $$f(b)-f(a)-\dfrac{1}{2}(b-a)\{f'(a)-f'(b)\}=-\dfrac{(b-a)^3}{12}f'''(d)$$
    where $d \in R$ such that $d\in]a, b[$.

15. Prove that
    $$\sin ax= ax-\dfrac{a^3 x^3}{3!}+\dfrac{a^5 x^5}{5!}-...+\dfrac{a^{n-1}x^{n-1}}{(n-1)!}$$
    $$\sin\left(\dfrac{n-1}{2}.\pi\right)+\dfrac{a^n x^n}{n!}\sin\left(a\theta x+\dfrac{n\pi}{2}\right)$$

16. If $f(x)= f(0)+xf'(0)+\dfrac{x^2}{2!}f''(\theta x)$
    find the value of θ as $x\to 1$, $f(x)$ being $(1-x)^{5/2}$.

17. Show that the number θ which occurs in the Taylor's Theorem with Lagrange's form of remainder after n terms approaches the limit $\dfrac{f^{n+1}(a)}{(n+1)}$ as $h\to 0$ provided that $f^{n+1}(x)$ is continuous and different from zero as $x\to a$.

18. Show that the function x^3-3x^2+3x+2 is monotonically increasing in every interval.

19. Obtain by Maclaurin's theorem the expansion of $e^{\sin x}$.

20. If $f(x)=\exp\left[-\dfrac{1}{x^2}\right]$, for $x\ne 0$ and $f(0)=0$, then show that :
    (i) $f^n(0)=0 \ \forall n=0,1,2,...$ and
    (ii) The Taylor's series for f about 0 agrees with $f(x)$ only at $x=0$.

21. Expand "log sec x" by Maclaurin's series expansion, upto the term containing x^6.   (VTU–2009, MUMBAI–2000)

22. If $x>0$, show that
    $$x-\dfrac{x^2}{2}+\dfrac{x^3}{3(1+x)}< \log(1+x)< x-\dfrac{x^2}{2}+\dfrac{x^3}{3}.$$

Hints to Selected Problems

1. (a) Since $f'(x)$ does not exist at $x=1$, the second condition of Rolle's theorem is not satisfied.

2. Let a,b be two distinct roots of $e^x\cos x -1=0$.
   Then $e^a\cos a=1$ and $e^b\cos b=1$
   Define a function $f(x)=e^{-x}-\cos x$.

5. $f(x)$ can be written as
   $(\cos\alpha\tan\beta-\cos\beta\tan\alpha)\sin x$
   $f(x)=-(\sin\alpha\tan\beta-\sin\beta\tan\alpha)\cos x$
   $+(\sin\alpha\cos\beta-\sin\beta\cos\alpha)\tan x$

Since $\sin x, \cos x, \tan x$ have finite derivatives in $]0,\pi/2[$
$\Rightarrow f'(x)$ exists.

Also, $f(\alpha)=f(\beta)$. Hence, all the conditions of Rolle's theorem are satisfied.

7. Let us suppose $f(x)=\log(1+x)-x$
   $\Rightarrow \qquad f(0)=0$
   $$f'(x)=\dfrac{1}{1+x}-1=\dfrac{-x}{1+x}\le 0$$

$\Rightarrow f(x)$ is a decreasing function.

$\Rightarrow \qquad\qquad f(x) \leq f(0) \;\; \forall x \geq 0$

$\Rightarrow \qquad\qquad \log(1+x) - x \leq 0$

$\Rightarrow \qquad\qquad \log(1+x) \leq x$

9. $f'(x) = nx^{n-1}$. Clearly $f(x)$ is an increasing function.

Let $x_1, x_2 \in\;]0, \infty[$ and $0 < x < r < x_2$ such that $f(r) = 0$.

Then $f(x_1) < f(r) < f(x_2) \Rightarrow f(x_1) < 0 < f(x_2)$

$\Rightarrow$ If $x \neq r, f(x) \neq 0$ on $(0, \infty)$.

$\Rightarrow x^n - a$ has at most one real positive root.

14. Define two functions $g(x)$ and $h(x)$ such that

$$g(x) = f(x) - f(a) - \frac{1}{2}(x-a)\{f'(a) + f'(x)\} + A(x-a)^3$$

and

$$h(x) = \frac{1}{2}\left[f'(x) - f'(a)\right]^{-1/2}$$

$$(x-a)f''(x) + 3A(x-a)^2$$

Clearly, $g(x)$ and $h(x)$ satisfying all conditions of Rolle's theorem. Then use Rolle's theorem for both the above functions.

18. Since $f'(x) \geq 0$ in $]-\infty, 1]$. Hence, it is monotonically increasing.

22. $f'(x) = \dfrac{1}{1+x} - 1 + x - \dfrac{3x^2 + 2x^3}{3(1+x^2)} = \dfrac{x^3}{3(1+x)^2} > 0$

f is increasing $\Rightarrow f(x) > f(0) = 0$ for $x > 0$

$$x - \frac{x^2}{2} + \frac{x^3}{3(1+x)} < \log(1+x) \text{ if } x > 0$$

Now, $\quad g'(x) = 1 - x + x^2 - \dfrac{1}{1+x} = \dfrac{x^3}{1+x} > 0$

g is increasing $\Rightarrow g(x) > g(0)$

$$\log(1+x) < x - \frac{x^2}{2} + \frac{x^3}{3}$$

1. (a) Not applicable (b) Not applicable (c) Not applicable (d) Verified (e) Verified (f) Verified (g) Not applicable (h) Verified

(i) Verified   **4.** (a) Verified   (b) Verified   **8.** $\theta = \dfrac{1}{2}$

11. (a) Verified   (b) Verified (c) Verified   (d) Verified   **12.** (a) $c = \sqrt{6}$ (b) $c = 3/2$

16. $\theta = \dfrac{9}{25}$ **19.** $y = 1 + x + \dfrac{x^2}{2} - \dfrac{x^4}{8} + \ldots$ **21.** $y = \dfrac{x^2}{2} + \dfrac{x^4}{12} + \dfrac{x^6}{45} + \ldots$

2.2.8 SOME MORE EXPANSIONS

EXAMPLE 1. *Expand* $\tan^{-1} x$.

SOLUTION. Let $f(x) = \tan^{-1} x \Rightarrow f(0) = 0$

$$f'(x) = \frac{1}{1+x^2} \Rightarrow f'(0) = 1$$

$$= (1+x^2)^{-1} = 1 - x^2 + x^4 - x^6 + \ldots$$

(By binomial expansion)

$$f''(x) = -2x + 4x^3 - 6x^5 + \ldots \Rightarrow f''(0) = 0$$

$f'''(x) = -2 + 12x^2 - 30x^4 + \ldots \qquad \Rightarrow f'''(0) = -2$

$f^{iv}(x) = 24x - 120x^3 + \ldots \qquad \Rightarrow f^{iv}(0) = 0$

$f^{v}(x) = 24 - 360x^2 + \ldots \qquad\qquad \Rightarrow f^{v}(0) = 24$

Put all these values in Maclaurin's series, we get

$$\tan^{-1} x = x - \frac{x^3}{3} + \frac{x^5}{5} - \frac{x^7}{7} + \ldots$$

REMARKS

- To expand an alone inverse function, find its first derivative, expand by binomial theorem and then find other derivatives.
- The expansion of $\tan^{-1} x$ is valid only if $-1 < x < 1$.
- This expansion for $\tan^{-1} x$ known as Gregory's series, which is very useful in finding the value of π.
- In a like manner, we may get $\sin^{-1} x = x + \dfrac{1}{2}\cdot\dfrac{x^3}{3} + \dfrac{1.3}{2.4}\cdot\dfrac{x^5}{5} + \dfrac{1.3.5}{2.4.6}\cdot\dfrac{x^7}{7} + \ldots$

EXAMPLE 2. *If* $y = \sin(m \sin^{-1} x)$, *then show that*

$$\left(1-x^2\right)\frac{d^2y}{dx^2} - x\frac{dy}{dx} + m^2 y = 0$$

Hence, or otherwise expand $\sin m\theta$ *in powers of* $\sin \theta$.   (SVTU–2008)

SOLUTION. Here, we have

$$y = f(x) = \sin(m\sin^{-1} x) \qquad \ldots(1)$$

$$\Rightarrow \quad y_1 = \cos\left(m\sin^{-1} x\right)\cdot\frac{m}{\sqrt{1-x^2}} \qquad \ldots(2)$$

$$\Rightarrow (1-x^2)y_1^2 = m^2\cos^2(m\sin^{-1} x)$$

$$\Rightarrow (1-x^2)y_1^2 = m^2[1 - \sin^2(m\sin^{-1} x)]$$

$$\Rightarrow (1-x^2)y_1^2 = m^2(1-y^2) \;[\because y = \sin(m\sin^{-1} x)]$$

$$\Rightarrow \quad (1-x^2)y_1^2 + m^2y^2 - m^2 = 0 \qquad \ldots(3)$$

Differentiating w.r.t. x, we get

$$(1-x^2)2y_1y_2 - 2xy_1^2 + 2m^2yy_1 = 0$$

$$\Rightarrow \quad 2y_1[(1-x^2)y_2 - xy_1 + m^2y] = 0$$

$$\Rightarrow \quad (1-x^2)y_2 - xy_1 + m^2y = 0 \qquad \ldots(4)$$

Now, differentiating (4) n times, using Leibnitz's theorem, we get

$$\left(1-x^2\right)y_{n+2} + n\cdot y_{n+1}(-2x) + \frac{n(n-1)}{1.2}y_n(-2)$$

$$- xy_{n+1} - n\cdot y_n + m^2y_n = 0$$

$\Rightarrow \left(1-x^2\right)y_{n+2}-\left(2n+1\right)xy_{n+1}-\left(n^2-m^2\right)y_n=0$

...(5)

Now, put $x=0$ in (1), (2), (4) and (5), we get

$y(0)=0, y_1(0)=m, y_2(0)+m^2y(0)=0$

$\Rightarrow y_2(0)=0$

and $y_{n+2}(0)=(n^2-m^2)y_n(0).$ ...(6)

Putting $n=2,4,6,...$ in (6), we get

$y_4(0)=(2^2-m^2)y_2(0)=0$

$y_6(0)=(4^2-m^2)y_4(0)=0$

$y_8(0)=0$

.... and so on.

Here, we observe that $y_n(0)=0$ if n is even.

Now, putting $n=1,3,5,...$ in (6), we get

$y_3(0)=(1^2-m^2)y_1(0)=(1^2-m^2).m$

$y_5(0)=(3^2-m^2)y_3(0)=(3^2-m^2)(1^2-m^2).m$

$\cdots\cdots\cdots\cdots\cdots\cdots\cdots\cdots\cdots\cdots$

Putting all these values in Maclaurin's series, we get

$\sin\left(m\sin^{-1}x\right)=mx+\dfrac{m\left(1^2-m^2\right)}{3!}x^3$

$+\dfrac{m\left(1^2-m^2\right)\left(3^2-m^2\right)}{5!}x^5+...$

Let $\theta=\sin^{-1}x\Rightarrow x=\sin\theta$

Then, we get

$\sin m\theta=m\sin\theta+\dfrac{m(1^2-m^2)}{3!}\sin^3\theta$

$+\dfrac{m(1^2-m^2)(3^2-m^2)}{5!}\sin^5\theta+...$

EXAMPLE 3. *Expand tan x by Maclaurin's theorem as far as* x^5 *and hence find the value of tan 46°30' upto four decimal places.* (VTU–2006)

SOLUTION. Let $f(x)=\tan x$

$\Rightarrow f(0)=0$

$f'(x)=\sec^2x=1+\tan^2x$

$\Rightarrow f'(0)=1$

$f''(x)=2\tan x\sec^2x=2\tan x(1+\tan^2x)$

$=2\tan x+2\tan^3x$

$\Rightarrow f''(0)=0$

$f'''(x)=2\sec^2x+6\tan^2x\sec^2x$

$=2(1+\tan^2x)+6\tan^2x(1+\tan^2x)$

$=2+8\tan^2x+6\tan^4x$

$\Rightarrow f'''(0)=2$

$f^{iv}(x)=16\tan x\sec^2x+24\tan^3x\sec^2x$

$=8\sec^2x(2\tan x+3\tan^3x)$

$=8(1+\tan^2x)(2\tan x+3\tan^3x)$

$=16\tan x+40\tan^3x+24\tan^5x$

$\Rightarrow f^{iv}(0)=0$

and $f^{v}(x)=16\sec^2x+120\tan^2x\sec^2x$

$+120\tan^4x\sec^2x$

$=8\sec^2x(2+15\tan^2x+15\tan^4x)$

$\Rightarrow f^{v}(0)=16$

Now, putting all these values in Maclaurin's series'

$f(x)=f(0)+xf'(0)+\dfrac{x^2}{2!}f''(0)+\dfrac{x^3}{3!}f'''(0)$

$+\dfrac{x^4}{4!}f^{iv}(0)+\dfrac{x^5}{5!}f^{v}(0)+...$

we get $\tan x=0+x+\dfrac{x^3}{3!}.2+\dfrac{x^5}{5!}.16+...$

$\Rightarrow \tan x=x+\dfrac{x^3}{3}+\dfrac{2}{15}x^5+...$ ...(1)

Deduction. Here

$x=46°30'=\left(46\dfrac{1}{2}\right)°=\left(\dfrac{93}{2}\right)°=\dfrac{93}{2}\times\dfrac{\pi}{180}$ Radians

$=\dfrac{31}{120}\times\dfrac{22}{7}=\dfrac{31\times11}{60\times7}=\dfrac{314}{420}=0.812$

Now, putting $x=46°30'=0.812$ in (1), we get

$\tan 46°30'=0.812+\dfrac{(0.812)^3}{3}+\dfrac{2}{15}(0.812)^5$

$=0.812+0.1784+0.047=1.0374$

EXAMPLE 4. *Prove by Maclaurin's theorem, that*

$e^{\sin x}=1+x+\dfrac{x^2}{1.2}-\dfrac{3.x^4}{1.2.3.4}+\cdots$ (VTU-2011, BHOPAL-2009)

SOLUTION. Let $f(x)=e^{\sin x}$ $\Rightarrow f(0)=e^0=1$

$f'(x)=e^{\sin x}.\cos x$ $\Rightarrow f'(0)=e^0\cos 0=1$

$f''(x)=e^{\sin x}(-\sin x)+\cos x\,e^{\sin x}\cos x$

$=e^{\sin x}[\cos^2x-\sin x]$

$\Rightarrow f''(0)=e^0[1-0]=1$

$f'''(x)=e^{\sin x}[2\cos x(-\sin x)-\cos x]$

$+e^{\sin x}\cos x.[\cos^2x-\sin x]$

$=e^{\sin x}\cos x[-2\sin x-1+\cos^2x-\sin x]$

$=-e^{\sin x}\cos x[3\sin x+\sin^2x]\Rightarrow f'''(0)=0$

$f^{iv}(x)=-e^{\sin x}\cos x[3\cos x+2\sin x\cos x]$

$+e^{\sin x}\sin x[3\sin x+\sin^2x]$

$-[3\sin x+\sin^2x]\cos x\,e^{\sin x}\cos x$

$\Rightarrow f^{iv}(0)=-3$

Putting all these values in Maclaurin's theorem, given by

$f(x)=f(0)+xf'(0)+\dfrac{x^2}{2!}f''(0)+\dfrac{x^3}{3!}f'''(0)+\dfrac{x^4}{4!}f^{iv}(0)+...$

we get, $e^{\sin x}=1+x+\dfrac{x^2}{1.2}-\dfrac{3.x^4}{1.2.3.4}+...$

EXAMPLE 5. *Expand log (1+sin x) by Maclaurin's theorem in ascending power of x upto first five terms.*

(SVTU–2009, JNTU–2006)

SOLUTION. Let $y=f(x)=\log(1+\sin x)$

By Maclaurin's expansion for $f(x)$, we have

$y=f(x)$

$=(y)_0+\dfrac{x}{1!}(y_1)_0+\dfrac{x^2}{2!}(y_2)_0+\dfrac{x^3}{3!}(y_3)_0+\dfrac{x^4}{4!}(y_4)_0+...$

Now $\qquad y = \log(1+\sin x) \qquad \therefore \quad (y)_0 = 0$ $\qquad$ ...(1)

$y_1 = \dfrac{\cos x}{1+\sin x} \Rightarrow (y_1)_0 = 1$

$y_2 = \dfrac{-\sin x(1+\sin x)-\cos^2 x}{(1+\sin x)^2} = -\dfrac{(1+\sin x)}{(1+\sin x)^2} = -\dfrac{1}{1+\sin x}$

$\Rightarrow \quad (y_2)_0 = -1$

$y_3 = \dfrac{\cos x}{(1+\sin x)^2} = \dfrac{\cos x}{(1+\sin x)} \cdot \dfrac{1}{(1+\sin x)} = -y_1 y_2$

$\Rightarrow \quad (y_3)_0 = -1(-1) = 1$

$y_4 = -y_1 y_3 - y_2^2$

$\Rightarrow \quad (y_4)_0 = -1.1-(-1)^2 = -1-1 = -2$

$y_5 = -y_1 y_4 - y_2 y_3 - 2y_2 y_3 = -y_1 y_4 - 3y_2 y_3$

$\Rightarrow \quad (y_5)_0 = -1.(-2)-3(-1).1 = 2+3 = 5 \ $ and so on.

Therefore, $\log(1+\sin x)$

$= 0 + \dfrac{x}{1!}.1 + \dfrac{x^2}{2!}.(-1) + \dfrac{x^3}{3!}.1 + \dfrac{x^4}{4!}.(-2) + ...$

$= x - \dfrac{x^2}{2} + \dfrac{x^3}{6} - \dfrac{x^4}{12} + \dfrac{x^5}{24} ...$

EXAMPLE 6. *Expand sin $(\pi/4+9)$ in powers of* θ.

SOLUTION . Let $\qquad f(\theta) = \sin(\pi/4+\theta)$

$\Rightarrow \qquad f(0) = \sin \pi/4 = 1/\sqrt{2}$

$f'(\theta) = \cos(\pi/4+\theta)$

$\Rightarrow \qquad f'(0) = \cos \pi/4 = 1/\sqrt{2}$

$f''(\theta) = -\sin(\pi/4+\theta)$

$\Rightarrow \qquad f''(0) = -\sin \pi/4 = -1/\sqrt{2}$

$f'''(\theta) = -\cos(\pi/4+\theta)$

$\Rightarrow \qquad f'''(0) = -\cos \pi/4 = -1/\sqrt{2}$

$f^{iv}(\theta) = \sin(\pi/4+\theta)$

$\Rightarrow \qquad f^{iv}(0) = 1/\sqrt{2}$ and so on.

The nth derivative of $f(\theta)$ is given by

$$f^n(\theta) = \sin\left(\theta + \dfrac{\pi}{4} + \dfrac{n\pi}{4}\right)$$

The Maclaurin's expansion of $f(\theta)$ with Lagrange's form of remainder is

$$f(\theta) = f(0) + \dfrac{\theta}{1!}f'(0) + \dfrac{\theta^2}{2!}f''(0)$$

$$+ \dfrac{\theta^3}{3!}f'''(0) + ... + \dfrac{\theta^{n-1}}{(n-1)!}f^{n-1}(0) + R_n ...(1)$$

where $R_n = \dfrac{\theta^n}{n!}f^n(t.\theta) = \dfrac{\theta^n}{n!}\sin\left(t.\theta + \dfrac{\pi}{4} + \dfrac{n\pi}{2}\right),\ 0<t<1.$

$$|R_n| = \left|\dfrac{\theta^n}{n!}\sin\left(t.\theta + \dfrac{\pi}{4} + \dfrac{n\pi}{2}\right)\right|$$

Now

$$= \left|\dfrac{\theta^n}{n!}\right|.\left|\sin\left(t.\theta + \dfrac{\pi}{4} + \dfrac{n\pi}{2}\right)\right| \le \left|\dfrac{\theta^n}{n!}\right|$$

$$\therefore \quad \lim_{n\to\infty}|R_n| \le \lim_{n\to\infty}\left|\dfrac{\theta^n}{n!}\right| = 0 \qquad \left[\because \lim_{n\to\infty}\dfrac{\theta^n}{n!} = 0\right]$$

$$\therefore \quad \lim_{n\to\infty}R_n = 0$$

Thus all the conditions of Maclaurin's series expansion are satisfied. Hence, from (1), the expansion of sin $(\theta+\pi/4)$ is given by

$$\sin\left(\theta + \dfrac{\pi}{4}\right) = \dfrac{1}{\sqrt{2}} + \dfrac{\theta}{1!}\dfrac{1}{\sqrt{2}}$$

$$+ \dfrac{\theta^2}{2!}\left(-\dfrac{1}{\sqrt{2}}\right) + \dfrac{\theta^3}{3!}\left(-\dfrac{1}{\sqrt{2}}\right) + ...$$

Exercise-2.5

1. Expand the following functions by Maclaurin's theorem :

(i) $\sec x$ $\qquad\qquad$ (ii) $e^{x\cos x}$

(iii) $e^x \sec x$ $\qquad\qquad$ (iv) $\log_e(1+e^x)$ $\quad$ (BHOPAL-2008)

(v) $\log(1+\tan x)$

2. Apply Maclaurin's theorem to prove that

$\log \sec x = \dfrac{1}{2}x^2 + \dfrac{1}{12}x^4 + \dfrac{1}{45}x^6 + ...$

3. If $y = \sin^{-1}x = a_0 + a_1 x + a_2 x^2 + ...$ Prove that

$(n+1)(n+2)a_{n+2} = n^2 a_n$.

4. Show that :

(i) $ex \cos x = 1 + x$

$- \dfrac{2x^3}{3!} + \dfrac{2^2 x^4}{4!} + \dfrac{2^2 x^5}{5!} + \dfrac{2^3 x^7}{7!} + ... + \cos\left(\dfrac{n\pi}{4}\right)\dfrac{2^{n/2}}{n!}x^n + ...$

(ii) $e^x \sin x = x + x^2 - \dfrac{2x^3}{3!} + \dfrac{2^2 x^5}{5!} - ... + \sin\left(\dfrac{n\pi}{4}\right)\dfrac{2^{n/2}}{n!}x^n + ...$

(iii) $e^{ax}\sin bx = bx + abx^2 + \dfrac{3a^2 b - b^3}{3!}x^3 +$

$... + \dfrac{\left(a^2 + b^2\right)^{\frac{n}{2}}}{n!}x^n \sin\left(n\tan^{-1}\dfrac{b}{a}\right) + ...$

(iv) $e^{ax}\cos bx = 1 + ax + \dfrac{a^2 - b^2}{2!}x^2 + \dfrac{a\left(a^2 - 3b^2\right)}{3!}x^3$

$... + \dfrac{\left(a^2 + b^2\right)^{\frac{n}{2}}}{n!}x^n \cos\left(n\tan^{-1}\dfrac{b}{a}\right) + ...$

5. Expand the following :

(i) $\tan^{-1}x$ in powers of $\left(x - \dfrac{\pi}{4}\right)$.

(ii) $2x^3 + 7x^2 + x - 1$ in powers of $x-2$. $\qquad$ (BURDWAN-2003)

(iii) $\sin^{-1}(x+h)$ in power of x.

(iv) $\log \sin x$ in power of $(x-a)$.

6. Show that $\log(x+h) = \log h + \dfrac{x}{h} - \dfrac{x^2}{2h^2} + \dfrac{x^3}{3h^3} - \ldots$

7. Use Taylor's theorem to prove that

$$\tan^{-1}(x+h) = \tan^{-1}x + h\sin\theta\,\frac{\sin\theta}{1}$$
$$-(h\sin\theta)^2\frac{\sin 2\theta}{2} + (h\sin\theta)^3\frac{\sin 3\theta}{3} +$$
$$\ldots + (-1)^{n-1}(h\sin\theta)^n\frac{\sin n\theta}{n} + \ldots$$

where $\qquad \theta = \cot^{-1}x$.

8. If $y = e^{\tan^{-1}x}$, show that

$$(1+x^2)y_{n+2} + [2(n+1)x-1]y_{n+1} + n(n+1)y_n = 0.$$

Hence, or otherwise, find out the coefficient of x^5 if $e^{\tan^{-1}x}$ is expanded in powers of x.

9. Expand $(\sin^{-1}x)^2$ in ascending powers of x and deduce that

$$\theta^2 = 2.\frac{\sin^2\theta}{2!} + 2^2.\frac{2\sin^4\theta}{4!} + 2^2.4^2\frac{2\sin^6\theta}{6!} + \ldots$$

10. If $y = e^{m\tan^{-1}x} = a_0 + a_1x + a_2x^2 + \ldots + a_nx^n + \ldots$, show that

$$(n+1)a_{n+1} + (n-1)a_{n-1} = ma_n.$$

11. If $e^{e^x} = a_0 + a_1x + a_2x^2 + \ldots + a_nx^n + \ldots$ show that

$$a_{n+1} = \frac{1}{n+1}\left[a_n + \frac{a_{n-1}}{1!} + \frac{a_{n-2}}{2!} + \ldots + \frac{a_{n-r}}{r!} + \ldots \frac{a_0}{n!}\right]$$

12. Show that $f(mx) = f(x) + (m-1)xf'(x) + \dfrac{(m-1)^2}{2!}x^2f''(x)$

$$+ \frac{(m-1)^3}{3!}x^3f'''(x) + \ldots$$

13. By Maclaurin's theorem find the expansion of $y = \sin(e^x - 1)$ upto and including the term in x. Find also the first non-vanishing terms in the expansion of x as a series ascending powers of y.

14. Prove that

$$f\left(\frac{x^2}{1+x}\right) = f(x) - \frac{x}{1+x}f'(x)$$
$$+\left(\frac{x}{1+x}\right)^2\frac{1}{2!}f''(x) - \left(\frac{x}{1+x}\right)^3 f'''(x) + \ldots$$

15. Calculate the approximate value of :

(i) $\sqrt{17}$ to four decimal places.

(ii) $\sqrt{26}$ to three decimal places by Taylor's expansion.

▓ Hints to Selected Problems

1. (i) $y = \sec x \Rightarrow y_1 = \sec x \tan x = y\tan x$
$$\Rightarrow y_1^2 = y^2\tan^2 x = y^2(\sec^2 x - 1)$$
$$= y^2(y^2-1) = y^4 - y^2.$$
Again differentiating, we get
$$2y_1y_2 = 4y^3y_1 - 2yy_1 \Rightarrow y_2 = 2y^3 - y$$
Similarly $\quad y_3 = 6y^2y_1 - y_1$
$$y_4 = 12yy_1^2 + 6y^2y_2 - y_2$$
$$y_5 = 12y_1^3 - 36y_1y_2 - y_3$$
$$\ldots \quad \ldots \quad \ldots \quad \ldots$$
Now putting $x=0$ in the above equation and use Maclaurin's series.

2. $y = \log\sec x \Rightarrow y_1 = \tan x$
$$y_2 = \sec^2 x = 1 + \tan^2 x = 1 + y_1^2$$
$$y_3 = 2y_1y_2$$
$$y_4 = 2y_2^2 + 2y_1y_3$$
$$y_5 = 4y_2y_3 + 2y_2y_3 + 2y_1y_4$$
$$y_6 = 8y_2y_4 + 6y_3^2 + 2y_1y_5$$
$$\ldots \quad \ldots \quad \ldots \quad \ldots \quad \ldots$$
Now putting $x=0$ in the above equations and use Maclaurin's series.

3. $\qquad y = \sin^{-1}x$
$$\Rightarrow y_1 = \frac{1}{\sqrt{1-x^2}} \Rightarrow (1-x^2)y_1^2 = 1$$
Again differentiating, we get
$$(1-x^2)y_2 - xy_1 = 0$$
Now apply Leibnitz's theorem to differentiating n times.

4. $y = e^x\cos x$
$$\Rightarrow \quad y_1 = e^x\cos x - e^x\sin x = e^x(\cos x - \sin x)$$
$$= re^x(\cos\theta\cos x - \sin\theta\sin x),$$
where $r = \sqrt{2}$, $\theta = \dfrac{\pi}{4}$
$$\Rightarrow \quad y_1 = re^x\cos(x+\theta)$$
$$\Rightarrow \quad y_2 = r^2e^x\cos(x+2\theta)$$
$$\ldots \quad \ldots \quad \ldots$$
$$y_n = r^ne^x\cos(x+n\theta) = (2)^{n/2}e^x\cos\left(x+\frac{n\pi}{4}\right)$$
Now putting $x=0$ and use Maclaurin's series.

(ii) Here $\quad y_n = 2^{n/2}e^x\sin\left(x+\dfrac{n\pi}{4}\right)$

(iii) $y_n = (a^2+b^2)^{n/2}e^{ax}\sin\left[bx + n\tan^{-1}\left(\dfrac{b}{a}\right)\right]$

5. (i) $\quad y = f(x) = \tan^{-1}x$
$$\Rightarrow y_1 = f'(x) = \frac{1}{1+x^2} \Rightarrow y_2 = \frac{-2.x}{(1+x^2)^2}$$
Putting $x = \pi/4$ and find $f(\pi/4), f'(\pi/4), f''(\pi/4)\ldots$ and so on.
Now $f(x) = f\left(\dfrac{\pi}{4} + x - \dfrac{\pi}{4}\right)$
Then expand by Taylor's theorem.

6. $\qquad y = f(x) = \log x$
$$\Rightarrow f(x+h) = \log(x+h)$$
$$\Rightarrow \quad f'(x) = \frac{1}{x}, f''(x) = -\frac{1}{x^2}, f'''(x) = \frac{2}{x^3}\ldots$$
and so on.

$\Rightarrow$ Now putting $x=h$ in above derivatives and expand $f(x+h)$ by Taylor's theorem.

7. Take $\quad y=f(x)=\tan^{-1}x$

$$\Rightarrow f'(x)=\frac{1}{1+x^2}=\frac{1}{2i}\left(\frac{1}{x-i}-\frac{1}{x+i}\right)$$

$$\Rightarrow f''(x)=\frac{1}{2i}\left[(-1)(x-i)^{-2}-(-1)(x+i)^{-2}\right]$$

$$\cdots \qquad \cdots \qquad \cdots \quad \cdots \quad \cdots \quad \cdots \quad \cdots$$

$$f^n(x)=\frac{1}{2i}\left[(-1)^{n-1}(n-1)!(x-i)^{-n}-(-1)^{n-1}(n-1)!(x+i)^{-n}\right]$$

Put $\quad \theta=\cot^{-1}x$, we get

$$f^n(x)=(-1)^{n-1}(n-1)!\sin^n\theta\sin n\theta$$

Now use Taylor's series.

8. $y=f(x)=e^{\tan^{-1}x}$

$$\Rightarrow y_1=\frac{e^{\tan^{-1}x}}{1+x^2}=\frac{y}{1+x^2}\Rightarrow(1+x^2)y_1=y$$

$$(1+x^2)y_2+(2x-1)y_1=0$$

Now to find y_n, use Leibnitz's theorem.

9. Let $\quad y=f(x)=(\sin^{-1}x)^2$

$$\Rightarrow y_1=\frac{2\sin^{-1}x}{\sqrt{1-x^2}}\Rightarrow(1-x^2)y_1^2=4\left(\sin^{-1}x\right)^2=4y$$

Now differentiating n times by Leibnitz's rule.

10. $y=f(x)=e^{m\tan^{-1}x}$

$$\Rightarrow y_1=\frac{me^{\tan^{-1}x}}{1+x^2}\Rightarrow(1+x^2)y_1=my$$

Now differentiating n times by using Leibnitz's theorem.

11. Let $\quad y=f(x)=e^{e^x}\Rightarrow y_1=e^xe^{e^x}=e^x.y$

Now to find nth derivative, using Leibnitz's theorem.

12. Write $f(mx)=f[x+(m-1)x]$. Now expand by Taylor's theorem.

13. $y=f(x)=\sin(e^x-1)\Rightarrow y_1=e^x\cos(e^x-1)$ Now differentiating successively.

14. Write $f\left(\dfrac{x^2}{1+x}\right)=f\left(\dfrac{x^2}{1+x}-x+x\right)=f\left(x-\dfrac{x}{1+x}\right)$

Now expand by Taylor's theorem.

15. Let $\qquad y=f(17)=\sqrt{17}=\sqrt{x}$

$$y=f(16+1).$$

Now expand by Taylor's theorem.

Answers

1. (i) $1+\dfrac{x^2}{2!}+\dfrac{5x^4}{4!}+\dfrac{61x^6}{6!}+\cdots$ (ii) $1+x+\dfrac{x^2}{2}-\dfrac{x^3}{3}-\dfrac{11x^4}{24}-\dfrac{x^5}{5}+\cdots$ (iii) $1+x+\dfrac{2x^2}{2!}+\dfrac{4x^3}{3!}+\cdots$ (iv) $\log 2+\dfrac{x}{2}+\dfrac{x^2}{8}-\dfrac{x^4}{192}+\cdots$

(v) $x-\dfrac{x^2}{2}+\dfrac{2}{3}x^3-\dfrac{7x^4}{12}+\cdots$ 5. (i) $\tan^{-1}\left(\dfrac{\pi}{4}\right)+\left(x-\dfrac{\pi}{4}\right)\bigg/\left(1+\dfrac{\pi^2}{16}\right)-\pi\left(x-\dfrac{\pi}{4}\right)^2\bigg/\left[4\left(1+\dfrac{\pi^2}{16}\right)^2\right]+\cdots$

(ii) $45+53(x-2)+19(x-2)^2+2(x-2)^3+\cdots$ (iii) $\sin^{-1}h+x\left(1-h^2\right)^{-1/2}+\dfrac{x^2}{2!}h\left(1-h^2\right)^{-3/2}+\dfrac{x^3}{3!}\left[\left(1-h^2\right)^{-5/2}\left(1+2h^2\right)\right]+\cdots$

(iv) $\log\sin a+(x-a)\cot a-\dfrac{(x-a)^2}{2!}\text{cosec}^2 a+\dfrac{(x-a)^3}{3!}2\text{cosec}^2 a\cot a+\cdots$

8. $\dfrac{1}{24}$ 9. $2.\dfrac{x^2}{2!}+\dfrac{2.2^2}{4!}x^4+\dfrac{2.2^2.4^2}{6!}x^6+\cdots+\dfrac{2.2^2.4^2\dots(2n-2)^2}{(2n)!}x^{2n}+\cdots$ 13. $x+\dfrac{x^2}{2!}-\dfrac{5x^4}{24}+\cdots, y-\dfrac{y^2}{2}+\cdots$ 15. (i) 4.123 (ii) 5.099

2.3 INDETERMINATE FORMS

When a function involves the independent variable in such a manner that for a certain assigned value of that variable, its value cannot be found by simply substituting that value of the variable, the function is said to take an indeterminate form.

The most common cases occuring is that of a fraction whose numerator and denominator both vanish for the value of the variable involved.

As $f(x)\to 0$ and $g(x)\to 0$ when $x\to a$, then the quotient $\dfrac{f(x)}{g(x)}$ is said to have attained the indeterminate form $\dfrac{0}{0}$.

Similarly if $\lim\limits_{x\to a}f(x)=\infty$ and $\lim\limits_{x\to a}g(x)=\infty$, then the fraction $\dfrac{f(x)}{g(x)}$ is said to have attained the indeterminate form $\dfrac{\infty}{\infty}$.

The other important indeterminate forms are $0\times\infty$, $\infty-\infty$, 0^0, 1^∞ and ∞^0.

REMARKS

- The limiting value of the indeterminate forms is also called the true value.
- The most standard form among all indeterminate forms is $\dfrac{0}{0}$. We reduce all other cases of limits to this form.
- It will always be assumed that $f(x)$, $g(x)$, etc. and their respective derivatives are all continuous functions.
- The true value of the indeterminate form $\dfrac{0}{0}$ and $\dfrac{\infty}{\infty}$ is determined by the application of L' Hospital Rule.

2.3.1 L'Hospital rule For Indeterminate Form $\dfrac{0}{0}$

To find $\lim\limits_{x \to a} \dfrac{f(x)}{g(x)}$ when $\lim\limits_{x \to a} f(x) = 0 = \lim\limits_{x \to a} g(x)$.

Let us assume $f(x)$ and $g(x)$ be continuous at $x = a$, then, we have

$$f(a) = \lim_{x \to a} f(x) = 0, \; g(a) = \lim_{x \to a} g(x) = 0$$

By Taylor's theorem, we have

$$f(a + h) = f(a) + hf'(a + \theta_1 h) = hf'(a + \theta_1 h), \; 0 < \theta_1 < 1$$
$$g(a + h) = g(a) + hg'(a + \theta_2 h) = hg'(a + \theta_2 h), \; 0 < \theta_2 < 1$$

Therefore $\quad \lim\limits_{x \to a} \dfrac{f(x)}{g(x)} = \lim\limits_{h \to 0} \dfrac{f(a+h)}{g(a+h)} = \lim\limits_{h \to 0} \dfrac{hf'(a+\theta_1 h)}{hg'(a+\theta_2 h)} = \lim\limits_{h \to 0} \dfrac{f'(a+\theta_1 h)}{g'(a+\theta_2 h)} = \dfrac{f'(a)}{g'(a)} = \lim\limits_{x \to a} \dfrac{f'(x)}{g'(x)}$ (Provided $g'(a) \neq 0$)

$\Rightarrow \quad \lim\limits_{x \to a} \dfrac{f(x)}{g(x)} = \lim\limits_{x \to a} \dfrac{f'(x)}{g'(x)}$, provided $g'(a) \neq 0$.

If $g'(a) = 0$, then this argument fails. The case when $g'(a) = 0$ but $f'(0) \neq 0$.

$$\lim_{x \to a} \dfrac{f'(x)}{g'(x)} \to +\infty \text{ or } -\infty$$

If $f'(a) = 0 = g'(a)$, then by Taylor's theorem, we have

$$f(a+h) = f(a) + hf'(a) + \dfrac{h^2}{2!} f''(a + \theta_3 h) = \dfrac{h^2}{2!} f''(a + \theta_3 h), \; 0 < \theta_3 < 1$$

$$g(a+h) = g(a) + hg'(a) + \dfrac{h^2}{2!} g''(a + \theta_4 h) = \dfrac{h^2}{2!} g''(a + \theta_4 h), \; 0 < \theta_4 < 1$$

$\Rightarrow \quad \lim\limits_{x \to a} \dfrac{f(x)}{g(x)} = \lim\limits_{h \to a} \dfrac{f(a+h)}{g(a+h)} = \lim\limits_{h \to a} \dfrac{f''(a+\theta_3 h)}{g''(a+\theta_4 h)} = \dfrac{f''(a)}{g''(a)}$, provided $g''(a) \neq 0$.

The case of failure, when $g''(a) = 0$, the limit can be determined as before.

Now, in general if $\quad f(a) = f'(a) = f''(a) = \ldots = f^{n-1}(a) = 0$
$$g(a) = g'(a) = g''(a) = \ldots = g^{n-1}(a) = 0$$
and $\quad g^n(a) \neq 0$.

Then, by Taylor's theorem, we get

$$f(a+h) = f(a) + hf'(a) + \ldots + \dfrac{h^{n-1}}{(n-1)!} f^{n-1}(a) + \dfrac{h^n}{n!} f^n(a + \theta_n h), \qquad 0 < \theta_n < 1$$

$$= \dfrac{h^n}{n!} f^n(a + \theta_n h).$$

and

$$g(a+h) = g(a) + hg'(a) + \ldots + \dfrac{h^{n-1}}{(n-1)!} g^{n-1}(a) + \dfrac{h^n}{n!} g^n(a + \theta_n' h), \qquad 0 < \theta_n' < 1$$

$$= \dfrac{h^n}{n!} g^n(a + \theta_n' h).$$

Therefore,

$$\lim_{x \to a} \dfrac{f(x)}{g(x)} = \lim_{h \to a} \dfrac{f(a+h)}{g(a+h)} = \lim_{x \to a} \dfrac{f^n(a+\theta_n h)}{g^n(a+\theta_n' h)} = \dfrac{f^n(a)}{g^n(a)}, \text{ if } g^n(a) \neq 0$$

$\Rightarrow \quad \lim\limits_{x \to a} \dfrac{f(x)}{g(x)} = \lim\limits_{x \to a} \dfrac{f^n(x)}{g^n(x)}$, provided $g^n(a) \neq 0$.

2.3.2 L'Hospital rule For Indeterminate Form $\dfrac{\infty}{\infty}$

If $\lim\limits_{x \to a} f(x) = \infty$ and $\lim\limits_{x \to a} g(x) = \infty$, then

$$\lim_{x \to a} \dfrac{f(x)}{g(x)} = \lim_{x \to a} \dfrac{f'(x)}{g'(x)} \text{ provided } \lim_{x \to a} \dfrac{f'(x)}{g'(x)} \text{ exists.}$$

Proof. Consider $\quad \lim\limits_{x \to a} \dfrac{f(x)}{g(x)} = \lim\limits_{x \to a} \dfrac{\dfrac{1}{g(x)}}{\dfrac{1}{f(x)}} = \lim\limits_{x \to a} \left\{ \dfrac{-\dfrac{g'(x)}{[g(x)]^2}}{-\dfrac{f'(x)}{[f(x)]^2}} \right\}$ $\qquad \left[\dfrac{0}{0} \text{ form} \right]$

$\qquad\qquad\qquad\qquad\qquad\qquad\qquad\qquad\qquad$ [By L'Hospital rule]

$\Rightarrow$

$$\lim_{x \to a} \frac{f(x)}{g(x)} = \lim_{x \to a} \frac{g'(x)}{f'(x)} \cdot \lim_{x \to a} \left[\frac{f(x)}{g(x)} \right]^2 \qquad \text{\textbackslash...(1)}$$

Now, let

$$\lim_{x \to a} \frac{f(x)}{g(x)} = l. \qquad \qquad \qquad ...(2)$$

Then there are following three cases :

Case (i) If $l \neq 0$ and $l \neq \infty$.

In this case, (1) becomes

$$l = \lim_{x \to a} \frac{g'(x)}{f'(x)} \cdot l^2$$

Dividing by l^2, we get

$$\frac{1}{l} = \lim_{x \to a} \frac{g'(x)}{f'(x)}$$

$\Rightarrow$

$$\lim_{x \to a} \frac{f'(x)}{g'(x)} = l = \lim_{x \to a} \frac{f(x)}{g(x)} \qquad \qquad \text{[Using (2)]}$$

Case (ii) If $l = 0$.

In this case, adding 1 to each side of (2), we get

$$l + 1 = \lim_{x \to a} \frac{f(x)}{g(x)} + 1 = \lim_{x \to a} \frac{f(x) + g(x)}{g(x)} = \lim_{x \to a} \frac{f'(x) + g'(x)}{g'(x)} \qquad \text{[By case (i)]}$$

$$= \lim_{x \to a} \frac{f'(x)}{g'(x)} + 1$$

$\Rightarrow$

$$l = \lim_{x \to a} \frac{f'(x)}{g'(x)}$$

Case (iii) Let $\qquad\qquad l = \infty$

In this case, by reciprocating, we have $\quad \lim\limits_{x \to a} \dfrac{g(x)}{f(x)} = 0$

By case (ii)

$$0 = \lim_{x \to a} \frac{g(x)}{f(x)} = \lim_{x \to a} \frac{g'(x)}{f'(x)}$$

Therefore,

$$\lim_{x \to a} \frac{g'(x)}{f'(x)} = \infty$$

Hence, the result $\lim\limits_{x \to a} \dfrac{f(x)}{g(x)} = \lim\limits_{x \to a} \dfrac{g'(x)}{f'(x)}$ has been established in every case.

✎ REMARKS

- The above result can be extended to the case when $x \to \infty$, *i.e.*, we can show that $\lim\limits_{x \to \infty} \dfrac{f(x)}{g(x)} = \lim\limits_{x \to \infty} \dfrac{f'(x)}{g'(x)}$

  Let $x = \dfrac{1}{y}$ then $\lim\limits_{x \to \infty} \dfrac{f(x)}{g(x)} = \lim\limits_{y \to 0} \dfrac{f\left(\dfrac{1}{y}\right)}{g\left(\dfrac{1}{y}\right)} = \lim\limits_{y \to 0} \dfrac{f'\left(\dfrac{1}{y}\right)\left(-\dfrac{1}{y^2}\right)}{g'\left(\dfrac{1}{y}\right)\left(-\dfrac{1}{y^2}\right)} = \lim\limits_{y \to 0} \dfrac{f'\left(\dfrac{1}{y}\right)}{g'\left(\dfrac{1}{y}\right)} = \lim\limits_{x \to \infty} \dfrac{f'(x)}{g'(x)}$

- While evaluating $\lim\limits_{x \to \infty} \dfrac{f(x)}{g(x)}$ when it is of the form $\dfrac{\infty}{\infty}$, care must be taken to change over to the form $\dfrac{0}{0}$ as early as possible, otherwise process of differentiating the numerator and denominator may never terminate.

- While appplying L' Hospital rule, we are not to differentiate $\dfrac{f(x)}{g(x)}$ by the rule for finding the differential coefficient of the quotient of two functions, but we are to differentiate the numerator and denominator separately.

- It must be remember that $\log 1 = 0$, $\log 0 = -\infty$, and $\log \infty = \infty$.

📚 Solved Examples

EXAMPLE 1. *Find* $\lim\limits_{x \to 0} \dfrac{e^x - e^{\sin x}}{x - \sin x}$.

SOLUTION. We have $\lim\limits_{x \to 0} \dfrac{e^x - e^{\sin x}}{x - \sin x}$ $\left| \dfrac{0}{0} \right.$ form

$$= \lim_{x \to 0} \frac{e^x - e^{\sin x} \cdot \cos x}{1 - \cos x} \qquad \left| \frac{0}{0} \right. \text{ form}$$

$$= \lim_{x \to 0} \frac{e^x - [\cos x \cdot e^{\sin x} \cdot \cos x + e^{\sin x}(-\sin x)]}{\sin x}$$

$$= \lim_{x \to 0} \frac{e^x - e^{\sin x}[\cos^2 x - \sin x]}{\sin x} \qquad \left| \frac{0}{0} \right. \text{ form}$$

$$= \lim_{x \to 0} \frac{e^x - e^{\sin x}[2\cos x(-\sin x) - \cos x]}{-[(\cos^2 x - \sin x)e^{x \sin x} \cos x]}{\sin x}$$

$$= \lim_{x \to 0} \frac{e^x - e^{\sin x}[-\sin 2x - \cos x + \cos^3 x - \sin x \cos x]}{\cos x}$$

$$= \frac{1 - 1(-1 + 1)}{1} = \frac{1}{1} = 1$$

EXAMPLE 2. *Find* $\lim\limits_{x \to 0} \dfrac{x \cos x - \log(1+x)}{x^2}$

SOLUTION. We have $\lim\limits_{x \to 0} \dfrac{x \cos x - \log(1+x)}{x^2}$

$$= \lim_{x \to 0} \dfrac{x\left(1 - \dfrac{x^2}{2!} + \dfrac{x^4}{4!} - ....\right) - \left(x - \dfrac{x^2}{2} + \dfrac{x^3}{3} - ...\right)}{x^2}$$

$$\qquad\qquad\qquad \left|\dfrac{0}{0}\text{ form}\right.$$

$$= \lim_{x \to 0}\left(\dfrac{\dfrac{x^2}{2} - \dfrac{5}{6}x^3 + ...}{x^2}\right)$$

$$= \lim_{x \to 0}\left(\dfrac{1}{2} - \dfrac{5}{6}x + \text{terms containing } x\right) = \dfrac{1}{2}.$$

EXAMPLE 3. *Find* $\lim\limits_{x \to 0} \dfrac{(1+x)^{1/x} - e}{x}$.

SOLUTION. We have $\lim\limits_{x \to 0} \dfrac{(1+x)^{1/x} - e}{x}$ $\quad\left|\dfrac{0}{0}\text{ form}\right.$

Evaluating the limit of expansion for $(1+x)^{1/x}$ in ascending power of x, we get

$$\lim_{x \to 0}\dfrac{(1+x)^{1/x} - e}{x} = \lim_{x \to 0}\dfrac{e\left[1 - \dfrac{1}{2}x + \dfrac{11}{24}x^2 + ...\right] - e}{x}$$

$$= \lim_{x \to 0}\dfrac{e\left[-\dfrac{1}{2}x + \dfrac{11}{24}x^2 + ...\right]}{x} = -\dfrac{1}{2}e.$$

EXAMPLE 4. *Find* $\lim\limits_{x \to 0} \dfrac{\log\log(1-x^2)}{\log\log\cos x}$.

SOLUTION. We have $\lim\limits_{x \to 0} \dfrac{\log\log(1-x^2)}{\log\log\cos x}$ $\quad\left|\dfrac{\infty}{\infty}\text{ form}\right.$

$$= \lim_{x \to 0}\dfrac{\dfrac{1}{\log(1-x^2)} \cdot \dfrac{1}{(1-x^2)} \cdot (-2x)}{\dfrac{1}{\log\cos x} \cdot \dfrac{1}{\cos x} \cdot (-\sin x)}$$

$$= 2\lim_{x \to 0}\dfrac{x\cos x \log\cos x}{\sin x (1-x^2)\log(1-x^2)}$$

$$= \left(2\lim_{x \to 0}\dfrac{x}{\sin x}\right)\left(\lim_{x \to 0}\dfrac{\cos x}{1-x^2}\right) \cdot \left(\lim_{x \to 0}\dfrac{\log\cos x}{\log(1-x^2)}\right)$$

$$= 2 \times 1 \times 1 \times \lim_{x \to 0}\dfrac{\log\cos x}{\log(1-x^2)} \qquad \left|\dfrac{0}{0}\text{ form}\right.$$

$$= 2\lim_{x \to 0}\dfrac{\dfrac{1}{\cos x} \cdot (-\sin x)}{\dfrac{1}{(1-x^2)} \cdot (-2x)}$$

$$= 2 \times \dfrac{1}{2} \cdot \lim_{x \to 0}\left(\dfrac{\sin x}{x} \cdot \dfrac{1-x^2}{\cos x}\right) = 1.$$

EXAMPLE 5. *Find* $\lim\limits_{x \to \frac{\pi}{2}} \dfrac{\log\left(x - \dfrac{\pi}{2}\right)}{\tan x}$.

SOLUTION. We have $\lim\limits_{x \to \frac{\pi}{2}} \dfrac{\log\left(x - \dfrac{\pi}{2}\right)}{\tan x}$

$$= \lim_{x \to \frac{\pi}{2}}\left(\dfrac{\dfrac{1}{x - \pi/2}}{\sec^2 x}\right) \qquad \left|\dfrac{\infty}{\infty}\text{ form}\right.$$

$$= \lim_{x \to \frac{\pi}{2}}\left(\dfrac{\dfrac{1}{x - \pi/2}}{\dfrac{1}{\cos^2 x}}\right) = \lim_{x \to \frac{\pi}{2}}\left(\dfrac{\cos^2 x}{x - \pi/2}\right) \qquad \left|\dfrac{0}{0}\text{ form}\right.$$

$$= \lim_{x \to \frac{\pi}{2}}\left(\dfrac{-2\cos x \sin x}{1}\right) = -2\cos\dfrac{\pi}{2} \cdot \sin\dfrac{\pi}{2} = 0.$$

EXAMPLE 6. *Evaluate :* $\lim\limits_{x \to 1} \dfrac{x^x - x}{x - 1 - \log x}$.

SOLUTION. $\lim\limits_{x \to 1} \dfrac{x^x - x}{x - 1 - \log x}$ $\quad\left|\dfrac{0}{0}\text{ form}\right.$

$$= \lim_{x \to 1}\dfrac{x^x(1 + \log x) - 1}{1 - \dfrac{1}{x}} \quad \text{(By L Hospital Rule)}$$

$$\qquad\qquad\qquad \left|\dfrac{0}{0}\text{ form}\right.$$

$$= \lim_{x \to 1}\dfrac{x^x(1 + \log x)^2 + x^x\left(\dfrac{1}{x}\right)}{x^{-2}}$$

$$\left[\because \dfrac{d}{dx}x^x = x^x(1 + \log x)\right]$$

$$= \dfrac{1.(1+0)^2 + 1.1}{1} = 2.$$

Exercise-2.6

1. Evaluate the following limits :

(i) $\lim\limits_{x \to 0} \dfrac{x - \sin x}{x^3}$

(ii) $\lim\limits_{x \to 0} \dfrac{1 - \cos x}{x^2}$

(iii) $\lim\limits_{x \to 0} \dfrac{a^x - b^x}{x}$

(iv) $\lim\limits_{x \to 1} \dfrac{\log x}{x - 1}$

(v) $\lim\limits_{x \to 0} \dfrac{(1+x)^n - 1}{x}$

(vi) $\lim\limits_{x \to 0} \dfrac{xe^x - \log(1+x)}{x^2}$

(vii) $\lim\limits_{x \to a} \dfrac{a^x - x^a}{x^x - a^a}$

(viii) $\lim\limits_{x \to 0} \dfrac{5\sin x - 7\sin 2x + 3\sin 3x}{\tan x - x}$

(ix) $\lim\limits_{x \to 0} \dfrac{\sin 2x + a\sin x}{x^2}$

(x) $\lim\limits_{x \to 0} \dfrac{[\cosh x + \log(1-x) - 1 + x]}{x^2}$

(xi) $\lim\limits_{x \to 1} \dfrac{x^5 - 2x^3 - 4x^2 + 9x - 4}{x^4 - 2x^3 + 2x - 1}$

(VTU–2004, OSMANIA–2000)

2. Evaluate $\lim\limits_{x \to 0} \dfrac{\sin x \sin^{-1} x - x^2}{x^6}$.

3. Evaluate $\lim\limits_{x \to 0} \dfrac{(1+x)^{1/x} - e + \dfrac{1}{2}ex}{x^2}$.

4. Evaluate the following limits :

(i) $\lim\limits_{x \to \infty} \dfrac{a^{1/x} - b^{1/x}}{\log\left(\dfrac{x}{x-1}\right)}$

(ii) $\lim\limits_{x \to \pi/2} \dfrac{\left(\dfrac{\pi}{2} - x\right)^2 \sin x}{\cos^2 x}$

(iii) $\lim\limits_{x \to 0} \dfrac{e^x + \log\left(\dfrac{1-x}{e}\right)}{\tan x - x}$

(iv) $\lim\limits_{x \to 0+} \dfrac{3^x - 2^x}{\sqrt{x}}$

5. (i) If $\lim\limits_{y \to 0} \dfrac{re^y - q\cos y + pe^{-y}}{y \tan y} = 3$, find the vlaues of p, q, and r.

(MUMBAI–2009)

(ii) Find the values of a and b in order that

$\lim\limits_{x \to 0} \dfrac{x(1 + a\cos x) - b\sin x}{x^3}$ may be equatl to 1. (MUMBAI–2007)

(iii) If $\lim\limits_{x \to 0} \dfrac{\sin 2x + a\sin x}{x^3}$ be finite, find the value of a and the limit. (NAGPUR–2009)

6. Evaluate the following limits :

(i) $\lim\limits_{x \to 0} \dfrac{\log x^2}{\cot x^2}$

(ii) $\lim\limits_{x \to a} \dfrac{\log(x-a)}{\log(e^x - e^a)}$

(iii) $\lim\limits_{x \to 1-0} \dfrac{\log(1-x)}{\cot \pi x}$

(iv) $\lim\limits_{x \to \infty} \dfrac{\log x}{a^x}, a > 1$

(v) $\lim\limits_{x \to \pi/2} \dfrac{\tan x}{\tan 3x}$

Hints to Selected Problems

1. (i) $\lim\limits_{x \to 0} \dfrac{x - \sin x}{x^3} = \lim\limits_{x \to 0} \dfrac{x - \left(x - \dfrac{x^3}{3!} + \dfrac{x^5}{5!} - \ldots\right)}{x^3}$

$= \lim\limits_{x \to 0} \left\{\dfrac{\dfrac{x^3}{3!} - \dfrac{x^5}{5!} + \ldots}{x^3}\right\}$

$= \lim\limits_{x \to 0}\left[\dfrac{1}{3!} - \dfrac{x^2}{5!} + \ldots\right] = \dfrac{1}{3!} = \dfrac{1}{6}$.

(vii) $\lim\limits_{x \to a} \dfrac{a^x - x^a}{x^x - a^a} = \lim\limits_{x \to a} \dfrac{a^x \log a - a \cdot x^{a-1}}{x^x(1 + \log x)}$

$= \dfrac{a^a \log a - a^a}{a^a(1 + \log a)}$

$= \dfrac{\log a - 1}{\log a + 1}$.

5. (i) If $r - q + p = 0$, ...(1)

we obtained $\dfrac{0}{0}$ form, Then we have

$\lim\limits_{y \to 0} \dfrac{re^y + q\sin y + pe^{-y}}{y\sec^2 y + \tan y}$

Again if $r - p = 0$, ...(2)

We obtained $\dfrac{0}{0}$ form. Then we have

$\lim\limits_{y \to 0} \dfrac{re^y + q\cos y + pe^{-y}}{\sec^2 y + 2y\sec^2 y \tan y + \sec^2 y}$

$\Rightarrow \qquad r + q + p = 6$ ...(3)

Now solving (1), (2) and (3).

6. (iv) $\lim\limits_{x \to \infty} \dfrac{\log x}{a^x}, a > 1$ $\quad\bigg|$ form $\dfrac{\infty}{\infty}$

$= \lim\limits_{x \to \infty} \dfrac{1/x}{a^x \log a} = \dfrac{1}{\log a}\lim\limits_{x \to \infty} \dfrac{1}{xa^x} = \dfrac{1}{\log a}.0 = 0$.

Answers

1. (i) $\dfrac{1}{6}$ (ii) $\dfrac{1}{2}$ (iii) $\log\dfrac{a}{b}$ (iv) 1 (v) n (vi) $\dfrac{3}{2}$ (vii) $\dfrac{\log a - 1}{\log a + 1}$ (viii) -15 (ix) $\begin{cases}\infty \text{ if } a \neq -2 \\ 0 \text{ if } a = -2\end{cases}$ (x) 0 (xi) 4

2. $\dfrac{1}{18}$ **3.** $\dfrac{11e}{24}$ **4.** (i) $\log\dfrac{a}{b}$ (ii) 1 (iii) $-\dfrac{1}{2}$ (iv) 0

5. (i) $p = \dfrac{3}{2}, q = 3, r = \dfrac{3}{2}$ (ii) $a = -\dfrac{5}{2}, b = -\dfrac{3}{2}$ (iii) $\begin{cases}-1 \text{ if } a = -2 \\ \infty \text{ if } a \neq -2\end{cases}$ **6.** (i) 0 (ii) 1 (iii) 0 (iv) 0 (v) 3

2.3.4 THE INDETERMINATE FORM $0 \times \infty$

To find $\lim\limits_{x \to a}[f(x) \cdot g(x)]$, *when* $\lim\limits_{x \to a} f(x) = 0$ *and* $\lim\limits_{x \to a} g(x) = \infty$.

To determine this limit, the product may be transformed into the form $\dfrac{0}{0}$ or $\dfrac{\infty}{\infty}$, using any one of the following relations

$$f(x) \cdot g(x) = \dfrac{f(x)}{\dfrac{1}{g(x)}} \quad \text{or} \quad f(x) \cdot g(x) = \dfrac{g(x)}{\dfrac{1}{f(x)}}$$

and then apply previous method.

Solved Examples

EXAMPLE 1. *Evaluate* $\lim\limits_{x \to 0^+} (x \log x)$.

SOLUTION. $\lim\limits_{x \to 0^+} (x \log x) = \lim\limits_{x \to 0^+} \dfrac{\log x}{1/x}$ $\quad\left|\dfrac{\infty}{\infty}\right.$ form

$= \lim\limits_{x \to 0^+} \dfrac{1/x}{-1/x^2}$

$= \lim\limits_{x \to 0^+} (-x) = 0.$

EXAMPLE 2. *Evaluate* $\lim\limits_{x \to 0} x \log \sin x$.

SOLUTION. $\lim\limits_{x \to 0} x \log \sin x$ $\quad | \; 0 \times \infty$ from

$= \lim\limits_{x \to 0} \left(\dfrac{\log \sin x}{1/x}\right)$ $\quad\left|\dfrac{\infty}{\infty}\right.$ form

$= \lim\limits_{x \to 0} \dfrac{(1/\sin x)\cdot \cos x}{-1/x^2}$ $\quad\left|\dfrac{\infty}{\infty}\right.$ form

$= \lim\limits_{x \to 0} \dfrac{-x^2 \cos x}{\sin x}$ $\quad\left|\dfrac{0}{0}\right.$ form

$= \lim\limits_{x \to 0} \dfrac{x^2 \sin x - 2x \cos x}{\cos x} = 0.$

2.3.5 THE INDETERMINATE FORM $\infty - \infty$

To determine $\lim\limits_{x \to a} [f(x) - g(x)]$, *when* $\lim\limits_{x \to a} f(x) = \infty = \lim\limits_{x \to \infty} g(x)$.

Here, this can be reduced to the form $\dfrac{0}{0}$ by the relation $f(x) - g(x) = \left\{ \dfrac{\left[\dfrac{1}{g(x)} - \dfrac{1}{f(x)}\right]}{\dfrac{1}{f(x)\cdot g(x)}} \right\}$ and then evaluate by previous method.

WORKING PROCEDURE

Step 1. Change all trigonometric-ratio into $\sin x$ and $\cos x$ (if T-ratio are present)

Step 2. Take L.C.M.

Now the indeterminate form is reduced into $\dfrac{0}{0}$ form.

 Solved Examples

EXAMPLE 1. *Evaluate* $\lim\limits_{x \to 0}\left(\dfrac{1}{x^2} - \dfrac{1}{\sin^2 x}\right)$.

SOLUTION. $\lim\limits_{x \to 0}\left(\dfrac{1}{x^2} - \dfrac{1}{\sin^2 x}\right)$ $\quad | \; \infty - \infty$ form

$= \lim\limits_{x \to 0} \dfrac{\sin^2 x - x^2}{x^2 \sin^2 x}$ $\quad\left|\dfrac{0}{0}\right.$ form

$= \lim\limits_{x \to 0} \dfrac{\left(x - \dfrac{x^3}{3!} + ...\right)^2 - x^2}{x^2\left(x - \dfrac{x^3}{3!} + ...\right)}$

$= \lim\limits_{x \to 0} \dfrac{-\dfrac{2x^4}{3!} + \text{terms containing higher powers of } x}{x^4 + \text{terms containing higher power of } x}$

$= \lim\limits_{x \to 0} \dfrac{-\dfrac{2}{3!} + \text{terms containing } x \text{ in the numerator}}{1 + \text{terms containing } x \text{ in the numerator}}$

$= -\dfrac{2}{3!} = -\dfrac{1}{3}$

EXAMPLE 2. *Evaluate* $\lim\limits_{x \to \pi/2} (\sec x - \tan x)$.

SOLUTION. We have $\lim\limits_{x \to \pi/2} (\sec x - \tan x)$ $\quad | \infty - \infty$ form

$= \lim\limits_{x \to \pi/2}\left(\dfrac{1}{\cos x} - \dfrac{\sin x}{\cos x}\right)$ $\quad\left|\dfrac{0}{0}\right.$ form

$= \lim\limits_{x \to \pi/2}\left(\dfrac{1 - \sin x}{\cos x}\right)$

$= \lim\limits_{x \to \pi/2} \dfrac{-\cos x}{-\sin x}$

$= \lim\limits_{x \to \pi/2} \cot x = 0.$

EXAMPLE 3. *Evaluate* $\lim\limits_{\to \pi/2}\left(\sec \quad \dfrac{}{1 \quad \sin}\right)$.

SOLUTION. We have $\lim\limits_{x \to \pi/2}\left(\sec x - \dfrac{1}{1 - \sin x}\right)$ $\quad | \infty - \infty$ form

$= \lim\limits_{x \to \pi/2}\left(\dfrac{1}{\cos x} - \dfrac{1}{1 - \sin x}\right)$ $\quad | \infty - \infty$ form

$= \lim\limits_{x \to \pi/2}\left(\dfrac{1 - \sin x - \cos x}{\cos x (1 - \sin x)}\right)$ $\quad\left|\dfrac{0}{0}\right.$ form

$= \lim\limits_{x \to \pi/2} \dfrac{-\cos x + \sin x}{-\sin x + \sin^2 x - \cos^2 x}$

$= \dfrac{-0 + 1}{-1 + 1 - 0} = \infty.$

EXAMPLE 4. *Evaluate* $\lim\limits_{x\to 0}\left(\tan x \log x\right)$.   (VTU–2009)   (By Ĺ Hospital Rule)

SOLUTION. We can write

$$\lim_{x\to 0}\left(\tan x\log x\right)=\lim_{x\to 0}\left(\frac{\log x}{\cot x}\right)\qquad\Bigg|\text{form }\frac{\infty}{\infty}$$

$$=\lim_{x\to 0}\frac{1/x}{-\operatorname{cosec}^2 x}$$

$$=-\lim_{x\to 0}\left(\frac{\sin^2 x}{x}\right)$$

$$=\lim_{x\to 0}\frac{2\sin x\cos x}{1}=0.$$

2.3.6 THE INDETERMINATE FORM $0^\circ, 1^\infty, \infty^\circ$

To determinate $\lim\limits_{x\to a}[f(x)]^{g(x)}$ *when the limit is of the form* $0^\circ, 1^\infty, \infty^\circ$

Let $\qquad\qquad\qquad\qquad y=[f(x)]^{g(x)}$

Taking logs; $\qquad\qquad\qquad \log y=g(x)\log f(x)$

The RHS assumes the indeterminate forms $0\times\infty$ in each of these above cases. The limit can, therefore, be determined by the method used in the article 3.4.

Suppose $\qquad\qquad\qquad \lim\limits_{x\to a}[g(x)\log f(x)]=l$ (say)

$\Rightarrow\qquad\qquad\qquad \lim\limits_{x\to a}\log y=l\quad\Rightarrow\quad \lim\limits_{x\to a}[\log y]=l$

$\Rightarrow\qquad\qquad\qquad \lim\limits_{x\to a}y=e^l\quad\Rightarrow\quad \lim\limits_{x\to a}[f(x)]^{g(x)}=e^l.$

WORKING PROCEDURE

Step 1. Let the given limit = y.
Step 2. Take logs on both sides to get the forms $0\times\infty$ and proceed by the method of the type $0\times\infty$.

 Solved Examples

EXAMPLE 1. *Evaluate* $\lim\limits_{\theta\to\frac{\pi}{2}}(\cos\theta)^{\cos\theta}$.

SOLUTION. Let $\qquad y=(\cos\theta)^{\cos\theta}$ | 0° form

Taking logs,

$\qquad\qquad \log y=\cos\theta\log\cos\theta$

$\therefore\quad \lim\limits_{\theta\to\frac{\pi}{2}}(\log y)=\lim\limits_{\theta\to\frac{\pi}{2}}\cos\theta\log\cos\theta$ | $0\times\infty$ form

$$=\lim_{\theta\to\frac{\pi}{2}}\frac{\log\cos\theta}{\sec\theta}\qquad\Bigg|\frac{\infty}{\infty}\text{ form}$$

$$=\lim_{\theta\to\frac{\pi}{2}}\frac{\dfrac{1}{\cos\theta}\times-\sin\theta}{\sec\theta\tan\theta}$$

$$=\lim_{\theta\to\frac{\pi}{2}}(-\cos\theta)=0$$

$\Rightarrow\quad \lim\limits_{\theta\to\frac{\pi}{2}}(\log y)=0$

$\Rightarrow\quad \log\left(\lim\limits_{\theta\to\frac{\pi}{2}}y\right)=0\Rightarrow \lim\limits_{\theta\to\frac{\pi}{2}}y=e^0=1$

$\Rightarrow\quad \lim\limits_{\theta\to\frac{\pi}{2}}(\cos\theta)^{\cos\theta}=1.$

EXAMPLE 2. *Evaluate* $\lim\limits_{x\to 0}\left(\dfrac{\sin x}{x}\right)^{1/x}$.

SOLUTION. Let $\quad y=\lim\limits_{x\to 0}\left(\dfrac{\sin x}{x}\right)^{1/x}$

$$\log y=\lim_{x\to 0}\left(\frac{1}{x}\log\frac{\sin x}{x}\right)$$

$$\therefore\qquad =\lim_{x\to 0}\frac{1}{x}\log\left\{\frac{x-\dfrac{x^3}{3!}+\dfrac{x^5}{5!}-\dots}{x}\right\}$$

$$=\lim_{x\to 0}\frac{1}{x}\log\left(1-\frac{x^2}{3!}+\frac{x^4}{5!}-\dots\right)$$

$$=\lim_{x\to 0}\frac{1}{x}\log\left[1-\left(\frac{x^2}{6}-\frac{x^4}{120}+\dots\right)\right]$$

$$=\lim_{x\to 0}\frac{1}{x}\log(1-z)\quad\text{where }z=\frac{x^2}{6}-\frac{x^4}{120}+\dots$$

$$=\lim_{x\to 0}\frac{1}{x}\left(-z-\frac{z^2}{2}-\dots\right)$$

$$=\lim_{x\to 0}\frac{1}{x}\left[-\left(\frac{x^2}{6}-\frac{x^4}{120}+\dots\right)-\frac{1}{2}\left(\frac{x^2}{6}-\frac{x^4}{120}+\dots\right)^2-\dots\right]$$

$$=\lim_{x\to 0}\frac{1}{x}\left[-\frac{x^2}{6}+\left(\frac{x^4}{120}-\frac{x^4}{72}\right)+\dots\right]$$

$$= \lim_{x \to 0} \frac{1}{x}\left[-\frac{x^2}{6} + \frac{x^4}{180} + \ldots\right]$$

$$= \lim_{x \to 0}\left[-\frac{x}{6} + \frac{x^3}{180} + \ldots\right] = 0$$

Hence, $y = e^0 = 1$.

EXAMPLE 3. *Evaluate* $\lim\limits_{x \to 0}(\operatorname{cosec} x)^{1/\log x}$.

SOLUTION. We have $\quad y = \lim\limits_{x \to 0}(\operatorname{cosec} x)^{1/\log x} \qquad |\infty^0 \text{ form}$

$$\therefore \quad \log y = \lim_{x \to 0}\frac{1}{\log x}(\log \operatorname{cosec} x) \qquad \left|\frac{\infty}{\infty}\text{ form}\right.$$

$$= \lim_{x \to 0}\frac{\left(\dfrac{1}{\operatorname{cosec} x}\right)(-\operatorname{cosec} x \cot x)}{1/x}$$

$$= \lim_{x \to 0}\left(-\frac{x}{\tan x}\right) \qquad \left|\frac{0}{0}\text{ form}\right.$$

$$= \lim_{x \to 0}\left(-\frac{1}{\sec^2 x}\right) = -1$$

$$\Rightarrow \quad y = e^{-1} = \frac{1}{e}.$$

EXAMPLE 4. *Evaluate the following limits* :

(i) $\lim\limits_{x \to 0}(\cos x)^{\cot x}$

(ii) $\lim\limits_{x \to 0}\dfrac{e^x - e^{-x} - 2\log(1+x)}{x \sin x}$

SOLUTION. (i) Let $\quad y = \lim\limits_{x \to 0}(\cos x)^{\cot x} \qquad (1^\infty \text{ form})$

$$\log y = \lim_{x \to 0}\log(\cos x)^{\cot x}$$

$$= \lim_{x \to 0}\cot \log \cos x$$

$$\log y = \lim_{x \to 0}\frac{\log \cos x}{\tan x} \qquad \left(\frac{0}{0}\text{ form}\right)$$

$$= \lim_{x \to 0}\frac{(-\sin x)}{\cos x \cdot \sec^2 x} = \lim_{x \to 0}\frac{-\sin x}{\sec x}$$

$$= \lim_{x \to 0} -\sin x \cos x = \lim_{x \to 0}-\frac{\sin 2x}{2 \cdot x} \cdot x$$

$$= -\lim_{x \to 0}\left(\frac{\sin 2x}{2x}\right)\cdot \lim_{x \to 0}x = -1 \times 0.$$

$$\therefore \quad \log y = 0 \Rightarrow y = e^0 \Rightarrow y = 1.$$

(ii) We have

$$\lim_{x \to 0}\frac{e^x - e^{-x} - 2\log(1-x)}{x \sin x} \qquad \left|\frac{0}{0}\text{ form}\right.$$

$$= \lim_{x \to 0}\frac{e^x + e^{-x} - \dfrac{2}{1+x}}{x \cos x + \sin x} \qquad \left|\frac{0}{0}\text{ form}\right.$$

$$= \lim_{x \to 0}\frac{e^x - e^{-x} + \dfrac{2}{(1+x)^2}}{2\cos x - x \sin x}$$

$$= \frac{e^0 - e^0 + \dfrac{2}{(1+0)^2}}{2\cdot \cos 0 - 0} = \frac{1 - 1 + \dfrac{2}{(1+0)^2}}{2 \cdot \cos 0 - 0} = \frac{2}{2} = 1.$$

EXAMPLE 5. *Find* $\lim\limits_{x \to 0}\left(\dfrac{\tan x}{x}\right)^{1/x^3}$.

SOLUTION. Let $\quad y = \lim\limits_{x \to 0}\left(\dfrac{\tan x}{x}\right)^{1/x^3}$

$$\log y = \lim_{x \to 0}\frac{1}{x^3}\log_e\left(\frac{\tan x}{x}\right)$$

$$= \lim_{x \to 0}\frac{\log_e \tan x - \log_e x}{x^3} \quad \left|\frac{0}{0}\text{ form}\right.$$

$$= \lim_{x \to 0}\frac{\dfrac{\sec^2 x}{\tan x} - \dfrac{1}{x}}{3x^2} = \lim_{x \to 0}\frac{\dfrac{2x}{\sin 2x} - \dfrac{1}{x}}{3x^2}$$

$$= \lim_{x \to 0}\frac{2x - \sin 2x}{3x^2 \sin 2x} \qquad \left|\frac{0}{0}\text{ form}\right.$$

$$= \lim_{x \to 0}\frac{2 - 2\cos 2x}{6x^2 \sin 2x + 6x^3 \cos 2x} \qquad \left|\frac{0}{0}\text{ form}\right.$$

$$= \lim_{x \to 0}\frac{2\sin 2x}{15x^2 \cos 2x + 6x \sin 2x - 6x^3 \sin 2x} \quad \left|\frac{0}{0}\text{ form}\right.$$

$$= \lim_{x \to 0}\frac{4\cos 2x}{-30x^2 \sin 2x + 30x \cos 2x + 6\sin 2x}$$
$$+ 12x \cos 2x - 18x^2 \sin 2x - 12x^3 \cos 2x$$

$$= \frac{4}{0} = \infty$$

Hence, $\quad y = e^\infty = \infty.$

Exercise-2.7

1. Evaluate the following limits :

(i) $\lim\limits_{x \to 0} x \log \tan x$

(ii) $\lim\limits_{x \to 0}\tan\left(\dfrac{\pi}{2} - x\right)$

(iii) $\lim\limits_{x \to \infty} 2^x \sin\dfrac{a}{2^x}$

(iv) $\lim\limits_{x \to \infty}(a^{1/x} - 1)\cdot x$

(v) $\lim\limits_{x \to 1}\sec\dfrac{\pi}{2x}\log x$

(vi) $\lim\limits_{x \to 0} x^m(\log x)^n \quad m; n \in Z^+$.

2. Evaluate the following limits :

(i) $\lim\limits_{x \to 0}\left[\dfrac{1}{x} - \dfrac{1}{x^2}\log(1+x)\right]$

(ii) $\lim\limits_{x \to 2}\left[\dfrac{1}{x-2} - \dfrac{1}{\log(x-1)}\right]$

(iii) $\lim\limits_{x \to 0}\left[\dfrac{1}{x^2} - \operatorname{cosec}^2 x\right]$

(iv) $\lim\limits_{x \to 0}\left(\dfrac{1}{x^2} - \cot^2 x\right)$

(v) $\lim\limits_{x \to 0}\left(\dfrac{1}{x^2} - \dfrac{1}{x \tan x}\right)$

(vi) $\lim\limits_{x \to 0}\left(\operatorname{cosec} x - \dfrac{1}{x}\right)$.

3. Evaluate the following limits :

(i) $\lim\limits_{x \to 0}\left(\dfrac{1}{x}\right)^{\tan x}$

(ii) $\lim\limits_{x \to 0} x^x$

(iii) $\lim\limits_{x \to \infty}\left(\dfrac{\pi}{2} - \tan^{-1} x\right)^{1/x}$

(iv) $\lim\limits_{x \to 0}\left(\dfrac{\tan x}{x}\right)^{1/x}$ (v) $\lim\limits_{x \to \pi/2}(\sin x)^{\tan x}$ (x) $\lim\limits_{x \to \infty}\left(1+\dfrac{k}{x}\right)^{x}$

(vi) $\lim\limits_{x \to \pi/4}(\tan x)^{\tan 2x}$ (VTU–2004)   **4.** Evaluate the following limits :

(vii) $\lim\limits_{x \to 0}\left[\dfrac{2(\cosh x - 1)}{x^2}\right]^{1/x^2}$ (viii) $\lim\limits_{x \to 0}(\operatorname{cosec} x)^{1/\log x}$ (i) $\lim\limits_{x \to 0}\left[\dfrac{a^x + b^x}{2}\right]^{1/x}$ (VTU–2007)

(ix) $\lim\limits_{x \to a}\left(2 - \dfrac{x}{a}\right)^{\tan\left(\frac{\pi x}{2a}\right)}$ (VTU–2010, NAGPUR–2009)   (ii) $\lim\limits_{x \to 0}\left[\dfrac{a_1^x + a_2^x + ... + a_n^x}{n}\right]^{1/x}$ (VTU–2011)

(iii) $\lim\limits_{x \to \infty}\left(1 + \dfrac{a}{x}\right)^{x}$.

🗒️ Hints to Selected Problems

1. (iii) $\lim\limits_{x \to \infty} 2^x \cdot \sin\dfrac{a}{2^x} = \lim\limits_{x \to \infty}\dfrac{\sin\left(\frac{a}{2^x}\right)}{\frac{1}{2^x}}$

$\qquad = a \lim\limits_{x \to \infty}\left[\dfrac{\sin\left(\frac{a}{2^x}\right)}{\frac{a}{2^x}}\right] = a \cdot 1 = a.$

(v) $\lim\limits_{x \to 1}\sec\left(\dfrac{\pi}{2x}\right) \cdot \log x = \lim\limits_{x \to 1}\dfrac{\log x}{\cos\left(\frac{\pi}{2x}\right)}$

$= \lim\limits_{x \to 1}\dfrac{\frac{1}{x}}{\frac{\pi}{2x^2}\sin\left(\frac{\pi}{2}x\right)} = \lim\limits_{x \to 1}\dfrac{2x}{\pi\sin\left(\frac{\pi}{2x}\right)} = \dfrac{2 \times 1}{\pi \cdot \sin\left(\frac{\pi}{2}\right)} = \dfrac{2}{\pi}.$

$\lim\limits_{x \to 0}\left(\dfrac{1}{x^2} - \dfrac{1}{x\tan x}\right) = \lim\limits_{x \to 0}\dfrac{\tan x - x}{x^2 \tan x}$

2. (v)

$= \lim\limits_{x \to 0}\dfrac{\left(x + \frac{x^3}{3} + \frac{2}{15}x^5 + ...\right) - x}{x^2\left(x + \frac{x^3}{3} + \frac{2}{15}x^5 + ...\right)}$

$= \lim\limits_{x \to 0}\dfrac{x^3\left(\frac{1}{3} + \frac{2}{15}x^2 + ...\right)}{x^3\left(1 + \frac{x^2}{3} + \frac{2}{15}x^4 + ...\right)}$

$= \lim\limits_{x \to 0}\dfrac{\frac{1}{3} + \frac{2}{15}x^2 + ...}{1 + \frac{x^2}{3} + \frac{2}{15}x^4 + ...} = \dfrac{1}{3}$

3. (i) $\qquad \log y = \lim\limits_{x \to 0}\left(\dfrac{1}{x}\right)^{\tan x}$

$\Rightarrow \; \log y = \lim\limits_{x \to 0}\tan x \log\left(\dfrac{1}{x}\right)$

$= -\lim\limits_{x \to 0}\tan x \log x = -\lim\limits_{x \to 0}\dfrac{\log x}{\cot x}$

$= -\lim\limits_{x \to 0}\dfrac{1/x}{-\operatorname{cosec}^2 x} = \lim\limits_{x \to 0}\dfrac{\sin^2 x}{x}$

$= \lim\limits_{x \to 0}\left(\dfrac{\sin x}{x}\right) \cdot \sin x = 1 \times 0 = 0$

$\Rightarrow \qquad y = e^0 = 1.$

🗒️ Answers

1. (i) 0   (ii) ∞ (iii) a (iv) $\log a$ (v) $-\dfrac{2}{\pi}$ (vi) 0.

2. (i) $\dfrac{1}{2}$ (ii) $-\dfrac{1}{2}$ (iii) $-\dfrac{1}{3}$ (iv) $\dfrac{2}{3}$ (v) $\dfrac{1}{3}$ (vi) 0.

3. (i) 1   (ii) 1   (iii) 1   (iv) 1   (v) 1   (vi) $\dfrac{1}{e}$ (vii) $e^{1/12}$

(viii) $\dfrac{1}{e}$ (ix) $e^{2/\pi}$ (x) e^k.   **4.** (i) $\sqrt{ab}$ (ii) $(a_1 . a_2 .... a_n)^{1/n}$ (iii) e^a.

2.4 TANGENT AND NORMAL

Let P be a given point and Q be any other point on it. Let Q travel towards P along the curve.

Let Q travel towards P along the curve.

Then, the limiting position PT of the secant PQ is known as the tangent to the curve.

The line PS through P which is perpendicular to the tangent PT is called the normal of the curve.

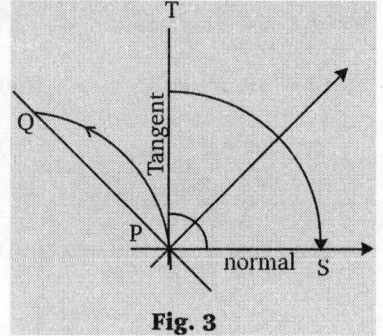

Fig. 3

2.4.1 Some Fundamental Concepts

(i) Slope of a line, $m = \tan\theta$, where θ is the angle which the line makes with the positive direction of x-axis.

(ii) Slope of the line $ax + by + c = 0$ is given by $m = -\dfrac{a}{b}$

(iii) Slope of the line joining the points (x_1, y_1) and (x_2, y_2) is $= \dfrac{y_2 - y_1}{x_2 - x_1}$

(iv) Slope of x-axis = 0, Slope of y-axis = ∞

(v) Two lines are parallel iff $m_1 = m_2$.

(vi) Two lines are perpendicular iff $m_1 m_2 = -1$.

(vii) Angle between two lines having slopes m_1 and m_2 is given by $\theta = \tan^{-1}\left(\dfrac{m_1 - m_2}{1 + m_1 m_2}\right)$

(viii) Equation of the line (one point form) $y - y_1 = m(x - x_1)$ passing through the point (

(ix) Perpendicular distance formula $= \dfrac{|ax_1 + by_1 + c|}{\sqrt{a^2 + b^2}}$

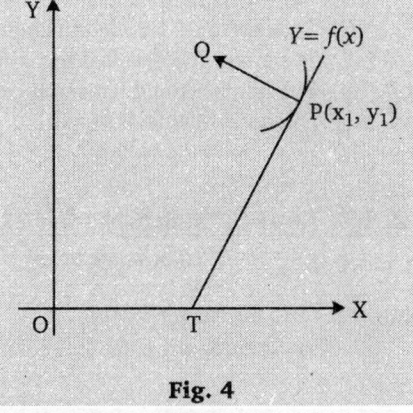

2.4.2 Equation of the Tangent

Let $y = f(x)$ be the equation of the curve, and $P(x_1, y_1)$ be any given point on this curve. Let $Q = Q(x + \delta x, y + \delta y)$ be any neighbouring point of P. Let PT be the tangent at the point (x_1, y_1).

The slope of the tangent at $(x_1, y_1) = \dfrac{dy_1}{dx_1}$.

Now, tangent is a line through the point $P(x_1, y_1)$ and its slope $m = \dfrac{dy_1}{dx_1}$.

Fig. 4

Hence, by Co-ordinate Geometry, the equation of the tangent is $y - y_1 = \dfrac{dy_1}{dx_1}(x - x_1)$.

✍ Remarks

- It should be clearly understood that by $\dfrac{dy_1}{dx_1}$ we mean the value of $\dfrac{dy}{dx}$ at (x_1, y_1) and not as derivative of y_1 with respect to x_1.

- The equation of the tangent at a point t_1 to the curve $x = f(t), y = g(t)$ is given by $y - g(t_1) = \dfrac{g'(t_1)}{f'(t_1)}[x - f(t_1)]$.

2.4.3 Geometrical Meaning of $\dfrac{dy}{dx}$

Let $y = f(x)$ be the given function and let it be represented by the curve AB. Take two neighbouring points $P(x, y)$ and $Q(x + \delta x, y + \delta y)$ on the curve AB. Join PQ and let PQ be produced to meet OX at the point R.

Slope of the secant $PQ = \dfrac{y + \delta y - y}{x + \delta x - x} = \dfrac{\delta y}{\delta x}$.   ...(i)

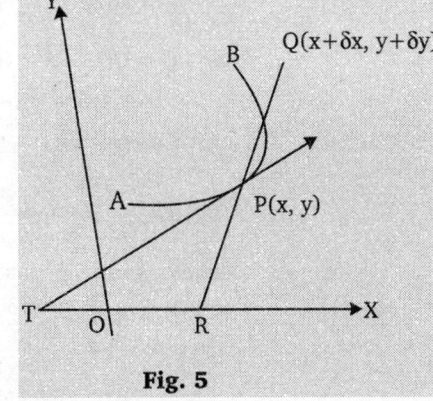

Now, let the point Q move along the curve and approach the point P in the limiting position. $\delta x \to 0$, $\delta y \to 0$ and the secant PQ becomes the tangent PT at P.

Therefore, from (1)

Slope of the tangent PT at $(x, y) = \lim\limits_{\substack{\delta x \to 0 \\ \delta y \to 0}} \dfrac{\delta y}{\delta x} = \dfrac{dy}{dx}$

Fig. 5

i.e., the value of the derivative at a point P of the curve is equal to the slope of tangent at that point to the curve.

✍ Remarks

- If the tangent at a point on the curve $y = f(x)$ is parallel to x-axis, its slope is zero *i.e.*, $\dfrac{dy}{dx}$ at the point = 0.

- If the tangent at a point on the curve is prependicular to x-axis, *i.e.*, parallel to y-axis. Its slope is ∞, *i.e.*, $\dfrac{dy}{dx}$ at the point = ∞.

2.4.4 Equation of the Normal

The normal to a curve at a given point is a line perpendicular to the tangent at that point and passes through the point. The slope of the normal at point $P(x_1, y_1)$ will be negative reciprocal of the slope of the tangent.

Hence,   the slope of the normal at $(x_1, y_1) = -\dfrac{1}{dy_1 / dx_1}$

$\therefore$ The equation of the normal at $P(x_1, y_1)$ is $y - y_1 = -\dfrac{1}{dy_1 / dx_1}(x - x_1)$

2.4.5 POLAR CO-ORDINATES

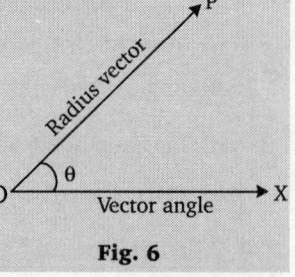

Let OX be a fixed straight line through fixed point O. The fixed point O is called the pole, or the origin and the fixed straight line OX is called initial line or the polar axis.

Let P be any point in the plane through the line OX. Join OP, then
 (i) The length OP is called the radius vector of the point P and is denoted by r.
 (ii) The angle XOP is called the vectorial angle of the point P and denoted by θ.
 (iii) The number r and θ taken together in this order and called P, the polar-co-ordinates of the point P and we write it as $P(r, \theta)$.
 (iv) If (x, y) are the co-ordinates of P referred to cartesian system, then it can be easily found that $x = r \cos \theta, y = r \sin \theta$.

Fig. 6

2.4.6 ANGLE BETWEEN RADIUS VECTOR AND TANGENT

Let (r, θ) be the co-ordinate of any point P' on the curve $r = f(\theta)$. Let the tangent at P makes an angle ψ with OX.

Let ϕ be the angle between the radius vector and the tangent at P, *i.e.*, $\angle MPN = \phi$ is the angle between the radius vector OP and the tangent at P to the curve $r = f(\theta)$.

To show that for any point (r, θ) of the curve $r = f(\theta)$, the angle ϕ between the radius vector and tangent is given by

$$\tan \phi = r \frac{d\theta}{dr}.$$

Let $P(r, \theta)$ be any point on the given curve $r = f(\theta)$ or $f(r, \theta) = 0$.

Let us suppose $Q(r + \delta r, \theta + \delta \theta)$ be the point in the neighbourhood of P on the curve.

Join OP, OQ, PQ, then $OP = r, OQ = r + \delta r$

$\angle XOP = \theta,$ $\angle XOQ = \theta + \delta \theta$ and $\angle POQ = \delta \theta.$

Draw $PR \perp OQ$ and $\angle PQR = \alpha.$

Now, let the angle between the radius vector OP and the tangent PT is ϕ *i.e.*,

      $\angle OPT = \phi$

Also, we have

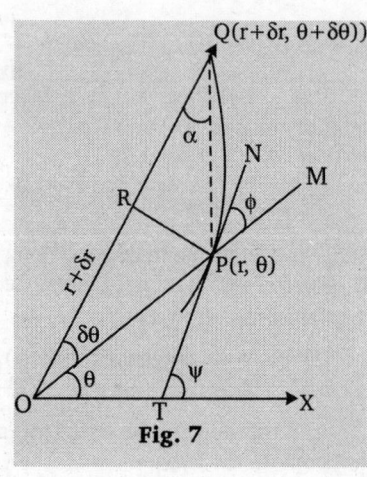

$$\frac{PR}{OP} = \sin \delta \theta \quad \Rightarrow \quad PR = r \sin \delta \theta$$

$$RQ = OQ - OR = (r + \delta r) - OP \cos \delta \theta$$

$$= r + \delta r - r \cos \delta \theta$$

$$= \delta r + r (1 - \cos \delta \theta) = \delta r + 2r \sin^2 \frac{\delta \theta}{2}.$$

$$\tan \alpha = \frac{PR}{QR} = \frac{r \sin \delta \theta}{\delta r + 2r \sin^2 \delta \theta / 2}$$

Dividing the numerator and denominator by $\delta \theta$.

Then, $\tan \alpha = \dfrac{r \cdot \dfrac{\sin \delta \theta}{\delta \theta}}{\dfrac{\delta r}{\delta \theta} + r \cdot \dfrac{\sin \delta \theta / 2}{\delta \theta / 2} \sin \dfrac{\delta \theta}{2}}$

Fig. 7

when $Q \to P$ along the curve $\alpha \to \phi$ ($\because PQ$ becomes the tangent PT and OQ coincides with OP).

$$\tan \phi = \lim_{Q \to P} \tan \alpha = \lim_{\delta \theta \to 0} \frac{r \cdot \dfrac{\sin \delta \theta}{\delta \theta}}{\dfrac{\delta r}{\delta \theta} + r \cdot \dfrac{\sin \delta \theta / 2}{\delta \theta / 2} \sin \dfrac{\delta \theta}{2}} = \frac{r \cdot 1}{dr / d\theta + r \cdot 1.0} = \frac{r}{dr / d\theta}$$

Hence, $\tan \phi = r \dfrac{d\theta}{dr}.$

- ϕ is the angle between the radius vector and tangent and taken to be positive when measured in the anticlockwise direction.
- Relation between θ, ϕ and ψ is $\psi = \theta + \phi$.

2.4.7 ANGLE OF INTERSECTION OF TWO CURVES

If the tangent to the two curves make angle ϕ_1 and ϕ_2 with the common radius vector to their point of intersection, then angle between the curves.

$$= \text{angle between tangents} = |\phi_1 - \phi_2|.$$

- The two curves intersect orthogonally if $\tan\phi_1 \tan\phi_2 = -1$.
- If $\dfrac{\tan\phi_1 - \tan\phi_2}{1 + \tan\phi_1 . \tan\phi_2}$ is positive, we shall get acute angle of intersection at P and if $\dfrac{\tan\phi_1 - \tan\phi_2}{1 + \tan\phi_1 . \tan\phi_2}$ is negative, we get the obtuse angle of intersection at P.

2.4.8 LENGTH OF SUBTANGENT AND SUBNORMAL

Let P be any point (r, θ) on a curve $f(r, \theta) = 0$. Let the tangent and normal at P meet the straight line through the pole O perpendicular to the radius vector OP in T and N respectively. Then OT and ON are called polar subtangent and polar subnormal at P.

Hence,

Polar subtangent $= r^2 \dfrac{d\theta}{dr}$

Polar subnormal $= \dfrac{dr}{d\theta}$

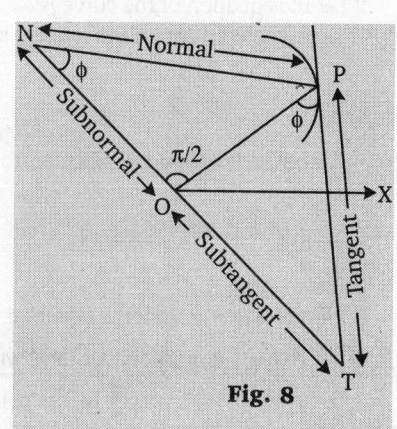

Fig. 8

2.4.9 LENGTH OF THE PERPENDICULAR FROM POLE TO THE TANGENT

Let p be the length of the perpendicular from the pole to the tangent at any point (r, θ) of a curve $r = f(\theta)$, then

(i) $p = r \sin\phi$

(ii) $\dfrac{1}{p^2} = \dfrac{1}{r^2} + \dfrac{1}{r^4} . \left(\dfrac{dr}{d\theta}\right)^2$

(iii) $\dfrac{1}{p^2} = u^2 + \left(\dfrac{du}{d\theta}\right)^2$ where $u = \dfrac{1}{r}$

PROOF.

(i) Let PT be the tangent at any point $P(r, \theta)$ on the curve $r = f(\theta)$ making an angle ψ with the initial line OX.

From the pole O, draw $OR \perp$ to the tangent PT.

$\therefore \qquad\qquad OR = p.$

Joint OP, also, $\angle OPT = \phi$.

Now from figure, we have

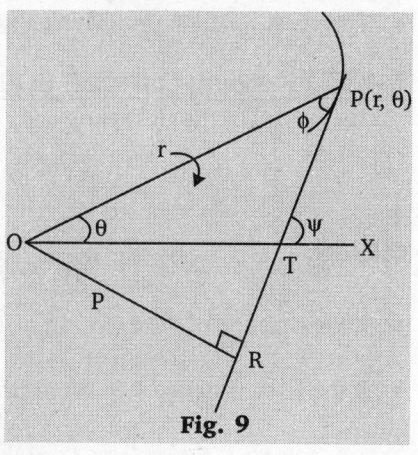

Fig. 9

$$\frac{OR}{OP} = \sin\phi \quad \Rightarrow \quad \frac{p}{r} = \sin\phi$$

$$\Rightarrow \qquad\qquad p = r \sin\phi$$

(ii) From (i), we have $\qquad \dfrac{1}{p^2} = \dfrac{1}{r^2 \sin^2\phi} = \dfrac{1}{r^2}\operatorname{cosec}^2\phi \qquad\qquad ...(1)$

Also, $\qquad\qquad\qquad \tan\phi = r\dfrac{d\theta}{dr}.$

$\therefore \qquad\qquad \operatorname{cosec}^2\phi = 1 + \cot^2\phi = 1 + \dfrac{1}{r^2}\left(\dfrac{dr}{d\theta}\right)^2$

Substitute it in (1), we get

$$\frac{1}{p^2} = \frac{1}{r^2}\left[1 + \frac{1}{r^2}\left(\frac{dr}{d\theta}\right)^2\right] \quad \Rightarrow \quad \frac{1}{p^2} = \frac{1}{r^2} + \frac{1}{r^4}\left(\frac{dr}{d\theta}\right)^2$$

(iii) Put $r = -$ in (ii), $\quad \dfrac{1}{p^2} = \dfrac{1}{r^2} + \dfrac{1}{r^4}\left(\dfrac{dr}{d\theta}\right)^2 \Rightarrow u^2 + u^4 \cdot \dfrac{1}{u^4}\left(\dfrac{du}{d\theta}\right)^2 \left(\because r = \dfrac{1}{u} \Rightarrow \dfrac{dr}{d\theta} = -\dfrac{1}{u^2} \cdot \dfrac{du}{d\theta}\right)$

$$\Rightarrow \qquad \frac{1}{p^2} = u^2 + \left(\frac{du}{d\theta}\right)^2$$

2.4.10 The Pedal Equation

Let r be the distance of any point on the curve from the origin (or pole), and p, is the length prependicular from the origin to the tangent at that point, then

The relation between p and r, where r is the distance of any point on the curve from the origin (or pole) and p is perpendicular from origin (or pole) to the tangent at that point is called the Pedal equation of the curve.

✒ REMARK

• The Pedal equation is also called per equation of the curve.

2.4.11 Pedal Equation of a Curve whose Cartesian Equation is Given

Let the equation of the curve is $\qquad f(x, y) = 0 \qquad\qquad\qquad\qquad\qquad$ ...(1)

Then, the equation of the tangent at any point (x, y) is

$$Y - y = \frac{dy}{dx}(X - x) = y_1(X - x) \text{ where } y_1 = \frac{dy}{dx}$$

$$\Rightarrow \qquad Xy_1 - Y + y - xy_1 = 0.$$

If p be the length prependicular from the origin to this tangent, then

$$p = \frac{y - xy_1}{\sqrt{1 + y_1^2}} \qquad\qquad\qquad\qquad\qquad ...(2)$$

Also, $\qquad\qquad\qquad r^2 = x^2 + y^2 \qquad\qquad\qquad\qquad\qquad ...(3)$

Eliminating x, y from the equation (1), (2) and (3), we get the required pedal equation of the curve (1).

2.4.12 Pedal Equation of a Curve whose Polar Equation is Given

Let $r = f(\theta) \qquad$ ...(i) be the polar curve. To find ϕ in terms of θ, eliminating θ and ϕ from both the above equations and $p = r \sin \phi$, we get the required pedal equation of curve (1).

✒ REMARK

• The pedal equation is sometimes more conveniently obtained by eliminating θ between (i) and $\dfrac{1}{p^2} = \dfrac{1}{r^2} + \dfrac{1}{r^4}\left(\dfrac{dr}{d\theta}\right)^2$.

2.4.13 Differential Coefficient Of Arc Length (Cartesian Form)

Let $y = f(x)$ be the given curve and s denote the length of the arc, then

$$\frac{ds}{dx} = \pm\sqrt{\left[1 + \left(\frac{dy}{dx}\right)^2\right]}$$

✒ REMARKS

• If the equation of the curve is $x = f(y)$, then $\quad \dfrac{ds}{dy} = \pm\sqrt{\left[1 + \left(\dfrac{dx}{dy}\right)^2\right]}$

• If the given equation is in parametric form *i.e.*, $x = f_1(t), y = f_2(t)$, then $\quad \dfrac{ds}{dt} = \pm\sqrt{\left[\left(\dfrac{dx}{dt}\right)^2 + \left(\dfrac{dy}{dt}\right)^2\right]}$

2.4.14 Differential Coefficient Of Arc Length (Polar Form)

To prove that $\dfrac{ds}{d\theta} = \sqrt{r^2 + \left(\dfrac{dr}{d\theta}\right)^2}$ *where* $r = f(\theta)$ *is the polar form of curve :*

Let $r = f(\theta)$ be the equation of the curve and s denote the length of arc AP. Obviously s is a function of θ. Let Q be the neighbouring point of P such that

$$AQ = s + \delta s \qquad\qquad \Rightarrow \qquad\qquad PQ = \delta s.$$

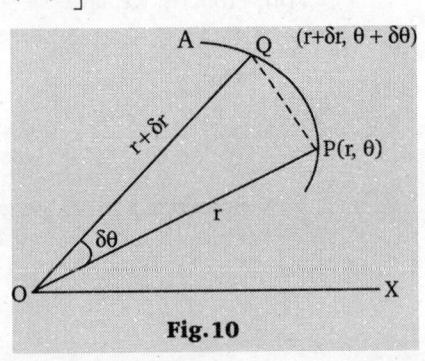

Fig. 10

As $Q \to P$, $\delta\theta \to \theta$ and $\delta r \to 0$

From ΔOPQ, we have $\quad$ (chord $PQ)^2 = OP^2 + OQ^2 - 2OP.OQ \cos(\angle QOP) = r^2 + (r + \delta r)^2 - 2r(r + \delta r) \cos \delta\theta$

$$= (\delta r)^2 + 2r\delta r(1 - \cos \delta\theta) + 2r^2(1 - \cos \delta\theta)$$

Dividing by $(\delta\theta)^2$, we get $\quad \left(\dfrac{\text{chord } PQ}{\delta\theta} \right)^2 = \left(\dfrac{\delta r}{\delta\theta} \right)^2 + r \left(\dfrac{\sin \frac{\delta\theta}{2}}{\frac{\delta\theta}{2}} \right)^2 . \delta r + r^2 \left(\dfrac{\sin \frac{\delta\theta}{2}}{\frac{\delta\theta}{2}} \right)^2$

$$\left(\dfrac{\text{chord } PQ}{\delta s} \right)^2 = \left(\dfrac{\delta x}{\delta\theta} \right)^2 + r \left(\dfrac{\sin \frac{\delta\theta}{2}}{\frac{\delta\theta}{2}} \right)^2 . \delta r + r^2 \left(\dfrac{\sin \frac{\delta\theta}{2}}{\frac{\delta\theta}{2}} \right)^2$$

Taking limit as $Q \to P$, we have

$$\left(\dfrac{ds}{d\theta} \right)^2 = \left(\dfrac{dr}{d\theta} \right)^2 + r.1.0 + r^2.1 \qquad \left[\because \lim_{Q \to P} \dfrac{\text{chord } PQ}{PQ(=\delta s)} = 1 \text{ and } \lim_{\delta\theta \to \theta} \dfrac{\delta r}{\delta\theta} = \dfrac{dr}{d\theta} \right]$$

$$\Rightarrow \qquad \left(\dfrac{ds}{d\theta} \right)^2 = r^2 + \left(\dfrac{dr}{d\theta} \right)^2 \Rightarrow \dfrac{ds}{d\theta} = \pm \sqrt{\left\{ r^2 + \left(\dfrac{dr}{d\theta} \right)^2 \right\}}$$

➡ **REMARKS**

- Here $+$ or $-$ sign is to be taken according as s increases or decreases as θ increases, we have

$$\dfrac{ds}{d\theta} = \pm \sqrt{\left\{ r^2 + \left(\dfrac{dr}{d\theta} \right)^2 \right\}}$$

- If $\theta = f(r)$ is the given equation of the curve, then

$$\dfrac{ds}{dr} = \pm \sqrt{\left\{ 1 + r^2 \left(\dfrac{d\theta}{dr} \right)^2 \right\}}$$

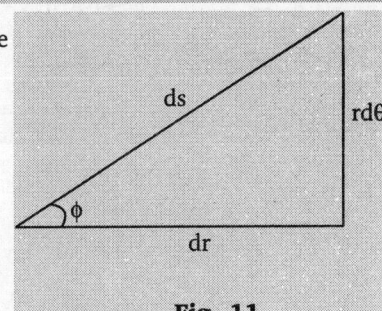

Fig. 11

- The result $\cos \phi = \dfrac{dr}{ds}$ and $\sin \phi = r \dfrac{d\theta}{ds}$ can be remember with the help of adjoining figure(11).

Solved Examples

EXAMPLE 1. *Find the equations on the tangent at the point t to the cycloid $x = a(t + \sin t)$, $y = a(1 - \cos t)$.*

SOLUTION. We have

$$x = a(t + \sin t) \Rightarrow \dfrac{dx}{dt} = a(1 + \cos t)$$

and $y = a(1 - \cos t) \Rightarrow \dfrac{dy}{dt} = a \sin t$

Therefore,

$$\dfrac{dy}{dx} = \dfrac{dy/dt}{dx/dt} = \dfrac{a \sin t}{a(1 + \cos t)} = \dfrac{2 \sin \frac{t}{2} \cdot \cos \frac{t}{2}}{2 \cos^2 t/2} = \tan \dfrac{t}{2}$$

Now, the equation of the tangent at 't' is

$$y - a(1 - \cos t) = \tan \dfrac{t}{2} [x - a(t + \sin t)]$$

$$\Rightarrow \quad y - 2a \sin^2 t/2 = (x - at) \tan \dfrac{t}{2} - a \sin t . \tan t/2$$

$$\Rightarrow \quad y - 2a \sin^2 t/2 = (x - at) \tan t/2 - 2a \sin^2 t/2$$

$$\Rightarrow \quad y = (x - at) \tan t/2.$$

EXAMPLE 2. *Show that the parabolas $r = \dfrac{a}{(1 + \cos\theta)}$ and $r = \dfrac{b}{(1 - \cos\theta)}$ intersect orthogonally.*

SOLUTION. Here we have

$$r = \dfrac{a}{(1 + \cos\theta)} \qquad \text{...(1)}$$

and $\qquad r = \dfrac{b}{(1 - \cos\theta)} \qquad \text{...(2)}$

Taking log of both sides of (1), we get

$$\log r = \log a - \log(1 + \cos\theta)$$

Differentiating with respect to θ, we get

$$\dfrac{1}{r} . \dfrac{dr}{d\theta} = \dfrac{-(-\sin\theta)}{(1 + \cos\theta)} = \dfrac{2 \sin \theta/2 \cos \theta/2}{2 \cos^2 \theta/2} = \tan \dfrac{\theta}{2}$$

$$\Rightarrow \qquad \cot \phi = \tan \dfrac{\theta}{2} = \cot \left(\dfrac{\pi}{2} - \dfrac{\theta}{2} \right)$$

$$\Rightarrow \qquad \phi_1 = \dfrac{\pi}{2} - \dfrac{\theta}{2}$$

Now, from (2), we get

$$\log r = \log b - \log(1 - \cos\theta)$$

Differentiating with respect to θ, we get

$$\dfrac{1}{r} . \dfrac{dr}{d\theta} = \dfrac{-\sin\theta}{1 - \cos\theta} = -\dfrac{2 \sin \theta/2 \cdot \cos \theta/2}{2 \sin^2 \theta/2} = -\cot \dfrac{\theta}{2}$$

$$\therefore \qquad \cot \phi = -\cot \dfrac{1}{2}\theta = \cot \left(\pi - \dfrac{1}{2}\theta \right)$$

$\Rightarrow \qquad \phi = \pi - \dfrac{1}{2}\theta \quad \Rightarrow \quad \phi_2 = \pi - \dfrac{1}{2}\theta$

Now, the angle of intersection $= \phi_1 \sim \phi_2$

$= \left(\pi - \dfrac{1}{2}\theta\right) - \left(\dfrac{1}{2}\pi - \dfrac{1}{2}\theta\right) = \dfrac{\pi}{2}$

Both curves intersect orthogonally.

EXAMPLE 3. *Show that the pedal equation of the ellipse*

$\dfrac{x^2}{a^2} + \dfrac{y^2}{b^2} = 1$ *is* $\dfrac{1}{p^2} = \dfrac{1}{a^2} + \dfrac{1}{b^2} - \dfrac{r^2}{a^2 b^2}.$

SOLUTION . Here, the equation of the curve is $\dfrac{x^2}{a^2} + \dfrac{y^2}{b^2} = 1$.

Let $x = a\cos t, y = b\sin t$.

$\therefore \qquad \dfrac{dx}{dt} = -a\sin t, \dfrac{dy}{dt} = b\cos t$

$\Rightarrow \qquad \dfrac{dy}{dx} = -\dfrac{b\cos t}{a\sin t}$

Therefore, the equation of the tangent at 't' is

$Y - b\sin t = -\dfrac{b\cos t}{a\sin t}(X - a\cos t)$

$\Rightarrow \qquad ab - b\cos t . X - a\sin t . Y = 0 \qquad \ldots(1)$

Since p denote the length prependicular from $(0, 0)$ to (1), therefore

$p = \dfrac{ab}{\sqrt{a^2 \sin^2 t + b^2 \cos^2 t}}$

$\dfrac{1}{p^2} = \dfrac{a^2 \sin^2 t + b^2 \cos^2 t}{a^2 b^2} \qquad \ldots(2)$

Now, $\quad r^2 = x^2 + y^2 = a^2 \cos^2 t + b^2 \sin^2 t$

$= a^2 + b^2 - a^2 \sin^2 t - b^2 \cos^2 t \ \ldots(3)$

From (3) $a^2 \sin^2 t + b^2 \cos^2 t = (a^2 + b^2) - r^2$.

Therefore, from (3), we get

$\dfrac{1}{p^2} = \dfrac{(a^2 + b^2) - r^2}{a^2 b^2} = \dfrac{1}{a^2} + \dfrac{1}{b^2} - \dfrac{r^2}{a^2 b^2}.$

EXAMPLE 4. *Find the pedal equation of* $r^n = a^n \sin n\theta$.

SOLUTION . Here, the given curve is

$r^n = a^n \sin n\theta \qquad \ldots(1)$

Taking logarithm of both the sides of (1), we get

$n \log r = n \log a + \log \sin n\theta. \qquad \ldots(2)$

Differentiating w.r.t. θ, we get

$\dfrac{n}{r} \cdot \dfrac{dr}{d\theta} = n\dfrac{\cos n\theta}{\sin n\theta} = n\cot n\theta$

$\Rightarrow \qquad \cot\phi = \dfrac{1}{r} \cdot \dfrac{dr}{d\theta} = \cot n\theta$

$\therefore \qquad \phi = n\theta$

Also, $\qquad p = r\sin\phi \Rightarrow p = r\sin n\theta \qquad \ldots(3)$

Now from (1) and (3), we have $\sin n\theta = \dfrac{p}{r}$

Putting the value in (1), we get $\quad pa^n = r^{n+1}.$

EXAMPLE 5. *For the cardiod* $r = a(1 - \cos\theta)$, *prove that*

(i) $\phi = \dfrac{1}{2}\theta$ (VTU–2004) **(ii)** $2ap^2 = r^3$

SOLUTION . Here the given curve is

$r = a(1 - \cos\theta) \qquad \ldots(1)$

$\Rightarrow \qquad \dfrac{dr}{d\theta} = a\sin\theta$

(i) Since, we have

$\tan\phi = r\dfrac{d\theta}{dr} = \dfrac{a(1 - \cos\theta)}{a\sin\theta}$

$= \dfrac{2a\sin^2\theta/2}{2a\sin\theta/2 \cdot \cos\theta/2} = \tan\dfrac{\theta}{2}$

$\Rightarrow \qquad \phi = \dfrac{\theta}{2}$

(ii) Since, we have $p = r\sin\phi = r\sin\theta/2$

$\Rightarrow \qquad r = 2a\sin^2\dfrac{\theta}{2} = 2a\dfrac{p^2}{r^2}$

$\therefore \ 2ap^2 = r^3$

EXAMPLE 6. *Find the pedal equation of the curve*

$x^{2/3} + y^{2/3} = a^{2/3}.$

SOLUTION . Here, the given curve is

$x^{2/3} + y^{2/3} = a^{2/3} \qquad \ldots(1)$

Let $\ x = a\cos^3 t, y = a\sin^3 t$

$\Rightarrow \qquad \dfrac{dy}{dx} = \dfrac{dy/dt}{dx/dt} = \dfrac{3a\sin^2 t\cos t}{-3a\cos^2 t\sin t} = -\dfrac{\sin t}{\cos t}.$

Hence, the equation of tangent of (1) is

$y - a\sin^3 t = -\dfrac{\sin t}{\cos t}(x - a\cos^3 t)$

$\Rightarrow x\sin t + y\cos t = a\sin t\cos t(\cos^2 t + \sin^2 t)$

$= a\sin t\cos t \qquad \ldots(2)$

$p = $ the length of the prependicular

from $(0, 0)$ to (2)

$= \dfrac{a\sin t\cos t}{\sqrt{\sin^2 t + \cos^2 t}} = a\sin t\cos t.$

Now, $\quad r^2 = x^2 + y^2 = a^2\cos^6 t + a^2\sin^6 t$

$= a^2[(\cos^2 t)^3 + (\sin^2 t)^3]$

$= a^2[(\cos^2 t + \sin^2 t)^3$

$- 3\cos^2 t\sin^2 t(\cos^2 t + \sin^2 t)]$

$= a^2[1 - 3(p^2/a^2).1] = a^2 - 3p^2.$

Exercise-2.8

1. Find the angle of intersection of the curve $r^2 = 16\sin 2\theta$ and $r^2 \sin 2\theta = 4$.

2. Show that in the curve $r = a\theta$, the polar subnormal is constant and in the curve $r\theta = a$, the polar subtangent is constant.

3. Show that the curves $r = a(1 + \cos\theta)$ and $r = b(1 - \cos\theta)$ intersect at right angles. (VTU–2011)

4. Show that the spiral $r^n = a^n \cos n\theta$ and $r^n = b^n \sin n\theta$ intersect orthogonally. (VTU–2010)

5. Find the angle ϕ for the curve $a\theta = (r^2 - a^2)^{1/2} - a\cos^{-1} a/r$.

6. Show that the curves $r = (1 + \sin\theta)$ and $r = a(1 - \sin\theta)$ cut orthogonally.

7. Show that the curves $r = 2\sin\theta$ and $r = 2\cos\theta$ intersect at right angles.

8. Find the angle of intersection between the pair of curves $r = 6\cos\theta$ and $r = 2(1 + \cos\theta)$.

9. Show that the pedal equation of the

   (i) conic $\dfrac{l}{r} = 1 + e\cos\theta$ is $\dfrac{1}{p^2} = \dfrac{1}{l^2}\left(\dfrac{2l}{r} - 1 + e^2\right)$

   (ii) curve $r = a\theta$ is $p^2 = \dfrac{r^4}{r^2 + a^2}$

   (iii) cosine spiral $r^n = a^n \cos n\theta$ is $pa^n = r^{n+1}$. (VTU–2009)

   (iv) cardiod $r = a(1 + \cos\theta)$ is $r^3 = 2ap^2$.

   (v) spiral $r = a\sec hn\theta$ is $\dfrac{1}{p^2} = \dfrac{A}{r^2} + B$.

   (vi) hyperbola $r^2\cos 2\theta = a^2$ is $pr = a^2$.

   (vii) lemniscate $r^2 = a^2\cos 2\theta$ is $r^3 = a^2 p$.

10. Show that the normal at any point (r, θ) to the curve $r^n = a^n \cos n\theta$ makes an angle $(n + 1)\theta$ with the initial line.

11. Show that in the equiangular spiral $r = ae^{\theta\cot\alpha}$, the tangent is inclined at a constant angle α to the radius vector.

12. For the curve $r = ae^{\theta\cot\alpha}$, prove that $\dfrac{s}{r} = $ constant, s being measured from the pole.

13. Show that

   (i) $\dfrac{ds}{d\theta} = \dfrac{r^2}{p}$ (ii) $\dfrac{ds}{dr} = \dfrac{r}{\sqrt{r^2 - p^2}}$

14. For the ellipse $x = a\cos t, y = b\sin t$, prove that $\dfrac{ds}{dt} = a(1 - e^2\cos^2 t)^{1/2}$.

15. For the curve $r^n = a^n \cos n\theta$, show that $a^{2n}\dfrac{d^2 r}{ds^2} + nr^{2n-1} = 0$.

16. For the cycloid $x = a(1 - \cos t), y = a(t + \sin t)$, show that

   (i) $\dfrac{ds}{dt} = 2a\cos\dfrac{t}{2}$ (ii) $\dfrac{ds}{dx} = \text{cosec}\dfrac{t}{2}$ (iii) $\dfrac{ds}{dy} = \sec\dfrac{t}{2}$

17. Show that for the curve $r^m = a^m \cos m\theta, \dfrac{ds}{d\theta} = \dfrac{a^m}{r^{m-1}}$.

18. Show that the pedal equation of the parabola $y^2 = 4a(x + a)$ is $p^2 = ar$.

19. Prove that for the ellipse $\dfrac{x^2}{a^2} + \dfrac{y^2}{b^2} = 1, f = \dfrac{a^2 b^2}{p^3}$, p being the perpendicular from centre upon the tangent (x, y).

Answers

1. $\dfrac{2\pi}{3}$ 5. $\cos^{-1}\dfrac{a}{r}$ 8. $\dfrac{\pi}{6}$

2.5 CURVATURE

In figure (12), curve PQ bends more sharply than the curve AB. Then measure of the sharpness of the bending of a curve at a particular point is called curvature of the curve at the point. In this chapter, we shall find mathematical expressions for the curvature of a curve at a given point.

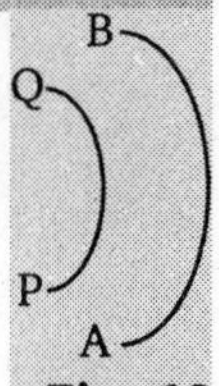

Let P, Q be two neighbouring points on a curve AB.

Also, let $AP = s$, arc $AQ = s + \delta s$ and arc $PQ = \delta s$.

Fig. 12

Let the tangent to the curve at points P and Q makes angle ψ and $\psi + \delta\psi$ respectively with a fixed line say X-axis, then

(i) The angle $\delta\psi$ through which the tangent turns as its points of contact travels along the arc PQ is called the total bending or total curvature of arc PQ.

(ii) The ratio $\dfrac{\delta\psi}{\delta s}$ is called the mean or average curvature of arc PQ.

(iii) The limiting value of the mean curvature when Q tends to P is called the curvature of the curve at the point P. Therefore, the curvature K at point P is

$$\lim_{Q \to P}\dfrac{\delta\psi}{\delta s} = \lim_{\delta s \to 0}\dfrac{\delta\psi}{\delta s} = \dfrac{d\psi}{ds}$$

(iv) The reciprocal of the curvature of the given curve at P. (provided this curvature is not equal to zero), is called the radius of curvature of the curve at P. This is denoted by ρ.

$$\rho = \dfrac{1}{K} = \dfrac{ds}{d\psi}$$

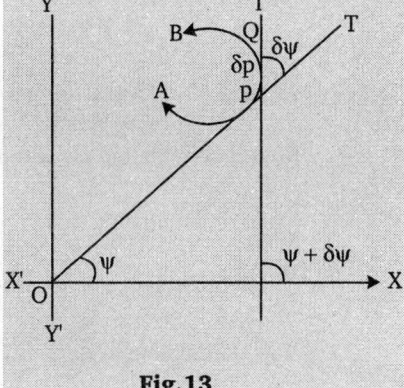

Fig.13

2.5.1 Formula for Radius of Curvature (Cartesian form)

Let $y = f(x)$ be the equation of curve. Then the slope of the tangent at any point $= \tan\psi = \dfrac{dy}{dx}$

Differentiating both sides, w.r.t. s, we get

$$\sec^2\psi\dfrac{d\psi}{ds} = \dfrac{d}{ds}\left(\dfrac{dy}{dx}\right) \Rightarrow \sec^2\psi.\dfrac{1}{\rho} = \dfrac{d}{dx}\left(\dfrac{dy}{dx}\right)\dfrac{dx}{ds}$$

$$\Rightarrow \qquad \sec^2\psi.\frac{1}{\rho} = \frac{d^2y}{dx^2}.\cos\psi \qquad\qquad \left(\because \frac{dx}{ds} = \cos\psi\right)$$

$$\text{Therefore} \qquad \rho = \frac{\sec^2\psi}{\cos\psi\dfrac{d^2y}{dx^2}} = \frac{\sec^3\psi}{\dfrac{d^2y}{dx^2}} = \frac{(1+\tan^2\psi)^{3/2}}{\dfrac{d^2y}{dx^2}} \Rightarrow \rho = \frac{\left[1+\left(\dfrac{dy}{dx}\right)^2\right]^{3/2}}{\dfrac{d^2y}{dx^2}}$$

REMARKS

- The positive root is taken in numerator of above formula, therefore, radius of curvature ρ, will be positive when $\dfrac{d^2y}{dx^2}$ is positive (*i.e.*, when the curve is concave upward) and negative when $\dfrac{d^2y}{dx^2}$ is negative (*i.e.*, when the curve is concave downward).

- At a point of inflexion, the curvature of a curve is not defined. $\qquad\left(\because \text{at the point of inflexion,} \dfrac{d^2y}{dx^2}=0\right)$

- When the equation of the curve is given in the form $x = f(y)$ then by interchanging x and y (It is justify because curvature is a length, and its value is independent of the choice of axis), we get $\rho = \dfrac{\left[1+(dx/dy)^2\right]^{3/2}}{d^2x/dy^2}$

- When the equation of curve is given in paraetric form *i.e.*, $x = f(t)$ and $y = g(t)$, then radius of curvature is given by $\rho = \dfrac{(x'^2 + y'^2)^{3/2}}{x'y'' - y'x''}$, where dash (') denote the derivative w.r.t., 't'.

$$\frac{1}{\rho^2} = \left(\frac{d^2x}{ds^2}\right)^2 + \left(\frac{d^2y}{ds^2}\right)^2$$

2.5.2 RADIUS OF CURVATURE AT THE ORIGIN

Let the curve $y = f(x)$ passes through the origin. Then, we may use the following methdos, to find the radius of curvature.

(i) Method of direct substitution. Since $y = f(x)$ be given. Calculate the values of $\dfrac{dy}{dx}$ and $\dfrac{d^2y}{dx^2}$ at origin and then use the following formula $\quad \rho = \dfrac{\left[1+\left(\dfrac{dy}{dx}\right)^2\right]^{3/2}}{d^2y/dx^2}$

(ii) Method of Expansion. Let $y = f(x)$ be the equation of curve. Since, it passes through the origin, therefore $f(0) = 0$.
Therefore, by Maclaurin's series expansion, we have

$$y = f(0) + xf'(0) + \frac{x^2}{2!}f''(0) + \frac{x^3}{3!}f'''(0) + ... \Rightarrow y = xf'(0) + \frac{x^2}{2!}f''(0) + \frac{x^3}{3!}f'''(0) + ... \qquad [\because f(0) = 0]$$

$$\Rightarrow \qquad y = p_1x + \frac{1}{2!}p_2x^2 + \frac{1}{3!}p_3x^3 + ... \quad ...(1) \text{ where } \quad p_1 = f'(0) = y_1(0), p_2 = f''(0) = y_2(0), \text{ etc.}$$

Now, differentiating (1) with respect to x, we get $\quad y_1 = p_1 + \dfrac{2p_2x}{2!} + \dfrac{3p_3x^2}{3!} + ...$

Again differentiating w.r.t. x. we get $\quad y_2 = \dfrac{2p_2}{2!} + \dfrac{6p_3x}{3!} + ...$

At the origin (*i.e.*, $x = 0$), we have $\quad y_1 = p_1$ and $\quad y_2 = \dfrac{2p_2}{2!} = p_2$

Now putting these values of y_1 and y_2 in the formula $\rho = \dfrac{(1+y_1^2)^{3/2}}{y_2}$, We have $\rho = \dfrac{(1+p_1^2)^{3/2}}{p_2}$

REMARKS

- We can find the values of p and q in the following manner:

Put the value of $y = p_1x + \dfrac{p_2x^2}{2!} + \dfrac{p_3x^3}{3!} + ...$ in the given equation of the curve and equating the coefficients of the powers of x.

(iii) Newton's Method. If a curve passes through the origin, and axis of x is the tangent at the origin, then radius of curvature ρ at origin

$$= \lim_{\substack{x\to 0 \\ y\to 0}} \frac{x^2}{2y}$$

Since the axis of x is the tangent at the origin, therefore, we have $y_1(0) = \left(\dfrac{dy}{dx}\right)_{(0,0)} = 0$

Here, we observed that $\dfrac{x^2}{2y}$ is of the indeterminate form $\left(\dfrac{0}{0}\right)$ as $x \to 0, y \to 0$.

Using L' Hospital rule, we have $\displaystyle\lim_{\substack{x\to 0\\ y\to 0}} \dfrac{x^2}{2y} = \lim_{\substack{x\to 0\\ y\to 0}} \dfrac{2x}{2y_1} = \lim_{\substack{x\to 0\\ y\to 0}} \dfrac{x}{y_1} = \lim_{\substack{x\to 0\\ y\to 0}} \dfrac{1}{y_2} = \dfrac{1}{y_2(0)}$ ...(1)

Now, $\qquad\qquad \rho$ at origin $= \dfrac{[1+y_1^2(0)]^{3/2}}{y_2(0)} = \dfrac{(1+0)^{3/2}}{y_2(0)} = \dfrac{1}{y_2(0)}$. ...(2)

From (1) and (2), we have $\quad \rho_{(\text{at origin})} = \displaystyle\lim_{\substack{x\to 0\\ y\to 0}} \dfrac{x^2}{2y}$

☛ REMARK

- If a curve passes through the origin and axis of y is the tangent, then radius of curvature at the origin is given by $\displaystyle\lim_{\substack{x\to 0\\ y\to 0}} \dfrac{y^2}{2x}$.

📂 Solved Examples

EXAMPLE 1. *In the cycloid $x = a(t + \sin t)$, $y = a(1- \cos t)$,*

prove that $\rho = 4a\cos\dfrac{t}{2}$. (SRM-2010)

SOLUTION . We have

$$x = a(t+\sin t) \Rightarrow \frac{dx}{dt} = a(1+\cos t)$$

and $y = a(1-\cos t) \Rightarrow \dfrac{dy}{dt} = a\sin t$

$$\Rightarrow \frac{dy}{dx} = \frac{dy/dt}{dx/dt} = \frac{a\sin t}{a(1+\cos t)} = \frac{2\sin t/2 \cos t/2}{2\cos^2 t/2} = \tan\frac{t}{2}$$

Also $\dfrac{d^2y}{dx^2} = \dfrac{d}{dx}\left(\dfrac{dy}{dx}\right) = \dfrac{d}{dx}\left(\tan\dfrac{t}{2}\right) = \dfrac{1}{2}\sec^2\dfrac{t}{2}\cdot\dfrac{dt}{dx}$

$$= \frac{1}{2}\sec^2\frac{t}{2}\cdot\frac{1}{a(1+\cos t)} = \frac{1}{4a}\sec^4\frac{t}{2}$$

Now, putting the values of $\dfrac{dy}{dx}$ and $\dfrac{d^2y}{dx^2}$ in

$$\rho = \frac{\left[1+\left(\dfrac{dy}{dx}\right)^2\right]^{3/2}}{\dfrac{d^2y}{dx^2}}$$

We get $\rho = \dfrac{[1+\tan^2 t/2]^{3/2}}{\dfrac{1}{4a}\sec^4 t/2} = \dfrac{4a\sec^3 t/2}{\sec^4 t/2} = 4a\cos t/2$

EXAMPLE 2. *Find the curvature of the curve $x^3 + y^3 = 3axy$ at the point $(3a/2, 3a/2)$.*

(ANNA–2009, VTU–2008, KURUKSHATRA–2009, KERELA–2005)

SOLUTION . The equation of the curve is

$$x^3 + y^3 = 3axy \qquad\qquad ...(1)$$

Differentiating w.r.t. x, we get

$$3x^2 + 3y^2\frac{dy}{dx} = 3ay + 3ax\frac{dy}{dx}$$

$$\Rightarrow \quad x^2 + y^2\frac{dy}{dx} = ay + ax\frac{dy}{dx}$$

$$\Rightarrow \quad \frac{dy}{dx} = \frac{x^2 - ay}{ax - y^2} \qquad\qquad ...(2)$$

$$\Rightarrow \quad \left(\frac{dy}{dx}\right)_{at\left(\frac{3}{2}a, \frac{3}{2}a\right)} = -1$$

From (2), we have

$$2x + 2y\left(\frac{dy}{dx}\right)^2 + y^2\frac{d^2y}{dx^2} = a\frac{dy}{dx} + a\frac{dy}{dx} + ax\frac{d^2y}{dx^2}$$

$$\Rightarrow \quad (ax - y^2)\frac{d^2y}{dx^2} = 2x + 2y\left(\frac{dy}{dx}\right)^2 - 2a\frac{dy}{dx} \quad ...(3)$$

Putting $x = \dfrac{3a}{2}, y = \dfrac{3a}{2}$ and $\left(\dfrac{dy}{dx}\right)_{\left(\frac{3a}{2},\frac{3a}{2}\right)} = -1$,

we get $\left[\dfrac{d^2y}{dx^2}\right]_{\left(\frac{3a}{2},\frac{3a}{2}\right)} = -\dfrac{32}{3}\cdot\dfrac{1}{a}$

Hence, the radius of curvature ρ at $\left(\dfrac{3a}{2}, \dfrac{3a}{2}\right)$, we get

$$\rho = \left[\frac{\left[1+\left(\dfrac{dy}{dx}\right)^2\right]^{3/2}}{\dfrac{d^2y}{dx^2}}\right]_{at\left(\frac{3a}{2},\frac{3a}{2}\right)} = \frac{(1+1)^{3/2}}{-\dfrac{32}{3}\cdot\dfrac{1}{a}} = -\frac{3a}{8\sqrt{2}}$$

Therefore, the curvature $\dfrac{1}{\rho} = +\dfrac{8\sqrt{2}}{3a}$.

(By ignoring the negative sign)

EXAMPLE 3. *If CP, CD be a pair of conjugate semi-diameters of an ellipse, show that the radius of curvature at P is $\dfrac{CD^3}{ab}$ where a and b are the lengths of the semi-axes of the ellipse.*

SOLUTION . Let CP and CD be a conjugate of semi-diameters of the ellipse $\dfrac{x^2}{a^2} + \dfrac{y^2}{b^2} = 1$

Let the coordinate of P are $x = a\cos t, y = b\sin t$. ...(1)

Also, the coordinate of D are

$$\left[a\cos\left(\frac{\pi}{2}+t\right), b\sin\left(\frac{\pi}{2}+t\right)\right] = (-a\sin t, b\cos t)$$

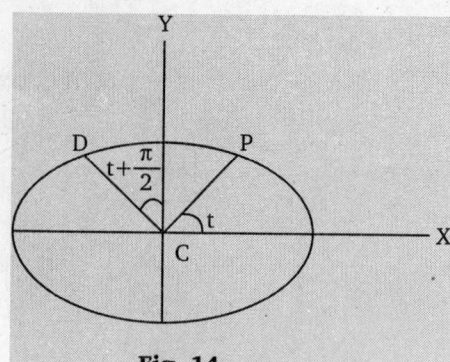

Fig. 14

From (1), we have

$$\frac{dx}{dt} = -a\sin t, \frac{dy}{dt} = b\cos t$$

$$\Rightarrow \frac{dy}{dx} = \frac{dy/dt}{dx/dt} = -\frac{b}{a}\cot t$$

$$\Rightarrow \frac{d^2y}{dx^2} = \frac{d}{dx}\left(\frac{dy}{dx}\right) = \frac{d}{dx}\left(-\frac{b}{a}\cot t\right) = \frac{d}{dt}\left(-\frac{b}{a}\cot t\right)\frac{dt}{dx}$$

$$= \left(\frac{b}{a}\csc^2 t\right)\left(-\frac{1}{a}\csc t\right) = -\frac{b}{a^2}\csc^3 t$$

Therefore, radius of curvature is given by

$$\rho = \frac{\left[1+\left(\frac{dy}{dx}\right)^2\right]^{3/2}}{-\frac{b}{a^2}\csc^3 t} = \frac{\left[1+\frac{b^2}{a^2}\cdot\frac{\cos^2 t}{\sin^2 t}\right]^{3/2}}{-\frac{b}{a^2}\csc^3 t}$$

$$= -\frac{(a^2\sin^2 t + b^2\cos^2 t)^{3/2}}{ab}$$

$$\Rightarrow \rho = \frac{(a^2\sin^2 t + b^2\cos^2 t)^{3/2}}{ab} \quad \ldots(2)$$

(By neglecting the negative sign)

From figure

$$CD = \sqrt{(-a\sin t - 0)^2 + (b\cos t - 0)^2}$$

$$= (a^2\sin^2 t + b^2\cos^2 t)^{1/2}$$

$$\therefore \frac{CD^3}{ab} = \frac{(a^2\sin^2 t + b^2\cos^2 t)^{3/2}}{ab} \quad \ldots(3)$$

Now from (2) and (3), we have $\rho = \frac{CD^3}{ab}$.

EXAMPLE 4. **For the curve** $y = \dfrac{ax}{a+x}$**, if** ρ **is the radius of curvature at any point** (x, y)**, show that**

$$\left(\frac{2\rho}{a}\right)^{2/3} = \left(\frac{y}{x}\right)^2 + \left(\frac{x}{y}\right)^2.$$ (VTU–2008)

SOLUTION. Let

$$y = \frac{ax}{a+x} \quad \ldots(1)$$

Therefore, $\dfrac{dy}{dx} = a\dfrac{a+x-x}{(a+x)^2} = a^2(a+x)^{-2}$

Now, again $\dfrac{d^2y}{dx^2} = \dfrac{d}{dx}\left(\dfrac{dy}{dx}\right) = -2a^2(a+x)^{-3} = \dfrac{-2a^2}{\left(\dfrac{ax}{y}\right)^3}$

$$\Rightarrow \frac{d^2y}{dx^2} = \frac{-2y^3}{ax^3}$$

$$\therefore 1+\left(\frac{dy}{dx}\right)^2 = 1+\frac{a^4}{(a+x)^4} = 1+\frac{a^4}{\left(\frac{ax}{y}\right)^4} = 1+\frac{y^4}{x^4}$$

$$\therefore \rho = \frac{\left[1+\left(\frac{dy}{dx}\right)^2\right]^{3/2}}{d^2y/dx^2} = \frac{[(x^4+y^4)/x^4]^{3/2}}{(-2y^3/ax^3)}$$

$$= -\frac{a(x^4+y^4)^{3/2}}{2x^6(y^3/x^3)} = -\frac{a}{2}\frac{(x^4+y^4)^{3/2}}{x^3 y^3}$$

Hence, $\left(\dfrac{2\rho}{a}\right)^{2/3} = \dfrac{x^4+y^4}{x^2 y^2} = \dfrac{x^2}{y^2} + \dfrac{y^2}{x^2}$

$$\Rightarrow \left(\frac{2\rho}{a}\right)^{2/3} = \left(\frac{x}{y}\right)^2 + \left(\frac{y}{x}\right)^2.$$

EXAMPLE 5. **Find the radius of curvature at origin for the curve** $x^3 + y^3 - 2x^2 + 6y = 0$**.** (BURDWAN 2003)

SOLUTION. The curve passes through origin. Equating to zero the lowest degree terms we get $y=0$ *i.e.*, x axis as tangent to the curve at origin.

$\therefore$ By Newtons method, ρ (at origin) $= \lim\limits_{\substack{x\to 0 \\ y\to 0}} \dfrac{x^2}{2y}$

Dividing by $2y$, the equation of the curve can be written as

$$x.\frac{x^2}{2y} + \frac{1}{2}y^2 - 2.\frac{x^2}{2y} + 3 = 0$$

Taking limit as $x\to 0, y\to 0$ and $\lim\limits_{\substack{x\to 0 \\ y\to 0}}\dfrac{x^2}{2y} = \rho$, we get

$$0.\rho + 0 - 2\rho + 3 = 0 \quad i.e., \quad \rho = 3/2.$$

EXAMPLE 6. **If** ρ_1 **and** ρ_2 **be the radii of curvature of the extremities of two conjugate diameters of an ellipse prove that**

$$(\rho_1^{2/3} + \rho_2^{2/3})(ab)^{2/3} = a^2 + b^2.$$

SOLUTION. Let the equation of an ellipse be

$$\frac{x^2}{a^2} + \frac{y^2}{b^2} = 1. \quad \ldots(1)$$

Let $P(a\cos\theta, b\sin\theta)$ and $Q(-a\sin\theta, b\cos\theta)$ be the extremities of two conjugate diameters of (1).
Differentiating both sides of (1) w.r.t x we get

$$\frac{2x}{a^2} + \frac{2y}{b^2}.\frac{dy}{dx} = 0 \quad \text{or} \quad \frac{dy}{dx} = -\frac{b^2 x}{a^2 y} \quad \ldots(2)$$

Again differentiating, we get

$$\frac{d^2y}{dx^2} = -\frac{b^2}{a^2}\left[\frac{y-x\dfrac{dy}{dx}}{y^2}\right]$$

$$= -\frac{b^2}{a^2}\left[\frac{y-x\left(-\dfrac{b^2 x}{a^2 y}\right)}{y^2}\right] = -\frac{b^2}{a^2}\left[\frac{\left(\dfrac{y^2}{b^2}+\dfrac{x^2}{a^2}\right)}{y^3}\right]b^2$$

$$= -\frac{b^4}{a^2 y^3} \quad \text{[Using (1)]}$$

We know that

$$\rho = \frac{\left[1+\left(\dfrac{dy}{dx}\right)^2\right]^{3/2}}{\dfrac{d^2y}{dx^2}} = \frac{\left[1+\left(-\dfrac{b^2x}{a^2y}\right)^2\right]^{3/2}}{-b^4/a^2y^3}$$

$$\rho = \frac{(a^4y^2+b^4x^2)^{3/2}}{-a^4b^4}$$

At $P(a\cos\theta, b\sin\theta)$, $\rho = \rho_1$

$$\therefore \quad \rho_1 = \frac{(a^4.b^2\sin^2\theta + b^4a^2\cos^2\theta)^{3/2}}{-a^4b^4}$$

or $$\rho_1 = \frac{(a^2\sin^2\theta + b^2\cos^2\theta)^{3/2}}{-ab}$$

or $$\rho_1(-ab) = (a^2\sin^2\theta + b^2\cos^2\theta)^{3/2}$$

or $$\rho_1^{2/3}(ab)^{2/3} = a^2\sin^2\theta + b^2\cos^2\theta \qquad \ldots(3)$$

At $Q(-a\sin\theta, b\cos\theta), \rho = \rho_2$

$$\therefore \quad \rho_2^{2/3}(ab)^{2/3} = a^2\cos^2\theta + b^2\sin^2\theta \qquad \ldots(4)$$

Adding (3) and (4), we get

$$(\rho_1^{2/3} + \rho_2^{2/3})(ab)^{2/3} = a^2 + b^2$$

EXAMPLE 7. *Prove that for the ellipse* $\dfrac{x^2}{a^2} + \dfrac{y^2}{b^2} = 1$, $\rho = \dfrac{a^2b^2}{p^3}$

p being the perpendicular from centre upon the tangent at (x, y).   (JNTU–2002)

SOLUTION . We have $\dfrac{x^2}{a^2} + \dfrac{y^2}{b^2} = 1 \Rightarrow \dfrac{dy}{dx} = -\dfrac{b^2x}{a^2y}$

and $\dfrac{d^2y}{dx^2} = -\dfrac{b^2}{a^2}\left[\dfrac{y - x\dfrac{dy}{dx}}{y^2}\right] = -\dfrac{b^4}{a^2y^3}$

Let $(a\cos\theta, b\sin\theta)$ be any point on the ellipse. The equation of the tangent at this point is

$$y - b\sin\theta = \frac{-b\cos\theta}{a\sin\theta}(x - a\cos\theta)$$

or $bx\cos\theta + ay\sin\theta - ab = 0$ ... (2)

We are given that

$p =$ Perpendicular from (0, 0) to the tangent (2)

or $$p = \frac{-ab}{\sqrt{b^2\cos^2\theta + a^2\sin^2\theta}} \qquad \ldots(3)$$

Now the radius of curvature ρ is

$$\rho = \frac{\left[1+\left(\dfrac{dy}{dx}\right)^2\right]^{\frac{3}{2}}}{\dfrac{d^2y}{dx^2}} = \frac{a^2y^3\left(1+\dfrac{b^4x^2}{a^4y^2}\right)^{\frac{3}{2}}}{-b^4} = \frac{(a^4y^2+b^4x^2)^{\frac{3}{2}}}{-a^4b^4}$$

The ρ at $(a\cos\theta, b\sin\theta)$ is given by

$$\rho = -\frac{(a^4b^2\sin^2\theta + b^4a^2\cos^2\theta)^{3/2}}{a^4b^4}$$

$$= -\frac{(a^2\sin^2\theta + b^2\cos^2\theta)^{3/2}}{ab}$$

$$= -\frac{(-ab/p)^3}{ab} \qquad \text{[Using (3)]}$$

$$\rho = a^2b^2/p^3.$$

EXAMPLE 8. *If* ρ_1 *and* ρ_2 *be the radii of curvature at the ends of a focal chord of the parabola* $y^2 = 4ax$, *then show that* $\rho_1^{-2/3} + \rho_2^{-2/3} = (2a)^{-2/3}$.

(KURUKSHETRA–2005, ROHTAK–2006)

SOLUTION . We have $y^2 = 4ax$ ...(1)

Parametric form of (1) is given by $x = at^2, y = 2at$

$\therefore \qquad x' = 2at, y' = 2a$

and $x'' = 2a, y'' = 0$

Therefore, radius of curvature ρ at $(at^2, 2at)$ is given by

$$\rho = \frac{(x'^2+y'^2)^{3/2}}{x'y''-x''y'} = \frac{(4a^2t^2+4a^2)^{3/2}}{0-4a^2} = 2a(1+t^2)^{3/2}$$

[Ignoring (–ve) sign]

If $P(t_1)$ and $Q(t_2)$ be the extremities of the focal chord of the parabola, then $t_1t_2 = -1 \Rightarrow t_2 = -\dfrac{1}{t_1}$

So, ρ_1 at $P(t_1) = 2a(1+t_1^2)^{3/2}$

ρ_2 at $Q(t_2) = 2a(1+t_2^2)^{3/2}$

$\therefore \quad \rho_1^{-2/3} + \rho_2^{-2/3} = (2a)^{-2/3}.[(1+t_1^2)^{-1}+(1+t_2^2)^{-1}]$

$$= (2a)^{-2/3}.\left[\frac{1}{1+t_1^2} + \frac{t_1^2}{1+t_1^2}\right]$$

$$= (2a)^{-2/3}.$$

EXAMPLE 9. *In the ellipse* $\dfrac{x^2}{a^2} + \dfrac{y^2}{b^2} = 1$, *show that the radius of curvature of an end of the major axis is equal to its semi-latusrectum of the ellipse.* (OSMANIA–2000)

SOLUTION . We have $\dfrac{x^2}{a^2} + \dfrac{y^2}{b^2} = 1$ ...(1)

$\Rightarrow \quad \dfrac{2x}{a^2} + \dfrac{2y}{b^2}\dfrac{dy}{dx} = 0 \Rightarrow \dfrac{dy}{dx} = -\dfrac{b^2}{a^2}\left(\dfrac{x}{y}\right)$

Therefore, $\dfrac{d^2y}{dx^2} = -\dfrac{b^2}{a^2}\left[\dfrac{y.1-x(dy/dx)}{y^2}\right]$

$$= -\frac{b^2}{a^2y^2}\left[y - x\left(-\frac{b^2x}{a^2y}\right)\right]$$

$$= -\frac{b^2}{a^2y^2}\left(\frac{a^2y^2+b^2x^2}{a^2y}\right)$$

$$= -\frac{b^2}{a^2y^2}\left(\frac{a^2b^2}{a^2y}\right) \qquad \text{[Using (1)]}$$

$$= -\frac{b^4}{a^2y^3}$$

Hence, ρ at $(x,y) = \dfrac{\left[1+\left(\dfrac{dy}{dx}\right)^2\right]^{\frac{3}{2}}}{d^2y/dx^2} = \dfrac{\left[1+\left(-b^2x/a^2y\right)^2\right]^{\frac{3}{2}}}{b^4/a^2y^3}$

[Ignoring (–ve) sign]

$$= \frac{\left(a^4y^2 + b^4.x^2\right)^{3/2}}{a^4b^4}$$

Now, the coordinate of one end of major axis are $(a, 0)$,

$\therefore \qquad \rho_{at(a,0)} = \dfrac{\left(a^4.0 + b^4.a^2\right)^{3/2}}{a^4b^4} = \dfrac{b^2}{a}$,

semi-latusrectum of the ellipse.

Exercise-2.9

1. Find the radius of curvature of the following curves:

(i) $x^{1/2} + y^{1/2} = a^{1/2}$ (ii) $a^2 y = x^3 - a^3$

(iii) $x^{2/3} + y^{2/3} = a^{2/3}$ (JNTU-2005)

(iv) $x^m + y^m = 1$

(v) $\sqrt{x} + \sqrt{y} = 1$ at $\left(\frac{1}{4}, \frac{1}{4}\right)$ (JNTU-2006)

(vi) $s = 4a \sin \psi$ at (s, ψ) (vii) $ay^2 = x^3$

(viii) $y = e^x$ at the point where it cuts the y-axis.

(ix) $x^{2/3} + y^{2/3} = a^{2/3}$ at $(a\cos^3 \theta, a\sin^3 \theta)$ (ANNA–2009)

(x) $y = 4\sin x - \sin 2x$ at $x = \dfrac{\pi}{2}$ (VTU–2009)

(xi) $y = x^3(x-a)$ at $(a, 0)$ (VTU–2010)

2. Find the radius of curvature at the origin of the following curves :

(i) $x^3 + y^3 = 3axy$ (ii) $y = x^3 + 5x^2 + 6x$

(iii) $5x^3 + 7y^3 + 4x^2 y + xy^2 + 2x^2 + 3xy + y^2 + 4x = 0$

(iv) $a(y^2 - x^2) = x^3$ (v) $y - x = x^2 + 2xy + y^2$

(vi) $2x^4 + 4x^3 + xy^2 + 6y^3 - 3x^2 - 2xy + y^2 - 4x = 0$

(vii) $\sqrt{x} + \sqrt{y} = a$ at $\left(\dfrac{a}{4}, \dfrac{a}{4}\right)$ (JNTU–2006)

3. Show that the curvature at a point of the curve $y = f(x)$ is given by $\dfrac{d^2 y}{dx^2} \cos^3 \psi$, where ψ is the inclination of the tangent at the point to the axis of x.

4. Show that for the curve $s = ae^{x/a}$, $a\rho = s(s^2 - a^2)^{1/2}$.

5. Show that if ρ be the radius of curvature at any point P on the parabola $y^2 = 4ax$ and S be its focus, then ρ varies as $(SP)^2$.

(KURUKSHETRA–2006)

6. Show that for any curve $\dfrac{1}{\rho} = \dfrac{d}{dx}\left(\dfrac{dy}{dx}\right)$.

Hints to Selected Problems

1. (i) Differentiating two times the given equation w.r.t. x, we get

$$\frac{d^2 y}{dx^2} = \frac{x^{1/2} + y^{1/2}}{2x^{3/2}} = \frac{a^{1/2}}{2x^{3/2}}.$$

Then put this value in the required formula.

2. (i) The given curve is passes through the origin. Therefore, equating the lowest degree term equal to zero, *i.e.*, $x = 0$ and $y = 0$ are the required trangent.

Then use $\rho = \lim\limits_{x \to 0}\left(\dfrac{y^2}{x^2}\right)$.

3. We have $\tan \psi = \left(\dfrac{dy}{dx}\right)$. Put this value in the formula of curvature.

4. Find $\dfrac{dy}{dx}$ and $\dfrac{d^2 y}{dx^2}$ by using $\dfrac{ds}{dx} = \sqrt{1 + \left(\dfrac{dy}{dx}\right)^2}$.

$$\therefore \quad \frac{d^2 y}{dx^2} = \frac{s^2}{a^2 \sqrt{s^2 - a^2}}.$$

Then use the formula of radius of curvature.

5. Let $P(at^2, 2at)$ be any point on $y^2 = 4ax$ and $S = (a, 0)$ be the coordinate of its focus. Then

$$SP = \sqrt{(at - a)^2 + (2at)^2} = \sqrt{(at^2 + a)^2} = t^2 + 1$$

Now find $\dfrac{d^2 y}{dx^2}$ by the equation of the parabola.

Then find the relation between ρ and SP, by using formula of radius of curvature.

6. Since we have $\dfrac{ds}{dy} = \sqrt{1 + \left(\dfrac{dx}{dy}\right)^2} \Rightarrow \dfrac{dy}{ds} = \dfrac{ds/dx}{\sqrt{1 + \left(\dfrac{dy}{dx}\right)^2}}$

$$\therefore \quad \frac{d}{dx}\left(\frac{dy}{ds}\right) = \frac{\left(\dfrac{d^2 y}{dx^2}\right)}{\left[1 + \left(\dfrac{dy}{dx}\right)^2\right]^{3/2}} = \frac{1}{\rho}.$$

Answers

1.(i) $\dfrac{2(x+y)^{3/2}}{\sqrt{y}}$ (ii) $\dfrac{(a^4 + 9x^4)^{3/2}}{6a^4 x}$ (iii) $3a^{1/3}x^{1/3}y^{1/3}$ (iv) $\dfrac{(x^{2m-2} + y^{2m-2})^{3/2}}{(1-m)x^{m-2}y^{m-2}}$ (v) $\dfrac{1}{\sqrt{2}}$ (vi) $4a \cos \psi$ (vii) —$(4a \quad 9x)^{3/2}x^{1/2}$ (viii) $\sqrt{8}$

(ix) $3a \sin \theta \cos \theta$ (x) $\dfrac{5\sqrt{5}}{4}$ (xi) $(1 + a^3)^{3b}/6a^2$ **2.**(i) $\dfrac{3a}{2}$ (ii) $\dfrac{37\sqrt{37}}{10}$ (iii) -2 (iv) $2a\sqrt{2}$ (v) $\dfrac{1}{2\sqrt{2}}$ (vi) 2 (vii) $\dfrac{a}{\sqrt{2}}$

2.5.4 Radius of Curvature for Pedal Equations

To prove: $\rho = r\dfrac{dr}{dp}$

Proof. Let the pedal equation of the curve be $p = f(r)$.
Form the adjoining figure, we have

$$\psi = \theta + \phi$$

$$\Rightarrow \quad \frac{d\psi}{ds} = \frac{d\theta}{ds} + \frac{d\phi}{ds} \Rightarrow \frac{1}{\rho} = \frac{d\theta}{ds} + \frac{d\phi}{ds} \quad \ldots(1)$$

Since, we know that

$$p = r \sin \phi$$

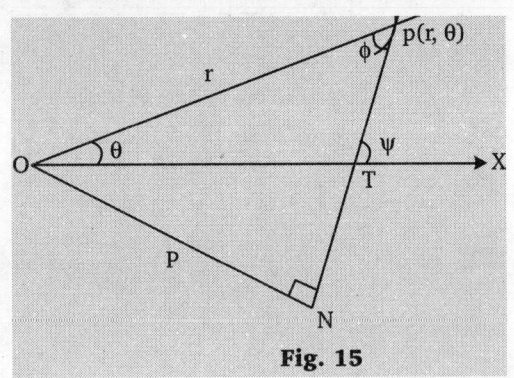

Fig. 15

$$\therefore \qquad \frac{dp}{dr} = \sin\phi + r\cos\phi \frac{d\phi}{dr}$$

$$= r.\frac{d\theta}{ds} + r\frac{dr}{ds}.\frac{d\phi}{dr} \left[\because \sin\phi = r.\frac{d\theta}{ds} \text{ and } \cos\phi = \frac{dr}{ds} \right]$$

$$= r\left[\frac{d\theta}{ds} + \frac{d\phi}{ds} \right] = r\frac{1}{\rho}$$

or $\qquad \dfrac{dp}{dr} = r\dfrac{1}{\rho}$ $\quad$ Hence, $\rho = \dfrac{r}{dp/dr} = r.\dfrac{dr}{dp}$ $\quad\Rightarrow\quad \rho = r\dfrac{dr}{dp}$.

2.5.4 RADIUS OF CURVATURE FOR TANGENTIAL POLAR EQUATIONS $p = f(\psi)$

To prove: $\rho = p + \dfrac{d^2p}{d\psi^2}$

Proof. Let p be the length of the perpendicular drawn from the origin on the tangent to curve at the point $P(x, y)$. Also, let ψ be the angle which the tangent makes with X-axis.

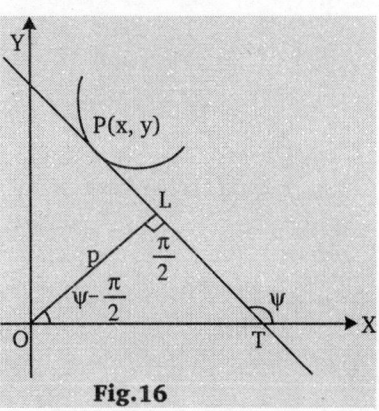

Fig.16

Here we observe that OL makes an angle $\psi - \dfrac{\pi}{2}$ with the positive direction of X-axis.

$\therefore$ Equation of the tangent PT is

$$p = X\cos\left(\psi - \frac{\pi}{2} \right) + Y\sin\left(\psi - \frac{\pi}{2} \right) \qquad \text{[Normal form, } x\cos\alpha + y\sin\alpha = p \text{]}$$

$$\Rightarrow \qquad p = X\sin\psi - Y\cos\psi$$

where X and Y are cartesian co-ordinates of any point on the tangent PT.

Since, $P(x, y)$ lies on PT, therefore

$$p = x\sin\psi - y\cos\psi \qquad\qquad\qquad ...(1)$$

$$\Rightarrow \qquad \frac{dp}{d\psi} = x\cos\psi + \sin\psi \frac{dx}{d\psi} + y\sin\psi - \cos\psi.\frac{dy}{d\psi}$$

$$= x\cos\psi + y\sin\psi + \sin\psi \frac{dx}{ds}.\frac{ds}{d\psi} - \cos\psi.\frac{dy}{ds}.\frac{ds}{d\psi}$$

$$= x\cos\psi + y\sin\psi + \sin\psi.\rho.\cos\psi - \cos\psi.\rho.\sin\psi \qquad\qquad \left(\frac{dx}{ds} = \cos\psi \text{ and } \frac{dy}{ds} = \sin\psi \right)$$

$$= x\cos\psi + y\sin\psi$$

Differentiating again w.r.t. ψ, we get

$$\frac{d^2p}{d\psi^2} = -x\sin\psi + \cos\psi.\frac{dx}{d\psi} + y\cos\psi + \sin\psi.\frac{dy}{d\psi}$$

$$= -x\sin\psi + y\cos\psi + \cos\psi.\frac{dx}{ds}.\frac{ds}{d\psi} + \sin\psi.\frac{dy}{ds}.\frac{ds}{d\psi}$$

$$= (-x\sin\psi + y\cos\psi) + \cos\psi.\cos\psi.\rho + \sin\psi.\sin\psi.\rho$$

$$= -p + \rho[\cos^2\psi + \sin^2\psi] \qquad\qquad\qquad \text{(Using (1))}$$

$$\Rightarrow \qquad \rho = p + \frac{d^2p}{d\psi^2}.$$

WORKING PROCEDURE

To transform polar equation to pedal equation, proceed as follows :

Step 1. Find ϕ, using formula $\tan\phi = r\dfrac{d\theta}{dr}$.

Step 2. Substittute the value of ϕ in $p = r\sin\phi$.

Step 3. Eliminate θ.

2.5.6 RADIUS OF CURVATURE IN POLAR FORM

To prove that $\rho = \dfrac{\left[r^2 + \left(\dfrac{dr}{d\theta} \right)^2 \right]^{3/2}}{r^2 + 2\left(\dfrac{dr}{d\theta} \right)^2 - r\dfrac{d^2r}{d\theta^2}}$

Proof. We know that $\qquad \dfrac{1}{p^2} = \dfrac{1}{r^2} + \dfrac{1}{r^4}\left(\dfrac{dr}{d\theta} \right)^2.$ $\qquad\qquad\qquad\qquad\qquad ...(1)$

Differentiating (1) w.r.t. r, we get

$$-\frac{2}{p^3}\frac{dp}{dr} = -\frac{2}{r^3} - \frac{4}{r^5}\left(\frac{dr}{d\theta}\right)^2 + \frac{1}{r^4}\left\{\frac{d}{dr}\left(\frac{dr}{d\theta}\right)^2\right\}$$

$$= -\frac{2}{r^3} - \frac{4}{r^5}\left(\frac{dr}{d\theta}\right)^2 + \frac{1}{r^4}\left[\frac{d}{d\theta}\left(\frac{dr}{d\theta}\right)^2\right]\cdot\frac{d\theta}{dr} = -\frac{2}{r^3} - \frac{4}{r^5}\left(\frac{dr}{d\theta}\right)^2 + \frac{2}{r^4}\frac{d^2r}{d\theta^4}$$

$$\frac{1}{p^3}\cdot\frac{dp}{dr} = \frac{1}{r^5}\left[r^2 + 2\left(\frac{dr}{d\theta}\right)^2 - r\frac{d^2r}{d\theta^2}\right]$$

Now

$$\rho = r\frac{dr}{dp} = \frac{r\cdot\dfrac{1}{p^3}}{\dfrac{1}{r^5}\left[r^2 + 2\left(\dfrac{dr}{d\theta}\right)^2 - r\dfrac{d^2r}{d\theta^2}\right]}$$

Form (1), we have

$$\frac{1}{p^3} = \left[\frac{1}{r^2} + \frac{1}{r^4}\left(\frac{dr}{d\theta}\right)^2\right]^{3/2} = \frac{1}{r^6}\left[r^2 + \left(\frac{dr}{d\theta}\right)^2\right]^{3/2}$$

Hence,

$$\rho = \frac{r^6\cdot\dfrac{1}{r^6}\left[r^2 + \left(\dfrac{dr}{d\theta}\right)^2\right]^{3/2}}{r^2 + 2\left(\dfrac{dr}{d\theta}\right)^2 - r\dfrac{d^2r}{d\theta^2}} \Rightarrow \rho = \frac{\left[r^2 + \left(\dfrac{dr}{d\theta}\right)^2\right]^{3/2}}{r^2 + 2\left(\dfrac{dr}{d\theta}\right)^2 - r\dfrac{d^2r}{d\theta^2}}.$$

 Solved Examples

EXAMPLE 1. *Find the radius of curvature for the curve* $r^n = a^n \cos n\theta$. (JNTU–2006, PTU–2010)

SOLUTION . We have $r^n = a^n \cos n\theta$
$\Rightarrow n \log r = n \log a + \log \cos n\theta$.
Now differentiating w.r.t. θ, we get

$$\frac{n}{r}\cdot\frac{dr}{d\theta} = 0 + \frac{1}{\cos n\theta}(-n\sin n\theta) = -n\tan n\theta \quad ...(1)$$

$$\Rightarrow \quad r = -r\tan n\theta$$

Again diiferentiating, we get

$$r_2 = -r.n.\sec^2 n\theta - r_1.\tan n\theta \quad ...(2)$$

$$= -rn\sec^2 n\theta + r\tan^2 n\theta.$$

Putting all these values in

$$\rho = \frac{[r^2 + r_1^2]^{3/2}}{r^2 + 2r_1^2 - rr_2}$$

$$= \frac{(r^2 + r^2\tan^2 n\theta)^{3/2}}{r^2 + 2r^2\tan^2 n\theta + r^2.n\sec^2 n\theta - r^2\tan^2 n\theta}$$

$$= \frac{r^3\sec^3 n\theta}{(n+1)r^2\sec^2 n\theta} = \frac{r\sec n\theta}{(n+1)} = \frac{r}{n+1}\cdot\frac{1}{\cos n\theta}$$

$$= \frac{r}{(n+1)\dfrac{r^n}{a^n}} = \frac{a^n}{(n+1)r^{n-1}}$$

EXAMPLE 2. *Prove that for any curve* $\dfrac{r}{\rho} = \sin\phi\left(1 + \dfrac{d\phi}{d\theta}\right)$, *where*

ρ *is the radius of curvature and* $\tan\phi = r\dfrac{d\theta}{dr}$.

SOLUTION . We know that the $\psi = \theta + \phi$. ...(1)
Differentiating (1) w.r.t. to s, we get

$$\frac{d\psi}{ds} = \frac{d\theta}{ds} + \frac{d\phi}{ds} = \frac{d\theta}{ds} + \frac{d\phi}{d\theta}\cdot\frac{d\theta}{ds} = \frac{d\theta}{ds}\left(1 + \frac{d\phi}{d\theta}\right).$$

$$\therefore \quad \frac{1}{\rho} = \frac{\sin\phi}{r}\left(1 + \frac{d\phi}{d\theta}\right) \left[\because \rho = \frac{ds}{d\psi} \text{ and } \sin\phi = r\frac{d\theta}{ds}\right]$$

$$\text{or} \quad \frac{r}{\rho} = \sin\phi\left(1 + \frac{d\phi}{d\theta}\right).$$

EXAMPLE 3. *Show that at any point on the equiangular spiral* $r = ae^{\theta\cot\alpha}$, $\rho = r\cosec\alpha$ *and that it subtends a right angle at the pole.*

SOLUTION . The given equation is $r = ae^{\theta\cot\alpha}$. ...(1)
Differentiating (1) w.r.t. θ, we have

$$\frac{dr}{d\theta} = ae^{\theta\cot\alpha}.\cot\alpha = r\cot\alpha.$$

$$\therefore \quad (1/r)\frac{dr}{d\theta} = \cot\alpha$$

$$\text{or} \quad \cot\phi = \cot\alpha \Rightarrow \phi = \alpha.$$

Now, $p = r\sin\phi$, thus the pedal equation of (1) is $p = r\sin\alpha$.

Therefore, $\dfrac{dp}{dr} = \sin\alpha$.

Now

$$\rho = r\frac{dr}{dp} = \frac{r}{\sin\alpha} = r\cosec\alpha.$$

Second part. Let $P(r, \theta)$ be any point on the given curve. PQ is the tangent and PR is the normal to the curve at P. Let R be center of curvatrure of the point P of the curve. Then $PR =$ the radius of curvature of the curve at $P = r\cosec\alpha$.

Intersect OP and OR, where O is the pole.
Let $\angle POR = \beta$. Then to show that $\beta = 90°$.
We have $\angle OPQ = \phi = \alpha$
$\angle OPR = 90° - \alpha$, (since PR is normal at P)
i.e., perpendicular to the tangent PQ.

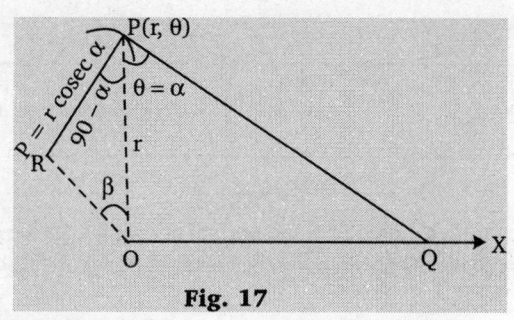

Fig. 17

Now in $\triangle OPR$, we have
$$\angle ORP = 180° - (90° - \alpha + \beta) = 90° + \alpha - \beta.$$

Therefore, applying the sine theorem for $\triangle OPR$, we get
$$\frac{OP}{\sin \angle ORP} = \frac{PR}{\sin \beta} \text{ or } \frac{r}{\sin(90 + \alpha - \beta)} = \frac{\rho}{\sin \beta}$$

or $\quad \frac{r}{\cos(\alpha - \beta)} = \frac{r \cosec \alpha}{\sin \beta} \quad (\because \rho = r \cosec \alpha)$

$\quad \sin \alpha \sin \beta = \cos(\alpha - \beta)$

or $\quad \sin \alpha \sin \beta = \cos \alpha \cos \beta + \sin \alpha \sin \beta$

or $\quad \cos \alpha \cos \beta = 0$ or $\cos \beta = 0$.

Hence, $\qquad \beta = 90°$.

Exercise-2.10

1. FInd the radius of curvature in polar form on each of the following curves :

  (i) $r = a(1 - \cos \theta)$ (VTU–2003)

  (ii) $r(1 + \cos \theta) = 2a$ (iii) $r^2 = a^2 \cos 2\theta$

  (iv) $r^m = a^m \sin m\theta$ (v) $r = ae^{\theta \cot \alpha}$

2. Find the radius of curvature at any point (p,r) on the following curves :

  (i) $p^2 = ar$ (ii) $r^2 = a^2 - b^2 + \dfrac{a^2 b^2}{p^2}$

  (iii) $2ap^2 = r^3$ (iv) $pa^2 = r^3$

3. Show that for the cardoid $r = a(1 + \cos \theta), \rho = -\sqrt{2ar}$.

4. Show that the radius of curvature of the cardoid $r = a(1 + \cos \theta)$ at the origin is 0.

5. Show that the radius of curvature at any point on the curve $r = a(1 \pm \cos \theta)$ varies as square root of the radius vector.

6. If ρ_1, ρ_2 be the radii of curvature at the extrimities of any chord of the cardoid $r = a(1 + \cos \theta)$, which passes through the pole, then $\rho_1^2 + \rho_2^2 = 16a^2/9$.

7. Show that the radius of curvature at the point (p, r) of the ellipse $\dfrac{1}{p^2} = \dfrac{1}{a^2} + \dfrac{1}{b^2} - \dfrac{r^2}{a^2 b^2}$ is $\dfrac{a^2 b^2}{p^3}$.   (VTU–2010)

8. Show that the radius of curvature for the hyperbola
$$p^2 = a^2 \cos^2 \psi + b^2 \sin^2 \psi \text{ is } \frac{a^2 b^2}{p^3}.$$

9. Show that the curvature of the curves $r = a\theta$ and $r\theta = a$ at their common point are in the ratio 3 : 1.

10. By Newton's method, show that the radius of curvature of the curve $r = a \sin n\theta$ at the origin is $\dfrac{na}{2}$.

11. Show that the radius of curvature at each point of the curve
$$x = a\left(\cos t + \log \tan \frac{t}{2} \right), y = a \sin t \text{ is inversely proportional to}$$
the length of the normal intercepted between the point on the curve and the x-axis.   (JNTU–2003)

Hints to Selected Problems

1. (i) Find $\dfrac{d^2 r}{d\theta^2} (= -a \cos \theta)$ to the given equation and use the

  formula $\rho = \dfrac{\left[r^2 + \left(\dfrac{dr}{d\theta} \right)^2 \right]^{3/2}}{r^2 + 2\left(\dfrac{dr}{d\theta} \right)^2 - r\dfrac{d^2 r}{d\theta^2}}.$

2. For the curve of the type $r = f(p)$, the radius of curvature is
$\rho = r\dfrac{dr}{dp}.$

3. $r = a(1 + \cos \theta) \Rightarrow \dfrac{dr}{d\theta} = -a \sin \theta \Rightarrow \dfrac{d^2 r}{d\theta^2} = -a \cos \theta.$

  Put these values in the formula for radius of curvature.

4. Proceed as question (3). Finally, put $r = 0$.

5. Proceed same as question (3).

6. Let $P(r, \theta)$ and $Q(r, \pi + \theta)$ be the extrimities of any chord of the cardoid $r = a(1 + \cos \theta)$

  Then for ρ_1, use $r = a(1 + \cos \theta)$

  $\Rightarrow \dfrac{dr}{d\theta} = -a \sin \theta$ and $\dfrac{d^2 r}{d\theta^2} = -a \cos \theta$

  and for ρ_2, use $r = a[1 + \cos(\pi + \theta)] = a(1 - \cos \theta)$

  $\Rightarrow \dfrac{dr}{d\theta} = a \sin \theta$ and $\dfrac{d^2 r}{d\theta^2} = a \cos \theta$

  Then find ρ_1 and ρ_2 and form a relation between ρ_1 and ρ_2.

7. Find $r\dfrac{dr}{dp}$ from the given equation

  $i.e., r\dfrac{dr}{dp} = \dfrac{a^2 b^2}{p^3}$. Then use the formula $\rho = r\dfrac{dr}{dp}$.

8. If the curve is $p = f(\psi)$, then the radius of curvature is
$$\rho = p + \frac{d^2 p}{d\psi^2}. \qquad \qquad \dots(1)$$

  Obtain the value of $\dfrac{d^2 p}{d\psi^2}$ form the given equation and substitute in (1).

9. Clearly, $(a, 1)$ and $(a, -1)$ are the common points.

  Now find the radius of curvature for both the above points.

10. $r = a \sin n\theta$ is the equation of the given curve.

  At $r = 0 \Rightarrow \theta = 0$. Then, use the formula given below
$$\rho_{\text{at origin}} = \lim_{\substack{x \to 0 \\ y \to 0}} \frac{x^2}{2y}.$$

1. (i) $\dfrac{2}{3}\sqrt{2ar}$ (ii) $2\sqrt{(r^3/a)}$ (iii) $\dfrac{a^2}{3r}$ (iv) $\dfrac{a^m}{(m+1)r^{m-1}}$ (v) $r\,\mathrm{cosec}\,\alpha$ **2.** (i) $\dfrac{2r^{3/2}}{\sqrt{a}}$ (ii) $\dfrac{a^2b^2}{p^3}$ (iii) $\dfrac{2}{3}\sqrt{2ar}$ (iv) $\dfrac{a^2}{3r}$

2.5.6 CENTRE OF CURVATURE

For any point P of a curve, the center of curvature is the point on the positive direction of the normal at P, at a distance ρ from it. Let PD be the normal curve at P and C be a point on it such that $PC = \rho$, then C is said to be the center of curvature at P.

EVOLUTE OF A CURVE

The locus of the center of curvature of the given curve is called the evolute of the curve.

CIRCLE OF CURVATURE

The circle with its center at the center of curvature c and radius equal to ρ is called the circle of curvature.

☞ REMARK
- The circle of curvature touches the curve at P and both the curve and the circle of curvature have the same curvature at this point.

2.5.7 CO-ORDINATES OF THE CENTRE OF CURVATURE

Let $y = f(x)$ be the given curve and $P(x, y)$ be any given point.

Let $C(\alpha, \beta)$ be the center of curvature corresponding to any point $P(x, y)$ on the given curve, then from above fig. (7), we have $PC = \rho$.

Suppose, the tangent TP makes an angle ψ with positive direction of x-axis. Draw PM and CN perpendicular to x-axis and draw perpendicular to CN. Then

$$\angle PCN = 90° - \angle CPR = 90° - (90° - \angle RPT) = \angle RPT = \angle PTX = \psi$$

$\therefore$ $\alpha = ON = OM - NM = OM - RP = x - CP\sin\psi = x - \rho\sin\psi$ ...(1)

Also, $\beta = NC = NR + RC = MP + RC = y + CP\cos\psi = y + \rho\cos\psi$ ...(2)

Since, we know that

$$y_1 = \tan\psi$$

$\Rightarrow$ $\sin\psi = \dfrac{y_1}{\sqrt{1+y_1^2}}$ and $\cos\psi = \dfrac{1}{\sqrt{1+y_1^2}}$.

Also, $\rho = \dfrac{(1+y_1^2)^{3/2}}{y_2}$

Putting all these values in (1) and (2), we get

$$\alpha = x - \dfrac{y_1(1+y_1^2)}{y_2} \text{ and } \beta = y + \dfrac{(1+y_1^2)}{y_2}.$$

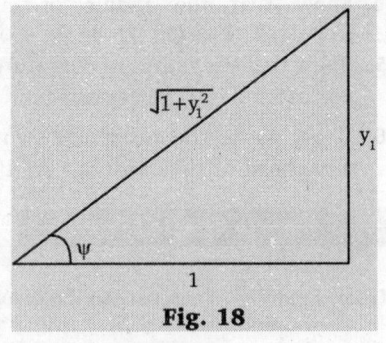

Fig. 18

☞ REMARKS
- From (1) and (2) we have $\alpha = x - \rho\sin\psi$ and $\beta = y + \rho\cos\psi$. Since x, y, ρ, ψ depends upon s, therefore the above equations may be treated as parametric equations of the evolute.
- The equation of the circle of curvature at the given point is $(x - \alpha)^2 + (y - \beta)^2 = \rho^2$.

2.5.8 CHORD OF CURVATURE

The length intercepted by the circle of curvature of the curve at P, on a straight line drawn through P in any given direction is called chord of curvature through P in that direction.

Let the chord of curvature PQ makes an angle α, with the normal PD, then its length PQ is given by

     $PQ = PD\cos\alpha$ ($\because \angle DQP$, being a semicircle is a right angle.)

     $= 2\rho\cos\alpha$, which is the chord of curvature perpendicular to radius vector.

☞ REMARK
- The chord of curvature through pole is given by $2\rho\sin\alpha$.

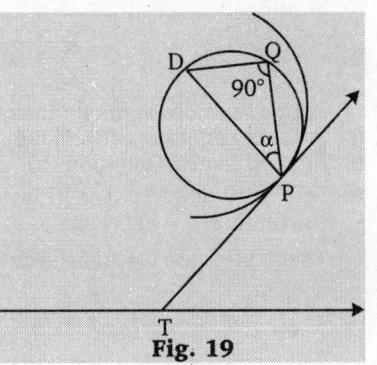

Fig. 19

2.5.9 LENGTH OF THE CHORD OF CURVATURE

(1) Cartesian form. Since, the tangernt at P makes an angle ψ with the x-axis therefore, the chord of curvature PA is parallel to x-axis, which makes an angle $90 - \psi$ with the normal PCD and chord of curvature PB parallel to y-axis makes angle ψ with the normal PCD.

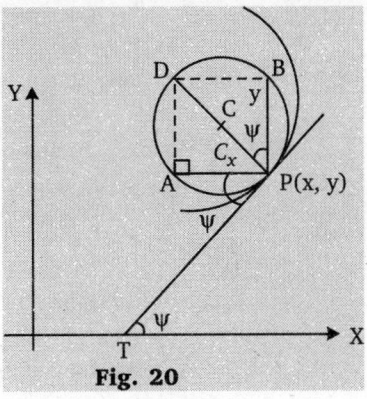

Fig. 20

$\therefore \qquad C_x =$ length of the chord of curvature PA, parallel to x-axis.

$= PD \cos(90 - \psi) = 2\rho \sin \psi$

$$= \frac{2(1 + y_1{}^2)^{3/2}}{y_2} \cdot \frac{y_1}{\sqrt{1 + y_1{}^2}} = \frac{2y_1(1 + y_1{}^2)}{y_2}$$

Similarly, $\qquad C_y = \dfrac{2(1 + y_1{}^2)^{3/2}}{y_2}$.

(2) Polar form. Let the chord of curvature PL makes an angle $90 - \phi$ with PCD, the normal of the curve at P, and PM, the chord of curvature perpendicular to the radius vector OP, makes an angle ϕ with the normal PCD.

$\therefore \qquad C_o =$ Length of the chord of curvature PL through origin (or pole)

$= PD(\cos 90 - \phi)$

$$= 2\rho \sin \phi = \frac{2(r^2 + r_1{}^2)^{3/2}}{r^2 + 2r_1{}^2 - rr_2} \cdot \frac{r}{\sqrt{r^2 + r_1{}^2}}$$

$$= 2\rho \sin \phi = \frac{2r(r^2 + r_1{}^2)}{r^2 + 2r_1{}^2 - rr_2}$$

and $\quad C_p =$ length of the chord of curvature PM perpendicular to radius vector.

$$= PD \cos \phi = 2\rho \cos \phi = \frac{2(r^2 + r_1{}^2)^{3/2}}{r^2 + 2r_1{}^2 - rr_2} \cdot \frac{r}{\sqrt{r^2 + r_1{}^2}} = \frac{2r(r^2 + r_1{}^2)}{r^2 + 2r_1{}^2 - rr_2}$$

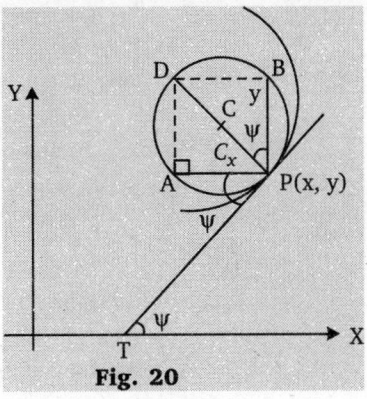

Fig. 21

(3) Pedal form. Let $p = f(r)$ be the given equation of the curve.

Let $\qquad C_o =$ length of the chord of curvature through pole along radius vector

$= PD \cos (90 - \phi) = 2\rho \sin \phi$ ...(1)

Now using $\qquad \rho = r \dfrac{dr}{dp}$ and $\sin \phi = \dfrac{p}{r}$ in (1), we get $C_o = 2r \dfrac{dr}{dp} \cdot \dfrac{p}{r} = 2p \cdot \dfrac{dr}{dp}$ ...(2)

Now $\qquad p = f(r) \Rightarrow \dfrac{dp}{dr} = f'(r)$ and $\sin \phi = \dfrac{p}{r} = \dfrac{f(r)}{r}$

$\therefore$ From (1), $\qquad C_o = 2\rho \sin \phi = 2.r \cdot \dfrac{dr}{dp} . \sin \phi = 2r \cdot \dfrac{1}{f'(r)} \cdot \dfrac{f(r)}{r} = \dfrac{2f(r)}{f'(r)}$

Also $\qquad C_p =$ length of the chord perpendiular to the radius vector

$$= DP \cos \phi = 2\rho \cos \phi = 2.r \cdot \frac{dr}{dp} \frac{\sqrt{r^2 - p^2}}{r} \qquad \left[\because \sin \phi = \frac{p}{r} \text{ and } \cos \phi = \frac{\sqrt{r^2 - p^2}}{r} \right]$$

$$= 2.\sqrt{r^2 - p^2} \cdot \frac{dr}{dp}.$$

 Solved Examples

EXAMPLE 1. *Find the chord of curvature through the pole of the cardioid $r = a(1 + \cos \theta)$.*

SOLUTION. We have $r = a(1 + \cos \theta)$

$\Rightarrow \qquad \dfrac{dr}{d\theta} = -a \sin \theta$

$\therefore \quad \tan \phi = r \dfrac{d\theta}{dr} = \dfrac{a(1 + \cos \theta)}{-a \sin \theta} = -\cot \dfrac{1}{2}\theta = \tan\left(\dfrac{\pi}{2} + \dfrac{\theta}{2}\right)$

Now $\quad p = r \sin \phi = r \sin\left(\dfrac{\pi}{2} + \dfrac{\theta}{2}\right) = r \cos \dfrac{\theta}{2}$

$\therefore \quad 2p^2 = r^2\left(2 \cos^2 \dfrac{\theta}{2}\right) = r^2(1 + \cos \theta) = r^2 - \dfrac{r}{a} = \dfrac{r^3}{a}$

$\Rightarrow \quad 2p^2 a = r^3$ is the pedal equation of the curve. On

differentiating w.r.t. r we get

$4ap \dfrac{dp}{dr} = 3r^2$

$\therefore \qquad \rho = r \dfrac{dr}{dp} = r.\dfrac{4ap}{3r^2} = \dfrac{4ap}{3r}$

Therefore, the chord of curvature through the pole

$= 2\rho \sin \phi = 2. \dfrac{4ap}{3r} . \dfrac{p}{r} \qquad [\because p = r \sin \phi]$

$= \dfrac{8ap^2}{3r^2} = \dfrac{8}{3r^2} . \dfrac{r^3}{2} = \dfrac{4r}{3} \qquad [\because 2ap^2 = r^3].$

EXAMPLE 2. *Show that the chord of curvature through the pole of the curve $r^n = a^n \cos n\theta$ is $\dfrac{2r}{n+1}$.*

SOLUTION . The given curve is
$$r^n = a^n \cos n\theta$$
$$\Rightarrow \quad n \log r = n \log a + \log \cos n\theta$$
Differentiating w.r.t. θ, we have
$$\frac{n}{r}\frac{dr}{d\theta} = -\frac{n}{\cos n\theta}.\sin n\theta$$
$$\Rightarrow \quad \cot\phi = -\tan n\theta = \cot\left(\frac{\pi}{2}+n\theta\right)$$
$$\therefore \qquad \phi = \frac{\pi}{2}+n\theta$$

Now $\quad p = r\sin\phi = r\sin\left(\frac{\pi}{2}+n\theta\right) = r\cos n\theta$

$\because$ Pedal equation of the curve is $p = \dfrac{r^{n+1}}{a^n}$.

$$\therefore \qquad \frac{dp}{dr} = \frac{(n+1)r^n}{a^n}$$

Also, $\quad \rho = r\dfrac{dr}{dp} = \dfrac{a^n}{(n+1)r^{n-1}}$

Therefore, the chord of curvature through pole is
$$= 2\rho\sin\phi = 2\rho\sin\left(\frac{\pi}{2}+n\theta\right) = 2\rho\cos n\theta$$
$$= 2\frac{a^n}{(n+1)r^{n-1}}.\frac{r^n}{a^n} = \frac{2r}{(n+1)}.$$

EXAMPLE 3. *Find the co-ordinate of the centre of curvature at any point of the parabola $y^2 = 4ax$. Hence, show that its evolute is $27ay^2 = 4(x-2a)^3$.* (VTU–2000)

SOLUTION . We have $\qquad y^2 = 4ax$
$$\Rightarrow 2yy_1 = 4a \ i.e., y_1 = \frac{2a}{y}$$
and $\quad y_2 = -\dfrac{2a}{y^2}.y_1 = -\dfrac{4a^2}{y^3}$

If $(\overline{x},\overline{y})$ be the centre of curvature, then
$$\overline{x} = x - \frac{y_1(1+y_1^2)}{y_2} = x - \frac{\dfrac{2a}{y}\left(1+\dfrac{4a^2}{y^2}\right)}{-4a^2/y^3} \qquad ...(1)$$
and $\overline{y} = y + \dfrac{1+y_1^2}{y_2} = y + \dfrac{1+4a^2/y^2}{-4a^2/y^3}$
$$= y - \frac{y(y^2+4a^2)}{4a^2} \qquad = \frac{-y^3}{4a^2} = -\frac{2x^{3/2}}{\sqrt{a}} \qquad ...(2)$$

Therefore, the required cenre of curvature is

$$\left\{(3x+2a), -2x\sqrt{\frac{x}{a}}\right\}$$

To find the required evolute, eliminate x from (1) and (2), we have
$$(\overline{y})^2 = \frac{4x^3}{a} = \frac{4}{a}\left(\frac{\overline{x}-2a}{3}\right)^3$$
$$\Rightarrow \quad 27a(\overline{y})^2 = 4(\overline{x}-2a)^3 \qquad\qquad ...(3)$$
Now, locus of $(\overline{x},\overline{y})$ is $27ay^2 = 4(x-2a)^3$ which is the required equation of evolute.

EXAMPLE 4. *Show that the evolute of the cycloid $x = a(\theta-\sin\theta)$, $y = a(1-\cos\theta)$ is another equal cycloid.* (MADRAS–2006)

SOLUTION. We have $x = a(\theta-\sin\theta)$ and $y = a(1-\cos\theta)$
$$\Rightarrow \quad y_1 = \frac{dy}{d\theta}.\frac{d\theta}{dx} = \frac{a\sin\theta}{a(1-\cos\theta)} = \cot\frac{\theta}{2}$$

Now $\qquad y_2 = \dfrac{d}{dx}(y_1) = \dfrac{d}{d\theta}\left(\cot\dfrac{\theta}{2}\right).\dfrac{d\theta}{dx}$
$$= -\mathrm{cosec}^2\frac{\theta}{2}.\frac{1}{2}.\frac{1}{a(1-\cos\theta)} = -\frac{1}{4a\sin^4\theta/2}$$

If $(\overline{x},\overline{y})$ be the center of curvature, then
$$\overline{x} = x - \frac{y_1(1+y_1^2)}{y_2}$$
$$= a(\theta-\sin\theta) + \cot\frac{\theta}{2}\left(4a\sin\frac{\theta}{2}\right)\left(1+\cot^2\frac{\theta}{2}\right)$$
$$= a(\theta-\sin\theta) + \frac{\cos\theta/2}{\sin\theta/2}.4a\sin^4\frac{\theta}{2}.\mathrm{cosec}^2\frac{\theta}{2}$$
$$= a(\theta-\sin\theta) + 4a\sin\frac{\theta}{2}.\cos\frac{\theta}{2}$$
$$= a(\theta-\sin\theta) + 2a\sin\theta$$
$$= a(\theta+\sin\theta)$$
and $\overline{y} = y + \dfrac{1+y_1^2}{y_2}$
$$= a(1-\cos\theta) + (1+\cot^2\frac{\theta}{2})(-4a\sin^4\frac{\theta}{2})$$
$$= a(1-\cos\theta) - 4a\sin^4\theta/2.\mathrm{cosec}^2\theta/2$$
$$= a(1-\cos\theta) - 4a\sin^2\frac{\theta}{2}$$
$$= a(1-\cos\theta) - 2a(1-\cos\theta) = -a(1-\cos\theta)$$

Hence, the required evolute is given by
$$x = a(\theta+\sin\theta), y = -a(1-\cos\theta) \qquad \text{which is}$$
another equal cycloid.

Exercise-2.11

1. In the curve $y = a\log\sec\left(\dfrac{x}{a}\right)$, show that the chord of curvature parallel to the axis of y is of constant length.

2. Prove that the centre of curvature (α,β) for the curve
$$x = 3t, y = t^2-6 \text{ is } \alpha = -\frac{4}{3}t^3, \beta = 3t^2-\frac{3}{2}.$$

3. If C_x and C_y be the chords of curvature parallel to the axis at any point of the curve $y = ae^{x/a}$, show that

$$\frac{1}{C_x^2} + \frac{1}{C_y^2} = \frac{1}{2aC_x}.$$

4. Show that the centre of curvature (α,β) at the point determined by t on the ellipse $x = a\cos t, y = b\sin t$, is given by
$$\alpha = \frac{a^2-b^2}{a}\cos^3 t, \beta = -\left(\frac{a^2-b^2}{b}\right)\sin^3 t.$$

5. Show that in any curve the chord of curvature prependicular to the radius vector is $2\rho\sqrt{(r^2 - p^2)}\,/\,r$.

6. Show that the chord of curvature through the pole of the equiangular spiral $r = ae^{m\theta}$ is $2r$.

7. Find the coordinates of the centre of curvatrue of ellipse $\dfrac{x^2}{a^2} + \dfrac{y^2}{b^2} = 1$ or $x = a\cos\theta, y = b\sin\theta$. Hence, show that the

equation of its evolute is $(ax)^{2/3} + (by)^{2/3} = (a^2 - b^2)^{2/3}$.

8. Find the chord of curvature through the pole of the curve $a\theta = \sqrt{r^2 - a^2} - a\cos^{-1}(a\,/\,r)$.

9. If C_r and C_θ be the chords of curvature of the curve $r = a(1 + \cos\theta)$ through the pole and perpendicular to the radius vector, then prove that $3(C_r^2 + C_\theta^2) = 8rC_r$.

Hints to Selected Problems

1. Given that $y = a\log\sec\dfrac{x}{a} \Rightarrow \dfrac{dy}{dx} = \tan\dfrac{x}{a}$ and $\dfrac{d^2y}{dx^2} = \dfrac{1}{a}\sec^2\dfrac{2}{a}$

   Put all these values in the radius of curvature, we get $\rho = a\sec\dfrac{x}{a}$.

   Then, chord of curvature parallel to y-axis is $= 2\rho\cos\psi$.

2. $\dfrac{dx}{dt} = 3, \dfrac{dy}{dt} = 2t \Rightarrow \dfrac{dy}{dx} = \dfrac{2t}{3}$ and $\dfrac{d^2y}{dx^2} = \dfrac{2}{3}\dfrac{dt}{dx} = \dfrac{2}{3}\cdot\dfrac{1}{3} = \dfrac{2}{9}$.

   Putting these values in the formula of centre of curvature.

3. Since $C_x = 2\rho\sin\psi$, $C_y = 2\rho\cos\psi$.

   Then, find the value of ρ using the given curve and by putting the value of ρ in the above expression, find a relation betwen C_x and C_y.

4. $x = a\cos t, y = b\sin t$

   $\Rightarrow \dfrac{dx}{dt} = -a\sin t, \dfrac{dy}{dt} = b\cos t \Rightarrow \dfrac{dy}{dx} = -\dfrac{b}{a}\cot t$

   $\therefore \dfrac{d^2y}{dx^2} = -\dfrac{b}{a}(-\text{cosec}^2 t)\dfrac{dt}{dx} = -\dfrac{b}{a^2}\text{cosec}^3 t$

   Then use the formulae for the centre of curvature.

5. The chord of curvature perpendicular to the radius vector is $2\rho\cos\phi$.

   Since $p = r\sin\phi \Rightarrow \dfrac{dp}{dr} = \sin\phi$ $\therefore$ $\rho = r\dfrac{dr}{dp} = \dfrac{r^2}{p}$

   Now $\cos\phi = \sqrt{1 - \sin^2\phi} = \sqrt{1 - \dfrac{p^2}{n}}$. Put this value in $2\rho\cos\phi$.

6. Here, $\dfrac{dr}{d\theta} = mr, \dfrac{d^2r}{d\theta^2} = m^2r$.

   Then we may get $\rho = r(1 + m^2)^{1/2}$.

   To find the chord of curvature through the pole is given by $2\rho\sin\phi$.

   where value of $\sin\phi$ can be obtained by using $r\dfrac{d\theta}{dr} = \tan\phi$.

7. $x = a\cos\theta, y = b\sin\theta \Rightarrow \dfrac{dx}{d\theta} = -a\sin\theta, \dfrac{dy}{d\theta} = b\cos\theta$.

   Therefore, $\dfrac{dy}{dx} = -\left(\dfrac{b}{a}\right)\cot\theta$ and $\dfrac{d^2y}{dx^2} = -\dfrac{b}{a^2}\text{cosec}^3\theta$

   Put these values in the formulae of centre of curvature (α, β). Also, to find the equation of evolute, find the locus of α and β.

Answers

7. $\left(\dfrac{a^2 - b^2}{a}\cos^3\theta, -\dfrac{a^2 - b^2}{b}\sin^3\theta\right)$ 8. $\dfrac{2(r^2 - a^2)}{r}$

2.6 ENVELOPES AND EVOLUTES

(i) **Family of curves with one parameter.** An equation in two variables x and y of the form $F(x, y, \lambda) = 0$ where λ is any constant, is known as a curve.

   If λ takes all real values, then the equation $F(x, y, \lambda) = 0$ is known as a family of curves with one parameter λ.

(ii) **Family of curves with two parameters.** An equation in two variables x and y of the form $F(x, y, \lambda, \mu) = 0$ is known as a family of curves with two parameter λ and μ if λ and μ take all real values.

For Example :

(1) The equation $x\cos\lambda + y\sin\lambda = p$ represents a family of straight line with one parameter λ.

(2) The equation $y = mx + a/m$ represents a family of straight lines which are the tangents to parabola $y^2 = 4ax$ with one parameter m.

(3) The equation $(x - \alpha)^2 + (y - \beta)^2 = a^2$ represents a family of circles with centred at (α, β) and radius a with two parameters α and β.

2.6.1 Some Standard Equations

(i) $\dfrac{x}{a} + \dfrac{y}{b} = 1$ (Equation of straight line)

(ii) System of Concentric Coaxial ellipses : $\dfrac{x^2}{a^2} + \dfrac{y^2}{b^2} = 1$

(iii) Family of parabola: $\left(\dfrac{x}{a}\right)^{1/2} + \left(\dfrac{y}{b}\right)^{1/2} = 1$

(iv) Equation of tractrix: $x = a\left(\cos t + \log\tan\dfrac{1}{2}t\right)$ $y = a\sin t$

(v) Pedal equation of equiangular spiral : $p = r\sin\alpha$

(vi) Equation of cardioid : $r = a(1 + \cos \theta)$

(vii) Other equations of straight line:

(a) $y = mx + \dfrac{a}{m}$ (b) $y = m^2 x + \dfrac{1}{m^2}$ (c) $\dfrac{x}{a} \cos \theta + \dfrac{y}{b} \sin \theta = 1$

2.6.2 ENVELOPE OF A FAMILY OF CURVES WITH ONE PARAMETER

Let $F(x, y, \lambda) = 0$ be a family of curves with parameter λ and let $F(x, y, \lambda) = 0$ and $F(x, y, \lambda + \delta\lambda) = 0$ be two members of a family of curves $F(x, y, \lambda) = 0$ corresponding to the parameter λ and $\lambda + \delta\lambda$, suppose P is a point of intersection of two members $F(x, y, \lambda) = 0$ and $F(x, y, \lambda + \delta\lambda) = 0$. As $\delta\lambda \to 0$, the point P tends to a definite point Q which depends upon λ. Thus the locus of such points Q gives an envelop of the family.

Definition. *The locus of the limiting positions of the points of intersection of any two members of the family of curves $F(x, y, \lambda) = 0$, when one of them tends to coincide with the other fixed point is called envelope.*

✎ REMARK

- The envelope of a family of curves is the locus of the points intersection of consecutive members of the family.

2.6.3 PROCEDURE FOR FINDING THE ENVELOPE

Let $F(x, y, \lambda) = 0$ be a family of the curve with one parameter λ.

Suppose $F(x, y, \lambda) = 0$ and $F(x, y, \lambda + \delta\lambda) = 0$ are two consecutive members of the family of curves corresponding to λ and $\lambda + \delta\lambda$. Thus the co-ordinates of the point of the intersection of these two members are obtained by the equations

$$F(x, y, \lambda) = 0 \qquad \qquad \text{...(1)}$$

and $\qquad F(x, y, \lambda) - F(x, y, \lambda + \delta\lambda) = 0 \qquad \qquad \text{...(2)}$

Divide the equation (2) by $\delta\lambda$, we get

$$\frac{F(x, y, \lambda) - F(x, y, \lambda + \delta\lambda)}{\delta\lambda} = 0 \quad \text{or} \quad \frac{F(x, y, \lambda + \delta\lambda) - F(x, y, \lambda)}{\delta\lambda} = 0$$

Taking limit as $\delta\lambda \to 0$, we get

$$\frac{\partial F(x, y, \lambda)}{\delta\lambda} = 0 \qquad \qquad \text{...(3)}$$

Now eliminating λ between $F(x, y, \lambda) = 0$ and $\dfrac{\partial F(x, y, \lambda)}{\partial \lambda} = 0$, we therefore, obtain the envelop of the family of curves $F(x, y, \lambda) = 0$.

WORKING PROCEDURE

To obtain an envelope of the family of curves $F(x, y, \lambda) = 0$, we use following steps :

Step 1. Differentiate partially $F(x, y, \lambda) = 0$ with respect to λ, to get $\dfrac{\partial F}{\partial \lambda} = 0$.

Step 2. Now eliminating λ between $F(x, y, \lambda) = 0$ and $\dfrac{\partial F}{\partial \lambda} = 0$, we therefore obtain envelope of the given family of curves.

📖 Solved Examples

EXAMPLE 1. ***Find the envelope of the family of straight lines*** $y = mx + \dfrac{a}{m}$, ***the parameter being m.***

SOLUTION. Here, the family of straight lines is

$$y = mx + \frac{a}{m}. \qquad \qquad \text{...(1)}$$

Differentiating (1) partially with respect to m, we get

$$0 = x - \frac{a}{m^2} \qquad \qquad \text{...(2)}$$

Eliminating m between (1) and (2), we get the required envelope.

From (2), we have

$$m^2 = \frac{a}{x}.$$

From (1), we have

$$ym = m^2 x + a \Rightarrow y^2 m^2 = (m^2 x + a)^2$$

$$\Rightarrow \quad y^2 \left(\frac{a}{x}\right) = \left(\frac{a}{x} \cdot x + a\right)^2. \qquad \left(\because m^2 = \left(\frac{a}{x}\right)\right)$$

$$\Rightarrow \qquad \frac{ay^2}{x} = (2a)^2 \qquad \Rightarrow \quad y^2 = 4ax.$$

This is the required envelope.

EXAMPLE 2. ***Find the envelope of the family of straight lines*** : $x \csc \theta - y \cot \theta = c$, ***the parameter being*** θ.

SOLUTION. Since the family of straight lines is

$$x \csc \theta - y \cot \theta = c. \qquad \qquad \text{...(1)}$$

Diff. (1) partially with respect to θ, we get

$$-x \csc \theta \cot \theta + y \csc^2 \theta = 0$$

or $\qquad x \cot \theta - y \csc \theta = 0. \qquad \qquad \text{...(2)}$

Eliminating θ between (1) and (2), we get

$$(x \csc \theta - y \cot \theta)^2 - (x \cot \theta - y \csc \theta)^2 = c^2$$

or $\quad x^2(\csc^2 \theta - \cot^2 \theta) - y^2(\csc^2 \theta - \cot^2 \theta)$

$$\qquad \qquad - 2xy \csc \theta \cot \theta + 2xy \csc \theta \cot \theta = c^2$$

or $\qquad x^2 - y^2 = c^2. \qquad (\because \csc^2 \theta - \cot^2 \theta = 1)$

This is the required envelope.

2.6.4 ENVELOPE OF THE FAMILY OF CURVES OF THE FORM $A\lambda^2 + B\lambda + C = 0$

The family of curve is $\qquad A\lambda^2 + B\lambda + C = 0$ ...(1)

where the parameter being λ.

Differentiating (1) partially w.r.t. to λ, we get $\qquad 2A\lambda + B = 0$ ...(2)

Eliminating λ between (1) and (2), we get

$$A\left[-\frac{B}{2A}\right]^2 + B\left[-\frac{B}{2A}\right] + C = 0 \quad \text{or} \quad \frac{B^2}{4A} - \frac{B^2}{2A} + C = 0 \quad \text{or} \quad B^2 - 4AC = 0.$$

This is the required equation of an envelope.

☞ REMARK

- If the equation of the family of curves is a quadratic equation in parameter, then its envelope is obtained by $D = 0$, where D is the discriminant of the quadratic.

2.6.5 ENVELOPE OF THE FAMILY OF CURVES WITH TWO PARAMETERS CONNECTED BY A RELATION

Let $F(x, y, \lambda, \mu) = 0$ be a family of curves with two parameters λ and μ. Let $f(\lambda, \mu) = 0$ be a relation between λ and μ.
To obtain the envelope, we proceed as follows :

WORKING PROCEDURE

Differentiating the equations $F(x, y, \lambda, \mu) = 0$ and $f(\lambda, \mu) = 0$ with respect to λ regarding x and y as constants and μ as a function of λ, we get two equations. Now elimainting λ, μ between the given equations and two obtained equations. We therefore obtain the envelope.

2.6.6 GEOMETRICAL INTERPRETATION OF THE ENVELOPE

Let the equation of the family of curves be $F(x, y, \lambda) = 0$; λ is the parameter.   ...(1)

Thus the envelope of (1) is obtained by eliminating λ between (1) and

$$\frac{\partial F}{\partial \lambda} = 0 \qquad ...(2)$$

Therefore, we can say that (2) is taken as the equation of the envelope of (1), if λ is a function of x and y but not constant.

Now from (1), we have $\quad \left(\dfrac{\partial F}{\partial x} + \dfrac{\partial F}{\partial \lambda} \cdot \dfrac{\partial \lambda}{\partial x}\right) + \left(\dfrac{\partial F}{\partial y} + \dfrac{\partial F}{\partial \lambda} \cdot \dfrac{\partial \lambda}{\partial y}\right)\dfrac{dy}{dx} = 0 \;\Rightarrow\; \dfrac{dy}{dx} = -\dfrac{\dfrac{\partial F}{\partial x} + \dfrac{\partial F}{\partial \lambda} \cdot \dfrac{\partial \lambda}{\partial x}}{\dfrac{\partial F}{\partial y} + \dfrac{\partial F}{\partial \lambda} \cdot \dfrac{\partial \lambda}{\partial y}}.$ ...(3)

This gives the slope of the tangent to the envelope of (1) at any point (x, y), where (x, y) is a common point to the member 'λ' of the family of curves and the envelope.

If $\dfrac{\partial F}{\partial x} \neq 0$ and $\dfrac{\partial F}{\partial y} \neq 0$ at (x, y), then the slope of the tangent to the member $F(x, y, \lambda) = 0$ is

$$\frac{dy}{dx} = -\frac{\partial F / \partial x}{\partial F / \partial y}. \qquad ...(4)$$

But $F(x, y, \lambda) = 0$ is also the equation of the envelope if λ is a function of x and y which is given by $\dfrac{\partial F}{\partial \lambda} = 0$

Since at every point of the envelope, $\dfrac{\partial F}{\partial \lambda} = 0$, then the slopes given by (3) and (4) are same.

Hence the curve of the family and its envelope have the same tangent lines at the common point. Consequently the envelope of a family of curves touch each member of the family.

☞ REMARK

- If $\dfrac{\partial F}{\partial x} = 0$ and $\dfrac{\partial F}{\partial y} = 0$ at any points on the curve, then the envelope may not touch curve at that points.

Solved Examples

EXAMPLE 1. *Find the envelope of the family of straight lines*

$$y = mx + a\sqrt{1 + m^2}, \text{ the parameter being } m.$$

(ANNA– 2006)

SOLUTION . The given equation of the family can be written as :

$$(y - mx)^2 = a^2(1 + m^2)$$

$$\text{or} \quad (x^2 - a^2)m^2 - 2mxy - a^2 + y^2 = 0 \qquad ...(1)$$

This equation is quadratic in m. Then the envelope of (1) is obtained by equating the discriminant of (1) to zero, we get $\qquad (-2xy)^2 - 4(x^2 - a^2)(y^2 - a^2) = 0$

$$(\because B^2 - 4AC = 0)$$

or $\qquad 4x^2y^2 - 4[x^2y^2 - x^2a^2 - a^2y^2 + a^4] = 0$

or $\qquad\qquad x^2a^2 + a^2y^2 = a^4$

or $\qquad\qquad x^2 + y^2 = a^2.$

This is the required equation of envelope.

EXAMPLE 2. *Find the envelope of the family of circles* $x^2 + y^2 - 2ax \cos \alpha - 2ay \sin \alpha = c^2$ *where* α *being the parameter. Also interpret the result.*

SOLUTION. The equation of the family of circles is
$$x^2 + y^2 - 2ax \cos \alpha - 2ay \sin \alpha = c^2. \qquad ...(1)$$
Differentiating (1) partially w.r.t. α, we get
$$2ax \sin \alpha - 2ay \cos \alpha = 0$$
or $\qquad x \sin \alpha - y \cos \alpha = 0. \qquad ...(2)$

Eliminating α between (1) and (2), we get
$$4a^2(x \sin \alpha - y \cos \alpha)^2 + 4a^2(x \cos \alpha + y \sin \alpha)^2$$
$$= 0 + (x^2 + y^2 - c^2)^2$$
or $\qquad 4a^2(x^2 + y^2) = (x^2 + y^2 - c^2)^2.$

This is the required envelope.

Intrepretation. The equation of envelope can be written as
$$(x^2 + y^2)^2 - (4a^2 + 2c^2)(x^2 + y^2) + c^4 = 0.$$

It is quadratic in $(x^2 + y^2)$ so solving, we get
$$x^2 + y^2 = \frac{2(2a^2 + c^2) \pm \sqrt{4(2a^2 + c^2)^2 - 4c^4}}{2}$$
$$= (2a^2 + c^2) \pm 2a\sqrt{c^2 + a^2}$$
$$= (\sqrt{a^2 + c^2} \pm a)^2$$

Thus, the equation of the envelope contains two circles with centred at $(0, 0)$ and radius $\sqrt{a^2 + c^2} \pm a$.

EXAMPLE 3. *Find the envelope of the circles drawn on the radii vectors of the parabola* $y^2 = 4ax$ *as diameter.* (MADRAS–2003)

SOLUTION. Let $(at^2, 2at)$ be any point on the parabola $y^2 = 4ax$. Then the equation of circles drawn on the line joining $(0, 0)$ and $(at^2, 2at)$ as diameter is
$$(x - 0)(x - at^2) + (y - 0)(y - 2at) = 0$$
or $\qquad x^2 + y^2 - axt^2 - 2aty = 0 \qquad ...(1)$
where t being the parameter.

Differentiating (1) partially with respect to t, we get
$$- 2axt - 2ay = 0$$
or $\qquad xt + y = 0. \qquad ...(2)$

Eliminating t between (1) and (2), we get
$$x^2 + y^2 - ax\left(-\frac{y}{x}\right)^2 - 2a\left(-\frac{y}{x}\right)y = 0$$
or $\qquad x^2 + y^2 - \frac{ay^2}{x} + \frac{2ay^2}{x} = 0$

or $\qquad\qquad x^2 + y^2 + \frac{ay^2}{x} = 0$

or $\qquad\qquad x(x^2 + y^2) + ay^2 = 0.$

This is the required envelope.

EXAMPLE 4. *Find the envelope of the straight lines drawn at right angles to the radii vectors of the cardioid* $r = a(1 + \cos \theta)$ *through their extremities.*

SOLUTION. Let P be any point on the cardioid $r = a(1 + \cos \theta)$, whose vectorial angle is α. Then $OP = a(1 + \cos \alpha)$. Let Q be any point (r, θ) on the straight line drawn through P and perpendicular to OP.

In $\triangle OPQ$, $\quad \angle P = 90°$, then
$$\frac{OP}{OQ} = \cos(\theta - \alpha)$$
or $\qquad OP = OQ \cos(\theta - \alpha)$
or $\qquad a(1 + \cos \alpha) = r \cos(\theta - \alpha)$
or $\qquad r \cos(\theta - \alpha) = a(1 + \cos \alpha) \qquad ...(1)$

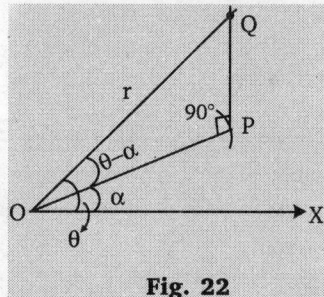

Fig. 22

This is the equation of family of straight lines drawn through P at right angle to OP.

Differentiating (1) partially with respect to α, we get
$$r \sin(\theta - \alpha) = - a \sin \alpha$$
or $\qquad \dfrac{a(1 + \cos \alpha) \sin(\theta - \alpha)}{\cos(\theta - \alpha)} = -a \sin \alpha \qquad$ [From (1)]

or $\tan(\theta - \alpha) = -\dfrac{\sin \alpha}{1 + \cos \alpha}$
$$= -\frac{2 \sin(\alpha/2) \cos(\alpha/2)}{2 \cos^2(\alpha/2)}$$
$$= -\tan(\alpha/2) = \tan(\pi - \alpha/2).$$
$\therefore \qquad \theta - \alpha = \pi - \alpha/2 \quad$ or $\quad \dfrac{\alpha}{2} = \theta - \pi$
or $\qquad \alpha = 2(\theta - \pi)$

Putting this value of α in (1), we get
$$r\cos(\theta - 2\theta + 2\pi) = a[1 + \cos(2\theta - 2\pi)]$$
$$r\cos(2\pi - \theta) = 2a\cos^2(\theta - \pi)$$
$$r\cos \theta = 2a\cos^2\theta$$
or $\qquad r = 2a\cos \theta.$

This is the required envelope.

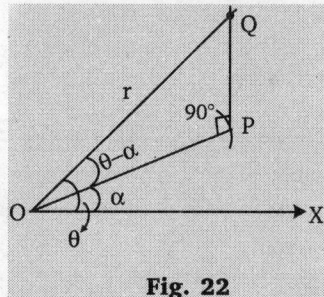

Exercise-2.12

1. Find the envelope of the family of straight lines
   $$ax \sec \theta - by \operatorname{cosec} \theta = a^2 - b^2$$
   where θ being the parameter.

2. Find the envelope of the following families of straight lines :
   (i) $y = mx + am^3$, the parameter being m.
   (ii) $y = mx + am^p$, the parameter being m.
   (iii) $x \cos^3 \alpha + y \sin^3 \alpha = a$, α being the parameter.
   (iv) $y = mx + \dfrac{a}{m}$ (MADRAS–2006)
   (v) $y = mx - 2am - m^3$

3. Find the envelope of the family of straight lines
   $$x \cos \alpha + y \sin \alpha = a$$
   where α being the parameter, and interpret the result.

4. Find the envelope of the family of circles
   $$x^2 + y^2 - 2ax \cos \alpha - 2ay \sin \alpha + c^2 = 0, \quad (a^2 > c^2)$$
   where α being the parameter and interpret the result.

5. Find the envelope of the straight lines
$$x \cos \alpha + y \sin \alpha = t \sin \alpha \cos \alpha,$$
where α being the parameter and interpret the result.

6. Find the envelope of the family of straight lines $\dfrac{x}{a} + \dfrac{y}{b} = 1$, where two parameters a and b are connected by a relation $a + b = c$, c being the constant.

7. Find the envelope of the family of curves $tx^2 + t^2 y = a$, t being the parameter.

8. Find the envelope of following families of circles :
   (i) $(x - \alpha)^2 + (y - \alpha)^2 = 2\alpha$, α being the parameter.
   (ii) $(x - c)^2 + y^2 = r^2$, where c being the parameter.

9. Find the envelope of the family of curves given by $\dfrac{x^2}{\alpha^2} + \dfrac{y^2}{k^2 - \alpha^2} = 1$, α being the parameter.

10. Find the envelope of the ellipse
$$x = a \sin(\theta - \alpha), \quad y = b \cos \theta$$
where α being the parameter.

11. $x^{2/3} + y^{2/3} = k^{2/3}$ is the envelope of the lines $\dfrac{x}{a} + \dfrac{y}{b} = 1$, then find the necessary relation between a, b and k.

12. Find the envelope of the family of the curves $x^2 \sin \alpha + y^2 \cos \alpha = a^2$, α being the parameter.

13. Show that the envelope of the family of straight lines $y = mx + \sqrt{a^2 m^2 + b^2}$, m being the parameter is $\dfrac{x^2}{a^2} + \dfrac{y^2}{b^2} = 1$.
   (ANNA–2006)

14. Find the envelope of the circles which pass through the origin and whose centres lie on the ellipse $x^2/a^2 + y^2/b^2 = 1$.

15. Find the envelope of the circles drawn upon the radii vectors of the ellipse $x^2/a^2 + y^2/b^2 = 1$ as diameter.

16. Circles are described on the double ordinates of the parabola $y^2 = 4ax$ as diameter; prove that their envelope is the parabola $y^2 = 4a(x + a)$.

17. Show that the envelope of the straight line joining the extremities of a pair of conjugate diameters of the ellipse $x^2/a^2 + y^2/b^2 = 1$ is the ellipse $x^2/a^2 + y^2/b^2 = \dfrac{1}{2}$.

18. Find the envelopes of the straight lines drawn at right angles to the radii vectors of the following curves through their extremities :
   (i) $r = ae^{\theta \cot \alpha}$ (ii) $r^n = a^n \cos n\theta$
   (iii) $r = a + b \cos \theta$.

19. Find the envelope of the straight line $x/a + y/b = 1$ where parameter a and b are connected by the relation $a^n + b^n = c^n$, c being the constant.

20. Find the envelope of the straight lines $\dfrac{x}{a} + \dfrac{y}{b} = 1$ when $a^m b^n = c^{m+n}$, c being a constant.

21. Find the envelope of the family of curves $\dfrac{x^m}{a^m} + \dfrac{y^m}{b^m} = 1$ where the parameters a and b are connected by the relation $a^p + b^p = c^p$.

22. Find the envelope of the family of parabola $y = x \tan \alpha - \dfrac{gx^2}{2u^2 \cos \alpha}$, α being the parameter.

Hints to Selected Problems

1. Differentiating the given equation w.r.t. θ, we get
$$\frac{ax \sin \theta}{\cos^2 \theta} + \frac{by \cos \theta}{\sin^2 \theta} = 0. \qquad \ldots(1)$$
On eliminating θ between (1) and the given equation, we get
$$\tan \theta = -\frac{(by)^{1/3}}{(ax)^{1/3}} \Rightarrow \sin \theta = \frac{(by)^{1/3}}{\sqrt{(ax)^{2/3} + (by)^{2/3}}},$$
and $\cos \theta = -\dfrac{(ax)^{1/3}}{\sqrt{(ax)^{2/3} + (by)^{2/3}}}$.
Put all these values in the given equation.

2. (ii) Differentiating the given equation partially w.r.t. m, we get
$$m^{p-1} = -\frac{x}{pa}.$$ Put this value in the given equation.

3. Differentiating the given equation partially w.r.t. α we get
$$-x \sin \alpha + y \cos \alpha = 0. \qquad \ldots(1)$$
Eliminating α between (1) and the given equation.

5. The given equation can be written as
$$x \operatorname{cosec} \alpha + y \sec \alpha = 0. \qquad \ldots(2)$$
Differentiating (1) partially w.r.t. α, we get
$$\tan \alpha = x^{1/3} \cdot y^{1/3}$$
$$\Rightarrow \operatorname{cosec} \alpha = \frac{\sqrt{x^{2/3} + y^{2/3}}}{x^{1/3}} \text{ and } \sec \alpha = \frac{\sqrt{x^{2/3} + y^{2/3}}}{y^{1/3}}$$
Put all these values in equation (1).

9. The given equation can be written as
$$\alpha^4 + \alpha^2 (y^2 - x^2 - k^2) + x^2 k^2 = 0$$
which is a quadratic equation in α^2. Then envelope of the family is given by $b^2 - 4ac = 0$
i.e., $(y^2 - x^2 - k^2)^2 = 4k^2 x^2$.

10. The given equation can be written as
$$x = a \left\{ \sqrt{\left(1 - \frac{y^2}{b^2}\right)} \cos \alpha - \left(\frac{y}{b}\right) \sin \alpha \right\}.$$

15. Consider the point $(a \cos \theta, b \sin \theta)$ on the ellipse.
Then we get $x^2 + y^2 = ax \cos \theta + by \sin \theta$.

16. The co-ordinate of the extremities of a double ordinate of the parabola may be written as
$$(at^2, 2at) \text{ and } (at^2, -2at).$$
Therefore, the given equation of the circle becomes
$$a^2 t^4 - 2at^2 (x + 2a) + x^2 + y^2 = 0$$
which is a quadratic equation in t^2. Hence, the required envelope is given by
$$4a^2 (x + 2a)^4 - 4a^2 (x^2 + y^2) = 0.$$

Answers

1. $(ax)^{2/3} + (by)^{2/3} = (a^2 - b^2)^{2/3}$ **2.**(i)$4x^3 + 27ay^2 = 0$ (ii) $(p-1)^{p-1}x^p + p^p ay^{p-1} = 0$ (iii) $a^2(x^2 + y^2) = x^2y^2$

(iv) $y^2 = 4ax$ (v) $\left(\dfrac{x}{a}\right)^2 + \left(\dfrac{y}{b}\right)^2 = 1$ **3.** $x^2 + y^2 = a^2$ **4.** $(x^2 + y^2 + c^2)^2 = 4a^2(x^2 + y^2)$, circles with centre at origin and radii $a \pm \sqrt{a^2 - c^2}$.

5. $x^{2/3} + y^{2/3} = t^{2/3}$ **6.** $x^{1/2} + y^{1/2} = c^{1/2}$ **7.** $x^4 + 4ay = 0$ **8.**   (i) $(x + y + 1)^2 = 2(x^2 + y^2)$ (ii) $y = \pm r$ **9.** $x \pm y = \pm k$

10. $x = \pm a$ **11.** $a^2 + b^2 = k^2$ **12.** $x^4 + y^4 = a^4$ **14.** $(x^2 + y^2)^2 = 4(a^2x^2 + b^2y^2)$ **15.** $(x^2 + y^2)^2 = a^2x^2 + b^2y^2$

18. (i) $r\sin\alpha = ae^{(\alpha - \pi/2)\cot\alpha}e^{\theta\cot\alpha}$ (ii) $r^{n/(1-n)} = a^{n/(1-n)}\cos[n\theta/(1-n)]$ (iii) $r^2 - 2br\cos\theta + b^2 - a^2 = 0$

19. $x^{n/(n+1)} + y^{n/(n+1)} = c^{n/(n+1)}$ **20.** $\{(m+n)^{m+n}x^m y^n\}/m^m n^n = c^{m+n}$ **21.** $x^{mp/(p+m)} + y^{mp/(p+m)} = c^{mp/(p+m)}$ **22.** $y = \dfrac{u^2}{2g} - \dfrac{gx^2}{2u^2}$.

2.6.7 EVOLUTE

Definition. *The evolute of a curve is the envelope of the normals to that curve.*
In other words, The locus of the centre of curvature of a curve is called evolute for that curve.

Since the centre of curvature of a curve for a given point P on it is the limiting position of the intersection of the normal at P and the normal at other point Q as Q tends to P. Thus the envelope of the normals to a given curve is called an evolute of that curve.

2.6.8 EVOLUTE OF PEDAL FORM OF CURVES

Let the pedal equation of the given curve be $p = f(r)$ ...(1)
and let C be the centre of curvature of (1) at the point P.

Then $PC = \rho$ (radius of curvature) and the equation joining P and C is the normal to the curve (1) at P. The point C will be on evolute corresponding to the point P on the curve. Since the evolute of the given curve $p = f(r)$ is the envelope of the normals at P of that curve, so that the normal PC of the given curve is a tangent to the evolute at C.

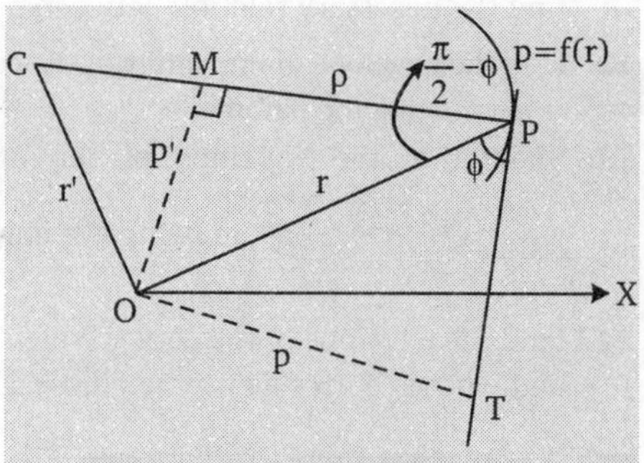

Here PT is the tangent at P to the given curve $p = f(r)$ and OT is perpendicular to PT such that $OT = p$ and $OP = r$. Now draw a perpendicular OM from O to PC such that $OM = p'$ and $CO = r'$. Then in triangle OPC, we have

$$\cos\angle OPC = \frac{r^2 + \rho^2 - r'^2}{2r\rho}$$

$\therefore$
$$r'^2 = r^2 + \rho^2 - 2r\rho\cos\angle OPC = r^2 + \rho^2 - 2r\rho\cos\left(-- \phi\right)$$
$$= r^2 + \rho^2 - 2r\rho\sin\phi$$
$$r'^2 = r^2 + \rho^2 - 2\rho p \qquad (\because p = r\sin\phi) \qquad ...(2)$$

Since $OTPM$ is a rectangle, so that $OM = TP = p'$, then in $\triangle PTO$, $r^2 = p^2 + p'^2$

$\Rightarrow$ $p'^2 = r^2 - p^2$ ...(3)

Also, we have $\rho = r\dfrac{dr}{dp}$.   ...(4)

Fig. 23

Now eliminating r, p and ρ between (1), (2), (3) and (4), we get the pedal equation of the evolute of the curve $p = f(r)$.

REMARK

- In above formulation the relation between p' and r' gives the evolute of the curve $p = f(r)$.

Solved Examples

EXAMPLE 1. *Find the evolute of the hyperbola* $x^2/a^2 - y^2/b^2 = 1$.

SOLUTION. Let $P(a\sec\theta, b\tan\theta)$ be any point on the hyperbola

$$x^2/a^2 - y^2/b^2 = 1.$$

The equation of the normal at P to the given hyperbola is

$$ax\cos\theta + by\cot\theta = a^2 + b^2. \qquad ...(1)$$

Differentiating (1) partially w.r.t. θ, we get

$$-ax\sin\theta - by\,\mathrm{cosec}^2\theta = 0$$

or $\sin^3\theta = -\dfrac{by}{ax}$ or $\sin\theta = \left(-\dfrac{by}{ax}\right)^{1/3}$

$\therefore$ $\cos\theta = \sqrt{1 - \sin^2\theta} = \sqrt{1 - \left(\dfrac{by}{ax}\right)^{2/3}}$

and $\cot\theta = \dfrac{\sqrt{\left(\dfrac{by}{ax}\right)^{2/3}}}{\left(\dfrac{by}{ax}\right)^{1/3}}$

Putting the values of $\cos\theta$ and $\cot\theta$ in (1), we get

$$ax\left[\sqrt{1-\left(\frac{by}{ax}\right)^{2/3}}\right]+by\frac{\sqrt{1-\left(\frac{by}{ax}\right)^{2/3}}}{\left(-\frac{by}{ax}\right)^{1/3}}=(a^2+b^2)$$

or

$$\frac{ax}{(ax)^{1/3}}\sqrt{ax^{2/3}-by^{2/3}}$$

$$-\frac{by}{(by)^{1/3}}\sqrt{ax^{2/3}-by^{2/3}}=(a^2+b^2)$$

or $\quad \sqrt{ax^{2/3}-by^{2/3}}[(ax)^{2/3}-(by)^{2/3}]=(a^2+b^2)$

or $\quad [(ax)^{2/3}-(by)^{2/3}]^{3/2}=(a^2+b^2)$
or $\quad (ax)^{2/3}-(by)^{2/3}=(a^2+b^2)^{2/3}$.

This is the required evolute of the given curve.

EXAMPLE 2. ***Show that the evolute of an equiangular spiral is an equiangular spiral.***

SOLUTION. The pedal equation of an equiangular spiral is

$$p=r\sin\alpha. \qquad \ldots(1)$$

So that $\qquad \dfrac{dp}{dr}=\sin\alpha.$

$\therefore \qquad \rho=r\dfrac{dr}{dp}=r.\dfrac{1}{\sin\alpha}=r\,\mathrm{cosec}\,\alpha$

or $\qquad \rho=r\,\mathrm{cosec}\,\alpha \qquad \ldots(2)$

Let (p',r') be any point on the evolute corresponding to the point (p,r) on the curve (1). Then we have,

$$r'^2=r^2+\rho^2-2\rho p$$
$$=r^2+r^2\,\mathrm{cosec}^2\,\alpha-2r\,\mathrm{cosec}\,\alpha.r\sin\alpha$$
$$\text{[Using (1) and (2)]}$$
$$=r^2\,\mathrm{cosec}^2\,\alpha-r^2$$
$$r'^2=r^2\cot^2\alpha.$$

Also, we have

$$p'^2=r^2-p^2=r^2-r^2\sin^2\alpha=r^2(1-\sin^2\alpha)$$
$$p'^2=r^2\cos^2\alpha. \qquad \ldots(4)$$

Dividing (4) by (3), we get

$$\frac{p'^2}{r'^2}=\frac{r^2\cos^2\alpha}{r^2\cot^2\alpha}=\sin^2\alpha.$$

$\therefore \qquad p'^2=r'^2\sin^2\alpha$

or $\qquad p'=r'\sin\alpha.$

Thus the locus of the point (p',r') is $p=r\sin\alpha$, which is an equiangular spiral.

Exercise-2.13

1. Find the equation of the evolute of the parabola $y^2=2ax$.

2. Show that the equation of the evolute of the ellipse $x^2/a^2+y^2/b^2=1$ is $(ax)^{2/3}+(by)^{2/3}=(a^2-b^2)^{2/3}$.
   (JNTU–2006, ANNA–2005)

3. Show that the evolute of the tractrix $x=a(\cos t+\log\tan(t/2)),\ y=a\sin t$ is the catenary $y=a\cosh(x/a)$.

4. Find the evolute of the curve $x^{2/3}+y^{2/3}=a^{2/3}$.

5. Prove that the evolute of the ellipse $x^2/a^2+y^2/b^2=1$ is the envelope of the family of the ellipses given by $a^2x^2\sec^4\alpha+b^2y^2\,\mathrm{cosec}^4\alpha=(a^2-b^2)^2$, α being the parameter.

6. Prove that the evolute of the hyperbola $2xy=a^2$ is
   $$(x+y)^{2/3}+(x-y)^{2/3}=2a^{2/3}.$$

7. Show that the evolute of the curve whose pedal equation is $r^2-a^2=mp^2$ is the curve whose pedal equation is
   $$r^2-(1-m)a^2=mp^2.$$

8. Show that the evolute of the cardiod $r=a(1+\cos\theta)$, is the cardiod $r=\dfrac{1}{3}a(1-\cos\theta)$, the pole of the latter equation being at the point $\left(\dfrac{2}{3}a,0\right)$.

9. Show that the whole length of the evolute of the ellipse $\dfrac{x^2}{a^2}+\dfrac{y^2}{b^2}=1$ is $4\left(\dfrac{a^2}{b}-\dfrac{b^2}{a}\right)$.

10. Find the evolute of the parabola $y^2=4ax$.

Hints to Selected Problems

1. Equation of any normal to the parabola $y^2=2ax$ is
   $$y=mx-am-\frac{1}{2}am^3,\quad\text{where m is a parameter.}$$
   Now differentiating partially w.r.t. m.

2. Let $x=a\cos\theta,\ y=b\sin\theta$ Then $\dfrac{dy}{dx}=\dfrac{b\cos\theta}{-a\sin\theta}$.

   Therefore, slope of the normal to the given ellipse at the point
   $$(x,y)=-\frac{dx}{dy}=\frac{a\sin}{b\cos}$$
   $\therefore$ The equation of the normal is given by
   $$y-b\sin\theta=\frac{a\sin\theta}{b\cos\theta}(x-a\cos\theta).\text{ Now proceed as usual.}$$

4. Take the point $x=a\cos^3\theta,\ y=a\sin^3\theta$, θ being the parameter. Then proceed as in (2).

6. Let $\qquad xy=\dfrac{a^2}{2}=c^2.\qquad\ldots(1)$

   Let $P\left(ct,\dfrac{c}{t}\right)$ be any point on the hyperbola (1).

   Then $\qquad \dfrac{dy}{dx}=-\dfrac{c^2}{x^2}$.

   $\therefore$ Slope of the normal $=-\dfrac{dx}{dy}=\dfrac{x^2}{c^2}=t^2$.

   Hence, the equation of the normal at P is given by
   $$y-\frac{c}{t}=t^2(x-ct)$$
   Then proceed as usual.

7. The pedal equation of the given curve is
   $$r^2-a^2=mp^2\Rightarrow\rho=r\frac{dr}{dp}=mp.$$
   Now use the relation $r'^2=r^2+\rho^2-2\rho p$.

Answers

1. $27ay^2 = 8(x-a)^3$ **4.** $(x+y)^{2/3} + (x-y)^{2/3} = 2a^{2/3}$ **10.** $27ay^2 = 4(x-2a)^3$

2.7 ASYMPTOTES

In calculus, there are some curves whose branches seem to go to infinity. It is not necessary that there always exists a definite straight line for all such curves which seems to touch the branch of the curves at infinite but more or less there are some certain curves for which this type of definite straight line exists, this straight line is therefore known as asymptote.

Definition. *A definite straight line whose distance from branch of the curve continuously decreases as we move away from the origin along the branch of the curve and seems to touch the branch at infinity, provided the distance of this line from origin should be finite initially, is called an asymptote of the curve.*

Suppose in the equalion of a curve, two or more than two values of y exists for every value of x, then we obtain different branches of the curve corresponding to these distinct values of y. If each branch have its own separate asymptote, then we can say that a curve may have more than one asymptote.

2.7.1 DETERMINATION OF ASYMPTOTES

Consider a curve $f(x,y) = 0$ ...(1)

and also consider that there are no asymptotes parallel to y-axis. Thus we shall take the equation which is not parallel to y-axis. in the form of

$$y = mx + c \qquad \qquad ...(2)$$

Let us take a point $P(x, y)$ on the curve (1), therefore this point as tends to infinity along the straight line (2), x must tend to infinity. Now find the tangent to the curve $f(x, y) = 0$ at the point $P(x, y)$.

∴ The equation of tangent at $P(x, y)$ is

$$Y - y = \frac{dy}{dx}(X - x) \quad \text{or} \quad Y = \frac{dy}{dx}X + \left(y - x\frac{dy}{dx}\right). \qquad ...(3)$$

The equation (3) is of the form $y = mx + c$, so in order to exist the asymptote of the curve there must both $\frac{dy}{dx}$ and $\left(y - x\frac{dy}{dx}\right)$ tend to finite limits as x tends to infinity. Therefore, if the equation (3) tends to the straight line given in (2) as x tends to infinity, then the line (2) will be an asymptote of the curve $f(x, y) = 0$ and also we have

$$m = \lim_{x \to \infty} \frac{dy}{dx} \quad \text{and} \quad c = \lim_{x \to \infty}\left(y - x\frac{dy}{dx}\right)$$

Since c is finite, then we have

$$\lim_{x \to \infty}\left(\frac{y - x\frac{dy}{dx}}{x}\right) = \lim_{x \to \infty}\frac{c}{x} = 0 \quad \text{or} \quad \lim_{x \to \infty}\left(\frac{y}{x} - \frac{dy}{dx}\right) = 0$$

or

$$\lim_{x \to \infty}\left(\frac{y}{x}\right) = \lim_{x \to \infty}\frac{dy}{dx} \quad \text{or} \quad \lim_{x \to \infty}\frac{y}{x} = m.$$

Also

$$c = \lim_{x \to \infty}\left(y - x\frac{dy}{xx}\right) \quad \text{or} \quad c = \lim_{x \to \infty}(y - mx).$$

Hence, if $y = mx + c$ is an asymptote to the curve $f(x, y) = 0$, then we obtain

$$m = \lim_{x \to \infty}\frac{dy}{dx} = \lim_{x \to \infty}\frac{y}{x} \quad \text{and} \quad c = \lim_{x \to \infty}(y - mx). \; .$$

2.7.2 ASYMPTOTES OF GENERAL EQUATION

Let the general rational algebraic equation of a curve be

$$\{a_0 y^n + a_1 y^{n-1}x + a_2 y^{n-2}x^2 + ... + a_{n-1}yx^{n-1} + a_n x^n\} + \{b_1 y^{n-1} + b_2 y^{n-2}x + ... + b_{n-1}yx^{n-2} + b_n x^{n-1}\}$$

$$+ \{c_2 y^{n-2} + c_3 y^{n-3} + ... + c_{n-1}yx^{n-3} + c_n x^{n-2}\} + ... = 0 \qquad ...(1)$$

or

$$x^n\left\{a_0\left(\frac{y}{x}\right)^n + a_1\left(\frac{y}{x}\right)^{n-1} + a_2\left(\frac{y}{x}\right)^{n-2} + ... + a_{n-1}\left(\frac{y}{x}\right) + a_n\right\} + x^{n-1}\left\{b_1\left(\frac{y}{x}\right)^{n-1} + b_2\left(\frac{y}{x}\right)^{n-2} + ... + b_n\right\}$$

$$+ x^{n-2}\left\{c_2\left(\frac{y}{x}\right)^{n-2} + c_3\left(\frac{y}{x}\right)^{n-3} + ... + c_n\right\} + ... = 0$$

or
$$x^n \phi_n\left(\frac{y}{x}\right) + x^{n-1}\phi_{n-1}\left(\frac{y}{x}\right) + x^{n-2}\phi_{n-2}\left(\frac{y}{x}\right) + ... + x\phi_1\left(\frac{y}{x}\right) + \phi_0\left(\frac{y}{x}\right) = 0 \qquad ...(2)$$

where $\phi_k\left(\dfrac{y}{x}\right)$ is a polynomial of degree k in $\left(\dfrac{y}{x}\right)$.

Divide (2) by x^n, we get
$$\phi_n\left(\frac{y}{x}\right) + \frac{1}{x}\phi_{n-1}\left(\frac{y}{x}\right) + \frac{1}{x^2}\phi_{n-2}\left(\frac{y}{x}\right) + ... + \frac{1}{x^{n-1}}\phi_1\left(\frac{y}{x}\right) + \frac{1}{x^n}\phi_0\left(\frac{y}{x}\right) = 0$$

Now taking limit as $x \to \infty$, and assuming there is no asymptote parallel to y-axis then $m = \lim\limits_{x\to\infty}\left(\dfrac{y}{x}\right)$, we get
$$\phi_n(m) = 0. \qquad ...(3)$$

This equation (3) is of degree n in m so it has at most n roots, real as well as imaginary. Out of these n roots some roots may be identical. Thus we get n values of m corresponding to the n branches of the curve (1). Since, we will have only real values of m so ignore all imaginary roots of (3) if they exists. Further if $y = mx + c$ is an asymptote of (1), then we have
$$c = \lim_{x\to\infty}(y - mx), \text{ for each specified value of } m.$$

Determination of c. For the determination of c corresponding to each distinct value of m, we put $y = mx + p$ in the equation of curve (2), where $p \to c$ as $x \to \infty$.

Now putting $y = mx + p$ i.e., $\dfrac{y}{x} = m + \dfrac{p}{x}$, in the (2), we get
$$x^n \phi_n\left(m + \frac{p}{x}\right) + x^{n-1}\phi_{n-1}\left(m + \frac{p}{x}\right) + x^{n-2}\phi_{n-2}\left(m + \frac{p}{x}\right) + ... + x\phi_1\left(m + \frac{p}{x}\right) + \phi_0\left(m + \frac{p}{x}\right) = 0.$$

Expand each term by Taylor's expansion, we get
$$x^n\left[\phi_n(m) + \frac{p}{x}\phi'_n(m) + \frac{p^2}{2!x^2}\phi''_n(m) + ...\right] + x^{n-1}\left[\phi_{n-1}(m) + \frac{p}{x}\phi'_{n-1}(m) + ...\right] + x^{n-2}\left[\phi_{n-2}(m) + \frac{p}{x}\phi'_{n-2}(m) + ...\right] + ... = 0$$

or
$$x^n\phi_n(m) + x^{n-1}[p\phi'_n(m) + \phi_{n-1}(m)] + x^{n-2}\left[\frac{p^2}{2!}\phi''_n(m) + \frac{p}{1!}\phi'_{n-1}(m) + \phi_{n-2}(m)\right] + ... = 0$$

Since we know that $\phi_n(m) = 0$, then
$$x^{n-1}[p\phi'_n(m) + \phi_{n-1}(m)] + x^{n-2}\left[\frac{p^2}{2!}\phi''_n(m) + \frac{p}{1!}\phi'_{n-1}(m) + \phi_{n-2}(m)\right] + ... = 0$$

Dividing by x^{n-1} and taking limit as $x \to \infty$, we get
$$\lim_{\to\infty}[p\phi'_n(m) + \phi_n \ (m)] = 0 \qquad \text{or} \qquad \left(\lim_{x\to\infty}p\right)\phi'_n(m) + \phi_{n-1}(m) = 0$$

or
$$c\phi'_n(m) + \phi_{n-1}(m) = 0 \qquad\qquad \left(\because \lim_{x\to\infty}p = c\right)$$

Hence, from above relation we can determine the value of c for each distinct value of m.

📧 REMARK

- To find the polynomial $\phi_n(m)$. We should put $y = m$ and $x = 1$ in the nth degree terms of the curve. Similarly to get $\phi_{n-1}(m)$ we put $y = m$ and $x = 1$ in the $(n-1)^{th}$ degree terms of the curve. Therefore in general, to get $\phi_k(m)$ we should put $y = m$ and $x = 1$ in the kth degree terms of the curves.

2.7.3 EXISTENCE OF ASYMPTOTES

From the equation $\phi_n(m) = 0$, if we obtain one or more than one values of m such that $\phi'_n(m) = 0$ and $\phi_{n-1}(m) \neq 0$, then from the equation for the determining of c. we obtain $0.c + \phi_{n-1}(m) = 0$

Thus we get c is either, $+\infty$ or $-\infty$. Hence, we can say that corresponding to such values of m no asymptotes will exists.

2.7.4 DETERMINATION OF c CORRESPONDING TO SOME IDENTICAL VALUES OF m

Let us suppose some of the roots of the equation $\phi_n(m) = 0$ are identical and let these identical values be r in number which will make $\phi'_n(m), \phi''_n(m), ... \phi_m^{r-1}(m)$ equal to zero. Now for the existence of the asymptotes $\phi_{n-1}(m)$ must be zero corresponding to the identical values of m. Also, if it will make $\phi'_{n-1}(m), \phi''_{n-1}(m), ...\phi_{n-1}^{r-2}(m); \phi_{n-2}(m), \phi'_{n-2}(m), ...\phi_{n-2}^{r-3}(m); \phi_{n-3}(m), \phi'_{n-3}(m), ...\phi_{n-3}^{r-4}(m)...; \ \phi_{n-r+2}(m), \phi'_{n-r+2}(m)$ and $\phi_{n-r+1}(m)$ equal to zero, then the equation to determine c will become
$$0.c^{r-1} + 0.c^{r-2} + ... + 0.c + 0 = 0$$

and thus we cannot find the value of c in this way.

So to determine c let us put $\phi_n(m), \phi'(m), ..., \phi_n^{r-1}(m); \phi_{n-1}(m), \phi'_{n-1}(m), ...\phi_{n-1}^{r-2}(m); \phi_{n-2}(m), \phi'_{n-2}(m), ...\phi_{n-2}^{r-3}(m); \phi_{n-3}(m),$ $(\phi'_{n-3}(m), ... \phi_{n-3}^{r-4}(m) ... \phi_{n-r+2}(m), \phi'_{n-r+2}(m)$ and $\phi_{n-r+1}(m)$ equal to zero in the following equation

$$x^n\phi_n(m) + x^{n-1}[p\phi'_n(m) + \phi_{n-1}(m)] + x^{n-2}\left[\frac{p^2}{2!}\phi''_n(m) + \frac{p}{1!}\phi'_{n-1}(m) + \phi_{n-2}(m)\right] + ... + x^{n-r+1}\left[\frac{p^{r-1}}{r-1!}\phi_n^{r-1}(m) + \right.$$

$$\left.\frac{p^{r-2}}{r-2!}\phi_{n-1}^{r-2}(m) + ... + \frac{p}{1!}\phi'_{n-r+2}(m) + \phi'_{n+r+1}(m)\right] + x^{n-r}\left[\frac{p^r}{r!}\phi_n^r(m) + \frac{p^{r-1}}{r-1!}\phi_{n-1}^{r-1}(m) + \frac{p^{r-2}}{r-2!}\phi_{n-2}^{r-2}(m) + ...\right.$$

$$\left. + \frac{p}{1!}\phi'_{n-r+1}(m) + \phi_{n-r}(m)\right] = 0$$

We have

$$x^{n-r}\left[\frac{p^r}{r!}\phi_n^r(m) + \frac{p^{r-1}}{r-1!}\phi_{n-1}^{r-1}(m) + ... + \frac{p}{1!}\phi'_{n-r+1}(m) + \phi_{n-r}(m)\right] + x^{n-r-1}\left[\frac{p^{r+1}}{r+1!}\phi_n^{r+1}(m) + \frac{p^r}{1!}\phi_{n-1}^r(m) + ...\right]. = 0$$

Now dividing above equation by x^{n-r} and taking the limit as $x \to \infty$, we get

$$\frac{c^r}{r!}\phi_n^r(m) + \frac{c^{r-1}}{r-1!}\phi_{n-1}^{r-1}(m) + ... + \frac{c}{1!}\phi'_{n-r+1}(m) + \phi_{n-r}(m) = 0 \text{ where } c = \lim_{x\to\infty} p.$$

Therefore this equation gives r values of c corresponding to the identical values of m. Hence, we obtain r parallel asymptotes.

2.7.5 NUMBER OF ASYMPTOTES OF A CURVE

Suppose the degree of an algebraic curve is n, then we find a polynomial $\phi_n(m)$ by putting $y = m$ and $x = 1$ in the nth degree terms of the curve. Thus the equation $\phi_n(m) = 0$ is of degree n in m and which gives almost n values of m real as well as imaginary. These n values of m are nothing but the slopes of the asymptotes, which are not parallel to y axis. If there are some asymptotes, parallel to y-axis, then the degree of $\phi_n(m)$ will be smaller than n by the same number of parallel asymptotes. Suppose all the roots of $\phi_n(m) = 0$ are distinct and real, then to each value of m we obtain one value of c. Hence, we obtain n asymptotes. In case, there some roots say r (out of n) of $\phi_n(m) = 0$ are same, then we can find the values of c for these same roots the following equation

$$\frac{c}{r!}\phi_n^r(m) + \frac{c^{r-1}}{r-1!}\phi_{n-1}^{r-1}(m) + ... + \phi_{n-r}(m) = 0$$

This equation in c is of degree r so we get r distinct values of c for the same roots, hence, again we obtain n asymptotes. Therefore we can say that the total number of asymptotes of a curve are equal to the degree of the curve, These asymptotes are real as well as imaginary but we have required only real asymptotes so we ignore all the imaginary asymptotes.

2.7.6 ASYMPTOTES PARALLEL TO CO-ORDINATES AXES

(a) Asymptotes parallel to x-axis. Let the general equation of an algebraic curve in decreasing powers of x be

$$x^n\phi(y) + x^{n-1}\phi_1(y) + x^{n-2}\phi_2(y) + ... = 0 \qquad ...(1)$$

where $\phi(y), \phi_1(y), \phi_2(y), ...$ are the function of y only.

Now divide (1) by x^n, we get $\phi(y) + \frac{1}{x}\phi_1(y) + \frac{1}{x^2}\phi_2(y) + ... = 0$. $\qquad ...(2)$

If $y = k$ is an asymptote parallel to x-axis, then we can say that x alone tends to infinity as a point $P(x, y)$ on the curve tends to infinity along the line $y = k$ and also we have $k = \lim_{x\to\infty} y$.

Now taking the limit of both sides of (2) as $x \to \infty$ and $y \to k$, we get $\phi(k) = 0$.

Thus k is a root of the equation $\phi(y) = 0$. If k_1, k_2, etc. are the roots of $\phi(y) = 0$, then the asymptotes parallel to x-axis are given by $y = k_1, y = k_2$, etc. Since k is a root of the equation $\phi(y) = 0$, then $(y - k)$ is a factor of the equation $\phi(y) = 0$. Also $\phi(y)$ is the coefficient of the highest power of x i.e., x^n in the equation of the curve. Hence, we obtain the asymptotes parallel to x-axis by taking the coefficient of highest power of x in the equation of the curve equal to zero.

(b) Asymptotes parallel to y-axis. Similarly, we may obtain the asymptotes parallel to y-axis by taking the coefficient of highest power of y in the equation of the curve equal to zero.

REMARK

- If the coefficient of highest power of x or y or both are constant, then no asymptotes parallel to either x or y or both axes exists respectively.

Solved Examples

EXAMPLE 1. *Find the asymptotes of the curve* $x^3 + y^3 - 3axy = 0$.

SOLUTION. Obviously, the degree of the curve is 3, so it will have 3 asymptotes real as well as imaginary. Here the coefficient of highest degree in x and y are constant so no asymptote parallel to co-ordinate axis exist. Let

$$y = mx + c \qquad ...(1)$$

be the asymptote of the curve.

So putting $y = m$ and $x = 1$ in the highest degree terms of the curve, we get

$$\phi_3(m) = 1 + m^3.$$

Solving the equation $\phi_3(m) = 0$

i.e., $\qquad\qquad 1 + m^3 = 0$

or $\qquad (1+m)(m^2-m+1)=0$
or $\qquad\qquad m=-1$
is only real root and other two roots are imaginary so ignore them.

Next, putting $y=m$ and $x=1$ is second degree terms in the equation of the curve (1), we get
$$\phi_2(m)=-3am.$$
Now we find value of c by the following equation
$$c\phi'_n(m)+\phi_{n-1}(m)=0$$
or $\qquad c\phi'_3(m)+\phi_2(m)=0$
or $\qquad c(3m^2)+(-3am)=0$
$$[\because \phi_3(m)=1+m^3,\Rightarrow \phi'_3(m)=3m^2]$$
If $m=-1$, then
$$c[3(-1)^2]+[-3a(-1)]=0$$
$$3c+3a=0$$
or $\qquad\qquad c=-a.$
Hence, the asymptote is $\qquad y=-x-a$
or $\qquad\qquad x+y+a=0.$

EXAMPLE 2. *Find all the asymptotes of the curve* $x^3+x^2y-xy^2-y^3-3x-y-1=0.$

SOLUTION. The degree of the curve is 3 so it has 3 asymptotes which are real as well as imaginary. Since the coefficients of highest degree i.e., 3rd degree of x and y are constant so there are no asymptotes parallel to co-ordinate axes. Thus there are oblique asymptotes of the form $y=mx+c$.

Now putting $y=m$ and $x=1$ in the third degree terms of the curve, we get
$$\phi_3(m)=1+m-m^2-m^3.$$
Solving the equation
$$\phi_3(m)=0 \text{ i.e, } 1+m-m^2-m^3=0,$$
we get $\qquad (1+m)(1-m^2)=0$
or $\qquad m=-1,-1,1.$

Determination of c. For $m=1$, we use the following equation
$$c\phi'_n(m)+\phi_{n-1}(m)=0$$
or $\qquad c\phi'_3(m)+\phi_2(m)=0 \qquad\qquad ...(1)$
Putting $y=m$ and $x=1$ in the second degree terms of the equation we get $\phi_2(m)=0$.
From (1), we get
$$c(1-2m-3m^2)+0=0$$
at $m=1, \qquad c(1-2-3)+0=0$
or $\qquad\qquad -4c=0$
or $\qquad\qquad c=0$
Thus one of the asymptote is $y=x$

Determination of c for m = -1, -1. Since two out of three roots of the equation $\phi_3(m)=0$ are same, then we use the following formula to determine c
$$\frac{c^2}{2!}\phi''_3(m)+\frac{c}{1!}\phi'_2(m)+\phi_1(m)=0. \qquad ...(2)$$
Putting $y=m$ and $x=1$ in the first degree terms of the quation we obtain $\phi_1(m)=-3-m$.
From (2), we have
$$\frac{c^2}{2!}(-2-6m)+\frac{c}{1!}.0+(-3-m)=0$$
at $m=-1$
$$\frac{c^2}{2}(-2+6)-3+1=0$$

or $\qquad 2c^2-2=0 \qquad$ or $\qquad c=\pm 1$
Thus other two asymptotes are
$$y=-x+1, \ y=-x-1.$$
Hence, all the asymptotes of the given curve are
$$y=x, x+y-1=0, x+y+1=0.$$

EXAMPLE 3. *Find all the asymptotes of the curve*
$$(x-2y)^2(x-y)-4y(x-2y)-(8x+7y)=0.$$

SOLUTION. Simplifying the equation of curve
$$(x^2+4y^2-4xy)(x-y)-4xy+8y^2-8x-7y=0$$
or $x^3+8xy^2-5x^2y-4y^3-4xy+8y^2-8x-7y=0. ...(1)$
The degree of the curve (1) is 3 so it has 3 asymptotes which are real as well as imaginary. Obviously there are no asymptotes parallel to co-ordinate axis. Thus there are only oblique asymptotes of the form $y=mx+c$.
Putting $y=m$ and $x=1$ in the highest degree i.e., third degree terms of the curve (1), we obtain
$$\phi_3(m)=1-5m+8m^2-4m^3.$$
Solving the equation $\phi_3(m)=0$
i.e., $\qquad 1-5m+8m^2-4m^3=0$
or $\qquad (1-m)(1-2m)^2=0 \quad$ or $\quad m=\frac{1}{2},\frac{1}{2},1.$

Determination of c for m = 1 :
Putting $y=m$ and $x=1$ in the second degree terms of the curve (1), we obtain
$$\phi_2(m)=-4m+8m^2.$$
Applying the formula
$$c.\phi'_3(m)+\phi_2(m)=0$$
or $c(-5+16m-12m^2)-4m+8m^2=0.$
Substitute $m=1$, we get
$$c(-5+16-12)-4+8=0$$
or $\qquad\qquad -c+4=0$
or $\qquad\qquad c=4.$
Thus the asymptote is $\qquad y=x+4$
or $\qquad\qquad x-y+4=0.$

Determination of c for $m=\dfrac{1}{2},\dfrac{1}{2}$:

Putting $y=m$ and $x=1$ in the first degree terms of the curve (1) we obtain
$$\phi_1(m)=-8-7m.$$

Since $m=\dfrac{1}{2},\dfrac{1}{2}$ are two repeated roots of $\phi_3(m)=0$, then apply the following formula to determine c,

$$\frac{c^2}{2!}[\phi''_3(m)]+\frac{c}{1!}\phi'_2(m)+\phi_1(m)=0$$

or $\dfrac{c^2}{2!}(16-24m)+c(-4+16m)-8-7m=0$

At $\quad m=\dfrac{1}{2}$

$$\frac{c^2}{2}(16-12)+c(-4+8)-8-\frac{7}{2}=0$$

or $\qquad\qquad 2c^2+4c-\dfrac{23}{2}=0$

or $\quad 4c^2+8c-23=0 \Rightarrow c=\dfrac{-2\pm 3\sqrt{3}}{2}.$

Thus the other asymptotes are

$$y = \frac{1}{2}x + \frac{-2 \pm 3\sqrt{3}}{2}$$

or $\qquad 2y = x - 2 \pm 3\sqrt{3}.$

Hence, all the three asymptotes of the curve are

$$x - y + 4 = 0, \quad 2y = x - 2 \pm 3\sqrt{3}.$$

EXAMPLE 4. *Find all the aysmptotes of the curve*

$$y^2(x^2 - a^2) = x^2(x^2 - 4a^2).$$

SOLUTION. Obviously the degree of the curve is 4 so its has at most 4 asymptotes. Since highest degree term of x is x^4 whose coefficient is a constant so there is no asymptote parallel to x-axis but highest degree term of y is y^2 and whose coefficient is $x^2 - a^2$. Thus the asymptotes parallel to y-axis are

$$x^2 - a^2 = 0$$

or $\qquad\qquad x = \pm a \Rightarrow x = -a, x = a.$

Putting $y = m$ and $x = 1$ in the highest *i.e.*, 4th degree terms of the curve we obtain

$$\phi_4(m) = m^2 - 1.$$

Solving the equation $\phi_4(m) = 0$ *i.e.*, $m^2 - 1 = 0$. then

$$m = \pm 1.$$

Now putting $y = m$ and $x = 1$ in the third degree terms of the curve we obtain $\phi_3(m) = 0$

Determination of c for $m = \pm 1$. To determine the values of c at $m = \pm 1$ we use the following formula

$$c.\phi_4'(m) + \phi_3(m) = 0$$

or $\qquad\qquad c(2m) + 0 = 0$

At $\qquad\qquad m = \pm 1, \ c = 0.$

Thus other two asymptotes are $y = \pm x.$

Hence, all the four asymptotes are

$$x = \pm a, y = \pm x.$$

 Exercise-2.14

Find all the asymptotes of the following curves :

1. $a^2/x^2 - b^2/y^2 = 1$
2. $a^2/x^2 + b^2/y^2 = 1$
3. $y^2(a^2 - x^2) = x^4$
4. $x^2y^2 = a^2(x^2 + y^2)$
5. $x^2y^2 - x^2y - xy^2 - y + 1 = 0$
6. $3x^3 + 2x^2y - 7xy^2 + 2y^3 + 14xy + 7y^2 + 4x + 5y = 0$
7. $2x^3 - x^2y - 2xy^2 + y^3 - 4x^2 + 8xy - 4x + 1 = 0$
8. $x^3 + 2x^2y + xy^2 - x^2 - xy + 2 = 0$
9. $y^3 - 5xy^2 + 8x^2y - 4x^3 - 3y^2 + 9xy - 6x^2 + 2y - 2x + 1 = 0$
10. $y^3 - x^2y - 2xy^2 + 2x^3 - 7xy + 3y^2 + 2x^2 + 2x + 2y + 1 = 0$
11. $y^3 - xy^2 - x^2y + x^3 + x^2 - y^2 - 1 = 0$ (MTU–2012)
12. $(x^2 - y^2)(y^2 - 4x^2) - 6x^3 + 5yx^2 + 3xy^2 - 2y^3 - x^2 + 3xy - 1 = 0$
13. $y^3 = x^3 + ax^2$
14. $x^2y^3 + x^3y^2 = x^3 + y^3.$

15. $(y - x)(y - 2x)^2 + (y + 3x)(y - 2x) + 2x + 2y - 1 = 0.$
16. $x^3 + 2x^2y - xy^2 - 2y^3 + 4y^2 + 2xy + y - 1 = 0.$
17. $(x + y)^2(x + 2y + 2) = x + 9y + 2$ (MDU–2005)
18. $x^2(x - y)^2 + a^2(x^2 - y^2) - a^2xy = 0$
19. $y^3 - 2y^2x - yx^2 + 2x^3 + y^2 - 6xy + 5x^2 - 2y + 2x + 1 = 0$
20. $x^3 + 3x^2y - 4y^3 - x + y + 3 = 0$
21. $x^3 - 5x^2y + 8xy^2 - 4y^3 + x^2 - 3xy + 2y^2 - 1 = 0$
22. $xy^2 = 4a^2(2a + x).$
23. $x^3 + 2x^2y - xy^2 - 2y^3 + xy - y^2 - 1 = 0$
24. $y^3 + x^2y + 2xy^2 - y + 1 = 0$
25. $(2x - 3y + 1)^2(x + y) - 8x + 2y - 9 = 0$
26. $(x^2 - y^2)^2 - 4y^2 + y = 0$
27. $y^2(x - 2a) = x^3 - a^3$
28. $(x^3 + a^3)y = bx^3$

Answers

1. $x = \pm a$ 2. $x = \pm a, y = \pm b$ 3. $x = \pm a$ 4. $x = \pm a, y = \pm a$ 5. $y = 0; y = 1; x = 0; x = 1$
6. $x + 2y = 1, 2x - 2y = -7, 6x - 2y = 15$ 7. $x + y - 2 = 0; x - y + 2 = 0; 2x - y - 4 = 0$
8. $x = 0; x + y = 0; x + y - 1 = 0$ 9. $x - y = 0; 2x - y + 2 = 0; 2x - y + 1 = 0$
10. $x - y - 1 = 0; x + y + 2 = 0; 2x - y = 0$ 11. $x + y = 0; x - y = 0; x - y + 1 = 0$
12. $x - y = 0; 2x - y = 0; x + y + 1 = 0; 2x + y + 1 = 0$ 13. $3x - 3y + a = 0$
14. $y = \pm 1; x = \pm 1; x + y = 0$ 15. $2x - y - 2 = 0; 2x - y - 3 = 0; x - y + 4 = 0$
16. $x - y + 1 = 0; x + y - 1 = 0; x + 2y = 0$ 17. $x + 2y + 2 = 0; x + y \pm 2\sqrt{2} = 0$
18. $y = \pm a; x - y = \pm a$ 19. $x - y = 0; 2x - y + 1 = 0; x + y + 2 = 0$
20. $x - y = 0; x + 2y - 1 = 0; x + 2y + 1 = 0$ 21. $x - y = 0; x - 2y = 0; x - 2y + 1 = 0$
22. $x = 0, y = \pm 2a$ 23. $x + 2y - 1 = 0; x - y = 0; x + y + 1 = 0$
24. $y = 0; x + y - 1 = 0; x + y + 1 = 0$ 25. $x + y = 0; 2x - 3y + 3 = 0; 2x - 3y - 1 = 0$
26. $x + y = \pm 1; x - y = \pm 1$ 27. $x = 2a, y = x + a, y = -x - a$
28. $x + a = 0, y - b = 0$

2.7.7 OTHER METHODS FOR FINDING THE ASYMPTOTE OF AN ALGEBRAIC CURVE

THEOREM 1.   *The asymptotes of an algebraic curve are parallel to the lines which obtained by equating to zero the linear factors of the highest degree terms of the equation of curve.*

PROOF.   Let us suppose the equation of the curve is of degree n and let $y - mx$ be a linear factor of the nth degree term in the equation of the curve. Since $\phi_n(m)$ is a polynomial of degree n in m and obtained by putting $y = m$ and $x = 1$ in the nth degree terms of the curve, then $(m - m_1)$ is a factor of $\phi_n(m)$. Thus m_1 is a root of the equation $\phi_n(m) = 0$ which gives the slope of the asymptote. Hence, there is an asymptote parallel to the line $y = m_1 x = 0$.

Conversely, let m_1 be a root of the equation $\phi_n(m) = 0$ so that there is an asymptote which is parallel to the line $y - m_1 x = 0$, then $(m_1 - m)$ must be a factor of $\phi_n(m)$ and therefore, $(y/x - m_1)$ will be a linear factor of $\phi_n(y/x)$. Hence $(y - m_1 x)$ is a linear factor of $x^n \phi_n(y/x)$ which is the highest degree terms in the equation of the curve.

Hence the theorem is proved.

Since we know that if $y = mx + c$ is an asymptote of the curve $f(x, y) = 0$, then we have

$$m = \lim_{x \to \infty} \frac{y}{x} \text{ and } c = \lim_{x \to \infty}(y - mx) = \lim_{x \to \infty, \frac{y}{x} \to \infty}(y - mx) \Bigg\} \qquad \ldots(1)$$

With the help of (1) and above theorem we may find the asymptotes of an algebraic curves.

WORKING PROCEDURE

Step 1. First we collect all the highest degree terms in the equation of the curve and then resolve into linear factors.

Step 2. After getting linear factors there may arise some cases.

Case I. If the linear factor $(y - m_1 x)$ of the highest degree *i.e.*, n^{th} degree terms in the equation of the curve is simple (non-repeated). Then the given equation of the curve can be written as

$$(y - m_1 x)F_{n-1} + P_{n-1} = 0. \qquad \ldots(2)$$

where F_{n-1} contains only terms of degree $n - 1$ and P_{n-1} contains the terms of various degree not exceeding $n - 1$. Therefore $y - m_1 x = c$ is an asymptote of the curve where c is to be determined. Let us take a point (x, y) on the curve (1), then we have

$$y - m_1 x = -\frac{P_{n-1}}{F_{n-1}}.$$

Now taking the limit as $x \to \infty$, $y/x \to m_1$, then we have

$$\lim_{x \to \infty, \frac{y}{x} \to m_1}(y - m_1 x) = \lim_{x \to \infty, \frac{y}{x} \to m_1}\left(-\frac{P_{n-1}}{F_{n-1}}\right) \text{ or } c = \lim_{x \to \infty, \frac{y}{x} \to m_1}\left(-\frac{P_{n-1}}{F_{n-1}}\right).$$

Now substitute this value of c in the equation $y = m_1 x + c$

we obtained the asymptote which is parallel to the line $y - m_1 x = 0$ corresponding to the linear factor $(y - m_1 x)$. Similarly we may obtain other asymptotes.

Case II. If $(y - m_1 x)$ is a linear factor of the nth degree terms of order two but $(y - m_1 x)$ is not a factor of the $(n-1)^{\text{th}}$ degree terms of the curve, then we have $\phi'_n(m_1) = 0$ and $\phi_{n-1}(m_1) \neq 0$. Therefore, no asymptotes corresponding to $(y - m_1 x)^2$ will exist. On the other hand if there are no terms of $(n-1)^{\text{th}}$ degree in the equation of the curve, then make them by adding with zero coefficient and thus we can say that $(y - m_1 x)$ is now a factor of $(n-1)^{\text{th}}$ degree terms, then we have the case III.

Case III. If $(y - m_1 x)^2$ is a linear factor of nth degree terms and $(y - m_1 x)$ is a factor of $(n-1)^{\text{th}}$ degree terms, then the equation of the curve can be written as

$$(y - m_1 x)^2 F_{n-2} + (y - m_1 x)G_{n-2} + P_{n-2} = 0 \qquad \ldots(3)$$

where F_{n-2} and G_{n-2} contain only the terms of degree $n - 2$, and P_{n-2} contains various degree terms not exceeding $n - 2$. Now divide (2) by F_{n-2} and taking the limit as $x \to \infty$ and $y/x \to m_1$, we get

$$\lim_{x \to \infty, y/x \to m_1}(y - m_1 x)^2 + \lim_{x \to \infty, y/x \to m_1}(y - m_1 x)\left(\frac{G_{n-2}}{F_{n-2}}\right) + \lim_{x \to \infty, y/x \to m_1}\left(\frac{P_{n-2}}{F_{n-2}}\right) \qquad \ldots(4)$$

Since we know that $c = \lim_{x \to \infty, y/x \to m_1}(y - m_1 x)$

and $A = \lim_{x \to \infty, y/x \to m_1}\left(\frac{G_{n-2}}{F_{n-2}}\right)$ and $B = \lim_{x \to \infty, y/x \to m_1}\left(\frac{P_{n-2}}{F_{n-2}}\right)$

then (4) becomes $c^2 + Ac + B = 0$.

This is a quadratic equation in c so it has two roots let c_1 and c_2 be these two roots. Then we obtain two asymptotes $y - m_1 x = c_1$ and $y - m_1 x = c_2$ corresponding to m_1.

☞ **REMARK**

- As a consequence we can say that the two asymptotes corresponding to the factor $(y - m_1x)^2$ may obtain by solving the quadratic equation $(y - m_1x)^2 + A(y - m_1x) + B = 0$.

  Similarly, we can also find the asymptotes corresponding to the factor $(y - m_1x)^3$, etc. of the nth degree terms in the equation of the curve.

Case IV. Suppose the equation of the curve is of the form $ax + by + c)P_{n-1} + Q_{n-1} = 0$ ...(5)

where P_{n-1} and Q_{n-1} contain various degree term not exceeding the degree $(n-1)^{th}$, and P_{n-1} contains atleast one term of degree $(n-1)$ such that (5) becomes of degree n. Therefore, we can say that $(ax + by)$ is a linear factor of nth degree terms in the equation (5). Thus (5) can also be written as

$$(ax + by) P_{n-1} + cP_{n-1} + Q_{n-1} = 0.$$

Divide this equation by P_{n-1} and taking the limit as $x \to \infty$ and $y/x \to -a/b$, we obtain

$$(ax + by + c) + \lim_{x \to \infty, y/x \to (-a/b)} (Q_{n-1}/P_{n-1}) = 0$$

This the required equation of the asymptote.

Case V. Let the equation of the curve of nth degree be of the form

$$F_n + P = 0$$ ...(1)

where F_n is of degree n and P is of degree $n - 2$ or lower and if $F_n = 0$ can be expressed as the product of n linear factors which give n straight lines such that no two of them are parallel or coincident, then all the asymptotes of the curve (1) are obtained by equating to zero the linear factors of F_n.

Solved Examples

EXAMPLE 1. *Find the asymptotes of*

$$(x-y)^2(x^2+y^2)-10(x-y)x^2+12y^2+2x+y = 0.$$

SOLUTION . We have

$$(x - y)^2 - 10(x - y) \lim_{\substack{x \to \infty \\ y/x \to 1}} \frac{x^2}{x^2 + y^2}$$

$$+ 12 + \lim_{\substack{x \to \infty \\ y/x \to 1}} \frac{y^2}{x^2 + y^2} = 0$$

or $$(x - y)^2 - 5(x - y) + 6 = 0$$

which gives parallel asymptotes $x - y = 2$ and $x - y = 3$.

The other two asymptotes are imaginary. Since the remaining linear factors of the four degree terms in the equation to the curve are imaginary.

EXAMPLE 2. *Find the asymptotes of* $(x - y - 1)^2(x^2 + y^2 + 2)$ $+ 6(x - y - 1)(xy + 7) - 8x^2 - 2x - 1 = 0.$

SOLUTION . Dividing by the coefficient of $(x - y - 1)^2$ and taking limits, we see that the asymptotes parallel to $x - y - 1 = 0$ are

$$(x-y-1)^2 + 6(x-y-1) \lim_{\substack{x \to \infty \\ \frac{y}{x} \to 1}} \frac{xy+7}{x^2+y^2+2}$$

$$+ \lim_{\substack{x \to \infty \\ \frac{y}{x} \to 1}} \frac{-8x^2 - 2x - 1}{x^2 + y^2 + 2} = 0$$

$$\Rightarrow \qquad (x-y-1)^2 + 3(x-y-1) - 4 = 0$$

$$\Rightarrow \qquad x - y - 1 = \frac{-3 \pm \sqrt{9 + 16}}{2} = 1, -4.$$

Hence, the two asymptotes are $x - y - 2 = 0$ and $x - y + 3 = 0$ the remaining two asymptotes are imaginary.

2.7.8 ASYMPTOTES BY EXPANSION

THEOREM. *Let the equation of the curve be of the form* $y = mx + c + \dfrac{A_1}{x} + \dfrac{A_2}{x^2} + \dfrac{A_3}{x^3} + ...$ ...(1)

then $y = mx + c$ *is the asymptote of* (1).

PROOF. Since the equation of the curve is

$$y = mx + c + \frac{A_1}{x} + \frac{A_2}{x^2} + \frac{A_3}{x^3} + ... \; ; \text{ where } \frac{A_1}{x} + \frac{A_2}{x^2} + \frac{A_3}{x^3} + ... \text{ is convergent for sufficiently large values of } x.$$

Differentiating (1) w.r.t. 'x', we get $\dfrac{dy}{dx} = m - \dfrac{A_1}{x^2} - \dfrac{2A_2}{x^3} - \dfrac{3A_3}{x^4} - ...$

Now the equation of the tangent to (1) at the point $P(x, y)$ is

$$Y - y = \left(m - \frac{A_1}{x^2} - \frac{2A_2}{x^3} - \frac{3A_3}{x^4} - ... \right)(X - x)$$

or $$Y = \left(m - \frac{A_1}{x^2} - \frac{2A_2}{x^3} - \frac{3A_3}{x^4} - ... \right)X + c + \frac{2A_1}{x} + \frac{3A_2}{x^2} + ... \qquad \text{[Using(1)]}$$

Now taking the limit as $x \to \infty$, we get

$$Y = mX + c.$$

Hence $y = mx + c$ is an asymptote of the curve $y = mx + c + \dfrac{A_1}{x} + \dfrac{A_2}{x^2} + \dfrac{A_3}{x^3} + ...$

Solved Examples

EXAMPLE 1. *Find the asymptotes of the hyperbola* $\dfrac{x^2}{a^2}-\dfrac{y^2}{b^2}=1.$

SOLUTION. The equation of the curve can be written as

$$y^2=b^2\left(-1+\frac{x^2}{a^2}\right)$$

or $\qquad y=\pm b\sqrt{\left(-1+\dfrac{x^2}{a^2}\right)}=\pm\dfrac{b}{a}x\sqrt{\left(1-\dfrac{a^2}{x^2}\right)}$

$$y=\pm\frac{b}{a}x\left[1-\frac{1}{2}\frac{a^2}{x^2}-\frac{1}{8}\frac{a^4}{x^4}+...\right]$$

[Using binomial expansion]

Since we know that $y = mx + c$ is an asymptote of the curve

$$y=mx+c+\frac{A_1}{x}+\frac{A_2}{x^2}+...$$

Hence, $y=\pm\dfrac{b}{a}x$ are the asymptotes of the given curve.

EXAMPLE 2. *Find all the asymptotes of the curve* $(y^2-x^2)(y-2x)-7xy+3y^2+2x^2+2x+2y+1=0.$

SOLUTION. The given equation can be written as
$$(y-x)(y+x)(y-2x)-7xy+3y^2+2x^2+2x+2y+1=0.$$
$$...(1)$$

The slope of the asymptote corresponding to the factor $y - x$ is 1. Thus the asymptote corresponding to this factor is

$$y-x=\lim_{x\to\infty,\frac{y}{x}\to1}\frac{7xy-3y^2-2x^2-2x-2y-1}{(y+x)(y-2x)}$$

$$=\lim_{x\to\infty,\frac{y}{x}\to1}\frac{7\left(\dfrac{y}{x}\right)-3\left(\dfrac{y}{x}\right)^2-2-\dfrac{2}{x}-2\dfrac{y}{x}\left(\dfrac{1}{x}\right)-\dfrac{1}{x^2}}{\left(\dfrac{y}{x}-1\right)\left(\dfrac{y}{x}-2\right)}$$

$$=\frac{7-3-2}{2(1-2)}=\frac{2}{-2}=-1.$$

$\therefore\quad y-x+1=0$

Similarly the second asymptote corresponding to the factor $(y + x)$ is

$$x+y=\lim_{x\to\infty,\frac{y}{x}\to-1}\frac{7xy-3y^2-2x^2-2x-2y-1}{(y-x)(y-2x)}$$

$$=\lim_{x\to\infty,\frac{y}{x}\to-1}\frac{7\left(\dfrac{y}{x}\right)-3\left(\dfrac{y}{x}\right)^2-2-\dfrac{2}{x}-2\left(\dfrac{y}{x}\right)\left(\dfrac{1}{x}\right)-\dfrac{1}{x^2}}{\left(\dfrac{y}{x}-1\right)\dfrac{y}{x}-2}$$

$$=\frac{7(-1)-3(-1)^2-2}{(-1-1)(-1-2)}=\frac{-7-3-2}{(-2)(-3)}=-2$$

$\therefore\qquad x+y+2=0$

and the third asymptote corresponding to the factor $y - 2x$ is

$$y-2x=\lim_{x\to\infty,\frac{y}{x}\to2}\frac{7xy-3y^2-2x^2-2x-2y-1}{(y-x)(y+x)}$$

$$=\lim_{x\to\infty,\frac{y}{x}\to2}\frac{7\left(\dfrac{y}{x}\right)-3\left(\dfrac{y}{x}\right)^2-2-2\left(\dfrac{1}{x}\right)-2\left(\dfrac{y}{x}\right)\left(\dfrac{1}{x}\right)-\dfrac{1}{x^2}}{\left(\dfrac{y}{x}+1\right)\left(\dfrac{y}{x}+1\right)}$$

$$=\frac{7(2)-3(2)^2-2}{(2-1)(2+1)}=\frac{14-12-2}{3}=0.$$

$\Rightarrow y-2x=0$

Hence, all the asymptotes are $y - x + 1 = 0$, $x + y + 2 = 0$ and $y - 2x = 0$.

EXAMPLE 3. *Find all the asymptotes of the curve* $(y-x)(y-2x)^2+(y+3x)(y-2x)+2x+2y-1=0.$

SOLUTION. The equation of the curve is
$$(y-x)(y-2x)^2+(y+3x)(y-2x)+2x+2y-1=0$$

The asymptotes corresponding to the factor $(y-2x)^2$ are

$$(y-2x)^2+(y-2x)\lim_{x\to\infty,y/x\to2}\frac{y+3x}{y-x}$$

$$+\lim_{x\to\infty,y/x\to2}\frac{2x+2y-1}{(y-x)}=0$$

or $(y-2x)^2+(y-2x)\lim_{x\to\infty,y/x\to2}\left(\dfrac{y/x+3}{y/x-1}\right)$

$$+\lim_{x\to\infty,y/x\to2}\frac{2+2(y/x)-1/x}{(y/x-1)}=0$$

or $\qquad (y-2x)^2+5(y-2x)+6=0$

or $\qquad (y-2x)=\dfrac{-5\pm\sqrt{(25-24)}}{2}=\dfrac{-5\pm1}{2}$

or $\qquad y-2x=-2$ and $\qquad y-2x=-3$

or $\quad y-2x+2=0$ and $y-2x+3=0$

And the asymptote corresponding to the factor $(y-x)$ is

$$(y-x)+\lim_{x\to\infty,y/x\to1}\frac{(y+3x)(y-2x)}{(y-2x)^2}$$

$$+\lim_{x\to\infty,y/x\to1}\frac{2x+2y-1}{(y-2x)^2}=0$$

or
$$(y-x)+\lim_{x\to\infty,y/x\to1}\frac{(y/x+3)(y/x-2)}{(y/x-2)^2}$$

$$+\lim_{x\to\infty,y/x\to1}\frac{2+2(y/x)-1/x}{x(y/x-2)^2}=0$$

or $\qquad (y-x)+\dfrac{(1+3)(1-2)}{(1-2)^2}+0=0$

or $\qquad\qquad y-x-4=0$

Hence, all the asymptotes of the given curve are $y - 2x + 2 = 0$, $y - 2x + 3 = 0$ and $y - x - 4 = 0$

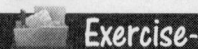 **Exercise-2.15**

Find all the asymptotes of the following curves :

1. $(x^2 - y^2)(x + 2y + 1) + x + y + 1 = 0$
2. $x^5 - y^5 = a^3 xy$
3. $(x^2 - y^2)(y^2 - 4x^2) - 6x^3 + 5x^2 y + 3xy^2 - 2y^3 - x^2 + 3xy - 1 = 0$
4. $x^2(x^2 - y^2)(x - y) + 2x^3(x - y) - 4y^3 = 0$
5. $xy(x^2 - y^2)(x^2 - 4y^2) + xy(x^2 - y^2) + x^2 + y^2 - 7 = 0$
6. $(x - 2y)^2(x - y) - 4y(x - 2y) - (8x + 7y) = 0$
7. $(x - y)^2(x^2 + y^2) - 10(x - y)x^2 + 12y^2 + 2x + y = 0$

8. $(x - y - 1)^2(x^2 + y^2 + 2) + 6(x - y - 1)(xy + 7) - 8x^2 - 2x - 1 = 0$
9. $(\alpha_1 x + \beta_1 y + \gamma_1)(\alpha_2 x + \beta_2 y + \gamma_2) + \gamma_3 = 0$
10. $(x - y + 2)(2x - 3y + 4)(4x - 5y + 6) + 5x - 6y + 7 = 0$
11. $(x - y + 1)(x - y - 2)(x + y) = 8x - 1$
12. $(x^2 - 3x + 2)(x + y - 2) + 1 = 0$
13. $x(y - 3)^3 - 4y(x - 1)^3 = 0$
14. $x^2(x + y)(x - y)^2 + ax^3(x - y) - a^2 y^3 = 0$
15. $(y - a)^2(x^2 - a^2) = x^4 + a^4$

Hints to Selected Problems

1. Asymptotes corresponding to the factor $(x - y)$ is

$$(x - y) + \lim_{\substack{x \to \infty \\ y/x \to 1}} \frac{(x + y + 1)}{(x + y)(x + 2y + 1)} = 0$$

which gives $x - y = 0$.

Similarly, asymptotes corresponding to the factor $(x + y)$ is given by $x + y = 0$ and so on.

Note. Apply the same procedure to all other questions.

Answers

1. $x - y = 0, x + y = 0, x + 2y + 1 = 0$
2. $y - x = 0$
3. $x - y = 0, 2x - y = 0, x + y + 1 = 0, 2x + y + 1 = 0$
4. $x - y + 2 = 0, x - y - 1 = 0, x + y + 1 = 0, x + 2 = 0$
5. $x = 0, y = 0, x - y = 0, x + y = 0, x - 2y = 0$ and $x + 2y = 0$
6. $x - y + 4 = 0, x - 2y = 2 \pm 3\sqrt{3}$
7. $x - y - 2 = 0, x - y - 3 = 0$
8. $x - y - 2 = 0, x - y + 3 = 0$
9. $\alpha_1 x + \beta_1 x + \gamma_1 = 0, \alpha_2 x + \beta_2 y + \gamma_2 = 0$
10. $x - y + 2 = 0, 2x - 3y + 4 = 0, 4x - 5y + 6 = 0$
11. $y + x = 0, x - y - 2 = 0, x - y + 1 = 0$
12. $x = 1, x = 2, x + y - 2 = 0$
13. $x = 0, y = 0, 4x - 2y + 3 = 0, 4x + 2y - 15 = 0$
14. $x \pm a = 0, x - y + a = 0, y = \pm x - \frac{1}{2}a$
15. $x \pm a = 0, x - y + a = 0, x + y - a = 0$

2.7.9 Intersection of a Curve with its Asymptotes

Let the equation $\quad\quad\quad\quad\quad\quad\quad\quad y = mx + c \quad\quad\quad\quad\quad\quad\quad\quad\quad$ ...(1)

be an asymptote of the curve $\quad x^2 \phi_n\left(\dfrac{y}{x}\right) + x^{n-1}\phi_{n-1}\left(\dfrac{y}{x}\right) + x^{n-2}\phi_{n-2}\left(\dfrac{y}{x}\right) + \ldots = 0.$ ...(2)

Solving (1) and (2) to find the intersection points so eliminating y between (1) and (2), we get

$$x^n \phi_n\left(m + \frac{c}{x}\right) + x^{n-1}\phi_{n-1}\left(m + \frac{c}{x}\right) + x^{n-2}\phi_{n-2}\left(m + \frac{c}{x}\right) + \ldots = 0.$$

Now expand each term of above equation by Taylor's theorem, we have

$$x^n \left[\phi_n(m) + \frac{c}{x}\phi'_n(m) + \frac{c^2}{x^2}\cdot\frac{1}{2!}\phi''_n(m) + \ldots\right] + x^{n-1}\left[\phi_{n-1}(m) + \frac{c}{x}\phi'_{n-1}(m) + \ldots\right] + x^{n-2}\left[\phi_{n-2}(m) + \frac{c}{x}\phi'_{n-2}(m) + \ldots\right] = 0$$

or $\quad x^n \phi_n(m) + [c\phi'_n(m) + \phi_{n-1}(m)x^{n-1}] + \left[\dfrac{c^2}{2!}\phi''_n(m) + \dfrac{c}{1!}\phi'_{n-1}(m) + \phi_{n-2}(m)\right]x^{n-2} + \ldots = 0.$ ...(3)

Since $y = mx + c$ is an asymptotes of the curve (2), then we have $\phi_n(m) = 0$ and $c\phi'_n(m) + \phi_{n-1}(m) = 0$.

Thus (3) becomes $\quad\quad\quad\quad\quad\quad\quad \left[\dfrac{c^2}{2!}\phi''_n(m) + \dfrac{c}{1!}\phi'_{n-1}(m) + \phi_{n-2}(m)\right]x^{n-2} + \ldots = 0.$ ...(4)

This is a equation of degree $n - 2$ in x so it will have atmost $n - 2$ values of x provided there is no asymptote parallel to $y = mx + c$ of the given curve.

Hence, in general we can say that any asymptote of a curve of the nth degree cuts the curve in $(n - 2)$ points.

Remarks

- Since one asymptote of the curve of nth degree cuts the curve in $(n - 2)$ points so n asymptotes of that curve will cut in $n(n - 2)$ points.
- If the equation of the curve of degree n can be written as $F_n + P = 0$, where F_n contains n non-repeated linear factors and P contains the terms almost of degree $n - 2$, then $n(n - 2)$ points of intersection of the curve will lie on the curve $P = 0$.

Solved Examples

EXAMPLE 1. *Show that the four asymptotes of the curve*
$$(x^2 - y^2)(y^2 - 4x^2) + 6x^3 - 5x^2y$$
$$- 3xy^2 + 2y^3 - x^2 + 3xy - 1 = 0.$$
cut the curve in eight points which lie on the circle $x^2 + y^2 = 1$.

SOLUTION. The given equation of the curve can be written as
$$(x - y)(x + y)(y - 2x)(y + 2x) + 6x^3 - 5x^2y$$
$$- 3xy^2 + 2y^3 - x^2 + 3xy - 1 = 0 \ ...(1)$$
The asymptote corresponding to the factor $x - y$ is
$$6x^3 - 5x^2y - 3xy^2$$
$$x - y + \lim_{x\to\infty, y/x\to1} \frac{+2y^3 - x^2 + 3xy - 1}{(x + y)(y - 2x)(y + 2x)} = 0$$
or
$$x - y + \lim_{x\to\infty, \frac{y}{x}\to1} \frac{6 - 5\left(\frac{y}{x}\right) - 3\left(\frac{y}{x}\right)^2 + 2\left(\frac{y}{x}\right)^3 - \frac{1}{x} + 3\left(\frac{y}{x}\right)\left(\frac{1}{x}\right) - \frac{1}{x^3}}{\left(1 + \frac{y}{x}\right)\left(\frac{y}{x} - 2\right)\left(\frac{y}{x} + 2\right)} = 0$$
or $$x - y + \lim_{x\to\infty, y/x\to1} \frac{6 - 5 - 3 + 2}{(1 + 1)(1 - 2)(1 + 2)} = 0$$
or $$x - y = 0.$$
The asymptote corresponding to the factor $x + y$ is
$$6x^3 - 5x^2y - 3xy^2$$
$$x + y + \lim_{x\to\infty, y/x\to-1} \frac{+2y^3 - x^2 + 3xy - 1}{(x - y)(y - 2x)(y + 2x)} = 0$$
or
$$x + y + \lim_{x\to\infty, y/x\to-1} \frac{6 - 5(y/x) - 3(y/x)^2 + 2(y/x)^3 - (1/x) + 3(y/x)(1/x) - (1/x^3)}{(1 - y/x)(y/x - 2)(y/x + 2)} = 0$$
or $$x + y + \frac{6 - 5(-1) - 3(-1)^2 + 2(-1)^3}{(1 + 1)(-1 - 2)(-1 + 2)} = 0$$
or $$x + y - 1 = 0.$$
Now the asymptote corresponding to the factor $y - 2x$ is
$$6x^3 - 5x^2y - 3xy^2$$
$$y - 2x + \lim_{x\to\infty, y/x\to2} \frac{+2y^3 - x^2 + 3xy - 1}{(x - y)(x + y)(y + 2x)} = 0$$

or
$$6 - 5(y/x) - 3(y/x)^2 + 2(y/x)^3 - (1/x)$$
$$y - 2x + \lim_{\substack{x\to\infty \\ y/x\to2}} \frac{+3(y/x)(1/x) - (1/x^3)}{(1 - y/x)(1 + y/x)(y/x + 2)} = 0$$
or $$y - 2x + \frac{6 - 5(2) - 3(2)^2 + 2(2)^3}{(1 - 2)(1 + 2)(2 + 2)} = 0$$
or $$y - 2x = 0.$$
The asymptote corresponding to the factor $y + 2x$ is
$$y + 2x + \lim_{\substack{x\to\infty, \\ y/x\to-2}} \frac{6x^3 - 5x^2y - 3xy^2 + 2y^3 - x^2 + 3xy - 1}{(x - y)(x + y)(y - 2x)} = 0$$

or $$y + 2x + \lim_{\substack{x\to\infty, \\ y/x\to-2}} \frac{6 - 5\left(\frac{y}{x}\right) - 3\left(\frac{y}{x}\right)^2 + 2\left(\frac{y}{x}\right)^3 - \left(\frac{1}{x}\right) + 3\left(\frac{y}{x}\right)\left(\frac{1}{x}\right) - \left(\frac{1}{x^3}\right)}{(1 - y/x)(1 + y/x)(y/x - 2)} = 0$$
or $$y + 2x + \frac{6 - 5(-2) - 3(-2)^2 + 2(-2)^3}{(1 + 2)(1 - 2)(-2 - 2)} = 0$$
or $$y + 2x - 1 = 0.$$
Hence, all the four asymptotes are $x - y = 0$, $x + y - 1 = 0$, $y - 2x = 0$ and $y + 2x - 1 = 0$.
Since one asymptote cuts the curve in $(4 - 2) = 2$ points so all the four asymptotes cut the curve in $4 \times 2 = 8$ points. Now combine all the asymptotes, we get
$$(x - y)(x + y - 1)(y - 2x)(y + 2x - 1) = 0$$
or $[x^2 - y^2 - (x - y)][y^2 - 4x^2 - (y - 2x)] = 0$
or $(x^2 - y^2)(y^2 - 4x^2) - (x^2 - y^2)(y - 2x)$
$$- (x - y)(y^2 - 4x^2) + (x - y)(y - 2x) = 0$$
or $(x^2 - y^2)(y^2 - 4x^2) - (x^2y - 2x^3 - y^3 + 2xy^2)$
$$- (xy^2 - 4x^3 - y^3 + 4x^2y) + xy - 2x^2 - y^2 - 2xy = 0$$
or $$(x^2 - y^2)(y^2 - 4x^2) + 6x^3 - 5x^2y$$
$$- 3xy^2 + 2y^3 - 2x^2 - y^2 + 3xy = 0. \qquad ...(2)$$
Now subtract (2) from (1), we get
$$x^2 + y^2 = 1.$$
Hence, all the eight points of intersection lie on the circle $x^2 + y^2 = 1$.

 Exercise-2.16

1. Show that the asymptotes of the curve
$$4(x^4 + y^4) - 17x^2y^2 - 4x(4y^2 - x^2) + 2(x^2 - 2) = 0$$
cut the curve in eight points which lie on the ellipse $x^2 + 4y^2 = 4$.

2. Find the asymptotes of the curve $x^2y - xy^2 + xy + y^2 + x - y = 0$ and show that they cut the curve again in three points which lie on the straight line $x + y = 0$.

3. Show that the eight points of intersection of the curve
$$x^4 - 5x^2y^2 + 4y^4 + x^2 - y^2 + x + y + 1 = 0$$
and its asymptotes lie on a rectangular hyperbola.

4. Show that the asymptotes of the cubic
$$x^3 - 2y^3 + xy(2x - y) + y(x - y) + 1 = 0$$
cut the curve in three points which lie on the straight line $x - y + 1 = 0$.

5. Find the equation of the cubic which has the same asymptotes as the curve $x^3 - 6x^2y + 11xy^2 - 6y^3 + x + y + 1 = 0$ and which passes through the points $(0, 0)$, $(1, 0)$ and $(0, 1)$.

6. Show that the asymptotes of the curve $y^2(x^2 - a^2) = x^2(x^2 - 4a^2)$ form two right angle triangles with the x-axis. $(y > 0)$.

Answers

2. $y = 0, x = 1, x - y + 2 = 0$ **5.** $x^3 - 6x^2y + 11xy^2 - 6y^3 - x + 6y = 0.$

2.7.10 ASYMPTOTES OF NON-ALGEBRAIC CURVES

Definition. *A curve in which there are some terms involving cosine, sine, etc. is called non-algebraic curve.*

The method for finding the asymptotes of non-algebraic curves can be explained by following example.

Let the equation of the curve be $y = \sec x$, then differentiating this w.r.t. 'x', we get

$$\frac{dy}{dx} = \sec x \tan x.$$

Therefore, the tangent at $P(x, y)$ on the curve is

$$Y - \sec x = \frac{dy}{dx}(X - x)$$

or $Y - \sec x = \sec x \tan x(X - x)$

or $Y \cos^2 x - \cos x = (X - x)\sin x.$ ...(1)

Now taking the distance of $P(x, y)$ from $(0, 0)$ infinity as $x \to \pi/2$ and $y \to \infty$, we get

$$Y.0 - 0 = (X - \pi/2).1 \quad \text{or} \quad X = \pi/2.$$

This is one asymptote and the other asymptotes are $X = -\pi/2, \pm 3/2\pi,...$

2.7.11 ASYMPTOTES OF POLAR CURVES

(i) Equation of a line in polar form. Let O be the pole and OX the initial line and let $P(r, \theta)$ be any point on the line whose equation is to be required as shown in fig. 24.

Draw a perpendicular OM from O to the line such that $OM = p$ and $\angle MOX = \alpha$ (say).

$\therefore$ In $\triangle OPM$, $\angle POM = \theta - \alpha$

then, $\dfrac{OM}{OP} = \cos \angle POM$

or $\dfrac{p}{r} = \cos (\theta - \alpha)$

or $p = r \cos (\theta - \alpha).$

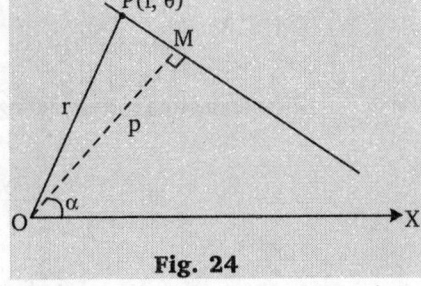

Fig. 24

This is the equation of line in polar form, where p is the perpendicular length from pole to this line and α is an angle which the perpendicular makes with initial line.

(ii) Asymptotes of polar curves.

THEOREM 1. *If $\theta = \alpha$ is a root of the equation $f(\theta) = 0$, then $r \sin (\theta - \alpha) = 1/f'(\alpha)$ is an asymptote of the curve $1/r = f(\theta)$.*

PROOF.   Since the equation of a curve in polar form is $\dfrac{1}{r} = f(\theta).$ ...(1)

Let $P(r, \theta)$ be any point on this curve and draw a line through O perpendicular to OP, then radius vector which meets that tangent at P in T as show in fig. 25.

Then OT is a polar subtangent of the curve at P.

$$OT = r^2 \frac{d\theta}{dr} \qquad \text{(From calculus)}$$

Now differentiating (1) w.r.t. 'θ', we get

$$-\frac{1}{r^2}\frac{dr}{d\theta} = f'(\theta).$$

$\therefore$ $OT = r^2 \dfrac{d\theta}{dr} = -\dfrac{1}{f'(\theta)}.$

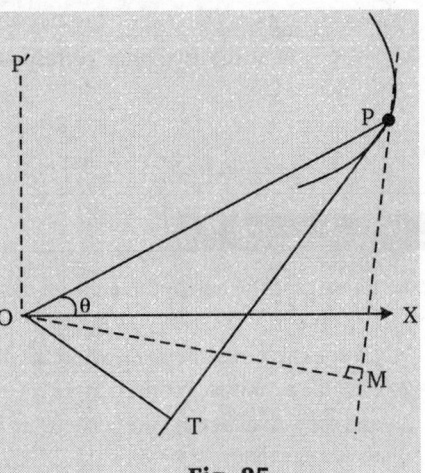

Fig. 25

Since α is a root of $f(\theta) = 0$ as $\theta \to \alpha$, then $r \to \infty$ from (1) and the tangent PT

tends to the asymptote and $OT \to \left[-\dfrac{1}{f'(\theta)}\right]_{\theta=\alpha}, f'(\alpha) \neq 0.$

And OP, PT will become parallel to lines shown dotted in the above fig. Thus $\angle OTP \to \pi/2$ and $OT \to OM$, where OM is a perpendicular distance from O to the asymptote.

$\therefore$ $OM = -\dfrac{1}{f'(\alpha)}$

when $\theta \to \alpha$ i.e., $OP \to OP'$. Then $\angle XOP' = \alpha$

$\therefore \qquad \angle MOX = -\left(\dfrac{\pi}{2} - \alpha\right)$ (In the clockwise direction)

Therefore the equation of the asymptote is

$$r\cos\left[\theta - \left\{-\left(\dfrac{\pi}{2} - \alpha\right)\right\}\right] = -\dfrac{1}{f'(\alpha)} \qquad\qquad [\text{using } p = r\cos(1 - \alpha)]$$

or $\qquad\qquad r\cos\left(\dfrac{\pi}{2} + \theta - \alpha\right) = -\dfrac{1}{f'(\alpha)} \qquad$ or $\qquad -r\sin(\theta - \alpha) = -\dfrac{1}{f'(\alpha)}$

or $\qquad\qquad r\sin(\theta - \alpha) = \dfrac{1}{f'(\alpha)}$

WORKING PROCEDURE

To find the asymptotes of polar curves, we use the follows steps :

Step 1. Convert the equation of the given curve in the form $\dfrac{1}{r} = f(\theta)$.

Step 2. Find the roots of the equation $f(\theta) = 0$ i.e., values of θ. Suppose α, β, etc. are the roots of $f(\theta) = 0$.

Step 3. Now the asymptote corresponding to $\theta = \alpha$ is $r\sin(\theta - \alpha) = \dfrac{1}{f'(\alpha)}$
where $f'(\alpha) = $ value of $f'(\theta)$ at $\theta = \alpha$.

Solved Examples

EXAMPLE 1. *Find the asymptotes of the curve $r \sin n\theta = a$.*

SOLUTION. **Step I.** Convert the given curve into the form

$$\dfrac{1}{r} = f(\theta).$$

$\therefore \qquad \dfrac{1}{r} = \dfrac{\sin n\theta}{a} = f(\theta).$ \qquad ...(1)

Step II. Solve the equation $f(\theta) = 0$.

i.e., $\qquad \dfrac{\sin n\theta}{a} = 0.$

or $\qquad \sin n\theta = \sin r\pi, \quad r = 0, 1, 2, ...,$

or $\qquad n\theta = r\pi$ or $\theta = \dfrac{r\pi}{n}, \quad r = 0, 1, 2, 3, ...$

Let $\qquad \alpha = \dfrac{r\pi}{n}.$

Now differentiating (1) w.r.t. 'θ', we get

$$f'(\theta) = +\dfrac{n\cos n\theta}{a}.$$

$\therefore \qquad f'(\alpha) = \dfrac{n\cos n\alpha}{a} = \dfrac{n}{a}\cos r\pi = \dfrac{n}{a}(-1)^r.$

Step III. Therefore, the asymptotes of the curve are

$$r\sin(\theta - \alpha) = \dfrac{1}{f'(\alpha)}$$

or $\qquad r\sin\left(\theta - \dfrac{r\pi}{n}\right) = \dfrac{a}{n(-1)^r}$, where r is any integer.

EXAMPLE 2. *Find the asymptotes of the curve $r \sin\theta = a \cos 2\theta$.*

SOLUTION. First put the equation in the form of $\dfrac{1}{r} = f(\theta)$.

i.e., $\qquad \dfrac{1}{r} = \dfrac{\sin\theta}{a\cos 2\theta}.$

$\therefore \qquad f(\theta) = \dfrac{\sin\theta}{a\cos 2\theta}.$ \qquad ...(1)

Now solve the equation $f(\theta) = 0$. Then

$$\dfrac{\sin\theta}{a\cos 2\theta} = 0$$

or $\qquad \sin\theta = \sin n\pi$ or $\theta = n\pi.$

Let $\alpha = n\pi$ be the root of the equation $f(\theta) = 0$.

Now differentiating (1) w.r.t. 'θ', we get

$$f'(\theta) = \dfrac{1}{a}\left[\dfrac{\cos 2\theta.\cos\theta + 2\sin 2\theta\sin\theta}{\cos^2 2\theta}\right]$$

$\therefore f'(\alpha) = \dfrac{1}{a}\left[\dfrac{\cos 2\alpha.\cos\alpha + 2\sin 2\alpha\sin\alpha}{\cos^2 2\alpha}\right]$

$\qquad = \dfrac{1}{2a}\left[\dfrac{\cos 2n\pi.\cos n\pi + 2\sin 2n\pi\sin n\pi}{\cos^2 2n\pi}\right]$

$\qquad\qquad\qquad\qquad\qquad\qquad (\because \alpha = n\pi)$

$\qquad = \dfrac{1}{a}\cos n\pi.$

The asymptote corresponding to $\alpha = n\pi$ is

$$r\sin(\theta - n\pi) = \dfrac{1}{f'(\alpha)} = \dfrac{a}{\cos n\pi}$$

or $\quad r(\sin\theta\cos n\pi - \cos\theta\sin n\pi) = \dfrac{a}{\cos n\pi}$

or $\qquad r\sin\theta\cos n\pi = \dfrac{a}{\cos n\pi} \qquad (\because \sin n\pi = 0)$

or $\qquad r\sin\theta\cos^2 n\pi = a$

or $\qquad\qquad r\sin\theta = a \qquad (\because \cos n\pi = 1)$

Exercise-2.17

Find the asymptotes of the following curves :

1. $y = \tan x.$

2. $r = a\,\mathrm{cosec}\,\theta + b$

3. $r\sin 2\theta = a$

4. $r\sin\theta = 2\cos 2\theta$

5. $r \sin \theta = 2 \cos \theta$

6. $r\theta \cos \theta = a \cos 2\theta$

7. $r(1 - 2\cos \theta) = 2a$

8. $r = 4(\sec \theta + \tan \theta)$

9. $r \cos \theta = 4 \sin^2\theta$

10. $r(e^\theta - 1) = a(e^\theta + 1)$

11. $r \cos \theta = a \sin \theta$

12. $r(1 + 2\sin \theta) = 2$

13. $r \sin \theta = 2\theta$

Hints to Selected Problems

1. (i) $y = \tan x \Rightarrow \dfrac{dy}{dx} = \sec^2 x$

Tangent at (x, y)

$Y - \tan x = \sec^2 x(Y - x) \Rightarrow Y \cos^2 x - \sin x \cos x = (X - x).$

Now as $x \to \pi/2, y \to \infty$ and the distance of (x, y) from $(0, 0) \to \infty$.

$\therefore Y.0 - 0 = (X - \pi/2) \Rightarrow X = \pi/2.$

2. $\dfrac{1}{r} = f(\theta) = \dfrac{\sin \theta}{a + b \sin \theta}$. Solving, $f(0) = 0$. we get $\theta = n\pi = \alpha$ (say)

$\Rightarrow f'(\alpha) = \dfrac{1}{a} \cos n\pi.$

Now required asymptotes are given by $r \sin(\theta - \alpha) = \dfrac{1}{f'(\alpha)}.$

3. $\dfrac{1}{r} = f(\theta) = \dfrac{\sin 2\theta}{a}.$

Now on solving $f(0) = 0$ we get $\theta = \dfrac{n\pi}{2} = \alpha$ (say).

Also, $f'(\alpha) = \dfrac{2\cos n\pi}{a}.$

Therefore, the asymptotes of the given curve is

$$r \sin(\theta - \alpha) = \dfrac{1}{f'(\alpha)}.$$

4. $\dfrac{1}{r} = f(\theta) = \dfrac{\sin \theta}{2 \cos 2\theta}$

Now, $f'(\alpha) = \dfrac{\cos n\pi}{2}$. Therefore, the asymptotes of the given curve is given by $r \sin(\theta - \alpha) = \dfrac{1}{f'(\alpha)}.$

5. Here $\theta = n\pi = \alpha$, $f'(\alpha) = \dfrac{1}{2}$. Then use the required formula.

6. $\theta = 0, \left(k\pi + \dfrac{\pi}{2}\right)$, $f'(\alpha) = \dfrac{1}{a}$

7. $\theta = \pm\dfrac{\pi}{3}, f'(\theta_1) = \dfrac{-\sqrt{3}}{2a}$, $f'(\theta_2) = \dfrac{\sqrt{3}}{2a}$

8. $\theta = \left(2n\pi + \dfrac{\pi}{2}\right) = \alpha$ (say), $f'(\alpha) = -\dfrac{1}{8}$

9. $\theta = \left(2n\pi + \dfrac{\pi}{2}\right) = \alpha$ (say), $f'(\alpha) = -\dfrac{1}{4}$

10. $\theta = 2n\pi = \alpha$, $f'(\alpha) = \dfrac{1}{2a}$

11. $\theta = \left(n\pi + \dfrac{\pi}{2}\right) = \alpha$, $f'(\alpha) = \dfrac{-1}{a}$

12. $\theta = \dfrac{-\pi}{6} = \alpha$, $f'(\alpha) = \dfrac{\sqrt{3}}{2}$

13. $\theta = n\pi = \alpha$, $f'(\alpha) = \dfrac{1}{2}\left(\dfrac{\cos n\pi}{n\pi}\right)$

Answers

1. $x = \pm\pi/2, \pm3\pi/2...$ **2.** $r \sin \theta = a$ **3.** $r \sin \theta = \pm\dfrac{1}{2}a, r \cos \theta = \pm\dfrac{1}{2}a$ **4.** $r \sin \theta = 2$ **5.** $r \sin \theta = \pm2$

6. $r \sin \theta = a, r \cos \theta = \dfrac{a}{\left(k + \dfrac{1}{2}\right)\pi}, k$ is any integer   **7.** $r \sin\left(\theta - \dfrac{\pi}{3}\right) = \dfrac{2a}{\sqrt{3}}, r \sin\left(\theta + \dfrac{\pi}{3}\right) = -\dfrac{2a}{\sqrt{3}}$ **8.** $r \cos \theta = 8$ **9.** $r \cos \theta = 4$

10. $r \sin \theta = 2a$ **11.** $r \cos \theta = \pm a$ **12.** $r \sin\left(\theta \pm \dfrac{\pi}{6}\right) = \dfrac{2}{\sqrt{3}}$ **13.** $r \sin \theta = 2n\pi, n = \pm1, \pm2, ...$

2.8 CURVE TRACING

If P is any point on a curve and CD is any given line which does not passes through this point P. Then the curve is said to be concave at P with respect to the line CD if the small arc of the curve containing P lies entirely within the acute angle between the tangent at P to the curve and the line CD and the curve is said to be convex at P if the arc of the curve containing P lies wholly outside the acute angle between that tangent at P and the line CD which are shown in figures below :

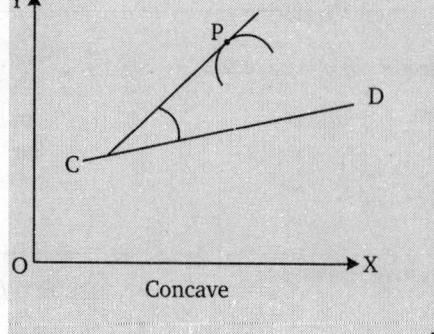
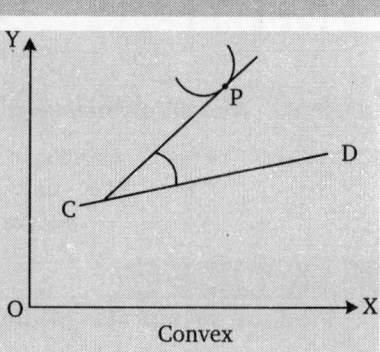

Fig. 26

2.8.1 POINT OF INFLEXION

A point P on the curve is said to be the point of inflexion, if the curve in one side of P is concave and other side of P is convex with respect to the line CD which does not passes through the point P as shown in fig. 27.

Inflexion tangent. The tangent at the point of inflexion of a curve is said to be inflexion tangent. In the fig. 2 the line PQ is the inflexion tangent.

2.8.2 DETERMINATION OF THE POINTS OF INFLEXION

Let $y = f(x)$ be the equation of a curve and let $P(x, y)$ be any point on the curve and assuming that the tangent at P is not parallel to y-axis as shown in fig. 28.

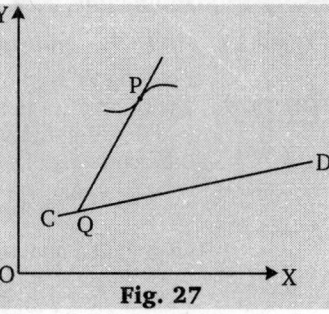

Fig. 27

Since the tangent is taken not to be parallel to y-axis, then $\dfrac{dy}{dx} = f'(x)$ must be finite. Let $Q(x + h,$ $y + k)$ be any point on the curve in the neighbourhood of P. We may take this point Q either side of P. Suppose the ordinate OM of Q intersects the tangent line at Q'.

$$Y - y = f'(x)(X - x) \qquad \qquad ...(1)$$

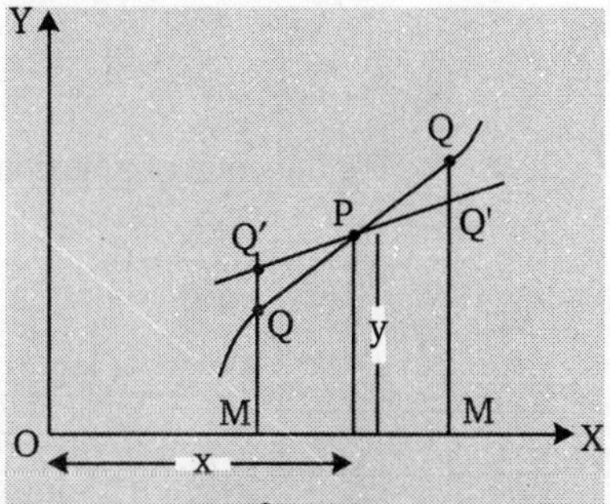

Since at point $Q(x + h, x + k)$ we have $X = x + h$ so putting $X = x + h$ in (1), we get

$$Q'M - y = f'(x)(x + h - x) \qquad \qquad [\because Y = Q'M]$$

or
$$Q'M = y + hf'(x)$$

or
$$Q'M = f(x) + hf'(x). \qquad \qquad [\because y = f(x)]$$

But we know that
$$QM = f(x + h) = f(x) + hf'(x) + \frac{h^2}{2!}f''(x) + \frac{h^3}{3!}f'''(x) + ...$$

(Using Taylor's theorem)

$$\therefore \qquad QM - Q'M = \frac{h^2}{2!}f''(x) + \frac{h^3}{3!}f'''(x) + ... + \frac{h^n}{n!}f^{(n)}(x + \theta h) \text{ where } 0 < \theta < 1. \qquad ...(2)$$

Fig. 28

Let us suppose $f''(x) \neq 0$ and taking h sufficiently small, then $(QM - Q'M)$ will have the same sign as $\dfrac{h^2}{2!}f''(x)$. But $\dfrac{h^2}{2!}f''(x)$ will have invariable sign because h^2 will always be positive. This means that on both sides of P the curve will be either concave or convex. Hence, we can say that the necessary condition for the existence of a point of inflexion at P is given by

$$f''(x) = 0 \text{ or } \frac{d^2y}{dx^2} = 0.$$

Thus (2) now becomes
$$QM - Q'M = \frac{h^2}{3!}f'''(x) + \frac{h^4}{4!}f^{iv}(x) + ... + \frac{h^n}{n!}f^{(n)}(x + \theta h) \qquad \qquad ...(3)$$

Further, if $f'''(x) \neq 0$ and taking h to be very small, then $(QM - Q'M)$ will have the same sign as $\dfrac{h^3}{3!}f'''(x)$ and this changes sign when h changes sign. Thus we can say that the curve with respect to the x-axis is concave on one side of P and convex on other side of P. Hence, there will exist a point of inflexion at P.

Consequently, we can have a point of inflexion at P, if $\dfrac{d^2y}{dx^2} = 0$ but $\dfrac{d^3y}{dx^3} \neq 0$.

☞ REMARKS

- The position of a point of inflexion is independent of the choice of co-ordinate axes so we can say that a point of inflexion at P exists if $\dfrac{d^2y}{dx^2} = 0$ but $\dfrac{d^3y}{dx^3} \neq 0$.

- If $f''(x) = 0 = f'''(x) = ... = f^{(n-1)}(x)$ and $f^{(n)}(x) \neq 0$, then there will be a point of inflexion if n is odd and if n is even and greater than 2, then the point is called point of undulation.

- If the tangent at P is parallel to y-axis, then $\dfrac{dy}{dx}$ will be infinite at P so change the curve to the form $x = f(y)$ and then find the point of inflexion.

Solved Examples

EXAMPLE 1. *Find the points of inflexion of the curve*
$$x = (\log y)^3.$$

SOLUTION . The equation of the curve is
$$x = (\log y)^3 \qquad \ldots(1)$$

Differentiating (1) with respect to 'y', we get
$$\frac{dx}{dy} = 3(\log y)^2 . \frac{1}{y}$$

Again differentiating w.r.t. y
$$\frac{d^2 x}{dy^2} = 3\left[\frac{2\log y}{y^2} - \frac{(\log y)^2}{y^2} \right]. \qquad \ldots(2)$$

Again differentiating w.r.t. 'y', we get
$$\frac{d^3 x}{dy^3} = 3\left[\frac{2}{y^3} - \frac{4\log y}{y^3} - \frac{2\log y}{y^3} - \frac{2(\log y)^2}{y^2} \right]. \qquad \ldots(3)$$

For the point of inflexion, we have
$$\frac{d^2 x}{dy^2} = 0.$$

$$\therefore \quad 3\left[\frac{2\log y - (\log y)^2}{y^2} \right] = 0$$

or $\quad 3(\log y)(2 - \log y) = 0$

or $\quad \log y = 0, \log y = 2$

or $\quad\quad\quad\quad\quad y = 1, y = e^2$

From (3) it is obvious that at $y = 1, y = e^2$,
$$\frac{d^3 x}{dy^3} \neq 0.$$

Hence, the points of inflexion are $(0, 1)(8, e^2)$.

EXAMPLE 2. *Find the points of inflexion of the curve*
$$y^2 = x(x + 1)^2.$$

SOLUTION . The equation of the curve can be written as
$$y = (x+1)\sqrt{x}. \qquad \ldots(1)$$

Differentiating (1) w.r.t. 'x', we get
$$\frac{dy}{dx} = \frac{3}{2}.x^{1/2} + \frac{1}{2\sqrt{x}}.$$

Again differentiating w.r.t. 'x'
$$\frac{d^2 y}{dx^2} = \frac{3}{4\sqrt{x}} - \frac{1}{4x^{3/2}}. \qquad \ldots(2)$$

and again differentiating w.r.t. 'x', we get
$$\frac{d^3 y}{dx^3} = -\frac{3}{8x^{3/2}} + \frac{3}{8x^{5/2}}. \qquad \ldots(3)$$

For the point of inflexion, we have
$$\frac{d^2 y}{dx^2} = 0.$$

$$\therefore \quad \frac{3}{4\sqrt{x}} - \frac{1}{4x\sqrt{x}} = 0$$

or $\quad \left(3 - \frac{1}{x}\right) = 0 \quad$ or $\quad x = 1/3.$

From (3) it is obvious that at $x = 1/3$, $\frac{d^3 y}{dx^3} \neq 0$.

Thus, the point of inflexion are given by $(1/3, \pm 4/3\sqrt{3})$.

Exercise-2.18

1. Find the points of inflexion of the curve $x = \log(y/x)$.

2. Find the points of inflexion of the curve $y(a^2 + x^2) = x^3$.

3. Find the points of inflexion of the curve $y = (x-1)^4 (x-2)^3$.

4. Find the points of inflexion of the curve $xy = a^2 \log(y/a)$.

5. Show that the points of inflexion of the curve $y^2 = (x-a)^2(x-b)$ lie on the line $3x + a = 4b$.

6. Show that the origin is a point of inflexion of the curve $a^{m-1}.y = x^m$, if m is odd and greater than 2.

7. Show that the points of inflexion of the curve $x^2 y = a^2(x-y)$ are given by $x = 0, x = \pm a\sqrt{3}$.

8. Prove that the curve $y = (1-x)/(1+x^2)$ has three points of inflexion which lie on a straight line.

9. Show that the abscissae of the points of inflexion on the curve $y^2 = f(x)$ satisfy the equation
$$[f'(x)]^2 = 2f(x)f''(x).$$

10. Show that the points of inflexion on the curve $y = be^{-(x/a)^2}$ are given by $x = \pm a / \sqrt{2}$.

11. Find the points of inflexion on the curve $r(\theta^2 - 1) = a\theta^2$.

12. Show that the points of inflexion of the curve $r = b\theta^n$ are given by $r = b\{-n(n+1)\}^{n/2}$.

13. Find the points of inflexion of the curve
$$x = a(2\theta - \sin \theta), y = a(2 - \cos \theta).$$

14. Find the points of inflexion of the curve $y = 3x^4 - 4x^3 + 1$.

Hints to Selected Problems

1. Given that $x = \log\left(\frac{y}{x}\right) \Rightarrow y = xe^x$

$\Rightarrow \frac{dy}{dx} = xe^x + e^x, \frac{d^2 y}{dx^2} = xe^x + 2e^x$ and $\frac{d^3 y}{dx^3} = xe^x + 3e^x$

For the point of inflexion, putting $\frac{d^2 y}{dx^2} = 0$ and $\frac{d^3 y}{dx^3} \neq 0$.

5. $y = (x-a)\sqrt{x-b}$

$\Rightarrow \quad \frac{dy}{dx} = \frac{3x - a - 2b}{2\sqrt{x-b}}$ and $\frac{d^2 y}{dx^2} = \frac{3x + a - 4b}{4(x-b)^{3/2}}$

Also, $\frac{d^3 y}{dx^3} = \frac{-3x - 3a + 6b}{4(x-b)^{5/2}}$

By putting $\frac{d^2 y}{dx^2} = 0$, we get $3x + a = 4b$ at which $\frac{d^3 y}{dx^3} \neq 0$.

9. $y^2 = f(x) \Rightarrow \frac{dy}{dx} = \frac{f'(x)}{2y} = \frac{f'(x)}{2\sqrt{f(x)}}$

$\frac{d^2 y}{dx^2} = \frac{2f(x).f''(x) - [f'(x)]^2}{4[f(x)]^{3/2}}$

and $\dfrac{d^3y}{dx^3} = \dfrac{4[f(x)]^2 \cdot f'''(x) - 6f(x)f'(x)f''(x) + 3[f'(x)]^3}{8[f(x)]^{5/2}}$.

For point of inflexion, put $\dfrac{d^2y}{dx^2} = 0$.

11. Given that $r(\theta^2 - 1) = a\theta^2$

$\Rightarrow \dfrac{dr}{d\theta} = -\dfrac{2a\theta}{(\theta^2 - 1)^2}$ and $\dfrac{d^2r}{d\theta^2} = \dfrac{2a(1 + 3\theta^2)}{(\theta^2 - 1)^3}$.

At the point of inflexion, use

$$r^2 + 2\left(\dfrac{dr}{d\theta}\right)^2 - r\dfrac{d^2r}{d\theta^2} = 0.$$

Answers

1. $(-2, -2/e^2)$ **2.** $(0,0), \left(\sqrt{3}a, \dfrac{3\sqrt{3}}{4}a\right), \left(-\sqrt{3}a, \dfrac{-3\sqrt{3}}{4}a\right)$ **3.** Point of inflection at $x = 2$, $(11 \pm \sqrt{2})/7$

4. $\left(\dfrac{3}{2}ae^{-3/2}, ae^{3/2}\right)$ **11.** $\theta = \pm\sqrt{3}$ **13.** $\left[\left(4n\pi \pm \dfrac{2\pi}{3} \mp \dfrac{\sqrt{3}}{2}\right)a, \dfrac{3a}{2}\right]$ **14.** $\left(\dfrac{2}{3}, \dfrac{11}{27}\right), (0,1)$

2.8.3 Multiple and Singular Points

Definition 1. *A point on the curve is said to be multiple points if through this point more than one branches of a curve passes.*

Definition 2. *A point on the curve is called a double point if through it two branches of the curve passes.*

Definition 3. *If three branches of the curve passes through a point, then this point is called triple point.*

Definition 4. *If n branches passes through a point on the curve, then this point is called a multiple point of nth order.*

Definition 5. *The point of inflexion and multiple points are also called the singular points. Or An unusual point on the curve is basically called a singular point.*

2.8.4 Types of Double Points

(i) Node. A double point on a curve is said to be a node, if through this double point two branches of the curve passes which are real and having two different tangents at that point (Fig. 29).

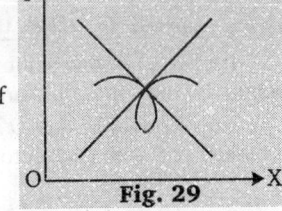

Fig. 29

(ii) Cusp. A double point on a curve is called a cusp if through this double point two real branches of the curve passes and have real coincident tangents at that point (Fig. 30).

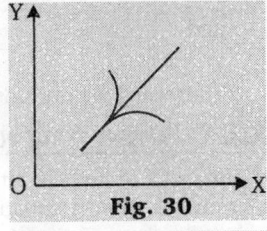

Fig. 30

(iii) Conjugate point. A point P on the curve is said to be conjugate point if there are no real points on the curve in the neighbourhood of that point and having no real tangent at that point (Fig. 31).

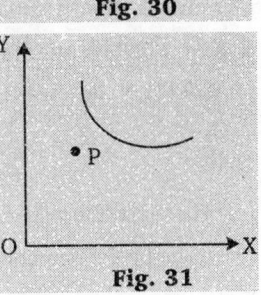

Fig. 31

2.8.5 Species of Cusp

Definition 1. *A cusp is said to be single if the curve lies entirely on one side of the common tangent (Fig. 32(ii)).*

Definition 2. *A cusp is said to be double if the curve lies on both sides of the common tangent (Fig. 32(i)).*

Definition 3. *A cusp is said to be of first species if the two branches of the curve lie on opposite sides of common tangent (Fig. 32(iii)).*

Definition 4. *A cusp is said to be of second species if the two branches of the curve lie on same side of the common tangent (Fig. 32(ii)).*

There are five different types of cusp :

(i) Single cusp of first species
(ii) Single cusp of second species
(iii) Double cusp of first species
(iv) Double cusp of second species
(v) Double cusp with change of species.

These all five types of cusp are shown below respectively :

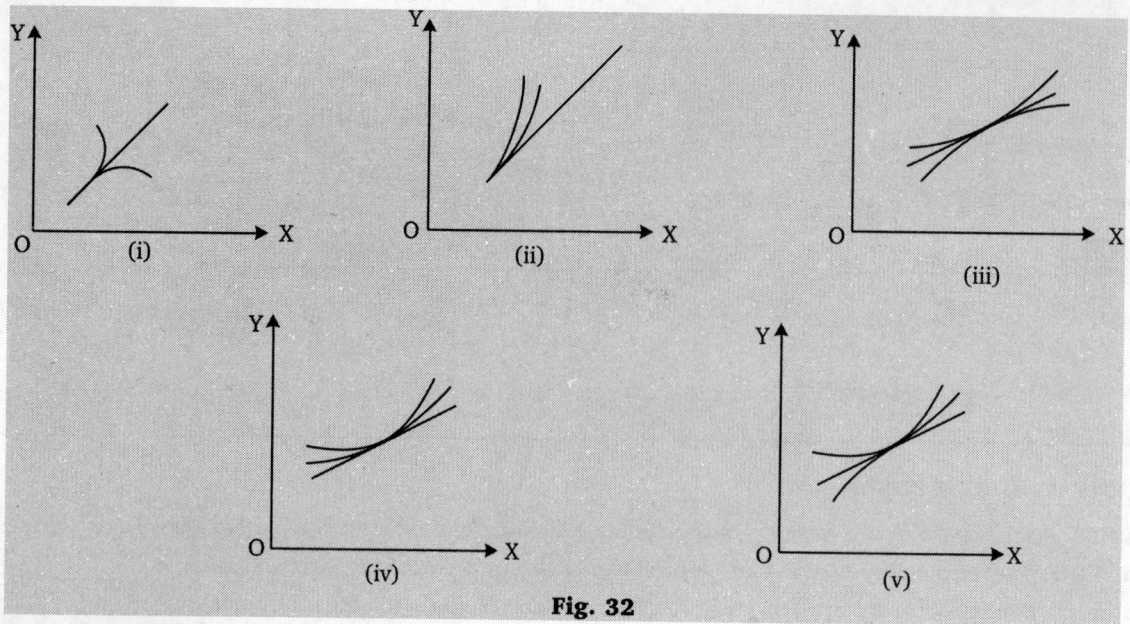

Fig. 32

2.8.6 Tangents at the Origin

The nature of a double point depends on the tangents so we find the tangent or tangents there. If a curve passes through the origin, then the equation of the tangent or tangents at the origin are obtained by equating to zero the lowest degree terms in the equation of the curve.

2.8.7 Change of Origin (Shift of Origin)

Let $P(x, y)$ be any point with respect to the co-ordinate axes OX and OY and let $O'(h, k)$ be any other point with respect to the same co-ordinate system with origin O'.

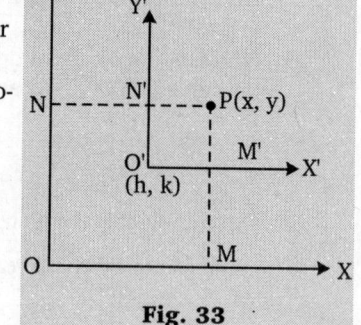

Fig. 33

Now draw $O'X$ and $O'Y$ parallel to the OX and OY axis respectively through $O'(h, k)$ and let co-ordinates of P with respect to the axes OX and OY be (X, Y). Then

$$M'P = PM - M'M.$$
$$\therefore \qquad Y = y - k \text{ or } y = Y + k$$
$$\text{and} \qquad N'P = PN - NN'.$$
$$\therefore \qquad X = x - h \text{ or } x = X + h.$$

Thus using the transformations $x = X + h$, and $y = Y + k$, the origin O is shifted to $O'(h, k)$.

2.8.8 Tangent at the Point (h, k) to a Curve

In order to find the tangent at (h, k) to the given curve, we first shift the origin at (h, k) and then find the tangent at the origin to the transformed curve by equating to zero the lowest degree terms.

2.8.9 Position and Nature of Double Points

Let $P(x, y)$ be any point on the curve $f(x, y) = 0$, we have

$$\frac{dy}{dx} = -\frac{\partial f / \partial x}{\partial f / \partial y} \qquad \text{or} \qquad \frac{\partial f}{\partial x} + \frac{\partial f}{\partial y}\frac{dy}{dx} = 0. \qquad \qquad ...(1)$$

Therefore, the slope of the tangent at $P(x, y)$ is equal to dy/dx which is given above.

Since by the definition of a multiple point we know that the curve has atleast two tangents so $\dfrac{dy}{dx}$ has atleast two values at a multiple point. But the equation (1) is of first degree in dy/dx. Therefore dy/dx will have two values or more than one value, if and only if

$$\frac{\partial f}{\partial x} = 0, \frac{\partial f}{\partial y} = 0$$

Thus the necessary and sufficient condition for any point of the curve $f(x, y) = 0$ to be a multiple point are that

$$\frac{\partial f}{\partial x} = 0 = \frac{\partial f}{\partial y}.$$

Hence, to find the multiple point of the curve $f(x, y) = 0$ we shall simultaneously solve the following equations

$$f(x, y) = 0, \frac{\partial f}{\partial x} = 0, \frac{\partial f}{\partial y} = 0.$$

Next, differentiating (1) w.r.t. 'x', we get $\dfrac{d}{dx}\left(\dfrac{\partial f}{\partial x}\right)+\dfrac{d}{dx}\left(\dfrac{\partial f}{\partial y}\cdot\dfrac{dy}{dx}\right)=0$

$$\dfrac{\partial}{\partial x}\left(\dfrac{\partial f}{\partial x}\right)+\dfrac{\partial}{\partial y}\left(\dfrac{\partial f}{\partial x}\right)\dfrac{dy}{dx}+\dfrac{d}{dx}\left(\dfrac{\partial f}{\partial y}\right)\dfrac{dy}{dx}+\dfrac{\partial f}{\partial y}\cdot\dfrac{d^2y}{dx^2}=0$$

or $\qquad \dfrac{\partial^2 f}{\partial x^2}+\dfrac{\partial}{\partial y}\left(\dfrac{\partial f}{\partial x}\right)\dfrac{dy}{dx}+\left[\dfrac{\partial}{\partial x}\left(\dfrac{\partial f}{\partial y}\right)+\dfrac{\partial}{\partial y}\left(\dfrac{\partial f}{\partial y}\right)\cdot\dfrac{dy}{dx}\right]\dfrac{dy}{dx}+\dfrac{\partial f}{\partial y}\cdot\dfrac{d^2y}{dx^2}=0.$

Since at the multiple point $\dfrac{\partial f}{\partial y}=0$. Therefore, $\qquad \dfrac{\partial^2 f}{\partial x^2}+\dfrac{\partial^2 f}{\partial y\partial x}\cdot\dfrac{dy}{dx}+\dfrac{\partial^2 f}{\partial x\partial y}\dfrac{dy}{dx}+\dfrac{\partial^2 f}{\partial y^2}\left(\dfrac{dy}{dx}\right)^2=0$

or $\qquad\qquad \dfrac{\partial^2 f}{\partial x^2}+2\dfrac{\partial^2 f}{\partial x\partial y}\dfrac{dy}{dx}+\dfrac{\partial^2 f}{\partial y^2}\left(\dfrac{dy}{dx}\right)^2=0$ ...(2)

$$\left(\because \dfrac{\partial^2 f}{\partial x\partial y}=\dfrac{\partial^2 f}{\partial y\partial x}\right)$$

This is a quadratic equation in $\dfrac{dy}{dx}$ and the multiple point will be double point if the equation (2) will remain quadratic in $\dfrac{dy}{dx}$, and for the

quadratic in $\dfrac{dy}{dx}$ it is assumed that $\dfrac{\partial^2 f}{\partial x^2},\dfrac{\partial^2 f}{\partial x\partial y},\dfrac{\partial^2 f}{\partial y^2}$ are not all zero. From the equation (2) it is obvious that the two values of dy/dx will be real and

distinct, coincident, or imaginary according as $\left[\left(\dfrac{\partial^2 f}{\partial x\partial y}\right)^2-\dfrac{\partial^2 f}{\partial x^2}\dfrac{\partial^2 f}{\partial y^2}\right]>,=\text{or}<0.$

Therefore, the two tangents will be real and distinct, coincident or imaginary according as

$$\left[\left(\dfrac{\partial^2 f}{\partial x\partial y}\right)^2-\dfrac{\partial^2 f}{\partial x^2}\dfrac{\partial^2 f}{\partial y^2}\right]>,=\text{or}<0.$$

Hence we obtained that the double point will be node, cusp or conjugate point according as

$$\left(\dfrac{\partial^2 f}{\partial x\partial y}\right)^2>\text{or}=\text{or}<\dfrac{\partial^2 f}{\partial x^2}\cdot\dfrac{\partial^2 f}{\partial y^2}.$$

📑 Remark

- If $\dfrac{\partial^2 f}{\partial x^2},\dfrac{\partial^2 f}{\partial x\partial y},\dfrac{\partial^2 f}{\partial y^2}$ are all zero, then the point $P(x,y)$ will be a multiple point of order greater that two.

2.8.10 Nature of a Cusp at the Origin

Let $(0, 0)$ be a cusp of the curve. Then there will be two coincident tangents at $(0, 0)$. Therefore, the curve will be of the form
$$(ax+by)^2+\text{terms of degree greater then two}=0 \qquad ...(1)$$
Thus the common tangent to the curve (1) at the origin is
$$ax+by=0. \qquad ...(2)$$
Let us suppose p is perpendicular from any point $P(x, y)$ to the equation (2), then
$$p=\dfrac{ax+by}{\sqrt{a^2+b^2}} \qquad ...(3)$$
where $P(x, y)$ is any point in the neighbourhood of $(0,0)$.

From the equation (3) it is obvious that p is proportional to $ax+by$ so let us take
$$p=ax+by. \qquad ...(4)$$
Now eliminating either x or y between (1) and (4), we get the equation involving p and x. Since p is small and there are two branches of the curve passes through the origin, therefore, neglecting all those terms having the degree of p greater than two. Thus we obtain a quadratic in p of the form $\qquad\qquad Ap^2+Bp+C=0 \qquad ...(5)$
where A, B, C are the functions of x only.

Now solving (5), we get

$$p = -\frac{B \pm \sqrt{(B^2 - 4AC)}}{2A} \quad \text{also } p_1 p_2 = C/A$$

where p_1 and p_2 are the roots of (5).

Now there arises following cases :

Case I. If for all numerically small values of x either negative or positive, the values of p obtained from (5) are imaginary, then the origin will be a conjugate point.

Case II. If the values of p are real for all numerically small values of x, then there will be a double cusp at origin.

Case III. If the reality of p depends on the sign of x, then origin will be a single cusp.

Case IV. If p is real for numerically small values of x and if $p_1 p_2 > 0$, then p_1 and p_2 will have same sign. Therefore the origin will be a cusp of second species because the two perpendiculars p_1 and p_2 lie on the same side of the common tangent. On the other hand if $p_1 p_2 < 0$, then p_1 and p_2 are of opposite signs. Then the origin will be a cusp of the first species because the two perpendicular line on the opposite sides of the common tangent.

2.8.11 NATURE OF A CUS AT ANY POINT

In order to find the nature of the cusp at any point (h, k). We first shift the origin at (h, k) and then apply above process discussed in § 2.8.10.

Solved Examples

EXAMPLE 1. *Show that the origin is a node on the curve*
$$x^3 + y^3 - 3axy = 0.$$

SOLUTION. The tangent at the origin are obtained by equating to zero the lowest degree terms *i.e.*, second degree term in the given equation of the curve.

$$\therefore \quad -3axy = 0 \text{ or } x = 0, y = 0.$$

Thus at the origin there are two real and distinct tangents. Hence $(0, 0)$ is a node.

EXAMPLE 2. *Find the double point of the curve*
$$(x - 2)^2 = y(y - 1)^2.$$

SOLUTION. Let $f(x, y) \equiv (x - 2)^2 - y(y - 1)^2 = 0$ ...(1)

Differentiating (1) partially w.r.t. x and y, we get

$$\frac{\partial f}{\partial x} = 2(x - 2) \qquad \text{...(2)}$$

and $\quad \dfrac{\partial f}{\partial y} = -(y - 1)^2 - 2y(y - 1).$ ...(3)

Since the necessary and sufficient condition for a double points are

$$\frac{\partial f}{\partial x} = 0, \frac{\partial f}{\partial y} = 0, \Rightarrow 2(x - 2) = 0 \qquad \text{...(4)}$$

$$-(y - 1)^2 - 2y(y - 1) = 0. \qquad \text{...(5)}$$

Now solving $f(x, y) = 0, \dfrac{\partial f}{\partial x} = 0$ and $\dfrac{\partial f}{\partial y} = 0$ simultaneously.

From (4), we get $x = 2$ and from (5), we get

$$-(y - 1)(y - 1 + 2y) = 0$$

or $-(y - 1)(3y - 1) = 0$ or $y = 1$ and $y = 1/3$.

$\therefore$ Possible double points are $(2, 1)$ and $(2, 1/3)$

But $(2, 1/3)$ does not satisfy $f(x, y) = 0$. Hence only double point is $(2, 1)$.

Exercise-2.19

1. Find the equation of the tangents at the origin to the following curves :

   (a) $(x^2 + y^2)(2a - x) = b^2 x$ (b) $a^4 y^2 = x^4(x^2 - a^2)$

   (c) $x^4 + 3x^3 y + 2xy - y^2 = 0$ (d) $x^3 + y^3 = 3axy$

2. Examine the nature of the origin on the curve
   $$(2x + y)^2 - 6xy(2x + y) - 7x^3 = 0.$$

3. Show that the origin is a conjugate point on the curve
   $$a^2 x^2 + b^2 y^2 = (x^2 + y^2)^2.$$

4. Show that the origin is a conjugate point on the curve
   $$y^2 = 2x^2 y + x^4 y - 2x^4.$$

5. Find the position and nature of double points of the curve
   $$y^3 = x^3 + ax^2.$$

6. Examine the nature of the double points of the curve
   $$2(x^3 + y^3) - 3(3x^2 + y^2) + 12x = 4.$$

7. Find the position and nature of the double points of the curve
   $$a^4 y^2 = x^4(2x^2 - 3a^2).$$

8. Find the position and nature of the double points of the curve
   $$x^4 - 2y^3 - 3y^2 - 2x^2 + 1 = 0.$$

9. Determine the existence and nature of the double points on the curve $y^2 = (x - 2)^2(x - 1)$.

10. Prove that the curve $ay^2 = (x - a)^2(x - b)$ has at $x = a$, a conjugate point if $a < b$, a node if $a > b$ and a cusp if $a = b$.

11. Examine the curve $x^3 + 2x^2 + 2xy - y^2 + 5x - 2y = 0$ for singular points and show that it has a cusp of the first kind at the point $(-1, -2)$.

12. Show that the curve $y^2 = bx \tan(x/a)$ has a node or a conjugate point at the origin according as a and b have like or unlike signs.

13. Determine the position and nature of the double points of the curves :

    (a) $y(y - 1)^2 = (x - 2)^2$.   (b) $x^3 - y^2 - 7x^2 + 4y + 15x - 13 = 0$.

    (c) $y^2 = x(x - a)^2, a > 0$ (d) $y^2 = x^2(a - x^2)$

    (e) $y(y - 6) = x^2(x - 2)^3 - 9$.

14. Discuss the nature of the double points of the curve
    $$(x + y)^3 - \sqrt{2}(y - x + 2)^2 = 0.$$

15. Show that the origin is a conjugate point on the curve
    $$x^4 - ax^2 y + axy^2 + a^2 y^2 = 0.$$

16. Show that curve $(xy + 1)^2 + (x - 1)^3(x - 2) = 0$ has a single cusp of the first species at the point $(1, -1)$.

17. Show that the curve $y^3 = (x - a)^2(2x - a)$ has a single cusp of the first species at the point $(a, 0)$.

18. Find the nature and position of double points of the curve $a^4 y^2 = x^4(a^2 - x^2)$.

1. (a)$x = 0$ (b) $y = 0, y = 0$ (c) $y = 0, 2x - y = 0$ (d) $x = 0, y = 0$ **2.** Origin is a single cusp of first species
5. A cusp at $(0, 0)$ **6.** Node at $(2, 0)$ **7.** Cusp at $(0, 0)$ **8.** Double points $(0, -1)$, $(1, 0)$ and $(-1, 0)$ are nodes
9. Node at $(2, 0)$ **13.** (a) Node at $(2, 1)$ (b) Node at $(3, 2)$ (c) Node at $(a, 0)$ (d) Node at $(0, 0)$
   (e) Conjugate at $(0, 3)$ and a single cusp of the first species at $(2, 3)$ **14.** A single cusp of first species at $(-1, 1)$
18. Double cusp of the first species at $(0, 0)$.

2.8.12 Curve Tracing : Cartesian Form

To trace any curve of cartesian form we should apply following process :

(a) Symmetricity. In order to find the symmetry of the curve we should apply following rules :
 (i) If the powers of y in the equation of the curve are all even, then curve is symmetrical about x-axis.
 (ii) If the powers of x in the equation of the curve are all even, then the curve is symmetrical about y-axis.
 (iii) If the powers of x as well as y in the equation of the curve are all even, the curve is symmetrical about both axes.
 (iv) If the equation of curve remains unchanged when x is replaced by $-x$ and y is replaced by $-y$, then the curve is symmetrical in opposite quadrants.
 (v) If the equation of the curve remains unchanged when x and y are interchanged, then the curve is symmetrical about the line $y = x$.

(b) Nature of the origin on the curve. If the curve passes through the origin, then find the tangent at $(0, 0)$ by equating to zero the lowest degree terms of the curve. If we obtain two tangent at the origin, then origin will be a double point and then find the nature of this double point.

(c) Intersection of curve with co-ordinate axes. We should check whether the curve cuts the co-ordinate axes or not, for this put $y = 0$ in the equation of the curve and find the values of x, then we get the points at which the curve cuts the x-axis. Similarly if the curve cuts the y-axis, then put $x = 0$ in the equation of the curve and obtain the points on the y-axis. Hence in this way we obtain the points of intersection of the curve with co-ordinate axis. Thereafter we should find the tangents at these points of intersection. For this first we shift the origin at these points and then obtain the tangent at these new origin by equating to zero the lowest degree terms in the new equation of the curve. On the other hand the value of dy/dx at these points of intersection can also be used to find the slope of the tangent at that point.

(d) Nature of y or x in the curve. We should now solve the equation of the curve either for y or for x whichever is convenient. Suppose we solve for y and see that nature of y as x increases from 0 to $+ \infty$. Similarly see the nature of y as x decreases from 0 to $- \infty$ and finally collect those values of x for which $y = 0$ or $y \to \infty$ or $- \infty$.

REMARK

● If the curve is symmetrical about x-axis in opposite quadrants then there is no need to take the values of x of both positive and negative. We can take only positive values of x to see the variation in y.

(e) Regions in which curve does not exist. In order to find the regions where the curve does not exist we should solve the equation of curve for one variable in terms of the other. Therefore, the curve will not exist for those values of one variable which make the other variable imaginary.

(f) Asymptotes. Next, we should find all the asymptotes of the curve because the branches of the curve approach to the asymptotes if they exist.

(g) Sign of dy/dx. Next, we should find the value of dy/dx from the equation of the curve and find the points on the curve at which $dy/dx = 0$ or $dy/dx = \infty$. Therefore at these points we obtain the nature of tangents. Suppose in any region $a < x < b$, dy/dx remains positive throughout, then in this region y increase continuously as x increases. On the other hand if dy/dx remains negative, then y decreases continuously as x increases.

(h) Special points. If necessary, we should find the some special point on the curve.

(i) Points of inflexion. If necessary, we should find the point of inflexion to know the position of the curves at that point.

Now taking all above considerations in mind, draw an approximate shape of the curve.

Solved Examples

EXAMPLE 1. *Trace the curve $y^2(2a - x) = x^3$.*
(UPTU–2007, 2011, PTU–2010, VTU–2008, RAJASTHAN–2006)

SOLUTION . (i) Obviously the given curve is symmetrical about x-axis.
(ii) The curve passes through the origin and the

tangents at the origin are obtained by equating to zero the lowest degree terms *i.e.*, $2ay^2$ in the equation of the curve.

$\therefore \qquad 2ay^2 = 0$ or $y = 0, y = 0$.

Thus at the origin we obtained two coincident

tangents $y = 0$, $y = 0$ *i.e.*, x-axis. Therefore $(0, 0)$ is a cusp.

(iii) From the equation of the curve it is obvious that the curve does not cut the co-ordinate axes.

(iv) Now solving the equation of the curve for y, we get $y^2 = x^3/(2a - x)$ when $x=0$, $y^2 = 0$ and when $x = 2a$, thus $y^2 \to \infty$ thus $x = 2a$ is an asymptotes of the curve.

It is observed that y increases as x increases from 0 to $2a$.

(v) When x lies between 0 and $2a$, y^2 will be positive and the curve will exist in this region. When $x > 2a$, y^2 will be negative so the curve will not exist beyond the line $x = 2a$. When $x < 0$ again y^2 will be negative and thus the curve will also not exist for $x < 0$. Hence we can say that the curve only exists in the region $0 < x < 2a$.

(vi) In order to find the asymptotes, putting $y = m$ and $x = 1$ in the third degree terms in the equation of the curve, we get

$$\phi_3(m) = m^2 + 1.$$

Therefore the equation $\phi_3(m) = 0$ gives both its roots imaginary so ignore them. Consequently $x = 2a$ is only the asymptote of the curve.

(vii) Differentiating the equation of the curve

$$y = \frac{x^{3/2}}{\sqrt{(2a - x)}}$$

We get $\quad \dfrac{dy}{dx} = \dfrac{(3a - x)x^{1/2}}{(2a - x)^{3/2}}.$

In the region $0 < x < 2a$; $\dfrac{dy}{dx}$ will be positive, so in this region y increase continuously as x increases. Now taking all the above points of consideration in the mind and draw the curve whose shape is shown in fig. 34.

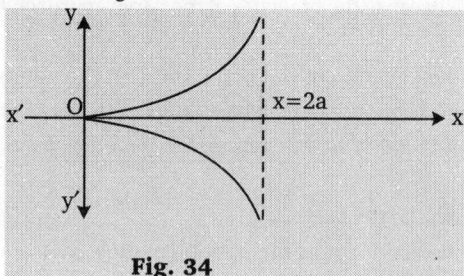

Fig. 34

EXAMPLE 2. *Trace the curve* $y^2 (1 - x^2) = x^2(1 + x^2)$.

SOLUTION . (i) In the equation of the curve the powers of both x and y are all even so the curve is symmetrical about both axes.

(ii) The curve passes through the origin. The tangents at the origin are obtained by equating to zero the lowest degree terms in the equation of the curve.

SOME STANDARD FIGURES AND THEIR EQUATIONS

S. No.	Name	Equation
1.	Cubical Parabola	$y = x^3$
2.	Semi cubical parabola	$ay^2 = x^3$
3.	Cissoid	$y^2(2a - x) = x^3$
4.	Folium of Descartes	$x^3 + y^3 = 3axy$
5.	Circle	$x^2 + y^2 = a^2$ $r = 2a \cos \theta$
6.	Cardioid	$r = a(1 - \cos \theta)$ $r = a(1 + \cos \theta)$
7.	Limacon	$r = a + b \cos \theta$
8.	Equiangular spiral	$r = ae^{m\theta}$
9.	Cycloid	$x = a(t + \sin t)$ $y = a(t + \cos t)$
10.	Tractrix	$x = a \cos t + \dfrac{1}{2}a \log \tan^2 \dfrac{t}{2}$ $y = a \sin t$
11.	Astroid	$x^{2/3} + y^{2/3} = a^{2/3}$
12.	Inverted Cycloid	$x = a(t + \sin t)$ $y = a(1 - \cos t)$
13.	Strophoid	$y^2(a - x) = x^2(a + x)$
14.	Four leaved rose	$r = a \sin 2\theta$
15.	Spiral of Archimedes	$r\theta = a$

$$\therefore \qquad y^2 - x^2 = 0 \text{ or } y = \pm x.$$

Thus there are two real and distinct tangent at the origin so $(0, 0)$ is a node.

(iii) From the equation of the curve it is clear that curve does not cut any co-ordinate axes.

(iv) Solving the equation of the curve for y, we get

$$y^2 = \frac{x^2(1 + x^2)}{(1 - x^2)}.$$

When $x = 0$, $y = 0$ and when $x = \pm 1$, $y \to \infty$ so $x = \pm 1$ are two asymptotes parallel to y-axis.

(v) When $-1 < x < 1$, y^2 is positive, so the curve exists in this region. When $x > 1$, y will be negative thus curve will not exist beyond the line $x = 1$. Also when $x < -1$, y^2 will be negative so that curve will not exist for $x < -1$.

(vi) In order to find the asymptotes, putting $y = m$ and $x = 1$ in the fourth degree terms of the curve, we get $\phi_4(m) = m^2 + 1$.

Solving $\phi_4(m) = 0$, we get both values of m imaginary so ignore them. Consequently $x = \pm 1$ are only two real asymptotes.

(vii) Since, we have

$$y = x\sqrt{\frac{1 + x^2}{1 - x^2}}.$$

$$\sqrt{1-x^2}\left(\sqrt{1+x^2}+\frac{x^2}{\sqrt{1+x^2}}\right)$$

$$\therefore \frac{dy}{dx}=\frac{-x\sqrt{(1+x^2)}\left[\dfrac{-x}{\sqrt{1-x^2}}\right]}{(1-x^2)}$$

$$=\frac{(1-x^2)(1+x^2+x^2)-x(1+x^2)(-x)}{(1-x^2)^{3/2}(1+x^2)^{1/2}}$$

$$=\frac{2x^2+1-x^4}{(1-x^2)^{3/2}(1+x^2)^{1/2}}.$$

When $-1 < x < 0, \dfrac{dy}{dx}$ is negative this means that when x decreases from -1 to 0, y decreases. When $0 < x < 1$; $\dfrac{dy}{dx}$ is positive this implies that when x increases from 0 to 1, y increases.

Now taking all the above facts in mind and draw the shape of the curve we get the following figure:

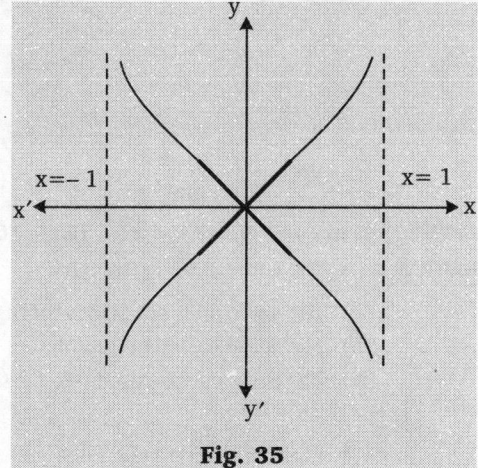

Fig. 35

EXAMPLE 3. *Trace the curve $ay^2 = x^2 (a - x)$.*

(SVTU–2004, KURUKSHETRA–2009)

SOLUTION. (i) The curve is symmetrical about x-axis because the powers of y are all even.

(ii) This curve passes through the origin. The tangents at the origin are obtained by equating to zero the lowest degree terms in the equation of the curve, we get $ay^2 - ax^2 = 0$ or $y = \pm x$.

Thus there are two real and distinct tangents at $(0, 0)$. Therefore $(0, 0)$ is a node.

(iii) The curve cuts the x-axis only at the point where $y = 0$.
$\therefore x^2(a - x) = 0$ or $x = 0, x = a$.
Thus the curve cuts the x-axis at $(a, 0)$.

Now $y = x\sqrt{\dfrac{a-x}{a}}$

$\therefore$ $\dfrac{dy}{dx} = \dfrac{1}{\sqrt{a}}\left[\sqrt{a-x} - \dfrac{x}{2\sqrt{a-x}}\right]$

$$= \frac{2(a-x)-x}{2\sqrt{a(a-x)}} = \frac{2a-3x}{2\sqrt{a(a-x)}}.$$

At $(a, 0)$, $\dfrac{dy}{dx} = \infty$. Therefore the tangent at $(a, 0)$ is perpendicular to x-axis.

(iv) Since we have $y = x\sqrt{\dfrac{a-x}{a}}$

when $x = 0$, $y^2 = 0$ and when $x = a$, y^2 also equals zero. Also when x increases from 0 to $a/2$ y increases and when x increases from $a/2$ to a, y decreases.

(v) When $0 < x < a$, y^2 is always positive so the curve will exist in this region. When $x < 0$, y^2 is also positive so that the curve will also exist for $x < 0$. When $x > a$, y^2 will be negative and therefore in this region the curve will not exist. Taking all above facts into consideration and draw the shape of the curve we obtain as shown below.

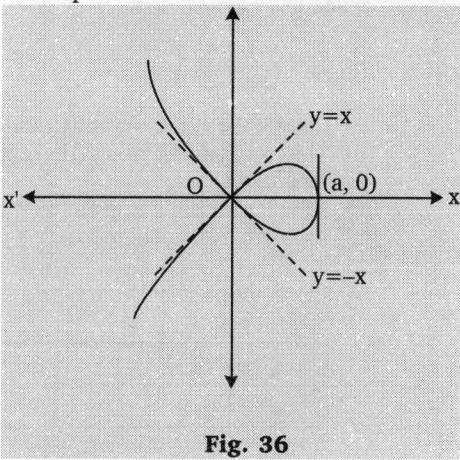

Fig. 36

EXAMPLE 4. *Trace the curve $x^3 + y^3 = 3axy$.*

(UPTU–2007, KURUKSHETRA–2005)

SOLUTION. (i) The given curve is symmetrical about the line $y = x$.

(ii) Curve is passing through the origin so the tangents at $(0, 0)$ are
$$xy = 0 \quad \Rightarrow \quad x = 0 \text{ and } y = 0.$$
Thus $(0, 0)$ is a node.

(iii) The curve does not intersect the axes.

(iv) If x is replaced by $-x$ and that of y by $-y$, the equation of the curve is changed. Thus the curve does not exist in third quadrant.

(v) The given curve has only one real asymptote which is $x + y + a = 0$.

(vi) The curve cuts the line $y = x$ at the point $\left(\dfrac{3a}{2}, \dfrac{3a}{2}\right)$.

From the curve, $\dfrac{dy}{dx} = \dfrac{ay - x^2}{y^2 - ax}$.

so at $\left(\dfrac{3a}{2}, \dfrac{3a}{2}\right)$ $\dfrac{dy}{dx} = -1$

Thus, the tangent at $\left(\dfrac{3a}{2}, \dfrac{3a}{2}\right)$ makes an angle

$135°$ with the positive axis of x.

Now taking all above facts into consideration and draw the curve, we get figure (37).

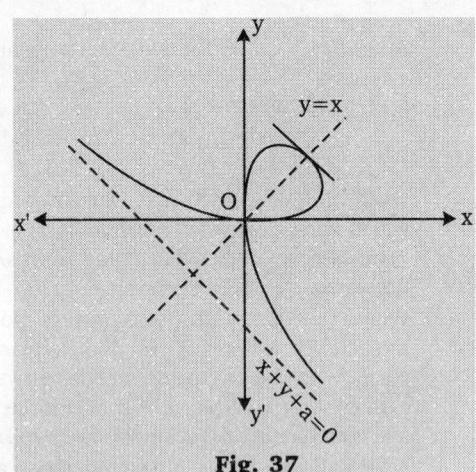

Fig. 37

EXAMPLE 5. *Trace the curve* $y = 1 - \dfrac{1}{1+x^2}$.   (UPTU–2008)

SOLUTION. The given curve is $y = 1 - \dfrac{1}{1+x^2}$.   ...(1)

(i) Clearly, the curve is symmetric about y-axis.
(ii) The curve passes through the origin, because (0, 0) satisfies (1).
(iii) Tangent at origin is given by $y = 0$.
(iv) Asymptotes parallel to x-axis is $y - 1 = 0 \Rightarrow y = 1$.
(v) Clearly the given curve meets the axes only at the origin. Special points

x	–2	–1	0	1	2
y	4/5	1/2	0	1/2	4/5

Now taking all the above facts in find, we draw the curve as in figure (38).

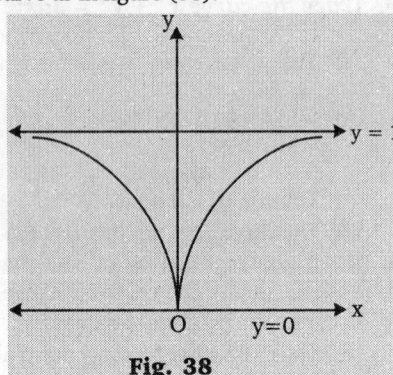

Fig. 38

EXAMPLE 6. *Trace the curve* $x^{2/3} + y^{2/3} = a^{2/3}$.
(UPTU–2006, VTU–2003, PTU–2009)

SOLUTION. The given curve is $x^{2/3} + y^{2/3} = a^{2/3}$ ...(1)

(i) The curve is symmetric about both the axes. Also, the curve is symmetric about the line $y = x$ and $y = -x$.
(ii) The curve does not passes through the origin.
(iii) The curve meets x-axis at $(a, 0)$ and $(-a, 0)$. Also, the curve meets y-axis at $(0, a)$ and $(0, -a)$.
(iv) The curve has no asymptotes.
(v) From (1) $\dfrac{dy}{dx} = -\dfrac{y^{1/3}}{x^{1/3}}$

$\dfrac{dy}{dx} = 0$, when $y = 0$

Again from (1) $x = \pm a$, when $y = 0$
Thus, the tangents to the curve are parallel to x-axis at the points $(\pm a, 0)$. Further, $\dfrac{dy}{dx} \to 0$ when $x = 0$.
From (1) when $x = 0, y = \pm a$.
Therefore, the tangents to the curve are parallel to y-axis at the points $(0, \pm a)$
(vi) From (1) we can write $y^{2/3} = a^{2/3} - x^{2/3}$.
If $|x| > a, y^{2/3}$ is negative $\Rightarrow y^2$ is negative.
$\Rightarrow y$ is imaginary.
$\Rightarrow$ Curve does not lie beyond the lines $x = \pm a$.
Similarly, the curve does not lie beyond the lines $y = \pm a$.
Further, when $x = 0, y = a$. As x-increases from 0 to a, y decreases from a to 0.

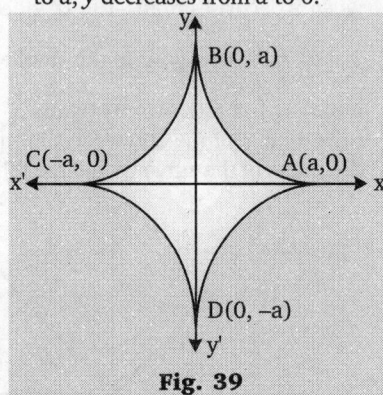

Fig. 39

EXAMPLE 7. *Trace the curve* $y = x^3 - 12x - 16$.   (PTU–2008)

SOLUTION. (i) The curve has no symmetry.
(ii) The curve does not passes through the origin.
(iii) The curve has no asymptotes.
(iv) The curve cuts x-axis at $(-2, 0)$, $(4, 0)$ and y-axis at $(0, -16)$.
(v) We have $\dfrac{dy}{dx} = 3x^2 - 12$.

At $(-2, 0), \dfrac{dy}{dx} = 0 \Rightarrow$ tangent is parallel to x-axis

At $(4, 0), \dfrac{dy}{dx} = 36 \Rightarrow$ tangent makes an acute angle $\tan^{-1}36$ with x-axis at $(4, 0)$

Also, $\dfrac{dy}{dx} = 0 \Rightarrow 3x^2 - 12 = 0 \Rightarrow x = \pm 2$

$\Rightarrow$ tangent is parallel to x-axis at $(2, -32)$.
(vi) $y \to \infty$ as $x \to \infty$ and $y \to -\infty$ as $x \to -\infty$: y is positive for $x > 4$ and y is –ve for $x < 4$.

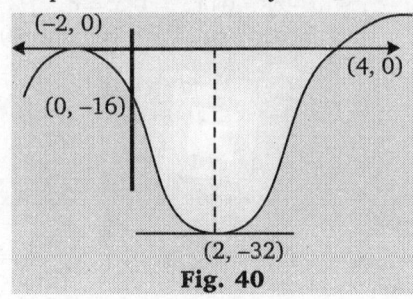

Fig. 40

EXAMPLE 8. *Trace the curve* $9ay^2 = (x - 2a)(x - 5a)^2$. (JNTU–2008)

SOLUTION. (i) The curve is symmetric about x-axis.

(ii) The curve does not passes through the origin.

(iii) The curve has no asymptotes.

(iv) The curve cuts the x-axis at $x = 2a$ and $x = 5a$ i.e., at $A(2a, 0)$ and $B(5a, 0)$. It cuts the y-axis at

$$y^2 = -50\frac{a^2}{9}$$

$\Rightarrow y$ is imaginary. Therefore curve does not cut the x-axis

(v) $y = \dfrac{(x - 5a)\sqrt{x - 2a}}{3\sqrt{a}}$

$\Rightarrow y$ is imaginary for $x < 2a$.

$\Rightarrow$ curve exists only for $x \geq 2a$.

and $\dfrac{dy}{dx} = \pm\dfrac{(x - 3a)}{2\sqrt{a}\sqrt{x - 2a}}$

At $A(2a, 0)$, $\dfrac{dy}{dx} = \infty$ i.e., tangent is parallel to y-axis.

At $B(5a, 0)$, $\dfrac{dy}{dx} = \pm\dfrac{1}{\sqrt{3}}$ i.e., there are two distinct tangents.

$\Rightarrow$ There is a node at $B(5a, 0)$.

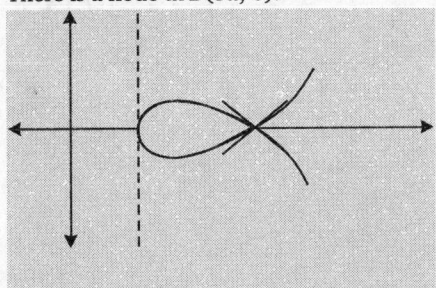

Fig. 41

EXAMPLE 9. *Trace the curve :* $y^2(x^2 + y^2) + a^2(x^2 - y^2) = 0$

or $x^2(y^2 + a^2) + y^2(y^2 - a^2) = 0$ (GBTU–2010)

SOLUTION. (i) The curve is symmetric about both the axes.

(ii) The curve passes through the origin and $a^2(x^2 - y^2) = 0$ i.e., $y = \pm x$ are two tangents at origin. So, origin is a node.

(iii) The curve intersects the x-axis only at origin. It intersects the y-axis at $(0, 0)$, $(0, a)$ and $(0, -a)$.

(iv) Shifting the origin at $(0, a)$ the equation of the curve becomes

$$(y+a)^2\{x^2 + (y+a)^2\} + a^2\{x^2 - (y+a)^2\} = 0$$

$$\Rightarrow (y^2 + 2ay + a^2)(x^2 + y^2 + 2ay + a^2) + a^2(x^2 - y^2 - 2ay - a^2) = 0$$

Equating to zero the the lowest degree terms, we get $2a^3y + 2a^3y - 2a^3y = 0 \Rightarrow y = 0$, which is the tangent at new origin. Here we need not find the tangent at $(0, -a)$ as the curve is symmetric about x-axis.

(v) On solving the given equation for x, we get $x^2 = y^2(a^2 - y^2)/(a^2 + y^2)$ when $y = 0$, $x^2 = 0$ and when $y = a$, $x^2 = 0$.

When $0 < y < a$, x^2 is positive. Therefore, the curve exists in the region $0 < y < a$. When $y > a$, x^2 is negative, so the curve does not exist in the region $y > a$.

(vi) The asymptotes parallel to x-axis are given by $a^2 + y^2 = 0$ i.e., $y = \pm ai$.

Also, $\phi_4(m) = m^2(1 + m^2)$. Its roots are $m = 0, 0, i, -i$.

The asymptote are imaginary.

(vii) In the positive quadrant we have

$$x = y(a^2 - y^2)^{1/2}/(a^2 + y^2)^{1/2}, y > 0$$

$$x = y\left(1 - \frac{y^2}{a^2}\right)^{1/2} \Big/ \left(1 + \frac{y^2}{a^2}\right)^{1/2}$$

When $0 < y < a$, we observe that x is less than y. Hence, the curve lies above the line $y = x$ which is tangent at origin.

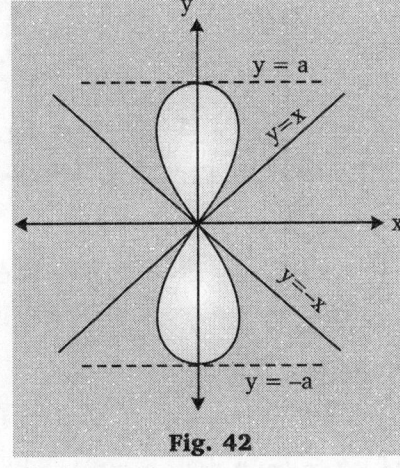

Fig. 42

Exercise-2.20

Trace the following curves :

1. $ay^2 = x^3$
2. $a^2y = x^3$
3. $y = x(x^2 - 1)$
4. $xy^2 = 4a^2(2a - x)$
5. $y^2(a + x) = x^2(a - x)$ (VTU–2004, UPTU–2006, SVTU–2008)
6. $x^2(x^2 - 4a^2) = y^2(x^2 - a^2)$.   (UPTU–2009)
7. $x^3 + y^3 = x$
8. $x^2y^2 = a^2(x^2 + y^2)$
9. $y^2(a^2 + x^2) = x^2(a^2 - x^2)$ (VTU–2010)
10. $y^2(x + 3a) = x(x - a)(x - 2a)$
11. $y^2(x^2 + y^2) + a^2(x^2 - y^2) = 0$
12. $a^2y^2 = x^3(2a - x)$
13. $9ay^2 = x(x - 3a)^2$
14. $x^2y^2 = (1 + y)^2(4 - y^2)$
15. $y^3 + x^3 = a^2x$
16. $x^4 + y^4 = 4a^2xy$
17. $y^2 = (x - a)(x - b)(x - c), a > b > c$
18. $y^2(x - a) = x^2(x + a)$ (UPTU–2007, VTU–2010, BPTU–2005)
19. $y^2(x^2 - 1) = x$
20. $y(x^2 - 1) = (x^2 + 1)$
21. $a^2y^2 = x^2(a^2 - x^2)$ (PTU–2009, VTU–2008)
22. $y(x^2 + 4a^2) = 8a^3$
23. $x^3y = x + 1$
24. $a^3y^2 = (x - a)^4(x - b), a > b$

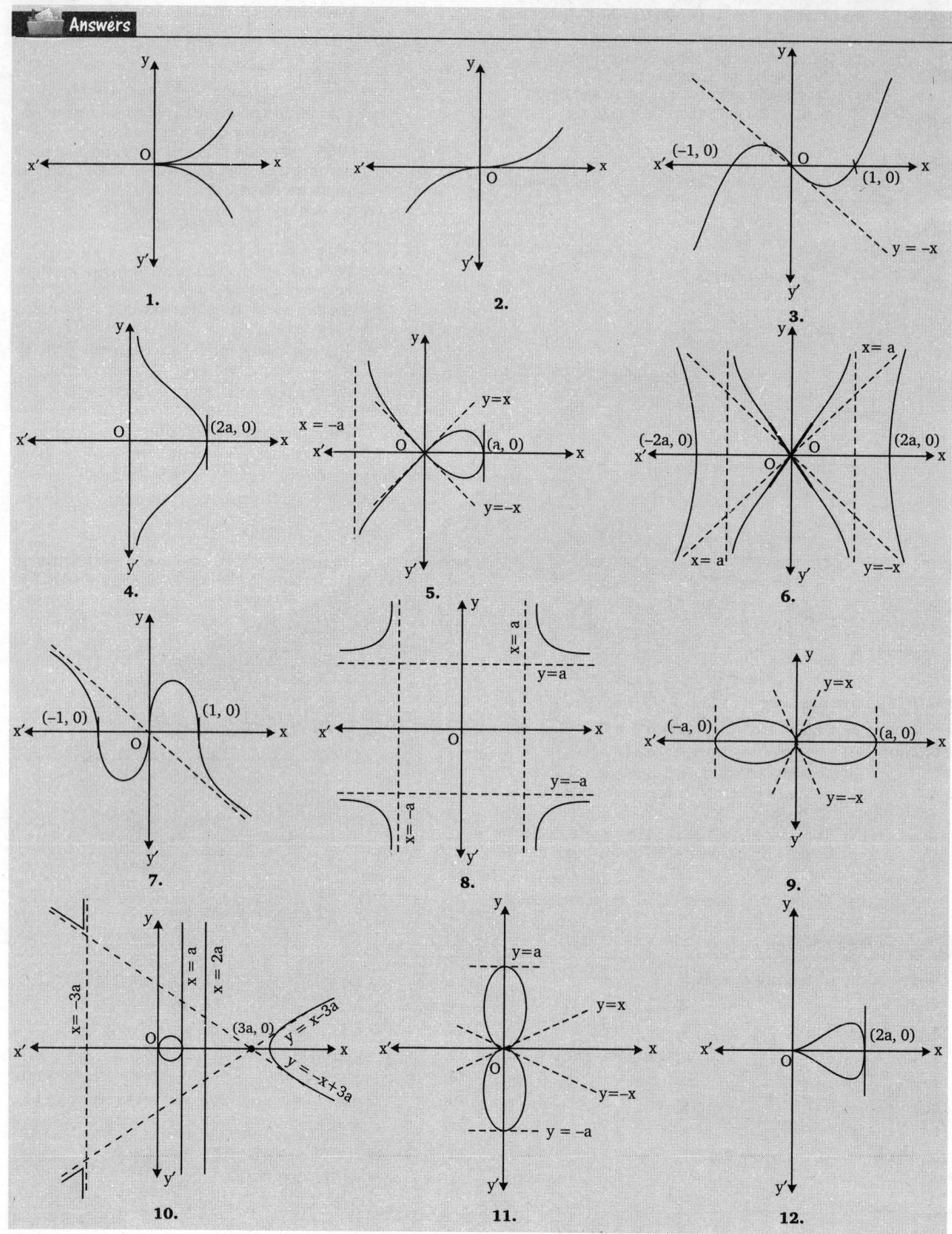

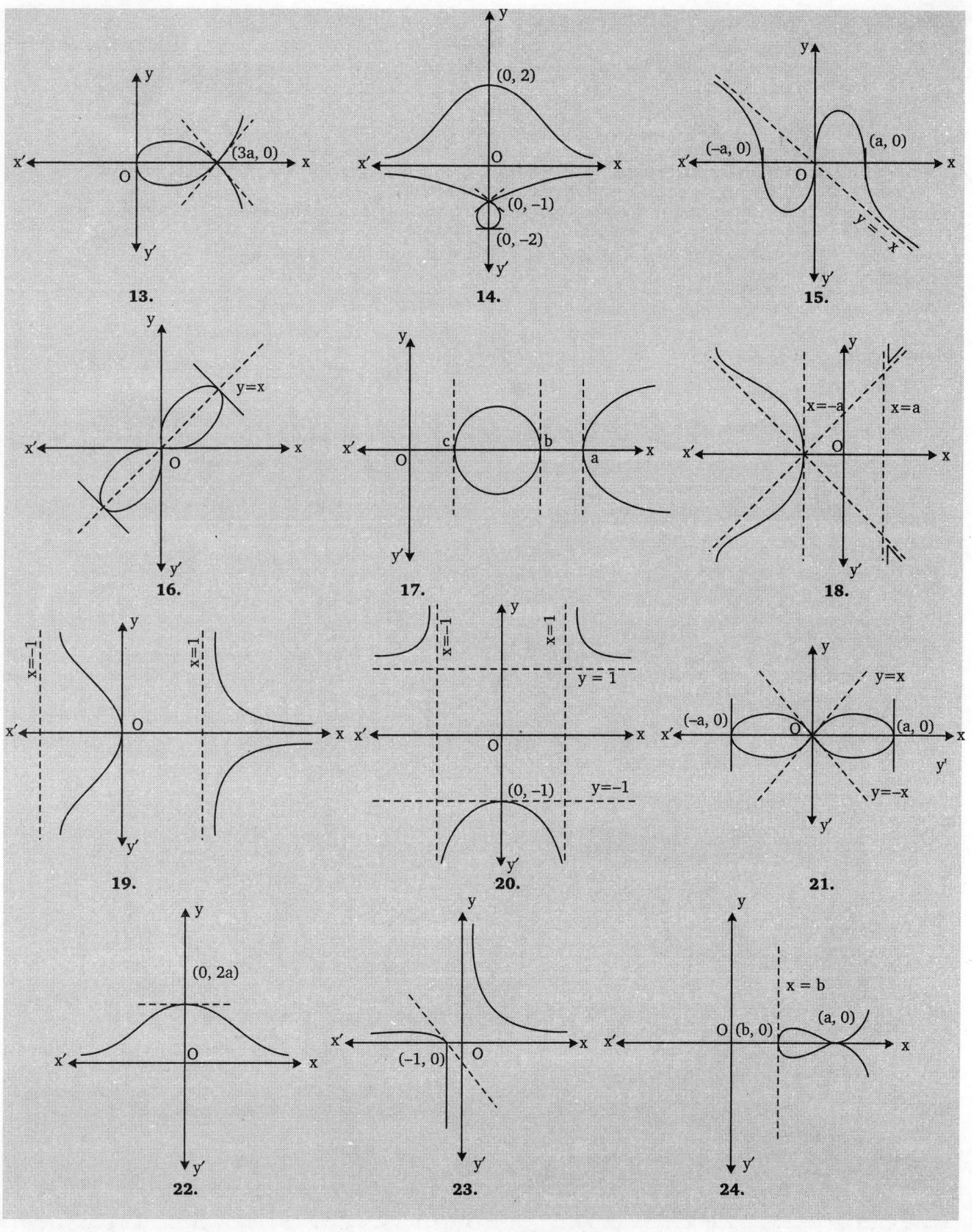

2.8.13 Tracing of a Curve Given by Parametric Equations

If the equations of the curve are in parametric form *i.e.*, $x = f(t), y = g(t)$. If conveniently possible, the parameter is eliminated and the corresponding cartesian equation is obtained. But if it is not convenient to eliminate t, a series of values are given to t and the corresponding values of $x, y,$ and dy/dx are found and proceed as follows :

(1) Symmetry

    (i) Curve is symmetric about x-axis if on replacing t by $-t$, $f(t)$ remains unchanged and $g(t)$ changes to $-g(t)$.

    (ii) Curve is symmetric about y-axis if on replacing t by $-t$, $f(t)$ changes to $-f(t)$ and $g(t)$ remains unchanged.

    (iii) The curve is symmetric in the opposite quadrants if on replacing t by $-t$, both $f(t)$ and $g(t)$ remains unchanged.

(2) Find the least and greatest value of x and y to find the region where the curve lies.

(3) Find the points where the curve cuts the axes. The point of intersection of the curve with x-axis given by the roots of $g(t) = 0$ and the point of intersection of the curve with y-axis are given by the roots of $f(t) = 0$.

(4) Find the points where the tangent is parallel or perpendicular to the x-axis (*i.e.*, where $dy/dy = 0$ or $\rightarrow \infty$)

Solved Examples

EXAMPLE 1. *Trace the curve* $x = a(t+\sin t)$ *and* $y = a(1+\cos t)$

SOLUTION . (I) Given $x = a(t+\sin t)$

$$\Rightarrow \frac{dx}{dt} = a(1+\cos t)$$

$$y = a(1 + \cos t)$$

$$\Rightarrow \frac{dy}{dt} = -a\sin t$$

$$\therefore \frac{dy}{dx} = \frac{dy/dt}{dx/dt} = \frac{-a\sin t}{a(1+\cos t)}$$

$$= \frac{-2a\sin t/2\cos t/2}{2a\cos^2 \frac{t}{2}} = -\tan\frac{t}{2}$$

(ii) We have $y = 0$,
when $\cos t = -1$, *i.e*, $t = -\pi, \pi$
When $t = \pi$, $x = a\pi$, $dy/dx = -\infty$
At the point $(a\pi, 0)$, the tangent to the curve is perpendicular to the x-axis. Also when $t = -\pi$,
$x = -a\pi$, $dy/dx = \infty$

(iii) y is maximum when $\cos t = 1$, *i.e.*, when $t = 0$
Also, when $t = 0$, $x = 0$, $y = 2a$ and $dy/dx = 0$.
So at the point $(0, 2a)$, the tangent to the curve is parallel to x-axis.

(iv) y can be negative. Also no part of the curve lies in the region $y > 2a$.

t	$-\pi$	$-\pi/2$	0	$\pi/2$	π
x	$-a\pi$	$-a\left(\frac{\pi}{2}+1\right)$	0	$a\left(\frac{\pi}{2}+1\right)$	$a\pi$
y	0	a	$2a$	a	0
dy/dx	∞	1	0	-1	$-\infty$

At $(-a\pi, 0)$ the tangent inclined to x-axis at the angle $\psi = \frac{\pi}{2}$
Also curve is symmetric about the y-axis.

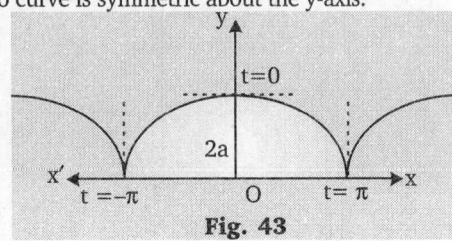

Fig. 43

EXAMPLE 2. *Trace the curve*

$$x = a\cos t + \frac{1}{2}a\log\tan^2\frac{t}{2}, y = a\sin t$$

SOLUTION . (i) Put $-t$ for t in the given equation of the curve, we get

$$x = a\cos t + \frac{1}{2}a\log\tan^2\frac{t}{2} \text{ and } y = -a\sin t$$

Therefore, for every value of x there are two equal and opposite value of $y \Rightarrow$ curve is summetric about x-axis.

Further, put $\pi - t$ for t in the given equation of the curve, we get

and

$$x = -a\cos t + \frac{1}{2}a\log\cot^2\frac{t}{2}$$

$$= -a\cos t - \frac{1}{2}a\log\tan^2\frac{t}{2}$$

$$y = a\sin t$$

$\Rightarrow$ For every value of y, there are two equal and opposite values of x, so curve is symmetric about y-axis.

(ii) Differentiating the given equation w.r.t. t we get

$$\frac{dx}{dt} = -a\sin t + \frac{1}{2}a\frac{1}{\tan^2\frac{t}{2}}(2\tan\frac{t}{2}\sec^2\frac{t}{2}).\frac{1}{2}$$

$$= -a\sin t + \frac{a}{2\sin\frac{t}{2}\cos\frac{t}{2}}$$

$$= -a\sin t + \frac{a}{\sin t} = \frac{a(1-\sin^2 t)}{\sin t} = \frac{a\cos^2 t}{\sin t}$$

and

$$\frac{dy}{dt} = a\cos t$$

$$\therefore \quad \frac{dy}{dx} = \frac{dy/dt}{dx/dt} = \frac{a\cos t.\sin t}{a\cos^2 t} = \tan t$$

(iii) We have $y = 0$ when $\sin t = 0$, *i.e.*, $t = 0$, when $t \rightarrow 0$, $x \rightarrow -\infty$.

Therefore, $x \rightarrow -\infty$ when $y \rightarrow 0$ showing that the line $y = 0$ is an asymptote of the curve.

(iv) Clearly y is maximum when $\sin t = 1$ *i.e.*, $t = \pi/2$. When $t = \pi/2, x = 0, y = a$

and $\frac{dy}{dx} = \tan\frac{\pi}{2} = \infty$

$\Rightarrow$ Curve passes through the point $(0, a)$ and the tangent at this point is the x-axis.

(v) Clearly, the numerical value of y cannot be greater than a. Therefore, curve does exist in the region $y > a$ and $y < -a$.

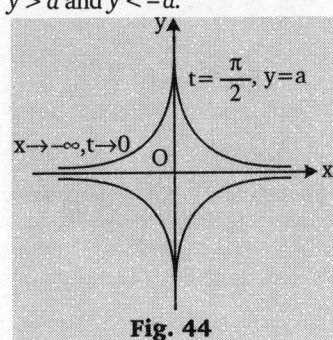

Fig. 44

EXAMPLE 3. *Trace the curve $x = a(t + \sin t), y = a(1 - \cos t)$ when $-\pi \le t \le \pi$.*

SOLUTION. We have $x = a(t + \sin t) \Rightarrow \dfrac{dx}{dt} = a(1 + \cos t)$

$$y = a(1 - \cos t) \Rightarrow \frac{dy}{dt} = a \sin t$$

$$\therefore \quad \frac{dy}{dx} = \frac{dy/dt}{dx/dt} = \frac{a \sin t}{a(1 + \cos t)} = \tan \frac{t}{2}$$

(i) Clearly, $y = 0$ when $\cos t = 0$ *i.e.*, $t = 0$

When $t = 0, x = 0, \dfrac{dy}{dx} = \tan 0 = 0$

$\Rightarrow$ Curve passes through the origin and the axis of x is tangent at the origin.

(ii) y is maximum when $\cos t = -1$ *i.e.*, $t = \pi$ and $-\pi$.

When $t = \pi, x = a\pi, y = 2a$ and $\dfrac{dy}{dx} = \infty$. So at the point $t = \pi$, whose coordinates are $(a\pi, 2a)$, the tangent is perpendicular to the x-axis. When $t = -\pi, x = -a\pi, y = 2a, \dfrac{dy}{dx} = -\infty$.

(iii) Here y can not be negative, so curve lies entirely above the axis of x and no portion of the curve lies in the region $y > 2a$.

(iv)

t	$-\pi$	$-\pi/2$	0	$\pi/2$	π
x	$-a\pi$	$-a\left(\dfrac{\pi}{2}+1\right)$	0	$a\left(\dfrac{\pi}{2}+1\right)$	$a\pi$
y	$2a$	a	0	a	$2a$
dy/dx	$-\infty$	-1	0	1	∞

Replace t by $-t$ in the given curve, we get $x = -a(t + \sin t)$ and $y = a(1 - \cos t)$. Therefore, for every value of y, there are two equal and opposite value of x. So, curve is symmetric about y-axis.

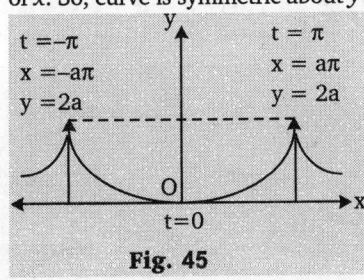

Fig. 45

- $x = a \cos t, y = b \sin t$

  then $\dfrac{x^2}{a^2} + \dfrac{y^2}{b^2} = 1$ (Ellipse)

- $x = a \cos t, y = b \sin t$

  Then $x^2 + y^2 = a^2$ (Circle)

- $x = a \cos^3 t, y = b \sin^3 t$

  Then $\left(\dfrac{x}{a}\right)^{2/3} + \left(\dfrac{y}{b}\right)^{2/3} = 1$ (Hypo-Cycloid)

- $x = a \cos^3 t, y = a \sin^3 t$

  Then $x^{2/3} + y^{2/3} = a^{2/3}$ (Astroid)

- $x = t^2, y = t - \dfrac{t^3}{3}$

  $y^2 = x(1 - x/3)^2$

- $x = a \sin^3 t, y = a \dfrac{\sin^3 t}{\cos t}$

  Then $y^2(a - x) = x^3$ (Cissoid)

- $x = \dfrac{1 - t^2}{1 + t^2}, y = \dfrac{2t}{1 + t^2}$

  Then $x^2 + y^2 = 1$ (Circle)

- $x = \dfrac{3at}{1 + t^3}, y = \dfrac{3at^2}{1 + t^3}$

  Then $x^3 + y^3 = 3axy$ (Follium of Descarte's)

EXAMPLE 4. *Trace the curve $x = a(t - \sin t), y = a(t - \cos t)$.*

SOLUTION. We have $x = a(t - \sin t) \Rightarrow \dfrac{dx}{dt} = a(1 - \cos t)$

$$y = a(t - \cos t) \Rightarrow \frac{dy}{dt} = a \sin t$$

$$\therefore \quad \frac{dy}{dx} = \frac{dy/dt}{dx/dt} = \frac{a \sin t}{a(1 - \cos t)} = \frac{2 \sin t/2 \cos t/2}{2 \sin^2 t/2} = \cot t/2$$

Here, $y = 0$ when $\cos t = 1$ *i.e.*, $t = 0, 2\pi$. When $t = 0, x = 0, y = 0$ and $\dfrac{dy}{dx} = \cot 0 = \infty$. Therefore, the curve passes through the origin and axis of y is tangent to the curve at this point.

Also, y is maximum when $\cos t = -1$ *i.e.*, $t = \pi$

When $t = \pi, x = a(\pi - \sin \pi) = a\pi, y = 2a, \dfrac{dy}{dx} = \cot \dfrac{\pi}{2} = 0$

Therefore, at $t = \pi$, whose cartesian coordinates are $(a\pi, 2a)$ the tangent to the curve is parallel to x-axis and curve does not lie in the region $y > 2a$.

In this curve y cannot be negative because $\cos t$ cannot be greater than 1. Hence, one complete arc of the given cycloid lying between $0 \le t \le 2\pi$.

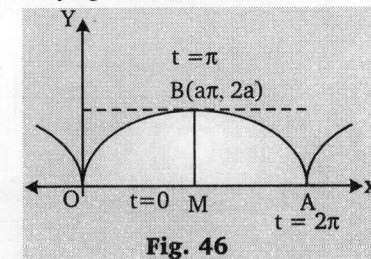

Fig. 46

2.8.14 TRACING OF EQUATION IN POLAR FORM

To trace the curve in polar form, we use the following procedure :

(a) Symmetry. In order to find the symmetry we use the following rules :
 (i) If the curve $r = f(\theta)$ remains unchanged when θ is replaced by $-\theta$, then the curve is symmetrical about the initial line.
 (ii) If the curve $r = f(\theta)$ remains unchanged when r is replaced by $-r$ then the curve is symmetrical about the pole. (origin).

(b) Special points on the curve. If r becomes zero for some values of θ, then the curve will pass through the pole. Therefore, if $r = 0$ when $\theta = \alpha$ (say), then $\theta = \alpha$ is the tangent to the curve at the pole.
Next we should find the maximum and minimum values of r which will exist for some values of θ.

(c) Solve these equations for r and observe the variation of r as θ varies form 0 to $+\infty$ and also we have to observe the variation as θ decreases form 0 to $-\infty$. Therefore, we should from the table for the values of r corresponding to the values of θ.

(d) Regions where curve does not exist. If we obtain the values of r, imaginary for $\alpha < \theta < \beta$ then the curve will not exist in this region.

(e) Asymptotes. Next we should find the asymptotes if exist. For this if $r \to \infty$ for $\theta = \alpha$, then $\theta = \alpha$ is an asymptotes of the curve.

(f) Direction of tangents. Find $r\dfrac{d\theta}{dr}$ from $r = f(\theta)$. But we have $\tan \phi = r\dfrac{d\theta}{dr}$. If for some $\theta = \alpha$, ϕ comes to be zero at any point then the line $\theta = \alpha$ will be the tangent at that point. Therefore, if ϕ comes out to be zero $\theta = \pi/2$ then $\theta = \pi/2$ is the tangent perpendicular to the radius vector $\theta = \alpha$.
Now taking all above facts into consideration draw the shape of the curve.

📝 **REMARK**

• Sometimes we face some problem to trace the curve of the form $r = f(\theta)$. Then for conveniently, change the polar form of the curve into cartesian form by the transformation : $x = r \cos \theta$, $y = r \sin \theta$ and then trace the curve.

Solved Examples

EXAMPLE 1. *Trace the following curve $r^2 = a^2 \cos 2\theta$.*

(VTU–2007, GBTU–2012, MTU–2011, UPTU–2009,
BPTU–2005,
KURUKSHETRA–2006)

SOLUTION . (i) The curve is symmetrical about both the initial line and the pole.

(ii) $r = 0$ when $\cos 2\theta = 0$ *i.e.*, when $\theta = \pm\pi/4$. Thus $\theta = -\pi/4$ and $\theta = \pi/4$ are two real and distinct tangent at the pole so that the pole is a node.

(iii) The maximum value of r is a when $\cos 2\theta = 1$ *i.e.*, when $\theta = 0$ and π.

(iv) $\tan \phi = r\dfrac{d\theta}{dr} = -\cot 2\theta \Rightarrow \tan \phi = \tan\left(\dfrac{\pi}{2} + 2\theta\right)$.

$\therefore \qquad \phi = \dfrac{\pi}{2} + 2\theta$.

At the points $(a, 0)$ and (a, π), ϕ comes out to be $\pi/2$ and $2\pi + \pi/2$. Thus at these points tangents the perpendicular to the initial line.

(v) The following table gives the corresponding values of r and θ.

θ	0	$\pi/6$	$\pi/4$	$\pi/3$	$3\pi/4$	π
r	a	$a/\sqrt{2}$	0	imag.	0	$-a$

(vi) **Region.** When $-\pi/4 < \theta < \pi/4$, r^2 is negative so the curve will not exist in this region and when $\dfrac{5\pi}{4} < \theta < \dfrac{7\pi}{4}$, r^2 is negative. Also the curve will not exist in this region.

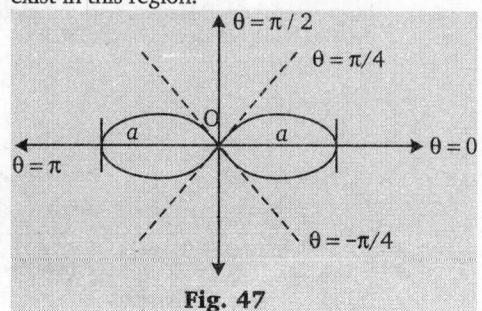

Fig. 47

EXAMPLE 2. *Trace the curve $r = a \sin 4\theta$.*

SOLUTION . (i) The curve is symmetrical about the initial line, since on putting $(\pi - \theta)$ in place of θ and $-r$ in place of r, the equation of the curve remains unchanged.

(ii) The curve is symmetrical about the pole because on replacing θ by $(\pi + \theta)$ the equation of the curves remains unchanged.

(iii) Since the curve is finite, there are no asymptotes.

(iv) Draw a table for values of r and θ.
It is evident that the greatest numerical value of r is a. Hence, the curve, between $\theta = 0$ and $\theta = \pi$, is as shown in fig. 23.
If θ increases from π to 2π, the corresponding branches of the curve are known because of symmetry about the pole. Since the curve is periodic, the values of θ outside the range $(0, 2\pi)$ need not be considered.

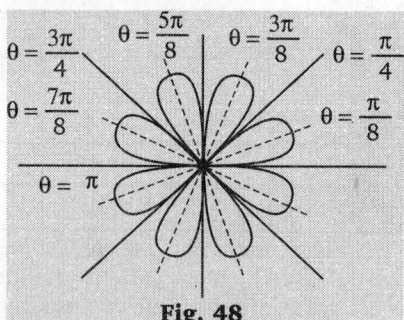

Fig. 48

EXAMPLE 3. *Trace the curve $r = a + b \cos \theta$, $a < b$.*

SOLUTION . (i) The curve is symmetrical about the initial line.

(ii) $r = 0$, when $a + b \cos \theta = 0$ *i.e.*,
$\cos \theta = \left(-\dfrac{a}{b}\right)$ or $\theta = \cos^{-1}\left(-\dfrac{a}{b}\right)$ but $a < b$ *i.e.*, $\dfrac{a}{b}$
< 1, therefore $\cos^{-1}\left(-\dfrac{a}{b}\right)$ comes out to be real so that $\theta = \cos^{-1}\left(-\dfrac{a}{b}\right)$ is the trangent at the pole.

(iii) r is maximum when $\cos\theta = 1$ *i.e.*, $\theta = 0$.

Then the maximum value of $r = a + b$ and the minimum value of $r = a - b$ when $\cos\theta = -1$ *i.e.*, $\theta = \pi$.

(iv) Since we have $r = a + b\cos\theta$.

$$\therefore \qquad \frac{dr}{d\theta} = -b\sin\theta$$

then $\tan\phi = \dfrac{d\theta}{dr} = -\dfrac{(a + b\cos\theta)}{b\sin}$.

Now if $\theta = 0$ and π, $\phi = 90°$, thus at the points $(a + b, 0)$, $(a - b, \pi)$ the tangents are perpendicular to the initial line.

(v) The following table gives the corresponding value of r and θ

θ	0	$\pi/2$	$\cos^{-1}\left(-\dfrac{a}{b}\right)$	$\cos\left(-\dfrac{a}{b}\right) < \theta < \pi$	π
r	$a+b$	a	0	r is negative	$a-b$

Thus from above facts the shape of the curve is shown below :

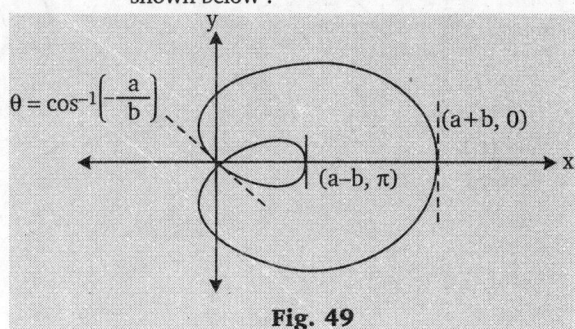

Fig. 49

Exercise-2.21

Trace the following curves :

1. $r = a(1 + \cos\theta)$
2. $r = 2a\cos\theta$
3. $r = a + b\cos\theta, a > b$
4. $r = a(\sec\theta + \cos\theta)$
5. $r\cos\theta = 2a\sin^2\theta$
6. $r^2 = a^2\sin 2\theta$
7. $r = a\sin 3\theta$ (GBTU–2011, UPTU–2002)
8. $r = ae^{m\theta}$
9. $2a/r = 1 + \cos\theta$.
10. $r = a\cos 3\theta$.
11. $r = a(1 - \cos\theta)$.

12. $r = a\sin 2\theta$. (VTU–2009)   13. $2r = 1 + 2\cos 2\theta$.
14. $x = a(\theta - \sin\theta), y = a(1 - \cos\theta)$.   (SRM-2008)

15. $x = a\cos\theta + \dfrac{1}{2}a\log\tan^2\dfrac{\theta}{2}, y = a\sin\theta$.

16. $x = a(\theta + \sin\theta), y = a(1 - \cos\theta)$.   (JNTU–2009)
17. $r = a\cos 2\theta$.
18. $x = a(t - \sin t), y = a(1 - \cos t)$.

Answers

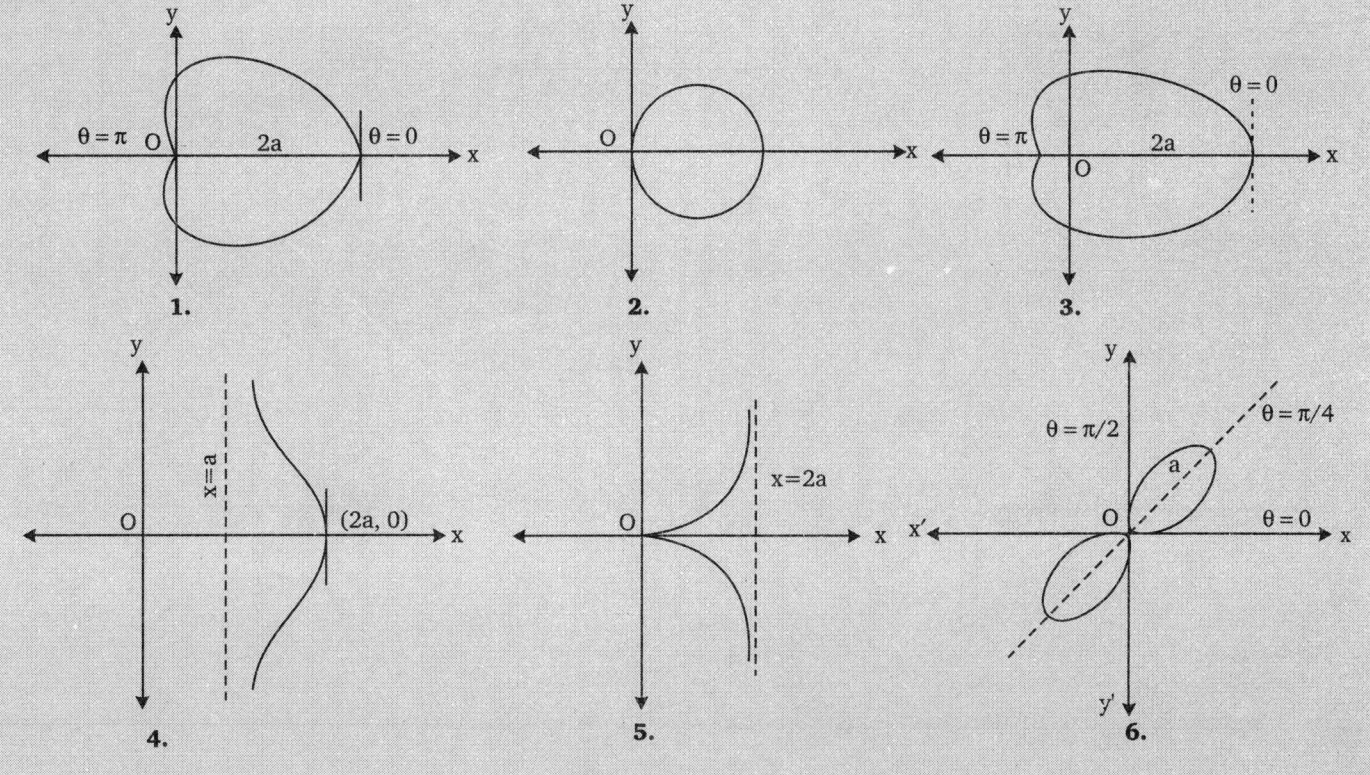

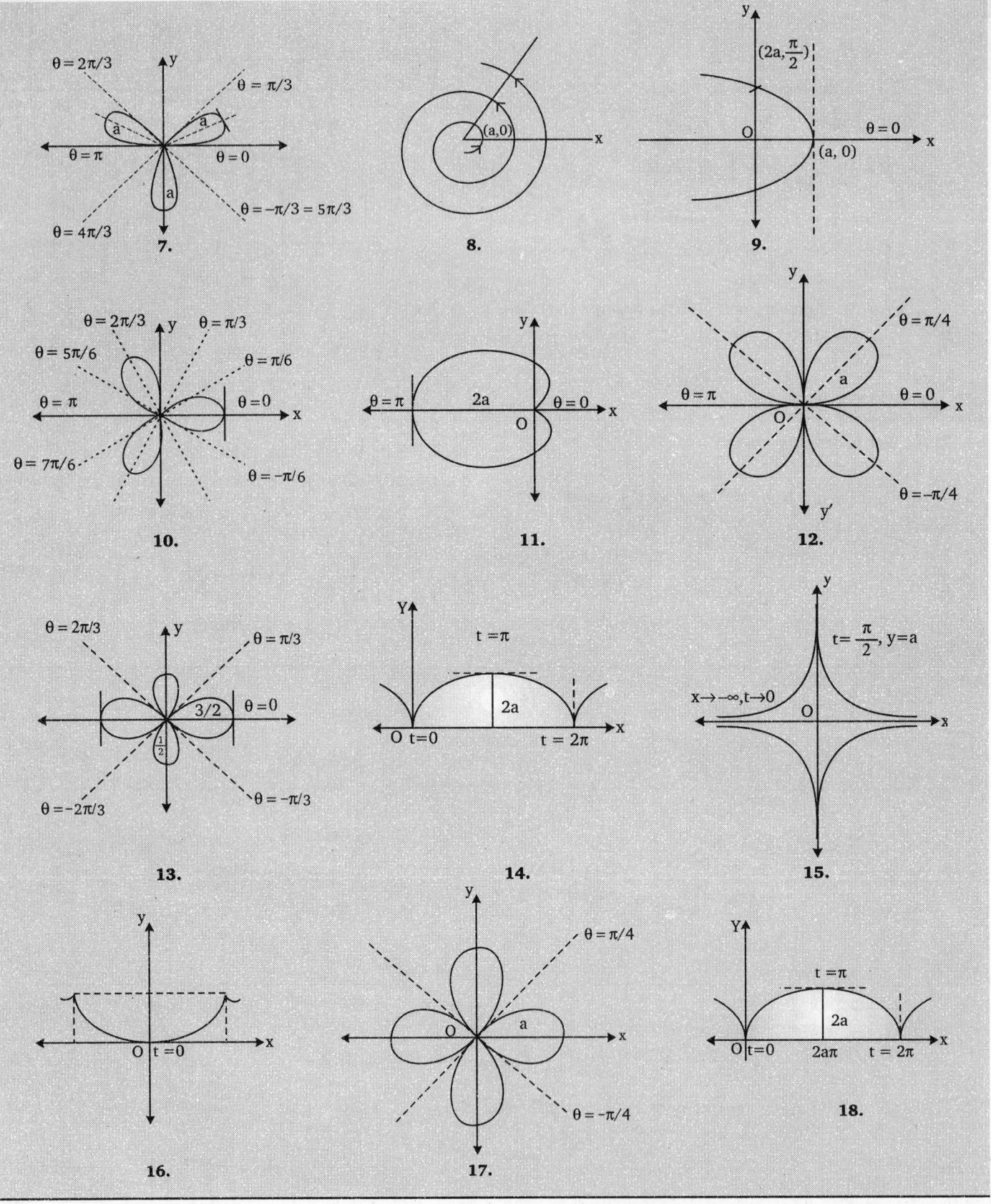

ARCHIVE

1. Find the nth derivative of the following functions :
   (i) $\sinh 2x \sin 4x$ (VTU–2010)
   (ii) $\log(4x^2 - 1)$ (VTU–2010)
   (iii) $\dfrac{x+3}{(x-1)(x+2)}$ (VTU–2009)
   (iv) $\dfrac{x^2}{2x^2+7x+6}$ (VTU–2005)
   (v) $\dfrac{x^n-1}{x-1}$ (MTU–2012)
   (vi) $x^{n-1}\log x$ (GBTU–2012)
   (vii) $e^{\sin^{-1}x}$ at $x = 0$. (UPTU–2009)
   (viii) $\sin^3 x$ (UPTU Model Paper–2009)

2. Find the nth derivative of the following functions :
   (i) $2^x\cos^9 x$ (MUMBAI–2009)
   (ii) $x^2\sin x$ at $x = 0$ (UPTU–2009)

3. If $y^{1/m} + y^{-1/m} = 2x$, prove that
   $(x^2-1)y_{n+2} + (2n+1)xy_{n+1} + (n^2-m^2)y_n = 0$
   (UPTU–2008, SVTU–2007, MUMBAI–2007)

4. If $y = \log\left[x+\sqrt{1+x^2}\right]^2$, prove that
   $(1+x^2)y_{n+2} + (2n+1)xy_{n+1} + n^2 y_n = 0$ (VTU–2007)

5. Verify the Rolle's theorem for the following functions :
   (i) $f(x) = (x-a)^m(x-b)^n$, m, n are positive integers in $[a, b]$. (NAGARJUNA–2008, VTU–2010)
   (ii) $f(x) = \dfrac{\sin x}{e^x}$ in $[0, \pi[$ (GNTU–2010)

6. If $f(x) = f(0) + kf_1(0) + \dfrac{k^2}{2!}f_2(\theta k)$, $0 < \theta < 1$, then find the value of θ where $k = 1$ and $f(x) = (1-x)^{5/2}$. (GBTU–2010)

7. If $0 < a < b < 1$, prove that
   $\dfrac{b-a}{1+b^2} < \tan^{-1}b - \tan^{-1}a < \dfrac{b-a}{1+a^2}$ (VTU–2006, MUMBAI–2009)

8. By applying mean value theorem to
   $f(x) = \log 2 . \sin\dfrac{\pi x}{2} + \log x$, prove that $\dfrac{\pi}{2}\log 2 . \cos\dfrac{\pi x}{2} + \dfrac{1}{x} = 0$ for some x between 1 and 2.

9. If x is positive, show that $x > \log(1+x) > x - \dfrac{x^2}{2}$. (VTU–2000)

10. If $f(x) = \sin^{-1}x$, $0 < a < b < 1$, use mean value theorem to prove that $\dfrac{b-a}{\sqrt{1-a^2}} < \sin^{-1}b - \sin^{-1}a < \dfrac{b-a}{\sqrt{1-b^2}}$.

11. Prove that $\dfrac{b-a}{b} < \log\left(\dfrac{b}{a}\right) < \dfrac{b-a}{b}$, for $0 < a < b$. Hence, show that $\dfrac{1}{4} < \log\dfrac{4}{3} < \dfrac{1}{3}$.

12. Expand $\log_e x$ in powers of $(x-1)$ and hence evaluate $\log_e 1.1$ correct to four decimal places. (BHOPAL–2007, KURUKSHETRA–2006)

13. Prove that $x \csc x = 1 + \dfrac{x^2}{6} + \dfrac{7x^4}{360} + ...$ (MUMBAI -2007)

14. Show that the coordinates for the line $x\cos\alpha + y\sin\alpha = p$ to touch the curve $\left(\dfrac{x}{a}\right)^m + \left(\dfrac{y}{b}\right)^m = 1$ is
   $(a\cos\alpha)^{m/m-1} + (b\sin\alpha)^{m/m-1} = p^{m/m-1}$.

15. If x, y be the parts of the axes x and y intercepted by the tangent at any point (x, y) on the curve $\left(\dfrac{x}{a}\right)^{2/3} + \left(\dfrac{y}{b}\right)^{2/3} = 1$ then show that $\left(\dfrac{x_1}{a}\right)^2 + \left(\dfrac{y_1}{b}\right)^2 = 1$. (BHOPAL–2008)

16. Show that in the exponential curve $y = be^{x/a}$, the subtangent is of constant length and that the subnormal varies as the square of the ordinate. (MADRAS–2000)

17. Show that the tangent to the cardoid $r = a(1 + \cos\theta)$ at the point $\theta = \dfrac{\pi}{3}$ and $\theta = \dfrac{2\pi}{3}$ are respectively parallel and perpendicular to the initial line. (VTU–2006)

18. With the usual meaning for r, s, θ and ϕ for the polar curve $r = f(\theta)$, show that $\dfrac{d\phi}{d\theta} + r\csc^2\theta\dfrac{d^2r}{ds^2} = 0$. (VTU–2000)

19. Show that the radius of curvature at $(a, 0)$ on the curve $y^2 = \dfrac{a^2(a-x)}{x}$ is $\dfrac{a}{2}$. (VTU–2000)

20. Show that the radius of curvature at the point t on the curve $x = e^t\cos t$, $y = e^t\sin t$ is $\sqrt{2}e^t$. (CALICUT–2005)

21. Show that the circle of convergence at the point $\left(\dfrac{3}{2}, \dfrac{3}{2}\right)$ of the curve $x^3 + y^3 = 3xy$ is $x^2 + y^2 - \dfrac{21}{8}(x+y) + \dfrac{432}{128} = 0$ (ANNA–2009, MADRAS–2006, CALICUT–2005)

22. Show that the circle of curvature at the origin for the curve $x + y = ax^2 + by^2 + cx^3$ is $(a+b)(x^2+y^2) = 2(x+y)$. (NAGPUR–2009)

23. Find the asymptotes of the following curves :
   (i) $(x^2-a^2)(y^2-b^2) = a^2b^2$ (OSMANIA–2002)
   (ii) $x^2(x-y)^2 - a^2(x^2+y^2) = 0$ (KURUKSHETRA–2006)
   (iii) $r = a\tan\theta$ (ROHTAK–2006)
   (iv) $r\sin\theta = 2\cos 2\theta$ (KURUKSHETRA–2009)

24. Choose the most appropriate answer :
   (i) The curve $r = a/(1 + \cos\theta)$ intersect orthogonally with the curve :
   a. $r = \dfrac{b}{1-\cos\theta}$ b. $r = \dfrac{b}{1+\sin\theta}$
   c. $r = \dfrac{b}{1+\sin^2\theta}$ c. $r = \dfrac{b}{1+\cos^2\theta}$ (VTU–2010)
   (ii) If $f(x)$ is continuous in the closed interval $[a, b]$, differentiable in $]a, b[$ and $f(a) = f(b)$ then $\exists c \in]a, b[$ such that $f'(c) =$
   a. 1 b. 0
   c. –1 d. none of these (VTU–2009)
   (iii) If the angle between the radius vector and the tangent is constant then the curve is
   a. $r = a\cos\theta$ b. $r^2 = a^2\cos^2\theta$
   c. $r = ae^{5\theta}$ d. none of these (VTU–2009)

Answers

1. (i) $\dfrac{20^{n/2}}{2}(e^{2x}\sin(2x+n\tan^{-1}2))-e^{-2x}(4x-n\tan^{-1}2)$ 　　　(ii) $(-1)^{n-1}(n-1)!\,2^n[(2x+1)^{-n}+(2x-1)^{-n}]$

(iii) $\dfrac{(-1)^n n!}{3}\left[\dfrac{4}{(x-1)^{n+1}}-\dfrac{1}{(x+2)^{n+1}}\right]$ 　(iv) $(-1)^n n!\left[\dfrac{9(2)^{n-1}}{(2x+3)^{n+1}}-\dfrac{4}{(x+2)^{n+1}}\right]$ 　(v) 0 　　(vi) $\dfrac{(n-1)!}{x}$

(vii) $y_n(0)=\begin{cases}1^2.(1^2+2^2).(1^2+4^2)\ldots[1^2+(n-2)^2] &,\ \text{if }n\text{ is even}\\ 1^2.(1^2+1^2).(1^2+3^2)\ldots[1^2+(n-2)^2] &,\ \text{if }n\text{ is odd}\end{cases}$ 　(viii) $\dfrac{3}{4}\sin\left(x+\dfrac{n\pi}{2}\right)-\dfrac{1}{4}(3)^n\sin\left(3x+\dfrac{n\pi}{2}\right)$

2. (i) $\dfrac{1}{256}\Big[(\log 2)^n.2^x(\cos 9\theta+9\cos 7\theta+3\cos 5\theta+84\cos 3\theta+126\cos\theta)+{}^nC_1(\log 2)^{n-1}.2^x\Big(\cos 9\theta+\dfrac{\pi}{2}\Big)+9\cos\Big(7\theta+\dfrac{n\pi}{2}\Big)+36\cos\Big(5\theta+\dfrac{\pi}{2}\Big)$

$+84\cos\Big(3\theta+\dfrac{\pi}{2}\Big)+126\cos\Big(\theta+\dfrac{\pi}{2}\Big)+\ldots+2^x\Big(\cos 9\theta+\dfrac{n\pi}{2}\Big)+9\cos\Big(7\theta+\dfrac{n\pi}{2}\Big)+36\cos\Big(5\theta+\dfrac{n\pi}{2}\Big)+84\cos\Big(3\theta+\dfrac{n\pi}{2}\Big)+126\cos\Big(\theta+\dfrac{n\pi}{2}\Big)\Big]$

(ii) $(n-n^2)\sin\dfrac{n\pi}{2}$ 　　**12.** $\log x=(x-1)-\dfrac{(x-1)^2}{2}+\dfrac{(x-1)^3}{3}-\dfrac{(x-1)^4}{4}+\ldots;0.0953$

23. (i) $x=\pm a,y=\pm b$ 　　(ii) $x+a=0,x-a=0,\ x-y+\sqrt{2}a=0$ 　　(iii) $r\cos\theta=a,r\cos\theta=-a$ 　　(iv) $r\sin\theta=2$

24. (i) a 　　(ii) b 　　(iii) c

✳✳✳✳✳✳✳

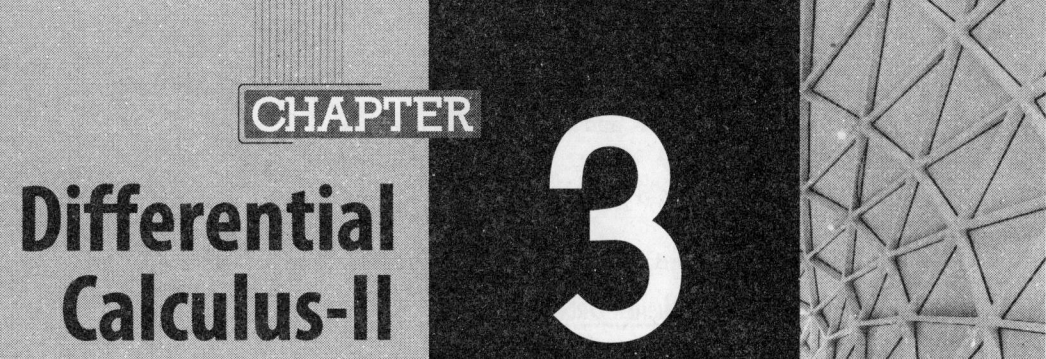

Differential Calculus-II

3.1 PARTIAL DIFFERENTIATION

We know that the differential coefficient of $f(x)$ with respect to x is $\lim\limits_{\delta x \to 0} \dfrac{f(x + \delta x) - f(x)}{\delta x}$, provided this limit exists, and it is denoted by

$$f'(x) \qquad \text{or} \qquad \frac{d}{dx}[f(x)]$$

If $u = f(x, y)$ be a continuous function of two independent variables x and y, then the differential coefficient of u w.r.t. x (regarding y as constant) is called the partial derivative or partial differential co-efficient of u w.r.t. x and is denoted by various symbols such as

$$\frac{\partial u}{\partial x}, \frac{\partial f}{\partial x}, f_x(x, y), f_x$$

Symbolically, if $u = f(x, y)$, then $\qquad \lim\limits_{\delta x \to 0} \dfrac{f(x + \delta x, y) - f(x, y)}{\delta x}$ if it exists,

is called the partial derivative or partial differential co-efficient of u w.r.t. x and is denoted by

$$\frac{\partial u}{\partial x} \quad \text{or} \quad \frac{\partial f}{\partial x} \quad \text{or} \quad f_x \quad \text{or} \quad u_x.$$

Similarly, by keeping x constant and allowing y alone to vary, we can define the partial derivative or partial differential coefficient of u w.r.t. y. It is denoted by any one of the symbols $\dfrac{\partial u}{\partial y}, \dfrac{\partial f}{\partial y}, f_y(x, y), f_y$.

Symbolically, $\qquad\qquad\qquad\qquad \dfrac{\partial u}{\partial y} = \lim\limits_{\delta y \to 0} \dfrac{f(x, y + \delta y) - f(x, y)}{\delta y}$

provided this limit exists.

For Example :

If $\quad u = ax^2 + 2hxy + by^2 \quad$ then $\qquad \dfrac{\partial u}{\partial x} = 2ax + 2hy \qquad$ and $\qquad \dfrac{\partial u}{\partial y} = 2hx + 2by.$

3.1.1 RULES OF PARTIAL DIFFERENTIATION

Rule (1) :

(a) If u is a function of x, y and we are to differentiate partially w.r.t. x then, y is treated as constant.

(b) Similarly, if we are to differentiate u partially w.r.t. y then x is treated as constant.

(c) If u is a function of x, y, z and we are to differentiate partially w.r.t. x, then y and z are treated as constant.

Rule (2) : If $z = u \pm v$, where u and v are functions of x and y, then

$$\frac{\partial z}{\partial x} = \frac{\partial u}{\partial x} \pm \frac{\partial v}{\partial x} \qquad\qquad \text{and} \qquad \frac{\partial z}{\partial y} = \frac{\partial u}{\partial y} \pm \frac{\partial v}{\partial y}.$$

Rule (3) : If $z = uv$, where u and v are functions of x and y, then

$$\frac{\partial z}{\partial x} = \frac{\partial}{\partial x}(uv) = u\frac{\partial v}{\partial x} + v\frac{\partial u}{\partial x} \qquad \text{and} \qquad \frac{\partial z}{\partial y} = \frac{\partial}{\partial y}(uv) = u\frac{\partial v}{\partial y} + v\frac{\partial u}{\partial y}.$$

Rule (4) : If $z = \dfrac{u}{v}$, where u, v are functions of x and y, then

$$\frac{\partial z}{\partial x} = \frac{\partial}{\partial x}\left(\frac{u}{v}\right) = \frac{v\dfrac{\partial u}{\partial x} - u\dfrac{\partial v}{\partial x}}{v^2} \qquad \text{and} \qquad \frac{\partial z}{\partial y} = \frac{\partial}{\partial y}\left(\frac{u}{v}\right) = \frac{v\dfrac{\partial u}{\partial y} - u\dfrac{\partial v}{\partial y}}{v^2}.$$

Rule (5) : If $z = f(u)$, where u is a function of x and y, then

$$\frac{\partial z}{\partial x} = \frac{\partial z}{\partial u} \cdot \frac{\partial u}{\partial x} \qquad\qquad \text{and} \qquad \frac{\partial z}{\partial y} = \frac{\partial z}{\partial u} \cdot \frac{\partial u}{\partial y}.$$

REMARKS

- Partial means a 'part of'.
- If z is a function of one variable x, then $\dfrac{\partial z}{\partial x} = \dfrac{dz}{dx}$.
- If z is a function of two variables x_1 and x_2, we get $\dfrac{\partial z}{\partial x_1}$ and $\dfrac{\partial z}{\partial x_2}$.
- If z is a function of n variables $x_1, x_2, \ldots, x_n$ we can find $\dfrac{\partial z}{\partial x_1}, \dfrac{\partial z}{\partial x_2}, \ldots, \dfrac{\partial z}{\partial x_n}$.

Symmetric Function of x and y. A function $u = u(x, y)$ is said to be symmetric if, on interchanging x and y, u remains unchanged.

3.1.2 PARTIAL DERIVATIVES OF THE HIGHER ORDER

We can find partial derivative of $\dfrac{\partial u}{\partial x}$ and $\dfrac{\partial u}{\partial y}$ just as we found those of u for $\dfrac{\partial u}{\partial x}$ and $\dfrac{\partial u}{\partial y}$ are itself functions of x and y.

The four derivatives, thus obtained, called the second order partial derivatives of u or $f(x, y)$ are

$$\frac{\partial}{\partial x}\left(\frac{\partial u}{\partial x}\right), \frac{\partial}{\partial y}\left(\frac{\partial u}{\partial x}\right), \frac{\partial}{\partial x}\left(\frac{\partial u}{\partial y}\right), \frac{\partial}{\partial y}\left(\frac{\partial u}{\partial y}\right)$$

and are denoted as

$$\frac{\partial^2 u}{\partial x^2}, \frac{\partial^2 u}{\partial y \partial x}, \frac{\partial^2 u}{\partial x \partial y}, \frac{\partial^2 u}{\partial y^2}$$

or

$$f_{xx}, f_{yx}, f_{xy}, f_{yy}.$$

REMARKS

- $\dfrac{\partial^2 u}{\partial x \partial y} = \dfrac{\partial}{\partial x}\left(\dfrac{\partial u}{\partial y}\right)$ and $\dfrac{\partial^2 u}{\partial y \partial x} = \dfrac{\partial}{\partial y}\left(\dfrac{\partial u}{\partial x}\right)$, Also $\dfrac{\partial^2 u}{\partial x \partial y} \neq \dfrac{\partial u}{\partial x} \cdot \dfrac{\partial u}{\partial y}$

- The partial derivatives $\dfrac{\partial^2 u}{\partial x \partial y}$ and $\dfrac{\partial^2 u}{\partial y \partial x}$ are distinguished by the order in which u is successively differentiated by the order in which u is successively differntiated w.r.t. x and y, but it will be seen that , in general, that are equal.

Solved Examples

EXAMPLE 1. *If $u = x^2y + y^2z + z^2x$, then show that*
$$\frac{\partial u}{\partial x} + \frac{\partial u}{\partial y} + \frac{\partial u}{\partial z} = (x + y + z)^2.$$
(UPTU(AG)–2006)

SOLUTION . Given that $\quad u = x^2y + y^2z + z^2x.$ ...(1)
Differentiating partially both sides of (1) *w.r.t.* x, y and z respectively, we get

$$\frac{\partial u}{\partial x} = 2xy + z^2 \qquad \text{...(2)}$$

$$\frac{\partial u}{\partial y} = x^2 + 2yz \qquad \text{...(3)}$$

and $\qquad \dfrac{\partial u}{\partial z} = y^2 + 2zx.$ ...(4)

Adding (2), (3) and (4), we get
$$\frac{\partial u}{\partial x} + \frac{\partial u}{\partial y} + \frac{\partial u}{\partial z} = 2xy + z^2 + x^2 + 2yz + y^2 + 2zx$$
$$= x^2 + y^2 + z^2 + 2xy + 2yz + 2zx$$
$$= (x + y + z)^2.$$

EXAMPLE 2. *If $u = f\left(\dfrac{y}{x}\right)$, show that $x\dfrac{\partial u}{\partial x} + y\dfrac{\partial u}{\partial y} = 0$.*

SOLUTION . We have $\quad u = f\left(\dfrac{y}{x}\right)$ ...(1)

Differentiating (1) partially w.r.t. x and y respectively, we get

$$\frac{\partial u}{\partial x} = f'\left(\frac{y}{x}\right) \cdot \left(-\frac{y}{x^2}\right)$$

$$\Rightarrow \qquad x\frac{\partial u}{\partial x} = -\frac{y}{x}f'\left(\frac{y}{x}\right) \qquad \text{...(2)}$$

and $\qquad \dfrac{\partial u}{\partial y} = f'\left(\dfrac{y}{x}\right) \cdot \dfrac{1}{x}$

$$\Rightarrow \qquad y\frac{\partial u}{\partial y} = \frac{y}{x}f'\left(\frac{y}{x}\right) \qquad \text{...(3)}$$

Adding (2) and (3), we get $x\dfrac{\partial u}{\partial x} + y\dfrac{\partial u}{\partial y} = 0.$

EXAMPLE 3. *If $z = f(x + ay) + \phi(x - ay)$, prove that*
$$\frac{\partial^2 z}{\partial y^2} = a^2 \frac{\partial^2 z}{\partial x^2}.$$
(SRM-2006, 10, 12)

SOLUTION . Given that $\quad z = f(x + ay) + \phi(x - ay).$ ...(1)
Differentiating partially both sides of (1) w.r.t. x and y respectively, we get

$$\frac{\partial z}{\partial x} = f'(x + ay) + \phi'(x - ay) \qquad \text{...(2)}$$

and $\qquad \dfrac{\partial z}{\partial y} = af'(x + ay) - a\phi'(x - ay).$ ...(3)

Again differentiating partially both sides of (2) w.r.t. x and (3) w.r.t. y, we get

$$\frac{\partial^2 z}{\partial x^2} = f''(x + ay) + \phi''(x - ay) \qquad \text{...(4)}$$

and $\qquad \dfrac{\partial^2 z}{\partial y^2} = a^2 f''(x+ay) + a^2 \phi''(x-ay).$ ...(5)

Form (4) and (5), we get

$$\dfrac{\partial^2 z}{\partial y^2} = a^2 \dfrac{\partial^2 z}{\partial x^2}.$$

EXAMPLE 4. *If* $u = \log(x^3 + y^3 + z^3 - 3xyz),$ *show that*

$$\left(\dfrac{\partial}{\partial x} + \dfrac{\partial}{\partial y} + \dfrac{\partial}{\partial z}\right)^2 u = -\dfrac{9}{(x+y+z)^2}.$$

(UKTU–2011, PTU–2010, ANNA–2007, UPTU–2006, BHOPAL–2008, VTU–2003)

SOLUTION . We have $\quad u = \log(x^3 + y^3 + z^3 - 3xyz)$

Differentiating partially with respect to x, we have

$$\dfrac{\partial u}{\partial x} = \dfrac{1}{x^3 + y^3 + z^3 - 3xyz}(3x^2 - 3yz)$$

$\Rightarrow \qquad \dfrac{\partial u}{\partial x} = \dfrac{3(x^2 - yz)}{x^3 + y^3 + z^3 - 3xyz}.$ ...(1)

Similarly, $\dfrac{\partial u}{\partial y} = \dfrac{3(y^2 - zx)}{x^3 + y^3 + z^3 - 3xyz}$ ...(2)

and $\qquad \dfrac{\partial u}{\partial z} = \dfrac{3(z^2 - xy)}{x^3 + y^3 + z^3 - 3xyz}$ ...(3)

Adding (1), (2) and (3), we get

$$\dfrac{\partial u}{\partial x} + \dfrac{\partial u}{\partial y} + \dfrac{\partial u}{\partial z} = \dfrac{3(x^2 + y^2 + z^2 - yz - zx - xy)}{x^3 + y^3 + z^3 - 3xyz}$$

$$= \dfrac{3(x^2 + y^2 + z^2 - yz - zx - xy)}{(x+y+z)(x^2 + y^2 + z^2 - yz - zx - xy)}$$

$$= \dfrac{3}{(x+y+z)}.$$

Also,

$$\left(\dfrac{\partial}{\partial x} + \dfrac{\partial}{\partial y} + \dfrac{\partial}{\partial z}\right)^2 u = \left(\dfrac{\partial}{\partial x} + \dfrac{\partial}{\partial y} + \dfrac{\partial}{\partial z}\right)\left(\dfrac{\partial}{\partial x} + \dfrac{\partial}{\partial y} + \dfrac{\partial}{\partial z}\right)u$$

$$= \left(\dfrac{\partial}{\partial x} + \dfrac{\partial}{\partial y} + \dfrac{\partial}{\partial z}\right)\left(\dfrac{\partial u}{\partial x} + \dfrac{\partial u}{\partial y} + \dfrac{\partial u}{\partial z}\right)$$

$$= \left(\dfrac{\partial}{\partial x} + \dfrac{\partial}{\partial y} + \dfrac{\partial}{\partial z}\right)\left(\dfrac{3}{x+y+z}\right)$$

$$= 3\left[\dfrac{\partial}{\partial x}\left(\dfrac{1}{x+y+z}\right) + \dfrac{\partial}{\partial y}\left(\dfrac{1}{x+y+z}\right) + \dfrac{\partial}{\partial z}\left(\dfrac{1}{x+y+z}\right)\right]$$

$$= 3\left[-\dfrac{1}{(x+y+z)^2} - \dfrac{1}{(x+y+z)^2} - \dfrac{1}{(x+y+z)^2}\right]$$

$$= -\dfrac{9}{(x+y+z)^2}$$

EXAMPLE 5. *If* $u = \sin^{-1}\dfrac{x}{y} + \tan^{-1}\dfrac{y}{x},$ *show that* $x\dfrac{\partial u}{\partial x} + y\dfrac{\partial u}{\partial y} = 0.$

(UPTU–2007)

SOLUTION . We have $u = \sin^{-1}\dfrac{x}{y} + \tan^{-1}\dfrac{y}{x}$

$\Rightarrow \qquad \dfrac{\partial u}{\partial x} = \dfrac{1}{\sqrt{1 - \left(\dfrac{x}{y}\right)^2}} \cdot \dfrac{1}{y} + \dfrac{1}{1 + \left(\dfrac{y}{x}\right)^2} \cdot \left(-\dfrac{y}{x^2}\right)$

$$= \dfrac{1}{\sqrt{y^2 - x^2}} - \dfrac{y}{(x^2 + y^2)}$$

$\Rightarrow \quad x\dfrac{\partial u}{\partial x} = \dfrac{x}{\sqrt{(y^2 - x^2)}} - \dfrac{xy}{x^2 + y^2}$ ...(1)

Also, $\dfrac{\partial u}{\partial y} = \dfrac{1}{\sqrt{1 - \left(\dfrac{x}{y}\right)^2}} \cdot \left(-\dfrac{x}{y^2}\right) + \dfrac{1}{1 + \left(\dfrac{y}{x}\right)^2} \cdot \left(\dfrac{1}{x}\right)$

$$= -\dfrac{x}{y\sqrt{y^2 - x^2}} + \dfrac{x}{x^2 + y^2}$$

$\Rightarrow \quad y\dfrac{\partial u}{\partial y} = -\dfrac{x}{\sqrt{(y^2 - x^2)}} + \dfrac{xy}{x^2 + y^2}$ ...(2)

On adding (1) and (2), we get $x\dfrac{\partial u}{\partial x} + y\dfrac{\partial u}{\partial y} = 0.$

EXAMPLE 6. *If* $u = f(r),$ *where* $r^2 = x^2 + y^2,$ *show that*

$$\dfrac{\partial^2 u}{\partial x^2} + \dfrac{\partial^2 u}{\partial y^2} = f''(r) + \dfrac{1}{r}f'(r).$$

(UPTU–2005, SVTU–2008, RAJASTHAN–2006)

SOLUTION . We have $r^2 = x^2 + y^2$

$\Rightarrow \quad \left.\begin{array}{l} 2r\dfrac{\partial r}{\partial x} = 2x \text{ or } \dfrac{\partial r}{\partial x} = \dfrac{x}{r} \\[2mm] 2r\dfrac{\partial r}{\partial y} = 2y \text{ or } \dfrac{\partial r}{\partial y} = \dfrac{y}{r} \end{array}\right]$

and ...(1)

Since, $u = f(r)$

$\Rightarrow \qquad \dfrac{\partial u}{\partial x} = [f'(r)] \cdot \dfrac{\partial r}{\partial x} = \dfrac{x}{r}f'(r)$

and $\dfrac{\partial^2 u}{\partial x^2} = \dfrac{\partial}{\partial x}\left(\dfrac{\partial u}{\partial x}\right) = \dfrac{\partial}{\partial x}\left[x \cdot \dfrac{1}{r}f'(r)\right]$

$$= 1 \cdot \dfrac{1}{r} \cdot f'(r) + [xf'(r)]\left[-\dfrac{1}{r^2}\dfrac{\partial r}{\partial x}\right] + \dfrac{x}{r}[f''(r)]\dfrac{\partial r}{\partial x}$$

$$= \dfrac{1}{r} \cdot f'(r) - \dfrac{x}{r^2} \cdot \dfrac{x}{r}f'(r) + \dfrac{x^2}{r^2}f''(r)$$

$$= \dfrac{1}{r} \cdot f'(r) - \dfrac{x^2}{r^3}f'(r) + \dfrac{x^2}{r^2}f''(r).$$ ...(2)

Similarly, we may get

$$\dfrac{\partial^2 u}{\partial y^2} = \dfrac{1}{r} \cdot f'(r) - \dfrac{y^2}{r^3}f'(r) + \dfrac{y^2}{r^2}f''(r)$$ ...(3)

Adding (2) and (3), we get

$$\dfrac{\partial^2 u}{\partial x^2} + \dfrac{\partial^2 u}{\partial y^2} = \dfrac{2}{r} \cdot f'(r) - \dfrac{x^2 + y^2}{r^3}f'(r) + \dfrac{x^2 + y^2}{r^2}f''(r)$$

$$= \dfrac{2}{r} \cdot f'(r) - \dfrac{r^2}{r^3}f'(r) + \dfrac{r^2}{r^2}f''(r)$$

$$= \dfrac{2}{r} \cdot f'(r) - \dfrac{1}{r}f'(r) + f''(r)$$

$$= f''(r) + \dfrac{1}{r} \cdot f'(r).$$

EXAMPLE 7. *If* $x^x y^y z^z = c.$ *Show that at* $x = y = z,$

$$\dfrac{\partial^2 z}{\partial x \partial y} = -[x \log ex]^{-1}$$

(BHOPAL–2008)

SOLUTION. We have $x^x y^y z^z = c.$ ...(1)

We observe that z can be regarding as a function of two independent variables x and y.

Taking log of both sides of (1), we have

$x \log x + y \log y + z \log z = \log c.$ ...(2)

Differentiating (2) partially w.r.t. x, we get

$$x.\frac{1}{x} + 1.\log x + \left[z.\frac{1}{z} + 1.\log z \right]\frac{\partial z}{\partial x} = 0$$

$$\Rightarrow \quad \frac{\partial z}{\partial x} = -\frac{(1 + \log x)}{(1 + \log z)}. \quad ...(3)$$

Similarly differentiating (2) partially, w.r.t. y, we get

$$\frac{\partial z}{\partial y} = -\frac{(1 + \log y)}{(1 + \log z)}. \quad ...(4)$$

Also, $\dfrac{\partial^2 z}{\partial x \partial y} = \dfrac{\partial}{\partial x}\left(\dfrac{\partial z}{\partial y}\right) = \dfrac{\partial}{\partial x}\left[-\left(\dfrac{1 + \log y}{1 + \log z}\right) \right]$

$$= -(1 + \log y)\frac{\partial}{\partial x}[(1 + \log z)^{-1}]$$

$$= -(1 + \log y).\left[-(1 + \log z)^{-2}\frac{1}{z}.\frac{\partial z}{\partial x} \right]$$

$$= \frac{(1 + \log y)}{z(1 + \log z)^2}.\left[-\left(\frac{1 + \log x}{1 + \log z}\right) \right].$$

For $x = y = z$, we have

$$\frac{\partial^2 z}{\partial x \partial y} = -\frac{(1 + \log x)^2}{x(1 + \log x)^3} = -\frac{1}{x(1 + \log x)}$$

$$= \frac{-1}{x[\log e + \log x]} \qquad [\because \log e = 1]$$

$$= \frac{-1}{x \log(ex)} = -[x \log(ex)]^{-1}.$$

EXAMPLE 8. *If* $u = (1 - 2xy + y^2)^{-1/2}$, **prove that**

$$\frac{\partial}{\partial x}\left\{ (1 - x^2)\frac{\partial u}{\partial x} \right\} + \frac{\partial}{\partial y}\left\{ y^2 \frac{\partial u}{\partial y} \right\} = 0. \qquad \text{(ROHTAK–2006)}$$

SOLUTION. We have $u = (1 - 2xy + y^2)^{-1/2}$ ...(1)

Differentiating (1) partially with respect to x, we get

$$\frac{\partial u}{\partial x} = -\frac{1}{2}(1 - 2xy + y^2)^{-3/2}(-2y)$$

or $\quad \dfrac{\partial u}{\partial x} = y(1 - 2xy + y^2)^{-3/2}$

$$\Rightarrow \quad (1 - x^2)\frac{\partial u}{\partial x} = y(1 - x^2)(1 - 2xy + y^2)^{-3/2}$$

Again differentiating partially w.r.t. x, we get

$$\frac{\partial}{\partial x}\left\{ (1 - x^2)\frac{\partial u}{\partial x} \right\} = y[y - 2x(1 - 2xy + y^2)^{-3/2}$$

$$+ (1 - x^2)\left(-\frac{3}{2}\right)(-2y)(1 - 2xy + y^2)^{-5/2}$$

$$= -2xy(1 - 2xy + y^2)^{-3/2}$$

$$+ 3y^2(1 - x^2)(1 - 2xy + y^2)^{-5/2}$$

$$\therefore \quad \frac{\partial}{\partial x}\left\{ (1 - x^2)\frac{\partial u}{\partial x} \right\} = -2xyu^3 + 3y^2(1 - x^2)u^5$$

$$[\text{Using (1)}]$$

...(2)

Differentiating (1) partially w.r.t. y, we get

$$\frac{\partial u}{\partial y} = -\frac{1}{2}(1 - 2xy + y^2)^{-3/2}(-2x + 2y)$$

or $\quad \dfrac{\partial u}{\partial y} = (x - y)(1 - 2xy + y^2)^{-3/2}$

$$\Rightarrow \quad y^2 \frac{\partial u}{\partial y} = (x - y)y^2(1 - 2xy + y^2)^{-3/2}$$

Again differentiating partially w.r.t. y, we get

$$\frac{\partial}{\partial y}\left(y^2 \frac{\partial u}{\partial y} \right) = (2xy - 3y^2)(1 - 2xy + y^2)^{-3/2} + (xy^2 - y^3)\left(-\frac{3}{2}\right)$$

$$(-2x + 2y)(1 - 2xy + y^2)^{-5/2}$$

$$= 2xy(1 - 2xy + y^2)^{-3/2} - 3y^2(1 - 2xy + y^2)^{-3/2}$$

$$+ 3y^2(x - y)^2(1 - 2xy + y^2)^{-5/2}$$

$$= 2xy(1 - 2xy + y^2)^{-3/2} - 3y^2(1 - 2xy + y^2)^{-5/2}$$

$$\{(1 - 2xy + y^2) - (x - y)^2\}$$

$$= 2xy(1 - 2xy + y^2)^{-3/2}$$

$$- 3y^2(1 - x^2)(1 - 2xy + y^2)^{-5/2}$$

$$\therefore \quad \frac{\partial}{\partial y}\left\{ y^2 \frac{\partial u}{\partial y} \right\} = 2xyu^3 - 3y^2(1 - x^2)u^5 \text{ [Using (1)]} \qquad ...(3)$$

Adding (2) and (3), we get

$$\frac{\partial}{\partial x}\left\{ (1 - x^2)\frac{\partial u}{\partial x} \right\} + \frac{\partial}{\partial y}\left\{ y^2 \frac{\partial u}{\partial y} \right\} = 0.$$

EXAMPLE 9. *If* $u = (x^2 + y^2 + z^2)^{-1/2}$, *show that*

(i) $x \dfrac{\partial u}{\partial x} + y \dfrac{\partial u}{\partial y} + z \dfrac{\partial u}{\partial z} = -u$ **(ii)** $\dfrac{\partial^2 u}{\partial x^2} + \dfrac{\partial^2 u}{\partial y^2} + \dfrac{\partial^2 u}{\partial z^2} = 0$

(GBTU–2011, VTU–2006, OSMANIA–2003)

SOLUTION. (i) We have $u = (x^2 + y^2 + z^2)^{-1/2}$ ...(1)

Differentiating (1) partially w.r.t. x, y and z respectively, we get

$$\frac{\partial u}{\partial x} = \left(-\frac{1}{2}\right)(x^2 + y^2 + z^2)^{-3/2}(2x)$$

or $\quad \dfrac{\partial u}{\partial x} = \dfrac{-x}{(x^2 + y^2 + z^2)^{3/2}}$

$$\Rightarrow \quad x\frac{\partial u}{\partial x} = \frac{-x^2}{(x^2 + y^2 + z^2)^{3/2}} \qquad ...(2)$$

Similarly,

$$y\frac{\partial u}{\partial y} = \frac{-y^2}{(x^2 + y^2 + z^2)^{3/2}} \qquad ...(3)$$

and $z\dfrac{\partial u}{\partial z} = \dfrac{-z^2}{(x^2 + y^2 + z^2)^{3/2}} \qquad ...(4)$

Adding (2), (3) and (4), we get

$$x\frac{\partial u}{\partial x} + y\frac{\partial u}{\partial y} + z\frac{\partial u}{\partial z} = \frac{-(x^2 + y^2 + z^2)}{(x^2 + y^2 + z^2)^{3/2}}$$

$$= -(x^2 + y^2 + z^2)^{-1/2}$$

$$\therefore \quad x\frac{\partial u}{\partial x} + y\frac{\partial u}{\partial y} + z\frac{\partial u}{\partial z} = -u$$

(ii) We have $\dfrac{\partial^2 u}{\partial x^2} = \dfrac{\partial}{\partial x}\left(\dfrac{\partial u}{\partial x}\right) = \dfrac{\partial}{\partial x}\left\{ \dfrac{-x}{(x^2 + y^2 + z^2)^{3/2}} \right\}$

$$= -\left[\frac{1}{(x^2 + y^2 + z^2)^{3/2}} + x\left\{ \left(-\frac{3}{2}\right)(2x)(x^2 + y^2 + z^2)^{-5/2} \right\} \right]$$

$$= -\left[\frac{1}{(x^2+y^2+z^2)^{3/2}} - \frac{3x^2}{(x^2+y^2+z^2)^{5/2}}\right]$$

$$= -\frac{(y^2+z^2-2x^2)}{(x^2+y^2+z^2)^{5/2}}$$

$$\frac{\partial^2 u}{\partial x^2} = \frac{2x^2-y^2-z^2}{(x^2+y^2+z^2)^{5/2}} \qquad \dots(5)$$

Similarly,

$$\frac{\partial^2 u}{\partial y^2} = \frac{2y^2-x^2-z^2}{(x^2+y^2+z^2)^{5/2}} \qquad \dots(6)$$

and $$\frac{\partial^2 u}{\partial z^2} = \frac{2z^2-y^2-x^2}{(x^2+y^2+z^2)^{5/2}} \qquad \dots(7)$$

Adding (5), (6) and (7), we get

$$\frac{\partial^2 u}{\partial x^2} + \frac{\partial^2 u}{\partial y^2} + \frac{\partial^2 u}{\partial z^2} = 0$$

EXAMPLE 10. *If* $\theta = t^n e^{-r^2/4t}$, *find the value of n for which*

$$\frac{1}{r^2}\cdot\frac{\partial}{\partial r}\left(r^2\frac{\partial\theta}{\partial r}\right) = \frac{\partial\theta}{\partial t}.$$

(UPTU–2006, KURUKSHETRA–2006, NAGPUR–2009)

SOLUTION. We have $\theta = t^n e^{-r^2/4t}$ ...(1)

Then $$\frac{\partial\theta}{\partial r} = t^n\left[e^{-r^2/4t}\left(-\frac{2r}{4t}\right)\right] = -\frac{r}{2}t^{n-1}e^{-r^2/4t}$$

$$\Rightarrow \quad r^2\frac{\partial\theta}{\partial r} = -\frac{r^3}{2}t^{n-1}e^{-r^2/4t}$$

Now

$$\frac{\partial}{\partial r}\left(r^2\frac{\partial\theta}{\partial r}\right) = -\frac{1}{2}t^{n-1}\left[3r^2 e^{-r^2/4t} + r^3 e^{-r^2/4t}\left(\frac{-2r}{4t}\right)\right]$$

$$= -\frac{3}{2}r^2 t^{n-1}e^{-r^2/4t} + \frac{1}{4}r^4 t^{n-2}e^{-r^2/4t}$$

$$\therefore \frac{1}{r^2}\cdot\frac{\partial}{\partial r}\left(r^2\frac{\partial\theta}{\partial r}\right) = -\frac{3}{2}t^{n-1}e^{-r^2/4t} + \frac{1}{4}r^2 t^{n-2}e^{-r^2/4t} \quad \dots(2)$$

Again from (1), we get

$$\frac{\partial\theta}{\partial t} = nt^{n-1}e^{-r^2/4t} + t^n\cdot e^{-r^2/4t}\cdot\left(\frac{r^2}{4t^2}\right)$$

or $$\frac{\partial\theta}{\partial t} = nt^{n-1}e^{-r^2/4t} + \frac{1}{4}r^2 t^{n-2}\cdot e^{-r^2/4t} \quad \dots(3)$$

Since, $$\frac{1}{r^2}\cdot\frac{\partial}{\partial r}\left(r^2\frac{\partial\theta}{\partial r}\right) = \frac{\partial\theta}{\partial t}$$

Then form (2) and (3), we have

$$-\frac{3}{2}t^{n-1}e^{-r^2/4t} + \frac{1}{4}r^2 t^{n-2}e^{-r^2/4t}$$

$$= nt^{n-1}e^{-r^2/4t} + \frac{1}{4}r^2 t^{n-2}\cdot e^{-r^2/4t}$$

$$\Rightarrow \quad n = -\frac{3}{2}.$$

EXAMPLE 11. *If* $u(x, y, z) = \log(\tan x + \tan y + \tan z)$, *show that* $\sin 2x \dfrac{\partial u}{\partial x} + \sin 2y \dfrac{\partial u}{\partial y} + \sin 2z \dfrac{\partial u}{\partial z} = 2.$

(UPTU–2007, MTU–2011, GBTU–2012)

SOLUTION. Differentiating the given function partially w.r.t. x, we get

$$\frac{\partial u}{\partial x} = \frac{1}{\tan x + \tan y + \tan z}(\sec^2 x) \qquad \dots(1)$$

Similarly,

$$\frac{\partial u}{\partial y} = \frac{1}{\tan x + \tan y + \tan z}(\sec^2 y) \qquad \dots(2)$$

and $$\frac{\partial u}{\partial z} = \frac{1}{\tan x + \tan y + \tan z}(\sec^2 z) \qquad \dots(3)$$

Now multiply (1) by $\sin 2x$, (2) by $\sin 2y$, (3) by $\sin 2z$ and then adding we get

$$\sin 2x\frac{\partial u}{\partial x} + \sin 2y\frac{\partial u}{\partial y} + \sin 2z\frac{\partial u}{\partial z}$$

$$= \frac{\sin 2x \sec^2 x + \sin 2y \sec^2 y + \sin 2z \sec^2 z}{\tan x + \tan y + \tan z}$$

$$= \frac{2(\tan x + \tan y + \tan z)}{\tan x + \tan y + \tan z} = 2.$$

EXAMPLE 12. *If* $z = \tan(y + ax) + (y - ax)^{3/2}$, *show that*

$$\frac{\partial^2 z}{\partial x^2} = a^2\frac{\partial^2 z}{\partial y^2}. \qquad \text{(MUMBAI–2009)}$$

SOLUTION. We have $z = \tan(y + ax) + (y - ax)^{3/2}$

$$\Rightarrow \quad \frac{\partial z}{\partial x} = (\sec^2(y + ax)).a + \frac{3}{2}(y - ax)^{1/2}(-a)$$

and

$$\frac{\partial^2 z}{\partial x^2} = 2a^2\tan(y+ax)\sec^2(y+ax) + \frac{3}{4}a^2(y-ax)^{-1/2} \qquad \dots(1)$$

Also, $\dfrac{\partial z}{\partial y} = \sec^2(y + ax) + \dfrac{3}{2}(y - ax)^{1/2}$

and $$\frac{\partial^2 z}{\partial y^2} = 2\sec^2(y+ax)\tan(y+ax) + \frac{3}{4}(y-ax)^{-1/2} \qquad \dots(2)$$

From (1) and (2) we conclude that

$$\frac{\partial^2 z}{\partial x^2} = a^2\left(\frac{\partial^2 z}{\partial y^2}\right).$$

Exercise-3.1

1. Find $\dfrac{\partial u}{\partial x}$ and $\dfrac{\partial u}{\partial y}$ when:

(i) $u = \log(x^2 + y^2)$

(ii) $u = \cos^{-1}\left(\dfrac{x}{y}\right)$

(iii) $u = \dfrac{x^2}{a^2} + \dfrac{y^2}{b^2}$

(iv) $u = \tan^{-1}\left(\dfrac{x^2+y^2}{x+y}\right)$

2. Find the second order partial derivatives of $\log(e^x + e^y)$.

3. Verify that $\dfrac{\partial^2 u}{\partial x\partial y} = \dfrac{\partial^2 u}{\partial y\partial x}$, where

(i) $u = \log(y \sin x + x \sin y)$

(ii) $u = \log\left(\dfrac{x^2+y^2}{xy}\right)$ (UPTU–2008)

(iii) $u = \log\left(\dfrac{x^2+y^2}{x+y}\right)$

(iv) $u = \sin^{-1}\dfrac{x}{y}$

(v) $u = x^y$ (ANNA–2009)

(vi) $u = \log \tan\left(\dfrac{y}{x}\right)$ (vii) $u = x^4 + x^2y^2 + y^4$

(viii) $u = \log\left(\dfrac{xy}{x^2 + y^2}\right)$ (ix) $u = x \log y$

4. If $x = r \cos\theta$, $y = r \sin\theta$, show that $\dfrac{\partial r}{\partial x} = \dfrac{\partial x}{\partial r}$, $\dfrac{\partial x}{r\partial\theta} = r\dfrac{\partial\theta}{\partial x}$.

5. If $u = \log(\tan x + \tan y)$, prove that $\sin 2x \dfrac{\partial u}{\partial x} + \sin 2y \dfrac{\partial u}{\partial y} = 2$.

6. If $u = x^2 \tan^{-1}\dfrac{y}{x} - y^2\tan^{-1}\dfrac{x}{y}$, prove that $\dfrac{\partial^2 u}{\partial x \partial y} = \dfrac{x^2 - y^2}{x^2 + y^2}$.

(UKTU–2012, MUMBAI–2008, MADRAS–2000)

7. If $u = 2(ax + by)^2 - (x^2 + y^2)$ and $a^2 + b^2 = 1$, prove that $\dfrac{\partial^2 u}{\partial x^2} + \dfrac{\partial^2 u}{\partial y^2} = 0$.

8. If $u = \log(x^3 + y^3 - x^2y - xy^2)$, prove that

(i) $\dfrac{\partial u}{\partial x} + \dfrac{\partial u}{\partial y} = 2(x + y)^{-1}$

(ii) $\dfrac{\partial^2 u}{\partial x^2} + 2\dfrac{\partial^2 u}{\partial x \partial y} + \dfrac{\partial^2 u}{\partial y^2} = -4(x + y)^{-2}$

9. If $u = f(x + 2y) + g(x - 2y)$, show that $4\dfrac{\partial^2 u}{\partial x^2} = \dfrac{\partial^2 u}{\partial y^2}$.

10. If $u = e^{xyz}$, show that $\dfrac{\partial^3 u}{\partial x \partial y \partial z} = (1 + 3xyz + x^2y^2z^2)e^{xyz}$.

(UPTU–2007, 2009)

11. If $u(x + y) = x^2 + y^2$, show that $\left(\dfrac{\partial u}{\partial x} - \dfrac{\partial u}{\partial y}\right)^2 = 4\left(1 - \dfrac{\partial u}{\partial x} - \dfrac{\partial u}{\partial y}\right)$.

(VTU–2003, 06, 12)

12. If $\tan u = \dfrac{\cos x}{\sinh y}$ and $\tanh v = \dfrac{\sinh x}{\cosh y}$ show that $\dfrac{\partial u}{\partial x} = \dfrac{\partial v}{\partial y}$ and $\dfrac{\partial u}{\partial y} = -\dfrac{\partial v}{\partial x}$.

13. Show that $\dfrac{\partial^2 u}{\partial x^2} + \dfrac{\partial^2 u}{\partial y^2} = 0$, if

(i) $u = e^{my} \cos mx$ (ii) $u = \tan^{-1}\dfrac{y}{x}$.

14. If $\dfrac{x^2}{a^2 + u} + \dfrac{y^2}{b^2 + u} + \dfrac{z^2}{c^2 + u} = 1$, show that

$\left(\dfrac{\partial u}{\partial x}\right)^2 + \left(\dfrac{\partial u}{\partial y}\right)^2 + \left(\dfrac{\partial u}{\partial z}\right)^2 = 2\left(x\dfrac{\partial u}{\partial x} + y\dfrac{\partial u}{\partial y} + z\dfrac{\partial u}{\partial z}\right)$ (UPTU–2003)

15. Find the value of $\dfrac{1}{a^2}\dfrac{\partial^2 z}{\partial x^2} + \dfrac{1}{b^2}\dfrac{\partial^2 z}{\partial y^2}$, when $a^2x^2 + b^2y^2 - c^2z^2 = 0$.

16. If $z = e^{ax + by}f(ax - by)$, show that $b\dfrac{\partial z}{\partial x} + a\dfrac{\partial z}{\partial y} = 2abz$.

(UPTU–2006, VTU–2010)

17. If $u = \sqrt{x^2 + y^2 + z^2}$, show that $\left(\dfrac{\partial u}{\partial x}\right)^2 + \left(\dfrac{\partial u}{\partial y}\right)^2 + \left(\dfrac{\partial u}{\partial z}\right)^2 = 1$.

18. If $x = r \cos\theta$, $y = r \sin\theta$, prove that

(i) $\dfrac{\partial^2\theta}{\partial x^2} + \dfrac{\partial^2\theta}{\partial y^2} = 0$ except when $x = 0$, $y = 0$,

(ii) $\left(\dfrac{\partial r}{\partial x}\right)^2 + \left(\dfrac{\partial r}{\partial y}\right)^2 = 1$ (BURDWAN–2003, 13)

(iii) $\dfrac{\partial^2 r}{\partial x^2} + \dfrac{\partial^2 r}{\partial y^2} = \dfrac{1}{r}\left\{\left(\dfrac{\partial r}{\partial x}\right)^2 + \left(\dfrac{\partial r}{\partial y}\right)^2\right\}$.

19. If $u = \log(x^2 + y^2 + z^2)$, then prove that

$x\dfrac{\partial^2 u}{\partial y \partial z} = y\dfrac{\partial^2 u}{\partial z \partial x} = z\dfrac{\partial^2 u}{\partial x \partial y}$.

20. If $x^2(y - z) + y^2(z - x) + z^2(x - y)$, prove that $\dfrac{\partial u}{\partial x} + \dfrac{\partial u}{\partial y} + \dfrac{\partial u}{\partial z} = 0$.

21. (i) If $u = \sqrt{x^2 + y^2 + z^2}$, then prove that $\dfrac{\partial^2 u}{\partial x^2} + \dfrac{\partial^2 u}{\partial y^2} + \dfrac{\partial^2 u}{\partial z^2} = \dfrac{2}{u}$.

(ii) If $u = \log\sqrt{x^2 + y^2 + z^2}$, show that

$(x^2 + y^2 + z^2)\left(\dfrac{\partial^2 u}{\partial x^2} + \dfrac{\partial^2 u}{\partial y^2} + \dfrac{\partial^2 u}{\partial z^2}\right) = 1$.   (UKTU–2011)

22. (i) If $u = \sin^{-1}\left(\dfrac{x^{1/3} + y^{1/3}}{x^{1/2} - y^{1/2}}\right)^{1/2}$, show that

$x\dfrac{\partial u}{\partial x} + y\dfrac{\partial u}{\partial y} = -\dfrac{1}{12}\tan u$.   (MTU–2012)

(ii) If $u = x\sin^{-1}\left(\dfrac{x}{y}\right) + y\sin^{-1}\left(\dfrac{y}{x}\right)$, show that

$x^2\dfrac{\partial^2 u}{\partial x^2} + 2xy\dfrac{\partial^2 u}{\partial x \partial y} + y^2\dfrac{\partial^2 u}{\partial y^2} = 0$.   (UPTU–2008, 15)

(iii) If $u = x^2\tan^{-1}\left(\dfrac{y}{x}\right) - y^2\tan^{-1}\left(\dfrac{x}{y}\right)$, show that

$x^2\dfrac{\partial^2 u}{\partial x^2} + 2xy\dfrac{\partial^2 u}{\partial x \partial y} + y^2\dfrac{\partial^2 u}{\partial y^2} = 2u$.

(UPTU–2009, UKTU–2010, HISAR–2003)

Answers

1. (i) $\dfrac{2x}{x^2 + y^2}, \dfrac{2y}{x^2 + y^2}$ (ii) $-\dfrac{1}{\sqrt{y^2 - x^2}}, \dfrac{x}{y\sqrt{y^2 - x^2}}$ (iii) $\dfrac{2x}{a^2}, \dfrac{2y}{b^2}$ (iv) $\dfrac{(x^2 + 2xy - y^2)}{(x + y)^2 + (x^2 + y^2)^2}, \dfrac{(y^2 + 2xy - x^2)}{(x + y)^2 + (x^2 + y^2)^2}$

2. (i) $\dfrac{e^{x+y}}{(e^x + e^y)^2}, -\dfrac{e^{x+y}}{(e^x + e^y)^2}, \dfrac{e^{x+y}}{(e^x + e^y)^2}$ **15.** $\dfrac{1}{c^2 z}$

3.1.3 HOMOGENEOUS FUNCTIONS

A function $f(x, y)$ is said to be homogeneous function of degree n, if the degree of each of its terms in x and y is equal to n. Thus

$$a_0 x^n + a_1 x^{n-1}y + a_2 x^{n-2}y^2 + \dots + a_{n-1}xy^{n-1} + a_n y^n \qquad \dots(1)$$

is homogeneous function in x and y of order n.

☞ REMARKS

- This definition of homogeneity applies to polynomial functions only. To widen the concept of homogeneity so as to bring even transcendental functions within its scope, we define u as a homogeneous function in x and y of order or degree n, if it can be expressed in the form of $x^n f\left(\dfrac{y}{x}\right)$.

- This definition also covers the polynomial function (1), which can be written as

$$x^n\left[a_0 + a_1\frac{y}{x} + a_2\left(\frac{y}{x}\right)^2 + \ldots + a_n\left(\frac{y}{x}\right)^n\right] = x^n f\left(\frac{y}{x}\right).$$

∴ It is a homogeneous function of order n.

- To test whether a given function $f(x, y)$, is homogeneous or not we put $x = hx$ and $y = hy$ in it.

  If we get $f(hx, hy) = h^n f(x, y)$, the function $f(x, y)$ is homogeneous of degree n, otherwise $f(x, y)$ is not a homogeneous function.

- A homogeneous function in x and y of degree n can also be written as $y^n f\left(\dfrac{x}{y}\right)$.

- A function u of three variables x, y, z is said to be homogeneous function of degree n, if it can be expressed in the form

$$u = x^n f_1\left(\frac{y}{x}, \frac{z}{x}\right) \quad \text{or} \quad y^n f_2\left(\frac{x}{y}, \frac{z}{y}\right) \quad \text{or} \quad z^n f_3\left(\frac{x}{z}, \frac{y}{z}\right).$$

In general, a function u of several variables $x_1, x_2, \ldots, x_n$ is said to be homogeneous function of degree m if it can be expressed in the form

$$u = x_1^m f_1\left(\frac{x_2}{x_1}, \frac{x_3}{x_1}, \ldots, \frac{x_n}{x_1}\right) \quad \text{or} \quad x_2^m f_2\left(\frac{x_1}{x_2}, \frac{x_3}{x_2}, \ldots, \frac{x_n}{x_2}\right) \text{ or etc.}$$

THEOREM 1. *If u is a homogeneous function of x and y of degree n, then $\dfrac{\partial u}{\partial x}$ and $\dfrac{\partial u}{\partial y}$ are homogeneous function of degree $(n-1)$ each.*

PROOF.   Since, u is a homogeneous function of x and y of degree n therefore, u can be expressed as

$$u = x^n f\left(\frac{y}{x}\right). \qquad \ldots(1)$$

Now from (1)

$$\frac{\partial u}{\partial x} = nx^{n-1}f\left(\frac{y}{x}\right) + x^n f'\left(\frac{y}{x}\right)\left(-\frac{y}{x^2}\right) = x^{n-1}\left[nf\left(\frac{y}{x}\right) + f'\left(\frac{y}{x}\right)\left(-\frac{y}{x}\right)\right]$$

$$= x^{n-1} \times \text{a function of } \frac{y}{x} = x^{n-1}g\left(\frac{y}{x}\right) \text{ (say)}.$$

which is a homogeneous function of degree $(n-1)$.

Also,

$$\frac{\partial u}{\partial y} = x^n f'\left(\frac{y}{x}\right).\left(\frac{1}{x}\right) = x^{n-1}f'\left(\frac{y}{x}\right) = x^{n-1} \times \text{a function of } \frac{y}{x} = x^{n-1}g\left(\frac{y}{x}\right) \text{ (say)}.$$

which is a homogeneous function of x and y of degree $(n-1)$.

THEOREM 2. [Euler's Theorem on Homogeneous Functions].

If u be a homogeneous function of x and y of degree n, then $x\dfrac{\partial u}{\partial x} + y\dfrac{\partial u}{\partial y} = nu$.   (UPTU–2006, 07, GBTU–2010, UKTU–2011)

PROOF.   Since, u is a homogeneous function of x and y of degree n therefore, u can be expressed as

$$u = x^n f\left(\frac{y}{x}\right).$$

∴

$$\frac{\partial u}{\partial x} = nx^{n-1}f\left(\frac{y}{x}\right) + x^n f'\left(\frac{y}{x}\right)\left(-\frac{y}{x^2}\right) = nx^{n-1}f\left(\frac{y}{x}\right) - yx^{n-2}f'\left(\frac{y}{x}\right).$$

Also,

$$\frac{\partial u}{\partial y} = x^n f'\left(\frac{y}{x}\right).\left(\frac{1}{x}\right) = x^{n-1}f'\left(\frac{y}{x}\right).$$

Now,

$$\text{L.H.S.} = x\frac{\partial u}{\partial x} + y\frac{\partial u}{\partial y} = x\left[nx^{n-1}f\left(\frac{y}{x}\right) - yx^{n-2}f'\left(\frac{y}{x}\right)\right] + yx^{n-1}f'\left(\frac{y}{x}\right)$$

$$= nx^n f\left(\frac{y}{x}\right) - yx^{n-1}f'\left(\frac{y}{x}\right) + yx^{n-1}f'\left(\frac{y}{x}\right) = nx^n f\left(\frac{y}{x}\right) = nu = \text{R.H.S.}$$

☞ REMARK

- Euler's theorem can be extended to a homogeneous functions of several variables. Thus, if u be the function of m independent variables $x_1, x_2, \ldots, x_m$ of degree n then, Euler's theorem states that

$$x_1\frac{\partial u}{\partial x_1} + x_2\frac{\partial u}{\partial x_2} + \ldots + x_m\frac{\partial u}{\partial x_m} = nu.$$

THEOREM 3. *If u is a homogeneous function in x and y of degree n, then* $x^2\dfrac{\partial^2 u}{\partial x^2} + 2xy\dfrac{\partial^2 u}{\partial x \partial y} + y^2\dfrac{\partial^2 u}{\partial y^2} = n(n-1)u.$

(UPTU–2006, 07, VTU–2007, 12, ANNA–2009, 15)

PROOF.   Since, u is a homogeneous function in x and y of degree n therefore, by Euler's theorem

$$x\frac{\partial u}{\partial x} + y\frac{\partial u}{\partial y} = nu \qquad\qquad\qquad \ldots(1)$$

Differentiating (1) partially *w.r.t. x*, we get

$$\frac{\partial}{\partial x}\left(x\frac{\partial u}{\partial x}\right) + \frac{\partial}{\partial x}\left(y\frac{\partial u}{\partial y}\right) = \frac{\partial}{\partial x}(nu) \qquad \left(\because \text{ Each of } \frac{\partial u}{\partial x} \text{ and } \frac{\partial u}{\partial y} \text{ is a function of both } x \text{ and } y\right)$$

$$\Rightarrow \qquad x\frac{\partial^2 u}{\partial x^2} + \frac{\partial u}{\partial x}\cdot 1 + y\frac{\partial^2 u}{\partial x \partial y} = n\frac{\partial u}{\partial x}$$

$$\Rightarrow \qquad x\frac{\partial^2 u}{\partial x^2} + y\frac{\partial^2 u}{\partial x \partial y} = (n-1)\frac{\partial u}{\partial x} \qquad\qquad \ldots(2)$$

Again differentiating (2) partially w.r.t. y, we get

$$y\frac{\partial^2 u}{\partial y^2} + x\frac{\partial^2 u}{\partial x \partial y} = (n-1)\frac{\partial u}{\partial y} \qquad\qquad \ldots(3)$$

Now, multiply (2) by x , (3) by y and then adding, we get

$$x^2\frac{\partial^2 u}{\partial x^2} + 2xy\frac{\partial^2 u}{\partial x \partial y} + y^2\frac{\partial^2 u}{\partial y^2} = (n-1)\left[x\frac{\partial u}{\partial y} + y\frac{\partial u}{\partial y}\right] = (n-1)nu = n(n-1)u.$$

☞ REMARK

- If z is a homogeneous function of x and y of degree n and if $z = f(u)$, then we have the following results :

(i) $\ x\dfrac{\partial u}{\partial x} + y\dfrac{\partial u}{\partial y} = n\dfrac{f(u)}{f'(u)} = G(u)$ (ii) $\ x^2\dfrac{\partial^2 u}{\partial x^2} + 2xy\dfrac{\partial^2 u}{\partial x \partial y} + y^2\dfrac{\partial^2 u}{\partial y^2} = G(u)[G'(u)-1]$

Solved Examples

EXAMPLE 1. **If** $u = \sin^{-1}\left[\dfrac{x^2+y^2}{x+y}\right]$, **show that** $x\dfrac{\partial u}{\partial x} + y\dfrac{\partial u}{\partial y} = \tan u.$

(UPTU–2007, VTU–2003, BHOPAL–2009)

SOLUTION.   We have $\sin u = \left[\dfrac{x^2+y^2}{x+y}\right]$

Let $v = \dfrac{x^2+y^2}{x+y}$

$\Rightarrow$ v is a homogeneous of x and y of degree 1.
Then, by Euler's theorem, we have

$$x\frac{\partial v}{\partial x} + y\frac{\partial v}{\partial y} = v \qquad\qquad \ldots(1)$$

$$v = \sin u \Rightarrow \frac{\partial v}{\partial x} = \cos u\,\frac{\partial u}{\partial x}$$

and $\dfrac{\partial v}{\partial y} = \cos u\,\dfrac{\partial u}{\partial y}.$

Put these values in (1), we get

$$x\cos u\,\frac{\partial u}{\partial x} + y\cos u\,\frac{\partial u}{\partial y} = v$$

$$\Rightarrow \quad x\frac{\partial u}{\partial x} + y\frac{\partial u}{\partial y} = \frac{v}{\cos u} = \frac{\sin u}{\cos u} = \tan u.$$

EXAMPLE 2. **If** $u = \tan^{-1}\dfrac{x^3+y^3}{x-y}$, **prove that**

$$x^2\frac{\partial^2 u}{\partial x^2} + 2xy\frac{\partial^2 u}{\partial x \partial y} + y^2\frac{\partial^2 u}{\partial y^2} = (1 - 4\sin^2 u)\sin 2u.$$

SOLUTION .   We have $u = \tan^{-1}\dfrac{x^3+y^3}{x-y}$

$$\therefore \quad \tan u = \frac{x^3+y^3}{x-y} = \frac{x^3\left[1+\left(\dfrac{y}{x}\right)^3\right]}{x\left[1-\dfrac{y}{x}\right]} = x^2 f\left(\frac{y}{x}\right)$$

$\tan u$ is of the form $x^n f\left(\dfrac{y}{x}\right)$ with $n = 2.$

$\therefore$ $\tan u$ is a homogeneous function in x, y of degree 2.
Then, by Euler's theorem

$$x\frac{\partial}{\partial x}(\tan u) + y\frac{\partial}{\partial y}(\tan u) = 2\tan u$$

$$\Rightarrow \quad x\sec^2 u\,\frac{\partial u}{\partial x} + y\sec^2 u\,\frac{\partial u}{\partial y} = 2\tan u$$

$$\Rightarrow \quad x\frac{\partial u}{\partial x} + y\frac{\partial u}{\partial y} = \frac{2\tan u}{\sec^2 u} = 2\sin u\cos u = \sin 2u \quad \ldots(1)$$

Differentiate (1) partially *w.r.t. x*, we get

$$\left(x\frac{\partial^2 u}{\partial x^2} + \frac{\partial u}{\partial x}\right) + y\frac{\partial^2 u}{\partial x \partial y} = 2\cos 2u\,\frac{\partial u}{\partial x}$$

$$\therefore \quad x\frac{\partial^2 u}{\partial x^2} + y\frac{\partial^2 u}{\partial x \partial y} = (2\cos 2u - 1)\frac{\partial u}{\partial x} \quad \ldots(2)$$

Interchanging x and y in (2), we get

$$y\frac{\partial^2 u}{\partial y^2} + x\frac{\partial^2 u}{\partial x \partial y} = (2\cos 2u - 1)\frac{\partial u}{\partial y} \quad \ldots(3)$$

Now multiplying (2) by x, (3) by y and then adding, we get

$$x^2 \frac{\partial^2 u}{\partial x^2} + 2xy \frac{\partial^2 u}{\partial x \partial y} + y^2 \frac{\partial^2 u}{\partial y^2} = (2\cos 2u - 1)\left[x \frac{\partial u}{\partial x} + y \frac{\partial u}{\partial y} \right]$$

$$= (2\cos 2u - 1).\sin 2u$$

$$= [2(1 - 2\sin^2 u) - 1]\sin 2u$$

$$= (1 - 4\sin^2 u)\sin 2u.$$

EXAMPLE 3. *If* $u = \sin^{-1}\left(\dfrac{x}{y}\right) + \tan^{-1}\left(\dfrac{y}{x}\right)$, *show that*

$$x\frac{\partial u}{\partial x} + y\frac{\partial u}{\partial y} = 0 \times u = 0$$

(UPTU–2007, HAZARIBAGH–2009, OSMANIA–2003, 12)

SOLUTION. We have $u = \sin^{-1}\left(\dfrac{x}{y}\right) + \tan^{-1}\left(\dfrac{y}{x}\right)$

$$= x^0\left[\sin^{-1}\left(\frac{1}{y/x}\right) + \tan^{-1}\left(\frac{y}{x}\right) \right]$$

$\Rightarrow$ u is a homogeneous function of order 0.
Then, by Euler's theorem, we have

$$x\frac{\partial u}{\partial x} + y\frac{\partial u}{\partial y} = 0 \times u = 0$$

EXAMPLE 4. *If* $u = \left(x^{1/4} + y^{1/4}\right)\left(x^{1/5} + y^{1/5}\right)$. *Apply Euler's theorem to find the value of* $x\dfrac{\partial u}{\partial x} + y\dfrac{\partial u}{\partial y}$. (MTU–2011)

SOLUTION. Here, we have

$$u(x, y) = \left(x^{1/4} + y^{1/4}\right)\left(x^{1/5} + y^{1/5}\right)$$

$$\Rightarrow \quad u(tx, ty) = t^{1/4}\left(x^{1/4} + y^{1/4}\right)t^{1/5}\left(x^{1/5} + y^{1/5}\right)$$

$$= t^{9/20}\left(x^{1/4} + y^{1/4}\right)\left(x^{1/5} + y^{1/5}\right)$$

$$= t^{9/20} u(x, y)$$

Clearly, u is a homogeneous function of degree $\dfrac{9}{20}$.

Hence, by Euler's theorem we have $x\dfrac{\partial u}{\partial x} + y\dfrac{\partial u}{\partial y} = \dfrac{9}{20}u.$

EXAMPLE 5. *Verify Euler's theoerm for*
$$f(x, y, z) = 3x^2yz + 5xy^2z + 4z^4.$$
(JNTU–1999)

SOLUTION. Let $f(x, y, z) = 3x^2yz + 5xy^2z + 4z^4$.

$$\therefore \quad \frac{\partial f}{\partial x} = 6xyz + 5y^2z; \frac{\partial f}{\partial y} = 3x^2z + 10xyz$$

and $\dfrac{\partial f}{\partial z} = 3x^2y + 5xy^2 + 16z^3$

$$\therefore \ x\frac{\partial f}{\partial x} + y\frac{\partial f}{\partial y} + z\frac{\partial f}{\partial z} = x(6xyz + 5y^2z) + y(3x^2z + 10xyz)$$

$$+ z(3x^2y + 5xy^2 + 16z^3)$$

$$= 4(3x^2yz + 5xy^2z + 4z^4) = 4f \ ...(1)$$

Also, $f(x, y, z) = x^4\left[3.\dfrac{y}{x}.\dfrac{z}{x} + 5\left(\dfrac{y}{x}\right)^2\left(\dfrac{z}{x}\right) + 4\left(\dfrac{z}{x}\right)^4 \right]$

is a homogeneous function of x, y, z of degree 4.
Hence, by Euler's theorem

$$x\frac{\partial f}{\partial x} + y\frac{\partial f}{\partial y} + z\frac{\partial f}{\partial z} = 4f. \qquad ...(2)$$

From (1) and (2) we conclude that Euler's theorem is verified.

EXAMPLE 6. *If* $u = f\left(\dfrac{y}{x}\right) + \sqrt{x^2 + y^2}$, *show that*

$$x\frac{\partial u}{\partial x} + y\frac{\partial u}{\partial y} = \sqrt{x^2 + y^2}.$$
(MUMBAI–2008)

SOLUTION. Let us write $u = v + w$

where $v = f\left(\dfrac{y}{x}\right) = x^0 f\left(\dfrac{y}{x}\right)$

and $w = \sqrt{x^2 + y^2} = x\sqrt{1 + \left(\dfrac{y}{x}\right)^2}$

Therefore, v and w are homogeneous function of degree 0 and 1 in x and y respectively. Hence, by Euler's theorem

$$x\frac{\partial v}{\partial x} + y\frac{\partial v}{\partial y} = 0.v = 0 \qquad ...(1)$$

and $x\dfrac{\partial w}{\partial x} + y\dfrac{\partial w}{\partial y} = 1.w = \sqrt{x^2 + y^2} \qquad ...(2)$

On adding (1) and (2), we get

$$x\left(\frac{\partial v}{\partial x} + \frac{\partial w}{\partial x}\right) + y\left(\frac{\partial v}{\partial y} + \frac{\partial w}{\partial y}\right) = \sqrt{x^2 + y^2} \qquad ...(3)$$

Now, since $u = v + w$, then using (3) we get

$$x\frac{\partial u}{\partial x} + y\frac{\partial u}{\partial y} = \sqrt{x^2 + y^2}.$$

EXAMPLE 7. *If* $z = x^n f_1\left(\dfrac{y}{x}\right) + y^{-n} f_2\left(\dfrac{x}{y}\right)$, *then show that*

$$x^2\frac{\partial^2 z}{\partial x^2} + 2xy\frac{\partial^2 z}{\partial x \partial y} + y^2\frac{\partial^2 z}{\partial y^2} + x\frac{\partial z}{\partial x} + y\frac{\partial z}{\partial y} = n^2 z$$

(ROHTAK(MDU)–2003, KURUKSHETRA–2009)

SOLUTION. Let $u = x^n f_1\left(\dfrac{y}{x}\right), v = y^{-n} f_2\left(\dfrac{x}{y}\right) \qquad ...(1)$

$\therefore \qquad z = u + v \qquad ...(2)$

Clearly, u and v are homogeneous functions of degree n and $-n$ respectively. Then by Euler's theorem, we get

$$x\frac{\partial u}{\partial x} + y\frac{\partial u}{\partial y} = nu \qquad ...(3)$$

$$x\frac{\partial v}{\partial x} + y\frac{\partial v}{\partial y} = (-n).v \qquad ...(4)$$

$$x^2\frac{\partial^2 u}{\partial x^2} + 2xy\frac{\partial^2 u}{\partial x \partial y} + y^2\frac{\partial^2 u}{\partial y^2} = n(n-1)u \qquad ...(5)$$

and

$$x^2\frac{\partial^2 v}{\partial x^2} + 2xy\frac{\partial^2 v}{\partial x \partial y} + y^2\frac{\partial^2 v}{\partial y^2} = (-n)(-n-1)v = n(n+1)v$$
$$...(6)$$

Since $z = u + v \Rightarrow \dfrac{\partial z}{\partial x} = \dfrac{\partial u}{\partial x} + \dfrac{\partial v}{\partial x}$ and $\dfrac{\partial z}{\partial y} = \dfrac{\partial u}{\partial y} + \dfrac{\partial v}{\partial y} \ ...(7)$

Adding (3) and (4) and using (7) we get

$$x\frac{\partial z}{\partial x} + y\frac{\partial z}{\partial y} = n(u - v) \qquad ...(8)$$

Similarly, adding (5) and (6) and using (7) we get

$$x^2\left[\frac{\partial^2 u}{\partial x^2} + \frac{\partial^2 v}{\partial x^2}\right] + 2xy\left[\frac{\partial^2 u}{\partial x \partial y} + \frac{\partial^2 v}{\partial x \partial y}\right] + y^2\left[\frac{\partial^2 u}{\partial y^2} + \frac{\partial^2 v}{\partial y^2}\right]$$

$$= n(n-1)u + n(n+1)v$$

$$\Rightarrow x^2\frac{\partial^2 z}{\partial x^2} + 2xy\frac{\partial^2 z}{\partial x \partial y} + y^2\frac{\partial^2 z}{\partial y^2} = n^2(u+v) - n(u-v)$$

$$= n^2 z - \left(x \frac{\partial z}{\partial x} + y \frac{\partial z}{\partial y} \right) \qquad \text{(Using 8)}$$

$$\Rightarrow \quad x^2 \frac{\partial^2 z}{\partial x^2} + 2xy \frac{\partial^2 z}{\partial x \partial y} + y^2 \frac{\partial^2 z}{\partial y^2} + x \frac{\partial z}{\partial x} + y \frac{\partial z}{\partial y} = n^2 z$$

EXAMPLE 8. *If* $u = \mathrm{cosec}^{-1} \sqrt{\dfrac{x^{1/2} + y^{1/2}}{x^{1/3} + y^{1/3}}}$, *then prove that*

$$x^2 \frac{\partial^2 u}{\partial x^2} + 2xy \frac{\partial^2 u}{\partial x \partial y} + y^2 \frac{\partial^2 u}{\partial y^2} = \frac{\tan u}{12} \left(\frac{13}{12} + \frac{\tan^2 u}{12} \right)$$

(ROHTAK–2006, MUMBAI–2008)

SOLUTION. We have $z = \mathrm{cosec}\, u = x^{1/12} \sqrt{\dfrac{1 + (y/x)^{1/2}}{1 + (y/x)^{1/3}}}$

$\Rightarrow \quad z = f(u) = \mathrm{cosec}\, u$ is a homogeneous function of degree $\dfrac{1}{12}$.

Then by result (ii) of remark of theorem-3, we have

$$x^2 \frac{\partial^2 u}{\partial x^2} + 2xy \frac{\partial^2 u}{\partial x \partial y} + y^2 \frac{\partial^2 u}{\partial y^2} = G(u)[G'(u) - 1] \quad \dots(1)$$

where $G(u) = \dfrac{n f(u)}{f'(u)} \Rightarrow G(u) = \dfrac{1}{12} \dfrac{f(u)}{f'(u)}$

$\because \qquad f(u) = \mathrm{cosec}\, u$ therefore

$$G(u) = \frac{1}{12} \frac{\mathrm{cosec}\, u}{(-\mathrm{cosec}\, u . \cot u)} = -\frac{1}{12} \tan u$$

$$\Rightarrow \quad G'(u) = -\frac{1}{12} \sec^2 u$$

Putting all these values in equation (1), we get

$$x^2 \frac{\partial^2 u}{\partial x^2} + 2xy \frac{\partial^2 u}{\partial x \partial y} + y^2 \frac{\partial^2 u}{\partial y^2} = -\frac{1}{12} \tan u \left[-\frac{1}{12} \sec^2 u - 1 \right]$$

$$= \frac{1}{12} \tan u \left[\frac{1}{12}(1 + \tan^2 u) + 1 \right]$$

$$= \frac{1}{12} \tan u \left[\frac{1}{12} \tan^2 u + \frac{13}{12} \right]$$

EXAMPLE 9. *If* $u = \sin^{-1} \dfrac{x + 2y + 3z}{x^8 + y^8 + z^8}$, *find the value of*

$$x \frac{\partial u}{\partial x} + y \frac{\partial u}{\partial y} + z \frac{\partial u}{\partial z}.$$

(UKTU–2010, UPTU–2004)

SOLUTION. We can write

$$w = \sin u = \frac{x + 2y + 3z}{x^8 + y^8 + z^8} = x^{-7} \frac{1 + 2\left(\dfrac{y}{x}\right) + 3\left(\dfrac{z}{x}\right)}{1 + \left(\dfrac{y}{x}\right)^8 + \left(\dfrac{z}{x}\right)^8}$$

$\Rightarrow \quad w$ is a homogeneous function of degree –7.

Then by Euler's theorem, we get

$$x \frac{\partial w}{\partial x} + y \frac{\partial w}{\partial y} + z \frac{\partial w}{\partial z} = (-7)w \qquad \dots(1)$$

Here, $\dfrac{\partial w}{\partial x} = \cos u \dfrac{\partial u}{\partial x}, \dfrac{\partial w}{\partial y} = \cos u \dfrac{\partial u}{\partial y}, \dfrac{\partial w}{\partial z} = \cos u \dfrac{\partial u}{\partial z}$

Then from (1)

$$x \cos u \frac{\partial u}{\partial x} + y \cos u \frac{\partial u}{\partial y} + z \cos u \frac{\partial u}{\partial z} = -7 \sin u$$

$$\Rightarrow \qquad x \frac{\partial u}{\partial x} + y \frac{\partial u}{\partial y} + z \frac{\partial u}{\partial z} = -7 \tan u.$$

EXAMPLE 10. *If* $u = \dfrac{x^3 y^3 z^3}{x^3 + y^3 + z^3} + \log\left(\dfrac{xy + yz + zx}{x^2 + y^2 + z^2} \right)$, *find*

the value of $x \dfrac{\partial u}{\partial x} + y \dfrac{\partial u}{\partial y} + z \dfrac{\partial u}{\partial z}$. (MUMBAI–2009)

SOLUTION. Let $v = \dfrac{x^3 y^3 z^3}{x^3 + y^3 + z^3}$

and $w = \log\left(\dfrac{xy + yz + zx}{x^2 + y^2 + z^2} \right)$

Clearly, $v = x^6 \left[\dfrac{\left(\dfrac{y}{x}\right)^3 \left(\dfrac{z}{x}\right)^3}{1 + \left(\dfrac{y}{x}\right)^3 + \left(\dfrac{z}{x}\right)^3} \right]$

is a homogeneous function of degree 6.

$\therefore$ By Euler's theorem

$$x \frac{\partial v}{\partial x} + y \frac{\partial v}{\partial y} + z \frac{\partial v}{\partial z} = 6v \qquad \dots(1)$$

Further, $w = \log\left[\dfrac{\dfrac{y}{x} + \dfrac{y}{x} . \dfrac{z}{x} + \dfrac{z}{x}}{1 + \left(\dfrac{y}{x}\right)^2 + \left(\dfrac{z}{x}\right)^2} \right]$

is a homogeneous function of degree zero.

Then, by Euler's theorem

$$x \frac{\partial w}{\partial x} + y \frac{\partial w}{\partial y} + z \frac{\partial w}{\partial z} = 0 \qquad \dots(2)$$

Adding (1) and (2), we get

$$x \left(\frac{\partial v}{\partial x} + \frac{\partial w}{\partial x} \right) + y \left(\frac{\partial v}{\partial y} + \frac{\partial w}{\partial y} \right) + z \left(\frac{\partial v}{\partial z} + \frac{\partial w}{\partial z} \right) = 6v$$

$$\Rightarrow \quad x \left(\frac{\partial u}{\partial x} \right) + y \left(\frac{\partial u}{\partial y} \right) + z \left(\frac{\partial u}{\partial z} \right) = 6 . \frac{x^3 y^3 z^3}{x^3 + y^3 + z^3}$$

Exercise-3.2

1. Verify the Euler's theorem for the following functions :

(i) $u = \dfrac{x(x^3 - y^3)}{x^3 + y^3}$

(ii) $u = x^n \sin\left(\dfrac{y}{x} \right)$

(iii) $u = x^n \log\left(\dfrac{y}{x} \right)$

(iv) $u = \dfrac{1}{\sqrt{x^2 + y^2}}$

(v) $u = x^n \sin \dfrac{y}{x}$

(vi) $x^4 \log \dfrac{y}{x}$

(vii) $u = \log\left(\dfrac{x^2 + y^2}{xy} \right)$ (UPTU–2006, UKTU–2011)

(viii) $u = \dfrac{x^{1/3} + y^{1/3}}{x^{1/2} + y^{1/2}}$ (GBTU–2010)

2. (i) If $u = x f\left(\dfrac{y}{x} \right)$, prove that $x \dfrac{\partial u}{\partial x} + y \dfrac{\partial u}{\partial y} = u$.

(ii) If $u = f\left(\dfrac{y}{x}\right)$, prove that $x\dfrac{\partial u}{\partial x} + y\dfrac{\partial u}{\partial y} = 0$. (GBTU–2011)

(iii) If $u = xyf\left(\dfrac{y}{x}\right)$, prove that $x\dfrac{\partial u}{\partial x} + y\dfrac{\partial u}{\partial y} = 2u$.

(iv) If $u = \log\left(\dfrac{x^2 + y^2}{x + y}\right)$, show by Euler's theorem :

$$x\dfrac{\partial u}{\partial x} + y\dfrac{\partial u}{\partial y} = 1.$$ (UPTU–2009)

3. If $u = \tan^{-1}\left(\dfrac{x^3 + y^3}{x + y}\right)$, show that $x\dfrac{\partial u}{\partial x} + y\dfrac{\partial u}{\partial y} = \sin 2u$

and $x^2\dfrac{\partial^2 u}{\partial x^2} + 2xy\dfrac{\partial^2 u}{\partial x \partial y} + y^2\dfrac{\partial^2 u}{\partial y^2} = 2\cos 3u \sin u$.

(PTU–2009, SVTU–2009, BHOPAL–2008, MUMBAI–2009)

4. If $u = \tan^{-1}\dfrac{y}{x}$, show that (using Euler's theorem)

$$x\dfrac{\partial u}{\partial x} + y\dfrac{\partial u}{\partial y} = 0.$$

5. If $u = \sin^{-1}\dfrac{x + y}{\sqrt{x} + \sqrt{y}}$, show that

(i) $x\dfrac{\partial u}{\partial x} + y\dfrac{\partial u}{\partial y} = \dfrac{1}{2}\tan u$ (RAJASTHAN–2006, CALICUT–2005)

(ii) $x^2\dfrac{\partial^2 u}{\partial x^2} + 2xy\dfrac{\partial^2 u}{\partial x \partial y} + y^2\dfrac{\partial^2 u}{\partial y^2} = -\dfrac{\sin u \cos 2u}{4\cos^3 u}$ (PTU-2006, 15, 18)

6. If $u = \sin^{-1}\dfrac{\sqrt{x} - \sqrt{y}}{\sqrt{x} + \sqrt{y}}$, show that $x\dfrac{\partial u}{\partial x} + y\dfrac{\partial u}{\partial y} = 0$.

7. (i) If $u = \log\dfrac{x^4 + y^4}{x + y}$, show that $x\dfrac{\partial u}{\partial x} + y\dfrac{\partial u}{\partial y} = 3$.

(UPTU–2009, UKTU–2012)

(ii) If $u = \log\dfrac{x^3 + y^3}{x + y}$, show that $x\dfrac{\partial u}{\partial x} + y\dfrac{\partial u}{\partial y} = 2$.

8. If $\sin u = \dfrac{x^2 y^2}{x + y}$, show that $x\dfrac{\partial u}{\partial x} + y\dfrac{\partial u}{\partial y} = 3\tan u$.

(VTU–2003, KOTTAYAM–2005)

9. (i) If $u = \dfrac{x^2 y^2}{x + y}$, show that $y\dfrac{\partial^2 u}{\partial y^2} + x\dfrac{\partial^2 u}{\partial x \partial y} = 2\dfrac{\partial u}{\partial y}$.

(ii) If $u = \dfrac{xy}{x + y}$, show that $x\dfrac{\partial^2 u}{\partial x^2} + 2xy\dfrac{\partial^2 u}{\partial x \partial y} + y^2\dfrac{\partial^2 u}{\partial y^2} = 0$.

(iii) If $u = \dfrac{x^2 y^2}{x + y}$, show that $x\dfrac{\partial^2 u}{\partial x^2} + y\dfrac{\partial^2 u}{\partial y \partial x} = 2\dfrac{\partial u}{\partial x}$.

10. If $u = xf_1\left(\dfrac{y}{x}\right) + f_2\left(\dfrac{y}{x}\right)$, show that

$$x^2\dfrac{\partial^2 u}{\partial x^2} + 2xy\dfrac{\partial^2 u}{\partial x \partial y} + y^2\dfrac{\partial^2 u}{\partial y^2} = 0.$$ (UPTU–2006, SVTU–2009)

11. (i) If $u = \log\left(\sqrt{x} + \sqrt{y}\right)$, show that $x\dfrac{\partial u}{\partial x} + y\dfrac{\partial u}{\partial y} = \dfrac{1}{2}$.

(ii) If $u = \log\dfrac{x^4 + y^4 + x^2 y^2}{x + y + \sqrt{xy}}$, show that $x\dfrac{\partial u}{\partial x} + y\dfrac{\partial u}{\partial y} = 3$.

12. If z be a homogeneous function of degree n, show that

$$x\dfrac{\partial^2 z}{\partial x^2} + y\dfrac{\partial^2 z}{\partial x \partial y} = (n-1)\dfrac{\partial z}{\partial x}.$$

13. If $u = \cos^{-1}\dfrac{x + y}{\sqrt{x} + \sqrt{y}}$, prove that $x\dfrac{\partial u}{\partial x} + y\dfrac{\partial u}{\partial y} = -\dfrac{1}{2}\cot u$.

(VTU–2004, GBTU–2010)

14. If $\sin u = \dfrac{x + 2y + 3z}{\sqrt{x^8 + y^8 + z^8}}$, show that $x\dfrac{\partial u}{\partial x} + y\dfrac{\partial u}{\partial y} + z\dfrac{\partial u}{\partial z} = -3\tan u$.

(VTU–2009, SVTU–2009)

15. If $u = \dfrac{x}{y + z} + \dfrac{y}{z + x} + \dfrac{z}{x + y}$, show that $x\dfrac{\partial u}{\partial x} + y\dfrac{\partial u}{\partial y} + z\dfrac{\partial u}{\partial z} = 0$.

(VTU–2000)

16. If $u = \tan^{-1}\left(\dfrac{y^2}{x}\right)$, show that

$$x^2\dfrac{\partial^2 u}{\partial x^2} + 2xy\dfrac{\partial^2 u}{\partial x \partial y} + y^2\dfrac{\partial^2 u}{\partial y^2} = -\sin^2 u . \sin 2u.$$

(PTU–2005, BHILLAI–2005)

17. Show that $x\dfrac{\partial u}{\partial x} + y\dfrac{\partial u}{\partial y} = 2u \log u$ where $u = e^{x^2 + y^2}$. (PTU–2010)

18. If $\log u = \dfrac{x^3 + y^3}{3x + 4y}$, show that $x\dfrac{\partial u}{\partial x} + y\dfrac{\partial u}{\partial y} = 2u \log u$ (UKTU–2011)

19. If $u = x^3 + y^3 + z^3 + 3xyz$, show that $x\dfrac{\partial u}{\partial x} + y\dfrac{\partial u}{\partial y} + z\dfrac{\partial u}{\partial z} = 3u$.

(GBTU–2010)

20. If $u = \sec^{-1}\left(\dfrac{x^3 - y^3}{x + y}\right)$, show that $x\dfrac{\partial u}{\partial x} + y\dfrac{\partial u}{\partial y} = 2\cot u$.

(UPTU–2008, 18)

3.1.4 Total Differential

Let $$u = f(x, y)$$ ...(1)

be the given function of x and y, which have continuous partial derivatives of first order *w.r.t.* x and y.

Let δx and δy be the increments in x and y respectively and let δu be the consequent change in u, then we have

$$u + \delta u = f(x + \delta x, y + \delta y)$$

$\therefore$ $$\delta u = f(x + \delta x, y + \delta y) - f(x, y)$$ ...(2)

$$= [f(x + \delta x, y + \delta y) - f(x, y + \delta y)] + [f(x, y + \delta y) - f(x, y)]$$

$\Rightarrow$ $$\dfrac{\delta u}{\delta t} = \dfrac{[f(x + \delta x, y + \delta y) - f(x, y + \delta y)]}{\delta t} + \dfrac{[f(x, y + \delta y) - f(x, y)]}{\delta t}$$

Now, $$\dfrac{du}{dt} = \lim_{\delta t \to 0} \dfrac{\delta u}{\delta t}$$

$$= \lim_{\delta t \to 0}\left[\dfrac{f(x + \delta x, y + \delta y) - f(x, y + \delta y)}{\delta x}\dfrac{\delta x}{\delta t} + \dfrac{f(x, y + \delta y) - f(x, y)}{\delta y}\dfrac{\delta y}{\delta t}\right]$$ ...(3)

Since δx and δy tends to zero, when $\delta t \to 0$ so we have

$$\lim_{\delta x \to 0} \frac{f(x+\delta x, y+\delta y) - f(x, y+\delta y)}{\delta x} = \frac{\partial f}{\partial x} = \frac{\partial u}{\partial x}.$$

Similarly, $\lim_{\delta y \to 0} \frac{f(x, y+\delta y) - f(x, y)}{\delta y} = \frac{\partial f}{\partial y} = \frac{\partial u}{\partial y}$ and $\lim_{\delta t \to 0} \frac{\delta x}{\delta t} = \frac{dx}{dt}, \lim_{\delta t \to 0} \frac{\delta y}{\delta t} = \frac{dy}{dt}.$

Therefore, from (3), we get $\frac{du}{dt} = \frac{\partial u}{\partial x}\cdot\frac{dx}{dt} + \frac{\partial u}{\partial y}\cdot\frac{dy}{dt}.$

REMARKS

- This result can be extended as follows :

If $u = f(x_1, x_2, ..., x_m)$ and $x_1, x_2, ..., x_m$ all are functions of t, then $\frac{du}{dt} = \frac{\partial u}{\partial x_1}\cdot\frac{dx_1}{dt} + \frac{\partial u}{\partial x_2}\cdot\frac{dx_2}{dt} + ... + \frac{\partial u}{\partial x_m}\cdot\frac{dx_m}{dt}.$

- The differentials dx and dy of the independent variables x and y are the actual changes δx and δy but the differential du of the dependent variable u is not the same as the change δu, it being the principal part of the increment δu.

3.1.5 IMPLICIT RELATION OF X AND Y

In most of the cases, we are mainly concerned with the case in which y is expressed explicity *i.e.*, directly in terms of x. There are so many cases in which y is not expreesed directly in terms of x, but functionally it is implied by an algebraic relation $f(x, y) = 0$ connecting x and y.

The relation of the type $f(x, y) = c$, where y is not explicity in terms of x are called implicit function.

3.1.6 DIFFERENTIATION OF IMPLICIT FUNCTIONS

Let $f(x, y)$ be a function of two variables x and y and y itself is a function of x i.e., $f(x, y)$ may be consider as a composite function of x. Then, we have

$$\frac{df}{dx} = \frac{\partial f}{\partial x}\cdot\frac{dx}{dx} + \frac{\partial f}{\partial y}\cdot\frac{dy}{dx}$$

$$\Rightarrow \qquad \frac{df}{dx} = \frac{\partial f}{\partial x} + \frac{\partial f}{\partial y}\cdot\frac{dy}{dx} \qquad ...(1)$$

Since $f(x, y) = 0$, therefore $\frac{df}{dx} = 0.$

Now from (1), we have $\frac{\partial f}{\partial x} + \frac{\partial f}{\partial y}\cdot\frac{dy}{dx} = 0$

$$\Rightarrow \qquad \frac{dy}{dx} = -\frac{\partial f}{\partial x}\Big/\frac{\partial f}{\partial y} = -\frac{f_x}{f_y}, \text{ provided } f_y \neq 0.$$

Solved Examples

EXAMPLE 1. If $x^y + y^x = a^b$. Find $\frac{dy}{dx}$.

SOLUTION. Let $f(x, y) = x^y + y^x - a^b$

$\Rightarrow \quad f(x, y) = 0$

Therefore $\frac{dy}{dx} = -\frac{\partial f/\partial x}{\partial f/\partial y} = -\frac{yx^{y-1} + y^x \log y}{x^y \log x + xy^{x-1}}.$

EXAMPLE 2. If $u = \log[(x^2 + y^2)/xy]$, find du.

SOLUTION. Let $u = \log(x^2 + y^2) - \log x - \log y.$

$\therefore \quad \frac{\partial u}{\partial x} = \frac{2x}{x^2+y^2} - \frac{1}{x} = \frac{2x^2 - x^2 - y^2}{x(x^2+y^2)} = \frac{x^2 - y^2}{x(x^2+y^2)}$

and $\frac{\partial u}{\partial y} = \frac{2y}{x^2+y^2} - \frac{1}{y} = \frac{2y^2 - x^2 - y^2}{y(x^2+y^2)} = \frac{y^2 - x^2}{y(x^2+y^2)}$

Now,

$du = \frac{\partial u}{\partial x}dx + \frac{\partial u}{\partial y}dy = \frac{(x^2-y^2)}{x(x^2+y^2)}dx + \frac{(y^2-x^2)}{y(x^2+y^2)}dy$

$= \frac{(x^2-y^2)}{xy(x^2+y^2)}(y\,dx - x\,dy).$

EXAMPLE 3. If $f(x, y) = 0$ and $g(y, z) = 0$, show that
$$\frac{\partial f}{\partial y}\cdot\frac{\partial g}{\partial z}\cdot\frac{dz}{dx} = \frac{\partial f}{\partial x}\cdot\frac{\partial g}{\partial y}.$$

SOLUTION. Let $f(x, y) = 0$, then we have

$$\frac{dy}{dx} = -\frac{\partial f/\partial x}{\partial f/\partial y}. \qquad ...(1)$$

Also, let $g(y, z) = 0$

$$\Rightarrow \qquad \frac{dz}{dy} = -\frac{\partial g/\partial y}{\partial g/\partial z}. \qquad ...(2)$$

Now, from (1) and (2), we have

$$\frac{dy}{dx}\cdot\frac{dz}{dy} = \left(\frac{\partial f}{\partial x}\cdot\frac{\partial g}{\partial y}\right)\Big/\left(\frac{\partial f}{\partial y}\cdot\frac{\partial g}{\partial z}\right)$$

$$\Rightarrow \qquad \frac{dz}{dx}\cdot\frac{\partial f}{\partial y}\cdot\frac{\partial g}{\partial z} = \frac{\partial f}{\partial x}\cdot\frac{\partial g}{\partial y}$$

EXAMPLE 4. If $u = x^2y$, where $x^2 + xy + y^2 = 1$. Find $\frac{du}{dx}$.

SOLUTION. We know that $\frac{du}{dx} = \frac{\partial u}{\partial x} + \frac{\partial u}{\partial y}\cdot\frac{dy}{dx}.$ $\qquad ...(1)$

Given that $u = x^2 y$

$$\frac{\partial u}{\partial x} = 2xy \text{ and } \frac{\partial u}{\partial y} = x^2$$

$$\therefore \quad f(x, y) = x^2 + xy + y^2 - 1$$

Then $\dfrac{dy}{dx} = -\dfrac{\partial f/\partial x}{\partial f/\partial y} = -\dfrac{2x+y}{x+2y}$

Putting all these values in (1), we get

$$\frac{du}{dx} = 2xy + x^2 \cdot \left(-\frac{2x+y}{x+2y}\right) = 2xy - \frac{x^2(2x+y)}{x+2y}$$

EXAMPLE 5. *If $u = x \log (xy)$, where $x^3 + y^3 + 3xy = 1$. Find $\dfrac{du}{dx}$.*

(VTU–2009, UPTU–2006)

SOLUTION . We have $u = x \log (xy)$. ...(1)

$$\Rightarrow \quad \frac{\partial u}{\partial x} = x\left(\frac{1}{xy}.y\right) + \log xy = 1 + \log xy$$

and $\quad \dfrac{\partial u}{\partial y} = x\left(\dfrac{1}{xy}.x\right) = \dfrac{x}{y}$

Also it is given that

$$x^3 + y^3 + 3xy = 1 \qquad ...(2)$$

Differentiating (2) we get

$$3x^2 + 3y^2 \frac{dy}{dx} + 3\left(x\frac{dy}{dx} + y\right) = 0$$

$$\Rightarrow \quad \frac{dy}{dx} = -\left(\frac{x^2+y}{x+y^2}\right)$$

Now, $\dfrac{du}{dx} = \dfrac{\partial u}{\partial x} + \dfrac{\partial u}{\partial y} . \dfrac{dy}{dx} = 1 + \log(xy) + \dfrac{x}{y}\left\{-\dfrac{(x^2+y)}{(y^2+x)}\right\}$

$$= 1 + \log (xy) - \frac{x(x^2+y)}{y(y^2+x)}$$

EXAMPLE 6. *Show that $\dfrac{\partial^2 z}{\partial u^2} + \dfrac{\partial^2 z}{\partial v^2} = \dfrac{\partial^2 z}{\partial x^2} + \dfrac{\partial^2 z}{\partial y^2}$ where*

$x = u \cos \alpha - v \sin \alpha, \; y = u \sin \alpha + v \cos \alpha.$

(UPTU–2008)

SOLUTION . We have z is a composite function of u and v. Therefore,

$$\frac{\partial z}{\partial u} = \frac{\partial z}{\partial x} . \frac{\partial x}{\partial u} + \frac{\partial z}{\partial y} . \frac{\partial y}{\partial u} = \cos\alpha\frac{\partial z}{\partial x} + \sin\alpha\frac{\partial z}{\partial y}$$

$$\Rightarrow \quad \frac{\partial}{\partial u} = \cos\alpha\frac{\partial}{\partial x} + \sin\alpha\frac{\partial}{\partial y} \qquad ...(1)$$

Also, $\dfrac{\partial z}{\partial v} = \dfrac{\partial z}{\partial x} . \dfrac{\partial x}{\partial v} + \dfrac{\partial z}{\partial y} . \dfrac{\partial y}{\partial v} = -\sin\alpha\dfrac{\partial z}{\partial x} + \cos\alpha\dfrac{\partial z}{\partial y}$

$$\frac{\partial}{\partial v} = -\sin\alpha\frac{\partial}{\partial x} + \cos\alpha\frac{\partial}{\partial y} \qquad ...(2)$$

Now, $\dfrac{\partial^2 z}{\partial u^2} = \dfrac{\partial}{\partial u}\left(\dfrac{\partial z}{\partial u}\right)$

$$= \left(\cos\alpha\frac{\partial}{\partial x} + \sin\alpha\frac{\partial}{\partial y}\right)\left(\cos\alpha\frac{\partial z}{\partial x} + \sin\alpha\frac{\partial z}{\partial y}\right)$$

$$= \cos^2\alpha\frac{\partial^2 z}{\partial x^2} + \cos\alpha\sin\alpha\frac{\partial^2 z}{\partial x\partial y}$$

$$+ \sin\alpha\cos\alpha\frac{\partial^2 z}{\partial y\partial x} + \sin^2\alpha\frac{\partial^2 z}{\partial y^2}$$

$$= \cos^2\alpha\frac{\partial^2 z}{\partial x^2} + 2\cos\alpha\sin\alpha\frac{\partial^2 z}{\partial x\partial y} + \sin^2\alpha\frac{\partial^2 z}{\partial y^2} \qquad ...(3)$$

Also,

$$\frac{\partial^2 z}{\partial v^2} = \frac{\partial}{\partial v}\left(\frac{\partial z}{\partial v}\right) = \left(-\sin\alpha\frac{\partial}{\partial x} + \cos\alpha\frac{\partial}{\partial y}\right)\left(-\sin\alpha\frac{\partial z}{\partial x} + \cos\alpha\frac{\partial z}{\partial y}\right)$$

$$= \sin^2\alpha\frac{\partial^2 z}{\partial x^2} - \sin\alpha\cos\alpha\frac{\partial^2 z}{\partial x\partial y} - \cos\alpha\sin\alpha\frac{\partial^2 z}{\partial x\partial y} + \cos^2\alpha\frac{\partial^2 z}{\partial y^2}$$

$$= \sin^2\alpha\frac{\partial^2 z}{\partial x^2} - 2\cos\alpha\sin\alpha\frac{\partial^2 z}{\partial x\partial y} + \cos^2\alpha\frac{\partial^2 z}{\partial y^2} \qquad ...(4)$$

Adding (3) and (4), we get

$$\frac{\partial^2 z}{\partial u^2} + \frac{\partial^2 z}{\partial v^2} = \frac{\partial^2 z}{\partial x^2} + \frac{\partial^2 z}{\partial y^2}$$

EXAMPLE 7. *If $u = u\left(\dfrac{y-x}{xy}, \dfrac{z-x}{xz}\right)$, show that*

$$x^2 \frac{\partial u}{\partial x} + y^2 \frac{\partial u}{\partial y} + z^2 \frac{\partial u}{\partial z} = 0.$$

(UPTU–2005)

SOLUTION . Suppose $\quad v = \dfrac{y-x}{xy} = \dfrac{1}{x} - \dfrac{1}{y}$

and $\quad w = \dfrac{z-x}{xz} = \dfrac{1}{x} - \dfrac{1}{z} \qquad ...(1)$

Then clearly, $\quad u = u(v, w)$

$$\therefore \quad \frac{\partial u}{\partial x} = \frac{\partial u}{\partial v} . \frac{\partial v}{\partial x} + \frac{\partial u}{\partial w} . \frac{\partial w}{\partial x}$$

$$= \frac{\partial u}{\partial v}\left(-\frac{1}{x^2}\right) + \frac{\partial u}{\partial w}\left(-\frac{1}{x^2}\right)$$

$$\Rightarrow \quad x^2 \frac{\partial u}{\partial x} = -\frac{\partial u}{\partial v} - \frac{\partial u}{\partial w} \qquad ...(2)$$

Further, $\dfrac{\partial u}{\partial y} = \dfrac{\partial u}{\partial v} . \dfrac{\partial v}{\partial y} + \dfrac{\partial u}{\partial w} . \dfrac{\partial w}{\partial y} = \dfrac{\partial u}{\partial v}\left(\dfrac{1}{y^2}\right) + \dfrac{\partial u}{\partial w}(0)$

$$\Rightarrow \quad y^2 \frac{\partial u}{\partial y} = \frac{\partial u}{\partial v} \qquad ...(3)$$

Similarly, $\dfrac{\partial u}{\partial z} = \dfrac{\partial u}{\partial v} . \dfrac{\partial v}{\partial z} + \dfrac{\partial u}{\partial w} . \dfrac{\partial w}{\partial z} = \dfrac{\partial u}{\partial v}(0) + \dfrac{\partial u}{\partial w}\left(\dfrac{1}{z^2}\right)$

$$\Rightarrow \quad z^2 \frac{\partial u}{\partial z} = \frac{\partial u}{\partial w} \qquad ...(4)$$

Finally, adding (2), (3) and (4), we get

$$x^2 \frac{\partial u}{\partial x} + y^2 \frac{\partial u}{\partial y} + z^2 \frac{\partial u}{\partial z} = 0.$$

EXAMPLE 8. *If $f(x, y) = 0$, show that $\dfrac{\partial^2 y}{\partial x^2} = -\dfrac{q^2 r - 2pqs + p^2 t}{q^3}$.*

(KURUKSHETRA–2006)

SOLUTION . We have $\dfrac{dy}{dx} = -\dfrac{\partial f/\partial x}{\partial f/\partial y} = \dfrac{-p}{q}$

$$\Rightarrow \quad \frac{d^2 y}{dx^2} = \frac{d}{dx}\left(\frac{dy}{dx}\right) = -\frac{d}{dx}\left(\frac{p}{q}\right) = \frac{-q\frac{dp}{dx} + p\frac{dq}{dx}}{q^2} \; ...(1)$$

Now, $\dfrac{dp}{dx} = \dfrac{\partial p}{\partial x} + \dfrac{\partial p}{\partial y} . \dfrac{dy}{dx} = r + s\left(-\dfrac{p}{q}\right) = \dfrac{qr - ps}{q}$

and $\dfrac{dq}{dx} = \dfrac{\partial q}{\partial x} + \dfrac{\partial q}{\partial y} \cdot \dfrac{dy}{dx} = s + t\left(-\dfrac{p}{q}\right) = \dfrac{qs - pt}{q}$

Putting all these value in (1), we get

$\dfrac{d^2 y}{dx^2} = -\dfrac{1}{q^2}\left[q\left(\dfrac{qr - ps}{q}\right) - p\left(\dfrac{qs - pt}{q}\right)\right] = -\dfrac{q^2 r - 2pqs + p^2 t}{q^3}$

Here, $p = \dfrac{\partial f}{\partial x}, q = \dfrac{\partial f}{\partial y}, r = \dfrac{\partial^2 f}{\partial x^2} = \dfrac{\partial p}{\partial x}$

$s = \dfrac{\partial^2 f}{\partial x \partial y} = \dfrac{\partial q}{\partial x}, \quad t = \dfrac{\partial^2 f}{\partial y^2} = \dfrac{\partial q}{\partial y}$

EXAMPLE 9. *If $u = f(r, s, t)$ and $r = \dfrac{x}{y}, s = \dfrac{y}{z}, t = \dfrac{z}{x}$, prove that*

$x\dfrac{\partial u}{\partial x} + y\dfrac{\partial u}{\partial y} + z\dfrac{\partial u}{\partial z} = 0$.   (JNTU–1990, 2007, UKTU–2011, 12)

SOLUTION. We have $u = f(r, s, t)$ ...(1)

then $\dfrac{\partial u}{\partial x} = \dfrac{\partial u}{\partial r} \cdot \dfrac{\partial r}{\partial x} + \dfrac{\partial u}{\partial s} \cdot \dfrac{\partial s}{\partial x} + \dfrac{\partial u}{\partial t} \cdot \dfrac{\partial t}{\partial x}$

$= \dfrac{1}{y}\dfrac{\partial u}{\partial r} + 0 \cdot \dfrac{\partial u}{\partial s} - \dfrac{z}{x^2} \cdot \dfrac{\partial u}{\partial t}$ ...(2)

Also, $\dfrac{\partial u}{\partial y} = \dfrac{\partial u}{\partial r} \cdot \dfrac{\partial r}{\partial y} + \dfrac{\partial u}{\partial s} \cdot \dfrac{\partial s}{\partial y} + \dfrac{\partial u}{\partial t} \cdot \dfrac{\partial t}{\partial y}$

$= \dfrac{x}{y^2} \cdot \dfrac{\partial u}{\partial r} + \dfrac{1}{z}\dfrac{\partial u}{\partial s} + 0 \cdot \dfrac{\partial u}{\partial t}$ ...(3)

and $\dfrac{\partial u}{\partial z} = \dfrac{\partial u}{\partial r} \cdot \dfrac{\partial r}{\partial z} + \dfrac{\partial u}{\partial s} \cdot \dfrac{\partial s}{\partial z} + \dfrac{\partial u}{\partial t} \cdot \dfrac{\partial t}{\partial z}$

$= 0 \cdot \dfrac{\partial u}{\partial r} + \left(\dfrac{-y}{z^2}\right) \cdot \dfrac{\partial u}{\partial s} + \dfrac{1}{x}\dfrac{\partial u}{\partial t}$ ...(4)

Now multiplying (2) by x, (3) by y and (4) by z and then adding we get

$x\dfrac{\partial u}{\partial x} + y\dfrac{\partial u}{\partial y} + z\dfrac{\partial u}{\partial z} = 0$

Exercise-3.3

1. If $(\tan x)^y + (y)^{\cot x} = a$. Find the value of $\dfrac{dy}{dx}$.

2. If $u = \sin(x^2 + y^2)$, where $a^2 x^2 + b^2 y^2 = c^2$. Find the value of $\dfrac{du}{dx}$.

3. If $u = f(y - z, z - x, x - y)$, prove that $\dfrac{\partial u}{\partial x} + \dfrac{\partial u}{\partial y} + \dfrac{\partial u}{\partial z} = 0$.

(UKTU–2010, GBTU–2010)

4. If z is a function of x and y; where $x = e^u + e^{-v}$ and $y = e^{-u} - e^v$, show that $\dfrac{\partial z}{\partial u} - \dfrac{\partial z}{\partial v} = x\dfrac{\partial z}{\partial x} - y\dfrac{\partial z}{\partial y}$.   (VTU–2003, 06)

5. Find the total derivative of u with respect to t, when

(i) $u = \cosh\left(\dfrac{y}{x}\right)$, where $x = t^2, y = e^t$

(ii) $u = e^x \sin y$, where $x = \log t, y = t^2$

6. If $u = \sqrt{(x^2 + y^2)}$ and $x^3 + y^3 + 3axy = 5a^2$. Find the value of $\dfrac{du}{dx}$ at $x = a, y = a$.

7. Find $\dfrac{dy}{dx}$ and $\dfrac{d^2 y}{dx^2}$ from the following implicit relations.

(i) $x^2 + y^2 = a^2$ (ii) $x^{2/3} + y^{2/3} = a^{2/3}$

8. If $f(x, y, z) = 0$, show that

$\left(\dfrac{\partial y}{\partial z}\right)_{x \text{ const.}} \left(\dfrac{\partial z}{\partial x}\right)_{y \text{ const.}} \left(\dfrac{\partial x}{\partial y}\right)_{z \text{ const.}} = -1$.

9. If $x + y = 2e^\theta \cos\phi, x - y = 2ie^\theta \sin\phi$, where $i = \sqrt{(-1)}$,

show that $\dfrac{\partial^2 u}{\partial \theta^2} + \dfrac{\partial^2 u}{\partial \phi^2} = 4xy\dfrac{\partial^2 u}{\partial x \partial y}$.   (LUCKNOW–2005)

Answers

1. $-\dfrac{y(\tan x)^{y-1}\sec^2 x - y^{\cot x} \cdot \log y \cdot \mathrm{cosec}^2 x}{(\tan x)^y \log \tan x + \cot x \, y^{\cot x - 1}}$ **2.** $2x[\cos(x^2 + y^2)]\left(1 - \dfrac{a^2}{b^2}\right)$ **5.** (i) $\dfrac{du}{dt} = \dfrac{1}{x^2}(xe^t - 2yt)\sinh\dfrac{y}{x}$

5. (ii) $\dfrac{du}{dt} = \dfrac{e^x}{t}(\sin y + 2t^2 \cos y)$, where $x = \log t, y = e^t$ **6.** 0 **7.** (i) $-\dfrac{x}{y}, \dfrac{-a^2}{y^3}$ (ii) $\dfrac{dy}{dx} = -\dfrac{y^{1/3}}{x^{1/3}}, \dfrac{d^2 y}{dx^2} = \dfrac{a^{1/3}}{3x^{4/3} \cdot y^{1/3}}$

3.2 CHANGE OF VARIABLES

3.2.1 CHANGE OF INDEPENDENT VARIABLE INTO DEPENDENT VARIABLES

Let $y = f(x)$ be a function with x independent and y is dependent variable. Then

$$\dfrac{dy}{dx} = 1 \bigg/ \left(\dfrac{dx}{dy}\right) = \left(\dfrac{dx}{dy}\right)^{-1} \qquad \qquad \text{...(1)}$$

$\therefore \qquad \dfrac{d^2 y}{dx^2} = \dfrac{d}{dx}\left(\dfrac{dy}{dx}\right) = \dfrac{d}{dx}\left[\left(\dfrac{dx}{dy}\right)^{-1}\right] = \dfrac{d}{dy}\left[\left(\dfrac{dx}{dy}\right)^{-1}\right]\dfrac{dy}{dx}$

$= -\left(\dfrac{dx}{dy}\right)^{-2} \cdot \dfrac{d^2 x}{dy^2} \cdot \dfrac{dy}{dx} = -\left(\dfrac{dx}{dy}\right)^{-2} \cdot \dfrac{d^2 x}{dy^2}\left(\dfrac{dx}{dy}\right)^{-1}$ [From (1)]

or $\qquad \dfrac{d^2 y}{dx^2} = -\left(\dfrac{dx}{dy}\right)^{-3} \cdot \dfrac{d^2 x}{dy^2}$ ...(2)

and
$$\frac{d^3y}{dx^3} = \frac{d}{dx}\left(\frac{d^2y}{dx^2}\right) = \frac{d}{dx}\left[\left(-\frac{dx}{dy}\right)^{-3}\frac{d^2x}{dy^2}\right] \qquad \text{[From (2)]}$$

$$= \frac{d}{dy}\left[-\left(\frac{dx}{dy}\right)^{-3}\frac{d^2x}{dy^2}\right]\frac{dy}{dx} = \left\{3\left(\frac{dx}{dy}\right)^{-4}\frac{d^2x}{dy^2}\right\}\frac{d^2x}{dy^2}\cdot\frac{dy}{dx} - \left(\frac{dx}{dy}\right)^{-3}\frac{d^3x}{dy^3}\frac{dy}{dx}$$

$$= 3\left(\frac{dx}{dy}\right)^{-4}\left(\frac{d^2x}{dy^2}\right)^2\left(\frac{dx}{dy}\right)^{-1} - \left(\frac{dx}{dy}\right)^{-3}\frac{d^3x}{dy^3}\left(\frac{dx}{dy}\right)^{-1} \qquad \text{[From (1)]}$$

or
$$\frac{d^3y}{dx^3} = 3\left(\frac{dx}{dy}\right)^{-5}\left(\frac{d^2x}{dy^2}\right)^2 - \left(\frac{dx}{dy}\right)^{-4}\frac{d^3x}{dy^3} \qquad \text{...(3)}$$

Similarly, we can find $\dfrac{d^4y}{dx^4}, \dfrac{d^5y}{dx^5}$, etc.

Solved Examples

EXAMPLE. *Show that the equation* $\dfrac{dy}{dx}\cdot\dfrac{d^3y}{dx^3} - 3\left(\dfrac{d^2y}{dx^2}\right)^2 = 0$

can be written in the form $\dfrac{d^3x}{dy^3} = 0$.

SOLUTION. Here, we have $\dfrac{dy}{dx} = \left(\dfrac{dx}{dy}\right)^{-1}, \dfrac{d^2y}{dx^2} = -\left(\dfrac{dx}{dy}\right)^{-3}\dfrac{d^2x}{dy^2}$

and $\dfrac{d^3y}{dx^3} = 3\left(\dfrac{dx}{dy}\right)^{-5}\left(\dfrac{d^2x}{dy^2}\right)^2 - \left(\dfrac{d^3x}{dy^3}\right)\left(\dfrac{dx}{dy}\right)^{-4}$

Making these substitutions in the given equation, we have

$$\left(\frac{dx}{dy}\right)^{-1}\left[3\left(\frac{dx}{dy}\right)^{-5}\left(\frac{d^2x}{dy^2}\right)^2 - \left(\frac{d^3x}{dy^3}\right)\left(\frac{dx}{dy}\right)^{-4}\right]$$

$$-3\left[-\left(\frac{dx}{dy}\right)^{-3}\frac{d^2x}{dy^2}\right]^2 = 0$$

$$\Rightarrow \qquad -\frac{d^3x}{dy^3}\left(\frac{dx}{dy}\right)^{-3} = 0$$

$$\Rightarrow \qquad \frac{d^3x}{dy^3} = 0 \text{ since } \frac{dx}{dy} \neq 0.$$

3.2.2 CHANGE OF INDEPENDENT VARIABLE INTO ANOTHER VARIABLE z, GIVEN x = f(z)

We have
$$\frac{dy}{dx} = \frac{dy}{dz}\cdot\frac{dz}{dx} = \frac{dy}{dz}\left(\frac{dx}{dz}\right)^{-1} \qquad \text{...(1)}$$

or
$$\frac{d}{dx}(y) = \left(\frac{dx}{dz}\right)^{-1}\frac{d}{dz}(y) \qquad \text{...(2)}$$

i.e., the operator $\dfrac{d}{dx}$ is equivalent to the operator $\left(\dfrac{dx}{dz}\right)^{-1}\dfrac{d}{dz}$ or $\dfrac{d}{dx} \equiv \left(\dfrac{dx}{dz}\right)^{-1}\dfrac{d}{dz}$.

Therefore,
$$\frac{d^2y}{dx^2} = \frac{d}{dx}\left(\frac{dy}{dx}\right) = \left(\frac{dx}{dz}\right)^{-1}\frac{d}{dz}\left(\frac{dy}{dx}\right)$$

$$= \left(\frac{dx}{dz}\right)^{-1}\frac{d}{dz}\left[\frac{dy}{dz}\left(\frac{dx}{dz}\right)^{-1}\right] \qquad \text{[From (1)]}$$

$$= \left(\frac{dx}{dz}\right)^{-1}\left[\frac{dy}{dz}\left\{-\left(\frac{dx}{dz}\right)^{-2}\cdot\frac{d^2x}{dz^2}\right\} + \left(\frac{dx}{dz}\right)^{-1}\frac{d^2y}{dz^2}\right]$$

or
$$\frac{d^2y}{dx^2} = \left(\frac{dx}{dz}\right)^{-3}\left[\frac{dx}{dz}\cdot\frac{d^2y}{dz^2} - \frac{dy}{dz}\cdot\frac{d^2x}{dz^2}\right] = \frac{\dfrac{dx}{dz}\cdot\dfrac{d^2y}{dz^2} - \dfrac{dy}{dz}\cdot\dfrac{d^2x}{dz^2}}{\left(\dfrac{dx}{dz}\right)^3} \qquad \text{...(3)}$$

and
$$\frac{d^3y}{dx^3} = \frac{d}{dx}\left[\frac{d^2y}{dx^2}\right] = \left(\frac{dx}{dz}\right)^{-1}\frac{d}{dz}\left(\frac{d^2y}{dx^2}\right), \qquad \text{[From (2)]}$$

$$= \left(\frac{dx}{dz}\right)^{-1}\frac{d}{dz}\left[\frac{\dfrac{dx}{dz}\cdot\dfrac{d^2y}{dz^2} - \dfrac{dy}{dz}\cdot\dfrac{d^2x}{dz^2}}{\left(\dfrac{dx}{dz}\right)^3}\right], \qquad \text{[From (3)]}$$

$$= \frac{\left(\frac{dx}{dz}\right)^{-1}\left[\left(\frac{dx}{dz}\right)^3\left\{\frac{dx}{dz}\cdot\frac{d^3y}{dz^3}+\frac{d^2x}{dz^2}\cdot\frac{d^2y}{dz^2}-\frac{d^2y}{dz^2}\cdot\frac{d^2x}{dz^2}-\frac{dy}{dz}\cdot\frac{d^3x}{dz^3}\right\}-3\left(\frac{dx}{dz}\right)^2\cdot\frac{d^2x}{dz^2}.N_r z\right]}{\left(\frac{dx}{dz}\right)^6}$$

where

$$N_r = \frac{dx}{dz}\cdot\frac{d^2y}{dz^2}-\frac{dy}{dz}\cdot\frac{d^2x}{dz^2}$$

or

$$\frac{d^3y}{dx^3} = \left(\frac{dx}{dz}\right)^{-5}\left[\left(\frac{dx}{dz}\frac{d^3y}{dz^3}-\frac{dy}{dz}\frac{d^3x}{dz^3}\right)\frac{dx}{dz}-3\frac{d^2x}{dz^2}\left(\frac{dx}{dz}\frac{d^2y}{dz^2}-\frac{dy}{dz}\frac{d^2x}{dz^2}\right)\right]$$

Solved Examples

EXAMPLE 1. *Transform the equation*

$$\frac{d}{dx}\left\{(1-x^2)\frac{dy}{dx}\right\}+n(n+1)y=0,$$

by the substitution $x=\frac{1}{2}[z+(1/z)]$.

SOLUTION . Given $x=\frac{1}{2}[z+(1/z)]$

$$\therefore \qquad \frac{dx}{dz}=\frac{1}{2}\left(1-\frac{1}{z^2}\right)=\frac{z^2-1}{2z^2} \qquad \text{...(1)}$$

$$\therefore \qquad \frac{dy}{dx}=\frac{dy}{dz}\cdot\frac{dz}{dx}=\frac{2z^2}{z^2-1}\cdot\frac{dy}{dz} \qquad \text{[From (1)]}$$

$$\text{or} \qquad (1-x^2)\frac{dy}{dx}=\left[1-\frac{1}{4}\left(z+\frac{1}{z}\right)^2\right]\frac{2z^2}{z^2-1}\frac{dy}{dz},$$

(substituting value of x)

$$=\left[1-\frac{1}{4}z^2-\frac{1}{4z^2}-\frac{1}{2}\right]\frac{2z^2}{z^2-1}\frac{dy}{dz}$$

$$=\frac{2z^2-z^4-1}{4z^2}\cdot\frac{2z^2}{z^2-1}\frac{dy}{dz}=-\frac{1}{2}(z^2-1)\frac{dy}{dz}$$

$\therefore$ Putting this value of $(1-x^2)\frac{dy}{dx}$ in given equation,

we get $\qquad \frac{d}{dx}\left\{-\frac{1}{2}(z^2-1)\frac{dy}{dz}\right\}+n(n+1)y=0$

$$\text{or} \qquad -\frac{1}{2}\frac{d}{dz}\left\{(z^2-1)\frac{dy}{dz}\right\}\cdot\frac{dz}{dx}+n(n+1)y=0$$

$$\text{or} \qquad -\frac{1}{2}\left[(z^2-1)\frac{d^2y}{dz^2}+2z\cdot\frac{dy}{dz}\right]\cdot\frac{2z^2}{z^2-1}+n(n+1)y=0$$

$$\text{or} \qquad z^2(z^2-1)\frac{d^2y}{dz^2}+2z^3\frac{dy}{dz}-n(n+1)(z^2-1)y=0.$$

EXAMPLE 2. *Transform the equation*

$$(1+x^2)^2 y_2 + 2x(1+x^2)y_1 + y = 0 \textbf{ by}$$

the substitution $x = \tan z$.

SOLUTION. Given $x=\tan z$. $\qquad$ ...(1)

$$\therefore \qquad \frac{dx}{dz}=\sec^2 z$$

$$\text{or} \qquad \frac{dz}{dx}=\cos^2 z \qquad \text{...(2)}$$

Now $y_1 = \dfrac{dy}{dx}=\dfrac{dy}{dz}\cdot\dfrac{dz}{dx}=\cos^2 z.\dfrac{dy}{dz}$ $\qquad$ [From (1)]

$$y_2 = \frac{d^2y}{dx^2}=\frac{d}{dx}\left(\frac{dy}{dx}\right)=\frac{d}{dx}\left[\cos^2 z\frac{dy}{dz}\right] \qquad \text{[From (2)]}$$

$$=\frac{d}{dz}\left[\cos^2 z\frac{dy}{dz}\right]\frac{dz}{dx}$$

$$=\left[\cos^2 z\frac{d^2y}{dz^2}+2\cos z(-\sin z)\frac{dy}{dz}\right]\cos^2 z$$

[From (1)]

$$\text{or} \qquad \frac{d^2y}{dx^2}=\cos^4 z\frac{d^2y}{dz^2}-2\sin z\cos^3 z\frac{dy}{dz} \qquad \text{...(3)}$$

Substituting the value of x, y_1 and y_2 in the given equation, we get

$$(1+\tan^2 z)^2\left[\cos^4 z\frac{d^2y}{dz^2}-2\sin z\cos^3 z\frac{dy}{dz}\right]$$

$$+2\tan z(1+\tan^2 z)\cos^2 z\frac{dy}{dz}+y=0$$

$$\text{or} \qquad \frac{d^2y}{dz^2}-2\tan z\frac{dy}{dz}+2\tan z\frac{dy}{dz}+y=0$$

$$\text{or} \qquad \frac{d^2y}{dz^2}+y=0.$$

3.2.3 Transformation involving Change of Dependent as well as Independent Variables

Such transformations will be clear from the examples given below.

EXAMPLE 1. *Transform into cartesian the polar formula*

$$\tan\phi=\frac{rd\theta}{dr}.$$

SOLUTION . We know $x=r\cos\theta$, $y=r\sin\theta$.

Hence we get $r^2=x^2+y^2$ and $\theta=\tan^{-1}(y/x)$.

$$\therefore \qquad 2r\frac{dr}{dx}-2x+2y\frac{dy}{dx}$$

$$\text{or} \qquad r\frac{dr}{dx}=x+y\frac{dy}{dx} \qquad \text{...(1)}$$

$$\text{and} \qquad \frac{d\theta}{dx}=\frac{1}{1+(y/x)^2}\cdot\frac{x(dy/dx)-y.1}{x^2}$$

$$=\frac{1}{x^2+y^2}\left(x\frac{dy}{dx}-y\right) \qquad \text{...(2)}$$

Now, $\tan\phi = r\dfrac{d\theta}{dr} = \dfrac{r\dfrac{d\theta}{dx}}{\dfrac{dr}{dx}} = \dfrac{r\left[1/(x^2+y^2)\right]\left(x\dfrac{dy}{dx}-y\right)}{(1/r)\left(x+y\dfrac{dy}{dx}\right)}$

[From (1) and (2)]

or $\quad \tan\phi = \dfrac{x(dy/dx)-y}{x+y(dy/dx)}$, using $r^2 = x^2 + y^2$.

This is the required formula in cartesian form.

EXAMPLE 2. *Transform cartesian formula*

$$\rho = \frac{\left[1+(dy/dx)^2\right]^{3/2}}{d^2y/dx^2} \text{ into polar form.}$$

SOLUTION. We know $x = r\cos\theta, y = r\sin\theta$.

$\therefore \qquad \dfrac{dx}{d\theta} = r(-\sin\theta) + \cos\theta.\dfrac{dr}{d\theta}$ ...(1)

and $\qquad \dfrac{dy}{d\theta} = r\cos\theta + \sin\theta\dfrac{dr}{d\theta}$ ...(2)

Now $\dfrac{dy}{dx} = \dfrac{dy/d\theta}{dx/d\theta} = \dfrac{r\cos\theta + \sin\theta(dr/d\theta)}{\cos\theta(dr/d\theta) - r\sin\theta}$ ...(3)

Again $\dfrac{d^2y}{dx^2} = \dfrac{d}{dx}\left(\dfrac{dy}{dx}\right) = \dfrac{d}{d\theta}\left(\dfrac{dy}{dx}\right)\dfrac{d\theta}{dx}$

$= \dfrac{d}{d\theta}\left[\dfrac{r\cos\theta + \sin\theta.\dfrac{dr}{d\theta}}{\cos\theta(dr/d\theta) - r\sin\theta}\right]\dfrac{1}{d\theta/dx}.$

$\dfrac{\left\{\dfrac{dr}{d\theta}\cos\theta - r\sin\theta\right\}}{\left\{\dfrac{dr}{d\theta}\cos\theta - r\sin\theta + \dfrac{dr}{d\theta}\cos\theta + \sin\theta\dfrac{d^2r}{d\theta^2}\right\}}$ [From (3)]

$= \dfrac{-\left\{r\cos\theta + \sin\theta.\dfrac{dr}{d\theta}\right\}}{\times\left\{\cos\theta\dfrac{d^2r}{d\theta^2} - \sin\theta\dfrac{dr}{d\theta} - \dfrac{dr}{d\theta}\sin\theta - r\cos\theta\right\}}{\left[(dr/d\theta)\cos\theta - r\sin\theta\right]^3}$

$= \dfrac{\left\{\dfrac{dr}{d\theta}\cos\theta - r\sin\theta\right\}\left\{\sin\theta\dfrac{d^2r}{d\theta^2} + 2\dfrac{dr}{d\theta}\cos\theta - r\sin\theta\right\}}{}$

$\dfrac{-\left\{r\cos\theta + \sin\theta.\dfrac{dr}{d\theta}\right\}}{\times\left\{\cos\theta\dfrac{d^2r}{d\theta^2} - 2\sin\theta\dfrac{dr}{d\theta} - r\cos\theta\right\}}{\left[(dr/d\theta)\cos\theta - r\sin\theta\right]^3}$

$= \left[2\left(\dfrac{dr}{d\theta}\right)^2 - r\dfrac{d^2r}{d\theta^2} + r^2\right]\bigg/\left[\dfrac{dr}{d\theta}\cos\theta - r\sin\theta\right]^3.$

After simplification putting these values in cartesian formula, we get

$\rho = \dfrac{\left[1+(dy/dx)^2\right]^{3/2}}{d^2y/dx^2}$

$= \dfrac{\left[1+\left\{\dfrac{r\cos\theta + \sin\theta(dr/d\theta)}{\cos\theta(dr/d\theta) - r\sin\theta}\right\}^2\right]^{3/2}}{\left[2\left(\dfrac{dr}{d\theta}\right)^2 - r\dfrac{d^2r}{d\theta^2} + r^2\right]\bigg/\left[\dfrac{dr}{d\theta}\cos\theta - r\sin\theta\right]^3}$

$= \left[\left(\dfrac{dr}{d\theta}\right)^2 + r^2\right]^{3/2}\bigg/\left[2\left(\dfrac{dr}{d\theta}\right)^2 - r\dfrac{d^2r}{d\theta^2} + r^2\right]$

which is the required polar form.

3.2.4 TRANSFORMATION IN THE CASE OF TWO INDEPENDENT VARIABLES

Here we use the following important results :

1. If $z = f(x,y)$, then $dz = \dfrac{\partial z}{\partial x}dx + \dfrac{\partial z}{\partial y}dy$. ...(1)

2. If $z = f(x,y)$, where $x = \phi(t)$ and $y = \psi(t)$, then $\dfrac{dz}{dt} = \dfrac{\partial z}{\partial x}\dfrac{dx}{dt} + \dfrac{\partial z}{\partial y}\dfrac{dy}{dt}$. ...(2)

3. If $z = f(x,y)$, where $x = \phi(t_1, t_2)$ and $y = \psi(t_1, t_2)$, then $\dfrac{\partial z}{\partial t_1} = \dfrac{\partial z}{\partial x}\dfrac{\partial x}{\partial t_1} + \dfrac{\partial z}{\partial y}\dfrac{\partial y}{\partial t_1}$ and $\dfrac{\partial z}{\partial t_2} = \dfrac{\partial z}{\partial x}\dfrac{\partial x}{\partial t_2} + \dfrac{\partial z}{\partial y}\dfrac{\partial y}{\partial t_2}$ ...(3)

4. The case $x = \phi(t_1, t_2)$ and $y = \psi(t_1, t_2)$ can easily be solved for t_1 and t_2 in terms of x and y, say $t_1 = F_1(x,y)$ and $t_2 = F_2(x,y)$, then the following formulae are used $\dfrac{\partial z}{\partial x} = \dfrac{\partial z}{\partial t_1}\dfrac{\partial t_1}{\partial x} + \dfrac{\partial z}{\partial t_2}\dfrac{\partial t_2}{\partial x}$ and $\dfrac{\partial z}{\partial y} = \dfrac{\partial z}{\partial t_1}\dfrac{\partial t_1}{\partial y} + \dfrac{\partial z}{\partial t_2}\dfrac{\partial t_2}{\partial y}$. ...(4)

Solved Examples

EXAMPLE 1. *If $z = f(x,y), x^2 = uv$ and $y^2 = u/v$, change the independent variables to u, v in the equation*

$$x^2\frac{\partial^2 z}{\partial x^2} - 2xy\frac{\partial^2 z}{\partial x\partial y} + y^2\frac{\partial^2 z}{\partial y^2} + 2y\frac{\partial z}{\partial y} = 0.$$

SOLUTION. Solving $x^2 = uv$ and $y^2 = u/v$, we get

$u = xy$ and $v = x/y$. ...(1)

$\therefore \qquad \dfrac{\partial u}{\partial x} = y, \dfrac{\partial u}{\partial y} = x, \dfrac{\partial v}{\partial x} = \dfrac{1}{y}, \dfrac{\partial v}{\partial y} = -\dfrac{x}{y^2}.$ ...(2)

Now, $\dfrac{\partial z}{\partial x} = \dfrac{\partial z}{\partial u}.\dfrac{\partial u}{\partial x} + \dfrac{\partial z}{\partial v}.\dfrac{\partial v}{\partial x} = \dfrac{\partial z}{\partial u}(y) + \dfrac{\partial z}{\partial v}\left(\dfrac{1}{y}\right)$

[From (2)]

or $x\dfrac{\partial z}{\partial x} = (xy)\dfrac{\partial z}{\partial u} + \left(\dfrac{x}{y}\right)\dfrac{\partial z}{\partial v} = u\dfrac{\partial z}{\partial u} + v\dfrac{\partial z}{\partial v}$ [From (1)]

$$\therefore \qquad x\frac{\partial}{\partial x} \equiv u\frac{\partial}{\partial u} + v\frac{\partial}{\partial v} \qquad \qquad \ldots(3)$$

and similarly $\quad y\frac{\partial}{\partial y} \equiv u\frac{\partial}{\partial u} - v\frac{\partial}{\partial v} \qquad \qquad \ldots(4)$

Now $\quad \left(x\dfrac{\partial}{\partial x} - y\dfrac{\partial}{\partial y}\right)^2 z$

$$= \left(x\frac{\partial}{\partial x} - y\frac{\partial}{\partial y}\right)\left(x\frac{\partial z}{\partial x} - y\frac{\partial z}{\partial y}\right)z$$

$$= x\frac{\partial}{\partial x}\left(x\frac{\partial z}{\partial x} - y\frac{\partial z}{\partial y}\right)z - y\frac{\partial}{\partial y}\left(x\frac{\partial z}{\partial x} - y\frac{\partial z}{\partial y}\right)z$$

$$= x\left[x\frac{\partial^2 z}{\partial x^2} + \frac{\partial z}{\partial x} - y\frac{\partial^2 z}{\partial x \partial y}\right] - y\left[x\frac{\partial^2 z}{\partial x \partial y} - y\frac{\partial^2 z}{\partial y^2} - \frac{\partial z}{\partial y}\right]z$$

or

$$\left(x\frac{\partial}{\partial x} - y\frac{\partial}{\partial y}\right)^2 z = x^2\frac{\partial^2 z}{\partial x^2} - 2xy\frac{\partial^2 z}{\partial x \partial y} + y^2\frac{\partial^2 z}{\partial y^2} + x\frac{\partial z}{\partial x} + y\frac{\partial z}{\partial y}$$

$\therefore$ The given equation

$$\left(x\frac{\partial}{\partial x} - y\frac{\partial}{\partial y}\right)^2 z + \left(y\frac{\partial z}{\partial y} - x\frac{\partial z}{\partial x}\right) = 0$$

which with the help of (3) and (4) reduces to

$$\left[\left(u\frac{\partial}{\partial u} + v\frac{\partial}{\partial v}\right) - \left(u\frac{\partial}{\partial u} - v\frac{\partial}{\partial v}\right)\right]^2 z$$

$$+ \left[\left(u\frac{\partial z}{\partial u} - v\frac{\partial z}{\partial v}\right) - \left(u\frac{\partial z}{\partial u} + v\frac{\partial z}{\partial v}\right)\right] = 0$$

or $\qquad\qquad 4\left(v\dfrac{\partial}{\partial v}\right)^2 z - 2v\dfrac{\partial z}{\partial v} = 0$

or $\qquad\qquad 2v\dfrac{\partial}{\partial v}\left(v\dfrac{\partial z}{\partial v}\right) - v\dfrac{\partial z}{\partial v} = 0$

or $\qquad\qquad 2\left[v\dfrac{\partial^2 z}{\partial v^2} + \dfrac{\partial z}{\partial v}\right] - \dfrac{\partial z}{\partial v} = 0$

or $\qquad\qquad 2v\dfrac{\partial^2 z}{\partial v^2} + \dfrac{\partial z}{\partial v} = 0$

3.2.5 TRANSFORMATION FROM CARTESIAN TO POLAR CO-ORDINATES AND VICE-VERSA

Transform the Laplace equation $\dfrac{\partial^2 u}{\partial x^2} + \dfrac{\partial^2 u}{\partial y^2} = 0$ *to polars.*

We know that $\qquad\qquad x = r\cos\theta, y = r\sin\theta. \qquad\qquad\qquad\qquad\qquad \ldots(1)$

$\therefore \qquad\qquad\qquad\qquad\quad r^2 = x^2 + y^2. \qquad\qquad\qquad\qquad\qquad\qquad \ldots(2)$

and $\qquad\qquad\qquad\qquad \theta = \tan^{-1}(y/x) \qquad\qquad\qquad\qquad\qquad\qquad \ldots(3)$

From (2), we get $\quad 2r\dfrac{\partial r}{\partial x} = 2x \quad$ or $\quad \dfrac{\partial r}{\partial x} = \dfrac{x}{r} = \dfrac{r\cos\theta}{r} \qquad\qquad\qquad$ [From (1)]

or $\qquad\qquad\qquad\qquad \dfrac{\partial r}{\partial x} = \cos\theta \qquad\qquad\qquad\qquad\qquad\qquad\qquad \ldots(4)$

Similarly, $\qquad\qquad\qquad \dfrac{\partial r}{\partial y} = \dfrac{y}{r} = \dfrac{r\sin\theta}{r} = \sin\theta \qquad\qquad\qquad\qquad \ldots(5)$

Also, from (3), $\qquad \dfrac{\partial\theta}{\partial x} = \dfrac{1}{1+(y/x)^2}\left(-\dfrac{y}{x^2}\right) = -\dfrac{y}{x^2+y^2} = -\dfrac{r\sin\theta}{r^2}$ or $\dfrac{\partial\theta}{\partial x} = -\dfrac{\sin\theta}{r} \quad \ldots(6)$

and $\qquad\qquad \dfrac{\partial\theta}{\partial y} = \dfrac{1}{1+(y/x)^2}\left(\dfrac{1}{x}\right) = \dfrac{x}{x^2+y^2} = \dfrac{r\cos\theta}{r^2} = \dfrac{\cos\theta}{r} \qquad\qquad \ldots(7)$

Now $\qquad\qquad\qquad \dfrac{\partial u}{\partial x} = \dfrac{\partial u}{\partial r}\cdot\dfrac{\partial r}{\partial x} + \dfrac{\partial u}{\partial\theta}\cdot\dfrac{\partial\theta}{\partial x} \qquad\qquad\qquad\qquad\qquad \ldots(8)$

$$= \frac{\partial u}{\partial r}\cdot(\cos\theta) + \frac{\partial u}{\partial\theta}\left(-\frac{\sin\theta}{r}\right) \qquad\qquad\qquad \text{[From (4) and (6)]}$$

or $\qquad\qquad \dfrac{\partial}{\partial x}(u) = \cos\theta\dfrac{\partial}{\partial r}(u) - \dfrac{\sin\theta}{r}\dfrac{\partial}{\partial\theta}(u) \qquad\qquad\qquad\qquad \ldots(9)$

Again $\qquad\qquad \dfrac{\partial u}{\partial y} = \dfrac{\partial u}{\partial r}\cdot\dfrac{\partial r}{\partial y} + \dfrac{\partial u}{\partial\theta}\cdot\dfrac{\partial\theta}{\partial y} \qquad\qquad\qquad \text{[From (5) and (7)]}$

or $\qquad\qquad \dfrac{\partial}{\partial y}(u) = \sin\theta\dfrac{\partial}{\partial r}(u) + \dfrac{\cos\theta}{r}\dfrac{\partial}{\partial\theta}(u) \qquad\qquad\qquad\qquad \ldots(10)$

$\therefore \qquad \dfrac{\partial^2 u}{\partial x^2} = \dfrac{\partial}{\partial x}\left(\dfrac{\partial u}{\partial x}\right) = \cos\theta\dfrac{\partial}{\partial r}\left(\dfrac{\partial u}{\partial x}\right) - \dfrac{\sin\theta}{r}\dfrac{\partial}{\partial\theta}\left(\dfrac{\partial u}{\partial x}\right) \qquad$ replacing u by $\dfrac{\partial u}{\partial x}$ in (9)

$$= \cos\theta\frac{\partial}{\partial r}\left[\cos\theta\frac{\partial u}{\partial r} - \frac{\sin\theta}{r}\frac{\partial u}{\partial\theta}\right] - \frac{\sin\theta}{r}\frac{\partial}{\partial\theta}\left[\cos\theta\frac{\partial u}{\partial r} - \frac{\sin\theta}{r}\frac{\partial u}{\partial\theta}\right]$$

substituting from (9) the polar equivalent of $\dfrac{\partial u}{\partial x}$

$$= \cos\theta\left[\cos\theta\frac{\partial^2 u}{\partial r^2} - \sin\theta\frac{\partial}{\partial r}\left(\frac{1}{r}\frac{\partial u}{\partial\theta}\right)\right] - \frac{\sin\theta}{r}\left[\left(\cos\theta\frac{\partial^2 u}{\partial\theta\partial r} - \sin\theta\frac{\partial u}{\partial r}\right) - \frac{1}{r}\frac{\partial}{\partial\theta}\left(\sin\theta\frac{\partial u}{\partial\theta}\right)\right]$$

$$= \cos\theta\left[\cos\theta\frac{\partial^2 u}{\partial r^2} - \sin\theta\left\{\frac{1}{r}\frac{\partial^2 u}{\partial r\partial\theta} - \frac{1}{r^2}\frac{\partial u}{\partial\theta}\right\}\right] - \frac{\sin\theta}{r}\left[\left(\cos\theta\frac{\partial^2 u}{\partial r\partial\theta} - \sin\theta\frac{\partial u}{\partial r}\right) - \frac{1}{r}\left(\sin\theta\frac{\partial^2 u}{\partial\theta^2} + \cos\theta\frac{\partial u}{\partial\theta}\right)\right]$$

or

$$\frac{\partial^2 u}{\partial x^2} = \cos^2\theta\frac{\partial^2 u}{\partial r^2} - \frac{2\sin\theta\cos\theta}{r}\frac{\partial^2 u}{\partial r\partial\theta} + \frac{\sin^2\theta}{r^2}\frac{\partial^2 u}{\partial\theta^2} + \frac{\sin^2\theta}{r}\frac{\partial u}{\partial r} + \frac{2\cos\theta\sin\theta}{r^2}\frac{\partial u}{\partial\theta} \qquad \dots(11)$$

Similarly, from

$$\frac{\partial^2 u}{\partial y^2} = \frac{\partial}{\partial y}\left(\frac{\partial u}{\partial y}\right) = \sin\theta\frac{\partial}{\partial r}\left(\frac{\partial u}{\partial y}\right) + \frac{\cos\theta}{r}\frac{\partial}{\partial\theta}\left(\frac{\partial u}{\partial y}\right), \text{ from (10), we get}$$

$$\frac{\partial^2 u}{\partial y^2} = \sin^2\theta\frac{\partial^2 u}{\partial r^2} + \frac{2\sin\theta\cos\theta}{r}\frac{\partial^2 u}{\partial r\partial\theta} + \frac{\cos^2\theta}{r^2}\frac{\partial^2 u}{\partial\theta^2} + \frac{\cos^2\theta}{r}\frac{\partial u}{\partial r} - \frac{2\cos\theta\sin\theta}{r^2}\frac{\partial u}{\partial\theta}. \qquad \dots(12)$$

Adding (11) and (12), we get $\quad \dfrac{\partial^2 u}{\partial x^2} + \dfrac{\partial^2 u}{\partial y^2} = (\cos^2\theta + \sin^2\theta)\dfrac{\partial^2 u}{\partial r^2} + \dfrac{1}{r^2}(\sin^2\theta + \cos^2\theta)\dfrac{\partial^2 u}{\partial\theta^2} + \dfrac{1}{r}(\sin^2\theta + \cos^2\theta)\dfrac{\partial u}{\partial r} = \dfrac{\partial^2 u}{\partial r^2} + \dfrac{1}{r^2}\dfrac{\partial^2 u}{\partial\theta^2} + \dfrac{1}{r}\dfrac{\partial u}{\partial r}$

Hence, the given differential equation $\dfrac{\partial^2 u}{\partial x^2} + \dfrac{\partial^2 u}{\partial y^2} = 0$ transforms into $\dfrac{\partial^2 u}{\partial r^2} + \dfrac{1}{r^2}\dfrac{\partial^2 u}{\partial\theta^2} + \dfrac{1}{r}\dfrac{\partial u}{\partial r} = 0$.

Case II. Transformation from Polar to Cartesian.

Now let us consider the converse of Case I, *i.e.*, let us transform the polar differential equation $\dfrac{\partial^2 u}{\partial r^2} + \dfrac{1}{r^2}\dfrac{\partial^2 u}{\partial\theta^2} + \dfrac{1}{r}\dfrac{\partial u}{\partial r} = 0$ to cartesian.

As before, $\qquad\qquad\qquad\qquad\qquad x = r\cos\theta, y = r\sin\theta. \qquad\qquad\qquad\qquad\qquad \dots(1)$

Then, $\qquad\qquad\qquad\qquad\qquad \dfrac{\partial u}{\partial r} = \dfrac{\partial u}{\partial x}\cdot\dfrac{\partial x}{\partial r} + \dfrac{\partial u}{\partial y}\cdot\dfrac{\partial y}{\partial r} \qquad\qquad\qquad\qquad\qquad \dots(2)$

Now, from (1) $\qquad\qquad\left.\begin{array}{l} \dfrac{\partial x}{\partial r} = \cos\theta = \dfrac{x}{r}; \dfrac{\partial x}{\partial\theta} = -r\sin\theta = -y \\[3mm] \dfrac{\partial y}{\partial r} = \sin\theta = \dfrac{y}{r}; \dfrac{\partial y}{\partial\theta} = r\cos\theta = x \end{array}\right\} \qquad\qquad \dots(3)$

Hence, from (2), we get $\qquad\qquad \dfrac{\partial u}{\partial r} = \dfrac{\partial u}{\partial x}\left(\dfrac{x}{r}\right) + \dfrac{\partial u}{\partial y}\left(\dfrac{y}{r}\right) \qquad\qquad\qquad$ [Using (3)]

or $\qquad\qquad\qquad\qquad\qquad r\dfrac{\partial}{\partial r}(u) = x\dfrac{\partial}{\partial x}(u) + y\dfrac{\partial}{\partial y}(u) \qquad\qquad\qquad\qquad \dots(4)$

Also we have $\qquad\qquad \dfrac{\partial u}{\partial\theta} = \dfrac{\partial u}{\partial x}\dfrac{\partial x}{\partial\theta} + \dfrac{\partial u}{\partial y}\dfrac{\partial y}{\partial\theta} = \dfrac{\partial u}{\partial x}(-y) + \dfrac{\partial u}{\partial y}(x) \qquad\qquad$ [From (3)]

or $\qquad\qquad\qquad\qquad\qquad \dfrac{\partial}{\partial\theta}(u) = x\dfrac{\partial}{\partial y}(u) - y\dfrac{\partial}{\partial x}(u) \qquad\qquad\qquad\qquad\qquad \dots(5)$

Now $\qquad\qquad r\dfrac{\partial}{\partial r}\left(r\dfrac{\partial u}{\partial r}\right) = x\dfrac{\partial}{\partial x}\left(r\dfrac{\partial u}{\partial r}\right) + y\dfrac{\partial}{\partial y}\left(r\dfrac{\partial u}{\partial r}\right) \qquad$ (replacing u by $r\dfrac{\partial u}{\partial r}$ in (4))

$$= x\dfrac{\partial}{\partial x}\left[x\dfrac{\partial u}{\partial x} + y\dfrac{\partial u}{\partial y}\right] + y\dfrac{\partial}{\partial y}\left(x\dfrac{\partial u}{\partial x} + y\dfrac{\partial u}{\partial y}\right) \text{ Substituting the value of } r\dfrac{\partial u}{\partial r} \text{ from (4)}$$

or $\qquad r\left[r\dfrac{\partial^2 u}{\partial r^2} + \dfrac{\partial u}{\partial r}\right] = x\left[x\dfrac{\partial^2 u}{\partial x^2} + \dfrac{\partial u}{\partial x} + y\dfrac{\partial^2 u}{\partial x\partial y}\right] + y\left[x\dfrac{\partial^2 u}{\partial y\partial x} + y\dfrac{\partial^2 u}{\partial y^2} + \dfrac{\partial u}{\partial y}\right]$

or $\qquad r^2\dfrac{\partial^2 u}{\partial r^2} + r\dfrac{\partial u}{\partial r} = x^2\dfrac{\partial^2 u}{\partial x^2} + 2xy\dfrac{\partial^2 u}{\partial x\partial y} + y^2\dfrac{\partial^2 u}{\partial y^2} + x\dfrac{\partial u}{\partial x} + y\dfrac{\partial u}{\partial y}$

or $\qquad\qquad r^2\dfrac{\partial^2 u}{\partial r^2} = x^2\dfrac{\partial^2 u}{\partial x^2} + 2xy\dfrac{\partial^2 u}{\partial x\partial y} + y^2\dfrac{\partial^2 u}{\partial y^2} \qquad\qquad\qquad\qquad \dots(6)$

Since $r\dfrac{\partial u}{\partial r} = x\dfrac{\partial u}{\partial x} + y\dfrac{\partial u}{\partial y}, \quad$ (from (4))

and $\qquad\qquad \dfrac{\partial^2 u}{\partial\theta^2} = \dfrac{\partial}{\partial\theta}\left(\dfrac{\partial u}{\partial\theta}\right) = x\dfrac{\partial}{\partial y}\left(\dfrac{\partial u}{\partial\theta}\right) - y\dfrac{\partial}{\partial x}\left(\dfrac{\partial u}{\partial\theta}\right) \qquad$ replacing u by $\dfrac{\partial u}{\partial\theta}$ in (5)

$$= x\dfrac{\partial}{\partial y}\left(x\dfrac{\partial u}{\partial y} - y\dfrac{\partial u}{\partial x}\right) - y\dfrac{\partial}{\partial x}\left(x\dfrac{\partial u}{\partial y} - y\dfrac{\partial u}{\partial x}\right)$$

$$\left(\text{Substituting the value of } \dfrac{\partial u}{\partial\theta} \text{ from (5)}\right)$$

$$= x\left(x\frac{\partial^2 u}{\partial y^2} - y\frac{\partial^2 u}{\partial x \partial y} - \frac{\partial u}{\partial x}\right) - y\left(x\frac{\partial^2 u}{\partial x \partial y} + \frac{\partial u}{\partial y} - y\frac{\partial^2 u}{\partial x^2}\right)$$

$$= x\frac{\partial^2 u}{\partial y^2} - 2xy\frac{\partial^2 u}{\partial x \partial y} + y^2\frac{\partial^2 u}{\partial x^2} - \left(x\frac{\partial u}{\partial x} + y\frac{\partial u}{\partial y}\right)$$

or

$$\frac{\partial^2 u}{\partial \theta^2} + r\frac{\partial u}{\partial r} = x^2\frac{\partial^2 u}{\partial y^2} - 2xy\frac{\partial^2 u}{\partial x \partial y} + y^2\frac{\partial^2 u}{\partial x^2} \qquad\qquad \ldots(7)$$

Adding (6) and (7), we get

$$r^2\frac{\partial^2 u}{\partial r^2} + \frac{\partial^2 u}{\partial \theta^2} + r\frac{\partial u}{\partial r} = (x^2 + y^2)\frac{\partial^2 u}{\partial x^2} + (x^2 + y^2)\frac{\partial^2 u}{\partial y^2} = r^2\left(\frac{\partial^2 u}{\partial x^2} + \frac{\partial^2 u}{\partial y^2}\right) \qquad [\because r^2 = x^2 + y^2]$$

or

$$\frac{\partial^2 u}{\partial r^2} + \frac{1}{r^2}\frac{\partial^2 u}{\partial \theta^2} + \frac{1}{r}\frac{\partial u}{\partial r} = \frac{\partial^2 u}{\partial x^2} + \frac{\partial^2 u}{\partial y^2}$$

Hence, the given differential equation $\dfrac{\partial^2 u}{\partial r^2} + \dfrac{1}{r^2}\dfrac{\partial^2 u}{\partial \theta^2} + \dfrac{1}{r^2}\dfrac{\partial u}{\partial r} = 0$, transforms into $\dfrac{\partial^2 u}{\partial x^2} + \dfrac{\partial^2 u}{\partial y^2} = 0$.

Solved Examples

EXAMPLE 1. *If x = r cos θ, y = r sin θ, prove that*

$$\frac{\partial^2 r}{\partial x^2}\frac{\partial^2 r}{\partial y^2} = \left(\frac{\partial^2 r}{\partial x \partial y}\right)^2.$$

SOLUTION. We know that $\dfrac{\partial r}{\partial x} = \dfrac{x}{r}, \dfrac{\partial r}{\partial y} = \dfrac{y}{r}$.

Now $\dfrac{\partial r}{\partial x^2} = \dfrac{\partial}{\partial x}\left(\dfrac{\partial r}{\partial x}\right) = \dfrac{\partial}{\partial x}\left(\dfrac{x}{r}\right) = \dfrac{r.1 - x(\partial r/\partial x)}{r^2}$

$$= \frac{r - x(x/r)}{r^2} = \frac{r^2 - x^2}{r^3} = \frac{y^2}{r^3}.$$

$$[\because x^2 + y^2 = r^2] \qquad \ldots(1)$$

Similarly, we can get

$$\frac{\partial^2 r}{\partial y^2} = \frac{x^2}{r^3} \qquad\qquad \ldots(2)$$

Also, $\dfrac{\partial^2 r}{\partial x \partial y} = \dfrac{\partial}{\partial x}\left(\dfrac{\partial r}{\partial y}\right) = \dfrac{\partial}{\partial x}\left(\dfrac{y}{r}\right) = \dfrac{r.0 - y(\partial r/\partial x)}{r^2}$

$$= -\frac{y(x/r)}{r^2} = -\frac{xy}{r^3}.$$

$$\therefore \left(\frac{\partial^2 r}{\partial x \partial y}\right)^2 = \left(-\frac{xy}{r^3}\right)^2 = \frac{x^2}{r^3}.\frac{y^2}{r^3} = \frac{\partial^2 r}{\partial y^2}.\frac{\partial^2 r}{\partial x^2}.$$

[From (1) an d (2)]

EXAMPLE 2. *Transform* $\dfrac{\partial^2 u}{\partial x^2} + \dfrac{\partial^2 u}{\partial y^2} = 0$ *into polars and show that*

$u = \left(Ar^n + Br^{-n}\right)\sin n\theta$ *satisfies the above equation.*

SOLUTION. We can transform the given equation

$$\frac{\partial^2 u}{\partial x^2} + \frac{\partial^2 u}{\partial y^2} = 0 \text{ into } \frac{\partial^2 u}{\partial r^2} + \frac{1}{r^2}\frac{\partial^2 u}{\partial \theta^2} + \frac{1}{r}\frac{\partial u}{\partial r} = 0,.$$

Now $u = \left(Ar^n + Br^{-n}\right)\sin n\theta$

$$\therefore \quad \frac{\partial u}{\partial r} = \left(nAr^{n-1} - Bnr^{-n-1}\right)\sin n\theta$$

$$\frac{\partial^2 u}{\partial r^2} = n[(n-1)Ar^{n-2} + Bn(n+1)r^{-n-2}]\sin n\theta$$

$$\frac{\partial u}{\partial \theta} = n[Ar^n + Br^{-n}]\cos n\theta;$$

$$\frac{\partial^2 u}{\partial \theta^2} = -n^2[Ar^n + Br^{-n}]\sin n\theta.$$

$$\therefore \quad \frac{\partial^2 u}{\partial r^2} + \frac{1}{r^2}\frac{\partial^2 u}{\partial \theta^2} + \frac{1}{r}\frac{\partial u}{\partial r}$$

$$= n[(n-1)Ar^{n-2} + B(n+1)r^{-n-2}]\sin n\theta$$

$$+ \frac{1}{r^2}[-n^2(Ar^n + Br^{-n})\sin n\theta]$$

$$+ \frac{1}{r}n(Ar^{n-1} - Br^{-n-1})\sin n\theta$$

$$= [A\{n(n-1) - n^2 + n\}r^{n-2}$$

$$+ \{Bn(n+1) - n^2 - n\}r^{-n-2}]\sin n\theta = 0$$

Hence, the equation $\dfrac{\partial^2 u}{\partial r^2} + \dfrac{1}{r^2}\dfrac{\partial^2 u}{\partial \theta^2} + \dfrac{1}{r}\dfrac{\partial u}{\partial r} = 0$, *i.e.,*

$\dfrac{\partial^2 u}{\partial x^2} + \dfrac{\partial^2 u}{\partial y^2} = 0$ is satisfied by $(Ar^n + Br^{-n})\sin n\theta$.

3.2.6 To Transform $\nabla^2 V$ into Polar Co-ordinates, where the operator ∇^2 stands for $\dfrac{\partial^2}{\partial x^2} + \dfrac{\partial^2}{\partial y^2} + \dfrac{\partial^2}{\partial z^2}$

For polar transformation (in three dimensions), we have $x = r\sin\theta\cos\phi, y = r\sin\theta\sin\phi, z = r\cos\theta$.

Let $r\sin\theta = u$, then $x = u\cos\phi, y = u\sin\phi$.

Then, as in § 9.8.5 Case I, we can have

$$\frac{\partial^2 V}{\partial x^2} + \frac{\partial^2 V}{\partial y^2} = \frac{\partial^2 V}{\partial u^2} + \frac{1}{u}.\frac{\partial V}{\partial u} + \frac{1}{u^2}.\frac{\partial^2 V}{\partial \phi^2} \qquad\qquad \ldots(1)$$

Again, we have $z = r\cos\theta$, $u = r\sin\theta$.

$\therefore$
$$\frac{\partial^2 V}{\partial z^2} + \frac{\partial^2 V}{\partial u^2} = \frac{\partial^2 V}{\partial u^2} + \frac{1}{r}\cdot\frac{\partial V}{\partial r} + \frac{1}{r^2}\cdot\frac{\partial^2 V}{\partial \theta^2} \qquad \ldots(2)$$

Also,
$$\frac{\partial V}{\partial u} = \frac{\partial V}{\partial r}\cdot\frac{\partial r}{\partial u} + \frac{\partial V}{\partial \theta}\cdot\frac{\partial \theta}{\partial u} \qquad \ldots(3)$$

Now $u = r\sin\theta$, $z = r\cos\theta$, wherence we get
$$r^2 = u^2 + z^2 \quad\text{and}\quad \theta = \tan^{-1}(u/z).$$

$\therefore \qquad 2r\dfrac{\partial r}{\partial u} = 2u \qquad$ or $\qquad \dfrac{\partial r}{\partial u} = \dfrac{u}{r} = \dfrac{r\sin\theta}{r} = \sin\theta$

and $\qquad \dfrac{\partial \theta}{\partial u} = \dfrac{1}{1+(x/z)^2}\cdot\dfrac{1}{z} = \dfrac{z}{u^2+z^2} = \dfrac{r\cos\theta}{r^2} = \dfrac{\cos\theta}{r}$

Substituting these values of $\partial r/\partial u$ and $\partial\theta/\partial u$ in (3), we get

$$\frac{\partial V}{\partial u} = \frac{\partial V}{\partial r}(\sin\theta) + \frac{\partial V}{\partial \theta}\left(\frac{\cos\theta}{r}\right)$$

or $\quad \dfrac{1}{u}\left(\dfrac{\partial V}{\partial u}\right) = \dfrac{1}{r\sin\theta}\left(\dfrac{\partial V}{\partial u}\right) = \dfrac{1}{r\sin\theta}\left[\sin\theta\,\dfrac{\partial V}{\partial r} + \dfrac{\cos\theta}{r}\dfrac{\partial V}{\partial\theta}\right]$ or $\dfrac{1}{u}\dfrac{\partial V}{\partial u} = \dfrac{1}{r}\dfrac{\partial V}{\partial r} + \dfrac{\cot\theta}{r^2}\dfrac{\partial V}{\partial\theta}.$ $\ldots(4)$

Now adding (1) and (2), we get
$$\nabla^2 V = \frac{\partial^2 V}{\partial x^2} + \frac{\partial^2 V}{\partial y^2} + \frac{\partial^2 V}{\partial z^2} = \frac{1}{u}\frac{\partial V}{\partial u} + \frac{1}{u^2}\frac{\partial^2 V}{\partial \phi^2} + \frac{\partial^2 V}{\partial r^2} + \frac{1}{r}\frac{\partial V}{\partial r} + \frac{1}{r^2}\frac{\partial^2 V}{\partial \theta^2}$$

$$= \frac{1}{r}\frac{\partial V}{\partial r} + \frac{\cot\theta}{r^2}\frac{\partial V}{\partial \theta} + \frac{1}{r^2\sin^2\theta}\cdot\frac{\partial^2 V}{\partial \phi^2} + \frac{\partial^2 V}{\partial r^2} + \frac{1}{r}\frac{\partial V}{\partial r} + \frac{1}{r^2}\frac{\partial^2 V}{\partial \theta^2}$$

i.e.,
$$\nabla^2 V = \frac{\partial^2 V}{\partial r^2} + \frac{2}{r}\frac{\partial V}{\partial r} + \frac{1}{r^2}\frac{\partial^2 V}{\partial \theta^2} + \frac{\cot\theta}{r^2}\frac{\partial V}{\partial \theta} + \frac{1}{r^2\sin^2\theta}\cdot\frac{\partial^2 V}{\partial \phi^2}$$

Solved Examples

EXAMPLE 1. *If $x = r\cos\theta$, $y = r\sin\theta$ and $r = e^t$, prove that*

$$x^2\frac{\partial^2 u}{\partial x^2} + 2xy\frac{\partial^2 u}{\partial y\partial x} + y^2\frac{\partial^2 u}{\partial y^2} = r\frac{\partial}{\partial r}\left(r\frac{\partial}{\partial r}-1\right)u = \frac{\partial}{\partial z}\left(\frac{\partial}{\partial z}-1\right)u$$

and $\quad x^2\dfrac{\partial^2 u}{\partial y^2} - 2xy\dfrac{\partial^2 u}{\partial x\partial y} + y^2\dfrac{\partial^2 u}{\partial x^2} = \dfrac{\partial^2 u}{\partial\theta^2} + r\dfrac{\partial u}{\partial r} = \dfrac{\partial^2 u}{\partial\theta^2} + \dfrac{\partial u}{\partial z}.$

SOLUTION. As in § 9.8.5 Case II, we can show that
$$r\frac{\partial u}{\partial r} = x\frac{\partial u}{\partial x} + y\frac{\partial u}{\partial y}$$

Therefore,
$$r\frac{\partial}{\partial r}\left(r\frac{\partial}{\partial r}-1\right)u = \left(x\frac{\partial}{\partial x}+y\frac{\partial}{\partial y}\right)\left[x\frac{\partial u}{\partial x}+y\frac{\partial u}{\partial y}-u\right]$$

$$= x\left[x\frac{\partial^2 u}{\partial x^2} + \frac{\partial u}{\partial x} + y\frac{\partial^2 u}{\partial x\partial y} - \frac{\partial u}{\partial x}\right]$$

$$+ y\left[x\frac{\partial^2 u}{\partial y\partial x} + y\frac{\partial^2 u}{\partial y^2} + \frac{\partial u}{\partial y} - \frac{\partial u}{\partial y}\right]$$

$$= x^2\frac{\partial^2 u}{\partial x^2} + 2xy\frac{\partial^2 u}{\partial x\partial y} + y^2\frac{\partial^2 u}{\partial y^2}$$

Also as $r = e^z$ or $z = \log r$.

$\therefore$ Using $\quad \dfrac{\partial u}{\partial r} = \dfrac{\partial u}{\partial z}\cdot\dfrac{\partial z}{\partial r} = \dfrac{\partial u}{\partial z}\cdot\dfrac{1}{r}$

$\therefore \qquad r\dfrac{\partial u}{\partial r} = \dfrac{\partial u}{\partial z}$

or $\qquad r\dfrac{\partial}{\partial r} = \dfrac{\partial}{\partial z}$ $\qquad\qquad \ldots(1)$

$\therefore \quad r\dfrac{\partial}{\partial r}\left(r\dfrac{\partial}{\partial r}-1\right)u = \dfrac{\partial}{\partial z}\left(\dfrac{\partial}{\partial z}-1\right)u.$

Again, we can easily prove that
$$\frac{\partial u}{\partial \theta} = x\frac{\partial u}{\partial y} - y\frac{\partial u}{\partial x}.$$

$$\therefore \quad \frac{\partial^2 u}{\partial \theta^2} = \frac{\partial}{\partial\theta}\left(\frac{\partial u}{\partial\theta}\right) = \left(x\frac{\partial}{\partial y}-y\frac{\partial}{\partial x}\right)\left(x\frac{\partial u}{\partial y}-y\frac{\partial u}{\partial x}\right)$$

$$= x\frac{\partial}{\partial y}\left(x\frac{\partial u}{\partial y}-y\frac{\partial u}{\partial x}\right) - y\frac{\partial}{\partial x}\left(x\frac{\partial u}{\partial y}-y\frac{\partial u}{\partial x}\right)$$

$$= x^2\frac{\partial^2 u}{\partial y^2} - 2xy\frac{\partial^2 u}{\partial x\partial y} + y^2\frac{\partial^2 u}{\partial x^2} - \left(x\frac{\partial u}{\partial x}+y\frac{\partial u}{\partial y}\right)$$

$$= x\left[x\frac{\partial^2 u}{\partial y^2} - \frac{\partial u}{\partial x} - y\frac{\partial^2 u}{\partial y\partial x}\right] - y\left[x\frac{\partial^2 u}{\partial x\partial y} + \frac{\partial u}{\partial y} - y\frac{\partial^2 u}{\partial x^2}\right]$$

or $\quad \dfrac{\partial^2 u}{\partial\theta^2} + r\dfrac{\partial u}{\partial r} = x^2\dfrac{\partial^2 u}{\partial y^2} - 2xy\dfrac{\partial^2 u}{\partial x\partial y} + y^2\dfrac{\partial^2 u}{\partial x^2}.$

[From (1)]

Also, from (1), we get $\dfrac{\partial^2 u}{\partial\theta^2} + r\dfrac{\partial u}{\partial r} = \dfrac{\partial^2 u}{\partial\theta^2} + \dfrac{\partial u}{\partial z}.$

EXAMPLE 2. *If V be a function of r along where $r^2 = x^2 + y^2 + z^2$,*

show that $\dfrac{\partial^2 V}{\partial x^2} + \dfrac{\partial^2 V}{\partial y^2} + \dfrac{\partial^2 V}{\partial z^2} = \dfrac{\partial^2 V}{\partial r^2} + \dfrac{2}{r}\dfrac{dV}{dr}.$

SOLUTION . As V is given to be a function of r alone, so we have

$$\frac{\partial V}{\partial x} = \frac{dV}{dr} \frac{\partial r}{\partial x} \qquad \qquad ...(1)$$

Also, from $r^2 = x^2 + y^2 + z^2$,

we get $\qquad 2r \frac{\partial r}{\partial x} = 2x.$

$\therefore$ From (1), $\quad \dfrac{\partial V}{\partial x} = \dfrac{dV}{dr} \dfrac{x}{r}$

$\therefore \quad \dfrac{\partial^2 V}{\partial x^2} = \dfrac{\partial}{\partial x}\left(\dfrac{\partial V}{\partial x}\right) = \dfrac{\partial}{\partial x}\left[\dfrac{dV}{dr}\dfrac{x}{r}\right] = \dfrac{dV}{dr}\dfrac{\partial}{\partial x}\left(\dfrac{x}{r}\right) + \dfrac{x}{r}\dfrac{\partial}{\partial x}\left(\dfrac{dV}{dr}\right)$

$\qquad\qquad = \dfrac{dV}{dr}\left[\dfrac{1}{r} + x\left(-\dfrac{1}{r^2}\dfrac{\partial r}{\partial x}\right)\right] + \dfrac{x}{r}\left(\dfrac{d^2V}{dr^2}\dfrac{\partial r}{\partial x}\right)$

$\qquad\qquad = \dfrac{1}{r}\dfrac{dV}{dr} - \dfrac{x}{r^2}\dfrac{dV}{dr}\left(\dfrac{x}{r}\right) + \dfrac{x}{r}\dfrac{d V}{dr^2}\left(\dfrac{x}{r}\right) \quad \left(\because \dfrac{\partial r}{\partial x} = \dfrac{x}{r}\right)$

or $\quad \dfrac{\partial^2 V}{\partial x^2} = \dfrac{1}{r}\dfrac{dV}{dr} - \dfrac{x^2}{r^3}\dfrac{dV}{dr} + \dfrac{x^2}{r^2}\dfrac{d^2V}{dr^2}$

Similarly, $\quad \dfrac{\partial^2 V}{\partial y^2} = \dfrac{1}{r}\dfrac{dV}{dr} - \dfrac{y^2}{r^3}\dfrac{dV}{dr} + \dfrac{y^2}{r^2}\dfrac{d^2V}{dr^2}$

and $\quad \dfrac{\partial^2 V}{\partial z^2} = \dfrac{1}{r}\dfrac{dV}{dr} - \dfrac{z^2}{r^3}\dfrac{dV}{dr} + \dfrac{z^2}{r^2}\dfrac{d^2V}{dr^2}$

On adding, we get

$$\frac{\partial^2 V}{\partial x^2} + \frac{\partial^2 V}{\partial y^2} + \frac{\partial^2 V}{\partial z^2} = \frac{3}{r}\frac{dV}{dr} - \frac{1}{r}\frac{dV}{dr} + \frac{d^2V}{dr^2} = \frac{2}{r}\frac{dV}{dr} + \frac{d^2V}{dr^2}$$

Exercise-3.4

1. Reduce the equation $\dfrac{d^2x}{dy^2} = a$ to the form $\dfrac{d^2y}{dx^2} + a\left(\dfrac{dy}{dx}\right)^3 = 0.$

2. Transform the equation $x^4\left(\dfrac{d^2y}{dx^2}\right) + a^2y = 0$ by the substitution $x = 1/z.$

3. Transform the equation $\sin^2 2z\dfrac{d^2y}{dz^2} + \sin 4z\dfrac{dy}{dz} + 4y = 0$ the substitution $\tan z = e^z.$

4. If $x^2 + z^2 = 1$, show that the equation

$\dfrac{d}{dx}\left\{(1-x^2)\dfrac{dy}{dx}\right\} + n(n+1)y = 0$ becomes

$z(z^2 - 1)\dfrac{d^2y}{dz^2} + (2z^2 - 1)\dfrac{dy}{dz} - n(n+1)zy = 0.$

5. Show that the equation $x^2\dfrac{d^2y}{dx^2} + x\dfrac{dy}{dx} + y = 0$ becomes $\dfrac{d^2y}{dz^2} + y = 0$ by substituting e^z for $x.$

6. Transform $\dfrac{d^2y}{dx^2}$ to new variables u and v by taking u as independent variable such that $y = uv, xy = 1.$

7. Show that $\dfrac{\partial^2 u}{\partial x^2} + \dfrac{\partial^2 u}{\partial y^2} = \dfrac{\partial^2 u}{\partial \xi^2} + \dfrac{\partial^2 u}{\partial \eta^2}$ where $x = \xi \cos \alpha - \eta \sin \alpha,$ $y = \xi \sin \alpha + \eta \cos \alpha.$

8. If $x = e^\theta, y = e^\phi$, show that

$e^{2\theta}\dfrac{\partial^2 v}{\partial x^2} + e^{2\phi}\dfrac{\partial^2 v}{\partial y^2} + e^\theta\dfrac{\partial v}{\partial x} + e^\phi\dfrac{\partial v}{\partial y} = \dfrac{\partial^2 v}{\partial \theta^2} + \dfrac{\partial^2 v}{\partial \phi^2}.$

9. If $f(x, y)$ has continuous partial derivatives of first two orders and $x + y = (u + v)^3, (x - y) = (u - v)^3$, then show that

$9(x^2 - y^2)\left(\dfrac{\partial^2 f}{\partial x^2} - \dfrac{\partial^2 f}{\partial y^2}\right) = (u^2 - v^2)\left\{\dfrac{\partial^2 f}{\partial u^2} - \dfrac{\partial^2 f}{\partial v^2}\right\}.$

10. If z is a function of u and v, where $u = x^2 - y^2 - 2xy$ and $v = y$, show that the equation $(x + y)\dfrac{\partial z}{\partial x} + (x - y)\left(\dfrac{\partial z}{\partial y}\right) = 0$ is transformed into $\dfrac{\partial v}{\partial z} = 0.$

11. Show that the equation $xy\left(\dfrac{\partial^2 u}{\partial x^2} - \dfrac{\partial^2 u}{\partial y^2}\right) - (x^2 - y^2)\dfrac{\partial^2 u}{\partial x \partial y} = 0$ becomes $r\cdot\dfrac{\partial^2 u}{\partial r \partial \theta} - \dfrac{\partial u}{\partial \theta} = 0$, when transformed to polar.

12. If $x = r\cos\theta, y = r\sin\theta$ and $z = f(x, y)$, show that

$\dfrac{\partial z}{\partial x} = \dfrac{\partial z}{\partial r}\cos\theta - \dfrac{1}{r}\dfrac{\partial z}{\partial \theta}\sin\theta.$

Also show that $\dfrac{\partial^2(r^n\cos n\theta)}{\partial x \partial y} = -n(n-1)r^{n-2}\sin(n-2)\theta.$

13. If $x + y = 2e^\theta\cos\phi$ and $x - y = 2e^\theta\sin\phi$, show that

$$\frac{\partial^2 V}{\partial \theta^2} + \frac{\partial^2 V}{\partial \phi^2} = 4xy\frac{\partial^2 V}{\partial x \partial y}. \qquad \text{(UPTU–2002)}$$

Answers

2. $\dfrac{d^2y}{dz^2} + \dfrac{2}{z}\dfrac{dy}{dz} + a^2y = 0$ **3.** $\dfrac{d^2y}{dx^2} + y = 0$ **6.** $4v^4\left(\dfrac{dv}{du}\right)^{-1} + 2uv^3 - v^5\left(\dfrac{d^2v}{du^2}\right)\left(\dfrac{dv}{du}\right)^{-3}$

3.3 EXPANSION OF FUNCTION OF SEVERAL VARIABLES

We know that the polynomial approximation of the function $z = f(x, y)$ is of great importance to engineering problems. We can find the approximate value of a function of several variables at a point by using mathematical tools. In this sector we shall discuss the expansion of functions of several variables by using Taylor's and Maclaurin's series.

3.3.1 TAYLOR'S THEOREM FOR TWO VARIABLES

Let $f(x, y)$ be a function of two independent variables and h, k are small increment in x and y respectively, then

$$f(x+h,y+k) = f(x,y) + \left(h\frac{\partial}{\partial x} + k\frac{\partial}{\partial y}\right)f(x,y) + \frac{1}{2!}\left[h\frac{\partial}{\partial x} + k\frac{\partial}{\partial y}\right]^2 f(x,y) + ...$$

PROOF.   Consider the function of two variables $f(x+h, y+k)$.

Expand $f(x+h, y+k)$ using Taylor's series by considering $f(x+h, y+k)$ as a function of single variable x, we get

$$f(x+h,y+k) = f(x,y+k) + h\frac{\partial f(x,y+k)}{\partial x} + \frac{h^2}{2!}\frac{\partial^2 f(x,y+k)}{\partial x^2} + ... \qquad ...(1)$$

Now expand $f(x, y+k)$ using Taylor's series by considering $f(x, y+k)$ as a function of single variable y. Then

$$f(x,y+k) = f(x,y) + k\frac{\partial f(x,y)}{\partial y} + \frac{k^2}{2!}\frac{\partial^2 f(x,y)}{\partial y^2} + ... \qquad ...(2)$$

Using (2) in (1), we get

$$f(x+h,y+k) = f(x,y) + k\frac{\partial f(x,y)}{\partial y} + \frac{k^2}{2!}\frac{\partial^2 f(x,y)}{\partial y^2} + ... + h\frac{\partial}{\partial x}\left[f(x,y) + k\frac{\partial f(x,y)}{\partial y} + \frac{k^2}{2!}\frac{\partial^2 f(x,y)}{\partial y^2} + ...\right]$$

$$+ \frac{h^2}{2!}\frac{\partial^2}{\partial x^2}\left[f(x,y) + k\frac{\partial f(x,y)}{\partial y} + \frac{k^2}{2!}\frac{\partial^2 f(x,y)}{\partial y^2} + ...\right] + ...$$

$$= f(x,y) + \left(h\frac{\partial}{\partial x} + k\frac{\partial}{\partial y}\right)f(x,y) + \frac{1}{2!}\left(h^2\frac{\partial^2}{\partial x^2} + 2hk\frac{\partial^2}{\partial x\partial y} + k^2\frac{\partial^2}{\partial y^2}\right)f(x,y) + ...$$

$$\Rightarrow f(x+h,y+k) = f(x,y) + \left(h\frac{\partial}{\partial x} + k\frac{\partial}{\partial y}\right)f(x,y) + \frac{1}{2!}\left(h\frac{\partial}{\partial x} + k\frac{\partial}{\partial y}\right)^2 f(x,y) + ... \qquad ...(3)$$

which is the required Taylor's series for functions of two variables.

DEDUCTIONS

(1) On putting $x = a$ and $y = b$ in (3), we get

$$f(a+h,b+k) = f(a,b) + hf_x(a,b) + kf_y(a,b) + \frac{1}{2!}\left[h^2 f_{xx}(a,b) + 2hk f_{xy}(a,b) + k^2 f_{yy}(a,b)\right]$$

$$+ \frac{1}{3!}\left[h^3 f_{xxx}(a,b) + 3h^2 k f_{xxy}(a,b) + 3hk^2 f_{xyy}(a,b) + k^3 f_{yyy}(a,b)\right] + ... \qquad ...(4)$$

(2) On putting $h = x - a$ and $k = y - b$ in (4), we get

$$f(x,y) = f(a,b) + \left[(x-a)f_x(a,b) + (y-b)f_y(a,b)\right]$$

$$+ \frac{1}{2!}\left[(x-a)^2 f_{xx}(a,b) + 2(x-a)(y-b)f_{xy}(a,b) + (y-b)^2 f_{yy}(a,b)\right] + ... \qquad ...(5)$$

(3) On putting $a = 0$ and $b = 0$ in (5), we get

$$f(x,y) = f(0,0) + \left[xf_x(0,0) + yf_y(0,0)\right] + \frac{1}{2!}\left[x^2 f_{xx}(0,0) + 2xy f_{xy}(0,0) + y^2 f_{yy}(0,0)\right] + ... \qquad ...(6)$$

This is called Maclaurin's series for two variables.

Solved Examples

EXAMPLE 1. *Expand $e^x \cos y$ about the point $(1, \pi/4)$.*

SOLUTION .   Let $f(x,y) = e^x \cos y$ (UPTU–2008)

By Taylor's theorem, we have

$$f(x+h,y+k) = f(x,y) + \left(h\frac{\partial}{\partial x} + k\frac{\partial}{\partial y}\right)f(x,y)$$

$$+ \frac{1}{2!}\left(h\frac{\partial}{\partial x} + k\frac{\partial}{\partial y}\right)^2 f(x,y) + ... \quad ...(1)$$

Now $e^x \cos y = f(x,y) = f\left[1 + (x-1).\frac{\pi}{4} + \left(y - \frac{\pi}{4}\right)\right]$

$$= f\left(1 + h, \frac{\pi}{4} + k\right)$$

where $h = x - 1$, $k = y - \pi/4$

Therefore, $f(x,y) = e^x \cos y \Rightarrow f\left(1, \frac{\pi}{4}\right) = e.\frac{1}{\sqrt{2}}$

$$\frac{\partial f}{\partial x} = e^x \cos y \Rightarrow \left(\frac{\partial f}{\partial x}\right)_{(1,\pi/4)} = \frac{e}{\sqrt{2}}$$

$$\frac{\partial f}{\partial y} = -e^x \sin y \Rightarrow \left(\frac{\partial f}{\partial y}\right)_{(1,\pi/4)} = -\frac{e}{\sqrt{2}}$$

$$\frac{\partial^2 f}{\partial x^2} = e^x \cos y \Rightarrow \left(\frac{\partial^2 f}{\partial x^2}\right)_{\left(1,\frac{\pi}{4}\right)} = \frac{e}{\sqrt{2}}$$

$$\frac{\partial^2 f}{\partial y^2} = -e^x \cos y \Rightarrow \left(\frac{\partial^2 f}{\partial y^2}\right)_{\left(1,\frac{\pi}{4}\right)} = -\frac{e}{\sqrt{2}}$$

$$\frac{\partial^2 f}{\partial x\partial y} = -e^x \sin y \Rightarrow \left(\frac{\partial^2 f}{\partial x\partial y}\right)_{\left(1,\frac{\pi}{4}\right)} = -\frac{e}{\sqrt{2}}$$

Putting all these values in (1), we get $e^x \cos y$

$$= \frac{e}{\sqrt{2}} + \left[(x-1)\frac{e}{\sqrt{2}} + \left(y - \frac{\pi}{4}\right)\left(-\frac{e}{\sqrt{2}}\right)\right] + \frac{1}{2!}\left[(x-1)^2 \frac{e}{\sqrt{2}}\right.$$

$$\left. + 2(x-1)\left(y - \frac{\pi}{4}\right)\left(-\frac{e}{\sqrt{2}}\right) + \left(y - \frac{\pi}{4}\right)^2\left(-\frac{e}{\sqrt{2}}\right)\right] + ...$$

$$= \frac{e}{\sqrt{2}} + \left[1 + (x-1) + \left(y - \frac{\pi}{4}\right) + \frac{(x-1)^2}{2}\right.$$

$$\left. + (x-1)\left(y - \frac{\pi}{4}\right) + \ldots\right]$$

EXAMPLE 2. *Expand* $\tan^{-1}\frac{y}{x}$ *in the neighbourhood of (1, 1) upto and inclusive of second degree term. Hence, compute f(1.1, 0.9) approximately.*

(UPTU–2005,06,07, GBTU–2010, UKTU–2011,12, VTU–2010, JNTU–2006)

SOLUTION. We have

$$f(x,y) = \tan^{-1}\frac{y}{x} \quad \Rightarrow \quad f(1,1) = \tan^{-1}1 = \frac{\pi}{4}$$

$$\frac{\partial f}{\partial x} = \frac{1}{1 + \frac{y^2}{x^2}}\left(-\frac{y}{x^2}\right) = -\frac{y}{x^2 + y^2}$$

$$\Rightarrow \quad \left(\frac{\partial f}{\partial x}\right)_{(1,1)} = -\frac{1}{2}$$

$$\frac{\partial f}{\partial y} = \frac{1}{1 + \frac{y^2}{x^2}} \cdot \frac{1}{x} = \frac{x}{x^2 + y^2}$$

$$\Rightarrow \quad \left(\frac{\partial f}{\partial y}\right)_{(1,1)} = \frac{1}{2}$$

$$\frac{\partial^2 f}{\partial x^2} = -y(-1)(x^2 + y^2)^{-2}.2x = \frac{2xy}{(x^2 + y^2)^2}$$

$$\Rightarrow \quad \left(\frac{\partial^2 f}{\partial x^2}\right)_{(1,1)} = \frac{1}{2}$$

$$\frac{\partial^2 f}{\partial x \partial y} = \frac{(x^2 + y^2) - x.2x}{(x^2 + y^2)^2} = \frac{y^2 - x^2}{(x^2 + y^2)^2}$$

$$\Rightarrow \quad \left(\frac{\partial^2 f}{\partial x \partial y}\right)_{(1,1)} = 0$$

$$\frac{\partial^2 f}{\partial y^2} = x(-1)(x^2 + y^2)^{-2}.2y = -\frac{2xy}{(x^2 + y^2)^2}$$

$$\Rightarrow \quad \left(\frac{\partial^2 f}{\partial y^2}\right)_{(1,1)} = -\frac{1}{2}$$

Similarly, we may get

$$\left(\frac{\partial^3 f}{\partial x^3}\right)_{(1,1)} = -\frac{1}{2}, \left(\frac{\partial^3 f}{\partial x^2 \partial y}\right)_{(1,1)} = -\frac{1}{2}$$

$$\left(\frac{\partial^3 f}{\partial x \partial y^2}\right)_{(1,1)} = \frac{1}{2}$$

We know that

$$f(x,y) = f(1,1) + [(x-1)f_x(1,1) + (y-1)f_y(1,1)]$$

$$+ \frac{1}{2!}\Big[(x-1)^2 f_{xx}(1,1)$$

$$+ 2(x-1)(y-1)f_{xy}(1,1) + (y-1)^2 f_{yy}(1,1)\Big]$$

$$+ \frac{1}{3!}\Big[(x-1)^3 f_{xxx}(1,1) + 3(x-1)^2(y-1)f_{xxy}(1,1)$$

$$+ 3(x-1)(y-1)^2 f_{xyy}(1,1) + (y-1)^3 f_{yyy}(1,1)\Big] + \ldots$$

Hence $\tan^{-1}\frac{y}{x}$

$$= \frac{\pi}{4} + \left[(x-1)\left(-\frac{1}{2}\right) + (y-1)\frac{1}{2}\right]$$

$$+ \frac{1}{2!}\left[(x-1)^2\frac{1}{2} + 2(x-1)(y-1).0 + (y-1)^2\left(-\frac{1}{2}\right)\right]$$

$$- \frac{1}{12}[(x-1)^3 + 3(x-1)^2(y-1) - 3(x-1)(y-1)^2 - (y-1)^3] + \ldots$$

$$= \frac{\pi}{4} - \frac{1}{2}(x-1) + \frac{1}{2}(y-1) + \frac{1}{4}(x-1)^2$$

$$- \frac{1}{4}(y-1)^2 - \frac{1}{12}[(x-1)^3 + 3(x-1)^2(y-1)$$

$$- 3(x-1)(y-1)^2 - (y-1)^3] + \ldots$$

Further, $f(1.1, 0.9)$

$$= \frac{\pi}{4} - \frac{1}{2}(.1) + \frac{1}{2}(-.1) + \frac{1}{4}(.1)^2 - \frac{1}{4}(-.1)^2$$

$$- \frac{1}{12}[(.1)^3 + 3(.1)^2(-.1) - 3(.1)(-.1)^2 - (-.1)^3] + \ldots$$

$$= \frac{\pi}{4} - \frac{(.2)}{2} - \frac{1}{12}[.001 - .003 - .003 + .001] + \ldots$$

$$\approx 0.6857$$

EXAMPLE 3. *Find Taylor's series expansion of function on* $f(x,y) = e^{-x^2 - y^2}\cos xy$ *about the point* $x_0 = 0, y_0 = 0$ *upto three terms.*   (UPTU–2007)

SOLUTION. We have $f(x,y) = e^{-x^2 - y^2}\cos xy \quad \Rightarrow \quad f(0,0) = 1$

$$\frac{\partial f}{\partial x} = -e^{-x^2 - y^2} y \sin xy + e^{-x^2 - y^2}(-2x)\cos xy$$

$$\Rightarrow \quad \left(\frac{\partial f}{\partial x}\right)_{(0,0)} = 0$$

and $\quad \frac{\partial f}{\partial y} = -xe^{-x^2 - y^2}\sin xy - 2ye^{-x^2 - y^2}\cos xy$

$$\Rightarrow \quad \left(\frac{\partial f}{\partial y}\right)_{(0,0)} = 0$$

$$\frac{\partial^2 f}{\partial x^2} = -y\left[e^{-x^2 - y^2}y\cos xy - 2xe^{-x^2 - y^2}\sin xy\right]$$

$$- 2\left[e^{-x^2 - y^2}\cos xy\right]$$

$$- 2x\left[-ye^{-x^2 - y^2}\sin xy - 2xe^{-x^2 - y^2}\cos xy\right]$$

$$\Rightarrow \quad \left(\frac{\partial^2 f}{\partial x^2}\right)_{(0,0)} = -2$$

$$\frac{\partial^2 f}{\partial y^2} = -x\left[-2ye^{-x^2 - y^2}\sin xy + e^{-x^2 - y^2}x\cos xy\right]$$

$$- 2e^{-x^2 - y^2}\cos xy$$

$$- 2y\left[-2ye^{-x^2 - y^2}\cos xy - e^{-x^2 - y^2}x\sin xy\right]$$

$$\Rightarrow \quad \left(\frac{\partial^2 f}{\partial y^2}\right)_{(0,0)} = -2$$

$$\frac{\partial^2 f}{\partial x \partial y} = e^{-x^2 - y^2}\sin xy$$

$$- x\left[-2xe^{-x^2 - y^2}\sin xy + e^{-x^2 - y^2}y\cos xy\right]$$

$$- 2y\left[-2xe^{-x^2 - y^2}\cos xy - e^{-x^2 - y^2}y\sin xy\right]$$

$$\Rightarrow \left(\frac{\partial^2 f}{\partial x \partial y}\right)_{(0,0)} = 0$$

Putting all these values in Taylor's theorem

$$f(x,y) = f(0,0) + \left(x\frac{\partial}{\partial x} + y\frac{\partial}{\partial y}\right)f(0,0)$$

$$+ \frac{1}{2!}\left(x\frac{\partial}{\partial x} + y\frac{\partial}{\partial y}\right)^2 f(0,0) + \ldots$$

we get $f(x,y) = 1 + 0 + \frac{1}{2!}[x(-2) + 2xy(0) + y(-2)] + \ldots$

$$= 1 - (x + y) + \ldots$$

EXAMPLE 4. *Expand $x^2 y + 3y - 2$ in powers of $(x - 1)$ and $(y + 2)$ using Taylor's theorem.*

(PTU–2010, VTU–2008, ANNA–2005, UPTU(AG)–2006, AMIETE–2003, AMIE–1999)

SOLUTION. We have $f(x,y) = x^2 y + 3y - 2$

$a + h = x$ and $h = x - 1 \Rightarrow a = 1$
$b + k = y$ and $k = y + 2 \Rightarrow b = -2$

Now

$f(x,y) = x^2 y + 3y - 2$	$\Rightarrow$	$f(1,-2) = -10$
$f_x(x,y) = 2xy$	$\Rightarrow$	$f_x(1,-2) = -4$
$f_y(x,y) = x^2 + 3$	$\Rightarrow$	$f_y(1,-2) = 4$
$f_{xx}(x,y) = 2y$	$\Rightarrow$	$f_{xx}(1,-2) = -4$
$f_{xy}(x,y) = 2x$	$\Rightarrow$	$f_{xy}(1,-2) = 2$
$f_{yy}(x,y) = 0$	$\Rightarrow$	$f_{yy}(1,-2) = 0$
$f_{xxx}(x,y) = 0$	$\Rightarrow$	$f_{xxx}(1,-2) = 0$
$f_{xxy}(x,y) = 2$	$\Rightarrow$	$f_{xxy}(1,-2) = 2$
$f_{xyy}(x,y) = 0$	$\Rightarrow$	$f_{xyy}(1,-2) = 0$
$f_{yyy}(x,y) = 0$	$\Rightarrow$	$f_{yyy}(1,-2) = 0$

Putting all these values in Taylor's theorem given by

$$f(a+h, b+k) = f(a,b) + \left(h\frac{\partial f}{\partial x} + k\frac{\partial f}{\partial y}\right)_{(a,b)}$$

$$+ \frac{1}{2!}\left(h^2\frac{\partial^2 f}{\partial x^2} + 2hk\frac{\partial^2 f}{\partial x \partial y} + k^2\frac{\partial^2 f}{\partial y^2}\right)_{(a,b)}$$

$$+ \frac{1}{3!}\left(h^3\frac{\partial^3 f}{\partial x^3} + 3h^2 k\frac{\partial f}{\partial x^2 \partial y}\right.$$

$$\left. + 3hk^2\frac{\partial^3 f}{\partial x \partial y^2} + k^3\frac{\partial^3 f}{\partial y^3}\right)_{(a,b)} + \ldots$$

We get $x^2 y + 3y - 2$

$$= -10 + \big((x-1)(-4) + (y+2)4\big)$$

$$+ \frac{1}{2!}\big((x-1)^2(-4) + 2(x-1)y(y+2)2 + (y+2)^2(0)\big)$$

$$+ \frac{1}{3!}\big((x-1)^3 \cdot 0 + 3(x-1)^2(y+2) \cdot 2$$

$$+ 3(x-1)(y+2)^2 \cdot 0 + (y+2)^3 \cdot 0\big) + \ldots$$

$$= -10 - 4(x-1) + 4(y+2) - 2(x-1)^2$$

$$+ 2(x-1)(y+2) + (x-1)^2(y+2)$$

EXAMPLE 5. *Expand $e^x \sin y$ in the powers of x and y in the neighbourhood of $(0, \pi/4)$ upto the third degree terms.*

(ANNA–2009)

SOLUTION. We have

$f(x,y) = e^x \sin y$	$\Rightarrow$	$f\left(0,\frac{\pi}{4}\right) = \frac{1}{\sqrt{2}}$
$f_x(x,y) = e^x \sin y$	$\Rightarrow$	$f_x\left(0,\frac{\pi}{4}\right) = \frac{1}{\sqrt{2}}$
$f_y(x,y) = e^x \cos y$	$\Rightarrow$	$f_y\left(0,\frac{\pi}{4}\right) = \frac{1}{\sqrt{2}}$
$f_{xx}(x,y) = e^x \sin y$	$\Rightarrow$	$f_{xx}\left(0,\frac{\pi}{4}\right) = \frac{1}{\sqrt{2}}$
$f_{xy}(x,y) = e^x \cos y$	$\Rightarrow$	$f_{xy}\left(0,\frac{\pi}{4}\right) = \frac{1}{\sqrt{2}}$
$f_{yy}(x,y) = -e^x \sin y$	$\Rightarrow$	$f_{yy}\left(0,\frac{\pi}{4}\right) = -\frac{1}{\sqrt{2}}$
$f_{xxx}(x,y) = e^x \sin y$	$\Rightarrow$	$f_{xxx}\left(0,\frac{\pi}{4}\right) = \frac{1}{\sqrt{2}}$
$f_{xxy}(x,y) = e^x \cos y$	$\Rightarrow$	$f_{xxy}\left(0,\frac{\pi}{4}\right) = \frac{1}{\sqrt{2}}$
$f_{xyy}(x,y) = -e^x \sin y$	$\Rightarrow$	$f_{xyy}\left(0,\frac{\pi}{4}\right) = -\frac{1}{\sqrt{2}}$
$f_{yyy}(x,y) = -e^x \cos y$	$\Rightarrow$	$f_{yyy}\left(0,\frac{\pi}{4}\right) = -\frac{1}{\sqrt{2}}$

By Taylor's theorem, we get

$$f(x,y) = f\left(0,\frac{\pi}{4}\right) + \left(x\frac{\partial}{\partial x} + \left(y - \frac{\pi}{4}\right)\frac{\partial}{\partial y}\right)f\left(0,\frac{\pi}{4}\right)$$

$$+ \frac{1}{2!}\left(x\frac{\partial}{\partial x} + \left(y - \frac{\pi}{4}\right)\frac{\partial}{\partial y}\right)^2 f\left(0,\frac{\pi}{4}\right)$$

$$+ \frac{1}{3!}\left(x\frac{\partial}{\partial x} + \left(y - \frac{\pi}{4}\right)\frac{\partial}{\partial y}\right)^3 f\left(0,\frac{\pi}{4}\right) + \ldots \quad \ldots(1)$$

Putting all the above values in (1), we get

$$f(x,y) = \frac{1}{\sqrt{2}} + x \cdot \frac{1}{\sqrt{2}} + \left(y - \frac{\pi}{4}\right)\frac{1}{\sqrt{2}}$$

$$+ \frac{1}{2!}\left[x^2\frac{1}{\sqrt{2}} + 2x\left(y - \frac{\pi}{4}\right)\frac{1}{\sqrt{2}} - \left(y - \frac{\pi}{4}\right)^2\frac{1}{\sqrt{2}}\right]$$

$$+ \frac{1}{3!}\left[x^3\frac{1}{\sqrt{2}} + 3x^2\left(y - \frac{\pi}{4}\right)\frac{1}{\sqrt{2}}\right.$$

$$\left. + 3x\left(y - \frac{\pi}{4}\right)^2\left(-\frac{1}{\sqrt{2}}\right) + \left(y - \frac{\pi}{4}\right)^3\left(-\frac{1}{\sqrt{2}}\right)\right] + \ldots$$

$$= \frac{1}{\sqrt{2}}\left[1 + x + \left(y - \frac{\pi}{4}\right) + \frac{1}{2!}\left\{x^2 + 2x\left(y - \frac{\pi}{4}\right)\right.\right.$$

$$\left.- \left(y - \frac{\pi}{4}\right)^2\right\} + \frac{1}{3!}\left\{x^3 + 3x^2\left(y - \frac{\pi}{4}\right)\right.$$

$$\left.\left. - 3x\left(y - \frac{\pi}{4}\right)^2 - \left(y - \frac{\pi}{4}\right)^3 - \right\} + \ldots\right]$$

EXAMPLE 6. *Expand e^x log $(1 + y)$ in the neighbourhood of the point (0, 0).*   (VTU–2010, PTU–2009, JNTU–2006)

SOLUTION . We have

$$f(x, y) = e^x \log(1+y) \Rightarrow f(0, 0) = 0$$

$$f_x(x, y) = e^x \log(1+y) \Rightarrow f_x(0, 0) = 0$$

$$f_y(x, y) = \frac{e^x}{1+y} \qquad \Rightarrow f_y(0, 0) = 1$$

$$f_{xx}(x, y) = e^x \log(1+y) \Rightarrow f_{xx}(0, 0) = 0$$

$$f_{xy}(x, y) = \frac{e^x}{1+y} \qquad \Rightarrow f_{xy}(0, 0) = 1$$

$$f_{yy}(x, y) = -\frac{e^x}{(1+y)^2} \qquad \Rightarrow f_{yy}(0, 0) = -1$$

$$f_{xxx}(x, y) = e^x \log(1+y) \Rightarrow f_{xxx}(0, 0) = 0$$

$$f_{xxy}(x, y) = \frac{e^x}{1+y} \qquad \Rightarrow f_{xxy}(0, 0) = 1$$

$$f_{xyy}(x, y) = -\frac{e^x}{(1+y)^2} \qquad \Rightarrow f_{xyy}(0, 0) = -1$$

$$f_{yyy}(x, y) = \frac{2e^x}{(1+y)^3} \qquad \Rightarrow f_{yyy}(0, 0) = 2$$

Putting all these values in Taylor's series given by

$$f(x, y) = f(0,0) + [xf_x(0,0) + yf_y(0,0)]$$
$$+ \frac{1}{2!}\Big[x^2 f_{xx}(0,0) + 2xy f_{xy}(0,0) + y^2 f_{yy}(0,0)\Big]$$
$$+ \frac{1}{3!}\Big[x^3 f_{xxx}(0,0) + 3x^2 y f_{xxy}(0,0)$$
$$+ 3xy^2 f_{xyy}(0,0) + y^3 f_{yyy}(0,0)\Big] + ...$$

we get $e^x \log (1+y)$

$$= 0 + [x.0 + y.1] + \frac{1}{2}\Big[x^2.0 + 2xy.1 + y^2(-1)\Big]$$
$$+ \frac{1}{6}\Big[x^3.0 + 3x^2y.1 + 3xy^2(-1) + y^3.2\Big] + ...$$
$$= y + xy - \frac{1}{2}y^2 + \frac{1}{2}x^2y - \frac{1}{2}xy^2 + \frac{1}{3}y^3 + ...$$

EXAMPLE 7. *Expand x^y in powers of $(x-1)$ and $(y-1)$ upto the third degree terms.* (UPTU–2004, UKTU–2010, VTU–2009)

SOLUTION . We have

$$f(x, y) = x^y \qquad\qquad \Rightarrow \quad f(1, 1) = 1$$
$$f_x(x, y) = yx^{y-1} \qquad\qquad \Rightarrow \quad f_x(1, 1) = 1$$
$$f_y(x, y) = x^y \log x \qquad\qquad \Rightarrow \quad f_y(1, 1) = 0$$
$$f_{xx}(x, y) = y(y-1)x^{y-2} \qquad \Rightarrow f_{xx}(1, 1) = 0$$
$$f_{xy}(x, y) = x^{y-1} + yx^{y-1}\log x \quad \Rightarrow f_{xy}(1, 1) = 1$$
$$f_{yy}(x, y) = x^y(\log x)^2 \qquad\quad \Rightarrow f_{yy}(1, 1) = 0$$
$$f_{xxx}(x, y) = y(y-1)(y-2)x^{y-3} \Rightarrow f_{yyy}(1, 1) = 0$$
$$f_{xxy}(x, y) = (y-1)x^{y-2} + y(y-1)x^{y-2}\log_e x + yx^{y-2}$$
$$\Rightarrow f_{xxy}(1, 1) = 1$$

$$f_{xyy}(x, y) = yx^{y-1}(\log x)^2 + x^4.\frac{2\log x}{x}$$
$$= yx^{y-1}(\log x)^2 + 2x^{y-1}\log x$$

$$\Rightarrow \quad f_{xyy}(1, 1) = 0$$
$$f_{yyy}(x, y) = x^4(\log x)^3$$
$$\Rightarrow \quad f_{yyy}(1, 1) = 0$$

Putting all these values in Taylor's series given by $f(x, y)$

$$f(x, y) = f_x(a,b) + (x-a)f_x(a,b) + (y-b)f_y(a,b)$$
$$+ \frac{1}{2!}\Big[(x-a)^2 f_{xx}(a,b) + 2(x-a)$$
$$(y-b)f_{xy}(a,b) + (y-b)^2 f_{yy}(a,b)\Big]$$
$$+ \frac{1}{3!}\Big[(x-a)^3 f_{xxx}(a,b) + 3(x-a)^2(y-b)f_{xxy}(a,b)$$
$$+ 3(x-a)(y-b)^2 f_{xyy}(a,b) + (y-b)^3 f_{yyy}(a,b)\Big] + ...$$

we get $f(x, y) = x^y$

$$= 1 + (x-1).1 + (y-1).0 + \frac{1}{2!}\Big[(x-1)^2.0$$
$$+ 2(x-1)(y-1)^2.1 + (y-1)^2.0\Big]$$
$$+ \frac{1}{3!}\Big[(x-1)^3.0 + 3(x-1)^2(y-1).1$$
$$+ 3(x-1)(y-1)^2.0 + (y-1)^3.0\Big] + ...$$
$$= 1 + (x-1) + (x-1)(y-1) + \frac{1}{2}(x-1)^2(y-1) + ...$$

EXAMPLE 8. *Expand $\frac{(x+h)(y+k)}{x+h+y+k}$ in powers of h and k upto and inclusive of the second degree terms.*
(AMIE–1999, 2001)

SOLUTION . We have $f(x+h, y+k) = \frac{(x+h)(y+k)}{x+h+y+k}$

Putting $h = k = 0 \Rightarrow f(x, y) = \frac{xy}{x+y}$

Now $f_x(x, y) = \frac{(x+y).y - xy.1}{(x+y)^2} = \frac{y^2}{(x+y)^2}$,

$$f_y(x, y) = \frac{x^2}{(x+y)^2}, f_{xx}(x, y) = -\frac{2y^2}{(x+y)^3},$$

$$f_{yy}(x, y) = -\frac{2x^2}{(x+y)^3},$$

$$f_{xy}(x, y) = \frac{(x+y)^2.2x - x^2.2(x+y)}{(x+y)^4} = \frac{2xy}{(x+y)^3}$$

Therefore,

$$\frac{(x+h)(y+k)}{x+h+y+k} = f(x+h, y+k)$$
$$= f(x, y) + (hf_x + kf_y)$$
$$+ \frac{1}{2!}\Big[h^2 f_{xx} + 2hk f_{xy} + k^2 f_{yy}\Big] + ...$$
$$= \frac{xy}{x+y} + \Big[h.\frac{y^2}{(x+y)^2} + k\frac{x^2}{(x+y)^2}\Big]$$
$$+ \frac{1}{2}\Big[h^2 \frac{(-2y^2)}{(x+y)^3} + 2hk \frac{2xy}{(x+y)^3} + k^2 \frac{(-2x^2)}{(x+y)^3}\Big] + ...$$
$$= \frac{xy}{x+y} + \frac{y^2}{(x+y)^2}.h + \frac{x^2}{(x+y)^2}.k$$
$$- \frac{y^2}{(x+y)^3}h^2 + 2xy.hk - \frac{x^2}{(x+y)^3}k^2 + ...$$

EXAMPLE 9. *Expand e^{ax} sin by in powers of x and y upto third degree term.*   (GBTU–2011)

SOLUTION . Since, the point is not given, so we expand it about the point (0, 0).

Let

$$f(x, y) = e^{ax} \sin by \quad \Rightarrow \quad f(0, 0) = 0$$

$$f_x(x, y) = ae^{ax} \sin by \quad \Rightarrow \quad f_x(0, 0) = 0$$

$$f_{xx}(x, y) = a^2 e^{ax} \sin by \quad \Rightarrow \quad f_{xx}(0, 0) = 0$$

$$f_{xxx}(x, y) = a^3 e^{ax} \sin by \quad \Rightarrow \quad f_{xxx}(0, 0) = 0$$

$$f_y(x, y) = be^{ax} \cos by \quad \Rightarrow \quad f_y(0, 0) = b$$

$$f_{yy}(x, y) = -b^2 e^{ax} \sin by \quad \Rightarrow \quad f_{yy}(0, 0) = 0$$

$$f_{yyy}(x, y) = -b^3 e^{ax} \cos by \quad \Rightarrow f_{yyy}(0, 0) = -b^3$$

$$f_{xy}(x, y) = abe^{ax} \cos by \quad \Rightarrow \quad f_{xy}(0, 0) = ab$$

$$f_{xxy}(x, y) = a^2 be^{ax} \cos by \quad \Rightarrow f_{xxy}(0, 0) = a^2 b$$

$$f_{xyy}(x, y) = -ab^2 e^{ax} \sin by \quad \Rightarrow f_{xyy}(0, 0) = 0$$

Putting all these values in Maclaurin's theorem given by

$$f(x, y)$$
$$= f(0, 0) + [xf_x(0, 0) + yf_y(0, 0)]$$
$$+ \frac{1}{2!}\left[x^2 f_{xx}(0, 0) + 2xyf_{xy}(0, 0) + y^2 f_{yy}(0, 0)\right]$$
$$+ \frac{1}{3!}\left[x^3 f_{xxx}(0, 0) + 3x^2 yf_{xxy}(0, 0)\right.$$
$$\left. + 3xy^2 f_{xyy}(0, 0) + y^3 f_{yyy}(0, 0)\right] + ...$$

We get $e^{ax} \sin by$

$$= by + \frac{1}{2!}ab(2xy) + \frac{1}{3!}\left[3x^2 y(a^2 b - b^3 y^3)\right] + ...$$

$$= by + abxy + \frac{1}{2}a^2 bx^2 y - \frac{b^3}{6}y^3 + ...$$

EXAMPLE 10. *Find the Taylor's series expansion of* $f(x, y) = x^3 + xy^2$ *about the point (2, 1).* (UPTU–2012)

SOLUTION. We have $f(x, y) = x^3 + xy^2$
$$a = 2, b = 1$$

Let

$$f(x, y) = x^3 + xy^2 \quad \Rightarrow \quad f(2, 1) = 10$$

$$f_x(x, y) = 3x^2 + y^2 \quad \Rightarrow \quad f_x(2, 1) = 13$$

$$f_{xx}(x, y) = 6x \quad \Rightarrow \quad f_{xx}(2, 1) = 12$$

$$f_{xxx}(x, y) = 0 \quad \Rightarrow \quad f_{xxx}(2, 1) = 0$$

$$f_y(x, y) = 2xy \quad \Rightarrow \quad f_y(2, 1) = 4$$

$$f_{yy}(x, y) = 2x \quad \Rightarrow \quad f_{yy}(2, 1) = 4$$

$$f_{yyy}(x, y) = 0 \quad \Rightarrow \quad f_{yyy}(2, 1) = 0$$

$$f_{xy}(x, y) = 2y \quad \Rightarrow \quad f_{xy}(2, 1) = 2$$

Putting all these values in Taylor's series expansion given by

$$f(x, y) = f(a, b) + \left[(x - a)f_x(a, b) + (y - b)f_y(a, b)\right]$$
$$+ \frac{1}{2!}\left[(x - a)^2 f_{xx}(a, b) + (y - b)^2 f_{yy}(a, b)\right.$$
$$\left. + 2(x - a)(y - b)f_{xy}(a, b)\right] + ...$$

we get $f(x, y) = 10 + \left[(x - 2).13 + (y - 1).4\right]$
$$+ \frac{1}{2}\left[(x - 2)^2.12 + (y - 1)^2.4\right.$$
$$\left. + 2(x - 2)(y - 1).2\right] + ...$$

Hence, $x^3 + xy^2 = 10 + 13(x - 2) + 4(y - 1) + 6(x - 2)^2$
$$+ 2(y - 1)^2 + 2(x - 2)(y - 1) + ...$$

EXAMPLE 11. *If $f(x, y) = \tan^{-1}xy$, compute an approximate value of $f(0.9, -1.2)$* (AMIE–1996, 2007)

SOLUTION. Let $f(x, y) = \tan^{-1}xy$

Let us expand it near the point $(1, -1)$

$$f(0.9, -1.2) = f(1 - 0.1, -1 - 0.2)$$
$$= f(1, -1) + \left[(-0.1)\frac{\partial f}{\partial x} + (-0.2)\frac{\partial f}{\partial y}\right]$$
$$+ \frac{1}{2!}\left[(-0.1)^2\frac{\partial^2 f}{\partial x^2} + 2(-0.1)(-0.2)\frac{\partial^2 f}{\partial x \partial y} + (0.2)^2\frac{\partial^2 f}{\partial y^2}\right] + ...$$
$$...(1)$$

Now $f(x, y) = \tan^{-1}xy$

$$\Rightarrow \quad f(1, -1) = \tan^{-1}(-1) = -\frac{\pi}{4}$$

$$f_x(x, y) = \frac{y}{1 + x^2 y^2}$$

$$\Rightarrow f_x(1, -1) = \frac{-1}{1 + 1} = -\frac{1}{2}$$

$$f_y(x, y) = \frac{x}{1 + x^2 y^2}$$

$$\Rightarrow f_y(1, -1) = \frac{1}{1 + 1} = \frac{1}{2}$$

$$f_{xx}(x, y) = -\frac{2x.y}{(1 + x^2 y^2)^2}$$

$$\Rightarrow f_{xx}(1, -1) = \frac{(-2)(-1)}{(1 + 1)^2} = \frac{1}{2}$$

$$f_{xy}(x, y) = \frac{1 - x^2 y^2}{(1 + x^2 y^2)^2} \quad \Rightarrow f_{xy}(1, -1) = 0$$

$$f_{yy}(x, y) = \frac{-x(2x^2 y)}{(1 + x^2 y^2)^2}$$

$$\Rightarrow f_{yy}(1, -1) = \frac{2}{(1 + 1)^2} = \frac{1}{2}$$

Putting all these values in (1), we get

$$f(0.9, -1.2)$$
$$= -\frac{\pi}{4} + (-0.1)\left(-\frac{1}{2}\right) + (-0.2)\left(\frac{1}{2}\right)$$
$$+ \frac{1}{2}\left[(-0.1)^2\left(\frac{1}{2}\right) + 2(-0.1)(-0.2).0 + (-0.2)^2\left(\frac{1}{2}\right)\right] + ...$$
$$= -\frac{\pi}{4} + 0.05 - 0.1 + \frac{1}{2}(0.005 + 0.02) + ...$$
$$= -\frac{\pi}{4} + 0.05 - 0.1 + 0.0125 = -0.823$$

 **Exercise-3.5**

1. Expand $(1 + x + y^2)^{1/2}$ at $(1, 0)$.

2. Show that $e^y \log(1 + x) = x + xy - \dfrac{x^2}{2} + ...$

3. Expand $\sin xy$ in powers of $(x - 1)$ and $\left(y - \dfrac{\pi}{2}\right)$ upto the second degree terms. (UPTU–2008, UKTU–2011)

5. Expand $\sin y$ about the origin upto third order terms.

6. Expand $x^2 + 3y^2 - 9x - 9y + 26$ in powers of $(x - 2)$ and $(y - 2)$ using Taylor's series expansion.

7. Evaluate $\log[(1.03)^{1/3} + (0.98)^{1/4} - 1]$ approximately using Taylor's series expansion.

Answers

1. $\sqrt{2}\left[1 + \dfrac{x-1}{4} - \dfrac{(x-1)^2}{32} + \dfrac{y^2}{4} + ...\right]$

3. $1 - \dfrac{\pi^2}{8}(x-1)^2 - \dfrac{\pi}{2}(x-1)\left(y - \dfrac{\pi}{2}\right) - \dfrac{1}{2}\left(y - \dfrac{\pi}{2}\right)^2$

4. $xy + ...$

6. $6 - 5(x-2) + 3(y-2) + (x-2)^2 + 3(y-2)^2 + ...$

7. 0.005

3.4 JACOBIAN

Sometimes a function is not exclusively defined in terms of the independent variable and we assumed that a functional equation $f(x,y) = 0$ gives y as a function of x. Some times y cannot be expressed in terms of x then we say that function is implicit.

Definition. *Let $f(x,y)$ be a function of two variables and $y = g(x)$ be a function of x such that for every value of x for which $g(x)$ is defined, $f(x,g(x))$ vanish identically, i.e, $y = g(x)$ is a root of the equation $f(x,y) = 0$. Then $y = g(x)$ is called the implicit function defined by the functional equation $f(x,y) = 0$.*

REMARK

- A functional equation in general may or may not define an implicit function.

3.4.1 EXISTENCE AND DERIVABILITY OF IMPLICIT FUNCTIONS

If $f_x(x,y), f_y(x,y)$ are continuous in a nbd of (a,b) and $f(a,b) \neq 0$ then $\exists$ a rectangle $R : [a–h, a+h, b–k, b+k]$ about (a,b) such that

(i) for each $x \in [a–h, a+h]$, the equation $f(x,y) = 0$ determine unique solution $y = g(x)$ in $(b–k, b+k)$

and (ii) $g'(x)$ is continuous in $[a–h, a+h]$ and $f_y(x,g(x)) \neq 0$ and $g'(x) = \dfrac{-f_x(x, g(x))}{f_y(x, g(x))}$.

GENERAL CASE

If $f(x_1, x_2, ..., x_n; y)$ be a function of $(n+1)$ variables $x_1, x_2, ..., x_n, y$ and $(a_1, a_2, ..., a_n, b)$ be a point of its domain such that

(i) $f(a_1, a_2, ..., a_n, b) = 0$

(ii) the partial derivatives w.r.t. all $(n+1)$ variables exists and are continuous in a nbd of $(a_1, a_2, ..., a_n, b)$

(iii) $f_y(a_1, a_2, ..., a_n, b) \neq 0$

then $\exists$ a nbd $(a_1–h_1, a_1+h_1; a_2–h_2, a_2+h_2; ...; a_n–h_n, a_n+h_n; b–k, b+k)$ of $(a_1, a_2, ..., a_n, b)$ such that for every point $(x_1, x_2, ..., x_n)$ of the nbd $R:(a_1–h_1, a_1+h_1; a_2–h_2, a_2+h_2; ...; a_n–h_n, a_n+h_n)$.

The equation $f(x_1, x_2, ..., x_n, y) = 0$ gives only one value $y = g(x_1, x_2, ..., x_n)$ in $[b–k, b+k]$ satisfying the following conditions :

(i) $b = g(a_1, a_2, ..., a_n)$

(ii) $f(x_1, x_2, ..., x_n, g) = 0$ for every point $(x_1, x_2, ..., x_n)$ in R.

(iii) g is continuous and having continuous first order partial derivatives w.r.t. $x_1, x_2, ... x_n$ in R.

3.4.2 DERIVATIVE OF IMPLICIT FUNCTION

If the equation $f(x,y) = 0$ defines y as a function of x. Then derivative $\dfrac{dy}{dx}$ can be obtained simply by differentiating the equation w.r.t. x assuming y as a function of $g(x)$.

Therefore, $$f_x + f_y \frac{dy}{dx} = 0$$

$$f_{xx} + f_{xy}\frac{dy}{dx} + \left(f_{xy} + f_{yy}\frac{dy}{dx}\right)\frac{dy}{dx} + f_y \frac{d^2y}{dx^2} = 0$$

or $$f_{xx} + 2f_{xy}\frac{dy}{dx} + f_{yy}\left(\frac{dy}{dx}\right)^2 + f_y \frac{d^2y}{dx^2} = 0 \quad \text{provided } f_y \neq 0$$

In a similar manner, we may find the other higher order derivatives.

DEFINITIONS AND PROPERTIES

Here we shall discuss some important definitions related to Jacobian and its properties.

(i) If u and v are the functions of two independent variables x and y, then the determinant

$$\begin{vmatrix} \dfrac{\partial u}{\partial x} & \dfrac{\partial u}{\partial y} \\[2mm] \dfrac{\partial v}{\partial x} & \dfrac{\partial v}{\partial y} \end{vmatrix}$$

is called the Jacobian of u and v with respect to x and y. It is denoted by $\dfrac{\partial(u,v)}{\partial(x,y)}$ or $J(u,v)$.

i.e.,
$$\dfrac{\partial(u,v)}{\partial(x,y)} = \begin{vmatrix} \dfrac{\partial u}{\partial x} & \dfrac{\partial u}{\partial y} \\[2mm] \dfrac{\partial v}{\partial x} & \dfrac{\partial v}{\partial y} \end{vmatrix}$$

(ii) If u, v and w are the functions of three independent variables x, y and z, then the determinant

$$\begin{vmatrix} \dfrac{\partial u}{\partial x} & \dfrac{\partial u}{\partial y} & \dfrac{\partial u}{\partial z} \\[2mm] \dfrac{\partial v}{\partial x} & \dfrac{\partial v}{\partial y} & \dfrac{\partial v}{\partial z} \\[2mm] \dfrac{\partial w}{\partial x} & \dfrac{\partial w}{\partial y} & \dfrac{\partial w}{\partial z} \end{vmatrix}$$

is called the Jacobian of u, v and w with respect to x, y and z. It is denoted by $\dfrac{\partial(u,v,w)}{\partial(x,y,z)}$ or $J(u,v,w)$.

(iii) If $u_1, u_2, ..., u_n$ are n functions of independent variables $x_1, x_2, ..., x_n$, then the determinant

$$\begin{vmatrix} \dfrac{\partial u_1}{\partial x_1} & \dfrac{\partial u_1}{\partial x_2} & \dfrac{\partial u_1}{\partial x_3} & \cdots & \dfrac{\partial u_1}{\partial x_n} \\[2mm] \dfrac{\partial u_2}{\partial x_1} & \dfrac{\partial u_2}{\partial x_2} & \dfrac{\partial u_2}{\partial x_3} & \cdots & \dfrac{\partial u_2}{\partial x_n} \\[2mm] \vdots & \vdots & \vdots & & \vdots \\[2mm] \dfrac{\partial u_n}{\partial x_1} & \dfrac{\partial u_n}{\partial x_2} & \dfrac{\partial u_n}{\partial x_3} & \cdots & \dfrac{\partial u_n}{\partial x_n} \end{vmatrix}$$

is called the Jacobian of $u_1, u_2, ..., u_n$ with respect to $x_1, x_2, ..., x_n$. It is denoted by $\dfrac{\partial(u_1, u_2, ..., u_n)}{\partial(x_1, x_2, ..., x_n)}$ or $J(u_1, u_2, ..., u_n)$.

(iv) If the functions $u_1, u_2, ..., u_n$ of n independent variables $x_1, x_2, ..., x_n$ are of the following form

$$u_1 = f_1(x_1), u_2 = f_2(x_1, x_2), ..., u_n = f_n(x_1, x_2, ..., x_n)$$

Then
$$\dfrac{\partial(u_1, u_2, ..., u_n)}{\partial(x_1, x_2, ..., x_n)} = \begin{vmatrix} \dfrac{\partial u_1}{\partial x_1} & \dfrac{\partial u_1}{\partial x_2} & \dfrac{\partial u_1}{\partial x_3} & \cdots & \dfrac{\partial u_1}{\partial x_n} \\[2mm] \dfrac{\partial u_2}{\partial x_1} & \dfrac{\partial u_2}{\partial x_2} & \dfrac{\partial u_2}{\partial x_3} & \cdots & \dfrac{\partial u_2}{\partial x_n} \\[2mm] \vdots & \vdots & \vdots & & \vdots \\[2mm] \dfrac{\partial u_n}{\partial x_1} & \dfrac{\partial u_n}{\partial x_2} & \dfrac{\partial u_n}{\partial x_3} & \cdots & \dfrac{\partial u_n}{\partial x_n} \end{vmatrix}$$

3.4.3 IMPORTANT THEOREMS ON JACOBIAN

THEOREM 1. $\dfrac{\partial(u,v)}{\partial(x,y)} \cdot \dfrac{\partial(x,y)}{\partial(u,v)} = 1$ [UPTU -2006; UPTU (SUM) 2008, 2009]

PROOF. Let $u = u(x, y)$ and $v = v(x, y)$ be the given functions.

Differentiating partially each of these functions, we get

$$1 = \dfrac{\partial u}{\partial x} \cdot \dfrac{\partial x}{\partial u} + \dfrac{\partial u}{\partial y} \cdot \dfrac{\partial y}{\partial u}$$

$$0 = \dfrac{\partial u}{\partial x} \cdot \dfrac{\partial x}{\partial v} + \dfrac{\partial u}{\partial y} \cdot \dfrac{\partial y}{\partial v}$$

and
$$0 = \frac{\partial v}{\partial x}\cdot\frac{\partial x}{\partial u} + \frac{\partial v}{\partial y}\cdot\frac{\partial y}{\partial u}$$

$$1 = \frac{\partial v}{\partial x}\cdot\frac{\partial x}{\partial v} + \frac{\partial v}{\partial y}\cdot\frac{\partial y}{\partial v}$$

Now

$$\frac{\partial(u,v)}{\partial(x,y)}\cdot\frac{\partial(x,y)}{\partial(u,v)} = \begin{vmatrix} \dfrac{\partial u}{\partial x} & \dfrac{\partial u}{\partial y} \\[2mm] \dfrac{\partial v}{\partial x} & \dfrac{\partial v}{\partial y} \end{vmatrix}\cdot\begin{vmatrix} \dfrac{\partial x}{\partial u} & \dfrac{\partial x}{\partial v} \\[2mm] \dfrac{\partial y}{\partial u} & \dfrac{\partial y}{\partial v} \end{vmatrix} = \begin{vmatrix} \dfrac{\partial u}{\partial x}\cdot\dfrac{\partial x}{\partial u} + \dfrac{\partial u}{\partial y}\cdot\dfrac{\partial y}{\partial u} & \dfrac{\partial u}{\partial x}\cdot\dfrac{\partial x}{\partial v} + \dfrac{\partial u}{\partial y}\cdot\dfrac{\partial y}{\partial v} \\[2mm] \dfrac{\partial v}{\partial x}\cdot\dfrac{\partial x}{\partial u} + \dfrac{\partial v}{\partial y}\cdot\dfrac{\partial y}{\partial u} & \dfrac{\partial v}{\partial x}\cdot\dfrac{\partial x}{\partial v} + \dfrac{\partial v}{\partial y}\cdot\dfrac{\partial y}{\partial v} \end{vmatrix}$$

$$= \begin{vmatrix} \dfrac{\partial u}{\partial x} & \dfrac{\partial u}{\partial y} \\[2mm] \dfrac{\partial v}{\partial x} & \dfrac{\partial v}{\partial y} \end{vmatrix}\cdot\begin{vmatrix} \dfrac{\partial x}{\partial u} & \dfrac{\partial y}{\partial u} \\[2mm] \dfrac{\partial x}{\partial v} & \dfrac{\partial y}{\partial v} \end{vmatrix} = \begin{vmatrix} 1 & 0 \\ 0 & 1 \end{vmatrix} = 1$$

THEOREM 2. *If the function $u_1, u_2, ..., u_n$ of n independent variables $x_1, x_2, .., x_n$ are of the following form*

$$u_1 = f_1(x_1)$$

$$u_2 = f_2(x_1, x_2)$$

$$\vdots \qquad \vdots \qquad \vdots$$

$$u_n = f_n(x_1, x_2, ..., x_n)$$

Then
$$\frac{\partial(u_1, u_2, ..., u_n)}{\partial(x_1, x_2, ..., x_n)} = \frac{\partial u_1}{\partial x_1}\cdot\frac{\partial u_2}{\partial x_2}\cdot\frac{\partial u_3}{\partial x_3}\cdots\frac{\partial u_n}{\partial x_n}$$

PROOF. We know that $\dfrac{\partial(u_1, u_2, ..., u_n)}{\partial(x_1, x_2, ..., x_n)} = \begin{vmatrix} \dfrac{\partial u_1}{\partial x_1} & \dfrac{\partial u_1}{\partial x_2} & \dfrac{\partial u_1}{\partial x_3} & \cdots & \dfrac{\partial u_1}{\partial x_n} \\[2mm] \dfrac{\partial u_2}{\partial x_1} & \dfrac{\partial u_2}{\partial x_2} & \dfrac{\partial u_2}{\partial x_3} & \cdots & \dfrac{\partial u_2}{\partial x_n} \\[2mm] \vdots & \vdots & \vdots & & \vdots \\[2mm] \dfrac{\partial u_n}{\partial x_1} & \dfrac{\partial u_n}{\partial x_2} & \dfrac{\partial u_n}{\partial x_3} & \cdots & \dfrac{\partial u_n}{\partial x_n} \end{vmatrix}$ ...(1)

(i) u_1 is a function of x_1 only, therefore $\dfrac{\partial u_1}{\partial x_1}$ exists and $\dfrac{\partial u_1}{\partial x_2} = 0, \dfrac{\partial u_1}{\partial x_3} = 0, ..., \dfrac{\partial u_1}{\partial x_n} = 0$

(ii) u_2 is a function of x_1 and x_2 only, therefore $\dfrac{\partial u_2}{\partial x_1}$ and $\dfrac{\partial u_2}{\partial x_2}$ exist and $\dfrac{\partial u_2}{\partial x_3} = 0, \dfrac{\partial u_2}{\partial x_4} = 0, ..., \dfrac{\partial u_2}{\partial x_n} = 0$

(iii) u_3 is a function of x_1, x_2 and x_3 only,

therefore, $\dfrac{\partial u_3}{\partial x_1}, \dfrac{\partial u_3}{\partial x_2}$ and $\dfrac{\partial u_3}{\partial x_3}$ exist and $\dfrac{\partial u_3}{\partial x_4} = 0, \dfrac{\partial u_3}{\partial x_5} = 0, ..., \dfrac{\partial u_3}{\partial x_n} = 0$

Preceding in the same manner, we have u_n is a function of $x_1, x_2, ..., x_n$,

therefore $\dfrac{\partial u_n}{\partial x_1}, \dfrac{\partial u_n}{\partial x_2}, ..., \dfrac{\partial u_n}{\partial x_n}$ all exist.

Putting all these values in (1), we get

$$\frac{\partial(u_1, u_2, ..., u_n)}{\partial(x_1, x_2, ..., x_n)} = \begin{vmatrix} \dfrac{\partial u_1}{\partial x_1} & 0 & 0 & \cdots & 0 \\[2mm] \dfrac{\partial u_2}{\partial x_1} & \dfrac{\partial u_2}{\partial x_2} & 0 & \cdots & 0 \\[2mm] \vdots & \vdots & \vdots & \vdots & \vdots \\[2mm] \dfrac{\partial u_n}{\partial x_1} & \dfrac{\partial u_n}{\partial x_2} & \dfrac{\partial u_n}{\partial x_3} & \cdots & \dfrac{\partial u_n}{\partial x_n} \end{vmatrix}$$

Now expanding the determinant along the first row, we get

$$\frac{\partial(u_1, u_2, ..., u_n)}{\partial(x_1, x_2, ..., x_n)} = \frac{\partial u_1}{\partial x_1}\cdot\frac{\partial u_2}{\partial x_2}\cdot\frac{\partial u_3}{\partial x_3}\cdots\frac{\partial u_n}{\partial x_n}$$

THEOREM 3. *If u_1, u_2 are functions of y_1, y_2 and y_1, y_2 are functions of x_1, x_2 then $\dfrac{\partial(u_1, u_2)}{\partial(x_1, x_2)} = \dfrac{\partial(u_1, u_2)}{\partial(y_1, y_2)}\cdot\dfrac{\partial(y_1, y_2)}{\partial(x_1, x_2)}$*

PROOF. Since u_1, u_2 are functions of y_1, y_2. Also y_1, y_2 are functions of x_1, x_2; therefore, we get

$$\left.\begin{array}{l}\dfrac{\partial u_1}{\partial x_1} = \dfrac{\partial u_1}{\partial y_1}\cdot\dfrac{\partial y_1}{\partial x_1} + \dfrac{\partial u_1}{\partial y_2}\cdot\dfrac{\partial y_2}{\partial x_1} \\[2mm] \dfrac{\partial u_1}{\partial x_2} = \dfrac{\partial u_1}{\partial y_1}\cdot\dfrac{\partial y_1}{\partial x_2} + \dfrac{\partial u_1}{\partial y_2}\cdot\dfrac{\partial y_2}{\partial x_2} \\[2mm] \dfrac{\partial u_2}{\partial x_1} = \dfrac{\partial u_2}{\partial y_1}\cdot\dfrac{\partial y_1}{\partial x_1} + \dfrac{\partial u_2}{\partial y_2}\cdot\dfrac{\partial y_2}{\partial x_1} \\[2mm] \dfrac{\partial u_2}{\partial x_2} = \dfrac{\partial u_2}{\partial y_1}\cdot\dfrac{\partial y_1}{\partial x_2} + \dfrac{\partial u_2}{\partial y_2}\cdot\dfrac{\partial y_2}{\partial x_2}\end{array}\right] \qquad \ldots(1)$$

We have

$$\frac{\partial(u_1,u_2)}{\partial(y_1,y_2)}\cdot\frac{\partial(y_1,y_2)}{\partial(x_1,x_2)} = \begin{vmatrix} \dfrac{\partial u_1}{\partial y_1} & \dfrac{\partial u_1}{\partial y_2} \\[2mm] \dfrac{\partial u_2}{\partial y_1} & \dfrac{\partial u_2}{\partial y_2} \end{vmatrix} \times \begin{vmatrix} \dfrac{\partial y_1}{\partial x_1} & \dfrac{\partial y_1}{\partial x_2} \\[2mm] \dfrac{\partial y_2}{\partial x_1} & \dfrac{\partial y_2}{\partial x_2} \end{vmatrix} = \begin{vmatrix} \dfrac{\partial u_1}{\partial y_1}\dfrac{\partial y_1}{\partial x_1}+\dfrac{\partial u_1}{\partial y_2}\dfrac{\partial y_2}{\partial x_1} & \dfrac{\partial u_1}{\partial y_1}\dfrac{\partial y_1}{\partial x_2}+\dfrac{\partial u_1}{\partial y_2}\dfrac{\partial y_2}{\partial x_2} \\[2mm] \dfrac{\partial u_2}{\partial y_1}\dfrac{\partial y_1}{\partial x_1}+\dfrac{\partial u_2}{\partial y_2}\dfrac{\partial y_2}{\partial x_1} & \dfrac{\partial u_2}{\partial y_1}\dfrac{\partial y_1}{\partial x_2}+\dfrac{\partial u_2}{\partial y_2}\dfrac{\partial y_2}{\partial x_2} \end{vmatrix}$$

Now, using relation (1), we get

$$\frac{\partial(u_1,u_2)}{\partial(y_1,y_2)}\cdot\frac{\partial(y_1,y_2)}{\partial(x_1,x_2)} = \begin{vmatrix} \dfrac{\partial u_1}{\partial x_1} & \dfrac{\partial u_1}{\partial x_2} \\[2mm] \dfrac{\partial u_2}{\partial x_1} & \dfrac{\partial u_2}{\partial x_2} \end{vmatrix} = \frac{\partial(u_1,u_2)}{\partial(x_1,x_2)}$$

THEOREM 4. *If u_1, u_2, u_3 are functions of y_1, y_2, y_3 and y_1, y_2, y_3 are functions of x_1, x_2, x_3, then*

$$\frac{\partial(u_1,u_2,u_3)}{\partial(x_1,x_2,x_3)} = \frac{\partial(u_1,u_2,u_3)}{\partial(y_1,y_2,y_3)}\cdot\frac{\partial(y_1,y_2,y_3)}{\partial(x_1,x_2,x_3)}$$

PROOF. Since u_1, u_2 and u_3 are functions of y_1, y_2 and y_3. Also y_1, y_2 and y_3 are functions of x_1, x_2 and x_3 therefore, we get

$$\frac{\partial u_1}{\partial x_1} = \frac{\partial u_1}{\partial y_1}\cdot\frac{\partial y_1}{\partial x_1} + \frac{\partial u_1}{\partial y_2}\cdot\frac{\partial y_2}{\partial x_1} + \frac{\partial u_1}{\partial y_3}\cdot\frac{\partial y_3}{\partial x_1} = \sum_{i=1}^{3}\frac{\partial u_1}{\partial y_i}\cdot\frac{\partial y_i}{\partial x_1}$$

$$\frac{\partial u_1}{\partial x_2} = \frac{\partial u_1}{\partial y_1}\cdot\frac{\partial y_1}{\partial x_2} + \frac{\partial u_1}{\partial y_2}\cdot\frac{\partial y_2}{\partial x_2} + \frac{\partial u_1}{\partial y_3}\cdot\frac{\partial y_3}{\partial x_2} = \sum_{i=1}^{3}\frac{\partial u_1}{\partial y_i}\cdot\frac{\partial y_i}{\partial x_2}$$

Similarly,

$$\frac{\partial u_1}{\partial x_3} = \sum_{i=1}^{3}\frac{\partial u_1}{\partial y_i}\cdot\frac{\partial y_i}{\partial x_3}, \quad \frac{\partial u_2}{\partial x_1} = \sum_{i=1}^{3}\frac{\partial u_2}{\partial y_i}\cdot\frac{\partial y_i}{\partial x_1}, \quad \frac{\partial u_2}{\partial x_2} = \sum_{i=1}^{3}\frac{\partial u_2}{\partial y_i}\cdot\frac{\partial y_i}{\partial x_2}, \quad \frac{\partial u_2}{\partial x_3} = \sum_{i=1}^{3}\frac{\partial u_2}{\partial y_i}\cdot\frac{\partial y_i}{\partial x_3},$$

$$\frac{\partial u_3}{\partial x_1} \quad \sum\frac{\partial u_3}{\partial y}\frac{\partial y}{\partial x_1}, \quad \frac{\partial u_3}{\partial x_2} = \sum_{i=1}^{3}\frac{\partial u_3}{\partial y_i}\cdot\frac{\partial y_i}{\partial x_2}, \text{ and } \frac{\partial u_3}{\partial x_3} \quad \sum\frac{\partial u_3}{\partial y}\frac{\partial y}{\partial x_3}$$

Now, $\dfrac{\partial(u_1,u_2,u_3)}{\partial(y_1,y_2,y_3)}\cdot\dfrac{\partial(y_1,y_2,y_3)}{\partial(x_1,x_2,x_3)} =$

$$\begin{vmatrix} \dfrac{\partial u_1}{\partial y_1} & \dfrac{\partial u_1}{\partial y_2} & \dfrac{\partial u_1}{\partial y_3} \\[2mm] \dfrac{\partial u_2}{\partial y_1} & \dfrac{\partial u_2}{\partial y_2} & \dfrac{\partial u_2}{\partial y_3} \\[2mm] \dfrac{\partial u_3}{\partial y_1} & \dfrac{\partial u_3}{\partial y_2} & \dfrac{\partial u_3}{\partial y_3} \end{vmatrix} \begin{vmatrix} \dfrac{\partial y_1}{\partial x_1} & \dfrac{\partial y_1}{\partial x_2} & \dfrac{\partial y_1}{\partial x_3} \\[2mm] \dfrac{\partial y_2}{\partial x_1} & \dfrac{\partial y_2}{\partial x_2} & \dfrac{\partial y_2}{\partial x_3} \\[2mm] \dfrac{\partial y_3}{\partial x_1} & \dfrac{\partial y_3}{\partial x_2} & \dfrac{\partial y_3}{\partial x_3} \end{vmatrix} = \begin{vmatrix} \sum\dfrac{\partial u_1}{\partial y_i}\cdot\dfrac{\partial y_i}{\partial x_1} & \sum\dfrac{\partial u_1}{\partial y_i}\cdot\dfrac{\partial y_i}{\partial x_2} & \sum\dfrac{\partial u_1}{\partial y_i}\cdot\dfrac{\partial y_i}{\partial x_3} \\[2mm] \sum\dfrac{\partial u_2}{\partial y_i}\cdot\dfrac{\partial y_i}{\partial x_1} & \sum\dfrac{\partial u_2}{\partial y_i}\cdot\dfrac{\partial y_i}{\partial x_2} & \sum\dfrac{\partial u_2}{\partial y_i}\cdot\dfrac{\partial y_i}{\partial x_3} \\[2mm] \sum\dfrac{\partial u_3}{\partial y_i}\cdot\dfrac{\partial y_i}{\partial x_1} & \sum\dfrac{\partial u_3}{\partial y_i}\cdot\dfrac{\partial y_i}{\partial x_2} & \sum\dfrac{\partial u_3}{\partial y_i}\cdot\dfrac{\partial y_i}{\partial x_3} \end{vmatrix}$$

Putting the values of each element of the determinant from the above relations, we get

$$\frac{\partial(u_1,u_2,u_3)}{\partial(y_1,y_2,y_3)}\cdot\frac{\partial(y_1,y_2,y_3)}{\partial(x_1,x_2,x_3)} = \begin{vmatrix} \dfrac{\partial u_1}{\partial x_1} & \dfrac{\partial u_1}{\partial x_2} & \dfrac{\partial u_1}{\partial x_3} \\[2mm] \dfrac{\partial u_2}{\partial x_1} & \dfrac{\partial u_2}{\partial x_2} & \dfrac{\partial u_2}{\partial x_3} \\[2mm] \dfrac{\partial u_3}{\partial x_1} & \dfrac{\partial u_3}{\partial x_2} & \dfrac{\partial u_3}{\partial x_3} \end{vmatrix} = \frac{\partial(u_1,u_2,u_3)}{\partial(x_1,x_2,x_3)}$$

Generalization. If $u_1, u_2, ..., u_n$ are functions of $y_1, y_2, ..., y_n$ and $y_1, y_2, ..., y_n$ are functions of $x_1, x_2, ..., x_n$, then

$$\frac{\partial(u_1,u_2,u_3,...,u_n)}{\partial(x_1,x_2,x_3,...,x_n)} = \frac{\partial(u_1,u_2,u_3,...,u_n)}{\partial(y_1,y_2,y_3,...y_n)}\cdot\frac{\partial(y_1,y_2,y_3,...,y_n)}{\partial(x_1,x_2,x_3,...,x_n)}$$

The proof may be easily extended as in the case of two and three variables.

THEOREM 5. *If the functions u, v, w of three independent variables x, y and z are not independent, then the Jacobian of u, v, w with respect to x, y, z vanishes.*

PROOF. Since, the functions u, v and w (of three independent variables x, y and z) are not independent. Then there will be a relation

$$F(u,v,w) = 0 \qquad\qquad \text{...(A)}$$

which will connect these independent variables.

Differentiating (A), with respect to x, y and z, we get

$$\frac{\partial F}{\partial u}\cdot\frac{\partial u}{\partial x} + \frac{\partial F}{\partial v}\cdot\frac{\partial v}{\partial x} + \frac{\partial F}{\partial w}\cdot\frac{\partial w}{\partial x} = 0 \qquad\qquad \text{...(1)}$$

$$\frac{\partial F}{\partial u}\cdot\frac{\partial u}{\partial y} + \frac{\partial F}{\partial v}\cdot\frac{\partial v}{\partial y} + \frac{\partial F}{\partial w}\cdot\frac{\partial w}{\partial y} = 0 \qquad\qquad \text{...(2)}$$

$$\frac{\partial F}{\partial u}\cdot\frac{\partial u}{\partial z} + \frac{\partial F}{\partial v}\cdot\frac{\partial v}{\partial z} + \frac{\partial F}{\partial w}\cdot\frac{\partial w}{\partial z} = 0 \qquad\qquad \text{...(3)}$$

Eliminating $\dfrac{\partial F}{\partial u}$, $\dfrac{\partial F}{\partial v}$ and $\dfrac{\partial F}{\partial w}$ from (1), (2) and (3), we get

$$\begin{vmatrix} \dfrac{\partial u}{\partial x} & \dfrac{\partial v}{\partial x} & \dfrac{\partial w}{\partial x} \\[2mm] \dfrac{\partial u}{\partial y} & \dfrac{\partial v}{\partial y} & \dfrac{\partial w}{\partial y} \\[2mm] \dfrac{\partial u}{\partial z} & \dfrac{\partial v}{\partial z} & \dfrac{\partial w}{\partial z} \end{vmatrix} = 0 \Rightarrow \begin{vmatrix} \dfrac{\partial u}{\partial x} & \dfrac{\partial u}{\partial y} & \dfrac{\partial u}{\partial z} \\[2mm] \dfrac{\partial v}{\partial x} & \dfrac{\partial v}{\partial y} & \dfrac{\partial v}{\partial z} \\[2mm] \dfrac{\partial w}{\partial x} & \dfrac{\partial w}{\partial y} & \dfrac{\partial w}{\partial z} \end{vmatrix} = 0 \text{ (Interchanging rows and columns)}$$

Hence, $\qquad\qquad \dfrac{\partial(u,v,w)}{\partial(x,y,z)} = 0$

THEOREM 6. $\quad \dfrac{\partial(u,v,w)}{\partial(x,y,z)} \times \dfrac{\partial(x,y,z)}{\partial(u,v,w)} = 1$

PROOF.   Let us suppose $\qquad u = f_1(x,y,z); \ v = f_2(x,y,z) \qquad$ and $\qquad w = f_3(x,y,z).$

Then we may write these equations as

$$x = \phi_1(u,v,w); \ y = \phi_2(u,v,w) \text{ and } \qquad z = \phi_3(u,v,w)$$

Now, differentiating $u = f_1(x,y,z)$ partially w.r.t. u, v and w respectively, we get

$$\frac{\partial u}{\partial u} = \frac{\partial u}{\partial x}\cdot\frac{\partial x}{\partial u} + \frac{\partial u}{\partial y}\cdot\frac{\partial y}{\partial u} + \frac{\partial u}{\partial z}\cdot\frac{\partial z}{\partial u} \quad\Rightarrow\quad 1 = \frac{\partial u}{\partial x}\cdot\frac{\partial x}{\partial u} + \frac{\partial u}{\partial y}\cdot\frac{\partial y}{\partial u} + \frac{\partial u}{\partial z}\cdot\frac{\partial z}{\partial u} \qquad \text{...(1)}$$

and

$$\frac{\partial u}{\partial v} = \frac{\partial u}{\partial x}\cdot\frac{\partial x}{\partial v} + \frac{\partial u}{\partial y}\cdot\frac{\partial y}{\partial v} + \frac{\partial u}{\partial z}\cdot\frac{\partial z}{\partial v} \quad\Rightarrow\quad 0 = \frac{\partial u}{\partial x}\cdot\frac{\partial x}{\partial v} + \frac{\partial u}{\partial y}\cdot\frac{\partial y}{\partial v} + \frac{\partial u}{\partial z}\cdot\frac{\partial z}{\partial v}. \qquad \text{..(2)}$$

Similarly $\qquad 0 = \dfrac{\partial u}{\partial x}\cdot\dfrac{\partial x}{\partial w} + \dfrac{\partial u}{\partial y}\cdot\dfrac{\partial y}{\partial w} + \dfrac{\partial u}{\partial z}\cdot\dfrac{\partial z}{\partial w} \qquad\qquad \text{...(3)}$

Now differentiating $v = f_2(x,y,z)$ and $w = f_3(x,y,z)$ partially with respect to u, v and w respectively, we get

$$\left.\begin{aligned} 0 &= \frac{\partial v}{\partial x}\cdot\frac{\partial x}{\partial u} + \frac{\partial v}{\partial y}\cdot\frac{\partial y}{\partial u} + \frac{\partial v}{\partial z}\cdot\frac{\partial z}{\partial u} \\[1mm] 1 &= \frac{\partial v}{\partial x}\cdot\frac{\partial x}{\partial v} + \frac{\partial v}{\partial y}\cdot\frac{\partial y}{\partial v} + \frac{\partial v}{\partial z}\cdot\frac{\partial z}{\partial v} \\[1mm] 0 &= \frac{\partial v}{\partial x}\cdot\frac{\partial x}{\partial w} + \frac{\partial v}{\partial y}\cdot\frac{\partial y}{\partial w} + \frac{\partial v}{\partial z}\cdot\frac{\partial z}{\partial w} \end{aligned}\right\} \qquad \text{...(4)}$$

and

$$\left.\begin{aligned} 0 &= \frac{\partial w}{\partial x}\cdot\frac{\partial x}{\partial u} + \frac{\partial w}{\partial y}\cdot\frac{\partial y}{\partial u} + \frac{\partial w}{\partial z}\cdot\frac{\partial z}{\partial u} \\[1mm] 0 &= \frac{\partial w}{\partial x}\cdot\frac{\partial x}{\partial v} + \frac{\partial w}{\partial y}\cdot\frac{\partial y}{\partial v} + \frac{\partial w}{\partial z}\cdot\frac{\partial z}{\partial v} \\[1mm] 1 &= \frac{\partial w}{\partial x}\cdot\frac{\partial x}{\partial w} + \frac{\partial w}{\partial y}\cdot\frac{\partial y}{\partial w} + \frac{\partial w}{\partial z}\cdot\frac{\partial z}{\partial w} \end{aligned}\right\} \qquad \text{...(5)}$$

We have $\dfrac{\partial(u,v,w)}{\partial(x,y,z)} \times \dfrac{\partial(x,y,z)}{\partial(u,v,w)} = \begin{vmatrix} \dfrac{\partial u}{\partial x} & \dfrac{\partial u}{\partial y} & \dfrac{\partial u}{\partial z} \\[2mm] \dfrac{\partial v}{\partial x} & \dfrac{\partial v}{\partial y} & \dfrac{\partial v}{\partial z} \\[2mm] \dfrac{\partial w}{\partial x} & \dfrac{\partial w}{\partial y} & \dfrac{\partial w}{\partial z} \end{vmatrix} \begin{vmatrix} \dfrac{\partial x}{\partial u} & \dfrac{\partial x}{\partial v} & \dfrac{\partial x}{\partial w} \\[2mm] \dfrac{\partial y}{\partial u} & \dfrac{\partial y}{\partial v} & \dfrac{\partial y}{\partial w} \\[2mm] \dfrac{\partial z}{\partial u} & \dfrac{\partial z}{\partial v} & \dfrac{\partial z}{\partial w} \end{vmatrix} = \begin{vmatrix} \sum\dfrac{\partial u}{\partial x}\cdot\dfrac{\partial x}{\partial u} & \sum\dfrac{\partial u}{\partial x}\cdot\dfrac{\partial x}{\partial v} & \sum\dfrac{\partial u}{\partial x}\cdot\dfrac{\partial x}{\partial w} \\[2mm] \sum\dfrac{\partial v}{\partial x}\cdot\dfrac{\partial x}{\partial u} & \sum\dfrac{\partial v}{\partial x}\cdot\dfrac{\partial x}{\partial v} & \sum\dfrac{\partial v}{\partial x}\cdot\dfrac{\partial x}{\partial w} \\[2mm] \sum\dfrac{\partial w}{\partial x}\cdot\dfrac{\partial x}{\partial u} & \sum\dfrac{\partial w}{\partial x}\cdot\dfrac{\partial x}{\partial v} & \sum\dfrac{\partial w}{\partial x}\cdot\dfrac{\partial x}{\partial w} \end{vmatrix}$

$$= \begin{vmatrix} 1 & 0 & 0 \\ 0 & 1 & 0 \\ 0 & 0 & 1 \end{vmatrix} = 1 \qquad \text{(Using the relations (1) to (5))}$$

Hence $\qquad \dfrac{\partial(u,v,w)}{\partial(x,y,z)} \times \dfrac{\partial(x,y,z)}{\partial(u,v,w)} = 1$

3.4.4 Jacobian of Implicit Functions

THEOREM 1. *If u_1, u_2, are implicit functions of x_1, x_2 that is $F_1(u_1, u_2, x_1, x_2) = 0$ and $F_2(u_1, u_2, x_1, x_2) = 0$, then*

$$\frac{\partial(u_1, u_2)}{\partial(x_1, x_2)} = (-1)^2 \left[\frac{\partial(F_1, F_2)}{\partial(x_1, x_2)} \bigg/ \frac{\partial(F_1, F_2)}{\partial(u_1, u_2)} \right]$$

PROOF. We have
$$\left. \begin{array}{l} F_1(u_1, u_2, x_1, x_2) = 0 \\ F_2(u_1, u_2, x_1, x_2) = 0 \end{array} \right] \qquad \ldots(1)$$

Differentiating relation (1), partially *w.r.t.* x_1 and x_2 respectively, we get

$$\left. \begin{array}{l} \dfrac{\partial F_1}{\partial x_1} + \dfrac{\partial F_1}{\partial u_1} \cdot \dfrac{\partial u_1}{\partial x_1} + \dfrac{\partial F_1}{\partial u_2} \cdot \dfrac{\partial u_2}{\partial x_1} = 0 \\[2mm] \dfrac{\partial F_1}{\partial x_2} + \dfrac{\partial F_1}{\partial u_1} \cdot \dfrac{\partial u_1}{\partial x_2} + \dfrac{\partial F_1}{\partial u_2} \cdot \dfrac{\partial u_2}{\partial x_2} = 0 \\[2mm] \dfrac{\partial F_2}{\partial x_1} + \dfrac{\partial F_2}{\partial u_1} \cdot \dfrac{\partial u_1}{\partial x_1} + \dfrac{\partial F_2}{\partial u_2} \cdot \dfrac{\partial u_2}{\partial x_1} = 0 \\[2mm] \dfrac{\partial F_2}{\partial x_2} + \dfrac{\partial F_2}{\partial u_1} \cdot \dfrac{\partial u_1}{\partial x_2} + \dfrac{\partial F_2}{\partial u_2} \cdot \dfrac{\partial u_2}{\partial x_2} = 0 \end{array} \right] \qquad \ldots(2)$$

We have $\dfrac{\partial(F_1, F_2)}{\partial(u_1, u_2)} \times \dfrac{\partial(u_1, u_2)}{\partial(x_1, x_2)} = \begin{vmatrix} \dfrac{\partial F_1}{\partial u_1} & \dfrac{\partial F_1}{\partial u_2} \\[2mm] \dfrac{\partial F_2}{\partial u_1} & \dfrac{\partial F_2}{\partial u_2} \end{vmatrix} \times \begin{vmatrix} \dfrac{\partial u_1}{\partial x_1} & \dfrac{\partial u_1}{\partial x_2} \\[2mm] \dfrac{\partial u_2}{\partial x_1} & \dfrac{\partial u_2}{\partial x_2} \end{vmatrix} = \begin{vmatrix} \dfrac{\partial F_1}{\partial u_1} \cdot \dfrac{\partial u_1}{\partial x_1} + \dfrac{\partial F_1}{\partial u_2} \cdot \dfrac{\partial u_2}{\partial x_1} & \dfrac{\partial F_1}{\partial u_1} \cdot \dfrac{\partial u_1}{\partial x_2} + \dfrac{\partial F_1}{\partial u_2} \cdot \dfrac{\partial u_2}{\partial x_2} \\[2mm] \dfrac{\partial F_2}{\partial u_1} \cdot \dfrac{\partial u_1}{\partial x_1} + \dfrac{\partial F_2}{\partial u_2} \cdot \dfrac{\partial u_2}{\partial x_1} & \dfrac{\partial F_2}{\partial u_1} \cdot \dfrac{\partial u_1}{\partial x_2} + \dfrac{\partial F_2}{\partial u_2} \cdot \dfrac{\partial u_2}{\partial x_2} \end{vmatrix}$

$$= \begin{vmatrix} -\dfrac{\partial F_1}{\partial x_1} & -\dfrac{\partial F_1}{\partial x_2} \\[2mm] -\dfrac{\partial F_2}{\partial x_1} & -\dfrac{\partial F_2}{\partial x_2} \end{vmatrix} = (-1)^2 \frac{\partial(F_1, F_2)}{\partial(x_1, x_2)}$$

$\Rightarrow \qquad \dfrac{\partial(u_1, u_2)}{\partial(x_1, x_2)} = (-1)^2 \left[\dfrac{\partial(F_1, F_2)}{\partial(x_1, x_2)} \bigg/ \dfrac{\partial(F_1, F_2)}{\partial(u_1, u_2)} \right]$ \qquad Using relation (2),

THEOREM 2. *If u_1, u_2 and u_3 be the implicit functions of x_1, x_2, x_3 that is*
$$F_1(u_1, u_2, u_3, x_1, x_2, x_3) = 0, \ F_2(u_1, u_2, u_3, x_1, x_2, x_3) = 0, \ F_3(u_1, u_2, u_3, x_1, x_2, x_3) = 0$$

then $\qquad \dfrac{\partial(u_1, u_2, u_3)}{\partial(x_1, x_2, x_3)} = (-1)^3 \left[\dfrac{\partial(F_1, F_2, F_3)}{\partial(x_1, x_2, x_3)} \bigg/ \dfrac{\partial(F_1, F_2, F_3)}{\partial(u_1, u_2, u_3)} \right]$

PROOF. We have
$$\left. \begin{array}{l} F_1(u_1, u_2, u_3, x_1, x_2, x_3) = 0 \\ F_2(u_1, u_2, u_3, x_1, x_2, x_3) = 0 \\ F_3(u_1, u_2, u_3, x_1, x_2, x_3) = 0 \end{array} \right] \qquad \ldots(1)$$

Differentiating (1), partially w.r.t. x_1, x_2 and x_3 respectively, we get $\dfrac{\partial F_1}{\partial x_1} + \dfrac{\partial F_1}{\partial u_1} \cdot \dfrac{\partial u_1}{\partial x_1} + \dfrac{\partial F_1}{\partial u_2} \cdot \dfrac{\partial u_2}{\partial x_1} + \dfrac{\partial F_1}{\partial u_3} \cdot \dfrac{\partial u_3}{\partial x_1} = 0$

$\Rightarrow \qquad \sum\limits_{r=1}^{3} \dfrac{\partial F_1}{\partial u_r} \cdot \dfrac{\partial u_r}{\partial x_1} = -\dfrac{\partial F_1}{\partial x_1}, \quad \sum\limits_{r=1}^{3} \dfrac{\partial F_1}{\partial u_r} \cdot \dfrac{\partial u_r}{\partial x_2} = -\dfrac{\partial F_1}{\partial x_2}, \quad \sum\limits_{r=1}^{3} \dfrac{\partial F_1}{\partial u_r} \cdot \dfrac{\partial u_r}{\partial x_3} = -\dfrac{\partial F_1}{\partial x_3},$

Similarly $\qquad \sum\limits_{r=1}^{3} \dfrac{\partial F_2}{\partial u_r} \cdot \dfrac{\partial u_r}{\partial x_1} = -\dfrac{\partial F_2}{\partial x_1}, \quad \sum\limits_{r=1}^{3} \dfrac{\partial F_2}{\partial u_r} \cdot \dfrac{\partial u_r}{\partial x_2} = -\dfrac{\partial F_2}{\partial x_2}, \quad \sum\limits_{r=1}^{3} \dfrac{\partial F_2}{\partial u_r} \cdot \dfrac{\partial u_r}{\partial x_3} = -\dfrac{\partial F_2}{\partial x_3},$

and $\qquad \sum\limits_{r=1}^{3} \dfrac{\partial F_3}{\partial u_r} \cdot \dfrac{\partial u_r}{\partial x_1} = -\dfrac{\partial F_3}{\partial x_1}, \quad \sum\limits_{r=1}^{3} \dfrac{\partial F_3}{\partial u_r} \cdot \dfrac{\partial u_r}{\partial x_2} = -\dfrac{\partial F_3}{\partial x_2}, \quad \sum\limits_{r=1}^{3} \dfrac{\partial F_3}{\partial u_r} \cdot \dfrac{\partial u_r}{\partial x_3} = -\dfrac{\partial F_3}{\partial x_3},$

Now consider

$$\frac{\partial(F_1, F_2, F_3)}{\partial(u_1, u_2, u_3)} \times \frac{\partial(u_1, u_2, u_3)}{\partial(x_1, x_2, x_3)} = \begin{vmatrix} \dfrac{\partial F_1}{\partial u_1} & \dfrac{\partial F_1}{\partial u_2} & \dfrac{\partial F_1}{\partial u_3} \\[2mm] \dfrac{\partial F_2}{\partial u_1} & \dfrac{\partial F_2}{\partial u_2} & \dfrac{\partial F_2}{\partial u_3} \\[2mm] \dfrac{\partial F_3}{\partial u_1} & \dfrac{\partial F_3}{\partial u_2} & \dfrac{\partial F_3}{\partial u_3} \end{vmatrix} \times \begin{vmatrix} \dfrac{\partial u_1}{\partial x_1} & \dfrac{\partial u_1}{\partial x_2} & \dfrac{\partial u_1}{\partial x_3} \\[2mm] \dfrac{\partial u_2}{\partial x_1} & \dfrac{\partial u_2}{\partial x_2} & \dfrac{\partial u_2}{\partial x_3} \\[2mm] \dfrac{\partial u_3}{\partial x_1} & \dfrac{\partial u_3}{\partial x_2} & \dfrac{\partial u_3}{\partial x_3} \end{vmatrix} = \begin{vmatrix} \sum \dfrac{\partial F_1}{\partial u_r} \cdot \dfrac{\partial u_r}{\partial x_1} & \sum \dfrac{\partial F_1}{\partial u_r} \cdot \dfrac{\partial u_r}{\partial x_2} & \sum \dfrac{\partial F_1}{\partial u_r} \cdot \dfrac{\partial u_r}{\partial x_3} \\[2mm] \sum \dfrac{\partial F_2}{\partial u_r} \cdot \dfrac{\partial u_r}{\partial x_1} & \sum \dfrac{\partial F_2}{\partial u_r} \cdot \dfrac{\partial u_r}{\partial x_2} & \sum \dfrac{\partial F_2}{\partial u_r} \cdot \dfrac{\partial u_r}{\partial x_3} \\[2mm] \sum \dfrac{\partial F_3}{\partial u_r} \cdot \dfrac{\partial u_r}{\partial x_1} & \sum \dfrac{\partial F_3}{\partial u_r} \cdot \dfrac{\partial u_r}{\partial x_2} & \sum \dfrac{\partial F_3}{\partial u_r} \cdot \dfrac{\partial u_r}{\partial x_3} \end{vmatrix}$$

Now, using (2), we get

$$
= -\begin{vmatrix} -\dfrac{\partial F_1}{\partial x_1} & -\dfrac{\partial F_1}{\partial x_2} & -\dfrac{\partial F_1}{\partial x_3} \\[2mm] -\dfrac{\partial F_2}{\partial x_1} & -\dfrac{\partial F_2}{\partial x_2} & -\dfrac{\partial F_2}{\partial x_3} \\[2mm] -\dfrac{\partial F_3}{\partial x_1} & -\dfrac{\partial F_3}{\partial x_2} & -\dfrac{\partial F_3}{\partial x_3} \end{vmatrix} = (-1)^3 \begin{vmatrix} \dfrac{\partial F_1}{\partial x_1} & \dfrac{\partial F_1}{\partial x_2} & \dfrac{\partial F_1}{\partial x_3} \\[2mm] \dfrac{\partial F_2}{\partial x_1} & \dfrac{\partial F_2}{\partial x_2} & \dfrac{\partial F_2}{\partial x_3} \\[2mm] \dfrac{\partial F_3}{\partial x_1} & \dfrac{\partial F_3}{\partial x_2} & \dfrac{\partial F_3}{\partial x_3} \end{vmatrix} = (-1)^3 \frac{\partial(F_1, F_2, F_3)}{\partial(x_1, x_2, x_3)}
$$

Hence,
$$
\frac{\partial(u_1, u_2, u_3)}{\partial(x_1, x_2, x_3)} = (-1)^3 \left[\frac{\partial(F_1, F_2, F_3)}{\partial(x_1, x_2, x_3)} \Big/ \frac{\partial(F_1, F_2, F_3)}{\partial(u_1, u_2, u_3)} \right]
$$

Generalization. Let $u_1, u_2, ..., u_n$ be the implicit functions of $x_1, x_2, ..., x_n$, that is

$$
\begin{aligned}
F_1(u_1, u_2, ..., u_n, x_1, x_2, ..., x_n) &= 0 \\
F_2(u_1, u_2, ..., u_n, x_1, x_2, ..., x_n) &= 0 \\
\vdots \qquad \vdots \qquad \vdots \qquad \vdots \qquad \vdots \\
F_n(u_1, u_2, ..., u_n, x_1, x_2, ..., x_n) &= 0
\end{aligned}
$$

Then,
$$
\frac{\partial(u_1, u_2, ..., u_n)}{\partial(x_1, x_2, ..., x_n)} = (-1)^n \left[\frac{\partial(F_1, F_2, ..., F_n)}{\partial(x_1, x_2, ..., x_n)} \Big/ \frac{\partial(F_1, F_2, ..., F_n)}{\partial(u_1, u_2, ..., u_n)} \right]
$$

The proof may be easily extended as in case of two and three implicit functions. (Theorem 7 and Theorem 8).

3.4.5 Necessary and Sufficient Conditions for a Jacobian to be Vanish

THEOREM 1.   *If $v_1, v_2, ..., v_n$ be the functions of n independent variables $x_1, x_2, ..., x_n$. For $F(v_1, v_2, ..., v_n) = 0$, it is necessary and sufficient that the Jacobian $\dfrac{\partial(v_1, v_2, ..., v_n)}{\partial(x_1, x_2, ..., x_n)}$ should vanish identically.*

PROOF.   **Necessary Condition.** Suppose that there exists a relation of $v_1, v_2, ..., v_n$ such that
$$
F(v_1, v_2, ..., v_n) = 0 \tag{...1}
$$
To show that Jacobian is necessarily zero.

Differentiating (1), partially w.r.t. $x_1, x_2, ..., x_n$ respectively, we get

$$
\begin{aligned}
\frac{\partial F}{\partial v_1} \cdot \frac{\partial v_1}{\partial x_1} + \frac{\partial F}{\partial v_2} \cdot \frac{\partial v_2}{\partial x_1} + ... + \frac{\partial F}{\partial v_n} \cdot \frac{\partial v_n}{\partial x_1} &= 0 \\
\frac{\partial F}{\partial v_1} \cdot \frac{\partial v_1}{\partial x_2} + \frac{\partial F}{\partial v_2} \cdot \frac{\partial v_2}{\partial x_2} + ... + \frac{\partial F}{\partial v_n} \cdot \frac{\partial v_n}{\partial x_2} &= 0 \\
... \qquad ... \qquad ... \\
\frac{\partial F}{\partial v_1} \cdot \frac{\partial v_1}{\partial x_n} + \frac{\partial F}{\partial v_2} \cdot \frac{\partial v_2}{\partial x_n} + ... + \frac{\partial F}{\partial v_n} \cdot \frac{\partial v_n}{\partial x_n} &= 0
\end{aligned}
$$

Now eliminating $\dfrac{\partial F}{\partial v_1}, \dfrac{\partial F}{\partial v_2}, ..., \dfrac{\partial F}{\partial v_n}$ from these equations, we get

$$
\begin{vmatrix} \dfrac{\partial v_1}{\partial x_1} & \dfrac{\partial v_2}{\partial x_1} & ... & \dfrac{\partial v_n}{\partial x_1} \\[2mm] \dfrac{\partial v_1}{\partial x_2} & \dfrac{\partial v_2}{\partial x_2} & ... & \dfrac{\partial v_n}{\partial x_2} \\[2mm] ... & ... & ... & ... \\[2mm] \dfrac{\partial v_1}{\partial x_n} & \dfrac{\partial v_2}{\partial x_n} & ... & \dfrac{\partial v_n}{\partial x_n} \end{vmatrix} = 0
$$

$\Rightarrow$
$$
\frac{\partial(v_1, v_2, ..., v_n)}{\partial(x_1, x_2, ..., x_n)} = 0
$$

Sufficient Condition. If the Jacobian $J(v_1, v_2, ..., v_n)$ is zero, then to show that there must exist a relation between $v_1, v_2, ..., v_n$. The equation connecting the functions $v_1, v_2, ..., v_n$ and the variables $x_1, x_2, ..., x_n$ can be written as

$$
\begin{aligned}
g_1(x_1, x_2, ..., x_n, v_1) &= 0 \\
g_2(x_2, x_3, ..., x_n, v_1, v_2) &= 0 \\
... \qquad ... \qquad ... \qquad ... \qquad ... \\
g_k(x_k, x_{k+1}, ..., x_n, v_1, v_2, ..., v_k) &= 0 \\
... \qquad ... \qquad ... \qquad ... \qquad ... \\
g_n(x_n, v_1, v_2, ..., v_n) &= 0
\end{aligned}
$$

Then, we have
$$J = \frac{\partial(v_1, v_2, ..., v_n)}{\partial(x_1, x_2, ..., x_n)} = (-1)^n \frac{\left[\dfrac{\partial(g_1, g_2, ..., g_n)}{\partial(x_1, x_2, ..., x_n)}\right]}{\left[\dfrac{\partial(g_1, g_2, ..., g_n)}{\partial(v_1, v_2, ..., v_n)}\right]} = (-1)^n \frac{\left(\dfrac{\partial g_1}{\partial x_1} \cdot \dfrac{\partial g_2}{\partial x_2} ... \dfrac{\partial g_n}{\partial x_n}\right)}{\left(\dfrac{\partial g_1}{\partial v_1} \cdot \dfrac{\partial g_2}{\partial v_2} ... \dfrac{\partial g_n}{\partial v_n}\right)}$$

If $J = 0$, then
$$\frac{\partial g_1}{\partial x_1} \cdot \frac{\partial g_2}{\partial x_2} ... \frac{\partial g_r}{\partial x_r} ... \frac{\partial g_n}{\partial x_n} = 0$$

$\Rightarrow$ At least one of $\dfrac{\partial v_1}{\partial x_1}, \dfrac{\partial v_2}{\partial x_2}, ..., \dfrac{\partial v_n}{\partial x_n}$ is zero.

$\Rightarrow$ $\dfrac{\partial g_k}{\partial x_k} = 0$ for some value of k between 1 and n.

$\Rightarrow$ For that particular value of k, the function g_k must not contain x_k and hence
$$g_k(x_{k+1}, ..., x_n, v_1, v_2, ..., v_k) = 0 \qquad ...(2)$$

Now we may easily eliminate the variables $x_{k+1}, x_{k+2}, ..., x_n$ between (2) and $g_{r+1} = 0, g_{r+2} = 0, ..., g_n = 0$ and an equation between $v_1, v_2, ..., v_n$ alone, can be obtained.

Solved Examples

EXAMPLE 1. *If $x = r \cos\theta$, $y = r \sin\theta$, show that*

(i) $\dfrac{\partial(x, y)}{\partial(r, \theta)} = r$ **(ii)** $\dfrac{\partial(r, \theta)}{\partial(x, y)} = \dfrac{1}{r}$

(GBTU (SUM) 2010)

SOLUTION. (a) We have $\dfrac{\partial(x, y)}{\partial(r, \theta)} = \begin{vmatrix} \partial x/\partial r & \partial x/\partial\theta \\ \partial y/\partial r & \partial y/\partial\theta \end{vmatrix}$

$= \begin{vmatrix} \cos\theta & -r\sin\theta \\ \sin\theta & r\cos\theta \end{vmatrix}$

$= r\cos^2\theta + r\sin^2\theta = r$

(b) From the given relations, we get
$$r^2 = x^2 + y^2 \text{ and } \tan\theta = y/x$$

Now differentiating partially w.r.t. x and y, we obtain

$2r\dfrac{\partial r}{\partial x} = 2x$ or $\dfrac{\partial r}{\partial x} = \dfrac{x}{r}$

$2r\dfrac{\partial r}{\partial y} = 2y$ or $\dfrac{\partial r}{\partial y} = \dfrac{y}{r}$

and $\tan\theta = y/x \Rightarrow \sec^2\theta \dfrac{\partial\theta}{\partial x} = -\dfrac{y}{x^2}$

or $\dfrac{\partial\theta}{\partial x} = -\dfrac{y}{x^2\sec^2\theta} = -\dfrac{y}{r^2\cos^2\theta\sec^2\theta} = -\dfrac{y}{r^2}$

and $\sec^2\theta\dfrac{\partial\theta}{\partial y} = \dfrac{1}{x}$

or $\dfrac{\partial\theta}{\partial y} = \dfrac{1}{x\sec^2\theta} = \dfrac{\cos^2\theta}{x} = \dfrac{x^2}{r^2} \cdot \dfrac{1}{x} = \dfrac{x}{r^2}$

$\dfrac{\partial(r, \theta)}{\partial(x, y)} = \begin{vmatrix} \dfrac{\partial r}{\partial x} & \dfrac{\partial r}{\partial y} \\ \dfrac{\partial\theta}{\partial x} & \dfrac{\partial\theta}{\partial y} \end{vmatrix} = \begin{vmatrix} x/r & y/r \\ -y/r^2 & x/r^2 \end{vmatrix}$

$= \dfrac{x^2}{r^3} + \dfrac{y^2}{r^3} = \dfrac{x^2 + y^2}{r^3} = \dfrac{r^2}{r^3} = \dfrac{1}{r}$

EXAMPLE 2. *If $x = r\sin\theta\cos\phi$, $y = r\sin\theta\sin\phi$, $z = r\cos\theta$, show that*
$$\frac{\partial(x, y, z)}{\partial(r, \theta, \phi)} = r^2\sin\theta \qquad \text{(GBTU (C.O) 2010)}$$

SOLUTION. We know that

$$\frac{\partial(x, y, z)}{\partial(r, \theta, \phi)} = \begin{vmatrix} \dfrac{\partial x}{\partial r} & \dfrac{\partial x}{\partial\theta} & \dfrac{\partial x}{\partial\phi} \\ \dfrac{\partial y}{\partial r} & \dfrac{\partial y}{\partial\theta} & \dfrac{\partial y}{\partial\phi} \\ \dfrac{\partial z}{\partial r} & \dfrac{\partial z}{\partial\theta} & \dfrac{\partial z}{\partial\phi} \end{vmatrix}$$

$$= \begin{vmatrix} \sin\theta\cos\phi & r\cos\theta\cos\phi & -r\sin\theta\sin\phi \\ \sin\theta\sin\phi & r\cos\theta\sin\phi & r\sin\theta\cos\phi \\ \cos\theta & -r\sin\theta & 0 \end{vmatrix}$$

[expanding the determinant along the third row]

$= \cos\theta\,(r^2\sin\theta\cos\theta\cos^2\phi + r^2\sin\theta\cos\theta\sin^2\phi)$
$\qquad + r\sin\theta(r\sin^2\theta\cos^2\phi + r\sin^2\theta\sin^2\phi)$
$= r^2\sin\theta\cos^2\theta + r^2\sin^3\theta = r^2\sin\theta(\cos^2\theta + \sin^2\theta)$
$= r^2\sin\theta.$

EXAMPLE 3. *If $y_1 = r\sin\theta_1\sin\theta_2$, $y_2 = r\sin\theta_1\cos\theta_2$, $y_3 = r\cos\theta_1\sin\theta_3$, $y_4 = r\cos\theta_1\cos\theta_3$, find the value of Jacobian.*

SOLUTION. We have
$$y_1 = r\sin\theta_1\sin\theta_2 \qquad ...(1)$$
$$y_2 = r\sin\theta_1\cos\theta_2 \qquad ...(2)$$
$$y_3 = r\cos\theta_1\sin\theta_3 \qquad ...(3)$$
$$y_4 = r\cos\theta_1\cos\theta_3 \qquad ...(4)$$

Squaring and adding the given four relations, we get
$$y_1^2 + y_2^2 + y_3^2 + y_4^2 = r^2$$

$\therefore$ $\left. \begin{array}{l} y_1\dfrac{\partial y_1}{\partial r} + y_2\dfrac{\partial y_2}{\partial r} + y_3\dfrac{\partial y_3}{\partial r} + y_4\dfrac{\partial y_4}{\partial r} = r \\ \text{and} \quad y_1\dfrac{\partial y_1}{\partial\theta_i} + y_2\dfrac{\partial y_2}{\partial\theta_i} + y_3\dfrac{\partial y_3}{\partial\theta_i} + y_4\dfrac{\partial y_4}{\partial\theta_i} = 0, \\ \qquad\qquad\qquad\qquad\qquad\qquad\qquad i = 1, 2, 3 \end{array} \right\} ...(5)$

Also $y_3^2 + y_4^2 = r^2\cos^2 q_1$, so that

$\left. \begin{array}{l} y_3\dfrac{\partial y_3}{\partial\theta_1} + y_4\dfrac{\partial y_4}{\partial\theta_1} = -r^2\cos\theta_1\sin\theta_1 \\ y_3\dfrac{\partial y_3}{\partial\theta_j} + y_4\dfrac{\partial y_4}{\partial\theta_j} = 0, j = 2, 3 \end{array} \right\} ...(6)$

Now the required Jacobian

$$J = \begin{vmatrix} \partial y_1/\partial r & \partial y_1/\partial\theta_1 & \partial y_1/\partial\theta_2 & \partial y_1/\partial\theta_3 \\ \partial y_2/\partial r & \partial y_2/\partial\theta_1 & \partial y_2/\partial\theta_2 & \partial y_2/\partial\theta_3 \\ \partial y_3/\partial r & \partial y_3/\partial\theta_1 & \partial y_3/\partial\theta_2 & \partial y_3/\partial\theta_3 \\ \partial y_4/\partial r & \partial y_4/\partial\theta_1 & \partial y_4/\partial\theta_2 & \partial y_4/\partial\theta_3 \end{vmatrix}$$

Operating, $y_1 R_1 + (y_2 R_2 + y_3 R_3 + y_4 R_4)$ and using result (5), we obtain

$$J = \frac{1}{y_1} \begin{vmatrix} r & 0 & 0 & 0 \\ \partial y_2/\partial r & \partial y_2/\partial\theta_1 & \partial y_2/\partial\theta_2 & \partial y_2/\partial\theta_3 \\ \partial y_3/\partial r & \partial y_3/\partial\theta_1 & \partial y_3/\partial\theta_2 & \partial y_3/\partial\theta_3 \\ \partial y_4/\partial r & \partial y_4/\partial\theta_1 & \partial y_4/\partial\theta_2 & \partial y_4/\partial\theta_3 \end{vmatrix}$$

$$= \frac{r}{y_1} \begin{vmatrix} \dfrac{\partial y_2}{\partial\theta_1} & \dfrac{\partial y_2}{\partial\theta_2} & \dfrac{\partial y_2}{\partial\theta_3} \\ \dfrac{\partial y_3}{\partial\theta_1} & \dfrac{\partial y_3}{\partial\theta_2} & \dfrac{\partial y_3}{\partial\theta_3} \\ \dfrac{\partial y_4}{\partial\theta_1} & \dfrac{\partial y_4}{\partial\theta_2} & \dfrac{\partial y_4}{\partial\theta_3} \end{vmatrix} = \frac{r}{y_1 y_3} \begin{vmatrix} \dfrac{\partial y_2}{\partial\theta_1} & \dfrac{\partial y_2}{\partial\theta_2} & \dfrac{\partial y_2}{\partial\theta_3} \\ -r^2\cos\theta_1\sin\theta_1 & 0 & 0 \\ \dfrac{\partial y_4}{\partial\theta_1} & \dfrac{\partial y_4}{\partial\theta_2} & \dfrac{\partial y_4}{\partial\theta_3} \end{vmatrix}$$

[Adding $y_4 R_3$ to $y_3 R_2$ and using the result (6)]

$$= \frac{r}{y_1 y_3} . r^2 \cos\theta_1 \sin\theta_1 \left[\frac{\partial y_2}{\partial\theta_2} . \frac{\partial y_4}{\partial\theta_3} - \frac{\partial y_4}{\partial\theta_2} . \frac{\partial y_2}{\partial\theta_3} \right]$$

$$= \frac{r^3 \cos\theta_1 \sin\theta_1}{y_1 y_3} \left[(-r\sin\theta_1\sin\theta_2)(-r\cos\theta_1\sin\theta_3) - 0 \right]$$

$$= \frac{r^5 \sin^2\theta_1 \cos^2\theta_1 \sin\theta_2 \sin\theta_3}{r^2 \sin\theta_1 \cos\theta_1 \sin\theta_2 \sin\theta_3}$$

$$= r^3 \sin\theta_1 \cos\theta_1.$$

EXAMPLE 4. *If $y_1 = 1 - x_1$, $y_2 = x_1(1 - x_2)$, $y_3 = x_1 x_2 (1 - x_3)$...*
$y_n = x_1 x_2 ... x_{n-1}(1 - x_n)$. Prove that
$J(y_1, y_2, ... y_n) = (-1)^n x_1^{n-1} . x_2^{n-2} ... x_{n-1}.$

SOLUTION. In the above relations

y_1 is a function of x_1
y_2 is a function of x_1, x_2
y_3 is a function of $x_1 x_2 x_3$
...　　　...　　　...　　...　　...
and y_n is a function of $x_1 x_2, ..., x_n$.

$$\therefore \frac{\partial(y_1, y_2, ..., y_n)}{\partial(x_1, x_2, ..., x_n)} = \begin{vmatrix} \dfrac{\partial y_1}{\partial x_1} & 0 & 0 & \cdots & 0 \\ \dfrac{\partial y_2}{\partial x_1} & \dfrac{\partial y_2}{\partial x_2} & 0 & \cdots & 0 \\ \vdots & \vdots & \vdots & \vdots & \vdots \\ \dfrac{\partial y_n}{\partial x_1} & \dfrac{\partial y_n}{\partial x_2} & \dfrac{\partial y_n}{\partial x_3} & \cdots & \dfrac{\partial y_n}{\partial x_n} \end{vmatrix}$$

$$= \frac{\partial y_1}{\partial x_1} . \frac{\partial y_2}{\partial x_2} . \frac{\partial y_3}{\partial x_3} ... \frac{\partial y_n}{\partial x_n}$$

$$= (-1)(-x_1)(-x_1 x_2)...(-x_1 x_2 x_3 ... x_{n-1})$$

$$= (-1)^n x_1^{n-1} . x_2^{n-2} ... x_{n-1}.$$

EXAMPLE 5. *If $u = x + 2y + z$, $v = x - 2y + 3z$ and $w = 2xy - xz + 4yz - 2z^2$, then prove that they are not independent. Find the relation between u, v and w.*

[UPTU (SUM) 2007; G.B.T.U(C.O.) 2011]

SOLUTION. We have

$$\frac{\partial(u, v, w)}{\partial(x, y, z)} = \begin{vmatrix} 1 & 2 & 1 \\ 1 & -2 & 3 \\ 2y - z & 2x + 4z & -x + 4y - 4z \end{vmatrix}$$

$$= \begin{vmatrix} 1 & 0 & 0 \\ 1 & -4 & 2 \\ 2y - z & -x + 2y - 6z & -x + 2y - 3z \end{vmatrix}$$

by $c_2 - 2c_1$ and $c_3 - c_1$

$$= -2 \begin{vmatrix} 1 & 0 & 0 \\ 1 & 2 & 2 \\ 2y - z & -x + 2y - 3z & -x + 2y - 3z \end{vmatrix} = 0$$

Here last two columns are identical. So the Jacobian of the functions u, v, w is zero, therefore these functions are not independent so there must exist a relation between them.

We have $u^2 - v^2 = (x + 2y + z)^2 - (x - 2y + 3z)^2$

$$= (2x + 4z)(4y - 2z) = 4(x + 2z)(2y - z)$$

By simplification $= 4(2xy - xz + 4yz - 2z^2) = 4w$

Therefore $u^2 - v^2 = 4w$, which is the required relation between u, v and w.

Exercise-3.6

1. If $u = \dfrac{y^2}{2x}$ and $v = \dfrac{x}{2} + \dfrac{y^2}{2x}$, find $\dfrac{\partial(u, v)}{\partial(x, y)}$.

2. (a) If $u_1 = \dfrac{x_2 x_3}{x_1}$, $u_2 = \dfrac{x_3 x_1}{x_2}$, $u_3 = \dfrac{x_1 x_2}{x_3}$,

then show that $J(u_1, u_2, u_3) = 4$.

(GBTU 2011, 2012; GBTU(AG) SUM 2010; UKTU 2011)

(b) If $x = u(1 + v)$, $y = v(1 + u)$, then show that

$$\frac{\partial(x, y)}{\partial(u, v)} = 1 + u + v. \qquad \text{(UPTU (SUM) 2009)}$$

3. If $x = \sin\theta\sqrt{1 - a^2\sin^2\phi}$, $y = \cos\theta\cos\phi$, then show that

$$\frac{\partial(x, y)}{\partial(\theta, \phi)} = -\sin\phi \frac{\left[(1 - a^2)\cos^2\theta + a^2\cos^2\phi \right]}{\sqrt{1 - a^2\sin^2\phi}}$$

4. If $y_1 = x_1(1 - x_1)$, $y_2 = x_1 x_2(1 - x_3)$,...,

$y_{n-1} = x_1 x_2 ... x_{n-1}(1 - x_n)$, $y_n = x_1 x_2 ... x_n$.

Then prove that $\dfrac{\partial(y_1, y_2, ..., y_n)}{\partial(x_1, x_2, ..., x_n)} = x_1^{n-1} . x_2^{n-2} .... . x_{n-1}.$

5. If $y_1 = \cos x_1$, $y_2 = \sin x_1 \cos x_2$, $y_3 = \sin x_1 \sin x_2 \cos x_3$,..., $y_n = \sin x_1 \sin x_2 \sin x_3 ... \sin x_{n-1} \cos x_n$. Then find the Jacobian of $y_1, y_2, ..., y_n$ with respect to $x_1, x_2, ..., x_n$.

6. Show that $\dfrac{\partial(u, v)}{\partial(x, y)} . \dfrac{\partial(x, y)}{\partial(u, v)} = 1$.

7. If $u^3 = xyz$, $\dfrac{1}{v} = \dfrac{1}{x} + \dfrac{1}{y} + \dfrac{1}{z}$, $w^2 = x^2 + y^2 + z^2$, show that

$$\frac{\partial(u, v, w)}{\partial(x, y, z)} = \frac{-v(y - z)(z - x)(x - y)(x + y + z)}{3u^2 w(yz + zx + xy)}$$

8. If $u=2xy$, $v=x^2-y^2$, and $x=r\cos\theta$, $y=r\sin\theta$, show that

$$\frac{\partial(u,v)}{\partial(r,\theta)}=-4r^3.$$

9. (i) If $u_1=x_1+x_2+x_3+x_4$, $u_1u_2=x_2+x_3+x_4$, $u_1u_2u_3=x_3+x_4$,

$u_1u_2u_3u_4=x_4$. Show that $\dfrac{\partial(x_1,x_2,x_3,x_4)}{\partial(u_1,u_2,u_3,u_4)}=u_1^3u_2^2u_3$.

(ii) If $x+y+z=u$, $y+z=uv$, $z=uvw$, show that

$$\frac{\partial(x,y,z)}{\partial(u,v,w)}=u^2v.$$

10. If $y_1.(x_1-x_2)=0$, $y_2.(x_1^2+x_1x_2+x_2^2)=0$, show that

$$\frac{\partial(y_1,y_2)}{\partial(x_1,x_2)}=3y_1y_2\frac{x_1+x_2}{x_1^3-x_2^3}.$$

11. If l,m,n are the roots of the equation in k.

$$\frac{x}{a+k}+\frac{y}{b+k}+\frac{z}{c+k}=1$$

Prove that $\dfrac{\partial(x,y,z)}{\partial(l,m,n)}=-\left[\dfrac{(m-n)(n-l)(l-m)}{(b-c)(c-a)(a-b)}\right]$.

12. If the roots of the equation

$(\lambda-x)^3+(\lambda-y)^3+(\lambda-z)^3=0$ are u,v,w. Prove that

$$\frac{\partial(u,v,w)}{\partial(x,y,z)}=-2\frac{(y-z)(z-x)(x-y)}{(v-w)(w-u)(u-v)}. \qquad \text{[UKTU 2011, 18]}$$

13. Show that the functions $u=x+y-z$, $v=x-y+z$, $w=x^2+y^2+z^2-2yz$ are not independent of one another. Also find the relation between them.

14. If $u=x^2+y^2+z^2$, $v=x+y+z$, $w=xy+yz+zx$. Show that the Jacobian $\dfrac{\partial(u,v,w)}{\partial(x,y,z)}$ vanish identically. Also find the relation between u,v and w. [UKTU 2010; UPTU 2009]

15. If $u=x+y+z+t$, $v=x+y-z-t$, $w=xy-zt$, $r=x^2+y^2-z^2-t^2$. Show that $\dfrac{\partial(u,v,w,r)}{\partial(x,y,z,t)}=0$.

16. If $f(0)=0$ and $f'(x)=\dfrac{1}{1+x^2}$. Then prove that

$$f(x)+f(y)=f\left(\frac{(x+y)}{(1-xy)}\right) \text{ (without using the method of integration)}.$$

17. $u=x\left(1-r^2\right)^{-\frac{1}{2}}$, $v=y\left(1-r^2\right)^{-\frac{1}{2}}$, $w=z\left(1-r^2\right)^{-\frac{1}{2}}$ where

$r^2=x^2+y^2+z^2$, then show that $\dfrac{\partial(u,v,w)}{\partial(x,y,z)}=\left(1-r^2\right)^{-5/2}$.

[UPTU(SUM) 2009]

18. If $u=\dfrac{x+y}{z}$, $v=\dfrac{y+z}{x}$, $w=y\dfrac{(x+y+z)}{xz}$

then show that u,v,w are not independent. (GBTU 2010)

19. If $x_1+x_2+...+x_n=y_1$

$x_2+x_3+...+x_n=y_1y_2$

$x_3+x_4+...+x_n=y_1y_2y_3$

$... \quad ... \quad ... \quad ... \quad ... \quad ...$

$x_n=y_1y_2y_3...y_n$

then show that $\dfrac{\partial(x_1,x_2,...,x_n)}{\partial(y_1,y_2,...,y_n)}=y_1^{n-1}y_2^{n-2}...y_{n-2}^2y_{n-1}$.

20. If $u_1=\dfrac{x_1}{x_n}$, $u_2=\dfrac{x_2}{x_n}$, $u_3=\dfrac{x_3}{x_n}$, $..,u_{n-1}=\dfrac{x_{n-1}}{x_n}$ and

$x_1^2+x_2^2+...x_{n-1}^2+x_n^2=1$, find the value of $\dfrac{\partial(u_1,u_2,...,u_n)}{\partial(x_1,x_2,...,x_n)}$.

21. (i) Calculate the Jacobian $\dfrac{\partial(x,y,z)}{\partial(u,v,w)}$ of the following :

$u=x+2y+z$, $v=x+2y+3z$, $w=2x+3y+5z$. (UPTU 2008)

(ii) If $u=xyz$, $v=xy+yz+zx$, $w=x+y+z$, then compute the jacobian $\dfrac{\partial(x,y,z)}{\partial(u,v,w)}$. (UPTU (SUM) 2008)

22. (i) Verify the chain rule for Jacobians if $x=u$, $y=u\tan v$, $z=w$. (UPTU 2009)

(ii) If $x=\sqrt{vw}$, $y=\sqrt{wu}$, $z=\sqrt{uv}$ and $u=r\sin\theta\cos\phi$, $v=r\sin\theta\sin\phi$, $w=r\cos\theta$ then calculate the Jacobian $\dfrac{\partial(x,y,z)}{\partial(u,v,w)}$ (UPTU, MODEL PAPER 2008)

23. If $u=xyz$, $v=x^2+y^2+z^2$, $w=x+y+z$, find the Jacobian $\dfrac{\partial(x,y,z)}{\partial(u,v,w)}$ (UKTU 2012, 16)

24. If $x=e^v\sec u$, $y=e^v\tan u$, then evaluate $\dfrac{\partial(x,y)}{\partial(u,v)}$. (GBTU 2010)

Hints to Selected Problems

1. $\dfrac{\partial u}{\partial x}=\dfrac{-y^2}{2x^2}$, $\dfrac{\partial u}{\partial y}=\dfrac{y}{x}$, $\dfrac{\partial v}{\partial x}=\dfrac{1}{2}-\dfrac{y^2}{2x^2}$, $\dfrac{\partial v}{\partial y}=\dfrac{y}{x}$

2. $\dfrac{\partial u_1}{\partial x_1}=-\dfrac{x_2x_3}{x_1^2}$, $\dfrac{\partial u_1}{\partial x_2}=\dfrac{x_3}{x_1}$, $\dfrac{\partial u_1}{\partial x_3}=\dfrac{x_2}{x_1}$

$\dfrac{\partial u_2}{\partial x_1}=\dfrac{x_3}{x_2}$, $\dfrac{\partial u_2}{\partial x_2}=-\dfrac{x_3x_1}{x_2}$, $\dfrac{\partial u_2}{\partial x_3}=\dfrac{x_1}{x_2}$, $\dfrac{\partial u_3}{\partial x_1}=\dfrac{x_2}{x_3}$, $\dfrac{\partial u_3}{\partial x_2}=\dfrac{x_1}{x_3}$.

3. $\dfrac{\partial x}{\partial\theta}=\cos\theta\sqrt{1-a^2\sin^2\phi}$, $\dfrac{\partial x}{\partial\phi}=\dfrac{-a^2\sin\theta\sin\phi\cos\phi}{\sqrt{1-a^2\sin^2\phi}}$

$\dfrac{\partial y}{\partial\theta}=-\sin\theta\cos\phi$, $\dfrac{\partial y}{\partial\phi}=-\cos\theta\sin\phi$

9. $u_1=x_1+x_3+x_4$, $u_2=x_2+x_3+x_4$, $u_1u_2u_3=x_3+x_4$, $u_1u_2u_3u_4=x_4$. Therefore $x_3=u_1u_2u_3(1-u_4)$, $x_2=u_1u_2(1-u_3)$, $x_1=u_1(1-u_2)$.

Now find required partial derivative of x_1, x_2 and x_3 w.r.t. u_1, u_2 and u_3.

13. If $\dfrac{\partial(u,v,w)}{\partial(x,y,z)}=0$ Then u,v,w are not independent.

Answers

1. $-\dfrac{y}{2x}$ **5.** $(-1)^n\sin^nx_1\sin^{n-1}x_2\sin^{n-2}x_3...\sin x_n$ **13.** $u^2+v^2=2w$ **14.** $v^2=u+2w$ **15.** $uv=r+2w$ **20.** $\dfrac{1}{x_n^n}$

21. (i) 2 (ii) $(x-y)(y-z)(z-x)$ **22.**(ii) $\dfrac{1}{4}r^2\sin\theta$ **23.** $\dfrac{-1}{2(x-y)(y-z)(z-x)}$ **24.** $-e^{2v}\sec u$

3.5 CONCEPTS OF MAXIMA AND MINIMA

If $y = f(x)$ be a continuous function. At a point $x = x_1$, if the function $f(x)$ does not increase and begins to decrease then $f(x)$ has its maximum value at $x = x_1$ and if at a point $x = x_2$, $f(x)$ does not decrease and begins to increase, then $f(x)$ has its minimum value at $x = x_2$.

If $f(x)$ is maximum at a point $x = x_1$ then $f(x)$ is an increasing function for the preceding values of x_1 and is a decreasing function for those value of x just below x_1 or we can say derivative of the function $\left(i.e., \dfrac{dy}{dx}\right)$ will be positive before $x = x_1$ and will be negative after $x = x_1$. But $\dfrac{dy}{dx}$ is a continuous function and $\dfrac{dy}{dx}$ changes the sign from positive to negative. So, $\dfrac{dy}{dx}$ will be zero at any point.

Therefore, for a maximum value of $y = f(x)$ at a point, we have $\dfrac{dy}{dx} = 0$ and $\dfrac{dy}{dx}$ changes the sign from positive to negative. On the other hand, for a minimum value of $y = f(x)$ at point we have $\dfrac{dy}{dx} = 0$ and $\dfrac{dy}{dx}$ changes the sign negative to positive.

☛ REMARKS

- If $\dfrac{dy}{dx}$ changes the sign positive to negative; it means that $f(x)$ is a decreasing function of x i.e., $\dfrac{d^2y}{dx^2} < 0$.

- If $\dfrac{dy}{dx}$ changes the sign from negative to positive, it means that the $f(x)$ is an increasing function of x, i.e., $\dfrac{d^2y}{dx^2} > 0$.

- A function may have more than one maximum and minimum value.
- Any minimum value of the function $f(x)$ can be greater than any maximum value.
- Maximum and minimum values of the function occur alternately.
- Maximum and minimum values of the function are sometimes known as extreme value.
- From the definition of maxima and minima, it is clear that $\dfrac{dy}{dx} = 0$ is the necessary condition for maximum or minimum.

- $\dfrac{d^2y}{dx^2} < 0$ is sufficient condition for maximum and $\dfrac{d^2y}{dx^2} > 0$ is sufficient condition for minimum.

WORKING PROCEDURE

Step 1. Find the derivative of the given function *i.e.*, $\dfrac{dy}{dx}$.

Step 2. Put $\dfrac{dy}{dx} = 0$ and find all the real values of x. (say $x_1, x_2, x_3 \dots$).

Step 3. Find $\dfrac{d^2y}{dx^2}$.

Step 4. Put $x = x_i$ in $\dfrac{d^2y}{dx^2}$ and find the result. If result is negative then the function $f(x)$ is maximum at $x = x_i$ and max. $f(x) = f(x_i)$. On the other hand, if result is positive then the function $f(x)$ is minimum at $x = x_i$ and minimum $f(x) = f(x_i)$.

☛ REMARKS

- In a continuous function, maxima and minima values occur alternately, *i.e.*, between two successive maxima there is one minimum and between two successive minima, there is one maximum.

- If $\dfrac{d^2y}{dx^2}$ is equal to 0 at any point $x = x_i$ then find $\dfrac{d^3y}{dx^3}, \dfrac{d^4y}{dx^4}$, and find the values of these derivatives at $x = x_i$ successively and check the sign.

Solved Examples

EXAMPLE 1. *Find the value of x for which $f(x) = y = x^4 + 2x^3 - 3x^2 - 4x + 4$ is maximum or minimum and also find those value of $f(x)$.*

SOLUTION. Here, the given function is

$$y = f(x) = x^4 + 2x^3 - 3x^2 - 4x + 4 \qquad \dots(1)$$

So $\dfrac{dy}{dx} = 4x^3 + 6x^2 - 6x - 4 = 2(x+2)(2x+1)(x-1)$

Now, put $\dfrac{dy}{dx} = 0$, we have

$$2(x+2)(2x+1)(x-1) = 0$$

So, $\qquad x = -2, -\dfrac{1}{2}, 1$

Again differentiating (2) w.r.t. to x, we get

$$\dfrac{d^2y}{dx^2} = 12x^2 + 12x - 6$$

At $\quad x = -2$, we have

$$\dfrac{d^2y}{dx^2} = 12(-2)^2 + 12(-2) - 6 = 48 - 24 - 6 = 18 > 0$$

Since, $\dfrac{d^2y}{dx^2} > 0$ (*i.e.*, positive). So $f(x)$ is minimum at

$x=-2$. The minimum value of $f(x)$ at $x=-2$ is given by

$$f(-2) = (-2)^4 + 2(-2)^3 - 3(-2)^2 - 4(-2) + 4 = 0$$

Now, at $x = -\dfrac{1}{2}$, we have

$$\frac{d^2y}{dx^2} = 12\left(-\frac{1}{2}\right)^2 + 12\left(-\frac{1}{2}\right) - 6 = 3 - 6 - 6 = -9 < 0$$

Since, $\dfrac{d^2y}{dx^2} < 0$ (i.e., negative). So, $f(x)$ is maximum at

$x = -\dfrac{1}{2}$ and maximum value of $f(x)$ at $x = -\dfrac{1}{2}$ is

$$f\left(-\frac{1}{2}\right) = \left(-\frac{1}{2}\right)^4 + 2\left(-\frac{1}{2}\right)^3 - 3\left(-\frac{1}{2}\right)^2 - 4\left(-\frac{1}{2}\right) + 4$$

$$= \frac{1}{16} - \frac{1}{4} - \frac{3}{4} + 2 + 4 = \frac{81}{16}$$

Similarly, at $x = 1$, we have

$$\frac{d^2y}{dx^2} = 12(1)^2 + 12(1) - 6 = 12 + 12 - 6 = 18 > 0$$

Since, $\dfrac{d^2y}{dx^2} > 0$ (i.e., positive). So $f(x)$ is minimum at x

$= 1$ and minimum value of $f(x)$ at $x = 1$ is

$$f(1) = (1)^4 + 2(1)^3 - 3(1)^2 - 4(1) + 4$$
$$= 1 + 2 - 3 - 4 + 4 = 0.$$

EXAMPLE 2. *Find the maximum and minimum value of the function* $y = f(x) = x^3 - 12x^2 + 36x + 21$.

SOLUTION. Here, the given function is

$$y = x^3 - 12x^2 + 36x + 21$$

Now, differentiating w.r.t. x, we get

$$\frac{dy}{dx} = 3x^2 - 24x + 36$$

Puting $\dfrac{dy}{dx} = 0$, we get

$$3x^2 - 24x + 36 = 0 \quad \text{or} \quad x^2 - 8x + 12 = 0$$

or $(x-2)(x-6) = 0$ or $x = 2, 6$

Again, differentiating w.r.t. x, we get

$$\frac{d^2y}{dx^2} = 6x - 24$$

At $x = 2$, we have $\dfrac{d^2y}{dx^2} = 6(2) - 24 = -12 < 0$

Since, $\dfrac{d^2y}{dx^2} < 0$ so $f(x)$ is maximum at $x = 2$. The

maximum value of $f(x)$ at $x = 2$ is given by

$$f(2) = (2)^3 - 12(2)^2 + 36(2) + 21$$
$$= 8 - 48 + 72 + 21 = 53.$$

Similarly, at $x = 6$, we have

$$\frac{d^2y}{dx^2} = 6 \times 6 - 24 = 36 - 24 = 12 > 0$$

Since, $\dfrac{d^2y}{dx^2} > 0$ so, $f(x)$ is minimum at $x = 6$ and

minimum value of $f(x)$ at $x = 6$ is

$$f(6) = (6)^3 - 12(6)^2 + 36(6) + 21$$
$$= 216 - 432 + 216 + 21 = 453 - 432$$
$$= 21.$$

3.5.1 MAXIMA AND MINIMA OF A FUNCTION OF SEVERAL INDEPENDENT VARIABLES

Let $f(x, y, z, ...)$ be a function of several independent variables $x, y, z....$ If f is continuous and finite for all values of $x, y, z, ...$ in the neighbourhood of $x = a, y = b, z = c, ...$ respectively, then the value of $(a, b, c, ...)$ is said to be a maximum or minimum if $f(a+h, b+k, c+l, ...)$ is less than or greater than $f(a, b, c, ...)$ for all values of $h, k, l, ...$ (where $h, k, l, ...$) are sufficiently small, may be positive or negative provided they are not all zero.

In other words we can say, the value of $f(a, b, c, ...)$ is said to be a maximum or minimum if $f(a+h, b + k, c + l, ...)$ – $f(a, b, c, ...)$ maintain an invariant sign (may be positive or negative) for all values of $h, k, l, ...$ positive or negative provided they are taken sufficiently small and finite.

STATIONARY AND EXTREME POINTS

A point $(a_1, a_2, ..., a_n)$ is called a stationary point, if all the first order partial derivatives of the function $f(x_1, x_2, ..., x_n)$ vanish at the point. A stationary point, if it is maximum or minimum is known as extreme point and the value of the function at an extreme point is known as an extreme value.

☞ REMARK
- A stationary point may be a maximum or minimum or neither of these two.

3.5.2 NECESSARY CONDITION FOR THE EXISTENCE OF MAXIMA OR MINIMA

Let $f(x, y, z, ...)$ be a function of several independent variables $x, y, z, ...$ It is clear from the definition of maxima and minima that maximum or minimum of $f(x, y, z, ..)$ will occur for those values of $x, y, z, ...$, for which the expression $f(x+h, y +k, z+l, ...)$ $-f(x, y, z, ...)$ maintain an invariant sign for all sufficiently small and finite values of $h, k, l, ...$ positive or negative.

Now, expanding $f(x+h, y+k, z+l, ...)$ by Taylor's theorem, we have

$$f(x+h, y+k, z+l...) = f(x,y,z) + \left(h\frac{\partial f}{\partial x} + k\frac{\partial f}{\partial y} + l\frac{\partial f}{\partial z} + ... \right) + \text{terms of second and higher order.}$$

$\Rightarrow \quad f(x+h, y+k, z+l...) - f(x, y, z,...) = \left(h\dfrac{\partial f}{\partial x} + k\dfrac{\partial f}{\partial y} + l\dfrac{\partial f}{\partial z} + ...\right) +$ terms of second and higher orders.   ...(1)

Now, since $h, k, l, ...$ are sufficiently small, the first degree expression $\left(h\dfrac{\partial f}{\partial x} + k\dfrac{\partial f}{\partial y} + l\dfrac{\partial f}{\partial z} + ...\right)$ of the equation (1) can be made to govern the sign of right hand side and hence, of the left hand side as well as. Thus, by changing the sign of the left hand side of the equation (1) will also change.

Since, left hand side is to preserve an invariable sign for maxima or minima, therefore, as a necessary condition for maximum and minimum values, we must have

$$h\dfrac{\partial f}{\partial x} + k\dfrac{\partial f}{\partial y} + l\dfrac{\partial f}{\partial z} + ... = 0 \qquad\qquad ...(2)$$

Now, since $h, k, l, ...$ are arbitrary and independent of each other, we must have

$$\dfrac{\partial f}{\partial x} = 0, \dfrac{\partial f}{\partial y} = 0, \dfrac{\partial f}{\partial z} = 0, \text{ etc.} \qquad\qquad ...(3)$$

If the number of independent variables be n, we shall get n simultaneous equations in these n variables, which will give the values $a, b, c, ...$ of the n variables $x, y, z, ...$ respectively for which $f(x, y, z, ...)$ will have a maximum or a minimum values.

✎ REMARKS

- The necessary condition for a function $f(x, y, z, ...)$ of the independent variables $x, y, z, ...$ to be maximum or minimum is given by

  $$\dfrac{\partial f}{\partial x} = 0, \dfrac{\partial f}{\partial y} = 0, \dfrac{\partial f}{\partial z} = 0,$$

- The conditions given above is only a necessary condition for the maxima and minima of the function $f(x, y, z, ...)$. These conditions are not sufficient.

3.5.3 MAXIMA AND MINIMA FOR A FUNCTION OF TWO INDEPENDENT VARIABLES

(1) *To find the condition which governs the sign of a quadratic expression.*
Consider, a binary expression $I = ax^2 + 2hxy + by^2$

of two variables x and y. Then I can be written as $I = ax^2 + 2hxy + by^2 = \dfrac{1}{a}[(ax + hy)^2 + (ab - h^2)y^2]$.

If $(ab - h^2)$ is positive, the sign of I will be the same as that of a.
But if $(ab - h^2)$ is negative, then, the expression within the brackets may be positive or negative and therefore we cannot say anything about the sign of expression I.

(2) *Stationary and extreme points (For the function of two independent variables)*:
Let $f(x, y)$ be a function of two independent variables x and y. A point (a, b) is called a stationary point, if both the first order partial derivatives $\left(\dfrac{\partial f}{\partial a} \text{ and } \dfrac{\partial f}{\partial b}\right)$ of the function $f(x, y)$ at (a, b) vanish.

A stationary point which is either a maximum or minimum is called an extreme point.

✎ REMARKS

- A stationary point is not necessarily an extreme point, hence a stationary point may be a maximum or a minimum or neither of these two.
- The value of the function at extreme point is called extreme value.
- A point at which function is neither maximum nor minimum, is known as saddle point.

3.5.4 NECESSARY CONDITION FOR MAXIMA AND MINIMA

Let $f(x, y)$ be a function of two independent variables x and y. Then, we have the maximum or minimum of $f(x, y)$ at $x = a$ and $x = b$ if the expression $f(a + h, b + k) - f(a, b)$ is of invariable sign for all sufficiently small independent variables h and k provided both of them are not equal to zero.

We observe that,

(i) If the sign of $f(a+h, b+k) - f(a, b)$ is negative, then we have a maximum of $f(x, y)$ at $x = a, y = b$.

(ii) If the sign of $f(a+h, b+k) - f(a, b)$ is positive, we have a minimum of $f(x, y)$ at $x = a, y = b$.

Expand $f(a+h, b+k)$ by Taylor's theorem, we have

$$f(a+h, b+k) = f(a, b) + \left(h\dfrac{\partial f}{\partial x} + k\dfrac{\partial f}{\partial y}\right)_{\substack{x=a \\ y=b}} + \dfrac{1}{2!}\left(h^2\dfrac{\partial^2 f}{\partial x^2} + 2hk\dfrac{\partial^2 f}{\partial x \partial y} + k^2\dfrac{\partial^2 f}{\partial y^2}\right)_{\substack{x=a \\ y=b}} + ... \qquad ....(1)$$

$\Rightarrow \qquad f(a+h,b+k)-f(a,b) = h\left(\frac{\partial f}{\partial x}\right)_{\substack{x=a \\ y=b}} + k\left(\frac{\partial f}{\partial y}\right)_{\substack{x=a \\ y=b}}$ + term of the second and higher orders in h and k.

Now, since h and k are sufficiently small, the expression $h\left(\frac{\partial f}{\partial x}\right)_{\substack{x=a \\ y=b}} + k\left(\frac{\partial f}{\partial y}\right)_{\substack{x=a \\ y=b}}$ of the equation (1) can be made to govern the sign of right hand side and hence of the left hand side as well. Thus by changing the sign of h and k, the sign of the left hand side of the equation (1) will also change.

Since L.H.S. is to preserve an invariable sign for maximum or minimum, therefore as a necessary condition for maximum and minimum values, we must have

$$h\left(\frac{\partial f}{\partial x}\right)_{\substack{x=a \\ y=b}} + k\left(\frac{\partial f}{\partial y}\right)_{\substack{x=a \\ y=b}} = 0. \qquad \qquad ...(2)$$

If $k = 0$, we find that if $\left(\frac{\partial f}{\partial x}\right)_{\substack{x=a \\ y=b}} \neq 0$, the R.H.S. of (2) changes sign when h changes sign. Therefore $f(x, y)$ cannot have a maximum or minimum at $x = a, y = b$ if $\left(\frac{\partial f}{\partial x}\right)_{\substack{x=a \\ y=b}} \neq 0$.

Similarly, taking $h = 0$, we see that $f(x, y)$ cannot have a maximum or a minimum at $x = a, y = b$ if $\left(\frac{\partial f}{\partial y}\right)_{\substack{x=a \\ y=b}} \neq 0$.

Thus, a set of necessary conditions that $f(x, y)$ should have a maximum or minimum at $x=a, y=b$ is that

$$\left(\frac{\partial f}{\partial x}\right)_{\substack{x=a \\ y=b}} = 0 \text{ and } \left(\frac{\partial f}{\partial y}\right)_{\substack{x=a \\ y=b}} = 0.$$

3.5.5 Sufficient Condition for Maxima and Minima: The Lagrange's Condition

Let $f(x, y)$ be a function of two variables x and y.

Let $$r = \frac{\partial^2 f}{\partial x^2}, s = \frac{\partial^2 f}{\partial x \partial y}, t = \frac{\partial^2 f}{\partial y^2} \text{ at } x = a \text{ and } y = b.$$

As a set of necessary conditions for a maximum or minimum at (a, b) we have

$$\frac{\partial f}{\partial x} = 0 \text{ and } \frac{\partial f}{\partial y} = 0 \text{ at } (a, b)$$

then $$f(a + h, b + k) - f(a, b) = \frac{1}{2!}[rh^2 + 2shk + tk^2] + R \qquad \qquad ...(1)$$

Where R consists of terms of third and higher order of small quantities h and k.

Now, by taking h and k sufficiently small, the second degree terms in R.H.S. of (1) may be made to govern the sign of R.H.S. and therefore of the L.H.S. also i.e., for sufficiently small values of h and k, the sign of $\frac{1}{2}(rh^2 + 2shk + tk^2) + R$ is same as that of $rh^2 + 2shk + tk^2$.

If the sign is negative, then the function is maximum at (a, b) and if the sign is positive, then the function is minimum at (a, b).

Case (i) If $(rt - s^2) > 0$.

Then, neither r nor t can be zero. Hence, we can write $rh^2 + 2shk + tk^2 = \frac{1}{r}[r^2h^2 + 2rshk + rtk^2] = \frac{1}{r}[(rh + sk)^2 + (rt - s^2)k^2]$

Since $rt - s^2 > 0$, therefore $(rh + sk)^2 + (rt - s^2)k^2 > 0$ for all values of h and k except when $rh + sk = 0$, $k = 0$ i.e., at $h = 0, k = 0$, which is not possible.

Hence, in this case the expression $rh^2 + 2shk + tk^2$ will have the same sign for all values of h and k, and the sign is determined by the sign of r.

Thus, the function $f(x, y)$ will have a maximum or minimum at $x = a$ and $y = b$. If $rt - s^2 > 0$. The function $f(x, y)$ is maximum or minimum according as r is negative or positive.

Case (ii) If $(rt - s^2) < 0$.

If $rt - s^2$ is negative, we are not sure about the sign of second degree term of R.H.S. of (1) and hence there is neither a maximum nor a minimum value.

Case (iii) If $rt - s^2 = 0$.

If $rt = s^2$, then quadratic expression $rh^2 + 2shk + tk^2$ becomes $-(hr \quad ks)$.

So that, the quadratic expression will be of the same sign as that of r or t unless

$$\frac{h}{k} = -\frac{s}{r} = \alpha \text{ (say) } i.e., \quad rh + sk = 0.$$

If this condition is satisfied, then the second degree expression in R.H.S. of (1) vanishes and hence, the sign of the R.H.S. of (1) depends upon third degree expression in h and k, which change sign with the change of sign of h and k and hence, the sign of L.H.S. of (1) will also change and hence, there will be neither maximum nor minimum.

Thus, the necessary condition for the existence of maxima and minima now is that the cubic terms must vanish collectively in R.H.S. of (1) when $\frac{h}{k} = -\frac{s}{r} = \alpha$; and then the biquadratic

terms of R.H.S. of (1) must collectively be of the same sign as r and t, when

$$\frac{h}{k} = -\frac{s}{r} = \alpha \quad i.e., \quad hr + ks = 0$$

Hence, the case is doubtful.

Thus, if $rt - s^2 = 0$, the case is doubtful and further, investigation is needed to determine the maxima and minima of $f(x, y)$ at (a, b).

WORKING PROCEDURE

To discuss the maxima and minima at $x = a, y = b$, we must find

$$r = \left(\frac{\partial^2 u}{\partial x^2}\right)_{\substack{x=a \\ y=b}}, \ s = \left(\frac{\partial^2 u}{\partial x \partial y}\right)_{\substack{x=a \\ y=b}}, \ t = \left(\frac{\partial^2 u}{\partial y^2}\right)_{\substack{x=a \\ y=b}}$$

Then, calculate $rt - s^2$.

Now following cases arise :

(i) If $rt - s^2 > 0$, then

   (A) If r is negative then, $f(x, y)$ is maximum at $x = a, y = b$.

   (B) If r is positive then, $f(x, y)$ is minimum at $x = a, y = b$.

(ii) If $rt - s^2 < 0$, $f(x, y)$ is neither maximum nor minimum at $x = a, y = b$.

(iii) If $rt - s^2 = 0$, the case is doubtful, and further investigation will be required.

REMARK

● While solving problems, we frequently used the identity, given by Lagrange.

$$\{(a^2 + b^2 + c^2)(p^2 + q^2 + r^2) - (ap + bq + cr)^2\} = \{(br - cq)^2 + (cp + ar)^2 + (aq - bp)^2\}.$$

Solved Examples

EXAMPLE 1. *Find the maximum or minimum values of the function* $x^3 y^2 (1 - x - y)$.

(ANNA-2009, JNTU-2006, 08, BHOPAL-2012)

SOLUTION. Let $\quad u = x^3 y^2 (1 - x - y)$

$\Rightarrow \quad \dfrac{\partial u}{\partial x} = 3x^2 y^2 (1 - x - y) - x^3 y^2$

and $\quad \dfrac{\partial u}{\partial y} = 2x^3 y (1 - x - y) - x^3 y^2$.

For a maximum or minimum of u, we must have

$$\frac{\partial u}{\partial x} = 0 \text{ and } \frac{\partial u}{\partial y} = 0$$

$\Rightarrow \quad 3x^2 y^2 (1 - x - y) - x^3 y^2 = 0 \qquad ...(1)$

and $\quad 2x^3 y (1 - x - y) - x^3 y^2 = 0. \qquad ...(2)$

Now, subtracting (2) from (1), we have

$$x^2 y (1 - x - y)(3y - 2x) = 0$$

which gives $\qquad\qquad y = \dfrac{2}{3} x.$

Putting the value of y in (1), we get $x = \dfrac{1}{2}$

So $\left(\dfrac{1}{2}, \dfrac{2}{3}\right)$ be the point of maxima or minima.

Now $\qquad r = \dfrac{\partial^2 u}{\partial x^2} = 6xy^2 - 12x^2 y^2 - 6xy^3$

$$= -\frac{1}{9}, \text{ at } \left(\frac{1}{2}, \frac{1}{3}\right)$$

$$t = \frac{\partial^2 u}{\partial y^2} = 2x^3 - 2x^4 - 6x^3 y$$

$$= -\frac{1}{8}, \text{ at } \left(\frac{1}{2}, \frac{1}{3}\right)$$

$$s = \frac{\partial^2 u}{\partial x \partial y} = 6x^2 y - 8x^3 y - 9x^2 y^2$$

$$= -\frac{1}{12} \text{ at } \left(\frac{1}{2}, \frac{1}{3}\right).$$

Now, $\qquad rt - s^2 = $ positive.

Also, r is negative, hence the function u has a maximum at $x = \dfrac{1}{2}, y = \dfrac{1}{3}$.

The maximum value is $= \left(\dfrac{1}{2}\right)^3 \left(\dfrac{1}{3}\right)^2 \left(1 - \dfrac{1}{2} - \dfrac{1}{3}\right) = \dfrac{1}{432}.$

EXAMPLE 2. *Discuss the maximum or minimum values of u, where* $u = 2a^2 xy - 3ax^2 y - ay^3 + x^3 y + xy^3$.

SOLUTION. We have $u = 2a^2xy - 3ax^2y - ay^3 + x^3y + xy^3$

which gives $\dfrac{\partial u}{\partial x} = 2a^2y - 6axy + 3x^2y + y^3$

and $\dfrac{\partial u}{\partial y} = 2a^2x - 3ax^2 - 3ay^2 + x^3 + 3xy^2$

For a maximum and minima of u, we have

$$\dfrac{\partial u}{\partial x} = 0, \dfrac{\partial u}{\partial y} = 0$$

which gives, $y(2a^2 - 6ax + 3x^2 + y^2) = 0$ ...(1)

and $2a^2x - 3ax^2 - 3ay^2 + x^3 + 3xy^2 = 0$ ...(2)

Equation (1) and (2) gives the following values of x and y :

$x = 0,\ y = 0;\ x = a,\ y = 0;$

$x = 2a,\ y = 0;\ x = \dfrac{3}{2}a,\ y = \pm\dfrac{1}{2}a;$

$x = a, y = a,\ x = \dfrac{1}{2}a,\ y = \dfrac{1}{2}a;$

$x = a, y = -a;\ x = \dfrac{1}{2}a,\ y = -\dfrac{1}{2}a.$

Then, we get the following pairs of values of x and y which make the function u stationary.

$(0,0),\ (a,0),\ (2a,0),\ \left(\dfrac{3}{2}a, \dfrac{1}{2}a\right),\ \left(\dfrac{3}{2}a, -\dfrac{1}{2}a\right)$

$(a,a),\ \left(\dfrac{1}{2}a, \dfrac{1}{2}a\right),\ (a,-a),\ \left(\dfrac{1}{2}a, -\dfrac{1}{2}a\right).$

Also $r = \dfrac{\partial^2 u}{\partial x^2} = -6ay + 6xy,$

$s = \dfrac{\partial^2 u}{\partial x\, \partial y} = 2a^2 - 6ax + 3x^2 + 3y^2,$

and $t = \dfrac{\partial^2 u}{\partial y^2} = -6ay + 6xy.$

For $(0, 0)$.

$$r = 0, s = 2a^2, t = 0$$

$\Rightarrow$ $rt - s^2$, is negative.

Therefore, we have neither maximum nor a minimum of u at $(0, 0)$.

Similarly, we can easily shown that u has neither a maximum nor a minimum at $(a, 0)$, $(2a, 0)$, (a, a), $(a, -a)$.

For $\left(\dfrac{3a}{2}, \dfrac{a}{2}\right)$.

$$r = \dfrac{3}{2}a^2, s = \dfrac{1}{2}a^2, t = \dfrac{3}{2}a^2,$$

$\Rightarrow$ $rt - s^2$ is positive.

Here, since r is positive, therefore u has minimum at $\left(\dfrac{3a}{2}, \dfrac{a}{2}\right).$

Similarly, we can check the maxima and minima at all other points.

✒ REMARK

- The point $\left(\dfrac{x_1 + x_2 + x_3}{3}, \dfrac{y_1 + y_2 + y_3}{3}\right)$ is the centroid of the given triangle.

EXAMPLE 3. *Determine the points where a function* $x^3 + y^3 - 3axy$ *has maximum or minimum.*

(UPTU -2009, GBTU-2010, 12, MTU -2011)

SOLUTION. Here, we have $u = x^3 + y^3 - 3axy$

$\Rightarrow$ $\dfrac{\partial u}{\partial x} = 3x^2 - 3ay$ and $\dfrac{\partial u}{\partial y} = 3y^2 - 3ax.$

For a maximum or minimum of u, we must have

$$\dfrac{\partial u}{\partial x} = 0 \text{ and } \dfrac{\partial u}{\partial y} = 0$$

which gives, $x^2 - ay = 0$ ... (1)

and $y^2 - ax = 0$ ... (2)

Solving (1) and (2), we get

$$x = 0, y = 0; x = a, y = a.$$

Thus $(0, 0)$ and (a, a) are the stationary points of u.

Now $r = \dfrac{\partial^2 u}{\partial x^2} = 6x, s = \dfrac{\partial^2 u}{\partial x\, \partial y} = -3a, t = \dfrac{\partial^2 u}{\partial y^2} = 6y.$

For $x = 0, y = 0$

$$r = 0, s = -3a \text{ and } t = 0$$

$\therefore$ $rt - s^2 = -9a^2 < 0$, for all values of a.

$\Rightarrow$ u is neither maximum nor minimum at $x = 0, y = 0.$

For $x = a, y = a$

$$r = 6a, s = -3a \text{ and } t = 6a$$

$\Rightarrow$ $rt - s^2 = 27a^2 > 0$, for all values of a.

Also $r = 6a$, which is positive if $a > 0$.

Thus (i) u is maximum at $x = a, y = a$ if $a < 0$

and (ii) u is minimum at $x = a, y = a$ if $a > 0$.

EXAMPLE 4. *Discuss the maxima and minima of the function* $u = \sin x \sin y \sin (x+y)$. (UPTU-2009)

SOLUTION. Here, we have

$$u = \sin x \sin y \sin (x + y)$$

$\Rightarrow$ $\dfrac{\partial u}{\partial x} = \sin y [\sin x \cos(x+y) + \cos x \sin(x+y)]$

and $\dfrac{\partial u}{\partial y} = \sin x [\sin y \cos(x+y) + \cos y \sin(x+y)].$

For a maxima and minima of u, we must have

$$\dfrac{\partial u}{\partial x} = 0 \text{ and } \dfrac{\partial u}{\partial y} = 0.$$

$\Rightarrow \sin y [\sin x \cos (x + y) + \cos x \sin (x+y)] = 0$

and $\sin x [\sin y \cos (x + y) + \cos y \sin (x+y)] = 0.$

Equation (1) and (2) gives

$\tan (x + y) = -\tan x$ ...(1)

$\Rightarrow$ $\tan x = \tan y$

and $\tan (x + y) = -\tan y$ ...(2)

$\Rightarrow$ $x = y$

From (1) and (2), we have

$\tan 2x = -\tan x = \tan (\pi - x)$

$\Rightarrow$ $2x = \pi - x$

$\Rightarrow$ $3x = \pi \Rightarrow x = \dfrac{\pi}{3} = y.$

Moreover, $\dfrac{\partial u}{\partial x} = 0$, gives $\sin y = 0 \Rightarrow y = 0$

and $\dfrac{\partial u}{\partial y} = 0$, gives $\sin x = 0 \Rightarrow x = 0.$

Thus, we get the following pair of values, which makes the function u stationary

$(0,0), \left(\dfrac{\pi}{3}, \dfrac{\pi}{3}\right).$

Now $\qquad r = \dfrac{\partial^2 u}{\partial x^2} = 2\sin y \cos(2x + y),$

$\qquad s = \dfrac{\partial^2 u}{\partial x \partial y} = \sin 2(x + y),$

and $\qquad t = \dfrac{\partial^2 u}{\partial y^2} = 2\sin x \cos(2y + x).$

For (0, 0).

$\qquad r = 0, s = 0, t = 0$

$\Rightarrow \qquad rt - s^2 = 0.$

$\therefore$ this case is doubtful and need further investigation.

For $\left(\dfrac{\pi}{3}, \dfrac{\pi}{3}\right).$

$\qquad r = 2\sin\dfrac{1}{3}\pi.\cos\pi = -\sqrt{3},$

$\qquad s = \sin\left(\dfrac{4\pi}{3}\right) = -\sin\dfrac{\pi}{3} = -\dfrac{\sqrt{3}}{2},$

and $\qquad t = 2\sin\dfrac{1}{3}\pi\cos\pi = -\sqrt{3}.$

$\therefore \qquad rt - s^2 = \dfrac{9}{4} = \text{positive}.$

Also $\qquad r = -\sqrt{3}.$

Hence, u has a maximum value at $\left(\dfrac{\pi}{3}, \dfrac{\pi}{3}\right).$

EXAMPLE 5. *Find the minimum value of $x^2 + y^2 + z^2$ when $ax + by + cz = p$.*

SOLUTION. Here, $\qquad u = x^2 + y^2 + z^2 \qquad$ ...(1)

Also $ax + by + cz = p$

$\Rightarrow \qquad z = \dfrac{p - ax - by}{c}.$

Put this value of z in equation (1), we get

$\qquad u = x^2 + y^2 + \dfrac{(p - ax - by)^2}{c^2}$

$\Rightarrow \qquad \dfrac{\partial u}{\partial x} = 2x - \dfrac{2a}{c^2}(p - ax - by)$

$\Rightarrow \qquad \dfrac{\partial u}{\partial y} = 2y - \dfrac{2b}{c^2}(p - ax - by).$

For a maxima and minima of u, we must have

$\qquad \dfrac{\partial u}{\partial x} = 0 \text{ and } \dfrac{\partial u}{\partial y} = 0$

$\Rightarrow \qquad x = \dfrac{ap}{a^2 + b^2 + c^2} \text{ and } y = \dfrac{bp}{a^2 + b^2 + c^2}.$

Now, $\qquad r = \dfrac{\partial^2 u}{\partial x^2} = 2 + \dfrac{2a^2}{c^2}, \; s = \dfrac{\partial^2 u}{\partial x \partial y} = \dfrac{2ab}{c^2}$

and $\qquad t = \dfrac{\partial^2 u}{\partial y^2} = 2 + \dfrac{2b^2}{c^2}$

$\Rightarrow \qquad rt - s^2 = 4\left(1 + \dfrac{a^2}{c^2}\right)\left(1 + \dfrac{b^2}{c^2}\right) - \dfrac{4a^2 b^2}{c^4}$

$\qquad = 4\left(1 + \dfrac{a^2}{c^2} + \dfrac{b^2}{c^2}\right) = \text{positive}.$

Since r is positive and $rt - s^2 > 0$, therefore u is minimum for the above values of x and y.

The minimum value is $\dfrac{p^2}{a^2 + b^2 + c^2}.$

EXAMPLE 6. *Find the stationary point of $x^4 + y^4 - 2x^2 + 4xy - 2y^2$ and determine their nature.* (JNTU-2009)

SOLUTION. Here, we have $\quad u = x^4 + y^4 - 2x^2 + 4xy - 2y^2$

$\Rightarrow \qquad \dfrac{\partial u}{\partial x} = 4x^3 - 4x + 4y$

and $\qquad \dfrac{\partial u}{\partial y} = 4y^3 + 4x - 4y.$

For, a maxima and minima of u, we must have

$\qquad \dfrac{\partial u}{\partial x} = 0 \Rightarrow 4x^3 - 4x + 4y = 0 \qquad$ ...(1)

and $\qquad \dfrac{\partial u}{\partial y} = 0 \Rightarrow 4y^3 + 4x - 4y = 0. \qquad$ ...(2)

Solving (1) and (2), we get

$\qquad 4x^3 + 4y^3 = 0 \Rightarrow x^3 + y^3 = 0$

$\Rightarrow \quad (x + y)(x^2 - xy + y^2) = 0$

$\Rightarrow \quad$ either $x + y = 0$ or $x^2 - xy + y^2 = 0 \qquad$ ...(3)

$\qquad x + y = 0 \Rightarrow y = -x.$

Put $y = -x$ in equation (1), we get

$\qquad 4x^3 - 8x = 0$

$\Rightarrow \qquad x(x^2 - 2) = 0$

$\Rightarrow \qquad x = 0, \sqrt{2}, -\sqrt{2}.$

Then $\qquad y = 0, -\sqrt{2}, \sqrt{2}$

Also from (3) $\quad x = 0, y = 0$(only real solution)

Hence, the stationary points u are given by $(0,0), (\sqrt{2}, -\sqrt{2}), (-\sqrt{2}, \sqrt{2}).$

Now, we have $\quad r = \dfrac{\partial^2 u}{\partial x^2} = 12x^2 - 4,$

$\qquad s = \dfrac{\partial^2 u}{\partial x \partial y} = 4,$

and $\qquad t = \dfrac{\partial^2 u}{\partial y^2} = 12y^2 - 4.$

For (0, 0). $\quad r = -4, s = 4, t = -4$

$\Rightarrow \qquad rt - s^2 = 0.$

$\Rightarrow \quad$ At the point (0, 0) the case is doubtful, and there is a need of further investigation.

For $(\sqrt{2}, -\sqrt{2})$.

$\qquad r = 20, s = 4, t = 20$

$\Rightarrow \qquad rt - s^2 = 400 - 16 = 384 > 0.$

Also $\qquad r > 0.$

$\Rightarrow \quad u$ has a minimum value at $(\sqrt{2}, -\sqrt{2}).$

Similarly u has a minimum value at $(-\sqrt{2}, \sqrt{2}).$

REMARK

- If there is a function of two variables x and y connected by a relation $g(x, y) = 0$. Then we find the maxima and minima of the function in the following manner.

Let
$$u = f(x, y) \qquad \ldots(1)$$
and
$$g(x, y) = 0. \qquad \ldots(2)$$

Generally, it is possible to eliminate one of the variables x and y from (1) and (2), then u is expressed in terms of a single **variable** and we can proceed in the usual way. But if it is not convenient to take the value of one variable in terms of the other from (2), then we should proceed as follows :

From (2), we get
$$\frac{dg}{dx} = -\frac{\partial g / \partial x}{\partial g / \partial y} \qquad \ldots(3)$$

Now, differentiating (1) with respect to x, we get
$$\frac{du}{dx} = \frac{\partial f}{\partial x} + \frac{\partial f}{\partial y}\frac{dg}{dx} \qquad \ldots(4)$$

Now, from (3) and (4),we get
$$\frac{du}{dx} = 0$$

Solve (5) with the help of (2), and get the required values of x and y for which u will have maximum or minimum values.

EXAMPLE 7. *Show that distance l of any point (x, y, z) on the plane $2x + 3y - z = 12$ from the origin is given by*
$$l = \sqrt{[x^2 + y^2 + (2x + 3y - 12)^2]}.$$
Hence, find the point on the plane that is nearest to the origin.

SOLUTION. If l is the distance from $(0, 0, 0)$ of any point (x, y, z) then $l = \sqrt{(x^2 + y^2 + z^2)}$. If the point (x, y, z) lies on the plane $2x + 3y - z = 12$, then

$$l = \sqrt{[x^2 + y^2 + (2x + 3y - 12)^2]}$$

$[\because z = 2x + 3y - 12$, from the equation of the plane$]$

$$\therefore \quad l^2 = x^2 + y^2 + (2x + 3y - 12)^2$$
$$= 5x^2 + 10y^2 + 12xy$$
$$- 48x + 72y + 144 = u \,(\text{say}).$$

Now l is maximum or minimum according as l^2 i.e., u is maximum or minimum.

For a maximum or minimum of u, we get

$$\frac{\partial u}{\partial x} = 10x + 12y - 48 = 0$$

and $\quad \dfrac{\partial u}{\partial y} = 20y + 12x - 72 = 0$

Solving these equations, we get

$$x = \frac{12}{7} \text{ and } y = \frac{18}{7}.$$

Also $\quad r = \dfrac{\partial^2 u}{\partial x^2} = 10, \ s = \dfrac{\partial^2 u}{\partial x \, \partial y} = 12 \text{ and } t = \dfrac{\partial^2 u}{\partial y^2} = 20.$

Therefore $rt - s^2 = 10 \times 20 - (12)^2 = +\text{ve}$, since $rt - s^2 > 0$ and $r > 0$, then u is minimum and hence l is minimum.

When $x = \dfrac{12}{7}$ and $y = \dfrac{18}{7}$. Putting these values of x and y in the equation of the plane, we get

$$z = 2 \cdot \left(\frac{12}{7}\right) + 3 \cdot \left(\frac{18}{7}\right) - 12 = -\frac{6}{7}.$$

Hence, the required point is $\left(\dfrac{12}{7}, \dfrac{18}{7}, -\dfrac{6}{7}\right)$.

EXAMPLE 8. *Find the points on $z^2 = xy + 1$ nearest to the origin.*

SOLUTION. Let l be the distance from the origin $(0, 0, 0)$ of any point (x, y, z) on the surface

$$z^2 = xy + 1 \qquad \ldots(1)$$

Then $\quad l = \sqrt{x^2 + y^2 + z^2} = \sqrt{(x^2 + y^2 + xy + 1)}$

[Using equation (1)]

Since l is always greater than zero, therefore l is maximum or minimum according as l^2, i.e., u is maximum or minimum, where $u = l^2$.

For a maximum or minimum of u, we must have

$$\frac{\partial u}{\partial x} = 2x + y = 0 \qquad \ldots(2)$$

and $\quad \dfrac{\partial u}{\partial y} = 2y + x = 0. \qquad \ldots(3)$

Solving the equation (2) and (3), we get
$$x = 0, y = 0$$

Also $\quad r = \dfrac{\partial^2 u}{\partial x^2} = 2, \ s = \dfrac{\partial^2 u}{\partial x \, \partial y} = 1, \ t = \dfrac{\partial^2 u}{\partial y^2} = 2.$

$\therefore \quad rt - s^2 = 2 \cdot 2 - 1 = 3 > 0.$

Since at $x = 0, y = 0$, then $rt - s^2 > 0$ and $r > 0$.

Therefore u is minimum at $x = 0, y = 0$. Hence l is minimum, when $x = 0, y = 0$.

Putting $x = 0, y = 0$ in the equation (1), we get $z^2 = 1$ i.e., $z = \pm 1$.

Hence, the required points are $(0, 0, 1)$ and $(0, 0, -1)$.

EXAMPLE 9. *Prove that the maxima or minima of the function* $u = \left[\dfrac{ax^2 + by^2 + 2hxy + 2gx + 2fy + c}{a'x^2 + b'y^2 + 2h'xy + 2g'x + 2f'y + c'}\right]$

are given by the roots of the equation

$$\begin{vmatrix} a - a'u & h - h'u & g - g'u \\ h - h'u & b - b'u & f - f'u \\ g - g'u & f - f'u & c - c'u \end{vmatrix} = 0.$$

SOLUTION. Here, we have

$$u = \left[\frac{ax^2 + by^2 + 2hxy + 2gx + 2fy + c}{a'x^2 + b'y^2 + 2h'xy + 2g'x + 2f'y + c'}\right]$$

$\Rightarrow \quad u[a'x^2 + b'y^2 + 2h'xy + 2g'x + 2f'y + c']$
$$= [ax^2 + by^2 + 2hxy + 2gx + 2fy + c]. \qquad \ldots(1)$$

Differentiating (1) partially w.r.t. x and y, we have

$$\frac{\partial u}{\partial x}[a'x^2 + b'y^2 + 2h'xy + 2g'x + 2f'y + c'] \qquad \ldots(2)$$
$$+ u[2a'x + 2h'y + 2g'] = 2ax + 2hy + 2g$$

and $\dfrac{\partial u}{\partial y}[a'x^2+b'y^2+2h'xy+2g'x+2f'y+c']$
$+u[2b'y+2h'x+2f'] = 2by+2hx+2f$ ....(3)

For the maxima and minima of u, we must have

$$\dfrac{\partial u}{\partial x}=0 \Rightarrow u[a'x+h'y+g']=ax+hy+g \quad ...(4)$$

and $\quad \dfrac{\partial u}{\partial y}=0 \Rightarrow u[h'x+b'y+f']=hx+by+f. \quad ...(5)$

Now, multiplying (4) by x, (5) by y and adding, we have

$$u[a'x^2+b'y^2+2h'xy+g'x+f'y]$$
$$=ax^2+by^2+2hxy+gx+fy \qquad ...(6)$$

Subtracting (6) from (1), we get

$$u(g'x+f'y+c')=gx+fy+c \qquad ...(7)$$

Now, from (4), (5) and (7), we have

$$(a-a'u)x+(h-h'u)y+(g-g'u)=0 \qquad ...(8)$$
$$(h-h'u)x+(b-b'u)y+(f-f'u)=0 \qquad ...(9)$$
$$(g-g'u)x+(f-f'u)y+(c-c'u)=0. \qquad ...(10)$$

By eliminating x and y from (8), (9) and (10), we get

$$\begin{vmatrix} a-a'u & h-h'u & g-g'u \\ h-h'u & b-b'u & f-f'u \\ g-g'u & f-f'u & c-c'u \end{vmatrix}=0.$$

which is a cubic equation in u. The roots of this equation gives the required maxima and minima.

Exercise-3.7

1. Discuss the maxima and minima of the function
   $f(x,y)=x^2+y^2+\dfrac{2}{x}+\dfrac{2}{y}$.

2. Find the values of x and y for which the expression
   $(a_1x+b_1y+c_1)^2+(a_2x+b_2y+c_2)^2+...+(a_nx+b_ny+c_n)^2$
   is minimum.

3. Examine for maximum and minimum values of the function
   $f(x,y)=x^2-3xy+y^2+2x$.

4. Examine the function $f(x,y)=x^2y-y^2x-x+y$ for maxima and minima.

5. Discuss the maxima and minima of the function
   $$f(x,y)=2\sin\frac{1}{2}(x+y)\cos\frac{1}{2}(x-y)+\cos(x+y).$$

6. Find the maximum and minimum values of
   $u=6xy+(47-x-y)(4x+3y)$.

7. Examine for extreme values
   (i) $x^2+y^2+6x+12$ (GBTU 2012) (ii) $x^3+y^3-63(x+y)+12xy$

   (UKTU 2011)

Hints to Selected Problems

1. $\dfrac{\partial f}{\partial x}=0 \Rightarrow y-2xy-y^2=0$

   $\dfrac{\partial f}{\partial y}=0 \Rightarrow x-2xy-x^2=0.$

   On solving above two equations, we get $(0, 0)$, $(1, 0)$, $(0, 1)$ and $(1/3, 1/3)$ are the extreme points.

   At (0, 0), $rt-s^2$ is negative $\Rightarrow f(x,y)$ is neither maximum nor minimum.

   At (1, 0), $rt-s^2$ is negative $\Rightarrow f(x,y)$ is neither maximum nor minimum.

   At (0, 1), $rt-s^2$ is negative $\Rightarrow f(x,y)$ is neither maximum nor minimum.

   At (1/3, 1/3), $rt-s^2$ is positive and $r\left(=-\dfrac{2}{3}\right)$ is negative. Hence, at $\left(\dfrac{1}{3},\dfrac{1}{3}\right)$, $f(x,y)$ is maximum.

7. $\dfrac{\partial f}{\partial x}=\cos x-\sin(x+y)$, $\dfrac{\partial f}{\partial y}=\cos y-\sin(x+y)$

   $\dfrac{\partial f}{\partial x}=0$, $\dfrac{\partial f}{\partial y}=0$, we get $\cos x=\sin(x+y)$, and $\cos y=\sin(x+y)$.

   The extreme points are given by
   $$\left(-\frac{\pi}{2},\frac{\pi}{2}\right),\left(\frac{3\pi}{2},\frac{\pi}{2}\right) \text{ and } \left(\frac{\pi}{2},\frac{\pi}{2}\right).$$

Answers

1. $f(x,y)$ is minimum at $(1, 1)$.
2. $f(x,y)$ is minimum for the value of x and y which are obtained by $\Sigma(a_1^2)x+(a_1b_1)y+a_1c_1=0$ and $\Sigma(a_1b_1)x+(b_1^2)y+b_1c_1=0$.
3. Stationary point is $x=\dfrac{4}{5}$, $y=\dfrac{6}{5}$. The function $f(x,y)$ is neither maximum nor minimum at $\left(\dfrac{4}{5},\dfrac{6}{5}\right)$.
4. At $(1, 1)$ and $(-1, -1)$ function is neither maximum nor minimum.
5. $x=y=2n\pi\pm\pi/2$; neither maximum nor minimum ; $x=y=n\pi+(-1)^n\pi/6$; f is maximum.
6. Maximum value is 3384.
7. (i) At $x=-3$, $y=0$, minimum (ii) max at $(-7, -7)$ min. at $(3, 3)$ neither max nor min. at $(5, -1)$ and $(-1, 5)$.

3.5.6 Maxima and Minima of the Function of Three Independent Variables

(1) *To find the condition, which governs the sign of the quadratic equation of three independent variables.*

Let I be the expression of three independent variables x, y and z given by

$$I=ax^2+by^2+cz^2+2fyz+2gzx+2hxy$$

I can be written as

$$I = \frac{1}{a}\left[a^2x^2 + aby^2 + acz^2 + 2afyz + 2agzx + 2ahxy\right] (a \neq 0)$$

$$= \frac{1}{a}\left[a^2x^2 + 2ax(gz + hy) + aby^2 + acz^2 + 2afyz\right]$$

$$= \frac{1}{a}\left[(ax + hy + gz)^2 + aby^2 + acz^2 + 2afyz - (gz + hy)^2\right]$$

$$= \frac{1}{a}\left[(ax + hy + gz)^2 + (ab - h^2)y^2 + 2yz(af - gh) + (ac - g^2)z^2\right]$$

Here, we observe that I be of the same sign as provided the expression within the square brackets is positive which will of course be so if $ab-h^2$ and $\{(ah-h^2)(ac-g^2) - (ab-gh)^2\}$ are positive *i.e.*, if $ab-h^2$ and $a[abc + 2fgh - af^2 - bg^2 - ch^2]$ are both positive.

Hence, I will be positive if

$$a, \quad \begin{vmatrix} a & h \\ h & b \end{vmatrix}, \quad \begin{vmatrix} a & h & g \\ h & b & f \\ g & f & c \end{vmatrix}$$

be all positive and will be negative if these three expression are alternately negative and positive.

3.4.6 MAXIMA AND MINIMA FOR A FUNCTION OF THREE INDEPENDENT VARIABLES : THE LAGRANGE'S CONDITION

Let $f(x,y,z)$ be a given function of three independent variables x, y and z.

Let A, B, C, F, G, H stand for $\dfrac{\partial^2 f}{\partial x^2}, \dfrac{\partial^2 f}{\partial y^2}, \dfrac{\partial^2 f}{\partial z^2}, \dfrac{\partial^2 f}{\partial y \partial z}, \dfrac{\partial^2 f}{\partial z \partial x}, \dfrac{\partial^2 f}{\partial x \partial y}$ respectively.

Let a set of the values of x, y, z obtained by solving the equations

$$\frac{\partial f}{\partial x} = \frac{\partial f}{\partial y} = \frac{\partial f}{\partial z} = 0 \text{ be } a, b, c.$$

By Taylor's theorem, we have $f(a+h, b+k, c+l), -f(a,b,c) = \dfrac{1}{2!}\left[Ah^2 + Bk^2 + Cl^2 + 2Fkl + 2Glh + 2Hhk\right] + R$ ...(1)

where, remainder term R consist of third and higher order of same quantity (*i.e.*, h, k, l).

Now, by taking h, k, l sufficiently small the second term of R.H.S. of (1) can be made to govern the sign of R.H.S. and therefore of L.H.S. also.

If for all such values of h, k and l, these terms be of permanent sign, then we shall have a maximum or minimum of $f(x,y,z)$ according as that sign is negative or positive.

Hence, the function will be minimum if the expression $A, \quad \begin{vmatrix} A & H \\ H & B \end{vmatrix}, \quad \begin{vmatrix} A & H & G \\ H & B & F \\ G & F & C \end{vmatrix}$ be all positive.

The function will have a maximum value, if the above three quantities are alternately negative and positive. If these conditions are not satisfied, we have neither a maximum nor a minimum.

WORKING PROCEDURE

Let $f(x, y, z)$ be a function of three independent variables x, y and z. Find the values of triads (a,b,c) of the value x, y and z by putting $\dfrac{\partial f}{\partial x} = 0, \dfrac{\partial f}{\partial y} = 0, \dfrac{\partial f}{\partial z} = 0$. The values of triads (a,b,c) will give the stationary values of $f(x, y, z)$.

Now, to discuss maximum and minimum values, at (a, b, c) we find the following six partial derivatives of second order

$$A = \frac{\partial^2 f}{\partial x^2}, B = \frac{\partial^2 f}{\partial y^2}, C = \frac{\partial^2 f}{\partial z^2}, F = \frac{\partial^2 f}{\partial y \partial z}, G = \frac{\partial^2 f}{\partial z \partial x}, and\ H = \frac{\partial^2 f}{\partial x \partial y}$$

Now, we have the following cases :

Case (i) The function $f(x,y,z)$ will be minimum at (a,b,c) if the expressions

$$A, \quad \begin{vmatrix} A & H \\ H & B \end{vmatrix}, \quad \begin{vmatrix} A & H & G \\ H & B & F \\ G & F & C \end{vmatrix}$$ be all positive at (a, b, c).

Case (ii) The function $f(x, y, z)$ will be maximum at (a, b, c) if the expressions

$$A, \quad \begin{vmatrix} A & H \\ H & B \end{vmatrix}, \quad \begin{vmatrix} A & H & G \\ H & B & F \\ G & F & C \end{vmatrix}$$ be alternately negative and positive.

Case (iii) If the expression, using in case (i) and (ii) neither be all positive nor having alternately negative and positive sign at (a,b,c). Then $f(x, y, z)$ is neither maximum nor minimum at (a,b,c).

☞ **REMARK**

- To find the maximum and minimum of the function at stationary point, it is sufficient to find the value of a second order partial derivative of function with respect to any of the independent variables. Then, the value of the function is maximum or minimum according as the value of this second order partial derivative at the stationary point under consideration is negative or positive.

Solved Examples

EXAMPLE 1. *Find the maximum value of u, where* $u = \dfrac{xyz}{(a+x)(x+y)(y+z)(z+b)}$.

SOLUTION. We have $u = \dfrac{xyz}{(a+x)(x+y)(y+z)(z+b)}$

Taking, log of both the sides, we have

$\log u = \log x + \log y + \log z - \log(a+x)$
$\qquad - \log(x+y) - \log(y+z) - \log(z+b).$

Differentiating w.r.t. x, we have

$\dfrac{1}{u}\dfrac{\partial u}{\partial x} = \dfrac{1}{x} - \dfrac{1}{a+x} - \dfrac{1}{x+y} = \dfrac{ay-x^2}{x(a+x)(x+y)}$

$\Rightarrow \quad \dfrac{\partial u}{\partial x} = \dfrac{(ay-x^2)u}{x(a+x)(x+y)}$

Similarly

$\dfrac{\partial u}{\partial y} = \dfrac{(xz-y^2)u}{y(x+y)(y+z)}$ and $\dfrac{\partial u}{\partial z} = \dfrac{(by-z^2)u}{z(y+z)(z+b)}$

For, a maxima and minima of u, we must have

$\dfrac{\partial u}{\partial x} = 0 \Rightarrow ay - x^2 = 0$; $\dfrac{\partial u}{\partial y} = 0$

$\Rightarrow xz - y^2 = 0$ and $\dfrac{\partial u}{\partial z} = 0 \Rightarrow by - z^2 = 0$

Here, we observe that $x^2 = ay$, $y^2 = xz$, $z^2 = by$ which implies that a, x, y, z and b are in G.P. Let r be the common ratio of this G.P.

Then $\qquad ar^4 = b$ or $r = \left(\dfrac{b}{a}\right)^{1/4}$

Also $\qquad x = ar, y = ar^2, z = ar^3.$

Hence, we have

$u = \dfrac{ar.ar^2.ar^3}{a(1+r)ar(1+r)ar^2(1+r)ar^3(1+r)}$

$= \dfrac{1}{a(1+r)^4} = \dfrac{1}{a\left[1+\left(\dfrac{b}{a}\right)^{1/4}\right]^4} = \dfrac{1}{\left(a^{1/4}+b^{1/4}\right)^4}$

which gives a stationary value of u. Now, to decide whether this value of u is a maximum or a minimum, we proceed to find the second order partial derivative of u. Here

$\dfrac{\partial^2 u}{\partial x^2} = \dfrac{-2ux}{x(a+x)(x+y)} + (ay-x^2)\dfrac{\partial}{\partial x}\left[\dfrac{u}{x(a+x)(x+y)}\right]$

When $x = ar, y = ar^2, z = ar^3$, we have

$A = \dfrac{\partial^2 u}{\partial x^2} = -\dfrac{2u}{a^2 r(1+r)^2} < 0$

Hence, the above stationary value of u is maximum.

EXAMPLE 2. *Find the maxima and minima value of the function $u = \sin x \sin y \sin z$ where x, y and z are the vertex angles of a triangle.*

SOLUTION. Here, we have

$u = \sin x \sin y \sin z$; where $x+y+z = \pi$ ...(1)

$\therefore \quad u = \sin x \sin y \sin[\pi-(x+y)]$
$\qquad = \sin x \sin y \sin(x+y)$

$\therefore \quad \dfrac{\partial u}{\partial x} = \cos x \sin y \sin(x+y)$
$\qquad\qquad\qquad + \sin x \sin y \cos(x+y)$
$\qquad = \sin y \sin(2x+y).$ ...(2)

Similarly $\dfrac{\partial u}{\partial y} = \sin x \sin(2y+x)$ ...(3)

For a maxima and minima, we must have

$\dfrac{\partial u}{\partial x} = 0, \dfrac{\partial u}{\partial y} = 0$

So, $\dfrac{\partial u}{\partial x} = 0 \Rightarrow \sin y \sin(2x+y) = 0$

$\Rightarrow \sin y = 0$ or $\sin(2x+y) = 0$

$\Rightarrow y = 0$ or $\sin(x+x+y) = 0$

$\Rightarrow y = 0$

or $\sin x \cos(x+y) + \cos x \sin(x+y) = 0$

$\Rightarrow \qquad \tan(x+y) = -\tan x$

$\Rightarrow \qquad \tan(x+y) = \tan(-x) = \tan(\pi-x)$ ...(4)

$\Rightarrow \qquad x+y = \pi - x$

$\Rightarrow \qquad 2x+y = \pi$ ...(5)

Similarly, from (3)
$\qquad\qquad x = 0$

or $\qquad \tan(x+y) = -\tan y$ ...(6)

Now, by (4) and (6), we have
$\qquad \tan x = \tan y \Rightarrow x = y.$

Hence, by (5), we have

$\qquad 3y = \pi \Rightarrow y = \dfrac{\pi}{3}$ and $x = \dfrac{\pi}{3}$

Therefore, the stationary points are $\left(\dfrac{\pi}{3}, \dfrac{\pi}{3}\right)$ and $(0, 0)$.

For (0,0): $u = 0.$

For $\left(\dfrac{\pi}{3}, \dfrac{\pi}{3}\right)$

$r = \dfrac{\partial^2 u}{\partial x^2} = 2\sin y \cos(2x+y)$

$= 2\sin\dfrac{\pi}{3}\cos\left(\dfrac{2\pi}{3} + \dfrac{\pi}{3}\right) = -\sqrt{3} < 0$

and $s = \dfrac{\partial^2 u}{\partial x \partial y} = \sin(2x+2y) = \sin\left(\dfrac{2\pi}{3}+\dfrac{2\pi}{3}\right)$

$= \sin\left(\dfrac{4\pi}{3}\right) = -\dfrac{\sqrt{3}}{2} < 0$

$t = \dfrac{\partial^2 u}{\partial y^2} = 2\sin x \cos(x+2y) = 2\sin\dfrac{\pi}{3}\cos\pi = -\sqrt{3} < 0$

Now $rt - s^2 = (-\sqrt{3})(-\sqrt{3}) - \left(\dfrac{\sqrt{3}}{2}\right)^2 = \dfrac{9}{4} > 0$

Thus $rt - s^2 > 0$ and $r < 0$.

Hence, the function u will be maximum at $\left(\dfrac{\pi}{3},\dfrac{\pi}{3}\right)$.

Exercise-3.8

1. Prove that the function $u = x^2 + y^2 + x - 2z - xy$ is minimum at $\left(-\dfrac{2}{3}, -\dfrac{1}{3}, 1\right)$.

2. Find the maximum and minimum values of $u = y^2 + 2z^2 - 5x^4 + 4x^5$.

3. Find the maximum or minimum values of the function u, where $u = axy^2 z^3 - x^2 y^2 z^3 - xy^3 z^3 - xy^2 z^4$.

4. Find the maximum value of $(ax+by+cz)\, e^{-\left(\alpha^2 x^2 + \beta^2 y^2 + \gamma^2 z^2\right)}$.

5. A rectangle box is placed on x-y plane. The one end of the box is at the origin. If the vertex opposite to the origin be on the plane $6x+4y+3z=24$, then find the maximum value of this box.

6. In a plane triangle xyz, find the maximum value of $\sin x \sin y \sin z$.

7. A rectangular box, open at the top is to have a given capacity. Show that the domain of the box requiring least material for its construction $x = y = (2v)^{1/3}$, where $v = xyz$.

(UPTU 2006, GBTU 2010)

Answers

2. Minimum at $(1,0,0)$, neither maximum nor minimum at $(0,0,0)$. **3.** Maximum at $\left(\dfrac{a}{7},\dfrac{2a}{7},\dfrac{3a}{7}\right)$, max. value $= \dfrac{108a^7}{7^7}$

4. Maximum at $\left(\dfrac{a}{2\alpha^2 k},\dfrac{b}{2\beta^2 k},\dfrac{c}{2\gamma^2 k}\right)$ where $k = \sqrt{\left\{\dfrac{1}{2}\left(\dfrac{a^2}{\alpha^2}+\dfrac{b^2}{\beta^2}+\dfrac{c^2}{\gamma^2}\right)\right\}}$, Maximum value $= \sqrt{\left\{\dfrac{1}{2e}\left(\dfrac{a^2}{\alpha^2}+\dfrac{b^2}{\beta^2}+\dfrac{c^2}{\gamma^2}\right)\right\}}$

5. Maximum at $\left(\dfrac{4}{3},2\right)$. maximum value $= \dfrac{64}{9}$ cube units. Neither maximum nor minimum at $(0,0)$.

6. Maximum at $\left(\dfrac{\pi}{3},\dfrac{\pi}{3},\dfrac{\pi}{3}\right)$, value $= \dfrac{3\sqrt{3}}{8}$

3.5.8 Lagrange's Method of Undetermined Multipliers

Let $u = f(x_1, x_2, ..., x_n)$ be a function of n variables $x_1, x_2, ..., x_n$.

Let us suppose these variables $x_1, x_2, ..., x_n$ are connected by k equations

$$g_1(x_1, x_2, ..., x_n) = 0$$
$$g_2(x_1, x_2, ..., x_n) = 0$$
$$... ... ...$$
$$g_k(x_1, x_2, ..., x_n) = 0$$

so, that there are $n-k$ independent variables out of these n variables. For the maxima and minima of u, we find

$$du = \dfrac{\partial u}{\partial x_1}dx_1 + \dfrac{\partial u}{\partial x_2}dx_2 + ... + \dfrac{\partial u}{\partial x_n}dx_n = 0 \qquad ...(1)$$

Also

$$dg_1 = \dfrac{\partial g_1}{\partial x_1}dx_1 + \dfrac{\partial g_1}{\partial x_2}dx_2 + ... + \dfrac{\partial g_1}{\partial x_n}dx_n = 0 \qquad ...(2)$$

$$dg_2 = \dfrac{\partial g_2}{\partial x_1}dx_1 + \dfrac{\partial g_2}{\partial x_2}dx_2 + ... + \dfrac{\partial g_2}{\partial x_n}dx_n = 0 \qquad ...(3)$$

$$dg_k = \dfrac{\partial g_k}{\partial x_1}dx_1 + \dfrac{\partial g_k}{\partial x_2}dx_2 + ... + \dfrac{\partial g_k}{\partial x_n}dx_n = 0 \qquad ...(k+1)$$

Multiplying equation (1),(2),(3)...(k+1) by $1, l_1, l_2, ..., l_k$ respectively and adding, we get the result, which can be written as

$$P_1 dx_1 + P_2 dx_2 + P_3 dx_3 + ... + P_n dx_n = 0 \qquad ...(4)$$

where

$$P_k = \dfrac{\partial u}{\partial x_k} + l_1\dfrac{\partial g_1}{\partial x_k} + l_2\dfrac{\partial g_2}{\partial x_k} + ... + l_k\dfrac{\partial g_k}{\partial x_k}$$

Now we have at our choice k multiple *viz* $l_1, l_2, ..., l_k$ and can be chosen such that

$$P_1 = 0, P_2 = 0, ..., P_k = 0$$

Then, the equation (4) reduces to $\quad P_{k+1} dx_{k+1} + P_{k+2} dx_{k+2} + P_{k+3} dx_{k+3} + ... + P_n dx_n = 0 \qquad ...(5)$

Now, let us suppose that out of n variables, the $(n-k)$ variables $x_{k+1}, x_{k+2}, ..., x_n$ are independent.

Then, since $n-k$ quantities $dx_{k+1}, dx_{k+2}, ..., dx_n$ are independent so their coefficients must be separately zero. Hence, we have

$$P_{k+1} = 0, P_{k+2} = 0, ..., P_n = 0$$

Thus, we have $k + n$ equations $\qquad\qquad P_1 = 0, P_2 = 0, ..., P_n = 0$

and $\qquad\qquad\qquad\qquad\qquad\qquad g_1 = 0, g_2 = 0, ..., g_k = 0.$

Hence, we get $(n+k)$ equations which determine the k multipliers $l_1, l_2, ..., l_k$ and get the possible value of u.

☞ REMARKS

- The Lagrange's method of undetermined multipliers is very convenient to apply. It gives the maximum and minimum values of the function without actually determining the values of the multipliers $l_1, l_2, ..., l_k$.

- It does not determine the nature of stationary point, which is the only drawback of this method.

3.5.9 APPLICATIONS OF THE METHOD OF UNDETERMINED MULTIPLIERS

The Lagrange's method of undetermined multipliers can be applied to determine the extreme values of the given functions, it does not detemine the nature of stationary point. Now, it is more convenient to find out the extreme values of a function F with the help of new function, given by

$$V = g + l_1 f_1 + l_2 f_2 + ... + l_m f_m$$

and use the following method. Here, we give the method for four variables x, y, u, v connected by the following two relations.

Let $F = g(x, y, u, v)$ be subjected to the conditions $\qquad f_1(x, y, u, v) = 0 \qquad\qquad\qquad ...(1)$

and $\qquad\qquad\qquad\qquad\qquad\qquad\qquad\qquad f_2(x, y, u, v) = 0. \qquad\qquad\qquad ...(2)$

For the maxima and minima of F, we have $\qquad dF = \dfrac{\partial g}{\partial x} dx + \dfrac{\partial g}{\partial y} dy + \dfrac{\partial g}{\partial u} du + \dfrac{\partial g}{\partial v} dv = 0 \qquad ...(3)$

Now, from (1) and (2), we have $\qquad df_1 = \dfrac{\partial f_1}{\partial x} dx + \dfrac{\partial f_1}{\partial y} dy + \dfrac{\partial f_1}{\partial u} du + \dfrac{\partial f_1}{\partial v} dv = 0 \qquad ...(4)$

and $\qquad\qquad\qquad\qquad\qquad df_2 = \dfrac{\partial f_2}{\partial x} dx + \dfrac{\partial f_2}{\partial y} dy + \dfrac{\partial f_2}{\partial u} du + \dfrac{\partial f_2}{\partial v} dv = 0 \qquad ...(5)$

Multiplying (4) by l_1, (5) by l_2 and adding their sum to (3), we get

$$\left(\frac{\partial g}{\partial x} + l_1 \frac{\partial f_1}{\partial x} + l_2 \frac{\partial f_2}{\partial x} \right) dx + \left(\frac{\partial g}{\partial y} + l_1 \frac{\partial f_1}{\partial y} + l_2 \frac{\partial f_2}{\partial y} \right) dy$$

$$+ \left(\frac{\partial g}{\partial u} + l_1 \frac{\partial f_1}{\partial u} + l_2 \frac{\partial f_2}{\partial u} \right) du + \left(\frac{\partial g}{\partial v} + l_1 \frac{\partial f_1}{\partial v} + l_2 \frac{\partial f_2}{\partial v} \right) dv = 0 \qquad ...(6)$$

Here, we have l_1 and l_2 are arbitrary, therefore we can choose them to satisfy the two linear equations

$$\frac{\partial g}{\partial x} + l_1 \frac{\partial f_1}{\partial x} + l_2 \frac{\partial f_2}{\partial x} = 0 \qquad ...(7)$$

and $\qquad\qquad\qquad\qquad\qquad\qquad \dfrac{\partial g}{\partial y} + l_1 \dfrac{\partial f_1}{\partial y} + l_2 \dfrac{\partial f_2}{\partial y} = 0 \qquad ...(8)$

Using (7) and (8), equation (6) reduces to

$$\left(\frac{\partial g}{\partial u} + l_1 \frac{\partial f_1}{\partial u} + l_2 \frac{\partial f_2}{\partial u} \right) du + \left(\frac{\partial g}{\partial v} + l_1 \frac{\partial f_1}{\partial v} + l_2 \frac{\partial f_2}{\partial v} \right) dv = 0$$

Since, the given function contains four variables (namely x, y, u and v) and we are given two equations of conditions, therefore, only two of the variables are independent and it is immaterial which two of the four variables are regarded as independent. Let them be u and v then du and dv are also independent, therefore, their coefficients must be zero separately. Thus

$$\frac{\partial g}{\partial u} + l_1 \frac{\partial f_1}{\partial u} + l_2 \frac{\partial f_2}{\partial u} = 0 \qquad ...(9)$$

$$\frac{\partial g}{\partial v} + l_1 \frac{\partial f_1}{\partial v} + l_2 \frac{\partial f_2}{\partial v} = 0 \qquad ...(10)$$

Now, we have six equations namely (1), (2), (7), (8), (9) and (10) to determine the two multipliers l_1, l_2 and values of the four variables x, y, u and v for which maximum and minimum values of F are possible.

Now, defined a new function $V(x, y, u, v)$ such that

$$V(x, y, u, v) = g(x, y, u, v) + l_1 f_1(x, y, u, v) + l_2 f_2(x, y, u, v).$$

Assuming that x, y, u, v are now all independent variables. Hence, for the maxima and minima of V, we must have

$$\frac{\partial V}{\partial x} = \frac{\partial g}{\partial x} + l_1 \frac{\partial f_1}{\partial x} + l_2 \frac{\partial f_2}{\partial x} = 0 \qquad \text{...(11)}$$

$$\frac{\partial V}{\partial y} = \frac{\partial g}{\partial y} + l_1 \frac{\partial f_1}{\partial y} + l_2 \frac{\partial f_2}{\partial y} = 0 \qquad \text{...(12)}$$

$$\frac{\partial V}{\partial u} = \frac{\partial g}{\partial u} + l_1 \frac{\partial f_1}{\partial u} + l_2 \frac{\partial f_2}{\partial u} = 0 \qquad \text{...(13)}$$

and $$\frac{\partial V}{\partial v} = \frac{\partial g}{\partial v} + l_1 \frac{\partial f_1}{\partial v} + l_2 \frac{\partial f_2}{\partial v} = 0 \qquad \text{...(14)}$$

Equations (11), (12), (13) and (14) are exactly the same as the equations (7). (8), (9) and (10). Hence, the maxima and minima of $V(x, y, u, v)$ are same as those of $F(x, y, u, v)$ assuming that $V(x, y, u, v)$ the variables x, y, u, v are now all independent.

Now, we proceed to find whether the values of F obtained with the help of above equations are maximum or minimum. For this, adopt the procedure, which is discussed ahead.

From (3), we get
$$d^2 F = \left(\frac{\partial}{\partial x} dx + \frac{\partial}{\partial y} dy + \frac{\partial}{\partial u} du + \frac{\partial}{\partial y} dy \right)^2 g + \left(\frac{\partial g}{\partial x} d^2 x + \frac{\partial g}{\partial y} d^2 y + \frac{\partial g}{\partial u} d^2 u + \frac{\partial g}{\partial y} d^2 v \right) \ldots \qquad \text{...(15)}$$

Also
$$d^2 f_1 = \left(\frac{\partial}{\partial x} dx + \frac{\partial}{\partial y} dy + \frac{\partial}{\partial u} du + \frac{\partial}{\partial v} dv \right)^2 f_1 + \frac{\partial f_1}{\partial x} d^2 x + \frac{\partial f_1}{\partial y} d^2 y + \frac{\partial f_1}{\partial u} d^2 u + \frac{\partial f_1}{\partial v} d^2 v = 0 \qquad \text{...(16)}$$

and
$$d^2 f_2 = \left(\frac{\partial}{\partial x} dx + \frac{\partial}{\partial y} dy + \frac{\partial}{\partial u} du + \frac{\partial}{\partial v} dv \right)^2 f_2 + \frac{\partial f_2}{\partial x} d^2 x + \frac{\partial f_2}{\partial y} d^2 y + \frac{\partial f_2}{\partial u} d^2 u + \frac{\partial f_2}{\partial v} d^2 v = 0 \qquad \text{...(17)}$$

Multiplying (16) by l_1 and (17) by l_2 and adding their sum to (15) and using the result (11), (12),(13) and (14), we have

$$\left(\frac{\partial}{\partial x} + \frac{\partial}{\partial y} + \frac{\partial}{\partial u} + \frac{\partial}{\partial v} \right)^2 (+ 1\,1 + 2\,2) = \left(\frac{\partial}{\partial x} dx + \frac{\partial}{\partial y} dy + \frac{\partial}{\partial u} du + \frac{\partial}{\partial v} dv \right)^2 V = d^2 V.$$

Hence $d^2 F$ is equal to $d^2 V$, where $d^2 V$ is obtained by assuming all the variables x, y, u and v as independent. Therefore, it is clear that $d^2 F$ and $d^2 V$ have the same sign. Hence, F will be minimum or maximum according as V is minimum or maximum.

☞ **REMARK**
• This method has the advantage over the Lagrange's methods that it enables us to decide whether the values are maximum or minimum.

Solved Examples

EXAMPLE 1. *Find the maxima and minima of $x^2+y^2+z^2$ subject to the conditions :*
$$ax^2+by^2+cz^2 = 1 \text{ and } lx+my+nz = 0$$
(UKTU- 2011)

SOLUTION. Here, we have $u = x^2 + y^2 + z^2$ ...(1)
where, the relations between the variables x, y and z are given by
$$ax^2+by^2+cz^2 = 1 \qquad \text{...(2)}$$
and $$lx+my+nz = 0 \qquad \text{...(3)}$$
For the maxima and minima of u, we must have
$$du = 0$$
$$\Rightarrow \quad 2x\,dx+2y\,dy+2z\,dz = 0$$
$$\Rightarrow \quad x\,dx+y\,dy+z\,dz = 0 \qquad \text{...(4)}$$
From (2) and (3), we get
$$ax\,dx+by\,dy+cz\,dz = 0 \qquad \text{...(5)}$$
$$l\,dx+m\,dy+n\,dz = 0 \qquad \text{...(6)}$$
Now, multiplying (4) by 1, (5) by l_1 and (6) by l_2 and adding, we get
$$(x\,dx+y\,dy+z\,dz)+l_1(ax\,dx+by\,dy+cz\,dz)$$
$$+l_2(l\,dx+m\,dy+n\,dz) = 0$$
$$\Rightarrow \quad (x+al_1x+ll_2)dx+(y+bl_1y+ml_2)$$
$$dy+(z+cl_1z+nl_2)dz = 0$$

Now equating the coefficient of dx, dy, dz to zero, we get
$$x+l_1ax+l_2l = 0 \qquad \text{...(7)}$$
$$y+bl_1y+ml_2 = 0 \qquad \text{...(8)}$$
and $$z+cl_1z+nl_2 = 0 \qquad \text{...(9)}$$
Multiplying the equations (7), (8) and (9) by x, y and z respectively, and adding we get
$$x^2+y^2+z^2+l_1(ax^2+by^2+cz^2) +l_2(lx+my+nz) = 0$$
or $$u+l_1.1+l_2.0 = 0 \text{ [By using (1), (2) and (3)]}$$
$$\Rightarrow \quad l_1 = -u$$
Substituting for l_1 in the equations (7), (8) and (9), we get
$$x = \frac{l_2 l}{au-1}, y = \frac{l_2 m}{bu-1}, z = \frac{l_2 n}{cu-1} \text{ ...(10)}$$
Now from (10) and (3), we get
$$\frac{l_2 l^2}{au-1} + \frac{l_2 m^2}{bu-1} + \frac{l_2 n^2}{cu-1} = 0$$
or $$\frac{l^2}{au-1} + \frac{m^2}{bu-1} + \frac{n^2}{cu-1} = 0 \qquad \text{...(11)}$$
which gives the maximum and minimum of $u = x^2+y^2+z^2$.

☞ REMARKS
- Equation (11) is a quadratic in u. So it gives two stationary values of u.
- Geometrically, the surface $ax^2 + by^2 + cz^2 = 1$ represents an ellipsoid whose centre is origin, and $lx + my + nz = 0$ represents a plane passing through the origin. The points (x, y, z) satisfying both the conditions (2) and (3) lies on the conic in which (2) and (3) intersect. $x^2 + y^2 + z^2$ gives the square of the distance (x, y, z) from the origin, which is also the centre of the conic of intersection. The maximum value of this distance is the major axis of this conic, and the minimum value of this distance is the minor axis of this conic. Hence, equation (11) gives the squares of the lengths of the semi-axis of the conic of intersection.

EXAMPLE 2. *Find the maxima and minima of $x^2 + y^2 + z^2$, where $ax^2 + by^2 + cz^2 + 2fyz + 2gzx + 2hxy = 1$.*

SOLUTION. Let
$$u = x^2 + y^2 + z^2 \qquad \ldots(1)$$
where the relation between the variables x, y and z is
$$ax^2 + by^2 + cz^2 + 2fyz + 2gzx + 2hxy = 1. \qquad \ldots(2)$$
For a maximum or minima of u, we must have
$$du = 0$$
$$\Rightarrow \qquad x\,dx + y\,dy + z\,dz = 0. \qquad \ldots(3)$$
From (2), we have
$$2ax\,dx + 2by\,dy + 2cz\,dz + 2fy\,dz + 2fz\,dy + 2gz$$
$$dx + 2gx\,dz + 2hx\,dy + 2hy\,dx = 0$$
$$\Rightarrow \quad (ax + hy + gz)dx + (hx + by + fz)$$
$$dy + (gx + fy + cz)dz = 0 \ldots(4)$$
Now, multiplying (3) by 1 and (4) by l_1, adding, and then equating the coefficient of dx, dy, dz to zero, we have
$$x + l_1(ax + hy + gz) = 0. \qquad \ldots(5)$$
$$y + l_1(hx + by + fz) = 0. \qquad \ldots(6)$$
$$z + l_1(gx + fy + cz) = 0. \qquad \ldots(7)$$
Multiplying (5) by x, (6) by y, (7) by z and adding, we get
$$x^2 + y^2 + z^2 + l_1(ax^2 + by^2 + cz^2 + 2fyz + 2gzx + 2hxy) = 0$$
$$\Rightarrow \qquad\qquad u + l_1 \cdot 1 = 0 \quad \text{[From (1) and (2)]}$$
$$\therefore \qquad\qquad l_1 = -u.$$
Hence, from (5), we have
$$x - u(ax + hy + gz) = 0$$
$$\Rightarrow \qquad \left(a - \frac{1}{u}\right)x + hy + gz = 0 \qquad \ldots(8)$$
Similarly from (6) and (7), we get
$$hx + \left(b - \frac{1}{u}\right)y + fz = 0 \qquad \ldots(9)$$
and
$$gx + fy + \left(c - \frac{1}{u}\right)z = 0 \qquad \ldots(10)$$
Eliminating x, y, z from (8), (9) and (10), we get
$$\begin{vmatrix} \left(a - \dfrac{1}{u}\right) & h & g \\[2mm] h & \left(b - \dfrac{1}{u}\right) & f \\[2mm] g & f & \left(c - \dfrac{1}{u}\right) \end{vmatrix} = 0 \qquad \ldots(11)$$
Hence, the maximum or minimum values of u are the roots of the equation (11).

EXAMPLE 3. *Find the maximum value of $u = x^m y^n z^p$ subject to the condition $x + y + z = a$.* (ANNA-2009)

SOLUTION. Here, we have $u = x^m y^n z^p$ $\qquad \ldots(1)$
and x, y, z connected by the relation given by
$$x + y + z = a \qquad \ldots(2)$$
Taking log of both the sides of (1), we get
$$\log u = m \log x + n \log y + p \log z.$$

On differentiating, we get
$$\frac{1}{u}du = \frac{m}{x}dx + \frac{n}{y}dy + \frac{p}{z}dz$$
For the maxima and minima of u, we must have
$$du = 0$$
$$\Rightarrow \qquad \frac{m}{x}dx + \frac{n}{y}dy + \frac{p}{z}dz = 0 \qquad \ldots(3)$$
Now, differentiating (2), we get
$$dx + dy + dz = 0. \qquad \ldots(4)$$
Now, multiplying (3) by 1 and (4) by l, and equating the coefficient of dx, dy, dz to zero (after adding), we get
$$\frac{m}{x} + l = 0, \quad \frac{n}{y} + l = 0 \quad \text{and} \quad \frac{p}{z} + l = 0$$
which implies $\qquad x = -\frac{m}{l}, y = -\frac{n}{l}, z = -\frac{p}{l}$
Putting the values of x, y and z in (2), we get
$$l = -\left(\frac{m + n + p}{a}\right)$$
therefore, we can say that, u is stationary when
$$x = \frac{am}{m + n + p}, y = \frac{an}{m + n + p}, z = \frac{ap}{m + n + p}$$
Now, we find the nature of this stationary value of u.

Let us regard x and y as independent variable and z is a function of x and y given by (2) [It is justify, because the variables x, y and z are connected by the relation (2), any two of them may be regarded as independent].

Now from (1), we get
$$\log u = m \log x + n \log y + p \log z$$
$$\therefore \quad \frac{1}{u}\frac{\partial u}{\partial x} = \frac{m}{x} + \frac{p}{z}\frac{\partial z}{\partial x} \qquad \ldots(5)$$
Now, differentiating (2) partially *w.r.t* x (treating y as constant), we get
$$1 + \frac{\partial z}{\partial x} = 0 \quad \Rightarrow \quad \frac{\partial z}{\partial x} = -1$$
Put this value in (5), we get
$$\frac{1}{u}\frac{\partial u}{\partial x} = \frac{m}{x} - \frac{p}{z}$$
$$\Rightarrow \frac{1}{u}\frac{\partial^2 u}{\partial x^2} - \frac{1}{u^2}\left(\frac{\partial u}{\partial x}\right)^2 = -\frac{m}{x^2} + \frac{p}{z^2}\frac{\partial z}{\partial x} = -\frac{m}{x^2} - \frac{p}{z^2}$$
At stationary point, $\frac{\partial u}{\partial x} = 0$

Therefore, $\frac{1}{u}\frac{\partial^2 u}{\partial x^2} = \frac{-m}{x^2} - \frac{p}{z^2}$

$$\Rightarrow \quad \frac{\partial^2 u}{\partial x^2} = u\left[-\frac{m}{x^2} - \frac{p}{z^2}\right] = -x^m y^n z^p \left[-\frac{m}{x^2} - \frac{p}{z^2}\right]$$

which is negative for the obtained values of x, y and z.

Hence, at the stationary point, u is maximum and maximum value is

$$= \left(\frac{am}{m+n+p}\right)^m \left(\frac{an}{m+n+p}\right)^n \left(\frac{ap}{m+n+p}\right)^p$$

EXAMPLE 4. *In a plane triangle ABC, find the maximum value*

of u=cos A cos B cos C.   (VTU-2010, ANNA-2006)

SOLUTION . Here, we have

$$u = \cos A \cos B \cos C \qquad \qquad ...(1)$$

Since, we know that the sum of the angles of a triangle is always 180°.

∴ The variables A, B and C are connected by the relation $A + B + C = \pi$ ...(2)

From (1), we get

$$\log u = \log \cos A + \log \cos B + \log \cos C$$

$$\Rightarrow \quad \frac{1}{u}\,du = -\tan A\,dA - \tan B\,dB - \tan C\,dC.$$

For the maxima and minima of u, we must have $du=0$

$$\Rightarrow \tan A\,dA + \tan B\,dB + \tan C\,dC = 0 \quad ...(3)$$

Also from (2),

$$dA + dB + dC = 0 \qquad \qquad ...(4)$$

Now, multiply (3) by 1, (4) by l , adding, and equating the coefficients of dA, dB and dC to zero, we get

$$\tan A + l = 0$$
$$\tan B + l = 0$$
$$\tan C + l = 0$$

$$\Rightarrow \quad l = -\tan A = -\tan B = -\tan C$$
$$\Rightarrow \quad A = B = C.$$

Now from (2), $A = B = C = \dfrac{\pi}{3}$ *i.e.,* the triangle is equilateral.

Now to show that the stationary value of u given

by $A = B = C = \dfrac{\pi}{3}$ is maximum.

Let C be a function of A and B, regarding A and B as independent variables. From (1),

$$\log u = \log \cos A + \log \cos B + \log \cos C$$

$$\Rightarrow \quad \frac{1}{u}\frac{\partial u}{\partial A} = -\tan A - \tan C \frac{\partial C}{\partial A}$$

Now, differentiating (2), partially w.r.t. A, we get

$$1 + \frac{\partial C}{\partial A} = 0 \quad \Rightarrow \quad \frac{\partial C}{\partial A} = -1$$

$$\therefore \qquad \frac{1}{u}\frac{\partial u}{\partial A} = -\tan A + \tan C$$

$$\Rightarrow \quad \frac{1}{u}\frac{\partial^2 u}{\partial A^2} - \frac{1}{u^2}\left(\frac{\partial u}{\partial A}\right)^2 = -\sec^2 A + \sec^2 C \cdot \frac{\partial C}{\partial A}$$

$$= -\left(\sec^2 A + \sec^2 C\right)$$

At stationary point $\dfrac{\partial u}{\partial A} = 0$

$$\therefore \qquad \frac{\partial^2 u}{\partial A^2} = -u\left(\sec^2 A + \sec^2 C\right) = -\text{ve}$$

for $A = B = C = \dfrac{\pi}{3}$.

Hence, u is maximum at $A = B = C = \dfrac{\pi}{3}$ and the maximum value is given by

$$u = \left(\cos\frac{\pi}{3}\right)^3 = \left(\frac{1}{2}\right)^3 = \frac{1}{8}.$$

Exercise-3.9

Using Lagrange's method of undetermined multiplirers:

1. Find the maximum and minimum values of $\dfrac{x^2}{a^4} + \dfrac{y^2}{b^4} + \dfrac{z^2}{c^4}$

   where $lx + my + nz = 0$ and $\dfrac{x^2}{a^2} + \dfrac{y^2}{b^2} + \dfrac{z^2}{c^2} = 1$.

2. Find the maximum and minimum values of $f = a^2 x^2 + b^2 y^2 + c^2 z^2$ where $x^2 + y^2 + z^2 = 1$ and $lx + my + nz = 0$.

3. Show that the maximum and minimum values of $u = x^2 + y^2 + z^2$

   subject to the conditions $px + qy + rz = 0$ and $\dfrac{x^2}{a^2} + \dfrac{y^2}{b^2} + \dfrac{z^2}{c^2} = 1$

   are given by $\dfrac{a^2 p^2}{u - a^2} + \dfrac{b^2 q^2}{u - b^2}$.

4. Find the minimum value of $u = x + y + z$ subject to the condition

   $\dfrac{a}{x} + \dfrac{b}{y} + \dfrac{c}{z} = 1$.

5. Find the minimum value of $u = x^2 + y^2 + z^2$, subject to the condition $ax + by + cz = p$.   (UKTU-2012, UPTU-2009)

6. Find the minimum value of $x + y + z$ where $xyz = c^3$.

7. Find the extreme values of $x^p y^q z^r$ subject to the condition $\dfrac{a}{x} + \dfrac{b}{y} + \dfrac{c}{z} = 1$.

8. Show that the maximum and minimum values of the radii vectors of the sections of the surface

   $$(x^2 + y^2 + z^2)^2 = \frac{x^2}{a^2} + \frac{y^2}{b^2} + \frac{z^2}{c^2}$$

   by the plane $\lambda x + \mu y + \nu z = 0$

   are given by $\dfrac{a^2 \lambda^2}{1 - a^2 r^2} + \dfrac{b^2 \mu^2}{1 - b^2 r^2} + \dfrac{c^2 \nu^2}{1 - c^2 r^2} = 0$

9. Find the stationary points of the function $u = ax^p + by^q + cz^r$ subject to the condition $x^l + y^m + z^n = k$.

10. If two variables x and y are connected by the relation $ax^2 + by^2 = ab$, show that the maximum and minimum values of the function $u = x^2 + y^2 + xy$ will be the roots of the equation $4(u-a)(u-b) = ab$.

11. Prove that of all rectangular parallelopipeds of the same volume, the cube has the least surface.

           (KURUKSHETRA-2006, UPTU-2004)

12. Prove that if $x+y+z=1$, $ayz+bzx+cxy$ has an extreme value equal to $\dfrac{abc}{2bc+2ca+2ab-a^2-b^2-c^2}$

Also, prove if a, b, c are all positive and c lies between $a+b-2\sqrt{ab}$ and $a+b+2\sqrt{ab}$ this value is true maximum and that if a, b, c are all negative and c lies between $a+b\pm2\sqrt{ab}$. It is true minimum.

13. Find the maximum value of u, when $u=\sin x \sin y \sin z$ and x, y, z are the angles of a triangle.

14. Find the triangle of maximum area inscribed in a circle.

15. Prove that the rectangular solid of maximum volume which can be inscribed in a sphere is a cube.

16. Find a plane triangle ABC such that $u=\sin^a A \sin^b B \sin^c C$ has maximum value.

17. Find the rectangular parallelopiped of maximum volume that can be inscribed in the ellipsoid
$$\frac{x^2}{a^2}+\frac{y^2}{b^2}+\frac{z^2}{c^2}=1$$
(UKTU-2010, ANNA-2009, MADRAS-2006)

18. Divide a number n into three parts x, y, z such that $ayz+bzx+cxy$ shall have maximum or minimum and determine which it is.

19. Prove that a rectangular solid of maximum volume which can be inscribed in a sphere is a cube.

20. Find the maximum or minimum value of $x^p y^q z^r$ subject to the condition $ax+by+cz=p+q+r$.

21. Show that the maximum and minimum value of
$$u=ax^2+by^2+cz^2+2fyz+2gzx+2hxy$$
subject to the conditions $lx+my+nz=0$
and $x^2+y^2+z^2=1$
are given by the equation
$$\begin{vmatrix} a-u & h & g & l \\ h & b-u & f & m \\ g & f & c-u & n \\ l & m & n & o \end{vmatrix}=0$$

22. Show that of the perimeter of a triangle is constant, its area is maximum when it is equilateral.

23. Show that the volume of the largest rectangular parallelopiped that can be inscribed in the ellipsoid $\dfrac{x^2}{a^2}+\dfrac{y^2}{b^2}+\dfrac{z^2}{c^2}=1$ is $\dfrac{8abc}{3\sqrt{3}}$.
(UKTU-2010)

24. Show that the maximum and minimum distances from the origin to the curve $x^2+4xy+6y^2=140$ are respectively given by 21.6589 and 4.5706.
(MTU-2012)

25. A rectangular box which is open at the top, has a capacity of 32 c.c. Find the dimension of the box such that the least material is required for the construction of the box.
(UPTU-2008, GBTU-2011, PTU-2006, KURUKSHETRA-2006)

Answers

1. The maximum and minimum values of the given function is given by the equation $\dfrac{l^2 a^4}{a^2 u-1}+\dfrac{m^2 b^4}{b^2 u-1}+\dfrac{n^2 c^4}{c^2 u-1}=0$

2. The maximum and minimum values of the given function is given by the equation $\dfrac{l^2}{u-a^2}+\dfrac{m^2}{u-b^2}+\dfrac{m^2}{u-c^2}=0$

4. Stationary points are $x=\sqrt{a}\left(\sqrt{a}+\sqrt{b}+\sqrt{c}\right), y=\sqrt{b}\left(\sqrt{a}+\sqrt{b}+\sqrt{c}\right), z=\sqrt{c}\left(\sqrt{a}+\sqrt{b}+\sqrt{c}\right)$, minimum value is $\left(\sqrt{a}+\sqrt{b}+\sqrt{c}\right)^2$.

5. Minimum value is $\dfrac{p^2}{\left(a^2+b^2+c^2\right)}$ 6. u is minimum at the point $x=y=z=c$. Value is $=3c^4$.

7. u is stationary when $\dfrac{px}{a}=\dfrac{qy}{b}=\dfrac{rc}{c}=p+q+r$, Minimum value is $\dfrac{a^p b^q c^r}{p^p q^q r^r}\left(p+q+r\right)^{p+q+r}$.

9. Stationary points are given by $\dfrac{x^{p-1}}{l/pa}=\dfrac{y^{q-m}}{m/qb}=\dfrac{z^{r-n}}{n/rc}$ 13. u is maximum, when $x=y=z=\dfrac{\pi}{3}$. Maximum value is $\dfrac{3\sqrt{3}}{8}$.

14. Equilateral.   16. u is maximum when, A, B, C are given by $\dfrac{\tan A}{a}=\dfrac{\tan B}{b}=\dfrac{\tan C}{c}$.

17. Stationary points are $x=\dfrac{a}{\sqrt{3}}, y=\dfrac{b}{\sqrt{3}}, z=\dfrac{c}{\sqrt{3}}$, Maximum value $=\dfrac{8abc}{3\sqrt{3}}$.   25. $x=4, y=4, z=2$

3.6   ERROR AND APPROXIMATIONS

It is a well known fact that error play an important role in daily life as well as in computational mathematics. In this chapter, we shall discuss the methods to estimate errors in any measurement process by applying differentiation method.

3.6.1 ERROR AND APPROXIMATIONS

Let $y = f(x)$ be a function and Δx be the small error in x. Then Δy be the corresponding error in y. Therefore, we have
$$y + \Delta y = f(x + \Delta x)$$
$$\Rightarrow \qquad \Delta y = f(x + \Delta x) - y = f(x + \Delta x) - f(x)$$
$$= \left[f(x) + \Delta x.f'(x) + \frac{(\Delta x)^2}{2!}f''(x) + ...\right] - f(x) \qquad \text{(on expanding by Taylor's series)}$$

On neglecting the second and higher powers of Δx, we get

$$\Delta y = \Delta x. f'(x) \qquad \ldots(1)$$

$$= \Delta x. \frac{df(x)}{dx} \qquad \ldots(2)$$

Similarly, when y is a function of more than one variable, *i.e.*, $y = f(x_1, x_2)$. Then $\Delta y = f(x_1 + \Delta x_1, x_2 + \Delta x_2)$

where Δx_1 and Δx_2 are the errors in x_1 and x_2 respectively.

Now using Taylor's theorem, we get $\quad \Delta y = \left(\left(f(x_1, x_2) + \frac{\partial f}{\partial x_1} \Delta x_1 + \frac{\partial f}{\partial x_2} \Delta x_2 + \ldots \right) - f(x_1, x_2) \right)$

On neglecting the second and higher degree terms we get

$$\Delta y = \frac{\partial f}{\partial x_1} \Delta x_1 + \frac{\partial f}{\partial x_2} \Delta x_2 \qquad \ldots(3)$$

and so on.

In general, if $y = f(x_1, x_2, \ldots, x_n)$ and $\Delta x_1, \Delta x_2, \ldots, \Delta x_n$ be the errors in $x_1, x_2, \ldots, x_n$ respectively and Δy is the corresponding error in y.

Then we have $\qquad \Delta y = \frac{\partial f}{\partial x_1} \Delta x_1 + \frac{\partial f}{\partial x_2} \Delta x_2 + \ldots + \frac{\partial f}{\partial x_n} \Delta x_n. \qquad \ldots(4)$

which is the required error relation of the function $y = f(x_1, x_2, \ldots, x_n)$.

Here, we observe that

(1) Δy is the absolute error in y.   (2) $\dfrac{\Delta y}{y}$ is the relative error in y.   (3) $\dfrac{\Delta y}{y} \times 100$ is percentage error in y.

Solved Examples

EXAMPLE 1. *Find the percentage error in the area of an ellipse when an error of one percent is made in measuring the major and minor axes.* (GBTU–2011)

SOLUTION. Let x and y are semi-major and semi-minor axes of the ellipse. Then area of the ellipse is

$A = \pi x y$

Taking log of both the sides, we get

$\log A = \log \pi + \log x + \log y$

On differentiating, we get

$\dfrac{\Delta A}{A} = 0 + \dfrac{\Delta x}{x} + \dfrac{\Delta y}{y}$

$\dfrac{\Delta A}{A} \times 100 = \dfrac{\Delta x}{x} \times 100 + \dfrac{\Delta y}{y} \times 100 = 1 + 1 = 2$

which is the required error in area.

EXAMPLE 2. *If the base radius and height of a cone are measured as 4 and 8 inches with a possible error of 0.04 and 0.08 inches respectively. Calculate the percentage error in calculating volume of the cone.* (GBTU–2012)

SOLUTION. We know that

Volume, $V = \dfrac{1}{3} \pi r^2 h$

$\Rightarrow \qquad \log V = \log \dfrac{1}{3} + \log \pi + 2 \log r + \log h$

Differentiating, we get

$\dfrac{\Delta V}{V} = 2 \dfrac{\Delta r}{r} + \dfrac{\Delta h}{h}$

$\Rightarrow \qquad \dfrac{\Delta V}{V} = 2 \left(\dfrac{0.04}{4} \right) + \left(\dfrac{0.08}{8} \right) = 0.03$

Hence, percentage error in volume $= 0.03 \times 100 = 3\%$

EXAMPLE 3. *Let A be the area of a triangle, prove that the error A resulting from a small error in c is given by* $\Delta A = \dfrac{A}{4} \left[\dfrac{1}{s} + \dfrac{1}{s-a} + \dfrac{1}{s-b} - \dfrac{1}{s-c} \right] \Delta c.$

(UPTU–2007, 08)

SOLUTION. Let a, b, c be the sides of a triangle, then

$A = \sqrt{s(s-a)(s-b)(s-c)}$

where $s = \dfrac{a+b+c}{2}$

Taking log of both sides, we get

$\log A = \dfrac{1}{2} \log(s(s-a)(s-b)(s-c))$

$= \dfrac{1}{2} \left(\log s + \log(s-a) + \log(s-b) + \log(s-c) \right)$

Differentiating, we get

$\dfrac{1}{A} \cdot \dfrac{dA}{dc} = \dfrac{1}{2} \left[\dfrac{1}{s} \dfrac{ds}{dc} + \dfrac{1}{s-a} \dfrac{d(s-a)}{dc} \right.$

$\left. + \dfrac{1}{s-b} \dfrac{d(s-b)}{dc} + \dfrac{1}{s-c} \dfrac{d(s-c)}{dc} \right] \qquad \ldots(1)$

We have $\dfrac{ds}{dc} = \dfrac{1}{2} \qquad \left[\because s = \dfrac{a+b+c}{2} \right]$

$\dfrac{d(s-c)}{dc} = \dfrac{d}{dc} \left(\dfrac{a+b-c}{2} \right) = -\dfrac{1}{2}$

$\dfrac{d(s-a)}{dc} = \dfrac{1}{2}, \dfrac{d(s-b)}{dc} = \dfrac{1}{2}$

Putting all these values in (1), we get

$\dfrac{1}{A} \dfrac{dA}{dc} = \dfrac{1}{2} \left[\dfrac{1}{s} \times \dfrac{1}{2} + \dfrac{1}{s-a} \times \dfrac{1}{2} + \dfrac{1}{s-b} \times \dfrac{1}{2} + \dfrac{1}{s-c} \times \left(\dfrac{-1}{2} \right) \right]$

$\Rightarrow \qquad \dfrac{dA}{dc} = \dfrac{A}{4} \left[\dfrac{1}{s} + \dfrac{1}{s-a} + \dfrac{1}{s-b} - \dfrac{1}{s-c} \right]$

$\Rightarrow \qquad \dfrac{\Delta A}{\Delta c} = \dfrac{A}{4} \left[\dfrac{1}{s} + \dfrac{1}{s-a} + \dfrac{1}{s-b} - \dfrac{1}{s-c} \right]$

$\Rightarrow \qquad \Delta A = -\left[\dfrac{1}{s} + \dfrac{1}{s-a} + \dfrac{1}{s-b} - \dfrac{1}{s-c} \right] \Delta c$

EXAMPLE 4. *The angles of a triangle are calculated from the sides a, b, c if small changes δa, δb and δc are made in the sides. Find δA, δB and δC approximately, where Δ is the area of the triangle and A, B, C are angles opposite to sides A, B, C respectively. Also, show that $\delta A + \delta B + \delta C = 0$.* (GBTU–2012)

SOLUTION. From trigonometry, we have

$$a^2 = b^2 + c^2 - 2bc \, \text{Cos} \, A \qquad \ldots(1)$$

Differentiating, we get

$$2a\delta a = 2b\delta b + 2c\delta c - 2b\delta c \, \text{Cos} \, A$$
$$\qquad - 2c\delta b \, \text{Cos} \, A + 2bc \, \text{sin} A \delta A$$

$$\Rightarrow bc \text{Sin} \, A . \delta A = a\delta a - (b - c \, \text{Cos} \, A)\delta b - (c - b \, \text{Cos} \, A)\delta c$$

$$\Rightarrow \quad 2 \, \Delta \, \delta A = a\delta a - (a \text{Cos} \, C + c \text{Cos} \, A - c \text{Cos} \, A) \, \delta b$$
$$\qquad - (a \text{Cos} \, B + b \text{Cos} \, A - b \text{Cos} \, A) \, \delta c$$

$$\Rightarrow \quad \delta A = \frac{a}{2\Delta}(\delta a - \delta b \cos C - \delta c \cos B) \qquad \ldots(2)$$

In a similar manner, we may get

$$\delta B = \frac{b}{2\Delta}(\delta b - \delta c \cos A - \delta a \cos C) \qquad \ldots(3)$$

and $\qquad \delta C = \frac{c}{2\Delta}(\delta c - \delta a \cos B - \delta b \cos A) \qquad \ldots(4)$

On adding (2), (3) and (4), we get

$$\delta A + \delta B + \delta C = \frac{1}{2\Delta}[(a - b \cos C - c \cos B)\delta a$$
$$+ (b - c \cos A - a \cos C)\delta b$$
$$+ (c - a \cos B - b \cos A)\delta c]$$
$$= \frac{1}{2\Delta}[(a - a)\delta a + (b - b)\delta b + (c - c)\delta c]$$
$$= \frac{1}{2\Delta}.0 = 0 .$$

EXAMPLE 5. *Find the possible percentage error in computing the parallel resistance r of three resistances r_1, r_2, r_3 from the formula $\frac{1}{r} = \frac{1}{r_1} + \frac{1}{r_2} + \frac{1}{r_3}$.*

If r_1, r_2, r_3 are each in error by +1.2%.

(UKTU–2011, 13)

SOLUTION. Given that $\frac{1}{r} = \frac{1}{r_1} + \frac{1}{r_2} + \frac{1}{r_3}$

$$\Rightarrow \quad -\frac{1}{r^2}\Delta r = -\frac{1}{r_1^2}\Delta r_1 - \frac{1}{r_2^2}\Delta r_2 - \frac{1}{r_3^2}\Delta r_3$$

$$\Rightarrow \frac{1}{r}\left(\frac{\Delta r}{r} \times 100\right) = \frac{1}{r_1}\left(\frac{\Delta r_1}{r_1} \times 100\right)$$
$$+ \frac{1}{r_2}\left(\frac{\Delta r_2}{r_2} \times 100\right) + \frac{1}{r_3}\left(\frac{\Delta r_3}{r_3} \times 100\right)$$
$$= \frac{1}{r_1}(1.2) + \frac{1}{r_2}(1.2) + \frac{1}{r_3}(1.2)$$
$$= 1.2\left(\frac{1}{r_1} + \frac{1}{r_2} + \frac{1}{r_3}\right) = 1.2\left(\frac{1}{r}\right)$$

$$\Rightarrow \quad \frac{\Delta r}{r} \times 100 = 1.2, \text{which is the required error in } r.$$

EXAMPLE 6.(i) *In estimating the number of bricks in a pile which is measured to be (5m × 10m × 5m), the count of bricks is taken as 100 bricks per m³. Find the error in the cost when the tape is stretched 2% beyond its standard length. The cost of bricks is Rs 2000 per thousand bricks.*

(UPTU–2001, 07)

(ii) *In estimating the cost of a pile of bricks as 6m × 50m × 4m, the tape is stretched 1% beyond the standard length. If the count is 12 bricks per 1 m³ and bricks cost is Rs 1000, find the approximate error in the cost.* (UKTU–2010)

SOLUTION. (i) Let length, breadth and height of the pile be x, y and z respectively.

The volume of the pile, $V = xyz$

$$\Rightarrow \quad \log V = \log x + \log y + \log z$$

Differentiating, we get

$$\frac{\Delta V}{V} \times 100 = \left(\frac{\Delta x}{x} \times 100\right) + \left(\frac{\Delta y}{y} \times 100\right) + \left(\frac{\Delta z}{z} \times 100\right)$$
$$= 2 + 2 + 2 = 6$$

Therefore, $\quad \Delta V = \frac{6}{100}V = \frac{6}{100}(250) = 15 \, \text{m}^3$

$$[\because V = 5 \times 10 \times 5 = 250 \, \text{m}^3]$$

Hence, number of bricks in $\Delta V = 15 \times 100 = 1500$

and error in the cost $= 1500 \times \frac{2000}{1000} = \text{Rs} \, 3000$

(ii) Proceed same as in part (i), we get

$$\frac{\Delta V}{V} \times 100 = \left(\frac{\Delta x}{x} \times 100\right) + \left(\frac{\Delta y}{y} \times 100\right) + \left(\frac{\Delta z}{z} \times 100\right)$$
$$= 1 + 1 + 1 = 3$$

$$\therefore \quad \Delta V = \frac{3}{100}V = \frac{3}{100} \times 1200 = 36 \, \text{m}^3$$

$$[\because V = 6 \times 50 \times 4 = 1200 \, \text{m}^3]$$

$$\therefore \quad \Delta V = 36 \times 12 = 432$$

and error in the cost $= 432 \times \frac{100}{1000} = \text{Rs} \, 43.20$

EXAMPLE 7. *The period of a simple pendulum is $T = 2\pi\sqrt{\dfrac{l}{g}}$. Find the maximum percentage error in T due to the possible errors upto 1% in l and 2.5% in g.*

(UPTU–2004)

SOLUTION. We have $\quad T = 2\pi\sqrt{\dfrac{l}{g}}$

$$\Rightarrow \quad \log T = \log 2\pi + \frac{1}{2}\log l - \frac{1}{2}\log g$$

$$\Rightarrow \quad \frac{1}{T}\Delta T = \frac{1}{2}\frac{\Delta l}{l} - \frac{1}{2}\frac{\Delta g}{g}$$

$$\Rightarrow \quad \frac{\Delta T}{T} \times 100 = \frac{1}{2}\left[\frac{\Delta l}{l} \times 100 - \frac{\Delta g}{g} \times 100\right] = \frac{1}{2}[1 \pm 2.5]$$

Hence, maximum error in $T = \frac{1}{2}(1 + 2.5) = 1.75\%$.

EXAMPLE 8. *The height h and the semi-vertical angle α of a cone are measured and from them, the total surface area of the cone including the base is calculated. If h and α are in error by small quantities δh and $\delta \alpha$ respectively. Find the corresponding error in the area. Show further that if $\alpha = \dfrac{\pi}{6}$ an error of +1% in h will be approximately compensated by an error of – 0.33° in α.*

(UPTU–2009)

SOLUTION. We have

Base radius, $r = h \tan \alpha$

Slant height, $l = h \sec \alpha$

Total area, $A = \pi r^2 + \pi r l = \pi r(r + l)$

$= \pi h \tan \alpha(h \tan \alpha + h \sec \alpha)$

$= \pi h^2(\tan^2\alpha + \sec \alpha \tan \alpha)$

So, $\delta A = \dfrac{\partial A}{\partial h}\delta h + \dfrac{\partial A}{\partial \alpha}\delta \alpha$

$= 2\pi h(\tan^2\alpha + \sec \alpha \tan \alpha)\,\delta h$
$\quad + \pi h^2\,(2\tan \alpha \sec^2\alpha + \sec^3\alpha$
$\quad + \sec \alpha \tan^2\alpha)\delta\alpha \qquad \dots(1)$

Let us set $\alpha = \dfrac{\pi}{6}, \delta h = \dfrac{h}{100}$ in (1), we get

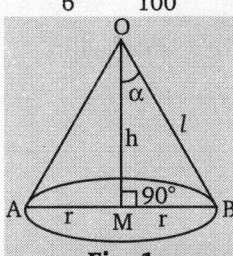

Fig. 1

$\delta A = 2\pi h\left(\dfrac{1}{3} + \dfrac{2}{3}\right)\dfrac{h}{100} + \pi h^2\left[\dfrac{2}{\sqrt{3}}\left(\dfrac{4}{3}\right) + \dfrac{8}{3\sqrt{3}} + \dfrac{2}{3\sqrt{3}}\right]\delta \alpha$

$= \dfrac{\pi h^2}{50} + 2\sqrt{3}\pi h^2\delta\alpha$

Now, since the error in h is to be compensated by the error in α.

Therefore, $\delta A = 0$

$\therefore \quad \delta\alpha = \dfrac{-1}{100\sqrt{3}}\text{ radians} = -\dfrac{57.3°}{173.2} = -0.33°.$

$(\because 1 \text{ radian} \approx 57.3°)$

EXAMPLE 9. *If* $f(x, y) = x^2y^{1/10}$. *Compute the value of* f *when* $x = 1.99$ *and* $y = 3.01$. (UPTU–2008)

SOLUTION. We have $f(x,y) = x^2y^{1/10}$.

We have to calculate $f(1.99, 3.01)$

Let $x = 2, y = 3$.

Then $x + \delta x = 1.99 \Rightarrow \quad \delta x = -0.01$

$\quad y + \delta y = 3.01 \qquad \Rightarrow \quad \delta y = 0.01$

Now $\quad f = x^2y^{1/10}$

$\Rightarrow \quad \delta f = \dfrac{\partial f}{\partial x}\delta x + \dfrac{\partial f}{\partial y}\delta y$

$= 2xy^{1/10}\delta x + \dfrac{1}{10}x^2y^{-9/10}\delta y$

$= 2(2)3^{1/10}(-0.01) + \dfrac{1}{10}(2)^2(3)^{-9/10}(0.01)$

$= 3^{1/10}\left[-0.04 + \dfrac{1}{10}(4)(3)^{-1}(0.01)\right]$

$= 3^{1/10}\left(-0.04 + \dfrac{0.004}{3}\right)$

$= 3^{-9/10}(-0.12 + 0.004)$

$= 3^{-9/10}(-0.116)$

Hence, $f(1.99, 3.01) = f(2, 3) + \delta f$

$= 2^2(3)^{1/10} + (-0.116)3^{-9/10}$

$= 3^{1/10}\left[4 - \dfrac{0.116}{3}\right]$

$= 3^{-9/10}(12 - 0.116)$

$= 3^{-9/10} \times 11.884$

$= 0.3720 \times 11.884 = 4.4213$

Exercise-3.10

1. The time T of a complete oscillation of a simple pendulum of length l is governed by the equation $T = 2\pi\sqrt{\dfrac{l}{g}}$, g is constant. Find the approximate error in the calculated value of T corresponding to an error of 2% in the value of l. (UPTU–2009)

2. The diameter and height of a right circular cylinder are found by measurement to be 8.0 cm and 12.5 cm respectively with possible errors of 0.05 in each measurement. Find the maximum possible approximate error in the computed volume. (GBTU–2011)

3. The work that must be done to propel a ship of displacement D for a distance S in time t is proportional to $\dfrac{S^2D^{2/3}}{t^2}$. Find approximately the increase of work necessary when the displacement is increased by 1%, the time diminished by 1% and the distance diminished by 2%.

4. The power P required to propel a steamer of length l at a speed u is given by $P = \lambda u^3 l^3$ where λ is constant. If u is increased by 3% and l is decreased by 1%, find the corresponding increase in P. (GBTU–2010, UKTU–2012)

5. The diameter and altitude of a can in the shape of a right circualr cylinder are measured as 4 cm and 6 cm respectively. The possible error in each measurement is 0.1 cm. Find approximately the maximum possible error in the value computed for the volume and lateral surface.

6. What error in the common logarithm of a number will be produced by an error of 1% in the number.

7. Find the possible percentage error in computing the parallel resistance r of two resistances r_1 and r_2 from the formula $\dfrac{1}{r} = \dfrac{1}{r_1} + \dfrac{1}{r_2}$ where the error in both r_1 and r_2 is +2% each. (UKTU–2011)

8. Compute an approximate value of $[(3.82)^2 + 2(2.1)^3]^{1/5}$. (MTU–2012)

9. Evaluate $\log[(1.01)^{1/3} + (0.99)^{1/4} - 1]$. (UPTU–2009, MTU–2011)

10. The resistance R of a circuit was found by formula $I = \dfrac{E}{R}$. If there is an error of 0.1 amp in reading I, 0.5 volts in E. Find the corresponding possible percentage error in R when reading are $I = 15$ amp and $E = 100$ volts. (MTU–2011)

11. Prove that the relative error of a quotient does not exceed the sum of the relative errors of dividend and the divisor.

12. The side a and the opposite angle A of a $\triangle ABC$ remain constant. Show that when the other sides and angles are slightly varied $\dfrac{\delta b}{\cos B} + \dfrac{\delta c}{\cos C} = 0$.

13. In determining the specific gravity by the formula $S = \dfrac{A}{A-W}$

where A is the weight in air and W is the weight in water; A can be read within 0.01 gm and W within 0.02 gm. Find approximately the maximum error in S if the readings are $A = 1.1$ gm $W = 0.6$ gm. Find also the maximum relative error.

14. If the kinetic energy T is given by $T = \dfrac{1}{2}mv^2$, find approximate

the change in T as the mass m changes from 49 to 49.5 and the velocity v changes from 1600 to 1590.

15. The deflection at the centre of a rod of length l and diameter d supported at its ends and loaded at the centre with a weight w varies as wl^2d^{-4}. What is the increase in the deflection corresponding to $p\%$ increase in w, $q\%$ decrease in l and $r\%$ increase in d.

Answers

1. 1% **2.** $3.3\,\pi$ cu cm **3.** $-\dfrac{4}{3}\%$ **4.** 6% **5.** $1.6\,\pi$ cu cm, π sq cm **6.** 0.0043429 **7.** 2% **8.** 2.012

9. 0.00083 **10.** -16.66% **11.** $\left|\dfrac{\delta z}{z}\right| \le \left|\dfrac{\delta x}{x}\right| + \left|\dfrac{\delta y}{y}\right|$ **13.** 0.112, 0.05091 **14.** 144000 units **15.** $(p - 3q - 4r)\%$

ARCHIVE

1. If $u = \log(x^2 + y^2) + \tan^{-1}\left(\dfrac{y}{x}\right)$, prove that $\dfrac{\partial^2 u}{\partial x^2} + \dfrac{\partial^2 u}{\partial y^2} = 0$

(ANNA-2009)

2. If $r^2 = x^2 + y^2 + z^2$ and $u = r^m$, prove that

$u_{xx} + u_{yy} + u_{zz} = m(m+1)r^{m-2}$ (SRM, 2009; RAIPUR -2005, 11)

3. If $u = e^{xyz}$ show that $\dfrac{\partial^3 u}{\partial x \partial y \partial z} = e^{xyz}(x^2 y^2 z^2 + 3xyz + 1)$.

(RAJSTHAN-2005, 07; OSMANIA-2003)

4. If $x = r\cos\theta, y = r\sin\theta$, prove that $\left(\dfrac{\delta r}{\delta y}\right)^2 + \left(\dfrac{\delta r}{\delta y}\right)^2 = 1$

(BURDWAN-2003)

5. If $z = x\log(x+r) - r$ where $r = x^2 + y^2$, prove that

$\dfrac{\partial^2 z}{\partial x^2} + \dfrac{\partial^2 z}{\partial yz} = \dfrac{1}{x+y}, \dfrac{\partial^3 z}{\partial x^3} = \dfrac{-x}{r^3}$.   (MUMBAI-2008)

6. If $u = y^2 - 4ax, x = at^2, y = 2at$, show that $\dfrac{du}{dt} = 0$.

(ANNA-2009)

7. If $x + y = 2e^\theta \cos\phi$ and $x - y = 2ie^\theta \sin\phi$, show that

$\dfrac{\partial^2 u}{\partial\theta^2} + \dfrac{\partial^2 u}{\partial\theta^2} = 4xy\dfrac{\partial^2 u}{\partial x \partial y}$ (UPTU-2002, 09; NAGPUR-2009, 16)

8. If $u = f(x, y)$ and $x = r\cos\theta, y = r\sin\theta$, prove that

$\left(\dfrac{\partial u}{\partial x}\right)^2 + \left(\dfrac{\partial u}{\partial y}\right)^2 = \left(\dfrac{\partial u}{\partial r}\right)^2 + \dfrac{1}{r^2}\left(\dfrac{\partial u}{\partial \theta}\right)^2$

(VTU–2010, ROHTAK–2005, MUMBAI-2006)

9. If $u = f(2x - 3y), 3y - 4z, 4z - 2x$, prove that

$\dfrac{1}{2}\dfrac{\partial u}{\partial x} + \dfrac{1}{3}\dfrac{\partial u}{\partial y} + \dfrac{1}{4}\dfrac{\partial u}{\partial z} = 0$ (UPTU-2006; RAIPUR-2005)

10. If $u = x^2 - y^2, v = 2xy$ and $x = r\cos\theta, y = r\sin\theta$

show that $\dfrac{\partial(u, v)}{\partial(r, \theta)} = 4r^3$ (VTU-2009, MADRAS-2006)

11. If $u = x + 3y^2 - z^3, V = 4x^2yz, w = 2z^2 - xy$, show that at

$(1, -1, 0)$; $\dfrac{\partial(u, v, w)}{\partial(x, y, z)} = 20$ (VTU-2006)

12. If $u = x + y + z, uv = y + z, uvw = z$, show that $\dfrac{\partial(x, y, z)}{\partial(u, v, w)} = u^2 v$

(PTU–2009, VTU–2003, KURUKSHETRA-2009)

13. In estimating the cost of a pile of bricks measured as $2m \times 15m \times 1.2m$, the tape is streched 1% beyond the standard length. If the count is 450 bricks to 1 m^3 and bricks cost Rs. 530 per 1000. Show that the approximate error in the cost is Rs. 257.58.

14. In a place triangle, show that the maximum value of $\cos A \cos B \cos C$ is $\dfrac{1}{8}$.   (VTU-2010, NAGPUR-2009, 14)

15. If $u = a^3x^2 + b^3y^2 + c^3z^2$, where $\dfrac{1}{x} + \dfrac{1}{y} + \dfrac{1}{z} = 1$, show that the stationary value of u is given by

$x = \dfrac{\Sigma a}{a}, y = \dfrac{\Sigma a}{b}, z = \dfrac{\Sigma c}{c}$ (SRM-2011, KERALA–2006, 17)

16. Show that, if the perimeter of a triangle is constant, the triangle has maximum area when it is equivalent.

17. The temperature T at any point (x, y, z) in space is $T = 400xyz^2$. Show that the highest temperature on the surface of the unit sphere $x^2 + y^2 + z^2 = 1$ is 50.

(VTU-2009, HISSAR-2005)

❋❋❋❋❋❋

Integral Calculus-I

4

4.1 REDUCTION FORMULAE

The process relating one integral to one or more integrals of the same type but simpler is called a reduction and this relation, between integrals is called a reduction formula.

The reduction formula is derived by two different methods; one is integration by parts and other is differentiation of a suitable function, says $P(x)$. Here the function $P(x)$ is chosen in such a way that $\dfrac{dP}{dx}$ has atleast one function, which is the integral of the given integral, whose reduction formula is required. It consists of the following steps :

   (i) Selection of $P(x)$ (ii)   Differentiation of $P(x)$ (iii) Integral of (ii).

4.1.1 REDUCTION FORMULAE FOR $\int \sin^m x \cos^n x\, dx$

  (i) $I_{m,n} = \int \sin^m x \cos^n x\, dx = \dfrac{\cos^{n-1} x \sin^{m+1} x}{m+n} + \dfrac{n-1}{m+n} I_{m,n-2}$ $(m+n \neq 0)$

  (ii) $I_{m,n} = \int \sin^m x \cos^n x\, dx = -\dfrac{\sin^{m-1} x \cos^{n+1} x}{m+n} + \dfrac{m-1}{m+n} I_{m-2,n}$ $(m+n \neq 0)$

PROOF.   (i)   Here $I_{m,n} = \int \sin^m x \cos^n x\, dx = \cos^{n-1} x \int \sin^m x \cos x\, dx - \int \left[\dfrac{d}{dx}(\cos^{n-1} x) \int \sin^m x \cos x\, dx \right].dx$

$$= \dfrac{\cos^{n-1} x \sin^{m+1} x}{m+1} + \dfrac{n-1}{m+1} \int \cos^{n-2} x \sin^{m+2} x\, dx$$

$$= \dfrac{\cos^{n-1} x \sin^{m+1} x}{m+1} + \dfrac{n-1}{m+1} \int \cos^{n-2} x \sin^m x (1 - \cos^2 x)\, dx$$

$$= \dfrac{\cos^{n-1} x \sin^{m+1} x}{m+1} + \dfrac{n-1}{m+1} . I_{m,n-2} - \dfrac{n-1}{m+1} . I_{m,n}$$

or $\left(1 + \dfrac{n-1}{m+1}\right) I_{m,n} = \dfrac{\cos^{n-1} x \sin^{m+1} x}{m+1} + \dfrac{n-1}{m+1} . I_{m,n-2}$

$\Rightarrow$ $I_{m,n} = \dfrac{\cos^{n-1} x \sin^{m+1} x}{m+n} + \dfrac{n-1}{m+n} I_{m,n-2}, m+n \neq 0.$

☞ REMARKS

- The above formula reduces the power or exponent m of cosine in each successive steps by 2. So, to establish the relation, one cosine is separated from the product so that $\sin^m x \cos x$ can be integrated.

- Here, one sine is separated from the product, so that $\cos^n x \sin x$ can be integrated and can be treated as a second function of the integration by parts. Therefore, following the same procedure as above, we can easily find

$$I_{m,n} = -\dfrac{\sin^{m-1} x \cos^{n+1} x}{m+n} + \dfrac{m-1}{m+n} I_{m-2,n}, m+n \neq 0.$$

- Put $m = 0$ in reduction formula (i), we may find $I_n = \int \cos^n x\, dx = \dfrac{\sin x \cos^{n-1} x}{n} + \dfrac{n-1}{n} I_{n-2}, n \neq 0.$

- Put $n = 0$ in (ii), we may get $I_m = \int \sin^m x\, dx = -\dfrac{\sin^{m-1} x \cos x}{m} + \dfrac{m-1}{m} I_{m-2}, m \neq 0.$

4.1.2 REDUCTION FORMULA FOR $\int \dfrac{dx}{\sin^m x \cos^n x} = \int \operatorname{cosec}^m x \sec^n x\, dx$

Replace n by $n+2$ in § 4.1.1 (i), we get

$$I_{m,n+2} = \int \sin^m x \cos^{n+2} x\, dx = \frac{\sin^{m+1} x \cos^{n+1} x}{m+n+2} + \frac{n+1}{m+n+2} I_{m,n}$$

Let us replace m by $-m$ and n by $-n$, we get

$$I_{-m,-n+2} = \int \frac{dx}{\sin^m x \cos^{n-2} x} = -\frac{1}{m+n-2} \cdot \frac{1}{\sin^{m-1} x \cos^{n-1} x} + \frac{n-1}{m+n-2} I_{-m,-n}$$

or

$$I_{-m,-n} = \int \frac{dx}{\sin^m x \cos^n x} = \frac{1}{(n-1)\sin^{m-1} x \cos^{n-1} x} + \frac{m+n-2}{n-1} I_{-m,-n+2}, n \neq 1 \qquad \dots(1)$$

Similarly from § 14.2 (ii), we get

$$I_{-m,-n} = \int \frac{dx}{\sin^m x \cos^n x} = \frac{1}{(m-1)\sin^{m-1} x \cos^{n-1} x} + \frac{m+n-2}{m-1} I_{-m+2,-n}, m \neq 1 \qquad \dots(2)$$

Here, from (1) and (2), we observed that powers of cosines and sines are respectively reduced by 2 in each step of reduction.

Now, putting $m = 0$ in (1), and $n = 0$ in (2), we can obtain

$$I_{-n} = \int \frac{1}{\cos^n x}\, dx = \int \sec^n x\, dx = \frac{\sin x}{(n-1)\cos^{n-1} x} + \frac{n-2}{n-1} I_{-n+2}, n \neq 1 \qquad \dots(3)$$

and

$$I_{-m} = \int \frac{1}{\sin^m x}\, dx = \int \operatorname{cosec}^m x\, dx = -\frac{\cos x}{(m-1)\sin^{m-1} x} + \frac{m-2}{m-1} I_{-m+2}, m \neq 1 \qquad \dots(4)$$

From equations (1), (2), (3) and (4), we may get

$$I_{m,n} = \int \operatorname{cosec}^m x \sec^n x\, dx = \frac{\operatorname{cosec}^{m-1} x \sec^{n-1} x}{n-1} + \frac{m+n-2}{n-1} I_{m,n-2}, n \neq 1$$

$$I_{m,n} = \int \operatorname{cosec}^m x \sec^n x\, dx = \frac{\operatorname{cosec}^{m-1} x \sec^{n-1} x}{m-1} + \frac{m+n-2}{m-1} I_{m-2n}, m \neq 1$$

$$I_n = \int \sec^n x\, dx = \frac{\tan x \sec^{n-2} x}{n-1} + \frac{n-2}{n-1} I_{n-2}, n \neq 1$$

and

$$I_m = \int \operatorname{cosec}^m x\, dx = -\frac{\cot x \operatorname{cosec}^{m-2} x}{m-1} + \frac{m-2}{m-1} I_{m-2}, m \neq 1$$

4.1.3 REDUCTION FORMULA FOR $\int \sin^m x \sec^n x\, dx$ AND $\int \cos^n x \operatorname{cosec}^m x\, dx$

Let

$$I_{m,-n} = \int \frac{\sin^m x}{\cos^n x}\, dx = \int \frac{\sin^{m-1} x \sin x}{\cos^n x}\, dx = \frac{\sin^{m-1} x}{(n-1)\cos^{n-1} x} - \frac{m-1}{n-1} \int \frac{\sin^{m-2} x}{\cos^{n-2} x}\, dx$$

$$= \frac{\sin^{m-1} x}{(n-1)\cos^{n-1} x} - \frac{m-1}{n-1} I_{m-2,-n+2}, n \neq 1$$

$$\Rightarrow \qquad \int \sin^m x \sec^n x\, dx = \frac{\sin^{m-1} x}{(n-1)\cos^{n-1} x} - \frac{m-1}{n-1} I_{m-2,-n+2}, n \neq 1 \qquad \dots(1)$$

Similarly we can find

$$I_{-m,n} = \int \frac{\cos^n x}{\sin^m x}\, dx = -\frac{\cos^{m-1} x}{(m-1)\sin^{m-1} x} - \frac{n-1}{m-1} I_{-m+2,n-2}, m \neq 1 \qquad \dots(2)$$

Here, we observed that, in the above two cases powers of the numerator have been reduced. In the former case, a sine and in the latter case a cosine of the numerator has been separated and integrated separately. Finally, the powers of both numerator and denominator have been reduced by two in both cases.

The formuale (1) and (2) can also be written as

$$I_{m,n} = \int \sin^m x \sec^n x\, dx = \frac{\sin^{m-1} x \sec^{n-1} x}{n-1} - \frac{m-1}{n-1} I_{m-2,n-2}, n \neq 1 \qquad \dots(3)$$

and

$$I_{m,n} = \int \operatorname{cosec}^m x \cos^n x\, dx = -\frac{\operatorname{cosec}^{m-1} x \cos^{n-1} x}{m-1} - \frac{n-1}{m-1} I_{m-2,n-2}, m \neq 1 \qquad \dots(4)$$

4.1.4 REDUCTION FORMULAE FOR $\int \tan^n x\, dx$ AND $\int \cot^n x\, dx$

(i)

$$I_n = \int \tan^n x\, dx = \int \tan^{n-2} x \tan^2 x\, dx = \int \tan^{n-2} x(\sec^2 x - 1)\, dx$$

$$= \int \tan^{n-2} x \sec^2 x\, dx - \int \tan^{n-2} x\, dx \qquad \dots(1)$$

Now $\qquad \int \tan^{n-2} x \sec^2 x\, dx = \int t^{n-2} dt$, putting $\tan x = t$ and $\sec^2 t\, dt = dx$

$$\int \tan^{n-2} x \sec^2 x dx = \frac{t^{n-1}}{n-1} = \frac{\tan^{n-1} x}{n-1} \qquad \ldots(2)$$

From equations (1) and (2), we get the reduction formula :

$$\therefore \qquad \int \tan^n x dx = \frac{\tan^{n-1} x}{n-1} - \int \tan^{n-2} x dx, n \neq 1$$

(ii) Similarly,
$$I_n = \int \cot^n x dx = \int \cot^{n-2} x \cot^2 x dx = \int \cot^{n-2} x(\operatorname{cosec}^2 x - 1) dx$$

$$= \int \cot^{n-2} x \operatorname{cosec}^2 x \, dx - \int \cot^{n-2} x \, dx$$

$$\therefore \qquad \int \cot^n x dx = -\frac{\cot^{n-1} x}{n-1} - \int \cot^{n-2} x dx \ .$$

4.1.5 Reduction Formulae for $\int \sec^n x dx$ and $\int \operatorname{cosec}^n x dx$

(i) Let
$$I_n = \int \sec^n x dx = \int \sec^{n-2} x \sec^2 x dx$$

$$= \sec^{n-2} x \int \sec^2 x dx - \int \left\{ \frac{d}{dx}(\sec^{n-2} x) \int \sec^2 x \, dx \right\} dx$$

$$= \tan x \sec^{n-2} x - (n-2)\int \sec^{n-2} x \tan^2 x dx$$

$$= \tan x \sec^{n-2} x - (n-2)\int \sec^{n-2} x(\sec^2 x - 1) dx$$

or
$$I_n = \tan x \sec^{n-2} x - (n-2)\int \sec^n x dx + (n-2)\int \sec^{n-2} x dx$$

$$= \tan x \sec^{n-2} x - (n-2)I_n + (n-2)I_{n-2}$$

or
$$I_n(1+n-2) = \tan x \sec^{n-2} x + (n-2)I_{n-2}$$

$$\therefore \qquad I_n = \frac{1}{n-1} \tan x \sec^{n-2} x + \frac{n-2}{n-1} I_{n-2}, n \neq 1$$

Hence
$$\int \sec^n x dx = \frac{\sec^{n-2} x \tan x}{n-1} + \frac{n-2}{n-1} \int \sec^{n-2} x dx$$

Similarly, we obtain
$$\int \operatorname{cosec}^n x dx = -\frac{\operatorname{cosec}^{n-2} x \cot x}{n-1} + \frac{n-2}{n-1} \int \operatorname{cosec}^{n-2} x dx$$

4.1.6 Reduction Formulae for $\int \tan^m x \sec^n x dx$ and $\int \cot^m x \operatorname{cosec}^n x dx$

(i) Let
$$I_{m,n} = \int \tan^m x \sec^n x dx = \int \tan^{m-1} x(\sec^{n-1} x \sec x \tan x) dx$$

$$= \tan^{m-1} x \frac{\sec^n x}{n} - \frac{m-1}{n} \int \tan^{m-2} x \sec^2 x \sec^n x dx$$

$$= \frac{\tan^{m-1} x \sec^n x}{n} - \frac{m-1}{n} I_{m-2,n} - \frac{m-1}{m+n-1} I_{m,n}$$

$$\Rightarrow \qquad I_{m,n} = \int \tan^m x \sec^n x dx = \frac{\tan^{m-1} x \sec^n x}{m+n-1} - \frac{m-1}{m+n-1} I_{m-2,n}, m+n \neq 1$$

(ii) Similarly, the power of $\sec x$ is reduced by two during integrating $\int \tan^m x \sec^n x dx$ by parts and then, we get

$$I_{m,n} = \int \tan^m x \sec^n x dx = \frac{\tan^{m+1} x \sec^{n-2} x}{m+n-1} - \frac{n-2}{m+n-1} I_{m,n-2}, (m+n \neq 1)$$

Remarks

- $I_{m,n} = \int \cot^m x \operatorname{cosec}^n x dx = \frac{\cot^{m-1} x \operatorname{cosec}^n x}{m-n+1} - \left(\frac{m-1}{m-n+1}\right) I_{m-2,n}, m+1 \neq n$

- $I_{m,n} = \int \cot^m x \operatorname{cosec}^n x dx = \frac{\cot^{m+1} x \operatorname{cosec}^{n-2} x}{n-m+1} - \frac{n-2}{n-m+1} I_{m,n-2}, n+1 \neq m$

4.1.7 Reduction Formulae for $\int \cos^m x \cos nx dx$, $\int \cos^m x \sin nx dx$ and $\int \sin^m x \cos nx dx$ etc.

(i) Let
$$I_{m,n} = \int \cos^m x \cos nx dx = \cos^m x \frac{\sin nx}{n} + \frac{m}{n} \int \cos^{m-1} x \sin x \sin nx dx \qquad \ldots(1)$$

$$= \frac{\cos^m x \sin nx}{n} + \frac{m}{n} \int \cos^{m-1} x \{\cos(n-1)x - \cos nx \cos x\} dx$$

$$[\because \cos(n-1)x = \cos nx \cos x + \sin nx \sin x]$$

$$= \frac{\cos^m x \sin nx}{n} + \frac{m}{n} I_{m-1,n-1} - \frac{m}{n} I_{m,n}$$

$$\Rightarrow \qquad \left(1 + \frac{m}{n}\right) I_{m,n} = \frac{\cos^m x \sin nx}{n} + \frac{m}{n} I_{m-1,n-1}$$

$$\Rightarrow \qquad I_{m,n} = \frac{\cos^m x \sin nx}{n} + \frac{m}{n} I_{m-1,n-1}, m+n \neq 0 \qquad \ldots(2)$$

Intetgrate (1) by parts and using (2), we get

$$I_{m,n} = \frac{\cos^m x \sin nx}{n} + \frac{m}{n} \left\{ \cos^{m-1} x \sin x \left(-\frac{\cos nx}{n}\right) + \int [\cos^m x - (m-1)\cos^{m-2} x \sin^2 x] \frac{\cos nx}{n} dx \right\}$$

$$= \frac{\cos^m x \sin nx}{n} - \frac{m}{n^2} \cos^{m-1} x \sin x \cos nx + \frac{m^2}{n^2} I_{m,n} - \frac{m(m-1)}{n^2} I_{m-2,n}$$

$$\Rightarrow \qquad \left(1 - \frac{m^2}{n^2}\right) I_{m,n} = \frac{\cos^m x \sin nx}{n} - \frac{m}{n^2} \cos^{m-1} x \sin x \cos nx - \frac{m(m-1)}{n^2} I_{m-2,n}$$

$$\Rightarrow \qquad I_{m,n} = -\frac{n \cos^m x \sin nx}{m^2 - n^2} + \frac{m}{m^2 - n^2} \cos^{m-1} x \sin x \cos nx + \frac{m(m-1)}{m^2 - n^2} I_{m-2,n}$$

$$= -\frac{\cos^{m-1} x}{m^2 - n^2} [n \cos x \sin nx - m \sin x \cos nx] + \frac{m(m-1)}{m^2 - n^2} I_{m-2,n}$$

$$\Rightarrow \qquad I_{m,n} = -\frac{\cos^2 nx}{m^2 - n^2} \frac{d}{dx}\left(\frac{\cos^m x}{\cos nx}\right) + \frac{m(m-1)}{m^2 - n^2} I_{m-2,n}$$

REMARK

- Here the power of cosine is reduced by two only.

(ii) Let $$I_{m,n} = \int \cos^m x \sin nx dx = \cos^m x \left(-\frac{\cos nx}{n}\right) - \int m \cos^{m-1} x (-\sin x)\left(-\frac{\cos nx}{n}\right) dx$$

$$= -\frac{\cos^m x \cos nx}{n} - \frac{m}{n} \int \cos^{m-1} x \sin x \cos nx \, dx \qquad \ldots(1)$$

$$= -\frac{\cos^m x \cos nx}{n} - \frac{m}{n} \int \cos^{m-1} x [\sin nx \cos x - \sin(n-1)x] dx$$

$$[\because \sin(n-1)x = \sin nx \cos x - \cos nx \sin x]$$

$$= -\frac{\cos^m x \cos nx}{n} - \frac{m}{n} I_{m,n} + \frac{m}{n} I_{m-1,n-1}$$

$$\Rightarrow \qquad \left(1 + \frac{m}{n}\right) I_{m,n} = -\frac{\cos^m x \cos nx}{m+n} + \frac{m}{m+n} I_{m-1,n-1}$$

$$\Rightarrow \qquad I_{m,n} = -\frac{\cos^m x \cos nx}{m+n} + \frac{m}{m+n} I_{m-1,n-1}, m+n \neq 0$$

On intetgrating by parts the right hand integral of (1), we get

$$I_{m,n} = -\frac{\cos^m x \cos nx}{n} - \frac{m}{n}\left[(\cos^{m-1} x \sin x)\frac{\sin nx}{n} - \int \{(m-1)\cos^{m-2} x(-\sin^2 x) + \cos^m x\} \frac{\sin nx}{n} dx \right]$$

$$= -\frac{\cos^m x \cos nx}{n} - \frac{m}{n^2} \cos^{m-1} x \sin x \sin nx$$

$$\qquad + \frac{m}{n^2} \int \{-(m-1)\cos^{m-2} x(1 - \cos^2 x) + \cos^m x\} \sin nx dx$$

$$= -\frac{\cos^m x \cos nx}{n} - \frac{m}{n^2} \cos^{m-1} x \sin x \sin nx - \frac{m(m-1)}{n^2} I_{m-2,n} + \frac{m^2}{n^2} I_{m,n}$$

$$\Rightarrow \qquad \left(1 - \frac{m^2}{n^2}\right) I_{m,n} = -\frac{\cos^m x \cos nx}{n} - \frac{m}{n^2} \cos^{m-1} x \sin x \sin nx - \frac{m(m-1)}{n^2} I_{m-2,n}$$

$$\Rightarrow \qquad I_{m,n} = \frac{n \cos^m x \cos nx}{m^2 - n^2} + \frac{m \cos^{m-1} x \sin x \sin nx}{m^2 - n^2} - \frac{m(m-1)}{m^2 - n^2} I_{m-2,n}$$

(iii) Here, let $I_{m,n} = \int \sin^m x \cos nx\, dx = \sin^m x \cdot \dfrac{\sin nx}{n} - \int m \sin^{m-1} x \cos x \cdot \dfrac{\sin nx}{n}\, dx$

$$= \frac{\sin^m x \sin nx}{n} - \frac{m}{n} \int \sin^{m-1} x \cos x \sin nx\, dx$$

$$= \frac{\sin^m x \sin nx}{n} - \frac{m}{n}\left[(\sin^{m-1} x \cos x)\left(-\frac{\cos nx}{n}\right) - \int \{(m-1)\sin^{m-2} x \cos^2 x - \sin^m x\}\left(-\frac{\cos nx}{n}\right)dx \right]$$

[Integrating by parts]

$$= \frac{\sin^m x \sin nx}{n} + \frac{m}{n^2} \sin^{m-1} x \cos x \cos nx - \frac{m}{n^2} \int \{(m-1)\sin^{m-2} x (1-\sin^2 x) - \sin^m x\} \cos nx\, dx$$

$$= \frac{\sin^m x \sin nx}{n} + \frac{m}{n^2} \sin^{m-1} x \cos x \cos nx - \frac{m(m-1)}{n^2} I_{m-2,n} + \frac{m^2}{n^2} I_{m,n}$$

$\Rightarrow \qquad I_{m,n} = -\dfrac{n \sin^m x \sin nx}{m^2 - n^2} - \dfrac{m \sin^{m-1} x \cos x \cos nx}{m^2 - n^2} + \dfrac{m(m-1)}{m^2 - n^2} I_{m-2,n}$

(iv) Let $\qquad I_{m,n} = \int \sin^m x \sin nx\, dx$

Proceed as above, we may easily get

$\Rightarrow \qquad I_{m,n} = -\dfrac{n \sin^m x \cos nx}{m^2 - n^2} - \dfrac{m \sin^{m-1} x \cos x \sin nx}{m^2 - n^2} + \dfrac{m(m+1)}{m^2 - n^2} I_{m-2,n}.$

4.1.8 REDUCTION FORMULAE FOR $\int x^n \sin mx\, dx$ AND $\int x^n \cos mx\, dx$

(i) Let $\qquad I(m,n) = \int x^n \sin mx\, dx = x^n \int \sin mx\, dx - \int \left\{ \frac{d}{dx}(x^n) \int \sin mx\, dx \right\} dx$ (KANPUR-2002)

$$= -\frac{x^n \cos mx}{m} + \frac{n}{m} \int x^{n-1} \cos mx\, dx$$

$$= -\frac{x^n \cos mx}{m} + \frac{n}{m}\left[\frac{x^{n-1} \sin mx}{m} - \frac{n-1}{m} \int x^{n-2} \sin mx\, dx \right]$$

or $\qquad I(m,n) = -\dfrac{x^n \cos mx}{m} + \dfrac{n x^{n-1} \sin mx}{m^2} - \dfrac{n(n-1)}{m^2} \int x^{n-2} \sin mx\, dx$

Hence, $\qquad \int x^n \sin mx\, dx = -\dfrac{x^n \cos mx}{m} + \dfrac{n x^{n-1} \sin mx}{m^2} - \dfrac{n(n-1)}{m^2} \int x^{n-2} \sin mx\, dx$

Similarly, $\qquad \int x^n \cos mx\, dx = \dfrac{x^n \sin mx}{m} + \dfrac{n x^{n-1} \cos mx}{m^2} - \dfrac{n(n-1)}{m^2} \int x^{n-2} \cos mx\, dx$.

4.1.9 REDUCTION FORMULAE FOR $\int x \sin^n x\, dx$ AND $\int x \cos^n x\, dx$

(i) Let $\qquad I_n = \int x \sin^n x\, dx$

$I_n = \int (x \sin^{n-1} x) \sin x\, dx$ (Integrating by parts)

$$= x \sin^{n-1} x \int \sin x\, dx - \int \{ \sin^{n-1} x + x(n-1)\sin^{n-2} x \cos x \}(-\cos x)\, dx$$

$$= -x \sin^{n-1} x \cos x + \int \sin^{n-1} x \cos x\, dx + (n-1)\int x \sin^{n-2} x \cos^2 x\, dx$$

$$= -x \sin^{n-1} x \cos x + \int \sin^{n-1} x \cos x\, dx + (n-1)\int x \sin^{n-2} x (1-\sin^2 x)\, dx$$

$\therefore \qquad I_n = -x \sin^{n-1} x \cos x + \dfrac{\sin^n x}{n} + (n-1)\int x \sin^{n-2} x\, dx - (n-1)\int x \sin^n x\, dx$

or $\qquad I_n = -x \sin^{n-1} x \cos x + \dfrac{\sin^n x}{n} + (n-1)\int x \sin^{n-2} x\, dx - (n-1)I_n$

or $\qquad I_n(1 + n - 1) = -x \sin^{n-1} x \cos x + \dfrac{\sin^n x}{n} + (n-1)\int x \sin^{n-2} x\, dx$

Hence, $\qquad \int x \sin^n x\, dx = \dfrac{-x \sin^{n-1} x \cos x}{n} + \dfrac{\sin^n x}{n^2} + \dfrac{n-1}{n} \int x \sin^{n-2} x\, dx$

Similarly, $\qquad \int x \cos^n x\, dx = \dfrac{x \cos^{n-1} x \sin x}{n} + \dfrac{\cos^n x}{n^2} + \dfrac{n-1}{n} \int x \cos^{n-2} x\, dx$.

Similarly, $\qquad \int x \cos^n x\, dx = \dfrac{x \cos^{n-1} x \sin x}{n} + \dfrac{\cos^n x}{n^2} + \dfrac{n-1}{n} \int x \cos^{n-2} x\, dx$.

4.1.10 REDUCTION FORMULAE FOR $\int e^{ax} \cos^n x\, dx$ AND $\int e^{ax} \sin^n x\, dx$

(i) Let

$$I_n = \int e^{ax} \cos^n x\, dx = \frac{e^{ax}}{a} \cos^n x + \frac{n}{a} \int e^{ax} \cos^{n-1} x \sin x\, dx$$

$$= \frac{e^{ax}}{a} \cos^n x + \frac{n}{a} \left\{ \frac{e^{ax}}{a} \cos^{n-1} x \sin x - \int \frac{e^{ax}}{a} [\cos^n x - (n-1)\cos^{n-2} x \sin^2 x]dx \right\}$$

$$= \frac{e^{ax}}{a} \cos^n x + \frac{n}{a^2} e^{ax} \cos^{n-1} x \sin x - \frac{n}{a^2} I_n + \frac{n(n-1)}{a^2} \int e^{ax} \cos^{n-2} x (1 - \cos^2 x)dx$$

$$= \frac{e^{ax}}{a} \cos^n x + \frac{n}{a^2} e^{ax} \cos^{n-1} x \sin x - \frac{n}{a^2} I_n + \frac{n(n-1)}{a^2} I_{n-2} - \frac{n(n-1)}{a^2} I_n$$

$\Rightarrow \qquad \left(1 + \dfrac{n}{a^2} + \dfrac{n(n-1)}{a^2}\right) I_n = e^{ax} \cos^{n-1} x \dfrac{(a \cos x + n \sin x)}{a^2} + \dfrac{n(n-1)}{a^2} I_{n-2}$

$\Rightarrow \qquad I_n = \dfrac{e^{ax} \cos^{n-1} x(a \cos x + n \sin x)}{(n^2 + a^2)} + \dfrac{n(n-1)}{n^2 + a^2} I_{n-2}.$

(ii) Similarly, we can find that

$$I_n = \int e^{ax} \sin^n x\, dx = \frac{e^{ax} \sin^{n-1} x}{n^2 + a^2}(a \sin x - n \cos x) + \frac{n(n-1)}{n^2 + a^2} I_{n-2}.$$

4.1.11 REDUCTION FORMULAE FOR $\int \cos nx\, \csc x\, dx$ AND $\int \sin nx\, \sec x\, dx$

(i) Let

$$I_n = \int \cos nx\, \csc x\, dx = \frac{\sin nx}{n} \csc x + \int \frac{\sin nx}{n} \csc x \cot x\, dx$$

$$= \frac{\sin nx}{n} \csc x + \int \frac{1}{n} \frac{\sin nx \cos x}{\sin^2 x} dx$$

$$= \frac{\sin nx}{n} \csc x + \int \frac{1}{n} \frac{\sin(n-1)x + \cos nx \sin x}{\sin^2 x} dx$$

$$= \frac{\sin nx \csc x}{n} + \frac{1}{n} \int \sin(n-1)x \csc^2 x\, dx + \frac{1}{n} I_n$$

$\Rightarrow \quad \left(1 - \dfrac{1}{n}\right) I_n = \dfrac{\sin nx \csc x}{n} + \dfrac{1}{n}[\sin(n-1)x(-\cot x) - (n-1)\int \cos(n-1)x(-\cot x)dx]$

$\Rightarrow \quad \dfrac{n-1}{n} I_n = \dfrac{\sin nx \csc x}{n} - \dfrac{\sin(n-1)x \cot x}{n} + \dfrac{(n-1)}{n} \int \dfrac{\cos(n-1)x \cos x}{\sin x} dx$

$\Rightarrow \quad I_n = \dfrac{\sin nx \csc x}{n-1} - \dfrac{\sin(n-1)x \cot x}{n-1} + \int \dfrac{\cos(n-2)x - \sin(n-1)x \sin x}{\sin x} dx$

$$= \frac{\sin nx \csc x}{n-1} - \frac{\sin(n-1)x \cot x}{n-1} + I_{n-2} - \int \sin(n-1)x\, dx$$

$$= \frac{\sin nx \csc x}{n-1} - \frac{\sin(n-1)x \cot x}{n-1} + I_{n-2} + \frac{1}{n-1} \cos(n-1)x$$

$$I_n - I_{n-2} = \frac{1}{n-1} \left[\frac{\sin nx}{\sin x} - \frac{\sin(n-1)x \cos x}{\sin x} + \cos(n-1)x \right]$$

$$= \frac{1}{n-1} \left[\frac{\sin nx - \sin(n-1)x \cos x}{\sin x} + \cos(n-1)x \right]$$

$$= \frac{1}{n-1} \left[\frac{\sin[(n-1)+1]x - \sin(n-1)x \cos x}{\sin x} + \cos(n-1)x \right] = \frac{2\cos(n-1)x}{n-1}$$

$$I_n = \frac{2\cos(n-1)x}{n-1} + I_{n-2}$$

(ii) Similarly,

$$I_n = \int \sin nx\, \sec x\, dx = -\frac{2\cos(n-1)x}{n-1} - I_{n-2}.$$

4.1.12 REDUCTION FORMULAE FOR $\int (\sin^{-1}x)^n\, dx$ AND $\int (\cos^{-1}x)^n\, dx$

(i) Let

$$I_n = \int (\cos^{-1}x)^n\, dx = x(\cos^{-1}x)^n - \int n(\cos^{-1}x)^{n-1} \left(\frac{-1}{\sqrt{1-x^2}} \right) x\, dx$$

$$= x(\cos^{-1}x)^n + n \int (\cos^{-1}x)^{n-1} \frac{x}{\sqrt{1-x^2}} dx$$

$$= x(\cos^{-1} x)^n - n(\cos^{-1} x)^{n-1}\sqrt{1-x^2} - n\int (n-1)(\cos^{-1} x)^{n-2}\left(-\frac{1}{\sqrt{1-x^2}}\right)\sqrt{1-x^2}\,dx$$

$$\Rightarrow \qquad I_n = x(\cos^{-1} x)^n - n(\cos^{-1} x)^{n-1}\sqrt{1-x^2} + n(n-1)I_{n-2}$$

(ii) Proceeding as same manner we get

$$I_n = \int(\sin^{-1} x)^n dx = x(\sin^{-1} x)^n + n(\sin^{-1} x)^{n-1}\sqrt{1-x^2} - n(n-1)I_{n-2}.$$

4.1.13 Reduction Formulae for $\int \dfrac{dx}{(a+b\cos x)^n}$ and $\int \dfrac{dx}{(a+b\sin x)^n}$

(i) Let

$$I_n = \int \frac{dx}{(a+b\cos x)^n}$$

Also, let

$$P(x) = \frac{\sin x}{(a+b\cos x)^{n-1}} = \frac{\sin x}{t^{n-1}}, \text{ where } t = a + b\cos x$$

$$\Rightarrow \qquad \cos x = \frac{t-a}{b}$$

Differentiating with respect to x, we get

$$\frac{dP(x)}{dx} = \frac{\cos x}{t^{n-1}} - (n-1)\frac{\sin x}{t^n}\frac{dt}{dx} = \frac{\cos x}{t^{n-1}} - (n-1)\frac{\sin x}{t^n}(-b\sin x)$$

$$= \frac{t-a}{bt^{n-1}} + \frac{b(n-1)}{t^n}(1-\cos^2 x) = \frac{t-a}{bt^{n-1}} + \frac{b(n-1)}{t^n}\left[1-\left(\frac{t-a}{b}\right)^2\right]$$

$$= \frac{t-a}{bt^{n-1}} + \frac{b(n-1)}{t^n}\left[1-\frac{t^2-2at+a^2}{b^2}\right] = \frac{(n-1)(b^2-a^2)}{bt^n} + \frac{a(2n-3)}{bt^{n-1}} - \frac{n-2}{bt^{n-2}}$$

$$= \frac{(n-1)(b^2-a^2)}{b}\cdot\frac{1}{(a+b\cos x)^n} + \frac{a(2n-3)}{b}\cdot\frac{1}{(a+b\cos x)^{n-1}} - \frac{n-2}{b}\cdot\frac{1}{(a+b\cos x)^{n-2}}$$

On integrating *w.r.t.* x, we get

$$P(x) = \frac{(n-1)(b^2-a^2)}{b}I_n + \frac{a(2n-3)}{b}I_{n-1} - \frac{n-2}{b}I_{n-2}$$

$$\Rightarrow \qquad I_n = \frac{b\sin x}{(n-1)(b^2-a^2)(a+b\cos x)^{n-1}} - \frac{a(2n-3)}{(n-1)(b^2-a^2)}I_{n-1} + \frac{n-2}{(n-1)(b^2-a^2)}I_{n-2}$$

(ii) Similarly, by setting $P(x) = \dfrac{\cos x}{(a+b\sin x)^{n-1}}$, we may easily get

$$I_n = \int\frac{dx}{(a+b\sin x)^n} = \frac{b}{(n-1)(a^2-b^2)}\cdot\frac{\cos x}{(a+b\sin x)^{n-1}} + \frac{(2n-3)a}{(n-1)(a^2-b^2)}I_{n-1} - \frac{n-2}{(n-1)(a^2-b^2)}I_{n-2}.$$

4.1.14 Reduction Formula for $\int x^m(\log x)^n\,dx$

Let

$$I(m,n) = \int x^m(\log x)^n dx \qquad\qquad\qquad\qquad\qquad \text{(Integrating by parts)}$$

$$= (\log x)^n\int x^m dx - \int\left\{\frac{n(\log x)^{n-1}}{x}\cdot\frac{x^{m+1}dx}{m+1}\right\}$$

$$= \frac{x^{m+1}(\log x)^n}{m+1} - \frac{n}{m+1}\int x^m(\log x)^{n-1}dx$$

Hence,

$$\int x^m(\log x)^n dx = \frac{x^{m+1}(\log x)^n}{m+1} - \frac{n}{m+1}\int x^m(\log x)^{n-1}dx.$$

4.1.15 Reduction Formula for $\int \dfrac{x^n}{\sqrt{ax^2+bx+c}}\,dx$

Let

$$I_n = \int\frac{x^n}{\sqrt{ax^2+bx+c}}dx = \frac{1}{2a}\int\frac{[(2ax+b)-b]x^{n-1}}{\sqrt{ax^2+bx+c}}dx$$

$$= \frac{1}{2a} \int \frac{(2ax+b)x^{n-1}}{\sqrt{ax^2+bx+c}} dx - \frac{b}{2a} \int \frac{x^{n-1}}{\sqrt{ax^2+bx+c}} dx$$

$$= \frac{1}{2a} \left[x^{n-1}.2\sqrt{ax^2+bx+c} - \int (n-1)x^{n-2} 2\sqrt{ax^2+bx+c}\, dx \right] - \frac{b}{2a} I_{n-1}$$

$$= \frac{1}{a} x^{n-1}\sqrt{ax^2+bx+c} - \frac{(n-1)}{a} \int \frac{x^{n-2}(ax^2+bx+c)}{\sqrt{ax^2+bx+c}} dx - \frac{b}{2a} I_{n-1}$$

$$= \frac{1}{a} x^{n-1}\sqrt{ax^2+bx+c} - (n-1)I_n - \frac{b(n-1)}{a} I_{n-1} - \frac{(n-1)}{a} I_{n-2} - \frac{b}{2a} I_{n-1}$$

$$\Rightarrow \qquad I_n = \frac{1}{na} x^{n-1}\sqrt{ax^2+bx+c} - \frac{b(2n-1)}{2an} I_{n-1} - \frac{c(n-1)}{an} I_{n-2}.$$

4.1.16 Reduction Formula for $\int \dfrac{px+q}{(ax^2+bx+c)^n} dx$

Let
$$I_n = \int \frac{px+q}{(ax^2+bx+c)} dx = \frac{p}{2a} \int \frac{2ax}{(ax^2+bx+c)^n} dx + \left(q - \frac{bp}{2a}\right) \int \frac{dx}{(ax^2+bx+c)^n}$$

$$= -\frac{p}{2a(n-1)} \frac{1}{(ax^2+bx+c)^{n-1}} + \left(q - \frac{bp}{2a}\right).\frac{1}{a^n} \int \frac{dx}{\left\{\left(x+\dfrac{b}{2a}\right)^2 + \dfrac{4ac-b^2}{4a^2}\right\}^n}$$

$$= -\frac{p}{2a(n-1)}.\frac{1}{(ax^2+bx+c)^{n-1}} + \frac{2qa-pb}{2a^{n+1}} \int \frac{dt}{(t^2+r)^n} \qquad \qquad …(1)$$

where $t = x + \dfrac{b}{2a}$ and $r = \dfrac{4ac-b^2}{4a^2}$

Integrating right hand integral of (1), by parts, we have

$$P_n = \int -\frac{dt}{(t^2+r)^n} = \frac{t}{(t^2+r)^n} + \int \frac{2t^2}{(t^2+r)^{n+1}} dt = \frac{t}{(t^2+r)^n} + 2n \int \frac{(t^2+r)-r}{(t^2+r)^{n+1}} dt$$

$$= \frac{t}{(t^2+r)^n} + 2n \int \frac{dt}{(t^2+r)^n} - 2nr \int \frac{dt}{(t^2+r)^{n+1}}$$

Now, replacing n by $n-1$ in this relation and solving for $\int \dfrac{dt}{(t^2+r)^n}$, we get

$$\int \frac{dt}{(t^2+r)^n} = \frac{1}{2(n-r)}.\frac{t}{(t^2+r)^{n-1}} + \frac{2n-3}{2(n-1)r} \int \frac{dt}{(t^2+r)^{n-1}} \qquad \qquad …(2)$$

From (1) and (2), we conclude that $I_n = -\dfrac{p}{2a(n-1)}.\dfrac{1}{(ax^2+bx+c)^{n-1}} + \dfrac{2aq-pb}{2a^{n+1}} \left[\dfrac{1}{2(n-r)r}.\dfrac{1}{(t^2+r)^{n-1}} + \dfrac{2n-3}{2(n-1)r} P_{n-1} \right]$

$$\Rightarrow \qquad I_n = \frac{-p}{2a(n-1)}.\frac{1}{(ax^2+bx+c)^{n-1}} + \frac{2aq-qb}{4(n-1)r} \left[\frac{1}{t^2+r)^{n-1}} + (2n-3)P_{n-1} \right]$$

where $P_n = \int \dfrac{dt}{(t^2+r)^n}, t = x + \dfrac{b}{2a}$ and $r = \dfrac{4ac-b^2}{4a^2}$.

4.1.17 Some More Important Reduction Formulae

The reduction formulae for $I_{m,p} = \int x^m (a+bx^n)^p dx$ are :

(i) $I_{m,p} = \dfrac{x^{m-n+1}.(a+bx^n)^{p+1}}{(np+m+1)b} - \dfrac{(m-n+1)}{(np+m+1)b} I_{m-n,p}, (np+m+1 \neq 0)$

(ii) $I_{m,p} = \dfrac{x^{m+1}(a+bx^n)^p}{np+m+1} + \dfrac{anp}{np+m+1} I_{m,p-1}, (np+m+1 \neq 0)$

(iii) $I_{m,p} = \dfrac{x^{m+1}(a+bx^n)^{p+1}}{(m+1)a} - \dfrac{(np+n+m+1)b}{(m+1)a} I_{m+n,p}, (m+1 \neq 0)$

(iv) $I_{m,p} = \dfrac{-x^{m+1}(a+bx^n)^{p+1}}{n(p+1)a} + \dfrac{(np+n+m+1)}{n(p+1)a} I_{m,p+1}, (n, p+1 \neq 0)$

Solved Examples

EXAMPLE 1. *Use reduction formula to integrate* $\sin^{1/2} x \cos^{7/2} x$.

SOLUTION. Since, we have

$$I_{m,n} = \int \sin^m x \cos^n x\,dx$$

$$= \frac{\cos^{n-1} x \sin^{m+1} x}{m+n} + \frac{n-1}{m+n} I_{m,n-2}, \ m+n \neq 0$$

Put $m = 1/2, n = 7/2$, we get

$$I_{\frac{1}{2},\frac{7}{2}} = \int \sin^{1/2} x \cos^{7/2} x \ dx$$

$$= \frac{\cos^{5/2} x \sin^{3/2} x}{4} + \frac{5}{8} I_{1/2,3/2}.$$

Also, $I_{\frac{1}{2},\frac{3}{2}} = \dfrac{\cos^{1/2} x \sin^{3/2} x}{2} + \dfrac{1}{4} I_{1/2,-1/2}$

$$\Rightarrow \quad I_{\frac{1}{2},\frac{7}{2}} = \frac{\cos^{5/2} x \sin^{3/2} x}{4}$$

$$+ \frac{5}{16} \cos^{1/2} x \sin^{3/2} x + \frac{5}{32} I_{1/2,-1/2}.$$

Here, further reduction is not possible because

$$m+n = \frac{1}{2} - \frac{1}{2} = 0.$$

But $I_{\frac{1}{2},-\frac{1}{2}} = \int \dfrac{\sqrt{\sin x}}{\sqrt{\cos x}}\,dx = \int \sqrt{\tan x}\,dx$

$$= \frac{1}{2} \tan^{-1}\left(\frac{\tan x - 1}{\sqrt{2\tan x}}\right) + \frac{1}{2\sqrt{2}} \log\frac{\tan x - \sqrt{2\tan x}}{\tan x + \sqrt{2\tan x}}$$

Therefore,

$$I_{\frac{1}{2},\frac{7}{2}} = \frac{1}{4}\sqrt{\cos x \sin^{3/2} x}\left(\cos^2 x + \frac{5}{4}\right)$$

$$+ \frac{5}{32}\left[\frac{1}{2} \tan^{-1}\left(\frac{\tan x - 1}{\sqrt{2\tan x}}\right)\right.$$

$$\left. + \frac{1}{2\sqrt{2}} \log\frac{\tan x - \sqrt{2\tan x}}{\tan x + \sqrt{2\tan x}}\right] + c.$$

EXAMPLE 2. *Compute* $\int \sin^4 x \cos^5 x\,dx$. (COCHIN–2005)

SOLUTION. We know that

$$I_{m,n} = \int \sin^m x \cos^n x\,dx$$

$$= -\frac{\sin^{m-1} x \cos^{n+1} x}{m+n} + \frac{m-1}{m+n} I_{m-2,n}, m+n \neq 0.$$

Put $m = 4, n = 5$, we get

$$I_{4,5} = \int \sin^4 x \cos^5 x\,dx = -\frac{\sin^3 x \cos^6 x}{9} + \frac{1}{3} I_{2,5}$$

Now $I_{2,5} = \int \sin^2 x \cos^5 x\,dx = -\frac{\sin x \cos^6 x}{5} + \frac{1}{7} I_{0,5}$

Here, $I_{0,5} = I_5 = \int \cos^5 x\,dx = \frac{\sin x \cos^4 x}{5} + \frac{4}{5} I_3$

and $I_3 = \int \cos^3 x\,dx = \frac{\sin x \cos^2 x}{3} + \frac{2}{3} I_1$

$I_1 = \int \cos x\,dx = \sin x$

Therefore,

$$I_{4,5} = -\frac{\sin^3 x \cos^6 x}{9} - \frac{1}{21}\sin x \cos^6 x$$

$$+ \frac{1}{105}\sin x \cos^4 x + \frac{4}{315}\sin x \cos^2 x + \frac{8}{315}\sin x + c$$

$$= \frac{\sin x}{315}[35\cos^8 x - 50\cos^6 x$$

$$+ 3\cos^4 x + 4\cos^2 x + 8] + c$$

EXAMPLE 3. *Compute* $\int \dfrac{dx}{(2+\cos x)^5}$.

SOLUTION. Since, we know that

$$I_n = \int \frac{dx}{(a+b\cos x)^n}$$

$$= \frac{b\sin x}{(n-1)(b^2-a^2)(a+b\cos x)^{n-1}}$$

$$- \frac{a(2n-3)}{(n-1)(b^2-a^2)} I_{n-1} + \frac{(n-2)}{(n-1)(b^2-a^2)} I_{n-2}.$$

Here $a = 2, b = 1, n = 5$

Therefore, $I_5 = \int \dfrac{dx}{(2+\cos x)^5}$

$$= -\frac{\sin x}{4.3(2+\cos x)^4} + \frac{2(10-3)}{4.3}.I_4 - \frac{3}{4.3} I_3.$$

Also, $I_4 = \int \dfrac{dx}{(2+\cos x)^4}$

$$= -\frac{\sin x}{3.3(2+\cos x)^3} + \frac{2(8-3)}{3.3}.I_3 - \frac{2}{3.3} I_2$$

$I_3 = \int \dfrac{dx}{(2+\cos x)^3}$

$$= -\frac{\sin x}{2.3(2+\cos x)^2} + \frac{2(6-3)}{2.3}.I_2 - \frac{1}{2.3} I_1$$

$I_2 = \int \dfrac{dx}{(2+\cos x)^2}$

$$= -\frac{\sin x}{1.3(2+\cos x)} + \frac{2(4-3)}{1.3}.I_1$$

and $I_1 = \int \dfrac{dx}{(2+\cos x)} = 2\tan^{-1}\left(\dfrac{1}{\sqrt{3}}\tan\dfrac{x}{2}\right) + c_1.$

Therefore,

$$I_5 = -\frac{1}{12}\frac{\sin x}{(2+\cos x)^4} + \frac{7}{6}\left[-\frac{1}{9}\frac{\sin x}{(2+\cos x)^3}\right.$$

$$\left. + \frac{10}{9} I_3 - \frac{2}{9} I_2\right] - \frac{1}{3} I_3$$

$$= -\frac{1}{12}\frac{\sin x}{(2+\cos x)^4} - \frac{7}{54}.\frac{\sin x}{(2+\cos x)^3} + \frac{113}{108} I_3 - \frac{7}{27} I_2$$

$$= -\frac{1}{12}\frac{\sin x}{(2+\cos x)^4} - \frac{7}{54}.\frac{\sin x}{(2+\cos x)^3}$$

$$+ \frac{113}{108}\left[-\frac{1}{6}.\frac{\sin x}{(2+3\cos x)^2} + I_2 - \frac{I_1}{6}\right] - \frac{7}{27} I_2$$

$$= -\frac{1}{12}\frac{\sin x}{(2+\cos x)^4} - \frac{7}{54}.\frac{\sin x}{(2+\cos x)^3} - \frac{113}{648}.\frac{\sin x}{(2+\cos x)^2}$$

$$+ \frac{85}{324}.\frac{\sin x}{(2+\cos x)} + \frac{170}{324} I_1 - \frac{113}{648} I_0$$

$$= -\frac{1}{12}\frac{\sin x}{(2+\cos x)^4} - \frac{7}{54}\cdot\frac{\sin x}{(2+\cos x)^3} - \frac{113}{660}\cdot\frac{\sin x}{(2+3\cos x)^2}$$

$$+ \frac{20}{27}\left[-\frac{1}{3}\cdot\frac{\sin x}{2+\cos x} + \frac{2}{3}I_1 - \frac{1}{3}I_0\right] - \frac{1}{6}I_1$$

$$= -\frac{1}{12}\frac{\sin x}{(2+\cos x)^4} - \frac{7}{54}\cdot\frac{\sin x}{(2+\cos x)^3}$$

$$-\frac{113}{648}\cdot\frac{\sin x}{(2+\cos x)^2} - \frac{85}{324}\cdot\frac{\sin x}{(2+\cos x)}$$

$$+ \frac{227}{324}\tan^{-1}\left(\frac{1}{\sqrt{3}}\tan\frac{x}{2}\right).$$

EXAMPLE 4. *If $I_n = \int_0^{\pi/4}\tan^n x\,dx$, show that*

$$I_n + I_{n-2} = \frac{1}{n-1}. \text{ Hence, deduce the value of } I_5.$$

SOLUTION . We know that
$$\int\tan^n x\,dx = \frac{\tan^{n-1}x}{n-1} - \int\tan^{n-2}x\,dx$$

Then
$$I_n = \int_0^{\pi/4}\tan^n x\,dx$$
$$= \left[\frac{\tan^{n-1}x}{n-1}\right]_0^{\pi/4} - \int_0^{\pi/4}\tan^{n-2}x\,dx$$

or
$$I_n = \frac{1}{n-1} - I_{n-2}$$

$$\therefore \quad I_n + I_{n-2} = \frac{1}{n-1} \qquad\qquad \dots(1)$$

Next, putting $n = 3$ and 5 successively, we get
$$I_3 + I_1 = \frac{1}{2} \quad, \quad I_5 + I_3 = \frac{1}{4}$$

Form these equations, we get
$$I_5 = \frac{1}{4} - I_3 = \frac{1}{4} - \left(\frac{1}{2} - I_1\right)$$

or
$$I_5 = -\frac{1}{4} + I_1$$

or
$$I_5 = -\frac{1}{4} + \int_0^{\pi/4}\tan x\,dx$$

$$= -\frac{1}{4} + \left[\log\sec x\right]_0^{\pi/4}$$

$$= -\frac{1}{4} + \left[\log\sec\frac{\pi}{4} - \log\sec 0\right]$$

$$= -\frac{1}{4} + \left[\log\sqrt{2} - \log 1\right]$$

$$= -\frac{1}{4} + \log\sqrt{2} - 0$$

Hence,
$$I_5 = \frac{1}{2}\left[\log 2 - \frac{1}{2}\right].$$

EXAMPLE 5. *If $u_n = \int_0^{\pi/2}x^n\sin x\,dx$ and $n > 1$, prove that*

$$u_n + n(n-1)u_{n-2} = n\left(\frac{\pi}{2}\right)^{n-1}$$

Hence evaluate $\int_0^{\pi/2}x^5\sin x\,dx$. (MADRAS–2000)

SOLUTION . By reduction formula, we have

$$\int x^n\sin mx\,dx = -\frac{x^n\cos mx}{m} + \frac{nx^{n-1}\sin mx}{m^2}$$
$$-\frac{n(n-1)}{m^2}\int x^{n-2}\sin mx\,dx$$

Putting $m = 1$ in both sides, we get
$$\int x^n\sin x\,dx = -x^n\cos x + nx^{n-1}\sin x$$
$$-n(n-1)\int x^{n-2}\sin x\,dx$$

Then
$$u_n = \int_0^{\pi/2}x^n\sin nx\,dx$$
$$= \left[-x^n\cos x + nx^{n-1}\sin x\right]_0^{\pi/2}$$
$$-n(n-1)\int_0^{\pi/2}x^{n-2}\sin x\,dx$$

or
$$u_n = \left[u\left(\frac{\pi}{2}\right)^{n-1}\right] - n(n-1)u_{n-2}$$

$$\therefore u_n + n(n-1)u_{n-2} = n\left(\frac{\pi}{2}\right)^{n-1} \qquad \dots(1)$$

Next, putting $n = 3$ and 5 successively in (1), we get
$$u_{5+5(5-1)}u_3 = 5\left(\frac{\pi}{2}\right)^4$$

and
$$u_{3+3(3-1)}u_1 = 3\left(\frac{\pi}{2}\right)^2$$

From these equations, we get
$$u_5 + 20\left[3\left(\frac{\pi}{2}\right)^2 - 6u_1\right] = 5\left(\frac{\pi}{2}\right)^4$$

Now,
$$u_1 = \int_0^{\pi/2}x\sin x\,dx$$
$$= \left[-x\cos x\right]_0^{\pi/2} - \int_0^{\pi/2}(-\cos x)dx = 1$$
$$= [0] + \left[\sin x\right]_0^{\pi/2}$$

$$\therefore u_5 + 20\left[3\left(\frac{\pi}{2}\right)^2 - 6(1)\right] = 5\left(\frac{\pi}{2}\right)^4$$

or
$$u_5 = \frac{5}{16}\pi^4 - 15\pi^2 + 120$$

EXAMPLE 6. *Evaluate $\int_0^1 x^m(\log x)^n\,dx$, when $m \geq 0$ and n is positive integer.* (SVTU–2009, BHILLAI–2005)

SOLUTION . We know by reduction formula
$$\int x^m(\log x)^n\,dx = \frac{x^{m+1}(\log x)^n}{m+1}$$
$$-\frac{n}{m+1}\int x^m(\log x)^{n-1}\,dx$$

Let $I(m,n) = \int_0^1 x^m(\log x)^n\,dx$, then we have
$$I(m,n) = \left[\frac{x^{m+1}(\log x)^n}{m+1}\right]_0^1 - \frac{n}{m+1}I(m,n-1)$$

or $I(m,n) = [0 - 0] - \frac{n}{m+1}I(m,n-1)$

or $I(m,n) = -\frac{n}{m+1}I(m,n-1) \qquad \dots(1)$

Replace n by $n-1$ in (1), we get

$$I(m, n-1) = -\frac{n-1}{m+1}I(m, n-2) \qquad \ldots(2)$$

From (2) and (1), we get

$$I(m, n) = (-1)^2\left(\frac{n}{m+1}\right)\left(\frac{n-1}{m+1}\right)I(m, n-2) \qquad \ldots(3)$$

Again by repeated application of (1), we get

$$I(m, n) = (-1)^n\left(\frac{n}{m+1}\right)\left(\frac{n-1}{m+1}\right)\left(\frac{n-2}{m+1}\right)$$
$$\ldots\left(\frac{1}{m+1}\right)I(m, 0)$$

or $\quad I(m, n) = (-1)^n\dfrac{n!}{(m+1)^n}I(m, 0)$

Now $I(m, 0) = \int_0^1 x^m dx = \left[\dfrac{x^{m+1}}{m+1}\right]_0^1 = \dfrac{1}{m+1}$

$\therefore \quad I(m, n) = (-1)^n\dfrac{n!}{(m+1)^n}\cdot\dfrac{1}{m+1} = \dfrac{(-1)^n n!}{(m+1)^{n+1}}$

Hence, $\int_0^1 x^m(\log x)^n dx = \dfrac{(-1)^n n!}{(m+1)^{n+1}}$.

Exercise-4.1

Use reduction formulae, compute the following :

1. $\int \cos^3 x \operatorname{cosec}^2 x\, dx$

2. $\int \sqrt{\cos\theta}\cdot\sin^3\theta\, d\theta$

3. $\int \cos^{-3}\theta \sin^{-1}\theta\, d\theta$

4. $\int \dfrac{x^4}{\sqrt{1-x^2}}dx$

5. $\int (1+x^2)^{3/2}dx$

6. $\int \dfrac{x^4}{(a^2+x^2)^2}dx$

7. $\int \dfrac{x^2}{\sqrt{2ax-x^2}}dx$ (MADRAS–2000)

8. $\int \dfrac{x^5}{(a+bx^2)^4}dx$

9. $\int \dfrac{dx}{x^{1/2}(1+x^2)^{5/4}}$

10. $\int \dfrac{x^3}{\sqrt{4x-x^2}}dx$

11. $\int \dfrac{x^3+5x^2-3x+4}{\sqrt{x^2+x+1}}dx$

12. $\int x^3\cos 3x\, dx$

13. $\int x^3 e^{ax}dx$

14. Prove that : $I_n = \int x^n e^{-x}dx = -x^n e^{-x} + nI_{n-1}$.

15. Find the following integrals using reduction formula
    (i) $\int \dfrac{x+1}{(x^2+1)^3}dx$
    (ii) $\int \dfrac{x^2-a^2}{(x^2+a^2)^3}dx$

16. (i) If $I_n = \int \dfrac{dx}{(x^2+a^2)^n}$, then show that
    $$I_{n+1} = \dfrac{1}{2na^2}\cdot\dfrac{x}{(x^2+a^2)^n} + \dfrac{2n-1}{2n}\cdot\dfrac{1}{a^2}I_n .$$

    (ii) If $I_n = \int(\log x)^n dx$, then show that $I_n = x(\log x)^n - nI_{n-1}$.

    (iii) If $I_n = \int x^n e^x dx$, then show that $I_n = x^n e^x - nI_{n-1}$.

    (iv) If $I_n = \int e^{ax}\sin^n x\, dx$, then show that
    $$I_n = \dfrac{e^{ax}}{a^2+n^2}\sin^{n-1}x(a\sin x - n\cos x) + \dfrac{n(n-1)}{a^2+n^2}I_{n-2}.$$
    (MADRAS–2000, 17)
    (GORAKHPUR–1999, 2014, 18)

17. Evaluate the following integrals :
    (i) $\int_0^{\pi/4}\tan^5\theta\, d\theta$
    (ii) $\int_0^{\pi/4}\tan^7 x\, dx$

18. If $I_n = \int_0^{\pi/4}\tan^n x\, dx$, prove that $n(I_{n-1}+I_{n+1}) = 1$. (VTU–2009)

19. If $I_n = \int_0^{\pi/3}\tan^n x\, dx$, prove that $(n-1)(I_n+I_{n-2}) = \left(\sqrt{3}\right)^{n-1}$.

20. Evaluate $\int \dfrac{d\theta}{\sin^4\frac{\theta}{2}}$

21. Prove that $\int_0^{\pi/4}\sec^3 x\, dx = \dfrac{1}{3}\left\{\sqrt{2}+\log(\sqrt{2}+1)\right\}$.

22. Prove that $\int_0^\pi\sin^m x\sin nx\, dx = \dfrac{m(m-1)}{m^2-n^2}\int_0^\pi\sin^{m-2}x\sin nx\, dx$.

23. If n is a positive integer greater than 1, prove that
    $$\int_0^{\pi/2}\cos^{n-2}x\sin nx\, dx = \dfrac{1}{n-1}.$$

24. If $u_n = \int_0^{\pi/2}x^n\sin mx\, dx$, prove that
    $$u_n = \dfrac{n}{m^2}\left(\dfrac{\pi}{2}\right)^{n-1} - \dfrac{n(n-1)}{m^2}u_{n-2}$$
    where m is of the form $4r+1$. (MARATHWADA–2008)

25. Evaluate the following integrals :
    (i) $\int_0^{\pi/2}x^3\sin 3x\, dx$
    (ii) $\int_0^\pi x\sin^3 x\, dx$
    (iii) $\int_0^{\pi/2}x^5\sin x\, dx$
    (iv) $\int_0^\pi\theta\sin^2\theta\cos\theta\, d\theta$

26. If $I_m = \int_0^\infty e^{-x}\sin^m x\, dx$, where $m\geq 2$, prove that
    $$(1+m^2)I_m = m(m-1)I_{m-2}.$$
    Hence deduce I_4.

27. If $I_n = \int_{-\pi/2}^{\pi/2}e^{ax}\cos^n x\, dx$, prove that $I_n = \dfrac{n(n-1)}{a^2+n^2}I_{n-2}$.

28. Evaluate $\int_0^1 x^{n-1}(\log x)^m dx$, when m and n are positive integers. (SVTU–2009, 14)

29. Evaluate the following integrals :
    (i) $\int_0^1 x^m(\log x)^5 dx$
    (ii) $\int_0^1 x^m(\log x)^4 dx$

30. If $I_n = \int x^n(a-x)^{1/2}dx$, prove that
    $$(2n+3)I_n = 2anI_{n-1} - 2x^n(a-x)^{3/2}.$$

31. If $I_n = \int_0^a(a^2-x^2)^n dx$, prove that $I_n = \dfrac{2na^2}{(2n+1)}I_{n-1}$. Hence deduce I_3.

32. If $I_n = \int_0^1 x^p(1-x^q)^n dx$, where p, q and n are positive, prove that $(qn+p+1)I_n = qnI_{n-1}$.

33. Prove that $\int_0^1 x^{-1/4}(1-x^{1/2})^{5/2}dx = \dfrac{5}{16}\int_0^1 x^{-1/4}(1-x^{1/2})^{1/2}dx$.

34. Prove that $\int_0^\pi\dfrac{\sin nx}{\sin x}dx = \begin{cases} 0, & \text{when } n \text{ is even positive integer} \\ \pi, & \text{when } n \text{ is odd positive integer} \end{cases}$

35. If $I_n = \int_0^1 e^{-1}x^n dx$, find the reduction formula and hence show that $I_n = n!$.

Answers

1. $-\dfrac{1}{3}\sin x[\cot^2 x + 2\cos^2 x + 4] + c$ **2.** $\dfrac{2}{7}\cos^{7/2}\theta - \dfrac{2}{3}\cos^{3/2}\theta + c$ **3.** $\dfrac{1}{2}\tan^2\theta + \log\tan\theta + c$ **4.** $\dfrac{1.3}{2.4}\sin^{-1}x - \dfrac{x\sqrt{1-x^2}}{8}(3+2x^2) + c$

5. $\dfrac{\sin\theta}{4\cos^4\theta} + \dfrac{3\sin\theta}{8\cos^2\theta} + \dfrac{3}{8}\log(\sec\theta + \tan\theta) + c \ (x = \tan\theta)$ **6.** $c - \dfrac{x^3}{2(a^2+x^4)} + \dfrac{3}{2}\left(x - a\tan^{-1}\dfrac{x}{a}\right)$ **7.** $c - \sqrt{2ax - x^2}\left(\dfrac{x}{2} + \dfrac{3a}{2}\right) + 3a^2\sin^{-1}\dfrac{x-a}{a}$

8. $\dfrac{1}{6a}\cdot\dfrac{x^4}{(a+bx^2)^3} + \dfrac{1}{12a^2}\cdot\dfrac{x^4}{(a+bx^2)^2} + c$ **9.** $2\sqrt{x}(1+x^2)^{-1/4} + c$ **10.** $c - \dfrac{1}{3}(x^2 + 5x + 30)\sqrt{4x - x^2} + 10\cos^{-1}\left(1 - \dfrac{x-2}{2}\right)$

11. $\left(\dfrac{1}{3}x^3 + \dfrac{25}{12}x - \dfrac{163}{24}\right)\sqrt{x^2+x+1} + \dfrac{85}{16}\sin^{-1}\left(\dfrac{2x+1}{\sqrt{3}}\right) + c$ **12.** $\dfrac{1}{27}(9x^2 - 2)\cos 3x + \dfrac{1}{9}(3x^2 - 2x)\sin 3x + c$

13. $\dfrac{e^{ax}}{a^4}(a^3x^3 - 3a^2x^2 + 6ax - 6) + c$ **15.** (i) $\dfrac{x-1}{4(x^2+1)^2} + \dfrac{3x}{8(x^2+1)} + \dfrac{3}{8}\tan^{-1}x + c$ (ii) $-\dfrac{x}{4a^2(x^2+a^2)} - \dfrac{x}{2(x^2+a^2)^2} - \dfrac{1}{4a^3}\tan^{-1}\dfrac{x}{a} + c$

17. (i) $\dfrac{1}{2}\left(\log 2 - \dfrac{1}{2}\right)$ (ii) $\dfrac{5}{12} - \dfrac{1}{2}\log 2$ **20.** $-\dfrac{2}{3}\left[\csc^2\dfrac{\theta}{2}\cot\dfrac{\theta}{2} + 2\cot\dfrac{\theta}{2}\right]$

25. (i) $\dfrac{2}{27} - \dfrac{\pi^2}{12}$ (ii) $\dfrac{2\pi}{3}$ (iii) $\dfrac{5}{16}\pi^4 - 15\pi^2 + 120$ (iv) $-\dfrac{4}{9}$ **26.** $I_4 = \dfrac{24}{85}$ **28.** $\dfrac{(-1)^m m!}{n^{m+1}}$

29. (i) $-\dfrac{120}{(m+1)^6}$ (ii) $\dfrac{24}{(m+1)^5}$ **34.** $\dfrac{16a^7}{35}$ **35.** $I_n = nI_{n-1}$

4.2 DEFINITE INTEGRALS

If $f(x)$ is continuous and non-negative function over a closed interval $[a, b]$ then $\int_a^b f(x)dx$ is called the definite integral of $f(x)$ between the limits a and b $(b > a)$.

If $\int f(x)dx = F(x) + c$, then $\int_a^b f(x)dx = \left[F(x) + c\right]_a^b = F(b) - F(a)$ is a definite value.

Here, a is called lower limit and b is called the upper limit and the interval $[a, b]$ is called the range of integration.

◄ REMARKS

* $\int_a^b f(x)dx$ represents the area bounded by the lines $x = a$ and $x = b$.
*   If $F(b) - F(a)$ is not a definite value, then the integral $\int_a^b f(x)dx$ is indefinite.

4.2.1 PROPERTIES OF DEFINITE INTEGRALS

PROPERTY 1. $\int_a^a f(x)dx = 0$.

PROOF.   Let $\int f(x)dx = F(x)$ then $\int_a^a f(x)dx = \left[F(x)\right]_a^a = F(a) - F(a) = 0$.

PROPERTY 2. *The value of definite integral is independent of the variable of integration. i.e.,* $\int_a^b f(x)dx = \int_a^b f(u)du$.

PROOF.   Let $\int f(x)dx = F(x)$ then $\int_a^b f(x)dx = \left[F(x)\right]_a^b = F(b) - F(a) = \int_a^b f(u)du$.

PROPERTY 3. $\int_a^b f(x)dx = -\int_b^a f(x)dx$.

PROOF.   Let $\int f(x)dx = F(x)$

then $\int_a^b f(x)dx = \left[F(x)\right]_a^b = F(b) - F(a) = -\left[F(x)\right]_b^a = \int_b^a f(x)dx$.

PROPERTY 4. $\int_a^c f(x)dx + \int_c^b f(x)dx = \int_a^b f(x)dx$, *where* $a < c < b$.

PROOF.   Let $\int f(x)dx = F(x)$

then $\int_a^c f(x)dx + \int_c^b f(x)dx = \left[F(x)\right]_a^c + \left[F(x)\right]_c^b = [F(c) - F(a)] + [F(b) - F(c)] = F(b) - F(a) = \left[F(x)\right]_a^b = \int_a^b f(x)dx$.

PROPERTY 5. $\int_0^a f(a-x)dx = \int_0^a f(x)dx$.

PROOF.   We have $\int_0^a f(a-x)dx$.

Let $a - x = t$, then $-dx = dt$. If $x = 0 \Rightarrow t = a$ and $x = a, t = 0$.

So, $\int_0^a f(a-x)dx = -\int f(t)dt = \int_0^a f(t)dt = \int_0^a f(x)dx$ (By property 2)

PROPERTY 6. *If* $f(x)$ *is an even function of* x, *then* $\int_{-a}^a f(x)dx = 2\int_0^a f(x)dx$ *and if* $f(x)$ *is an odd function then* $\int_{-a}^a f(x)dx = 0$.

(BHOPAL–2008)

PROOF.   Consider $\int_{-a}^a f(x)dx$.   Then, $\int_{-a}^a f(x)dx = \int_{-a}^0 f(x)dx + \int_0^a f(x)dx$.

Now, let $I = \int_{-a}^0 f(x)dx$

put $x = -t$, so $x = -a \Rightarrow t = a$ and $x = 0 \Rightarrow t = 0$.

So, $\int_{-a}^{0} f(x)dx = -\int_{a}^{0} f(-t)dt = \int_{0}^{a} f(-t)dt = \int_{0}^{a} f(-x)dx$ (by 2)

Thus, we have $\int_{-a}^{a} f(x)dx = \int_{0}^{a} f(-x)dx + \int_{0}^{a} f(x)dx = \int_{0}^{a}[f(-x)] + f(x)dx$.

Now, if $f(x)$ is an even function i.e., $f(-x) = f(x)$, then we get $\int_{-a}^{a} f(x)dx = \int_{0}^{a}[f(x) + f(x)]dx = 2\int_{0}^{a} f(x)dx$

and, if $f(x)$ is an odd function i.e., $f(-x) = -f(x)$, then we get $\int_{-a}^{a} f(x)dx = \int_{0}^{a}[-f(x) + f(x)]dx = 0$.

PROPERTY 7. $\int_{0}^{2a} f(x)dx = 2\int_{0}^{a} f(x)dx$ if $f(2a - x) = f(x)$ and $\int_{0}^{2a} f(x)dx = 0$ if $f(2a - x) = -f(x)$.

PROOF. This integral can be written as $\int_{0}^{2a} f(x)dx = \int_{0}^{a} f(x)dx + \int_{a}^{2a} f(x)dx$ ...(1)

Now, consider the integral $\int_{a}^{2a} f(x)dx$.
Put $x = 2a - t$, then $dx = -dt$ and if $x = a$ then $t = a$ and if $x = 2a$ then $t = 0$.
So, $\int_{a}^{2a} f(x)dx = -\int_{a}^{0} f(2a - t)dt = \int_{0}^{a} f(2a - t)dt = \int_{0}^{a} f(2a - x)dx$ (by 2)
Therefore, from (1), we have

$$\int_{0}^{2a} f(x)dx = \int_{0}^{a} f(x)dx + \int_{0}^{a} f(2a - x)dx = \int_{0}^{a}[f(x) + f(2a - x)]dx \quad ...(2)$$

Now, if $f(2a - x) = f(x)$ then from (2), we get $\int_{0}^{2a} f(x)dx = 2\int_{0}^{a} f(x)dx$

and if $f(2a - x) = -f(x)$ then from (2), we get $\int_{0}^{2a} f(x)dx = \int_{0}^{a}[f(x) - f(x)]dx = 0$

PROPERTY 8. $\int_{0}^{na} f(x)dx = n\int_{0}^{a} f(x)dx$, if $f(x + ma) = f(x)$ for all integral values of m.

PROOF. The given integral can be written as

$$\int_{0}^{na} f(x)dx = \int_{0}^{a} f(x)dx + \int_{a}^{2a} f(x)dx + ... + \int_{(m-1)a}^{ma} f(x)dx + ... + \int_{(n-1)a}^{na} f(x)dx \quad ...(1)$$

Now consider the integral $\int_{(m-1)a}^{ma} f(x)dx$.

put $x = y + ma \Rightarrow dx = dy$.
but $f(y + ma) = f(y)$ is given, so we have

$$\int_{(m-1)a}^{ma} f(x)dx = \int_{0}^{a} f(y)dy = \int_{0}^{a} f(x)dx.$$

Now from (1), we have

$$\int_{0}^{na} f(x)dx = \int_{0}^{a} f(x)dx + \int_{0}^{a} f(x)dx + ... + \int_{0}^{a} f(x)dx = n\int_{0}^{a} f(x)dx.$$

Hence, $\int_{0}^{na} f(x)dx = n\int_{0}^{a} f(x)dx$, if $f(x + ma) = f(x)$ for all integral values of m.

Solved Examples

EXAMPLE 1. *Evaluate the following integrals.*

(i) $\int_{0}^{\pi/2} \log \tan x \, dx$

(ii) $\int_{0}^{\pi/4} \log(1 + \tan \theta)d\theta$ (MADRAS–2000)

(iii) $\int_{0}^{\pi} \dfrac{x \sin x}{1 + \sin x} dx$

(iv) $\int_{0}^{\pi/2} \log \sin x \, dx$ (ANNA–2005)

SOLUTION. (i) Consider, $I = \int_{0}^{\pi/2} \log \tan x \, dx$.

Now, $I = \int_{0}^{\pi/2} \log \tan\left(\dfrac{\pi}{2} - x\right)dx$

$\left(\because \int_{0}^{a} f(x)dx = \int_{0}^{a} f(a - x)dx\right)$

$I = \int_{0}^{\pi/2} \log \cot x \, dx$
On adding, we get

$2I = \int_{0}^{\pi/2} \log \tan x \, dx + \int_{0}^{\pi/2} \log \cot x \, dx$

$= \int_{0}^{\pi/2} \log(\tan x \cdot \cot x)dx$

$= \int_{0}^{\pi/2} \log 1 = 0$

Hence $2I = 0 \Rightarrow I = 0$.

(ii) Consider, $I = \int_{0}^{\pi/4} \log(1 + \tan \theta)d\theta$

Now, $I = \int_{0}^{\pi/4} \log\left[1 + \tan\left(\dfrac{\pi}{4} - \theta\right)\right]d\theta$

$\left[\because \int_{0}^{a} f(x)dx = \int_{0}^{a} f(a - x)dx\right]$

$= \int_{0}^{\pi/4} \log\left[1 + \dfrac{1 - \tan \theta}{1 + \tan \theta}\right]d\theta$

$= \int_{0}^{\pi/4} \log\left(\dfrac{2}{1 + \tan \theta}\right)d\theta$

$= \int_{0}^{\pi/4}[\log 2 - \log(1 + \tan \theta)]d\theta$

$= \int_{0}^{\pi/4} \log 2 d\theta - \int_{0}^{\pi/4} \log(1 + \tan \theta)d\theta$

$= \dfrac{\pi}{4} \log 2 - I$

So, $2I = \dfrac{\pi}{4} \log 2 \Rightarrow I = \dfrac{\pi}{8} \log 2$.

(iii) Here, $I = \int_{0}^{\pi} \dfrac{x \sin x}{1 + \sin x} dx$

$\Rightarrow I = \int_{0}^{\pi} \dfrac{(\pi - x)\sin(\pi - x)}{1 + \sin(\pi - x)} dx$

$\left(\because \int_{0}^{a} f(x)dx = \int_{0}^{a} f(a - x)dx\right)$

$$= \int_0^\pi \frac{(\pi - x)\sin x}{1 + \sin x}dx$$

$$= \int_0^\pi \frac{\pi \sin x}{1 + \sin x}dx - \int_0^\pi \frac{x \sin x}{1 + \sin x}dx$$

$$= \int_0^\pi \frac{\pi \sin x}{1 + \sin x}dx - I$$

So, $\quad 2I = \int_0^\pi \frac{\pi \sin x}{1 + \sin x}dx = \pi \int_0^\pi \frac{\sin x}{1 + \sin x}dx$

$$= \pi \int_0^\pi \left(1 - \frac{1}{1 + \sin x}\right)dx$$

$$= \pi \int_0^\pi \left(1 - \frac{1 - \sin x}{\cos^2 x}\right)dx$$

$$= \pi \int_0^\pi [1 - \sec^2 x + \sec x \tan x]dx$$

$$= \pi \left[x - \tan x + \sec x\right]_0^\pi = \pi(\pi - 2)$$

$\therefore \qquad 2I = \pi(\pi - 2)$

Hence, $\qquad I = \pi\left(\dfrac{\pi}{2} - 1\right)$.

(iv) Here, $\quad I = \int_0^{\pi/2} \log \sin x\, dx$.

Also, $\quad I = \int_0^{\pi/2} \log \sin\left(\dfrac{\pi}{2} - x\right)dx$

$$\left(\because \int_0^a f(x)dx = \int_0^a f(a - x)dx\right)$$

$\therefore \qquad I = \int_0^{\pi/2} \log \cos x\, dx$.

On adding, we get

$$2I = \int_0^{\pi/2} (\log \sin x)dx$$

$$+ \int_0^{\pi/2} (\log \cos x)dx$$

$$= \int_0^{\pi/2} \log\left(\frac{\sin 2x}{2}\right)dx$$

$$= \int_0^{\pi/2} \log \sin 2x\, dx - \int_0^{\pi/2} \log 2\, dx$$

Let $2x = t$ for first integral, then on differentiating, we get $2dx = dt$

Now, $\quad 2I = \dfrac{1}{2}\int_0^\pi \log \sin t\, dt - \left[(x \log 2)\right]_0^{\pi/2}$

$$= \frac{1}{2}\int_0^\pi \log \sin t\, dt - \frac{\pi}{2}\log 2$$

$$= \int_0^{\pi/2} \log \sin t\, dt - \frac{\pi}{2}\log 2$$

$\therefore \qquad 2I = I - \dfrac{\pi}{2}\log 2$

Hence, $\qquad I = -\dfrac{\pi}{2}\log 2$.

EXAMPLE 2. *Evaluate the following integrals* :

   (i) $\int_0^3 |3x - 1|\, dx$ **(ii)** $\int_0^\pi |\cos x|\, dx$

SOLUTION. (i) Given integral is $\int_0^3 |3x - 1|\, dx$.

Now, $|3x - 1| = \begin{cases} 3x - 1, & \text{when } x \geq 1/3 \\ -(3x - 1), & \text{when } x < 1/3 \end{cases}$

So,
$\int_0^3 |3x - 1|\, dx = \int_0^{1/3} -(3x - 1)dx + \int_{1/3}^3 (3x - 1)dx$

$$= \left[-\frac{3x^2}{2} + x\right]_0^{1/3} + \left[\frac{3x^2}{2} - x\right]_{1/3}^3 = \frac{65}{6}.$$

(ii) Here, the given intregral is $\int_0^\pi |\cos x|\, dx$.

Now, $\quad |\cos x| = \begin{cases} \cos x, & 0 \leq x \leq \pi/2 \\ -\cos x, & \pi/2 \leq x \leq \pi \end{cases}$

So, $\int_0^\pi |\cos x|\, dx = \int_0^{\pi/2} \cos x\, dx + \int_{\pi/2}^\pi (-\cos x)dx$

$$= \int_0^{\pi/2} \cos x\, dx - \int_{\pi/2}^\pi \cos x\, dx$$

$$= \left[\sin x\right]_0^{\pi/2} - \left[\sin x\right]_{\pi/2}^\pi = 2.$$

(iii) Let $\int_0^6 |x + 2|\, dx = \int_0^6 (x + 2)dx = \int_0^6 x\, dx + \int_0^6 2\, dx$

$$= \left(\frac{x^2}{2}\right)_0^6 + \left(2x\right)_0^6 = \frac{36}{7} - 0 + 2 \times 6$$

$$= 18 + 12 = 30.$$

EXAMPLE 3. *Evaluate the integral* $\int_0^\pi \log(1 + \cos x)dx$.

SOLUTION. We have, $I = \int_0^\pi \log(1 + \cos x)dx$

Now, $\quad I = \int_0^\pi \log\{1 + \cos(\pi - x)\}dx$

$$= \int_0^\pi \log(1 - \cos x)dx$$

On adding,

$$2I = \int_0^\pi \log(1 + \cos x)dx + \int_0^{\pi/2} \log(1 - \cos x)dx$$

$$= \int_0^\pi \log(1 + \cos x)(1 - \cos x)dx$$

$$= \int_0^\pi \log \sin^2 x\, dx = 2\int_0^\pi \log \sin x\, dx$$

$$= 4\int_0^{\pi/2} \log \sin x\, dx \qquad \text{(Using property 7)}$$

$\therefore \qquad 2I = 4\left(-\dfrac{\pi}{2}\log 2\right)$

So, $\qquad I = -\pi \log 2$

Hence, $\quad I = \pi \log\left(\dfrac{1}{2}\right)$.

EXAMPLE 4. *Evaluate the integral* $\int_0^\infty \log\left(\dfrac{1 + x^2}{x}\right)\dfrac{dx}{1 + x^2}$.

SOLUTION. We have $I = \int_0^\infty \log\left(\dfrac{1 + x^2}{x}\right)\dfrac{dx}{1 + x^2}$.

Now, putting $x = \tan\theta \Rightarrow \theta = \tan^{-1}x$, $d\theta = \dfrac{dx}{1 + x^2}$.

So, $\quad I = \int_0^{\pi/2} \log\left(\dfrac{\sec^2\theta}{\tan\theta}\right)d\theta$

$$= \int_0^{\pi/2} \log\left(\frac{1}{\sin\theta . \cos\theta}\right)d\theta$$

$$= \int_0^{\pi/2} \log\left(\frac{2}{2\sin\theta . \cos\theta}\right)d\theta$$

$$= \int_0^{\pi/2} \log 2\, d\theta - \int_0^{\pi/2} \log \sin 2\theta\, d\theta$$

$$= \frac{\pi}{2}\log 2 - \int_0^{\pi/2} \log \sin 2\theta\, d\theta$$

Let $2\theta = t \quad \Rightarrow \quad 2d\theta = dt$,

then $\quad I = \dfrac{\pi}{2}\log 2 - \dfrac{1}{2}\int_0^\pi \log \sin t\, dt$

$$= \frac{\pi}{2}.\log 2 - \int_0^{\pi/2} \log \sin x\, dx \text{(Using property 7)}$$

$$= \frac{\pi}{2}.\log 2 - \left(-\frac{\pi}{2}\log 2\right) = \pi \log 2$$

$$\left(\because \int_0^{\pi/2} \log \sin x\, dx = -\frac{\pi}{2}\log 2\right)$$

EXAMPLE 5. *Evaluate the integral* $\int_{-1}^{1} f(x)dx$, *where*

$$f(x) = \begin{cases} e^x & , \quad -1 \le x \le 0 \\ 1 & , \quad 0 \le x \le \dfrac{1}{2} \\ 3^x & , \quad \dfrac{1}{2} \le x \le 1 \end{cases}$$

SOLUTION. We can write

$$\int_{-1}^{1} f(x)dx = \int_{-1}^{0} f(x)dx + \int_{0}^{1/2} f(x)dx + \int_{1/2}^{1} f(x)dx$$

$$= \int_{-1}^{0} e^x dx + \int_{0}^{1/2} 1\,dx + \int_{1/2}^{1} 3^x dx$$

$$= \left[e^x\right]_{-1}^{0} + \left[x\right]_{0}^{\frac{1}{2}} + \left[\frac{3^x}{\log 3}\right]_{\frac{1}{2}}^{1}$$

$$= \left[e^0 - e^{-1}\right] + \frac{1}{2} + \frac{1}{\log 3}(3 - \sqrt{3})$$

$$= 1 - \frac{1}{e} + \frac{1}{2} + \frac{1}{\log 3}(3 - \sqrt{3}).$$

EXAMPLE 6. *Evaluate the integral* $\int_{0}^{\pi} \dfrac{x \tan x}{\sec x + \tan x} dx$.

SOLUTION. Here, $\quad I = \int_{0}^{\pi} \dfrac{x \tan x}{\sec x + \tan x} dx \qquad \ldots(1)$

Now, $\quad I = \int_{0}^{\pi} \dfrac{(\pi - x)\tan(\pi - x)}{\sec(\pi - x) + \tan(\pi - x)} dx$

$$\left(\because \int_{0}^{a} f(x)dx = \int_{0}^{a} f(a - x)dx\right)$$

$$= \int_{0}^{\pi} \dfrac{(\pi - x)\tan x}{\sec x + \tan x} dx \qquad \ldots(2)$$

On adding (1) and (2), we get

$$2I = \int_{0}^{\pi} \dfrac{x \tan x}{\sec x + \tan x} dx + \int_{0}^{\pi} \dfrac{(\pi - x)\tan x}{\sec x + \tan x} dx$$

$$= \int_{0}^{\pi} \dfrac{\pi \tan x}{\sec x + \tan x} dx$$

$$= \int_{0}^{\pi} \dfrac{\pi \tan x (\sec x - \tan x)}{\sec^2 x - \tan^2 x} dx$$

$$\left(\because \sec^2 x - \tan^2 x = 1\right)$$

$$= \pi \int_{0}^{\pi} [\sec x \tan x - \sec^2 x + 1] dx$$

$$= \pi \left[\sec x - \tan x + x\right]_{0}^{\pi}$$

$$= \pi [\{\sec \pi - \tan \pi + \pi\} - \{\sec 0 - \tan 0 + 0\}]$$

$$= \pi(-1 + \pi - 1) = \pi(\pi - 2)$$

Hence, $\quad I = \pi\left(\dfrac{\pi}{2} - 1\right).$

EXAMPLE 7. *Evaluate* $\int_{0}^{\pi/2} \dfrac{dx}{5 + 4 \sin x}$.

SOLUTION. Let $\quad I = \int_{0}^{\pi/2} \dfrac{dx}{5 + 8 \sin x/2 \cdot \cos x/2}$

Divide Nr and Dr by $\cos^2 x/2$ we have

$$I = \int_{0}^{\pi/2} \dfrac{\sec^2(x/2)dx}{5 \sec^2 \dfrac{x}{2} + 8 \tan \dfrac{x}{2}}$$

$$= \int_{0}^{\pi/2} \dfrac{\sec^2(x/2)dx}{5\left(1 + \tan^2 \dfrac{x}{2}\right) + 8 \tan \dfrac{x}{2}}$$

Put $\tan \dfrac{x}{2} = t \Rightarrow \dfrac{1}{2} \sec^2 \dfrac{x}{2} dx = dt$

$$\Rightarrow \sec^2 \dfrac{x}{2} dx = 2dt$$

When $x = 0 \Rightarrow t = 0$

$$x = \dfrac{\pi}{2} \Rightarrow t = 1$$

Then $\quad I = \int_{0}^{1} \dfrac{2dt}{5t^2 + 8t + 5} = \dfrac{2}{5}\int_{0}^{1} \dfrac{dt}{t^2 + \dfrac{8}{5}t + 1}$

$$= \dfrac{2}{5}\int_{0}^{1} \dfrac{dt}{\left(t + \dfrac{4}{5}\right)^2 + \left(\dfrac{3}{4}\right)^2}$$

$$= \dfrac{2}{5} \cdot \dfrac{4}{3}\left[\tan^{-1}\left(\dfrac{t + \dfrac{4}{5}}{\dfrac{3}{4}}\right)\right]_{0}^{1}$$

$$= \dfrac{8}{15}\left[\tan^{-1}\dfrac{36}{15} - \tan^{-1}\dfrac{16}{15}\right]$$

EXAMPLE 8. *Evaluate* $\int_{0}^{1} \sin^{-1} x\, dx$.

SOLUTION. Let $\quad I = \int_{0}^{1} \sin^{-1} x\, dx$

Putting $\quad x = \sin\theta \Rightarrow dx = \cos\theta\, d\theta$

When $\quad x = 0 \quad \Rightarrow \quad \theta = 0$

$$x = 1 \quad \Rightarrow \quad \theta = \dfrac{\pi}{2}$$

Then $\quad I = \int_{0}^{\pi/2} \theta \cos\theta\, d\theta$;

Integrating by parts, we get

$$I = \left(\theta \sin\theta\right)_{0}^{\pi/2} - \int_{0}^{\pi/2} \sin\theta\, d\theta$$

$$= \dfrac{\pi}{2} + [\cos\theta]_{0}^{\pi/2} = \dfrac{\pi}{2} - 1 = \dfrac{\pi - 2}{2}$$

Exercise-4.2

1. Evaluate the integral $\int_{0}^{\pi/2} \dfrac{\sin^4 x\, dx}{\sin^4 x + \cos^4 x}$.

2. Evaluate the integral $\int_{0}^{\pi/2} \dfrac{\sin x - \cos x}{1 + \sin x \cos x} dx$.

3. Evaluate the integral $\int_{0}^{4} f(x)dx$

where $f(x) = \begin{cases} 2x + 3 & , \quad 0 \le x \le 3 \\ 3x & , \quad 3 \le x \le 4 \end{cases}$

4. Evaluate $\int_{0}^{\pi/2} \dfrac{x \sin x \cos x}{\cos^4 x + \sin^4 x} dx$.

5. Evaluate the integral $\int_{0}^{\pi} \dfrac{x \sin x}{1 + \cos^2 x} dx$.

6. Show that $\int_{0}^{\pi} \dfrac{x\, dx}{a^2 \cos^2 x + b^2 \sin^2 x} = \dfrac{\pi^2}{2ab}$.

7. Evaluate the integral $\int_{0}^{\pi/2} \log(\tan x + \cot x)dx$.

8. Evaluate $\int_{0}^{1} e^{\sin^{-1} x} dx$.

Hints to Selected Problems

1.
$$I = \int_0^{\pi/2} \frac{\sin^4 x \, dx}{\sin^4 x + \cos^4 x} \qquad \ldots(1)$$

$$= \int_0^{\pi/2} \frac{\sin^4\left(\frac{\pi}{2} - x\right)}{\sin^4\left(\frac{\pi}{2} - x\right) + \cos^4\left(\frac{\pi}{2} - x\right)} dx$$

$$I = \int_0^{\pi/2} \frac{\cos^4 x}{\sin^4 x + \cos^4 x} dx \qquad \ldots(2)$$

Adding (1) and (2) we get

$$2I = \int_0^{\pi/2} dx = \frac{\pi}{2} \quad \Rightarrow \quad I = \frac{\pi}{4}$$

2.
$$I = \int_0^{\pi/2} \frac{\sin x - \cos x}{1 + \cos x \sin x} dx \qquad \ldots(1)$$

$$= \int_0^{\pi/2} \frac{\sin\left(\frac{\pi}{2} - x\right) - \cos\left(\frac{\pi}{2} - x\right)}{1 + \cos\left(\frac{\pi}{2} - x\right)\sin\left(\frac{\pi}{2} - x\right)} dx$$

$$= \int_0^{\pi/2} \frac{\cos x - \sin x}{1 + \cos x \sin x} dx \qquad \ldots(2)$$

Adding (1) and (2) $2I = 0 \quad \Rightarrow \quad I = 0.$

3.
$$I = \int_0^4 f(x) dx = \int_0^3 (2x + 3) dx + \int_3^4 3x \, dx$$

$$= [x^2 + 3x]_0^3 + \left[\frac{3x^2}{2}\right]_3^4$$

$$= 18 + 24 - \frac{27}{2} = 42 - \frac{27}{2} = \frac{57}{2}.$$

4.
$$I = \int_0^{\pi/2} \frac{x \sin x \cos x \, dx}{\cos^4 x + \sin^4 x}$$

$$= \int_0^{\pi/2} \frac{\left(\frac{\pi}{2} - x\right)\sin\left(\frac{\pi}{2} - x\right)\cos\left(\frac{\pi}{2} - x\right)}{\cos^4\left(\frac{\pi}{2} - x\right) + \sin^4\left(\frac{\pi}{2} - x\right)} dx$$

$$= \int_0^{\pi/2} \frac{\pi}{2} \frac{\sin x \cos x}{\sin^4 x + \cos^4 x} dx$$

$$\quad - \int_0^{\pi/2} \frac{x \sin x \cos x}{\sin^4 x + \cos^4 x} dx$$

$$\Rightarrow \quad 2I = \frac{\pi}{2}\int_0^{\pi/2} \frac{\sin x \cos x}{\sin^4 x + \cos^4 x} dx$$
divide Nr and Dr by $\cos^4 x$

$$2I = \frac{\pi}{2}\int_0^{\pi/2} \frac{\tan x \sec^2 x}{1 + \tan^4 x} dx$$

Put $\tan^2 x = t \quad \Rightarrow \quad 2\tan x \sec^2 x \, dx = dt$

$$\Rightarrow \quad \tan x \sec^2 x \, dx = \frac{dt}{2}$$

$$\therefore \quad 2I = \frac{\pi}{2}\int_0^\infty \frac{1}{2} \frac{dt}{1 + t^2}$$

When $x = 0 \quad \Rightarrow \quad t = 0$
When $x = \pi/2 \quad \Rightarrow \quad t = \infty$

$$= \frac{\pi}{4}\left[\tan^{-1} t\right]_0^\infty = \frac{\pi}{4}\left(\frac{\pi}{2} - 0\right) = \frac{\pi^2}{8}.$$

So, $I = \frac{\pi^2}{16}.$

5.
$$I = \int_0^\pi \frac{x \sin x}{1 + \cos^2 x} dx \qquad \ldots(1)$$

$$= \int_0^\pi \frac{(\pi - x)\sin(\pi - x)}{1 + \cos^2(\pi - x)} dx \qquad \ldots(2)$$

Adding (1) and (2)

$$2I = \int_0^\pi \frac{(x + \pi - x)\sin x}{1 + \cos^2 x} dx = \pi\int_0^\pi \frac{\sin x}{1 + \cos^2 x} dx$$

Let $\cos x = t \quad \Rightarrow \quad \sin x \, dx = -dt$
$$\begin{aligned} x = 0 \quad &\Rightarrow \quad t = 1 \\ x = \pi \quad &\Rightarrow \quad t = -1 \end{aligned}$$

Now $2I = -\pi\int_1^{-1} \frac{dt}{1 + t^2} = -\pi\left[\tan^{-1} t\right]_1^{-1} = -\pi\left[-\frac{\pi}{4} - \frac{\pi}{4}\right]$

$$2I = \frac{\pi^2}{2} \quad \Rightarrow \quad I = \frac{\pi^2}{4}.$$

6.
$$I = \int_0^\pi \frac{x \, dx}{a^2 \cos^2 x + b^2 \sin^2 x} \qquad \ldots(1)$$

$$= \int_0^\pi \frac{(\pi - x) dx}{a^2 \cos^2(\pi - x) + b^2 \sin^2(\pi - x)}$$

$$= \int_0^\pi \frac{(\pi - x) dx}{a^2 \cos^2 x + b^2 \sin^2 x} \qquad \ldots(2)$$

Adding (1) and (2) we get

$$2I = \int_0^\pi \frac{(x + \pi - x) dx}{a^2 \cos^2 x + b^2 \sin^2 x}$$

$$= \pi\int_0^\pi \frac{dx}{a^2 \cos^2 x + b^2 \sin^2 x}$$

$$= 2\pi\int_0^{\pi/2} \frac{dx}{a^2 \cos^2 x + b^2 \sin^2 x}$$

$$\left[\int_0^{2a} f(x) dx = 2\int_0^a f(x) dx; f(2a - x) = f(x)\right]$$

Divide Nr and Dr by $\cos^2 x$

$$I = \pi\int_0^{\pi/2} \frac{\sec^2 x \, dx}{a^2 + b^2 \tan^2 x}$$

Put $\tan x = t$
$$\sec^2 x \, dx = dt$$

When $\begin{aligned} x = 0 \quad &\Rightarrow \quad t = 0 \\ x = \pi/2 \quad &\Rightarrow \quad t = \infty \end{aligned}$

Then, $I = \pi\int_0^\infty \frac{dt}{a^2 + b^2 t^2} = \frac{\pi}{b^2}\int_0^\infty \frac{dt}{\left(\frac{a}{b}\right)^2 + t^2}$

$$= \frac{\pi}{b^2} \frac{1}{a/b}\left[\tan^{-1} \frac{t}{a/b}\right]_0^\infty$$

$$I = \frac{\pi}{ab}\left[\tan^{-1} \frac{bt}{a}\right]_0^\infty = \frac{\pi}{ab}\left(\frac{\pi}{2} - 0\right) = \frac{\pi^2}{2ab}$$

7.
$$I = \int_0^{\pi/2} \log(\tan x + \cot x) dx$$

$$= \int_0^{\pi/2} \log\left(\frac{\sin^2 x + \cos^2 x}{\sin x \cos x}\right) dx$$

$$= \int_0^{\pi/2} \log\left(\frac{1}{\sin x \cos x}\right) dx$$

$$= \int_0^{\pi/2} -(\log \cos x + \log \sin x) dx$$

$$= -\left[\int_0^{\pi/2} \log \sin x\, dx + \int_0^{\pi/2} \log \cos x\, dx\right]$$

$$= -2\int_0^{\pi/2} \log \sin x\, dx$$

$$\left[\because \int_0^{\pi/2} \log \sin x\, dx = \int_0^{\pi/2} \log \cos x\, dx = -\frac{\pi}{2}\log 2\right]$$

$$= -2\left(-\frac{\pi}{2}\log 2\right) = \pi \log 2 .$$

8. $\qquad I = \int_0^1 e^{\sin^{-1} x}\, dx$

Put $\sin^{-1} x = t \Rightarrow \sin t = x \Rightarrow \cos t\, dt = dx$

When $\qquad x = 0 \Rightarrow t = 0$

$x = 1 \Rightarrow t = \pi/2$

$I = \int_0^{\pi/2} e^t \cos t\, dt$

$$= \left[\frac{1}{1^2 + t^2}[e^t \cos t + e^t \sin t]\right]_0^{\pi/2}$$

$$= \frac{1}{2}[e^{\pi/2}(0+1) - 1]$$

$$\left[\text{Since } \int_0^{\pi/2} e^{ax} \cos bx\, dx\right.$$

$$\left. = \frac{1}{a^2 + b^2}[ae^{ax}\cos bx + be^{ax}\sin bx = \frac{1}{2}(e^{\pi/2}-1)]\right]$$

Answers

1. $\dfrac{\pi}{4}$ $\qquad$ **2.** 0 $\qquad$ **3.** $\dfrac{57}{2}$ $\qquad$ **4.** $\dfrac{\pi^2}{16}$ $\qquad$ **5.** $\dfrac{\pi^2}{4}$ $\qquad$ **7.** $\pi \log 2$ $\qquad$ **8.** $\dfrac{e^{\pi/2}-1}{2}$

4.2.2 DEFINITE INTEGRAL AS THE LIMIT OF SUM

It is always possible to regard a definite integral as the limit of the sum of certain number of terms, when the number of terms tends to infinity and each term tends to zero.

Here, we define the definite integral as follows :

$$\int_0^a f(x)dx = \lim h[f(a) + f(a+h) + f(a+2h) + ... + f\{a+(n-1)h\}]$$

when $n \to \infty$, $h \to 0$ and $nh \to b - a$.

Solved Examples

EXAMPLE 1. *Evaluate $\int_a^b x^2 dx$, directly from the definition of integral as the limit of a sum.*

SOLUTION. We know that

$$\int_0^a f(x)dx = \lim_{n \to \infty} h[f(a) + f(a+h) + f(a+2h) +$$

$$... + f\{a+(n-1)h\}] \qquad ...(1)$$

Here $f(x) = x^2 \qquad f(a) = a^2$

$\qquad\qquad\qquad f(a+h) = (a+h)^2$ and so on.

Put all these values in (1), we get

$$\int_a^b x^2 dx = \lim_{n \to \infty} h[a^2 + (a+h)^2 + (a+2h)^2 + ... + \{a+(n-1)h\}^2]$$

when $h \to 0$, $n \to \infty$ and $nh \to b - a$.

$$= \lim_{n \to \infty} h[na^2 + 2ah\{1+2+3+...+(n-1)\}$$

$$+ h^2[1^2 + 2^2 + ... + (n-1)^2]$$

Using $\Sigma n = \dfrac{n(n+1)}{2}$ and $\Sigma n^2 = \dfrac{n(n+1)(2n+1)}{6}$

$$\therefore \int_a^b x^2 dx = \lim_{n \to \infty} h\left[na^2 + 2ah\frac{(n-1)n}{2} + \frac{h^2}{6}(n-1)n(2n-1)\right]$$

$$= \lim_{n \to \infty}\left[(nh)a^2 + a(nh)(n-1)h\right.$$

$$\left. + \frac{1}{6}(nh)(n-1)h(2n-1)h\right]$$

$$= \lim_{n \to \infty}\left[(nh)a^2 + a(nh)^2\left(1 - \frac{1}{n}\right)\right.$$

$$\left. + \frac{1}{6}2(nh)^3\left(1 - \frac{1}{n}\right)\left(1 - \frac{1}{2n}\right)\right]$$

$$= (b-a)a^2 + a(b-a)^2 + \frac{1}{3}(b-a)^3$$

$$(\because \text{as } n \to \infty, h \to 0, nh \to b - a)$$

$$= \frac{1}{3}(b-a)[3a^2 + 3(b-a)a + b^2 - 2ab + a^2]$$

$$= \frac{1}{3}(b-a)(a^2 + ab + b^2) = \frac{1}{3}(b^3 - a^3).$$

EXAMPLE 2. *From the definition of a definite integral as the limit of a sum, evaluate $\int_a^b e^x dx$.*

SOLUTION. Here we have $f(x) = e^x$

Therefore, $\qquad f(a) = e^a$

$\qquad\qquad\qquad f(a+h) = e^{a+h}$

$\qquad\qquad\qquad ...\quad ...\quad ...\quad ...$ etc.

Now $\int_a^b e^x dx = \lim_{h \to 0} h[e^a + e^{a+h} + e^{a+2h} + ... + e^{a+(n-1)h}]$

where, $nh \to b - a$ and $n \to \infty$ as $h \to 0$.

$$= \lim_{h \to \infty} he^a[1 + e^h + e^{2h} + ... + e^{(n-1).h}]$$

$$= \lim_{h \to 0} he^a\left\{\frac{(e^h)^n - 1}{e^h - 1}\right\} = \lim_{h \to 0} he^a\left\{\frac{e^{nh} - 1}{e^h - 1}\right\}$$

$$= \lim_{h \to 0} he^a\left[\frac{e^{b-a} - 1}{e^h - 1}\right] = \lim_{h \to 0} e^a\left[\frac{e^{b-a} - 1}{\dfrac{e^h - 1}{h}}\right]$$

$$= e^b - e^a \qquad \left(\because \lim_{h \to 0}\frac{e^h - 1}{h} = 1\right)$$

4.2.3 SUMMATION OF SERIES WITH THE HELP OF DEFINITE INTEGRAL

We know that $\qquad \int_a^b f(x)dx = \lim_{n \to \infty} h[f(a) + f(a+h) + f(a+2h) + ... + f\{a+(n-1)h\}]$

$$= \lim_{n \to \infty} n \sum_{r=0}^{n-1} f(a+rh). \qquad\qquad \text{where, } nh = b - a$$

Now putting $a = 0$ and $b = 1$ so that $h = 1/n$, we get

$$\int_0^1 f(x)dx = \lim_{n \to \infty} \frac{1}{n} \sum_{r=0}^{n-1} f\left(\frac{r}{n}\right).$$

WORKING PROCEDURE

Step 1. Write the rth terms of the series.

Step 2. Write the rth term in the form of $\frac{1}{n} f\left(\frac{r}{n}\right)$.

Step 3. Replace $\frac{r}{n}$ by x, $\frac{1}{n}$ by dx and $\lim_{x \to \infty}$ by $\int$.

Then lower limit of the definite integral will be the value of $\frac{r}{n}$ for the first term as $n \to \infty$ and the upper limit will be the value of $\frac{r}{n}$ for the last term as $n \to \infty$.

Solved Examples

EXAMPLE 1. Evaluate the following

$$\lim_{\to \infty}\left[\frac{1}{n+1} + \frac{1}{n+2} + \dots + \frac{1}{2n}\right].$$

SOLUTION. The general term r^{th} term $= \frac{1}{n+r}$.

We have to find $\lim_{n \to \infty} \sum_{r=1}^{n} \frac{1}{n+r} = \lim_{n \to \infty} \sum_{r=1}^{n} \frac{1}{n[1+r/n]}$.

$$= \lim_{n \to \infty} \frac{1}{n} \sum_{r=1}^{n} \frac{1}{1+(r/n)}$$

Since the limit of r in the summation are 1 to n, therefore, the lower limit of integration

$$= \lim_{n \to \infty} \frac{1}{n} = 0.$$

Also, the upper limit of integration $= \lim_{n \to \infty} \frac{n}{n} = 1$.
Hence, the required limit

$$\int_0^1 \frac{1}{1+x}dx = \left[\log(1+x)\right]_0^1 = \log 2.$$

EXAMPLE 2. Evaluate

$$\lim_{n \to \infty}\left[\left(1+\frac{1}{n^2}\right)\left(1+\frac{2^2}{n^2}\right)\left(1+\frac{3^2}{n^2}\right)\dots\left(1+\frac{n^2}{n^2}\right)\right]^{\frac{1}{n}}$$

(BHOPAL–2008)

SOLUTION. Let $A = \lim_{n \to \infty}\left[\left(1+\frac{1}{n^2}\right)\left(1+\frac{2^2}{n^2}\right)\left(1+\frac{3^2}{n^2}\right)\dots\left(1+\frac{n^2}{n^2}\right)\right]^{\frac{1}{n}}$

$\log A = \lim_{n \to \infty} \frac{1}{n}\left[\log\left(1+\frac{1}{n^2}\right) + \log\left(1+\frac{2^2}{n^2}\right)\right.$

$\left. + \log\left(1+\frac{3^2}{n^2}\right) + \dots + \log\left(1+\frac{n^2}{n^2}\right)\right]$

$= \lim_{n \to \infty} \frac{1}{n} \sum_{r=1}^{n} \log\left(1+\frac{r^2}{n^2}\right)$

$= \int_0^1 \log(1+x^2)dx = \int_0^1 \log(1+x^2)1.dx$

$= \left[x\log(1+x^2)\right]_0^1 - \int_0^1 \frac{2x.xdx}{1+x^2}$

$= \log 2 - 2\int_0^1 \frac{(1+x^2)-1}{1+x^2}dx$

$= \log 2 - 2\int_0^1\left[1 - \frac{1}{(1+x^2)}\right]dx$

$= \log 2 - 2\left[x - \tan^{-1}x\right]_0^1$

$= \log 2 - 2\left(1 - \frac{\pi}{4}\right)$

Therefore, $\log A = \log 2 + \frac{1}{2}(\pi - 4)$

$\Rightarrow \quad \log \frac{A}{2} = \frac{1}{2}(\pi - 4)$

$\Rightarrow \quad A = 2e^{(\pi-4)/2}$.

EXAMPLE 3. Find the limit of $\left[\frac{n!}{n^n}\right]^{1/n}$ **when** $n \to \infty$.

SOLUTION. Let $A = \lim_{n \to \infty}\left[\frac{n!}{n^n}\right]^{1/n}$

$= \lim_{n \to \infty}\left[\frac{1.2.3.4\dots n}{n.n.n\dots n}\right]^{1/n}$

$\Rightarrow \quad \log A = \lim_{n \to \infty} \frac{1}{n}\left[\log\left(\frac{1}{n}\right) + \log\left(\frac{2}{n}\right)\right.$

$\left. + \log\left(\frac{3}{n}\right) + \dots + \log\left(\frac{n}{n}\right)\right]$

$= \lim_{n \to \infty} \sum_{r=1}^{n} \frac{1}{n}\log\left(\frac{r}{n}\right)$

$= \int_0^1 \log x dx = \int_0^1 \log x.1 dx$

$= \left[(\log x).x\right]_0^1 - \int_0^1 \frac{1}{x}.xdx$

$= 0 - [x]_0^1 = -1$

Hence, $A = e^{-1} = \frac{1}{e}$.

Exercise-4.3

1. Show that the limit of the sum $\frac{1}{n} + \frac{1}{n+1} + \frac{1}{n+2} + \dots + \frac{1}{6n}$ when n is indefinitely increased is $\log 6$.

2. Evaluate $\int_a^b x^2 dx$ directly from the definition of the integral as the limit of the sum.

3. Evaluate by summation $\int_1^2 x\,dx$.

4. Evaluate by summation $\int_a^b \sin x\,dx$.

5. Evaluate by summation $\int_0^{\pi/2} \sin x\,dx$.

6. Show that the limit (when $n \to \infty$) of the series

$$\frac{n}{(n+1)^2}+\frac{n}{(n+2)^2}+...+\frac{n}{(n+n)^2} \text{ is } \frac{1}{2}.$$

7. Show that

$$\lim_{n\to\infty}\left[\frac{n}{n^2}+\frac{n}{n^2+1^2}+\frac{n}{n^2+2^2}+...+\frac{n}{n^2+(n+1)^2}\right]=\frac{\pi}{4}.$$

8. Show that $\displaystyle\lim_{n\to\infty}\left[\frac{n}{n^2+1^2}+\frac{n}{n^2+2^2}+...+\frac{1}{2n}\right]=\frac{\pi}{4}.$

9. Show that $\displaystyle\lim_{n\to\infty}\left[\frac{1}{n^3}(1+4+9+...+n^2)\right]=\frac{1}{3}.$

10. Show that $\displaystyle\lim_{n\to\infty}\left[\frac{1}{n}+\frac{n^2}{(n+1)^3}+\frac{n^2}{(n+2)^2}+...+\frac{1}{8n}\right]=\frac{3}{8}.$

11. Show that

$$\lim_{\to\infty}\left[\frac{1}{\sqrt{n^2-1^2}}+\frac{1}{\sqrt{n^2-2^2}}+...+\frac{1}{\sqrt{n^2-(n-1)^2}}\right]=-.$$

12. Show that $\displaystyle\lim_{n\to\infty}\left[\frac{1}{n^2}\sec^2\frac{1}{n^2}+\frac{2}{n^2}\sec^2\frac{4}{n^2}\right.$

$$\left.+\frac{3}{n^2}\sec^2\frac{9}{n^2}+...+\frac{1}{n}\sec^2 1\right]=\frac{1}{2}\tan 1.$$

13. Show that

$$\lim_{n\to\infty}\left[\frac{1}{\sqrt{n^2-1^2}}+\frac{1}{\sqrt{n^2-2^2}}+...+\frac{1}{\sqrt{n^2-(n-1)^2}}\right]=\frac{\pi}{2}.$$

14. Show that

$$\lim_{n\to\infty}\left[\frac{n^{1/2}}{n^{3/2}}+\frac{n^{1/2}}{(n+3)^{3/2}}+\frac{n^{1/2}}{(n+6)^{3/2}}+...+\frac{n^{1/2}}{[n+3(n+1)]^{3/2}}\right]=\frac{1}{3}.$$

Hints to Selected Problems

1. $\displaystyle\lim_{n\to\infty}\left[\frac{1}{n}+\frac{1}{n+1}+\frac{1}{n+2}+...+\frac{1}{6n}\right]=\lim_{n\to\infty}\sum_{r=0}^{5n}\left[\frac{1}{n+r}\right]$

$$=\lim_{n\to\infty}\frac{1}{n}\sum_{r=0}^{5n}\left[\frac{1}{1+\frac{r}{n}}\right]=\int_0^5\frac{1}{1+x}dx$$

$$=\left[\log(1+x)\right]_0^5=\log 6-\log 1=\log 6$$

4. Let $I=\int_a^b\sin x\,dx$. Here $f(x)=\sin x$

Let $h=\dfrac{b-a}{n}, n\in N$

$I=\lim_{n\to\infty}h[f(a)+f(a+h)+...+f(a+(n-1)h)]$

$=\lim_{n\to\infty}h[\sin a+\sin(a+h)+...+\sin(a+(n-1)h)]$

Now $\sin a+\sin(a+h)+...+\sin(a+(n-1)h)$

$=\dfrac{1}{2\sin h/2}\left[2\sin a\sin\dfrac{h}{2}+2\sin(a+h)\cdot\sin\dfrac{h}{2}\right.$

$$\left.+...+2\sin(a+(n-1)\sin\dfrac{h}{2}\right]$$

$=\dfrac{1}{2\sin h/2}\left[\left\{\cos\left(a-\dfrac{h}{2}\right)-\cos\left(a+\dfrac{h}{2}\right)\right\}\right.$

$$+\left\{\cos\left(a+\dfrac{h}{2}\right)-\cos\left(a+\dfrac{3h}{2}\right)\right\}$$

$$\left.+...+\left\{\cos\left(a+\left(n-\dfrac{3}{2}\right)h\right)-\cos\left(a+\left(n-\dfrac{1}{2}\right)h\right)\right\}\right]$$

$=\dfrac{1}{2\sin h/2}\left[\cos\left(a-\dfrac{h}{2}\right)-\cos\left(a+\left(n-\dfrac{1}{2}\right)h\right)\right]$

$=\dfrac{1}{2\sin h/2}\left[\cos\left(a-\dfrac{h}{2}\right)-\cos\left(b-\dfrac{h}{2}\right)\right]$

$b=a+nh$

$I=\lim_{h\to 0}h\cdot\dfrac{1}{2\sin h/2}\left[\cos\left(a-\dfrac{h}{2}\right)-\cos\left(b-\dfrac{h}{2}\right)\right]$

$=\lim_{h\to 0}\dfrac{h/2}{\sin h/2}\lim_{h\to 0}\left[\cos\left(a-\dfrac{h}{2}\right)-\cos\left(b-\dfrac{h}{2}\right)\right]$

$=1.[\cos(a-0)-\cos(b-0)]$

$=\cos a-\cos b$

6. $\displaystyle\lim_{n\to\infty}\left[\frac{n}{(n+1)^2}+\frac{n}{(n+2)^2}+\frac{n}{(n+3)^2}+...+\frac{n}{(n+n)^2}\right]$

$=\lim_{n\to\infty}\sum_{r=0}^{n+1}\frac{n}{(n+r)^2}=\lim_{n\to\infty}\frac{1}{n}\sum_{r=0}^{n+1}\frac{1}{\left(1+\frac{r}{n}\right)^2}$

$=\int_0^1\frac{1}{(1+x)^2}dx=\left[-\frac{1}{1+x}\right]_0^1=\frac{1}{2}$

7. $\displaystyle\lim_{n\to\infty}\left[\frac{n}{n^2}+\frac{n}{n^2+1^2}+\frac{n}{n^2+2^2}+...+\frac{n}{n^2+(n+1)^2}\right]$

$=\lim_{n\to\infty}\sum_{r=0}^{n}\frac{n}{n^2+r^2}=\lim_{n\to\infty}\frac{1}{n}\sum_{r=0}^{n+1}\frac{1}{1+\left(\frac{r}{n}\right)^2}$

$=\int_0^1\frac{1}{1+x^2}dx=\left(\tan^{-1}x\right)_0^1=\tan^{-1}1-\tan^{-1}0=\pi/4.$

9. $\displaystyle\lim_{n\to\infty}\left[\frac{1}{n^3}(1+4+9+...+n^2)\right]$

$=\lim_{n\to\infty}\left[\frac{1}{n^3}(1^2+2^2+3^2+...+n^2)\right]=\lim_{n\to\infty}\sum_{r=0}^{n}\frac{1}{n^3}r^2$

$=\lim_{n\to\infty}\frac{1}{n}\sum_{r=0}^{n}\left(\frac{r}{n}\right)^2=\int_0^1 x^2dx=\left(\frac{x^3}{3}\right)_0^1=\frac{1}{3}.$

11. $\displaystyle\lim_{n\to\infty}\left[\frac{1}{n}+\frac{1}{\sqrt{n^2-1^2}}+\frac{1}{\sqrt{n^2-2^2}}+...+\frac{1}{\sqrt{n^2-(n-1)^2}}\right]$

$=\lim_{n\to\infty}\sum_{r=0}^{n-1}\frac{1}{\sqrt{n^2-r^2}}=\lim_{n\to\infty}\frac{1}{n}\sum_{r=0}^{n-1}\frac{1}{\sqrt{1-\left(\frac{r}{n}\right)^2}}$

$=\int_0^1\frac{1}{\sqrt{1-x^2}}dx=\left(\sin^{-1}x\right)_0^1$

$=\sin^{-1}(1)-\sin^{-1}(0)=\pi/2.$

12. $\lim_{n\to\infty}\left[\dfrac{1}{n^2}\sec^2\dfrac{1}{n^2}+\dfrac{2}{n^2}\sec^2\dfrac{4}{n^2}+\dfrac{3}{n^2}\sec^2\dfrac{9}{n^2}+\ldots+\dfrac{1}{n}\sec^2 1\right]$

$= \lim_{n\to\infty}\sum_{r=0}^{n}\dfrac{r}{n^2}\sec^2\dfrac{r^2}{n^2}=\lim_{n\to\infty}\dfrac{1}{n}\sum_{r=0}^{n}\left(\dfrac{r}{n}\right)\sec^2\left(\dfrac{r}{n}\right)^2$

$= \int_0^1 x\sec^2 x\,dx$

$= \dfrac{1}{2}\int_0^1\sec^2 t\,dt=\dfrac{1}{2}(\tan t)_0^1=\dfrac{1}{2}\tan 1$

$\left(\text{Let } x^2=t\Rightarrow 2x\,dx=dt\right)$

14. $\lim_{n\to\infty}\left[\dfrac{n^{1/2}}{n^{3/2}}+\dfrac{n^{1/2}}{(n+3)^{3/2}}+\dfrac{n^{1/2}}{(n+6)^{3/2}}+\ldots+\dfrac{n^{1/2}}{\{n+3(n+1)\}^{3/2}}\right]$

$= \lim_{n\to\infty}\sum_{r=0}^{n}\dfrac{n^{1/2}}{(n+3r)^{3/2}}=\lim_{n\to\infty}\dfrac{1}{n}\sum_{r=0}^{n}\dfrac{1}{\left(1+\dfrac{3r}{n}\right)^{3/2}}$

$= \int_0^1\dfrac{1}{(1+3x)^{3/2}}dx=\int_0^1(1+3x)^{-3/2}dx$

$= \left[\dfrac{(1+3x)^{-1/2}}{(-1/2)\times 3}\right]_0^1=-\dfrac{2}{3}\left[(1+3x)^{-1/2}\right]_0^1$

$= -\dfrac{2}{3}[4^{-1/2}-1^{-1/2}]=-\dfrac{2}{3}\left[\dfrac{1}{2}-1\right]=-\dfrac{2}{3}\left[-\dfrac{1}{2}\right]=\dfrac{1}{3}$.

Answers

1. $\dfrac{1}{3}(b^3-a^3)$ **3.** $\dfrac{3}{2}$ **4.** $\cos b-\cos a$ **5.** 1

4.3 RECTIFICATION OF CURVES

Rectification is a process for finding the length of an arc of a plane curve between two given points on a curve.

4.3.1 FORMULAE FOR FINDING THE LENGTH OF THE CURVES

(a) Let the equation of a curve be $y=f(x)$ and let A and B be two points on this curve. Between A and B, the length of curve is to be required. Let s be the length of an arc from a fixed point on the curve to any point on it.

Therefore, we have $\dfrac{ds}{dx}=\pm\sqrt{\left[1+\left(\dfrac{dy}{dx}\right)^2\right]}$ or $ds=\pm\sqrt{\left[1+\left(\dfrac{dy}{dx}\right)^2\right]}dx$

where positive and negative sign will have to take according as x increases and decreases as s increases. Thus, the length of an arc between the points A and B where at A, $x=a$ and at B, $x=b$ is given by

$$s=\int_a^b\sqrt{\left[1+\left(\dfrac{dy}{dx}\right)^2\right]}dx \qquad\qquad (a<b)$$

(b) If the equation of the curve is $x=f(y)$, then the length of an arc between c and d is given by

$$s=\int_c^d\sqrt{\left[1+\left(\dfrac{dx}{dy}\right)^2\right]}dy \qquad\qquad (c<d)$$

(c) If the equation of the curve is in parametric form, *i.e.*, $x=f(t), y=g(t)$, then we have

$$\dfrac{ds}{dt}=\sqrt{\left[\left(\dfrac{dx}{dt}\right)^2+\left(\dfrac{dy}{dt}\right)^2\right]} \quad\text{or}\quad ds=\sqrt{\left(\dfrac{dx}{dt}\right)^2+\left(\dfrac{dy}{dt}\right)^2}\,dt$$

Thus the length of an arc between A and B where at pt. A, $t=t_1$ and at pt. B, $t=t_2$ is given by

$$s=\int_{t_1}^{t_2}\sqrt{\left[\left(\dfrac{dx}{dt}\right)^2+\left(\dfrac{dy}{dt}\right)^2\right]}dt$$

(d) If the equation of the curve is $r=f(\theta)$ (in polar form), then

$$\dfrac{ds}{d\theta}=\sqrt{\left[r^2+\left(\dfrac{dr}{d\theta}\right)^2\right]} \quad\text{or}\quad ds=\sqrt{r^2+\left(\dfrac{dr}{d\theta}\right)^2}\,d\theta$$

and s is measured in the direction of θ increasing. Let at point A, $\theta=\theta_1$ and at B, $\theta=\theta_2$. Therefore, the length of an arc between A and B is given by

$$s=\int_{\theta_1}^{\theta_2}\sqrt{\left[r^2+\left(\dfrac{dr}{d\theta}\right)^2\right]}d\theta$$

(e) If the equation of the curve is $\theta=f(r)$, then the length of an arc between A and B is given by

$$s=\int_{r_1}^{r_2}\sqrt{\left[1+\left(r\dfrac{d\theta}{dr}\right)^2\right]}dr$$

(f) If the equaton of the curve is in pedal form *i.e.*, $p = f(r)$. Since we know that

$$\frac{ds}{dr} = \frac{r}{\sqrt{(r^2 - p^2)}}$$

Thus the length of an arc between $A(r = r_1)$ and $B(r = r_2)$ is given by $s = \int_{r_1}^{r_2} \frac{rdr}{\sqrt{(r^2 - p^2)}}$

 REMARK

- If the curve is symmetrical about some lines, then in order to find the length of an arc, we first find the length of one of the symmetrical part and multiply this length by the number of symmetrical parts.

Solved Examples

EXAMPLE 1. *Find the length of the arc of the parabola* $y^2 = 4ax$ *cut off by its latus rectum.*
(VTU–2008, MUMBAI–2006)

SOLUTION. We know that line which passes through the focus of the given parabola and perpendicular to the axis of that parabola is called its latus rectum..

Here the equation of the parabola is $y^2 = 4ax$ whose trace is shown above in the figure, LL' is the latus rectum, the co-ordinates of L and L' are respectively $(a, 2a)$ and $(a, -2a)$. Since

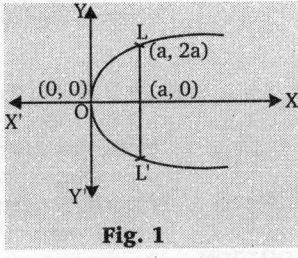

Fig. 1

$y^2 = 4ax$ is symmetrical about the line OX. Therefore the required arc length

$$= 2 \times \text{arc length } OL.$$

Since $y^2 = 4ax$.

$\therefore \qquad y = 2\sqrt{a}\sqrt{x}$

$\therefore \qquad \frac{dy}{dx} = \frac{\sqrt{a}}{\sqrt{x}}.$

Now arc length

$$OL = \int_0^a \sqrt{1 + \left(\frac{dy}{dx}\right)^2} \, dx$$

$(\because$ At point O, $x = 0$ and at point L, $x = a)$

$$= \int_0^a \sqrt{\left(1 + \frac{a}{x}\right)} \, dx = \int_0^a \frac{\sqrt{x+a}}{\sqrt{x}} \, dx$$

$$= \int_0^a \frac{x+a}{\sqrt{x^2+ax}} \, dx = \frac{1}{2}\int_0^a \frac{2x+2a}{\sqrt{(x^2+ax)}} \, dx$$

$$= \frac{1}{2}\int_0^a \frac{(2x+a)dx}{\sqrt{x^2+ax}} + \frac{a}{2}\int_0^a \frac{dx}{\sqrt{(x^2+ax)}}$$

$$= \frac{1}{2}\int_0^a \frac{(2x+a)dx}{\sqrt{x^2+ax}} + \frac{a}{2}\int_0^a \frac{dx}{\sqrt{\left[\left(x+\frac{a}{2}\right)^2 - \left(\frac{a}{2}\right)^2\right]}}$$

$$= \frac{1}{2}\left(2\sqrt{x^2+ax}\right)_0^a + \frac{a}{2}\left[\log\left\{\left(x+\frac{a}{2}\right)+\sqrt{x^2+ax}\right\}\right]_0^a$$

$$= a\sqrt{2} + \frac{a}{2}\log(3+2\sqrt{2}) = a\sqrt{2} + \frac{a}{2}\log(1+\sqrt{2})^2$$

Arc length $OL = a\sqrt{2} + a\log(1+\sqrt{2})$.

Hence the required arc length

$$= 2 \times \text{arc length } OL$$

$$= 2\sqrt{2}a + 2a\log(1+\sqrt{2}).$$

EXAMPLE 2. *Find the length of the curve* $y = \log\sec x$ *between the points* $x = 0$ *and* $x = \pi/3$. (VTU-2010, PTU-2007)

SOLUTION. Since the equation of the curve is

$$y = \log\sec x$$

$$\therefore \frac{dy}{dx} = \frac{1}{\sec x}.\sec x \tan x = \tan x.$$

Now $\sqrt{\left[1 + \left(\frac{dy}{dx}\right)^2\right]} = \sqrt{(1+\tan^2 x)} = \sqrt{\sec^2 x} = \sec x.$

Therefore the length of the given curve between $x = 0$ and $x = \pi/3$ is

$$s = \int_0^{\pi/3} \sqrt{\left[1 + \left(\frac{dy}{dx}\right)^2\right]} dx = \int_0^{\pi/3} \sec x \, dx$$

$$= \left[\log\left\{\tan\left(\frac{\pi}{4} + \frac{x}{2}\right)\right\}\right]_0^{\pi/3}$$

$$s = \log\left[\tan\left(\frac{\pi}{4} + \frac{\pi}{6}\right)\right] - \log\left[\tan\left(\frac{\pi}{4}\right)\right]$$

$$= \log\left[\frac{\tan\frac{\pi}{4} + \tan\frac{\pi}{6}}{1 - \tan\frac{\pi}{4}\tan\frac{\pi}{6}}\right] - \log 1$$

$$\left[\because \tan\left(\frac{\pi}{4} + \theta\right) = \frac{1 + \tan\theta}{1 - \tan\theta}\right]$$

$$= \log\left[\frac{1 + 1/\sqrt{3}}{1 - 1/\sqrt{3}}\right] - 0 \qquad (\because \log 1 = 0)$$

$$= \log\left(\frac{\sqrt{3}+1}{\sqrt{3}-1}\right)$$

Hence, $s = \log(2+\sqrt{3})$

EXAMPLE 3. *Find the whole length of the astroid* $x^{2/3} + y^{2/3} = a^{2/3}.$

or $x = a\cos^3\theta, y = a\sin^3\theta$
(VTU–2010, RAJASTHAN–2006, MARATHWADA–2008)

SOLUTION. The equation of the curve is

$$x^{2/3} + y^{2/3} = a^{2/3} \qquad \dots(1)$$

Since the curve is symmetrical about both the axis, *i.e.*, curve lies in all the four quadrants as shown in fig. 2. Therefore the whole length of the given curve

$$= 4 \times \text{arc length of the curve in first quadrant.}$$

Since the co-ordinates of A and B are $(a, 0)$ and $(0, a)$ respectively. Thus in the first quadrant x varies from 0 to a.

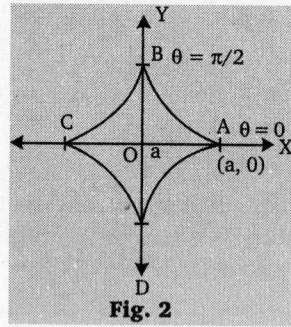

Fig. 2

Now differentiating (1) *w.r.t. x*, we get

$$\frac{2}{3}x^{-1/3} + \frac{2}{3}y^{-1/3}\frac{dy}{dx} = 0$$

or

$$\frac{dy}{dx} = -\left(\frac{y}{x}\right)^{1/3}$$

Therefore, the length of the curve in the first quadrant

$$= \int_0^a \sqrt{\left[1 + \left(\frac{dy}{dx}\right)^2\right]}\,dx = \int_0^a \sqrt{\left(1 + \frac{y^{2/3}}{x^{2/3}}\right)}\,dx$$

$$= \int_0^a \sqrt{\left(\frac{x^{2/3} + y^{2/3}}{x^{2/3}}\right)}\,dx = \int_0^a \sqrt{\left(\frac{a^{2/3}}{x^{2/3}}\right)}\,dx$$

[Using (1)]

$$= a^{1/3}\int_0^a x^{-1/3}\,dx = a^{1/3}\left[\frac{3}{2}x^{2/3}\right]_0^a$$

$$= a^{1/3}\left[\frac{3}{2}a^{2/3}\right] = \frac{3}{2}a$$

Hence, the whole length of the astroid

$$= 4 \times \frac{3a}{2} = 6a$$

EXAMPLE 4. *Find the entire length of the cardioid*
$$r = a(1 + \cos\theta).$$

(PTU–2010, KURUKSHETRA–2005, BHOPAL–2008, BHILLAI–2005)

SOLUTION. From the equation of the cardioid $r = a(1 + \cos\theta)$ it is obvious, that the curve is symmetrical about the initial line (x-axis) and r will become zero, when $\theta = \pi$ and maximum, *i.e.*, $r = 2a$, when $\theta = 0$. Thus r varies from 0 to $2a$. As θ varies from π to 0. Therefore curve is shown in the fig.3.

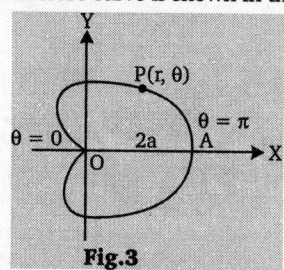

Fig.3

Let s be the arc length from O to any point $P(r, \theta)$.

$\therefore$ Entire length of the cardioid

$= 2 \times$ Arc length of the upper half of the cardioid.

Now the arc length of the upper half

$$= \int_\pi^0 \sqrt{\left[r^2 + \left(\frac{dr}{d\theta}\right)^2\right]}\,d\theta \qquad \ldots(1)$$

Since we have
$$r = a(1 + \cos\theta).$$

$\therefore$
$$\frac{dr}{d\theta} = -a\sin\theta$$

and
$$\frac{ds}{d\theta} = \sqrt{\left[r^2 + \left(\frac{dr}{d\theta}\right)^2\right]}$$

$$= \sqrt{a^2(1 + \cos\theta)^2 + a^2\sin^2\theta}$$

$$= \sqrt{[2a^2(1 + \cos\theta)]} = \sqrt{\left[2a^2\left(2\cos^2\frac{\theta}{2}\right)\right]}$$

$$\frac{ds}{d\theta} = 2a\cos\theta/2.$$

Here we have to measure the arc length 's' from the cusp O where $\theta = \pi$ to any point $P(r, \theta)$ in the direction of θ decreasing, then the arc length 's' increases as θ decreases. Therefore, we will take $\frac{ds}{d\theta}$ to be negative. Thus from (1), we obtain

$$s = \int_\pi^0 \left(-2a\cos\frac{\theta}{2}\right)d\theta = -2a\int_\pi^0 \cos\theta/2\,d\theta$$

$$= -2a\left[2\sin\frac{\theta}{2}\right]_\pi^0 = -4a\left[0 - \sin\frac{\pi}{2}\right] = 4a.$$

Hence, the entire length of the given cardioid
$$= 2s = 8a.$$

EXAMPLE 5. *Find the length of an arc of the curve*
$$x = a(t + \sin t),\ y = a(1 - \cos t) \quad \text{(PTU–2009, VTU–2004)}$$

SOLUTION. The given equation of the curve is
$$x = a(t + \sin t),\ y = a(1 - \cos t).$$

$\therefore$
$$\frac{dx}{dt} = a(1 + \cos t),\ \frac{dy}{dt} = a\sin t.$$

Since,
$$\left(\frac{ds}{dt}\right)^2 = \left(\frac{dx}{dt}\right)^2 + \left(\frac{dy}{dt}\right)^2$$

$$= a^2(1 + \cos t)^2 + a^2\sin^2 t$$

$$= 2a^2(1 + \cos t)$$

$\therefore$
$$\left(\frac{ds}{dt}\right)^2 = 4a^2\cos^2 t/2$$

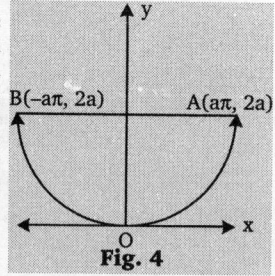

Fig. 4

From the equation of the curve it is obvious when $\cos t = 1$, $y = 0$ *i.e.*, when $t = 0$, $y = 0$. This implies when $t = 0$, $x = 0$ and $y = 0$. Therefore the curve will passes through the point $(0, 0)$ and at the point $(0,0)$ tangent is the x-axis. Also y will always positive. Thus the tracing of the given curve is shown in fig. 4.

From the fig. 4 it is obvious that the curve is symmetrical about the y-axis. Therefore, the entire length of the arc of the given curve will be twice of the arc OA. At O, $t = 0$ and at A, $t = \pi$.

Let s be the arc length measured from O to any point P on the curve towards the point A, then s increases as t increases so that $\frac{ds}{dt}$ will be taken positive.

$\therefore$ Arc length OA

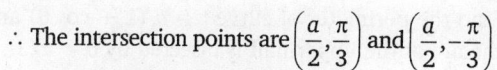

$$\int_0^\pi \frac{ds}{dt}.dt = \int_0^\pi 2a\cos t/2\,dt \qquad \text{[Using(1)]}$$

$$= 2a\left[2\sin t/2\right]_0^\pi = 4a .$$

Hence, the entire length of the given curve

$$= 2 \times \text{arc } OA = 2 \times 4a = 8a.$$

EXAMPLE 6. *Find the length of the arc of the equiangular spiral $r = ae^{\theta \cot \alpha}$ between the points for which radii vectors are r_1 and r_2.*

SOLUTION . Since we have $\quad r = ae^{\theta \cot \alpha}$ ...(1)

Now differentiating (1) *w.r.t.* θ, we get

$$\frac{dr}{d\theta} = a\cot\alpha\, e^{\theta\cot\alpha} = r\cot\alpha .$$

If s is the arc length of the spiral in the direction of r increasing, then

$$\frac{ds}{dr} = \sqrt{1+\left(r\frac{d\theta}{dr}\right)^2} = \sqrt{(1+\tan^2\alpha)}$$

$$\left(\because r\frac{d\theta}{dr} = \tan\alpha\right)$$

$$= \sqrt{(\sec^2\alpha)}$$

$$\frac{ds}{dr} = \sec\alpha \Rightarrow ds = \sec\alpha\, dr$$

Now integrating *w.r.t.* r from $r = r_1$ to $r = r_2$, we get

$$s = \int_{r_1}^{r_2}\sec\alpha\, dr = \sec\alpha\int_{r_1}^{r_2} dr$$

$$= \sec\alpha\left[r\right]_{r_1}^{r_2} = (r_2 - r_1)\sec\alpha$$

EXAMPLE 7. *Find the length of the cardioid $r = a\,(1 - \cos\theta)$ lying outside the circle $r = a\cos\theta$.*

SOLUTION . Both curves intersect. Therefore, we have

$$a(1-\cos\theta) = a\cos\theta$$

$$\text{or } 2\cos\theta = 1 \text{ or } \cos\theta = \frac{1}{2} \text{ or } \theta = \pm\frac{\pi}{3}$$

$\therefore$ The intersection points are $\left(\dfrac{a}{2}, \dfrac{\pi}{3}\right)$ and $\left(\dfrac{a}{2}, -\dfrac{\pi}{3}\right)$:

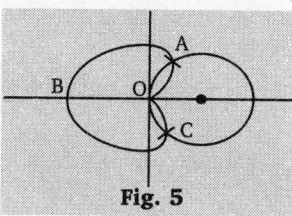

Fig. 5

The equation of the cardioid is

$$r = a(1-\cos\theta)$$

$$\therefore \quad \frac{dr}{d\theta} = a\sin\theta$$

$$\Rightarrow \quad \frac{ds}{d\theta} = \sqrt{r^2 + \left(\frac{dr}{d\theta}\right)^2} = \sqrt{[a^2(1-\cos\theta)^2 + a^2\sin^2\theta]}$$

$$\text{or } \quad \frac{ds}{d\theta} = 2a\sin\theta/2 .$$

Since s is the arc length of the cardioid to any point $P(r, \theta)$ in the direction of θ increasing so we will take $\dfrac{ds}{d\theta}$ positive. But we have to find the length of the cardioid lying outside the circle. Therefore if s_1 is the required length, then

$s_1 = 2 \times$ upper portion of the cardioid from A to B.

At the point A, $\theta = \pi/3$ and at B, $\theta = \pi$.

$$\therefore \quad s_1 = 2\int_{\pi/3}^\pi\left(\frac{ds}{d\theta}\right).d\theta = 2\int_{\pi/3}^\pi 2a\sin\theta/2\,d\theta$$

$$= 4a\left[-2\cos\theta/2\right]_{\pi/3}^\pi$$

$$= 4a\left[-2\cos\frac{\pi}{2} + 2\cos\frac{\pi}{6}\right] = 4a\left[2.\frac{\sqrt{3}}{2}\right]$$

Hence, $\quad s_1 = 4a\sqrt{3}$.

Exercise-4.4

1. Find the whole length of the curve $x^2 + y^2 = b^2$.

2. Find the length of the arc of the curve $x^2 = 8y$ from the vertex to an extremity of the latusrectum.

3. Find the arc length of the curve $y = \dfrac{1}{2}x^2 - \dfrac{1}{4}\log x$ from $x = 1$ to $x = 2$.

4. Find the length of the arc from $\theta = 0$ to $\theta = 2\pi$ of the curve
$$x = a\,(\cos\theta + \theta\sin\theta), y = a\,(\sin\theta - \theta\cos\theta).$$

5. Find the length of the arc of the curve $ay^2 = x^3$ between the points $(0, 0)$ and (a, a).

6. Show that the length of the curve
$$x = e^\theta\left(\sin\frac{\theta}{2} + 2\cos\frac{\theta}{2}\right), y = e^\theta\left(\cos\frac{\theta}{2} - 2\sin\frac{\theta}{2}\right)$$
between $\theta = 0$ to $\theta = \pi$ is $\dfrac{5}{2}(e^\pi - 1)$.

7. Show that the length of the arc of the curve $y^2 = 4ax$ which is intercepted between the points of intersection of the curve and the line $3y = 8x$ is $a(\log 2 + 15/16)$.

8. Find the length of the loop of the curve
$$9ay^2 = (x - 2a)(x - 5a)^2.$$

9. Find the length of an arc of the curve
$$x = a\left(\sin t + \frac{1}{3}\sin 3t\right), y = a\left(\cos t - \frac{1}{3}\cos 3t\right)$$
between $t = 0$ and $t = \pi/4$.

10. Find the whole length of the hypo-cycloid $x = a\cos^3 t, y = b\sin^3 t$.

11. Find the length of the loop of the curve $x = t^2, y = t - \dfrac{1}{3}t^3$.

12. Show that the whole length of the curve $x^2(a^2 - x^2) = 8a^2y^2$ is $\pi a^2\sqrt{2}$.

13. Find the length of an arc of the cycloid
$$x = a(\theta - \sin\theta), y = a(1 - \cos\theta). \qquad \text{(VTU–2004, PTU–2009)}$$

14. Find the length of the arc of the curve $x = e^\theta\sin\theta, y = e^\theta\cos\theta$ between $\theta = 0$ to $\theta = \pi/2$.

15. Find the entire length of the cardioid $r = a(1 - \cos\theta)$.

16. Show that the arc of the upper half of the curve $r = a(1 - \cos\theta)$ is bisected by $\theta = 2\pi/3$.

17. In the ellipse $x = a\cos\theta, y = b\sin\theta$, show that $ds = a\sqrt{(1 - e^2\cos^2\theta)}d\theta$, and hence show that the whole length of the ellipse is
$$2\pi a\left[1 - \left(\frac{1}{2}\right)^2 \cdot \frac{e^2}{1} - \left(\frac{1.3}{2.4}\right)^2 \cdot \frac{e^4}{3} - \left(\frac{1.3.5}{2.4.6}\right)^2 \cdot \frac{e^6}{5} - ...\right].$$

18. Find the perimeter of curve $r = a(1 + \cos\theta)$ and show that the arc of the upper half is bisected by $\theta = \pi/3$.   (JNTU–2003)

19. Prove that the perimeter of the limacon $r = a + b\cos\theta$, $a > b$ is approximately $2\pi a(1 + b^2/4a^2)$.

20. Find the length of the arc of the curve $r = ae^{\theta\cot\alpha}$ taking $s = 0$ when $\theta = 0$.

21. Show that $\theta = \pi/3$ divides the length of the cycloid
$$x = a(\theta - \sin\theta), y = a(1 - \cos\theta)$$
in the ratio 1:3.

22. Find the length of the arc of the curve
$$x = a(3\sin\theta - \sin^3\theta), y = a\cos^3\theta$$
between $\theta = 0$ and $\theta = \pi/2$.

23. Find the length of the arc of the curve

$$x = a\sin 2\theta(1 + \cos 2\theta), y = a\cos 2\theta(1 - \cos 2\theta)$$
between $\theta = 0$ and $\theta = \pi/2$.

24. Show that $\theta = \pi/6$ divides the arc in the first quadrant of the curve $x = a\cos^3\theta, y = a\sin^3\theta$ in the ratio 1:3.

25. Find the length of any arc of the cissoid
$$r = \frac{a\sin^2\theta}{\cos\theta}.$$

26. Show that the ratio of the lengths of the cardioid $r = a(1 - \cos\theta)$ lying inside and outside the circle $r = a\cos\theta$ is $(2 - \sqrt{3}):\sqrt{3}$.

27. Find the perimeter of the loop of the curve $3ay^2 = x^2(a - x)$.

28. Show that the length of the arc of the curve $x\cos\theta - y\sin\theta = f''(\theta)$, $x\sin\theta + y\cos\theta = f(\theta)$ is given by $S = f(\theta) = $ constant.

Hints to Selected Problems

1. The given equation of the curve $x^2 + y^2 = b^2$, which is a circle.

∴ Therefore, $s = 4\int_0^b \dfrac{ds}{dx}dx$.

Here $\dfrac{dy}{dx} = -\left(\dfrac{x}{y}\right)$ which implies $\dfrac{ds}{dx} = \sqrt{1 + \left(\dfrac{dy}{dx}\right)^2} = \dfrac{b}{\sqrt{b^2 - x^2}}$.

2. Let s be the length from the vertex $(0, 0)$ to $L(4, 2)$. Then, we have
$$s = \int_0^4 \frac{ds}{dx}dx \text{ and using } \int\sqrt{16 + x^2}dx = \frac{x}{2}\sqrt{16 + x^2}$$
$$= 8\log\left(x + \sqrt{16x^2}\right)$$
Then proceed as in (1).

4. $\dfrac{dx}{d\theta} = a\theta\cos\theta, \dfrac{dy}{d\theta} = a\theta\sin\theta$

$\Rightarrow \dfrac{dy}{dx} = \tan\theta$. Therefore, $\dfrac{ds}{dx} = \sqrt{1 + \left(\dfrac{dy}{dx}\right)^2} = \sqrt{1 + \tan^2\theta}$
$$= \sec\theta.$$
Then obtained the arc using the following formula
$$s = \int_0^{2\pi}\frac{ds}{dx}dx.$$

5. $\dfrac{dy}{dx} = \dfrac{3x^2}{2ay}$

∴ $\dfrac{ds}{dx} = \sqrt{1 + \left(\dfrac{dy}{dx}\right)^2} = \sqrt{1 + \dfrac{9x}{4a}}$

Now using the following formula $s = \int_0^a \dfrac{ds}{dx}.dx$.

6. $\dfrac{dx}{d\theta} = \dfrac{5}{2}e^\theta\cos\dfrac{\theta}{2}, \dfrac{dy}{d\theta} = -\dfrac{5}{2}e^\theta\sin\dfrac{\theta}{2}$

$\Rightarrow \dfrac{dy}{dx} = \dfrac{dy/d\theta}{dx/d\theta} = -\tan\dfrac{\theta}{2}$

∴ $\dfrac{ds}{dx} = \sqrt{1 + \tan^2\theta/2} = \sec\dfrac{\theta}{2}$.

Now $\dfrac{ds}{d\theta} = \dfrac{ds}{dx}.\dfrac{dx}{d\theta} = \dfrac{5}{2}e^\theta$. Now using the formula $s = \int_0^\pi \dfrac{ds}{d\theta}d\theta$.

7. The points of intersection of the given curves are $(0, 0)$ and

$$\left(\frac{9a}{16}, \frac{3a}{2}\right).$$

Also $\dfrac{dy}{dx} = \dfrac{2a}{y}$. Now find $\dfrac{ds}{dx}$ and then use the formula, given below
$$s = \int_0^{9a/16}\frac{ds}{dx}.dx.$$

10. $\dfrac{dx}{dt} = -3a\cos^2 t\sin^2 t, \dfrac{dy}{dt} = 3b\sin^2 t\cos t$

∴ $\dfrac{ds}{dt} = \sqrt{\left(\dfrac{dx}{dt}\right)^2 + \left(\dfrac{dy}{dt}\right)^2}$ ∴ $s = 4\int_0^{\pi/2}\dfrac{ds}{dt}.dt$

11. Eliminate t between the given curve, we get $9y^2 = x(3 - x)^2$.

Now $\dfrac{dx}{dt} = 2t, \dfrac{dy}{dt} = (1 - t)$.

Then use $\dfrac{ds}{dt} = \sqrt{\left(\dfrac{dx}{dt}\right)^2 + \left(\dfrac{dy}{dt}\right)^2}$ and $s = 2\int_0^{\sqrt{3}}\dfrac{ds}{dt}.dt$

15. $\dfrac{dr}{d\theta} = a\sin\theta$.

Since the given curve is symmetrical about the initial line, therefore, the entire length is twice the arc measure from 0 to π.

Here, $\dfrac{ds}{d\theta} = \sqrt{r^2 + \left(\dfrac{dr}{d\theta}\right)^2} = 2a\sin\dfrac{\theta}{2}$. Then using the formula, we

get $s = 2\int_0^\pi \dfrac{ds}{d\theta}.d\theta$.

26. The points of the intersection of the given curve are $\theta = \pm\dfrac{\pi}{3}$.

Let s_1 be the arc length of the cardioid inside the circle and s_2 be the arc length of the cardioid outside of the circle.

Therefore, $s_1 = 2\int_0^{\pi/3}\dfrac{ds}{d\theta}.d\theta$ and $s_2 = 2\int_{\pi/3}^{\pi}\dfrac{ds}{d\theta}.d\theta$

27. Do same as ex. 9 by taking $a = 1$.

28. Do same as ex. 6.

Answers

1. $2\pi b$ **2.** $2[\sqrt{2} + \log(1 + \sqrt{2})]$ **3.** $\dfrac{3}{2} + \dfrac{1}{4}\log 2$ **4.** $2a\pi^2$ **5.** $\dfrac{1}{27}a[13\sqrt{13} \quad 8]$

8. $4a\sqrt{3}$ **9.** a **10.** $4(a^2 + ab + b^2)/(a + b)$ **11.** $4\sqrt{3}$ **13.** $8a$

14. $\sqrt{2}[e^{\pi/2} - 1]$ **15.** $8a$ **18.** $8a$ **20.** $a \sec \alpha \, (e^{\theta \cot \alpha} - 1)$ **22.** $3a\pi/4$ **23.** $4a/3$

25. $s_1(\theta_2) - s_1(\theta_1)$ where $s_1(\theta) = a\sqrt{(\sec^2 \theta + 3)} - a\sqrt{3}[\log\{\cos\theta + \sqrt{(\cos^2\theta + \frac{1}{3})}\}]$ **27.** $\dfrac{4a}{\sqrt{3}}$

4.3.2 INTRINSIC EQUATION OF CURVES

The relation between s and ψ is called an intrinsic equation of any curve, where s is the length of an arc of the curve measured from a fixed point on the curve to any point P on it and ψ is the angle which the tangent to the curve at P makes with the positive x-axis. Thus the co-ordinates (s, ψ) is known as Intrinsic co-ordinates.

4.3.3 DERIVATION OF INTRINSIC EQUATION OF THE CURVES

(a) If the equation of a curve is in cartesian form. Let the equation of the curve be $y = f(x)$ as shown in fig. 6.

Let us consider a point A as fixed on this curve and let $P(x, y)$ be any point on this curve such that $AP = s$. Suppose the tangent at P makes the angle ψ with the fixed straight line (*i.e.*, x-axis). Since, we have

$$\frac{dy}{dx} = \tan \psi \, .$$

$\therefore \qquad \tan \psi = \frac{d}{dx} f(x) = f'(x) .$ ...(1)

Further since we know that $\qquad \dfrac{ds}{dx} = \sqrt{\left[1 + \left(\dfrac{dy}{dx}\right)^2\right]}.$ ...(2)

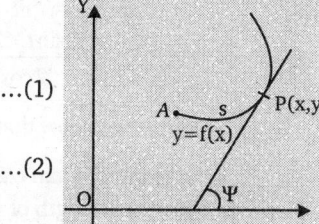

Fig. 6

Let x_1 be the x-co-ordinate of the fixed point A. Therefore, we have

$$s = \int_a^{x_1}\left(\frac{ds}{dx}\right)dx \quad \text{or} \quad s = \int_a^{x_1}\sqrt{\left[1 + \left(\frac{dy}{dx}\right)^2\right]}dx \qquad \text{[Using (2)]}$$

or $\qquad s = \int_a^{x_1}\sqrt{1 + [f'(x)]^2}dx \qquad$ [Using (1)]   ...(3)

Now eliminate x and $f'(x)$ between (1) and (3), we obtain the relation between s and ψ and thus obtain the intrinsic equations.

REMARK

- If $x = f(t), y = g(t)$, then $\dfrac{dy}{dx} = \dfrac{dy}{dt} \Big/ \dfrac{dx}{dt}$.

(b) Equations of a curve in polar form. Let the equations of a curve be $r = f(\theta)$ and consider a point A as fixed on this curve. Let $P(r, \theta)$ be any point on this curve such that arc $AP = s$ and the tangent at P makes an angle ψ with the positive x-axis as shown in fig. 7.

Let ϕ be the angle between the radius vector and the tangent at P. Thus, we have

$$\psi = \theta + \phi \qquad \text{...(1)}$$

Since we know that $\qquad \tan\phi = r\dfrac{d\theta}{dr}.$ ...(2)

But, the equation of a curve is $\qquad \left. \begin{aligned} r &= f(\theta) \\ \frac{dr}{d\theta} &= f'(\theta) \end{aligned} \right\}$ ...(3)

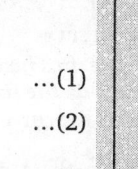

Fig. 7

Now using (2) and (3), we get

$$\tan\phi = \frac{f(\theta)}{f'(\theta)}. \qquad \text{...(4)}$$

Now $\qquad \dfrac{ds}{d\theta} = \sqrt{\left[r^2 + \left(\dfrac{dr}{d\theta}\right)^2\right]}. \Rightarrow \quad \dfrac{ds}{d\theta} = \sqrt{[[f(\theta)]^2 + [f'(\theta)]^2]}.$ ...(5)

Let the vectorial angle of the point A be α. Then, we have

$$s = \int_\alpha^\theta\left(\frac{ds}{d\theta}\right)d\theta \qquad \Rightarrow \qquad s = \int_\alpha^\theta\sqrt{[\{f(\theta)\}^2 + \{f'(\theta)\}^2]}d\theta \qquad \text{...(6)}$$

[Using (5)]

Now eliminate θ and ϕ between (1), (4) and (6), we obtain the required intrinsic equation of the curve.

(c) Equation of curve in Pedal form. Let the equation of a curve in pedal form be $p = f(r)$ and let A be the fixed point on the curve such that at A, $r = a$ (say), then, we have

$$s = \int_a^r \frac{rdr}{\sqrt{(r^2 - p^2)}} \cdot \Rightarrow \quad s = \int_a^r \frac{rdr}{\sqrt{[r^2 - \{f(r)\}^2]}} \qquad \text{...(1)}$$

where s is an arc measured from A to any point $P(p, r)$. Let ρ be the radius of curvature of the given curve at P, then we have

$$\rho = \frac{ds}{d\psi} = r\frac{dr}{dp} = r / f'(r).\qquad \ldots(2)$$

Now eliminating r between (1) and (2), we obtain the required intrinsic equation.

Solved Examples

EXAMPLE 1. *Show that the intrinsic equation of the curve*
$x = a(t + \sin t), y = a(1 - \cos t)$ *is* $s = 4a \sin \psi$.

SOLUTION . Since, we have

$$x = a(t + \sin t), y = a(1 - \cos t)$$

$$\therefore\quad \frac{dx}{dt} = a(1 + \cos t), \frac{dy}{dt} = a\sin t$$

Now, $\dfrac{dy}{dx} = \dfrac{dy}{dt} \Big/ \dfrac{dx}{dt} = \dfrac{\sin t}{1 + \cos t}$

$$= \frac{2\sin t / 2\cos t / 2}{2\cos^2 t / 2} = \tan t / 2$$

Since, we know that $\dfrac{dy}{dx} = \tan \psi$.

$$\tan \psi = \tan t/2 \quad \text{or} \quad \psi = t/2 \quad \text{or} \quad t = 2\psi.$$

If s is the length of the arc of the curve measured from the vertex $A(0, 0)$ to any point $P(x, y)$ in the direction of t increasing, then

$$s = \int_0^t \left(\frac{ds}{dt}\right).dt \quad \text{or} \quad s = \int_0^t \sqrt{\left[\left(\frac{dx}{dt}\right)^2 + \left(\frac{dy}{dt}\right)^2\right]}.dt$$

where $\left(\dfrac{ds}{dt}\right)^2 = \left(\dfrac{dx}{dt}\right)^2 + \left(\dfrac{dy}{dt}\right)^2$

$$\therefore\quad s = \int_0^t \sqrt{[a^2(1 + \cos t)^2 + a^2 \sin^2 t]}.dt$$

$$= \int_0^t 2a\cos t / 2\, dt$$

$$s = 2a\left[2\sin t / 2\right]_0^t = 4a\sin t / 2$$

But $\quad t = 2\psi.$

$$\therefore\quad s = 4a \sin \psi.$$

EXAMPLE 2. *Show that the intrinsic equation of* $3ay^2 = 2x^3$ *taking its cusp as the fixed point is* $9s = 4a(\sec^3 \psi)$.

SOLUTION . The equation of the curve is

$$3ay^2 = 2x^3 \quad \text{or} \quad y = \frac{\sqrt{2}x^{3/2}}{\sqrt{3a}}$$

$$\therefore\quad \frac{dy}{dx} = \sqrt{\frac{2}{3a}}.\frac{3}{2}x^{1/2} = \sqrt{\frac{3x}{2a}}.$$

Further since, we have

$$\frac{dy}{dx} = \tan \psi$$

$$\therefore\quad \tan \psi = \sqrt{\frac{3x}{2a}} \quad \text{or} \quad x = \frac{2a}{3}\tan^2 \psi \qquad \ldots(1)$$

But $\quad \dfrac{ds}{dx} = \sqrt{\left[1 + \left(\dfrac{dy}{dx}\right)^2\right]} = \sqrt{\left[1 + \dfrac{3x}{2a}\right]}.$

If s is the arc length of the given curve measure from the cusp at which $x = 0$ to any point $P(x, y)$ in the direction of x decreasing. Then

$$\frac{ds}{dx} = \sqrt{\left(1 + \frac{3x}{2a}\right)} \qquad \ldots(2)$$

$$\therefore\quad s = \int_0^x \left(\frac{ds}{dx}\right)dx \text{ or } s = \int_0^x \sqrt{\left(1 + \frac{3x}{2a}\right)}dx$$

[Using (2)]

$$= \left[\frac{\frac{2}{3}\left(1 + \frac{3x}{2a}\right)^{3/2}}{3/2a}\right]_0^x$$

$$\Rightarrow\quad s = \frac{4a}{9}\left[\left(1 + \frac{3x}{2a}\right)^{3/2} - 1\right] \qquad \ldots(3)$$

Now eliminating x between (1) and (3), we get

$$s = \frac{4a}{9}\left[(1 + \tan^2 \psi)^{3/2} - 1\right]$$

$$\Rightarrow\quad s = \frac{4a}{9}[\sec^3 \psi - 1] \text{ or } 9s = 4a[\sec^3 \psi - 1].$$

EXAMPLE 3. *Find the intrinsic equation of the cardioid* $r = a(1 - \cos \theta)$.

SOLUTION . The equation of the curve is

$$r = a(1 - \cos \theta). \qquad \ldots(1)$$

$$\therefore\quad \frac{dr}{d\theta} = a\sin \theta.$$

Now, $\dfrac{ds}{d\theta} = \sqrt{\left[r^2 + \left(\dfrac{dr}{d\theta}\right)^2\right]}$

$$= \sqrt{[a^2(1 - \cos \theta)^2 + a^2 \sin^2 \theta]}$$

$$= \sqrt{2a^2(1 - \cos \theta)} = \pm 2a\sin \theta / 2.$$

If s is the length of the arc of the curve measured from pole $(0, 0)$ to any point $P(r, \theta)$ in the direction of θ increasing, then we will take the sign of $\dfrac{ds}{d\theta}$ positive.

$$s = \int_0^\theta \left(\frac{ds}{d\theta}\right)d\theta = \int_0^\theta 2a\sin(\theta / 2)d\theta$$

$$= 2a\left[-2\cos(\theta / 2)\right]_0^\theta$$

$$s = 4a(1 - \cos(\theta / 2)) \qquad \ldots(2)$$

Further since, we know that

$$\tan \phi = r\frac{d\theta}{dr} = a(1 - \cos \theta).\frac{1}{a\sin \theta}$$

$$= \frac{1 - \cos \theta}{\sin \theta} = \frac{2\sin^2(\theta / 2)}{2\sin(\theta / 2)\cos(\theta / 2)}$$

$$\therefore\quad \tan \phi = \tan \theta / 2$$

$$\therefore\quad \phi = \theta/2 \qquad \text{or} \qquad \theta = 2\phi.$$

But we have

$$\psi = \theta + \phi = \theta + \theta / 2 \qquad (\because \phi = \theta/2)$$

$$\psi = \frac{3}{2}\theta \text{ or } \theta = \frac{2}{3}\psi .$$

Substitute the value of θ in (2) , we get

$$s = 4a\left(1 - \cos\frac{2}{6}\psi\right) = 4a\left(2\sin^2\frac{1}{6}\psi\right)$$

Hence, $s = 8a\sin^2\left(\frac{\psi}{6}\right)$.

EXAMPLE 4. *Find the intrinsic equation of the curve p= r sin α.*

SOLUTION. Since the equation of the curve is in pedal form

i.e., $p = r\sin\alpha$

$\therefore \quad \dfrac{dp}{dr} = \sin\alpha$

Now, $\quad \rho = \dfrac{ds}{d\psi} = r\dfrac{dr}{dp}$

$\therefore \quad \dfrac{ds}{d\psi} = r.\dfrac{1}{\sin\alpha} = r\,\text{cosec}\,\alpha \qquad ...(1)$

If s is the length of arc measured from the point $r = 0$

to any point P in the direction of r increasing. Then, we have $\quad s = \int_0^r \dfrac{r\,dr}{\sqrt{r^2 - p^2}}$

$= \int_0^r \dfrac{r\,dr}{\sqrt{r^2 - r^2\sin^2\alpha}} \qquad (\because p = r\sin\alpha)$

$= \int_0^r \sec\alpha\, dr = \sec\alpha[r]_0^r$

$s = r\sec\alpha \qquad ...(2)$

Now eliminate r between (1) and (2), we get

$$\frac{ds}{d\psi} = s\cot\alpha \quad \text{or} \quad \frac{ds}{s} = \cot\alpha\, d\psi\,.$$

Integrating, we get

$\log s = \psi\cot\alpha + \log c \quad \text{or} \quad s = ce^{\psi\cot\alpha}$

where c is the constant of integration.

This is the required intrinsic equation of the curve.

Exercise-4.5

1. Show that the intrinsic equation of the curve $y^2 = 4ax$ is
$s = a\cot\psi\,\text{cosec}\,\psi + a\log(\cot\psi + \text{cosec}\,\psi)$.

2. Find the intrinsic equation of the curve $y = c\cosh(x/c)$.

3. Find the intrinsic equation of the curve $x^2 = 4ay$.

4. Find the intrinsic equation of the curve $x = a(1 + \sin t)$, $y = a(1 + \cos t)$.

5. Find the intrinsic equation of the curve $x^{2/3} + y^{2/3} = a^{2/3}$.
   (i) If s is measured from the vertex.
   (ii) If s is measured from the cusp on x-axis.

6. Find the intrinsic equation of the cardioid $r = a(1 + \cos\theta)$, $\theta = 0$ being the fixed point.

7. Find the intrinsic equation of the curve $r = ae^{\theta\cot\alpha}$, where s is measured from the point $(a, 0)$.

8. Find the intrinsic equation of the curve $r = a\theta$, s being measured from $(0, 0)$.

9. Find the intrinsic equation of the curve $p = \sqrt{(r^2 - a^2)}$.

10. Prove that the intrinsic equation of the curve $x^2 = 4ay$ is
$s = a\tan\psi\sec\psi + a\log(\tan\psi + \sec\psi)$.

11. Find the intrinsic equation of the curve $x = e^t\sin t, y = e^t\cos t$, where $t = \pi/4$ being the fixed point.

12. Find the intrinsic equation of the curve $y = a\log\sec\dfrac{x}{a}$.

Hints to Selected Problems

1. Since $y^2 = 4ax$. Therefore, $\dfrac{dy}{dx} = \dfrac{2a}{y}$. $\quad \left[\because \dfrac{dy}{dx} = \tan\psi\right]$

$\therefore \quad y = 2a\cot\psi$

Now using $\dfrac{ds}{dy} = \sqrt{1 + \left(\dfrac{dx}{dy}\right)^2} = \dfrac{1}{2a}\sqrt{4a^2 + y^2}$

Then integrating after separating the variables.

2. $\dfrac{dy}{dx} = \sinh\left(\dfrac{x}{c}\right) \therefore \dfrac{ds}{dx} = \cosh\dfrac{x}{c}$.

Now using the formula given below

$\int_0^s ds = \int_0^x \cosh\left(\dfrac{x}{c}\right) dx$.

3. Do same as (1)

4. $\dfrac{dx}{dt} = a\cos t, \dfrac{dy}{dt} = -a\sin t$

$\therefore \quad \dfrac{dy}{dx} = -\tan t = \tan\psi \Rightarrow t = -\psi$.

Now $\dfrac{ds}{dt} = \sqrt{\left(\dfrac{dx}{dt}\right)^2 + \left(\dfrac{dy}{dt}\right)^2} = a$. Then use $\int_0^s ds = a\int_0^t dt$.

6. $\dfrac{dr}{d\theta} = -a\sin\theta$. Also, $\tan\phi = r\dfrac{d\theta}{dr} = -\cot\dfrac{\theta}{2} = \tan\left(\dfrac{\pi}{2} + \dfrac{\theta}{2}\right)$.

Now using $\psi = \theta + \phi = \dfrac{3\theta}{2} + \dfrac{\pi}{2} \Rightarrow \theta = \dfrac{2}{3}\left(\psi - \dfrac{\pi}{2}\right)$

$\dfrac{ds}{d\theta} = \sqrt{r^2 + \left(\dfrac{dr}{d\theta}\right)^2} = 2a\cos\dfrac{\theta}{2}$.

$\therefore \quad \int_0^s ds = \int_0^\theta 2a\cos\dfrac{\theta}{2}d\theta$.

7. $\dfrac{dr}{d\theta} = r\cot\alpha$. Also, we have $\tan\phi = r\dfrac{d\theta}{dr} = r.\dfrac{1}{r\cot\alpha} = \tan\alpha$

$\Rightarrow \tan\phi = \tan\alpha \Rightarrow \phi = \alpha$.

Now $\psi = \theta + \phi \Rightarrow \phi = \psi - \alpha$

$\therefore \quad \dfrac{ds}{d\theta} = \sqrt{r^2 + \left(\dfrac{dr}{d\theta}\right)^2}$

$\Rightarrow \quad ds = a\,\text{cosec}\,\alpha e^{\theta\cot\alpha}d\theta$

then integrating *w.r.t.* θ.

8. Do same as (7).

11. $\dfrac{dx}{dt} = e^t\sin t + e^t\cos t, \dfrac{dy}{dt} = e^t\cos t - e^t\sin t$.

$\therefore \quad \dfrac{dy}{dx} = \dfrac{1 - \tan t}{1 + \tan t} = \tan\left(\dfrac{\pi}{4} - t\right) \Rightarrow \psi = \dfrac{\pi}{4} - t$ *i.e.*, $t = \dfrac{\pi}{4} - \psi$.

Now proceed as above.

12. $y = a\log\sec(x/a) \quad \Rightarrow \quad \dfrac{dy}{dx} = \tan(x/a)$

$\Rightarrow \quad \tan\psi = \tan(x/a) \Rightarrow \quad x = a\psi$

$\Rightarrow \quad \dfrac{ds}{dx} = \sqrt{1 + \left(\dfrac{dy}{dx}\right)^2} = \sqrt{1 + \tan^2\dfrac{x}{a}} \Rightarrow \dfrac{ds}{dx} = \sec\dfrac{x}{a}$

$\Rightarrow \quad s = \int_0^x \dfrac{ds}{dx}.dx$.

Answers

2. $s = c \tan \psi$ **3.** $s = a \tan \psi \sec \psi + a \log (\tan \psi + \sec \psi)$ **4.** $s + a\psi = 0$

5. (i) $4s = 3a \cos 2\psi$ (ii) $2s = 3a \sin^2 \psi$ **6.** $s = 4a \sin\left(\dfrac{\psi}{3} - \dfrac{\pi}{6}\right)$ **7.** $s = a \sec \alpha \,[e^{(\psi - \alpha)\cot \alpha} - 1]$

8. $s = \dfrac{1}{2} a[\theta\sqrt{(1+\theta)^2} + \log\{\theta + \sqrt{(1+\theta)^2}\}]$ **9.** $s = \dfrac{1}{2} a(\psi^2 + 1)$ **11.** $s = \sqrt{2} e^{\pi/4}[e^{-\psi} - 1]$

12. $s = a \log [\tan \psi + \sec \psi]$ **13.** $s = a \tan \psi$

4.4 AREA OF CURVE

4.4.1 AREA OF CURVE IN CARTESIAN FORM

Let $y = f(x)$ be a continuous curve in cartesian form and let A be the area of the region bounded by the curve $y = f(x)$, the axis of x and the two ordinates $x = a$ and $y = b$. Then

$$A = \int_a^b y\,dx = \int_a^b f(x)\,dx$$

Proof. Let BC be the arc of the curve $y = f(x)$ cut by the lines $x = a$ and $y = b$ as shown in fig. 8.

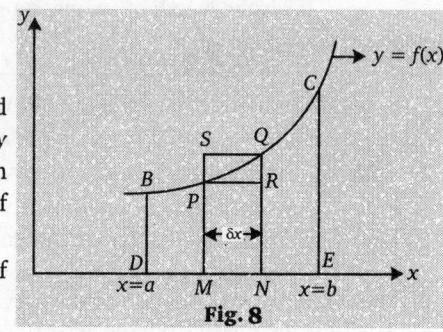

Let $P(x, y)$ and $Q(x + \delta x, y + \delta y)$ be two neighbouring points on the cuve between B and C. Now draw the perpendicular PM and QN on the axis of x such that $PM = y$ and $QN = y + dy$ and $MN = \delta x$. Since we observe that the area of $DMPB$ increases as P moves along arc BC from B to C. Draw the perpendicular PR on QN and QS to MP produced to S. Let δA be the area of $MNQP$, this area lies between the area $MNRP$ and the area $MNQS$.

Since $MNRP$ and $MNQS$ both are rectangles and the area of $MNRP$ is $y\,\delta x$ and the area of $MNQS$ is $(y + \delta y)\,\delta x$.

Fig. 8

$\therefore$ Area of $MNRP < \delta A <$ area of $MNQS$

or $y\,\delta x < \delta A < (y + \delta y)\,\delta x$ or $\;y < \dfrac{\delta A}{\partial x} < y + \delta y$.

Taking the limit as $\delta x \to 0$ and $\delta y \to 0$, where $Q \to P$.

$\therefore$ $y < \lim\limits_{\delta x \to 0} \dfrac{\delta A}{\partial x} < y \;$ or $\; y < \dfrac{dA}{dx} < y$

$\Rightarrow$ $\dfrac{dA}{dx} = y \;$ or $\; \dfrac{dA}{dx} = f(x)$ $[\because y = f(x)]$

or $dA = f(x)\,dx$.

Now integrating *w.r.t.* 'x' from $x = a$ to $x = b$, we get

$$\int_{x=a}^{x=b} dA = \int_a^b f(x)\,dx \qquad\qquad \Rightarrow \qquad\qquad [A]_{x=a}^{x=b} = \int_a^b f(x)\,dx$$

or   Area $DECB = \int_a^b f(x)\,dx$

or $A = \int_a^b f(x)\,dx = \int_a^b y\,dx$.

Similarly, the area bounded by the curve $x = f(y)$, the axis of y and the ordinates $y = a$ and $y = b$ is given by

$$A = \int_a^b f(y)\,dy = \int_a^b x\,dy.$$

REMARKS

- If the given curve is symmetrical either about x-axis or about y-axis or both, then find the area of one of the symmetrical part and multiply this area by the number of symmetrical parts, we get the whole area of the bounded region.
- Area bounded by two curves = | Area bounded by one curve – Area bounded by other curve |.

4.4.2 AREA OF CURVE IN POLAR FORM

Let $r = f(\theta)$ be the equation of a curve in polar form where $f(\theta)$ is a continuous function of θ, then the area of the sector bounded by the curve $r = f(\theta)$ and two radii vectors $\theta = \theta_1$ and $\theta = \theta_2$ such that $\theta_2 > \theta_1$ is given by

$$A = \frac{1}{2}\int_{\theta_1}^{\theta_2} r^2\,d\theta.$$

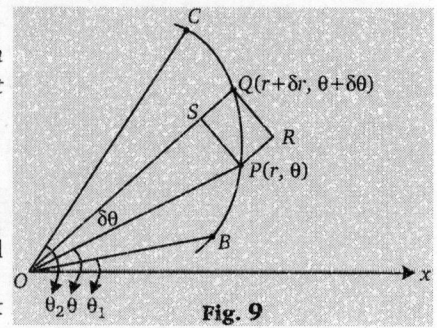

Proof. Since the equation of curve in polar form is $r = f(\theta)$.

Let A be the area of the sector OBC bounded by the curve and two radii vector $\theta = \theta_1$ and $\theta = \theta_2$ as shown in fig. 9.

Let $P(r, \theta)$ and $Q(r + \delta r, \theta + \delta \theta)$ be two neighbouring points on the curve $r = f(\theta)$ such that

Fig. 9

$OP = r$, $OQ = r + \delta r$ and $\angle POQ = \delta\theta$. Draw the perpendicular PS to OQ and QR to OR where OP produced to R. Let δA be the area of the sector OPQ. Obviously, this area δA lies between the area of isosceles triangle OPS and the area of isosceles triangle ORQ.

Now, the area of the isosceles triangle $OPS = \dfrac{1}{2}OP \times OS \sin\delta\theta = \dfrac{1}{2}r \times r \cdot \delta\theta$

$$= \dfrac{1}{2}r^2\delta\theta \quad (\delta\theta \text{ is very small})$$

and the area of isosceles triangle, $\quad ORQ = \dfrac{1}{2}OR \times OQ \sin\delta\theta$

$$= \dfrac{1}{2}(r + \delta r)^2\delta\theta \qquad\qquad (\because \delta\theta \text{ is very small so } \sin\delta\theta \approx \delta\theta)$$

Since δA lies between the areas of triangle OPS and ORQ, then

$$\dfrac{1}{2}r^2\delta\theta < \delta A < \dfrac{1}{2}(r + \delta r)^2\,\delta\theta\,.$$

Divide by $\delta\theta$, we have

$$\dfrac{1}{2}r^2 < \dfrac{\delta A}{\delta\theta} < \dfrac{1}{2}(r + \delta r)^2.$$

As $Q \to P$, $\delta\theta > 0$ and $\delta r \to 0$ so taking the limit as $\delta\theta \to 0$, we get

$$\dfrac{1}{2}r^2 < \lim_{\delta\theta \to 0}\dfrac{\delta A}{\delta\theta} < \dfrac{1}{2}r^2$$

or $\qquad\qquad\qquad\qquad \dfrac{1}{2}r^2 < \dfrac{dA}{d\theta} < \dfrac{1}{2}r^2$

or $\qquad\qquad\qquad\qquad \dfrac{dA}{d\theta} = \dfrac{1}{2}r^2 \quad \text{or} \quad dA = \dfrac{1}{2}r^2 d\theta.$

Integrating *w.r.t.* θ from $\theta = \theta_1$ to $\theta = \theta_2$, we get

$$\int_{\theta=\theta_1}^{\theta=\theta_2} dA = \int_{\theta_1}^{\theta_2}\dfrac{1}{2}r^2\,d\theta$$

Hence, $\qquad\qquad\qquad\qquad A = \dfrac{1}{2}\int_{\theta_1}^{\theta_2} r^2 d\theta \qquad\qquad \left(\because \int_{\theta_1}^{\theta_2} dA = \int_B^C dA = A\right)$

REMARKS

- In case of $r = a\cos n\theta$ or $r = a\sin n\theta$, the number of loops are n and $2n$ according as n is odd and n is even.
- Area by double integration bounded by $r = f(\theta)$ and $\theta = \theta_1$ and $\theta = \theta_2$ is given by $\int_{\theta_1}^{\theta_2}\int_{r=0}^{r=f(\theta)} r\,d\theta\,dr$.

Solved Examples

EXAMPLE 1. *Find the area of the region bounded by the line* $x = 2$ *and the parabola* $y^2 = 8x$.

SOLUTION. The equation of the parabola is

$$y^2 = 8x$$

which is symmetrical about x-axis and the line $x=2$ intersects the parabola $y^2 = 8x$ in two points $(2,4)$ and $(2, -4)$ as shown in fig. 10.

∴ The required area

$$= 2\int_0^2 y\,dx$$

$$= 2\int_0^2 \sqrt{8x}\,dx$$

$$= 4\sqrt{2}\left[\dfrac{2}{3}x^{3/2}\right]_0^2$$

$$= \dfrac{32}{3} \text{ sq. units.}$$

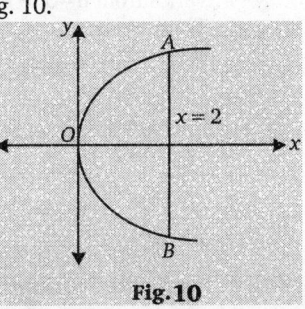

Fig.10

SOLUTION. We know that a line through the focus of the parabola and perpendicular to its axis is called latus-rectum. Since the equation of the parabola is

$$y^2 = 4ax.$$

∴ Extremities of latusrectum are $(a, 2a)$ and $(a, -2a)$ which is shown in fig. 11.

∴ The required area

$$= 2\int_0^a y\,dx = 2\int_0^a \sqrt{4ax}\,dx$$

$$= 4\sqrt{a}\left[\dfrac{2}{3}x^{3/2}\right]_0^a = 4\sqrt{a}\left[\dfrac{2}{3}a\sqrt{a}\right]$$

$$= \dfrac{8}{3}a^2 \text{ sq. units.}$$

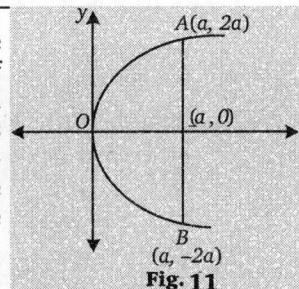

Fig. 11

EXAMPLE 2. *Find the area bounded by the parabola* $y^2 = 4ax$ *and its latusrectum.*

EXAMPLE 3. *Find the whole area of the ellipse* $\dfrac{x^2}{a^2} + \dfrac{y^2}{b^2} = 1$.

SOLUTION. The ellipse $\frac{x^2}{a^2}+\frac{y^2}{b^2}=1$ is symmetrical about both axes and whose trace is shown in fig. 12.

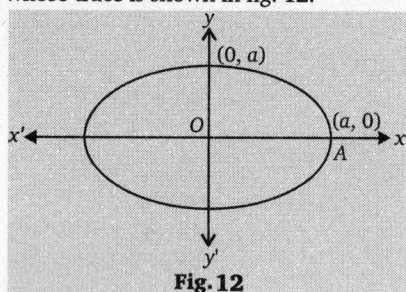

Fig.12

∴ Required area $=4\int_0^a y\,dx$

$=4\int_0^a b\sqrt{\left(1-\frac{x^2}{a^2}\right)}\,dx = 4b\int_0^a\sqrt{\left(1-\frac{x^2}{a^2}\right)}\,dx.$

Let us put $x = a\sin\theta$.

∴ $dx = a\cos\theta\,d\theta$ and θ varies from 0 to $\pi/2$.

∴　A. $=4b\int_0^{\pi/2}\cos\theta.a\cos\theta\,d\theta$

$=4ab\int_0^{\pi/2}\cos^2\theta\,d\theta$

$=4ab\left[\frac{(2-1)}{2}.\frac{\pi}{2}\right]$　(By Walli's formula)

$=\pi ab$ sq. units.

EXAMPLE 4. *Find the area of the loop of the curve*
$$ay^2 = x^2(a-x).$$　(SVTU-2009, OSMANIA-2000)

SOLUTION. Since the curve is symmetrical about x-axis and $y = 0$ when $x = 0$ and $x = a$ so the loop exists between $x = 0$ and $x = a$ as shown in fig. 13.

∴ Required area

$=2\int_0^a y\,dx$

$=2\int_0^a x\sqrt{\frac{a-x}{a}}\,dx.$

Let us put

$a-x = t^2,$

then $-dx = 2t\,dt$ and t varies from $\sqrt{a}$ to 0.

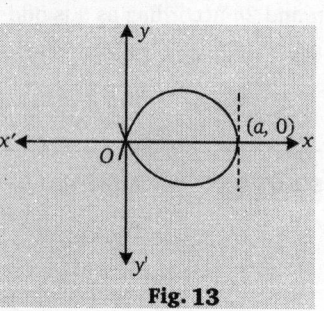

Fig. 13

∴　$=\frac{2}{\sqrt{a}}\int_{\sqrt{a}}^0 (a-t^2)t(-2t\,dt)$

$=\frac{2}{\sqrt{a}}.2\int_0^{\sqrt{a}}(at^2-t^4)dt$

$=\frac{4}{\sqrt{a}}\left[\frac{at^3}{3}-\frac{t^5}{5}\right]_0^{\sqrt{a}}$

$=\frac{4}{\sqrt{a}}\left[\frac{a^2\sqrt{a}}{3}-\frac{a^2\sqrt{a}}{5}\right]$

$=\frac{8}{15}a^2$ sq. units.

EXAMPLE 5. *Find the area of the one loop of the curve*
$$y^2 = x^2(a^2-x^2).$$

SOLUTION. Obviously the given curve is symmetrical about both

axes and $y = 0$ when $x = 0$, $x = \pm a$. Thus the curve has two loops one of them lies between $x = 0$ and $x = a$ and other lies between $x = -a$ and $x=0$ as shown in fig. 14.

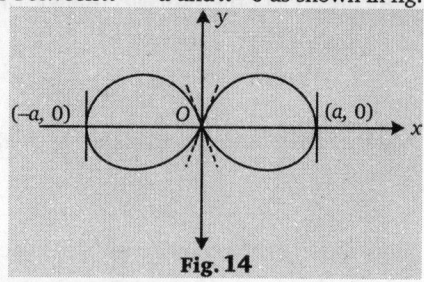

Fig. 14

∴ The required area

$=2\int_0^a y\,dx$

$=2\int_0^a x\sqrt{(a^2-x^2)}\,dx$

Let us put $a^2 - x^2 = t^2$, then $x\,dx = -t\,dt$ and t lies between a and 0. Thus the required area is

$=2\int_a^0 t(-t\,dt) = -2\int_a^0 t^2dt$

$=2\int_0^a t^2dt = 2\left[\frac{t^3}{3}\right]_0^a$

$=\frac{2}{3}a^3$ sq. units.

EXAMPLE 6. *Find the whole area of the curve* $a^2y^2 = x^3(2a-x)$.

SOLUTION. Clearly, the curve is symmetrical about x-axis and $y=0$ when $x = 0$ and $x = 2a$ so the curve has a loop between $x=0$ and $x=2a$ and curve does not exist in the regions $x < 0$ and $x>2a$. The tracing of the curve is shown in fig. 15.

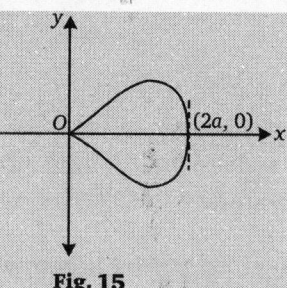

Fig. 15

∴ The required area

$= 2\times$ area in the first quadrant

$=2\int_0^{2a} y\,dx = 2\int_0^{2a}\frac{x}{a}\sqrt{(2ax-x^2)}\,dx$

$=\frac{2}{a}\int_0^{2a} x\sqrt{(2ax-x^2)}\,dx.$

Let us put $x=2a\sin^2\theta$, then $dx=4a\sin\theta\cos\theta\,d\theta$ and θ varies from $\theta=0$ to $\theta=\pi/2$.

Thus, we have

$=\frac{2}{a}\int_0^{\pi/2}2a\sin^2\theta.2a\sin\theta\cos\theta.4a\sin\theta\cos\theta\,d\theta$

$=32a^2\int_0^{\pi/2}\sin^4\theta\cos^2\theta\,d\theta$

$=32a^2\left[\frac{(4-1)(4-3)}{6.4.(6-4)}.(2-1).\frac{\pi}{2}\right]$

$=\pi a^2$ sq. units.　[By Walli's formula]

EXAMPLE 7. *Find the common area between the curves* $y^2=4ax$ *and* $x^2 = 4ay$.　(SVTU-2008, KURUKSHETRA-2005)

SOLUTION. The curve $y^2 = 4ax$ is symmetrical about x-axis and the curve $x^2 = 4ay$ is symmetrical about y-axis. Both curves intersects at two points (0,0) and (4a, 4a) which are obtained by solving both equations of curves. The tracing of the curves is shown in fig. 16.

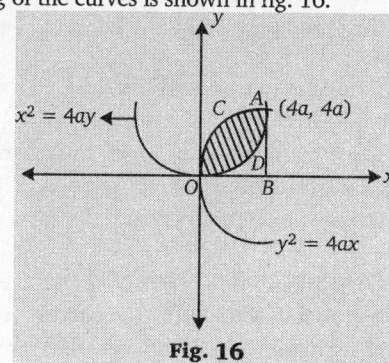

Fig. 16

The required area

= area of *OBACO* – area of *OBADO*   ...(1)

Now area of *OBACO*

$$= \int_0^{4a} y\,dx, \text{ where } y^2 = 4ax$$

$$= \int_0^{4a} \sqrt{4ax}\,dx = 2\sqrt{a}\left[\frac{2}{3}x^{3/2}\right]_0^{4a} = \frac{32}{3}a^2$$

and area of *OBADO* $= \int_0^{4a} y\,dx,$

where $y = \dfrac{x^2}{4a} = \int_0^{4a} \dfrac{x^2}{4a}dx$

$$= \frac{1}{4a}\left[\frac{x^3}{3}\right]_0^{4a} = \frac{16a^2}{3}.$$

From (1), we get

Required area $= \dfrac{32}{3}a^2 - \dfrac{16}{3}a^2 = \dfrac{16}{3}a^2$ sq. units.

EXAMPLE 8. *Find the area included between the curve* $x = a(t+\sin t), y = a(1-\cos t)$ *and its base.*

(VTU- 2000)

SOLUTION . Since the equation of the curve is

$$x = a(t + \sin t)$$
$$y = a(1 - \cos t)$$

Obviously, $y = 0$ when $\cos t = 1$ *i.e.*, $t = 0$ and $x = 0$. Thus the curve passes through the point (0,0). Also

$$\frac{dy}{dx} = \tan t/2, \text{ at } t = 0, \frac{dy}{dx} = 0.$$

Therefore, the tangent at (0,0) is the *x*-axis. The tracing of the curve is shown in fig 17.

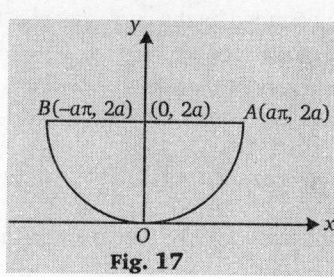

Fig. 17

Obviously, curve is symmetrical about *y*-axis so the required area is given by

$A = 2\times$ area bounded by the axis of *y* and *y*=2*a*

$$A = 2\int_{y=0}^{y=2a} x\,dy.$$

Since $x = a(t+\sin t), y = a(1-\cos t)$

$\therefore \quad dy = a \sin t\,dt.$

$\therefore \quad A = 2\int_0^\pi a(t + \sin t).a\sin t\,dt$

$$= 2a^2\int_0^\pi (t\sin t + \sin^2 t)\,dt$$

$$= 2a^2\left[\int_0^\pi t\sin t\,dt + \int_0^\pi \sin^2 t\,dt\right]$$

$$= 2a^2\left[(-t\cos t + \sin t)_0^\pi + \frac{1}{2}\left(t - \frac{\sin 2t}{2}\right)_0^\pi\right]$$

$$= 2a^2\left[\pi + \frac{\pi}{2}\right] = 3\pi a^2 \text{ sq. units.}$$

EXAMPLE 9. *Find the whole area between the curve* $x^2y^2=a^2(y^2-x^2)$ *and its asymptotes.* (VTU-2007)

SOLUTION . The given curve is symmetrical about both axes and curve passes through only (0,0). The tangents at (0, 0) are given by
$y^2 - x^2 = 0$ or $y = \pm x$.
Thus we obtained two distinct tangents at (0, 0) so (0, 0) is a node and its two real asymptotes are
$x = \pm a$. The tracing of the curve is shown in fig. 18.

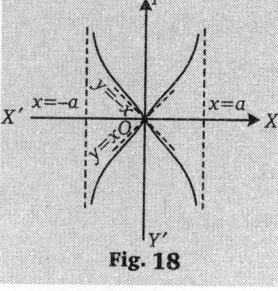

Fig. 18

The required area = 4 × area between curve and asymptote in the first quadrant.

$\therefore \qquad A = 4\int_0^a y\,dx \text{ or } A = 4\int_0^a \dfrac{ax\,dx}{\sqrt{(a^2 - x^2)}}$

Let us put $a^2 - x^2 = t^2$, so $x\,dx = -t\,dt$ and t takes the values from a to 0. Then

$$A = 4a\int_a^0 \frac{-t\,dt}{t} = 4a\int_0^a dt$$

$$= 4a[t]_0^a = 4a^2 \text{ sq. units.}$$

Exercise-4.6

1. Find the area of the region bounded by the following curves, and the axis of *x* and the given ordinates :
   (i) $y = \log x$; $x = a, \ x = b$
   (b) $y = c\cosh(x/c)$; $x = 0, \ x = a$
   (c) $y = \sin^2 x$; $x = 0, \ x = \pi/2$

2. Find the area of the region bounded by the parabola $y^2 = 4x$ and the line $y = 2x$.

3. Find the area of a quadrant of the ellipse $x^2/a^2 + y^2/b^2 = 1$.
   (VTU-2003, KERALA-2005)

4. Find the area of a loop of the curve $xy^2 + (x+a)^2(x+2a) = 0$.

5. Find the area of the loop of the curve $3ay^2 = x(x-a)^2$.
   (RAJASTHAN-2005)

6. Find the whole area of the curve $a^2x^2 = y^3(2a-y)$.
   (NAGPUR-2009)

7. Prove that the area of a loop of the curve $a^4y^2 = x^4(a^2-x^2)$ is $\pi a^2/8$.

8. Find the area bounded by the curve $xy^2 = a^2(a-x)$ and y-axis.

9. Find the area of the loop of the curve $y^2 = x(x-1)^2$.

10. Find the whole area of the curve $y^2 = x^2(a^2 - x^2)$.

11. Find the area between the curve $y^2(a-x) = x^2$ and its asymptote.

12. Find the area between the curve $y^2(2a-x) = x^3$ and its asymptote.   (VTU 2003)

13. Find the area between the curve $xy^2 = 4a^2(2a-x)$ and its asymptote.

14. Find the area between the $y^2(a-x) = x^3$ and its asymptote. Also find the ratio in which the ordinate $x = a/2$ divides this area.

15. Find the area of the region bounded by the parabola $y^2 = 4ax$ and $x^2 = 4by$.

16. Find the area bounded by the curves $x^2 + y^2 \le 2ax$ and $y^2 \ge ax$, $a > 0, x > 0, y > 0$.

17. Find the area bounded by the curves $y \ge x^2$ and $y \le |x|$.

18. Find the area of the segment cut off from the parabola $y^2 = 4x$ by the line $y = 8x - 1$.

19. Find the area bounded by the parabola $4y = 3x^2$ and the line $3x - 2y + 12 = 0$.

20. Find the area enclosed by the curves $y^2 \le 3x$ and $3x^2 + 3y^2 \ge 16$.

21. Find the area included between the cycloid $x = a(\theta - \sin\theta)$, $y = a(1 - \cos\theta)$ and its base.   (GORAKHPUR-1999)

22. Find the area enclosed between the curve $y = x^3$ and the line $y = x$.

23. Find whole area of the curve $x^{2/3} + y^{2/3} = a^{2/3}$.   (VTU- 2005)

24. Find the area of the loop of the curve $x^3 + y^3 = 3axy$.

25. Find the area enclosed between the parabola $y^2 = 4a(x+a)$ and $y^2 = -4a(x-a)$.

26. Find the area of the smaller portion enclosed by the curves $x^2 + y^2 = 9$ and $y^2 = 8x$.

27. Find the area between the parabola $y = 4x - x^2$ and the line $y = x$.   (VTU-2010, SVTU-2008, UPTU-2008)

Hints to Selected Problems

1. Required area $A = \int_a^b y\, dx$.

2. Required area $A = \int_0^1 (\sqrt{4x})\, dx - \int_0^1 (2x)\, dx$.

3. Required area $A = \int_0^a y\, dx$.

4. $A = 2\int_{-2a}^a y\, dx = \pm 2\int_{-2a}^a \dfrac{\sqrt{(x+a)^2(x+2a)}}{-x}\, dx$.

   Put $-x = t$. Then integrate.

5. $3ay^2 = x(x-a)^2 \Rightarrow y = (x-a)\sqrt{\dfrac{x}{3a}}$. Then $A = 2\int_0^a y\, dx$.

6. Here $x = 0, 0$ be two real tangents. Therefore $A = 2\int_0^{2a} x\, dy$.

8. $A = 2\int_0^\infty x\, dy$.

9. $A = 2\int_0^1 y\, dx$.

12. $x = 2a$ is the asymptote of the given curve, therefore, the required area $A = 2\int_0^{2a} y\, dx$.

13. $x = 0$ is the asympotote. Therefore, $A = 2\int_0^{2a} y\, dx$.

14. $A = 2\int_0^a y\, dx \Rightarrow A = 2\int_0^a x\sqrt{\dfrac{x}{a-x}}\, dx$

   $A_1 = 2\int_0^{a/2} y\, dx \Rightarrow A_1 = 2\int_0^{a/2} x\sqrt{\dfrac{x}{a-x}}\, dx$

   Put $x = a\sin^2\theta$ and integrate.

16. $A = \int_0^a (\sqrt{2ax - x^2})\, dx - \int_0^a (\sqrt{ax})\, dx$.

17. $A = 2\int_0^1 x\, dx - \int_0^1 x^2\, dx$.

21. Do same as example 10.

24. Put $x = r\cos\theta, y = r\sin\theta$, in the equation of the given curve, we get

$$r = \frac{3a\sin\theta\cos\theta}{(\cos^3\theta + \sin^3\theta)}.$$

Required area $= \int_0^{\pi/2} \dfrac{r^2}{2}\, d\theta$.

Answers

1.(i) $b\log(b/e) - a\log(a/e)$ (ii) $c^2\sinh(a/c)$ (iii) $\pi/4$ **2.** 1/3   **3.** $\dfrac{1}{4}\pi ab$ **4.** $2a^2(1-\pi/4)$ **5.** $8a^2/(15\sqrt{3})$ **6.** πa^2 **8.** πa^2

9. 8/15   **10.** $4a^3/3$ **11.** $(8/3)a^3$ **12.** $3\pi a^2$ **13.** $4\pi a^2$ **14.** $3\pi a^2/4$; $(3\pi - 8):(3\pi + 8)$ **15.** $(16/3)\,ab$ **16.** $(a^2/12)(3\pi - 8)$ **17.** 1/3

18. 9/64   **19.** 27   **20.** $\dfrac{4}{3}a^{3/2} + \dfrac{8\pi}{3} - \dfrac{a}{2}\sqrt{\left(\dfrac{16}{3}\right) - a^2} - \dfrac{8}{3}\sin^{-1}\left(\dfrac{a}{4\sqrt{3}}\right)$, where $a = (-a + \sqrt{273})/6$ **21.** $3\pi a^2$ **22.** 1/2   **23.** $\dfrac{3}{8}\pi a^2$

24. $3a^2/2$ **25.** $(16/3)a^2$ **26.** $2\left[\dfrac{\sqrt{2}}{3} + \dfrac{9\pi}{4} - \dfrac{9}{2}\sin^{-1}\left(\dfrac{1}{3}\right)\right]$ **27.** 9/2

4.4.3 Problem Based on Polar Form

EXAMPLE 1. Find the area of the cardioid $r = a(1 + \cos\theta)$.
     (VTU-2008)

SOLUTION . The equation of the cardioid $r = a(1 + \cos\theta)$ is symmetrical about the initial line and $r = 0$ when $\cos\theta = -1$ *i.e.*, $\theta = \pi$ and r is maximum when $\cos\theta = 1$

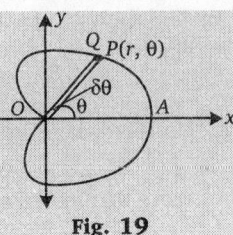

Fig. 19

i.e., $\theta = 0$. Thus the tracing of the curve is shown in fig. 19.

Let A be the area of the cardiod $r = a(1 + \cos\theta)$. This area A is the twice the area of the upper half of the curve between $\theta = 0$ and $\theta = \pi$.

     Now the required area

$$A = 2\int_0^\pi \frac{1}{2}r^2\, d\theta = \int_0^\pi a^2(1 + \cos\theta)^2\, d\theta$$

$$= a^2 \int_0^\pi (2)^2 \cos^4 \frac{\theta}{2} d\theta.$$

$$= 4a^2 \int_0^\pi \cos^4 \theta / 2 \, d\theta.$$

Let us put $\theta/2 = \phi$ so $d\theta = 2 \, d\phi$ and ϕ runs from 0 to $\pi/2$.

$$\therefore \quad A = 8a^2 \int_0^{\pi/2} \cos^4 \phi \, d\phi$$

$$= 8a^2 \left[\frac{(4-1)(4-3)}{4.2} \cdot \frac{\pi}{2} \right] \quad \text{[By Walli's formula]}$$

$$= \frac{3}{2} \pi a^2.$$

EXAMPLE 2. *Find the area of a loop of the curve $r^2 = a^2 \cos 2\theta$.*

(VTU- 2006)

SOLUTION . The curve is symmetrical about pole and initial line both and $r = 0$ when $\cos 2\theta = 0$ *i.e.*, $\theta = \pm \pi/4$. Thus a loop of the curve lies between $\theta = -\pi/4$ and $\theta = \pi/4$.
Let A be the area of this loop. Then

$$A = \int_{-\pi/4}^{\pi/4} \frac{1}{2} r^2 d\theta = \frac{1}{2} \int_{-\pi/4}^{\pi/4} a^2 \cos 2\theta \, d\theta$$

$$= \frac{a^2}{2} \int_{-\pi/4}^{\pi/4} \cos 2\theta \, d\theta$$

$$= \frac{a^2}{2} \left[\frac{1}{2} \sin 2\theta \right]_{-\pi/4}^{\pi/4} = \frac{a^2}{2} \left[\frac{1}{2} + \frac{1}{2} \right] = \frac{a^2}{2}.$$

EXAMPLE 3. *Find the area of common to the cardioids $r = a(1 + \cos \theta)$ and $r = a(1 - \cos \theta)$.*

(VTU-2006, KURUKSHETRA-2006)

SOLUTION . Clearly both the cardioids are symmetrical about the initial line OX and intersect at B and B' as shown in the adjoining figure.

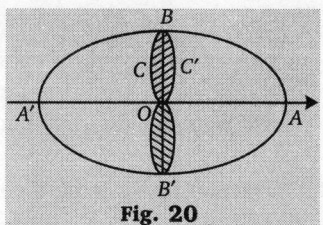

Fig. 20

Required area = 2 . Area OC'BCO
= 2[area OC'BO + area OBCO]

$$= 2 \left\{ \left[\int_0^{\pi/2} \frac{1}{2} r^2 d\theta \right]_{r=a(1-\cos\theta)} \right.$$

$$\left. + \left[\int_{\pi/2}^{\pi} \frac{1}{2} r^2 d\theta \right]_{r=a(1+\cos\theta)} \right\}$$

$$= a^2 \int_0^{\pi/2} (1 - \cos\theta)^2 d\theta + a^2 \int_{\pi/2}^{\pi} (1 + \cos\theta)^2 d\theta$$

$$= a^2 \left\{ \int_0^{\pi/2} (1 - 2\cos\theta + \cos^2\theta) d\theta \right.$$

$$\left. + \int_{\pi/2}^{\pi} (1 + 2\cos\theta + \cos^2\theta) d\theta \right\}$$

$$= a^2 \left\{ \int_0^{\pi} (1 + \cos^2\theta) d\theta - 2\int_0^{\pi/2} \cos\theta \, d\theta \right.$$

$$\left. + 2\int_{\pi/2}^{\pi} \cos\theta \, d\theta \right\}$$

$$= a^2 \left\{ \int_0^{\pi} \left(1 + \frac{1 + \cos 2\theta}{2} \right) d\theta - 2|\sin\theta|_0^{\pi/2} \right.$$

$$\left. + 2|\sin\theta|_{\pi/2}^{\pi} \right\}$$

$$= a^2 \left\{ \left| \frac{3}{2}\theta + \frac{\sin 2\theta}{4} \right|_0^{\pi} - 2(1 - 0) + 2(0 - 1) \right\}$$

$$= \left(\frac{3\pi}{2} - 4 \right) a^2.$$

Exercise-4.7

1. Find the area of the parabola $r(1 + \cos \theta) = l$ between $\theta = 0$, $\theta = \alpha$.
2. Find the area of one loop of the curve $r = a \cos 4\theta$.
3. Find the whole area of the curve $r^2 = a^2 \cos^2\theta + b^2 \sin^2\theta$.
4. Find the area of a loop of the curve $x^3 + y^3 = 3axy$.
5. Show that the area of the limacon $r = a + b \cos \theta$, $(a > b)$ is $\pi \left(a^2 + \frac{1}{2} b^2 \right)$.
6. Prove that the sum of the areas of the two loops of the limacon $r = a + b \cos \theta$ $(a < b)$ is $\pi(2a^2 + b^2)/2$.
7. Find the ratio of the two parts into which the parabola $2a = r(1 + \cos \theta)$ divides the area of the cardioid $r = 2a(1 + \cos\theta)$.
8. Find the area outside the circle $r = 2a \cos \theta$ and inside the cardioid $r = a (1 + \cos \theta)$. (KURUKSHETRA-2006)
9. Find the area between the curve $r = a (\sec \theta + \cos \theta)$ and its asymptote.
10. Find the area of a loop of the curve $x^4 + y^4 = 4a^2 xy$.
11. Prove that the area of a loop of the curve $x^6 + y^6 = a^2 x^2 y^2$ is $\pi a^2 / 12$.
12. Find the area lying between the cardioid $r = (1 - \cos \theta)$ and its double tangent. (VTU -2004)
13. Find the area common to the circles $r = a\sqrt{2}$ and $r = 2a \cos \theta$.
14. If O is the pole of the lemniscate $r^2 = a^2 \cos 2\theta$ and PQ is a common tangent to its two loops. Find the area bounded by the line PQ and the arcs OP and OQ of the curve.
15. Find the area of a loop of the curve $x^4 + 3x^2 y^2 + 2y^4 = a^2 xy$.
16. Find the total area inside $r = \sin \theta$ and outside $r = 1 - \cos \theta$. (ANNA-2009)
17. Find the area of a loop of the curve $r = a \cos 3\theta + b \sin 3\theta$.

Hints to Selected Problems

1. $A = \frac{1}{2} \int_0^\alpha r^2 d\theta$.

2. $A = \frac{1}{2} \int_{-\pi/8}^{\pi/8} r^2 d\theta$.

3. The curve is symmetrical about the initial line and line $\theta = \pi/2$. Also it is symmetric about the pole. $A = 4 \times$ Area lying in the first quadrant $= 4 \times \frac{1}{2} \int_0^{\pi/2} r^2 d\theta$.

4. In polar form the given equation becomes $r = \dfrac{3a \cos\theta \sin\theta}{\cos^3\theta + \sin^3\theta}$

Then $A = \int_0^{\pi/2} \frac{1}{2} r^2 d\theta$.

5. $A = 2 \int_0^\pi \frac{1}{2} r^2 d\theta$.

6. $A = 2 \left[\int_0^{\cos^{-1}\left(-\frac{a}{b}\right)} \frac{1}{2} r^2 d\theta + \int_{\cos^{-1}\left(-\frac{a}{b}\right)}^{\pi} \frac{1}{2} r^2 d\theta \right]$.

7. Smaller area

$$A_1 = \frac{1}{2}\int_0^{\pi/2} \frac{4a^2}{(1+\cos\theta)^2} d\theta + \frac{1}{2}\int_{\pi/2}^{\pi} 4a^2(1+\cos\theta)^2 d\theta.$$

Larger area A_2 = whole area $- A_1$. Then find $\dfrac{A_1}{A_2}$.

13. The points of intersection are given by $\theta = \pm\dfrac{\pi}{4}$.

Required area $= 2 (A_1 + A_2)$

where $A_1 = \int_{\pi/4}^{\pi/2} \frac{1}{2} r^2 d\theta$ for $r = 2a\cos\theta$

and $A_2 = \int_0^{\pi/4} \frac{1}{2} r^2 d\theta$ for $r = a\sqrt{2}$.

16. The both curve intersect at $(0, 0)$ and $(1, \pi/2)$.

Therefore, the required area $= A_1 - A_2$

where $A_1 = \frac{1}{2}\int_0^{\pi/2} r^2 d\theta$ or $r = \sin\theta$

and $A_2 = \frac{1}{2}\int_0^{\pi/2} r^2 d\theta$ for $r = 1 - \cos\theta$.

Answers

1. $\dfrac{1}{4}l^2\left[\tan\dfrac{\alpha}{2} + \dfrac{1}{3}\tan^3\dfrac{\alpha}{2}\right]$ **2.** $\dfrac{1}{16}\pi a^2$ **3.** $\dfrac{1}{2}\pi(a^2+b^2)$ **4.** $\dfrac{3a^2}{2}$ **7.** $(9\pi+16):(9\pi-16)$ **8.** $\dfrac{1}{2}\pi a^2$ **9.** $\dfrac{5}{4}\pi a^2$

10. $\dfrac{1}{2}\pi a^2$ **12.** $\dfrac{1}{16}(15\sqrt{3}-8\pi)a^2$ **13.** $a^2(\pi-1)$ **14.** $\dfrac{1}{8}a^2(3\sqrt{3}-4)$ **15.** $\dfrac{1}{4}a^2\log 2$ **16.** $\left(1-\dfrac{\pi}{4}\right)$ **17.** $(a^2+b^2)\dfrac{\pi}{12}$.

4.5 SURFACE AND VOLUME OF SOLID OF REVOLUTIONS

When a plane curve is revolved about a certain fixed line lying in its own plane, a surface is generated. This surface is called a surface of revolution. Also the fixed line is called the axis of revolution.

4.5.1 REVOLUTION ABOUT X-AXIS

Let S be the surface area (curved surface) of a solid which is generated by the revolution of the curve $y = f(x)$ about x-axis between the ordinates $x = a$ and $x = b$ and let s be the arc length measured from the point $(a, f(a))$ to any point $P(x, y)$. Then

$$S = \int_a^b 2\pi y \, ds = \int_a^b 2\pi y \frac{ds}{dx}.dx.$$

Proof. Let $A(a, f(a))$ and $B(b, f(b))$ be the points on the curve $y = f(x)$ and assuming that the curve $y = f(x)$ is continuous in (a, b) and does not intersect the axis of x. Let $P(x, y)$ be any point on the curve and s be the arc length of the curve measured from A as shown in fig. 1.

Let $Q(x + \delta x, y + \delta y)$ be any other point very near to $P(x, y)$. Then $PQ = \delta s$, because $AP = s$ and $AQ = s + \delta s$. Draw the perpendiculars PM and QN to the axis of x from P and Q respectively. As the curve revolves about x-axis, the arc length $PQ = \delta s$ also revolves and form a right circular cydinder of thickness δs of radii y and $y + \delta y$. Let δS be the surface area of this cylinderical element which lies between the surface areas $2\pi y \delta s$ and $2\pi(y + \delta y)\delta s$. That is,

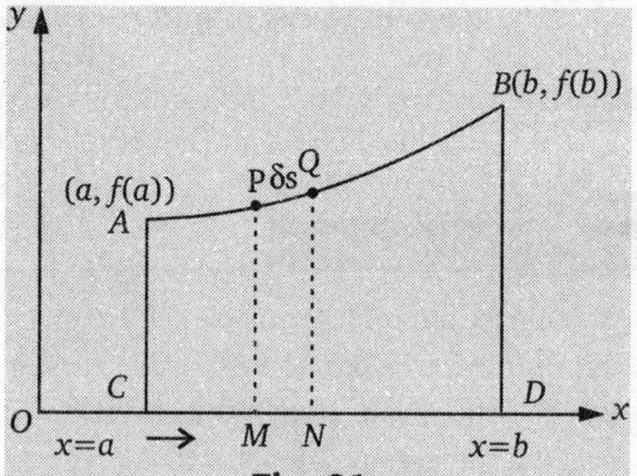

$$2\pi y \, \delta s < \delta S < 2\pi(y + \delta y) \, \delta s.$$

Fig. 21

Divide by δs, we get

$$2\pi y < \frac{\delta S}{\delta s} < 2\pi(y + \delta y).$$

As $Q \to P$, $\delta s \to 0$ and $\delta y \to 0$, then taking the limit as $\delta s \to 0$, we obtain

$$2\pi y < \lim_{\delta s \to 0} \frac{\delta S}{\delta s} < 2\pi y.$$

$\therefore$

$$\lim_{\delta s \to 0} \frac{\delta S}{\delta s} = 2\pi y \text{ or } \frac{dS}{ds} = 2\pi y \text{ or } dS = 2\pi y \, ds.$$

Now integrating, we get

$$\int_{x=a}^{x=b} dS = \int_{x=a}^{x=b} 2\pi y \, ds$$

or

$$S = \int_a^b 2\pi y \, ds = \int_a^b 2\pi y \frac{ds}{dx}.dx.$$

where $\dfrac{ds}{dx} = \sqrt{1 + \left(\dfrac{dy}{dx}\right)^2}$ and S is the surface area of the solid of revolution of the curve $y = f(x)$ about x-axis between $x = a$ and $x = b$.

4.5.2 REVOLUTION ABOUT Y-AXIS

Let S be the surface area of a solid generated by the revolution of the curve $x = f(y)$ about y-axis between $y = a$ and $y = b$. Then

$$S = \int_a^b 2\pi x \, ds = \int_a^b 2\pi x \frac{ds}{dy}.dy$$

where $\dfrac{ds}{dy} = \sqrt{1 + \left(\dfrac{dx}{dy}\right)^2}$ and s the arc length being measured from the point $(f(a), a)$.

Proof. Similar as before given in § 4.5.2.

4.5.3 REVOLUTION ABOUT ANY LINE

Let S be the surface area of a solid generated by the curve about any line between certain points. Let s be the arc length of the curve measured from one of the two given points to any point P on the curve and let Q be any point very near to P such that $PQ = \delta s$. Now draw a perpendicular PM from the point P to the line of axis of the revolution. Then

$$S = \int 2\pi (PM)\, ds$$

4.5.4 SURFACE FORMULAE FOR DIFFERENT FORM OF EQUATIONS

(a) Equation of a curve in parametric form. Suppose the equation of a curve is given in parametric form $x = f(t)$, $y = g(t)$ where t is the parameter, then the surface area of a solid generated by the revolution of the given curve about x-axis between the suitable limits is

$$S = \int 2\pi y \left(\frac{ds}{dt} \right) dt$$

where

$$\frac{ds}{dt} = \sqrt{\left[\left(\frac{dx}{dt} \right)^2 + \left(\frac{dy}{dt} \right)^2 \right]}.$$

Similarly for y-axis as the axis of revolution we may find the surface area.

(b) Equation of a curve in polar form. Suppose the equation of a curve is given in polar form $r = f(\theta)$. Then the formula for finding the surface area between the proper limits is given by

$$S = \int 2\pi (r \sin \theta) \frac{ds}{d\theta}. d\theta \qquad \qquad \dots(1)$$

where

$$\frac{ds}{d\theta} = \sqrt{\left[r^2 + \left(\frac{dr}{d\theta} \right)^2 \right]}.$$

The formula given in (1) can also be taken as $S = \int 2\pi (r \sin \theta) \frac{ds}{dr}. dr \qquad \dots(2)$

where

$$\frac{ds}{dr} = \sqrt{\left[1 + \left(r \frac{d\theta}{dr} \right)^2 \right]}.$$

WORKING PROCEDURE

To find the surface area of a solid generated by the revolution of the curve about any line, use the following steps :

Step 1. Take any point P on the curve between the given points.

Step 2. Draw the perpendicular from P to the line of axis which meets the axis at the point M (say).

Step 3. Find the perpendicular PM.

Step 4. Use the given formula.

Solved Examples

EXAMPLE 1. *Find the surface of a sphere of radius a.*

SOLUTION. The sphere is generated, if a semi-circle is revolved about its diameter. Let the equation of a circle of radius a is

$$x^2 + y^2 = a^2. \qquad \dots(1)$$

$$\therefore \qquad 2x + 2y \frac{dy}{dx} = 0 \text{ or } \frac{dy}{dx} = -\frac{x}{y}.$$

$$\Rightarrow \quad \frac{ds}{dx} = \sqrt{\left[1 + \left(\frac{dy}{dx} \right)^2 \right]} = \sqrt{\left(1 + \frac{x^2}{y^2} \right)}$$

$$= \sqrt{\left(\frac{x^2 + y^2}{y^2} \right)} = \frac{a}{y} \qquad \text{[Using (1)]}$$

Let $A(-a, 0)$ and $B(a, 0)$ be the bounding points of the semi-circle as shown in fig. 22.

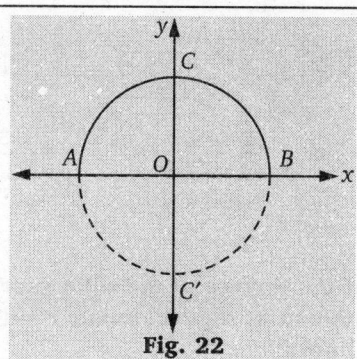

Fig. 22

Here the diameter is taken as x-axis. Let S be the surface area of the sphere, then

$$S = \int_{-a}^{a} 2\pi y . \frac{ds}{dx}. dx = \int_{-a}^{a} 2\pi y . \frac{a}{y}. dx \qquad \left(\because \frac{ds}{dx} = \frac{a}{y} \right)$$

$$= 2\pi a \int_{-a}^{a} dx = 2\pi a [x]_{-a}^{a}$$

$$\therefore \qquad S = 4\pi a^2.$$

EXAMPLE 2. *Find the surface of the solid generated by the* *revolution of the ellipse $x^2 + 4y^2 = 16$ about its major axis.*

__SOLUTION.__ The equation of the curve is

$$x^2 + 4y^2 = 16 \qquad ...(1)$$

or

$$\frac{x^2}{16} + \frac{y^2}{4} = 1.$$

The end points of major axis are $A(-4, 0)$ and $B(4, 0)$ which are on the x-axis so the major axis is the axis of x. Thus the curve is revolved about the x-axis.

Now differentiating (1) *w.r.t. 'x'*, we get

$$2x + 8y\frac{dy}{dx} = 0 \text{ or } \frac{dy}{dx} = -\frac{x}{4y}.$$

$$\therefore \quad \frac{ds}{dx} = \sqrt{1 + \left(\frac{dy}{dx}\right)^2}$$

$$= \sqrt{\left(1 + \frac{x^2}{16y^2}\right)} = \frac{\sqrt{(16y^2 + x^2)}}{4y}$$

$$= \frac{\sqrt{4(16 - x^2) + x^2]}}{4y} \qquad \text{[Using (1)]}$$

$$\therefore \quad \frac{ds}{dx} = \frac{\sqrt{(64 - 3x^2)}}{4y}.$$

Let S be the surface area of the solid so formed by the revolution of the ellipse given in (1) about its major axis (x-axis is)

$$S = \int_{-4}^{4} 2\pi y \frac{ds}{dx}.dx = \int_{-4}^{4} 2\pi y \cdot \frac{\sqrt{(64 - 3x^2)}}{4y} dx$$

$$= \frac{\pi}{2}\int_{-4}^{4} \sqrt{(64 - 3x^2)}\,dx = \pi\int_{0}^{4}\sqrt{(64 - 3x^2)}\,dx$$

$$= \sqrt{3}\pi\int_{0}^{4} \sqrt{\left[\left(\frac{8}{\sqrt{3}}\right)^2 - x^2\right]}\,dx$$

$$= \sqrt{3}\pi\left[\frac{x}{2}\sqrt{\left[\left(\frac{8}{\sqrt{3}}\right)^2 - x^2\right]} + \frac{32}{3}\sin^{-1}\frac{\sqrt{3}x}{8}\right]_{0}^{4}$$

$$= \sqrt{3}\pi\left[2\sqrt{\left(\frac{64}{3} - 16\right)} + \frac{32}{3}\sin^{-1}\left(\frac{\sqrt{3}}{2}\right)\right]$$

$$= \sqrt{3}\pi\left[2.\frac{4}{\sqrt{3}} + \frac{32}{(\sqrt{3})^2}.\frac{\pi}{3}\right] = \pi\left[8 + \frac{32}{3\sqrt{3}}\pi\right]$$

$$\therefore \quad S = 8\pi\left[1 + \frac{4\pi}{3\sqrt{3}}\right].$$

__EXAMPLE 3.__ *Find the surface of the solid generated by the revolution of the lemniscate $r^2 = a^2\cos2\theta$ about the initial line.*

__SOLUTION.__ The equation of the curve is

$$r^2 = a^2\cos 2\theta. \qquad ...(1)$$

From (1) $r = 0$, when $\cos 2\theta = 0$ *i.e.*, $2\theta = \pm\,\pi/2$ or $\theta = \pm\,\pi/4$ and maximum value of r is a, when $\cos 2\theta = 1$ *i.e.*, $\theta = 0$. Thus the tracing of this curve is as fig. 23.

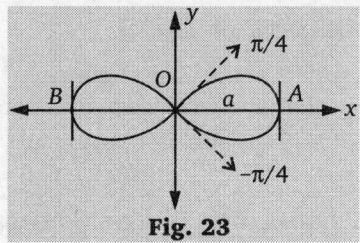

Fig. 23

Now differentiating (1) w.r.t. θ, we get

$$2r\frac{dr}{d\theta} = -2a^2\sin 2\theta$$

or

$$\frac{dr}{d\theta} = -\frac{a^2\sin 2\theta}{r}. \qquad ...(2)$$

$$\therefore \quad \frac{ds}{d\theta} = \sqrt{\left[r^2 + \left(\frac{dr}{d\theta}\right)^2\right]}$$

$$= \sqrt{\left[a^2\cos 2\theta + \frac{a^4\sin^2 2\theta}{r^2}\right]}$$

$$\qquad\qquad\qquad\qquad \text{[Using (1) and (2)]}$$

$$= \frac{\sqrt{(r^2 a^2\cos 2\theta + a^4\sin^2 2\theta)}}{r}$$

$$= \frac{\sqrt{(a^4\cos^2 2\theta + a^4\sin^2 2\theta)}}{r} \quad \text{[Using (1)]}$$

$$= \frac{a^2}{r}.$$

Since there are two loops in the curve and one loop of the curve lies between $\theta = -\pi/4$ and $\theta = \pi/4$. Also the curve is symmetrical about the pole as well as about the initial line. Let S be the surface of the solid generated by the revolution of the given curve. Then

$$S = 2\int_{0}^{\pi/4} 2\pi y\frac{ds}{d\theta}.d\theta, \text{ where } y = r\sin\theta$$

$$= 2\int_{0}^{\pi/4} 2\pi(r\sin\theta).\frac{a^2}{r}.d\theta \quad \text{[Using (2)]}$$

$$= 4\pi a^2\int_{0}^{\pi/4}\sin\theta\,d\theta = 4\pi a^2\left[-\cos\theta\right]_{0}^{\pi/4}$$

$$= 4\pi a^2\left[-\cos\frac{\pi}{4} + \cos 0\right]$$

$$= 4\pi a^2\left(-\frac{1}{\sqrt{2}} + 1\right)$$

$$\therefore \quad S = 4\pi a^2\left(1 - \frac{1}{\sqrt{2}}\right).$$

__EXAMPLE 4.__ *The lemniscate $r^2 = a^2\cos2\theta$ revolves about a tangent at the pole. Find the surface of the solid thus generated.*

__SOLUTION.__ We have $\qquad r^2 = a^2\cos 2\theta.$

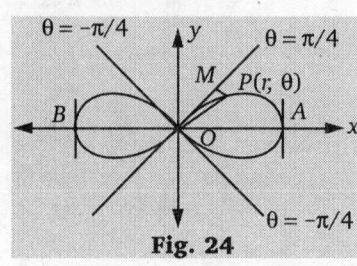

Fig. 24

Clearly, $\theta = \pi/4$ is one of the tangent at the pole.

Let $P(r, \theta)$ be any point on the curve. Draw PM as perpendicular from P to the line $\theta = \pi/4$. Then

$$\angle POM = \frac{\pi}{4} - \theta.$$

So in ΔPMO, $\qquad \dfrac{PM}{PO} = \sin\left(\dfrac{\pi}{4} - \theta\right)$

$\therefore \qquad\qquad PM = r \sin\left(\dfrac{\pi}{4} - \theta\right).$

Now differentiating (1) w.r.t. θ, we get

$$2r \frac{dr}{d\theta} = -2a^2 \sin 2\theta$$

or $\qquad\qquad \dfrac{dr}{d\theta} = -\dfrac{a^2 \sin 2\theta}{r}$

$\therefore \qquad \dfrac{ds}{d\theta} = \sqrt{r^2 + \left(\dfrac{dr}{d\theta}\right)^2}$

$$= \sqrt{a^2 \cos 2\theta + \frac{a^4 \sin^2 2\theta}{r^2}}$$

$$= \frac{1}{r}\sqrt{a^4 \cos^2 2\theta + a^4 \sin^2 2\theta}$$

$$= \frac{a^2}{r}$$

There are two loops in the curve and loop lies between $\theta = -\pi/4$ and $\theta = \pi/4$. Also the curve is symmetrical about the pole as well as about the initial line.

Let S be the surface of the solid generated by the revolution of the given curve about the line $\theta = \pi/4$. Then

$$S = 2\int_{-\pi/4}^{\pi/4} 2\pi(PM)\frac{ds}{d\theta}d\theta$$

$$= 4\pi \int_{-\pi/4}^{\pi/4} r\sin\left(\frac{\pi}{4} - \theta\right)\frac{a^2}{r}d\theta$$

$$= 4\pi a^2 \int_{-\pi/4}^{\pi/4} \sin\left(\frac{\pi}{4} - \theta\right)d\theta$$

$$= 4\pi a^2 \left[\cos\left(\frac{\pi}{4} - \theta\right)\right]_{-\pi/4}^{\pi/4}$$

$$= 4\pi a^2 \left[1 - \cos\frac{\pi}{2}\right] = 4\pi a^2(1-0) = 4\pi a^2.$$

EXAMPLE 5. *Find the surface of the solid generated by the revolution of the curve $x = a \cos^3 t$ and $y = a \sin^3 t$ about the x-axis.*

SOLUTION. We have $\qquad x = a\cos^3 t$ and $y = a\sin^3 t$.
This curve is symmetrical about both the axes.

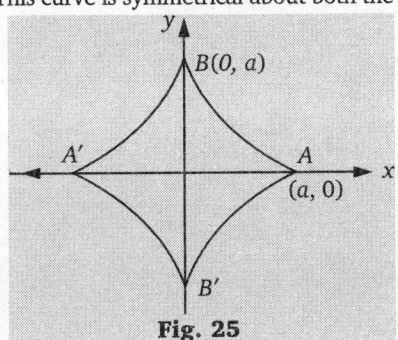

Fig. 25

At $A(a, 0)$, $t = 0$ and at $B(0, a)$, $t = \pi/2$.

Now $\quad \dfrac{dx}{dt} = -3a\cos^2 t \sin t$

$\qquad\quad \dfrac{dy}{dt} = 3a\sin^2 t\cos t$

$\therefore \quad \dfrac{ds}{dt} = \sqrt{\left(\dfrac{dx}{dt}\right)^2 + \left(\dfrac{dy}{dt}\right)^2}$

$$= 3a\sqrt{\cos^4 t\sin^2 t + \sin^4 t\cos^2 t}$$

$$= 3a\sin t\cos t = \frac{3a}{2}\sin 2t.$$

Let S be the surface of a solid of revolution of the given curve about x-axis. Then

$$S = 2\int_0^{\pi/2} 2\pi y\frac{ds}{dt}dt$$

$$= 4\pi\int_0^{\pi/2} a\sin^3 t \cdot \frac{3a}{2}\sin 2t\,dt$$

$$= 12\pi a^2\int_0^{\pi/2}\sin^4 t\cos t\,dt$$

$$= 12\pi a^2\left[\frac{\sin^5 t}{5}\right]_0^{\pi/2} = \frac{12\pi a^2}{5}.$$

EXAMPLE 6. *Find the surface of the solid generated by revolving the arc of the parabola $y^2 = 4ax$ bounded by its latusrectum about x-axis.*
(ROHTAK 2003)

SOLUTION. Let S be the surface of solid of revolution of arc LOL' of the parabola $y^2 = 4ax$ about x-axis. Then

$$S = \int_0^a 2\pi y\,ds$$

or $\qquad S = \int_0^a 2\pi y\,dx \cdot \dfrac{ds}{dx} \qquad\qquad …(1)$

We have

$$y^2 = 4ax \implies y = 2\sqrt{ax} \implies \frac{dy}{dx} = \frac{\sqrt{a}}{\sqrt{x}}$$

$\therefore \quad \dfrac{ds}{dx} = \sqrt{1 + \left(\dfrac{dy}{dx}\right)^2} = \sqrt{1 + \dfrac{a}{x}} = \sqrt{\dfrac{x+a}{x}}$

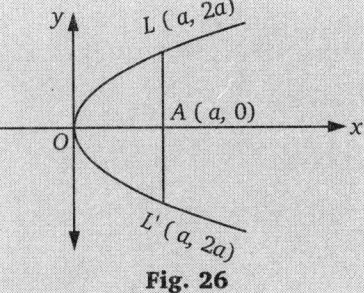

Fig. 26

Therefore,

$$S = 2\pi\int_0^a 2\sqrt{ax}\left[\sqrt{\frac{x+a}{x}}\right]dx$$

$$= 4\pi\sqrt{a}\int_0^a \sqrt{x+a}\,dx = 4\pi\sqrt{a}\left[\frac{2}{3}(x+a)^{3/2}\right]_0^a$$

$$= \frac{8\pi\sqrt{a}}{3}\left[(2a)^{3/2} - a^{3/2}\right]$$

$$= \frac{8\pi}{3}(2^{3/2} - 1)a^2 = \frac{8\pi}{3}(2\sqrt{2} - 1)a^2$$

Exercise-4.8

1. Find the curved surface of a hemi-sphere of radius a.

2. For a catenary $y = a \cosh (x/a)$, prove that
$$aS = \pi a (ax + sy).$$
where s is the length of the arc measured from the vertex, S is the area of curved surface of the solid generated by the revolution of the arc about x-axis.

3. Find the surface of the solid generated by the revolution of the astroid $x^{2/3} + y^{2/3} = a^{2/3}$ aboul x-axis.

4. Find the surface of the solid formed by the revolution, about the axis of y, of the part of the curve $ay^2 = x^3$ from $x = 0$ to $x = 4a$ which is above the x-axis.

5. Prove that the surface of the prolate spheroid formed by the revolution of the ellipse of essentricity e about its major axis is equal to $2\pi ab\left[\sqrt{(1-e^2)} + \dfrac{1}{e}\sin^{-1}e\right]$.

6. Prove that the surface of the oblate spheroid formed by the revolution of the ellipse of essentricity e about its minor axis is $2\pi a^2\left[1 + \dfrac{1-e^2}{2e}\log\left(\dfrac{1+e}{1-e}\right)\right]$.

7. Find the surface area of the solid generated by revolution of the cycloid $x = a (\theta - \sin \theta), y = a(1 - \cos \theta)$ about x-axis.
   (VTU-2003)

8. The portion between two consecutive cusps of the cycloid, $x = a (\theta + \sin \theta), y = a (1 + \cos \theta)$ is revolved about x-axis. Prove that the area of the surface so formed is to the area of the cycloid as $64 : 9$.

9. Prove that the surface area of the solid generated by the revolution of the loop of the curve $x = t^2$, $y = t - \dfrac{1}{3}t^3$ about x-axis is 3π.

10. Find the surface of the solid formed by the revolution of the cardioid $r = a (1 + \cos \theta)$ about the initial line.
    (VTU-2009, JNTU-2003, RAJASTHAN-2006)

11. The arc of the cardioid $r = a (1 + \cos \theta)$ included between $-\dfrac{\pi}{2} \le \theta \le \dfrac{\pi}{2}$ is rotated about the line $\theta = \pi / 2$. Find the area of the surface thus generated.

12. Find the area of the surface of revolution formed by revolving the curve $r = 2a \cos \theta$ about the initial line. (VTU 2009)

13. A circular arc revolves aboul its chord. Find the area of the surface generated, when 2α is the angle subtended by the arc at the centre.

Hints to Selected Problems

1. $S = 2\pi\int_0^a 2\pi x \, ds = 2\pi\int_0^a \sqrt{a^2 - y^2}\cdot\dfrac{ds}{dy}\cdot dy.$

2. $y = a\cosh\dfrac{x}{a} \Rightarrow \dfrac{dy}{dx} = \sinh\left(\dfrac{x}{a}\right)$
   $\Rightarrow \dfrac{ds}{dx} = \cosh\dfrac{x}{a} \Rightarrow s = a\sinh\dfrac{x}{a}.$
   Now $S = 2\pi\int_0^x y\dfrac{ds}{dx}\cdot dx = 2\pi\int_0^x a\cosh\dfrac{x}{a}\cdot\cosh\dfrac{x}{a}\cdot dx.$

3. From the given curve, we obtained $\dfrac{dy}{dx} = -\left(\dfrac{y}{x}\right)^{1/3}$
   $\therefore \qquad \dfrac{ds}{dx} = \sqrt{1 + \left(\dfrac{dy}{dx}\right)^2} = \dfrac{a^{1/3}}{x^{1/3}}.$
   Now use $\quad S = 4\pi\int_0^a y\cdot\dfrac{ds}{dx}\cdot dx$
   $\qquad\qquad = 4\pi\int_0^a (a^{2/3} - x^{2/3})^{3/2}\dfrac{a^{1/3}}{x^{1/3}}dx.$
   Then put $x^{2/3} = a^{2/3}\sin^2\theta.$

4. $S = \int_0^{8a} 2\pi x\cdot\dfrac{ds}{dy}\cdot dy.$ ...(1)
   Also $\dfrac{ds}{dy} = \sqrt{1 + \left(\dfrac{dx}{dy}\right)^2} = \dfrac{\sqrt{9x^4 + 4a^2y^2}}{3x^2}.$
   Put in (1) and then solve.

5. Given that $\dfrac{x^2}{a^2} + \dfrac{y^2}{b^2} = 1$ ...(1)
   and $\qquad b^2 = a^2 (1 - e^2)$
   Then $\qquad S = 2\int_0^a 2\pi y\, ds.$

From (1), $\dfrac{ds}{dx} = -\left(\dfrac{b^2 x}{a^2 y}\right).$

Then find $\dfrac{ds}{dx}$ and put in $S = 2\int_0^a 2\pi y\, ds.$

9. $x = t^2, y = t - \dfrac{t^3}{3} \Rightarrow \dfrac{dx}{dt} = 2t, \dfrac{dy}{dt} = 1 - t^2.$
   $\therefore \qquad \dfrac{dy}{dx} = \dfrac{1-t^2}{2t}.$
   Now by using the formula
   $\left(\dfrac{ds}{dt}\right)^2 = \left(\dfrac{dx}{dt}\right)^2 + \left(\dfrac{dy}{dt}\right)^2.$
   We obtained $\dfrac{ds}{dt} = 1 + t^2.$
   Then $\qquad S = \int_0^{\sqrt{3}} 2\pi y\cdot\dfrac{ds}{dt}\cdot dt.$

10. $r = a(1 + \cos\theta) \Rightarrow \dfrac{dr}{d\theta} = -a\sin\theta$
    $\Rightarrow \dfrac{ds}{d\theta} = \sqrt{r^2 + \left(\dfrac{dr}{d\theta}\right)^2} = 2a\cos\dfrac{\theta}{2}$
    Then use $S = \int_0^\pi 2\pi(r\sin\theta)\cdot\dfrac{ds}{d\theta}d\theta.$

11. Here, we have $\dfrac{dr}{d\theta} = -a\sin\theta$
    $\Rightarrow \dfrac{ds}{d\theta} = \sqrt{r^2 + \left(\dfrac{dr}{d\theta}\right)^2} = 2a\cos\dfrac{\theta}{2}$

12. Here, we have $\dfrac{dr}{d\theta} = -2a\sin\theta.$

$$\therefore \qquad \frac{ds}{d\theta} = \sqrt{r^2 + \left(\frac{dr}{d\theta}\right)^2} = 2a. \qquad\qquad \text{Then we use} \quad S = \int_0^{\pi/2} 2\pi y \frac{ds}{d\theta} d\theta.$$

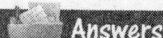 **Answers**

1. $2\pi a^2$ **3.** $\dfrac{12\pi a^2}{5}$ **4.** $\dfrac{128}{1215}\pi a^2[125\sqrt{10}+1]$ **7.** $\dfrac{64}{3}\pi a^2$ **10.** $\dfrac{32}{5}\pi a^2$ **11.** $\dfrac{48}{5}\sqrt{2}\,\pi a^2$ **12.** $4\pi a^2$ **13.** $4\pi a^2(\sin\alpha - \alpha\cos\alpha)$.

4.5.5 VOLUME OF SOLID OF REVOLUTIONS

When a plane area is revolved about any fixed line lying in the same plane, a solid (body) is generated. This solid (body) is called a solid of revolution. Also about the x-axis, two right circular cylinders are formed of volumes $\pi y^2 \delta x$ and $\pi(y+\delta y)^2 \delta x$ respectively. Since the plane area $PMNQ$ lies between $PMNP'$ and $Q'MNQ$. Therefore δV_1 lies between $\pi y^2 \delta x$ and $\pi(y+\delta y)^2 \delta x$. Then

$$\pi y^2 \,\delta x < \delta V_1 < \pi(y+\delta y)^2\,\delta x.$$

Divide by δx, we get

$$\pi y^2 < \frac{\delta V_1}{\delta x} < \pi(y+\delta y)^2.$$

As $Q \to P$, $\delta x \to 0$ and $\delta y \to 0$ so taking the limit as $\delta x \to 0$, we get

$$\pi y^2 < \lim_{\delta x \to 0}\frac{\delta V_1}{\delta x} < \pi y^2 \quad \text{or} \quad \lim_{\delta x \to 0}\frac{\delta V_1}{\delta x} = \pi y^2$$

or

$$\frac{dV_1}{dx} = \pi y^2 \qquad\qquad \text{or} \qquad dV_1 = \pi y^2\,dx.$$

Now integrating w.r.t. x between the limits $x = a$ to $x = b$, we get $\int_a^b dV_1 = \int_a^b \pi y^2\,dx$.

$\therefore$ Volume generated by the plane area $ABDC = \int_a^b \pi y^2 dx$

or

$$V_1 = \int_a^b \pi y^2\,dx$$

Where V_1 is the required volume of a solid formed by the plane area bounded by the curve $y = f(x)$, the ordinates $x = a$ and $x = b$ and x-axis about x-axis. Here, fixed line is called axis of revolution.

(1) REVOLUTION ABOUT x-AXIS

Let V be the volume of a solid which is generated by the revolution of a plane area bounded by the curve $y = f(x)$, the ordinates $x = a$, $x = b$ and x-axis about the x-axis. Then

$$V = \int_a^b \pi y^2\,dx$$

where $y = f(x)$ is a continuous and single valued function defined on $[a, b]$.

Proof. Let us assume that the curve $y = f(x)$ does not cut the x-axis and let AB be the arc of $y = f(x)$ between the ordinates $x = a$ and $x = b$ as shown in fig. 27.

Let $P(x, y)$ and $Q(x + \delta x, y + \delta y)$ be two neighbouring points on the curve $y = f(x)$. Draw the perpendiculars PM and QN to the x-axis about which the plane area $ACDB$ is revolved. Also PP' is the perpendicular to QN and QQ' is the perpendicular to PM, where (Q' is a point on MP when P produced to Q'. Let V_1 be the volume of the solid formed by the revolution of plane area $ACMP$ about x-axis and $(V_1 + \delta V_1)$ be the volume of the solid formed by the revolution of the plane area $ACNQ$. Then δV_1 is the volume of a solid formed by the revolution of the plane area $PMNQ$ about x-axis.

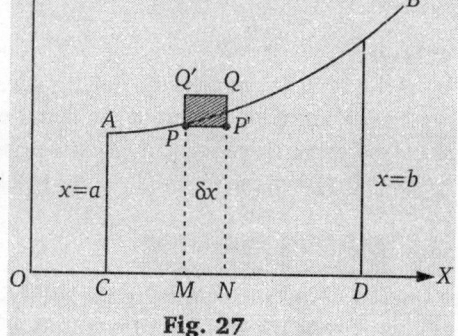

Fig. 27

(2) REVOLUTION ABOUT y-AXIS

The volume of a solid formed by the revolution of a plane area bounded by the curve $x = f(y)$ and the lines $y = a$ and $y = b$ and y-axis about y-axis is

$$V = \int_a^b \pi x^2\,dy.$$

(3) REVOLUTION ABOUT ANY LINE

The volume of a solid formed by the revolution of a plane area bounded by the arc AB and the lines AC and BD and the axis CD about any line CD (different from x-axis and y-axis) is

$$V = \int_{OC}^{OD} \pi(PM)^2\,d(OM)$$

where PM is the length of perpendicular from any point P on the arc AB to the axis CD and O be any fixed point on the axis CD.

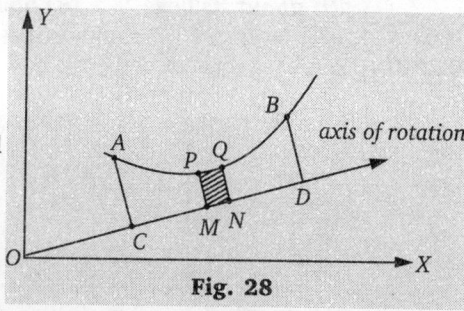

Fig. 28

4.5.6 Volume of a Solid of Revolution when the Equation of the Curve are in Different Forms

(1) EQUATION OF CURVE IN PARAMETRIC FORMS

Suppose the equation of a generating curve are in parametric form $x = f(t)$ and $y = g(t)$, then the volume of the solid generated by the revolution of the plane area bounded by the given curve, axis of x and the ordinates at the points where $t = a$ and $t = b$, about x-axis is

$$V = \int_a^b \pi y^2 \left(\frac{dx}{dt} \right).dt = \int_a^b \pi[g(t)]^2 \frac{dx}{dt}.dt \quad \text{where} \quad \frac{dx}{dt} = \frac{d}{dt}[f(t)].$$

Similarly, the volume of a solid formed by the revolution of a plane area bounded by $x = f(t)$, $y = g(t)$ axis of y and the two absciassae where $t = a$ and $t = b$ about y-axis is

$$V = \int_a^b \pi x^2 \left(\frac{dy}{dt} \right).dt = \int_a^b \pi[f(t)]^2 \frac{dy}{dt}.dt \quad \text{where} \quad \frac{dy}{dt} = \frac{d}{dt}[g(t)].$$

(2) EQUATION OF CURVE IN POLAR FORM

Suppose the equation of a curve in polar form is $r = f(\theta)$ where $x = r \cos \theta$, $y = r \sin \theta$, then the volume of a solid generated by the revolution of the plane area of the curve about the initial line (x-axis) between the lines $\theta = \alpha$ and $\theta = \beta$ is given by

$$V = \int_\alpha^\beta \pi y^2 \frac{dx}{d\theta}.d\theta = \int_\alpha^\beta \pi(r \sin \theta)^2 \frac{d}{d\theta}(r \cos \theta)d\theta \qquad \text{where} \qquad r = f(\theta).$$

Similarly, the volume of a solid formed by the revolution of the plane area bounded by the curve $r = f(\theta)$ and the lines $\theta = \alpha$ and $\theta = \beta$ about the line $\theta = \pi/2$ (y-axis) is given by

$$V = \int_\alpha^\beta \pi x^2 \frac{dy}{d\theta}.d\theta = \int_\alpha^\beta \pi(r \cos \theta)^2 \frac{d}{d\theta}(r \sin \theta)d\theta \qquad \text{where} \qquad r = f(\theta).$$

(3) FORMULAE FOR FINDING THE VOLUME IN CASE OF POLAR FORM.

(i) The volume of a solid formed by the revolution of the plane area bounded by the curve $r = f(\theta)$ and the radii vectors $\theta = \alpha$ and $\theta = \beta$ about the initial line *i.e.*, $\theta = 0$ (x-axis) is

$$V = \int_\alpha^\beta \frac{2}{3}\pi r^3 \sin \theta \, d\theta.$$

(ii) The volume of a solid formed by the revolution of a plane area bounded by $r = f(\theta)$ and $\theta = \alpha$, $\theta = \beta$ about the line $\theta = \pi/2$ (y-axis) is given by

$$V = \int_\alpha^\beta \frac{2}{3}\pi r^3 \cos \theta \, d\theta.$$

(iii) The volume of a solid formed by the revolution of a plane area bounded by $r = f(\theta)$ and $\theta = \alpha$, $\theta = \beta$ about any line $\theta = \gamma$ is given by

$$V = \int_\alpha^\beta \frac{2}{3}\pi r^3 \sin(\theta - \gamma)d\theta.$$

✒ Remarks

- If the curve is symmetrical about x-axis. then the portion of the curve above x-axis overlaps the other portion of the curve below x-axis during the revolution. So that the volume shall not double between the bounding points.
- If the curve is symmetrical about x-axis and the volume of a solid generated by the revolution of the plane area about y-axis is required, then the required volume will be double the volume which is obtained by the revolution of half of the symmetrical curve.

Solved Examples

EXAMPLE 1. *Find the volume of the solid formed by revolving the cycloid $x = a(\theta - \sin \theta)$, $y = (1 - \cos \theta)$*

    (i) about its base   *(ii) about y-axis.*

(VTU- 2003, 05; KURUKSHETRA- 2006)

SOLUTION . Since the equation of the cyloid is

$$x = a(\theta - \sin \theta), \ y = a(1 - \cos \theta) \ ...(1)$$

The tracing of this curve is given below :

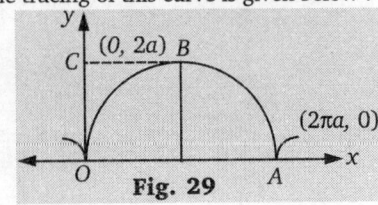

Fig. 29

(i) In the above fig. base is the axis of x so the volume of the solid formed by revolving the cycloid given in (1) about its base (x-axis) between $x = 0$ to $x = 2\pi a$ where $\theta = 0$ to $\theta = 2\pi$, is given by

$$\Rightarrow \quad V = \int_0^{2\pi a} \pi y^2 dx = \pi \int_0^{2\pi} y^2 \left(\frac{dx}{d\theta} \right).d\theta$$

$$= \pi \int_0^{2\pi} a^2(1 - \cos \theta)^2.a(1 - \cos \theta)d\theta$$

$$[\text{Using (1)}]$$

$$= \pi a^3 \int_0^{2\pi} (1 - \cos \theta)^3 \, d\theta$$

$$\Rightarrow \quad V = \pi a^3 \int_0^{2\pi} 8\sin^6(\theta/2)d\theta. \qquad ...(2)$$

Let us put $\theta/2 = \phi$ so $d\theta = 2d\phi$ and ϕ varies from 0 to π, then (2) becomes

$$V = 8\pi a^3 \int_0^\pi \sin^6 \phi (2d\phi) = 16\pi a^3 \int_0^\pi \sin^6 \phi \, d\phi$$

$$= 16\pi a^3 . 2\int_0^{\pi/2} \sin^6 \phi \, d\phi$$

$$= 32\pi a^3 \int_0^{\pi/2} \sin^6 \phi \, d\phi$$

$$= 32\pi a^3 \left[\frac{(6-1)(6-3)(6-5)}{6.4.2} . \frac{\pi}{2} \right]$$

$$V = 5\pi^2 a^3. \qquad \text{(By Walli's formula)}$$

(ii) Let V be the volume of the solid formed by revolving of the cycloid about y-axis. This volume is the difference of the volume generated by the revolution of the area $OABCO$ and the volume generated by the revolution of the area $OBCO$. Since we have that at A, $\theta = 2\pi$, at O, $\theta = 0$ and at B, $\theta = \pi$. Therefore, the volume generated by the revolution of the area $OABCO$ about y-axis is given by V_1 (say)

$$V_1 = \int_{2\pi}^\pi \pi x^2 \frac{dy}{d\theta} . d\theta$$

$$= \pi \int_{2\pi}^\pi a^2 (\theta - \sin\theta)^2 . a\sin\theta \, d\theta$$

$$= \pi a^3 \int_{2\pi}^\pi (\theta^2 + \sin^2\theta - 2\theta\sin\theta)\sin\theta \, d\theta$$

$$= \pi a^3 \int_{2\pi}^\pi (\theta^2 \sin\theta + \sin^3\theta - 2\theta\sin^2\theta) \, d\theta$$

$$= \pi a^3 \int_{2\pi}^\pi \left(\theta^2 \sin\theta + \frac{3}{4}\sin\theta - \frac{1}{4}\sin 3\theta - \theta + \theta\cos 2\theta \right) d\theta$$

$$= \pi a^3 \left[-\theta^2 \cos\theta + 2\theta\sin\theta + 2\cos\theta - \frac{3}{4}\cos\theta \right.$$
$$\left. + \frac{1}{12}\cos 3\theta - \frac{\theta^2}{2} + \frac{1}{2}\theta\sin 2\theta + \frac{1}{4}\cos 2\theta \right]_{2\pi}^\pi$$

(where $\int \theta^2 \sin\theta \, d\theta$ and $\int \theta\cos 2\theta \, d\theta$ are solved by integration by parts.)

$$= \pi a^3 \left[\left(\pi^2 - 2 + \frac{3}{4} - \frac{1}{12} - \frac{\pi^2}{2} + \frac{1}{4} \right) \right.$$
$$\left. - \left(-4\pi^2 + 2 - \frac{3}{4} + \frac{1}{12} - 2\pi^2 + \frac{1}{4} \right) \right]$$

$$= \pi a^3 \left(\frac{13}{2}\pi^2 - \frac{8}{3} \right)$$

and let V_2 be the volume generated by the revolution of the area $OBCO$ about y-axis, then

$$V_2 = \int_0^\pi \pi x^2 \frac{dy}{d\theta} d\theta$$

$$= \pi \int_0^\pi a^2 (\theta - \sin\theta)^2 a\sin\theta \, d\theta$$

$$= \pi a^3 \int_0^\pi (\theta^2 \sin\theta + \sin^3\theta - 2\theta\sin^2\theta) \, d\theta$$

$$= \pi a^3 \left[-\theta^2 \cos\theta + 2\theta\sin\theta + 2\cos\theta - \frac{3}{4}\cos\theta \right.$$
$$\left. + \frac{1}{12}\cos 3\theta - \frac{\theta^2}{2} + \frac{1}{2}\theta\sin 2\theta + \frac{1}{4}\cos 2\theta \right]_0^\pi$$

$$= \pi a^3 \left[\left(\pi^2 - 2 + \frac{3}{4} - \frac{1}{12} - \frac{\pi^2}{2} + \frac{1}{4} \right) \right.$$
$$\left. - \left(2 - \frac{3}{4} + \frac{1}{12} + \frac{1}{4} \right) \right]$$

$$V_2 = \pi a^3 \left(\frac{1}{2}\pi^2 - \frac{8}{3} \right).$$

$$\therefore \quad V = V_1 - V_2$$

$$= \pi a^3 \left(\frac{13}{2}\pi^2 - \frac{8}{3} \right) - \pi a^3 \left(\frac{1}{2}\pi^2 - \frac{8}{3} \right) = \pi a^3 (6\pi^2)$$

$$V = 6\pi^3 a^3.$$

EXAMPLE 2. *Find the volume of a solid formed by the revolution of the loop of the curve $y^2 (a+x) = x^2(a-x)$ about x-axis.*

SOLUTION . Clearly, the given curve is symmetrical about x-axis and the curve cuts the x-axis only at the points $(0, 0)$ and $(a, 0)$ so the loop exists between these points. The tracing of this curve is shown in fig. 32.

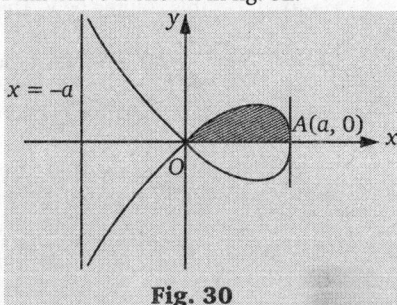

Fig. 30

Therefore, the required volume is the volume of a solid formed by the revolution of upper half of the loop of the curve about x-axis where x varies from 0 to a. Then

$$V = \int_0^a \pi y^2 \, dx$$

$$= \pi \int_0^a \frac{x^2(a-x)}{a+x} dx \qquad [\because y^2(a+x) = x^2(a-x)]$$

$$= \pi \int_0^a \frac{x^2(2a-a-x)}{a+x} dx = \pi \int_0^a \frac{2ax^2}{a+x} dx - \pi \int_0^a x^2 \, dx$$

$$= 2a\pi \int_0^a \frac{(x^2 - a^2 + a^2)}{a+x} dx - \pi \int_0^a x^2 dx$$

$$= 2a\pi \left[\int_0^a (x-a) \, dx + a^2 \int_0^a \frac{dx}{a+x} \right] - \pi \int_0^a x^2 dx$$

$$= 2a\pi \left[\frac{x^2}{2} - ax + a^2 \log(a+x) \right]_0^a - \pi \left[\frac{x^3}{3} \right]_0^a$$

$$= 2a\pi \left[\frac{a^2}{2} - a^2 + a^2 \log 2a - a^2 \log a \right] - \frac{\pi a^3}{3}$$

$$= 2a\pi \left[-\frac{a^2}{2} + a^2 \log \frac{2a}{a} \right] - \frac{\pi a^3}{3}$$

$$= -\pi a^3 + 2a^3\pi \log 2 - \frac{\pi a^3}{3} = 2a^3\pi \log 2 - \frac{4\pi}{3}a^3$$

$$\Rightarrow \quad V = 2\pi a^3 \left[\log 2 - \frac{2}{3} \right].$$

EXAMPLE 3. *Find the volume of the solid generated by the revolution of the cardioid $r = a(1+\cos\theta)$ about the initial line.* (VTU-2010; KURUKSHATRA-2009)

SOLUTION . Obviously, the curve is symmetrical about the initial line and $r = 0$ when $\cos\theta = -1$ i.e., $\theta = \pi$ and the maximum value of $r = 2a$ when $\cos\theta = 1$ i.e., $\theta = 0$. Thus the tracing of the curve is as under :

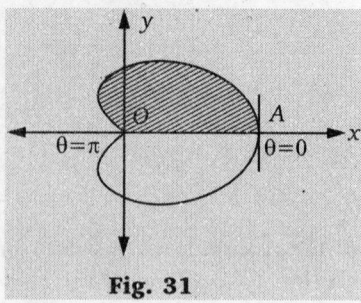

Fig. 31

Therefore, the required volume is the volume of solid generated by the revolution of the upper half of the curve between $\theta = 0$ and $\theta = \pi$ about initial line (*x*-axis). Let this volume be *V*. Then

$$V = \int_{\theta=0}^{\theta=\pi} \pi y^2 \frac{dy}{d\theta} . d\theta \quad \text{(as } \theta \text{ increases } x \text{ decreases so } \frac{dx}{d\theta}$$

will have to take negative.)

$$= \pi \int_0^\pi (r \sin\theta)^2 . \frac{d}{d\theta}(r \cos\theta) d\theta \quad (\because x = r \cos\theta, y = r \sin\theta)$$

$$= -\pi \int_0^\pi a^2 (1 + \cos\theta)^2 \sin^2\theta \frac{d}{d\theta}[a(1 + \cos\theta)\cos\theta] d\theta$$

$$[\because r = a(1 + \cos\theta)]$$

$$= -\pi a^3 \int_0^\pi (1 + \cos\theta)^2 \sin^2\theta .(-\sin\theta - 2\cos\theta\sin\theta) d\theta$$

$$= +\pi a^3 \int_0^\pi (1 + \cos^2\theta + 2\cos\theta)(1 + 2\cos\theta).\sin^3\theta d\theta$$

$$= +\pi a^3 \int_0^\pi (\sin^3\theta + 4\cos\theta\sin^3\theta + 5\cos^2\theta\sin^3\theta$$

$$+ 2\cos^3\theta\sin^3\theta) d\theta$$

$$= +\pi a^3 \left[\int_0^\pi \sin^3\theta d\theta + 4\int_0^\pi \cos\theta\sin^3\theta d\theta \right.$$

$$\left. + 5\int_0^\pi \cos^2\theta\sin^3\theta d\theta + 2\int_0^\pi \cos^3\theta\sin^3\theta d\theta \right]$$

$$= \pi a^3 \left[\int_0^\pi \sin^3\theta d\theta + 5\int_0^\pi \cos^2\theta\sin^3\theta d\theta \right]$$

(The second and fourth integral vanish by the property of definite integral.)

$$= \pi a^3 \left[2\int_0^{\pi/2} \sin^3\theta d\theta + 10\int_0^{\pi/2} \cos^2\theta\sin^3\theta d\theta \right]$$

(By the property of definite integral)

$$= \pi a^3 \left[2.\frac{(3-1)}{3.1}.1 + 10.\frac{(2-1)(3-1)}{5.3.1}.1 \right]$$

(By Walli's formula)

$$= \pi a^3 \left[\frac{4}{3} + \frac{4}{3} \right] = \frac{8}{3}\pi a^3.$$

$$\therefore \quad V = \frac{8}{3}\pi a^3.$$

EXAMPLE 4. *Find the volume of the solid generated by the revolution of the tractrix* $x = a\cos t + \frac{a}{2}\log\tan^2\left(\frac{t}{2}\right), y = a\sin t$ *about its*

asymptote.

SOLUTION . We have $\quad x = a\cos t + \frac{a}{2}\log\tan^2\left(\frac{t}{2}\right)$

and $\qquad y = a\sin t.$

Now $\qquad \frac{dx}{dt} = -a\sin t + \frac{a}{\sin t} = \frac{a\cos^2 t}{\sin t}.$

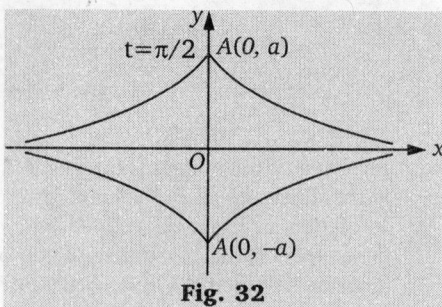

Fig. 32

Here, *x*-axis is the asymptote of the given curve so that for *x*-axis *i.e.*, $y = 0$, $t = \pi/2$ and at $A(0, a)$, $t = 0$.

Let *V* be the volume of the solid generated by the revolution of the tractrix. Then

$$V = 2\int_{\pi/2}^0 \pi y^2 \frac{dx}{dt} dt$$

$$= 2\pi \int_{\pi/2}^0 a^2 \sin^2 t \frac{a\cos^2 t}{\sin t} dt$$

$$= 2\pi a^3 \int_{\pi/2}^0 \cos^2 t \sin t \, dt$$

$$= 2\pi a^3 \left[\frac{\cos^3 t}{3} \right]_{\pi/2}^0 = 2\pi a^3 \left[\frac{1}{3} - 0 \right] = \frac{2}{3}\pi a^3.$$

EXAMPLE 5. *Find the volume of the solid generated by revolution of one loop of the lemniscate* $r^2 = a^2\cos 2\theta$ *about the line* $\theta = \pi/2.$ (UKTU 2006)

SOLUTION . We have $\qquad r^2 = a^2\cos 2\theta.$ ...(1)

Clearly one loop of the curve lies between

$$\theta = -\pi/4 \quad \text{and} \quad \theta = \pi/4.$$

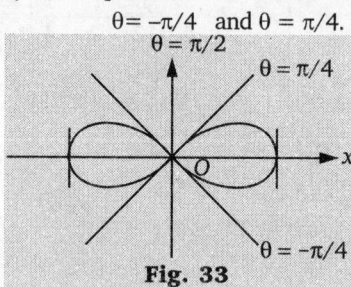

Fig. 33

Let *V* be the volume of the solid generated by revolving one loop of the curve (1) about the line $\theta = \pi/2$. Then

$$V = \int_{-\pi/4}^{\pi/4} \frac{2}{3}\pi r^3 \cos\theta d\theta$$

$$= \frac{2}{3}\pi \int_{-\pi/4}^{\pi/4} a^3 (\cos 2\theta)^{3/2} \cos\theta d\theta$$

$$= \frac{2\pi a^3}{3} \int_{-\pi/4}^{\pi/4} (1 - 2\sin^2\theta)^{3/2} \cos\theta d\theta$$

$$= \frac{4\pi a^3}{3} \int_0^{\pi/4} (1 - 2\sin^2\theta)^{3/2} \cos\theta d\theta$$

$$= \frac{4\pi a^3}{3} \int_0^{\pi/4} (1-\sin^2\phi)^{3/2} \frac{1}{\sqrt{2}} \cos\phi \, d\phi$$

$$\text{put } \sqrt{2}\sin\theta = \sin\phi$$

$$= \frac{4\pi a^3}{3\sqrt{2}} \int_0^{\pi/2} \cos^4\phi \, d\phi$$

$$= \frac{4\pi a^3}{3\sqrt{2}} \left[\frac{(4-1)(4-3)}{4.(4-2)} \cdot \frac{\pi}{2} \right] = \frac{\pi^2 a^3}{4\sqrt{2}}.$$

EXAMPLE 8. *Find the volume of the solid generated by the revolution of the cissoid $y^2(2a-x) = x^3$ about its asymptotes.* (VTU -2000)

SOLUTION . Let $P(x, y)$ be any point on the curve and let V be the volume of the solid generated by revolution of $y^2(2a-x)=x^3$ about its asymptote $x = 2a$ then

$$V = \int_{-\infty}^{\infty} \pi(PM)^2 dy$$

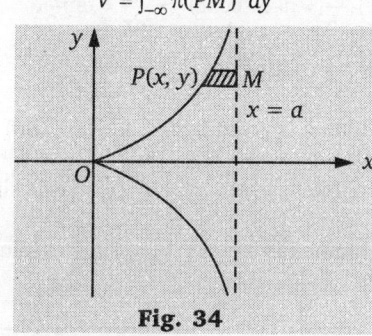

Fig. 34

$$= 2\int_0^\infty \pi(PM)^2 dy$$

$$V = 2\pi\int_0^\infty (2a-x)^2 dy \qquad \ldots(1)$$

Now, $y^2(2a-x) = x^3$

$$\Rightarrow \quad y = \frac{x^{3/2}}{\sqrt{2a-x}} \Rightarrow \frac{dy}{dx} = \frac{\sqrt{x}.(3a-x)}{(2a-x)^{3/2}}$$

Then $V = 2\pi\int_0^{2a} \frac{(2a-x)^2(3a-x)\sqrt{x}}{(2a-x)^{3/2}} dx$

$$[\text{ as } y \to 0 \text{ to } \infty, x \to 0 \text{ to } 2a]$$

$$= 2\pi\int_0^{2a} \sqrt{2ax-x^2}(3a-x)dx$$

$$= \pi\left[\int_0^{2a} 2(a-x)\sqrt{2ax-x^2} \, dx \right.$$

$$\left. + 4a\int_0^{2a} \sqrt{a^2-(x-a)^2}dx \right]$$

$$= \pi\left[\frac{2}{3}(2ax-x^2) + 4a\left\{ \frac{(x-a)}{2}\sqrt{a^2-(x-a)^2} \right. \right.$$

$$\left. \left. + \frac{a^2}{2}\sin^{-1}\frac{(x-a)}{a} \right\} \right]_0^{2a}$$

$$= \pi\left\{ 0 + 4a\left\{ \frac{a^2\pi}{2} \right\} \right\} = 2\pi^2 a^3.$$

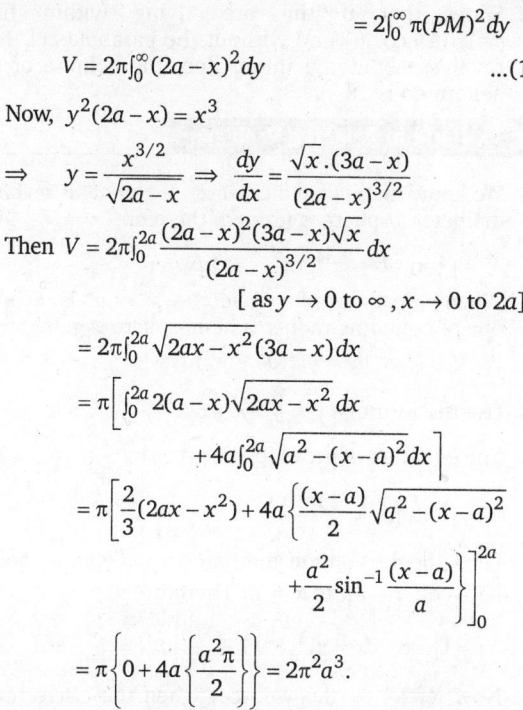

Exercise-4.9

1. Show that the volume of sphere of radius a is $\frac{4}{3}\pi a^3$. (SVTV- 2007)

2. Find the volume of a hemi-sphere,

3. (i) The part of a parabola $y^2 = 4ax$ cut off by the latus rectum revolves about the tangent at the vertex. Find the volume of a solid thus generated. (VTU- 2009)

   (ii) Find the volume of the paraboloid generated by the revolution about the axis of x, of the parabola $y^2 = 4ax$ from $x = 0$ to x to $x= h$.

4. Find the volume of the solid generated by the revolution of an arc of the catenary $y = c \cosh (x/c)$ about x-axis.

5. Find the volume of solid generated by the revolution of the loop of the curve $y^2 = x^2 (a - x)$ about the axis of x.

6. Prove that the volume of this solid generated by the revolution of an ellipse around its minor axis is a mean proportional between those generated by the revolution of the ellipse and of the auxiliary circle about the major axis.

7. Find the volume of a solid generated by the loop of the curve $y^2(a+ x)= x^2(3a - x)$ about x-axis. (MARATHWADA-2008)

8. The area of the curves $x^{2/3}+ y^{2/3} = a^{2/3}$ lying in the first quadrant revolves about x-axis. Find the volume of the solid generated. (UPTU-2010, SVTU-2008)

9. Show that the volume of the solid generated by the revolution of the upper half of the loop of the curve $y^2 = x^2(2-x)$ about x-axis is $\frac{4}{3}\pi$. .

10. Find the volume of the solid generated by revolving the loop of the curve $a^2y^2 = x^2(2a-x)(x-a)$ about x-axis.

11. Show that the volume of the solid generated by the revolution of the curve $(a-x)y^2 = a^2 x$, about its asymptote is $\frac{1}{2}\pi^2 a^3$.

12. The area cut off from the parabola $y^2 = 4ax$ by the chord joining the vertex to an end of the latusrectum is rotated through four right angles about the chord. Find the volume of the solid thus generated.

13. Prove that the volume of the reel formed by the revolution of the cycloid $x = a (t + \sin t)$, $y = a (1 - \cos t)$ about the tangent at the vertex is $\pi^2 a^3$.

14. Find the volume of the solid generated by the revolution of the cycloid $x = a(t + \sin t)$, $y = a(t - \cos t)$, $0 \le t \le n$.
    (i) about the x-axis (ii) about its base.

15. Find the volume of the solid generated by the revolution of the loop of the curves $x= t^2$, $y= t - \frac{1}{2}t^3$ about x-axis.

16. Find the volume of the solid generated by the revolution of the cissoid $x=2a\sin^2 t$, $y=2a\sin^3 t / \cos t$ about its asymptote. (PTU-2001)

17. Find the volume of the solid generated by the revolution of the cardioid $r= a(1-\cos\theta)$ about the initial line. (PTU-2006)

18. (i) Find the volume of the solid formed by revolving one loop of the curve $r^2 = a^2 \cos 2\theta$ about the initial line.

    (ii) The lemniscate $r^2 = a^2 \cos 2\theta$ revolves about a tangent at the pole. Show that the volume generated is $\frac{\pi^2 a^3}{4}$.

19. Show that the volume of the solid formed by the revolution of the curve $r = a + b \cos\theta(a > b)$ about the initial line is $\frac{4}{3}\pi a(a^2+b^2)$.

20. Show that if the area lying within the cardioid $r = 2a(1 + \cos\theta)$ and without the parabola $r(1 + \cos\theta) = 2a$ revolves about the initial line, the volume of a solid thus generated is $18\,\pi a^3$.

21. Find the volume of the solid generated by the revolution of the curve $r = 2a\cos\theta$ about the initial line.

22. Find the volume of the solid generated by the revolution of the curve $x(b^2 + y^2) = b^3$ about its asymptote.

Hints to Selected Problems

1. We know that, when a circle $x^2 + y^2 = a^2$ is revolved about its diameter a sphere is formed, therefore
$$V = \int_{-a}^{a} \pi y^2\,dx = \pi \int_{-a}^{a}(a^2 - x^2)\,dx.$$

2. When a quadrant of a circle $x^2 + y^2 = a^2$ is revolved about its one of bounding radius, a hemi-sphere is generated.
$$\therefore \quad V = \int_{0}^{a} \pi x^2\,dy = \pi \int_{0}^{a}(a^2 - y^2)\,dy.$$

4. Use the formula $V = \int_{0}^{x} \pi y^2\,dx$.

6. The equation of the ellipse is given by $\dfrac{x^2}{a^2} + \dfrac{y^2}{b^2} = 1$. Therefore,
$$V = \int_{-b}^{b} \pi x^2\,dy = \frac{4\pi a^2 b}{3}.$$

Let V_1 be the volume generated by an ellipse about major axis between $x = -a$ to $x = a$. Therefore
$$V_1 = \int_{-a}^{a} \pi y^2\,dx = \pi \int_{-a}^{a} b^2\left(1 - \frac{x^2}{a^2}\right)dx = \sqrt{\frac{4\pi}{3}b^2 a}.$$

Now let V_2 be the volume, when the circle revolves about major axis.

7. Let V be the volume of a solid generated by the revolution of the loop about x-axis between $x = 0$ and $x = 3a$. Then, we have $V = \int_{0}^{3a} \pi y^2\,dx = \pi \int_{0}^{3a} \dfrac{x^2(3a - x)}{a + x}\,dx$.

8. Here $V = \int_{0}^{a} \pi y^2\,dx$. Change the given equation into polar form by assuming $x = a\cos^3\theta$ and $y = a\sin^3\theta$.

13. $\dfrac{dx}{dt} = a(1 + \cos t), \dfrac{dy}{dt} = a(1 - \cos t) \Rightarrow \dfrac{dy}{dx} = \tan\dfrac{t}{2}$.

Then, the required volume $V = \int_{-a\pi}^{a\pi} \pi y^2\,dx$.

17. The required volume is given by $V = \int_{0}^{\pi} \dfrac{2}{3}\pi r^3 \sin\theta\,d\theta$.

18. The required volume is given by $V = 2\int_{0}^{\pi/4} \dfrac{2}{3}\pi r^3 \sin\theta\,d\theta$.

21. The required volume is given by $V = \int_{0}^{\pi/2} \dfrac{2}{3}\pi r^3 \sin\theta\,d\theta$.

$$\therefore \qquad V_2 = \int_{-a}^{a} \pi y^2\,dx = \frac{4\pi}{3a^2}.$$

Answers

2. $\dfrac{2}{3}\pi a^3$ **3.** (i) $\dfrac{4}{5}\pi a^3$ (ii) $2ah^2$ **4.** $\dfrac{\pi c^2}{2}\left[x + \dfrac{c}{2}\sinh\left(\dfrac{2x}{c}\right)\right]$ **5.** $\dfrac{1}{12}\pi a^4$ **7.** $\pi a^3[8\log 2 - 3]$ **8.** $\dfrac{16}{105}\pi a^3$ **10.** $\dfrac{23}{60}\pi a^3$

12. $\dfrac{2}{75}\sqrt{5}\,\pi a^3$ **14.** (i) $\dfrac{1}{2}\pi^2 a^3$ (ii) $\dfrac{5}{2}\pi^2 a^3$ **15.** $\dfrac{4}{3}\pi$ **16.** $2\pi^2 a^3$ **17.** $\dfrac{8}{3}\pi a^3$ **18.** $\dfrac{\pi a^3}{24}\sqrt{2}[3\log(\sqrt{2} + 1) - \sqrt{2}]$ **21.** $\dfrac{4}{3}\pi a^3$ **22.** $\dfrac{1}{2}\pi^2 b^3$

4.5.7 PAPPUS AND GULDIN'S THEOREMS
(I) FOR THE VOLUME OF A SOLID OF REVOLUTION

Statement. *When a closed plane curve revolves about any straight line lying in the same plane but not intersecting the closed curve, then the volume of a ring shape solid thus generated, is equal to the product of the area of a closed curve and the perimeter of a cirlce described by the centroid of the closed curve.*

Proof. Let *ABCDA* be a closed curve which revolves about *x*-axis. Let the lines $x = a$ and $x = b$ touch the closed plane curve at the points *A* and *C* as shown in fig. 19.

Now draw a line through the centroid of the closed curve and parallel to the line $x = a$ and $x = b$. This line intersects the given curve at two points P_1 and P_2 such that $MP_1 = y_1$ and $MP_2 = y_2$ provided y_1 and y_2 both are the functions of x only. Therefore the required volume of a solid of revolution of the closed plane curve about x-axis is equal to the difference of the volume generated by the plane area *AEFCDA* and the volume generated by the plane area *AEFCBA* about x-axis.

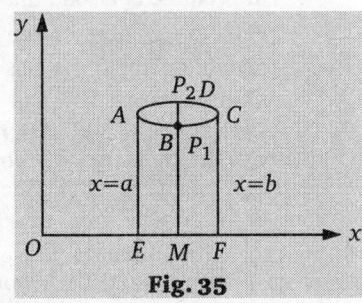

Fig. 35

Now the volume generated by *AEFCDA* $= \int_{a}^{b} \pi y_2^2\,dx$

and the volume generated by *AEFCBA* $= \int_{a}^{b} \pi y_1^2\,dx$.

$$\therefore \qquad \text{Required volume, } V = \int_{a}^{b} \pi y_2^2\,dx - \int_{a}^{b} \pi y_1^2\,dx = \pi \int_{a}^{b}(y_2^2 - y_1^2)\,dx. \qquad \text{...(1)}$$

Let $(\bar{x}, \bar{y})$ be the co-ordinates of the centroid of the closed plane curve *ABCDA*. Then by the method of finding the centre of gravity, we have

$$\bar{y} = \frac{\int_{a}^{b} \frac{1}{2}(y_2 + y_1)(y_2 - y_1)\,dx}{A} \quad \text{where } A \text{ is the area of the closed plane curve.}$$

$$\therefore \qquad \bar{y}A = \frac{1}{2}\int_{a}^{b}(y_2^2 - y_1^2)\,dx \qquad \text{...(2)}$$

Now from (1) and (2), we get

Required volume $= 2\pi\,\bar{y}\,A = A \times 2\pi\,\bar{y} = $ (area of the closed curve) $\times$ (perimeter of the circle whose radius is $\bar{y}$)

$V = $ (area of the closed curve) $\times$ (perimeter of the circle described by the centroid of the closed curve)

(2) FOR THE SURFACE OF A SOLID OF REVOLUTION

Statement. *When an arc of a plane curve revolves about any straight line lying in the plane of curve but not intersecting it, then the surface area of a solid of revolution thus generated is equal to the product of the length of the arc and the perimeter of a circle described by the centroid of that arc.*

Proof. Let *ACB* be an arc of a plane curve cut off by the lines $x = a$ and $x = b (a < b)$. Let $P(x, y)$ be any point on the arc. Now draw a line parallel to the lines $x = a$ and $x = b$ through P which meets the *x*-axis at *M* such that $PM = y$. Suppose the arc revolves about the *x*-axis as shown in fig. 20.

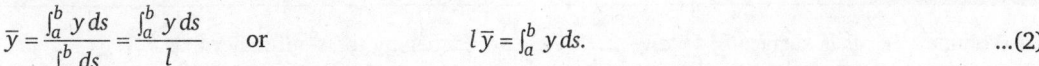

Let *l* be the length of an arc *ACB*. Therefore the surface area is given by

$$S = \int_a^b 2\pi y \, ds \qquad \qquad ...(1)$$

where *s* is the length of the arc measured from *A* to any point *P*.

Let $(\bar{x}, \bar{y})$ be the co-ordinates of the centroid of the arc *ACB*, then we know that

$$\bar{y} = \frac{\int_a^b y \, ds}{\int_a^b ds} = \frac{\int_a^b y \, ds}{l} \qquad \text{or} \qquad l\bar{y} = \int_a^b y \, ds. \qquad ...(2)$$

From (1) and (2), we get $\quad S = 2\pi \bar{y} \, l = l \times 2\pi \bar{y} = $ length of the arc *ACB* × perimeter of the circle of radius *y*

∴ $\quad\quad\quad\quad\quad$ *S* = (length of an arc *ACB*) × (perimeter of the circle described by the centroid of that arc)

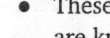 **REMARKS**

- These theorems are only applicable when the closed curve or arc donot intersect the line of revolution.
- These theorems are also used to find the centroid of the closed curve or an arc only when volume and surface area of revolution are known.

Solved Examples

EXAMPLE 1. ***The volume generated by the revolution of an ellipse having semi-axes a and b about a tangent at vertex.***

SOLUTION. Let the equation of an ellipse be $\dfrac{x^2}{a^2} + \dfrac{y^2}{b^2} = 1$

The centroid of this ellipse is $(0, 0)$ and the area is πab. There are four vertices $(\pm a, 0)$ and $(0, \pm b)$. Now first we revolve the ellipse about tangent at $(a, 0)$, then the distance of the centroid $(0, 0)$ from this tangent is *a*. Thus the generated volume

= (area of ellipse) × (perimeter of the circle of radius *a*)

= $\pi ab \times 2\pi a = 2\pi^2 a^2 b$.

Similarly, if we revolve the ellipse about the tangent at $(0, b)$, then the distance of the centroid $(0, 0)$ from this tangent is *b*. Thus the required volume is

= (area of an ellipse)

$\quad\quad$ × (perimeter of the circle of radius *b*)

= $\pi ab \times 2\pi b = 2\pi^2 ab^2$.

EXAMPLE 2. ***Find the position of the centroid of a semi-circular area.***

SOLUTION. Let the equation of a circle whose radius is *a* and centre is $(0,0)$ be $x^2 + y^2 = a^2$.

Let the semi-circular area be obtained by the circle $x^2 + y^2 = a^2$ and the *x*-axis as shown in fig. 37.

Let *V* be the volume of a solid generated by the semi-circular area about the *x*-axis, between $x = -a$ to $x = a$.

Then

$$V = \int_{-a}^a \pi y^2 \, dx = \pi \int_{-a}^a (a^2 - x^2) \, dx$$

$$= \pi \left[a^2 x - \frac{x^3}{3} \right]_{-a}^a = \frac{4}{3}\pi a^3$$

and let *A* be the area of this semi-circle. Then

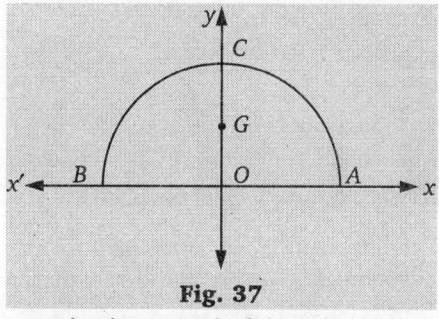

Fig. 37

Let *G* be the centroid of this semi-circular area whose position be $\bar{y}$ from *x*-axis (axis of revolution).

Then by Pappus theorem, we have

$$V = (\text{area of the semi-circle}) \times 2\pi \bar{y}$$

$$\Rightarrow \quad \frac{4}{3}\pi a^3 = \frac{\pi a^2}{2} \times 2\pi \bar{y}$$

∴ $\quad\quad\quad \bar{y} = \frac{4a}{3\pi}.$

Exercise-4.10

1. Find the volume and surface area of the anchor-ring generated by the revolution of a circle of radius *a* about an axis in its own plane distance *b* from its centre $(b < a)$.

2. The volume of the ring generated by the revolution of an ellipse of eccentricity $1/\sqrt{2}$ about a straight line parallel to the minor axis and situated at a distance from the centre equal to three times the major axis.

3. The loop of the curve $2ay^2 = x(x - a)^2$ revolves about the straight line $y = a$. Find the volume of the solid generated.

4. The volume of a ring generated by the revolution of the cardioid $r = a(1 + \cos \theta)$ about the line $r \cos \theta + a = 0$, given that the centroid of the cardioid is at a distance $5a/6$ from the pole.

5. Using Pappus theorem, determine the position of the centre of gravity of the quadrant of a uniform circular lamina of radius *a*, where the volume of the solid generated by the revolution of the quadrant of circular lamina about the tangent at either of its extremities is $\dfrac{\pi(3\pi - 4)}{6} a^3$.

Hints to Selected Problems

1. The area of the circle of radius A is given by $A = \pi a^2$.
   Circular circumference $= 2\pi b$.
   Then by Pappus and Guldin's theorem, we have
   Volume, $V = A \times 2\pi b = 2\pi^2 a^2 b$
   and surface $= 2\pi a \times 2\pi b = 4\pi^2 ab$.

2. Here $A = \pi a\left(\dfrac{a}{\sqrt{2}}\right) = \dfrac{1}{\sqrt{2}}\pi a^2$.

   Let l be the distance of straight line parallel to the minor axis

from C.G. of the ellipse.

Then $l = 3$ (Major axis) $= 6a$

length of the circular path $= 12\,\pi a$

$\therefore$ Required volume $V = \dfrac{1}{\sqrt{2}}\pi a^2 \times 12\pi a$.

4. $A = 2\int_0^\pi \dfrac{1}{2} r^2 d\theta = \dfrac{11\pi a}{3}$. The volume $V = A \times \dfrac{11\pi a}{3}$.

 **Answers**

1. Volume $= 2\pi^2 a^2 b$, surface $= 4\pi^2 ab$ 2. $6\sqrt{2}\,\pi^2 a^3$, a beging the semi-major axis. 3. $\dfrac{8}{15}\sqrt{2}\,\pi a^3$ 4. $\dfrac{11}{2}\pi^2 a^3$

5. $d = \dfrac{(3\pi-4)a}{3\pi}$, d being the distance of centroid from the tangent.

ARCHIVE

1. Find the reduction formulae for
   (i) $\displaystyle\int\tan^n x\,dx$ (ii) $\displaystyle\int\cot^n x\,dx$

2. Prove that $\displaystyle\int_0^{\pi/2} \dfrac{\sqrt{\sin x}}{\sqrt{\sin x}+\sqrt{\cos x}}\,dx = \dfrac{\pi}{2}$.

3. If $I_n = \int_0^{\pi/2} x\cos^n x\,dx, n>1$ prove that $I_n = \dfrac{n-1}{n}I_{n-2} - \dfrac{1}{n^2}$.

4. Prove that $\int_0^\pi \theta\sin^2\theta\cos^4\theta\,d\theta = \dfrac{\pi^2}{32}$.

5. Prove the following :
   (i) $\int_0^\pi \dfrac{x\,dx}{a^2\cos^2 x + b^2\sin^2 x} = \dfrac{\pi^2}{2ab}$.
   (ii) $\int_0^\pi \dfrac{x\,dx}{a^2-\cos^2 x} = \dfrac{\pi^2}{2a\sqrt{a^2-1}}, a>1$.

6. Show that the area of the tangent cut off from the parabola $x^2 = 8y$ by the line $x-2y+8=0$ is 36 square units.

7. Show that the area of a loop of the curve $x^3+y^3=3axy$ is $\dfrac{3a^2}{2}$.

8. Show that the area included between the folium $x^3+y^3=3axy$ and its asymptotes is equal to the area of the loop.

9. Show that the whole length of the curve $x^2(a^2-x^2)=8a^2y^2$ is $\neq a\sqrt{2}$.

10. Show that the volume of the solid formed by the revolution of the curve $(a^2+x^2)=a^2$ about its asymptotes is $\dfrac{1}{2}\neq^2 a^3$.

11. Show that the surface area of the solid generated by the revolution of the curve $x=a\cos^3 t, y=a\sin^3 t$ about x-axis is $12\dfrac{\neq^2}{5}$.

12. Show that the surface of the solid generated by the revolution of the tractrix $x=a\cos t + \dfrac{a}{2}\log\tan^2 t/2, y=a\sin t$ about x-axis is $4\pi a^2$.

13. If $\int_0^\pi \dfrac{dx}{a+b\cos x} = \dfrac{\pi}{\sqrt{a^2-b^2}}$ $(a>b)$ then show that $\int_0^\pi \dfrac{\cos x}{(a+b\cos x)^2}\,dx = \dfrac{\pi b}{(a^2-b^2)^{3/2}}$. (MADRAS-2006)

14. Prove that $\int_0^1 \dfrac{\log(1+x)}{1+x^2}\,dx = \dfrac{\pi}{8}\log_e 2$. (HISSAR-2005)

15. Prove that $\int_0^\infty \dfrac{\tan^{-1}ax}{x(1+x^2)}\,dx = \dfrac{\pi}{2}\log(1+a), a\geq 0$ (VTU-2010, SVTU-2009, ROHTAK-2006, ANNA-2005, 09)

16. Prove that $\int_0^{\pi/2} \dfrac{\log(1+y\sin^2 x)}{\sin^2 x}\,dx = \pi[\sqrt{1+y}-1]$. (SVTU-2008)

17. Using reduction formula, prove that
   $\int_0^a \dfrac{x^7}{\sqrt{a^2-x^2}}\,dx = \dfrac{1}{a^{2n-1}}\dfrac{(2n-3)(2n-5)...3.1}{(2n-2)(2n-4)...4.2}\left(\dfrac{\pi}{2}\right)$. (VTU-2006)

18. Prove that $\int_0^\infty \dfrac{x^2}{(1+x^2)^{7/2}}\,dx = \dfrac{2}{15}$. (VTU-2010)

19. If $I_n = \int x^n\sqrt{a-x}\,dx$, prove that
   $(2n+3)I_n = 2anI_{n-1} - 2x^n(a-x)^{3/2}$. (MARATHWADA-2008)

20. Prove that $\int_0^{\pi/2} \cos^n x\cos nx\,dx = \dfrac{\pi}{2^{n+1}}$. (SVTU-2008)

21. Prove that $\int_0^\pi \theta\sin^2\theta\cos^4\theta\,d\theta = \dfrac{\pi^2}{32}$. (VTU-2009)

22. Prove that the area of cardioid $r = a(1-\cos\theta) = \dfrac{3\pi a^2}{2}$. (VTU-2004)

23. Show that the length of the arc of the parabola $x^2=4ay$ measured from the vertex to one extremity of the latusrectum is given by $a(\sqrt{2}+\log(1+\sqrt{2}))$. (DELHI-2002)

24. Show that the surface of the solid generated by the revolution of the ellipse $2\pi ab\left\{\dfrac{b}{a}+\dfrac{a}{\sqrt{a^2-b^2}}\sin^{-1}\sqrt{\dfrac{(a^2-b^2)}{a}}\right\}$. (SRM-2008, 17)

25. Show that the surface area of the solid generated by revolving the cyloid $x=a(t-\sin t), y=a(1-\cos t)$ about the base is given by $\dfrac{64}{3}\neq a^2$. (MARATHWADA-2008, COCHIN-2005, 12, KURUKSHETRA-2005)

✱✱✱✱✱✱✱

CHAPTER 5

Integral Calculus-II

5.1 MULTIPLE INTEGRALS

Double integral is an extension of a definite integral in two-dimensional space.

Let $f(x, y)$ be a single valued function of x and y, bounded and defined in the region R of XY-plane, and A be the area of region R and let R be divided in any manner into n-sub regions α_1, $\alpha_2,...,\alpha_n$, whose areas are δs_1, $\delta s_2,...\delta s_n$ respectively. If $P_r(\xi_r, \eta_r)$ is any point inside the region α_n. $\beta_n = f(\xi_1)$. Let $B_n = \sum_{r=1}^{n} f(\xi_r, \eta_r)\delta s_r$ then the limits of B_n which is assumed to exists as $n \to \infty$ such that every $\alpha_r \to 0$ in all its dimensions is known as double integral of $f(x, y)$ over the region R and is denoted by

$$\iint_R f(x, y)ds \quad \text{or} \quad \iint_R f(x, y)dx\, dy .$$

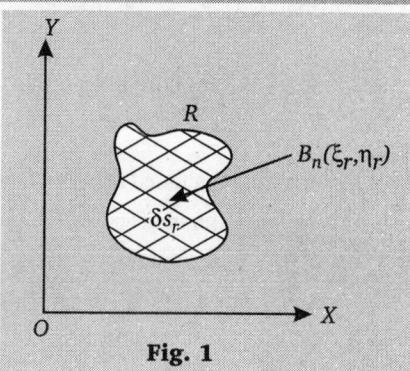

Fig. 1

Hence, the area R is called the region or field of integration for the double integral and ds is called element of area.

☞ REMARK

- Let the region A be divided into the rectangular partitions and dx be the length of a sub-rectangular and dy be its width, so that the $dx\, dy$ is an element of area in cartesian co-ordinates, then the integral $\iint f(x, y)\, ds$ is written as $\iint_A f(x, y)dx\, dy$ and is called the double integral of $f(x, y)$ over the region R.

5.1.1 PROPERTIES OF DOUBLE INTEGRALS

(1) When the region R is partitioned into two parts say R_1 and R_2 then

$$\iint_R f(x, y)dx\, dy = \iint_{R_1} f(x, y)dx\, dy + \iint_{R_2} f(x, y)dx\, dy$$

Similarly, we divide the region into three or more parts.

(2) The double integral of a algebraic sum of a fixed number of functions is equal to the algebraic sum of double integrals taken for each term separately. Thus

$$\iint_R [f_1,(x, y) + f_2(x, y) + f_3(x, y) + ...]dx\, dy = \iint_R f_1(x, y)\, dx\, dy + \iint_R f_2(x, y)\, dx\, dy + \iint_R f_3(x, y)\, dx\, dy + ...$$

(3) A constant factor may be taken outside the integral sign. Thus

$$\iint_R mf(x, y)dx\, dy = m\iint_R f(x, y)dx\, dy \text{ where } m \text{ is a constant.}$$

5.1.2 EVALUATION OF DOUBLE INTEGRALS

(i) *Over a rectangular region R.* If the region R be given by the inequalities $a \le x \le b, c \le y \le d$, then the double integral

$$\iint_R f(x, y)dx\, dy = \int_a^b \int_c^d f(x, y)dx\, dy$$
$$= \int_a^b \left[\int_c^d f(x, y)dy \right]dx . \qquad ...(1)$$

We first evaluate $\int_c^d f(x, y)dy$ *i.e.*, integrate $f(x, y)$ with respect to y regarding x as constant and then resulting function of x is to be integrated with respect to x between the limits a and b

or
$$\iint_R f(x, y)dx\, dy = \int_c^d \int_a^b f(x, y)\, dx\, dy$$
$$= \int_c^d \left[\int_a^b f(x, y)dx \right]dy . \qquad ...(2)$$

Now, we integrate $\int_a^b f(x, y)dx$ first and then integrate with respect to y.

(ii) *Over the regions which are not rectangular.* Let the region R be described by $a \le x \le b$ and $\phi_1(x) \le y \le \phi_2(x)$ so that $y = \phi_1(x)$ and $y = \phi_2(x)$ respectively, the boundary of R then

$$\iint_R f(x,y)dx\,dy = \int_a^b \left[\int_{\phi_1(x)}^{\phi_2(x)} f(x,y)dy \right] dx$$

Here, the inner integral $\int_{\phi_1(x)}^{\phi_2(x)} f(x,y)dy$ is integrated

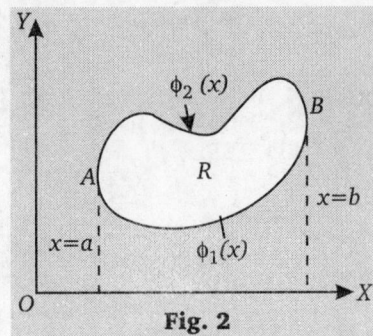

Fig. 2

first and in this integral the result of integration is a function of x, say $\phi_1(x)$. Then $\phi_1(x)$ is integrated with respect to x between the limits a and b to obtain the value of double integral.

In a similar way, if R can be described by

$$c \le y \le d, \quad \phi_3(y) \le x \le \phi_4(y)$$

then we get

$$\iint_R f(x,y)dx\,dy = \int_c^d \left[\int_{\phi_3(y)}^{\phi_4(y)} f(x,y)dx \right] dy .$$

Here, the result of integration

$$\int_{\phi_3(y)}^{\phi_4(y)} f(x,y)dx .$$

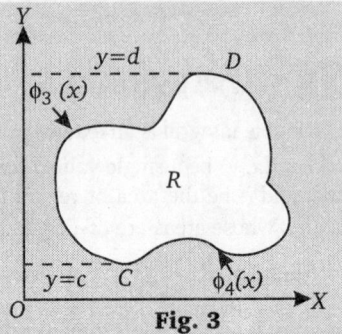

Fig. 3

which is evaluated first, is a function of y say $\phi_2(y)$, then $\phi_2(y)$ is integrated with respect to y between the limits c to d.

WORKING PROCEDURE

- While evaluating double integrals, first integrate with respect to variable having variable limits and treating the other variable as constant and then integrate with respect to variable with constant limits.

5.1.3 CONVERSION OF CARTESIAN TO POLAR CO-ORDINATES

The transformation formula required is $x = r\cos\theta, y = r\sin\theta$ and elementary area $\delta A = r\delta\theta . \delta r$ so that

$$\iint f(x,y)dx\,dy = \iint f(x,y)dA = \iint f(r,\theta)r\,d\theta\,dr.$$

 Solved Examples

EXAMPLE 1. *Evaluate* $\int_1^2 \int_0^{y/2} y\,dy\,dx$.

SOLUTION. We have

$$\int_1^2 \int_0^{y/2} y\,dy\,dx = \int_1^2 y(x)_0^{y/2}\,dy = \int_1^2 y\left(\frac{1}{2}y\right)dy$$

$$= \frac{1}{2}\int_1^2 y^2 dy = \frac{1}{2}\left[\frac{1}{3}y^3\right]_1^2 = \frac{1}{6}(2^3 - 1^3) = 7/6$$

EXAMPLE 2. *Evaluate* $\int_1^2 \int_0^x \dfrac{1}{x^2 + y^2}\,dx\,dy$.

SOLUTION. We have

$$\int_1^2 \int_0^x \frac{dx\,dy}{x^2+y^2} = \int_1^2 \left[\int_0^x \frac{dy}{x^2+y^2} \right] dx$$

$$= \int_1^2 \left[\frac{1}{x}\tan^{-1}\frac{y}{x} \right]_{y=0}^x dx$$

$$= \int_1^2 \left[\frac{1}{x}(\tan^{-1}1 - \tan^{-1}0) \right] dx$$

$$= \frac{\pi}{4}\int_1^2 \frac{dx}{x} = \frac{\pi}{4}[\log x]_1^2$$

$$= \frac{\pi}{4}.[\log 2 - \log 1] = \frac{1}{4}\pi \log 2.$$

EXAMPLE 3. *Evaluate* $\iint_A (x^2 + y^2)dx\,dy$, *where A is the region bounded by* $x=0, y=0, x+y=1$.

SOLUTION. Let R be the region of integration $x+y=1$ and

the limit of itegration can be expressed as $0 \le x \le 1, 0 < y < 1-x$.

From the equation $x+y =1$, we have $x = 1$ for $y = 0$ and for the positive quadrant x varies from 0 to 1 and for y which varies from $y = 0$ to $y = 1-x$. First integrate with respect to y, treated x as constant and then integrate with respect to 'x'.

Hence, the integral

$$= \int_0^1 \int_0^{1-x}(x^2+y^2)\,dx\,dy = \int_0^1 \left(x^2 y + \frac{1}{3}y^3 \right)_0^{1-x} dx$$

$$= \int_0^1 \left[x^2(1-x) + \frac{1}{3}(1-x)^3 \right] dx$$

$$= \int_0^1 (1-x)\left\{ x^2 + \frac{1}{3}(1-x)^2 \right\} dx$$

$$= \int_0^1 \frac{1}{3}[1 - 3x + 6x^2 - 4x^3]dx$$

$$= \frac{1}{3}\left[x - \frac{3}{2}x^2 + 2x^3 - x^4 \right]_0^1$$

$$= \frac{1}{3}\left[1 - \frac{3}{2} + 2 - 1 \right] = \frac{1}{6}$$

EXAMPLE 4. *Evaluate* $\iint (x+y)^2 dx\,dy$ *over the region bounded by ellipse* $\dfrac{x^2}{a^2} + \dfrac{y^2}{b^2} = 1$. *Hence find the*

mass of an elliptic plate whose density per unit area is given by $\rho = k(x+y)$. (UKTU-2011)

SOLUTION. Since the region is bounded by ellipse $\dfrac{x^2}{a^2}+\dfrac{y^2}{b^2}=1$, we expressed it as:

$$x = -a \text{ and } x = a$$

$$y = -b\sqrt{(1-x^2/a^2)},\ y = b\sqrt{(1-x^2/a^2)}.$$

$$\therefore \iint (x+y)^2 dx\,dy = \int_{-a}^{a}\int_{-b\sqrt{(1-x^2/a^2)}}^{b\sqrt{(1-x^2/a^2)}}(x^2+y^2+2xy)\,dx\,dy$$

$$= \int_{-a}^{a} 2\int_{0}^{b\sqrt{(1-x^2/a^2)}}(x^2+y^2)\,dx\,dy$$

[$\because 2xy$ being an odd function of f, its integration under the given limits of y is 0]

$$= 2\int_{-a}^{a}\left[x^2 y+\frac{y^3}{3}\right]_{0}^{b\sqrt{1-x^2/a^2}} dx$$

$$= 2\int_{-a}^{a}\left\{x^2 b\sqrt{\left(1-\frac{x^2}{a^2}\right)}+\frac{b^3}{3}\left(1-\frac{x^2}{a^2}\right)^{3/2}\right\} dx$$

$$= 2\times 2\int_{0}^{a}\left\{x^2 b\sqrt{\left(1-\frac{x^2}{a^2}\right)}+\frac{b^3}{3}\left(1-\frac{x^2}{a^2}\right)^{3/2}\right\} dx$$

$$= 4b\int_{0}^{\pi/2}\left\{a^2\sin^2\theta\cos\theta+\frac{b^2}{3}\cos^3\theta dx\right\} a\cos\theta\,d\theta$$

(By putting $x = a\sin\theta$ so that $dx = a\cos\theta\,d\theta$)

$$= 4ab\int_{0}^{\pi/2}\left[a^2\sin^2\theta\cos^2\theta+\frac{b^2}{3}\cos^4\theta\right]d\theta$$

$$= 4ab\left[a^2\int_{0}^{\pi/2}\sin^2\theta\cos^2\theta\,d\theta+\frac{b^2}{3}\int_{0}^{\pi/2}\cos^4\theta\,d\theta\right]$$

$$= 4ab\left[\frac{1}{16}\pi a^2+\frac{1}{16}\pi b^2\right]=\frac{1}{4}\pi ab(a^2+b^2)$$

the mass of elliptic plate whose density is given by

$\rho = k(x+y)^2 = \iint_A k(x+y)^2 dx\,dy$

(where integration is to be performed over the area A of ellipse.)

$$= k.\frac{1}{4}\pi ab(a^2+b^2)$$

EXAMPLE 5. **Evaluate** $\iint xy(x+y)\,dx\,dy$ **over the region between** $y = x^2$ **and** $y = x$. (UKTU-2011, 12)

SOLUTION. When we draw the given curve, the parabola $y = x^2$ and line $y = x$ intersect at the point $(0, 0)$ and $(1, 1)$,

Here $x^2 = x$ or $x(x-1) = 0$ *i.e.*, $x=0$ or 1, when $x =0$, $y = 0$, and $x=1, y=1$].

So the area of integration for x is from $x=0$ to $x = 1$ and for y from x^2 to x.

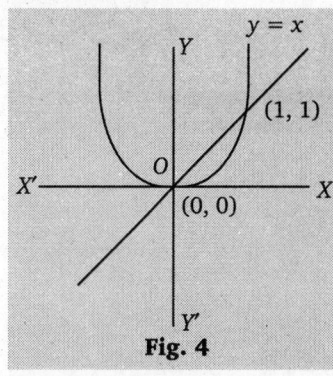

Fig. 4

$\therefore$ Given integral

$$= \int_{0}^{1}\int_{x^2}^{x} xy(x+y)\,dx\,dy = \int_{0}^{1}\int_{x^2}^{x} x(yx+y^2)\,dx\,dy$$

$$= \int_{0}^{1} x\left[x.\frac{y^2}{2}+\frac{1}{3}y^3\right]_{x^2}^{x} dx$$

$$= \int_{0}^{1} x.\left[\left(x.\frac{x^2}{2}+\frac{1}{3}x^3\right)-\left(x.\frac{x^4}{2}+\frac{1}{3}x^6\right)\right]dx$$

$$= \int_{0}^{1} x\left[\frac{5x^3}{6}-\frac{1}{2}x^5-\frac{1}{3}x^6\right]dx$$

$$= \int_{0}^{1}\left[\frac{5}{6}x^4-\frac{1}{2}x^6-\frac{1}{3}x^7\right]dx$$

$$= \left[\frac{1}{6}x^5-\frac{1}{14}x^7-\frac{1}{24}x^8\right]_{0}^{1}=\frac{1}{6}-\frac{1}{14}-\frac{1}{24}=\frac{3}{56}$$

Exercise-5.1

1. Evaluate $\int_{2}^{3} dx\int_{0}^{1}(x^2+3y^2)dy$.

2. Evaluate $\int_{0}^{2}\int_{0}^{\sqrt{4+x^2}}\dfrac{dx\,dy}{(4+x^2+y^2)}$.

3. Evaluate $\int_{0}^{\pi/2}\int_{\pi/2}^{\pi}\cos(x+y)dx\,dy$.

4. Evaluate $\int_{0}^{2}\int_{0}^{\sqrt{2x-x^2}} x\,dx\,dy$. (SRM-2010)

5. Evaluate $\int_{0}^{1}\int_{0}^{x^2} e^{y/x}\,dx\,dy$.

6. Evaluate $\int_{0}^{1}\int_{0}^{1}\dfrac{dx\,dy}{\sqrt{(1-x^2)(1-y^2)}}$ (UPTU (AG)-2006)

7. Evaluate $\iint e^{2x+3y}dx\,dy$ over the triangle bounded by $x= 0$, $y =0$ and $x+y =1$.

8. Evaluate $\iint_{p} x\sin(x+y)dx\,dy$, where p is a rectangle $[0 \le x \le \pi, 0 \le y \le \pi/2]$.

9. Show that $\int_{1}^{2}\int_{3}^{4}(xy+e^y)dx\,dy = \int_{3}^{4}\int_{1}^{2}(xy+e^y)dy\,dx$.

10. Evaluate $\iint x^2 y^2 dx\,dy$ over the region bounded by $x =0$, $y =0$, where A is the region bounded by $x^2+y^2 =1$.

11. Find the area of the ellipse $\dfrac{x^2}{a^2}+\dfrac{y^2}{b^2}=1$ by double integration.

12. Show that by double integration that the area between the parabolas $y^2 = 4ax$ and $x^2 = 4by$ is $(16/3)\,ab$.

13. Find by double integration the region included between the parabola $x^2 = 4ay$ and the curve $y = 8a^3/(x^2+4a^2)$.

14. Evaluate $\iint y\,dx\,dy$ over the region between the parabolas $y^2 = 4x$ and $x^2 = 4y$.

15. Find the double integration the region lying between the parabola $y = 4x - x^2$, and the line $y = x$.

16. Find by double integration the area of the region lying between the parabola $y^2 = 4ax$, and line $y = mx$.

17. Find by double integration the area of the region lying between the semi-cubical parbola $y^2 = x^3$, and line $y = mx$.

18. Find by double integration the area of the region lying between the circle $x^2 + y^2 = a^2$, and line $x + y = a$ (in first quadrant)

19. Find by double integration the area of the region lying between the curves $(x^2 + 4a^2) y = 8a^3$, $2y = x$ and $x = 0$.

20. Evaluate

(i) $\int_0^1 \int_0^{\sqrt{1+x^2}} \dfrac{dx\,dy}{1 + x^2 + y^2}$ (UPTU-2006)

(ii) $\int_0^a \int_0^{\sqrt{a^2+y^2}} (a^2 - x^2 - y^2)\,dx\,dy$

21. (i) Evaluate $\iint_R \left(1 - \dfrac{x^2}{a^2} + \dfrac{y^2}{b^2}\right) dx\,dy$ over the first quadrant

of the ellipse $\dfrac{x^2}{a^2} + \dfrac{y^2}{b^2} = 1$.   (GBTU-2010)

(ii) Evaluate $\iint xy\,dx\,dy$ where A is the domain bounded by x-axis, ordinate $x = 2a$ and the curve $x^2 = 4ay$.

                           (MTU-2012, GBTU-2010)

Hints to Selected Problems

5. $I = \int_0^1 \int_0^{x^2} e^{y/x}\,dx\,dy = \int_0^1 x [e^{y/x}]_{y=0}^{x^2}\,dx$

$= \int_0^1 x(e^x - 1)dx = \int_0^1 xe^x\,dx - \int_0^1 x\,dx$

$= \left(xe^x\right)_0 - \left[e^x\right]_0 - \left(\dfrac{1}{2}\right) = (e-0) - (e-1) - \dfrac{1}{2} = \dfrac{1}{2}$

10. $I = \iint_A x^2 y^2\,dx\,dy = \int_0^1 \int_0^{\sqrt{1-x^2}} x^2 y^2\,dx\,dy$

$= \int_0^1 x^2 \left[\dfrac{y^3}{3}\right]_0^{\sqrt{1-x^2}} dx = \dfrac{1}{3}\int_0^1 x^2(1-x^2)^{3/2}\,dx$

Now put $x^2 = t$.

11. $I = \iint_A dx\,dy = \int_{-a}^a \int_{-b/a\sqrt{a^2-x^2}}^{b/a\sqrt{a^2-x^2}} dx\,dy$

12. $A = \int_0^{4a^{1/3}b^{2/3}} \int_{x^2/4b}^{\sqrt{4ax}} dx\,dy$.

13. $A = \int_{-2a}^{2a} \int_{x^2/4a}^{8a^3/x^2+4a^2} dx\,dy$.

14. The curves $y^2 = 4a$ and $x^2 = 4y$ intersect at the points where $x = 0$ and $x = 4$. Also, when

$$0 < x < 4, \sqrt{4x} > \dfrac{x^2}{4}.$$

$$\therefore \quad I = \int_0^4 \int_{x^2/4}^{\sqrt{4x}} y\,dx\,dy.$$

16. Since, the two corners cut at the point where $x = 0$ and $x = \dfrac{4a}{m^2}$

$$\therefore \quad I = \int_0^{4a/m^2} \int_{mx}^{\sqrt{4ax}} dx\,dy$$

Answers

1. $\dfrac{22}{3}$ **2.** $\dfrac{\pi}{4}\log(1+\sqrt{2})$ **3.** -2 **4.** $\dfrac{\pi}{2}$ **5.** $\dfrac{1}{2}$ **6.** $\dfrac{\pi^2}{4}$ **7.** $\dfrac{1}{6}(e-1)^2(2e+1)$ **8.** $\pi+2$ **9.** $\dfrac{21}{4}+e^4-e^3$

10. $\pi/96$ **11.** πab **13.** $\left(2\pi - \dfrac{4}{3}\right)a^2$ **14.** $48/5$ **15.** $9/2$ **16.** $8a^2/3m^2$ **17.** $1/10\,m^5$ **18.** $\dfrac{1}{4}(\pi-2)a^2$

19. $(\pi-1)a^2$ **20.** (i) $\dfrac{\pi}{4}\log(1+\sqrt{2})$ (ii) $\dfrac{\pi a^4}{8}$ **21.** (i) $\dfrac{\pi ab}{4}$ (ii) $\dfrac{a^4}{3}$.

5.1.4 Double Integral in Polar Co-ordinates

Let us consider a function $f(r, \theta)$ of polar co-ordinates r, θ over a certain area A with whose boundary is also given in terms of polar co-ordinates. We divide the area into n parts of elementary areas $\delta A_1, \delta A_2, \delta A_3, \dots \delta A_n$ and let $s_n = \sum_{r=1}^{n} f(r, \theta)\delta A$

where (r_1, θ_1) is a point inside the elementary area δA_1, the dobule integral of $f(r, \theta)$ is then defined as

$$\iint_A f(r, \theta)dA = \lim_{\substack{n \to \infty \\ \delta A_i \to 0}} \sum_{i=1}^{n} f(r_i, \theta_i)\delta A_i ,$$

provided limit toward right hand side exists.

REMARK

- In case of cartesian co-ordinates when the double integral $\iint_A f(x, y)dA$ is expressed in the form of repeated integral, dA representes the area of the rectangle with sides dx and dy and hence $dA = dx\,dy$.

  If the radius vector of OS and OP

  are r and $r + \delta r$ respectively and $\angle POQ = d\theta \Rightarrow RS = r\,d\theta$

  as RS and PQ are arcs of circles. Then

  $$dA = RS \times SP$$
  $$= r\,d\theta . dr = r\,dr\,d\theta$$

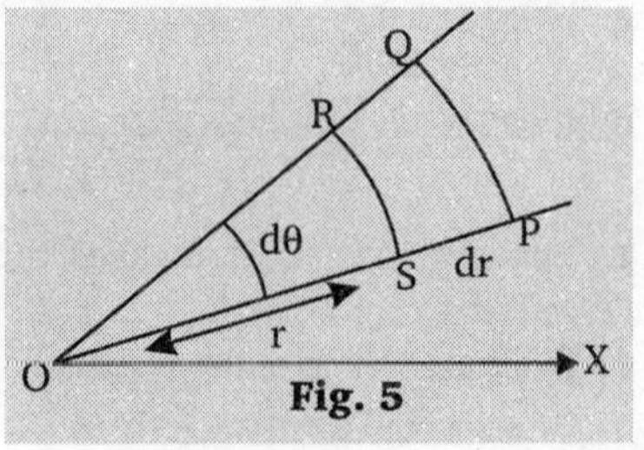

Fig. 5

Solved Examples

EXAMPLE 1. *Evaluate the double integral*
$$\int_0^{\pi/2}\int_0^{2a\cos\theta} r^2\sin\theta.\cos\theta.d\theta\,dr\,.$$

SOLUTION. We have $\int_0^{\pi/2}\int_0^{2a\cos\theta} r^2\sin\theta.\cos\theta.d\theta\,dr$

$$=\int_0^{\pi/2}\int_0^{2a\cos\theta}(r^2\sin\theta.\cos\theta)\,dr\,d\theta$$

$$=\int_0^{\pi/2}\left[\frac{r^3}{3}.\sin\theta.\cos\theta\right]_0^{2a\cos\theta}d\theta$$

$$=\frac{1}{3}\int_0^{\pi/2}(2a\cos\theta)^3.\sin\theta.\cos\theta\,d\theta$$

$$=\frac{8a^3}{3}\int_0^{\pi/2}\sin\theta.\cos^4\theta\,d\theta$$

$$=-\frac{8a^3}{3}\int_0^{\pi/2}\cos^4\theta\,d(\cos\theta)$$

$$=-\frac{8a^3}{3}\left[\frac{\cos^5\theta}{5}\right]_0^{\pi/2}$$

$$=-\frac{8a^3}{3}\left[0-\frac{1}{5}\right]=\frac{8a^3}{15}.$$

EXAMPLE 2. *Evaluate* $\iint\dfrac{r\,d\theta\,dr}{\sqrt{a^2+r^2}}$ *over one loop of the lemniscate* $r^2=a^2\cos2\theta$

SOLUTION. In lemniscate, there are two loops.
We see that when $-\pi/4<\theta<\pi/4$ or $3\pi/4<\theta<5\pi/4$, where r is real.

We want to evaluate the given integral over the right loop of the leminscate. [Fig. (6)]

Therefore, $\iint\dfrac{r\,d\theta\,dr}{\sqrt{a^2+r^2}}=\int_{-\pi/4}^{\pi/4}\int_{-a\sqrt{\cos2\theta}}^{a\sqrt{\cos2\theta}}\dfrac{r}{\sqrt{r^2+a^2}}\,dr\,d\theta$

as $r^2=a^2\cos2\theta.$

$\therefore \qquad r=\pm a\sqrt{\cos2\theta}$

Thus, r varies from $-a\sqrt{\cos2\theta}$ to $a\sqrt{\cos2\theta}$.

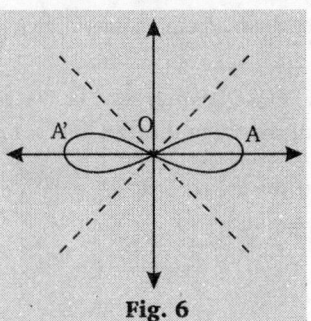

Fig. 6

Since, there is a symmetry about *X*-axis, we should evaluate the double integral over half of the right loop as follows:

$$\iint\frac{r\,d\theta\,dr}{\sqrt{a^2+r^2}}=\int_0^{\pi/4}\int_0^{a\sqrt{\cos2\theta}}\frac{r}{\sqrt{a^2+r^2}}\,dr\,d\theta$$

$$=\frac{1}{2}\int_0^{\pi/4}\int_0^{a\sqrt{\cos2\theta}}\frac{2r}{\sqrt{a^2+r^2}}\,dr\,d\theta$$

$$=\frac{1}{2}\int_0^{\pi/4}\int_0^{a\sqrt{\cos2\theta}}\frac{d(a^2+r^2)}{\sqrt{a^2+r^2}}\,d\theta$$

$$=\frac{1}{2}\int_0^{\pi/4}\left[2\sqrt{a^2+r^2}\right]_0^{a\sqrt{\cos2\theta}}d\theta$$

$$=\int_0^{\pi/4}\left[\sqrt{a^2+a^2\cos2\theta}-\sqrt{a^2+0}\right]d\theta$$

$$=\int_0^{\pi/4}[\sqrt{2}.a\cos\theta-a]d\theta$$

$$=\sqrt{2}.a\int_0^{\pi/4}\cos\theta\,d\theta-a\int_0^{\pi/4}d\theta$$

$$=\sqrt{2}.a[\sin\theta]_0^{\pi/4}-a[\theta]_0^{\pi/4}$$

$$=\sqrt{2}.a.\frac{1}{\sqrt{2}}-a.\frac{\pi}{4}=a-\frac{a}{4}\pi$$

∴ Value of double integral over the complete right loop

$$=2a-\frac{2a\pi}{4}=2a(1-\pi/4).$$

📝 **REMARK**

• If we evaluate the double integral as $\int_{-\pi/4}^{\pi/4}\int_{-a\sqrt{\cos2\theta}}^{a\sqrt{\cos2\theta}}\dfrac{r}{\sqrt{a^2+r^2}}\,d\theta\,dr$, then it will become zero due to oddness of function $\dfrac{r}{\sqrt{a^2+r^2}}$ therefore we must not calculate the double integral over the complete loop.

 Exercise-5.2

1. Integrate $r\sin\theta$ over the area of cardiod $r=a(1+\cos\theta)$ lying above the initial line.
2. Find by double integration that the area lying inside the cardiod $r=a(1+\cos\theta)$ and outside the circle $r=a$.
3. Find by double integration the area lying inside the cardiod $r=1+\cos\theta$ and outside the parabola $r(1+\cos\theta)=1$.
4. Find by double integration the area lying inside the circle $r=a\sin\theta$ and outside the cardioid $r=a(1-\cos\theta)$.

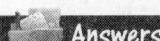 **Answers**

1. $\dfrac{4a^3}{3}$ 2. $\dfrac{1}{4}a^2(\pi+8)$ 3. $\dfrac{9\pi+16}{12}$ 4. $\dfrac{a^2}{4}(4-\pi)$.

5.1.5 Applications of Double Integration

Double integration is generally used to find the area of curves, volume and surface of solids of revolution.

(1) *Area of curves.* Let AD be an arc of the curve $y=f(x)$.

Let area $ABCD$ be divided into sub-area by drawing lines parallel to X and Y axis respectively such that distance between two adjoining lines drawn parallel to Y-axis be δx and those drawn parallel to X-axis be δy.

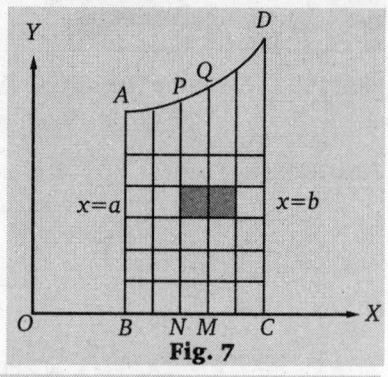

(i) Let $P(x, y)$ and $Q(x+\delta x, y+\delta y)$ be two neighbouring points on the curve AD. Then the area of element shown by shadded lines is $\delta x\,\delta y$.

Therefore, the area of strip $PN = \int_{y=0}^{f(x)} dx\,dy$ where $y = f(x)$.

The required area

$$ABCD = \int_{x=a}^{b} \int_{y=0}^{f(x)} dx\,dy .$$

Fig. 7

(ii) We can find the area bounded by the two curves $y = f_1(x)$ and $y = f_2(x)$ and the ordinates $x=a$ and $x=b$

$$\int_{x=a}^{b} \int_{y=f_1(x)}^{f_2(x)} dx\,dy .$$

(iii) *In polar co-ordinates.* The area bounded by curve $r= f(\theta)$ where $f(\theta)$ is a single valued function of θ in the domain (α, β) and the radii vector $\theta= \alpha$ and $\theta= \beta$ is

$$\int_{\theta=\alpha}^{\beta} \int_{r=0}^{f(\theta)} r\,d\theta\,dr .$$

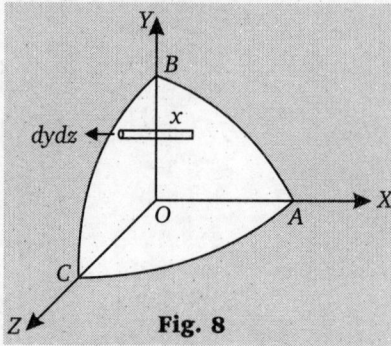

(2) *Volume of a solid.* Consider the area $dy\,dz$ on the plane $x=0$ through each point on the boundary of this small area. Draw the lines parallel to X-axis and thus construct a small cylinder whose base is area to X-axis. This cylinder cuts the given surface, and volume of this cylinder $= x\,dy\,dz$.

$\therefore$ Volume of solid $= \iint x\,dy\,dz$.

Fig. 8

🖝 **REMARKS**

- By considering area $dx\,dy$ on plane $z=0$ the volume of solid $= \iint z\,dx\,dy$.
- By considering area $dx\,dz$ on plane $y=0$ the volume of solid $= \iint y\,dx\,dz$.

(3) *Area of surface of a solid.* Let the equation of surface be $z= f(x, y)$. Consider a point $P(x, y, z)$ on this surface surrounding this point P. Consider an element of area δs of the surface. Let $\delta x\,\delta y$ be the projection of this area δs on the plane $z = 0$, then we have

$$\delta x\,\delta y =\delta s \cos \alpha \qquad \qquad ... (1)$$

where α is the angle between the tangent plane to the given surface at $P(x, y, z)$ and the plane $z=0$ then by co-ordinate geometry, we have

$$\sec \alpha = \sqrt{\left[1+\left\{\frac{\partial z}{\partial x}\right\} +\left\{\frac{\partial z}{\partial y}\right\}\right]} \qquad \qquad ...(2)$$

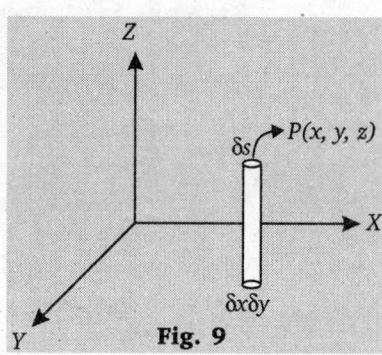

From (1) we have $\qquad \delta s = \delta x\,\delta y \sec \alpha = \delta x \delta y \sqrt{\left[+\left\{\frac{\partial z}{\partial x}\right\} +\left\{\frac{\partial z}{\partial y}\right\}\right]}$ [From (2)]

Fig. 9

$\therefore$ The required area of surface $= \iint \sqrt{\left[1+\left\{\frac{\partial z}{\partial x}\right\}^2 +\left\{\frac{\partial z}{\partial y}\right\}^2\right]}\, dx\,dy$.

Solved Examples

EXAMPLE 1. *Find the whole area of curve* $a^2 x^2 = y^3(2a - y)$.

SOLUTION. The shape of curve is shown in fig. 10.

The required area $= 2\times \text{area } OAB = 2\int_{y=0}^{2a} \int_{x=0}^{f(y)} dy\,dx$

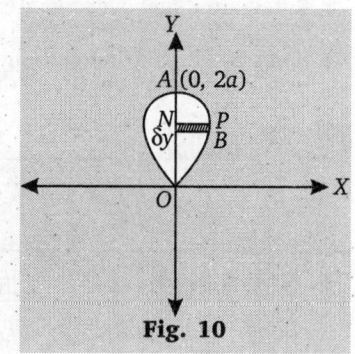

where $x= f(y)$ i.e., $x = y^{3/2}\dfrac{\sqrt{2a - y}}{a}$ is the equation of curve.

$\therefore$ The required area $= 2\int_{y=0}^{2a}[x]_0^{f(y)} dy = 2\int_0^{2a} f(y)\,dy$

Fig. 10

$$= 2\int_0^{2a} \frac{y^{3/2}\sqrt{2a-y}}{a}\,dy$$

$$\left[\because f(y) = x = y^{3/2}\frac{\sqrt{2a-y}}{a}\right]$$

Put $\qquad\qquad y = 2a\sin^2\theta$

$\Rightarrow\qquad\qquad dy = 4a\sin\theta\cos\theta\,d\theta$

at $\qquad\qquad y = 0, \theta = 0$ and $\quad y = 2a, \theta = \pi/2$

$\therefore$ Required area

$$= \frac{2}{a}\int_0^{\pi/2}(2a\sin^2\theta)^{3/2}\sqrt{(2a-2a\sin^2\theta)}\ 4a\sin\theta\cos\theta\,d\theta$$

$$= 32a^2\int_0^{\pi/2}\sin^4\theta\cos^2\theta\,d\theta$$

$$= \frac{32a^2\Gamma(5/2)\Gamma(3/2)}{2\Gamma 4}$$

$$= \frac{32a^2.(3/2).(1/2).\sqrt{\pi}.(1/2).\sqrt{\pi}}{2.3.2.1} = \pi a^2$$

EXAMPLE 2. *Find the area of curve r=a (1+cos θ).*

SOLUTION. The required area $= 2\times$ area $OABO = 2\int_{\theta=0}^{\pi}\int_{r=0}^{f(\theta)} r\,d\theta\,dr,$

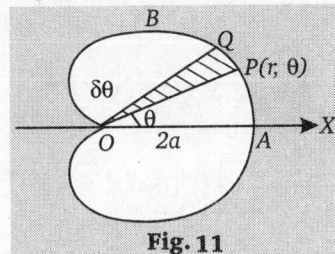

Fig. 11

(where $r = f(\theta)$ and $r = a(1+\cos\theta)$ is the equation of the curve.)

$$= 2\int_{\theta=0}^{\pi}\left[\frac{1}{2}r^2\right]_{r=0}^{f(\theta)}d\theta = \int_0^{\pi}[f(\theta)]^2\,d\theta$$

$$= \int_0^{\pi}a^2(1+\cos\theta)^2\,d\theta = a^2\int_0^{\pi}(2\cos^2\theta/2)^2\,d\theta$$

$$= 4a^2\int_0^{\pi}\cos^4\frac{\theta}{2}\,d\theta$$

$$= 8a^2\int_0^{\pi/2}\cos^4\phi\,d\phi, \qquad\qquad \text{(Putting } \theta = 2\phi)$$

$$= 8a^2.\frac{3}{4}.\frac{1}{2}.\frac{1}{2}\pi = (3/2)a^2\pi.$$

EXAMPLE 3. *Find the volume bounded by co-ordinates planes and the plane $\frac{x}{a}+\frac{y}{b}+\frac{z}{c} = 1$.*

SOLUTION. The plane cuts X, Y and Z- axis at point $(a,0,0)$, $(0,b,0)$ and $(0,0,c)$ respectively. The surface $ABCD$ of co-ordinates planes will be equal to

$$\int_0^a\int_0^{b(1-x/a)}\int_0^{c(1-x/a-y/b)} dx\,dy\,dz$$

$$= \int_0^a\int_0^{b(1-x/a)}c\left(1-\frac{x}{a}-\frac{y}{b}\right)dy\,dx$$

$$= c\int_0^a\int_0^{b(1-x/a)}\left(1-\frac{x}{a}-\frac{y}{b}\right)dy\,dx$$

$$= c\int_0^a\left[y-\frac{x}{a}.y-\frac{y^2}{2b}\right]_0^{b(1-x/a)} dx$$

$$= c\int_0^a\left[b\left(1-\frac{x}{a}\right)-\frac{x}{a}.b\left(1-\frac{x}{a}\right)-\frac{1}{2b}b^2\left(1-\frac{x}{a}\right)^2\right]dx$$

EXAMPLE 4. *Transform the integral $\int_0^2\int_0^{\sqrt{2x-x^2}}\frac{x\,dx\,dy}{\sqrt{(x^2+y^2)}}$ by changing into polar co-ordinates and hence evaluate it.*

SOLUTION. We have the limit of integration be

$$y = 0, y = \sqrt{(2x-x^2)} \text{ and } x = 0, x = 2.$$

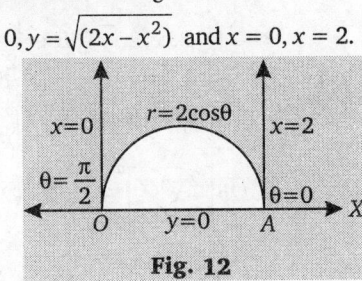

Fig. 12

$x^2+y^2-2x = 0$ which is change into

$$r^2(\cos^2\theta+\sin^2\theta)-2r\cos\theta = 0$$

or $\qquad\qquad r = 2\cos\theta.$

Now r varies from 0 to $2\cos\theta$ and θ varies from 0 to $\pi/2$. Note that at the point A of the circle, $\theta = 0$ and at point O, $r = 0$ and so from $r = 2\cos\theta$, we get

$$\theta = \frac{\pi}{2} \text{ at } O$$

the polar equivalent of the elementary area $dx\,dy$ is $r\,d\theta\,dr$.

$\therefore \iint_A f(x,y)dx\,dy = \iint_A f(r\cos\theta,r\sin\theta)r\,d\theta\,dr$

where A is the region of integration.

Therefore, transforming to polar co-ordinates, the given double integral

$$= \int_{\theta=0}^{\pi/2}\int_{r=0}^{2\cos\theta}\frac{r\cos\theta}{r}r\,d\theta\,dr = \int_0^{\pi/2}\cos\theta\left[\frac{r^2}{2}\right]_0^{2\cos\theta} d\theta$$

$$= \int_0^{\pi/2}\frac{1}{2}\cos\theta.4\cos^2\theta\,d\theta = 2\int_0^{\pi/2}\cos^3\theta\,d\theta$$

$$= 2.\frac{2}{3} = \frac{4}{3}.$$

EXAMPLE 5. *Find the area of the surface $z^2 = 2xy$ included between planes x=0, x=a, y=0, y=b.*

SOLUTION. The given surface is $z^2 = 2xy$.

$\therefore\qquad\qquad 2z\frac{\partial z}{\partial x} = 2y$ or $\frac{\partial z}{\partial x} = \frac{y}{z}$

Similarly $\qquad \frac{\partial z}{\partial y} = \frac{x}{z}$

Then required area of the surface

$$= \iint\sqrt{\left[1+\left(\frac{\partial z}{\partial x}\right)^2+\left(\frac{\partial z}{\partial y}\right)^2\right]}\,dx\,dy$$

$$= \int_{x=0}^a\int_{y=0}^b\sqrt{\left\{1+\left(\frac{y}{z}\right)^2+\left(\frac{x}{z}\right)^2\right\}}\,dx\,dy$$

$$= \int_{x=0}^a\int_{y=0}^b\sqrt{\left(\frac{z^2+y^2+x^2}{2xy}\right)}\,dx\,dy$$

$$= \int_{x=0}^a\int_{y=0}^b\sqrt{\frac{x^2+y^2+z^2}{2xy}}\,dx\,dy$$

$$= \int_{x=0}^{a} \int_{y=0}^{b} \frac{(x+y)}{\sqrt{2}\sqrt{(xy)}} \, dx \, dy$$

$$= \frac{1}{\sqrt{2}} \int_{x=0}^{a} \int_{y=0}^{b} \left(\sqrt{x} \frac{1}{\sqrt{y}} + \sqrt{y} \cdot \frac{1}{\sqrt{x}} \right) dx \, dy$$

$$= \frac{1}{\sqrt{2}} \int_{x=0}^{a} \sqrt{x} (2\sqrt{y})_0^b dx + \frac{1}{\sqrt{2}} \int_{x=0}^{a} \frac{1}{\sqrt{x}} \left(\frac{2}{3} y^{3/2} \right)_0^b dx$$

$$= \sqrt{(2b)} \int_0^a \sqrt{x} \, dx + \frac{\sqrt{2}}{3} b^{3/2} \int_0^a \frac{1}{\sqrt{x}} dx$$

$$= \sqrt{2b} \left[\frac{2}{3} x^{3/2} \right]_0^a + \frac{1}{3} \sqrt{2b^3} (2x^{1/2})_0^a$$

$$= \frac{2}{3} \sqrt{2} \sqrt{(ab)}(a+b) \ .$$

EXAMPLE 6. *Show that the area of the surface of paraboloid* $x^2 + y^2 = a^2$ *which lies between the planes* $z = 0$ *and* $z = a$ *is* $(\pi/6)(5.\sqrt{5}-1)$.

SOLUTION. The projection of given surface between the planes $z=0$ and $z=a$ on the x-y planes is circle $x^2 + y^2 = a^2, z = 0$.

Also, $\dfrac{\partial z}{\partial x} = \dfrac{2x}{a}, \dfrac{\partial z}{\partial y} = \dfrac{2y}{a}$

$$\therefore \quad S = \iint_A \sqrt{1 + \left(\frac{\partial z}{\partial x} \right)^2 + \left(\frac{\partial z}{\partial y} \right)^2} \, dx \, dy$$

$$= \iint_A \sqrt{1 + \frac{4x^2}{a^2} + \frac{4y^2}{a^2}} \, dx \, dy$$

$$= \frac{1}{a} \iint_A \sqrt{a^2 + 4x^2 + 4y^2} \, dx \, dy$$

where A is the circle $x^2 + y^2 = a^2$ in the xy-plane.

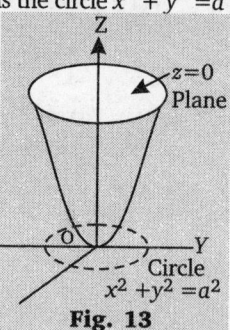

Fig. 13

The equation of the circle $x^2 + y^2 = a^2$ in polar co-ordinates is $r = a$. Hence, transforming the above double integral into polar co-ordinates, we have

$$S = \frac{1}{a} \int_0^{2\pi} \int_0^a \sqrt{a^2 + 4r^2} (r \, d\theta \, dr)$$

$$= \frac{1}{a} \int_0^{2\pi} \left[\int_0^a r\sqrt{a^2 + 4r^2} dr \right] d\theta$$

$$= \frac{1}{8a} \int_0^{2\pi} \left[\int_0^a \sqrt{a^2 + 4r^2} d(a^2 + 4r^2) \right] d\theta$$

$$= \frac{1}{8a} \int_0^{2\pi} \left[\frac{2}{3} \cdot (a^2 + 4r^2)^{3/2} \right]_0^a d\theta$$

$$= \frac{1}{8} \times \frac{2}{3} \times \frac{1}{a} \int_0^{2\pi} [(5a^2)^{3/2} - (a^2)^{3/2}] d\theta$$

$$= \frac{1}{12a} (5\sqrt{5} - 1) a^3 . 2\pi = \frac{\pi}{6}(5\sqrt{5} - 1) a^2$$

Exercise-5.3

1. Find the volume of the region bounded by $z = x^2 + y^2$ and $z = 2x$.

2. Find the volume cut off the sphere $x^2 + y^2 + z^2 = a^2$ by the cone $x^2 + y^2 = z^2$.

3. Transforms the following double integrals to polar co-ordinates and hence, evaluate them

    (a) $\int_{y=0}^{a} \int_{x=0}^{\sqrt{a^2-y^2}} (a^2 - x^2 - y^2) dx \, dy$

    (b) $\int_0^1 \int_x^{\sqrt{2x-x^2}} (x^2 + y^2) dx \, dy$

    (c) $\int_0^a \int_0^{\sqrt{a^2-x^2}} y^2 \sqrt{(x^2+y^2)} dx \, dy$

4. Evaluate $\iint r^2 d\theta \, dr$ over the area of circle $r = a \cos\theta$.

Hints to Selected Problems

1. $I = \int_0^\pi \int_0^{a(1+\sin\theta)} r \sin\theta . r \, d\theta \, dr$

2. Since the two curves intersect at the points where $\cos\theta = 0$ i.e., $\theta = \dfrac{\pi}{2}$ $\therefore$ $I = \int_{\theta=-\pi/2}^{\pi/2} \int_{r=a}^{a(1+\cos\theta)} r \, d\theta \, dr$.

4. The required volume is given by

$$V = \int_0^{2\pi} \int_0^{a/\sqrt{2}} \int_r^{\sqrt{a^2-r^2}} dz(r \, d\theta \, dr) .$$

Answers

1. $\dfrac{\pi^3}{-2}$ 2. $\dfrac{(2-\sqrt{2})\pi a^3}{3}$ 3. (a) $\dfrac{\pi a^4}{8}$ (b) $\dfrac{3\pi}{8} - 1$ (c) $\dfrac{\pi a^5}{20}$ 4. $\dfrac{4a^3}{9}$.

5.1.6 TRIPLE INTEGRAL

Let $f(x, y, z)$ be a single-valued function of the independent variables x, y, z in finite region V. Divide the region V into n subregions $\delta V_1, \delta V_2, \delta V_3, \ldots$ Let P be any point on the boundary or inside.

Take a point in each part and form the sum $s_n = f(x_1, y_1, z_1)\delta V_1 + f(x_2, y_2, z_2)\delta V_2 + \ldots + f(x_n, y_n, z_n)\delta V_n$

$$= \sum_{r=1}^{n} f(x_r, y_r, z_r)\delta V_r \qquad \ldots(1)$$

when n tends to infinity, the limit of sum (1) tends to zero is called the triple integral of function $f(x, y, z)$ over the region V and is denoted by

$$\iiint_V f(x,y,z)\,dv .$$

The triple integral can be utilised in evaluating a number of physical quantities like, $f(x,y,z) = 1$

We find volume, $V = \iiint_V dV$ and putting $f(x,y,z) = \rho$

We get, $\qquad mass = \iiint_V \rho\,dV$.

5.1.7 EVALUATION OF TRIPLE INTEGRALS

The region V divide into elementary cuboids by drawing parallel co-ordinate planes. The volume V can then be considered as the sum of number of columns parallel to z-axis extending from the lower surface of V say $z = z_1(x,y)$ to the upper surface of V say $z = z_2(x,y)$ the bases of these as column (only one column has been shown in fig. 14) are the elementary area δs_r, which cover a certain area S in x-y plane $i.e.$ plane $z = 0$.

$\therefore$ Summing up over the elementary cuboids in the same column

first and then taking the sum of all such columns

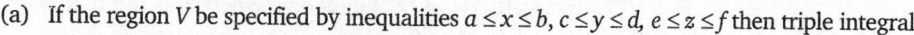

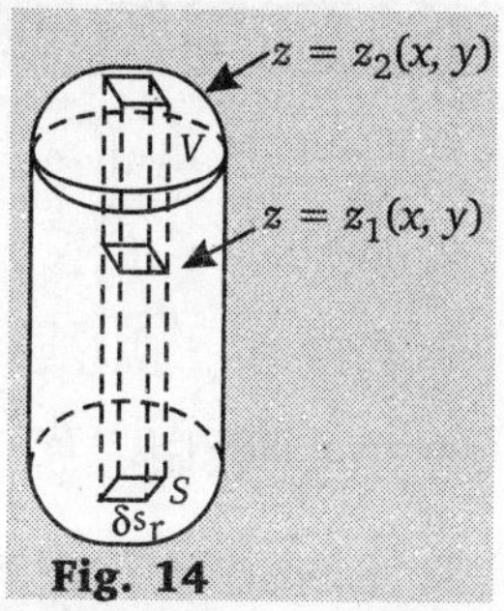

we can write $\sum\limits_{r=1}^{n} f(x_r, y_r, z_r)$ as $\sum\limits_{r=1}^{n} f(x_r, y_r, z_r)\delta z]\delta s_r$

where (x_r, y_r, z_r) is a point in the mth cuboid.

When δS_r and δz tend to zero this becomes equal to $\iint_S \left\{ \int_{z=z_1(x,y)}^{z_2(x,y)} f(x,y,z)\,dz \right\} ds$

(a) If the region V be specified by inequalities $a \le x \le b, c \le y \le d, e \le z \le f$ then triple integral

$\iiint_V f(x,y,z)\,dx\,dy\,dz = \int_a^b \int_c^d \int_e^f f(x,y,z)\,dx\,dy\,dz = \int_a^b dx \int_c^d dy \int_e^f f(x,y,z)\,dz.$

Here, we integrate first with respect to z keeping x and y constant and then the remaining integration is done as in the case of double integrals.

The integration with respect to z is performed first regarding x and y as constant then integration w.r to y regarding x as a constant and then integrate w.r to x.

(b) If the limits of z are function of x and y and y as function of x and x takes the constant values as from $x = a$ to $x = b$.

$\iiint_V f(x,y,z)\,dx\,dy\,dz = \int_a^b dx \int_{y_1(x)}^{y_2(x)} dy \int_{z(x,y)}^{z(x,y)} f(x,y,z)\,dz$

The integration with respect to z perform first regarding x and y as constant then integral w.r.t. y regarding x as a constant and then integrate w.r.t. x.

$z = z_2(x,y)$

$z = z_1(x,y)$

Fig. 14

 Solved Examples

EXAMPLE 1. *Evaluate* $\int_0^1 \int_{y^2}^1 \int_0^{1-x} x\,dy\,dx\,dz$.

SOLUTION. We have

$I = \int_0^1 \int_{y^2}^1 (z)_0^{1-x} x\,dy\,dx = \int_0^1 \int_{y^2}^1 x(1-x)\,dy\,dx$

$= \int_0^1 \int_{y^2}^1 (x - x^2)\,dy\,dx = \int_0^1 \left[\frac{1}{2}x^2 - \frac{1}{3}x^3 \right]_{y^2}^1 dy$

$= \int_0^1 \left[\left\{ \frac{1}{2}(1)^2 - \frac{1}{3}(1)^3 \right\} - \left\{ \frac{1}{2}(y^2)^2 - \frac{1}{3}(y^2)^3 \right\} \right] dy$

$= \int_0^1 \left(\frac{1}{6} - \frac{1}{2}y^4 + \frac{1}{3}y^6 \right) dy = \left(\frac{1}{6}y - \frac{1}{10}y^5 + \frac{1}{21}y^7 \right)_0^1$

$= \frac{1}{6} - \frac{1}{10} + \frac{1}{21} = \frac{4}{35}$.

EXAMPLE 2. *Evaluate* $\int_{x=0}^1 \int_{y=0}^{\sqrt{1-x^2}} \int_{z=0}^{\sqrt{1-x^2-y^2}} xyz\,dx\,dy\,dz.$

SOLUTION. The given integral

$I = \int_{x=0}^1 \int_0^{\sqrt{1-x^2}} xy\left(\frac{1}{2}z^2 \right)_0^{\sqrt{1-x^2-y^2}} dx\,dy$

$= \frac{1}{2}\int_{x=0}^1 \int_{y=0}^{\sqrt{1-x^2}} xy(1-x^2-y^2)\,dx\,dy$

$= \frac{1}{2}\int_{x=0}^1 \int_{y=0}^{\sqrt{1-x^2}} x[y(1-x^2) - y^3)]\,dx\,dy$

$= \frac{1}{2}\int_{x=0}^1 x\left[\frac{1}{2}(1-x^2)y^2 - \frac{1}{4}y^4 \right]_0^{\sqrt{1-x^2}} dx$

$= \frac{1}{2}\int_0^1 x\left[\frac{1}{2}(1-x^2)(1-x^2) - \frac{1}{4}(1-x^2)^2 \right] dx$

$= \frac{1}{2}\int_0^1 x\left(\frac{1}{2} - \frac{1}{4} \right)(1-x^2)^2 dx$

$= \frac{1}{8}\int_0^1 (x - 2x^3 + x^5)\,dx = \frac{1}{8}\left[\frac{1}{2}x^2 - \frac{1}{2}x^4 + \frac{1}{6}x^6 \right]_0^1$

$= \frac{1}{8}\left(\frac{1}{2} - \frac{1}{2} + \frac{1}{6} \right) = \frac{1}{48}$.

EXAMPLE 3. *Evaluate* $\int_0^{\log a} \int_0^x \int_0^{x+y} e^{x+y+z}\,dx\,dy\,dz.$

SOLUTION. Let

$I = \int_0^{\log a} \int_0^x \int_0^{x+y} e^{x+y} e^z\,dx\,dy\,dz$

$= \int_0^{\log a} \int_0^x e^{x+y}(e^z)_0^{x+y}\,dx\,dy$

$= \int_0^{\log a} \int_0^x e^{x+y}[e^{x+y} - 1]\,dx\,dy$

$= \int_0^{\log a} \int_0^x e^{2(x+y)}\,dx\,dy - \int_0^{\log a} \int_0^x e^{x+y}\,dx\,dy$

$= \int_0^{\log a} \int_0^x e^{2x}\cdot e^{2y}\,dx\,dy - \int_0^{\log a} \int_0^x e^{x+y}\,dx\,dy$

$= \int_0^{\log a} e^{2x}\left(\frac{1}{2}e^{2y} \right)_0^x dx - \int_0^{\log a} e^x(e^y)_0^x\,dx$

$= \frac{1}{2}\int_0^{\log a} e^{2x}(e^{2x} - e^0)\,dx - \int_0^{\log a} e^x(e^x - e^0)\,dx$

$= \frac{1}{2}\int_0^{\log a} (e^{4x} - e^{2x})\,dx - \int_0^{\log a} (e^{2x} - e^x)\,dx$

$= \frac{1}{2}\int_0^{\log a} (e^{4x} - 3e^{2x} + 2e^x)\,dx = \frac{1}{2}\left[\frac{1}{4}e^{4x} - \frac{3}{2}e^{2x} + 2e^x \right]_0^{\log a}$

$= \frac{1}{8}(e^{4\log a} - e^0) - \frac{3}{4}(e^{2\log a} - e^0) + (e^{\log a} - e^0)$

$= \frac{1}{8}(a^4 - 1) - \frac{3}{4}(a^2 - 1) + (a - 1)$

$$= \frac{1}{8}a^4 - \frac{3}{4}a^2 + a - \frac{3}{8} = \frac{1}{8}[a^4 - 6a^2 + 8a - 3] \ .$$

EXAMPLE 4. *Evaluate the volume of tetrahedron bounded by the co-ordinate planes and the planes* $x+y+z=1$.

SOLUTION. The volume of tetrahedron can be expressed as

$$0 \le x \le 1, 0 \le y \le 1-x, 0 \le z \le 1-x-y.$$

∴ The integral

$$I = \iiint dx\,dy\,dz = \int_0^1 \int_0^{1-x} \int_0^{1-x-y} dx\,dy\,dz$$

$$= \int_0^1 \int_0^{1-x} [z]_0^{1-x-y} dx\,dy = \int_0^1 \int_0^{1-x} (1-x-y)dx\,dy$$

$$= \int_0^1 \left[(1-x)y - \frac{y^2}{2}\right]_0^{1-x} dx$$

$$= \int_0^1 \left[(1-x)^2 - \frac{(1-x)^2}{2}\right] dx = \int_0^1 \frac{1}{2}(1-x)^2 dx$$

$$= \frac{1}{2}\left[\frac{(1-x)^3}{3.(-1)}\right]_0^1 = -\frac{1}{6}(0-1) = \frac{1}{6}.$$

Exercise-5.4

1. Evaluate $\int_{x=0}^1 \int_{y=0}^2 \int_{z=1}^2 x^2 yz\,dx\,dy\,dz$.

2. Evaluate $\int_{-a}^a \int_{-b}^b \int_{-c}^c (x^2 + y^2 + z^2)dx\,dy\,dz$.

3. Evaluate $\int_{-1}^1 \int_0^z \int_{x-z}^{x+z}(x+y+z)dy\,dx\,dz$.　　(GBTU-2010)

4. Evaluate $\int_0^1 \int_0^{1-x} \int_0^{1-x-y} \frac{dy\,dx\,dz}{(1+x+y+z)^3}$.

5. Evaluate $\int_0^{\pi/2} d\theta \int_0^{a\sin\theta} dr \int_0^{(a^2-r^2)/a} r\,dz$.

6. Evaluate $\int_0^a \int_0^{a-x} \int_0^{a-x-y} x^2\,dx\,dy\,dz$.

7. Evaluate $\int_0^2 \int_0^x \int_0^{x+y} e^x (y+2z)dx\,dy\,dz$.

8. Evaluate $\int_0^{\log 2} \int_0^x \int_0^{x+\log y} e^{x+y+z}\,dx\,dy\,dz$.

9. Evaluate the integral $\iiint xyz\,dx\,dy\,dz$ over the the volume enclosed by three co-ordinates plane and the plane $x+y+z=1$.

10. Evaluate $\iiint \frac{dx\,dy\,dz}{(x+y+z+1)^2}$ over the region z. Evaluate $\iiint (z^5+z)dx\,dy\,dz$ over the sphere $x^2+y^2+z^2=1$.

12. Evaluate $\iiint_R u^2 v^2 w\,du\,dv\,dw$, where R is the region $u^2+v^2 \le 1$, $0 \le w \le 1$.　　(VTU-2011; SRM-2009)

13. Find the volume of the tetrahedron bounded by the plane $\frac{x}{a}+\frac{y}{b}+\frac{z}{c}=1$ and $(x+z=a)$ and coordinate plane.

14. Evaluate $\int_0^2 \int_0^x \int_0^{x+y} e^x (y+2z)dx\,dy\,dz$.

15. Evaluate $\iiint_R (x-2y+z)$ where R is the region determined by $0 \le x \le 1, 0 < y \le x^2, 0 \le z \le x+y$　　(UPTU-2009)

Answers

1. 1　**2.** $\frac{2}{3}abc(a^2+b^2+c^2)$　**3.** 0　**4.** $\frac{1}{2}\left(\log 2 - \frac{5}{8}\right)$　**5.** $\frac{5a^3\pi}{64}$　**6.** $\frac{a^5}{60}$　**7.** $19[(1/3)e^2+1]$

8. $\frac{8}{3}\log 2 - \frac{19}{9}$　**9.** 0　**10.** $\frac{1}{2}\left(\log 2 - \frac{5}{8}\right)$　**11.** 0　**12.** $\pi/48$　**13.** $abc/5$　**14.** $\frac{19}{3}(e^2+3)$　**15.** 8/35

5.2 DIRICHLET'S THEOREMS

5.2.1 DIRICHLET'S THEOREM FOR THREE VARIABLES

Let V be the region given by $x \ge 0, y \ge 0, z \ge 0, x+y+z \le 1, l,m,n$ are positive. Then

$$\int_V x^{l-1}y^{m-1}z^{n-1}dx\,dy\,dz = \frac{\Gamma(l)\Gamma(m)\Gamma(n)}{\Gamma(l+m+n+1)}.$$

Proof. We evaluate the given integral over the volume enclosed by the three co-ordinates planes and the plane $x+y+z=1, x=0, y=0, z=0$. The limits of integration for this region can be expressed as $0 \le x \le 1, 0 \le y \le 1-x, 0 < z \le 1-x-y$.

Hence we may write the given triple integral as

$$\int_0^1 \int_0^{1-x} \int_0^{1-x-y} x^{l-1}y^{m-1}z^{n-1}dx\,dy\,dz = \int_0^1 \int_0^{1-x} x^{l-1}y^{m-1}[z^n/n]_0^{1-x-y}dx\,dy = \frac{1}{n}\int_0^1 \int_0^{1-x} x^{l-1}y^{m-1}(1-x-y)^n dx\,dy$$

$$= \frac{1}{n}\int_0^1 \int_0^1 x^{l-1}\{(1-x)t\}^{m-1}[1-x-(1-x)t]^n(1-x)dx\,dt$$
　　　　　(Putting $y=(1-x)t, \Rightarrow dy=(1-x)dt$)

$$= \frac{1}{n}\int_0^1 \int_0^1 x^{l-1}(1-x)^{m-1}t^{m-1}(1-x)^n(1-t)^n(1-x)dx\,dt$$

$$= \frac{1}{n}\int_0^1 \int_0^1 x^{l-1}(1-x)^{m+n}t^{m-1}(1-t)^n dx\,dt = \frac{1}{n}\int_0^1 x^{l-1}(1-x)^{m+n}dx \times \int_0^1 (t)^{m-1}(1-t)^n dt$$

$$= \frac{1}{n}B(l,m+n+1)B(m,n+1) \quad \text{(By definition of Beta function)}$$

$$= \frac{1}{n}\cdot\frac{\Gamma(l)\Gamma(m+n+1)}{\Gamma(l+m+n+1)}\cdot\frac{\Gamma(m)\Gamma(n+1)}{\Gamma(m+n+1)} \qquad \left[\because B(m,n)=\frac{\Gamma(m)\Gamma(n)}{\Gamma(m+n)}\right]$$

$$= \frac{\Gamma(l)\Gamma(m)}{\Gamma(l+m+n+1)}\cdot\frac{n\Gamma(n)}{n} = \frac{\Gamma(l)\Gamma(m)\Gamma(n)}{\Gamma(l+m+n+1)}. \qquad [\because \Gamma(n+1)=n\,\Gamma(n)]$$

📖 REMARKS

- Dirichlet's theorem holds good even if the conditions is taken as $x+y+z<1$ in place of $x + y + z \leq 1$.
- The triple integral $\iiint x^{l-1} y^{m-1} z^{n-1} dx\, dy\, dz = h^{l+m+n} \dfrac{\Gamma(l)\Gamma(m)\Gamma(n)}{\Gamma(l+m+n+1)}$

where the integral is extended to all positive values of the variables x, y and z, when $x+y+z \leq h$.

5.2.2 DIRICHLET'S THEOREM FOR n VARIABLES

If the integral is extended to all positive values of the variables x_1, x_2,.., x_n subject to the condition $x_1 + x_2 + ... + x_n \leq 1$. Then

$$\iint ... \int x_1^{l_1-1} x_2^{l_2-1} ... x_n^{l_n-1} dx_1 dx_2 ... dx_n = \frac{\Gamma(l_1)\Gamma(l_2)...\Gamma(l_n)}{\Gamma(1+l_1+l_2+...+l_n)}$$

Proof. We shall prove this theorem by mathematical induction.

First we prove the theorem for 2-variables *i.e.*, $n=2$

Let us consider the integral $\qquad I_2 = \iint x_1^{l_1-1} x_2^{l_2-1} dx_1 dx_2$ such that $x_1 + x_2 \leq 1$.

Now, using previous theorem, we have $\qquad I_2 = \dfrac{\Gamma(l_1)\Gamma(l_2)}{\Gamma(1+l_1+l_2)}$ ... (2)

Equation (1) is true for two variables. Now assume that theorem is true for n variables. Therefore

$$I_n = \iint ... \int x_1^{l_1-l} x_2^{l_2-l} ... x_n^{l_n-l} dx_1 . dx_2 .... dx_n$$

$$= \frac{\Gamma(l_1)\Gamma(l_2)...\Gamma(l_n)}{\Gamma(1+l_1+l_2+...+l_n)} \qquad \qquad ...(2)$$

with condition $x_1 + x_2 + ... + x_n \leq 1$.

If the condition $x_1 + x_2 + ... + x_n \leq h$, then putting $\dfrac{x_1}{h} = u_1, \dfrac{x_2}{h} = u_2 ... \dfrac{x_n}{h} = u_n$ so that

$$dx_1 = h\, du_1, dx_2 = h\, du_2,... dx_n = h\, du_n$$

We have $\quad \iint ... \int x_1^{l_1-1} x_2^{l_2-1} ... x_n^{l_n-1} dx_1 dx_2 ... dx_n = h^{l_1+l_2+...+l_n} \iint ... \int u_1^{l_1-1} u_2^{l_2-1} ... u_n^{l_n-1} du_1 du_2 ... du_n$

$$\text{subject to the condition } u_1 + u_2 + ... + u_n \leq 1$$

$$= h^{l_1+l_2+...+l_n} \frac{\Gamma(l_1)\Gamma(l_2)...\Gamma(l_n)}{\Gamma(1+l_1+l_2+...+l_n)} \qquad \qquad ... (3)$$

(Using the assumed result (2))

Now for $n+1$ variables the conditions are $x_1 + x_2 + ... + x_n + x_{n+1} \leq 1$ *i.e.*, $x_2 + x_3 + ... + x_n + x_{n+1} \leq 1 - x_1$ and $0 \leq x_1 \leq 1$.

We have

$$\iint ... \int x_1^{l_1-1} x_2^{l_2-1} ... x_n^{l_n-1} x_{n+1}^{l_{(n+1)}-1} dx_1 dx_2 ... dx_n dx_{n+1} \quad \text{where } x_1 + x_2 + ... x_{n+1} \leq 1$$

$$= \int_{x_1=0}^{1} x_1^{l_1-1} \left[\iint ... \int x_2^{l_2-1} ... x_{n+1}^{l_{n+1}-1} dx_2 ... dx_{n+1} \right] dx_1$$

$$= \frac{\Gamma(l_2)\Gamma(l_3)...\Gamma(l_{n+1})}{\Gamma(l_1+1+l_2+...l_n+l_{n+1})} . \int_0^1 x_1^{l_1-1} (1-x_1)^{(1+l_2+l_3+...+l_{n+1})-1} dx_1 \qquad \text{Using (3)}$$

$$= \frac{\Gamma(l_2)\Gamma(l_3)...\Gamma(l_{n+1})}{\Gamma(1+l_2+...+l_{n+1})} . \frac{\Gamma(1+l_2+...+l_{n+1})}{\Gamma(1+l_1+l_2+...+l_n+l_{n+1})}$$

$$= \frac{\Gamma(l_1)\Gamma(l_2)...\Gamma(l_{n+1})}{\Gamma(1+l_1+l_2+...+l_{n+1})} \qquad \qquad ...(4)$$

The result (4) shows that the theorem hold for $(n+1)$ variables. Hence, by principle of mathematical induction, theorem is true for all values of n.

Solved Examples

EXAMPLE 1. *Evaluate $\iiint x^{l-1} y^{m-1} z^{n-1} dx\, dy\, dz$ in which $x \geq 0$, $y \geq 0, z = 0$ and $(x/a)^{1/2} + (y/b)^{1/2} + (z/c)^{1/2} \leq 1$.*

SOLUTION. Let $\quad (x/a)^{1/2} = u$, $(y/b)^{1/2} = v$

and $\quad (z/c)^{1/2} = w$

Then $\quad x = au^2, y = bv^2, z = cw^2$

$dx = 2au\, du;\ dy = 2bv\, dv;\ dz = 2cw\, dw$,

$u \geq 0, v \geq 0, w \geq 0$ and $u + v + w \leq 1$

Hence, $\iiint (au^2)^{l-1} (bv^2)^{m-1} (cw^2)^{n-1}$
$\qquad\qquad . 2au . 2bv . 2cw\, du\, dv\, dw$

$= 8 a^l b^m c^n \iiint u^{2l-1} v^{2m-1} w^{2n-1}\, du\, dv\, dw$

$= 8 a^l b^m c^n \dfrac{\Gamma(2l)\Gamma(2m)\Gamma(2n)}{\Gamma(2l+2m+2n+1)}$

EXAMPLE 2. *Show that the integral $\iiint x^{l-1} y^{m-1} z^{n-1}\, dx\, dy\, dz$ integrand over the region in the first octant below the surface $(x/a)^p + (y/b)^q + (z/c)^r = 1$.*

is $\dfrac{a^l b^m c^n}{pqr} . \dfrac{\Gamma(l/p)\Gamma(m/q)\Gamma(n/r)}{\Gamma[(l/p) + (m/q) + (n/r) + 1]}$.

(UPTU-2006, 2009, GBTU-2011, 16)

SOLUTION. Putting $\left(\dfrac{x}{a}\right)^p = u$ or $x = au^{1/p}$

$\Rightarrow \qquad dx \quad a(1/p)u^{(1/\ )\ 1}.du.$

Similarly putting $(y/b)^q = v$ and $(z/c)^r = w$, we get

$$dy = b(1/q)v^{(1/q)-1}dv$$

and $\qquad dz = c(1/r)w^{(1/r)-1}dw$.

$\therefore \qquad x^{l-1}dx = a^{l-1}u^{(l-1)/p}a(1/p)u^{(1-p)/p}du$

$$= a^l(1/p)u^{(l/p)-1}du$$

Similarly, $y^{m-1}dy = b^m(1/q)v^{(m/q)-1}dv$;

$$z^{n-1}dz = c^n(1/r)w^{(n/r)-1}dw$$

Hence, subject to the condition $u+v+w \le 1$, the given integral

$$= \frac{a^l b^m c^n}{pqr}\iiint u^{(l/p)-1}v^{(m/q)-1}w^{(n/r)-1}du\,dv\,dw$$

$$= \frac{a^l b^m c^n}{pqr}\cdot\frac{\Gamma(l/p)\Gamma(m/q)\Gamma(n/r)}{\Gamma[(l/p)+(m/q)+(n/r)+1]}$$

EXAMPLE 3. *Evaluate* $\iiint dx\,dy\,dz$ *where* $\dfrac{x^2}{a^2}+\dfrac{y^2}{b^2}+\dfrac{z^2}{c^2} \le 1$
or find the volume of
$(x^2/a^2)+(y^2/b^2)+(z^2/c^2)=1$.

SOLUTION. Let $\dfrac{x^2}{a^2} = u, x = au^{1/2}$ so that $dx = \dfrac{1}{2}au^{-1/2}du$.

Similarly, putting $\dfrac{y^2}{b^2} = v$ and $\dfrac{z^2}{c^2} = w$, we get

$$dy = \frac{1}{2}bv^{-1/2}dv \quad \text{and} \quad dz = \frac{1}{2}cw^{-1/2}dw.$$

$\therefore \iiint dx\,dy\,dz = \iiint \dfrac{1}{2}au^{-1/2}du.\dfrac{1}{2}bv^{-1/2}.dv.\dfrac{1}{2}cw^{-1/2}dw$

$$= \frac{1}{8}abc\iiint u^{1/2-1}v^{1/2-1}w^{1/2-1}du\,dv\,dw$$

$$= \frac{1}{8}abc\frac{\Gamma(1/2)\Gamma(1/2)\Gamma(1/2)}{\Gamma(1/2+1/2+1/2+1)}$$

$$= \frac{1}{8}abc\frac{\sqrt{\pi}\sqrt{\pi}\sqrt{\pi}}{\Gamma(5/2)} = \frac{\pi\sqrt{\pi}\,bca}{8\cdot 3/2\cdot 1/2\sqrt{\pi}}$$

$$= -\pi abc$$

EXAMPLE 4. *Find the volume of the tetrahedron bounded by the plane* $\dfrac{x}{a}+\dfrac{y}{b}+\dfrac{z}{c}=1$ *and the co-ordinate planes.*   (GBTU-2012, MTU-2012)

SOLUTION. The volume of a small element at a point $(x, y, z) = dx\,dy\,dz$

$\therefore$ the volume of the given tetrahedron $= \iiint dx\,dy\,dz$ where the integral is extended to all positive values of variables x, y, z.

Put $x/a = u, y/b = v, z/c = w$ subject to the condition so that $\dfrac{x}{a}+\dfrac{y}{b}+\dfrac{z}{c} \le 1$

$$dx = a.du, dy = b.dv \text{ and } dz = c.dw$$

then the required volume

$$= \iiint abc\,du\,dv\,dw \quad \text{where } u+v+w \le 1$$

$$= abc\iiint u^{1-1}v^{1-1}w^{1-1}\,du\,dv\,dw$$

$$= abc\frac{[\Gamma(1)]^3}{\Gamma(1+1+1+1)} \quad \text{[By Dirichlet's theorem]}$$

$$= abc\frac{1}{\Gamma(4)} = \frac{abc}{3.2.1} = \frac{abc}{6}$$

5.3   LIOUVILLE'S EXTENSION OF DIRICHLET'S THEOREM

Statement. *If x, y, z are all positive and such that $h_1 < x+y+z \le h_2$ then*

$$\iiint f(x+y+z)x^{l-1}y^{m-1}z^{n-1}dx\,dy\,dz = \frac{\Gamma(l)\Gamma(m)\Gamma(n)}{\Gamma(l+m+n)}\int_{h_1}^{h_2} f(u)u^{l+m+n-1}du.$$

Proof. From Dirichlet's theorem, we have

$$I = \iiint x^{l-1}y^{m-1}z^{n-1}dx\,dy\,dz = \frac{\Gamma(l)\Gamma(m)\Gamma(n)}{\Gamma(l+m+n+1)}u^{(l+m+n)} \qquad \ldots (1)$$

subject to the condition that $x, y, z \ge 0$ and $x+y+z \le u$.

Now if $x, y, z \ge 0$ and $x+y+z \le u+\delta u$, then we have

$$I = \iiint x^{l-1}y^{m-1}z^{n-1}dx\,dy\,dz = \frac{\Gamma(l)\Gamma(m)\Gamma(n)}{\Gamma(l+m+n+1)}(u+\delta u)^{(l+m+n)} \qquad \ldots (2)$$

So the value of integral given above extended to all such positive value of x, y, z such that $x+y+z$ lies between u and $u+\delta u$, is given by

$$I = \iiint x^{l-1}y^{m-1}z^{n-1}dx\,dy\,dz$$

$$= \frac{\Gamma(l)\Gamma(m)\Gamma(n)}{\Gamma(l+m+n+1)}[(u+\delta u)^{l+m+n}-u^{l+m+n}] = \frac{\Gamma(l)\Gamma(m)\Gamma(n)}{\Gamma(l+m+n+1)}u^{l+m+n}\left[\left(1+\frac{\delta u}{u}\right)^{l+m+n}-1\right]$$

$$= \frac{\Gamma(l)\Gamma(m)\Gamma(n)}{\Gamma(l+m+n+1)}u^{l+m+n}\left[1+(l+m+n)\frac{\delta u}{u}+\ldots-1\right] \qquad \text{[On expanding by Taylor's series]}$$

$$= \frac{\Gamma(l)\Gamma(m)\Gamma(n)}{\Gamma(l+m+n+1)}(l+m+n)u^{(l+m+n-1)}\delta u \qquad \text{[Neglecting the second and higher degree terms of } \delta u]$$

$$= \frac{\Gamma(l)\Gamma(m)\Gamma(n)}{\Gamma(l+m+n)}u^{(l+m+n-1)}\delta u \qquad [\because \Gamma(l+m+n+1) = (l+m+n)\Gamma(l+m+n)]$$

Now, consider the intergral $\iiint f(x+y+z)x^{l-1}y^{m-1}z^{n-1}dx\,dy\,dz$.

Since $u \le x+y+z \le \delta u,$ so the function $f(x+y+z)$ will differ by a small quantity of same order of solution. Hence, the integral

$$\iiint f(x+y+z)x^{l-1}y^{m-1}z^{n-1}dx\,dy\,dz = \frac{\Gamma(l)\Gamma(m)\Gamma(n)}{\Gamma(l+m+n)}f(u)u^{(l+m+n-l)}\delta u$$

subject to the condition that $x, y, z \ge 0$ and $u \le x+y+z \le u+\delta u$, to the first approximation.

So finally for the given condition that for positive x, y, z such that $h_1 \le x+y+z \le h_2$,

we get $\iiint f(x+y+z)x^{l-1}y^{m-1}z^{n-1}dx\,dy\,dz = \frac{\Gamma(l)\Gamma(m)\Gamma(n)}{\Gamma(l+m+n)}\int_{h_1}^{h_2} f(u)u^{(l+m+n-1)}du$

 Solved Examples

EXAMPLE 1. *Evaluate $\iiint e^{x+y+z}dx\,dy\,dz$ taken over the positive octant such that $x+y+z \le 1$.* (UPTU-2008)

SOLUTION. In the positive octant x, y, z are all positive and therefore $0 < (x+y+z) \le 1$.
Therefore, we have
$$\iiint e^{x+y+z}dx\,dy\,dz$$
$$= \frac{\Gamma(1)\Gamma(1)\Gamma(1)}{\Gamma(1+1+1)}\int_0^1 e^h h^{1+1+1-1}dh$$
$$\hspace{4cm}\text{[By Liouville's theorem]}$$
$$= \frac{1}{\Gamma(3)}\int_0^1 h^2 e^h dh = \frac{1}{2!}\left[(h^2 e^h)_0^1 - \int_0^1 2he^h dh\right]$$
$$= \frac{1}{2}\left[e - 2\left\{(he^h)_0^1 - \int_0^1 e^h dh\right\}\right]$$
$$= \frac{1}{2}\left[e - 2\left\{e - (e^h)_0^1\right\}\right] = \frac{1}{2}[e - 2\{e-e+1\}]$$
$$= \frac{1}{2}(e-2) .$$

EXAMPLE 2. *Evaluate $\iiint \log(x+y+z)\,dx\,dy\,dz$ taken over all positive values of x, y, z subject to the condition $x+y+z \le 1$.*

SOLUTION. Since x, y, z are to be taken positive value only, we have
$$0 < (x+y+z) \le 1.$$
Therefore, we have
$$\iiint \log(x+y+z)\,dx\,dy\,dz$$
$$= \iiint \log(x+y+z)x^{1-1}y^{1-1}z^{1-1}\,dx\,dy\,dz$$
$$= \frac{\Gamma(1)\Gamma(1)\Gamma(1)}{\Gamma(1+1+1)}\int_0^1 (\log h)h^{1+1+1-1}dh,$$
$$\hspace{4cm}\text{[By Liouville's theorem]}$$
$$= \frac{1}{\Gamma(3)}\int_0^1 h^2 (\log h)dh$$
$$= \frac{1}{2!}\left[\left\{(\log h)\frac{1}{3}h^3\right\}_0^1 - \int_0^1 \frac{1}{h}\cdot\frac{1}{3}h^3 dh\right]$$
$$= \frac{1}{6}\left[h^3 \log h - \frac{h^3}{3}\right]_0^1 = -\frac{1}{18}$$

EXAMPLE 3. *Evaluate $\iiint x^\alpha y^\beta z^\gamma (1-x-y-z)^\lambda\,dx\,dy\,dz$ over the interior of tetrahedron formed by the co-ordinate plane and the plane $x+y+z=1$.*

SOLUTION. The region of integration is bounded by the plane $x = 0$, $y = 0, z = 0$ and $x+y+z=1$. So, the variable x, y, z take all positive values subject to the condition
$$0 < x+y+z < 1.$$
Therefore the given integral

$$= \iiint x^{(\alpha+1)-1}y^{(\beta+1)-1}z^{(\gamma+1)-1}$$
$$\hspace{2cm}[1-(x+y+z)]^\lambda dx\,dy\,dz$$
$$= \frac{\Gamma(\alpha+1)\Gamma(\beta+1)\Gamma(\gamma+1)}{\Gamma(\alpha+\beta+\gamma+3)}$$
$$\hspace{1cm}\int_0^1 u^{\alpha+1+\beta+1+\gamma+1-1}(1-u)^\lambda du$$
[By Liouville's extension of Dirichlet's theorem]
$$= \frac{\Gamma(\alpha+1)\Gamma(\beta+1)\Gamma(\gamma+1)}{\Gamma(\alpha+\beta+\gamma+3)}$$
$$\hspace{1cm}\int_0^1 u^{(\alpha+\beta+\gamma+3)-1}(1-u)^{(\lambda+1)-1}du$$
$$= \frac{\Gamma(\alpha+1)\Gamma(\beta+1)\Gamma(\gamma+1)}{\Gamma(\alpha+\beta+\gamma+3)}B(\alpha+\beta+\gamma+3,\lambda+1)$$
$$= \frac{\Gamma(\alpha+1)\Gamma(\beta+1)\Gamma(\gamma+1)}{\Gamma(\alpha+\beta+\gamma+3)}\cdot\frac{\Gamma(\alpha+\beta+\gamma+3)\Gamma(\lambda+1)}{\Gamma(\alpha+\beta+\gamma+4)}$$
$$= \frac{\Gamma(\alpha+1)\Gamma(\beta+1)\Gamma(\gamma+1)\Gamma(\lambda+1)}{\Gamma(\alpha+\beta+\gamma+\lambda+4)}$$

EXAMPLE 4. *Evaluate*
$$\iiint \sqrt{(a^2b^2c^2 - b^2c^2x^2 - c^2a^2y^2 - a^2b^2z^2)}dx\,dy\,dz$$
taken throughout the ellipsoid $\frac{x^2}{a^2}+\frac{y^2}{b^2}+\frac{z^2}{c^2} = 1$.

SOLUTION. Let us first evaluate the given integral over the region of ellipsoid which lie in the positive octants the given ellipsoid $\frac{x^2}{a^2}+\frac{y^2}{b^2}+\frac{z^2}{c^2} = 1$ is symmetrical in all the eight octants.

Put $\frac{x^2}{a^2} = u, \frac{y^2}{b^2} = v, \frac{z^2}{c^2} = w$ then $x = au^{1/2}$

$\Rightarrow \quad dx = 1/2au^{-1/2}du$

$\quad y = bv^{1/2} \Rightarrow \quad dy = \frac{1}{2}bv^{-1/2}dv$

and $z = cw^{1/2} \Rightarrow \quad dz = \frac{1}{2}cw^{-1/2}dw$
The given integral
$$I = abc\iiint \sqrt{\left(1 - \frac{x^2}{a^2} - \frac{y^2}{b^2} - \frac{z^2}{c^2}\right)}dx\,dy\,dz$$
where $0 < x^2/a^2 + y^2/b^2 + z^2/c^2 \le 1$
$$= abc\iiint \sqrt{1-u-v-w}(1/8)$$
$$\hspace{1cm}abcu^{-1/2}v^{-1/2}w^{-1/2}du\,dv\,dw$$
$$\hspace{3cm}\text{where } 0 < u+v+w \le 1$$
$$= \frac{a^2b^2c^2}{8}\iiint u^{(1/2)-1}v^{(1/2)-1}$$
$$\hspace{1cm}w^{(1/2)-1}\sqrt{1-(u+v+w)}\,du\,dv\,dw$$

$$= \frac{a^2b^2c^2}{8} \frac{[\Gamma(1/2)]^3}{\Gamma(3/2)} \int_0^1 \sqrt{1-t} \cdot t^{1/2+1/2+1/2-1} dt$$

[By Liouville's theorem]

$$= \frac{a^2b^2c^2}{8} \frac{(\sqrt{\pi})^3}{1/2\sqrt{\pi}} \int_0^1 (1-t)^{(3/2)-1} t^{(3/2)-1} dt$$

$$= \frac{a^2b^2c^2}{8} \cdot 2\pi \frac{\Gamma(3/2)\Gamma(3/2)}{\Gamma(3)} = \frac{\pi^2 a^2 b^2 c^2}{32}$$

Hence, if the integration is extended throughout the ellipsoid then the given integral

$$= 8I = 8 \cdot \frac{\pi^2 a^2 b^2 c^2}{32} = \frac{\pi^2 a^2 b^2 c^2}{4}$$

EXAMPLE 5. *Show that* $\iint \left(\dfrac{1-x^2-y^2}{1+x^2+y^2}\right)^{1/2} dx\, dy = \dfrac{\pi}{8}(\pi - 2)$

over the positive quadrant of circle $x^2 + y^2 = 1$.

(UPTU-2008)

SOLUTION. The given integral is to be extended to all positive values of x and y such that

$$0 \le x^2 + y^2 \le 1 \qquad \dots (1)$$

Put $\qquad x^2 = u, y^2 = v \Rightarrow x = u^{1/2}, y = v^{1/2}$

so that $\qquad dx = \dfrac{1}{2} u^{-1/2} du,\ dy = \dfrac{1}{2} v^{-1/2} dv$

With these substitution, the condition (1) become $0 \le u + v \le 1$

Therefore the integral

$$= \iint \left[\frac{1-(u+v)}{1+(u+v)}\right]^{1/2} \frac{1}{4} u^{-1/2} v^{-1/2} du\, dv$$

$$= \frac{1}{4} \iint \left[\frac{1-(u+v)}{1+(u+v)}\right]^{1/2} u^{(1/2)-1} v^{(1/2)-1} du\, dv$$

where $0 \le u + v \le 1$

$$= \frac{1}{4} \frac{\Gamma(1/2)\Gamma(1/2)}{\Gamma(1/2+1/2)} \int_0^1 \left[\frac{1-h}{1+h}\right]^{1/2} \cdot h^{(1/2)+(1/2)-1} dh$$

[By Liouville's extension of Dirichlet's theorem]

$$= \frac{1}{4} \frac{\sqrt{\pi}\sqrt{\pi}}{\Gamma(1)} \int_0^1 \frac{1-h}{\sqrt{(1-h^2)}} dh = \frac{\pi}{4} \int_0^1 \frac{(1-\sin\theta)}{\cos\theta} \cos\theta\, d\theta$$

Putting $h = \sin\theta$, so that $dh \Rightarrow \cos\theta\, d\theta$

$$= \frac{\pi}{4}[\theta + \cos\theta]_0^{\pi/2} = \frac{\pi}{4}\left[\frac{\pi}{2} - 1\right] = \frac{\pi}{8}(\pi - 2).$$

EXAMPLE 6. *Prove that* $I = \iiint dx\, dy\, dz\, dw$, *for all positive values of the variables for which* $x^2 + y^2 + z^2 + w^2$ *is not less than* a^2 *and not greater than* b^2, *is* $\pi^2(b^4 - a^4)/32$.

SOLUTION. As per given, we have
$$a^2 < x^2 + y^2 + z^2 + w^2 < b^2.$$

Put $x^2 = u_1$ or $x = u_1^{1/2}, \Rightarrow dx = 1/2u_1^{-1/2}du_1$.

Similarly, putting $y^2 = u_2, z^2 = u_3, w^2 = u_4$, we get
$$dy = \frac{1}{2}u_2^{-1/2}du_2, dz = \frac{1}{2}u_3^{-1/2}du_3, dw = \frac{1}{2}u_4^{-1/2}du_4$$

∴ Then

$$I = \iiint \frac{1}{2}u_1^{-1/2}\frac{1}{2}u_2^{-1/2}\cdot\frac{1}{2}u_3^{-1/2}\frac{1}{2}u_4^{-1/2} du_1 du_2 du_3 du_4$$

$$= \frac{1}{16}\iiint u_1^{1/2-1}u_2^{1/2-1}u_3^{1/2-1}u_4^{1/2-1} du_1\, du_2\, du_3\, du_4$$

$$= \frac{1}{16} \frac{\Gamma(1/2)\Gamma(1/2)\Gamma(1/2)\Gamma(1/2)}{\Gamma(1/2+1/2+1/2+1/2)}\int_{a^2}^{b^2} h^{\frac{1}{2}+\frac{1}{2}+\frac{1}{2}+\frac{1}{2}-1} dh$$

$$= \frac{1}{16}\frac{(\sqrt{\pi})^4}{\Gamma(2)}\int_{a^2}^{b^2} h\, dh \qquad [\because \Gamma(1/2) = \sqrt{\pi}]$$

$$= \frac{\pi^2}{16}\left(\frac{1}{2}h^2\right)_{a^2}^{b^2} = \frac{\pi^2}{32}(b^4 - a^4).$$

Exercise-5.5

1. Show that if l, m, n are all positive, then

$$\iiint x^{l-1}y^{m-1}z^{n-1}dx\, dy\, dz = \frac{a^l b^m c^n}{8} \cdot \frac{\Gamma(l/2)\Gamma(m/2)\Gamma(n/2)}{\Gamma(l/2+m/2+n/2+1)}$$

where the triple integral is taken throughout the part of the ellipsoid $\dfrac{x^2}{a^2} + \dfrac{y^2}{b^2} + \dfrac{z^2}{c^2} = 1$ which lies in the positive octant.

2. Evaluate $\iint x^{2l-1}y^{2m-1}dx\, dy$ such that $x^2 + y^2 \le c^2$ for all positive values of x and y.

3. Find the volume of solid surrounded by the surface $\left(\dfrac{x}{a}\right)^{2/3} + \left(\dfrac{y}{b}\right)^{2/3} + \left(\dfrac{z}{c}\right)^{2/3} = 1$.

4. Evaluate the double integral $\iint_p x^{1/2}y^{1/2}(1-x-y)^{2/3}dx\, dy$ over the domain D bounded by lines $x = 0, y = 0, x + y = 1$.

5. Evaluate $\iint_T x^{1/2}y^{1/2}(1-x-y)^{3/2}dx\, dy$, where T is the region bounded by $x \ge 0, y \ge 0, x+y \le 1$.

6. Evaluate $\iiint \sqrt{\left(\dfrac{1-x^2-y^2-z^2}{1+x^2+y^2+z^2}\right)} dx\, dy\, dz$, integral being taken

over all positive values of x, y, z such that $x^2 + y^2 + z^2 \le 1$.

7. Evaluate $\iint \sqrt{\left\{\dfrac{1-(x^2/a^2-y^2/b^2)}{1+(x^2/a^2+y^2/b^2)}\right\}} dx\, dy$ where $\dfrac{x^2}{a^2} + \dfrac{y^2}{b^2} \le 1$.

8. Evaluate $\iint_R \sqrt{(x^2+y^2)}dx\, dy$, where R is the region in the xy plane bounded by $x^2+y^2 = 4$ and $x^2+y^2 = 9$.

9. Prove that $\iiint \dfrac{dx\, dy\, dz}{(x+y+z+1)^2} = \dfrac{1}{2}\left[\log 2 - \dfrac{5}{8}\right]$ throughout the volume bounded by the co-ordinates planes and plane $x+y+z=1$.

10. Evaluate the integral $\iiint_R (ax^2+by^2+cz^2)dx\, dy\, dz$ where R is the region given by $x^2+y^2+z^2 \le d^2$

11. Find the value of $\iiint xyz \sin(x+y+z)dx\, dy\, dz$, the integral being extended to all positive values of variables subject to the condition $x+y+z \le \pi/2$. (UKTU-2011)

12. Evaluate $\iiint_R x^2y^2z^2 dx.dy.dz$ where R is the region given by $x^2+y^2 < 1, 0 \le z \le 1$.

13. Find the value of $\iint x^{l-1}y^{-1}e^{x+y}\,dx,dy, 0 < l < 1$ to all positive values subject to $x+y < h$.

14. A triangular prism is formed by planes whose equations are $ay=bx$, $y=0$ and $x=a$. Show that the volume of the prism between the planes $z=0$ and surface $z=c+xy$ is $\dfrac{ab}{8}(4c+ab)$.

(GBTU-2015)

15. Show that the volume of the paraboloid of revolution $x^2+y^2=4z$ cut off by the plane $z=4$ is 32π. (UPTU-2006)

16. Show that the volume common to the cylinder

$x^2+y^2 = a^2$ and $x^2+z^2=a^2$ is $\dfrac{16a^3}{3}$. (UPTU-2009)

17. Show that the volume of the solid which is bounded by the surface $2z = x^2+y^2$ and $z = x$ is $\pi/4$. (MTU-2012)

18. Show that the volume enclosed between the two surfaces $z = 8-x^2-y^2$ and $z = x^2 +3y^2$ is $8\pi\sqrt{2}$. (GBTU-2011)

19. Show that the volume bounded by the elliptic paraboloids $z = x^2 + 9y^2$ and $z = 18-x^2-9y^2$ is 27π. (GBTU-2012)

20. Apply Dirichlet's integral to find the mass of an octant of the ellipsoid $\dfrac{x^2}{a^2}+\dfrac{y^2}{b^2}+\dfrac{z^2}{c^2}=1$, the density at any point being $\rho = kxyz$. (UPTU-2006, 07, GBTU-2011)

Hints to Selected Problems

1. Put $\dfrac{x^2}{a^2}=u$ i.e., $x=au^{1/2}$, $\dfrac{y^2}{b^2}=v$ i.e., $y=bv^{1/2}$ and $z=cw^{1/2}$

Then apply Dirichlet's theorem.

2. Put $x^2=c^2u, y^2=c^2v$

3. Put $\left(\dfrac{x}{a}\right)^{2/3}=u, \left(\dfrac{y}{b}\right)^{2/3}=v, \left(\dfrac{z}{c}\right)^{2/3}=w$.

4. The given integral can be written as

$I = \iint x^{(3/2)-1}y^{(3/2)-1}[1-(x+y)]^{2/3}dx\,dy$.

Then apply Liouville's extension of Dirichlet's theorem.

6. Put $x^2=u, y^2=v, z^2<w$ and apply Liouville's extension of Dirichlet's theorem.

7. Put $\dfrac{x^2}{a^2}=u$ and $\dfrac{y^2}{b^2}=v$ and apply Liouville's extension of Dirichlet's theorem.

8. Put $x^2=u, y^2=v$ and Liouville's extension of Dirichlet's theorem.

10. Put $x^2=d^2u, y^2=d^2v, z^2=d^2w$.

Answers

2. $\dfrac{c^{2l+2m}}{4}\dfrac{\Gamma(l)\Gamma(m)}{\Gamma(l+m+1)}$ **3.** $\dfrac{4}{35}\pi abc$ **4.** $\dfrac{27\pi}{1760}$ **5.** $\dfrac{2\pi}{315}$ **6.** $\dfrac{abc}{6}$ **7.** $\dfrac{\pi}{8}\left[B\left(\dfrac{3}{4},\dfrac{1}{2}\right)-B\left(\dfrac{5}{4},\dfrac{1}{2}\right)\right]$ **8.** $\pi ab\left[\dfrac{\pi}{2}-1\right]$ **9.** $\dfrac{38\pi}{3}$

10. $\dfrac{4}{15}\pi(a+b+c)d^5$ **11.** $\dfrac{1}{384}[\pi^4 - 48\pi^2 + 384]$ **12.** $\dfrac{\pi}{48}$ **13.** $\dfrac{\pi}{\sin l\pi}(e^n -1)$ **14.** $\dfrac{ka^2b^2c^2}{48}$.

5.4 CHANGE OF VARIABLES

Some time we change the variables from one system to another system for more convenient way to find the double integrals. The variables x, y in $\iint_R f(x,y)dx\,dy$ are changed to u, v by means of the relations $x=f_1(u,v), y=f_2(u,v)$ then the double integral is transformed into

$$\iint f\{f_1(u,v), f_2(u,v)\}\,|J|\,du\,dv$$

where $J = \begin{vmatrix} \dfrac{\partial x}{\partial u} & \dfrac{\partial x}{\partial v} \\ \dfrac{\partial y}{\partial u} & \dfrac{\partial y}{\partial v} \end{vmatrix}$ and R' is the region in the u-v plane corresponding to region R in the x-y plane.

WORKING PROCEDURE

- Replace x, y by their equivalent in terms of u and v, the element of area $dx\,dy$ by $|J|\,du\,dv$ and the region R of integration in xy plane by the region R', in the uv plane.

5.4.1 CHANGE TO POLAR CO-ORDINATES

To change the variable from cartesian to polar form we put $x = r\cos\theta, y = r\sin\theta$.

Then $J = \begin{vmatrix} \dfrac{\partial x}{\partial r} & \dfrac{\partial x}{\partial \theta} \\ \dfrac{\partial y}{\partial r} & \dfrac{\partial y}{\partial \theta} \end{vmatrix} = \begin{vmatrix} \cos\theta & -r\sin\theta \\ \sin\theta & r\cos\theta \end{vmatrix} = r$

$\int_R f(x,y)dx\,dy = \iint_{R'} f(r\cos\theta, r\sin\theta)\,|J|\,dr\,d\theta = \iint_{R'} f(r\cos\theta, r\sin\theta)\,|r\,dr\,d\theta$

Solved Examples

EXAMPLE 1. *Transform* $\iint f(x,y)dx\,dy$, *by the substitution* $x+y= u, y=vu.$

SOLUTION. We have $x+y= u$ and $y=uv$ therefore,

$$x=u-y=u-uv \text{ and } y=uv.$$

$$\therefore \frac{\partial x}{\partial u}=1-v, \frac{\partial x}{\partial v}=-u, \frac{\partial y}{\partial u}=v \text{ and } \frac{\partial y}{\partial v}=u \qquad ...(1)$$

$$\therefore \quad J=\frac{\partial(x,y)}{\partial(u,v)}=\begin{vmatrix} \dfrac{\partial x}{\partial u} & \dfrac{\partial x}{\partial v} \\ \dfrac{\partial y}{\partial u} & \dfrac{\partial y}{\partial v} \end{vmatrix}=\begin{vmatrix} 1-v & -u \\ v & u \end{vmatrix}=u$$

$$\therefore \qquad dx\,dy = J\,du\,dv = u\,du\,dv.$$

Hence, the given integral transforms to $\iint F(u,v)u\,du\,dv$

EXAMPLE 2. *By using the transformation* $x+y = u, y= vu,$ *show that* $\int_0^1\int_0^{1-x} e^{y/(x+y)}dx\,dy=\dfrac{1}{2}(e-1)$

SOLUTION. We have $\qquad dx\,dy = u\,du\,dv$

The region of integration is bounded by the lines $y = 0,$ $y = 1-x, x=0$ and $x=1$

Changing these equations into new variables u and v by using the relation $x = u-y = u-uv = u(1-v)$ and $y= uv$, we have $uv = 0, uv = 1 - u(1-v), u(1-v) = 0$ and $u(1-v)=1$

giving $\qquad v=0$ to $v =1, u = 0$ to $u=1.$

Therefore for the given region v varies from 0 to 1 and u varies from 0 to 1.

and $\qquad e^{y/(x+y)} = e^{uv/u} = e^v.$

Changing the variables to u, v the given integral becomes

$$I = \int_0^1\int_0^1 e^v u\,du\,dv = \int_0^1 [e^v]_0^1 u\,du$$

$$= \int_0^1 (e^1-e^0)u\,du$$

$$= (e-1)\int_0^1 u\,du = (e-1)\left[\frac{u^2}{2}\right]_0^1 = \frac{1}{2}(e-1) \;.$$

EXAMPLE 3. *Evaluate the integral* $\int_0^a\int_0^{\sqrt{a^2-y^2}} (x^2+y^2)dy\,dx$ *by changing into polar co-ordinates.* (UPTU-2009)

SOLUTION. Putting $x= r\cos\theta, y = r\sin\theta$, we have

$$J=\begin{vmatrix} \dfrac{\partial x}{\partial r} & \dfrac{\partial x}{\partial \theta} \\ \dfrac{\partial y}{\partial r} & \dfrac{\partial y}{\partial \theta} \end{vmatrix}=\begin{vmatrix} \cos\theta & -r\sin\theta \\ \sin\theta & r\cos\theta \end{vmatrix}=r$$

$\therefore \quad dx\,dy$ is to be replaced by $J\,dr\,d\theta$.

$$x^2+y^2=r^2\cos^2\theta+r^2\sin^2\theta=r^2 \;.$$

Again we find that in the upper limit $x=\sqrt{a^2-y^2}$, y varies from 0 to a and x varies from 0 to any point on the circle $x^2+y^2=a^2$.

In the polar form of the circle $r^2=a^2$ i.e., $r=a$ we find that r varies from 0 to a and θ varies from 0 to $\pi/2$.

$$\therefore \int_0^a\int_0^{\sqrt{a^2-y^2}} (x^2+y^2)dy\,dx = \int_{\theta=0}^{\pi/2}\int_{r=0}^a r^2 \cdot r\,d\theta\,dr$$

$$= \int_{\theta=0}^{\pi/2}\left[\frac{r^4}{4}\right]_0^a d\theta = \frac{1}{4}a^4(\theta)_0^{\pi/2} = \left(\frac{1}{8}\right)\pi a^4 \;.$$

 Exercise-5.6

1. Transform $\int_0^a\int_0^{a-x} f(x,y)dx\,dy$, by the substitution $x+y = u, y=uv.$

2. By using the transformation $x+y = u, y = uv$ show that
$$\iint\{xy(1-x-y)\}^{1/2}dx\,dy$$
taken over the area of the triangle bounded by lines $x=0,$ $y=0, x+y =1$ is $\dfrac{2\pi}{105}$. (GBTU-2011)

3. Transform the integral $\int_0^a\int_0^{\sqrt{a^2-x^2}} y\sqrt{x^2+y^2}dx\,dy$ by changing to polar co-ordinates and hence solve it.

4. Evaluate $\int_0^2\int_0^{\sqrt{2x-x^2}} \dfrac{x\,dx\,dy}{\sqrt{x^2+y^2}}$, by changing to polar co-ordinates.

5. Evaluate $\iint(x^2+y^2)^{7/2}dx\,dy$, over the circle $x^2+y^2=1$.

6. Evaluate $\iint xy(x^2+y^2)^{3/2}dx\,dy$, over the positive axes of circle $x^2+y^2=1$.

7. Transform the integral $\int_0^{\pi/2}\int_0^{\pi/2}\sqrt{\dfrac{\sin\phi}{\sin\theta}}d\phi\,d\theta$ by the substitutions $x = \sin\phi\cos\theta, y = \sin\phi\sin\theta$ and show that its value is π. (UPTU-2007)

8. Evaluate $\int_0^a\int_0^{\sqrt{a^2-y^2}} y^2\sqrt{x^2+y^2}\,dx\,dy$. (UPTU-2008)

Hints to Selected Problems

1. $x+y = u$ and $y = uv \Rightarrow x= u-uv$ and $y = uv$

Now, $J=\dfrac{\partial(x,y)}{\partial(x,v)}=\begin{vmatrix} \dfrac{\partial x}{\partial u} & \dfrac{\partial x}{\partial v} \\ \dfrac{\partial y}{\partial u} & \dfrac{\partial y}{\partial v} \end{vmatrix}=\begin{vmatrix} 1-v & -u \\ v & u \end{vmatrix}=u$

$\Rightarrow dx\,dy = u\,du\,dv.$

2. Proceed same as (1), we have $dx\,dy = u\,du\,dv$

$\therefore \{xy\,(1-x-y)\}^{1/2} = [u(1-v)uv(1-u)]^{1/2}$

$$= u(1-u)^{1/2}v^{1/2}(1-v)^{1/2} \;.$$

3. $x = r\cos\theta, y = r\sin\theta, J = r$

$\Rightarrow \qquad dx\,dy = r\,d\theta\,dr$

$\therefore \int_0^a\int_0^{\sqrt{(a^2-x^2)}} y\sqrt{(x^2+y^2)}dx\,dy = \int_{\theta=0}^{\pi/2}\int_{r=0}^a r\sin\theta\cdot r\,d\theta\,dr \;.$

1. $\int_0^a \int_0^1 F(u,v) u \, du \, dv$ **3.** $\dfrac{a^4}{4}$ **4.** $\dfrac{4}{3}$ **5.** $\dfrac{2\pi}{9}$ **6.** $\dfrac{1}{14}$ **7.** $\int_0^1 \int_0^{\sqrt{1-y^2}} \dfrac{dx\,dy}{\sqrt{y - y(x^2 + y^2)}}$ **8.** $\dfrac{\pi a^5}{20}$

5.5 CHANGE OF ORDER OF INTEGRATIONS

If the limits of integration are constants in the double integration then the value of integration can be obtained by integrating with respect to any independent variable.

When the limits are not constant but are the function of x and y then firstly we integrate with respect to first independent variable and then with respect to second.

In this case the limits of integration are determined in the given region by drawing the strips parallel to Y-axis or X-axis.

In this case limits of y are function of x then we find the new limits of x as function of y and new constant.

WORKING PROCEDURE

Step 1. If we perform the integration first with respect to y, we take the elementary strip parallel to y-axis and determine the limits of y and add up the vertical strip from extreme left to the extreme right of the region.

Step 2. If the order of integration is performed first with respect to x, we take the elementary strip parallel to x-axis and proceed.

Solved Examples

EXAMPLE 1. *Evaluate the following integral by changing the order of integration*

$$\int_0^{2a} \int_0^{\sqrt{2ax-x^2}} a - \sqrt{(a^2 - y^2)} \, dx \, dy.$$

SOLUTION. Here, we have the figure (17).

The limits of y are from $y = 0$ to $y = \sqrt{2ax - x^2}$ or between x-axis and semicircle

$$x^2 + y^2 - 2ax = 0$$

or $\qquad (x - a)^2 + y^2 = a^2$

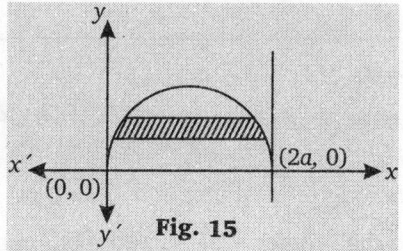

Fig. 15

In the figure, the region x varies from one end of the circle to other end

i.e., $(x - a)^2 = a^2 - y^2 \implies x = a \pm \sqrt{a^2 - y^2}$.

The strips are taken parallel to x-axis and y varies from 0 to a. So the order of integration is changed as

$$I = \int_0^{2a} \int_0^{\sqrt{2ax-x^2}} (a - \sqrt{a^2 - y^2}) \, dx \, dy$$

$$= \int_0^a \int_{a - \sqrt{a^2 - y^2}}^{a + \sqrt{a^2 - y^2}} (a - \sqrt{a^2 - y^2}) \, dx \, dy.$$

EXAMPLE 2. *Change the order of integration in the integral*

$$\int_0^a \int_0^x f(x,y) \, dx \, dy.$$

SOLUTION. The given limits shows that the region of integration is bounded by the curve $y = 0, y = x, x = 0, x = a$.

Hence $y = 0$ represent X-axis and $y = x$ represent a straight line through the origin. Also $x = 0$ and $x = a$, represent straight lines parallel to y-axis therefore the region of the integration is the triangle OAB in the figure 16 and B is (a, a).

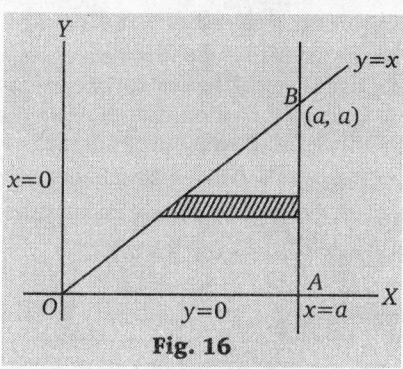

Fig. 16

In the given integral, the limits of integration of y being variable, we are required to integrate first w.r. to y regarding x as constant and then w.r. to x. To change the order of integration drawn parallel strip along x-axis, straight from the line OB and terminating on the line AB. Thus in region OBA, x varies from y to a and y varies from 0 to a.

Hence, by changing the order of integration we have

$$\int_0^a \int_0^x f(x,y) \, dx \, dy = \int_0^a \int_y^a f(x,y) \, dy \, dx.$$

EXAMPLE 3. *Change the order of integration and evaluate*

$$\int_0^1 \int_{e^x}^e \frac{dx \, dy}{\log y}.$$

SOLUTION. The region of integration is bounded by $e^x = y, y = e, x = 0$ and $x = 1$.

Here $y = e^x$ represents a curve. Putting $x = 0$ and $x = 1$ in $y = e^x$, then we get $y = 1$ and $y = e$. So $A(0, 1)$ and $B(1, e)$ are the points on this curve.

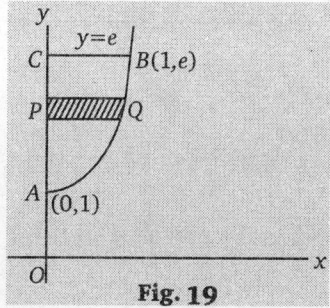

Fig. 19

When we integrate with respect to x first drawn a strip parallel to x-axis.

The strip starts from $x = 0$ and extends upto the curve $y = e^x$ i.e., $x = \log y$. Also for the given region y varies from $y = 1$ to $y = e$. On changing the order of integration, the given integral

$$I = \int_1^e \int_0^{\log y} \frac{dy\,dx}{(\log y)} = \int_1^e \frac{1}{\log y}(x)_0^{\log y}\,dy$$

$$= \int_1^e \frac{1}{\log y}(\log y - 0)\,dy = \int_1^e dy = (y)_1^e = e - 1 \quad .$$

EXAMPLE 4. *Change the order of integration in the integral*

$$\int_0^{\pi/2} \int_0^{2a\cos\theta} f(r,\theta)\,d\theta\,dr.$$

SOLUTION. The limits are given by $\theta = 0$ to $\theta = \pi/2$ and $r = 0$ to $r = 2a\cos\theta$. Also the curve $r = 2a\cos\theta$ is the circle.

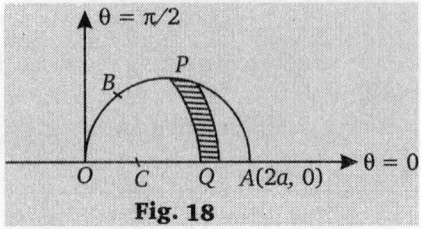

Fig. 18

The region of integration is the area $OABO$ of circle. In the given integral the limits of integration of r is variable while limit of θ are constant. Now draw a strip parallel to θ (pole) such strip extends from the points O to the point A i.e., $r = 0$ to $r = 2a$ and for a particular circular strip of this type we observe that θ varies from $\theta = 0$ to θ of curve i.e., $\theta = \cos^{-1}(r/2a)$.

Hence, the given integral $= \int_{r=0}^{2a}\int_{\theta=0}^{\cos^{-1}(r/2a)} f(r,\theta)\,dr\,d\theta.$

EXAMPLE 5. *Change the order of integration in*

$$\int_0^a \int_x^{a^2/x} \phi(x,y)\,dx\,dy.$$

SOLUTION. We observe that the region is bounded by $y = x$ and $y = a^2/x$ and $x = 0$ to $x = a$. Clearly the region is OAC.

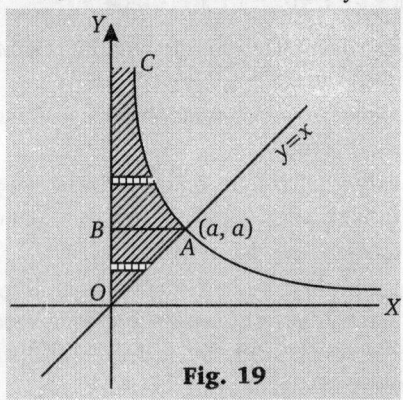

Fig. 19

Now draw a line AB parallel to x-axis, which divides the given region into two parts, OAB and BAC, Therefore, we draw a strip parallel to x-axis in OAB, where the left end of the strip is at y-axis and the right end is at $y = x$ and y takes the values from 0 to a. Also in region BAC, the left end of this strip is at y-axis whereas the right end is at the curve $y = a^2/x$, and y takes the values from $y = a$ to $y = \infty$. Hence, the given integral becomes

$$\int_0^a \int_x^{a^2/x} \phi(x,y)\,dx\,dy = \int_0^a \int_0^y \phi(x,y)\,dx\,dy$$
$$+ \int_a^\infty \int_0^{a^2/y} \phi(x,y)\,dx\,dy.$$

Exercise-5.7

Change the order of integration in the following integrals (Ques. 1-9)

1. $\int_0^{a/2}\int_{x^2/2}^{x-x^2/a} dy\,dx.$

2. $\int_0^a \int_x^{a^3/x} f(x,y)\,dx\,dy.$

3. $\int_0^1 \int_x^{x(2-x)} f(x,y)\,dx\,dy.$

4. $\int_0^{a\cos\alpha}\int_{x\tan\alpha}^{\sqrt{a^2-x^2}} f(x,y)\,dx\,dy.$

5. $\int^a \int_{mx}^{lx} f(x,y)\,dx\,dy.$

6. $\int_0^a \int_{(b/a)\sqrt{a^2-x^2}}^b f(x,y)\,dx\,dy,\text{ where } c < a.$

7. $\int_0^{2a} \int_{\sqrt{2a-x^2}}^{\sqrt{2ax}} f(x,y)\,dx\,dy.$

8. $\int_0^{\pi/2}\int_0^{2a\cos\theta} f(r,\theta)\,rd\theta\,dr.$

9. $\int_0^a \int_0^{b/(b+x)} f(x,y)\,dx\,dy.$

10. Change the order of integration in $\int_0^\infty \int_0^\infty \frac{e^{-y}}{y}\,dx\,dy$ and hence find its value.

11. Change the order of integration in $\int_0^\infty \int_x^\infty f(x,y)\,dx\,dy.$

12. Change the order of integration in $\int_0^{2a}\int_{x(2a-x)/2a}^{\sqrt{2a-x^2}} f(x,y)\,dx\,dy.$

13. Change the order of integration and evaluate

$$\int_0^\infty \int_0^x xe^{-x^2/y}\,dx\,dy.$$

14. Change the order of integration in

$$\int_0^a \int_{\sqrt{ax-x^2}}^{\sqrt{ax}} f(x,y)\,dx\,dy.$$

11. Change the order of integration to evaluate $\int_0^1 \int_{2y}^2 e^{x^2}\,dx\,dy.$

(GBTU-2010)

Answers

1. $\int_0^{a/4}\int_{\frac{1}{2}\left[a-\sqrt{a^2-4ay}\right]}^{\sqrt{ay}} f(x,y)\,dy\,dx.$ **2.** $\int_0^a \int_0^y f(x,y)\,dy\,dx + \int_a^\infty \int_0^{a^2/y} f(x,y)\,dy\,dx$ **3.** $\int_0^1 \int_{1-\sqrt{1-y}}^y f(x,y)\,dy\,dx.$

4. $\int_0^{a\sin\alpha}\int_0^{y\cot\alpha} f(x,y)\,dy\,dx + \int_{a\sin\alpha}^a \int_0^{\sqrt{a^2-y^2}} f(x,y)\,dy\,dx$ **5.** $\int_0^{am} \int_{y/l}^{y/m} f(x,y)\,dy\,dx + \int_{am}^{al} \int_{y/l}^a f(x,y)\,dy\,dx$

6. $\int_0^{b\sqrt{1-(c^2/a^2)}}\int_{a\sqrt{1-y^2/b^2}}^a f(x,y)\,dy\,dx + \int_{b\sqrt{1-c^2/a^2}}^b \int_c^a f(x,y)\,dy\,dx$

7. $\int_0^a \int_{y^2/2a}^{a-\sqrt{a^2-y^2}} f(x,y)\,dy\,dx + \int_0^a \int_{a+\sqrt{a^2-y^2}}^{2a} f(x,y)\,dy\,dx + \int_a^{2a}\int_{y^2/2a}^{2a} f(x,y)\,dy\,dx$ **8.** $\int_0^{2a}\int_0^{\cos^{-1}(r/2a)} f(r,\theta)\,dr\,d\theta$

9. $\int_0^{b/(a+b)}\int_0^a f(x,y)\,dy\,dx + \int_{b/(a+b)}^1 \int_0^{b(1-y)/y} f(x,y)\,dy\,dx$ **10.** 1   **11.** $\int_0^\infty \int_0^y f(x,y)\,dy\,dx$

12. $\int_0^{a/2}\int_{a-\sqrt{a^2-y^2}}^{a-\sqrt{a^2-2ay}} f(x,y)\,dx\,dy + \int_{a/2}^a \int_{a-\sqrt{a^2-y^2}}^{a+\sqrt{a^2-y^2}} f(x,y)\,dx\,dy + \int_a^{a/2}\int_{a+\sqrt{a^2-2ay}}^{a+\sqrt{a^2-y^2}} f(x,y)\,dx\,dy$ **13.** $\int_0^\infty \int_0^y xe^{-x^2/y}\,dx\,dy$

14. $\int_0^{a/2}\int_{y^2/a}^{a/2-\sqrt{(a^2/4)-y^2}} f(x,y)\,dx\,dy + \int_{a/2}^a \int_{y^2/a}^a f(x,y)\,dx\,dy + \int_0^{a/2}\int_{a/2+\sqrt{a^2/4-y^2}}^a f(x,y)\,dx\,dy$ **15.** e^2-3

5.6 GAMMA FUNCTIONS

The definite integral $\int_0^\infty e^{-x}x^{n-1}\,dx$, for $n>0$ is known as the gamma function and is denoted by $\Gamma(n)$ ['read as Gamma n']. Gamma function is also called the Eulerian integral of second kind. Weierstrass defined it as infinite product as

$$\frac{1}{\Gamma(z)} = z\,e^{2/z}\prod_{n=1}^{\infty}\left[\left(1+\frac{z}{n}\right)e^{-z/n}\right]$$

for non-zero and non-negative number z and n is an Euler's constant.   (UPTU-2006, 2008, GBTU-2010)

☞ REMARK
- The integral is valid only for $n>0$ because it is for just those values of m and n that the above integral are convergent.

5.6.1 PROPERTIES OF GAMMA FUNCTION

(1) $\Gamma(1)=1$.

   Proof. We have $\qquad\qquad \Gamma(n) = \int_0^\infty e^{-x}x^{n-1}dx,\ n>0.$ ...(1)

   Put $n=1$ in equation (1), we get

   $$\Gamma(1) = \int_0^\infty e^{-x}\,dx = \left[-e^{-x}\right]_0^\infty = 1.$$

   $\Rightarrow \qquad\qquad \Gamma(1)=1$

(2) $\Gamma(n+1) = n\,\Gamma(n),\ n>0$.

   Proof. We have $\qquad\qquad \Gamma(n) = \int_0^\infty e^{-x}x^{n-1}\,dx,$ for $n>0$

   Replacing n by $(n+1)$, we have

   $$\Gamma(n+1) = \int_0^\infty e^{-x}x^{n+1-1}dx = \int_0^\infty e^{-x}x^n\,dx$$

   $$= \left[x^n.(-e^{-x})\right]_0^\infty - \int_0^\infty (nx^{n-1})(-e^{-x})\,dx$$

   $\therefore \qquad\qquad \Gamma(n+1) = -\lim_{x\to\infty}\frac{x^n}{e^x} + 0 + n\int_0^\infty e^{-x}x^{n-1}dx$ ...(1)

   But $\qquad \lim_{x\to\infty}\frac{x^n}{e^x} = \lim_{x\to\infty}\dfrac{x^n}{1+\dfrac{x}{1!}+\dfrac{x^2}{2!}+...+\dfrac{x^n}{n!}+\dfrac{x^{n+1}}{(n+1)!}+...}$

   $$= \lim_{x\to\infty}\dfrac{1}{\dfrac{1}{x^n}+\dfrac{1}{1!x^{n-1}}+...+\dfrac{1}{n!}+\dfrac{x}{(n+1)!}+...}$$

   $$= 0 \qquad\qquad\qquad\qquad ...(2)$$

   Also, by definition, we have $\quad \Gamma(n) = \int_0^\infty e^{-x}x^{n-1}\,dx.$ ...(3)

   Using (2) and (3), (1) redcuces to $\Gamma(n+1)=n\,\Gamma(n)$.

☞ REMARKS
- The formula $\Gamma(n+1)=n\,\Gamma(n)$ is known as a recurrence formula for gamma function.
- The gamma function can be generalized to $n<0$ by using recurrence formula in the form of

  $\Gamma(n) = \dfrac{\Gamma(n+1)}{n}$. This process is known as analytic continuation.

(3) *If n is a non-negative integer, then $\Gamma(n+1)=n$!*

   Proof. We know that for $n>0$.

   $$\Gamma(n+1) = n\,\Gamma(n) = n\,\Gamma(n-1+1)$$

$$= n(n-1)\Gamma(n-1) \qquad\qquad\qquad\qquad\qquad\text{[By property 2]}$$
$$= n(n-1)(n-2)\,\Gamma(n-2)$$
$$= n(n-1)(n-2)\,...\,3.\,2.\,1.\,\Gamma(1)$$
$$= n! \qquad\qquad\qquad\qquad\qquad\qquad\qquad\qquad [\because\ \Gamma(1)=1]$$

☞ REMARK

- Gauss's Pi-function is denoted by $\pi(n)$ and is defined by $\pi(n) = \Gamma(n+1)$, when n is +ve integer.

(4) $\Gamma(1/2) = \sqrt{\pi}$.   (UPTU-2006, GBTU-2010)

Proof. By definition, we have

$$\Gamma(n) = \int_0^\infty e^{-t}t^{n-1}\,dt,\ n > 0 \qquad\qquad\qquad\qquad\qquad\qquad ...(1)$$

Replacing n by $1/2$ in equation (1), we get

$$\Gamma(1/2) = \int_0^\infty e^{-t}t^{-1/2}dt = 2\int_0^\infty e^{-u^2}du \qquad\qquad\qquad\qquad ...(2)$$

$$\text{[Putting } t = u^2,\ i.e.,\ dt = 2u\,du]$$

$$\therefore \qquad\qquad \Gamma(1/2) = 2\int_0^\infty e^{-x^2}dx \text{ and } \Gamma(1/2) = 2\int_0^\infty e^{-y^2}dy \qquad\qquad ...(3)$$

$$\text{(Limits remaining same)}$$

Multiplying the corresponding sides of two equations of (3), we get

$$[\Gamma(1/2)]^2 = \left(2\int_0^\infty e^{-x^2}dx\right)\left(2\int_0^\infty e^{-y^2}dy\right) = 4\int_0^\infty\int_0^\infty e^{-(x^2+y^2)}dx\,dy.$$

Now, changing the variables to polar co-ordinates (r, θ) where $x = r\cos\theta, y = r\sin\theta$
$$\Rightarrow \qquad\qquad x^2+y^2 = r^2 \text{ and } dx\,dy = r\,d\theta\,dr$$

we have
$$[\Gamma(1/2)]^2 = 4\int_{\theta=0}^{\pi/2}\int_{r=0}^\infty e^{-r^2}\,r\,d\theta\,dr .$$

The area of integration in the positive quadrant of plane is

$$= 2\int_0^{\pi/2}\left\{\int_0^\infty 2e^{-r^2}r\,.\,dr\right\}d\theta . \qquad\qquad \text{(Putting } r^2 = v, \text{ so that } 2r\,dr = dv)$$

$$= 2\int_0^{\pi/2}\left[-e^{-v}\right]_0^\infty d\theta = 2\int_0^{\pi/2}d\theta = 2\left[\theta\right]_0^{\pi/2} = \pi$$

Therefore, $\qquad\qquad [\Gamma(1/2)]^2 = \pi$ so that $\Gamma(1/2) = \sqrt{\pi}$.

(5) $\Gamma(n) = \int_0^1 (\log 1/y)^{n-1}dy$.

Proof. By definition of gamma function, we have

$$\Gamma(n) = \int_0^\infty e^{-x}x^{n-1}\,dx, n > 0$$

Putting $x = \log(1/y)$ in gamma function, we get

$$\Gamma(n) = -\int_1^0 (\log 1/y)^{n-1}dy = \int_0^1 (\log 1/y)^{n-1}dy.$$

5.6.2 SOME TRANSFORMATIONS OF GAMMA FUNCTION

Gamma function is given by $\qquad\qquad \Gamma(n) = \int_0^\infty x^{n-1}e^{-x}\,dx$.   ...(1)

(1) $\dfrac{\Gamma(n)}{a^n} = \int_0^\infty e^{-ay}\,y^{n-1}\,dy,\ n > 0,\ a > 0$

Proof. We have $\qquad\qquad \Gamma(n) = \int_0^\infty x^{n-1}e^{-x}\,dx,\ n > 0.$

Put $x = ay$, so that $\qquad\qquad dx = a\,dy.$ Also, when $x = 0, y = 0$ and when $x \to \infty, y \to \infty$.

$$\therefore \qquad\qquad\qquad \Gamma(n) = \int_0^\infty e^{-ay}(ay)^{n-1}.a\,dy .$$

Hence, $\qquad\qquad \int_0^\infty e^{-ay}y^{n-1}dy = \dfrac{\Gamma(n)}{a^n} .$

(2) $\Gamma(n) = \dfrac{1}{n}\int_0^\infty e^{-x^{1/n}}dx, n > 0$

Proof. We have $\qquad\qquad \Gamma(n) = \int_0^\infty e^{-x}x^{n-1}dx,\ n > 0$ ...(1)

Put $\qquad x^n = t.$ *i.e.,* $\qquad nx^{n-1}\,dx = dt$, then (1) gives

$$\Gamma(n) = \dfrac{1}{n}\int_0^\infty e^{-t^{1/n}}dt$$

$$\Rightarrow \qquad\qquad \Gamma(n) = \dfrac{1}{n}\int_0^\infty e^{-x^{1/n}}dx \qquad\qquad\qquad \text{[By the property of definite integral]}$$

(3) $\Gamma(n) = 2 \int_0^\infty e^{-x^2} x^{2n-1} dx, n > 0$

Proof. We have $\qquad \Gamma(n) = \int_0^\infty e^{-x} x^{n-1} dx,$ ...(1)

Put $\qquad x = t^2,$ so that $\qquad dx = 2t \, dt,$

Therefore, $\qquad \Gamma(n) = \int_0^\infty e^{-t^2} (t^2)^{n-1} 2t \, dt$

or $\qquad \Gamma(n) = 2 \int_0^\infty e^{-t^2} t^{2n-1} dt$

$\Rightarrow \qquad \Gamma(n) = 2 \int_0^\infty e^{-x^2} x^{2n-1} dx.$

Solved Examples

EXAMPLE 1. *Evaluate :*

(i) $\int_0^\infty e^{-x} x^4 \, dx$ **(ii)** $\int_0^\infty x^6 e^{-2x} \, dx$

SOLUTION. (i) We have

$$\int_0^\infty e^{-x} x^4 \, dx = \int_0^\infty e^{-x} x^{5-1} dx$$

[By definition of gamma function]

$$= \Gamma(5) = (4)! = 24.$$

(ii) Let $\qquad I = \int_0^\infty x^6 e^{-2x} dx$ ...(1)

Put $2x = t$, so that $dx = 1/2 \, dt$ then

$$I = \int_0^\infty \left(\frac{t}{2}\right)^6 e^{-t} \cdot \frac{1}{2} dt = \frac{1}{2^7} \int_0^\infty e^{-t} t^{7-1} dt$$

$$= \frac{1}{2^7} \Gamma(7)$$

[By definition of gamma function]

$$= \frac{1}{2^7} \times (6!) = \frac{45}{8}.$$

EXAMPLE 2. *Show that* $\int_0^1 \dfrac{dx}{\sqrt{(-\log x)}} = \sqrt{\pi}.$

SOLUTION. We know that $\quad \Gamma(n) = \int_0^1 (-\log x)^{n-1} dx$

Putting $n = 1/2$, we have

$$\Gamma(1/2) = \int_0^1 (-\log x)^{(1/2)-1} dx$$

or $\qquad \sqrt{\pi} = \int_0^1 (-\log x)^{-1/2} dx$

or $\qquad \sqrt{\pi} = \int_0^1 \dfrac{dx}{\sqrt{(-\log x)}}.$

EXAMPLE 3. *If n is a positive integer, prove that*
$$2^n \, \Gamma(n+1/2) = 1. \, 3. \, 5. \, ...(2n-1) \sqrt{\pi}$$

SOLUTION. We know that

$$\Gamma(n+1) = n\Gamma(n)$$ ...(1)

Now $\Gamma(n + 1/2) = \Gamma(n - 1/2 + 1)$

$$= (n-1/2) \, \Gamma(n-1/2) \text{ [Using (1)]}$$

$$= (n-1/2) \, \Gamma(n-3/2+1)$$

$$= (n-1/2)(n-3/2) \, \Gamma(n-3/2)$$

$$= \frac{2n-1}{2} \cdot \frac{2n-3}{2} \cdot \Gamma\left(\frac{2n-3}{2}\right)$$

$$= \frac{2n-1}{2} \cdot \frac{2n-3}{2} \cdots \frac{5}{2} \cdot \frac{3}{2} \cdot \frac{1}{2} \cdot \Gamma\left(\frac{1}{2}\right)$$

$$= \frac{(2n-1)(2n-3)\dots 5.3.1}{2^n} \sqrt{\pi}$$

$[\because \Gamma(1/2) = \sqrt{\pi}]$

Hence,

$$2^n \Gamma(n+1/2) = (2n-1)(2n-3)\dots 5.3.1. \sqrt{\pi}.$$

EXAMPLE 4. *Prove that*

(i) $\int_0^\infty x e^{-\alpha x} \cos\beta x \, dx = \dfrac{\alpha^2 - \beta^2}{(\alpha^2 + \beta^2)^2}, \alpha > 0$

(ii) $\int_0^\infty x e^{-\alpha x} \sin\beta x \, dx = \dfrac{2\alpha\beta}{(\alpha^2 + \beta^2)^2}, \alpha > 0$

(Remember)

SOLUTION. We know that

$$\int_0^\infty e^{-kx} x^{n-1} dx = \frac{\Gamma(n)}{k^n}, n > 0, k > 0.$$ ...(1)

Putting $k = \alpha - i\beta$ and $n = 2$ in (1), we get

$$\int_0^\infty e^{-(\alpha - i\beta)x} x \, dx = \frac{\Gamma(2)}{(\alpha - i\beta)^2}$$

or $\int_0^\infty x e^{-\alpha x} e^{i\beta x} dx = \dfrac{(\alpha + i\beta)^2}{(\alpha - i\beta)^2 (\alpha + i\beta)^2}$ [as $\Gamma(2) = 1$]

$$\int_0^\infty x e^{-\alpha x} e^{i\beta x} dx = \frac{\alpha^2 - \beta^2 + 2i\alpha\beta}{[(\alpha + i\beta)(\alpha - i\beta)]^2}$$

$\Rightarrow \int_0^\infty x e^{-\alpha x} (\cos\beta x + i\sin\beta x) dx = \dfrac{\alpha^2 - \beta^2 + 2i\alpha\beta}{(\alpha^2 + \beta^2)^2}$

or $\int_0^\infty x e^{-\alpha x} \cos\beta x \, dx + i \int_0^\infty x e^{-\alpha x} \sin\beta x$

$$= \frac{\alpha^2 - \beta^2}{(\alpha^2 + \beta^2)^2} + i \frac{2\alpha\beta}{(\alpha^2 + \beta^2)^2}.$$

Equating real and imaginary parts of both sides, we get

$$\int_0^\infty x e^{-\alpha x} \cos\beta x \, dx = \frac{\alpha^2 - \beta^2}{(\alpha^2 + \beta^2)^2}$$

and $\int_0^\infty x e^{-\alpha x} \sin\beta x \, dx = \dfrac{2\alpha\beta}{(\alpha^2 + \beta^2)^2}.$

5.7 BETA FUNCTION

The definite integral $\int_0^1 x^{m-1} (1-x)^{n-1} dx,$ for $m > 0$, $n > 0$ is known as the Beta function and denoted by $B(m, n)$ which is read as "Beta m, n", where m, n are positive number or integers. Thus $B(m,n) = \int_0^1 x^{m-1}(1-x)^{n-1} dx$. (UPTU-2006, 2008, GBTU(AG)-2010)

REMARK
- Beta function is also called the Eulerian integral of first kind.

5.7.1 PROPERTIES OF BETA FUNCTION

(1) *Symmetry of Beta function i.e., B(m, n) = B(n, m)*   (UPTU-2006)

Proof. By definition of beta function, we have

$$B(m,n) = \int_0^1 x^{m-1}(1-x)^{n-1}\,dx$$

$$= \int_0^1 (1-x)^{m-1}[1-(1-x)]^{n-1}\,dx \qquad \left[\because \int_0^a f(x)\,dx = \int_0^a f(a-x)\,dx\right]$$

$$= \int_0^1 (1-x)^{m-1}x^{n-1}\,dx \;=\; \int_0^1 x^{n-1}(1-x)^{m-1}\,dx$$

$$= B(n,m) \qquad\qquad\qquad \text{[By definition of Beta function]}$$

$$B(m,n) = B(n,m)$$

i.e., the interchange of position of m and n does not change the value of beta function.

☞ REMARK

- This is the fundamental property of Beta function and also called symmetric property of Beta function.

(2) *Beta function B(m, n) can be evaluated in an explicit form if m or n is a positive integer.*

Proof. Case I. *When 'n' is a positive integer.*

If $n = 1$, then by definition of Beta function, we have

$$B(m,n) = \int_0^1 x^{m-1}(1-x)^{n-1}\,dx \qquad\qquad\qquad\qquad \text{...(1)}$$

$$\Rightarrow \qquad B(m,1) = \int_0^1 x^{m-1}(1-x)^{1-1}\,dx$$

$$= \int_0^1 x^{m-1}\,dx = \left[\frac{x^m}{m}\right]_0^1 = \frac{1}{m}. \qquad\qquad\qquad \text{...(2)}$$

Now, let $n > 1$, then from (1), we have

$$B(m,n) = \int_0^1 (1-x)^{n-1}x^{m-1}\,dx \;=\; \left[(1-x)^{n-1}\cdot\frac{x^m}{m}\right]_0^1 - \int_0^1 (n-1)(1-x)^{n-2}\cdot(-1)\frac{x^m}{m}\,dx.$$

Integrating by parts taking x^{m-1} as second function, we have

$$B(m,n) = 0 + \frac{n-1}{m}\int_0^1 x^m(1-x)^{n-2}\,dx \qquad \left[\because n>1 \quad \text{and} \quad \lim_{x\to 0}(1-x)^{n-1}\frac{x^m}{m}=0\right]$$

$$= \frac{n-1}{m}\int_0^1 x^{(m+1)-1}(1-x)^{(n-1)-1}\,dx = \frac{n-1}{m}B(m+1,n-1)$$

Thus $$B(m,n) = \frac{n-1}{m}B(m+1,n-1) \qquad\qquad\qquad\qquad \text{...(3)}$$

Now replacing m by $m+1$ and n by $n-1$ in (3), we get

$$B(m+1,n-1) = \frac{n-1-1}{m+1}B(m+2,n-2) \qquad\qquad\qquad\qquad \text{...(4)}$$

Using equation (4), the equation (3) becomes

$$B(m,n) = \frac{n-1}{m}\cdot\frac{n-2}{m+1}B(m+2,n-2) \qquad\qquad\qquad\qquad \text{...(5)}$$

After applying the above process successively, we get

$$B(m,n) = \frac{n-1}{m}\cdot\frac{n-2}{m+1}\cdot\frac{n-3}{m+2}\cdots\frac{1}{m+n-2}B(m+n-1,1) \qquad\qquad \text{...(6)}$$

$$= \frac{n-1}{m}\cdot\frac{n-2}{m+1}\cdot\frac{n-3}{m+2}\cdots\frac{1}{m+n-2}\int_0^1 x^{m+n-2}(1-x)^0\,dx$$

$$= \frac{n-1}{m}\cdot\frac{n-2}{m+1}\cdot\frac{n-3}{m+2}\cdots\frac{1}{m+n-2}\left[\frac{x^{m+n-1}}{m+n-1}\right]_0^1$$

$$= \frac{n-1}{m}\cdot\frac{n-2}{m+1}\cdot\frac{n-3}{m+2}\cdots\frac{1}{m+n-2}\cdot\frac{1}{m+n-1}$$

$$\Rightarrow \qquad B(m,n) = \frac{n-1}{m}\cdot\frac{n-2}{m+1}\cdot\frac{n-3}{m+2}\cdots\frac{1}{m+n-2}\cdot\frac{1}{m+n-1}$$

$$\therefore \qquad B(m,n) = \frac{(n-1)!}{m(m+1)(m+2)\cdots(m+n-2)(m+n-1)} \qquad\qquad \text{...(7)}$$

Case II. *When m is a positive integer.*

Since the beta function is symmetrical in m and n i.e., $B(m, n) = B(n, m)$ therefore by interchanging m and n in Case I we get

$$B(m,n) = \frac{(m-1)!}{n(n+1)(n+2)\ldots(n+m-2)(n+m-1)}. \qquad \ldots(8)$$

Case III. *When both m and n are positive integers.*

We have, by Case I

$$B(m,n) = \frac{(n-1)!}{m(m+1)(m+2)\ldots(m+n-2)(m+n-1)}.$$

Multiplying both numerator and denominator by 1. 2. 3. .. .(m–1) !, we get

$$B(m,n) = \frac{[1.2.3\ldots(m-1)](n-1)!}{1.2.3\ldots.m(m+1)(m+2)\ldots(m+n-2)(m+n-1)}$$

$$B(m,n) = \frac{(m-1)!(n-1)!}{(m+n-1)!}$$

5.7.2 Transformation of Beta Function

The Beta function $\qquad B(m,n) = \int_0^1 x^{m-1}(1-x)^{n-1}\,dx \qquad \ldots(A)$

can be transformed into many forms given below :

(I) $B(m,n) = \int_0^\infty \dfrac{x^{n-1}}{(1+x)^{m+n}}\,dx = \int_0^\infty \dfrac{x^{m-1}}{(1+x)^{m+n}}\,dx$.

Proof. Put $x = \dfrac{1}{(1+y)}$ and $dx = -\dfrac{dy}{(1+y)^2}$ and $y \to 0$ when $x=1$, $y \to \infty$, when $x = 0$.

$$\therefore \qquad B(m,n) = \int_\infty^0 \left(\frac{1}{1+y}\right)^{m-1}\left[1 - \frac{1}{1+y}\right]^{n-1}\left[\frac{-dy}{(1+y)^2}\right]$$

$$= \int_0^\infty \frac{(y)^{n-1}}{(1+y)^{m+1}}\left(\frac{1}{1+y}\right)^{n-1}. dy \; = \int_0^\infty \frac{y^{n-1}}{(1+y)^{m+n}}\,dy$$

$$\Rightarrow \qquad B(m,n) = \int_0^\infty \frac{x^{n-1}\,dx}{(1+x)^{m+n}}. \qquad \ldots(1)$$

Since m and n are interchangeable in beta function by symmetric property therefore (1) gives

$$B(m,n) = \int_0^\infty \frac{x^{m-1}}{(1+x)^{m+n}}\,dx$$

thus $\qquad B(m,n) = \int_0^\infty \dfrac{x^{n-1}\,dx}{(1+x)^{m+n}} = \int_0^\infty \dfrac{x^{m-1}\,dx}{(1+x)^{m+n}}$.

(II) $B(m,n) = 2\int_0^{\pi/2} \cos^{2m-1}\theta \sin^{2n-1}\theta\,d\theta$.

Proof. Put $x = \sin^2\theta$ and $dx = 2\sin\theta\cos\theta\,d\theta$ and $\theta = 0$ when $x = 0$, $\theta = \pi/2$ when $x = 1$ in (A) we get

$$B(m,n) = 2\int_0^{\pi/2} \sin^{2m-1}\theta\cos^{2n-1}\theta\,d\theta$$

$$= 2\int_0^{\pi/2}\cos^{2m-1}\theta\sin^{2n-1}\theta\,d\theta \qquad \text{[By symmetric property of beta function]}$$

(III) $B(m,n) = \dfrac{1}{a^{m+n-1}}\int_0^a x^{m-1}(a-x)^{n-1}dx$.

Proof. Put $\quad x = \dfrac{y}{a}$, i.e., $dx = \dfrac{1}{a}dy$ and when $x \to 0$, then $y \to 0$, when $x=1$ then $y \to a$.

So $\qquad B(m,n) = \dfrac{1}{a^{m+n-1}}\int_0^a y^{m-1}(a-y)^{n-1}dy \; = \dfrac{1}{a^{m+n-1}}\int_0^a x^{m-1}(a-x)^{n-1}dx$.

(IV) $\dfrac{B(m,n)}{a^n(1+a)^m} = \int_0^1 \dfrac{x^{m-1}(1-x)^{n-1}dx}{(x+a)^{m+n}}$.

Proof. Let $\quad \dfrac{x}{1+a} = \dfrac{t}{t+a} \quad \Rightarrow \quad dx = a(1+a)\dfrac{dt}{(t+a)^2}$

then we have $\qquad B(m,n) = \int_0^1 (1+a)^{m-1}\left(\dfrac{t}{t+a}\right)^{m-1}a^{n-1}\left(\dfrac{1-t}{a+t}\right)^{n-1}\dfrac{a(a+1)}{(t+a)^2}dt$

$$= a^n(1+a)^m \int_0^1 \frac{t^{m-1}(1-t)^{n-1}}{(t+a)^{m+n}} dt \quad = a^n(1+a)^m \int_0^1 \frac{x^{m-1}(1-x)^{n-1}}{(x+a)^{m+n}} dx$$

Hence, $$\frac{B(m,n)}{a^n(1+a)^m} = \int_0^1 \frac{x^{m-1}(1-x)^{n-1}}{(x+a)^{m+n}} dx$$

(V) $B(m,n)(a-b)^{m+n-1} = \int_b^a (x-b)^{m-1}(a-x)^{n-1} dx$.

Proof. Put $x = \dfrac{t-b}{a-b}$ so that $dx = \dfrac{dt}{a-b}$. in (A), we get

$$B(m,n) = \int_b^a \left(\frac{t-b}{a-b}\right)^{m-1} \left(\frac{a-t}{a-b}\right)^{n-1} \frac{dt}{a-b} = \frac{1}{(a-b)^{m+n-1}} \int_b^a (t-b)^{m-1}(a-t)^{n-1} dt$$

$$= \frac{1}{(a-b)^{m+n-1}} \int_b^a (x-b)^{m-1}(a-x)^{n-1} dx.$$

$\therefore \qquad B(m,n)(a-b)^{m+n-1} = \int_b^a (x-b)^{m-1}(a-x)^{n-1} dx.$

(VI) $\dfrac{1}{a^n b^m} B(m,n) = \int_0^1 \dfrac{x^{m-1}(1-x)^{n-1} dx}{\{a+(b-a)x\}^{m+n}}$

Proof. We put $$\frac{a}{y} - \frac{b}{x} = a - b. \quad \text{(Remember)} \qquad \qquad \qquad \qquad ...(1)$$

$\therefore \qquad \qquad \dfrac{b}{x} = \dfrac{a}{y} + (b-a) = \dfrac{a+(b-a)y}{y}$

$\Rightarrow \qquad \qquad x = \dfrac{by}{a+(b-a)y} \qquad \qquad \qquad \qquad \qquad ...(2)$

$\Rightarrow \qquad \qquad dx = \dfrac{b[a+(b-a)y] - by(b-a)}{\{a+(b-a)y\}^2} dy$

i.e., $\qquad \qquad dx = \dfrac{ab\, dy}{[a+(b-a)y]^2}. \qquad \qquad \qquad \qquad ...(3)$

Again from (1), we see that when $x = 1, y = 1$ and $x = 0, y = 0$.

and from (2), we have $\qquad 1-x = 1 - \dfrac{by}{a+by-ay} = \dfrac{a(1-y)}{a+(b-a)y}. \qquad \qquad ...(4)$

Using (2), (3) and (4), (1) gives

$$B(m,n) = \int_0^1 \left\{\frac{by}{a+(b-a)y}\right\}^{m-1} \left\{\frac{a(1-y)}{a+(b-a)y}\right\}^{n-1} \frac{ab\, dy}{\{a+(b-a)y\}^2}$$

$$= a^n b^m \int_0^1 \frac{y^{m-1}(1-y)^{n-1}}{\{a+(b-a)y\}^{m+n}} dy = a^n b^m \int_0^1 \frac{x^{m-1}(1-x)^{n-1} dx}{\{a+(b-a)x\}^{m+n}}$$

$\Rightarrow \qquad \dfrac{1}{a^n b^m} B(m,n) = \int_0^1 \dfrac{x^{m-1}(1-x)^{n-1} dx}{\{a+(b-a)x\}^{m+n}}.$

Solved Examples

EXAMPLE 1. *Evaluate the following integrals by expressing them in terms of Beta function*

    (i) $\int_0^1 x^m(1-x^2)^n\, dx, \; m > 1, n > -1$

    (ii) $\int_0^1 \dfrac{x^2 dx}{\sqrt{(1-x^5)}}.$

SOLUTION. (i) We have

$\int_0^1 x^m(1-x^2)^n dx = \int_0^1 x^{m-1}(1-x^2)^n x \cdot dx$

$= \int_0^1 y^{\frac{m-1}{2}}(1-y)^n \cdot \dfrac{dy}{2}$

(Putting $x^2 = y$ so that $2x\, dx = dy$)

$= \dfrac{1}{2} \int_0^1 y^{\frac{(m-1)}{2}} (1-y)^n dy$

$= \dfrac{1}{2} \int_0^1 y^{\frac{m+1}{2}-1} (1-y)^{(n+1)-1} dy$

$= \dfrac{1}{2} B\left[\dfrac{1}{2}(m+1), n+1\right].$

(ii) We have

$\int_0^1 \dfrac{x^2}{\sqrt{(1-x^5)}} dx = \int_0^1 x^2(1-x^5)^{-1/2} dx$

$$= \int_0^1 x^2 \cdot \frac{1}{x^4} (1-x^5)^{-1/2} x^4 dx$$

$$= \int_0^1 x^{-2} (1-x^5)^{-1/2} x^4 dx$$

$$= \int_0^1 y^{-2/5} (1-y)^{-1/2} \frac{1}{5} dy$$

(Putting $x^5 = y$, i.e., $5x^4 dx = dy$)

$$= \frac{1}{5} \int_0^1 y^{-2/5} (1-y)^{-1/2} dy$$

$$= \frac{1}{5} \int_0^1 y^{(3/5)-1} (1-y)^{(1/2)-1} dy$$

$$= \frac{1}{5} B\left(\frac{3}{5}, \frac{1}{2}\right).$$

EXAMPLE 2. **Show that**

$$\int_0^1 \frac{x^{m-1}(1-x)^{n-1}}{(a+bx)^{m+n}} dx = \frac{1}{(a+b)^m a^n} B(m,n).$$

SOLUTION. Let $I = \int_0^1 \frac{x^{m-1}(1-x)^{n-1}}{(a+bx)^{m+n}} dx$

$$= \int_0^1 \left(\frac{x}{a+bx}\right)^{m-1} \cdot \left(\frac{1-x}{a+bx}\right)^{n-1} \frac{1}{(a+bx)^2} dx.$$

Put $\dfrac{x}{a+bx} = \dfrac{y}{a+b}$

i.e. $\dfrac{(a+bx) - x \cdot b}{(a+bx)^2} dx = \dfrac{dy}{a+b}$

$$\Rightarrow \frac{1}{(a+bx)^2} dx = \frac{dy}{a(a+b)}$$

$$\Rightarrow \frac{1-x}{a+bx} = \frac{1}{a}\left(\frac{a-ax}{a+bx}\right) = \frac{1}{a}\left[\frac{a+bx-ax-bx}{a+bx}\right]$$

$$= \frac{1}{a}\left[1 - \frac{x(a+b)}{a+bx}\right] = \frac{1-y}{a}.$$

Also when $x = 0$, $y = 0$, and when $x = 1$, $y = 1$. Therefore,

$$I = \int_0^1 \left(\frac{y}{a+b}\right)^{m-1} \left(\frac{1-y}{a}\right)^{n-1} \cdot \frac{dy}{a(a+b)}$$

$$= \frac{1}{(a+b)^m \cdot a^n} \int_0^1 y^{m-1}(1-y)^{n-1} dy$$

$$= \frac{B(m,n)}{(a+b)^m \cdot a^n}.$$

EXAMPLE 3. **Evaluate** $\int_0^\infty x^m e^{-ax^n} dx$, **when m, n and a are all positive constant.**

SOLUTION. Let $ax^n = y \Rightarrow nax^{n-1} = dy$
and at $x = 0, y = 0$ and at $x = \infty, y = \infty$.

$$\therefore \int_0^\infty x^m e^{-ax^n} dx = \frac{1}{na} \int_0^\infty \left(\frac{y}{a}\right)^{m/n} \left(\frac{y}{a}\right)^{1/n-1} e^{-y} dy$$

$$= \frac{1}{na^{(m+1)/n}} \int_0^\infty y^{\left(\frac{m+1}{n}\right)-1} \cdot e^{-y} dy$$

$$= \frac{1}{na^{(m+1)/n}} \Gamma\left(\frac{m+1}{n}\right).$$

5.8 RELATION BETWEEN BETA AND GAMMA FUNCTION

We have
$$B(m,n) = \frac{\Gamma(m)\Gamma(n)}{\Gamma(m+n)}, m > 0, n > 0 = \int_0^\infty \frac{y^{n-1} dy}{(1+y)^{m+n}}.$$
(UPTU-2009, 10 ; VKTU-2010)

Proof. We have
$$\int_0^\infty y^{n-1} e^{-xy} dy = \frac{\Gamma(n)}{x^n}$$

or
$$\Gamma(n) = \int_0^\infty x^n y^{n-1} e^{-xy} dy \qquad \qquad ...(1)$$

Also
$$\Gamma(m) = \int_0^\infty x^{m-1} e^{-x} dx \qquad \qquad ...(2)$$

Multiplying both sides of (1) by $x^{m-1} e^{-x}$, we have
$$\Gamma(n) \cdot x^{m-1} e^{-x} = \int_0^\infty x^{n+m-1} y^{n-1} e^{-(y+1)x} dy.$$

Integrating both sides with respect to x within limits $x = 0$ to $x = \infty$, we have
$$\Gamma(n) \int_0^\infty x^{m-1} e^{-x} dx = \int_0^\infty \left[\int_0^\infty x^{n+m-1} e^{-(y+1)x} dx\right] y^{n-1} dy \qquad ...(3)$$

But
$$\int_0^\infty x^{(n+m)-1} e^{-(y+1)x} dx = \frac{\Gamma(n+m)}{(1+y)^{m+n}}.$$

Hence with the help of this result and (2), we get from (3)
$$\Gamma(n)\Gamma(m) = \int_0^\infty \Gamma(n+m) \frac{y^{n-1}}{(1+y)^{n+m}} dy$$

$$= \Gamma(n+m) \int_0^\infty \frac{y^{n-1}}{(1+y)^{n+m}} dy = \Gamma(n+m) B(m,n)$$

or
$$B(m,n) = \frac{\Gamma(m)\Gamma(n)}{\Gamma(n+m)}.$$

DEDUCTION 1.
$$\Gamma(n)\Gamma(1-n) = \frac{\pi}{\sin n\pi}, \text{ where } 0 < n < 1.$$

Proof. We have
$$B(m,n) = \int_0^\infty \frac{x^{n-1} dx}{(1+x)^{m+n}}, m > 0, n > 0.$$

Therefore the relation between beta and gamma functions becomes

$$\int_0^\infty \frac{x^{n-1}dx}{(1+x)^{m+n}} = \frac{\Gamma(m)\,\Gamma(n)}{\Gamma(m+n)}.$$

Taking $m + n = 1$, so that $m = 1 - n$, we get

$$\int_0^\infty \frac{x^{n-1}}{1+x}dx = \frac{\Gamma(1-n)\,\Gamma(n)}{\Gamma(1)}, \; 0 < n < 1. \qquad [\because m > 0 \Rightarrow 1-n > 0 \Rightarrow n < 1. \text{ Also } n > 0]$$

But also we know that $\int_0^\infty \frac{x^{n-1}}{1+x}dx = \dfrac{\pi}{\sin n\pi}$ and $\Gamma(1) = 1$

$$\therefore \qquad\qquad \frac{\pi}{\sin n\pi} = \Gamma(1-n)\,\Gamma(n), \; 0 < n < 1.$$

DEDUCTION 2. $\Gamma(1/2) = \sqrt{\pi}$.

Proof. We have just proved that

$$\Gamma(n)\,\Gamma(1-n) = \frac{\pi}{\sin n\pi} \qquad\qquad\qquad\qquad \text{...(1)}$$

Putting $n = 1/2$ in (1), we obtain

$$\Gamma\left(\frac{1}{2}\right)\Gamma\left(1-\frac{1}{2}\right) = \frac{\pi}{\sin \pi/2} \text{ or } \left[\Gamma\left(\frac{1}{2}\right)\right]^2 = \pi$$

$$\Rightarrow \qquad\qquad\qquad \Gamma\left(\frac{1}{2}\right) = \sqrt{\pi}.$$

Aliter. We know $\qquad\qquad B(m,n) = \dfrac{\Gamma(m)\,\Gamma(n)}{\Gamma(m+n)}.$

Putting $m = n = 1/2$, we get $\quad B(1/2, 1/2) = \dfrac{\Gamma(1/2)\,\Gamma(1/2)}{\Gamma(1/2+1/2)} = \dfrac{\{\Gamma(1/2)\}^2}{\Gamma(1)} \qquad\qquad [\because \Gamma(1) = 1]$

or $\qquad \{\Gamma(1/2)\}^2 = B(1/2, 1/2) = \int_0^1 x^{(1/2)-1}(1-x)^{(1/2)-1}dx$

$$= \int_0^1 x^{-1/2}(1-x)^{-1/2}dx = \int_0^1 \frac{dx}{\sqrt{x}\sqrt{1-x}}$$

$$= \int_0^{\pi/2} \frac{2\sin\theta\cos\theta\, d\theta}{\sin\theta\sqrt{(1-\sin^2\theta)}} \qquad\qquad \text{(By putting } x = \sin^2\theta)$$

$$= 2\int_0^{\pi/2} d\theta = 2\big[\theta\big]_0^{\pi/2} = 2\left(\frac{\pi}{2}\right) = \pi$$

$$\Rightarrow \qquad\qquad \left\{\Gamma\left(\frac{1}{2}\right)\right\}^2 = \pi \Rightarrow \Gamma\left(\frac{1}{2}\right) = \sqrt{\pi}.$$

DEDUCTION 3. $\int_0^1 e^{-x^2}dx = \dfrac{1}{2}\sqrt{\pi}.$

Proof. We have $\int_0^1 e^{-x^2}dx = \int_0^1 e^{-y}\cdot\dfrac{1}{2\sqrt{y}}dy$, putting $x^2 = y$, $2x\,dx = dy$

$$= \frac{1}{2}\int_0^\infty e^{-y}\,y^{-1/2}\,dy \;= \frac{1}{2}\int_0^1 e^{-y}\,y^{(1/2)-1}\,dy$$

$$= \frac{1}{2}\Gamma\left(\frac{1}{2}\right) \qquad\qquad\qquad \left[\because \int_0^\infty e^{-x}\cdot x^{n-1}dx = \Gamma n\right]$$

$$= \frac{1}{2}\sqrt{\pi} \qquad\qquad\qquad\qquad \left[\because \Gamma\left(\frac{1}{2}\right) = \sqrt{\pi}\right]$$

$$\Rightarrow \qquad\qquad \int_0^\infty e^{-x^2}dx = \frac{1}{2}\sqrt{\pi}.$$

DEDUCTION 4. $\int_0^{\pi/2} \cos^m\theta \sin^n\theta\, d\theta = \dfrac{\Gamma\left(\dfrac{m+1}{2}\right)\Gamma\left(\dfrac{n+1}{2}\right)}{2\Gamma\left(\dfrac{m+n+2}{2}\right)}$ *for all values of m and n such that* $m > -1, n > -1$.

Proof. We put $\qquad\qquad \sin^2\theta = x, \qquad\qquad \Rightarrow \quad 2\sin\theta\cos\theta\, d\theta = dx$

$$\Rightarrow \qquad\qquad 2\sin\theta\cdot\sqrt{(1-\sin^2\theta)}\, d\theta = dx \quad \Rightarrow \quad 2x^{1/2}\sqrt{1-x}\, d\theta = dx$$

$$\Rightarrow \qquad d\theta = \frac{dx}{2x^{1/2}(1-x)^{1/2}}.$$

Also, when $\theta = \pi/2$, $x = 1$ and $\theta = 0$, $x = 0$.

Putting these values in L.H.S. of the given equation, we get

$$\int_0^{\pi/2} \cos^m \theta \sin^n \theta \, d\theta = \int_0^{\pi/2} (1 - \sin^2 \theta)^{m/2}.\sin^n \theta \, d\theta$$

$$= \int_0^1 (1-x)^{m/2}.x^{n/2}.\frac{dx}{2x^{1/2}(1-x)^{1/2}} = \frac{1}{2}\int_0^1 x^{\frac{(n-1)}{2}}(1-x)^{\frac{(m-1)}{2}}dx$$

$$= \frac{1}{2}\int_0^1 x^{\left\{\frac{(n+1)}{2}\right\}-1}(1-x)^{\left\{\frac{(m+1)}{2}\right\}-1}dx = \frac{1}{2}B\left(\frac{m+1}{2},\frac{n+1}{2}\right)$$

$$= \frac{\frac{1}{2}\Gamma\frac{1}{2}(m+1)\Gamma\frac{1}{2}(n+1)}{\Gamma\frac{1}{2}(m+n+1+1)} \qquad \left(\text{Because } B(m,n) = \frac{\Gamma(m)\Gamma(n)}{\Gamma(m+n)}\right)$$

$$= \frac{\Gamma\frac{1}{2}(m+1)\Gamma\frac{1}{2}(n+1)}{2\Gamma\frac{1}{2}(m+n+2)}.$$

<u>**DEDUCTION 5.**</u> $\int_0^{\pi/2} \sin^{p-1}\theta \cos^{q-1}\theta \, d\theta = \dfrac{\Gamma(p/2)\Gamma(q/2)}{2\Gamma\left(\dfrac{p+q}{2}\right)}.$

Proof. By definition of Beta function, we have

$$B(m,n) = \int_0^1 x^{m-1}(1-x)^{n-1}dx = 2\int_0^{\pi/2}\cos^{2m-1}\theta\sin^{2n-1}\theta\,d\theta = \frac{\Gamma(m).\Gamma(n)}{\Gamma(m+n)} \qquad \ldots(1)$$

Let $2m = p$ and $2n = q$. So that $m = p/2$ and $n = q/2$

Put in equation (1), we get $\int_0^{\pi/2} \sin^{p-1}\theta\cos^{q-1}\theta\,d\theta = \dfrac{\Gamma(p/2)\Gamma(q/2)}{2\Gamma\left(\dfrac{p+q}{2}\right)}.$

<u>**DEDUCTION 6.**</u> $\int_0^{\pi/2}\sin^{p-1}\theta\,d\theta = \int_0^{\pi/2}\cos^{p-1}\theta\,d\theta = \dfrac{\Gamma\left(\dfrac{p}{2}\right)\Gamma\left(\dfrac{1}{2}\right)}{2\Gamma\left(\dfrac{p+1}{2}\right)} = \dfrac{\sqrt{\pi}}{2}\dfrac{\Gamma(p/2)}{\Gamma\left(\dfrac{p+1}{2}\right)}.\left(\because \Gamma\left(\dfrac{1}{2}\right) = \sqrt{\pi}\right)$

Proof. Replacing q by 1 in deduction 5, we get

$$\int_0^{\pi/2}\sin^{p-1}\theta\,d\theta = \frac{\Gamma\left(\dfrac{p}{2}\right)\Gamma\left(\dfrac{1}{2}\right)}{2\Gamma\left(\dfrac{p+1}{2}\right)}.$$

Next, replacing p by 1 and q by p in equation of deduction 5, we have

$$\int_0^{\pi/2}\cos^{p-1}\theta\,d\theta = \frac{\sqrt{\pi}}{2}.\frac{\Gamma\left(\dfrac{p}{2}\right)}{\Gamma\left(\dfrac{p+1}{2}\right)} = \frac{\Gamma\left(\dfrac{1}{2}\right)\Gamma\left(\dfrac{p}{2}\right)}{2\Gamma\left(\dfrac{1+p}{2}\right)}$$

 Solved Examples

EXAMPLE 1. *Evaluate the following integrals:*

    (i) $\int_0^1 x^4(1-x^2)\,dx$ *(ii)* $\int_0^a y^4\sqrt{a^2-y^2}\,dy$

SOLUTION. (i) We have

$$\int_0^1 x^4(1-x^2)dx = \int_0^1 x^{5-1}(1-x^2)^{3-1}dx$$

$$= \frac{\Gamma(5)\,\Gamma(3)}{\Gamma(5+3)} = \frac{4!\,2!}{7!}$$

$$= \frac{4!\times 2}{7\times 5\times 4!\times 6} = \frac{1}{105}.$$

(ii) $\int_0^a y^4\sqrt{a^2-y^2}\,dy$.

Let $y^2 = a^2\,t$ so that $dy = \dfrac{a^2\,dt}{2y} = \dfrac{a\,dt}{2\sqrt{t}}$, then

$$I = \int_0^1 (a^2 t)^2\sqrt{(a^2 - ta^2)}\left(\frac{a\,dt}{2\sqrt{t}}\right)$$

$$= \frac{a^6}{2}\int_0^1 t^{3/2}(1-t)^{1/2}dt$$

$$= \frac{a^6}{2} \int_0^1 t^{(5/2)-1}(1-t)^{(3/2)-1} dt$$

$$= \frac{a^6}{2} \frac{\Gamma\left(\frac{5}{2}\right)\Gamma\left(\frac{3}{2}\right)}{\Gamma\left(\frac{5}{2}+\frac{3}{2}\right)} = \frac{a^6}{2} \frac{\frac{3}{2}\cdot\frac{1}{2}\sqrt{\pi}\cdot\frac{1}{2}\sqrt{\pi}}{3!} = \frac{\pi a^6}{32}.$$

EXAMPLE 2. *Prove that* $\int_0^2 (8-x^3)^{-1/3} dx = \frac{2\pi}{3\sqrt{3}}$

SOLUTION. Let $\quad x^3 = 8t$, then $x = 2t^{1/3}$

$$\Rightarrow \quad dx = \frac{2}{3} t^{-2/3} dt$$

and when $x = 0$ to $x = 2$, $t = 0$ to $t = 1$

So $\int_0^2 (8-x^3)^{-1/3} dx = \int_0^1 (8-8t)^{-1/3} \cdot \frac{2}{3} t^{-2/3} dt$

$$= (8)^{-1/3} \cdot \frac{2}{3} \int_0^1 t^{-2/3}(1-t)^{-1/3} dt$$

$$= \frac{1}{3} \int_0^1 t^{(1/3)-1}(1-t)^{(2/3)-1} dt$$

$$= \frac{1}{3} B\left(\frac{1}{3},\frac{2}{3}\right) = \frac{1}{3} \frac{\Gamma\left(\frac{1}{3}\right)\Gamma\left(\frac{2}{3}\right)}{\Gamma\left(\frac{1}{3}+\frac{2}{3}\right)}$$

$$= \frac{1}{3} \frac{\Gamma\left(\frac{1}{3}\right)\Gamma\left(1-\frac{1}{3}\right)}{\Gamma(1)} = \frac{1}{3} \frac{\pi}{\sin\frac{\pi}{3}} = \frac{2\pi}{3\sqrt{3}}.$$

EXAMPLE 3. *Show that* $\int_0^1 \frac{dx}{(1-x^n)^{1/2}} = \frac{\sqrt{\pi}\,\Gamma(1/n)}{n\Gamma\left(\frac{1}{n}+\frac{1}{2}\right)}.$

SOLUTION. Let $\quad x^n = \sin^2\theta \Rightarrow x = \sin^{2/n}\theta$

so that $dx = 2.\frac{1}{n}.\sin^{\left(\frac{2}{n}-1\right)}\theta\cos\theta\, d\theta$

then $\int_0^1 \frac{dx}{\sqrt{(1 \quad)}} \quad \frac{2}{n}\int_0^{/2} \frac{\sin^{(2/\ )\ 1}\theta\cos\theta\, d\theta}{\cos }$

$$= \frac{2}{n} \int_0^{\pi/2} \sin^{(2/n)-1}\theta\cos^0\theta\, d\theta$$

$$= \frac{2}{n} \cdot \frac{\Gamma\left(\frac{1}{n}\right)\Gamma\left(\frac{1}{2}\right)}{2\Gamma\left(\frac{1}{n}+\frac{1}{2}\right)} = \frac{\sqrt{\pi}}{n} \cdot \frac{\Gamma\left(\frac{1}{n}\right)}{\Gamma\left(\frac{1}{n}+\frac{1}{2}\right)}.$$

$$= \frac{\pi}{\sin n\pi}.$$

EXAMPLE 4. *Prove that*

(a) $\int_0^{\pi/2} \sqrt{\tan\theta}\, d\theta = \frac{1}{2}\Gamma\left(\frac{1}{4}\right)\Gamma\left(\frac{3}{4}\right) = \frac{\pi\sqrt{2}}{2}.$

(UKTU-2011)

(b) $\int_0^{\pi/2} \tan^n x\, dx = \frac{\pi}{2}\sec\frac{n\pi}{2}, -1 < n < 1.$

SOLUTION. (a) We have

$$\int_0^{\pi/2} \sqrt{\tan\theta}\, d\theta = \int_0^{\pi/2} \left(\frac{\sin\theta}{\cos\theta}\right)^{1/2} d\theta$$

$$= \int_0^{\pi/2} \sin^{1/2}\theta\cos^{-1/2}\theta\, d\theta$$

$$= \frac{\Gamma\left(\frac{1+\frac{1}{2}}{2}\right)\Gamma\left(\frac{1-\frac{1}{2}}{2}\right)}{2\Gamma\left(\frac{\frac{1}{2}-\frac{1}{2}+2}{2}\right)} \quad \because \int_0^{\pi/2} \sin^n\theta\cos^m\theta\, d\theta$$

$$\left[= \frac{\Gamma\left(\frac{n+1}{2}\right)\Gamma\left(\frac{m+1}{2}\right)}{2\Gamma\left(\frac{n+m+2}{2}\right)}, n > -1, m > -1\right]$$

$$= \frac{\Gamma\left(\frac{3}{4}\right)\Gamma\left(\frac{1}{4}\right)}{2\Gamma(1)} = \frac{1}{2}\Gamma\left(\frac{3}{4}\right)\Gamma\left(\frac{1}{4}\right)$$

$$= \frac{1}{2}\Gamma\left(\frac{1}{4}\right)\Gamma\left(1-\frac{1}{4}\right) = \frac{1}{2}\frac{\pi}{\sin\left(\frac{\pi}{4}\right)} = \frac{\pi\sqrt{2}}{2}.$$

(b) Consider L.H.S.

$$\int_0^{\pi/2} \tan^n x\, dx = \int_0^{\pi/2} \sin^n x\cos^{-n} x\, dx$$

$$= \frac{\Gamma\left(\frac{1+n}{2}\right)\Gamma\left(\frac{1-n}{2}\right)}{2\Gamma\left(\frac{n-n+2}{2}\right)}$$

$$\left[\begin{array}{l} \text{Here } \frac{1+n}{2} > 0, \ \frac{1-n}{2} > 0 \\ \Rightarrow \qquad n > -1 \text{ and } n < 1 \end{array}\right]$$

$$= \frac{1}{2}\Gamma\left(\frac{1+n}{2}\right)\Gamma\left(1-\frac{1+n}{2}\right)$$

$$= \frac{1}{2}\frac{\pi}{\sin\left(\frac{1+n}{2}\right)\pi} = \frac{\pi}{2\sin\left(\frac{\pi}{2}+\frac{n\pi}{2}\right)} = \frac{\pi}{2\cos\frac{n\pi}{2}}$$

$$= \frac{\pi}{2}\sec\frac{n\pi}{2}, \quad \text{where } -1 < n < 1.$$

EXAMPLE 5. *Evaluate* $\int_{-1}^1 \left(\frac{1+x}{1-x}\right)^{1/2} dx.$

SOLUTION. Let $I = \int_{-1}^1 \left(\frac{1+x}{1-x}\right)^{1/2} dx$ ...(1)

Putting $t = \frac{1}{2}(1+x)$ so that $x = 2t-1$, $dx = 2\, dt$ in (1),

we get $I = \int_0^1 \left(\frac{1+2t-1}{1-2t+1}\right)^{1/2} 2\, dt = 2\int_0^1 \left(\frac{t}{1-t}\right)^{1/2} dt$

$$= 2\int_0^1 t^{(3/2)-1}(1-t)^{(1/2)-1} dt = 2B\left(\frac{3}{2},\frac{1}{2}\right)$$

$$= \frac{2\Gamma\left(\frac{3}{2}\right)\Gamma\left(\frac{1}{2}\right)}{\Gamma\left(\frac{3}{2}+\frac{1}{2}\right)} = 2\frac{\frac{1}{2}\sqrt{\pi}\sqrt{\pi}}{\Gamma(2)} = \pi.$$

EXAMPLE 6. *Show that* $B(n,n+1) = \frac{1}{2}\frac{\Gamma(n)^2}{\Gamma(2n)}$ *and hence deduce that*

$$\int_0^{\pi/2} \left(\frac{1}{\sin^3\theta} - \frac{1}{\sin^2\theta}\right)^{1/4} \cos\theta\, d\theta = \frac{\left\{\Gamma\left(\frac{1}{4}\right)\right\}^2}{2\sqrt{\pi}}.$$

SOLUTION. We have

$$B(n, n+1) = \frac{\Gamma(n)\Gamma(n+1)}{\Gamma(n+n+1)} = \frac{\Gamma(n) \cdot n\Gamma n}{(2n)\,\Gamma(2n)}$$

$$\therefore \quad B(n, n+1) = \frac{1}{2}\frac{\{\Gamma(n)\}^2}{\Gamma(2n)} \qquad [\because \ \Gamma(p+1) = p\Gamma(p)]$$

Let $\ I = \int_0^{\pi/2} \left(\frac{1}{\sin^3\theta} - \frac{1}{\sin^2\theta}\right)^{1/4} \cos\theta\, d\theta \qquad \dots(2)$

Putting $x = \sin\theta$, so that $dx = \cos\theta\, d\theta$ in (2), we get

$$I = \int_0^1 \left(\frac{1}{x^3} - \frac{1}{x^2}\right)^{1/4} dx = \int_0^1 \left(\frac{1-x}{x^3}\right)^{1/4} dx$$

$$= \int_0^1 x^{-3/4}(1-x)^{1/4} dx = \int_0^1 x^{(1/4)-1}(1-x)^{(5/4)-1} dx$$

$$= B\left(\frac{1}{4}, \frac{5}{4}\right)$$

$$= B\left(\frac{1}{4}, \frac{1}{4}+1\right) = \frac{1}{2} \frac{\left\{\Gamma\left(\frac{1}{4}\right)\right\}^2}{\Gamma\left(\frac{1}{2}\right)} = \frac{\left\{\Gamma\left(\frac{1}{4}\right)\right\}^2}{2\sqrt{\pi}}.$$

EXAMPLE 7. *Prove that*

(a) $\int_0^\pi \dfrac{\sin^{n-1} x\, dx}{(a+b\cos x)^n} = \dfrac{2^{n-1}}{(a^2-b^2)^{n/2}} B\left(\dfrac{n}{2}, \dfrac{n}{2}\right).$

(b) $\int_0^\pi \dfrac{\sqrt{\sin x}}{[5+3\cos x]^{3/2}} = \dfrac{\left[\Gamma\left(\dfrac{3}{4}\right)\right]^2}{2\sqrt{2}\,\pi}.$

SOLUTION. (a) Let

$$I = \int_0^\pi \frac{\sin^{n-1} x\, dx}{(a+b\cos x)^n} = \int_0^\pi \frac{(\sin x)^{n-1} dx}{(a+b\cos x)^n} \qquad \dots(1)$$

$$= \int_0^\pi \frac{\left(2\sin\dfrac{x}{2}\cos\dfrac{x}{2}\right)^{n-1} dx}{\left[a\left\{\cos^2\left(\dfrac{x}{2}\right)+\sin^2\left(\dfrac{x}{2}\right)\right\} +b\left\{\cos^2\left(\dfrac{x}{2}\right)-\sin^2\left(\dfrac{x}{2}\right)\right\}\right]^n} \qquad \text{[by (1)]}$$

$$= 2^{n-1}\int_0^\pi \frac{\sin^{n-1}\left(\dfrac{x}{2}\right)\cos^{n-1}\left(\dfrac{x}{2}\right) dx}{\left[(a+b)\cos^2\left(\dfrac{x}{2}\right)+(a-b)\sin^2\left(\dfrac{x}{2}\right)\right]^n}$$

$$= \frac{2^{n-1}}{(a+b)^n}\int_0^\pi \frac{\sin^{n-1}\left(\dfrac{x}{2}\right)\cos^{n-1}\left(\dfrac{x}{2}\right) dx}{\cos^{2n}\left(\dfrac{x}{2}\right)\left[1+\dfrac{a-b}{a+b}\tan^2\dfrac{x}{2}\right]^n}$$

$$= \frac{2^{n-1}}{(a+b)^n}\int_0^\infty \frac{\tan^{n-1}\left(\dfrac{x}{2}\right)\sec^2\left(\dfrac{x}{2}\right) dx}{\left[1+\dfrac{a-b}{a+b}\tan^2\dfrac{x}{2}\right]^n}$$

$$= \frac{2^{n-1}}{(a+b)^n}\int_0^\infty \frac{\left[\dfrac{a+b}{a-b}t\right]^{\frac{(n-2)}{2}} \cdot \dfrac{a+b}{a-b}\, dt}{(1+t)^n}$$

$$\left[\begin{array}{l} \text{Put } \dfrac{a-b}{a+b}\tan^2\dfrac{x}{2} = t, \\[2mm] \text{i.e., } \ 2\dfrac{a-b}{a+b}\tan\dfrac{x}{2}\sec^2\dfrac{x}{2}\cdot\dfrac{dx}{2} = dt \end{array}\right]$$

$$I = \frac{2^{n-1}}{[(a+b)(a-b)]^{n/2}}\int_0^\infty \frac{t^{(n/2)-1}}{(1+t)^{n/2+n/2}}\, dt$$

$$= \frac{2^{n-1}}{(a^2-b^2)^{n/2}} B\left(\frac{n}{2}, \frac{n}{2}\right).$$

(b) Taking $n = 3/2$, $a = 5$ and $b = 3$, in part (a), we get

$$\int_0^\pi \frac{\sin^{(3/2)-1} x\, dx}{(5+3\cos x)^{3/2}} = \frac{(2)^{(3/2)-1}}{(25-9)^{3/4}} B\left(\frac{3}{4}, \frac{3}{4}\right)$$

$$= \frac{\sqrt{2}}{2^3}\cdot\frac{\Gamma\left(\dfrac{3}{4}\right)\Gamma\left(\dfrac{3}{4}\right)}{\Gamma\left(\dfrac{3}{4}+\dfrac{3}{4}\right)} = \frac{\sqrt{2}\left\{\Gamma\left(\dfrac{3}{4}\right)\right\}^2}{8\Gamma\left(\dfrac{3}{2}\right)}$$

$$= \frac{\sqrt{2}\left\{\Gamma\left(\dfrac{3}{4}\right)\right\}^2}{8.\dfrac{1}{2}\sqrt{\pi}} = \frac{\left\{\Gamma\left(\dfrac{3}{4}\right)\right\}^2}{2\sqrt{2\pi}}.$$

Exercise-5.8

1. Show that $\int_0^\infty e^{-4x} x^{3/2}\, dx = \dfrac{3\sqrt{\pi}}{128}.$

2. Show that $\int_0^\infty e^{-x^2}\cdot x^2\, dx = \dfrac{\sqrt{\pi}}{4}.$

3. Show that $\int_0^1 \dfrac{dx}{\sqrt{(-\log x)}} = \sqrt{\pi}.$

4. Show that $\Gamma\left(-\dfrac{15}{2}\right) = \dfrac{2^8\sqrt{\pi}}{1.3.5.7.9.11.13.15}.$

5. Show that $\int_0^1 x^{n-1}\left(\log\dfrac{1}{x}\right)^{m-1} dx = \dfrac{\Gamma(m)}{n^m}\ \ m>0,\, n>0.$

6. Show that

$$\int_0^{\pi/2} \sin^m\theta\cos^n\theta\, d\theta = \frac{\Gamma\left(\dfrac{m+1}{2}\right)\Gamma\left(\dfrac{n+1}{2}\right)}{2\Gamma\left(\dfrac{m+n+2}{2}\right)} = \frac{1}{2} B\left(\frac{m+1}{2}, \frac{n+1}{2}\right).$$

7. Show that $\int_0^\infty 3^{-4x^2} dx = \dfrac{\sqrt{\pi}}{4\sqrt{(\log 3)}}.$

8. Show that $\int_0^a (a-x)^{m-1} x^{n-1} dx = \dfrac{a^{m+n-1}\Gamma(m)\Gamma(n)}{\Gamma(m+n)}.$

9. Show that $\int_0^{\pi/2} \sin^p\theta\cos^q\theta\, d\theta = \dfrac{1}{2} B\left(\dfrac{p+1}{2}, \dfrac{q+1}{2}\right),\quad p>-1,$

$q>-1$. Deduce that $\int_0^2 x^4(8-x^3)^{-1/3} dx = \dfrac{16}{3} B\left(\dfrac{5}{3}, \dfrac{2}{3}\right).$

10. Show that $\int_0^1 \left(\dfrac{1}{x}-1\right)^{1/4} dx = B\left(\dfrac{5}{4}, \dfrac{3}{4}\right) = \dfrac{\pi}{2\sqrt{2}}.$

11. Prove that

(i) $B(l, m)\, B(l+m, n) = B(m, n)B(m+n, l) = B(n, l)\, B(n+l, m)$

(ii) $B(l, m)B(l+m, n)\, B(l+m+n, p) = \dfrac{\Gamma(l)\Gamma(m)\Gamma(n)\Gamma(p)}{\Gamma(l+m+n+p)}.$

12. Show that $\int_0^1 x^m(1-x^n)^p dx = \frac{1}{n} B\left(\frac{m+1}{n}, p+1\right)$.

13. Show that $\int_0^1 \left(1-x^n\right)^{1/n} dx = \frac{1}{n}\frac{[\Gamma(1/n)]^2}{2\Gamma(2/n)}$.

14. Show that $\int_0^1 \frac{x^2 dx}{(1-x^4)^{1/2}} \times \int_0^1 \frac{dx}{(1+x^4)^{1/2}} = \frac{\pi}{4\sqrt{2}}$.

15. Show that $B(m,n) = B(m+1, n) + B(m, n+1)$ for $m > 0, n > 0$.

16. Prove the following

    (i) $\int_0^\infty e^{-x} x^{n-1} dx$, for $n > 0$ (UPTU-2008)

    (ii) $\frac{1}{\Gamma(z)} = z\, e^{2/z} \prod_{n=1}^{\infty} \left[\left(1+\frac{z}{n}\right) e^{-z/n}\right]$ (UPTU-2007)

    (iii) $\Gamma(n) = \int_0^\infty e^{-x} x^{n-1} dx$, $n > 0$.   (UPTU-2008)

Hints to Selected Problems

1. $I = \int_0^\infty e^{-4x} x^{3/2} dx = \int_0^\infty e^{-4x} x^{(5/2)-1} dx$

$$= \frac{\Gamma\left(\frac{5}{2}\right)}{4^{5/2}} = \frac{\frac{3}{2} \cdot \frac{1}{2}\sqrt{\pi}}{2^5} = \frac{3\sqrt{\pi}}{128}.$$

3. $I = \int_0^1 \frac{dx}{\sqrt{\log\dfrac{1}{x}}} = \int_0^1 \left[\log\left(\frac{1}{x}\right)\right]^{-1/2} dx$.

Now put $\log(1/x) = t$.

5. $I = \int_0^1 x^{n-1} \left(\log\frac{1}{x}\right)^{m-1} dx$.

Put $\frac{1}{x} = t$, i.e., $x = e^{-t} \Rightarrow dx = -e^{-t} dt$.

6. The given integral can be written as

$I = \int_0^{\pi/2} (\sin^2\theta)^{(m-1)/2}(1-\sin^2\theta)^{(n-1)/2} \sin\theta \cos\theta\, d\theta$

Now put $\sin^2\theta = x$ i.e., $dx = 2\sin\theta \cos\theta\, d\theta$.

7. Put $3^{-4x^2} = e^{-t}$ i.e., $x = \frac{\sqrt{t}}{2\sqrt{\log 3}}$.

8. Put $x = at$ in the LHS.

9. Do same as (6).

12. Put $x^n = t$ i.e., $x = (t)^{1/n} \Rightarrow dx = \frac{1}{n}(t)^{1/n-1} dt$.

5.8.1 DUPLICATION FORMULA

We have $\Gamma(n)\Gamma\left(n+\frac{1}{2}\right) = \frac{\sqrt{\pi}}{2^{2n-1}}\Gamma(2n),\ n > 0.$

Proof. We know that $B(m,n) = \frac{\Gamma(m)\Gamma(n)}{\Gamma(m+n)}$ where $m > 0, n > 0$.   ...(1)

Now putting $m = n$ in equation (1), we get

$$B(n,n) = \frac{[\Gamma(n)]^2}{\Gamma(2n)}. \qquad\qquad ...(2)$$

By definition of Beta function, we get

$$B(n,n) = \int_0^1 x^{n-1}(1-x)^{n-1} dx.$$

Putting $x = \sin^2\theta$ so that $dx = 2\sin\theta \cos\theta\, d\theta$ in (1), we get

$$B(n,n) = \int_0^{\pi/2} (\sin^2\theta)^{n-1}(1-\sin^2\theta)^{n-1} \cdot 2\sin\theta \cos\theta\, d\theta$$

$$= 2\int_0^{\pi/2} (\sin\theta \cos\theta)^{2n-1} d\theta = 2\int_0^{\pi/2} \left(\frac{\sin 2\theta}{2}\right)^{2n-1} d\theta$$

$$= \frac{1}{2^{2n-2}}\int_0^{\pi/2} \sin^{2n-1} 2\theta\, d\theta$$

$$= \frac{1}{2^{2n-2}}\int_0^{\pi} \sin^{2n-1}\phi\, \frac{d\phi}{2} \ \text{(By putting } 2\theta=\phi \Rightarrow d\theta = \frac{1}{2}d\phi)$$

$$= \frac{1}{2^{2n-1}}\int_0^{\pi} \sin^{2n-1}\phi\, d\phi = \frac{1}{2^{2n-2}}\int_0^{\pi/2} \sin^{2n-1}\phi\, d\phi$$

$$\left[\because \int_0^{2a} f(x)\, dx = 2\int_0^a f(x)\, dx \text{ when } f(2a-x) = f(x)\right]$$

$$= \frac{1}{2^{2n-2}}\int_0^{\pi/2} \sin^{2n-1}\phi(\cos\phi)^0 d\phi$$

$$= \frac{1}{2^{2n-2}} \frac{\Gamma\left(\dfrac{2n-1+1}{2}\right)\Gamma\left(\dfrac{0+1}{2}\right)}{2\Gamma\left(\dfrac{2n-1+0+2}{2}\right)}$$

$$B(n,n) = \frac{1}{2^{2n-1}} \cdot \frac{\Gamma(n)\sqrt{\pi}}{\Gamma\left(n+\dfrac{1}{2}\right)} \text{ as } \Gamma\left(\frac{1}{2}\right) = \sqrt{\pi}. \qquad ...(3)$$

Equating two values of $B(n, n)$ given by (2) and (3), we obtain

$$\Rightarrow \qquad \frac{[\Gamma(n)]^2}{\Gamma(2n)} = \frac{1}{2^{2n-1}} \frac{\Gamma(n)\sqrt{\pi}}{\Gamma\left(n+\frac{1}{2}\right)}$$

or $\qquad\qquad \Gamma(n)\,\Gamma\left(n+\frac{1}{2}\right) = \frac{\sqrt{\pi}}{2^{2n-1}}\,\Gamma(2n)\,.$ $\hspace{4cm}$ ...(4)

DEDUCTION 1. *For all positive real value of p, we have* $2^p\,\Gamma\left(\frac{p+1}{2}\right)\Gamma\left(\frac{p+2}{2}\right) = \sqrt{\pi}\,\Gamma(p+1)$

Proof. We know that

$$\Gamma(n)\,\Gamma\left(n+\frac{1}{2}\right) = \frac{\sqrt{\pi}}{2^{2n-1}}\,\Gamma(2n). \hspace{4cm} ...(1)$$

Putting $2n - 1 = p$, so that $n = \frac{1}{2}(p+1)$ in equation (1), we get

$$\Gamma\left(\frac{p+1}{2}\right)\Gamma\left(\frac{p+1}{2}+\frac{1}{2}\right) = \frac{\sqrt{\pi}}{2^p}\,\Gamma(p+1) \;\Rightarrow\; 2^p\,\Gamma\left(\frac{p+1}{2}\right)\Gamma\left(\frac{p+2}{2}\right) = \sqrt{\pi}\,\Gamma(p+1).$$

DEDUCTION 2. *For any positive integer n, we have* $\Gamma\left(n+\frac{1}{2}\right) = \frac{(2n)!}{2^{2n}.n!}\sqrt{\pi}.$

Proof. Let n be positive integer, then we have

$$\frac{\Gamma(2n)}{\Gamma(n)} = \frac{(2n-1)!}{(n-1)!} = \frac{(2n)(2n-1)!}{2.n.(n-1)!} = \frac{(2n)!}{2.(n)!} \hspace{3cm} ...(1)$$

Now, from the duplication formula (4), we have

$$\Gamma\left(n+\frac{1}{2}\right) = \frac{\sqrt{\pi}}{2^{2n-1}}\cdot\frac{\Gamma(2n)}{\Gamma(n)} = \frac{\sqrt{\pi}}{2^{2n-1}}\cdot\frac{(2n)!}{2.n!} \hspace{3cm} \text{[By (1)]}$$

$$= \frac{(2n)!}{2^{2n}.n!}\sqrt{\pi}.$$

DEDUCTION 3. *For any integer n, we have* $\Gamma\left(\frac{1}{n}\right)\Gamma\left(\frac{2}{n}\right)\Gamma\left(\frac{3}{n}\right)...\Gamma\left(\frac{n-1}{n}\right) = \frac{(2\pi)^{(n-1)/2}}{n^{1/2}}\,.$

Proof. Let $\qquad X = \Gamma\left(\frac{1}{n}\right)\Gamma\left(\frac{2}{n}\right)\Gamma\left(\frac{3}{n}\right)...\Gamma\left(\frac{n-2}{n}\right)\Gamma\left(\frac{n-1}{n}\right).$ $\hspace{2.5cm}$ ...(1)

Writing the above expression in the reversed order, we get

$$X = \Gamma\left(\frac{n-1}{n}\right)\Gamma\left(\frac{n-2}{n}\right)...\Gamma\left(\frac{2}{n}\right)\Gamma\left(\frac{1}{n}\right)$$

$$X = \Gamma\left(1-\frac{1}{n}\right)\Gamma\left(1-\frac{2}{n}\right)...\Gamma\left(1-\frac{n-2}{2}\right)\Gamma\left(1-\frac{n-1}{n}\right). \hspace{2cm} ...(2)$$

Multiplying (1) and (2) and arranging in products of terms in the $\Gamma(n)\,\Gamma(1{-}n)$, we have

$$X^2 = \left[\Gamma\left(\frac{1}{n}\right)\Gamma\left(1-\frac{1}{n}\right)\right]\cdot\left[\Gamma\left(\frac{2}{n}\right)\Gamma\left(1-\frac{2}{n}\right)\right]....$$

$$...\left[\Gamma\left(\frac{n-2}{n}\right)\Gamma\left(1-\frac{n-2}{n}\right)\right]\left[\Gamma\left(\frac{n-1}{n}\right)\Gamma\left(1-\frac{n-1}{n}\right)\right]$$

$$= \frac{\pi\,.\,\pi}{\sin\dfrac{\pi}{n}\,.\,\sin\dfrac{2\pi}{n}}...\frac{\pi}{\sin\dfrac{n-2}{n}\pi\,\sin\dfrac{n-1}{n}\pi} \qquad \left[\because \Gamma(m)\,\Gamma(1-m) = \frac{\pi}{\sin m\pi}\right]$$

$$\therefore \qquad X^2 = \frac{\pi^{n-1}}{\sin\left(\dfrac{\pi}{n}\right)\sin\left(\dfrac{2\pi}{n}\right)...\sin\left(\dfrac{(n-1)\pi}{n}\right)} \hspace{3cm} ...(3)$$

Now, using the following trigonometrical identity :

$$2^{n-1}\sin\left(\theta+\frac{\pi}{n}\right)\sin\left(\theta+\frac{2\pi}{n}\right)\sin\left(\theta+\frac{n-1}{n}\pi\right) = \frac{\sin n\theta}{\sin\theta} \hspace{2cm} ...(4)$$

and, taking limit as $\theta \to 0$, equation (4) gives

$$2^{n-1}\sin\frac{\pi}{n}\sin\frac{2\pi}{n}...\sin\left(\frac{n-1}{n}\pi\right) = \lim_{\theta\to 0}\frac{\sin n\theta}{\sin\theta} = n\lim_{\theta\to 0}\left[\frac{\sin n\theta}{n\theta}\cdot\frac{\theta}{\sin\theta}\right] = n.$$

$$\therefore \qquad \sin\left(\frac{\pi}{n}\right)\sin\left(\frac{2\pi}{n}\right)\dots\sin\left\{\frac{(n-1)\pi}{n}\right\} = \frac{n}{2^{n-1}}. \qquad \dots(5)$$

Using (5), (3) reduces to

$$X^2 = \frac{\pi^{n-1}}{n/2^{n-1}} = \frac{(2\pi)^{n-1}}{n} \text{ or } X = \frac{(2\pi)^{(n-1)/2}}{n^{1/2}} \qquad \dots(6)$$

From (1) and (6), we get

$$\Gamma\left(\frac{1}{n}\right)\Gamma\left(\frac{2}{n}\right)\dots\Gamma\left(\frac{n-1}{n}\right) = \frac{(2\pi)^{\frac{(n-1)}{2}}}{n^{1/2}}$$

DEDUCTION 4. *To prove that*

(i) $\int_0^\infty e^{-ax}\cos bx\, x^{m-1}dx = \frac{\Gamma(m)}{r^m}\cos m\theta, \qquad r^2 = a^2 + b^2.$

(ii) $\int_0^\infty e^{-ax}\sin bx\, x^{m-1}dx = \frac{\Gamma(m)}{r^m}\sin m\theta \qquad$ *where* $r = (a^2 + b^2)^{1/2}$ *and* $\theta = \tan^{-1}\left(\dfrac{b}{a}\right).$

Proof. We know that $\int_0^\infty e^{-kx}x^{m-1}dx = \dfrac{\Gamma(m)}{k^m},\ m > 0,\ k > 0 \qquad \dots(1)$

Putting $k = a - ib$ in both sides of (1), we get

$$\int_0^\infty e^{-(a-ib)x}x^{m-1}dx = \frac{\Gamma(m)}{(a-ib)^m}$$

or

$$\int_0^\infty e^{-ax}e^{ibx}x^{m-1}dx = \frac{\Gamma(m)(a+ib)^m}{[(a+ib)(a-ib)]^m}$$

$$\Rightarrow \qquad \int_0^\infty e^{-ax}x^{m-1}(\cos bx + i\sin bx)\,dx = \frac{\Gamma(m)(a+ib)^m}{(a^2+b^2)^m} \qquad \dots(2)$$

Let $\qquad\qquad\qquad\qquad a + ib = r(\cos\theta + i\sin\theta) \qquad \dots(3)$

Equating real and imaginary parts of both sides, we get

$$r^2 = a^2 + b^2, \qquad\qquad \tan\theta = \frac{b}{a} \qquad \dots(4)$$

Now $\qquad\qquad\qquad (a+ib)^m = [r(\cos\theta + i\sin\theta)]^m \qquad\qquad$ [By (3)]

$$(a+ib)^m = r^m(\cos m\theta + i\sin m\theta) \qquad \dots(5)$$

[By De'Moivre's theorem]

Using (4), (5) and (2) reduces to

$$\int_0^\infty e^{-ax}x^{m-1}(\cos bx + i\sin bx)\,dx = \frac{\Gamma(m)r^m(\cos m\theta + i\sin m\theta)}{r^{2m}}.$$

Equating real and imaginary parts of both sides, we get

$$\int_0^\infty e^{-ax}x^{m-1}\cos bx\,dx = \frac{\Gamma(m)}{r^m}\cos m\theta \qquad \dots(6)$$

and $\qquad\qquad \int_0^\infty e^{-ax}x^{m-1}\sin bx\,dx = \dfrac{\Gamma(m)}{r^m}\sin m\theta. \qquad \dots(7)$

DEDUCTION 5. Let $m = 1$, then $\Gamma(m) = \Gamma(1) = 1$, so (6), (7) reduces to

$$\int_0^\infty e^{-ax}\cos bx\,dx = \frac{\cos\theta}{r} \qquad \dots(8)$$

$$\int_0^\infty e^{-ax}\sin bx\,dx = \frac{\sin\theta}{r} \qquad \dots(9)$$

But $\tan\theta = \dfrac{b}{a}$, $\qquad$ so that $\qquad \sin\theta = \dfrac{b}{\sqrt{a^2+b^2}}$ and $\cos\theta = \dfrac{a}{\sqrt{(a^2+b^2)}}$

Also, $r^2 = (a^2 + b^2)$. Hence (8) and (9) becomes

$$\int_0^\infty e^{-ax}\cos bx\,dx \qquad \frac{}{a^2 \quad b^2} \qquad \dots(10)$$

and $\qquad\qquad \int_0^\infty e^{-ax}\sin bx\,dx = \dfrac{b}{a^2+b^2} \qquad \dots(11)$

Solved Examples

EXAMPLE 1. *Express* $\Gamma(1/6)$ *in terms of* $\Gamma(1/3)$.

SOLUTION. By duplication formula, we have

$$\Gamma(n)\Gamma\left(n+\frac{1}{2}\right) = \frac{\sqrt{\pi}}{2^{n-1}}\Gamma(2n) \qquad \dots(1)$$

Put $n = 1/6$ in (1), we get

$$\Gamma\left(\frac{1}{6}\right)\Gamma\left(\frac{2}{3}\right) = \frac{\sqrt{\pi}\,\Gamma\left(\frac{1}{3}\right)}{2^{-2/3}} \qquad \dots(2)$$

$$\Rightarrow \quad \Gamma\left(\frac{1}{6}\right) = \frac{\sqrt{\pi}\,\Gamma\left(\frac{1}{3}\right)}{2^{-2/3}\Gamma\left(\frac{2}{3}\right)}$$

Also, we know that

$$\Gamma(n)\Gamma(1-n) = \frac{\pi}{\sin n\pi}. \qquad \ldots(3)$$

Putting $n = 1/3$ in (3), we get

$$\Gamma\left(\frac{1}{3}\right)\Gamma\left(\frac{2}{3}\right) = \frac{\pi}{\sin(\pi/3)} = \frac{2\pi}{\sqrt{3}}$$

$$\Gamma\left(\frac{2}{3}\right) = \frac{2\pi}{\sqrt{3}\,\Gamma(1/3)} \qquad \ldots(4)$$

Substituting the value of $\Gamma(2/3)$ given by (4) in (2), we get

$$\Gamma\left(\frac{1}{6}\right) = \frac{\sqrt{\pi}\,\Gamma\left(\frac{1}{3}\right)}{2^{-2/3}} \cdot \frac{\sqrt{3}\,\Gamma\left(\frac{1}{3}\right)}{2\pi} = \frac{\sqrt{3}}{2^{1/3}\sqrt{\pi}}\left[\Gamma\left(\frac{1}{3}\right)\right]^2$$

EXAMPLE 2. *Find the value of* $\Gamma\left(\frac{1}{9}\right)\Gamma\left(\frac{2}{9}\right)\Gamma\left(\frac{3}{9}\right)\ldots\Gamma\left(\frac{8}{9}\right).$

SOLUTION. We know that

$$\Gamma\left(\frac{1}{n}\right)\Gamma\left(\frac{2}{n}\right)\Gamma\left(\frac{3}{n}\right)\ldots\Gamma\left(\frac{n-1}{n}\right) = \frac{(2\pi)^{(n-1)/2}}{n^{1/2}}.$$

Putting $n = 9$, in the above relation, we get

$$\Gamma\left(\frac{1}{9}\right)\Gamma\left(\frac{2}{9}\right)\Gamma\left(\frac{3}{9}\right)\ldots = \frac{(2\pi)^{(9-1)/2}}{9^{1/2}} = \frac{(2\pi)^4}{3} = \frac{16}{3}\pi^4.$$

EXAMPLE 3. *Prove that* $B(m,m)\,B\left(m+\frac{1}{2},m+\frac{1}{2}\right) = \frac{\pi m^{-1}}{2^{4m-1}}.$

SOLUTION. L.H.S. $= \dfrac{\Gamma(m)\Gamma(m)}{\Gamma(m+m)} \cdot \dfrac{\Gamma\left(m+\frac{1}{2}\right)\Gamma\left(m+\frac{1}{2}\right)}{\Gamma\left(m+\frac{1}{2}+m+\frac{1}{2}\right)}$

$$= \frac{[\Gamma(m)\Gamma(m+1/2)]^2}{\Gamma(2m)\Gamma(2m+1)} = \frac{[\Gamma(m)\Gamma(m+1/2)]^2}{\Gamma(2m).2m\Gamma(2m)}$$

$$[\because \Gamma(p+1)=p\Gamma(p)]$$

$$= \frac{1}{2m}\left[\frac{\Gamma(m)\Gamma(m+1/2)}{\Gamma(2m)}\right]^2 = \frac{1}{2m}\cdot\left(\frac{\sqrt{\pi}}{2^{2m-1}}\right)^2$$

$$\text{(By duplication formula)}$$

$$= \frac{\pi}{2m.2^{4m-2}} = \frac{\pi m^{-1}}{2^{4m-1}}.$$

EXAMPLE 4. *Show that* $\Gamma\left(\frac{3}{2}-x\right)\Gamma\left(\frac{3}{2}+x\right) = \left(\frac{1}{4}-x^2\right)\pi\sec\pi x,$ *provided* $-1 < 2x < 1.$

SOLUTION. We have

$$\Gamma\left(\frac{3}{2}-x\right)\Gamma\left(\frac{3}{2}+x\right) = \left(\frac{1}{2}-x\right)\Gamma\left(\frac{1}{2}-x\right)\cdot\left(\frac{1}{2}+x\right)\Gamma\left(\frac{1}{2}+x\right)$$

$$= \left(\frac{1}{4}-x^2\right)\Gamma\left(\frac{1-2x}{2}\right)\Gamma\left(\frac{1+2x}{2}\right)$$

$$= \left(\frac{1}{4}-x^2\right)\Gamma\left(\frac{1-2x}{2}\right)\Gamma\left(1-\frac{1-2x}{2}\right)$$

$$= \left(\frac{1}{4}-x^2\right)\frac{\pi}{\sin\left(\frac{1-2x}{2}\pi\right)} = \left(\frac{1}{4}-x^2\right)\frac{\pi}{\sin\left(\frac{\pi}{2}-\pi x\right)}$$

$$= \left(\frac{1}{4}-x^2\right)\cdot\frac{\pi}{\cos\pi x} = \left(\frac{1}{4}-x^2\right)\sec\pi x.\pi.$$

EXAMPLE 5. *Prove that* $\int_0^\infty \cos(bx^{1/n})\,dx = \frac{\Gamma(n+1)}{b^n}\cos\frac{n\pi}{2}.$

SOLUTION. Let $I = \int_0^\infty \cos(bx^{1/n})\,dx.$ $\qquad \ldots(1)$

Putting $x = t^n$ so that $dx = nt^{n-1}dt$ then (1) gives

$$I = n\int_0^\infty \cos(bt).t^{n-1}\,dt = \frac{n\Gamma(n)}{b^n}\cos\frac{n\pi}{2} \qquad \frac{\Gamma(n+1)}{b^n}\cos\frac{n\pi}{2}.$$

ALITER. $\int_0^\infty \cos(bz^{1/n})\,dz = \frac{1}{b^n}\Gamma(n+1)\cos\frac{n\pi}{2}.$

Put $\quad z^{1/n} = x \Rightarrow z = x^n.$

so that $\quad dz = nx^{n-1}\,dx.$

$\therefore \quad \int_0^\infty \cos(bz^{1/n})\,dz = \int_0^\infty \cos(bx).nx^{n-1}dx$

$$= n\int_0^\infty x^{n-1}\cos(bx)\,dx$$

$$= \text{real part of } n\int_0^\infty e^{-bxi}x^{n-1}dx$$

$$= \text{real part of } n\frac{\Gamma(n)}{(bi)^n}$$

$$= \text{real part of } \frac{n\Gamma(n)}{b^n}\left(\cos\frac{\pi}{2}+i\sin\frac{\pi}{2}\right)^{-n}$$

$$= \text{real part of } \frac{\Gamma(n+1)}{b^n}\left(\cos\frac{n\pi}{2}-i\sin\frac{n\pi}{2}\right)$$

$$= \frac{1}{b^n}\Gamma(n+1)\cos\left(\frac{n\pi}{2}\right).$$

Exercise-5.9

1. Show that $\Gamma(n)\Gamma\left(\dfrac{1-n}{2}\right) = \dfrac{\sqrt{\pi}\,\Gamma(n/2)}{2^{1-n}\cos\left(\dfrac{n\pi}{2}\right)}, 0 < n < 1$.

2. Show that $\int_0^1 \dfrac{dx}{\sqrt{(1-x^6)}} = \dfrac{\sqrt{3}}{2}\int_0^1 \dfrac{dx}{\sqrt{1-x^3}} = \dfrac{1}{2^{7/3}\pi}\left[\Gamma\left(\dfrac{1}{3}\right)\right]^3.$

3. Show that $\Gamma(0.1).\Gamma(0.2).\Gamma(0.3)\ldots\Gamma(0.9) = \dfrac{(2\pi)^{9/2}}{\sqrt{10}}.$

4. Show that $\int_0^1 \dfrac{dx}{\sqrt{(1-x^4)}} = \dfrac{\sqrt{2}}{8\sqrt{\pi}}\left[\Gamma\left(\dfrac{1}{4}\right)\right]^2.$

5. Show that if $m > -1$ then

$$\int_0^\infty x^m e^{-n^2x^2}\,dx = \frac{1}{2n^{m+1}}\Gamma\left(\frac{m+1}{2}\right).$$

6. Prove the following

(i) $B(l,m).B(l+m,n).B(l+m+n,p) = \dfrac{\Gamma(l)\Gamma(m)\Gamma(n)}{\Gamma(l+m+n+p)}$

(UPTU-2008)

(ii) $B(m,n) = 2^{1-2m}\left(m,\dfrac{1}{2}\right)$

(UPTU-2007)

7. Prove that $B(m,n) = \int_0^1 \dfrac{x^{m-1}+x^{n-1}}{(1+x)^{m+n}}\,dx$

(MTU-2012)

Hints to Selected Problems

2. Let $A = \int_0^1 \dfrac{dx}{\sqrt{1-x^6}}$. Now putting $x^6 = \sin^2\theta$ *i.e.*, $x = \sin^{1/3}\theta \Rightarrow dx = -\sin^{-2/3}\theta\cos\theta\,d\theta$.

$\therefore \quad A = \dfrac{1}{3}\int_0^{\pi/2}\sin^{-2/3}\theta\cos^0\theta\,d\theta = \dfrac{1}{3}\dfrac{\Gamma\left(\frac{1}{6}\right)\Gamma\left(\frac{1}{2}\right)}{2\Gamma\left(\frac{2}{3}\right)} = \dfrac{1}{6}\dfrac{\Gamma\left(\frac{1}{6}\right)\cdot\sqrt{\pi}}{\Gamma\left(\frac{2}{3}\right)}$. Now find the values of $\Gamma\left(\dfrac{1}{6}\right)$ and $\Gamma\left(\dfrac{2}{3}\right)$ separately.

4. Let $I = \int_0^1 \dfrac{dx}{\sqrt{1-x^4}}$. Then solve after putting $x^4 = t$.   **5.** Let $I = \int_0^\infty x^m e^{-n^2x^2}dx$. Then solve after putting $n^2x^2 = t$.

ARCHIVE

1. Prove that $\int_0^{a/\sqrt{2}}\int_y^{\sqrt{a^2-y^2}}\log\left(x^2+y^2\right)dx\,dy\,(a>0)$
$$= \frac{\pi a^2}{4}\left(\log a - \frac{1}{2}\right).$$

2. Prove that $\int_0^a\int_{x^2/a}^{2a-x}xy\,dy\,dx = \dfrac{3}{8}a^4$.

3. Prove that $\iiint_V \sqrt{1-\left(x^2+y^2+z^2\right)}dx\,dy\,dz = \dfrac{\pi^2}{3^2}$, where V is the region interior to the sphere $x^2+y^2+z^2=1$.

4. Prove that $\iiiint dx\,dy\,dz\,dw = \dfrac{\pi^2}{32}\left(b^4-a^4\right)$, where $a^2 < x^2+y^2+z^2+w^2 < b^2, a < b$.

5. Prove that the volume of the solid bounded by the surface .
$$\left(\frac{x}{a}\right)^{2/3}+\left(\frac{y}{b}\right)^{2/3}+\left(\frac{z}{c}\right)^{2/3} = 1 \text{ is } \frac{4\pi abc}{35}$$

6. Prove that the volume determined by the surface $x^n+y^n+z^n=a^n, n>0$ in the positive octant is $\dfrac{a^3\left[\Gamma\left(1+\frac{1}{n}\right)\right]^3}{\sqrt{\Gamma\left(1+\frac{3}{n}\right)}}$.

7. Prove that the volume bounded by the surface $\dfrac{x^2}{a^2}+\dfrac{y^2}{b^2}+\dfrac{z^2}{c^2}=1$ is $\dfrac{8\pi abc}{5}$.

8. Prove that the volume of solid whose surface is represent by the equation $\dfrac{x^4}{a^4}+\dfrac{y^4}{b^4}+\dfrac{z^4}{c^4}=1$ is $\dfrac{abc}{6\sqrt{2}.\pi}\left[\Gamma\left(\frac{1}{4}\right)\right]^4$.

9. Prove that the volume of the ellipsoid $\dfrac{x^2}{a^2}+\dfrac{y^2}{b^2}+\dfrac{z^2}{c^2}=1$ is $\dfrac{4}{3}\pi abc$.

10. Prove that the area of the ellipse $\dfrac{x^2}{9}+\dfrac{y^2}{4}=1$ is 6π.

11. Prove that the volume of the tetrahedron bounded by the co-ordinate planes and the plane $x+y+z=1$ is __

12. Prove that the area of ellipse $\dfrac{x^2}{a^2}+\dfrac{y^2}{b^2}=1$ is πab.

13. Prove that the $\iiint(x+y+z)dx\,dy\,dz = \dfrac{1}{8}$ over the tetrahedron bounded by the planes $x=0$, $y=0$, $z=0$ and $x+y+z=1$.

14. Prove that the value of $\iint \sqrt{a^2-x^2-y^2}dx\,dy$ over the semicircle $x^2+y^2=ax$ in the positive quadrant is $\dfrac{a^3}{3}\left(\dfrac{\pi}{2}-\dfrac{2}{3}\right)$.

15. Prove that if R is a region bounded by the curves $x=f_1(y)$, $x=f_2(y), y=c$ $y=d$ then
$$\iint_R f(x,y)dA = \int_c^d\left[\int_{f_1(y)}^{f_2(y)}f(x,y)dx\right]dy.$$

16. Prove that if $p > 0$, $B(p,p) = B\left(p+\dfrac{1}{2},p+\dfrac{1}{2}\right) = \dfrac{\pi}{2^{4p-1}.p}$.

17. Prove that if $\int_0^1 x^{-1/3}(1-x)^{-2/3}(1-2x)^{-1}dx = \dfrac{1}{9^{1/3}}B\left(\dfrac{2}{3},\dfrac{1}{3}\right)$

18. Prove that if $\int_{-\infty}^\infty \dfrac{e^{pt}}{1+e^t}dt = \Gamma(p)\Gamma(1-p), 1 > p > 0$

19. Prove that if $p > 0$, $q > 1$ then $\sum\limits_{n=0}^\infty B(p+r,q)$ converges to $B(p, q-1)$.

20. Prove that if $\sin\dfrac{\pi}{2n}.\sin\dfrac{2\pi}{2n}.\sin\dfrac{3\pi}{2n}...\sin\dfrac{(n-1)\pi}{2n} = \sqrt{n}.2^{-n+1}$.

21. Prove that if $p > 0, q > 0, a > 0, b > 0$, then
$$\int_0^{\pi/2}\frac{\cos^{2p-1}\theta\sin^{2q-1}\theta}{(a\cos^2\theta+b\sin^2\theta)^{p+q}}d\theta = \frac{B(p,q)}{2a^p.b^q}.$$

22. Prove that if $n > 0$, then $\dfrac{\Gamma\left(n+\frac{1}{2}\right)}{\Gamma(n)} = \dfrac{(2n)!}{2^{2n}.n!}$.

23. Prove that $\int_0^{\pi/2}\dfrac{d\theta}{\sqrt{\sin\theta}} \times \int_0^{\pi/2}\sqrt{\sin\theta}\,d\theta = \pi$.

24. Prove that $\int_0^1 \sqrt{(1-x^4)}\,dx = \dfrac{1}{12}\sqrt{\dfrac{2}{\pi}}\left[\Gamma\left(\dfrac{1}{4}\right)\right]^2$.

25. Prove that $\int_0^\infty x^{m-1}\cos bx = \dfrac{\Gamma(m)}{b^m}\cos\left(\dfrac{m\pi}{2}\right)$.

26. Change the order of integration in $I = \int_0^1\int_{x^2}^{2-x}xy\,dx\,dy$.

    (BHOPAL-2008, VTU-2008, SVTU-2007, PTU-200, UPTU-2005)

27. Change the order of integration $I = \int_0^{4a}\int_{x^2/4a}^{2\sqrt{ax}}dy\,dx$ and hence show that $I = \dfrac{16a^2}{3}$.   (NAGPUR-2009, 12, PTU-2009)

28. Show that $\int_0^1\int_x^{\sqrt{2-x^2}}\dfrac{x}{\sqrt{x^2+y^2}}dy\,dx = 1 - \dfrac{1}{\sqrt{2}}$

    (PTU-2010, 14, MARATHWADA-2008)

29. Show that the value of $\int_0^\infty\int_0^x xe^{-x^2/y}dy\,dx = 1$

    (SVTU-2006; VTU-2004, 10)

30. Show that the value of $\iint r\sin dr\,d\theta$ over $r = a(1-\cos\theta)$ above the initial line is $\dfrac{4a^2}{3}$.   (KERALA-2005)

❋❋❋❋❋❋

6.1 DIFFERENTIATION AND INTEGRATION OF VECTORS

6.1.1 SCALAR FUNCTION

Since we know that the quantity which is associated with the magnitude but not associated with direction is known as scalar quantity. Therefore every real number is a scalar quantity.

Let D be a subset of a set of real numbers. Then a function f defined over the subset D such that for all $t \in D$, $f(t)$ is obtained as a scalar quantity, is called a scalar function.

6.1.2 VECTOR FUNCTION

If the scalar fucntion $f(t)$ for all $t \in D$ is associated with some direction then this function is called a vector function and is therefore, denoted by $\boldsymbol{f}(t)$ or $\boldsymbol{f}$.

Let $f_1(t), f_2(t), f_3(t)$ be three components of a vector function $\boldsymbol{f}(t)$, then this function can be uniquely expressed as a linear combination of these three fixed non-coplanar vectors $f_1(t)\hat{i}, f_2(t)\hat{j}, f_3(t)\hat{k}$.

$$\therefore \qquad \boldsymbol{f}(t) = f_1(t)\hat{i} + f_2(t)\hat{j} + f_3(t)\hat{k}$$

where $\hat{i}, \hat{j}, \hat{k}$ are three mutually perpendicular non-coplanar unit vectors.

6.1.3 SCALAR AND VECTOR FIELDS

Scalar fields. A scalar point function f defined over some region R such that to each point $P(x, y, z)$ in space, there corresponds a unique scalar $f(P)$, is called a scalar field. For example

$$f(x, y, z) = x^2 + y^2 + z^2 - 3xyz.$$

Vector fields. A vector point function f defined over a region R such that to each **point** $P(x, y, z)$ there exists a unique vector $\mathbf{f}(P)$, is called vector field. For example

$$\boldsymbol{f}(x, y, z) = x^2 y\hat{i} + x^3 z\hat{j} - y^3 z\hat{k}.$$

6.1.4 LIMIT AND CONTINUITY OF A VECTOR FUNCTION

Limit. A vector function $f(t)$ is said to have the limit l as t tends to t_0, for given $\varepsilon > 0$ there exists a positive number δ such that $\left| \boldsymbol{f}(t) - l \right| < \varepsilon$ whenever $0 < \left| t - t_0 \right| < \delta$ i.e., $\lim\limits_{t \to t_0} \boldsymbol{f}(t) = l.$

Continuity. A vector function $\boldsymbol{f}(t)$ is said to be continuous at t_0, if for given $\varepsilon > 0$ there must exists a positive number δ such that $\left| \boldsymbol{f}(t) - \boldsymbol{f}(t_0) \right| < \varepsilon$ whenever $\left| t - t_0 \right| < \delta$, provided $\boldsymbol{f}(t_0)$ is defined.

📝 **REMARK**

- A vector function $\boldsymbol{f}(t)$ is said to be continuous for every value of t in the domain over which $\boldsymbol{f}(t)$ is defined.

6.1.5 SOME RESULTS RELATED TO THE LIMIT AND CONTINUITY OF A VECTOR FUNCTION

1. The necessary and sufficient condition for a vcector function $\boldsymbol{f}(t)$ to be continuous at t_0 is that $\lim\limits_{t \to t_0} \boldsymbol{f}(t) = \boldsymbol{f}(t_0).$

2. If $\boldsymbol{f}(t) = f_1(t)\hat{i} + f_2(t)\hat{j} + f_3(t)\hat{k}$, then $\boldsymbol{f}(t)$ is continuous iff $f_1(t), f_2(t), f_3(t)$ are continuous.

3. If $\boldsymbol{f}(t) = f_1(t)\hat{i} + f_2(t)\hat{j} + f_3(t)\hat{k}$ and $\boldsymbol{l} = l_1\hat{i} + l_2\hat{j} + l_3\hat{k}$, then $\lim\limits_{t \to t_0} \boldsymbol{f}(t) = \boldsymbol{l}$ iff $\lim\limits_{t \to t_0} f_1(t) = l_1, \lim\limits_{t \to t_0} f_2(t) = l_2$ and $\lim\limits_{t \to t_0} f_3(t) = l_3.$

4. If $\boldsymbol{f}(t)$ and $\boldsymbol{g}(t)$ are vector functions of scalar variable t and $\phi(t)$ is a scalar function, then

   (i) $\lim\limits_{t \to t_0} [\boldsymbol{f}(t) \pm \boldsymbol{g}(t)] = \lim\limits_{t \to t_0} \boldsymbol{f}(t) \pm \lim\limits_{t \to t_0} \boldsymbol{g}(t)$ (ii) $\lim\limits_{t \to t_0} [\boldsymbol{f}(t).\boldsymbol{g}(t)] = \left[\lim\limits_{t \to t_0} \boldsymbol{f}(t) \right].\left[\lim\limits_{t \to t_0} \boldsymbol{g}(t) \right]$

(iii) $\lim\limits_{t \to t_0} [\boldsymbol{f}(t) \times \boldsymbol{g}(t)] = \left[\lim\limits_{t \to t_0} \boldsymbol{f}(t)\right] \times \left[\lim\limits_{t \to t_0} \boldsymbol{g}(t)\right]$ (iv) $\lim\limits_{t \to t_0} |\boldsymbol{f}(t)| = \left|\lim\limits_{t \to t_0} \boldsymbol{f}(t)\right|$

(v) $\lim\limits_{t \to t_0} [\phi(t)\boldsymbol{f}(t)] = \left[\lim\limits_{t \to t_0} \phi(t)\right]\left[\lim\limits_{t \to t_0} \boldsymbol{f}(t)\right].$

6.1.6 Differentiation of a Vector Function with respect to a Scalar

Definition (1) *Let $\boldsymbol{f}(t)$ be a vector function of scalar variable t. The function $\boldsymbol{f}(t)$ is differentiable with respect to t if*

$$\lim_{\delta t \to 0} \frac{\boldsymbol{f}(t + \delta t) - \boldsymbol{f}(t)}{\delta t} \; exists.$$

and its derivative is denoted by $\dfrac{d\boldsymbol{f}(t)}{dt}$.

Definition (2). *If $\dfrac{d\boldsymbol{f}(t)}{dt}$ exists, then $\boldsymbol{f}(t)$ is differentiable and $\dfrac{d\boldsymbol{f}(t)}{dt}$ is also a vector function of variable t. If $\dfrac{d\boldsymbol{f}(t)}{dt}$ is differentiable, then $\dfrac{d^2\boldsymbol{f}(t)}{dt^2}$ is called second derivative of $\boldsymbol{f}(t)$. Similarly we can find third, fourth, etc. derivaties of $\boldsymbol{f}(t)$.*

◀ Remark

- If $\boldsymbol{r} = \boldsymbol{f}(t)$, then $\dfrac{d\boldsymbol{r}}{dt}, \dfrac{d^2\boldsymbol{r}}{dt^2}$ etc. are the first, second etc. derivaties of $r = \boldsymbol{f}(t)$ and also denoted by $\dot{\boldsymbol{r}}, \ddot{\boldsymbol{r}}$ respectively etc.

6.1.7 Differentiation Formulae for the Vector Function

Let $\boldsymbol{a}, \boldsymbol{b}, \boldsymbol{c}$ be differentiable vector function of a scalar variable f and ϕ be a differentiable scalar function of t, then

(i) $\dfrac{d}{dt}(\boldsymbol{a} \pm \boldsymbol{b}) = \dfrac{d\boldsymbol{a}}{dt} \pm \dfrac{d\boldsymbol{b}}{dt}$ (ii) $\dfrac{d}{dt}(\boldsymbol{a}.\boldsymbol{b}) = \boldsymbol{a}.\dfrac{d\boldsymbol{b}}{dt} + \dfrac{d\boldsymbol{a}}{dt}.\boldsymbol{b}$

(iii) $\dfrac{d}{dt}(\boldsymbol{a} \times \boldsymbol{b}) = \boldsymbol{a} \times \dfrac{d\boldsymbol{b}}{dt} + \dfrac{d\boldsymbol{a}}{dt} \times \boldsymbol{b}$ (iv) $\dfrac{d}{dt}(\phi\boldsymbol{a}) = \phi\dfrac{d\boldsymbol{a}}{dt} + \dfrac{d\phi}{dt}\boldsymbol{a}$

(v) $\dfrac{d}{dt}[\boldsymbol{a}\,\boldsymbol{b}\,\boldsymbol{c}] = \left[\dfrac{d\boldsymbol{a}}{dt}\,\boldsymbol{b}\,\boldsymbol{c}\right] + \left[\boldsymbol{a}\,\dfrac{d\boldsymbol{b}}{dt}\,\boldsymbol{c}\right] + \left[\boldsymbol{a}\,\boldsymbol{b}\,\dfrac{d\boldsymbol{c}}{dt}\right]$ (vi) $\dfrac{d}{dt}\{\boldsymbol{a} \times (\boldsymbol{b} \times \boldsymbol{c})\} = \dfrac{d\boldsymbol{a}}{dt} \times (\boldsymbol{b} \times \boldsymbol{c}) + \boldsymbol{a} \times \left[\dfrac{d\boldsymbol{b}}{dt} \times \boldsymbol{c}\right] + \boldsymbol{a} \times \left[\boldsymbol{b} \times \dfrac{d\boldsymbol{c}}{dt}\right].$

PROOF.   (i) Let $\qquad\qquad\qquad \boldsymbol{p} = \boldsymbol{a} + \boldsymbol{b}.$

$\Rightarrow \qquad\qquad\qquad\qquad \boldsymbol{p} + \delta\boldsymbol{p} = \boldsymbol{a} + \boldsymbol{b} + \delta\boldsymbol{a} + \delta\boldsymbol{b}$

$\therefore \qquad\qquad\qquad\qquad \delta\boldsymbol{p} = (\boldsymbol{p} + \delta\boldsymbol{p}) - \boldsymbol{p}$

$\qquad\qquad\qquad\qquad\qquad \delta\boldsymbol{p} = \delta\boldsymbol{a} + \delta\boldsymbol{b}.$

Divide by δt and taking the limit of both sides as $\delta t \to 0$, we get

$$\lim_{\delta t \to 0} \frac{\delta\boldsymbol{p}}{\delta t} = \lim_{\delta t \to 0}\left(\frac{\delta\boldsymbol{a} + \delta\boldsymbol{b}}{\delta t}\right) = \lim_{\delta t \to 0} \frac{\delta\boldsymbol{a}}{\delta t} + \lim_{\delta t \to 0} \frac{\delta\boldsymbol{b}}{\delta t}$$

$\therefore \qquad\qquad\qquad\qquad \dfrac{d\boldsymbol{p}}{dt} = \dfrac{d\boldsymbol{a}}{dt} + \dfrac{d\boldsymbol{b}}{dt}$

or $\qquad\qquad\qquad\qquad \dfrac{d}{dt}(\boldsymbol{a} + \boldsymbol{b}) = \dfrac{d\boldsymbol{a}}{dt} + \dfrac{d\boldsymbol{b}}{dt}$

Similarly, $\qquad\qquad\quad \dfrac{d}{dt}(\boldsymbol{a} - \boldsymbol{b}) = \dfrac{d\boldsymbol{a}}{dt} - \dfrac{d\boldsymbol{b}}{dt}$

Hence, $\qquad\qquad\qquad \dfrac{d}{dt}(\boldsymbol{a} \pm \boldsymbol{b}) = \dfrac{d\boldsymbol{a}}{dt} \pm \dfrac{d\boldsymbol{b}}{dt}$

(ii) Let $\qquad\qquad\qquad \boldsymbol{p} = \boldsymbol{a}.\boldsymbol{b}$ ...(1)

$\therefore \qquad\qquad\qquad\qquad \boldsymbol{p} + \delta\boldsymbol{p} = (\boldsymbol{a} + \delta\boldsymbol{a}).(\boldsymbol{b} + \delta\boldsymbol{b})$ ...(2)

From (1) and (2), we get

$$\delta\boldsymbol{p} = (\boldsymbol{a} + \delta\boldsymbol{a}).(\boldsymbol{b} + \delta\boldsymbol{b}) - \boldsymbol{a}.\boldsymbol{b} = \boldsymbol{a}.\boldsymbol{b} + \boldsymbol{a}.\delta\boldsymbol{b} + \delta\boldsymbol{a}.\boldsymbol{b} + \delta\boldsymbol{a}.\delta\boldsymbol{b} - \boldsymbol{a}.\boldsymbol{b}$$

$$\delta\boldsymbol{p} = \boldsymbol{a}.\delta\boldsymbol{b} + \delta\boldsymbol{a}.\boldsymbol{b} + \delta\boldsymbol{a}.\delta\boldsymbol{b}$$

$\therefore \qquad\qquad\qquad \dfrac{\delta\boldsymbol{p}}{\delta t} = \boldsymbol{a} \cdot \dfrac{\delta\boldsymbol{b}}{\delta t} + \dfrac{\delta\boldsymbol{a}}{\delta t} \cdot \boldsymbol{b} + \left(\dfrac{\delta\boldsymbol{a}}{\delta t} \cdot \dfrac{\delta\boldsymbol{b}}{\delta t} \cdot \delta t\right).$

Now taking the limit of both sides as $\delta t \to 0$, we get

$$\lim_{\delta t \to 0} \frac{\delta\boldsymbol{p}}{\delta t} = \boldsymbol{a} \cdot \lim_{\delta t \to 0} \frac{\delta\boldsymbol{b}}{\delta t} + \lim_{\delta t \to 0} \frac{\delta\boldsymbol{a}}{\delta t} \cdot \boldsymbol{b} + \left(\lim_{\delta t \to 0} \frac{\delta\boldsymbol{a}}{\delta t} \cdot \lim_{\delta t \to 0} \frac{\delta\boldsymbol{b}}{\delta t}\right)\lim_{\delta t \to 0} \delta t$$

or
$$\frac{dp}{dt} = a \cdot \frac{db}{dt} + \frac{da}{dt} \cdot b \qquad\qquad \left[\because \lim_{\delta t \to 0} \delta t = 0\right]$$

or
$$\frac{d}{dt}(a \cdot b) = a \cdot \frac{db}{dt} + \frac{da}{dt} \cdot b \qquad\qquad (\because p = a.b)$$

(iii) Let
$$p = a \times b \qquad\qquad ...(1)$$

and
$$p + \delta p = (a + \delta a) \times (b + \delta b)$$
$$= a \times b + a \times \delta b + \delta a \times b + \delta a \times \delta b \qquad\qquad ...(2)$$

Subtract (1) from (2), we get
$$\delta p = a \times \delta b + \delta a \times b + \delta a \times \delta b \qquad\qquad ...(3)$$

$$\Rightarrow \quad \frac{\delta p}{\delta t} = a \times \frac{\delta b}{\delta t} + \frac{\delta a}{\delta t} \times b + \frac{\delta a}{\delta t} \times \frac{\delta b}{\delta t} \delta t$$

$$\therefore \quad \frac{\delta p}{\delta t} = a \times \frac{\delta b}{\delta t} + \frac{\delta a}{\delta t} \times b + \frac{\delta a}{\delta t} \times \frac{\delta b}{\delta t} \delta t.$$

Now taking the limit of both sides as $\delta t \to 0$, we get

$$\lim_{\delta t \to 0} \frac{\delta p}{\delta t} = \lim_{\delta t \to 0} \left[a \times \frac{\delta b}{\delta t} + \frac{\delta a}{\delta t} \times b + \frac{\delta a}{\delta t} \times \frac{\delta b}{\delta t} \delta t \right]$$

$$= \lim_{\delta t \to 0} \left(a \times \frac{\delta b}{\delta t} \right) + \lim_{\delta t \to 0} \left(\frac{\delta a}{\delta t} \times b \right) + \lim_{\delta t \to 0} \left(\frac{\delta a}{\delta t} \times \frac{\delta b}{\delta t} \delta t \right)$$

$$= a \times \lim_{\delta t \to 0} \frac{\delta b}{\delta t} + \lim_{\delta t \to 0} \frac{\delta a}{\delta t} \times b + \left(\lim_{\delta t \to 0} \frac{\delta a}{\delta t} \times \lim_{\delta t \to 0} \frac{\delta b}{\delta t} \right) \lim_{\delta t \to 0} \delta t$$

$$\therefore \quad \frac{dp}{dt} = a \times \frac{db}{dt} + \frac{da}{dt} \times b + \left(\frac{da}{dt} \times \frac{db}{dt} \right)(0) = a \times \frac{db}{dt} + \frac{da}{dt} \times b$$

or
$$\frac{d}{dt}(a \times b) = a \times \frac{db}{dt} + \frac{da}{dt} \times b \qquad\qquad (\because p = a \times b)$$

(iv)
$$\frac{d}{dt}(\phi a) = \phi \frac{da}{dt} + \frac{d\phi}{dt} a$$

Let
$$p = \phi a \qquad\qquad ...(1)$$

and
$$p + \delta p = (\phi + \delta\phi)(a + \delta a) = \phi a + \phi \delta a + \delta\phi a + \delta\phi \delta a \qquad\qquad ...(2)$$

Subtract (1) from (2), we get
$$\delta p = \phi \delta a + \delta\phi a + \delta\phi \delta a \qquad\qquad ...(3)$$

$$\Rightarrow \quad \frac{\delta p}{\delta t} = \phi \frac{\delta a}{\delta t} + \frac{\delta\phi}{\delta t} a + \frac{\delta\phi}{\delta t} \frac{\delta a}{\delta t} \delta t .$$

Now taking the limit of both sides as $\delta t \to 0$, we get

$$\lim_{\delta t \to 0} \frac{\delta p}{\delta t} = \phi \lim_{\delta t \to 0} \frac{\delta a}{\delta t} + \lim_{\delta t \to 0} \frac{\delta\phi}{\delta t} a + \lim_{\delta t \to 0} \frac{\delta\phi}{\delta t} \lim_{\delta t \to 0} \frac{\delta a}{\delta t} \lim_{\delta t \to 0} \delta t$$

$$\therefore \quad \frac{dp}{dt} = \phi \frac{da}{dt} + \frac{d\phi}{dt} a + 0$$

or
$$\frac{d}{dt}(\phi a) = \phi \frac{da}{dt} + \frac{d\phi}{dt} a \qquad\qquad (\because p = \phi a)$$

(v) Let
$$p = [a\ b\ c] = a \cdot (b \times c) \qquad\qquad ...(1)$$

and let $b \times c = q$, then $p = a \cdot q$

$$\therefore \quad \frac{dp}{dt} = \frac{d}{dt}(a.q) = a \cdot \frac{dq}{dt} + \frac{da}{dt} \cdot q \qquad\text{[Using (ii)]} \qquad ...(2)$$

Now
$$\frac{dq}{dt} = \frac{d}{dt}(b \times c) = b \times \frac{dc}{dt} + \frac{db}{dt} \times c \qquad\text{[Using (iii)]} \qquad ...(3)$$

From (2) and (3), we get

$$\frac{dp}{dt} = a \cdot \left(b \times \frac{dc}{dt} + \frac{db}{dt} \times c \right) + \frac{da}{dt} \cdot b \times c$$

$$= a.b \times \frac{dc}{dt} + a. \frac{db}{dt} \times c + \frac{da}{dt} . b \times c$$

$$\frac{dp}{dt} = \left[a\ b\ \frac{dc}{dt} \right] + \left[a\ \frac{db}{dt}\ c \right] + \left[\frac{da}{dt}\ b\ c \right]$$

or $\quad \dfrac{d}{dt}[a\ b\ c] = \left[a\ b\ \dfrac{dc}{dt}\right] + \left[a\ \dfrac{db}{dt}\ c\right] + \left[\dfrac{da}{dt}\ b\ c\right]$ $\qquad\qquad (\because p = [a\ b\ c])$

(vi) $\quad \dfrac{d}{dt}\{a \times (b \times c)\} = \dfrac{da}{dt} \times (b \times c) + a \times \left(\dfrac{db}{dt} \times c\right) + a \times \left(b \times \dfrac{dc}{dt}\right)$

Let $\qquad\qquad p = a \times (b \times c)$ $\qquad\qquad\qquad\qquad\qquad\qquad ...(1)$

and let $\qquad\qquad b \times c = q$

$\therefore \qquad\qquad\qquad p = a \times q$

$\therefore \qquad\qquad \dfrac{dp}{dt} = \dfrac{d}{dt}(a \times q) = a \times \dfrac{dq}{dt} + \dfrac{da}{dt} \times q$ $\qquad$ [Using (iii)] $\qquad\qquad ...(2)$

Now $\qquad\qquad \dfrac{dq}{dt} = \dfrac{d}{dt}(b \times c)$

$\qquad\qquad\qquad \dfrac{dq}{dt} = b \times \dfrac{dc}{dt} + \dfrac{db}{dt} \times c$ $\qquad$ [Using (iii)] $\qquad\qquad ...(3)$

From (2) and (3), we get

$$\dfrac{dp}{dt} = a \times \left(b \times \dfrac{dc}{dt} + \dfrac{db}{dt} \times c\right) + \dfrac{da}{dt} \times (b \times c)$$

$$\dfrac{dp}{dt} = a \times \left(b \times \dfrac{dc}{dt}\right) + a \times \left(\dfrac{db}{dt} \times c\right) + \dfrac{da}{dt} \times (b \times c)$$

or $\quad \dfrac{d}{dt}\{a \times (b \times c)\} = a \times \left(b \times \dfrac{dc}{dt}\right) + a \times \left(\dfrac{db}{dt} \times c\right) + \dfrac{da}{dt} \times (b \times c)$ $\qquad [\because p = a \times (b \times c)]$

6.1.8 Derivative of a Constant Vector

Definition. *A vector is said to be constant vector if its magnitude as well as direction are fixed.*

Let r be a constant vector, then $\qquad r = c$ (a constant vector) $\qquad\qquad\qquad ...(1)$

$\therefore \qquad\qquad\qquad\qquad r + \delta r = c.$ $\qquad\qquad\qquad\qquad\qquad\qquad\qquad ...(2)$

Subtract (1) from (2), we get $\qquad \delta r = 0.$

Divide by δt and taking the limit as $\delta t \to 0$, we get $\quad \lim\limits_{\delta t \to 0} \dfrac{\delta r}{\delta t} = 0$ or $\dfrac{dr}{dt} = 0.$

Hence the derivative of a constant vector is a zero vector.

6.1.9 Derivative of a Vector Function in terms of its Components

Let $P(x, y, z)$ be any point in space and its position vector with respect to the origin O be $\mathbf{r}$ and let x, y, z be the function of scalar variable t, then we have

$$r = x\hat{i} + y\hat{j} + z\hat{k} \qquad\qquad\qquad ...(1)$$

where $\hat{i}, \hat{j}, \hat{k}$ are constant vectors.

$\therefore \qquad\qquad r + \delta r = (x + \delta x)\hat{i} + (y + \delta y)\hat{j} + (z + \delta z)\hat{k} \qquad\qquad\qquad ...(2)$

Subtract (1) from (2), we get

$$\delta r = \delta x\hat{i} + \delta y\hat{j} + \delta z\hat{k}$$

Now divide this equation by δt and taking the limit as $\delta t \to 0$, we have

$$\lim\limits_{\delta t \to 0} \dfrac{\delta r}{\delta t} = \lim\limits_{\delta t \to 0}\left(\dfrac{\delta x}{\delta t}\hat{i} + \dfrac{\delta y}{\delta t}\hat{j} + \dfrac{\delta z}{\delta t}\hat{k}\right)$$

$$\dfrac{dr}{dt} = \left(\lim\limits_{\delta t \to 0} \dfrac{\delta x}{\delta t}\right)\hat{i} + \left(\lim\limits_{\delta t \to 0} \dfrac{\delta y}{\delta t}\right)\hat{j} + \left(\lim\limits_{\delta t \to 0} \dfrac{\delta z}{\delta t}\right)\hat{k}$$

$\therefore \qquad\qquad \dfrac{dr}{dt} = \dfrac{dx}{dt}\hat{i} + \dfrac{dy}{dt}\hat{j} + \dfrac{dz}{dt}\hat{k}.$

Similarly, we can find $\dfrac{d^2 r}{dt^2}, \dfrac{d^3 r}{dt^3}$ etc.

6.1.10 Derivative of a Vector Function of Function

Let $\mathbf{r}$ be a function of a scalar variable u, and u is also a scalar function of scalar variable t.

$\therefore \qquad\qquad\qquad r = f(u) \qquad\qquad\qquad\qquad\qquad\qquad ...(1)$

and $\qquad\qquad\qquad u = g(t) \qquad\qquad\qquad\qquad\qquad\qquad ...(2)$

$\therefore \qquad\qquad\qquad r + \delta r = f(u + \delta u) \qquad\qquad\qquad\qquad ...(3)$

and $\qquad\qquad\qquad u + \delta u = g(t + \delta t) \qquad\qquad\qquad\qquad ...(4)$

Subtract (1) from (3), we get

$$\delta \boldsymbol{r} = \boldsymbol{f}(u + \delta u) - \boldsymbol{f}(u) \qquad \qquad \dots(5)$$

and subtract (2) from (4), we get

$$\delta u = g(t + \delta t) - g(t) \qquad \qquad \dots(6)$$

Now divide (5) by δt, we have

$$\frac{\delta \boldsymbol{r}}{\delta t} = \frac{\boldsymbol{f}(u + \delta u) - \boldsymbol{f}(u)}{\delta t} = \frac{\boldsymbol{f}(u + \delta u) - \boldsymbol{f}(u)}{\delta u} \cdot \frac{\delta u}{\delta t}$$

$$\frac{\delta \boldsymbol{r}}{\delta t} = \frac{\boldsymbol{f}(u + \delta u) - \boldsymbol{f}(u)}{\delta u} \cdot \frac{g(t + \delta t) - g(t)}{\delta t} \qquad \qquad \text{[Using (6)]}$$

Taking the limit $\delta t \to 0$, when $\delta t \to 0$, $\delta \boldsymbol{r} \to 0$ and $\delta u \to 0$, we get

$$\lim_{\delta t \to 0} \frac{\delta \boldsymbol{r}}{\delta t} = \lim_{\delta u \to 0} \frac{\boldsymbol{f}(u + \delta u) - \boldsymbol{f}(u)}{\delta u} \cdot \lim_{\delta t \to 0} \frac{g(t + \delta t) - g(t)}{\delta t}$$

$$\frac{d\boldsymbol{r}}{dt} = \frac{d\boldsymbol{f}}{du} \frac{dg}{dt} \qquad \text{or} \qquad \frac{d\boldsymbol{r}}{dt} = \frac{d\boldsymbol{r}}{du} \frac{du}{dt} \qquad \qquad [\because \boldsymbol{r} = \boldsymbol{f}(u), u = g(t)]$$

THEOREM 1. *The vector $\boldsymbol{a}(t)$ has a constant magnitude if and only if $\boldsymbol{a} \cdot \dfrac{d\boldsymbol{a}}{dt} = 0$.*

PROOF. Let us suppose $\boldsymbol{a}(t)$ has a constant magnitude. Therefore, $|\boldsymbol{a}(t)| = a(\text{constant})$

or $$\boldsymbol{a} \cdot \boldsymbol{a} = a^2 (\text{constant})$$

$\therefore$ $$\frac{d}{dt}(\boldsymbol{a} \cdot \boldsymbol{a}) = \boldsymbol{a} \cdot \frac{d\boldsymbol{a}}{dt} + \frac{d\boldsymbol{a}}{dt} \cdot \boldsymbol{a} = 2\boldsymbol{a} \cdot \frac{d\boldsymbol{a}}{dt} \qquad \qquad (\because \boldsymbol{a} \cdot \boldsymbol{b} = \boldsymbol{b} \cdot \boldsymbol{a})$$

Since $$\boldsymbol{a} \cdot \boldsymbol{a} = a^2$$

$$\frac{d}{dt}(\boldsymbol{a} \cdot \boldsymbol{a}) = \frac{d}{dt}(a^2) = 0.$$

$\therefore$ $$2\boldsymbol{a} \cdot \frac{d\boldsymbol{a}}{dt} = 0 \qquad \text{or} \qquad \boldsymbol{a} \cdot \frac{d\boldsymbol{a}}{dt} = 0$$

Conversely, suppose $\boldsymbol{a} \cdot \dfrac{d\boldsymbol{a}}{dt} = 0$, then we get

$$\frac{d}{dt}(\boldsymbol{a} \cdot \boldsymbol{a}) = \boldsymbol{a} \cdot \frac{d\boldsymbol{a}}{dt} + \frac{d\boldsymbol{a}}{dt} \cdot \boldsymbol{a} = 2\boldsymbol{a} \cdot \frac{d\boldsymbol{a}}{dt} \qquad \qquad (\boldsymbol{a}.\boldsymbol{b} = \boldsymbol{b}.\boldsymbol{a})$$

$$\frac{d}{dt}(\boldsymbol{a}.\boldsymbol{a}) = 0 \qquad \text{or} \qquad \boldsymbol{a}.\boldsymbol{a} = \text{constant}$$

or $$|\boldsymbol{a}|^2 = \text{constant} \qquad \text{or} \qquad |\boldsymbol{a}| = \text{constant}$$

THEOREM 2. *The vector function $\boldsymbol{a}(t)$ is constant vector if and only if $\dfrac{d\boldsymbol{a}}{dt} = 0$.*

PROOF. Let us suppose first $\boldsymbol{a}(t)$ is a constant vector such that $\boldsymbol{a}(t) = \boldsymbol{c}$ where c is a constant vector, then

$$\boldsymbol{a}(t + \delta t) = \boldsymbol{c}$$

$\therefore$ $$\boldsymbol{a}(t + \delta t) - \boldsymbol{a}(t) = \boldsymbol{c} - \boldsymbol{c} = \boldsymbol{0}.$$

Divide by δt and taking the limit as $\delta t \to 0$, we get

$$\lim_{\delta t \to 0} \frac{\boldsymbol{a}(t + \delta t) - \boldsymbol{a}(t)}{\delta t} = \boldsymbol{0} \implies \frac{d\boldsymbol{a}}{dt} = \boldsymbol{0}.$$

Conversely, suppose $\dfrac{d\boldsymbol{a}}{dt} = \boldsymbol{0}$.

Let $$\boldsymbol{a}(t) = a_1(t)\hat{i} + a_2(t)\hat{j} + a_3(t)\hat{k}$$

$\therefore$ $$\frac{d\boldsymbol{a}}{dt} = \frac{da_1(t)}{dt}\hat{i} + \frac{da_2(t)}{dt}\hat{j} + \frac{da_3(t)}{dt}\hat{k}$$

$\therefore$ $$\frac{da_1(t)}{dt}\hat{i} + \frac{da_2(t)}{dt}\hat{j} + \frac{da_3(t)}{dt}\hat{k} = \boldsymbol{0} \qquad \qquad \left(\because \frac{d\boldsymbol{a}}{dt} = 0\right)$$

This implies $$\frac{da_1}{dt} = 0, \frac{da_2}{dt} = 0, \frac{da_3}{dt} = 0.$$

Therefore, a_1, a_2, a_3 are all constants.

Hence $\boldsymbol{a}(t) = a_1\hat{i} + a_2\hat{j} + a_3\hat{k}$ is a constant vector.

THEOREM 3. *If the vector $\boldsymbol{a}$ has a constant magnitude $|\boldsymbol{a}| = a$, then $\boldsymbol{a}$ and $\dfrac{d\boldsymbol{a}}{dt}$ are perpendicular, provided $\left|\dfrac{d\boldsymbol{a}}{dt}\right| \neq 0$.*

PROOF.   Since, we have that $|\boldsymbol{a}| = a$ (constant), then

$$\boldsymbol{a}\cdot\boldsymbol{a} = |\boldsymbol{a}|^2 = a^2 \text{ (constant)}$$

$$\therefore \qquad \frac{d}{dt}(\boldsymbol{a}\cdot\boldsymbol{a}) = \boldsymbol{a}\cdot\frac{d\boldsymbol{a}}{dt} + \frac{d\boldsymbol{a}}{dt}\cdot\boldsymbol{a} = 2\boldsymbol{a}\cdot\frac{d\boldsymbol{a}}{dt}$$

and $$\frac{d}{dt}(\boldsymbol{a}\cdot\boldsymbol{a}) = \frac{d}{dt}(a^2) = 0$$

$$\therefore \qquad 2\boldsymbol{a}\cdot\frac{d\boldsymbol{a}}{dt} = 0 \qquad \text{or} \qquad \boldsymbol{a}\cdot\frac{d\boldsymbol{a}}{dt} = 0.$$

This implies that vector $\boldsymbol{a}$ is perpendicular to $\frac{d\boldsymbol{a}}{dt}$, provided $\left|\frac{d\boldsymbol{a}}{dt}\right| \neq 0$.

THEOREM 4.   *If a vector $\boldsymbol{a}$ is a differentiable vector function of t, then* $\frac{d}{dt}\left(\boldsymbol{a}\times\frac{d\boldsymbol{a}}{dt}\right) = \boldsymbol{a}\times\frac{d^2\boldsymbol{a}}{dt^2}$.

PROOF.   Since, we have $\frac{d}{dt}(\boldsymbol{a}\times\boldsymbol{b}) = \boldsymbol{a}\times\frac{d\boldsymbol{b}}{dt} + \frac{d\boldsymbol{a}}{dt}\times\boldsymbol{b}$.

$$\therefore \quad \frac{d}{dt}\left(\boldsymbol{a}\times\frac{d\boldsymbol{a}}{dt}\right) = \boldsymbol{a}\times\frac{d}{dt}\left(\frac{d\boldsymbol{a}}{dt}\right) + \frac{d\boldsymbol{a}}{dt}\times\frac{d\boldsymbol{a}}{dt} = \boldsymbol{a}\times\frac{d^2\boldsymbol{a}}{dt^2} + \frac{d\boldsymbol{a}}{dt}\times\frac{d\boldsymbol{a}}{dt} = \boldsymbol{a}\times\frac{d^2\boldsymbol{a}}{dt^2}$$

$$\left(\because \frac{d\boldsymbol{a}}{dt}\times\frac{d\boldsymbol{a}}{dt} = 0 \text{ i.e., cross product of two same vector is zero.}\right)$$

THEOREM 5.   *The vector $\boldsymbol{a}(t)$ has a constant direction if and only if $\boldsymbol{a}\times\frac{d\boldsymbol{a}}{dt} = \boldsymbol{0}$.*

PROOF.   Suppose $\boldsymbol{a}(t)$ has a constant direction. Let $\hat{a}$ be the unit vector along $\boldsymbol{a}(t)$ and $|\boldsymbol{a}(t)| = a$, then

$$\boldsymbol{a}(t) = a\hat{a}.$$

$$\therefore \qquad \frac{d\boldsymbol{a}}{dt} = \frac{d}{dt}(a\hat{a}) = \frac{da}{dt}\hat{a} + a\frac{d\hat{a}}{dt}$$

$$\therefore \qquad \boldsymbol{a}\times\frac{d\boldsymbol{a}}{dt} = \boldsymbol{a}\times\left(\frac{da}{dt}\hat{a} + a\frac{d\hat{a}}{dt}\right) = \frac{da}{dt}\boldsymbol{a}\times\hat{a} + a\boldsymbol{a}\times\frac{d\hat{a}}{dt}$$

$$\text{or} \qquad \boldsymbol{a}\times\frac{d\boldsymbol{a}}{dt} = a\left(\boldsymbol{a}\times\frac{d\hat{a}}{dt}\right). \qquad\qquad \left(\because \boldsymbol{a}\times\hat{a} = \boldsymbol{a}\times\frac{\boldsymbol{a}}{a} = 0\right) \qquad\qquad …(1)$$

Since, $\boldsymbol{a}$ has a constant direction, then $\hat{a}$ is a constant vector, and thus we have $\frac{d\hat{a}}{dt} = \boldsymbol{0}$.

$$\therefore \qquad \boldsymbol{a}\times\frac{d\boldsymbol{a}}{dt} = a(\boldsymbol{a}\times\boldsymbol{0}) = \boldsymbol{0}.$$

Conversely, suppose $\boldsymbol{a}\times\frac{d\boldsymbol{a}}{dt} = 0$, then from (1)

$$a\left(\boldsymbol{a}\times\frac{d\hat{a}}{dt}\right) = 0 \qquad \text{or} \quad \boldsymbol{a}\times\frac{d\hat{a}}{dt} = 0 \quad \text{or} \quad \hat{a}\times\frac{d\hat{a}}{dt} = 0 \qquad\qquad …(2)$$
$$[\because \boldsymbol{a} = a\hat{a}]$$

Since $\hat{a}$ has a constant magnitude, then by theorem (1)

$$\hat{a}\times\frac{d\hat{a}}{dt} = 0. \qquad\qquad …(3)$$

From (2) and (3), we get $\frac{d\hat{a}}{dt} = \boldsymbol{0}$.

This implies $\hat{a}$ is a constant vector hence $\boldsymbol{a}$ has a constant directions.

6.1.11 Curves in Three Dimensional Space

Let $f(x, y, z) = 0$ and $\phi(x, y, z) = 0$ be two surfaces, then a curve in three dimensional space is obtained by the intersection of the surfaces $f(x, y, z) = 0$ and $\phi(x, y, z) = 0$. Therefore, the equation is of the form

$$x = f_1(t), y = f_2(t), z = f_3(t) \qquad\qquad …(1)$$

and it also represents a curve in three dimensional space. Where t takes the value between a and b, i.e. $a \leq t \leq b$. Let (x, y, z) be any point on the curve (1) and let $\boldsymbol{r}$ be its position vector, then we have

$$\boldsymbol{r} = x\hat{i} + y\hat{j} + z\hat{k} \qquad \text{and} \qquad \boldsymbol{f}(t) = f_1(t)\hat{i} + f_2(t)\hat{j} + f_3(t)\hat{k}.$$

$\therefore$ From (1), we have

$$x\hat{i} + y\hat{j} + z\hat{k} = f_1(t)\hat{i} + f_2(t)\hat{j} + f_3(t)\hat{k} \qquad \text{or} \qquad \boldsymbol{r} = \boldsymbol{f}(t). \qquad\qquad …(2)$$

Thus equation (2) represents a curve in three dimensional space.

GEOMETRICAL INTERPRETATION OF $\dfrac{dr}{dt}$

Let $r = f(t)$ be a curve in three dimensional space and let P and Q be two neighbouring points on this curve and r and $r + \delta r$ be the position vectors of P and Q respectively as shown in fig. 1.

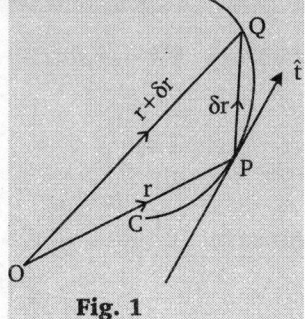

Fig. 1

$$\therefore \qquad r = \overline{OP} = f(t) \qquad \text{and} \qquad \overline{OQ} = r + \delta r = f(t + \delta t)$$

$$\therefore \qquad \overline{PQ} = \overline{OQ} - \overline{OP} = r + \delta r - r = \delta r.$$

Thus $\dfrac{\delta r}{\delta t}$ is a vector parallel to the vector $\overline{PQ}$ (which is a chord).

As $\delta t \to 0$ *i.e.*, $Q \to P$, then the chord PQ tending to a line at P to the curve. This line is known as tangent.

$$\therefore \qquad \lim_{\delta t \to 0} \frac{\delta r}{\delta t} = \frac{dr}{dt}.$$

Hence, $\dfrac{dr}{dt}$ is a vector which is parallel to the line (tangent) at P to the curve $r = f(t)$.

SIGNIFICANCE OF $\dfrac{dr}{ds}$

Let C be any fixed point on the curve as shown in fig. 1 at $r = f(s)$, where s is the arc length measured from the point C. Then

$$CP = s, \; CQ = s + \delta s.$$

Thus $\dfrac{dr}{ds}$ is a vector along the tangent at P to the curve $r = f(s)$ in the direction of s increasing.

$$\therefore \qquad \frac{dr}{ds} = \lim_{\delta x \to 0} \frac{\delta r}{\delta s} = \lim_{Q \to P} \frac{\delta r}{\text{arc } PQ} \qquad (\because \text{ as } \delta s \to 0, Q \to P \text{ and } \delta s = \text{arc length } PQ)$$

$$\text{or} \qquad \left| \frac{dr}{ds} \right| = \lim_{Q \to P} \frac{|\delta r|}{\text{arc } PQ} = \lim_{Q \to P} \frac{\text{chord } PQ}{\text{arc } PQ} \qquad (\because \; \delta r = \overline{PQ})$$

Thus $\dfrac{dr}{ds}$ is a unit vector along the tangent at P to the curve $r = f(s)$. Thus unit vector is known as unit tangent vector and is denoted by $\hat{t}$.

Hence, $$\hat{t} = \frac{dr}{ds}.$$

6.2 VELOCITY AND ACCELERATION VECTORS (GBTU–2010)

Let a particle be moving along the curve $r = f(t)$. At any instant t the moving particle is at P whose position vector is r. In time interval δt the moving particle reached to the point Q whose position vector is $r + \delta r$.

$\therefore$ δr is the displacement of the moving particle in the time interval δt.

Thus $\dfrac{\delta r}{\delta t}$ gives an average velocity of the particle during the interval δt. If the vector v represents the velocity vector at P, then

$$v = \lim_{\delta t \to 0} \frac{\delta r}{\delta t} = \frac{dr}{dt}.$$

Since $\dfrac{dr}{dt}$ is a vector along the tangent at P to the curve. Hence, the vector v of the particle always along the tangent.

Further, if δv is the change in velocity during the time interval δt, then $\dfrac{\delta v}{\delta t}$ represents a vector which gives an average acceleration of the particle. Let a be the acceleration vector of the particle, then we have

$$a = \lim_{\delta t \to 0} \frac{\delta v}{\delta t} = \frac{dv}{dt}.$$

Since $\quad v = \dfrac{dr}{dt} \quad$ therefore $\quad a = \dfrac{d}{dt}\left(\dfrac{dr}{dt}\right) = \dfrac{d^2 r}{dt^2}.$

Hence $$a = \frac{dv}{dt} = \frac{d^2 r}{dt^2}.$$

REMARKS

- The equation $\mathbf{r} = (a\cos t)\hat{i} + (b\sin t)\hat{j} + 0.\hat{k}$ represents the equation of an ellipse.
- The equation $\mathbf{r} = (a\sec t)\hat{i} + (b\tan t)\hat{j} + 0.\hat{k}$ represents the equation of hyperbola.
- The equation $\mathbf{r} = (at^2)\hat{i} + (2at)\hat{j} + 0.\hat{k}$ represents the equation of a parabola.

Solved Examples

EXAMPLE 1. *If* $\mathbf{r} = (2\sin t)\hat{i} + (3\cos t)\hat{j} + t\hat{k}$, *find*

(i) $\dfrac{d\mathbf{r}}{dt}$

(ii) $\left|\dfrac{d\mathbf{r}}{dt}\right|$

(iii) $\dfrac{d^2\mathbf{r}}{dt^2}$

(iv) $\left|\dfrac{d^2\mathbf{r}}{dt^2}\right|$

SOLUTION. Since we know that $\hat{i}, \hat{j}, \hat{k}$ are constant vectors, so

$$\frac{d\hat{i}}{dt} = \mathbf{0}, \frac{d\hat{j}}{dt} = \mathbf{0} \text{ and } \frac{d\hat{k}}{dt} = \mathbf{0}.$$

(i) $\quad \mathbf{r} = (2\sin t)\hat{i} + (3\cos t)\hat{j} + t\hat{k},$

$$\therefore \quad \frac{d\mathbf{r}}{dt} = (2\cos t)\hat{i} - (3\sin t)\hat{j} + \hat{k}$$

(ii) $\quad \left|\dfrac{d\mathbf{r}}{dt}\right| = \sqrt{4\cos^2 t + 9\sin^2 t + 1}$

$$= \sqrt{5(1 + \sin^2 t)}.$$

(iii) $\quad \dfrac{d^2\mathbf{r}}{dt^2} = \dfrac{d}{dt}\left(\dfrac{d\mathbf{r}}{dt}\right)$

$$= \frac{d}{dt}(2\cos t\hat{i} - 3\sin t\hat{j} + \hat{k})$$

$$= -2\sin t\hat{i} - 3\cos t\hat{j}.$$

(iv) $\quad \left|\dfrac{d^2\mathbf{r}}{dt^2}\right| = \sqrt{(4\sin^2 t + 9\cos^2 t)}$

$$= \sqrt{(4 + 5\cos^2 t)}.$$

EXAMPLE 2. *If* $\hat{r}$ *be a unit vector in the direction of* $\mathbf{r}$, *prove that*
$$\hat{r} \times \frac{d\hat{r}}{dt} = \frac{1}{r^2}\mathbf{r} \times \frac{d\mathbf{r}}{dt} \text{ where } |\mathbf{r}| = r.$$

SOLUTION. Since $\hat{r}$ is a unit vector along the vector $\mathbf{r}$, so we have

$$\mathbf{r} = r\hat{r} \qquad \dots(1)$$

$$\because \quad |\mathbf{r}| = r.$$

Differentiating w.r.t. t of both sides, we get

$$\frac{d\mathbf{r}}{dt} = \frac{d}{dt}(r\hat{r})$$

$$\therefore \quad \frac{d\mathbf{r}}{dt} = r\frac{d\hat{r}}{dt} + \frac{dr}{dt}\hat{r} \qquad \dots(2)$$

Now $\mathbf{r} \times \dfrac{d\mathbf{r}}{dt} = \mathbf{r} \times \left(r\dfrac{d\hat{r}}{dt} + \dfrac{dr}{dt}\hat{r}\right)$

$$= r\mathbf{r} \times \frac{d\hat{r}}{dt} + \frac{dr}{dt}\mathbf{r} \times \hat{r}$$

$$= r(r\hat{r}) \times \frac{d\hat{r}}{dt} + \frac{dr}{dt}r\hat{r} \times \hat{r} \qquad (\because \mathbf{r} = r\hat{r})$$

$$= r^2\hat{r} \times \frac{d\hat{r}}{dt} + 0$$

($\because$ Cross product of same vector is zero, *i.e.*, $\hat{r} \times \hat{r} = \mathbf{0}$)

$$= r^2\hat{r} \times \frac{d\hat{r}}{dt}$$

$$\therefore \quad \hat{r} \times \frac{d\hat{r}}{dt} = \frac{1}{r^2}\mathbf{r} \times \frac{d\mathbf{r}}{dt}$$

EXAMPLE 3. *If* $\mathbf{r} = \mathbf{a}\sin\omega t + \mathbf{b}\cos\omega t + \dfrac{\mathbf{c}t}{\omega^2}\sin\omega t$, *prove that*

$$\frac{d^2\mathbf{r}}{dt^2} + \omega^2\mathbf{r} = \frac{2\mathbf{c}}{\omega}\cos\omega t,$$

where $\mathbf{a}, \mathbf{b}, \mathbf{c}$, *are constant vectors and* ω *is a constant scalar.*

SOLUTION. Since $\mathbf{a}, \mathbf{b}, \mathbf{c}$, are constant vectors so $\dfrac{d\mathbf{a}}{dt} = 0, \dfrac{d\mathbf{b}}{dt} = 0$ and

$$\frac{d\mathbf{c}}{dt} = 0.$$

and $\quad \mathbf{r} = \mathbf{a}\sin\omega t + \mathbf{b}\cos\omega t + \dfrac{\mathbf{c}t}{\omega^2}\sin\omega t \qquad \dots(1)$

$$\therefore \quad \frac{d\mathbf{r}}{dt} = \omega\mathbf{a}\cos\omega t - \omega\mathbf{b}\sin\omega t$$

$$+ \frac{\mathbf{c}}{\omega^2}\sin\omega t + \frac{\mathbf{c}t}{\omega}\cos\omega t$$

and $\dfrac{d^2\mathbf{r}}{dt^2} = \dfrac{d}{dt}\left(\dfrac{d\mathbf{r}}{dt}\right)$

$$= -\omega^2\mathbf{a}\sin\omega t - \omega^2\mathbf{b}\cos\omega t$$

$$+ \frac{\mathbf{c}}{\omega}\cos\omega t + \frac{\mathbf{c}}{\omega}\cos\omega t - \mathbf{c}t\sin\omega t$$

$$= -\omega^2\left(\mathbf{a}\sin\omega t + \mathbf{b}\cos\omega t + \frac{\mathbf{c}t}{\omega^2}\sin\omega t\right)$$

$$+ \frac{2\mathbf{c}}{\omega}\cos\omega t$$

$$= -\omega^2\mathbf{r} + \frac{2\mathbf{c}}{\omega}\cos\omega t$$

$$\therefore \quad \frac{d^2\mathbf{r}}{dt^2} + \omega^2\mathbf{r} = \frac{2\mathbf{c}}{\omega}\cos\omega t.$$

EXAMPLE 4. *If* $\dfrac{d\mathbf{a}}{dt} = \mathbf{c} \times \mathbf{a}, \dfrac{d\mathbf{b}}{dt} = \mathbf{c} \times \mathbf{b}$ *show that*

$$\frac{d}{dt}(\mathbf{a} \times \mathbf{b}) = \mathbf{c} \times (\mathbf{a} \times \mathbf{b}).$$

SOLUTION. Since we know that

$$\frac{d}{dt}(\mathbf{a} \times \mathbf{b}) = \mathbf{a} \times \frac{d\mathbf{b}}{dt} + \frac{d\mathbf{a}}{dt} \times \mathbf{b}$$

$$= \mathbf{a} \times (\mathbf{c} \times \mathbf{b}) + (\mathbf{c} \times \mathbf{a}) \times \mathbf{b}$$

$$\left(\because \frac{d\mathbf{a}}{dt} = \mathbf{c} \times \mathbf{a}, \frac{d\mathbf{b}}{dt} = \mathbf{c} \times \mathbf{b}\right)$$

$$= [(\mathbf{a} \cdot \mathbf{b})\mathbf{c} - (\mathbf{a} \cdot \mathbf{c})\mathbf{b}] - [(\mathbf{b} \cdot \mathbf{a})\mathbf{c} - (\mathbf{b} \cdot \mathbf{c})\mathbf{a}]$$

$$= (\mathbf{a} \cdot \mathbf{b})\mathbf{c} - (\mathbf{a} \cdot \mathbf{c})\mathbf{b} - (\mathbf{a} \cdot \mathbf{b})\mathbf{c} + (\mathbf{b} \cdot \mathbf{c})\mathbf{a}$$

$$(\because \mathbf{a} \cdot \mathbf{b} = \mathbf{b} \cdot \mathbf{a})$$

$$= (\mathbf{b} \cdot \mathbf{c})\mathbf{a} - (\mathbf{a} \cdot \mathbf{c})\mathbf{b}$$

$$= (c \cdot b)a - (c \cdot a)b$$

$$(\because \text{ dot products are commutative.})$$

$$= c \times (a \times b).$$

$$\therefore \quad \frac{d}{dt}(a \times b) = c \times (a \times b).$$

EXAMPLE 5. *Show that* $\dfrac{d}{dt}\left[a.\left(\dfrac{da}{dt} \times \dfrac{d^2a}{dt^2}\right)\right] = a.\left(\dfrac{da}{dt} \times \dfrac{d^3a}{dt^3}\right).$

SOLUTION. Let $\quad A = \left[a.\left(\dfrac{da}{dt} \times \dfrac{d^2a}{dt^2}\right)\right]$

Then A is the scalar triple product of three vectors a, $\dfrac{da}{dt}$ and $\dfrac{d^2a}{dt^2}$. Therefore using the rule for finding the derivative of a scalar triple product, we have

$$\frac{dA}{dt} = \frac{da}{dt}.\left(\frac{da}{dt} \times \frac{d^2a}{dt^2}\right)$$

$$+ a.\left(\frac{d^2a}{dt^2} \times \frac{d^2a}{dt^2}\right) + a.\left(\frac{da}{dt} \times \frac{d^3a}{dt^3}\right)$$

$$= a.\left(\frac{da}{dt} \times \frac{d^3a}{dt^3}\right),$$

since scalar triple products having two equal vectors vanish.

EXAMPLE 6. *A particle moves along the curve*
$$x = 2t^2,\ y = t^2 - 4t,\ z = 3t - 5,$$
where t is the time. Find the magnitude of the velocity and acceleration at t = 0. (VTU–2008)

SOLUTION. Since $\quad r = x\hat{i} + y\hat{j} + z\hat{k}$

$$= 2t^2\hat{i} + (t^2 - 4t)\hat{j} + (3t - 5)\hat{k}$$

Velocity vector

$$v = \frac{dr}{dt} = 4t\hat{i} + (2t - 4)\hat{j} + 3\hat{k}$$

Acceleration vector

$$a = \frac{dv}{dt} = 4\hat{i} + 2\hat{j} + 0\hat{k}$$

$$\therefore \quad \text{Velocity} = |v| = \sqrt{(4t)^2 + (2t-4)^2 + 3^2}$$
and
$$\text{Acceleration} = |a| = \sqrt{(4)^2 + (2)^2 + 0} = \sqrt{20}$$

$$\therefore \text{ at } t = 0, \text{ velocity} = \sqrt{0 + (0-4)^2 + 9} = 5$$

and acceleration $= \sqrt{20} = 2\sqrt{5}.$

EXAMPLE 7. *If* $A = x^2yz\hat{i} - 2xz^3\hat{j} + xz^2\hat{k}$ *and*

$B = 2z\hat{i} + y\hat{j} - x^2\hat{k}$ *find the value of the*

$$\frac{\partial^2}{\partial x \partial y}(A \times B) \text{ at the point (1, 0, 2).}$$

SOLUTION. We have $A \times B = \begin{vmatrix} \hat{i} & \hat{j} & \hat{k} \\ x^2yz & -2xz^3 & xz^2 \\ 2z & y & -x^2 \end{vmatrix}$

$$= \hat{i}(2x^3z^3 - xyz^2) - \hat{j}(-x^4yz - 2xz^3)$$
$$+ \hat{k}(x^2y^2z + 2xz^4)$$

Then

$$\frac{\partial}{\partial y}(A \times B) = \hat{i}(-xz^2) - \hat{j}(-x^4z) + \hat{k}(2x^2yz)$$

and $\dfrac{\partial^2}{\partial x \partial y}(A \times B) = \hat{i}(-z^2) - \hat{j}(-4x^3z) + \hat{k}(4xyz)$

$\therefore$ At the point (1, 0, 2)

$$\frac{\partial^2}{\partial x \partial y}(A \times B) = -4\hat{i} + 8\hat{j}$$

EXAMPLE 8. *Show that the radial and transverse acceleration of a particle moving in a plane are*

$$\frac{d^2r}{dt^2} - r\left(\frac{d\theta}{dt}\right)^2 \text{ and } 2\frac{dr}{dt}.\frac{d\theta}{dt} + r\frac{d^2\theta}{dt^2} \text{ respectively.}$$

(UPTU–2008, KURUKSHETRA–2006, RAJASTHAN–2005, 2006)

SOLUTION. Let $\hat{e}_r$ and $\hat{e}_\theta$ be two unit vectors along and perpendicular to radius vector OP where $\hat{i}$ and $\hat{j}$ be two mutually perpendicular unit vectors in the plane.

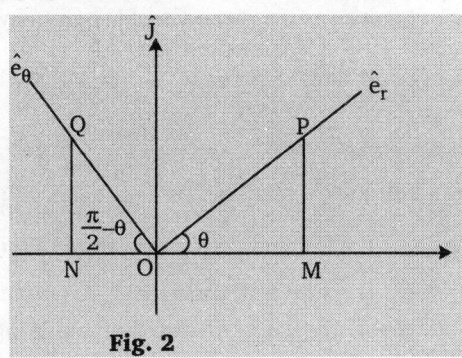

Fig. 2

Then $\quad \hat{e}_r = \overrightarrow{OP} = \overrightarrow{OM} + \overrightarrow{MP}$

$$= \cos\theta\hat{i} + \sin\theta\hat{j} \qquad \ldots(1)$$

and $\quad \hat{e}_\theta = \cos\left(\dfrac{\pi}{2} + \theta\right)\hat{i} + \sin\left(\dfrac{\pi}{2} + \theta\right)\hat{j}$

$$= -\sin\theta\hat{i} + \cos\theta\hat{j} \qquad \ldots(2)$$

Also $\quad r = r\hat{e}_r = r\cos\theta\hat{i} + r\sin\theta\hat{j} \qquad \ldots(3)$

$$\frac{d\hat{e}_r}{dt} = (-\sin\theta\hat{i} + \cos\theta\hat{j})\frac{d\theta}{dt} \qquad \ldots(4)$$

$$= \hat{e}_\theta\frac{d\theta}{dt} = \dot{\theta}\hat{e}_\theta$$

$$\frac{d\hat{e}_\theta}{dt} = (-\cos\theta\hat{i} - \sin\theta\hat{j})\frac{d\theta}{dt} \qquad \ldots(5)$$

$$= -\hat{e}_r\frac{d\theta}{dt} = -\dot{\theta}\hat{e}_r$$

Now, $\quad \dfrac{dr}{dt} = \dfrac{d}{dt}(r\cos\theta\hat{i} + r\sin\theta\hat{j})$

$$= \left(\frac{dr}{dt}\cos\theta - r\sin\theta\frac{d\theta}{dt}\right)\hat{i}$$

$$+ \left(\frac{dr}{dt}\sin\theta + r\cos\theta\frac{d\theta}{dt}\right)\hat{j}$$

$$= \frac{dr}{dt}\left(\cos\theta\hat{i} + \sin\theta\hat{j}\right)$$

$$+ r\frac{d\theta}{dt}\left(-\sin\theta\hat{i} + \cos\theta\hat{j}\right)$$

$$= \dot{r}\hat{e}_r + r\dot{\theta}\hat{e}_\theta$$

∴ Acceleration

$$\frac{d^2\boldsymbol{r}}{dt^2} = \ddot{r}\hat{e}_r + \dot{r}\frac{d\hat{e}_r}{dt} + \dot{r}\dot{\theta}\hat{e}_\theta + r\ddot{\theta}\hat{e}_\theta + r\dot{\theta}\frac{d\hat{e}_\theta}{dt}$$

$$= \ddot{r}\hat{e}_r + 2\dot{r}\dot{\theta}\hat{e}_\theta + r\ddot{\theta}\hat{e}_\theta - r\dot{\theta}^2\hat{e}_r$$

$$= (\ddot{r} - r\dot{\theta}^2)\hat{e}_r + (2\dot{r}\dot{\theta} + r\ddot{\theta})\hat{e}_\theta$$

$$= \left[\frac{d^2\boldsymbol{r}}{dt^2} - \boldsymbol{r}\left(\frac{d\theta}{dt}\right)^2\right]\hat{e}_r + \frac{1}{\boldsymbol{r}}\frac{d}{dt}\left(\boldsymbol{r}^2\frac{d\theta}{dt}\right)\hat{e}_\theta$$

Hence, Radial acceleration

= Acceleration along $\hat{e}_r$

$$= \frac{d^2\boldsymbol{r}}{dt^2} - \boldsymbol{r}\left(\frac{d\theta}{dt}\right)^2.$$

Transverse acceleration

= Acceleration along $\hat{e}_\theta$

$$= \frac{1}{\boldsymbol{r}}\frac{d}{dt}\left(\boldsymbol{r}^2\frac{d\theta}{dt}\right)$$

$$= 2\frac{d\boldsymbol{r}}{dt}\cdot\frac{d\theta}{dt} + \boldsymbol{r}\frac{d^2\theta}{dt^2}.$$

Exercise-6.1

1. If $\mathbf{r} = (t+1)\hat{i} + (t^2+t+1)\hat{j} + (t^3+t^2+t+1)\hat{k}$, find $\dfrac{d\boldsymbol{r}}{dt}, \dfrac{d^2\boldsymbol{r}}{dt^2}$.

2. If $\boldsymbol{a}, \boldsymbol{b}$ are constant vectors, ω is a constant, and $\boldsymbol{r}$ is a vector function of the scalar variable t given by
$$\boldsymbol{r} = \cos\omega t\,\boldsymbol{a} + \sin\omega t\,\boldsymbol{b}$$
Show that
   (i) $\dfrac{d^2\boldsymbol{r}}{dt^2} + \omega^2\boldsymbol{r} = \boldsymbol{0}$

   (ii) $\boldsymbol{r} \times \dfrac{d\boldsymbol{r}}{dt} = \omega\boldsymbol{a} \times \boldsymbol{b}$ (UPTU–2007, BHOPAL–2007)

3. (i) If $\boldsymbol{r}$ is a unit vector, then show that $\left|\boldsymbol{r} \times \dfrac{d\boldsymbol{r}}{dt}\right| = \left|\dfrac{d\boldsymbol{r}}{dt}\right|$.

   (ii) If $\boldsymbol{r} \times d\boldsymbol{r} = \boldsymbol{0}$, show that $\hat{r} =$ constant.

4. If $\boldsymbol{r}$ is the position vector of a moving point and r is the modulus of $\boldsymbol{r}$, show that $\boldsymbol{r} \cdot \dfrac{d\boldsymbol{r}}{dt} = r\dfrac{dr}{dt}$.

5. If $\boldsymbol{r}$ is a vector function of a scalar variable t and $\boldsymbol{a}$ is a constant vector, differentiate the following with respect to t :
   (i) $\boldsymbol{r} \times \boldsymbol{a}$ (ii) $\boldsymbol{r} \times \dfrac{d\boldsymbol{r}}{dt}$

   (iii) $\dfrac{\boldsymbol{r} \times \boldsymbol{a}}{\boldsymbol{r}.\boldsymbol{a}}$ (iv) $r^3\boldsymbol{r} + \boldsymbol{a} \times \dfrac{d\boldsymbol{r}}{dt}$

6. (i) If $\boldsymbol{r} = \sin t\hat{i} + \cos t\hat{j} + t\hat{k}$, find $\left|\dfrac{d\boldsymbol{r}}{dt} \times \dfrac{d^2\boldsymbol{r}}{dt^2}\right|, \left|\dfrac{d^2\boldsymbol{r}}{dt^2}\right|$.

   (ii) If $\boldsymbol{r} = a\cos t\hat{i} + a\sin t\hat{j} + at\tan\alpha\hat{k}$, find $\left|\dfrac{d\boldsymbol{r}}{dt} \times \dfrac{d^2\boldsymbol{r}}{dt^2}\right|$ and
   $\dfrac{d\boldsymbol{r}}{dt} \cdot \left(\dfrac{d^2\boldsymbol{r}}{dt^2} \times \dfrac{d^3\boldsymbol{r}}{dt^3}\right)$.   (ROHTAK–2005)

7. If $\boldsymbol{r} = r^3\hat{i} + \left(2t^3 - \dfrac{1}{5t^2}\right)\hat{j}$, show that $\boldsymbol{r} \times \dfrac{d\boldsymbol{r}}{dt} = \hat{k}$.

8. Show that if $\boldsymbol{a}, \boldsymbol{b}, \boldsymbol{c}$ are constant vectors, then $\boldsymbol{r} = \boldsymbol{a}t^2 + \boldsymbol{b}t + \boldsymbol{c}$ is the path of a particle moving with constant acceleration.

9. If $\boldsymbol{r} = e^{nt}\boldsymbol{a} + e^{-nt}\boldsymbol{b}$, where $\boldsymbol{a}, \boldsymbol{b}$ are constant vectors, show that $\dfrac{d^2\boldsymbol{r}}{dt^2} - n^2\boldsymbol{r} = \boldsymbol{0}$.

10. Show that $\boldsymbol{r} = \boldsymbol{a}e^{nt} + \boldsymbol{b}e^{nt}$ is the solution of the differential equation
$$\frac{d^2\boldsymbol{r}}{dt^2} - (m+n)\frac{d\boldsymbol{r}}{dt} + mn\boldsymbol{r} = 0.$$
Hence solve the equation $\dfrac{d^2\boldsymbol{r}}{dt^2} - \dfrac{d\boldsymbol{r}}{dt} - 2\boldsymbol{r} = 0.$

where $\boldsymbol{r} = \hat{i}$ and $\dfrac{d\boldsymbol{r}}{dt} = \hat{j}$ at $t = 0$.

11. If $\boldsymbol{a} = 5t^2\hat{i} + t\hat{j} - t^3\hat{k}$ and $\boldsymbol{b} = \sin t\hat{i} - \cos t\hat{j}$, then find
   (i) $\dfrac{d}{dt}(\boldsymbol{a}.\boldsymbol{b})$ (ii) $\dfrac{d}{dt}(\boldsymbol{a} \times \boldsymbol{b})$
   (iii) $\dfrac{d}{dt}(\boldsymbol{a}.\boldsymbol{a})$

12. A particle moves along the curve $x = 4\cos t$, $y = 4\sin t$, $z = 6t$. Find the velocity and acceleration at time $t = 0$ and $t = \dfrac{\pi}{2}$.

13. A particle moves along the curve $x = e^{-t}$, $y = 2\cos 3t$, $z = 2\sin 3t$. Find the velocity and acceleration at any time t and their magnitudes at $t = 0$.   (PTU–2003, VTU–2003)

14. Find the unit tangent vector to any point on the curve $x = a\cos t$, $y = a\sin t$, $z = bt$.

15. A particle P is moving on a circle of radius r with constant angular velocity $\omega = \dfrac{d\theta}{dt}$ show that the acceleration is $-\omega^2\boldsymbol{r}$.

16. The position vector of a moving particle at a time t is $\boldsymbol{r} = 3\cos t\hat{i} + 3\sin t\hat{j} + 4t\hat{k}$. Find the tangent and normal components of its acceleration at $t = 1$.   (MARATHWADA–2008)

Hints to Selected Problems

1. $\boldsymbol{r} = (t+1)\hat{i} + (t^2+t+1)\hat{j} + (t^3+t^2+t+1)\hat{k}$

   ∴ $\dfrac{d\boldsymbol{r}}{dt} = \hat{i} + (2t+1)\hat{j} + (3t^2+2t+1)\hat{k}$

       $\dfrac{d^2\boldsymbol{r}}{dt^2} = 2\hat{j} + (6t+2)\hat{k}$.

2. $\boldsymbol{r} = (\cos\omega t)\boldsymbol{a} + (\sin\omega t)\boldsymbol{b}$

        $\dfrac{d\boldsymbol{r}}{dt} = -\omega\sin\omega t\boldsymbol{a} + \omega\cos\omega t\boldsymbol{b}$

        $\dfrac{d^2\boldsymbol{r}}{dt^2} = -\omega^2[\cos\omega t\boldsymbol{a} + \sin\omega t\boldsymbol{b}] = -\omega^2\boldsymbol{r}$

   ∴ $\dfrac{d^2\boldsymbol{r}}{dt^2} + \omega^2\boldsymbol{r} = \boldsymbol{0}$.

Similarly

$$r \times \frac{dr}{dt} = (\cos\omega t\, a + \sin\omega t\, b) \times (-\omega\sin\omega t\, a + \omega\cos\omega t\, b)$$

$$= \omega a \times b.$$

3. (i) Since **r** is a unit vector, then

$$|r| = 1 \qquad \Rightarrow \qquad r.r = 1$$

$$\Rightarrow \quad \frac{d}{dt}(r.r) = 0 \quad \Rightarrow \quad r.\frac{dr}{dt} = 0$$

$$\Rightarrow \quad r \text{ is perpendicular to } \frac{dr}{dt}$$

$$\therefore \quad \left|r \times \frac{dr}{dt}\right| = |r|\left|\frac{dr}{dt}\right|\sin\frac{\pi}{2} = \left|\frac{dr}{dt}\right|.$$

(ii) $r = r\hat{r}$

$$\Rightarrow \quad dr = d(r\hat{r}) = rd\hat{r} + \hat{r}dr$$

$$\therefore \quad r \times dr = r\hat{r} \times (rd\hat{r} + \hat{r}dr) = r^2\hat{r} + d\hat{r}$$

$$\Rightarrow \quad 0 = r^2\hat{r} + d\hat{r} \qquad \Rightarrow \quad \hat{r} \times d\hat{r} = 0$$

$$\therefore \quad r \times dr = 0 \qquad \text{and} \qquad \hat{r} \times d\hat{r} = 0$$

$$\Rightarrow \quad \hat{r} = \text{constant}.$$

4. Since $\qquad r = r\hat{r}$

$$\Rightarrow \quad \frac{dr}{dt} = \frac{d}{dr}(r\hat{r}) = \frac{rd\hat{r}}{dt} + \hat{r}\frac{dr}{dt}$$

$$\therefore \quad r.\frac{dr}{dt} = r.\left(r\frac{d\hat{r}}{dt} + \hat{r}\frac{dr}{dt}\right) = rr.\frac{d\hat{r}}{dt} + r.\hat{r}\frac{dr}{dt}$$

$$= r^2\hat{r}.\frac{d\hat{r}}{dt} + r(r\hat{r})\frac{dr}{dt} = 0 + r\frac{dr}{dt} \qquad [\because \hat{r}.\hat{r} = 1]$$

$$= r\frac{dr}{dt}.$$

6. (i) $\qquad r = \sin t\,\hat{i} + \cos t\,\hat{j} + t\hat{k}$

$$\therefore \quad \frac{dr}{dt} = \cos t\,\hat{i} - \sin t\,\hat{j} + \hat{k}$$

$$\frac{d^2r}{dt^2} = -\sin t\,\hat{i} - \cos t\,\hat{j}$$

$$\therefore \quad \frac{dr}{dt} \times \frac{d^2r}{dt^2} = |\cos t\,\hat{i} - \sin t\,\hat{j} - \hat{k}| = \sqrt{2}.$$

11. (i) $\qquad a = 5t^2\hat{i} + t\hat{j} - t^3\hat{k}$ and $b = \sin t\,\hat{i} - \cos t\,\hat{j}$

$$a.b = 5t^2\sin t - t\cos t$$

$$\therefore \quad \frac{d}{dt}(a.b) = \frac{d}{dt}(5t^2\sin t - t\cos t)$$

$$= 10t\sin t + 5t^2\cos t - \cos t + t\sin t$$

$$= 5t^2\cos t + 11t\sin t - \cos t$$

Similarly, we can find $\frac{d}{dt}(a \times b)$ and $\frac{d}{dt}(a \cdot b)$

13. $\qquad x = e^{-t}, y = 2\cos 3t, z = 2\sin 3t$

$$\therefore \qquad r = e^{-t}\hat{i} + 2\cos 3t\hat{j} + 2\sin 3t\hat{k}$$

$$\text{so} \qquad \frac{dr}{dt} = -e^{-t}\hat{i} - 6\sin 3t\hat{j} + 6\cos 3t\hat{k}$$

$$\text{and} \qquad \frac{d^2r}{dt^2} = e^{-t}\hat{i} - 18\cos 3t\hat{j} - 18\sin 3t\hat{k}$$

$$\text{Velocity} = v = \left(\frac{dr}{dt}\right)_{\text{at } t=0} = -\hat{i} + 6\hat{k}$$

$$\therefore \qquad |v| = \sqrt{1+36} = \sqrt{37}$$

$$\text{and} \qquad \text{acceleration} = \left(\frac{d^2r}{dt^2}\right)_{\text{at } t=0} = \hat{i} - 18\hat{j}$$

$$\therefore \qquad \left|\frac{d^2r}{dt^2}\right| = \sqrt{1+324} = \sqrt{325}.$$

15. Let i and j be the unit vector two perpendicular radii of the circle.

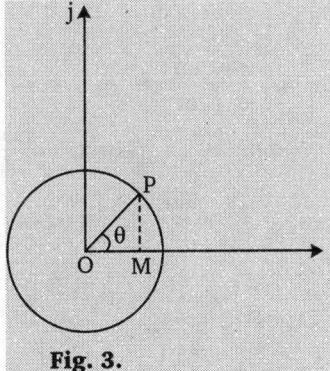

Fig. 3.

If P be any point on the circle such that OP makes an angle θ with the direction of i, then the position vector of P is given by

$$r = \overrightarrow{OP} = \overrightarrow{OM} + \overrightarrow{MP}$$

$$= (r\cos\theta)i + (r\sin\theta)j$$

where r is the radius of circle and hence constant and further

$$\frac{dr}{dt}, \frac{d^2r}{dt^2} = -\omega^2 r.$$

Answers

1. $\frac{dr}{dt} = \hat{i} + (2t+1)\hat{j} + (3t^2 + 2t + 1)\hat{k}$, $\frac{d^2r}{dt^2} = 2\hat{j} + (6t+2)\hat{k}$ **5.** (i) $\frac{dr}{dt} \times a$ (ii) $r \times \frac{d^2r}{dt^2}$ (iii) $\frac{\frac{dr}{dt} \times a}{r.a} - \frac{\frac{dr}{dt}.a}{(r.a)^2}(r \times a)$

(iv) $3r^2\frac{dr}{dt}r + r^3\frac{dr}{dt} + a \times \frac{d^2r}{dt^2}$ **6.**(i) $\sqrt{2}, 1$ (ii)$a2\sec\alpha, a3\tan\alpha$ **10.** $r = \frac{1}{3}(e^{2t} + 2e^{-t})\hat{i} + \frac{1}{3}(e^{2t} - e^{-t})\hat{j}$

11. (i) $(5t^2 - 1)\cos t + 11t\sin t$ (ii) $(t^3\sin t - 3t^2\cos t)\hat{i} - (t^3\cos t + 3t^2\sin t)\hat{j} + (5t^2\sin t - 11t\cos t - \sin t)\hat{k}$ (iii) $100t^3 + 2t + 6t^5$

12. $v = 4\hat{j} + 6\hat{k}, a = -4\hat{i}$ and $v = -4\hat{i} + 6\hat{k}, a = -4\hat{j}$ **13.** $|v| = \sqrt{37}, a = \sqrt{(325)}$ **14.** $\frac{1}{\sqrt{(a^2+b^2)}}[-a\sin t\,\hat{i} + a\cos\hat{j} + b\hat{k}]$ **16.** 0, 3

6.3 INTEGRATION OF A VECTOR FUNCTION

Let $F(t)$ be a differentiable vector function and let $f(t)$ be its differential coefficient, then

$$\frac{d}{dt}(\boldsymbol{F}(t)) = \boldsymbol{f}(t) .$$ ...(1)

Therefore the integral of $\boldsymbol{f}(t)$ is $\boldsymbol{F}(t)$. Consequently we can say that integration is the reverse process of differentiation.

∴ $$\int \boldsymbol{f}(t)dt = \boldsymbol{F}(t) .$$ ...(2)

This is an indefinite integral and the function $\boldsymbol{f}(t)$ which is being integrated is known as integrand.

Moreover, let $\boldsymbol{c}$ be a constant vector which is independent of t, then (1) can also be written as

$$\frac{d}{dt}[\boldsymbol{F}(t) + \boldsymbol{c}] = \boldsymbol{f}(t)$$ ...(3)

∴ $$\int \boldsymbol{f}(t)dt = \boldsymbol{F}(t) + \boldsymbol{c} .$$ ...(4)

This constant vector $\boldsymbol{c}$ is called constant of integration, since this vector $\boldsymbol{c}$ is taken to be arbitrary so the integral given by (4) is therefore known as indefinite integral.

If $\boldsymbol{f}(t)$ is defined over the closed interval $[a, b]$, then the integral given in (5)

$$\int_a^b \boldsymbol{f}(t)dt = [\boldsymbol{F}(t) + \boldsymbol{c}]_a^b = \boldsymbol{F}(b) - \boldsymbol{F}(a)$$ ...(5)

is called the definite integral and a and b are called limits of integration.

☛ REMARK

- If $\boldsymbol{f}(t) = f_1(t)\hat{i} + f_2(t)\hat{j} + f_3(t)\hat{k}$, then $\int \boldsymbol{f}(t)dt = \hat{i}\int f_1(t)dt + \hat{j}\int f_2(t)dt + \hat{k}\int f_3(t)dt$.

SOME IMPORTANT RESULTS

1. Since $\dfrac{d}{dt}(\boldsymbol{a.b}) = \boldsymbol{a}.\dfrac{d\boldsymbol{b}}{dt} + \dfrac{d\boldsymbol{a}}{dt}.\boldsymbol{b}$, therefore $\int\left(\boldsymbol{a}.\dfrac{d\boldsymbol{b}}{dt} + \dfrac{d\boldsymbol{a}}{dt}.\boldsymbol{b}\right)dt = \boldsymbol{a.b} + \boldsymbol{c}$ where, $\boldsymbol{c}$ is a constant of integration.

2. $\dfrac{d}{dt}(\boldsymbol{a} \times \boldsymbol{b}) = \boldsymbol{a} \times \dfrac{d\boldsymbol{b}}{dt} + \dfrac{d\boldsymbol{a}}{dt} \times \boldsymbol{b}$, so $\int\left(\boldsymbol{a} \times \dfrac{d\boldsymbol{b}}{dt} + \dfrac{d\boldsymbol{a}}{dt} \times \boldsymbol{b}\right)dt = (\boldsymbol{a} \times \boldsymbol{b}) + \boldsymbol{c}$, where, $\boldsymbol{c}$ is a constant vector.

3. Since $\quad \boldsymbol{a.a} = \boldsymbol{a}^2 \Rightarrow \dfrac{d}{dt}(\boldsymbol{a.a}) = \boldsymbol{a}.\dfrac{d\boldsymbol{a}}{dt} + \dfrac{d\boldsymbol{a}}{dt}.\boldsymbol{a} = 2\boldsymbol{a}.\dfrac{d\boldsymbol{a}}{dt}$

   ∴ $\qquad \int\left(2\boldsymbol{a}.\dfrac{d\boldsymbol{a}}{dt}\right)dt = (\boldsymbol{a.a}) + c$ or $\int\left(2\boldsymbol{a}.\dfrac{d\boldsymbol{a}}{dt}\right)dt = \boldsymbol{a}^2 + c$, where, c is a scalar quantity.

4. Since $\qquad \dfrac{d}{dt}\left(\boldsymbol{a} \times \dfrac{d\boldsymbol{a}}{dt}\right) = \boldsymbol{a} \times \dfrac{d^2\boldsymbol{a}}{dt^2} + \dfrac{d\boldsymbol{a}}{dt} \times \dfrac{d\boldsymbol{a}}{dt} = \boldsymbol{a} \times \dfrac{d^2\boldsymbol{a}}{dt^2}$ $\left(\because \dfrac{d\boldsymbol{a}}{dt} \times \dfrac{d\boldsymbol{a}}{dt} = 0\right)$

   ∴ $\qquad \int\left(\boldsymbol{a} \times \dfrac{d^2\boldsymbol{a}}{dt^2}\right)dt = \left(\boldsymbol{a} \times \dfrac{d\boldsymbol{a}}{dt}\right) + \boldsymbol{c}$ where $\boldsymbol{c}$ is a constant vectror of integration.

5. Since $\qquad \dfrac{d}{dt}\left(\boldsymbol{a} \times \dfrac{d\boldsymbol{b}}{dt}\right) = \boldsymbol{a} \times \dfrac{d^2\boldsymbol{b}}{dt^2} + \dfrac{d\boldsymbol{a}}{dt} \times \dfrac{d\boldsymbol{b}}{dt}$

   If $\boldsymbol{a}$ is a constant vector, then $\dfrac{d\boldsymbol{a}}{dt} = 0$ and $\dfrac{d}{dt}\left(\boldsymbol{a} \times \dfrac{d\boldsymbol{b}}{dt}\right) = \boldsymbol{a} \times \dfrac{d^2\boldsymbol{b}}{dt^2}$

   ∴ $\qquad \int\left(\boldsymbol{a} \times \dfrac{d^2\boldsymbol{b}}{dt^2}\right)dt = \left(\boldsymbol{a} \times \dfrac{d\boldsymbol{b}}{dt}\right) + \boldsymbol{c}$ where $\boldsymbol{c}$ is a constant vector of integration.

6. If $\boldsymbol{a}$ is a constant vector, we have

   $$\frac{d}{dt}(\boldsymbol{a} \times \boldsymbol{b}) = \boldsymbol{a} \times \frac{d\boldsymbol{b}}{dt}$$ $\left(\because \dfrac{d\boldsymbol{a}}{dt} = 0\right)$

   ∴ $\qquad \int\left(\boldsymbol{a} \times \dfrac{d\boldsymbol{b}}{dt}\right)dt = (\boldsymbol{a} \times \boldsymbol{b}) + \boldsymbol{c}$ where, $\boldsymbol{c}$ is a constant vector which is constant of integration.

7. If c is constant scalar and $\boldsymbol{a}$ is a vector function of t, then $\int c\boldsymbol{a}dt = c\int \boldsymbol{a}dt$.

Solved Examples

EXAMPLE 1. *If* $\boldsymbol{f}(t) = (t+1)\hat{i} + (t^2 + t + 1)\hat{j} + (t^3 + t^2 + t + 1)\hat{k}$

$\qquad$ *find* $\int_0^1 \boldsymbol{f}(t)dt$.

SOLUTION . Since $\boldsymbol{f}(t) = (t+1)\hat{i} + (t^2 + t + 1)\hat{j} + (t^3 + t^2 + t + 1)\hat{k}$,

$\qquad$ then $\int_0^1 \boldsymbol{f}(t)dt$

$= \int_0^1 [(t+1)\hat{i} + (t^2 + t + 1)\hat{j} + (t^3 + t^2 + t + 1)\hat{k}]dt$

$= \hat{i}\int_0^1 (t+1)dt + \hat{j}\int_0^1 (t^2 + t + 1)dt$

$\qquad + \hat{k}\int_0^1 (t^3 + t^2 + t + 1)dt$

$$= \hat{i}\left(\frac{t^2}{2}+t\right)_0^1 + \hat{j}\left(\frac{t^3}{3}+\frac{t^2}{2}+t\right)_0^1$$

$$+ \hat{k}\left(\frac{t^4}{4}+\frac{t^3}{3}+\frac{t^2}{2}+t\right)_0^1$$

$$= \frac{3}{2}\hat{i}+\frac{11}{6}\hat{j}+\frac{25}{12}\hat{k}.$$

EXAMPLE 2. *If $r = 5t^2\hat{i} + t\hat{j} - t^3\hat{k}$, then prove that*

$$\int_1^2\left(r\times\frac{d^2r}{dt^2}\right)dt = -14\hat{i}+75\hat{j}-15\hat{k}.$$

SOLUTION. Since $\quad r = 5t^2\hat{i}+t\hat{j}-t^3\hat{k}$, then

$$\frac{dr}{dt}=10t\hat{i}+\hat{j}-3t^2\hat{k}$$

again $\quad \dfrac{d^2r}{dt^2}=10\hat{i}-6t\hat{k}$.

$$\therefore \quad r\times\frac{d^2r}{dt^2}=(5t^2\hat{i}+t\hat{j}-t^3\hat{k})\times(10\hat{i}-6t\hat{k})$$

$$= -30t^3\hat{i}\times\hat{k}+10t\hat{j}\times\hat{i}$$
$$-6t^2\hat{j}\times\hat{k}-10t^3\hat{k}\times\hat{i}$$

$$= 30t^3\hat{j}-10t\hat{k}-6t^2\hat{i}-10t^3\hat{j}$$

$$= -6t^2\hat{i}+20t^3\hat{j}-10t\hat{k}.$$

Now

$$\int_1^2\left(r\times\frac{d^2r}{dt^2}\right)=\int_1^2(-6t^2\hat{i}+20t^3\hat{j}-10t\hat{k})dt$$

$$=\left[-2t^3\hat{i}+5t^4\hat{j}-5t^2\hat{k}\right]_1^2$$

$$= -14\hat{i}+75\hat{j}-15\hat{k}.$$

EXAMPLE 3. *Find the value of r satisfying the equation*

$$\frac{d^2r}{dt^2}=6t\hat{i}-24t^2\hat{j}+4\sin t\hat{k}$$

given that $r = 2\hat{i}+\hat{j}$ and $\dfrac{dr}{dt}=0$ at $t = 0$.

SOLUTION. We know that

$$\frac{d^2r}{dt^2}=\frac{d}{dt}\left(\frac{dr}{dt}\right). \text{ so } \frac{dr}{dt}=\int\left(\frac{d^2r}{dt^2}\right)dt+c$$

(where c is constant vector taken to be as constant of integration.)

$$\frac{dr}{dt}=\int(6t\hat{i}-24t^2\hat{j}+4\sin t\hat{k})dt+c$$

$$\frac{dr}{dt}=3t^2\hat{i}-8t^3\hat{j}-4\cos t\hat{k}+c \qquad ...(1)$$

Initially at $t = 0, \dfrac{dr}{dt}=0 \Rightarrow \; 0=-4\hat{k}+c$

$$\therefore \qquad\qquad c = 4\hat{k} \qquad\qquad ...(2)$$

Form (1) and (2), we get

$$\frac{dr}{dt}=3t^2\hat{i}-8t^3\hat{j}-4\cos t\hat{k}+4\hat{k}$$

Again integrating, we get

$$r = \int(3t^2\hat{i}-8t^3\hat{j}-4\cos t\hat{k}+4\hat{k})+d$$

Here, d is constant vector of integration.

$$\therefore \qquad r = (t^3\hat{i}-2t^4\hat{j}-4\sin t\hat{k}+4t\hat{k})+d \quad ...(3)$$

Again initially, at $t = 0, r = 2\hat{i}+\hat{j}$.

$$\therefore \qquad 2\hat{i}+\hat{j}=d \qquad\qquad ...(4)$$

Thus, form (3) and (4), we get the required result

$$r = t^3\hat{i}-2t^4\hat{j}+4(t-\sin t)\hat{k}+2\hat{i}+\hat{j}$$

or $\qquad r = (t^3+2)\hat{i}+(1-2t^4)\hat{j}+4(t-\sin t)\hat{k}$

EXAMPLE 4. *If $F(t) = 3t^2\hat{i}+t\hat{j}+2\hat{k}$ and*

$G(t) = 6t^2\hat{i}+(t-1)\hat{j}+3t\hat{k}$ then find

$$\int_0^1\left(\frac{dF}{dt}.G+F.\frac{dG}{dt}\right)dt \text{ and } \int_0^1\left(F\times\frac{dG}{dt}+\frac{dF}{dt}\times G\right)dt$$

SOLUTION. We have

$$F.G = 18t^4+t(t-1)+6t$$

$$F\times G = \begin{vmatrix} \hat{i} & \hat{j} & \hat{k} \\ 3t^2 & t & 2 \\ 6t^2 & (t-1) & 3t \end{vmatrix}$$

$$= \hat{i}(3t^2-2t+2)-\hat{j}(9t^3-12t^2)$$
$$+\hat{k}(3t^3-3t-6t^3)$$

$$= (3t^2-2t+2)\hat{i}-(9t^3-12t^2)\hat{j}$$
$$-(3t^3+3t)\hat{k}$$

Now, $\int_0^1\left(\dfrac{dF}{dt}.G+F.\dfrac{dG}{dt}\right)dt$

$$=\left[F.G\right]_0^1=\left[18t^4+t(t-1)+6t\right]_0^1$$

$$=18+6=24$$

$$\int_0^1\left(F\times\frac{dG}{dt}+\frac{dF}{dt}\times G\right)dt$$

$$=\left[F\times G\right]_0^1$$

$$=\left[\begin{array}{c}(3t^2-2t+2)\hat{i}-(9t^3-12t^2)\hat{j}\\-(3t^3+3t)\hat{k}\end{array}\right]_0^1$$

$$=(3\hat{i}+3\hat{j}-6\hat{k})-(2\hat{i})$$

$$=\hat{i}+3\hat{j}-6\hat{k}.$$

Exercise-6.2

1. If $f(t) = (t-t^2)\hat{i}+2t^3\hat{j}-3\hat{k}$, find

(i) $\int f(t)dt$ (ii) $\int_1^2 f(t)dt$

2. Integrate $a\times\dfrac{d^2r}{dt^2}=b$, where a and b are constant vectors.

3. Find the value of r satisfying the equation $\dfrac{d^2r}{dt^2}=ta+b$ where a and b are constant vectors.

4. Given that $r(t) = \begin{cases} 2\hat{i} - \hat{j} + 2\hat{k} & , \quad t = 2 \\ 4\hat{i} - 2\hat{j} + 3\hat{k} & , \quad t = 3 \end{cases}$

Show that $\int_2^3 \left(r.\dfrac{dr}{dt} \right) dt = 10$.

5. Find $\int_0^1 \left(e^t \hat{i} + e^{-2t} \hat{j} + t\hat{k} \right) dt$

6. If $r = t\hat{i} - t^2\hat{j} + (t-1)\hat{k}$ and $s = 2t^2\hat{i} + 6t\hat{k}$, evaluate

(i) $\int_0^2 r.s\, dt$ (ii) $\int_0^2 r \times s\, dt$.

7. Solve the equation $\dfrac{d^2 r}{dt^2} = a$ where a is a constant vector given

that $r = 0$ and $\dfrac{dr}{dt} = 0$ when $t = 0$.

8. If $f(t) = t\hat{i} + (t^2 - 2t)\hat{j} + (3t^2 + 3t^3)\hat{k}$, find $\int_0^1 f(t)dt$.

9. If $\quad a = t\hat{i} - 3\hat{j} + 2t\hat{k}, b = \hat{i} - 2\hat{j} + 2\hat{k}$ and $c = 3\hat{i} + t\hat{j} - \hat{k}$, then

evaluate $\int_1^2 a \cdot (b \times c)dt$.

10. The acceleration of a particle at any time $t \geq 0$ is given by

$$a = \frac{dv}{dt} = 12\cos 2t\hat{i} - 8\sin 2t\hat{j} + 16t\hat{k}$$

if the velocity v and displacement r, are zero at $t = 0$, find v and r at any time t.

11. The acceleration of a particle at any time t is given by

$$a = \frac{dv}{dt} = e^t\hat{i} + e^{2t}\hat{j} + \hat{k}, \text{find } v \text{ if } v = \hat{i} + \hat{j} \text{ at } t = 0.$$

12. Find the value of r satisfying the equation $\dfrac{d^2 r}{dt^2} = a$, where a is

a constant vector. Also it is given that when $t = 0$, $r = 0$ and

$\dfrac{dr}{dt} = u$.

13. If $A = \hat{i} + u^2\hat{j} - 2u\hat{k}$ and $B = e^u\hat{i} - u\hat{j} - \hat{k}$, find $\int (A \times B)du$.

Hints to Selected Problems

1.
$$f(t) = (t - t^2)\hat{i} + 2t^3\hat{j} - 3\hat{k}$$

$$\therefore \quad \int f(t)dt = \int [(t - t^2)\hat{i} + 2t^3\hat{j} - 3\hat{k}]dt$$

$$= \left(\frac{t^2}{2} - \frac{t^3}{3} \right)\hat{i} + \frac{t^4}{2}\hat{j} - 3t\hat{k}$$

and $\int_1^2 f(t)dt = \left[\left(\frac{t^2}{2} - \frac{t^3}{3} \right)\hat{i} + \frac{t^4}{2}\hat{j} - 3t\hat{k} \right]_1^2 = \dfrac{-5}{6}\hat{i} + \dfrac{15}{2}\hat{j} - 3\hat{k}$

4.
$$r(t) = \begin{cases} 2\hat{i} - \hat{j} + 2\hat{k} & , \quad t = 2 \\ 4\hat{i} - 2\hat{j} + 3\hat{k} & , \quad t = 3 \end{cases}$$

Since $\int \left(r.\dfrac{dr}{dt} \right)dt = \dfrac{1}{2}\int \dfrac{d}{dt}(r.r)dt = \dfrac{1}{2}(r.r)$

$\therefore \quad \int_2^3 \left(r.\dfrac{dr}{dt} \right)dt = \dfrac{1}{2}[r(t).r(t)]_2^3 = \dfrac{1}{2}[r(3).r(3) - r(2).r(2)]$

$= \dfrac{1}{2}[(4\hat{i} - 2\hat{j} + 3\hat{k}).(4\hat{i} - 2\hat{j} + 3\hat{k})$

$\qquad - (2\hat{i} - \hat{j} + 2\hat{k}).(2\hat{i} - \hat{j} + 2\hat{k})]$

$= \dfrac{1}{2}[(16 + 4 + 9) - (4 + 1 + 4)]$

$= \dfrac{1}{2}[29 - 9] = \dfrac{1}{2}(20) = 10.$

9. $a = t\hat{i} - 3\hat{j} + 2t\hat{k}, b = \hat{i} - 2\hat{j} + 2\hat{k}, c = 3\hat{i} + t\hat{j} - \hat{k},$

then $\quad a.(b \times c) = \begin{vmatrix} t & -3 & 2t \\ 1 & -2 & 2 \\ 3 & t & -1 \end{vmatrix}$

$= t(2 - 2t) + 3(-1 - 6) + 2t(t + 6)$

$= 2t - 2t^2 - 21 = 2t^2 + 12t = 14t - 21$

$\therefore \int_1^2 a.(b \times c)dt = \int_1^2 (14t - 21)dt = \left[7t^2 - 21t \right]_1^2$

$= (28 - 42) - (7 - 21) = 0.$

Answers

1. (i) $\left(\dfrac{t^2}{2} - \dfrac{t^3}{3} \right)\hat{i} + \dfrac{t^4}{2}\hat{j} - 3t\hat{k} + c$ (ii) $-\dfrac{5}{6}\hat{i} + \dfrac{15}{2}\hat{j} - 3\hat{k}$ **2.** $a \times r = -t\, b + tc + d$ **3.** $r = \dfrac{1}{6}t^3 a + \dfrac{1}{2}t^2 b + tc + d$

5. $(e - 1)\hat{i} - \dfrac{1}{2}(e^{-2} - 1)\hat{j} + \dfrac{1}{2}\hat{k}$ **6.** (i) 12   (ii) $-24\hat{i} - \dfrac{40}{3}\hat{j} + \dfrac{64}{5}\hat{k}$ **7.** $r = \dfrac{1}{2}t^2 a$ **8.** $\dfrac{1}{2}\hat{i} - \dfrac{2}{3}\hat{j} + \dfrac{7}{4}\hat{k}$ **9.** 0

10. $v = 6\sin 2t\hat{i} + (4\cos 2t - 4)\hat{j} + 8t^2\hat{k}$ **11.** $v = e^t\hat{i} + \dfrac{1}{2}(e^{2t} + 1)\hat{j} + t\hat{k}$ **12.** $r = \dfrac{1}{2}t^2 a + tu$

$r = (3 - 3\cos 2t)\hat{i} + (2\sin 2t - 4t)\hat{j} + \dfrac{8}{3}t^3\hat{k}$

13. $-\dfrac{u^3}{3}\hat{i} - \hat{j}\left(-u + 2ue^u - 2e^u \right) + \hat{k}\left(-\dfrac{u^2}{2} - u^2 e^u + 2ue^u + 2e^u \right)$

6.4 PARTIAL DERIVATIVE OF VECTORS

Let $r = f(x, y, z)$ be a vector function of three scalar variables x, y, z. The first order partial derivative of r with respect to x is given by

$$\frac{\partial r}{\partial x} = \lim_{\delta x \to 0} \frac{f(x + \delta x, y, z) - f(x, y, z)}{\delta x}, \text{ if this limit exists.}$$

Similarly we can find first order partial derivatives of r with respect to y and z respectively and are denoted by $\dfrac{\partial r}{\partial y}, \dfrac{\partial r}{\partial z}$.

During the differentiation if y and z are treating as constant, then $\dfrac{\partial \mathbf{r}}{\partial x}$ is regarded as ordinary derivative. Likewise we can find higher order partial derivatives.

6.4.1 Vector Differential Operator

The vector differential operator is defined by the formula $\nabla = \dfrac{\partial}{\partial x}\hat{i} + \dfrac{\partial}{\partial y}\hat{j} + \dfrac{\partial}{\partial z}\hat{k}.$

Obviously, ∇ is a vector quantity. This vector ∇ is read as **nabla** or **del**.

6.5 GRADIENT OF A SCALAR FIELD

(UPTU–2006, 07, 08)

Let $f(x, y, z)$ be a scalar point function which is defined over some region R in space and also differentiable at each point (x, y, z) in R, then the gradient of $f(x, y, z)$ is defined as

$$\text{grad } f = \frac{\partial f}{\partial x}\hat{i} + \frac{\partial f}{\partial y}\hat{j} + \frac{\partial f}{\partial z}\hat{k} \quad \text{or} \quad \text{grad } f = \left(\frac{\partial}{\partial x}\hat{i} + \frac{\partial}{\partial y}\hat{j} + \frac{\partial}{\partial z}\hat{k}\right)f = \nabla f$$

Thus gradient of f can also be written in terms of vector differential operator(∇). Since ∇ is a vector quantity, thus ∇f is a vector whose components are $\dfrac{\partial f}{\partial x}, \dfrac{\partial f}{\partial y}, \dfrac{\partial f}{\partial z}$. Hence, gradient of a scalar field is a vector field.

6.5.1 Some Formulae related to Gradient

1. If f and g are two scalar point functions, then grad $(f + g)$ = grad f + grad g or $\nabla(f + g) = \nabla f + \nabla g$.

PROOF. Since we know that

$$\nabla f = \frac{\partial f}{\partial x}\hat{i} + \frac{\partial f}{\partial y}\hat{j} + \frac{\partial f}{\partial z}\hat{k}$$

$$\therefore \quad \nabla(f + g) = \frac{\partial}{\partial x}(f+g)\hat{i} + \frac{\partial}{\partial y}(f+g)\hat{j} + \frac{\partial}{\partial z}(f+g)\hat{k} = \frac{\partial f}{\partial x}\hat{i} + \frac{\partial g}{\partial x}\hat{i} + \frac{\partial f}{\partial y}\hat{j} + \frac{\partial g}{\partial y}\hat{j} + \frac{\partial f}{\partial z}\hat{k} + \frac{\partial g}{\partial z}\hat{k}$$

$$= \left(\frac{\partial f}{\partial x}\hat{i} + \frac{\partial f}{\partial y}\hat{j} + \frac{\partial f}{\partial z}\hat{k}\right) + \left(\frac{\partial g}{\partial x}\hat{i} + \frac{\partial g}{\partial y}\hat{j} + \frac{\partial g}{\partial z}\hat{k}\right) = \nabla f + \nabla g.$$

Hence, $\nabla(f + g) = \nabla f + \nabla g.$

2. If f and g are two scalar point functions, then $\nabla(fg) = f\nabla g + g\nabla f$. or grad (fg) = f(grad g) + g(grad f).

PROOF. Since we know that

$$\nabla f = \frac{\partial f}{\partial x}\hat{i} + \frac{\partial f}{\partial y}\hat{j} + \frac{\partial f}{\partial z}\hat{k}$$

$$\therefore \quad \nabla(fg) = \frac{\partial}{\partial x}(fg)\hat{i} + \frac{\partial}{\partial y}(fg)\hat{j} + \frac{\partial}{\partial z}(fg)\hat{k}$$

$$= \left(f\frac{\partial g}{\partial x} + g\frac{\partial f}{\partial x}\right)\hat{i} + \left(f\frac{\partial g}{\partial y} + g\frac{\partial f}{\partial y}\right)\hat{j} + \left(f\frac{\partial g}{\partial z} + g\frac{\partial f}{\partial z}\right)\hat{k}$$

$$= f\left(\frac{\partial g}{\partial x}\hat{i} + \frac{\partial g}{\partial y}\hat{j} + \frac{\partial g}{\partial z}\hat{k}\right) + g\left(\frac{\partial f}{\partial x}\hat{i} + \frac{\partial f}{\partial y}\hat{j} + \frac{\partial f}{\partial z}\hat{k}\right) = f\nabla g + g\nabla f.$$

Hence, $\nabla(fg) = f\nabla g + g\nabla f.$

3. If f and g are scalar point functions and $g \neq 0$ for all point in the region R, then $\nabla\left(\dfrac{f}{g}\right) = \dfrac{g\nabla f - f\nabla g}{g^2}.$

PROOF. Since

$$\nabla f = \frac{\partial f}{\partial x}\hat{i} + \frac{\partial f}{\partial y}\hat{j} + \frac{\partial f}{\partial z}\hat{k}$$

$$\therefore \quad \nabla\left(\frac{f}{g}\right) = \frac{\partial}{\partial x}\left(\frac{f}{g}\right)\hat{i} + \frac{\partial}{\partial y}\left(\frac{f}{g}\right)\hat{j} + \frac{\partial}{\partial z}\left(\frac{f}{g}\right)\hat{k}$$

$$= \frac{1}{g^2}\left(g\frac{\partial f}{\partial x} - f\frac{\partial g}{\partial x}\right)\hat{i} + \frac{1}{g^2}\left(g\frac{\partial f}{\partial y} - f\frac{\partial g}{\partial y}\right)\hat{j} + \frac{1}{g^2}\left(g\frac{\partial f}{\partial z} - f\frac{\partial g}{\partial z}\right)\hat{k}$$

$$= \frac{1}{g^2}\left[g\left(\frac{\partial f}{\partial x}\hat{i} + \frac{\partial f}{\partial y}\hat{j} + \frac{\partial f}{\partial z}\hat{k}\right) - f\left(\frac{\partial g}{\partial x}\hat{i} + \frac{\partial g}{\partial y}\hat{j} + \frac{\partial g}{\partial z}\hat{k}\right)\right] = \frac{1}{g^2}[g\nabla f - f\nabla g]$$

$$= \frac{g(\nabla f) - f(\nabla g)}{g^2}.$$

Hence,
$$\nabla\left(\frac{f}{g}\right) = \frac{g\nabla f - f\nabla g}{g^2}.$$

4. If f is a scalar point function, then f is constant if and only if $\nabla f = 0$.

PROOF. Suppose f is constant, then
$$\frac{\partial f}{\partial x} = 0, \frac{\partial f}{\partial y} = 0, \frac{\partial f}{\partial z} = 0 \qquad\qquad (\because f(x,y,z) = c)$$

$$\therefore \qquad \nabla f = \frac{\partial f}{\partial x}\hat{i} + \frac{\partial f}{\partial y}\hat{j} + \frac{\partial f}{\partial z}\hat{k} = 0\hat{i} + 0\hat{j} + 0\hat{k} = \mathbf{0}.$$

Conversely, suppose $\nabla f = \mathbf{0}$. Then we have $\nabla f = \dfrac{\partial f}{\partial x}\hat{i} + \dfrac{\partial f}{\partial y}\hat{j} + \dfrac{\partial f}{\partial z}\hat{k} = \mathbf{0}$. So, $\dfrac{\partial f}{\partial x} = 0, \dfrac{\partial f}{\partial y} = 0, \dfrac{\partial f}{\partial z} = 0$.

Hence, $\qquad\qquad f(x,y,z) = c$ (constant)

☞ REMARKS

- $\nabla(f - g) = \nabla f - \nabla g$
- $\nabla(cf) = c\nabla f$, where c is a constant.
- $\nabla\left(\dfrac{1}{f}\right) = -\dfrac{\nabla f}{f^2}$, where $f \neq 0 \ \forall \ (x, y, z) \in \mathbf{R}.$

6.6 DIRECTIONAL DERIVATIVES

Let us consider a scalar field given by a scalar point function $f(P) = f(x, y, z)$ where P is any point in space whose co-ordinates are (x, y, z). Since we know that the first order partial derivatives of f are the rates of change of f in the direction of co-ordinate axes. Now we shall have to discuss the rate of change of f in any direction, this leads the notion of a directional derivative.

Let us choose a point P in space and a direction at P, given by a unit vector $\hat{a}$. Let C be the ray from P in the direction of $\hat{a}$ and let Q be any point on this ray C such that PQ is as shown in fig. 4.

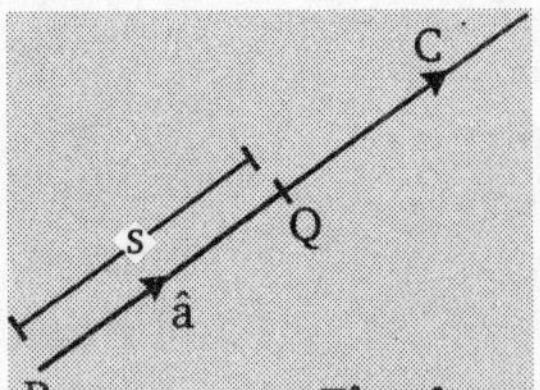

Then the limit
$$\frac{\partial f}{\partial s} = \lim_{s \to 0} \frac{f(Q) - f(P)}{s}, \text{ where } s = PQ$$

Fig. 4

if exists is called the directional derivative of f at P in the direction of $\hat{a}$. In fact there are infinitely many directional derivatives of f at P, each corresponding to a certain direction. But if a cartesian co-ordinates system is given, then we may represent any such derivative in terms of the first order partial derivatives of f at P. If the position vector P is $\mathbf{p}$, then the ray C can be written as
$$\mathbf{r}(s) = x(s)\hat{i} + y(s)\hat{j} + z(s)\hat{k} \qquad\qquad\qquad \dots(1)$$
$$= \mathbf{p} + s\hat{a} \qquad (s \geq 0)$$

and $\dfrac{\partial f}{\partial s}$ is the derivative of the function $f[x(s), y(s), z(s)]$ with respect to the arc length s of C. Hence, assuming that f has continuous partial derivative of first order, we have

$$\frac{\partial f}{\partial s} = \frac{\partial f}{\partial x}\frac{dx}{ds} + \frac{\partial f}{\partial y}\frac{dy}{ds} + \frac{\partial f}{\partial z}\frac{dz}{ds} \qquad\qquad \dots(2)$$

Form (1)
$$\frac{d\mathbf{r}}{ds} = \frac{dx}{ds}\hat{i} + \frac{dy}{ds}\hat{j} + \frac{dz}{ds}\hat{k} = \hat{a} \qquad\qquad \dots(3)$$

Since we have
$$\text{grad } f = \frac{\partial f}{\partial x}\hat{i} + \frac{\partial f}{\partial y}\hat{j} + \frac{\partial f}{\partial z}\hat{k}. \qquad\qquad \dots(4)$$

Thus, equation (2) becomes
$$\frac{\partial f}{\partial s} = \left(\frac{\partial f}{\partial x}\hat{i} + \frac{\partial f}{\partial y}\hat{j} + \frac{\partial f}{\partial z}\hat{k}\right).\left(\frac{dx}{ds}\hat{i} + \frac{dy}{ds}\hat{j} + \frac{dz}{ds}\hat{k}\right)$$
$$= (\text{grad } f).\hat{a} \qquad\qquad\qquad \text{[From (3) and (4)]}$$

or
$$\frac{\partial f}{\partial s} = \hat{a} \cdot \text{grad } f = \hat{a} \cdot \nabla f.$$

Hence, the directional derivative $\dfrac{\partial f}{\partial s}$ is given as $\hat{a} \cdot \nabla f$.

☞ REMARKS

- If $\hat{a} = \hat{i}$, then $\dfrac{\partial f}{\partial s} = \hat{i}.\nabla f = \hat{i}.\left(\dfrac{\partial f}{\partial x}\hat{i} + \dfrac{\partial f}{\partial y}\hat{j} + \dfrac{\partial f}{\partial z}\hat{k}\right) = \dfrac{\partial f}{\partial x}$. Similarly, if $\hat{a} = \hat{j}, \hat{k}$, then $\dfrac{\partial f}{\partial s} = \dfrac{\partial f}{\partial y}, \dfrac{\partial f}{\partial s} = \dfrac{\partial f}{\partial z}$.

- Maximum value of the directional derivative is $|\text{grad } f|$.

Solved Examples

EXAMPLE 1. *If $r = |r|$ where $r = x\hat{i} + y\hat{j} + z\hat{k}$, prove that*

   (i) $\nabla f(r) = f'(r)\nabla r$ (GBTU–2011)

   (ii) $\nabla r = \dfrac{r}{r}$ (GBTU–2011)

   (iii) $\nabla f(r) \times r = 0$

   (iv) $\nabla r^n = nr^{n-2}r$.   (UPTU–2008, GBTU–2011, ANNA–2003, BHOPAL–2007, VTU–2000)

   (v) $\nabla r^{-3} = -3r^{-5}r$.

SOLUTION. (i) Since we know that

$$\nabla f = \frac{\partial f}{\partial x}\hat{i} + \frac{\partial f}{\partial y}\hat{j} + \frac{\partial f}{\partial z}\hat{k} \qquad \dots(1)$$

$$\therefore \ \nabla f(r) = \frac{\partial}{\partial x}(f(r))\hat{i} + \frac{\partial}{\partial y}(f(r))\hat{j} + \frac{\partial}{\partial z}(f(r))\hat{k}$$

$$\text{or } \nabla f(r) = f'(r)\frac{\partial r}{\partial x}\hat{i} + f'(r)\frac{\partial r}{\partial y}\hat{j} + f'(r)\frac{\partial r}{\partial z}\hat{k}$$

$$\nabla f(r) = f'(r)\left[\frac{\partial r}{\partial x}\hat{i} + \frac{\partial r}{\partial y}\hat{j} + \frac{\partial r}{\partial z}\hat{k}\right]$$

$$\nabla f(r) = f'(r)\nabla r. \text{ [Using (i)]}$$

   (ii) We have $\nabla r = \dfrac{\partial r}{\partial x}\hat{i} + \dfrac{\partial r}{\partial y}\hat{j} + \dfrac{\partial r}{\partial z}\hat{k}$

   Since $r = x\hat{i} + y\hat{j} + z\hat{k}$

   $\therefore \ |r|^2 = x^2 + y^2 + z^2$

   or $\quad r^2 = x^2 + y^2 + z^2 \qquad (\because |r| = r)$

   $\therefore \ \dfrac{\partial r}{\partial x} = \dfrac{x}{r}, \dfrac{\partial r}{\partial y} = \dfrac{y}{r}, \dfrac{\partial r}{\partial z} = \dfrac{z}{r}$

   $\therefore \ \nabla r = \dfrac{x}{r}\hat{i} + \dfrac{y}{r}\hat{j} + \dfrac{z}{r}\hat{k}$

   $$= \frac{1}{r}(x\hat{i} + y\hat{j} + z\hat{k})$$

   or $\quad \nabla r = \dfrac{r}{r}$.

   (iii) $\quad \nabla f(r) = f'(r)\nabla r \qquad$ [From (i)]

   $$= f'(r)\frac{r}{r} \qquad \text{[From (ii)]}$$

   Now

   $$\nabla f(r) \times r = \frac{f'(r)}{r}r \times r = 0 \qquad (\because r \times r = 0)$$

   (iv) Since

   $$\nabla f(r) = f'(r)\nabla r \qquad \text{(From (i))}$$

   Let $f(r) = r^n$.

   $\therefore \ \nabla r^n = nr^{n-1}\nabla r = nr^{n-1}\left(\dfrac{r}{r}\right) \qquad \left(\because \nabla r = \dfrac{r}{r}\right)$

$$= nr^{n-2}r$$

   or $\quad \nabla r^n = nr^{n-2}r$.

   (v) From part (iv), $\nabla r^n = nr^n$

   $\Rightarrow \nabla r^{-3} = -3r^{-3-2}r = -3r^{-5}r$.

EXAMPLE 2. *If $f(x, y, z) = 3x^2y - y^3z^2$, find grad f and $|grad f|$ at $(1, -2, -1)$.* (UPTU–2007)

SOLUTION. Since we know that

$$\text{grad } f = \nabla f = \frac{\partial f}{\partial x}\hat{i} + \frac{\partial f}{\partial y}\hat{j} + \frac{\partial f}{\partial z}\hat{k}$$

$$= \frac{\partial}{\partial x}(3x^2y - y^3z^2)\hat{i}$$

$$+ \frac{\partial}{\partial y}(3x^2y - y^3z^2)\hat{j}$$

$$+ \frac{\partial}{\partial z}(3x^2y - y^3z^2)\hat{k}$$

$$= 6xy\hat{i} + (3x^2 - 3y^2z^2)\hat{j}$$

$$+ (-2y^3z)\hat{k}$$

At $(1, -2, -1)$

$$\text{grad } f = -12\hat{i} - 9\hat{j} - 16\hat{k}$$

and $\quad |\text{grad } f| = \sqrt{144 + 81 + 256} = \sqrt{481}$.

EXAMPLE 3. *If $\phi(x, y, z) = xy^2z$ and $f(x, y, z) = xz\hat{i} - xy\hat{j} + yz\hat{k}$, show that $\dfrac{\partial^3}{\partial x^2 \partial z}(\phi f)$ at $(2, -1, 1)$ is $4\hat{i} + 2\hat{j}$.* (BHOPAL–2008)

SOLUTION. We have $\phi f = x^2y^2z^2\hat{i} - x^2y^3z\hat{j} + xy^3z^2\hat{k}$.

$\therefore \quad \dfrac{\partial}{\partial z}(\phi f) = 2x^2y^2z\hat{i} - x^2y^3\hat{j} + 2xy^3z\hat{k}$

$\dfrac{\partial^2}{\partial x \partial z}(\phi f) = 4xy^2z\hat{i} - 2xy^3\hat{j} + 2y^3z\hat{k}$

$\dfrac{\partial^3}{\partial x^2 \partial z}(\phi f) = 4y^2z\hat{i} - 2y^3\hat{j}$

At $(2, -1, 1)$

$\dfrac{\partial^3(\phi f)}{\partial x^2 \partial z} = 4(-1)^2(1)\hat{i} - 2(-1)^3\hat{j} = 4\hat{i} + 2\hat{j}$.

EXAMPLE 4. *Find $\nabla\phi$ and $|\nabla\phi|$ when*

$$\phi = (x^2 + y^2 + z^2)e^{-(x^2+y^2+z^2)^{1/2}}.$$

SOLUTION. Let $r^2 = x^2 + y^2 + z^2$, then ϕ can be written as

$$\phi = r^2e^{-r} \qquad \dots(1)$$

Now $\quad \nabla\phi = \dfrac{\partial\phi}{\partial x}\hat{i} + \dfrac{\partial\phi}{\partial y}\hat{j} + \dfrac{\partial\phi}{\partial z}\hat{k} \qquad \dots(2)$

By (1), $\dfrac{\partial\phi}{\partial x} = \dfrac{\partial\phi}{\partial r}\dfrac{\partial r}{\partial x} = (2re^{-r} - r^2e^{-r})\dfrac{\partial r}{\partial x}$.

Again by $r^2 = x^2 + y^2 + z^2$,

$$2r\frac{\partial r}{\partial x} = 2x \Rightarrow \frac{\partial r}{\partial x} = \frac{x}{r}.$$

$$\therefore \quad \frac{\partial \phi}{\partial x} = r(2-r)e^{-r}\frac{x}{r} = (2-r)e^{-r}x$$

Similarly, $\dfrac{\partial \phi}{\partial y} = (2-r)e^{-r}y$ and $\dfrac{\partial \phi}{\partial z} = (2-r)e^{-r}z$

$\therefore$ By (2), we have

$$\nabla \phi = (2-r)e^{-r}(x\hat{i} + y\hat{j} + z\hat{k})$$

$$= (2-r)e^{-r}\boldsymbol{r}.$$

Also, $\quad |\nabla \phi| = \left|(2-r)e^{-r}\boldsymbol{r}\right| = (2-r)e^{-r}|\boldsymbol{r}|$

$$= (2-r)e^{-r}r = (2-r)re^{-r}.$$

EXAMPLE 5. *Prove that*
    (i) $\nabla(\boldsymbol{r}.\boldsymbol{a}) = \boldsymbol{a}$ (UPTU–2008)
    (ii) $\nabla[\boldsymbol{r}\ \boldsymbol{a}\ \boldsymbol{b}] = \boldsymbol{a} \times \boldsymbol{b}.$
where a and b are constant vectors.

SOLUTION. Suppose $\quad \boldsymbol{a} = a_1\hat{i} + a_2\hat{j} + a_3\hat{k}$ and

$$\boldsymbol{r} = x\hat{i} + y\hat{j} + z\hat{k}, \boldsymbol{b} = b_1\hat{i} + b_2\hat{j} + b_3\hat{k},$$

then $\quad\quad \boldsymbol{r}.\boldsymbol{a} = xa_1 + a_2y + a_3z$

and $\boldsymbol{r}.(\boldsymbol{a} \times \boldsymbol{b}) = \begin{vmatrix} x & y & z \\ a_1 & a_2 & a_3 \\ b_1 & b_2 & b_3 \end{vmatrix}$

$$= x(a_2 b_3 - a_3 b_2) + y(a_3 b_1 - a_1 b_3)$$
$$+ z(a_1 b_2 - a_2 b_1)$$

(i) $\nabla(\boldsymbol{r}.\boldsymbol{a}) = \nabla(xa_1 + a_2y + a_3z)$

$$= a_1\nabla(x) + a_2\nabla(y) + a_3\nabla(z)$$

$$= a_1\hat{i} + a_2\hat{j} + a_3\hat{k}$$

$$(\because \nabla(x) = \hat{i}, \nabla(y) = \hat{j}, \nabla(z) = \hat{k})$$

$$= \boldsymbol{a}$$

(ii) $\nabla[\boldsymbol{r}\ \boldsymbol{a}\ \boldsymbol{b}] = \nabla(\boldsymbol{r}.(\boldsymbol{a} \times \boldsymbol{b}))$

$$= \nabla[x(a_2b_3 - a_3b_2) + y(a_3b_1 - a_1b_3)$$
$$+ z(a_1b_2 - a_2b_1)]$$

$$= \nabla[x(a_2b_3 - a_3b_2)] + \nabla[y(a_3b_1 - a_1b_3)]$$
$$+ \nabla[z(a_1b_2 - a_2b_1)]$$

$$= (a_2b_3 - a_3b_2)\nabla(x) + (a_3b_1 - a_1b_3)\nabla(y)$$
$$+ (a_1b_2 - a_2b_1)\nabla(z)$$

$$= (a_2b_3 - a_3b_2)\hat{i} + (a_3b_1 - a_1b_3)\hat{j}$$
$$+ (a_1b_2 - a_2b_1)\hat{k}$$

$$= \begin{vmatrix} \hat{i} & \hat{j} & \hat{k} \\ a_1 & a_2 & a_3 \\ b_1 & b_2 & b_3 \end{vmatrix} = \boldsymbol{a} \times \boldsymbol{b}.$$

$\therefore \quad \nabla[\boldsymbol{r}\ \boldsymbol{a}\ \boldsymbol{b}] = \boldsymbol{a} \times \boldsymbol{b}.$

EXAMPLE 6. *Prove that* $\nabla \phi.\overline{d\boldsymbol{r}} = d\phi.$

SOLUTION. $\quad \nabla \phi = \dfrac{\partial \phi}{\partial x}\hat{i} + \dfrac{\partial \phi}{\partial y}\hat{j} + \dfrac{\partial \phi}{\partial z}\hat{k}$

$$\overline{d\boldsymbol{r}} = dx\hat{i} + dy\hat{j} + dz\hat{k} \quad\quad [\because \boldsymbol{r} = x\hat{i} + y\hat{j} + z\hat{k}]$$

$$\therefore \ \nabla \phi.\overline{d\boldsymbol{r}} = \left(\left(\frac{\partial \phi}{\partial x}\hat{i} + \frac{\partial \phi}{\partial y}\hat{j} + \frac{\partial \phi}{\partial z}\hat{k}\right).(dx\hat{i} + dy\hat{j} + dz\hat{k})\right)$$

$$= \frac{\partial \phi}{\partial x}dx + \frac{\partial \phi}{\partial y}dy + \frac{\partial \phi}{\partial z}dz = d\phi.$$

EXAMPLE 7. *If* $\phi(x, y) = \log\sqrt{x^2 + y^2}$ *show that*

$$\boldsymbol{grad}\ \phi = \frac{\boldsymbol{r} - (\hat{\boldsymbol{k}} \cdot \boldsymbol{r})\hat{\boldsymbol{k}}}{\{\boldsymbol{r} - (\hat{\boldsymbol{k}} \cdot \boldsymbol{r} \cdot \hat{\boldsymbol{k}})\}.\{\boldsymbol{r} - (\hat{\boldsymbol{k}} \cdot \boldsymbol{r})\vec{\boldsymbol{k}}\}}.$$

SOLUTION. We have $\quad \boldsymbol{r} = x\hat{i} + y\hat{j} + z\hat{k}$

$$\phi = \frac{1}{2}\log(x^2 + y^2)$$

$$\frac{\partial \phi}{\partial x} = \frac{1}{2(x^2 + y^2)}.2x = \frac{x}{x^2 + y^2}$$

Similarly, $\dfrac{\partial \phi}{\partial y} = \dfrac{y}{x^2 + y^2}, \dfrac{\partial \phi}{\partial z} = 0$

$$\text{grad}\ \phi = \hat{i}\frac{\partial \phi}{\partial x} + \hat{j}\frac{\partial \phi}{\partial y} + \hat{k}\frac{\partial \phi}{\partial z}$$

$$= \frac{x}{x^2 + y^2}\hat{i} + \frac{y}{x^2 + y^2}\hat{j} + 0\hat{k}$$

$$= \frac{x\hat{i} + y\hat{j}}{x^2\ y^2} = \frac{-z\hat{k}}{(x\hat{i} + y\hat{j}).(x\hat{i} + y\hat{j})}$$

$$= \frac{\boldsymbol{r} - z\hat{k}}{(\boldsymbol{r} - z\hat{k}).(\boldsymbol{r} - z\hat{k})} \quad\quad [\text{By (1)}]$$

Now replacing z by $\hat{k} \cdot \boldsymbol{r}$, we get

$$\text{grad}\ \phi = \frac{\boldsymbol{r} - (\hat{k}.\boldsymbol{r})\hat{k}}{\{\boldsymbol{r} - (\hat{k}.\boldsymbol{r})\hat{k}\}.\{\boldsymbol{r} - (\hat{k}.\boldsymbol{r})\hat{k}\}}.$$

EXAMPLE 8. *Find the directional derivative of* $f(x, y, z) = x^2yz + 4xz^2$ *at the point* (1, –2, –1) *in the direction of the vector* $2\hat{i} - \hat{j} - 2\hat{k}.$

           (UPTU–2006, JNTU–2006, VTU–2007, ROHTAK–2006)

SOLUTION. Let $\boldsymbol{a} = 2\hat{i} - \hat{j} - 2\hat{k}$, then

$$\hat{a} = \frac{\boldsymbol{a}}{|\boldsymbol{a}|} = \frac{2\hat{i} - \hat{j} - 2\hat{k}}{\sqrt{4 + 1 + 4}} = \frac{1}{3}(2\hat{i} - \hat{j} - 2\hat{k}).$$

Since $f(x, y, z) = x^2yz + 4xz^2$

$$\therefore \quad \frac{\partial f}{\partial x} = 2xyz + 4z^2$$

$$\frac{\partial f}{\partial y} = x^2z, \frac{\partial f}{\partial z} = x^2y + 8xz$$

$$\therefore \quad \nabla f = \frac{\partial f}{\partial x}\hat{i} + \frac{\partial f}{\partial y}\hat{j} + \frac{\partial f}{\partial z}\hat{k}$$

$$= (2xyz + 4z^2)\hat{i} + x^2z\hat{j}$$
$$+ (x^2y + 8xz)\hat{k}$$

At (1, –2, –1)

$$\nabla f = 8\hat{i} - \hat{j} - 10\hat{k}.$$

Now directional derivative of f at (1, –2, –1) in the direction of $2\hat{i} - \hat{j} - 2\hat{k}$ is

$$\nabla f \cdot \hat{a} = (8\hat{i} - \hat{j} - 10\hat{k}).\left(\frac{1}{3}(2\hat{i} - \hat{j} - 2\hat{k})\right)$$

$$= \frac{1}{3}(16 + 1 + 20) = \frac{37}{3}.$$

EXAMPLE 9. *Find the maximum value of the directional derivative of* $\phi = x^2yz$ *at the point (1, 4, 1).x*

SOLUTION. Since $\phi = x^2yz$

$\therefore \quad \nabla\phi = \dfrac{\partial\phi}{\partial x}\hat{i} + \dfrac{\partial\phi}{\partial y}\hat{j} + \dfrac{\partial\phi}{\partial z}\hat{k}$

$\quad = 2xyz\hat{i} + x^2z\hat{j} + x^2y\hat{k}x$

At (1, 4, 1),

$\nabla\phi = 8\hat{i} + \hat{j} + 4\hat{k}$.

Maximum value of directional derivative of ϕ at (1, 4, 1)

$= |\nabla\phi| = \sqrt{64+1+16}$

$= \sqrt{81} = 9$.

EXAMPLE 10. *Find the directional derivative of* $f(x, y, z) = xy^2 + yz^3$ *at the point (2, –1, 1) in the direction of the vector* $\hat{i} + 2\hat{j} + 2\hat{k}$. (ROHTAK–2003, BHOPAL–2008, KURUKSHETRA–2006, UPTU–2011, SVTU–2009, GBTU–2011)

SOLUTION. Given $f(x, y, z) = xy^2 + yz^3$.

Therefore, $\nabla f = \hat{i}\dfrac{\partial}{\partial x}(xy^2 + yz^3) + \hat{j}\dfrac{\partial}{\partial y}(xy^2 + yz^3)$

$\qquad + \hat{k}\dfrac{\partial}{\partial z}(xy^2 + yz^3)$

$= \hat{i}(y^2) + \hat{j}(2xy + z^3) + \hat{k}(3yz^2)$

$\nabla f_{\text{at }(2,-1,1)} = \hat{i} - 3\hat{j} - 3\hat{k}$

$\therefore$ Directional derivative of f in the direction

$\hat{i} + 2\hat{j} + 2\hat{k} = (\hat{i} - 3\hat{j} - 3\hat{k})\dfrac{\hat{i} + 2\hat{j} + 2\hat{k}}{\sqrt{1^2 + 2^2 + 2^2}}$

$= (1.1 - 3.2 - 3.2)/3 = -3\dfrac{2}{3}$.

EXAMPLE 11. *Find the directional derivative of* $\phi = 5x^2y - 5y^2z + \dfrac{5}{2}z^2x$ *at the point P(1, 1, 1) in the direction of the line* $\dfrac{x-1}{2} = \dfrac{y-3}{2} = \dfrac{z}{1}$.

(GBTU–2010, UPTU–2004, BHOPAL–2008)

SOLUTION. We have $\phi = 5x^2y - 5y^2z + \dfrac{5}{2}z^2x$

$\therefore \quad \text{grad }\phi = \hat{i}\dfrac{\partial\phi}{\partial x} + \hat{j}\dfrac{\partial\phi}{\partial y} + \hat{k}\dfrac{\partial\phi}{\partial z}$

$= \left(10xy + \dfrac{5}{2}z^2\right)\hat{i} + (5x^2 - 10yz)\hat{j}$

$\qquad + (-5y^2 + 5zx)\hat{k}$

$= \dfrac{25}{2}\hat{i} - 5\hat{j}$ at the point (1, 1, 1)

$\therefore$ Required direction derivative

$= \left(\dfrac{25}{2}\hat{i} - 5\hat{j}\right)\cdot\left(\dfrac{2}{3}\hat{i} - \dfrac{2}{3}\hat{j} + \dfrac{1}{3}\hat{k}\right)$

$= \dfrac{25}{3} + \dfrac{10}{3} = \dfrac{35}{3}$.

EXAMPLE 12. *Find the directional derivative of* $\phi = (x^2 + y^2 + z^2)^{-\frac{1}{2}}$ *at the point (3, 1, 2) in the*

direction of the vector $yz\hat{i} + zx\hat{j} + xy\hat{k}$.

(UPTU–2007)

SOLUTION. We have $\phi = (x^2 + y^2 + z^2)^{-\frac{1}{2}}$

Therefore,

$\text{grad }\phi = \hat{i}\dfrac{\partial\phi}{\partial x} + \hat{j}\dfrac{\partial\phi}{\partial y} + \hat{k}\dfrac{\partial\phi}{\partial z}$

$= \hat{i}\left(-\dfrac{1}{2}(x^2 + y^2 + z^2)^{-3/2}.2x\right)$

$\quad + \hat{j}\left(-\dfrac{1}{2}(x^2 + y^2 + z^2)^{-3/2}.2y\right)$

$\quad + \hat{k}\left(-\dfrac{1}{2}(x^2 + y^2 + z^2)^{-3/2}.2z\right)$

$= -\dfrac{(x\hat{i} + y\hat{j} + z\hat{k})}{(x^2 + y^2 + z^2)^{3/2}}$

$= -\dfrac{3\hat{i} + \hat{j} + 2\hat{k}}{14\sqrt{14}}$ at (3, 1, 2).

Let $\hat{a}$ be the unit vector in the given direction then

$\hat{a} = \dfrac{yz\hat{i} + zx\hat{j} + xy\hat{k}}{\sqrt{y^2z^2 + z^2x^2 + x^2y^2}}$

$= \dfrac{2\hat{i} + 6\hat{j} + 3\hat{k}}{7}$ at (3, 1, 2).

Hence $\dfrac{d\phi}{ds} = \hat{a}.\text{grad }\phi$

$= \dfrac{2\hat{i} + 6\hat{j} + 3\hat{k}}{7}\left(-\dfrac{3\hat{i} + \hat{j} + 2\hat{k}}{14\sqrt{14}}\right)$

$= -\dfrac{2(3) + 6\cdot1 + 3\cdot2}{7\cdot14\cdot\sqrt{14}} = -\dfrac{9}{49\sqrt{14}}$

EXAMPLE 13. *If the directional derivative of* $\phi = ax^2y + by^2z + cz^2x$ *at the point (1, 1, 1) has maximum magnitude 15 in the direction parallel to the line* $\dfrac{x-1}{2} = \dfrac{y-3}{-2} = \dfrac{z}{1}$. *Find the values of a, b and c.*

SOLUTION. We have $\phi = ax^2y + by^2z + cz^2x$

$\Rightarrow \quad \text{grad }\phi = \hat{i}\dfrac{\partial\phi}{\partial x} + \hat{j}\dfrac{\partial\phi}{\partial y} + \hat{k}\dfrac{\partial\phi}{\partial z}$

$= (2axy + cz^2)\hat{i} + (ax^2 + 2byz)\hat{j}$

$\qquad + (by^2 + 2czx)\hat{k}$

$= (2a + c)\hat{i} + (a + 2b)\hat{j} + (b + 2c)\hat{k}$

at (1, 1, 1)

Now the directional derivative is maximum along the normal to the surface, *i.e.* along grad ϕ

$|\text{grad }\phi| = \sqrt{(2a+c)^2 + (a+2b)^2 + (b+2c)^2}$

$\Rightarrow \quad 15 = \sqrt{(2a+c)^2 + (a+2b)^2 + (b+2c)^2}$

$\Rightarrow (2a+c)^2 + (a+2b)^2 + (b+2c)^2 = 225 \quad \text{...(1)}$

Since the directional derivative is maximum in the

direction parallel to the line $\frac{x-1}{2} = \frac{y-3}{-2} = \frac{z}{1}$, i.e.,

parallel to the vector $2\hat{i} - 2\hat{j} + \hat{k}$, therefore,

$$\frac{2a+c}{2} = \frac{a+2b}{-2} = \frac{2c+b}{1}$$

$\Rightarrow \quad 2a + c = -a - 2b \Rightarrow 3a + 2b + c = 0 \quad \ldots(2)$

and $2b + a = -4c - 2b \Rightarrow a + 4b + 4c = 0 \quad \ldots(3)$

On solving (1), (2) and (3) by cross multiplication we get

$$\frac{a}{4} = \frac{b}{-11} = \frac{c}{10} = k \text{ (say)}$$

$\Rightarrow \quad a = 4k, b = -11k, c = 10k$

Then from (1)

$(8k + 10k)^2 + (4k - 22k)^2$

$$+ (-11k + 20k)^2 = 225$$

$\Rightarrow \quad k = \pm\dfrac{5}{9}$

Hence $a = \pm\dfrac{20}{9}, b = \mp\dfrac{55}{9}$ and $c = \pm\dfrac{50}{9}$.

Exercise-6.3

1. If $\phi(x, y, z) = x^2 y + y^2 x + z^2$, find $\nabla\phi$ at the point $(1, 1, 1)$.

2. If $f(x, y, z) = x^2 yz\hat{i} - 2xz^3\hat{j} + xz^2\hat{k}, \phi(x, y, z) = 2z\hat{i} + y\hat{j} - x^2\hat{k}$, find the value of $\dfrac{\partial^2}{\partial x \partial y}(f \times \phi)$ at $(1, 0, -2)$.

3. If $|r| = r$ where $r = x\hat{i} + y\hat{j} + z\hat{k}$, prove that

(i) $\nabla\left(\dfrac{1}{r}\right) = -\dfrac{r}{r^3}$ (GBTU–2011)

(ii) $\nabla \log r = \dfrac{r}{r^2}$ (GBTU–2011)

4. Prove that $f(r)\nabla r = \nabla \int f(r) dr$.

5. (i) Interpret the symbol $\mathbf{a}.\nabla$. (ii) Prove that $(\mathbf{a}.\nabla)\phi = \mathbf{a}.\nabla\phi$.
(iii) Prove that $(\mathbf{a}.\nabla)\mathbf{r} = \mathbf{a}$.

6. Find the grad f, where f is given by $f(x, y, z) = x^3 - y^3 + xz^2$, at the point $(1, -1, 2)$.

7. If $u = x + y + z, v = x^2 + y^2 + z^2, w = yz + zx + xy$, prove that (grad u) . [(grad v) × (grad w)] = 0.
(UKTU–2010, UPTU–2002)

8. f and p are two scalar point functions such that f is a function of p, show that

$$\nabla f = \frac{df}{dp}\nabla p.$$

9. If $\mathbf{F} = \left(y\dfrac{\partial f}{\partial z} - z\dfrac{\partial f}{\partial y}\right)\hat{i} + \left(z\dfrac{\partial f}{\partial x} - x\dfrac{\partial f}{\partial z}\right)\hat{j} + \left(x\dfrac{\partial f}{\partial y} - y\dfrac{\partial f}{\partial x}\right)\hat{k}$. Prove that

(i) $\mathbf{F} = \mathbf{r} \times \nabla f$ (ii) $\mathbf{F} . \mathbf{r} = 0$
(iii) $\mathbf{F} . \nabla f = 0$

10. Prove that the directional derivative of a scalar field f at a point $P(x, y, z)$ in the direction of a unit vector $\hat{a}$ is given by

$$\frac{\partial f}{\partial s} = \nabla f.\hat{a}$$

11. Find the directional derivative of the function

$$f(x, y, z) = x^2 - y^2 + 2z^2.$$

at the point $P(1, 2, 3)$ in the direction of the line PQ where Q is the point $(5, 0, 4)$. (UKTU–2011, GBTU–2010)

12. In what direction from the point $(1, 1, -1)$ is the directional derivative of $f = x^2 - 2y^2 + 4z^2$ a maximum? Also find the value of this maximum directional derivative.

13. Find the directional derivative of the function $f = xy + yz + zx$ in the direction of the vector $2\hat{i} + 3\hat{j} + 6\hat{k}$ at the point $(3, 1, 2)$.

14. Find the greatest value of the derivative of the function $f = 2x^2 - y - z^4$ at the point $(2, -1, 1)$.

15. Find the directional derivative $\partial f/\partial s$ of $f(x, y, z) = 2x^2 + 3y^2 + z^2$ at the point $P(2, 1, 3)$ in the direction of the vector $\mathbf{a} = \hat{i} - 2\hat{k}$. (UPTU–2009)

16. Find the directional derivative of $f = x^2 + y^2 + z^2$ at $(1, 2, 3)$ in the direction of the line $\dfrac{x}{3} = \dfrac{y}{4} = \dfrac{z}{5}$.

17. Find the directional derivative of the function
$$f(x, y, z) = 4e^{x + 5y - 13z} \text{ at the point } (1, 2, 3)$$
in the direction towards the point $(-3, 5, 7)$. (UPTU–2009)

18. In what direction from $(3, 1, -2)$ is the directional derivative of $\phi = x^2 y^2 z^4$ maximum and what is its magnitude.
(ROHTAK–2003)

19. Show that grad $(e^{r^2}) = 2e^{r^2} \cdot r$.

20. Show that grad $f(r) \times r = 0$. (GBTU–2011)

21. Show that the directional derivative of $\dfrac{1}{r}$ in the direction of $\mathbf{r}$ where $\vec{r} = x\hat{i} + y\hat{j} + z\hat{k}$ is $-\dfrac{1}{r^2}$. (UKTU–2011)

22. Show that the directional derivative of $\dfrac{1}{r^2}$ in the direction of $\mathbf{r}$ where $\vec{r} = x\hat{i} + y\hat{j} + z\hat{k}$ is $-\dfrac{2}{r^3}$. (UPTU–2006)

Hints to Selected Problems

1. $\phi(x, y, z) = x^2 y + y^2 x + z^2$

$\therefore \quad \nabla\phi = \dfrac{\partial\phi}{\partial x}\hat{i} + \dfrac{\partial\phi}{\partial y}\hat{j} + \dfrac{\partial\phi}{\partial z}\hat{k}$

$\qquad = (2xy + y^2)\hat{i} + (x^2 + 2xy)\hat{j} + 2z\hat{k}$

$\therefore \quad [\nabla\phi]_{(1,1,1)} = 3\hat{i} + 3\hat{j} + 2\hat{k}$.

3. (ii) Since $\quad \nabla f(r) = f(r)\nabla r = f'(r)\dfrac{r}{r}$.

$\therefore \quad \nabla \log r = \dfrac{d}{dt}(\log r)\dfrac{r}{r} = \dfrac{1}{r}\dfrac{r}{r} = \dfrac{r}{r^2}$.

(iv) $\qquad = x\hat{i} + y\hat{j} + z\hat{k} \quad \therefore \quad dr = dx\hat{i} + dy\hat{j} + dz\hat{k}$.

$\therefore \quad \nabla\phi.dr = \left(\dfrac{\partial\phi}{\partial x}\hat{i} + \dfrac{\partial\phi}{\partial y}\hat{j} + \dfrac{\partial\phi}{\partial z}\hat{k}\right).(dx\hat{i} + dy\hat{j} + dz\hat{k})$

$\qquad = \dfrac{\partial\phi}{\partial x}dx + \dfrac{\partial\phi}{\partial y}dy + \dfrac{\partial\phi}{\partial z}dz = d\phi.$

5. Let $\quad \boldsymbol{a} = a_1\hat{i} + a_2\hat{j} + a_3\hat{k}, \boldsymbol{r} = x\hat{i} + y\hat{j} + z\hat{k}.$

$\therefore \qquad \boldsymbol{a}\cdot\nabla = a_1\dfrac{\partial}{\partial x} + a_2\dfrac{\partial}{\partial y} + a_3\dfrac{\partial}{\partial z}$

Now $\qquad (\boldsymbol{a}\cdot\nabla)\boldsymbol{r} = a_1\dfrac{\partial \boldsymbol{r}}{\partial x} + a_2\dfrac{\partial \boldsymbol{r}}{\partial y} + a_3\dfrac{\partial \boldsymbol{r}}{\partial z}$

$\qquad\qquad = a_1\hat{i} + a_2\hat{j} + a_3\hat{k} = \boldsymbol{a}$

7. $\quad u = x + y + z, v = x^2 + y^2 + z^2, w = yz + zx + xy$

$\qquad \text{grad } v = \dfrac{\partial v}{\partial x}\hat{i} + \dfrac{\partial v}{\partial y}\hat{j} + \dfrac{\partial v}{\partial z}\hat{k} = 2x\hat{i} + 2y\hat{j} + 2z\hat{k}$

$\qquad \text{grad } w = \dfrac{\partial w}{\partial x}\hat{i} + \dfrac{\partial w}{\partial y}\hat{j} + \dfrac{\partial w}{\partial z}\hat{k}$

$\qquad\qquad = (z+y)\hat{i} + (z+x)\hat{j} + (y+x)\hat{k}$

$(\text{grad } v)\times(\text{grad } w) = \begin{vmatrix} \hat{i} & \hat{j} & \hat{k} \\ 2x & 2y & 2z \\ z+y & z+x & y+x \end{vmatrix}$

$\qquad\qquad = \hat{i}(2y^2 + 2xy - 2z^2 - 2zx)$

$\qquad\qquad + \hat{j}(2z^2 + 2yz - 2xy - 2x^2)$

$\qquad\qquad + \hat{k}(2xz + 2x^2 - 2yz - 2y^2)$

$\qquad \text{grad } u = \dfrac{\partial u}{\partial x}\hat{i} + \dfrac{\partial u}{\partial y}\hat{j} + \dfrac{\partial u}{\partial z}\hat{k} = \hat{i} + \hat{j} + \hat{k}$

$\therefore \ (\text{grad } u)\cdot[(\text{grad } v)\times(\text{grad } w)]$

$\qquad\qquad = (2y^2 + 2xy - 2z^2 - 2zx)$

$\qquad\qquad + (2z^2 + 2yz - 2xy - 2x^2)$

$\qquad\qquad + (2xz + 2x^2 - 2yz - 2y^2)$

8. Since $\qquad\qquad\qquad = 0.$
$\qquad\qquad\qquad f = f(p)$

$\therefore \qquad \nabla f = \nabla(f(p))$

$\qquad\qquad = \dfrac{\partial}{\partial x}(f(p))\hat{i} + \dfrac{\partial}{\partial y}(f(p))\hat{j} + \dfrac{\partial}{\partial z}(f(p))\hat{k}$

$\qquad\qquad = \dfrac{df}{dp}\dfrac{\partial p}{\partial x}\hat{i} + \dfrac{df}{dp}\dfrac{\partial p}{\partial y}\hat{j} + \dfrac{df}{dp}\dfrac{\partial p}{\partial z}\hat{k}$

$\qquad\qquad = \dfrac{df}{dp}\left(\dfrac{\partial p}{\partial x}\hat{i} + \dfrac{\partial p}{\partial y}\hat{j} + \dfrac{\partial p}{\partial z}\hat{k}\right) = \dfrac{df}{dp}\nabla p.$

11. $\qquad f(x,y,z) = x^2 - y^2 + 2z^2$

$\therefore \qquad \dfrac{\partial f}{\partial x} = 2x, \dfrac{\partial f}{\partial y} = -2y, \dfrac{\partial f}{\partial z} = 4z.$

$\therefore \qquad \nabla f = \dfrac{\partial f}{\partial x}\hat{i} + \dfrac{\partial f}{\partial y}\hat{j} + \dfrac{\partial f}{\partial z}\hat{k} = 2x\hat{i} - 2y\hat{j} + 4z\hat{k}$

at $P(1, 2, 3)$ $\qquad \nabla f = 2\hat{i} - 4\hat{j} + 12\hat{k}$

Now $\qquad \overline{PQ} = 4\hat{i} - 2\hat{j} + \hat{k}$

$\therefore \qquad$ directional derivative along $\overline{PQ}$ is

$(\nabla f)\cdot\dfrac{\overline{PQ}}{|\overline{PQ}|} = \dfrac{(8 + 8 + 12)}{\sqrt{16 + 4 + 1}}$

$\qquad\qquad = \dfrac{28}{\sqrt{21}} = \dfrac{28\sqrt{21}}{21} = \dfrac{4}{3}\sqrt{21}.$

Answers

1. $3\hat{i} + 3\hat{j} + 2\hat{k}$ $\qquad$ **2.** $-4\hat{i} - 8\hat{j}$ $\qquad$ **6.** $7\hat{i} - 3\hat{j} + 4\hat{k}$ $\qquad$ **11.** $\dfrac{4}{3}\sqrt{21}$ $\quad$ **12.** $2\hat{i} - 4\hat{j} - 8\hat{k}, 2\sqrt{(21)}$

13. $45/7$ $\quad$ **14.** 9 $\quad$ **15.** $-4/\sqrt{5}$ $\qquad$ **16.** $\dfrac{52}{\sqrt{50}}$ $\quad$ **17.** $-4\sqrt{41}e^{-28}$ $\qquad$ **18.** $96(\hat{i} + 3\hat{j} - 3\hat{k}), 96\sqrt{19}$

6.7 LEVEL SURFACES

Let us consider a scalar function $f(x, y, z)$ and suppose that for each constant c, the equation

$$f(x, y, z) = c, \text{ constant}$$

represents a surface in space (in three dimensional space). Then assuming c takes all values, we obtain a family of surfaces, which are known as level surfaces.

THEOREM 1. *Let $f(x, y, z)$ be a scalar point function over some region R, then show that through any point on R there passes one and only one, level surface of f.*

PROOF. Let (x_1, y_1, z_1) be any point in space (on R). Then the level surface $f(x, y, z)$ will pass through this point

$\therefore \qquad\qquad f(x, y, z) = f(x_1, y_1, z_1).$ $\qquad\qquad$ ...(1)

Let us suppose that the level surfaces $f(x, y, z) = c_1$ and $f(x, y, z) = c_2$ passes through the point (x_1, y_1, z_1), then

$\qquad f(x_1, y_1, z_1) = c_1 \qquad$ and $\qquad f(x_1, y_1, z_1) = c_2.$

Using (1), we get

$$c_1 = c_2.$$

Hence, through (x_1, y_1, z_1) there passes one and only one level surface.

THEOREM 2. *grad $(f) = \nabla f$ is a normal vector to the surface $f(x, y, z) = c$, where c is a constant.*

PROOF. The equation of a curve in space can be represented in the form

$$\boldsymbol{r}(t) = x(t)\hat{i} + y(t)\hat{j} + z(t)\hat{k}.$$ $\qquad\qquad$ ...(1)

If the curve lies on the surface $f(x, y, z) = c$, then we have $f[x(t), y(t), z(t)] = c$.

Differentiating with respect to t, we get

$$\dfrac{\partial f}{\partial x}\cdot\dfrac{dx}{dt} + \dfrac{\partial f}{\partial y}\cdot\dfrac{dy}{dt} + \dfrac{\partial f}{\partial z}\cdot\dfrac{dz}{dt} = 0$$

or $\qquad \left(\dfrac{\partial f}{\partial x}\hat{i}+\dfrac{\partial f}{\partial y}\hat{j}+\dfrac{\partial f}{\partial z}\hat{k}\right)\cdot\left(\dfrac{dx}{dt}\hat{i}+\dfrac{dy}{dt}\hat{j}+\dfrac{dz}{dt}\hat{k}\right)=0$ ·

or $\qquad\qquad (\mathrm{grad}\,f).\dfrac{d\boldsymbol{r}}{dt}=0 \qquad\qquad\qquad$ [Using (1)] $\qquad$ ...(2)

where the vector $\quad \dfrac{d\boldsymbol{r}}{dt}=\dfrac{dx}{dt}\hat{i}+\dfrac{dy}{dt}\hat{j}+\dfrac{dz}{dt}\hat{k}\ $ is a vector parallel to the tangent at the point $P(x,y,z)$ to the surface.

Equation (2) implies that $(\mathrm{grad}\,f)$ is perpendicular to the tangent plane at $P(x,y,z)$ to the surface $f(x,y,z)=c$. Hence, $(\mathrm{grad}\,f)$ or ∇f is a vector normal to the surface $f(x,y,z)=c$.

6.7.1 Tangent and Normal to the Level Surface

(i) **Tangent Plane.** Let $f(x,y,z)=c$ be the equation of a level surfaces and let $P(x,y,z)$ be any point on this surface whose position vector be $\boldsymbol{r}$.

$\therefore \qquad\qquad\qquad\qquad\qquad \boldsymbol{r}=x\hat{i}+y\hat{j}+z\hat{k}. \qquad\qquad\qquad\qquad\qquad$ ...(1)

Since ∇f is perpendicular to the tangent plane at $P(x,y,z)$. Let Q be any variable point on the surface tangent plane to the whose co-ordinates are (X,Y,Z) and whose position vector is $\boldsymbol{R}$.

$\therefore \qquad\qquad\qquad \overrightarrow{PQ}=\boldsymbol{R}-\boldsymbol{r}=(X-x)\hat{i}+(Y-y)\hat{j}+(Z-z)\hat{k}.$

Since $\overrightarrow{PQ}$ is along the tangent plane at P, then ∇f is perpendicular to $\overrightarrow{PQ}$.

$\therefore \qquad\qquad\qquad\qquad\qquad \nabla f.\overrightarrow{PQ}=0$

$$\left(\dfrac{\partial f}{\partial x}\hat{i}+\dfrac{\partial f}{\partial y}\hat{j}+\dfrac{\partial f}{\partial z}\hat{k}\right)((X-x)\hat{i}+(Y-y)\hat{j}+(Z-z)\hat{k})=0$$

or $\qquad\qquad (X-x)\dfrac{\partial f}{\partial x}+(Y-y)\dfrac{\partial f}{\partial y}+(Z-z)\dfrac{\partial f}{\partial z}=0$

This is the equation of a tangent plane at $P(x,y,z)$ to the level surface $f(x,y,z)=c$.

(ii) **Normal.** In this case the point Q is taken on the normal to the surface $f(x,y,z)=c$. So that the direction ratios of the line PQ are $X-x,Y-y,Z-z$

or $\qquad\qquad\qquad \overrightarrow{PQ}=(X-x)\hat{i}+(Y-y)\hat{j}+(Z-z)\hat{k}$

Thus the vector $\overrightarrow{PQ}$ is now parallel to the ∇f.

$\therefore \qquad\qquad\qquad\qquad\qquad \nabla f\times\overrightarrow{PQ}=\mathbf{0}$

or $\qquad \left(\dfrac{\partial f}{\partial x}\hat{i}+\dfrac{\partial f}{\partial y}\hat{j}+\dfrac{\partial f}{\partial z}\hat{k}\right)\times((X-x)\hat{i}+(Y-y)\hat{j}+(Z-z)\hat{k})=\mathbf{0}$

$$\left[\dfrac{\partial f}{\partial y}(Z-z)-\dfrac{\partial f}{\partial z}(Y-y)\right]\hat{i}+\left[\dfrac{\partial f}{\partial z}(X-x)-\dfrac{\partial f}{\partial x}(Z-z)\right]\hat{j}+\left[\dfrac{\partial f}{\partial x}(Y-y)-\dfrac{\partial f}{\partial y}(X-x)\right]\hat{k}=\mathbf{0}$$

or $\quad \dfrac{\partial f}{\partial y}(Z-z)-\dfrac{\partial f}{\partial z}(Y-y)=0,\quad \dfrac{\partial f}{\partial z}(X-x)-\dfrac{\partial f}{\partial x}(Z-z)=0\ $ and $\ \dfrac{\partial f}{\partial x}(Y-y)-\dfrac{\partial f}{\partial y}(X-x)=0.$

From these three equations we obtain the equation to the normal which is given as follows :

$$\dfrac{X-x}{\dfrac{\partial f}{\partial x}}=\dfrac{Y-y}{\dfrac{\partial f}{\partial y}}=\dfrac{Z-z}{\dfrac{\partial f}{\partial z}}.$$

✒ REMARK

- If $f(x,y,z)=c$, then $\dfrac{\partial f}{\partial x},\dfrac{\partial f}{\partial y},\dfrac{\partial f}{\partial z}$ are the direction ratios of the normal to the surface.

Solved Examples

EXAMPLE 1. *Find a unit vector normal to the surface* $xy^3z^2=0$
at the point $(-1,-1,2)$. (UPTU–2008, MUMBAI–2008)

SOLUTION . We have $\phi=xy^3z^2$

$\therefore \qquad \mathrm{grad}\,\phi=\hat{i}\dfrac{\partial\phi}{\partial x}+\hat{j}\dfrac{\partial\phi}{\partial y}+\hat{k}\dfrac{\partial\phi}{\partial z}$

$= \hat{i}\dfrac{\partial}{\partial x}(xy^3z^2)+\hat{j}\dfrac{\partial}{\partial y}(xy^3z^2)$

$\qquad\qquad +\hat{k}\dfrac{\partial}{\partial z}(xy^3z^2)$

$= \hat{i}(y^3z^2)+\hat{j}(3xy^2z^2)+\hat{k}(2xy^3z)$

$$= -4\hat{i} - 12\hat{j} + 4\hat{k} \text{ at the point } (-1, -1, 2)$$

Hence, required unit normal vector to the surface is

$$= \frac{-4\hat{i} - 12\hat{j} + 4\hat{k}}{\sqrt{(-4)^2 + (-12)^2 + 4^2}}$$

$$= -\frac{1}{\sqrt{11}}(\hat{i} + 3\hat{j} - \hat{k}).$$

EXAMPLE 2. *Find a unit normal vector $\hat{n}$ of the cone of revolution $z^2 = 4(x^2 + y^2)$ at the point $(1, 0, 2)$.*

(UPTU-2010)

SOLUTION. We have $\quad \phi = z^2 - 4x^2 - 4y^2$.

Then $\quad \nabla\phi = \hat{i}\dfrac{\partial\phi}{\partial x} + \hat{j}\dfrac{\partial\phi}{\partial y} + \hat{k}\dfrac{\partial\phi}{\partial z}$

$$= -8x\hat{i} - 8y\hat{j} + 2z\hat{k}$$

$$= -8\hat{i} + 4\hat{k} \text{ at the point } (1, 0, 2)$$

$$\Rightarrow \quad |\nabla\phi| = \sqrt{64 + 16} = \sqrt{80}$$

Hence, unit normal vetor $\hat{n}$ to the given cone at $(1, 0, 2)$ is

$$\hat{n} = \frac{\nabla\phi}{|\nabla\phi|} = \frac{-8\hat{i} + 4\hat{k}}{\sqrt{80}} = \frac{-2\hat{i} + \hat{k}}{\sqrt{5}}.$$

EXAMPLE 3. *Find the equation of the normal to the surface $2xz^2 - 3xy - 4x = 7$ at the point $(1, -1, 2)$.*

SOLUTION. Let $f(x, y, z) \equiv 2xz^2 - 3xy - 4x - 7 = 0$

$$\therefore \qquad \nabla f = \frac{\partial f}{\partial x}\hat{i} + \frac{\partial f}{\partial y}\hat{j} + \frac{\partial f}{\partial z}\hat{k}$$

$$= (2z^2 - 3y - 4)\hat{i} - 3x\hat{j} + 4xz\hat{k}$$

At $(1, -1, 2)$

$$\nabla f = 7\hat{i} - 3\hat{j} + 8\hat{k}$$

Now the position vector of the point $(1, -1, 2)$ is

$$\therefore \qquad \boldsymbol{r} = \hat{i} - \hat{j} + 2\hat{k}$$

Let $\boldsymbol{R} = X\hat{i} + Y\hat{j} + Z\hat{k}$ be the position vector of any variable point (X, Y, Z) on the normal, then the vector $\boldsymbol{R} - \boldsymbol{r}$ is parallel to ∇f.

$$(\boldsymbol{R} - \boldsymbol{r}) \times \nabla f = 0$$

or the equation of the normal is

$$\frac{X-1}{7} = \frac{Y+1}{-3} = \frac{Z-2}{8}.$$

EXAMPLE 4. *Find the angle between the surface $x^2 + y^2 + z^2 = 9$ and $z = x^2 + y^2 - 3$ at the point $(2, -1, 2)$.*

(VTU-2010, UPTU-2003, KOTTAYAM-2005)

SOLUTION. Let the given surfaces be

$$f_1(x, y, z) \equiv x^2 + y^2 + z^2 = 9 \text{ as } f_1(x, y, z) = c_1$$
$$\qquad \qquad \dots(1)$$

$$f_2(x, y, z) \equiv x^2 + y^2 - z = 3 \text{ as } f_2(x, y, z) = c_2$$
$$\qquad \qquad \dots(2)$$

Normal vector to surface (1) is

$$\boldsymbol{n_1} = \text{grad } f_1$$

$$= \left(\hat{i}\frac{\partial}{\partial x} + \hat{j}\frac{\partial}{\partial y} + \hat{k}\frac{\partial}{\partial z}\right)(x^2 + y^2 + z^2)$$

$$= 2x\hat{i} + 2y\hat{j} + 2z\hat{k}$$

At point $(2, -1, 2)$,

$$\boldsymbol{n_1} = 2.2\hat{i} + 2(-1)\hat{j} + 2.2\hat{k} = 4\hat{i} - 2\hat{j} + 4\hat{k}.$$

Normal vector to surface (2) is

$$\boldsymbol{n_2} = \text{grad } f_2$$

$$= \left(\hat{i}\frac{\partial}{\partial x} + \hat{j}\frac{\partial}{\partial y} + \hat{k}\frac{\partial}{\partial z}\right)(x^2 + y^2 - z)$$

$$= 2x\hat{i} + 2y\hat{j} - \hat{k}.$$

At point $(2, -1, 2)$,

$$\boldsymbol{n_2} = 2.2\hat{i} + 2(-1)\hat{j} - \hat{k} = 4\hat{i} - 2\hat{j} - \hat{k}.$$

Now let θ be the angle between surfaces (1) and (2), then the angle between their normals $\boldsymbol{n_1}$ and $\boldsymbol{n_2}$ is also θ.

$$\therefore \qquad \cos\theta = \frac{\boldsymbol{n_1}.\boldsymbol{n_2}}{|\boldsymbol{n_1}||\boldsymbol{n_2}|}$$

$$= \frac{4.4 + (-2)(-2) + 4(-1)}{\sqrt{(4)^2 + (-2)^2 + (4)^2}\sqrt{(4)^2 + (-2)^2 + (-1)^2}}$$

$$= \frac{16 + 4 - 4}{\sqrt{36}\sqrt{21}} = \frac{16}{6\sqrt{21}} = \frac{8}{3\sqrt{21}}$$

$$\therefore \qquad \theta = \cos^{-1}\left(\frac{8}{3\sqrt{21}}\right).$$

Exercise-6.4

1. Find the unit vector normal to the surfce $x^2 - y^2 + z = 2$ at the point $(1, -1, 2)$.

2. Find the vector normal to the surface $z = x^2 + y^2$ at the point $(-1, -2, 5)$.

3. Find the unit normal to the surface $x^2 + y - z = 4$ at the point $(2, 0, 0)$.

4. Find the equation of the tangent plane and normal to the surface $xyz = 4$ at the point $(1, 2, 2)$.

5. Find the equation of the tangent plane and normal to the surface $x^2 + y^2 + z^2 = 25$ at the point $(4, 0, 3)$.

6. Find the equation of the tangent plane and normal to the surface $z = x^2 + y^2$ at the point $(2, -1, 5)$.

7. If $\hat{n}$ be a unit vector normal to the level surface $f(x, y, z) = c$ at a point $P(x, y, z)$ and n be the distance of P from some fixed point A in the direction of $\hat{n}$ so that δn represents element of normal at P in the direction of $\hat{n}$, then

$$\text{grad } f = \frac{df}{dn}\hat{n}.$$

8. Prove that grad f is a vector in the direction of which the maximum value of the directional derivative of f.

9. Find the angle between the normals to the surface $xy = z^2$ at the points $(4, 1, 2)$ and $(3, 3, -3)$.

(UKTU-2012)

Hints to Selected Problems

1. $f \equiv x^2 - y^2 + z - 2 = 0.$

     $\therefore \quad \nabla f = \dfrac{\partial f}{\partial x}\hat{i} + \dfrac{\partial f}{\partial y}\hat{j} + \dfrac{\partial f}{\partial z}\hat{k} = 2x\hat{i} - 2y\hat{j} + \hat{k}$

     At $(1, -1, 2)$

         $\nabla f = 2\hat{i} + 2\hat{j} + \hat{k}$

     $\therefore$ Unit normal to the surface $f = \dfrac{\nabla f}{|\nabla f|} = \dfrac{2\hat{i} + 2\hat{j} + \hat{k}}{3}.$

4. $f(x, y, z) \equiv xyz - 4$

     $\therefore \quad \dfrac{\partial f}{\partial x} = yz, \dfrac{\partial f}{\partial y} = xz, \dfrac{\partial f}{\partial z} = xy$

     $\therefore$ at $(1, 2, 2)$ $\nabla f = 4\hat{i} + 2\hat{j} + 2\hat{k}$.

The equation of the tangent plane at $(1, 2, 2)$ is

     $[(x-1)\hat{i} + (y-2)\hat{j} + (z-2)\hat{k}].(4\hat{i} + 2\hat{j} + 2\hat{k}) = 0$

or $4(x-1) + 2(y-2) + 2(z-2) = 0$

or $4x + 2y + 2z - 12 = 0$ or $2x + y + z - 6 = 0$

and the equation of the normal at $(1, 2, 2)$ is

     $\dfrac{x-1}{4} = \dfrac{y-2}{2} = \dfrac{z-2}{2}$ or $\dfrac{x-1}{2} = \dfrac{y-2}{1} = \dfrac{z-2}{1}$

6. Same as (4).

Answers

1. $\dfrac{1}{3}(2\hat{i} + 2\hat{j} + \hat{k})$ **2.** $2\hat{i} + 4\hat{j} + \hat{k}$ **3.** $\dfrac{1}{3\sqrt{2}}(4\hat{i} + \hat{j} - \hat{k})$ **4.** $2x + y + z = 6; \dfrac{x-1}{2} = \dfrac{y-2}{1} = \dfrac{z-2}{1}$

5. $4x + 3z = 25; \dfrac{x-4}{4} = \dfrac{y}{0} = \dfrac{z-3}{3}$ **6.** $4x - 2y - z = 5; \dfrac{x-2}{4} = \dfrac{y+1}{-2} = \dfrac{z-5}{-1}$ **9.** $\cos^{-1}\left(\dfrac{1}{\sqrt{22}}\right)$

6.8 DIVERGENCE OF A VECTOR FIELD   (UPTU–2006, 07, 08)

Let $V(x, y, z)$ be a differentiable vector function, where x, y, z are cartesian co-ordinates in space and let V_1, V_2, V_3 be the components of V, then the function

$$\text{div } V = \frac{\partial V_1}{\partial x} + \frac{\partial V_2}{\partial y} + \frac{\partial V_3}{\partial z} \qquad \qquad \ldots(1)$$

is called the divergence of V.

Since we have that the differential operator $\nabla \equiv \dfrac{\partial}{\partial x}\hat{i} + \dfrac{\partial}{\partial y}\hat{j} + \dfrac{\partial}{\partial z}\hat{k}$ and the vector $V = V_1\hat{i} + V_2\hat{j} + V_3\hat{k}$.

Then $\nabla.V = \dfrac{\partial V_1}{\partial x} + \dfrac{\partial V_2}{\partial y} + \dfrac{\partial V_3}{\partial z}$ $\ldots(2)$

From equation (1) and (2), we get

               div $V = \nabla.V$

Hence, divergence of a vector function V can also be written as $\nabla.V$. Consequently divergence of a vector function is scalar because dot product of ∇ and V gives a scalar quantity.

✎ REMARKS

- Though dot product is cummulative but ∇ being operator which operates right side function only, we have $\nabla.f \neq f.\nabla$.
- If div $V = 0$, then the vector V is called solenoidal vector.
- If the vector V is a velocity vector of a fluid and if div $V = 0$, then the fluid is incompressible.
- div $V = \nabla.V = \Sigma \hat{i}.\dfrac{\partial V}{\partial x}$.

6.9 CURL OF A VECTOR FIELD   (UPTU–2006, 07,08; GBTU–2012)

Let $V(x, y, z)$ be a vector function of x, y, z where (x, y, z) are right handed cartesian co-ordinates in space and let

     $V(x, y, z) = V_1(x, y, z)\hat{i} + V_2(x, y, z)\hat{j} + V_3(x, y, z)\hat{k}$

be a differentiable vector function. Then the function

$$\text{curl } V = \nabla \times V = \begin{vmatrix} \hat{i} & \hat{j} & \hat{k} \\ \dfrac{\partial}{\partial x} & \dfrac{\partial}{\partial y} & \dfrac{\partial}{\partial z} \\ V_1 & V_2 & V_3 \end{vmatrix} = \left(\frac{\partial V_3}{\partial y} - \frac{\partial V_2}{\partial z}\right)\hat{i} + \left(\frac{\partial V_1}{\partial z} - \frac{\partial V_3}{\partial x}\right)\hat{j} + \left(\frac{\partial V_2}{\partial x} - \frac{\partial V_1}{\partial y}\right)\hat{k}$$

is called the curl of the vector function V or the curl of the vector field definded by Curl V is a vector quantity.

☞ REMARKS

- If curl $V = 0$, then the vector V is called irrotational.
- Curl $V = \nabla \times V = \Sigma \hat{i} \times \dfrac{\partial V}{\partial x}$.
- Curl V is perpendicular to V.
- In the case of a rigid body rotation, the curl of the velocity field has the direction of the axis of rotation and its magnitude equals twice the angular speed of the rotation.

6.10 LAPLACIAN OPERATOR

If the function $f(x, y, z)$ is a twice differentiable scalar function, then we have

$$\text{grad } f = \frac{\partial f}{\partial x}\hat{i} + \frac{\partial f}{\partial y}\hat{j} + \frac{\partial f}{\partial z}\hat{k}$$

Since grad f is a vector function, then

$$\text{div (grad } f) = \left(\frac{\partial}{\partial x}\hat{i} + \frac{\partial}{\partial y}\hat{j} + \frac{\partial}{\partial z}\hat{k}\right) \cdot \left(\frac{\partial f}{\partial x}\hat{i} + \frac{\partial f}{\partial y}\hat{j} + \frac{\partial f}{\partial z}\hat{k}\right) = \frac{\partial^2 f}{\partial x^2} + \frac{\partial^2 f}{\partial y^2} + \frac{\partial^2 f}{\partial z^2} = \left(\frac{\partial^2}{\partial x^2} + \frac{\partial^2}{\partial y^2} + \frac{\partial^2}{\partial z^2}\right) f$$

$$\therefore \qquad \text{div (grad } f) = \nabla^2 f. \qquad \qquad \qquad \dots(1)$$

Thus, R.H.S. of (1) is the Laplacian of f. Consequently the Laplacian is defined as

$$\nabla^2 = \left(\frac{\partial^2}{\partial x^2} + \frac{\partial^2}{\partial y^2} + \frac{\partial^2}{\partial z^2}\right).$$

Hence, ∇^2 is a Laplacian operator.

☞ REMARKS

- The equation $\nabla^2 f = 0$ is called Laplace's equation.
- If f is a scalar point function, then $\nabla^2 f$ is a scalar quantity.
- If f is a vector point function, then $\nabla^2 f$ is a vector quantity.
- If a function f satisfies the Laplace's equation then f is called harmonic function.

6.11 PHYSICAL INTERPRETATION OF DIVERGENCE AND CURL

(i) Physical interpretation of divergence. Let us consider the motion of a fluid in a region R having no sources or sinks in R.

(GBTU–2010)

Let $V(x, y, z)$ be the velocity of the fluid at any point. Now consider the flow through a small rectangular parallelopiped of edges δx, δy, δz parallel to the co-ordinate axes as shown in fig. 5
The volume of this parallelopiped is $\delta x \, \delta y \, \delta z$. Let

$$V = V_1 \hat{i} + V_2 \hat{j} + V_3 \hat{k}$$

and assuming that V is continuous differentiable vector function of x, y, z. Let us calculate the change in the mass in the parallelopiped by considering the flux across the boundary, that is the total loss of mass leaving the parallelopiped per unit time. Let the co-ordinates of A be (x, y, z). Now consider a flow through the face $ABCD$ whose area is $\delta x \delta z$.

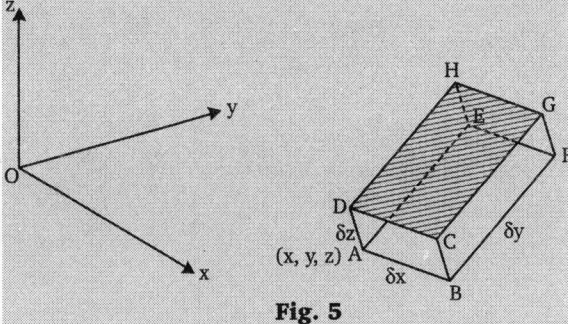

Fig. 5

In this case, the components V_1 and V_2 of V are parallel to that face, so no contribution to that flow. Hence, the mass of the fluid entering through the face $ABCD$ per unit time is given approximately by $V_2(y)\delta x \delta z$ and the mass of fluid leaving the face $EFGH$ per unit time is approximately $V_2(y + \delta y)\delta x \delta z$.

The loss of mass along y-axis is given by

$$V_2(y + \delta y)\delta x \delta z - V_2(y)\delta x \delta z = [V_2(y + \delta y) - V_2(y)]\delta x \delta z$$

$$= \frac{\delta V_2}{\delta y}\delta x \delta y \delta z.$$

where $V_2(y + \delta y)$ and $V_2(y)$ are the components of V along y-axis at point A and E respectively.

Similarly, loss in mass per unit time in x-direction is $\dfrac{\partial V_1}{\partial x}\delta x \delta y \delta z$ and loss in mass per unit time in z-direction is $\dfrac{\partial V_3}{\partial z}\delta x \delta y \delta z$.

$\therefore$ The total loss in mass per unit time in the parallelopiped is given by

$$\left(\frac{\partial V_1}{\partial x} + \frac{\partial V_2}{\partial y} + \frac{\partial V_3}{\partial z}\right)\delta x \delta y \delta z.$$

Hence, the total loss in the mass per unit time and per unit volume is

$$\left(\frac{\partial V_1}{\partial x} + \frac{\partial V_2}{\partial y} + \frac{\partial V_3}{\partial z}\right) = \text{div } \mathbf{V} = \nabla.\mathbf{V}.$$

Hence, the divergence of $\mathbf{V}$ is nothing but the loss in mass of fluid per unit volume per unit time in a small parallelopiped whose edges are parallel to the co-ordinate axes.

(ii) Physical interpretation of the curl.

(GBTU–2010, MTU–2011)

Circulation. The integral $\oint_C \mathbf{V}.d\mathbf{r}$ is called circulation where C is a closed curve and $\mathbf{V}$ a velocity vector.

Let S be a circular disc of radius r and enclosed by a closed curve C (a circle). Let $\mathbf{V}(x, y, z)$ be the velocity vector and assuming that $\mathbf{V}$ is continuously differentiable in S. Then by Stoke's theorem

$$\oint_C \mathbf{V}.d\mathbf{r} = \iint_S (\text{curl } \mathbf{V}).\mathbf{n}\, dS$$

Let λ be the intermediate value between the maximum and minimum of $(\text{curl } \mathbf{V}).\mathbf{n}$ over S. Then by mean value theorem of integral calculus, we have

$$\oint_C \mathbf{V}.d\mathbf{r} = \lambda \iint_S dS = \lambda S$$

or

$$\lambda = \frac{\oint_C \mathbf{V}.d\mathbf{r}}{S}$$

Now taking limit as $r \to 0$, we get

$$(\text{curl } \mathbf{V}).\mathbf{n} = \lim_{r \to 0} \frac{\oint_C \mathbf{V}.d\mathbf{r}}{S}$$

Now $(\text{curl } \mathbf{V}).\mathbf{n}$ is a normal component of curl $\mathbf{V}$ at the centre of circular disc S and $\oint_C \mathbf{V}.d\mathbf{r}$ is a circulation of $\mathbf{V}$ about C. Hence the normal component of the curl is nothing but the limit of the circulation per unit area.

REMARK

- Stoke's theorem states that $\oint_C \mathbf{A}.d\mathbf{r} = \iint_S (\text{curl } \mathbf{A}).\mathbf{n}\, dS$. This theorem will discuss later on.

Solved Examples

EXAMPLE 1. *Prove that the followings :*
  (i) *div* $\mathbf{r}$ = 3   (ii) *curl* $\mathbf{r}$ = 0
  where $\mathbf{r} = x\hat{i} + y\hat{j} + z\hat{k}$.

SOLUTION. (i) Since div $\mathbf{r} = \nabla.\mathbf{r}$

$$= \left(\frac{\partial}{\partial x}\hat{i} + \frac{\partial}{\partial y}\hat{j} + \frac{\partial}{\partial z}\hat{k}\right).(x\hat{i} + y\hat{j} + z\hat{k})$$

$$= \frac{\partial x}{\partial x} + \frac{\partial y}{\partial y} + \frac{\partial z}{\partial z} = 1 + 1 + 1 = 3.$$

(ii) Curl $\mathbf{r} = \nabla \times \mathbf{r}$

$$= \begin{vmatrix} \hat{i} & \hat{j} & \hat{k} \\ \frac{\partial}{\partial x} & \frac{\partial}{\partial y} & \frac{\partial}{\partial z} \\ x & y & z \end{vmatrix}$$

$$= \hat{i}\left(\frac{\partial z}{\partial y} - \frac{\partial y}{\partial z}\right) + \hat{j}\left(\frac{\partial x}{\partial z} - \frac{\partial z}{\partial x}\right) + \hat{k}\left(\frac{\partial y}{\partial x} - \frac{\partial x}{\partial y}\right)$$

$$= \hat{i}(0 - 0) + \hat{j}(0 - 0) + \hat{k}(0 - 0) = \mathbf{0}.$$

EXAMPLE 2. *If a is a constant vector, find*
  (i) *div* $(\mathbf{r} \times \mathbf{a})$ (ii) *curl* $(\mathbf{r} \times \mathbf{a})$
  where $\mathbf{r} = x\hat{i} + y\hat{j} + z\hat{k}$.

SOLUTION. Let $\mathbf{a} = a_1\hat{i} + a_2\hat{j} + a_3\hat{k}$ be a constant vector, then

$$\mathbf{r} \times \mathbf{a} = \begin{vmatrix} \hat{i} & \hat{j} & \hat{k} \\ x & y & z \\ a_1 & a_2 & a_3 \end{vmatrix}$$

$$= \hat{i}(ya_3 - za_2) + \hat{j}(za_1 - xa_3) + \hat{k}(xa_2 - ya_1)$$

(i) $\text{div}(\mathbf{r} \times \mathbf{a}) = \nabla.(\mathbf{r} \times \mathbf{a})$

$$= \frac{\partial}{\partial x}(ya_3 - za_2) + \frac{\partial}{\partial y}(za_1 - xa_3)$$

$$+ \frac{\partial}{\partial z}(xa_2 - ya_1)$$

$$= 0 + 0 + 0 = 0$$

(ii) $\text{curl}(\mathbf{r} \times \mathbf{a}) = \nabla \times (\mathbf{r} \times \mathbf{a})$

$$= \begin{vmatrix} \hat{i} & \hat{j} & \hat{k} \\ \frac{\partial}{\partial x} & \frac{\partial}{\partial y} & \frac{\partial}{\partial z} \\ (ya_3 - za_2) & (za_1 - xa_3) & (xa_2 - ya_1) \end{vmatrix}$$

$$= \hat{i}\left[\frac{\partial}{\partial y}(xa_2 - ya_1) - \frac{\partial}{\partial z}(za_1 - xa_3)\right]$$

$$+ \hat{j}\left[\frac{\partial}{\partial z}(ya_3 - za_2) - \frac{\partial}{\partial x}(xa_2 - ya_1)\right]$$

$$+ \hat{k}\left[\frac{\partial}{\partial x}(za_1 - xa_3) - \frac{\partial}{\partial y}(ya_3 - za_2)\right]$$

$$= \hat{i}[-a_1 - a_1] + \hat{j}[-a_2 - a_2] + \hat{k}[-a_3 - a_3]$$

$$= -2(a_1\hat{i} + a_2\hat{j} + a_3\hat{k}) = -2\mathbf{a}$$

Similarly, we can show that

$$\frac{1}{2}\text{curl}(\mathbf{a} \times \mathbf{r}) = \mathbf{a}.$$

EXAMPLE 3. *If V is differentiable vector function and f is a scalar point function, then show that*
  (i) *div (f V) = f div V + V.(grad f)*   (UPTU–2006)
  (ii) *curl (f V) = (∇f) × V + f(∇ × V)*

SOLUTION. (i) Since we know that

$$\operatorname{div}(f\mathbf{V}) = \nabla.(f\mathbf{V})$$

$$= \left(\hat{i}\frac{\partial}{\partial x} + \hat{j}\frac{\partial}{\partial y} + \hat{k}\frac{\partial}{\partial z}\right).(f\mathbf{V})$$

$$= \hat{i}\cdot\frac{\partial}{\partial x}(f\mathbf{V}) + \hat{j}\cdot\frac{\partial}{\partial y}(f\mathbf{V}) + \hat{k}\cdot\frac{\partial}{\partial z}(f\mathbf{V})$$

$$= \hat{i}\cdot\left(\frac{\partial f}{\partial x}\mathbf{V} + f\frac{\partial \mathbf{V}}{\partial x}\right) + \hat{j}\cdot\left(\frac{\partial f}{\partial y}\mathbf{V} + f\frac{\partial \mathbf{V}}{\partial y}\right)$$

$$+ \hat{k}\cdot\left(\frac{\partial f}{\partial z}\mathbf{V} + f\frac{\partial \mathbf{V}}{\partial z}\right)$$

$$= \left[\hat{i}\cdot\left(\frac{\partial f}{\partial x}\mathbf{V}\right) + \hat{j}\cdot\left(\frac{\partial f}{\partial y}\mathbf{V}\right) + \hat{k}\cdot\left(\frac{\partial f}{\partial z}\mathbf{V}\right)\right]$$

$$+ \left[\hat{i}\cdot\left(f\frac{\partial \mathbf{V}}{\partial x}\right) + \hat{j}\cdot\left(f\frac{\partial \mathbf{V}}{\partial y}\right) + \hat{k}\cdot\left(\frac{\partial \mathbf{V}}{\partial z}f\right)\right]$$

$$= \left[\left(\frac{\partial f}{\partial x}\hat{i}\right)\cdot\mathbf{V} + \left(\frac{\partial f}{\partial y}\hat{j}\right)\cdot\mathbf{V} + \left(\frac{\partial f}{\partial z}\hat{k}\right)\cdot\mathbf{V}\right]$$

$$+ \left[f\left(\hat{i}\cdot\frac{\partial \mathbf{V}}{\partial x}\right) + f\left(\hat{j}\cdot\frac{\partial \mathbf{V}}{\partial y}\right) + f\left(\hat{k}\cdot\frac{\partial \mathbf{V}}{\partial z}\right)\right]$$

$$= \left[\left(\frac{\partial f}{\partial x}\cdot\hat{i} + \frac{\partial f}{\partial z}\cdot\hat{k} + \frac{\partial f}{\partial y}\cdot\hat{j}\right)\mathbf{V}\right]$$

$$+ \left[f\left(\hat{i}\cdot\frac{\partial \mathbf{V}}{\partial x} + \hat{j}\cdot\frac{\partial \mathbf{V}}{\partial y} + \hat{k}\cdot\frac{\partial \mathbf{V}}{\partial z}\right)\right]$$

$$= (\nabla f).\mathbf{V} + f(\nabla.\mathbf{V})$$

$$\therefore \operatorname{div}(f\mathbf{V}) = f(\operatorname{div}\mathbf{V}) + \mathbf{V}\cdot(\operatorname{grad} f)$$

(ii) Curl (f **V**) = $\nabla \times$ (f **V**)

$$= \left(\hat{i}\frac{\partial}{\partial x} + \hat{j}\frac{\partial}{\partial y} + \hat{k}\frac{\partial}{\partial z}\right)\times(f\mathbf{V})$$

$$= \left(\hat{i}\frac{\partial}{\partial x}\right)\times(f\mathbf{V}) + \left(\hat{j}\frac{\partial}{\partial y}\right)\times(f\mathbf{V})$$

$$+ \left(\hat{k}\frac{\partial}{\partial z}\right)\times(f\mathbf{V})$$

$$= \hat{i}\times\frac{\partial}{\partial x}(f\mathbf{V}) + \hat{j}\times\frac{\partial}{\partial y}(f\mathbf{V})$$

$$+ \hat{k}\times\frac{\partial}{\partial z}(f\mathbf{V})$$

$$= \hat{i}\times\left(\frac{\partial f}{\partial x}\mathbf{V} + f\frac{\partial \mathbf{V}}{\partial x}\right)$$

$$+ \hat{j}\times\left(\frac{\partial f}{\partial y}\mathbf{V} + f\frac{\partial \mathbf{V}}{\partial y}\right)$$

$$+ \hat{k}\times\left(\frac{\partial f}{\partial z}\mathbf{V} + f\frac{\partial \mathbf{V}}{\partial z}\right)$$

$$= \left[\left(\frac{\partial f}{\partial x}\hat{i}\right)\times\mathbf{V} + \left(\frac{\partial f}{\partial y}\hat{j}\right)\times\mathbf{V} + \left(\frac{\partial f}{\partial z}\hat{k}\right)\times\mathbf{V}\right]$$

$$+ \left[f\left(\hat{i}\times\frac{\partial \mathbf{V}}{\partial x}\right) + f\left(\hat{j}\times\frac{\partial \mathbf{V}}{\partial y}\right) + f\left(\hat{k}\times\frac{\partial \mathbf{V}}{\partial z}\right)\right]$$

$$= \left(\frac{\partial f}{\partial x}\hat{i} + \frac{\partial f}{\partial y}\hat{j} + \frac{\partial f}{\partial z}\hat{k}\right)\times\mathbf{V}$$

$$+ f\left(\hat{i}\times\frac{\partial \mathbf{V}}{\partial x} + \hat{j}\times\frac{\partial \mathbf{V}}{\partial y} + \hat{k}\times\frac{\partial \mathbf{V}}{\partial z}\right)$$

$$= (\nabla f)\times\mathbf{V} + f(\nabla\times\mathbf{V}).$$

$$\therefore \operatorname{curl}(f\mathbf{V}) = (\nabla f)\times\mathbf{V} + f(\nabla\times\mathbf{V}).$$

EXAMPLE 4. *Find the divergence of the following vector functions:*
  (i) $\mathbf{f} = x^2\hat{i} + y^2\hat{j} - z\hat{k}$ (ii) $\mathbf{f} = xyz(\hat{i} + \hat{j} + \hat{k})$

SOLUTION. (i) We have

$$\operatorname{div}\mathbf{f} = \nabla.\mathbf{f}$$

$$= \left(\hat{i}\frac{\partial}{\partial x} + \hat{j}\frac{\partial}{\partial y} + \hat{k}\frac{\partial}{\partial z}\right)$$

$$.(x^2\hat{i} + y^2\hat{j} - z\hat{k})$$

$$= \frac{\partial}{\partial x}(x^2) + \frac{\partial}{\partial y}(y^2) + \frac{\partial}{\partial z}(-z)$$

$$= 2x + 2y - 1$$

(ii) $\mathbf{f} = xyz(\hat{i} + \hat{j} + \hat{k})$

$$\therefore \operatorname{div}\mathbf{f} = \nabla.\mathbf{f}$$

$$= \left(\hat{i}\frac{\partial}{\partial x} + \hat{j}\frac{\partial}{\partial y} + \hat{k}\frac{\partial}{\partial z}\right)$$

$$\cdot xyz.(\hat{i} + \hat{j} + \hat{k})$$

$$= \frac{\partial}{\partial x}(xyz) + \frac{\partial}{\partial y}(xyz) + \frac{\partial}{\partial z}(xyz)$$

$$= yz + zx + xy$$

EXAMPLE 5. *Find curl f if*
  (i) $\mathbf{f} = z^2\hat{i} + x^2\hat{j} + y^2\hat{k}$ (ii) $\mathbf{f} = e^{xyz}(\hat{i} + \hat{j} + \hat{k})$
  (iii) $\mathbf{f} = e^x\sin y\,\hat{i} + e^x\cos y\,\hat{j}$

SOLUTION. (i) $\mathbf{f} = z^2\hat{i} + x^2\hat{j} + y^2\hat{k}$

We have curl $\mathbf{f} = \nabla \times \mathbf{f}$

$$= \begin{vmatrix} \hat{i} & \hat{j} & \hat{k} \\ \dfrac{\partial}{\partial x} & \dfrac{\partial}{\partial y} & \dfrac{\partial}{\partial z} \\ z^2 & x^2 & y^2 \end{vmatrix}$$

$$= \hat{i}\left(\frac{\partial}{\partial y}(y)^2 - \frac{\partial}{\partial z}(x)^2\right)$$

$$+ \hat{j}\left(\frac{\partial}{\partial z}(z)^2 - \frac{\partial}{\partial x}(y)^2\right)$$

$$+ \hat{k}\left(\frac{\partial}{\partial x}(x)^2 - \frac{\partial}{\partial y}(z)^2\right)$$

$$= 2y\hat{i} + 2z\hat{j} + 2x\hat{k} = 2(y\hat{i} + z\hat{j} + x\hat{k}).$$

(ii) $\mathbf{f} = e^{xyz}(\hat{i} + \hat{j} + \hat{k})$

$$\therefore \quad \operatorname{curl}\mathbf{f} = \nabla \times \mathbf{f}$$

$$= \begin{vmatrix} \hat{i} & \hat{j} & \hat{k} \\ \dfrac{\partial}{\partial x} & \dfrac{\partial}{\partial y} & \dfrac{\partial}{\partial z} \\ e^{xyz} & e^{xyz} & e^{xyz} \end{vmatrix}$$

$$= \hat{i}\left(\frac{\partial}{\partial y}(e^{xyz}) - \frac{\partial}{\partial z}(e^{xyz})\right)$$

$$+ \hat{j}\left(\frac{\partial}{\partial z}(e^{xyz}) - \frac{\partial}{\partial x}(e^{xyz})\right)$$

$$+ \hat{k}\left(\frac{\partial}{\partial x}(e^{xyz}) - \frac{\partial}{\partial y}(e^{xyz})\right)$$

$$= \hat{i}[e^{xyz}(zx - xy)] + \hat{j}[e^{xyz}(xy - yz)]$$

$$+ \hat{k}[e^{xyz}(yz - zx)]$$

$$= e^{xyz}[(zx - xy)\hat{i} + (xy - yz)\hat{j}$$

$$+ (yz - zx)\hat{k}]$$

(iii) $\qquad f = e^x \sin y\hat{i} + e^x \cos y\hat{j}$

$\therefore \qquad \text{curl} f = \nabla \times f$

$$= \begin{vmatrix} \hat{i} & \hat{j} & \hat{k} \\ \dfrac{\partial}{\partial x} & \dfrac{\partial}{\partial y} & \dfrac{\partial}{\partial z} \\ e^x \sin y & e^x \cos y & 0 \end{vmatrix}$$

$$= \hat{i}\left(-\frac{\partial}{\partial z}(e^x \cos y)\right) + \hat{j}\left(\frac{\partial}{\partial z}(e^x \sin y)\right)$$

$$+ \hat{k}\left(\frac{\partial}{\partial x}(e^x \cos y) - \frac{\partial}{\partial y}(e^x \sin y)\right)$$

$$= 0\hat{i} + 0\hat{j} + (e^x \cos y - e^x \cos y)\hat{k}$$

$$= 0\hat{i} + 0\hat{j} + 0\hat{k} = \mathbf{0}$$

EXAMPLE 6. ***Prove that***
div $(\mathbf{a} \times \mathbf{b}) = \mathbf{b} \cdot \text{curl } \mathbf{a} - \mathbf{a}. \text{curl } \mathbf{b}$ (GBTU–2012)
or $\nabla \cdot (\mathbf{a} \times \mathbf{b}) = \mathbf{b} \cdot (\nabla \times \mathbf{a}) - \mathbf{a} \cdot (\nabla \times \mathbf{b})$

SOLUTION. We have
$\text{div } (\mathbf{a} \times \mathbf{b}) = \nabla \cdot (\mathbf{a} \times \mathbf{b})$

$$= \left(\hat{i}\frac{\partial}{\partial x} + \hat{j}\frac{\partial}{\partial y} + \hat{k}\frac{\partial}{\partial z}\right).(\mathbf{a} \times \mathbf{b})$$

$$= \left(\hat{i}\frac{\partial}{\partial x}\right).(\mathbf{a} \times \mathbf{b}) + \left(\hat{j}\frac{\partial}{\partial y}\right).(\mathbf{a} \times \mathbf{b})$$

$$+ \left(\hat{k}\frac{\partial}{\partial z}\right).(\mathbf{a} \times \mathbf{b})$$

$$= \hat{i}.\frac{\partial}{\partial x}(\mathbf{a} \times \mathbf{b}) + \hat{j}.\frac{\partial}{\partial y}(\mathbf{a} \times \mathbf{b})$$

$$+ \hat{k}.\frac{\partial}{\partial z}(\mathbf{a} \times \mathbf{b})$$

$$= \hat{i}.\left(\frac{\partial \mathbf{a}}{\partial x} \times \mathbf{b} + \mathbf{a} \times \frac{\partial \mathbf{b}}{\partial x}\right)$$

$$+ \hat{j}.\left(\frac{\partial \mathbf{a}}{\partial y} \times \mathbf{b} + \mathbf{a} \times \frac{\partial \mathbf{b}}{\partial y}\right)$$

$$+ \hat{k}.\left(\frac{\partial \mathbf{a}}{\partial z} \times \mathbf{b} + \mathbf{a} \times \frac{\partial \mathbf{b}}{\partial z}\right)$$

$$= \hat{i}.\left(\frac{\partial \mathbf{a}}{\partial x} \times \mathbf{b}\right) + \hat{i}.\left(\mathbf{a} \times \frac{\partial \mathbf{b}}{\partial x}\right)$$

$$+ \hat{j}.\left(\frac{\partial \mathbf{a}}{\partial y} \times \mathbf{b}\right) + \hat{j}.\left(\mathbf{a} \times \frac{\partial \mathbf{b}}{\partial y}\right)$$

$$+ \hat{k}.\left(\frac{\partial \mathbf{a}}{\partial z} \times \mathbf{b}\right) + \hat{k}.\left(\mathbf{a} \times \frac{\partial \mathbf{b}}{\partial z}\right)$$

$$= \left[\hat{i}.\left(\frac{\partial \mathbf{a}}{\partial x} \times \mathbf{b}\right) + \hat{j}.\left(\frac{\partial \mathbf{a}}{\partial y} \times \mathbf{b}\right) +\right.$$

$$\left. + \hat{k}.\left(\frac{\partial \mathbf{a}}{\partial z} \times \mathbf{b}\right)\right] + \left[\hat{i}.\left(\mathbf{a} \times \frac{\partial \mathbf{b}}{\partial x}\right)\right.$$

$$\left. + \hat{j}.\left(\mathbf{a} \times \frac{\partial \mathbf{b}}{\partial y}\right) + \hat{k}.\left(\mathbf{a} \times \frac{\partial \mathbf{b}}{\partial z}\right)\right]$$

Using $\mathbf{a}.(\mathbf{b} \times \mathbf{c}) = \mathbf{a} \times (\mathbf{b}.\mathbf{c})$ and $\mathbf{a}.(\mathbf{b} \times \mathbf{c})$
$\qquad = -\mathbf{a}.(\mathbf{c} \times \mathbf{b})$

$$= \left[\left(\hat{i} \times \frac{\partial \mathbf{a}}{\partial x}\right).\mathbf{b} + \left(\hat{j} \times \frac{\partial \mathbf{a}}{\partial y}\right).\mathbf{b} +\right.$$

$$\left. + \left(\hat{k} \times \frac{\partial \mathbf{a}}{\partial z}\right).\mathbf{b}\right] - \left[\hat{i}.\left(\frac{\partial \mathbf{b}}{\partial x} \times \mathbf{a}\right)\right.$$

$$\left. + \hat{j}.\left(\frac{\partial \mathbf{b}}{\partial y} \times \mathbf{a}\right) + \hat{k}.\left(\frac{\partial \mathbf{b}}{\partial z} \times \mathbf{a}\right)\right]$$

$$= \left(\hat{i} \times \frac{\partial \mathbf{a}}{\partial x} + \hat{j} \times \frac{\partial \mathbf{a}}{\partial y} + \hat{k} \times \frac{\partial \mathbf{a}}{\partial z}\right).\mathbf{b}$$

$$- \left(\hat{i} \times \frac{\partial \mathbf{b}}{\partial x} + \hat{j} \times \frac{\partial \mathbf{b}}{\partial y} + \hat{k} \times \frac{\partial \mathbf{b}}{\partial z}\right).\mathbf{a}$$

$$= (\nabla \times \mathbf{a}).\mathbf{b} - (\nabla \times \mathbf{b}).\mathbf{a}$$

$\therefore \quad \nabla \cdot (\mathbf{a} \times \mathbf{b}) = \mathbf{b} \cdot (\nabla \times \mathbf{a}) - \mathbf{a} \cdot (\nabla \times \mathbf{b})$ $(\because \mathbf{a}.\mathbf{b} = \mathbf{b}.\mathbf{a})$

EXAMPLE 7. ***Prove that***
$\text{curl}(\mathbf{a} \times \mathbf{b}) = (\mathbf{b} \cdot \nabla)\mathbf{a} - \mathbf{b} \text{ div } \mathbf{a} - (\mathbf{a} \cdot \nabla)\mathbf{b} + \mathbf{a} \text{ div } \mathbf{b}$

SOLUTION. We have $\text{curl } (\mathbf{a} \times \mathbf{b}) = \nabla \times (\mathbf{a} \times \mathbf{b})$

$$= \left(\hat{i}\frac{\partial}{\partial x} + \hat{j}\frac{\partial}{\partial y} + \hat{k}\frac{\partial}{\partial z}\right) \times (\mathbf{a} \times \mathbf{b})$$

$$= \hat{i} \times \frac{\partial}{\partial x}(\mathbf{a} \times \mathbf{b}) + \hat{j} \times \frac{\partial}{\partial y}(\mathbf{a} \times \mathbf{b})$$

$$+ \hat{k} \times \frac{\partial}{\partial z}(\mathbf{a} \times \mathbf{b})$$

$$= \hat{i} \times \left(\frac{\partial \mathbf{a}}{\partial x} \times \mathbf{b} + \mathbf{a} \times \frac{\partial \mathbf{b}}{\partial x}\right)$$

$$+ \hat{j} \times \left(\frac{\partial \mathbf{a}}{\partial y} \times \mathbf{b} + \mathbf{a} \times \frac{\partial \mathbf{b}}{\partial y}\right)$$

$$+ \hat{k} \times \left(\frac{\partial \mathbf{a}}{\partial z} \times \mathbf{b} + \mathbf{a} \times \frac{\partial \mathbf{b}}{\partial z}\right)$$

$$= \hat{i} \times \left(\frac{\partial \mathbf{a}}{\partial x} \times \mathbf{b}\right) + \hat{i} \times \left(\mathbf{a} \times \frac{\partial \mathbf{b}}{\partial x}\right)$$

$$+ \hat{j} \times \left(\frac{\partial \mathbf{a}}{\partial y} \times \mathbf{b}\right) + \hat{j} \times \left(\mathbf{a} \times \frac{\partial \mathbf{b}}{\partial y}\right)$$

$$+ \hat{k} \times \left(\frac{\partial \mathbf{a}}{\partial z} \times \mathbf{b}\right) + \hat{k} \times \left(\mathbf{a} \times \frac{\partial \mathbf{b}}{\partial z}\right)$$

$$= \left[\hat{i} \times \left(\frac{\partial \mathbf{a}}{\partial x} \times \mathbf{b}\right) + \hat{j} \times \left(\frac{\partial \mathbf{a}}{\partial y} \times \mathbf{b}\right) +\right.$$

$$\left. + \hat{k} \times \left(\frac{\partial \mathbf{a}}{\partial z} \times \mathbf{b}\right)\right] + \left[\hat{i} \times \left(\mathbf{a} \times \frac{\partial \mathbf{b}}{\partial x}\right)\right.$$

$$\left. + \hat{j} \times \left(\mathbf{a} \times \frac{\partial \mathbf{b}}{\partial y}\right) + \hat{k} \times \left(\mathbf{a} \times \frac{\partial \mathbf{b}}{\partial z}\right)\right]$$

$$= \left[(\hat{i} \cdot \boldsymbol{b}) \frac{\partial \boldsymbol{a}}{\partial x} - \left(\hat{i} \cdot \frac{\partial \boldsymbol{a}}{\partial x} \right) \boldsymbol{b} + (\hat{j} \cdot \boldsymbol{b}) \frac{\partial \boldsymbol{a}}{\partial y} \right.$$

$$- \left(\hat{j} \cdot \frac{\partial \boldsymbol{a}}{\partial y} \right) \boldsymbol{b} + (\hat{k} \cdot \boldsymbol{b}) \frac{\partial \boldsymbol{a}}{\partial z} - \left(\hat{k} \cdot \frac{\partial \boldsymbol{a}}{\partial z} \right) \boldsymbol{b} \right]$$

$$- \left[(\hat{i} \cdot \boldsymbol{a}) \frac{\partial \boldsymbol{b}}{\partial x} - \left(\hat{i} \cdot \frac{\partial \boldsymbol{b}}{\partial x} \right) \boldsymbol{a} + (\hat{j} \cdot \boldsymbol{a}) \frac{\partial \boldsymbol{b}}{\partial y} \right.$$

$$- \left(\hat{j} \cdot \frac{\partial \boldsymbol{b}}{\partial y} \right) \boldsymbol{a} + (\hat{k} \cdot \boldsymbol{a}) \frac{\partial \boldsymbol{b}}{\partial z} - \left(\hat{k} \cdot \frac{\partial \boldsymbol{b}}{\partial z} \right) \boldsymbol{a} \right]$$

Using $\boldsymbol{a} \cdot \boldsymbol{b} = \boldsymbol{b} \cdot \boldsymbol{a}$, we get

$$= \left[\left(\boldsymbol{b} \cdot \hat{i} \frac{\partial \boldsymbol{a}}{\partial x} + \boldsymbol{b} \cdot \hat{j} \frac{\partial \boldsymbol{a}}{\partial y} + \boldsymbol{b} \cdot \hat{k} \frac{\partial \boldsymbol{a}}{\partial z} \right) \right.$$

$$- \left(\hat{i} \cdot \frac{\partial \boldsymbol{a}}{\partial x} + \hat{j} \cdot \frac{\partial \boldsymbol{a}}{\partial y} + \hat{k} \cdot \frac{\partial \boldsymbol{a}}{\partial z} \right) \boldsymbol{b} \right]$$

$$- \left[\left(\boldsymbol{a} \cdot \hat{i} \frac{\partial \boldsymbol{b}}{\partial x} + \boldsymbol{a} \cdot \hat{j} \frac{\partial \boldsymbol{b}}{\partial y} + \boldsymbol{a} \cdot \hat{k} \frac{\partial \boldsymbol{b}}{\partial z} \right) \right.$$

$$- \left(\hat{i} \cdot \frac{\partial \boldsymbol{b}}{\partial x} + \hat{j} \cdot \frac{\partial \boldsymbol{b}}{\partial y} + \hat{k} \cdot \frac{\partial \boldsymbol{b}}{\partial z} \right) \boldsymbol{a} \right]$$

$$= \left[\left(\boldsymbol{b} \cdot \hat{i} \frac{\partial}{\partial x} + \boldsymbol{b} \cdot \hat{j} \frac{\partial}{\partial y} + \boldsymbol{b} \cdot \hat{k} \frac{\partial}{\partial z} \right) \boldsymbol{a} - (\nabla \cdot \boldsymbol{a}) \boldsymbol{b} \right]$$

$$- \left[\left(\boldsymbol{a} \cdot \hat{i} \frac{\partial}{\partial x} + \boldsymbol{a} \cdot \hat{j} \frac{\partial}{\partial y} + \boldsymbol{a} \cdot \hat{k} \frac{\partial}{\partial z} \right) \boldsymbol{b} - (\nabla \cdot \boldsymbol{b}) \boldsymbol{a} \right]$$

$$= (\boldsymbol{b} \cdot \nabla) \boldsymbol{a} - (\nabla \cdot \boldsymbol{a}) \boldsymbol{b} - (\boldsymbol{a} \cdot \nabla) \boldsymbol{b} + (\nabla \cdot \boldsymbol{b}) \boldsymbol{a}$$

Hence, curl $(\boldsymbol{a} \times \boldsymbol{b}) = (\boldsymbol{b} \cdot \nabla) \boldsymbol{a} - \boldsymbol{b} \operatorname{div} \boldsymbol{a} - (\boldsymbol{a} \cdot \nabla) \boldsymbol{b} + \boldsymbol{a} \operatorname{div} \boldsymbol{b}$.

EXAMPLE 8. *Prove that grad $(\boldsymbol{a} \cdot \boldsymbol{b}) = (\boldsymbol{b} \cdot \nabla) \boldsymbol{a} + (\boldsymbol{a} \cdot \nabla) \boldsymbol{b} + \boldsymbol{b} \times curl\, \boldsymbol{a} + \boldsymbol{a} \times curl\, \boldsymbol{b}$.*

SOLUTION . Since we have

$$\operatorname{grad}(\boldsymbol{a} \cdot \boldsymbol{b}) = \left(\hat{i} \frac{\partial}{\partial x} + \hat{j} \frac{\partial}{\partial y} + \hat{k} \frac{\partial}{\partial z} \right) (\boldsymbol{a} \cdot \boldsymbol{b})$$

$$= \hat{i} \frac{\partial}{\partial x} (\boldsymbol{a} \cdot \boldsymbol{b}) + \hat{j} \frac{\partial}{\partial y} (\boldsymbol{a} \cdot \boldsymbol{b}) + \hat{k} \frac{\partial}{\partial z} (\boldsymbol{a} \cdot \boldsymbol{b})$$

$$= \hat{i} \left(\boldsymbol{a} \cdot \frac{\partial \boldsymbol{b}}{\partial x} + \frac{\partial \boldsymbol{a}}{\partial x} \cdot \boldsymbol{b} \right) + \hat{j} \left(\boldsymbol{a} \cdot \frac{\partial \boldsymbol{b}}{\partial y} + \frac{\partial \boldsymbol{a}}{\partial y} \cdot \boldsymbol{b} \right)$$

$$+ \hat{k} \left(\boldsymbol{a} \cdot \frac{\partial \boldsymbol{b}}{\partial z} + \frac{\partial \boldsymbol{a}}{\partial z} \cdot \boldsymbol{b} \right)$$

$$\operatorname{grad}(\boldsymbol{a} \cdot \boldsymbol{b}) = \left[\hat{i} \left(\boldsymbol{a} \cdot \frac{\partial \boldsymbol{b}}{\partial x} \right) + \hat{j} \left(\boldsymbol{a} \cdot \frac{\partial \boldsymbol{b}}{\partial y} \right) + \hat{k} \left(\boldsymbol{a} \cdot \frac{\partial \boldsymbol{b}}{\partial z} \right) \right]$$

$$+ \left[\hat{i} \left(\frac{\partial \boldsymbol{a}}{\partial x} \cdot \boldsymbol{b} \right) + \hat{j} \left(\frac{\partial \boldsymbol{a}}{\partial y} \cdot \boldsymbol{b} \right) + \hat{k} \left(\frac{\partial \boldsymbol{a}}{\partial z} \cdot \boldsymbol{b} \right) \right] \dots (1)$$

Further , since we know that

$$\boldsymbol{a} \times (\boldsymbol{b} \times \boldsymbol{c}) = (\boldsymbol{a} \cdot \boldsymbol{c}) \boldsymbol{b} - (\boldsymbol{a} \cdot \boldsymbol{b}) \boldsymbol{c}$$

$$\therefore \quad (\boldsymbol{a} \cdot \boldsymbol{b}) \boldsymbol{c} \equiv (\boldsymbol{a} \cdot \boldsymbol{c}) \boldsymbol{b} - \boldsymbol{a} \times (\boldsymbol{b} \times \boldsymbol{c})$$

$$\therefore \left(\boldsymbol{a} \cdot \frac{\partial \boldsymbol{b}}{\partial x} \right) \hat{i} = (\boldsymbol{a} \cdot \hat{i}) \frac{\partial \boldsymbol{b}}{\partial x} - \boldsymbol{a} \times \left(\frac{\partial \boldsymbol{b}}{\partial x} \times \hat{i} \right)$$

$$= (\boldsymbol{a} \cdot \hat{i}) \frac{\partial \boldsymbol{b}}{\partial x} + \boldsymbol{a} \times \left(\hat{i} \times \frac{\partial \boldsymbol{b}}{\partial x} \right)$$

$$[\because \boldsymbol{a} \times \boldsymbol{b} = -\boldsymbol{b} \times \boldsymbol{a}]$$

Similarly, $\left(\boldsymbol{a} \cdot \frac{\partial \boldsymbol{b}}{\partial y} \right) \hat{j} = (\boldsymbol{a} \cdot \hat{j}) \frac{\partial \boldsymbol{b}}{\partial y} + \boldsymbol{a} \times \left(\hat{j} \times \frac{\partial \boldsymbol{b}}{\partial y} \right)$

and $\quad \left(\boldsymbol{a} \cdot \frac{\partial \boldsymbol{b}}{\partial z} \right) \hat{k} = (\boldsymbol{a} \cdot \hat{k}) \frac{\partial \boldsymbol{b}}{\partial z} + \boldsymbol{a} \times \left(\hat{k} \times \frac{\partial \boldsymbol{b}}{\partial z} \right)$

$$\therefore \hat{i} \left(\boldsymbol{a} \cdot \frac{\partial \boldsymbol{b}}{\partial x} \right) + \hat{j} \left(\boldsymbol{a} \cdot \frac{\partial \boldsymbol{b}}{\partial y} \right) + \hat{k} \left(\boldsymbol{a} \cdot \frac{\partial \boldsymbol{b}}{\partial z} \right)$$

$$= \left(\boldsymbol{a} \cdot \hat{i} \frac{\partial}{\partial x} + \boldsymbol{a} \cdot \hat{j} \frac{\partial}{\partial y} + \boldsymbol{a} \cdot \hat{k} \frac{\partial}{\partial z} \right) \boldsymbol{b} + \boldsymbol{a} \times (\nabla \times \boldsymbol{b})$$

$$= (\boldsymbol{a} \cdot \nabla) \boldsymbol{b} + \boldsymbol{a} \times (\nabla \times \boldsymbol{b}) \qquad \dots (2)$$

Similarly, $\hat{i} \left(\frac{\partial \boldsymbol{a}}{\partial x} \cdot \boldsymbol{b} \right) + \hat{j} \left(\frac{\partial \boldsymbol{a}}{\partial y} \cdot \boldsymbol{b} \right) + \hat{k} \left(\frac{\partial \boldsymbol{a}}{\partial z} \cdot \boldsymbol{b} \right)$

$$= (\boldsymbol{b} \cdot \nabla) \boldsymbol{a} + \boldsymbol{b} \times (\nabla \times \boldsymbol{a}) \qquad \dots (3)$$

From (1), (2) and (3), we get

grad $(\boldsymbol{a} \cdot \boldsymbol{b}) = (\boldsymbol{a} \cdot \nabla) \boldsymbol{b} + \boldsymbol{a} \times (\nabla \times \boldsymbol{b}) \times (\boldsymbol{b} \cdot \nabla) \boldsymbol{a}$

$$+ \boldsymbol{b} \times (\nabla \times \boldsymbol{a})$$

$$(\boldsymbol{a} \cdot \nabla) \boldsymbol{b} + (\boldsymbol{b} \cdot \nabla) \boldsymbol{a} + \boldsymbol{b} \times (\nabla \times \boldsymbol{a}) + \boldsymbol{a} \times (\nabla \times \boldsymbol{b})$$

Hence, grad $(\boldsymbol{a} \cdot \boldsymbol{b}) = (\boldsymbol{b} \cdot \nabla) \boldsymbol{a} + (\boldsymbol{a} \cdot \nabla) \boldsymbol{b}$

$$+ \boldsymbol{b} \times \operatorname{curl} \boldsymbol{a} + \boldsymbol{a} \times \operatorname{curl} \boldsymbol{b}.$$

EXAMPLE 9. *Prove that the curl of the gradient of f (scalar function) is zero, i.e. $\nabla \times (\nabla f) = 0$* (GBTU–2011)

or

If $f = r^n$, then $\nabla \times (\nabla r^n) = 0$

SOLUTION . Since we have

$$\nabla f = \frac{\partial f}{\partial x} \hat{i} + \frac{\partial f}{\partial y} \hat{j} + \frac{\partial f}{\partial z} \hat{k}$$

$$\therefore \quad \nabla \times (\nabla f) = \begin{vmatrix} \hat{i} & \hat{j} & \hat{k} \\ \dfrac{\partial}{\partial x} & \dfrac{\partial}{\partial y} & \dfrac{\partial}{\partial z} \\ \dfrac{\partial f}{\partial x} & \dfrac{\partial f}{\partial y} & \dfrac{\partial f}{\partial z} \end{vmatrix}$$

$$= \hat{i} \left[\frac{\partial^2 f}{\partial y \partial z} - \frac{\partial^2 f}{\partial z \partial y} \right] + \hat{j} \left[\frac{\partial^2 f}{\partial z \partial x} - \frac{\partial^2 f}{\partial x \partial z} \right]$$

$$+ \hat{k} \left[\frac{\partial^2 f}{\partial x \partial y} - \frac{\partial^2 f}{\partial y \partial x} \right]$$

Since $\dfrac{\partial^2 f}{\partial y \partial z} = \dfrac{\partial^2 f}{\partial z \partial y}$ etc.

$$\therefore \nabla \times (\nabla f) = 0 \hat{i} + 0 \hat{j} + 0 \hat{k} = \boldsymbol{0}$$

EXAMPLE 10. *Prove that the div(curl V) = 0.* (GBTU–2010, 11)
i.e. $\nabla \cdot (\nabla \times \boldsymbol{V}) = 0$.

SOLUTION . Since we have

$$\nabla \times V = \begin{vmatrix} \hat{i} & \hat{j} & \hat{k} \\ \dfrac{\partial}{\partial x} & \dfrac{\partial}{\partial y} & \dfrac{\partial}{\partial z} \\ V_1 & V_2 & V_3 \end{vmatrix}$$

where $\boldsymbol{V} = V_1 \hat{i} + V_2 \hat{j} + V_3 \hat{k}$ (say).

$$\therefore \quad \nabla \times \boldsymbol{V} = \hat{i} \left(\frac{\partial V_3}{\partial y} - \frac{\partial V_2}{\partial z} \right) + \hat{j} \left(\frac{\partial V_1}{\partial z} - \frac{\partial V_3}{\partial x} \right)$$

$$+ \hat{k} \left(\frac{\partial V_2}{\partial x} - \frac{\partial V_1}{\partial y} \right)$$

Now $\nabla \cdot (\nabla \times \mathbf{V})$

$$= \frac{\partial}{\partial x}\left(\frac{\partial V_3}{\partial y} - \frac{\partial V_2}{\partial z}\right) + \frac{\partial}{\partial y}\left(\frac{\partial V_1}{\partial z} - \frac{\partial V_3}{\partial x}\right)$$

$$+ \frac{\partial}{\partial z}\left(\frac{\partial V_2}{\partial x} - \frac{\partial V_1}{\partial y}\right)$$

$$= \frac{\partial^2 V_3}{\partial x \partial y} - \frac{\partial^2 V_2}{\partial x \partial z} + \frac{\partial^2 V_1}{\partial y \partial z} - \frac{\partial^2 V_3}{\partial y \partial x}$$

$$+ \frac{\partial^2 V_2}{\partial z \partial x} - \frac{\partial^2 V_1}{\partial z \partial y}$$

Since $\dfrac{\partial^2 V_1}{\partial y \partial z} = \dfrac{\partial^2 V_1}{\partial z \partial y}$ etc.

$\therefore \nabla \cdot (\nabla \times \mathbf{V}) = 0$.

EXAMPLE 11. $\nabla^2 f(r) = f''(r) + \dfrac{2}{r} f'(r)$

(UKTU–2010, GBTU–2012, BHOPAL–2008, SVTU–2008, VTU–2006)

SOLUTION . Since we have

$$\nabla^2 f(r) = \nabla \cdot (\nabla f(r)) = \nabla \cdot (f'(r)\nabla r)$$

$$(\because \nabla f(r) = f'(r)\nabla r)$$

$$= \nabla \cdot \left(f'(r)\frac{\mathbf{r}}{r}\right) \qquad \left(\because \nabla r = \frac{\mathbf{r}}{r}\right)$$

$$= \nabla \cdot \left(\frac{f'(r)}{r}\mathbf{r}\right) = \frac{f'(r)}{r}\nabla \cdot \mathbf{r} + \mathbf{r} \cdot \nabla\left\{\frac{f'(r)}{r}\right\}$$

$$[\because \nabla f(\mathbf{V}) = f\nabla \cdot \mathbf{V} + \mathbf{V} \cdot (\nabla f)]$$

$$= \frac{3}{r}f'(r) + \mathbf{r} \cdot \left\{\frac{f'(r)}{r}\right\}' \nabla r$$

$$= \frac{3}{r}f'(r) + \mathbf{r} \cdot \left\{\frac{rf''(r) - f'(r)}{r^2}\right\}\frac{\mathbf{r}}{r}$$

$$= \frac{3}{r}f'(r) + \frac{rf''(r) - f'(r)}{r^2}\frac{\mathbf{r} \cdot \mathbf{r}}{r}$$

$$= \frac{3}{r}f'(r) + \frac{rf''(r) - f'(r)}{r^2}\frac{r^2}{r}$$

$$= \frac{3}{r}f'(r) + \frac{rf''(r) - f'(r)}{r}$$

$$= f''(r) + \frac{2}{r}f'(r)$$

Hence, $\nabla^2 f(r) = f''(r) + \dfrac{2}{r}f'(r)$.

EXAMPLE 12. *Solve* $\nabla^2 f(r) = 0$.

SOLUTION . We have

$$\nabla^2 f(r) = f''(r) + \frac{2}{r}f'(r)$$

Since $\nabla^2 f(r) = 0$ given

$$\therefore f''(r) + \frac{2}{r}f'(r) = 0 \qquad \text{or} \qquad \frac{f''(r)}{f'(r)} = -\frac{2}{r}$$

Integrating w.r.t. 'r', we get

$$\log f'(r) = -2\log r + \log c_1$$

where c_1 is a constant of integration.

$$\therefore f'(r) = \frac{c_1}{r^2}.$$

Again integrating w.r.t. 'r', we get

$$f(r) = -\frac{c_1}{r} + c_2$$

where c_1 and c_2 are constant of integration.

EXAMPLE 13. *Prove that* $\nabla^2\left(\dfrac{1}{r}\right) = 0$.   (UPTU–2003, PTU–2003)

SOLUTION . Since we have

$$\nabla^2\left(\frac{1}{r}\right) = \nabla \cdot \left(\nabla\left(\frac{1}{r}\right)\right) = \nabla \cdot \left(-\frac{1}{r^2}\nabla r\right)$$

$$[\because \text{grad} f(r) = f'(r)\,\text{grad}\,r]$$

$$= \nabla \cdot \left(-\frac{1}{r^3}\mathbf{r}\right) \qquad \left[\because \nabla r = \frac{\mathbf{r}}{r}\right]$$

$$= \left(-\frac{1}{r^3}\right)\nabla \cdot \mathbf{r} + \mathbf{r} \cdot \text{grad}\left(-\frac{1}{r^3}\right)$$

$$= -\frac{3}{r^3} + \mathbf{r} \cdot \left(\frac{3}{r^4}\text{grad}\,r\right) = -\frac{3}{r^3} + \mathbf{r} \cdot \left(\frac{3}{r^4}\frac{\mathbf{r}}{r}\right)$$

$$= -\frac{3}{r^3} + \frac{3}{r^5}\mathbf{r} \cdot \mathbf{r} = -\frac{3}{r^3} + \frac{3}{r^5}r^2 \qquad (\because \mathbf{r} \cdot \mathbf{r} = r^2)$$

$$= -\frac{3}{r^3} + \frac{3}{r^3} = 0.$$

$$\therefore \nabla^2\left(\frac{1}{r}\right) = 0.$$

EXAMPLE 14. *Prove that*

(i) $\nabla \times (\nabla \times \mathbf{a}) = \nabla(\nabla \cdot \mathbf{a}) - \nabla^2 \mathbf{a}$.

(ii) $\mathbf{a} \times (\nabla \times \mathbf{r}) = \nabla(\mathbf{a} \cdot \mathbf{r}) - (\mathbf{a} \cdot \nabla)\mathbf{r}$.

where $\mathbf{a} = x\hat{\mathbf{i}} + y\hat{\mathbf{j}} + z\hat{\mathbf{k}}$ (UPTU–2008)

SOLUTION . (i) Let $\mathbf{a} = a_1\hat{i} + a_2\hat{j} + a_3\hat{k}$. Then

$$\nabla \times \mathbf{a} = \begin{vmatrix} \hat{i} & \hat{j} & \hat{k} \\ \dfrac{\partial}{\partial x} & \dfrac{\partial}{\partial y} & \dfrac{\partial}{\partial z} \\ a_1 & a_2 & a_3 \end{vmatrix}$$

$$= \left(\frac{\partial a_3}{\partial y} - \frac{\partial a_2}{\partial z}\right)\hat{i} + \left(\frac{\partial a_1}{\partial z} - \frac{\partial a_3}{\partial x}\right)\hat{j}$$

$$+ \left(\frac{\partial a_2}{\partial x} - \frac{\partial a_1}{\partial y}\right)\hat{k}$$

$$\therefore \nabla \times (\nabla \times \mathbf{a})$$

$$= \begin{vmatrix} \hat{i} & \hat{j} & \hat{k} \\ \dfrac{\partial}{\partial x} & \dfrac{\partial}{\partial y} & \dfrac{\partial}{\partial z} \\ \dfrac{\partial a_3}{\partial y} - \dfrac{\partial a_2}{\partial z} & \dfrac{\partial a_1}{\partial z} - \dfrac{\partial a_3}{\partial x} & \dfrac{\partial a_2}{\partial x} - \dfrac{\partial a_1}{\partial y} \end{vmatrix}$$

$$= \Sigma\left[\left\{\frac{\partial}{\partial y}\left(\frac{\partial a_2}{\partial x} - \frac{\partial a_1}{\partial y}\right) - \frac{\partial}{\partial z}\left(\frac{\partial a_1}{\partial z} - \frac{\partial a_3}{\partial x}\right)\right\}\hat{i}\right]$$

$$= \Sigma\left[\left\{\left(\frac{\partial^2 a_2}{\partial x \partial y} + \frac{\partial^2 a_3}{\partial x \partial z}\right) - \left(\frac{\partial^2 a_1}{\partial y^2} - \frac{\partial^2 a_1}{\partial z^2}\right)\right\}\hat{i}\right]$$

$$= \Sigma \left[\left\{ \frac{\partial}{\partial x} \left(\frac{\partial a_1}{\partial x} + \frac{\partial a_2}{\partial y} + \frac{\partial a_3}{\partial z} \right) \right. \right.$$

$$\left. \left. - \left(\frac{\partial^2 a_1}{\partial x^2} + \frac{\partial^2 a_1}{\partial y^2} + \frac{\partial^2 a_1}{\partial z^2} \right) \right\} \hat{i} \right]$$

$$= \Sigma \left[\left\{ \frac{\partial}{\partial x} (\nabla \cdot \boldsymbol{a}) - (\nabla^2 \cdot a_1) \right\} \hat{i} \right]$$

$$= \left(\Sigma \hat{i} \frac{\partial}{\partial x} \right) (\nabla \cdot \boldsymbol{a}) - \nabla^2 (\Sigma a_1 \hat{i})$$

$$= \nabla (\nabla \cdot \boldsymbol{a}) - \nabla^2 \boldsymbol{a}.$$

(ii) LHS $= \boldsymbol{a} \times (\nabla \times \boldsymbol{r}) = \boldsymbol{a} \times \boldsymbol{0} = 0$ $[\because \operatorname{curl} \boldsymbol{r} = 0]$

RHS $= \nabla (\boldsymbol{a} \cdot \boldsymbol{r}) - (\boldsymbol{a} \cdot \nabla) \boldsymbol{r}$

$$= \nabla (a_1 x + a_2 y + a_3 z)$$

$$- \left(a_1 \frac{\partial}{\partial x} + a_2 \frac{\partial}{\partial y} + a_3 \frac{\partial}{\partial z} \right) \boldsymbol{r}$$

$$= \hat{i}(a_1) + \hat{j}(a_2) + \hat{k}(a_3) - (a_1 \hat{i} + a_2 \hat{j} + a_3 \hat{k})$$

$$= \boldsymbol{0} = \text{LHS}$$

EXAMPLE 15. *If f and g are two scalar point functions, show that*

  (i) $div (f \nabla g) = f \nabla^2 g + \nabla f \cdot \nabla g$

  (ii) $div (f \nabla g) - div(g \nabla f) = f \nabla^2 g - g \nabla^2 f$

SOLUTION. (i) Since $\nabla g = \dfrac{\partial g}{\partial x} \hat{i} + \dfrac{\partial g}{\partial y} \hat{j} + \dfrac{\partial g}{\partial z} \hat{k}$

$$\therefore \quad f \nabla g = f \frac{\partial g}{\partial x} \hat{i} + f \frac{\partial g}{\partial y} \hat{j} + f \frac{\partial g}{\partial z} \hat{k}$$

and

$$\operatorname{div}(f \nabla g) = \frac{\partial}{\partial x} \left(f \frac{\partial g}{\partial x} \right) + \frac{\partial}{\partial y} \left(f \frac{\partial g}{\partial y} \right)$$

$$+ \frac{\partial}{\partial z} \left(f \frac{\partial g}{\partial z} \right)$$

$$= f \frac{\partial^2 g}{\partial x^2} + \frac{\partial f}{\partial x} \frac{\partial g}{\partial x} + f \frac{\partial^2 g}{\partial y^2}$$

$$+ \frac{\partial f}{\partial y} \frac{\partial g}{\partial y} + f \frac{\partial^2 g}{\partial z^2} + \frac{\partial f}{\partial z} \frac{\partial g}{\partial z}$$

$$= f \left(\frac{\partial^2 g}{\partial x^2} + \frac{\partial^2 g}{\partial y^2} + \frac{\partial^2 g}{\partial z^2} \right)$$

$$+ \frac{\partial f}{\partial x} \frac{\partial g}{\partial x} + \frac{\partial f}{\partial y} \frac{\partial g}{\partial y} + \frac{\partial f}{\partial z} \frac{\partial g}{\partial z}$$

$$= f \nabla^2 g + \left(\frac{\partial f}{\partial x} \hat{i} + \frac{\partial f}{\partial y} \hat{j} + \frac{\partial f}{\partial z} \hat{k} \right)$$

$$\cdot \left(\frac{\partial g}{\partial x} \hat{i} + \frac{\partial g}{\partial y} \hat{j} + \frac{\partial g}{\partial z} \hat{k} \right)$$

$$= f \nabla^2 g + (\nabla f) \cdot (\nabla g)$$

Hence, $\operatorname{div} (f \nabla g) = f \nabla^2 g + (\nabla f) \cdot (\nabla g)$ ...(1)

(ii) Similarly, we may get

$\operatorname{div} (g \nabla f) = g \nabla^2 f + (\nabla g) \cdot (\nabla f)$ ...(2)

Form (1) and (2), we get

$\operatorname{div} (f \nabla g) - \operatorname{div} (g \nabla f) = f \nabla^2 g - g \nabla^2 f$

EXAMPLE 16. *Prove that div grad $(r^n) = n(n + 1)r^{n-2}$.*

  (UKTU–2012, UPTU–2005, BHOPAL–2008, JNTU–2006, SVTU–2006)

SOLUTION. Since we have

$$\operatorname{grad} (r^n) = (nr^{n-1} \operatorname{grad} r)$$

$$[\because \operatorname{grad} f(r) = f'(r)(\nabla r)]$$

$$= nr^{n-1} \frac{\boldsymbol{r}}{r} \qquad \left[\because \operatorname{grad} r = \frac{\boldsymbol{r}}{r} \right]$$

$$= nr^{n-2} \boldsymbol{r}$$

Now

$$\operatorname{div} \operatorname{grad} (r^n) = \operatorname{div} (nr^{n-2} \boldsymbol{r})$$

$$= n \operatorname{div}(r^{n-2} \boldsymbol{r})$$

$$= n[r^{n-2} \nabla \cdot \boldsymbol{r} + \boldsymbol{r} \cdot \operatorname{grad} (r^{n-2})]$$

$$= n[3r^{n-2} + \boldsymbol{r} \cdot ((n-2)r^{n-3} \operatorname{grad} r)]$$

$$(\because \nabla \cdot \boldsymbol{r} = 3)$$

$$= n \left[3r^{n-2} + \boldsymbol{r} \cdot \left((n-2)r^{n-3} \frac{\boldsymbol{r}}{r} \right) \right]$$

$$= n[3r^{n-2} + (n-2)r^{n-4} \boldsymbol{r} \cdot \boldsymbol{r}]$$

$$= n[3r^{n-2} + (n-2)r^{n-4} r^2]$$

$$[\because \boldsymbol{r} \cdot \boldsymbol{r} = r^2]$$

$$= n(n+1)r^{n-2}$$

$$\therefore \quad \operatorname{div} \operatorname{grad} r^n = n(n+1)r^{n-2}.$$

EXAMPLE 17. *Prove that $r^n r$ is an irrotational vector for any value of n but is solenodial if $n + 3 = 0$.*

  (GBTU–2010, VTU–2006, UPTU–2006, PTU–2005, 06, KOTTAYAM–2005)

SOLUTION. Since we know that if $\boldsymbol{a}$ is irrotational, then

$$\nabla \times \boldsymbol{a} = 0$$

$$\therefore \quad \nabla \times (r^n \boldsymbol{r}) = (\operatorname{grad} r^n) \times \boldsymbol{r} + r^n \operatorname{curl} \boldsymbol{r}$$

$$= (nr^{n-1} \operatorname{grad} r) \times \boldsymbol{r} + \boldsymbol{0} \; [\because \operatorname{curl} \boldsymbol{r} = 0]$$

$$= \left(nr^{n-1} \frac{\boldsymbol{r}}{r} \right) \times \boldsymbol{r} \qquad \left[\because \operatorname{grad} r = \frac{\boldsymbol{r}}{r} \right]$$

$$= nr^{n-2} \boldsymbol{r} \times \boldsymbol{r} = \boldsymbol{0}$$

$[\because$ Vector product of two same vectors is zero.$]$

Hence $r^n \boldsymbol{r}$ is irrotational for any value of n.

Further since if $\boldsymbol{a}$ is solenoidal, then $\nabla \cdot \boldsymbol{a} = 0$

$$\therefore \quad \nabla \cdot (r^n \boldsymbol{r}) = (\operatorname{grad} r^n) \cdot \boldsymbol{r} + r^n \nabla \cdot \boldsymbol{r}$$

$$= (nr^{n-1} \operatorname{grad} r) \cdot \boldsymbol{r} + 3r^n \; [\because \nabla \cdot \boldsymbol{r} = 3]$$

$$= \left(nr^{n-1} \frac{\boldsymbol{r}}{r} \right) \cdot \boldsymbol{r} + 3r^n \qquad \left[\because \operatorname{grad} r = \frac{\boldsymbol{r}}{r} \right]$$

$$= nr^{n-2} \boldsymbol{r} \cdot \boldsymbol{r} + 3 r^n$$

$$= nr^{n-2} r^2 + 3 r^n \qquad [\because \boldsymbol{r} \cdot \boldsymbol{r} = r^2]$$

$$= nr^n + 3 r^n = r^n (n+3).$$

If $n + 3 = 0$, then $\nabla \cdot (r^n \boldsymbol{r}) = 0$, and hence $r^n \boldsymbol{r}$ is solenoidal.

EXAMPLE 18. *If a and b are irrotational, prove that $a \times b$ is solenoidal.*   (GBTU–2010, MADRAS–2003, VTU–2001)

SOLUTION. Since $\boldsymbol{a}$ and $\boldsymbol{b}$ are irrotational, then

$$\nabla \times \boldsymbol{a} = 0, \; \nabla \times \boldsymbol{b} = 0.$$

Now we have to prove that $\boldsymbol{a} \times \boldsymbol{b}$ is solenoidal.

$$\therefore \quad \nabla \cdot (\boldsymbol{a} \times \boldsymbol{b}) = \boldsymbol{b} \cdot \operatorname{curl} \boldsymbol{a} - \boldsymbol{a} \cdot \operatorname{curl} \boldsymbol{b}$$

$$= \boldsymbol{b} \cdot (\nabla \times \boldsymbol{a}) - \boldsymbol{a} \cdot (\nabla \times \boldsymbol{b})$$

$$= 0 - 0 = 0. \; [\because \nabla \times \boldsymbol{a} = 0, \nabla \times \boldsymbol{b} = 0]$$

Thus, $\nabla \cdot (\boldsymbol{a} \times \boldsymbol{b}) = 0.$

Hence $(\boldsymbol{a} \times \boldsymbol{b})$ is solenoidal.

EXAMPLE 19. *Show that the vector field* $F = \dfrac{r}{r^3}$ *is irrotaitonal*

as well as solenoidal. Find the scalar potential.

(UPTU–2006)

SOLUTION . We know that

$$\text{curl } (u.\boldsymbol{a}) = u \text{ curl } \boldsymbol{a} + (\text{grad } u) \times \boldsymbol{a}$$

Therefore,

$$\text{curl}\left(\frac{1}{r^3}\cdot\boldsymbol{r}\right) = \frac{1}{r^3}\text{ curl } \boldsymbol{r} + \left(\text{grad}\frac{1}{r^3}\right)\times \boldsymbol{r}$$

$$= \frac{1}{r^3}\boldsymbol{0} + \left(-\frac{3}{r^4}\hat{r}\right)\times \boldsymbol{r}$$

$$= \boldsymbol{0} - \frac{3}{r^5}(\boldsymbol{r}\times\boldsymbol{r}) = \boldsymbol{0} - \boldsymbol{0} = \boldsymbol{0}$$

$\Rightarrow$ $\boldsymbol{F}$ is irrotational.

Also, we know that for the vector field $\boldsymbol{F}$ to be solenoidal, div $\boldsymbol{F} = 0$

We know that

$$\text{div } (u\boldsymbol{a}) = u \text{ div } \boldsymbol{a} + \boldsymbol{a}.\text{grad } u$$

$\therefore$ $\text{div}\left(\dfrac{\boldsymbol{r}}{r^3}\right) = \dfrac{1}{r^3}\text{div } \boldsymbol{r} + \boldsymbol{r}.\text{grad}\left(\dfrac{1}{r^3}\right)$

$$= \frac{3}{r^3} + r\left(-\frac{3}{r^4}\frac{\boldsymbol{r}}{r}\right) \qquad (\because \text{ div } \boldsymbol{r} = 3)$$

$$= \frac{3}{r^3} - \frac{3}{r^5}r^2 = \frac{3}{r^3} - \frac{3}{r^3} = 0$$

$\Rightarrow$ $\vec{F}$ is solenoidal.

Now, let $\vec{F} = \nabla\phi$, where ϕ is scalar potential.

$\therefore$ $F.d\boldsymbol{r} = \nabla\phi.d\boldsymbol{r}$

      $\vec{F}.d\boldsymbol{r} = d\phi$

$\therefore$ $d\phi = \dfrac{x\hat{i} + y\hat{j} + z\hat{k}}{(x^2 + y^2 + z^2)^{3/2}}(dx\hat{i} + dy\hat{j} + dz\hat{k})$

$$= \frac{xdx + ydy + zdz}{(x^2 + y^2 + z^2)^{3/2}}$$

$$= d[-(x^2 + y^2 + z^2)^{-1/2}]$$

$\Rightarrow$ $\phi = -\dfrac{1}{\sqrt{x^2 + y^2 + z^2}} + c$.

$\therefore$ $\phi = -\dfrac{1}{r} + c$.

Exercise-6.5

1. If $\boldsymbol{f} = x^2 y\hat{i} - 2xz\hat{j} + 2yz\hat{k}$, find
  (i) div $\boldsymbol{f}$ (ii) curl $\boldsymbol{f}$
  (iii) curl curl $\boldsymbol{f}$

2. If $\boldsymbol{f} = xy^2\hat{i} + 2x^2 yz\hat{j} - 3yz^2\hat{k}$, then at the point $(1, -1, 1)$ find
  (i) div $\boldsymbol{f}$ (ii) curl $\boldsymbol{f}$

3. If $\boldsymbol{f} = \text{grad}(x^3 + y^3 + z^3 - 3xyz)$, find
  (i) div $\boldsymbol{f}$ (ii) curl $\boldsymbol{f}$

4. (i) Determine the constant λ so that the vector
     $\boldsymbol{f} = (x + 3y)\hat{i} + (y - 2z)\hat{j} + (x + \lambda z)\hat{k}$
     is solenoidal.
  (ii) Find the constants a, b, c so that the vector
     $\boldsymbol{f} = (x + 2y + az)\hat{i} + (bx - 3y - z)\hat{j} + (4x + cy + 2z)\hat{k}$
     is irrotational, *i.e.* curl $\boldsymbol{f} = \boldsymbol{0}$.

5. Show that the vector $\boldsymbol{f} = (\sin y + z)\hat{i} + (x\cos y - z)\hat{j} + (x - y)\hat{k}$ is irrotational.

6. Show that $\nabla^2\left(\dfrac{x}{r^2}\right) = -\dfrac{2x}{r^4}$.

7. If $\boldsymbol{v} = (x\hat{i} + y\hat{j} + z\hat{k}) / (x^2 + y^2 + z^2)^{3/2}$, find
  (i) $\nabla\cdot\boldsymbol{v}$ (ii) $\nabla\times\boldsymbol{v}$

8. If $\boldsymbol{v} = (x + y + 1)\hat{i} + \hat{j} + (-x - y)\hat{k}$, prove that
     $\boldsymbol{v}\cdot(\nabla\times\boldsymbol{v}) = 0$

9. If $\boldsymbol{v} = (y^2 + z^2 - x^2)\hat{i} + (z^2 + x^2 - y^2)\hat{j} + (x^2 + y^2 - z^2)\hat{k}$, find
  (i) div $\boldsymbol{v}$ (ii) curl $\boldsymbol{v}$

10. Prove that
  (i) div $(\boldsymbol{a} + \boldsymbol{b})$ = div $\boldsymbol{a}$ + div $\boldsymbol{b}$
  (ii) curl $(\boldsymbol{a} + \boldsymbol{b})$ = curl $\boldsymbol{a}$ + curl $\boldsymbol{b}$

11. Prove that div $\nabla\phi = \nabla^2\phi$, *i.e.* $\nabla\cdot\nabla\phi = \nabla^2\phi$, where ϕ is a scalar point function.

12. If $\boldsymbol{f} = x^2 y\hat{i} + xz\hat{j} + 2yz\hat{k}$, prove that div curl $\boldsymbol{f} = 0$.

13. Prove that div $\hat{r} = 2/r$, where $\hat{r}$ is a unit vector.

14. Prove that the vector $f(r)\boldsymbol{r}$ is irrotational.

15. Prove that :
  (i) $\nabla^2(fg) = f\nabla^2 g + 2\nabla f\cdot\nabla g + g\nabla^2 f$
  (ii) div $(\nabla f \times \nabla g) = 0$
  where f and g are two scalar point function.

16. Prove that : $\nabla\cdot\left\{r\nabla\left(\dfrac{1}{r^3}\right)\right\} = \dfrac{3}{r^4}$.

17. If $\boldsymbol{a}$ is a constant vector, prove that
  (i) div $\{r^n(\boldsymbol{a}\times\boldsymbol{r})\} = 0$
  (ii) curl $\left(\dfrac{\boldsymbol{a}\times\boldsymbol{r}}{r^3}\right) = -\dfrac{\boldsymbol{a}}{r^3} + \dfrac{3r}{r^5}(\boldsymbol{a}\cdot\boldsymbol{r})$

18. If $\boldsymbol{f} = f_1\hat{i} + f_2\hat{j} + f_3\hat{k}$, show that
  (i) $\nabla\cdot\boldsymbol{f} = \nabla f_1\cdot\hat{i} + \nabla f_2\cdot\hat{j} + \nabla f_3\cdot\hat{k}$
  (ii) $\nabla\times\boldsymbol{f} = \nabla f_1\times\hat{i} + \nabla f_2\times\hat{j} + \nabla f_3\times\hat{k}$

19. Prove that :
  (i) $\boldsymbol{a}.\nabla\left(\dfrac{1}{r}\right) = -\dfrac{\boldsymbol{a}.\boldsymbol{r}}{r^3}$
  (ii) $\boldsymbol{b}\cdot\nabla\left[\boldsymbol{a}\cdot\nabla\left(\dfrac{1}{r}\right)\right] = -\dfrac{\boldsymbol{a}\cdot\boldsymbol{b}}{r^3} + \dfrac{3(\boldsymbol{a}\cdot\boldsymbol{r})(\boldsymbol{b}\cdot\boldsymbol{r})}{r^5}$

20. Prove that div $\left\{\dfrac{f(r)}{r}\boldsymbol{r}\right\} = \dfrac{1}{r^2}\dfrac{d}{dr}r^2 f(r)$.

21. Prove that curl $(g\nabla f) = \nabla g \times \nabla f = -$ curl $(f\nabla g)$.

22. Prove that curl $(\boldsymbol{a}\times r)r^n = (n + 2)r^n\boldsymbol{a} - nr^{n-2}(\boldsymbol{r}\cdot\boldsymbol{a})\boldsymbol{r}$.

23. If $\hat{a}$ is a constant unit vector, prove that
     $\hat{a}\cdot\{\nabla(\boldsymbol{v}\cdot\hat{a}) - \nabla\times(\boldsymbol{v}\times\hat{a})\} = $ div $\boldsymbol{v}$

24. Prove that $\nabla^2\left[\nabla\cdot\left(\dfrac{\boldsymbol{r}}{r^2}\right)\right] = 2r^{-4}$.

25. Prove that $\dfrac{1}{2}\nabla a^2 = (\boldsymbol{a}\cdot\nabla)\boldsymbol{a} + \boldsymbol{a}\times$ curl $\boldsymbol{a}$

26. Prove that curl grad $r^n = \boldsymbol{0}$.

27. If $\boldsymbol{a}$ and $\boldsymbol{b}$ are constant vectors, prove that $\nabla[(\boldsymbol{a}\cdot\boldsymbol{b})r] = \boldsymbol{a}\cdot\boldsymbol{b}$.

28. If $\boldsymbol{a}$ is a constant vector, prove that
(i) $\nabla(\boldsymbol{a \cdot u}) = (\boldsymbol{a} \cdot \nabla)\boldsymbol{u} + \boldsymbol{a} \times \text{curl } \boldsymbol{u}$
(ii) $\nabla \cdot (\boldsymbol{a} \times \boldsymbol{u}) = -\boldsymbol{a} \cdot \text{curl } \boldsymbol{u}$
(iii) $\nabla \times (\boldsymbol{a} \times \boldsymbol{u}) = \boldsymbol{a} \text{ div } \boldsymbol{u} - (\boldsymbol{a} \cdot \nabla)\boldsymbol{u}$

29. If $\boldsymbol{u} = \left(\dfrac{1}{r}\right)\boldsymbol{r}$, then prove that

(i) $\nabla \times \boldsymbol{u} = 0$ (ii) $\text{grad (div } \boldsymbol{u}) = -\left(\dfrac{2}{r^3}\right)\boldsymbol{r}$.

30. Prove that
(i) $\text{div }(\boldsymbol{a} \times \boldsymbol{r}) = \boldsymbol{r} \cdot \text{curl } \boldsymbol{a}$ (ii) $\text{div }(\boldsymbol{r} \times \boldsymbol{a}) = 0$
(iii) $\text{curl }(\boldsymbol{r} \times a) = -2a$ (VTU–2010, BURDWAN–2009)

31. If $\boldsymbol{V} = \dfrac{x\hat{i} + y\hat{j} + z\hat{k}}{\sqrt{x^2+y^2+z^2}}$, show that $\nabla \boldsymbol{V} = \dfrac{2}{\sqrt{x^2+y^2+z^2}}$ and

$\nabla \times \boldsymbol{V} = 0$.   (MTU–2011)

32. Show that $\boldsymbol{A} = (6xy + z^3)\hat{i} + (3x^2 - z)\hat{j} + (3xz^2 - y)\hat{k}$ is irrotational.   (UKTU–2011)

33. Show that the fluid motion given by
$$\boldsymbol{V} = (y\sin z + \sin x)\hat{i} + (x\sin z + 2yz)\hat{j} + (xy\cos z + y^2)\hat{k}$$
is irrotational.   (UKTU–2011)

34. Show that the directional derivative of $\nabla(\nabla\phi)$ at the point $(1, -2, 1)$ in the direction of the normal to the surface $xy^2z = 3x + z^2$, where $\phi = 2x^3y^2z^4$ is given by $\dfrac{1724}{\sqrt{21}}$.   (UPTU–2009)

35. Show that
$$(y^2 - z^2 + 3yz - 2x)\hat{i} + (3xz + 2xy)\hat{j} + (3xy - 2xz + 2z)\hat{k}$$
is both solenoidal and irrotational.   (UPTU–2009)

Hints to Selected Problems

1. $\boldsymbol{f} = x^2y\hat{i} - 2xz\hat{j} + 2yz\hat{k}$

(i) $\nabla \cdot \boldsymbol{f} = \dfrac{\partial}{\partial x}(x^2y)\hat{i} + \dfrac{\partial}{\partial y}(-2xz)\hat{j} + \dfrac{\partial}{\partial z}(2yz)\hat{k}$
$= 2xy + 0 + 2y = 2y(x+1)$.

3. $\boldsymbol{f} = \text{grad}(x^3 + y^3 + z^3 - 3xyz)$
Let $\phi = x^3 + y^3 + z^3 - 3xyz$.
$\therefore \boldsymbol{f} = \nabla\phi$.
(i) $\nabla \cdot \boldsymbol{f} = \nabla \cdot (\nabla\phi) = \nabla^2\phi = \dfrac{\partial^2\phi}{\partial x^2} + \dfrac{\partial^2\phi}{\partial y^2} + \dfrac{\partial^2\phi}{\partial z^2}$
$= 6x + 6y + 6z = 6(x + y + z)$.
(ii) $\nabla \times (\nabla\phi) = 0$.

4. (ii) $\boldsymbol{f} = (x + 2y + az)\hat{i} + (bx - 3y - z)\hat{j} + (4x + cy + 2z)\hat{k}$

$\nabla \times \boldsymbol{f} = \begin{vmatrix} \hat{i} & \hat{j} & \hat{k} \\ \dfrac{\partial}{\partial x} & \dfrac{\partial}{\partial y} & \dfrac{\partial}{\partial z} \\ x+2y+az & bx-3y-z & 4x+cy+2z \end{vmatrix}$

$= (c+1)\hat{i} + (a-4)\hat{j} + (b-2)\hat{k}$
For irrotaional, we have $\nabla \times \boldsymbol{f} = 0$.
$\therefore \qquad c = -1, a = 4, b = 2$.

8. $\boldsymbol{v} = (x + y + 1)\hat{i} + \hat{j} + (-x - y)\hat{k}$, then

$\nabla \times \boldsymbol{v} = \begin{vmatrix} \hat{i} & \hat{j} & \hat{k} \\ \dfrac{\partial}{\partial x} & \dfrac{\partial}{\partial y} & \dfrac{\partial}{\partial z} \\ (x+y+1) & 1 & (-x-y) \end{vmatrix}$

$= \hat{i}[-1-0] - \hat{j}[-1-0] + \hat{k}[0-1] = -\hat{i} + \hat{j} - \hat{k}$
$\therefore \boldsymbol{v} \cdot \nabla \times \boldsymbol{v} = -(x + y + 1) + 1 + x + y = 0$.

11. $\text{div}(\nabla\phi) = \nabla \cdot (\nabla\phi) = \nabla \cdot \left[\dfrac{\partial\phi}{\partial x}\hat{i} + \dfrac{\partial\phi}{\partial y}\hat{j} + \dfrac{\partial\phi}{\partial z}\hat{k}\right]$

$= \dfrac{\partial}{\partial x}\left(\dfrac{\partial\phi}{\partial x}\right) + \dfrac{\partial}{\partial y}\left(\dfrac{\partial\phi}{\partial y}\right) + \dfrac{\partial}{\partial z}\left(\dfrac{\partial\phi}{\partial z}\right)$

$= \dfrac{\partial^2\phi}{\partial x^2} + \dfrac{\partial^2\phi}{\partial y^2} + \dfrac{\partial^2\phi}{\partial z^2} = \nabla^2\phi$.

16. Since $\nabla f(r) = f'(r) \cdot \nabla r = f'(r)\left(\dfrac{\boldsymbol{r}}{r}\right)$.

$\therefore \nabla\left(\dfrac{1}{r^3}\right) = -\dfrac{3}{r^5}\boldsymbol{r}$.

$\therefore \nabla\left\{r\nabla\left(\dfrac{1}{r^3}\right)\right\} = \nabla \cdot \left(-\dfrac{3}{r^4}\boldsymbol{r}\right) = \left(-\dfrac{3}{r^4}\right)\nabla \cdot \boldsymbol{r} + r\nabla\left(-\dfrac{3}{r^4}\right)$

$= -\dfrac{9}{r^4} + \boldsymbol{r} \cdot \left(\dfrac{12}{r^6}\boldsymbol{r}\right) = -\dfrac{9}{r^4} + \dfrac{12}{r^6}(\boldsymbol{r} \cdot \boldsymbol{r})$.

$= -\dfrac{9}{r^4} + \dfrac{12}{r^6}(r^2) = -\dfrac{9}{r^4} + \dfrac{12}{r^4} = \dfrac{3}{r^4}$.

17. (ii) $\text{curl}\left\{\dfrac{\boldsymbol{a} \times \boldsymbol{r}}{r^3}\right\} = \nabla\left(\dfrac{1}{r^3}\right) \times (\boldsymbol{a} \times \boldsymbol{r}) + \dfrac{1}{r^3}\nabla \times (\boldsymbol{a} \times \boldsymbol{r})$

$= -\dfrac{3}{r^3}\left(\dfrac{\boldsymbol{r}}{r}\right) \times (\boldsymbol{a} \times \boldsymbol{r}) + \dfrac{1}{r^3}(0 - 0 - \boldsymbol{a} + 3\boldsymbol{a})$

$= -\dfrac{3}{r^5}[(\boldsymbol{r} \cdot \boldsymbol{r})\boldsymbol{a} - (\boldsymbol{r} \cdot \boldsymbol{a})\boldsymbol{r}] + \dfrac{2\boldsymbol{a}}{r^3}$

$= -\dfrac{3\boldsymbol{a}}{r^3} + \dfrac{3\boldsymbol{r}}{r^5}(\boldsymbol{r} \cdot \boldsymbol{a}) + \dfrac{2\boldsymbol{a}}{r^3} = -\dfrac{\boldsymbol{a}}{r^3} + \dfrac{3\boldsymbol{r}}{r^5}(\boldsymbol{r} \cdot \boldsymbol{a})$

19. (i) $\boldsymbol{a} \cdot \nabla\left(\dfrac{1}{r}\right) = \boldsymbol{a} \cdot \left[-\dfrac{1}{r^2}\dfrac{\boldsymbol{r}}{r}\right] = -\dfrac{\boldsymbol{a} \cdot \boldsymbol{r}}{r^3}$

(ii) $\boldsymbol{b} \cdot \nabla\left[\boldsymbol{a} \cdot \nabla\left(\dfrac{1}{r}\right)\right] = \boldsymbol{b} \cdot \nabla\left[-\dfrac{\boldsymbol{a} \cdot \boldsymbol{r}}{r^3}\right]$

$= \boldsymbol{b} \cdot \left[-\dfrac{\boldsymbol{a}}{r^3} + \dfrac{3(\boldsymbol{a} \cdot \boldsymbol{r})}{r^5}\boldsymbol{r}\right] = -\dfrac{\boldsymbol{a} \cdot \boldsymbol{b}}{r^3} + \dfrac{3(\boldsymbol{a} \cdot \boldsymbol{r})(\boldsymbol{b} \cdot \boldsymbol{r})}{r^5}$

24. $\nabla \cdot \left(\dfrac{\boldsymbol{r}}{r^2}\right) = \nabla\left(\dfrac{1}{r^2}\right) \cdot \boldsymbol{r} + \dfrac{1}{r^2}(\nabla \cdot \boldsymbol{r})$

$= -\dfrac{2}{r^3}\left(\dfrac{\boldsymbol{r}}{r}\right) \cdot \boldsymbol{r} + \dfrac{3}{r^2} = -\dfrac{2}{r^2} + \dfrac{3}{r^2} = \dfrac{1}{r^2}$

$\therefore \nabla^2\left[\nabla \cdot \left(\dfrac{\boldsymbol{r}}{r^2}\right)\right] = \nabla^2 \cdot \left(\dfrac{1}{r^2}\right) = \nabla\left(\nabla\left(\dfrac{1}{r^2}\right)\right) = \nabla \cdot \left[-\dfrac{2\boldsymbol{r}}{r^4}\right]$

$= \nabla\left(-\dfrac{2}{r^4}\right) \cdot \boldsymbol{r} + \left(-\dfrac{2}{r^4}\right)\nabla \cdot \boldsymbol{r} = \dfrac{8}{r^4} - \dfrac{6}{r^4} = \dfrac{2}{r^4} = 2r^{-4}$.

Answers

1. (i) $2y(x + 1)$ (ii) $(2x + 2z)\hat{i} - (x^2 + 2z)\hat{k}$ (iii) $(2x + 2)\hat{j}$ **2.** (i) $\text{div }\boldsymbol{f} = 9$ (ii) $\text{curl }\boldsymbol{f} = -\hat{i} - 2\hat{k}$

3. (i) $\text{div }\boldsymbol{f} = 6(x + y + z)$ (ii) $\text{curl }\boldsymbol{f} = 0$ **4.** (i) $\lambda = -2$ (ii) $a = 4, b = 2\ c = -1$

7. (i) $\dfrac{3}{2} \cdot \dfrac{1}{(x^2 + y^2 + z^2)^{3/2}}$ (ii) **0**   **9.** (i) $-2x - 2y - 2z$ (ii) $2(y - z)\hat{i} + 2(z - x)\hat{j} + 2(x - y)\hat{k}$

6.12 GAUSS'S, STOKE'S AND GREEN'S THEOREMS

In this section we shall discuss line integral, surface integrals and volume intetgrals and shall consider some important applications of such integrals which deal with physical and engineering problems. We shall see that line integral is a natural generalization of a definite integral, a surface integral is a generalization of double integral and volume intgral is generalization of triple integral.

Line integrals can be transformed into double integral and conversely. Triple integrals can be transformed into double integral and these transformations has great importance. Therefore, we shall discuss some formulas of Gauss, Green and Stokes which are powerful in many applications as well as in theoretical problems. These formulas give the physical meaning of the divergence and the curl of a vector function.

6.12.1 ORIENTED CURVES

Let us consider a curve C in space and oriented the curve C by choosing one of the two directions along C as the positive direction and the opposite direction along C is then called the negative direction.

Let A be the initial point and B be the terminal point of C under the chosen orientation. Therefore, we may now represent the curve C by the parametric equation

$$r(s) = x(s)\hat{i} + y(s)\hat{j} + z(s)\hat{k}$$

where s is the arc length of C and for the point A, $s = a$ and for the point B, $s = b$, hence $a \leq s \leq b$.

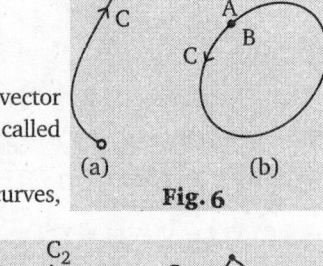

(i) Closed curve. If the point A and B coincide as shown in fig. 6(b), then the curve is closed.

(ii) Smooth curve. If $r(s)$ is continuously differentiable and its first derivative is different from zero vector for all values of s and the curve C has a unique tangent at each of its points, then the curve C is called smooth curve.

(iii) Piecewise smooth curve. A curve C which is the composition of a finite number of smooth curves, is called piecewise smooth curve.

In the adjoining fig. 7 the curve is composed of four smooth curves C_1, C_2, C_3 and C_4, hence the curve is piecewise smooth.

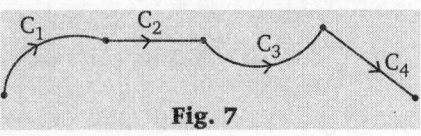

Fig. 7

(iv) Smooth surface. A surface S over each of its points a unique normal may drawn and the direction of each normal depends only on the point at which it is drawn, is called smooth surface.

(v) Piecewise smooth surface. A surface which is composed of a finite number of smooth surfaces, is called piecewise smooth surface.

(vi) Simply connected domain. A region (or domain) in which every closed curve can be shrink to a point without crossover the boundary of the region, is called simply connected domain. Otherwise the region is called multiply connected domain.

(a) Simply connected   (b) Multiply connected

Fig. 8

6.12.2 LINE, SURFACE AND VOLUME INTEGRALS

(i) Line integrals. Let $f(x, y, z)$ be a given function which is defined at each point of the curve C and $f(x, y, z)$ is continuous function of s and let P be a point on C with co-ordintaes $(x(s), y(s), z(s))$. Thus $f(x, y, z)$ is written as $f(P)$. Now divide the curve C into n parts in an arbitrary way and letting $P_0 = A, P_1, P_2, ..., P_{n-1}, P_n = B$ where A and B are the end points of the curve C.

Let us divide in the interval $a \leq s \leq b$ such that

$$a = s_0 < s_1 < s_2 < ... < s_n = b.$$

Now choose an arbitrary point between each portion, *i.e.* between A and P_1, P_1 and P_2 and so on. Let Q_1 be that point between A and P_1, Q_2 between P_1 and P_2 etc. and form the sum

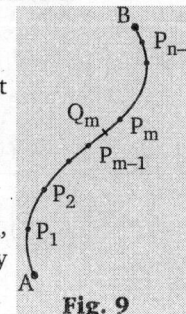

$$S_n = \sum_{m=1}^{n} f(Q_m)\Delta s_m, \text{ where } \Delta s_m = s_m - s_{m-1}.$$

Now for $n = 2, 3, 4, ...,$ and the greatest $\Delta s_m \to 0$ as $n \to \infty$, we get a sequence of real numbers $S_2, S_3, S_4, ...$. The limit of this squence $\langle s_n \rangle$ is called the line integral of f along the curve C from A to B is denoted by $\int_C f(x, y, z)ds$.

Fig. 9

In most cases the representation of C will be of the form

$$r(t) = x(t)\hat{i} + y(t)\hat{j} + z(t)\hat{k}, t_0 \leq t \leq t_1$$

then we have

$$\int_C f(x, y, z)dS = \int_a^b f[x(s), y(s), z(s)]ds \qquad ...(1)$$

and $$\int_a^b f[x(s), y(s), z(s)]ds = \int_{t_0}^{t_1} f[x(t), y(t), z(t)]\frac{ds}{dt}dt \qquad ...(2)$$

In particular, suppose $\mathbf{r}(t)$ is the position vector of (x, y, z), then $\mathbf{r}(t) = x\hat{i} + y\hat{j} + z\hat{k}$

and let $t = t_0$ at A and $t = t_1$ at B and suppose $\mathbf{F}(x, y, z) = f_1\hat{i} + f_2\hat{j} + f_3\hat{k}$
is a vector function and continuous along C. Let s be the arc length of the curve C i.e., $s = $ arc AP, then

$$\frac{d\mathbf{r}}{ds} = \mathbf{t}$$

is a unit tangent vector at the point $P(x, y, z)$. Thus the component of $\mathbf{f}$ along this tangent is $\mathbf{f} \cdot \dfrac{d\mathbf{r}}{ds}$. Therefore, we have

$$\int_A^B \left(\mathbf{f} \cdot \frac{d\mathbf{r}}{ds}\right) ds = \int_A^B \mathbf{f} \cdot d\mathbf{r} = \int_C \mathbf{f} \cdot d\mathbf{r}.$$

$\therefore \qquad\qquad \int_C \mathbf{f} \cdot d\mathbf{r} = \int_C (f_1 dx + f_2 dy + f_3 dz).$ ...(3)

Since $x = x(t), y = y(t), z = z(t)$, then

$$\int_C \mathbf{f} \cdot d\mathbf{r} = \int_{t_0}^{t_1} \left[f_1 \frac{dx}{dt} + f_2 \frac{dy}{dt} + f_3 \frac{dz}{dt} \right] dt.$$ ...(4)

Hence (2) and (4) are equivalent.

REMARK

- If the curve C is simple closed curve, then the integral $\int_C \mathbf{f} \cdot d\mathbf{r}$ is known as the circulation.

(ii) Surface Integral (double integral). Let S be a surface of finite area and let $f(x, y, z)$ be defined over this surface S which is single valued function. Now divide the whole surface S into n surface elements of areas $\Delta S_1, \Delta S_2, ..., \Delta S_m, ..., \Delta S_n$. Let us take an surface element of area ΔS_m and choose an arbitrary point P_m inside ΔS_m and form the sum

$$J_n = \sum_{m=1}^{n} f(P_m) \Delta S_m.$$

Now taking the limit as $n \to \infty$ in such a way that $\Delta S_m \to 0$, then this limit if exists is called the surface integral of f over S and is denoted by

$$\iint_S f(x, y, z) dS.$$

It can be shown that the sequence $\langle J_n \rangle$ converges and its limit is independent of the choice of subdivisions and corresponding point P_m.

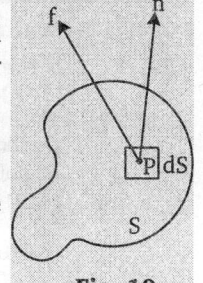

In particular, let S be a piecewise smooth surface and $\mathbf{f}(x, y, z)$ is a vector function which is continuous and defined over S. Let us consider a surface element of area dS enclosing a point P and let $\mathbf{n}$ be the unit vector drawn at P outward to the element dS and normal to it which is shown in fig. 10.

Thus $\mathbf{f} \cdot \mathbf{n}$ is the normal component of $\mathbf{f}$ at P. Therefore, the integral of $\mathbf{f} \cdot \mathbf{n}$ over S can be written as

$$\iint_S \mathbf{f} \cdot \mathbf{n} \, dS = \iint_S \mathbf{f} \cdot d\mathbf{S}, \text{ where } d\mathbf{S} = \mathbf{n} dS$$ ...(1)

If $\mathbf{n} = l\hat{i} + m\hat{j} + n\hat{k}$, where l, m, n are the direction cosines of normal which makes the angles α, β and γ with the positive axis i.e., $l = \cos\alpha$, $m = \cos\beta$, $n = \cos\gamma$.

Let $\mathbf{f}(x, y, z) = f_1\hat{i} + f_2\hat{j} + f_3\hat{k}$, then

$$\iint_S \mathbf{f} \cdot \mathbf{n} \, dS = \iint_S (f_1 \cos\alpha + f_2 \cos\beta + f_3 \cos\gamma) dS.$$

Since we have $\cos\alpha \, dS = dydz$, $\cos\beta \, dS = dzdx$, $\cos\gamma \, dS = dxdy$

$\therefore \qquad\qquad \iint_S \mathbf{f} \cdot \mathbf{n} \, dS = \iint_S (f_1 dzdy + f_2 dzdx + f_3 dxdy).$

Fig. 10

Evaluation of double integral. To find the value of double integral, it is taken over the orthogonal projection of S on one of the co-ordinate planes. This projection is obtained only if a line perpendicular to the closed co-ordinate planes meet the given surface S in only one point. If it is not so, then subdivide the surface S into surface which do satisfy this condition. This method can be understood by the following process : For this let us consider the surface S in such a way that a line perpendicular to xy-plane meets S in only one point. The equation of the surface S is then taken

$$z = f(x, y).$$

Let R be the orthogonal projection of S on xy-plane and let $\hat{n}$ be the unit vector drawn at some point P on S and normal to S and let this normal $\hat{n}$ makes the acute angle γ with the z-axis then

$$\cos\gamma \, dS = dxdy$$

where dS is the area of a small patch at P on S.

$\therefore \qquad\qquad dS = \frac{dxdy}{\cos\gamma}.$

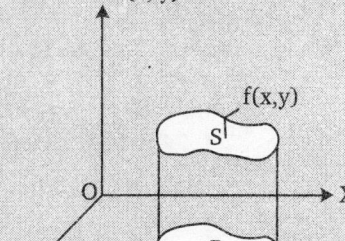

Fig. 11

Since $\quad \boldsymbol{n} = \cos\alpha\hat{i} + \cos\beta\hat{j} + \cos\gamma\hat{k}$, therefore $\cos\gamma = \boldsymbol{n}\cdot\hat{k}$.

Since γ is acute, then $\cos\gamma = \left|\boldsymbol{n}\cdot\hat{k}\right|$, so $\quad dS = \dfrac{dxdy}{\left|\boldsymbol{n}\cdot\hat{k}\right|}$

Thus $\qquad \iint_S \boldsymbol{f}\cdot\boldsymbol{n}dS = \iint_R \boldsymbol{f}\cdot\boldsymbol{n}\dfrac{dxdy}{\left|\boldsymbol{n}\cdot\hat{k}\right|}.$ $\qquad\qquad\qquad$ ...(1)

Hence the surface integral over S can be evaluated using (1)

☛ REMARK

- If S can be described by inequalities of the form $a \le x \le b$ and $g(x) \le y \le h(x)$ where $S : z = f(x, y)$, then

$$\iint_S f dxdy = \int_a^b\left[\int_{g(x)}^{h(x)} f dy\right]dx.$$

(iii) **Volume integral.** Let V be a volume enclosed by a surface S and let $f(x, y, z)$ be a point function defined over V. Now divide the volume V into n subvolume element of volumes $\Delta V_1, \Delta V_2, ..., \Delta V_{n-1}, \Delta V_n$ and choose an arbitrary point $P_m(x_m, y_m, z_m)$ in each of elements ΔV_m such that $f(P_m) = f(x_m, y_m, z_m)$ and form the sum

$$J_n = \sum_{m=1}^{n} f(P_m)\Delta V_m.$$

for $n = 2, 3, 4, ...$, and taking greatest $\Delta V_m \to 0$ as $n \to \infty$. Then we get the sequence $J_2, J_3, J_4, ... $. If the limit of this sequence $\langle J_n\rangle$ exists, then this limit is called the volume integral of f over the volume V which is denoted by

$$\iiint_V f(x, y, z)dV.$$

This limit is independent of the choice of the subdivision of V, if V is piecewise smooth volume. Therefore, we can take the volume elememts in the form of urboids whose edges are parallel to the co-ordinates axis. Then $dV = dxdydz$, hence

$$\iiint_V f(x, y, z)dV = \iiint_V f(x, y, z)dxdydz.$$

☛ REMARK

- If $f(x, y, z)$ is a vector function, then the volume integral of $f(x, y, z$) over V is $\quad \iiint_V \boldsymbol{f}(x, y, z)dV$.

Solved Examples

EXAMPLE 1. *Evaluate* $\int_C \boldsymbol{F}\cdot d\boldsymbol{r}$, *where* $\boldsymbol{F} = 3xy\hat{i} - y^2\hat{j}$ *and C is the curve* $y = 2x^2$ *with xy-plane from* (0, 0) *to* (1, 2).

SOLUTION . The parametric equations of the given curve *i.e.*, the parabola $y = 2x^2$ can be taken as $x = t, y = 2t^2$.

At point (0, 0), $x = 0$ and so $t = 0$ and at the point (1, 2), $x = 1$ and so $t = 1$.

Again $\dfrac{dx}{dt} = 1$ and $\dfrac{dy}{dt} = 4t$

and $\quad dr = dx\hat{i} + dy\hat{j} \qquad [\because \boldsymbol{r} = x\hat{i} + y\hat{j}]$

$\therefore \int_C \boldsymbol{F}\cdot d\boldsymbol{r} = \int_C (3xy\hat{i} - y^2\hat{j})\cdot(dx\hat{i} + dy\hat{j})$

$= \int_C (3xydx - y^2dy)$

$= \int_{t=0}^{1}\left(3xy\dfrac{dx}{dt} - y^2\dfrac{dy}{dt}\right)dt$

$= \int_{t=0}^{1}(3.t.2t^2.1 - 4t^4.4t)dt$

$= \int_{t=0}^{1}(6t^3 - 16t^5)dt = \left[6.\dfrac{t^4}{4} - 16.\dfrac{t^6}{6}\right]_{t=0}^{1}$

$= \dfrac{6}{4} - \dfrac{16}{6} = \dfrac{3}{2} - \dfrac{8}{3} = -\dfrac{7}{6}.$

EXAMPLE 2. *Evaluate* $\int_C \boldsymbol{F}\cdot d\boldsymbol{r}$ *where* $\boldsymbol{F} = xy\hat{i} + yz\hat{j} + zx\hat{k}$ *and curve C is* $\boldsymbol{r} = t\hat{i} + t^2\hat{j} + t^3\hat{k}$ *and* $-1 \le t \le 1$.

SOLUTION . Since $\boldsymbol{r} = t\hat{i} + t^2\hat{j} + t^3\hat{k}$ is given but

$\boldsymbol{r} = x\hat{i} + y\hat{j} + z\hat{k}$, then $x = t, y = t^2, z = t^3$

and $\quad \dfrac{d\boldsymbol{r}}{dt} = \hat{i} + 2t\hat{j} + 3t^2\hat{k}$

$\therefore \quad \int_C \boldsymbol{F}\cdot d\boldsymbol{r} = \int_{-1}^{1}\left(\boldsymbol{F}\cdot\dfrac{d\boldsymbol{r}}{dt}\right)dt \qquad$ ...(1)

and $\qquad \boldsymbol{F} = xy\hat{i} + yz\hat{j} + zx\hat{k}$

$\qquad\qquad = t^3\hat{i} + t^5\hat{j} + t^4\hat{k}$

$\therefore \qquad \boldsymbol{F}\cdot\dfrac{d\boldsymbol{r}}{dt} = (t^3\hat{i} + t^5\hat{j} + t^4\hat{k})\cdot(\hat{i} + 2t\hat{j} + 3t^2\hat{k})$

$\qquad\qquad = t^3 + 2t^6 + 3t^6 = t^3 + 5t^6.$

$\therefore$ From (1)

$\qquad \int_C \boldsymbol{F}\cdot d\boldsymbol{r} = \int_{-1}^{1}(t^3 + 5t^6)dt$

$\qquad = \left[\dfrac{t^4}{4} + \dfrac{5t^7}{7}\right]_{-1}^{1} = \dfrac{1}{4} - \dfrac{1}{4} + \dfrac{10}{7} = \dfrac{10}{7}.$

EXAMPLE 3. *Evaluate* $\int_C \boldsymbol{F}\cdot d\boldsymbol{r}$ *where* $\boldsymbol{F} = (x^2 - y^2)\hat{i} + xy\hat{j}$ *and curve C is the arc of the curve* $y = x^3$ *from* (0, 0) *to* (2, 8).

SOLUTION . Taking the given arc of the following form $x = t$ and $y = t^3$ and takes the value from $t = 0$ to $t = 2$ (*i.e.*, from (0, 0) to (2, 8)).

$\therefore \qquad \boldsymbol{r}(t) = x\hat{i} + y\hat{j} = t\hat{i} + t^3\hat{j}$ so $\dfrac{d\boldsymbol{r}}{dt} = \hat{i} + 3t^2\hat{j}.$

Also, $\quad \boldsymbol{F} = (x^2 - y^2)\hat{i} + xy\hat{j} = (t^2 - t^6)\hat{i} + t^4\hat{j}$

$\therefore \int_C \boldsymbol{F}\cdot d\boldsymbol{r} = \int_C\left(\boldsymbol{F}\cdot\dfrac{d\boldsymbol{r}}{dt}\right)dt$

$$= \int_0^2 [(t^2 - t^6)\hat{i} + t^4\hat{j}].[\hat{i} + 3t^2\hat{j}]dt$$

$$= \int_0^2 (t^2 - t^6 + 3t^6)dt = \int_0^2 (t^2 + 2t^6)dt$$

$$= \left[\frac{t^3}{3} + \frac{2}{7}t^7\right]_0^2 = \frac{8}{3} + \frac{256}{7} = \frac{824}{21}.$$

EXAMPLE 4. *Evaluate $\int_C F \cdot dr$ where $F = z\hat{i} + x\hat{j} + y\hat{k}$ and C is the arc of the curve*

$$r = \cos t\hat{i} + \sin t\hat{j} + t\hat{k} \text{ from } t = 0 \text{ to } t = 2\pi.$$

SOLUTION. Here

$$\therefore \quad r(t) = \cos t\hat{i} + \sin t\hat{j} + t\hat{k}$$

$$\therefore \quad x = \cos t, y = \sin t, z = t$$

$$\therefore \quad F = t\hat{i} + \cos t\hat{j} + \sin t\hat{k}$$

$$\therefore \quad dr = (-\sin t\hat{i} + \cos t\hat{j} + \hat{k})dt$$

$$\therefore \quad F \cdot dr = (-t\sin t + \cos^2 t + \sin t)dt$$

Now $\int_C F \cdot dr = \int_0^{2\pi}(-t\sin t + \cos^2 t + \sin t)dt$

$$= -\int_0^{2\pi} t\sin t\, dt + \int_0^{2\pi}\cos^2 t\, dt + \int_0^{2\pi}\sin t\, dt$$

$$= -[-t\cos t + \sin t]_0^{2\pi}$$

$$+ \int_0^{2\pi}\frac{1}{2}(1 + \cos 2t)dt + [-\cos t]_0^{2\pi}$$

$$= -[-2\pi + \sin 2\pi] + \frac{1}{2}\left[t + \frac{1}{2}\sin 2t\right]_0^{2\pi}$$

$$+ [-\cos 2\pi + \cos 0]$$

$$= 2\pi + \frac{1}{2}[2\pi + 0] + [-1 + 1] = 3\pi.$$

EXAMPLE 5. *Evaluate $\int_C F \cdot dr$ where $F = (x^2 + y^2)\hat{i} - 2xy\hat{j}$, curve C is the rectangle in the xy-plane bounded by $y = 0$, $x = a$, $y = b$, $x = 0$.* (UPTU–2006)

SOLUTION. The curve C is shown in fig. 12.

Since C is in xy-plane, then

$$r = x\hat{i} + y\hat{j}$$

$$\therefore \quad dr = dx\hat{i} + dy\hat{j}$$

$$\therefore \quad F \cdot dr = [(x^2 + y^2)\hat{i} - 2xy\hat{j}].(dx\hat{i} + dy\hat{j})$$

$$= (x^2 + y^2)dx - 2xy\, dy$$

$$\therefore \quad \int_C F \cdot dr = \int_{OABD} F \cdot dr$$

$$= \int_O^A F \cdot dr + \int_A^B F \cdot dr + \int_B^D F \cdot dr + \int_D^O F \cdot dr \quad ...(1)$$

Along the line OA, $y = 0$, x varies from 0 to a.

$$\therefore \quad dy = 0$$

and $\int_O^A F \cdot dr = \int_O^A [(x^2 + y^2)dx - 2xy\, dy]$

$$= \int_0^a x^2 dx = \left[\frac{x^3}{3}\right]_0^a = \frac{a^3}{3}.$$

Along the line AB, $x = a$, y varies from 0 to b.

$$\therefore \quad dx = 0$$

and $\int_A^B F \cdot dr = \int_A^B [(x^2 + y^2)dx - 2xy\, dy]$

$$= \int_0^b (-2ay)dy = -2a\left[\frac{y^2}{2}\right]_0^b = -ab^2$$

Along the line BD, $y = b$, x varies from a to 0, then $dy = 0$

and $\int_B^D F \cdot dr = \int_B^D [(x^2 + y^2)dx - 2xy\, dy]$

$$= \int_a^0 (x^2 + b^2)dx = \left[\frac{x^3}{3} + b^2 x\right]_a^0$$

$$= -\frac{a^3}{3} - ab^2.$$

And along the line DO, $x = 0$ and y varies from b to 0, then $dx = 0$

and $\int_D^O F \cdot dr = \int_D^O [(x^2 + y^2)dx - 2xy\, dy]$

$$= \int_b^0 y^2.0 - 0.dy = 0$$

Substitute the values of these integral in (1), we get

$$\int_C F \cdot dr = \frac{a^3}{3} - ab^2 - \frac{a^3}{3} - ab^2 + 0$$

$$= -2ab^2$$

EXAMPLE 6. *If $F = (3x^2 + 6y)\hat{i} - 14yz\hat{j} + 20xz^2\hat{k}$, evaluate $\int_C F \cdot dr$ where C is a straight line joining $(0,0,0)$ and $(1, 1, 1)$.*

SOLUTION. Since curve C is a straight line joining $(0, 0, 0)$ and $(1, 1, 1)$. Then

$$C: \frac{x}{1} = \frac{y}{1} = \frac{z}{1} = t \text{ (say)}$$

$$\therefore \quad x = t, y = t, z = t \text{ such that } 0 \le t \le 1$$

and $\quad r = x\hat{i} + y\hat{j} + z\hat{k} = t\hat{i} + t\hat{j} + t\hat{k}$

$$\therefore \quad dr = (\hat{i} + \hat{j} + \hat{k})dt$$

and $\quad F = (3t^2 + 6t)\hat{i} - 14t^2\hat{j} + 20t^3\hat{k}$

$$\therefore \quad F \cdot dr = [(3t^2 + 6t)\hat{i} - 14t^2\hat{j} + 20t^3\hat{k}]$$

$$.(\hat{i} + \hat{j} + \hat{k})dt$$

$$= (3t^2 + 6t - 14t^2 + 20t^3)dt$$

$$= (20t^3 - 11t^2 + 6t)dt.$$

Now $\int_C F \cdot dr = \int_0^1 F \cdot dr = \int_0^1 (20t^3 - 11t^2 + 6t)dt$

$$= \left[20\frac{t^4}{4} - 11\frac{t^3}{3} + \frac{6t^2}{2}\right]_0^1 = \frac{13}{3}.$$

EXAMPLE 7. *Evaluate $\int_C F \cdot dr$ where $F = xy\hat{i} + (x^2 + y^2)\hat{j}$ and C is the x-axis from $x = 2$ to $x = 4$ and the line $x = 4$ from $y = 0$ to $y = 12$.*

SOLUTION. Here the curve C consists of two lines, one of them is x-axis from $x = 2$ to $x = 4$ and other is the line $x = 4$ from $y = 0$ to $y = 12$ has shown in fig. 13.

Since $\quad r = x\hat{i} + y\hat{j}, z = 0$

$$dr = dx\hat{i} + dy\hat{j}$$

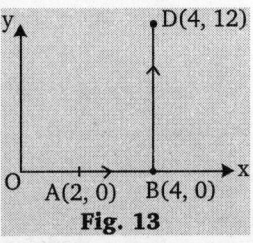

Fig. 13

and $\qquad F = xy\hat{i} + (x^2 + y^2)\hat{j}$

$\therefore \qquad F \cdot dr = [xy\hat{i} + (x^2 + y^2)\hat{j}] \cdot (dx\hat{i} + dy\hat{j})$

$$= xy\,dx + (x^2 + y^2)\,dy.$$

Now $\qquad \int_C F \cdot dr = \int_{ABD} F \cdot dr = \int_A^B F \cdot dr + \int_B^D F \cdot dr$

$\qquad\qquad\qquad\qquad\qquad\qquad\qquad \dots(1)$

Along the line $AB, y = 0$ and x varies from 2 to 4 and $dy = 0$

$\therefore \quad \int_A^B F \cdot dr = \int_A^B [xy\,dx + (x^2 + y^2)\,dy]$

$$= \int_2^4 0 \cdot dx = 0$$

Along the line $BD, x = 4$ and y varies from 0 to 12 and $dx = 0$

$\therefore \quad \int_B^D F \cdot dr = \int_B^D [xy\,dx + (x^2 + y^2)\,dy]$

$$= \int_0^{12} (16 + y^2)\,dy = \left[16y + \frac{y^3}{3}\right]_0^{12}$$

$$= 192 + 576 = 768.$$

EXAMPLE 8. *Let* $F(x, y, z) = x^3\hat{i} + y\hat{j} + z\hat{k}$ *is the force field. Find the workdone F along the line from* (1, 2, 3) *to* (3, 5, 7).   (UPTU–2006)

SOLUTION. Equation of the line from (1, 2, 3) to (3, 5, 7) is given by

$$\frac{x-1}{3-1} = \frac{y-2}{5-2} = \frac{z-3}{7-3} = k \ \text{(say)}$$

$\Rightarrow \qquad x = 2k + 1, y = 3k + 2, z = 4k + 3$

Now $\qquad r = x\hat{i} + y\hat{j} + z\hat{k}$

$$= (2k+1)\hat{i} + (3k+2)\hat{j} + (4k+3)\hat{k}$$

$\Rightarrow \qquad dr = (2\hat{i} + 3\hat{j} + 4\hat{k})\,dk$

Also at (1, 2, 3), $k = 0$ and at (3, 5, 7), $k = 1$

$\therefore$ Required workdone

$$= \int_C F \cdot dr$$

$$= \int_0^1 [(2k+1)^3\hat{i} + (3k+2)\hat{j}$$

$$+ (4k+3)\hat{k})] \cdot (2\hat{i} + 3\hat{j} + 4\hat{k})\,dk$$

$$= \int_0^1 [2(2k+1)^3 + 3(3k+2)$$

$$+ 4(4k+3)]\,dk$$

$$= \int_0^1 (16k^3 + 24k^2 + 37k + 20)\,dk$$

$$= \left(4k^4 + 8k^3 + \frac{37}{2}k^2 + 20k\right)_0^1$$

$$= \frac{101}{2}.$$

EXAMPLE 9. *Find the workdone by the force*

$$F = (2y + 3)\hat{i} + xz\hat{j} + (yz - x)\hat{k}$$

when it moves a particle from the point (0, 0, 0) *to the point* (2, 1, 1) *along the curve* $x = 2t^2, y = t$ *and* $z = t^3$.   (UPTU–2011, MADRAS–2010)

SOLUTION. We have $F = (2y + 3)\hat{i} + xz\hat{j} + (yz - x)\hat{k}$

and $r = x\hat{i} + y\hat{j} + z\hat{k} \Rightarrow dr = dx\hat{i} + dy\hat{j} + dz\hat{k}$

$\Rightarrow \quad F \cdot dr = (2y + 3)dx + xz\,dy + (yz - x)dz$

$\therefore$ Workdone $= \int_C F \cdot dr$

$$= \int_C (2y + 3)dx + xz\,dy + (yz - x)dz \quad \dots(1)$$

Along $C, \ x = 2t^2, y = t, z = t^3$

$\Rightarrow \qquad dx = 4t\,dt, dy = dt$ and $dz = 3t^2\,dt$

Also t varies from 0 to 1

$\therefore \qquad W = \int_0^1 (2t + 3)4t\,dt + \int_0^1 2t^2 \cdot t^3\,dt$

$$+ \int_0^1 (t \cdot t^3 - 2t^2)3t^2\,dt$$

$$= \int_0^1 (8t^2 + 12t + 2t^5 + 3t^6 - 6t^4)\,dt$$

$$= \left[\frac{8t^3}{3} + \frac{12t^2}{2} + \frac{2t^6}{6} + \frac{3t^7}{7} - \frac{6t^5}{5}\right]_0^1$$

$$= \frac{8}{3} + \frac{12}{2} + \frac{2}{6} + \frac{3}{7} - \frac{6}{5} = \frac{288}{35}.$$

EXAMPLE 10. *A vector field is given by* $F = \sin y\hat{i} + x(1 + \cos y)\hat{j}$ *Evaluate the line integral over a circular path given by* $x^2 + y^2 = a^2$.   (PTU–2003, ROHTAK–2006)

SOLUTION. Let $\qquad r = x\hat{i} + y\hat{j} + z\hat{k}$

Since $\qquad z = 0$ (given)

$\therefore \qquad r = x\hat{i} + y\hat{j} \ \Rightarrow \ dr = dx\hat{i} + dy\hat{j}$

Also, the circular path is $x = a \cos t, y = a \sin t, z = 0$ where t varies from 0 to 2π.

$\therefore \int_C F \cdot dr = \int_C (\sin y\hat{i} + x(1 + \cos y)\hat{j}) \cdot (dx\hat{i} + dy\hat{j})$

$$= \int_C [\sin y\,dx + x(1 + \cos y)\,dy]$$

$$= \int \ [\sin y\,dx + x \cos y\,dy + x\,dy]$$

$$= \int_C [d(x \sin y) + x\,dy]$$

$$= \int_0^{2\pi} [d(a \cos t \sin(a \sin t))] + a^2 \cos^2 t\,dt$$

$$= \left|a \cos t \sin(a \sin t)\right|_0^{2\pi}$$

$$+ \frac{a^2}{2}\int_0^{2\pi} (1 + \cos 2t)\,dt$$

$$= \frac{a^2}{2}\left|t + \frac{\sin 2t}{2}\right|_0^{2\pi} = \pi a^2.$$

EXAMPLE 11. *Find the workdone in moving a particle in a force field* $F = 3x^2\hat{i} + (2xy - y)\hat{j} + 3\hat{k}$ *along the curve* $x^2 = 4y, 3x^2 = 8z$ *from* $x = 0$ *to* $x = 2$.   (MTU–2012)

SOLUTION. Required work done,

$$W = \int_C F \cdot dr$$

$$= \int_C [3x^2\hat{i} + (2xy - y)\hat{j} + 3\hat{k}] \cdot (dx\hat{i} + dy\hat{j} + dz\hat{k})$$

$$= \int_C (3x^2\,dx + (2xy - y))\,dy + 3\,dz \quad \dots(1)$$

Put $x = t$, in $x^2 = 4y$ and $3x^2 = 8z$. The parametric equation of C are $x = t, y = t^2/4, z = (3/8)\,t^2$ (t varies from 0 to 2).

Then from (1)

$$W = \int_0^2 3t^2\,dt + \left[2t\left(\frac{t^2}{4}\right) - \frac{t^2}{4}\right]\frac{2t}{4}\,dt + 3\left(\frac{6t}{4}\right)dt$$

$$= \int_0^2 \left(3t^2 + \frac{t^4}{4} - \frac{t^3}{8} + \frac{9t}{4}\right)dt$$

$$= \left[t^3 + \frac{t^5}{20} - \frac{t^4}{32} + \frac{9t^2}{8}\right]_0^2$$

$$= \left[8 + \frac{32}{20} - \frac{16}{32} + \frac{9}{2}\right] = \frac{136}{10} = 13.6.$$

EXAMPLE 12. *Find the total workdone in moving a particle in a force field given by* $F = 3xy\hat{i} - 5z\hat{j} + 10x\hat{k}$ *along the curve* $x = t^2 + 1, y = 2t^2, z = t^3$ *from* $t = 1$ *to 2.* (MTU-2013)

SOLUTION. Required workdone $W = \int_C F.dr$

$$= \int_C [3xy\hat{i} - 5z\hat{j} + 10x\hat{k}].(dx\hat{i} + dy\hat{j} + dz\hat{k})$$

$$= \int_C (3xydx - 5zdy + 10xdz)$$

$$= \int_1^2 (3xy\frac{dx}{dt} - 5z\frac{dy}{dt} + 10x\frac{dz}{dt})dt$$

$$= \int_1^2 [3(t^2+1)(2t^2)2t - (5t^3)(4t)$$
$$\qquad + 10(t^2+1)(3t^2)]dt$$

$$= \int_1^2 (12t^5 + 12t^3 - 20t^4 + 30t^4 + 30t^2)dt$$

$$= \int_1^2 (12t^5 + 10t^4 + 12t^3 + 30t^2)dt$$

$$= \left[\frac{12t^6}{6} + \frac{10t^5}{5} + \frac{12t^4}{4} + \frac{30t^3}{3}\right]_1^2 = 303.$$

EXAMPLE 13. *If* $F = (2y+3)\hat{i} + xz\hat{j} + (yz-x)\hat{k}$, *evaluate* $\int_C F \cdot dr$ *where C is the path consisting of the straight lines from* (0, 0, 0) *to* (0, 0, 1) *then to* (0, 1, 1) *and then to* (2, 1, 1). (GBTU–2011)

SOLUTION. We have

$$F \cdot dr = [(2y+3)\hat{i} + xz\hat{j} + (yz-x)\hat{k}]$$
$$\qquad .(dx\hat{i} + dy\hat{j} + dz\hat{k})$$

$$= (2y+3)dx + xzdy + (yz-x)dz$$

Let C_1 denote the straight line joining (0, 0, 0) to (0, 0, 1), C_2 denote the straight line joining (0, 0, 1) to (0, 1, 1) and C_3 denote the straight line joining (0, 1, 1) to (2, 1, 1).

Along C_1, $x = 0, y = 0 \Rightarrow dx = 0, dy = 0$ and z varies from 0 to 1.
Along C_2, $x = 0, z = 1 \Rightarrow dx = 0, dz = 0$ and y varies from 0 to 1.
Along C_3, $y = 1, z = 1 \Rightarrow dy = 0, dz = 0$ and x varies from 0 to 2.
Therefore,

$$\int_C F \cdot dr = \int_{C_1} F \cdot dr + \int_{C_2} F \cdot dr + \int_{C_3} F \cdot dr$$

$$= \int_{z=0}^1 (0.z-0)dz + \int_{y=0}^1 (0.1)dy$$

$$\qquad + \int_{x=0}^2 (2.1+3)dx$$

$$= 0 + 0 + 5[x]_0^2 = 10.$$

EXAMPLE 14. *Show that the vector field F defined by*
$$F = (\sin y + z)\hat{i} + (x\cos y - z)\hat{j} + (x - y)\hat{k}$$
is conservative and find a function ϕ *such that* $F = \nabla\phi$.

SOLUTION. We have $\nabla \times F = \begin{vmatrix} \hat{i} & \hat{j} & \hat{k} \\ \frac{\partial}{\partial x} & \frac{\partial}{\partial y} & \frac{\partial}{\partial z} \\ \sin y + z & x\cos y - z & x - y \end{vmatrix}$

$$= \hat{i}\left[\frac{\partial}{\partial y}(x-y) - \frac{\partial}{\partial z}(x\cos y - z)\right]$$

$$+ \hat{j}\left[\frac{\partial}{\partial z}(\sin y + z) - \frac{\partial}{\partial x}(x-y)\right]$$

$$+ \hat{k}\left[-\frac{\partial}{\partial y}(\sin y + z) + \frac{\partial}{\partial x}(x\cos y - z)\right]$$

$$= \hat{i}[-1+1] + \hat{j}[1-1]$$
$$\qquad + \hat{k}[-\cos y + \cos y]$$

$$= \mathbf{0}$$

Hence, the field F is conservative.

Since $F = \nabla\phi = \frac{\partial\phi}{\partial x}\hat{i} + \frac{\partial\phi}{\partial y}\hat{j} + \frac{\partial\phi}{\partial z}\hat{k}$

$$\Rightarrow F_x = \frac{\partial\phi}{\partial x}, F_y = \frac{\partial\phi}{\partial y}, F_z = \frac{\partial\phi}{\partial z} \text{ if } F = (F_x, F_y, F_z)$$

$$\therefore F_x dx + F_y dy + F_z dz = \frac{\partial\phi}{\partial x}dx + \frac{\partial\phi}{\partial y}dy + \frac{\partial\phi}{\partial z}dz$$
$$= d\phi$$

$$\Rightarrow d\phi = (\sin y + z)dx + (x\cos y - z)dy$$
$$\qquad + (x-y)dz$$

$$\Rightarrow d\phi = \sin y dx + x\cos y dy + z dx + x dz$$
$$\qquad - (zdy - ydz)$$

$$\Rightarrow d\phi = d(x\sin y) + d(zx) - d(yz)$$

Integrating, we get

$$\phi = x\sin y + zx - yz + c, c \text{ is a constant}$$

EXAMPLE 15. *Evaluate* $\iint_S F \cdot ndS$, *where* $F = z\hat{i} + x\hat{j} - 3y^2z\hat{k}$ *and S is the surface of the cylinder* $x^2 + y^2 = 16$ *included in the first octant between* $z = 0$ *and* $z = 5$.

SOLUTION. Since $S : x^2 + y^2 = 16$
Let $f \equiv x^2 + y^2 - 16$
then the vector normal to the surface S is the gradient of f and let n be the unit normal to S.
Then $n = (\nabla f) / |\nabla f|$

$$= \frac{2x\hat{i} + 2y\hat{j}}{\sqrt{(4x^2 + 4y^2)}} = \frac{2x\hat{i} + 2y\hat{j}}{\sqrt{4(x^2 + y^2)}}$$

$$= \frac{2x\hat{i} + 2y\hat{j}}{\sqrt{4 \times 16}} \qquad (\because x^2 + y^2 = 16)$$

$$= \frac{2x\hat{i} + 2y\hat{j}}{8}$$

$$\therefore \quad n = \frac{x\hat{i} + y\hat{j}}{4}.$$

Here the surface S is perpendicular to xy-plane so we will take the projection of S on zx-plane. Let R be that projection

$$\iint_S F \cdot ndS = \iint_R F \cdot n \frac{dxdz}{|n \cdot \hat{j}|} \qquad ...(1)$$

Now $F \cdot n = (z\hat{i} + x\hat{j} - 3y^2z\hat{k}) \cdot \left(\frac{x\hat{i} + y\hat{j}}{4}\right)$

$$= \frac{zx + xy}{4}$$

and $\quad \left| \boldsymbol{n} \cdot \hat{j} \right| = \left| \left(\dfrac{x\hat{i} + y\hat{j}}{4} \right) \cdot \hat{j} \right| = \dfrac{y}{4}$

From (1), we get

$$\iint_S \boldsymbol{F} \cdot \boldsymbol{n}\, dS = \iint_R \dfrac{zx + xy}{4} \cdot \dfrac{4}{y}\, dx\, dz$$

$$= \iint_R \dfrac{(xz + xy)}{y}\, dx\, dz.$$

Since z varies from 0 to 5 and $y = \sqrt{16 - x^2}$ on S.

$$\therefore \quad \iint_R \dfrac{(xz + xy)}{y}\, dx\, dz$$

$$= \int_{z=0}^5 \int_{x=0}^4 \left[\dfrac{xz}{\sqrt{16 - x^2}} + x \right] dx\, dz$$

$$= \int_0^5 \left[\dfrac{x^2}{2} - z\sqrt{16 - x^2} \right]_0^4 dz$$

$$= \int_0^5 (4z + 8)\, dz = \left[2z^2 + 8z \right]_0^5$$

$$= 50 + 40 = 90 \cdot$$

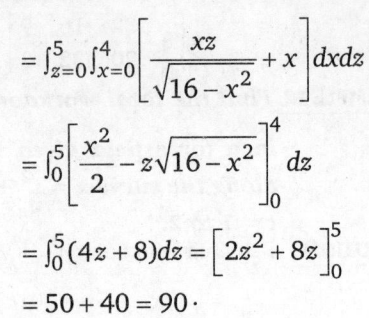

Exercise-6.6

1. Evaluate $\int_C \boldsymbol{F} \cdot \boldsymbol{dr}$, where $\boldsymbol{F} = x^2\hat{i} + y^3\hat{j}$ and curve C is the arc of the parabola $y = x^2$ in the xy-plane from $(0, 0)$ to $(1, 1)$.

2. If $\boldsymbol{F} = (3x^2 + 6y)\hat{i} - 14yz\hat{j} + 20xz^2\hat{k}$, then evaluate $\int_C \boldsymbol{F} \cdot \boldsymbol{dr}$ from $(0, 0, 0)$ to $(1, 1, 1)$ along the curve C
$$x = t, y = t^2, z = t^3. \qquad \text{(GBTU–2010, VTU–2001)}$$

3. If $\boldsymbol{F} = y\hat{i} - x\hat{j}$, evaluate $\int_C \boldsymbol{F} \cdot \boldsymbol{dr}$ from $(0, 0)$ to $(1, 1)$ along the following paths C:
   (i) The parabola $y = x^2$
   (ii) The straight lines form $(0, 0)$ to $(1, 0)$ and then to $(1, 1)$
   (iii) The straight line joining $(0, 0)$ and $(1, 1)$.

4. Find the workdone in moving a particle in a force field
$$\boldsymbol{F} = 3x^2\hat{i} + (2xz - y)\hat{j} + z\hat{k}$$
along the line joining $(0, 0, 0)$ to $(2, 1, 3)$.
   (MTU–2012, DELHI–2002, JNTU–2002, SVTU–2007)

   [**Hint :** Workdone$= \int_C \boldsymbol{F} \cdot \boldsymbol{dr}$]

5. Evaluate $\int_C \boldsymbol{F} \cdot \boldsymbol{dr}$, where $\boldsymbol{F} = yz\hat{i} + zx\hat{j} + xy\hat{k}$ and the curve C is the position of the curve $\boldsymbol{r} = a\cos t\hat{i} + b\sin t\hat{j} + ct\hat{k}$ from $t = 0$ to $t = \pi/2$.

6. Evaluate the integral
$$\int_C [(2xy^3 - y^2 \cos x)dx + (1 - 2y\sin x + 3x^2y^2)dy],$$
where C is the arc of the parabola $2x = \pi y^2$ from $(0, 0)$ to $\left(\dfrac{\pi}{2}, 1 \right)$.

7. Evaluate $\int_C x^{-1}(y + z)ds$ where C is the arc of the circle $x^2 + y^2 = 4$, $z = 0$ from $(2, 0, 0)$ to $(\sqrt{2}, \sqrt{2}, 0)$.

8. If $\boldsymbol{F} = (2x + y)\hat{i} + (3y - x)\hat{j}$, evaluate $\int_C \boldsymbol{F} \cdot \boldsymbol{dr}$ where C is the curve in the xy-plane consisting of the straight lines from $(0, 0)$ to $(2, 0)$ and then to $(3, 2)$.

9. Evaluate $\iint_S \boldsymbol{F} \cdot \boldsymbol{n}\, dS$, where $\boldsymbol{F} = yz\hat{i} + zx\hat{j} + xy\hat{k}$ and S is that part of the surface of the sphere $x^2 + y^2 + z^2 = 1$ which lies in the first octant.

10. Evaluate $\iint_S \boldsymbol{F} \cdot \boldsymbol{n}\, dS$, where $\boldsymbol{F} = xy\hat{i} - x^2\hat{j} + (x + z)\hat{k}$, S is the portion of the plane $2x + 2y + z = 6$ included in the first octant.

11. Evaluate $\iint_S \boldsymbol{F} \cdot \boldsymbol{n}\, dS$, where $\boldsymbol{F} = y\hat{i} + 2x\hat{j} - z\hat{k}$ and S is the surface of the plane $2x + y = 6$ in the first octant cut off by the plane $z = 4$.

12. If $\boldsymbol{F} = 2y\hat{i} - 3\hat{j} + x^2\hat{k}$ and S is the surface of the parabolic cylinder $y^2 = 8x$ in the first octant bounded by the planes $y = 4$ and $z = 6$, then evaluate $\iint_S \boldsymbol{F} \cdot \boldsymbol{n}\, dS$.

13. If $\boldsymbol{F} = (2x^2 - 3z)\hat{i} - 2xy\hat{j} - 4x\hat{k}$, then evaluate $\iiint_V \nabla \cdot \boldsymbol{F}\, dV$ where V is the closed region bounded by the planes $x = 0$, $y = 0$, $z = 0$ and $2x + 2y + z = 4$.

14. Evaluate $\int_C \boldsymbol{F} \cdot \boldsymbol{dr}$, where $\boldsymbol{F} = xy\hat{i} + yz\hat{j} + zx\hat{k}$ and C is the arc of the curve $\boldsymbol{r} = (a\cos\theta)\hat{i} + (a\sin\theta)\hat{j} + a\theta\hat{k}$ from $\theta = 0$ to $\theta = \dfrac{\pi}{2}$.

15. If $\boldsymbol{F} = yz\hat{i} + zx\hat{j} - xy\hat{k}$, find $\int_C \boldsymbol{F} \cdot \boldsymbol{dr}$ where C is given by $x = t$, $y = t^2$, $z = t^3$ from $P(0, 0, 0)$ to $Q(2, 4, 8)$.

16. Evaluate $\int_C \boldsymbol{F} \cdot \boldsymbol{dr}$, where $\boldsymbol{F} = (2x + y)\hat{i} + (3y - x)\hat{j} + yz\hat{k}$ and C is the curve $x = 2t^2$, $y = t$, $z = t^3$ from $t = 0$ to $t = 1$.

17. Find the circulation of $\boldsymbol{F}$ round the curve C where $\boldsymbol{F} = y\hat{i} + z\hat{j} + x\hat{k}$ and C is the circle $x^2 + y^2 = 1$, $z = 0$.

18. Show that $\int_C \left[-\dfrac{y}{x^2 + y^2}\hat{i} + \dfrac{x}{x^2 + y^2}\hat{j} \right] \cdot \boldsymbol{dr} = 2\pi$ where C is the circle $x^2 + y^2 = 1$ in the xy-plane described in counter-clockwise sense.

19. Evaluate $\int_C \boldsymbol{F} \cdot \boldsymbol{dr}$, where
$$\boldsymbol{F} = c[-3a\sin^2 t\cos t\hat{i} + a(2\sin t - 3\sin^3 t)\hat{j} + b\sin 2t\hat{k}]$$
and C is given by $\boldsymbol{r} = a\cos t\hat{i} + a\sin t\hat{j} + bt\hat{k}$ from $t = \pi/4$ to $\pi/2$.

20. If $\boldsymbol{F} = (x^2 + y^3)\hat{i} + (x^3 - y^2)\hat{j}$, evaluate line integral $\int \boldsymbol{F} \cdot \boldsymbol{dr}$ along the path $y^2 = x$, joining $(0, 0)$ to $(1, 0)$.

21. If $\boldsymbol{A} = (x - y)\hat{i} + (x + y)\hat{j}$ show that around the curve C consistintg of $y = x^2$ and $y^2 = x$, $\int_C \boldsymbol{A} \cdot \boldsymbol{dr} = \dfrac{2}{3}$.   (GBTU–2012)

22. If $\boldsymbol{F} = e^{xyz}(yz\hat{i} + zx\hat{j} + xy\hat{k})$ and $\boldsymbol{r} = x\hat{i} + y\hat{j} + z\hat{k}$ and C is the boundary of $0 \le x \le 1$, $0 \le y \le 1$ and $z = 1$ clockwise then show that $\int_C \boldsymbol{F} \cdot \boldsymbol{dr} = 0$.   (UPTU–2008)

23. Show that the surface area of the plane $x + 2y + 2z = 12$ cut off by $x = 0$, $y = 0$ and $x^2 + y^2 = 16$ is given by 6π square units.   (GBTU–2012)

24. Show that the integral $\int_C (x^2 + xy)dx + (x^2 + y^2)dy$ where C is the square formed by the lines $y = \pm 1$, and $x = \pm 1$ is equal to zero.   (DELHI–2002)

25. If $\boldsymbol{F} = (5xy - 6x^2)\hat{i} + (2y - 4x)\hat{j}$, show that $\int_C \boldsymbol{F} \cdot \boldsymbol{dr} = 35$ along the curve C in the xy plane, $y = x^3$ from the point $(1, 1)$ to $(2, 8)$.   (JNTU–2006)

26. Show that the total workdone by the force $\boldsymbol{F} = 3xy\hat{i} - y\hat{j} + 2zx\hat{k}$ in moving a particle around the circle $x^2 + y^2 = 4$ is 0.   (VTU–2010)

Hints to Selected Problems

1. Let $x = t, y = t^2$ for $0 \le t \le 1$

$\mathbf{F} = x^2\hat{i} + y^3\hat{j} = t^2\hat{i} + t^6\hat{j}$ and $d\mathbf{r} = dx\hat{i} + dy\hat{j} = (\hat{i} + 2t\hat{j})dt$

$\therefore \qquad \mathbf{F} \cdot d\mathbf{r} = (t^2 + 2t^7)dt$

$\therefore \quad \int_C \mathbf{F} \cdot d\mathbf{r} = \int_0^1 (t^2 + 2t^7)dt = \left[\dfrac{t^3}{3} + \dfrac{2t^8}{8}\right]_0^1 = \dfrac{7}{12}$.

3. (i) $C : y = x^2$ so $x = t, y = t^2$ and $0 \le t \le 1$

$\mathbf{F} = y\hat{i} - x\hat{j} = t^2\hat{i} - t\hat{j}$ and $d\mathbf{r} = dx\hat{i} + dy\hat{j} = (\hat{i} + 2t\hat{j})dt$

$\therefore \qquad \mathbf{F} \cdot d\mathbf{r} = -t^2 dt$

$\therefore \quad \int_C \mathbf{F} \cdot d\mathbf{r} = -\int_0^1 t^2 dt = -\left[\dfrac{t^3}{3}\right]_0^1 = -\dfrac{1}{3}$.

(ii) C : The straight line from $(0, 0)$ to $(1, 0)$ and to $(1, 1)$.

Let OA and OB be the lines where $O(0, 0)$, $A(1, 0)$ and $B(1, 1)$

$\int_C \mathbf{F} \cdot d\mathbf{r} = \int_{OA} \mathbf{F} \cdot d\mathbf{r} + \int_{AB} \mathbf{F} \cdot d\mathbf{r}$

Along OA, $y = 0, x = 1$, then $dy = 0$

$\mathbf{F} \cdot d\mathbf{r} = y\,dx - x\,dy$

$\therefore \quad \int_{OA} \mathbf{F} \cdot d\mathbf{r} = \int_0^1 0\,dx = 0$

Along AB, $x = 1, 0 \le y \le 1$, then $dx = 0$

$\therefore \quad \int_{AB} \mathbf{F} \cdot d\mathbf{r} = \int_0^1 (-1)dy = (-y)_0^1 = -1$.

Hence $\int_C \mathbf{F} \cdot d\mathbf{r} = 0 - 1 = -1$.

(iii) Same as part (ii).

6. Since $2x = \pi y^2$ so $dx = \pi y\,dy$ and $0 \le x \le \dfrac{\pi}{2}, 0 \le y \le 1$

$\int_C (2xy^3 - y^2 \cos x)dx + \int_C (1 - 2y \sin x + 3x^2 y^2)dy$

$= \int_0^{\pi/2}\left[2x\left(\dfrac{2x}{\pi}\right)^{3/2} - \dfrac{2x}{\pi}\cos x\right]dx$

$+ \int_0^{\pi/2}\left[1 - 2\left(\dfrac{2x}{\pi}\right)^{1/2}\sin x + 3x^2\left(\dfrac{2x}{\pi}\right)\right]\dfrac{1}{\pi}\left(\dfrac{\pi}{2x}\right)^{1/2}dx$

$= \dfrac{7}{2}\left(\dfrac{2}{\pi}\right)^{3/2}\int_0^{\pi/2}x^{5/2}dx - \dfrac{2}{\pi}\int_0^{\pi/2}x \cos x\,dx$

$+ \dfrac{1}{\sqrt{2\pi}}\int_0^{\pi/2}x^{-1/2}dx - \dfrac{2}{\pi}\int_0^{\pi/2}\sin x\,dx = \dfrac{\pi^2}{4}$.

9. The parametric equation of S are

$S = \sin\theta\cos\phi, y = \sin\theta\sin\phi, z = \cos\theta$

where $0 \le \theta \le \pi/2, 0 \le \phi \le \pi/2$

$\mathbf{F} = yz\hat{i} + zx\hat{j} + xy\hat{k}$

$\iint_S \mathbf{F} \cdot \mathbf{n}\,dS = \iint_S (yz\,dydz + zx\,dzdx + xy\,dxdy)$

Also, $dydz = \sin^2\theta\cos\phi\,d\theta\,d\phi, dzdx = \sin^2\theta\sin\phi\,d\theta\,d\phi$

$dxdy = \sin\theta\cos\theta\,d\theta\,d\phi$

$\therefore \quad \iint_S \mathbf{F} \cdot \mathbf{n}\,dS = 3\int_0^{\pi/2}\int_0^{\pi/2}\sin^3\theta\cos\theta\cos\phi\sin\phi\,d\theta\,d\phi$

$= 3\int_0^{\pi/2}\sin^3\theta\cos\theta\,d\theta.\int_0^{\pi/2}\cos\phi\sin\phi\,d\phi$

$= 3\left[\dfrac{\sin^4\theta}{4}\right]_0^{\pi/2}.\left[\dfrac{\sin^2\phi}{2}\right]_0^{\pi/2} = 3\left(\dfrac{1}{4}\right)\left(\dfrac{1}{2}\right) = \dfrac{3}{8}$

12. $f \equiv -y^2 + 8x = 0$, then $\mathbf{n} = \dfrac{\nabla f}{|\nabla f|} = \dfrac{8\hat{i} - 2y\hat{j}}{\sqrt{64 + 4y^2}} = \dfrac{8\hat{i} - 2y\hat{j}}{2\sqrt{16 + 8x}}$

$\therefore \qquad \mathbf{F} \cdot \mathbf{n} = \dfrac{11y}{\sqrt{16 + 8x}}$ and $|\mathbf{n} \cdot \hat{j}| = \dfrac{y}{\sqrt{16 + 8x}}$

$\iint_S \mathbf{F} \cdot \mathbf{n}\,dS = \iint_R \dfrac{11y}{\sqrt{16 + 8x}}\dfrac{dzdx}{|\mathbf{n} \cdot \hat{j}|}$

$= 11\int_R dzdx = 11\int_0^2\int_0^6 dzdx = 132$.

14. Since $\mathbf{r} = a(\cos\theta)\hat{i} + a(\sin\theta)\hat{j} + a\theta\hat{k}$

$\therefore \qquad x = a\cos\theta, y = a\sin\theta, z = a\theta$

so $\mathbf{F} = (a^2\cos\theta\sin\theta)\hat{i} + (a^2\theta\sin\theta)\hat{j} + (a^2\theta\cos\theta)\hat{k}$

and $d\mathbf{r} = [(-a\sin\theta)\hat{i} + (a\cos\theta)\hat{j} + a\hat{k}]d\theta$.

$\therefore \qquad \mathbf{F} \cdot d\mathbf{r} = (-a^3\cos\theta\sin^2\theta + a^3\theta\sin\theta\cos\theta + a^3\theta\cos\theta)d\theta$.

$\int_C \mathbf{F} \cdot d\mathbf{r} = -a^3\int_0^{\pi/2}\sin^2\theta\cos\theta\,d\theta$

$+ \dfrac{a^3}{2}\int_0^{\pi/2}\theta\sin 2\theta\,d\theta + a^3\int_0^{\pi/2}\theta\cos\theta\,d\theta$

$= -\dfrac{1}{3}a^3 + \dfrac{1}{2}a^3\left(\dfrac{\pi}{4}\right) + a^3\left(\dfrac{1}{2}\pi - 1\right) = a^3\left(\dfrac{5}{8}\pi - \dfrac{4}{3}\right)$.

15. $x = t, y = t^2, z = t^3$ and $0 \le t \le 2$. $\therefore \mathbf{r} = t\hat{i} + t^2\hat{j} + t^3\hat{k}$

$\dfrac{d\mathbf{r}}{dt} = \hat{i} + 2t\hat{j} + 3t^2\hat{k}$ and $\mathbf{F} = t^5\hat{i} + t^4\hat{j} - t^3\hat{k}$

$\int_C \mathbf{F}.d\mathbf{r} = \int_0^2\left(\mathbf{F}.\dfrac{d\mathbf{r}}{dt}\right)dt = \int_0^2(t^5 + 2t^5 - 3t^5)dt = \int_0^2 0\,dt = 0$.

16. Same as 15

18. The parametric equation of the circle are $x = \cos t, y = \sin t, z = 0$, and $0 \le t \le 2\pi$

So, $\mathbf{r} = x\hat{i} + y\hat{j} = (\cos t)\hat{i} + (\sin t)\hat{j}$

$\therefore \qquad \dfrac{d\mathbf{r}}{dt} = (-\sin t)\hat{i} + (\cos t)\hat{j}$

$\therefore \int_C\left[-\dfrac{y}{x^2 + y^2}\hat{i} + \dfrac{x}{x^2 + y^2}\hat{j}\right].d\mathbf{r}$

$= \int_0^{2\pi}(-\sin t\hat{i} + \cos t\hat{j}).(-\sin t\hat{i} + \cos t\hat{j})dt$

$= \int_0^{2\pi}(\sin^2 t + \cos^2 t)dt = \int_0^{2\pi}dt = [t]_0^{2\pi} = 2\pi$.

Answers

1. $7/12$ **2.** 5 **3.** (i) $-\dfrac{1}{3}$ (ii) -1 (iii) 0 **4.** 16 **5.** 0 **6.** $\pi^2/4$ **7.** $\log_e 2$ **8.** 11 **9.** $3/8$

10. $27/4$ **11.** 108 **12.** 132 **13.** $8/3$ **14.** $a^3\left(\dfrac{5\pi}{8} - \dfrac{4}{3}\right)$ **15.** 0 **16.** $277/42$ **17.** $-\pi$ **18.** 2π **19.** $\dfrac{1}{2}c[a^2 + b^2]$ **20.** $\dfrac{97}{105}$

6.12.3 Green's Theorem in the Plane

George Green (1793 – 1841), the English mathematician, discovered a method to transform a double integral over a plane region into line integral over the boundary of the region and conversely. This method (transformation) is of practical as well as theoretical interest. This transformation is as follows :

THEOREM.   *Let* **R** *be a closed and bounded region in xy-plane whose boundary C consists of finitely many smooth curves. Let P(x, y) and Q(x, y) be the continuous functions and have continuous partial derivatives $\dfrac{\partial P}{\partial y}$ and $\dfrac{\partial Q}{\partial x}$ everywhere in* **R.** *Then*

$$\iint_R \left(\frac{\partial Q}{\partial x} - \frac{\partial P}{\partial y}\right) dxdy = \oint_C (Pdx + Qdy)$$

the integration being taken along the entire boundary C of **R** *such that* **R** *is on the left as one advances in the direction of integration.*

(UPTU–2008, GBTU–2012)

PROOF.   We shall first prove the theorem for a special region R which is given as follows :

$$a \le x \le b, u(x) \le y \le v(x) \quad \text{and} \quad c \le y \le d, p(y) \le x \le q(y).$$

This region **R** has been shown in the adjoining fig. 14.

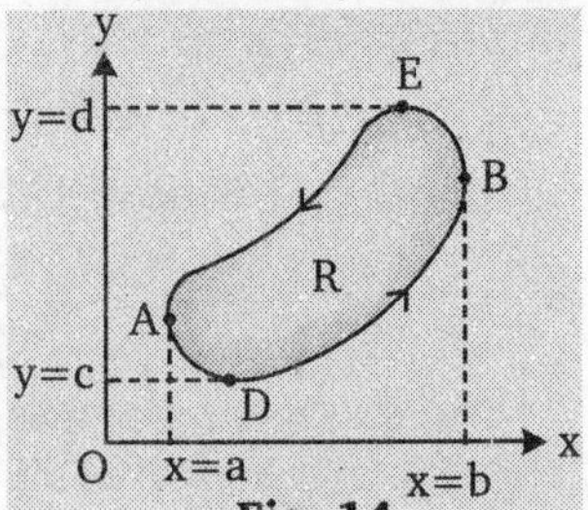

In the above fig. the equation of the curves ADB and BEA are respectively $y = u(x)$ and $y = v(x)$ and the equation of the curves EAD and DBE are represented by $x = p(y)$ and $x = q(y)$ respectively. Therefore, we have

$$\iint_R \frac{\partial P}{\partial y} dxdy \quad \int_{x=a}^{x=b}\left[\int_{y=u(x)}^{y=v(x)} \frac{\partial P}{\partial y} dy\right]dx \qquad \dots(1)$$

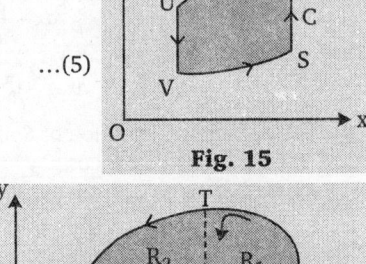

Fig. 14

Now evaluate the integral

$$\int_{u(x)}^{v(x)} \frac{\partial P}{\partial y} dy = \left[P(x,y)\right]_{u(x)}^{v(x)} = P[x, v(x)] - P[x, u(x)]. \qquad \dots(2)$$

From (1) and (2), we have

$$\iint_R \frac{\partial P}{\partial y} dxdy = \int_a^b (P[x, v(x)] - P[x, u(x)])dx = \int_a^b P(x, v(x))dx - \int_b^a P(x, u(x))dx.$$

$$\therefore \quad \iint_R \frac{\partial P}{\partial y} dxdy = -\int_a^b P(x, u(x))dx - \int_b^a P(x, v(x))dx. \qquad \dots(3)$$

Since $y = u(x)$ represents the oriented curve ADB and $y = v(x)$ represents oriented curve BEA. Thus the integrals on R.H.S. of (3) may be written as the line integral over ADB and BEA. Therefore, we obtain

$$\iint_R \frac{\partial P}{\partial y} dxdy = -\oint_C P(x, y)dx \quad \text{or} \quad -\iint_R \frac{\partial P}{\partial y} dxdy = \oint_C P(x, y)dx \qquad \dots(4)$$

If the portions of the curve C are the segments parallel to y-axis as shown in fig. 15.

Then the result in (4) does not change therefore the value of the integral $\int P(x, y)dx$ along the segments ST and UV are zero. The reason for the value of the above integral to be zero are that along ST and UV, x are constant and thus $dx = 0$ and so the value of integral becomes zero.

Similarly, we obtain

$$\iint_R \frac{\partial Q}{\partial x} dxdy = \int_{x=c}^{x=d}\left[\int_{x=p(y)}^{x=q(y)} \frac{\partial Q}{\partial x} dx\right]dy = \oint_C Q(x, y)dy. \qquad \dots(5)$$

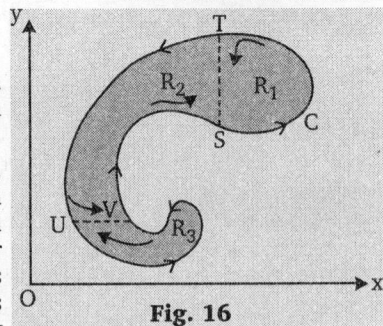

Fig. 15

From (4) and (5), we have

$$\iint_R \left(\frac{\partial Q}{\partial x} - \frac{\partial P}{\partial y}\right) dxdy = \oint_C (Pdx + Qdy).$$

Hence this is the required formula for a special region.

The proof of the above theorem can now be extended to a region R which is itself not a special region but can be subdivided into finitely many special regions as shown below in fig 16.

The special regions are obtained by drawing the lines parallel to the co-ordinate axes. In fig. 16 ST and UV are the two lines drawn parallel to y and x-axis respectively which divide the region R_1, R_2 and R_3 and apply above theorem to each subregion R_1, R_2 and R_3 and then add the results. The sum of the left hand members will give the integral over the region R and the sum of the right hand members will give the line integral over C plus integrals over the curves introduced for subdividing R. Each of the latter integrals occurs twice, taken in each direction, hence these two integrals will cancel each other and finally we obtain the value of the integral over C.

Fig. 16

6.12.4 Applications of Green's Theorem

Green's theorem has various applications and important consequences, some of which may be illustrated by the subsequent examples.

(i) Area of plane region as a line integral over the boundary. Let A be the area of a plane region R and $P(x, y) = 0$ and $Q(x, y) = x$. Then from Green's theorem

$$\iint_R \left(\frac{\partial Q}{\partial x} - \frac{\partial P}{\partial y}\right) dxdy = \oint_C (Pdx + Qdy). \qquad \qquad \text{...(1)}$$

Putting $P(x, y) = 0$ and $Q(x, y) = x$ into above formula, we get

$$\iint_R dxdy = \oint_C xdy.$$

Since $\qquad A = \iint_R dxdy \qquad \qquad \therefore \qquad A = \oint_C xdy. \qquad \text{...(2)}$

Similarly, let $P(x, y) = -y$ and $Q(x, y) = 0$, then from (1), we have

$$\iint_R dxdy = -\oint_C ydx. \qquad \qquad \therefore \qquad A = -\oint_C ydx. \qquad \text{...(3)}$$

Adding (2) and (3), we get

$$A = \frac{1}{2}\oint_C (xdy - ydx). \qquad \qquad \text{...(4)}$$

The integration in (4) being taken as indicated in Green's theorem. This formula gives the area of R in terms of a line integral over the boundary of C.

REMARK

- The theory of planimeters is based upon this formula.

(ii) Transformation of a double integral of the Laplacian of a function into a line integral of its normal derivative.

Let $\phi(x, y)$ be a function which is continuous and has continuous first and second derivatives in a region **R** of the xy-plane where **R** is as same as taken in Green's theorem. Let $P(x, y) = -\dfrac{\partial \phi}{\partial y}$ and $Q(x, y) = \dfrac{\partial \phi}{\partial x}$. Then $\dfrac{\partial P}{\partial y}$ and $\dfrac{\partial Q}{\partial x}$ are continuous in R and

$$\frac{\partial Q}{\partial x} - \frac{\partial P}{\partial y} = \frac{\partial^2 \phi}{\partial x^2} + \frac{\partial^2 \phi}{\partial y^2} = \left(\frac{\partial^2}{\partial x^2} + \frac{\partial^2}{\partial y^2}\right)\phi = \nabla^2\phi. \qquad \text{...(1)}$$

where ∇^2 is a Laplacian operator. The region **R** is shown in fig. 17.

Now we have

$$\int_C (Pdx + Qdy) = \int_C \left(P\frac{dx}{ds} + Q\frac{dy}{ds}\right)ds = \int_C \left(-\frac{\partial \phi}{\partial y}\frac{dx}{ds} + \frac{\partial \phi}{\partial x}\frac{dy}{ds}\right)ds$$

$$\text{...(2)}$$

where s is the arc length of C and

$$-\frac{\partial \phi}{\partial y}\frac{dx}{ds} + \frac{\partial \phi}{\partial x}\frac{dy}{ds} = (\text{grad } \phi) \cdot \mathbf{n} \qquad \text{...(3)}$$

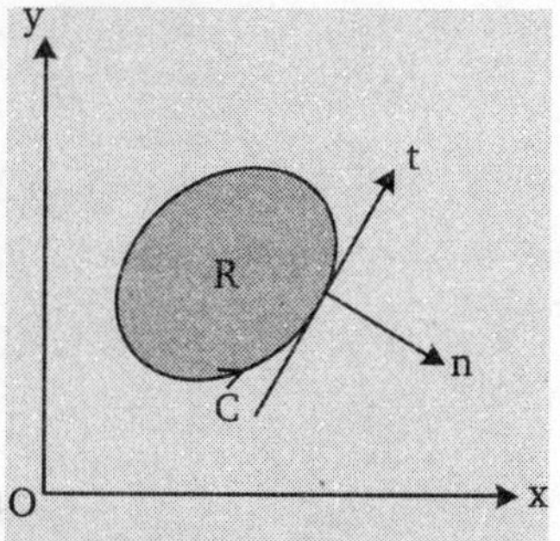

Let t be the unit tangent vector to C which is given by

$$\mathbf{t} = \frac{d\mathbf{r}}{ds} = \frac{dx}{ds}\hat{i} + \frac{dy}{ds}\hat{j} \qquad (\because \mathbf{r} = x\hat{i} + y\hat{j})$$

Fig. 17

From (3) $\qquad \mathbf{n} = \dfrac{dy}{ds}\hat{i} - \dfrac{dx}{ds}\hat{j}$, then $\mathbf{t} \cdot \mathbf{n} = 0$

this implies that the vector **n** is a unit outward drawn normal to C. Thus $(\text{grad } \phi) \cdot \mathbf{n} = \dfrac{\partial \phi}{\partial n}$

That is the expression on R.H.S. of (3) is the derivative of ϕ in the direction of the outward normal to C. Therefore (3) becomes

$$-\frac{\partial \phi}{\partial y}\frac{dx}{ds} + \frac{\partial \phi}{\partial x}\frac{dy}{ds} = \frac{\partial \phi}{\partial n}$$

so, equation (2) becomes

$$\int_C (Pdx + Qdy) = \int_C \frac{\partial \phi}{\partial n}ds \qquad \text{...(4)}$$

From Green's theorem we have

$$\iint_R \left(\frac{\partial Q}{\partial x} - \frac{\partial P}{\partial y}\right)dxdy = \int_C (Pdx + Qdy). \qquad \text{...(5)}$$

From (1) and (4) and using (5), we get

$$\iint_C \nabla^2\phi \, dxdy = \int_C \frac{\partial \phi}{\partial n}ds \,.$$

This is the required transformation and is important application of Green's theorem.

Solved Examples

EXAMPLE 1. *Verify Green's theorem in the plane for*

$$\int_C [(2xy - x^2)dx + (x^2 + y^2)dy]$$

where C is the boundary of the region enclosed by $y = x^2$ and $y^2 = x$ described in the positive sense.

SOLUTION. Let **R** be the region enclosed by $y = x^2$ and $y^2 = x$ whose boundary C is traversed in the positive direction as shown in the fig. 18.

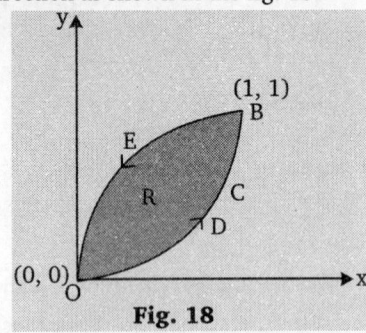

Fig. 18

The curves $y = x^2$ and $y^2 = x$ intersect at $(0, 0)$ and $(1, 1)$ and have

$$P(x, y) = 2xy - x^2$$

and $Q(x, y) = x^2 + y^2$.

$$\therefore \quad \frac{\partial P}{\partial y} = 2x \quad \text{and} \quad \frac{\partial Q}{\partial x} = 2x$$

By Green's theorem, we have

$$\iint_R \left(\frac{\partial Q}{\partial x} - \frac{\partial P}{\partial y} \right) dxdy = \int_C (Pdx + Qdy). \quad ...(1)$$

$$\therefore \quad \text{L.H.S.} = \iint_R \left(\frac{\partial Q}{\partial x} - \frac{\partial P}{\partial y} \right) dxdy$$

$$= \iint_R (2x - 2x)dxdy = \iint_R 0.dxdy = 0$$

and R.H.S. $= \int_C (Pdx + Qdy)$

$$= \int_C [(2xy - x^2)dx + (x^2 + y^2)dy]$$

$$= \int_{ODB} [(2xy - x^2)dx + (x^2 + y^2)dy]$$

$$+ \int_{BEO} [(2xy - x^2)dx + (x^2 + y^2)dy] \quad ...(2)$$

($\because$ C consists of two curves ODB and BEO.)

Along the curve BEO, we have
$$y^2 = x \text{ and } x \text{ varies from 0 to 1.}$$

$$\therefore \quad 2ydy = dx$$

$$\therefore \quad \int_{BEO} [(2xy - x^2)dx + (x^2 + y^2)dy]$$

$$= \int_0^1 (2x^{3/2} - x^2)dx + \int_0^1 (x^2 + x)\frac{dx}{2\sqrt{x}}$$

$$= \left[2\frac{x^{5/2}}{5/2} - \frac{x^3}{3} \right]_0^1 + \frac{1}{2}\left[\frac{x^{5/2}}{5/2} + \frac{x^{3/2}}{3/2} \right]_0^1$$

$$= \left[\frac{4}{5} - \frac{1}{3} \right] + \frac{1}{2}\left[\frac{2}{5} + \frac{2}{3} \right] = 1$$

and along the curve ODB, we have
$$y = x^2 \text{ and } x \text{ varies from 1 to 0.}$$

$$\therefore \quad dy = 2xdx$$

$$\therefore \quad \int_{ODB} [(2xy - x^2)dx + (x^2 + y^2)dy]$$

$$= \int_1^0 [2x^3 - x^2 + 2x^3 + 2x^5]dx$$

$$= \int_1^0 (4x^3 - x^2 + 2x^5)dx$$

$$= -\int_0^1 (4x^3 - x^2 + 2x^5)dx$$

(By the property of definite integral)

$$= -\left[x^4 - \frac{x^3}{3} + \frac{x^6}{3} \right]_0^1$$

$$= \left[1 - \frac{1}{3} + \frac{1}{3} \right] = -1$$

$\therefore$ R.H.S. $= -1 + 1 = 0.$

Thus, L.H.S. = R.H.S.

Hence, Green's theorem is verified.

EXAMPLE 2. *Apply Green's theorem in the plane to evaluate*
$$\int_C [(y - \sin x)dx + \cos xdy]$$
where C is the triangle enclosed by the lines
$y = 0$, $x = \pi$, $\pi y = 2x$. (ANNA–2003, JNTU–2005)

SOLUTION. By Green's theorem, we have

$$\iint_R \left(\frac{\partial Q}{\partial x} - \frac{\partial P}{\partial y} \right) dxdy = \int_C (Pdx + Qdy). \quad ...(1)$$

Here $P = y - \sin x, Q = \cos x$.

$$\therefore \quad \frac{\partial P}{\partial y} = 1, \frac{\partial Q}{\partial x} = -\sin x.$$

Now from (1), we have
$$\int_C [(y - \sin x)dx + \cos xdy]$$

$$= \iint_R [(-\sin x - 1)dxdy]$$

$$= \int_{x=0}^{\pi} \int_{y=0}^{(2x)/\pi} (-\sin x - 1)dxdy$$

$$= -\int_0^{\pi} (1 + \sin x)[y]_0^{2x/\pi} dx$$

$$= -\frac{2}{\pi} \int_0^{\pi} x(1 + \sin x)dx$$

$$= -\frac{2}{\pi} \int_0^{\pi} xdx - \frac{2}{\pi} \int_0^{\pi} x\sin xdx$$

$$= -\frac{2}{\pi} \left[\frac{x^2}{2} \right]_0^{\pi} - \frac{2}{\pi}\left[-x\cos x + \sin x \right]_0^{\pi}$$

$$= -\frac{2}{\pi}\left[\frac{\pi^2}{2} \right] - \frac{2}{\pi}\left[\pi \right] = -\pi - 2.$$

EXAMPLE 3. *Evaluate by Green's theorem :*
$$\oint_C [(x^2 - \cosh y)dx + (y + \sin x)dy)],$$
where C is the rectangle with vertices $(0, 0)$,
$(\pi, 0)$, $(\pi, 1)$, $(0, 1)$. (NAGPUR–2009, PTU–2006)

SOLUTION. By Green's theorem in the plane, we have

$$\oint_C (Pdx + Qdy) = \iint_R \left(\frac{\partial Q}{\partial x} - \frac{\partial P}{\partial y} \right) dxdy. \quad ...(1)$$

Here, $P = x^2 - \cosh y, Q = y + \sin x$.

$$\therefore \quad \frac{\partial P}{\partial y} = -\sinh y, \frac{\partial Q}{\partial x} = \cos x.$$

Thus from (1), we get
$$\oint_C [(x^2 - \cosh y)dx + (y + \sin x)dy)]$$

$$= \iint_R (\cos x + \sinh y)dxdy$$

$$= \int_{x=0}^{\pi} \int_{y=0}^{1} (\cos x + \sinh y)dxdy$$

$$= \int_{x=0}^{\pi} [y\cos x + \cosh y]_{y=0}^{1} dx$$

$$= \int_{x=0}^{\pi}(\cos x + \cosh 1 - 1)dx$$

$$= \left[\sin x + x\cosh 1 - x\right]_{x=0}^{1} = \pi(\cosh 1 - 1)$$

EXAMPLE 4. *Using Green's theorem, show that area bounded by a simple closed curve C is given by $\frac{1}{2}\int(xdy - ydx)$. Hence, find the area of an ellipse.*   (KERALA–2005, VTU–2000)

SOLUTION. We have $M = -y$ and $N = x$.

$$\Rightarrow \quad \frac{\partial M}{\partial y} = -1 \text{ and } \frac{\partial N}{\partial x} = 1 \qquad ...(1)$$

From Green's theorem, we have

$$\int_C (Mdx + Ndy) = \int\int_S \left(\frac{\partial N}{\partial x} - \frac{\partial M}{\partial y}\right)dxdy. \qquad ...(2)$$

Using (1) in (2), we get

$$\int_C (-ydx + xdy) = \int\int_S (1+1)dxdy$$

$$= 2\int\int_S dxdy = 2A$$

where A is the required area given by

$$A = \frac{1}{2}\int_C (xdy - ydx)$$

Any point (x, y) on the ellipse is given by
$$x = a\cos\phi, y = b\sin\phi, \phi \text{ is a parameter}$$
Hence, area of the ellipse

$$= \frac{1}{2}\int_0^{2\pi}(a\cos\phi)(b\cos\phi)d\phi$$

$$- (b\sin\phi)(-a\sin\phi)d\phi$$

$$= \frac{1}{2}ab\int_0^{2\pi}(\cos^2\phi + \sin^2\phi)d\phi$$

$$= \frac{1}{2}ab(2\pi) = \pi ab.$$

EXAMPLE 5. *Use Green's theorem to evaluate*

$$\int_C (x^2 + xy)dx + (x^2 + y^2)dy$$

where C is the square formed by the lines $y = \pm 1$, $x = \pm 1$.   (GBTU–2010, MTU–2011, SVTU–2008, SRM-2006, MARATHWADA–2008)

SOLUTION. We have

$$\int_C (x^2 + xy)dx + (x^2 + y^2)dy$$

$$= \int\int_S \left[\frac{\partial}{\partial x}(x^2 + y^2) - \frac{\partial}{\partial y}(x^2 + xy)\right]dxdy$$

$$= \int\int_S (2x - x)dxdy$$

$$= \int_{x=-1}^{1}\int_{y=-1}^{1} xdxdy = \int_{-1}^{1} x[y]_{-1}^{1} dx$$

$$= \int_{-1}^{1} 2xdx = 0.$$

EXAMPLE 6. *Apply Green's theorem to evaluate*

$$\int_C [(2x^2 - y^2)dx + (x^2 + y^2)dy]$$

where C is the boundary of the area enclosed by the x-axis and upper half of the circle $x^2 + y^2 = a^2$.   (UPTU–2005, GBTU(AG)–2010)

SOLUTION. Let $\mathbf{F} = M\hat{i} + N\hat{j}$

Then, $\mathbf{F}.d\mathbf{r} = (M\hat{i} + N\hat{j}).(dx\hat{i} + dy\hat{j})$

$$= Mdx + Ndy$$

Here we have

$$M = 2x^2 - y^2 \quad \Rightarrow \quad \frac{\partial M}{\partial y} = -2y$$

$$N = x^2 + y^2 \quad \Rightarrow \quad \frac{\partial N}{\partial x} = 2x$$

Therefore,

$$\int_C \mathbf{F}.d\mathbf{r} = \int_C [(2x^2 - y^2)dx + (x^2 + y^2)dy]$$

$$= \int\int_S (2x + 2y)dxdy \qquad \text{(By Green's theorem)}$$

$$= 2\int_{x=-a}^{a}\int_{y=0}^{\sqrt{a^2-x^2}}(x + y)dxdy$$

$$= 2\int_{-a}^{a}\left(xy + \frac{y^2}{2}\right)_0^{\sqrt{a^2-x^2}} dx$$

$$= 2\int_{-a}^{a}\left(x\sqrt{a^2 - x^2} + \frac{a^2 - x^2}{2}\right)dx$$

$$= 0 + 2\int_0^a (a^2 - x^2)dx$$

$$= 2\left(a^2 x - \frac{x^3}{3}\right)_0^a = 2\left(a^3 - \frac{a^3}{3}\right) = \frac{4}{3}a^3$$

EXAMPLE 7. *Using Green's theorem, evaluate $\int_C (x^2y + x^2dy)$ where C is the boundary described counter clockwise of the triangle with vertices (0, 0), (1, 0), (1, 1).*   (UKTU–2010)

SOLUTION. We have

$$\int_C (x^2y\,dx + x^2dy)$$

$$= \int\int_S \left[\frac{\partial}{\partial x}(x^2) - \frac{\partial}{\partial y}(x^2y)\right]dxdy$$

(By Green's theorem)

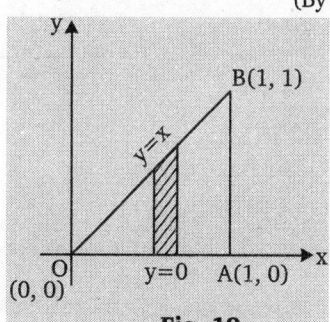

Fig. 19

$$= \int_{x=0}^{1}\int_{y=0}^{x}(2x - x^2)dxdy$$

$$= \int_0^1 (2x - x^2)(y)_0^x dx$$

$$= \int_0^1 (2x^2 - x^3)dx = \left(\frac{2}{3}x^3 - \frac{x^4}{4}\right)_0^1$$

$$= \frac{2}{3} - \frac{1}{4} = \frac{5}{12}$$

EXAMPLE 8. *Using Green's theorem, find the area of the region in the first quadrant bounded by the curves $y = x, y = \frac{1}{x}, y = \frac{x}{4}$.*   (UPTU–2009)

SOLUTION. Using Green's theorem, we have

$$A = \frac{1}{2}\int_C (x\,dy - y\,dx)$$

$$A = \frac{1}{2}\Big[\int_{C_1}(x\,dy - y\,dx) + \int_{C_2}(x\,dy - y\,dx)$$

$$+ \int_{C_3}(x\,dy - y\,dx)\Big] = \frac{1}{2}(I_1 + I_2 + I_3) \quad …(1)$$

Along C_1:

We have $y = \dfrac{x}{4} \Rightarrow dy = \dfrac{1}{4}dx$, $x = 0$ to 2

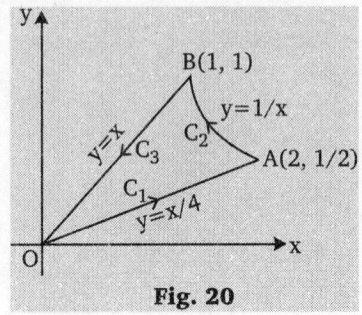

Fig. 20

$$\therefore \quad I_1 = \int_{C_1}(x\,dy - y\,dx)$$

$$= \int_{C_1}\left(x\cdot\frac{dx}{4} - \frac{x}{4}dx\right) = 0 \quad …(2)$$

Along C_2:

We have $y = \dfrac{1}{x} \Rightarrow dy = -\dfrac{1}{x^2}dx$, $x = 2$ to 1

$$\therefore \quad I_2 = \int_{C_2}(x\,dy - y\,dx)$$

$$= \int_2^1\left(x\left(\frac{-1}{x^2}\right)dx - \frac{1}{x}dx\right)$$

$$= -2\int_2^1\frac{1}{x}dx = -2\log x\Big|_2^1 = 2\log 2 \quad …(3)$$

Along C_3:
We have $y = x \Rightarrow dy = dx$, $x = 1$ to 0

$$\therefore \quad I_3 = \int_{C_3}(x\,dy - y\,dx)$$

$$= \int x\,dx - x\,dx = 0 \quad …(4)$$

Using (2), (3) and (4) in (1), we get

$$A = \frac{1}{2}(I_1 + I_2 + I_3) = \log 2$$

EXAMPLE 9. *If C is a simple closed curve in the xy-plane not enclosing the origin, show that*

$$\int_C F.dr = 0 \quad where \quad F = \frac{y\hat{i} - x\hat{j}}{x^2 + y^2}. \qquad \text{(PTU–2005)}$$

SOLUTION. We have

$$\int_C F.dr = \int_C \frac{y\hat{i} - x\hat{j}}{x^2 + y^2}(dx\,\hat{i} + dy\,\hat{j})$$

$$= \int_C \frac{y\,dx - x\,dy}{x^2 + y^2} = \int_C M\,dx + N\,dy$$

where $\quad M = \dfrac{y}{x^2 + y^2}, N = \dfrac{-x}{x^2 + y^2}$

$$= \iint_S\left(\frac{\partial N}{\partial x} - \frac{\partial M}{\partial y}\right)dx\,dy$$

$$\text{(By Green's theorem)}$$

$$= \iint_S\left[\frac{-(x^2 + y^2) + x(2x)}{(x^2 + y^2)^2}\right.$$

$$\left. - \frac{(x^2 + y^2) - y(2y)}{(x^2 + y^2)^2}\right]dx\,dy$$

$$= \iint\left[\frac{x^2 - y^2}{(x^2 + y^2)^2} - \frac{x^2 - y^2}{(x^2 + y^2)^2}\right]dx\,dy$$

$$= 0$$

Exercise-6.7

1. Verify Green's theorem in the plane for

$$\int_C[(xy + y^2)dx + x^2 dy]$$

where C is the closed curve of the region bounded by $y = x$ and $y = x^2$.

(UPTU–2008, GBTU–2010, VTU–2011, SVTU–2009, ROHTAK–2003)

2. Verify Green's threorem in the plane for

$$\int_C[(x^2 - xy^3)dx + (y^2 - 2xy)dy]$$

where C is the square with vertices $(0, 0)$, $(2, 0)$, $(2, 2)$, $(0, 2)$.

(UPTU–2008)

3. Apply Green's theorem to evaluate :

$$\int_C[(e^{-x}\sin y\,dx + e^{-x}\cos y\,dy)$$

where C is the rectangle with vertices $(0, 0)$, $(\pi, 0)$, $(\pi, \pi/2)$, $(0, \pi/2)$.

(UPTU–2007)

4. Using Green's theorem evaluate

$$\int_C[(3x^2 + y)dx + 4y^2 dy]$$

where C is the triangle with vertices $(0, 0)$, $(1, 0)$, $(0, 2)$.

5. Using Green's theorem evaluate

$$\int_C[(x^2 - \cosh y)dx + (y + \sin x)dy]$$

where C is the boundary of the rectangle $0 \le x \le \pi$, $0 \le y \le 1$.

6. Using Green's theorem evaluate

$$\int_C[(\cos x\sin y - xy)dx + \sin x\cos y\,dy]$$

where C is the circle $x^2 + y^2 = 1$.

7. Using Green's theorem evaluate

$$\int_C[x^{-1}e^y dx + (e^y \log x + 2x)dy]$$

where C is the boundary of the region bounded by $y = x^4 + 1$ and $y = 2$.

8. Verify the Green's theorem for

$$\int_C(y^2 dx + x^2 dy)$$

where C is the boundary of the square $-1 \le x \le 1$ and $-1 \le y \le 1$.

9. Using Green's theorem in the plane, evaluate

$$\int_C[2\tan^{-1}(y/x)dx + \log(x^2 + y^2)dy]$$

where C is the boundary of the circle $(x - 1)^2 + (y + 1)^2 = 4$.

10. Using the following formula

$$\iint_R \nabla^2\phi\,dx\,dy = \int_C\frac{\partial\phi}{\partial n}ds$$

evaluate $\int_C \frac{\partial \phi}{\partial n} ds$, where $\phi = x^2 + 3y^2$ and C is the boundary of the circle $x^2 + y^2 = 4$.

11. If $A = Q\hat{i} - P\hat{j}$, show that the formula in Green's theorem may be written as $\iint_R \text{div } A \, dxdy = \int_C A.n ds$ where n is the outward unit normal vector to C and s is the arc length of C.

12. Show that the formula in Green's theorem may be written as

$$\iint_R (\text{curl } A).\hat{k} dxdy = \int_C A.t ds \quad \text{where } \hat{k} \text{ is a unit vector}$$ perpendicular to the xy-plane, t is the unit tangent vector to C and s is the arc length of C.

13. If $\phi(x, y)$ satisfies Laplace equation $\nabla^2 \phi = 0$ in a region R, then using Green's theorem show that

$$\iint_R \left[\left(\frac{\partial \phi}{\partial x}\right)^2 + \left(\frac{\partial \phi}{\partial y}\right)^2\right] dxdy = \int_C \phi.\frac{\partial \phi}{\partial n} ds.$$

Hints to Selected Problems

2. Here $\quad P = x^2 - xy^3, Q = y^2 - 2xy$

$$\therefore \qquad \frac{\partial P}{\partial y} = -3xy , \frac{\partial Q}{\partial x} = -2y$$

Then by Green's theorem

$$\iint_R \left(\frac{\partial Q}{\partial x} - \frac{\partial P}{\partial y}\right) dxdy = \int_C Pdx + Qdy.$$

$$\text{L.H.S.} = \iint_R \left(\frac{\partial Q}{\partial x} - \frac{\partial P}{\partial y}\right) dxdy = \iint_R (-2y + 3xy^2)dxdy$$

$$= \int_0^2 \int_{x=0}^{x=2} (-2y + 3xy^2)dxdy = \int_0^2 \left[-2xy + \frac{3x^2y^2}{2}\right]_0^2 dy$$

$$= \int_0^2 (-4y + 6y^2)dy = 8 .$$

Since C is the boundary of a square with vertices $O(0, 0)$, $A(2, 0)$, $B(2, 2)$ and $C(0, 2)$. Therefore

$$\text{R.H.S.} = \int_C Pdx + Qdy$$

$$= \int_{OA} Pdx + Qdy + \int_{AB} Pdx + Qdy$$
$$+ \int_{BC} Pdx + Qdy + \int_{CO} Pdx + Qdy.$$

Along $OA: y = 0, 0 \leq x \leq 2$, then

$$\int_{OA} Pdx + Qdy = \int_0^2 x^2 dx = \left[\frac{x^3}{3}\right]_0^2 = \frac{8}{3}.$$

Along $AB: x = 2, 0 \leq y \leq 2$, then

$$\int_{AB} Pdx + Qdy = \int_0^2 (y^2 - 4y)dy = \left[\frac{y^3}{3} - 2y^2\right]_0^2 = -\frac{16}{3}.$$

Along $BC: y = 2$ and x varies from 2 to 0.

$$\int_{BC} Pdx + Qdy = \int_2^0 (x^2 - 8x)dx = \frac{40}{3}.$$

Along $CO: x = 0$ and y varies from 2 to 0, then

$$\int_{CO} Pdx + Qdy = \int_2^0 y^2 dy = -\frac{8}{3}.$$

$$\text{L.H.S.} \qquad = \int_C Pdx + Qdy = \frac{8}{3} - \frac{16}{3} + \frac{40}{3} - \frac{8}{3} = 8.$$

$$\therefore \qquad \text{L.H.S.} = \text{R.H.S.}$$

4. Let R be the region enclosed by the triangle with vertices $O(0, 0)$, $A(1, 0)$ and $B(0, 2)$.

By Green's theorem

$$\int_C Pdx + Qdy = \iint_R \left(\frac{\partial Q}{\partial x} - \frac{\partial P}{\partial y}\right) dxdy$$

Here $P = 3x^2 + y$, $Q = 4y^2$

so, $\quad \frac{\partial P}{\partial y} = 1, \frac{\partial Q}{\partial x} = 0.$

$$\therefore \quad \int_C (3x^2 + y)dx + 4y^2 dy = \iint_R (0 - 1)dxdy = -\iint_R dxdy$$

$$= -(\text{Area of the triangle } OAB)$$

$$= -\left(\frac{1}{2} \times 1 \times 2\right) = -1$$

7. Let R be the region enclosed by $y = x^4 + 1$ and $y = 2$.

The intersection points of $y = 2$ and $y = x^4 + 1$ are $(1, 2)$ and $(-1, 2)$ and $y = x^4 + 1$ cuts only y-axis at $(0, 1)$, therefore y varies from 1 to 2 and x varies from -1 to 1.

By Green's theorem

$$\iint_R \left(\frac{\partial Q}{\partial x} - \frac{\partial P}{\partial y}\right) dxdy = \int_C Pdx + Qdy$$

Here $P = x^{-1}e^y$, $Q = e^y \log x + 2x$.

$$\therefore \quad \frac{\partial P}{\partial y} = x^{-1}e^y, \frac{\partial Q}{\partial x} = \frac{e^y}{x} + 2$$

$$\int_C x^{-1}e^y dx + (e^y \log x + 2x)dy$$

$$= \iint_R \left(e^y x^{-1} + 2 - x^{-1}e^y\right) dxdy$$

$$= 2\iint_R dxdy = 2\int_{-1}^1 \int_{y=x^4+1}^2 dxdy$$

$$= 2\int_{-1}^1 [y]_{x^4+1}^2 dx = 2\int_{-1}^1 (2 - x^4 - 1)dx$$

$$= 2\int_{-1}^1 (1 - x^4)dx = 4\int_0^1 (1 - x^4)dx$$

$$= 4\left(x - \frac{x^5}{5}\right)_0^1 = \frac{16}{5}.$$

10. Since R is region enclosed by the circle $x^2 + y^2 = 4$ whose boundary is C and $\phi = x^2 + 3y^2$.

$$\therefore \quad \nabla^2\phi = \frac{\partial^2\phi}{\partial x^2} + \frac{\partial^2\phi}{\partial y^2} = \frac{\partial}{\partial x}(2x) + \frac{\partial}{\partial y}(6y) = 2 + 6 = 8$$

$$\therefore \quad \int_C \frac{\partial\phi}{\partial n} ds = \iint_R \nabla^2\phi dxdy = \iint_R 8dxdy = 8\iint_R dxdy$$

$$= 8 (\text{Area of the circle } R) = 8[\pi(2)^2] = 32\pi$$

13. $\qquad\qquad A = Q\hat{i} - P\hat{j}$

$$\nabla \cdot A = \frac{\partial Q}{\partial x} - \frac{\partial P}{\partial y}$$

By Green's theorem,

$$\int_C Pdx + Qdy = \iint_R \left(\frac{\partial Q}{\partial x} - \frac{\partial P}{\partial y}\right) dxdy$$

$$\Rightarrow \qquad \iint_R \nabla \cdot A dxdy = \int_C \left(P\frac{dx}{ds} + Q\frac{dy}{ds}\right) ds$$

$$\Rightarrow \qquad \iint_R \nabla \cdot A dxdy = \int_C (Q\hat{i} - P\hat{j})\left(\frac{dy}{ds}\hat{i} - \frac{dx}{ds}\hat{j}\right) ds = \int_C A \cdot n ds$$

Now putting $\quad A = \phi(\nabla\phi)$

$$\therefore \qquad \nabla \cdot A = \nabla\phi \cdot \nabla\phi$$

$$\therefore \iint_R \nabla\phi \cdot \nabla\phi dxdy = \int_C \phi\nabla\phi \cdot n ds$$

$$\Rightarrow \iint_R |\nabla\phi|^2 dxdy = \int_C \phi\frac{\partial\phi}{\partial n} ds$$

$$\Rightarrow \iint_R \left[\left(\frac{\partial\phi}{\partial x}\right)^2 + \left(\frac{\partial\phi}{\partial y}\right)^2\right] dxdy = \int_C \phi\frac{\partial\phi}{\partial n} ds.$$

3. $2(e^{-\pi} - 1)$ **4.** -1 **5.** $\pi(\cosh 1 - 1)$ **6.** 0 **7.** $\dfrac{16}{5}$ **9.** 0 **10.** 32π

6.12.5 Gauss's Divergence Theorem

THEOREM. *Let V be the volume enclosed by a closed and bounded piecewise smooth surface S and let* $\mathbf{F}(x, y, z)$ *be a vector function which is continuous and has continuous first partial derivatives on V. Then*

$$\iiint_V \operatorname{div} \mathbf{F}\, dV = \iint_S \mathbf{F} \cdot \mathbf{n}\, dS \qquad \ldots(1)$$

where $\mathbf{n}$ *is the outward unit normal vector the surface S.*   (UPTU–2006, 07, GBTU–2011, 12)

Cartesian form of (1). *Let* $\mathbf{F} = F_1 \hat{i} + F_2 \hat{j} + F_3 \hat{k}$ *and suppose the outward unit normal vector* $\hat{\mathbf{n}}$ *makes the angle α, β and g with the positive axes of x, y, z respectively. Then* $\cos \alpha$, $\cos \beta$ *and* $\cos \gamma$ *are the direction-cosines of* $\hat{\mathbf{n}}$, *we have*

$$\hat{\mathbf{n}} = \cos\alpha\, \hat{i} + \cos\beta\, \hat{j} + \cos\gamma\, \hat{k}.$$

$$\therefore \qquad \mathbf{F} \cdot \hat{\mathbf{n}} = F_1 \cos\alpha + F_2 \cos\beta + F_3 \cos\gamma.$$

and $\qquad\qquad\qquad \operatorname{div} \mathbf{F} = \dfrac{\partial F_1}{\partial x} + \dfrac{\partial F_2}{\partial y} + \dfrac{\partial F_3}{\partial z}$ *and* $\qquad dV = dxdydz.$

Thus (1) becomes

$$\iiint_V \left(\frac{\partial F_1}{\partial x} + \frac{\partial F_2}{\partial y} + \frac{\partial F_3}{\partial z} \right) dxdydz = \iint_S (F_1 \cos\alpha + F_2 \cos\beta + F_3 \cos\gamma)\, dS. \qquad \ldots(2)$$

Proof. We shall first prove the theorem for a special volume V which is bounded by a piecewise smooth oriented surface S and has the property that any straight line drawn parallel to any one of the co-ordinate axes and intersecting V has only one point (or one segment) in common with V.

Then V can be represented by $\qquad f(x, y) \le z \le g(x, y) \qquad \ldots(3)$

where $(x, y) \in \mathbf{R}$. This R is the orthogonal projection of V in the xy-plane. Obviously $z = f(x, y)$ represents the lower part S_2 of S and $z = g(x, y)$ represents the upper part S_1 of S and there may be a remaining vertical S_3 of S has shown in fig. 21.

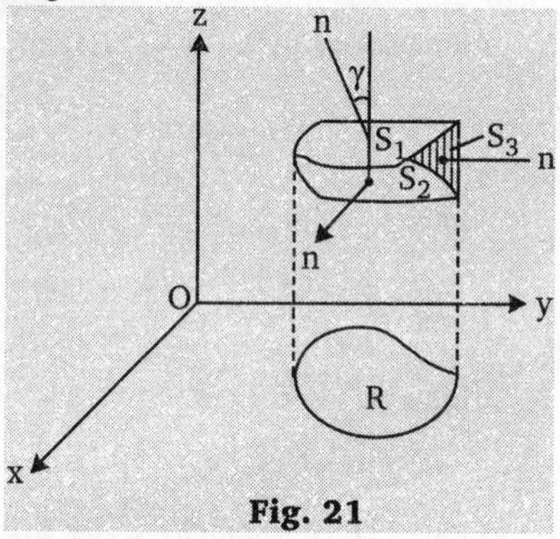

First we prove that

$$\iiint_V \frac{\partial F_3}{\partial z} dxdydz = \iint_S F_3 \cos\gamma\, dS. \qquad \ldots(4)$$

Since $\mathbf{F}(x, y, z)$ is continuously differentiable in V and using (3), we have

$$\iiint_V \frac{\partial F_3}{\partial z} dxdydz = \iint_R \left[\int_{z=f(x,y)}^{z=g(x,y)} \frac{\partial F_3}{\partial z} dz \right] dxdy$$

$$= \iint_R \left[F_3(x, y, z) \right]_{z=f(x,y)}^{z=g(x,y)} dxdy.$$

$$\therefore \qquad \iiint_V \frac{\partial F_3}{\partial z} dxdydz = \iint_R F_3[x, y, g(x, y)] dxdy - \iint_R F_3[x, y, f(x, y)] dxdy \qquad \ldots(5)$$

Now we have

$$\iint_S F_3 \cos\gamma\, dS = \iint_{S_1} F_3 \cos\gamma\, dS + \iint_{S_2} F_3 \cos\gamma\, dS + \iint_{S_3} F_3 \cos\gamma\, dS. \qquad \ldots(6)$$

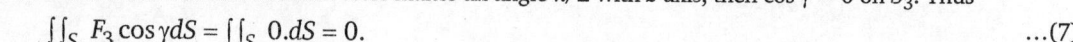

Fig. 21

Since on the portion S_3 of S the outward drawn unit normal vector makes an angle $\pi/2$ with z-axis, then $\cos\gamma = 0$ on S_3. Thus

$$\iint_{S_3} F_3 \cos\gamma\, dS = \iint_{S_3} 0.dS = 0. \qquad \ldots(7)$$

On the portion S_1 of S the outward drawn unit normal makes an acute angle γ with positive z-axis and the equation of S_1 is $z = g(x, y)$. Then

$$\cos\gamma\, dS = dxdy$$

$$\therefore \qquad \iint_{S_1} F_3 \cos\gamma\, dS = \iint_R F_3[x, y, g(x, y)] dxdy \qquad \ldots(8)$$

and on the portion S_2 of S the outward drawn unit normal vector makes obtuse angle γ with positive z-axis and the equation of S_2 is $z = f(x, y)$. Then

$$\cos\gamma\, dS = -dxdy$$

$$\therefore \qquad \iint_{S_2} F_3 \cos\gamma\, dS = -\iint_R F_3[x, y, f(x, y)] dxdy. \qquad \ldots(9)$$

Using (7), (8) and (9) the equation (6) becomes

$$\iint_S F_3 \cos\gamma\, dS = \iint_R F_3[x, y, g(x, y)] dxdy - \iint_R F_3[x, y, f(x, y)] dxdy. \qquad \ldots(10)$$

From (5) and (10), we obtain

$$\iiint_V \frac{\partial F_3}{\partial z} dxdydz = \iint_S F_3 \cos\gamma\, dS. \qquad \ldots(11)$$

Similarly taking the projection of S on the other co-ordinate planes, we have

$$\iiint_V \frac{\partial F_1}{\partial x} dxdydz = \iint_S F_1 \cos\alpha \, dS. \qquad \ldots(12)$$

and

$$\iiint_V \frac{\partial F_2}{\partial y} dxdydz = \iint_S F_2 \cos\beta \, dS. \qquad \ldots(13)$$

Now adding (11), (12) and (13), we get

$$\iiint_V \left(\frac{\partial F_1}{\partial x} + \frac{\partial F_2}{\partial y} + \frac{\partial F_3}{\partial z}\right) dxdydz = \iint_S (F_1 \cos\alpha + F_2 \cos\beta + F_3 \cos\gamma) dS. \qquad \ldots(14)$$

or

$$\iiint_V \operatorname{div} \boldsymbol{F} dV = \iint_S \boldsymbol{F} \cdot \boldsymbol{n} dS.$$

Hence proved the theorem for special region V.

6.12.6 GAUSS' DIVERGENCE THEOREM FOR ANY REGION

Let V be any volume which is not a special volume but can be subdivided into finitely many special volumes by drawing auxiliary surfaces. Now apply above theorem to each special volume and adding the result for each part. On the left hand side of this result we obtain the sum of volume integral over parts of V and which gives the volume integral over V. On the right hand side we obtain the sum of surface itegral over auxiliary surfaces plus the sum of the remaining surface integral. In this side the surface integral over auxiliary surfaces cancel in pairs and the remaining surface integrals give the surface integral over the whole boundary S of V.

REMARK

- The divergence theorem of Gauss can also be stated as the surface integral of the normal component of a vector $\boldsymbol{F}$ taken over a closed surface is equal to the volume integral of the divergence of $\boldsymbol{F}$ taken over the volume V enclosed by the surface.

6.12.7 APPLICATIONS OF GAUSS' DIVERGENCE THEOREM

The divergence theorem has various applications, some of which may be illustrated by the examples.

1. Representation of the divergence independent of the coordinates. By the divergence theorem, we have

$$\iiint_V \operatorname{div} \boldsymbol{F} dV = \iint_S \boldsymbol{F} \cdot \boldsymbol{n} dS. \qquad \ldots(1)$$

Dividing by the volume V of both sides of (1), we get

$$\frac{1}{V}\iiint_V \operatorname{div} \boldsymbol{F} dV = \frac{1}{V}\iint_S \boldsymbol{F} \cdot \boldsymbol{n} dS. \qquad \ldots(2)$$

Since for any continuous function $f(x, y, z)$, then by the mean value theorem for triple integral, we have

$$\iiint_V f(x, y, z) dV = f(x_0, y_0, z_0) V \qquad \ldots(3)$$

where (x_0, y_0, z_0) is any point in V. Thus from (3), we have

$$\frac{1}{V}\iiint_V \operatorname{div} \boldsymbol{F} dV = \operatorname{div} \boldsymbol{F}(x_0, y_0, z_0). \qquad \ldots(4)$$

Now let $P(x_1, y_1, z_1)$ be any fixed point in V and suppose V shrinks to the point P, so that the maximum distance $d(V)$ of the points of V from $P \to 0$, then $Q \to P$ and from (1) and (4), we have

$$\operatorname{div} \boldsymbol{F}(x_1, y_1, z_1) = \lim_{d(V)\to 0} \frac{1}{V}\iint_S \boldsymbol{F} \cdot \boldsymbol{n} dS.$$

This formula is independent of the co-ordinate system while the definition of divergence involves co-ordintaes.

2. Heat flow. Since we know that in a body heat will flow from high temperature to lower temperature region and the ratio of flow is proportional to the gradient to the temperature. Let $\boldsymbol{F}$ be the velocity of the heat flow in a body. Then we have

$$\boldsymbol{F} = -k \operatorname{grad} U \qquad \ldots(1)$$

where $U(x, y, z, t)$ is the temperature at the time t and k is the thermal conductivity of the body which is a constant.

Let V be a volume in the body and S be its boundary surface. Then the amount of heat leaving V per unit time is

$$\iint_S \boldsymbol{F} \cdot \boldsymbol{n} dS.$$

Now using Gauss's divergence theorem

$$\iint_S \boldsymbol{F} \cdot \boldsymbol{n} dS = \iiint_V \operatorname{div} \boldsymbol{F} dV.$$

Using (1), we have $\qquad \operatorname{div} \boldsymbol{F} = -k \operatorname{div} (\operatorname{grad} U) = -k\nabla^2 U.$

$$\therefore \qquad \iint_S \boldsymbol{F} \cdot \boldsymbol{n} dS = -k\iiint_V \nabla^2 U dxdydz. \qquad (\because dV = dxdydz) \qquad \ldots(2)$$

Let H be the total amount of heat in V which is given by

$$H = \iiint_V \sigma\rho U dxdydz.$$

where the constant σ is the specific heat of the body and ρ is the density. Therefore the rate of decrease of H is

$$-\frac{\partial H}{\partial t} = -\iiint_V \sigma\rho\frac{\partial U}{\partial t} dxdydz.$$

Since $\qquad\qquad\qquad\qquad -\dfrac{\partial H}{\partial t} = -\iint_S \mathbf{F} \cdot \mathbf{n}\, dS.$

$\therefore \qquad\qquad -\iiint_V \sigma\rho \dfrac{\partial U}{\partial t} dxdydz = -k\iiint_V \nabla^2 U dxdydz.$ $\qquad$ or $\qquad$ $\iiint_V \left(\sigma\rho \dfrac{\partial U}{\partial t} - k\nabla^2 U \right) dxdydz = 0.$

This equation holds for any volume V in the body, hence

$$\sigma\rho\dfrac{\partial U}{\partial t} - k\nabla^2 U = 0 \qquad \text{or} \qquad \dfrac{\partial U}{\partial t} = \dfrac{k}{\sigma\rho}\nabla^2 U \qquad \text{or} \qquad \dfrac{\partial U}{\partial t} = c^2 \nabla^2 U, \quad c^2 = \dfrac{k}{\sigma\rho}$$

This is the heat equation.

✒ REMARKS

- The formula given by (1) is called the first Green's formula or first form of Green's theorem.
- The formula given by (2) is called the second Green's formula or second form of Green's theorem.
- The formula obtained in (3) is called Green's theorem in symmetrical form.

⚓ Solved Examples

EXAMPLE 1. *Verify Gauss's divergence theorem for*

$$\mathbf{F} = (2x - z)\hat{i} + x^2 y\hat{j} - xz^2\hat{k}$$

taken over the region bounded by $x = 0$, $x = 1$, $y = 0$, $y = 1$, $z = 0$, $z = 1$.

(UPTU–2009, GBTU–2011)

SOLUTION. By Gauss's divergence theorem, we have

$$\iiint_V \text{div } \mathbf{F}\, dV = \iint_S \mathbf{F} \cdot \mathbf{n}\, dS. \qquad ...(1)$$

Here $\qquad \mathbf{F} = (2x - z)\hat{i} + x^2 y\hat{j} - xz^2\hat{k}$

$\therefore \qquad \text{div } \mathbf{F} = \dfrac{\partial}{\partial x}(2x - z) + \dfrac{\partial}{\partial y}x^2 y + \dfrac{\partial}{\partial z}(-xz^2).$

$\qquad\qquad = 2 + x^2 - 2xz.$

$\therefore \qquad \iiint_V \text{div } \mathbf{F}\, dV = \iiint_V (2 + x^2 - 2xz)\, dV$

Here V is a cube bounded by $x = 0, x = 1, y = 0, y = 1,$ $z = 0, z = 1$ is as shown in fig. 22.

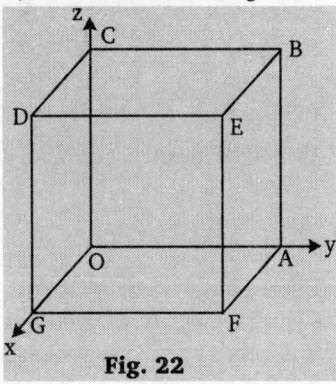

Fig. 22

$\therefore \iiint_V (2 + x^2 - 2xz)\, dV$

$= \iiint_V (2 + x^2 - 2xz)\, dxdydz$

$= \int_{x=0}^1 \int_{y=0}^1 \int_{z=0}^1 (2 + x^2 - 2xz)\, dxdydz$

$= \int_{x=0}^1 \int_{y=0}^1 \left[2z + x^2 z - xz^2 \right]_0^1 dydx$

$= \int_{x=0}^1 \int_{y=0}^1 (2 + x^2 - x)\, dydx$

$= \int_{x=0}^1 \left[(2 + x^2 - x)y \right]_0^1 dx$

$= \int_{x=0}^1 (2 + x^2 - x)\, dx$

$= \left(2x + \dfrac{x^3}{3} - \dfrac{x^2}{2} \right)_0^1 = \left(2 + \dfrac{1}{3} - \dfrac{1}{2} \right) = \dfrac{11}{6}.$

Now we find $\iint_S \mathbf{F} \cdot \mathbf{n}\, dS$ over six faces of the cube.

Over the face $DEFG : \mathbf{n} = \hat{i}, x = 1$ and $dS = dydz$

$\therefore \iint_{DEFG} \mathbf{F} \cdot \mathbf{n}\, dS$

$= \int_{z=0}^{z=1} \int_{y=0}^{y=1} [(2x - z)\hat{i} + x^2 y\hat{j} - xz^2\hat{k}] . \hat{i}\, dydz.$

$= \int_{z=0}^{z=1} \int_{y=0}^{y=1} (2x - z)\, dydz$

$= \int_{z=0}^{z=1} \left[(2x - z)y \right]_0^1 dz = \int_{z=0}^{z=1} (2x - z)\, dz$

$= \int_0^1 (2 - z)\, dz \qquad\qquad (\because x = 1)$

$= \left(2z - \dfrac{z^2}{2} \right)_0^1 = 2 - \dfrac{1}{2} = \dfrac{3}{2}$

Over the face $ABCO : \mathbf{n} = -\hat{i}, x = 0,$ then

$\therefore \iint_{ABCO} \mathbf{F} \cdot \mathbf{n}\, dS$

$= \iint_{ABCO} (-z)\hat{i} . (-\hat{i})\, dydz.$

$= \int_{z=0}^{z=1} \int_{y=0}^{y=1} z\, dydz = \int_{z=0}^{z=1} z \left[y \right]_0^1 dz = \int_{z=0}^{z=1} z\, dz$

$= \left(\dfrac{z^2}{2} \right)_0^1 = \dfrac{1}{2}$

Over the face $ABEF : \mathbf{n} = \hat{j}, y = 1$ and $dS = dzdx$

$\therefore \iint_{ABEF} \mathbf{F} \cdot \mathbf{n}\, dS$

$= \iint_{ABEF} [(2x - z)\hat{i} + x^2\hat{j} - xz^2\hat{k}] . \hat{j}\, dzdx.$

$= \int_{z=0}^{z=1} \int_{x=0}^{x=1} x^2\, dzdx$

$= \int_{z=0}^{z=1} \left[\dfrac{x^3}{3} \right]_0^1 dz = \dfrac{1}{3}\int_0^1 dz = \dfrac{1}{3}(z)_0^1 = \dfrac{1}{3}$

Over the face $OCDG : \mathbf{n} = -\hat{j}, y = 0$ then

$\therefore \iint_{OCDG} \mathbf{F} \cdot \mathbf{n}\, dS$

$= \iint_{OCDG} [(2x - z)\hat{i} - xz^2\hat{k}] . (-\hat{j})\, dzdx$

$= \iint_{OCDG} 0\, dzdx = 0$

Over the face $BCDE : \boldsymbol{n} = \hat{k}, z = 1$ and $dS = dxdy$

$\therefore \iint_{BCDE} \boldsymbol{F.n}dS$

$= \iint_{BCDE} [(2x-1)\hat{i} + x^2\hat{j} - x\hat{k}].\hat{k}dxdy.$

$= \int_{y=0}^{1} \int_{x=0}^{1} (-x)dxdy$

$= -\int_{y=0}^{1} \left[\frac{x^2}{2} \right]_0^1 dy = -\frac{1}{2}\int_{y=0}^{1} dy$

$= -\frac{1}{2}(y)_0^1 = -\frac{1}{2}$

Over the face $AOGF : \quad = -\hat{k}, z = 0$ then

$\therefore \iint_{AOGF} \boldsymbol{F} \cdot \boldsymbol{n}dS$

$= \iint_{AOGF} (2x\hat{i} + x^2 y\hat{j}).(-\hat{k})dxdy$

$= \iint_{AOGF} 0.dxdy = 0$

Thus $\iint_S \boldsymbol{F} \cdot \boldsymbol{n}dS = \iint_{DEFG} \boldsymbol{F} \cdot \boldsymbol{n}dS + \iint_{ABCO} \boldsymbol{F} \cdot \boldsymbol{n}dS$

$\qquad + \iint_{ABEF} \boldsymbol{F} \cdot \boldsymbol{n}dS + \iint_{OCDG} \boldsymbol{F} \cdot \boldsymbol{n}dS$

$\qquad + \iint_{BCDE} \boldsymbol{F} \cdot \boldsymbol{n}dS + \iint_{AOGF} \boldsymbol{F} \cdot \boldsymbol{n}dS$

$\qquad = \frac{3}{2} + \frac{1}{2} + \frac{1}{3} + 0 - \frac{1}{2} + 0 = \frac{11}{6}$

$\iiint_V \operatorname{div} \boldsymbol{F}dV = \iint_S \boldsymbol{F} \cdot \boldsymbol{n}dS = \frac{11}{6}.$

Hence Gauss's divergence theorem is verified.

EXAMPLE 2. **Prove that** $\iiint_V \dfrac{dV}{r^2} = \iint_S \dfrac{\boldsymbol{r} \cdot \boldsymbol{n}}{r^2} dS.$

SOLUTION . Since we have $\iint_S \dfrac{\boldsymbol{r} \cdot \boldsymbol{n}}{r^2} dS = \iint_S \dfrac{\boldsymbol{r}}{r^2} \cdot \boldsymbol{n}dS$ ...(1)

Using Gauss's divergence theorem

$\iint_S \dfrac{\boldsymbol{r}}{r^2} \cdot \boldsymbol{n}dS = \iiint_V \operatorname{div}\left(\dfrac{\boldsymbol{r}}{r^2}\right) dV.$ ...(2)

Since

$\operatorname{div}\left(\dfrac{\boldsymbol{r}}{r^2}\right) = \operatorname{grad}\left(\dfrac{1}{r^2}\right).\boldsymbol{r} + \dfrac{1}{r^2}\operatorname{div}\boldsymbol{r}$

$\qquad = -\dfrac{2}{r^3}\operatorname{grad}r.\boldsymbol{r} + \dfrac{3}{r^2}$ $(\because \operatorname{div}\boldsymbol{r} = 3)$

$\qquad = -\dfrac{2}{r^3}.\dfrac{\boldsymbol{r}}{r}.\boldsymbol{r} + \dfrac{3}{r^2}$ $\left(\because \operatorname{grad}r = \dfrac{\boldsymbol{r}}{r}\right)$

$\qquad = \dfrac{-2}{r^2} + \dfrac{3}{r^2}$ $[\because \boldsymbol{r}^2 = r^2]$

$\therefore \quad \operatorname{div}\left(\dfrac{\boldsymbol{r}}{r}\right) = \dfrac{1}{r^2}$

$\therefore \iiint_V \operatorname{div}\left(\dfrac{\boldsymbol{r}}{r^2}\right) dV = \iiint_V \dfrac{dV}{r^2}.$ ...(3)

From (1), (2) and (3), we get

$\iint_S \dfrac{\boldsymbol{r.n}}{r^2} dS = \iiint_V \dfrac{dV}{r^2}.$

EXAMPLE 3. **Using Gauss's divergence theorem evaluate**

$\iint_S [(x+z)dydz + (y+z)dzdx + (x+y)dxdy]$

where S is the surface of the sphere $x^2+y^2+z^2= 4.$

SOLUTION . By Gauss's divergence theorem, we have

$\iint_S [(x+z)dydz + (y+z)dzdx + (x+y)dxdy]$

$= \iiint_V \left[\dfrac{\partial}{\partial x}(x+z) + \dfrac{\partial}{\partial y}(y+z) \right.$

$\qquad \left. + \dfrac{\partial}{\partial z}(x+y) \right]dxdydz$

$= \iiint_V 2dxdydz = 2\iiint_V dV,$

where V is the volume of the sphere $x^2 + y^2 + z^2 = 4.$

$= 2\left[\dfrac{4}{3}\pi(2)^3 \right] = \dfrac{64}{3}\pi$

EXAMPLE 4. **Using divergence theorem evaluate**

$\iint_S xyzdydz$

where S is surface of the parallelopiped $0 \le x \le 3,$ $0 \le y \le 2,\ 0 \le z \le 1.$

SOLUTION. By Gauss's divergence theorem, we have

$\iint_S (F_1 dydz + F_2 dzdx + F_3 dxdy)$

$\qquad = \iiint_V \left(\dfrac{\partial F_1}{\partial x} + \dfrac{\partial F_2}{\partial y} + \dfrac{\partial F_3}{\partial z} \right) dxdydz.$

Here $F_1 = xyz, F_2 = 0, F_3 = 0$

$\therefore \qquad \dfrac{\partial F_1}{\partial x} = \dfrac{\partial}{\partial x}(xyz) = yz$

$\therefore \iint_S xyzdydz = \iiint_V yzdxdydz$

$\qquad = \int_{z=0}^{1} \int_{y=0}^{2} \int_{x=0}^{3} yzdxdydz$

$\qquad = \int_{z=0}^{1} \int_{y=0}^{2} yz[x]_0^3 dydz$

$\qquad = 3\int_{z=0}^{1} \int_{y=0}^{2} yzdydz$

$\qquad = 3\int_{z=0}^{1} z\left[\dfrac{y^2}{2} \right]_0^2 dz = 6\int_{z=0}^{1} zdz$

$\qquad = 6\left[\dfrac{z^2}{2} \right]_0^1 = 6\left(\dfrac{1}{2} \right) = 3.$

EXAMPLE 5. **Evaluate** $\iint_S (\nabla \times \boldsymbol{F}) \cdot \boldsymbol{n}dS$ **where**

$\boldsymbol{F} = (x-z)\hat{i} + (x^3 + yz)\hat{j} - 3xy^2\hat{k}$

and S is the surface of the cone $x^2 + y^2 = (2-z)^2$ **above the xy-plane.**

SOLUTION . The cone $x^2 + y^2 = (2-z)^2$ meets the xy-plane in a circle whose equation is $x^2 + y^2 = 4, z = 0$. Let S_1 be the plane region bounded by the circle. If S_2 is the surface consisting of the surfaces S and S_1, then S_2 is closed surface. Then we have

$\iint_{S_2} (\nabla \times \boldsymbol{F}) \cdot \boldsymbol{n}dS = 0$

$\therefore \iint_S (\nabla \times \boldsymbol{F}) \cdot \boldsymbol{n}dS + \iint_{S_1} (\nabla \times \boldsymbol{F}) \cdot \boldsymbol{n}dS = 0$ ...(1)

Since outward drawn unit normal to S_1 is $-\hat{k}$

Thus $\boldsymbol{n} = -k$ on S_1.

$\therefore$ From (1), we get

$\iint_S (\nabla \times \boldsymbol{F}) \cdot \boldsymbol{n}dS = \iint_{S_1} (\nabla \times \boldsymbol{F}) \cdot \hat{k}dS.$ ...(2)

Now find $\nabla \times \boldsymbol{F}$

$\nabla \times \boldsymbol{F} = \begin{vmatrix} \hat{i} & \hat{j} & \hat{k} \\ \dfrac{\partial}{\partial x} & \dfrac{\partial}{\partial y} & \dfrac{\partial}{\partial z} \\ x-z & x^3+yz & -3xy^2 \end{vmatrix}$

$$= \hat{i}\left[\frac{\partial}{\partial y}(-3xy^2) - \frac{\partial}{\partial z}(x^3 + yz)\right]$$

$$+ \hat{j}\left[\frac{\partial}{\partial z}(x - z) - \frac{\partial}{\partial x}(-3xy^2)\right]$$

$$+ \hat{k}\left[\frac{\partial}{\partial x}(x^3 + yz) - \frac{\partial}{\partial y}(x - z)\right]$$

$$= \hat{i}(-6xy - y) + \hat{j}(-1 + 3y^2) + \hat{k}(3x^2)$$

and $(\nabla \times \boldsymbol{F}).\hat{k} = 3x^2$.

Now from (2), we get

$$\iint_S (\nabla \times \boldsymbol{F}).\boldsymbol{n}dS = \iint_{S_1} 3x^2 dS.$$

$$= \int_{\theta=0}^{2\pi}\int_{r=0}^{2} 3r^2 \cos^2\theta r d\theta dr$$

$$(\because x = r\cos\theta, y = r\sin\theta, dxdy = rd\theta dr)$$

$$= 3\int_{\theta=0}^{2\pi}\left[\frac{r^4}{4}\right]_0^2 \cos^2\theta d\theta$$

$$= 12\int_0^{2\pi}\cos^2\theta d\theta$$

$$= \frac{12}{2}\left[\theta + \frac{\sin 2\theta}{2}\right]_0^{2\pi} = \frac{12}{2}(2\pi) = 12\pi$$

EXAMPLE 6. *By Gauss divergence theorem, show that*

$$\iint_S (x^2\hat{i} + y^2\hat{j} + z^2\hat{k}) \cdot \boldsymbol{n}dS = 0,$$

where S is the surface of the ellipsoid

$$\frac{x^2}{a^2} + \frac{y^2}{b^2} + \frac{z^2}{c^2} = 1.$$

SOLUTION . By Gauss divergence theorem, we have

$$\iint_S (x^2\hat{i} + y^2\hat{j} + z^2\hat{k}) \cdot \boldsymbol{n}dS$$

$$= \iiint_V \text{div}(x^2\hat{i} + y^2\hat{j} + z^2\hat{k})dV$$

where V is the volume enclosed by S.

$$= \iiint_V (2x + 2y + 2z)dxdydz \quad (\because dV = dxdydz)$$

$$= 2\iiint_V (x + y + z)dxdydz$$

$$= 2\int_{z=-c}^{c}\int_{y=-b\sqrt{1-z^2/c^2}}^{y=b\sqrt{1-z^2/c^2}}\int_{x=-a\sqrt{1-y^2/b^2-z^2/c^2}}^{x=a\sqrt{1-y^2/b^2-z^2/c^2}}$$
$$(x + y + z)dxdydz$$

$$= 2\int_{z=-c}^{c}\int_{y=-b\sqrt{1-z^2/c^2}}^{y=b\sqrt{1-z^2/c^2}}$$

$$\left[\frac{x^2}{2} + x(y + z)\right]_{-a\sqrt{1-y^2/b^2-z^2/c^2}}^{a\sqrt{1-y^2/b^2-z^2/c^2}} dydz$$

$$= 4a\int_{z=-c}^{c}\int_{y=-b\sqrt{1-z^2/c^2}}^{y=b\sqrt{1-z^2/c^2}}(y + z)\sqrt{1 - \frac{y^2}{b^2} - \frac{z^2}{c^2}}$$
$$dydz$$

$$= 8a\int_{z=-c}^{c}\int_{0}^{y=b\sqrt{1-z^2/c^2}} z\left(\sqrt{\left(1 - \frac{z^2}{c^2}\right) - \frac{y^2}{b^2}}\right)dydz$$

$$= \frac{8a}{b}\int_{z=-c}^{c} z\left[\frac{y}{b}\sqrt{b^2\left(1 - \frac{z^2}{c^2}\right) - y^2}\right.$$

$$\left. + \frac{b^2}{2}\left(1 - \frac{z^2}{c^2}\right)\sin^{-1}\left\{\frac{y}{b\sqrt{1 - \frac{z^2}{c^2}}}\right\}\right]_0^{b\sqrt{1-z^2/c^2}} dz$$

$$= \frac{8a}{b}\int_{z=-c}^{c} z\left[\frac{b^2}{2}\left(1 - \frac{z^2}{c^2}\right)\sin^{-1}1\right]dz$$

$$= \frac{8ab^2\pi}{4b}\int_{z=-c}^{c} z\left(1 - \frac{z^2}{c^2}\right)dz = 0$$

Exercise-6.8

1. Verify divergence theorem for

$$\boldsymbol{F} = (x^2 - yz)\hat{i} + (y^2 - zx)\hat{j} + (z^2 - xy)\hat{k}$$

taken over the region bounded by $0 \le x \le a$, $0 \le y \le b$, $0 \le z \le c$.
(ROHTAK–2006, MADRAS–2000)

2. Verify Gauss's divergence theorem for $\boldsymbol{F} = 4xz\hat{i} - y^2\hat{j} + yz\hat{k}$ taken over the cube bounded by $x = 0$, $x = 1$, $y = 0$, $y = 1$, $z = 0$, $z = 1$.
(MADRAS–2006)

3. Verify divergence theorem for $\boldsymbol{F} = 4x\hat{i} - 2y^2\hat{j} + z^2\hat{k}$ taken over the region bounded by the surfaces $x^2 + y^2 = 4$, $z = 0$, $z = 3$.
(JNTU–2006, SVTU–2007, MUMBAI–2006)

4. Using Gauss's divergence theorem, evaluate

$$\iint_S (x\hat{i} + y\hat{j} + z\hat{k}).\boldsymbol{n}dS$$

where S is the closed surface bounded by the cone $x^2 + y^2 = z^2$ and the plane $z = 1$.

5. For any closed surface S, prove that

$$\iint_S \text{curl } \boldsymbol{F} \cdot \boldsymbol{n}dS = 0.$$

6. If $\boldsymbol{F} = \nabla\phi$ and $\nabla^2\phi = 0$, show that for a closed surface S

$$\iiint_V \boldsymbol{F}^2 dV = \iint_S \phi\boldsymbol{F} \cdot \boldsymbol{n}dS.$$

7. If ϕ and ψ are harmonic in V and $\frac{\partial\phi}{\partial n} = \frac{\partial\psi}{\partial n}$ on S, then $\phi = \psi + c$ in V where C is a constant.

8. For any closed surface S, show that

(i) $\iint_S \boldsymbol{n}dS = \boldsymbol{0}$.   (ii) $\iint_S \boldsymbol{r} \times \boldsymbol{n}dS = \boldsymbol{0}$.

9. Using the divergence theorem, show that the volume V of a region bounded by a surface S is

$$V = \iint_S xdydz = \iint_S ydzdx = \iint_S zdxdy$$

$$= \frac{1}{3}\iint_S (xdydz + ydzdx + zdxdy). \text{ (KURUKSHETRA–2008)}$$

10. By Gauss's divergence theorem evaluate

$$\iint_S (x^2dydz + y^2dzdx + z^2dxdy)$$

where S is a cube bounded by $0 \le x \le 1$, $0 \le y \le 1$, $0 \le z \le 1$.

11. Using divergence theorem evaluate
$$\iint_S (yzdydz + zxdzdx + xydxdy)$$
where S is the boundary of the sphere $x^2 + y^2 + z^2 = 4$.
(UPTU–2004, 09)

12. Apply Gauss's divergence theorem, evaluate
$$\iint_S [\sin x dydz + (2 - \cos x) y dzdx]$$
where S is the parallelopiped bounded by $0 \le x \le 3$, $0 \le y \le 2$ and $0 \le z \le 1$.

13. Evaluate $\iint_S [x^2 dydz + y^2 dzdx + 2z(xy - x - y)dxdy]$ where S is the surface of the cube $0 \le x \le 1$, $0 \le y \le 1$ and $0 \le z \le 1$.

14. If $F = x\hat{i} - y\hat{j} + (z^2 - 1)\hat{k}$, find the value of $\iint_S F \cdot n dS$ where S is the closed surface bounded by the planes $z = 0$, $z = 1$ and the cylinder $x^2 + y^2 = 4$.

15. Evaluate $\iint_S (ax^2 + by^2 + cz^2)dS$
over the sphere $x^2 + y^2 + z^2 = 1$ using divergence theorem.

16. If $F = y\hat{i} + (x - 2xz)\hat{j} - xy\hat{k}$, evaluate $\iint_S (\nabla \times F) \cdot n dS$ where S is the surface of the sphere $x^2 + y^2 + z^2 = a^2$ above the xy-plane.

17. Evaluate $\iint_S (x^2 + y^2)dS$ where S is surface of the cone $z^2 = 3(x^2 + y^2)$ bounded by $z = 0$ and $z = 3$.

18. If $F = (x^2 + y - 4)\hat{i} + 3xy\hat{j} + (2xz + z^2)\hat{k}$, evaluate
$$\iint_S (\nabla \times F) \cdot n dS$$
where S is the surface of the sphere $x^2 + y^2 + z^2 = 16$ above the xy-plane.

19. Apply Gauss's divergence theorem, evaluate
$$\iint_S [(x^3 - yz)dydz - 2x^2 ydzdx + zdxdy]$$
over the surface of a cube bounded by the co-ordinate planes and the planes $x = y = z = a$.

20. Prove that $\iint_S n \times (a \times r)dS = 2Va$
where a is a constant vector and V is the volume enclosed by the closed surface S.

21. If $F = \nabla \phi, \nabla^2 \phi = -4\pi\rho$, show that $\iint_S F \cdot n dS = -4\pi \iiint_V \rho dV$.

22. Prove that $\iiint_V \nabla \phi \cdot F dV = \iint_S \phi F \cdot n dS - \iiint_V \nabla \phi \cdot F dV$.

23. If $C = \frac{1}{2} \nabla \times B, B = \nabla \times A$, show that
$$\iiint_V B^2 dV = \iint_S A \times B.n dS + 2\iiint_V A.C dV.$$

24. A vector B is always normal to a given closed surface S. Show that $\iiint_V \nabla \times B dV = 0$ wherer V is the region bounded by S.

25. Let r denoted the position vector of the point $P(x, y, z)$ with respect to the origin 0 and $|r| = r$, evaluate $\iint_S \frac{r}{r^3}.n dS$, where S is the sphere $x^2 + y^2 + z^2 = a^2$.

26. If V is the volume enclosed by a closed surface S and $F = x\hat{i} + 2y\hat{j} + 3z\hat{k}$, show that $\int_S F.\hat{n} dS = 6V$.

27. Verify Gauss's Divergence theorem and show that
$$\iint_S [(x^3 - yz)\hat{i} - 2x^2 y\hat{j} + 2\hat{k}] \cdot \hat{n} dS = \frac{a^5}{2}$$
where S denotes the surface of the cube bounded by the planes $x = 0$, $x = a$, $y = 0$, $y = a$, $z = 0$, $z = a$. (UPTU–2006)

Hints to Selected Problems

5. Since S is closed and let V be the volume enclosed by S. Then by Gauss's divergence theorem,
$$\iint_S F \cdot n dS = \iiint_V \nabla \cdot F dV$$
$$\Rightarrow \iint_S (\text{curl} F) \cdot n dS = \iiint_V \text{div}(\text{curl} F)dV = 0 \quad (\because \text{div}(\text{curl} F) = 0)$$

7. Since ϕ and ψ are harmonic so $\nabla^2 \phi = 0$, $\nabla^2 \psi = 0$
$$\Rightarrow \nabla^2 \phi - \nabla^2 \psi = 0 \Rightarrow \nabla^2 (\phi - \psi) = 0$$
$$\Rightarrow (\phi - \psi) \text{ is also harmonic on V.}$$
Also $\frac{\partial \phi}{\partial n} = \frac{\partial \psi}{\partial n}$ on S $\Rightarrow \frac{\partial}{\partial n}(\phi - \psi) = 0$
$$\Rightarrow \phi - \psi = \text{constant} \Rightarrow \phi = \psi + c, c \text{ being constant}$$

10. $\iint_S (x^2 dydz + y^2 dzdx + z^2 dxdy) = \iint_S (x^2\hat{i} + y^2\hat{j} + z^2\hat{k}).n dS$
$$= \iiint_V (2x + 2y + 2z)dxdydz \text{ (By Gauss's divergence theorem)}$$
$$= 2\int_{x=0}^1 \int_{y=0}^1 \int_{z=0}^1 (x + y + z)dxdydz$$
$$= 2\int_{x=0}^1 \int_{y=0}^1 \left[xz + yz + \frac{z^2}{2} \right]_{z=0}^1 dydx$$
$$= 2\int_{x=0}^1 \int_{y=0}^1 \left(x + y + \frac{1}{2} \right) dydx = 2\int_{x=0}^1 \left[xy + \frac{y^2}{2} + \frac{1}{2}y \right]_{y=0}^1 dx$$
$$= 2\int_0^1 \left(x + \frac{1}{2} + \frac{1}{2} \right) dx = 2\left[\frac{x^2}{2} + x \right]_{x=0}^1 = 3.$$

12. $\iint_S [\sin x dydz + (2 - \cos x) y dzdx]$
$$= \iint_S (\sin x\hat{i} + (2 - \cos x)\hat{j}) \cdot n dS = \iiint_V (\cos x + 2 - \cos x)dx$$
(By Gauss's divergence theorem)
$$= 2\iiint_V dV = 2(\text{Volume of cuboid}) = 2[(3)(2)(1)] = 12$$

15. Let $f \equiv x^2 + y^2 + z^2 - 1 = 0$
$$n = \frac{\nabla f}{|\nabla f|} = (x\hat{i} + y\hat{j} + z\hat{k})$$
$$\therefore \iint_S (ax^2 + by^2 + cz^2)dS = \iint_S F \cdot (x\hat{i} + y\hat{j} + z\hat{k})dS.$$
$$\therefore F = ax\hat{i} + by\hat{j} + cz\hat{k} \text{ so div } F = a + b + c.$$
$$\therefore \iint_S F \cdot n dS = \iiint_V \text{div } F dV \quad [\text{By Gauss's divergence theorem}]$$
$$= \iiint_V (a + b + c)dV = (a + b + c)\iiint_V dV$$
$$= (a + b + c) \text{ (Volume of sphere)}$$
$$= \frac{4\pi}{3}(a + b + c).$$

19. $\iint_S [(x^3 - yz)dydz - 2x^2 ydzdx + zdxdy]$
$$= \iint_S [(x^3 - yz)\hat{i} - 2x^2 y\hat{j} + z\hat{k}] \cdot n dS$$
$$= \iiint_V (3x^2 - 2x^2 + 1).dxdydz$$
(By Gauss's divergence theorem)
$$= \int_{z=0}^a \int_{y=0}^a \int_{x=0}^a (x^2 + 1)dxdydz.$$

21. Since $F = \nabla \phi, \nabla^2 \phi = -4\pi\rho$
$$\iint_S F \cdot n dS = \iiint_V \text{div } F dV = \iiint_V \nabla \cdot (\nabla\phi)dV$$
$$= \iiint_V \nabla^2 \phi dV = \iiint_V (-4\pi\rho)dV = -4\pi\rho\iiint_V dV.$$

24. Let n be the unit vector normal to the surface S and also B is always normal to S, so n and B are parallel
$$\therefore n \times B = 0$$
$$\iiint_V \nabla \times B dV = \iint_S n \times B)dS = \iint_S 0 dS = 0.$$

26. Using Gauss's divergence theorem
$$\iint_S F \cdot \hat{n} dS = \iiint_V \text{div} \cdot F dV = \iiint_V (1 + 2 + 3)dV = 6\iiint_V dV = 6V.$$

4. $\dfrac{7\pi}{6}$ **10.** 3   **11.** 0   **12.** 12   **13.** $\dfrac{1}{2}$ **14.** 4π **15.** $\dfrac{4\pi}{3}(a+b+c)$ **16.** 0   **17.** 9π **18.** -16π **19.** $a^2\left(\dfrac{a^3}{3}+a\right)$ **25.** 4π

6.12.8 STOKE'S THEOREM

Let S be a piecewise smooth oriented surface in space bounded by a piecewise smooth simple closed curve C. Let **F**(x, y, z) *be a continuous vector function having continuous first order partial derivatives in a region of space in which S lies interior. Then*

$$\iint_S (\operatorname{curl} \mathbf{F})\cdot \mathbf{n}\, dS = \int_C \mathbf{F}\cdot d\mathbf{r} \qquad\qquad …(1)$$

where C is taken as counterclockwise direction and $\mathbf{n}$ is a outward drawn unit normal vector to S.   (UPTU–2007, 08, 09, UKTU–2012)

PROOF.   We shall first prove Stoke's theorem for a surface S which represented simultaneously in the forms of

$$z = f(x, y),\, y = g(x, z),\, x = h(y, z)$$

where f, g, h are continuous functions and having continuous first order partial derivatives. Let

$\mathbf{n} = \cos\alpha\, \hat{i} + \cos\beta\, \hat{j} + \cos\gamma\, \hat{k}$ be outward drawn unit normal to the surface S which makes the angles α, β, γ with positive co-ordinate axes respectively, and let

$$\mathbf{F} = F_1 \hat{i} + F_2 \hat{j} + F_3 \hat{k}.$$

$$\therefore \qquad \nabla \times \mathbf{F} = \begin{vmatrix} \hat{i} & \hat{j} & \hat{k} \\ \dfrac{\partial}{\partial x} & \dfrac{\partial}{\partial y} & \dfrac{\partial}{\partial z} \\ F_1 & F_2 & F_3 \end{vmatrix} = \hat{i}\left(\dfrac{\partial F_3}{\partial y} - \dfrac{\partial F_2}{\partial z}\right) + \hat{j}\left(\dfrac{\partial F_1}{\partial z} - \dfrac{\partial F_3}{\partial x}\right) + \hat{k}\left(\dfrac{\partial F_2}{\partial x} - \dfrac{\partial F_1}{\partial y}\right).$$

and $\qquad (\nabla \times \mathbf{F})\cdot \mathbf{n} = \left(\dfrac{\partial F_3}{\partial y} - \dfrac{\partial F_2}{\partial z}\right)\cos\alpha + \left(\dfrac{\partial F_1}{\partial z} - \dfrac{\partial F_3}{\partial x}\right)\cos\beta + \left(\dfrac{\partial F_2}{\partial x} - \dfrac{\partial F_1}{\partial y}\right)\cos\gamma.$

Let $P(x, y, z)$ be any point on C whose position vector is

$$\mathbf{r} = x\hat{i} + y\hat{j} + z\hat{k}.$$

$$\therefore \qquad\qquad d\mathbf{r} = dx\hat{i} + dy\hat{j} + dz\hat{k}.$$

Thus $\qquad\qquad \mathbf{F}\cdot d\mathbf{r} = F_1 dx + F_2 dy + F_3 dz.$

Now the equation (1) becomes

$$\iint_S \left[\left(\dfrac{\partial F_3}{\partial y} - \dfrac{\partial F_2}{\partial z}\right)\cos\alpha + \left(\dfrac{\partial F_1}{\partial z} - \dfrac{\partial F_3}{\partial x}\right)\cos\beta + \left(\dfrac{\partial F_2}{\partial x} - \dfrac{\partial F_1}{\partial y}\right)\cos\gamma\right] = \int_C (F_1 dx + F_2 dy + F_3 dz). \qquad …(2)$$

First, we shall prove that

$$\iint_S \left(\dfrac{\partial F_1}{\partial z}\cos\beta - \dfrac{\partial F_1}{\partial y}\cos\gamma\right) = \int_C F_1 dx. \qquad\qquad …(3)$$

Let R be the orthogonal projection of S in the xy-plane and C^* be its boundary which is oriented in positive direction as shown in fig. 23.

Using the representation $z = f(x, y)$ of S, we may write the line integral over C as a line integral over C^* as follows :

$$\int_C F_1(x, y, z)dx = \int_{C^*} F_1[x, y, f(x, y)]dx = \int_{C^*}[F_1[x, y, f(x, y)]dx + 0\, dy].$$

We now apply Green's theorem in the plane to the functions $F_1[x, y, f(x, y)]$ and 0. Then we have

$$\int_C F_1(x, y, z)dx = -\iint_R \dfrac{\partial F_1}{\partial y}dx\, dy.$$

But $\qquad \dfrac{\partial F_1[x, y, f(x, y)]}{\partial y} = \dfrac{\partial F_1(x, y, z)}{\partial y} + \dfrac{\partial F_1(x, y, z)}{\partial z}\cdot\dfrac{\partial f}{\partial y}$ $\qquad [\because z = f(x, y)]$

$$\therefore \qquad \int_C F_1(x, y, z)dx = -\iint_R \left(\dfrac{\partial F_1}{\partial y} + \dfrac{\partial F_1}{\partial z}\cdot\dfrac{\partial f}{\partial y}\right)dx\, dy. \qquad …(4)$$

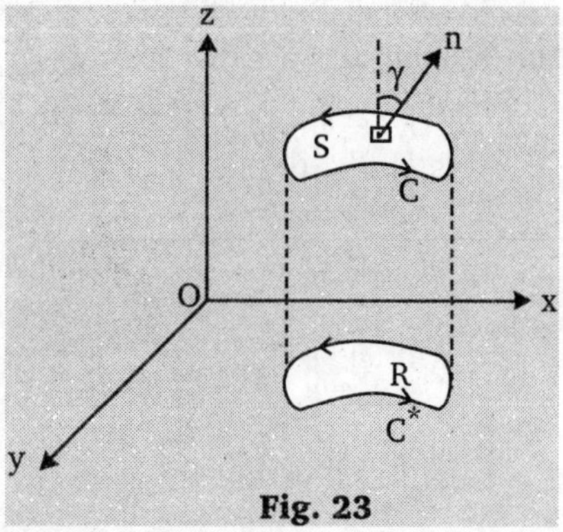

Now we shall prove that the integral on R.H.S. of (4) is equal to integral on L.H.S. of (3). For this let us consider

$$\phi(x, y, z) = z - f(x, y) = 0.$$

$$\therefore \qquad \operatorname{grad}\phi = \dfrac{\partial\phi}{\partial x}\hat{i} + \dfrac{\partial\phi}{\partial y}\hat{j} + \dfrac{\partial\phi}{\partial z}\hat{k} = -\dfrac{\partial f}{\partial x}\hat{i} - \dfrac{\partial f}{\partial y}\hat{j} + \hat{k}. \qquad \left(\because \dfrac{\partial\phi}{\partial x} = -\dfrac{\partial f}{\partial x}, \dfrac{\partial\phi}{\partial y} = -\dfrac{\partial f}{\partial y}, \dfrac{\partial\phi}{\partial z} = 1\right)$$

Fig. 23

Let the length of grad ϕ be a

$\therefore$ $\qquad\qquad\qquad\qquad a = |\text{grad }\phi|.$

Since we know that grad ϕ is perpendicular to the surface S. Therefore, we have

$$n = \pm \frac{\text{grad }\phi}{|\text{grad }\phi|} = \pm \frac{\text{grad }\phi}{a}$$

But the components of both n and grad ϕ in the positive direction of z-axis are positive. Thus

$$n = + \frac{\text{grad }\phi}{a} = \frac{1}{a}\left(-\frac{\partial f}{\partial x}\hat{i} - \frac{\partial f}{\partial y}\hat{j} + \hat{k}\right).$$

Since $= \cos\alpha\,\hat{i} + \cos\beta\,\hat{j} + \cos\gamma\,\hat{k}$, therefore on comparing these twos, we get

$$\cos\alpha = -\frac{1}{a}\frac{\partial f}{\partial x}, \cos\beta = -\frac{1}{a}\frac{\partial f}{\partial y}, \cos\gamma = \frac{1}{a}$$

Since $\qquad\qquad \cos\gamma\,dS = dxdy \quad \therefore \quad dS = \dfrac{dxdy}{\cos\gamma} = a\,dxdy$ $\qquad\qquad\left(\because \cos\gamma = \dfrac{1}{a}\right)$

$$\therefore \iint_S \left(\frac{\partial F_1}{\partial z}\cos\beta - \frac{\partial F_1}{\partial y}\cos\gamma\right)dS = \iint_R \left[\frac{\partial F_1}{\partial z}\left(-\frac{1}{a}\frac{\partial f}{\partial y}\right) - \frac{\partial F_1}{\partial y}\frac{1}{a}\right]a\,dxdy = -\iint_R \left(\frac{\partial F_1}{\partial y} + \frac{\partial F_1}{\partial z}\frac{\partial f}{\partial y}\right)dxdy. \qquad\ldots(5)$$

Thus from (4) and (5), we get

$$\int_C F_1 dx = \iint_S \left(\frac{\partial F_1}{\partial z}\cos\beta - \frac{\partial F_1}{\partial y}\cos\gamma\right)dS. \qquad\qquad\ldots(6)$$

Similarly using the representation $y = g(x, z)$ and $x = h(y, z)$ and having the projection on the other co-ordinates planes, we obtain

$$\int_C F_2 dy = \iint_S \left(\frac{\partial F_2}{\partial x}\cos\gamma - \frac{\partial F_2}{\partial z}\cos\alpha\right)dS. \qquad\qquad\ldots(7)$$

and

$$\int_C F_3 dz = \iint_S \left(\frac{\partial F_3}{\partial y}\cos\alpha - \frac{\partial F_3}{\partial x}\cos\beta\right)dS. \qquad\qquad\ldots(8)$$

Adding (6), (7) and (8), we get

$$\int_C (F_1 dx + F_2 dy + F_3 dz) = \iint_S \left[\left(\frac{\partial F_3}{\partial y} - \frac{\partial F_2}{\partial z}\right)\cos\alpha + \left(\frac{\partial F_1}{\partial z} - \frac{\partial F_3}{\partial x}\right)\cos\beta + \left(\frac{\partial F_2}{\partial x} - \frac{\partial F_1}{\partial y}\right)\cos\gamma\right]dS.$$

or $\qquad\qquad\qquad \int_C \mathbf{F}\cdot d\mathbf{r} = \iint_S (\nabla \times \mathbf{F})\cdot \mathbf{n}\,dS.$

This proves Stoke's theorem for the surface S which can be represented simultaneously by $z = f(x, y), y = g(x, y)$ and $x = h(y, z)$. The proof of this theorem can be extended to a surface S which does not satisfy above conditions but can be decomposed into finitely many surfaces $S_1, S_2, ..., S_n$ whose boundary are $C_1, C_2, ..., C_n$. To each surface this theorem is applied as follows :

$$\int_{C_1} \mathbf{F}\cdot d\mathbf{r} = \iint_{S_1} (\nabla \times \mathbf{F})\cdot \mathbf{n}\,dS.$$

$$\int_{C_2} \mathbf{F}\cdot d\mathbf{r} = \iint_{S_2} (\nabla \times \mathbf{F})\cdot \mathbf{n}\,dS.$$

$$\cdots \qquad \cdots \qquad \cdots \qquad \cdots \qquad \cdots$$
$$\cdots \qquad \cdots \qquad \cdots \qquad \cdots \qquad \cdots$$

$$\int_{C_n} \mathbf{F}\cdot d\mathbf{r} = \iint_{S_n} (\nabla \times \mathbf{F})\cdot \mathbf{n}\,dS.$$

On adding, we get

$$\int_{C_1} \mathbf{F}\cdot d\mathbf{r} + \int_{C_2} \mathbf{F}\cdot d\mathbf{r} + ... + \int_{C_n} \mathbf{F}\cdot d\mathbf{r} = \iint_{S_1} (\nabla \times \mathbf{F})\cdot \mathbf{n}\,dS + \iint_{S_2} (\nabla \times \mathbf{F})\cdot \mathbf{n}\,dS + ... + \iint_{S_n} (\nabla \times \mathbf{F})\cdot \mathbf{n}\,dS.$$

or $\qquad\qquad\qquad \int_C \mathbf{F}\cdot d\mathbf{r} = \iint_S (\nabla \times \mathbf{F})\cdot \mathbf{n}\,dS.$

Hence Stoke's theorem is proved for any surface S enclosed by a closed curve C.

REMARK

- Stoke's theorem can also be stated as the line integral of the tangential component of the vector function $\mathbf{F}$ taken around a closed curve C is equal to the surface integral of the normal component of the curl $\mathbf{F}$ taken over any surface S enclosed by C. That is

$$\int_C \mathbf{F}\cdot \mathbf{t}\,ds = \iint_S (\nabla \times \mathbf{F})\cdot \mathbf{n}\,dS$$

where $\mathbf{t} = \dfrac{d\mathbf{r}}{ds}$ and s is the arc length of C.

6.12.9 APPLICATIONS OF STOKE'S THEOREM

The applications of Stoke's theorem may be illustrated by some examples.

1. Green's theorem in the plane as a special case of Stoke's theorem.

If $F = F_1\hat{i} + F_2\hat{j}$ is a vector function which is continuously differentiable in a simply connected bounded closed surfaces in the xy-plane whose boundary is the piecewise smooth simple closed curve, then

$$(\nabla \times F) \cdot n = \left(\frac{\partial F_2}{\partial x} - \frac{\partial F_1}{\partial y}\right)$$

and

$$F \cdot dr = F_1 dx + F_2 dy.$$

By Stoke's theorem, we have

$$\iint_S (\nabla \times F) \cdot n \, dS = \int_C F \cdot dr.$$

$$\therefore \qquad \iint_S \left(\frac{\partial F_2}{\partial x} - \frac{\partial F_1}{\partial y}\right) dS = \int_C (F_1 dx + F_2 dy).$$

This is a Green's theorem in the plane which is a special case of Stoke's theorem.

2. Stoke's theorem gives the physical interpretation of curl.

Let S be circular disc of radius r centred at P bounded by the circle C as shown in fig. 24.

Let $F(Q) = F(x, y, z)$ be a continuously differentiable vector function in S. Then by Stoke's theorem, we have

Fig. 24

$$\int_C F \cdot t \, ds = \iint_S (\nabla \times F) \cdot n \, dS. \qquad \dots(1)$$

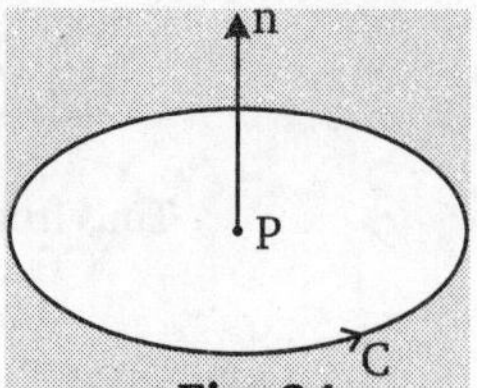

By mean value theorem for integrals, we have

$$\iint_S (\nabla \times F) \cdot n \, dS = [\nabla \times F(P)] \cdot n. \qquad \dots(2)$$

where A is the area of S and P is a suitable point of S. From (1) and (2), we have

$$[\nabla \times F(P)] \cdot n = \frac{1}{A} \int_C F \cdot dr = \frac{1}{A} \int_C F \cdot t \, ds.$$

If F is the velocity vector of the fluid motion, then the integral $\int_C F \cdot dr$ is called the circulation for the flow around C. If we now let $r \to 0$, then

$$[\nabla \times F(P)] \cdot n = \lim_{r \to 0} \frac{1}{A} \int_C F \cdot dr.$$

Thus the component of the curl in the positive normal direction can be regarded as the specific circulation (circulation per unit area) of the flow in the surface at the corresponding point.

Solved Examples

EXAMPLE 1. *Using Stoke's theorem prove that :*

(i) *div curl F = 0*

(ii) *curl grad ϕ = 0.* (KERALA–2005)

SOLUTION . (i) Let V be a any volume enclosed by a closed surface S. Then by Gauss's divergence theorem,

$$\iiint_V \text{div}(\text{curl } F) dV = \iint_S (\text{curl } F) \cdot n \, dS. \qquad \dots(1)$$

Now divide the surface S into S_1 and S_2 in a closed curve C. Then

$$\iint_S (\text{curl } F) \cdot n \, dS = \iint_{S_1} (\text{curl } F) \cdot n \, dS_1$$
$$+ \iint_{S_2} (\text{curl } F) \cdot n \, dS_2.$$

Using Stoke's theorem, we get

$$\iint_S (\text{curl } F) \cdot n \, dS = \int_C F \cdot dr + \int_C F \cdot dr.$$
$$= 0.$$

(Negative sign is taken because positive direction along the boundaries of two surfaces are opposite) Thus the equation (1) becomes

$$\iiint_V \text{div}(\text{curl } F) dV = 0. \qquad \dots(2)$$

Since the equation (2) is true for all volume V, hence

$$\text{div curl } F = 0.$$

(ii) Let S be any surface enclosed by a simple closed curve C. Then by Stoke's theorem

$$\iint_S (\text{curl } F) \cdot n \, dS = \int_C F \cdot dr.$$

Let $F = \text{grad } \phi$

$$\therefore \iint_S (\text{curl grad } \phi) \cdot n \, dS = \int_C \text{grad } \phi \cdot dr. \qquad \dots(1)$$

Since $r = x\hat{i} + y\hat{j} + z\hat{k}.$

$$\therefore \qquad dr = dx\hat{i} + dy\hat{j} + dz\hat{k}.$$

Now $\text{grad } \phi \cdot dr = \left(\frac{\partial \phi}{\partial x}\hat{i} + \frac{\partial \phi}{\partial y}\hat{j} + \frac{\partial \phi}{\partial z}\hat{k}\right)$
$$\cdot (dx\hat{i} + dy\hat{j} + dz\hat{k}).$$
$$= \frac{\partial \phi}{\partial x}dx + \frac{\partial \phi}{\partial y}dy + \frac{\partial \phi}{\partial z}dz = d\phi.$$

$$\therefore \int_C \text{grad } \phi \cdot dr = \int_C d\phi = 0. \quad (\because C \text{ is closed curve})$$

Thus the equation (1) becomes

$$\iint_S (\text{curl grad } \phi) \cdot n \, dS = 0. \qquad \dots(2)$$

Since S is an arbitrary surface, and the equation (2) holds for any S. Then we have

$$\text{curl grad } \phi = \mathbf{0}.$$

EXAMPLE 2. *Verify Stoke's theorem for*

$$F = (2x - y)\hat{i} - yz^2\hat{j} - y^2 z\hat{k}$$

where S is the upper half surface of the supere $x^2 + y^2 + z^2 = 1$ and C is its boundary.

(SVTU–2006, BHOPAL–2008, MADRAS–2006)

SOLUTION. The boundary C of the surface S in the xy-plane is a circle whose equation is $x^2 + y^2 + z^2 = 1$, $z = 0$, therefore its parametric equations are

$$x = \cos\theta, y = \sin\theta, z = 0, 0 \le \theta \le 2\pi$$

Now $\mathbf{F}\cdot d\mathbf{r} = [(2x-y)\hat{i} - yz^2\hat{j} - y^2z\hat{k}]$

$$.[dx\hat{i} + dy\hat{j} + dz\hat{k}]$$

$$= (2x-y)dx - yz^2dy - y^2zdz$$

$$= (2x-y)dx \qquad [\because z = 0]$$

$\therefore \quad \oint_C \mathbf{F}.d\mathbf{r} = \oint_C (2x-y).dx.$

$$= \int_0^{2\pi}[2\cos\theta - \sin\theta](-\sin\theta d\theta)$$

$$= -\int_0^{2\pi}\sin 2\theta d\theta + \int_0^{2\pi}\sin^2\theta d\theta)$$

$$= -\left[-\frac{\cos 2\theta}{2}\right]_0^{2\pi} + \frac{1}{2}\left[\theta - \frac{\sin 2\theta}{2}\right]_0^{2\pi}$$

$$= -\left[-\frac{\cos 4}{2} + \frac{\cos 0}{2}\right]$$

$$+ \frac{1}{2}\left[2\pi - \frac{\sin 4}{2} - 0 + \frac{\sin 0}{2}\right]$$

$$= -\left[-\frac{1}{2} + \frac{1}{2}\right] + \frac{1}{2}[2\pi - 0] = \pi$$

and $\quad \nabla \times \mathbf{F} = \begin{vmatrix} \hat{i} & \hat{j} & \hat{k} \\ \frac{\partial}{\partial x} & \frac{\partial}{\partial y} & \frac{\partial}{\partial z} \\ 2x-y & -yz^2 & -y^2z \end{vmatrix}$

$$= (-2yz + 2yz)\hat{i} - (0-0)\hat{j} + (0+1)\hat{k} = \hat{k}.$$

Let S' be the surface consisting of surface S and S_1 and S_1 is the plane region by the circle C, then S' is closed. Then we have

$$\iint_{S'}(\nabla\times\mathbf{F})\cdot\mathbf{n}dS = 0$$

$$\iint_S(\nabla\times\mathbf{F})\cdot\mathbf{n}dS + \iint_{S_1}(\nabla\times\mathbf{F})\cdot\mathbf{n}dS = 0$$

Since the outward drawn unit normal vector $\mathbf{n}$ to S_1 is $-\hat{k}$, then

$$\iint_S(\nabla\times\mathbf{F})\cdot\mathbf{n}dS - \iint_{S_1}(\nabla\times\mathbf{F})\cdot\hat{k}dS = 0$$

or $\quad \iint_S(\nabla\times\mathbf{F})\cdot\mathbf{n}dS = \iint_{S_1}(\hat{k}\cdot\hat{k})dS \qquad [\because \nabla\times\mathbf{F} = \hat{k}]$

$$= \iint_{S_1}dS \qquad [\because \hat{k}\cdot\hat{k} = 1]$$

$$= S_1 \text{(the area of the circle } C \text{ of radius 1)}$$

$$= \pi[1]^2 = \pi$$

$\therefore \quad \oint_C\mathbf{F}\cdot d\mathbf{r} = \iint_S(\nabla\times\mathbf{F})\cdot\mathbf{n}dS.$

Hence Stoke's theorem is verified.

EXAMPLE 3. *Verify Stroke's theorem for the vector function*

$$\mathbf{F} = 3y\hat{i} - xz\hat{j} + yz^2\hat{k}$$

where S is the surface of the paraboloid $2z = x^2 + y^2$ bounded by $z = 2$ and C its boundary. (UPTU–2007)

SOLUTION. C is the boundary of the surface given by $2z = x^2 + y^2$ and $z = 2$. Thus C is the boundary of the circle $x^2 + y^2 = 4$, $z = 2$ whose parametric equations are $x = 2\cos\theta, y = 2\sin\theta, z = 2$, where $0 \le \theta \le 2\pi$.

$\therefore \quad \mathbf{F}\cdot d\mathbf{r} = (3y\hat{i} - xz\hat{j} + yz^2\hat{k})\cdot(dx\hat{i} + dy\hat{j} + dz\hat{k})$

$$= 3ydx - xzdy + yz^2dz$$

Now $\int_C\mathbf{F}\cdot d\mathbf{r} = \int_C(3ydx - xzdy + yz^2dz)$.

$$= \int_C 3ydx - 2xdy. \qquad (\because z = 2, \therefore dz = 0)$$

$$= \int_0^{2\pi}[3(2\sin\theta)(-2\sin\theta)d\theta - 2(2\cos\theta)(2\cos\theta)d\theta]$$

$$(\because x = 2\cos\theta, y = 2\sin\theta)$$

$$= -12\int_0^{2\pi}\sin^2\theta d\theta - 8\int_0^{2\pi}\cos^2\theta d\theta.$$

$$= -\frac{12}{2}\left[\theta - \frac{\sin 2\theta}{2}\right]_0^{2\pi} - \frac{8}{2}\left[\theta + \frac{\sin 2\theta}{2}\right]_0^{2\pi}.$$

$$= -6[2\pi] - 4[2\pi] = -12\pi - 8\pi = -20\pi.$$

Now $\nabla\times\mathbf{F} = \begin{vmatrix} \hat{i} & \hat{j} & \hat{k} \\ \frac{\partial}{\partial x} & \frac{\partial}{\partial y} & \frac{\partial}{\partial z} \\ 3y & -xz & yz^2 \end{vmatrix}$

$$= \hat{i}(z^2 + x) + \hat{j}(0-0) + \hat{k}(-z-3)$$

$$= (z^2 + x)\hat{i} - (z+3)\hat{k}.$$

Let S' be the surface consisting of surface S and S_1 where S_1 is the surface whose boundary is the curve C, then S' is a closed surface. Therefore we have

$$\iint_{S'}(\nabla\times\mathbf{F})\cdot\mathbf{n}dS = 0$$

or $\iint_S(\nabla\times\mathbf{F})\cdot\mathbf{n}dS + \iint_{S_1}(\nabla\times\mathbf{F})\cdot\mathbf{n}dS = 0$

Since the unit normal vector drawn outward to S_1 is $-\hat{k}$, then

$$\iint_S(\nabla\times\mathbf{F})\cdot\mathbf{n}dS = \iint_{S_1}(\nabla\times\mathbf{F})\cdot\hat{k}dS$$

$$= \iint_{S_1}[(z^2+x)\hat{i} - (z+3)\hat{k}].\hat{k}dS$$

$$= -\iint_{S_1}(z+3)dS$$

$$(\because \hat{k}.\hat{i} = 0, \hat{k}.\hat{k} = 1)$$

$$= -\iint_{S_1}(2+3)dS \qquad (\because z = 2)$$

$$= -5\iint_{S_1}dS$$

$$= -5 S_1$$

(where S_1 is the area of the circle $x^2 + y^2 = 4$)

$$= -5[\pi(2)^2] = -20\pi.$$

Thus $\iint_S(\nabla\times\mathbf{F})\cdot\mathbf{n}dS = \int_C\mathbf{F}\cdot d\mathbf{r}.$

Hence Stoke's theorem is verified.

EXAMPLE 4. *Using Stoke's theorem evaluate*

$$\int_C(\sin zdx - \cos xdy + \sin ydz)$$

where C is the boundary of the rectangle $0 \le x \le \pi$, $0 \le y \le 1$ and $z = 3$. (UKTU-2013; ROHTAK–2005)

SOLUTION. Since $\sin zdx + (-\cos x)dy + \sin ydz$

$$= (\sin z\hat{i} - \cos x\hat{j} + \sin y\hat{k})\cdot(dx\hat{i} + dy\hat{j} + dz\hat{k})$$

$\therefore \quad \mathbf{F} = \sin z\hat{i} - \cos x\hat{j} + \sin y\hat{k}$

Now $\nabla\times\mathbf{F} = \begin{vmatrix} \hat{i} & \hat{j} & \hat{k} \\ \frac{\partial}{\partial x} & \frac{\partial}{\partial y} & \frac{\partial}{\partial z} \\ \sin z & -\cos x & \sin y \end{vmatrix}$

$$= \hat{i}(\cos y - 0) + \hat{j}(\cos z - 0) + \hat{k}(\sin x - 0)$$

$$= \cos y \hat{i} + \cos z \hat{j} + \sin x \hat{k}.$$

Since the rectangle is parallel to xy-plane at the distance $z = 3$ from the origin so the outward drawn unit normal to S bounded by C is $\hat{k}$. By Stoke's theorem

$$\iint_S (\nabla \times F) \cdot n \, dS = \int_C F \cdot dr$$

$$\int_C F \cdot dr = \iint_S (\cos y \hat{i} + \cos z \hat{j} + \sin x \hat{k}) \cdot \hat{k} \, dS$$

$$= \iint_S \sin x \, dS = \iint_S \sin x \, dx \, dy$$

$$= \int_{x=0}^{\pi} \int_{y=0}^{1} \sin x \, dx \, dy$$

$$(\because \ 0 \le x \le \pi, \ 0 \le y \le 1 \text{ on } S)$$

$$= \int_{x=0}^{\pi} \sin x \left[y \right]_0^1 dx = \int_{x=0}^{\pi} \sin x \, dx$$

$$= \left[-\cos x \right]_0^{\pi} = -\cos \pi + \cos 0 = 2$$

$$\therefore \quad \int_C (\sin z \, dx - \cos x \, dy + \sin y \, dz) = 2.$$

EXAMPLE 5. *Using Stoke's theorem to prove that*

$$\int_C (y \, dx + z \, dy + x \, dz) = -2\sqrt{2}\pi a^2.$$

where C is the curve by $x^2 + y^2 + z^2 - 2ax - 2ay = 0$, $x + y = 2a$ and begins at the point $(2a, 0, 0)$ and goes at first below the z-plane. (BHOPAL–2008)

SOLUTION. Since the intersection of the sphere $x^2 + y^2 + z^2 - 2ax - 2ay = 0$ and the plane $x + y = 2a$ is a circle. The centre of the sphere is $(a, a, 0)$ and this centre also lies on the plane $x + y = 2a$ so the obtained circle is a great circle whose radius is the radius of the sphere.

$$\therefore \text{ Radius of the circle} = \sqrt{a^2 + a^2} = a\sqrt{2}.$$

Since $y \, dx + z \, dy + x \, dz$

$$= (y \hat{i} + z \hat{j} + x \hat{k}) \cdot (dx \hat{i} + dy \hat{j} + dz \hat{k}).$$

$$\therefore \qquad F = y \hat{i} + z \hat{j} + x \hat{k}.$$

$$dr = dx \hat{i} + dy \hat{j} + dz \hat{k}.$$

By Stoke's theorem, we have

$$\int_C F \cdot dr = \iint_S (\nabla \times F) \cdot n \, dS. \qquad \ldots(1)$$

where S is the surface whose boundary C is the circle.

Now $$\nabla \times F = \begin{vmatrix} \hat{i} & \hat{j} & \hat{k} \\ \dfrac{\partial}{\partial x} & \dfrac{\partial}{\partial y} & \dfrac{\partial}{\partial z} \\ y & z & x \end{vmatrix}$$

$$= \hat{i}(0-1) + \hat{j}(0-1) + \hat{k}(0-1)$$

$$= -\hat{i} - \hat{j} - \hat{k}$$

Since S is the surface of the plane $x + y = 2a$ bounded by the circle C. Then

$$n = \frac{\nabla(x + y - 2a)}{|\nabla(x + y - 2a)|} = \frac{\hat{i} + \hat{j}}{\sqrt{2}}.$$

From (1)

$$\int_C F \cdot dr = \iint_S -(\hat{i} + \hat{j} + \hat{k}) \cdot \frac{\hat{i} + \hat{j}}{\sqrt{2}} \, dS.$$

$$= -\frac{2}{\sqrt{2}} \iint_S dS = -\frac{2}{\sqrt{2}} S$$

where S is the area of the circle of radius $a\sqrt{2}$

$$= -\frac{2}{\sqrt{2}} (\pi(a\sqrt{2})^2) = -\frac{4\pi a^2}{\sqrt{2}} = -2\sqrt{2}a^2.$$

Hence, $\int_C (y \, dx + z \, dy + x \, dz) = -2\sqrt{2}\pi a^2.$

EXAMPLE 6. *Use Stoke's theorem evaluate*

$$\int [(x + y)dx + (2x - z)dy + (y + z)dz]$$

where C is the boundary of the triangle with vertices $(2, 0, 0)$, $(0, 3, 0)$ and $(0, 0, 6)$.

(NAGPUR–2009, KURUKSHETRA–2009, KERALA–2005)

SOLUTION. We have $F = (x + y)\hat{i} + (2x - z)\hat{j} + (y + z)\hat{k}$

Therefore,

$$\text{curl } F = \begin{vmatrix} \hat{i} & \hat{j} & \hat{k} \\ \dfrac{\partial}{\partial x} & \dfrac{\partial}{\partial y} & \dfrac{\partial}{\partial z} \\ x + y & 2x - y & y + z \end{vmatrix} = 2\hat{i} + \hat{k}$$

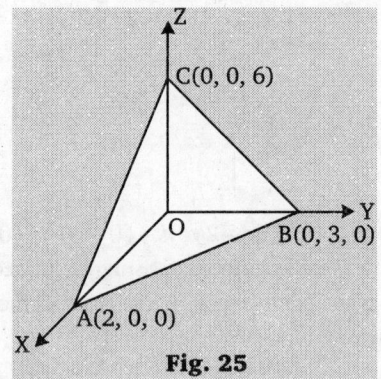

Fig. 25

Equation of the plane through A, B and C

$$\frac{x}{2} + \frac{y}{3} + \frac{z}{6} = 1$$

$$\Rightarrow \quad 3x + 2y + z = 6$$

Now vector $\hat{n}$ normal to this plane is

$$\hat{n} = \frac{3\hat{i} + 2\hat{j} + \hat{k}}{\sqrt{9 + 4 + 1}} = \frac{1}{\sqrt{14}} (3\hat{i} + 2\hat{j} + \hat{k}).$$

Therefore,

$$\int_C [(x + y)dx + (2x - z)dy + (y + z)dz] = \int_C F \cdot dr$$

$$= \int_S \text{curl } F \cdot \hat{n} \, dS, \text{ where } S \text{ is the triangle ABC}$$

$$= \int_S (2\hat{i} + \hat{k}) \cdot \left(\frac{3\hat{i} + 2\hat{j} + \hat{k}}{\sqrt{14}} \right) dS$$

$$= \frac{1}{\sqrt{14}} (6 + 1) \int_S dS = \frac{7}{\sqrt{14}} (\text{Area of } \Delta \, ABC)$$

$$= \frac{7}{\sqrt{14}} \cdot 3\sqrt{14} = 21$$

Exercise-6.9

1. Verify Stoke's theorem for $F = z\hat{i} + x\hat{j} + y\hat{k}$ taken over the half of the sphere $x^2 + y^2 + z^2 = a^2$ lying above the xy-plane.

2. Verify Stoke's theorem for $F = (x^2 + y^2)\hat{i} - 2xy\hat{j}$ taken round the rectangle bounded by $x = \pm a, y = 0, y = b$.

(VTU–2007, BHOPAL–2008, JNTU–2003, UPTU–2003, MUMBAI–2007, BPTU–2006,)

3. Verify Stoke's theorem fot the vector function $F = xy\hat{i} + xy^2\hat{j}$ taken over round the square with vertices (1, 0, 0), (1, 1, 0), (0, 1, 0) and (0, 0, 0).

4. Verify Stoke's theorem for $F = x^2\hat{i} + xy\hat{j}$ taken over round the square in the plane $z = 0$, whose sides are along the lines $x = 0, y = 0, x = a, y = a$.

5. Verify Stoke's theorem for $F = -y^3\hat{i} + x^3\hat{j}$, where S is the circular disc $x^2 + y^2 \le 1, z = 0$.

6. Using Stoke's theorem, evaluate $\int_C F \cdot dr$ where $= y^2\hat{i} + x^2\hat{j} - (x+z)\hat{k}$ and C is the boundary of the triangle with vertices at (0, 0, 0), (1, 0, 0), (1, 1, 0).   (UKTU–2012)

7. If $F = 2y\hat{i} + 2\hat{j} + 3y\hat{k}$, evaluate by Stoke's theorem $\int_C F \cdot dr$, where C is the boundary of the intersection $x^2 + y^2 + z^2 = 6z$ and $z = x + 3$.

8. Using Stoke's theorem evaluate
$$\int_C (yzdx + xzdy + xydz)$$
where C is the boundary of the intersection of $x^2 + y^2 = 1$ and $z = y^2$.   (JNTU–2005)

9. Using Stoke's theorem evaluate
$$\iint_S (\nabla \times F) \cdot n dS$$
where $F = (x^2 + y - 4)\hat{i} + 3xy\hat{j} + (2xz + z^2)\hat{k}$ and S is the surface of
 (i) The hemisphere $x^2 + y^2 + z^2 = 16$ above the xy-plane.
 (ii) The paraboloid $z = 4 - (x^2 + y^2)$ above the xy-plane.

10. Using Stoke's theorem evaluate
$$\iint_S (\nabla \times F) \cdot n dS$$
where $F = (x - z)\hat{i} + (x^3 + yz)\hat{j} - 3xy^2\hat{k}$ and S is the surface of the cone $z = 2 - \sqrt{x^2 + y^2}$ above the xy-plane.

11. Prove that the necessary and sufficient condition that $\int_C F \cdot dr = 0$ for every closed curve C lying in a simply connected region R is that $\nabla \times F = 0$ identically.

12. Show that $\iint_S \phi$ curl $F . dS = \int_C \phi F . dr - \iint_S (\text{grad } \phi \times F) . dS$.

13. If $F = \nabla\phi$ and $G = \nabla\psi$ such that $\nabla^2\phi = 0, \nabla^2\psi = 0$, then show that
$$\iint_S (G.\nabla)F.dS = \int_C (F \times G).dr + \iint_S (F.\nabla)G.dS.$$

14. Verify Stoke's theorem for the function
$$F = (x - y - z)\hat{i} + (y - z - x)\hat{j} + (z - x - y)\hat{k}$$
over the unclosed surface of cylinder $\dfrac{x^2}{a^2} + \dfrac{y^2}{b^2} = 1$ bounded by the plane $z = h$ and open at the end $z = 0$.   (MTU–2011)

15. Use Stoke's theorem to evaluate
$$\int_C [(x + 2y)dx + (x - z)dy + (y - z)dz]$$
where C is the boundary of the triangle with vertices (2, 0, 0), (0, 3, 0) and (0, 0, 6) oriented in the anticlockwise direction
(UPTU–2009)

16. Verify Stoke's theorem by evaluating the line integral
$$\int_C xe^z dx + ye^x dy + ze^y dz$$
along the boundary of the triangle with vertices (0, 0, 1), (1, 0, 1), (1, 1, 1).   (UPTU–2008)

17. Verify Stoke's theorem for $F = (y - z + 2)\hat{i} + (yz + 4)\hat{j} - xz\hat{k}$ where S is the surface of the cube $x = 0, y = 0, z = 0, x = 2, y = 2, z = 2$ above the xy-plane.   (ANDHRA–2000)

18. If S be the surface of the sphere $x^2 + y^2 + z^2 = 1$, prove that $\int_C$ curl $. F dS = 0$.   (JNTU–2009)

19. Use Stoke's theorem to evaluate $(\nabla \times F) \cdot \hat{n} dS$ where
$$F = y\hat{i} + (x - 2xz)\hat{j} - xy\hat{k}$$
and S is the surface of the sphere $x^2 + y^2 + z^2 = a^2$ above xy-plane.   (KOTTAYAM–2005)

20. Verify Stoke's theorem first by evaluating
$$\int_C (3x^2 - 6yz)dx + (2y + 3zx)dy + (1 - 4yz^2)dz$$
from (0, 0, 0) to (1, 1, 1) along the path C given by the straight line from (1, 0, 0) to (1, 1, 0) then from (1, 1, 0) to (1, 1, 1) then from (1, 1, 1) to (1, 0, 0) in counter clockwise and secondly evaluating the integral as surface integral, the surface being enclosed by the path C and the co-ordinate plane.   (UPTU–2009)

Hints to Selected Problems

7. Let S be the surface of the intersection of the sphere $x^2 + y^2 + z^2 = 6z$ and the plane $z = x + 3$ with boundary C, then S is obtained as $2x^2 + y^2 = 9$.

$\therefore \qquad n = \hat{k}$

$$\Rightarrow \quad \nabla \times F = \begin{vmatrix} \hat{i} & \hat{j} & \hat{k} \\ \dfrac{\partial}{\partial x} & \dfrac{\partial}{\partial y} & \dfrac{\partial}{\partial z} \\ 4y & 2 & 3y \end{vmatrix} = 3\hat{i} - 4\hat{k}.$$

By Stoke's theorem
$$\int_C F \cdot dr = \iint_S (\nabla \times F) \cdot n dS$$

$$= \iint_S (3\hat{i} - 4\hat{k}) \cdot \hat{k} dS = -4 \iint_S dS$$
$$= -4 \text{ (area of an ellipse : } 2x^2 + y^2 = 9)$$
$$= -4 \left[\pi \left(\dfrac{\sqrt{3}}{2} \right) \right] = -18\pi\sqrt{2}.$$

9. (i) Let S be the surface of the intersection of hemisphere $x^2 + y^2 + z^2 = 16$ and $z = 0$ with boundary C, therefore C is a circle $x^2 + y^2 = 16, z = 0$.
So its parametric equations are
$$x = 4 \cos \theta, y = 4 \sin \theta, z = 0.$$
Also $F \cdot dr = (x^2 + y - 4)dx + xydy + (2xz + z^2)dz$.

By Stoke's theorem

$$\iint_S (\nabla \times \boldsymbol{F}) \cdot \boldsymbol{n}\, dS = \int_C \boldsymbol{F} \cdot d\boldsymbol{r}$$

$$= \int_0^{2\pi} (-16\sin^2\theta + 16\sin\theta)d\theta = -16\pi$$

(ii) Same as part (i)

13. By Stoke's theorem

$$\int_C (\boldsymbol{F} \times \boldsymbol{G}) \cdot d\boldsymbol{r} = \iint_S [\nabla \times (\boldsymbol{F} \times \boldsymbol{G})] \cdot d\boldsymbol{S}. \qquad \dots(1)$$

Since

$$\nabla \times (\boldsymbol{F} \times \boldsymbol{G}) = (\boldsymbol{G} \cdot \nabla)\boldsymbol{F} - \boldsymbol{G} \operatorname{div} \boldsymbol{F} - (\boldsymbol{F} \cdot \nabla)\boldsymbol{G} + \boldsymbol{F} \operatorname{div} \boldsymbol{G} \qquad \dots(2)$$

But $\quad \boldsymbol{F} = \nabla\phi$, $\boldsymbol{G} = \nabla\psi$, $\operatorname{div} \boldsymbol{F} = 0$, $\operatorname{div} \boldsymbol{G} = 0$. $\qquad \dots(3)$

Using (1), (2), (3), we obtain the required result.

Answers

6. $\dfrac{1}{3}$

7. $-18\pi\sqrt{2}$

8. 0

9. (i) -16π　　(ii) -4π

10. 12π

15. 15

19. 0

20. $-\dfrac{\pi a^2}{\sqrt{2}}$

ARCHIVE

1. Find the tangent vector to the curve $\boldsymbol{H}(t) = t^2\hat{i} + \sin t\hat{j} - t^2\hat{k}$ at $t = 0$ and $t = 1$.

2. If $\boldsymbol{r} = \cos nt\hat{i} + \sin nt\hat{j}$, where n is a constant and t varies, show that $\boldsymbol{r} \times \dfrac{d\boldsymbol{r}}{dt} = n\hat{k}$.

3. If $u = x + y + z$, $v = x^2 + y^2 + z^2$ and $w = xy + yz + zx$, show that ∇u, ∇v and ∇w are coplanar.

4. If $\boldsymbol{r} = x\hat{i} + y\hat{j} + z\hat{k}$, show that

(i) $\nabla(\boldsymbol{a} \cdot \boldsymbol{r}) = \boldsymbol{a}$, where $\boldsymbol{a}$ is a constant vector.

(ii) $\operatorname{grad}|\boldsymbol{r}| = \dfrac{\boldsymbol{r}}{|\boldsymbol{r}|}$

(iii) $\operatorname{grad}\dfrac{1}{|\boldsymbol{r}|} = -\dfrac{\boldsymbol{r}}{|\boldsymbol{r}|^3}$

(iv) $\operatorname{grad}|\boldsymbol{r}|^n = n|\boldsymbol{r}|^{n-2} \cdot \boldsymbol{r}$

5. Prove that $f(r) \times \boldsymbol{r} = 0$.

6. If $\boldsymbol{a}$ is a constant vector and $\boldsymbol{r}$ is a point function, prove that
$$(\boldsymbol{a} \cdot \nabla)\boldsymbol{r} = \boldsymbol{a}$$

7. Show that $\nabla^2\left(\dfrac{x}{r^3}\right) = 0$ where r is a magnitude of the position vector $\boldsymbol{r} = x\hat{i} + y\hat{j} + z\hat{k}$.

8. Find the divergence and curl of the vectot field
$$\boldsymbol{V} = x^2y^2\hat{i} + 2xy\hat{j} + (y^2 - xy)\hat{k}.$$

9. Show that the vector field $\boldsymbol{F} = \dfrac{\boldsymbol{r}}{|\boldsymbol{r}|^3}$ is rotational as well as solenoidal.

10. Prove : $\operatorname{curl} \cdot \operatorname{curl} \boldsymbol{f} = \nabla \cdot \operatorname{div} \boldsymbol{f} - \nabla^2\boldsymbol{f}$.

11. Prove : $\operatorname{grad} \operatorname{div} \boldsymbol{V} = \operatorname{curl} \operatorname{curl} \boldsymbol{F} + \Sigma \dfrac{\partial^2 \boldsymbol{F}}{\partial x^2}$.

12. Show that vector field $\boldsymbol{F}$ defined by
$$\boldsymbol{F} = (\sin y + z)\hat{i} + (x\cos y - z)\hat{j} + (x - y)\hat{k}$$
is conservative.

13. Show that $\iint_S (yz\hat{i} + zx\hat{j} + xy\hat{k})\, dS$, where S is the surface of the sphere $x^2 + y^2 + z^2 = 1$ in the first octant is equal to $\dfrac{3}{8}$.

14. Verify the Green's theorem for the functions $e^{-x}\sin y$ and $e^{-x}\cos y$.

15. Verify Green's theorem in the xy-plane for $\int_C [(3x^2 - 8y^2)dz + (4y - 6xy)dy]$ where C is the region bounded by the paraboles $y = \sqrt{x}$ and $y = x^2$.

16. Verify Gauss's divergence theorem for the function
$$\boldsymbol{F} = (x^2 - yz)\hat{i} + (y^2 - zx)\hat{j} + (z^2 - xy)\hat{k}$$
taken over the rectangular parallelopiped $0 \le x \le a$, $0 \le y \le b$, $0 \le z \le c$.

17. If $\boldsymbol{F} = 3y\hat{i} - xz\hat{j} + yz^2\hat{k}$ and S is the surface of the paraboloid $2z = x^2 + y^2$ bounded by $z = 2$, show using Stoke's theorem that $\iint_S (\nabla \cdot \boldsymbol{F}) \cdot dS = -20\pi$.

18. Verify Stoke's theorem for the function
$$\boldsymbol{F} = (x^2 + y - 4)\hat{i} + 3xy\hat{j} + (2xz + z^2)\hat{k}$$
where S is the upper half of the sphere $x^2 + y^2 + z^2 = 16$ and C is its boundary.

❋❋❋❋❋❋

CHAPTER 7
Series

7.1 INTRODUCTION

Infinite series are essential in the calculation of values of many functions, and can frequently be used for the evaluation of definite integrals. They can also serve to define new and useful functions that are fundamental in many investigations in advanced mathematics and its applications.

Problems connected with the concept of series attracted the Indian mathematician as early as the third century A.D. Their work on series continued till late fourteenth century, but they never took up a critical study of series. In Europe, it was during the sixth century A.D., that the wider significance of finite and infinite series was realized.

The English mathematicians, Brook Taylor (1685-1731) and James Sterling (1692-1770), and the Scotch mathematician Colin Maclaurin (1698-1746), made important contributions to the study of infinite series. The question of convergent of infinite series was first subjected to rigorous investigation by the German Mathematician Carl Friedrich Gauss (1777-1855).

In this chapter, we are going to discuss the convergence behaviour of infinite series of real numbers and shall obtain a few tests for ascertaining the convergence of the infinite series. Some writer use the word Progression instead of word series. But here the word Series, which is due to the writers of the 17th century and is most commonly used in preferred. We shall also discuss the concepts of Fourier series and its applications in engineering problems.

7.1.1 INFINITE SERIES

Let $<u_n>$ be a sequence of real numbers, then an expression of the form

$$u_1 + u_2 + ... + u_n + ... \qquad \qquad ...(1)$$

is called an infinite series. In symbols it is generally written as $\sum\limits_{n=1}^{\infty} u_n$ or $\sum u_n$

If all the terms of $<u_n>$ after a certain number are zero then the expression written as $u_1 + u_2 + ... + u_m$, written as $\sum\limits_{n=1}^{m} u_n$ is called a finite series.

The term u_n is called the nth term or general term of the series (1). The sum of first n terms of the series is denoted by s_n. Thus,

$$s_n = u_1 + u_2 + ... + u_n = \sum\limits_{r=1}^{n} u_r$$

7.1.2 SEQUENCE OF PARTIAL SUM OF AN INFINITE SERIES

An expression of the form $u_1 + u_2 + ... + u_n ...$ which involves addition of infinitely many terms has in itself no meaning. In order to give a meaning to the value of such as infinite sum, we form a sequence of partial sums. It is the limit of such a sequence which gives meaning to the infinite series.

Let us associate to the infinite series $u_1 + u_2 + ... + u_n ....$; a sequence $<s_n>$ defined by $s_n = u_1 + u_2 + ... + u_n$.

Then the sequence $<s_n>$ is called the sequence of partial sums of the given series $u_1 + u_2 + ... + u_n ...$

7.1.3 CONVERGENCE, DIVERGENCE OR OSCILLATION OF AN INFINITE SERIES

An infinite series $\sum\limits_{n=1}^{\infty} u_n$ is said to be

(i) convergent if the sequence $<s_n>$ of its partial sums converges to a real number l and in that case l is called the sum of the series $\sum\limits_{n=1}^{\infty} u_n$ and we write $\sum\limits_{n=1}^{\infty} u_n = l$. In this case, we also say that the series is convergent to l.

(ii) converges absolutely if $\sum\limits_{n=1}^{\infty} |u_n|$ converges.

(iii) converges conditionally if $\sum\limits_{n=1}^{\infty} u_n$ converges but $\sum\limits_{n=1}^{\infty} |u_n|$ does not converge.

(iv) diverges to ∞ (or $-\infty$) if the sequence $<s_n>$ diverges to ∞ (or $-\infty$) and in that case $\sum\limits_{n=1}^{\infty} u_n = \infty \left(\text{or } \sum\limits_{n=1}^{\infty} u_n = -\infty \right)$

(v) oscillate finitely if the sequence $<s_n>$ oscillates finitely.

(vi) oscillate infinitely if the sequence $<s_n>$ oscillates infinitely.

(vii) oscillatory if s_n, the sum of its first n terms, neither tends to a definite finite limit nor to $+\infty$ or $-\infty$ as $n \rightarrow \infty$.

REMARKS

- Divergent and oscillatory series are often called non-convergent series.
- The value of s_n of oscillate finitely series fluctuate within a finite range as $n \rightarrow \infty$.
- The value of s_n of oscillate infinitely series, tends to infinity as $n \rightarrow \infty$ and its sign is alternatively positive and negative.

ILLUSTRATIONS

(1) The series $1 + \dfrac{2}{3} + \left(\dfrac{2}{3}\right)^2 + \dots + \left(\dfrac{2}{3}\right)^{n-1} +$ is convergent.

(2) The series $\dfrac{1}{2} + \dfrac{1}{2^2} + \dfrac{1}{2^3} + \dots$ is convergent.

(3) The series $1 + 2 + 3 + \dots + n + \dots$ is divergent.

(4) The series $3 - 3 + 3 - 3 + \dots$ is oscillatory.

THEOREM 1   **(Necessary condition for convergence).** *For a series Σu_n to be convergent, it is necessary that $\lim u_n = 0$.*   (PTU–2009)

PROOF.   Let us suppose, the series Σu_n be convergent. Let s_n denote the sum of n terms of the series

Therefore $\left. \begin{array}{l} s_n = u_1 + u_2 + \dots + u_n \\ s_{n-1} = u_1 + u_2 + \dots + u_{n-1} \end{array} \right] \Rightarrow u_n = s_n - s_{n-1}$ ...(1)

The series Σu_n is convergent, therefore s_n and s_{n-1} both will tend to the same finite limit, say l as $n \rightarrow \infty$.

Now, from (1) $\lim u_n = \lim s_n - \lim s_{n-1} = l - l = 0$

Hence, for a convergent series, it is necessary that $\lim u_n = 0$.

REMARKS

- The converse of the above theorem is not necessarily true :

  For example, in the series $\Sigma u_n = 1 + \dfrac{1}{2} + \dfrac{1}{3} + \dots ;\ u_n = \dfrac{1}{n} \rightarrow 0$ as $n \rightarrow \infty$ but the series is not convergent.

- If a series Σu_n be such that u_n does not tend to zero as $n \rightarrow \infty$ then the series does not converge.

THEOREM 2.   **(Cauchy's General principle of convergence for series).** *A necessary and sufficient condition for a series Σu_n to be convergent is that to each $\varepsilon > 0$, there exists a positive integer m such that $|u_{n+1} + u_{n+2} + \dots + u_{n+p}| < \varepsilon\ \forall\ n > m,\ p \geq 1$.*

PROOF.   Let $<s_n>$ be the sequence of partial sums of the series Σu_n. The series Σu_n will converge if and only if the sequence $<s_n>$ of its partial sums converges. But by Cauchy's general principle of convergence for sequences, we know that a necessary and sufficient condition for the convergence of $<s_n>$ is that for each $\varepsilon > 0$, there exists $m \in N$ such that

$$|s_n - s_m| < \varepsilon,\ \forall\ n > m$$

$\Rightarrow$ $|u_{n+1} + u_{n+2} + \dots + u_{n+p}| < \varepsilon,\ \forall\ n > m$ and $p \geq 1$.

REMARKS

- The nature of a series remains unaltered if :
  (a) the sign of all terms are changed.
  (b) a finite number of terms are added or omitted.
  (c) each term of the series is multiplied or divided by the same fixed number k ($k \neq 0$).
- If Σu_n converges to l and $k \in R$, then $\Sigma k u_n$ converges to lk.
- If Σu_n converges to l_1 and Σv_n converges to l_2 then $\Sigma(u_n + v_n)$ converges to $(l_1 + l_2)$.
- If Σu_n diverges and $k \in R,\ k \neq 0$, then $\Sigma k u_n$ diverges.
- If Σu_n and Σv_n are two divergent series having all terms positive, then $\Sigma(u_n + v_n)$ also diverges.

THEOREM 3.   *A series of positive terms is convergent if s_n, the sum of n terms is less than a fixed number for all values of n.*

PROOF.   Let $u_1 + u_2 + \dots u_n + \dots$ be the series of positive terms.

Then $s_n = u_1 + u_2 + \dots u_n$.

Obviously if n increases, then s_n increases and may tend to a finite limit or to $+\infty$. The series cannot oscillate. If s_n remains less than a fixed number for all values of n it cannot tend to infinity and so it must tend to a finite limit. Hence, the series is convergent.

7.2 FUNDAMENTAL RESULTS FOR THE CONVERGENCE OF POSITIVE TERM SERIES

THEOREM 1. *A series Σu_n of positive term is convergent if and only if the sequence $<s_n>$ (where $s_n = u_1 + u_2 + \ldots + u_n$) of its partial sum is bounded above.*

PROOF. Since $u_n > 0 \; \forall \; n$, the sequence $<s_n>$ of partial sums of the series is monotonically increasing. Now the series Σu_n is convergent iff the sequence $<s_n>$ is convergent, *i.e.*, iff the sequence $<s_n>$ is bounded above.

($\therefore$ a monotonically increasing sequence is convergent iff it is bounded above.)

☞ REMARKS

- To show that a series of positive term is convergent it is enough to show the sequence of its partial sum is bounded above and to show that the series of positive term is divergent, we have to show that the sequence of its partial sum is unbounded above.
- A series Σu_n where the terms are not necessarily positive, may fail to the convergent even if the sequence of its partial sums is bounded above.

For example, consider the series $\sum\limits_{n=1}^{\infty} u_n$ where $u_n = (-1)^n$. Then $s_n = \begin{cases} -1 & \text{if } n \text{ is odd} \\ 0 & \text{if } n \text{ is even.} \end{cases}$

The sequence $<s_n>$ is bounded above but not convergent and as such, the series is not convergent. Hence, boundedness of the sequence of partial sums of a series Σu_n is only a necessary condition and not a sufficient one. However, it is a sufficient condition for a positive term series.

THEOREM 2. **(Pringsheim theorem).** *If a series Σu_n of positive monotonic decreasing terms converges, then not only u_n tends to zero but also nu_n tends to zero as n tends to infinity.*

PROOF. By Cauchy's general principle of convergence, we have for a convergent series, that for given $\varepsilon > 0$ there exists a positive integer k such that $|u_{m+1} + u_{m+2} + \ldots + u_{m+p}| < \varepsilon/2 \; \forall \; m \geq k; \; p \geq 1$

Choose $m + p = n > 2k$ and $m = \left[\dfrac{n}{2}\right]$ where $\left[\dfrac{n}{2}\right]$ denote the greatest integer not greater than $\left(\dfrac{n}{2}\right)$.

Then $\qquad |u_{m+1} + u_{m+2} + \ldots + u_n| < \dfrac{\varepsilon}{2}$

But Σu_n is monotonic decreasing sequence of positive terms. Therefore,

$$(n - m)u_n < |u_{m+1} + u_{m+2} + \ldots + u_n| < \dfrac{\varepsilon}{2}$$

$$\Rightarrow \qquad \dfrac{1}{2} n u_n < \dfrac{\varepsilon}{2} \qquad \Rightarrow \qquad n u_n < \varepsilon, \forall n \in N \qquad \Rightarrow \qquad \lim_{n \to \infty} n u_n = 0.$$

THEOREM 3. *If each term of a series Σu_n of positive terms does not exceed the corresponding terms of a convergent series Σv_n of positive terms, then Σu_n is convergent. On the other hand, if each term of Σu_n exceed (or equals) the corresponding terms of a divergent series of positive terms, then Σu_n is divergent.*

PROOF. Let us suppose $u_n < v_n \; \forall \; n \in N$. Now let $s_n = u_1 + u_2 + \ldots + u_n$ and $s'_n = v_1 + v_2 + \ldots + v_n$

Since $u_n \leq v_n \; \forall \; n$, therefore $s_n \leq s'_n$.

Now Σv_n is convergent, therefore $\lim s_n \leq \lim s'_n = s'$ (a finite quantity).

Thus s_n tends to a finite limit as $n \to \infty$. Hence, the series Σu_n is convergent.

Now, if $u_n > v_n$, $\forall \; n$ then $s_n > s'_n$.

But Σv_n is divergent, therefore $s'_n \to \infty$ as $n \to \infty$ and hence $s_n \to \infty$ as $n \to \infty$ which gives that Σu_n is divergent.

THEOREM 4. **(Convergence of geometric series).** *The geometric series $1 + r + r^2 + \ldots + r^{n-1} + \ldots$*

$\qquad$ *(i) converges to $\dfrac{1}{1-r}$ if $|r| < 1$* $\qquad$ *(ii) diverges to $+\infty$ if $r \geq 1$*

$\qquad$ *(iii) oscillate finitely if $r = -1$.* $\qquad$ *(iv) oscillate infinitely if $r < -1$.*

PROOF. Here $\qquad s_n = 1 + r + r^2 + \ldots + r^{n-1} = \begin{cases} \dfrac{1 - r^n}{1 - r} & \text{if } r \neq 1 \\ n & \text{if } r = 1 \end{cases}$

Now, there are following cases :

Case (I). If $|r| < 1$.

Then $\qquad \lim\limits_{n \to \infty} r^n = 0$ so that $\qquad \lim\limits_{n \to \infty} s_n = \dfrac{1}{1-r} \qquad \Rightarrow$ the series is convergent to $\dfrac{1}{1-r}$.

Case (II). If $r > 1$.

Then $\qquad \lim\limits_{n \to \infty} r^n = \infty$ so that $\qquad s_n = \dfrac{1 - r^n}{1 - r} = \dfrac{1}{1-r} + \dfrac{r^n}{r-1} \to \infty$ as $n \to \infty$.

Hence, the series diverges to ∞.

if $r = 1$, then $\qquad s_n = 1 + 1 + \ldots + 1 + \ldots$ to n times

$\qquad\qquad\qquad\qquad = n$

Thus, the sequence $< s_n >$ diverges and hence the series diverges.

Case (III). If $r = -1$.

Then, $\qquad\qquad s_n = \begin{cases} 0 \text{ if } n \text{ is even} \\ 1 \text{ if } n \text{ is odd} \end{cases}$

Therefore the sequence $< s_n >$ oscillate between 0 and 1. $\Rightarrow$ The series oscillates finitely between 0 and 1.

Case (IV). If $r < -1$

Let $r = -a$ where $a > 1$ Then $s_n = \dfrac{1}{1+a} - \dfrac{(-1)^n \cdot a^n}{1+a}$ so that $s_{2n} \to \infty$ and $s_{2n+1} \to \infty$

Therefore, the sequence $< s_n >$ oscillate infinitely between $-\infty$ and $+\infty$.

Hence, the series oscillate infinitely.

THEOREM 5.   *A positive terms series Σu_n either converges to a finite limit or diverges to ∞.*

PROOF.   Let $\qquad\qquad s_n = u_1 + u_2 + \ldots + u_n$

$\Rightarrow \qquad\qquad\qquad s_{n+1} = u_1 + u_2 + \ldots u_{n+1}$

Therefore, $\quad s_{n+1} - s_n = u_{n+1} > 0 \qquad\qquad \Rightarrow \qquad\qquad s_{n+1} > s_n, \forall\, n$

$\Rightarrow\ < s_n >$ is monotonically increasing sequence.

Since, a monotonically increasing sequence is either convergent to a finite limit or divergent to ∞, the sequence $< s_n >$ of partial sums of the series Σu_n is either convergent a finite limit or divergent to ∞.

Hence, the series Σu_n is either converges or diverges to ∞.

✒ REMARKS

- In view of the above theorem, a positive term series has only two possible behaviours, *i.e.* convergence or divergence while a general term has got five behaviour (*i.e.*, convergent, divergent to ∞, divergent to $-\infty$, oscillate finitely and oscillate infinitely).
- If, in a positive terms series Σu_n, u_n does not tend to 0 as $n \to \infty$, the series is divergent.
- Similarly, it can be proved that a negative term series either converges to a finite limit or diverges to $-\infty$.

THEOREM 6. $\left(\textbf{The Auxillary series } \Sigma \dfrac{1}{n^p} \right)$. *The infinite series* $\Sigma\left(\dfrac{1}{n^p}\right) = \dfrac{1}{1^p} + \dfrac{1}{2^p} + \ldots + \dfrac{1}{n^p} + \ldots$ *is convergent if $p > 1$ and divergent if $p \le 1$.*

(PTU–2009, VTU–2006, ROHTAK–2003)

PROOF.   **Case (I).** When $p \le 1$.

Since each term of the given series is positive so that the given series can be written as :

$$\Sigma\left(\frac{1}{n^p}\right) = \frac{1}{1^p} + \left(\frac{1}{2^p} + \frac{1}{3^p}\right) + \left(\frac{1}{4^p} + \frac{1}{5^p} + \frac{1}{6^p} + \frac{1}{7^p}\right) + \left(\frac{1}{8^p} + \frac{1}{9^p} + \frac{1}{10^p} + \frac{1}{11^p} + \frac{1}{12^p} + \frac{1}{13^p} + \frac{1}{14^p} + \frac{1}{15^p}\right) + \ldots$$

Since $p > 1$, then $\qquad\qquad\qquad\qquad\qquad\qquad\qquad\qquad\qquad\qquad\qquad\qquad\qquad\qquad\qquad\qquad$ ...(1)

$3^p > 2^p \Rightarrow \dfrac{1}{3^p} < \dfrac{1}{2^p} \qquad\qquad\qquad\qquad \Rightarrow \qquad \dfrac{1}{2^p} + \dfrac{1}{3^p} < \dfrac{1}{2^p} + \dfrac{1}{2^p} = \dfrac{2}{2^p} = \dfrac{1}{2^{p-1}}$

Also, $5^p > 4^p, 6^p > 4^p, 7^p > 4^p \qquad\qquad \Rightarrow \qquad \dfrac{1}{5^p} < \dfrac{1}{4^p}, \dfrac{1}{6^p} < \dfrac{1}{4^p}, \dfrac{1}{7^p} < \dfrac{1}{4^p}$

$\Rightarrow \qquad\qquad\qquad \dfrac{1}{4^p} + \dfrac{1}{5^p} + \dfrac{1}{6^p} + \dfrac{1}{7^p} < \dfrac{1}{4^p} + \dfrac{1}{4^p} + \dfrac{1}{4^p} + \dfrac{1}{4^p} = \left(\dfrac{1}{2^{p-1}}\right)^2$

Similarly $\qquad \dfrac{1}{8^p} + \dfrac{1}{9^p} + \dfrac{1}{10^p} + \ldots + \dfrac{1}{15^p} < \dfrac{8}{8^p} = \left(\dfrac{1}{2^{p-1}}\right)^3$ ... and so on.

Now using above inequalities equation (1) becomes

$$\Sigma\left(\frac{1}{n^p}\right) < 1 + \frac{1}{2^{p-1}} + \left(\frac{1}{2^{p-1}}\right)^2 + \left(\frac{1}{2^{p-1}}\right)^3 + \ldots \qquad\qquad\qquad\qquad\qquad\qquad ...(2)$$

The R.H.S. of (2) is a geometric series with common ratio less than 1 as $p > 1$, which is therefore convergent thus the series on L.H.S of (2) is convergent, hence, $\Sigma\left(\dfrac{1}{n^p}\right)$ is convergent, when $p > 1$.

Case (II). When $p = 1$. Then the given series becomes

$$\Sigma \frac{1}{n^p} = 1 + \frac{1}{2} + \frac{1}{3} + \ldots$$

Now, this series may be written as follows

$$\Sigma \frac{1}{n^p} = 1 + \frac{1}{2} + \left(\frac{1}{3} + \frac{1}{4}\right) + \ldots > 1 + \frac{1}{2} + \left(\frac{1}{4} + \frac{1}{4}\right) + \ldots = 1 + \frac{1}{2} + \frac{2}{4} + \ldots = 1 + \frac{1}{2} + \frac{1}{2} \ldots.$$

Now since $\lim u_n = \frac{1}{2} \neq 0$, the series is divergent.

Case (III). *When* $p < 1$. Then $2^p < 2$, $3^p < 3$, $4^p < 4$ and so on.

Hence, the given series reduces to $\Sigma \frac{1}{n^p} > 1 + \frac{1}{2} + \frac{1}{3} + \frac{1}{4} + \ldots$

Clearly, the series on the right hand side is divergent. [By case (II)]

Hence, the given series is divergent when $p < 1$.

7.3 COMPARISON TESTS

The most important technique for deciding whether a series is convergent or not is to compare it with another suitable chosen series which is already known to be convergent or divergent.

1. **First form.** Let Σu_n and Σv_n be two series of positive terms such that $u_n < k v_n$, $\forall\, n$ and $k > 0$

   Then,

   (i) Σv_n converges $\Rightarrow \Sigma u_n$ converges   (ii) Σu_n diverges $\Rightarrow \Sigma v_n$ diverges.

2. **Second form.** Let Σu_n and Σv_n be two series of positive terms and let k_1 and k_2 be positive real numbers such that $k_1 v_n < u_n < k_2 v_n$, $\forall\, n$ then series Σu_n and Σv_n converge or diverge together.

3. **Third form.** If Σu_n and Σv_n be two given positive terms series such that $u_n < k v_n$, $\forall\, n > m$, $k > 0$ and $m \in \mathbf{N}$

   Then

   (i) Σv_n *is* convergent $\Rightarrow \Sigma u_n$ is convergent   (ii) Σu_n is divergent $\Rightarrow \Sigma v_n$ is also divergent.

4. **Fourth form.** Let Σu_n and Σv_n be two series of positive terms and let k_1, k_2 be positive real numbers such that $k_1 v_n < u_n < k_2 v_n$ $\forall\, n > m$, m being a fixed positive integer. Then the series Σu_n and Σv_n converge or diverge together.

5. **Fifth form.** Let Σu_n and Σv_n be two sereis of positive terms such that

   $$\lim_{\to \infty} \text{——} \quad \text{(finite and non-zero)}$$

   then both the series converge or diverge together.

REMARKS

- In the above form of the comparison test, the condition $\lim\limits_{n \to \infty} \dfrac{u_n}{v_n}$ be finite and non-zero cannot be dropped. For example if

  $u_n = \dfrac{1}{n}$ and $v_n = \dfrac{1}{n^2}$ then $\lim \dfrac{u_n}{v_n} = +\infty$, Σu_n is divergent and Σv_n is convergent.

  In this case, neither the hypothesis nor the conclusion of the comparison test happens to be true.

- The comparison test is usually applied when the nth term u_n of the given series Σu_n contains the powers of n only which may be positive or negative, integral or fractional.

- v_n can be choosen such that $v_n = \dfrac{1}{n^{p-q}}$ where p and q are respectively the highest indices of n in the denominator and numerator of u_n when it is in the form of fraction, if u_n can be expanded in ascending powers of $\dfrac{1}{n}$ then to get v_n we should retain only the lowest power of $\dfrac{1}{n}$ the numerical factor being disregarded.

- We always denote the given series by Σu_n and the series which is used for comparison by Σv_n

- The series Σv_n is known as auxiliary series. We select the auxiliary series in such a way that $\lim\limits_{n \to \infty} \left(\dfrac{u_n}{v_n}\right)$ exists finitely and non-zero.

6. **Sixth form.** Let Σu_n and Σv_n be two series of positive terms and let $\exists$ a positive integer m such that

   $$\frac{u_n}{u_{n+1}} > \frac{v_n}{v_{n+1}}, \forall n \geq m \text{ then } \Sigma u_n \text{ and } \Sigma v_n \text{ both converge or diverge together.}$$

Solved Examples

EXAMPLE 1. *Examine the following series for convergence :*

(i) $\Sigma \dfrac{1}{(\log n)^{\log n}}$ (ii) $\Sigma \dfrac{1}{(\log \log n)^{\log n}}$

SOLUTION. (i) Since we have $\lim\limits_{n \to \infty} \log(\log n) = \infty$

$\Rightarrow$ There exists a large positive integer n such that

$$\log(\log n) > 2$$

$$\Rightarrow \qquad [\log(\log n)]\log n > 2\log n$$

$$\Rightarrow \qquad (\log n)[\log(\log n)] > \log n^2$$

$$\Rightarrow \qquad \log[(\log n)^{\log n}] > \log n^2$$

$$\Rightarrow \qquad (\log n)^{\log n} > n^2$$

$$\Rightarrow \qquad \frac{1}{(\log n)^{\log n}} < \frac{1}{n^2}$$

since $\sum \dfrac{1}{n^2}$ is convergent hence by comparison test

$\Sigma (\log n)^{\log n}$ is also convergent.

(ii) Similarly,

$$\lim_{n \to \infty} \log(\log \log n) = \infty$$

$\Rightarrow$ there exists a positive integer n is so large such that

$$\log(\log \log n) > 2$$

$$\Rightarrow \qquad \log n \,[\log(\log \log n)] > 2\log n$$

$$\Rightarrow \qquad \log [\{\log(\log n)\}^{\log n}] > \log n^2$$

$$\Rightarrow \qquad [\log(\log n)]^{\log n} > n^2$$

$$\Rightarrow \qquad \frac{1}{[\log(\log n)]^{\log n}} < \frac{1}{n^2}$$

Since $\sum \dfrac{1}{n^2}$ is convergent, hence by comparison

test $\sum \dfrac{1}{[\log(\log n)]^{\log n}}$ is convergent.

EXAMPLE 2. *Show that the series* $1 + \dfrac{1}{2!} + \dfrac{1}{3!} + \ldots$ *is convergent.*

SOLUTION. Since,

$$\frac{1}{2!} = \frac{1}{2}$$

$$\frac{1}{3!} < \frac{1}{2^2}$$

$$\ldots \quad \ldots \quad \ldots$$

$$\ldots \quad \ldots \quad \ldots$$

Therefore, $\qquad \dfrac{1}{n!} < \dfrac{1}{2^{n-1}}$

$$1 + \frac{1}{2!} + \frac{1}{3!} + \ldots + \frac{1}{n!} + \ldots < 1 + \frac{1}{2} + \frac{1}{2^2} + \ldots$$

The series on the right hand side is a geometric series with common ratio $\dfrac{1}{2}$ and hence convergent. So the series on the left hand side will also be convergent.

EXAMPLE 3. *Test the convergence of the series*

$$\frac{1}{1 \cdot 2} + \frac{1}{2 \cdot 3} + \ldots + \frac{1}{n(n+1)} + \ldots \qquad \text{(VTU–2006)}$$

SOLUTION. Here $u_n = \dfrac{1}{n(n+1)} = \dfrac{1}{n} - \dfrac{1}{n+1}$

If s_n is the partial sum of n terms of the series Σu_n, then

$$s_n = u_1 + u_2 + \ldots + \ldots + u_n$$

$$= \left(1 - \frac{1}{2}\right) + \left(\frac{1}{2} - \frac{1}{3}\right) + \ldots + \left(\frac{1}{n} - \frac{1}{n+1}\right)$$

$$= 1 - \frac{1}{n+1}$$

Now, $\displaystyle\lim_{n \to \infty} s_n = \lim_{n \to \infty}\left[1 - \frac{1}{n+1}\right] = 1$, which is finite and non-zero.

Hence, the given series is convegent.

EXAMPLE 4. *Test the convergence or divergence of*

$$\frac{1}{1 \cdot 2 \cdot 3} + \frac{3}{2 \cdot 3 \cdot 4} + \frac{5}{3 \cdot 4 \cdot 5} + \ldots + \frac{2n-1}{n(n+1)(n+2)} + \ldots$$

(PTU–2009)

SOLUTION. Here, $\qquad u_n = \dfrac{2n-1}{n(n+1)(n+1)}$

Take $\qquad v_n = \dfrac{n}{n(n)(n)} = \dfrac{1}{n^2}$

Then, $\qquad \dfrac{u_n}{v_n} = \dfrac{2n-1}{n(n+1)(n+2)} \cdot \dfrac{n^2}{1} = \dfrac{\left(2 - \dfrac{1}{n}\right)}{\left(1 + \dfrac{1}{n}\right) \cdot \left(1 + \dfrac{2}{n}\right)}$

$$\Rightarrow \quad \lim_{n \to \infty} \frac{u_n}{v_n} = \lim_{n \to \infty} \frac{\left(2 - \dfrac{1}{n}\right)}{\left(1 + \dfrac{1}{n}\right)\left(1 + \dfrac{2}{n}\right)} = 2, \text{ which is finite.}$$

Now, the auxiliary series $\sum v_n = \sum \dfrac{1}{n^2}$ is conver- gent ($\because p = 2 > 1$). Hence the given series is convergent.

EXAMPLE 5. *Test the convergence of the series whose general term is* $[n^3 + 1]^{1/3} - n]$.   (PTU–2007, ROHTAK–2003)

SOLUTION. Here, we have

$$u_n = (n^3 + 1)^{1/3} - n$$

$$= n\left[\left(1 + \frac{1}{n^3}\right)^{1/3} - 1\right]$$

$$= n\left[\left(1 + \frac{1}{3n^3} + \frac{\dfrac{1}{3}\left(\dfrac{1}{3} - 1\right)}{2!} \cdot \frac{1}{n^6} + \ldots\right) - 1\right]$$

$$= \frac{1}{n^2}\left[\frac{1}{3} - \frac{1}{9n^3} + \ldots\right]$$

Let $v_n = \dfrac{1}{n^2}$, then the auxiliary series

$$\sum v_n = \sum \frac{1}{n^2}$$

Now, $\displaystyle\lim \frac{u_n}{v_n} = \frac{1}{3} - \frac{1}{9n^3} + \ldots = \frac{1}{3}$ which is finite and non-zero.

Since the series $\sum v_n = \sum \dfrac{1}{n^2}$ is convergent ($\because p = 2 > 1$), therefore, the given series is also convergent.

Exercise-7.1

Check the convergence of the following series:

1. $\sum u_n = 1 + \dfrac{1}{3} + \dfrac{1}{5} + \dfrac{1}{7} + \ldots$

2. $\sum u_n = 1 + \dfrac{1}{\sqrt{2}} + \dfrac{1}{\sqrt{3}} + \dfrac{1}{\sqrt{4}} + \ldots$

3. $\sum u_n = 1 + \dfrac{4}{5} + \dfrac{6}{10} + \dfrac{8}{17} + \ldots + \dfrac{2n}{n^2 + 1} \ldots$

4. $\sum u_n = \sqrt{\dfrac{1}{2^3}} + \sqrt{\dfrac{2}{3^3}} + \sqrt{\dfrac{3}{4^3}} + \ldots$

5. $\sum u_n = \dfrac{1}{2} + \dfrac{\sqrt{2}}{5} + \dfrac{\sqrt{3}}{10} + \ldots \dfrac{\sqrt{n}}{n^2 + 1} + \ldots$

6. $\sum u_n = \dfrac{\sqrt{1}}{1 + \sqrt{1}} + \dfrac{\sqrt{2}}{2 + \sqrt{2}} + \dfrac{\sqrt{3}}{3 + \sqrt{3}} + \ldots$

7. $\sum u_n = \dfrac{1}{a + b} + \dfrac{1}{a + 2b} + \dfrac{1}{a + 3b} + \ldots + \dfrac{1}{a + nb} + \ldots$

8. $\sum u_n = \dfrac{1}{a(a + b)} + \dfrac{1}{(a + 2b)(a + 3b)} + \dfrac{1}{(a + 4b)(a + 5b)} + \ldots$

9. $\sum u_n = \sum \dfrac{n}{n^2 + \sqrt{n}}$

10. $\sum u_n = \sum \dfrac{n}{(a + nb)^2}$

11. $\sum u_n = \sum \dfrac{\sqrt{n + 1} + \sqrt{n - 1}}{n}$

12. $\sum u_n = \sum \dfrac{1}{n} \sin \dfrac{1}{n}$

13. $\sum u_n = \sum \tan^{-1} \dfrac{1}{n}$

14. $\sum u_n = \sum \dfrac{n^p}{(n + 1)^q}$

15. $\sum u_n = \sum \dfrac{n^2 - 1}{n^2 + 1}$

16. $\sum u_n = \sum \dfrac{1}{n} \sqrt{n^2 + n + 1} - \sqrt{n^2 - n + 1}$

17. $\dfrac{1}{4.7.10} + \dfrac{4}{7.10.13} + \dfrac{9}{10.13.16} + \ldots + \infty$ (VTU–2010)

18. $\displaystyle\sum_{n=1}^{\infty} \dfrac{1}{\sqrt{n} + \sqrt{(n + 1)}}$ (VTU–2008)

19. $\displaystyle\sum_{n=1}^{\infty} \sqrt{\dfrac{3^n - 1}{2^n + 1}}$ (VTU–2000S)

20. $1 - \dfrac{1}{3} + \dfrac{1}{3^2} - \dfrac{1}{3^3} + \dfrac{1}{3^4} - \ldots \infty$ (JNTU–2000)

21. $f(x) = \dfrac{a_0}{2} + \displaystyle\sum_{n=1}^{\infty} a_n \cos nx + \sum_{n=1}^{\infty} b_n \sin nx$ (COCHIN–2001)

22. $\dfrac{1}{1.3} + \dfrac{2}{3.5} + \dfrac{3}{5.7} + \ldots \infty$ (PTU–2009, 14)

23. $\displaystyle\sum_{n=1}^{\infty} [\sqrt{(n^2 + 1)} - n]$ (VTU–2010, PTU–2009)

24. $\dfrac{1}{1.3.5} + \dfrac{2}{3.5.7} + \dfrac{3}{5.7.9} + \ldots \infty$ (VTU–2009S)

25. $\sum \dfrac{\sqrt{n}}{n^2 + 1}$ (OSMANIA–2000S)

26. $\sum \dfrac{(n + 1)(n + 2)}{n^2 \sqrt{n}}$ (JNTU–2006S, 12)

27. $\displaystyle\sum_{n=1}^{\infty} \dfrac{\sqrt{(n + 1)} - 1}{(n + 2)^3 - 1}$ (JNTU–2003, 10)

Hints to Selected Problems

1. $\sum u_n = \displaystyle\sum_{n=1}^{\infty} \left(\dfrac{1}{2n - 1} \right) \Rightarrow \lim_{n \to \infty} u_n = \lim_{n \to \infty} \left(\dfrac{1}{2n - 1} \right) = 0$

Now apply comparison test.

4. $u_n = \dfrac{\sqrt{n}}{(n + 1)\sqrt{n + 1}}$

Then $\displaystyle\lim_{n \to \infty} \dfrac{u_n}{v_n} = \lim_{n \to \infty} \sqrt{\dfrac{1}{1 + \dfrac{1}{n}} \cdot \dfrac{1}{\left(1 + \dfrac{1}{n}\right)}} = n \ne 0$

7. $u_n = \dfrac{1}{a + nb}$ Let $v_n = \dfrac{1}{n}$.

Then $\displaystyle\lim_{n \to \infty} \dfrac{u_n}{v_n} = \lim_{n \to \infty} \dfrac{n}{a + nb} = \dfrac{1}{b} \ne 0.$

8. Here, we have $u_n = \dfrac{1}{[a + (2n - 2)b][a + (2n - 1)b]}$

Let $v_n = \dfrac{1}{n^2}$

Then $\displaystyle\lim_{n \to \infty} \dfrac{u_n}{v_n} = \dfrac{1}{4b^2} \ne 0$

12. $u_n = \dfrac{1}{n} \sin \dfrac{1}{n} = \dfrac{1}{n} \left[\dfrac{1}{n} - \dfrac{1}{6n^3} + \dfrac{1}{120n^5} - \ldots \right]$

Let $v_n = \dfrac{1}{n^2}$. Then $\displaystyle\lim_{n \to \infty} \dfrac{u_n}{v_n} = 1 \ne 0$

14. Here we have $u_n = \dfrac{n^p}{(n + 1)^q}$

Let $v_n = \dfrac{1}{n^{q - p}}$

Then $\displaystyle\lim_{n \to \infty} \dfrac{v_n}{u_n} = \lim_{n \to \infty} n^{q - p} \left[\dfrac{n^p}{(n + 1)^q} \right] = \lim_{n \to \infty} \dfrac{1}{\left(1 + \dfrac{1}{n}\right)^q} = 1 \ne 0$

15. $\displaystyle\lim_{n \to \infty} u_n = \lim_{n \to \infty} \dfrac{\left(1 - \dfrac{1}{n^2}\right)}{\left(1 + \dfrac{1}{n^2}\right)} = 1 \ne 0$

Answers

1. Divergent	**2.** Divergent	**3.** Divergent	**4.** Divergent	**5.** Convergent
6. Divergent	**7.** Divergent	**8.** Convergent	**9.** Divergent	**10.** Divergent
11. Divergent	**12.** Convergent	**13.** Divergent		
14. Convergent if $p - q + 1 < 0$ and divergent if $p - q + 1 \ge 0$	**15.** Divergent	**16.** Divergent		**17.** Divergent
18. Divergent	**19.** Divergent	**20.** Convergent	**21.** Divergent	**22.** Divergent
23. Divergent	**24.** Convergent	**25.** Convergent	**26.** Divergent	**27.** Convergent

7.4 CAUCHY'S ROOT TEST

Let Σu_n be a series of positive terms and let $\lim_{n \to \infty} u_n^{1/n} = l.$

If,

(i) $l < 1$, *then Σu_n converges;*   (ii) $l > 1$, *then Σu_n diverges;*

(iii) $l = 1$, *then the test fails and the series may either converge or diverge.*

Proof. Case (I). Let $u_n^{1/n} = l < 1.$

Since $l < 1$, we can choose an $\varepsilon > 0$ such that $l + \varepsilon < 1.$

Let $l + \varepsilon = r$ such that $0 < r < 1.$

Since $\lim_{n \to \infty} u_n^{1/n} = l$, therefore there exists a positive integer $m1$ such that.

$$|u_n^{1/n} - l| < \varepsilon, \forall\, n > m_1 \Rightarrow \quad l - \varepsilon < u_n^{1/n} < l + \varepsilon \,\forall\, n > m_1$$

$$\Rightarrow \qquad\qquad (l - \varepsilon)^n < u_n < (l + \varepsilon)^n \,\forall\, n > m_1$$

Since $u_n < r^n \,\forall\, n > m_1$ and since Σr^n converges (being a geometric series with common ratio less than one). Then by comparison test Σu_n converges.

Case (II). Let $u_n^{1/n} = l > 1.$

Since $l > 1$, we can choose an $\varepsilon > 0$ such that $l - \varepsilon > 1.$

Let $l - \varepsilon = R$ then $R > 1.$

Since $R^n < u_n \,\forall\, n > m_2$ and since ΣR^n diverges (being a G.P. with common ratio greater than one). Then by comparison test Σu_n diverges.

Case (III).   Let $u_n = \dfrac{1}{n}$ then $\lim_{n \to \infty} u_n^{1/n} = 1.$

Since $\Sigma \left(\dfrac{1}{n}\right)$ diverges, therefore we find that if $\lim_{n \to \infty} u_n^{1/n} = 1$, then the series Σu_n may diverge.

Again, let $u_n = \dfrac{1}{n^2}$. In this case also $\lim_{n \to \infty} u_n^{1/n} = 1$ but the series Σu_n converges. Thus we find that if $\lim_{n \to \infty} u_n^{1/n} = 1$, then the series Σu_n may converge. The above two examples show that if

Then the test fail. $\lim_{n \to \infty} (u_n)^{1/n} = 1$

REMARKS

- Cauchy's root test can be applied with advantage to series in which the nth term happens to be an exponential fraction of n.
- In this test, it is understood that $u_n^{1/n}$ stands for the positive n^{th} root of u_n.
- The Cauchy's root test can also be stated as follows :
- "A series Σu_n of positive terms is convergent if for every value of $n \geq m$, m being finite, $(u_n)^{1/n}$ less than a fixed number, which is less than unity, and the series is divergent if $(u_n)^{1/n} \geq 1$ for every value of $n \geq m$."

7.5 D'ALEMBERT'S RATIO TEST

Let Σu_n be a series of positive terms and let

(a) $\lim_{n \to \infty} \dfrac{u_n}{u_{n+1}} = l$

Then if,

(i) $l > 1$, *the series converges,*   (ii) $l < 1$, *the series diverges*

(iii) $l = 1$, *the series may converge or diverge and therefore the test fails.*

(b) $\dfrac{u_n}{u_{n+1}} = \infty$ *as $n \to \infty$. Then Σu_n converges.*

Proof. (a) Case(I). When $l > 1$, Let $\varepsilon > 0$ be a positive number such that $l - \varepsilon > 1$.

Now since $\lim_{n \to \infty} \dfrac{u_n}{u_{n+1}} = l$, therefore $\exists$ a positive integer m such that $l - \varepsilon < \dfrac{u_n}{u_{n+1}} < l + \varepsilon$, whenever $n > m$

Now, putting $n = m + 1, m + 2, \ldots p_{-1}$, in succession in the above inequality, we get

$$l - \varepsilon < \frac{u_{m+1}}{u_{m+2}} < l + \varepsilon$$

$$l - \varepsilon < \frac{u_{m+2}}{u_{m+3}} < l + \varepsilon$$

$$\ldots \qquad \ldots \qquad \ldots \qquad \ldots$$

$$l - \varepsilon < \frac{u_{p-1}}{u_p} < l + \varepsilon$$

Multiplying the corresponding sides of the first part of the above inequalities, we get

$$(l-\varepsilon)^{p-1-m} < \frac{u_{m+1}}{u_{m+2}} \cdot \frac{u_{m+2}}{u_{m+3}} \cdots \frac{u_{p-1}}{u_p} \qquad \Rightarrow \qquad (l-\varepsilon)^{p-1-m} < \frac{u_{m+1}}{u_p}$$

$$\Rightarrow \qquad u_p < u_{m+1}(l-\varepsilon)^{m+1} \cdot (l-\varepsilon)^{-p} \qquad \Rightarrow \qquad u_p < k(l-\varepsilon)^{-p}, \forall p \ge m+2 \text{ and } k = u_{m+1}(l-\varepsilon)^{m+1}$$

Since, the series $\Sigma(l-\varepsilon)^{-p}$ converges (being a geometric series with common ratio $(l-\varepsilon)^{-1}$, which is certainly less than unity), then by comparison test it follows that Σu_n converges,

Case (II). When $l < 1$, let $\varepsilon > 0$ be a positive number such that $l + \varepsilon < 1$.

Now since $\lim\limits_{n \to \infty} \dfrac{u_n}{u_{n+1}} = l$, therefore, $\exists$ a positive intger m such that $l - \varepsilon < \dfrac{u_n}{u_{n+1}} < l + \varepsilon, \forall n > m$

Putting $n = m+1, m+2, ..., p-1$ in succession in the second part of the above inequality, we get

$$\frac{u_{m+1}}{u_{m+2}} < l+\varepsilon, \frac{u_{m+2}}{u_{m+3}} < l+\varepsilon, \ldots\ldots\ldots \frac{u_{p-1}}{u_p} < l+\varepsilon.$$

Multiplying the corresponding sides of the above inequalities, we have

$$\frac{u_{m+1}}{u_p} < (l+\varepsilon)^{p-1-m} \qquad \Rightarrow \qquad u_p > u_{m+1}(l+\varepsilon)^{m+1}(l+\varepsilon)^{-p}$$

$$\Rightarrow \qquad u_p > A(l+\varepsilon)^{-p} \, \forall \, p \ge m+2 \text{ and } A = u_{m+1}(l+\varepsilon)^{m+1}$$

Since, $\Sigma(l+\varepsilon)^{-p}$ is a divergent series (being a geometric series with common ratio $(l+\varepsilon)^{-1}$, which is certainly greater than unity), then by comparison test, it follows that Σu_n diverges.

Case (III). Let $l = 1$.

Now, first consider the harmonic series $1 + \dfrac{1}{2} + \dfrac{1}{3} + \dfrac{1}{5} + ... + \dfrac{1}{n} + ...$

Then $\qquad \dfrac{u_n}{u_{n+1}} = \dfrac{n+1}{n} = 1 + \dfrac{1}{n} \Rightarrow \lim\limits_{n \to \infty} \dfrac{u_n}{u_{n+1}} = 1$

Since, the harmonic series is divergent, we find that if $l = 1$, a series may diverge.

Now, consider the series $\dfrac{1}{1^2} + \dfrac{1}{2^2} + ... + \dfrac{1}{n^2} + ...$

Then $\qquad \dfrac{u_n}{u_{n+1}} = \dfrac{(n+1)^2}{n^2} = \left(1 + \dfrac{1}{n}\right)^2 \Rightarrow \lim\limits_{n \to \infty} \dfrac{u_n}{u_{n+1}} = 1$

since, the series $\Sigma \dfrac{1}{n^2}$ converges, we find that if $l = 1$, a series may converge.

(b) Let us suppose $\lim\limits_{n \to \infty} \dfrac{u_n}{u_{n+1}} = +\infty$ then there exists positive integers m and p such

$$\frac{u_n}{u_{n+1}} > p \, \forall \, n \ge m, p > 1$$

Replacing n by $m, m+1, m+2, ..., n-1$, we have

$$\frac{u_m}{u_{m+1}} > p, \frac{u_{m+1}}{u_{m+2}} > p \, \ldots\ldots\ldots \, \frac{u_{n-1}}{u_n} > p.$$

Multiplying the corresponding sides of the above inequalities, we have

$$\frac{u_m}{u_n} > p^{n-m} \qquad \Rightarrow \qquad u_n < p^{m-n} \cdot u_m,$$

$$\Rightarrow \qquad u_n < A.p^{-n} \, \forall \, n \ge m \quad \text{and} \quad A = p^m u_m.$$

Since Σp^{-n} is convergent, then by comparison test, the series Σu_n is convergent.

☑ REMARKS

- The ratio test is generally applied when the nth term of the series involves factorials, products of several factors, or combination of powers and factorials.
- The ratio test can also be stated as follows:
  "An inifinite series of positive terms is convergent if from and after some terms the ratio of each term to the preceding term is less than a fixed number which is less than unity and series is divergent if the ratio, defined above is greater than or equal to unity".
- The ratio test is easier to apply than the root test. However, the root test is stronger than the ratio test.
- The ratio test does not tell us anything about the convergence of the series Σu_n if we only have $\dfrac{u_n}{u_{n+1}} > 1 \, \forall n$.

Solved Examples

EXAMPLE 1. *Test the convergence of the series*

$$1 + \frac{1}{2^2} + \frac{1}{3^3} + \frac{1}{4^4} + \dots$$

SOLUTION. Here, we have $u_n = \frac{1}{n^n}$

$$\Rightarrow \lim_{n \to \infty} (u_n)^{1/n} = \lim_{n \to \infty} \frac{1}{n} = 0 < 1$$

Hence by Cauchy's root test the given series is convergent.

EXAMPLE 2. *Examine the convergene of the following series :*

(i) $\sum \left(1 + \frac{1}{\sqrt{n}}\right)^{-n^{3/2}}$

                        (PTU–2009, KURUKSHETRA–2005)

(ii) $\sum \frac{(n - \log n)^n}{2^n \cdot n^n}$ **(iii)** $\sum \left(1 + \frac{1}{n}\right)^{-n^2}$ (PTU–2010)

(iv) $\sum_{n=2}^{\infty} \frac{1}{(\log n)^n}$ (PTU–2005)

SOLUTION. **(i)** We have $u_n = \left(1 + \frac{1}{\sqrt{n}}\right)^{-n^{3/2}}$

$$\therefore \lim_{n \to \infty} {}^{1/} = \lim_{n \to \infty} \left(1 + \frac{}{\sqrt{}}\right)^{\sqrt{}}$$

$$= \lim_{n \to \infty} \left[\left(1 + \frac{1}{\sqrt{n}}\right)^{\sqrt{n}}\right]^{-1}$$

$$= e^{-1} = \frac{1}{e} < 1$$

Hence, the given series is convergent.

(ii) We have $u_n = \frac{(n - \log n)^n}{2^n \cdot n^n}$

$$\therefore \quad u_n^{1/n} = \frac{n - \log n}{2n} = \frac{1}{2}\left(1 - \frac{\log n}{n}\right)$$

$$\therefore \quad u_n^{1/n} = \frac{n - \log n}{2n} = \frac{1}{2}\left(1 - \frac{\log n}{n}\right)$$

$$= \frac{1}{2}(1 - 0) = \frac{1}{2} < 1$$

$$\left[\because \lim_{n \to \infty} \frac{\log n}{n} = 0\right]$$

Hence, by Cauchy's root test the given series is convergent.

(iii) We have $u_n = \left(1 + \frac{1}{n}\right)^{-n^2}$

$$\therefore \quad u_n^{1/n} = \left(1 + \frac{1}{n}\right)^{-n}$$

$$\therefore \quad \lim_{n \to \infty} u_n^{1/n} = \lim_{n \to \infty} \left(1 + \frac{1}{n}\right)^{-n}$$

$$= \lim_{n \to \infty} \left[\left(1 + \frac{1}{n}\right)^n\right]^{-1}$$

$$= e^{-1} = \frac{1}{e} < 1$$

Hence, by Cauchy's root test the given series is convergent.

(iv) We have $u_n = \frac{1}{(\log n)^n}$

$$\therefore \quad u_n^{1/n} = \frac{1}{(\log n)}$$

$$\therefore \quad \lim_{n \to \infty} u_n^{1/n} = \lim_{n \to \infty} \frac{1}{(\log n)} = 0 < 1$$

Hence, by Cauchy's root test the given series is convergent.

EXAMPLE 3. *Test the convergence of the series*

$$\left(\frac{2^2}{1^2} - \frac{2}{1}\right)^{-1} + \left(\frac{3^3}{2^3} - \frac{3}{2}\right)^{-2} + \left(\frac{4^4}{3^4} - \frac{4}{3}\right)^{-3} + \dots$$
                           (VTU–2006)

SOLUTION. Here we have $u_n = \left[\frac{(n+1)^{n+1}}{n^{n+1}} - \frac{(n+1)}{n}\right]^{-n}$

Therefore

$$\lim_{n \to \infty} u_n^{1/n} = \lim_{n \to \infty} \left[\frac{(n+1)^{n+1}}{n^{n+1}} - \frac{n+1}{n}\right]^{-1}$$

$$= \lim_{n \to \infty} \left[\left(1 + \frac{1}{n}\right)^{n+1} - \left(1 + \frac{1}{n}\right)\right]^{-1}$$

$$= \lim_{n \to \infty} \left(1 + \frac{1}{n}\right)^{-1} \left[\left(1 + \frac{1}{n}\right)^n - 1\right]^{-1}$$

$$= (1 + 0)^{-1} [e - 1]^{-1}$$

$$= \frac{1}{e - 1} < 1.$$

Hence, by Cauchy's root test the given series is convergent.

EXAMPLE 4. *Test the convergence of the series*

$$\frac{1}{2} + \left(\frac{2}{3}\right)x + \left(\frac{3}{4}\right)^2 x^2 + \left(\frac{4}{5}\right)^3 x^3 + \dots \infty, \ x > 0.$$

                           (JNTU–2006)

SOLUTION. Omitting the first term of the series (because it will not effect the convergence or divergence of the series),

we have $u_n = \left(\frac{n+1}{n+2}\right)^n \cdot x^n$

Therefore $\lim_{n \to \infty} u_n^{1/n} = \lim_{n \to \infty} \left[\frac{\left(1 + \frac{1}{n}\right)x}{1 + \left(\frac{2}{n}\right)}\right] = x.$

Therefore by Cauchy's root test, the given series Σu_n converges if $x < 1$, divegent if $x > 1$.

For $x = 1$, test fails

$$\therefore \quad \lim_{n \to \infty} u_n = \lim_{n \to \infty} \frac{\left(1 + \frac{1}{n}\right)^n}{\left(1 + \frac{2}{n}\right)^n} = \frac{e}{e^2} = \frac{1}{e} > 0.$$

$\therefore$ The series Σu_n diverges if $x = 1$.

Hence, the given series is convergent if $x < 1$ and divergent if $x \geq 1$.

EXAMPLE 5. *Test for convergence the series*

$$1 + \frac{2^p}{2!} + \frac{3^p}{3!} + \frac{4^p}{4!} + \dots \quad \text{(KURUKSHETRA–2005)}$$

SOLUTION. Here, we have

Now
$$u_n = \frac{n^p}{n!} \Rightarrow u_{n+1} = \frac{(n+1)^p}{(n+1)!}$$

$$\lim_{n \to \infty} \frac{}{u} \quad \lim_{n \to \infty} \frac{(n\ 1)\ n!}{(n\ 1)!}$$

$$= \lim_{\to \infty} \left[1 + \frac{1}{n} \right] \cdot \frac{1}{(n\ 1)}$$

$$= e^p \times 0 = 0 < 1$$

Hence, by ratio test, the given series is divergent.

EXAMPLE 6. *Test the series* $x + \frac{x^3}{3!} + \frac{x^5}{5!} + \frac{x^7}{7!} + \dots$ *for*

convergence, for all positive value of x.

SOLUTION. Since x is positive. Hence the given series is of positive term series.

Here
$$u_n = \frac{x^{2n-1}}{(2n-1)!}, u_{n+1} = \frac{x^{2n+1}}{(2n+1)!}$$

$$\Rightarrow \lim_{n \to \infty} \frac{u_n}{u_{n+1}} = \lim_{n \to \infty} \frac{x^{2n-1}}{(2n-1)!} \frac{(2n+1)!}{x^{2n+1}}$$

$$= \lim_{n \to \infty} \frac{2n(2n+1)}{x^2}$$

$$= +\infty, \forall \text{ positive value of } x.$$

Then, by ratio test the given series converges for all positive value of x.

Exercise-7.2

Based on Cauchy's Root Test :

1. Test the convergence of the following series :

(i) $\displaystyle\sum_{n=1}^{\infty} \left(1 + \frac{2}{n}\right)^{-n^2}$

(ii) $\displaystyle\sum_{n=1}^{\infty} \frac{n^{n^2}}{(n+1)^{n^2}}$

(iii) $\displaystyle\sum_{n=1}^{\infty} 2^{-n-(-1)^n}$

(iv) $\displaystyle\sum_{n=1}^{\infty} 5^{-n-(-1)^n}$

(v) $\displaystyle\sum_{n=1}^{\infty} (n^{1/n} + x)$ for all positive values of x

(vi) $\displaystyle\sum_{n=1}^{\infty} \frac{n^3}{3^n}$

(vii) $\displaystyle\sum_{n=1}^{\infty} \frac{x^n}{n^n}, x > 0$

2. Test the convergence of the following series :

(i) $\displaystyle\sum \left(\frac{n}{n+1}\right)^{n^2}$

(ii) $\displaystyle\sum n^n x^n, x > 0$

(iii) $\displaystyle\sum \left(\frac{n+1}{3n}\right)^n$

(iv) $\displaystyle\sum \left(\frac{nx}{n+1}\right)^n$

(v) $\displaystyle\sum \frac{(1+nx)^n}{n^n}$

(vi) $\displaystyle\sum (n^{1/n} - 1)^n$

3. Test the convergence of the following series :

(i) $\dfrac{1^3}{} + \dfrac{2^3}{3^2} + \dfrac{3^3}{3^3} + \dfrac{4^3}{3^4} + \dots$

(ii) $\dfrac{2}{1^2}x + \dfrac{3^2}{2^3}x^2 + \dfrac{4^3}{3^4}x^3 + \dots + \dfrac{(n+1)^n x^n}{n^{n+1}} + \dots$ if $x > 0$

(iii) $\displaystyle\sum q^{n^2} r^2, q, r > 0$

(iv) $\displaystyle\sum_{n=2}^{\infty} \frac{1}{[\log(\log n)]^n}$

Based on D'Alembert's Ratio test

4. Test the convergence of the following series :

(i) $\displaystyle\sum_{n=1}^{\infty} \frac{2^{n-1}}{3^n + 1}$

(ii) $\displaystyle\sum_{n=1}^{\infty} \frac{n!}{n^n}$

(iii) $\displaystyle\sum_{n=1}^{\infty} \frac{x^n}{n!}, x > 0$

(iv) $\displaystyle\sum_{n=1}^{\infty} \frac{x^n}{n^n}, x > 0$

(v) $\displaystyle\sum_{n=1}^{\infty} \frac{2^n n!}{n^n}$

(vi) $\displaystyle\sum_{n=1}^{\infty} \frac{n^n x^n}{n!}$

(vii) $\displaystyle\sum_{n=1}^{\infty} \frac{5^n}{n^2 + 5}$

(viii) $\displaystyle\sum_{n=1}^{\infty} \frac{n^3 + a}{2^n + a}$

(ix) $\displaystyle\sum_{n=1}^{\infty} \frac{\sqrt{n}}{\sqrt{n^2+1}} x^n, x > 0$

(PTU–2006)

(x) $\displaystyle\sum_{n=1}^{\infty} \sqrt{\frac{n-1}{n^3+1}} x^n, x > 0$

5. Test the convergence of the series with n^{th} term :

(i) $\dfrac{1}{x^n + x^{-n}}$

(ii) $\left[\sqrt{n^2+1} - n\right] x^{2n}$

(iii) $\dfrac{1}{2^n + x}, x \geq 0$

(iv) $\dfrac{x^n}{n^2 + 1}$

(v) $\dfrac{a^n}{x^n + a^n}$

(vi) $\sqrt{\dfrac{2^n - 1}{3^n - 1}}$

6. Test the convergence of the following series :

(i) $\dfrac{2!}{3} + \dfrac{3!}{3^2} + \dfrac{4!}{3^3} + \dots + \dfrac{(n+1)!}{3^n} + \dots$

(ii) $\dfrac{1^2 \cdot 2^2}{1!} + \dfrac{2^2 \cdot 3^2}{2!} + \dfrac{3^2 \cdot 4^2}{3!} + \dots$

(iii) $\dfrac{1}{1+2} + \dfrac{2}{1+2^2} + \dfrac{3}{1+2^3} + \dots$

(iv) $1 + 3x + 5x^2 + 7x^3 + \dots$ (v) $1 + \dfrac{x}{2^2} + \dfrac{x^2}{3^2} + \dfrac{x^3}{4^2} + \dots$

(vi) $2x + \dfrac{3x^2}{8} + \dfrac{4x^3}{27} + \dots + \dfrac{(n+1)x^n}{n^3} + \dots$

(vii) $\dfrac{1}{\sqrt{1}+\sqrt{2}} + \dfrac{1}{\sqrt{2}+\sqrt{3}} + \dfrac{1}{\sqrt{3}+\sqrt{4}} + \dots$

(viii) $\dfrac{\sqrt{2}-1}{3^3-1} + \dfrac{\sqrt{3}-1}{4^3-1} + \dfrac{\sqrt{4}-1}{5^3-1} + \dots$

(ix) $\dfrac{1}{2} + \dfrac{2!}{8} + \dfrac{3!}{32} + \dfrac{4!}{128} + \dots$

(x) $1 + \dfrac{1}{2 \cdot 2^{1/100}} + \dfrac{1}{3 \cdot 3^{1/100}} + \dfrac{1}{4 \cdot 4^{1/100}} + \dots$

7. Test for convergence the following series :

(i) $\dfrac{1}{2 \cdot 3} + \dfrac{1}{3 \cdot 4} + \dfrac{1}{4 \cdot 5} + \dfrac{1}{5 \cdot 6} + \dots$

(ii) $\dfrac{1}{1 \cdot 2 \cdot 3} + \dfrac{3}{2 \cdot 3 \cdot 4} + \dfrac{5}{3 \cdot 4 \cdot 5} + \dots$

(iii) $\dfrac{1 \cdot 2}{3^2 \cdot 4^2} + \dfrac{3 \cdot 4}{5^2 \cdot 6^2} + \dfrac{5 \cdot 6}{7^2 \cdot 8^2} + \dots$

(iv) $\dfrac{1}{3} + \dfrac{1\cdot2}{3\cdot5} + \dfrac{1\cdot2\cdot3}{3\cdot5\cdot7} + \dfrac{1\cdot2\cdot3\cdot4}{3\cdot5\cdot7\cdot9} + ...$

8. Test the series : $1 + \dfrac{x^2}{2} + \dfrac{x^4}{4} + \dfrac{x^6}{6} + ...$

for convergence for all positive values of x.

9. Test for convergence the series :

$$\dfrac{x}{1\cdot2} + \dfrac{x^2}{2\cdot3} + \dfrac{x^3}{3\cdot4} + \dfrac{x^4}{4\cdot5} + ..., x > 0$$

10. Show that the series $(\alpha > 0, \beta > 0)$

$$1 + \dfrac{\alpha+1}{\beta+1} + \dfrac{(\alpha+1)(2\alpha+1)}{(\beta+1)(2\beta+1)} + \dfrac{(\alpha+1)(2\alpha+1)(3\alpha+1)}{(\beta+1)(2\beta+1)(3\beta+1)} + ...$$

converges if $\beta > \alpha > 0$
and diverges if $\alpha \geq \beta > 0$

11. Test for convergence the series :

$$\dfrac{x}{1\cdot3} + \dfrac{x^2}{2\cdot4} + \dfrac{x^3}{3\cdot5} + \dfrac{x^4}{4\cdot6} + ...$$

12. Test for convergence the following series :

(i) $1 + \dfrac{x}{2} + \dfrac{x^2}{3^2} + \dfrac{x^3}{4^3} + ..., x > 0$

(ii) $x + 2x^2 + 3x^3 + 4x^4 + ...$

(iii) $2 + \dfrac{3}{2}x + \dfrac{4}{3}x^2 + \dfrac{5}{4}x^3 + ..., x > 0$

(iv) $\dfrac{(1+a)(1+b)}{1\cdot2\cdot3} + \dfrac{(2+a)(2+b)}{2\cdot3\cdot4} + \dfrac{(3+a)(3+b)}{3\cdot4\cdot5} + ...$

(v) $x \log x + x^2 \log 2x + x^3 \log 3x + ... + x^n \log nx + ...$

(vi) $\sum\limits_{n=1}^{\infty} \dfrac{n!}{(n^n)^2}$ (PTU–2010)

(vii) $1 + \dfrac{2!}{2^2} + \dfrac{3!}{3^3} + \dfrac{4!}{4^4} + ...\infty$ (VTU–2008S)

(viii) $\sum\limits_{n=1}^{\infty} \dfrac{n!\,3^n}{n^n}$ (KERALA–2005)

(ix) $\dfrac{2}{3\cdot4} + \dfrac{2\cdot4}{3\cdot5\cdot6} + \dfrac{2\cdot4\cdot6}{3\cdot5\cdot7\cdot8} + ...$ (VTU–2010)

(x) $\sum\limits_{n=2}^{\infty} \dfrac{x^n}{n(n-1)(n-2)}$ (JNTU–2006)

(xi) $\sum\limits_{n=1}^{\infty} \left(\dfrac{n^2}{2^n} + \dfrac{1}{n^2} \right)$ (ROHTAK–2005)

(xii) $\sum\limits_{1}^{\infty} \dfrac{n^3 - n + 1}{n!}$ (MADRAS–2000)

(xiii) $1 + \dfrac{1^2 \cdot 2^2}{1\cdot3\cdot5} + \dfrac{1^2 \cdot 2^2 \cdot 3^3}{1\cdot3\cdot5\cdot7\cdot9} + ...\infty$ (DELHI–2002)

(xiv) $\dfrac{4}{18} + \dfrac{4\cdot12}{18\cdot27} + \dfrac{4\cdot12\cdot20}{18\cdot27\cdot36} + ...\infty$ (MADRAS–2000, 12)

(xv) $\dfrac{1}{1^P} + \dfrac{x}{3^P} + \dfrac{x^2}{5^P} + ... \dfrac{x^{n-1}}{(2n-1)^P} + ...\infty$ (JNTU–2006)

(xvi) $\sum\limits_{n=1}^{\infty} \dfrac{3\cdot6\cdot9...3n}{4\cdot7\cdot10...(3n+1)} \cdot \dfrac{5^n}{3n+2}$ (VTU–2004)

(xvii) $\dfrac{3}{4}x + \left(\dfrac{4}{5} \right)^2 x^2 + \left(\dfrac{5}{6} \right)^3 x^3 + ... + \infty \; (x > 0)$ (VTU–2007)

13. Test for convergence the series with n^{th} term :

(i) $\dfrac{n^3 - 1}{n^3 + 1} x^n, \; x > 0$ (ii) $\dfrac{x^n}{a + \sqrt{n}}$ (iii) $\dfrac{x^n}{x+n}$

(iv) $\dfrac{3n+1}{4n+3} x^n, \; x > 0$ (v) $\dfrac{x^n}{(2n+1)^P}$

(vi) $\dfrac{3^n - 2}{3^n + 1} x^{n-1}, \; x > 0$ (vii) $\dfrac{1}{n} \sin \dfrac{1}{n}$

Answers

1. (i) Convergent (ii) Convergent (iii) Convergent (iv) Convergent (v) Divergent (vi) Convergent   (vii) Convergent

2. (i) Convergent (ii) Divergent (iii) Convergent (iv) Convergent if $x < 1$, divergent if $x \geq 1$
   (v) Convergent if $x < 1$, divergent if $x \geq 1$ (vi) Convergent   **3.** (i) Convergent (ii) Convergent if $x < 1$ and divergent if $x \geq 1$
   (iii) Convergent if $0 < q < 1$ and divergent if $q > 1$, Convergent if $0 < r < 1$, when $q = 1$, divergent if $q > 1$ or $q = 1, r \geq 1$ (iv) Convergent

4. (i) Convergent (ii) Convergent (iii) Convergent (iv) Convergent (v) Convergent (vi) Convergent if $x < 1$, divergent if $x \geq 1$
   (vii) Divergent (viii) Convergent (ix) Convergent if $x < 1$, divergent if $x \geq 1$ (x) Convergent if $x < 1$, divergent if $x \geq 1$

5. (i) Convergent if $x > 1$ or $x < 1$, and divergent if $x = 1$ (ii) Convergent if $x < 1$, divergent if $x \geq 1$
   (iii) Convergent (iv) Convergent if $x \leq 1$, divergent if $x > 1$ (v) Convergent if $x > a$, divergent if $x \leq a$, (vi) Convergent.

6. (i) Divergent (ii) Convergent (iii) Convergent (iv) Convergent if $x < 1$, divergent if $x \geq 1$ (v) Convergent if $x \leq 1$, divergent if $x > 1$
   (vi) Convergent if $x \leq 1$, divergent if $x > 1$ (vii) Divergent (viii) Convergent (ix) Divergent (x) Convergent

7. (i) Convergent (ii) Convergent (iii) Convergent (iv) Convergent

8. Convergent if $x < 1$, divergent if $x \geq 1$ **9.** Convergent if $x \leq 1$, divergent if $x > 1$ **11.**   Convergent if $x \leq 1$, divergent $x > 1$

12. (i) Convergent (ii) Convergent if $x < 1$, divergent if $x \geq 1$ (iii) Convergent if $x < 1$, divergent if $x \geq 1$ (iv) Divergent
   (v) Convergent if $x < 1$, divergent if $x \geq 1$ (vi) Convergent (vii) Convergent (viii) Convergent (ix) Convergent
   (x) Convergent for $x \geq 1$, divergent for $x < 1$ (xi) Convergent (xii) Convergent (xiii) Divergent (xiv) Convergent
   (xv) Convergent $x < 1$, divergent for $x > 1$; Covergent for $P > 1$ and divergent for $P \leq 1$ (xvi) Divergent (xvii) Convergent

13. (i) Convergent if $x < 1$, divergent if $x \geq 1$ (ii) Convergent if $x < 1$, divergent if $x \geq 1$
   (iii) Convergent if $x < 1$, divergent if $x \geq 1$ (iv) Convergent if $x < 1$, divergent if $x \geq 1$
   (v) Convergent if $x < 1$, divergent if $x > 1$, when $x = 1$, then convergent if $p > 1$ and divergent if $p \leq 1$
   (vi) Convergent if $x < 1$, divergent if $x \geq 1$ (vii) Convergent.

7.6 RAABE'S TEST

If $\Sigma\, u_n$ be a series of positive terms such that $\lim\limits_{n\to\infty}\left\{n\left(\dfrac{u_n}{u_{n+1}}-1\right)\right\}=l$.
Then, if
  (i)   *$l > 1$, the series converges,*   (ii)   *$l < 1$, the series diverges,*
  (iii)   *$l = 1$, the series may either converge or diverge and therefore the test fails.*

Proof. Case (I) When $l > 1$. We can write $l = 1 + r$, where $r > 0$. Choosing $\varepsilon = r/2$, we can find a positive integer m such that

$$l-\varepsilon < n\left(\frac{u_n}{u_{n+1}}-1\right)<l+\varepsilon \ \forall\ n \geq m$$

Now, from the first part of the above inequality, we have

$$(1+r)-\frac{1}{2}r < n\left(\frac{u_n}{u_{n+1}}-1\right)\ \forall\ n\geq m \qquad \Rightarrow \qquad \frac{1}{2}ru_{n+1} < nu_n -(n+1)u_{n+1}\ \forall n\geq m \qquad \ldots(1)$$

Putting $n = m+1, m+2, \ldots, p-1$ in succession in (1), we have

$$\frac{1}{2}ru_{m+2} < (m+1)u_{m+1}-(m+2)u_{m+2}$$
$$\cdots \quad \cdots \qquad \cdots \qquad \cdots \qquad \cdots$$
$$\frac{1}{2}ru_p < (p-1)u_{p-1}-pu_p.$$

Now, adding the corresponding sides of the above inqualities, we have

$$\frac{1}{2}r[u_{m+2}+u_{m+3}+\ldots+u_p] < (m+1)u_{m+1}-pu_p, \ \Rightarrow\ \frac{1}{2}r[u_{m+2}+\ldots+u_p]<(m+1)u_{m+1},$$

or $\qquad\qquad\qquad u_1 + u_2 +\ldots+u_p < \dfrac{2(m+1)}{r}u_{m+1} < u_1+u_2+\ldots+u_{m+1},\ \forall\ p\geq m+2.$

The above inequality shows that the sequence $\langle s_n\rangle$ of the partial sums of the series Σu_n is bounded and therefore Σu_n converges.

Case (II) When $l < 1$. Let us choose $\varepsilon = 1-l$, then we can find a positive integer m such that

$$l-\varepsilon < n\left(\frac{u_n}{u_{n+1}}-1\right)<1(=l+\varepsilon)\,\forall\,n\geq m \qquad \text{or} \qquad nu_n < (n+1)\,u_{n+1}\ \forall\ n\geq m$$

Putting $n = m+1, m+2, \ldots, p-1\ (p \geq m+2)$, in succession, we get

$$(m+1)u_{m+1} < (m+2)u_{m+2},$$
$$(m+2)u_{m+2} < (m+3)u_{m+3},$$
$$\cdots\quad\cdots\quad\cdots\quad\cdots\quad\cdots$$
$$(p-1)\,u_{p-1} < pu_p.$$

From the above inequality, we have by transitivity

$$(m+1)u_{m+1} < pu_p\ \forall\ p\geq m+2 \qquad \text{or} \qquad u_p > k(1/p)\ \forall\ p\geq m+2 \text{ and } k=(m+1)u_{m+1}.$$

Now, since the series $\Sigma\left(\dfrac{1}{p}\right)$ diverges, then by comparison test the given series diverges.

Case (III) When $l = 1$. In this case the test fails to give any definite information. For example, consider the series $\Sigma\dfrac{1}{n}$ and $\Sigma\dfrac{1}{n(\log n)^2}$ then, we have

$$\lim_{n\to\infty}n\left[\frac{u_n}{u_{n+1}}-1\right]=1.$$

But the former series is divergent, while the latter is convergent.

✒ REMARKS

- Raabe's test is to be applied when D'Alembert's ratio test fails.
- Raabe's test is stronger than D'Alembert ratio test.
- It can be shown as in the proof of case (I) above that if $\lim\limits_{n\to\infty}\left\{n\left[\dfrac{u_n}{u_{n+1}}-1\right]\right\}=+\infty.$, Then, Σu_n converges.
- The case in which $\lim\limits_{n\to\infty}\left\{n\left[\dfrac{u_n}{u_{n+1}}-1\right]\right\}=-\infty.$ the given series Σu_n diverges.

7.7 LOGARITHMIC TEST

If Σu_n be a series of positive terms such that $\lim\limits_{n\to\infty}\left(n\log\dfrac{u_n}{u_{n+1}}\right)=l$.

then Σu_n converges if $l > 1$ and diverges when $l < 1$.

Proof. Case (I) When $l > 1$. In this case, we can choose $\varepsilon > 0$ such that $l - \varepsilon > 1$. Let $l-\varepsilon = p$ (say).

Since $\lim\limits_{n\to\infty}\left(n\log\dfrac{u_n}{u_{n+1}}\right)=l$. Therefore, we can find a positive integer m such that

$$l-\varepsilon < n\log\frac{u_n}{u_{n+1}}<l+\varepsilon\ \forall\ n\geq m.$$

Consider the first part of the above inequality, we have

$$n \log \frac{u_n}{u_{n+1}} > p \quad \forall \, n \geq m. \qquad \Rightarrow \frac{u_n}{u_{n+1}} > e^{p/n} \quad \forall \, n \geq m. \qquad \qquad ...(1)$$

Since, $a_n = \left(1 + \frac{1}{n}\right)^n$ defines a monotonically increasing sequence converging to e, therefore,

$$e \geq \left(1 + \frac{1}{n}\right)^n \quad \forall \, n. \qquad \qquad ...(2)$$

From (1) and (2), we have

$$\frac{u_n}{u_{n+1}} > \left(1 + \frac{1}{n}\right)^p \quad \forall \, n \geq m. \quad \Rightarrow \frac{u_n}{u_{n+1}} > \frac{v_n}{v_{n+1}} \quad \forall \, n \geq m. \qquad \qquad ...(3)$$

where

$$v_n = \frac{1}{n^p}.$$

Now since $p > 1$, therefore Σv_n converges and from (3) it then follows by comparison test that Σu_n converges.

Case (II) When $l < 1$. Let the comparison series $\Sigma v_n = \Sigma \frac{1}{n^p}$ be divergent, *i.e.*, $p < 1$.

$$\therefore \quad \Sigma u_n \text{ will be divergent if } \frac{v_n}{v_{n+1}} > \frac{u_n}{u_{n+1}} \quad \Rightarrow \quad \frac{u_n}{u_{n+1}} < \left(1 + \frac{1}{n}\right)^p \Rightarrow \log\left(\frac{u_n}{u_{n+1}}\right) < p \log\left(1 + \frac{1}{n}\right)$$

$$= p \left[\frac{1}{n} - \frac{1}{2n^2} + \frac{1}{3n^3} + ...\right]$$

$$\therefore \qquad n \log\left(\frac{u_n}{u_{n+1}}\right) = p \left[1 - \frac{1}{2n} + \frac{1}{3n^2} + ...\right]$$

$$\therefore \qquad \lim_{n \to \infty}\left[n \log \frac{u_n}{u_{n+1}}\right] = p < 1$$

$$\therefore \quad \Sigma u_n \text{ will be divergent if } l < 1.$$

☞ REMARKS

- Logarithmic test is to be applied only when :
  (a) ratio test fails
  (b) the ratio test involves the exponent 'e'
- This test is an alternative to Raabe's test.

7.8 SOME MODIFIED FORMS

Various test of convergence, involving limits can be modified in terms of the upper and lower limits. For example, a few modification are given below :

1. Cauchy's Root test.
The series of non-negative term Σu_n converges or diverges according as

$$\underline{\lim}\, u_n^{1/n} < 1 \qquad \qquad \text{or} \qquad \qquad \overline{\lim}\, u_n^{1/n} > 1.$$

2. D' Alembert's Ratio test.
The series Σu_n of positive terms converges or diverges according as

$$\underline{\lim}\, \frac{u_n}{u_{n+1}} > 1 \qquad \qquad \text{or} \qquad \qquad \overline{\lim}\, \frac{u_n}{u_{n+1}} < 1.$$

3. Raabe's test.
The series Σu_n of positive terms converges or diverges according as

$$\underline{\lim}\left\{n\left(\frac{u_n}{u_{n+1}} - 1\right)\right\} > 1 \text{ or} \qquad \qquad \overline{\lim}\left\{n \log \frac{u_n}{u_{n+1}}\right\} < 1.$$

4. Logarithmic test.
The series Σu_n of positive terms converges or diverges according as

$$\underline{\lim}\left\{n \log \frac{u_n}{u_{n+1}}\right\} > 1 \text{ or} \qquad \qquad \overline{\lim}\left\{n \log \frac{u_n}{u_{n+1}}\right\} < 1.$$

7.8.1 SOME OTHER IMPORTANT TESTS

(1) De Morgan's and Bertrand's test : The series Σu_n of positive terms is convergent or divergent according as

$$\lim\left[n\left(\frac{u_n}{u_{n+1}} - 1\right) - 1\right]\log n > 1 \quad \text{or} \quad < 1.$$

(2) **Alternative to Bertrand's test :** The series Σu_n of positive terms is convergent or divergent according as

$$\lim\left[\left(n\log\frac{u_n}{u_{n+1}}-1\right)\log n\right]>1 \quad \text{or} \quad <1.$$

 Solved Examples

EXAMPLE 1. *Test the convergence of the series*

$$1+\frac{3}{7}x+\frac{3\cdot6}{7\cdot10}x^2+\frac{3\cdot6\cdot9}{7\cdot10\cdot13}x^3+...$$

SOLUTION. After leaving the first term we have

$$u_n=\frac{3\cdot6\cdot9\cdot...\cdot3n}{7\cdot10\cdot13\cdot...\cdot(3n+4)}x^n$$

$$\Rightarrow \quad u_{n+1}=\frac{3\cdot6\cdot9\cdot...\cdot3n(3n+3)}{7\cdot10\cdot13\cdot...\cdot(3n+4)(3n+7)}x^{n+1}$$

Now $\lim_{n\to\infty}\dfrac{u_{n+1}}{u_n}=\lim_{n\to\infty}\left(\dfrac{3n+3}{3n+7}\right)x$

$$\frac{x^2}{2\sqrt{1}}+\frac{x^3}{3\sqrt{2}}+\frac{x^4}{4\sqrt{3}}+....=x$$

Then, by D'Alembert ratio test the series is convergent if $x<1$, divergent if $x>1$ and the test fails if $x=1$.
For $x=1$, we have

$$\frac{u_n}{u_{n+1}}=\frac{3n+7}{3n+3}$$

or $\quad n\left(\dfrac{u_n}{u_{n+1}}-1\right)=n\left(\dfrac{3n+7}{3n+3}-1\right)=\dfrac{4n}{3n+3}$

$$\Rightarrow \lim_{n\to\infty}n\left[\left(\frac{u_n}{u_{n+1}}-1\right)\right]=\lim_{n\to\infty}\frac{4n}{3n+3}=\lim_{n\to\infty}\frac{4}{3+3/n}$$

$$=\frac{4}{3}>1$$

Therefore, by Raabe's test the series is convergent when $x=1$.
Hence, the given series is convergent when $x\le1$ and divergent when $x>1$.

EXAMPLE 2. *Test the convergence of the following series*

$$\sum_{n=1}^{\infty}\frac{1.3.5....(2n-1)}{2.4.6....(2n)}\cdot\frac{x^{2n}}{2n}, \ (x>0).$$

SOLUTION. Here, we have

$$u_n=\frac{1.3.5....(2n-1)}{2.4.6....(2n)}\cdot\frac{x^{2n}}{2n}$$

and $\quad u_{n+1}=\dfrac{1.3.5....(2n-1)(2n+1)}{2.4.6....(2n)(2n+2)}\cdot\dfrac{x^{2n+2}}{(2n+2)}$

$$\Rightarrow \lim_{n\to\infty}\frac{u_n}{u_{n+1}}=\lim_{n\to\infty}\left(\frac{2n+2}{2n+1}\cdot\frac{2n+2}{2n}\cdot\frac{1}{x^2}\right)=\frac{1}{x^2}.$$

∴ By D'Alembert's ratio test, the series is convergent if $x^2<1$ and divergent if $x^2>1$.
Now since $x>0$ this gives that the series is convergent if $x<1$ and divergent if $x>1$.
If $x=1$. Then D'Alembert's ratio test fails.
Now consider

$$\lim_{n\to\infty}n\left[\frac{u_n}{u_{n+1}}-1\right]=\lim_{n\to\infty}n\left(\frac{2n+2}{2n+1}\cdot\frac{2n+2}{2n}-1\right)$$

$$=\lim_{n\to\infty}\frac{n(6n+4)}{2n(2n+1)}=\frac{3}{2}>1.$$

Then by Raabe's test, the series is convergent for $x=1$.
Hence, the series is convergent if $x\le1$ and divergent if

$x>1$.

EXAMPLE 3. *Test the convergence of the series*

$$\frac{a}{b}+\frac{(1+a)}{(1+b)}+\frac{(1+a)(2+a)}{(1+b)(2+b)}+...$$

SOLUTION. Here, we have

$$u_n=\frac{(1+a)(2+a)...(n-1+a)}{(1+b)(2+b)...(n-1+b)}$$

$$\Rightarrow \quad u_{n+1}=\frac{(1+a)(2+a)...(n+a)}{(1+b)(2+b)...(n+b)}$$

$$\therefore \lim_{n\to\infty}\frac{u_n}{u_{n+1}}=\lim_{n\to\infty}\left[\frac{n+b}{n+a}\right]=\lim_{n\to\infty}\left[\frac{1+\dfrac{b}{n}}{1+\dfrac{a}{n}}\right]=1.$$

Hence, the D'Alembert's ratio test fails.
Now, consider

$$\lim_{n\to\infty}n\left[\frac{u_n}{u_{n+1}}-1\right]=\lim_{n\to\infty}n\left[\frac{n+b}{n+a}-1\right]$$

$$=\lim_{n\to\infty}n\left[\frac{b-a}{n+b}\right]=\lim_{n\to\infty}\left[\frac{b-a}{1+b/n}\right]$$

$$=(b-a).$$

Then by Raabe's test the given series is convergent if $b-a>1$, i.e., $b>a+1$ and divergent if $b<a+1$.
The test fails for $b=a+1$.
Now, for $b=a+1$, the given series becomes

$$\frac{a}{a+1}+\frac{1+a}{2+a}+...=\Sigma\frac{1+a}{n+a}.$$

Taking $v_n=\dfrac{1}{n}$, by comparison test, we can easily

shown that the series is divergent.
Hence, the given series is convergent if $b>a+1$ and divergent if $b\le a+1$.

EXAMPLE 4. *Test the convergence of the series*

$$1+a+\frac{a(a+1)}{1\cdot2}+\frac{a(a+1)(a+2)}{1\cdot2\cdot3}+...$$

SOLUTION. On leaving the first term we have

$$u_n=\frac{a(a+1)(a+2)...(a+n-1)}{1.2....n}$$

$$\Rightarrow \quad u_{n+1}=\frac{a(a+1)...(a+n)}{1.2....n(n+1)}$$

$$\therefore \quad \lim_{n\to\infty}\frac{u_n}{u_{n+1}}=\lim_{n\to\infty}\frac{(n+1)}{(a+n)}=\lim_{n\to\infty}\frac{1+\dfrac{1}{n}}{1+\dfrac{a}{n}}=1.$$

$\Rightarrow$ The D' Alembert's ratio test fails.

Now $\lim_{n\to\infty}n\left[\dfrac{u_n}{u_{n+1}}-1\right]=\lim_{n\to\infty}n\left[\dfrac{n+1}{a+n}-1\right]$

$$=\lim_{n\to\infty}n\left[\frac{1-a}{a+n}\right]=\lim_{n\to\infty}\frac{(1-a)}{(1+a/n)}=(1-a).$$

Hence, by Raabe's test the given series is convergent if $1-a>1$, i.e., $a<0$ and divergent if $a>0$ and test fails if $a=0$.

In case $a = 0$, the given series becomes $1 + 0 + 0 + ...$ The sum of n terms is always 1. Therefore, the series is convergent if $a = 0$. Thus the given series Σu_n is convergent if $a \le 0$ and divergent if $a > 0$.

EXAMPLE 5. *Test the convergence of the series*

$$x + \frac{2^2 x^2}{2!} + \frac{3^3 x^3}{3!} + \frac{4^4 x^4}{4!} + ...$$

(PTU–2008, COCHIN–2005, ROHTAK–2003)

SOLUTION. Here, we have $u_n = \dfrac{n^n x^n}{n!} \Rightarrow u_{n+1} = \dfrac{(n+1)^{n+1} \cdot x^{n+1}}{(n+1)!}$

Therefore, $\lim\limits_{n \to \infty} \dfrac{u_n}{u_{n+1}} = \lim\limits_{n \to \infty} \dfrac{(n+1)! n^n x^n}{(n+1)^{n+1} x^{n+1} \cdot n!}$

$$= \lim\limits_{n \to \infty} \frac{1}{\left(1 + \dfrac{1}{n}\right)^n x} = \frac{1}{ex}.$$

Thus, by D'Alembert's ratio test the series is convergent if $ex < 1$ *i.e.*, $x < \dfrac{1}{e}$,

divergent if $x > \dfrac{1}{e}$ and the test fails if $\dfrac{1}{ex} = 1$, *i.e.*, $x = \dfrac{1}{e}$. In this case

$$\lim\limits_{n \to \infty} n \left[\log \frac{u_n}{u_{n+1}} \right] = \lim\limits_{n \to \infty} n \log \left[\frac{e}{\left(1 + \dfrac{1}{n}\right)^n} \right]$$

$$= \lim\limits_{n \to \infty} n \left[\log e - n \log \left(1 + \frac{1}{n}\right) \right]$$

$$= \lim\limits_{n \to \infty} n \left[1 - n \left(\frac{1}{n} - \frac{1}{2n^2} + \frac{1}{3n^2} - ... \right) \right]$$

$$= \lim\limits_{n \to \infty} \left[\frac{1}{2} - \frac{1}{3n} + ... \right] = \frac{1}{2} < 1.$$

Hence, by Logarithmic test, the series is divergent if $x = \dfrac{1}{e}$.

Thus the given series Σu_n is convergent if $x < \dfrac{1}{e}$ and divergent if $x \ge \dfrac{1}{e}$.

EXAMPLE 6. *Test the convergence of the series*

$$1^p + \left(\frac{1}{2}\right)^p + \left(\frac{1 \cdot 3}{2 \cdot 4}\right)^p + \left(\frac{1 \cdot 3 \cdot 5}{2 \cdot 4 \cdot 6}\right)^p + ...$$

SOLUTION. Leaving the first term 1^p, we have

$$u_n = \left[\frac{1 \cdot 3 \cdot 5 ... (2n-1)}{2 \cdot 4 \cdot 6 ... (2n)} \right]^p$$

$$\Rightarrow \quad u_{n+1} = \left[\frac{1 \cdot 3 \cdot 5 ... (2n-1)(2n+1)}{2 \cdot 4 \cdot 6 ... (2n)(2n+2)} \right]^p$$

Now $\dfrac{u_n}{u_{n+1}} = \left[\dfrac{(2n+2)}{(2n+1)} \right]^p = \left(\dfrac{1 + \dfrac{1}{n}}{1 + \dfrac{1}{2n}} \right)^p$

$$\Rightarrow \lim\limits_{n \to \infty} \frac{u_n}{u_{n+1}} = \left(\frac{1}{1} \right)^p \Rightarrow \text{ ratio test fails.}$$

Now, applying logarithmic test, we have

$$\log \frac{u_n}{u_{n+1}} = \log \left(\frac{2n+2}{2n+1} \right)^p = \log \left(\frac{1 + 1/n}{1 + 1/2n} \right)^p$$

$$= p \left[\log \left(1 + \frac{1}{n}\right) - \log \left(1 + \frac{1}{2n}\right) \right]$$

$$= p \left[\left(\frac{1}{n} - \frac{1}{2n^2} + \frac{1}{3n^3} - ... \right) \right.$$

$$\left. - \left(\frac{1}{2n} - \frac{1}{2 \cdot 2^2 n^2} + \frac{1}{3 \cdot 2^3 \cdot n^3} - ... \right) \right]$$

$$= p \left[\left\{ 1 - \frac{1}{2} \right\} \frac{1}{n} - \frac{1}{2} \cdot \left\{ 1 - \frac{1}{4} \right\} \frac{1}{n^2} + \frac{1}{3} \left\{ 1 - \frac{1}{8} \right\} \frac{1}{n^3} - ... \right]$$

$$= p \left[\frac{1}{2n} - \frac{3}{8n^2} + \frac{7}{24n^3} - ... \right]$$

$$\therefore \quad n \log \frac{u_n}{u_{n+1}} = p \left[\frac{1}{2} - \frac{3}{8n} + \frac{7}{24n^3} - ... \right]$$

Therefore $\lim\limits_{n \to \infty} \left[n \log \dfrac{u_n}{u_{n+1}} \right] = \dfrac{p}{2}$.

So that, the series is convergent if $p/2 > 1$, *i.e.*, if $p > 2$, and divergent if $p \le 2$.

Exercise-7.3

Test the convergence of the following series

1. $1 + \dfrac{2}{3}\left(\dfrac{1}{4}\right) + \dfrac{2 \cdot 4}{3 \cdot 5}\left(\dfrac{1}{6}\right) + \dfrac{2 \cdot 4 \cdot 6}{3 \cdot 5 \cdot 7}\left(\dfrac{1}{8}\right) + ...$

2. $\dfrac{1^2}{4^2} + \dfrac{1^2 \cdot 5^2}{4^2 \cdot 8^2} + \dfrac{1^2 \cdot 5^2 \cdot 9^2}{4^2 \cdot 8^2 \cdot 12^2} + \dfrac{1^2 \cdot 5^2 \cdot 9^2 \cdot 13^2}{4^2 \cdot 8^2 \cdot 12^2 \cdot 16^2} + ...$

3. $1 + \dfrac{1}{2}x + \dfrac{1 \cdot 3}{2 \cdot 4}x^2 + \dfrac{1 \cdot 3 \cdot 5}{2 \cdot 4 \cdot 6}x^3, ..., (x > 0)$ (RAIPUR–2005)

4. $x^2 + \dfrac{2^2}{3 \cdot 4}x^4 + \dfrac{2^2 \cdot 4^2}{3 \cdot 4 \cdot 5 \cdot 6}x^6 + ...$

5. $1 + \dfrac{1}{2}\dfrac{x^2}{4} + \dfrac{1 \cdot 3 \cdot 5}{2 \cdot 4 \cdot 6} \cdot \dfrac{x^4}{8} + + \dfrac{1 \cdot 3 \cdot 5 \cdot 7 \cdot 9}{2 \cdot 4 \cdot 6 \cdot 8 \cdot 10} \cdot \dfrac{x^6}{12} + ...$

6. $\sum\limits_{n=1}^{\infty} \dfrac{n!}{(n+1)^n} x^n, x > 0$ 7. $\sum\limits_{n=1}^{\infty} \left[\dfrac{1}{1 + \log n} \right]$

8. $1 + \dfrac{2}{3 \cdot 5} + \dfrac{2 \cdot 4}{3 \cdot 5 \cdot 7} + \dfrac{2 \cdot 4 \cdot 6}{3 \cdot 5 \cdot 7 \cdot 9} + ...$

9. Test the convergence of the series

$$x + x^{1 + \frac{1}{2}} + x^{1 + \frac{1}{2} + \frac{1}{3}} + x^{1 + \frac{1}{2} + \frac{1}{3} + \frac{1}{4}} + ...$$

10. Test the convergence of the following series:

   (i) $\dfrac{1^2}{2^2} + \dfrac{1^2 \cdot 3^2}{2^2 \cdot 4^2}x + \dfrac{1^2 \cdot 3^2 \cdot 5^2}{2^2 \cdot 4^2 \cdot 6^2}x^2 + ...$

   (ii) $1 + \dfrac{2^2}{3^2} + \dfrac{2^2 \cdot 4^2}{3^2 \cdot 5^2} + \dfrac{2^2 \cdot 4^2 \cdot 6^2}{3^2 \cdot 5^2 \cdot 7^2} + ...$

11. Test for convergence, the following series :

   (i) $1 + \dfrac{x}{1} + \dfrac{1}{2} \cdot \dfrac{x^3}{3} + \dfrac{1 \cdot 3}{2 \cdot 4} \cdot \dfrac{x^5}{5} + \dfrac{1 \cdot 3 \cdot 5}{2 \cdot 4 \cdot 6} \cdot \dfrac{x^7}{7} + ...$

   (ii) $\dfrac{x}{1} + \dfrac{1}{2} \dfrac{x^2}{3} + \dfrac{1 \cdot 3}{2 \cdot 4} \cdot \dfrac{x^3}{5} + \dfrac{1 \cdot 3 \cdot 5}{2 \cdot 4 \cdot 6} \cdot \dfrac{x^4}{7} + ... (x > 0)$

(iii) $\displaystyle\sum_{n=1}^{\infty} \frac{1.3.5....(4n-5)(4n-3)}{2.4.6....(4n-4)(4n-2)} \frac{x^{2n}}{4n}, x > 0$

(iv) $\displaystyle\sum_{n=1}^{\infty} \frac{2.4.6....2n}{1.3.5...(2n+1)}$

12. Test for convergence, the following series :

(i) $1 + \dfrac{x}{1!} + \dfrac{2^2 x^2}{2!} + \dfrac{3^3 x^3}{3!} + ...$ for $x > 0$

(ii) $\dfrac{1}{2}x + \dfrac{1.3}{2.4}x^2 + \dfrac{1.3.5}{2.4.6}x^3 + ..., x > 0$

(iii) $1 + \dfrac{2!}{2^2}x + \dfrac{3!}{3^3}x^2 + ..., x > 0$

13. Test for convergence, the following series :

(i) $1 + \dfrac{a(1-a)}{1^2} + \dfrac{(1+a)a(1-a)(2-a)}{1^2 . 2^2} +$

$\qquad + \dfrac{(2+a)(1+a)a(1-a)(2-a)(3-a)}{1^2 . 2^2 . 3^2} + ...$

(ii) $\dfrac{(1+a)(1+b)}{1.2.3} + \dfrac{(2+a)(2+b)}{2.3.4} + \dfrac{(3+a)(3+b)}{3.4.5.} + ...$

14. Test for convergence the following series :

(i) $1 + \dfrac{\alpha}{1.\beta}x + \dfrac{\alpha(\alpha+1)^2}{1.2\beta(\beta+1)^2}x^2 + \dfrac{\alpha(\alpha+1)^2(\alpha+2)^2}{1.2.3\beta(\beta+1)(\beta+2)}x^3 + ...$

(ii) $1 + \dfrac{\alpha.\beta}{1.\gamma}x + \dfrac{\alpha(\alpha+1)\beta(\beta+1)}{1.2.\gamma(\gamma+1)}x^2$

$\qquad + \dfrac{\alpha(\alpha+1)(\alpha+2)\beta(\beta+1)(\beta+2)}{1.2.3.\gamma(\gamma+1)(\gamma+2)}x^3 + ...$ (KURUKSHETRA–2005)

15. Test for convergence the following series :

$\dfrac{a}{a+3} + \dfrac{a(a+2)}{(a+3)(a+5)}x + \dfrac{a(a+2)(a+4)}{(a+3)(a+5)(a+7)}x^2 + ...$

16. Test for convergence the following series :

$\left(\dfrac{1}{2.4}\right)^{2/3} + \left(\dfrac{1.3}{2.4.6}\right)^{2/3} + \left(\dfrac{1.3.5}{2.4.6.8}\right)^{2/3} + ...$

17. Test for convergence the following series :

(i) $\displaystyle\sum_{n=1}^{\infty} \frac{1 \cdot 3 \cdot 5....(2n-1)}{2 \cdot 4 \cdot 6....2n} \cdot \frac{1}{n}$

(ii) $\displaystyle\sum_{n=1}^{\infty} \frac{4 \cdot 7 \cdot 10....(3n+1)}{1 \cdot 2 \cdot 3....n}x^n$ (VTU–2009, PTU–2006S)

(iii) $\displaystyle\sum_{n=1}^{\infty} \frac{3 \cdot 6 \cdot 9....(3n)}{7 \cdot 10 \cdot 13....(3n+4)}x^n, x > 0$

(iv) $\displaystyle\sum_{n=1}^{\infty} \frac{(2n)!}{(n!)^2}x^n, x > 0$

18. Test for convergence the following series :

$\dfrac{1^2}{2^2} + \dfrac{1^2 . 3^2}{2^2 . 4^2} + \dfrac{1^2 . 3^2 . 5^2}{2^2 . 4^2 . 6^2} + ...$

19. Test for convergence the following series :

(i) $\dfrac{1}{(\log 2)^p} + \dfrac{1}{(\log 3)^p} + ... + \dfrac{1}{(\log n)^p} + ...$

(ii) $x^2(\log 2)^p + x^3(\log 3)^p + x^4(\log 4)^p + ...$

20. Test for convergence the following series :

(i) $\dfrac{x}{1.2} + \dfrac{x^2}{3.4} + \dfrac{x^3}{5.6} + \dfrac{x^4}{7.8} + ...\infty \ (x > 0)$ (MUMBAI–2009)

(ii) $\dfrac{x}{1.2} + \dfrac{x^2}{2.3} + \dfrac{x^3}{3.4} + \dfrac{x^4}{4.5} + ...\infty$ (VTU–2008, JNTU–2003)

(iii) $1 + \dfrac{2}{3}x + \dfrac{2.3}{3.5}x^2 + \dfrac{2.3.4}{3.5.7}x^3 + ...\infty$ (VTU–2009S)

(iv) $\dfrac{x}{1} + \dfrac{1}{2}\dfrac{x^3}{3} + \dfrac{1.3}{2.4}\dfrac{x^5}{5} + \dfrac{1.3.5}{2.4.6}\dfrac{x^7}{7} + ...\infty \ (x > 0)$

(VTU–2007, RAIPUR–2005)

(v) $1 + \dfrac{1}{2}\dfrac{x^2}{4} + \dfrac{1.3.5}{2.4.6}\dfrac{x^4}{8} + \dfrac{1.3.5.7.9}{2.4.6.8.10}\dfrac{x^6}{12} + ...\infty$

(ROHTAK–2006S, ROORKEE–2000)

(vi) $\dfrac{1}{1^2} + \dfrac{1+2}{1^2 + 2^2} + \dfrac{1+2+3}{1^2 + 2^2 + 3^2} + ...$ (VTU–2000)

Answers

1. Convergent   **2.** Convergent   **3.** $\begin{cases} \text{Convergent if } x < 1, \\ \text{Divergent if } x \geq 1 \end{cases}$ **4.** Convergent if $x^2 \leq 1$, divergnet if $x^2 > 1$

5. Convergent if $x \leq 1$, divergent if $x > 1$ **6.** Convergent if $x < e$, divergent if $x \geq e$ **7.** Convergent   **8.** Convergent

9. Convergent if $x < \dfrac{1}{e}$, divergent if $x \geq \dfrac{1}{e}$ **10.** (i) Convergent if $x < 1$, divergent if $x \geq 1$ (ii) Divergent

11. (i) Convergent if $x^2 \leq 1$, divergent if $x^2 > 1$ (ii) Convergent if $0 < x \leq 1$, divergent if $x > 1$
(iii) Convergent if $x \leq 1$, divergent if $x > 1$ (iv) Divergent   **12.** (i) Convergent if $x < \dfrac{1}{e}$, divergent if $x \geq \dfrac{1}{e}$

(ii) Convergent if $x < 1$, divergent if $x \geq 1$,   (iii) Convergent if $x < e$, divergent if $x \geq e$ **13.** (i) Divergent (ii) Divergent

14. (i) Convergent if $x < 1$, divergent if $x > 1$, When $x = 1$, then convergent if $\beta > 2\alpha$, divergent if $\beta \leq 2\alpha$

(ii) Convergent if $x < 1$, divergent if $x > 1$, When $x = 1$, then convergent if $\gamma > \alpha + \beta$, divergent if $\gamma \leq \alpha + \beta$.

15. Convergent if $x \leq 1$, divergent if $x > 1$ **16.** Divergent   **17.** (i) Convergent (ii) Convergent if $x < \dfrac{1}{3}$, divergent if $x \geq \dfrac{1}{3}$

(iii) Convergent if $x \leq 1$, divergentif $x > 1$ (iv) Convergent if $x < \dfrac{1}{4}$, divergent if $x \geq \dfrac{1}{4}$.   **18.** Divergent

19. (i) Divergent for all values of p,   (ii) Convergent if $x < 1$, divergent if $x \geq 1$

20. (i) Convergent for $x \leq 1$; divergent for $x > 1$ (ii) Convergent for $x \leq 1$; divergent for $x > 1$
(iii) Convergent for $x < 2$; divergent for $x \geq 2$ (iv) Convergent for $x \leq 1$; divergent for $x > 1$
(v) Convergent for $x^2 \leq 1$; divergent for $x^2 > 1$ (vi) Diverges

7.9 GAUSS'S TEST

If Σu_n be a series of positive terms such that $\dfrac{u_n}{u_{n+1}} = \alpha + \dfrac{\beta}{n} + \dfrac{\gamma_n}{n^p}$,

where $\alpha > 0, p > 1$ and $<\gamma_n>$ is a bounded sequence. Then

(i) Σu_n converges for $\alpha > 1$, diverges for $\alpha < 1$, whatever β may be.

(ii) If $\alpha = 1$, Σu_n converges whenever $\beta > 1$, and diverges whenever $\beta \leq 1$.

Proof. We have $\qquad\qquad \lim\limits_{n\to\infty} \dfrac{u_n}{u_{n+1}} = \alpha.$

Then by D'Alembert's ratio test Σu_n is convergent if $\alpha > 1$ and divergent if $\alpha < 1$.

For $\alpha = 1$, we have $\qquad n\left[\dfrac{u_n}{u_{n+1}} - 1\right] = \beta + \dfrac{\gamma_n}{n^{p-1}},$

where $p > 1$ and $< \gamma_n >$ is a bounded sequence.

$\therefore \qquad\qquad\qquad \lim\limits_{n\to\infty} n\left[\dfrac{u_n}{u_{n+1}} - 1\right] = \beta.$

Then, by Raabe's test Σu_n is convergent if $\beta > 1$ and divergent if $\beta < 1$.

Now for $\alpha = \beta = 1$, we compare the series with the divergent series Σv_n where $v_n = \dfrac{1}{n \log n}$.

Now, consider $\quad \dfrac{u_n}{u_{n+1}} - \dfrac{v_n}{v_{n+1}} = 1 + \dfrac{1}{n} + \dfrac{\gamma_n}{n^p} - \dfrac{(n+1)\log(n+1)}{n\log n} = \dfrac{\gamma_n}{n^p} - \dfrac{(n+1)}{n}\left[\dfrac{\log(n+1)}{\log n} - 1\right]$

$$= \dfrac{1}{n^p}\left[\gamma_n - (n+1)\log\left(1 + \dfrac{1}{n}\right)\cdot\dfrac{n^{p-1}}{\log n}\right].$$

But $\quad \lim\limits_{n\to\infty}(n+1)\log\left(1+\dfrac{1}{n}\right) = \lim\limits_{n\to\infty}\left[\log\left(1+\dfrac{1}{n}\right)^n + \log\left(1+\dfrac{1}{n}\right)\right]$

Also, $\qquad \lim\limits_{n\to\infty}\dfrac{n^{p-1}}{\log n} = \infty, p > 1$ and $<\gamma_n>$ is bounded.

Therefore, for large value of n, $\gamma_n - (n+1)\log\left(1+\dfrac{1}{n}\right)\dfrac{n^{p-1}}{\log n}$ remains negative.

$\therefore \qquad\qquad \dfrac{u_n}{u_{n+1}} - \dfrac{v_n}{v_{n+1}} < \qquad$ or $\qquad \dfrac{u_n}{u_{n+1}} < \dfrac{v_n}{v_{n+1}}.$

Now, since Σv_n is divergent, by comparison test Σu_n is divergent.

Hence, the series Σu_n is convergent if $\alpha > 1$ or $\alpha = 1$ and $\beta > 1$ and divergent if $\alpha < 1$ or $\alpha = 1$ and $\beta \leq 1$.

7.10 CAUCHY'S INTEGRAL TEST

Let $f(x)$ be non-negative monotonically decreasing integrable function on $[1, \infty[$ then the series $\sum\limits_{n=1}^{\infty} f(n)$ and the improper integral $\int_1^{\infty} f(x)\,dx$ converge or diverge together.

7.11 CAUCHY'S CONDENSATION TEST

If $f(n)$ is a monotonically decreasing function of n for all $n \in N$ such that each $f(n)$ is positive,

then two infinite series $\sum\limits_{n=1}^{\infty} f(n)$ and $\sum\limits_{n=1}^{\infty} a^n f(a^n)$ converge or diverge together, where a is a positive integer greater than unity.

7.12 REARRANGEMENT OF TERMS

A series Σv_n is said to be rearrangement of a series Σu_n if there exists one-one correspondence between the terms of the two series and if v_n corresponds to u_n then $v_n = u_n$.

In other words, we can say that a series Σu_n is said to be rearrangement of a series Σv_n if every term of Σu_n is a term of Σv_n and *vice-versa*.

7.12.1 ALTERNATING SERIES

A series, whose terms are alternatively positive and negative is called an alternating series.

Thus, a series of the form $u_1 - u_2 + u_3 - u_4 + \ldots + (-1)^{n-1} u_n + \ldots$ where $u_n > 0 \ \forall \ n$, is an alternating series.

7.13 DIFFERENT TYPES OF CONVERGENCE

(1) ABSOLUTE CONVERGENCE

A series Σu_n is said to be absolutely convergent if the series $\Sigma |u_n|$ is convergent.

(2) UNCONDITIONALLY CONVERGENT SERIES

A series Σu_n is said to be unconditionally convergent if every rearrangement converge to the same sum Σu_n, i.e, Σu_n is conditionally convergent iff it is absolutely convergent.

(3) CONDITIONAL CONVERGENCE

A series Σu_n is said to be conditionally convergent if Σu_n is convergent but $\Sigma |u_n|$ is divergent.

☞ REMARK

- The conditional convergence of a series is also known as semi-convergent or non-absolutely convergent.

▪ ILLUSTRATIONS

(1) The series $\Sigma u_n = 1 - \dfrac{1}{2} + \dfrac{1}{2^2} - \dfrac{1}{2^3} + ...$ is absolutely convergent.

(2) The series $\dfrac{1}{1^2} - \dfrac{1}{2^2} + \dfrac{1}{3^2} - \dfrac{1}{4^2} + ...$ is absolutely convergent.

THEOREM 1. *An absolutely convergent series is convergent.*

PROOF. Let us suppose, the series Σu_n is absolutely convergent. Then by definition $\Sigma |u_n|$ is convergent.

Now $\qquad u_n + |u_n| = \begin{cases} 2u_n, & \text{if } u_n \text{ is positive} \\ 0, & \text{if } u_n \text{ is negative.} \end{cases}$

Therefore, every term of the series $\Sigma(u_n + |u_n|)$ is ≥ 0 and less than equal to the corresponding term of the convergent series $\Sigma 2|u_n|$.

Hence, $\Sigma(u_n + |u_n|)$ is convergent. Hence Σu_n is convergent.

☞ REMARKS

- The converse of the above theorem is not necessarily true :

  For example : The series $\Sigma u_n = 1 - \dfrac{1}{2} + \dfrac{1}{3} - ....$ is convergent, but the series $\Sigma |u_n| = 1 + \dfrac{1}{2} + \dfrac{1}{3} + ...$ is divergent. Hence a convergent series need not be absolutely convergent.

- The usefulness of absolute convergence is partly due to the fact that it is often easier to establish absolute convergence than convergence :

  For example : Consider the series $\Sigma \dfrac{a^n}{2^n}$, where $a_n = 1$ if n is prime number and $a_n = -1$ otherwise. Here, $\Sigma |a_n| = \Sigma \dfrac{1}{2^n}$ is convergent. Accordingly $\Sigma a_n/2^n$ is absolutely convergent, and hence convergent.

THEOREM 2. If the terms of a convergent series of positive terms are rearranged, the series remains convergent and its sum is unaltered.

PROOF. Let us suppose Σu_n be a convergent series, and let the terms be rearranged in any manner. Denote the new series by Σv_n, so that every u is a v and every v is a u.

Let $\quad s_n = u_1 + u_2 + ... + u_n \quad$ and $\qquad t_n = v_1 + v_2 + ... + v_n$.

Then, for any definite value of n, s_n contains n terms each of which occurs, sooner or later, in the v series and so we can find a corresponding m such that t_m contains all the terms of s_n (and possibly other not contained in s_n).

Now, since each term is positive, therefore $s_n \leq t_m$.

Also, suppose that the first m terms of Σv_n are among the first $(n+p)$ terms of Σu_n. Therefore,

$$s_n \leq t_m \leq s_{n+p}.$$

and m tends to infinity with n.

Let Σu_n converges to s, so that $\lim s_n = \lim s_{n+p} = s$

$\therefore \qquad\qquad \lim t_m = s.$

Hence, Σv_n is convergent and has the same sum as Σu_n.

☞ REMARK

- The arrangement fails for a dearrangement such as $u_1 + u_3 + u_5 + ... + u_2 + u_4 + u_6 + ...$ where Σu_n is broken up into two (or any finite no. of) infinite series.

  Here, we cannot find an m so that the first n terms of Σu_n occur among the first m terms of Σv_n.

  For instance, u_2 does not occur even if infinitely many of the terms $u_1, u_3, u_5, ...$ have been placed.

THEOREM 3. **(Dirichlet's Theorem).** If the terms of an absolutely convergent series are rearranged, the series remains convergent and its sum is unaltered.

PROOF. Let Σu_n be an absolutely convergent series, and let its terms be rearranged in a different order. Let, the new series be denoted by Σv_n so that every v occurs somewhere in the u series and every u occurs somewhere in the v series.

Now, we have $u_n + |u_n| = 2u_n$ or 0 according as u_n is positive or negative. Now $\Sigma |u_n|$ is a convergent series of positive terms, so also in the series $\Sigma(u_n + |u_n|)$, because its terms are less than equal to be corresponding terms of the series $\Sigma 2|u_n|$.

Let $\Sigma|u_n| = s$ and $\Sigma(u_n + |u_n|) = s'$ so that $\Sigma u_n = s' - s$.

Also, since $\Sigma|u_n|$ and $\Sigma(u_n + |u_n|)$ are convergent series of positive terms, their sum remains unchanged by any rearrangement of term (By theorem 2).

Accordingly, $\Sigma|v_n| = s$ and $\Sigma(v_n + |v_n|) = s'$.

Hence, $\Sigma v_n = s' - s = \Sigma u_n$.

REMARKS
- If we rearrange the order of terms of a semi-convergent series, we may or may not changed the sum of the series.
- The sum will be changed if we interfere too much with the balance between positive and negative terms.
- By a suitable rearrangement of the terms a semi-convergent series may be made to diverge. The reason is that in a semi-convergent series the positive and negative terms taken separately from two divergent series.

THEOREM 4. **(Riemann's Rearrangement theorem).** By a suitable rearrangement of terms of a conditionally convergent series can be made to converge to any number λ or to diverge to ∞ or $-\infty$ even to oscillate.

In other words, this theorem can be stated as follows

To a given conditionally convergent series and to any given number there corresponds a rearrangement of the given series which is convergent and whose sum is the given number.

THEOREM 5. **(Pringsheim theorem).** *Let $f(x)$ be a sequence of positive terms which* monotonically converges to zero and let the series

$$\sum_{n=1}^{\infty} (-1)^{n-1} f(x)$$ be rearranged so that in the first $p+n$ terms there are p-positive and n negative terms, *i.e.,* $\lim_{n \to \infty} n\, f(x) = \lambda$

and $\lim_{n \to \infty} \dfrac{p}{n} = k$ then the sum of the series is increased by $\dfrac{1}{2}\lambda \log k$.

7.14 LEIBNITZ'S TEST

If the alternative series $u_1 - u_2 + u_3 - ...(u_n > 0, \forall\, n \in N)$ is such that

(i) $u_{n+1} \le u_n$, $\forall\, n \in N$ (ii) $\lim_{n \to \infty} u_n = 0$

Then the series converges.

Proof. Let $s_n = u_1 - u_2 + u_3 - ... + (-1)^{n-1} u_n$ so that $<s_n>$ is a sequence of partial sums of the given series.

Now for all n

$$s_{2n+2} - s_{2n} = u_{2n+1} - u_{2n+2} \ge 0 \qquad \text{[By (i)]}$$

which gives that $<s_{2n}>$ is a monotonically increasing sequence.

Further, $s_{2n} = u_1 - u_2 + u_3 - .... u_{2n-1} - u_{2n} = u_1 - (u_2 - u_3) - (u_4 - u_5) - ... - u_{2n}$

$$= u_1 - [(u_2 - u_3) + ... + u_{2n}] = u_1 - \text{some positive number} \le u_1.$$

Therefore, the monotonically increasing sequence $<s_{2n}>$ is bounded above and consequently it is convergent.

Let $\lim_{n \to \infty} s_{2n} = s$.

Now $s_{2n+1} = s_{2n} + u_{2n+1}$ $\Rightarrow$ $\lim_{n \to \infty} s_{2n+1} = \lim_{n \to \infty} s_{2n} + \lim_{n \to \infty} u_{2n+1}$ $\left[\because \lim_{n \to \infty} u_n = 0 \right]$

$$= s + 0 = s$$

Thus, the subsequences $<s_{2n}>$ and $<s_{2n+1}>$ both converge to the same limit. Now we shall show that the sequence $<s_n>$ also converges to s.

Let $\varepsilon > 0$ be given. Since, the sequences $<s_{2n}>$ *and* $<s_{2n+1}>$ both converges to s, there exists positive integers m_1, m_2 such that

$$|s_{2n} - s| < \varepsilon \,\forall\, n \ge m_1,$$

and $|s_{2n+1} - s| < \varepsilon \,\forall\, n \ge m_2.$

Let $m = \max \{m_1, m_2\}.$

Then $|s_n - s| < \varepsilon \,\forall\, n \ge m$

which gives that the sequence $<s_n>$ converges to s.

Hence, the given series $\Sigma(-1)^{n-1} u_n$ converges.

REMARKS
- This test gives us a set of sufficient conditions for the convergence of an alternating series.
- If the test does not show a series to be convergent, we may not immediately say that the series is divergent.

Solved Examples

EXAMPLE 1. *Show that* $\lim_{n \to \infty}\left[1 + \dfrac{1}{2} + ... + \dfrac{1}{n} - \log n\right]$ *exists.*

SOLUTION. Let $f(x) = \dfrac{1}{x}, x \in [1, \infty[$.

Then $f(x) > 0$ and monotonically decreasing on $[1, \infty[$.

Let $S_n = f(1) + f(2) + ... + f(n)$

$$= 1 + \frac{1}{2} + \frac{1}{3} + ... + \frac{1}{n}$$

and $I_n = \int_1^n f(x)\,dx = \int_1^n \dfrac{1}{x}\,dx = [\log x]_1^n = \log n$.

It can be easily shown that

$$f(n) \le S_n - I_n \le f(1) \; \forall \, n \in \mathbb{N}$$

or $\quad 0 < \dfrac{1}{n} \le S_n - I_n \le 1 \; \forall \, n \in \mathbb{N}$

which gives that the sequence $<u_n>$, where $u_n = S_n - I_n$, is bounded below.

Now, it can also be shown easily that the sequence $<u_n>$ is a monotonically decreasing. Therefore it converges.

Hence, $\lim_{n \to \infty}\left(1 + \dfrac{1}{2} + ... + \dfrac{1}{n} - \log n\right)$ exist.

REMARK

- The limit of the above sequence is called Euler's constant and is denoted by γ.

EXAMPLE 2. *Show by integral test that* $\Sigma \dfrac{1}{n^p}$ *converges if* $p > 1$ *and diverges if* $p \le 1$.

SOLUTION. Let $f(x) = \dfrac{1}{x^p}, p > 0$. Then $f(x)$ is positive valued and monotonically decreasing.

Therefore by Cauchy's integral test $\Sigma \dfrac{1}{n^p}$ and $\int_1^\infty f(x)\,dx$ converges and diverges together.

Let $I_n = \int_1^n \dfrac{1}{x^p}\,dx = \int_1^n x^{-p}\,dx$

$$= \begin{cases} \left(\dfrac{n^{1-p}}{1-p} - \dfrac{1}{1-p}\right), & \text{if } p \ne 1 \\[2mm] \log n, & \text{if } p = 1. \end{cases}$$

If $n \to \infty$, $n^{1-p} = \dfrac{1}{n^{p-1}} \to 0$ as $p > 1$ and tends to ∞ if $p < 1$ and $\log n \to \infty$

$\therefore \quad \lim_{n \to \infty} I_n = -\dfrac{1}{1-p} = \dfrac{1}{p-1}$, if $p > 1$

and $\lim_{n \to \infty} I_n = \infty$, if $p \le 1$.

Hence, $\int_1^\infty f(x)\,dx$ converges if $p > 1$ and diverges if $p \le 1$. Then by Cauchy's integral test the series $\Sigma \dfrac{1}{n^p}$ is convergent if $p > 1$ and divergent if $p \le 1$.

EXAMPLE 3. *Show by Cauchy's integral test that the series* $\sum_{n=2}^{\infty} \dfrac{1}{n(\log n)^p}$ *converges if* $p > 1$ *and diverges if* $0 < p \le 1$.

(PTU–2010)

SOLUTION. Let us suppose $f(x) = \dfrac{1}{x(\log x)^p}, p > 0$

and $x \in [2, \infty[$; then obviously $f(x)$ is monotoni-cally decreasing in $[2, \infty[$ and positive valued.

Let $I_n = \int_2^n \dfrac{dx}{x(\log x)^p}$

Then $I_n = \left[\dfrac{(\log x)^{1-p}}{1-p}\right]_2^n, p \ne 1$

$$= \frac{1}{(1-p)}[(\log n)^{1-p} - (\log 2)^{1-p}], p \ne 1$$

and $I_n = [\log \log x]_2^n, p = 1$

$$= [\log \log n - \log \log 2], p = 1.$$

Therefore, we have

$$\lim_{n \to \infty} I_n = \lim_{n \to \infty} \int_2^n f(x)\,dx = \infty, \text{ if } p < 1$$

and $\lim_{n \to \infty} I_n = -\dfrac{1}{(1-p)}(\log 2)^{1-p}$, if $p > 1$.

Thus the integral $\int_2^\infty f(x)\,dx$ converges if $p > 1$ and diverges if $0 < p \le 1$.

Hence, by Cauchy's integral test, the series

$$\sum_{n=2}^{\infty} f(x) = \sum_{n=2}^{\infty} \frac{1}{n(\log n)^p}$$

converges if $p > 1$ and diverges if $0 < p \le 1$.

EXAMPLE 4. *Apply the Cauchy's condensation test to discuss the convergence of the series*

$$\sum_{n=2}^{\infty} \frac{1}{(n \log n)(\log \log n)^p}.$$

SOLUTION. Here, we have

$$f(n) = \frac{1}{(n \log n)(\log \log n)^p}$$

$\therefore \quad a^n f(a^n) = \dfrac{a^n}{(a^n \log a^n)(\log \log a^n)^p}$

$$= \frac{1}{(n \log a)[\log(n \log a)]^p}$$

Since, a is a positive integer greater than 1 and can be chosen that $\log a > 1$ so that $n \log a > n$.

Then $a^n f(a^n) < \dfrac{1}{(n \log a)(\log n)^p}$

Since, the series $\dfrac{1}{\log a} \Sigma \dfrac{1}{n(\log n)^p}$ is convergent when $p > 1$, therefore $\Sigma a^n f(a^n)$ is also convergent and consequently the given series is convergent when $p > 1$.

Now let $p \le 1$. If we take $a = 2$, then $\log_e a < 1$ so that $n \log_e a < n$

$\therefore \quad a^n f(a^n) > \dfrac{1}{(n \log a)(\log n)^p}$

But the series $\dfrac{1}{\log a} \Sigma \dfrac{1}{n(\log n)^p}$ is divergent when $p \le 1$

and therefore $\Sigma a^n f(a^n)$

is also divergent. Then by Cauchy condensation test, the

given series is divergent when $p \leq 1$.

EXAMPLE 5. *Test the convergence of the series*

$$\frac{1^2}{2^2} + \frac{1^2 \cdot 3^2}{2^2 \cdot 4^2} + \frac{1^2 \cdot 3^2 \cdot 5^2}{2^2 \cdot 4^2 \cdot 6^2} + \dots$$

SOLUTION. Here, we have $u_n = \dfrac{1^2 \cdot 3^2 \cdot 5^2 \dots (2n-1)^2}{2^2 \cdot 4^2 \cdot 6^2 \dots (2n)^2}$

$\therefore \qquad u_{n+1} = \dfrac{1^2 \cdot 3^2 \dots (2n-1)^2 (2n+1)^2}{2^2 \cdot 4^2 \dots (2n)^2 (2n+2)^2}$

$\therefore \qquad \dfrac{u_n}{u_{n+1}} = \dfrac{(2n+2)^2}{(2n+1)^2}$

$\Rightarrow \quad \lim\limits_{n \to \infty} \dfrac{u_n}{u_{n+1}} = \lim\limits_{n \to \infty} \dfrac{\left(2 + \dfrac{2}{n}\right)^2}{\left(2 + \dfrac{1}{n}\right)^2} = 1$

which gives that, the ratio test is fail.
Now, we can easily see that

$$\lim\limits_{n \to \infty} n\left[\dfrac{u_n}{u_{n+1}} - 1\right] = 1.$$

$\Rightarrow$ Raabe's test also fails.
Now applying Gauss test,
Consider

$\dfrac{u_n}{u_{n+1}} = \dfrac{(2n+2)^2}{(2n+1)^2} = \left(1 + \dfrac{1}{n}\right)^2 \left(1 + \dfrac{1}{2n}\right)^{-2}$

$= \left(1 + \dfrac{2}{n} + \dfrac{1}{n^2}\right)\left(1 - 2 \cdot \dfrac{1}{2n} + 3 \cdot \dfrac{1}{4n^2} \dots\right)$

$= 1 + \dfrac{1}{n} - \dfrac{1}{4n^2} + \dots$

$= \alpha + \dfrac{\beta}{n} + \dfrac{\gamma_n}{n^2}$, where $\gamma_n \to -\dfrac{1}{4}$ as $n \to \infty$.

Here, $\alpha = 1$, $\beta = 1$. Therefore by Gauss test the series Σu_n is divergent.

EXAMPLE 6. *Test the convergence of the series*

$$\frac{\log 2}{2^2} - \frac{\log 3}{3^2} + \frac{\log 4}{4^2} - \dots$$

SOLUTION. The given series is an alternating series
Here, the nth term

$$t_n = (-1)^n u_n, \text{ where } u_n = \frac{\log(n+1)}{(n+1)^2} > 0$$

$\lim\limits_{n \to \infty} u_n = \lim\limits_{n \to \infty} \dfrac{\log(n+1)}{(n+1)^2}$

$= \lim\limits_{n \to \infty} \dfrac{\log(n+1)}{(n+1)} \cdot \dfrac{1}{(n+1)} = 0.$

Now, we shall show that $u_{n+1} \leq u_n \; \forall \; n$.

Let $\quad f(x) = \dfrac{\log x}{x^2}, x > 0$

Then $\quad f'(x) = \dfrac{x^2 \cdot \dfrac{1}{x} - 2x \log x}{x^4}$

$= \dfrac{1 - 2\log x}{x^3} < 0 \text{ when } x > e^{1/2}.$

Therefore, the function $f(x)$ is monotonically decreasing for all $x > e^{1/2}$. We know that

$$2 < e < 3 \Rightarrow 2^{1/2} < e^{1/2} < 3^{1/2}$$

so $\qquad f(n+2) \leq f(n+1)$ for all n.

i.e, $\qquad u_{n+1} \leq u_n \; \forall \; n.$

Hence, by Leibnitz test the given series is convergent.

📑 **REMARK**

- Since for a convergent series $\sum\limits_{n=1}^{\infty} u_n$, $\lim\limits_{n \to \infty} u_n = 0$. Therefore, $\lim\limits_{n \to \infty} \dfrac{x^n}{n!} = 0$ is a useful result.

7.15 MORE ABOUT CONDITIONAL AND ABSOLUTE CONVERGENCE

(i) If Σu_n is an absolute convergent series, then the series of its positive and the series of its negative terms are both convergent.

(ii) The divergence of $\Sigma |u_n|$ does not imply the divergence of Σu_n. For example, if $u_n = \dfrac{(-1)^{n-1}}{n}$ then $\Sigma |u_n|$ is divergent, whereas Σu_n is convergent.

(iii) Since, the series $\Sigma |u_n|$ is of positive terms, therefore, all the tests established for testing the convergence of series of positive terms, will also be the tests for determining the absolute convergence of the series Σu_n.

(iv) If Σu_n is conditionally convergent, then the series of its positive terms and the series of its negative terms are both divergent.

(v) A series with mixed signs cannot converge, if the series of its positive terms is convergent (divergent) and the series of its negative terms is divergent (convergent).

7.16 SUMMARY OF THE TESTS

For the guidance of the students we given below a working procedure for determining the convergence of a series.

(i) If in a series of positive terms, n^{th} term does not tend to zero, the series is divergent.

(ii) If n^{th} terms tends to zero, then a comparison test may be applied when its nth term neither involves any power of n nor involve factorials.

(iii) If the nth term is the nth power of some expression, then Cauchy's root test may be applied.

(iv) When the series involves increasing power of x or involves factorials, one should start with the ratio test.

(v) If the $\lim\limits_{n\to\infty}\dfrac{u_n}{u_{n+1}}$ turns out to be 1, then the ratio test fails and Raabe's test or Gauss's test is applied provided $\dfrac{u_n}{u_{n+1}}$ does not involves e, otherwise logarithmic test is applied.

(vi) For an arbitrary terms series, try with the ratio test for absolute convergence. If the limit turns out to be 1, then try some other tests. When the terms have alternating signs, then Leibnitz's test is suggested.

Exercise-7.4

1. Test the convergence of the following series $1-\dfrac{1}{2}+\dfrac{1}{3}-\dfrac{1}{4}+...$
(PTU–2009)

2. Prove that the following series is absolute convergent.

$$\left(\dfrac{\sqrt{2}-1}{1}\right)-\left(\dfrac{\sqrt{3}-\sqrt{2}}{2}\right)+\left(\dfrac{\sqrt{4}-\sqrt{3}}{3}\right)-...$$

3. Show that the series $\Sigma\dfrac{\sin n\theta}{n^2}$ is absolutely convergent.

4. Show that the series $\dfrac{1^2}{4^2}+\dfrac{1^2.5^2}{4^2.8^2}+\dfrac{1^2.5^2.9^2}{4^2.8^2.12^2}+...$ is convergent.

5. Examine the convergence of the series $1+a+b^2+a^3+b^4+...$

6. Test the convergence of the series $\dfrac{x}{1}+\dfrac{1}{2}.\dfrac{x^2}{3}+\dfrac{1.3}{2.4}.\dfrac{x^3}{5}+...$

7. Show that the series $\Sigma(-1)^{n-1}\sin\dfrac{1}{n}$ is conditionally convergent.

8. Test for convergence the series $\Sigma\left(\dfrac{n^{n-1}.x^{n-1}}{n!}\right)$.

9. Test the convergence of the series

$$1+\dfrac{a(1-a)}{1^2}+\dfrac{(1+a)a(1-a)(2-a)}{1^2.2^2}+...$$

10. Show that the series $\dfrac{2}{1^2}-\dfrac{3}{2^2}+\dfrac{4}{3^2}-\dfrac{5}{4^2}+...$ converge conditionally.

11. Show that the series $\dfrac{1}{x+1}-\dfrac{1}{x+2}+\dfrac{1}{x+3}-...$ is convergent except when x is a negative integer.

12. Show that the series $\Sigma(-1)^n[\sqrt{n^2+1}-n]$ is conditionally convergent.

13. Show that the series $\sum\limits_{n=1}^{\infty}\dfrac{(-1)^{n+1}.n}{n^2+1}$ is not absolutely convergent.

14. Show that the binomial series

$$1+nx+\dfrac{n(n-1)}{2!}x^2+...\dfrac{n(n-1)...(n-r+1)}{r!}x^r+...$$

is absolutely convergent when $|x|<1$.

15. Test the convergence of the series $\sum\limits_{n=1}^{\infty}\left[\dfrac{1}{n}+\dfrac{(-1)^{n+1}}{\sqrt{n}}\right]$.

16. Show that the series

$$x+\dfrac{a-b}{2!}x^2+\dfrac{(a-b)(a-2b)}{3!}x^3+\dfrac{(a-b)(a-2b)(a-3b)}{4!}x^4+...$$

is absolutely convergent if $|x|<\dfrac{1}{|b|}$.

17. Show that the series

$$1-\dfrac{1}{2^3}-\dfrac{1}{4^3}+\dfrac{1}{3^3}-\dfrac{1}{6^3}-\dfrac{1}{8^3}+....+\dfrac{1}{(2n-1)^3}-\dfrac{1}{(4n-2)^3}-\dfrac{1}{(4n)^3}+...$$

is absolutely convergent.

18. Show that the series $2\sin\dfrac{x}{3}+4\sin\dfrac{x}{9}+8\sin\dfrac{x}{27}+...$

converges absolutely for all finite values of x.

19. Discuss the convergence of the series

$$x^2(\log 2)^q+x^3(\log 3)^q+x^4(\log 4)^q+...$$

20. Discuss the convergence of the following series :

(i) $\dfrac{1}{\log 2}-\dfrac{1}{\log 3}+\dfrac{1}{\log 4}-\dfrac{1}{\log 5}+...$ (PTU–2010)

(ii) $1-\dfrac{1}{5}+\dfrac{1}{9}-\dfrac{1}{13}+...\infty$ (VTU–2010)

(iii) $\sum\limits_{n=0}^{\infty}\dfrac{(-1)^n}{n!}$ (DELHI–2002)

(iv) $\dfrac{1}{1.2}-\dfrac{1}{3.4}+\dfrac{1}{5.6}-\dfrac{1}{7.8}+...\infty$ (OSMANIA–2003)

(v) $1-2x+3x^2-4x^3+...\infty\left(x<\dfrac{1}{2}\right)$ (COCHIN–2005)

(vi) $\dfrac{x}{1+x}-\dfrac{x^2}{1+x^2}+\dfrac{x^3}{1+x^3}-\dfrac{x^4}{1+x^4}+...\infty\ (0<x<1)$

(VTU–2004, DELHI–2002)

Hints to Selected Problems

1. $u_n=\dfrac{1}{n},\ u_{n+1}=\dfrac{1}{n+1}$

$u_n-u_{n+1}=\dfrac{1}{n+1}>0\ \forall\ n\in N\Rightarrow u_n>u_{n+1}.$

Also $\lim u_n=0.$

3. Since $\Sigma\left|\dfrac{\sin n\theta}{n^2}\right|\le\Sigma\dfrac{1}{n^2}.$

The series $\Sigma\dfrac{1}{n^2}$ is convergent.

7. $u_n=\sin\left(\dfrac{1}{n}\right)>0$

$\sin\left(\dfrac{1}{n+1}\right)<\dfrac{1}{\sin n}\Rightarrow u_{n+1}<u_n.$

Also $\lim\limits_{n\to\infty}u_n=\lim\limits_{n\to\infty}\sin\dfrac{1}{n}=0$

8. By D'Alembert's test, series is convergent for $x<\dfrac{1}{e}$ and divergent for $x>\dfrac{1}{e}$.

Then by logarithmic series, for $x=\dfrac{1}{e}$ the series is convergent.

18. $\sum\limits_{n=1}^{\infty}u_n=\sum\limits_{n=1}^{\infty}2^n\sin\left(\dfrac{x}{3^n}\right)$

$\therefore\quad u_n=2^n\sin\dfrac{x}{3^n}>0,\forall\ n\in N$

$\Rightarrow \; |u_n| = u_n \Rightarrow \Sigma |u_n| = \Sigma u_n$

$u_{n+1} = 2^{n+1} \sin\left(\dfrac{x}{3^{n+1}}\right) \Rightarrow \lim\limits_{n \to \infty} \left|\dfrac{u_n}{u_{n+1}}\right| = \dfrac{3}{2} > 1.$

Then by D'Alembert's ratio test. The given series is absolutely convergent.

 **Answers**

1. Convergent   **5.** Convergent.   **6.** Convergent if $x \le 1$ and divergent if $x > 1$.
20. (i) Convergent   (ii) Convergent   (iii) Convergent   (iv) Convergent   (v) Convergent

7.17   MULTIPLICATION OF SERIES

Consider two convergent infinite series given by

$$\sum_{n=1}^{\infty} a_n = a_1 + a_2 + \cdots + \cdots \quad \text{and} \quad \sum_{n=1}^{\infty} b_n = b_1 + b_2 + \cdots$$

Then

$$\sum_{n=1}^{\infty} c_n = a_1 b_n + a_2 b_{n-1} + \cdots + a_n b_1 = \sum_{r=1}^{n} a_r b_{n-r+1}$$

is called the Cauchy product or simply product of two given series.

THEOREM 1.   (Cauchy's theorem). *Let Σa_n and Σb_n be two absolutely convergent series then their Cauchy product Σc_n is also absolutely convergent and sum of the cauchy product series is the product of the sums.*

PROOF.   By definition of Cauchy product of two infinite series Σa_n and Σb_n we can write $\Sigma c_n = \Sigma a_n \cdot \Sigma b_n$ where

$\Sigma c_n = \sum\limits_{r=1}^{\infty} a_r b_{n-r+1}$

Let us suppose $\alpha_n = \sum\limits_{r=1}^{n} a_r$, $\beta_n = \sum\limits_{r=1}^{n} b_r$, $\sum\limits_{n=1}^{\infty} a_n = l$, and $\sum\limits_{n=1}^{\infty} b_n = m$

Then $<\alpha_n>$ and $<\beta_n>$ are the sequences of partial sums of Σa_n and Σb_n respectively. Also l and m are the sums of Σa_n and Σb_n respectively.

Consider $|a_1 b_1| + |a_1 b_2| + |a_2 b_2| + |a_2 b_1| + \ldots$ to n terms

$$\le (|a_1| + |a_2| + \ldots |a_n|)(|b_1| + |b_2| + \ldots + |b_n|) \le l.m, \; \forall \, n \in \mathbb{N}$$

Thus, the series $a_1 b_1 + a_1 b_2 + a_2 b_2 + a_2 b_1 + \ldots$ ...(1)

must be absolulety convergent. Then by Dirichlet's theorem, (1) is absolutely convergent.

Hence, by grouping, we have $\sum\limits_{n=1}^{\infty}\left(\sum\limits_{r=1}^{n} a_r b_{n-r+1}\right)$ *i.e.* $\sum\limits_{n=1}^{\infty} c_n$ ...(2)

is absolutely convergent.

Finally, since the sum of first n^2 terms of (1) is $\alpha_n \beta_n$ and $\alpha_n \beta_n \to l.m$ as $n \to \infty$, therefore, the sum of the series (1) is lm. Hence, the series (2) must converge to the same sum.

Hence $\Sigma c_n = (\Sigma a_n)(\Sigma b_n)$

THEOREM 2.   (Merten's theorem). *Let $\sum\limits_{n=1}^{\infty} a_n$ and $\sum\limits_{n=1}^{\infty} b_n$ be two convergent series and $\sum\limits_{n=1}^{\infty} a_n$ converges absolutely. Then the Cauchy product*

series and $\sum\limits_{n=1}^{\infty} c_n$ converges to lm where $l = \sum\limits_{n=1}^{\infty} a_n$ and $m = \sum\limits_{n=1}^{\infty} b_n$.

PROOF.   Let $<A_n>$, $<B_n>$ and $<s_n>$ denote the sequence of partial sums of $\sum\limits_{n=1}^{\infty} a_n$, $\sum\limits_{n=1}^{\infty} b_n$ and $\sum\limits_{n=1}^{\infty} c_n$ respectively. As per

given, the series $\sum\limits_{n=1}^{\infty} a_n$ and $\sum\limits_{n=1}^{\infty} b_n$ converges to l and m respectively, therefore

$$\lim\limits_{n \to \infty} A_n = l \quad \text{and} \quad \lim\limits_{n \to \infty} B_n = m \qquad\qquad\qquad …(1)$$

Suppose that $p_n = s_n - m \; \forall \, n$ so that $\lim\limits_{n \to \infty} p_n = \lim\limits_{n \to \infty}(B_n - m) = m - m = 0$ ...(2)

Now $s_n = n^{th}$ partial sum of $\sum\limits_{n=1}^{\infty} c_n$

$\therefore$ $s_n = c_1 + c_2 + \ldots + c_n = a_1 b_1 + a_1 b_2 + a_2 b_1 + a_1 b_3 + a_2 b_2 + a_3 b_1 + \ldots + a_1 b_n + a_2 b_{n-1} + \ldots + a_n b_1$

$= a_1(b_1 + b_2 + \ldots + b_n) + a_2(b_1 + b_2 + \ldots + b_{n-1}) + a_3(b_1 + b_2 + \ldots + b_{n-2}) + \ldots + \ldots + a_n b_1$

$= a_1 B_n + a_2 B_{n-1} + a_2 B_{n-2} + \ldots + a_n B_1$

$= a_1(p_n + m) + a_2(p_{n-1} + m) + a_3(p_{n-2} + m) + \ldots + a_n(p_1 + m)$

$$= a_1 p_n + a_2 p_{n-1} + a_3 p_{n-2} + \dots + a_n p_1 + m(a_1 + a_2 + \dots + a_n)$$

$$= q_n + m A_n \text{ where } q_n = a_1 p_n + a_2 p_{n-1} + \dots + a_n p_1$$

$$\therefore \quad \lim_{n \to \infty} s_n = \lim_{n \to \infty} q_n + m \lim_{n \to \infty} A_n = \lim_{n \to \infty} q_n + l_m \qquad \dots(3)$$

Now, we have to show that $q_n \to 0$ as $n \to \infty$.

Here, since $\sum\limits_{n=1}^{\infty} a_n$ converges absolutely $\Rightarrow \sum\limits_{n=1}^{\infty} |a_n|$ converges.

Let $\sum\limits_{n=1}^{\infty} |a_n|$ converges to l'.

From (2) $\lim\limits_{n \to \infty} p_n = 0 \Rightarrow <p_n>$ converges $\Rightarrow <p_n>$ is bounded $\qquad \dots(4)$

$\Rightarrow \exists k > 0$ such that $|p_n| < k \; \forall \, n \qquad \dots(5)$

By definition of convergence of sequence, for a given $\varepsilon > 0 \; \exists \, a$ positive integer m_1 such that

$$|p_n| < \frac{\varepsilon}{2A'+1} \quad \forall \, n \geq m_1 \qquad \dots(6)$$

Since, $\sum\limits_{n=1}^{\infty} |a_n|$ converge, so by Cauchy's general principle of convergence there exists a positive integer m_2 such that

$$|a_{m_2}+1| + |a_{m_2}+2| + \dots + |a_n| < \frac{\varepsilon}{2k+1} \forall \, n > m_1 \quad \Rightarrow \quad |a_{m_2}+1| + |a_{m_2}+2| + \dots + |a_n| < \frac{\varepsilon}{2k+1} \forall \, n > m_2 \qquad \dots(7)$$

Let $m^* = \max\{m_1, m_2\}$

then (6) and (7) are true for $n > m^*$

When $n > 2m^*$ then $n - m^* > m^*$, we have

$$|q_n| = |a_1 p_n + a_2 p_{n-1} + \dots + a_n p_1|$$

or $\quad |q_n| = |a_n p_1 + a_{n-1} p_2 + p_{m+1} a_{n-m} + p_{m+2} a_{n-m-1} + \dots + a_1 p_n)$

$$\leq |p_1| |a_n| + |p_2| |a_{n-1}| + |p_{m+1}| |a_{n-m}| + |p_{m+2}| |a_{n-m-1}| + \dots + |p_n| |a_1|$$

$$< k(|a_n| + |a_{n-1}| + \dots + |a_{n-m}|) + \frac{\varepsilon}{2A'+1}(|a_{n-m-1}| + \dots + |a_1|)$$

$$< k \cdot \frac{\varepsilon}{2k+1} + \frac{\varepsilon}{2A'+1}(|a_1| + |a_2| + \dots + |a_{n-m-1}|)$$

$$< k \cdot \frac{\varepsilon}{2k+1} + \frac{\varepsilon}{2A'+1} A' \qquad \left[\because \sum_{n=1}^{\infty} |a_n| = A' \Rightarrow \sum_{n=1}^{n-m-1} |a_n| < A' \right]$$

$$< \frac{k\varepsilon}{2k+1} + \frac{\varepsilon}{2A'+1} \cdot A' \qquad = \frac{\varepsilon}{2\left(1+\dfrac{1}{k}\right)} + \frac{\varepsilon}{2\left(1+\dfrac{1}{A'}\right)}$$

$\Rightarrow \quad |q_n| < \varepsilon$ whenever $n > 2m^* \qquad \Rightarrow \lim\limits_{n \to \infty} q_n = 0$

$\Rightarrow \quad \lim\limits_{n \to \infty} S_n = lm \qquad \Rightarrow \sum\limits_{n=1}^{\infty} c_n$ converges to $l.m.$

THEOREM 3. **(Abel's theorem).** Let $\sum\limits_{n=1}^{\infty} a_n$ and $\sum\limits_{n=1}^{\infty} b_n$ be two convergent series such that $\sum\limits_{n=1}^{\infty} a_n$ and $\sum\limits_{n=1}^{\infty} b_n$ converge to l and m respectively.

If their Cauchy product $\sum\limits_{n=1}^{\infty} c_n$ converges, then it converges to $l.m.$

PROOF. Let $<A_n>$, $<B_n>$ and $<s_n>$ be the sequences of partial sums of $\sum\limits_{n=1}^{\infty} a_n$, $\sum\limits_{n=1}^{\infty} b_n$ and $\sum\limits_{n=1}^{\infty} c_n$ respectively, then

$$\lim_{n \to \infty} A_n = l \quad \text{and} \quad \lim_{n \to \infty} B_n = m \qquad \dots(1)$$

Now $s_n = c_1 + c_2 + \dots + c_n$

$$= a_1 b_1 + a_1 b_2 + a_2 b_1 + a_1 b_3 + a_2 b_2 + \dots + a_1 b_n + a_2 b_{n-1} + \dots + a_n b_1$$

$$= a_1(b_1 + b_2 + \dots + b_n) + a_2(b_1 + b_2 + \dots + b_{n-1}) + a_3(b_1 + b_2 + \dots + b_{n-2}) + \dots + a_n b_1$$

$$= a_1 B_n + a_2 B_{n-1} + a_3 B_{n-2} + \dots + a_n B_1 \qquad \dots(2)$$

which imples

$$s_{n-1} = a_1 B_{n-1} + a_2 B_{n-2} + a_3 B_{n-3} + \dots + a_{n-1} B_1$$

$$s_{n-2} = a_1 B_{n-2} + a_2 B_{n-3} + a_3 B_{n-4} + \dots + a_{n-2} B_1$$

....

$$s_1 = a_1 B_1$$

On adding all these we get,

$$s_1 + s_2 + \dots + s_n = a_1 B_n + (a_1 + a_2) B_{n-1} + (a_1 + a_2 + a_3) B_{n-2} + \dots + (a_1 + a_2 + \dots + a_n) B_1$$

$$= A_1 B_n + A_2 B_{n-1} + A_3 B_{n-2} + \dots + A_n B_1$$

$$\Rightarrow \qquad \frac{s_1 + s_2 + \dots + s_n}{n} = \frac{A_1 B_n + A_2 B_{n-1} + A_3 B_{n-2} + \dots + A_n B_1}{n}$$

As per given, $\sum\limits_{n=1}^{\infty} c_n$ converges, Let it converges to s, then $\lim\limits_{n \to \infty} s_n = s$

$$\Rightarrow \qquad \lim_{n \to \infty} \frac{s_1 + s_2 + \dots + s_n}{n} = s \qquad \text{(By Cauchy's first theorem on limits)}$$

Since $<A_n> \to l$ and $<B_n> \to m$ then by Ceasaro's theorem

$$\lim_{n \to \infty} \frac{A_1 B_n + A_2 B_{n-1} + \dots + A_n B_1}{n} = l.m$$

$$\therefore \qquad \lim_{n \to \infty} \frac{s_1 + s_2 + \dots + s_n}{n} = \lim_{n \to \infty} \frac{A_1 B_n + A_2 B_{n-1} + \dots + A_n B_1}{n} = l.m \Rightarrow s = l.m \Rightarrow \sum_{n=1}^{\infty} c_n \text{ conveges to } l.m.$$

 **Solved Examples**

EXAMPLE 1. *Show that the Cauchy product of two divergent series given by*

$$\sum_{n=0}^{\infty} a_n = 1 - \left(\frac{3}{2}\right) - \left(\frac{3}{2}\right)^2 - \left(\frac{3}{2}\right)^3 - \dots \text{ and}$$

$$\sum_{n=0}^{\infty} b_n = 1 + \left(2 + \frac{1}{2^2}\right) + \frac{3}{2}\left(2^2 + \frac{1}{2^3}\right) + \left(\frac{3}{2}\right)^2 \left(2^3 + \frac{1}{2^4}\right) + \dots$$

is convergent.

SOLUTION. Clearly, we can write

$$\sum_{n=0}^{\infty} a_n = 1 - \sum_{n=1}^{\infty} \left(\frac{3}{2}\right)^n$$

and $\quad \sum\limits_{n=0}^{\infty} b_n = 1 + \sum\limits_{n=1}^{\infty} \left(\frac{3}{2}\right)^{n-1} \left(2^n + \frac{1}{2^{n+1}}\right)$

If we leave the first term of both the series, we see that remaining series form G.P. with common ratio greater than 1 and therefore, both are divergent.

If $\sum\limits_{n=0}^{\infty} c_n$ be the Cauchy product, then $c_0 = a_0 b_0 = 1$

$c_n = a_0 b_n + a_1 b_{n-1} + a_2 b_{n-2} + \dots + a_n b_0, n \ge 1$

Therefore

$$c_n = 1 . \left(\frac{3}{2}\right)^{n-1} \left(2^n + \frac{1}{2^{n+1}}\right) + \left(\frac{-3}{2}\right)\left(\frac{3}{2}\right)^{n-2}$$

$$\left\{2^{n-1} + \frac{1}{2^n}\right\} + \dots + \left(\frac{-3}{2}\right)^n \times 1$$

$$= \left(\frac{3}{2}\right)^{n-1} \left[\left(2^n + \frac{1}{2^{n+1}}\right) - \left(2^{n-1} + \frac{1}{2^n}\right)\right.$$

$$\left. - \left(2^{n-2} + \frac{1}{2^{n-1}}\right) - \left(2 + \frac{1}{2^2}\right)\right] - \left(\frac{3}{2}\right)^n$$

$$= \left(\frac{3}{2}\right)^{n-1} \left[2^n + \frac{1}{2^{n+1}} - \{2^{n-1} + 2^{n-2} + \dots + 2\}\right.$$

$$\left. - \left(\frac{1}{2^n} + \frac{1}{2^{n-1}} + \dots + \frac{1}{2^2}\right)\right] - \left(\frac{3}{2}\right)^n$$

$$= \left(\frac{3}{2}\right)^{n-1} \left[2^n + \frac{1}{2^{n+1}} - 2\left(\frac{2^{n-1} - 1}{2 - 1}\right)\right.$$

$$\left. - \frac{1}{2^2}\left(\frac{1 - \frac{1}{2^{n-1}}}{1 - \frac{1}{2}}\right)\right] - \left(\frac{3}{2}\right)^n$$

$$= \left(\frac{3}{2}\right)\left[2^n + \frac{1}{2^{n+1}} - 2^n + 2 - \frac{1}{2} + \frac{1}{2^n}\right] - \left(\frac{3}{2}\right)^n$$

$$= \left(\frac{3}{2}\right)^{n-1} \left[\frac{3}{2} + \frac{1}{2^n} + \frac{1}{2^{n+1}}\right] - \left(\frac{3}{2}\right)^n$$

$$= \left(\frac{3}{2}\right)^{n-1} \left[\frac{3}{2} + \frac{3}{2^{n+1}}\right] - \left(\frac{3}{2}\right)^n$$

$$= \left(\frac{3}{2}\right)^n + \left(\frac{3}{2}\right)^{n-1} . \frac{3}{2} . \frac{1}{2^n} - \left(\frac{3}{2}\right)^n$$

$$= \left(\frac{3}{2}\right)^n \frac{1}{2^n} = \frac{3^n}{2^{2n}} = \left(\frac{3}{4}\right)^n$$

$$\Rightarrow \sum_{n=1}^{\infty} c_n = \sum_{n=1}^{\infty} \left(\frac{3}{4}\right)^4 \text{ which is a G.P with common}$$

ratio $\frac{3}{4} < 1$ and hence convergent.

EXAMPLE 2. *Prove that*

$$\frac{1}{1-x} \log \frac{1}{1-x} = \sum_{n=1}^{\infty} \left(1 + \frac{1}{2} + \frac{1}{3} + \dots + \frac{1}{n}\right) x^n,$$

for $|x| < 1.$

SOLUTION. Consider

$$(1+x)^{-1} = 1 + x + x^2 + \dots = \sum_{n=0}^{\infty} x^n = \sum_{n=0}^{\infty} a_n$$

where $a_n = x^n, n > 0$

and $\log \dfrac{1}{1-x} = \log(1-x)^{-1} = -\log(1-x)$

$$= \dfrac{x}{1} + \dfrac{x^2}{2} + \dfrac{x^3}{3} + \cdots + \dfrac{x^n}{n} + \cdots$$

$$= \sum_{n=0}^{\infty} \dfrac{x^{n+1}}{n+1} = \sum_{n=0}^{\infty} b_n \quad \text{(say)}$$

where $b_n = \dfrac{x^{n+1}}{n+1}, n > 0$

Now, $\lim_{n\to\infty} |a_n|^{1/n} = \lim_{n\to\infty} (|x|^n)^{1/n} = |x|$

Hence, by root test $\sum_{n=0}^{\infty} |a_n|$ is convegent for $|x| < 1$

and therefore $\sum_{n=0}^{\infty} a_n$ is absolutely convergent for $|x| < 1$.

Further

$$\lim_{n\to\infty} |b_{n-1}|^{1/n} = \lim_{n\to\infty} \left(\left|\dfrac{x^n}{n}\right| \right)^{1/n} = \lim_{n\to\infty} \dfrac{|x|}{n^{1/n}} = |x|$$

By root test, $\sum_{n=0}^{\infty} |b_n|$ is convergent for $|x| < 1$ and

therefore $\sum_{n=0}^{\infty} b_n$ is absolutely convergent for $|x| < 1$.

If $\sum_{n=0}^{\infty} c_n$ be the Cauchy product of $\sum_{n=0}^{\infty} a_n$ and $\sum_{n=0}^{\infty} b_n$

then $c_n = \sum_{k=0}^{n} a_{n-k} \cdot b_k$

$$= \sum_{k=0}^{n} x^{n-k} \left(\dfrac{x^{k+1}}{k+1} \right) = x^{n+1} \sum_{k=0}^{n} \dfrac{1}{k+1}$$

$$= x^{n+1} \left(1 + \dfrac{1}{2} + \dfrac{1}{3} + \cdots + \dfrac{1}{n+1} \right)$$

Now, since both $\sum_{n=0}^{\infty} a_n$ and $\sum_{n=0}^{\infty} b_n$ are absolutely

convergent for $|x| < 1$, so their Cauchy product $\sum_{n=0}^{\infty} c_n$ is convergent for $|x| < 1$.

Also, $\sum_{n=0}^{\infty} a_n \cdot \sum_{n=0}^{\infty} b_n = \sum_{n=0}^{\infty} c_n$

$$\Rightarrow \dfrac{1}{1-x} \log \dfrac{1}{1-x}$$

$$= \sum_{n=0}^{\infty} x^{n+1} \left(1 + \dfrac{1}{2} + \dfrac{1}{3} + \cdots + \dfrac{1}{n+1} \right)$$

$$= \sum_{n=0}^{\infty} x^n \left(1 + \dfrac{1}{2} + \dfrac{1}{3} + \cdots + \dfrac{1}{n} \right)$$

Exercise-7.5

1. Show that the Cauchy product of the convergent series.

$\sum_{n=1}^{\infty} \dfrac{(-1)^n}{\sqrt{n+1}}$ with itself is divergent.

2. Show that the Cauchy product of the series $\sum_{n=1}^{\infty} \dfrac{(-1)^n}{(n+1)^p}, p > 0$

with itself converges for $p > \dfrac{1}{2}$ and diverges for $p \le \dfrac{1}{2}$.

3. Show that the Cauchy product of two series $3 + \sum_{n=1}^{\infty} 3^n$ and

$-2 + \sum_{n=1}^{\infty} 2^n$ is absolutely convergent, although both the series

are divergent.

4. Show that

$$\dfrac{1}{2} \left(1 - \dfrac{1}{3} + \dfrac{1}{5} - \dfrac{1}{7} + \cdots \right)^2 = \dfrac{1}{2} - \dfrac{1}{4} \left(1 + \dfrac{1}{3} \right) + \dfrac{1}{6} \left(1 + \dfrac{1}{3} + \dfrac{1}{5} \right) \cdots$$

5. Given, $\log 2 = 1 - \dfrac{1}{2} + \dfrac{1}{3} - \dfrac{1}{4} + \cdots + \dfrac{(-1)^n}{n+1}$ prove that

$$\sum_{n=0}^{\infty} (-1)^n \left[\dfrac{1}{(n+1).1} + \dfrac{1}{n-2} + \dfrac{1}{(n-1).3} + \cdots + \dfrac{1}{1.(n-1)} \right] = (\log 2)^2$$

7.18 FOURIER SERIES

In this section, we shall study a special type of functional series extensively studied by Joseph Fourier. Joseph Fourier represented expansions in trigonometrical series in connection with boundary value problem in conduction of heat. Although such expansions had been studied earlier, these series bear the name 'Fourier series' because of the major contributions of Fourier in this field.

7.18.1 PERIODIC FUNCTIONS   (UPTU–2002)

A function $f(x)$ which satisfies the relation $f(x + T) = f(x)$ for all real x and some fixed T is called a periodic function. The smallest positive number T, for which this relation holds, is called the period of $f(x)$.

If T is the period of $f(x)$. Then

$$f(x) = f(x + T) = f(x + 2T) = \ldots = f(x + nT) = \ldots$$

Also, $\qquad f(x) = f(x - T) = f(x - 2T) = \ldots = f(x - nT) = \ldots$

$\therefore \qquad f(x) = f(x \pm nT)$, where n is a positive integer.

For example: Consider the function $f(x) = \sin x$. We have

$$\sin x = \sin (x + 2\pi) = \sin (x + 4\pi) = \ldots.$$

Here, $f(x) = \sin x$ is a periodic function with period 2π. This function is also called sinusoidal periodic function.

We have studied about the Macluarian's theorem which is used to expand a function provided the function's derivative are continuous. Now, the need arise to expand functions which have discontinuities in their derivatives. By Fourier series, we can expand both types of functions under certain conditions as an infinite series of sine and cosine of x and it's integral multiple of a function $f(x)$ is defined in the interval $c < x < c + 2\pi$.

Then, Fourier series of $f(x)$ is given by

$$f(x) = \frac{a_0}{2} + \sum_{n=1}^{\infty} a_n \cos nx + \sum_{n=1}^{\infty} b_n \sin nx \qquad \text{...(1)}$$

where a_0, a_n and b_n are called Fourier coefficient of $f(x)$ and their values are given as :

$$a_0 = \frac{1}{\pi} \int_c^{c+2\pi} f(x)dx \qquad \text{...(2)}$$

$$a_n = \frac{1}{\pi} \int_c^{c+2\pi} f(x)\cos nx\, dx \qquad \text{...(3)}$$

$$b_n = \frac{1}{\pi} \int_c^{c+2\pi} f(x)\sin nx dx \qquad \text{...(4)}$$

The series (1) with coefficients a_0, a_n and b_n given by (2), (3) and (4) respectively is called the Fourier series of $f(x)$ and the coefficients a_0, a_n and b_n are called the Fourier coefficients corresponding to $f(x)$.

(i) When $c = 0$, the interval becomes $0 < x < 2\pi$ and formula for a_0, a_n, b_n is obtained by putting $c = 0$.

(ii) When $c = -\pi$, then interval becomes $-\pi < x < \pi$. In this interval, the formula for a_0, a_n and b_n becomes as under :

(a) When $f(x)$ is an odd function, then

$$a_0 = \frac{1}{\pi} \int_{-\pi}^{\pi} f(x)dx = 0 \qquad a_n = \frac{1}{\pi} \int_{-\pi}^{\pi} f(x)\cos nx\, dx = 0 \quad \text{[By property of definite integral]}$$

$$b_n = \frac{1}{\pi} \int_{-\pi}^{\pi} f(x)\sin nx\, dx = \frac{2}{\pi} \int_0^{\pi} f(x)\sin x dx$$

Hence, if function $f(x)$ is odd, its Fourier expansion contains only sine series,

i.e., $f(x) = \sum_{n=1}^{\infty} b_n \sin nx$, where $b_n = \frac{2}{\pi} \int_0^{\pi} f(x)\sin nx\, dx.$

(b) When $f(x)$ is even function, then formula for a_0, a_n and b_n are given by

$$a_0 = \frac{1}{\pi} \int_{-\pi}^{\pi} f(x)dx = \frac{2}{\pi} \int_0^{\pi} f(x)dx, \quad a_n = \frac{1}{\pi} \int_{-\pi}^{\pi} f(x)\cos nx\, dx = \frac{2}{\pi} \int_0^{\pi} f(x)\cos nx\, dx$$

and $\qquad b_n = \frac{1}{\pi} \int_{-\pi}^{\pi} f(x)\sin nx\, dx = 0 \qquad\qquad\qquad\qquad\qquad\qquad$ [$\because f(x) \sin nx$ is odd.]

Hence, if a periodic function $f(x)$ is even, its Fourier expansion contains only cosine terms, *i.e.,* $f(x) = \frac{a_0}{2} + \sum_{n=1}^{\infty} \int_0^{\pi} f(x)dx$, where

$$a_0 = \frac{2}{\pi} \int_0^{\pi} f(x)dx \quad \text{and} \quad a_n = \frac{2}{\pi} \int_0^{\pi} f(x).\cos nx\, dx$$

7.18.2 SOME IMPORTANT RESULTS

The following results are useful in the Fourier series :

(i) $\sin n\pi = 0, \cos n\pi = (-1)^n, \cos\left(n + \frac{1}{2}\right)\pi = 0$, where $n \in \mathbf{Z}$.

(ii) $\int uv = uv_1 - u'v_2 + u''v_3 - u'''v_4 + \cdots$, where $u' = \dfrac{du}{dx}, u'' = \dfrac{d^2u}{dx^2}, \cdots$

$\qquad v_1 = \int v dx, v_2 = \int v_1 dx, \cdots$

(iii) $\int_0^{2\pi} \sin nx\, dx = 0$ (iv) $\int_0^{2\pi} \cos nx\, dx = 0$ (v) $\int_0^{2\pi} \sin^2 nx\, dx = \pi$

(vi) $\int_0^{2\pi} \cos^2 nx\, dx = \pi$ (vii) $\int_0^{2\pi} \sin nx.\sin mx\, dx = 0$ (viii) $\int_0^{2\pi} \cos nx.\cos mx\, dx = 0$

(ix) $\int\limits_{0}^{2\pi} \sin nx . \cos mx\, dx = 0$ (x) $\int\limits_{0}^{2\pi} \sin mx . \cos nx\, dx = 0$

(xi) $\int e^{ax} \sin bx\, dx = \dfrac{e^{ax}}{a^2+b^2}(a\sin bx - b\cos bx) + c$ (xii) $\int e^{ax} \cos bx\, dx = \dfrac{e^{ax}}{a^2+b^2}(a\cos bx + b\sin bx) + c$

7.18.3 DETERMINATION OF FOURIER COEFFICIENTS: EULER'S FORMULAE

The fourier series is given by

$$f(x) = \frac{a_0}{2} + a_1 \cos x + a_2 \cos 2x + \dots + a_n \cos nx + b_1 \sin x + \dots + b_2 \sin 2x + \dots + b_n \sin nx + \dots$$

or $$f(x) = \frac{a_0}{2} + \sum_{n=1}^{\infty} a_n \cos nx + \sum_{n=1}^{\infty} b_n \sin nx.$$ ...(i)

To find a_0 : Integrating both sides of equation (1) from $x = c+0, x = c+2\pi$

$$\int\limits_{c}^{c+2\pi} f(x)dx = \frac{a_0}{2} \int\limits_{c}^{c+2\pi} dx + \int\limits_{c}^{c+2\pi} \left(\sum_{n=1}^{\infty} a_n \cos nx \right) dx + \int\limits_{c}^{c+2\pi} \left(\sum_{n=1}^{\infty} b_n \sin nx \right) dx = \frac{a_0}{2}(c + 2\pi - c) + 0 + 0 = a_0 \pi$$

$\Rightarrow$ $$a_0 = \frac{1}{\pi} \int\limits_{c}^{c+2\pi} f(x)\,dx .$$

To find a_n : Multipling each side of equation (1) by $\cos nx$ and integrate w.r.t. x., between the limit c to $c+2\pi$.

$$\int\limits_{c}^{c+2\pi} f(x)\cos nx\, dx = \frac{a_0}{2} \int\limits_{c}^{c+2\pi} \cos nx\, dx + \int\limits_{c}^{c+2\pi} \left(\sum_{n=1}^{\infty} a_n \cos nx \right) \cos nx\, dx + \int\limits_{c}^{c+2\pi} \left(\sum_{n=1}^{\infty} b_n \sin nx \right) \cos nx\, dx$$

$$= 0 + a_n \pi + 0 = a_n \pi$$

$\Rightarrow$ $$a_n = \frac{1}{\pi} \int\limits_{c}^{c+2\pi} f(x)\cos nx\, dx .$$

To find b_n : Multiplying each side of equation (1) by $\sin nx$ and integrate w.r.t. x between the limit c to $c + 2\pi$.

$$\int\limits_{c}^{c+2\pi} f(x)\sin nx\, dx = \frac{a_0}{2} \int\limits_{c}^{c+2\pi} \sin nx\,dx + \int\limits_{c}^{c+2\pi} \left(\sum_{n=1}^{\infty} a_n \cos nx \right) \sin nx\, dx + \int\limits_{c}^{c+2\pi} \left(\sum_{n=1}^{\infty} b_n \sin nx \right) \sin nx\, dx$$

$$= 0 + 0 + b_n \pi = b_n \pi$$

$\Rightarrow$ $$b_n = \frac{1}{\pi} \int\limits_{c}^{c+2\pi} f(x)\sin nx\, dx$$

These values of a_0, a_n and b_n are called Euler's formulae.

7.18.4 DIRICHLET'S CONDITIONS

Any function $f(x)$ can be expressed as a Fourier series $\dfrac{a_0}{2} + \sum\limits_{n=1}^{\infty} a_n \cos nx + \sum\limits_{n=1}^{\infty} b_n \sin nx$, where a_0, a_n and b_n are constants.

(i) $f(x)$ is finite and single valued in the interval $c < x < c + 2\pi$.

(ii) $f(x)$ is periodic with period 2π.

(iii) $f(x)$ and $f'(x)$ are piecewise continuous in the interval $c < x < c + 2\pi$.

The Fourier series with its coefficients converge to

(a) $f(x)$ if x is a point of continuity.

(b) $\dfrac{f(x+0) + f(x-0)}{2}$, if x is a point of discontinuity.

The conditions (i), (ii) and (iii) imposed on $f(x)$ are sufficient but not necessary, *i.e.*, if the conditions are satisfied, the convergence is guranteed. However, if they are not satisfied the series may or may not converge.

Solved Examples

EXAMPLE 1. *Expand the function f(x) = x sin x as a Fourier series in interval* $-\pi \le x \le \pi$ *. Deduce that*

$$\frac{1}{1.3} - \frac{1}{3.5} + \frac{1}{5.7} - \frac{1}{7.9} + ... = \frac{\pi - 2}{4}$$

(UPTU–2001, 2005, 2008, Q.BANK–2001, SVTU–2009, BHOPAL–2009, ROHTAK–2006)

SOLUTION . Since $x \sin x$ is an even function of x, so $b_n = 0$, then Fourier series is given by

$$f(x) = x \sin x = \frac{a_0}{2} + \sum_{n=1}^{\infty} a_n \cos nx,$$

where $a_0 = \frac{2}{\pi} \int_0^\pi x \sin x \, dx = \frac{2}{\pi} \left[-x \cos x + \sin x \right]_0^\pi$

$$= \frac{2}{\pi} (-\pi \cos \pi) = 2$$

$$a_n = \frac{2}{\pi} \int_0^\pi x \sin x \cos nx \, dx = \frac{1}{\pi} \int_0^\pi x . 2 \cos nx \sin x \, dx$$
$$= \frac{1}{\pi} \int_0^\pi x \{ \sin(n+1)x - \sin(n-1)x \} \, dx$$

$$= \frac{1}{\pi} \left[x \left\{ \frac{-\cos(n+1)x}{n+1} + \frac{\cos(n+1)x}{n-1} \right\} \right.$$

$$\left. -1 \left\{ \frac{-\sin(n+1)x}{(n+1)^2} + \frac{\sin(n-1)x}{(n+1)^2} \right\} \right]_0^\pi$$

$$= \frac{1}{\pi} \left[\pi \left\{ \frac{-\cos(n+1)\pi}{n+1} + \frac{\cos(n-1)\pi}{n-1} \right\} \right]$$

$$= \frac{\cos(n-1)\pi}{n-1} - \frac{\cos(n+1)\pi}{n+1} ; n \ne 1$$

$$= \begin{cases} \frac{1}{n-1} - \frac{1}{n+1} = \frac{2}{n^2-1} & \text{if } n \text{ is odd } n \ne 1 \\ \frac{-1}{n-1} + \frac{1}{n+1} = \frac{-2}{n^2-1} & \text{if } n \text{ is even} \end{cases}$$

When $n = 1$, then

$$a_1 = \frac{2}{\pi} \int_0^\pi x \sin x \cos x \, dx = \frac{1}{\pi} \int_0^\pi x \sin 2x \, dx$$

$$= \frac{1}{\pi} \left[x \left(\frac{-\cos 2x}{2} \right) - \left(\frac{-\sin 2x}{4} \right) \right]_0^\pi$$

$$= \frac{1}{\pi} \left[\frac{-\pi \cos 2\pi}{2} \right] = -\frac{1}{2}$$

$$\therefore \quad x \sin x = 1 - \frac{1}{2} \cos x - 2 \left[\frac{\cos 2x}{2^2 - 1} - \frac{\cos 3x}{3^2 - 1} \right.$$

$$\left. + \frac{\cos 4x}{4^2 - 1} - \frac{\cos 5x}{5^2 - 1} + ... \right]$$

Putting $x = \frac{\pi}{2}$, we get

$$\frac{\pi}{2} = 1 - 2 \left(\frac{-1}{2^2 - 1} + \frac{1}{4^2 - 1} - \frac{1}{6^2 - 1} + ... \right)$$

$$\Rightarrow \quad \frac{\pi}{2} - 1 = 2 \left(\frac{1}{3} - \frac{1}{15} + \frac{1}{35} - ... \right)$$

$$\Rightarrow \quad \frac{\pi - 2}{4} = \left(\frac{1}{1.3} - \frac{1}{3.5} + \frac{1}{5.7} - ... \right)$$

EXAMPLE 2. *Obtain the Fourier series for the function f(x) = x², $-\pi < x < \pi$. Sketch the graph f function f(x). Hence, show that*

(PTU–2009, BHOPAL–2008, BPTU–2006)

(i) $\frac{1}{1^2} + \frac{1}{2^2} + \frac{1}{3^2} + \frac{1}{4^2} + ... = \sum_{n=1}^{\infty} \frac{1}{n^2} = \frac{\pi^2}{6}$

(UPTU(Q. BANK)–2001, ANNA–2009, PTU–2009, OSMANIA–2003, MUMBAI–2009, SVTU–2008)

(ii) $\frac{1}{1^2} - \frac{1}{2^2} + \frac{1}{3^2} - \frac{1}{4^2} + ... = \frac{\pi^2}{12}$

(UPTU–2004, 08, SVTU–2008)

(iii) $\frac{1}{1^2} + \frac{1}{3^2} + \frac{1}{5^2} + ... = \sum_{n=1}^{\infty} \frac{1}{(2n-1)^2} = \frac{\pi^2}{8}$

SOLUTION. $f(x) = x^2$ is an even function, therefore $b_n = 0$

Now $f(x) = x^2 = \frac{a_0}{2} + \sum_{n=1}^{\infty} a_n \cos nx$. Then

$$a_0 = \frac{2}{\pi} \int_0^\pi f(x) \, dx = \frac{2}{\pi} \int_0^\pi x^2 \, dx = \frac{2}{\pi} \left[\frac{x^3}{3} \right]_0^\pi = \frac{2}{3} \pi^2$$

$$a_n = \frac{2}{\pi} \int_0^\pi f(x) \cos nx \, dx = \frac{2}{\pi} \int_0^\pi x^2 \cos nx \, dx$$

$$= \frac{2}{\pi} \left[x^2 \left(\frac{\sin nx}{n} \right) - 2x \left(\frac{-\cos nx}{n^2} \right) + 2 \left(\frac{-\sin nx}{n^2} \right) \right]_0^\pi$$

$$= \frac{2}{\pi} \left[2\pi \frac{\cos n\pi}{n^2} \right] = 4 \frac{(-1)^n}{n^2}$$

$$\therefore \quad x^2 = \frac{\pi^2}{3} - 4 \left(\frac{\cos x}{1^2} - \frac{\cos 2x}{2^2} \right.$$

$$\left. + \frac{\cos 3x}{3^2} - \frac{\cos 4x}{4^2} + ... \right)$$

$$\Rightarrow \quad x^2 = \frac{\pi^2}{3} + 4 \sum_{n=1}^{\infty} \frac{(-1)^n}{n^2} \cos nx \qquad ... (1)$$

Put $x = \pi$ in (1), we get

$$\pi^2 = \frac{\pi^2}{3} - 4 \left(-\frac{1}{1^2} - \frac{1}{2^2} - \frac{1}{3^2} - \frac{1}{4^2} - ... \right)$$

$$\Rightarrow \quad \frac{2\pi^2}{3} = -4 \left(-\frac{1}{1^2} - \frac{1}{2^2} - \frac{1}{3^2} - \frac{1}{4^2} - ... \right)$$

$$\therefore \quad \frac{1}{1^2} + \frac{1}{2^2} + \frac{1}{3^2} + \frac{1}{4^2} ... = \frac{\pi^2}{6} \qquad ... (2)$$

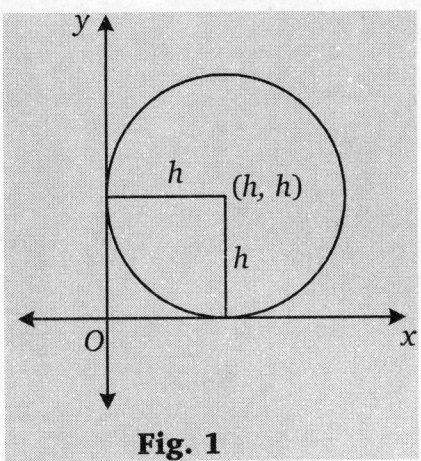

Fig. 1

Put $x = 0$ in (1), we get

$$0 = \frac{\pi^2}{3} - 4\left(-\frac{1}{1^2} - \frac{1}{2^2} - \frac{1}{3^2} - \frac{1}{4^2} - \ldots\right)$$

$$\therefore \quad \frac{1}{1^2} - \frac{1}{2^2} - \frac{1}{3^2} - \frac{1}{4^2} - \ldots = \frac{\pi^2}{12} \quad \ldots(3)$$

Adding (2) and (3), we get

$$\frac{\pi^2}{4} = 2\left(\frac{1}{1^2} + \frac{1}{3^2} + \frac{1}{5^2} + \ldots\right)$$

$$\therefore \quad \frac{1}{1^2} + \frac{1}{3^2} + \frac{1}{5^2} + \ldots = \frac{\pi^2}{8}.$$

EXAMPLE 3. *Obtain the Fouries series for $f(x) = e^{-x}$ in the interval $0 < x < 2\pi$.* (UPTU(Q.BANK)–2001, SVTU–2007)

SOLUTION. Let $f(x) = e^{-x}$. The Fourier series of $f(x)$ can be written as

$$f(x) = e^{-x} = \frac{a_0}{2} + \sum_{n=1}^{\infty} a_n \cos nx + \sum_{n=1}^{\infty} b_n \sin nx$$

Then, $\quad a_0 = \frac{1}{\pi} \int_0^{2\pi} f(x)dx = \frac{1}{\pi} \int_0^{2\pi} e^{-x} dx$

$$= \frac{1}{\pi} \cdot \left[-e^{-x}\right]_0^{2\pi} = \frac{1 - e^{-2\pi}}{\pi}$$

$$a_n = \frac{1}{\pi} \int_0^{2\pi} f(x) \cos nx \, dx = \frac{1}{\pi} \int_0^{2\pi} e^{-x} \cos nx \, dx$$

$$= \frac{1}{\pi(1+n^2)} [e^{-x}(-\cos nx + n \sin nx)]_0^{2\pi}$$

$$= \frac{1 - e^{-2\pi}}{\pi(1+n^2)}$$

$$b_n = \frac{1}{\pi} \int_0^{2\pi} f(x) \sin nx \, dx = \frac{1}{\pi} \int_0^{2\pi} e^{-x} \sin nx \, dx$$

$$= \left[\frac{e^{-x}}{\pi(1+n^2)} - \sin nx - n \cos nx\right]_0^{2\pi}$$

$$= \frac{1 - e^{-2\pi}}{\pi} \cdot \frac{n}{1+n^2}$$

$$\therefore \quad e^{-x} = \frac{1 - e^{-2\pi}}{\pi}$$

$$\left[\frac{1}{2} + \left(\frac{1}{2}\cos x + \frac{1}{5}\cos 2x + \frac{1}{10}\cos 3x + \ldots\right)\right]$$

$$+ \left(\frac{1}{2}\sin x + \frac{2}{5}\sin 2x + \frac{3}{10}\sin 3x + \ldots\right)$$

$$= \frac{1 - e^{-2\pi}}{2\pi} + \frac{1 - e^{-2\pi}}{\pi} \sum_{n=1}^{\infty} \frac{\cos nx}{1+n^2}$$

$$+ \frac{1 - e^{-2\pi}}{\pi} \cdot \sum_{n=1}^{\infty} \frac{n \sin nx}{1+n^2}$$

EXAMPLE 4. *Obtain Fourier series for the function $f(x)$, given by* $f(x) = \begin{cases} 1 + \dfrac{2x}{\pi}; & -\pi \leq x \leq 0 \\[2mm] 1 - \dfrac{2x}{\pi}; & 0 \leq x \leq \pi \end{cases}$

(VTU–2010, MUMBAI–2007)

Hence, deduce that $\dfrac{1}{1^2} + \dfrac{1}{3^2} + \dfrac{1}{5^2} + \ldots = \dfrac{\pi^2}{8}$.

SOLUTION. When $-\pi \leq x \leq 0 \Rightarrow 0 \leq -x \leq \pi$

$$\therefore \quad f(-x) = 1 - \frac{2(-x)}{\pi} = 1 + \frac{2x}{\pi} = f(x)$$

When $0 \leq x \leq \pi \Rightarrow -\pi \leq -x \leq 0$

$$\therefore \quad f(-x) = 1 + \frac{2(-x)}{\pi} = 1 - \frac{2x}{\pi} = 1 - \frac{2x}{\pi} = f(x)$$

Therefore, $f(x)$ is an even function of x in the interval $[-\pi, \pi]$. Hence $b_n = 0$.

Now, Fourier series of $f(x)$ is given by

$$f(x) = \frac{a_0}{2} + \sum_{n=1}^{\infty} a_n \cos nx$$

Then,

$$a_0 = \frac{2}{\pi} \int_0^{\pi} f(x)dx = \frac{2}{\pi} \int_0^{\pi} \left(1 - \frac{2x}{\pi}\right) dx = \frac{2}{\pi}\left[x - \frac{x^2}{\pi}\right]_0^{\pi} = 0$$

$$a_n = \frac{2}{\pi} \int_0^{\pi} f(x) \cos nx \, dx = \frac{2}{\pi} \int_0^{\pi} \left(1 - \frac{2x}{\pi}\right) \cos nx \, dx$$

$$= \frac{2}{\pi}\left[\left(1 - \frac{2x}{\pi}\right)\frac{\sin nx}{n} - \left(-\frac{2}{\pi}\right)\left(-\frac{\cos nx}{n^2}\right)\right]_0^{\pi}$$

$$= \frac{2}{\pi}\left[-\frac{2\cos n\pi}{\pi n^2} + \frac{2}{\pi n^2}\right] = \frac{4}{\pi^2 n^2}[1 - (-1)^n]$$

$$\Rightarrow f(x) = \frac{4}{\pi^2} \sum_{n=1}^{\infty} [1 - (-1)^n]\frac{\cos nx}{n^2}$$

$$= \frac{4}{\pi^2}\left(\frac{2\cos x}{1^2} + \frac{2\cos 3x}{3^2} + \frac{2\cos 5x}{5^2} + \ldots\right)$$

$$= \frac{8}{\pi^2}\left(\frac{\cos x}{1^2} + \frac{\cos 3x}{3^2} + \frac{\cos 5x}{5^2} + \ldots\right).$$

Putting $x = 0$, we get $\dfrac{1}{1^2} + \dfrac{1}{3^2} + \dfrac{1}{5^2} + \ldots = \dfrac{\pi^2}{8}$.

[Since $f(0) = 1$]

EXAMPLE 5. *Find a Fourier series to represent* $x - x^2$ *from* $x = -\pi$ *to* $x = \pi$.

Deduce that $\dfrac{1}{1^2} - \dfrac{1}{2^2} + \dfrac{1}{3^2} - \dfrac{1}{4^2} + \ldots = \dfrac{\pi^2}{12}$.

(VTU–2011, MADRAS–2006)

SOLUTION. The Fourier series for $f(x)$ in $(-\pi, \pi)$ is

$$f(x) = a_0 + \sum_{n=1}^{\infty} a_n \cos nx + \sum_{n=1}^{\infty} b_n \sin nx$$

Here,

$$a_0 = \frac{1}{2\pi} \int_{-\pi}^{\pi} (x - x^2)\,dx = \frac{1}{2\pi}\left[\frac{x^2}{2} - \frac{x^3}{3}\right]_{-\pi}^{\pi} = -\frac{\pi^2}{3}$$

$$a_n = \frac{1}{\pi} \int_{-\pi}^{\pi} (x - x^2) \cos nx\,dx$$

$$= \frac{1}{\pi}\left[(x - x^2)\frac{\sin nx}{n} - (1 - 2x)\left(-\frac{\cos nx}{n^2}\right)\right.$$
$$\left. + (-2)\left(-\frac{\sin nx}{n^3}\right)\right]_{-\pi}^{\pi}$$

$$= \frac{-4(-1)^n}{n^2}$$

and $\quad b_n = \dfrac{1}{\pi}\int_{-\pi}^{\pi} (x - x^2) \sin nx\,dx$

$$= \frac{1}{\pi}\left[(x - x^2)\left(-\frac{\cos nx}{n}\right) - (1 - 2x)\cdot\left(-\frac{\sin nx}{n^2}\right)\right.$$
$$\left. + (-2)\left(\frac{\cos nx}{n^3}\right)\right]_{-\pi}^{\pi}$$

$$= \frac{(-2)(-1)^n}{n}$$

∴ The required Fourier series is

$$x - x^2 = -\frac{\pi^2}{3} + 4\left[\frac{\cos x}{1^2} - \frac{\cos 2x}{2^2}\right.$$
$$\left. + \frac{\cos 3x}{3^2} - \frac{\cos 4x}{4^2} + \ldots\right]$$
$$+ 2\left[\frac{\sin x}{1} - \frac{\sin 2x}{2} + \frac{\sin 3x}{3} - \frac{\sin 4x}{4} + \ldots\right] \quad \ldots(1)$$

Deduction. Putting $x = 0$ in (1), we get

$$0 = -\frac{\pi^2}{3} + 4\left(\frac{1}{1^2} - \frac{1}{2^2} + \frac{1}{3^2} - \frac{1}{4^2} + \ldots\right)$$

or $\quad \dfrac{1}{1^2} - \dfrac{1}{2^2} + \dfrac{1}{3^2} - \dfrac{1}{4^2} + \ldots = \dfrac{\pi^2}{12}$.

EXAMPLE 6. *Find the Fourier series of the function defined as*

$$f(x) = \begin{cases} x + \pi & ; \quad 0 \le x \le \pi \\ -x - \pi & ; \quad -\pi \le x \le 0 \end{cases}$$

and $f(x + 2\pi) = f(x)$. (UPTU–2006, Q. BANK–2001)

SOLUTION. Let $\quad f(x) = \dfrac{a_0}{2} + \displaystyle\sum_{n=1}^{\infty} a_n \cos nx + \sum_{n=1}^{\infty} b_n \sin nx$

Then,

$$a_0 = \frac{1}{\pi} \int_{-\pi}^{\pi} f(x)\,dx = \frac{1}{\pi} \int_{-\pi}^{0} f(x)\,dx + \frac{1}{\pi} \int_{0}^{\pi} f(x)\,dx$$

$$= \frac{1}{\pi} \int_{-\pi}^{0} (-x - \pi)\,dx + \frac{1}{\pi} \int_{0}^{\pi} (x + \pi)\,dx$$

$$= \frac{1}{\pi}\left[\left(-\frac{x^2}{2} - \pi x\right)_{-\pi}^{0} + \left(\frac{x^2}{2} + \pi x\right)_{0}^{\pi}\right]$$

$$= \frac{1}{\pi}\left\{\left(\frac{\pi^2}{2} - \pi^2\right) + \left(\frac{\pi^2}{2} + \pi^2\right)\right\} = \pi$$

$$a_n = \frac{1}{\pi} \int_{-\pi}^{\pi} f(x) \cos nx\,dx$$

$$= \frac{1}{\pi} \int_{-\pi}^{0} f(x)\cdot\cos nx\,dx + \frac{1}{\pi} \int_{0}^{\pi} f(x)\cdot\cos nx\,dx$$

$$= \frac{1}{\pi} \int_{-\pi}^{0} (-x - \pi)\cos nx\,dx + \frac{1}{\pi} \int_{0}^{\pi} (x + \pi)\cos nx\,dx$$

$$= \frac{1}{\pi}\left[(-x - \pi)\frac{\sin nx}{n} - (-1)\left\{-\frac{\cos nx}{n^2}\right\}\right]_{-\pi}^{0}$$

$$+ \frac{1}{\pi}\left[(x + \pi)\frac{\sin nx}{n} - (-1)\left\{-\frac{\cos nx}{n^2}\right\}\right]_{0}^{\pi}$$

$$= \frac{1}{\pi}\left[-\frac{1}{n^2} + \frac{(-1)^n}{n^2}\right] + \frac{1}{\pi}\left[\frac{(-1)^n}{n^2} - \frac{1}{n^2}\right]$$

$$= \frac{2}{n^2\pi}[(-1)^n - 1] = \begin{cases} -\dfrac{4}{n^2\pi} & ; \quad \text{if } n \text{ is odd} \\ 0 & ; \quad \text{if } n \text{ is even} \end{cases}$$

Also $\quad b_n = \dfrac{1}{\pi} \displaystyle\int_{-\pi}^{\pi} f(x) \sin nx\,dx$

$$= \frac{1}{\pi}\left\{\int_{-\pi}^{0} f(x)\cdot\sin nx\,dx + \int_{0}^{\pi} f(x)\cdot\sin nx\,dx\right\}$$

$$= \frac{1}{\pi}\left\{\int_{-\pi}^{0} (-x - \pi)\sin nx\,dx + \int_{0}^{\pi} (x + \pi)\sin nx\,dx\right\}$$

$$= \frac{1}{\pi}\left[(-x - \pi)\left(-\frac{\cos nx}{n}\right) - (-1)\left\{-\frac{\sin nx}{n^2}\right\}\right]_{-\pi}^{0}$$

$$+ \frac{1}{\pi}\left[(x + \pi)\left(-\frac{\cos nx}{n}\right) - (-1)\left\{-\frac{\sin nx}{n^2}\right\}\right]_{0}^{\pi}$$

$$= \frac{1}{\pi}\left[\frac{\pi}{n}\right] + \frac{1}{\pi}\left[\frac{-2\pi}{n}(-1)^n + \frac{\pi}{n}\right]$$

$$= \frac{1}{n}[1 - 2(-1)^n + 1] = \frac{2}{n}[1 - (-1)^n]$$

$$= \begin{cases} \dfrac{4}{n} & , \quad \text{if } n \text{ is odd} \\ 0 & , \quad \text{if } n \text{ is even} \end{cases}$$

The required Fourier series is given by

$$f(x) = \frac{a_0}{2} + a_1 \cos x + a_2 \cos 2x$$
$$+ ... + b_1 \sin x + b_2 \sin 2x + ...$$

$$= \frac{\pi}{2} - \frac{4}{\pi}\left(\frac{\cos x}{1^2} + \frac{\cos 3x}{3^2} + ...\right)$$
$$+ 4\left(\frac{\sin x}{1} + \frac{\sin 3x}{3} + ...\right)$$

EXAMPLE 7. *Find the Fourier series for the function f(x) = x*

+ x^2, $-\pi < x < \pi$. Hence, show that

(KURUKSHETRA–2005, UPTU–2003)

(i) $\dfrac{\pi^2}{6} = 1 + \dfrac{1}{2^2} + \dfrac{1}{3^2} + \dfrac{1}{4^2} + ...$ (UPTU–2003)

(ii) $\dfrac{\pi^2}{12} = \dfrac{1}{1^2} - \dfrac{1}{2^2} + \dfrac{1}{3^2} - \dfrac{1}{4^2} + ...$

SOLUTION. Let the Fourier series be

$$x + x^2 = \frac{a_0}{2} + \sum_{n=1}^{\infty} a_n \cos nx + \sum_{n=1}^{\infty} b_n \sin nx \quad ... (1)$$

Here, $a_0 = \dfrac{1}{\pi}\displaystyle\int_{-\pi}^{\pi}(x + x^2)dx$

$$= \frac{1}{\pi}\left[\int_{-\pi}^{\pi} x\,dx + \int_{-\pi}^{\pi} x^2\,dx\right]$$

$$= \frac{2}{\pi}\int_0^{\pi} x^2 dx = \frac{2}{3}\pi^2$$

$$a_n = \frac{1}{\pi}\int_{-\pi}^{\pi}(x + x^2)\cos nx\,dx$$

$$= \frac{1}{\pi}\left[\int_{-\pi}^{\pi} x\cos nx\,dx + \int_{-\pi}^{\pi} x^2\cos nx\,dx\right]$$

$$= \frac{2}{\pi}\int_0^{\pi} x^2\cos nx\,dx$$

$$= \frac{2}{\pi}\left[\left(x^2\frac{\sin nx}{n}\right)_0^{\pi} - \int_0^{\pi} 2x.\frac{\sin nx}{n}dx\right]$$

$$= -\frac{4}{n\pi}\int_0^{\pi} x\sin nx\,dx$$

$$= -\frac{4}{n\pi}\left[\left\{x\left(-\frac{\cos nx}{n}\right)\right\}_0^{\pi} - \int_0^{\pi} 1.\left(-\frac{\cos nx}{n}\right)dx\right]$$

$$= -\frac{4}{n\pi}\left(-\frac{\pi}{n}\cos nx\right) = \frac{4}{n^2}\cos n\pi = \frac{4}{n^2}(-1)^n$$

and $b_n = \dfrac{1}{\pi}\displaystyle\int_{-\pi}^{\pi}(x + x^2)\sin nx\,dx$

$$= \frac{2}{\pi}\int_0^{\pi} x\sin nx\,dx + \frac{2}{\pi}\int_0^{\pi} x^2\sin nx\,dx$$

$$\left[\because \int_0^{\pi} x^2 \sin nx\,dx = 0\right]$$

$$= \frac{2}{\pi}\left(-\frac{\pi}{n}\cos n\pi\right) = -\frac{2}{n}(-1)^n.$$

From (1), $x + x^2 = \dfrac{\pi^2}{3} + 4\displaystyle\sum_{n=1}^{\infty}\frac{(-1)^n}{n^2}$

$$\cos nx - 2\sum_{n=1}^{\infty}\frac{(-1)^n}{n}\sin nx$$

$$f(x) = \frac{\pi^2}{3} + 4\left[-\frac{1}{1^2}\cos x + \frac{1}{2^2}\cos 2x\right.$$

$$\left. -\frac{1}{3^2}\cos 3x + ...\right]$$

$$-2\left[-\frac{1}{1}\sin x + \frac{1}{2}\sin 2x - \frac{1}{3}\sin 3x + ...\right]. \quad ...(2)$$

We observe that the series on the R.H.S. given by equation (2), always represents $x + x^2$ for all values of x except the end points $-\pi$ or π.

At the point of discontinuity

$$f(-\pi) = \frac{1}{2}(\text{L.H.L.} + \text{R.H.L.})$$

$$= \frac{1}{2}[f(-\pi - 0) + f(-\pi + 0)]$$

$$= \frac{1}{2}[f(\pi - 0) + f(-\pi + 0)]$$

$$= \frac{1}{2}[\pi + \pi^2 + (-\pi) + (-\pi)^2] = \pi^2.$$

Putting $x = -\pi$ in equation (2), we get

$$\pi^2 = \frac{\pi^2}{3} + 4\left[\frac{1}{1^2} + \frac{1}{2^2} + \frac{1}{3^2} + \frac{1}{4^2} + ...\right]$$

Therefore, $\dfrac{\pi^2}{6} = 1 + \dfrac{1}{2^2} + \dfrac{1}{3^2} + \dfrac{1}{4^2} + ...$ ... (3)

Again, putting $x = 0$ in equaton (2), we get

$$0 = \frac{\pi^2}{3} + 4\left[-\frac{1}{1^2} + \frac{1}{2^2} - \frac{1}{3^2} + \frac{1}{4^2} - ...\right]$$

$$\Rightarrow \frac{\pi^2}{12} = \frac{1}{1^2} - \frac{1}{2^2} + \frac{1}{3^2} - \frac{1}{4^2}...$$

EXAMPLE 8. *Express f(x) = |x|, $-\pi < x < \pi$, as Fourier series.*

Hence, show that $\dfrac{1}{1^2} + \dfrac{1}{3^2} + \dfrac{1}{5^2} + ... = —$.

(UPTU(Q.BANK)–2001, SVTU–2009, KERALA–2005, PTU–2005)

SOLUTION. Here, $f(-x) = |-x| = |x| = f(x)$

$\therefore$ $f(x)$ is an even function and hence $b_n = 0$.

Let $f(x) = |x| = \dfrac{a_0}{2} + \displaystyle\sum_{n=1}^{\infty} a_n \cos nx$

Then, $a_0 = \dfrac{2}{\pi}\displaystyle\int_0^{\pi} f(x)dx = \frac{2}{\pi}\int_0^{\pi}|x|\,dx$

$$= \frac{2}{\pi}\int_0^{\pi} x\,dx = \frac{2}{\pi}\left[\frac{x^2}{2}\right]_0^{\pi} = \pi$$

and $a_n = \dfrac{2}{\pi}\displaystyle\int_0^{\pi} f(x)\cos nx\,dx$

$$= \frac{2}{\pi}\int_0^\pi |x|.\cos nx\,dx = \frac{2}{\pi}\int_0^\pi x\cos nx\,dx$$

$$= \frac{2}{\pi}\left[x\left(\frac{\sin nx}{n}\right) - 1\left(-\frac{\cos nx}{n^2}\right)\right]_0^\pi$$

$$= \frac{2}{\pi}\left[\frac{\cos nx}{n^2} - \frac{1}{n^2}\right]$$

$$= \frac{2}{\pi n^2}[(-1)^n - 1] = \begin{cases} 0 & , \text{ if } n \text{ is even} \\ -\frac{4}{\pi n^2} & , \text{ if } n \text{ is odd} \end{cases}$$

Hence,

$$|x| = \frac{\pi}{2} - \frac{4}{\pi}\left(\cos x + \frac{\cos 3x}{3^2} + \frac{\cos 5x}{5^2} + ...\right)$$

... (1)

DEDUCTION.   Putting $x = 0$, in equation (1), we get

$$\frac{1}{1^2} + \frac{1}{3^2} + \frac{1}{5^2} + ... = \frac{\pi^2}{8}.$$

Exercise-7.6

1. Express $f(x) = \frac{1}{2}(\pi - x)$ in a Fourier series in the interval $0 < x < 2\pi$.

2. Find the Fourier series to represent the function $f(x) = |\sin x|$, $-\pi < x < \pi$.   (UPTU(Q.BANK)–2001)

3. Obtain the Fourier series to represent $f(x) = \frac{1}{4}(\pi - x)^2, 0 < x < 2\pi$.   Hence, obtain the following results :   (UPTU(Q.BANK)–2001)

(i) $\frac{1}{1^2} + \frac{1}{2^2} + \frac{1}{3^2} + \frac{1}{4^2} + ... = \frac{\pi^2}{6}$

(ii) $\frac{1}{1^2} - \frac{1}{2^2} + \frac{1}{3^2} - \frac{1}{4^2} + ... = \frac{\pi^2}{12}$

(iii) $\frac{1}{1^2} + \frac{1}{3^2} + \frac{1}{5^2} + ... = \frac{\pi^2}{8}$

4. Expand in a Fourier series the function $f(x) = x$ in the interval $0 < x < 2\pi$, sketch its graph from $x = -4\pi$ to $x = 4\pi$.   (UPTU(Q.BANK)–2001)

5. Show that for $-\pi < x < \pi$

$$\sin ax = \frac{2\sin a\pi}{\pi}\left(\frac{\sin x}{1^2 - a^2} - \frac{2\sin 2x}{2^2 - a^2} + \frac{3\sin 3x}{3^2 - a^2} - ...\right).$$

6. Obtain a Fourier expansion for $\sqrt{1 - \cos x}$ in the interval $-\pi < x < \pi$.

7. Obtain a Fourier series to represent e^{-ax} from $x = -\pi$ to $x = \pi$. Hence derive the series for $\frac{\pi}{\sinh \pi}$.

8. Find the Fourier series to represent the periodic function:

$$f(x) = \begin{cases} x & , -\pi/2 < x < \pi/2 \\ \pi - x & , \pi/2 < x < 3\pi/2 \end{cases}$$ (UPTU(Q.BANK)–2001)

9. Find a series of sines and cosines to multiples of x which will represent $\frac{\pi}{\sinh \pi}e^x$ in the interval $-\pi < x < \pi$.

10. Prove that $x^2 = \frac{\pi^2}{3} + 4\sum_{n=1}^{\infty}(-1)^n \frac{\cos nx}{n^2}, -\pi < x < \pi$.   (UPTU–2004)

11. Prove that in the interval $x\cos x = -\frac{1}{2}\sin x + 2\sum_{n=2}^{\infty}\frac{n(-1)^n}{n^2 - 1}\sin nx$ (UPTU(SP)–2001)

12. If $f(x) = \cos \omega x, -\pi < x < \pi$, where ω is a fraction as a fourier series, prove that $\cot \theta = \frac{1}{\theta} + \frac{2\theta}{\theta^2 - \pi^2} + \frac{2\theta}{\theta^2 - 4\pi^2} + ...$ (UPTU(Q.BANK)–2001)

Hints to Selected Problems

3. $a_0 = \frac{1}{\pi}\int_0^{2\pi}\frac{1}{4}(\pi - x)^2 dx = \frac{1}{4\pi}\left[\frac{(\pi - x)^3}{-3}\right]_0^{2\pi}$

$$= -\frac{1}{12\pi}[-\pi^3 - \pi^3] = \frac{\pi^2}{6}$$

$$a_n = \frac{1}{\pi}\int_0^{2\pi}f(x)\cos nx\,dx = \frac{1}{\pi}\int_0^{2\pi}\frac{1}{4}(\pi - x)^2\cos nx\,dx$$

$$= \frac{1}{4\pi}\left[(\pi - x)^2\frac{\sin nx}{n} - \{-2(\pi - x)\}\right.$$

$$\left.\left(-\frac{\cos nx}{n^2}\right) + 2\left(\frac{-\sin nx}{n^3}\right)\right]_0^{2\pi}$$

$$= \frac{1}{4\pi}\left[\frac{2\pi}{n^2} + \frac{2\pi}{n^2}\right] = \frac{1}{n^2}$$

and $b_n = \frac{1}{\pi}\int_0^{2\pi}\frac{1}{4}(\pi - x)^2\sin nx\,dx$

$$= \frac{1}{4\pi}\left[(\pi - x)^2\left(-\frac{\cos nx}{n}\right) - \{-2(\pi - x)\}\right.$$

$$\left.\left(-\frac{\sin nx}{n^2}\right) + 2\left(\frac{\cos nx}{n^3}\right)\right]_0^{2\pi}$$

$$= \frac{1}{4\pi}\left[\left(-\frac{\pi^2}{n} + \frac{2}{n^3}\right) - \left(-\frac{\pi^2}{n} + \frac{2}{n^3}\right)\right] = 0$$

$$\therefore\ f(x) = \frac{\pi^2}{12} + \sum_{n=1}^{\infty}\frac{\cos nx}{n^2}$$

$$= \frac{\pi^2}{12} + \frac{\cos x}{1^2} + \frac{\cos 2x}{2^2} + \frac{\cos 3x}{3^2} + ...$$ ...(1)

(i) Putting $x = 0$ in equation (1), we get

$$\frac{\pi^2}{4} = \frac{\pi^2}{12} + \left(\frac{1}{1^2} + \frac{1}{2^2} + \frac{1}{3^2} + \frac{1}{4^2} + ...\right)$$

$$\frac{\pi^2}{6} = \frac{1}{1^2} + \frac{1}{2^2} + \frac{1}{3^2} + \frac{1}{4^2} + ...$$ ...(2)

(ii) Putting $x=\pi$ in equation (1), we get

$$0 = \frac{\pi^2}{12} + \left[\left(\frac{-1}{1^2}\right) + \frac{1}{2^2} + \left(-\frac{1}{3^2}\right) + \frac{1}{4^2} + \ldots\right]$$

$$\frac{\pi^2}{12} = \frac{1}{1^2} - \frac{1}{2^2} + \frac{1}{3^2} - \frac{1}{4^2} + \ldots \qquad \ldots(3)$$

(iii) Adding equations (2) and (3), we get

$$\Rightarrow \frac{\pi^2}{6} + \frac{\pi^2}{12} = 2\left(\frac{1}{1^2} + \frac{1}{3^2} + \frac{1}{5^2} + \ldots\right)$$

$$\Rightarrow \frac{\pi^2}{4} = 2\left(\frac{1}{1^2} + \frac{1}{3^2} + \frac{1}{5^2} + \ldots\right) \qquad \ldots(4)$$

$$\Rightarrow \frac{\pi^2}{8} = \frac{1}{1^2} + \frac{1}{3^2} + \frac{1}{5^2} + \ldots$$

5. Here, $a_0 = 0$, $a_n = 0$ $\qquad (\because f(x)$ is an odd function$)$

$$b_n = \frac{2}{\pi}\int_0^\pi \sin ax \sin nx\, dx = \frac{1}{\pi}\int_0^\pi [\cos(n-a)x - \cos(n+a)x]dx$$

$$= \frac{1}{\pi}\left[\frac{\sin(n-a)x}{(n-a)} - \frac{\sin(n+a)x}{n+a}\right]_0^\pi$$

$$= \frac{1}{\pi}\left[\frac{\sin(n-a)\pi}{n-a} - \frac{\sin(n+a)\pi}{n+a}\right]$$

$$= \frac{1}{\pi}\left[\frac{(-1)^n(-\sin a\pi)}{n-a} - \frac{(-1)^n \sin a\pi}{n+a}\right]$$

$$= \frac{(-1)^n \sin a\pi}{\pi}\left[\frac{1}{n-a} + \frac{1}{n+a}\right] = (-1)^{n+1}\frac{2n\sin a\pi}{\pi(n^2-a^2)}$$

$$\therefore \sin ax = \frac{2\sin a\pi}{\pi}\sum_{n=1}^\infty \frac{(-1)^{n+1} n}{n^2-a^2}\sin nx$$

$$= \frac{2\sin a\pi}{\pi}\left[\frac{\sin x}{1^2-a^2} - \frac{2\sin 2x}{2^2-a^2} + \frac{3\sin 3x}{3^2-a^2} - \ldots\right]$$

9. $f(n) = \frac{1}{2l}\int_{-l}^l f(x)dx$

$$+ \frac{1}{l}\sum_{n=1}^\infty\left[\cos\frac{n\pi x}{l} - \int_{-l}^l f(x)\cdot\cos\frac{n\pi x}{l}dx\right.$$

$$\left. + \sin\frac{n\pi x}{l}\int_{-l}^l f(x)\sin\frac{n\pi x}{l}dx\right]$$

$$\frac{\pi}{2\sin n\pi}e^x = \frac{1}{2\pi}\int_{-\pi}^\pi \frac{\pi}{2\sin n\pi}e^x dx$$

$$+ \frac{1}{\pi}\sum_{n=1}^\infty \cos nx\int_{-\pi}^\pi \frac{\pi}{2\sin n\pi}e^x \cos nu\, du$$

$$+ \frac{1}{\pi}\sum_{n=1}^\infty \sin nx\int_{-\pi}^\pi \frac{\pi}{2\sin n\pi}e^x \sin nu\, du.$$

We have $\int_{-\pi}^\pi e^u du = \left[e^u\right]_{-\pi}^\pi = 2\sin n\pi$

$$\int_{-\pi}^\pi e^u \cos nu\, du$$

$$= \left[e^u \frac{\sin nu}{n}\right]_{-\pi}^\pi - \frac{1}{n}\int_{-\pi}^\pi e^u \sin nu\, du$$

$$= \frac{1}{n^2}\left[e^u \cos nu\right]_{-\pi}^\pi - \frac{1}{n^2}\int_{-\pi}^\pi e^u \cos nu\, du$$

$$\left(1 + \frac{1}{n^2}\right)\int_{-\pi}^\pi e^u \cos nu\, du = \frac{1}{n^2}(e^\pi - e^{-\pi})\cos n\pi$$

$$\int_{-\pi}^\pi e^u \cos nu\, du = \frac{2}{1+n^2}\sin n\pi \cos n\pi$$

$$\int_{-\pi}^\pi e^u \sin nu\, du = \left[\frac{-e^u \cos nu}{n}\right]_{-\pi}^\pi + \frac{1}{n}\int_{-\pi}^\pi e^u \cos nu\, du$$

$$= -\frac{1}{n}(e^\pi - e^{-\pi})\cos n\pi + \frac{1}{n}\left\{\left[\frac{e^u \sin nu}{n}\right]_{-\pi}^\pi - \int_{-\pi}^\pi \frac{e^u \sin nu}{n}du\right\}$$

$$\left(1 + \frac{1}{n^2}\right)\int_{-\pi}^\pi e^u \sin nu\, du = -\frac{1}{n}(e^\pi - e^{-\pi})\cos n\pi$$

$$\int_{-\pi}^\pi e^u \sin nu\, du = \frac{2}{1+n^2}\sin n\pi \cos n\pi$$

$$\therefore \frac{\pi}{2\sin n\pi}e^x = \frac{1}{2} + \sum_{n=1}^\infty \frac{\cos n\pi}{1+n^2}\cos nx - \sum_{n=1}^\infty\left\{\frac{n}{1+n^2}\cos n\pi \sin nx\right\}$$

$$= \frac{1}{2} - \left(\frac{1}{2}\cos x - \frac{1}{2}\cos 2x + \frac{1}{10}\cos 3x - \frac{1}{17}\cos 4x + \ldots\right)$$

$$+ \left(\frac{1}{2}\sin x - \frac{2}{5}\sin 2x + \frac{3}{10}\sin 3x - \frac{4}{17}\sin 4x + \ldots\right)$$

Answers

1. $f(x) = \sum_{n=1}^\infty \frac{\sin nx}{n}$

2. $|\sin x| = \frac{2}{\pi} \cdot \frac{4}{\pi}\left(\frac{\cos 2x}{3} + \frac{\cos 4x}{15} + \ldots + \frac{\cos 2nx}{4n^2-1} + \ldots\right)$

3. $f(x) = \frac{\pi^2}{12} + \sum_{n=1}^\infty \frac{\cos nx}{n^2} = \frac{\pi^2}{12} + \frac{\cos x}{1^2} + \frac{\cos 2x}{2^2} + \frac{\cos 3x}{3^2} + \ldots$

4. $f(x) = \pi - 2 \cdot \sum_{n=1}^\infty \frac{\sin nx}{n}$

5. $\sin ax = \frac{2\sin a\pi}{\pi}\sum_{n=1}^\infty \frac{(-1)^{n+1}}{n^2-a^2}\sin nx$

6. $\sqrt{1-\cos x} = \frac{2\sqrt{2}}{\pi} - \frac{4\sqrt{2}}{\pi}\sum_{n=1}^\infty \frac{\cos nx}{4n^2-1}$

7. $e^{-ax} = 2\frac{\sinh a\pi}{\pi}\left[\left(\frac{1}{2a} - \frac{a\cos x}{1^2+a^2} + \frac{a\cos 2x}{2^2+a^2} - \ldots\right) - \left(\frac{\sin x}{1^2+a^2} - \frac{2\sin 2x}{2^2+a^2} + \frac{3\sin 3x}{3^2+a^2} \ldots\right)\right]$; $\frac{\pi}{\sinh\pi} = 2\left[\frac{1}{2^2+1} - \frac{1}{3^2+1} + \frac{1}{4^2+1} - \ldots\right]$

8. $f(x) = \frac{4}{\pi}\left[\frac{\sin x}{1^2} - \frac{\sin 3x}{3^2} + \frac{\sin 5x}{5^2} - \ldots\right]$

9. $\frac{\pi}{2\sin n\pi}e^x = \frac{1}{2} + \sum_{n=1}^\infty \frac{\cos n\pi}{1+n^2}\cos nx - \sum_{n=1}^\infty\left\{\frac{n}{1+n^2}\cos nx \sin n\pi\right\} = \frac{1}{2} - \left(\frac{1}{2}\cos x - \frac{1}{2}\cos 2x + \frac{1}{10}\cos 3x - \frac{1}{17}\cos 4x + \ldots\right)$

7.19 FOURIER SERIES FOR DISCONTINUOUS FUNCTIONS

At the point of discontinuity, the value of function for Fourier series is obtained by the average of left hand limit and right hand limit of function at that point of discontinuity.

Solved Examples

EXAMPLE 1. *Obtain Fourier series for the function*

$$f(x) = \begin{cases} x & ; & -\pi < x < 0 \\ -x & ; & 0 < x < \pi \end{cases}$$

and hence show that $\dfrac{1}{1^2} + \dfrac{1}{3^2} + \dfrac{1}{5^2} + \ldots = \dfrac{\pi^2}{8}$.

(UPTU–2002)

SOLUTION .   We know that

$$f(x) = \frac{a_0}{2} + \sum_{n=1}^{\infty} a_n \cos nx + \sum_{n=1}^{\infty} b_n \sin nx \quad \ldots(1)$$

$$a_0 = \frac{1}{\pi} \int_{-\pi}^{\pi} f(x)\, dx = \frac{1}{\pi}\left[\int_{-\pi}^{0} x\, dx + \int_{0}^{\pi} -x\, dx \right]$$

$$= \frac{1}{\pi}\left[\left(\frac{x^2}{2}\right)_{-\pi}^{0} - \left(\frac{x^2}{2}\right)_{0}^{\pi} \right] = \frac{1}{\pi}\left[0 - \frac{\pi^2}{2} - \frac{\pi^2}{2} \right] = -\pi$$

$$a_n = \frac{1}{\pi} \int_{-\pi}^{\pi} f(x)\cos nx\, dx$$

$$= \frac{1}{\pi}\left[\int_{-\pi}^{0} x\cos nx\, dx + \int_{0}^{\pi} -x\cos nx\, dx \right]$$

$$= \frac{1}{\pi}\left[\left(\frac{x\sin nx}{n} \right)_{-\pi}^{0} - \int_{-\pi}^{0} \frac{\sin nx}{n}\, dx \right.$$

$$\left. + \left(-x\frac{\sin nx}{n} \right)_{0}^{\pi} - \int_{0}^{\pi}(-1)\frac{\sin nx}{n}\, dx \right]$$

$$= \frac{1}{\pi}\left[\frac{1}{n^2}(\cos nx)_{-\pi}^{0} - \frac{1}{n^2}(\cos nx)_{0}^{\pi} \right]$$

$$= \frac{1}{\pi}\left[\left\{ \frac{1-(-1)^n}{n^2} \right\} - \left\{ \frac{(-1)^n - 1}{n^2} \right\} \right]$$

$$= \frac{1}{\pi}\left[\frac{2\{1-(-1)^n\}}{n^2} \right] = \frac{2}{\pi n^2}[1-(-1)^n]$$

$$= \begin{cases} 0 & ; \text{ if } n \text{ is even} \\ \dfrac{4}{\pi n^2} & ; \text{ if } n \text{ is odd} \end{cases}$$

and $b_n = \dfrac{1}{\pi} \int_{-\pi}^{\pi} f(x)\sin nx\, dx$

$$= \frac{1}{\pi}\left[\int_{-\pi}^{0} x\sin nx\, dx + \int_{0}^{\pi} -x\sin nx\, dx \right]$$

$$= \frac{1}{\pi}\left[\left(x\frac{-\cos nx}{n} \right)_{-\pi}^{0} - \int_{-\pi}^{0} \frac{-\cos nx}{n}\, dx \right.$$

$$\left. + \left(x\frac{\cos nx}{n} \right)_{0}^{\pi} - \int_{0}^{\pi}(-1)\frac{-\cos nx}{n}\, dx \right]$$

$$= \frac{1}{\pi}\left[\frac{-\pi}{n}(-1)^n + \frac{1}{n}(-1)^n \right] = 0$$

From (1)

$$f(x) = -\frac{\pi}{2} + \frac{4}{\pi}\left(\frac{\cos x}{1^2} + \frac{\cos 3x}{3^2} + \frac{\cos 5x}{5^2} + \ldots \right)$$

At the point of discontinuity

$$f(0) = \frac{1}{2}[f(0^-) + f(0^+)] = \frac{1}{2}[0-0] = 0$$

Putting, $x = 0$ in (2), we get

$$0 = -\frac{\pi}{2} + \frac{4}{\pi}\left(\frac{1}{1^2} + \frac{1}{3^2} + \frac{1}{5^2} + \ldots \right)$$

Hence, $\dfrac{1}{1^2} + \dfrac{1}{3^2} + \dfrac{1}{5^2} + \ldots = \dfrac{\pi^2}{8}$.

EXAMPLE 2. *Obtain the Fourier series to represent f(x) given*

as follows : $f(x) = \begin{cases} x & ; & \text{for } 0 \le x \le \pi \\ 2\pi - x & ; & \text{for } \pi \le x \le 2\pi \end{cases}$

(SVTU–2008, BPTU–2005S)

SOLUTION.   Let $f(x) = \dfrac{a_0}{2} + \sum\limits_{n=1}^{\infty} a_n \cos nx + \sum\limits_{n=1}^{\infty} b_n \sin nx$, $0 \le x \le 2\pi$

$$\ldots (1)$$

where $a_0 = \dfrac{1}{\pi} \int_{0}^{2\pi} f(x)\, dx$

$$= \frac{1}{\pi}\left[\int_{0}^{\pi} x\, dx + \int_{\pi}^{2\pi}(2\pi - x)\, dx \right]$$

$$= \frac{1}{\pi}\left[\left(\frac{x^2}{2} \right)_{0}^{\pi} + \left(2\pi x - \frac{x^2}{2} \right)_{\pi}^{2\pi} \right]$$

$$= \frac{1}{\pi}\left[\frac{\pi^2}{2} + 2\pi(2\pi - x) - \frac{1}{2}(4\pi^2 - \pi^2) \right]$$

$$= \frac{1}{\pi}(\pi^2) = \pi$$

$$a_n = \frac{1}{\pi} \int_{0}^{2\pi} f(x)\cos nx\, dx$$

$$= \frac{1}{\pi}\left[\int_{0}^{\pi} x\cos nx\, dx + \int_{0}^{2\pi}(2\pi - x)\cos nx\, dx \right]$$

$$= \frac{1}{\pi}\left[\left\{ \frac{x\sin nx}{n} + \frac{\cos nx}{n^2} \right\}_{0}^{\pi} \right.$$

$$\left. + \left\{ (2\pi - x)\frac{\sin nx}{n} - \frac{\cos nx}{n^2} \right\}_{\pi}^{2\pi} \right]$$

$$= \frac{1}{\pi}\left[\left(\frac{\cos n\pi - 1}{n^2} \right) - \left(\frac{1 - \cos n\pi}{n^2} \right) \right]$$

$$= \frac{2}{n^2\pi}[(-1)^n - 1] = \begin{cases} 0 & , \text{ if } n \text{ is even} \\ -\dfrac{4}{n\pi^2} & , \text{ if } n \text{ is odd} \end{cases}$$

Again $b_n = \frac{1}{\pi}\int_0^{2\pi} f(x).\sin nx\, dx$

$$= \frac{1}{\pi}\left[\int_0^\pi x \sin nx\, dx + \int_\pi^{2\pi}(2\pi - x)\sin nx\, dx\right]$$

$$= \frac{1}{\pi}\left[\left\{-\frac{x\cos nx}{n} + \frac{\sin nx}{n^2}\right\}_0^\pi\right.$$

$$\left.+ \left\{-(2\pi - x)\frac{\cos nx}{n} - \frac{\sin nx}{n^2}\right\}_\pi^{2\pi}\right]$$

$$= \left(\frac{-\pi\cos n\pi}{n} + \frac{\pi\cos n\pi}{n}\right) = 0$$

Therefore, $f(x) = \dfrac{\pi}{2} - \dfrac{4}{\pi}\left[\cos x + \dfrac{\cos 3x}{3^2}\right.$

$$\left.+ \frac{\cos 5x}{5^2} + ...\right], 0 \le x \le 2\pi$$

which is the required Fourier series for $f(x)$.

EXAMPLE 3. *If* $f(x) = \begin{cases} 0 & , \quad -\pi \le x \le 0 \\ \sin x & , \quad 0 \le x \le \pi \end{cases}$

Prove that $f(x) = \dfrac{1}{\pi} + \dfrac{1}{2}\sin x - \dfrac{2}{\pi}\sum_{n=1}^{\infty}\dfrac{\cos 2nx}{4n^2 - 1}$.

(UPTU(Q.BANK)–2001)

Hence, show that

(i) $\dfrac{1}{1.3} + \dfrac{1}{3.5} + \dfrac{1}{5.7} + ... = \dfrac{1}{2}$

(ii) $\dfrac{1}{1.3} - \dfrac{1}{3.5} + \dfrac{1}{5.7} - ... = \dfrac{\pi - 2}{4}$

(BHOPAL–2008, MUMBAI–2005S, ROHTAK–2005, 15)

SOLUTION. Let $f(x) = \dfrac{a_0}{2} + \sum_{n=1}^\infty a_n \cos nx + \sum_{n=1}^\infty b_n \sin nx$

Then, $a_0 = \dfrac{1}{\pi}\int_{-\pi}^\pi f(x)\, dx$

$$= \frac{1}{\pi}\left[\int_{-\pi}^0 0.dx + \int_0^\pi \sin x\, dx\right] = \frac{2}{\pi}$$

$$a_n = \frac{1}{\pi}\int_{-\pi}^\pi f(x).\cos nx\, dx$$

$$= \frac{1}{\pi}\left[\int_{-\pi}^0 0.dx + \int_0^\pi \sin x \cos nx\, dx\right]$$

$$= \frac{1}{2\pi}\int_0^\pi 2\cos nx.\sin x\, dx$$

$$= \frac{1}{2\pi}\int_0^\pi[\sin(n+1)x - \sin(n-1)x]dx$$

$$= \frac{1}{2\pi}\left[-\frac{\cos(n+1)x}{n+1} + \frac{\cos(n-1)x}{n-1}\right]_0^\pi, n \ne 1$$

$$= \frac{1}{2\pi}\left[-\frac{\cos(n+1)\pi}{n+1} + \frac{\cos(n-1)\pi}{n-1}\right.$$

$$\left.+ \frac{1}{n+1} - \frac{1}{n-1}\right]$$

$$= \frac{1}{2\pi}\left[-\frac{(-1)^{n+1}}{n+1} + \frac{(-1)^{n-1}}{n-1} + \frac{1}{n+1} - \frac{1}{n-1}\right]$$

$$= \begin{cases} \dfrac{1}{2\pi}\left(-\dfrac{1}{n+1} + \dfrac{1}{n-1} + \dfrac{1}{n+1} - \dfrac{1}{n-1}\right) & , \text{ when } n \text{ is odd} \\ \dfrac{1}{2\pi}\left(\dfrac{1}{n+1} - \dfrac{1}{n-1} + \dfrac{1}{n+1} - \dfrac{1}{n-1}\right) & , \text{ when } n \text{ is even} \end{cases}$$

$$= \begin{cases} 0 & , \text{ when } n \text{ is odd}, i.e., n = 3,5,7,... \\ -\dfrac{2}{\pi(n^2 - 1)} & , \text{ when } n \text{ is even} \end{cases}$$

When $n = 1$, we have

$$a_1 = \frac{1}{\pi}\int_0^\pi \sin x \cos x\, dx = \frac{1}{2\pi}\int_0^\pi \sin 2x\, dx$$

$$= \frac{1}{2\pi}\left[-\frac{\cos 2x}{2}\right]_0^\pi = 0$$

and $b_n = \dfrac{1}{\pi}\int_{-\pi}^\pi f(x)\sin nx\, dx$

$$= \frac{1}{\pi}\left[\int_{-\pi}^0 0.dx + \int_0^\pi \sin x \sin nx\, dx\right]$$

$$= \frac{1}{2\pi}\int_0^\pi 2\sin nx \sin x\, dx$$

$$= \frac{1}{2\pi}\int_0^\pi[\cos(n-1)x - \cos(n+1)x]dx$$

$$= \frac{1}{2\pi}\left[\frac{\sin(n-1)x}{(n-1)} - \frac{\sin(n+1)x}{(n+1)}\right]_0^\pi = 0, n \ne 1$$

When $n = 1$, we have $b_1 = \dfrac{1}{\pi}\int_0^\pi \sin x \sin x\, dx$

$$= \frac{1}{2\pi}\int_0^\pi(1 - \cos 2x)dx = \frac{1}{2\pi}\left[x - \frac{\sin 2x}{2}\right]_0^\pi = \frac{1}{2}$$

$$\therefore f(x) = \frac{1}{\pi} - \frac{2}{\pi}\left[\frac{\cos 2x}{2^2 - 1} + \frac{\cos 4x}{4^2 - 1} + \frac{\cos 6x}{6^2 - 1} + ...\right]$$

$$+ \frac{1}{2}\sin x$$

$$= \frac{1}{\pi} + \frac{1}{2}\sin x - \frac{2}{\pi}\sum_{n=1}^\infty \frac{\cos 2nx}{(2n)^2 - 1}$$

(i) Putting $x = 0$ in equation (1), we have

$$0 = \frac{1}{\pi} - \frac{2}{\pi}\sum_{n=1}^\infty \frac{1}{4n^2 - 1}$$

$$\frac{1}{2} = \sum_{n=1}^\infty \frac{1}{4n^2 - 1} = \sum_{n=1}^\infty \frac{1}{(2n-1)(2n+1)}$$

$$= \frac{1}{1.3} + \frac{1}{3.5} + \frac{1}{5.7} + ...$$

(ii) Putting $x = \pi/2$ in equation (1), we have,

$$1 = \frac{1}{\pi} + \frac{1}{2} - \frac{2}{\pi}\sum_{n=1}^{\infty}\frac{\cos n\pi}{4n^2 - 1}$$

$$\Rightarrow \frac{1}{2} - \frac{1}{\pi} = -\frac{2}{\pi}\sum_{n=1}^{\infty}\frac{(-1)^n}{4n^2 - 1}$$

$$\Rightarrow \frac{\pi - 2}{4} = -\sum_{n=1}^{\infty}\frac{(-1)^n}{(2n-1)(2n+1)}$$

$$= -\left(-\frac{1}{1.3} + \frac{1}{3.5} - \frac{1}{5.7} + ...\right)$$

$$\Rightarrow \frac{1}{1.3} - \frac{1}{3.5} + \frac{1}{5.7} - ... = \frac{\pi - 2}{4}$$

 Exercise-7.7

1. Find the Fourier series for the following function:

$$f(x) = \begin{cases} x^2 , & 0 \le x \le \pi \\ -x^2 , & -\pi \le x \le 0 \end{cases}$$ (MUMBAI–2009, HISSAR–2007)

2. Find the Fourier series to represent the function:

$$f(x) = \begin{cases} -k , & \text{when } -\pi < x < 0 \\ k , & \text{when } 0 < x < \pi \end{cases}$$

Also deduce that $\frac{\pi}{4} = 1 - \frac{1}{3} + \frac{1}{5} - \frac{1}{7} + ...$

3. Find the Fourier series for the function:

$$f(x) = \begin{cases} -1 , & -\pi < x < -\pi/2 \\ 0 , & -\pi/2 < x < \pi/2 \\ 1 , & \pi/2 < x < \pi \end{cases}$$ (UPTU–2004, 2005)

4. Find the Fourier series expansion for $f(x)$ if

$$f(x) = \begin{cases} -\pi , & -\pi < x < 0 \\ x , & 0 < x < \pi \end{cases}$$ (BHOPAL–2008S)

Deduce that $\frac{1}{1^2} + \frac{1}{3^2} + \frac{1}{5^2} + ... = \frac{\pi^2}{8}$ (KOTTAYAM–2005)

5. Find the Fourier expansion of the function defined in one period by the relations:

$$f(x) = \begin{cases} 1 , & 0 < x < \pi \\ 2 , & \pi < x < 2\pi \end{cases}$$

and deduce that $\frac{\pi}{4} = 1 - \frac{1}{3} + \frac{1}{5} - \frac{1}{7} + ...$.

6. An alternating current after passing through a rectifier has the form $i = \begin{cases} I_0 \sin x & \text{for } 0 \le x < \pi \\ 0 & \text{for } \pi \le x \le 2\pi \end{cases}$

(VTU–2007, CALICUT–2005, UPTU(Q.BANK)–2001)

where I_0 is the maximum current and the period is 2π. Express i as a Fourier series.

 **Hints to Selected Problems**

3. $a_0 = \frac{1}{\pi}\int_{-\pi}^{-\pi/2}(-1)dx + \frac{1}{\pi}\int_{-\pi/2}^{\pi/2}0dx + \frac{1}{\pi}\int_{\pi/2}^{\pi}1\,dx = 0$

$a_n = \frac{1}{\pi}\int_{-\pi}^{-\pi/2}(-1)\cos nxdx + \frac{1}{\pi}\int_{-\pi/2}^{\pi/2}(0)\cos nxdx$

$+ \frac{1}{\pi}\int_{\pi/2}^{\pi}(1)\cos nx\,dx = 0$

$b_n = \frac{1}{\pi}\int_{-\pi}^{\pi/2}(-1)\sin nxdx + \frac{1}{\pi}\int_{-\pi}^{\pi/2}(0)\sin nxdx$

$+ \frac{1}{\pi}\int_{\pi/2}^{\pi}(1)\sin nx\,dx = 0 = -\left[\cos\frac{}{} - \cos\frac{}{}\pi\right]$

$b_1 = \frac{2}{\pi}, b_2 = -\frac{2}{\pi}, b_3 = \frac{2}{3\pi}$

$f(x) = \frac{1}{\pi}\left[2\sin x - 2\sin 2x + \frac{2}{3}\sin 3x + ...\right].$

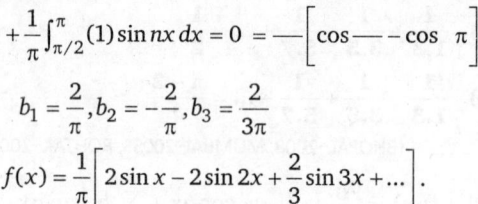 **Answers**

1. $f(x) = 2\left(\pi - \frac{4}{\pi}\right)\sin x - \pi\sin 2x + \frac{2}{3}\left(\pi - \frac{4}{9\pi}\right)\sin 3x - \frac{\pi}{2}\sin 4x + ..$ **2.** $f(x) = \frac{4k}{\pi}\left(\sin x + \frac{\sin 3x}{3} + \frac{\sin 5x}{5} + ...\right)$

3. $f(x) = \frac{2}{\pi}\left[\sin x - \sin 2x + \frac{\sin 3x}{3} + ...\right]$ **4.** $f(x) = -\frac{\pi}{4} - \frac{2}{\pi}\left(\cos x + \frac{\cos 3x}{3^2} + \frac{\cos 5x}{5^2} + ...\right) + \left(3\sin x - \frac{\sin 2x}{2} + \sin 3x - \frac{\sin 4x}{4} + ...\right)$

5. $f(x) = \frac{3}{2} - \frac{2}{\pi}\left(\sin x + \frac{\sin 3x}{3} + \frac{\sin 5x}{5} + ...\right)$ **6.** $i = \frac{I_0}{\pi} + \frac{I_0}{2}\sin x - \frac{2I_0}{\pi}\left(\frac{\cos 2x}{2^2 - 1} + \frac{\cos 4x}{4^2 - 1} + \frac{\cos 6x}{6^2 - 1} + ...\right)$

7.20 **CHANGE OF INTERVAL**

In many problems, the interval of Fourier expansion is $2l$ and not 2π. In order to apply this theory, this interval must be transformed into an interval of length 2π.

Consider a periodic function $f(x)$ defined in the interval $c < x < c + 2l$. To change the interval into one of length 2π, we put

$$\frac{x}{l} = \frac{z}{\pi} \text{ or } z = \frac{\pi x}{l} \text{ so that at } x = c, z = \frac{\pi c}{l} = d(\text{say})$$

When $\qquad x = c + 2l, z = \dfrac{\pi(c+2l)}{l} = \dfrac{\pi c}{l} + 2\pi = d + 2\pi$

Thus, the function $f(x)$ of period $2l$ in $(c, c+2l)$ is transformed to the function $f\left(\dfrac{lz}{\pi}\right) = F(z)$ say, or period in $(d, d+2\pi)$ and then function $F(z)$ can be expressed as a Fourier series such that

$$F(z) = \frac{a_0}{2} + \sum_{n=1}^{\infty} a_n \cos nz + \sum_{n=1}^{\infty} b_n \sin nz \qquad \qquad \ldots(1)$$

where $\qquad a_0 = \dfrac{1}{\pi}\int_d^{d+2\pi} F(z)dz; a_n = \dfrac{1}{\pi}\int_d^{d+2\pi} F(z)\cos nz\, dz$

and $\qquad b_n = \dfrac{1}{\pi}\int_d^{d+2\pi} F(z)\sin nz\, dz$

Now, making the inverse substitution $z = \dfrac{\pi x}{l}, dz = \dfrac{\pi}{l}dx$, when $z = d, x = c$ and when $z = d + 2\pi, x = c + 2l$. The expression (1) becomes

$$F(z) = F\left(\frac{\pi x}{l}\right) = F(x) = \frac{a_0}{2} + \sum_{n=1}^{\infty} a_n \cos\frac{n\pi x}{l} + \sum_{n=1}^{\infty} b_n \sin\frac{n\pi x}{l} \qquad \ldots(2)$$

The coefficient a_0, a_n, b_n in (2) becomes

$$a_0 = \frac{1}{l}\int_c^{c+2l} f(x)dx, \quad a_n = \frac{1}{l}\int_c^{c+2l} f(x)\cos\frac{n\pi x}{l}dx, \quad b_n = \frac{1}{l}\int_c^{c+2\pi} f(x)\sin\frac{n\pi x}{l}dx$$

⚓ REMARKS

- If $c = 0$, the interval become $0 < x < 2l$ and the a_0, a_n, b_n are given by

$$a_0 = \frac{1}{l}\int_0^{2l} f(x)dx, a_n = \frac{1}{l}\int_0^{2l} f(x)\cos\frac{n\pi x}{l}dx, \quad b_n = \frac{1}{l}\int_0^{2l} f(x)\sin\frac{n\pi x}{l}dx \; .$$

- If $c = -l$, the interval become $-l < x < l$ and a_0, a_n, b_n are given by

$$a_0 = \frac{1}{l}\int_{-l}^{l} f(x)dx, \quad a_n = \frac{1}{l}\int_{-l}^{l} f(x)\cos\frac{n\pi x}{l}dx, \quad b_n = \frac{1}{l}\int_{-l}^{l} f(x)\sin\frac{n\pi x}{l}dx.$$

📂 Solved Examples

EXAMPLE 1. *Find the Fourier series to represent $f(x) = x^2 - 2$ when $-2 \le x \le 2$.*

SOLUTION. Here, $b_n = 0$ because $f(x)$ is an even function

Let $f(x) = x^2 - 2 = \dfrac{a_0}{2} + \displaystyle\sum_{n=1}^{\infty} a_n \cos\frac{n\pi x}{2}$

$$[\because 2l = 4 \Rightarrow l = 2]$$

Then,

$$a_0 = \frac{2}{2}\int_0^2 (x^2 - 2)dx = \left[\frac{x^3}{3} - 2x\right]_0^2 = \frac{8}{3} - 4 = -\frac{4}{3}$$

and $\quad a_n = \dfrac{2}{2}\displaystyle\int_0^2 (x^2 - x)\cos\frac{n\pi x}{2}dx$

$$= \left[(x^2 - 2)\frac{\sin n\pi x/2}{(n\pi/2)}\right.$$

$$\left. - 2x\left(-\frac{\cos\frac{n\pi x}{2}}{(n^2\pi^2/4)} + 2\left(\frac{\sin\frac{n\pi x}{2}}{(n^3\pi^3/8)}\right)\right)\right]_0^2$$

$$= \frac{16\cos n\pi}{n^2\pi^2} = \frac{16(-1)^n}{n^2\pi^2}\; .$$

$$\therefore\; f(x) = (x^2 - 2) = -\frac{2}{3} + \frac{16}{\pi^2}\sum \frac{(-1)^n}{n^2}\cos\frac{n\pi x}{2}$$

$$= -\frac{2}{3} - \frac{16}{\pi^2}\left(\cos\frac{\pi x}{2} - \frac{1}{4}\cos\pi x + \frac{1}{9}\cos\frac{3\pi x}{2} - \ldots\right)$$

EXAMPLE 2. *Obtain the Fourier series for the function*

$$f(x) = \begin{cases} \pi x & ; \quad 0 \le x \le 1 \\ \pi(2-x) & ; \quad 1 \le x \le 2 \end{cases}$$

(UPTU–2001, VTU–2011, BHOPAL–2008, MUMBAI–2007)

SOLUTION. Here, $2l = 2 \Rightarrow l = 1$.

Let $f(x) = \dfrac{a_0}{2} + \displaystyle\sum_{n=1}^{\infty} a_n \cos n\pi x + \sum_{n=1}^{\infty} b_n \sin n\pi x$

where $a_0 = \displaystyle\int_0^2 f(x)dx = \int_0^1 \pi x\, dx + \int_1^2 \pi(2-x)dx$

$$= \pi\left[\frac{x^2}{2}\right]_0^1 + \pi\left[2x - \frac{x^2}{2}\right]_1^2$$

$$= \pi\left(\frac{1}{2}\right) + \pi\left[(4-2) - \left(2 - \frac{1}{2}\right)\right] = \pi$$

$$a_n = \int_0^2 f(x)\cos n\pi x\, dx$$

$$= \int_0^1 \pi x\cos n\pi x\, dx + \int_1^2 \pi(2-x)\cos n\pi x\, dx$$

$$= \left[\pi x\frac{\sin n\pi x}{n\pi} - \pi\left(-\frac{\cos n\pi x}{n^2\pi^2}\right)\right]_0^1$$

$$+ \left[\pi(2-x)\frac{\sin n\pi x}{n\pi} - (-\pi)\left(-\frac{\cos n\pi x}{n^2\pi^2}\right)\right]_1^2$$

$$= \left(\frac{\cos n\pi}{n^2\pi} - \frac{1}{n^2\pi}\right) + \left[-\frac{\cos 2n\pi}{n^2\pi} + \frac{\cos n\pi}{n^2\pi}\right]$$

$$= \frac{2}{n^2\pi}(\cos n\pi - 1)$$

$$= \frac{2}{n^2\pi}[(-1)^n - 1] = \begin{cases} 0 & ; \text{ if } n \text{ is even} \\ -\dfrac{4}{n^2\pi} & ; \text{ if } n \text{ is odd} \end{cases}$$

and $b_n = \int_0^2 f(x)\sin n\pi x\,dx$

$$= \int_0^1 \pi x \sin n\pi x\,dx + \int_1^2 \pi(2-x)\sin n\pi x\,dx$$

$$= \left[\pi x\left(\frac{-\cos n\pi x}{n\pi}\right) - \pi\left(-\frac{\sin n\pi x}{n^2\pi^2}\right)\right]_0^1$$

$$+ \left[-\pi(2-x)\frac{\cos n\pi x}{n\pi} - (-\pi)\left(-\frac{\sin n\pi x}{n^2\pi^2}\right)\right]_1^2$$

$$= \left[-\frac{\cos n\pi}{n}\right] + \left[\frac{\cos n\pi}{n}\right] = 0$$

Hence,

$$f(x) = \frac{\pi}{2} - \frac{4}{\pi}\left(\frac{\cos \pi x}{1^2} + \frac{\cos 3\pi x}{3^2} + \frac{\cos 5\pi x}{5^2} + \dots\right)$$

EXAMPLE 3. *Expand* $f(x) = e^{-x}$ *as a Fourier series in the interval* $(-l, l)$. (KERALA–2005, VTU–2004)

SOLUTION.   Let $f(x) = e^{-x} = \dfrac{a_0}{2} + \sum_{n=1}^{\infty} a_n \dfrac{\cos n\pi x}{l}$

$$+ \sum_{n=1}^{\infty} b_n \frac{\sin n\pi x}{l}$$

Then, $a_0 = \dfrac{1}{l}\int_{-l}^{l} e^{-x}dx = \dfrac{1}{l}\left[-e^{-x}\right]_{-l}^{l}$

$$= -\frac{1}{l}(e^l - e^{-l}) = \frac{2\sinh l}{l}$$

$$a_n = \frac{1}{l}\int_{-l}^{l} e^{-x}\cos\frac{n\pi x}{l}dx$$

$$= \frac{1}{l}\left[\frac{e^{-x}}{1+\left(\frac{n\pi}{l}\right)^2}\left(-\cos\frac{n\pi x}{l} + \frac{n\pi}{l}\sin\frac{n\pi x}{l}\right)\right]_{-l}^{l}$$

$$= \frac{l}{l^2+(n\pi)^2}[-e^{-l}\cos n\pi + e^l\cos n\pi]$$

$$= -\frac{2l\cos n\pi}{l^2+(n\pi)^2}\left(\frac{e^l - e^{-l}}{2}\right) = \frac{2l(-1)^n\sinh l}{l^2+(n\pi)^2}$$

$$b_n = \frac{1}{l}\int_{-l}^{l} e^{-x}\sin\frac{n\pi x}{l}dx$$

$$= \frac{1}{l}\left[\frac{e^{-x}}{1+\left(\frac{n\pi}{l}\right)^2}\left(-\sin\frac{n\pi x}{l} - \frac{n\pi}{l}\cos\frac{n\pi x}{l}\right)\right]_{-l}^{l}$$

$$= -\frac{l}{l^2+(n\pi)^2}\left[\frac{n\pi}{l}(e^{-l} - e^l)\cos n\pi\right]$$

$$= \frac{2n\pi\cos n\pi}{l^2+(n\pi)^2}\left(\frac{e^l - e^{-l}}{2}\right) = \frac{2n\pi(-1)^n\sinh l}{l^2+(n\pi)^2}$$

Hence, $e^{-x} = \sinh l\left[\frac{1}{l} - 2l\left(\frac{1}{l^2+\pi^2}\cos\frac{\pi x}{l}\right.\right.$

$$-\frac{1}{l^2+2^2\pi^2}\cos\frac{2\pi x}{l} + \frac{1}{l^2+3^2\pi^2}\cos\frac{3\pi x}{l} - \dots\Big)$$

$$-2\pi\left(\frac{1}{l^2+\pi^2}\sin\frac{\pi x}{l} - \frac{2}{l^2+2^2\pi^2}\sin\frac{2\pi x}{l}\right.$$

$$\left.\left. + \frac{3}{l^2+3^2\pi^2}\sin\frac{3\pi x}{l} - \dots\right)\right].$$

EXAMPLE 4. *Find the Fourier expansion for the function* $f(x) = x - x^2; -1 < x < 1$. (UPTU(Q.BANK)–2001)

SOLUTION. Let $f(x) = \dfrac{a_0}{2} + \sum_{n=1}^{\infty} a_n \cos n\pi x + \sum_{n=1}^{\infty} b_n \sin n\pi x$

Then, $a_0 = \int_{-1}^{1}(x - x^2)dx = \int_{-1}^{1} x\,dx - \int_{-1}^{1} x^2 dx$

$$= 0 - 2\int_0^1 x^2 dx = -2\left[\frac{x^3}{3}\right]_0^1 = -\frac{2}{3}$$

$$a_n = \int_{-1}^{1}(x - x^2)\cos n\pi x\,dx$$

$$= \int_{-1}^{1} x\cos n\pi x\,dx - \int_{-1}^{1} x^2\cos n\pi x\,dx$$

$$= 0 - 2\int_0^1 x^2\cos n\pi x\,dx$$

$$= -2\left[x^2\frac{\sin n\pi x}{n\pi} - 2x\left(-\frac{\cos n\pi x}{n^2\pi^2}\right)\right.$$

$$\left. + 2\left(-\frac{\sin n\pi x}{n^3\pi^3}\right)\right]_0^1$$

$$= -2\left[\frac{2\cos n\pi}{n^2\pi^2}\right] = -\frac{4(-1)^n}{n^2\pi^2}$$

and $b_n = \int_{-1}^{1}(x - x^2)\sin n\pi x\,dx = \int_{-1}^{1} x\sin n\pi x\,dx$

$$-1\int_{-1}^{1} x^2 n\pi x\,dx$$

$$= 2\int_0^1 x\sin n\pi x\,dx - 0$$

$$= 2\left[x\left(-\frac{\cos n\pi x}{n\pi}\right) - 1\left(-\frac{\sin n\pi x}{n^2\pi^2}\right)\right]_0^1$$

$$= 2\left[-\frac{\cos n\pi}{n\pi}\right] = -2\frac{(-1)^n}{n\pi}$$

$$\therefore\ x - x^2 = -\frac{1}{3} + \frac{4}{\pi^2}$$

$$\left(\frac{\cos\pi x}{1^2} - \frac{\cos 2\pi x}{2^2} + \frac{\cos 3\pi x}{3^2} - \ldots\right)$$

$$+\frac{2}{\pi}\left(\frac{\sin\pi x}{1} - \frac{\sin 2\pi x}{2} + \frac{\sin 3\pi x}{3} - \ldots\right).$$

 Exercise-7.8

1. Expand $f(x)$ in a Fourier series in the interval $(0, 2)$ if
$$f(x) = \begin{cases} x & , & 0 < x < 1 \\ 0 & , & 1 < x < 2 \end{cases}$$

2. Given $f(x) = \begin{cases} 0 & , & 0 < x < c \\ 1 & , & c < x < 2c \end{cases}$ expand $f(x)$ in a Fourier series of period $2c$.

3. Expand $f(x)$ in Fourier series in the interval $(-2, 2)$ when
$$f(x) = \begin{cases} 0 & , & -2 < x < 0 \\ 1 & , & 0 < x < 2 \end{cases}$$

4. Find a Fourier series for the function given by
$$f(t) = \begin{cases} t & , & 0 < t < 1 \\ 1 - t & , & 1 < t < 2 \end{cases}$$

5. Find a Fourier series corresponding to the function $f(x)$ defined in $(-2, 2)$ as follows;
$$f(x) = \begin{cases} 2 & , & \text{if } -2 \le x \le 0 \\ x & , & \text{if } 0 < x < 2 \end{cases}$$

6. Find a Fourier series for the function
$$f(x) = \begin{cases} 0 & , & \text{when } -2 < x < -1 \\ k & , & \text{when } -1 < x < 1 \\ 0 & , & \text{when } 1 < x < 2 \end{cases}$$

7. Find the Fourier series expansion of $f(x) = 2x - x^2$ in $(0, 3)$ and hence show that
$$\frac{1}{1^2} - \frac{1}{2^2} + \frac{1}{3^3} - \frac{1}{4^2} + \ldots - \infty = \frac{\pi}{12} \qquad \text{(MUMBAI–2005)}$$

Hints to Selected Problems

2. $a_0 = \frac{1}{c}\int_0^{2c} f(x)dx = \frac{1}{c}\int_0^c 0.dx + \frac{1}{c}\int_c^{2c} 1.dx = \frac{1}{c}[x]_c^{2c} = 1$

$a_n = \frac{1}{c}\int_0^{2c} f(x)\cos\frac{n\pi x}{c}dx$

$= \frac{1}{c}\int^c 0.\cos\frac{n\pi x}{c}dx + \frac{1}{c}\int^c 1.\cos\frac{n\pi x}{c}dx = \frac{1}{c}\left[\frac{c}{n\pi}\sin\frac{n\pi x}{c}\right]_c^{2c}$

$= \frac{1}{n\pi}[\sin 2n\pi - \sin n\pi] = 0$

$b_n = \frac{1}{c}\int_0^{2c} f(x).\sin\frac{n\pi x}{c}dx$

$= \frac{1}{c}\int_0^c 0.\sin\frac{n\pi x}{c}dx + \frac{1}{c}\int_c^{2c} 1.\sin\frac{n\pi x}{c}dx$

$= \frac{1}{c}\left[-\frac{c}{n\pi}\cos\frac{n\pi x}{c}\right]_c^{2c}$

$= -\frac{1}{n\pi}[\cos 2n\pi - \cos n\pi] = -\frac{1}{n\pi}[1 - (-1)^n]$

$= \begin{cases} -\dfrac{2}{n\pi} & , & \text{when } n \text{ is odd} \\ 0 & , & \text{when } n \text{ is even} \end{cases}$

Then, $f(x) = \frac{1}{2} - \frac{2}{\pi}\left(\frac{1}{1}\sin\frac{\pi x}{c} + \frac{1}{3}\sin\frac{3\pi x}{c} + \ldots\right)$

5. $a_0 = \frac{1}{l}\int_{-l}^{l} f(x)dx = \frac{1}{2}\left[\int_{-2}^0 2dx + \int_0^2 x\, dx\right]$

$= \frac{1}{2}\left[(2x)_{-2}^0 + \left(\frac{x^2}{2}\right)_0^2\right] = 3$

$a_n = \frac{1}{l}\int_{-l}^l f(x)\cos\left(\frac{n\pi x}{l}\right)dx$

$= \frac{1}{2}\left[\int_{-2}^0 2\cos\frac{n\pi x}{2}dx + \int_0^2 x\cos\frac{n\pi x}{2}dx\right]$

$= \frac{1}{2}\left[\frac{4}{n\pi}\left(\sin\frac{n\pi x}{2}\right)_{-2}^0 + \left(x\frac{2}{n\pi}\sin\frac{n\pi x}{2} + \frac{4}{n^2\pi^2}\cos\frac{n\pi x}{2}\right)_0^2\right]$

$= \frac{1}{2}\left[\frac{4}{n^2\pi^2}\cos n\pi - \frac{4}{n^2\pi^2}\right] = \frac{2}{n^2\pi^2}[(-1)^n - 1]$

$= \begin{cases} -\dfrac{4}{n^2\pi^2} & , & \text{when } n \text{ is odd} \\ 0 & , & \text{when } n \text{ is even} \end{cases}$

$b_n = \frac{1}{l}\int_{-l}^l f(x)\sin\left(\frac{n\pi x}{l}\right)dx$

$= \frac{1}{2}\left[2\int_{-2}^0 \sin\frac{n\pi x}{2}dx + \int_0^2 x\sin\frac{n\pi x}{2}dx\right]$

$= \frac{1}{2}\left[2\left(-\frac{2}{n\pi}\cos\frac{n\pi x}{2}\right)\right]_{-2}^0$

$+ \frac{1}{2}\left[x\left(-\frac{2}{n\pi}\cos\frac{n\pi x}{2}\right) + (1)\frac{4}{n^2\pi^2}\sin\frac{n\pi x}{2}\right]_0^2$

$= \frac{1}{2}\left[-\frac{4}{n\pi} + \frac{4}{n\pi}\cos n\pi\right]$

$+ \frac{1}{2}\left[-\frac{4}{n\pi}\cos n\pi + \frac{4}{n^2\pi^2}\sin n\pi\right]$

$= \frac{1}{2}\left[-\frac{4}{n\pi}\right] = -\frac{2}{n\pi}$

$f(x) = \frac{3}{2} - \frac{4}{\pi^2}\left\{\frac{1}{1^2}\cos\frac{\pi x}{2} + \frac{1}{3^2}\cos\frac{3\pi x}{2} + \ldots\right\}$

$- \frac{2}{\pi}\left\{\frac{1}{1}\sin\frac{\pi x}{2} + \frac{1}{2}\sin\frac{2\pi x}{2} + \frac{1}{3}\sin\frac{3\pi x}{3} + \ldots\right\}.$

Answers

1. $f(x) = \dfrac{1}{4} - \dfrac{2}{\pi^2}\left(\cos \pi x + \dfrac{\cos 3\pi x}{3^2} + \dfrac{\cos 5\pi x}{5^2} + ...\right) + \dfrac{1}{\pi}\left(\sin \pi x - \dfrac{\sin 2\pi x}{2} + \dfrac{\sin 3\pi x}{3} + ...\right)$

2. $f(x) = \dfrac{1}{2} - \dfrac{2}{\pi}\left\{\sin \dfrac{\pi x}{c} + \dfrac{1}{3}\sin \dfrac{3\pi x}{c} + ...\right\}$ 3. $f(x) = \dfrac{1}{2} + \dfrac{2}{\pi^2}\left(\sin \dfrac{\pi x}{2} + \dfrac{1}{3}\sin \dfrac{3\pi x}{2} + \dfrac{1}{5}\sin \dfrac{5\pi x}{2} + ...\right)$

4. $f(t) = -\dfrac{4}{\pi^2}\left(\cos \pi t + \dfrac{\cos 3\pi t}{3^2} + \dfrac{\cos 5\pi t}{5^2} + ...\right) + \dfrac{2}{\pi}\left(\sin \pi t + \sin \dfrac{3\pi t}{3} + ...\right)$

5. $f(x) = \dfrac{3}{2} - \dfrac{4}{\pi^2}\left\{\dfrac{1}{1^2}\cos \dfrac{\pi x}{2} + \dfrac{1}{3^2}\cos \dfrac{3\pi x}{2} + ...\right\} - \dfrac{2}{\pi}\left\{\sin \dfrac{\pi x}{2} + \dfrac{1}{2}\sin \dfrac{2\pi x}{2} + \dfrac{1}{3}\sin \dfrac{3\pi x}{2} + ...\right\}$

6. $f(x) = \dfrac{k}{2} + \dfrac{2R}{\pi}\left(\cos \dfrac{\pi x}{2} - \dfrac{1}{3}\cos \dfrac{3\pi x}{2} + \dfrac{1}{5}\cos \dfrac{5\pi x}{5} - ...\right)$ 7. $f(x) = -\sum_{n=1}^{\infty} \dfrac{9}{n^2\pi^2}\cos \dfrac{2n\pi x}{3} + \sum_{n=1}^{\infty} \dfrac{3}{n\pi}\sin \dfrac{2n\pi x}{3}$

7.21 HALF RANGE SERIES

When we require to expand a function $f(x)$ in the range $(0, \pi)$ in a Fourier series of period 2π or more generally in the range $(0, l)$ in a Fourier series of period $2l$, a function $f(x)$ defined over the interval $0 < x < l$ is capable of two distinct half range series.

The half range cosine series is $\quad f(x) = \dfrac{a_0}{2} + \sum\limits_{n=1}^{\infty} a_n \cos \dfrac{n\pi x}{l}$

where, $\qquad\qquad\qquad\qquad a_0 = \dfrac{2}{l}\int_0^l f(x).dx$, where $a_n = \dfrac{2}{l}\int_0^l f(x)\cos \dfrac{n\pi x}{l}dx$

The half range sine series is

$$f(x) = \sum_{n=1}^{\infty} b_n \sin \dfrac{n\pi x}{l}, \quad \text{where} \quad b_n = \dfrac{2}{l}\int_0^l f(x)\sin \dfrac{n\pi x}{l}dx$$

Solved Examples

EXAMPLE 1. If $f(x) = \begin{cases} x & ; & 0 < x < \pi/2 \\ \pi - x & ; & \pi/2 < x < \pi \end{cases}$ **Show that**

(i) $\quad f(x) = \dfrac{4}{\pi}\left[\sin x - \dfrac{\sin 3x}{3^2} + \dfrac{\sin 5x}{5^2} - ...\right]$

(MUMBAI–2008, SVTU–2008, VTU–2004)

(ii) $\quad f(x) = \dfrac{\pi}{4} - \dfrac{2}{\pi}\left[\dfrac{\cos 2x}{1^2} + \dfrac{\cos 6x}{3^2} + \dfrac{\cos 10x}{5^2} + ...\right]$

(VTU–2011)

SOLUTION. (i) Half range sine series, we have $l = \pi$ so

$$f(x) = \sum_{n=1}^{\infty} b_n \sin \dfrac{n\pi x}{\pi} = \sum_{n=1}^{\infty} b_n \sin nx$$

$$b_n = \dfrac{2}{\pi}\int_0^\pi f(x)\sin nx\, dx$$

$$= \dfrac{2}{\pi}\left[\int_0^{\pi/2} x\sin nx\, dx + \int_{\pi/2}^\pi (\pi - x)\sin nx\, dx\right]$$

$$= \dfrac{2}{\pi}\left[x\left(-\dfrac{\cos nx}{n}\right) - 1\left(-\dfrac{\sin nx}{n^2}\right)\right]_0^{\pi/2}$$

$$+ \dfrac{2}{\pi}\left[(\pi - x)\left(-\dfrac{\cos nx}{nx}\right) - (-1)\left(-\dfrac{\sin nx}{n^2}\right)\right]_0^\pi$$

$$= \dfrac{2}{\pi}\left[-\dfrac{\pi}{2n}\cos \dfrac{n\pi}{2} + \dfrac{1}{n^2}\sin \dfrac{n\pi}{2}\right]$$

$$+ \dfrac{2}{\pi}\left[\dfrac{\pi}{2n}\cos \dfrac{n\pi}{2} + \dfrac{1}{n^2}\sin \dfrac{n\pi}{2}\right]$$

$$= \dfrac{2}{\pi}\left[\dfrac{2}{n^2}\sin \dfrac{n\pi}{2}\right] = \dfrac{4}{\pi n^2}\sin \dfrac{n\pi}{2}$$

Hence, $f(x) = \dfrac{4}{\pi}\left[\sin x - \dfrac{\sin 3x}{3^2} + \dfrac{\sin 5x}{5^2} - ...\right]$.

(ii) Half range cosine series

Let $\quad f(x) = \dfrac{a_0}{2} + \sum\limits_{n=1}^{\infty} a_n \cos nx$

Then, $a_0 = \dfrac{2}{\pi}\int_0^\pi f(x)dx$

$$= \dfrac{2}{\pi}\left[\int_0^{\pi/2} xdx + \int_{\pi/2}^\pi (\pi - x)dx\right]$$

$$= \dfrac{2}{\pi}\left[\dfrac{x^2}{2}\right]_0^{\pi/2} + \left[\pi x - \dfrac{x^2}{2}\right]_{\pi/2}^\pi$$

$$= \dfrac{2}{\pi}\left[\dfrac{\pi^2}{8} + \left(\pi^2 - \dfrac{\pi^2}{2}\right) - \left(\dfrac{\pi^2}{2} - \dfrac{\pi^2}{8}\right)\right]$$

$$= \dfrac{2}{\pi}\left[\dfrac{\pi^2}{4}\right] = \dfrac{\pi}{2}$$

and

$$a_n = \dfrac{2}{\pi}\int_0^\pi f(x)\cos nx\, dx$$

$$= \dfrac{2}{\pi}\left[\int_0^{\pi/2} x\cos nx\, dx + \int_{\pi/2}^\pi (\pi - x)\cos nx\, dx\right]$$

$$= \dfrac{2}{\pi}\left[\dfrac{x\sin nx}{n} - 1\left(-\dfrac{\cos nx}{n^2}\right)\right]_0^{\pi/2}$$

$$+ \dfrac{2}{\pi}\left[(\pi - x)\dfrac{\sin nx}{n} - (-1)\left(-\dfrac{\cos nx}{n^2}\right)\right]_{\pi/2}^\pi$$

$$= \frac{2}{\pi}\left[\frac{\pi}{2n}\sin\frac{n\pi}{2}+\frac{1}{n^2}\cos\frac{n\pi}{2}-\frac{1}{n^2}\right]$$

$$+\frac{2}{\pi}\left[-\frac{\cos n\pi}{n^2}-\frac{\pi}{2n}\sin\frac{n\pi}{2}+\frac{1}{n^2}\cos\frac{n\pi}{2}\right]$$

$$=\frac{2}{\pi}\left[\frac{2}{n^2}\cos\frac{n\pi}{2}-\frac{\cos n\pi}{n^2}-\frac{1}{n^2}\right]$$

$$=\frac{2}{\pi n^2}\left[2\cos\frac{n\pi}{2}-\cos n\pi-1\right]$$

Put $n = 0, 1, 2, 3, \ldots$ in equation (1), we get

$$a_1 = 0, a_2 = \frac{2}{\pi.2^2}(2\cos\pi-\cos 2\pi-1)=\frac{-2}{1^2.\pi}$$

$$a_3 = 0, a_4 = 0, a_5 = 0,$$

$$a_6 = \frac{2}{6^2\pi}(2\cos 3\pi-\cos 6\pi-1)=\frac{-2}{3^2\pi}$$

$$a_7 = a_8 = a_9 = 0,$$

$$a_{10} = \frac{2}{10^2.\pi}(2\cos 5\pi-\cos 10\pi-1)=\frac{-2}{5^2\pi}$$

Hence,

$$f(x)=\frac{\pi}{4}-\frac{2}{\pi}\left[\frac{\cos 2x}{1^2}+\frac{\cos 6x}{3^2}+\frac{\cos 10x}{5^2}+\ldots\right].$$

EXAMPLE 2. *Develop* $\sin\frac{\pi x}{l}$ *in half range cosine series in range* $0 < x < l$. *(UPTU–2001)*

SOLUTION. Let $\sin\frac{\pi x}{l}=\frac{a_0}{2}+\sum_{n=1}^{\infty}a_n\cos\frac{n\pi x}{l}$

where, $a_0 = \frac{2}{l}\int_0^l \sin\frac{\pi x}{l}dx=\frac{2}{l}\left[-\frac{\cos(\pi x/l)}{\pi/l}\right]_0^l$

$$=\frac{2}{\pi}[\cos\pi-1]=\frac{4}{\pi}$$

and $a_n = \frac{2}{l}\int_0^l\sin\frac{\pi x}{l}\cos\frac{n\pi x}{l}dx$

$$=\frac{1}{l}\int_0^l\left[\sin(n+1)\frac{\pi x}{l}-\sin(n-1)\frac{\pi x}{l}\right]dx$$

$$=\frac{1}{l}\left[-\frac{\cos(n+1)\frac{\pi x}{l}}{(n+1)\pi/l}+\frac{\cos(n-1)\frac{\pi x}{l}}{(n-1)\pi/l}\right]_0^l$$

$$=\frac{1}{\pi}\left[-\frac{(-1)^{n+1}}{n+1}+\frac{(-1)^{n-1}}{n-1}+\frac{1}{n+1}-\frac{1}{n-1}\right]$$

(i) When n is odd

$$a_n = \frac{1}{\pi}\left[-\frac{1}{n+1}+\frac{1}{n-1}+\frac{1}{n+1}-\frac{1}{n-1}\right]=0$$

(ii) When n is even

$$a_n = \frac{1}{\pi}\left[\frac{1}{n+1}-\frac{1}{n-1}+\frac{1}{n+1}-\frac{1}{n-1}\right]$$

$$=\frac{2}{\pi}\left[\frac{1}{n+1}-\frac{1}{n-1}\right]$$

$$=\frac{-4}{\pi(n+1)(n-1)}, n\neq 1$$

$$\therefore \quad \sin\frac{\pi x}{l}=\frac{2}{\pi}$$

$$-\frac{4}{\pi}\left[\frac{\cos\frac{2\pi x}{l}}{1.3}+\frac{\cos\frac{4\pi x}{l}}{3.5}+\frac{\cos\frac{6\pi x}{l}}{5.7}+\ldots\right].$$

EXAMPLE 3. *Obtain the half range sine series for function* $f(x) = x^2$ *in the interval* $0 < x < 3$.

(UPTU–2001(SP), 2002)

SOLUTION. The Fourier half range sine series in the interval $(0, c)$ is given by

$$f(x)=\sum_{n=1}^{\infty}b_n\sin nx \qquad \ldots(1)$$

where, $b_n = \frac{2}{c}\int_0^c f(x)\sin\frac{n\pi x}{c}dx$

Here, $c = 3$ and $f(x)=x^2$

$$\therefore \quad b_n = \frac{2}{3}\int_0^3 x^2\sin\frac{n\pi x}{3}dx$$

$$=\frac{2}{3}\left[x^2\left(\frac{-3}{n\pi}\right)\left(\cos\frac{n\pi x}{3}\right)+2x\left(\frac{3}{n\pi}\right)\left(\frac{3}{n\pi}\right)\sin\frac{n\pi x}{3}\right.$$

$$\left.-2\left(\frac{3}{n\pi}\right)\left(\frac{3}{n\pi}\right)\left(\frac{3}{n\pi}\right)\cos\frac{n\pi x}{3}\right]_0^3$$

$$=\frac{2}{3}\left[\left\{-\frac{27}{n\pi}(-1)^n-\frac{54}{n^3\pi^3}(-1)^n\right\}+\frac{54}{n^3\pi^3}\right]$$

$$=\frac{2}{3}\left[\frac{54}{n^3\pi^3}\{1-(-1)^n\}-\frac{27}{n\pi}(-1)^n\right]$$

$$=\begin{cases}\frac{2}{3}\left(\frac{108}{n^3\pi^3}+\frac{27}{n\pi}\right), & \text{if } n \text{ is odd}\\ -\frac{18}{n\pi}, & \text{if } n \text{ is even}\end{cases}$$

Hence, the required half range sine series is given by

$$f(x) = b_1\sin x+b_2\sin 2x+b_3\sin 3x+\ldots$$

$$=\frac{2}{3}\left[\frac{108}{\pi^3}\left(\frac{\sin x}{1^3}+\frac{\sin 3x}{3^3}+\frac{\sin 5x}{5^3}+\ldots\right)\right.$$

$$+\frac{27}{\pi}\left(\frac{\sin x}{1}+\frac{\sin 3x}{3}+\frac{\sin 5x}{5}+\ldots\right)$$

$$\left.-\frac{18}{\pi}\left(\frac{\sin 2x}{2}+\frac{\sin 4x}{4}+\ldots\right)\right].$$

EXAMPLE 4. (i) *Express* $f(x) = x$ *as a half range sine series in* $0 < x < 2$, *(UPTU–2004)*

(ii) *Express* $f(x) = x$ *as a half-range cosine series in* $0 < x < 2$.

(SVTU–2009, BHOPAL–2007, MUMBAI–2006)

SOLUTION. (i) The Fourier sine series for $F(x)$ in $(0, 2)$ is

$$f(x)=\sum_{n=1}^{\infty}b_n\sin\frac{n\pi x}{2}$$

where

$$b_n = \frac{2}{2}\int_0^2 f(x)\sin\frac{n\pi x}{2}dx = \int_0^2 x\sin\frac{n\pi x}{2}dx$$

$$= \left[-\frac{2x}{n\pi}\cos\frac{n\pi x}{2} + \frac{4}{n^2\pi^2}\sin\frac{n\pi x}{2}\right]_0^2 = \frac{-4(-1)^n}{n\pi}$$

$$\Rightarrow b_1 = 4/\pi_1, b_2 = -4/2\pi, b_3 = 4/3\pi,$$
$$b_4 = -4/4\pi, \text{ etc.}$$

Required half range Fourier sine series is

$$f(x) = \frac{4}{\pi}\left[\sin\frac{\pi x}{2} - \frac{1}{2}\sin\frac{2\pi x}{2} + \frac{1}{3}\sin\frac{3\pi x}{2}\right.$$
$$\left. - \frac{1}{4}\sin\frac{4\pi x}{2} + ...\right]$$

(ii) The Fourier cosine series for $f(x)$ in (0, 2) is

$$f(x) = \frac{a_0}{2} + \sum_{n=1}^{\infty} a_n \cos\frac{n\pi x}{2}$$

where $a_0 = \frac{2}{2}\int_0^2 f(x)dx = \int_0^2 x\,dx = 2$

and $a_n = \frac{2}{2}\int_0^2 f(x)\cos\frac{n\pi x}{2}dx = \int_0^2 x\cos\frac{n\pi x}{2}dx$

$$= \left[\frac{2x}{n\pi}\sin\frac{n\pi x}{2} + \frac{4}{n^2\pi^2}\cos\frac{n\pi x}{2}\right]_0^2$$

$$= \frac{4}{n^2\pi^2}[(-1)^n - 1]$$

$$\Rightarrow a_1 = -8/\pi^2, a_2 = 0, a_3 = -8/3^2\pi^2, a_4 = 0,$$
$$a_5 = -8/5^2\pi^2, \text{ etc.}$$

Required half range Fourier cosine series is given by

$$f(x) = 1 - \frac{8}{\pi^2}\left[\frac{\cos\pi x/2}{1^2} + \frac{\cos 3\pi x/2}{3^2}\right.$$
$$\left. + \frac{\cos 5\pi x/2}{5^2} + ...\right].$$

EXAMPLE 5. *Find a series of cosines of multiples of x which will represent x sin x in the interval (0, π) and show that*

$$\frac{1}{1.3} - \frac{1}{3.5} + \frac{1}{5.7} - ... = \frac{\pi - 2}{4}.$$

(UPTU–2002, VTU–2003, ANNA–2001)

SOLUTION. Let $\quad x\sin x = \frac{a_0}{2} + \sum_{n=1}^{\infty} a_n \cos nx$

Then

$$a_0 = \frac{2}{\pi}\int_0^\pi x\sin x\,dx = \frac{2}{\pi}[x(-\cos x) - 1.(-\sin x)]_0^\pi$$

$$= \frac{2}{\pi}[-\pi\cos x] = 2$$

and

$$a_n = \frac{2}{\pi}\int_0^\pi x\sin x\cos nx\,dx$$

$$= \frac{1}{\pi}\int_0^\pi x(2\cos nx\sin x)dx$$

$$= \frac{1}{\pi}\int_0^\pi x[\sin(n+1)x - \sin(n-1)x]dx$$

$$= \frac{1}{\pi}\left[x\left\{-\frac{\cos(n+1)x}{n+1} + \frac{\cos(n-1)x}{n-1}\right\}\right.$$
$$\left. -1\left\{-\frac{\sin(n+1)x}{(n+1)^2} - \frac{\sin(n-1)\pi}{(n-1)^2}\right\}\right]_0^\pi$$

$$= \frac{1}{\pi}\left[-\frac{\pi\cos(n+1)\pi}{n+1} + \frac{\pi\cos(n-1)\pi}{(n-1)}\right],$$
$$\text{when } n \neq 1$$

$$= \frac{-(-1)^{n+1}}{n+1} + \frac{(-1)^{n-1}}{n-1}$$

$$= (-1)^n\left[\frac{1}{n-1} - \frac{1}{n+1}\right] = \frac{2(-1)^{n-1}}{(n-1)(n+1)}$$

When $n=1$, we have

$$a_1 = \frac{2}{\pi}\int_0^\pi x\sin x\cos x\,dx = \frac{1}{\pi}\int_0^\pi x\sin 2x\,dx$$

$$= \frac{1}{\pi}\left[x\left(-\frac{\cos 2x}{2}\right) - 1\left(-\frac{\sin 2x}{2^2}\right)\right]_0^\pi$$

$$= \frac{1}{\pi}\left[-\frac{\pi\cos 2x}{2}\right] = -\frac{1}{2}$$

$$\therefore \quad x\sin x = 1 - \frac{1}{2}\cos x$$

$$-2\left(\frac{\cos 2x}{1.3} - \frac{\cos 3x}{2.4} + \frac{\cos 4x}{3.5} - ...\right)$$

Putting $x = \frac{\pi}{2}$, we get

$$\frac{\pi}{2} = 1 - 2\left(-\frac{1}{1.3} + \frac{1}{3.5} - \frac{1}{5.7} - ...\right)$$

$$\therefore \quad 1 + \frac{2}{1.3} - \frac{2}{3.5} + \frac{2}{5.7} - ... = \frac{\pi}{2}$$

$$\Rightarrow \frac{2}{1.3} - \frac{2}{3.5} + \frac{2}{5.7} - ... = \frac{\pi}{2} - 1$$

Hence, $\frac{1}{1.3} - \frac{1}{3.5} + \frac{1}{5.7} - ... = \frac{\pi - 2}{4}.$

EXAMPLE 6. *Expand* $f(x) = \begin{cases} \frac{1}{4} - x & ,\text{if } 0 < x < \frac{1}{2} \\ x - \frac{3}{4} & ,\text{if } \frac{1}{2} < x < 1 \end{cases}$ *as the*

Fourier series of sine terms.

(UPTU–2001, VTU–2011, ANDHRA–2000)

SOLUTION. The Fourier sine series for $f(x)$ in (0, 1) is

$$f(x) = \sum_{n=1}^{\infty} b_n \sin n\pi x$$

where, $b_n = \frac{2}{1}\int_0^1 f(x)\sin n\pi x\,dx$

$$=2\left[\int_0^{1/2}\left(\frac{1}{4}-x\right)\sin n\pi x\,dx+\int_{1/2}^1\left(x-\frac{3}{4}\right)\sin n\pi x\,dx\right]$$

$$=2\left|-\left(\frac{1}{4}-x\right)\frac{\cos n\pi x}{n\pi}-\frac{\sin n\pi x}{n\pi}\right|_0^{1/2}$$

$$+2\left|-\left(x-\frac{3}{4}\right)\frac{\cos n\pi x}{n\pi}+\frac{\sin n\pi x}{n^2\pi^2}\right|_{1/2}^1$$

$$=2\left[\frac{1}{4n\pi}\cos\frac{n\pi}{2}+\frac{1}{4n\pi}-\frac{\sin n\pi/2}{n^2\pi^2}\right]$$

$$+2\left[-\frac{1}{4n\pi}\cos n\pi-\frac{1}{4n\pi}\cos\frac{n\pi}{2}-\frac{\sin n\pi/2}{n^2\pi^2}\right]$$

$$=\frac{1}{2n\pi}[1-(-1)^n]-\frac{4\sin n\pi/2}{n^2\pi^2}$$

$$\Rightarrow b_1=\frac{1}{\pi}-\frac{4}{\pi^2},\,b_2=0,\,b_3=\frac{1}{3\pi}+\frac{4}{3^2\pi^2},$$

$$b_4=0,\,b_5=\frac{1}{5}-\frac{4}{5^2\pi^2},\,b_6=0\text{ etc.}$$

Hence, the required Fourier series is

$$f(x)=\left(\frac{1}{\pi}-\frac{4}{\pi^2}\right)\sin\pi x+\left(\frac{1}{3\pi}+\frac{4}{3^2\pi^2}\right)\sin 3\pi x$$

$$+\left(\frac{1}{5\pi}-\frac{4}{5^2\pi^2}\right)\sin 5\pi x+...$$

Exercise-7.9

1. Find the Fourier half range series expansion of the function
   $$f(x)=(-x/l)+1, 0\le x\le l.$$

2. Find a series of sines of multiples of x which will represent $f(x)$ in the interval $(0,\pi)$, where

$$f(x)=\begin{cases}\dfrac{1}{3}\pi, & 0<x<\dfrac{1}{3}\pi\\[2mm] 0, & \dfrac{1}{3}\pi<x<\dfrac{2}{3}\pi\\[2mm] -\dfrac{1}{3}\pi, & \dfrac{2}{3}\pi<x<\pi\end{cases}$$

   Also, represent this function by a series of cosines of multiples of x as well. Draw graph of these series and find the sine and cosine series where $x=-\dfrac{1}{3}\pi,-\dfrac{2}{3}\pi,-\pi$.

3. Find the half range cosine series for function $f(x)=(x-1)^2$ in the interval $0<x<1$.   (VTU–2010, JNTU–2006)
   Hence show that

   (i) $\dfrac{1}{1^2}+\dfrac{1}{2^2}+\dfrac{1}{3^2}+\dfrac{1}{4^2}+...=\dfrac{\pi^2}{6}$,

   (ii) $\dfrac{1}{1^2}-\dfrac{1}{2^2}+\dfrac{1}{3^2}-\dfrac{1}{4^2}+...=\dfrac{\pi^2}{12}$,

   (iii) $\dfrac{1}{1^2}+\dfrac{1}{3^2}+\dfrac{1}{5^2}+\dfrac{1}{7^2}+...=\dfrac{\pi^2}{8}$.   (ANNA–2003)

4. If $\begin{aligned}f(x)&=mx, & 0\le x\le\pi/2\\ &=m(\pi-x), & \pi/2\le x\le\pi\end{aligned}$

   Then show that $f(x)=\dfrac{4m}{\pi}\left[\dfrac{\sin x}{1^2}-\dfrac{\sin 3x}{3^2}+\dfrac{\sin 5x}{5^2}-...\right]$.

5. If $f(x)=\begin{cases}\dfrac{hx}{a}, & 0<x<a\\[2mm]\dfrac{h(l-x)}{l-a}, & a<x<l\end{cases}$

   Prove that for all values of x between 0 and l

   $$f(x)=\dfrac{2hl^2}{a(l-a)\pi^2}\left[\sin\dfrac{\pi a}{l}\sin\dfrac{\pi x}{l}\sin\dfrac{\pi x}{l}\right.$$
   $$\left.+\dfrac{1}{2^2}\sin\dfrac{2\pi a}{l}\sin\dfrac{2\pi x}{l}+...\right]$$ (UPTU(Q.BANK)–2001)

6. Expand $\pi x-x^2$ as a half range sine series in the interval $(0,\pi)$ upto first three terms.

7. Obtain a half range cosine series for

   $$f(x)=\begin{cases}kx, & \text{for } 0\le x\le l/2\\ k(l-x), & \text{for } l/2\le x\le l\end{cases}$$
   (BHOPAL–2008, VTU–2008)

   Show that $f(x)=\dfrac{4Kl}{\pi^2}\sum_{n=0}^{\infty}\dfrac{(-1)^n}{(2n+1)^2}\sin\dfrac{(2n+1)\pi x}{l}$

   Deduce the sum of the series $\dfrac{1}{1^2}+\dfrac{1}{3^2}+\dfrac{1}{5^2}+....$
   (UPTU–2003, ROHTAK–2006)

8. Find the half range sine series for the function $f(t)=t-t^2$ in the interval $0<t<1$.   (UPTU–2004)

Hints to Selected Problems

1.

$$a_0=\frac{2}{l}\int_0^l f(x)dx=\frac{2}{l}\int_0^l\left(-\frac{x}{l}+1\right)dx$$

$$=\frac{2}{l}\left[-\frac{x^2}{2l}+x\right]_0^l=\frac{2}{l}\left[-\frac{l^2}{2l}+l\right]=1$$

$$a_n=\frac{2}{l}\int_0^l f(x)\cos\frac{n\pi x}{l}dx=\frac{2}{l}\int_0^l\left(-\frac{x}{l}+1\right)\cos\frac{n\pi x}{l}dx$$

$$=\frac{2}{l}\left[\left(-\frac{x}{l}+1\right)\left(\frac{l}{n\pi}\sin\frac{n\pi x}{l}\right)-\left(-\frac{1}{l}\right)\left(-\frac{l^2}{n^2\pi^2}\cos\frac{n\pi x}{l}\right)\right]_0^l$$

$$=\frac{2}{l}\left[0-\frac{l}{n^2\pi^2}\cos n\pi+\frac{l}{n^2\pi^2}\right]=\frac{2}{n^2\pi^2}[1-(-1)^n]$$

$$=\begin{cases}\dfrac{4}{n^2\pi^2}, & \text{when }n\text{ is odd}\\ 0, & \text{when }n\text{ is odd}\end{cases}$$

and $b_n=0$

$$\Rightarrow f(x)=\frac{1}{2}+\frac{4}{\pi^2}\left[\frac{1}{1^2}\cos\frac{\pi x}{l}+\frac{1}{3^2}\cos\frac{3\pi x}{l}+\frac{1}{5^2}\cos\frac{5\pi x}{l}+...\right]$$

2. $b_n=\dfrac{2}{\pi}\int_0^\pi f(v)\sin nv\,dv$

$$= \frac{2}{\pi}\left[\int_0^{\pi/2}\frac{\pi}{3}\sin nv\, dv + \int_{\pi/3}^{2\pi/3}0.\sin nv\, dv + \int_{2\pi/3}^{\pi}-\frac{\pi}{3}.\sin nv\, dv\right]$$

$$= \frac{2}{3}\left[-\frac{\cos nv}{n}\right]_0^{\pi/3} - \frac{2}{3}\left[-\frac{\cos nv}{n}\right]_{2\pi/3}^{\pi}$$

$$= \frac{2}{3n}\left[1 - \cos\frac{n\pi}{3} + \cos n\pi - \cos\frac{2n\pi}{3}\right]$$

$$= -\frac{8}{3n}\sin\frac{n\pi}{6}\sin\frac{n\pi}{3}.\cos\frac{n\pi}{2}$$

$$f(x) = -\frac{8}{3}\sum_{n=1}^{\infty}\frac{1}{n}\sin\frac{n\pi}{6}\sin\frac{n\pi}{3}\cos\frac{n\pi}{2}\sin nx$$

$$= \frac{1}{2}\left[\frac{1}{2}\sin 2x + \frac{1}{2}\sin 4x + \frac{1}{3}\sin 3x + \frac{1}{10}\sin 10x + ...\right]$$

$$a_0 = \frac{1}{\pi}\int_0^{\pi}f(v)dv = \frac{1}{\pi}\int_0^{\pi/3}\frac{\pi}{3}dv + \int_{\pi/3}^{2\pi/3}0.dv + \int_{2\pi/3}^{\pi}-\frac{\pi}{3}dv = 0$$

$$a_n = \frac{2}{\pi}\int_0^{\pi}f(v)\cos nv\, dv$$

$$= \frac{2}{\pi}\left[\int_0^{\pi/3}\frac{\pi}{3}.\cos nv\, dv + \int_{\pi/3}^{2\pi/3}0.dv + \int_{2\pi/3}^{\pi/3}-\frac{\pi}{3}.\cos nv\, dv\right]$$

$$= \frac{2}{3n}\left[\sin\frac{n\pi}{3} + \sin\frac{2n\pi}{3}\right] = \frac{4}{3n}\sin\frac{n\pi}{2}\cos\frac{n\pi}{6}$$

$$\Rightarrow f(x) = \frac{4}{3}\sum_{n=1}^{\infty}\frac{1}{n}\cos\frac{n\pi}{6}\sin\frac{n\pi}{2}.\cos nx$$

$$= \frac{2}{\sqrt{3}}\left[\cos x - \frac{1}{5}\cos 5x + \frac{1}{7}\cos 7x - \frac{1}{11}\cos 11x + ...\right]$$

6. $b_n = \dfrac{2}{\pi}\int_0^{\pi}(\pi x - x^2)\sin x\, dx$

$$= \frac{2}{\pi}\left[(\pi x - x^2)\left(-\frac{\cos nx}{n}\right) - (\pi - 2x).\right.$$

$$\left.\left(-\frac{\sin nx}{n^2}\right) + (-2)\left(\frac{\cos nx}{n^3}\right)\right]_0^{\pi}$$

$$= \frac{2}{\pi}\left[-\frac{2\cos n\pi}{n^3} + \frac{2}{n^3}\right] = \frac{4}{\pi n^3}[1-(-1)^n] = 0 \text{ or } \frac{8}{n\pi^3}$$

according as n is even or odd

$$\Rightarrow \pi x - x^2 = \frac{8}{\pi}\left(\sin x + \frac{\sin 3x}{3^3} + \frac{\sin 5x}{5^3} + ...\right).$$

7. $a_0 = \dfrac{2}{l}\int_0^{l}f(x)dx = \dfrac{2}{l}\left[\int_0^{l/2}kx\,dx + \int_{l/2}^{l}k(l-x)dx\right] = \dfrac{kl}{2}$

$$= \frac{2}{l}\left[\int_0^{l/2}k.x\cos\frac{n\pi x}{l}dx + \int_{l/2}^{l}k(l-x)\frac{\cos n\pi x}{l}dx\right]$$

$$= \frac{2kl}{n^2\pi^2}\left[2\cos\frac{n\pi}{2} - 1 - \cos n\pi\right]$$

When n is odd, $\cos\dfrac{n\pi}{2} = 0$ and $\cos n\pi = -1$,

$$\Rightarrow a_n = 0 \Rightarrow a_1 = a_3 = a_5 = ... = 0$$

When n is even,

$$a_2 = \frac{2kl}{2^2\pi^2}[2\cos\pi - 1 - \cos 2\pi] = -\frac{8kl}{2^2\pi^2}$$

$$a_4 = \frac{2kl}{4^2\pi^2}[2\cos 2\pi - 1 - \cos 4\pi] = 0$$

$$a_6 = \frac{2kl}{6^2\pi^2}[2\cos 3\pi - 1 - \cos 6\pi]$$

$$= \frac{2kl}{6^2\pi^2}(-2-1-1) = -\frac{8kl}{6^2\pi^2} \text{ and so on}$$

$$\Rightarrow f(x) = \frac{kl}{4} - \frac{8kl}{\pi^2}\left[\frac{1}{2^2}\cos\frac{2\pi x}{l} + \frac{1}{6^2}\cos\frac{6\pi x}{l} + ...\right] \qquad ...(1)$$

Putting $x = 1$, $f(x) = 0$

From (1), we have

$$0 = \frac{kl}{4} - \frac{8kl}{\pi^2}\left(\frac{1}{2^2} + \frac{1}{6^2} + ...\right)$$

$$\frac{1}{2^2} + \frac{1}{6^2} + ... = \frac{\pi^2}{32}$$

$$\Rightarrow \frac{1}{2^2}\left(\frac{1}{1^2} + \frac{1}{3^2} + ...\right)... = \frac{\pi^2}{32}.$$

Hence $\dfrac{1}{1^2} + \dfrac{1}{3^2} + ... = \dfrac{\pi^2}{8}.$

Answers

1. $f(x) = \dfrac{1}{2} + \dfrac{4}{\pi^2}\left[\dfrac{1}{1^2}\cos\dfrac{\pi x}{l} + \dfrac{1}{3^2}\cos\dfrac{3\pi x}{l} + \dfrac{1}{5^2}\cos\dfrac{5\pi x}{l} + ...\right]$

2. $f(x) = \dfrac{1}{2}\left[\dfrac{1}{2}\sin 2x + \dfrac{1}{2}\sin 4x + \dfrac{1}{8}\sin 8x + \dfrac{1}{10}\sin 10x...\right]$ and $f(x) = \dfrac{2}{\sqrt{3}}\left[\cos x - \dfrac{1}{5}\cos 5x + \dfrac{1}{7}\cos 7x - \dfrac{1}{11}\cos 11x + ...\right]$

3. $\dfrac{1}{3} + \dfrac{4}{\pi^2}\left(\cos\pi x + \dfrac{\cos 2\pi x}{2^2} + \dfrac{\cos 3\pi x}{3^2} + ...\right)$ **6.** $\pi x - x^2 = \dfrac{8}{\pi}\left(\sin x + \dfrac{\sin 3x}{3^2} + \dfrac{\sin 5x}{5^3} + ...\right)$

7. $\dfrac{2kl}{n^2\pi^2}\left[2\cos\dfrac{n\pi}{2} - 1 - \cos n\pi\right]$ **8.** $f(t) = \dfrac{a}{2} + \dfrac{2a}{\pi}\left[\sin x + \dfrac{1}{3}\sin 3x + \dfrac{1}{5}\sin 5x + \dfrac{1}{7}\sin 7x + ...\right]$

7.22   PARSEVEL'S IDENTITY FOR FOURIER SERIES

Consider the Fourier series $\dfrac{a_0}{2} + \sum_{n=1}^{\infty}(a_n\cos nx + b_n\sin nx)$ *.If f(x) converges uniformly to f(x) at every point of the interval* $(0, 2\pi)$*, then*

$$\frac{1}{\pi}\int_0^{2\pi}\{f(x)\}^2 dx = \frac{a_0^2}{2} + \sum_{n=1}^{\infty}(a_n^2 + b_n^2).$$

Proof. Let the series $\dfrac{a_0}{2} + \sum_{n=1}^{\infty}(a_n\cos nx + b_n\sin nx)$ represents the Fourier series of $f(x)$. Also, let this series converges uniformly to $f(x)$ at every point of the interval $(0, 2\pi)$ so that

$$f(x) = \frac{a_0}{2} + \sum_{n=1}^{\infty} (a_n \cos nx + b_n \sin nx)$$ ...(1)

and that term by term integration is possible.

To prove that $\frac{1}{\pi} \int_0^{2\pi} \{f(x)\}^2 dx = \frac{a_0^2}{2} + \sum (a_n^2 + b_n^2)$

We have $a_n = \frac{1}{\pi} \int_0^{2\pi} f(x) . \cos nx \, dx$ $(n = 0, 1, 2, 3, ...)$

$b_n = \frac{1}{\pi} \int_0^{2\pi} f(x) . \sin nx \, dx$ $(n = 0, 1, 2, 3 ...)$

Multiplying (1) by $f(x)$ and then integrating from $x = 0$ to $x = 2\pi$, we get

$$\int_0^{2\pi} \{f(x)\}^2 dx = \frac{a_0}{2} + \int_0^{2\pi} f(x) dx + \sum_{n=1}^{\infty} \left(a_n \int_0^{2\pi} f(x) . \cos nx \, dx + b_n \int_0^{2\pi} f(x) \sin nx \, dx \right) = \frac{a_0}{2} . \pi a_0 + \sum_{n=1}^{\infty} (\pi a_n^2 + \pi b_n^2)$$

Dividing by π, we get $\frac{1}{\pi} \int_0^{2\pi} \{f(x)\}^2 dx = \frac{a_0^2}{2} + \sum_{n=1}^{\infty} (a_n^2 + b_n^2)$.

Solved Examples

EXAMPLE 1. *Obtain the Fourier series expansion of $f(x) = x^2$ in $-\pi < x < \pi$ and prove that $\sum_{n=1}^{\infty} \frac{1}{n^4} = \frac{\pi^4}{90}$ by using Parsevel's theorem.*

SOLUTION. Since $f(x) = x^2$ is even function so $b_n = 0$

Let the Fourier series expansion of $f(x)$ is given by

$$f(x) = x^2 = \frac{a_0}{2} + \sum_{n=1}^{\infty} a_n \cos nx$$ ...(1)

where $a_0 = \frac{2}{\pi} \int_0^{\pi} f(x) dx = \frac{2}{\pi} \int_0^{\pi} x^2 dx = \frac{2\pi^2}{3}$

$a_n = \frac{2}{\pi} \int_0^{\pi} f(x) \cos nx \, dx = \frac{2}{\pi} \int_0^{\pi} x^2 \cos nx \, dx$

$\Rightarrow a_n = \frac{2}{\pi} \left[x^2 \frac{\sin nx}{n} + 2x . \frac{\cos nx}{n^2} - 2 \frac{\sin nx}{n^2} \right]_0^{\pi}$

$= \frac{4(-1)^2}{n^2}$

$\therefore$ (1) becomes,

$$x^2 = \frac{\pi^2}{3} + 4 \sum_{n=1}^{\infty} \frac{(-1)^n \cos nx}{n^2}$$ ...(2)

which is the required Fourier expansion.

Now, by Parsevel's theorem, we have

$\int_{-\pi}^{\pi} \{f(x)\}^2 dx = \pi \left[\frac{a_0^2}{2} + \sum_{n=1}^{\infty} (a_n^2 + b_n^2) \right]$

Hence, $\int_{-\pi}^{\pi} x^4 dx = \pi \left[\frac{4\pi^4}{2.9} + \sum_{n=1}^{\infty} \frac{16}{n^4} \right]$

$\Rightarrow \left(\frac{x^5}{5} \right)_{-\pi}^{\pi} = \frac{2\pi^5}{9} + \pi \sum_{n=1}^{\infty} \frac{16}{n^4}$

or $\frac{2\pi^5}{5} - \frac{2\pi^5}{9} = \pi \sum_{n=1}^{\infty} \frac{16}{n^4} \Rightarrow \frac{\pi^4}{90} = \sum_{n=1}^{\infty} \frac{1}{n^4}$.

EXAMPLE 2. *If $f(x) = \begin{cases} \pi x & , \quad 0 < x < 1 \\ \pi(2 - x) & , \quad 1 < x < 2 \end{cases}$*

Using half range cosine series, show that

$$\frac{1}{1^4} + \frac{1}{3^4} + \frac{1}{5^4} + ... = \frac{\pi^4}{96}.$$

SOLUTION. The half range cosine series for $f(x)$ in $(0, c)$ is

$$f(x) = \frac{a_0}{2} + \sum_{n=1}^{\infty} a_n \cos \frac{n\pi x}{c}$$

Here, $a_0 = \frac{2}{c} \int_0^c f(x) dx$

$= \frac{2}{2} \left[\int_0^1 \pi x \, dx + \int_1^2 \pi(2 - x) dx \right]$

$= \pi \left[\frac{x^2}{2} \right]_0^1 + \pi \left[2x - \frac{x^2}{2} \right]_0^1$

$= \frac{\pi}{2} + \pi \left[(4 - 2) - \left(2 - \frac{1}{2} \right) \right] = \pi$

$a_n = \frac{2}{c} \int_0^c f(x) . \cos \frac{n\pi x}{c} dx$

$= \frac{2}{2} \left[\int_0^1 \pi x \cos \frac{n\pi x}{2} dx + \int_1^2 \pi(2 - x) \cos \frac{n\pi x}{2} dx \right]$

$= \pi \left[\frac{x \sin \frac{n\pi x}{2}}{\frac{n\pi}{2}} - \left(- \frac{\cos \frac{n\pi x}{2}}{\frac{n^2\pi^2}{4}} \right) \right]_0^1$

$+ \pi \left[(2 - x) \frac{\sin \frac{n\pi x}{2}}{\frac{n\pi}{2}} - (-1) \left(- \frac{\cos \frac{n\pi x}{2}}{\frac{n^2\pi^2}{4}} \right) \right]_1^2$

$= \pi \left[\frac{2}{n\pi} \sin \frac{n\pi}{2} + \frac{4}{n^2\pi^2} \cos \frac{n\pi}{2} - \frac{4}{n^2\pi^2} \right]$

$+ \pi \left[0 - \frac{4}{n^2\pi^2} \cos n\pi - \frac{2}{n\pi} \sin \frac{n\pi}{2} + \frac{4}{n^2\pi^2} \cos \frac{n\pi}{2} \right]$

$= \pi \left[\frac{8}{n^2\pi^2} \cos \frac{n\pi}{2} - \frac{4}{n^2\pi^2} - \frac{4}{n^2\pi^2} \cos n\pi \right]$

$= \frac{4}{n^2\pi} \left[2 \cos \frac{n\pi}{2} - 1 - \cos n\pi \right]$

Putting $n = 1, 2, 3, \ldots$, we get

$$a_1 = 0, a_2 = \frac{-4}{\pi}, a_3 = 0, a_4 = 0, a_5 = 0, a_6 = -\frac{4}{9\pi}$$

By Parsevel's formula, we get

$$\int_0^c \{f(x)\}^2 dx = \frac{c}{2}\left[\frac{a_0^2}{2} + a_1^2 + a_2^2 + a_3^2 + \ldots\right]$$

$$\int_0^1 (\pi x)^2 dx + \int_1^2 \pi^2 (2-x)^2 dx = \frac{2}{2}\left[\frac{\pi^2}{2} + \frac{16}{\pi^2} + \frac{16}{81\pi^2} + \ldots\right]$$

$$\pi^2\left[\frac{x^3}{3}\right]_0^1 - \pi^2\left[\frac{(2-x)^3}{3}\right]_1^2 = \frac{\pi^2}{2} + \frac{16}{\pi^2} + \frac{16}{81\pi^2} + \ldots$$

$$\Rightarrow \frac{\pi^2}{3} - \pi^2\left(0 - \frac{1}{3}\right) = \frac{\pi^2}{3^2} + \frac{16}{\pi^2}\left[1 + \frac{1}{81} + \ldots\right]$$

$$\Rightarrow \frac{2\pi^2}{3} - \frac{\pi^2}{2} = \frac{16}{\pi^2}\left[1 + \frac{1}{3^4} + \frac{1}{5^4} + \ldots\right]$$

$$\Rightarrow \frac{\pi^2}{6} = \frac{16}{\pi^2}\left[1 + \frac{1}{3^4} + \frac{1}{5^4} + \ldots\right] \Rightarrow \frac{\pi^4}{96} = 1 + \frac{1}{3^4} + \frac{1}{5^4} + \ldots$$

7.23   COMPLEX FORM OF FOURIER SERIES

In complex notations, the Fourier series is written as

$$f(x) = \sum_{n=-\infty}^{+\infty} C_n e^{inx} \qquad\qquad (n = 0, 1, 2, \ldots)$$

$$f(x) = C_0 + \sum_{n=1}^{\infty} C_n e^{inx} + \sum_{n=1}^{\infty} C_{-n} e^{-inx} \qquad\qquad \ldots(1)$$

With $\qquad C_n = \frac{1}{2\pi}\int_C^{C+2\pi} f(x) e^{-inx} dx$ and $C_{-n} = \frac{1}{2\pi}\int_C^{C+2\pi} f(x) e^{inx} dx \qquad (n = 0, 1, 2, \ldots)$

$f(x)$ being defined in the interval $[C, C+2\pi]$.

Here, $\qquad C_0 = \frac{1}{2\pi}\int_C^{C+2\pi} f(x)\, dx = \frac{a_0}{2}$.

$$C_n = \frac{1}{2\pi}\int_C^{C+2\pi} f(x) e^{-inx} dx = \frac{1}{2\pi}\int_C^{C+2\pi} f(x)(\cos nx - i\sin nx) dx$$

$$= \frac{1}{2\pi}\int_C^{C+2\pi} f(x)\cos nx\, dx - i\frac{1}{2\pi}\int_C^{C+2\pi} f(x)\sin nx\, dx = \frac{a_n - ib_n}{2}$$

Similarly, $\qquad C_{-n} = \frac{1}{2\pi}\int_C^{C+2\pi} f(x) e^{inx} dx = \frac{a_n + ib_n}{2}$.

Now, (1) becomes

$$f(x) = \frac{a_0}{2} + \sum_{n=1}^{\infty}\left\{\left(\frac{a_n - ib_n}{2}\right)\cos nx + \left(\frac{a_n + ib_n}{2}\right)\sin nx\right\} \qquad\qquad \ldots(2)$$

with $\qquad a_n = \frac{1}{\pi}\int_C^{C+2\pi} f(x).\cos nx\, dx$ and $b_n = \frac{1}{\pi}\int_C^{C+2\pi} f(x)\sin nx\, dx$

Solved Examples

EXAMPLE 1. *Obtain the complex form of the Fourier series of*

$f(x) = e^{-x}$ *in* $-1 \le x \le 1$. (MUMBAI-2005S, MADRAS-2000S)

SOLUTION. The complex form of Fourier series for the given function $f(x)$ is

$$f(x) = \sum_{n=-\infty}^{\infty} C_n e^{inx}.$$

Here, $C_n = \frac{1}{2}\int_{-1}^1 e^{-x} e^{-inx\pi} dx$

$$= \frac{1}{2}\int_{-1}^1 e^{-(1+in\pi)x} dx$$

$$= \frac{1}{2}\left[\frac{e^{-(1+in\pi)x}}{-(1+in\pi)}\right]_{-1}^1 = \frac{e^{1+in\pi} - e^{-(1+in\pi)}}{2(1+in\pi)}$$

$$= \frac{e(\cos n\pi + i\sin n\pi) - e^{-1}(\cos n\pi - i\sin n\pi)}{2(1+in\pi)}$$

$$= \frac{e - e^{-1}}{2}(-1)^n \frac{1 - in\pi}{1 + n^2\pi^2} = \frac{(-1)^n(1 - in\pi)\sinh 1}{1 + n^2\pi^2}$$

Hence, the required complex form of the Fourier series is

$$e^{-x} = \sum_{n=-\infty}^{\infty} \frac{(-1)^n(1 - in\pi)}{1 + n^2\pi^2}\sinh 1.e^{in\pi x}.$$

Exercise-7.10

1. Prove that in $0 < x < C$,

$$x = \frac{C}{2} - \frac{4C}{\pi^2}\left(\cos\frac{\pi x}{C} + \frac{1}{3^2}\cos\frac{3\pi x}{C} + \frac{1}{5^2}\cos\frac{5\pi x}{C} + \ldots\right)$$

and deduce that

(i) $\frac{1}{1^4} + \frac{1}{3^4} + \frac{1}{5^4} + \ldots = \frac{\pi^4}{96}$ (ii) $\frac{1}{1^4} + \frac{1}{2^4} + \frac{1}{3^4} + \frac{1}{4^4} + \ldots = \frac{\pi^4}{90}$

2. Find the complex form of the Fourier series of

$$f(x) = \cos ax, \ -\pi < x < \pi. \qquad \text{(ANNA-2009, MUMBAI-2009)}$$

3. Find the complex form of the Fourier series of

$$f(x) = e^{ax}, \quad -l < x < l. \qquad \text{(MADRAS–2003)}$$

4. Find the complex form of the Fourier series of

$$f(x) = \begin{cases} 0, & \text{if } 0 < x < l \\ a, & \text{if } 0 < x < 2l \end{cases}.$$

5. Find the complex form of the Fourier series of

$$f(x) = \cosh 3x + \sinh 3x \text{ in } (-3, 3). \qquad \text{(MUMBAI–2008)}$$

Answers

2. $\dfrac{a}{\pi} \sin a\pi \displaystyle\sum_{n=-\infty}^{\infty} \dfrac{(-1)^n e^{inx}}{a^2 - n^2}$

3. $\dfrac{2}{\pi} - \dfrac{2}{\pi}\left[\dfrac{e^{2it} + e^{-2it}}{1.3} + \dfrac{e^{4it} + e^{-4it}}{3.5} + \dfrac{e^{6it} + e^{-6it}}{5.7} + \ldots\right]$

4. $\dfrac{a}{2} - \dfrac{a}{\pi}\left[(e^u - e^{-u}) + \dfrac{1}{3}(e^{3u} - e^{-3u}) + \dfrac{1}{5}(e^{5u} - e^{-5u}) + \ldots\right]$, where $u = \dfrac{i\pi x}{l}$

5. $\sinh 9 \displaystyle\sum_{-\infty}^{\infty} \dfrac{(-1)^n (9 + n\pi i)}{81 + (n\pi)^2} e^{n\pi i x/3}$

ARCHIVE

Verify the following:

1. If $|r| < 1$ then the series $\displaystyle\sum_{n=1}^{\infty} r^n$ converges to $\dfrac{r}{1-r}$.

2. If Σa_n is convergent then $\Sigma a_n x^n$ is absolutely convergent when $|x| < 1$.

3. If $u_n \geq u_{n+1} \geq 0 \; \forall \, n$ then $\Sigma (u_n - u_{n+1})$ converges.

4. If the series of positive terms Σa_n diverges and $<s_n>$ be its sequence of partial sums then $\displaystyle\sum \dfrac{a_n}{s_n^2}$ converges.

5. Positive and decreasing terms of convergent series Σu_n implies $n u_n \to 0$.

6. $\displaystyle\sum\left(\sin\dfrac{1}{n} + \dfrac{1}{n^2}\right)$ is divergent.

7. If Σu_n is a convergent series of non-negative numbers and v_n is bounded then $\Sigma u_n . v_n$ converges absolutely.

8. The series $\displaystyle\sum a_n^2$ and $\Sigma \,|\, (1+a_n)\, e^{-a_n} - 1\,|$ converge or diverge together.

9. If $f(n)$ be monotonic and non-negative then $\Sigma f(n)$ and $\Sigma 2^n f(2^n)$ are either both convergent or both divergent to ∞.

10. If $u_n > 0$ and $c > 0$ then $\displaystyle\sum \dfrac{1}{u_n}$ and $\displaystyle\sum \dfrac{1}{(c + u_n)}$ converge or diverge together.

11. If $u_n > 0 \; \forall \, n$ then Σu_n and $\displaystyle\sum \dfrac{1}{u_n}$ shall diverge but cannot converge together.

12. If for a positive term series Σu_n, $\lim n u_n = 0$ then Σu_n may or may not converge.

13. If Σa_n, Σb_n be the convergent series of non-negative terms then Σ min $[\, a_n, b_n]$, Σ max $[a_n, b_n]$ and also $\displaystyle\sum \sqrt{a_n . b_n}$ converges.

14. The series $\displaystyle\sum \dfrac{n! e^n}{n^n}$ is divergent.

15. The series $\displaystyle\sum_{n=1}^{\infty} \dfrac{2^2 . 3^2 \ldots n^2}{(1 + 2 + 2^2)(1 + 3 + 3^2) \ldots (1 + n + n^2)}$ diverges.

16. The series $\displaystyle\sum_{n=2}^{\infty} \dfrac{(\log n)^p}{n^q}$ converges if $q > 1$ or if $q = 1$ and $p < -1$ and otherwise diverges.

17. For $x > 0$, the series $\displaystyle\sum x^{1 + \frac{1}{2} + \ldots + \frac{1}{n}}$ converges if $x < \dfrac{1}{e}$ and diverges if $x \geq \dfrac{1}{e} \ldots$.

18. The series $1 - \dfrac{1}{2^s} + \dfrac{1}{3^s} - \dfrac{1}{4^s} + \ldots$ converges but its rearranged series

$$1 + \dfrac{1}{3^s} - \dfrac{1}{2^s} + \dfrac{1}{5^s} - \dfrac{1}{7^s} - \dfrac{1}{4^s} + \ldots \text{ diverges to } + \infty \text{ if } 0 < s < 1.$$

19. The series

$$1 - \dfrac{1}{2} - \dfrac{1}{4} + \dfrac{1}{3} - \dfrac{1}{6} - \dfrac{1}{8} + \dfrac{1}{5} - \dfrac{1}{10} - \dfrac{1}{12} + \ldots$$

converges to the sum $\dfrac{1}{2} \log 2$.

20. Every conditionally convergent series shall be rearranged so as to converge to a desired sum or diverge or oscillate finitely or infinitety.

21. $\lim \int_0^a g \dfrac{\sin nx}{\sin x} dx = \lim \int_0^a g \dfrac{\sin nx}{x} dx \; ; 0 \le a \le \pi$

22. $\dfrac{1}{2} a_0 + \displaystyle\sum_{n=1}^{\infty} a_n = \dfrac{f(0^-) + f(0^+)}{2}$ where $\int_0^{\infty} \dfrac{\sin x}{x} dx = \dfrac{\pi}{2}$

23. $f(x) = |x| = \dfrac{\pi}{2} - \dfrac{4}{\pi} \left(\dfrac{\cos x}{1^2} + \dfrac{\cos 3x}{3^2} + \dfrac{\cos 5x}{5^2} + ... \right)$

24. The function x^2 is periodic with period $2l$ on the interval $(-l, l)$ and its Fourier series is given by

$$\dfrac{l^2}{3} + \dfrac{4l^2}{\pi^2} \sum_{n=1}^{\infty} \dfrac{(-1)^n}{n^2} \cos\left(\dfrac{n\pi x}{l} \right).$$

25. $\left| \cos\left(\dfrac{\pi x}{l} \right) \right| = \dfrac{4}{\pi} + \displaystyle\sum_{n=1}^{\infty} \dfrac{2(-1)^{n+1}}{\pi(4n^2 - 1)} \cos\left(\dfrac{2n\pi x}{l} \right)$

26. $x - [x] - \dfrac{1}{2} = \displaystyle\sum_{n=1}^{\infty} \left(-\dfrac{1}{n\pi} \right) \sin(2n\pi x) \text{ in } \left[-\dfrac{1}{2}, \dfrac{1}{2} \right]$

27. For all values of x in $[-\pi, \pi]$, k is not an integer

$$\cos kx = \dfrac{\sin k\pi}{\pi} \left[\dfrac{1}{k} + \sum_{n=1}^{\infty} \dfrac{(-1)^n 2k \cos nx}{k^2 - n^2} \right]$$

28. The Fourier series of e^x in $[-l, l]$ is given by

$$\dfrac{\sinh l}{l} + 2l \sinh l \sum \dfrac{(-1)^n \cos \dfrac{n\pi x}{l}}{l^2 + n^2 \pi^2} - 2\pi \sinh l \sum \dfrac{(-1)^n n \sin \dfrac{n\pi x}{l}}{l^2 + n^2 \pi^2}$$

29. Test the series for convergence

$1 + \dfrac{1}{2^2} - \dfrac{1}{3^2} - \dfrac{1}{4^2} + \dfrac{1}{5^2} + \dfrac{1}{6^2} - \dfrac{1}{7^2} - \dfrac{1}{8^2} + ...\infty$ (VTU–2006)

30. Test the series $\dfrac{x}{\sqrt{3}} - \dfrac{x^2}{\sqrt{5}} + \dfrac{x^3}{\sqrt{7}} - ...$ for absolute convergence and conditional convergence.   (VTU–2010)

31. Prove that the series $\dfrac{\sin x}{1^3} - \dfrac{\sin 2x}{2^3} + \dfrac{\sin 3x}{3^3} - ...$ converges absolutely.   (ROHTAK–2006S)

32. Discuss the absolute convergence of $\displaystyle\sum_{n=0}^{\infty} \dfrac{(-1)^n x^n}{n+1}$.   (HISSAR–2005S)

33. Find the nature of the series

$\dfrac{x}{1.2} - \dfrac{x^2}{2.3} + \dfrac{x^3}{3.4} - \dfrac{x^4}{4.5} + ...\infty$ (VTU–2009)

34. For what values of x the following series convergent

$x - \dfrac{x^2}{\sqrt{2}} + \dfrac{x^3}{\sqrt{3}} - \dfrac{x^4}{\sqrt{4}} + ...\infty$ (PTU–2009S, VTU–2008)

35. Find the readius of convergence of the series $\sum \dfrac{n!}{n^n} x^n$.   (CALICUT–2005)

36. Test the series $1 - \dfrac{1}{2\sqrt{2}} + \dfrac{1}{3\sqrt{3}} - \dfrac{1}{4\sqrt{4}}$ for :

   (i) absolute convergence   (ii) conditional convergence   (VTU–2007, ROHTAK–2005)

37. Test the convergence of the series

$\displaystyle\sum_{n=1}^{\infty} \dfrac{2^n - 2}{2^n + 1} x^{n-1} \quad (x > 0)$ (OSMANIA–1999)

38. Examine the following series for uniform convergence :

   (i) $\displaystyle\sum_{n=1}^{\infty} \dfrac{\sin(nx + x^2)}{x(n+2)}$ (PTU–2009)

   (ii) $\displaystyle\sum_{n=1}^{\infty} \dfrac{1}{n^p + n^q x^2}$ (PTU–2005S)

39. Test for uniform convergence the series :

$\dfrac{\sin x}{1^2} + \dfrac{\sin 2x}{2^2} + \dfrac{\sin 3x}{3^2} + ...\infty$ (PTU–2003, ANDHRA–2000)

Answers

1. Convergent   **4.** Absolute convergent for $0 < x < 1$ **5.** Convergent for $x \le 1$ and not convergent for $x > 1$

6. $-1 < x \le 1$ **7.** $-e < x \le e$ **8.** Absolutely convergent   **9.** Convergent for $x < 1$; Divergent for $x \ge 1$

❋❋❋❋❋❋

Differential Equations and Integral Transform Methods

CHAPTER 8

If $y = f(x)$ be a given function, then its derivative dy / dx can be interpreted as the rate of change of y with respect to x. In any natural process, the variables involved and their rates of changes are connected with one another by means of the basic scientific principles that govern the process. When this connection is expressed in mathematical symbols, the result is often called a differential equation.

Differential equations arise from many problems in Algebra, Geometry, Mechanics, Physics and Chemistry.

1. **Differential Equation.** An equation involving one dependent variable and its derivatives with respect to one or more independent variables is called a differential equation.

   For example :   (i) $e^x dx + e^y dy = 0$,   (ii) $y = x \dfrac{dy}{dx}$

   (iii) $\left[1 + \left(\dfrac{dy}{dx}\right)^2\right]^{3/2} = c \dfrac{d^2 y}{dx^2}$ (iv) $\dfrac{\partial^2 y}{\partial x^2} = \dfrac{1}{c}\dfrac{\partial^2 y}{\partial t^2}$

   (v) $\dfrac{dx}{dt} - \omega y = a \cos pt, \dfrac{dy}{dt} + \omega x = a \sin pt$

2. **Ordinary Differential Equation.** A differential equation which contains only one independent variable is called an ordinary differential equation.

   For example :   (i) $xdx + ydy = 0$,   (ii) $1 + \left(\dfrac{dy}{dx}\right)^2 = \dfrac{d^3 y}{dx^3}$ are both ordinary differential equations as they have one independent variable.

3. **Partial Differential Equation.** A differential equation which contains more than one independent variable is called partial differential equation.

   For example :   (i) $\dfrac{\partial^2 u}{\partial x^2} + \dfrac{\partial^2 u}{\partial y^2} + \dfrac{\partial^2 u}{\partial z^2} = 0$,   (ii) $\dfrac{\partial^2 u}{\partial x^2} = \dfrac{1}{c}\dfrac{\partial^2 u}{\partial y^2}$ are both partial differential equations as they have more than one independent variable such as x, y, z.

4. **Order of Differential Equations.** The order of a differential equation is the order of the highest derivative appearing in it.

5. **Degree of Differential Equations.** The degree of a differential equation is the degree of the highest derivatives occurring in it, after the equation has been expressed in a form free from radicals and fractions as far as the derivative are concerned.

   For example :

(i) $e^x dx + e^y dy = 0$, is of the first order and first degree.   (ii) $\dfrac{d^2 y}{dx^2} + x^2 y = 0$, is of second order and first degree.

(iii) The differential equation $\left[1 + \left(\dfrac{dy}{dx}\right)^2\right]^{3/2} = c\left(\dfrac{d^2 y}{dx^2}\right)$ can be written as $\left[1 + \left(\dfrac{dy}{dx}\right)^2\right]^3 = c^2\left(\dfrac{d^2 y}{dx^2}\right)^2$, it is of second order and second degree.

Solved Examples

EXAMPLE 1. *Determine the order and degree of the differential equation*

$$x\left(\dfrac{d^2 y}{dx^2}\right)^3 + y\left(\dfrac{dy}{dx}\right)^4 + y^2 = 0.$$

SOLUTION. The given differential equation contains second order derivative which is the highest derivative, so the order of the differential equation is 2. The power of second order derivative is 3, so the degree of the differential equation is 3.

Hence, order = 2 and degree = 3.

EXAMPLE 2. *Determine the order and degree of the differential equation*

$$\left[1+\left(\frac{dy}{dx}\right)^2\right]^{1/2} = \left(\frac{d^2y}{dx^2}\right)^{1/3}.$$

SOLUTION. Making the differential equation free from radicals, we get

$$\left[1+\left(\frac{dy}{dx}\right)^2\right]^3 = \left(\frac{d^2y}{dx^2}\right)^2$$

Now, in this equation, the order of the highest derivative is 2, so its order is 2. The power of the highest derivative is 2, so its degree is 2.

Hence, the given differential equation is of order 2 and degree 2.

EXAMPLE 3. *Determine the order and degree of the differential equation*

$$x\left(\frac{dy}{dx}\right) + \frac{2}{(dy/dx)} = y^2.$$

SOLUTION. Making the differential equation free from radicals, we get

$$x\left(\frac{dy}{dx}\right)^2 + 2 = y^2\left(\frac{dy}{dx}\right).$$

Now, in this equation, the order of the highest derivative is 1 and its power is 2. Hence, the differential equation is of order 1 and degree 2.

EXAMPLE 4. *Determine the order and degree of the differential equation*

$$y = px + \sqrt{a^2p^2 + b^2}, \text{ where } p = dy/dx.$$

SOLUTION. Making the differential equation free from radicals, we get

$$y = px + \sqrt{a^2p^2 + b^2}$$
$$\Rightarrow \quad y - px = \sqrt{a^2p^2 + b^2}$$
$$\Rightarrow \quad (y - px)^2 = a^2p^2 + b^2$$
$$\Rightarrow y^2 + p^2x^2 - 2ypx = a^2p^2 + b^2$$
$$\text{or } (x^2 - a^2)p^2 - 2xyp + y^2 - b^2 = 0$$
$$\text{or } (x^2 - a^2)\left(\frac{dy}{dx}\right)^2 - 2xy\frac{dy}{dx} + y^2 - b^2 = 0.$$

Now, in this equation, the order of the highest derivative is 1 and its power is 2. Hence, the given differential equation is of order 1 and degree 2.

Exercise-8.1

Determine the order and degree of each of the following differential equations :

1. $\left(\frac{dy}{dx}\right)^2 + 5y = \sin x$

2. $\frac{d^2y}{dx^2} + 3\left(\frac{dy}{dx}\right)^3 + 2y = 0$

3. $(x^3 - y^3)\,dx + (xy^2 - x^2y)dy = 0$

4. $x^2\left(\frac{dy}{dx}\right) + 2xy - 6x^3 = 0$

5. $\frac{d^2y}{dx^2} + 5xy = -3xe^{-x}$

6. $\left(\frac{d^2y}{dx^2}\right)^3 + 2\left(\frac{dy}{dx}\right)^4 + 9 = \sin x$

7. $(2x - 2y + 5)dx = (x - y + 3)dy$

8. $8y^2 = 4xy\frac{dy}{dx} - \left(\frac{dy}{dx}\right)^3$

9. $y = xp \pm c\sqrt{p^2 + 1}$ where $p = \frac{dy}{dx}$

10. $\frac{d^2y}{dx^2} = \sqrt{1 + \left(\frac{dy}{dx}\right)^2}$

11. $\sqrt{a + x}\left(\frac{dy}{dx}\right) + x = 0$

12. $x\sqrt{1 + y^2}\,dx + y\sqrt{1 - x^2}\,dy = 0$

Hints to Selected Problems

3. $\frac{dy}{dx} = \frac{x^3 - y^3}{x^2y - xy^2}$ **7.** $\frac{dy}{dx} = \frac{(2x - 2y + 5)}{(x - y + 3)}$

9. $y = xp \pm c\sqrt{p^2 + 1} \Rightarrow y - xp = \pm c\sqrt{p^2 + 1}$

(squaring and then simplifying)

10. Square both sides and then simplify.

11. $\left(\frac{dy}{dx}\right)^2 = \frac{x^2}{a + x}$ **12.** $\left(\frac{dy}{dx}\right)^2 = \frac{x^2(1 - y^2)}{y^2(1 - x^2)}$

Answers

1. Order = 1, Degree = 2   **2.** Order = 2, Degree = 1   **3.** Order = 1, Degree = 1
4. Order = 1, Degree = 1   **5.** Order = 2, Degree = 1   **6.** Order = 2, Degree = 3
7. Order = 1, Degree = 1   **8.** Order = 1, Degree = 3   **9.** Order = 1, Degree = 2
10. Order = 2, Degree = 2   **11.** Order = 1, Degree = 2   **12.** Order = 1, Degree = 2

8.1.2 SOLUTION OF THE ORDINARY DIFFERENTIAL EQUATION

Any relation between the dependent and independent variables *i.e.*, a function of the form $y = f(x) + C$, which satisfies the given differential equation is called its solution or primitive.

For example : Let $\frac{dy}{dx} = \cos x$ be a given differential equation. Then a function $y = \sin x + C$ is its solution.

1. **General Solution :** If the solution of nth order differential equation contains n arbitrary constants, then it is called general solution or complete primitive.

   For example : $y = C_1 \cos x + C_2 \sin x$ (inviolving two arbitrary constants C_1 and C_2) is the general solution of the differential equation $\dfrac{d^2 y}{dx^2} + y = 0$ of second order.

2. **Particular Solution :** A solution obtained from the general solution by giving particular values to the arbitrary constants in the general solution is called particular solution or particular integral.

   For example : $y = C_1 e^x + C_2 e^{-x}$ is the general solution of the differential equation $\dfrac{d^2 y}{dx^2} - y = 0$ whereas $y = e^x - e^{-x}$ or $y = e^x$ are its particular solution.

3. **Singular Solution :** The solution which cannot be obtained from general solution by assigning particular values to the arbitrary constants, is called singular solution.

Solved Examples

EXAMPLE 1. *Verify that* $y = A \cos x - B \sin x$ *is a solution of the differential equation* $\dfrac{d^2 y}{dx^2} + y = 0$

SOLUTION. We have $y = A \cos x - B \sin x$.   ...(1)

Differentiating (1) w.r.t. x , we get
$$\frac{dy}{dx} = -A \sin x - B \cos x .$$
Differentiating again, we get
$$\frac{d^2 y}{dx^2} = -A \cos x + B \sin x$$
$$= -(A \cos x - B \sin x) = -y \qquad \text{(Using (1))}$$
$$\therefore \quad \frac{d^2 y}{dx^2} + y = 0$$

Hence, $y = A \cos x - B \sin x$ is a solution of $\dfrac{d^2 y}{dx^2} + y = 0$.

EXAMPLE 2. *Show that* $y = Ae^x + Be^{-x}$ *is a solution of the differential equation*
$$\frac{d^2 y}{dx^2} - y = 0 .$$

SOLUTION. We have $y = Ae^x + Be^{-x}$.   ...(1)

Differentiating (1) w.r.t. x, we get
$$\frac{dy}{dx} = Ae^x - Be^{-x} .$$
Differentiating again, we get
$$\frac{d^2 y}{dx^2} = Ae^x + Be^{-x} = y \qquad \text{(Using (1))}$$
$$\Rightarrow \quad \frac{d^2 y}{dx^2} - y = 0 .$$

Hence, $y = Ae^x + Be^{-x}$ is a solution of $\dfrac{d^2 y}{dx^2} - y = 0$.

Exercise-8.2

1. Show that $y = ae^x$ is a solution of the differential equation $\dfrac{dy}{dx} - y = 0$.

2. Show that $y = ae^{-x}$ is a solution of the differential equation $\dfrac{dy}{dx} + y = 0$.

3. Show that $y = c(x-c)^2$ is a solution of the differential equation $8y^2 = 4xy \dfrac{dy}{dx} - \left(\dfrac{dy}{dx}\right)^3$.

4. Show that $y = 4 \sin 3x$ is a solution of the differential equation $\dfrac{d^2 y}{dx^2} + 9y = 0$.

5. Verify that $y = A \cos 2x + B \sin 2x$ is a solution of the differential equation $\dfrac{d^2 y}{dx^2} + 4y = 0$.

6. Verify that $y = e^x (A \cos x + B \sin x)$ is a solution of the differential equation $\dfrac{d^2 y}{dx^2} - 2\dfrac{dy}{dx} + 2y = 0$.   (ANDHRA–1998)

7. Verify that $y = \dfrac{a}{x} + b$ is a solution of the differential equation
$$\frac{d^2 y}{dx^2} + \frac{2}{x}\left(\frac{dy}{dx}\right) = 0 .$$

8. Verify that $y = Ae^{Bx}$ is a solution of the differential equation $\dfrac{d^2 y}{dx^2} = \dfrac{1}{y}\left(\dfrac{dy}{dx}\right)^2$.

9. Show that $y = x^n (A + B \log x)$ is a solution of the differential equation $x^2 \dfrac{d^2 y}{dx^2} - (2n-1)x\dfrac{dy}{dx} + n^2 y = 0$.

10. Show that $y = e^{3x}(A + Bx)$ is a solution of the differential equation $\dfrac{d^2 y}{dx^2} - 6\dfrac{dy}{dx} + 9y = 0$.

11. Verify that $y = ax^3 + bx^2 + c$ is a solution of the differential equation $\dfrac{d^3 y}{dx^3} = 6a$.

12. Verify that $y^2 = 4a(x+a)$ is a solution of the differential equation $y\left[1 - \left(\dfrac{dy}{dx}\right)^2\right] = 2x\dfrac{dy}{dx}$.

13. Verify that $y = Ae^{mx} + Be^{nx}$ is a solution of the differential equation $\dfrac{d^2y}{dx^2} - (m+n)\dfrac{dy}{dx} + mny = 0$.

14. Show that $x^2 + y^2 = cx$ is a solution of the differential equation $2xy\dfrac{dy}{dx} = y^2 - x^2$.

15. Show that $y^2 = a(b - x^2)$ is a solution of the differential equation $y\dfrac{dy}{dx} = x\left[y\dfrac{d^2y}{dx^2} + \left(\dfrac{dy}{dx}\right)^2 \right]$.

16. Show that $y = \log\left[x + \sqrt{x^2 + a^2}\right]^2$ is a solution of the differential equation $(a^2 + x^2)\dfrac{d^2y}{dx^2} + x\dfrac{dy}{dx} = 0$.

17. Show that $y = e^{m\cos^{-1}x}$ is a solution of the differential equation $(1 - x^2)\dfrac{d^2y}{dx^2} - x\dfrac{dy}{dx} - m^2 y = 0$.

18. Show that $y = ae^{\tan^{-1}x}$ is a solution of the differential equation $(1 + x^2)\dfrac{d^2y}{dx^2} + (2x - 1)\dfrac{dy}{dx} = 0$.

Hints to Selected Problems

3. $\qquad y = C(x - C)^2 \Rightarrow \qquad \dfrac{dy}{dx} = 2C(x - C)$

To prove

$$8y^2 = 4xy\dfrac{dy}{dx} - \left(\dfrac{dy}{dx}\right)^3$$

R.H.S. $= 4xy.2C(x - C) - 8C^3(x - C)^3$

$\qquad = 4x.C(x - C)^2.2C(x - C) - 8C^3(x - C)^3$

$\qquad = 8C^2(x - C)^3(x - C) - 8C^2(x - C)^4$

$\qquad = 8[C(x - C)^2]^2 = 8y^2 = $ L.H.S.

10. $\qquad y = e^{3x}(A + Bx)$

$\Rightarrow \dfrac{dy}{dx} = 3e^{3x}(A + Bx) + e^{3x} \cdot B$

and $\dfrac{d^2y}{dx^2} = 3e^{3x}(A + Bx)\cdot 3 + 3e^{3x}\cdot B + 3Be^{3x}$

L.H.S. : $\dfrac{d^2y}{dx^2} - 6\dfrac{dy}{dx} + 9y$

$\qquad = \left[9e^{3x}(A + Bx) + 3Be^{3x} + 3Be^{3x} \right]$

$\qquad\quad - 6\left[3e^{3x}(A + Bx) + e^{3x}B \right] + 9e^{3x}(A + Bx)$

17. $\qquad y = e^{m\cos^{-1}x}$

$\Rightarrow \dfrac{dy}{dx} = e^{m\cos^{-1}x}.m\left(\dfrac{-1}{\sqrt{1 - x^2}}\right) = \dfrac{-me^m \cos^{-1}x}{\sqrt{1 - x^2}}$

Let $\dfrac{dy}{dx} = y_1$ and $\dfrac{d^2y}{dx^2} = y_2$

$\Rightarrow y_1 = \dfrac{-my}{\sqrt{1 - x^2}} \Rightarrow y_1\sqrt{1 - x^2} = -my$

Differentiate again

$y_2\sqrt{1 - x^2} + y_1\dfrac{1}{2}\dfrac{(-2x)}{\sqrt{1 - x^2}} = -my_1$

$\Rightarrow y_2\sqrt{1 - x^2} - \dfrac{xy_1}{\sqrt{1 - x^2}} = -my_1$

$y_2(1 - x^2) - xy_1 = -my_1\sqrt{1 - x^2}$

Putting all values of y_1, y_2

$\dfrac{d^2y}{dx^2}(1 - x^2) - x\dfrac{dy}{dx} + m(my) = 0$

$\Rightarrow \dfrac{d^2y}{dx^2}(1 - x^2) - x\dfrac{dy}{dx} - m^2 y = 0$

$(1 - x^2)\dfrac{d^2y}{dx^2} - x\dfrac{dy}{dx} - m^2 y = 0$

8.1.3 FORMATION OF A DIFFERENTIAL EQUATION

The general solution of a first order differential equation contains one arbitrary constant, which is called a parameter. If this parameter takes various values, then we get a family of curve of one parameter.

For example : The equation $x^2 + y^2 = c^2$ represents a one parameter family of circles if c takes all real values.

Suppose there is an equation, representing a family of curves, containing n arbitrary constants. Then, in order to find its differential equation, we proceed as follows :

WORKING PROCEDURE

Step 1. Differentiate the given equation of family of curve n times to get n more equations containing n arbitrary constants and derivatives.

Step 2. Eliminate all the n constants from all of these $(n + 1)$ equations to get an equation containing a nth order derivative, which is the required differential equation of the given family of curves.

Solved Examples

EXAMPLE 1. ***Find the differential equation of family of all straight lines passing through the origin.***

SOLUTION. The general equation of the family of all straight lines passing through the origin is

$$y = mx . \qquad\qquad\qquad ...(1)$$

where m is a parameter.

Differentiating (1) w.r.t. x, we get

$$\dfrac{dy}{dx} = m . \qquad\qquad\qquad ...(2)$$

Eliminating m between (1) and (2), we get

$$y = x\frac{dy}{dx}$$

which is the required differential equation.

EXAMPLE 2. *Find the differential equation of the family of all straight lines making equal intercepts on the co-ordinate axes.*

SOLUTION. The general equation of the family of all straight lines making equal intercepts on the axes is

$$\frac{x}{c} + \frac{y}{c} = 1$$

or $x + y = c$, where c is a parameter.   ...(1)

Differentiating (1) w.r.t. x, we get

$$1 + \frac{dy}{dx} = 0$$

which is the required differential equation.

EXAMPLE 3. *Find the differential equation that will represent the family of all circles having their centres on the x-axis and radii equal to unity.*

SOLUTION. Let $(h,0)$ be any arbitrary point on the x-axis and suppose $(h,0)$ be the co-ordinates of the centre of a circle of radius equal to unity.

Then, the general equation of the family of all circles having their centres on x-axis and radii equal to unity is

$$(x-h)^2 + y^2 = 1, \; h \text{ is a parameter.} \quad ...(1)$$

Differentiating (1) w.r.t. x, we get

$$2(x-h) + 2y\frac{dy}{dx} = 0$$

or $\quad (x-h) + y\frac{dy}{dx} = 0$ ...(2)

Eliminating h between (1) and (2), we get

$$(x-h) = -y\frac{dy}{dx}.$$

Putting the value of $(x-h)$ in (1), we get

$$\left[-y\left(\frac{dy}{dx}\right)\right]^2 + y^2 = 1$$

$y^2\left(\frac{dy}{dx}\right)^2 + y^2 = 1$, which is the required differential equation.

EXAMPLE 4. *Find the differential equation of the family of curves $y^2 - 2ay + x^2 = a^2$, where a is an arbitrary constant.*

SOLUTION. We have

$$y^2 - 2ay + x^2 = a^2. \quad ...(1)$$

Differentiating (1) w.r.t. x, we get

$$2y\frac{dy}{dx} - 2a\frac{dy}{dx} + 2x = 0$$

or $\quad a = y + \frac{x}{(dy/dx)}$

Putting the value of a in (1), we get

$$y^2 - 2y\left[y + \frac{x}{(dy/dx)}\right] + x^2$$

$$= \left[y + \frac{x}{(dy/dx)}\right]^2$$

or $\quad y^2 - 2y\left[y + \frac{x}{(dy/dx)}\right] + x^2$

$$= y^2 + \frac{x^2}{(dy/dx)^2} + \frac{2xy}{(dy/dx)}$$

or $\quad -2y\frac{dy}{dx}\left[y\frac{dy}{dx} + x\right] + x^2\left(\frac{dy}{dx}\right)^2$

$$= x^2 + 2xy\frac{dy}{dx}$$

or $\quad (x^2 - 2y^2)\left(\frac{dy}{dx}\right)^2 - 4xy\frac{dy}{dx} - x^2 = 0$

which is the required differential equation.

EXAMPLE 5. *Find the differential equation of the family of all circles in the first quadrant which touch the co-ordinate axes.*

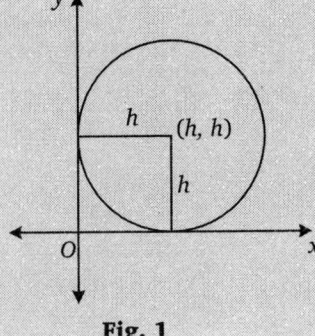

Fig. 1

SOLUTION. The general equation of the family of all circles in the first quadrant which touch the co-ordinate axes is given by

$$(x-h)^2 + (y-h)^2 = h^2 \quad ...(1)$$

Differentiating (1) w.r.t. x, we get

$$2(x-h) + 2(y-h)\frac{dy}{dx} = 0$$

or $2x + 2y\frac{dy}{dx} - 2h - 2h\frac{dy}{dx} = 0$

or $h = \dfrac{\left(x + y\dfrac{dy}{dx}\right)}{\left(1 + \dfrac{dy}{dx}\right)} = \left(\dfrac{x + yp}{1 + p}\right)$, where $p = \dfrac{dy}{dx}$

Putting the value of h in (1), we get

$$\left[x - \frac{x+yp}{1+p}\right]^2 + \left[y - \frac{x+yp}{1+p}\right]^2 = \left[\frac{x+yp}{1+p}\right]^2$$

or $\quad \left(\dfrac{xp - yp}{1+p}\right)^2 + \left(\dfrac{y-x}{1+p}\right)^2 = \left(\dfrac{x+yp}{1+p}\right)^2$

or $\quad (x-y)^2 p^2 + (y-x)^2 = (x+yp)^2$

or $\quad (x-y)^2(p^2 + 1) = (x+yp)^2$

$$(x-y)^2\left[\left(\frac{dy}{dx}\right)^2 + 1\right] = \left[x + y\frac{dy}{dx}\right]^2$$

which is the required differential equation.

EXAMPLE 6. *Find the differential equation of the family of all circles of radius r.*

(ANDHRA–1999, MYSORE–1997, 2009)

SOLUTION. The general equation of the family of all circles with centres (a, b) and radius r is given by

$$(x-a)^2 + (y-b)^2 = r^2 \quad ...(1)$$

Differentiating (1) w.r.t. x, we get

$$2(x-a) + 2(y-b)\frac{dy}{dx} = 0$$

or $\quad (x-a) + (y-b)\frac{dy}{dx} = 0 \qquad \qquad ...(2)$

Differentiating again, we get

$$1 + (y-b)\frac{d^2y}{dx^2} + \left(\frac{dy}{dx}\right)^2 = 0 \qquad \qquad ...(3)$$

Eliminating a and b between (1), (2) and (3), we get

$$y - b = \frac{1 + (dy/dx)^2}{-\dfrac{d^2y}{dx^2}}$$

From (2), we get

$$(x-a) = -(y-b)\left(\frac{dy}{dx}\right) = \frac{\dfrac{dy}{dx}\left[1+\left(\dfrac{dy}{dx}\right)^2\right]}{\dfrac{d^2y}{dx^2}}$$

Putting the values of $(x-a)$ and $(y-b)$ in (1), we get

$$\frac{\left(\dfrac{dy}{dx}\right)^2\left[1+\left(\dfrac{dy}{dx}\right)^2\right]^2}{\left(\dfrac{d^2y}{dx^2}\right)^2} + \frac{\left[1+\left(1+\dfrac{dy}{dx}\right)^2\right]^2}{\left(\dfrac{d^2y}{dx^2}\right)^2} = r^2$$

or $\quad \left[1+\left(\dfrac{dy}{dx}\right)^2\right]^2 = r^2\left(\dfrac{d^2y}{dx^2}\right)^2$

which is the required differential equation.

EXAMPLE 7. *Find the differential equation of the coaxial circles of system* $x^2 + y^2 + 2ax + c^2 = 0$ *where c is a constant and a is a variable.* (JNTU–2003)

SOLUTION. We have, $x^2 + y^2 + 2ax + c^2 = 0 \qquad ...(1)$
Differentiating w.r.t. x, we get

$$2x + 2y\frac{dy}{dx} + 2a = 0$$

or $\qquad \qquad 2a = -2\left(x + y\frac{dy}{dx}\right)$

Substituting in equation (1), we get

$$x^2 + y^2 - 2\left(x + y\frac{dy}{dx}\right)x + c^2 = 0$$

or $\qquad \qquad 2xy\frac{dy}{dx} = y^2 - x^2 + c^2$

which is the required differential equation.

EXAMPLE 8. *Determine the differential equation whose set of independent solution is* $\{e^x, xe^x, x^2e^x\}$.

(UPTU–2002)

SOLUTION. Here, we have

$$y = C_1 e^x + C_2 x e^x + C_3 x^2 e^x \qquad ...(1)$$

Differentiating w.r.t. x, we get

$$y' = C_1 e^x + C_2 x e^x + C_2 e^x + C_3 x^2 e^x + C_3 2x e^x$$

$$= (C_1 e^x + C_2 x e^x + C_3 x^2 e^x) + C_2 e^x + C_3 2x e^x$$

$$= y + C_2 e^x + C_3 2x e^x \qquad \text{(Using (1))}$$

$$\Rightarrow y' - y = C_2 e^x + C_3 2x e^x \qquad ...(2)$$

Again differentiating w.r.t. x, we get

$$y'' - y' = C_2 e^x + 2C_3 e^x + 2x C_3 e^x$$

or $\quad y'' - y' = (C_2 e^x + 2x C_3 e^x) + 2C_3 e^x \quad \text{(Using (2))}$

$$\Rightarrow \quad y'' - y' = y' - y + 2C_3 e^x$$

$$\Rightarrow \quad y'' - 2y' + y = 2C_3 e^x \qquad ...(3)$$

Differentiating w.r.t. x, we get

$$y''' - 2y'' + y' = 2C_3 e^x$$

$$\Rightarrow \quad y''' - 2y'' + y' = y'' - 2y' + y \quad \text{(Using (3))}$$

$$\Rightarrow \quad y''' - 3y'' + 3y' - y = 0$$

$$\Rightarrow \quad (D-1)^3 y = 0.$$

EXAMPLE 9. *A particle falls under gravity in a resisting medium whose resistance varies with velocity. Find the relation between distance and velocity if initially the particle starts from rest.*

(UPTU–2003, 04)

SOLUTION. Let v be the velocity when the particle has fallen distance s in time t from rest. If the resistance is mkv, equation of motion is $m\dfrac{dv}{dt} = mg - mkv$.

Since the forces acting on the particle are the weight mg downwards and resistance mkv upwards.

$$\frac{dv}{dt} = g - kv \quad \Rightarrow \quad \frac{dv}{g - kv} = dt$$

Integration gives $-\dfrac{1}{k}\log(g - kv) = t + C_1$

$v = 0$ when $t = 0$

$$C_1 = -\frac{1}{k}\log g \quad \Rightarrow \quad \frac{g - kv}{g} = e^{-kt}$$

$$v = \frac{g}{k}(1 - e^{-kt}) \qquad ...(1)$$

$$\frac{ds}{dt} = \frac{g}{k}(1 - e^{-kt})$$

Integration yields, $s = \dfrac{g}{k}t + \dfrac{g}{k^2}e^{-kt} + C_2$

Since $s = 0$ when $t = 0$

$$C_2 = -\frac{g}{k^2}$$

$$s = \frac{g}{k}t + \frac{g}{k^2}(e^{-kt} - 1) \qquad ...(2)$$

Eliminating t between (1) and (2), we get

$$s = \frac{g}{k^2}\log\left(\frac{g}{g - kv}\right) - \frac{v}{k}$$

which gives a relation between distance and velocity.

Exercise-8.3

Find the differential equations of the following families of the curves :

1. $y = 3x + c$, where c is a parameter.

2. $(2x + a)^2 + y^2 = a^2$, where a is a parameter.

3. $x^2 + y^2 = a^2$, where a is a parameter.

4. $x^2 + y^2 = 2ax$, where a is a parameter.

5. $y = a\sin(x + b)$, where a and b are parameters.

6. $y = a\cos(x + b)$, where a and b are parameters.

7. $y = a\cos nx + b\sin nx$, where a and b are parameters.

8. $y = ax + \dfrac{b}{x}$, where a and b are parameters.

9. $y = ae^{2x} + be^{-x}$, where a and b are parameters.

10. $y = ae^{2x} + be^{-2x}$, where a and b are parameters.

11. $y = ae^{3x} + be^{5x}$, where a and b are parameters.

12. $y^2 = b(a^2 - x^2)$, where a and b are parameters.

13. $(y - b)^2 = 4(x - a)$, where a and b are parameters.

14. $xy = ae^x + be^{-x} + x^2$, where a and b are parameters.

15. Find the differential equation of the family of all circles which passes through the origin and whose centre lie on the y-axis.

Hints to Selected Problems

3. $x^2 + y^2 = a^2 \Rightarrow 2x + 2y\dfrac{dy}{dx} = 0 \Rightarrow x + y\dfrac{dy}{dx} = 0$

7. $y = a\cos nx + b\sin nx \Rightarrow \dfrac{dy}{dx} = -a.n\sin nx + b.n\cos nx$

$\dfrac{d^2y}{dx^2} = -a.n^2\cos nx - b.n^2\sin nx = -n^2[a\cos nx + b\sin nx]$

$= -n^2 y$

$\dfrac{d^2y}{dx^2} + n^2 y = 0$.

11. $y = ae^{3x} + be^{5x}$ ...(1)

$\dfrac{dy}{dx} = a.3.e^{3x} + b.5\,e^{5x}$ ...(2)

$\dfrac{d^2y}{dx^2} = 9ae^{3x} + 25\,be^{5x}$ ...(3)

From (2) and (3) on removing a and b, we get

$\dfrac{d^2y}{dx^2} - 8\dfrac{dy}{dx} + 15y = 0$.

Answers

1. $\dfrac{dy}{dx} = 3$ **2.** $2xy\dfrac{dy}{dx} - y^2 + 4x^2 = 0$ **3.** $x + y\dfrac{dy}{dx} = 0$ **4.** $2xy\dfrac{dy}{dx} + x^2 - y^2 = 0$ **5.** $\dfrac{d^2y}{dx^2} + y = 0$

6. $\dfrac{d^2y}{dx^2} + y = 0$ **7.** $\dfrac{d\,y}{dx} + n\,y =$ **8.** $x^2\dfrac{d^2y}{dx^2} + x\dfrac{dy}{dx} - y = 0$ **9.** $\dfrac{d^2y}{dx^2} - \dfrac{dy}{dx} - 2y = 0$ **10.** $\dfrac{d^2y}{dx^2} - 4y = 0$

11. $\dfrac{d^2y}{dx^2} - 8\dfrac{dy}{dx} + 15y = 0$ **12.** $xy\dfrac{d^2y}{dx^2} + x\left(\dfrac{dy}{dx}\right)^2 - y\dfrac{dy}{dx} = 0$ **13.** $2\dfrac{d^2y}{dx^2} + \left(\dfrac{dy}{dx}\right)^3 = 0$ **14.** $x\dfrac{d^2y}{dx^2} + 2\dfrac{dy}{dx} = xy - x^2 + 2$

15. $(x^2 - y^2)\dfrac{dy}{dx} - 2xy = 0$

8.1.4 Method of Solving Differential Equation by Separation of Variables

If any differential equation can be expressed as

$$f(x)\,dx = g(y)\,dy$$ ...(1)

Then we say that variables are separable.

WORKING PROCEDURE

In order to solve the given differential equation, using variable separable,

Step 1. Separate the variables as $f(x)dx = g(y)dy$.

Step 2. Integrate both sides as $\int f(x)dx = \int g(y)dy$.

Step 3. Add an arbitrary constant to any of the sides.

Solved Examples

EXAMPLE 1. *Solve* $\dfrac{dy}{dx} = xy + x + y + 1$.

SOLUTION. We have

$\dfrac{dy}{dx} = xy + x + y + 1$

$\Rightarrow \dfrac{dy}{dx} = (1 + x)(1 + y)$.

Separating the variables, we get

$\dfrac{dy}{1 + y} = (1 + x)\,dx$.

Integrating, we get

$\log|1 + y| = x + \dfrac{x^2}{2} + C$

which is the required solution.

EXAMPLE 2. *Solve* $\dfrac{dy}{dx} = \log(x + 1)$.

SOLUTION. We have $\dfrac{dy}{dx} = \log(x + 1)$

Separating the variables, we get

$dy = \log(x + 1)dx$.

Integrating both sides, we get

$y = \int \log(x + 1)dx + C$

$= \log(x + 1)\int dx - \int\left\{\dfrac{d}{dx}(\log(x + 1)).\int dx\right\}dx + C$

$$= x \log(x+1) - \int \frac{x}{x+1} dx + C$$

$$= x \log(x+1) - \int \frac{x+1-1}{x+1} dx + C$$

$$= x \log(x+1) - \int dx + \int \frac{dx}{x+1} + C$$

$$\therefore \quad y = x \log(x+1) - x + \log|x+1| + C$$

which is the required solution.

EXAMPLE 3. *Solve* $x\sqrt{1-y^2}\, dx + y\sqrt{1-x^2}\, dy = 0$.

SOLUTION. We have $x\sqrt{1-y^2}\, dx + y\sqrt{1-x^2}\, dy = 0$

Separating the variables, we get

$$\frac{x}{\sqrt{1-x^2}} dx + \frac{y}{\sqrt{1-y^2}} dy = 0$$

Integrating both sides, we get

$$\int \frac{x}{\sqrt{1-x^2}} dx + \int \frac{y}{\sqrt{1-y^2}} dy = C$$

$$\Rightarrow \quad \frac{1}{2}\int \frac{2x}{\sqrt{1-x^2}} dx + \frac{1}{2}\int \frac{2y}{\sqrt{1-y^2}} dy = C$$

$$\Rightarrow \qquad -\frac{1}{2}\int \frac{dt}{\sqrt{t}} - \frac{1}{2}\int \frac{du}{\sqrt{u}} = C,$$

where $t = 1-x^2, u = 1-y^2$

$$\Rightarrow \qquad -\frac{1}{2}\left[2\sqrt{t}\right] - \frac{1}{2}\left[2\sqrt{u}\right] = C$$

$$\Rightarrow \qquad -\sqrt{t} - \sqrt{u} = C$$

$$\Rightarrow \qquad -\sqrt{1-x^2} - \sqrt{1-y^2} = C$$

$$\therefore \quad \sqrt{1-x^2} + \sqrt{1-y^2} = k \text{ , where } k = -C$$

which is the required solution.

EXAMPLE 4. *Solve* $\quad y - x\dfrac{dy}{dx} = a\left(y^2 + \dfrac{dy}{dx}\right)$.   (BHOPAL 1991)

SOLUTION. We have $y - x\dfrac{dy}{dx} = a\left(y^2 + \dfrac{dy}{dx}\right)$

$$\Rightarrow \qquad y - ay^2 = (x+a)\frac{dy}{dx}$$

Separating the variables, we get

$$\int \frac{dx}{x+a} = \frac{1}{(y-ay^2)} dy.$$

Integrating both sides, we get

$$\int \frac{dx}{x+a} = \int \frac{dy}{y-ay^2} + \log C$$

$$\Rightarrow \quad \log|x+a| = \int \frac{dy}{y(1-ay)} + \log C$$

$$= \int \left(\frac{1}{y} + \frac{a}{1-ay}\right) dy + \log C$$

$$= \int \frac{dy}{y} + a\int \frac{dy}{1-ay} + \log C$$

$$\log|x+a| = \log|y|$$
$$- \log|1-ay| + \log C$$

$$\Rightarrow \quad \log\left|\frac{(x+a)(1-ay)}{y}\right| = \log C$$

$$\Rightarrow \qquad \left|\frac{(x+a)(1-ay)}{y}\right| = C$$

$$\Rightarrow \qquad \frac{(x+a)(1-ay)}{y} = k, \text{ where } k = \pm C.$$

$$\therefore \quad (x+a)(1-ay) = ky, \text{ which is the required solution.}$$

EXAMPLE 5. *Solve* $(1+x)(1+y^2)dx + (1+y)(1+x^2)dy = 0$.

SOLUTION. We have

$$(1+x)(1+y^2)dx + (1+y)(1+x^2)dy = 0$$

Separating the variables, we get

$$\frac{(1+x)}{(1+x)^2} dx + \frac{(1+y)}{(1+y^2)} dy = 0.$$

Integrating both sides, we get

$$\int \frac{(1+x)}{(1+x^2)} dx + \int \frac{(1+y)}{(1+y^2)} dy = C$$

$$\Rightarrow \quad \int \frac{dx}{1+x^2} + \int \frac{xdx}{1+x^2}$$

$$+ \int \frac{dy}{1+y^2} + \int \frac{ydy}{1+y^2} = C$$

$$\Rightarrow \quad \tan^{-1} x + \frac{1}{2}\int \frac{2x}{1+x^2} dx +$$

$$\tan^{-1} y + \frac{1}{2}\int \frac{2y}{1+y^2} dy = C$$

$$\Rightarrow \quad \tan^{-1} x + \frac{1}{2}\log|1+x^2| +$$

$$\tan^{-1} y + \frac{1}{2}\log|1+y^2| = C$$

$$\therefore \quad \tan^{-1} x + \tan^{-1} y$$

$$+ \frac{1}{2}[\log(1+x^2) + \log(1+y^2)] = C$$

which is the required solution.

EXAMPLE 6. *Solve* $\dfrac{dy}{dx} = \dfrac{1-\cos x}{1+\cos x}$.

SOLUTION. We have $\dfrac{dy}{dx} = \dfrac{1-\cos x}{1+\cos x}$

Separating the variables, we get

$$dy = \left(\frac{1-\cos x}{1+\cos x}\right) dx.$$

Integrating both sides, we get

$$\int dy = \int \frac{1-\cos x}{1+\cos x} dx + C$$

$$\Rightarrow y = \int \frac{2\sin^2 \frac{x}{2}}{2\cos^2 \frac{x}{2}} dx + C \Rightarrow y = \int \tan^2 \frac{x}{2} dx + C$$

$$\Rightarrow \qquad y = \int \left(\sec^2 \frac{x}{2} - 1\right) dx + C$$

$$\Rightarrow \qquad y = \int \sec^2 \frac{x}{2} dx - \int dx + C$$

$$\Rightarrow \qquad y = 2\tan \frac{x}{2} - x + C$$

$$\therefore \quad x + y = 2\tan \frac{x}{2} + C, \text{ which is the required solution.}$$

EXAMPLE 7. *Solve* $(1 + y^2)(1 + \log x)dx + xdy = 0$, *it being given that* $y = 1$ *when* $x = 1$.

SOLUTION. We have $(1 + y^2)(1 + \log x)dx + xdy = 0$

Separating the variables, we get

$$\frac{1 + \log x}{x}dx + \frac{dy}{1 + y^2} = 0.$$

Integrating both sides, we get

$$\int \frac{(1 + \log x)}{x}dx + \int \frac{dy}{1 + y^2} = C$$

$$\Rightarrow \int t\,dt + \tan^{-1}y = C, \text{ where } t = 1 + \log x$$

$$\Rightarrow \frac{t^2}{2} + \tan^{-1}y = C$$

$$\Rightarrow \frac{(1 + \log x)^2}{2} + \tan^{-1}y = C.$$

Putting $x = 1$ and $y = 1$, we get

$$\frac{1}{2} + \tan^{-1}1 = C$$

$$\Rightarrow \qquad C = \frac{1}{2} + \frac{\pi}{4}.$$

$$\therefore \quad \frac{(1 + \log\ )}{2} + \tan\quad = \frac{1}{2} + \frac{\pi}{4}$$

or $(1 + \log x)^2 + 2\tan^{-1}y = 1 + \frac{\pi}{2}$.

EXAMPLE 8. *Solve* $(1 + x^2)\sec^2 ydy + 2x\tan y\,dx = 0$, *given*

that $y = \frac{\pi}{4}$ *when* $x = 1$.

SOLUTION. We have $(1 + x^2)\sec^2 ydy + 2x\tan y\,dx = 0$

Separating the variables, we get

$$\frac{\sec^2 y}{\tan y}dy + \frac{2x}{1 + x^2}dx = 0.$$

Integrating both sides, we get

$$\int \frac{\sec^2 y}{\tan y}dy + \int \frac{2x}{1 + x^2}dx = C$$

$$\Rightarrow \int \frac{dt}{t} + \int \frac{du}{u} = C, \text{ where } t = \tan y, u = 1 + x^2$$

$$\Rightarrow \log|t| + \log|u| = C$$

$$\Rightarrow \log|\tan y| + \log|1 + x^2| = C.$$

Putting $x = 1$ and $y = \frac{\pi}{4}$, we get

$$C = \log\left|\tan\frac{\pi}{4}\right| + \log|1 + 1|$$

$$\Rightarrow \quad C = \log 1 + \log 2 = \log 2$$

$$\therefore \log|\tan y| + \log(1 + x^2) = \log 2$$

$$\therefore \qquad \log|(1 + x^2)\tan y| = \log 2$$

or $\qquad 2\cot y = (1 + x^2)$

8.1.5 DIFFERENTIAL EQUATION REDUCIBLE TO VARIABLE SEPARABLE FORM

Sometimes, we come across with some differential equation, in which the variables cannot be separated. In such type of differential equation, some suitable substitution reduces it to a form in which the variables are separable.

Solved Examples

EXAMPLE 1. *Solve* $\dfrac{dy}{dx} = \sin(x + y)$.

SOLUTION. Let $v = x + y$.

Then, $\dfrac{dv}{dx} = 1 + \dfrac{dy}{dx}$ so that $\dfrac{dy}{dx} = \dfrac{dv}{dx} - 1$.

Now, the given differential equation reduces to

$\dfrac{dv}{dx} - 1 = \sin v$ or $\dfrac{dv}{dx} = 1 + \sin v$.

Separating the variables, we get

$$\frac{dv}{1 + \sin v} = dx.$$

Integrating both sides, we get

$$\int \frac{dv}{1 + \sin v} = \int dx + C$$

$$\Rightarrow \int \frac{(1 - \sin v)}{(1 + \sin v)(1 - \sin v)}dv = x + C$$

$$\Rightarrow \int \frac{(1 - \sin v)}{1 - \sin^2 v}dv = x + C$$

$$\Rightarrow \int \frac{1 - \sin v}{\cos^2 v}dv = x + C$$

$$\Rightarrow \int \sec^ v\,dv - \int \sec v\tan v\,dv = x + C.$$

$$\Rightarrow \qquad \tan v - \sec v = x + C$$

$\therefore \tan(x + y) - \sec(x + y) = x + C$, which is the required solution.

EXAMPLE 2. *Solve* $\dfrac{dy}{dx} = (4x + y + 1)^2$. (MANGLORE–1999)

SOLUTION. Let $v = 4x + y + 1$.

Then, $\dfrac{dv}{dx} = 4 + \dfrac{dy}{dx}$ so that $\dfrac{dy}{dx} = \dfrac{dv}{dx} - 4$

Now, the given differential equation reduces to

$$\frac{dv}{dx} - 4 = v^2 \text{ or } \frac{dv}{dx} = 4 + v^2.$$

Separating the variables, we get

$$\frac{dv}{4 + v^2} = dx.$$

Integrating both sides, we get

$$\int \frac{dv}{4 + v^2} = \int dx + C$$

$$\Rightarrow \int \frac{dv}{2^2 + v^2} = \int dx + C.$$

$$\Rightarrow \frac{1}{2}\tan^{-1}\frac{v}{2} = x + C$$

$$\therefore \frac{1}{2}\tan^{-1}\left(\frac{4x + y + 1}{2}\right) = x + C$$

or $4x + y + 1 = 2\tan(2x + 2C)$, which is the required solution.

EXAMPLE 3. *Solve* $\dfrac{dy}{dx} = \sin(x+y) + \cos(x+y)$.   (VTU–2005)

SOLUTION. Let $x+y = v$.

Then, $1 + \dfrac{dy}{dx} = \dfrac{dv}{dx}$ so that $\dfrac{dy}{dx} = \dfrac{dv}{dx} - 1$.

Using above substitution, the given differential equation reduces to

$\dfrac{dv}{dx} - 1 = \sin v + \cos v$ or $\dfrac{dv}{dx} = 1 + \sin v + \cos v$.

Separating the variables, we get

$$\dfrac{dv}{1 + \sin v + \cos v} = dx.$$

Integrating both sides, we get

$$\int \dfrac{dv}{1 + \sin v + \cos v} = \int dx + C$$

$$\Rightarrow \int \dfrac{dv}{2\sin\dfrac{v}{2}\cos\dfrac{v}{2} + 2\cos^2\dfrac{v}{2}} = x + C$$

$$\Rightarrow \dfrac{1}{2}\int \dfrac{\sec^2\dfrac{v}{2}\,dv}{\left(\tan\dfrac{v}{2} + 1\right)} = x + C$$

$$\Rightarrow \int \dfrac{dt}{t} = x + C, \text{ where } t = 1 + \tan\dfrac{v}{2}$$

$$\Rightarrow \log|t| = x + C$$

$$\Rightarrow \log\left|1 + \tan\dfrac{v}{2}\right| = x + C$$

$$\therefore \log\left|1 + \tan\left(\dfrac{x+y}{2}\right)\right| = x + C \qquad [\because v = x+y]$$

which is the required solution.

EXAMPLE 4. *Solve* $\dfrac{dy}{dx} = \dfrac{2x+3y+4}{4x+6y+5}$.

SOLUTION. Let $2x + 3y = v$.

Then, $2 + 3\dfrac{dy}{dx} = \dfrac{dv}{dx} \Rightarrow \dfrac{dy}{dx} = \dfrac{1}{3}\left(\dfrac{dv}{dx} - 2\right)$.

Using above substitution, the given differential equation reduces to

$$\dfrac{1}{3}\left(\dfrac{dv}{dx} - 2\right) = \dfrac{v+4}{2v+5}$$

or $\qquad \dfrac{dv}{dx} - 2 = \dfrac{3v+12}{2v+5}$

or $\qquad \dfrac{dv}{dx} = 2 + \dfrac{3v+12}{2v+5} = \dfrac{7v+22}{2v+5}$.

Separating the variables, we get

$$\dfrac{2v+5}{7v+22}dv = dx.$$

Integrating both sides, we get

$$\int \dfrac{2v+5}{7v+22}dv = \int dx + C$$

$$\Rightarrow \quad 2\int \dfrac{v}{7v+22}dv + 5\int \dfrac{dv}{7v+22} = x + C$$

$$\Rightarrow \quad \dfrac{2}{7}\int \dfrac{7v}{7v+22}dv + 5\int \dfrac{dv}{7v+22} = x + C$$

$$\Rightarrow \dfrac{2}{7}\int \dfrac{7v+22-22}{7v+22}dv + 5\int \dfrac{dv}{7v+22} = x + C$$

$$\Rightarrow \dfrac{2}{7}\int dv - \dfrac{44}{7}\int \dfrac{dv}{7v+22} + 5\int \dfrac{dv}{7v+22} = x + C$$

$$\Rightarrow \quad \dfrac{2}{7}v - \dfrac{9}{7}\int \dfrac{dv}{7v+22} = x + C$$

$$\Rightarrow \quad \dfrac{2}{7}v - \dfrac{9}{49}\log|7v+22| = x + C$$

$$\Rightarrow \quad 14v - 9\log|7v+22| = 49x + 49C$$

$$\Rightarrow 14(2x+3y) - 9\log|14x+21y+22| = 49x + 49C$$

$$\Rightarrow 9\log|14x+21y+22| = 42y - 21x + k, k = 49C,$$

which is the required solution.

EXAMPLE 5. *Solve the equation*

$$(2x^2 + 3y^2 - 7)x\,dx - (3x^2 + 2y^2 - 8)y\,dy = 0.$$

SOLUTION. The given equation can be written as   (UPTU–2005)

$$\dfrac{x}{y}\dfrac{dx}{dy} = \dfrac{3x^2 + 2y^2 - 8}{2x^2 + 3y^2 - 7}$$

Applying componendo and dividendo rule, we get

$$\dfrac{xdx + ydy}{xdx - ydy} = \dfrac{5x^2 + 5y^2 - 15}{x^2 - y^2 - 1}$$

or $\dfrac{xdx + ydy}{x^2 + y^2 - 3} = 5\left(\dfrac{xdx - ydy}{x^2 - y^2 - 1}\right)$

Multiplying by 2, we get

$$\left(\dfrac{2xdx + 2ydy}{x^2 + y^2 - 3}\right) = 5\left(\dfrac{2xdx - 2ydy}{x^2 - y^2 - 1}\right)$$

On integrating, we get

$$\log(x^2 + y^2 - 3) = 5\log(x^2 - y^2 - 1) + \log C$$

or $\quad x^2 + y^2 - 3 = C(x^2 - y^2 - 1)^5$.

EXAPMPLE 6. *Solve* $\dfrac{dy}{dx} = \dfrac{x(2\log x + 1)}{\sin y + y\cos y}$ (VTU–2008)

SOLUTION. Given equation is

$x(2\log x + 1)dx = (\sin y + y\cos y)dy$

Integrating both sides, we get

$$2\int (\log x.x + x)dx = \int \sin y\,dy + \int y\cos y\,dy + c$$

$$2\left[\left(\log x.\dfrac{x^2}{2} - \int \dfrac{1}{x}.\dfrac{x^2}{2}dx\right) + \dfrac{x^2}{2}\right]$$

$$= -\cos y + [y\sin y - \int \sin y.1dy + c]$$

$$2x^2\log x - \dfrac{x^2}{2} + \dfrac{x^2}{2}$$

$$= -\cos y + y\sin y + \cos y + c$$

or $\quad 2x^2\log x - y\sin y = c$.

EXAPMPLE 7. *Solve* $\dfrac{y}{x}\dfrac{dy}{dx} + \dfrac{x^2 + y^2 - 1}{2(x^2 + y^2) + 1} = 0$.   (VTU–2003)

SOLUTION. Putting $x^2 + y^2 = t$, we get

$$2x + 2y\dfrac{dy}{dx} = \dfrac{dt}{dx} \text{ or } \dfrac{y}{x}\dfrac{dy}{dx} = \dfrac{1}{2x}\dfrac{dt}{dx} - 1$$

Therefore, the given equation becomes,

$$\dfrac{1}{2x}\dfrac{dt}{dx} - 1 + \dfrac{t-1}{2t+1} = 0 \Rightarrow \dfrac{1}{2x}\dfrac{dt}{dx} = 1 - \dfrac{t-1}{2t+1}$$

or $\dfrac{1}{2x}\dfrac{dt}{dx} = \dfrac{t+2}{2t+1}$ or $2xdx = \dfrac{2t+1}{t+2}dt$

or $2xdx = \left(2 - \dfrac{3}{t+2}\right)dt$

On integrating, we get

$$x^2 = 2t - 3\log(t+2) + c$$

or $x^2 + 2y^2 - 3\log(x^2 + y^2 + 2) + c = 0$

$$(\because x^2 + y^2 = t)$$

EXAPMPLE 8. *Solve* $\sec^2 x \tan y\, dx + \sec^2 y \tan x\, dy = 0$.

(UPTU(B.PHARMA)SUM–2009, PTU–2003, VTU–2009)

SOLUTION. Separating the variables, we get

$$\dfrac{\sec^2 x}{\tan x}dx + \dfrac{\sec^2 y}{\tan^2 y}dy = 0$$

On integrating, we get

$$\log \tan x + \log \tan y = \log C$$

or $\log \tan x \tan y = \log C$

or $\tan x \tan y = C$

where C is an arbitrary constant.

Exercise-8.4

Solve the following differential equations :

1. $x\dfrac{dy}{dx} = y$

2. $\dfrac{dy}{dx} + y = 1$

3. $\dfrac{dy}{dx} = \sqrt{4 - y^2}$

4. $\dfrac{dy}{dx} = e^{x+y}$

5. $\dfrac{dy}{dx} = (e^x + 1)y$

6. $\dfrac{dy}{dx} = \dfrac{(1+y^2)}{(1+x^2)}$

7. $\dfrac{dy}{dx} + \sqrt{\dfrac{1-y^2}{1-x^2}} = 0$ (ANDHRA–1999)

8. $(x^2 + 1)\dfrac{dy}{dx} = xy$

9. $(x\log x)\dfrac{dy}{dx} = y$

10. $x\dfrac{dy}{dx} + y = y^2$

11. $\dfrac{dy}{dx} = 1 - x + y - xy$

12. $\dfrac{dy}{dx} = (1+x)(1+y^2)$

13. $(e^x + 1)ydy = (y+1)e^x dx$

14. $(x+2)\dfrac{dy}{dx} = 4x^2 y$

15. $\dfrac{dy}{dx} + \dfrac{(xy+y)}{(xy+x)} = 0$

16. $x\sqrt{1+y^2}\,dx + y\sqrt{1+x^2}\,dy = 0$

17. $\dfrac{dy}{dx} + \dfrac{\cos x \sin y}{\cos y} = 0$

18. $\dfrac{dy}{dx} + \dfrac{(1+\cos 2y)}{(1-\cos 2x)} = 0$

19. $(y + xy)dx + (x - xy^2)dy = 0$

20. $(x^2 y - x^2)dx + (xy^2 - y^2)dy = 0$

21. $(1+x)ydx + (1-y)x\,dy = 0$

22. $(1-x^2)(1-y)dx = xy(1+y)dy$

23. $\sqrt{1+x^2+y^2+x^2 y^2} + xy\dfrac{dy}{dx} = 0$

24. $x\cos^2 y\,dx = y\cos^2 x\,dy$

25. $3e^x \tan y\,dx + (1-e^x)\sec^2 y\,dy = 0$

26. $(e^y + 1)\cos x\,dx + e^y \sin x\,dy = 0$ (MANGLORE–1999)

27. $x(e^{2y} - 1)dy + (x^2 - 1)e^y dx = 0$

28. $e^{2x-3y}dx + e^{2y-3x}dy = 0$

29. $\sin y\dfrac{dy}{dx} = \sin^3 x$

30. $\sin^2 x\,dy + \cos^2 y\,dx = 0$

31. $x\cos^2 y\,dx - y\cos^2 x\,dy = 0$

32. $x\cos y\,dy = e^x(x\log x + 1)dx$

33. $\dfrac{dy}{dx} = \dfrac{\sin x + x\cos x}{y(2\log y + 1)}$

34. $\dfrac{dy}{dx} = \dfrac{x(2\log x + 1)}{(\sin y + y\cos y)}$ (VTU–2000)

35. $\tan y\dfrac{dy}{dx} = \sin(x+y) + \sin(x-y)$

36. $\dfrac{dy}{dx} = x^2 e^{-3y}$, given that $y = 0$ when $x = 0$.

37. $\dfrac{dy}{dx} = e^{x+y}$, given that $y = 1$ when $x = 1$.

38. $e^x\dfrac{dy}{dx} = 3y^3$, given that $y(0) = \dfrac{1}{2}$.

39. $(x-1)\dfrac{dy}{dx} = 2xy$, given that $y(2) = 1$

40. $xy\dfrac{dy}{dx} = y + 2$, given that $y(2) = 0$

41. $(1+x^2)\dfrac{dy}{dx} + (1+y^2) = 0$, given that $y = 1$ when $x = 0$.

42. $\dfrac{dy}{dx} = \sec y$, given that $y = 0$ when $x = 0$.

43. $\dfrac{dy}{dx} = y\sin 2x$, given that $y = 0$ when $x = 0$.

44. $\dfrac{dy}{dx} = y\cot 2x$, given that $y\left(\dfrac{\pi}{4}\right) = 2$.

45. Find the equation of the curve that passes through the point $(1, 2)$ and satisfies the differential equation $\dfrac{dy}{dx} = -\dfrac{2xy}{(x^2+1)}$.

46. Find the equation of the curve that passes through the point $(1, 0)$ and satisfies the differential equation $(1+y^2)dx - xydy = 0$.

47. $\dfrac{dy}{dx} = (x+y)^2$

48. $\dfrac{dy}{dx} = \sec(x+y)$

49. $(x+y)^2\dfrac{dy}{dx} = a^2$

50. $(x-y)^2\dfrac{dy}{dx} = a^2$ (AMIE 1999)

51. $\dfrac{dy}{dx} = \tan^2(x+y)$

52. $\dfrac{dy}{dx} = \dfrac{x+y-1}{x+y+1}$

53. $\dfrac{dy}{dx} = \dfrac{x-y+3}{2x-2y+5}$

54. $\dfrac{dy}{dx} = \dfrac{8x+6y+12}{4x+3y+2}$

Hints to Selected Problems

11. $\dfrac{dy}{dx} = 1 - x + y - xy = (1-x)(1+y)$

$\Rightarrow \dfrac{dy}{(1+y)} = (1-x)dx \Rightarrow \int \dfrac{dy}{1+y} = \int (1-x)dx + C$

18. $\dfrac{dy}{dx} + \left(\dfrac{1+\cos 2y}{1-\cos 2x}\right) = 0 \Rightarrow \dfrac{dy}{dx} + \dfrac{\cos^2 y}{\sin^2 x} = 0$

$\Rightarrow \sec^2 y\,dy + \text{cosec}^2 x\,dx = 0$

23. $\sqrt{1+x^2+y^2+x^2y^2}+xy\dfrac{dy}{dx}=0$

$\Rightarrow \sqrt{1+x^2}\sqrt{1+y^2}+xy\dfrac{dy}{dx}=0$

$\Rightarrow \dfrac{\sqrt{1+x^2}}{x}dx+\dfrac{y}{\sqrt{1+y^2}}dy=0$

$\Rightarrow \displaystyle\int \dfrac{\sqrt{1+x^2}}{x}dx+\int \dfrac{y}{\sqrt{1+y^2}}dy=C$

$\Rightarrow \displaystyle\int \dfrac{t^2}{t^2-1}dt+\sqrt{1+y^2}=C$, where $t^2=1+x^2$

$\Rightarrow \displaystyle\int dt+\int \dfrac{dt}{t^2-1}+\sqrt{1+y^2}=C$.

29. $\sin y\dfrac{dy}{dx}=\sin^3 x \Rightarrow \sin y\,dy=\sin^3 x\,dx$

$\Rightarrow \sin y\,dy=\dfrac{1}{4}(3\sin x-\sin 3x)dx$

$\Rightarrow \displaystyle\int \sin y\,dy=\dfrac{3}{4}\int \sin x\,dx-\dfrac{1}{4}\int \sin 3x\,dx+C$

32. $x\cos y\,dy=e^x(x\log x+1)dx \Rightarrow \cos y\,dy=e^x\left(\log x+\dfrac{1}{x}\right)dx$

$\Rightarrow \displaystyle\int \cos y\,dy=\int e^x\left(\log x+\dfrac{1}{x}\right)dx+C$

$\Rightarrow \sin y=\displaystyle\int e^x\left[f(x)+f'(x)\right]dx+C$

$\Rightarrow \sin y=e^x f(x)+C=e^x\log x+C$

35. $\tan y\dfrac{dy}{dx}=\sin(x+y)+\sin(x-y)$

$\Rightarrow \tan y\dfrac{dy}{dx}=2\sin x\cos y \Rightarrow \dfrac{\tan y}{\cos y}dy=2\sin x\,dx$

$\Rightarrow \displaystyle\int \dfrac{\tan y}{\cos y}dy=2\int \sin x\,dx+C$

45. $\dfrac{dy}{dx}=\dfrac{-2xy}{(x^2+1)} \Rightarrow \dfrac{dy}{y}+\dfrac{2x}{x^2+1}dx=0$

$\Rightarrow \displaystyle\int \dfrac{dy}{y}+\int \dfrac{2x}{x^2+1}dx=C \Rightarrow \log|y|+\log|x^2+1|=C$

Putting $x=1$ and $y=2$, we get $C=2\log 2=\log 4$

$\therefore \quad \log|y|+\log|x^2+1|=\log 4 \Rightarrow y(x^2+1)=4$.

50. Let $x-y=v$ so that $\dfrac{dy}{dx}=1-\dfrac{dv}{dx}$

Then, $(x-y)^2\dfrac{dy}{dx}=a^2$

$\Rightarrow v^2\left(1-\dfrac{dv}{dx}\right)=a^2 \Rightarrow 1-\dfrac{dv}{dx}=\dfrac{a^2}{v^2}$

$\Rightarrow \dfrac{dv}{dx}=-\dfrac{a^2}{v^2}+1=\dfrac{v^2-a^2}{v^2} \Rightarrow \dfrac{v^2}{v^2-a^2}dv=dx$

$\Rightarrow \displaystyle\int \dfrac{v^2}{v^2-a^2}dv=\int dx+C \Rightarrow a^2\int \dfrac{dv}{v^2-a^2}+\int dv=x+C$

$\Rightarrow a^2\dfrac{1}{2a}\log\left|\dfrac{v-a}{v+a}\right|+v=x+C$

$\Rightarrow \dfrac{a}{2}\log\left|\dfrac{-a+x-y}{a+x-y}\right|+x-y=x+C$

$\Rightarrow y+C=\dfrac{a}{2}\log\left|\dfrac{x-y-a}{x-y+a}\right|$

54. Let $4x+3y=v$ so that $\dfrac{dy}{dx}=\dfrac{1}{3}\left(\dfrac{dv}{dx}-4\right)$

Then $\dfrac{dy}{dx}=\dfrac{8x+6y+12}{4x+3y+2}$

$\Rightarrow \dfrac{1}{3}\left(\dfrac{dv}{dx}-4\right)=\dfrac{2v+12}{v+2} \Rightarrow \dfrac{dv}{dx}=\dfrac{10v+44}{v+2}$

Answers

1. $y=Cx$ **2.** $\log|1-y|+x=C$ **3.** $y=2\sin(x+C)$ **4.** $e^x+e^{-y}=C$ **5.** $\log|y|=e^x+x+C$

6. $\tan^{-1}y-\tan^{-1}x=C$ **7.** $\sin^{-1}y+\sin^{-1}x=C$ **8.** $y=C\sqrt{x^2+1}$ **9.** $y=C\log x$

10. $(y-1)=Cxy$ **11.** $\log|1+y|=x-\dfrac{x^2}{2}+C$ **12.** $\tan^{-1}y=x+\dfrac{x^2}{2}+C$ **13.** $(e^x+1)(y+1)=Ce^y$

14. $\log|y|=2x^2-8x+16\log|2+x|+C$ **15.** $x+y+\log|xy|=C$ **16.** $\sqrt{1+y^2}+\sqrt{1+x^2}=C$ **17.** $\log|\sin y|+\sin x=C$

18. $\tan y=\cot x+C$ **19.** $\log|xy|+x-\dfrac{1}{2}y^2=C$ **20.** $\dfrac{1}{2}(x^2+y^2)+(x+y)+\log|(x-1)(y-1)|=C$

21. $\log|xy|+x-y=C$ **22.** $\log|x(1-y)^2|=\dfrac{x^2}{2}-\dfrac{y^2}{2}-2y+C$ **23.** $\log|x|-\log|1+\sqrt{1+x^2}|+\sqrt{1+x^2}+\sqrt{1+y^2}=C$

24. $y\tan y+\log\cos y=x\tan x+\log\cos x+C$ **25.** $\tan y=C(e^x-1)^3$ **26.** $(1+e^y)\sin x=C$

27. $e^y+e^{-y}+\dfrac{x^2}{2}-\log|x|=C$ **28.** $e^{5x}+e^{5y}=C$ **29.** $\dfrac{3}{4}\cos x-\cos y-\dfrac{\cos 3x}{12}=C$ **30.** $\tan y=\cot x+C$

31. $x\tan x-y\tan y-\log|\sec x|+\log|\sec y|=C$ **32.** $\sin y=e^x\log x+C$ **33.** $y^2\log y=x\sin x+C$ **34.** $y\sin y=x^2\log x+C$

35. $\sec y+2\cos x=C$ **36.** $e^{3y}=x^3+1$ **37.** $e^x+e^{-y}=e+\dfrac{1}{e}$ **38.** $\dfrac{1}{6y^2}=e^{-x}-\dfrac{1}{3}$ **39.** $y=(x-1)^2\,e^{2(x-2)}$

40. $y-2\log(y+2)=\log x-3\log 2$ **41.** $x+y=1-xy$ **42.** $x=\sin y$ **43.** $y^2e^{\cos 2x}=0$

44. $y^2=4\sin 2x$ **45.** $y(x^2+1)=4$ **46.** $x^2=(1+y^2)$ **47.** $x+y=\tan(x+C)$ **48.** $y=\tan\left(\dfrac{x+y}{2}\right)+C$

49. $y-a\tan^{-1}\left(\dfrac{x+y}{a}\right)=C$ **50.** $y+C=\dfrac{a}{2}\log\left|\dfrac{x-y-a}{x-y+a}\right|$ **51.** $2(y-x)+\sin 2(x+y)+C$

52. $x-y+C=\log|x+y|$ **53.** $x-2y+\log|x-y+2|=C$ **54.** $15y-30x-12\log|20x+15y+22|=C$

8.1.6 HOMOGENEOUS DIFFERENTIAL EQUATIONS

Definition 1. *A function f(x, y) in x and y is said to be homogeneous function of degree n, if the degree of each term of f(x, y) is n.*

For example : $f(x, y) = ax^2 + 2hxy + by^2$ is a homogeneous function of degree 2.

REMARK

- In general, a homogeneous function of degree n can be expressed as $f(x, y) = x^n F(y / x)$

Definition 2. *A differential equation of the form $\dfrac{dy}{dx} = \dfrac{f_1(x, y)}{f_2(x, y)}$ is said to be homogeneous, if $f_1(x, y)$ and $f_2(x, y)$ are homogeneous functions of same degree.*

For example : $\dfrac{dy}{dx} = \dfrac{x^2 + y^2}{xy}$ is a homogeneous differential equation.

WORKING PROCEDURE

In order to solve a homogeneous differential equation, we proceed through the following steps :

Step 1. Substitute $y = vx$ (or $x = vy$.

Step 2. Reduce the given differential equation in terms of v and x.

Step 3. Solve the reduced equation by the method of separation of variables.

Step 4. Replace v by $\dfrac{y}{x}$ (or v by $\dfrac{x}{y}$) after integration.

Solved Examples

EXAMPLE 1. Solve $(x^2 + y^2)\dfrac{dy}{dx} = xy$. (AMIETE–1999)

SOLUTION. The given differential equation can be expressed as

$$\frac{dy}{dx} = \frac{xy}{x^2 + y^2} \qquad \text{...(1)}$$

Clearly, (1) is a homogeneous differential equation.

Putting $y = vx$ and $\dfrac{dy}{dx} = v + x\dfrac{dv}{dx}$ in (1), we get

$$v + x\frac{dv}{dx} = \frac{vx^2}{x^2 + v^2 x^2} = \frac{v}{1 + v^2}$$

$$\Rightarrow \quad x\frac{dv}{dx} = \frac{v}{1 + v^2} - v = \frac{-v^3}{1 + v^2}.$$

Separating the variables, we get

$$-\frac{1 + v^2}{v^3}\, dv = \frac{dx}{x}.$$

Integrating both sides, we get

$$-\int \frac{dv}{v^3} - \int \frac{dv}{v} = \int \frac{dx}{x} + \log|C|$$

$$\Rightarrow \quad \frac{1}{2v^2} - \log|v| = \log|x| + \log|C|$$

$$\Rightarrow \quad \frac{1}{2v^2} = \log|xvC| \qquad \Rightarrow \quad xvC = e^{1/2v^2}$$

$$\Rightarrow \quad yC = e^{1/2y^2}, \text{ which is the required solution.}$$

EXAMPLE 2. Solve $(x + y)dy + (x - y)dx = 0$.

SOLUTION. The given differential equation can be expressed as

$$\frac{dy}{dx} = \frac{y - x}{y + x} \qquad \text{...(1)}$$

Clearly, (1) is homogeneous.

Putting $y = vx$ and $\dfrac{dy}{dx} = v + x\dfrac{dv}{dx}$ in (1), we get

$$v + x\frac{dv}{dx} = \frac{vx - x}{vx + x} = \frac{v - 1}{v + 1}$$

$$\Rightarrow \quad x\frac{dv}{dx} = \frac{v - 1}{v + 1} - v = \frac{-1 - v^2}{v + 1}.$$

Separating the variables, we get

$$\frac{v + 1}{1 + v^2}\, dv + \frac{dx}{x} = 0$$

$$\Rightarrow \quad \frac{v}{1 + v^2}\, dv + \frac{1}{1 + v^2}\, dv + \frac{dx}{x} = 0$$

Integrating both sides, we get

$$\int \frac{v}{1 + v^2}\, dv + \int \frac{dv}{1 + v^2} + \int \frac{dx}{x} = C, \text{ where } C \text{ is the}$$

constant of integration.

$$\Rightarrow \quad \frac{1}{2}\log|1 + v^2| + \tan^{-1} v + \log|x| = C$$

$$\Rightarrow \quad \frac{1}{2}\log\left(1 + \frac{y^2}{x^2}\right) + \tan^{-1}\frac{y}{x} + \log|x| = C$$

$$\Rightarrow \quad \frac{1}{2}\log\left(\frac{x^2 + y^2}{x^2}\right) + \tan^{-1}\frac{y}{x} + \log|x| = C$$

$$\Rightarrow \quad \frac{1}{2}\log(x^2 + y^2) + \tan^{-1}\frac{y}{x} = C, \text{ which is the required}$$

solution.

EXAMPLE 3. Solve $xdy - ydx = \sqrt{x^2 + y^2}\, dx$.

SOLUTION. The given differential equation can be expressed as

$$\frac{dy}{dx} = \frac{y + \sqrt{x^2 + y^2}}{x} \qquad \text{...(1)}$$

Clearly, (1) is homogeneous as $y + \sqrt{x^2 + y^2}$ and x are homogeneous functions of degree 1.

Putting $y = vx$ and $\dfrac{dy}{dx} = v + x\dfrac{dv}{dx}$ in (1), we get

$$v + x\frac{dv}{dx} = \frac{vx + \sqrt{x^2 + v^2 x^2}}{x} = \frac{v + \sqrt{1 + v^2}}{1}$$

$\Rightarrow \qquad x\dfrac{dv}{dx} = \sqrt{1+v^2}$.

Separating the variables, we get

$$\dfrac{dv}{\sqrt{1+v^2}} = \dfrac{dx}{x} .$$

Integrating both sides, we get

$$\int \dfrac{dv}{\sqrt{1+v^2}} = \int \dfrac{dx}{x} + \log C$$

$$\log\left| v + \sqrt{1+v^2} \right| = \log|x| + \log C$$

$\Rightarrow \quad \log\left| \dfrac{y}{x} + \sqrt{1+\dfrac{y^2}{x^2}} \right| = \log|x| + \log C$

$\Rightarrow \quad \log\left| \dfrac{y+\sqrt{x^2+y^2}}{x^2} \right| = \log C$

$\Rightarrow \quad \left| \dfrac{y+\sqrt{x^2+y^2}}{x^2} \right| = C$

$\Rightarrow \qquad y + \sqrt{x^2+y^2} = \pm Cx^2$

$\therefore \; y + \sqrt{x^2+y^2} = kx^2$, where $k = \pm C$

EXAMPLE 4. **Solve** $(1 + e^{x/y})dx + e^{x/y}\left(1 - \dfrac{x}{y}\right)dy = 0$.

(VTU–2001,   2003,   TIRUPATI–1998,   BHOPAL–1998,   PTU–2006, RAJASTHAN–2005)

SOLUTION. The given differential equation can be expressed as

$$\dfrac{dx}{dy} = \dfrac{e^{x/y}\left(\dfrac{x}{y} - 1\right)}{1 + e^{x/y}} \qquad \qquad \dots(1)$$

Putting $x = vy$ and $\dfrac{dx}{dy} = v + y\dfrac{dv}{dy}$ in (1), we get

$$v + y\dfrac{dv}{dy} = \dfrac{e^v(v-1)}{1+e^v}$$

$\Rightarrow \qquad y\dfrac{dv}{dy} = \dfrac{e^v(v-1)}{1+e^v} - v = -\dfrac{v+e^v}{1+e^v}$.

Separating the variables, we get

$$\dfrac{(1+e^v)}{(v+e^v)}dv + \dfrac{dy}{y} = 0 .$$

Integrating both sides, we get

$\int \dfrac{(1+e^v)}{(v+e^v)}dv + \int \dfrac{dy}{y} = \log C$, where C is the constant of integration.

$\Rightarrow \qquad \log|v + e^v| + \log|y| = \log C$

$\Rightarrow \qquad \log|y(v+e^v)| = \log C$

$\Rightarrow \qquad |y(v+e^v)| = C$

$\Rightarrow \qquad y(v+e^v) = \pm C$

$\Rightarrow \qquad y\left(\dfrac{x}{y} + e^{x/y}\right) = \pm C$

$\Rightarrow \qquad x + ye^{x/y} = k$, where $k = \pm C$

which is the required solution.

EXAMPLE 5. **Solve** $(x\sqrt{x^2+y^2} - y^2)dx + xydy = 0$.

(AMIETE–2000)

SOLUTION. The given differential equation can be expressed as

$$\dfrac{dy}{dx} = \dfrac{y^2 - x\sqrt{x^2+y^2}}{xy} \qquad \qquad \dots(1)$$

Clearly (1) is homogeneous.

Putting $y = vx$ and $\dfrac{dy}{dx} = v + x\dfrac{dv}{dx}$ in (1), we get

$$v + x\dfrac{dv}{dx} = \dfrac{v^2x^2 - x\sqrt{x^2+v^2x^2}}{x^2v} = \dfrac{v^2 - \sqrt{1+v^2}}{v}$$

$\Rightarrow \quad x\dfrac{dv}{dx} = \dfrac{v^2 - \sqrt{1+v^2}}{v} - v = -\dfrac{\sqrt{1+v^2}}{v}$.

Separating the variables, we get

$$\dfrac{v\,dv}{\sqrt{1+v^2}} + \dfrac{dx}{x} = 0 .$$

Integrating both sides, we get

$$\int \dfrac{v\,dv}{\sqrt{1+v^2}} + \int \dfrac{dx}{x} = C$$

$\Rightarrow \qquad \sqrt{1+v^2} + \log|x| = C$

$\Rightarrow \qquad \sqrt{1+\dfrac{y^2}{x^2}} + \log|x| = C$

$\Rightarrow \qquad \sqrt{x^2+y^2} + x\log|x| = Cx$

which is the required solution.

EXAMPLE 6. **Solve** $\left(x\sin\dfrac{y}{x}\right)dy = \left(y\sin\dfrac{y}{x} - x\right)dx$.

SOLUTION. The given differential equation can be expressed as

$$\dfrac{dy}{dx} = \dfrac{y\sin\dfrac{y}{x} - x}{x\sin\dfrac{y}{x}} = \dfrac{\dfrac{y}{x}\sin\dfrac{y}{x} - 1}{\sin\dfrac{y}{x}} \qquad \dots(1)$$

Clearly (1) is homogeneous, as it is of the form $x^nF\left(\dfrac{y}{x}\right)$.

Putting $y = vx$ and $\dfrac{dy}{dx} = v + x\dfrac{dv}{dx}$ in (1), we get

$$v + x\dfrac{dv}{dx} = \dfrac{v\sin v - 1}{\sin v}$$

$\Rightarrow x\dfrac{dv}{dx} = \dfrac{v\sin v - 1}{\sin v} - v = -\dfrac{1}{\sin v}$.

Separating the variables, we get

$$\sin v\,dv + \dfrac{dx}{x} = 0 .$$

Integrating both sides, we get

$$\int \sin v\,dv + \int \dfrac{dx}{x} = C$$

$-\cos v + \log|x| = C \;\Rightarrow\; \cos v = \log|x| - C$

$\cos v = \log|x| + C_1$, where $C_1 = -C$

$\cos\left(\dfrac{y}{x}\right) = \log|x| + C_1$, which is the required solution.

EXAMPLE 7. **Solve** $\left(x\tan\dfrac{y}{x} - y\sec^2\dfrac{y}{x}\right)dx - x\sec^2\dfrac{y}{x}dy = 0$.

SOLUTION. The given differential equation is,   (VTU–2006)

$$\dfrac{dy}{dx} = \left(\dfrac{y}{x}\sec^2\dfrac{y}{x} - \tan\dfrac{y}{x}\right)\cos^2\dfrac{y}{x} \qquad \dots(i)$$

which is a homogeneous equation.

Putting $y = vx$ in equation (i), we get

$$v + x\frac{dv}{dx} = \left(v\sec^2 v - \tan v\right)\cos^2 v$$

or $\quad x\frac{dv}{dx} = v - \tan v\cos^2 v - v$

Separating the variables, we get

$$\frac{\sec^2 v}{\tan v}dv = -\frac{dx}{x}$$

Integrating both sides, we get

$$\log\tan v = -\log x + \log c$$

or $\quad x\tan v = c \quad$ or $\quad x\tan\dfrac{y}{x} = c.$

Exercise-8.5

Solve the following differential equations :

1. $\dfrac{dy}{dx} = \dfrac{y^2 - x^2}{2xy}$

2. $\dfrac{dy}{dx} = \dfrac{x^2 + y^2}{2xy}$

3. $x + y\dfrac{dy}{dx} = 2y$

4. $x\dfrac{dy}{dx} = x + y$

5. $\dfrac{dy}{dx} = \dfrac{x+y}{x-y}$

6. $(x-y)\dfrac{dy}{dx} = x + 3y$

7. $\dfrac{dy}{dx} + \dfrac{x-2y}{2x-y} = 0$

8. $\dfrac{dy}{dx} = \dfrac{2xy}{x^2 - y^2}$

9. $2xy\dfrac{dy}{dx} = x^2 + 3y^2$

10. $x^2\dfrac{dy}{dx} = 2xy + y^2$

11. $y^2 + x^2\dfrac{dy}{dx} = xy\dfrac{dy}{dx}$

12. $x^2\dfrac{dy}{dx} = x^2 + xy + y^2$

13. $y^2 dx + (x^2 - xy + y^2)dy = 0$

14. $x^2\left(\dfrac{dy}{dx}\right) = (x^2 - 2y^2 + xy)$

15. $(x^2 + xy)dy = (x^2 + y^2)dx$

16. $(x^2 + y^2)dx + 3xy dy = 0$

17. $x(x-y)dy + y^2 dx = 0$

18. $x\dfrac{dy}{dx} + \dfrac{y^2}{x} = y$

19. $x\dfrac{dy}{dx} = y - \sqrt{x^2 + y^2}$

20. $x\dfrac{dy}{dx} - y = 2\sqrt{y^2 - x^2}$

21. $(x - \sqrt{xy})\,dy = y dx$

22. $y - x\dfrac{dy}{dx} = x + y\dfrac{dy}{dx}$

23. $(x^3 + 3xy^2)dx + (y^3 + 3x^2 y)dy = 0$ (AMIETE–1997)

24. $x\dfrac{dy}{dx} = y(\log y - \log x + 1)$

25. $\dfrac{dy}{dx} = \dfrac{y}{x} + \sin\dfrac{y}{x}$ (VTU–2000)

26. $x\dfrac{dy}{dx} = y - x\cos^2\dfrac{y}{x}$

27. $\left(x\cos\dfrac{y}{x}\right)(y dx + x dy) = \left(y\sin\dfrac{y}{x}\right)(x dy - y dx)$

28. $y dx + \left(x\log\dfrac{y}{x}\right)dy - 2x dy = 0$

29. $2x^2\dfrac{dy}{dx} - 2xy + y^2 = 0,\ \ y(e) = e$

30. $(y + \sqrt{x^2 + y^2})dx - x dy,\ y(1) = 0$ (UPTU–2008)

31. $(x^2 y - 2xy^2)dx - (x^3 - 3x^2 y)dy = 0$ (BHOPAL–2008)

32. $x^2 y dx - (x^3 + y^3)dy = 0$ (VTU–2010)

33. $y dx - x dy = \sqrt{x^2 + y^2}\,dx$ (RAIPUR–2005)

34. $(3xy - 2ay^2)dx + (x^2 - 2axy)dy = 0$ (SVTU–2009)

35. $\dfrac{dy}{dx} = \dfrac{y}{x} + \sin\dfrac{y}{x}$ (VTU–2005)

36. $ye^{x/y}dx = (xe^{x/y} + y^2)dy$ (VTU–2006)

Hints to Selected Problems

24. $x\dfrac{dy}{dx} = y(\log y - \log x + 1) \Rightarrow \dfrac{dy}{dx} = \dfrac{y}{x}\left(\log\dfrac{y}{x} + 1\right)$

Putting $y = vx$ and $\dfrac{dy}{dx} = v + x\dfrac{dv}{dx}$ in (1), we get

$v + x\dfrac{dv}{dx} = v(\log v + 1) \Rightarrow x\dfrac{dv}{dx} = v\log v$

$\Rightarrow \dfrac{dv}{v\log v} = \dfrac{dx}{x} \Rightarrow \int\dfrac{dv}{v\log v} = \int\dfrac{dx}{x} + \log C$.

28. $y dx + \left(x\log\dfrac{y}{x}\right)dy - 2x dy = 0$

$\Rightarrow \dfrac{dy}{dx} = \dfrac{-y}{x\log\dfrac{y}{x} - 2x} = \dfrac{-y/x}{\log(y/x) - 2}$

Putting $y = vx$ and $\dfrac{dy}{dx} = v + x\dfrac{dv}{dx}$ in (1).

Answers

1. $(x^2 + y^2) = Cx$ **2.** $x = C(x^2 - y^2)$ **3.** $\log(y-x) = C + \dfrac{x}{y-x}$ **4.** $y = x\log|x| + Cx$

5. $\tan^{-1}\dfrac{y}{x} = \dfrac{1}{2}\log(x^2 + y^2) + C$ **6.** $\log|x+y| + \dfrac{2x}{x+y} = C$ **7.** $y - x = C(x+y)^3$ **8.** $y = C(x^2 + y^2)$

9. $x^2 + y^2 = Cx^3$ **10.** $y = Cx(x+y)$ **11.** $y = x[C + \log|y|]$ **12.** $\tan^{-1}\dfrac{y}{x} = \log|x| + C$

13. $|y| = Ce^{\tan^{-1}(y/x)}$ **14.** $\dfrac{1}{2\sqrt{2}}\log\left|\dfrac{x+\sqrt{2}y}{x-\sqrt{2}y}\right| - \log|x| = C$ **15.** $C(x-y)^2 = xe^{y/x}$ **16.** $x^2(x^2 + 4y^2)^3 = C$

17. $y = Ce^{y/x}$ **18.** $\log|x| - \dfrac{x}{y} = C$ **19.** $y + \sqrt{x^2 + y^2} = C$ **20.** $y + \sqrt{y^2 - x^2} = Cx^3$ **21.** $2\sqrt{\dfrac{x}{y}} + \log|y| = C$

22. $\tan^{-1}\dfrac{y}{x} + \dfrac{1}{2}\log(x^2 + y^2) = C$ **23.** $y^4 + 6x^2 y^2 + x^4 = C$ **24.** $y = xe^{Cx}$ **25.** $\tan\left(\dfrac{y}{2x}\right) = Cx$

26. $\tan\dfrac{y}{x} = C - \log|x|$ **27.** $xy\cos\left(\dfrac{y}{x}\right) = C$ **28.** $y = C\left[\log\left(\dfrac{y}{x} - 1\right)\right]$ **29.** $y = \dfrac{2x}{1 + \log|x|}$ **30.** $y + \sqrt{x^2 + y^2} = Cx^2$

31. $Cy^3 = x^2 e^{-x/y}$ **32.** $\left(\dfrac{x}{y}\right)^3 = 3\log cy$ **33.** $y + \sqrt{(x^2 + y^2)} = c$ **34.** $x(c+y) = ay^2$ **35.** $y = 2x\tan^{-1}(cx)$

36. $e^{x/y} = y + c$

8.1.7 EQUATION REDUCIBLE TO THE HOMOGENEOUS FORM

The differential equation of the type

$$\frac{dy}{dx} = \frac{a_1 x + b_1 y + c_1}{a_2 x + b_2 y + c_2} \qquad \qquad \ldots(1)$$

can be reduced to the homogeneous form as follows. Put $X = x - h$ and $Y = y - k$.

Therefore, $\qquad \qquad \dfrac{dY}{dX} = \dfrac{dy}{dx}$

Then, (1) becomes $\qquad \dfrac{dY}{dX} = \dfrac{a_1 X + b_1 Y + (a_1 h + b_1 k + c_1)}{a_2 X + b_2 Y + (a_2 h + b_2 k + c_2)}$

where h and k can be chosen such that $a_1 h + b_1 k + c_1 = 0$ and $a_2 h + b_2 k + c_2 = 0$.

Then, we get $\dfrac{dY}{dX} = \dfrac{a_1 X + b_1 Y}{a_2 X + b_2 Y}$, which is homogeneous differential equation and can be solved by the method discussed earlier.

WORKING PROCEDURE

(A) **When** $\dfrac{a_1}{a_2} \neq \dfrac{b_1}{b_2}$ **:**

Step 1. Put $x = X + h$, $y = Y + k$, in the given differential equation.

Step 2. Equate the constant terms in the numerator and denominator to zero, and find the value of h and k.

Step 3. Solve the resulting homogeneous equation in X and Y.

Step 4. Replace X by $x - h$ and Y by $y - k$.

Step 5. Put the value of h and k.

(B) **When** $\dfrac{a_1}{a_2} = \dfrac{b_1}{b_2}$ **:**

Step 1. Put $a_1 x + b_1 y = v$.

Step 2. Solve the resulting equation.

Solved Examples

EXAMPLE 1. **Solve** $(2x + y + 3)dx = (2y + x + 1)dy$ **.**

SOLUTION. The given equation can be written as

$$\frac{dy}{dx} = \frac{2x + y + 3}{2y + x + 1}$$

This is the equation reducible to homogeneous form.
To solve it, put $x = X + h, y = Y + k$

We get $\qquad \dfrac{dY}{dX} = \dfrac{2X + Y + (2h + k + 3)}{2Y + X + (h + 2k + 1)}$

Now putting
$2h + k + 3 = 0$ and $h + 2k + 1 = 0$
which gives $h = -5/3, k = 1/3$.

We get $\dfrac{dY}{dX} = \dfrac{2X + Y}{2Y + X}$ which is the reduced homogeneous equation.

Put $Y = vX \Rightarrow \dfrac{dY}{dX} = v + X\dfrac{dv}{dX}$

and solve by the method discussed earlier, we get
$$(X + Y)(X - Y)^3 = c$$

Now putting the value of X, Y, h and k, we get

$$\left(x + y + \frac{4}{3}\right)(x - y + 2)^3 = 0\,.$$

EXAMPLE 2. **Solve** $(x + y)(dx - dy) = dx + dy$ **.**

SOLUTION. The given equation can be written as

$$\frac{dy}{dx} = \frac{x + y - 1}{x + y + 1}$$

Put $x + y = v \quad \Rightarrow \quad 1 + \dfrac{dy}{dx} = \dfrac{dv}{dx}$

$$\therefore \frac{dv}{dx} - 1 = \frac{v - 1}{v + 1} \Rightarrow \frac{dv}{dx} = \frac{2v}{v + 1}$$

Now, separating the variables, we get

$$\left(1 + \frac{1}{v}\right) dv = 2dx\,.$$

On integrating, we get
$v + \log v = 2x + c$ or $x + y + \log(x + y) = 2x + c$
$\Rightarrow \log(x + y) = x - y + c$

Exercise-8.6

Solve the following ordinary differential equations :

1. $\dfrac{dy}{dx} = \dfrac{y - x + 1}{y + x - 5}$

2. $(2x + y - 3)dy = (x + 2y - 3)dx$
(MADRAS–2000, VTU–2000, 2009S, ANDHRA–1998)

3. $\dfrac{dy}{dx} = \dfrac{2x + 2y - 2}{3x + y - 5}$

4. $(x + 2y - 2)dx + (2x - y + 3) = 0$

5. $\dfrac{dy}{dx} = \dfrac{x + y + 3}{2x + 2y + 1}$

6. $\dfrac{dy}{dx} = \dfrac{x - y + 3}{2x - 2y + 5}$

7. $(6x + 2y - 10)\dfrac{dy}{dx} - 2x - 9y + 20 = 0$

8. $(x - y)dy = (x + y + 1)dx$

9. $(x + y - 1)dy = (x + y)dx$

10. $(2x - 2y + 5)dy - (x - y + 3)dx = 0$

11. $\dfrac{dy}{dx} = \dfrac{x + y + 7}{2x + 2y + 3}$

12. $\dfrac{dy}{dx} = \dfrac{2y + x + 1}{2x + 4y + 1}$

13. $\dfrac{dy}{dx} = \left(\dfrac{2x - y}{4x - 2y + 1}\right)^2$

14. $(7y - 3)dx + (2x + 1)dy = 0$

15. $(2x + 3y - 1)dx + (4x + 6y + 2)dy = 0$

16. $\dfrac{dy}{dx} = \dfrac{y + x - 2}{y - x - 4}$ (RAIPUR–2005)

17. $(3y + 2x + 4)dx - (4x + 6y + 5)dy = 0$ (MADRAS–2000S)

Hints to Selected Problems

1. Put $x = X + h, y = Y + k$ in the given equation, we get

$\dfrac{dY}{dX} = \dfrac{Y - X}{Y + X}$, where $h = 3, k = 2$.

4. Proceeding as usual, we get

$\dfrac{dY}{dX} = -\dfrac{X + 2Y}{2X - Y}, h = -\dfrac{4}{5}, k = \dfrac{7}{5}$

12. Put $x + 2y = v$.

15. Put $2x + 3y = v$.

Answers

1. $\tan^{-1}\left(\dfrac{\beta}{\alpha}\right) + \dfrac{1}{2}\log(\alpha^2 + \beta^2) = c$, where $\beta = y - 2, \alpha = x - 3$ **2.** $(x + y - 2) = c^2(x - y)^3$ **3.** $(y - x + 3)^4 = c(2x + y - 3)$

4. $x^2 + 4xy - y^2 - 4x + 6y = c$ **5.** $x + c = \dfrac{2}{3}(x + y) - \dfrac{5}{9}\log(3x + 3y + 4)$ **6.** $x - 2y + \log(x - y + 2) = c$

7. $(y - 2x)^2 = c(x + 2y - 5)$ **8.** $2\tan^{-1}\{(2y + 1)/(2x + 1)\} = \log\{c^2(x^2 + y^2 + x + y + \tfrac{1}{2})\}$

9. $2(y - x) - \log(2x + 2y - 1) = c$ **10.** $x - 2y + \log(x - y + 2) = c$ **11.** $\dfrac{2}{3}(x + y) - \left(\dfrac{11}{9}\right)\log(3x + 3y + 10) = x + c$

12. $4(2y - x) - \log(4x + 8y + 3) = c$ **13.** $7x - 28y + 2\log(28x^2 - 28xy + 7y^2 - 16x + 8y + 2) + \dfrac{9}{2\sqrt{2}}\log\dfrac{14x - 7y - 4 - \sqrt{2}}{14x - 7y - 4 + \sqrt{2}}$

14. $7\log(2x + 1) + 2\log(7y - 3) = c$ **15.** $x + 2y - 4\log(2x + 3y + 7) = c$ **16.** $x^2 + 2xy - y^2 - 4x + 8y - 14 = c$

17. $21x - 42y + 9\log(14x + 21y + 22) = c$

8.1.8 LINEAR DIFFERENTIAL EQUATION

A differential equation in which the dependent variable and all its derivatives are of first degree only and not multiplied together, is called a linear differential equation.

For example : $x\dfrac{dy}{dx} + y = x$ is a linear differential equation.

GENERAL FORMS OF LINEAR DIFFERENTIAL EQUATION

(i) The equation of the form $\dfrac{dy}{dx} + Py = Q$ where P and Q are either constants or functions of x only, is most general form of linear differential equation.

WORKING PROCEDURE

To find the solution of such type of equation use the following steps :

Step 1. Find I.F. $= e^{\int P\,dx}$.

Step 2. The solution is $y \cdot (\text{I.F.}) = \int Q.(\text{I.F.})dx + C$.

From (ii) The equation of the form $\dfrac{dx}{dy} + Px = Q$ where P and Q are either constants or functions of y only, is most general form of linear differential equation.

WORKING PROCEDURE

To find the solution of such type of equation use the following steps :

Step 1. Find I.F. $= e^{\int P\,dy}$.

Step 2. The solution is $x.(\text{I.F.}) = \int Q.(\text{I.F.})dy + C$.

Solved Examples

EXAMPLE 1. *Solve* $(1 + x^2)\dfrac{dy}{dx} + y = \tan^{-1} x$. (UKTU–2012)

SOLUTION. The given differential equation can be written as

$$\dfrac{dy}{dx} + \dfrac{1}{1 + x^2}y = \dfrac{\tan^{-1} x}{1 + x^2} \qquad \ldots(1)$$

Comparing (1) with equation $\dfrac{dy}{dx} + Py = Q$, we get

$P = \dfrac{1}{1 + x^2}$ and $Q = \dfrac{\tan^{-1} x}{1 + x^2}$.

So, I.F. $= e^{\int P\,dx} = e^{\int \frac{1}{1 + x^2}dx} = e^{\tan^{-1} x}$.

Thus, the solution of (1) is

$$y.(\text{I.F.}) = \int Q.(\text{I.F.})dx + C$$

$$\Rightarrow \quad y.(e^{\tan^{-1}x}) = \int \frac{\tan^{-1}x}{1+x^2} e^{\tan^{-1}x}dx + C$$

$$= \int te^t dt + C, \text{ where } t = \tan^{-1}x$$

$$= t\int e^t dt - \int \left\{\frac{d}{dt}(t).\int e^t dt\right\}dt + C$$

(Integrating by parts)

$$= te^t - \int e^t dt + C = te^t - e^t + C$$

$$= (\tan^{-1}x - 1)e^{\tan^{-1}x} + C$$

$$\therefore \quad y = (\tan^{-1}x - 1) + Ce^{-\tan^{-1}x}, \text{ which is the required solution.}$$

EXAMPLE 2. **Solve** $(x^2 - 1)\dfrac{dy}{dx} + 2xy = \dfrac{2}{x^2 - 1}$.

SOLUTION. The given differential equation can be written as

$$\frac{dy}{dx} + \frac{2x}{x^2 - 1}y = \frac{2}{(x^2 - 1)^2} \qquad ...(1)$$

Comparing (1) with $\dfrac{dy}{dx} + Py = Q$, we get

$$P = \frac{2x}{x^2 - 1} \text{ and } Q = \frac{2}{(x^2 - 1)^2}.$$

$$\text{I.F.} = e^{\int P dx} = e^{\int \frac{2x}{x^2-1}dx} = e^{\log(x^2 - 1)} = x^2 - 1.$$

Thus, the equation of (1) is

$$y(\text{I.F.}) = \int Q.(\text{I.F.})\,dx + C$$

$$\Rightarrow \quad y.(x^2 - 1) = \int \frac{2}{(x^2-1)^2}(x^2 - 1)dx + C$$

$$= 2\int \frac{1}{x^2 - 1}dx + C = 2.\left[\frac{1}{2}\log\left|\frac{x-1}{x+1}\right|\right] + C$$

$$= \log\left|\frac{x-1}{x+1}\right| + C$$

$$\Rightarrow \quad y(x^2 - 1) = \log\left|\frac{x-1}{x+1}\right| + C$$

which is the required solution.

EXAMPLE 3. **Solve** $(x\log x)\dfrac{dy}{dx} + y = \dfrac{2}{x}\log x$.

SOLUTION. The given differential equation can be written as

$$\frac{dy}{dx} + \frac{1}{x\log x}y = \frac{2}{x^2} \qquad ...(1)$$

Comparing (1) with the equation $\dfrac{dy}{dx} + Py = Q$, we get

$$P = \frac{1}{x\log x} \text{ and } Q = \frac{2}{x^2}.$$

$$\text{So, I.F.} = e^{\int P dx} = e^{\int \frac{1}{x\log x}dx} = e^{\log(\log x)} = \log x.$$

Thus, the solution of (1) is

$$y(\text{I.F.}) = \int Q.(\text{I.F.})\,dx + C$$

$$\Rightarrow \quad y.(\log x) = \int \frac{2}{x^2}\log x\,dx + C$$

$$\Rightarrow \quad y\log x$$

$$= 2\left\{\log x\int \frac{1}{x^2}dx - \int\left(\frac{d}{dx}(\log x)\int \frac{1}{x^2}dx\right)dx\right\} + C$$

(Integrating by parts)

$$= 2\left\{-\frac{1}{x}\log x + \int \frac{1}{x^2}dx\right\} + C$$

$$= 2\left[-\frac{1}{x}\log x - \frac{1}{x}\right] + C$$

$$\therefore \quad y\log x + \frac{2}{x}(\log x + 1) = C, \text{ which is the required solution.}$$

EXAMPLE 4. **Solve** $\dfrac{dy}{dx} + y\cot x = 2x + x^2\cot x$, **given that** $y(0) = 0$.

SOLUTION. The given differential equation is

$$\frac{dy}{dx} + y\cot x = 2x + x^2\cot x \qquad ...(1)$$

Comparing (1) with the equation $\dfrac{dy}{dx} + Py = Q$, we get

$$P = \cot x \text{ and } Q = 2x + x^2\cot x.$$

$$\text{So, I.F.} = e^{\int P dx} = e^{\int \cot x dx} = e^{\log\sin x} = \sin x.$$

Thus, the solution is

$$y(\text{I.F.}) = \int Q.(\text{I.F.})dx + C$$

$$\Rightarrow \quad y\sin x = \int (2x + x^2\cot x)\sin x\,dx + C$$

$$= 2\int x\sin x\,dx + \int x^2\cos x\,dx + C$$

$$= 2\sin x\int x\,dx - 2\int\left\{\frac{d}{dx}(\sin x)\int x\,dx\right\}dx$$

$$\qquad + \int x^2\cos x\,dx + C$$

$$= x^2\sin x - \int x^2\cos x\,dx + \int x^2\cos x\,dx + C$$

$$\therefore \quad y\sin x = x^2\sin x + C.$$

Putting $x = 0$ and $y = 0$, we get $C = 0$.

$$\therefore \quad y\sin x = x^2\sin x \text{ or } y = x^2, \text{ which is the required solution.}$$

EXAMPLE 5. **Solve** $(1 + y^2)dx = (\tan^{-1}y - x)dy$.

(ASSAM-1998, KARNATAKA-1995, BHOPAL-1991, 2008, VTU-2008, UPTU-2005)

SOLUTION. The given differential equation can be written as

$$\frac{dx}{dy} + \frac{1}{1+y^2}x = \frac{\tan^{-1}y}{1+y^2} \qquad ...(1)$$

Comparing (1) with the equation $\dfrac{dx}{dy} + Px = Q$, we get

$$P = \frac{1}{1+y^2} \text{ and } Q = \frac{\tan^{-1}y}{1+y^2}.$$

$$\text{So, I.F.} = e^{\int P dy} = e^{\int \frac{1}{1+y^2}dy} = e^{\tan^{-1}y}.$$

Thus, the solution of (1) is

$$x.(\text{I.F.}) = \int Q.(\text{I.F.})dy + C$$

$$\Rightarrow \quad xe^{\tan^{-1}y} = \int \frac{\tan^{-1}y}{1+y^2}e^{\tan^{-1}y}dy + C$$

$$= \int te^t dt + C, \text{ where } t = \tan^{-1}y$$

$$= t\int e^t dt - \int\left\{\frac{d}{dt}(t).\int e^t dt\right\}dt + C$$

$$= te^t - e^t + C = (t - 1)e^t + C$$

$$= (\tan^{-1}y - 1)e^{\tan^{-1}y} + C$$

$$\therefore \quad x = (\tan^{-1}y - 1) + Ce^{-\tan^{-1}y}, \text{ which is the required solution.}$$

EXAMPLE 6. **Solve** $y\log y\dfrac{dx}{dy} + x - \log y = 0$.

(UPTU-2000, 04, UPTU(B.PHARM.)-2009)

SOLUTION.

$$y\log y\frac{dx}{dy} + x - \log y = 0$$

$$y\log y\frac{dx}{dy} + x = \log y \quad \Rightarrow \quad \frac{dx}{dy} + \frac{x}{y\log y} = \frac{1}{y}$$

$$\text{I.F.} = e^{\int \frac{dy}{y \log y}} = e^{\log(\log y)} = \log y$$

The solution is

$$x.\log y = \int \frac{1}{y}(\log y)\,dy + C$$

$$\Rightarrow \quad x.\log y = \frac{1}{2}(\log y)^2 + C.$$

EXAMPLE 7. Solve $\dfrac{dy}{dx} + \dfrac{3y}{x} = \dfrac{1}{x^4}$ (UPTU(CO)–2009)

SOLUTION. Comparing the given differential equation with

$$\frac{dy}{dx} + Py = Q, \text{ we get } P = \frac{3}{x} \text{ and } Q = \frac{1}{x^4}.$$

So, I.F. $= e^{\int P dx} = e^{3\int \frac{1}{x}dx} = e^{3\log x} = x^3$.

Thus, the solution is

$$y(I.F.) = \int Q(I.F.)dx + C$$

$$\Rightarrow \quad yx^3 = \int \frac{1}{x^4}(x^3)dx + c$$

$$\Rightarrow \quad yx^3 = \log x + c$$

where c is a constant of integration.

Exercise-8.7

Solve the following differential equations :

1. $\dfrac{dy}{dx} - y = e^{2x}$
2. $\dfrac{dy}{dx} + y = e^{x}$
3. $\dfrac{dy}{dx} - 4y = e^{2x}$
4. $\dfrac{dy}{dx} + y = e^{-2x}$
5. $\dfrac{dy}{dx} + 2y = e^{-x}$
6. $\dfrac{dy}{dx} + 2y = 6e^{x}$
7. $\dfrac{dy}{dx} - \dfrac{y}{x} = 2x^2$
8. $\dfrac{dy}{dx} + 2y = xe^{4x}$
9. $\dfrac{dy}{dx} + \dfrac{y}{x} = x^n$
10. $x\dfrac{dy}{dx} + 3y = x^2$
11. $\dfrac{dy}{dx} + 2y = 4x$
12. $x\dfrac{dy}{dx} - y = x+1$
13. $x\dfrac{dy}{dx} - y = \log x$
14. $(1+x^2)\dfrac{dy}{dx} + 2xy = \sqrt{x^2+4}$
15. $(1+x^2)\dfrac{dy}{dx} - 2xy = (x^2+2)(x^2+1)$
16. $\dfrac{dy}{dx} + \dfrac{y}{x} = e^{x}$
17. $(1+x)\dfrac{dy}{dx} + 2xy = \cos x$
18. $\dfrac{dy}{dx} + y = \cos x$
19. $\dfrac{dy}{dx} + 2y = \sin x$
20. $\dfrac{dy}{dx} - y\tan x = e^{x}\sec x$
21. $\dfrac{dy}{dx} + (\sec x)y = \tan x$
22. $(1+x^2)\dfrac{dy}{dx} + y = e^{\tan^{-1}x}$
23. $\dfrac{dy}{dx} + y\tan x = \sec x$
24. $\dfrac{dy}{dx} + y = \cos x - \sin x$
25. $x\dfrac{dy}{dx} + 2y = \sin x$

26. $\dfrac{dy}{dx} + y\cot x = x$
27. $\dfrac{dy}{dx} + y\cot x = \sin 2x$
28. $x\dfrac{dy}{dx} - y = 2x^2\sec x$
29. $\dfrac{dy}{dx} + 2y\cot x = 3x^2\csc^2 x$
30. $(\sec x)\dfrac{dy}{dx} = y + \tan x$
31. $\dfrac{dy}{dx} + y\cos x = \sin x\cos x$
32. $\dfrac{dy}{dx} - y\tan x = -2\sin x$
33. $\dfrac{dy}{dx} - 2y = \cos 3x$
34. $(1+x^2)\dfrac{dy}{dx} + 2xy = 4x^2$, given that $y(0) = 0$
35. $\dfrac{dy}{dx} - 3y\cot x = \sin 2x$, given that $y = 2$ when $x = \dfrac{\pi}{2}$.
36. $(x+2y^2)\dfrac{dy}{dx} = y$
37. $(x-y^3)\dfrac{dy}{dx} + y = 0$
38. $(x-y^2)\dfrac{dy}{dx} + y = 0$
39. $y^2 + \left(x - \dfrac{1}{y}\right)\dfrac{dy}{dx} = 0$
40. $\dfrac{dx}{dy} + 2xy = e^{-y^2}$
41. $3x(1-x^2)y^2\dfrac{dy}{dx} + (2x^2-1)y^3 = ax^3$ (RAJASTHAN–2006)
42. $x\log x\dfrac{dy}{dx} + y = \log x^2$ (VTU–2011)
43. $(1-x^2)\dfrac{dy}{dx} - xy = 1$ (VTU–2010)

Hints to Selected Problems

14. $(1+x^2)\dfrac{dy}{dx} + 2xy = \sqrt{x^2+4} \Rightarrow \dfrac{dy}{dx} + \dfrac{2x}{1+x^2}y = \dfrac{\sqrt{x^2+4}}{1+x^2}$

So, I.F. $= e^{\int \frac{2x}{1+x^2}dx} = e^{\log(1+x^2)} = 1+x^2$

$\therefore \ y(1+x^2) = \int \dfrac{\sqrt{x^2+4}}{1+x^2}.(1+x^2)dx + C = \int \sqrt{x^2+4}dx + C$

30. $\sec x\dfrac{dy}{dx} = y + \sin x \Rightarrow \dfrac{dy}{dx} - y\cos x = \sin x\cos x$

So, I.F. $= e^{-\int \cos x\,dx} = e^{-\sin x}$.

$y(e^{-\sin x}) = \int \sin x\cos x\,e^{-\sin x}dx + C = \int te^{-t}dt + C$,

where $t = \sin x$.

36. $(x+2y^2)\dfrac{dy}{dx} = y \Rightarrow \dfrac{dx}{dy} - \dfrac{x}{y} = 2y$

So, I.F. $= e^{-\int \frac{1}{y}dy} = e^{-\log y} = \dfrac{1}{y}$

$\therefore \ x\left\{\dfrac{1}{y}\right\} = \int 2y.\dfrac{1}{y}dy + C = 2y + C$

$\Rightarrow \quad x = 2y^2 + Cy$

Answers

1. $y = e^{2x} + Ce^{x}$
2. $y = \dfrac{1}{2}e^{x} + Ce^{-x}$
3. $y = -\dfrac{1}{2}e^{2x} + Ce^{4x}$
4. $y = -e^{-2x} + Ce^{-x}$
5. $y = e^{-x} + Ce^{-2x}$
6. $y = 2e^{x} + Ce^{-2x}$
7. $y = x^3 + Cx$
8. $y = \dfrac{1}{6}xe^{4x} - \dfrac{1}{36}e^{4x} + Ce^{-2x}$
9. $(n+2)xy = x^{n+2} + (n+2)C$

10. $5x^3 y - x^5 = C$ **11.** $y = 2x - 1 + Ce^{-2x}$ **12.** $y = x\log x - 1 + Cx$ **13.** $y = Cx - (\log x + 1)$

14. $y(1 + x^2) = \dfrac{x\sqrt{x^2 + 4}}{2} + 2\log|x + \sqrt{x^2 + 4}| + C$ **15.** $y = (1 + x^2)(x + \tan^{-1} x + C)$ **16.** $y = e^x - \dfrac{1}{x}e^x + \dfrac{C}{x}$

17. $y(1 + x^2) = \sin x + C$ **18.** $y = \dfrac{1}{2}(\cos x + \sin x) + Ce^{-x}$ **19.** $y = \dfrac{1}{5}(-4\cos x + 2\sin x) + Ce^{-2x}$ **20.** $y\cos x = e^x + C$

21. $y(\sec x + \tan x) = \sec x + \tan x - x + C$ **22.** $y = \dfrac{1}{2}e^{\tan^{-1} x} + Ce^{-\tan^{-1} x}$ **23.** $y\sec x = \tan x + C$

24. $y = \cos x + Ce^{-x}$ **25.** $x^2 y = (2 - x^2)\cos x + 2x\sin x + C$ **26.** $(y - 1)\sin x + x\cos x = C$ **27.** $y\sin x = \dfrac{2}{3}\sin^3 x + C$

28. $y = Cx + 2x\log|\sec x + \tan x|$ **29.** $y\sin^2 x = x^3 + C$ **30.** $y = Ce^{\sin x} - (1 + \sin x)$ **31.** $y = \sin x - 1 + Ce^{-\sin x}$

32. $2y\cos x = \cos 2x + C$ **33.** $y = \dfrac{3\sin 3x - 2\cos 3x}{13} + Ce^{2x}$ **34.** $3y(1 + x^2) = 4x^3$ **35.** $y = 4\sin^3 x - 2\sin^2 x$

36. $x = 2y^2 + Cy$ **37.** $4xy = y^4 + C$ **38.** $3xy = y^3 + C$ **39.** $xy = 1 + y + Cye^{1/y}$ **40.** $xe^{y^2} = y + C$

41. $y^3 = ax + cx\sqrt{1 - x^2}$ **42.** $y = \log x + c/\log x$ **43.** $y\sqrt{(1 - x^2)} = \sin^{-1} x + c$

8.1.9 EQUATION REDUCIBLE TO LINEAR FORM (BERNOULLI'S EQUATION)

A differential equation of the form $\dfrac{dy}{dx} + Py = Qy^n$ is called an equation reducible to the linear form or Bernoulli's equation.

Solution of the equation. The given equation can be written as

$$y^{-n}\frac{dy}{dx} + Py^{-n+1} = Q \qquad \dots(1)$$

Put $y^{-n+1} = v$ $\Rightarrow$ $(-n + 1)y^{-n}\dfrac{dy}{dx} = \dfrac{dv}{dx}$.

Then, (1) becomes

$$\frac{-1}{(n-1)}\frac{dv}{dx} + Pv = Q \qquad \text{or} \qquad \frac{dv}{dx} - (n-1)Pv = Q(1 - n)$$

which is a linear equation of first order, and can be solved in a usual manner.

WORKING PROCEDURE

Step 1. Write the given equation in the form $\dfrac{dy}{dx} + Py = Qy^n$.

Step 2. Divide by y^n, substitute $y^{-n+1} = v$ and get the equation of the form

$$\frac{dv}{dx} - (n-1)Pv = Q(1 - n)$$

Step 3. Then, apply the method of solution of linear equations.

Solved Examples

EXAMPLE 1. *Solve* $\dfrac{dy}{dx} + \dfrac{y}{x}\log y = \dfrac{y}{x^2}(\log y)^2$.

(AMIETE 2001, S. PATEL 1997, UKTU–2011)

SOLUTION. The given equation is Bernoulli's type because of presence of non-linear part on the right side, so dividing by $y(\log y)^2$, we get

$$\frac{1}{y(\log y)^2}\cdot\frac{dy}{dx} + \frac{1}{x}\cdot\frac{1}{\log y} = \frac{1}{x^2} \qquad \dots(1)$$

Now, substituting $u = \dfrac{1}{\log y}$

or $\dfrac{du}{dx} = -\dfrac{1}{y(\log y)^2}\cdot\dfrac{dy}{dx}$ in (1), we get

$$-\frac{du}{dx} + \frac{1}{x}u = \frac{1}{x^2} \Rightarrow \frac{du}{dx} - \frac{1}{x}u = -\frac{1}{x^2}$$

which is a linear equation in u, therefore

$$\text{I.F.} = e^{-\int \frac{1}{x}dx} = \frac{1}{x} \qquad \dots(2)$$

Multiplying equation (2) by I.F. and after integrating, we get

$$\frac{u}{x} = \frac{1}{2x^2} + c, \text{ where } c \text{ is the constant of integration.}$$

$\Rightarrow \dfrac{1}{x\log y} = \dfrac{1}{2x^2} + c$, which is the required solution.

EXAMPLE 2. *Solve* $x^2 dy - y(x - y)dx = 0$.   (UPTU(B.TECH.)–2006)

SOLUTION. The given equation can be written as

$$\frac{1}{y^2}\frac{dy}{dx} - \frac{1}{xy} = -\frac{1}{x^2}$$

Put $-\dfrac{1}{y} = u \Rightarrow \dfrac{1}{y^2}\dfrac{dy}{dx} = \dfrac{du}{dx}$

The given equation reduces to a linear differential equation in u.

$$\frac{du}{dx} - \frac{u}{x} = -\frac{1}{x^2}$$

$$\text{I.F.} = e^{-\int \frac{1}{x}dx} = e^{-\log x} = e^{\log 1/x} = \frac{1}{x}$$

Hence the solution is

$$u.\frac{1}{x} = \int -\frac{1}{x^2}.\frac{1}{x}\,dx + C = -\int x^{-3}dx + C$$

$$= -\frac{x^{-2}}{-2} + C$$

$$\Rightarrow \quad \frac{1}{xy} = -\frac{1}{2x^2} - C \text{, which is the required solution.}$$

EXAMPLE 3. *Solve* $xy(1 + xy^2)\dfrac{dy}{dx} = 1$.   (NAGPUR–2006)

SOLUTION. The given differential equation is

$$\frac{dx}{dy} - yx = y^3x^2$$

and dividing by x^2, we have

$$x^{-2}\frac{dx}{dy} - yx^{-1} = y^3 \qquad \ldots(1)$$

Putting $x^{-1} = z$ so that $-x^{-2}\dfrac{dx}{dy} = \dfrac{dz}{dy}$

Equation (1) becomes,

$$\frac{dz}{dy} + yz = -y^3 \text{ which is linear in } z.$$

Now, I.F. $= e^{\int ydy} = e^{y^2/2}$

∴ Solution is $z(\text{I.F.}) = \int (-y^3)(\text{I.F.})dy + c$

or $ze^{y^2/2} = -\int y^2.e^{y^2/2}.ydy + c$

Put $\dfrac{1}{2}y^2 = t$ so that $ydy = dt$

$$= -2\int t.e^t dt + c \qquad \text{[Integrate by parts]}$$

$$= -2\left[t.e^t - \int 1.e^t dt\right] + c$$

$$= -2[t.e^t - e^t] + c = (2 - y^2)e^{y^2/2} + c$$

or $\quad z = (2 - y^2) + ce^{-\frac{1}{2}y^2}$

or $\quad \dfrac{1}{x} = (2 - y^2) + ce^{-\frac{1}{2}y^2}$.

EXAMPLE 4. *Solve* $\dfrac{dy}{dx} + x\sin 2y = x^3\cos^2 y$.

(VTU–2011, MARATHWADA–2008, JNTU–2005)

SOLUTION. Dividing by $\cos^2 y$,

$$\sec^2 y\frac{dy}{dx} + 2x\frac{\sin y\cos y}{\cos^2 y} = x^3$$

or $\quad \sec^2 y\dfrac{dy}{dx} + 2x\tan y = x^3$

which is of the form (A) above. $\qquad \ldots(1)$

∴ Put $\tan y = z$ so that $\sec^2 y\dfrac{dy}{dx} = \dfrac{dz}{dx}$

So equation (1), becomes $\dfrac{dz}{dx} + 2xz = x^3$

This is linear equation in z

∴ I.F. $= e^{\int 2xdx} = e^{x^2}$

∴ The solution is

$$ze^{x^2} = \int e^{x^2}x^3 dx + c$$

$$= \frac{1}{2}(x^2 - 1)e^{x^2} + c$$

Replace z by $\tan y$, we get

$$\tan y = \frac{1}{2}(x^2 - 1) + ce^{-x^2}$$

Exercise-8.8

Solve the following ordinary differential equations :

1. $\dfrac{dy}{dx} + \dfrac{1}{x}y = x^2y^6$ (ANDHRA–1998, DELHI–1997)

2. $x\left(\dfrac{dy}{dx}\right) + y = x^2y^4$

3. $\dfrac{dy}{dx} + \left(\dfrac{x}{1-x^2}\right)y = x\sqrt{y}$

4. $\dfrac{dy}{dx} + \dfrac{y}{x} = y^3$

5. $y(1 + xy)\,dx - xdy = 0$

6. $\dfrac{dy}{dx} + \dfrac{1}{x} = \dfrac{e^y}{x^2}$

7. $(x^3y^2 + xy)dx = dy$ (BPTU–2005)

8. $\dfrac{dy}{dx} = x^3y^3 - xy$

9. $\dfrac{dy}{dx} + \left\{\dfrac{y}{x-1}\right\} = xy^{1/3}$

10. $\dfrac{dy}{dx} = e^{x-y}(e^x - e^y)$

11. $\dfrac{dy}{dx} - y\tan x = -y^2\sec x$

12. $x\dfrac{dy}{dx} + y\log y = xy\,e^x$ (AMIE–2000)

13. $\cos x\,dy = (\sin x - y)\,y\,dx$

14. $\dfrac{dy}{dx} + \dfrac{1}{x}\tan y = \dfrac{1}{x^2}\tan y\sin y$

15. $\sin y\dfrac{dy}{dx} = \cos y(1 - x\cos y)$

16. $\dfrac{dy}{dx} - \dfrac{\tan y}{(1+x)} = (1+x)\,e^x\sec y$

(ISM–2001, MARATHWADA–1993, BHOPAL–2009)

17. $\dfrac{dy}{dx} + y\cos x = y^n\sin 2x$

18. $\dfrac{dy}{dx} - 2y\tan x = y^2\tan^2 x$

19. $nx\dfrac{dy}{dx} + 2y = xy^{n+1}$

20. $x\dfrac{dy}{dx} + y = y^3$

Hints to Selected Problems

1. Put $\dfrac{1}{y^6} = v$, we get $\dfrac{dv}{dx} - \dfrac{5}{x}v = -5x^2$

2. Put $v = -\dfrac{1}{y^3}$

3. Put $y^{1/2} = v$

4. Put $\dfrac{1}{y^2} = v$

5. Put $v = -\dfrac{1}{y}$

6. Put $v = e^{-y}$

7. Put $-\dfrac{1}{y} = v$

8. Put $\dfrac{1}{y^2} = v$

9. Put $y^{2/3} = v$

10. Put $e^y = v$

11. Put $v = -\dfrac{1}{y}$

12. Put $\log y = v$

13. Put $1/y = v$

14. Put $-\csc y = v$

15. Put $\sec y = v$

16. Put $\sin y = v$

17. Put $\dfrac{1}{y^{n-1}} = v$

18. Put $-\dfrac{1}{y} = v$

19. Put $\dfrac{1}{y^n} = v$

Answers

1. $\dfrac{1}{x^5 y^5} = \dfrac{5}{2}(1/x^2) + c$ **2.** $\dfrac{1}{y^3} = x^2(3 - cx)$ **3.** $\sqrt{y}(1 - x^2)^{1/4} = -\dfrac{1}{3}(1 - x^2)^{3/4} + c$ **4.** $2xy^2 + cx^2 y^2 = 1$

5. $-\dfrac{x}{y} = \dfrac{1}{2}x^2 + c$ **6.** $2x = (2cx^2 + 1)\,e^y$ **7.** $1 = 2y\left(1 - \dfrac{x^2}{2}\right) - cye^{-x^2/2}$ **8.** $\dfrac{1}{y^2} = ce^{x^2} + x^2 + 1$

9. $y^{2/3}(x - 1)^{2/3} = \dfrac{2}{3}x(x - 1)^{5/3} - \dfrac{3}{20}(x - 1)^{8/3} + c$ **10.** $e^y = e^x - 1 + ce^{-(e^x)}$ **11.** $\sec x = y(\tan x - c)$

12. $x \log y = e^x(x - 1) + c$ **13.** $\sec x = y(\tan x - c)$ **14.** $2x = \sin y(1 - 2cx^2)$ **15.** $\sec y = x + 1 + ce^x$

16. $\sin y = (1 + x)(e^x + c)$ **17.** $\dfrac{1}{y^{n-1}} = 2\sin x - \left\{\dfrac{2}{1 - n}\right\} + ce^{(n-1)\sin x}$ **18.** $\sec^2 x + \dfrac{1}{3}y\tan^3 x + cy = 0$

19. $1/xy^n = cx + 1$ **20.** $1/y^2 = cx^2 + 1$

8.1.10 Exact Differential Equation

The differential equation which can be derived from its primitive by direct differentiation without any further transformation (such as elimination or reduction), is called an exact differential equation.

For example : The equation $ydx + xdy = 0$ is exact, as it is derived from its primitive $yx = c$.

WORKING PROCEDURE

Step 1. Compare the given equation with $Mdx + Ndy = 0$ and find out M and N.

Step 2. Show that $\dfrac{\partial M}{\partial y} = \dfrac{\partial N}{\partial x}$, which conclude the exactness of the given equation.

Step 3. Integrate the coefficient of dx (i.e., M) with respect to x, regarding y to be constant.

Step 4. Omit the terms containing x in N and find the integral of the coefficient of dy with respect to y.

Step 5. Add the above two results and equate this sum to an arbitrary constant i.e.,

$$\int_{(y\,\text{constant})} Mdx + \int_{\text{those terms which do not contain } x} Ndy = c.$$

REMARKS

- It is clear from the definition that the exact differential equation is formed from its general solution by direct differentiation and without any further operation of elimination or reduction.

- The necessary and sufficient condition for a differential equation (of first order and first degree) $Mdx + Ndy = 0$, where M and N are the functions of x and y to be exact equation is that $\dfrac{\partial M}{\partial y} = \dfrac{\partial N}{\partial x}$

- The equation $\dfrac{\partial M}{\partial y} = \dfrac{\partial N}{\partial x}$ is true whenever both sides exist and are continuous. Here we take the hypothesis that all the functions, which we discuss are sufficiently continuous and differentiable to guarantee the validity of the operations we perform on them.

Solved Examples

EXAMPLE 1. *Solve* $x\,dx + ydy = \dfrac{a^2(xdy - ydx)}{x^2 + y^2}$.

(UPTU(B.TECH)–2005, UKTU–2011)

SOLUTION. We have $x\,dx + ydy = \dfrac{a^2(xdy - ydx)}{x^2 + y^2}$

$\Rightarrow \left(x + \dfrac{a^2 y}{x^2 + y^2}\right) dx + \left(y - \dfrac{a^2 x}{x^2 + y^2}\right) dy = 0.$

Here, $M = x + \dfrac{a^2 y}{x^2 + y^2}, N = y - \dfrac{a^2 x}{x^2 + y^2}$

$\dfrac{\partial M}{\partial y} = \dfrac{a^2(x^2 - y^2)}{(x^2 + y^2)^2}, \dfrac{\partial N}{\partial x} = \dfrac{a^2(x^2 - y^2)}{(x^2 + y^2)^2}$

$\Rightarrow \dfrac{\partial M}{\partial y} = \dfrac{\partial N}{\partial x}$

Therefore, equation is exact. Hence,

$$\int \left(x + \dfrac{a^2 y}{x^2 + y^2}\right) dx + \int y\,dy = C$$

$$\Rightarrow \dfrac{x^2}{2} + a^2 y \dfrac{1}{y}\tan^{-1}\left(\dfrac{x}{y}\right) + \dfrac{y^2}{2} = C$$

$$\left(\dfrac{x^2 + y^2}{2}\right) + a^2 \tan^{-1}\dfrac{x}{y} = C$$

EXAMPLE 2. *Solve* $(y^2 e^{xy^2} + 4x^3)dx + (2xye^{xy^2} - 3y^2)dy = 0$

(VTU–2006)

SOLUTION. Compare the given equation with $Mdx + Ndy = 0$, we get

$$M = y^2 e^{xy^2} + 4x^3 \quad \text{and} \quad N = 2xye^{xy^2} - 3y^2$$

$$\dfrac{\partial M}{\partial y} = 2ye^{xy^2} + y^2 e^{xy^2} \cdot 2xy = \dfrac{\partial N}{\partial x}$$

Thus the equation is exact.
Hence, the solution is

$$\int_{(y\,\text{constant})} M dx + \int_{\text{those terms which do not contain } x} N dy = c$$

$$\Rightarrow \quad \int (y^2 e^{xy^2} + 4x^3) dx + \int (-3y^2) dy = C$$

$$\Rightarrow \qquad\qquad e^{xy^2} + x^4 - y^3 = C$$

EXAMPLE 3. Solve $\left\{ y\left(1 + \dfrac{1}{x}\right) + \cos y \right\} dx +$

$$(x + \log x - x \sin y) dy = 0$$

(MARATHWADA–2008S, VTU–2006)

SOLUTION. Compare the given equation with $M dx + N dy = 0$, we get

$$M = y\left(1 + \frac{1}{x}\right) + \cos y$$

and $\quad N = x + \log x - x \sin y$

$$\frac{\partial M}{\partial y} = 1 + \frac{1}{x} - \sin y = \frac{\partial N}{\partial x}$$

$\Rightarrow$ Equation is exact.
Hence, the solution is

$$\int_{(y\,\text{constant})} M dx + \int_{\text{those terms which do not contain } x} N dy = c$$

$$\Rightarrow \quad \int \left\{\left(1 + \frac{1}{x}\right) y + \cos y\right\} dx = C$$

$$\Rightarrow \qquad (x + \log x) y + x \cos y = C$$

EXAMPLE 4. Solve $\dfrac{dy}{dx} + \dfrac{y \cos x + \sin y + y}{\sin x + x \cos y + x} = 0$

(KURUKSHETRA–2005)

SOLUTION. The given equaton is

$$(y \cos x + \sin y + y) dx + (\sin x + x \cos y + x) dy = 0$$

Compare the given equation with $M dx + N dy = 0$, we get

$$M = y \cos x + \sin y + y$$

and $\quad N = \sin x + x \cos y + x$

$$\frac{\partial M}{\partial y} = \cos x + \cos y + 1 = \frac{\partial N}{\partial x}$$

$\Rightarrow$ Equation is exact.
Hence, the solution is

$$\int_{(y\,\text{constant})} M dx + \int_{\text{those terms which do not contain } x} N dy = c$$

$$\Rightarrow \quad \int (y \cos x + \sin y + y) dx + \int (0) dy = C$$

or $\qquad y \sin x + (\sin y + y) x = C$

Exercise-8.9

Solve the following ordinary differential equations :

1. $(4x + 3y + 1) dx + (3x + 2y + 1) dy = 0$

2. $\dfrac{dy}{dx} = \dfrac{(2x - y)}{x + 2y - 5}$

3. $x dx + y dy + \dfrac{x dy - y dx}{x^2 + y^2} = 0$

4. $(1 + 4xy + 2y^2) dx + (1 + 4xy + 2x^2) dy = 0$

5. $[1 + e^{x/y}] dx + e^{x/y}[1 - (x/y)] dy = 0$

6. $(y \sin 2x) dx - (1 + y^2 + \cos^2 x) dy = 0$ (VTU–2000)

7. $\dfrac{x dy}{x^2 + y^2} = \left(\dfrac{y}{x^2 + y^2} - 1\right) dx$

8. $(\cos x - x \cos y) dy - (\sin y + y \sin x) dx = 0$ (UPTU–2008)

Answers

1. $2x^2 + 3xy + y^2 + x + y = c$ **2.** $x^2 - y^2 - xy + 5y = c$ **3.** $x^2 - 2\tan^{-1}(x/y) + y^2 = c$ **4.** $x + 2x^2 y + 2xy^2 + y = c$

5. $x + y e^{x/y} = c$ **6.** $y \cos 2x + 2y + \dfrac{2}{3} y^3 = c$ **7.** $\tan^{-1}(x/y) - x = c$ **8.** $-x \sin y - y \cos x = C$

8.1.11 Integrating Factor

If the given equation $M dx + N dy = 0$ is not exact, then it can be made exact by multiplying some function of x and y. This multiplier is called an integrating factor (I.F.).

Remarks

- The number of integrating factor for the equation $M dx + N dy = 0$, is infinite.
- If μ is an integrating factor, then $\mu(M dx + N dy) = 0$ is an exact differential equation.
- Although an equation of the form $M dx + N dy = 0$, always has integrating factor, there is no general method of finding them.

8.1.12 Rules for Finding out Integrating Factor

Rule 1. By Inspection: Here, we explain some rules for finding I.F :
The following list of exact differentials should be noted very carefully.

(1) $d\left(\dfrac{x}{y}\right) = \dfrac{y dx - x dy}{y^2}$ **(2)** $d\left(\dfrac{y}{x}\right) = \dfrac{x dy - y dx}{x^2}$ **(3)** $d\left(\tan^{-1}\dfrac{y}{x}\right) = \dfrac{x dy - y dx}{x^2 + y^2}$

(4) $d\left(\tan^{-1}\dfrac{x}{y}\right) = \dfrac{y dx - x dy}{x^2 + y^2}$ **(5)** $d\left(\log \dfrac{x}{y}\right) = \dfrac{y dx - x dy}{xy}$ **(6)** $d\left(\log \dfrac{y}{x}\right) = \dfrac{x dy - y dx}{xy}$

(7) $d(x.y) = x dy + y dx$ **(8)** $d\left(\dfrac{1}{xy}\right) = -\left[\dfrac{x dy + y dx}{x^2 y^2}\right]$ **(9)** $d\left(\dfrac{x^2}{y}\right) = \dfrac{2xy\,dx - x^2 dy}{y^2}$

(10) $d\left(\dfrac{y^2}{x}\right) = \dfrac{2xy\,dy - y^2dx}{x^2}$ **(11)** $d\left(\dfrac{y^2}{x^2}\right) = \dfrac{2x^2ydy - 2y^2x\,dx}{x^4}$ **(12)** $\left(\dfrac{x^2}{y^2}\right)\ \dfrac{2xy^2dx}{y^4}\ \dfrac{2yx^2dy}{}$

(13) $d\left(\dfrac{e^x}{y}\right) = \dfrac{ye^x dx - e^x dy}{y^2}$ **(14)** $d\,[\log(x^2 + y^2)] = \dfrac{2xdx + 2ydy}{x^2 + y^2}$

 Solved Examples

EXAMPLE.   **Solve $y(2xy + e^x)dx = e^x\,dy$.**   (KURUKSHETRA–2005)

SOLUTION. $ye^x dx - e^x dy + 2xy^2 dx = 0$

Since the term $2xy^2 dx$ should not involve y^2. So, $1/y^2$ may be I.F. Multiplying both side by $1/y^2$, we get

$$\frac{ye^x dx - e^x dy}{y^2} + 2xdx = 0$$

or $d\left(\dfrac{e^x}{y}\right) + 2xdx = 0$

On integrating, we get

$$\frac{e^x}{y} + x^2 = c$$

Rule 2.   When $Mx + Ny \neq 0$ and the equation $Mdx + Ndy = 0$ is homogeneous, then an integrating factor of this differential equation is

$$\frac{1}{Mx + Ny}$$

 Solved Examples

EXAMPLE 1.   **Solve $y(y^2 - 2x^2)dx + x(2y^2 - x^2)dy = 0$.**

SOLUTION.   The given equation is

$$(y^3 - 2x^2y)dx + (2xy^2 - x^3)dy = 0 \qquad ...(1)$$

Compare with $Mdx + Ndy = 0$, we get

$$M = y^3 - 2x^2y, N = 2xy^2 - x^3$$

$\therefore$ $\dfrac{\partial M}{\partial y} = 3y^2 - 2x^2, \dfrac{\partial N}{\partial x} = 2y^2 - 3x^2$

$\because$ $\dfrac{\partial M}{\partial y} \neq \dfrac{\partial N}{\partial x} \Rightarrow$ Eqn. (1) is not exact.

But it is homogeneous in x and y.

Now, $Mx + Ny = x(y^3 - 2x^2y) + y(2xy^2 - x^3)$

$$= 3xy(y^2 - x^2)$$

$\therefore$ I.F. $= \dfrac{1}{Mx + Ny} = \dfrac{1}{3xy(y^2 - x^2)}$

Multiplying eqn. (1) by I.F, we get

$$\frac{y(y^2 - 2x^2)}{3xy(y^2 - x^2)}dx + \frac{x(2y^2 - x^2)}{3xy(y^2 - x^2)}dy = 0$$

$\Rightarrow$ $\dfrac{(y^2 - 2x^2)}{x(y^2 - x^2)}dx + \dfrac{(2y^2 - x^2)}{y(y^2 - x^2)}dy = 0$

$\Rightarrow$ $\dfrac{1}{x}dx - \dfrac{x}{y^2 - x^2}dx + \dfrac{1}{y}dy + \dfrac{y}{y^2 - x^2}dy = 0$

$\Rightarrow$ $\dfrac{1}{x}dx + \dfrac{1}{y}dy + \dfrac{1}{2}\left(\dfrac{2ydy - 2xdx}{y^2 - x^2}\right) = 0$

$\Rightarrow$ $d(\log x) + d(\log y) + \dfrac{1}{2}d\{\log(y^2 - x^2)\} = 0$

$\Rightarrow d(\log x) + d(\log y) + d\{\log\sqrt{(y^2 - x^2)}\} = d(\log c)$

On integrating, we get

$$\log x + \log y + \log\sqrt{(y^2 - x^2)} = \log c$$

$\Rightarrow$ $xy\sqrt{(y^2 - x^2)} = c$

where c is an arbitrary constant of integration.

EXAMPLE 2.   **Solve $x^2ydx - (x^3 + y^3)dy = 0$.**

SOLUTION.   The given equation is

$$x^2ydx - (x^3 + y^3)dy = 0 \qquad ...(1)$$

Compare with $Mdx + Ndy = 0$, we get

$$M = x^2y, N = -(x^3 + y^3)$$

$\therefore$ $\dfrac{\partial M}{\partial y} = x^2, \dfrac{\partial N}{\partial x} = -3x^2$

$\because$ $\dfrac{\partial M}{\partial y} \neq \dfrac{\partial N}{\partial x} \Rightarrow$ Eqn. (1) is not exact.

But it is homogeneous in x and y.

Now, $Mx + Ny = (x^2y).x + \{-(x^3 + y^3)\}y$

$$= x^3y - x^3y - y^4 = -y^4 \neq 0$$

$\therefore$ I.F. $= \dfrac{1}{Mx + Ny} = -\dfrac{1}{y^4}$

Multiplying eqn. (1) by $-\dfrac{1}{y^4}$, we get

$$\frac{-x^2}{y^3}dx + \left(\frac{x^3}{y^4} + \frac{1}{y}\right)dy = 0 \qquad ...(2)$$

Comparing with $Mdx + Ndy = 0$, we get

$$M = -\frac{x^2}{y^3}, N = \frac{x^3}{y^4} + \frac{1}{y}$$

$\therefore$ $\dfrac{\partial M}{\partial y} = \dfrac{3x^2}{y^4}, \dfrac{\partial N}{\partial x} = \dfrac{3x^2}{y^4}$

$\because$ $\dfrac{\partial M}{\partial y} = \dfrac{\partial N}{\partial x} \Rightarrow$ Eqn. (2) is exact.

Hence, the solution is

$\int_{(y\,constant)} Mdx + \int_{those\,terms\,which\,do\,not\,contain\,x} Ndy = c$

$\Rightarrow$ $\displaystyle\int\left(\frac{-x^2}{y^3}\right)dx + \int\frac{1}{y}dy = c$

$\Rightarrow$ $\dfrac{-x^3}{3y^3} + \log y = c$

where c is an arbitrary constant of integration.

Exercise-8.10

Solve the following ordinary differential equations :

1. $(x^2y - 2xy^2)dx - (x^3 - 3x^2y)dy = 0$ (OSMANIA–2003S)

2. $xy^2dy - (x^3 + y^3)dx = 0$

3. $(3xy^2 - y^3)dx - (2x^2y - xy^2)dy = 0$

 Answers

1. $\dfrac{x}{y} - 2\log x + 3\log y = c$ **2.** $\log x - \dfrac{y^3}{3x^3} = c$ **3.** $3\log x + \dfrac{y}{x} - 2\log y = c$

Rule 3. If the equation $Mdx + Ndy = 0$ is of form $f_1(xy)ydx + f_2(xy)xdy = 0$, then $\dfrac{1}{Mx - Ny}$ is an integrating factor, provided $Mx - Ny \neq 0$.

 Solved Examples

EXAMPLE 1. **Solve $(y - xy^2)dx - (x + x^2y)dy = 0$.**

SOLUTION. The given equation is

$$(y - xy^2)dx - (x + x^2y)dy = 0 \qquad \text{...(1)}$$

Comparing with $Mdx + Ndy = 0$, we get

$$M = y - xy^2, N = -(x + x^2y)$$

$$\therefore \quad \frac{\partial M}{\partial y} = 1 - 2xy, \frac{\partial N}{\partial x} = -1 - 2xy$$

$$\because \quad \frac{\partial M}{\partial y} \neq \frac{\partial N}{\partial x} \Rightarrow \text{Eqn. (1) is not exact.}$$

Now, $M = y - xy^2 = y(1 - xy) = yf_1(xy)$

and $N = -(x + x^2y) = -x(1 + xy) = xf_2(xy)$

$$\therefore \quad Mx - Ny = (y - xy^2)x + (x + x^2y)y$$

$$= 2xy \neq 0$$

$$\therefore \quad \text{I.F.} = \frac{1}{Mx - Ny} = \frac{1}{2xy}$$

Multiplying (1) by $\dfrac{1}{2xy}$, we get

$$\frac{y - xy^2}{2xy}dx - \frac{x + x^2y}{2xy}dy = 0$$

$$\frac{1}{2}\left(\frac{1}{x} - y\right)dx - \frac{1}{2}\left(\frac{1}{y} + x\right)dy = 0 \qquad \text{...(2)}$$

Comparing with $Mdx + Ndy = 0$, we get

$$M = \frac{1}{2}\left(\frac{1}{x} - y\right), N = -\frac{1}{2}\left(\frac{1}{y} + x\right)$$

$$\therefore \quad \frac{\partial M}{\partial y} = -\frac{1}{2}, \frac{\partial N}{\partial x} = -\frac{1}{2}$$

$$\because \quad \frac{\partial M}{\partial y} = \frac{\partial N}{\partial x} \Rightarrow \text{Eqn. (2) is exact.}$$

Hence, the solution is

$$\int_{(y\text{ constant})} Mdx + \int_{\text{those terms which do not contain }x} Ndy = c$$

$$\Rightarrow \quad \int \frac{1}{2}\left(\frac{1}{x} - y\right)dx + \int \frac{-1}{2y}dy = c$$

$$\Rightarrow \quad \frac{1}{2}(\log x - yx) - \frac{1}{2}\log y = c$$

$$\Rightarrow \quad \log x - xy - \log y = 2c$$

$$\Rightarrow \quad \log\left(\frac{x}{y}\right) - xy = A,$$

where $A = 2c$

 **Exercise-8.11**

Solve the following ordinary differential equations :

1. $(1 + xy)ydx + (1 - xy)xdy = 0$ (SVTU–2008)

2. $(x^2y^2 + xy + 1)ydx + (x^2y^2 - xy + 1)xdy = 0$

3. $(xy \sin xy + \cos xy)ydx + (xy \sin xy - \cos xy)xdy = 0$

4. $(xy^2 + 2x^2y^3)dx + (x^2y - x^3y^2)dy = 0$

Answers

1. $\log\dfrac{x}{y} - \dfrac{1}{xy} = c$ **2.** $xy - \dfrac{1}{xy} + \log\dfrac{x}{y} = c$ **3.** $x \sec xy = cy$ **4.** $-\dfrac{1}{xy} + 2\log x - \log y = c$

Rule 4. If in the equation $Mdx + Ndy = 0$, $\dfrac{1}{N}\left(\dfrac{\partial M}{\partial y} - \dfrac{\partial N}{\partial x}\right)$ is a function of x alone, say $f(x)$, then $e^{\int f(x)dx}$ is an integrating factor of the equation $Mdx + Ndy = 0$.

Rule 5. If in the equation $Mdx + Ndy = 0$, $\dfrac{1}{M}\left(\dfrac{\partial N}{\partial x} - \dfrac{\partial M}{\partial y}\right)$ is a function of y alone, say $f(y)$, then $e^{\int f(y)dy}$ is an integrating factor of the equation $Mdx + Ndy = 0$.

Solved Examples

EXAMPLE 1. *Solve* $\left(y + \dfrac{1}{3}y^3 + \dfrac{1}{2}x^2\right)dx + \dfrac{1}{4}(1 + y^2)xdy = 0.$

SOLUTION. The given equation is

$$\left(y + \frac{1}{3}y^3 + \frac{1}{2}x^2\right)dx + \frac{1}{4}(1 + y^2)xdy = 0. \qquad ...(1)$$

Comparing with $Mdx + Ndy = 0$, we get

$$M = y + \frac{1}{3}y^3 + \frac{1}{2}x^2, N = \frac{1}{4}(1 + y^2)x$$

$$\therefore \quad \frac{\partial M}{\partial y} = 1 + y^2, \frac{\partial N}{\partial x} = \frac{1}{4}(1 + y^2)$$

$$\because \quad \frac{\partial M}{\partial y} \neq \frac{\partial N}{\partial x} \Rightarrow \text{Eqn. (1) is not exact.}$$

Now, $\dfrac{1}{N}\left(\dfrac{\partial M}{\partial y} - \dfrac{\partial N}{\partial x}\right) = \dfrac{(1 + y^2) - \dfrac{1}{4}(1 + y^2)}{\dfrac{1}{4}(1 + y^2)x}$

$$= \frac{3}{x} = f(x)$$

$$\therefore \quad \text{I.F.} = e^{\int f(x)dx} = e^{\int \frac{3}{x}dx} = e^{3\log x} = x^3$$

Multiplying (1) by x^3, we get

$$\left(yx^3 + \frac{1}{3}y^3x^3 + \frac{1}{2}x^5\right)dx + \frac{1}{4}(1 + y^2)x^4dy = 0 \quad ...(2)$$

Comparing with $Mdx + Ndy = 0$, we get

$$M = yx^3 + \frac{1}{3}y^3x^3 + \frac{1}{2}x^5$$

and $\quad N = \dfrac{1}{4}(1 + y^2)x^4$

$$\therefore \quad \frac{\partial M}{\partial y} = x^3 + y^2x^3 = x^3(1 + y^2),$$

$$\frac{\partial N}{\partial x} = \frac{1}{4}(1 + y^2).4x^3 = x^3(1 + y^2)$$

$$\therefore \quad \frac{\partial M}{\partial y} = \frac{\partial N}{\partial x} \Rightarrow \text{Eqn. (2) is exact.}$$

Hence, the solution is

$$\int_{(y \text{ constant})} Mdx + \int_{\text{those terms which do not contain } x} Ndy = c$$

$$\Rightarrow \quad \int \left(yx^3 + \frac{1}{3}y^3x^3 + \frac{1}{2}x^5\right)dx + \int 0.dy = c$$

$$\Rightarrow \quad \frac{yx^4}{4} + \frac{y^3x^4}{12} + \frac{x^6}{12} = c$$

EXAMPLE 2. *Solve* $(y \log y)dx + (x - \log y)dy = 0.$ (UPTU-2004)

SOLUTION. The given equation is

$$(y \log y)dx + (x - \log y)dy = 0. \qquad ...(1)$$

Comparing with $Mdx + Ndy = 0$, we get

$$M = y \log y, N = x - \log y$$

$$\therefore \quad \frac{\partial M}{\partial y} = y.\frac{1}{y} + \log y = 1 + \log y$$

and $\quad \dfrac{\partial N}{\partial x} = 1$

$$\because \quad \frac{\partial M}{\partial y} \neq \frac{\partial N}{\partial x} \Rightarrow \text{Eqn. (1) is not exact.}$$

Now, $\dfrac{1}{M}\left(\dfrac{\partial N}{\partial x} - \dfrac{\partial M}{\partial y}\right) = \dfrac{1}{y \log y}(1 - 1 - \log y)$

$$= -\frac{1}{y} = f(y)$$

$$\therefore \quad \text{I.F.} = e^{\int f(y)dy} = e^{-\int \frac{1}{y}dy} = e^{-\log y} = \frac{1}{y}$$

Multiplying (1) by $\dfrac{1}{y}$, we get

$$(\log y)dx + \left(\frac{x - \log y}{y}\right)dy = 0 \qquad ...(2)$$

Comparing with $Mdx + Ndy = 0$, we get

$$M = \log y, \ N = \frac{x}{y} - \frac{\log y}{y}$$

$$\therefore \quad \frac{\partial M}{\partial y} = \frac{1}{y} \text{ and } \frac{\partial N}{\partial x} = \frac{1}{y}$$

$$\because \quad \frac{\partial M}{\partial y} = \frac{\partial N}{\partial x} \Rightarrow \text{Eqn. (2) is exact.}$$

Hence, the solution is

$$\int_{(y \text{ constant})} Mdx + \int_{\text{those terms which do not contain } x} Ndy = c$$

$$\Rightarrow \quad \int \log y dx + \int \left(-\frac{\log y}{y}\right)dy = c$$

$$\Rightarrow \quad x \log y - \frac{1}{2}(\log y)^2 = c$$

EXAMPLE 3. *Solve* $(xy^3 + y)dx + 2(x^2y^2 + x + y^4)dy = 0.$

SOLUTION. The given equation is

$$(xy^3 + y)dx + 2(x^2y^2 + x + y^4)dy = 0. \qquad ...(1)$$

Comparing with $Mdx + Ndy = 0$, we get

$$M = xy^3 + y, N = 2(x^2y^2 + x + y^4)$$

$$\therefore \quad \frac{\partial M}{\partial y} = 3xy^2 + 1, \frac{\partial N}{\partial x} = 2(2xy^2 + 1)$$

$$\because \quad \frac{\partial M}{\partial y} \neq \frac{\partial N}{\partial x} \Rightarrow \text{Eqn. (1) is not exact.}$$

Now, $\dfrac{1}{M}\left(\dfrac{\partial N}{\partial x} - \dfrac{\partial M}{\partial y}\right) = \dfrac{2.(2xy^2 + 1) - (3xy^2 + 1)}{xy^3 + y}$

$$= \frac{xy^2 + 1}{y(xy^2 + 1)} = \frac{1}{y} = f(y)$$

$$\therefore \quad \text{I.F.} = e^{\int f(y)dy} = e^{\int \frac{1}{y}dy} = e^{\log y} = y$$

Multiplying (1) by y, we get

$$(xy^4 + y^2)dx + 2(x^2y^3 + xy + y^5)dy = 0 \qquad ...(2)$$

Comparing with $Mdx + Ndy = 0$, we get

$$M = xy^4 + y^2, \ N = 2(x^2y^3 + xy + y^5)$$

$$\therefore \quad \frac{\partial M}{\partial y} = 4xy^3 + 2y, \ \frac{\partial N}{\partial x} = 4xy^3 + 2y$$

$$\because \quad \frac{\partial M}{\partial y} = \frac{\partial N}{\partial x} \Rightarrow \text{Eqn. (2) is exact.}$$

Hence, the solution is

$$\int_{(y \text{ constant})} Mdx + \int_{\text{those terms which do not contain } x} Ndy = c$$

$$\Rightarrow \quad \int (xy^4 + y^2)dx + \int 2y^5dy = c$$

$$\Rightarrow \quad \frac{x^2y^4}{2} + xy^2 + \frac{2y^6}{6} = c$$

$$\Rightarrow \quad 3x^2y^4 + 6xy^2 + 2y^6 = k, \text{ where } k = 6c.$$

Exercise-8.12

Solve the following ordinary differential equations :

1. $(xy^2 - e^{1/x^3})dx - x^2ydy = 0$ (SVTU–2009, MUMBAI–2007)

2. $(x^2 + y^2 + 2x)dx + 2ydy = 0$

3. $(y^4 + 2y)dx + (xy^3 + 2y^4 - 4x)dy = 0$

4. $(x\sec^2 y - x^2\cos y)dy = (\tan y - 3x^4)dx$

 **Answers**

1. $\dfrac{1}{3}e^{x^{-3}} - \dfrac{1}{2}\dfrac{y^2}{x^2} = c$ **2.** $e^x(x^2 + y^2) = c$ **3.** $\left(y + \dfrac{2}{y^2}\right)x + y^2 = c$ **4.** $-\dfrac{1}{x}\tan y - x^3 + \sin y = c$

Rule 6. If an equation is of the form $x^a y^b[mydx + nxdy] + x^r y^s[pydx + qxdy] = 0$ where a, b, m, n, r, s, p and q are all constants, then the integrating factor will be $x^h y^k$, where h and k are such that after multiplying by the integrating factor, the condition of exactness is satisfied.

 Solved Examples

EXAMPLE 1. *Solve* $(2ydx + 3xdy) + 2xy(3ydx + 4xdy) = 0$.

SOLUTION. The given equation is
$$(2ydx + 3xdy) + 2xy(3ydx + 4xdy) = 0. \qquad ...(1)$$
$$\Rightarrow \quad (2y + 6xy^2)dx + (3x + 8x^2y)dy = 0. \qquad ...(2)$$
Comparing (2) with $Mdx + Ndy = 0$, we get
$$M = 2y + 6xy^2, N = 3x + 8x^2y$$
$$\therefore \quad \frac{\partial M}{\partial y} = 2 + 12xy, \frac{\partial N}{\partial x} = 3 + 16xy$$
$$\because \quad \frac{\partial M}{\partial y} \neq \frac{\partial N}{\partial x} \Rightarrow \text{Eqn. (2) is not exact.}$$

Eqn. (1) is of the form
$$x^a y^b(mydx + nxdy) + x^r y^s(pydx + qxdy) = 0$$
$\therefore$ $x^h y^k$ is an integrating factor of eqn (1).
Multiplying (2) by $x^h y^k$, we get
$$(2x^h y^{k+1} + 6x^{h+1}y^{k+2})dx + (3x^{h+1}y^k + 8x^{h+2}y^{k+1})dy = 0. \qquad ...(3)$$
Comparing (3) with $Mdx + Ndy = 0$, we get
$$M = 2x^h y^{k+1} + 6x^{h+1}y^{k+2}, N = 3x^{h+1}y^k + 8x^{h+2}y^{k+1}$$
Now eqn. (3) must be exact, the condition for which is
$$\frac{\partial M}{\partial y} = \frac{\partial N}{\partial x}$$

$$2(k+1)x^h y^k + 6(k+2)x^{h+1}y^{k+1}$$
$$= 3(h+1)x^h y^k + 8(h+2)x^{h+1}y^{k+1}$$
Now, equating the coefficients of $x^h y^k$ and $x^{h+1}y^{k+1}$ from both sides, we get
$$2(k+1) = 3(h+1) \Rightarrow 3h - 2k + 1 = 0 \qquad ...(4)$$
and $6(k+2) = 8(h+2) \Rightarrow 4h - 3k + 2 = 0 \qquad ...(5)$
On solving (4) and (5), we get $h = 1, k = 2$
Hence, xy^2 is an integrating factor.
Multiplying eqn. (2) by xy^2, we get
$$(2xy^3 + 6x^2y^4)dx + (3x^2y^2 + 8x^3y^3)dy = 0 \qquad ...(6)$$
Comparing (6) with $Mdx + Ndy = 0$, we get
$$M = 2xy^3 + 6x^2y^4, N = 3x^2y^2 + 8x^3y^3$$
$$\therefore \quad \frac{\partial M}{\partial y} = 6xy^2 + 24x^2y^3, \frac{\partial N}{\partial x} = 6xy^2 + 24x^2y^3$$
$$\because \quad \frac{\partial M}{\partial y} = \frac{\partial N}{\partial x} \Rightarrow \text{Eqn. (6) is exact.}$$
Hence, the solution is
$$\int_{(y\,constant)} Mdx + \int_{\text{those terms which do not contain } x} Ndy = c$$
$$\Rightarrow \quad \int (2xy^3 + 6x^2y^4)dx + \int 0 \cdot dy = c$$
$$\Rightarrow \quad x^2y^3 + 2x^3y^4 = c$$

 Exercise-8.13

Solve the following ordinary differential equations :

1. $y(xy + 2x^2y^3)dx + x(xy - x^2y^2)dy = 0$

(HISSAR–2005, KURUKSHETRA–2005)

2. $x(3ydx + 2xdy) + 8y^4(ydx + 3xdy) = 0$

 Answers

1. $2\log x - \log y - \dfrac{1}{xy} = c$ **2.** $x^3y^2 + 4x^2y^6 = c$

 Miscellaneous Solved Examples

EXAMPLE 1. *Solve* $(y^2e^x + 2xy)dx - x^2dy = 0$.

SOLUTION. The given equation is
$$(y^2e^x + 2xy)dx - x^2dy = 0$$
Compare with $Mdx + Ndy = 0$, we get
$$M = y^2e^x + 2xy \text{ and } N = -x^2$$
$$\frac{\partial M}{\partial y} = 2ye^x + 2x, \quad \frac{\partial N}{\partial x} = -2x$$

$$\Rightarrow \quad \frac{\partial M}{\partial y} \neq \frac{\partial N}{\partial x}$$
The given equation is not exact.
Now, if in the equation e^x is multiplied by some other function, then it must occur twice in the differential equation. But since it is occurring only once, therefore, we should divide by y^2.

$$\therefore \quad \left(e^x + \frac{2x}{y}\right)dx + \left(\frac{-x^2}{y^2}\right)dy = 0$$

Now, $M = e^x + \dfrac{2x}{y}$ and $N = -\dfrac{x^2}{y^2}$

$$\Rightarrow \quad \frac{\partial M}{\partial y} = -\frac{2x}{y^2} = \frac{\partial N}{\partial x} \Rightarrow \text{Equation is exact.}$$

Hence, the solution is

$\int_{(y\, constant)} M dx + \int_{those\ terms\ which\ do\ not\ contain\ x} N dy = c$

$$\int \left(e^x + \frac{2x}{y}\right)dx = c \Rightarrow e^x + \frac{x^2}{y} = c.$$

EXAMPLE 2. *Solve* $(y^2 + 2x^2y)dx + (2x^3 - xy)dy = 0$.

(RAJASTHAN–2005)

SOLUTION. Here, the given equation is

$$(y^2 + 2x^2y)dx + (2x^3 - xy)dy = 0 \quad \ldots(1)$$

which can be written as

$$y(ydx - xdy) + 2x^2(ydx + xdy) = 0.$$

Let us suppose that the possible integrating factor be $x^h y^k$.

Multiplying (1) by I.F. (i.e., $x^h y^k$), we have

$$(x^h y^{k+2} + 2x^{h+2} y^{k+1})dx + (2x^{h+3}y^k - x^{h+1}y^{k+1})dy = 0 \quad \ldots(2)$$

Now, $M = x^h y^{k+2} + 2x^{h+2}y^{k+1}$

and $N = 2x^{h+3}y^k - x^{h+1}y^{k+1}$

$$\Rightarrow \quad \frac{\partial M}{\partial y} = (k+2)x^h y^{k+1} + 2(k+1)x^{h+2}y^k$$

and $\dfrac{\partial N}{\partial x} = 2(h+3)x^{h+2}y^k - (h+1)x^h y^{k+1}$.

If the equation (2) is exact, then $\dfrac{\partial M}{\partial y} = \dfrac{\partial N}{\partial x}$.

$\therefore$ Equating the coefficients of $x^h y^{k+1}$ and $x^{h+2}y^k$ on both sides, we have

$$k + 2 = -h - 1 \Rightarrow h + k + 3 = 0$$

and $2k + 2 = 2h + 6 \Rightarrow h - k + 2 = 0$

Solving these equations, we get $h = -5/2$ and $k = -1/2$

The I.F. $= x^h y^k = x^{-5/2} y^{-1/2}$

Now, equation (2) becomes

$$(x^{-5/2}y^{3/2} + 2x^{-1/2}y^{1/2})dx$$
$$+ (2x^{1/2}y^{-1/2} - x^{-3/2}y^{1/2})dy = 0.$$

We can easily verify that this equation is exact by condition $\dfrac{\partial M}{\partial y} = \dfrac{\partial N}{\partial x}$.

Hence, the solution is

$\int_{(y\, constant)} M dx + \int_{those\ terms\ which\ do\ not\ contain\ x} N dy = c$

$$\Rightarrow \quad -\frac{2}{3}x^{-3/2}y^{3/2} + 4x^{1/2}y^{1/2} = c$$

is required solution.

Miscellaneous Exercise

Solve the following ordinary differential equations :

1. $(1 + xy)y\,dx + (1 - xy)x\,dy = 0$

    (AMIE–1996, MADURAI–1990, MDU–2000)

2. $ydx - xdy + (1 + x^2)dx + x^2 \sin y\,dy = 0$

3. $(x^3y^3 + x^2y^2 + xy + 1)y\,dx + (x^3y^3 - x^2y^2 - xy + 1)x\,dy = 0$

4. $x^2y\,dx - (x^3 + y^3)dy = 0$

5. $y(2x^2y + e^x)dx - (e^x + y^3)dy = 0$ (BANGALORE–1990)

6. $xdy - ydx = -2x^3dx$

7. $xdx + ydy + (x^2 + y^2)dy = 0$

8. $(x^2 + y^2)dx - 2xy\,dy = 0$

9. $x(3y\,dx + 2x\,dy) + 8y^4(y\,dx + 3x\,dy) = 0$

10. $(xy \sin xy + \cos xy)y\,dx + (xy \sin xy - \cos xy)x\,dy = 0$

11. $(3x^2y^4 + 2xy)dx + (2x^3y^3 - x^2)dy = 0$

12. $(2ydx + 3x\,dy) + 2xy\,(3y\,dx + 4x\,dy) = 0$

13. $(xy^2 + 2x^2y^3)dx + (x^2y - x^3y^2)dy = 0$

14. $(xy^2 - x^2)dx + (3x^2y^2 + x^2y - 2x^3 + y^2)dy = 0$

15. $xdx + ydy = \dfrac{a\ (xdy\ ydx)}{x^2\ y^2}$ (UPTU–2005)

16. $(4xy + 3y^2 - x)dx + x(x + 2y)dy = 0$ (MUMBAI–2006)

17. $(y - xy^2)dx + (x + x^2y)dy = 0$ (MUMBAI–2006)

18. $ydx - xdy + 3x^2y^2e^{x^3}dx = 0$ (KURUKSHETRA–2006)

19. $2ydx + x(2 \log x - y)dy = 0$ (PTU–2005)

Hints to Selected Problems

1. The given equation can be written as $d(yx) + xy^2dx - x^2ydy = 0$

   Divide by x^2y^2 and then put $xy = v$.

2. The given equation can be written as $d(e^x) + d\left(\dfrac{x^2}{y}\right) = 0$. Now integrate it.

4. I.F. $= -\dfrac{1}{y^4}$

5. I.F. $= \dfrac{1}{y^2}$

6. Dividing throughout by x^6, we get $d\left(\dfrac{y}{x}\right) + 2x\,dx = 0$. Then integrate it.

7. The given equation can be written as $\dfrac{d(x^2 + y^2)}{x^2 + y^2} + 2dy = 0$.

8. I.F. $= \dfrac{1}{N}\left(\dfrac{\partial M}{\partial y} - \dfrac{\partial N}{\partial x}\right) = \dfrac{1}{x^2}$

9. I.F. $= xy$

10. I.F. $= \dfrac{1}{Mx - Ny} = \dfrac{1}{2yx \cos xy}$

11. I.F. $= \dfrac{1}{M}\left(\dfrac{\partial N}{\partial x} - \dfrac{\partial M}{\partial y}\right) = \dfrac{1}{y^2}$

12. I.F. $= xy^2$

13. I.F. $= \dfrac{1}{Mx - Ny} = \dfrac{1}{3x^3y^3}$

14. I.F. $= \dfrac{1}{M}\left(\dfrac{\partial N}{\partial x} - \dfrac{\partial M}{\partial y}\right) = e^{6y}$

Answers

1. $\log(x/y) = c + (1/xy)$ **2.** $\dfrac{y}{x} + \dfrac{1}{x} - x + \cos y = c$ **3.** $xy + (-1/xy) - 2\log y = c$

4. $y = ce^{x^3/3y^3}$ **5.** $\dfrac{2}{3}x^3 - \dfrac{1}{2}y^2 + \dfrac{e^x}{y} = c$ **6.** $y + x^3 = cx$ **7.** $x^2 + y^2 = e^{c-2y}$ **8.** $x^2 - y^2 = cx$

9. $x^3y^2 + 4y^6x^2 = c$ **10.** $x\sec(xy) = cy$ **11.** $x^3y^2 + \left(x^2/y\right) = c$ **12.** $x^2y^3 + 2x^3y^2 = c$ **13.** $x^2 = cy\, e^{1/xy}$

14. $e^{6y}\left[x^2\left(\dfrac{y^2}{2} - \dfrac{x}{3}\right) + \left(\dfrac{y^2}{6} - \dfrac{y}{18} + \dfrac{1}{108}\right)\right] = c$ **15.** $x^2 + y^2 - 2a^2\tan^{-1}\left(\dfrac{y}{x}\right) = c$ **16.** $4x^4y + 4x^3y^2 - x^4 = c$

17. $\log\left(\dfrac{x}{y}\right) = c + xy$ **18.** $\left(\dfrac{x}{y}\right) + e^{x^3} = c$ **19.** $4y\log x = y^2 + c$

8.2 LINEAR DIFFERENTIAL EQUATION WITH CONSTANT COEFFICIENTS

The equation $\dfrac{d^n y}{dx^n} + A_1\dfrac{d^{n-1}y}{dx^{n-1}} + A_2\dfrac{d^{n-2}y}{dx^{n-2}} + \ldots + A_n y = B$ ...(1)

having $A_1, \ldots, A_n$ and B either constant or function of x, is called the linear differential equation of nth order.

If $A_1, A_2, \ldots, A_n$ are all constants and B may not be constant, then equation (1) is said to be linear differential equation of nth degree with constant coefficients.

If we take $B = 0$, then the corresponding equation is called homogeneous equation.

Using the symbols $D, D^2, \ldots, D^n$ for $\dfrac{d}{dx}, \dfrac{d^2}{dx^2}, \ldots, \dfrac{d^n}{dx^n}$ respectively in (1), then we get

$$D^n y + A_1 D^{n-1}y + A_2 D^{n-2}y + \ldots + A_n y = B$$

$\Rightarrow \qquad (D^n + A_1 D^{n-1} + A_2 D^{n-2} + \ldots + A_n)y = B \Rightarrow f(D)y = B$ ...(2)

where, $f(D) = D^n + A_1 D^{n-1} + A_2 D^{n-2} + \ldots + A_n$.

Now, consider the homogeneous differential equation

$$f(D)y = 0$$ ...(3)

(Obtained by putting right hand side, *i.e.*, B equal to zero).

Now, we shall show that if $y_1, y_2, \ldots, y_n$ are n linearly independent solutions of (3), then $(C_1 y_1 + C_2 y_2 + \ldots + C_n y_n)$ is also a solution of (3), where $C_1, C_2, \ldots, C_n$ are arbitrary constants.

Since, we assumed that $y_1, y_2, \ldots, y_n$ are solution of (3) $\Rightarrow y_1, y_2, \ldots, y_n$ must satisfy (3).

which gives $\qquad \left.\begin{array}{l} f(D)y_1 = 0 \\ f(D)y_2 = 0 \\ \cdots\cdots\cdots\cdots \\ f(D)y_n = 0 \end{array}\right]$ ...(4)

Now consider

$$f(D)(C_1 y_1 + C_2 y_2 + \ldots + C_n y_n) = f(D)(C_1 y_1) + f(D)(C_2 y_2) + \ldots + f(D)(C_n y_n)$$

$$= C_1 f(D)y_1 + C_2 f(D)y_2 + \ldots + C_n f(D)y_n$$

$$= C_1.0 + C_2.0 + \ldots + C_n.0 \qquad \text{(By using (4))}$$

Therefore, we have $\qquad f(D)[C_1 y_1 + C_2 y_2 + \ldots C_n y_n] = 0$ ...(5)

$\Rightarrow \qquad (C_1 y_1 + C_2 y_2 + \ldots + C_n y_n)$ satisfies (3).

$\Rightarrow \qquad (C_1 y_1 + C_2 y_2 + \ldots + C_n y_n)$ is also a solution of (3).

Hence, we can say that if $y_1, y_2, \ldots, y_n$ are n linearly independent solutions of (3), then $(C_1 y_1 + C_2 y_2 + \ldots + C_n y_n)$ is also a solution of (3) known as complete or general solution of (3), containing n arbitrary constants $C_1, C_2, \ldots, C_n$.

Now, let us suppose $\qquad (C_1 y_1 + C_2 y_2 + \ldots + C_n y_n) = u$ (say).

Then, from (5), we have $\qquad f(D)u = 0$ ...(6)

Again, let v be any particular solution of (2). Therefore, we have

$$f(D)v = B$$ ...(7)

Now, $\qquad f(D)(u + v) = f(D)u + f(D)v \qquad \text{(Using (6) and (7))}$

which shows that $(u + v)$, *i.e.*, $\{(C_1 y_1 + C_2 y_2 + \ldots + C_n y_n) + v\}$ is the general solution of (2).

WORKING PROCEDURE

Step 1.   Firstly, we find the general solution of (2), which is called the complimentary function (C.F), contains as many arbitrary constants as is the order of the given differential equation.

Step 2.   Next, find the solution of (1), with no arbitrary constant which is called the particular integral (P.I.).

Step 3.   To find the general solution of (1), add C.F. and P.I. obtained in (1) and (2), *i.e.*, $y = u + v = $ C.F. + P.I.

REMARKS

- Here, the operator D stands for d/dx, D^2 for d^2/dx^2 and so on.
- The operator D^{-1} stands for integration.
- Since, the symbol D satisfies the fundamental laws of algebra, therefore it can be regarded as an algebraic quantity.
- The general solution of (1) is $y = $ C.F. + P.I., where C.F. involves n arbitrary constants and P.I. does not involve any arbitrary constant.
- Since P.I. appears due to B in (1), therefore, if a linear differential equation with constant coefficients is given with $B = 0$, then its general solution will not involve P.I. and hence the general solution of the differential equation is given by $y = $ C.F.
- The equations, discussed in this chapter are most important in the study of vibrations of all kinds, mechanical, acoustical and electrical.
- The method (discussed above) of solving these type of equations, is given by Euler and D'Alembert.

8.2.1 AUXILIARY EQUATION

Consider the differential equation (1) with $B = 0$, *i.e.*,

$$(D^n + A_1 D^{n-1} + A_2 D^{n-2} + ... + A_n) y = 0 \qquad \text{or} \qquad f(D) y = 0. \qquad \text{...(1)}$$

Substitute $y = e^{mx}$ on the trial basis, then we get $e^{mx}(m^n + A_1 m^{n-1} + A_2 m^{n-2} + ... + A_n) = 0$

which holds if $m^n + A_1 m^{n-1} + A_2 m^{n-2} + ... + A_n = 0$ or $f(m) = 0$. $\qquad \text{...(2)}$

Equation (2) is called the auxiliary equation.

From (1) and (2), we observe that the auxiliary equation $f(m) = 0$ will give the same value of m as the equation $f(D) = 0$ gives the value of D.

8.2.2 METHOD OF FINDING THE COMPLEMENTARY FUNCTION (C.F.)

To find the C.F., the roots of the auxiliary equation (2) are to be considered. Three different cases arise :

 (i) The roots of auxiliary equation (2) are real.

 (ii) The roots of auxiliary equation (2) are complex, *i.e.*, $\alpha \pm i\beta$ type.

 (iii) The roots of auxiliary equation (2) are surds, *i.e.*, $\alpha \pm \sqrt{\beta}$ type.

Case (i) : (a) Suppose that the auxiliary equation (2) has n distinct roots $m_1, m_2, ..., m_n$, then C.F. is given by

$$C_1 e^{m_1 x} + C_2 e^{m_2 x} + ... + C_n e^{m_n x}$$

where $C_1, C_2, ..., C_n$ are arbitrary constants.

 (b) If the auxiliary equation having r roots are equal to m_1(say) and remaining roots are distinct, then the C.F. is given by

$$[C_1 + C_2 x + C_3 x^2 + ... + C_r x^{r-1}] e^{m_1 x} + C_{r+1} e^{m_{r+1} x} + ... + C_n e^{m_n x}.$$

Case (ii):   If some of the roots of the auxiliary equation are complex, then we shall use the following procedure.

Let $\alpha \pm i\beta$ be the roots of the auxiliary equation, then the corresponding part becomes

$$= C_1 e^{(\alpha + i\beta)x} + C_2 e^{(\alpha - i\beta)x} = C_1 e^{\alpha x} . e^{i\beta x} + C_2 e^{\alpha x} . e^{-i\beta x}$$

$$= e^{\alpha x}[C_1 \cos\beta x + iC_1 \sin\beta x] + e^{\alpha x}[C_2 \cos\beta x - iC_2 \sin\beta x] = e^{\alpha x}[(C_1 + C_2)\cos\beta x + (iC_1 - iC_2)\sin\beta x]$$

$$\text{C.F.} = e^{\alpha x}[B_1 \cos\beta x + B_2 \sin\beta x] \qquad \text{...(1)}$$

where, B_1, B_2 are arbitrary constants.

The expression (1) can also be written as

 (a) $B_1 e^{\alpha x} \cos(\beta x + B_2)$ 　　　　　　　　　　　　　 (b) $B_1 e^{\alpha x} \sin(\beta x + B_2)$.

If, the equation has two equal pair of complex roots $\alpha + i\beta$ and $\alpha - i\beta$, say, occur twice, then the corresponding part of C.F. is written as

$$e^{\alpha x}[(B_1 + B_2 x)\cos\beta x + (B_3 + B_4 x)\sin\beta x].$$

In general, if $\alpha \pm i\beta$ occur k times, then the corresponding part of the C.F. can be written as

$$e^{\alpha x}\{(B_1 + B_2 x + ... + B_k x^{k-1})\cos\beta x + (B_{k+1} + B_{k+2} x + ... + B_{2k} x^{k-1})\} \sin\beta x$$

where $B_1, B_2, ..., B_k, B_{k+1}, ..., B_{2k}$ are arbitrary constants.

Case (iii) :   If a pair of the roots of the auxiliary equation involves surds, say $\alpha \pm \sqrt{\beta}$, where $\beta > 0$, then the corresponding part of C.F. in one of the following three forms

 (a) $e^{\alpha x}[B_1 \cosh(x\sqrt{\beta}) + B_2 \sinh(x\sqrt{\beta})]$ 　　　　 (b) $B_1 e^{\alpha x} \cosh(x\sqrt{\beta} + B_2)$ 　　　 (c) $B_1 e^{\alpha x} \sinh(x\sqrt{\beta} + B_2)$

REMARKS

- The results obtained in case (iii), are exactly similar to those of case (ii) except that sin and cos replaced by sinh and cosh respectively.
- The method of finding the complimentary function (C.F.) of the following differential equation of the form

$$(D^n + A_1 D^{n-1} + A_2 D^{n-2} + ... + A_n)y = 0$$

can be concluded as follows :

S. No.	Nature of the Roots	Solution
1.	Real and distinct, say $m_1, m_2, ..., m_n$	$y = B_1 e^{m_1 x} + B_2 e^{m_2 x} ... + B_n e^{m_n x}$
2.	Real and equal, say m_1	$y = (B_1 + B_2 x + B_3 x^2 + ... + B_n x^{n-1})e^{m_1 x}$
3.	Non-repeated roots : $\alpha \pm i\beta$	(a) $y = (B_1 \cos \beta x + B_2 \sin \beta x)e^{\alpha x}$ (b) $y = B_1 e^{\alpha x} \cos(\beta x + B_2)$
4.	Repeated roots : $\alpha \pm i\beta$, r times	$y = \{(B_1 + B_2 x + ... + B_r x^{r-1})\cos \beta x + (B_1' + B_2' x + ... + B_r' x^{r-1})\sin \beta x\}e^{\alpha x}$
5.	Irrational roots : $\alpha \pm \sqrt{\beta}$	(a) $y = B_1 e^{\alpha x} \cosh(x\sqrt{\beta} + B_2)$ (b) $y = B_1 e^{\alpha x} \sinh(x\sqrt{\beta} + B_2)$

Solved Examples

EXAMPLE 1. **Solve** $[D^3 + 6D^2 + 11D + 6] y = 0$.

SOLUTION. Here, the given differential equation is
$$[D^3 + 6D^2 + 11D + 6] y = 0$$
To find the auxiliary equation, replace D by m, then (1) becomes,
$$m^3 + 6m^2 + 11m + 6 = 0$$
$$\Rightarrow (m+1)(m^2 + 5m + 6) = 0$$
$$\Rightarrow (m+1)(m+2)(m+3) = 0$$
$$\Rightarrow m = -1, -2, -3$$
i.e., Roots are real and unequal. Hence, the general solution is
$$y = C_1 e^{-x} + C_2 e^{-2x} + C_3 e^{-3x} .$$

EXAMPLE 2. **Solve** $[D^4 + 2D^3 - 3D^2 - 4D + 4] y = 0$.

SOLUTION. Here, the auxiliary equation is
$$m^4 + 2m^3 - 3m^2 - 4m + 4 = 0$$
or $\quad (m-1)(m^3 + 3m^2 - 4) = 0$
$$\Rightarrow (m-1)(m-1)(m^2 + 4m + 4) = 0$$
$$\Rightarrow (m-1)(m-1)(m+2)^2 = 0$$
$$\Rightarrow m = +1, +1, -2, -2$$
$\Rightarrow$ Repeated real roots exist.
Hence, general solution is
$$y_1 = (C_1 + C_2 x)e^x + (C_3 + C_4 x) e^{-2x} .$$

EXAMPLE 3. **Solve** $(D^4 + k^4)y = 0$.   [UPTU(Q. Bank)–2001]

SOLUTION. Here, the auxiliary equation is
$$m^4 + k^4 = 0 \quad \text{or} \quad (m^2 + k^2)^2 - 2k^2 m^2 = 0$$

$$\Rightarrow (m^2 + k^2)^2 - (\sqrt{2} . km)^2 = 0$$
$$\Rightarrow (m^2 + k^2 - \sqrt{2}.km)(m^2 + k^2 + \sqrt{2}.km) = 0$$
$$\Rightarrow m^2 - \sqrt{2}.km + k^2 = 0$$
and $\quad m^2 + k^2 + \sqrt{2} . km = 0$
$$\Rightarrow \quad m = \frac{\sqrt{2}k \pm \sqrt{(2k^2 - 4k^2)}}{2}$$
and $\quad m = \frac{-\sqrt{2}k \pm \sqrt{(2k^2 - 4k^2)}}{2}$
$$\Rightarrow m = \frac{k}{\sqrt{2}} \pm i \frac{k}{\sqrt{2}} \quad \text{and} \quad m = -\frac{k}{\sqrt{2}} \pm i \frac{k}{\sqrt{2}}$$
Hence, the solution is
$$y = e^{kx/\sqrt{2}} \{C_1 \cos(kx / \sqrt{2}) + C_2 \sin(kx / \sqrt{2})$$
$$+ e^{-kx/\sqrt{2}} \{C_3 \cos(kx / \sqrt{2}) + C_4 \sin(kx / \sqrt{2})\}$$

EXAMPLE 4. **Solve** $(D^4 - n^4)y = 0$, **where** $D \equiv \dfrac{d}{dx}$.

[UPTU(Q. Bank)–2001]

SOLUTION. The auxiliary equation is
$$m^4 - n^4 = 0 \quad \Rightarrow \quad (m^2 - n^2)(m^2 + n^2) = 0$$
$$\Rightarrow \quad m = \pm n, \pm ni$$
C.F. $= C_1 e^{nx} + C_2 e^{-nx} + e^{0x}(C_3 \cos nx + C_4 \sin nx)$
$$= C_1 e^{nx} + C_2 e^{-nx} + C_3 \cos nx + C_4 \sin nx$$
Hence, the solution is
$$y = C_1 e^{nx} + C_2 e^{-nx} + C_3 \cos nx + C_4 \sin nx .$$

Exercise-8.14

Solve the following equations :

1. $\dfrac{d^2 y}{dx^2} + 3\dfrac{dy}{dx} + 2y = 0$

2. $(D^3 - 9D^2 + 23D - 15)y = 0$

3. $(D^4 - D^3 - 9D^2 - 11D - 4)y = 0$

4. $(D^2 + 1)^2 (D-1)^2 y = 0$

5. $(D^3 - D^2 - 12D)y = 0$

6. $(D^4 + 2n^2 D^2 + n^4)y = 0$

7. $[D^2 - 2\lambda D + (\lambda^2 + \mu^2)]y = 0$

8. $(D^4 + D^3 + 2D^2 - D + 3) y = 0$

9. $(D^5 - 13D^3 + 26D^2 + 82D + 104)y = 0$

10. $\dfrac{d^4 y}{dx^4} + y = 0$

11. $(D^3 - 3D^2 + 4)y = 0$ [UPTU(Q. Bank)–2001]

12. $(D^2 - 2D + 4)^2 y = 0$ [UPTU(Q. Bank)–2001]

13. $\dfrac{d^2 x}{dt^2} + 5\dfrac{dx}{dt} + 6x = 0$, given $x(0) = 0, \dfrac{dx(0)}{dt} = 15$ (VTU–2010)

14. $(D^4 - 4D + 4)y = 0$ (Bhopal–2008)

15. $(D^2 + 1)^3 y = 0$, where $D \equiv d/dx$

16. $\dfrac{d^2 x}{dt^2} - 4\dfrac{dx}{dt} + 13x = 0, x(0), \dfrac{dx(0)}{dt} = 2$ (VTU–2008)

17. $\dfrac{d^3 y}{dx^3} + y = 0$ (VTU–2000S)

18. $\dfrac{d^4 y}{dx^4} + 8\dfrac{d^2 y}{dx^2} + 16y = 0$ (JNTU–2005)

19. $(4D^4 - 8D^3 - 7D^2 + 11D + 6)y = 0$ (VTU–2008)

Hints to Selected Problems

1. $m = -1, -2$ **2.** $m = 1, 5, 3$

3. $m = -1, -1, -1, 4$ **4.** $m = \pm i, \pm i, 1, 1$

5. $m = 0, 4, 3$ **6.** $m = \pm ni, \pm ni$

7. $m = \lambda \pm \mu i$ **8.** $m = -1 \pm i\sqrt{2}, \dfrac{1}{2} \pm i\dfrac{\sqrt{3}}{2}$

9. $m = -1 \pm i, -3 \pm 2i, -4$ **10.** $m = \dfrac{-1 \pm i}{\sqrt{2}}, \dfrac{1 \pm i}{\sqrt{2}}$

11. $m = -1, 2, 2$ **12.** $m = 1 \pm \sqrt{3}\,i, 1 \pm \sqrt{3}\,i$

Answers

1. $y = C_1 e^{-x} + C_2 e^{-2x}$ **2.** $y = C_1 e^x + C_2 e^{3x} + C_3 e^{5x}$ **3.** $y = e^{-x}(C_1 + C_2 x + C_3 x^2) + C_4 e^{4x}$

4. $y = (C_1 + C_2 x)\sin x + (C_3 + C_4 x)\cos x + (C_5 + C_6 x)e^x$ **5.** $y = C_1 + C_2 e^{4x} + C_3 e^{-3x}$

6. $y = (C_1 - C_2 x)\cos nx + (C_3 + C_4 x)\sin nx$ **7.** $y = e^{\lambda x}(C_1 \cos \mu x + C_2 \sin \mu x)$

8. $y = e^{-x}\left[C_1 \cos(\sqrt{2}x) + C_2 \sin(\sqrt{2}x) + e^{x/2}\left(C_3 \cos\dfrac{\sqrt{3}}{2}x + C_4 \sin\dfrac{\sqrt{3}}{2}x \right) \right]$

9. $y = C_1 e^{-x}\cos(x + \alpha) + C_2 e^{-3x}\cos(2x + \beta) + C_3 e^{-4x}$ **10.** $y = C_1 e^{x/\sqrt{2}}\cos\left(\dfrac{x}{\sqrt{2}} + C_2 \right) + C_3 e^{-x/\sqrt{2}}\cos\left(\dfrac{x}{\sqrt{2}} + C_4 \right)$

11. $y = C_1 e^{-x} + (C_2 + C_3 x)e^{2x}$ **12.** $y = e\left[(C_1 + C_2 x)\cos\sqrt{3}x + (C_3 + C_4 x)\sin\sqrt{3}\,x \right]$ **13.** $x = 15(e^{-2t} - e^{-3t})$

14. $y = ((C_1 + C_2 x)e^{\sqrt{2}x} + (C_3 + C_4 x)e^{-\sqrt{2}x})$ **15.** $y = (C_1 + C_2 x + C_3 x^2)\cos x + (C_4 + C_5 x + C_6 x^2)\sin x$ **16.** $\dfrac{2}{3}e^{2t}\sin 3t$

17. $y = C_1 e^{-x} + e^{x/2}\left(C_2 \cos\dfrac{\sqrt{3x}}{2} + C_3 \sin\dfrac{\sqrt{3x}}{2} \right)$ **18.** $y = (C_1 + C_2 x)\cos 2x + (C_3 + C_4 x)\sin 2x$

19. $y = C_1 e^{-x} + C_2 e^{2x} + e^{x/2}\left(C_3 \cos\dfrac{x}{\sqrt{2}} + C_4 \sin\dfrac{x}{\sqrt{2}} \right)$

8.2.3 Particular Integral

Consider the differential equation

$$f(D)y = B \implies y = \frac{1}{f(D)} \cdot B$$

Let $\dfrac{1}{f(D)} \cdot B$ denote some function of x, which operated upon by $f(D)$ produces B. Hence, P.I. $= \dfrac{1}{f(D)} \cdot B$.

(I) GENERAL METHOD OF FINDING P.I.

THEOREM 1. *If B is a function of x, then* $\dfrac{1}{D-a}B = e^{ax}\int B e^{-ax}\,dx$.

PROOF.   Let $y = \dfrac{1}{D-a}B \implies (D-a)y = B \implies \left(\dfrac{d}{dx} - a \right)y = B \implies \dfrac{dy}{dx} - ay = B$

which is the linear differential equation. I.F. $= e^{-\int a\,dx} = e^{-ax}$.

Hence, solution is given by $ye^{-ax} = \int B e^{-ax}dx$.   (Since we find the P.I., therefore we omit the constant of integration.)

$$\therefore \quad y = e^{ax}\int B e^{-ax}\,dx \qquad\qquad \implies \qquad \frac{1}{D-a} \cdot B = e^{ax}\int B e^{-ax}dx$$

REMARKS

- P.I. never contains any arbitrary constant.
- The method discussed above can be used to evaluate P.I. in any problem. It does not depend upon the form of B.
- The method discussed above must be used when B is of the form $\sec ax$, $\csc ax$, $\tan ax$, etc.
- Here, the operator $\dfrac{1}{f(D)}$ (known as increase operator) having the following properties :

(a) If $B = u_1 + u_2 + ... + u_n$, then $\dfrac{1}{f(D)}.B = \dfrac{1}{f(D)}.u_1 + \dfrac{1}{f(D)}.u_2 + ... \dfrac{1}{f(D)}.u_n$

(b) $\dfrac{1}{f(D)}(KB) = \dfrac{K}{f(D)}.B$

(c) $\dfrac{1}{f(D)}$ can be resolved into factors.

(d) $\dfrac{1}{f(D)}$ can be broken into partial fractions.

(e) $\dfrac{1}{f(D)}.B$ is a particular integration.

Solved Examples

EXAMPLE 1. Solve $D^2 - 5D + 6 = e^{3x}$.

SOLUTION. The given equation can be written as

$$(D-3)(D-2)y = e^{3x}$$

$$\text{C.F.} = C_1 e^{3x} + C_2 e^{2x}$$

and

$$\text{P.I.} = \frac{1}{D-3}.\frac{1}{D-2}e^{3x} = \frac{1}{D-3}e^{2x}\int e^{3x}e^{-2x}dx$$

$$= \frac{1}{D-3}e^{2x}.e^x = e^{3x}\int e^{3x}.e^{-3x}dx = xe^{3x}.$$

Now, general solution = C.F. + P.I.

$$\Rightarrow \quad y = C_1 e^{3x} + C_2 e^{2x} + xe^{3x}.$$

EXAMPLE 2. Solve $(D^2 + 1)y = \sec^2 x$.

SOLUTION. Here, the given equation is

$$(D^2 + 1)y = \sec^2 x \qquad ...(1)$$

To find the C.F. of (1).

The auxiliary equation of (1) is given by

$$m^2 + 1 = 0 \Rightarrow m = \pm i$$

$$\Rightarrow \text{C.F.} = C_1 \cos x + C_2 \sin x$$

$$\text{P.I.} = \frac{1}{D^2 + 1}\sec^2 x$$

$$= \frac{1}{(D+i)(D-i)}\sec^2 x = \frac{1}{2i}\left[\frac{1}{D-i} - \frac{1}{D+i}\right]\sec^2 x$$

$$= \frac{1}{2i}\left[e^{xi}\int e^{-ix}\sec^2 x\, dx - e^{-ix}\int e^{ix}\sec^2 x\, dx\right]$$

$$= \frac{1}{2i}\left\{e^{ix}\int \frac{\cos x - i\sin x}{\cos^2 x}dx - e^{-ix}\int \frac{\cos x + i\sin x}{\cos^2 x}dx\right\}$$

$$= \frac{1}{2i}\left\{e^{ix}\int (\sec x - i\sec x\tan x)dx\right.$$

$$\left. -e^{-ix}\int (\sec x + i\sec x\tan x)dx\right\}$$

$$= \frac{1}{2i}\left\{(e^{ix} - e^{-ix})\int \sec x\, dx - i(e^{ix} + e^{-ix})\right.$$

$$\left. \int \tan x\sec x\, dx\right\}$$

$$= \frac{1}{2i}\{2i\sin x\log(\sec x + \tan x) - 2i\cos x\sec x\}$$

$$= \sin x\log(\sec x + \tan x) - 1.$$

Hence, the general solution is $y = $ C.F. + P.I.

$$\Rightarrow \quad y = C_1 \cos x + C_2 \sin x$$

$$+ \sin x\log(\sec x + \tan x) - 1.$$

EXAMPLE 3. Solve $(D^2 + 9)y = \sec 3x$.

SOLUTION. Auxiliary equation is $m^2 + 9 = 0 \Rightarrow m = \pm 3i$

$$\therefore \qquad \text{C.F.} = c_1 \cos 3x + c_2 \sin 3x$$

$$\text{P.I.} = \frac{\sec 3x}{D^2 + 9} = \frac{\sec 3x}{(D+3i)(D-3i)}$$

$$= \frac{1}{6i}\left[\frac{1}{D-3i} - \frac{1}{D+3i}\right]\sec 3x$$

$$= \frac{1}{6i}\left[e^{3ix}\int e^{-3ix}\sec 3x\, dx - e^{-3ix}\int e^{3ix}\sec 3x\, dx\right]$$

$$= \frac{1}{6i}\left[e^{3ix}\left\{\int\left(1 - i\frac{\sin 3x}{\cos 3x}\right)dx\right\} - e^{-3ix}\left\{\int\left(1 + i\frac{\sin 3x}{\cos 3x}\right)dx\right\}\right]$$

$$= \frac{1}{6i}\left[e^{3ix}\left\{x + \frac{i}{3}\log\cos 3x\right\} - e^{-3ix}\left\{x - \frac{i}{3}\log\cos 3x\right\}\right]$$

$$= \frac{1}{6i}\left[(\cos 3x + i\sin 3x)\left(x + \frac{i}{3}\log\cos 3x\right)\right.$$

$$\left. -(\cos 3x - i\sin 3x)\left(x - \frac{i}{3}\log\cos 3x\right)\right]$$

$$= \frac{1}{6i}\left[\frac{2i}{3}\cos 3x\log\cos 3x + 2i\sin 3x\right].$$

$$= \frac{1}{9}\cos 3x\log\cos 3x + \frac{x}{3}\sin 3x$$

Hence, $y = $ C.F. + P.I.

$$= c_1 \cos 3x + c_2 \sin 3x$$

$$+ \frac{x}{3}\sin 3x + \frac{1}{9}\cos 3x\log\cos 3x.$$

Exercise-8.15

Solve the following differential equations :

1. $(D^2 + a^2)y = \sec ax$ (MTU(B.PHARMA.)–2011, UKTU–2011)

2. $(D^2 + a^2)y = \tan ax$

3. $(D^2 + 1)y = \operatorname{cosec} x$

4. $(D^2 + n^2)y = \cot nx$

5. $(D^2 + n^2)y = \tan nx$

 **Hints to Selected Problems**

1. $m = \pm ai \Rightarrow$ C.F. $= C_1 \cos ax + C_2 \sin ax$

$$\text{P.I.} = \frac{1}{D^2 + a^2} \sec ax = \frac{1}{(D+ai)(D-ai)} \sec ax$$

$$= \frac{1}{2ai}\left[\frac{1}{(D-ai)} - \frac{1}{(D+ai)}\right] \sec ax$$

$$= \frac{1}{2ai}\left\{e^{iax}\int e^{-iax} \sec ax\, dx - e^{-iax}\int e^{iax} \sec ax\, dx\right\}$$

2. P.I. $= \dfrac{1}{D^2 + a^2} \tan ax = \dfrac{1}{2ai}\left[\dfrac{1}{D-ia} - \dfrac{1}{D+ia}\right] \tan ax$.

Then proceed as above.

3. P.I. $= \dfrac{1}{D^2 + 1} \operatorname{cosec} x$

$$= \frac{1}{(D+i)(D-i)} \operatorname{cosec} x = \frac{1}{2i}\left[\frac{1}{D-i} + \frac{1}{D+i}\right] \operatorname{cosec} x \ .$$

Now proceed as above.

4. C.F. $= C_1 \cos nx + C_2 \sin nx$

$$\text{P.I.} = \frac{1}{D^2 + n^2} \cot nx = \frac{1}{(D-in)(D+in)} \cot nx$$

$$= \frac{1}{2in}\left[\frac{1}{D-in} \cot nx - \frac{1}{D+in} \cot nx\right]$$

Answers

1. $y = C_1 \cos ax + C_2 \sin ax + \dfrac{x}{a} \sin ax + \dfrac{1}{a^2} \cos ax \log \cos ax$

2. $y = C_1 \cos ax + C_2 \sin ax - \dfrac{1}{a^2} \cos ax \log \tan\left(\dfrac{\pi}{4} + \dfrac{ax}{2}\right)$

3. $y = C_1 \cos x + C_2 \sin x + \sin x \log \sin x - x \cos x$

4. $y = C_1 \cos nx + C_2 \sin nx + \dfrac{1}{n^2} \sin nx \log(\operatorname{cosec} nx - \cot nx)$

5. $y = C_1 \cos nx + C_2 \sin nx - \dfrac{1}{n^2} \cos nx \log(\sec nx + \tan nx)$

(2) SHORT METHODS OF GETTING P.I.

The general method for getting P.I. discussed above requires lot of calculations. In certain cases, the P.I. can be obtained by methods which are shorter than the general method.

(1) To evaluate P.I., when B is of the form e^{ax} :

Here, we want to evaluate $\dfrac{1}{f(D)} e^{ax}$ where, $\qquad f(D) = A_0 D^n + A_1 D^{n-1} + ... + A_n$ with $f(a) \neq 0$.

Here, $\qquad B = e^{ax}$, we have

$$D(e^{ax}) = ae^{ax}$$

$$D^2(e^{ax}) = a^2 e^{ax}$$

$$.................$$

$$.................$$

$$D^n(e^{ax}) = a^n e^{ax}$$

$\Rightarrow \qquad f(D)e^{ax} = (A_0 D^n + A_1 D^{n-1} + ... + A_n)e^{ax} = A_0 D^n e^{ax} + A_1 D^{n-1} e^{ax} + ... + A_n e^{ax}$

$\qquad\qquad\qquad = A_0 a^n e^{ax} + A_1 a^{n-1} e^{ax} + ... + A_n e^{ax} = (A_0 a^n + A_1 a^{n-1} + ... + A_n)\, e^{ax}$

$\Rightarrow \qquad f(D)\, e^{ax} = f(a)\, e^{ax}$.

Operating upon both sides with $\dfrac{1}{f(D)}$, we get $\dfrac{1}{f(D)} . f(D).e^{ax} = \dfrac{1}{f(D)} . f(a)\, e^{ax}$ $\qquad \Rightarrow \qquad e^{ax} = f(a)\, \dfrac{1}{f(D)}\, e^{ax}$

$\Rightarrow \qquad \dfrac{1}{f(D)} e^{ax} = \dfrac{e^{ax}}{f(a)}$, provided $f(a) \neq 0$.

 Solved Examples

EXAMPLE 1. *Solve* $(D^2 - 3D + 2)y = e^{5x}$.

SOLUTION. The given equation is

$$(D^2 - 3D + 2)y = e^{5x}$$

Auxiliary equation is $m^2 - 3m + 2 = 0$

$\Rightarrow \quad (m-1)(m-2) = 0 \Rightarrow m = 1, 2$.

$\therefore \quad$ C.F. $= C_1 e^x + C_2 e^{2x}$

Now, P.I. $= \dfrac{1}{D^2 - 3D + 2} \cdot e^{5x} = \dfrac{1}{25 - 3 \times 5 + 2} e^{5x}$

$= \dfrac{1}{12} e^{5x}$

Hence, the general solution is

$y = $ C.F. + P.I.

$\Rightarrow \quad y = C_1 e^x + C_2 e^{2x} + \dfrac{1}{12} \cdot e^{5x}$.

EXAMPLE 2. *Solve* $(D^3 + 1) y = (e^x + 1)^2$.

SOLUTION. The given equation is

$(D^3 + 1) y = (e^x + 1)^2$

The auxiliary equation is $m^3 + 1 = 0$

$\Rightarrow (m + 1)(m^2 - m + 1) = 0 \Rightarrow m = -1, \dfrac{1}{2} \pm \dfrac{i\sqrt{3}}{2}$

Therefore,

C.F. $= C_1 e^{-x} + e^{x/2} \left[C_2 \cos\left(\dfrac{x\sqrt{3}}{2} \right) + C_3 \sin\left(\dfrac{x\sqrt{3}}{2} \right) \right]$.

Now,

P.I. $= \dfrac{1}{(D^3 + 1)} [e^x + 1]^2 = \dfrac{1}{(D^3 + 1)} (e^{2x} + 2e^x + 1)$

$= \dfrac{1}{D^3 + 1} (e^{2x} + 2e^x + e^{0x})$

$= \dfrac{1}{D^3 + 1} e^{2x} + 2 \dfrac{1}{D^3 + 1} e^x + \dfrac{1}{D^3 + 1} e^{0x}$

$= \dfrac{1}{2^3 + 1} e^{2x} + 2 \dfrac{1}{1^3 + 1} e^x + \dfrac{1}{0 + 1} e^{0x} = \dfrac{1}{9} e^{2x} + e^x + 1$

Here, the general solution is

$y = $ C.F. + P.I.

$\Rightarrow \quad y = C_1 e^{-x} + e^{x/2} \left[C_2 \cos\left(\dfrac{x\sqrt{3}}{2} \right) + C_3 \sin\left(\dfrac{x\sqrt{3}}{2} \right) \right]$

$+ \dfrac{1}{9} e^{2x} + e^x + 1$

EXAMPLE 3. *Solve* $(D^3 - 2D^2 + 4D - 8) y = 8$.

(UPTU(B.Pharm)SUM–2009)

SOLUTION. The given equation is

$(D^3 - 2D^2 + 4D - 8) y = 8$

Auxiliary equation is

$m^3 - 2m^2 + 4m - 8 = 0$

$\Rightarrow \quad (m^2 + 4)(m - 2) = 0$

$\Rightarrow \quad m = 2, \pm 2i$

$\therefore$ C.F. $= C_1 e^{2x} + C_2 \cos 2x + C_3 \sin 2x$

Now P.I. $= \dfrac{1}{D^3 - 2D^2 + 4D - 8} (8 e^{0x}) \qquad (\because e^{0x} = 1)$

$= \dfrac{1}{(0)^3 - 2(0)^2 + 4(0) - 8} (8 e^{0x}) = -1$

Hence, the general solution is

$y = $ C.F. + P.I.

$\Rightarrow \quad y = C_1 e^{2x} + C_2 \cos 2x + C_3 \sin 2x - 1$

EXAMPLE 4. *Solve* $(D - 2)^3 y = 17 e^{2x}$ (MTU–2011)

SOLUTION. The given equation is

$(D - 2)^3 y = 17 e^{2x}$

Auxiliary equation is

$(m - 2)^3 = 0$

$\Rightarrow \quad m = 2, 2, 2$

$\therefore$ C.F. $= (C_1 + C_2 x + C_3 x^2) e^{2x}$

Now P.I. $= \dfrac{1}{(D - 2)^3} 17 e^{2x}$ |Case of failure

$= 17 x \left[\dfrac{1}{3(D - 2)^2} e^{2x} \right]$

|Again case of failure

$= \dfrac{17}{3} x^2 \left[\dfrac{1}{2(D - 2)} e^{2x} \right]$

|Again case of failure

$= \dfrac{17}{6} x^3 e^{2x}$

EXAMPLE 5. *Solve* $2\dfrac{d^3 y}{dx^3} - \dfrac{d^2 y}{dx^2} + 4\dfrac{dy}{dx} - 2y = e^x$ (UPTU–2007)

SOLUTION. The given equation can be written as

$(2D^3 - D^2 + 4D - 2) y = e^x$

Auxiliary equation is

$2m^3 - m^2 + 4m - 2 = 0$

$\Rightarrow \quad (2m - 1)(m^2 + 2) = 0$

$\Rightarrow \quad m = \dfrac{1}{2}, \pm \sqrt{2} i$

$\therefore$ C.F. $= C_1 e^{x/2} + C_2 \cos \sqrt{2} x + C_3 \sin \sqrt{2} x$

Now P.I. $= \dfrac{1}{2D^3 - D^2 + 4D - 2} e^x$

$= \dfrac{1}{2(1)^3 - (1)^2 + 4(1) - 2} e^x = \dfrac{1}{3} e^x$

Hence, the complete solution is

$y = $ C.F. + P.I.

$= C_1 e^{x/2} + C_2 \cos \sqrt{2} x + C_3 \sin \sqrt{2} x + \dfrac{1}{3} e^x$

Exercise-8.16

Solve the following differential equations :

1. $(D^2 - 4D + 1) y = e^{2x} - e^{-x}$

2. $(D^2 + 5D + 6) y = e^{2x}$

3. $(4D^2 + 4D - 3) y = e^{2x}$

4. $(D^2 - 2D + 1) y = 2 e^{5x/2}$

5. $(D^2 + D + 1) y = e^{-x}$

6. $D^2 (D + 1)^2 (D^2 + D + 1)^2 y = e^x$

7. $[D^2 + 2pD + (p^2 + q^2)] y = e^{ax}$

8. $(4D^2 + 12D + 9) y = 144 e^{-3x}$

9. $(D^2 - 4D + 3) y = e^{3x}$ [UPTU(Q. Bank)–2001]

10. $(D^2 - a^2) y = e^{ax} - e^{-ax}$ [UPTU(Q. Bank)–2001]

11. $(D^2 + D + 1) y = (1 + e^x)^2$ [UPTU(Q. Bank)–2001]

12. $\dfrac{d^3 y}{dx^3} - 3\dfrac{d^2 y}{dx^2} + 3\dfrac{dy}{dx} - y = e^x + 2$ [UPTU(Q. Bank)–2001]

Hints to Selected Problems

1. $m = 2 \pm \sqrt{3} \Rightarrow$ C.F. $= e^{2x}[C_1 \cosh x\sqrt{3} + C_2 \sinh x\sqrt{3}]$

P.I. $= \dfrac{1}{D^2 - 4D + 1}[e^{2x} - e^{-x}]$

$= \dfrac{1}{D^2 - 4D + 1}e^{2x} - \dfrac{1}{(D^2 - 4D + 1)}e^{-x}$

$= \dfrac{e^{2x}}{2^2 - 4 \times 2 + 1} - \dfrac{e^{-x}}{(-1)^2 - 4(-1) + 1} = -\dfrac{e^{2x}}{3} - \dfrac{e^{-x}}{6}$

4. $m = 1, 1 \Rightarrow$ C.F. $= (C_1 + C_2 x)e^x$

P.I. $= \dfrac{1}{D^2 - 2D + 1}(2e^{5x/2}) = 2 \cdot \dfrac{e^{5x/2}}{\frac{25}{4} - 4}$.

6. $m = 0, 0, -1, -1, -\dfrac{1}{2} \pm \dfrac{i\sqrt{3}}{2}, -\dfrac{1}{2} \pm \dfrac{i\sqrt{3}}{2}$.

7. $m = -p \pm iq$.

8. $m = -\dfrac{3}{2}, -\dfrac{3}{2} \Rightarrow$ C.F. $= (C_1 + C_2 x)e^{-3x/2}$

P.I. $= 144\left(\dfrac{1}{4D^2 + 12D + 9}\right)e^{-3x} = \dfrac{144e^{-3x}}{9}$

9. P.I. $= \dfrac{1}{D^2 - 4D + 3}e^{3x} = \dfrac{1}{2D - 4}e^{3x} = \dfrac{x}{2} \cdot e^{3x}$

10. $m = \pm a$

11. $m = -\dfrac{1}{2} \pm \dfrac{\sqrt{3}}{2}i$

12. $m = 1, 1, 1$

 Answers

1. $y = e^{2x}(C_1 \cosh x\sqrt{3} + C_2 \sinh x\sqrt{3}) - \dfrac{1}{3}e^{2x} - \dfrac{1}{6}e^{-x}$

2. $y = C_1 e^{-2x} + C_2 e^{-3x} + \dfrac{1}{20}e^{2x}$

3. $y = C_1 e^{x/2} + C_2 e^{-3x/2} + \dfrac{1}{21}e^{2x}$ **4.** $y = (C_1 + C_2 x)e^x + \dfrac{8}{9}e^{5x/2}$ **5.** $y = e^{-x/2}\left[C_1 \cos\left(\dfrac{1}{2}x\sqrt{3}\right) + C_2 \sin\left(\dfrac{1}{2}x\sqrt{3}\right)\right] + e^{-x}$

6. $y = (C_1 + C_2 x)e^{0x} + (C_3 + C_4 x)e^{-x} + e^{-x/2}\left[(C_5 + C_6 x)\cos\left(\dfrac{1}{2}\sqrt{3}x\right) + (C_7 + C_8 x)\sin\left(\dfrac{1}{2}\sqrt{3}x\right)\right] + \dfrac{1}{36}e^x$

7. $y = e^{-px}(C_1 \cos qx + C_2 \sin qx) + \dfrac{e^{ax}}{[(p+a)^2 + q^2]}$

8. $y = (C_1 + C_2 x)e^{-3x/2} + 16e^{-3x}$

9. $y = C_1 e^x + C_2 e^{3x} + \dfrac{x}{2}e^{3x}$

10. $y = C_1 e^{ax} + C_2 e^{-ax} + \dfrac{x}{9}\cosh ax$

11. $y = e^{-x/2}\left[C_1 \cos\dfrac{\sqrt{3}}{2}x + C_2 \sin\dfrac{\sqrt{3}}{2}x\right] + 1 + \dfrac{1}{7}e^{2x} + \dfrac{2}{3}e^x$

12. $y = (C_1 + C_2 x + C_2 x^2)e^x + \dfrac{x^3}{6}e^x - 2$

(2) To evaluate P.I., when B is of the form $\sin ax$ or $\cos ax$:

Case (I) : If $f(D)$ contains even power of D :

Let us suppose

$$f(D^2) = A_0(D^2)^n + A_1(D^2)^{n-1} + \dots + A_n.$$

Here, we observe that

$$D^2 \sin ax = -a^2 \sin ax$$
$$D^4 \sin ax = (-a^2)^2 \sin ax$$
$$D^6 \sin ax = (-a^2)^3 \sin ax$$
$$\dots\dots\dots\dots\dots\dots$$
$$(D^2)^n \sin ax = (-a^2)^n \sin ax$$

Consider $f(D^2)\sin ax = [A_0(D^{2n}) + A_1(D^{2n-2}) + \dots + A_n]\sin ax$

$= A_0 D^{2n} \sin ax + A_1 D^{2n-2} \sin ax + \dots + A_n \sin ax$

$= A_0(-a^2)^n \sin ax + A_1(-a^2)^{n-1}\sin ax + \dots + A_n \sin ax = f(-a^2)\sin ax$

Now, operating on both sides with $\dfrac{1}{f(D^2)}$, we get $\dfrac{1}{f(D^2)} \cdot f(D^2)\sin ax = f(-a^2)\dfrac{1}{f(D^2)}\sin ax$

$\Rightarrow$ $\sin ax = f(-a^2)\left[\dfrac{1}{f(D^2)}\sin ax\right] \Rightarrow \dfrac{1}{f(D^2)}\sin ax = \dfrac{1}{f(-a^2)}\sin ax$.

Case (II) : If $f(D)$ contains odd power of D :

Let us suppose, it be put in the form $f_1(D^2) + f_2(D^2)D$, then

$$\dfrac{1}{f(D)}\sin ax = \dfrac{1}{f_1(D^2) + f_2(D^2)D}\sin ax = \dfrac{1}{f_1(-a^2) + f_2(-a^2)D}\sin ax$$

$$= \frac{1}{p+qD}\sin ax \ \text{(say)} \qquad\qquad \text{(Where } p = f_1(-a^2), q = f_2(-a^2)]$$

$$= (p-qD)\left[\frac{1}{(p-qD)(p+qD)}\sin ax\right] = (p-qD)\left[\frac{1}{p^2 - q^2 D^2}\sin ax\right]$$

$$= (p-qD)\left[\frac{1}{p^2 + q^2 a^2}\sin ax\right] \qquad\qquad \text{(By putting } D^2 = -a^2\text{)}$$

$$= \frac{(p-qD)\sin ax}{(p^2 + a^2 q^2)} = \frac{p\sin ax - qa\cos ax}{p^2 + a^2 q^2}$$

$$\Rightarrow \qquad \frac{1}{f(D)}\sin ax = \frac{f_1(-a^2)\sin ax - f_2(-a^2)a\cos ax}{\{f_1(-a^2)\}^2 + a^2\{f_2(-a^2)\}^2}$$

☞ REMARKS

- To find P.I. $= \dfrac{1}{f(D)}\sin ax$, replace D^2 by $-a^2$ provided $f(-a^2) \neq 0$.

- If the linear factors of D contains the odd powers of D, then first multiplying the numerator and denominator by the conjugate factors $(P \pm qD)$ and then replace D^2 by $(-a^2)$.

- Similar results are true for $\dfrac{1}{f(D)}\cos ax$.

Solved Examples

EXAMPLE 1. Solve $\dfrac{d^2 y}{dx^2} - 3\dfrac{dy}{dx} + 2y = \cos 3x$.

SOLUTION. The given differential equation can be written as

$$(D^2 - 3D + 2)y = \cos 3x \qquad\qquad \text{...(1)}$$

To find C.F., the auxiliary equation is

$$m^2 - 3m + 2 = 0$$
$$\Rightarrow \quad (m-1)(m-2) = 0$$

which gives $m = 1$ and $m = 2$.

Therefore, C.F. $= C_1 e^x + C_2 e^{2x}$.

Now, P.I. $= \dfrac{1}{D^2 - 3D + 2}\cos 3x = \dfrac{1}{-9 - 3D + 2}\cos 3x$

$$[\because D^2 = -a^2 = -9]$$

$$= \frac{1}{-7 - 3D}\cos 3x = -\frac{(7 - 3D)}{(7^2 - 9D^2)}\cos 3x$$

$$= -\frac{(7 - 3D)}{7^2 - 9(-9)}\cos 3x = -\frac{1}{130}[7\cos 3x - 3D\cos 3x]$$

$$= -\frac{7}{130}\cos 3x - \frac{9}{130}\sin 3x = -\frac{1}{130}(7\cos 3x + 9\sin 3x).$$

Hence, the general solution of (1) is given by

$$y = \text{C.F.} + \text{P.I.}$$
$$\Rightarrow \quad y = C_1 e^x + C_2 e^{2x} - \frac{1}{130}[7\cos 3x + 9\sin 3x].$$

EXAMPLE 2. Solve $(D^2 + 4)y = \sin 3x + \cos 2x$. (UPTU(SUM)-2008)

SOLUTION. The given differential equation is

$$(D^2 + 4)y = \sin 3x + \cos 2x \qquad\qquad \text{...(1)}$$

To find the C.F. of (1), the auxiliary equation is

$$m^2 + 4 = 0$$
$$\Rightarrow \qquad m = \pm 2i$$

Therefore, C.F. $= C_1 \cos 2x + C_2 \sin 2x$.

Now, P.I. $= \dfrac{1}{D^2 + 4}\sin 3x + \dfrac{1}{D^2 + 4}(\cos 2x)$

$$= \frac{1}{-(3)^2 + 4}\sin 3x + x \cdot \frac{1}{2D}(\cos 2x)$$

$$= -\frac{1}{5}\sin 3x + \frac{x}{2}\left(\frac{\sin 2x}{2}\right)$$

$$= -\frac{1}{5}\sin 3x + \frac{x}{4}\sin 2x$$

Hence, the general solution of (1) is given by

$$y = \text{C.F.} + \text{P.I.}$$
$$= C_1 \cos 2x + C_2 \sin 2x - \frac{1}{5}\sin 3x + \frac{x}{4}\sin 2x.$$

EXAMPLE 3. Solve $\dfrac{d^2 y}{dx^2} + a^2 y = \sin ax$. (UPTU–2008)

SOLUTION. Here, the given equation can be written as

$$(D^2 + a^2)y = \sin ax \qquad\qquad \text{...(1)}$$

To find the C.F. of (1), the auxiliary equation is

$$m^2 + a^2 = 0 \quad \Rightarrow \quad m = \pm ai$$

Therefore C.F. $= C_1 \cos ax + C_2 \sin ax$.

Now, P.I. $= \dfrac{1}{D^2 + a^2}\sin ax = \dfrac{x}{2D}\sin ax$

$$= \frac{x}{2}\left[\frac{-\cos ax}{a}\right] = -\frac{x}{2a}\cos ax$$

Hence, the general solution of (1) is given by

$$y = \text{C.F.} + \text{P.I.}$$
$$= C_1 \cos ax + C_2 \sin ax - \frac{x}{2a}\cos ax.$$

EXAMPLE 4. Solve $(D^2 + 4)y = \cos^2 x$. (MTU(B.Pharm)–2011)

SOLUTION. Here, the given differential equation is

$$(D^2 + 4)y = \cos^2 x \qquad\qquad \text{...(1)}$$

To find the C.F. of (1), the auxiliary equation is

$$m^2 + 4 = 0 \quad \Rightarrow \quad m = \pm 2i$$
$$\therefore \qquad \text{C.F.} = C_1 \cos 2x + C_2 \sin 2x.$$

Now, P.I. $= \dfrac{1}{D^2 + 4}\cos^2 x$

$$= \frac{1}{2}\left[\frac{1}{D^2+4}(1+\cos 2x)\right]$$

$$= \frac{1}{2}\left[\frac{1}{4}+\frac{x}{4}\sin 2x\right] = \frac{1}{8}(1+x+\sin 2x)$$

Hence, the general solution of (1) is given by

$$y = \text{C.F.} + \text{P.I.}$$

$$= \frac{1}{2}\left[\frac{1}{D^2+4}(e^{0x}) + \frac{1}{D^2+4}(\cos 2x)\right]$$

$$= \frac{1}{2}\left[\frac{1}{4}+x.\frac{1}{2D}(\cos 2x)\right]$$

$$= C_1\cos 2x + C_2\sin 2x + \frac{1}{8}(1+x+\sin 2x)$$

 Exercise-8.17

Solve the following differential equations :

1. $(D^2+9)y = \cos 4x$ **2.** $(D^2-2D+5)y = \sin 3x$

3. $(D^2-3D+2)y = \sin 3x$

4. $(D^4+2D^3-3D^2)y = 3e^{2x}+4\sin x$

5. $(D^3-2D^2+3)y = \cos x$

6. $(D^2+16)y = \sin 2x$, given that $y = 0$ and $\frac{dy}{dx} = \frac{5}{6}$ when $x = 0$.

7. $(D^4-2D^2+1)y = \cos x$ **8.** $(D^2+2D+2)y = \cos 2x$

9. $(D^2-9)y = \sin x + \cos x$

10. Solve $\frac{d^2y}{dx^2} + 2\frac{dy}{dx} + 10y + 37\sin 3x = 0$ and find the value of y

when $x = \frac{\pi}{2}$ being given that $y = 3, \frac{dy}{dx} = 0$ when $x = 0$.

(GBTU–2011)

11. $\frac{d^2y}{dx^2} + 4y = e^x + \sin 2x$ (UPTU(B.Pharm)–2009, 2010)

12. $(D^2+5D-6)y = \sin 3x + \cos 2x$ (GBTU–2010, GBTU(CO)–2011)

13. $(D^2+5D-6)y = \sin 4x \sin x$ (UPTU(SUM)–2009)

 Hints to Selected Problems

1. $m = \pm 3i \Rightarrow \text{C.F.} = C_1\cos 3x + C_2\sin 3x$

$$\text{P.I.} = \frac{1}{D^2+9}\cos 4x = \frac{1}{-4^2+9}\cos 4x = -\frac{1}{7}\cos 4x .$$

5. $m = -1, \frac{3}{2} \pm \frac{i\sqrt{3}}{2}$

$$\Rightarrow \text{C.F.} = C_1e^{-x} + e^{3/2.x}\left[C_2\cos\left(\frac{\sqrt{3}}{2}x\right) + C_3\sin\left(\frac{\sqrt{3}}{2}x\right)\right]$$

$$\text{P.I.} = \frac{1}{(D+1)(D^2-3D+3)}\cos x .$$

9. $\text{C.F.} = C_1e^{3x} + C_2e^{-3x}$

$$\text{P.I.} = \frac{1}{D^2-9}(\sin x + \cos x) = \frac{1}{D^2-9}\sin x + \frac{1}{D^2-9}\cos x$$

$$= \frac{1}{-1-9}\sin x + \frac{1}{-1-9}\cos x = -\frac{1}{10}\sin x - \frac{1}{10}\cos x$$

Answers

1. $y = C_1\cos 3x + C_2\sin 3x - \frac{1}{7}\cos 4x$

2. $y = e \ [C_1\cos 2x + C_2\sin 2x] + \frac{1}{26}(3\cos 3x - 2\sin 3x)$

3. $y = C_1e^x + C_2e^{2x} + \frac{1}{130}(9\cos 3x - 7\sin 3x)$

4. $y = (C_1 + C_2x) + C_3e^x + C_4e^{-3x} + \frac{3}{20}e^{2x} + \frac{4}{5}\sin x + \frac{2}{5}\cos x$

5. $y = C_1e^{-x} + \left\{C_2\cos\left(\frac{x\sqrt{3}}{2}\right) + C_3\sin\left(\frac{x\sqrt{3}}{2}\right)\right\}e^{3x/2} + \frac{1}{26}[5\cos x - \sin x]$

6. $y = \frac{1}{6}\sin 4x + \frac{1}{12}\sin 2x$

7. $y = (C_1 + C_2x)e^x + (C_3 + C_4x)e^{-x} + \frac{1}{4}\cos x$

8. $y = e^{-x}[C_1\cos x + C_2\sin x] - \frac{1}{10}(\cos 2x - 2\sin 2x)$

9. $y = C_1e^{3x} + C_2e^{-3x} - \frac{1}{10}[\sin x + \cos x]$ **10.** $y = e^{-x}(C_1\cos 3x + C_2\sin 3x) + 6\cos 3x - \sin 3x$ and $y = 1$ at $x = \pi/2$

11. $y = C_1\cos 2x + C_2\sin 2x + \frac{1}{5}e \ - \frac{1}{4}\cos 2x$

12. $y = C_1e^x + C_2e^{-6x} - \frac{1}{30}(\cos 3x + \sin 3x) + \frac{1}{20}(\sin 2x - \cos 2x)$

13. $y = C_1e^x + C_2e^{-6x} + \frac{1}{2}\left[\frac{\sin 3x - \cos 3x}{30} + \frac{31\cos 5x - 25\sin 5x}{1586}\right]$

(3) To evaluate P.I., when B is of the form x^m, when m is positive integer :

i.e., to evaluate $\frac{1}{f(D)}x^m$, $m \in Z^+$ and $f(D) = A_0D^n + A_1D^{n-1} + ... + A_n$

Let us consider $\frac{1}{D-a}x^m$

i.e., $\frac{1}{(D-a)}x^m = e^{ax}\int e^{-ax}x^m\,dx = e^{ax}\left\{\frac{e^{-ax}x^m}{a} - \frac{mx^{m-1}e^{-ax}}{a^2} - \frac{m(m-1)x^{m-2}e^{-ax}}{a^3} - ... - \frac{m(m-1)...2.1\,e^{-ax}}{a^{m+1}}\right\}$ ...(1)

If we expand $\dfrac{1}{D-a}$ in powers of D, we get

$$\frac{1}{(D-a)}x^m = -\frac{1}{a(1-D/a)}x^m = -\frac{1}{a}\left[1 + \frac{D}{a} + \frac{D^2}{a^2} + \dots\right]x^m$$

$$\frac{1}{D-a}x^m = -\frac{1}{a}\left[x^m + \frac{mx^{m-1}}{a} + \frac{m(m-1)x^{m-2}}{a^2} + \dots + \frac{m(m-1)\dots 2.1}{a^m}\right] \qquad \dots(2)$$

Here, we observe that (1) and (2) are the same.

WORKING PROCEDURE

Take the lowest degree term from $f(D)$ and remaining factor will be of the form $[1 + f(D)]$ or $[1 - f(D)]$. Now, this factor can be taken in the numerator with a negative index, which can be expanded by Binomial theorem. Here, it should be noted that the expansion is to be carried upto the term D^m, since we always have $D^{m+1}x^m = 0$, $D^{m+2}x^m = 0$ and all other higher differential coefficients of x^m are zero.

SOME IMPORTANT EXPANSIONS (TO BE USED DIRECTLY)

1. $[1+x]^n = 1 + nx + \dfrac{n(n-1)}{2!}x^2 + \dfrac{n(n-1)(n-2)}{3!}x^3 + \dots$

2. $(1+x)^{-1} = 1 - x + x^2 - x^3 + x^4 - x^5 + \dots$ **3.** $(1-x)^{-1} = 1 + x + x^2 + x^3 + x^4 + \dots$

4. $(1-x)^{-2} = 1 + 2x + 3x^2 + 4x^3 + \dots$ **5.** $(1+x)^{-2} = 1 - 2x + 3x^2 - 4x^3 + \dots$

Solved Examples

EXAMPLE 1. *Solve* $(D^2 + D - 2)y = x + \sin x$.

SOLUTION. The given equation is

$$(D^2 + D - 2)y = x + \sin x \qquad \dots(1)$$

To find C.F., the auxiliary equation is

$$m^2 + m - 2 = 0$$

$$\Rightarrow \quad (m-1)(m+2) = 0 \Rightarrow m = 1, -2$$

$$\therefore \quad \text{C.F.} = C_1 e^x + C_2 e^{-2x}$$

Now, P.I. $= \dfrac{1}{(D^2 + D - 2)}(x + \sin x)$

$$= \frac{1}{(D^2 + D - 2)}x + \frac{1}{(D^2 + D - 2)}\sin x$$

$$= \frac{1}{-2\left(1 - \frac{1}{2}D - \frac{1}{2}D^2\right)}x + \frac{1}{-1 + D - 2}\sin x$$

$$= -\frac{1}{2}\left[1 - \left(\frac{1}{2}D + \frac{1}{2}D^2\right)\right]^{-1}x + \frac{(D+3)}{(D-3)(D+3)}\sin x$$

$$= -\frac{1}{2}\left(1 + \frac{1}{2}D + \dots\right)x + \frac{(D+3)}{D^2 - 9}\sin x$$

$$= -\frac{1}{2}\left(x + \frac{1}{2}\right) + \frac{D+3}{-1-9}\sin x$$

$$= -\frac{1}{2}\left(x + \frac{1}{2}\right) - \left(\frac{1}{10}\right)[D(\sin x) + 3\sin x]$$

$$= -\frac{1}{2}x - \frac{1}{4} - \frac{1}{10}(\cos x + 3\sin x).$$

Hence, the complete solution is given by

$$y = \text{C.F.} + \text{P.I.}$$

$$\therefore \quad y = C_1 e^x + C_2 e^{-2x} - \frac{1}{2}x - \frac{1}{4} - \frac{1}{10}(\cos x + 3\sin x)$$

EXAMPLE 2. *Solve* $(D^2 - 4D + 4)y = x^2 + e^x + \cos 2x$.

(UPTU(B.Pharma)SUM–2009)

SOLUTION. The given differential equation is

$$(D^2 - 4D + 4)y = x^2 + e^x + \cos 2x \qquad \dots(1)$$

To find C.F., the auxiliary equation is given by

$$m^2 - 4m + 4 = 0 \Rightarrow (m-2)^2 = 0 \Rightarrow m = 2, 2$$

$$\therefore \quad \text{C.F.} = (C_1 + C_2 x)e^{2x}$$

Now, P.I. $= \dfrac{1}{(D^2 - 4D + 4)}(x^2 + e^x + \cos 2x)$

$$= \frac{1}{(D-2)^2}x^2 + \frac{1}{(D-2)^2}e^x + \frac{1}{(D^2 - 4D + 4)}\cos 2x$$

$$= \frac{1}{4\left(1 - \frac{D}{2}\right)^2}x^2 + \frac{1}{(1-2)^2}e^x + \frac{1}{(-2^2 - 4D + 4)}\cos 2x$$

$$= \frac{1}{4}\left(1 - \frac{D}{2}\right)^{-2}x^2 + \frac{e^x}{1} - \frac{1}{4D}\cos 2x$$

$$= \frac{1}{4}\left[1 + D + \frac{3}{4}D^2 + \dots\right]x^2 + e^x - \frac{1}{4}\int \cos 2x \, dx$$

$$= \frac{1}{4}\left(x^2 + D(x^2) + \frac{3}{4}D^2(x^2)\right) + e^x - \frac{1}{4}.\frac{1}{2}\sin 2x$$

$$= \frac{1}{4}\left[x^2 + 2x + \frac{3}{2}\right] + e^x - \frac{1}{8}\sin 2x.$$

Hence, the complete solution is given by

$$y = \text{C.F.} + \text{P.I.}$$

$$\Rightarrow \quad y = (C_1 + C_2 x)e^{2x} + \frac{1}{4}\left(x^2 + 2x + \frac{3}{2}\right) + e^x - \frac{1}{8}\sin 2x.$$

EXAMPLE 3. *Find the solution of the following differential equation:* $(D^2 - 4D - 5)y = e^{2x} + 3\cos(4x + 3)$, *where* $D \equiv \dfrac{d}{dx}$.

(UPTU–2008)

SOLUTION. Here, we have

$$(D^2 - 4D - 5)y = e^{2x} + 3\cos(4x + 3)$$

A.E. is $(m^2 - 4m - 5) = 0$

$$\Rightarrow \quad m^2 - 5m + m - 5 = 0$$

$$\Rightarrow \quad (m-5)(m+1) = 0 \Rightarrow m = -1, 5.$$

$\text{C.F.} = C_1 e^{-x} + C_2 e^{5x}$

$= \dfrac{1}{(D^2 - 4D - 5)}\left[e^{2x} + 3\cos(4x+3)\right]$

$= \dfrac{1}{D^2 - 4D - 5}e^{2x} + 3\dfrac{1}{D^2 - 4D - 5}\cos(4x+3)$

$= \dfrac{e^{2x}}{2^2 - 4(2) - 5} + 3.\dfrac{1}{-16 - 4D - 5}\cos(4x+3)$

$= \dfrac{e^{2x}}{4 - 8 - 5} + 3\dfrac{1}{-4D - 21}\cos(4x+3)$

$= -\dfrac{e^{2x}}{9} - 3\dfrac{1}{4D + 21}\cos(4x+3)$

$= -\dfrac{e^{2x}}{9} - 3\dfrac{4D - 21}{16D^2 - 441}\cos(4x+3)$

$= -\dfrac{e^{2x}}{9} - 3\dfrac{(4D - 21)}{16(-16) - 441}\cos(4x+3)$

$= \dfrac{e^{2x}}{9} + \dfrac{3}{697}(4D - 21)\cos(4x+3)$

$= -\dfrac{e^{2x}}{9} + \dfrac{3}{697}\left[4D\cos(4x+3) - 21\cos(4x+3)\right]$

$= -\dfrac{e^{2x}}{9} + \dfrac{3}{697}\left[-16\sin(4x+3) - 21\cos(4x+3)\right]$

$= -\dfrac{e^{2x}}{9} - \dfrac{3}{697}\left[16\sin(4x+3) + 21\cos(4x+3)\right]$

Complete solution is given by
$y = \text{C.F.} + \text{P.I.}$

$= C_1 e^{-x} + C_2 e^{5x} - \dfrac{e^{2x}}{9}$

$\qquad - \dfrac{3}{697}\left[16\sin(4x+3) + 21\cos(4x+3)\right]$

EXAMPLE 4. Solve $\dfrac{d^3 y}{dx^3} - 3\dfrac{d^2 y}{dx^2} + 4\dfrac{dy}{dx} - 2y = e^x + \cos x$.

(UPTU–2001, 2006)

SOLUTION. The given equation can be written as
$$(D^3 - 3D^2 + 4D - 2)y = e^x + \cos x$$

The auxiliary equation is $m^3 - 3m^2 + 4m - 2 = 0$

or $(m - 1)(m^2 - 2m + 2) = 0$, *i.e.*, $m = 1,\ 1 \pm i$

$\therefore \quad \text{C.F.} = C_1 e^x + e^x(C_2 \cos x + C_3 \sin x)$

$\qquad \text{P.I.} = \dfrac{1}{(D - 1)(D^2 - 2D + 2)}e^x$

$\qquad\qquad + \dfrac{1}{(D^3 - 3D^2 + 4D - 2)}\cos x$

$= \dfrac{1}{(D - 1)(1 - 2 + 2)}e^x$

$\qquad + \dfrac{1}{(-1)D - 3(-1) + 4D - 2}\cos x$

$= \dfrac{1}{(D - 1)}e^x + \dfrac{1}{3D + 1}\cos x$

$= e^x \dfrac{1}{D}.1 + \dfrac{(-3\sin x - \cos x)}{-9 - 1}$

$= e^x.x + \dfrac{1}{10}(3\sin x + \cos x)$

Hence, complete solution is
$\quad y = \text{C.F.} + \text{P.I.}$

$\therefore \quad y = C_1 e^x + e^x(C_2 \cos x + C_3 \sin x) + xe^x$

$\qquad\qquad + \dfrac{1}{10}(3\sin x + \cos x)$

EXAMPLE 5. *A body executes damped forced vibrations given by the equation* $\dfrac{d^2 x}{dt^2} + 2k\dfrac{dx}{dt} + b^2 x = e^{-kt}\sin wt$. *Solve the differential equation for both the cases where* $w^2 \neq b^2 - k^2$ *and when* $w^2 = b^2 - k^2$.

(UPTU–2002)

SOLUTION. The given equation can be written as
$$(D^2 + 2kD + b^2)x = e^{-kt}\sin wt$$

The auxiliary equation is $m^2 + 2km + b^2 = 0$

$\Rightarrow \quad m = \dfrac{-2k \pm \sqrt{4k^2 - 4b^2}}{2}$

$\qquad\quad = -k \pm \sqrt{k^2 - b^2}$

As the given problem is on vibration, we must have
$$k^2 < b^2.$$

$\therefore \quad m = -k \pm \sqrt{-(b^2 - k^2)} = -k \pm i\sqrt{(b^2 - k^2)}$

$\therefore \ \text{C.F.} = e^{-kt}\left\{C_1 \cos\sqrt{(b^2 - k^2)}t + C_2 \sin\sqrt{(b^2 - k^2)}t\right\}$

$\therefore \ \text{P.I.} = \dfrac{1}{D^2 + 2kD + b^2}e^{-kt}\sin wt$

$= e^{-kt}\dfrac{1}{(D - k)^2 + 2k(D - k) + b^2}\sin wt$

$= e^{-kt}\dfrac{1}{b^2 + (b^2 - k^2)}\sin wt = e^{-kt}\dfrac{1}{-w^2 + (b^2 - k^2)}\sin wt,$

$\qquad\qquad\qquad \text{if } w^2 \neq b^2 - k^2$

$= e^{-kt}\,t\,\dfrac{1}{2D}\sin wt = e^{-kt}\left(-\dfrac{t}{2w}\cos wt\right)$

$\qquad\qquad\qquad \text{if } w^2 = b^2 - k^2$

Exercise-8.18

Solve the following differential equations :

1. $(D^3 - D^2 - 6D)y = x^2 + 1$ (UPTU(Q. Bank)–2001)

2. $(D^4 - a^4)y = x^4$

3. $(D^3 + 2D^2 + D)y = e^{2x} + x^2 + x$

4. $(D^3 - 3D - 2)y = x^3$

5. $(D^3 - 3D^2 + 2D)y = 4 + 60e^{5x}$

6. $(D^3 + 1)y = \sin 3x - \cos^2\dfrac{x}{2}$

7. $(D^2 - 2D + 3)y = \cos x + x^2$

8. $(D^2 - 5D + 6)y = x + e^{mx}$

9. $(D^2 + 16)y = \cos 3x + e^{3x} + x^4$.

10. $(D^2 + 4)y = \sin 3x + x^2$

11. $\dfrac{d^2y}{dx^2} - \dfrac{dy}{dx} + 4y = x^2 + e^x$ (UPTU(B.Pharma)SUM–2010)

12. If $\dfrac{d^2x}{dt^2} + \dfrac{g}{b}(x-a) = 0; a, b$ and g are positive numbers and $x = a'$,

$\dfrac{dx}{dt} = 0$ when t = 0, show that

$$x = a + (a' - a)\cos\sqrt{\dfrac{g}{b}}\, t$$ (UPTU(SUM)–2007)

Hints to Selected Problems

1. $m = 0, -2, 3 \Rightarrow$ C.F. $= C_1 e^{0x} + C_2 e^{-2x} + C_3 e^{3x}$

P.I. $= \dfrac{1}{D^3 - D^2 - 6D}(1 + x^2) = -\dfrac{1}{6D}\left[1 + \dfrac{D}{6} - \dfrac{D^2}{6}\right]^{-1}(1 + x^2)$

$= -\dfrac{1}{6D}\left[1 - \dfrac{1}{6}(-D + D^2)\right]^{-1}(1 + x^2)$

Now expand by binomial theorem and D for differentiation and $1/D$ for integration.

7. $m = 1 \pm i\sqrt{2} \Rightarrow$ C.F. $= e^x[C_1 \cos\sqrt{2}x + C_2 \sin\sqrt{2}x]$

P.I. $= \dfrac{1}{D^2 - 2D + 3}(\cos x + x^2)$

$= \dfrac{1}{D^2 - 2D + 3}\cos x + \dfrac{1}{D^2 - 2D + 3}x^2$

$= \dfrac{1}{-1 - 2D + 3}\cos x + \dfrac{1}{3}\left[1 - \left(\dfrac{2D}{3} - \dfrac{D^2}{3}\right)\right]^{-1}x^2 .$

10. $m = \pm 2i \Rightarrow$ C.F. $= C_1 \cos 2x + C_2 \sin 2x$

P.I. $= \dfrac{1}{D^2 + 4}(\sin 3x + x^2) = \dfrac{1}{D^2 + 4}\sin 3x + \dfrac{1}{D^2 + 4}.x^2$

$= \dfrac{1}{-9 + 4}\sin 3x + \dfrac{1}{4\left(1 + \dfrac{D^2}{4}\right)}.x^2$

$= -\dfrac{1}{5}\sin 3x + \dfrac{1}{4}\left(1 + \dfrac{D^2}{4}\right)^{-1}.x^2$

Now expand by Binomial expansion.

Answers

1. $y = C_1 + C_2 e^{3x} + C_3 e^{-2x} - \dfrac{25}{108}x - \dfrac{1}{18}x^3 + \dfrac{1}{36}x^2$

2. $y = C_1 e^{ax} + C_2 e^{-ax} + C_3 \cos ax + C_4 \sin ax - \dfrac{x^4}{a^4} - \dfrac{24}{a^8}$

3. $y = C_1 + (C_2 + C_3 x)e^{-x} + \dfrac{1}{18}e^{2x} + \dfrac{1}{3}x^3 - \dfrac{3}{2}x^2 + 4x$

4. $y = (C_1 + C_2 x)e^{-x} + C_3 e^{2x} - \dfrac{1}{2}x^3 + \dfrac{9}{4}x^2 - \dfrac{27}{4}x + 15$

5. $y = C_1 + C_2 e^x + C_3 e^{2x} + 2x + e^{5x}$

6. $y = C_1 e^{-x} + e^{x/2}\left\{C_2 \cos\dfrac{x\sqrt{3}}{2} + C_3 \sin\dfrac{x\sqrt{3}}{2}\right\} + \dfrac{1}{730}[\sin 3x + 27\cos 3x] - \dfrac{1}{2} - \dfrac{1}{4}(\cos x - \sin x)$

7. $y = e^x[C_1 \cos(x\sqrt{2}) + C_2 \sin(x\sqrt{2})] + \dfrac{1}{4}(\cos x - \sin x) + \dfrac{x^2}{3} + \dfrac{4}{9}x + \dfrac{2}{27}$

8. $y = C_1 e^{2x} + C_2 e^{3x} + \dfrac{1}{6}\left[x + \dfrac{5}{6}\right] + [e^{mx}/m^2 - 5m + 6)]$

9. $y = C_1 \cos 4x + C_2 \sin 4x + \dfrac{1}{7}\cos 3x + \dfrac{1}{25}e^{3x} + \dfrac{1}{16}x^4 - \dfrac{3}{64}x^2 + \dfrac{3}{512}$

10. $y = C_1 \cos 2x + C_2 \sin 2x - \dfrac{1}{5}\sin 3x + \dfrac{1}{4}x^2 - \dfrac{1}{8}$

11. $y = e^{x/2}\left(C_1 \cos\dfrac{\sqrt{15}}{2}x + C_2 \sin\dfrac{\sqrt{15}}{2}x\right) + \dfrac{1}{4}\left(e^x + x^2 + \dfrac{x}{2} - \dfrac{3}{8}\right)$

12. $x = (a' - a)\cos\sqrt{\dfrac{g}{b}}\, t + a$

(4) To evaluate $\dfrac{1}{f(D)}e^{ax}.X$, where X is any function of x :

Let us consider any function X_1 of x. Then, by simple differentiation, we get

$D(e^{ax}.X_1) = e^{ax}D(X_1) + X_1 a e^{ax} = e^{ax}(D + a)X_1 .$ ...(1)

Now, let us assume

$D^n[e^{ax}.X_1] = e^{ax}(D + a)^n.X_1$ ...(2)

Then, consider $D^{n+1}[e^{ax}.X_1] = D[D^n(e^{ax}.X_1)] = D[e^{ax}(D + a)^n.X_1] = ae^{ax}(D + a)^n.X_1 + e^{ax}.D(D + a)^n.X_1$

$= e^{ax}(D + a)^{n+1}.X_1$

Therefore, by the method of induction, we have $D^n[e^{ax}.X_1] = e^{ax}(D + a)^n X_1$, for all positive integer n

$\therefore$ $f(D)e^{ax}.X_1 = e^{ax}f(D + a)X_1 .$ ...(3)

Now, operating on equation (3) with $\dfrac{1}{f(D)}$, we get

$\dfrac{1}{f(D)}.f(D)e^{ax}.X_1 = \dfrac{1}{f(D)}e^{ax}f(D + a).X_1$ $\Rightarrow$ $e^{ax}.X_1 = \dfrac{1}{f(D)}e^{ax}f(D + a).X_1 .$ ...(4)

Let $X = f(D + a).X_1 \Rightarrow X_1 = \dfrac{X}{f(D\ a)} .$

Now, (4) becomes $e^{ax}.\dfrac{X}{f(D + a)} = \dfrac{1}{f(D)}e^{ax}.\dfrac{X}{f(D + a)}.f(D + a) \Rightarrow \dfrac{1}{f(D)}[e^{ax}.X] = e^{ax}\left[\dfrac{1}{f(D + a)}.X\right]$

- Here, we observe that if e^{ax} is brought to the left from the right of $\dfrac{1}{f(D)}$, then D should be replaced by $(D+a)$.
- This method will be used if X is $\cos ax$, $\sin ax$ or x^m or a polynomial of degree m.
- This method is also capable to find $\left\{\dfrac{1}{f(D)} e^{ax}\right\}$, when $f(a) = 0$.

WORKING PROCEDURE

Replace D by $(D+a)$ and brought e^{ax} before the operator $\dfrac{1}{f(D)}$. After that, determine $\dfrac{1}{f(D+a)}.X$ as usual.

Solved Examples

EXAMPLE 1. *Solve* $(D^2 + 4D - 12)y = (x-1)e^{2x}$.

SOLUTION. The given differential equation is
$$(D^2 + 4D - 12)y = (x-1)e^{2x} \qquad \ldots(1)$$
To find C.F. of (1), the auxiliary equation is
$$m^2 + 4m - 12 = 0 \Rightarrow (m-2)(m+6) = 0$$
which gives $m = 2$ and $m = -6$.
$$\therefore \quad \text{C.F.} = C_1 e^{2x} + C_2 e^{-6x}$$
Now, $\text{P.I.} = \dfrac{1}{(D^2 + 4D - 12)} e^{2x}(x-1)$
$$= e^{2x} \dfrac{1}{[(D+2)^2 + 4(D+2) - 12]}(x-1)$$
$$= e^{2x} \dfrac{1}{(D^2 + 8D)}(x-1) = e^{2x} \dfrac{1}{8D\left(1 + \dfrac{D}{8}\right)}(x-1)$$
$$= \dfrac{1}{8} e^{2x} \dfrac{1}{D}\left(1 + \dfrac{1}{8}D\right)^{-1}(x-1)$$
$$= \dfrac{1}{8} e^{2x} \dfrac{1}{D}\left(1 - \dfrac{1}{8}D + \ldots\right)(x-1)$$
$$= \dfrac{1}{8} e^{2x} \dfrac{1}{D}\left(x - 1 - \dfrac{1}{8}\right) = \dfrac{1}{8} e^{2x} \dfrac{1}{D}\left(x - \dfrac{9}{8}\right)$$
$$= \dfrac{1}{8} e^{2x} \int \left(x - \dfrac{9}{8}\right) dx$$
$$= \dfrac{1}{8} e^{2x}\left(\dfrac{x^2}{2} - \dfrac{9}{8}x\right).$$

Hence, the general solution of (1), is given by
$$y = \text{C.F.} + \text{P.I.}$$
$$\Rightarrow \quad y = C_1 e^{2x} + C_2 e^{-6x} + \dfrac{1}{8} e^{2x}\left[\dfrac{x^2}{2} - \dfrac{9}{8}x\right].$$

EXAMPLE 2. *Solve* $(D^2 - 2D + 4)y = e^x \cos x$.

SOLUTION. The differential equation is
$$(D^2 - 2D + 4)y = e^x \cos x \qquad \ldots(1)$$
To find the C.F. of (1), the auxiliary equation is
$$m^2 - 2m + 4 = 0 \Rightarrow m = 1 \pm i\sqrt{3}.$$
Therefore, $\text{C.F.} = e^x(C_1 \cos \sqrt{3}.x + C_2 \sin \sqrt{3}.x)$

Now, $\text{P.I.} = \dfrac{1}{(D^2 - 2D + 4)} e^x \cos x$
$$= e^x \dfrac{1}{[(D+1)^2 - 2(D+1) + 4]} \cos x$$

$$= e^x \dfrac{1}{(D^2 + 3)} \cos x$$
$$= e^x \dfrac{1}{-1^2 + 3} \cos x = \dfrac{1}{2} e^x \cos x.$$
Hence, the complete solution of (1) is given by
$$y = \text{C.F.} + \text{P.I.}$$
$$\Rightarrow \quad y = e^x[C_1 \cos \sqrt{3}.x + C_2 \sin \sqrt{3}.x] + \dfrac{1}{2} e^x \cos x.$$

EXAMPLE 3. *Solve* $(D^2 - 5D + 6)y = e^{2x} \sin 2x$.

SOLUTION. The given differential equation is
$$(D^2 - 5D + 6)y = e^{2x} \sin 2x \qquad \ldots(1)$$
To find the C.F. of (1), the auxiliary equation is given by
$$m^2 - 5m + 6 = 0 \Rightarrow (m-2)(m-3) = 0$$
which gives, $m = 2$ and $m = 3$.
$$\therefore \quad \text{C.F.} = C_1 e^{2x} + C_2 e^{3x}$$

Now, $\text{P.I.} = \dfrac{1}{D^2 - 5D + 6} e^{2x} \sin 2x$
$$= e^{2x} \dfrac{1}{[(D+2)^2 - 5(D+2) + 6]} \sin x$$
$$= e^{2x} \dfrac{1}{D^2 - D} \sin 2x = e^{2x} \dfrac{1}{-2^2 - D} \sin 2x$$
$$= e^{2x} \dfrac{1}{-4 - D} \sin 2x = -e^{2x} \dfrac{1}{(4+D)} \sin 2x$$
$$= -e^{2x} \dfrac{(D-4)}{(D+4)(D-4)} \sin 2x$$
$$= -e^{2x}\left[\dfrac{D-4}{D^2 - 16}\right] \sin 2x = -e^{2x}\left[\dfrac{D-4}{-4-16}\right] \sin 2x$$
$$= \dfrac{e^{2x}}{20}(D-4) \sin 2x = \dfrac{e^{2x}}{20}[D \sin 2x - 4 \sin 2x]$$
$$= \dfrac{e^{2x}}{20}[2 \cos 2x - 4 \sin 2x]$$
Hence, the complete solution of (1) is given by
$$y = \text{C.F.} + \text{P.I.}$$
$$\Rightarrow \quad y = C_1 e^{2x} + C_2 e^{3x} + \dfrac{e^{2x}}{20}[2 \cos 2x - 4 \sin 2x]$$

EXAMPLE 4. *Solve* $\dfrac{d^2y}{dx^2} - 2\dfrac{dy}{dx} + y = xe^x \cos x$. (UPTU-2009)

SOLUTION. The differential equation can be written as
$$(D^2 - 2D + 1)y = xe^x \cos x \qquad \ldots(1)$$

To find the C.F., the auxiliary equation is

$$m^2 - 2m + 1 = 0 \implies m = 1,1$$

Therefore, C.F. $= (C_1 + C_2 x)e^x$

Now, P.I. $= \dfrac{1}{D^2 - 2D + 1} xe^x \cos x$

$$= \dfrac{1}{(D-1)^2} xe^x \cos x$$

$$= e^x \dfrac{1}{(D+1-1)^2} x \cos x = e^x \dfrac{1}{D^2} x \cos x$$

$$= e^x \dfrac{1}{D}(x \sin x + \cos x) = e^x (-x \cos x + 2 \sin x)$$

Hence, the complete solution of (1) is given by

$$y = \text{C.F.} + \text{P.I.}$$

$$= (C_1 + C_2 x)e^x + e^x(-x \cos x + 2 \sin x)$$

Exercise-8.19

Solve the following differential equations :

1. $(D^2 - 2D + 1)y = e^x \cdot x^2$
2. $(D^2 - 5D + 6)y = x^3 \cdot e^{2x}$
3. $(D^2 - 1)y = e^x(1 + x^2)$
4. $(D^2 - 4D + 1)y = e^{2x} \sin x$
5. $(D^2 - 2D + 1)y = x^2 e^3$
6. $(D^2 - 1)y = e^x \cos x$
7. $(D^2 - 2D + 5)y = e^{2x} \sin x$ (MTU(AG)–2011)
8. $(D^2 - 2D + 6)y = e^x \cos x$
9. $(D^2 - 1)y = \cosh x \cos x + a^x$

10. $(D^2 - 4D - 5)y = xe^{-x}$ given that $y = 0$ and $\dfrac{dy}{dx} = 0$ at $x = 0$.
11. $(D^2 - 4D + 4)y = e^x \cos x$ (GBTU(CO)–2010)
12. $\dfrac{d^2 y}{dx^2} - 2\dfrac{dy}{dx} + 4y = e^{2x} \cos x$ (MTU(B.Pharm.)–2011)
13. $(D^2 - 3D + 2)y = xe^x + \sin 2x$ (UPTU–2008)
14. $(D^2 - 1)y = xe^x + \cos^2 x$ (UPTU(SUM)–2007)
15. $(D^2 - 1)y = x \sin x + x^2 e^x$ (UPTU(SUM)–2009)
16. $(D^2 - 2D + 1)y = x \sin x$ (UKTU–2012)
17. $\dfrac{d^2 y}{dx^2} + 2\dfrac{dy}{dx} + y = x^2 e^{-x} \cos x$ (GBTU–2012)

Hints to Selected Problems

1. $m = 1,1, \therefore$ C.F. $= (C_1 + C_2 x)e^x$

P.I. $= \dfrac{1}{D^2 - 2D + 1} e^x \cdot x^2 = \left[\dfrac{1}{(D-1)^2} e^x \cdot x^2\right]$

$$= e^x \left[\dfrac{1}{[(D+1)-1]^2} \cdot x^2\right] = e^x \cdot \dfrac{1}{D^2} \cdot x^2 = \dfrac{e^x \cdot x^4}{12}$$

3. $m = \pm 1, \therefore$ C.F. $= C_1 e^x + C_2 e^{-x}$

P.I. $= \dfrac{1}{D^2 - 1} e^x (1 + x^2) = e^x \dfrac{1}{(D+1)^2 - 1}(1 + x^2)$

$$= e^x \left[\dfrac{1}{D^2 + 2D + 1 - 1}\right] \cdot (1 + x^2)$$

$$= e^x \cdot \dfrac{1}{D^2 + 2D}(1 + x^2) = \dfrac{e^x}{2D}\left[1 + \dfrac{D}{2}\right]^{-1}[1 + x^2]$$

Expand by binomial expansion.

7. $m = 1 \pm 2i, \therefore$ C.F. $= e^x(C_1 \cos 2x + C_2 \sin 2x)$

P.I. $= \dfrac{1}{D^2 - 2D + 5} e^{2x} \sin x = e^{2x} \dfrac{1}{(D+2)^2 - 2(D+2) + 5} \cdot \sin x$

$$= e^{2x} \cdot \dfrac{1}{D^2 + 2D + 5} \sin x.$$

9. C.F. $= C_1 e^x + C_2 e^{-x}$

P.I. $= \dfrac{1}{D^2 - 1} \cosh x \cos x + \dfrac{1}{D^2 - 1} a^x$

$$= \dfrac{1}{D^2 - 1}\left(\dfrac{e^x + e^{-x}}{2}\right) \cos x + \dfrac{1}{(D^2 - 1)} e^{\log a^x}$$

$$= \dfrac{1}{2} e^x \left\{ \dfrac{1}{(D+1)^2 - 1} \cos x + \dfrac{1}{2} e^{-x} \dfrac{1}{(D-1)^2 - 1} \cos x \right.$$

$$\left. + \dfrac{1}{(\log a)^2 - 1} e^{x \log a} \right\}.$$

Answers

1. $y = (C_1 + C_2 x)e^x + \dfrac{1}{12} e^x \cdot x^4$

2. $y = C_1 e^{2x} + C_2 e^{3x} - e^{2x}\left[\dfrac{x^4}{4} + x^3 + 3x^2 + 6x\right]$

3. $y = C_1 e^x + C_2 e^{-x} + \dfrac{1}{12} e^x[9x + 2x^3 - 3x^2]$

4. $y = C_1 e^{(2+\sqrt{3})x} + C_2 e^{(2-\sqrt{3})x} - \dfrac{1}{4} e^{2x} \sin x$

5. $y = (C_1 + C_2 x)e^x + \dfrac{1}{8} e^{3x}(2x^2 - 4x + 3)$

6. $y = C_1 e^x + C_2 e^{-x} - \dfrac{1}{5} e^x(\cos x - 2\sin x)$

7. $y = e^x[C_1 \cos 2x + C_2 \sin 2x] - \dfrac{1}{10} e^{2x}(\cos x - 2\sin x)$

8. $y = e^x[C_1 \cos \sqrt{5}.x + C_2 \sin \sqrt{5}.x] + \dfrac{1}{4} e^x \cos x$

9. $y = C_1 e^x + C_2 e^{-x} + \dfrac{1}{10} e^x[2\sin x - \cos x] - \dfrac{1}{10} e^{-x}(2\sin x + \cos x) + \dfrac{a^x}{(\log a)^2 - 1}$

10. $y = -\dfrac{1}{216} e^{-x} + \dfrac{1}{216} e^{5x} - \dfrac{1}{36} xe^{-x} - \dfrac{1}{12} x^2 e^{-x}$

11. $y = (C_1 + C_2 x)e^{2x} - \dfrac{e^x}{2} \sin x$

12. $y = e^x(C_1 \cos \sqrt{3}x + C_2 \sin \sqrt{3}x) + \dfrac{1}{13} e^{2x}(2\sin x + 3\cos x)$

13. $y = C_1 e^x + C_2 e^{2x} - e^x\left(\dfrac{x^2}{2} + x\right) + \dfrac{1}{20}(3\cos 2x - \sin 2x)$

14. $y = C_1 e^x + C_2 e^{-x} + \dfrac{1}{4} e^x(x^2 - x) - \dfrac{1}{2} - \dfrac{1}{10} \cos 2x$

15. $y = C_1 e^x + C_2 e^{-x} - \dfrac{1}{2}(x \sin x + \cos x) + \dfrac{xe^x}{12}(2x^2 - 3x + 3)$

16. $y = (C_1 + C_2 x)e^x + \dfrac{1}{2}[(x + 1)\cos x - \sin x]$

17. $y = (C_1 + C_2 x)e^{-x} + e^{-x}(-x^2 \cos x + 4x \sin x + 6\cos x)$

(5) To evaluate $\dfrac{1}{f(D)}e^{ax}.X$, **when** $f(a) = 0$:

Let us suppose $f(a) = 0$. In this case $(D - a)$ is at least one factor of $f(D)$.

Let $f(D) = (D - a)^r\, g(D)$, where $g(a) \neq 0$.

Then, $\dfrac{1}{f(D)}e^{ax} = \dfrac{1}{(D-a)^r} \cdot \dfrac{1}{g(a)} e^{ax} = \dfrac{1}{g(a)} \cdot \dfrac{1}{(D-a)^r} e^{ax} = \dfrac{1}{g(a)} \cdot \dfrac{1}{(D-a)^{r-1}} e^{ax} \int e^{ax} . e^{-ax} dx$

$= \dfrac{1}{g(a)} \cdot \dfrac{1}{(D-a)^{r-1}} x e^{ax} = \dfrac{1}{g(a)} \cdot \dfrac{1}{(D-a)^{r-2}} e^{ax} \int x e^{ax} . e^{-ax} dx = \dfrac{1}{g(a)} \cdot \dfrac{1}{(D-a)^{r-2}} \cdot \dfrac{x^2}{2!} e^{ax} .$

Proceeding in the same way, finally, we get $\dfrac{1}{f(D)} e^{ax} = \dfrac{1}{g(a)} \cdot \dfrac{x^r}{r!} e^{ax} .$

📝 Remarks

- Substitute $D = a$ in those factors of $f(D)$ which do not vanish for $D = a$ and then make the question as P.I. of a product of e^{ax} and 1, which is calculated by previous section and reduce to the calculation of $\dfrac{1}{D}.1$ or $\dfrac{1}{D^2}.1$ or $\dfrac{1}{D^3}.1$ and so on.

- Here, $\dfrac{1}{D^n}$ implies n times integral of 1, with respect to x.

Solved Examples

EXAMPLE 1. *Solve* $(D^2 + D - 6)y = e^{2x}$.

SOLUTION. The given equation is

$$(D^2 + D - 6)y = e^{2x} \qquad \ldots(1)$$

To find C.F. of (1), the auxiliary equation is

$$m^2 + m - 6 = 0$$

$\Rightarrow (m+3)(m-2) = 0 \Rightarrow m = 2, -3$

$\therefore$ C.F. $= C_1 e^{2x} + C_2 e^{-3x}$

Now, P.I. $= \dfrac{1}{D^2 + D - 6} e^{2x} = \dfrac{1}{(D+3)(D-2)} e^{2x}$

$= \dfrac{1}{(2+3)(D-2)} e^{2x} = \dfrac{1}{5(D-2)} e^{2x}.1$

$= \dfrac{1}{5} e^{2x} \dfrac{1}{(D+2)-2}.1 = \dfrac{1}{5} e^{2x} \dfrac{1}{D}.1 = \dfrac{1}{5} x e^{2x} .$

Hence, the complete solution of (1) is given by
$y = $ C.F. + P.I.

$\Rightarrow \quad y = C_1 e^{2x} + C_2 e^{-3x} + \dfrac{1}{5} x e^{2x} .$

EXAMPLE 2. *Solve* $\dfrac{d^2 y}{dx^2} - 3\dfrac{dy}{dx} + 2y = e^x$.

SOLUTION. The given differential equation can be written as

$$(D^2 - 3D + 2)y = e^x \qquad \ldots(1)$$

To find the C.F. of (1), the auxiliary equation is

$$m^2 - 3m + 2 = 0$$

$\Rightarrow (m-1)(m-2) = 0 \Rightarrow m = 1, 2$

$\therefore$ C.F. $= C_1 e^x + C_2 e^{2x}$

Now, P.I. $= \dfrac{1}{(D^2 - 3D + 2)} e^x$

$= \dfrac{1}{(D-2)(D-1)} e^x = \dfrac{1}{(1-2)(D-1)} e^x$

(By putting 1 for D in $(D-2)$, because at $D = 1$ $(D-2) \neq -1$)

$= -\dfrac{1}{D-1} e^x = -\dfrac{1}{D-1} e^x.1$

$= -e^x \dfrac{1}{(D+1)-1}.1 = -e^x.\dfrac{1}{D}.1 = -e^x.x$

Hence, the complete solution of (1) is given by
$y = C_1 e^x + C_2 e^{2x} - x e^x$.

EXAMPLE 3. *Solve* $(D^3 + 3D^2 + 3D + 1)\, y = e^{-x}$.

SOLUTION. The given differential equation is

$$(D^3 + 3D^2 + 3D + 1)\, y = e^{-x} \qquad \ldots(1)$$

To find the C.F. of (1), the auxiliary equation is given by

$(m + 1)^3 = 0 \Rightarrow m = -1, -1, -1$

$\therefore \quad$ C.F. $= (C_1 + C_2 x + C_3 x^2)\, e^{-x}$

Now, P.I. $= \dfrac{1}{(D+1)^3} e^{-x} = e^{-x} \dfrac{1}{(D-1+1)^3}.1$

$= e^{-x} . \dfrac{1}{D^3}.1 = e^{-x} . \dfrac{x^3}{3!} .$

Hence, the complete solution of (1) is given by
$y = $ C.F. + P.I.

$\Rightarrow \qquad y = (C_1 + C_2 x + C_3 x^2)e^{-x} + e^{-x}.\dfrac{x^3}{3!} .$

Exercise-8.20

Solve the following differential equations :

1. $(D^2 + 4D + 3)\, y = e^{-3x}$

2. $(D^2 + 6D + 9)y = 2e^{-3x}$

3. $(D^4 + D^3 + D^2 - D - 2)y = e^x$

4. $(D^2 - 9D + 18)y = \cosh 3x$

5. $(D-1)^2(D^2+1)^2 y = e^x$

6. $(D^2 - 3D + 2)y = e^x$ when $y = 3$, $\dfrac{dy}{dx} = 3$ at $x = 0$

7. $(D-1)^3(D+1)y = e^x + e^{-x}$

8. $(D^2 - 6D + 9)y = 4e^{3x}$

9. $(D^2 - 1)y = \cosh x$

10. $(D^2 - 4D + 4)y = 8(x^2 + e^{2x} + \sin 2x)$

Hints to Selected Problems

1. C.F. $= C_1 e^{-x} + C_2 e^{-3x}$

P.I. $= \dfrac{1}{D^2 + 4D + 3} e^{-3x} = \dfrac{1}{(D+1)(D+3)} e^{-3x}$

$= \dfrac{1}{(-3+1)(D+3)} e^{-3x} = -\dfrac{1}{2(D+3)} e^{-3x} \cdot 1$

$= -\dfrac{1}{2} e^{-3x} \dfrac{1}{[(D-3)+3]} \cdot 1 = -\dfrac{1}{2} e^{-3x} \cdot \dfrac{1}{D} \cdot 1 = -\dfrac{1}{2} e^{-3x} \cdot x$

4. C.F. $= C_1 e^{3x} + C_2 e^{6x}$

P.I. $= \dfrac{1}{D^2 - 9D + 18} \cosh 3x = \dfrac{1}{D^2 - 9D + 18} \left(\dfrac{e^{3x} + e^{-3x}}{2} \right)$

$= \dfrac{1}{2(D-3)(D-6)}(e^{3x} + e^{-3x})$

7. C.F. $= (C_1 + C_2 x + C_3 x^2) e^x + C_1 e^{-x}$

P.I. $= \dfrac{1}{(D-1)^3(D+1)}(e^x + e^{-x})$

$= \dfrac{1}{2} \dfrac{1}{(D-1)^3} e^x \cdot 1 - \dfrac{1}{8} \dfrac{1}{(D+1)} e^{-x} \cdot 1$

10. P.I. $= \dfrac{1}{(D^2 - 4D + 4)}(8x^2 + 8e^{2x} + 8\sin 2x)$

$= \dfrac{1}{(D-2)^2} 8x^2 + \dfrac{1}{(D-2)^2} 8e^{2x} + \dfrac{1}{(D-2)^2} \cdot 8\sin 2x$

Answers

1. $y = C_1 e^{-x} + C_2 e^{-3x} - \dfrac{x}{2} e^{-3x}$

2. $y = (C_1 + C_2 x) e^{-3x} + x^2 e^{-3x}$

3. $y = C_1 e^x + C_2 e^{-x} + e^{-x/2} \left[C_3 \cos\left(\dfrac{\sqrt{7}}{2} x \right) + C_4 \sin\left(\dfrac{\sqrt{7}}{2} x \right) \right] + \dfrac{1}{8} x e^x$

4. $y = C_1 e^{3x} + C_2 e^{6x} - \dfrac{1}{6} x e^{3x} + \dfrac{1}{108} e^{-3x}$

5. $y = (C_1 + C_2 x)e^x + (C_3 + C_4 x)\cos x + (C_5 + C_6 x)\sin x + \dfrac{1}{8} x^2 e^x$

6. $y = 2e^x + e^{2x} - xe^x$

7. $y = (C_1 + C_2 x + C_3 x^2)e^x + C_4 e^{-x} + \dfrac{1}{12} x^3 e^x - \dfrac{x}{8} e^{-x}$

8. $y = (C_1 + C_2 x) e^{3x} + 2x^2 e^{3x}$

9. $y = C_1 e^x + C_2 e^{-x} + \dfrac{1}{2} x \sinh x$

10. $(C_1 + C_2 x) e^{2x} + 2x^2 + 3 + 4x + 4x^2 e^{2x} + \cos 2x$

---

(6) To evaluate $\dfrac{1}{f(D^2)} \sin ax$ **or** $\cos ax$, **when** $f(-a^2)=0$:

To find the particular integral of such cases, we shall calculate P.I. for e^{iax} instead of $\sin ax$ or $\cos ax$.

Here, we have $e^{iax} = \cos ax + i \sin ax$.

Thus, P.I. for e^{iax} = P.I. for $(\cos ax + i \sin ax)$

$\Rightarrow$ P.I. for $\cos ax$ = Real part of P.I. for e^{iax} and P.I. for $\sin ax$ = imaginary part of P.I. for e^{iax}.

Therefore, $\dfrac{\cos ax}{D^2 + a^2}$ and $\dfrac{\sin ax}{D^2 + a^2}$ are respectively real and imaginary part of $\dfrac{e^{iax}}{D^2 + a^2}$

$= \dfrac{e^{iax}}{(D + ai)(D - ai)} = \dfrac{e^{iax}}{(ai + ai)(D - ai)}$ (By putting ai for in $(D + ai)$ because at $D = ai$ it does not vanish.)

$= \dfrac{e^{iax}}{2ai} \left[\dfrac{1}{D + ai - ai} \cdot 1 \right] = \dfrac{e^{iax}}{2ai} \cdot \dfrac{1}{D} \cdot 1 = \dfrac{x}{2ai}(e^{aix}) = -\dfrac{ix(\cos ax + i\sin ax)}{2a} = -\dfrac{ix}{2a}\cos ax + \dfrac{x}{2a}\sin ax$

$\Rightarrow \dfrac{1}{D^2 + a^2} \sin ax = -\dfrac{x}{2a} \cos ax = \dfrac{x}{2} \int \sin ax \, dx$ and $\dfrac{1}{D^2 + a^2} \cos ax = \dfrac{x}{2a} \sin ax = \dfrac{x}{2} \int \cos ax \, dx$.

Solved Examples

EXAMPLE 1. *Solve* $(D^2 + a^2)y = \sin ax$.

SOLUTION. The given equation is

$(D^2 + a^2)y = \sin ax$ ...(1)

To find the C.F of (1), the auxiliary equation is

$m^2 + a^2 = 0 \Rightarrow m = 0 \pm ai$

$\therefore$ C.F. $= e^{0x}[C_1 \cos ax + C_2 \sin ax]$

$= [C_1 \cos ax + C_2 \sin ax]$

Now, P.I. $= \dfrac{1}{D^2 + a^2} \sin ax$

$=$ Imaginary part of $\left[\dfrac{1}{D^2 + a^2}(\cos ax + i \sin ax) \right]$

$=$ Imaginary part of $\left[\dfrac{1}{D^2 + a^2} e^{iax} \right]$

$=$ Imaginary part of $\left[\dfrac{1}{(D + ai)(D - ai)} e^{iax} \right]$

$$= \text{Imaginary part of } \left[\frac{1}{(ai+ai)(D-ai)}e^{iax}\right]$$

$$= \text{Imaginary part of } \left[\frac{1}{2ai}\cdot\frac{1}{(D-ai)}e^{iax}\right]$$

$$= \text{Imaginary part of } \frac{1}{2ai}\frac{1}{(D-ai)}e^{iax}.1$$

$$= \text{Imaginary part of } \frac{1}{2ai}e^{iax}\frac{1}{[(D+ia)-ia]}.1$$

$$= \text{Imaginary part of } \frac{1}{2ai}e^{iax}\frac{1}{D}.1$$

$$= \text{Imaginary part of } \frac{1}{2ai}e^{iax}.x$$

$$= \text{Imaginary part of } \frac{1}{2ai}.x\,(\cos ax + i\sin ax)$$

$$= \text{Imaginary part of } \frac{1}{2a}x\left[\frac{1}{i}\cos ax + \sin ax\right]$$

$$= \text{Imaginary part of } \frac{1}{2a}x\left[\frac{i}{i^2}\cos ax + \sin ax\right]$$

$$= \text{Imaginary part of } \frac{1}{2a}x[-i\cos ax + \sin ax]$$

$$= -\frac{x}{2a}\cos ax.$$

Hence, the complete solution of (1) is given by

$$y = \text{C.F.} + \text{P.I.}$$

$$\Rightarrow \qquad y = C_1\cos ax + C_2\sin ax - \frac{x}{2a}\cos ax.$$

EXAMPLE 2. *Solve* $(D^2+1)y = \sin x \sin 2x$.

SOLUTION. The given equation is

$$(D^2+1)y = \sin x \sin 2x \qquad \dots(1)$$

To find the C.F. of (1), the auxiliary equation is

$$(m^2+1) = 0 \Rightarrow m = \pm i$$

$$\therefore \quad \text{C.F.} = C_1\cos x + C_2\sin x$$

Now, $\text{P.I.} = \dfrac{1}{(D^2+1)}(\sin x \sin 2x)$

$$= \frac{1}{(D^2+1)}\cdot\frac{1}{2}[2\sin x \sin 2x]$$

$$= \frac{1}{2}\cdot\frac{1}{D^2+1}[\cos x - \cos 3x]$$

$$= \frac{1}{2}\left[\frac{1}{D^2+1}\cos x - \frac{1}{D^2+1}\cos 3x\right].$$

Now, $\dfrac{1}{D^2+1}\cos 3x = \dfrac{1}{-3^2+1}\cos 3x = -\dfrac{1}{8}\cos 3x$

Again, $\dfrac{1}{D^2+1}\cos x = \text{Real part of }\left[\dfrac{1}{D^2+1}e^{ix}\right]$

$$= \text{Real part of }\left[\frac{1}{(D+i)(D-i)}e^{ix}\right]$$

$$= \text{Real part of }\left[\frac{1}{(i+i)(D-i)}e^{ix}\right]$$

$$= \text{Real part of }\left[\frac{1}{2i}\cdot\frac{1}{(D-i)}e^{ix}\right]$$

$$= \text{Real part of }\left[\frac{1}{2i}\frac{1}{D-i}e^{ix}.1\right]$$

$$= \text{Real part of }\left[\frac{1}{2i}\cdot\frac{1}{(D+i-i)}.1\right]$$

$$= \text{Real part of }\left[\frac{1}{2i}e^{ix}\frac{1}{D}.1\right]$$

$$= \text{Real part of }\left[\frac{1}{2i}.e^{ix}.x\right]$$

$$= \text{Real part of }\left[\frac{x}{2i}(\cos x + i\sin x)\right]$$

$$= \text{Real part of }\left[-i.\frac{1}{2}x\cos x + \frac{1}{2}x\sin x\right]$$

$$= \frac{1}{2}x\sin x$$

Hence, the complete solution of (1) is given by

$$y = \text{C.F.} + \text{P.I.}$$

$$\Rightarrow y = C_1\cos x + C_2\sin x + \frac{x}{4}\sin x - \frac{1}{16}\cos 3x.$$

EXAMPLE 3. *Solve the given differential equation*

$$\frac{d^2y}{dx^2} + 9y = 2\sin 3x + \cos 3x.$$

SOLUTION. The given differential equation can be written as

$$(D^2+9)y = 2\sin 3x + \cos 3x \qquad \dots(1)$$

To find the C.F. of (1), the auxiliary equation is given by

$$m^2+9 = 0 \Rightarrow m = \pm 3i$$

$$\therefore \quad \text{C.F.} = C_1\cos 3x + C_2\sin 3x$$

Now, $\text{P.I.} = \dfrac{1}{D^2+9}(2\sin 3x + \cos 3x)$

$$= 2\frac{1}{D^2+9}\sin 3x + \frac{1}{D^2+9}\cos 3x.$$

Now, $\dfrac{1}{D^2+9}\cos 3x + i.\dfrac{1}{D^2+9}\sin 3x$

$$= \frac{1}{D^2+9}(\cos 3x + i\sin 3x) = \frac{1}{D^2+9}e^{i3x}$$

$$= \frac{1}{(D+3i)(D-3i)}e^{i3x} = \frac{1}{(3i+3i)(D-3i)}e^{3ix}.1$$

$$= \frac{e^{i3x}}{6i}\cdot\frac{1}{[(D+3i)-3i]}.1 = \frac{1}{6i}.e^{i3x}.\frac{1}{D}.1 = \frac{1}{6i}e^{i3x}.x$$

$$= \frac{x}{6i}[\cos 3x + i\sin 3x] = \frac{x}{6}\sin 3x - i\frac{x}{6}\cos 3x$$

Now, equating real and imaginary part of both sides, we get

$$\frac{1}{D^2+9}\cos 3x = \frac{x}{6}\sin 3x$$

and $\dfrac{1}{D^2+9}\sin 3x = -\dfrac{x}{6}\cos 3x.$

$$\therefore \text{Required P.I.} = 2\left[-\frac{x}{6}\cos 3x\right] + \frac{x}{6}\sin 3x$$

$$= -\frac{x}{3}\cos 3x + \frac{x}{6}\sin 3x.$$

Hence, the complete solution of (1) is given by

$$y = \text{C.F.} + \text{P.I.}$$

$$\Rightarrow \qquad y = C_1\cos 3x + C_2\sin 3x - \frac{1}{3}x\cos x + \frac{1}{6}x\sin 3x$$

Exercise-8.21

Solve the following differential equations :

1. $(D^2 + a^2)y = \cos ax$

2. $(D^2 + 4)y = \cos 2x$

3. $(D^2 + 4)y = e^x + \sin 2x$

4. $(D^3 + a^2 D)y = \sin ax$

5. $(D^3 + 1)y = \cos 2x$

6. $(D^4 + D^2 + 1)y = e^{-x/2}\cos\dfrac{x\sqrt{3}}{2}$ (Rajasthan–2006)

7. $(D^2 + 2D + 2)y = 2e^{-x}\sin x$

8. $(D^4 + 2D^2 + 1)y = \cos x$

9. $(D^2 + 4)y = 4 + \sin 2x$

Hints to Selected Problems

1. $m = \pm ai \Rightarrow$ C.F. $= C_1 \cos ax + C_2 \sin ax$

P.I. $= \dfrac{1}{D^2 + a^2}\cos ax = $ Real part of $\left[\dfrac{1}{D^2 + a^2}e^{iax}\right]$.

3. $m = \pm 2i$, C.F. $= C_1 \cos 2x + C_2 \sin 2x$

P.I. $= \dfrac{1}{(D^2 + 4)}(e^x + \sin 2x) = \dfrac{e^x}{D^2 + 4} + \dfrac{1}{D^2 + 4}.\sin 2x$

$= \dfrac{e^x}{1 + 4} + $ Imag. part of $\left[\dfrac{1}{D^2 + 4}.e^{2ix}\right]$

4. P.I. $= \dfrac{1}{D^2 + a^2}\sin ax = $ Imaginary part of $\left[\dfrac{1}{(D + ai)(D - ai)}e^{iax}\right]$

Answers

1. $y = C_1 \cos ax + C_2 \sin ax + \dfrac{x}{2a}\sin ax$ 2. $y = C_1 \cos 2x + C_2 \sin 2x + \dfrac{x}{4}\sin 2x$ 3. $y = C_1 \cos 2x + C_2 \sin 2x + \dfrac{1}{5}e^x - \dfrac{1}{4}x\cos 2x$

4. $y = C_1 + C_2 \cos ax + C_3 \sin ax - \dfrac{1}{2a^2}x\sin ax$ 5. $y = C_1 e^{-x} + e^{x/2}\left[C_2 \cos\dfrac{\sqrt{3}}{2}x + C_3 \sin\dfrac{\sqrt{3}}{2}x\right] + \dfrac{1}{65}(\cos 2x - 8\sin 2x)$

6. $y = e^{-x/2}\left[C_1 \cos\dfrac{1}{2}x\sqrt{3} + C_2 \sin\dfrac{1}{2}x\sqrt{3}\right] + e^{x/2}\left[C_3 \cos\dfrac{1}{2}x\sqrt{3} + C_4 \sin\dfrac{1}{2}x\sqrt{3}\right] - \dfrac{1}{12}\sqrt{3}\,x\,e^{-x/2}\sin\dfrac{\sqrt{3}x}{2} + \dfrac{x}{4}e^{-x/2}\cos\dfrac{\sqrt{3}}{2}x$

7. $y = e^{-x}(C_1 \cos x + C_2 \sin x) - xe^{-x}\cos x$ 8. $y = (C_1 + C_2 x)\cos x + (C_3 + C_4 x)\sin x - \dfrac{1}{8}x^2 \cos x$

9. $y = C_1 \cos 2x + C_2 \sin 2x + 1 - \dfrac{x}{4}\cos 2x$

(7) To evaluate $\dfrac{1}{f(D)}x.X$, **where X is any function of x (except e^{ax}) :**

Consider $D^n(x.X) = xD^n.X + {}^nC_1 D^{n-1}.X$ (By Leibnitz's theorem)

We have $f(D)(xX) = x\,f(D)X + f'(D).X$

Now, taking the inverse operator, we have $\dfrac{1}{f(D)}(xX) = x.\dfrac{1}{f(D)}X + \left[\dfrac{d}{dD}\dfrac{1}{f(D)}\right]X$

But we have $\dfrac{d}{dD}\left[\dfrac{1}{f(D)}\right] = -\dfrac{f'(D)}{\{f(D)\}^2}$. Therefore, $\dfrac{1}{f(D)}(x.X) = x.\dfrac{1}{f(D)}X - \dfrac{f'(D)}{\{f(D)\}^2}X$.

REMARK

- If we want to find P.I. when B is of the form $x^m.X$, where X is any function of , then there are two cases

  (a) If $X = x^n$, then $x^m.X = x^{m+n}$.

  Then B is of the form x^{m+n} (Polynomial). Here, we should apply the method of finding P.I. for polynomial discussed earlier.

  (b) If $X = e^{ax}$, then $x^m.X = x^m.e^{ax}$ and we should apply the method, discussed earlier.

  (c) If $X = \cos ax$, then $x^m.X = x^m \cos ax$

  Then P.I. $= \dfrac{1}{f(D)}x^m \cos ax = \dfrac{1}{f(D)}$ (Real part of $x^m.e^{iax}$) $= $ Real part of $\dfrac{1}{f(D)}x^m.e^{iax}$, which can be easily calculated.

  Similar results hold if $X = \sin ax$, then taking imaginary part.

Solved Examples

EXAMPLE 1. Solve $(D^2 + 2D + 1)y = x\cos x$.   (Rajasthan–2006)

SOLUTION. The given equation is

$((D^2 + 2D + 1)y = x\cos x$ ...(1)

To find the C.F. of (1), the auxiliary equation is

$m^2 + 2m + 1 = 0 \Rightarrow m = -1, -1$

$\therefore$ C.F. $= (C_1 + C_2 x)e^{-x}$

Now, P.I. $= \dfrac{1}{(D^2 + 2D + 1)}.x\cos x$

$= x.\dfrac{1}{D^2 + 2D + 1}\cos x - \dfrac{2D + 2}{(D^2 + 2D + 1)^2}\cos x$

$= x.\dfrac{1}{2D}\cos x - \dfrac{2D + 2}{4D^2}\cos x = \dfrac{x}{2}\sin x + \dfrac{(D+1)}{2}\cos x$

$$= \frac{x}{2}\sin x + \frac{\cos x}{2} - \frac{\sin x}{2}$$

Hence, the complete solution of (1) is given by

$$y = \text{C.F.} + \text{P.I.}$$

$$\Rightarrow \quad y = (C_1 + C_2 x)e^{-x} + \frac{x}{2}\sin x$$

$$+ \frac{\cos x}{2} - \frac{\sin x}{2}.$$

EXAMPLE 2. *Solve* $(D^2 - 4D + 4) = 8x^2 e^{2x} \sin 2x$.

(JNTU–2006, UPTU 2004, 05, 09)

SOLUTION. The given equation is

$$(D^2 - 4D + 4) = 8x^2 e^{2x} \sin 2x \qquad \ldots(1)$$

To find the C.F. of (1), the auxiliary equation is

$$m^2 - 4m + 4 = 0 \Rightarrow m = 2, 2$$

$$\therefore \quad \text{C.F.} = (C_1 + C_2 x)\, e^{2x}$$

Now, $\quad \text{P.I.} = 8 \cdot \frac{1}{(D-2)^2} e^{2x}(x^2 \sin 2x)$

$$= 8e^{2x} \cdot \frac{1}{(D+2-2)^2} \cdot (x^2 \sin 2x)$$

$$= 8e^{2x} \cdot \frac{1}{D^2}(x^2 \sin 2x) = 8\, e^{2x} \cdot I_1$$

where, $\quad I_1 = \frac{1}{D^2}(x^2 \sin 2x)$

$$= \text{Imaginary part of } \frac{1}{D^2} x^2 e^{2ix}$$

$$= \text{Imaginary part of } e^{2ix} \frac{1}{(D+2i)^2} x^2$$

$$= \text{Imaginary part of } \frac{e^{2ix}}{4i^2}\left(1 + \frac{D}{2i}\right)^{-2} x^2$$

$$= \text{Imaginary part of } \frac{e^{2ix}}{-4}\left(1 - \frac{iD}{2}\right)^{-2} x^2$$

$$= \text{Imaginary part of } \frac{e^{2ix}}{-4}\left[1 + 2\left(\frac{iD}{2}\right) + 3\left(\frac{iD}{2}\right)^2 + \ldots\right] x^2$$

$$= \text{Imaginary part of } \frac{e^{2ix}}{-4}\left[1 + Di - \frac{3}{4}D^2 + \ldots\right] x^2$$

$$= \text{Imaginary part of } \frac{e^{2ix}}{-4}\left[x^2 + 2ix - \frac{3}{2}\right]$$

$$= \text{Imaginary part of}$$

$$\left\{-\frac{1}{4}(\cos 2x + i \sin 2x)\left(x^2 + 2ix - \frac{3}{2}\right)\right\}$$

$$= -\frac{1}{4}\left[\left(x^2 - \frac{3}{2}\right)\sin 2x + 2x \cos 2x\right]$$

$$= -\frac{1}{8}\left[(2x^2 - 3)\sin 2x + 4x \cos 2x\right]$$

$$\therefore \quad \text{P.I.} = 8e^{2x} \cdot I_1 = 8e^{2x}\left[-\frac{1}{8}\{(2x^2 - 3)\sin 2x + 4x \cos 2x\}\right]$$

$$= -e^{2x}\left[(2x^2 - 3)\sin 2x + 4x \cos 2x\right].$$

Hence, the complete solution of (1) is given by

$$y = \text{C.F.} + \text{P.I.}$$

$$\Rightarrow y = e^{2x}[C_1 + C_2 x + 3\sin 2x$$

$$- 2x^2 \sin 2x - 4x \cos 2x].$$

EXAMPLE 3. *Solve* $(D^2 - 2D + 1)\, y = xe^x \sin x$.

(SVTU–2007, JNTU–2006, UPTU–2005, MTU(B.Pharma.)–2011)

SOLUTION. The given differential equation can be written as

$$(D^2 - 2D + 1)\, y = xe^x \sin x \qquad \ldots(1)$$

To find the C.F. of (1), the auxiliary equation is given by

$$m^2 - 2m + 1 = 0 \Rightarrow m = 1, 1$$

$$\therefore \quad \text{C.F.} = (C_1 + C_2 x)\, e^x$$

Now, $\text{P.I.} = \frac{1}{(D^2 - 2D + 1)} xe^x \sin x$

$$= e^x \frac{1}{(D+1)^2 - 2(D+1) + 1} x \sin x$$

$$= e^x \frac{1}{D^2}(x \sin x) = e^x \cdot \frac{1}{D} \int x \sin x \, dx$$

$$= e^x \left(\frac{1}{D}\right)(-x \cos x + \sin x)$$

$$= e^x \left[\int -x \cos x\, dx + \int \sin x\, dx\right]$$

$$= e^x \left[-x \sin x - 2\cos x\right].$$

Hence, the complete solution of (1) is given by

$$y = \text{C.F.} + \text{P.I.}$$

$$\Rightarrow \quad y = (C_1 + C_2 x)\, e^x - e^x(x \sin x + 2\cos x).$$

EXAMPLE 4. *Solve* $\dfrac{d^2 y}{dx^2} + 4y = x \sin x$. (Madras–2004)

SOLUTION. The given differential equation can be written as

$$(D^2 + 4)y = x \sin x \qquad \ldots(1)$$

To find the C.F. of (1), the auxiliary equation is given by

$$m^2 + 4 = 0 \Rightarrow m = \pm 2i$$

$$\therefore \quad \text{C.F.} = C_1 \cos 2x + C_2 \sin 2x$$

Now, $\text{P.I.} = \dfrac{1}{D^2 + 4} x \sin x$

$$= x \frac{1}{D^2 + 4}\sin x - \frac{2D}{(D^2 + 4)^2}\sin x$$

$$= \frac{x \sin x}{-1^2 + 4} - \frac{2D}{(-1^2 + 4)^2}\sin x$$

$$= \frac{1}{3}x \sin x - \frac{2}{9}D(\sin x)$$

$$= \frac{1}{3}x \sin x - \frac{2}{9}\cos x$$

Hence, the complete solution of (1) is given by

$$y = \text{C.F.} + \text{P.I.}$$

$$\Rightarrow \quad y = C_1 \cos 2x + C_2 \sin 2x + \frac{x}{3}\sin x - \frac{2}{9}\cos x.$$

Exercise-8.22

Solve the following differential equations :

1. $(D^2 - 2D + 1)y = x \sin x$

2. $(D^2 + m^2)y = x \cos mx$

3. $(D^2 - 1) = x^2 \sin x$

4. $(D^2 + a^2)^2 y = \sin ax$

(SRM–2013)

5. $(D^4 - 1)y = x \sin x$

6. $(D^4 - 1)y = e^x \cos x$

7. $(D^4 + 2D^2 + 1)y = x^2 \cos x$ (Nagarjuna–2008, Rajasthan–2005)

8. $(D^2 + 1)y = x^2 \sin 2x$

9. $(D^2 + 1)^2 y = 24x \cos x$, given that $x = 0, y = 0, Dy = 0,$ $D^2 y = 0, D^3 y = 0$.

Hints to Selected Problems

1. C.F. $= (C_1 + C_2 x) e^x$

P.I. $= \dfrac{1}{D^2 - 2D + 1} x \sin x$

$= x \dfrac{1}{D^2 - 2D + 1} \sin x - \dfrac{(2D - 2)}{(D^2 - 2D + 1)^2} \sin x$

$= x . \dfrac{1}{-1 - 2D + 1} \sin x - \dfrac{(2D - 2)}{(-1 - 2D + 1)^2} \sin x$

$= x . \dfrac{1}{-2D} \sin x - \dfrac{(2D - 2)}{4D^2} \sin x$.

2. P.I. $= \dfrac{1}{D^2 + m^2} x \cos mx = $ Real part of $\dfrac{1}{(D^2 + m^2)} x e^{imx}$

$= $ Real part of $e^{imx} \left[\dfrac{1}{(D + im)^2 + m^2} . x \right]$

8. P.I. $= \dfrac{1}{D^2 + 1} x^2 \sin 2x = $ Imaginary part of $\dfrac{1}{D^2 + 1} . xe^{2ix}$

$= $ Imag. part of $e^{2ix} \left[\dfrac{1}{(D + 2i)^2 + 1} \right] . x$.

10. P.I. $= \dfrac{1}{(D^2 + 1)^2} 24x \cos x = 24 \dfrac{1}{(D^2 + 1)^2} x \cos x$

$= $ Real part of $24 e^{ix} \dfrac{1}{[(D + i)^2 + 1]^2} . x$

Answers

1. $y = (C_1 + C_2 x) e^x + \dfrac{1}{2}(x \cos x + \cos x - \sin x)$

2. $y = C_1 \cos mx + C_2 \sin mx + \dfrac{x^2}{4m} \sin mx + \dfrac{x}{4m^2} \cos mx$

3. $y = C_1 e^x + C_2 e^{-x} - x \cos x - \dfrac{1}{2}(x^2 - 1) \sin x$

4. $y = (C_1 + C_2 x) \cos ax + (C_3 + C_4 x) \sin ax - \dfrac{1}{8a^2}(x^2 \sin ax)$

5. $y = C_1 e^x + C_2 e^{-x} + C_3 \cos x + C_4 \sin x + \dfrac{1}{8}(x^2 \cos x - 3x \sin x)$

6. $y = C_1 e^x + C_2 e^{-x} + C_3 \cos x + C_4 \sin x - \dfrac{1}{5} e^x \cos x$

7. $y = (C_1 + C_2 x) \cos x + (C_3 + C_4 x) \sin x - \dfrac{1}{48}(x^4 - 9x^2) \cos x + \dfrac{1}{12} x^3 \sin x$

8. $y = C_1 \cos x + C_2 \sin x - \dfrac{1}{27}[24x \cos 2x + (9x^2 - 26) \sin 2x]$

9. $y = 3x^2 \sin x - x^3 \cos x$

8.3 HOMOGENEOUS LINEAR DIFFERENTIAL EQUATIONS

Any differential equation of the form

$$x^n \frac{d^n y}{dx^n} + A_1 x^{n-1} \frac{d^{n-1} y}{dx^{n-1}} + \ldots + A_n y = X$$

is called a homogeneous linear differential equation of nth order, where $A_1, A_2, \ldots, A_n$ are constants and X is a function of x or a constant.

For example : Consider the following differential equations :

(i) $x^2 \dfrac{d^2 y}{dx^2} + 3x \dfrac{dy}{dx} + 4y = e^x$ (ii) $x^2 \dfrac{d^2 y}{dx^2} + 2x \dfrac{dy}{dx} + 2y = e^x$

The above two differential equations are linear as the dependent variable y and its derivatives appear in their first degree and are not multiplied together.

REMARKS

- The differential equation in which the powers of x in the coefficients are equal to the orders of the derivative associated with them, is called the homogeneous linear differential equation.
- In linear homogeneous differential equation, the dependent variable y and its derivatives with respect to independent variable x, appears in their first degree and are not multiplied together.
- The homogeneous linear equations, discussed above, are also known as Cauchy-Euler equation.

8.3.2 SOLUTION OF HOMOGENEOUS LINEAR EQUATION

Consider the homogeneous linear differential equation

$$x^n \frac{d^n y}{dx^n} + A_1 x^{n-1} \frac{d^{n-1} y}{dx^{n-1}} + A_2 x^{n-2} \frac{d^{n-2} y}{dx^{n-2}} + \ldots + A_n y = X \qquad \ldots(1)$$

where, $A_1, A_2, \ldots, A_n$ are constants and X is a function of x or constant.

Now, equation (1) can be transformed to an equivalent equation (linear differential equation) with constant coefficients by changing the independent variable by the relation

$$x = e^z, \text{ i.e., } z = \log x$$

$\Rightarrow$
$$\frac{dz}{dx} = \frac{1}{x}$$

Now
$$\frac{dy}{dx} = \frac{dy}{dz}\frac{dz}{dx} = \frac{1}{x}\cdot\frac{dy}{dz} \quad \Rightarrow \quad x\frac{dy}{dx} = \frac{dy}{dz}$$

Again,
$$\frac{d^2y}{dx^2} = \frac{d}{dx}\left(\frac{dy}{dx}\right) = \frac{d}{dx}\left(\frac{1}{x}\cdot\frac{dy}{dz}\right) = \frac{1}{x}\frac{d}{dx}\left(\frac{dy}{dz}\right) + \frac{d}{dx}\left(\frac{1}{x}\right)\frac{dy}{dz}$$

$$= \frac{1}{x}\frac{d^2y}{dz^2}\cdot\frac{dz}{dx} - \frac{1}{x^2}\frac{dy}{dz} = \frac{1}{x^2}\cdot\frac{d^2y}{dz^2} - \frac{1}{x^2}\cdot\frac{dy}{dz}$$

$\Rightarrow$
$$x^2\frac{d^2y}{dx^2} = \frac{d^2y}{dz^2} - \frac{dy}{dz}$$

Proceeding likewise, we get

$$x^n\frac{d^ny}{dx^n} = \left[\frac{d^ny}{dz^n} - \frac{n(n-1)}{2!}\cdot\frac{d^{n-1}y}{dz^{n-1}} + \ldots + (-1)^{n-1}n!\frac{dy}{dz}\right]$$

If we write $D \equiv \dfrac{d}{dz}$, then we get

$$x\frac{dy}{dx} = Dy, x^2\frac{d^2y}{dx^2} = D(D-1)y, x^3\frac{d^3y}{dx^3} = D(D-1)(D-2)y$$

and so on
$$x^n\frac{d^ny}{dx^n} = D(D-1)(D-2)\ldots(D-n+1)y$$

Let us consider, the transformed equation
$$f(D)y = X \qquad\qquad\qquad\qquad \ldots(2)$$

The general solution of (2) is the sum of a particular solution of (2) and complementary function of (2). Now, it can be easily solved by the usual method given in section 8.2.

WORKING PROCEDURE

Step 1. Put $x = e^z$, $x\dfrac{d}{dx} = D = \dfrac{d}{dz}$, $x^2\dfrac{d^2}{dx^2} = D(D-1)$ and so on.

Step 2. Obtain the equation in terms of D (linear equation).

Step 3. To find the C.F. and P.I. used the usual method given in section 8.2.

Step 4. Find general solution by adding C.F. and P.I.

Step 5. Finally, substitute $z = \log x$.

REMARK

- To solve the homogeneous linear differential equation, we change the independent variable x to z by substitution $x = e^z$. The substitution can be easily justified as follows : "Since the exponential function e^{mz} has the property that its derivative are all constant multiples of the function itself. This leads us to consider $x = e^{mz}$ as a possible solution of the given equation."

Solved Examples

EXAMPLE 1. **Solve** $x^2\dfrac{d^2y}{dx^2} + x\dfrac{dy}{dx} - \lambda^2 y = 0$ (UPTU-2007)

SOLUTION. Put $x = e^z \Rightarrow \log x = z$

Also
$$x\frac{dy}{dx} = Dy$$

$$x^2\frac{d^2y}{dx^2} = D(D-1)y$$

The transformed equation is

$$D(D-1)y + Dy - \lambda^2 y = 0$$

$\Rightarrow \quad [D(D-1) + D - \lambda^2]y = 0$

$\Rightarrow \quad\quad\quad (D^2 - \lambda^2)y = 0$

A.E. is $\quad\quad m^2 - \lambda^2 = 0 \Rightarrow m = \pm\lambda$

C.F. $= C_1 e^{\lambda z} + C_2 e^{-\lambda z}$

P.I. $= 0$

Complete solution is

$$y = \text{C.F.} + \text{P.I.}$$
$$= C_1 e^{\lambda z} + C_2 e^{-\lambda z} = C_1 x^\lambda + C_2 x^{-\lambda}$$

EXAMPLE 2. **Solve** $x^2\dfrac{d^2y}{dx^2} - 4x\dfrac{dy}{dx} + 6y = x$ ...(1)

SOLUTION. Putting $x = e^z$

$\Rightarrow \quad z = \log x$ and $D \equiv \dfrac{d}{dz}$

Thus, the given equation becomes

$$[D(D-1) - 4D + 6]y = e^z$$

$\Rightarrow \quad\quad (D^2 - 5D + 6)y = e^z$ ...(2)

which is a linear equation in y .

To find the C.F. of (2), the auxiliary equation is

$$m^2 - 5m + 6 = 0$$

$\Rightarrow \quad (m-2)(m-3) = 0$ which gives $m = 2, 3$

$\therefore \quad\quad$ C.F. $= C_1 e^{2z} + C_2 e^{3z}$

Now, P.I. $= \dfrac{1}{D^2 - 5D + 6}e^z = \dfrac{1}{1-5+6}e^z = \dfrac{1}{2}e^z$

Hence, the general solution is given by
$$y = \text{C.F.} + \text{P.I.}$$
$$\Rightarrow \qquad y = C_1 e^{2z} + C_2 e^{3z} + \frac{1}{2}e^z$$
Now, put $\quad e^z = x.$
$$y = C_1 x^2 + C_2 x^3 + \frac{1}{2}x$$

EXAMPLE 3. *Solve* $x^2 \dfrac{d^2y}{dx^2} - 3x\dfrac{dy}{dx} + 4y = 2x^2.$

SOLUTION. Here, the given equation is
$$x^2 \frac{d^2y}{dx^2} - 3x\frac{dy}{dx} + 4y = 2x^2 \qquad ...(1)$$

Let $x = e^z$ then
$$z = \log x, \ x\frac{dy}{dx} = Dy, \ x^2\frac{d^2y}{dx^2} = D(D-1)y$$
Put in (1), we get
$$(D(D-1) - 3D + 4)y = 2e^{2z}$$
$$(D^2 - 4D + 4)\,y = 2e^{2z} \qquad ...(2)$$
To find the C.F. of (1), the auxiliary equation is
$$m^2 - 4m + 4 = 0$$
$$\Rightarrow \qquad (m-2)^2 = 0 \Rightarrow m = 2,2$$
$$\therefore \quad \text{C.F.} = (C_1 + C_2 z)e^{2z} = (C_1 + C_2 \log x)\, x^2$$
Now, P.I. $= \dfrac{1}{(D-2)^2} 2e^{2z}$
$$= 2e^{2z}\cdot\frac{1}{D^2}\cdot 1 \ = \ 2\frac{1}{(D-2)^2}e^{2z}$$
$$= 2e^{2z}\frac{1}{[(D+2)-2]^2}\cdot 1 = 2e^{2z}\cdot\frac{1}{D^2}\cdot 1$$
$$= 2e^{2z}\cdot\frac{z^2}{2} = z^2 e^{2z} = (\log x)^2 \cdot x^2$$
Hence, the complete solution of (1) is given by
$$y = \text{C.F.} + \text{P.I.}$$
$$\Rightarrow \qquad y = (C_1 + C_2 \log x)\, x^2 + x^2\,(\log x)^2$$

EXAMPLE 4. *Solve* $x^2 \dfrac{d^2y}{dx^2} + 2x\dfrac{dy}{dx} - 20y = (x+1)^2$

SOLUTION. The given equation is
$$x^2 \frac{d^2y}{dx^2} + 2x\frac{dy}{dx} - 20y = (x+1)^2 \qquad ...(1)$$
Putting $x = e^z$, $D \equiv \dfrac{d}{dz}$ in (1), we get
$$[D(D-1) + 2D - 20]\,y = (e^z+1)^2$$
$$\Rightarrow \qquad [D^2 + D - 20]y = (e^z+1)^2$$
or $\quad (D+5)(D-4)y = e^{2z} + 2e^z + 1 \qquad ...(2)$
To find the C.F. of (1), the auxiliary equation is
$$(m+5)(m-4) = 0 \ \Rightarrow \ m = 4, -5$$
$$\therefore \quad \text{C.F.} = C_1 e^{4z} + C_2 e^{-5z} = C_1 x^4 + C_2 x^{-5}$$
Now, P.I. $= \dfrac{1}{(D+5)(D-4)} e^{2z}$
$$+ \frac{1}{(D+5)(D-4)} 2e^z + \frac{1}{(D+5)(D-4)} e^{0z}$$
$$= \frac{e^{2z}}{(2+5)(2-4)} + 2\frac{1\cdot e^z}{(1+5)(1-4)} + \frac{1}{5(-4)} e^{0z}$$
$$= \frac{e^{2z}}{-14} + \frac{2e^z}{-18} - \frac{1}{20}$$
$$= \frac{e^{2z}}{-14} - \frac{1}{9}e^z - \frac{1}{20} = -\frac{1}{14}x^2 - \frac{1}{9}x - \frac{1}{20}$$

Hence, the general solution is given by
$$y = \text{C.F.} + \text{P.I.}$$
$$\Rightarrow \quad y = C_1 x^4 + C_2 x^{-5} - \frac{1}{14}x^2 - \frac{1}{9}x - \frac{1}{20}$$

EXAMPLE 5. *Solve* $(x+1)^2 \dfrac{d^2y}{dx^2} + (x+1)\dfrac{dy}{dx} = (2x+3)(2x+4)$

[MTU (SUM)-2011, GBTU (CO)-2011]

SOLUTION. Put $x+1 = e^z \ \Rightarrow \ z = \log(x+1)$
Also $D \equiv \dfrac{d}{dz}$
The transformed equation is
$$[D(D-1) + D]y = (2e^z + 1)(2e^z + 2)$$
$$D^2 y = 4e^{2z} + 6e^z + 2$$
A.E. is $m^2 = 0 \ \Rightarrow \ m = 0,0$
$$\therefore \quad \text{C.F.} = C_1 + C_2 z$$
$$\text{P.I.} = \frac{1}{D^2}(4e^{2z} + 6e^z + 2) = e^{2z} + 6e^z + z^2$$
Complete solution is
$$y = \text{C.F.} + \text{P.I.}$$
$$= C_1 + C_2 z + e^{2z} + 6e^z + z^2$$
$$= C_1 + C_2 \log(x+1) + (x+1)^2$$
$$+ 6(x+1) + [\log(x+1)]^2$$

EXAMPLE 6. *Solve* $x^3 \dfrac{d^3y}{dx^3} + 3x^2 \dfrac{d^2y}{dx^2} + x\dfrac{dy}{dx} + y = x + \log x.$

(UPTU–2001, BHOPAL–2008)

SOLUTION. Putting $x = e^z$, in the given equation, we get
$$(D(D-1)(D-2) + 3D(D-1) + (D+1)y$$
$$= e^z + z$$
where $D \equiv \dfrac{d}{dz}$
$$\Rightarrow \qquad (D^3 + 1)y = e^z + z \qquad ...(1)$$
To find the C.F. of (1), the auxiliary equation is
$$m^3 + 1 = 0 \ \Rightarrow \ (m+1)(m^2 - m + 1) = 0$$
$$\Rightarrow \quad m = -1, \frac{1 \pm i\sqrt{3}}{2}$$
$$\therefore \quad \text{C.F.} = C_1 e^{-z} + e^{z/2}$$
$$\left[C_2 \cos\left(\frac{\sqrt{3}}{2}\right)z + C_3 \sin\left(\frac{\sqrt{3}}{2}\right)z\right]$$
$$= C_1 x^{-1} + x^{1/2}\left[C_2 \cos\left(\frac{\sqrt{3}}{2}\log x\right)\right.$$
$$\left. + C_3 \sin\left(\frac{\sqrt{3}}{2}\log x\right)\right]$$
Now, P.I. $= \dfrac{1}{D^3+1}(e^z + z) = \dfrac{1}{D^3+1}e^z + \dfrac{1}{D^3+1}z$
$$= \frac{e^z}{1^3 + 1} + (1+D^3)^{-1}z = \frac{1}{2}e^z + (1 - D^3 + ...)z$$
$$= \frac{1}{2}e^z + z = \frac{1}{2}x + \log x$$
Hence, the general solution of (1) is given by
$$y = \text{C.F.} + \text{P.I.}$$

$$\Rightarrow \quad y = \frac{C_1}{x} + \sqrt{x}\left[C_2\cos\left(\frac{\sqrt{3}}{2}\log x\right)\right.$$

$$\left. + C_3\sin\left(\frac{\sqrt{3}}{2}\log x\right)\right] + \frac{1}{2}x + \log x$$

EXAMPLE 7. *Solve* $x^2\dfrac{d^2y}{dx^2} - x\dfrac{dy}{dx} + 4y$

$$= \cos(\log x) + x\sin(\log x)$$

SOLUTION. Putting, $x = e^z \Rightarrow z = \log x$ in the given equation, we get

$$[D(D-1) - D + 4]\,y = \cos z + e^z\sin z$$

where $\qquad D \equiv \dfrac{d}{dz}$

$$\Rightarrow \quad (D^2 - 2D + 4)\,y = \cos z + e^z\sin z \quad ...(1)$$

To find the C.F. of (1), the auxiliary equation is

$$m^2 - 2m + 4 = 0 \Rightarrow m = 1 \pm i\sqrt{3}$$

$$\text{C.F.} = e^z[C_1\cos\sqrt{3}z + C_2\sin\sqrt{3}z]$$

$$= x[C_1\cos(\sqrt{3}\log x) + C_2\sin(\sqrt{3}\log x)]$$

Now, P.I. $= \dfrac{1}{(D^2 - 2D + 4)}\cos z$

$$+ \frac{1}{(D^2 - 2D + 4)}e^z.\sin z$$

$$= \frac{1}{-1^2 - 2D + 4}\cos z$$

$$+ e^z\frac{1}{(D+1)^2 - 2(D+1) + 4}\sin z$$

$$= \frac{1}{(3 - 2D)}\cos z + e^z\frac{1}{D^2 + 3}\sin z$$

$$= \frac{(3 + 2D)}{9 - 4D^2}\cos z + e^z.\frac{1}{-1^2 + 3}\sin z$$

$$= \frac{3\cos z - 2\sin z}{9 + 4} + \frac{e^z\sin z}{2}$$

$$= \frac{1}{13}[3\cos(\log x) - 2\sin(\log x)]$$

$$+ \frac{1}{2}x\sin(\log x)$$

Hence, the general solution of the given equation is
$$y = \text{C.F.} + \text{P.I.}$$

$$\Rightarrow \quad y = x[C_1\cos(\sqrt{3}\log x) + C_2\sin(\sqrt{3}\log x)]$$

$$+ \frac{1}{13}[3\cos(\log x) - 2\sin(\log x)] + \frac{1}{2}x\sin(\log x)$$

EXAMPLE 8. *Solve* $x^3\dfrac{d^3y}{dx^3} + 2x^2\dfrac{d^2y}{dx^2} + 2y = 10\left[x + \dfrac{1}{x}\right].$

[SAMBHALPUR-1998, PTU-2003, SVTU-2006, UPTU(CO)-2009]

SOLUTION. Putting $x = e^z$, in the given equation, we get

$$[D(D-1)(D-2) + 2D(D-1) + 2]\,y$$

$$= 10(e^z + e^{-z})\ ,\ \text{where}\ \frac{d}{dz} \equiv D$$

$$\Rightarrow \quad (D^3 - D^2 + 2)\,y = 10(e^z + e^{-z}) \qquad ...(1)$$

To find the C.F. of (1), the auxiliary equation is

$$m^3 - m^2 + 2 = 0$$

$$\Rightarrow \quad (m+1)(m^2 - 2m + 2) = 0 \Rightarrow m = -1, 1 \pm i$$

$$\therefore \quad \text{C.F.} = C_1e^{-z} + e^z(C_2\cos z + C_3\sin z)$$

$$= \frac{C_1}{x} + x[C_2\cos(\log x) + C_3\sin(\log x)]$$

Now, P.I. $= \dfrac{10}{(D+1)(D^2 - 2D + 2)}e^z$

$$+ \frac{10}{(D+1)(D^2 - 2D + 2)}e^{-z}$$

$$= \frac{10e^z}{(1+1)(1-2+2)} + \frac{10}{(1+2+2)}.\frac{1}{(D+1)}e^{-z}.1$$

$$= 5e^z + 2e^{-z}\frac{1}{D - 1 + 1}.1$$

$$= 5e^z + 2e^{-z}.z = 5x + \frac{2}{x}\log x$$

Hence, the general solution of the given equation is
$$y = \text{C.F.} + \text{P.I.}$$

$$\Rightarrow \quad y = \frac{C_1}{x} + x\,[C_2\cos(\log x)$$

$$+ C_3\sin(\log x)] + 5x + \frac{2}{x}\log x$$

EXAMPLE 9. *Solve* $x^2\dfrac{d^2y}{dx^2} + 4x\dfrac{dy}{dx} + 2y = e^x.$

[UPTU-2005, KURUKSHETRA-2005]

SOLUTION. Here, the given equation is

$$x^2\frac{d^2y}{dx^2} + 4x\frac{dy}{dx} + 2y = e^x \qquad ...(1)$$

Putting $x = e^z$, $z = \log x$ and $\dfrac{d}{dz} \equiv D$, we get

$$[D(D-1) + 4D + 2]\,y = e^{e^z}$$

or $\qquad (D^2 + 3D + 2)\,y = e^{e^z} \qquad ...(2)$

To find the C.F. of (2), the auxiliary equation is

$$m^2 + 3m + 2 = 0 \Rightarrow m = -1, -2$$

$$\therefore \quad \text{C.F.} = C_1e^{-z} + C_2e^{-2z} = \frac{C_1}{x} + \frac{C_2}{x^2} \qquad ...(3)$$

Now, P.I. $= \dfrac{1}{(D+2)}\left[\dfrac{1}{(D+1)}e^{e^z}\right]$

Let $\qquad \dfrac{1}{(D+1)}e^{e^z} = u$

$$\therefore \qquad \frac{du}{dz} + u = e^{e^z}$$

This is a linear equation, with I.F. $= e^z$.

$$\therefore \qquad ue^z = \int e^{e^z}.e^z\,dz = e^{e^z}$$

$$\therefore \qquad u = e^{e^z}.e^{-z}$$

$$\therefore \qquad \text{P.I.} = \frac{1}{(D+2)}(e^{e^z}.e^{-z}) = v \ \ (\text{say})$$

$$\therefore \qquad \frac{dv}{dz} + 2v = e^{e^z}.e^{-z}$$

This is again a linear equation with I.F. $= e^{2z}$

$$\therefore \qquad v.e^{2z} = \int e^{e^z}.e^{-z}.e^{2z}\,dz$$

$$= \int e^{e^z}.e^z\,dz = e^{e^z}$$

$$\Rightarrow \qquad v = e^{e^z}.e^{-2z} = \frac{e^x}{x^2}$$

Hence, the general solution of the given equation is given by

$$y = \frac{C_1}{x} + \frac{C_2}{x^2} + \frac{e^x}{x^2}$$

EXAMPLE 10. *Solve the homogeneous linear differential equation* $x^2 \dfrac{d^2y}{dx^2} + x \dfrac{dy}{dx} + y = (\log x)\sin(\log x)$

[UPTU-2002, KURUKSHETRA-2006, MADRAS-2006, KERALA-2005]

SOLUTION. Since given equation is homogeneous

Put $\qquad x = e^z \Rightarrow \log x = z$

Also, $\qquad x \dfrac{dy}{dx} = Dy$

$$x^2 \frac{d^2y}{dx^2} = D(D-1)y \qquad \left[\because D \equiv \frac{d}{dz} \right]$$

The transformed equation is

$$D(D-1)y + Dy + y = z \sin z$$

$$\Rightarrow \qquad (D^2 - D + D + 1)y = z \sin z$$

$$\Rightarrow \qquad (D^2 + 1)y = z \sin z$$

A.E. is $m^2 + 1 = 0 \Rightarrow m = \pm i$

C.F. $= C_1 \cos z + C_2 \sin z$

P.I. $= \dfrac{1}{D^2 + 1} z \sin z$

$$= \text{Imaginary part of } \frac{1}{D^2 + 1} z e^{iz}$$

$$= \text{Imaginary part of } e^{iz} \frac{1}{(D+i)^2 + 1} z .$$

$$= \text{Imaginary part of } e^{iz} \frac{1}{D^2 + 2iD - 1 + 1} z$$

$$= \text{Imaginary part of } e^{iz} \frac{1}{D^2 + 2iD} z$$

$$= \text{Imaginary part of } e^{iz} \frac{1}{2iD} \frac{1}{\left(1 + \dfrac{D}{2i}\right)} z$$

$$= \text{Imaginary part of } e^{iz} \frac{1}{2iD} \left(1 - \frac{D}{2i}\right) z$$

$$= \text{Imaginary part of } e^{ir} \frac{1}{2iD} \left(z - \frac{1}{2i}\right)$$

$$= \text{Imaginary part of } e^{iz} \frac{1}{2i} \left(\frac{z^2}{2} - \frac{z}{2i}\right)$$

$$= \text{Imaginary part of } \frac{1}{2i}(\cos z + i \sin z)\left(\frac{z^2}{2} - \frac{z}{2i}\right)$$

$$= \text{Imaginary part of } (\cos z + i \sin z)\left(\frac{z^2}{4i} + \frac{z}{4}\right)$$

$$= -\frac{z^2}{4} \cos z + \frac{z}{4} \sin z$$

Complete solution is
$$y = \text{C.F.} + \text{P.I.}$$

$$y = C_1 \cos z + C_2 \sin z - \frac{z^2}{4} \cos z + \frac{z}{4} \sin z$$

$$y = C_1 \cos(\log x) + C_2 \sin(\log x)$$

$$- \frac{1}{4}(\log x)^2 \cos(\log x) + \frac{1}{4}(\log x)\sin(\log x)$$

Exercise-8.23

Solve the following differential equations :

1. $x^2 \dfrac{d^2y}{dx^2} - x \dfrac{dy}{dx} + y = x$

2. $x^2 \dfrac{d^2y}{dx^2} - 4x \dfrac{dy}{dx} + 6y = x^4$

3. $x^2 \dfrac{d^2y}{dx^2} + x \dfrac{dy}{dx} - 4y = x^2$

4. $x^4 \dfrac{d^3y}{dx^3} + 2x^3 \dfrac{d^2y}{dx^2} - x^2 \dfrac{dy}{dx} + xy = 1$

5. $x^3 \dfrac{d^3y}{dx^3} + 3x^2 \dfrac{d^2y}{dx^2} + x \dfrac{dy}{dx} + y = x \log x$

6. $x^3 \dfrac{d^3y}{dx^3} + 2x^2 \dfrac{d^2y}{dx^2} + 3x \dfrac{dy}{dx} - 3y = x^2 + x$

7. $x^3 \dfrac{d^3y}{dx^3} - x^2 \dfrac{d^2y}{dx^2} + 2x \dfrac{dy}{dx} - 2y = x^3 + 3x$

8. $x^2 \dfrac{d^2y}{dx^2} + x \dfrac{dy}{dx} - y = x^m$

9. $(x^3 D^3 + 3x^2 D^2 - 2xD + 2)y = 0$

10. $x^2 \dfrac{d^2y}{dx^2} - x \dfrac{dy}{dx} + y = 2 \log x$

11. $(x^4 D^4 + 6x^3 D^3 + 9x^2 D^2 + 3xD + 1)y = (1 + \log x)^2$

12. $x^3 \dfrac{d^3y}{dx^3} + 2x^2 \dfrac{d^2y}{dx^2} + x \dfrac{dy}{dx} - y = \cos(\log x)$

13. $x^2 \dfrac{d^2y}{dx^2} - 2x \dfrac{dy}{dx} - 4y = x^4$ [AMIE-2000]

14. $x^2 \dfrac{d^2y}{dx^2} + 6x \dfrac{dy}{dx} + 6y = (\log x)^2$

15. $x^3 \dfrac{d^3y}{dx^3} + 2x^2 \dfrac{d^2y}{dx^2} + 3x \dfrac{dy}{dx} - 3y = 0$

16. $(3x + 2)^2 \dfrac{d^2y}{dx^2} - (3x + 2) \dfrac{dy}{dx} - 12y = 6x$ (GBTU-2011)

17. $x^2 \dfrac{d^2y}{dx^2} - x \dfrac{dy}{dx} + y = \log x$ (VTU-2010)

18. $x^2 \dfrac{d^2y}{dx^2} - 3x \dfrac{dy}{dx} + y = \log x \dfrac{\sin(\log x + 1)}{x}$ (ISM-2001)

Hints to Selected Problems

1. Put $x = e^z$, $D \equiv \dfrac{d}{dz}$.

Then, given equation reduces to $(D^2 - 2D + 1)y = e^z$

C.F. $= (C_1 + C_2 z)e^z$

P.I. $= \dfrac{1}{(D-1)^2} e^z = e^z \left(\dfrac{1}{D^2}\right) . 1$

3. C.F. $= C_1 e^{2z} + C_2 e^{-2z}$

P.I. $= \dfrac{1}{(D^2 - 4)} e^{2z} = \dfrac{1}{(D+2)(D-2)} e^{2z} = \dfrac{1}{4(D-2)} e^{2z}$

$= \dfrac{1}{4} e^{2z} \dfrac{1}{(D+2)-2} \cdot 1 = \dfrac{1}{4} e^{2z} \cdot \dfrac{1}{D} \cdot 1 = \dfrac{1}{4} z e^{2z}$

5. C.F. $= C_1 e^{-z} + C_2 e^{z/2} \cos\left(\dfrac{\sqrt{3}}{2} z + C_3\right)$

P.I. $= \dfrac{1}{(D^3 + 1)} z e^z = e^z \dfrac{1}{(D^3 + 3D^2 + 3D + 2)} \cdot z$

12. C.F. $= C_1 x + C_2 \cos(\log x) + C_3 \sin(\log x)$

P.I. $= \dfrac{1}{(D-1)(D^2 + 1)} \cdot \cos z = \dfrac{D+1}{D^4 - 1} \cos z$

$= \dfrac{1}{D^2 + 1}\left[\dfrac{(D+1)\cos z}{(D^2 - 1)}\right] = \dfrac{1}{D^2 + 1}\left[\dfrac{-\sin z + \cos z}{-1 - 1}\right]$

Answers

1. $y = x(C_1 + C_2 \log x) + \dfrac{x}{2} (\log x)^2$

2. $y = C_1 x^2 + C_2 x^3 + \dfrac{1}{2} x^4$

3. $y = C_1 x^2 + \dfrac{C_2}{x^2} + \dfrac{1}{4} x^2 \log x$

4. $y = (C_1 + C_2 \log x) x + C_3 x^{-1} + \dfrac{1}{4x} \log x$

5. $y = C_1 x^{-1} + C_2 \sqrt{x} \cos\{(\sqrt{3}/2) \log x + C_3\} + \dfrac{3}{2} x \log x - \dfrac{3}{4} x$

6. $y = C_1 x + C_2 \cos\{(\sqrt{3} \log x) + C_3 \sin(\sqrt{3} \log x)\} + \dfrac{1}{7} x^2 + \dfrac{1}{4} x \log x$

7. $y = (C_1 + C_2 \log x) x + C_3 x^2 + \dfrac{x^3}{4} - \dfrac{3}{2} x (\log x)^2$

8. $y = C_1 x + C_2 x^{-1} + \dfrac{x^m}{(m^2 - 1)}$

9. $y = x[C_1 + C_2 \log x] + C_3 x^{-2}$

10. $y = x[C_1 + C_2 \log x] + 2\log x + 4$

11. $y = (C_1 + C_2 \log x) \cos(\log x) + (C_3 + C_4 \log x) \sin(\log x) + (\log x)^2 + 2(\log x) - 3$

12. $y = C_1 x + C_2 \cos(\log x) + C_3 \sin(\log x) - \dfrac{1}{4} \log x [\cos(\log x) + \sin(\log x)]$

13. $y = C_1 x^4 + \dfrac{C_2}{x} + \dfrac{x^4 \log x}{5}$

14. $y = C_1 x^{-2} + C_2 x^{-3} + \dfrac{1}{108}\{18(\log x)^2 - 30(\log x) + 19\}$

15. $y = C_1 x + C_2 \cos(\sqrt{3} \log x) + C_3 \sin(\sqrt{3} \log x)$

16. $y = C_1(3x + 2)^2 + C_2(3x + 2)^{-2/3} - \dfrac{2}{15}(3x + 2) + \dfrac{1}{3}$

17. $y = (C_1 + C_2 \log x) x + \log x + 2$

18. $y = x^2(C_1 x^{\sqrt{3}} + C_2 x^{-\sqrt{3}}) + \dfrac{1}{x}\left[\dfrac{1}{6}(\log x + 1) + \dfrac{\log x}{61}\{5\sin(\log x) + 6\cos(\log x) + \dfrac{2}{3721}[27\sin(\log x) + |9|\cos(\log x)\}\right]$

8.3.2 An Alternative Approach for getting P.I. when the R.H.S. is kept unchanged

If the given equation can be written as $\qquad F(D_1) y = f(x) \qquad$ ...(1)

Then, use the following results :

(a) $\dfrac{1}{D_1 - \alpha} f(x) = x^\alpha \int x^{-\alpha - 1} f(x)\, dx$

(b) $\dfrac{1}{D_1 + \alpha} f(x) = x^{-\alpha} \int x^{\alpha - 1} f(x)\, dx$

Here, to find the P.I., first factorize $F(D_1)$ into linear factors and then use any of the following method :

(i) The operator $\dfrac{1}{F(D_1)}$ can be broken up into partial fractions.

Then, $\quad$ P.I. $= \dfrac{1}{f(D_1)} f(x) = \left[\dfrac{A_1}{D_1 - \alpha_1} + \dfrac{A_2}{D_1 - \alpha_2} + ... + \dfrac{A_n}{D_1 - \alpha_n}\right] f(x)$

which can be obtained easily, by using (a) and (b).

(ii) P.I. $= \dfrac{1}{(D_1 - \alpha_1)(D_1 - \alpha_2)...(D_1 - \alpha_n)} f(x)$

Now, using (a) and (b), by taking the factors in succession beginning with the first on the right.

Solved Examples

EXAMPLE 1. *Solve* $x^2 \dfrac{d^2 y}{dx^2} + 4x \dfrac{dy}{dx} + 2y = e^x.$

SOLUTION. Putting $x = e^z \Rightarrow z = \log x$ and $\dfrac{d}{dz} \equiv D$, in given equation, we get (without changing R.H.S.)

$[D(D-1) + 4D + 2] y = e^x$

$\Rightarrow \quad (D^2 + 3D + 2) y = e^x \qquad$ ... (1)

To find the C.F. of (1), the auxiliary equation is

$m^2 + 3m + 2 = 0 \Rightarrow m = -1, -2$

$\therefore \quad$ C.F. $= C_1 e^{-z} + C_2 e^{-2z} = C_1 x^{-1} + C_2 x^{-2}$

Now, $\qquad$ P.I. $= \dfrac{1}{(D+2)(D+1)} e^x$

$= \left[\dfrac{1}{D+1} - \dfrac{1}{D+2}\right] e^x$

[By breaking up into partial fractions]

$= \dfrac{1}{D+1} e^x - \dfrac{1}{D+2} e^x$

$= x^{-1} \int x^{1-1} e^x dx - x^{-2} \int x^{2-1} e^x dx$

[By using (a)]

$= x^{-1} \int e^x dx - x^{-2} \int x\, e^x\, dx$

$= x^{-1} e^x - x^{-2}[x e^x - \int 1 . e^x dx]$

$= x^{-1} e^x - x^{-2}[x e^x - e^x] = x^{-2} e^x$

Hence, the general solution is given by

$y = $ C.F. + P.I.

$\Rightarrow \qquad y = C_1 x^{-1} + C_2 x^{-2} + x^{-2} e^x$

EXAMPLE 2. *Solve* $x^2 \dfrac{d^2 y}{dx^2} + 3x \dfrac{dy}{dx} + y = \dfrac{1}{(1-x)^2}.$ (PTU-2003)

SOLUTION. Putting $x = e^z$, i.e., $z = \log x$, we get

$$[D(D-1)+(3D+1)]y = \frac{1}{(1-x)^2}$$

[Without changing RHS]

$$\Rightarrow \qquad (D+1)^2 y = \frac{1}{(1-x)^2} \qquad \text{...(1)}$$

To find the C.F. of (1), the auxiliary equation is

$$(m+1)^2 = 0 \;\Rightarrow\; m = -1, -1$$

$$\therefore \quad \text{C.F.} = (C_1 + C_2 z)e^{-z} = (C_1 + C_2 \log x)\, x^{-1}$$

Now, P.I. $= \dfrac{1}{D+1} \cdot \dfrac{1}{D+1}(1-x)^{-2}$

$$= \frac{1}{D+1} x^{-1} \int x^{1-1}(1-x)^{-2}\,dx$$

$$= \frac{1}{D+1} x^{-1}(1-x)^{-1}$$

$$= x^{-1} \int x^{1-1} x^{-1}(1-x)^{-1}\,dx$$

$$= x^{-1} \int \frac{dx}{x(1-x)} = x^{-1} \int \left(\frac{1}{x} + \frac{1}{1-x}\right)dx$$

$$= x^{-1}\,[\log x - \log(1-x)]$$

$$= x^{-1}\log\left(\frac{x}{1-x}\right)$$

Hence, the required general solution is given by

$$y = \text{C.F.} + \text{P.I.}$$

$$\Rightarrow \qquad y = (C_1 + C_2 \log x)\,x^{-1} + x^{-1}\log\left(\frac{x}{1-x}\right)$$

8.3.3 EQUATION REDUCIBLE TO HOMOGENEOUS FORM

Any differential equation of the form

$$(a+bx)^n \frac{d^n y}{dx^n} + A_1(a+bx)^{n-1}\frac{d^{n-1}y}{dx^{n-1}} + \dots + A_{n-1}(a+bx)\frac{dy}{dx} + A_n y = X(x) \qquad \text{...(1)}$$

where the coefficients $A_1, A_2, \dots, A_n$ are constants, can be transformed into homogeneous linear equation with constant coefficients by changing independent variable from x to z, by a suitable substitution $z = a + bx$.

Let $\quad z = a + bx \;\Rightarrow\; \dfrac{dz}{dx} = b$

Now we have

$$\frac{dy}{dx} = \frac{dy}{dz}\cdot\frac{dz}{dx} = \frac{dy}{dz}\cdot b$$

$$\frac{d^2 y}{dx^2} = \frac{d}{dx}\left(\frac{dy}{dx}\right) = \frac{d}{dz}\left(b\frac{dy}{dz}\right)\frac{dz}{dx} = b^2\frac{d^2 y}{dz^2}$$

Similarly, $\qquad \dfrac{d^n y}{dx^n} = b^n \dfrac{d^n y}{dz^n}$

Putting all these values in equation (1), and dividing throughout by b^n, we get

$$z^n \frac{d^n y}{dz^n} + \frac{A_1}{b}z^{n-1}\frac{d^{n-1}y}{dz^{n-1}} + \frac{A_2}{b^2}z^{n-2}\frac{d^{n-2}y}{dz^{n-2}} + \dots + \frac{A_{n-1}}{b^{n-1}}z\frac{dy}{dz} + \frac{A_n}{b^n}y = \frac{1}{b^n}X\left(\frac{z-a}{b}\right) \qquad \text{...(2)}$$

Now, this is the standard homogeneous equation.

REMARK

- Sometimes, the given equation can be solved easily by making the substitution $e^z = a + bx \;\Rightarrow\; z = \log(a+bx)$

Solved Examples

EXAMPLE 1. *Solve* $(1+x)^2 \dfrac{d^2 y}{dx^2} + (1+x)\dfrac{dy}{dx} + y = 4\cos\log(1+x)$

SOLUTION. Let $(1+x) = e^z$ and $\dfrac{d}{dz} \equiv D$.

Then, the given equation becomes

$$[D(D-1)+D+1]\,y = 4\cos z$$

$$\Rightarrow \qquad (D^2+1)\,y = 4\cos z \qquad \text{...(1)}$$

To find the C.F. of (1), the auxiliary equation is

$$m^2 + 1 = 0 \;\Rightarrow\; m = \pm i$$

$$\therefore \quad \text{C.F.} = e^{0z}(C_1 \cos z + C_2 \sin z)$$

$$= C_1 \cos\log(1+x) + C_2 \sin\log(1+x)$$

Now, P.I. $= \dfrac{1}{D^2+1}\,4\cos z = 4\dfrac{1}{D^2+1}\cos z$

$$= \text{Real part of } 4\cdot\frac{1}{D^2+1}(\cos z + i\sin z)$$

$$= \text{Real part of } 4\cdot\frac{1}{D^2+1}\cdot e^{iz}$$

Consider $\dfrac{1}{D^2+1}e^{iz} = \dfrac{1}{(D+i)(D-i)}e^{iz} = \dfrac{1}{(i+i)(D-i)}e^{iz}$

$$= \frac{1}{2i(D-i)}e^{iz}\cdot 1 = \frac{1}{2i}e^{iz}\frac{1}{D+i-i}\cdot 1$$

$$= \frac{e^{iz}}{2i}\cdot\frac{1}{D}\cdot 1 = \frac{ze^{iz}}{2i} = \frac{z}{2i}(\cos z + i\sin z)$$

$$= -\frac{iz}{2}(\cos z + i\sin z) = -i\frac{z}{2}\cos z + \frac{z}{2}\sin z$$

$$\therefore \quad \text{P.I.} = \text{Real part of } 4\left(-\frac{iz}{2}\cos z + \frac{z}{2}\sin z\right)$$

$$= 2z\sin z = 2\log(1+x)\sin\log(1+x)$$

Hence, the general solution is given by

$$y = \text{C.F.} + \text{P.I.}$$

$$\Rightarrow \qquad y = C_1 \cos\log(1+x) + C_2 \sin\log(1+x)$$

$$+ 2\log(1+x)\sin\log(1+x)$$

EXAMPLE 2. **Solve** $2x^2 y \dfrac{dy}{dx} + 4y^2 = x^2 \left(\dfrac{dy}{dx}\right)^2 + 2xy \dfrac{dy}{dx}$
after making it homogeneous by the substitution $y = z^2$.

SOLUTION. Let $\quad y = z^2 \Rightarrow \dfrac{dy}{dx} = 2z \cdot \dfrac{dz}{dx}$

and $\quad \dfrac{d^2 y}{dx^2} = 2z \dfrac{d^2 z}{dx^2} + 2\left(\dfrac{dz}{dx}\right)^2$

Putting all these values in the given equation, we get

$$2x^2 z^2 \left[2z \dfrac{d^2 z}{dx^2} + 2\left(\dfrac{dz}{dx}\right)^2\right] + 4z^2$$

$$= x^2 \left(2z \dfrac{dz}{dx}\right)^2 + 2xz^2 \cdot 2z \dfrac{dz}{dx}$$

$$\Rightarrow \quad x^2 \dfrac{d^2 z}{dx^2} - x \dfrac{dz}{dx} + z = 0 \qquad \ldots(1)$$

which is standard homogeneous equation.
Putting $x = e^t$ or $\log x = t$ in (1), we get

$$(D(D-1) - D + 1)z = 0, \text{ where } D \equiv \dfrac{d}{dt}$$

$$\Rightarrow \qquad (D-1)^2 z = 0 \qquad \ldots(2)$$

To find the C.F. of (2), the auxiliary equation is
$(m-1)^2 = 0 \Rightarrow m = 1,1$

$\text{C.F.} = (C_1 + C_2 t)e^t = (C_1 + C_2 \log x) x$

$\Rightarrow$ the general solution is $z = \text{C.F.}$

$$[\because \text{R.H.S. of (1) is zero}]$$

$$\Rightarrow \qquad z = (C_1 + C_2 \log x)x$$

Squaring both sides, we get,

$$z^2 = x^2 (C_1 + C_2 \log x)^2$$

$$\Rightarrow \qquad y = x^2 (C_1 + C_2 \log x)^2$$

which is the required complete solution.

EXAMPLE 3. **Solve** $(1+x)^2 \dfrac{d^2 y}{dx^2} + (1+x)\dfrac{dy}{dx} + y = 2\sin[\log(1+x)]$

$$[(VTU)\text{-}2009 \text{ JNTU-}2005, \text{ KEWALS-}2005]$$

SOLUTION. Let $\quad 1 + x = e^z$ i.e., $z = \log(1+x)$

so that $\quad (1+x)\dfrac{dy}{dx} = Dy$

and $\quad (1+x)^2 \dfrac{d^2 y}{dx^2} = D(D-1)y$,

where $\qquad D \equiv \dfrac{d}{dz}$

Then, the given equation becomes

$$[D(D-1) + D + 1]y = 2\sin z$$

$$(D^2 + 1)y = 2\sin Z \qquad \ldots(1)$$

To find the C.F. of (1)
A.E. is $m^2 + 1 = 0 \Rightarrow m = \pm i$

$\therefore \quad \text{C.F.} = e^{oz}(c_1 \cos z + c_2 \sin z)$

$$= c_1 \cos\log(1+x) + c_2 \sin\log(1+x)$$

Now, P.I. $= 2\dfrac{1}{D^2 + 1}\sin z = 2z \cdot \dfrac{1}{2D}\sin z$

$$= z \int \sin z \, dz = -z \cos z$$

$$= -\log(1+x)\cos\log(1+x)$$

$$(\because \text{ On replacing } D^2 \text{ by } -1^2, D^2 + 1 = 0)$$

Hence, the general solution is given by
$y = \text{C.F.} + \text{P.I.}$

$$\Rightarrow \qquad y = c_1 \cos\log(1+x) + c_2 \sin\log(1+x)$$

$$-\log(1+x)\cos\log(1+x)$$

EXAMPLE 5. **Solve** $(2x-1)\dfrac{d^2 y}{dx} + (2x-1)\dfrac{dy}{dx} - 2y$

$$= 8x^2 - 2x + 3 \qquad (VTU\ 2006)$$

SOLUTION. Let $\quad 2x - 1 = e^2$, i.e., $z = \log(2x-1)$

So that $\quad (2x-1)\dfrac{dy}{dx} = 2Dy$

and $\quad (2x-1)^2 \dfrac{d^2 y}{dx^2} = 4D(D-1)y$,

where $\qquad D \equiv \dfrac{d}{dz}$

Then, the given equation becomes.
$4D(D-1)y + 2Dy - 2y$

$$= 8\left(\dfrac{1+e^z}{2}\right)^2 - 2\left(\dfrac{1+e^z}{2}\right) + 3$$

or $2D^2 y - Dy - y = e^{2z} + \dfrac{3}{2}e^z + 2 \qquad \ldots(1)$

To find the C.F. of (1)
A.E. is $2D^2 - D - 1 = 0 \Rightarrow D = 1, -1/2$

$\therefore \quad \text{C.F.} = c_1 e^z + c_2 e^{-z/2}$

Now, P.I. $= \dfrac{1}{2D^2 - D - 1}\left(e^{2z} + \dfrac{3}{2}e^z + 2\right)$

$$= \dfrac{1}{2 \cdot 4 - 2 - 1}e^{2z} + \dfrac{3}{2}\dfrac{z}{4D-1}e^z$$

$$+ 2 \cdot \dfrac{1}{2 \cdot 0^2 - 0 - 1}e^{oz}$$

$$[\because \text{ On putting } D = 1, 2D^2 - D - 1 = 0]$$

$$= \dfrac{1}{5}e^{2z} + \dfrac{3z}{2} \cdot \dfrac{1}{4-1}e^z - 2$$

$$= \dfrac{1}{5}e^{2z} + \dfrac{z}{2}e^z - 2$$

Hence, the general solution is given by
$y = \text{C.F.} + \text{P.I.}$

$$= C_1 e^z + C_2 e^{-z/2} + \dfrac{1}{5}e^{2z} + \dfrac{z}{2}e^z - 2$$

$$= C_1 (2x-1) + C_2 (2x-1)^{-1/2}$$

$$+ \dfrac{1}{5}(2x-1)^2 + \dfrac{1}{2}(2x-1)\log(2x-1) - 2$$

$$[\text{On replacing } z \text{ by } \log(2x-1)]$$

Exercise-8.24

Solve the following differential equations :

1. $(3x+2)^2 \dfrac{d^2 y}{dx^2} + 3(3x+2)\dfrac{dy}{dx} - 36y = 3x^2 + 4x + 1$

2. $(3x+2)^2 \dfrac{d^2 y}{dx^2} + 5(3x+2)\dfrac{dy}{dx} - 3y = x^2 + x + 1$ (MUMBAI-2006)

3. $(5+2x)^2 \dfrac{d^2 y}{dx^2} - 6(5+2x)\dfrac{dy}{dx} + 8y = 0$

4. $(1+2x)^2 \dfrac{d^2 y}{dx^2} - 6(1+2x)\dfrac{dy}{dx} + 16y = 8(1+2x)^2$

5. $16(x+1)^4\dfrac{d^4y}{dx^4} + 96(x+1)^3\dfrac{d^3y}{dx^3} + 104(x+1)^2\dfrac{d^2y}{dx^2}$

$\qquad\qquad + 8(x+1)\dfrac{dy}{dx} + y = x^2 + 4x + 3$

6. $(2x-1)^3\dfrac{d^3y}{dx^3} + (2x-1)\dfrac{dy}{dx} - 2y = 0$

7. $(x+1)^2\dfrac{d^2y}{dx^2} + (x+1)\dfrac{dy}{dx} = 4x^2 + 14x + 12$

Hints to Selected Problems

2. Putting $3x+2 = e^z$

$\text{C.F.} = C_1e^{-z} + C_2e^{z/3}$

$\text{P.I.} = \dfrac{1}{27}\dfrac{1}{(D+1)(3D-1)}(e^{2z} - e^z + 7)$

$= \dfrac{1}{27}\left[\dfrac{1}{(D+1)(3D-1)}e^{2z} - \dfrac{1}{(D+1)(3D-1)}e^z \right.$

$\qquad\qquad \left. + \dfrac{1}{(D+1)(3D-1)}.7e^{0z}\right]$

5. Putting $x+1 = e^z$

$\text{C.F.} = [C_1 + C_2\{\log(1+x)\}]e^{\log(1+x)/z} + [C_3 + C_4$

$\qquad\qquad\qquad\qquad \{\log(1+x)\}]e^{-\log\frac{(1-x)}{2}}$

$\text{P.I.} = \dfrac{1}{(4D^2-1)^2}[e^{2z} + 2e^z]$

6. $\text{C.F.} = C_1e^z + e^z\left[C_2\cosh\dfrac{\sqrt{3}}{2}z + C_3\sinh\dfrac{\sqrt{3}}{2}z\right],$

where $2x-1 = e^z$

Answers

1. $y = C_1(3x+2)^2 + C_2(3x+2)^{-2} + \dfrac{1}{108}[(3x+2)^2\log(3x+2)+1]$ **2.** $y = C_1(3x+2)^{1/3} + C_2(3x+2)^{-1} + \dfrac{(3x+2)^2}{405} - \dfrac{(3x+2)}{108} - \dfrac{7}{27}$

3. $y = (5+2x)^2[C_1(5+2x)^{\sqrt{2}} + C_2(5+2x)^{-\sqrt{2}}]$

4. $y = [C_1 + C_2\log(1+2x)](1+2x)^2 + (1+2x)^2[\log(1+2x)]^2$

5. $y = (C_1 + C_2t)e^{t/2} + (C_3 + C_4t)e^{-t/2} + \dfrac{e^{2t}}{225} + \dfrac{2}{9}e^t,$ where $t = \log(1+x)$

6. $y = C_1(2x-1) + (2x-1)C_2\cosh\left\{\left(\dfrac{\sqrt{3}}{2}\right)\log(2x-1)\right\} + C_3\sinh\left\{\left(\dfrac{\sqrt{3}}{2}\right)\log(2x-1)\right\}$

7. $y = C_1 + C_2\log(1+x) + x^2 + 8x + 7 + [\log(1+x)]^2$

8.4 ORDINARY SIMULTANEOUS LINEAR DIFFERENTIAL EQUATIONS

In this section, we shall discuss the ordinary differential equations involving two or more dependent variables. Here, we shall discuss the case when there are as many simultaneous equations as there are dependent variables. For solving such equations, obtain an equation involving one dependent variable with one independent variable, by the process of elimination. After solving the derived equation, we substitute back to get the other dependent variable.

Consider the simultaneous equation as

$$\left.\begin{array}{l} f_1(D)x + f_2(D)y = f(t) \\ \text{and} \quad g_1(D)x + g_2(D)y = g(t) \end{array}\right] \quad \text{with} \quad D \equiv \dfrac{d}{dt} \qquad \ldots(1)$$

where, x and y are functions of t and $f_1(D), f_2(D), g_1(D)$ and $g_2(D)$ are rational integral functions with constant coefficients, $f(t)$ and $g(t)$ are the functions of the independent variable t. Now define the determinant Δ such as

$$\Delta = \begin{vmatrix} f_1(D) & f_2(D) \\ g_1(D) & g_2(D) \end{vmatrix} \qquad \ldots(2)$$

then we can say Δ involves the operator coefficients of x and y in (1).

The equation (2) can be solved by the usual methods.

REMARKS

- To solve the simultaneous equations completely, we always require as many simultaneous equations as are the number of dependent variables.
- The method of solving the simultaneous differential equations with constant coefficients is similar to that of solving a set of simultaneous equations in Algebra.
- The number of arbitrary constants appearing in the general solution of the system (1) is equal to the degree in D of the determinant Δ given by (2), provided determinant is non-zero.
- The determinant Δ, defined by (2), involves the operator coefficients of x and y in (1).

8.4.1 METHOD OF SOLVING SIMULTANEOUS LINEAR DIFFERENTIAL EQUATION WITH CONSTANT COEFFICIENTS

Let x and y be the dependent variables and t be the independent variable. Generally, there are two methods for the solution of simultaneous linear differential equations with constant coefficients.

METHOD-1: SYMBOLIC METHOD WITH USE OF D

Consider the simultaneous equation such as

$$f_1(D)x + g_1(D)y = T_1 \qquad \dots (1)$$

and

$$f_2(D)x + g_2(D)y = T_2 \qquad \dots (2)$$

where T_1 and T_2 are the functions of independent variable t and f_1, f_2, g_1, g_2 are polynomial functions with constant coefficients.

Operate on both sides of equation (1) by $g_2(D)$ and equation (2) by $g_1(D)$, we get

$$g_2(D)f_1(D)x + g_2(D)g_1(D)y = g_2(D)T_1 \qquad \dots (3)$$

and

$$g_1(D)f_2(D)x + g_1(D)g_2(D)y = g_1(D)T_2 \qquad \dots (4)$$

Now, since $g_1(D)$ and $g_2(D)$ both have the constant coefficients then

$$g_1(D)g_2(D)y = g_2(D)g_1(D)y$$

therefore, from (3) and (4)

$$[g_2(D)f_1(D) - g_1(D)f_2(D)]x = g_2(D)T_1 - g_1(D)T_2 \qquad \dots (5)$$

Equation (5) is an ordinary differential equation with one dependent variable and can be solved by the usual methods. Thus, x can be obtained as a function of t. The value of y is then obtained by substituting the value of x in any of the given equations and integrating the resulting equation, if necessary. If however, y is obtained by an independent elimination as in the case of x, the values of x and y are to be substituted in given equation (1) and (2) and the arbitrary constants in x and y are to be so adjusted that the given equations are satisfied. Here, the number of independent arbitrary constants entering in the general solution is the index of the highest power of D.

METHOD-2: USE OF DIFFERENTIATION

If two equations containing $x, y, \dfrac{dx}{dt}$ and $\dfrac{dy}{dt}$ are given. Then we can obtain more equations containing $x, y, \dfrac{dx}{dt}, \dfrac{dy}{dt}, \dfrac{d^2x}{dt^2}$ and $\dfrac{d^2y}{dt^2}$ by differentiating the given equations with respect to t.

From these equations, we can obtain an equation containing x (or y) and its derivative, by eliminating x (or y) and its derivatives. Now, solve this new equation for x (or y) and substituting the value of x (or y) in any of the given equation and if necessary, solve the resulting equation.

REMARKS

- The method of differentiation will be used when found very necessary.
- Generally t will be the independent variable and x and y will be dependent variables. In some problems any other variable, say x, will be given as the independent variable and y and z as the dependent variables.

Solved Examples

EXAMPLE 1. **Solve** $\dfrac{dx}{dt} = -\omega y$ ...(1)

$$\dfrac{dy}{dt} = \omega x \qquad \dots(2)$$

[UPTU (SUM)-2007, UPTU-2008]

SOLUTION. Differentiating (1) with respect to t, we get

$$\frac{d^2x}{dt^2} + \omega \frac{dy}{dt} = 0 \implies \frac{d^2x}{dt^2} + \omega^2 x = 0 \quad \text{[By using (2)]}$$

$$\therefore \quad x = C_1 \cos \omega t + C_2 \sin \omega t$$

Putting this value of x in (1), we get

$$y = -(1/\omega)[-C_1\omega \sin \omega t + C_2\omega \cos \omega t]$$

$$= -C_2 \cos \omega t + C_1 \sin \omega t$$

EXAMPLE 2. **Solve** $\dfrac{dx}{dt} + 2\dfrac{dy}{dt} - 2x + 2y = 3e^t$

$$3\dfrac{dx}{dt} + \dfrac{dy}{dt} + 2x + y = 4e^{2t}.$$

SOLUTION. The given equation can be written as

$$(D-2)x + 2(D+1)y = 3e^t \qquad \dots (1)$$

and $\quad (3D+2)x + (D+1)y = 4e^{2t} \qquad \dots (2)$

Eliminating y between (1) and (2), we obtain

$$[2(3D+2) - (D-2)]x = 8e^{2t} - 3e^t$$

or $\qquad (5D+6)x = 8e^{2t} - 3e^t$

or $\quad \dfrac{dx}{dt} + \dfrac{6}{5}x = \dfrac{8}{5}e^{2t} - \dfrac{3}{5}e^t$

which is a linear differential equation with

$$\text{I.F.} = e^{\int 6/5\,dt} = e^{6t/5}$$

Now, solution becomes

$$x \cdot e^{6t/5} = \int e^{6t/5}\left[\frac{8}{5}e^{2t} - \frac{3}{5}e^t\right]dt + C_1$$

$$= \frac{8}{5}\int e^{16t/5} - \frac{3}{5}\int e^{11t/5}dt + C_1$$

$$= \frac{1}{2}e^{16t/5} - \frac{3}{11}e^{11t/5} + C_1$$

$$\therefore \quad x = \frac{1}{2}e^{2t} - \frac{3}{11}e^t + C_1 e^{-6t/5}$$

Now, $\qquad \dfrac{dx}{dt} = e^{2t} - \dfrac{3}{11}e^t - \dfrac{6}{5}C_1 e^{-6t/5}$

Putting this value in (1), we get

$$2Dy + 2y + e^{2t} - \frac{3}{11}e^t - \frac{6}{5}C_1e^{-6t/5}$$

$$-e^{2t} + \frac{6}{11}e^t - 2C_1e^{-6t/5} = 3e^t$$

or $\quad 2\dfrac{dy}{dt} + 2y = \dfrac{30}{11}e^t + \dfrac{16}{5}C_1e^{-6t/5}$

or $\quad \dfrac{dy}{dt} + y = \dfrac{15}{11}e^t + \dfrac{8}{5}C_1e^{-6t/5}$

which is a linear differential equation with

$$\text{I.F.} = e^{\int dt} = e^t$$

The solution is

$\therefore \quad y.e^t = \dfrac{15}{11}\int e^{2t}dt + \dfrac{8}{5}C_1\int e^{-t/5}dt$

$$+ C_2 = \frac{15}{22}e^{2t} - 8C_1e^{-t/5} + C_2$$

Hence, the solution is given by

$$x = \frac{1}{2}e^{2t} - \frac{3}{11}e^t + C_1e^{-6t/5}$$

$$y = \frac{15}{22}e^t - 8C_1e^{-6t/5} + C_1e^{-t}$$

EXAMPLE 3. **Solve** $\dfrac{dx}{dt} + \dfrac{dy}{dt} - 2y = 2\cos t - 7\sin t$

$$\dfrac{dx}{dt} - \dfrac{dy}{dt} + 2x = 4\cos t - 3\sin t \qquad \text{[UKTU-2001]}$$

SOLUTION. The given equation can be written as

$$Dx + (D-2)y = 2\cos t - 7\sin t \qquad \text{... (1)}$$

and $\;(D+2)x - Dy = 4\cos t - 3\sin t \qquad \text{... (2)}$

Eliminating y between (1) and (2), we obtain

$$[D^2 + (D-2)(D+2)]x$$

$$= D(2\cos t - 7\sin t) + (D-2)(4\cos t - 3\sin t)$$

or $\;(D^2 - 2)x = -9\cos t$

Auxiliary equation is $m^2 - 2 = 0 \Rightarrow m = \pm\sqrt{2}$

$$\text{C.F.} = C_1e^{\sqrt{2}t} + C_2e^{-\sqrt{2}t}$$

and P. I. $= -9\dfrac{1}{D^2 - 2}\cos t = \dfrac{-9\cos t}{D^2 - 2} = 3\cos t$

$$x = C_1e^{\sqrt{2}t} + C_2e^{-\sqrt{2}t} + 3\cos t$$

$$\frac{dx}{dt} = \sqrt{2}\,C_1\,e^{\sqrt{2}t} - C_2\sqrt{2}\,e^{-\sqrt{2}t} - 3\sin t$$

Now adding (1) and (2), we get

$$2Dx + 2x - 2y = 6\cos t - 10\sin t$$

or $\quad y = \dfrac{dx}{dt} + x - 3\cos t + 5\sin t$

$$= \sqrt{2}C_1e^{\sqrt{2}t} - C_2\sqrt{2}e^{-\sqrt{2}t} - 3\sin t + C_1e^{\sqrt{2}t}$$

$$+ C_2e^{-\sqrt{2}t} + 3\cos t - 3\cos t + 5\sin t$$

$$= (\sqrt{2}+1)C_1\,e^{\sqrt{2}t} + (1-\sqrt{2})C_2e^{-\sqrt{2}t} + 2\sin t$$

Hence, the solution is given as

$$x = C_1e^{\sqrt{2}t} + C_2e^{-\sqrt{2}t} + 3\cos t$$

$$y = (\sqrt{2}+1)C_1\,e^{\sqrt{2}t} + (1-\sqrt{2})C_2e^{-2t} + 2\sin t$$

EXAMPLE 4. **Solve** $\dfrac{dx}{dt} = ax + by,\; \dfrac{dy}{dt} = a'x + b'y$.

SOLUTION. The given equation can be written as

$$(D-a)x - by = 0 \qquad \text{... (1)}$$

and $-a'x + (D-b')y = 0 \qquad \text{... (2)}$

Eliminating y between (1) and (2), we get

$$[(D-a)(D-b') - a'b]\,x = 0$$

or $\quad [D^2 - (a+b')D + (ab' - a'b)]\,x = 0$

$\therefore \quad$ Auxiliary equation is

$$m^2 - (a+b')m + (ab' - a'b) = 0$$

$$m = \frac{(a+b') \pm \sqrt{[(a+b')^2 - 4(ab' - a'b)]}}{2}$$

$$= \frac{(a+b') \pm \sqrt{(a-b')^2 + 4a'b}}{2}$$

where roots m_1 and m_2 is

$$\left.\begin{array}{c} m_1 = \dfrac{(a+b') + \sqrt{(a-b')^2 + 4a'b}}{2} \\[4mm] m_2 = \dfrac{(a+b') - \sqrt{(a-b')^2 + 4a'b}}{2} \end{array}\right\} \qquad \text{... (3)}$$

$\therefore \quad x = C_1e^{m_1t} + C_2e^{m_2t}$

$$\frac{dx}{dt} = C_1m_1e^{m_1t} + C_2m_2e^{m_2t}$$

$\therefore \quad$ From (1), we get

$$y = \frac{1}{b}\left[\frac{dx}{dt} - ax\right]$$

$$= \frac{1}{b}[(m_1 - a)C_1e^{m_1t} + (m_2 - a)C_2e^{m_2t}]$$

Hence, the solution is

$$x = C_1e^{m_1t} + C_2e^{m_2t}$$

$$y = \frac{1}{b}[(m_1 - a)C_1e^{m_1t} + (m_2 - a)C_2e^{m_2t}]$$

EXAMPLE 5. **Solve** $\dfrac{dx}{dt} + 4x + 3y = t,\quad \dfrac{dy}{dt} + 2x + 5y = e^t$

[UPTU-2006, UPTU Q.BANK-2001]

SOLUTION. The given equation can be written as

$$(D+4)x + 3y = t \qquad \text{... (1)}$$

and $\;2x + (D+5)y = e^t \qquad \text{... (2)}$

Eliminating y between (1) and (2), we get

$$[(D+4)(D+5) - 6]x = (D+5)t - 3e^t$$

or $\;(D^2 + 9D + 14)x = 1 + 5t - 3e^t$

$\therefore \quad$ Auxiliary equation is $m^2 + 9m + 14 = 0$

$\therefore \quad m = -2, -7$

Complementary function $= C_1e^{-2t} + C_2e^{-7t}$

Particular integral

$$= \frac{1}{14 + 9D + D^2}(1 + 5t) - \frac{1}{14 + 9D + D^2}3e^t$$

$$= \frac{1}{14}\left[1 + \frac{9}{14}D + \frac{1}{14}D^2\right]^{-1}(1 + 5t) - \frac{3e^t}{14 + 9 + 1}$$

[On replacing D by 1]

$$= \frac{1}{14}\left[1 - \frac{9}{14}D - \frac{1}{14}D^2 + \ldots\right](1 + 5t) - \frac{1}{8}e^t$$

$$= \frac{1}{14}\left[1 + 5t - \frac{9}{14}.5\right] - \frac{1}{8}e^t$$

$$= \frac{1}{14}\left[5t - \frac{31}{14}\right] - \frac{1}{8}e^t$$

$$\therefore \quad x = C_1 e^{-2t} + C_2 e^{-7t} + \frac{5}{14}t - \frac{1}{8}e^t - \frac{31}{196}$$

$$\frac{dx}{dt} = -2C_1 e^{-2t} - 7C_2 e^{-7t} + \frac{5}{14} - \frac{1}{8}e^t$$

Putting above value in (1), we get

$$3y = -\frac{dx}{dt} - 4x + t = -2C_1 e^{-2t}$$

$$+3C_2 e^{-7t} - \frac{10}{7}t + t - \frac{5}{14} + \frac{31}{49} + \frac{1}{8}e^t + \frac{1}{2}e^t$$

or $\quad y = \dfrac{1}{3}\left[-2C_1 e^{-2t} + 3C_2 e^{-7t} - \dfrac{3}{7}t + \dfrac{27}{98} + \dfrac{5}{8}e^t\right]$

or $\quad x = C_1 e^{-2t} + C_2 e^{-7t} + \dfrac{5}{14}t - \dfrac{31}{196} - \dfrac{1}{8}e^t$

EXAMPLE 6. *Solve* $\dfrac{d^2x}{dt^2} + m^2 y = 0,\ \dfrac{d^2y}{dt^2} - m^2 x = 0.$

SOLUTION. The given equation can be written as

$$D^2 x + m^2 y = 0 \qquad \qquad ...(1)$$

$$-m^2 x + D^2 y = 0 \qquad \qquad ...(2)$$

Eliminating y between (1) and (2),

$$(D^4 + m^4)x = 0 .$$

Auxiliary equation is $M^4 + m^4 = 0$

or $\quad (M^2 + m^2)^2 - 2M^2 m^2 = 0$

or $(M^2 - \sqrt{2}Mm + m^2)(M^2 + \sqrt{2}Mm + m^2) = 0$

$\therefore \qquad\qquad M^2 - \sqrt{2}mM + m^2 = 0$

or $\qquad\qquad M^2 + \sqrt{2}\,mM + m^2 = 0$

$$\therefore \quad M = \frac{\sqrt{2}m \pm \sqrt{(2m^2 - 4m^2)}}{2},$$

$$M = \frac{-\sqrt{2}m \pm \sqrt{(2m^2 - 4m^2)}}{2}$$

$$= \frac{m}{\sqrt{2}} \pm \frac{m}{\sqrt{2}}i, \ -\frac{m}{\sqrt{2}} \pm \frac{m}{\sqrt{2}}i$$

$$\therefore \quad x = e^{mt/\sqrt{2}}\left[C_1 \cos\frac{mt}{\sqrt{2}} + C_2 \sin\frac{mt}{\sqrt{2}}\right]$$

$$+ e^{-mt/\sqrt{2}}\left[C_3 \cos\frac{mt}{\sqrt{2}} + C_4 \sin\frac{mt}{\sqrt{2}}\right] \quad ...(3)$$

so that

$$\frac{dx}{dt} = \frac{m}{\sqrt{2}}e^{mt/\sqrt{2}}\left[C_1 \cos\frac{mt}{\sqrt{2}} + C_2 \sin\frac{mt}{\sqrt{2}}\right]$$

$$+ e^{mt/\sqrt{2}}\frac{m}{\sqrt{2}}\left[-C_1 \sin\frac{mt}{\sqrt{2}} + C_2 \cos\frac{mt}{\sqrt{2}}\right]$$

$$+ \left(-\frac{m}{\sqrt{2}}\right)e^{-mt/\sqrt{2}}\left[C_3 \cos\frac{mt}{\sqrt{2}} + C_4 \sin\frac{mt}{\sqrt{2}}\right]$$

$$+ e^{-mt/\sqrt{2}}\frac{m}{\sqrt{2}}\left[-C_3 \sin\frac{mt}{\sqrt{2}} + C_4 \cos\frac{mt}{\sqrt{2}}\right]$$

and $\dfrac{d^2x}{dt^2} = \dfrac{m^2}{2}e^{mt/\sqrt{2}}\left[C_1 \cos\dfrac{mt}{\sqrt{2}} + C_2 \sin\dfrac{mt}{\sqrt{2}}\right]$

$$+ \frac{m^2}{2}e^{mt/\sqrt{2}}\left[-C_1 \sin\frac{mt}{\sqrt{2}} + C_2 \cos\frac{mt}{\sqrt{2}}\right]$$

$$+ \frac{m^2}{2}e^{mt/\sqrt{2}}\left[-C_1 \sin\frac{mt}{\sqrt{2}} + C_2 \cos\frac{mt}{\sqrt{2}}\right]$$

$$- \frac{m^2}{2}e^{mt/\sqrt{2}}\left[C_1 \cos\frac{mt}{\sqrt{2}} + C_2 \sin\frac{mt}{\sqrt{2}}\right]$$

$$+ \frac{m^2}{2}e^{-mt/\sqrt{2}}\left[C_3 \cos\frac{mt}{\sqrt{2}} + C_4 \sin\frac{mt}{\sqrt{2}}\right]$$

$$- \frac{m^2}{2}e^{-mt/\sqrt{2}}\left[-C_3 \sin\frac{mt}{\sqrt{2}} + C_4 \cos\frac{mt}{\sqrt{2}}\right]$$

$$- \frac{m^2}{2}e^{-mt/\sqrt{2}}\left[-C_3 \sin\frac{mt}{\sqrt{2}} + C_4 \cos\frac{mt}{\sqrt{2}}\right]$$

$$- \frac{m^2}{2}e^{-mt/\sqrt{2}}\left[C_3 \cos\frac{mt}{\sqrt{2}} + C_4 \sin\frac{mt}{\sqrt{2}}\right]$$

$$= m^2 e^{mt/\sqrt{2}}\left[-C_1 \sin\frac{mt}{\sqrt{2}} + C_2 \cos\frac{mt}{\sqrt{2}}\right]$$

$$- m^2 e^{-mt/\sqrt{2}}\left[-C_3 \sin\frac{mt}{\sqrt{2}} + C_4 \cos\frac{mt}{\sqrt{2}}\right]$$

$\therefore$ From (1)

$$y = -\frac{1}{m^2}\frac{d^2x}{dt^2}$$

$$= e^{mt/\sqrt{2}}\left[C_1 \sin\frac{mt}{\sqrt{2}} - C_2 \cos\frac{mt}{\sqrt{2}}\right]$$

$$+ e^{-mt/\sqrt{2}}\left[-C_3 \sin\frac{mt}{\sqrt{2}} + C_4 \cos\frac{mt}{\sqrt{2}}\right] \quad ...(4)$$

Hence, (3) and (4) be complete solution.

EXAMPLE 7. *Solve the following system of differential equations* $Dx + Dy + 3x = \sin t$ *and* $Dx + y - x = \cos t.$

[UPTU-2003]

SOLUTION. We have $\quad Dx + Dy + 3x = \sin t \qquad ...(1)$

and $\qquad\qquad Dx + y - x = \cos t \qquad ...(2)$

Differentiating equation (2), we get

$$D(D-1)x + Dy = -\sin t \qquad ...(3)$$

Subtracting (1) from (3), we get

$$\{D(D-1) - (D+3)\}\,x = -2\sin t$$

or $\quad \{D^2 - D - D - 3\}\,x = -2\sin t$

or $\quad (D^2 - 2D - 3)x = -2\sin t$

The auxiliary equation is

$$m^2 - 2m - 3 = 0 \ \text{ or } \ (m-3)(m+1) = 0,$$

i.e., $\qquad\qquad m = 3, -1$

$\therefore$ C.F. $= C_1 e^{3t} + C_2 e^{-t}$

P.I. $= \dfrac{1}{D^2 - 2D - 3}(-2\sin t) = -2\dfrac{1}{(-1) - 2D - 3}\sin t$

$$= 2.\frac{1}{2(D+2)}\sin t = \frac{(D-2)}{D^2 - 4}\sin t$$

$$= \frac{(D-2)}{-1-4}\sin t = \frac{\cos t - 2\sin t}{-5} = \frac{1}{5}(2\sin t - \cos t)$$

Hence, the general solution is

$$y = \text{C.F.} + \text{P.I.}$$

$$x = C_1 e^{3t} + C_2 e^{-t} + \frac{1}{5}(2\sin t - \cos t) \qquad \dots (4)$$

From (2), we get

$$(D-1)x + y = \cos t$$

$$\text{or } (D-1)\left\{ C_1 e^{3t} + C_2 e^{-t} + \frac{1}{5}(2\sin t - \cos t) \right\} + y = \cos t$$

$$\text{or } \quad y = \cos t - D\left\{ C_1 e^{3t} + C_2 e^{-t} + \frac{1}{5}(2\sin t - \cos t) \right\}$$

$$+ \left\{ C_1 e^{3t} + C_2 e^{-t} + \frac{1}{5}(2\sin t - \cos t) \right\}$$

$$= \cos t - 3C_1 e^{3t} + C_2 e^{-t} - \frac{1}{5}(2\cos t + \sin t)$$

$$+ C_1 e^{3t} + C_2 e^{-t} + \frac{1}{5}(2\sin t - \cos t)$$

$$= \cos t - 2C_1 e^{3t} + 2C_2 e^{-t} - \frac{1}{5}(3\cos t - \sin t)$$

$$= \frac{2}{5}\cos t + \frac{1}{5}\sin t + 2C_2 e^{-t} - 2C_1 e^{3t}$$

$$\therefore \qquad y = \frac{1}{5}(2\cos t + \sin t) - 2C_1 e^{3t} + 2C_2 e^{-t} \qquad \dots (5)$$

and $x = C_1 e^{3t} + C_2 e^{-t} + \frac{1}{5}(2\sin t - \cos t)$ $\qquad \dots (6)$

which is required solution.

EXAMPLE 8. *Solve $\dfrac{dx}{dt} = 2y$, $\dfrac{dy}{dt} = 2z$, $\dfrac{dz}{dt} = 2x$.*

[UPTU-2004, 07, SVTU-2006 S]

SOLUTION. We have $\dfrac{dx}{dt} = 2y \Rightarrow Dx = 2y$ $\qquad \dots (1)$

$$\frac{dy}{dt} = 2z \Rightarrow Dy = 2z \qquad \dots (2)$$

and $\dfrac{dz}{dt} = 2x \Rightarrow Dz = 2x$ $\qquad \dots (3)$

From (1), $\dfrac{dx}{dt} = 2y$

or $\dfrac{d^2 x}{dt^2} = 2\dfrac{dy}{dt} = 2(2z) = 4z$

$$\frac{d^3 x}{dt^3} = 4\frac{dz}{dt} = 4(2x) = 8x$$

$$\frac{d^3 x}{dt^3} - 8x = 0 \Rightarrow (D^3 - 8)x = 0$$

The auxiliary equation is $m^3 - 8 = 0$

or $\quad (m-2)(m^2 + 2m + 4) = 0$

$\Rightarrow \qquad\qquad m - 2 = 0 \Rightarrow m = 2$

and $\quad m^2 + 2m + 4 = 0 \Rightarrow m = \dfrac{-2 \pm \sqrt{4-16}}{2}$

$$= \frac{-2 \pm i\sqrt{12}}{2} = -1 \pm i\sqrt{3}$$

$\therefore$ The general solution is

$$x = C_1 e^{2t} + e^{-t}(A\cos\sqrt{3}t + B\sin\sqrt{3}t)$$

$$[A = C_2 \cos\alpha, B = C_2 \sin\alpha, \tan\alpha = \frac{B}{A}$$

$$\text{or } \alpha = \tan^{-1}\left(\frac{B}{A}\right)]$$

$$x = C_1 e^{2t} + e^{-t}[C_2 \cos\alpha \cos\sqrt{3}\,t + C_2 \sin\alpha \sin\sqrt{3}\,t]$$

$$x = C_1 e^{2t} + e^{-t}C_2 \cos(\sqrt{3}t - \alpha)$$

$$= C_1 e^{2t} + C_2 e^{-t}\cos(\sqrt{3}t - \alpha)$$

From (3), we have $\dfrac{dz}{dt} = 2x$

$$\Rightarrow \quad \frac{dz}{dt} = 2C_1 e^{2t} + 2C_2 e^{-t}\cos(\sqrt{3}t - \alpha)$$

$$z = C_1 e^{2t} + 2C_2 \frac{e^{-t}}{\sqrt{1+3}}\cos(\sqrt{3}t - \alpha - \beta)$$

$$\left[\because \int e^{ax}\cos bx\, dx = \frac{e^{ax}}{\sqrt{a^2 + b^2}}\cos(bx - \beta) \right.$$

$$\left. \text{where } \beta = \tan^{-1}\frac{\sqrt{3}}{-1} = \frac{2\pi}{3} \text{ and } -\frac{2\pi}{3} = \frac{4\pi}{3} \right]$$

$$z = C_1 e^{2t} + C_2 e^{-t}\cos\left(\sqrt{3}t - \alpha + \frac{4\pi}{3}\right)$$

From (2), $\dfrac{dy}{dt} = 2z$

$$\Rightarrow \quad \frac{dy}{dt} = 2C_1 e^{2t} + 2C_2 e^{-t}\cos\left(\sqrt{3}t - \alpha + \frac{4\pi}{3}\right)$$

$$y = \int 2C_1 e^{2t} dt + 2C_2 \int e^{-t}\cos\left(\sqrt{3}t - \alpha + \frac{4\pi}{3}\right) dt$$

$$\left(\because \gamma = \tan^{-1}\frac{\sqrt{3}}{-1} = \frac{2\pi}{3} \right)$$

$$= C_1 e^{2t} + 2C_2 \frac{e^{-x}}{\sqrt{1+3}}\cos\left(\sqrt{3}t - \alpha + \frac{4\pi}{3} - \gamma\right)$$

$$= C_1 e^{2t} + 2C_2 \frac{e^{-t}}{\sqrt{1+3}}\cos\left(\sqrt{3} - \alpha + \frac{4\pi}{3} - \frac{2\pi}{3}\right)$$

$$\therefore \qquad y = C_1 e^{2t} + C_2 e^{-t}\cos\left(\sqrt{3}t - \alpha + \frac{2\pi}{3}\right)$$

EXAMPLE 9. *Solve $\dfrac{d^2 x}{dt^2} + \dfrac{dy}{dt} + 3y = e^{-t}$,*

$$\dfrac{d^2 y}{dt^2} - 4\dfrac{dy}{dt} + 3y = \sin 2t.$$

[UPTU-2007, MTU (SUM)-2011]

SOLUTION. The given equation can be written as

$$D^2 x + Dy + 3x = e^{-t}$$

$$\Rightarrow \quad (D^2 + 3)x + Dy = e^{-t} \qquad \dots (1)$$

and $D^2 y - 4Dx + 3y = \sin 2t$

$$\Rightarrow -4Dx + (D^2 + 3)y = \sin 2t \qquad \dots (2)$$

From (1) and (2), we get

$$(D^4 + 6D^2 + 9)x + D(D^2 + 3)y = e^{-t} + 3e^{-t} \qquad \dots(3)$$

$$-4D^2 x + D(D^2 + 3)y = 2\cos 2t \qquad \dots (4)$$

Subtracting (4) from (3), we get

$$(D^4 + 10D^2 + 9) = 4e^{-t} - 2\cos 2t$$

The auxiliary equation is $m^4 + 10m^2 + 9 = 0$

$\Rightarrow (m^2 + 1)(m^2 + 9) = 0$, *i.e.,* $m = \pm i, \pm 3i$

C.F. $= C_1 \cos t + C_2 \sin t + C_3 \cos 3t + C_4 \sin 3t$

$$\text{P.I.} = \frac{1}{D^4 + 10D^2 + 9} 4e^{-t} - \frac{1}{D^4 + 10D^2 + 9}(2\cos 2t)$$

$$= \frac{4}{1 + 10 + 9}e^{-t} - \frac{1}{(-4)^2 + 10(-4) + 9}(2\cos 2t)$$

$$= \frac{e^{-t}}{5} + \frac{2}{15}\cos 2t$$

$$x = C_1 \cos t + C_2 \sin t + C_3 \cos 3t$$
$$+ C_4 \sin 3t + \frac{e^{-t}}{5} + \frac{2}{15}\cos 2t$$

Putting the value of x in (2), we get

$$-4D\left[C_1 \cos t + C_2 \sin t + C_3 \cos 3t \right.$$
$$\left. + C_4 \sin 3t + \frac{e^{-t}}{5} + \frac{2}{15}\cos 2t \right]$$
$$+ (D^2 + 3)y = \sin 2t$$

or $-4[-C_1 \sin t + C_2 \cos t - 3C_3 \sin 3t$
$$+ 3C_4 \cos 3t - \frac{e^t}{5} - \frac{4}{15}\sin 2t]$$
$$+ (D^3 + 3)y = \sin 2t$$

or $\quad (D^3 + 3)y = \sin 2t - 4C_1 \sin t + 4C_2 \cos t$

$$-12C_3 \sin 3t + 12C_4 \cos 3t - \frac{4}{5}e^{-t} - \frac{16}{15}\sin 2t$$

The auxiliary equation is $m^2 + 3 = 0 \Rightarrow \pm i\sqrt{3}$

$\therefore \quad$ C.F. $= C_1 \cos\sqrt{3}t + C_2 \sin\sqrt{3}t$

P.I. $= \dfrac{1}{D^2 + 3}[\sin 2t - 4C_1 \sin t + 4C_2 \cos t$

$$-12C_3 \sin 3t + 12C_4 \cos 3t - \frac{4}{5}e^{-t} - \frac{16}{15}\sin 2t]$$

$$= \frac{1}{D^2+3}\left(-\frac{1}{15}\sin 2t\right) + \frac{1}{D^2+3}(-4C_1\sin t)$$

$$+ \frac{1}{D^2+3}4C_2\cos t$$

$$+ \frac{1}{D^2+3}(-12C_3\sin 3t) + \frac{1}{D^2+3}(12C_4\cos 3t)$$

$$+ \frac{1}{D^2+3}\left(-\frac{4}{5}e^{-t}\right)$$

$$= \frac{1}{-4+3}\left(-\frac{1}{15}\sin 2t\right) + \frac{1}{-1+3}(-4C_1\sin t)$$

$$+ \frac{1}{-1+3}4C_2\cos t + \frac{1}{-9+3}(-12C_3\sin 3t)$$

$$+ \frac{1}{-9+3}(12C_4\cos 3t) + \frac{1}{1+3}\left(-\frac{4}{5}e^{-t}\right)$$

$$= \frac{1}{15}\sin 2t - 2C_1\sin t + 2C_2\cos t$$

$$+ 2C_3\sin 3t - 2C_4\cos 3t - \frac{1}{5}e^{-t}$$

$\therefore \quad y =$ C.F. + P.I.

$\therefore \quad y = C_1\cos\sqrt{3}t + C_2\sin\sqrt{3}t + \frac{1}{15}\sin 2t$

$$- 2C_1\sin t + 2C_2\cos t + 2C_3\sin 3t$$

$$-2C_4\cos 3t - \frac{1}{5}e^{-t}$$

and $x = C_1\cos 3t + C_2\sin t + C_3\cos 3t + C_4\sin 3t$

$$+ \frac{e^{-t}}{5} + \frac{2}{15}\cos 2t$$

Exercise-8.25

Solve the following simultaneous differential equations:

1. $\dfrac{d^2x}{dt^2} - 3x - 4y = 0, \quad \dfrac{d^2y}{dt^2} + x + y = 0$ [UPTU–2005]

2. $\dfrac{dx}{dt} = 3x + 2y, \quad \dfrac{dy}{dt} = 5x + 3y$

         [MTU (B. PHARMA)–2011, UPTU (SUM)–2008]

3. $\dfrac{d^2x}{dt^2} + 4x + y = te^{3t}, \quad \dfrac{d^2y}{dt^2} + y - 2x = \cos^2 t$

4. $\dfrac{dx}{dt} + 5x + y = e^t, \quad \dfrac{dy}{dt} - x + 3y = 0$

5. $\dfrac{dx}{dt} = 3x + 2y, \quad \dfrac{dy}{dt} + 5x + 3y = 0$

6. $\dfrac{dx}{dt} + 2x - 3y = t, \quad \dfrac{dy}{dt} - 3x + 2y = e^{2t}$ [NAGPUR–2009, UKTU–2012]

7. $(D - 17)y + (2D - 8)z = 0, (13D - 53)y - 2z = 0$

8. $2\dfrac{d^2y}{dx^2} - \dfrac{dz}{dx} - 4y = 2x, \quad 2\dfrac{dy}{dx} + 4\dfrac{dz}{dx} - 3z = 0$

9. $\dfrac{dx}{dt} + \dfrac{2}{t}(x - y) = 1, \quad \dfrac{dy}{dt} + \dfrac{1}{t}(x + 5y) = t$ [UPTU–2005]

10. $\dfrac{dx}{dt} + 5x - 2y = t, \quad \dfrac{dy}{dt} + 2x + y = 0$

         [UPTU-2008, SVTU-2009, KURUKSHETRA-2005]

11. $\dfrac{d^2x}{dt^2} + y = \sin t, \quad \dfrac{d^2y}{dt^2} + x = \cos t$ [UPTU-2004]

12. $(D^2 - 1)x + 8Dy = 16e^t, Dx + 3(D^2 + 1)y = 0$ [UPTU-2001]

13. $\dfrac{dx}{dt} + 7x - y = 0, \dfrac{dy}{dt} + 2x + 5y = 0$)   [GBTU (AG)SUM-2010]

14. $\dfrac{dx}{dt} + x - 2y = 0, \dfrac{dy}{dt} + x + 4y = 0$; $x(0) = y(0) = 1$ (MTU-2011)

15. $\dfrac{dx}{dt} = 3x + 8y, \dfrac{dy}{dt} = -x - 3y$; $x(0) = 6, y(0) = -2$ (GBTU-2010)

16. $\dfrac{dx}{dt} = y + 1, \dfrac{dy}{dt} = x + 1$ [UPTU (SUM)-2009]

17. $\dfrac{dx}{dt} - y = e^t, \dfrac{dy}{dt} + x = \sin t$; $x(0) = 1, y(0) = 0$

         [GBTU (SUM)-2010, GBTU (CO)-2011]

18. $\dfrac{dx}{dt} + 5x + y = e^t, \dfrac{dy}{dt} + x + 5y = e^{5t}$ [UPTU (CO)-2009]

19. $\dfrac{dx}{dt} + \dfrac{dy}{dt} + 2x + y = 0, \dfrac{dy}{dt} + 5x + 3y = 0$ [GBTU (CO)-2011]

20. $\dfrac{dx}{dt} = -4(x + y), \dfrac{dx}{dt} + 4\dfrac{dy}{dt} = -4y$ with conditions

     $x(0) = 1, y(0) = 0$ (GBTU-2011)

21. $\dfrac{dx}{dt} + y = \sin t, \dfrac{dy}{dt} + x = \cos t$ given that $x = 2$ and $y = 0$

     when $t = 0$ (BHOPAL-2009, JNTU-2006, KERALA-2005)

22. $\dfrac{dx}{dt} + 2x + 3y = 0, \quad 3x + \dfrac{dy}{dt} + 2y = 2e^{2t}$ (DELHI-2002)

23. $\dfrac{dx}{dt} + 2y = e^t, \dfrac{dy}{dt} - 2x = e^{-t}$ (BHOPAL-2005)

24. $\dfrac{d^2x}{dt^2} - 3x - 4y = 0, \dfrac{d^2y}{dt^2} + x + y = 0$ (UPTU-2005)

25. $\dfrac{d^2x}{dt^2} + y = \sin t, \dfrac{d^2y}{dt^2} + x = \cos t$ (UPTU-2004)

Hints to Selected Problems

1. Eliminating y, we get

$$[(D^2 + 1)(D^2 - 3) + 4]x = 0$$

$$\Rightarrow \qquad\qquad (D^2 - 1)^2 x = 0$$

The auxiliary equation is $(m^2 - 1)^2 = 0 \Rightarrow m = \pm 1, \pm 1$

$$x = (C_1 + C_2t)e^{-t} + (C_3 + C_4t)\,e^t$$

$$\frac{dx}{dt} = -(C_1 + C_2t)e^{-t} + C_2e^{-t} + (C_3 + C_4t)e^t + C_4e^t$$

$$\frac{d^2x}{dt^2} = (C_1 + C_2t)e^{-t} - 2C_2e^{-t} + (C_3 + C_4t)^t + 2C_4e^t$$

Now, for given equation

$$4y = D^2x - 3x$$

$$= (C_1 + C_2t)\,e^{-t} - 2C_2e^{-t} + (C_3 + C_4t)\,e^t$$

$$+ 2C_4e^t - 3(C_1 + C_2t)e^{-t} - 3(C_2t + C_4t)\,e^t$$

$$= -(2C_1 + 2C_2 + 2C_2t)\,e^{-t} + (-2C_3 + 2C_4 - 2C_4t)\,e^t$$

$$= -\frac{1}{2}(C_1 + C_2 + C_2t)\,e^{-t} + \frac{1}{2}(C_4 + C_3 - C_5t)\,e^t$$

3. The given equation can be written as

$$(D^2+4)x+y=t\,e^{3t} \qquad \ldots(1)$$

$$-2x+(D^2+1)y=\cos^2 t \qquad \ldots(2)$$

Eliminating y, we get

$$\left[(D^2+1)(D^2+4)+2\right]x=(D^2+1)t\,e^{3t}-\cos^2 t$$

$$(D^4+5D^2+6)x=10t\,e^{3t}-\cos^2 t+6e^{3t}$$

$$m^4+5m^2+6=0$$

$$(m^3+3)(m^2+2)=0$$

$$m=\pm\sqrt{3}\,i,\ \pm\sqrt{2}\,i$$

C.F. $= (C_1\cos\sqrt{3}\,t+C_2\sin\sqrt{3}\,t)+(C_3\cos\sqrt{2}\,t+C_4\sin\sqrt{2}\,t)$

$$\text{P.I.}=\frac{10}{D^4+5D^2+6}t\,e^{3t}+\frac{6}{D^4+5D^2+6}e^{3t}-\frac{1}{D^4-5D^2+6}\cos^2 t$$

$$=10\,e^{3t}\cdot\frac{1}{(D+3)^4+5(D+3)^2+6}t+\frac{6e^{3t}}{3^4+5.3^2+6}$$

$$\qquad\qquad -\frac{1}{(D^4+5D^2+6)}\cdot\frac{1}{2}(1+\cos 2t)$$

$$=10e^{3t}\cdot\frac{1}{131+138D+59D^2+\ldots}t+\frac{1}{22}e^{3t}$$

$$\qquad -\frac{1}{6+5D^2+D^4}\cdot\frac{1}{2}-\frac{1}{D^4+5D^2+6}\left(\frac{1}{2}\cos 2t\right)$$

$$=10^{3t}\frac{1}{132}\left(1+\frac{23}{22}D+\frac{59}{132}D^2+\ldots\right)^{-1}t$$

$$\qquad +\frac{e^{3t}}{22}-\frac{1}{6}\left(1+\frac{5D^2}{6}+\frac{D^4}{6}\right)^{-1}\frac{1}{2}-\frac{\frac{1}{2}\cos 2t}{(-2)^2+5(-2)^2+6}$$

$$=\frac{5}{66}te^{3t}-\frac{49}{1452}e^{3t}-\frac{1}{12}-\frac{1}{4}\cos 2t$$

$$x=(C_1\cos\sqrt{3}\,t+C_2\sin\sqrt{3}\,t)\ +(C_3\cos\sqrt{2}\,t+C_4\sin\sqrt{2}\,t)$$

$$\qquad +\frac{5}{66}t\,e^{3t}-\frac{49}{1452}e^{3t}-\frac{1}{12}-\frac{1}{4}\cos 2t \qquad \ldots(3)$$

$$\frac{dx}{dt}=\left(-C_1\sqrt{3}\sin\sqrt{3}\,t+C_2\sqrt{3}\cos\sqrt{3}\,t\right)$$

$$\qquad +\left(-C_3\sqrt{3}\sin\sqrt{2}\,t+C_4\sqrt{2}\cos\sqrt{2}\,t\right)$$

$$\qquad +\frac{5}{66}(3t\,e^{3t}+e^{3t})-\frac{49}{1452}3e^{3t}+\frac{1}{2}\sin 2t$$

$$\frac{d^2x}{dt^2}=-3(C_1\cos\sqrt{3}t+C_2\sin\sqrt{3}\,t)-2(C_3\cos\sqrt{2}\,t+C_4\sin\sqrt{2}\,t)$$

$$\qquad +\frac{5}{66}(9te^{3t}+6e^{3t})-\frac{49}{1452}9e^{3t}+\cos 2t$$

Substituting in (1), we get

$$y=-\frac{d^2x}{dt^2}-4x+te^{3t}$$

$$y=-(C_1\cos\sqrt{3}\,t+C_2\sin\sqrt{3}\,t)$$

$$\qquad -2(C_3\cos\sqrt{2}\,t+C_4\sin\sqrt{2}\,t)+\frac{1}{66}te^{2t}-\frac{23}{1452}e^{3t}+\frac{1}{3} \quad\ldots(4)$$

6. The given equation can be written as

$$(D+2)x-3y=t \qquad \ldots(1)$$

$$-3x+(D+2)y=e^{2t} \qquad \ldots(2)$$

Eliminating y, we get

$$[(D+2)^2-9]x=(D+2)t+3e^{2t}$$

$$(D^2+4+4D-9)x=(D+2)t+3e^{2t}=(1+2t)+3e^{2t}$$

A.E. $m^2+4m-5=0 \quad m=1,-5$

C.F. $= C_1e^{-5t}+C_2e^t$

$$\text{P.I.}=\frac{1+2t}{(D-1)(D+5)}+3\frac{e^{2t}}{(D-1)(D+5)}$$

$$=-\frac{1}{5}\left[1-\frac{4}{5}D-\frac{1}{5}D^2\right]^{-1}(1+2t)+\frac{3e^{2t}}{(2-1)(2+5)}$$

$$=-\frac{1}{5}\left[1+\frac{4}{5}D+\frac{1}{5}D^2+\ldots\right](1+2t)+\frac{3}{7}e^{2t}$$

$$=-\frac{13}{25}-\frac{2}{5}t+\frac{3}{7}e^{2t}$$

$$x=C_1e^{5t}+C_2e^t+\frac{3}{7}e^{2t}-\frac{2}{5}t-\frac{13}{25}$$

$$\frac{dx}{dt}=-5C_1e^{-5t}+C_2e^t+\frac{6}{7}e^{2t}-\frac{2}{5}$$

From (1) : $\quad 3y=-5C_1e^{-5t}+C_2e^t+\frac{6}{7}e^{2t}-\frac{2}{5}+2C_1e^{-5t}$

$$\qquad\qquad +2C_2e^t+\frac{6}{7}e^{2t}-\frac{4}{5}t-\frac{26}{25}-t$$

$$y=-C_1e^{-5t}+C_2e^t+\frac{4}{7}e^{2t}-\frac{12}{25}-\frac{3}{5}t$$

9. The given equation can be written as

$$t\frac{dx}{dt}+2(x-y)=t \qquad \ldots(1)$$

$$t\frac{dy}{dt}+x+5y=t^2 \qquad \ldots(2)$$

Differentiating (1) w.r.t. t, we have

$$t\frac{d^2x}{dt^2}+\frac{dx}{dt}+2\frac{dx}{dt}-2\frac{dy}{dt}=1$$

$$t^2\frac{d^2x}{dt^2}+t\frac{dx}{dt}+2t\frac{dx}{dt}-2t\frac{dy}{dt}=t \qquad \ldots(3)$$

Substituting the value of $t\dfrac{dy}{dt}$ from (2) in (1), we get

$$t^2\frac{d^2x}{dt^2}+3t\frac{dx}{dt}+2x+5\left(t\frac{dx}{dt}+2x-t\right)-2t^2=t$$

$$\Rightarrow \qquad\qquad t\frac{d^2x}{dt^2}+8t\frac{dx}{dt}+12x=2t^2+6t \qquad \ldots(4)$$

which is a homogeneous linear equation.

Put $\qquad\qquad\qquad t=e^z$

$$\frac{dx}{dt}=\frac{dx}{dz}\cdot\frac{dz}{dt}=\frac{dx}{dz}\cdot\frac{1}{t}$$

$$t\frac{d}{dt}\left(t\frac{dx}{dt}\right)=t^2\frac{d^2x}{dt^2}+t\frac{dx}{dt}$$

$$t^2D^2x=(D-1)Dx$$

Equation (4) gives

$$\left[(D-1)D+8D+12\right]x=2e^{2z}+6e^z$$

$$(D^2+7D+12)x=2e^{2z}+6e^z$$

A.E. is $m^2+7m+12=0 \Rightarrow m=-3,-4$

C.F. $= C_1e^{-3z}+C_2e^{-4z}$

$$\text{P.I.}=\frac{2}{(D^2+7D+12)}e^{2z}+\frac{6}{(D^2+7D+12)}e^z$$

$$=\frac{2}{30}e^{2z}+\frac{6}{20}e^z=\frac{1}{15}e^z+\frac{3}{10}e^z$$

$$x=C_1e^{-3z}+C_2e^{-4z}+\frac{1}{15}e^{2z}+\frac{3}{10}e^z$$

$$x=\frac{C_1}{t^3}+\frac{C_2}{t^4}+\frac{t^2}{15}+\frac{3}{10}t$$

$$\frac{dx}{dt}=-\frac{3C_1}{t^4}-\frac{4C_2}{t^5}+\frac{2t}{15}+\frac{3}{10}$$

From (1) :

$$2y = -\frac{3C_1}{t^3} - \frac{4C_2}{t^4} + \frac{2t^2}{15} + \frac{3t}{10} + \frac{2C_1}{t^3} + \frac{2C_2}{t^3} + \frac{2t^2}{15} + \frac{6}{10}t - t \qquad y = -\frac{C_1}{2t^3} - \frac{C_2}{t^4} + \frac{2t^2}{15} + \frac{3}{10}$$

Answers

1. $x = (C_1 + C_2 t)e^{-t} + (C_3 + C_4 t)e^t$, $y = -\frac{1}{2}(C_1 + C_2 + C_2 t)e^{-t} + \frac{1}{2}(C_4 - C_3 - C_4 t)e^t$

2. $x = C_1 e^{(3+\sqrt{10})t} + C_2 e^{(3-\sqrt{10})t}$, $y = \frac{1}{2}\sqrt{10}\,[C_1 e^{(3+\sqrt{10})t} - C_2 e^{(3-\sqrt{10})t}]$

3. $x = (C_1 \cos\sqrt{3}t + C_2 \sin\sqrt{3}t) + (C_3 \cos\sqrt{2}t + C_4 \sin\sqrt{2}t) + \frac{5}{66}te^{3t} - \frac{49}{1452}e^{3t} - \frac{1}{12} - \frac{1}{4}\cos 2t$

$y = (C_1 \cos\sqrt{3}t + C_2 \sin\sqrt{3}t) - (C_3 \cos\sqrt{2}t + C_4 \sin\sqrt{2}t) + \frac{1}{60}te^{3t} - \frac{23}{1452}e^{3t} + \frac{1}{3}$

4. $x = (C_1 + C_2 t)e^{-4t} + \frac{4}{25}e^t - \frac{1}{36}e^{2t}$, $y = -(C_1 + C_2 + C_2 t)e^{-4t} + \frac{7}{36}e^{2t} + \frac{1}{25}e^t$

5. $x = C_1 \cos t + C_2 \sin t$, $y = \frac{1}{2}(C_2 - 3C_1)\cos t - \frac{1}{2}(C_1 + 3C_2)\sin t$

6. $x = C_1 e^{-5t} + C_2 e^t + \frac{3}{7}e^{2t} - \frac{2}{5}t - \frac{13}{25}$, $y = -C_1 e^{-5t} + C_2 e^t + \frac{4}{7}e^{2t} - \frac{3}{5}t - \frac{12}{25}$ **7.** $y = C_1 e^{3x} + C_2 e^{5x}$, $z = -7C_1 e^{3x} + 6C_2 e^{5x}$

8. $y = (C_1 + C_2 x)e^x + C_3 e^{-3x/2} - \frac{1}{2}x$, $z = -2(C_1 + C_2 x - 3C_2)e^x - \frac{1}{3}C_3 e^{-3x/2} - \frac{1}{3}$

9. $x = \frac{C_1}{t^3} + \frac{C_2}{t^4} + \frac{t^2}{15} + \frac{3}{10}t$, $y = -\frac{C_1}{2t^3} - \frac{C_2}{t^4} + \frac{2t^2}{15} - \frac{t}{20}$ **10.** $x = -\frac{1}{27}(1+6t)e^{-3t} + \frac{1}{27}(1+3t)$, $y = -\frac{2}{27}(2+3t)e^{-3t} + \frac{2}{27}(2-3t)$

11. $x = C_1 e^t + C_2 e^{-t} + C_3 \cos t + C_4 \sin t + \frac{t}{4}(\sin t - \cos t)$, $y = -C_1 e^t - C_2 e^{-t} + C_3 \cos t + C_4 \sin t + \frac{1}{4}(2+t)(\sin t - \cos t)$

12. $y = C_1 \cos\frac{t}{\sqrt{3}} + C_2 \sin\frac{t}{\sqrt{3}} + C_3 \cosh\sqrt{3}.t + C_4 \sinh\sqrt{3}t + 2e^t$; $x = \sqrt{3}C_1 \sin\frac{t}{\sqrt{3}} - \sqrt{3}C_2 \cos\frac{t}{\sqrt{3}} - 3\sqrt{3}C_3 \sinh\sqrt{3}t - 3\sqrt{3}C_4 \cosh\sqrt{3}t - 6e^t - 3t$

13. $x = e^{-6t}(A\cos t + B\sin t)$, $y = e^{-6t}[(A+B)\cos t - (A-B)\sin t]$ **14.** $x = 4e^{-2t} - 3e^{-3t}$, $y = -2e^{-2t} + 3e^{-3t}$

15. $x = 4e^t + 2e^{-t}$, $y = -e^t - e^{-t}$ **16.** $x = C_1 e^t + C_2 e^{-t} - 1$, $y = C_1 e^t - C_2 e^{-t} - 1$

17. $x = 2\sin t + \frac{3}{2}\cos t + \frac{t}{2}\cos t - \frac{1}{2}e^t$, $y = \frac{1}{2}\cos t - \frac{3}{2}\sin t + \frac{t}{2}\sin t - \frac{1}{2}e^t$ **18.** $x = C_1 e^{-6t} + C_2 e^{-4t} + \frac{6e^t}{35} - \frac{e^{5t}}{99}$, $y = C_1 e^{-6t} - C_2 e^{-4t} - \frac{1}{35}e^t + \frac{10}{99}e^{5t}$

19. $x = \left(\frac{C_1 - 3C_2}{5}\right)\sin t - \left(\frac{C_2 + 3C_1}{5}\right)\cos t$, $y = C_1 \cos t + C_2 \sin t$ **20.** $x = (1-2t)e^{-2t}$, $y = te^{-2t}$ **21.** $x = e^t + e^{-t}$, $y = e^{-t} - e^t + \sin t$

22. $x = C_1 e^t + C_2 e^{-st} + \frac{6}{7}e^{2t}$; $y = C_2 e^{-st} - C_1 e^t + \frac{8}{7}e^{2t}$ **23.** $x = \frac{1}{5}e^t + \frac{2}{5}e^{-t} - C_1 \sin 2t + C_2 \cos 2t$, $y = \frac{2}{5}e^t + \frac{1}{5}e^{-t} + C_1 \cos 2t + C_2 \sin 2t$

24. $x = (C_1 + C_2 t)e^{-t} + (C_3 + C_4 t)e^t$, $y = -\frac{1}{2}[C_1 + C_2(1+t)]e^{-t} + \frac{1}{2}[C_4(1-t) - C_3]e^t$

25. $x = C_1 e^t + C_2 e^{-t} + C_3 \cos t + C_4 \sin t - \frac{t}{4}\cos t + \frac{t}{4}\sin t$, $y = -C_1 e^t - C_2 e^{-t} + C_3 \cos t + C_4 \sin t + \frac{1}{4}(2+t)(\sin t - \cos t)$

8.4.2 Simultaneous Equations in Different Form

Consider the equations of the type

$$P_1 dx + Q_1 dy + R_1 dz = 0$$
$$P_2 dx + Q_2 dy + R_2 dz = 0 \qquad\qquad \ldots (1)$$

where P_1, P_2, Q_1, Q_2, R_1 and R_2 are functions of x, y, z.

Equation (1) can be written as

$$P_1 \frac{dx}{dz} + Q_1 \frac{dy}{dt} + R_1 = 0 \ , P_2 \frac{dx}{dz} + Q_2 \frac{dy}{dz} + R_2 = 0 \ .$$

Solving the above equations for $\frac{dx}{dz}, \frac{dy}{dz}$, we get $\quad \frac{dx}{dz} = \frac{Q_1 R_2 - Q_2 R_1}{P_1 Q_2 - Q_1 P_2} \ , \frac{dy}{dz} = \frac{R_1 P_2 - P_1 R_2}{P_1 Q_2 - Q_1 P_2}$

Hence,
$$\frac{dx}{Q_1 R_2 - Q_2 R_1} = \frac{dy}{R_1 P_2 - R_2 P_1} = \frac{dz}{P_1 Q_2 - P_2 Q_1}$$

i.e., the equation can be put in the form $\qquad \frac{dx}{P} = \frac{dy}{Q} = \frac{dz}{R}$

where P, Q and R the functions of x, y and z.

WORKING PROCEDURE

Method-I

Step 1. Take any two member of an equation (1) say $\dfrac{dx}{P} = \dfrac{dy}{Q}$ *(say)*
After integrating it, we may get an equation.

Step 2. Again take two member of equation (1) say $\dfrac{dy}{Q} = \dfrac{dz}{R}$ *(say)*
After integrating it, we also get an equation.

Step 3. The solution obtained from (i) and (ii) give the required general solution.

Method-II

Step 1. The given equation is $\dfrac{dx}{P} = \dfrac{dy}{Q} = \dfrac{dz}{R}$

If we choose l, m and n such that $\dfrac{dx}{P} = \dfrac{dy}{Q} = \dfrac{dz}{R} = \dfrac{ldx + mdy + ndz}{lP + mQ + nR}$

If $lP + mQ + nR = 0$, then, $ldx + mdy + ndz = 0$

If it is an exact differential, say du, then $u = a$ is one equation of the complete solution.

REMARKS

- To find a solution of the given equation, we choose l, m, n such that $ldx + mdy + ndz$ is differential of $lP + mQ + nR$.
- If we have obtained one solution, then this solution can be used to simplify the other differential equations in the integrable form.
- Sometimes, we use only one set of multiples, but in some cases, we have a need of more than one set of multipliers.
- We can obtain one relation, say $u = a$ by the first method and the second relation by the second method.

Geometrical meaning of $\dfrac{dx}{P} = \dfrac{dy}{Q} = \dfrac{dz}{R}$

Since, we know that the direction cosines of the tangent to a curve at any point (x, y, z) are $\dfrac{dx}{ds}, \dfrac{dy}{ds}, \dfrac{dz}{ds}$ or proportional to dx, dy, dz. Therefore, geometrically the above situations represents a system of curves in such a way that the direction-ratios of the tangent from it at any point $A(x, y, z)$ are proportional to P, Q and R. If $u = a$ and $v = b$ are the complete solutions of $\dfrac{dx}{P} = \dfrac{dy}{Q} = \dfrac{dz}{R}$, then system of curves is intersection of the surfaces $u = a, v = b$.

Solved Examples

EXAMPLE 1. *Solve the simultaneous equations.*
$$\frac{a\,dx}{(b-c)\,yz} = \frac{b\,dy}{(c-a)\,zx} = \frac{c\,dz}{(a-b)\,xy}.$$

SOLUTION. Let us take the x, y, z as multipliers.

Each fraction $= \dfrac{ax\,dx + by\,dy + cz\,dz}{0}$

$\therefore \quad ax\,dx + by\,dy + cz\,dz = 0$

Integrating $ax^2 + by^2 + cz^2 = C_1$ ... (1)
Now taking ax, by, cz as multipliers.

Each fraction $= \dfrac{a^2 x\,dx + b^2 y\,dy + c^2 z\,dz}{0}$

$\therefore \quad a^2 x\,dx + b^2 y\,dy + c^2 z\,dz = 0$

On integrating,
$$a^2 x^2 + b^2 y^2 + c^2 z^2 = 0 \quad \text{... (2)}$$
Hence, complete solution is
$$\phi(ax^2 + by^2 + cz^2,\ a^2 x^2 + b^2 y^2 + c^2 z^2) = 0$$

EXAMPLE 2. *Solve* $\dfrac{xdx}{z^2 - 2yz - y^2} = \dfrac{dy}{y+z} = \dfrac{dz}{y-z}$.

SOLUTION. Let us take $1, y, z$ as multipliers, we get

Each fraction $= \dfrac{xdx + ydy + zdz}{0}$

$xdx + ydy + zdz = 0$

Integrating, $x^2 + y^2 + z^2 = C_1$ ... (1)
Again, last two members, we get
$$\frac{dy}{y+z} = \frac{dz}{y-z}$$
$$ydy - zdy = ydz + zdz$$
or $ydy - (ydz + zdy) - zdz = 0$

Integrating $y^2 - 2yz - z^2 = C_2$ ... (2)
Complete solution is
$$x^2 + y^2 + z^2 = C_1$$
$$y^2 - 2yz - z^2 = C_2$$

EXAMPLE 3. *Solve the simultaneous equation*
$$\frac{dx}{y^2 + z^2 - x^2} = \frac{dy}{-2xy} = \frac{dz}{-2xz}.$$

SOLUTION. From last two members, we get
$$\frac{dy}{y} = \frac{dz}{z}$$
$\therefore \quad y = C_1 z$ ... (1)
Now, taking x, y, z as multiplier, we get

Each fraction $= \dfrac{dz}{-2xz} = \dfrac{xdx + ydy + zdz}{-x(x^2 + y^2 + z^2)}$

$$\frac{dz}{z} = \frac{2xdx + 2ydy + 2zdz}{x^2 + y^2 + z^2}$$

Integrating $\log z + \log C = \log(x^2 + y^2 + z^2)$
$$x^2 + y^2 + z^2 = C_2 z \quad \text{... (2)}$$
Complete solution is
$$y = C_1 z$$
$$x^2 + y^2 + z^2 = C_2 z$$

EXAMPLE 4. *Solve* $\dfrac{dx}{x^2 + y^2 + yz} = \dfrac{dy}{x^2 + y^2 - xz} = \dfrac{dz}{z(x+y)}$.

SOLUTION. Given equation can change to be new form as
$$\frac{dx - dy}{z(x+y)} = \frac{dz}{z(x+y)} \quad \text{or} \quad dx - dy = dz$$
Integrating $x - y - z = C_1$ ... (1)

Again from the given equation, we get

$$\frac{xdx + ydy}{(x+y)(x^2+y^2)} = \frac{dz}{z(x+y)}$$

or

$$\frac{xdx + ydy}{x^2 + y^2} = \frac{dz}{z}.$$

Integrating, $\log(x^2 + y^2) = 2\log z + \log C_2$

$$\therefore \qquad\qquad x^2 + y^2 = z^2 C_2 \qquad\qquad ...(2)$$

From (1) and (2), we get complete solution

$$x - y - z = C_1$$
$$x^2 + y^2 = z^2 C_2$$

EXAMPLE 5. *Solve* $\dfrac{dx}{x(y^2 - z^2)} = \dfrac{dy}{-y(z^2 + x^2)} = \dfrac{dz}{z(x^2 + y^2)}$.

SOLUTION. Taking $\dfrac{1}{x}, -\dfrac{1}{y}, -\dfrac{1}{z}$ as multipliers, we get

Each fraction $= \dfrac{\dfrac{dx}{x} - \dfrac{dy}{y} - \dfrac{dz}{z}}{0}$

$\therefore \dfrac{dx}{x} - \dfrac{dy}{y} - \dfrac{dz}{z} = 0$ or $\dfrac{dy}{y} + \dfrac{dz}{z} = \dfrac{dx}{x}$

Integrating, $\log y + \log z = \log x + \log C_1$.

$\therefore \qquad\qquad yz = C_1 x \qquad\qquad ...(A)$

Again using x, y, z as multipliers, we get

Each fraction $= \dfrac{xdx + ydy + zdz}{0}$

$\therefore \qquad xdx + ydy + zdz = 0$

Integrating $x^2 + y^2 + z^2 = C_2$. $\qquad\qquad ...(B)$

From (A) and (B), we obtain complete solution.

Exercise-8.26

Solve the following simultaneous differential equations:

1. $\dfrac{dx}{xy} = \dfrac{dy}{y^2} = \dfrac{dz}{zyx - 2x^2}$ **2.** $\dfrac{dx}{x^2 + y^2} = \dfrac{dy}{2xy} = \dfrac{dz}{(x+y).z}$

3. $\dfrac{dx}{(x^2 - yz)} = \dfrac{dy}{y^2 - zx} = \dfrac{dz}{z^2 - xy}$ **4.** $\dfrac{dx}{yz} = \dfrac{dy}{zx} = \dfrac{dz}{xy}$

5. $\dfrac{dx}{y+z} = \dfrac{dy}{z+x} = \dfrac{dz}{x+y}$ **6.** $\dfrac{dx}{mz - ny} = \dfrac{dy}{nx - lz} = \dfrac{dz}{ly - mx}$

7. $\dfrac{dx}{z(x+y)} = \dfrac{dy}{z(x-y)} = \dfrac{dz}{x^2 + y^2}$ **8.** $\dfrac{dx}{z} = \dfrac{dy}{-z} = \dfrac{dz}{z^2 + (x+y)^2}$

9. $\dfrac{dx}{x(y-z)} = \dfrac{dy}{y(z-x)} = \dfrac{dz}{z(x-y)}$

10. $\dfrac{dx}{x^2 + y^2} = \dfrac{dy}{2xy} = \dfrac{dz}{(x+y)^2}$

11. $\dfrac{dx}{y^2 + yz + z^2} = \dfrac{dy}{z^2 + zx + x^2} = \dfrac{dz}{x^2 + xy + y^2}$

12. $\dfrac{dx}{\cos(x+y)} = \dfrac{dy}{\sin(x+y)} = \dfrac{dz}{z}$

Hints to Selected Problems

1. Taking the first two members, we get

$$\frac{dx}{xy} = \frac{dy}{y^2} \quad \text{or} \quad \frac{dx}{x} = \frac{dy}{y}$$

Integrating $\log x = \log y + \log C_1$

$$x = C_1 y \qquad\qquad ...(1)$$

Again taking the last two members, we have

$$\frac{dy}{y^2} = \frac{dz}{zxy - 2x^2}$$

$$\frac{dy}{y^2} = \frac{dz}{zC_1 y^2 - 2C_1^2 y^2} \quad \text{from (1)}$$

$$dy = \frac{dz}{zC_1 - 2C_1^2} \quad \text{or} \quad C_1 dy = \frac{dz}{z - 2C_1}$$

Integrating, we get

$$C_1 y = \log(z - 2C_1) + C_2$$

$$x = \log\left(z - \frac{2x}{y}\right) + C_2 \qquad [\because C_1 y = x]$$

$$x = \log(zy - 2x) - \log y + C_3 \qquad ...(2)$$

3. Obviously, each of the given ratios

$$= \frac{dx - dy}{x^2 yz - y^2 + zx} \qquad ...(1)$$

$$= \frac{dy - dz}{y^2 - zx - z^2 + xy} \qquad ...(2)$$

$$= \frac{dz - dx}{z^2 - xy - x^2 + yz} \qquad ...(3)$$

From (1) and (2)

$$\frac{dx - dy}{(x-y)(x+y+z)} = \frac{dy - dz}{(y-z)(x+y+z)}$$

$$\frac{dx - dy}{x - y} = \frac{dy - dz}{y - z}$$

Integrating, we get

$$\log(x - y) = \log(y - z) + \log a$$

$$\frac{x - y}{y - z} = a \qquad\qquad ...(4)$$

From (2) and (3), similarly

$$\frac{y - z}{z - x} = b \qquad\qquad ...(5)$$

From (4) and (5), the complete solution of the equation is

$$\phi\left(\frac{x-y}{y-z}, \frac{y-z}{z-x}\right) = 0$$

7. Using $x, -y, -z$ as multipliers, we have

Each fraction $= \dfrac{xdx - ydy - zdz}{xz(x+y) - yz(x-y) - z(x^2 + y^2)}$

$$= \frac{xdx - ydy - zdz}{0}$$

$$xdx - ydy - zdz = 0$$

Integrating, $x^2 - y^2 - z^2 = C \qquad\qquad ...(1)$

Similarly, using $y, x, -z$ as multipliers, we get

Each fraction $= \dfrac{ydx + xdy - zdz}{yz(x+y) + xz(x-y) - z(x^2 + y^2)}$

$$= \frac{ydx + xdy - zdz}{0}$$

$\therefore \quad ydx + xdy - zdz = 0$

Integrating, $\qquad\qquad 2xy - z^2 = C_2$

9. Obviously, $\dfrac{dx}{x(y-z)} = \dfrac{dy}{y(z-x)} = \dfrac{dz}{z(x-y)}$

$$= \frac{dx + dy + dz}{xy - xz + yz - yx + zx - zy}$$

$$dx + dy + dz = 0$$

Integrating, $x + y + z = C_1$ ... (1)

Now using $\dfrac{1}{x}, \dfrac{1}{y}, \dfrac{1}{z}$ as multipliers, we have

$$\dfrac{\dfrac{1}{x}dx}{y-z} = \dfrac{\dfrac{1}{y}dy}{z-x} = \dfrac{\dfrac{1}{z}dz}{x-y}$$

$$\quad\quad I \quad\quad\quad II \quad\quad\quad III$$

$$I = II = III = \dfrac{\dfrac{1}{x}dx + \dfrac{1}{y}dy + \dfrac{1}{z}dz}{y - z + z - x + x - y}$$

$$\dfrac{1}{x}dx + \dfrac{1}{y}dy + \dfrac{1}{z}dz = 0$$

On integrating, we get

$$\log x + \log y + \log z = \log C_2$$

$$xyz = C_2 \quad\quad ...(2)$$

10. Obviously, $\dfrac{dx+dy}{x^2+y^2+2xy} = \dfrac{dx-dy}{x^2+y^2-2xy} = \dfrac{dz}{(x+y)^2}$

$$\dfrac{dx+dy}{(x+y)^2} = \dfrac{dx-dy}{(x-y)^2} = \dfrac{dz}{(x+y)^2}$$

$$\quad\quad I \quad\quad\quad II \quad\quad\quad III$$

Taking first two members

$$\dfrac{dx+dy}{(x+y)^2} = \dfrac{dx-dy}{(x-y)^2}$$

Integrating, we get

$$-(x+y)^{-1} = -(x-y)^{-1} + C_1$$

$$\dfrac{1}{x-y} - \dfrac{1}{x+y} = C_1$$

$$\dfrac{2y}{x^2-y^2} = C_1 \quad\quad ...(1)$$

Now, taking first and last members

$$\dfrac{dx+dy}{(x+y)^2} = \dfrac{dz}{(x+y)^2}$$

$$dx + dy - dz = 0$$

Integrating, we get

$$x + y - z = C_2 \quad\quad ...(2)$$

From equation (1) and (2), the complete solution is given by

$$\phi\left(\dfrac{2y}{x^2-y^2},\ x+y-z\right) = 0$$

Answers

1. $\phi\left[\left(\dfrac{x}{y}\right),\ x - \log(zy-2) + \log y\right] = 0$ **2.** $\phi\left(\dfrac{x+y}{z},\ \dfrac{2y}{y^2-x^2}\right) = 0$ **3.** $\phi\left(\dfrac{x-y}{y-z},\ \dfrac{y-z}{z-x}\right) = 0$

4. $\phi(x^2-y^2,\ x^2-z^2) = 0$ **5.** $\phi\left[\left(\dfrac{y-x}{z-y}\right),\ (x-y)^2(x+y+z)\right] = 0$ **6.** $\phi(lx+my+nz,\ x^2+y^2+z^2) = 0$

7. $\phi(x^2-y^2-z^2,\ 2xy-z^2) = 0$ **8.** $\phi[x+y,\ \log\{z^2+(x+y)^2\} - 2x] = 0$ **9.** $\phi(x+y+z,\ xyz) = 0$

10. $\phi\left(\dfrac{2y}{x^2-y^2},\ x+y-z\right) = 0$ **11.** $\phi\left(\dfrac{y-x}{z-x},\ \dfrac{y-x}{z-y}\right) = 0$ **12.** $f\{(\cos(x+y)+\sin(x+y))\}\,e^{y-x}\,z^{\sqrt{2}}\cot\left(\dfrac{x+y}{2}+\dfrac{\pi}{8}\right) = 0$

8.5 LINEAR DIFFERENTIAL EQUATION OF SECOND ORDER WITH VARIABLE COEFFICIENTS

A differential equation of the form

$$\dfrac{d^2y}{dx^2} + P\dfrac{dy}{dx} + Qy = X \quad\quad ...(1)$$

where P, Q and X are function of x alone, are called linear equation of second order.

We have several methods of solving the equation (1).

For example: Linear equations with constant coefficients, homogeneous equations and exact equations.

When these methods are not applicable, we shall try the method, discussed in this chapter.

8.5.1 The Complete Solution in terms of a Known Solution

Consider the differential equation $\dfrac{d^2y}{dx^2} + P\dfrac{dy}{dx} + Qy = 0$

Let $y = u$ be a known solution of the complementary function of (1).

$\Rightarrow\quad y = u$ be a solution of $\dfrac{d^2y}{dx^2} + P\dfrac{dy}{dx} + Qy = 0$. Therefore,

$$\dfrac{d^2u}{dx^2} + P\dfrac{du}{dx} + Qu = 0. \quad\quad ...(2)$$

On substituting $y = uv$, we get $\dfrac{dy}{dx} = v\dfrac{du}{dx} + u\dfrac{dv}{dx}$ and $\dfrac{d^2y}{dx^2} = v\dfrac{d^2u}{dx^2} + 2\dfrac{du}{dx}\cdot\dfrac{dv}{dx} + u\dfrac{d^2v}{dx^2}$

Putting all these values in equation (1), we get

$$\left(v\frac{d^2u}{dx^2}+2\frac{du}{dx}.\frac{dv}{dx}+u\frac{d^2v}{dx^2}\right)+P\left(v\frac{du}{dx}+u\frac{dv}{dx}\right)+Qu.v=X \Rightarrow u\frac{d^2v}{dx^2}+\frac{dv}{dx}\left(2\frac{du}{dx}+Pu\right)+v\left(\frac{d^2u}{dx^2}+P\frac{du}{dx}+Qu\right)=X$$

$$\Rightarrow \qquad\qquad u\frac{d^2v}{dx^2}+\frac{dv}{dx}\left(2\frac{du}{dx}+Pu\right)=X \qquad\qquad \text{(By using (2))}$$

$$\Rightarrow \qquad\qquad \frac{d^2v}{dx^2}+\left(P+\frac{2}{u}\frac{du}{dx}\right)\frac{dv}{dx}=\frac{X}{u}. \qquad\qquad …(3)$$

Putting $\frac{dv}{dx}=p \Rightarrow \frac{d^2v}{dx^2}=\frac{dp}{dx}$ in (3), we get $\frac{dp}{dx}+\left(P+\frac{2}{u}\frac{du}{dx}\right)p=\frac{X}{u}.$ …(4)

Equation (4) is a linear differential equation with p as dependent variable.

$$\text{I.F.}=e^{\int\left(P+\frac{2}{u}\frac{du}{dx}\right)dx}=e^{\{2\log u+\int Pdx\}}=u^2e^{\int Pdx}.$$

The solution is $\qquad pu^2e^{\int Pdx}=\int\left[\frac{X}{u}.u^2e^{\int Pdx}\right]dx+C_1 \qquad\qquad …(5)$

$$\Rightarrow \qquad\qquad p=\frac{dv}{dx}=\frac{C_1e^{-\int Pdx}}{u^2}+\frac{e^{-\int Pdx}}{u^2}\int uXe^{\int Pdx}dx$$

On integrating we get $\qquad v=C_2+C_1\int\frac{e^{-\int Pdx}}{u^2}dx+\int\left[\frac{e^{-\int Pdx}}{u^2}\int uXe^{\int Pdx}.dx\right]dx.$

Hence, the solution of (1) is $y=u.v=C_2+C_1u\int\frac{e^{-\int Pdx}}{u^2}dx+u\int\left[\frac{e^{-\int Pdx}}{u^2}\int uXe^{\int Pdx}.dx\right]dx.$ …(6)

WORKING PROCEDURE

Step 1. Put in given equation into standard form $\frac{d^2y}{dx^2}+P\frac{dy}{dx}+Qy=X$. The coefficient of $\frac{d^2y}{dx^2}$ must be unity.

Step 2. Find an integral u.

Step 3. Assume that the complete solution is given by $y=uv$. Then the given equation reduces to

$$\frac{d^2v}{dx^2}+\left(P+\frac{2}{u}\frac{du}{dx}\right)\frac{dv}{dx}=\frac{X}{u} \qquad\qquad …(1)$$

Step 4. Put $\frac{dv}{dx}=p, \frac{d^2v}{dx^2}=\frac{dp}{dx}$ in (1) and then solve.

Step 5. Put the value of v in the assumed solution $y=uv$ and get the desired complete solution.

REMARKS

- The solution, given by (6) contains two arbitrary constants, hence it is the complete solution of the given equation in terms of the known integral.
- To find one integral belonging to the complementary function by inspection, the following points must be kept into mind.

(i) $u=e^x$ is a part of C.F if $P+Q+1=0$ (ii) $u=e^x$ is a part of C.F. if $1-P+Q=0$

(iii) $u=e^{ax}$ is a part of C.F. if $1+\frac{P}{a}+\frac{Q}{a^2}=0$ (iv) $u=x$ is a part of C.F. if $P+Q.x=0$

(v) $u=x^2$ is a part of C.F. if $2+2Px+Qx^2=0$

Solved Examples

EXAMPLE 1. **Solve** $x^2\frac{d^2y}{dx^2}-2x(1+x)\frac{dy}{dx}+2(1+x)y=x^3.$

SOLUTION. The given equation can be written as

$$\frac{d^2y}{dx^2}-2\left(1+\frac{1}{x}\right)\frac{dy}{dx}+2\left(\frac{1}{x^2}+\frac{1}{x}\right)y=x. \qquad …(1)$$

Compare with the standard equation, we get

$$P=-2\left(1+\frac{1}{x}\right), \quad Q=2\left(\frac{1}{x}+\frac{1}{x^2}\right), \quad X=x$$

Here, we observe that

$$P+Qx=-2\left(\frac{1}{x}+1\right)+2x\left(\frac{1}{x}+\frac{1}{x^2}\right)=0$$

$\Rightarrow \quad u=x$ is a part of C.F.

Putting $y=vx$

$$\therefore \quad \frac{dy}{dx}=x\frac{dv}{dx}+v, \quad \frac{d^2y}{dx^2}=x\frac{d^2v}{dx^2}+2\frac{dv}{dx}$$

Now, putting all those values in equation (1), we get

$$\frac{d^2v}{dx^2} - 2\frac{dv}{dx} = 1. \qquad \ldots(2)$$

Let $p = \dfrac{dv}{dx}$, then (2) gives $\dfrac{dp}{dx} - 2p = 1$

which is a linear equation

$$\text{I.F.} = e^{-2\int dx} = e^{-2x}.$$

$\therefore$ Solution is

$$p\, e^{-2x} = \int 1 \cdot e^{-2x}dx + C_1 = -\frac{1}{2}e^{-2x} + C_1$$

$$\therefore \qquad p = \frac{dv}{dx} = -\frac{1}{2} + C_1 e^{2x}.$$

On integrating, we get $v = -\dfrac{1}{2}x + \dfrac{C_1}{2}e^{2x} + C_2$.

Hence, the complete solution is given by

$$y = vx = -\frac{1}{2}x^2 + \frac{C_1}{2}xe^{2x} + C_2 x.$$

EXAMPLE 2. *Solve* $(x+2)\dfrac{d^2y}{dx^2} - (2x+5)\dfrac{dy}{dx} + 2y = (x+1)e^x.$

SOLUTION. The given equation can be written as

$$\frac{d^2y}{dx^2} - \left(\frac{2x+5}{x+2}\right)\frac{dy}{dx} + \frac{2}{(x+2)}y = \left(\frac{x+1}{x+2}\right)e^x. \qquad \ldots(1)$$

Compare with the standard equation, we get

$$P = -\left(\frac{2x+5}{x+2}\right), Q = \frac{2}{x+2} \text{ and } X = \left(\frac{x+1}{x+2}\right)e^x.$$

Here, we observe that

$$2^2 + 2P + Q = 4 - 2\left(\frac{2x+5}{x+2}\right) + \frac{2}{x+2} = 0$$

which implies $u = e^{2x}$ is a part of C.F. of (1).

Suppose the complete solution of (1) is given by

$$y = uv.$$

Then (1) reduces to $\dfrac{d^2v}{dx^2} + \left(P + \dfrac{2}{u}\dfrac{du}{dx}\right)\dfrac{dv}{dx} = \dfrac{X}{u}$

$$\Rightarrow \frac{d^2v}{dx^2} + \left(-\frac{2x+5}{x+2} + \frac{2}{e^{2x}}\cdot 2e^{2x}\right)\frac{dv}{dx} = \frac{x+1}{x+2}\cdot\frac{e^x}{e^{2x}}$$

$$\qquad \ldots(2)$$

$$\Rightarrow \frac{d^2v}{dx^2} + \frac{2x+3}{x+2}\frac{dv}{dx} = \frac{x+1}{x+2}e^{-x}$$

Let $\dfrac{dv}{dx} = p$ so that $\dfrac{d^2v}{dx^2} = \dfrac{dp}{dx}$, then (2) gives

$$\frac{dp}{dx} + \frac{2x+3}{x+2}p = \frac{x+1}{x+2}e^{-x}. \qquad \ldots(3)$$

Now, $\int \dfrac{2x+3}{x+2}dx = \int\left[2 - \dfrac{1}{x+2}\right]dx = 2x - \log(x+2)$

I.F. of (3) $= e^{2x - \log(x+2)} = e^{2x}(x+2)^{-1}$.

Therefore, the solution of (3) is given by

$$pe^{2x}\frac{1}{x+2} = \int\frac{x+1}{(x+2)^2}e^x dx + C_1$$

$$= C_1 + \int e^x\frac{x+2-1}{(x+2)^2}dx = C_1 + \frac{e^x}{(x+2)}$$

$$\text{(By using } \int e^x[f(x) + f'(x)]dx = e^x f(x))$$

$$\therefore \qquad p = e^{-x} + C_1(x+2)e^{-2x}$$

or $\dfrac{dv}{dx} = e^{-x} + C_1(x+2)e^{-2x}$

$\therefore \quad v = C_2 + \int e^{-x}dx + C_1\int (x+2)e^{-2x}dx$

$$\Rightarrow v = C_2 - e^{-x} + C_1\left[(x+2)(-\frac{1}{2}e^{-2x}) - 1\cdot\left(\frac{1}{4}e^{-2x}\right)\right]$$

$$= C_2 - e^{-x} - \frac{1}{4}C_1 e^{-2x}(2x+5) \qquad \ldots(4)$$

Hence, the complete solution is given by

$$y = uv = e^{2x}\left[C_2 - e^{-x} - \frac{1}{4}C_1 e^{-2x}(2x+5)\right]$$

$$y = C_2 e^{2x} - \frac{1}{4}C_1(2x+5) - e^x.$$

EXAMPLE 3. *Solve* $\sin^2 x\dfrac{d^2y}{dx^2} = 2y$, *given* $y = v\cot x$ *is a solution.* (BHOPAL–2007)

SOLUTION. The given equation is

$$\sin^2 x\frac{d^2y}{dx^2} = 2y. \qquad \ldots(1)$$

Put $y = v\cot x$

$$\frac{dy}{dx} = \frac{dv}{dx}\cdot\cot x - v\,\mathrm{cosec}^2 x$$

$$\frac{d^2y}{dx^2} = \frac{d^2v}{dx^2}\cot x - 2\,\mathrm{cosec}^2 x\frac{dv}{dx} + 2v\,\mathrm{cosec}^2 x\cot x.$$

In (1), we get

$$\cos x\sin x\frac{d^2v}{dx^2} - 2\frac{dv}{dx} = 0$$

or $\dfrac{d^2v}{dx^2} - \dfrac{2}{\sin x\cos x}\dfrac{dv}{dx} = 0$

or $\dfrac{dp}{dx} - \dfrac{2}{\sin x\cos x}\cdot p = 0 \qquad \left(p = \dfrac{dv}{dx}\right)$

On separating the variables, we get

$$\frac{dp}{p} = \frac{2}{\sin x\cos x}dx = \frac{2\sec^2 x}{\tan x}dx.$$

On integrating, we get

$$\log p = 2\log\tan x + \log c$$

$$p = C_1\tan^2 x$$

$$\Rightarrow \frac{dv}{dx} = C_1\tan^2 x = C_1(\sec^2 x - 1).$$

Integrating, we get $v = C_1(\tan x - x) + C_2$.

Hence, the complete solution is given by

$$y = v\cot x = C_1[1 - x\cot x] + C_2\cot x.$$

EXAMPLE 4. *Solve* $y'' - 4xy' + (4x^2 - 2)y = 0$ *given that* $y = e^{x^2}$ *is an integral included in the complementary function.* (UPTU–2004)

SOLUTION. We have $y'' - 4xy' + (4x^2 - 2)y = 0 \qquad \ldots(1)$

On putting $y = ve^{x^2}$ in (1), then

$$\frac{d^2v}{dx^2} + \left(P + \frac{2}{u}\right)\frac{dv}{dx} = 0$$

$$[\because P = -4x, Q = 4x^2 - 2, X = 0]$$

or $\dfrac{d^2v}{dx^2} + \left[-4x + \dfrac{2}{e^{x^2}}(2x\,e^{x^2})\right]\dfrac{dv}{dx} = 0$

or $\quad \dfrac{d^2v}{dx^2} + \left[-4x + 4x\right]\dfrac{dv}{dx} = 0$

or $\quad \dfrac{d^2v}{dx^2} = 0 \Rightarrow \dfrac{dv}{dx} = C_1 \Rightarrow v = C_1 x + C_2 \quad (\because u = e^{x^2})$

then $\quad y = uv$

$\therefore \quad y = e^{x^2}(C_1 x + C_2)$.

Exercise-8.27

Solve the following equations :

1. $x^2\dfrac{d^2y}{dx^2} - (x^2 + 2x)\dfrac{dy}{dx} + (x + 2)y = x^3 e^x$.

2. $(x\sin x + \cos x)\dfrac{d^2y}{dx^2} - x\cos x\dfrac{dy}{dx} + y\cos x$
$\qquad\qquad = \sin x(x\sin x + \cos x)^2$.

3. $(1 - x^2)\dfrac{d^2y}{dx^2} + x\dfrac{dy}{dx} - y = x(1 - x^2)^{3/2}$.

4. $\dfrac{d^2y}{dx^2} + \left(1 + \dfrac{2}{x}\cot x - \dfrac{2}{x^2}\right)y = x\cos x$ given that $\dfrac{\sin x}{x}$ is a C.F.

5. $x\dfrac{d^2y}{dx^2} - (2x - 1)\dfrac{dy}{dx} + (x - 1)y = 0$.   (BHOPAL−2008, 19S)

6. $\dfrac{d^2y}{dx^2} - 2(x + 1)\dfrac{dy}{dx} + (x + 2)y = (x - 2)e^{2x}$.

7. $(x + 1)\dfrac{d^2y}{dx^2} - 2(x + 3)\dfrac{dy}{dx} + (x + 5)\,y = e^x$

8. $x\dfrac{d^2y}{dx^2} + (x - 2)\dfrac{dy}{dx} - 2y = x^3$.

9. $x\dfrac{d^2y}{dx^2} - (x + 2)\dfrac{dy}{dx} + 2y = x^3$.

10. $x\dfrac{d^2y}{dx^2} + (1 - x)\dfrac{dy}{dx} - y = e^x$.

11. $x\dfrac{d^2y}{dx^2} + (x - 1)\dfrac{dy}{dx} - y = x^2$.

12. $x^2\dfrac{d^2y}{dx^2} + x\dfrac{dy}{dx} - 9y = 0$, given that $y = x^3$ is a C.F.

13. $x\dfrac{dy}{dx} - y = (x - 1)\left(\dfrac{d^2y}{dx^2} - x + 1\right)$.

14. Solve $x\dfrac{d^2y}{dx^2} - \dfrac{dy}{dx} - 4x^3y = -4x^5$. Given that $\quad y = e^{x^2}$ is a solution is the left hand side is equated to zero.

15. Verify that $f_1(x) = x^2$ is solution of the differential equation $\dfrac{d^2y}{dx^2} - \dfrac{2}{x^2}y = 0$, $0 < x < \infty$ and find a second independent solution. Also obtain the solution of the given equation.

16. Solve $\dfrac{d^2y}{dx^2} - \cot x\dfrac{dy}{dx} - (1 - \cot x)y = e^x\sin x$.

17. Solve $(1 - x^2)y'' - 2xy' + 2y = 0$ given that $y = x$ is a solution.
   (BPTU−2005S)

18. Solve $x\dfrac{d^2y}{dx^2} - (2x - 1)\dfrac{dy}{dx} + (x - 1)y = e^x$ given that $y = e^x$ is one integral.   (BHOPAL−2007, 12S)

Hints to Selected Problems

1. $P + Qx = 0 \Rightarrow y = x$ is a part of C.F. Take $y = vx$.
2. $P + Qx = 0 \Rightarrow y = x$ is a part of C.F.
4. Take $y = v.\dfrac{\sin x}{x}$.

6. Here $1 + P + Q = 0 \Rightarrow y = e^x$ is a part of C.F.
8. Here $1 - P + Q = 0 \Rightarrow y = e^{-x}$ is a part of C.F.

10. Here $1 + P + Q = 0 \Rightarrow y = e^x$ is a part of C.F.
11. Here $1 - P + Q = 0 \Rightarrow y = e^{-x}$ is a part of C.F.
13. Here $P + Qx = 0 \Rightarrow y = x$ is a part of C.F.
15. $y = x^2$ satisfying the given equation, therefore we can take $y = vx^2$.

Answers

1. $y = x^2e^x - xe^x + C_1 xe^x + C_2 x$.

2. $y = \dfrac{1}{4}x\cos 2x - \dfrac{1}{2}\sin 2x - C_1\dfrac{\cos x}{x} + C_x$

3. $y = -\dfrac{1}{9}x(1 - x^2)^{3/2} - C_1\{x\sin^{-1}x + \sqrt{(1 - x^2)}\} + C_2 x$.

4. $y = \dfrac{x^2\sin x}{6} + C_1\left[-x\cos x + 2\sin x\log\sin x - \dfrac{2\sin x}{x}\int\log\sin x\,dx\right] + C_2\dfrac{\sin x}{x}$ **5.** $y = e^x(C_1\log x + C_2)$.

6. $y = \dfrac{1}{3}C_1 x^3 e^x + C_2 e^x + e^{2x}$.   **7.** $y = \dfrac{1}{5}C_1 e^x(x + 1)^5 - \dfrac{1}{4}xe^x + C_2 e^x$.   **8.** $y = x^3 + (C_1 - 3)(x^2 - 2x + 2) + C_2 e^{-x}$.

9. $y = -x^3 - (C_1 - 3)(x^2 + 2x + 2) + C_2 e^x$.   **10.** $y = e^x\log x + C_1 e^x\int\dfrac{e^{-x}}{x}dx + C_2 e^x$.   **11.** $y = C_1(x - 1) + C_2 e^{-x} + x^2 - 2x + 2$.

12. $\dfrac{-4}{6}x^{-3} + C_2 + x^{-3}$.   **13.** $y = C_1 e^x + C_2 x - x^2 - x^2 - 1$.   **14.** $y = e^{x^2}x^2 e^{-x^2} - \dfrac{C_1}{4}e^{-2x^2} + C_2$.   **15.** $y = -\dfrac{1}{3}\left(\dfrac{C_1}{x}\right) + C_2 x^2$.

16. $y = -\dfrac{C_1}{2x} + C_2\left(x + \dfrac{1}{x}\right)$.   **17.** $y = x\left[C_1\left\{\log\left(\dfrac{x}{1 - x}\right)^2 - \dfrac{1}{x}\right\} + C_2\right]$ **18.** $y = e^x(C_1\log x + x + C_2)$

8.5.2 Method of Removal of the First Derivative

Transformation of the Equation into Normal Form

To obtain a suitable substitution for the dependent variable which transforms the equation $\dfrac{d^2y}{dx^2} + P\dfrac{dy}{dx} + Qy = X$ *into normal form, i.e., the form where the first derivative is absent :*

Consider the equation

$$\frac{d^2y}{dx^2} + P\frac{dy}{dx} + Qy = X. \qquad \qquad \dots(1)$$

Let us suppose $y = uv$ is the general solution of (1), where u is a function of x and is not a part of C.F.

Now, $\qquad \qquad y = uv \quad \Rightarrow \quad \dfrac{dy}{dx} = v\dfrac{du}{dx} + u\dfrac{dv}{dx} \quad$ and $\quad \dfrac{d^2y}{dx^2} = v\dfrac{d^2u}{dx^2} + 2\dfrac{dv}{dx}\cdot\dfrac{du}{dx} + u\dfrac{d^2v}{dx^2}$

Putting all these values in equation (1), we get

$$\left\{ v\frac{d^2u}{dx^2} + 2\frac{dv}{dx}\cdot\frac{du}{dx} + u\frac{d^2v}{dx^2} \right\} + P\left(v\frac{du}{dx} + u\frac{dv}{dx} \right) + Q\,uv = X \Rightarrow u\frac{d^2v}{dx^2} + u\frac{dv}{dx}\left[P + \frac{2}{u}\frac{du}{dx} \right] + v\left[\frac{d^2u}{dx^2} + P\frac{du}{dx} + Qu \right] = X \qquad \dots(2)$$

To remove the term of first derivative, we shall choose u such that

$$P + \frac{2}{u}\frac{du}{dx} = 0 \qquad \qquad \Rightarrow \qquad \qquad \frac{du}{u} = -\frac{P}{2}\,dx$$

On integrating, we get

$$\log u = -\int\frac{P}{2}\,dx \qquad \text{or} \qquad u = e^{\left\{ -\int\frac{P}{2}\,dx \right\}}. \qquad \dots(3)$$

Now equation (2) becomes

$$u\frac{d^2v}{dx^2} + v\left[\frac{d^2u}{dx^2} + P\frac{du}{dx} + Qu \right] = X \qquad \Rightarrow \qquad \frac{d^2v}{dx^2} + \frac{v}{u}\left[\frac{d^2u}{dx^2} + P\frac{du}{dx} + Qu \right] = \frac{X}{u} \qquad \dots(4)$$

From (3), we get

$$u = e^{-\int P/2\,dx} \quad \Rightarrow \qquad \qquad \frac{du}{dx} = -\frac{P}{2}u$$

and $\qquad \dfrac{d^2u}{dx^2} = -\dfrac{1}{2}\left(P\dfrac{du}{dx} + u\dfrac{dP}{dx} \right) = -\dfrac{1}{2}\left[P\left(-\dfrac{P}{2}u \right) + u\dfrac{dP}{dx} \right] = \dfrac{1}{4}P^2u - \dfrac{u}{2}\dfrac{dP}{dx}.$

Putting these values in (4), we get

$$\frac{d^2v}{dx^2} + v\left[\frac{1}{4}P^2 - \frac{P^2}{2} + Q - \frac{1}{2}\frac{dP}{dx} \right] = X_1 e^{\left\{ \int\frac{1}{2}P\,dx \right\}} \qquad \text{or} \qquad \frac{d^2v}{dx^2} + v\left[Q - \frac{1}{4}P^2 - \frac{1}{2}\frac{dP}{dx} \right] = X_1 e^{\left\{ \int\frac{1}{2}P\,dx \right\}}$$

which is known as the normal form of the equation (1) and be easily solved.

WORKING PROCEDURE

Step 1. Put the given equation into standard form $\dfrac{d^2y}{dx^2} + P\dfrac{dy}{dx} + Qy = X$ With coefficient of $\dfrac{d^2y}{dx^2}$ is unity.

Step 2. To remove the first derivative, choose $u \quad e^{\int -P\,dx}$.

Step 3. Assume that the complete solution of the given equation is $y = uv$. Then the equation reduces to normal form

$$\frac{d^2v}{dx^2} + P_1\,v = X_1, \text{ where } P_1 = Q - \frac{1}{4}P^2 - \frac{1}{2}\frac{dP}{dx} \text{ and } X_1 = \frac{X}{u}.$$

Step 4. Solve the equation (obtained in step 3) for v, then the complete solution is $y = uv$.

REMARKS

- Students are advised to remember the equation form, so that it may be written directly.

- To solve the equation $\dfrac{d^2v}{dx^2} + P_1 v = X_1$, we have following two cases :

  (a) If P_1 is constant, then equation being constant coefficient and can be solved by usual methods.

  (b) If $P_1 = \dfrac{\text{constant}}{x^2}$, then the resulting equation reduces to homogeneous form.

Solved Examples

EXAMPLE 1. *Solve*
$$\frac{d^2y}{dx^2} + \frac{1}{x^{1/3}}\frac{dy}{dx} + \left(\frac{1}{4x^{2/3}} - \frac{1}{6x^{4/3}} - \frac{6}{x^2}\right)y = 0.$$

SOLUTION. The given equation is
$$\frac{d^2y}{dx^2} + \frac{1}{x^{1/3}}\frac{dy}{dx} + \left(\frac{1}{4x^{2/3}} - \frac{1}{6x^{4/3}} - \frac{6}{x^2}\right)y = 0.$$

Comparing with the standard equation, we get
$$P = \frac{1}{x^{1/3}}, Q = \frac{1}{4x^{2/3}} - \frac{1}{6x^{4/3}} - \frac{6}{x^2} \text{ and } X = 0.$$

Let us take
$$u = e^{-\int \frac{1}{2}Pdx} = e^{-\int \frac{1}{2}x^{-1/3}dx} = e^{\left(-\frac{3}{4}x^{2/3}\right)}.$$

Now putting all these values, into the normal form
$$\frac{d^2v}{dx^2} + v\left[Q - \frac{1}{4}P^2 - \frac{1}{2}\frac{dP}{dx}\right] = Xe^{\int \frac{1}{2}P\,dx}$$

We get
$$\frac{d^2v}{dx^2} + v\left[\left(\frac{1}{4x^{2/3}} - \frac{1}{6x^{4/3}} - \frac{6}{x^2}\right)\right.$$
$$\left. - \frac{1}{4x^{2/3}} - \frac{1}{2}\left(-\frac{1}{3x^{4/3}}\right)\right] = 0$$

$$\Rightarrow \frac{d^2v}{dx^2} - \frac{6}{x^2}v = 0 \Rightarrow x^2\frac{d^2v}{dx^2} - 6v = 0$$

This is a homogeneous equation. To solve, put $x = e^z$
and let $\frac{d}{dz} \equiv D$, then we get
$$D(D-1)v - 6v = 0 \text{ or } (D+2)(D+3)v = 0.$$
Auxiliary equation is
$$(m+2)(m-3) = 0 \Rightarrow m = -2, +3.$$
$$\Rightarrow v = C_1e^{-2z} + C_2e^{3z} = C_1 \cdot \frac{1}{x^2} + C_2 \cdot x^3.$$

Hence, the complete solution is given by
$$y = uv = \left(\frac{C_1}{x^2} + C_2x^3\right)e^{\left(\frac{-3}{4}x^{2/3}\right)}.$$

EXAMPLE 2. **Remove the second term from the given equation and hence solve**
$$\frac{d^2y}{dx^2} - 4x\frac{dy}{dx} + (4x^2 - 1)y = -3e^{x^2}\sin 2x.$$
(UPTU–2004)

SOLUTION. The given equation is
$$\frac{d\,y}{dx} - 4x\frac{dy}{dx} + (4x - 1)y = -3e\ \sin 2x.$$
Comparing with the standard form, we get
$$P = -4x, Q = 4x^2 - 1, X = -3e^{x^2}\sin 2x$$
To remove second term, let us choose
$$u = e^{-1/2\int P\,dx} = e^{-1/2\int(-4x)\,dx} = e^{x^2}.$$
Putting all those values in the normal form, we get
$$\frac{d^2v}{dx^2} + v = -3\sin 2x.$$
The auxiliary equation is $m = \pm i$
$$\therefore \quad \text{C.F.} = C_1\cos x + C_2\sin x$$
$$\text{P.I.} = -3\frac{1}{D^2+1}\sin 2x$$
$$= -3 \cdot \frac{1}{-2^2+1}\sin 2x = \sin 2x$$
$$\Rightarrow v = \text{C.F.} + \text{P.I} = C_1\cos x + C_2\sin x + \sin 2x$$
Hence, the required complete solution of (1) is given by $y = uv$
$$\Rightarrow y = e^{x^2}(C_1\cos x + C_2\sin x + \sin 2x).$$

EXAMPLE 3. *Solve* $\frac{d^2y}{dx^2} - 2\tan x.\frac{dy}{dx} + 5y = 0.$

SOLUTION. Comparing the given equation with standard equation, we get $P = -2\tan x, Q = 5y, X = 0.$
To remove the term of first derivative, let us choose
$$u = e^{-1/2\int P\,dx} = e^{-1/2\int(-2\tan x)}$$
$$= e^{\log\sec x} = \sec x.$$
Put all these values into normal form, we get
$$\frac{d^2v}{dx^2} + 6v = 0.$$
Auxiliary equation is $m^2 + 6 = 0 \Rightarrow m \pm i\sqrt{6}$
$$\therefore \quad v = C_1\cos(x\sqrt{6}) + C_2\sin(x\sqrt{6})$$
Hence, the complete solution is given by $y = uv$
$$\Rightarrow y = \sec x\,[C_1\cos(x\sqrt{6}) + C_2\sin(x\sqrt{6})].$$

Exercise-8.28

Solve the following differential equations :

1. $x\frac{d}{dx}\left(x\frac{dy}{dx} - y\right) - 2x\frac{dy}{dx} + 2y + x^2y = 0.$

2. $\frac{d^2y}{dx^2} - 4x\frac{dy}{dx} + (4x^2 - 3)y = e^{x^2}.$

3. $\frac{d^2y}{dx^2} - \frac{2}{x}\frac{dy}{dx} + \left(1 + \frac{2}{x^2}\right)y = xe^x$

4. $\frac{d^2y}{dx^2} - \frac{1}{x^{1/2}}\frac{dy}{dx} + \frac{1}{4x^2}(-8 + x^{1/2} + x)y = 0$

5. $\frac{d^2y}{dx^2} - 2\tan x\frac{dy}{dx} + 5y = \sec x.e^x$

6. $\left(\frac{d^2y}{dx^2} + y\right)\cot x + 2\left(\frac{dy}{dx} + y\tan x\right) = \sec x$

7. $\frac{d^2y}{dx^2} - 4x\frac{dy}{dx} + (4x^2 - 1)y = -3e^{x^2}[\sin 2x + 5e^{-2x} + 6]$

8. $\frac{d^2y}{dx^2} + 2x\frac{dy}{dx} + (x^2 + 1)y = x^3 + 3x.$

9. $\frac{d}{dx}\left(\cos^2 x\frac{dy}{dx}\right) + y\cos^2 x = 0$

10. $x^2(\log x)^2\frac{d^2y}{dx^2} - 2x\log x\frac{dy}{dx} + [2 + \log x - 2(\log x)^2]y$
$$= (\log x)^3.x^2$$

11. $\dfrac{d^2y}{dx^2} + 2x\dfrac{dy}{dx} + (x^2+5)y = xe^{-1/2x^2}$

12. $\dfrac{d^2y}{dx^2} - 2x\dfrac{dy}{dx} + (x^2+2)y = e^{\frac{1}{2}(x^2+2x)}$ (GBTU(CO)–2011)

13. $\dfrac{d^2y}{dx^2} + 2x\dfrac{dy}{dx} + (x^2-8)y = x^2e^{-x^2/2}$ (GBTU–2012)

Hints to Selected Problems

1. $P = \dfrac{-2}{x}$, $Q = 2 + \dfrac{1}{x^2}$, $X = 0$,

$u = e^{(-1/2)\int Pdx} = e^{\int (1/x)dx} = e^{\log x} = x$

3. $P = -\dfrac{2}{x}$, $Q = 1 + \dfrac{2}{x}$, $X = xe^x$,

$u = e^{(-1/2)\int Pdx} = e^{\int (1/x)dx} = e^{\log x} = x$

4. $P = -\dfrac{1}{x^{-1/2}}$, $Q = \dfrac{1}{4x^2}(x + x^{1/2} - 8)$, $X = 0$,

$u = e^{(-1/2)\int Pdx} = e^{-1/2\int x^{-1/2}dx} = e^{x^{1/2}} = e^{\sqrt{x}}$

7. $P = -4x$, $Q = 4x^2 - 1$, $X = -3e^{x^2}[\sin 2x + 5e^{-2x} + 6]$,

$u = e^{-1/2\int Pdx} = e^{-1/2\int -4xdx} = e^{x^2}$

8. $P = 2x$, $Q = x^2 + 1$, $X = x(x^2 + 3)$, $u = e^{-1/2\int Pdx} = e^{-x^2/2}$

9. $P = -2\tan x$, $Q = 1$, $X = 0$,

$u = e^{-1/2\int Pdx} = e^{\int \tan xdx} = e^{\log \sec x} = \sec x$

Answers

1. $y = x[C_1\cos x + C_2\sin x]$

2. $y = e^{x^2}[C_1e^x + C_2e^{-x} - 1]$

3. $y = x\left[C_1\cos x + C_2\sin x + \dfrac{e^x}{2}\right]$

4. $y = e^{\sqrt{x}}[C_1x^2 + C_2x^{-1}]$

5. $y = \sec x\left[C_1\cos(\sqrt{6}x) + C_2\sin(\sqrt{6}x) + \dfrac{e^x}{7}\right]$

6. $y = \dfrac{1}{2}\sin x + (C_1x + C_2)\cos x$

7. $y = e^{x^2}(C_1\cos x + C_2\sin x + \sin 2x - 3e^{-2x} - 18)$

8. $y = x + (C_1x + C_2)e^{-x^2/2}$

9. $y = \sec x(C_1\cos\sqrt{2}x + C_2\sin\sqrt{2}x)$

10. $y = (\log x)(C_1x^2 + C_2x^{-1} + \dfrac{1}{3}x^2\log x)$

11. $y = e^{-x^2/2}\left[C_1\cos(2x + C_2) + \dfrac{1}{4}x\right]$

12. $y = e^{x^2/2}(C_1\cos\sqrt{3}x + C_2\sin\sqrt{3}x) + \dfrac{1}{4}e^{\frac{1}{2}(x^2+2x)}$

13. $y = e^{-\frac{x^2}{2}}\left[C_1e^{3x} + C_2e^{-3x} - \dfrac{1}{9}\left(x^2 + \dfrac{2}{9}\right)\right]$

8.5.3 TRANSFORMATION OF THE EQUATION BY CHANGING THE INDEPENDENT VARIABLE

Consider the equation $\qquad\qquad \dfrac{d^2y}{dx^2} + P\dfrac{dy}{dx} + Qy = X$ ...(1)

where P, Q and X are functions of z. Let the independent variable be changed from x to z, where $z = f(x)$ (say), we know that

$\dfrac{df}{dx} = \dfrac{df}{dz}\cdot\dfrac{dz}{dx} \Rightarrow \dfrac{dy}{dx} = \dfrac{dy}{dz}\cdot\dfrac{dz}{dx}$ $\dfrac{d^2y}{dx^2} = \dfrac{d}{dx}\left(\dfrac{dy}{dz}\cdot\dfrac{dz}{dx}\right) = \dfrac{d}{dz}\left(\dfrac{dy}{dz}\cdot\dfrac{dz}{dx}\right)\cdot\dfrac{dz}{dx} = \dfrac{d^2y}{dz^2}\left(\dfrac{dz}{dx}\right)^2 + \dfrac{dy}{dz}\cdot\dfrac{d^2z}{dx^2}$

Putting all those values in equation (1), we get

$\left\{\dfrac{d^2y}{dz^2}\left(\dfrac{dz}{dx}\right)^2 + \dfrac{dy}{dz}\cdot\dfrac{d^2z}{dx^2}\right\} + P\left\{\dfrac{dy}{dz}\cdot\dfrac{dz}{dx}\right\} + Q.y = X$

$\Rightarrow \qquad \dfrac{d^2y}{dz^2} + \left\{\dfrac{\dfrac{d^2z}{dx^2} + P\dfrac{dz}{dx}}{\left(\dfrac{dz}{dx}\right)^2}\right\}\dfrac{dy}{dz} + \dfrac{Q}{\left(\dfrac{dz}{dx}\right)^2}y = \dfrac{X}{\left(\dfrac{dz}{dx}\right)^2} \qquad \Rightarrow \qquad \dfrac{d^2y}{dz^2} + P_1\dfrac{dy}{dz} + Q_1y = X_1$ ...(2)

where, $\qquad P_1 = \left\{\dfrac{\dfrac{d^2z}{dx^2} + P\dfrac{dz}{dx}}{\left(\dfrac{dz}{dx}\right)^2}\right\}$, $Q_1 = \left\{\dfrac{Q}{\left(\dfrac{dz}{dx}\right)^2}\right\}$, and $X_1 = \left\{\dfrac{X}{\left(\dfrac{dz}{dx}\right)^2}\right\}$.

Here, P_1, Q_1, X_1 are functions of x but can be expressed in terms of z by the given relation between z and x.

WORKING PROCEDURE

Step 1. Put the given equation into standard form, and find the value of P, Q and X.

Step 2. Suppose $Q = \pm C f(x)$, then assume a relation between z and x given by $\left(\dfrac{dz}{dx}\right)^2 = Cf(x)$. (Here, we always omit –ve sign of Q).

Step 3. Now solve

$$\left(\frac{dz}{dx}\right)^2 = Cf(x) \Rightarrow \frac{dz}{dx} = \sqrt{C\,f(x)} \qquad \ldots(1)$$

(Omit the negative sign, to get a relation between z and x).

After solving (1), we get a relation between z and x (Don't use the constant of integration).

Step 4. Transform the equation into the form

$$\frac{d^2y}{dz^2} + P_1\frac{dy}{dz} + Q_1y = X_1 \qquad \ldots(2)$$

Step 5. Now solve (2) for z and then replace z by x by the relation between x and z.

REMARKS

- If we equate Q_1 to a constant quantity, then P_1 also becomes constant, then the equation (2) can be solved easily (since, then it will be linear equation with constant coefficients).
- Equation (2) can be solved in following two ways :

  (i) We choose z in such a way that $P_1 = 0$, i.e., $\dfrac{d^2z}{dx^2} + P\dfrac{dz}{dx} = 0 \Rightarrow z = \int e^{-\int P\,dx}dx$

  Then, the given equation is changed into $\dfrac{d^2y}{dz^2} + Q_1y = X_1$

  This is easily integrable, provided Q_1 comes out to be constant or of the form $\dfrac{\text{constant}}{z^2}$.

  (ii) Choose z in such a way that $Q_1 = \left\{\dfrac{Q}{(dz/dx)^2}\right\}$ is a constant, say C^2 , then $\dfrac{Q}{(dz/dx)^2} = C^2 \Rightarrow C\dfrac{dz}{dx} = \sqrt{Q}$
  $\therefore Cz = \int \sqrt{Q}\cdot dz$

  Put this value in the given equation, we get $\dfrac{d^2y}{dz^2} + P_1\dfrac{dy}{dz} + C^2y = X_1$

  This differential equation can be integrated provided P_1 also comes out to be a constant.
- The value of P_1, Q_1 and X_1 must be remembered for its direct use.

Solved Examples

EXAMPLE 1. *Solve* $x\dfrac{d^2y}{dx^2} - \dfrac{dy}{dx} + 4x^3y = x^5$. (UPTU-2002, 03, 05)

SOLUTION. The given equation can be written as

$$\frac{d^2y}{dx^2} - \frac{1}{x}\frac{dy}{dx} + 4x^2y = x^4 \qquad \ldots(1)$$

Comparing with the standard equation, we get

$$P = -\frac{1}{x}, \quad Q = 4x^2, \quad X = x^4 .$$

Let us choose z such that

$$\left(\frac{dz}{dx}\right)^2 = 4x^2 \Rightarrow \frac{dz}{dx} = 2x .$$

$\therefore z = x^2$

Then, the transformed equation is

$$\frac{d^2y}{dz^2} + P_1\frac{dy}{dz} + Q_1y = X_1 \qquad \ldots(2)$$

where, $P = \dfrac{\dfrac{d^2z}{dx^2} + P\dfrac{dz}{dx}}{(dz/dx)^2} = \dfrac{2 + \left(-\dfrac{1}{x}\right)2x}{4x^2} = 0,$

$$Q_1 = \frac{Q}{(dz/dx)^2} = 1$$

and $X_1 = \dfrac{X}{(dz/dx)^2} = \dfrac{x^4}{4x^2} = \dfrac{x^2}{4} = \dfrac{z}{4} .$

Put all these values in (2), we get $\dfrac{d^2y}{dz^2} + y = \dfrac{z}{4} .$

Auxiliary equation is $m^2 + 1 = 0$

$\Rightarrow \quad$ C.F. $= C_1\cos z + C_2\sin z$

$$\text{P.I.} = \frac{1}{D_1^2 + 1}\cdot\frac{z}{4} = \frac{z}{4}(D_1 + 1)^{-1} = \frac{z}{4} \quad \left(D_1 \equiv \frac{d}{dz}\right)$$

$\therefore y = $ C.F. + P.I. $= C_1\cos z + C_2\sin z + \dfrac{z}{4}$

$$\Rightarrow y = C_1\cos x^2 + C_2\sin x^2 + \frac{x^2}{4} .$$

EXAMPLE 2. *Solve* $(1 + x^2)^2\dfrac{d^2y}{dx^2} + 2x(1 + x^2)\dfrac{dy}{dx} + 4y = 0$.

(UPTU(B.TECH.)(Q. BANK)–2001)

SOLUTION. The given equation can be written as

$$\frac{d^2y}{dx^2} + \frac{2x}{(1+x^2)}\frac{dy}{dx} + \frac{4}{(1+x^2)^2}y = 0 \qquad \ldots(1)$$

Comparing with the standard equation, we get

$$P = \frac{2x}{(1+x^2)}, \quad Q = \frac{4}{(1+x^2)^2}, \quad X = 0 .$$

Let us choose z such that

$$\left(\frac{dz}{dx}\right)^2 = \frac{4}{(1+x^2)^2} \Rightarrow \frac{dz}{dx} = \frac{2}{1+x^2} .$$

On integrating, we get

$$z = 2\int \frac{dx}{1+x^2} \Rightarrow z = 2\tan^{-1}x .$$

Thus, the transformed equation is

$$\frac{d^2y}{dz^2} + P_1\frac{dy}{dz} + Q_1y = X_1 \qquad \ldots(2)$$

where, $P_1 = \dfrac{\dfrac{d^2z}{dx^2} + P\dfrac{dz}{dx}}{(dz/dx)^2}$

$$= \frac{-\dfrac{4x}{(1+x^2)^2} + \dfrac{2x}{1+x^2}\cdot\dfrac{2}{1+x^2}}{4/(1+x^2)^2} = 0$$

$Q_1 = \dfrac{Q}{(dz/dx)^2} = 1$ and $X_1 = \dfrac{X}{(dz/dx)^2} = 0$.

Putting all these in (2), we get $\dfrac{d^2y}{dz^2} + y = 0$

Auxiliary equation is $m^2 + 1 = 0 \Rightarrow m = \pm i$

$\therefore$ C.F. $= C_1 \cos z + C_2 \sin z$.

Hence, the complete solution is

$\Rightarrow y = C_1 \cos(2\tan^{-1} x) + C_2 \sin(2\tan^{-1} x)$.

EXAMPLE 3. *Transform the differential equation*

$$\cos x \dfrac{d^2y}{dx^2} + \sin x \dfrac{dy}{dx} - 2y \cos^3 x = 2\cos^5 x$$

into one having z as independent variable, where $z = \sin x$ and solve it.

(UPTU(Q. BANK)–2001, BHOPAL–2006S)

SOLUTION. Given that $z = \sin x \Rightarrow \dfrac{dz}{dx} = \cos x$ ...(1)

Now, $\dfrac{dy}{dx} = \dfrac{dy}{dz} \cdot \dfrac{dz}{dx} = \cos x \dfrac{dy}{dz}$ ...(2)

and $\dfrac{d^2y}{dx^2} = \dfrac{d}{dx}\left(\dfrac{dy}{dz}\right) = \dfrac{d}{dx}\left(\cos x \dfrac{dy}{dz}\right)$

$= -\sin x \dfrac{dy}{dz} + \cos x \dfrac{d}{dx}\left(\dfrac{dy}{dz}\right)$

$= -\sin x \dfrac{dy}{dz} + \cos x \dfrac{d}{dz}\left(\dfrac{dy}{dz}\right) \cdot \dfrac{dz}{dx}$

$\Rightarrow \dfrac{d^2y}{dx^2} = -\sin x \dfrac{dy}{dz} + \cos^2 x \dfrac{d^2y}{dz^2}$ (By using (1))

Now, using (2) and (3), given equation reduces to

$\cos x \left[-\sin x \dfrac{dy}{dz} + \cos^2 x \dfrac{d^2y}{dz^2} \right]$

$+ \sin x \cos x \dfrac{dy}{dz} - 2\cos^2 x \cdot x.y = 2\cos^5 x$

$\Rightarrow \dfrac{d^2y}{dz^2} - 2y = 2\cos^2 x$

$\Rightarrow \dfrac{d^2y}{dz^2} - 2y = 2(1 - \sin^2 x)$

$\Rightarrow (D_1^2 - 2)y = 2(1 - z^2)$ $\left(D_1 \equiv \dfrac{d}{dz}\right)$

Now, the auxiliary equation is

$m^2 - 2 = 0 \Rightarrow m = \pm\sqrt{2}$

$\Rightarrow$ C.F. $= C_1 e^{\sqrt{2}.z} + C_2 e^{-\sqrt{2}.z}$

P.I. $= 2 \cdot \dfrac{1}{D_1^2 - 2}(1 - z^2)$

$= 2 \cdot \dfrac{1}{-2\left(1 - \dfrac{D_1^2}{2}\right)}(1 - z^2)$.

$= -\left(1 - \dfrac{D_1^2}{2}\right)^{-1}(1 - z^2)$

$= -\left(1 + \dfrac{D_1^2}{2} + ...\right)(1 - z^2)$

$= -(1 - z^2 - 1) = z^2$.

Therefore, the complete solution is given by

$$y = C_1 e^{z\sqrt{2}} + C_2 e^{-z\sqrt{2}} + z^2.$$

Putting $z = \sin x$, we get the required general solution as

$$y = C_1 e^{\sqrt{2}.\sin x} + C_2 e^{-\sqrt{2}\sin x} + \sin^2 x.$$

Exercise-8.29

Solve the following differential equations :

1. $x^6 \dfrac{d^2y}{dx^2} + 3x^5 \dfrac{dy}{dx} + a^2y = \dfrac{1}{x^2}$

2. $\sin^2 x \dfrac{d^2y}{dx^2} + \sin x \cos x \dfrac{dy}{dx} + 4y = 0$

3. $\dfrac{d^2y}{dx^2} + \cot x \dfrac{dy}{dx} + 4\csc^2 x.y = 0$

4. $\dfrac{d^2y}{dx^2} + \dfrac{2}{x}\dfrac{dy}{dx} + \dfrac{a^2}{x^4}y = 0$

5. $\dfrac{d^2y}{dx^2} + \tan x \dfrac{dy}{dx} + y\cos^2 x = 0$ (BHOPAL–2005)

6. $\dfrac{d^2y}{dx^2} - \cot x \dfrac{dy}{dx} - y\sin^2 x = \cos x - \cos^3 x$ (GBTU–2011)

7. $x\dfrac{d^2y}{dx^2} + (4x^2 - 1)\dfrac{dy}{dx} + 4x^3y = 2x^3$ (UPTU–2006)

8. $\dfrac{d^2y}{dx^2} - (8e^{2x} + 2)\dfrac{dy}{dx} + 4e^{4x}.y = e^{6x}$

9. $\dfrac{d^2y}{dx^2}\left(1 - \dfrac{1}{x}\right)\dfrac{dy}{dx} + 4x^2 e^{-2x}y = 4(x^2 + x^3)e^{-3x}$

10. $(x^3 - x)\dfrac{d^2y}{dx^2} + \dfrac{dy}{dx} + n^2x^3y = 0$

11. $\dfrac{d^2y}{dx^2} - \dfrac{1}{x}\dfrac{dy}{dx} + 4x^2y = x^4$ (UPTU–2003, SPECIAL EXAM–2001)

12. $x^4 \dfrac{d^2y}{dx^2} + 2x^3 \dfrac{dy}{dx} + n^2y = 0$ (UKTU–2011)

Hints to Selected Problems

1. $P = \dfrac{3}{x}$, $Q = \dfrac{a^2}{x^6}$, $X = \dfrac{1}{x^8}$. Then take

$\dfrac{Q}{(dz/dx)^2} = \dfrac{a^2/x^6}{(dz/dx)^2} = a^2$ (say)

$\Rightarrow \left(\dfrac{dz}{dx}\right)^2 = \dfrac{1}{x^6} \Rightarrow \dfrac{dz}{dx} = \dfrac{1}{x^3} \Rightarrow z = -\dfrac{1}{2x^2}$

2. $P = \cot x$, $Q = 4\csc^2 x$, $X = 0$,

$P_1 = \dfrac{\dfrac{d^2z}{dx^2} + P\dfrac{dz}{dx}}{(dz/dx)^2}, Q = \dfrac{Q}{(dz/dx)^2}, X = \dfrac{X}{(dz/dx)^2}$

$\Rightarrow Q_1 = 1, P_1 = 0$.

Then transformed equation is $(D^2 + 1)y = 0$.

4. $P = \dfrac{2}{x}, Q = \dfrac{a^2}{x^4}, X = 0$.

Now, $\dfrac{Q}{(dz/dx)^2} = \dfrac{a^2/x^4}{(dz/dx)^2} = a^2 \Rightarrow \dfrac{dz}{dx} = -\dfrac{1}{x^2} \Rightarrow z = \dfrac{1}{x}$

6. $P = -\cot x, Q = -\sin^2 x, X = \cos x - \cos^3 x$.

Now, $\dfrac{Q}{(dz/dx)^2} = -\dfrac{\sin^2 x}{(dz/dx)^2} = -1$ (say) $\Rightarrow z = -\cos x$.

$\therefore \quad P_1 = 0, \ Q_1 = -1, \ X_1 = -z$.

Transformed equation is given by $(D^2 - 1)y = -z$.

10. $P = \dfrac{1}{x^3 - x}, \quad Q = \dfrac{n^2 x^2}{x^2 - 1}, \quad X = 0$.

Proceed as usual, we get $z = \sqrt{x^2 - 1}$, $P_1 = 0, Q = n^2, X_1 = 0$.

Then transformed equation is $\dfrac{d^2 y}{dx^2} + n^2 y = 0$.

Answers

1. $y = C_1 \cos\left(\dfrac{a}{2x^2}\right) - C_2 \sin\left(\dfrac{a}{2x^2}\right) + \dfrac{1}{a^2 x^2}$　　**2.** $y = C_1 \cos\left(2\log \tan\dfrac{x}{2}\right) + C_2 \sin\left(2\log \tan\dfrac{x}{2}\right)$　　**3.** Same as (2)

4. $y = C_1 \cos\dfrac{a}{x} + C_2 \sin\dfrac{a}{x}$　　**5.** $y = C_1 \cos(\sin x) + C_2 \sin(\sin x)$　　**6.** $y = C_1 e^{-\cos x} + C_2 e^{\cos x} - \cos x$

7. $y = e^{-x^2}[C_1 x^2 + C_2] + \dfrac{1}{2}$　　**8.** $y = C_1 e^{(2+\sqrt{3})e^{2x}} + C_2 e^{(2-\sqrt{3})e^{2x}} + \dfrac{1}{4}e^{2x} + 1$

9. $y = C_1 \cos\{2(x+1)e^{-x}\} - C_2 \sin[2(x+1)e^{-x}] + (x+1)e^{-x}$　　**10.** $y = C_1 \sin\left[n\sqrt{(x^2-1)} + C_2\right]$

11. $y = C_1 \cos(x^2) + C_2 \sin(x^2) + \dfrac{x^2}{4}$　　**12.** $y = C_1 \cos\left(-\dfrac{n}{x}\right) + C_2 \sin\left(-\dfrac{n}{x}\right)$

8.5.4 Method of Variation of Parameters

Consider the differential equation
$$\frac{d^2 y}{dx^2} + P\frac{dy}{dx} + Qy = X \qquad \qquad \ldots(1)$$

Let us suppose the C.F. of (1) is given by
$$y = A_1 u + B_1 v \qquad \qquad \ldots(2)$$

where A_1 and B_1 are constants and u, v are functions of x.

Now, since (2) is a C.F. of (1), therefore, u and v must be the solution of
$$\frac{d^2 y}{dx^2} + P\frac{dy}{dx} + Qy = 0 \qquad \qquad \ldots(3)$$

$\Rightarrow \qquad \qquad \dfrac{d^2 u}{dx^2} + P\dfrac{du}{dx} + Qu = 0 \quad \text{and} \quad \dfrac{d^2 v}{dx^2} + P\dfrac{dv}{dx} + Qv = 0 \qquad \ldots(4)$

Now, let us assume
$$y = A_2 u + B_2 v \qquad \qquad \ldots(5)$$

is the complete solution of (1), where A_2, B_2 are functions of x and to be so chosen that (1) shall be satisfied.

Now, we have two unknown quantities, A_2 and B_2 in terms of which y has been expressed by (5).

To determine A_2 and B_2, let us take $\qquad u\dfrac{dA}{dx} + v\dfrac{dB}{dx} = 0 \qquad \qquad \ldots(6)$

Differentiating (5) and using (6), we get $\quad \dfrac{dy}{dx} = A_1 \dfrac{du}{dx} + B_1 \dfrac{dv}{dx}$. $\qquad \qquad \ldots(7)$

Now, differentiating (6), we get $\qquad \dfrac{d^2 y}{dx^2} = A_1 \dfrac{d^2 u}{dx^2} + \dfrac{dA_1}{dx} \cdot \dfrac{du}{dx} + B_1 \dfrac{d^2 v}{dx^2} + \dfrac{dB_1}{dx} \cdot \dfrac{dv}{dx}$. $\qquad \ldots(8)$

Substituting all these values in (1), we get
$$A_1 \left[\frac{d^2 u}{dx^2} + P\frac{du}{dx} + Qu\right] + B_1 \left[\frac{d^2 v}{dx^2} + P\frac{dv}{dx} + Q.v\right] + \frac{dA_1}{dx} \cdot \frac{du}{dx} + \frac{dB_1}{dx} \cdot \frac{dv}{dx} = X.$$

Using (4), we get $\quad A_1.0 + B_1.0 + \dfrac{dA_1}{dx} \cdot \dfrac{du}{dx} + \dfrac{dB_1}{dx} \cdot \dfrac{dv}{dx} = X$

$$\frac{dA_1}{dx} \cdot \frac{du}{dx} + \frac{dB_1}{dx} \cdot \frac{dv}{dx} = X. \qquad \qquad \ldots(9)$$

Solving (6) and (9), we get $\qquad \dfrac{dA_1}{dx} = -\dfrac{vX}{u\dfrac{dv}{dx} - v\dfrac{du}{dx}}$ and $\dfrac{dB_1}{dx} = \dfrac{uX}{u\dfrac{dv}{dx} - v\dfrac{dy}{dx}}$.

On integrating, we get $\qquad A_1 = f(x) + C_1 \qquad$ and $\qquad B_1 = g(x) + C_2$.

Hence, from (5), the required complete solution is given by $y = C_1 u + C_2 v + u\, f(x) + v\, g(x)$ with C_1 and C_2 arbitrary constants.

Step 1. Reduce the given equation into normal form.

Step 2. Consider $\dfrac{d^2y}{dx^2} + P\dfrac{dy}{dx} + Qy = 0$ which can be solved by any method. ...(1)

Step 3. Let $y = Au + Bv$ ...(2)

be the solution of (1), where A and B are arbitrary constants and u and v are known functions of x. Then $Au + Bv$ is complementary function.

Step 4. Choose A_1 and B_1 such that $u\dfrac{dA_1}{dx} + v\dfrac{dB_1}{dx} = 0$ ...(3)

Step 5. Let $y = A_1 u + B_1 v$ be the general solution of the given equation. Then A_1 and B_1 are functions of x, which are to be determined.

Step 6. Differentiating (2) and using (3), we get $\dfrac{dy}{dx} = A_1\dfrac{du}{dx} + B_1\dfrac{dv}{dx}$ ...(4)

Now, differentiating (4), we get $\dfrac{d^2y}{dx^2} = \dfrac{dA_1}{dx}\dfrac{du}{dx} + A_1\dfrac{d^2u}{dx^2} + \dfrac{dB_1}{dx}\cdot\dfrac{dv}{dx} + B_1\cdot\dfrac{d^2v}{dx^2}$ ...(5)

Substituting all these values in the standard form of the given equation and after simplification, we get

$$\frac{dA_1}{dx}\cdot\frac{du}{dx} + \frac{dB_1}{dx}\cdot\frac{dv}{dx} = X \qquad ...(6)$$

Step 7. Solve (3) and (6) and get $\dfrac{dA_1}{dx}$ and $\dfrac{dB_1}{dx}$, integrate these to get A_1 and B_1. Then putting the values of A_1 and B_1 in (2) and get the required general solution.

✎ REMARKS

- Since, the form of y is the same for the equations (3) and (1), but the constants, which will occur, are different, therefore this method is known as variation of parameters.
- In this method, we must require a complete knowledge of the complementary functions.
- This method must be used only if instructed to do so.
- This method is more useful when the C.F. is known easily but particular integral of the differential equation cannot be obtained by any previous method (discussed in this section).

Solved Examples

EXAMPLE 1. *Solve the following equation by variation of parameters $\dfrac{d^2y}{dx^2} + a^2y = \sec ax$*

(UPTU(SUM)–2002, 08 Q. BANK, AMIE–2004, UKTU–2012)

SOLUTION. The solution of the given equation $\dfrac{d^2y}{dx^2} + a^2y = 0$ is given by

$$y = A\cos ax + B\sin ax \qquad ...(1)$$

Now, let us suppose that A and B are functions of x and let (1) satisfies the given equation.

Therefore, $\dfrac{dy}{dx} = -Aa\sin ax + B.a\cos ax$

$$+ \cos ax\frac{dA}{dx} + \sin ax\frac{dB}{dx}$$

Now, let us choose A and B such that

$$\cos ax\frac{dA}{dx} + \sin ax\frac{dB}{dx} = 0. \qquad ...(2)$$

Then, $\dfrac{dy}{dx} = -A.a\sin ax + B.a\cos ax$

$\Rightarrow \quad \dfrac{d^2y}{dx^2} = -Aa^2\cos ax - Ba^2\sin ax$

$$-\frac{dA}{dx}.a\sin ax + \frac{dB}{dx}.a\cos ax.$$

If equation (1) satisfy the given equation, when A and B are functions of x, we have

$$-\frac{dA}{dx}.a\sin ax + a\cos ax\frac{dB}{dx} = \sec ax \qquad ...(3)$$

Solving (2) and (3), we get

$$a\frac{dB}{dx} = 1 \quad \text{and} \quad a\frac{dA}{dx} = -\tan ax$$

$\Rightarrow \quad A = \dfrac{1}{a^2}\log\cos ax + C_1$ and $B = \dfrac{x}{a} + C_2$.

Hence, the complete solution of the given equation is

$$y = C_1\cos ax + C_2\sin ax$$
$$+\frac{1}{a^2}(\log\cos ax)\cos ax + \frac{x}{a}\sin ax.$$

EXAMPLE 2. *Apply the method of variation of parameters to solve $x^2\dfrac{d^2y}{dx^2} + x\dfrac{dy}{dx} - y = x^2 e^x$* (UPTU–2004, 06)

SOLUTION. Clearly, the given equation is homogeneous equation.

To solve it, put $x = e^z$ and $D \equiv \dfrac{d}{dz}$.

$$[D(D-1) + D - 1]y = 0 \quad \Rightarrow \quad (D^2-1)y = 0 \qquad ...(1)$$

Therefore, the solution is $y = Ae^z + Be^{-z}$

i.e., The complementary function of the given equation is

$$y = Ax + B(1/x) \qquad ...(2)$$

Now, if A and B are the functions of x, therefore

$$\frac{dy}{dx} = A - \frac{B}{x^2} + x\frac{dA}{dx} + \frac{1}{x}\frac{dB}{dx}.$$

Now, choosing A and B such that

$$x\frac{dA}{dx} + \frac{1}{x}\frac{dB}{dx} = 0. \qquad ...(3)$$

Then, $\dfrac{d^2y}{dx^2} = \dfrac{dA}{dx} - \dfrac{1}{x^2}\dfrac{dB}{dx} + \dfrac{2B}{x^3}$.

Putting all these values in the given equation, we get

$$x^2\dfrac{dA}{dx} - \dfrac{dB}{dx} = x^2 e^x \qquad \qquad ...(4)$$

Now, solving (3) and (4), we get

$$2x^2\dfrac{dA}{dx} = x^2 e^x \Rightarrow A = \dfrac{1}{2}e^x + C_1$$

and $2\dfrac{dB}{dx} = -x^2 e^x$

$$\Rightarrow B = -\dfrac{1}{2}(x^2 e^x - 2xe^x + 2e^x + C_2).$$

Hence, the solution of the given equation is

$$y = x\left(C_1 + \dfrac{1}{2}e^x\right) + \dfrac{1}{x}\left(-\dfrac{1}{2}x^2 e^x + xe^x - e^x + \dfrac{1}{2}C_2\right).$$

EXAMPLE 3. *Solve by method of variation of parameters*

$$\dfrac{d^2y}{dx^2} - y = \dfrac{2}{1 + e^x}.$$

(UPTU–2001, AMIETE–2001, GBTU(SUM)–2010)

SOLUTION. The C.F. of the given equation is

$$y = C_1 e^x + C_2 e^{-x} \qquad \qquad ...(1)$$

Now, if A and B are the functions of x, therefore

$$\dfrac{dy}{dx} = Ae^x - Be^{-x} + \dfrac{dA}{dx}e^x + \dfrac{dB}{dx}e^{-x}.$$

Now choosing A and B such that

$$e^x\dfrac{dA}{dx} + e^{-x}\dfrac{dB}{dx} = 0 \qquad \qquad ...(2)$$

$$\Rightarrow \quad \dfrac{dy}{dx} = Ae^x - Be^{-x}$$

$$\Rightarrow \quad \dfrac{d^2y}{dx^2} = \dfrac{dA}{dx}.e^x - \dfrac{dB}{dx}e^{-x} + Ae^x + Be^{-x}.$$

Putting all these values in the given equation, we get

$$e^x\dfrac{dA}{dx} - e^{-x}\dfrac{dB}{dx} = \dfrac{2}{1 + e^x}. \qquad \qquad ...(3)$$

Solving (2) and (3), we get

$$2e^x\dfrac{dA}{dx} = \dfrac{2}{1 + e^x} \Rightarrow \dfrac{dA}{dx} = \dfrac{e^{-x}}{1 + e^x}$$

$$\Rightarrow \quad A = \int \dfrac{e^{-x}dx}{1 + e^x} = \int \dfrac{dz}{z^2(1 + z)}$$

$$\text{(By putting } e^x = z \text{ and } e^x\, dx = dz)$$

$$= \int \left(\dfrac{1}{z^2} - \dfrac{1}{z} + \dfrac{1}{1 + z}\right)dz$$

$$= -\dfrac{1}{z} - \log z + \log(1 + z) + C_1$$

$$= \log\dfrac{(1 + e^x)}{e^x} - e^{-x} - C_1.$$

Similarly, we can find

$$B = -\log(1 + e^x) + C_2.$$

Hence, the complete solution of the given equation is

$$y = C_1 e^x + C_2 e^{-x} + e^x\log\dfrac{1 + e^x}{e^x}$$

$$-1 - e^{-x}\log(1 + e^x).$$

EXAMPLE 4. *By the method of variation of parameters, solve the differential equation*

$$\dfrac{d^2y}{dx^2} + (1 - \cot x)\dfrac{dy}{dx} - y\cot x = \sin^2 x.$$

(UPTU–2002, SPECIAL EXAM.–2001)

SOLUTION. Compare the given equation to standard form

$$\dfrac{d^2y}{dx^2} + P\dfrac{dy}{dx} + Qy = 0$$

Here, $P = 1 - \cot x$, $Q = -\cot x$

$\therefore\ 1 - P + Q = 1 - (1 - \cot x) - \cot x = 0$,

$\therefore \qquad y = e^{-x}$ is a part of C.F.

Putting all these above values in given equation, we get

$$\dfrac{d^2v}{dx^2} - (1 + \cot x)\dfrac{dv}{dx} = 0$$

or $\dfrac{dP}{dx} - (1 + \cot x)P = 0$, where $P = \dfrac{dv}{dx}$

or $\dfrac{dP}{P} = (1 + \cot x)dx$

On integrating, we get

$$\log P = x + \log\sin x + \log C_1$$

$$\Rightarrow \quad \log\left(\dfrac{P}{C_1\sin x}\right) = x$$

$$\Rightarrow \quad \dfrac{P}{C_1\sin x} = e^x \Rightarrow P = C_1 e^x\sin x$$

$$\Rightarrow \quad P = \dfrac{dv}{dx} = C_1 e^x\sin x \Rightarrow v = C_1\int e^x\sin x$$

$$\Rightarrow \quad v = C_1\dfrac{1}{2}e^x(\sin x - \cos x) + C_2$$

Hence, $y = v\, e^{-x}$

$$\therefore \qquad y = C_1.\dfrac{1}{2}(\sin x - \cos x) + C_2 e^{-x}$$

Let $y = A(\sin x - \cos x) + Be^{-x}$ be the complete solution of the given equation, where A and B are functions of x.

Now, $\dfrac{dy}{dx} = A(\cos x + \sin x) - Be^{-x}$

$$+ \dfrac{dA}{dx}(\sin x - \cos x) + \dfrac{dB}{dx}e^x$$

Let us suppose A and B such that

$$\dfrac{dA}{dx}(\sin x - \cos x) + \dfrac{dB}{dx}e^{-x} = 0 \qquad \qquad ...(3)$$

$$\dfrac{dy}{dx} = A(\cos x + \sin x) - Be^{-x}$$

and $\dfrac{d^2y}{dx^2} = \dfrac{dA}{dx}(\cos x + \sin x) - \dfrac{dB}{dx}e^{-x}$

$$+ A(-\sin x + \cos x) + Be^{-x}$$

Putting these above values in the given equation, we get

$$\dfrac{dA}{dx}(\cos x + \sin x) - \dfrac{dB}{dx}e^{-x} = \sin^2 x \qquad \qquad ...(4)$$

From (3) and (4), we get

$$\dfrac{dA}{dx} = \dfrac{1}{2}\sin x$$

and $\dfrac{dB}{dx} = \dfrac{1}{2}e^x(\sin x\cos x - \sin^2 x)$

$$= \dfrac{e^x}{4}(\sin 2x + \cos 2x - 1)$$

On integrating, we get

$$A = -\frac{1}{2}\cos x + C_1$$

and

$$B = \frac{1}{4}\int e^x(\sin 2x - 1 + \cos 2x)dx + C_2$$

$$= \frac{1}{4}\int e^x \sin 2x\, dx - \frac{1}{4}\int e^x dx$$

$$\quad + \frac{1}{4}\int e^x \cos 2x + C_2$$

$$= \frac{1}{4}\cdot\frac{e^x}{5}(\sin 2x - 2\cos 2x) - \frac{e^x}{4}$$

$$\quad + \frac{1}{4}\cdot\frac{e^x}{5}(\cos 2x + 2\sin 2x) + C_2$$

$$= \frac{e^x}{20}(3\sin 2x - \cos 2x) - \frac{e^{-x}}{4} + C_2$$

Putting the values of A and B in (2), the general solution is

$$y = \left(-\frac{1}{2}\cos x + C_1\right)(\sin x - \cos x)$$

$$\quad + \left\{\frac{e^x}{20}(3\sin 2x - \cos 2x) - \frac{e^x}{2} + C_2\right\}e^{-x}$$

$$\therefore \quad y = C_1(\sin x - \cos x) + C_2 e^{-x}$$

$$\quad - \frac{1}{10}(\sin 2x - 2\cos 2x).$$

EXAMPLE 5. *Solve by method of variation of parameters*

$$\frac{d^2y}{dx^2} + 2\frac{dy}{dx} + y = e^{-x}\log x. \quad \text{(UPTU–2008)}$$

SOLUTION. We have $\dfrac{d^2y}{dx^2} + 2\dfrac{dy}{dx} + y = e^{-x}\log x$

A.E. is $m^2 + 2m + 1 = 0$

$\Rightarrow \quad (m+1)^2 = 0 \Rightarrow m = -1, -1.$

C.F. $= C_1 e^{-x} + C_2 x e^{-x} = C_1 y_1 + C_2 y_2,$

where $y_1 = e^{-x}$ and $y_2 = xe^{-x}$

P.I. $= uy_1 + vy_2 \Rightarrow$ P.I. $= ue^{-x} + vxe^{-x}.$

where, $u = -\int \dfrac{y_2 X}{y_1 y_2' - y_1' y_2}dx$

$$= -\int \frac{xe^{-x}(e^{-x}\log x)}{e^{-x}(e^{-x} - xe^{-x}) - (e^{-x})(xe^{-x})}dx$$

$$= -\int \frac{xe^{-2x}\log x}{e^{-2x} - xe^{-2x} + xe^{-2x}}dx$$

$$= -\int \frac{e^{-2x}x\log x}{e^{-2x}}dx$$

$$= -\int x\log x\, dx$$

$$= -\left[\log x\,\frac{x^2}{2} - \int\left(\frac{1}{x}\right)\cdot\left(\frac{x^2}{2}\right)dx\right]$$

$$= -\frac{x^2}{2}\log x + \int \frac{x}{2}dx$$

$$= -\frac{x^2}{2}\log x + \frac{x^2}{4}.$$

$$v = \int \frac{y_1 X}{y_1 y_2' - y_1' y_2}dx$$

$$= \int \frac{e^{-x}e^{-x}\log x}{e^{-x}(e^{-x} - xe^{-x}) - (e^{-x})(xe^{-x})}dx$$

$$= \int \frac{e^{-2x}\log x}{e^{-2x}}dx$$

$$= \int \log x\, dx = (\log x)x - \int \frac{1}{x}\cdot x\, dx$$

$$= x\log x - \int dx = x\log x - x.$$

P.I. $= uy_1 + vy_2$

$$= \left(-\frac{x^2}{2}\log x + \frac{x^2}{4}\right)e^{-x} + (x\log x - x)xe^{-x}$$

$$= xe^{-x}\left[\left(-\frac{x}{2}\log x + \frac{x}{4}\right) + x\log x - x\right]$$

$$= xe^{-x}\left[\frac{x}{2}\log x - \frac{3x}{4}\right]$$

Complete solution is $y = $ C.F. + P.I.

$$y = C_1 e^{-x} + C_2 xe^{-x} + xe^{-x}\left[\frac{x}{2}\log x - \frac{3x}{4}\right].$$

EXAMPLE 6. *Using variation of parameters method, solve*

$$x^2\frac{d^2y}{dx^2} + 2x\frac{dy}{dx} - 12y = x^3\log x. \quad \text{(UPTU–2004)}$$

SOLUTION. The given equation is

$$x^2\frac{d^2y}{dx^2} + 2x\frac{dy}{dx} - 12y = x^3\log x \qquad \ldots(1)$$

Put $x = e^z$, so that $z = \log x$ and let $D \equiv \dfrac{d}{dz}$, then the given equation (1) reduces to

$$[D(D-1) + 2D - 12]y = ze^{3z}$$

$$\Rightarrow \qquad (D^2 + D - 12)y = z\,e^{3z}.$$

Auxiliary equation is

$$m^2 + m - 12 = 0 \Rightarrow m = 3, -4$$

C.F. $= C_1 e^{3z} + C_2 e^{-4z} = C_1 x^3 + C_2 x^{-4}.$

Hence, parts of C.F. are x^3 and x^{-4}

Let, $u = x^3$ and $v = x^{-4}$. Also, $R = x\log x$.

Let $y = Au + Bv$ be the complete solution, where A and B are some suitable functions of x. A and B are determined as follows :

$$A = -\int \frac{Rv}{uv_1 - u_1 v}dx + C_1$$

$$= -\int \frac{x\log x\cdot x^{-4}}{x^3(-4x^{-5}) - 3x^2(x^{-4})}dx + C_1$$

$$= -\int \frac{x^{-3}\log x}{-7x^{-2}}dx + C_1 = \frac{1}{7}\int \frac{\log x}{x}dx + C_1$$

$$= \frac{1}{7}\frac{(\log x)^2}{2} + C_1 = \frac{1}{14}(\log x)^2 + C_1$$

$$B = \int \frac{Ru}{uv_1 - u_1 v}dx + C_2 = \int \frac{x\log x\cdot x^3}{-7x^{-2}}dx + C_2$$

$$= -\frac{1}{7}\int x^6\log x\, dx + C_2 = -\frac{1}{7}\left[\log x\cdot\frac{x^7}{7} - \int \frac{1}{x}\cdot\frac{x^7}{7}dx\right] + C_2$$

$$= -\frac{1}{7}\left[\frac{x^7\log x}{7} - \frac{1}{7}\left(\frac{x^7}{7}\right)\right] + C_2$$

$$= \frac{x^7}{49}\left(\frac{1}{7} - \log x\right) + C_2$$

Hence, the complete solution is given by

$$y = Ax^3 + Bx^{-4}$$

$$= \left[\frac{1}{14}(\log x)^2 + C_1\right]x^3 + \left[\frac{x^7}{49}\left(\frac{1}{7} - \log x\right) + C_2\right]x^{-4}$$

$$= C_1 x^3 + C_2 x^{-4} + \frac{x^3}{98}\log x(7\log x - 2) + \frac{x^3}{343}$$

$$y = \left(C_1 + \frac{1}{343}\right)x^3 + C_2 x^{-4} + \frac{x^3}{98}\log x(7\log x - 2)$$

EXAMPLE 7. *Solve by the method of variation of parameters*
$$(D^2 - 1)y = 2(1 - e^{-2x})^{-1/2}.$$

(MTU–2011, UPTU(SUM)–2009)

SOLUTION. Proceeding same as above, we get

Parts of C.F. are $u = e^x$ and $v = e^{-x}$

$$\therefore \quad y = A\,e^x + B\,e^{-x}$$

where A and B are determined as follows :

$$A = -\int \frac{Rv}{uv_1 - u_1 v}dx + C_1$$

$$= -\int \frac{2(1 - e^{-2x})^{-1/2}.e^{-x}}{e^x(-e^{-x}) - e^x.e^{-x}}dx + C_1$$

$$= -2\int \frac{e^{-x}}{-2\sqrt{1 - e^{-2x}}}dx + C_1$$

$$= \int \frac{e^{-x}}{\sqrt{1 - e^{-2x}}}dx + C_1$$

Put $e^{-x} = t \therefore e^{-x}dx = -dt$

$$= -\int \frac{dt}{\sqrt{1 - t^2}} + C_1$$

$$\Rightarrow \quad A = -\sin^{-1}(e^{-x}) + C_1$$

and $B = \int \dfrac{Ru}{uv_1 - u_1 v}dx + C_2$

$$= -\int \frac{2(1 - e^{-2x})^{-1/2}.e^{-x}}{(-2)}dx + C_2$$

$$= -\int \frac{e^{-x}}{\sqrt{1 - e^{-2x}}}dx + C_2$$

$$= -\int \frac{e^{2x}}{\sqrt{e^{2x} - 1}}dx + C_2$$

Put $e^{2x} = t \therefore e^{2x}dx = dt/2$

$$= -\frac{1}{2}\int \frac{dt}{\sqrt{t - 1}} + C_2 = -\frac{1}{2}\frac{(t - 1)^{1/2}}{\left(\frac{1}{2}\right)} + C_2$$

$$\Rightarrow \quad B = -(e^{2x} - 1)^{1/2} + C_2$$

Hence, the complete solution is

$$y = A\,e^x + B\,e^{-x}$$

$$y = [-\sin^{-1}(e^{-x}) + C_1]e^x$$

$$+ [-(e^{2x} - 1)^{1/2} + C_2]e^{-x}$$

$$\Rightarrow \quad y = C_1 e^x + C_2 e^{-x}$$

$$- e^x \sin^{-1}(e^{-x}) - e^{-x}(e^{2x} - 1)^{1/2}$$

Exercise-8.30

Apply method of variation of parameters solve the following equations :

1. $x^2 \dfrac{d^2y}{dx^2} - 2x(1 + x)\dfrac{dy}{dx} + 2(x + 1)y = x^3$

2. $\dfrac{d^2y}{dx^2} + a^2 y = \operatorname{cosec} ax$

3. $(1 - x)\dfrac{d^2y}{dx^2} + x\dfrac{dy}{dx} - y = (1 - x)^2$

4. $(x + 2)\dfrac{d^2y}{dx^2} - (2x + 5)\dfrac{dy}{dx} + 2y = (x + 1)e^x$

5. $\dfrac{d\,y}{dx} + y = \operatorname{cosec} x$

6. $\dfrac{d^2y}{dx^2} + 9y = \sec 3x$

7. $\dfrac{d^2y}{dx^2} + (\tan x - 3\cos x)\dfrac{dy}{dx} + 2y\cos^2 x = \cos^4 x$

8. $\dfrac{d^3y}{dx^3} - 6\dfrac{d^2y}{dx^2} + 11\dfrac{dy}{dx} - 6y = e^{2x}$

9. If $y = x, y = x^2 - 1$ are linearly independent solution of

$$(x^2 + 1)\frac{d^2y}{dx^2} - 2x\frac{dy}{dx} + 2y = 0$$

find the general solution of

$$(x^2 + 1)\frac{d^2y}{dx^2} - 2x\frac{dy}{dx} + 2y = 6(x^2 + 1)^2.$$

10. $(1 - x^2)\dfrac{d^2y}{dx^2} + x\dfrac{dy}{dx} - y = x(1 - x^2)^{3/2}$

(UPTU(Q.BANK)–2001)

11. $\dfrac{d^2y}{dx^2} - 2\dfrac{dy}{dx} = e^x \sin x$ (UPTU(Q.BANK)–2001)

12. $\dfrac{d^2y}{dx^2} - 3\dfrac{dy}{dx} + 2y = \dfrac{e^x}{1 + e^x}$ (MTU(SUM)–2011)

13. $\dfrac{d^2y}{dx^2} - 3\dfrac{dy}{dx} + 2y = \sin e^{-x}$ (MTU–2012)

14. $x\dfrac{d\,y}{dx} + 4x\dfrac{dy}{dx} + 2y = e$ (GBTU–2012)

15. $x^2\dfrac{d^2y}{dx^2} + x\dfrac{dy}{dx} - 9y = 48x^5$ (GBTU(CO)–2010)

Hints to Selected Problems

1. We have $P + Q.x = 0 \quad \Rightarrow \quad y = x$ is a part of C.F.

On solving by usual procedure, we get $y = Ax + Bxe^{2x}$ is the general solution. Now, by letting A and B also the functions of x, use the technique of variation of parameters.

2. C.F. of the equation is given by $y = A\cos ax + B\sin ax$, where A and B both are functions of x. Now use method of variation of parameters.

4. Here, $y = e^{2x}$ is a part of C.F. Putting $y = ve^{2x}$ and on solving, we get
$y = A(2x + 5) + Be^{2x}$ is the solution.
Now by letting A and B, both functions of x, use the method of

variation of parameters.

9. It is given that $y = x$ and $y = x^2 - 1$ are two linearly independent solutions of the given differential equation.
Let $y = A(x^2 - 1) + Bx$ be the general solution of given equation. Then use method of variation of parameters by assuming A and B both are the functions of x.

Answers

1. $y = C_1 x + C_2 xe^{2x} - \dfrac{x^2}{2} - \dfrac{x}{4}$

2. $y = C_1 \cos ax + C_2 \sin ax - \dfrac{x}{a}\cos ax + \dfrac{1}{a^2}\sin ax \cdot \log\sin ax$

3. $y = C_1 e^x + C_2 x + x + 1 + x^2$

4. $y = C_1(2x + 5) + C_2 e^{2x} - e^x$

5. $y = C_1 \cos x + C_2 \sin x - x\cos x + \sin x \log\sin x$

6. $y = C_1 \cos 3x + C_2 \sin 3x + \dfrac{1}{9}\cos 3x \log\cos 3x + \dfrac{x}{3}\sin 3x$

7. $y = C_1 e^{\sin x} + C_2 e^{2\sin x} - \dfrac{5}{4} - \dfrac{3}{2}\sin x - \dfrac{1}{2}\sin^2 x$

8. $y = C_1 e^x + (C_2 - x)e^{2x} + C_3 e^{3x}$

9. $y = C_1(x^2 - 1) + C_2 x + x^4 + 3x^2$

10. $y = C_1(\sqrt{1 - x^2} + x\sin^{-1} x) + C_2 x - \dfrac{x}{9}(1 - x^3)^{3/2}$

11. $y = C_1 + C_2 e^{2x} - \dfrac{e^x}{2}\sin x$

12. $y = [\log(e^{-x} + 1) + C_1]e^x + [\log(1 + e^{-x}) - (1 + e^{-x}) + C_2]e^{2x}$

13. $y = C_1 e^x + C_2 e^{2x} - e^{2x}\sin e^{-x}$

14. $y = (e^x + C_1)\dfrac{1}{x} + [(1 - x)e^x + C_2]\dfrac{1}{x^2}$

15. $y = (4x^2 + C_1)x^3 + (C_2 - x^8)x^{-3}$

8.6 APPLICATIONS OF ORDINARY DIFFERENTIAL EQUATION TO ENGINEERING PROBLEMS

Differential equations are widely used in solving engineering problems. The physical principles forming the background of the problems are expressed mathematically by the formulation of one or more differential equations. Here, we study the applications concerned with the motion of a particle in a resistance medium, SHM, bending of beams, electric circuits, etc.

8.6.1 APPLICATION TO ELECTRIC CIRCUITS

If q be the electrical charge on a conductor of capacity C and i be the current, then

(a) $i = \dfrac{dq}{dt}$ or $q = \int i\, dt$

(b) the potential drop across the resistance R is Ri.

(c) the potential drop across the inductance L is $L\dfrac{di}{dt}$.

(d) the potential drop across the capacitance C is $\dfrac{q}{C}$.

KIRCHOFF'S LAW

1. **Voltage Law :** The algebraic sum of the voltage drop around any closed circuit is equal to the resultant electromotive force in the circuit.

2. **Current Law :** At a junction or node, current coming is equal to the current going.

(i) **L-C Electrical Circuit (without e.m.f):** Let q be the electrical charge on the condenser plate and i be the current in an electrical circuit containing an inductance L and capacitance C. Then

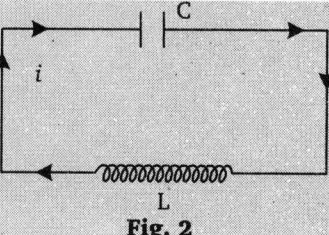

Fig. 2

(a) The voltage drop across L is $L\dfrac{di}{dt} = L\dfrac{d^2 q}{dt^2}$ $\left[\because i = \dfrac{dq}{dt}\right]$

(b) The voltage drop across C is $\dfrac{q}{C}$.

By Kirchoff's law, we have $L\dfrac{d^2 q}{dt^2} + \dfrac{q}{C} = 0$ [$\because$ No e.m.f. is applied.]

$$\dfrac{d^2 q}{dt^2} + \dfrac{1}{LC}q = 0$$

$$\dfrac{d^2 q}{dt^2} + \omega^2 q = 0 \qquad\qquad \left[\because \omega^2 = \dfrac{1}{LC}\right]$$

which is an equation of S.H.M. Thus it represents free electrical oscillations of current having period

$$T = \dfrac{2\pi}{\omega} = 2\pi\sqrt{LC}$$

Hence, the discharging of a condenser through an inductance L is the same as the motion of the mass m attached at the end of a spring.

(ii) L-R Series Circuit : Let i be the current flowing in the circuit containing resistance R and inductance L in series, with voltage source E at any time t.

By voltage law, $Ri + L\dfrac{di}{dt} = E$

$$\dfrac{di}{dt} + \dfrac{R}{L}i = \dfrac{E}{L} \qquad\qquad\qquad \text{... (1)}$$

This is the linear differential equation

$$\text{I.F.} = e^{\int \frac{R}{L}dt} = e^{\frac{R}{L}t}$$

Its solution is given by

$$i\,e^{\frac{R}{L}t} = \int \dfrac{E}{L}e^{\frac{R}{L}t}\,dt + C$$

$$i.e^{\frac{R}{L}t} = \dfrac{E}{L} \times \dfrac{L}{R}e^{\frac{R}{L}t} + C$$

$$i = - + Ce^{} + C \qquad\qquad \text{... (2)}$$

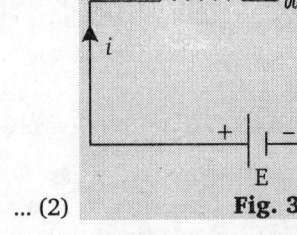

Fig. 3

At $t = 0,\ i = 0 \Rightarrow C = -\dfrac{E}{R}$

From (2), $i = \dfrac{E}{R}\left[1 - e^{-\frac{R}{L}t}\right]$

(iii) L-C Electrical Circuit with e.m.f. $Q\cos nt$: The equation for an L-C circuit with e.m.f. $Q\cos nt$ is

$$L\dfrac{d^2q}{dt^2} + \dfrac{q}{C} = Q\cos nt \qquad \Rightarrow \qquad \dfrac{d^2q}{dt^2} + \dfrac{1}{LC}q = \dfrac{Q}{L}\cos nt$$

$$\dfrac{d^2q}{dt^2} + \omega^2 q = E\cos nt\text{ , where, } \omega^2 = \dfrac{1}{LC} \text{ and } E = \dfrac{Q}{L}.$$

 **REMARK**

- Practically we see that we tune in a radio station. By changing C , the natural frequency of the tuning L-C circuit is made equal to the frequency of the desired receiving radio station, and so the output of the receiver becomes large at the desired receiving station.

(iv) L-C-R Electrical Circuit with e.m.f. $Q\cos nt$:

The differential equation of the L-C-R circuit which contains an alternating e.m.f. $Q\cos nt$ is

$$L\dfrac{d^2q}{dt^2} + R\dfrac{dq}{dt} + \dfrac{q}{C} = Q\cos nt$$

$$\dfrac{d^2q}{dt^2} + \dfrac{R}{L}\dfrac{dq}{dt} + \dfrac{1}{LC}q = \dfrac{Q}{L}\cos nt$$

$$\dfrac{d^2 q}{dt} + 2p\dfrac{dq}{dt} + \omega\, q = E\cos nt$$

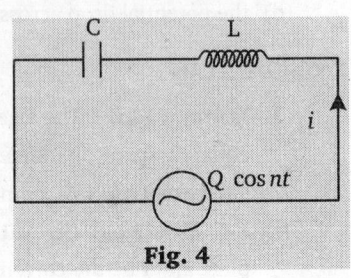

Fig. 4

where, $\dfrac{R}{L} = 2p,\ \dfrac{1}{LC} = \omega^2 \text{ and } \dfrac{Q}{L} = E$.

Solved Examples

EXAMPLE 1. *A condenser of capacity C Farades with V_0 is discharged through a resistance R Ohms. Show that if q Coulomb is the charge on the condenser, i Ampere the current and V the voltage at time t.*

$$q = CV,\ V = Ri \text{ and } \dfrac{dq}{dt}$$

Hence show that $V = V_0\, e^{\frac{t}{RC}}$.

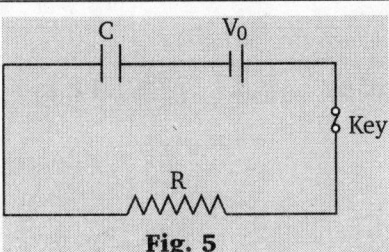

Fig. 5

SOLUTION. Voltage across resistance $R = Ri$.

Voltage drop across capacitance $= \dfrac{q}{C}$.

The equation of discharge of condenser can be written, when after release of key the condenser gets discharged and at that time, voltage across the battery gets zero so that $V_0 = 0$.

The differential equation of the above circuit is

$$Ri + \dfrac{q}{C} = 0$$

$$R\dfrac{dq}{dt} + \dfrac{q}{C} = 0$$

$$\frac{dq}{dt} + \frac{q}{RC} = 0 \quad \text{or} \quad \frac{dq}{dt} = -\frac{q}{RC}.$$

$$\frac{dq}{q} = -\frac{1}{RC} dt$$

Integrating both sides, we get

$$\int \frac{dq}{q} = -\frac{1}{RC} \int dt$$

$$\log q = -\frac{1}{RC} t + a \qquad \ldots (1)$$

At $t = 0$, the change at the condenser is q_0

such that $\quad \log q_0 = -\frac{1}{RC}(0) + a \qquad \ldots (2)$

$\Rightarrow \qquad\qquad a = \log q_0$

Now, from (1), we have

$$\log q = -\frac{1}{RC} t + \log q_0$$

$$\log q - \log q_0 = -\frac{1}{RC} t$$

$$\Rightarrow \quad \log \frac{q}{q_0} = -\frac{1}{RC} t \quad \Rightarrow \quad \frac{q}{q_0} = e^{-t/RC}$$

$$q = q_0 \, e^{-t/RC} \qquad \ldots (3)$$

Dividing both sides of (3) by C, we get

$$\Rightarrow \qquad \frac{q}{C} = \frac{q_0}{C} e^{-t/RC}$$

Hence, $\qquad V = V_0 \, e^{-t/RC}$

EXAMPLE 2. *Show that the frequency of free vibrations in a closed electrical circuit with inductance L and capacity C in series is* $\dfrac{30}{\pi\sqrt{LC}}$ *per minute.*

[UPTU(Q. BANK)–2002]

SOLUTION. Let i be the current and q be the charge in the condenser plate at any time t.

The voltage drop across $L = L\dfrac{di}{dt} = L\dfrac{d^2q}{dt^2}$

The voltage drop across $C = \dfrac{q}{C}$.

Since, there is no applied e.m.f. in the circuit, we have by Kirchoff's law :

$$L\frac{d^2q}{dt^2} + \frac{q}{C} = 0 \quad \text{or} \quad \frac{d^2q}{dt^2} + \frac{q}{CL} = 0.$$

$$\Rightarrow \qquad \frac{d^2q}{dt^2} = -\frac{1}{LC} q$$

Writing $\dfrac{1}{LC} = \omega^2$, it becomes $\dfrac{d^2q}{dt^2} = -\omega^2 q$

It represents oscillatory current with period

$$\frac{2\pi}{\omega} = 2\pi\sqrt{LC}.$$

Frequency $= \dfrac{1}{T}$ per second

$= \dfrac{60}{2\pi\sqrt{LC}}$ per minute

$= \dfrac{30}{\pi\sqrt{LC}}$ per minute.

EXAMPLE 3. *A condenser of capacity C is discharged through the inductance L and a resistance R in series and the charge q at any time t satisfies the equation*

$$L\frac{d^2q}{dt^2} + R\frac{dq}{dt} + \frac{q}{C} = 0.$$

Given that L = 0.25 Henry, R = 250 Ohms, $C = 2 \times 10^{-6}$ Farad and that when t = 0, the charge q is 0.02 Coulomb and the current $\dfrac{dq}{dt} = 0$. Obtain the value of q in terms of t.

SOLUTION. The given equation of charge is

$$L\frac{d^2q}{dt^2} + R\frac{dq}{dt} + \frac{q}{C} = 0 \qquad \ldots (1)$$

Substituting the given value in (1), we get

$$\frac{d^2q}{dt^2} + \frac{250}{0.25}\frac{dq}{dt} + \frac{q}{2\times10^{-6}\times0.25} = 0 \qquad \ldots (2)$$

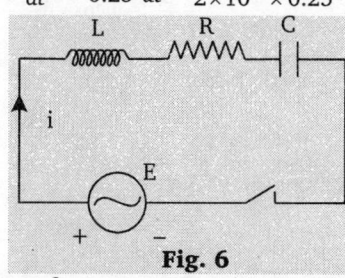

Fig. 6

$$\frac{d^2q}{dt^2} + 1000\frac{dq}{dt} + 2\times10^6 q = 0$$

A.E. is $\qquad m^2 + 1000m + 2\times10^6 = 0$

$\Rightarrow \qquad\qquad m = -500 \pm 1323i$

The solution is

$$q = e^{-500t}[A\cos1323t + B\sin1323t] \qquad \ldots(3)$$

Putting $t = 0$, $q = 0.002$ in equation (3), we get

$$A = 0.002.$$

From (3), we have

$$\frac{dq}{dt} = -500\, e^{-500t}[A\cos1323t + B\sin1323t]$$

$$+ 1323 e^{-500t}[-A\sin1323t + B\cos1323t] \ldots (4)$$

Putting $t = 0$, $\dfrac{dq}{dt} = 0$ in (4), we get $B = 0.0008$.

Hence, the required value of q is

$$q = e^{-0.500t}[0.002\cos1323t + 0.0003\sin1323t]$$

EXAMPLE 4. *An uncharged condenser of capacity C is charged by applying an e.m.f., $E\sin\dfrac{t}{\sqrt{LC}}$ through leads of self-inductance L and negligible resistance. Prove that at time t, the charge on one of the plate is*

$$\frac{EC}{2}\left[\sin\frac{t}{\sqrt{LC}} - \frac{t}{\sqrt{LC}}\cos\frac{t}{\sqrt{LC}}\right]$$

(UPTU–2003)

SOLUTION. Let q be the charge on the condenser at any time t.

The differential equation for the circuit is

$$L\frac{d^2q}{dt^2} + \frac{q}{C} = E\sin\frac{t}{\sqrt{LC}} \qquad \ldots (1)$$

It's A.E. is $Lm^2 + \dfrac{1}{C} = 0$ or $m^2 = -\dfrac{1}{LC}$

so that $m = \pm\dfrac{i}{\sqrt{LC}}$

C.F. $= C_1\cos\dfrac{t}{\sqrt{LC}} + C_2\sin\dfrac{t}{\sqrt{LC}}$

P.I. $= \dfrac{1}{LD^2 + \dfrac{1}{C}} E\sin\dfrac{t}{\sqrt{LC}} = Et \cdot \dfrac{1}{2LD}\sin\dfrac{t}{\sqrt{LC}}$

$$= \frac{Et}{2L}\left[-\sqrt{LC}\cos\frac{t}{\sqrt{LC}}\right] = -\frac{Et}{2}\sqrt{\frac{C}{L}}\cos\frac{t}{\sqrt{LC}}$$

The complete solution is

$$q = C_1\cos\frac{t}{\sqrt{LC}} + C_2\sin\frac{t}{\sqrt{LC}} - \frac{Et}{2}\sqrt{\frac{C}{L}}\cos\frac{t}{\sqrt{LC}} \quad ... (2)$$

When $t = 0$, $q = 0$, $\therefore C_1 = 0$.

Differentiating (2) w.r.t. t, we get

$$\frac{dq}{dt} = -\frac{q}{\sqrt{LC}}\sin\frac{t}{\sqrt{LC}} + \frac{C_2}{\sqrt{LC}}\cos\frac{t}{\sqrt{LC}}$$

$$- \frac{E}{2}\sqrt{\frac{C}{L}}\left[\cos\frac{t}{\sqrt{LC}} - \frac{t}{\sqrt{LC}}\sin\frac{t}{\sqrt{LC}}\right]$$

$$\frac{dq}{dt} = i = 0, \text{ when } t = 0$$

$$\Rightarrow \quad \frac{C_2}{\sqrt{LC}} - \frac{E}{2}\sqrt{\frac{C}{L}} = 0 \Rightarrow C_2 = \frac{EC}{2}$$

Now, from (2), we get

$$q = \frac{EC}{2}\sin\frac{t}{\sqrt{LC}} - \frac{Et}{2}\sqrt{\frac{C}{L}}\cos\frac{t}{\sqrt{LC}}$$

or $\quad q = \frac{EC}{2}\left[\sin\frac{t}{\sqrt{LC}} - \frac{t}{\sqrt{LC}}\cos\frac{t}{\sqrt{LC}}\right]$

EXAMPLE 5. *In an L-C-R circuit, the charge q on a plate of a condenser is given by* $L\dfrac{d^2q}{dt^2} + R\dfrac{dq}{dt} + \dfrac{q}{C} = E\sin pt$. *The circuit is tuned to resonance so that* $p^2 = 1/LC$. *If initially the current i and the charge q be zero, show that for small values of* R/L, *the current in the circuit at time t is given by* $\left\{\dfrac{Et}{2L}\right\}\sin pt$

(UPTU–2004, 05)

SOLUTION. The given equation is

$$\left(LD^2 + RD + \frac{1}{C}\right)q = E\sin pt; \quad D \equiv \frac{d}{dt} \quad ... (1)$$

A.E. is $\quad Lm^2 + Rm + \dfrac{1}{C} = 0$

$$m = \frac{-R \pm \sqrt{R^2 - \dfrac{4L}{C}}}{2L} = -\frac{R}{2L} \pm \frac{1}{2}\sqrt{\frac{R^2}{L^2} - \frac{4}{LC}}$$

$$= -\frac{R}{2L} \pm \frac{1}{2}\sqrt{\frac{-4}{LC}}$$

$$m = -\frac{R}{2L} \pm \frac{i}{\sqrt{LC}} = -\frac{R}{2L} \pm pi$$

C.F. $= e^{-\frac{Rt}{2L}}(C_1\cos pt + C_2\sin pt)$

$$= \left(1 - \frac{Rt}{2L}\right)(C_1\cos pt + C_2\sin pt)$$

P.I. $= \dfrac{1}{LD^2 + RD + \dfrac{1}{C}}(E\sin pt)$

$$= E.\frac{1}{-Lp^2 + RD + \dfrac{1}{C}}\sin pt$$

$$= E.\frac{1}{RD}\sin pt = -\frac{E}{Rp}\cos pt$$

Complete solution is given by

$$q = \left(1 - \frac{Rt}{2L}\right)(C_1\cos pt + C_2\sin pt) - \frac{E}{Rp}\cos pt \quad ... (2)$$

$$i = \frac{dq}{dt} = \left(1 - \frac{Rt}{2L}\right)(-pC_1\sin pt + pC_2\cos pt)$$

$$- \frac{R}{2L}(C_1\cos pt + C_2\sin pt) + \frac{E}{R}\sin pt \quad ...(3)$$

When $t = 0$, $q = 0$ and $i = 0$, then

from (2), $\quad 0 = C_1 - \dfrac{E}{Rp} \Rightarrow C_1 = \dfrac{E}{Rp}$.

from (3), $\quad 0 = C_2 p - \dfrac{R}{2L}C_1 \Rightarrow C_2 = \dfrac{E}{2Lp^2}$

Now, from (3),

$$i = \left(1 - \frac{Rt}{2L}\right)\left(-\frac{E}{R}\sin pt + \frac{E}{2Lp}\cos pt\right)$$

$$- \frac{R}{2L}\left(\frac{E}{Rp}\cos pt + \frac{E}{2Lp^2}\sin pt\right) + \frac{E}{R}\sin pt$$

$$= \frac{Et}{2L}\sin pt - \frac{ERt}{4pL^2}\cos pt - \frac{ER}{4L^2p^2}\sin pt$$

$$= \frac{Et}{2L}\sin pt; \quad \frac{R}{L} \text{ being small.}$$

EXAMPLE 6. *A 20 Ohms resistor is connected in series with a capacitor of 0.01 Farad and e.m.f. E Volts given by* $40e^{-3t} + 20e^{-6t}$. *If* $q = 0$, *at* $t = 0$. *Show that the maximum charge on the capacitor is 0.25 Coulomb.*

SOLUTION. Equation of charge and discharge can be written as follows :

$$R\frac{dQ}{dt} + \frac{Q}{C} = 40e^{-3t} + 20e^{-6t}$$

$$\Rightarrow \quad \frac{dQ}{dt} + \frac{Q}{RC} = \frac{40}{R}e^{-3t} + \frac{20}{R}e^{-6t}$$

$$\Rightarrow \quad \frac{dQ}{dt} + \frac{Q}{RC} = \frac{40}{20}e^{-3t} + \frac{30}{20}e^{-6t}$$

$$\Rightarrow \quad \frac{dQ}{dt} + \frac{Q}{20 \times 0.01} = 2e^{-3t} + e^{-6t}$$

$$\Rightarrow \quad \frac{dQ}{dt} + 5Q = 2e^{-3t} + e^{-6t}$$

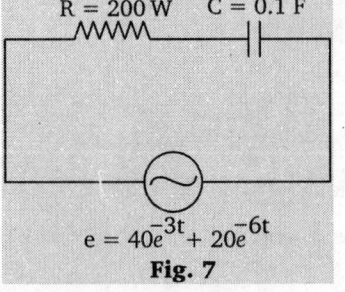

$$e = 40e^{-3t} + 20e^{-6t}$$

Fig. 7

I.F. $= e^{5\int dt} = e^{5t}$.

Its solution is

$$Qe^{5t} = \int e^{5t}(2e^{-3t} + e^{-6t})\,dt + a.$$

$$= \int (2e^{2t} + e^{-t})\,dt + a = e^{2t} - e^{-t} + a.$$

$$\Rightarrow \quad Q = e^{-3t} - e^{-6t} + ae^{-5t} \quad ... (1)$$

Putting $t = 0$, $Q = 0$ in (1), we get

$$0 = ae^{-3\times 0} + e^{-6\times 0} + ae^{-5\times 0}$$

$$\Rightarrow \quad 0 = 1 - 1 + a \Rightarrow a = 0$$

Putting value of a in (1)

$$Q = e^{-3t} - e^{-6t}$$

For maximum value $\dfrac{dQ}{dt} = 0$.

$-3e^{-3t} - (-6)e^{-6t} = 0$ or $3e^{-3t} = 6e^{-6t}$.

$$e^{-3t} = 2e^{-6t} \text{ or } \frac{1}{2} = e^{-3t}$$

or
$$2 = e^{3t}$$

$$\log 2 = 3t$$

$$t = \frac{1}{3}\log 2 = \log(2^{1/3})$$

Maximum charge (by putting the value of t)

$$Q = e^{-3t} - e^{-6t} = e^{-3(\log(2^{1/3}))} - e^{-6(\log 2^{1/3})}$$

$$= e^{\log 2^{(-3/3)}} - e^{\log 2^{(-6/2)}}$$

$$= e^{\log(1/2)} - e^{\log(1/4)} = \frac{1}{2} - \frac{1}{4} = \frac{2-1}{4}$$

$$= \frac{1}{4} = 0.25 \text{ Amp.}$$

EXAMPLE 7. *The voltage V and the current i at a distance x from the sending end of the transmission line satisfy the equation:*

$$-\frac{dV}{dx} = Ri, \quad -\frac{di}{dx} = GV$$

where R and G are constants. If $X = V_0$ at the sending end $(x = 0)$ and $V = 0$ at receiving end $(x = l)$. Show that $V = V_0\left\{\dfrac{\sinh n(l-x)}{\sinh nl}\right\}$, when $n^2 = RG$.

SOLUTION. We have $-\dfrac{dV}{dx} = Ri$ ... (1)

$$-\frac{di}{dx} = GV \quad \text{... (2)}$$

When $x = 0$, $V = V_0$, when $x = l$, $V = 0$

putting the value of i from (1) into (2), we get

$$-\frac{d}{dx}\left(-\frac{dV}{dx}\cdot\frac{1}{R}\right) = GV$$

or
$$\frac{d^2V}{dx^2} = RGV$$

$$\frac{d\,V}{dx} - (RG)V = 0$$

$$(D^2 - RG)V = 0$$

A.E. $(D^2 - n^2) = 0 \Rightarrow D = \pm n$

$$V = A\,e^{+nx} + Be^{-nx} \quad \text{... (3)}$$

On putting $x = 0$ and $V = V_0$ in (3), we get

$$V_0 = A + B \quad \text{... (4)}$$

On putting $x = l$ and $V = 0$ in (3), we get

$$0 = Ae^{+nl} + Be^{-nl} \quad \text{... (5)}$$

From (4) and (5), we have

$$A = \frac{V_0}{1 - e^{2nl}}, \quad B = -\frac{V_0 e^{2nl}}{1 - e^{2nl}}$$

Now, from (3), we have

$$V = \frac{V_0 e^{nx}}{1 - e^{2nl}} - V_0\frac{e^{2nl}e^{-nx}}{1 - e^{2nl}}$$

$$= \frac{V_0[e^{nx} - e^{2nl-nx}]}{1 - e^{2nl}} = \frac{V_0[e^{(nl-nx)} - e^{-(nl-nx)}]}{e^{nl} - e^{-nl}}$$

[Dividing numerator and denominator by e^{nl}]

$$= V_0\left\{\frac{\sinh n(l-x)}{\sinh nl}\right\}.$$

Exercise-8.31

1. In a condenser discharging electricity the voltage V satisfies the equation $\dfrac{1}{a}\dfrac{dy}{dx} = A\sinh ax + B\cosh ax$, where k is a constant and t is time measured in seconds. Given $k = 50$, find the time t, in which V decreases to one tenth of its original value.

2. A resistance of 100 Ohms, an inductance of 0.5 Henry are connected in series with a battery of 20 Volts. Find the current in the circuit as a function of time.

3. In an LCR circuit, the charge q on a plate of the condenser is given by $L\dfrac{d^2q}{dt^2} + R\dfrac{dq}{dt} + \dfrac{q}{C} = E\sin\omega t$, where $i = \dfrac{dq}{dt}$. The circuit is tuned to resonance so that $\omega^2 = \dfrac{1}{LC}$. If $R^2 < \dfrac{4L}{C}$ and $q = 0 = i$, when $t = 0$, show that

$$q = \frac{E}{R\omega}\left[-\cos\omega t + e^{-\frac{Rt}{2L}}\left(\cos pt + \frac{R}{2Lp}\sin pt\right)\right]$$

and $i = \dfrac{E}{R}\left[\sin\omega t - \dfrac{1}{p\sqrt{LC}}e^{-\frac{Rt}{2L}}\sin pt\right]$, where $p^2 = \dfrac{1}{LC} - \dfrac{R^2}{4L^2}$.

(UPTU(B.TECH)–2003)

4. The equations of electromotive force in terms of current i for an electrical circuit having resistance R and a condenser of capacity C_1 in series is $E = Ri + \int\dfrac{i}{C}dt$. Find the current i at any time t, when $E = E_0\sin\omega t$.

(UPTU–2006)

5. The damped LCR circuit is governed by the equation

$$L\frac{d^2q}{dt^2} + R\frac{dq}{dt} + \left(\frac{1}{C}\right)q = 0$$

where L, C, R are positive constants. Find the conditions under which the circuit is overdamped, underdamped and critically damped. Find also the critical resistance.

(UPTU–2005, (CO), 2008)

6. A condenser of capacity C is discharged through an inductance L and resistance R in series and the charge q at time t satisfies the equation $L\dfrac{d^2q}{dt^2} + R\dfrac{dq}{dt} + \dfrac{q}{C} = 0$ given that $L = 0.25$ Henry, $R = 250$ Ohm, $C = 2\times10^{-6}$ Farad and that when $t = 0$, charge q is 0.002 Coulomb and current $i = 0$. Obtain the value of q in terms of t.

(UPTU(Q. BANK)–2002)

7. A constant e.m.f. E at $t = 0$ is applied to a circuit consisting of an inductance L, resistance R and capacitance C in series. The initial values of the current and the charge being zero, find the current at any time t, if $CR^2 < 4L$. Show that the amplitudes of the successive vibrations are in geometrical progression.

(UPTU(Q. BANK)–2001)

8. The differential equation $\dfrac{d^2x}{dt^2} + 2k\dfrac{dx}{dt} + n^2x = 0$; $(k < n)$, represents the damped harmonic oscillations of apertures. Solve this equation and show that the ratio of the amplitude of any oscillation to that of the preceding one is constant, i.e., its amplitude forms a G.P.

9. An inductance (L) of 2.0 H and a resistance (R) of 20 Ω are connected in series with an e.m.f. E Volt. If the current i is zero, when $t = 0$, find the current (i) at the end of 0.01 seconds if $E = 100\,V$, using the following differential equation $L\dfrac{di}{dt} + iR = E$ [UPTU–2008]

10. A resistance R of 5 Ω and an inductance L of 0.1 H are conneated in series with a battery of 12 V. Find the current i in the circuit as a funation of time using the following differential equation $Ri + L\dfrac{di}{dt} = E$ [UPTU(SUM)–2008]

11. When a resistance R ohms is connected in series with an induatance L henries, an e.m.f. of E volts, the current i amperes at time t is given by $L\dfrac{di}{dt} + Ri = E$. If $E = 10\sin t$ volts and $i = 0$ when $t = 0$, find i as a function of t.

12. An R-L circuit has an e.m.f. given (in volts) by $4\sin t$, a resistance of 100 ohms, an inductance of 4 henries with no initial current. Find the current at any time t.

[GBTU–2012, GBTU(CO)–2011, MTU(SUM)–2011]

13. An inductance L of 5.0 H and a resistance R of 25 Ω are connected in series with an e.m.f. E volt. If the current I is zero when $t = 0$, find the current I at the end of 1 second if $E = 100\,V$ using the differential equation
$$L\dfrac{dI}{dt} + IR = E$$ [GBTU(CO)–2010]

14. Find the steady state solution in R-L-C circuit equation consisting of inductance $L = 0.05$ H, resistance $R = 5$ ohms and a condenser of capacitance 4×10^{-4} Farad, if $Q = I = 0$ when $t = 0$ and there is an alternating e.m.f. of $200\cos 100t$. Find $Q(t)$ and $I(t)$.   [MTU–2011]

15. An R-L-C circuit connected in series has $R = 90\,\Omega, C = \dfrac{1}{140}$ Farad, $L = 10$ henries and an applied voltage $E(t) = 10\cos t$. Assuming no initial charge on the capacitor but an initial current of 1 ampere at $t = 0$, when the voltage is first applied, find the subsequent charge on the capacitor and the amlplitude of the steady state charge.   [MTU–2012]

Answers

1. 115.13 secs   2. $-[1 \quad ^{200}\ ]$ 4. $i = E_0\omega\dfrac{C}{1 + \omega^2 R^2 C^2}\left[\cos\omega t + \omega RC\sin\omega t\right] + C\,e^{-t/RC}$

5. $R > 2\sqrt{\dfrac{L}{C}}$ [over damping], $R = 2\sqrt{\dfrac{L}{C}}$ [critically damped]   6. $q = e^{-500t}\left[0.002\cos 1323t + 0.0008\sin 1323t\right]$

10. $i = 2.4(1 - e^{-50t})$ ampere   11. $i = \dfrac{10}{L^2 + R^2}(R\sin t - L\cos t + Le^{-Rt/L})$ 12. $i(t) = \dfrac{1}{626}(e^{-25t} - \cos t + 25\sin t)$ 3. 3.973 ampere

14. $Q(t) = \dfrac{2}{85}(\sin 100t + 4\cos 100t);\ I(t) = \dfrac{40}{17}(\cos 100t - 4\sin 100t)$ 15. $q = \dfrac{3}{25}e^{-2t} - \dfrac{43}{250}e^{-7t} + \dfrac{1}{250}(9\sin t + 13\cos t);\ \dfrac{\sqrt{10}}{50}$

8.6.2 Applications as Rate of Cooling

EXAMPLE 1. *According to Newton's law of cooling, the rate at which a substance cools in moving air is proportional to the difference between the temperature of the substance and that of the air. If the temperature of the air is 30°C and the substance cools from 100°C to 70°C in 15 minutes, find when the temperature will be 40°C.*

SOLUTION. By Newton's law of cooling, we have
$$\frac{dT}{dt} = -k(T - 30) \quad \text{or} \quad \frac{dT}{T - 30} = -k\,dt$$

On integrating $\log(T - 30) = -kt + C$ ... (1)

Initially, when $t = 0$, $T = 100$
$$C = \log 70$$

Substituting the value of C in equation (1), we get
$$\log(T - 30) = -kt + \log 70 \qquad \text{... (2)}$$

Also, when $t = 15$, $T = 70$
$$15k = \log 70 - \log 40 \qquad \text{... (3)}$$

Dividing (2) by (3), we get
$$\frac{t}{15} = \frac{\log 70 - \log(T - 30)}{\log(70) - \log 40} \qquad \text{... (4)}$$

Now, when $T = 40$, we have from (4)
$$\frac{t}{15} = \frac{\log 70 - \log 10}{\log 70 - \log 40} = \frac{\log_e 7}{\log_e(7/4)}$$

$$= \frac{\log_{10} 7}{\log_{10}(7/4)} = 3.48$$
$$t = 15 \times 3.48 = 52.20$$

Hence, the temperature will be 40° after 52.20 minutes.

EXAMPLE 2. *The rate at which the ice melts is proportional to the amount of ice at the instant. Find the amount of ice after 2 hours if half the quantity melts in 30 minutes.*

SOLUTION. Let A be the amount of ice at any time t.
$$\frac{dA}{dt} = kA \quad \text{or} \quad \frac{dA}{A} = k\,dt$$
$$\int \frac{dA}{A} = k\int dt + C$$
$$\log A = kt + C \qquad \text{... (1)}$$

Putting $t = 0$, $A = A_0$ in equation (1),
we get $\log A_0 = 0 + C \Rightarrow C = \log A_0$

Substituting the value of c in equation (1), we get
$$\log A = kt + \log A_0 \qquad \text{... (2)}$$

$A = A_0/2$, when $t = 1/2$ hour
$$\log\frac{A_0}{2} = \frac{k}{2} + \log A_0 \Rightarrow \log\frac{A_0}{2A_0} = \frac{k}{2}$$
$$\Rightarrow k = 2.\log\frac{1}{2}$$

On putting the value of k in equation (2), we get
$$\log A = \left(2\log\frac{1}{2}\right)t + \log A_0 \qquad \text{... (3)}$$

On putting $t = 2$ hours in equation (3), we get

$$\log A = 4 \log \frac{1}{2} + \log A_0$$

$$\log \frac{A}{A_0} = \log \left(\frac{1}{2}\right)^4 \Rightarrow \frac{A}{A_0} = \left(\frac{1}{2}\right)^4$$

8.6.3 CHEMICAL ACTION

EXAMPLE 1. *Uranium disintegrates at a rate proportional to the amount present at any instant. If M_1 and M_2 grams of uranium are present at times T_1 and T_2 respectively, show that the half life of Uranium is*

$$\frac{(T_2 - T_1) \log 2}{\log(M_1 / M_2)}.$$

SOLUTION. Let M grams of Uranium be present at any time t. Then the equation of disintegration of Uranium is

$$\frac{dM}{dt} = -kM_1, \text{ where } k \text{ is a constant.}$$

$$\frac{dM}{dt} = -Rdt \qquad \dots (1)$$

Integrating (1) as

$$\frac{A}{A_0} = \frac{1}{16} \qquad \Rightarrow \qquad A = \frac{A_0}{16}$$

After 2 hours, amount of ice left $= \frac{1}{16}$ of amount of ice at the beginning.

$$\int_{M_1}^{M_2} \frac{dM}{M} = -k \int_{T_1}^{T_2} dt$$

$$\Rightarrow \log \frac{M_2}{M_1} = -k(T_2 - T_1)$$

$$\Rightarrow k = \frac{\log(M_1 / M_2)}{T_2 - T_1} \qquad \dots (2)$$

Let the half life of Uranium, *i.e.*, $t = T$, $M = \frac{1}{2}M_0$

From (1) again

$$\int_M^{M/2} \frac{dM}{M} = -k \int T \, dt$$

$$\log \frac{1}{2} = kT \Rightarrow T = \frac{\log 2}{k}$$

Hence, $\quad T = \dfrac{(T_2 - T_1) \log 2}{\log(M_1 / M_2)}.$

Exercise-8.32

1. If the temperature of the air is 30°C and the substance cools from 100°C to 70°C in 15 minutes, find when the temperature will be 40°C.

2. Water at temperature 100 °C cools in 10 minutes to 88 °C in a room of temperature 25°C. Find the temperature of water after 20 minutes.

3. A body originally at 80 °C cools down to 60 °C in 20 minutes, the temperature of the air being 40 °C. What will be the temperature of the body after 40 minutes from the original.

4. The rate at which a body cools in proportional to the difference between the temperature of the body and that of the surrounding air. If a body in air at 25°C will cool from 100°C to 75°C in one minutes. Find its temperature at the end of three minutes.

5. If a thermometer is taken outdoors where the temperature is 0°C from a room in which the temperature is 21°C and the reading drops at 10°C in 1 minute. How long after its removal will the reading be 5°C ?

6. Radium decomposes at a rate proportional to the amount present. If a fraction P of the original amount disappears in one year. How much will remain at the end of 21 years?

7. Under certain conditions cane sugar is converted into dextrose at a rate, which is proportional to the amount unconverted at any time. If out of 75 grams of sugar at $t = 0$, 8 grams are converted during the first 3 minutes, find the amount converted in 1½ hours.

8. A radioactive substance decomposes at a rate proportional to its mass. When the mass is 10 milligrams, the rate of disintegration is 0.051 milligram per day. How long will it take for the mass to be reduced from 10 milligrams to 5 milligrams?

Answers

1. 52.5 minutes   2. 77.9°C   3. 50 °C   4. 47.22   5. 2 minutes, 13 seconds

6. $\left(1 - \dfrac{1}{p}\right)^{21}$ times of the original amount   7. 21.53 grams   8. 136 days

8.6.4 SIMPLE HARMONIC MOTION (S.H.M.)

A motion in which a particle moves in a straight line in such a way that its acceleration is always directed towards a fixed point on the line (called centre of the force) and varies as the distance of the particle from the fixed point, is called simple harmonic motion.

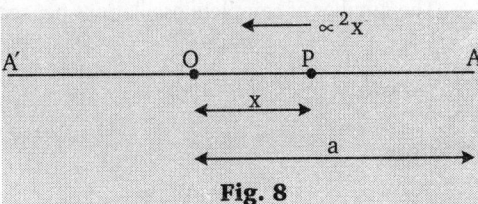

Fig. 8

Let O be the fixed point in the line $A'A$. Let P be the position of the particle at any time t, where $OP = x$.

Since the acceleration is always directed towards O, *i.e.*, the acceleration is in the direction opposite to that in which x increases, the equation of motion of the particle is

$$\frac{d^2 x}{dt^2} = -\mu^2 x.$$

$$(D^2 + \mu^2)x = 0, \text{ where } D = d/dx. \qquad \dots (1)$$

It is a linear differential equation with constant coefficients. Its auxiliary equation is $m^2 + \mu^2 = 0$ so that $m = \pm i\mu$

The solution of (1) is $\quad x = C_1 \cos \mu t + C_2 \sin \mu t \qquad \dots (2)$

Velocity of particle at $P = \dfrac{dx}{dt} = -C_1 \mu \sin \mu t + C_2 \mu \cos \mu t \qquad \dots (3)$

If the particle starts from rest at A, where $OA = a$,

then from (2), (at $t = 0$, $x = a$), $C_1 = a$ and from (3), (at $t = 0$, $\frac{dx}{dt} = 0$); $C_2 = 0$

$$x = a\cos\mu t \qquad \ldots (4)$$

$$\frac{dx}{dt} = -a\mu\sin\mu t \qquad \ldots (5)$$

$$= -a\mu\sqrt{1-\cos^2\mu t} = -a\mu\sqrt{1-\frac{x^2}{a^2}} = -\mu\sqrt{a^2-x^2} \qquad \ldots (6)$$

The equation (6) gives the velocity of the particle at any point P.

REMARKS

- We observe that velocity is maximum when $x = 0$. Thus in SHM, the velocity is maximum at the centre of force O.
- We observe from equation (6), that velocity = 0 when $x = a$.
  Thus, in S.H.M., the velocity is zero at points equidistant from the centre of force.

(1) NATURE OF MOTION

The particle starts from rest at A where its acceleration is maximum and is μa towards O. It begins to move towards the centre of attraction O and as it approaches the centre of force O, its velocity goes on increasing. When the particle reaches O its acceleration is zero and its velocity is maximum and is $a\mu$ in the direction OA'. Due to this velocity gained at O the particle moves towards the left of O. But on account of the centre of attraction at O, a force begins to act upon the particle against its direction of motion. So, its velocity goes on decreasing and it comes to instantaneous rest at A' where $OA' = OA$. The rest at A' is only instantaneous. The particle at once begins to move towards the centre of attraction O and retracing its path, it again comes to instantaneous rest at A. Thus the motion of particle is oscillatory and it continues to oscillate between A and A'. To start from A and to come back to A is called one complete oscillation.

(2) AMPLITUDE

In S.H.M., the distance from the centre to the position of maximum displacement is called the amplitude of the motion. OA is maximum distance and is called the amplitude.

From (6), $$-\frac{dx}{\sqrt{a^2-x^2}} = \mu dt$$

Integrating, we get $$\cos^{-1}\frac{x}{a} = \mu t + A \qquad \ldots (7)$$

Putting $t = 0$, $x = a$ in equation (7), we get $0 = 0 + A \Rightarrow A = 0$.

On putting the value of A, equation (7) becomes

$$\cos^{-1}\frac{x}{a} = \mu t \qquad \Rightarrow \qquad x = a\cos\mu t$$

Particle will reach O in time t_1, therefore,

$$0 = a\cos\mu t_1 \Rightarrow 0 = \cos\mu t_1$$

$$\cos\frac{\pi}{2} = \cos\mu t_1 \Rightarrow \frac{\pi}{2} = \mu t_1$$

Hence, $$t_1 = \frac{\pi}{2\mu}$$

(3) TIME PERIOD

In a S.H.M., the time taken to make a complete oscillation is called time period.

(4) FREQUENCY

The number of complete oscillations per second is called the frequency of motion. If n is the frequency and t is the time period.

$$\text{Time period } T = 4\left(\frac{\pi}{2}\mu\right) = \frac{2\pi}{\mu}$$

$$\text{Frequency } n = \frac{1}{T} = \frac{\mu}{2\pi}$$

REMARKS

- The equation of S.H.M. is $\frac{d\,x}{dt} = -\omega$
- The velocity V at a distance x from the centre at time t is $V^2 = \omega^2(a^2-x^2)$.
  $x = a\cos\omega t$ where a is the amplitude and ω is the angular velocity (at the extreme point).
- Maximum acceleration $= \omega^2 a$.
- Maximum velocity $= \omega a$.
- Time period $T = \frac{2\pi}{\omega}$.

(5) GEOMETRICAL REPRESENTATION OF S.H.M.

Let a particle moves with a uniform angular velocity ω round the circumference of a circle of radius a.

Suppose AA' is a fixed diameter of the circle. If the particle starts from A and P is its position at time t, then $\angle AOP = \omega t$. Draw PQ perpendicular to the diameter AA'.

If $OQ = x$, then $\quad x = a\cos\omega t$ ... (1)

Fig. 9

As the particle P moves round the circumference, the foot Q of the perpendicular on the diameter AA' oscillates on AA' from A to A' and from A' to A back. Thus the motion of the point Q is periodic.

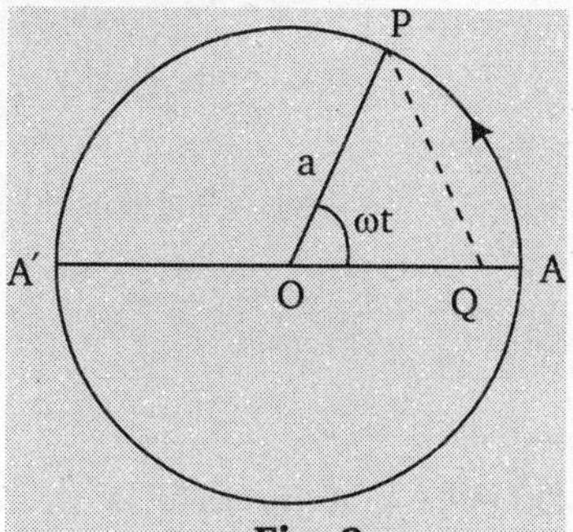

From (1), we have

$$\frac{dx}{dt} = -a\omega\sin\omega t \qquad ... (2)$$

and

$$\frac{d^2x}{dt^2} = -a\omega^2\cos\omega t = -\omega^2 x \qquad ... (3)$$

The equations (2) and (3) gives the velocity and acceleration of Q at any time t.

The equation (3) shows that Q executes a simple harmonic motion with centre at the origin O. From equation (1), we see that the amplitude of this S.H.M. is a because the maximum value of x is a.

The periodic time of Q = The time required by P to turn through an angle 2π with a uniform angular velocity ω.

$$= \frac{2\pi}{\omega}.$$

Thus, if a particle describes a circle with constant angular velocity, the foot of the perpendicular from it on any diameter executes a S.H.M.

Solved Examples

EXAMPLE 1. *A point moving in a straight line with S.H.M. has velocities V_1 and V_2 when its distances from the centre are x_1 and x_2. Show that the period of motion is $2\pi\sqrt{\dfrac{x_1^2 - x_2^2}{V_2^2 - V_1^2}}$.*

SOLUTION. Let the equation of S.H.M. with centre O as origin be

$\dfrac{d^2x}{dt^2} = -\mu x^2$. Then the time period $T = \dfrac{2\pi}{\mu}$. If a be the amplitude of the motion, we have

$$V^2 = \mu^2(a^2 - x^2)$$

where V is the velocity at a distance x from the centre.
But when $x = x_1$, $V = V_1$ and
when $x = x_2$, $V = V_2$.
Therefore, from (1), we have

$$V_1^2 = \mu^2(a^2 - x_1^2)$$

and $\quad V_2^2 = \mu^2(a^2 - x_2^2)$.

These gives

$$V_2^2 - V_1^2 = \mu^2\{(a^2 - x_2^2) - (a^2 - x_1^2)\}.$$
$$= \mu^2(x_1^2 - x_2^2)$$

$$\Rightarrow \quad \mu^2 = \frac{V_2^2 - V_1^2}{x_1^2 - x_2^2}$$

Hence, the time period

$$T = \frac{2\pi}{\mu} = 2\pi\sqrt{\frac{x_1^2 - x_2^2}{V_2^2 - V_1^2}}.$$

EXAMPLE 2. *At the end of the three successive seconds, the distance of a point moving with S.H.M. from its mean position are x_1, x_2, x_3 respectively. Show that the time of a complete oscillation is*

$$\frac{2\pi}{\cos^{-1}\left(\dfrac{x_1 + x_3}{2x_2}\right)}.$$

SOLUTION. Let the moving point be at distances x_1, x_2, x_3 from the mean position at the end of $t, t+1, t+2$ seconds respectively.
Using $\quad x = a\cos\mu t$

$$x_1 = a\cos\mu t \qquad ... (1)$$
$$x_2 = a\cos\mu(t+1) \qquad ... (2)$$
$$x_3 = a\cos\mu(t+2) \qquad ... (3)$$

Adding (1) and (3), we get

$x_1 + x_3 = a\left[\cos\mu(t+2) + \cos\mu t\right]$

$$= a.2\cos\left\{\frac{\mu(t+2)+\mu t}{2}\right\}\cos\left\{\frac{\mu(t+2)-\mu t}{2}\right\}$$

$$= 2a\cos\mu(t+1)\cos\mu = 2x_2\cos\mu \text{ [Using (2)]}$$

$$\mu = \cos^{-1}\left(\frac{x_1 + x_3}{2x_2}\right)$$

Hence, the time of a complete oscillation

$$= \frac{2\pi}{\mu} = \frac{2\pi}{\cos^{-1}\left(\dfrac{x_1+x_3}{2x_2}\right)}.$$

EXAMPLE 3. *A particle moves with S.H.M. of period 12 seconds, travels 8 cm from the position of rest in 2 seconds. Find the amplitude, the maximum velocity and the velocity at the end of 2 seconds.*

SOLUTION. $T = \dfrac{2\pi}{\sqrt{\mu}} = 12 \Rightarrow \sqrt{\mu} = \dfrac{\pi}{6}$

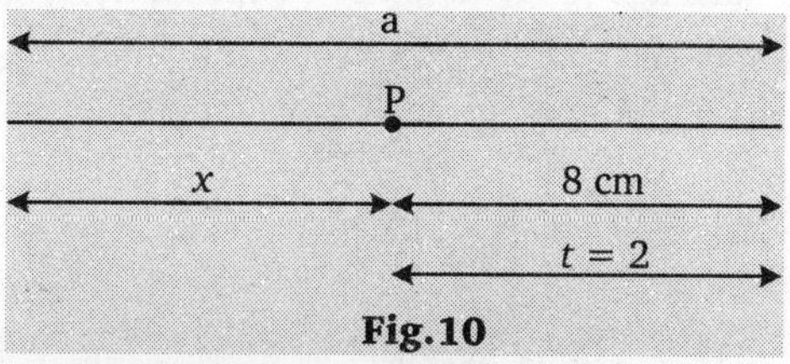

Fig.10

Let a be the amplitude OA
$AP = 8$ cm,
$OP = x = a - 8$, $t = 2$ seconds
We know that $x = a\cos\sqrt{\mu}\,t$

$a - 8 = a \cos 2\sqrt{\mu} = a \cos \dfrac{\pi}{3} = \dfrac{a}{2} \Rightarrow a = 16$

Maximum velocity

$= a\sqrt{\mu} = \dfrac{\pi}{6} \times 16 = \dfrac{8\pi}{3} = 4.619 \ cm/sec$

Velocity V at the end of two seconds

$= \sqrt{\mu} \sqrt{a^2 - x^2} = \dfrac{\pi}{6} \sqrt{256 - 64}$

$(\because x = a - 8 = 16 - 8 = 8)$

$= \dfrac{\pi}{6} \times \sqrt{192} = \dfrac{4\pi\sqrt{3}}{3} \ cm/sec.$

EXAMPLE 4. *A particle is performing a simple harmonic motion of period T about a centre O and it passes through a point P where OP = b with velocity V in the direction OP; prove that the time which elapses before it returns to P is*

$$\dfrac{T}{\pi} \tan^{-1}\left(\dfrac{VT}{2\pi b}\right).$$

SOLUTION. Let the equation of S.H.M. with centre O as origin as

$\dfrac{d^2x}{dt^2} = -\mu x.$

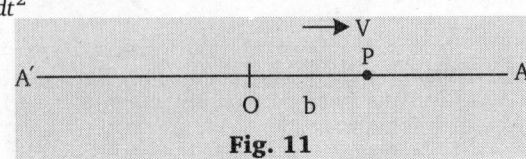

Fig. 11

The time period, $T = \dfrac{2\pi}{\sqrt{\mu}}.$

Let the amplitude be $a.$

Then, $\left(\dfrac{dx}{dt}\right)^2 = \mu(a^2 - x^2).$

When the particle passes through P, its velocity is given to be V in the direction OP. Also, OP= b. So putting $x = b$ and $dx/dt = V$ in (1), we get

$V^2 = \mu(a^2 - b^2)$ ... (2)

Let A be an extremity of the motion. From P the particle comes to instantaneous rest at A and then returns back to P. In S.H.M., the time from P to A is equal to the time from A to P.

$\therefore$ The required time $= 2 \times$ time from A to P.

Now, for the motion from A to P, we have

$\dfrac{dx}{dt} = -\sqrt{\mu} \sqrt{a^2 - x^2}$

$\Rightarrow \quad dt = -\dfrac{1}{\sqrt{\mu}} \dfrac{dx}{\sqrt{a^2 - x^2}}$

Let t_1 be the time from A to P. Then at $A, t = 0, x = a$ and at $P, t = t_1$ and $x = b$. Therefore, integrating (3), we get

$\int_0^{t_1} dt = \dfrac{1}{\sqrt{\mu}} \int_a^b \dfrac{-dx}{\sqrt{a^2 - x^2}}$

$t_1 = \dfrac{1}{\sqrt{\mu}}\left[\cos^{-1}\dfrac{x}{a}\right]_a^b$

$= \dfrac{1}{\sqrt{\mu}}\left[\cos^{-1}\dfrac{b}{a} - \cos^{-1}1\right]$

$= \dfrac{1}{\sqrt{\mu}}\cos^{-1}\left(\dfrac{b}{a}\right).$

Hence, the required time

$= 2t_1 = \dfrac{2}{\sqrt{\mu}} \cos^{-1}\dfrac{b}{a} = \dfrac{2}{\sqrt{\mu}} \tan^{-1}\left\{\dfrac{\sqrt{a^2 - b^2}}{b}\right\}$

$= \dfrac{2}{\sqrt{\mu}} \tan^{-1}\left(\dfrac{V}{b\sqrt{\mu}}\right) \quad$ [From (2), $\sqrt{a^2 - b^2} = \dfrac{V}{\sqrt{\mu}}$]

$= \dfrac{2}{2\pi/T} \tan^{-1}\left\{\dfrac{V}{b(2\pi/T)}\right\}$

$\left(\because T = \dfrac{2\pi}{\sqrt{l}} \text{ so that } \sqrt{\mu} = \dfrac{2\pi}{T} \right)$

$= \dfrac{T}{\pi} \tan^{-1}\left(\dfrac{VT}{2\pi b}\right).$

EXAMPLE 5. *A point moves in a straight line towards a centre of force* $\mu/(distance)^3$ *starting from rest at a distance 'a' from the centre of force. Show that the time of reaching a point distant 'b' from the centre of force is* $(a/\sqrt{\mu})\sqrt{a^2 - b^2}$ *and that its velocity is* $\dfrac{\sqrt{\mu}}{ab}\sqrt{a^2 - b^2}.$ [UPTU–2001]

SOLUTION. Let a point moves from A towards the centre of force O.

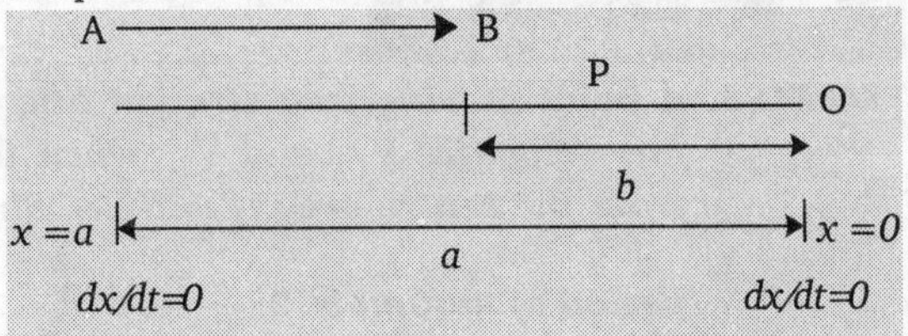

A ———————→ B
          P
                        O
          b
$x = a$ |←————— a —————→| $x = 0$
$dx/dt = 0$ $dx/dt = 0$

Fig. 12

$\dfrac{m\mu}{x^3} = -m\dfrac{d^2x}{dt^2} \Rightarrow \dfrac{d^2x}{dt^2} = -\dfrac{\mu}{x^3}.$

$2\dfrac{dx}{dt} \cdot \dfrac{d^2x}{dt^2} = -2\dfrac{\mu}{x^3} \cdot \dfrac{dx}{dt}$

Integrating, we get

$\left(\dfrac{dx}{dt}\right)^2 = -2\mu\dfrac{x^{-2}}{-2} + C$

$\Rightarrow \quad V^2 = \dfrac{\mu}{x^2} + C$ ... (1)

At $A, \quad V = 0, \quad$ and $x = a,$

$\therefore \ 0 = \dfrac{\mu}{a^2} + C \Rightarrow C = -\dfrac{\mu}{a^2}$

On putting the value of C, equation (1) becomes

$V^2 = \dfrac{\mu}{x^2} - \dfrac{\mu}{a^2} = \mu\left(\dfrac{a^2 - x^2}{x^2 a^2}\right)$ ... (2)

Therefore, velocity when $x = b$ is given by

$V^2 = \mu\left(\dfrac{a^2 - b^2}{a^2 b^2}\right) \Rightarrow V = \pm\sqrt{\mu}\dfrac{\sqrt{a^2 - b^2}}{ab}$

$\Rightarrow \ V = \sqrt{\mu}\dfrac{\sqrt{a^2 - b^2}}{ab}$ (Numerical value)

From (2), $\left(\dfrac{dx}{dt}\right)^2 = \mu\dfrac{(a^2 - x^2)}{x^2 a^2} \Rightarrow \dfrac{dx}{dt}$

$= -\sqrt{\mu}\dfrac{\sqrt{a^2 - x^2}}{xa}$

$dt = -\dfrac{1}{\sqrt{\mu}}\dfrac{xa}{\sqrt{a^2 - x^2}}dx$

Integrating, we get

$$t = -\frac{1}{\sqrt{\mu}} \int \frac{xa\,dx}{\sqrt{a^2 - x^2}}$$

(As the point 'A' is moving towards O)

Let $a^2 - x^2 = z^2 \Rightarrow -2x\,dx = 2z\,dz$

$$t = \frac{a}{\sqrt{\mu}} \int \frac{z\,dz}{z} = \frac{a}{\sqrt{\mu}} \int dz = \frac{a}{\sqrt{\mu}} z + C_1 \qquad \ldots (3)$$

$$= \frac{a}{\sqrt{\mu}} \sqrt{a^2 - x^2} + C_1$$

At A_1, $t = 0$, $x = a$

On putting $t = 0$, $x = a$ in equation (3), we get

$$0 = 0 + C_1 \Rightarrow C_1 = 0$$

On putting the value of C_1 in equation (3), we get

$$t = \frac{a}{\sqrt{\mu}} \sqrt{a^2 - x^2}. \text{ At } B_1, x = b, t = \frac{a}{\sqrt{\mu}} \sqrt{a^2 - b^2}$$

Exercise-8.33

1. A particle is executing simple harmonic motion with amplitude 20 cm and time 4 seconds. Find the time required by the particle in passing between points which are at distances 15 cm and 5 cm from the centre of force and are on the same side of it. [UPTU Q. BANK–2002]

2. A particle moves with S.H.M. if, when at a distance of 3 cm and 4 cm from the centre of the path, its velocities are 8 cm and 6 cm per sec respectively. Find its period, maximum velocity and acceleration when at its greatest distance from the centre.

3. A point executes S.H.M. such that in two of its positions, the velocities are u, v and the corresponding accelerations α, β.

   Show that the distance between the positions is $\dfrac{v^2 - u^2}{\alpha + \beta}$ and find the amplitude of the motion.

4. A particle moves with S.H.M. in a straight line under the action of force which is proportional to the distance of the particle from $x = 0$. If it starts at $x = 5$ cm with a velocity of 10 cm/sec and it reaches an extreme positions $x = 10$ cm, at what speed does it passes through the origin?

5. A particle whose mass is m, is acted upon by a force $m\mu(x + a^4 x^{-3})$ towards the origin. If it starts from rest at a distance a, show that it will arrive at the origin in time $\dfrac{\pi}{4\sqrt{\mu}}$.

6. At what distance from the centre, the velocity in a S.H.M. will be one forth of the maximum?

7. A particle begins to move from a distance 'a' towards a fixed centre which repels it with retardation μx. If its initial velocity is $a\sqrt{\mu}$, show that it continually approach the fixed centre but will never reach it. [UPTU(SUM)–2007]

8. A particle of mass m moves in a straight line under the action of force $mn^2 x$ which is always directed towards a fixed point O on the line. Determine the displacement $x(t)$ if the resistance to the motion is $2\lambda mnv$ given that initially $x = 0$,

   $$\frac{dx}{dt} = 0, (0 < \lambda < 1)$$

   (GBTU–2010)

Hints to Selected Problems

7. $\ddot{x} = \mu x; t = 0, x = a, \dot{x} = a\sqrt{\mu}$ so that $x = ae^{\sqrt{\mu}t}$

8. Equation of motion is $m\ddot{x} = -2\lambda mn\dot{x} - mn^2 x$

Answers

1. 0.38 sec

2. π sec, 10 cm/sec, 20 cm/sec^2

4. 11.546 cm/sec

6. $x = \pm\dfrac{\sqrt{15}\,a}{4}$

8. $x(t) = \dfrac{x_0}{n\omega} e^{-\lambda nt} \sin n\omega t$, where $\omega = \sqrt{1 - \lambda^2}$

8.6.7 Simple Pendulum

A light inextensible string and a heavy particle of negligible size tied to one end of the string whose other end is attached to a fixed point and oscillating in a vertical plane under gravity through a small angle, are said to form a simple pendulum.

(1) Oscillating of a Simple Pendulum

Let O be the fixed point, l the length of the string and m the mass of the bob (heavy mass).

Let P be the positions of the bob at any time t. Let arc $AP = s$ and $\angle AOP = \theta$, where OA is the vertical line through O.

The forces acting on the bob are

(i) Its weight mg acting vertically downward

(ii) The tension T in the string.

The components of weight along and perpendicular to the path of motion are $mg \sin\theta$ and $mg \cos\theta$ respectively. The component $mg \cos\theta$ is balanced by the tension in the string.

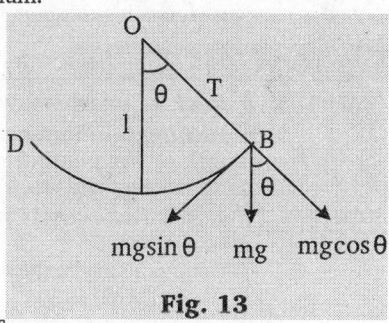

Fig. 13

$\therefore$ The equation of motion of the bob along the tangent is $m\dfrac{d^2 s}{dt^2} = -mg\sin\theta$

$$\frac{d^2(l\theta)}{dt^2} = -g\sin\theta \qquad (\because s = l\theta)$$

$$\frac{d^2\theta}{dt^2} = -\frac{g}{l}\left(\theta - \frac{\theta^3}{3!} + \dots\right) = -\frac{g}{l}\theta \text{ to a first approximation.}$$

or

$$\frac{d^2\theta}{dt^2} + \omega^2\theta = 0, \text{ where, } \omega^2 = \frac{g}{l}.$$

Its auxiliary equation is

$$m^2 + \omega^2 = 0 \Rightarrow m = \pm i\omega$$

Solution is given by

$$\theta = C_1 \cos\omega t + C_2 \sin\omega t$$

$$\theta = C_1 \cos\sqrt{\frac{g}{l}}\, t + C_2 \sin\sqrt{\frac{g}{l}}\, t$$

The motion of the bob is simple harmonic and the time of an oscillation is $\frac{2\pi}{\omega} = 2\pi\sqrt{\frac{l}{g}}$.

(2) BEAT OF A PENDULUM

A beat of a pendulum means its going from one extreme position of rest to the other position of rest, *i.e.*, half of the complete oscillation.

$$\therefore \qquad \text{The time of a beat} = \frac{1}{2}T = \pi\sqrt{\frac{l}{g}}.$$

(3) THE SECOND'S PENDULUM

If a simple pendulum oscillates from rest in one second, *i.e.*, if the time of one beat of a simple pendulum is one second, then it is called a second's pendulum and such a clock is said to be a correct clock.

Thus for a second pendulum,

$$1 = \pi\sqrt{\frac{l}{g}}, \quad l = \frac{g}{\pi^2} = \frac{g}{(3.1416)^2}$$

In F.P.S. system, $g = 32.2$, then $l = \dfrac{32.2}{(3.1416)^2} = 39.14$ inches (approx.)

and in C.G.S. system, $g = 981$, then $l = \dfrac{981}{(3.1416)^2} = 99.4$ cm (approx.)

(4) GAIN OR LOSS OF BEATS BY A CLOCK

Let t be the time of n beat.

$$t = n\pi\sqrt{\frac{l}{g}} \text{ or } n = \frac{t}{\pi}\sqrt{\frac{g}{l}} \qquad \dots (1)$$

Taking log on both sides of (1), we get

$$\log n = \log t - \log \pi + \frac{1}{2}\log g - \frac{1}{2}\log l \qquad \dots (2)$$

Taking differential of (2), we have

$$\frac{dn}{n} = \frac{1}{2}\frac{dg}{g} - \frac{1}{2}\frac{dl}{l} \qquad \dots (3)$$

If l changes, g remains constant, then

$$\frac{dn}{n} = -\frac{1}{2}\frac{dl}{l}$$

If g changes, l remains constant, then

$$\frac{dn}{n} = \frac{1}{2}\frac{dg}{g}$$

Solved Examples

EXAMPLE 1. *A clock is provided with a seconds pendulum is gaining 5 minutes a day. Find how much the length of the pendulum should be increased so as to correct the clock.*

SOLUTION. Here, n = Number of beats in a second pendulum
$\qquad\qquad = 86400$.
$\qquad \delta n$ = Change in the number of beats
$\qquad\qquad = 5 \times 60 = 300$
The length of second's pendulum $= 99.4$ cm

We have, $dn = -\dfrac{n}{2}\dfrac{\delta l}{l}$ $\qquad \dots (1)$

$$300 = -\frac{86400}{2}\frac{\delta l}{99.4}$$

$$\delta l = -\frac{300 \times 2 \times 99.4}{86400} = -0.690$$

As the pendulum gains 5 minutes a day, its length is less than its correct length by 0.690 cm. Hence, the length of the pendulum should be increased by 0.690 cm in order to correct the clock.

EXAMPLE 2. *A second's pendulum which gains 10 seconds per day at one place loses 10 seconds per day at another. Compare the accelerations due to gravity at the two places.* (KURUKSHETRA–2005)

SOLUTION. Let g be the acceleration due to gravity when the pendulum beats seconds.

Let $g + g_1$ be the acceleration due to gravity at the place, where it gains 10 seconds per day, then :
$$dn = 10, \quad n = 86400$$

Since $dn = \dfrac{n}{2} \times \dfrac{g_1}{g}$ (Here dg has been replaced by g_1)

$$10 = \frac{86400}{2} \cdot \frac{g_1}{g}$$

$$\frac{g_1}{g} = \frac{1}{4320}$$

Adding (1) to both sides

$$\frac{g + g_1}{g} = \frac{4321}{4320} \qquad \ldots (1)$$

Let $g + g_2$ be the acceleration due to gravity at the places where it loses 10 seconds per day, then

$$dn = -10$$

$$\Rightarrow \qquad -10 = \frac{86400}{2} \cdot \frac{g_2}{g}$$

$$\Rightarrow \qquad \frac{g_2}{g} = -\frac{1}{4320}$$

$$\Rightarrow \qquad \frac{g + g_2}{g} = \frac{4319}{4320} \qquad \ldots (2)$$

Dividing (1) by (2), we get $\dfrac{g + g_1}{g + g_2} = \dfrac{4321}{4319}$.

which is the required ratio.

EXAMPLE 3. *If a pendulum of length l makes n complete oscillations in a given time, show that if g is changed to (g+g´), the number of oscillations gained is ng´/(2g).*

SOLUTION. For a pendulum of length l, the time of one complete oscillation T is given by

$$T = 2\pi \sqrt{\frac{l}{g}}$$

n = the number of complete oscillation in a given time t

$$= \frac{t}{T} = \frac{t}{2\pi} \sqrt{\frac{g}{l}}$$

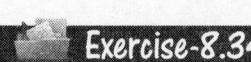
Exercise-8.34

$$\log n = \log \frac{t}{2\pi} + \frac{1}{2} \log g - \frac{1}{2} \log l$$

Differentiating, we get

$$\frac{1}{n} \delta n = \frac{1}{2g} \delta g - \frac{1}{2l} \delta l \qquad \ldots (1)$$

$$(\because t / 2\pi \text{ is constant})$$

If l is fixed then $\delta l = 0$ and if g is changed to $(g + g')$, then $\delta g - g'$

from (1), we get

$$\frac{1}{n} \delta n = \frac{1}{2g} g'$$

Hence, the number of oscillations gained

$$= \delta n = \frac{ng'}{2g}$$

EXAMPLE 4. *Find how many seconds a clock would lose per day, if the length of its pendulum were increased in the ratio 900 : 901.*

SOLUTION. Let the original length and $l + dl$, the increased length of the pendulum, then

$$\frac{l}{d + dl} = \frac{900}{901}$$

$$\frac{dl}{l} = \frac{901}{900} - 1 = \frac{1}{900}$$

Let n be the number of beats per day, then

$$n = 8.6400.$$

If dn is the change in the number of beats, then

$$dn = -\frac{n}{2} \frac{dl}{l} = -\frac{86400}{2} \times \frac{1}{900} = -48$$

Since dn is negative, the clock will lose 48 seconds per day.

1. Find how many seconds a clock would lose per day if the length of its pendulum were increased in the ratio 450 : 451.

2. A clock with a second's pendulum loses 10 seconds per day at a place where g = 980 cm/sec^2. What change in the gravity is necessary to make it accurate.

3. If a pendulum clock loses 9 minutes per week, find in mm, what change is required in the length of the pendulum in order that the clock may keep correct time.

4. A pendulum of length l hangs against a wall inclined at an angle θ to be horizontal. Show that the time of complete oscillation is $2\pi \sqrt{\dfrac{l}{g \sin \theta}}$.

5. If l_1 be the length of an imperfectly adjusted second's pendulum which gains n seconds in one hour and l_2 the length of one which loses n seconds in one hour, at the same place, show that the true length of second's pendulum is

$$\frac{4l_1 l_2}{l_1 + l_2 + 2\sqrt{l_1 l_2}}.$$

6. A pendulum of length l has one end of the string fastned to a peg on a smooth plane inclined to the horizon at an angle α. With the string and the weight on the plane, its time of oscillation in t sec. If the pendulum of length l' oscillates in one second when suspended vertically, prove that

$$\alpha = \sin^{-1}\left(\frac{1}{l't^2}\right) \qquad \text{(KURUKSHETRA–2006)}$$

7. It $I = \dfrac{d^2\theta}{dt^2} = -mgl \sin \theta$, where I, m, g, l are constants, given that at $t=0$, $\theta = 0$ and $\dfrac{d\theta}{dt} = \omega_0 = m\sqrt{mgl} / I$ then show that

$$t = \frac{2}{\omega_0} \log \frac{\pi + \theta}{4} \qquad \text{(NAGPUR–2009)}$$

8. A point moves in a straight line towards the centre of force $\mu / (\text{distance})^2$ starting from rest at a distance a from the centre of force, show that the time of reaching a point b from the centre of force is $a\sqrt{a^2 - b^2} / \sqrt{\mu}$ and that its velocity then is $\dfrac{\sqrt{\mu}}{ab} \sqrt{a^2 - b^2}$ (UPTU–2001)

Answers

1. 96 seconds per day   2. to be increased by 0.227 cm/sec^2

8.6.7 LINEAR MOTION OF A PARTICLE IN A RESISTING MEDIUM

When a body is moving vertically in a resisting medium, there are two forces acting on the body *i.e.*, its weight downward and the resistance of the air opposite to the direction of the motion. If it is moving horizontally, the only force acting on it is the resistance.

If the particle is falling under gravity in a resisting medium, the velocity will never exceed some definite quantity. When the resistance becomes equal to the weight, the acceleration is then zero, and the particle will continue to move with a constant velocity. This velocity for which the acceleration becomes zero is called the limiting or terminal velocity.

Solved Examples

EXAMPLE 1. *A particle of mass m is projected vertically upward under gravity, the resistance of the air being mk times the velocity. Show that the greatest height attained by the particle is $\dfrac{V^2}{g}[\lambda - \log(1+\lambda)]$, where V is the greatest velocity which the above mass will attain when it falls freely and λV is the initial velocity,*

SOLUTION. Let V be the velocity of the particle at time t. The forces acting on the particle are

(i) its weight mg acting vertically downwards.

(ii) the resistance mkV of the air acting vertically downwards.

Accelerating force on the particle $= -mg - mkV$

∴ By Newton's second law, the equation of motion of the particle is $mV\dfrac{dV}{dx} = -mg - mkV$.

or $\quad V\dfrac{dV}{dx} = -g - kV$ ... (1)

When the particle falls freely (under gravity), equation (1) becomes (changing g to $-g$)

$$V\dfrac{dV}{dx} = g - kV \qquad \ldots (2)$$

When the particle attains the greatest velocity V, its acceleration is zero.

From (2), $\quad 0 = g - kV$

$\Rightarrow \qquad k = g/V$

Putting this value of k in equation (1), we have

$$V\dfrac{dV}{dx} = -g - \dfrac{g}{V}v = -\dfrac{g}{V}(V+v)$$

$$\dfrac{v}{V+v}dv = -\dfrac{g}{V}dx$$

Integrating $\displaystyle\int \dfrac{v}{V+v}dv = -\dfrac{g}{V}\int dx + C$

or $\quad \displaystyle\int\left(1 - \dfrac{V}{V+v}\right)dv = -\dfrac{g}{V}x + C$

or $\quad v - V\log(V+v) = -\dfrac{g}{V}x + C$ ... (3)

Initially, when $x = 0$, $v = \lambda V$

From (3), we have $\lambda V - V\log(V + \lambda V) = C$

or $\quad C = V[\lambda - \log V(1+\lambda)]$

Substituting the value of C in (3), we get

$v - V\log(V+v) = -\dfrac{g}{V}x + V[\lambda - \log V(1+\lambda)]$ ... (4)

Let h be the greatest height attained by the particle, then $x = h$ when $V = 0$.

From (4), we have

$$-V\log V = -\dfrac{g}{V}h + V[\lambda - \log V(1+\lambda)]$$

$$\dfrac{g}{V}h = V\lambda - V[\log V(1+\lambda) - \log V]$$

$$= V\lambda - V\log\dfrac{V(1+\lambda)}{V}$$

Hence, $\quad h = \dfrac{V^2}{g}[\lambda - \log(1+\lambda)]$

EXAMPLE 2. *A particle falls under gravity in a resisting medium whose resistance varies with velocity. Find the relation between distance and velocity if initially the particle starts from rest.*

[UPTU–2003]

SOLUTION. By Newton's second law of motion, the equation of motion of the body is

$$mV\dfrac{dV}{dx} = mg - mkV$$

$\Rightarrow V\dfrac{dV}{dx} = g - kV \qquad \Rightarrow \qquad \dfrac{V\,dV}{g-kV} = dx$

[On separating the variables]

$\Rightarrow \quad -\dfrac{dV}{k} + \dfrac{g}{k}\dfrac{dV}{g-kV} = dx$

Integrating, we get

$$-\dfrac{V}{k} = \dfrac{g}{k}\left(-\dfrac{1}{k}\right)\log(g-kV) = x + C \qquad \ldots (1)$$

$$-\dfrac{V}{k} - \dfrac{g}{k^2}\log(g-kV) = x + C$$

Initially, $\quad x = 0$, $V = 0$

so, $\quad -\dfrac{g}{k^2}\log g = C$

Equation (1) becomes

$$-\dfrac{V}{k} - \dfrac{g}{k^2}\log(g-kV) = x - \dfrac{g}{k^2}\log g$$

$\Rightarrow \quad -\dfrac{V}{k} - \dfrac{g}{k^2}\log\dfrac{g-kV}{g} = x$

EXAMPLE 3. *A particle is projected with velocity V along a smooth horizontal plane in a medium whose resistance per unit mass is μ times the cube of the velocity. Show that the distance it was described in time t in $(1/\mu V)\left[\sqrt{1 + 2\mu V^2 t} - 1\right]$ and that its velocity then is $V/\sqrt{(1 + 2\mu V^2 t)}$.*

SOLUTION. Take the point of projection O as origin. Let v be the velocity of the particle at time t at a point distant x from the fixed point O. Then the resistance at this point will be $m\mu v^3$, acting in the direction of x decreasing. Here, the resistance is the only force acting on the particle during its motion.

∴ The equation of motion of the particle is

$$m\dfrac{dv}{dt} = -m\mu v^3$$

$\Rightarrow \qquad \dfrac{dv}{v^3} = -\mu\,dt$

Integrating, $-\dfrac{1}{2v^2} = -\mu t + A$, where A is a constant.

But initially, when $t = 0$, $v = V$ and

so $\quad A = -\dfrac{1}{2V^2}$

$-\dfrac{1}{2v^2} = -\mu t - \dfrac{1}{2V^2} \Rightarrow \dfrac{1}{v^2} = (2\mu V^2 t + 1)/V^2$

$$v = V/\sqrt{(1 + 2\mu V^2 t)} \qquad \ldots (1)$$

which gives the velocity of the particle at time t.

Since the particle is moving in the direction of x increasing, therefore from the equation (1), we have

$$\frac{dx}{dt} = v = V / \sqrt{(1+2\mu V^2 t)}$$

$$\Rightarrow \quad dx = V(1+2\mu V^2 t)^{-1/2} dt$$

Integrating, $x = \dfrac{1}{\mu V}(1+2\mu V^2 t)^{1/2} + B$,

where B is a constant.

But initially, when $t = 0$, $x = 0$, $\therefore B = -\dfrac{1}{\mu V}$

$$x = \frac{1}{\mu V}(1+2\mu V^2 t)^{1/2} - \frac{1}{\mu V}$$

Hence, $x = \dfrac{1}{\mu V}\left[\sqrt{(1+2\mu V^2 t)} - 1\right]$

which gives the distance described in time t.

EXAMPLE 4. *A moving body is opposed by a force per unit mass of value Cx and resistance per unit mass of value bv^2, where x and v are the displacement and velocity of the particle at that instant. Show that the velocity of the particle, if it starts from rest, is given by $v^2 = \dfrac{C}{2b^2}(1-e^{-2bx}) - \dfrac{Cx}{b}$.*

SOLUTION. By Newton's second law of motion, the equation of motion of the body is

$$v\frac{dv}{dx} = -Cx - bv^2$$

$$\Rightarrow \quad v\frac{dv}{dx} + bv^2 = -Cx \qquad \ldots(1)$$

Putting $v^2 = z$ and $2v\dfrac{dv}{dx} = \dfrac{dz}{dx}$

Equation (1) becomes $\dfrac{1}{2}\dfrac{dz}{dx} + bz = -Cx$.

$$\frac{dz}{dx} + 2bz = -2Cx \qquad \ldots(2)$$

which is Leibnitz's linear equation

$$\text{I.F.} = e^{\int 2b\,dx} = e^{2bx}.$$

The solution of (2) is

$$z.e^{2bx} = \int -2Cx.e^{2bx} dx + C_1$$

$$= -2C\left[x.\frac{e^{2bx}}{2b} - \int 1.\frac{e^{2bx}}{2b} dx\right] + C_1$$

$$= -\frac{Cx}{b}e^{2bx} + \frac{C}{2b^2}e^{2bx} + C_1$$

$$\Rightarrow v^2.e^{2bx} = -\frac{Cx}{b}e^{2bx} + \frac{C}{2b^2}e^{2bx} + C_1$$

$$\Rightarrow v^2 = -\frac{Cx}{b} + \frac{C}{2b^2} + C_1 e^{-2bx} \qquad \ldots(3)$$

Initially when $x = 0$, $v = 0$

$$\frac{C}{2b^2} + C_1 = 0 \Rightarrow C_1 = -\frac{C}{2b^2}$$

Substituting the value of C_1 in (3), we have

$$v^2 = -\frac{Cx}{b} + \frac{C}{2b^2} - \frac{C}{2b}e^{-2bx}$$

Hence, $v^2 = \dfrac{C}{2b^2}(1-e^{-2bx}) - \dfrac{Cx}{b}$

EXAMPLE 5. *The acceleration and velocity of a body falling in the air approximately satisfy the equation:*

Acceleration = $g - kV^2$, where V is the velocity

of the body at any time t, and g, k are constants. Find the distance traversed as a function of the time, if the body falls from rest. Show that value of v will never exceed $\sqrt{\dfrac{g}{k}}$

SOLUTION. Acceleration $= g - kV^2$

$$\frac{dV}{dt} = g - kV^2 \qquad \Rightarrow \qquad \frac{dV}{g - k^2} = dt$$

$$\Rightarrow \frac{1}{2\sqrt{g}}\left[\frac{1}{\sqrt{g} + \sqrt{k}V} + \frac{1}{\sqrt{g} - \sqrt{k}V}\right] dV = dt$$

On integrating, we get

$$\frac{1}{2\sqrt{g}}\frac{1}{\sqrt{k}}\log(\sqrt{g} + \sqrt{k}V)$$
$$- \frac{1}{2\sqrt{gk}}\log(\sqrt{g} - \sqrt{k}V) = t + C$$

$$\Rightarrow \frac{1}{2\sqrt{gk}}\log\frac{\sqrt{g} + \sqrt{k}.V}{\sqrt{g} - \sqrt{k}.V} = t + C \qquad \ldots(1)$$

On putting $t = 0$, $V = 0$ in equation (1), we get

$$\frac{1}{2\sqrt{gk}}\log 1 = 0 + C \Rightarrow C = 0$$

Equation (1) becomes

$$\frac{1}{2\sqrt{gk}}\log\frac{\sqrt{g} + \sqrt{k}.V}{\sqrt{g} - \sqrt{k}.V} = t$$

$$\Rightarrow \log\frac{\sqrt{g} + \sqrt{k}.V}{\sqrt{g} - \sqrt{k}.V} = 2\sqrt{gk}\,t$$

$$\Rightarrow \frac{\sqrt{g} + \sqrt{k}.V}{\sqrt{g} - \sqrt{k}.V} = e^{2\sqrt{gk}\,t}$$

By componendo and dividendo, we have

$$\frac{\sqrt{k}\,V}{\sqrt{g}} = \frac{e^{2\sqrt{gk}\,t} - 1}{e^{2\sqrt{gk}\,t} + 1} = \frac{e^{\sqrt{gk}\,t} - e^{\sqrt{gk}\,t}}{e^{\sqrt{gk}\,t} + e^{\sqrt{gk}\,t}} = \tanh\sqrt{gk}\,t$$

$$\Rightarrow V = \sqrt{\frac{g}{k}}\tanh\sqrt{gk}\,t$$

Whenever the value of t may be

$$\tanh\sqrt{gk}\,t \le 1.$$

Hence, the value of V will never exceed $\sqrt{\dfrac{g}{k}}$.

$$\therefore \qquad \frac{dx}{dt} = \sqrt{\frac{g}{k}}\tanh\sqrt{gk}\,t$$

Integrating again, $x = \sqrt{\dfrac{g}{k}}\int\tanh\sqrt{gk}\,t + B$

$$\Rightarrow x = \frac{1}{k}\log\cosh\sqrt{gk}.t + B$$

When $t = 0$, $x = 0$, then $B = 0$.

Hence, $x = \dfrac{1}{k}\log\cosh\sqrt{gk}\,t$

EXAMPLE 6. *A particle of mass m is falling under the influence of gravity through a medium whose resistance equals μ times the velocity. If the particle were released from rest, show that the distance fallen through in time t is*

$$\frac{gm^2}{\mu^2}\left[e^{-(\mu/m)t} - 1 + \frac{\mu t}{m}\right].$$

SOLUTION. Let a particle of mass m falling under gravity be at a distance x from the starting point after time t. If v is its velocity at this point, then the resistance on the particle is μv acting vertically upwards, *i.e.*, in the direction of x

decreasing. The weight mg of the particle acts vertically downwards, i.e., in the direction of x increasing.

∴ The equation of motion of the particle is

$$m\frac{d^2x}{dt^2} = mg - \mu v$$

$$\frac{dv}{dt} = g - \frac{\mu}{m}v \Rightarrow dt = \frac{dv}{g - (\mu/m)v}$$

Integrating, we get

$$t = -\frac{m}{\mu}\log\left(g - \frac{\mu}{m}v\right) + A,$$

where A is a constant.
But initially, when $t = 0$, $v = 0$ and
so $A = (m/\mu)\log g$

$$t = -\frac{m}{\mu}\log\left(g - \frac{\mu}{m}v\right) + \frac{m}{\mu}\log g$$

$$\Rightarrow t = -\frac{m}{\mu}\log\left\{\frac{g - (\mu/m)v}{g}\right\}$$

$$-\frac{\mu t}{m} = \log\left(1 - \frac{\mu}{gm}v\right)$$

$$1 - \frac{\mu v}{gm} = e^{-\mu t/m}$$

$$\Rightarrow v = \frac{dx}{dt} = \frac{gm}{\mu}\left(1 - e^{-\mu t/m}\right)$$

$$\Rightarrow dx = \frac{gm}{\mu}(1 - e^{-\mu t/m})\, dt$$

Integrating, we get

$$x = \frac{gm}{\mu}\left[t + \frac{m}{\mu}e^{-\mu t/m}\right] + B \qquad \dots (1)$$

where B is a constant.
But initially, when $t = 0$, $x = 0$. Therefore,

$$0 = \frac{gm}{\mu}\left[\frac{m}{\mu}\right] + B \qquad \dots (2)$$

Subtracting (2) from (1), we have

$$x = \frac{gm}{\mu}\left[\frac{m}{\mu}e^{-\mu t/m} - \frac{m}{\mu} + t\right]$$
$$= \frac{gm^2}{\mu^2}\left\{e^{-(\mu t/m)} - 1 + \frac{\mu t}{m}\right\}$$

Exercise-8.35

1. A particle falls in a vertical line under gravity (supposed constant) and the force of air resistance to its motion is proportional to its velocity. Show that its velocity cannot exceed a particular limit.

2. A body falling from rest is subjected to a force of gravity and an air resistance of n^2/g times of the square of velocity. Show that the distance travelled by the body in t seconds is $\frac{g}{n^2}\log\cosh nt$.

3. A particle of unit mass moves in a straight line under retardation which is k times its velocity. Initially, the particle is at a distance a from a given point O in the line and is moving towards O with velocity $u(>ak)$. Prove that it will reach O in time $\frac{1}{k}\log\frac{u}{u - ak}$.

4. A particle of unit mass is projected vertically upward with velocity V and the resistance of the air produces a retardation kv^2, where v is the velocity. Show that the velocity V', with which the particle will return to the point of projection, is given by $\frac{1}{V'^2} = \frac{1}{V^2} + \frac{k}{g}$.

5. A moving body is opposed by a force proportional to the displacement and by a resistance proportional to the square of the velocity. Prove that the velocity is given by
$$V^2 = aC - \frac{Cx}{b} + \frac{C}{ab^2}$$

Answers

1. $V = \frac{g}{k}$

8.6.8 HOOKE'S LAW

The tension of an elastic string is proportional to the extension of the string beyond its natural length.

If x is the stretched length of a string of natural length l, then by Hooke's law the tension T in the string is given by $T = E\frac{x - l}{l}$, where E is called the modulus of elasticity of the string that the direction of the tension is always opposite to the extension.

(1) PARTICLE ATTACHED TO ONE END OF A HORIZONTAL ELASTIC STRING

If one end of the elastic string be fixed at O on a table. The other end A of the elastic string of length l is attached to a particle of mass m.

The string is stretched to a point B and then released. The particle comes into motion. Let the particle be at a distance x from A at any time t. The weight mg of the particle is acting downward and is balanced by the reaction R of the table.

The only force acting upon the particle is the tension of the string

By Hooke's law, $\frac{\text{Stress}}{\text{Strain}} = $ Constant of elasticity

$$T = E\frac{x}{l}$$

where, x = Extension of the length of the string,
E = Modulus of elasticity.

Equation of motion is

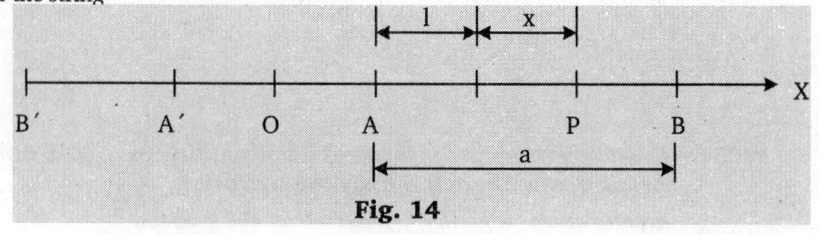

Fig. 14

$$m \frac{d^2x}{dt^2} = -\frac{Ex}{l}$$

$$\frac{d^2x}{dt^2} = -\left(\frac{E}{ml}\right)x \qquad \ldots (1)$$

The motion of the equation is S.H.M.

On multiplying (1) by $2\frac{dx}{dt}$, we get $\quad 2\frac{d^2x}{dt^2} \cdot \frac{dx}{dt} = -2\frac{Ex}{lm} \cdot \frac{dx}{dt}$

On integrating, we get $\qquad \left(\frac{dx}{dt}\right)^2 = \frac{-Ex^2}{lm} + C \quad \Rightarrow \quad V^2 = -\frac{Ex^2}{ml} + C \qquad \ldots (2)$

If the velocity of the particle is zero at B, amplitude $AB = a$. The particle moves from A to B and back B to A will be a S.H.M. The particle moves towards O. Then it moves with uniform velocity upto A'.

On putting $V = 0$, $x = a$ we get $\qquad 0 = -\frac{Ea^2}{ml} + C \qquad$ or $\qquad C = \frac{Ea^2}{ml}$

$$\Rightarrow \qquad V^2 = \frac{E}{lm}(a^2 - x^2)$$

At A, $x = 0$, $V = \sqrt{\frac{E}{l}} a$. This is the maximum velocity. The particle moves from A to A' with this velocity. After that the string again stretches and motion becomes S.H.M.

Periodic time of S.H.M. (from A to B, B to A, A' to B', B to A') + time taken by particle from A to A' and A' to A with constant velocity $\sqrt{\frac{E}{l}} a$.

$$\frac{2\pi}{\sqrt{E/l}} + \frac{4l}{\sqrt{(E/lm)}\,a} = \sqrt{\left(\frac{lm}{E}\right) \cdot \left(\pi + \frac{2l}{a}\right)}$$

(2) PARTICLE SUSPENDED BY AN ELASTIC STRING

Let one end of the string OA of natural length l be attached to the fixed point O and a particle of mass m be attached to the other end A . Due to the weight mg of the particle the string OA is stretched and if B is the position of equilibrium of the particle such that $AB = d$, then the tension T_B in the string will balance the weight of the particle, i.e.,

$$mg = T_B \quad \text{or} \quad mg = E\frac{AB}{OA} = E\frac{d}{l} \qquad \ldots (1)$$

The particle is pulled down to a point C such that $BC = C$ and then released. At the point C, the tension in the string is greater than weight of the particle and so the particle starts moving vertically upwards with velocity zero at C. Let P be the position of the particle at any time t, where $BP = x$. The tension in the string when the particle is at P is $T_P = E\frac{d+x}{l}$, acting vertically upwards.

The resultant force acting on the particle at P is the vertically upwards direction.

$$= T_P - mg = E\left(\frac{d+x}{l}\right) - mg = \frac{Ed}{l} + \frac{Ex}{l} - mg = \frac{Ex}{a} \qquad \text{[Using (1)]}$$

Also, the acceleration of the particle at P is d^2x / dx^2 in the direction of x increasing, i.e., in the vertically downwards direction.

$\therefore$ By Newton's law, the equation of motion of P is given by

$$m \frac{d^2x}{dt^2} = -\frac{Ex}{l} \quad \text{or} \quad \frac{d^2x}{dt^2} = -\frac{E}{lm}x \qquad \ldots (2)$$

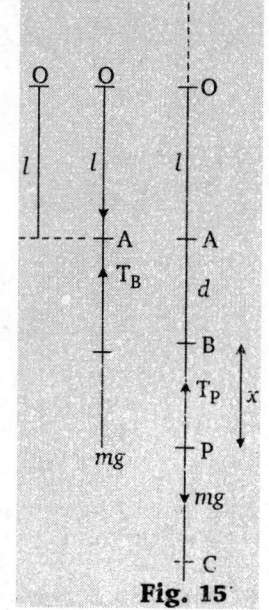

Fig. 15

This equation holds good so long as the tension operates, i.e., when the string is extended beyond its natural length.

Equation (2) is the standard equation of a S.H.M. with centre at the origin B and the amplitude of the motion is $BC = C$.

The periodic time T of the S.H.M. represented by the equation (2) is given by

$$T = \frac{2\pi}{\sqrt{E/ml}} = 2\pi\sqrt{\frac{lm}{E}} \qquad \ldots (3)$$

Solved Examples

EXAMPLE 1. *Two particles of mass m_1 and m_2 are tied to the ends of an elastic string of natural length a and modulus λ. They are placed on a smooth table so that the string is just taut and m_2 is projected with any velocity directly away from m_1. Prove that the string will become slack after the lapse of time :*

$$\pi\left[am_1m_2 / \lambda(m_1 + m_2)\right]^{1/2}.$$

SOLUTION. Let initially at $t = 0$, the particle m_1 be placed at O, the fixed point of reference. Given that the particle m_2 is at a distance a from m_1. Let at time t the particle m_1 be at a distance x from O and the particle m_2 at a distance y from O.

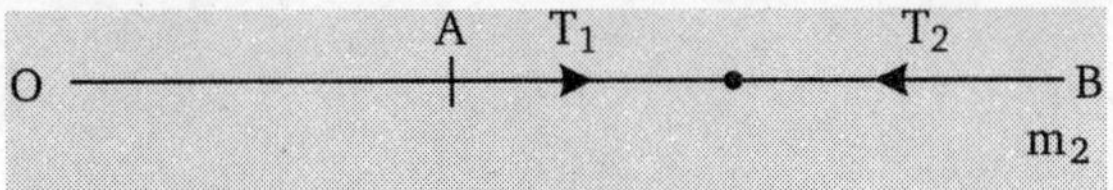

O ———————————— A T_1 ——————— T_2 —————— B
                                                        m_2

Fig. 16

Let T_1 be the tension in the string.
The equation of motion of m_1 is

$$m_1\frac{d^2x}{dt^2} = +T_1 \qquad \ldots (1)$$

The equation of motion of m_2 is

$$m_2\frac{d^2y}{dt^2} = -T_1 \qquad \ldots (2)$$

and $\qquad T_1 = \lambda\dfrac{y-x}{a} \qquad \ldots (3)$

From (1) and (2), we get

$$\frac{d^2y}{dt^2} - \frac{d^2x}{dt^2} = -\frac{T_1}{m_2} - \frac{T_1}{m_1}$$

$$\Rightarrow \qquad \frac{d^2(y-x)}{dt^2} = -\left(\frac{1}{m_1} + \frac{1}{m_2}\right)\lambda\frac{y-x}{a}$$

or $\qquad \dfrac{d^2z}{dt^2} = \dfrac{-\lambda(m_1+m_2)}{m_1m_2a}z$,

taking $\qquad y - x = z$.

This is the equation of motion of S.H.M. with periodic

time, $T = \dfrac{2\pi}{\sqrt{\dfrac{\lambda(m_1+m_2)}{a(m_1 . m_2)}}}$.

Now, when m_2 starts coming towards m_1 and when the string acquires its original length, the string becomes

slack after this time. This required time $= \dfrac{T}{4} + \dfrac{T}{4} = \dfrac{T}{2}$.

Hence, the string becomes slack after time

$$\frac{T}{2} = \pi\left[am_1m_2 / \lambda(m_1+m_2)^{1/2}\right]$$

EXAMPLE 2. *A light elastic string of original length l is hung by one end to the other end are tied successively particles of masses m, m´. If t_1 and t_2 be the periods of small oscillations corresponding to these weights are C_1 and C_2 the statical extensions, prove that $g(t_1^2 - t_2^2) = 4\pi^2(C_1 - C_2)$.*

SOLUTION. We have $\qquad mg = T_1 \qquad \ldots (1)$

$$mg = E\frac{C_1}{l} \qquad \ldots (2)$$

$$m'g = E\frac{C_2}{l} \qquad \ldots (3)$$

Equation of motion of first particle,

$$m\frac{d^2x}{dt^2} = mg - E\frac{(x+C_1)}{l}$$

$$\Rightarrow \quad m\frac{d^2x}{dt^2} = mg - E\frac{C_1}{l} - \frac{Ex}{l} = \frac{-Ex}{l}$$

[Using (2)]

Motion of S.H.M. with $\quad t_1 = \dfrac{2\pi}{\sqrt{E/lm}}$

Similarly, $\qquad t_2 = \dfrac{2\pi}{\sqrt{E/lm'}}$

Now, $\qquad t_1^2 - t_2^2 = 4\pi^2\dfrac{l}{E}(m-m') = 4\pi^2\left(\dfrac{C_1}{g} - \dfrac{C_2}{g}\right)$

[Using (2) and (3)]

Hence, $g(t_1^2 - t_2^2) = 4\pi^2(C_1 - C_2)$.

EXAMPLE 3. *An elastic string without weight of which the unstretched length is l and modulus of elasticity is the weight of n kg is suspended by one end and a mass m kg is attached to the other end. Show that the time of a small vertical oscillation*

is $2\pi\sqrt{\dfrac{ml}{ng}}$.

SOLUTION. Let $OA = l$ be the natural length of a string whose one end is fixed at O. B is the position of equilibrium of a particle of mass m kg attached to the other end of the string. Considering the equilibrium of the particle at B, we have $mg =$ the tension T_B in the string OB.

$\therefore \qquad mg = ng\dfrac{AB}{l} \qquad \ldots (1)$

because modulus of elasticity of the string is given to be ng. Now, suppose the particle is pulled slightly upto C (so that $BC < AB$ and then let go. It starts moving vertically upwards with velocity zero at C. Let P be its position at any point t, where $\quad BP = x$.

The direction BP is that of x increasing and the direction PB is that of x decreasing. At P, there are two forces acting on the particle :

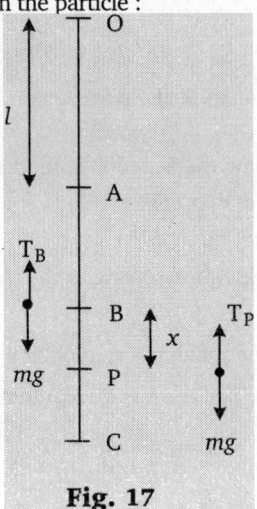

Fig. 17

(i) The weight mg acting vertically downwards, *i.e.*, in the direction of x increasing, and

(ii) the tension $T_P = ng \dfrac{AB + x}{l}$ in the string OP, acting vertically upwards, *i.e.*, in the direction of x decreasing.

Hence, by Newton's second law of motion, the equation of motion of the particle at P is

$$m \frac{d^2 x}{dt^2} = mg - ng \frac{AB + x}{l}$$

$$= mg - ng \frac{AB}{l} - ng \frac{x}{l} = -ng \frac{x}{l} \quad \text{[Using (1)]}$$

$$\therefore \quad \frac{d^2 x}{dt^2} = -\frac{ng}{ml} x \qquad \text{... (2)}$$

which is the equation of a simple harmonic motion with centre at the origin B and amplitude BC.

Since $BC < AB$, therefore during the entire motion of the particle the string will not become slack.

Thus the entire motion of the particle is governed by the equation (2) and the particle will make oscillations in simple harmonic motion about the centre B.

The time of oscillation

$$= \frac{2\pi}{\sqrt{\mu}} = \frac{2\pi}{\sqrt{ng/lm}} = 2\pi \sqrt{\frac{lm}{ng}}.$$

Exercise-8.36

1. Two light elastic strings are fastened to a particle of mass m and their other ends to fixed points so that the strings are taut. The modulus of each is λ, the tension T, and length a and b. Show that the period of an oscillation along the line of the strings is $2\pi \left[\dfrac{mab}{(T + \lambda)(a + b)} \right]^{1/2}$.

2. An elastic string of natural length $(a + b)$, where $a > b$ and modulus of elasticity λ has a particle of mass m attached to it at a distance a from one end, which is fixed to a point A of a smooth horizontal plane. The other end of the string is fixed to a point B so that the string is just unstretched. If the particle will be held at B and then released, show that it will oscillate to and fro through a distance $\dfrac{b(\sqrt{a} + \sqrt{b})}{\sqrt{a}}$ in a periodic time $\pi (\sqrt{a} + \sqrt{b}) \sqrt{m/\lambda}$.

3. A particle is performing S.H.M. in the line joining two points A and B on a smooth plane and is connected with these points by elastic strings of natural lengths a and a', the moduli of elasticity being λ and λ' respectively. Show that the periodic time is $2\pi \sqrt{m \left(\dfrac{\lambda}{a} + \dfrac{\lambda'}{a'} \right)^{-1}}$.

4. A mass m hangs from a light spring and is given a small vertical displacement. If l is the length of the spring when the system is in equilibrium and n the number of oscillations per second, show that the natural length of the spring is $l - (g/4\pi^2 n^2)$.

5. A heavy particle is attached to one point of a uniform elastic string. The ends of the string are attached to two points in a vertical line. Show that the period of a vertical oscillation in which the string remains taut is $2\pi \sqrt{mh/2\pi}$, where λ is the coefficient of elasticity of the string and h the harmonic mean of the unstretched lengths of the two parts of the string.

6. A light elastic string of natural length l has one extremity fixed at point O and the other attached to a stone, the weight of which in equilibrium would extend the string to a length l_1. Show that if the stone be dropped from rest at O, it will come to instantaneous rest at a depth $\sqrt{l_1^2 - l^2}$ below equilibrium position.

8.6.9 MECHANICAL OSCILLATORY SYSTEMS

(1) FREE OSCILLATIONS

Consider a spring OA suspended vertically from a fixed support at O. Let a body of mass m be suspended from the end A, the mass of the body being so large in comparison with the mass of the spring that the latter may be neglected. Let $e (= AB)$ be the elongation produced by the mass m hanging in equilibrium, then B is called the position of static equilibrium and e is called the static extension. Let k be the stiffness of the spring k, the restoring force per unit stretch of the spring due to elasticity.

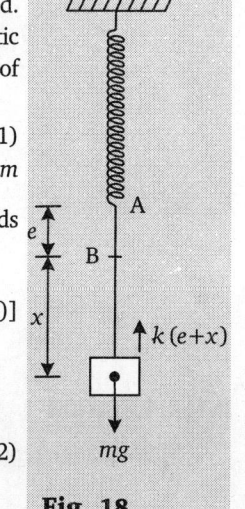

For equilibrium at B, $mg = T = ke$ ... (1)

Let the mass be displaced through a further distance x from the equilibrium position. The acceleration of the mass m at this position is $\dfrac{d^2 x}{dt^2}$ and the forces acting upon it are weight mg downwards and the restoring force $k(e + x)$ upwards

$\therefore$ The equation of motion of mass m is

$$m \frac{d^2 x}{dt^2} = mg - k(e + x) = -kx \qquad \text{[Using (1)]}$$

$$\Rightarrow \qquad \frac{d^2 x}{dt^2} + \frac{k}{m} x = 0$$

Writing $\omega^2 = \dfrac{k}{m}$, it becomes $\qquad \dfrac{d^2 x}{dt^2} + \omega^2 x = 0.$ ... (2)

Fig. 18

The solution of equation (2) is $x = C_1 \cos \omega t + C_2 \sin \omega t$, which can also be written as

$$x = A\cos(\omega t + B) \qquad \qquad ...(3)$$

where C_1, C_2 or A, B are constants to be determined from the initial conditions of the problem.

Equation (3) represents oscillatory variations of the variable x with amplitude A and period $T = \dfrac{2\pi}{\omega}$. It represents the S.H.M. of the mass m with period

$$2\pi\sqrt{\dfrac{m}{k}} = 2\pi\sqrt{\dfrac{e}{g}} \qquad \qquad \left(\text{Since } \omega^2 = \dfrac{k}{m} \text{ and } mg = ke\right)$$

(2) DAMPED FREE OSCILLATIONS

We have assumed so far that no frictional forces acting on the oscillatory body. This situation is impractical, for if it were true, a pendulum or a weight on a string would continue to oscillate indefinitely. Due to friction, the amplitude gradually decreases to zero. This motion is called damped harmonic motion. The damping force generally arises due to air resistance or internal forces. Its magnitude depends generally on the speed of the body. We will consider the case when the damping force is proportional to the velocity or equal to $\lambda \dfrac{dx}{dt}$ of the body and is directed opposite to it, where λ is damping constant.

Equation of damped harmonic motion is

$$m\dfrac{d^2x}{dt^2} = mg - ke - kx - \lambda\dfrac{dx}{dt} = -kx - \lambda\dfrac{dx}{dt} \qquad \qquad [\because mg = ke]$$

$\Rightarrow \qquad m\dfrac{d^2x}{dt^2} + kx + \lambda\dfrac{dx}{dt} = 0$ or $\dfrac{d^2x}{dt^2} + \left(\dfrac{\lambda}{m}\right)\dfrac{dx}{dt} + \left(\dfrac{k}{m}\right)x = C$

Let $\qquad \dfrac{\lambda}{m} = 2p$ and $\dfrac{k}{m} = \omega^2$, then $\dfrac{d^2x}{dt^2} + 2p\dfrac{dx}{dt} + \omega^2 x = 0 \qquad \qquad ...(1)$

which is a linear differential equation with constant coefficients.

A.E. is $\qquad \qquad m^2 + 2km + \omega^2 = 0 \quad\Rightarrow\qquad m = -k \pm \sqrt{k^2 - \omega^2}$

Hence, the general solution $\qquad x = A\, e^{(-p + \sqrt{p^2 - \omega^2})t} + B e^{(-p - \sqrt{p^2 - \omega^2})t}$

At $t = 0$, $x = x_0 = $ maximum and so $\quad A + B = x_0 \qquad \qquad ...(2)$

At $t = 0$, $\dfrac{dx}{dt} = 0$ and so $\qquad A - B = \dfrac{px_0}{\sqrt{p^2 - \omega^2}} \qquad \qquad ...(3)$

Solving (2) and (3), we get $A = \dfrac{x_0}{2}\left[1 + \dfrac{p}{\sqrt{p^2 - \omega^2}}\right]$, $B = \dfrac{x_0}{2}\left[1 - \dfrac{p}{\sqrt{p^2 - \omega^2}}\right]$

In damped harmonic motion there arise three cases :

Case I: When $\dfrac{\lambda}{m} < \sqrt{\dfrac{p}{m}}$, i.e., $2p < \omega$, i.e., $\sqrt{p^2 - \omega^2}$ is imaginary.

Here, $\qquad \qquad = -\left[\quad + \dfrac{}{\sqrt{{}^2 - \omega^2}}\right]$ or $\quad A = \dfrac{x_0}{2}\left[1 + \dfrac{p}{i\sqrt{\omega^2 - p^2}}\right]$

and $\qquad \qquad B = \dfrac{x_0}{2}\left[1 - \dfrac{p}{i\sqrt{\omega^2 - p^2}}\right]$

$\therefore \qquad x = \dfrac{x_0}{2}\left[1 + \dfrac{p}{i\sqrt{\omega^2 - p^2}}\right]e^{(-p + i\sqrt{\omega^2 - p^2})t} + \left[1 - \dfrac{p}{i\sqrt{\omega^2 - p^2}}\right]e^{(-p - i\sqrt{\omega^2 - p^2})t}$

$\qquad x = \dfrac{x_0\, e^{-pt}}{\sqrt{\omega^2 - p^2}}\left[\sqrt{\omega^2 - p^2}\,\sin\sqrt{(\omega^2 - p^2)}\,t + k\cos\sqrt{\omega^2 - k^2}\,t\right]$

$\qquad x = \dfrac{x_0\, e^{-pt}\omega}{\sqrt{\omega^2 - p^2}}\sin\left[\sqrt{\omega^2 - p^2}\,t + p\right]$, where, $\tan\phi = \dfrac{p}{\sqrt{\omega^2 - k^2}}$.

$\qquad x = Ae^{(-\lambda t/2m)}\sin\left[\sqrt{\left(\omega^2 - \dfrac{\lambda^2}{4m^2}\right)}\,t + \phi\right]$, where, $A = \dfrac{x_0\omega}{\sqrt{\omega^2 - \dfrac{\lambda^2}{4m^2}}}$.

INTERPRETATION

This equation represents damped harmonic motion with period $T = \dfrac{2\pi}{\sqrt{\omega^2 - \dfrac{\lambda^2}{4m^2}}}$

Period is longer and frequency is smaller when friction is present. The amplitude is not constant but it decays exponentially.

Case I : When $k = \omega$. Here $x = 0$, $\lambda = -k$. Hence, $x = x\,e^{-kt}$, $x = x_0\,e^{-\lambda t/2m}$

This is critical damping. There is no oscillatory term in the solution, hence the motion becomes non-oscillatory.

Case II : When $k > \omega$

In this case $\sqrt{k^2 - \omega^2}$ is real and $-k + \sqrt{k^2 - \omega^2}$ and $-k - \sqrt{k^2 - \omega^2}$ are both negative.

The displacement falls exponentially. Hence, in this case a body merely returns to the equilibrium position when released from its initial displacement A.

(3) FORCED OSCILLATIONS (WITHOUT DAMPING) AND RESONANCE

If an external force is applied on the point of support of the spring, it oscillates. The motion is called the forced oscillatory motion.

Let the external force be $q \cos nt$.

Equation of motion is
$$m\frac{d^2x}{dt^2} = mg - me - kx + q\cos nt \qquad\qquad [\because mg = ke]$$

$$\Rightarrow \qquad\qquad m\frac{d^2x}{dt^2} = -kx + q\cos nt$$

$$\Rightarrow \qquad\qquad \frac{d^2x}{dt^2} = -\frac{k}{m} + \frac{q}{m}\cos nt \qquad\qquad \ldots (1)$$

Let $\dfrac{k}{m} = \omega^2$ and $\dfrac{q}{m} = E$, then (1) becomes $\dfrac{d^2x}{dt^2} = -\omega^2 x + E\cos nt \quad\Rightarrow\quad \dfrac{d^2x}{dt^2} + \omega^2 x = E\cos nt$

$$\Rightarrow \qquad\qquad (p^2 + \omega^2)x = E\cos nt \qquad\qquad \ldots (2)$$

A.E. is $\quad^2 + \omega^2 = \quad$ or $D = \pm i\omega$

$$\text{C.F.} = C_1 \cos\omega t + C_2 \sin\omega t$$

and
$$\text{P.I.} = \frac{1}{D^2 + \omega^2} E\cos nt$$

Case I : If $\omega \neq n$, P.I. $= E\dfrac{1}{-n^2 + \omega^2}\cos nt$

Complete solution of (1) is
$$x = C_1\cos\omega t + C_2\sin\omega t + \frac{E}{\omega^2 - n^2}\cos nt$$

or
$$x = A\cos(\omega t + \alpha) + \frac{E}{\omega^2 - n^2}\cos nt \qquad\qquad \ldots (3)$$

Equation (2) shows that the motion is the resultant of two oscillatory motions, i.e., the first due to $A\cos(\omega t + \alpha)$ gives free oscillation of period $2\pi / \omega$ and the second due to $\dfrac{E}{\omega^2 - n^2}\cos nt$ gives forced oscillations of period $\dfrac{2\pi}{n}$. If ω is large, then the frequency of free oscillations is very high, then the amplitude $\dfrac{E}{\omega^2 - n^2}$ of forced oscillations is small.

Case II : If $\omega = n$,
$$\text{P.I.} = \frac{1}{D^2 + \omega^2} E\cos nt = E.t\,\frac{1}{2D}\cos nt = \frac{Et}{2}\int \cos nt\,dt$$

Now,
$$\text{P.I.} = \frac{Et}{2}\int \cos nt\,dt = \frac{Et}{2}\left(\frac{\sin nt}{n}\right)$$

So,
$$x = C_1\cos\omega t + C_2\sin\omega t + E\,t\,\frac{\sin nt}{2n}$$

$$\Rightarrow \qquad x = C_1\cos\omega t + C_2\sin\omega t + E\text{---}\sin\omega t \qquad\qquad [n = \omega]$$

or
$$x = C_1\cos\omega t + \left(C_2 + \frac{Et}{2\omega}\right)\sin\omega t$$

Let
$$C_1 = r\sin\phi \quad\text{and}\quad \left(C_2 + \frac{Et}{2\omega}\right) = r\cos\phi$$

Then,
$$x = r\sin\phi\cos\omega t + r\cos\phi\sin\omega t$$

$$\Rightarrow \qquad x = r\sin(\omega t + \phi)$$

The period of oscillations
$$= \frac{2\pi}{\omega}$$

and
$$\text{amplitude} = \sqrt{C_1^2 + \left(C_2 + \frac{Et}{2\omega}\right)}\ \text{ and it increases as } t \text{ increases.}$$

After long time, the amplitude of the oscillation may become abnormally large causing over strain and consequently break down the system. But it does not happen as there is always some resistance in the system.

(4) RESONANCE

If the frequency due to external periodic force becomes equal to the natural frequency of the system, the phenomenon is known as resonance.

In designing a machine or structure, occurrence of the resonance is always to be avoided so that the system may not breakdown. While marching over a bridge, the soldiers avoid their steps may not be in rhythm with the natural frequency of the bridge. Resonance may cause the bridge to collapse.

(5) FORCED OSCILLATIONS (WITH DAMPING)

In the above case, if in addition, there is a damping force which is proportional to the instantaneous velocity of the mass, say $\lambda \dfrac{dx}{dt}$, then the equation of motion of the mass m is $m \dfrac{d^2x}{dt^2} = mg - k(e+x) - \lambda \dfrac{dx}{dt} + q \cos nt$

Since $mg = ke$, it becomes

$$m \frac{d^2x}{dt^2} = -kx - \lambda \frac{dx}{dt} + q \cos nt$$

$$\Rightarrow \qquad \frac{d^2x}{dt^2} + \frac{\lambda}{m} \frac{dx}{dt} + \frac{k}{m} x = \frac{q}{m} \cos nt$$

Writing $\dfrac{\lambda}{m} = 2p$, $\dfrac{k}{m} = \omega^2$ and $\dfrac{q}{m} = E$, it becomes

$$\frac{d^2x}{dt^2} + 2p \frac{dx}{dt} + \omega^2 x = E \cos nt \qquad \ldots (1)$$

Equation (1) is a linear differential equation with constant coefficients.

It's A.E. is $\qquad m^2 + 2pm + \omega^2 = 0 \quad \Rightarrow \quad m = -p \pm \sqrt{p^2 - \omega^2}$

$$\therefore \qquad \text{C.F.} = e^{-pt}(C_1 e^{\sqrt{p^2-\omega^2}\,t} + C_2^{-\sqrt{p^2-\omega^2}\,t})$$

$$\text{P. I.} = E \frac{1}{D^2 + 2pD + \omega^2} \cos nt = E. \frac{1}{-n^2 + 2pD + \omega^2} \cos nt$$

$$= E. \frac{1}{(\omega^2 - n^2) + 2pD} \cos nt = E. \frac{(\omega^2 - n^2) - 2pD}{(\omega^2 - n^2)^2 - 4p^2 D^2} \cos nt$$

$$= \frac{E(\omega^2 - n^2) \cos nt + 2pn \sin nt}{(\omega^2 - n^2) + 4p^2 n^2}$$

Now, let $\omega^2 - n^2 = r \cos \phi$, $2pn = r \sin \phi$, so that $r = \sqrt{(\omega^2-n^2)^2 + 4p^2 n^2}$ and $\phi = \tan^{-1}\left(\dfrac{2pn}{\omega^2 - n^2}\right)$

$$\therefore \qquad \text{P.I.} = E \frac{r \cos(nt - \phi)}{r^2} = \frac{E}{r}. \cos(nt - \phi)$$

The complete solution of equation (1) is given by

$$x = e^{-pt}\left(C_1 e^{\sqrt{p^2-\omega^2}\,t} + C_2 e^{-\sqrt{p^2-\omega^2}\,t}\right) + \frac{E \cos\left\{nt - \tan^{-1}\left(\dfrac{2pn}{\omega^2 - n^2}\right)\right\}}{\sqrt{(\omega^2-n^2)^2 + 4p^2 n^2}}$$

The C.F. represents free oscillations of the system which die out as $t \to \infty$ due to the presence of the factory e^{-pt}.

The P.I. represents the forced oscillations of the system having a constant amplitude.

$$= \frac{E}{\sqrt{(\omega^2-n^2)^2 + 4p^2 n^2}}$$

and the period $= \dfrac{2\pi}{n}$, which is the same as that of the impressed force.

Thus, as t increases, the free oscillations (given by the C.F.) die out while the forced oscillations (given by the P.I.) persist giving the steady state of motion.

Solved Examples

EXAMPLE 1. *A mass M suspended from the end of a helical spring is subjected to a periodic force $f = F \sin \omega t$ in the direction of its length. The force f is measured positive vertically downwards and at zero time M is at rest. If the spring stiffness is S, prove that the displacement of M at time t from the commencement of motion is given by*

$$x = \frac{F}{M(p^2 - \omega^2)}\left[\sin \omega t - \frac{\omega}{p}\sin pt\right], \text{where}$$

$$p^2 = \frac{S}{M} \text{ and damping effects are neglected.}$$

[UPTU–2002]

SOLUTION. Let x be the displacement from the equilibrium position, the equation of motion is

$$M \frac{d^2x}{dt^2} = -Sx + F \sin \omega t$$

$$\frac{d^2x}{dt^2} + \frac{S}{M}x = \frac{F}{M}\sin \omega t$$

$$\Rightarrow \frac{d^2x}{dt^2} + P^2x = \frac{F}{M}\sin\omega t \qquad \left(\because \frac{S}{M} = P^2\right)$$

$$(D^2 + P^2)x = \frac{F}{M}\sin\omega t$$

A.E. is $D^2 + P^2 = 0$, $D = \pm iP$.

C.F. $= (C_1 \cos pt + C_2 \sin pt)$

P.I. $= \frac{1}{D^2 + P^2}\frac{F}{M}\sin\omega t = \frac{F}{M}\frac{1}{-\omega^2 + P^2}\sin\omega t$

$$\therefore \quad x = C_1 \cos pt + C_2 \sin pt + \frac{F}{M}\frac{1}{p^2 - \omega^2}\sin\omega t \quad ...(1)$$

Putting $t = 0$ and $x = 0$ in (1), we get $0 = C_1$

Equation (1) becomes

$$x = C_2 \sin pt + \frac{F}{M}\frac{1}{p^2 - \omega^2}\sin\omega t \quad ...(2)$$

Differentiating (2), we obtain

$$\frac{dx}{dt} = C_2 p \cos pt + \frac{F}{M}\frac{\omega}{p^2 - \omega^2}\cos\omega t \quad ...(3)$$

Putting $\frac{dx}{dt} = 0$ and $t = 0$ in (3), we have

$$0 = C_2 P = \frac{F}{M}\frac{\omega}{p^2 - \omega^2}$$

$$\Rightarrow \quad C_2 = -\frac{F\omega}{PM(p^2 - \omega^2)}.$$

On substituting the value of C_2 in (2), we get

$$x = -\frac{\omega}{p}\frac{F}{PM(p^2 - \omega^2)}\sin pt$$

$$+ \frac{F}{M}\frac{1}{p^2 - \omega^2}\sin\omega t$$

$$= \frac{F}{M(p^2 - \omega^2)}\left[\sin\omega t - \frac{\omega}{p}\sin pt\right].$$

EXAMPLE 2. *In a spring of T_1 and T_2 be the periods corresponding to two different weights attached and s_1 and s_2 the statical extensions due to these weights, prove that*

$$g = \frac{4\pi^2(s_1 - s_2)}{T_1^2 - T_2^2}.$$

SOLUTION. Let k be the stiffness of the spring and m_1, m_2 be the masses attached to the spring for which the periods are T_1 and T_2 respectively, we have

$$T_1 = 2\pi\sqrt{\frac{m_1}{k}} \Rightarrow T_1^2 = 4\pi^2 m_1/k \quad ...(1)$$

and $T_2 = 2\pi\sqrt{\frac{m_2}{k}} \Rightarrow T_2^2 = \frac{4\pi^2 m_2}{k} \quad ...(2)$

Subtracting (2) from (1), we get

$$T_1^2 - T_2^2 = \frac{4\pi^2}{k}(m_1 - m_2) \quad ...(3)$$

During static equilibrium spring force is equal to gravitational force, i.e.,

$$ks = mg$$
$$ks_1 = m_1 g \quad ...(4)$$
and $ks_2 = m_2 g \quad ...(5)$

Subtracting (4) from (5), we get

$$k(s_1 - s_2) = (m_1 - m_2)g$$

$$k^{-1}(m_1 - m_2) = \frac{1}{g}(s_1 - s_2)$$

Substituting the value of $(m_1 - m_2)$ in (2), we get

$$T_1^2 - T_2^2 = 4\pi^2\frac{1}{g}(s_1 - s_2)$$

$$g = \frac{4\pi^2(s_1 - s_2)}{T_1^2 - T_2^2}.$$

EXAMPLE 3. *A spring for which the spring constant $k = 700$ Nm^{-1} hangs in a vertical position with its upper end fixed to a support. A mass of 20 kg is attached to the lower end and system brought to rest. Find the position of the mass at time t, if a force 70 sin 2tN is applied to the support.* [UPTU Q. BANK–2002]

SOLUTION. Equation of motion is

$$m\frac{d^2x}{dt^2} = -kx + 70\sin 2t$$

$$\Rightarrow \quad 20\frac{d x}{dt} = -700x + 70\sin 2t$$

$$\Rightarrow \quad \frac{d^2x}{dt^2} + 35x = \frac{7}{2}\sin 2t$$

Auxiliary equation is

$$m^2 + 35 = 0$$
$$m = \pm i\sqrt{35}$$

C.F. $= C_1 \cos\sqrt{35}\,t + C_2 \sin\sqrt{35}\,t$

P.I. $= \frac{1}{D^2 + 35}\left(\frac{7}{2}\sin 2t\right) = \frac{7}{62}\sin 2t$

$$x = C_1 \cos\sqrt{35}\,t + C_2 \sin\sqrt{35}\,t + \frac{7}{62}\sin 2t \quad ...(1)$$

At $t = 0$, $x = 0$, $\therefore$ from (1), $C_1 = 0$.

From (1), $x = C_2 \sin\sqrt{35}\,t + \frac{7}{62}\sin 2t \quad ...(2)$

Differentiating w.r.t. t, we get

$$\frac{dx}{dt} = \sqrt{35}\,C_2 \cos\sqrt{35}\,t + \frac{7}{31}\cos 2t \quad ...(3)$$

Also, when $t = 0$, $v = \frac{dx}{dt} = 0$ in (3), we get

$$0 = \sqrt{35}\,C_2 + \frac{7}{31} \Rightarrow C_2 = -\frac{7}{31\sqrt{35}}$$

$$\therefore \text{ From (2), } x = -\frac{7}{31\sqrt{35}}\sin\sqrt{35}\,t + \frac{7}{62}\sin 2t.$$

Exercise-8.37

1. A body weighing 4.9 kg is hung from a spring. A pull of 10 kg will stretch the spring to 5 cm. The body is pulled down 6 cm below the static equilibrium position and then released. Find the displacement of the body from its equilibrium position at time t seconds, the maximum velocity and the period of oscillation. [UPTU Q. BANK–2002]

2. A mass of 200 gm is toed at the end of a spring which extends to 4 cm under a force 196,000 dynes. The spring is pulled 5 cm and released. Find the displacement, t seconds after release, if there be a damping force of 2000 dynes per cm per second. [UPTU Q. BANK–2002]

3. A spring for which stiffness $K = 700$ Newton/m hangs in a

vertical position with its upper end fixed. A mass of 7 kg is attached to the lower end. After coming to rest, the mass is pulled down 0.05 m and released. Discuss the resulting motion of the mass, neglecting air resistance.

4. A body executes damped forced vibrations given by the equation $\dfrac{d^2x}{dt^2} + 2K\dfrac{dx}{dt} + b^2x = e^{-kt}\sin\omega t$. Solve the equation for both the cases, when

$\omega^2 \neq b^2 - K^2$ and when $\omega^2 = b^2 - K^2$.

5. A spring of negligible weight which stretches l inch under tension of 2 lb is fixed at one end is attached to a weight of w lb at the other. It is found that resonance occurs when an axial periodic force $2\cos 2t$ lb acts on the weight. Show that when the free vibrations have dies out, the forced vibrations

are given by $x = Ct\sin 2t$, and find values of w and C.

6. A spring which stretches by an amount e under a force mk^2e is suspended from a support P and has a mass m at its lower end. At time $t = 0$, the mass is at rest in its equilibrium position at a point A below P. A vertical oscillation is now given to the support P such that at any time $t(>0)$, its displacement below its initial position is $a\sin nt$. Show that the displacement x of the mass below A is given by

$$\frac{d^2x}{dt^2} + K^2x = K^2a\sin nt$$

Hence, show that if $n \neq k$, the displacement is given by

$$x = \frac{Ka}{K^2 - n^2}(K\sin nt - n\sin nt)$$

What happens, when $n = k$? 　　[UPTU Q. BANK–2002]

8.6.10 BENDING OF BEAMS

Beam or Bar : A rod of a circular or rectangular cross-section with its length very much greater than its thickness is called a beam.

Supported Beam : If a beam may just rest on a support like a knife edge is called a supported beam.

Fixed Beam : If one of the ends of a beam is firmly fixed, then it is called a fixed beam.

Cantilever : If one end of the beam is fixed and the other end is loaded, it is called a cantilever.

Strut : A beam of homogeneous isotropic material subject to compressive stress is called strut.

(I) DEFLECTION OF BEAMS

A beam can be considered to be made up of fibres running lengthwise. When a beam is bent under the given loading, it is observed that the fibres in the upper of its face are compressed while those in the opposite face to the lower half are stretched. In between these surfaces of compression and tension, there must be a transition surface where the fibres are neither compressed nor stretched. This surface is called the neutral surface of the beam. The curve defined by any fibre in this surface known as the deflection curve or the elastic curve of the beam. The line in which any plane section of the beam cuts the neutral surface is called the neutral axis of that section.

Before we derive the equation of the deflection curve, we make the following assumptions :

(i) The beam is of uniform cross-section and its breadth and thickness are small compared to its length.

(ii) The axis of the beam is a horizontal line through the centre of the gravity of the cross-section.

(iii) The beam is perfectly elastic and obeys Hooke's law.

(iv) The deflections of the beam are small so that the slope at any point of the elastic curve is assumed small.

Consider a cross-section of the beam which cuts the neutral surface in the point P at a distance x from O, where OX and OY are the axes of coordinates. Let CD be the neutral axis of this section. We know from mechanics that in the equilibrium of any cross-section of the beam, the external bending moment M about CD, due to different loads acting on the beam, must balance the internal moment $\dfrac{EI}{R}$ which is the sum of the moment of the force about CD.

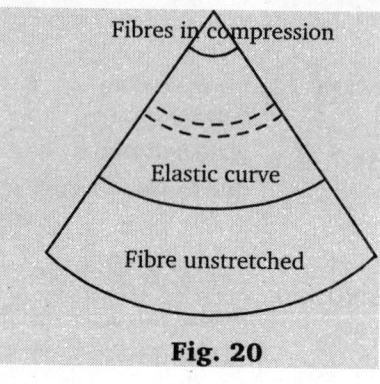

Fig. 20

Hence, we get 　　　　$M = \dfrac{EI}{R}$ 　　　　... (1)

where, E is the modulus of elasticity of the material of the beam I is the moment of inertia of the area of cross-section of the beam and R is the radius of curvature of the elastic curve at the point P.

Since $R = \dfrac{\left\{1 + (dy/dx)^2\right\}^{3/2}}{d^2y/dx^2}$ and $\dfrac{dy}{dx}$ is small; $\left(\dfrac{dy}{dx}\right)^2$ can be ignored.

Thus, (1) gives bending moment $M = EI\dfrac{d^2y}{dx^2}$ 　　　　... (2)

If the beam is subjected to transverse load only, we further obtain from equation (2) on differentiation

$$\frac{dM}{dx} = \text{Shear force} = EI\frac{d^3y}{dx^3} = F \qquad\qquad ...(3)$$

and $$\frac{d^2M}{dx^2} = \text{Intensity of loading} = \frac{dF}{dx} = EI\frac{d^4y}{dx^4} \qquad\qquad ...(4)$$

REMARKS

- Equation (1) is known as Bernoulli's-Euler formula.
- The product EI is called the flexural rigidity of the beam.
- The moment M about the neutral axis of all external forces acting on either side of the two portions of the beam separated by the cross-section is independent of the portion considered.
- The moment M is the algebraic sum of the moments of the external forces acting on the portion of the beam about the neutral axis. The upward forces (in anti-clockwise direction) gives negative moments.

BOUNDARY CONDITIONS

(i) End Freely Supported :

At the freely supported end O, there will be no deflection and no bending moment.

$$y = 0, \frac{d^2y}{dx^2} = 0.$$

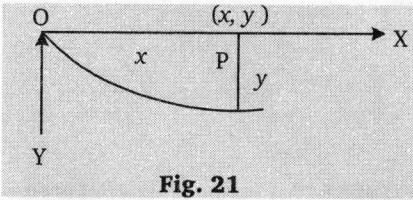

Fig. 21

(ii) Fixed End Horizontally :

Deflection and slope of the beam are zero

$$y = 0, \frac{dy}{dx} = 0.$$

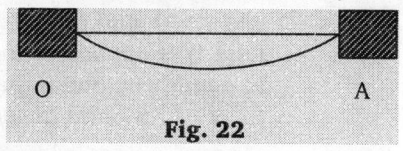

Fig. 22

(iii) Perfectly Free End :

At the free end, there is no bending moment of shear force.

$$\frac{d^2y}{dx^2} = 0, \frac{d^3y}{dx^3} = 0.$$

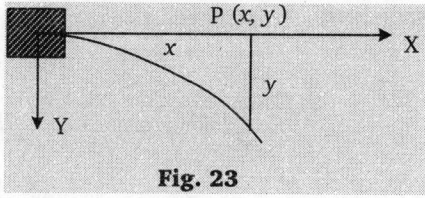

Fig. 23

 Solved Examples

EXAMPLE 1. *For a strut of length l, freely hinged at each end, prove that the bending moment at the centre is*

$$\frac{-Wl^2}{8} \frac{Q}{Q-P}, \text{ where } Q = \frac{EI\pi^2}{l^2}. \text{ The differential}$$

equation for a strut is

$$EI \frac{d^2y}{dx^2} + Py = -\frac{Wl^2}{8} \sin \frac{\pi x}{l}$$

SOLUTION. We have

$$EI \frac{d^2y}{dx^2} + Py = -\frac{Wl^2}{8} \sin \frac{\pi x}{l} \qquad \dots (1)$$

$$\Rightarrow \quad \frac{d^2y}{dx^2} + \frac{P}{EI} y = -\frac{W\pi^2}{8Q} \sin \frac{\pi x}{l}$$

$$\left[\because EI \frac{\pi^2}{l^2} \Rightarrow \frac{1}{EI} = \frac{\pi^2}{l^2 Q} \right]$$

$$\Rightarrow \quad \frac{d^2y}{dx^2} + \frac{P\pi^2}{Ql^2} y = \frac{-W\pi^2}{8Q} \sin \frac{\pi x}{l}$$

$$\left(D^2 + \frac{P\pi^2}{Ql^2} \right) y = \frac{-W\pi^2}{8Q} \sin \frac{\pi x}{l}$$

It's A.E. is $D^2 + \frac{P\pi^2}{Ql^2} = 0$

$$\Rightarrow \qquad D = \pm \frac{\pi}{l} \sqrt{\frac{P}{Q}}$$

$$\text{C.F.} = C_1 \cos \left(\frac{\pi}{l} \sqrt{\frac{P}{Q}} \right) x + C_2 \sin \left(\frac{\pi}{2} \sqrt{\frac{P}{Q}} \right) x$$

$$\text{P.I.} = \frac{1}{D^2 + \frac{P\pi^2}{Ql^2}} \left(-\frac{W\pi^2}{8Q} \sin \frac{\pi x}{l} \right)$$

$$= -\frac{W\pi^2}{8Q} \frac{1}{-\frac{\pi^2}{l^2} + \frac{P}{Q} \frac{\pi^2}{l^2}} \sin \frac{\pi x}{l}$$

$$= \frac{W\pi^2}{8Q} \cdot \frac{l^2}{\pi^2} \cdot \frac{1}{1 - \frac{P}{Q}} \sin \frac{\pi x}{l}$$

$$= \frac{Wl^2}{8} \cdot \frac{1}{Q-P} \sin \frac{\pi x}{l}$$

The general solution of (1) is

$$y = C_1 \cos \left(\frac{\pi}{l} \sqrt{\frac{P}{Q}} \right) x$$
$$+ C_2 \sin \left(\frac{\pi}{l} \sqrt{\frac{P}{Q}} \right) x + \frac{Wl^2}{8} \frac{1}{Q-P} \sin \frac{\pi x}{l} \quad \dots (2)$$

Since the ends are freely hinged.

$$y = 0, \frac{d^2y}{dx^2} \text{ at } x = 0$$

and $y = 0, \frac{d^2y}{dx^2}$ at $x = l$.

On putting $x = 0$, $y = 0$ in (2), we get $0 = C_1$.

Then, (2) becomes

$$y = C_2 \sin \left(\frac{\pi}{l} \sqrt{\frac{P}{Q}} \right) x + \frac{Wl^2}{8} \frac{1}{Q-P} \sin \frac{\pi x}{l} \qquad \dots (3)$$

On putting $x = l$ and $y = 0$ in (3), we get

$$0 = C_2 . \sin \pi \sqrt{\frac{P}{Q}} \Rightarrow C_2 = 0$$

(3) reduces to

$$y = \frac{Wl^2}{8} \frac{1}{Q-P} \sin \frac{\pi x}{l}$$

But bending moment is given by

$$M = EI \frac{d^2y}{dx^2}$$

$$M = EI \left[-\frac{Wl^2}{8} \frac{1}{Q-P} \frac{\pi^2}{l^2} \sin\frac{\pi x}{l} \right]$$

$$= EI \left[-\frac{W\pi^2}{8} \frac{1}{Q-P} \sin\frac{\pi x}{l} \right]$$

$$M = -\frac{Wl^2}{8} \frac{Q}{Q-P} \sin\frac{\pi x}{l} \quad \left(Q = \frac{EI \pi^2}{l^2} \right)$$

Hence, M at centre (at $x = l/2$)

$$= -\frac{Wl^2}{8} \frac{Q}{Q-P} \sin\frac{\pi}{2} = -\frac{Wl^2}{8} \frac{Q}{Q-P}.$$

EXAMPLE 2. *A beam of length l is clamped horizontally at its end $x = 0$ and is free at the end $x = l$. A point load W is applied at the end $x = l$, in addition to a uniform load w per unit length from $x = 0$ to $x = l/2$. Find the deflections at any point.*

[UPTU(B.TECH)–2002]

SOLUTION. From the equation of balance, we get

$$R = W + \frac{wl}{2} \qquad \dots(1)$$

Choose a random axis NN'. Let (x, y) be the coordinates of N. Taking moment about N, we get

$$EI \frac{d^2y}{dx^2} = \frac{wl}{2}\left(x - \frac{l}{4} \right) - Rx$$

$$= \frac{wl}{2}\left(x - \frac{l}{4} \right) - \left(W - \frac{wl}{2} \right)x$$

[From (1)]

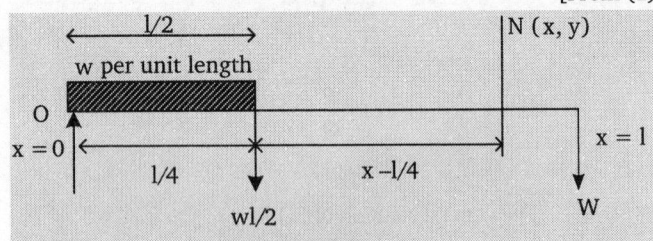

Fig. 24

or $\quad EI \frac{d^2y}{dx^2} = -\frac{wl^2}{8} - Wx \qquad \dots(2)$

Integrating (2) w.r.t. x, we get

$$EI\frac{dy}{dx} = -\frac{wl^2}{8}x - \frac{Wx^2}{2} + C.$$

Applying boundary conditions, we have $x = 0$

$$\Rightarrow \quad \text{slope} = \frac{dy}{dx} = 0 \; , \; \therefore C = 0$$

$$EI\frac{dy}{dx} = -\frac{wl^2}{8}x - \frac{Wx^2}{2} \qquad \dots(3)$$

Again integrating w.r.t. x, we get

$$EIy = \frac{-wl^2x^2}{16} - \frac{Wx^3}{6} + C_1$$

Applying boundary conditions, we have

$$x = 0, \; y = 0 \Rightarrow C_1 = 0.$$

$$EIy = \frac{-wl^2x^2}{16} - \frac{Wx^3}{6}$$

or $\quad y = -\frac{1}{EI}\left[\frac{Wx^3}{6} + \frac{wl^2x^2}{16} \right].$

EXAMPLE 3. *The deflection of a strut of length l with one end ($x = 0$) built in and the other supported and subjected to end thrust P, satisfies the equation $\frac{d^2y}{dx^2} + a^2y = \frac{a^2R}{P}(l-x)$. Prove that deflection curve in $y = \frac{R}{P}\left(\frac{\sin ax}{a} - l\cos ax + l - x \right)$, where $al = \tan al$.*

[UPTU–2001]

SOLUTION. The given equation is $\frac{d^2y}{dx^2} + a^2y = \frac{a^2R}{P}(l-x)$

$$(D^2 + a^2)y = \frac{a\,R}{P}(l-x) \qquad \dots(1)$$

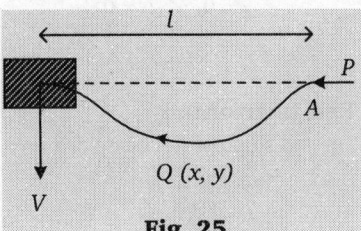

Fig. 25

It's A.E. is $m^2 + a^2 = 0$ so that
$$m^2 + a^2 = 0 \text{ so that } m = \pm ia$$

$\therefore$ C.F. $= C_1 \cos ax + C_2 \sin ax .$

$$\text{P. I.} = \frac{1}{D^2 + a^2}\left\{ \frac{a^2R}{P}(l-x) \right\}$$

$$= \frac{a^2R}{P} \frac{1}{a^2\left(1 + \frac{D^2}{a^2} \right)}(l-x) .$$

$$= \frac{R}{P}\left(1 + \frac{D^2}{a^2} \right)^{-1}(l-x)$$

$$= \frac{R}{P}\left(1 - \frac{D^2}{a^2} + \dots \right)(l-x) = \frac{R}{P}(l-x)$$

$\therefore$ The complete solution of (1) is

$$y = C_1 \cos ax + C_2 \sin ax + \frac{R}{P}(l-x) \qquad \dots(2)$$

Differentiating (2) w.r.t. x,

$$\frac{dy}{dx} = -aC_1 \sin ax + aC_2 \cos ax - \frac{R}{P} \qquad \dots(3)$$

Since the end O is build in, $y = 0$ and $\frac{dy}{dx} = 0$ at $x = 0$

$\therefore$ From (2), $0 = C_1 + \frac{Rl}{P} \Rightarrow C_1 = -\frac{Rl}{P}.$

From (4), $0 = aC_2 - \frac{R}{P} \Rightarrow C_2 = \frac{R}{aP}.$

Substituting the values of C_1 and C_2 in (2), we have

$$y = \frac{R}{P}\left(\frac{\sin ax}{a} - l\cos ax + l - x \right) \qquad \dots(4)$$

which is the equation of the deflection curve.

Also, at the end A, $y = 0$, when $x = l$

$\therefore$ From (1), $0 = \frac{R}{P}\left(\frac{\sin al}{a} - l\cos al \right)$

or $\quad \frac{\sin al}{a} = l\cos al$

$\therefore \quad al = \tan al.$

8.6.11 WHIRLING OF SHAFTS

Sometimes, rotating shafts rotate about its geometrical axes even in the absence of external load. The magnitude of the deflection depends upon the

(i) stiffness of shaft and its support

(ii) the total mass of shaft and attached parts

(iii) the imbalance of the mass with respect to the axis of rotation.

(iv) the amount of damping in the system.

At certain speeds the deflection of a rotating shaft tends to become large and the shaft will fracture unless the speed is lowered and the deflection becomes normal again. These dangerous speeds are called whirling speeds or critical speeds of the shaft.

The differential equation for the whirling of the shaft of weight W per unit length which is rotating with angular velocity ω (radian/sec) is

$$(D^4 - a^4)\, y = 0, \quad \text{where} \quad a^4 = \frac{W\omega^2}{gEl}$$

A.E. is $\qquad m^4 - a^4 = 0 \;\Rightarrow\; m = \pm a, \pm ai$

Hence, its complete solution is

$$y = C_1 e^{ax} + C_2 e^{-ax} + C_3 \cos ax + C_4 \sin ax$$
$$y = A\cosh ax + B\sinh ax + C\cos ax + D\sin ax.$$

To determine arbitrary constants, boundary conditions are

(1) In case of short or flexible bearings : There is no deflection and no bearing moment. Boundary conditions are

$$y = 0, \quad \frac{d^2 y}{dx^2} = 0$$

(2) In case of load bearings : The deflection and the slope of the shaft both will be zero. Boundary conditions are :

$$y = 0, \quad \frac{dy}{dx} = 0$$

(3) When end of shaft is perfectly free : There is no bending moment and no shear force. Boundary conditions are

$$\frac{d^2 y}{dx^2} = 0 \quad \text{and} \quad \frac{d^3 y}{dx^3} = 0$$

 Solved Examples

EXAMPLE 1. *The whirling speed of a shaft of length l is given by* $\dfrac{d^4 y}{dx^4} - a^4 y = 0$ *, where* $a^4 = \dfrac{W w^2}{gEl}$ *and y is the displacement. If* $\cos al.\cosh al = 1$ *, find the whirling speed of the shaft.*

SOLUTION. The given differential equation is

$$\frac{d^4 y}{dx^4} - a^4 y = 0$$

C.F. is $y = A\cosh ax + B\sinh ax$
$$\qquad\qquad + C\cos ax + D\sin ax \qquad \dots (1)$$

Differentiating (1) w.r.t. x, we get

$$\frac{dy}{dx} = Aa\sinh ax + Ba\cosh ax$$
$$\qquad\qquad - Ca\sin ax + Da\cos ax$$

$$\frac{1}{a}\frac{dy}{dx} = A\sinh ax + B\cosh ax$$
$$\qquad\qquad - C\sin ax + D\cos ax \qquad \dots (2)$$

As the ends of the shaft are fixed, the boundary conditions are

$$y = 0, \; \frac{dy}{dx} = 0, \; \text{when } x = 0 \qquad \dots (3)$$

and $\quad y = 0, \; \frac{dy}{dx} = 0, \; \text{when } x = 2 \qquad \dots (4)$

Using (3), from (1), we have $0 = A + C$,

or $\qquad C = -A \qquad\qquad\qquad\qquad \dots (5)$

and from (2), we have $0 = B + D$

$$D = -B \qquad\qquad\qquad\qquad \dots (6)$$

From (4) and (1), we have

$$0 = A\cosh al + B\sinh al + C\cos al + D\sin al \ \dots (7)$$

and from (2) and (1), we have

$$0 = A\sin al + B\cosh al - C\sin al + D\cos al \qquad \dots (8)$$

Substituting the values of C and D from (5) and (6) in (7) and (8), we get

$$A(\cosh al - \cos al) + B(\sinh al - \sin al) = 0$$

and $A(\sinh al + \sin al) + B(\cosh al - \cos al) = 0$

Solving for A and B, we get

$$\frac{\cosh al - \cos al}{\sinh al - \sin al} = -\frac{B}{A} = \frac{\sinh al + \sin al}{\cosh al - \cos al}$$

$$\cosh^2 al - 2\cosh al\cos al + \cos^2 al$$
$$\qquad\qquad = \sinh^2 al - \sin^2 al$$

$$-2\cosh al \ \cos al + 2 = 0$$

$$\cos al \cosh al = 1$$

When the shaft whirls, this equation must satisfy. The solution of the equation gives

$$al = 4.73 = \frac{3\pi}{2} \text{ radians approximately.}$$

Now, $\quad a^4 = \dfrac{W\omega^2}{gEl}$ or $\left(\dfrac{W\omega^2}{gEl}\right)^{1/2} = a^2$

$$\Rightarrow \quad \omega\sqrt{\left(\frac{W}{gEl}\right)} l^2 = a^2 l^2 = \frac{9\pi^2}{4}$$

∴ Whirling speed of a shaft with ends in long bearings

$$\omega = \frac{9\pi^2}{4}\sqrt{\left(\frac{gEl}{W}\right)} \text{ approximately.}$$

EXAMPLE 2. *If the shaft has one long bearing, find the whirling speed of the shaft.*

SOLUTION. $y = A\cosh ax + B\sinh ax + C\cos ax + D\sin ax$

Conditions : $y = 0$, $\dfrac{du}{dx} = 0$

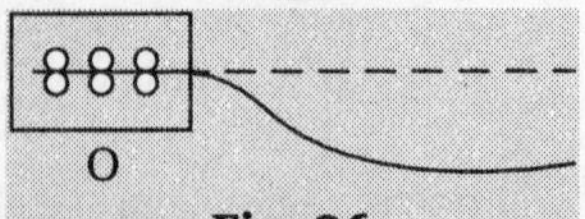

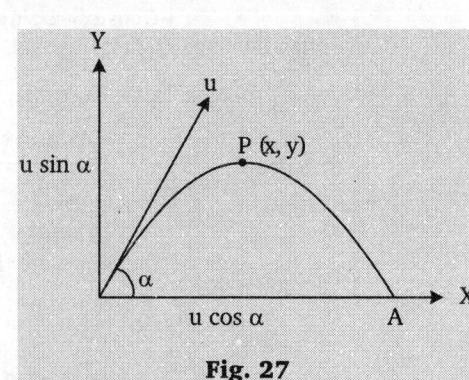

Fig. 26

Then $\cosh al \cos al = -1$

$\Rightarrow \qquad al \quad 1.865$

$$\omega \sqrt{\frac{W}{gEl}}\, l^2 = a^2 l^2$$

$$\omega \sqrt{\frac{W}{gEl}}\, l^2 = (1.865)^2$$

$\Rightarrow \qquad \omega = \dfrac{3.5}{l^2}\sqrt{\dfrac{gEl}{W}}.$

Exercise-8.38

1. A light horizontal strut AB of length l is freely pinned at A and B and is under the action of equal and opposite compressive forces P at each of its ends and carries a load W at its centre. Prove that the deflection at the centre is $\dfrac{W}{2P}\left(\dfrac{l}{n}\tan\dfrac{nl}{2}-\dfrac{l}{2}\right)$, where $n^2 = \dfrac{P}{EI}.$

2. Solve the differential equation

$$EI\frac{d^2y}{dx^2} + Py = -\frac{Wl^2}{8}\sin\frac{\pi x}{l}$$

for a strut of length l, freely hinged at each end, prove that bending moment at the centre is $-\dfrac{Wl^2}{8}\left(\dfrac{Q}{Q-P}\right)$, where $Q = EI\dfrac{\pi^2}{l^2}.$

3. A beam of length l metres is fixed horizontally at one end. Find the equation of the elastic curve and the maximum deflection, if there is uniform load W Newton/m along the length and load W Newton at the free end.

   [UPTU Q. BANK–2001]

4. The differential equation for the displacement y of a whirling shaft when the weight of the shaft is taken into account is

$$EI\frac{d^4y}{dx^4} - \frac{Ww^2}{g}y = W .$$

   Taking the shaft of length $2l$ with the origin at the centre and short bearings at both ends, show that the maximum deflection of the shaft is

$$\frac{g}{2w^2}(\operatorname{sech} al + \sec al - 2)$$

5. A horizontal tie rod of length $2l$ with concentrated load W at the centre and ends freely hinged, satisfies the differential equation $EI\dfrac{d^2y}{dx^2} = Py - \dfrac{W}{2}x$ with conditions $x = 0$, $y = 0$ and $x = l$, $\dfrac{dy}{dx} = 0$. Prove that the deflection δ and bending moment M at the centre $(x = l)$ are $\delta = \dfrac{W}{Pn}(nl - \tanh nl)$ and $M = -\dfrac{W}{2n}\tanh nl$, where $n^2 EI = P$.

6. The differential equation for the displacement of a heavy whirling shaft is $\dfrac{d^4y}{dx^4} = a^4\left(y + \dfrac{g}{w^2}\right)$, where $a^4 = \dfrac{Ww^2}{gEI}$.

   If both ends are short bearings, then ends being $x = 0$ and $x = l$, find the bending moment of the centre of the shaft.

Answers

2. $y = C_1\cos\dfrac{\pi}{l}\sqrt{\dfrac{P}{Q}}x + C_2\sin\dfrac{\pi}{l}\sqrt{\dfrac{P}{Q}}x + \dfrac{Wl^2}{8(Q-P)}\sin\dfrac{\pi x}{l}$ **3.** $y = \dfrac{W}{24EI}(-3l^4 - 8l^3) = \dfrac{Wl^3}{24EI}(3l+8)$ **6.** $\dfrac{W}{2a^2}\left(\operatorname{sech}\dfrac{al}{2} - \sec al\right)$

8.6.12 Applications of Simultaneous Linear Differential Equations

EXAMPLE 1. *A particle is projected with velocity u making an angle α with the horizontal. Neglecting air resistance, show that the equation of its path is the parabola $y = x\tan\alpha - \dfrac{9x^2}{2u^2\cos^2\alpha}$. Find the time of flight, the greatest height attained and range on the horizontal plane.*

SOLUTION. Let a particle of mass m be projected from a point O with velocity u in a direction making an angle α with the horizontal. Let the horizontal and the vertical lines through O is the plane of motion of the particle be taken as the axes of x and y respectively. Let $P(x, y)$ be the position of the particle at time t.

Fig. 27

The horizontal and vertical components of u are $u\cos \alpha$ and $u\sin \alpha$ respectively.

The only force acting on the particle is its weight mg acting vertically downwards.

∴   The equation of motion are

Parallel to x-axis

$$m\frac{d^2x}{dt^2} = 0 \Rightarrow \frac{d^2x}{dt^2} = 0 \qquad \dots (1)$$

Parallel to y-axis

$$m\frac{d^2x}{dt^2} = -mg \Rightarrow \frac{d^2y}{dt^2} \qquad \dots (2)$$

Integrating (1) w.r.t. t, we get $\frac{dx}{dt} = C_1$

Initially, when $t = 0$, $dx/dt = u\cos\alpha$,

$$\therefore \qquad C_1 = u\cos\alpha$$

$$\therefore \qquad \frac{dx}{dt} = u\cos\alpha \qquad \dots (3)$$

Integrating w.r.t. t, $x = (u\cos\alpha)t + C_2$

Initially, when $t = 0$, $x = 0$, $\therefore C_2 = 0$

$$x = (u\cos\alpha)t \qquad \dots (4)$$

Integrating (2) w.r.t. t, $\frac{dy}{dt} = -gt + C_3$

Initially, when $t = 0$, $\frac{dy}{dt} = u\sin\alpha$

so that $\qquad C_3 = u\sin\alpha$

$$\frac{dy}{dt} = u\sin\alpha - gt \qquad \dots (5)$$

Integrating w.r.t. t, we get

$$y = (u\sin\alpha)t - \frac{1}{2}gt^2 + C_4$$

Initially, when $t = 0$, $y = 0$, $\therefore C_4 = 0$

$$y = (u\sin\alpha)t - \frac{1}{2}gt^2 \qquad \dots (6)$$

Equation (4) and (6) give the position of the particle at any time t. The equation of the path described by the particle is obtained by eliminating the parameter t between equations (4) and (6).

From (4), $\quad t = \dfrac{x}{u\cos\alpha}$

Substituting this value of t in equation (6), we get

$$y = (u\sin\alpha).\frac{x}{u\cos\alpha} - \frac{1}{2}g\left(\frac{x^2}{u^2\cos^2\alpha}\right)$$

$$y = x\tan\alpha - \frac{gx^2}{2u^2\cos^2\alpha} \qquad \dots (7)$$

which is the equation of the path of the projectile. Clearly, the path is a parabola.

The time of flight of the projectile is the time taken by the particle to reach the horizontal plane through O.

At the point A, $\quad y = 0$

From (6), $0 = (u\sin\alpha)t - \dfrac{1}{2}gt^2$ or $t = \dfrac{2u\sin\alpha}{g}$

At the highest point, the vertical component of velocity vanishes, *i.e.*,

$$dy/dt = 0$$

so that $\quad u\sin\alpha - gt = 0$ or $t = \dfrac{u\sin\alpha}{g}$.

$\therefore$ The greatest height $= y$ (when $\dfrac{dy}{dx} = 0$).

$$= (u\sin\alpha)t - \frac{1}{2}gt^2 \text{ (when } t = \frac{u\sin\alpha}{g}\text{)}$$

$$= u\sin\alpha.\frac{u\sin\alpha}{g} - \frac{1}{2}g\frac{u^2\sin^2\alpha}{g}$$

$$= \frac{u^2\sin^2\alpha}{g} - \frac{u^2\sin^2\alpha}{2g} = \frac{u^2\sin^2\alpha}{2g}.$$

The range $R (= OA)$ is the horizontal distance covered by the particle in the time of flight.

$$R = u\cos\alpha \times \frac{2u\sin\alpha}{g} = \frac{u^2\sin 2\alpha}{g}.$$

EXAMPLE 2. *The equation of motion under certain conditions are*

$$m\frac{d^2x}{dt^2} + eh\frac{dy}{dt} = eE \qquad \dots (1)$$

$$m\frac{d^2y}{dt^2} - eh\frac{dx}{dt} = 0 \qquad \dots (2)$$

with condition $x = \dfrac{dx}{dt} = y = \dfrac{dy}{dt} = 0$, *when* $t = 0$, *find the path of electron.*

SOLUTION. Multiplying (2) by k and adding to (1), we get

$$m\frac{d^2x}{dt^2} + mk\frac{d^2y}{dt^2} + eh\frac{dy}{dt} - ehk\frac{dx}{dt} = eE$$

$$\Rightarrow m\frac{d^2}{dt^2}(x + ky) - ehk\frac{d}{dt}\left(-\frac{y}{k} + x\right) = e.E \qquad \dots (3)$$

Let us choose k such that $x + ky = x - \dfrac{y}{x}$.

$$k = -\frac{1}{k} \Rightarrow k^2 = -1 \Rightarrow k = \pm i$$

Putting $x + ky = u$ in (3), we have

$$m\frac{d^2u}{dt^2} - ehk\frac{du}{dt} = eE$$

$$\Rightarrow \frac{d^2u}{dt^2} - \omega k\frac{du}{dt} = \frac{eE}{m} \qquad \left(\omega = \frac{eh}{m}\right)$$

$$\Rightarrow D^2u - \omega k\,Du = \frac{eE}{m}$$

A.E. is $\quad m^2 - \omega km = 0$

$$m(m - \omega k) = 0$$

$$m = 0 \text{ or } m = \omega k$$

C.F. $= C_1 + C_2 e^{\omega kt}$

P.I. $= \dfrac{1}{D^2 - \omega kD}\dfrac{eE}{m} = \dfrac{eE}{m}\dfrac{1}{D^2 - \omega kD}e^{0t}$

$$= \frac{eEt}{m}\frac{1}{2D - \omega k}e^{0t} = \frac{\omega Et}{h}\frac{1}{0 - \omega k} = \frac{-Et}{hk}$$

$$\left(\omega = \frac{eh}{m}\right)$$

The complete solution is

$$v = C_1 + C_2 e^{\omega kt} - \frac{Et}{hk}$$

or $\quad x + ky = C_1 + C_2 e^{\omega kt} - \dfrac{Et}{hk}$. $\qquad \dots (4)$

Putting the value of k, *i.e.*, $i, -i$ in (4), we get

$$x + iy = C_1 + C_2 e^{i\omega t} - \frac{Et}{ih} \qquad \dots (5)$$

$$x - iy = C_3 + C_4 e^{-i\omega t} + \frac{Et}{ih} \qquad \dots (6)$$

Differentiating (5) and (6), we get

$$\frac{dx}{dt} + i\frac{dy}{dt} = C_2 i\omega e^{i\omega t} + \frac{iE}{h} \qquad \dots (7)$$

$$\frac{dx}{dt} - i\frac{dy}{dt} = -i\omega C_4 e^{-i\omega t} - \frac{iE}{h} \qquad \dots (8)$$

Initial conditions, $x = y = \dfrac{dx}{dt} = \dfrac{dy}{dt} = 0$, when $t = 0$.

Putting these values in (5), (6), (7) and (8), we get

$$0 = C_1 + C_2 \Rightarrow C_2 = -C_1$$

$$0 = C_3 + C_4 \Rightarrow C_4 = -C_3$$

$$0 = i\omega C_2 + \frac{iE}{h} \Rightarrow C_2 = -\frac{E}{\omega h}$$

$$0 = -i\omega C_4 - \frac{iE}{h} \Rightarrow C_4 = \frac{-E}{\omega h}$$

On substituting the values of C_1, C_2, C_3 and C_4 in (5) and (6), we get

$$x + iy = \frac{E}{\omega h} - \frac{E}{h\omega} e^{i\omega t} + i\frac{Et}{h} \qquad \dots (9)$$

$$x - iy = \frac{E}{\omega h} - \frac{E}{\omega h} e^{-i\omega t} - i\frac{Et}{h} \qquad \dots (10)$$

On adding (9) and (10), we get

$$2x = \frac{2E}{h\omega} - \frac{E}{h\omega}(e^{i\omega t} + e^{-i\omega t})$$

$$x = \frac{E}{h\omega} - \frac{E}{h\omega}\cos\omega t$$

$$x = \frac{E}{h\omega}(1 - \cos\omega t)$$

Substituting (10) from (9), we obtain

$$2iy = -\frac{E}{h\omega}(e^{i\omega t} - e^{-i\omega t}) + \frac{2iEt}{h}$$

$$y = \frac{-E}{\omega h}\left(\frac{e^{i\omega t} - e^{-i\omega t}}{2i}\right) + \frac{Et}{h}$$

$$= -\frac{E}{\omega h}\sin\omega t + \frac{Et}{h}$$

$$y = \frac{E}{h\omega}(\omega t - \sin\omega t).$$

EXAMPLE 3. *Two particles each of mass m gm are suspended from two strings of same stiffness k. After the system comes to rest, the lower mass is pulled l cm downwards and released. Discuss their motion.*

SOLUTION. Let the displacements of the upper and lower masses at time t from their respective positions of equilibrium be denoted by x and y.

The stretch of the upper string is x and the stretch of the lower spring is $(y - x)$.

Thus the restoring force acting on the upper mass

$$= -kx + k(y - x)$$
$$= k(y - 2x)$$

and the restoring force acting on the lower mass $= -k(y - x)$.

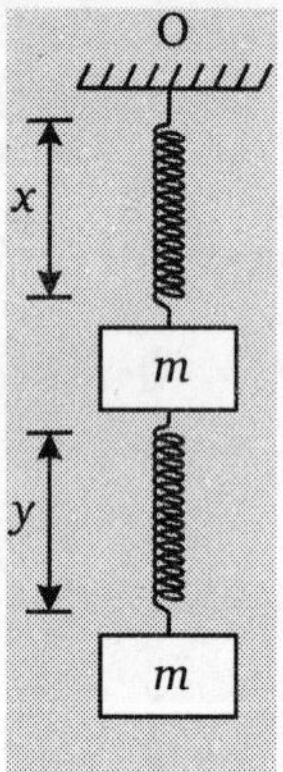

Thus the equations of motion are

$$m\frac{d^2x}{dt^2} = k(y - 2x)$$

and $\quad m\frac{d^2y}{dt^2} = -k(y - x)$.

$$(mD^2 + 2k)x - ky = 0 \qquad \dots (1)$$

and $\qquad (mD^2 + k)y - kx = 0 \qquad \dots (2)$

Operating (1) by $(mD^2 + k)$ and multiplying (2) by k and then adding, we get

$$[(mD^2 + k)(mD^2 + 2k) - k^2]x = 0$$

$$(D^4 + 3hD^2 + h^2)x = 0, \text{ where } h^2 = k/m.$$

The auxiliary equation is $m^4 + 3hm^2 + h^2 = 0$

$$m^2 = \frac{-3h \pm \sqrt{9h^2 - 4h^2}}{2} = -2.62h$$

or $-0.38h = -a^2, -b^2 \qquad\qquad\qquad$ (say)

Fig. 28

and so $\quad m = \pm ia, \pm ib$.

Thus, we have

$$x = C_1\cos at + C_2\sin at + C_3\cos bt + C_4\sin bt \quad \dots (3)$$

From (1), we have

$$y = \left(\frac{D^2}{h} + 2\right)x$$

$$= \left(2 - \frac{a^2}{h}\right)(C_1\cos at + C_2\sin at)$$

$$+ (2 - b^2/h)(C_3\cos bt + C_4\sin bt) \quad \dots (4)$$

Initially, we have $t = 0$, $x = y = l$

and $\quad \dfrac{dx}{dt} = \dfrac{dy}{dt} = 0$.

Therefore, from (3), we have

$$l = C_1 + C_3; \quad 0 = aC_2 + bC_4 \text{ and from (4)},$$

We have

$$l = (2 - a^2/h)C_1 + (2 - b^2/h)C_3$$

$$0 = (2 - a^2/h)C_2 + (2 - b^2/h)bC_4$$

Thus, $\quad C_1 = \dfrac{l(h - b^2)}{a^2 - b^2}$,

$$C_2 = \frac{l(h - a^2)}{b^2 - a^2}, \quad C_3 = C_4 = 0$$

Putting these values of C_1, C_2, C_3 and C_4 in (3) and (4), we get x and y. From the values of x and y, it is clear that the motion of the given system of springs in a combination of springs is a combination of two simple harmonic motion of periods $2\pi/a$ and $2\pi/b$.

EXAMPLE 4. *Assuming that a spherical rain drop evaporates at rate proportional to its surface area and if its radius originally is 3 mm and one hour later has been reduced to 2 mm, find an expression for the radius of the rain drop at any time.*

SOLUTION. Evaporation $\propto$ surface area

$$\frac{dV}{dt} \propto S$$

$\Rightarrow \qquad \dfrac{dV}{dt} = kS \qquad\qquad\qquad \dots (1)$

$$V = \frac{4}{3}\pi r^3 \qquad\qquad\qquad \dots (2)$$

$$\frac{dV}{dt} = 4\pi r^2 \frac{dr}{dt}$$

Putting the value of $\dfrac{dV}{dt}$ from (1) in (2), we have

$$4\pi r^2 \frac{dr}{dt} = kS$$

$$S\frac{dr}{dt} = kS \quad \text{or} \quad \frac{dr}{dt} = k.$$

$$r = kt + C \qquad\qquad\qquad \dots (3)$$

Putting $t = 0$, $r = 3$ in (3), we get $3 = C$.

(3) becomes $\qquad r = kt + 3 \qquad\qquad \dots (4)$

Putting $t = 1$ and $r = 2$ in (4),

we get $\qquad\qquad 2 = k + 3 \Rightarrow k = -1$

(4) becomes

$$r = -t + 3.$$

Exercise-8.39

1. A particle moving in a plane is subjected to a force directed towards a fixed point O and proportional to the distance of the particle from O. Show that the differential equations of motion are of the form

$$\frac{d^2x}{dt^2} = -K^2x, \quad \frac{d^2y}{dt^2} = -K^2y.$$

   Find the cartesian equation of path of the particle if $x = 1,\ y = 0,\ \frac{dx}{dt} = 0$ and $\frac{dy}{dt} = 2$, when $t = 0$.

2. A particle of unit mass is projected with velocity u at an inclination a above the horizon in a medium whose resistances is k times the velocity. Show that its direction will again make an angle a with horizontal after a time

$$\frac{1}{k}\log\left(1 + \frac{2ku}{g}\sin\alpha\right).$$

3. In a chemical transformation of certain substance following equation occur :

$$\frac{dx}{dt} = ax = 0, \quad \frac{dz}{dt} = by, \quad x + y + z = c$$

   where a, b, c are constants. Obtain a differential equation for z.

   Hence, prove that if $z = -\dfrac{dz}{dt} = 0$ when $t = 0$, then

$$z = c + \frac{c}{a-b}(be^{-at} - ae^{-bt})$$

4. Two particles each of mass m gm are suspended from two springs of same stiffness k. After the system comes to rest, the lower mass is pulled l cm downwards and released. Discuss their motion.

 Answers

1. $4x^2 + k^2y^2 = 4$

4. $x = \dfrac{l(\lambda - \beta^2)}{\alpha^2 - \beta^2}\cos\alpha t + \dfrac{l(\lambda - \alpha^2)}{\beta^2 - \alpha^2}\sin\alpha t$

8.7 SERIES SOLUTION OF SECOND ORDER DIFFERENTIAL EQUATIONS

We have observed that the linear homogeneous differential equation with constant coefficient can be solved by some algebraic methods, which give the solution as elementary function. If we consider the differential equations having variable coefficients, we get a more complicated situation and the solutions thus obtained may be non-elementary functions. Therefore, Legendre's, Bessel's and the hypergeometric differential equations and their solutions play an important role in the field of pure and applied mathematics and in engineering mathematics as well. We shall now discuss in this chapter a method for solving such type of differential equation of order two having their solutions in the forms of power series, and the method is therefore called a power series method. Besides the solutions, we shall also discuss some properties of the solutions.

8.7.1 POWER SERIES METHOD

This method is very effective with respect to linear homogeneous differential equation with variable coefficients. This method gives the solution of the differential equations in the form of a power series. Therefore, an infinite series of the form

$$\sum_{m=0}^{\infty} a_m x^m = a_0 + a_1 x + a_2 x^2 + \dots + a_m x^m + \dots$$

is called a power series. This power series is said to be convergent at a point x if $\lim\limits_{n \to \infty} \sum\limits_{m=0}^{n} a_m x^m$ exists. It is clear that the above series is always convergent at $x = 0$. To explain this method clear, let us consider a general homogeneous differential equation of second order $y'' + P(x)y' + Q(x)y = 0$.

The solution y of this given differential equation is assumed in the form of a power series as above with undetermined coefficient and these coefficients are determined by putting that series and the series for derivatives of y into the given differential equation.

(1) ORDINARY AND SINGULAR POINTS

Consider a general homogeneous linear differential equation of order two :

$$\frac{d^2y}{dx^2} + P(x)\frac{dy}{dx} + Qxy = 0 \qquad \text{or} \qquad y'' + P(x)y' + Q(x)y = 0. \qquad \dots(1)$$

The main concept about the solution of (1) is that the behaviour of the solutions near a point $x = x_0$ depends on the behaviour of $P(x)$ and $Q(x)$ near this point x_0. If $P(x)$ and $Q(x)$ are analytic at this point x_0, then power series method is applicable in some neighbourhood of x_0. Then this point x_0 is called an ordinary point of the differential equation (1). Thus we can say that every solution of (1) is analytic at x_0. If x_0 is not an ordinary point, then this point x_0 is called a singular point.

(2) REGULAR SINGULAR POINTS

In the above section, we have seen that if one of the coefficient functions $P(x)$ and $Q(x)$ is not differentiable at x_0 then this point is called a singular point. Thus a point x_0 of the differential equation (1) is called regular if the functions $(x - x_0)P(x)$ and $(x - x_0)^2 Q(x)$ are analytic at $x = x_0$.

If a singular point x_0 is located at the origin, then the general form of an analytic function at $x = x_0$ is $\sum\limits_{m=0}^{\infty} a_m x^m$.

This implies that the origin will definitely be a singular point of (1) of $P(x)$ and $Q(x)$ have at least one of the coefficients with negative subscripts non-zero. In this case we assume the solution of the differential equation (1) of the form

$$y = x^n \sum_{m=0}^{\infty} a_m x^m = \sum_{m=0}^{\infty} a_m x^{m+n}$$

where n may be a negative integer or may be a fraction or even an irrational number.

REMARKS

- The solution of the form $y = \sum\limits_{m=0}^{\infty} a_m x^m$ must be valid at least on the interval $|x| < 1$.

- The power series of the type $y = \sum\limits_{m=0}^{\infty} a_m x^{m+n}$ is known as "quasi power series".

8.7.2 POWER SERIES SOLUTION

(1) SOLUTION NEAR AN ORDINARY POINT

Consider the differential equation $\dfrac{d^2 y}{dx^2} + P(x)\dfrac{dy}{dx} + Q(x)y = 0$ ...(1)

Let us take a trial solution of the form $\quad y = \sum\limits_{n=0}^{\infty} C_n x^n$ ...(2)

$$\Rightarrow \qquad \left. \frac{dy}{dx} = \sum_{n=0}^{\infty} n\, C_n\, x^{n-1} \quad \text{and} \quad \frac{d^2 y}{dx^2} = \sum_{n=0}^{\infty} n(n-1)\, C_n\, x^{n-2} \right\}$$ ...(3)

Also, by letting $P(x)$ and $Q(x)$ are not polynomial in x, we can expand such that

$$P(x) = \sum_{n=1}^{\infty} p_n x^n \qquad \text{and} \qquad Q(x) = \sum_{n=0}^{\infty} q_n . x^n$$ ...(4)

Now, putting all these values in equation (1), we get the required solution.

(2) SOLUTION NEAR A REGULAR SINGULAR POINT

Here, we assume a trial series solution of the type

$$y = x^m (C_0 + C_1 x + C_2 x^2 + ...)$$ ...(1)

$$= x^m . \sum_{n=0}^{\infty} C_n x^n, \text{ where all } C_n\text{'s constant with } C_n \neq 0.$$

WORKING PROCEDURE

To find the values of m and C's, we proceed as follows :

Step 1. Put the values of $\dfrac{dy}{dx}$ and $\dfrac{d^2 y}{dx^2}$ in the given differential equation.

Step 2. By equating to zero the coefficients of the lowest power of x, get a quadratic equation in m, which is called indicial equation.

Step 3. To find the values of the equations $C_1, C_2, ...$, etc. in terms of C_0 equating to zero the coefficients of other powers of x.

Step 4. The nature of the root can be determined as follows :

(A) If roots of the indicial equation are equal :
Let $m = n$ be two equal roots. Then putting $m = n$ in y and in $\dfrac{\partial y}{\partial m}$ we may get the two independent solutions.

(B) If roots of the indicial equation unequal and not differing by an integer
If the indicial equation has two unequal roots $m = m_1$ and m_2 which do not differ by an integer, then by putting $m = m_1$ and m_2 in the series, we get two independent solutions.

(C) If the roots of the indicial equation differing by an integer and making the coefficients of some powers of x in the series for y infinity :
Let $m = m_1$ and m_2 be two roots of the indicial equation which differ by an integer and some of the coefficients of powers of k in the series for y infinity for $m = m_2$.
Here, put $C(m \quad m)$ for C_0, then we get two independent solutions for $m = m_2$. Then proceed as in case 1.

(D) If the roots of the indicial equation differing by an integer and making the coefficients of the series for y indeterminate :
If $m = m_1$ and $m_2 (m_1 > m_2)$ are two roots of the indicial equation which differ by an integer. If one of the coefficients of the series for y becomes indeterminate when $m = m_2$, the complete solution is given by putting $m = m_2$ in y, which have two arbitrary constants.

- The second solution always consists of a numerical multiple of the product of the first solution and log x added to another series.
- The series obtained by putting $m = m_1$ in y is merely a numerical multiple of one of the series contained in the first solution.

Solved Examples

EXAMPLE 1. *Solve* $x\dfrac{d^2y}{dx^2} + \dfrac{dy}{dx} + xy = 0$.

(UPTU-2003, VTU-2010, SVTU-2007)

SOLUTION. The given equation is

$$x\frac{d^2y}{dx^2} + \frac{dy}{dx} + xy = 0 \qquad \ldots(1)$$

Putting $y = x^m$ in the LHS of (1), we get

$$xm(m-1)x^{m-2} + mx^{m-1} + x.x^m$$

$$= x^{m+1} + m^2x^{m-1}.$$

Clearly, the common difference of the powers is $(m+1) - (m-1)$, *i.e.*, 2.

Let $y = \sum\limits_{r=0}^{\infty} C_r x^{m+2r}$

$$= C_0 x^m + C_1 x^{m+2} + C_2 x^{m+4} + \ldots \qquad \ldots(2)$$

is the solution of (1).

Then, we have

$$\frac{dy}{dx} = \sum_{r=0}^{\infty} C_r(m+2r)\, x^{m+2r-1}$$

$$\frac{d^2y}{dx^2} = \sum_{r=0}^{\infty} C_r(m+2r)(m+2r-1)^{m+2r-2}.$$

Put all these values in (1), we get

$$\sum_{r=0}^{\infty} C_r[(m+2r)(m+2r-1)x^{m+2r-1}$$

$$+ (m+2r)x^{m+2r-1} + x^{m+2r+1}] = 0$$

$$\Rightarrow \sum_{r=0}^{\infty} C_r[x^{m+2r+1} + (m+2r)^2\, x^{m+2r-1}] = 0.$$

Equating to zero, the coefficient of the lowest power of x, *i.e.*, of x^{m-1}, we have

$$C_0 m^2 = 0$$

which is the required indicial equation.

Since $C_0 \ne 0$, therefore $m = 0$, 0 are two equal roots.

Now, equating to zero the coefficient of the general term, *i.e.*, of x^{m+2p+1}, we get

$$C_p + (m+2p+2)^2 C_{p+1} = 0$$

$$\Rightarrow C_{p+1} = -\frac{1}{(m+2p+2)^2} C_p \qquad \ldots(3)$$

Putting $p = 0, 1, 2, \ldots$ in (3), we get

$$C_1 = -\frac{1}{(m+2)^2} C_0,$$

$$C_2 = -\frac{1}{(m+4)^2} C_1$$

$$= (-1)^2 \frac{1}{(m+2)^2(m+4)^2} C_0$$

$$C_3 = -\frac{1}{(m+6)^2} C_2$$

$$= (-1)^3 \frac{1}{(m+2)^2(m+4)^2(m+6)^2} C_0 \ldots$$

and so on.

Put all these values in (2), we get

$$y = C_0 x^m \left[1 - \frac{x^2}{(m+2)^2} + \frac{x^4}{(m+2)^2(m+4)^2} - \ldots \right]$$

$$\ldots(4)$$

Putting $m = 0$, we get

$$y = C_0 \left[1 - \frac{x^2}{2^2} + \frac{x^4}{2^2.4^2} - \frac{x^6}{2^2.4^2.6^2} \right] + \ldots \qquad \ldots(5)$$

$$= C_0.u \text{ (say)},$$

which is the first solution of the given equation (1).

when $u = 1 - \dfrac{x^2}{2^2} + \dfrac{x^2}{2^2.4^2} - \dfrac{x^6}{2^2.4^2.6^2} + \ldots$

Since, there are two equal values of m, therefore, second solution be obtained from (4).

Now, from (4)

$$\frac{dy}{dx} = C_0 \left[mx^{m-1} - \frac{(m+2)x^{m+1}}{(m+2)^2} \right.$$

$$\left. + \frac{(m+4)x^{m+3}}{(m+2)^2(m+4)^2} - \ldots \right]$$

$$\Rightarrow \frac{d^2y}{dx^2} = C_0 \left[m(m-1)x^{m-2} \right.$$

$$- \frac{(m+2)(m+1)}{(m+2)^2} x^m$$

$$\left. + \frac{(m+4)(m+3)x^{m+2}}{(m+2)^2(m+4)^2} - \ldots \right].$$

Put above two values in (1), we get

$$\text{LHS} = xC_0 \left[m(m-1)x^{m-2} \right.$$

$$- \frac{(m+2)(m+1)}{(m+2)^2} x^m$$

$$\left. + \frac{(m+4)(m+3)x^{m+2}}{(m+2)^2(m+4)^2} - \ldots \right]$$

$$+ C_0 \left[mx^{m-1} - \frac{(m+2)x^{m+1}}{(m+2)^2} \right.$$

$$\left. + \frac{(m+4)x^{m+3}}{(m+2)^2(m+4)^2} - \ldots \right]$$

$$+ xC_0 \left[x^m - \frac{x^{m+2}}{(m+2)^2} \right.$$

$$\left. + \frac{x^{m+4}}{(m+2)^2(m+4)^2} - \ldots \right]$$

$$= C_0 m^2 x^{m-1}$$

$$\therefore \left[x\frac{d^2}{dx^2} + \frac{d}{dx} + x \right] y = C_0 m^2 x^{m-1}.$$

Differentiating both sides partially, w.r.t. m, we get

$$\frac{\partial}{\partial m}\left[x\frac{d^2}{dx^2}+\frac{d}{dx}+x\right]y=\frac{\partial}{\partial m}(C_0 m^2 x^{m-1})$$

$$\Rightarrow\left[x\frac{d^2}{dx^2}+\frac{d}{dx}+x\right]\left(\frac{\partial y}{\partial m}\right)$$

$$=C_0.2mx^{m-1}+C_0 m^2 x^{m-1}\log x$$

Putting $m=0$, we get

$$\left[x\frac{d^2}{dx^2}+\frac{d}{dx}+x\right]\left[\frac{\partial y}{\partial m}\right]=0$$

$$\Rightarrow\left(\frac{\partial y}{\partial m}\right)_{m=0}$$ satisfy the equation (1), therefore it is also

a solution of (1).

Differentiating (4) partially w.r.t. m, we get

$$\frac{\partial y}{\partial m}=C_0 x^m \log x\left[1-\frac{x^2}{(m+2)^2}\right.$$

$$\left.+\frac{x^4}{(m+2)^2(m+4)^2}-...\right]$$

$$+C_0 x^m\left[\frac{2x^2}{(m+2)^3}+\left\{\frac{-2}{(m+2)^3(m+4)^2}\right.\right.$$

$$\left.\left.+\frac{-2}{(m+2)^2(m+4)^3}\right\}x^4+...\right]$$

Putting $m=0$, we get

$$\left(\frac{\partial y}{\partial m}\right)_{m=0}=C_0\log x\left[1-\frac{x^2}{2^2}+\frac{x^4}{2^2.4^2}-...\right]$$

$$+C_0\left[\frac{x^2}{2^2}+\left\{\frac{-2}{2^3.4^2}+\frac{-2}{2^2.4^3}\right\}x^4...\right]$$

$$=bu\log x+b\left[\frac{x^2}{2^2}-\frac{3}{2^3.4^2}x^4+...\right]$$

$$=bv\,(\text{say})$$

where, $v=u\log x+\left[\frac{x^2}{2^2}-\frac{3}{2^3.4^2}x^4+...\right]$ and b is any

arbitrary constant which replaces C_0. Hence, the required general solution of (1) is given by

$$y=au+bv,$$ where a and b are arbitrary constants.

EXAMPLE 2. *Solve the following Legendre's equation*

$$(1-x^2)y''-2xy'+p(p+1)y=0$$

in descending powers of x.

(UPTU(Q. BANK)–2002, GBTU–2010)

SOLUTION. The given equation can be written as

$$(1-x^2)\frac{d^2y}{dx^2}-2x\frac{dy}{dx}+p(p+1)y=0 \qquad ...(1)$$

Putting $y=x^m$ in the LHS of (1), we get

$$(1-x^2)m(m-1)x^{m-2}-2x.mx^{m-1}+p(p+1)x^m$$

or $(-m^2-m+p^2+p)x^m+m(m-1)x^{m-2}$.

Clearly, the common difference of the powers is $m-(m-2)$, *i.e.*, 2.

Let the solution of (1) in descending powers of x be

$$y=C_0 x^m+C_1 x^{m-2}+C_2 x^{m-4}+...$$

$$=\sum_{r=0}^{\infty}C_r x^{m-2r}. \qquad ...(2)$$

$$\Rightarrow\quad\frac{dy}{dx}=\sum_{r=0}^{\infty}C_r(m-2r)x^{m-2r-1}$$

and $$\frac{d^2y}{dx^2}=\sum_{r=0}^{\infty}C_r(m-2r)(m-2r-1)x^{m-2r-2}$$

Put all these values in (1), we get

$$\sum_{r=0}^{\infty}C_r[(1-x^2)(m-2r)$$

$$(m-2r-1)x^{m-2r-2}-2x(m-2r)x^{m-2r-1}$$

$$+p(p+1)x^{m-2r}]=0$$

$$\Rightarrow\sum_{r=0}^{\infty}C_r[(-(m-2r)(m-2r-1)$$

$$-2(m-2r)+p(p+1)x^{m-2r}$$

$$+(m-2r)(m-2r-1)x^{m-2r-2}]=0$$

$$\Rightarrow\sum_{r=0}^{\infty}C_r[\{p^2-(m-2r)^2+(p-m+2r)\}x^{m-2r}$$

$$+(m-2r)(m-2r-1)x^{m-2r-2}]=0$$

$$\Rightarrow\sum_{r=0}^{\infty}C_r[(p-m+2r)(p-m-2r+1)x^{m-2r}$$

$$+(m-2r)(m-2r-1)x^{m-2r-2}]=0$$

Equating to zero, the coefficients of the highest power of x, *i.e.*, x^m, we get the initial equation as

$$C_0(p-m)(p+m+1)=0.$$

Since $C_0\neq0$, therefore, we get $m=p,-(p+1)$.

Now, equating to zero the coefficients of x^{m-2r}, we get

$$C_r(p-m+2r)(p+m-2r+1)$$

$$+(m-2r+2)(m-2r+1)C_{r-1}=0$$

$$\Rightarrow\quad C_r=\frac{(m-2r+2)(m-2r+1)}{(p-m+2r)(p+m-2r+1)}C_{r-1}.$$

Putting $r=1,2,...$, we get

$$C_1=-\frac{m(m-1)}{(p-m+2)(p-m-1)}C_0$$

$$C_2=-\frac{(m-2)(m-3)}{(p-m+4)(p+m-3)}C_1$$

$$=(-1)^2\frac{m(m-1)}{(p-m+2)(p-m+4)}C_0$$

$$\frac{(m-2)(m-3)}{(p+m-1)(p+m-3)}$$

$$...... \quad ... \quad ...\text{ and so on.}$$

Put all these values in (2), we get

$$y=C_0\left[x^m-\frac{m(m-1)}{(p-m+2)(p+m-1)}x^{m-2}\right.$$

$$\left.+\frac{m(m-1)}{\begin{array}{c}(m-2)(m-3)\\(p-m+2)(p-m+4)\end{array}}x^{m-4}-...\right]$$

$$(p+m-1)(p+m-3)$$

Now putting $m = p, -(p+1)$ successively, we get

$$y = C_0 \left[x^p - \frac{p(p-1)}{2(2p-1)} x^{p-2} \right.$$

$$\left. + \frac{p(p-1)(p-2)(p-3)}{2.4.(2p-1)(2p-3)} x^{p-4} - \ldots \right]$$

$$= au \text{ (say)}$$

which is one solution of the given equation.

Also, $y = C_0 \left[x^{-p-1} + \frac{(p+1)(p+2)}{2(2p+3)} x^{-p-3} \right.$

$$\left. + \frac{(p+1)(p+2)(p+3)(p+4)}{2.4(2p+3)(2p+5)} x^{-p-5} + \ldots \right]$$

$$= bv \text{ (say)}$$

Here, the required solution of the given equation is $y = au + bv$, where a and b are arbitrary constants.

EXAMPLE 3. *Solve the following differential equation in series* $2x^2 \dfrac{d^2y}{dx^2} - x\dfrac{dy}{dx} + (x-5)y = 0$.

(UPTU–2008)

SOLUTION. We have

$$2x^2 \frac{d^2y}{dx^2} - x\frac{dy}{dx} + (x-5)y = 0 \qquad \ldots(1)$$

Here $xP(x)$ and $x^2P_2(x)$ are analytic (not infinite at $x = 0$). So x_0 is regular singular point. We assume the solution in the form :

$$y = \sum_{k=0}^{\infty} a_k x^{m+k}, \frac{dy}{dx} = \sum a_k(m+k)x^{m+k-1}$$

$$\frac{d^2y}{dx^2} = \sum a_k(m+k)(m+k-1)x^{m+k-2}$$

Putting the values of $y, \dfrac{dy}{dx}$ and $\dfrac{d^2y}{dx^2}$ in equation (1), we get

$$2x^2 \sum a^k(m+k)(m+k-1)x^{m+k-2}$$

$$- x\sum a_k(m+k)x^{m+k-1} + (x-5)\sum a_k x^{m+k} = 0$$

$$\Rightarrow \sum a_k[2(m+k)(m+k-1)-(m+k)-5]x^{m+k}$$

$$+ \sum a_k x^{m+k+1} = 0$$

$$\sum a_k[2(m+k)^2 - 3(m+k) - 5]x^{m+k}$$

$$+ \sum a_k x^{m+k+1} = 0$$

Equating the coefficient of the lowest degree term x^n to zero by putting $k = 0$ in first summation of (2), we get

$$a_0(m+1)(2m-5) = 0 \Rightarrow (m+1) = 0$$

or $\qquad\qquad 2m - 5 = 0$

$$\Rightarrow \qquad\qquad m = -1 \text{ or } \qquad m = 5/2$$

Now, equating the coefficient of next lower degree term x^{m+1} to zero by putting $k = 1$ in first summation and $k = 0$ in second summation, we get

$$a_1(m+2)(2m-3) + a_0 = 0$$

$$\Rightarrow \qquad a_1(m+2)(2m-3) = a_0$$

$$a_1 = \frac{a_0}{(m+2)(2m-3)}$$

Now, equating the coefficient of x^{m+k+1} to zero by putting $k = k+1$ in first summation

$$a_{k+1}(m+k+2)(2m+2k-3) + a_k = 0$$

$$a_{k+1} = \frac{-a_k}{(m+k+2)(2m+2k-3)}$$

If $k = 0, a_1 = \dfrac{-a_0}{(m+2)(2m-3)}$

If $k = 1$,

$$a_2 = \frac{-a_1}{(m+3)(2m-1)} = \frac{a_0}{(m+2)(m+3)(2m-1)(2m-3)}$$

If $k = 2$,

$$a_3 = \frac{-a_2}{(m+4)(2m+1)} = -\frac{a_0}{(m+2)(m+3)(m+4)}$$
$$(2m+1)(2m-1)(2m-3)$$

$m = -1$	$m = 5/2$
$a_1 = a_0/5$	$a_1 = -a_0/9$
$a_2 = a_0/30$	$a_2 = a_0/198$
$a_3 = -a_0/90$	$a_3 = -a_0/7722$

We have, $y = x^m(a_0 + a_1 x + a_2 x^2 + a_3 x^3 + \ldots)$

For $m = -1$, $y_1 = x^{-1}\left(a_0 + \dfrac{a_0}{5}x + \dfrac{a_0}{30}x^2 - \dfrac{a_0}{90}x^3 + \ldots\right)$

$$y_1 = a_0 x^{-1}\left(1 + \frac{x}{5} + \frac{x^2}{30} - \frac{x^3}{90} + \ldots\right)$$

For $m = 5/2$,

$$y_2 = x^{5/2}\left(a_0 - \frac{a_0}{9}x + \frac{a_0}{198}x^2 - \frac{a_0}{7722}x^2 + \ldots\right)$$

$$\Rightarrow \quad y_2 = a_0 x^{5/2}\left(1 - \frac{x}{9} + \frac{x^2}{198} - \frac{x^3}{7722} + \ldots\right)$$

Since two solutions are linearly independent, so the general solution of (1) may be represented as $y = Ay_1 + By_2$.

EXAMPLE 4. *Solve* $(1-x^2)\dfrac{d^2y}{dx^2} + 2x\dfrac{dy}{dx} + y = 0$.

(UPTU (Q. BANK) 2002, AMIETE-2000)

SOLUTION. The given equation is

$$(1-x^2)\frac{d^2y}{dx^2} + 2x\frac{dy}{dx} + y = 0 \qquad \ldots(1)$$

Let us assume the series solution of the equation (1) is

$$y = \sum_{r=0}^{\infty} C_r x^{m+r}$$

$$\therefore \quad \frac{dy}{dt} = \sum_{r=0}^{\infty} C_r(m+r)x^{m+r-1}$$

and $\dfrac{d^2y}{dx^2} = \sum_{r=0}^{\infty} C_r(m+r)(m+r-1)x^{m+r-2}$.

Putting all these values in (1), we get

$$\sum_{r=0}^{\infty} C_r[(1-x^2)(m+r)(m+r-1)x^{m+r-2}$$

$$+ 2x(m+r)x^{m+r-1} + x^{m+r}] = 0$$

$$\Rightarrow \quad \sum_{r=0}^{\infty} C_r[-\{(m+r)(m+r-3)-1\}x^{m+r}$$

$$+ (m+r)(m+r-1)x^{m+r-2}] = 0.$$

Now, equating to zero the coefficients of the lower power of x, i.e., of x^{m-2}, we get the indicial equation is

$$C_0 m(m-1) = 0$$

Now, $C_0 \neq 0 \Rightarrow m = 0, 1$

$\Rightarrow$ Roots are real and unequal differing by an integer.
Now equating to zero the coefficients of the next higher power of x, *i.e.*, of x^{m-1}, we get
$$C_1(m+1)m = 0 .$$

If $m = 0$, we get $C_1 = 0$.
Now, equating to zero the coefficients of the general term, *i.e.*, of x^{m+p}, we get
$$C_{p+2} = \frac{(m+p)(m+p-3)-1}{(m+p+1)(m+p+2)} C_p .$$

Putting $p = 0, 1, 2, \dots$, successively, we get

$$C_2 = \frac{m(m-3)-1}{(m+1)(m+2)} C_0 .$$

$$C_3 = \frac{(m+1)(m-2)-1}{(m+2)(m+3)} C_1 ,$$

$$C_4 = \frac{(m+2)(m-1)-1}{(m+3)(m+4)} C_2$$
$$= \frac{\{(m+2)(m-1)-1\}\{m(m-3)-1\}}{(m+1)(m+2)(m+3)(m+4)} C_0 ,$$

$$C_5 = \frac{(m+3)m-1}{(m+4)(m+5)} C_3$$
$$= \frac{\{m(m+3)-1\}\{(m+1)(m-2)-1\}}{(m+2)(m+3)(m+4)(m+5)} C_1$$

... ... and so on.
Put all these values in (2), we get

$$y = C_0 x^m \left[1 + \frac{m(m-3)-1}{(m+1)(m+2)} x^2 \right.$$
$$+ \frac{\{(m+2)(m-1)-1\}\{m(m-3)-1\}}{(m+1)(m+2)(m+3)(m+4)} x^4 + \dots \right]$$
$$+ C_1 x^m \left[x + \frac{(m+1)(m-2)-1}{(m+2)(m+3)} x^3 \right.$$
$$+ \left. \frac{\{m(m+3)-1\}\{(m+1)(m-2)-1\}}{(m+2)(m+3)(m+4)(m+5)} x^5 + \dots \right].$$

Putting $m = 0$, we get

$$y = C_0 \left[1 - \frac{1}{2}x^2 + \frac{1}{8}x^4 + \dots \right] + C_1 \left[x - \frac{1}{2}x^3 + \frac{1}{40}x^5 + \dots \right]$$

$$= a \left[1 - \frac{x^2}{2} + \frac{x^4}{8} + \dots \right] + b \left[x - \frac{x^3}{2} + \frac{x^5}{40} + \dots \right]$$

where a and b are arbitrary constants.
This is the required general solution of the given equation.

EXAMPLE 5. *Solved* $9x(1-x)\dfrac{d^2y}{dx^2} - 12\dfrac{dy}{dx} + 4y = 0$.

(ROORKEE-2000, MADRAS-2006)

SOLUTION. Here, the given equation is

$$9x(1-x)\frac{d^2y}{dx^2} - 12\frac{dy}{dx} + 4y = 0 \qquad \dots(1)$$

Now, putting $y = x^m$ on the LHS of (1), we get
$$\{-9m(m-1)+4\} x^m + \{9m(m-1)-12m\} x^{m-1} .$$
Clearly, the common difference of the powers is one.

Let us assume $y = \sum\limits_{r=0}^{\infty} C_r x^{m+r}$ be the solution of (1).

Therefore, $\dfrac{dy}{dx} = \sum\limits_{r=0}^{\infty} C_r (m+r) x^{m+r-1}$

and $\qquad \dfrac{d^2y}{dx^2} = \sum\limits_{r=0}^{\infty} C_r (m+r)(m+r-1) x^{m+r-2}$.

Putting all these values in (1) and after simplification, we get

$$\sum\limits_{r=0}^{\infty} C_r [(3m+3r-4)(3m+3r+1) x^{m+r}$$
$$- 3(m+r)(3m+3r-7)x^{m+r-1}]. \qquad \dots(2)$$

Equating to zero the coefficient of lower power of x, *i.e.*, of x^{m-1}, we have $-3C_0 m(3m-7) = 0$.
Since $C_0 \neq 0$. Therefore, $m = 0, 7/3$.
Now equating to zero the coefficient of general term, *i.e.* x^{m+p}, we have

$$C_p [(3m+3p-4)(3m+3p+1)$$
$$- 3(m+p+1)(3m+3p+3-7)]C_{p+1} = 0$$

$$\Rightarrow \quad C_{p+1} = \frac{(3m+3p+1)}{3(m+p+1)} C_p .$$

Putting $p = 0, 1, 2, \dots$ successively, we have

$$C_1 = \frac{(3m+1)}{3(m+1)} C_0$$

$$C_2 = \frac{(3m+4)}{3(m+2)} C_1 = \frac{(3m+4)(3m+1)}{3^2(m+2)(m+1)} C_0$$

$$C_3 = \frac{(3m+7)}{3(m+3)} C_2$$
$$= \frac{(3m+7)(3m+4)(3m+1)}{3(m+3)(m+2)(m+1)} C_0 \dots$$

and so on.
Therefore, we have

$$y = \sum\limits_{r=0}^{\infty} C_r x^{m+r} = C_0 x^m + C_1 x^{m+1} + C_2 x^{m+2} + \dots$$

$$= C_0 x^m \left[1 + \frac{3m+1}{3(m+1)} x + \frac{(3m+1)(3m+4)}{3^2(m+1)(m+2)} x^2 \right.$$

$$+ \left. \frac{(3m+1)(3m+4)(3m+7)}{3^3(m+1)(m+2)(m+3)} x^3 + \dots \right].$$

If $m = 0$, then taking $C_0 = a$.

$$\therefore \quad y = a \left[1 + \frac{1}{3}x + \frac{1.4}{3.6}x^2 + \frac{1.4.7}{3.6.9}x^3 + \dots \right]$$

$$= au \text{ (say)}$$

which is one solution of the given equation.
Also, if $m = \dfrac{7}{3}$, taking $C_0 = b$, we get

$$y = bx^{7/3} \left[1 + \frac{8}{10}x + \frac{8.11}{10.13}x^2 + \frac{8.11.14}{10.13.16}x^3 + \dots \right]$$

$$= bv \text{ (say)}.$$

Hence, the complete solution of the given equation is
$y = au + bv$, where a and b are arbitrary constants.

(3) METHOD OF DIFFERENTIATION

EXAMPLE 6. **Solve** $(1+x^2)\dfrac{d^2y}{dx^2} + 2x\dfrac{dy}{dx} = 0$,

where $x = 0, \dfrac{dy}{dx} = 1, y = 0$.

SOLUTION. The given equation can be written as

$$(1+x^2)y_2 + 2xy_1 = 0 \qquad \ldots(1)$$

Differentiating the equation (1) n times by Leibnitz's theorem, we have

$$(1+x^2)y_{n+2} + 2(n+1)xy_{n+1} + n(n+1)y_n = 0$$

when $x = 0$, we have

$$(y_{n+2})_0 = -n(n+1)(y_n)_0. \qquad \ldots(2)$$

Using the given equation, when $x = 0$, we have

$$(y_2)_0 = 0.$$

Putting $n = 2, 4, \ldots$ in (2), we have

$$(y_2)_0 = (y_4)_0 = \ldots = 0.$$

Again putting $n = 1, 2, \ldots,$ in (2), we have

$$(y_3)_0 = -1.2.(y_1)_0 = -2!$$

$$(y_5)_0 = -3.4(y_3)_0 = -4!$$

$$(y_7)_0 = -5.6(y_5)_0 = -6!$$

$$\ldots \quad \ldots \quad \ldots \quad \text{etc.}$$

Hence, $\quad y = (y)_0 + (y_1)_0 x + (y_2)_0\dfrac{x^2}{2!} + \ldots$

$$= x - 2!\dfrac{x^3}{3!} + 4!\dfrac{x^5}{5!} - 6!\dfrac{x^7}{7!}\ldots$$

$$= x - \dfrac{x^3}{3} + \dfrac{x^5}{5} - \dfrac{x^7}{7} + \ldots$$

Exercise-8.40

1. Solve $\dfrac{d^2y}{dx^2} - 2x^2\dfrac{dy}{dx} + 4xy = x^2 + 2x + 2$ in powers of x.

2. Solve $x\dfrac{d^2y}{dx^2} + \dfrac{dy}{dx} + xy = 0$.

3. Solve $x\dfrac{d^2y}{dx^2} + (1+x)\dfrac{dy}{dx} + 2y = 0$.

4. Transform the equation $\dfrac{d^2y}{dx^2} - y = 0$ by the substitution $x = \dfrac{1}{z}$ and show that it has no integrals that are regular in descending powers of x.

5. Solve completely in series the equation

$$x\dfrac{d^2y}{dx^2} + (x+n)\dfrac{dy}{dx} + (n+1)y = 0$$

6. Solve $(2x + x^3)\dfrac{d^2y}{dx^2} - \dfrac{dy}{dx} - 6xy = 0$. (BHOPAL-2006)

7. Solve $2x^2\dfrac{d^2y}{dx^2} - x\dfrac{dy}{dx} + (1-x^2)y = x^2$.

8. Solve $(x - x^2)\dfrac{d^2y}{dx^2} + (1-x)\dfrac{dy}{dx} - y = 0$.

9. Solve $x^4\dfrac{d^2y}{dx^2} + x\dfrac{dy}{dx} + y = \dfrac{1}{x}$.

10. Solve $\dfrac{d^2y}{dx^2} - y = x$.

11. Solve $\dfrac{d^2y}{dx^2} + x^2y = 0$. (AMIETE-1995)

12. Solve $(1+x^2)\dfrac{d^2y}{dx^2} + x\dfrac{dy}{dx} - y = 0$. (AMIETE-1997)

13. Solve $x^2\dfrac{d^2y}{dx^2} + x\dfrac{dy}{dx} + (x^2 - 4)y = 0$.

(AMIETE-1997, BHOPAL-2008, RAJASTHAN-2003)

14. Solve $x^2\dfrac{d^2y}{dx^2} + 4x\dfrac{dy}{dx} + (x^2 + 2)y = 0$. (AMIETE-1998)

15. Solve $x^2\dfrac{d^2y}{dx^2} + 6x\dfrac{dy}{dx} + (6 - x^2)y = 0$. (AMIETE-1997)

16. Solve $2x(1-x)\dfrac{d^2y}{dx^2} + (1-x)\dfrac{dy}{dx} + 3y = 0$.

(UPTU(B.TECH.)-2004, GBTU-2010)

17. Solve $2x^2\dfrac{d^2y}{dx^2} - x\dfrac{dy}{dx} + (x^2 + 1)y = 0$.

(AMIETE-1999, GBTU-2011)

18. Solve $x^2\dfrac{d^2y}{dx^2} + (x-1)\dfrac{dy}{dx} - y = 0$. (AMIETE-1996, 98)

19. Solve $x^2\dfrac{d^2y}{dx^2} + \dfrac{dy}{dx} + xy = 0$. (AMIETE-1997, 2000, 01)

20. Solve $x^2\dfrac{d^2y}{dx^2} + 5x\dfrac{dy}{dx} + x^2y = 0$. (UPTU-2002)

21. Solve $(1 - x^2)y'' - xy' + 4y = 0$ in series.

(UPTU(Q. BANK)-2002, BHOPAL-2008, UPTU-2006, GBTU-2012)

22. Find the power series solution of the following differential equation about $x = 0$

$$(1 - x^2)\dfrac{d^2y}{dx^2} - 2x\dfrac{dy}{dx} + 2y = 0 \qquad \text{(UPTU-2004)}$$

23. Solve $y'' + xy' + x^2y = 0$. (UPTU-2005, Q. BANK-2002)

24. Solve in series $x(2 + x^2)\dfrac{d^2y}{dx^2} - \dfrac{dy}{dx} - 6xy = 0$

(UPTU(Q. BANK)–2002)

25. Solve in series the equation $\dfrac{d^2y}{dx^2} + xy = 0$. (VTU-2010)

26. Solve in series the equation $\dfrac{d^2y}{dx^2} + y = 0, y(0) = 0$.

(BPTU-2005)

27. Solve in series the equation $(1+x^2)\dfrac{d^2y}{dx^2} + x\dfrac{dy}{dx} - y = 0$.

(SVTU-2008)

28. $4x\dfrac{d^2y}{dx^2} + 2\dfrac{dy}{dx} + y = 0$ (PTU-2005)

29. $8x^2\dfrac{d^2y}{dx^2}+10x\dfrac{dy}{dx}-(1+x)y=0$ (PTU-2009)   **30.** $3x\dfrac{d^2y}{dx^2}+(1-x)\dfrac{dy}{dx}-y=0$ (SVTU-2008)

Hints to Selected Problems

1. Let a trial solution of the given equation is

$$y=C_0+C_1x+C_2x^2+\dots$$

$$\Rightarrow\ \frac{dy}{dx}=C_1+2C_2x+3C_3x+\dots;$$

$$\frac{d^2y}{dx^2}=2C_2+6C_3x+\dots$$

Using these values in the given equation.

2. Let the series solution be

$$y=\sum_{r=0}^{\infty}C_r\,x^{m+2r}\ \Rightarrow\ \frac{dy}{dx}=\sum_{r=0}^{\infty}C_r(m+2r)\,x^{m+2r-1}$$

and $\dfrac{d^2y}{dx^2}=\sum_{r=0}^{\infty}C_r(m+2r)(m+2r-1)\,x^{m+2r-2}$. Put all these

values in the given equation.

3. Indicial equation has two equal roots, *i.e.*, 0, 0.

20. $y=\Sigma a_k(m+k)(m+k+4)x^{m+k}+\Sigma a_k x^{m+k-2}$.

Indical equation is $a_0m(m-4)=0$.

Answers

1. $y=C_0\left(1-\dfrac{2}{3}x^3-\dfrac{2}{45}x^6\dots\right)+C_1\left(x-\dfrac{1}{6}x^4-\dfrac{1}{63}x^7\right)+x^2+\dfrac{1}{3}x^3+\dfrac{1}{12}x^4+\dfrac{1}{45}x^6+\dots$

2. $y=au+bv$, where $u=1-\dfrac{x^2}{2^2}+\dfrac{x^4}{2^2.4^2}-\dfrac{x^6}{2^2.4^2.6^2}+\dots$ and $v=u\log x+\left[\dfrac{x^2}{2^2}-\dfrac{3}{2^3.4^2}x^4+\dots\right]$

3. $u=1-2x+\dfrac{3}{2!}x^2-\dfrac{4}{3!}x^3+\dots;\ v=bu\log x+b\left[2\left(2-\dfrac{1}{2}\right)x-\dfrac{3}{2!}\left(-\dfrac{1}{3}+2+\dfrac{1}{2}\right)x^2+\dots\right]$

4. Transformed equation is $z^4\dfrac{d^2y}{dz^2}+2z^2\dfrac{dy}{dz}-y=0$

5. $y=a\left[n-(n+1)x+(n+2)\dfrac{x^2}{2!}-(n+3)\dfrac{x^3}{3!}+\dots\right]+bx^{1-n}\left[1+\dfrac{2}{n-2}x+\dfrac{3}{(n-2)(n-3)}x^2+\dfrac{4}{(n-2)(n-3)(n-4)}x^3+\dots\right]$

6. $y=a\left[1+3x^2+\dfrac{3}{5}x^4-\dfrac{1}{15}x^6+\dots\right]+bx^{3/2}\left[1+\dfrac{3}{8}x^2-\dfrac{1}{8}.\dfrac{3}{16}x^4+\dfrac{1.3.5}{8.16.24}x^6\dots\right]$

7. $y=ax\left[1+\dfrac{x^2}{2.5}+\dfrac{x^4}{2.4.5.9}+\dfrac{x^6}{2.4.6.5.9.15}+\dots\right]+bx^{-1/2}\left[1+\dfrac{x^2}{2.3}+\dfrac{x^4}{2.4.3.7}+\dfrac{x^6}{2.4.6.3.7.11}+\dots\right]$

$$+\dfrac{1}{3}x^2+\dfrac{1}{3}.\dfrac{1}{3.7}x^4+\dfrac{1}{3}.\dfrac{1}{3.5.7.11}x^6+\dots$$

8. $y=a\left[1+x+\dfrac{2}{4}x^2+\dfrac{2.5}{4.9}x^3+\dots\right]\left[4\log x+\left(-2x-x^2-\dfrac{14}{27}x^3-\dots\right)\right]$

9. $y=a\left[1-\dfrac{1}{3!}.x^{-2}-\dfrac{1}{5!}x^{-4}-\dfrac{1.3}{7!}x^{-6}-\dots\right]+b\left[x-\dfrac{1}{x}\right]+2x^{-3}\left(\dfrac{1}{4!}+\dfrac{2}{6!}x^{-2}+\dfrac{2.4}{8!}x^{-4}+\dots\right)$

10. $y=a\left[1+\dfrac{1}{2}x^2+\dfrac{1}{24}x^4+\dots\right]+b\left[x+\dfrac{1}{6}x^3+\dfrac{1}{120}x^5+\dots\right]+\left(\dfrac{1}{6}x^3+\dfrac{1}{120}x^5+\dots\right)$

11. $y=a\left(1-\dfrac{1}{12}x^4+\dfrac{x^8}{12\times7\times8}-\dfrac{x^{12}}{12\times8\times7\times11\times12}+\dots\right)+b\left(x-\dfrac{x^5}{20}+\dfrac{x^9}{20\times8\times1}+\dots\right)$ **12.** $y=a\left(1+\dfrac{x^2}{2}-\dfrac{x^4}{8}\dots\right)+bx$

13. $y=ax^2\left[1-\dfrac{x^2}{2\times6}+\dfrac{x^4}{2.4.6.8}-\dfrac{x^6}{2.4.6.8.10}+\dots\right]+b\left[x^2\log x\left(-\dfrac{1}{2^2.4}+\dfrac{x^2}{2^3.4.6}-\dfrac{x^4}{2^3.4^3.6.8}+\dots\right)\right]+x^{-2}\left(1+\dfrac{x^2}{2^2}+\dfrac{x^4}{2^2.4^2}+\dots\right)$

14. $y_1=\dfrac{a}{x}\left[1-\dfrac{x^2}{6}+\dfrac{x^4}{12}+\dots\right]+b\left[1-\dfrac{x^2}{12}+\dfrac{x^4}{360}+\dots\right],\ y_2=x^{-2}(a\cos x+b\sin x)$

15. $y = Ay_1 + By_2$,

when $y_1 = ax^{-2}\left(1 + \dfrac{x^2}{3!} + \dfrac{x^4}{5!} + ...\right) + bx^{-2}\left(x + \dfrac{x^3}{3.4} + \dfrac{x^5}{3.4.5.6} + ...\right)$, $y_2 = ax^{-3}\left(1 + \dfrac{x^2}{2!} + \dfrac{x^4}{4!} + ...\right) + bx^{-3}\left(x + \dfrac{x^3}{3!} + \dfrac{x^5}{5!} + ...\right)$

16. $y = a\sqrt{x}.(1-x) + b\left(1 - 3x + \dfrac{3x^2}{1.3} + \dfrac{3.x^3}{3.5} + \dfrac{3x^4}{5.7} + ...\right)$.

17. $y = ax\left(1 - \dfrac{x^2}{10} + \dfrac{x^4}{360} + ...\right) + bx^{1/2}\left(1 + \dfrac{x^2}{6} + \dfrac{x^4}{168} + ...\right)$.

18. $y = a\left(1 - x + \dfrac{x^2}{2!} - \dfrac{x^3}{3!} + ...\right) + b\left(x^2 - \dfrac{2x^3}{3!} + \dfrac{2x^4}{4!} - \dfrac{2x^5}{5!} + ...\right)$

19. $y = a\left(1 - \dfrac{x^2}{2^2} + \dfrac{x^4}{2^2.4^2} - \dfrac{x^6}{2^2.4^2.6^2} + ...\right) + b\left\{y_1 \log x + C_0\left\{\dfrac{x^2}{2^2} + \dfrac{1}{2^2.4^2}\left(1 + \dfrac{1}{x}\right)x^4 + \dfrac{1}{2^2.4^2.6^2}\left(1 + \dfrac{1}{2} + \dfrac{1}{3}\right)x^6 + ...\right\}\right\}$

20. $y = a\left(1 - \dfrac{x^2}{12} + \dfrac{x^4}{364} - ...\right) + bx^{-4}\log x\left(1 - \dfrac{x^4}{16} - \dfrac{x^6}{16}...\right) + 6x^{-2}\left(\dfrac{1}{4} + \dfrac{x^2}{64}\right) + ...$.

21. $y = a_0(1 - 2x^2) + a_1 x\left(1 - \dfrac{x^2}{2} - \dfrac{x^4}{8} - ...\right)$

22. $y = a_0\left(1 - x^2 - \dfrac{x^4}{3} - \dfrac{x^6}{5} - ...\right) + a_1 x$

23. $y = a_0\left(1 - \dfrac{x^4}{12} + \dfrac{x^6}{90} - ...\right) + a_1\left(x - \dfrac{x^3}{6} - \dfrac{x^5}{40} - ...\right)$

24. $y = A\left(1 + 3x^2 + \dfrac{3}{5}x^4 - \dfrac{1}{15}x^6 + ...\right) + Bx^{3/2}\left(1 + \dfrac{3}{8}x^2 - \dfrac{3.1}{8.16}x^4 + \dfrac{5.3.1}{8.16.24}x^6 - ...\right)$

25. $y = a_0\left(1 - \dfrac{x^3}{3!} + \dfrac{1.4.x^6}{6!} - \dfrac{1.4.7.x^9}{9!} + ...\right) + a_1\left(x - \dfrac{2.x^4}{4!} + \dfrac{2.5x^7}{7!} - ...\right)$ **26.** $y = a_1\left(x - \dfrac{x^3}{3!} + \dfrac{x^5}{5!} - ...\right)$

27. $y = a_0\left(1 + \dfrac{x^2}{2} - \dfrac{x^4}{8} + \dfrac{x^6}{16} - \dfrac{5x^8}{128} + ...\right) + a_1 x$

28. $y = C_1 \cos\sqrt{x} + C_2 \sin\sqrt{x}$

29. $y = C_1 x^{-1/2}\left(1 + \dfrac{x}{2} + \dfrac{x^2}{40} + ...\right) + C_2 x^{1/4}\left(1 + \dfrac{x}{14} + \dfrac{x^2}{616} + ...\right) + C_2\sqrt{x}\left(x + \dfrac{x^2}{2.3} + \dfrac{x^4}{2.4.3.7} + \dfrac{x^6}{2.4.6.3.7.11} + ...\right)$

30. $y = C_1\left(1 + x + \dfrac{x^2}{4} + \dfrac{x^3}{4}.7 + ...\right) + C_2 x^{2/3}\left(1 + \dfrac{1}{3}x + \dfrac{x^2}{3.6} + \dfrac{x^3}{3.6.9} + ...\right)$

8.8 LEGENDRE AND BESSEL'S FUNCTIONS

It is known that the functions like $e^x, \sin x$ and $\cos x$ are categorised as elementary transcendental functions. The theory of higher transcendental functions constitutes the special functions of Mathematics. We can express the variety of the properties of these functions in terms of differential equation satisfied by them. The special function constitute the complex exponential functions, hypergeometric functions, Jacobi function, Legendre functions, Bessel's and Spherical Bessel's functions, etc.

8.8.1 LEGENDRE'S FUNCTIONS

Consider a homogeneous linear differential equation of order two of the form

$$(1 - x^2)\dfrac{d^2 y}{dx^2} - 2x\dfrac{dy}{dx} + n(n+1)y = 0 \qquad ...(1)$$

where n is a real number. This differential equation is known as Legendre's differential equation, and any solution of (1) is called a Legendre function.

(1) SOLUTION OF LEGENDRE'S EQUATION

Dividing (1) by $(1 - x^2)$, we get $\dfrac{d^2 y}{dx^2} - \dfrac{2x}{1-x^2}\dfrac{dy}{dx} + n(n+1).\dfrac{1}{1-x^2}y = 0$

Now compare this equation with the standard form $\dfrac{d^2 y}{dx^2} + P(x)\dfrac{dy}{dx} + Q(x)y = 0$

we get
$$P(x) = -\frac{2x}{1-x^2}, Q(x) = \frac{n(n+1)}{1-x^2}.$$

It is trivially obtained that $P(x)$ and $Q(x)$ are analytic at $x = 0$, so for finding the solution of (1) we apply the power series method. Let us assume the solution of (1)

$$y = \sum_{m=0}^{\infty} a_m x^m \qquad \text{...(2)}$$

Now differentiating (2) w.r.t. x one time and then two times, we get

$$\frac{dy}{dx} = \sum_{m=1}^{\infty} m a_m x^{m-1} \qquad \text{...(3)}$$

and

$$\frac{d^2 y}{dx^2} = \sum_{m=2}^{\infty} m(m-1) a_m x^{m-2} \qquad \text{...(4)}$$

Substitute the values of $y, \dfrac{dy}{dx}$ and $\dfrac{d^2 y}{dx^2}$ from (2), (3) and (4) into eq. (1), we get

$$(1-x^2) \sum_{m=2}^{\infty} m(m-1) a_m x^{m-2} - 2x \sum_{m=1}^{\infty} m a_m x^{m-1} + n(n+1) \sum_{m=0}^{\infty} a_m x^m = 0$$

or

$$\sum_{m=2}^{\infty} m(m-1) a_m x^{m-2} - \sum_{m=2}^{\infty} m(m-1) a_m x^m - 2 \sum_{m=1}^{\infty} m a_m x^m + n(n+1) \sum_{m=0}^{\infty} a_m x^m = 0$$

or

$$\{2.1 \, a_2 + 3.2 a_3 x + 4.3 a_4 x^2 + ... + (r+2)(r+1) a_{r+2} x^r + ...\}$$

$$- \{2.1 a_2 x^2 + 3.2 a_3 x^3 + ... + r(r-1) a_r x^r + ...\} - 2\{a_1 x + 2 a_2 x^2 + ... + r a_r x^r + ...\} + n(n+1)\{a_0 + a_1 x + ... + a_r x^r + ...\} = 0. \qquad \text{...(5)}$$

If equation (2) is a solution of (1), then equation (5) must be an identity in x. Thus in (5) the sum of the coefficients of each power of x must be zero. We therefore obtain

$$2 a_2 + n(n+1) a_0 = 0.$$
$$6 a_3 + \{-2 + n(n+1)\} a_1 = 0, \quad 12 a_4 + \{-2 a_2 - 4 a_2 + n(n+1) a_2\} = 0$$

$$\therefore \qquad a_{n-2} = -\frac{n(n-1)}{2(2n-1)} \cdot \frac{(2n)!}{2^n (n!)^2} \qquad \left(\because a_n = \frac{(2n)!}{2^n (n!)^2} \right)$$

$$= -\frac{n(n-1) 2n.(2n-1).(2n-2)!}{2(2n-1).2^n.n!.n(n-1).(n-2)!} = -\frac{n(n-1) 2n.(2n-1).(2n-2)!}{2(2n-1).2^n n.(n-1)!.n(n-1).(n-2)!} = -\frac{(2n-2)!}{2^n (n-1)!(n-2)!}.$$

Similarly,

$$a_{n-4} = -\frac{(n-2)(n-3)}{4(2n-3)} a_{n-2} = -\frac{(n-2)(n-3)}{4(2n-3)} \cdot \frac{-(2n-2)!}{2^n (n-1)!(n-2)!}$$

$$= \frac{(n-2)(n-3).(2n-2)(2n-3)(2n-4)!}{4(2n-3) 2^n (n-1).(n-2)!(n-2)(n-3).(n-4)!} = \frac{(2n-4)!}{2^n .(2)!(n-2)!(n-4)!}$$

Continuing in this way, we get in general,

$$a_{n-2m} = \frac{(-1)^m (2n-2m)!}{2^n (m)!(n-m)!(n-2m)!}, \quad n - 2m \geq 0$$

Thus we obtain the first kind of Legendre polynomial of degree n and it is denoted by $P_n(x)$ which is given by

$$P_n(x) = \sum_{m=0}^{N} a_{n-2m} x^{n-2m}$$

In general, we obtain

$$(r+2)(r+1) a_{r+2} + \{-r(r-1) - 2r + n(n+1)\} a_r = 0 \text{ for } r = 2, 3, 4, ...$$

or

$$(r+2)(r+1) a_{r+2} + (n-r)(n+r+1) a_r = 0$$

or

$$a_{r+2} = -\frac{(n-r)(n+r+1)}{(r+2)(r+1)} a_r, r = (0, 1, 2, ...) \qquad \text{...(6)}$$

This equation (6) is known as recursion formula. Now finding the coefficients successively for $r = 0, 1, 2, 3, ...$

$$a_2 = -\frac{n(n+1)}{2.1} a_0 = -\frac{n(n+1)}{(2)!} a_0 \, ; a_3 = -\frac{(n-1)(n+2)}{3.2} a_1 = -\frac{(n-1)(n+2)}{(3)!} a_1$$

$$a_4 = -\frac{(n-2)(n+3)}{4.3} a_2 = -\frac{(n-2)(n+3)}{4.3} \cdot \frac{-n(n+1)}{2.1} a_0 = \frac{(n-2)n(n+1)(n+3)}{(4)!} a_0$$

$$a_5 = -\frac{(n-3)(n+4)}{5.4} a_3 = \frac{(n-3)(n-1)(n+2)(n+4)}{(5)!} a_1$$

$$\vdots$$

$$\text{etc.}$$

We observed from above coefficients that all the even numbered coefficients are obtained in terms of a_0 while all odd numbered coefficients are obtained in terms of a_1. Thus we obtain the solution as

$$y = a_0 y_1(x) + a_1 y_2(x) \quad \text{where } y_1(x) = 1 - \frac{n(n+1)}{(2)!} x^2 + \frac{(n-2)n(n+1)(n+3)}{(4)!} x^4 - \ldots$$

and $$y_2(x) = x - \frac{(n-1)(n+2)}{(3)!} x^3 + \frac{(n-3)(n-1)n(n+2)(n+4)}{(5)!} x^5 - \ldots .$$

These both series are convergent if $|x| < 1$. Sometimes, we have observed that the parameter n in the Legendre's differential equation will be non-negative. Then, from recursion formula (6) we obtain

$$a_{r+2} = 0, \text{ when } r = n \qquad \qquad i.e., a_{n+2} = a_{n+4} = \ldots = 0$$

Hence we can say that if n is even. $y_1(x)$ becomes a polynomial of degree n whereas n is odd $y_2(x)$ becomes a polynomial of degree n. Therefore if $y_1(x)$ is multiplied by some constant, then this polynomial is called Legendre's polynomial of first kind and if $y_2(x)$ is multiplied by some constant, then $y_2(x)$ is called Legendre's polynomial of second kind. Now to obtain first kind of Legendre's polynomial we proceed as follows :

The recursion formula given in (6) may be written as

$$a_r = - \frac{(r+2)(r+1)}{(n-r)(n+r+1)} a_{r+2} \text{ for } r \leq n-2$$

Also, all a's may express in terms of the coefficient a_n which is the coefficient of the highest power of x of the polynomial. This a_n is an arbitrary so choose $a_n = 1$ when $n = 0$ and $= \dfrac{1.3.5\ldots(2n-1)}{2 \cdot (\ )!} = \dfrac{2n!}{2 \cdot (\ )!}$ for all $n = 1, 2, 3 \ldots$. This a_n is chosen in such a way that the values of all those polynomial will be 1 when $x = 1$. Now finding the coefficients as follows :

$$a_{n-2} = - \frac{n(n-1)}{2(2n-1)} a_n \qquad \text{or} \qquad P_n(x) = \sum_{m=0}^{N} \frac{(-1)^m (2n-2m)!}{2^n (m)!(n-m)!(n-2m)!} \cdot x^{n-2m}$$

where $N = \begin{cases} n/2 & ; \text{ if } n \text{ is even} \\ (n-1)/2 & ; \text{ if } n \text{ is odd.} \end{cases}$

Similarly, we can obtain second kind of Legendre polynomial. From recursion formula (6) we also have

$$a_{r+2} = 0 \text{ if } r = -(n+1) \quad i.e., \quad a_{-n+3} = a_{-n+5} = \ldots = 0$$

In this case we obtain second kind of Legendre polynomial of an infinite series. It is denoted by $Q_n(x)$. If we choose

$$a_{-n-1} = \frac{(n)!}{1.3.5\ldots(2n+1)}$$

Then from the following relation

$$a_r = - \frac{(r+2)(r+1)}{(n-r)(n+r+1)} a_{r+2}, \text{ we obtain } a_{-n-3} = \frac{(n+1)((n+2)}{2(2n+3)} \cdot a_{-n-1}$$

and $$a_{-n-5} = \frac{(n+3)(n+4)}{4(2n+5)} \cdot a_{-n-3} = \frac{(n+1)(n+2)(n+3)(n+4)}{2 \cdot 4(2n+3)(2n+5)} a_{-n-1} \text{ and so on.}$$

Hence $Q_n(x)$ is given as

$$Q_n(x) = a_{-n-1} \left[x^{-n-1} + \frac{(n+1)(n+2)}{2(2n+3)} x^{-n-3} + \frac{(n+1)(n+2)(n+3)(n+4)}{2 \cdot 4(2n+3)(2n+5)} x^{-n-5} + \ldots \right]$$

$$\Rightarrow \qquad Q = \frac{(n)!}{1 \cdot 3 \cdot 5 \ldots (2n+1)} \left[x^{-n-1} + \frac{(n+1)(n+2)}{2(2n+3)} x^{-n-3} + \frac{(n+1)(n+2)(n+3)(n+4)}{2 \cdot 4(2n+3)(2n+5)} x^{-n-5} + \ldots \right] \qquad \text{(VTU-2006)}$$

8.8.2 Generating Functions of Legendre's Polynomial $P_n(x)$

(UPTU-2005, KERALA-2005)

The functions of the type $\dfrac{1}{\sqrt{1 - 2xt + t^2}}$ *generates Legendre polynomial* $P_n(x)$, *is called generating functions. Thus we obtain*

$$\frac{1}{\sqrt{1 - 2xt + t^2}} = \sum_{n=0}^{\infty} P_n(x) t^n .$$

PROOF. Consider R.H.S. $= \dfrac{1}{\sqrt{1 - 2xt + t^2}} = \dfrac{1}{\sqrt{1-s}}$ $\qquad \qquad \qquad [\because s = 2xt - t^2]$

$$= (1-s)^{-1/2} = 1 + \frac{1}{2} s + \frac{1.3}{2.4} s^2 + \frac{1.3.5}{2.4.6} s^3 + \ldots \ldots + \frac{1.3.5\ldots(2n-3)}{2.4.6\ldots(2n-2)} s^{n-1} + \frac{1.3.5\ldots(2n-1)}{2.4.6\ldots(2n)} s^n + \ldots$$

(Expand by binomial theorem) $\qquad \ldots (1)$

Since $\qquad s = 2xt - t^2$

$\therefore \qquad s^n = (2xt - t^2)^n = t^n (2x-t)^n = t^n [{}^n C_0 (2x)^n - {}^n C_1 (2x)^{n-1} t + \ldots].$

Similarly $s^{n-1} = t^{n-1} \left[{}^{n-1} C_0 (2x)^{n-1} - {}^{n-1} C_1 (2x)^{n-2} t + \ldots \right]$

and $\quad s^{n-2} = t^{n-2}\left[{}^{n-2}C_0(2x)^{n-2} - {}^{n-2}C_1(2x)^{n-3}t + {}^{n-2}C_2(2x)^{n-4}t^2 \ldots\right]$

$$\vdots$$

etc.

Substitute these value in the above equation (1), we get

$$\text{L.H.S.} = 1 + \frac{1}{2}t(2x-t) + \frac{1.3}{2.4}t^2(2x-t)^2 + \frac{1.3.5}{2.4.6}t^3(2x-t)^3 + \ldots$$

$$\ldots + \frac{1.3.5\ldots(2x-5)}{2.4.6\ldots(2x-4)}t^{n-2}\left[{}^{n-2}C_0(2x)^{n-2} - {}^{n-2}C_1(2x)^{n-3}t + {}^{n-2}C_2(2x)^{n-4}t^2 + \ldots\right]$$

$$+ \frac{1.3.5\ldots(2n-3)}{2.4.6\ldots(2n-2)}t^{n-1}\left[{}^{n-1}C_0(2x)^{n-1} - {}^{n-1}C_1(2x)^{n-2}t + \ldots\right]$$

$$+ \frac{1.3.5\ldots(2n-1)}{2.4.6\ldots(2n)}t^n\left[{}^{n}C_0(2x)^n - {}^{n}C_1(2x)^{n-1}t + \ldots\right]$$

Now collecting the coefficients of t^n, we get

$$= \frac{1.3.5\ldots(2n-1)}{2.4.6\ldots(2n)}\,{}^{n}C_0(2x)^n - \frac{1.3.5\ldots(2n-3)}{2.4.6\ldots(2n-2)}\,{}^{n-1}C_1(2x)^{n-2} + \frac{1.3.5\ldots(2n-5)}{2.4.6\ldots(2n-4)}\,{}^{n-2}C_2(2x)^{n-4} - \ldots$$

$$= \frac{1.3.5\ldots(2n-1)}{2.4.6\ldots(2n)}\cdot 2^n x^n - \frac{1.3.5\ldots(2n-3)}{2.4.6\ldots(2n-2)}\cdot\frac{(n-1)}{(1)!}2^{n-2}\cdot x^{n-2}$$

$$+ \frac{1.3.5\ldots(2n-5)}{2.4.6\ldots(2n-4)}\cdot\frac{(n-2)(n-3)}{(2)!}2^{n-4}\cdot x^{n-4} - \ldots$$

$$= \frac{1.3.5\ldots(2n-1)}{2.4.6\ldots(2n)}2^n\left[x^n - \frac{2n(n-1)}{(2n-1)2^2}x^{n-2} + \frac{2n(2n-2)(n-2)(n-3)}{(2n-1)(2n-3)(2)!\cdot 2^4}\cdot x^{n-4} - \ldots\right]$$

$$= \frac{1.3.5\ldots(2n-1)}{(n)!}\left[x^n - \frac{n(n-1)}{2(2n-1)}x^{n-2} + \frac{n(n-1)(n-2)(n-3)}{2.4(2n-1)(2n-3)}\cdot x^{n-4} - \ldots\right]$$

Hence we obtain $\dfrac{1}{\sqrt{1-2xt+t^2}} = \sum\limits_{n=0}^{\infty} P_n(x)t^n$

(1) RODRIGUE'S FORMULA

(UPTU-2004, 07, VTU-2008, BHOPAL-2007, MADRAS-2003)

The expression for $P_n(x)$, given by $P_n(x) = \dfrac{1}{2^n(n)!}\dfrac{d^n}{dx^n}(x^2-1)^n$ is called Rodrigue's Formula.

PROOF. Since $P_n(x)$ is a Legendre polynomial whose expression is given as

$$P_n(x) = \sum_{m=0}^{[n/2]} \frac{(-1)^m(2n-2m)!}{2^n(m)!(n-m)!(n-2m)!}\cdot x^{n-2m} \qquad \ldots(1)$$

where $[n/2]$ is an integral value of $n/2$ not exceed $n/2$. Rearrange (1), we get

$$P_n(x) = \sum_{m=0}^{[n/2]} \frac{(-1)^m}{2^n(m)!(n-m)!}\left\{\frac{(2n-2m)!}{(n-2m)!}\cdot x^{n-2m}\right\} = \sum_{m=0}^{[n/2]} \frac{(-1)^m}{2^n(m)!(n-m)!}\cdot \frac{d^n}{dx^n}x^{2n-2m}$$

$$\left(\because \frac{d^r}{dx^r}x^{2n-2m} = \frac{(2n-2m)!}{(2n-2m-r)!}\cdot x^{2n-2m-r}\right)$$

$$= \frac{1}{2^n(n)!}\sum_{m=0}^{[n/2]} \frac{(n)!}{(m)!(n-m)!}\cdot \frac{d^n}{dx^n}(x^2)^{n-m}\cdot(-1)^m$$

Now extending the range of m from 0 to n. To do so no change will occur in the above expression, because n^{th} derivatives of those terms whose degree are less than n will be zero. Thus above expression can be written as

$$= \frac{1}{2^n(n)!}\frac{d^n}{dx^n}\sum_{m=0}^{n} \frac{(n)!}{(m)!(n-m)!}(x^2)^{n-m}(-1)^m = \frac{1}{2^n(n)!}\frac{d^n}{dx^n}\sum_{m=0}^{n} {}^{n}C_m(x^2)^{n-m}(-1)^m \qquad \left(\because {}^{n}C_m = \frac{(n)!}{(m)!(n-m)!}\right)$$

$$= \frac{1}{2^n(n)!}\frac{d^n}{dx^n}\left[{}^{n}C_0(x^2)^n - {}^{n}C_1(x^2)^{n-1} + {}^{n}C_2(x^2)^{n-2} + \ldots + {}^{n}C_n(-1)^n\right] = \frac{1}{2^n(n)!}\frac{d^n}{dx^n}(x^2-1)^n \quad \text{(By Binomial theorem)}$$

Hence $P_n(x) = \dfrac{1}{2^n(n)!}\dfrac{d^n}{dx^n}(x^2-1)^n$.

(2) LAPLACE INTEGRAL FOR $P_n(x)$

(i) *Laplace's First Integral for $P_n(x)$:*

$$P_n(x) = \frac{1}{\pi}\int_0^\pi \left[x \pm \sqrt{(x^2-1)}\cos\theta\right]^n d\theta, \text{ where } n \text{ is any positive integer.}$$

PROOF. Since we know that $\int_0^\pi \dfrac{d\theta}{a \pm b\cos\theta} = \dfrac{\pi}{\sqrt{a^2 - b^2}}$, where $a^2 > b^2$ ...(1)

Let us take $a = 1 - tx$ and $b = t\sqrt{x^2 - 1}$,

Then, $a^2 - b^2 = (1 - tx)^2 - t^2(x^2 - 1) = 1 + t^2x^2 - 2tx - t^2x^2 + t^2 = 1 - 2tx + t^2$.

Thus (1) becomes $\int_0^\pi \dfrac{d\theta}{(1 - tx) \pm t\sqrt{x^2 - 1}\cos\theta} = \dfrac{\pi}{\sqrt{1 - 2tx + t^2}}$ ...(2)

Since generating function gives $\dfrac{1}{\sqrt{1 - 2xt + t^2}} = \sum\limits_{n=0}^{\infty} P_n(x)t^n$

$\therefore$ (2) becomes

$$\pi \sum\limits_{n=0}^{\infty} P_n(x)t^n = \int_0^\pi \dfrac{d\theta}{1 - tx \pm t\sqrt{(x^2 - 1)}\cos\theta} = \int_0^\pi \dfrac{d\theta}{[1 - t\{x \mp \sqrt{(x^2 - 1)}\cos\theta\}]}$$

$$= \int_0^\pi [1 - t\{x \pm \sqrt{(x^2 - 1)}\cos\theta\}]^{-1}d\theta = \int_0^\pi (1 - ts)^{-1}d\theta, \quad \text{where} \qquad s = x \pm \sqrt{(x^2 - 1)}\cos\theta$$

$$= \int_0^\pi (1 + ts + t^2s^2 + \ldots + t^ns^n + \ldots)d\theta = \int_0^\pi \sum\limits_{n=0}^{\infty} t^ns^n d\theta = \sum\limits_{n=0}^{\infty} \int_0^\pi s^nt^n d\theta = \sum\limits_{n=0}^{\infty} t^n\int_0^\pi [x \mp \sqrt{(x^2 - 1)}\cos\theta]^n d\theta$$

$\therefore \quad \pi \sum\limits_{n=0}^{\infty} P_n(x)t^n = \sum\limits_{n=0}^{\infty} t^n\int_0^\pi [x \mp \sqrt{(x^2 - 1)}\cos\theta]^n \, d\theta$

$\therefore \qquad \pi P_n(x) = \int_0^\pi [x \pm \sqrt{(x^2 - 1)}\cos\theta]^n \, d\theta$

Hence, $\qquad P_n(x) = \dfrac{1}{\pi} \int_0^\pi [x \pm \sqrt{(x^2 - 1)}\cos\theta]^n \, d\theta$.

(ii) Laplace's Second integral for $P_n(x)$:

$$P_n(x) = \dfrac{1}{\pi} \int_0^\pi \dfrac{d\theta}{[x \pm \sqrt{(x^2 - 1)}\cos\theta]^{n+1}}, \textit{ where } n \textit{ is any positive integer.}$$

PROOF. Since we know that $\int_0^\pi \dfrac{d\theta}{a \pm b\cos\theta} = \dfrac{\pi}{\sqrt{a^2 - b^2}}$, where $a^2 > b^2$ ...(1)

Here taking $a = xt - 1$, and $b = t\sqrt{(x^2 - 1)}$, then $a^2 - b^2 = 1 - 2xt + t^2$

$\therefore \quad$ (1) becomes $\int_0^\pi \dfrac{d\theta}{(xt - 1) \pm t\sqrt{(x^2 - 1)}\cos\theta} = \dfrac{\pi}{\sqrt{1 - 2tx + t^2}}$ ...(2)

Since $\sum\limits_{n=0}^{\infty} P_n(x)t^n = \dfrac{1}{\sqrt{1 - 2xt + t^2}}$

$\therefore \quad$ (2) becomes

$$\pi \sum\limits_{n=0}^{\infty} P_n(x)t^n = \int_0^\pi \dfrac{d\theta}{[-1 + t\{x \pm \sqrt{(x^2 - 1)}\cos\theta\}]} = \int_0^\pi [t\{x \pm \sqrt{(x^2 - 1)}\cos\theta\} - 1]^{-1}d\theta$$

$$= \int_0^\pi (ts - 1)^{-1}d\theta, \, s = x \pm \sqrt{(x^2 - 1)}\cos\theta$$

$$= \int_0^\pi \dfrac{1}{ts}\left(1 - \dfrac{1}{ts}\right)^{-1}d\theta = \int_0^\pi \dfrac{1}{ts}\left(1 + \dfrac{1}{ts} + \dfrac{1}{t^2s^2} + \ldots + \dfrac{1}{t^ns^n} + \ldots\right)d\theta$$

$$= \int_0^\pi \left(\dfrac{1}{ts} + \dfrac{1}{t^2s^2} + \ldots + \dfrac{1}{t^{n+1}s^{n+1}} + \ldots\right)d\theta = \int_0^\pi \sum\limits_{n=0}^{\infty} \dfrac{1}{t^{n+1}s^{n+1}}d\theta$$

$\therefore \qquad \pi \sum\limits_{n=0}^{\infty} P_n(x)t^n = \sum\limits_{n=0}^{\infty} \dfrac{1}{t^{n+1}}\int_0^\pi \dfrac{d\theta}{[x \pm \sqrt{(x^2 - 1)}\cos\theta]^{n+1}}$

or $\qquad \pi \cdot \dfrac{1}{\sqrt{1 - 2xt + t^2}} = \sum\limits_{n=0}^{\infty} \dfrac{1}{t^{n+1}}\int_0^\pi \dfrac{d\theta}{[x \pm \sqrt{(x^2 - 1)}\cos\theta]^{n+1}}$

or $\qquad \dfrac{\pi}{t} \cdot \dfrac{1}{\sqrt{1 - 2x \cdot \dfrac{1}{t} + \dfrac{1}{t^2}}} = \sum\limits_{n=0}^{\infty} \dfrac{1}{t^{n+1}}\int_0^\pi \dfrac{d\theta}{[x \pm \sqrt{(x^2 - 1)}\cos\theta]^{n+1}}$

or $\qquad \dfrac{\pi}{t} \sum\limits_{n=0}^{\infty} \dfrac{1}{t^n}P_n(x) = \sum\limits_{n=0}^{\infty} \dfrac{1}{t^{n+1}}\int_0^\pi \dfrac{d\theta}{[x \pm \sqrt{(x^2 - 1)}\cos\theta]^{n+1}}$

or $\qquad \pi \sum\limits_{n=0}^{\infty} \dfrac{1}{t^{n+1}}P_n(x) = \sum\limits_{n=0}^{\infty} \dfrac{1}{t^{n+1}}\int_0^\pi \dfrac{d\theta}{[x \pm \sqrt{(x^2 - 1)}\cos\theta]^{n+1}}$

$$\therefore \quad \pi \, P_n(x) = \int_0^\pi \frac{d\theta}{[x \pm \sqrt{(x^2-1)}\cos\theta]^{n+1}}$$

Hence
$$P(x) \quad - \int \frac{}{[x \pm \sqrt{(x^2-1)}\cos\theta]^{1}}.$$

8.8.3 ORTHOGONAL PROPERTIES OF LEGENDRE'S POLYNOMIALS

(i) $\int_{-1}^{1} P_m(x)P_n(x)\,dx = 0$, when $m \ne n$.   (UPTU-2001, 02, 04, 08, SVTU-2008, MADRAS-2006, VTU-2006)

(ii) $\int_{-1}^{1} [P_n(x)]^2\,dx = \dfrac{2}{2n+1}$, when $m = n$.

PROOF.

(i) Legendre differential equation is given by

$$(1-x^2)\frac{d^2y}{dx^2} - 2x\frac{dy}{dx} + n(n+1)y = 0 \quad \text{or} \quad \frac{d}{dx}\left\{(1-x^2)\frac{dy}{dx}\right\} + n(n+1)y = 0 \qquad \ldots(1)$$

Since $P_m(x)$ and $P_n(x)$ are the solution of (1), so we have

$$\frac{d}{dx}\left\{(1-x^2)\frac{dP_m(x)}{dx}\right\} + m(m+1)P_m(x) = 0 \qquad \ldots(2)$$

and
$$\frac{d}{dx}\left\{(1-x^2)\frac{dP_n(x)}{dx}\right\} + n(n+1)P_n(x) = 0 \qquad \ldots(3)$$

Now, multiplying (2) by $P_n(x)$ and (3) by $P_m(x)$ and then subtract, we get

$$\frac{d}{dx}\left\{(1-x^2)\frac{dP_m(x)}{dx}\right\}P_n(x) - \frac{d}{dx}\left\{(1-x^2)\frac{dP_n(x)}{dx}\right\}P_m(x) + [m(m+1) - n(n+1)]P_m(x)P_n(x) = 0 \qquad \ldots(4)$$

Integrating (4) w.r.t. x from $x = -1$ to $x = 1$,, we get

$$\int_{-1}^{1} \frac{d}{dx}\left\{(1-x^2)\frac{dP_m(x)}{dx}\right\}P_n(x)\,dx - \int_{-1}^{1} \frac{d}{dx}\left\{(1-x^2)\frac{dP_n(x)}{dx}\right\}P_m(x)\,dx$$
$$+ (m-n)(m+n+1)\int_{-1}^{1} P_m(x)P_n(x)\,dx = 0 \qquad \ldots(5)$$

Let $I_1 = \int_{-1}^{1} \dfrac{d}{dx}\left\{(1-x^2)\dfrac{dP_m(x)}{dx}\right\}P_n(x)\,dx$ and $I_2 = \int_{-1}^{1} \dfrac{d}{dx}\left\{(1-x^2)\dfrac{dP_n(x)}{dx}\right\}P_m(x)\,dx.$

$\therefore$ (5) becomes $I_1 - I_2 + (m-n)(m+n+1)\int_{-1}^{1} P_m(x)P_n(x)\,dx = 0 \qquad \ldots(6)$

Now solving I_1 and I_2, we get

$$I_1 = \int_{-1}^{1} \frac{d}{dx}\left\{(1-x^2)\frac{dP_m(x)}{dx}\right\}P_n(x)\,dx = P_n(x)\left[(1-x^2)\frac{dP_m(x)}{dx}\right]_{-1}^{1} - \int_{-1}^{1} \frac{dP_n(x)}{dx}(1-x^2)\frac{dP_m(x)}{dx}\,dx$$

(Integration by parts)

$$= 0 - \int_{-1}^{1} (1-x^2)\frac{dP_n(x)}{dx}\cdot\frac{dP_m(x)}{dx}\,dx$$

$$\therefore \quad I_1 = -\int_{-1}^{1} (1-x^2)\frac{dP_n(x)}{dx}\cdot\frac{dP_m(x)}{dx}\,dx.$$

Also, $I_2 = \int_{-1}^{1} \dfrac{d}{dx}\left\{(1-x^2)\dfrac{dP_n(x)}{dx}\right\}P_m(x)\,dx = P_m(x)\left[(1-x^2)\dfrac{dP_n(x)}{dx}\right]_{-1}^{1} - \int_{-1}^{1} (1-x^2)\dfrac{dP_m(x)}{dx}\cdot\dfrac{dP_n(x)}{dx}\,dx$

$$= 0 - \int_{-1}^{1} (1-x^2)\frac{dP_m(x)}{dx}\cdot\frac{dP_n(x)}{dx}\,dx$$

$$\therefore \quad I_2 = -\int_{-1}^{1} (1-x^2)\frac{dP_m(x)}{dx}\cdot\frac{dP_n(x)}{dx}\,dx.$$

Thus $I_1 - I_2 = 0$. Now (6) becomes $0 + (m-n)(m+n+1)\int_{-1}^{1} P_m(x)P_n(x)\,dx = 0$

If $m \ne n$, then $\int_{-1}^{1} P_m(x)P_n(x)\,dx = 0$.

(ii) $\int_{-1}^{1} [P_n(x)]^2\,dx = \dfrac{2}{2n+1}$, if $m = n$

Since we know that $\dfrac{1}{\sqrt{1-2xt+t^2}} = \sum\limits_{n=0}^{\infty} P_n(x)t^n$

or $\dfrac{1}{\sqrt{1-2xt+t^2}} = P_0(x) + t\,P_1(x) + t^2 P_2(x) + \ldots + t^n P_n(x) + \ldots$

Squaring of both sides, we get

$$\frac{1}{1-2xt+t^2} = [P_0(x) + t\,P_1(x) + t^2\,P_2(x) + \dots + t^n\,P_n(x) + \dots]^2$$

$$= [P_0(x)]^2 + [t\,P_1(x)]^2 + [t^2\,P_2(x)]^2 + \dots + [t^n\,P_n(x)]^2 + \dots$$

$$+ 2[t P_0(x)\,P_1(x) + t^2\,P_0(x)\,P_2(x) + \dots + t^n\,P_0(x)\,P_n(x) + \dots$$

$$\dots + t^3\,P_1(x)\,P_2(x) + t^4\,P_1(x)\,P_3(x) + \dots + t^{n+1}\,P_1(x)\,P_n(x) + \dots]$$

$$= \sum_{n=0}^{\infty} t^{2n}[P_n(x)]^2 + 2 \sum_{\substack{m,n=0 \\ m \neq n}}^{\infty} t^{m+n} P_m(x) P_n(x)$$

$$\therefore \quad \frac{1}{1-2xt+t^2} = \sum_{n=0}^{\infty} t^{2n}[P_n(x)]^2 + 2 \sum_{\substack{m,n=0 \\ m \neq n}}^{\infty} t^{m+n} P_m(x) P_n(x).$$

Integrating both sides w.r.t. x from $x = -1$ to 1, we get

$$\int_{-1}^{1} \frac{1}{1-2xt+t^2} = \int_{-1}^{1} \sum_{n=0}^{\infty} t^{2n}[P_n(x)]^2\,dx + 2\int_{-1}^{1} \sum_{\substack{m,n=0 \\ m \neq n}}^{\infty} t^{m+n} P_m(x) P_n(x)\,dx$$

$$= \sum_{n=0}^{\infty} t^{2n}\int_{-1}^{1}[P_n(x)]^2\,dx + 2 \sum_{\substack{m,n=0 \\ m \neq n}}^{\infty} t^{m+n}\int_{-1}^{1} P_m(x) P_n(x)\,dx$$

$$= \sum_{n=0}^{\infty} t^{2n}\int_{-1}^{1}[P_n(x)]^2\,dx + 0 \qquad \left[\because \int_{-1}^{1} P_m(x)\,P_n(x)\,dx = 0 \ \text{when } m \neq n \right]$$

$$= \sum_{n=0}^{\infty} t^{2n}\int_{-1}^{1}[P_n(x)]^2\,dx = \int_{-1}^{1} \frac{dx}{1-2xt+t^2}$$

$$= -\frac{1}{2t}\Big[\log(1-2xt+t^2)\Big]_{-1}^{1} = -\frac{1}{2t}[\log(1-2t+t^2) - \log(1+2t+t^2)]$$

$$= -\frac{1}{2t}[\log(1-t)^2 - \log(1+t)^2] = -\frac{1}{2t}\left[\log\left(\frac{1-t}{1+t}\right)^2\right] = -\frac{1}{t}\left[\log\frac{1-t}{1+t}\right]$$

$$= \frac{1}{t}\left[\log\frac{1+t}{1-t}\right] = \frac{1}{t}[\log(1+t) - \log(1-t)]$$

$$= \frac{1}{t}\left[\left\{t - \frac{t^2}{2} + \frac{t^3}{3} - \frac{t^4}{4} + \dots\right\} - \left\{-t - \frac{t^2}{2} - \frac{t^3}{3} - \frac{t^4}{4} - \dots\right\}\right]$$

$$= \frac{1}{t}\left[2t + \frac{2t^3}{3} + \frac{2t^5}{5} + \dots\right] = 2\left[1 + \frac{t^2}{3} + \frac{t^4}{5} + \dots\right] = 2\sum_{n=0}^{\infty} \frac{t^{2n}}{2n+1}$$

$$\therefore \quad \sum_{n=0}^{\infty} t^{2n}\int_{-1}^{1}[P_n(x)]^2\,dx = \sum_{n=0}^{\infty} \frac{2}{2n+1} \cdot t^{2n}$$

Hence, $\displaystyle \int_{-1}^{1}[P_n(x)]^2\,dx = \frac{2}{2n+1}.$

8.8.5 Recurrence Relations for Legendre's Function

(I) $(2n+1)xP_n = (n+1)P_{n+1} + nP_{n-1}$ (MADRAS-2006)

PROOF. Since we know that $(1-2xt+t^2)^{-1/2} = \displaystyle\sum_{n=0}^{\infty} t^n\,P_n(x)$ ...(1)

Differentiating (1) both sides w.r.t. 't', we get

$$-\frac{1}{2}(1-2xt+t^2)^{-3/2} \cdot (-2x+2t) = \sum_{n=1}^{\infty} n\,t^{n-1}\,P_n(x) \ \text{ or } \ \frac{(x-t)(1-2xt+t^2)^{-1/2}}{(1-2xt+t^2)} = \sum_{n=1}^{\infty} n\,t^{n-1}\,P_n(x)$$

or $\quad (x-t)(1-2xt+t^2)^{-1/2} = (1-2xt+t^2)\displaystyle\sum_{n=1}^{\infty} n\,t^{n-1}\,P_n(x)$

or $\quad (x-t)\displaystyle\sum_{n=0}^{\infty} t^n\,P_n(x) = (1-2xt+t^2)\sum_{n=1}^{\infty} n t^{n-1}\,P_n(x)$ (From (1))

or $\quad x\displaystyle\sum_{n=0}^{\infty} t^n P_n(x) - \sum_{n=0}^{\infty} t^{n+1}P_n(x) = \sum_{n=1}^{\infty} n\,t^{n-1}P_n(x) - 2x\sum_{n=1}^{\infty} nt^n P_n(x) + \sum_{n=1}^{\infty} nt^{n+1}P_n(x)$

$$x[P_0(x) + tP_1(x) + \dots + t^n P_n(x) + \dots] - [tP_0(x) + t^2 P_1(x) + \dots + t^n P_{n-1}(x) + \dots]$$

$$= [P_1(x) + 2tP_2(x) + \dots + (n+1)\,t^n P_{n+1}(x) + \dots] - 2x[tP_1(x) + 2t^2 P_2(x) + \dots + nt^n P_n(x) + \dots]$$

$$+ [t^2 P_1(x) + 2t^3 P_2(x) + \dots + (n-1)\,t^n P_{n-1}(x) + \dots]$$

Taking the coefficient of t^n both sides, we get

$$xP_n(x) - P_{n-1}(x) = (n+1)\,P_{n+1}(x) - 2nxP_n(x) + (n-1)\,P_{n-1}(x)$$

or $\qquad (2n+1)\,xP_n(x) = (n+1)\,P_{n+1}(x) + nP_{n-1}(x)$

or $\qquad (2n+1)xP_n = (n+1)P_{n+1} + nP_{n-1}$

📑 REMARK

- Putting $(n-1)$ in place of n, we get $(2n-1)xP_{n-1} = nP_n + (n-1)P_{n-2}$

(II) $\quad nP_n = xP'_n - P'_{n-1}\ where\ P'_n \equiv \dfrac{dP_n}{dx}.$ (UPTU-2006)

PROOF.   Since we have $\dfrac{1}{\sqrt{1-2xt+t^2}} = \sum\limits_{n=0}^{\infty} t^n\,P_n(x)$ ...(1)

Differentiating (1) both sides w.r.t. 't' and w.r.t. x, respectively, we get

$$(x-t)\,(1-2xt+t^2)^{-3/2} = \sum\limits_{n=1}^{\infty} nt^{n-1}P_n(x)$$ ...(2)

and $\qquad t(1-2xt+t^2)^{-3/2} = \sum\limits_{n=0}^{\infty} t^n\,P'_n(x)$ ...(3)

From (2) and (3), we get

$$(x-t)\sum\limits_{n=0}^{\infty} t^n\,P'_n(x) = t\sum\limits_{n=1}^{\infty} nt^{n-1}P_n(x)\ \ \text{or}\ \ x\sum\limits_{n=0}^{\infty} t^n\,P'_n(x) - \sum\limits_{n=0}^{\infty} t^{n+1}P'_n(x) = \sum\limits_{n=1}^{\infty} nt^n P_n(x)$$

or $\ x[P'_0(x) + tP'_1(x) + ... + t^n P'_n(x) + ...] - [tP'_0(x) + t^2 P'_1(x) + ... + t^n P'_{n-1}(x) + ...]$

$$= tP_1(x) + 2t^2 P_2(x) + ... + nt^n P_n(x) + ...$$

Taking the coefficients of t^n of both sides, we get

$$xP'_n(x) - P'_{n-1}(x) = nP_n(x)$$

$\therefore \qquad\qquad nP_n = xP'_n - P'_{n-1}$

(III) $\quad (2n+1)P_n = P'_{n+1} - P'_{n-1}.$

PROOF.   From recurrence relation (I), we have

$$(2n+1)\,xP_n = (n+1)\,P_{n+1} + nP_{n-1}.$$

Differentiating this w.r.t. 'x' of both sides, we get

$$(2n+1)\,P_n + (2n+1)\,xP'_n = (n+1)\,P'_{n+1} + nP'_{n-1}$$ ...(1)

From recurrence relation (II), we have $nP_n = xP'_n - P'_{n-1}$

or $\qquad\qquad\qquad xP'_n = nP_n + P'_{n-1}$

Substitute this value of xP'_n into (1), we get

$$(2n+1)P_n + (2n+1)(nP_n + P'_{n-1}) = (n+1)\,P'_{n+1} + nP'_{n-1}$$

or $\quad (n+1)(2n+1)\,P_n = (n+1)P'_{n+1} - (2n+1)P'_{n-1} + nP'_{n-1} = (n+1)\,P'_{n+1} - (n+1)\,P'_{n-1}$

$\therefore \qquad\qquad (2n+1)\,P_n = P'_{n+1} - P'_{n-1}$

(IV) $\quad (n+1)P_n = P'_{n+1} - xP'_n.$

PROOF.   From recurrence relations (II) and (III), we have

$$nP_n = xP'_n - P'_{n-1}$$ ...(1)

and $\qquad\qquad (2n+1)\,P_n = P'_{n+1} - P'_{n-1}$ ...(2)

Subtract (1) from (2), we get

$$(2n+1)\,P_n - nP_n = P'_{n+1} - xP'_n \qquad\text{or}\qquad (n+1)\,P_n = P'_{n+1} - xP'_n.$$

(V) $\quad (1-x^2)P'_n = n(P_{n-1} - xP_n).$

PROOF.   From recurrence relations (II) and (IV), we have

$$nP_n = xP'_n - P'_{n-1}$$ ...(1)

and $\qquad\qquad (n+1)\,P_n = P'_{n+1} - xP'_n$ ...(2)

Putting $(n-1)$ in place of n in (2), we get

$$nP_{n-1} = P'_n - xP'_{n-1}$$ ...(3)

Now, multiplying (1) by x and subtract from (3), we get

$$nP_{n-1} - nxP_n = P'_n - x^2 P'_n \qquad\text{or}\qquad n(P_{n-1} - xP_n) = (1-x^2)\,P'_n$$

$\therefore \qquad (1-x^2)\,P'_n = n(P_{n-1} - xP_n)$

(VI) $(1-x^2)P_n' = (n+1)(xP_n - P_{n+1})$.

(UPTU-2007)

PROOF. From recurrence relations (I) and (V), we have

$$(2n+1)\, xP_n = (n+1)\, P_{n+1} + nP_{n-1} \qquad \qquad \dots(1)$$

and $\qquad (1-x^2)P_n' = n(P_{n-1} - xP_n) \qquad \qquad \dots(2)$

Substitute the value of nP_{n-1} from (1) into (2), we get

$$(1-x^2)P_n' = (2n+1)\, xP_n - (n+1)P_{n+1} - nxP_n = (n+1)xP_n - (n+1)\, P_{n+1}$$

$$(1-x^2)P_n' = (2n+1)\, xP_n - (n+1)P_{n+1} - nxP_n \Rightarrow (1-x^2)P_n' = (n+1)\, (xP_n - P_{n+1})$$

BELTRAMI'S RELATION

The following relation $(2n+1)(x^2-1)\, P_n' = n(n+1)\,)\, (P_{n+1} - P_{n-1})$ is known as Beltrami's Relation.

PROOF. From recurrence relations (V) and (VI), we have

$$(1-x^2)P_n' = n(P_{n-1} - xP_n) \qquad \qquad \dots(1)$$

and $\qquad (1-x^2)P_n' = (n+1)\, (xP_n - P_{n+1}) \qquad \qquad \dots(2)$

Eliminating xP_n from (1) and (2), we get

$$\frac{(1-x^2)P_n'}{n} + \frac{(1-x^2)P_n'}{n+1} = P_{n-1} - P_{n+1} \qquad \text{or} \qquad \frac{(n+1)(1-x^2)P_n' + n(1-x^2)P_n'}{n(n+1)} = P_{n-1} - P_{n+1}$$

or $\qquad (2n+1)(1-x^2)\, P_n' = n(n+1)\,)\, (P_{n-1} - P_{n+1})$

$\therefore \qquad (2n+1)(x^2-1)\, P_n' = n(n+1)\,)\, (P_{n+1} - P_{n-1})$

8.8.5 Christoffel's Expansion

The following series $P_n' = (2n-1)\, P_{n-1} + (2n-5)\, P_{n-3} + (2n-9)P_{n-5} + \dots + 1$

here, $1 = \begin{cases} 3P_1, & \text{if } n \text{ is even} \\ P_0, & \text{if } n \text{ is odd} \end{cases}$ is known as Christoffel's Expansion.

PROOF. From recurrence relation (III), we have

$$(2n+1)\, P_n = P_{n+1}' - P_{n-1}'$$

$\therefore \qquad P_{n+1}' = (2n+1)\, P_n + P_{n-1}' \qquad \qquad \dots(1)$

Now, putting $(n-1)$ in place of n in (1), we get

$$P_n' = (2n-1)\, P_{n-1} + P_{n-2}' \qquad \qquad \dots(2)$$

Now putting $(n-2), (n-4), (n-6)$ in place of n in (2), we get

$$P_{n-2}' = (2n-5)\, P_{n-3} + P_{n-4}' \qquad \qquad \dots(3)$$

$$P_{n-4}' = (2n-9)\, P_{n-5} + P_{n-6}' \qquad \qquad \dots(4)$$

$$P_{n-6}' = (2n-13)\, P_{n-7} + P_{n-8}' \qquad \qquad \dots(5)$$

$$\vdots$$

$$P_2' = 3P_1 + P_0' \text{ , if } n \text{ is even.}$$

Adding (2), (3), (4), (5), ..., we get

$$P_n' = (2n-1)\, P_{n-1} + (2n-5)P_{n-3} + (2n-9)P_{n-5} + (2n-13)P_{n-7} + \dots + 3P_1 + P_0'$$

$$= (2n-1)\, P_{n-1} + (2n-5)P_{n-3} + (2n-9)P_{n-5} + \dots + 3P_1 \qquad \left[\because P_0' = 0 \right]$$

If n is odd, then

$$P_n' = (2n-1)\, P_{n-1} + (2n-5)P_{n-3} + (2n-9)P_{n-5} + \dots + 5P_2 + P_1'$$

$$= (2n-1)\, P_{n-1} + (2n-5)P_{n-3} + (2n-9)P_{n-5} + \dots + 5P_2 + P_0 \qquad \left[\because P_0' = 1 = P_1' \right]$$

Hence, we obtained Christoffel's Expansion.

(I) CHRISTOFFEL'S SUMMATION FORMULA

The following summation

$$\sum_{K=0}^{n} (2k+1)P_k(x)P_k(y) = (n+1)\left[\frac{P_{n+1}(x)\, P_n(y) - P_{n+1}(y)P_n(x)}{(x-y)} \right]$$

is known as a Christoffel's summation.

PROOF. From recurrence relation I, we have

$$(2k+1)\,xP_k(x) = (k+1)P_{k+1}(x) + kP_{k-1}(x) \qquad \ldots(1)$$

and

$$(2k+1)\,yP_k(y) = (k+1)P_{k+1}(y) + kP_{k-1}(y) \qquad \ldots(2)$$

Multiplying (1) by $P_k(y)$ and (2) by $P_k(x)$ and then subtract, we get

$$(2k+1)(x-y)P_k(x)P_k(y) = (k+1)[P_{k+1}(x)P_k(y) - P_k(x)P_{k+1}(y)] + k[P_{k-1}(x)P_k(y) - P_k(x)P_{k-1}(y)]$$

Taking summation from $k = 0$ to $k = n$, we get

$$(x-y) \sum_{K=0}^{n} (2k+1)\,P_k(x)\,P_k(y)$$

$$= \sum_{k=0}^{n} (k+1)\,[P_{k+1}(x)\,P_k(y) - P_k(x)P_{k+1}(y)] + \sum_{k=0}^{n} k\,[P_{k-1}(x)\,P_k(y) - P_k(x)P_{k-1}(y)]$$

$$= \{[P_1(x)P_0(y) - P_0(x)P_1(y)] + 2[P_2(x)P_1(y) - P_1(x)P_2(y)] + 3[P_3(x)P_2(y) - P_2(x)P_3(y)] + \ldots$$

$$+ n[P_n(x)P_{n-1}(y) - P_{n-1}(x)P_n(y)] + (n+1)\,[P_{n+1}(x)P_n(y) - P_n(x)P_{n+1}(y)]\}$$

$$+ \{[P_0(x)P_1(y) - P_1(x)P_0(y)] + 2[P_1(x)P_2(y) - P_2(x)P_1(y)] + 3[P_2(x)P_3(y) - P_3(x)P_2(y)] + \ldots$$

$$+ (n-1)[P_{n-2}(x)P_{n-1}(y) - P_{n-1}(x)P_{n-2}(y)] + n\,[P_{n-1}(x)P_n(y) - P_n(x)P_{n-1}(y)]\}$$

$$= (n+1)\,[P_{n+1}(x)\,P_n(y) - P_n(x)P_{n+1}(y)] \qquad \text{(All the terms cancel except above)}$$

$$\therefore \quad \sum_{k=0}^{n} (2k+1)\,P_k(x)\,P_k(y) = (n+1)\left[\frac{P_{n+1}P_n(y) - P_n(x)P_{n+1}(y)}{(x-y)}\right]$$

 Solved Examples

EXAMPLE 1. ***Prove that*** $|P_n(x)| < 1$, ***when*** $-1 < x < 1$.

SOLUTION. From Laplace first integral for $P_n(x)$, we have

$$P_n(x) = \frac{1}{\pi}\int_0^{\pi} [x \pm \sqrt{(x^2-1)}\cos\theta]^n\,d\theta. \qquad \ldots(1)$$

Now taking

$$\left|[x \pm \sqrt{(x^2-1)}\cos\theta]\right| = \left|x \pm i\sqrt{(1-x^2)}\cos\theta\right|$$

$$= \sqrt{x^2 + (1-x^2)\cos^2\theta} = \sqrt{1 - (1-x^2)\sin^2\theta}$$

$$\therefore \; |x \pm \sqrt{(x^2-1)}\cos\theta| < 1 \text{ except } \theta = 0 \text{ and } \theta = \pi.$$

From (1), we have

$$|P_n(x)| = \left|\frac{1}{\pi}\int_0^{\pi} [x \pm \sqrt{(x^2-1)}\cos\theta]^n\,d\theta\right|$$

$$\leq \frac{1}{\pi}\int_0^{\pi} |x \pm \sqrt{(x^2-1)}\cos\theta|^n\,d\theta$$

$$< \frac{1}{\pi}\int_0^{\pi} 1.d\theta = \frac{1}{\pi}.\pi = 1$$

$$\therefore \; |P_n(x)| < 1.$$

EXAMPLE 2. ***Show that***

(i) $P_n(-x) = (-1)^n P_n(x)$ (BHOPAL-2008, VTU-2003)

and **(ii)** $P'_n(-x) = (-1)^{n+1} P'_n(x)$.

SOLUTION. (i) Since we have
$$P_n(x)$$

$$= \sum_{m=0}^{[n/2]} \frac{(-1)^m (2n-2m)!}{2^n (m)!(n-m)!(n-2m)!}.x^{n-2m}$$

Putting $-x$ in place of x, we get

$$P_n(-x) = \sum_{m=0}^{[n/2]} \frac{(-1)^m (2n-2m)!}{2^n (m)!(n-m)!(n-2m)!}.(-x)^{n-2m}$$

$$= \sum_{m=0}^{[n/2]} \frac{(-1)^m (2n-2m)!}{2^n (m)!(n-m)!(n-2m)!}.(-1)^{n-2m}.x^{n-2m}$$

$$= (-1)^n \sum_{m=0}^{[n/2]} \frac{(-1)^m (2n-2m)!}{2^n (m)!(n-m)!(n-2m)!}.x^{n-2m}$$

$$[\because (-1)^{-2m} = 1]$$

$$= (-1)^n P_n(x).$$

Hence $P_n(-x) = (-1)^n P_n(x).$

(ii) To show $P'_n(-x) = (-1)^{n+1} P'_n(x)$.

From above result we have
$$P_n(-x) = (-1)^n P_n(x).$$

Differentiate both sides w.r.t. x we get

$$-P'_n(-x) = (-1)^n P'_n(x)$$

or $\; P'_n(-x) = (-1)^{n+1} P'_n(x).$

Hence proved the result.

EXAMPLE 3. ***Show that*** $P_n(1) = 1$ ***and*** $(-1) = (-1)$.

(VTU-2003, DELHI-2002, BPTU-2005)

SOLUTION. Since we have

$$\sum_{n=0}^{\infty} P_n(x)t^n = (1 - 2xt + t^2)^{-1/2} \qquad \ldots(1)$$

putting $x = 1$ in (1), we get

$$\sum_{n=0}^{\infty} P_n(1)t^n = (1 - 2t + t^2)^{-1/2} = (1-t)^{-1}$$

or $\; [P_0(1) + tP_1(1) + \ldots + t^n P_n(1) + \ldots]$

$$= [1 + t + t^2 + \ldots + t^n + \ldots].$$

Taking the coefficient of t^n of both sides, we get $P_n(1) = 1$.
Hence proved.

Next putting $x = -1$ in (1), we get

$$\sum_{n=0}^{\infty} P_n(-1)t^n = (1 + 2t + t^2)^{-1/2} = (1+t)^{-1}$$

or $\; [P_0(-1) + tP_1(-1) + \ldots + t^n P_n(-1) + \ldots]$

$$= [1 - t + t^2 - \ldots + (-1)^n t^n + \ldots].$$

Comparing the coefficient of t^n of both sides, we get

$$P_n(-1) = (-1)^n.$$

EXAMPLE 4. *Show that*

(i) $P_n(0) = 0$, *if n is odd.*

(ii) $P_n(0) = \dfrac{(-1)^{n/2}(n)!}{2^n((n/2)!)^2}$, *if n is even.*

(UPTU-2008, SVTU-2008)

SOLUTION. Since we have

$$\sum_{n=0}^{\infty} P_n(x)t^n = (1 - 2xt + t^2)^{-1/2} \qquad \ldots(1)$$

Putting $x = 0$ in (1), we get

$$\sum_{n=0}^{\infty} P_n(0)t^n = (1 + t^2)^{-1/2}$$

or $[P_0(0) + t P_1(0) + \ldots + t^n P_n(0) + \ldots]$

$$= \left[1 - \frac{1}{2}t^2 + \frac{1.3}{2.4}t^4 - \ldots \right.$$

$$\left. + \frac{(-1)^n 1.3.5\ldots(2n-1)}{2^n(n)!}t^{2n} + \ldots \right].$$

(i) If n is odd, then comparing the coefficient of t^n, we get $P_n(0) = 0$

(ii) If n is even, then comparing the coefficient of t^n of both sides, *i.e.*, let $n = 2m$ then comparing the coefficient of t^{2m}, we get

$$P_{2m}(0) = \frac{(-1)^m 1.3.5\ldots(2m-1)}{2.4.6\ldots2m}$$

$$\therefore \quad P_{2m}(0) = \frac{(-1)^m(2m)!}{2^{2m}((m)!)^2}$$

or $\quad P_n(0) = \dfrac{(-1)^{n/2}(n)!}{2^n((n/2)!)^2} \qquad [\because n = 2m]$

EXAMPLE 5. *Show that*

(i) $P_{2n}(0) = (-1)^n \dfrac{1.3.5\ldots(2n-1)}{2.4.6\ldots2n}$

(ii) $P_{2n+1}(0) = 0$.   (UPTU-2005)

SOLUTION. We know that $\sum z^{2n} P_{2n}(x) = (1 - 2xz + z^2)^{-1/2}$

On putting $x = 0$, $\sum z^{2n} P_{2n}(0) = (1 + z^2)^{-1/2}$

$$= 1 + \left(-\frac{1}{2}\right)z^2 + \frac{\left(-\frac{1}{2}\right)\left(-\frac{3}{2}\right)}{2!}(z^2)^2$$

$$+ \frac{\left(-\frac{1}{2}\right)\left(-\frac{3}{2}\right)\left(-\frac{5}{2}\right)}{3!}(z^2)^3$$

$$+ \ldots + \frac{\left(-\frac{1}{2}\right)\left(-\frac{3}{2}\right)\left(-\frac{5}{2}\right)\ldots\left(-\frac{1}{2}-n+1\right)}{n!}$$

$$(z^2)^n + \ldots \quad \ldots(1)$$

Equating the coefficient of z^{2n} both sides, we get

$$P_{2n}(0) = \frac{\left(-\frac{1}{2}\right)\left(-\frac{3}{2}\right)\left(-\frac{5}{2}\right)\ldots\left(-\frac{1}{2}-n+1\right)}{n!}$$

$$= (-1)^n \frac{1.3.5\ldots(2n-1)}{2^n.n!}$$

$$= (-1)^n \frac{1.3.5\ldots(2n-1)}{2.4.6\ldots2n}$$

Coefficient of $z^{2n+1} = P_{2n+1}(0) = 0$.

EXAMPLE 6. *Express* $f(x) = x^4 + 2x^3 + 2x^2 - x - 3$ *in terms of Legendre's polynomials.*   (UPTU(Q. BANK)-2002)

SOLUTION. Degree of $f(x)$ is four. Therefore, we take Legendre's polynomials $P_0(x)$, $P_1(x)$, $P_2(x)$, $P_3(x)$, $P_4(x)$ as follows :

$$P_0(x) = 1, P_1(x) = x, P_2(x) = \frac{1}{2}(3x^2 - 1),$$

$$P_3(x) = \frac{1}{2}(5x^3 - 3x), P_4(x) = \frac{1}{8}(35x^4 - 30x^2 - 3).$$

From Legendre's polynomial, we have

$$1 = P_0(x), x = P_1(x),$$

$$x^2 = \frac{1}{3} + \frac{2}{3}P_2(x) = \frac{1}{3}P_0(x) + \frac{2}{3}P_2(x)$$

$$x^3 = \frac{3}{5}x + \frac{2}{5}P_3(x) = \frac{3}{5}P_1(x) + \frac{2}{5}P_3(x)$$

and $\quad x^4 = -\dfrac{3}{35} + \dfrac{30}{35}x^2 + \dfrac{8}{35}P_4(x)$

$$= -\frac{3}{35}P_0(x) + \frac{30}{35}\left(\frac{1}{3}P_0(x) + \frac{2}{5}P_2(x)\right)$$

$$+ \frac{8}{35}P_4(x)$$

$$= \frac{1}{5}P_0(x) + \frac{20}{35}P_2(x) + \frac{8}{35}P_4(x).$$

Now, substitute these values of $1, x, x^2, x^3$ and x^4 into

$$f(x) = x^4 + 2x^3 + 2x^2 - x - 3$$

$$= \left[\frac{1}{5}P_0(x) + \frac{20}{35}P_2(x) + \frac{8}{35}P_4(x)\right]$$

$$+ 2\left[\frac{3}{5}P_1(x) + \frac{2}{5}P_3(x)\right]$$

$$+ 2\left[\frac{1}{3}P_0(x) + \frac{2}{3}P_2(x)\right] - \left[P_1(x)\right] - 3P_0(x)$$

$$\therefore \quad f(x) = \frac{8}{35}P_4(x) + \frac{4}{5}P_3(x) + \frac{40}{21}P_2(x)$$

$$+ \frac{1}{5}P_1(x) - \frac{32}{15}P_0(x).$$

EXAMPLE 7. *Prove that*

$$\int_{-1}^{1} x^2 P_{n+1}P_{n-1} \, dx = \frac{2n(n+1)}{(2n-1)(2n+1)(2n+3)}.$$

(UPTU(Q. BANK)-2002, JNTU-2006, KERALA(M.TECH)-2005)

SOLUTION. From Recurrence relation I, we have

$$(2n+1)x P_n = (n+1)P_{n+1} + n P_{n-1} \qquad \ldots(1)$$

Putting $(n-1)$ and $(n+1)$ in place of n respectively, we get

$$(2n-1)xP_{n-1} = n P_n + (n-1)P_{n-2} \qquad \ldots(2)$$

$$(2n+3)xP_{n+1} = (n+2)P_{n+2} + (n+1)P_n \qquad \ldots(3)$$

Multiplying (2) and (3), we get

$$(2n-1)(2n+3)x^2 P_{n+1}P_{n-1}$$

$$= [n P_n + (n-1)P_{n-2}][(n+2)P_{n+2} + (n+1)P_n]$$

$$= n(n+2)P_n P_{n+2} + n(n+1)(P_n)^2$$

$$+ (n-1)(n+2)P_{n-2}P_{n+2} + (n^2-1)P_{n-2}P_n.$$

Now integrating from $x = -1$ to $x = 1$ w.r.t. x we get

$$(2n-1)(2n+3)\int_{-1}^1 x^2 P_{n+1} P_{n-1} dx$$

$$= n(n+2)\int_{-1}^1 P_n P_{n+2} dx$$

$$+ n(n+1)\int_{-1}^1 [P_n]^2 dx$$

$$+ (n-1)(n+2)\int_{-1}^1 P_{n-2} P_{n+2} dx$$

$$+ (n^2-1)\int_{-1}^1 P_{n-2} P_n dx$$

$$= n(n+1)\int_{-1}^1 [P_n]^2 dx + 0 + 0 + 0$$

$$= n(n+1)\left[\frac{2}{2n+1}\right] \qquad \text{(By orthogonal property)}$$

$$\therefore \int_{-1}^1 x^2 P_{n+1} P_{n-1} dx = \frac{2n(n+1)}{(2n-1)(2n+1)(2n+3)}.$$

EXAMPLE 8. *Prove that*

$$\frac{1+t}{t\sqrt{1-2xt+t^2}} - \frac{1}{t} = \sum_{n=0}^\infty [P_n(x) + P_{n+1}(x)]t^n.$$

SOLUTION. RHS $= \sum_{n=0}^\infty [P_n(x) + P_{n+1}(x)]t^n$

$$= \sum_{n=0}^\infty t^n P_n(x) + \sum_{n=0}^\infty t^n P_{n+1}(x)$$

$$= \frac{1}{\sqrt{1-2xt+t^2}} + \frac{1}{t}\sum_{n=0}^\infty t^{n+1} P_{n+1}(x)$$

$$= \frac{1}{\sqrt{1-2xt+t^2}} + \frac{1}{t}\left[\sum_{n=0}^\infty t^n P_n(x) - P_0(x)\right]$$

$$= \frac{1}{\sqrt{1-2xt+t^2}} + \frac{1}{t}\left[\sum_{n=0}^\infty t^n P_n(x) - P_0(x)\right]$$

$$= \frac{1}{\sqrt{1-2xt+t^2}} + \frac{1}{t}\left[\sum_{n=0}^\infty t^n P_n(x) - 1\right]$$

$$[\because P_0(x) = 1]$$

$$= \frac{1}{\sqrt{1-2xt+t^2}} + \frac{1}{t\sqrt{1-2xt+t^2}} - \frac{1}{t}$$

$$= \frac{1+t}{t\sqrt{1-2xt+t^2}} - \frac{1}{t} = \text{LHS}.$$

EXAMPLE 9. *Prove that* $\int_{-1}^1 (1-x^2) P_m' P_n' dx = 0, m \neq n$.

(UPTU-2006)

SOLUTION. From Legendre's differential equation, we have

$$(1-x^2)\frac{d^2y}{dx^2} - 2x\frac{dy}{dx} + n(n+1)y = 0 \qquad \dots(1)$$

Since $P_n(x)$ and $P_m(x)$ are the solution of (1), so

$$(1-x^2)\frac{d^2 P_m}{dx^2} - 2x\frac{dP_m}{dx} + m(m+1)P_m = 0$$

or $\quad \dfrac{d}{dx}\left\{(1-x^2)\dfrac{dP_m}{dx}\right\} + m(m+1)P_m = 0 \qquad \dots(2)$

Multiplying (2) by P_n and then integrating from $x = -1$ to 1, we get

$$\int_{-1}^1 \frac{d}{dx}\left\{(1-x^2)\frac{dP_m}{dx}\right\} P_n dx$$

$$+ m(m+1)\int_{-1}^1 P_m P_n dx = 0$$

or $\left[(1-x^2)\dfrac{dP_m}{dx} P_n\right]_{-1}^1 - \int_{-1}^1 (1-x^2)\dfrac{dP_m}{dx}\cdot\dfrac{dP_n}{dx} dx$

$$+ m(m+1)\int_{-1}^1 P_m P_n dx = 0$$

or $\quad 0 - \int_{-1}^1 (1-x^2)P_m' P_n' dx$

$$+ m(m+1)\int_{-1}^1 P_m P_n dx = 0$$

or $\int_{-1}^1 (1-x^2)P_m' P_n' dx = m(m+1)\int_{-1}^1 P_m P_n dx \quad \dots(3)$

Since $m \neq n$, then $\int_{-1}^1 P_m P_n dx = 0$

$$\Rightarrow \quad \int_{-1}^1 (1-x^2)P_m' P_n' dx = 0.$$

EXAMPLE 10. *Prove that* $\int_{-1}^1 (1-x^2)(P_n')^2 dx = \dfrac{2n(n+1)}{2n+1}$.

(UPTU(B.TECH.)2006, SVTU-2008, KERALA(ME)-2005)

SOLUTION. From equation (3) in Ex. 13, we have

$$\int_{-1}^1 (1-x^2)P_m' P_n' dx = m(m+1)\int_{-1}^1 P_m P_n dx$$

If $m = n$, then

$$\int_{-1}^1 (1-x^2)(P_n')^2 dx = n(n+1)\int_{-1}^1 (P_n)^2 dx$$

$$= n(n+1)\cdot\frac{2}{2n+1} = \frac{2n(n+1)}{2n+1}.$$

![Exercise-8.41]

1. Express the following function in terms of Legendre's polynomial.

   (i) $2x + 10x^3$

   (ii) $x^3 + 2x^2 - x - 3$

   (OSMANIA-2003)

   (iii) $x^4 + 3x^3 - x^2 + 5x - 2$

   (VTU-2010, SVTU-2007, BHOPAL-2008, MADRAS-2006)

   (iv) $4x^3 + 6x^2 + 7x + 2$

   (SVTU-2008)

2. Find $P_6(t)$.

3. Show that $\dfrac{1-t^2}{(1-2xt+t^2)^{3/2}} = \sum_{n=0}^\infty (2n+1)P_n(x)t^n$.

4. Prove that $\int_{-1}^1 (P_n')^2 dx = n(n+1)$.

5. Show that $2P_2(x) - 3P_1(x)P_1(x) + 1 = 0$.

6. Prove that $P_{n+1}' + P_n' = \sum_{r=0}^n (2r+1)P_r(x)$.

7. Prove that $(1-2xt+t^2)^{-1/2}$ is a solution of the equation

$$t\frac{\partial^2(ts)}{\partial t^2} + \frac{\partial}{\partial x}\left\{(1-x^2)\frac{\partial s}{\partial x}\right\} = 0$$

8. Prove that $\int_{-1}^1 (1-x^2)P_m' P_n' dx = \dfrac{2n(n+1)}{(2n+1)}\delta_{mn}, \delta_{mn} = \begin{cases} 1, m = n \\ 0, m \neq n \end{cases}$.

9. Prove that

   (i) $\int P_n(x) dx = \left[\dfrac{P_{n+1}(x) - P_{n-1}(x)}{2n+1}\right] + C$

   (ii) $\int_x^1 P_n(x) dx = \dfrac{P_{n-1}(x) - P_{n+1}(x)}{2n+1}$

10. Prove that $\int_x^1 x^n P_n(x) dx = \dfrac{2^{n+1}[(n!)]^2}{(2n+1)!}$

11. Prove that

(i) $\int_{-1}^{1} P_n(x)\, dx = 0,\ n \neq 0,$

(ii) $\int_{-1}^{1} P_0(x)\, dx = 2$

12. Find the value of integrals

(i) $\int_{-1}^{1} x^{99} P_{100}(x)\, dx$,   (ii) $\int_{-1}^{1} x^2 P_2(x)\, dx$

13. Prove that

(i) $P_n'(1) = \dfrac{1}{2}n(n+1)$,   (ii) $P_n'(-1) = (-1)^{n-1} \cdot \dfrac{1}{2}n(n+1)$

14. Prove that if $r_2 > 0$

$$\frac{1}{r} = \frac{1}{r_2}\left[P_0 + P_1(\cos\theta)\left(\frac{r_1}{r_2}\right) + \left(\frac{r_1}{r_2}\right)^2 P_2(\cos\theta) + ... \right],$$

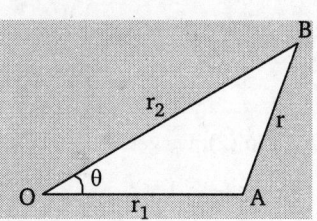

where in ΔOAB

15. Prove that $|P_n(x)| \leqslant \sqrt{\dfrac{\pi}{2n(1-x^2)}}$, when $-1 < x < 1$.

16. If $f(x) = \begin{cases} 0 & -1 < x < 0 \\ x & 0 < x < 1 \end{cases}$, show that

$$f(x) = \frac{1}{4}P_0(x) + \frac{1}{2}P_1(x) + \frac{5}{16}P_2(x) - \frac{3}{32}P_4(x) + ...$$

(UPTU-2003)

Hints to Selected Problems

1. (i) Use $P_n(x) = \sum\limits_{k=0}^{[n/2]} \dfrac{(-1)(2n-2k)!}{2^n . k! (n-2k)!(n-k)!} x^{n-2k}$, i.e.,

$P_0(x) = 1, P_1(x) = x, P_2(x) = \dfrac{1}{2}(3x^2 - 1), P_3(x) = \dfrac{1}{2}(5x^3 - 3x)$

3. Since we have $\dfrac{1}{\sqrt{1-2xt+t^2}} = \sum\limits_{n=0}^{\infty} P_n(x)\, t^n$ ...(1)

On differentiating, we get (after simplification)

$2t(x-t)(1-2xt-t^2)^{-3/2} = \sum\limits_{n=0}^{\infty} 2nP_n(x)\, t^n$ ...(2)

Then adding (1) and (2).

4. Use Christoffel's expansion.

5. Use $P_1(x) = x,\ P_2(x) = \dfrac{1}{2}(3x^2 - 1)$.

6. Put $n = 1, 2, 3, ..., n-1, n$ in the recurrence relation
$P_{n+1}' - P_{n-1}' = (2n+1)P_n$ and then adding all these expression.

7. The given equation can be written as

$$t^2 \frac{\partial^2 s}{\partial t^2} + 2t\frac{\partial s}{\partial t} + (1-x^2)\frac{\partial^2 s}{\partial x^2} - 2x\frac{\partial s}{\partial x} = 0.$$

Then, find the values of different derivatives by using $s = (1-2xt+t^2)^{-1/2}$.

8. Use Beltrami's result and Christoffel's expansion.

9. (i) We have $(2n+1)P_n = P_{n+1}' - P_{n-1}'$

$\Rightarrow P_n(x) = \dfrac{P_{n+1}'(x) - P_{n-1}'(x)}{(2n+1)}$. Then integrate.

10. From Rodrigue's formula, we have

$P_n(x) = \dfrac{1}{2^n . n!}\dfrac{d^n}{dx^n}(x^2-1)^n \Rightarrow x^n P_n(x) = \dfrac{1}{2^n . n!}x^n\dfrac{d^n}{dx^n}(x^2-1)^n.$

Now, integrating w.r.t. x from $x = -1$ to $x = +1$.

11. On integrating the Rodrigue's formula from $x = -1$ to $= +$.

12. Use $\int_{-1}^{1} x^m P_n(x_0) = 0$, if $m < n$.

13. Differentiating the generating function

$\sum\limits_{n=0}^{\infty} t^n P_n(x) = (1-2xt+t^2)^{-1/2}$ and then putting $x = 1$.

14. We have $\cos\theta = \dfrac{r_1^2 + r_2^2 - r^2}{2r_1 r_2}$

$$\Rightarrow \frac{1}{r} = \frac{1}{r_2}\left[\frac{1}{\sqrt{1 - 2\left(\dfrac{r_1}{r_2}\right)\cos\theta + \left(\dfrac{r_1}{r_2}\right)^2}} \right]$$

Then use the generating function $\dfrac{1}{\sqrt{1-2xt+t^2}} = \sum\limits_{n=0}^{\infty} t^n P_n(x)$

By putting $t = \dfrac{r_1}{r_2}$ and $x = \cos\theta$.

Answers

1. (i) $8P_1(x) + 4P_3(x)$ (ii) $\dfrac{2}{5}P_3 + \dfrac{4}{3}P_2 - \dfrac{2}{5}P_1 - \dfrac{7}{3}P_0$ (iii) $\dfrac{8}{35}P_4(x) + \dfrac{6}{5}P_3(x) - \dfrac{2}{21}P_2(x) + \dfrac{34}{5}P_1(x) - \dfrac{224}{105}P_0(x)$

(iv) $\dfrac{8}{5}P_3 - 4P_2 + \dfrac{47}{5}P_1 + 4$ 2. $\dfrac{1}{16}(231x^6 - 315x^4 + 105x^2 - 5)$ **12.** (i) 0   (ii) 4/15

8.8.6 Legendre's Function of the Second Kind

Here we have $Q_n(x) = \dfrac{n!}{1.3...(2n+1)}\left[x^{-n-1} + \dfrac{(n+1)(n+2)}{2(2n+3)}x^{-n-3} + \dfrac{(n+1)(n+2)(n+3)(n+4)}{2.4(2n+3)(2n+5)}x^{-n-5} + ... \right]$

$$= \frac{1}{1.3...(2n+1)}\sum_{r=0}^{\infty}\frac{(n+2r)! x^{-(n+2r+1)}}{2.4...2r(2n+3)(2n+5)...(2n+2r+1)}$$

$\Rightarrow \quad Q_n(x) = \dfrac{2^n . n!}{(2n+1)!}\sum\limits_{r=0}^{\infty}\dfrac{(n+2r)!\, x^{-(n+2r+1)}}{2^r\, r!\,(2n+3)(2n+5)...(2n+2r+1)}.$ ...(1)

Now, differentiating (1) w.r.t. x we have

$$Q_n'(x) = -\frac{2^n \cdot n!}{(2n+1)!} \sum_{r=0}^{\infty} \frac{(n+2r+1)! \; x^{-(n+2r+2)}}{2^r \cdot r! \, (2n+3)(2n+5)\ldots(2n+2r+1)} \qquad \ldots(2)$$

Putting $n-1$ for n in (2), we get

$$Q_{n-1}'(x) = \frac{-2^{n-1} \cdot (n-1)!}{(2n-1)!} \sum_{r=0}^{\infty} \frac{(n+2r)! \; x^{-(n+2r+1)}}{2^r \cdot r! \, (2n+1)(2n+3)\ldots(2n+2r+1)} \qquad \ldots(3)$$

Also, putting $n+1$ for n in (2), we get

$$Q_{n+1}'(x) = -\frac{2^{n+1} \cdot (n+1)!}{(2n+3)!} \sum_{r=0}^{\infty} \frac{(n+2r+2)! \; x^{-(n+2r+3)}}{2^r \cdot r! \, (2n+5)\ldots(2n+2r+1)(2n+2r+3)}$$

$$= -\frac{2^n \cdot n!(2n+2)}{(2n+3)(2n+2)(2n+1)(2n)!} \cdot \sum_{r=0}^{\infty} \frac{(n+2r+2)! \; x^{-(n+2r+3)}}{2^r \cdot r! \, (2n+5)\ldots(2n+2r+3)}$$

$$= -\frac{2^n \cdot n!}{(2n)!} \sum_{r=0}^{\infty} \frac{(n+2r+2)! \; x^{-(n+2r+3)}}{2^r \cdot (r!) \, (2n+1)(2n+3)\ldots(2n+2r+3)} \qquad \ldots(4)$$

8.8.7 IMPORTANT RECURRENCE RELATIONS

(i) $Q_{n+1}' - Q_{n-1}' = (2n+1)\,Q_n$

(ii) $nQ_{n+1}' + (n+1)\,Q_{n-1}' = (2n+1)\,xQ_n'$

(iii) $(2n+1)\,xQ_n = (n+1)\,Q_{n+1} + nQ_{n-1}$

(iv) $(2n+1)(1-x^2)\,Q_n' = n(n+1)\,(Q_{n-1} - nQ_{n+1})$

(v) $x\,Q_n' - Q_{n-1}' = n\,Q_n$

(vi) $Q_n' - xQ_{n-1}' = n\,Q_{n-1}$

(vii) $(x^2-1)\,Q_n' = nxQ_n - nQ_{n-1}$

(viii) $(x^2-1)\,Q_n' = (n+1)Q_{n+1} - (n+1)\,xQ_n$

(ix) $(x^2-1)\,(Q_nP_n' - P_nQ_n') = c$, a constant.

(x) $\dfrac{Q_n(x)}{P_n(x)} = \int_x^{\infty} \dfrac{dx}{(x^2-1)\,P_n^2(x)}$

(xi) $n[P_nQ_{n-1} - Q_nP_{n-1}] = 1$.

(xii) $P_nQ_{n-2} - Q_nP_{n-2} = \dfrac{(2n-1)}{n(n-1)}x$.

THEOREM 1. *If P_n is a solution of Legendre's equation, then the complete solution of the Legendre's equation is given by $aP_n + bQ_n$, where*

$$Q_n = cP_n \int \frac{dx}{(1-x^2)P_n^2}, \; c \text{ being a constant.}$$

PROOF. Consider the Legendre's equation

$$(1-x^2)\frac{d^2y}{dx^2} - 2x\frac{dy}{dx} + n(n+1)y = 0. \qquad \ldots(1)$$

Let $y = uP_n$ be the complete solution of (1), where u is a function of x.

Therefore, $\dfrac{dy}{dx} = u\dfrac{dP_n}{dx} + P_n\dfrac{du}{dx}$, $\dfrac{d^2y}{dx^2} = u\dfrac{d^2P_n}{dx^2} + 2\dfrac{dP_n}{dx}\cdot\dfrac{du}{dx} + P_n\dfrac{d^2u}{dx^2}$ $\qquad \ldots(2)$

Putting all these values in (1), we get

$$(1-x^2)\left\{u\frac{d^2P_n}{dx^2} + 2\frac{dP_n}{dx}\cdot\frac{du}{dx} + P_n\frac{d^2u}{dx^2}\right\} - 2x\left\{u\frac{dP_n}{dx} + P_n\frac{du}{dx}\right\} + n(n+1)uP_n = 0$$

$$\Rightarrow \quad (1-x^2)\left\{P_n\frac{d^2u}{dx^2} + 2\frac{dP_n}{dx}\frac{du}{dx}\right\} + (1-x^2)u\frac{d^2P_n}{dx^2} - 2xu\frac{dP_n}{dx} - 2uP_n\frac{du}{dx} + n(n+1)uP_n = 0$$

$$\Rightarrow (1-x^2)\left\{P_n\frac{d^2u}{dx^2} + 2\frac{du}{dx}\frac{dP_n}{dx}\right\} + u\left\{(1-x^2)\frac{d^2P_n}{dx^2} - 2x\frac{dP_n}{dx} + n(n+1)P_n\right\} - 2xP_n\frac{du}{dx} = 0. \qquad \ldots(3)$$

Now since $P_n(x)$ is the solution of (1), therefore, we have

$$(1-x^2)\frac{d^2P_n}{dx^2} - 2x\frac{dP_n}{dx} + n(n+1)P_n = 0.$$

Using this in (3), we get

$$(1-x^2)\left\{P_n\frac{d^2u}{dx^2} + 2\frac{du}{dx}\frac{dP_n}{dx}\right\} - 2xP_n\frac{du}{dx} = 0 \qquad \Rightarrow \qquad \frac{d^2u/dx^2}{du\,dx} + 2\frac{dP}{P}\frac{dx}{} - \frac{2x}{} = 0.$$

On integrating, we have

$$\log\frac{du}{dx} + 2\log P_n + \log(1-x^2) = \log k \qquad \Rightarrow \log\left\{\frac{du}{dx}\cdot P_n^2(1-x^2)\right\} = \log k$$

$$\Rightarrow \qquad \frac{du}{dx}\cdot P_n^2(1-x^2) = k \qquad \Rightarrow \quad \frac{du}{dx} = \frac{k}{(1-x^2)P_n^2}.$$

On integrating, we have $u = k\int \dfrac{1}{(1-x^2)P_n^2}\,dx + A$, where A and k are arbitrary constants of integration. Hence, the complete solution of Legendre's equation is

$$y = uP_n = \left[k\int \frac{dx}{(1-x^2)P_n^2} + A \right] P_n = aP_n + kP_n\int \frac{dx}{(1-x^2)P_n^2} = aP_n + \frac{k}{c}\cdot cP_n\int \frac{dx}{(1-x^2)P_n^2} = aP_n + bQ_n$$

where $b = k/c$.

Solved Examples

EXAMPLE 1. **Show that** $Q_2(x) = \dfrac{1}{2}P_2(x)\log\dfrac{x+1}{x-1} - \dfrac{3}{2}x.$

SOLUTION. Since we know that

$$(n+1)Q_{n+1} = (2n+1)xQ_n - nQ_{n-1}.$$

Putting $n = 1$, we get $2Q_2 = 3xQ_1 - Q_0$

$$= 3x\left[\frac{x}{2}\log\frac{x+1}{x-1} - 1\right] - \frac{1}{2}\log\frac{x+1}{x-1}$$

$$= \frac{1}{2}(3x^2 - 1)\log\frac{x+1}{x-1} - 3x$$

$$= P_2(x)\log\frac{x+1}{x-1} - 3x.$$

Hence, $Q_2(x) = \dfrac{1}{2}P_2(x)\log\dfrac{x+1}{x-1} - \dfrac{3}{2}x.$

EXAMPLE 2. **Show that**

(i) $n(Q_nP_{n-1} - Q_{n-1}P_n)$
$= (n-1)(Q_{n-1}P_{n-2} - Q_{n-2}P_{n-1})$

(ii) $n(Q_nP_{n-1} - Q_{n-1}P_n) = -1.$

SOLUTION. (i) Using recurrence relation of $P_n(x)$ and $Q_n(x)$, we have

$$(2n+1)xP_n = (n+1)P_{n+1} + nP_{n-1} \qquad \ldots(1)$$

$$(2n+1)xQ_n = (n+1)Q_{n+1} + nQ_{n-1}. \qquad \ldots(2)$$

Replacing n by $n-1$ in (1) and (2), we get

$$(2n-1)xP_{n-1} = nP_n + (n-1)P_{n-2} \qquad \ldots(3)$$

$$(2n-1)xQ_{n-1} = nQ_n + (n-1)Q_{n-2}. \qquad \ldots(4)$$

Multiplying (3) by Q_{n-1} and (4) by P_{n-1} and then subtracting, we have

$$0 = n(P_nQ_{n-1} - P_{n-1}Q_n)$$
$$+ (n-1)(P_{n-2}Q_{n-1} - P_{n-1}Q_{n-2})$$

$$\Rightarrow n(Q_nP_{n-1} - Q_{n-1}P_n)$$
$$= (n-1)(Q_{n-1}P_{n-2} - Q_{n-2}P_{n-1}). \qquad \ldots(5)$$

(ii) Let $u_n = n(Q_nP_{n-1} - Q_{n-1}P_n)$. Then (5) can be written as $u_n = u_{n-1}$

$$\Rightarrow u_{n-1} = u_{n-2} = u_{n-3} = \ldots = u_3 = u_2 = u_1$$

$$\therefore \quad u_n = u_1$$

$$\Rightarrow n(Q_nP_{n-1} - Q_{n-1}P_n) = Q_1P_0 - Q_0P_1$$

Hence, $n(Q_nP_{n-1} - Q_{n-1}P_n) = Q_1 - xQ_0$

$$= \frac{x}{2}\log\left(\frac{x+1}{x-1}\right) - 1 - \frac{x}{2}\log\frac{x+1}{x-1} = -1.$$

Exercise-8.42

1. Show that $\dfrac{Q_n}{P_n} = \int_x^\infty \dfrac{dx}{(x^2-1)P_n^2}.$

2. Show that $Q_1(x) = \dfrac{x}{2}\log\dfrac{x+1}{x-1} - 1.$

3. Show that

(a) $P_nQ_{n-1} - Q_nP_{n-1} = \dfrac{1}{n}.$

(b) $P_nQ_{n-2} - Q_nP_{n-2} = \dfrac{(2n-1)x}{n(n-1)}.$

4. Show that $Q_n(x) = 2^n n!\int_x^\infty dx\int_x^\infty dx\ldots\int_x^\infty(x^2-1)^{-n-1}dx.$

5. Show that

$$(2n+1)(1-x^2)Q_n'(x) = n(n+1)[Q_{n-1}(x) - Q_{n+1}(x)].$$

Hints to Selected Problems

1. Putting the values of P_n, Q_n, P_n' and Q_n' in the following relation

$$Q_nP_n' - P_nQ_n' = \frac{c}{x^2-1} = \frac{c}{x^2}\left(1 - \frac{1}{x^2}\right)^{-1} = c\left(\frac{1}{x^2} + \frac{1}{x^4} + \frac{1}{x^6} + \ldots\right)$$

Then equating the coefficient of $\dfrac{1}{x^2}$ of both the sides, we get $c = 1$.

Then write $\dfrac{Q_nP_n' - P_nQ_n'}{P_n^2} = -\dfrac{d}{dx}\left(\dfrac{Q_n}{P_n}\right) = \dfrac{1}{(x^2-1)P_n^2}$

Then taking the limit after integrating from x to ∞.

2. Putting $n = 1$ in eq. (4) of example 1, we get

$$\frac{Q_1(x)}{P_1(x)} = \int_x^\infty \frac{dx}{(x^2-1)P_1^2(x)}$$

$$\Rightarrow Q_1(x) = x\int_0^\infty \frac{dx}{(x^2-1)x^2}.$$ Now, integrating after breaking using partial fraction.

3. Using the following recurrence relation

$$(2n+1)xP_n = (n+1)P_{n+1} + nP_{n-1}$$

and $(2n+1)xQ_n = (n+1)Q_{n+1} + nQ_{n-1}.$

4. We have

$$(x^2-1)^{-n-1} = x^{-2n-1}\left(1 - \frac{1}{x^2}\right)^{-n-1}$$

$$= x^{-2n-2}\left[1 + (n+1)\frac{1}{x^2} + \frac{(n+1)(n+2)}{2!}\cdot\frac{1}{x^4}\ldots\right]$$

$$= x^{-(2n+2)} + (n+1)x^{-(2n+4)} + \frac{(n+1)(n+2)}{2!}x^{-(2n+6)} + \ldots$$

Now integrating both sides w.r.t. x between x to ∞, n times.

5. Since $Q_n(x)$ is the solution of $\dfrac{d}{dx}\left[(1-x^2)\dfrac{dy}{dx}\right] + n(n+1)\ldots$

So, $\dfrac{d}{dx}\left((1-x^2)Q_n'\right) = -n(n+1)Q_n.$ Then, integrating both sides from x to ∞.

8.8.8 Bessel's Function

The homogeneous linear differential equation of the form

$$x^2 \frac{d^2y}{dx^2} + x\frac{dy}{dx} + \left(x^2 - n^2\right)y = 0 \qquad \qquad ...(1)$$

is known as Bessel's differential equation, where n is a non-negative real number.

(I) SOLUTION OF THE BESSEL'S FUNCTION

Change the differential equation (1) into standard form by dividing (1) by x^2.

$$\frac{d^2y}{dx^2} + \frac{1}{x}\frac{dy}{dx} + \left(1 - \frac{n^2}{x^2}\right)y = 0 \qquad \qquad ...(2)$$

Now compare this differential equation with following equation

$$\frac{d^2y}{dx^2} + P(x)\frac{dy}{dx} + Q(x)y = 0$$

We get

$$P(x) = \frac{1}{x}, \quad Q(x) = \left(1 - \frac{n^2}{x^2}\right).$$

It is obvious from $P(x)$ and $Q(x)$, that $x = 0$ is a singular point which is located at the origin. Therefore we assume the solution of (1) in the form of a power series of the following type

$$y = \sum_{m=0}^{\infty} a_m x^{m+r} (a_0 \neq 0) \qquad \qquad ...(3)$$

Differentiating (3) w.r.t. x we get

$$\frac{dy}{dx} = \sum_{m=0}^{\infty} a_m (m+r) x^{m+r-1} \qquad \qquad ...(4)$$

Again differentiating (4) w.r.t. x we get

$$\frac{d^2y}{dx^2} = \sum_{m=0}^{\infty} a_m (m+r)(m+r-1) x^{m+r-2}. \qquad \qquad ...(5)$$

Now substitute the values of $y, \frac{dy}{dx}, \frac{d^2y}{dx^2}$ from (3), (4) and (5) into (1), we have

$$x^2 \sum_{m=0}^{\infty} a_m (m+r)(m+r-1) x^{m+r-2} + x \sum_{m=0}^{\infty} a_m (m+r) x^{m+r-1} + (x^2 - n^2) \sum_{m=0}^{\infty} a_m x^{m+r} = 0$$

$$\sum_{m=0}^{\infty} a_m (m+r)(m+r-1) x^{m+r} + \sum_{m=0}^{\infty} a_m (m+r) x^{m+r} - \sum_{m=0}^{\infty} a_m n^2 x^{m+r} + \sum_{m=0}^{\infty} a_m x^{m+r+2} = 0 \qquad ...(6)$$

Equation (6) will be an identity if the equation (3) is a solution of (1), then coefficient of each terms in (6) will be zero. Thus taking the coefficient of x^r, x^{r+1}

$$a_0 r(r-1) + a_0 r - n^2 a_0 = 0 \qquad \qquad ...(7)$$

$$a_1(r+1)r + a_1(r+1) - n^2 a_1 = 0. \qquad \qquad ...(8)$$

In general, taking the coefficients of x^{s+r}

$$a_s(s+r)(s+r-1) + a_s(s+r) - n^2 a_s + a_{s-2} = 0 \quad \text{for } s = 2,3,4,... \qquad \qquad ...(9)$$

From (7), we have $r(r-1) + r - n^2 = 0$ $[\because a_0 \neq 0]$

or $r^2 - n^2 = 0$ or $r = n, -n.$

From (8), we have $[(r+1)r + (r+1) - n^2]a_1 = 0$.

For any value of $r = n, -n$, we get $a_1 = 0$.

From (9), we have

$$a_s[(s+r)(s+r-1) + s + r - n^2] + a_{s-2} = 0 \qquad \text{or} \qquad a_s[(s+r)^2 - n^2] + a_{s-2} = 0$$

or $a_s(s+r-n)(s+r+n) + a_s = 0$ \qquad \qquad ...(10)

For case if $r = n$, then (1) becomes

$$a_s(s)(s+2n) + a_{s-2} = 0 \qquad \text{or} \qquad a_s = -\frac{1}{s(s+2n)}.a_{s-2}$$

Putting $s = 2,3,4,5,...$

$$a_2 = -\frac{1}{2(2+2n)}a_0 \; ; a_3 = -\frac{1}{3(3+2n)}a_1 = 0 \qquad \qquad [\because a_1 = 0]$$

$$a_4 = -\frac{1}{4(4+2n)}a_2 = -\frac{1}{4(4+2n)}\left(-\frac{1}{2(2+2n)}\right)a_0 = (-1)^2 \frac{1}{2 \cdot 4(2+2n)(4+2n)}a_0$$

$$\vdots$$

etc.

We observe that $a_1 = a_3 = a_5 = \dots = 0$. Since a_0 is arbitrary. Let us choose $a_0 = \dfrac{1}{2^n \Gamma(n+1)}$ where $\Gamma(n+1)$ is a gamma function, therefore, we know that $\Gamma(n+1) = n\Gamma(n)$ and if n is positive integer and $\Gamma(n+1) = (n)!$. Thus

$$a_2 = -\frac{1}{2(2+2n)}a_0 = -\frac{1}{2^2(1+n)} \cdot \frac{1}{2^n \Gamma(n+1)} \qquad \left(\because a_0 = \frac{1}{2^n \Gamma(n+1)} \right)$$

$$= -\frac{1}{2^{n+2}\Gamma(n+2)}$$

and

$$a_4 = (-1)^2 \frac{1}{2^4 \cdot (2)!(1+n)(2+n)} \cdot \frac{1}{2^n \Gamma(n+1)} = (-1)^2 \frac{1}{2^{n+4} \cdot (2)!\,\Gamma(n+3)}$$

and so on. Now From (3), we have

$$y = \sum_{m=0}^{\infty} a_m x^{m+r} = a_0 x^r + a_1 x^{1+r} + a_2 x^{2+r} + a_3 x^{3+r} + a_4 x^{4+r} + \dots$$

$$= a_0 x^r + a_2 x^{2+r} + a_4 x^{4+r} + \dots = a_0 x^n + \left\{ -\frac{1}{2^{n+2}\Gamma(n+2)} x^{n+2} \right\} + \frac{1}{2^{n+4}(2)!\,\Gamma(n+3)} x^{n+4} + \dots$$

$$= \frac{1}{2^n \Gamma(n+1)} x^n - \frac{1}{2^{n+2}(1)!\,\Gamma(n+2)} x^{n+2} + \frac{1}{2^{n+4}(2)!\,\Gamma(n+3)} x^{n+4} + \dots$$

$$= \sum_{m=0}^{\infty} \frac{(-1)^m x^{n+2m}}{2^{n+2m}(m)!\,\Gamma(n+m+1)}.$$

This solution is known as Bessel's function, which is denoted by $J_n(x)$. This function is also known as Bessel's function of first kind.

$$\therefore \qquad J_n(x) = \sum_{m=0}^{\infty} (-1)^m \frac{x^{n+2m}}{2^{n+2m}(m)!\,\Gamma(n+m+1)} \qquad \dots(11)$$

For case if $r = -n$, we have

$$J_{-n}(x) = \sum_{m=0}^{\infty} \frac{(-1)^m x^{-n+2m}}{2^{-n+2m}(m)!\,\Gamma(-n+m+1)} \qquad \dots(12)$$

(2) GENERAL SOLUTION

The solution of the Bessel's differential equation of the type $y(x) = A J_n(x) + B J_{-n}(x)$ where A and B are arbitrary constants, is called general solution.

(VTU-2006)

(3) LINEAR DEPENDENCE

For an integer $r = n$, the Bessel's function $J_n(x)$ and $J_{-n}(x)$ are linearly dependent because $J_{-n}(x) = (-1)^n J_n(x)$ for $n = 1, 2, \dots$

PROOF. Since

$$J_{-n}(x) = \sum_{m=0}^{\infty} \frac{(-1)^m x^{-n+2m}}{2^{-n+2m}(m)!\,\Gamma(-n+m+1)}, \qquad \dots(1)$$

if n is a positive integer, then the gamma functions in the coefficients of first n terms becomes infinite and coefficients of (1) becomes zero. Thus the summation will start at $m = n$ and in this case $\Gamma(-n+m+1) = (m-n)!$.

From (1), we now have,

$$J_{-n}(x) = \sum_{m=n}^{\infty} \frac{(-1)^m x^{-n+2m}}{2^{-n+2m}(m)!\,(m-n)!} = \sum_{k=0}^{\infty} \frac{(-1)^{n+k} x^{n+2k}}{2^{n+2k}(k)!\,(n+k)!} \qquad [\because m = n+k]$$

$$= (-1)^n \sum_{k=0}^{\infty} \frac{(-1)^k x^{n+2k}}{2^{n+2k}(k)!\,\Gamma(n+k+1)}$$

$$\therefore \qquad J_{-n}(x) = (-1)^n J_n(x)$$

(4) DEFINITION OF $J_n(x)$, WHEN n = 0

Putting $n = 0$ in the Bessel's differential equation, we get

$$x\frac{d^2y}{dx^2} + \frac{dy}{dx} + xy = 0 \qquad \dots(1)$$

Let us assume the solution

$$y = \sum_{m=0}^{\infty} a_m x^{m+r} \quad (a_0 \neq 0) \qquad \dots(2)$$

$$\therefore \qquad \frac{dy}{dx} = \sum_{m=0}^{\infty} a_m(m+r)x^{m+r-1} \qquad \text{and} \qquad \frac{d^2y}{dx^2} = \sum_{m=0}^{\infty} a_m(m+r)(m+r-1)x^{m+r-2}.$$

Substitute these values in (1), we get

$$x\sum_{m=0}^{\infty} a_m(m+r)(m+r-1)x^{m+r-2} + \sum_{m=0}^{\infty} a_m(m+r)x^{m+r-1} + x\sum_{m=0}^{\infty} a_m x^{m+r} = 0$$

or

$$\sum_{m=0}^{\infty} a_m(m+r)(m+r-1)x^{m+r-1} + \sum_{m=0}^{\infty} a_m(m+r)x^{m+r-1} + \sum_{m=0}^{\infty} a_m x^{m+r+1} = 0 \qquad \dots(3)$$

If (2) is the solution of (1), then (3) will be an identity, Thus coefficients of each terms will be zero. So that taking the coefficients of x^{r-1}, we get

$$a_0 r(r-1) + a_0 r = 0 \qquad \text{or} \qquad r^2 a_0 = 0 \qquad \text{or} \qquad r = 0 \qquad\qquad [\because a_0 \neq 0]$$

Now taking the coefficient of x^r, we have

$$a_1(1+r)r + a_1(1+r) = 0 \qquad \text{or} \qquad a_1(1+r)^2 = 0 \qquad \text{or} \qquad a_1 = 0 \qquad\qquad [\because r = 0]$$

In general, taking the coefficient of x^{m+r}

$$a_{m+1}(m+r+1)(m+r) + a_{m+1}(m+r+1) + a_{m-1} = 0$$

or $\qquad a_{m+1}(m+r+1)^2 + a_{m-1} = 0 \qquad\qquad$ or $\qquad a_{m+1} = -\dfrac{a_{m-1}}{(m+r+1)^2}$

For the case $r = 0$, $a_{m+1} = -\dfrac{a_{m-1}}{(m+1)^2}$.

Putting $m = 1, 2, 3, 4, 5, \ldots$, $a_3 = -\dfrac{a_1}{9} = 0 \qquad\qquad\qquad [\because a_1 = 0]$

$$a_2 = -\frac{a_0}{2^2} \, ; a_4 = -\frac{a_2}{4^2} = \frac{(-1)^2 a_0}{2^2 \cdot 4^2} \, ; a_5 = 0 \ldots \text{ etc.}$$

Thus we obtained $a_1 = a_3 = a_5 = \ldots = 0$. Hence, $y = a_0 \left(1 - \dfrac{x^2}{2^2} + \dfrac{x^4}{2^2 \cdot 4^2} - \dfrac{x^6}{2^2 \cdot 4^2 \cdot 6^2} + \ldots \right)$

If $a_0 = 1$, then $y = J_0(x)$.

$$\therefore \qquad J_0(x) = 1 - \frac{x^2}{2^2} + \frac{x^4}{2^2 \cdot 4^2} - \frac{x^6}{2^2 \cdot 4^2 \cdot 6^2} + \ldots$$

$J_0(x)$ is also known as Bessel's function of order zero.

(5) GENERATING FUNCTION FOR $J_n(x)$

The function of the form $e^{\left[\frac{1}{2}x\left(t - \frac{1}{t}\right)\right]}$ *generates* $J_n(x)$, *if taking coefficient of* t^n. *Thus this function is known as Generating function for* $J_n(x)$.

(VTU-2007)

PROOF.   Expand $e^{\left[\frac{1}{2}x\left(t - \frac{1}{t}\right)\right]} = e^{\frac{xt}{2}} \cdot e^{-\frac{x}{2t}}$

$$= \left[1 + \frac{xt}{2} + \frac{1}{(2)!}\left(\frac{xt}{2}\right)^2 + \ldots + \frac{1}{(n)!}\left(\frac{xt}{2}\right)^n + \frac{1}{(n+1)!}\left(\frac{xt}{2}\right)^{n+1} + \frac{1}{(n+2)!}\left(\frac{xt}{2}\right)^{n+2} + \ldots \right]$$

$$\cdot \left[1 - \frac{x}{2t} + \frac{1}{(2)!}\left(\frac{x}{2t}\right)^2 + \ldots + \frac{(-1)^n}{(n)!}\left(\frac{x}{2t}\right)^n + \frac{(-1)^{n+1}}{(n+1)!}\left(\frac{x}{2t}\right)^{n+1} + \frac{(-1)^{n+2}}{(n+2)!}\left(\frac{x}{2t}\right)^{n+2} + \ldots \right]$$

Now collecting the coefficient of t^n, in above expression (obtained after multiplication) is given by

$$= \frac{1}{(n)!}\left(\frac{x}{2}\right)^n - \frac{1}{(n+1)!}\left(\frac{x}{2}\right)^{n+2} + \frac{1}{(n+2)!} \cdot \frac{1}{(2)!}\left(\frac{x}{2}\right)^{n+4} + \ldots$$

$$= \sum_{m=0}^{\infty} (-1)^m \cdot \frac{1}{(m)!(n+m)!} \cdot \left(\frac{x}{2}\right)^{n+2m} = \sum_{m=0}^{\infty} \frac{(-1)^m x^{n+2m}}{2^{n+2m}(m)! \Gamma(m+n+1)} = J_n(x) \qquad [\because \Gamma(m+n+1) = (m+n!)]$$

$$\therefore \qquad e^{\left[\frac{1}{2}x\left(t - \frac{1}{t}\right)\right]} = \sum_{n=0}^{\infty} t^n J_n(x).$$

If taking the coefficient of t^{-n}, we get

$$= \frac{(-1)^n}{(n)!}\left(\frac{x}{2}\right)^n + \frac{(-1)^{n+1}}{(n+1)!}\left(\frac{x}{2}\right)^{n+2} + \frac{(-1)^{n+2}}{(n+2)!} \cdot \frac{1}{(2)!}\left(\frac{x}{2}\right)^{n+4} + \ldots$$

$$= (-1)^n \left[\frac{1}{(n)!}\left(\frac{x}{2}\right)^n - \frac{1}{(n+1)!}\left(\frac{x}{2}\right)^{n+2} + \frac{1}{(n+2)!} \cdot \frac{1}{(2)!}\left(\frac{x}{2}\right)^{n+4} - \ldots \right]$$

$$= (-1)^n \sum_{m=0}^{\infty} \frac{(-1)^m x^{n+2m}}{2^{n+2m}(m)! \Gamma(n+m+1)} = (-1)^n J_n(x) = J_{-n}(x) \qquad [\because J_{-n}(x) = (-1)^n J_n(x)]$$

Hence we obtain $e^{\left[\frac{1}{2}x\left(t - \frac{1}{t}\right)\right]} = \sum_{n=-\infty}^{\infty} t^n J_n(x)$.

8.8.9 ORTHOGONAL PROPERTIES OF BESSEL'S FUNCTION

$\int_0^1 x J_n(\alpha x) J_n(\beta x)\, dx = 0$, *where* α, β *are the roots of* $J_n(x) = 0$.

(VTU-2006, 13, UPTU(Q. BANK)-2002)

PROOF. We know that $x^2 \dfrac{d^2y}{dx^2} + x \dfrac{dy}{dx} + (\alpha^2 x^2 - n^2)y = 0$ ...(1)

and $x^2 \dfrac{d^2z}{dx^2} + x \dfrac{dz}{dx} + (\beta^2 x^2 - n^2)z = 0$ ...(2)

where, $y = J_n(\alpha x)$ and $z = J_n(\beta x)$

Multiplying (1) by z/x and (2) by $-y/x$ and then add, we get

$$x\left[z\dfrac{d^2y}{dx^2} - y\dfrac{d^2z}{dx^2}\right] + \left[z\dfrac{dy}{dx} - y\dfrac{dz}{dx}\right] + (\alpha^2 - \beta^2)xyz = 0 \;\Rightarrow\; \dfrac{d}{dx}\left[x\left(z\dfrac{dy}{dx} - y\dfrac{dz}{dx}\right)\right] + (\alpha^2 - \beta^2)xyz = 0 \quad ...(3)$$

On integrating from 0 to 1, we get

$$\left[x\left(z\dfrac{dy}{dx} - y\dfrac{dz}{dx}\right)\right]_0^1 + (\alpha^2 - \beta^2)\int_0^1 xyz\, dx = 0$$

$$\Rightarrow \quad (\beta^2 - \alpha^2)\int_0^1 xyz\, dx = \left[x\left(z\dfrac{dy}{dx} - y\dfrac{dz}{dx}\right)\right]_0^1 = \left[z\dfrac{dy}{dx} - y\dfrac{dz}{dx}\right]_{x=1} \quad ...(4)$$

Now, $y = J_n(\alpha x) \Rightarrow \dfrac{dy}{dx} = \alpha J_n'(\alpha x)$, $z = J_n(\beta x) \Rightarrow \dfrac{dz}{dx} = \beta J_n'(\beta x)$

Putting these values in (4), we get

$$(\beta^2 - \alpha^2)\int_0^1 x\, J_n(\alpha x)\, J_n(\beta x)dx$$

$$= \alpha J_n'(\alpha)\, J_n(\beta) ... \beta J_n'(\beta)\, J_n(\alpha) \quad ...(5)$$

Now, since α, β are the roots of $J_n(x) = 0 \Rightarrow J_n(\alpha) = J_n(\beta) = 0$.

Putting these values in (5), we get

$$(\beta^2 - \alpha^2)\int_0^1 x\, J_n(\alpha x).J_n(\beta x)\, dx = 0.$$

Hence, $\int_0^1 x\, J_n(\alpha x)\, J_n(\beta x)\, dx = 0$.

8.8.10 Recurrence Relations for $J_n(x)$

(I) $x J_n'(x) = n J_n(x) - x J_{n+1}(x)$, where $J_n'(x) = \dfrac{dJ_n(x)}{dx}$

PROOF. Since we have $J_n(x) = \sum\limits_{m=0}^{\infty} (-1)^m \dfrac{1}{(m)!\,\Gamma(n+m+1)} \cdot \left(\dfrac{x}{2}\right)^{n+2m}$ ...(1)

Differentiating (1) w.r.t. x, we get

$$J_n'(x) = \sum\limits_{m=0}^{\infty} \dfrac{(-1)^m (n+2m)}{(m)!\,\Gamma(n+m+1)} \cdot \dfrac{1}{2} \cdot \left(\dfrac{x}{2}\right)^{n+2m-1}$$

or $x J_n'(x) = \sum\limits_{m=0}^{\infty} \dfrac{(-1)^m (n+2m)}{(m)!\,\Gamma(n+m+1)} \cdot \left(\dfrac{x}{2}\right)^{n+2m} = \sum\limits_{m=0}^{\infty} \dfrac{(-1)^m (n+2m)}{(m)!\,\Gamma(n+m+1)} \cdot \left(\dfrac{x}{2}\right)^{n+2m}$

$$= \sum\limits_{m=0}^{\infty} \dfrac{(-1)^m\, n}{(m)!\,\Gamma(n+m+1)} \cdot \left(\dfrac{x}{2}\right)^{n+2m} + \sum\limits_{m=0}^{\infty} \dfrac{(-1)^m\, 2m}{(m)!\,\Gamma(n+m+1)} \cdot \left(\dfrac{x}{2}\right)^{n+2m}$$

$$= n J_n(x) + \sum\limits_{m=0}^{\infty} \dfrac{(-1)^m \cdot 2}{(m-1)!\,\Gamma(n+m+1)} \cdot \dfrac{x}{2} \cdot \left(\dfrac{x}{2}\right)^{n-1+2m}$$

$$= n J_n(x) + x \sum\limits_{m=0}^{\infty} \dfrac{(-1)^m}{(m-1)!\,\Gamma(n+m+1)} \cdot \left(\dfrac{x}{2}\right)^{n-1+2m}$$

$$= n J_n(x) + x \sum\limits_{m=1}^{\infty} \dfrac{(-1)^m}{(m-1)!\,\Gamma(n+m+1)} \cdot \left(\dfrac{x}{2}\right)^{n-1+2m} \qquad \left(\because \dfrac{1}{(-1)!} = 0\right)$$

$$= n J_n(x) + x \sum\limits_{k=0}^{\infty} \dfrac{(-1)^{k+1}}{(k)!\,\Gamma(n+1+k+1)} \cdot \left(\dfrac{x}{2}\right)^{n+1+2k} = n J_n(x) - x J_{n+1}(x)$$

$$\therefore \qquad x J_n'(x) = n J_n(x) - x J_{n+1}(x)$$

✎ Remark

- $\dfrac{d}{dx}(x^{-n} J_n) = -x^{-n} J_{n+1}$

(II) $x J_n'(x) = -n J_n(x) + x J_{n-1}(x)$. (UPTU-2004, 06)

PROOF. Since we have $J_n(x) = \sum\limits_{m=0}^{\infty} \dfrac{(-1)^m}{(m)! \, \Gamma(n+m+1)} \cdot \left(\dfrac{x}{2}\right)^{n+2m}$ ...(1)

Differentiating (1) w.r.t. x, we get

$$J_n'(x) = \sum_{m=0}^{\infty} \frac{(-1)^m (n+2m)}{(m)! \, \Gamma(n+m+1)} \cdot \frac{1}{2} \cdot \left(\frac{x}{2}\right)^{n+2m-1}$$

or $\quad x J_n'(x) = \sum\limits_{m=0}^{\infty} \dfrac{(-1)^m (n+2m)}{(m)! \, \Gamma(n+m+1)} \cdot \left(\dfrac{x}{2}\right)^{n+2m} = \sum\limits_{m=0}^{\infty} \dfrac{(-1)^m (2n+2m-n)}{(m)! \, \Gamma(n+m+1)} \cdot \left(\dfrac{x}{2}\right)^{n+2m}$

$$= -n \sum_{m=0}^{\infty} \frac{(-1)^m}{(m)! \, \Gamma(n+m+1)} \cdot \left(\frac{x}{2}\right)^{n+2m} + \sum_{m=0}^{\infty} \frac{(-1)^m \, 2(n+m)}{(m)! \, \Gamma(n+m+1)} \cdot \left(\frac{x}{2}\right)^{n+2m}$$

$$= -n J_n(x) + \sum_{m=0}^{\infty} \frac{(-1)^m \cdot 2}{(m)! \, \Gamma(n+m)} \cdot \frac{x}{2} \cdot \left(\frac{x}{2}\right)^{n+2m-1}$$

$$= -n J_n(x) + x \sum_{m=0}^{\infty} \frac{(-1)^m}{(m)! \, \Gamma(n-1+m+1)} \cdot \left(\frac{x}{2}\right)^{n-1+2m} = -n J_n(x) + x J_{n-1}(x)$$

$$\therefore \qquad x J_n'(x) = -n J_n(x) + x J_{n-1}(x).$$

REMARK

- $\dfrac{d}{dx}(x^n J_n) = x^n J_{n-1}$

(III) $2 J_n'(x) = J_{n-1}(x) - J_{n+1}(x)$. (SVTU-2015, MADRAS-2006, JNTU-2006, 16; PTU-2005, ANNA-2005, VTU-2005)

PROOF. From recurrence relations I and II, we have

$$x J_n'(x) = n J_n(x) - x J_{n+1}(x) \qquad \qquad \text{...(1)}$$

$$x J_n'(x) = -n J_n(x) + x J_{n-1}(x) \qquad \qquad \text{...(2)}$$

Adding (1) and (2), we get

$$2x J_n'(x) = x J_{n-1}(x) - x J_{n+1}(x)$$

$$\therefore \qquad 2 J_n'(x) = J_{n-1}(x) - J_{n+1}(x)$$

(IV) $2n J_n(x) = x[J_{n-1}(x) + J_{n+1}(x)]$.

PROOF. From recurrence relations I and II, we have

$$x J_n'(x) = n J_n(x) - x J_{n+1}(x) \qquad \qquad \text{...(1)}$$

$$x J_n'(x) = -n J_n(x) + x J_{n-1}(x) \qquad \qquad \text{...(2)}$$

From (1) and (2), we get

$$n J_n(x) - x J_{n+1}(x) = -n J_n(x) + x J_{n-1}(x)$$

$$\therefore \qquad 2n J_n(x) = x[\, J_{n-1}(x) + J_{n+1}(x)\,].$$

(V) $\dfrac{d}{dx}[x^{-n} J_n(x)] = -x^{-n} J_{n+1}(x)$. (UPTU-2003, BPTU-2005, PTU-2006)

PROOF. $\text{LHS} = \dfrac{d}{dx}[x^{-n} J_n(x)] = x^{-n} J_n'(x) - n x^{-n-1} J_n(x) = x^{-n-1}[x J_n'(x) - n J_n(x)]$

$$= x^{-n-1}[-x \, J_{n+1}(x)] \qquad \text{(from recurrence relation I)}$$

$$= -x^{-n} J_{n+1}(x) = \text{RHS}$$

$$\therefore \qquad \frac{d}{dx}[x^{-n} J_n(x)] = -x^{-n} J_{n+1}(x).$$

(VI) $\dfrac{d}{dx}[x^n J_n(x)] = x^n J_{n-1}(x)$. (UPTU-2005, VTU-2005, BHOPAL-2008,)

PROOF. Consider $\quad \text{LHS} = \dfrac{d}{dx}[x^n J_n(x)] = x^n J_n'(x) + n x^{n-1} J_n(x) = x^{n-1}[x J_n'(x) + n J_n(x)]$

$$= x^{n-1}[x \, J_{n-1}(x)] \qquad \text{(From recurrence relation II)}$$

$$= x^n J_{n-1}(x) = \text{RHS}$$

$$\therefore \qquad \frac{d}{dx}[x^n J_n(x)] = x^n J_{n-1}(x).$$

Solved Examples

EXAMPLE 1. *Prove the following using Generating function for $J_n(x)$.*

(i) $\cos(x\sin\theta)$
$= J_0 + 2J_2\cos 2\theta + 2J_4\cos 4\theta + ...$
(KERALA(M.TECH)-2005)

(ii) $\sin(x\sin\theta) = 2J_1\sin\theta + 2J_3\sin 3\theta + ...$
(ANNA-2005)

(iii) $\cos(x\cos\theta)$
$= J_0 - 2J_2\cos 2\theta + 2J_4\cos 4\theta - ...$

(iv) $\sin(x\cos\theta) = 2J_1\cos\theta - 2J_3\cos 3\theta + ...$

(v) $\cos x = J_0 - 2J_2 + 2J_4 - ...$

(vi) $\sin x = 2J_1 - 2J_3 + 2J_5 - ...$

SOLUTION. Since we have $e^{\left[\frac{1}{2}x\left(t-\frac{1}{t}\right)\right]} = \sum\limits_{n=-\infty}^{\infty} t^n J_n(x)$

$= J_0 + \left(t - \frac{1}{t}\right)J_1 + \left(t^2 + \frac{1}{t^2}\right)J_2$

$+ \left(t^3 - \frac{1}{t^3}\right)J_3 + ...$ ...(1)

Let us put $t = e^{i\theta}$, then

$t^n = e^{in\theta} = \cos n\theta + i\sin n\theta$

$t^{-n} = e^{-in\theta} = \cos n\theta - i\sin n\theta$

$\therefore \; t^n - \frac{1}{t^n} = 2i\sin n\theta$ and $t^n + \frac{1}{t^n} = 2\cos n\theta$,

$n = 1, 2, 3, ...$

From (1), we have

$e^{i(x\sin\theta)} = J_0 + (2i\sin\theta)J_1$

$+ (2\cos 2\theta)J_2 + (2i\sin 3\theta)J_3 + ...$

$\cos(x\sin\theta) + i\sin(x\sin\theta)$

or $= (J_0 + 2J_2\cos 2\theta + 2J_4\cos 4\theta + ...)$

$+ i(2J_1\sin\theta + 2J_3\sin 3\theta + ...)$

Separate real and imaginary parts, we get

(i) $\cos(x\sin\theta) = J_0 + 2J_2\cos 2\theta + 2J_4\cos 4\theta + ...$

and (ii) $\sin(x\sin\theta) = 2J_1\sin\theta + 2J_3\sin 3\theta + ...$

Now putting $\frac{\pi}{2} - \theta$ in place of θ in (i) and (ii), we get

(iii) $\cos\left(x\sin\left(\frac{\pi}{2} - \theta\right)\right) = J_0 + 2J_2\cos 2\left(\frac{\pi}{2} - \theta\right)$

$+ 2 \quad \cos 4\left(--\theta\right) + ...$

$\therefore \cos(x\cos\theta) = J_0 + 2J_2\cos(\pi - 2\theta)$

$+ 2J_4\cos(2\pi - 4\theta) + ...$

$= J_0 - 2J_2\cos 2\theta + 2J_4\cos 4\theta + ...$

(iv) $\sin\left(x\sin\left(\frac{\pi}{2} - \theta\right)\right)$

$= 2J_1\sin\left(\frac{\pi}{2} - \theta\right) + 2J_3\sin 3\left(\frac{\pi}{2} - \theta\right) + ...$

$\therefore \; \sin(x\cos\theta) = 2J_1\cos\theta - 2J_3\cos 3\theta + ...$

Now putting $\theta = \pi/2$ in (i) and (ii), we get

(v) $\cos\left(x\sin\frac{\pi}{2}\right)$

$= J_0 + 2J_2\cos\pi + 2J_4\cos 2\pi + ...$

$\therefore \; \cos x = J_0 - 2J_2 + 2J_4 - ...$

(vi) and $\sin\left(x\sin\frac{\pi}{2}\right) = 2J_1\sin\frac{\pi}{2} + 2J_3\sin\frac{3\pi}{2} + ...$

$\sin x = 2J_1 - 2J_3 + ...$

EXAMPLE 2. *Prove that* $J_2'(x) = \left(1 - \frac{4}{x^2}\right)J_1(x) + \frac{2}{x}J_0(x)$,
where $J_n(x)$ is the Bessel's function of first kind.
(UPTU-2001, 08)

SOLUTION. $xJ_n' = -nJ_n + xJ_{n-1}$ (Recurrence Formula) ...(1)

On putting $n = 2$ in (1), we have

$xJ_2' = -2J_2 + xJ_1$

$\Rightarrow \quad J_2' = -\frac{2}{x}J_2 + J_1$ ...(2)

$xJ_n' = nJ_n - xJ_{n+1}$ (Recurrence relation) ...(3)

From (1) and (3), we have

$-nJ_n + xJ_{n-1} = nJ_n - xJ_{n+1}$

On putting $n = 1$,

$-J_1 + xJ_0 = J_1 - xJ_2$

$-\frac{1}{x}J_1 + J_0 = \frac{1}{x}J_1 - J_2$

$\Rightarrow \quad J_2 = \frac{2}{x}J_1 - J_0$ ...(4)

Putting the value of J_2 from (4) in (2), we get

$J_2' = -\frac{2}{x}\left(\frac{2}{x}J_1 - J_0\right) + J_1$

$= -\frac{4}{x^2}J_1 + \frac{2}{x}J_0 + J_1$

$= \left(1 - \frac{4}{x^2}\right)J_1 + \frac{2}{x}J_0$

EXAMPLE 3. *Prove that*
$$\frac{d}{dx}[J_n^2 + J_{n+1}^2] = 2\left(\frac{n}{x}J_n^2 - \frac{n+1}{x}J_{n+1}^2\right).$$
(UPTU-2005, Q. BANK-2002, VTU-2000, 06)

SOLUTION. LHS $= \frac{d}{dx}[J_n^2 + J_{n+1}^2]$

$= 2J_nJ_n' + 2J_{n+1}J_{n+1}'$ ...(1)

From recurrence relation (1), we have

$xJ_n' = nJ_n - xJ_{n+1}$

$\therefore \quad J_n' = \frac{n}{x}J_n - J_{n+1}$ ...(2)

From recurrence relation (2), we have

$xJ_n' = -nJ_n + xJ_{n-1}$ or $J_n' = -\frac{n}{x}J_n + J_{n-1}$.

Putting $(n + 1)$ in place of n, we get

$J_{n+1}' = -\frac{n+1}{x}J_{n+1} + J_n$ ...(3)

Substitute the value of J_n' and J_{n+1}' from (2) and (3) into (1), we get

LHS $= 2J_n\left[\frac{n}{x}J_n - J_{n+1}\right] + 2\left[-\frac{n+1}{x}J_{n+1} + J_n\right]J_{n+1}$

$= 2\frac{n}{x}J_n^2 - 2J_nJ_{n+1} - 2\frac{(n+1)}{x}J_{n+1}^2 + 2J_nJ_{n+1}$

$= 2\left(\frac{n}{x}J_n^2 - \frac{n+1}{x}J_{n+1}^2\right) = $ R.H.S.

Hence $\dfrac{d}{dx}[J_n^2 + J_{n+1}^2] = 2\left(\dfrac{n}{x}J_n^2 - \dfrac{n+1}{x}J_{n+1}^2\right).$

EXAMPLE 4. **Prove that** $\dfrac{d}{dx}[x\,J_n J_{n+1}] = x\left(J_n^2 - J_{n+1}^2\right).$

SOLUTION. LHS $= \dfrac{d}{dx}(xJ_n J_{n+1})$

$$= xJ_n J'_{n+1} + xJ'_n J_{n+1} + J_n J_{n+1} \qquad \ldots(1)$$

From recurrence relations I and II, we have

$$xJ'_n = nJ_n - xJ_{n+1} \qquad \ldots(2)$$

and $\;\; xJ'_n = -nJ_n + xJ_{n-1} \qquad \ldots(3)$

Putting $(n + 1)$ in place of n in (3), we get

$$xJ'_{n+1} = -(n+1)J_{n+1} + xJ_n \qquad \ldots(4)$$

Substitute the values of xJ'_n and xJ'_{n+1} from (2) and (4) into (1), we get

LHS $= J_n[-(n+1)J_{n+1} + xJ_n]$

$$+ J_{n+1}[nJ_n - xJ_{n+1}] + J_n J_{n+1}$$

$$= -nJ_n J_{n+1} - J_n J_{n+1} + xJ_n^2$$

$$+ nJ_n J_{n+1} - xJ_{n+1}^2 + J_n J_{n+1}$$

$$= xJ_n^2 - xJ_{n+1}^2 = x(J_n^2 - J_{n+1}^2) = \text{RHS}$$

Hence, $\dfrac{d}{dx}[xJ_n J_{n+1}] = x[J_n^2 - J_{n+1}^2].$

EXAMPLE 5. **Prove the following relation :**

$$x^2 J_n''\,(x) = (n^2 - n - x^2)J_n(x) + xJ_{n+1}(x)$$

(UPTU-2006, 07, Q. BANK-2002)

SOLUTION. $x^2 \dfrac{d^2 y}{dx^2} + x\dfrac{dy}{dx} + (x^2 - n^2)y = 0 \qquad \ldots(1)$

(Bessel's equation)

Clearly, $J_n(x)$ is the solution of (1)

So, $\;\; x^2 J_n'' + xJ_n' + (x^2 - n^2)J_n = 0 \qquad \ldots(2)$

We know that

$$xJ_n' = nJ_n - xJ_{n+1} \quad \text{(Recurrence relation)} \quad \ldots(3)$$

Putting the value of xJ_n' from (3) in (2), we get

$$x^2 J_0'' = -nJ_n + xJ_{n+1} + (n^2 - x^2)J_n$$

$$x^2 J'' = (n^2 - n - x^2)J_n + xJ_{n+1}.$$

EXAMPLE 6. **Prove the followings :**

(i) $\;\; J_{1/2}(x) = \sqrt{\dfrac{2}{\pi x}}\cdot\sin x \qquad$ (JNTU-2003, VTU-2009)

(ii) $\;\; J_{-1/2}(x) = \sqrt{\dfrac{2}{\pi x}}\cdot\cos x$

(VTU-2003, ANNA-2005, WBTU-2005)

SOLUTION. (i) Since we have

$$J_n(x) = \dfrac{x^n}{2^n \Gamma(n+1)}\left[1 - \dfrac{x^2}{2(2n+2)}\right.$$

$$\left. + \dfrac{x^4}{2.4(2n+2)(2n+4)}\ldots\right] \qquad \ldots(1)$$

Putting $n = 1/2$ in (1), we get

$$J_{1/2}(x) = \dfrac{x^{1/2}}{2^{1/2}\Gamma\left(1+\dfrac{1}{2}\right)}$$

$$\left[1 - \dfrac{x^2}{2.3} + \dfrac{x^4}{2.4.3.5} - \ldots\right]$$

$$= \sqrt{\dfrac{x}{2}}\cdot\dfrac{1}{\dfrac{1}{2}\Gamma(1/2)}\left[1 - \dfrac{x^2}{(3)!} + \dfrac{x^4}{(5)!} - \ldots\right]$$

$$= \sqrt{\dfrac{2}{x}}\cdot\dfrac{1}{\Gamma(1/2)}\left[x - \dfrac{x^3}{(3)!} + \dfrac{x^5}{(5)!} - \ldots\right]$$

$$= \sqrt{\dfrac{2}{\pi x}}\cdot\sin x \quad \left(\because \Gamma\left(\dfrac{1}{2}\right) = \sqrt{\pi} \text{ and}\right.$$

$$\left. \sin\theta = \theta - \dfrac{\theta^3}{(3)!} + \dfrac{\theta^5}{(5)!} - \ldots\right)$$

(ii) Putting $n = -1/2$ in (1), we get

$$J_{-1/2}(x) = \dfrac{x^{-1/2}}{2^{-1/2}\Gamma\left(1-\dfrac{1}{2}\right)}$$

$$\left[1 - \dfrac{x^2}{1.2} + \dfrac{x^4}{1.2.3.4} - \ldots\right]$$

$$= \sqrt{\dfrac{2}{x}}\cdot\dfrac{1}{\Gamma\left(\dfrac{1}{2}\right)}\left[1 - \dfrac{x^2}{(2)!} + \dfrac{x^4}{(4)!} - \ldots\right]$$

$$= \sqrt{\dfrac{2}{\pi x}}\left[1 - \dfrac{x^2}{(2)!} + \dfrac{x^4}{(4)!} - \ldots\right]$$

$$= \sqrt{\dfrac{2}{\pi x}}\cdot\cos x \quad \left(\because \cos\theta = 1 - \dfrac{\theta^2}{(2)!} + \dfrac{\theta^4}{(4)!} - \ldots\right)$$

$$\therefore \;\; J_{-1/2}(x) = \sqrt{\dfrac{2}{\pi x}}\cdot\cos x\,.$$

EXAMPLE 7. **Prove the following :**

$$J_3(x) + 3J_0(x) + 4J_0'''(x) = 0\,.$$

(UPTU(B.TECH.)-2001, Q. BANK-2002)

SOLUTION. We know that $2J_n' = J_{n-1} - J_{n+1}.$

Differentiating and multiplying by 2, we get

$$2^2 J_n'' = 2J_{n-1}' - 2J_{n+1}'$$

$$= (J_{n-2} - J_n) - (J_n - J_{n+2})$$

$$= J_{n-2} - 2J_n + J_{n+2}.$$

Differentiating again and multiplying by 2, we get

$$2^3 J_n''' = 2J_{n-2}' - 4J_n' + 2J_{n+2}'$$

$$= (J_{n-3} - J_{n-1}) - 2(J_{n-1} - J_{n+1})$$

$$+ (J_{n+1} - J_{n+3})$$

$$= J_{n-3} - 3J_{n-1} + 3J_{n+1} - J_{n+3}.$$

Putting $n = 0$, we get

$$2^3 J_0''' = J_{-3} - 3J_{-1} + 3J_1 - J_3$$

$$= (-1)^3 J_3 - 3(-1)J_1 + 3J_1 - J_3$$

$$= -2J_3 + 6J_1$$

$$4J_0''' = -J_3 - 3J_1 = -J_3 + 3(-J_0')$$

$$= J_3 + 3J_0' + 4J_0''' = 0$$

$$\Rightarrow \;\; J_3(x) + 3J_0'(x) + 4J_0'''(x) = 0$$

EXAMPLE 8. **Prove that**

(i) $\;\; [J_{1/2}(x)]^2 + [J_{-1/2}(x)]^2 = \dfrac{2}{\pi x} \qquad$ (DELHI-2002)

(ii) $\;\; J_{-3/2}(x) = -\sqrt{\dfrac{2}{\pi x}}\left(\dfrac{1}{x}\cos x + \sin x\right)$

SOLUTION. (i) In Example 6, we have proved that

$$J_{1/2}(x) = \sqrt{\frac{2}{\pi x}} \cdot \sin x$$

and $J_{-1/2}(x) = \sqrt{\frac{2}{\pi x}} \cdot \cos x$

Squaring these and add, we get

$$[J_{1/2}(x)]^2 + [J_{-1/2}(x)]^2$$

$$= \frac{2}{\pi x}(\sin^2 x + \cos^2 x) = \frac{2}{\pi x}$$

(ii) Since we know that

$$2nJ_n(x) = x\,[J_{n-1}(x) + J_{n+1}(x)]$$

or $J_{n-1}(x) = \frac{2n}{x}J_n(x) - J_{n+1}(x)$

Now, putting $n = -1/2$, we get

$$J_{-3/2}(x) = \frac{2\left(-\dfrac{1}{2}\right)}{x}J_{-1/2} - J_{1/2}(x)$$

$$= -\frac{1}{x}J_{-1/2}(x) - J_{1/2}(x) \qquad \ldots(1)$$

Putting the value of $J_{1/2}(x) = \sqrt{\dfrac{2}{\pi x}} \cdot \sin x$ and

$J_{-1/2}(x) = \sqrt{\dfrac{2}{\pi x}} \cdot \cos x$ into (1), we get

$$J_{-3/2}(x) = -\sqrt{\frac{2}{\pi x}}\left(\frac{1}{x}\cos x + \sin x\right)$$

EXAMPLE 9. **Prove that** $\displaystyle \lim_{x \to 0} \frac{J_n(x)}{x^n} = \frac{1}{2^n \Gamma(n+1)}, n > -1.$

SOLUTION. Since we know that

$$J_n(x) = \frac{x^n}{2^n \Gamma(n+1)}$$

$$\left[1 - \frac{x^2}{2(2n+2)} + \frac{x^4}{2.4(2n+2)(2n+4)}\ldots\right]$$

or $\dfrac{J_n(x)}{x^n} = \dfrac{1}{2^n \Gamma(n+1)}$

$$\left[1 - \frac{x^2}{2(2n+2)} + \frac{x^4}{2.4(2n+2)(2n+4)}\ldots\right]$$

Taking limit of both sides as $x \to 0$

$$\therefore \lim_{x\to 0} \frac{J_n(x)}{x^n} = \frac{1}{2^n \Gamma(n+1)}$$

$$\lim_{x\to 0}\left[1 - \frac{x^2}{2(2n+2)} + \frac{x^4}{2.4(2n+2)(2n+4)}\ldots\right]$$

$$= \frac{1}{2^n\,\Gamma(n+1)} \cdot 1 = \frac{1}{2^n\,\Gamma(n+1)}$$

EXAMPLE 10. **Prove that** $J_0^2 + 2(J_1^2 + J_2^2 + J_3^2 + \ldots) = 1$ **and**

deduce that $|J(x)| \le 1, |J(x)| \le \dfrac{1}{\sqrt{}}, n \ge 1.$

(UPTU-2003, 08, KERALA(M.TECH)-2005, VTU-2003)

SOLUTION. Since we know that

$$e^{\left[\frac{1}{2}x\left(t-\frac{1}{t}\right)\right]}.e^{-\left[\frac{1}{2}x\left(t-\frac{1}{t}\right)\right]} = 1 \qquad \ldots(1)$$

$$\sum_{n=\infty}^{\infty} t^n J_n(x). \sum_{n=-\infty}^{\infty} J_n(-x).t^n = 1$$

(by generating function)

$$\therefore \left[\sum_{n=-\infty}^{\infty} t^n J_n(x)\right].\left[\sum_{n=-\infty}^{\infty} (-1)^n J_n(x).t^n\right] = 1$$

$$[\because J_n(-x) = (-1)^n J_n(x)]$$

or $[\ldots + J_2(x).t^{-2} - J_1(x).t^{-1}$

$$+ J_0(x) + tJ_1(x) + t^2.J_2(x)\ldots]$$

$$\times [\ldots + t^{-2}J_2(x) + t^{-1}J_1(x)$$

$$+ J_0(x) - tJ_1(x) + t^2 J_2(x) - \ldots] = 1$$

Comparing the common terms of both sides, we get

$$J_0^2(x) + 2J_1^2(x) + 2J_2^2(x) + \ldots = 1$$

or $J_0^2 + 2(J_1^2 + J_2^2 + J_3^2 + \ldots) = 1$

Further, since, if x is real and we have all the terms in the left hand side of above obtained result. Then,

$$|J_0(x)| \le 1 \text{ and } 2|J_n(x)|^2 \le 1 \text{ or } |J_n(x)| \le \frac{1}{\sqrt{2}}.$$

EXAMPLE 11. **Prove that**

$$x = 2J_0 J_1 + 6J_1 J_2 + \ldots + 2(2n+1)\,J_n J_{n+1} + \ldots$$

SOLUTION. Since we have a result

$$\frac{d}{dx}(xJ_n J_{n+1}) = x(J_n^2 - J_{n+1}^2) \qquad \ldots(1)$$

Putting $n = 0, 1, 2, 3, 4, \ldots$, we get

$$\frac{d}{dx}(xJ_0 J_1) = x(J_0^2 - J_1^2) \qquad \ldots(2)$$

$$\frac{d}{dx}(xJ_1 J_2) = x(J_1^2 - J_2^2) \qquad \ldots(3)$$

$$\frac{d}{dx}(xJ_2 J_3) = x(J_2^2 - J_3^2) \qquad \ldots(4)$$

$$\frac{d}{dx}(xJ_3 J_4) = x(J_3^2 - J_4^2)$$

and so on.

Now, multiplying (2), (3), (4), ... by 1, 3, 5,... respectively and adding, we get

$$\frac{d}{dx}[x(J_0 J_1 + 3J_1 J_2 + 5J_2 J_3 + \ldots)]$$

$$= x[J_0^2 + 2(J_1^2 + J_2^2 + \ldots)] = x[1]$$

$$[\because J_0^2 + 2(J_1^2 + J_2^2 + \ldots) = 1]$$

$$\therefore \frac{d}{dx}[x(J_0 J_1 + 3J_1 J_2 + 5J_2 J_3 + \ldots)] = x.$$

Integrating both sides, we get

$$x(J_0 J_1 + 3J_1 J_2 + 5J_2 J_3 + \ldots) = \frac{x^2}{2} + A$$

When $x = 0 \Rightarrow A = 0$

$$\therefore x(J_0 J_1 + 3J_1 J_2 + 5J_2 J_3 + \ldots) = \frac{x^2}{2}$$

$$\therefore x = 2J_0 J_1 + 6J_1 J_2 + 10J_2 J_3 + \ldots$$

or $x = 2J_0 J_1 + 6J_1 J_2 + \ldots + 2(2n+1)J_n J_{n+1} + \ldots$

8.8.11 BESSEL's INTEGRAL

We have

(a) $J_0(x) = \dfrac{1}{\pi} \int_0^\pi \cos(x \sin \theta) d\theta$ (MADRAS-2006)   (b) $J_n(x) = \dfrac{1}{\pi} \int_0^\pi \cos(n\theta - x \sin \theta) d\theta$ (VTU-2006)

PROOF.   It is known that $\cos(x \sin \theta) = J_0 + 2J_2 \cos 2\theta + 2J_4 \cos 4\theta + \ldots$ …(1)

and $\sin(x \sin \theta) = 2J_1 \sin \theta + 2J_3 \sin 3\theta + 2J_5 \sin 5\theta + \ldots$ …(2)

On integrating (1) between 0 to π, we get

$$\int_0^\pi \cos(x \sin \theta) d\theta = \int_0^\pi (J_0 + 2J_2 \cos 2\theta + 2J_4 \cos 4\theta + \ldots) d\theta = J_0 \int_0^\pi d\theta + 2J_2 \int_0^\pi \cos 2\theta d\theta + 2J_4 \int_0^\pi \cos 4\theta d\theta + \ldots$$

$$= (\pi) + 0 + 0 + \ldots$$

Hence, $J_0 = \dfrac{1}{\pi} \int_0^\pi \cos(x \sin \theta) d\theta$

Similarly, multiplying (1) by $\cos n\theta$ and integrating between 0 to π, we get

$$\int_0^\pi \cos(x \sin \theta) \cos n\theta \, d\theta = \int_0^\pi (J_0 \cos n\theta + 2J_2 \cos 2\theta + 2J_4 \cos 4\theta \cos n\theta + \ldots) d\theta$$

$$= 2J_0 \int_0^\pi \cos n\theta \, d\theta + 2J_2 \int_0^\pi \cos 2\theta \cos n\theta \, d\theta + \ldots = \begin{cases} 0 \, ; & \text{if } n \text{ is odd} \\ \pi J_n \, ; & \text{if } n \text{ is even} \end{cases} \quad \ldots(3)$$

Further, multiplying (2) by $\sin n\theta$ and integrating between 0 to π, we get

$$\int_0^\pi \sin(x \sin \theta) \sin n\theta \, d\theta = \int_0^\pi (2J_1 \sin \theta \sin n\theta + 2J_3 \sin 3\theta \sin n\theta + \ldots) d\theta$$

$$= 2J_1 \int_0^\pi \sin \theta \sin n\theta \, d\theta + 2J_3 \int_0^\pi \sin 3\theta \sin n\theta \, d\theta + \ldots = \begin{cases} 0 \, ; & \text{if } n \text{ is even} \\ \pi J_n \, ; & \text{if } n \text{ is odd} \end{cases} \quad \ldots(4)$$

Using (3) and (4), we get

$$\int_0^\pi (\cos(x \sin \theta) \cos n\theta + \sin(x \sin \theta) \sin n\theta) d\theta = \pi J_n$$

or $\int_0^\pi \cos(n\theta - x \sin \theta) d\theta = \pi J_n$ or $J_n = \dfrac{1}{\pi} \int_0^\pi \cos(n\theta - x \sin \theta) d\theta$

8.8.12 FOURIER-BESSEL EXPANSION

Let $f(x)$ be a function, which is continuous and has a finite number of oscillations in the interval $0 \le x \le a$, then $f(x)$ can be expanded as follows

$$f(x) = C_1 J_n(\alpha_1 x) + C_2 J_n(\alpha_2 x) + C_3 J_n(\alpha_3 x) + \ldots + C_n J_n(\alpha_n x) + \ldots$$

or $f(x) = \displaystyle\sum_{i=1}^\infty C_i J_n(\alpha_i x)$

where, $\alpha_1, \alpha_2, \ldots$ are the roots of the equation $J_n(x) = 0$.

PROOF.   It is given that $f(x) = \displaystyle\sum_{i=1}^\infty C_i J_n(\alpha_i x)$ …(1)

Multiplying (1) by $x J_n(\alpha_j x)$, we get $f(x) J_n(\alpha_j x) = \displaystyle\sum_{i=1}^\infty C_i x J_n(\alpha_j x) J_n(\alpha_i x)$ …(2)

On integrating from $x = 0$ to a, we get

$$\int_0^a x J_n(\alpha_i x) J_n(\alpha_j x) dx = \begin{cases} 0 \, ; & \text{if } i \ne j \\ \dfrac{a^2}{2} J_{n+1}^2(\alpha_i a) \, ; & \text{if } i = j \end{cases} \quad \ldots(4)$$

Using (4) in (3), we get

$$\int_0^a x \, f(x) J_n(\alpha_i x) dx = C_i \cdot \frac{a^2}{2} J_{n+1}^2(\alpha_i a) \quad \Rightarrow C_i = \frac{2 \int_0^a x \, f(x) J_n(\alpha_i x) dx}{a^2 \cdot J_{n+1}^2(\alpha_i a)}$$

Putting these values in (1), we get $f(x) = C_1 J_n(\alpha_1 x) + C_2 J_n(\alpha_2 x) + C_3 J_n(\alpha_3 x) + \ldots + C_n J_n(\alpha_n x) + \ldots$

8.8.13 BER AND BEI FUNCTION

Consider the differential equation

$$x \frac{d^2 y}{dx^2} + \frac{dy}{dx} - i x y = 0 \qquad \ldots(1)$$

which occurs in certain problems of electrical engineering.

We have $y = J_0(k x) = J_0[(-i)^{1/2} x] = J_0(i^{3/2} x)$

Replacing $i^{3/2}$ in the series for $J_0(x)$, we get

$$y = 1 - \frac{i^3 x^2}{2^2} + \frac{i^6 x^4}{(2!)^2 2^4} - \frac{i^9 x^6}{(3!)^2 2^6} + \frac{i^{12} x^8}{(4!)^2 2^8} - \ldots$$

$$= \left[1 - \frac{x^4}{2^2 . 4^2} + \frac{x^8}{2^2 . 4^2 . 6^2 . 8^2} - \dots\right] + i\left[\frac{x^2}{2^2} - \frac{x^6}{2^2 . 4^2 . 6^2} + \frac{x^{10}}{2^2 . 4^2 . 6^2 . 8^2 . 10^2} - \dots\right] \qquad \dots(2)$$

which is complex for x real. The series in the above brackets are taken to define Bessel-real (or ber) and Bessel-imaginary (or bei) functions.

Thus
$$\text{ber } x = 1 + \sum_{m=1}^{\infty} (-1)^m \frac{x^{4m}}{2^2 . 4^2 . 6^2 \dots (4m)^2} \qquad \dots(3)$$

and
$$\text{bei } x = -\sum_{m=1}^{\infty} (-1)^m \frac{x^{4m-2}}{2^2 . 4^2 . 6^2 \dots (4m-2)^2}$$

So that $y = \text{ber } x + i \text{ bei } x$ is a solution of (1).

Solved Examples

EXAMPLE 1. **Prove that** $\sum_{i=1}^{\infty} \frac{2J_0(a_i x)}{a_i J_1(a_1 i)} = 1$, **where** a_i **are the roots of** $J_0(x)$. (UPTU(Q. BANK)-2002)

SOLUTION. We have

$$f(x) = \sum_{i=1}^{\infty} C_i J_n(a_i x) \qquad \dots(1)$$

$$C_i = \frac{2}{J_{n+1}^2(a_i)} \int_0^1 x J_n(a_i x) f(x) \, dx \qquad \dots(2)$$

Putting $f(x) = 1$ and $n = 0$ in (1), we get

$$1 = \sum_{i=1}^{\infty} C_i J_0(a_i x)$$

$$\therefore \quad C_i = \frac{2}{J_1^2(a_i)} \int_0^1 x J_0(a_i x) dx$$

$$= \frac{2}{J_1^2(a_i)} \left(\frac{J_1(a_i)}{a_i}\right) = \frac{2}{a_i J_1(a_i)}$$

Then using these values in (1), we get

$$1 = \sum_{i=1}^{\infty} \frac{2}{a_i J_1(a_i)} J_0(a_i x).$$

Hence, $\Sigma \frac{2J_0(a_i x)}{a_i J_1(a_i)} = 1$

Exercise-8.43

1. Prove that

(i) $\int_0^x x^n J_{n-1}(x) dx = x^n J_n(x)$

(ii) $\int_0^x x^{n+1} J_n(x) \, dx = x^{n+1} J_{n+1}(x)$

2. Prove that $\int_0^{\pi/2} \sqrt{\pi x} \, J_{1/2}(2x) \, dx = 1$.

3. Prove that $J_0(x) = \frac{1}{\pi} \int_0^{\pi} \cos(x \sin \phi) \, d\phi$.

4. Show that if $n > -1, \int_0^x x^{-n} J_{n+1}(x) \, dx = \frac{1}{2^n \Gamma(n+1)} - x^{-n} J_n(x)$.

5. Prove that $4J_n''(x) = J_{n-2}(x) - 2J_n(x) + J_{n+2}(x)$.

6. Prove that $J_n J_{-n}' - J_{-n} J_n' = -\frac{2 \sin n\pi}{\pi x}$.

Hence deduce that $\frac{d}{dx}\left[\frac{J_{-n}}{J_n}\right] = -\frac{2 \sin n\pi}{\pi x J_n^2}$.

7. Prove that

(i) $J_2 = J_0'' - \frac{1}{x} J_0'$ (ii) $J_2 - J_0 = 2J_0''$

8. Prove that $J_{n+3} + J_{n+5} = \frac{2}{x}(n+4) J_{n+4}$.

9. Prove that $J_n' = \frac{2}{x}\left[\frac{n}{2} J_n - (n+2)J_{n+2} + (n+4) J_{n+4} - \dots\right]$.

10. Prove that $\int J_{n+1}(x) dx = \int J_{n-1}(x) \, dx - 2J_n(x) + A$.

11. Prove that $J_{n-1} = \frac{2}{x}[nJ_n - (n+2) J_{n+2} + (n+4) J_{n+4} - \dots]$ and hence deduce that

$\frac{1}{2} x J_n = (n+1) J_{n+1} - (n+3)J_{n+3} + (n+5) J_{n+5} - \dots$

12. Prove that

(i) $J_{3/2}(x) = \sqrt{\frac{2}{\pi x}} \left[\frac{1}{x} \sin x - \cos x\right]$

(ii) $J_{-5/2}(x) = \sqrt{\frac{2}{\pi x}} \left[\left(\frac{3-x^2}{x^2}\right) \cos x + \frac{3}{x} \sin x\right]$

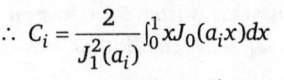 (UPTU(Q. BANK)-2002)

(iii) $J_{5/2}(x) = \sqrt{\frac{2}{\pi x}} \left[\left(\frac{3-x^2}{x^2}\right) \sin x - \frac{3}{x} \cos x\right]$

13. Show that $J_n(x) = \frac{1}{\pi} \int_0^{\pi} \cos(n\theta - x \sin \theta) d\theta$, where n is positive integer.

14. Prove that $\int_0^{\infty} e^{-ax} J_0(bx) dx = \frac{1}{\sqrt{a^2 + b^2}}, a > 0$.

15. Show that $\int x J_0^2(x) dx = \frac{1}{2} x^2 \left[J_0^2(x) + J_1^2(x)\right] + C$.

(UPTU-2003, 04)

16. Show that $4J_0'''(x) + 3J_0'(x) + J_3(x) = 0$.

(UPTU-2002, OSMANIA-2003, 11)

17. Show that $J_0(x) = \frac{1}{\pi} \int_0^{\pi} \cos(x \cos \phi) \, d\phi$.   (UPTU(Q. BANK)-2002)

18. Show that the solution of the differential equation

$$\frac{d^2 y}{dx^2} + \left(9x - \frac{20}{x^2}\right) y = 0$$

in terms of Bessel's function is given by

$$y = \left[\sqrt{x} \, (C_1 J_3(2x^{3/2}) + C_2 Y_3(2x^{3/2}))\right]$$ (UPTU-2002)

19. Show that the solution of the differential equation

$$y'' + \frac{y'}{x}\left(8 - \frac{1}{x^2}\right)y = 0 \text{ is given by}$$

$$y = C_1 J_2(4\sqrt{2}x) + C_2 Y_2(4\sqrt{2}.x) \qquad \text{(UPTU(Q. BANK)-2002)}$$

20. Show that the solution of the differential equation

$$y'' + \frac{y'}{x} + 4\left(x^2 - \frac{n^2}{x^2}\right)y = 0$$

is given by $y = C_1 J_n(x^2) + C_2 Y_n(x^2)$.

Hints to Selected Problems

1. Integrate both sides from 0 to x of the following relation

$$\frac{d}{dx}\int x^n J_n(x) = x^n J_{n-1}(x).$$

2. $J_{1/2}(x) = \sqrt{\frac{2}{\pi x}}.\sin x \Rightarrow J_{1/2}(2x) = \sqrt{\frac{2}{2\pi x}}.\sin 2x$

$\Rightarrow \sqrt{\pi x}. J_{1/2}(2x) = \sin 2x$. Now solve.

3. Integrating both sides w.r.t. ϕ from $\phi = 0$ to π of the following relation $\cos(x\sin\phi) = J_0 + 2J_2\cos 2\phi + 2J_4\cos 4\phi + ...$

4. Integrating the relation $\frac{d}{dx}[x^{-n}J(x)] = -x^{-n}J^{()}$

5. Differentiating $2J'_n(x) = J_{n-1}(x) - J_{n+1}(x)$ and then replace

$n-1$ and $n+1$ in place of n.

7. Put $n = 0$ in the recurrence relation $xJ'_n = nJ_n - xJ_{n+1}$.

8. Use recurrence relation $2nJ_n = x(J_{n-1} + J_{n+1})$.

9. Replacing n by $n+2, n+4, n+6, ...$ in $J_{n-1} + J_{n+1} = \frac{2n}{x}J_n$.

13. Using $\cos(x\sin\theta) = J_0 + 2J_2\cos 2\theta + ... + 2J_{2m}\cos 2m\theta...$ and $\sin(x\sin\theta) = 2\sin\theta.J_1 + 2\sin 3\theta J_3 + ... + 2J_{2m+1}\sin(2m+1)\theta + ...$

14. We have $J_0(x) = \frac{1}{\pi}\int_0^\pi \cos(x\sin\theta)d\phi \Rightarrow J_0(6x) = \frac{1}{\pi}$

Now multiplying by e^{-9x} and then integrate from $x = 0$ to ∞.

8.9 THE LAPLACE TRANSFORM

An integral of the type $\int_{-\infty}^\infty k(p,t)F(t)dt$ is defined as the integral transform of $F(t)$, provided it is convergent. It is denoted by $f(p)$ or $T[F(t)]$.

$$f(p) = TF(t) = \int_{-\infty}^\infty k(p,t)F(t)dt$$

✒ REMARK

- The function $k(p,t)$ appearing in the integral is called kernel of the transform. Here p is a parameter and is independent of t and p may be real or complex number.

Definition 1. *If $F(t)$ be a function of t defined for all values of t, then Laplace transform of $f(t)$, denoted by $L\{F(t)\}$ or $f(p)$ is defined by*

$$L\{F(t)\} = f(p) = \int_0^\infty e^{-pt} F(t)\, dt \qquad \qquad ...(1)$$

✒ REMARKS

- If the integral (1) converges for some value of p, then only the Laplace transform of $f(t)$ exists otherwise not.
- L is called Laplace transform operator.

Definition 2. *A function $f(x)$ is said to be of exponential order a as $x \to \infty$ if $\lim\limits_{x\to\infty} e^{-ax} f(x) = a$ finite quantity, i.e., for a given positive integer n, if a real number M such that $|e^{-ax} f(x)| < M, \forall x \geq n$ which can be written as $f(x) = O(e^{-ax}), x \to \infty$.*

Definition 3. *A function $f(x)$ is called sectionally continuous over the closed interval $x_1 \leq x \leq x_2$ if the closed interval can be divided into a finite number of subintervals $a \leq x \leq b$ such that*

(i) *$f(x)$ is continuous in the closed interval $[a, b]$.*

(ii) *$\lim\limits_{x\to a+0} f(x)$ and $\lim\limits_{x\to b-0} f(x)$ both exist.*

Definition 4. *A function which is sectionally (or piecewise) continuous over every finite interval in the range $t \geq 0$ and ω of exponential order as $t \to \infty$ is called a function of class A.*

8.9.1 LINEARITY PROPERTY

THEOREM. *The Laplace transformation is a linear transformation i.e. $L\{a_1 F_1(t) + a_2 F_2(t)\} = a_1 L\{F_1(t)\} + a_2 L\{F_2(t)\}$ if a_1 and a_2 are constants.*

PROOF. We know that $L = \{F(t)\} = \int_0^\infty e^{-pt}F(t)\,dt$.

Therefore,

$$L\{a_1F_1(t) + a_2F_1(t)\} = \int_0^\infty e^{-pt}\{a_1F_1(t) + a_2F_1(t)\}\,dt = a_1\int_0^\infty e^{-pt} F_1(t)\,dt + a_2\int_0^\infty e^{-pt}F_2(t)dt$$

$$= a_1 L\{F_1(t)\} + a_2 L\{F_2(t)\}.$$

8.9.2 Existence of Laplace Transform

THEOREM. *If $F(t)$ is a function which is piecewise continuous on every finite interval in the range $t \geq 0$ and satisfies $|F(t)| \leq M\, e^{at}$ for all $t \geq 0$ and for some constant a and M, then the Laplace transform of $f(t)$ exists for all $p > a$.*

PROOF. We know that $\quad L\{F(t)\} = \int_0^\infty e^{-pt} F(t)\, dt = \int_0^{t_0} F(t) e^{-pt} dt + \int_{t_0}^\infty F(t) e^{-pt} dt$ ...(1)

Now, $\int_0^{t_0} F(t)\, e^{-pt} dt$ exists since $F(t)$ is sectionally continuous on every finite interval $0 \leq t \leq t_0$

and $\quad \left| \int_{t_0}^\infty F(t)\, e^{-pt}\, dt \right| \leq \int_{t_0}^\infty |F(t)\, e^{-pt}|\, dt \leq \int_{t_0}^\infty e^{-pt}\, M\, e^{at}\, dt$ $\qquad [\because |F(t)| \leq Me^{at}]$

$$= \int_{t_0}^\infty e^{(a-p)t}\, M\, dt = M \left[\frac{e^{-(p-a)t}}{-(p-a)} \right]_{t_0}^\infty = \frac{M}{p-a} e^{-(p-a)t_0}, \text{ if } p > a$$

$$\Rightarrow \quad \left| \int_{t_0}^\infty F(t)\, e^{-pt}\, dt \right| \leq \frac{M}{p-a} e^{-(p-a)t_0}, \text{ if } p > a.$$

Now, $\dfrac{Me^{-(p-a)t_0}}{p-a}$ can be made small as we please by taking t_0 sufficiently large. Hence, from (1), we conclude that $L\{f(t)\}$ exists for all $p > a$.

✎ Remark

- The above conditions are sufficient but not necessary for the existence of the Laplace transform. If these conditions are satisfied, the Laplsce transform must exist. If these conditions are not satisfied, the Laplace transform may or may not exist.

8.9.3 Laplace Transform of some Elementary Functions

(i) $F(t) = 1$

SOLUTION. We have $\qquad L\{F(t)\} = \int_0^\infty e^{-pt} F(t)\, dt$ ...(1)

Here $F(t) = 1$.

Therefore, from (1) $\qquad L\{1\} = \int_0^\infty e^{-pt} \cdot 1\, dt = \left[-\frac{e^{-pt}}{p} \right]_0^\infty = \frac{1}{p},\ p > 0$

Hence, $\qquad L\{1\} = \dfrac{1}{p}.$

(ii) $F(t) = t^n$

SOLUTION. We have $\qquad L\{F(t)\} = \int_0^\infty e^{-pt} F(t)\, dt$

$$\Rightarrow \qquad L\{t^n\} = \int_0^\infty e^{-pt}\, t^n\, dt = \int_0^\infty e^{-pt} \cdot t^{(n+1)-1}\, dt$$

$$= \frac{\Gamma(n+1)}{p^{n+1}} = \frac{n!}{p^{n+1}},\ p > 0 \qquad \left[\because \int_0^\infty e^{-u}\, u^n\, du = \Gamma(n+1) \right]$$

Hence, $\qquad L\{t^n\} = \dfrac{n!}{p^{n+1}}.$

(iii) $F(t) = t$

SOLUTION. We have $\qquad L\{F(t)\} = \int_0^\infty e^{-pt} \cdot t\, dt = \left[-\frac{1}{p} t\, e^{-pt} \right]_0^\infty + \frac{1}{p} \int_0^\infty e^{-pt} dt = \frac{1}{p^2},\ p > 0$

(iv) $F(t) = e^{at}$

SOLUTION. We have $\qquad L\{F(t)\} = \int_0^\infty e^{-pt} e^{at}\, dt = \int_0^\infty e^{-(p-a)t}\, dt.$

If $p \leq a$, integral diverges. For $p > a$, the integral converges. Hence, for $p > a$,

$$L\{e^{at}\} = \int_0^\infty e^{-(p-a)t}\, dt = \left[-\frac{e^{-(p-a)t}}{p-a} \right]_0^\infty = 0 + \frac{1}{p-a} = \frac{1}{p-a},\ p > a.$$

(v) $F(t) = \sin at$

SOLUTION. $\quad L\{\sin at\} = \int_0^\infty e^{-pt} \sin at\, dt = \left[\frac{e^{-pt}(-p \sin at - a \cos at)}{p^2 + a^2} \right]_0^\infty \qquad \left[\because \int e^{ax} \sin bx\, dx = e^{ax} \frac{[a \sin bx - b \cos bx]}{a^2 + b^2} \right]$

$$= \frac{a}{p^2 + a^2},\ p > a$$

Hence, $\quad L\{\sin at\} = \dfrac{a}{p^2 + a^2}.$

(vi) $F(t) = \cos at$

SOLUTION. We know that

$$\int e^{ax} \cos bx \, dx = \frac{e^{ax}(a\cos bx + b\sin bx)}{a^2 + b^2}$$

Therefore, we have

$$L\{\cos at\} = \int_0^\infty e^{-pt} \cos at \, dt = \left[\frac{e^{-pt}(-p\cos at + a\sin at)}{a^2 + p^2}\right]_0^\infty = \frac{p}{p^2 + a^2}, \quad p > 0.$$

(vii) $F(t) = \sinh at$

SOLUTION. Consider $L\{\sinh at\} = L\left\{\dfrac{e^{at} - e^{-at}}{2}\right\} = \dfrac{1}{2}L\{e^{at}\} - \dfrac{1}{2}L\{e^{-at}\}$　　　　　(Using (iv))

$$= \frac{1}{2} \cdot \frac{1}{p-a} - \frac{1}{2} \cdot \frac{1}{p+a} = \frac{a}{p^2 - a^2}$$

Hence,　　$L\{\sinh at\} = \dfrac{a}{p^2 - a^2}, p > |a|$.

(viii) $F(t) = \cosh at$

SOLUTION. Consider

$$L\{\cosh at\} = L\left[\frac{1}{2}(e^{at} + e^{-at})\right] = \frac{1}{2}L\{e^{at}\} + \frac{1}{2}L\{e^{-at}\} = \frac{1}{2} \cdot \frac{1}{p-a} + \frac{1}{2} \cdot \frac{1}{p+a}, \quad p > a \text{ and } p > -a$$

$$= \frac{p}{p^2 - a^2}, \quad p > |a|$$

Hence,　　$L\{\cosh at\} = \dfrac{p}{p^2 - a^2}, p > |a|$.

Solved Examples

EXAMPLE 1. *Find the Laplace transform of the function*
$$F(t) = \frac{e^{at} - 1}{a}.$$

SOLUTION. We have

$$L\{F(t)\} = L\left\{\frac{e^{at} - 1}{a}\right\} = L\left\{\frac{1}{a}e^{at} - \frac{1}{a}\right\}$$

$$= \frac{1}{a}L\{e^{at}\} - \frac{1}{a}L\{1\}$$

$$= \frac{1}{a}\left(\frac{1}{p-a}\right) - \frac{1}{a}\left(\frac{1}{p}\right) = \frac{1}{p(p-a)}.$$

EXAMPLE 2. *Find* $L\{(t^2 + 1)^2\}$.

SOLUTION.　$L\{(t^2 + 1)^2\} = L\{t^4 + 2t^2 + 1\}$

$$= L\{t^4\} + 2L\{t^2\} + L(1)$$
　　　　　　　　(By linearity property)

$$= \frac{4!}{p^5} + 2 \cdot \frac{2!}{p^3} + \frac{1}{p}$$

$$= \frac{24 + 4p^2 + p^4}{p^5}, \quad p > 0.$$

EXAMPLE 3. *Find* $L\{F(t)\}$ *where* $F(t) = (\sin t - \cos t)^2$.

SOLUTION. Consider

$$L\{(\sin t - \cos t)^2\}$$

$$= L\{\sin^2 t + \cos^2 t - 2\sin t \cos t\}$$

$$= L\{1 - \sin 2t\} = L\{1\} - L\{\sin 2t\}$$

$$= \frac{1}{p} - \frac{2}{p^2 + 2^2}, \quad p > 0 \quad = \frac{p^2 - 2p + 4}{p(p^2 + 4)}, \quad p > 0$$

EXAMPLE 4. *Find* $L\{F(t)\}$, *if* $F(t) = \begin{cases} e^t, & 0 < t \leq 1 \\ 0, & t > 1 \end{cases}$.

SOLUTION.　$L\{f(t)\} = \int_0^\infty e^{-pt} F(t) \, dt$

$$= \int_0^1 e^{-pt} \cdot e^t \, dt + \int_1^\infty e^{-pt} \cdot 0 \, dt$$

$$= \int_0^1 e^{-(p-1)t} \, dt = \left[-\frac{e^{-(p-1)t}}{p-1}\right]_0^1$$

$$= \frac{1}{(p-1)}[1 - e^{-(p-1)}], \quad p \neq 1.$$

EXAMPLE 5. *Find* $L\{F(t)\}$, *where* $F(t) = \begin{cases} 0, & 0 < t < 1 \\ t, & 1 < t < 2 \\ 0, & t > 2 \end{cases}$.

　　　　　　　　　　(JNTU–2006, WBTU–2005)

SOLUTION. We have that $F(t)$ is not defined at $t = 0, 1$ and 2.

$$\therefore L\{F(t)\} = \int_0^\infty e^{-pt} F(t) \, dt$$

$$= \int_0^1 e^{-pt} \cdot 0 \, dt + \int_1^2 e^{-pt} \cdot t \, dt + \int_2^\infty e^{-pt} \cdot 0 \, dt$$

$$= \int_1^2 e^{-pt} \cdot t \, dt = \left[-t\frac{e^{-pt}}{p} - \frac{e^{-pt}}{p^2}\right]_1^2, p \neq 0$$

$$= -\left(\frac{2}{p} + \frac{1}{p^2}\right)e^{-2p} + \left(\frac{1}{p} + \frac{1}{p^2}\right)e^{-p}, p \neq 0.$$

EXAMPLE 6. *Find the Laplace transform of*

$$F(t) = \begin{cases} t^2, & 0 < t < 2 \\ t - 1, & 2 < t < 3 \\ 7, & t > 3 \end{cases}$$

　　　　　　　(UPTU–2007, MUMBAI–2007)

SOLUTION. $L[F(t)] = \int_0^\infty e^{-pt} F(t)dt$

$$= \int_0^2 t^2 e^{-pt}dt + \int_2^3 (t-1)e^{-pt}dt + \int_3^\infty 7e^{-pt}dt$$

$$= \left[t^2 \frac{e^{-pt}}{(-p)} - 2t \frac{e^{-pt}}{(-p)^2} + 2\frac{e^{-pt}}{(-p)^3} \right]_0^2$$

$$+ \left[(t-1)\left(\frac{e^{-pt}}{(-p)}\right) - \frac{e^{-pt}}{(-p)^2} \right]_2^3 + 7\left[\frac{e^{-pt}}{-p} \right]_3^\infty$$

$$= \left[-4\left(\frac{e^{-2p}}{p}\right) - 4\left(\frac{e^{-2p}}{p^2}\right) + \frac{2}{p^3} \right]$$

$$+ \left[2\left(\frac{e^{-3p}}{-p}\right) - \left(\frac{e^{-3p}}{p^2}\right) + \left(\frac{e^{-2p}}{p}\right) + \frac{e^{-2p}}{p^2} \right]$$

$$+ 7\left[0 + \frac{e^{-3p}}{p} \right]$$

$$= \frac{2}{p^3} + e^{-2p}\left[-\frac{4}{p} - \frac{4}{p^2} - \frac{2}{p^3} \right]$$

$$+ e^{-3p}\left[-\frac{2}{p} - \frac{1}{p^2} \right] + e^{-2p}\left[\frac{1}{p} + \frac{1}{p^2} \right] + e^{-3p}\left[\frac{7}{p} \right]$$

$$= \frac{2}{p^3} + e^{-2p}\left[-\frac{4}{p} - \frac{4}{p^2} - \frac{2}{p^3} + \frac{1}{p} + \frac{1}{p^2} \right]$$

$$+ e^{-3p}\left[-\frac{2}{p} - \frac{1}{p^2} + \frac{7}{p} \right]$$

$$= \frac{2}{p^3} - \frac{e^{-2p}}{p^3}(2 + 3p + 3p^2) + \frac{e^{-3p}}{p^2}(5p - 1)$$

EXAMPLE 7. *Show that* $L\left\{ \dfrac{1}{\sqrt{\pi t}} \right\} = \dfrac{1}{\sqrt{p}}$.

SOLUTION. We have $L\left\{ \dfrac{1}{\sqrt{\pi t}} \right\} = \int_0^\infty e^{-pt} \cdot \dfrac{1}{\sqrt{\pi t}} \cdot dt$

$$= \frac{1}{\sqrt{\pi}} \int_0^\infty e^{-pt} \cdot \frac{1}{\sqrt{t}}dt = \frac{1}{\sqrt{\pi}} \int_0^\infty e^{-pt} \cdot t^{-1/2}dt$$

$$= \frac{1}{\sqrt{\pi}} \int_0^\infty e^{-pt} \cdot t^{1/2-1}dt = \frac{1}{\sqrt{\pi}} \frac{\overline{1/2}}{p^{1/2}}, \ p > 0$$

$$\text{(Using gamma function)}$$

$$= \frac{1}{\sqrt{\pi}} \cdot \frac{\sqrt{\pi}}{p^{1/2}} = \frac{1}{\sqrt{p}}. \qquad \left[\because \Gamma(\tfrac{1}{2}) = \sqrt{\pi} \right]$$

EXAMPLE 8. *Show that* $L\left\{ \dfrac{\cos\sqrt{t}}{\sqrt{t}} \right\} = \sqrt{\left(\dfrac{\pi}{p}\right)} e^{-1/4p}$.

$$\text{(MUMBAI–2009)}$$

SOLUTION. Here, we have

$$\frac{\cos\sqrt{t}}{\sqrt{t}} = \frac{1}{\sqrt{t}}\left\{ 1 - \frac{1}{2!}(\sqrt{t})^2 \right.$$

$$\left. + \frac{1}{4!}(\sqrt{t})^4 - \frac{1}{6!}(\sqrt{t})^6 + ... \right\}$$

$$= t^{-1/2} - \frac{1}{2!}t^{1/2} + \frac{1}{4!}t^{3/2} - \frac{1}{6!}t^{5/2} + ...$$

Therefore,

$$L\left\{ \frac{\cos\sqrt{t}}{\sqrt{t}} \right\} = L\{t^{-1/2}\} - \frac{1}{2!}L\{t^{1/2}\}$$

$$+ \frac{1}{4!}L\{t^{3/2}\} - \frac{1}{6!}L\{t^{5/2}\} + ...$$

$$= \frac{\Gamma(\tfrac{1}{2})}{p^{1/2}} - \frac{1}{2!}\frac{\Gamma(\tfrac{3}{2})}{p^{3/2}} + \frac{1}{4!}\frac{\Gamma(\tfrac{5}{2})}{p^{5/2}} - \frac{1}{6!}\frac{\Gamma(\tfrac{7}{2})}{p^{7/2}} + ..., p > 0$$

$$= \frac{\sqrt{\pi}}{p^{1/2}} - \frac{1}{1.2}\cdot\frac{\tfrac{1}{2}\cdot\sqrt{\pi}}{p^{3/2}} + \frac{\tfrac{3}{2}\cdot\tfrac{1}{2}\cdot\sqrt{\pi}}{1.2.3.4}\cdot\frac{1}{p^{5/2}}$$

$$- \frac{\tfrac{5}{2}\cdot\tfrac{3}{2}\cdot\tfrac{1}{2}\cdot\sqrt{\pi}}{1.2.3.4.5.6}\cdot\frac{1}{p^{7/2}} + ...$$

$$= \sqrt{\left(\frac{\pi}{p}\right)}\left[1 - \frac{1}{1!}\left(\frac{1}{4p}\right) + \frac{1}{2!}\left(\frac{1}{4p}\right)^2 - \frac{1}{3!}\left(\frac{1}{4p}\right)^3 + ... \right]$$

$$= \sqrt{\left(\frac{\pi}{p}\right)} \cdot e^{-1/4p}.$$

Exercise-8.44

Find the Laplace transform of the following functions : (Ques. 1 to 11)

1. $\sin t \cos t$

2. $4\cos^2 2t$

3. $\sin^2 at$

4. $3\cosh 5t - 4\sinh 5t$ (NAGARJUNA–2006)

5. $3t^4 - 2t^3 + 4e^{-3t} - 2\sin 5t + 3\cos 2t$

6. $e^{-2t} - e^{-3t}$

7. $F(t) = \begin{cases} \sin t, & 0 < t < \pi \\ 0, & t > \pi \end{cases}$ (MADRAS–2000S)

8. $F(t) = \begin{cases} (t-1)^2, & t > 1 \\ 0, & 0 < t < 1 \end{cases}$

9. $F(t) = \begin{cases} e^t, & 0 < t < 5 \\ 3, & t > 5 \end{cases}$

10. $F(t) \begin{cases} t, & 0 < t < 4 \\ 5, & 4 \end{cases}$

11. $F(t) = \sin\sqrt{t}$

12. Show that t^2 is of exponential order 3.

13. Show that the function e^{t^2} is not of exponential order as $t \to \infty$.

14. Show that the Laplace transforms of the function $F(t) = t^n, -1 < n < 0$ exists, although it is not a function of class A.

15. Find the Laplace transform of $F(t) = \begin{cases} e^t, & 0 < t < 1 \\ 0, & t > 1 \end{cases}$.

$$\text{(UPTU–2004)}$$

Hints to Selected Problems

1. The given function can be written as $F(t) = \sin t . \cos t = \dfrac{1}{2}\sin 2t$.

2. $L\{4\cos^2 2t\} = L\{2(1+\cos 4t)\} = 2[L\{1\} + L\{\cos 4t\}]$.

7. $L\{F(t)\} = \int_0^\infty e^{-pt} F(t)dt = \int_0^\pi e^{-pt}.\sin t dt + \int_0^\infty e^{-pt}.0 dt$

8. $L\{F(t)\} = \int_0^\infty F(t)e^{-pt}dt = \int_0^1 0.e^{-pt}dt + \int_0^\infty (t-1)^2 e^{-pt}dt$.

11. $L\{\sin\sqrt{t}\} = L\left[\sqrt{t} - \dfrac{(\sqrt{t})^3}{3!} + \dfrac{(\sqrt{t})^5}{5!} - \dfrac{(\sqrt{t})^7}{7!} + ... \right]$

Answers

1. $\dfrac{1}{p^2+4}$, $p > 0$ **2.** $\dfrac{4(p^2+8)}{p(p^2+16)}$, $p > 0$ **3.** $\dfrac{2a^2}{p(p^2+4a^2)}$, $p > 0$ **4.** $\dfrac{3p-20}{p^2-25}$, $p > 5$

5. $\dfrac{72}{p^5} - \dfrac{12}{p^4} + \dfrac{4}{p+3} - \dfrac{10}{p^2+25} + \dfrac{3p}{p^2+4}$, $p > 0$ **6.** $\dfrac{1}{p^2+5p+6}$, $p > -2$ **7.** $\dfrac{e^{-p\pi}+1}{p^2+1}$ **8.** $\dfrac{2e^{-p}}{p^3}$, $p > 0$

9. $\dfrac{1-e^{-5(p-1)}}{p-1} + \dfrac{3}{p}e^{-5p}$, $p > 0$ **10.** $\dfrac{1+(p-1)e^{-4p}}{p^2}$, $p > 0$ **11.** $\dfrac{\sqrt{\pi}}{2p^{3/2}}e^{-1/4p}$ **15.** $\dfrac{e^{1-p}-1}{1-p}$

8.9.4 TRANSLATION OR SHIFTING THEOREMS

THEOREM 1.   **(First Translation or Shifting Theorem).** *If $f(p)$ is the Laplace transform of $F(t)$, then $f(p-a)$ is the Laplace transforms of $e^{at} F(t)$, i.e., if $L\{F(t)\} = f(P)$, when $p > a$, then $L\{e^{at}F(t)\} = f(p-a)$, $p > a$.*

PROOF.   We have, by definition of Laplace transform

$$L\{F(t)\} = f(p) = \int_0^\infty e^{-pt} F(t)\, dt$$

Therefore, $L\{e^{at} F(t)\} = \int_0^\infty e^{-pt}.e^{at}F(t)\, dt = \int_0^\infty e^{-(p-a)t}. F(t)\, dt = \int_0^\infty e^{-ut} F(t)\, dt$,

where $u = p - a > 0$

                 $= f(u)$ (By definition)

                 $= f(p - a)$.

THEOREM 2.   **(Second Translation or Heaviside's Shifting Theorem).**

     If $L\{F(t)\} = f(p)$ and $G(t) = \begin{cases} F(t-a), & t > a \\ 0, & t < a \end{cases}$ then, $L\{G(t)\} = e^{-ap} f(p)$.   (UPTU–2006, 08)

PROOF.   Let $L\{F(t)\} = f(p)$ and $G(t) = \begin{cases} F(t-a), & \text{if } t > a \\ 0, & \text{if } t < a \end{cases}$

Then, $L\{G(t)\} = \int_0^\infty e^{-pt} G(t)\, dt$

                      $= \int_0^a e^{-pt} G(t)\, dt + \int_a^\infty e^{-pt} G(t)\, dt = \int_0^a e^{-pt}.0\, dt + \int_a^\infty e^{-pt} F(t-a)\, dt$

                      $= 0 + \int_a^\infty e^{-pt} F(t-a)\, dt$.

Let $t - a = u$, therefore $dt = du$.

If $t = a$, then $u = t - a = a - a = 0$ and if $t = \infty$, then $u = \infty - a = \infty$.

Hence, $L\{G(t)\} = \int_0^\infty e^{-p(u+a)} F(u)\, du = e^{-pa} \int_0^\infty e^{-pu} F(u)\, du = e^{-pa} f(p)$.

THEOREM 3.   **(Change of Scale Property).** *If $L\{F(t)\} = f(p)$, then $L\{F(at)\} = \dfrac{1}{a}f\left(\dfrac{p}{a}\right)$.*

PROOF.   By definition

$$L\{F(at)\} = \int_0^\infty e^{-pt} F(at)\, dt = \int_0^\infty e^{-pu/a} F(u)\, \frac{du}{a} \qquad\qquad (\text{where } at = u)$$

$$= \frac{1}{a}\int_0^\infty e^{-pu/a} F(u)\, du = \frac{1}{a}\int_0^\infty e^{-su} F(u)\, du \text{, where } s = \frac{p}{a} = \frac{1}{a}f(s) = \frac{1}{a}f\left(\frac{p}{a}\right) .$$

 Solved Examples

EXAMPLE 1. *Find* $L\left\{\dfrac{e^{-at}\, t^{n-1}}{(n-1)!}\right\}$.

SOLUTION. We have $L\left\{\dfrac{t^{n-1}}{(n-1)!}\right\} = \dfrac{1}{(n-1)!}.\dfrac{(n-1)!}{p^n} = \dfrac{1}{p^n}$.

Therefore, using first shifting theorem, we have

$$L\left\{e^{-at}\,\frac{t^{n-1}}{(n-1)!}\right\} = f(p+a) = \frac{1}{(p+a)^n} .$$

EXAMPLE 2. *Find* $L\{e^t \cos^2 t\}$.

SOLUTION. We have $L\{\cos^2 t\} = L\left\{\dfrac{1}{2}(1+\cos 2t)\right\}$

$$= \frac{1}{2}[L\{1\} + L\{\cos 2t\}]$$

$$= \frac{1}{2}\left\{\frac{1}{p} + \frac{p}{p^2+2^2}\right\} = \frac{p^2+2}{p(p^2+4)} = f(p) \text{ (say)}$$

Using first shifting theorem, we have

$$L\{e^t \cos^2 t\} = f(p-1) = \frac{(p-1)^2 + 2}{(p-1)(p-1)^2 + 4}$$

$$= \frac{p^2 - 2p + 3}{(p-1)(p^2 - 2p + 5)}$$

EXAMPLE 3. **Find $L\{e^{-t}(3\sin 2t - 5\cosh 2t)\}$.**

SOLUTION. We have

$$L\{3\sin 2t - 5\cosh 2t\}$$

$$= 3 \cdot \frac{2}{p^2 + 2^2} - \frac{5p}{p^2 - 2^2} = f(p) \qquad \text{(say)}.$$

Using first shifting theorem, we have

$$L\{e^{-t}(3\sin 2t - 5\cosh 2t)\} = f(p+1)$$

$$= \frac{6}{(p+1)^2 + 4} - \frac{5(p+1)}{(p+1)^2 - 4}$$

$$= \frac{6}{p^2 + 2p + 5} - \frac{5(p+1)}{p^2 + 2p - 3}.$$

EXAMPLE 4. **If $L\{\cos^2 t\} = \dfrac{p^2 + 2}{p(p^2 + 4)}$, find $L[\cos^2 at]$.**

(UPTU–2006)

SOLUTION. We have $L\{\cos^2 t\} = \dfrac{p^2 + 2}{p(p^2 + 4)}$

By change of scale property, we have

$$L\{\cos^2 at\} = \frac{1}{a} \cdot \frac{\left(\dfrac{p}{a}\right)^2 + 2}{\left(\dfrac{p}{a}\right)\left[\left(\dfrac{p}{a}\right)^2 + 4\right]}$$

$$= \frac{1}{p}\left[\frac{p^2 + 2a^2}{\dfrac{p}{a}(p^2 + 4a^2)}\right] = \frac{p^2 + 2a^2}{p(p^2 + 4a^2)}.$$

EXAMPLE 5. **Find $L\{G(t)\}$, where**

$$G(t) = \begin{cases} \cos\left(t - \dfrac{2}{3}\pi\right), & t > \dfrac{2\pi}{3} \\ 0, & t < \dfrac{2\pi}{3} \end{cases}$$

SOLUTION. Let $\qquad F(t) = \cos t$

Then, $\qquad G(t) = \begin{cases} F\left(t - \dfrac{2\pi}{3}\right), & t > 2\pi/3 \\ 0, & t < 2\pi/3 \end{cases}$

We have $L\{F(t)\} = L\{\cos t\} = \dfrac{p}{p^2 + 1} = f(p)$ (say)

Using second shifting theorem, we have

$$L\{G(t)\} = e^{\left(-\frac{2\pi}{3}\right) \cdot p} \cdot f(p) = e^{-2\pi p/3} \cdot \frac{p}{p^2 + 1}.$$

EXAMPLE 6. **Find $L\{G(t)\}$, where $G(t) = \begin{cases} e^{t-a}, & t > a \\ 0, & t < a \end{cases}$.**

(UPTU–2008)

SOLUTION. By second shifting theorem, we have

$$L\{F(t)\} = f(p) \text{ and } G(t) = \begin{cases} F(t-a), & t > a \\ 0, & t < a \end{cases}$$

Then, $L\{G(t)\} = e^{-ap} f(p)$

Let $\qquad F(t) = e^t$

Then, $L\{F(t)\} = L\{e^t\} = \int_0^\infty e^{-pt} \cdot e^t dt$

$$= \frac{1}{p-1}, p > 1 = f(p) \text{ (say)}$$

Now, let $G(t) = \begin{cases} F(t-a) = e^{t-a}, & t > a \\ 0, & t < a \end{cases}$

Then, $L\{G(t)\} = e^{-ap} f(p) = \dfrac{e^{-ap}}{p-1}, \ p > 1$.

Exercise-8.45

1. Find $L\{t^3 e^{-3t}\}$.

2. Find $L\{e^{3t} \cos 5t\}$.

3. Find $L\{e^{-t} \sin^2 t\}$.   (MUMBAI–2009)

4. Find $L\{e^t \sin^2 t\}$

5. Find $L\{e^{-4t} \cosh 2t\}$

6. Find $L\{e^{-2t}(3\cos 6t - 5\sin 6t)\}$

7. Using first shifting theorem, find the value of $L\{e^{6t}(t+2)^2\}$

8. If $L\{F(t)\} = f(p)$, find $L\{F(t)\cos \omega t\}$

9. Applying change of scale property, find

    (i) $L\{\sinh 3t\}$,   (ii) $L\{\cos 5t\}$

10. Find $L\{F(t)\}$, where $F(t) = \begin{cases} \sin\left(t - \dfrac{\pi}{3}\right), & t > \pi/3 \\ 0, & t < \pi/3 \end{cases}$

11. Find $L\{F(t)\}$, where $F(t) = \begin{cases} \sin\left(t - \dfrac{2}{3}\pi\right), & t > 2\pi/3 \\ 0, & t < 2\pi/3 \end{cases}$

12. If $\{F(t)\} = \dfrac{1}{p} e^{-1/p}$, show that $L\{e^{-t}F(3t)\} = \dfrac{e^{-3/(p+1)}}{p+1}$.

13. Find the Laplace transform of $e^t t^{-1/2}$.

(UPTU(Q. BANK)–2001)

14. Find the Laplace transform of :

    (i) $t^2 e^t \sin 4t$ (UPTU(SP.)–2001)

    (ii) $t e^{-t} \sin 2t$ (UPTU(SP.)–2002)

15. Find the Laplace transform of

$$F(t) = \begin{cases} 1, & 0 \le t < 1 \\ t, & 1 \le t < 2 \\ t^2, & 2 \le t < \infty \end{cases} \qquad \text{(UPTU(Q. BANK)–2001)}$$

Hints to Selected Problems

1. $L\{t^3\} = \dfrac{3!}{p^4}$, then $L\{t^3 e^{-3t}\} = f(p+3) = \dfrac{6}{(p+3)^4}$

5. $L(\cosh 2t) = \dfrac{p}{p^2 - 2^2} = \dfrac{p}{p^2 - 4} = f(p)$, then apply first shifting theorem.

7. $L\{(t+3)^2\} = L\{t^2 + 6t + 9\} = \dfrac{2!}{p^3} + 6 \cdot \dfrac{1}{p^2} + \dfrac{9}{p} = f(p)$ (say), then applying first shifting theorem.

9. (i) $L\{\sinh t\} = \dfrac{1}{(p^2 - 1)} = f(p)$, then by change of scale property

$L\{\sinh 3t\} = \dfrac{1}{3} f\left(\dfrac{p}{3}\right) = \dfrac{1}{3} \cdot \dfrac{1}{(p/3)^2 - 1} = \dfrac{3}{p^2 - 9}$

10. Let $G(t) = \sin t$, then $F(t) = \begin{cases} G\left(t - \dfrac{\pi}{3}\right), & t > \pi/3 \\ 0, & t < \pi/3 \end{cases}$

Then, $L\{G(t)\} = L\{\sin t\} = \dfrac{1}{p^2 + 1} = f(p)$ (say) then apply second shifting theorem.

Answers

1. $\dfrac{6}{(p+3)^4}$ **2.** $\dfrac{p-3}{p^2 - 6p + 25}$ **3.** $\dfrac{2}{(p+1)(p^2 + 2p + 5)}$ **4.** $\dfrac{2}{(p-1)(p^2 - 2p + 5)}$ **5.** $\dfrac{p+4}{p^2 + 8p + 12}$

6. $\dfrac{3p - 24}{p^2 + 4p + 40}$ **7.** $\dfrac{4p^2 - 44p + 122}{(p-6)^3}$ **8.** $\dfrac{1}{2}[f(p - i\omega) + f(p + i\omega)]$ **9.** (i) $\dfrac{3}{p^2 - 9}$ (ii) $\dfrac{p}{p^2 + 25}$, $p > 0$

10. $e^{-\pi p/3} \cdot \dfrac{1}{p^2 + 1}$, $p > 0$ **11.** $\dfrac{e^{-2\pi p/3}}{p^2 + 1}$, $p > 0$ **13.** $\dfrac{\sqrt{\pi}}{\sqrt{p-1}}$ **14.** (i) $\dfrac{8(3p^2 - 6p - 13)}{(p^2 - 2p + 17)^3}$ (ii) $\dfrac{4p + 4}{(p^2 + 2p + 5)^2}$

15. $\dfrac{1}{p} + \dfrac{2}{p} e^{-2p} + \dfrac{e^{-p}}{p^2} + \dfrac{3}{p^2} e^{-2p} + \dfrac{2}{p^3} e^{-2p}$

8.9.5 LAPLACE TRANSFORM OF DERIVATIVES

THEOREM 1.   *Let $F(t)$ be continuous for all $t \geq 0$ and be of exponential order as $t \to \infty$ and if $F'(t)$ is of class A, the Laplace transforms of derivatives $F'(t)$ exists when $p > a$ and $L\{F'(t)\} = p\,L\{F(t)\} - F(0)$.*

PROOF.   By definition, we have

$$L\{F'(t)\} = \int_0^\infty e^{-pt} F'(t)\, dt = \left[e^{-pt} F(t)\right]_0^\infty + p \int_0^\infty e^{-pt} F(t)\, dt \qquad \text{(On integrating by parts)}$$

$$= -F(0) + p L\{F(t)\} \qquad \left[\because \lim_{t \to \infty} e^{-pt} F(t) = 0\right]$$

$$= p L\{F(t)\} - F(0).$$

REMARK

- Proceeding same as above, we get

$L\{F''(t)\} = p L\{F'(t)\} - F'(0) = p[p L\{F(t)\} - F(0)] - F'(0) = p^2 L\{F(t)\} - p F(0) - F'(0) = p^2 f(p) - p F(0) - F'(0).$

THEOREM 2.   *If $F(t), F'(t), ..., F^{n-1}(t)$ are continuous for $t \geq 0$ and be of exponential order as $t \to \infty$ and if $F^n(t)$ is of class A and if $L\{F(t)\} = f(p)$,*

then $L\{F^n(t)\} = p^n f(p) - p^{n-1} F(0) - p^{n-2} F(0) ... p F^{(n-2)}(0) - F^{(n-1)}(0) = p^n f(p) - \displaystyle\sum_{r=0}^{n-1} p^{n-1-r} F^r(0)$

PROOF.   Using above theorem, we have

$$L\{F'(t)\} = p L\{F(t)\} - F(0) \quad \text{and} \quad L\{F''(t)\} = p^2 L\{F(t)\} - p F(0) - F'(0)$$

Similarly, we can find

$$L\{F'''(t)\} = p L\{F''(t)\} - F''(0) = p[p^2 L\{F(t)\} - p F(0) - F'(0)] - F''(0)$$

$$= p^3 L\{F(t)\} - p^2 F(0) - p F'(0) - F''(0).$$

Proceeding, similarly, we get

$$L\{F^n(t)\} = p^n L\{F(t)\} - p^{n-1} F(0) - p^{n-2} F'(0) - ... - F^{n-1}(0) = p^n L\{F(t)\} - \displaystyle\sum_{r=0}^{n-1} p^{n-1-r} F^r(0).$$

THEOREM 3.   *If $F(t)$ is a function of class A and if $L\{F(t)\} = f(p)$, then $L\{t \cdot F(t)\} = -f'(p)$*

PROOF.   We know that

$$f(p) = L\{F(t)\} = \int_0^\infty e^{-pt} F(t)\, dt.$$

Therefore, $f'(p) = \dfrac{d}{dp} \int_0^\infty e^{-pt} F(t)\, dt = \int_0^\infty \dfrac{\partial}{\partial p} \{e^{-pt} F(t)\}\, dt$

(By Leibnitz's rule of differentiation under the sign of integral)

$$= -\int_0^\infty t\, e^{-pt}\, F(t)\, dt = -\int_0^\infty e^{-pt}\, \{t\, F(t)\}\, dt = -L\{t\, F(t)\}$$

$$\Rightarrow \qquad L\{t\, F(t)\} = -f'(p).$$

THEOREM 4. *If $F(t)$ is a function of class A and if $L\{F(t)\} = f(p)$ Then, $L\{t^n\, F(t)\} = (-1)^n \dfrac{d^n}{dp^n}\, f(p).$* \hfill (UPTU–2005)

PROOF.   We shall prove this theorem by the principle of Mathematical induction.

Step 1. Using previous theorem, we have

$$L\{t\, F(t)\} = (-1)^1 \frac{d}{dp}\, f(p) \qquad \Rightarrow \qquad \text{Theorem is true for } n = 1.$$

Step 2. Assume that the theorem is true for a particular value of n say k. Then, we have

$$L\{t^k\, F(t)\} = (-1)^k \frac{d^k}{dp^k}\, f(p) \qquad \Rightarrow \qquad \int_0^\infty e^{-pt}\, t^k\, F(t)\, dt = (-1)^k \frac{d^k}{dp^k}\, f(p).$$

Step 3. Differentiating both sides w.r.t. p, we have

$$\frac{d}{dp} \int_0^\infty e^{-pt}\, t^k\, F(t)\, dt = (-1)^k \frac{d^{k+1}}{dp^{k+1}}\, f(p).$$

Applying, Leibnitz's rule for differentiation under the sign of integration, we have

$$-\int_0^\infty e^{-pt}\, t^{k+1}\, F(t)\, dt = (-1)^{k+2} \frac{d^{k+1}}{dp^{k+1}}\, f(p) \quad \Rightarrow \quad \int_0^\infty e^{-pt}\, \{t^{k+1}\, F(t)\}\, dt = (-1)^{k+1} \frac{d^{k+1}}{dp^{k+1}}\, f(p)$$

$$\Rightarrow \qquad L\{t^{k+1}\, F(t)\} = (-1)^{k+1} \frac{d^{k+1}}{dp^{k+1}}\, f(p) \qquad \Rightarrow \qquad \text{Theorem is true for } n = k+1.$$

Hence, by the principle of mathematical induction, it is true for every positive integral value of n.

THEOREM 5. *Let a function $F(t)$ be periodic with period w, so that $F(t + nw) = F(t)$ for $n = 1, 2, 3, \ldots$, then $L\{F(t)\} = \dfrac{1}{1 - e^{-pw}} \int_0^w e^{-pt}\, F(t)\, dt.$*

PROOF.   We know that

$$L\{F(t)\} = \int_0^\infty e^{-pt}\, F(t)\, dt$$

$$= \int_0^w e^{-pt}\, F(t)\, dt + \int_w^{2w} e^{-pt}\, F(t)\, dt + \ldots + \int_{nw}^{(n+1)w} e^{-pt}\, F(t)\, dt + \ldots$$

$$= \sum_{n=0}^\infty \int_{nw}^{(n+1)w} e^{-pt}\, F(t)\, dt = \sum_{n=0}^\infty \int_0^w e^{-p(x+nw)}\, F(x + nw)\, dx \qquad [t = x + nw]$$

$$= \sum_{n=0}^\infty \int_0^w e^{-px}\, e^{-npw}\, F(x)\, dx \qquad [\because \text{By definition of periodic function } f(x + nw) = f(x)]$$

$$= \sum_{n=0}^\infty e^{-npw} \int_0^w e^{-px}\, F(x)\, dx = (1 + e^{-pw} + e^{-2pw} + \ldots) \int_0^w e^{-px}\, F(x)\, dx$$

$$= \frac{1}{1 - e^{-pw}} \int_0^w e^{-px}\, F(x)\, dx \qquad\qquad \left[\because e^{-pw} < 1\right]$$

THEOREM 6 **(Initial Value Theorem).** *Let $F(t)$ be continuous for all $t \geq 0$ and be of exponential order as $t \to \infty$ and if $F'(t)$ is of class A, then*

$$\lim_{t \to 0} F(t) = \lim_{p \to \infty} pL\{F(t)\}.$$

PROOF.   We know that

$$L\{F'(t)\} = \int_0^\infty e^{-pt}\, F'(t)\, dt = pL\{F(t)\} - F(0). \qquad \ldots(1)$$

Since $F'(t)$ is sectionally continuous and of exponential order.

Therefore, $\displaystyle \lim_{p \to \infty} \int_0^\infty e^{-pt}\, F'(t)\, dt = 0$

Now, taking limit as $p \to \infty$ in (1), we have

$$0 = \lim_{p \to \infty} pL\{F(t)\} - F(0) \quad \Rightarrow \quad F(0) = \lim_{p \to \infty} pL\{F(t)\} \quad \Rightarrow \quad \lim_{t \to 0} F(t) = \lim_{p \to \infty} pL\{F(t)\}.$$

THEOREM 7 **(Final Value Theorem).** *Let $F(t)$ be continuous for all $t \geq 0$ and be of exponential order as $t \to \infty$ and if $F'(t)$ is of class A, then*

$$\lim_{t \to \infty} F(t) = \lim_{p \to 0} pL\{F(t)\}.$$

PROOF.   We know that

$$L\{F'(t)\} = \int_0^\infty e^{-pt}\, F'(t)\, dt = pL\{F(t)\} - F(0). \qquad \ldots(1)$$

Taking limit as $p \to 0$ in (1), we get

$$\lim_{p \to 0} \int_0^\infty e^{-pt}\, F'(t)\, dt = \lim_{p \to 0} [pL\{F(t)\} - F(0)] \quad \Rightarrow \quad \int_0^\infty F'(t)\, dt = \lim_{p \to 0} pL\{F(t)\} - F(0)$$

$$\Rightarrow \qquad [F(t)]_0^\infty = \lim_{p \to 0} pL\{F(t)\} - F(0) \qquad \Rightarrow \qquad \lim_{t \to \infty} F(t) - F(0) = \lim_{p \to 0} pL\{F(t)\} - F(0)$$

$$\Rightarrow \qquad \lim_{t \to \infty} F(t) = \lim_{p \to 0} pL\{F(t)\}.$$

THEOREM 8   **(Laplace Transform of the Laplace Transform).** *We have*

$$L[L\{F(t)\}] = L\left\{\int_0^\infty e^{-pt} F(t) \, dt\right\} = \int_0^\infty e^{-up} \left\{\int_0^\infty e^{-pt} F(t) \, dt\right\} dp.$$

PROOF.   The area of integration being the whole positive quadrant. Now, changing the order of integration, we get

$$L[L\{F(t)\}] = \int_0^\infty F(t) \left\{\int_0^\infty e^{-p(t+u)} dp\right\} dt = \int_0^\infty F(t) \left\{\left[\frac{e^{-p(t+u)}}{-(t+u)}\right]_{p=0}^\infty\right\} dt = \int_0^\infty \frac{F(t)}{t+u} \, dt.$$

THEOREM 9   **(Laplace Transforms of Integrals).** *If $F(t)$ is piecewise continuous and satisfies $|F(t)| \le Me^{at}, \forall t \ge 0$ for some constant a and M, then*

$$L\left\{\int_0^t F(x) \, dx\right\} = \frac{1}{p} L\{F(t)\} \qquad\qquad (p > 0, \ p > a)$$

PROOF.   Let $F(t)$ be piecewise continuous such that

$$|F(t)| \le M e^{at} \qquad\qquad\qquad\qquad \dots(1)$$

for some constants a and M. •

If (1) holds for some negative value of a, then it also holds for positive value of a. Therefore, suppose that a is positive.

Let $\qquad\qquad\qquad G(t) = \int_0^t F(x) \, dx$.

Then $G(t)$ is continuous.   ($\because$ Integral of an integrable function is continuous)

Now, $|G(t)| \le \int_0^t |F(x)| \, dx \le \int_0^t Me^{ax} \, dx \qquad \Rightarrow \qquad |G(t)| \le \frac{M}{a}(e^{at} - 1), \ a > 0 \qquad \dots(2)$

Further, $G'(t) = F(t)$, except for points at which $F(t)$ is discontinuous. Therefore, $G'(t)$ is piecewise continuous on each finite interval.

$$\therefore \qquad\qquad L\{G'(t)\} = pL\{G(t)\} - G(0) = pL\{G(t)\} \qquad\qquad\qquad [\because G(0) = 0]$$

$$\Rightarrow \qquad\qquad L\{G(t)\} = \frac{1}{p} L\{G'(t)\} \qquad\qquad \Rightarrow \qquad L\left\{\int_0^t F(x) \, dx\right\} = \frac{1}{p} L\{F(t)\}.$$

THEOREM 10   **(Division by t).** *If $L\{F(t)\} = f(p)$, then $L\left\{\dfrac{1}{t} F(t)\right\} = \int_p^\infty f(x) \, dx$ provided $\lim_{t \to 0}\left\{\dfrac{1}{t} F(t)\right\}$ exists.*   (UPTU–2005, 07)

PROOF.   Let $\qquad\qquad G(t) = \dfrac{1}{t} F(t)$, *i.e.,* $F(t) = t \, G(t)$

Therefore, $\quad L\{F(t)\} = L\{t \, G(t)\} = -\dfrac{d}{dp} L\{G(t)\} \qquad \Rightarrow \qquad f(p) = -\dfrac{d}{dp} L\{G(t)\}.$

On integrating both sides with respect to p to ∞, we get

$$-\left[L\{G(t)\}\right]_p^\infty = \int_p^\infty f(p) \, dp \qquad \Rightarrow \qquad -\lim_{p \to \infty} L\{G(t)\} + L\{G(t)\} = \int_p^\infty f(p) \, dp$$

$$\Rightarrow \qquad 0 + L\{G(t)\} = \int_p^\infty f(p) \, dp, \qquad\qquad \left(\text{By using } \lim_{p \to \infty} L\{G(t)\} = \lim_{p \to \infty} \int_0^\infty e^{-pt} G(t) dt = 0 \right)$$

$$\Rightarrow \qquad L\left\{\frac{1}{t} F(t)\right\} = \int_p^\infty f(x) \, dx.$$

Solved Examples

EXAMPLE 1. ***Find $L\{t \cos at\}$.***   (RAIPUR–2005)

SOLUTION.   We know that

$$L\{\cos at\} = \frac{p}{p^2 + a^2}, \ p > 0.$$

Therefore,

$$L\{t \cos at\} = -\frac{d}{dp} L\{\cos at\} = -\frac{d}{dp}\left(\frac{p}{p^2 + a^2}\right).$$

$$= \frac{p^2 - a^2}{(p^2 + a^2)^2}.$$

EXAMPLE 2. ***Find $L\{t^2 \sin at\}$.***   (UPTU(Q. BANK)–2001)

SOLUTION.   We know that $L\{\sin at\} = \dfrac{a}{p^2 + a^2}$

Therefore, $L\{t^2 \sin at\} = (-1)^2 \dfrac{d^2}{dp^2} L\{\sin at\}$

$$= \frac{d^2}{dp^2}\left\{\frac{a}{p^2 + a^2}\right\}$$

$$= \frac{d}{dp}\left\{\frac{-2ap}{(p^2 + a^2)^2}\right\}$$

$$= \frac{2a(3p^2 - a^2)}{(p^2 + a^2)^3}, \ p > 0.$$

EXAMPLE 3. *Find L {(sin at – at cos at)}.*

SOLUTION. Consider $L\{\sin at - at \cos at\}$

$$= L\{\sin at\} - aL\{t \cos at\}$$

$$= \frac{a}{p^2 + a^2} - a.(-1)\frac{d}{dp}[L\{\cos at\}]$$

$$= \frac{a}{p^2 + a^2} + a\frac{d}{dp}\left(\frac{p}{p^2 + a^2}\right)$$

$$= \frac{a}{p^2 + a^2} + \frac{a(a^2 - p^2)}{(p^2 + a^2)^2}$$

$$= \frac{2a^3}{(p^2 + a^2)^2}.$$

EXAMPLE 4. *Find the Laplace transform of the function*

$$F(t) = t\, e^{-t}\, \sin 2t$$

(UPTU–2002, KURUKSHETRA–2005, 13)

SOLUTION. $L\{\sin 2t\} = \dfrac{2}{p^2 + 4}$,

$$L\{e^{-t}\sin 2t\} = \frac{2}{(p+1)^2 + 4} = f(p) \quad \text{(say)}$$

$$L\{te^{-t}\sin 2t\} = f'(p) = -\frac{d}{dp}\left[\frac{2}{(p+1)^2 + 4}\right]$$

$$= \frac{-2.2(p+1)}{[(p+1)^2 + 4]^2} = \frac{4(p+1)}{[(p+1)^2 + 4]^2}.$$

EXAMPLE 5. *Obtain the Laplace transform of $t^2 e^t \sin 4t$.*

(UPTU–2002)

SOLUTION.

$$L\{\sin 4t\} = \frac{4}{p^2 + 16},$$

$$L\{e^t \sin 4t\} = \frac{4}{(p-1)^2 + 16}$$

$$L\{te^t \sin 4t\} = -\frac{d}{dp}\left(\frac{4}{p^2 - 2p + 17}\right)$$

$$= \frac{4(2p - 2)}{(p^2 - 2p + 17)^2}$$

$$L\{t^2 e^t \sin 4t\} = -\frac{d}{dp}\left(\frac{4(2p - 2)}{(p^2 - 2p + 17)^2}\right)$$

$$= \frac{-4(2p^2 - 4p + 34 - 8p^2 + 16p - 8)}{(p^2 - 2p + 17)^3}$$

$$= \frac{-4(-6p^2 + 12p + 26)}{(p^2 - 2p + 17)^3}$$

$$= \frac{8(3p^2 - 6p - 13)}{(p^2 - 2p + 17)^3}.$$

EXAMPLE 6. *Given $L\{\sin\sqrt{t}\} = \dfrac{\sqrt{\pi}}{2p^{3/2}}\, e^{-1/4p}$, show that*

$$L\left\{\frac{\cos\sqrt{t}}{\sqrt{t}}\right\} = \sqrt{\left(\frac{\pi}{p}\right)}.\, e^{-1/4p}. \quad \text{(MUMBAI–2009)}$$

SOLUTION. Let $F(t) = \sin\sqrt{t}$

Then we have $F'(t) = \cos\dfrac{\sqrt{t}}{2\sqrt{t}}$ and $F(0) = 0$

Put all these values in

$$L\{F'(t)\} = pL\{F(t)\} - F(0)$$

We get $L\left\{\dfrac{\cos\sqrt{t}}{2\sqrt{t}}\right\} = pL\{\sin\sqrt{t}\}$

$$= p\left[\frac{\sqrt{\pi}}{2p^{3/2}}e^{-1/4p}\right] = \frac{1}{2}\sqrt{\left(\frac{\pi}{p}\right)}e^{-1/4p}.$$

Hence, $L\left\{\dfrac{\cos\sqrt{t}}{\sqrt{t}}\right\} = \sqrt{\left(\dfrac{\pi}{p}\right)}.\, e^{-1/4p}.$

EXAMPLE 7. *Show that $L\left\{\dfrac{\sin t}{t}\right\} = \tan^{-1}\dfrac{1}{p}$ and hence find $L\left\{\dfrac{\sin at}{t}\right\}$. Does the Laplace transform of $\dfrac{\cos at}{t}$ exists?*

(UPTU–2005, PTU–2010)

SOLUTION. Let $F(t) = \sin t$

Then, $\lim\limits_{t\to 0}\dfrac{F(t)}{t} = \lim\limits_{t\to 0}\dfrac{\sin t}{t} = 1$.

We know that $L\{\sin t\} = \dfrac{1}{p^2 + 1} = f(p)$ (say)

Then we have

$$L\left\{\frac{\sin t}{t}\right\} = \int_p^\infty f(x)\, dx$$

$$= \int_p^\infty \frac{dx}{x^2 + 1} = \left(\tan^{-1} x\right)_p^\infty$$

$$= \frac{\pi}{2} - \tan^{-1} p = \cot^{-1} p = \tan^{-1}\left(\frac{1}{p}\right).$$

Now, $L\left\{\dfrac{\sin at}{t}\right\} = aL\left\{\dfrac{\sin at}{at}\right\}$

$$= a.\frac{1}{a}\tan^{-1}\left(\frac{1}{p/a}\right)$$

$$\left[\because L\{f(at)\} = \frac{1}{a}f\left(\frac{p}{a}\right)\right]$$

$$= \tan^{-1}\left(\frac{a}{p}\right).$$

Also, since $L\{\cos at\} = \dfrac{p}{p^2 + a^2} = f(p)$ (say)

Then, $L\left\{\dfrac{\cos at}{t}\right\} = \int_p^\infty \dfrac{x}{x^2 + a^2}\, dx$

$$= \left[\frac{1}{2}\log(x^2 + a^2)\right]_p^\infty$$

$$= \frac{1}{2}\lim_{x\to\infty}\log(x^2 + a^2) - \frac{1}{2}\log(p^2 + a^2)$$

which does not exist since $\lim\limits_{x\to\infty}\log(x^2 + a^2)$ is infinite.

Therefore, $L\left\{\dfrac{\cos at}{t}\right\}$ does not exist.

EXAMPLE 8. *If $F(t) = \dfrac{e^{at} - \cos bt}{t}$, find the Laplace transform of $F(t)$.*

(UPTU–2003)

SOLUTION. $F(t) = \dfrac{e^{at} - \cos bt}{t} = \dfrac{e^{at}}{t} - \dfrac{\cos bt}{t}$

We know that

$$L\left(e^{at} - \cos bt\right) = \left(\dfrac{1}{p-a} - \dfrac{p}{p^2 + b^2}\right)$$

$$\therefore\ L\left(\dfrac{e^{at} - \cos bt}{t}\right) = \int_p^\infty \left(\dfrac{1}{p-a} - \dfrac{p}{p^2 + b^2}\right) dp$$

$$= \left[\log(p-a) - \dfrac{1}{2}\log(p^2 + b^2)\right]_p^\infty$$

$$= \left[\dfrac{2\log(p-a) - \log(p^2 + b^2)}{2}\right]_p^\infty$$

$$= \dfrac{1}{2}\left[\log(p-a)^2 - \log(p^2 + b^2)\right]_p^\infty$$

$$= \dfrac{1}{2}\left[\log\dfrac{(p-a)^2}{p^2 + b^2}\right]_p^\infty = \dfrac{1}{2}\left[\log\left[\dfrac{(1-(a/p))}{(1+(b^2/p^2))}\right]\right]_p^\infty$$

$$= \dfrac{1}{2}\left[0 - \log\dfrac{(1-(1/p))^2}{(1+(b^2/p^2))}\right] = \dfrac{1}{2}\left[\log\dfrac{p^2 + b^2}{(p-a)^2}\right]$$

Exercise-8.46

1. Show that $L\{-a\sin at\} = -\dfrac{a^2}{p^2 + a^2}$.

2. Evaluate
(i) $L\{t\cosh 3t\}$,   (ii) $L\{t\sinh at\}$

3. Show that $L\{t^2\cos at\} = \dfrac{2p(p^2 - 3a^2)}{(p^2 + a^2)^3}$, $p > 0$.

4. Show that $L\left(t^n e^{at}\right) = \dfrac{n!}{(p-a)^{n+1}}$, $p > a$.

5. Show that $L\{t\,(3\sin 2t - 2\cos 2t)\} = \dfrac{8 + 12p - 2p^2}{(p^2 + 4)^2}$.

6. Show that $L\{\sin\alpha t + t\cos\alpha t\} = \dfrac{(\alpha+1)p^2 + (\alpha-1)\alpha^2}{(p^2 + \alpha^2)^2}$.

7. Show that
$L\{t^2 - 3t + 2\}\sin 3t = \dfrac{6p^4 - 18p^3 + 126p^2 - 162p + 432}{(p^2 + 9)^3}$.

8. If $L\{F(t), t \to p\} = f(p)$, show that
$$L\left\{\int_0^t \dfrac{F(u)}{u}\,du,\ t \to p\right\} = \dfrac{1}{p}\int_p^\infty f(y)\,dy.$$

Hence, show that $L\left\{\int_0^t \dfrac{\sin u}{u}\,du,\ t \to p\right\} = \dfrac{\cot^{-1} p}{p}$.

9. Show that if $L\{F(t)\} = f(p)$, then

$$\int_0^\infty \dfrac{F(t)}{t}\,dt = \int_0^\infty f(x)\,dx\ ,\text{provided that the integral converges.}$$

10. If $L\{t\sin wt\} = \dfrac{2wp}{(p^2 + w^2)^2}$, evaluate $L\{wt\cos wt + \sin wt\}$.

(UPTU(Q. BANK)–2001)

11. Find the Laplace transform of
(i) $\int_0^t e^{-t}\cos t\,dt$,

(ii) $\int_0^t \dfrac{\sin t}{t}\,dt$ (UPTU(Q. BANK)–2001, JNTU–2005)

12. Find the Laplace transform of
(i) $te^{-t}\sin 2t$,   (UPTU–2002)

(ii) $t^2 e^t \sin 4t$ (UPTU(SP)–2001)

Hints to Selected Problems

1. $F(t) = -a\sin at, F'(t) = -a^2\cos at, F''(t) = a^3\sin at$
$F'(0) = -a^2$ and $F(0) = 0$, then using

$L\{F''(t)\} = p^2\,L\,F\{t\} - pF(0) - F'(0)$

9. Use Theorem 10.

Answers

2. (a) $\dfrac{p^2 + 9}{(p^2 - 9)^2}$ (b) $\dfrac{2ap}{(p^2 - a^2)^2}$ **10.** $\dfrac{2wp^2}{(p^2 + w^2)^2}$ **11.**(i) $\dfrac{p+1}{p(p^2 + 2p + 2)}$, (ii) $\dfrac{1}{p}\cot^{-1} p$

12. (i) $\dfrac{4p+4}{(p^2 + 2p + 5)^2}$, (ii) $\dfrac{8(3p^2 - 6p - 13)}{(p^2 - 2p + 17)^3}$

8.9.6 Evaluation of Integrals

If $L\{F(t)\} = f(p)$, i.e., $\int_0^\infty e^{-pt} F(t)\,dt = f(p)$

By taking limit as $p \to 0$, we have $\int_0^\infty F(t)\,dt = f(0)$, provided the integral is convergent.

8.9.7 SOME IMPORTANT SPECIAL FUNCTIONS

(i) The sine and cosine integrals. The sine the cosine integrals, which are denoted by $S_i(t)$ and $C_i(t)$ respectively are defined by

$S_i(t) = \int_0^t \dfrac{\sin u}{u}\,du$ and $C\,(t)\ \ \int^\infty \dfrac{\cos}{} du$.

(ii) Error Function. The error function denoted by $erf\,(t)$, is defined by $erf\,(t) = \dfrac{2}{\sqrt{\pi}}\int_0^t e^{-u^2}\,du$.

(iii) The Gamma function. If $n > 0$, the gamma function is defined by $\Gamma(n) = \int_0^\infty u^{n-1} e^{-u} du$.

(iv) Heaviside's unit function. The unit step function or heaviside's unit function denoted by $H(t-a)$ is defined by

$$H(t-a) = \begin{cases} 0, & t < a \\ 1, & t \geq a \end{cases}.$$

(v) Bessel's functions. $J_n(t) = \dfrac{t^n}{2^n \, \Gamma(n+1)} \left[1 - \dfrac{t^2}{2(2n+2)} + \dfrac{t^4}{2.4(2n+2)(2n+4)} \cdots \right]$.

Solved Examples

EXAMPLE 1. Find $\int_0^\infty \dfrac{(e^{-at} - e^{-bt})}{t} dt$.

(SVTU–2009, MUMBAI–2007, JNTU–2006)

SOLUTION. Let $F(t) = e^{-at} - e^{-bt}$.

Thus, we have

$$L\{F(t)\} = L\{e^{-at}\} - L\{e^{-bt}\}$$

$$= \frac{1}{p+a} - \frac{1}{p+b} = f(p) \text{ (say)}.$$

Therefore,

$$L\left\{\frac{F(t)}{t}\right\} = \int_p^\infty f(x)\, dx = \int_p^\infty \left(\frac{1}{x+a} - \frac{1}{x+b}\right) dx$$

$$= \left[\log\left(\frac{x+a}{x+b}\right)\right]_p^\infty = \lim_{x \to \infty} \log \frac{x+a}{x+b} - \log \frac{p+a}{p+b}$$

$$= \lim_{x \to \infty} \log \frac{1 + a/x}{1 + b/x} - \log \frac{p+a}{p+b}$$

$$= 0 - \log \frac{p+a}{p+b} = \log \frac{p+b}{p+a}$$

Therefore, $L\left\{\dfrac{F(t)}{t}\right\} = \int_0^\infty e^{-pt} \cdot \dfrac{e^{-at} - e^{-bt}}{t} dt$

$$= \log \frac{p+b}{p+a}$$

Hence, taking limit as $p \to 0$, we have

$$\int_0^\infty \frac{e^{-at} - e^{-bt}}{t} dt = \log \frac{b}{a}.$$

EXAMPLE 2. Show that $\int_0^\infty t\, e^{-2t} \cos t\, dt = \dfrac{3}{25}$.

SOLUTION. We have $L\{t \cos t\} = -\dfrac{d}{dp} L\{\cos t\}$

or $\int_0^\infty e^{-pt} \cdot t \cos t\, dt = -\dfrac{d}{dp}\left(\dfrac{p}{p^2+1}\right) = \dfrac{p^2 - 1}{(p^2+1)^2}$

Taking $p = 2$, we get

$$\int_0^\infty t\, e^{-2t} \cos t\, dt = \frac{3}{25}.$$

EXAMPLE 3. Show that

(i) $L\{\sinh at \cos at\} = \dfrac{a(p^2 - 2a^2)}{p^4 + 4a^4}$

(ii) $L\{\sinh at \sin at\} = \dfrac{2a^2 p}{p^4 + 4a^4}$.

SOLUTION. (i) We know that

$$L\{\sinh at\} = \frac{a}{p^2 - a^2} = f(p) \text{ (say)}$$

Therefore, $L\{e^{iat} \sin at\} = f(p - ia)$

$$= \frac{a}{(p - ia)^2 - a^2} = \frac{a}{(p^2 - 2a^2) - 2iap}$$

$$= \frac{a\{(p^2 - 2a^2) + 2iap\}}{(p^2 - 2a^2)^2 - (2ipa)^2}$$

$$\Rightarrow L\{\sinh at (\cos at + i \sin at)\}$$

$$= \frac{a(p^2 - 2a^2) + 2ia^2 p}{p^4 + 4a^4}$$

$$\Rightarrow L\{\sinh at \cos at\} + iL\{\sinh at \sin at\}$$

$$= \frac{a(p^2 - 2a^2)}{p^4 + 4a^4} + i\frac{2a^2 p}{p^4 + 4a^4}$$

Equating real and imaginary parts of both the sides, we get

$$L\{\sinh at \cos at\} = \frac{a(p^2 - 2a^2)}{p^4 + 4a^4}$$

and $L\{\sinh at \sin at\} = \dfrac{2a^2 p}{p^4 + 4a^4}$.

EXAMPLE 4. Find $L\left\{erf\sqrt{t}\right\}$ **and hence prove that**

$$L\left\{t \cdot erf(2\sqrt{t})\right\} = \frac{3p + 8}{p^2(p+4)^{3/2}}. \quad \text{(UPTU–2001)}$$

SOLUTION. We know that

$$erf\sqrt{t} = \frac{2}{\sqrt{\pi}} \int_0^{\sqrt{t}} e^{-x^2} dx$$

$$= \frac{2}{\sqrt{\pi}} \int_0^{\sqrt{t}} \left(1 - x^2 + \frac{x^4}{2!} - \frac{x^6}{6!} + \ldots\right) dx$$

$$= \frac{2}{\sqrt{\pi}} \left[x - \frac{x^3}{3} + \frac{x^5}{10} - \frac{x^7}{42} - \ldots\right]_0^{\sqrt{t}}$$

$$= \frac{2}{\sqrt{\pi}} \left[\sqrt{t} - \frac{t^{3/2}}{3} + \frac{t^{3/2}}{10} - \frac{t^{7/2}}{42} + \ldots\right]$$

$$L\left\{erf\sqrt{t}\right\} = \frac{2}{\sqrt{\pi}} \left[\frac{\Gamma(3/2)}{p^{3/2}} - \frac{\Gamma(5/2)}{3p^{5/2}}\right.$$

$$\left. + \frac{\Gamma(7/2)}{10 p^{7/2}} - \frac{\Gamma(9/2)}{42 p^{9/2}} + \ldots\right]$$

$$= \frac{2}{\sqrt{\pi}} \left[\frac{\frac{1}{2}\Gamma(1/2)}{p^{3/2}} - \frac{\frac{3}{2} \cdot \frac{1}{2}\Gamma(1/2)}{3p^{5/2}}\right.$$

$$\left. + \frac{\frac{5}{2} \cdot \frac{3}{2} \cdot \frac{1}{2}\Gamma(1/2)}{10 p^{7/2}} - \frac{\frac{7}{2} \cdot \frac{5}{2} \cdot \frac{3}{2} \cdot \frac{1}{2}\Gamma(1/2)}{42 p^{9/2}} + \ldots\right]$$

$$= \frac{1}{p^{3/2}} - \frac{1}{2}\frac{1}{p^{5/2}} + \frac{1.3}{2.4}\cdot\frac{1}{p^{7/2}}$$

$$- \frac{1}{2}\cdot\frac{3}{4}\cdot\frac{5}{6}\cdot\frac{1}{p^{9/2}} + \dots$$

$$= \frac{1}{p^{3/2}}\left[1 - \frac{1}{2}\cdot\frac{1}{p} + \frac{1}{2}\cdot\frac{3}{4}\cdot\frac{1}{p^2}\right.$$

$$\left. - \frac{1}{2}\cdot\frac{3}{4}\cdot\frac{5}{6}\cdot\frac{1}{p^3} + \dots\right]$$

$$= \frac{1}{p^{3/2}}\left[1 - \frac{1}{2}\cdot\frac{1}{p} + \frac{\left(-\frac{1}{2}\right)\left(-\frac{3}{2}\right)}{2!}\frac{1}{p^2}\right.$$

$$\left. + \frac{\left(-\frac{1}{2}\right)\left(-\frac{3}{2}\right)\left(-\frac{5}{2}\right)}{3!}\frac{1}{p^3} + \dots\right]$$

$$= \frac{1}{p^{3/2}}\left[1 + \frac{1}{p}\right]^{-1/2}$$

$$= \frac{1}{p^{3/2}}\left[\frac{p}{p+1}\right]^{1/2} = \frac{1}{p\sqrt{p+1}}$$

Now, $L\left\{erf\, 2\sqrt{t}\right\} = L\left\{erf\, \sqrt{4t}\right\}$

$$= \frac{1}{4}\frac{1}{p\sqrt{\frac{p}{4}+1}} = \frac{2}{p\sqrt{p+4}} = \frac{2}{p\sqrt{p+4}}$$

$$L\left\{t.erf\, (2\sqrt{t})\right\} = -\frac{d}{dp}\frac{2}{\sqrt{p^3\, p^2}}$$

$$= -2\left(-\frac{1}{2}\right)\left(p^3 + 4p^2\right)^{-3/2}(3p^2 + 8p)$$

$$= \frac{3p^2 + 8p}{(p^3 + 4p^2)^{3/2}} = \frac{3p + 8}{p^2(p+4)^{3/2}}.$$

EXAMPLE 5. **Show that** $L\left\{\dfrac{\sin^2 t}{t}\right\} = \dfrac{1}{4}\log\left(\dfrac{p^2+4}{p^2}\right).$

SOLUTION. We know that $\sin^2 t = \dfrac{1}{2}(1 - \cos 2t)$.

Now let $F(t) = \sin^2 t = \dfrac{1}{2}(1 - \cos 2t)$

$$\Rightarrow L\{F(t)\} = \frac{1}{2}[L\{1\} - L\{\cos 2t\}]$$

$$= \frac{1}{2}\left[\frac{1}{p} - \frac{p}{p^2+4}\right] = f(p) \text{ (say)}$$

Now, $\displaystyle\lim_{t\to 0}\left\{\frac{1}{t}F(t)\right\} = \lim_{t\to 0}\left(\frac{\sin t}{t}\right).\sin t = 1.0 = 0$

$\Rightarrow$ limit exists.

Therefore, $L\left\{\dfrac{1}{t}F(t)\right\} = \int_p^\infty f(x)\, dx$

$$= \frac{1}{2}\int_p^\infty\left(\frac{1}{x} - \frac{x}{x^2+4}\right)dx$$

$$= \frac{1}{2}\left[\log x - \frac{1}{2}\log(x^2+4)\right]_p^\infty$$

$$= \frac{1}{4}\left[\log\frac{x^2}{x^2+4}\right]_p^\infty$$

$$= \frac{1}{4}\left[\lim_{x\to\infty}\log\frac{x^2}{x^2+4} - \log\frac{p^2}{p^2+4}\right]$$

$$= \frac{1}{4}\left[\lim_{x\to\infty}\log\frac{1}{1+(4/x^2)} - \log\frac{p^2}{p^2+4}\right]$$

$$= \frac{1}{4}\left[\log 1 - \log\frac{p^2}{p^2+4}\right] = \frac{1}{4}\left[0 - \log\frac{p^2}{p^2+4}\right]$$

$$= -\frac{1}{4}\log\frac{p^2}{p^2+4} = \frac{1}{4}\log\frac{p^2+4}{p^2}.$$

EXAMPLE 6. **Show that** $\displaystyle\int_0^\infty\frac{\cos at - \cos bt}{t}\,dt = \frac{1}{2}\log\left(\frac{p^2+b^2}{p^2+a^2}\right).$

(UPTU–2004, 12)

SOLUTION. Let $F(t) = \cos at - \cos bt$

$$\Rightarrow L\{F(t)\} = L\{\cos at\} - L\{\cos bt\}$$

$$= \frac{p}{p^2+a^2} - \frac{p}{p^2+b^2} = f(p) \text{ (say)}$$

Now, $\displaystyle\lim_{t\to 0}\frac{F(t)}{t} = \lim_{t\to 0}\frac{\cos at - \cos bt}{t}$

$$= \lim_{t\to 0}\frac{-a\sin at + b\sin bt}{1} = 0$$

(By L-Hospital's rule)

$\Rightarrow$ limit exist.

Therefore,

$$L\left\{\frac{F(t)}{t}\right\} = \int_p^\infty F(x)\, dx$$

$$= \int_p^\infty\left[\frac{x}{x^2+a^2} - \frac{x}{x^2+b^2}\right]dx$$

$$= \frac{1}{2}\left[\log(x^2+a^2) - \log(x^2+b^2)\right]_p^\infty$$

$$= \frac{1}{2}\left[\log\frac{x^2+a^2}{x^2+b^2}\right]_p^\infty$$

$$= \frac{1}{2}\lim_{x\to\infty}\log\frac{x^2+a^2}{x^2+b^2} - \frac{1}{2}\log\frac{p^2+a^2}{p^2+b^2}$$

$$= \frac{1}{2}\lim_{x\to\infty}\log\frac{1+a^2/x^2}{1+b^2/x^2} - \frac{1}{2}\log\frac{p^2+a^2}{p^2+b^2}$$

$$= 0 - \frac{1}{2}\log\frac{p^2+a^2}{p^2+b^2} = -\frac{1}{2}\log\frac{p^2+a^2}{p^2+b^2}$$

$$= \frac{1}{2}\log\frac{p^2+b^2}{p^2+a^2}.$$

Exercise-8.47

1. Show that $L\{(1+te^{-t})^3\} = \frac{1}{p} + \frac{3}{(p+1)^2} + \frac{6}{(p+2)^3} + \frac{6}{(p+3)^4}$.

2. If $L\left\{2\sqrt{\frac{t}{\pi}}\right\} = \frac{1}{p^{3/2}}$, then show that $\frac{1}{p^{1/2}} = L\left\{\frac{1}{\sqrt{\pi t}}\right\}$.

(UPTU–2005, MADRAS–2003)

3. Find (i) $L\{F(t)\}$ and (ii) $L\{F'(t)\}$ for the function defined by

$$F(t) = \begin{cases} 2t, & 0 \le t \le 1 \\ t, & t > 1 \end{cases}.$$

4. Show that $\int_0^\infty \frac{\sin^2 t}{t^2} dt = \frac{\pi}{2}$.

5. Show that $\int_0^\infty \frac{\cos 6t - \cos 4t}{t} dt = \log\left(\frac{2}{3}\right)$.

(MUMBAI–2008, PTU–2006)

6. Show that

(i) $L\{J_0(t)\} = \frac{1}{\sqrt{1+p^2}}$ (ii) $L\{t J_0(at)\} = \frac{p}{(p^2+a^2)^{3/2}}$.

7. Show that $L\{J_1(t)\} = 1 - \frac{p}{\sqrt{p^2+1}}$, where $J_1(t)$ is the Bessel function of order one and hence deduce that

$$L\{t J_1(t)\} = \frac{1}{(p^2+1)^{3/2}}.$$

8. Prove that $L\{J_0(a\sqrt{t})\} = \frac{1}{p} e^{-a^2/4p}$.

9. Show that $L\{t.erf(2\sqrt{t})\} = \frac{3p+8}{p^2(p+4)^{3/2}}$.

10. Show that $L\{e^{3t} . erf\sqrt{t}\} = \frac{1}{(p-3)\sqrt{p-2}}$.

11. Show that $L\{c_i(t)\} = \frac{1}{2p}.\log(p^2+1)$, where $c_i(t) = \int_0^\infty \frac{\cos u}{u} du$.

12. If $F(t) = t^2$, $0 < t < 2$ and $F(t+2) = F(t)$. Then show that

$$L\{F(t)\} = \frac{-(4p^2+4p+2)e^{-2p+2}}{p^3(1-e^{-2p})}$$

13. If $F(t) = \begin{cases} 3t, & 0 < t < 2 \\ 6, & 2 < t < 4 \end{cases}$ and $F(t)$ is a periodic function of

period 4, then show that $L\{F(t)\} = \frac{3-3e^{-2p}-6pe^{-4p}}{p^2(1-e^{-4p})}$.

14. Find the Laplace transform of the Heaviside's unit step function $H(t-a)$.

15. Show that $\int_0^\infty \frac{\sin t}{t} dt = \frac{\pi}{2}$.

16. Show that $\int_0^\infty \frac{e^{-t}-e^{-3t}}{t} dt = \log 3$.

17. Show that $\int_0^\infty t^3 e^{-t} \sin t \, dt = 0$.

18. Show that $\int_0^\infty e^{-t} \frac{\sin t}{t} dt = \frac{1}{4} \log \frac{p^2+4}{p^2}$.

(UPTU–2008, VTU–2009S)

Hints to Selected Problems

1. Let $F(t) = 2\sqrt{\left(\frac{t}{\pi}\right)} \Rightarrow F'(t) = \frac{1}{\sqrt{\pi t}}$.

Then use $L\{F'(t)\} = pLF(t) - F(0)$.

5. (i) Use $J_0(t) = 1 - \frac{t^2}{2^2} + \frac{t^4}{2^2.4^2} - \frac{t^6}{2^2.4^2.6^2}$,

(ii) $L\{t J_0(t)\} = -\frac{d}{dp} L\{J_0(t)\}$.

6. Since $J_0'(t) = -J_1(t)$. Now using $L\{f'(t)\} = pL\{f(t)\} - f(0)$

$$L\{J_1(t)\} = L\{-J_0'(t)\} = -L\{J_0'(t)\}.$$

7. Use $J_0(t) = 1 - \frac{t^2}{2^2} + \frac{t^4}{2^2.4^2} - \frac{t^6}{2^2.4^2.6^2} + ...$

$\Rightarrow J_0(a\sqrt{t}) = 1 - \frac{a^2 t}{2^2} + \frac{a^4 t^2}{2^2.4^2} - \frac{a^6.t^3}{2^2.4^2.6^2} + ...$

8. Use $erf(\sqrt{t}) = \frac{2}{\sqrt{\pi}} \int_0^{\sqrt{t}} e^{-u^2} du$

$= \frac{2}{\sqrt{\pi}} \int_0^{\sqrt{t}} \left(1 - u^2 + \frac{u^4}{2!} - \frac{u^6}{3!} + ...\right) du$.

10. Using $L\{c_i(t)\} = L\left\{\int_t^\infty \frac{\cos u}{u} du\right\}$.

11. Using $L\{F(t)\} = \frac{\int_0^T e^{-pt} F(t)\, dt}{1-e^{-pT}} = \frac{\int_0^2 t^2 e^{-pt} dt}{1-e^{-2p}}$.

Answers

2. (i) $\frac{2}{p^2} - \left(\frac{1}{p} + \frac{1}{p^2}\right) e^{-p}$, $p > 0$ (ii) $\frac{1}{p}(2-e^{-p})$

8.9.8 THE UNIT STEP FUNCTION

The unit step function, denoted by $H(t-a)$ is defined by $H(t-a) = \begin{cases} 0, & t < a \\ 1, & t > a \end{cases}$

$\therefore \quad L\{H(t-a)\} = \int_0^\infty e^{-pt} H(t-a) dt = \int_a^\infty e^{-pt}.1\, dt = \left[\frac{e^{-pt}}{-p}\right]_a^\infty$

$= \left[\lim_{t\to\infty} \frac{e^{-pt}}{-p}\right] + \frac{e^{-ap}}{p} = 0 + \frac{e^{-ap}}{p}, \, p > 0 = \frac{e^{-ap}}{p}, \, p > 0$

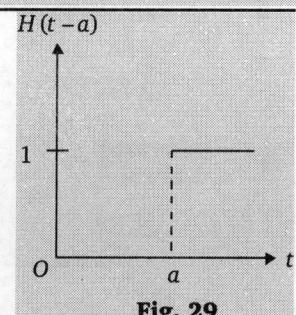

Fig. 29

THEOREM 1. *If* $L\{F(t)\} = f(p)$, *then* $L\{F(t-a).H(t-a)\} = e^{-ap}f(p)$.

PROOF.   We have $L\{F(t-a).H(t-a)\} = \int_0^\infty e^{-pt}[F(t-a).H(t-a)]dt = \int_0^a e^{-pt}F(t-a).0\,dt + \int_a^\infty e^{-pt}F(t-a).(1)dt$

$$= \int_a^\infty e^{-pt}F(t-a)dt = \int_0^\infty e^{-p(u+a)}\,F(u)du \qquad\qquad \text{[Putting } t-a = u]$$

$$= e^{-ap}\int_0^\infty e^{-pu}F(u)\,du = e^{-ap}f(p).$$

THEOREM 2. $L\{F(t).H(t-a)\} = e^{-ap}L\{F(t+a)\}dt$.

PROOF.   We have $L\{F(t).H(t-a)\} = \int_0^\infty e^{-pt}[F(t).H(t-a)]dt = \int_0^a e^{-pt}[F(t).H(t-a)]dt + \int_a^\infty e^{-pt}[F(t).H(t-a)]dt$

$$= 0 + \int_0^\infty e^{-pt}.F(t).(1)dt = \int_0^\infty e^{-p(u+a)}F(u+a)du, \qquad\qquad \text{[Putting } t-a = u]$$

$$= e^{-ap}\int_0^\infty e^{-pu}\,F(u+a)\,du = e^{-ap}\int_0^\infty e^{-pt}F(t+a)dt = e^{-ap}L\{F(t+a)\}$$

 Solved Examples

EXAMPLE 1. *Find the Laplace transform of* $t^2 H(t-3)$.

SOLUTION.   We have $L\{t^2H(t-3)\} = e^{-3p}L\{(t+3)^2\}$

$$= e^{-3p}L\{t^2 + 6t + 9\}$$

$$= \left[\frac{2}{p^3} + \frac{6}{p^2} + \frac{9}{p}\right]$$

EXAMPLE 2. *Express the following function in terms of unit step function and find its Laplace transform :*

$$F(t) = \begin{cases} t-1, & 1 < t < 2 \\ 3-t, & 2 < t < 3 \end{cases}.$$

SOLUTION.   We have

$$F(t) = \begin{cases} t-1, & 1 < t < 2 \\ 3-t, & 2 < t < 3 \end{cases}$$

$$= (t-1)[H(t-1) - H(t-2)]$$
$$\quad + (3-t)[H(t-2) - H(t-3)]$$
$$= (t-1)H(t-1) - (t-1)H(t-2)$$
$$\quad + (3-t)H(t-2) + (t-3)H(t-3)$$
$$= (t-1)H(t-1) - 2(t-2)H(t-2)$$
$$\quad + (t-3)H(t-3).$$

$$\therefore\ L\{F(t)\} = \frac{e^{-p}}{p^2} - 2\frac{e^{-2p}}{p^2} + \frac{e^{-3p}}{p^2}.$$

EXAMPLE 3. *Express the following function in terms of unit step function and find its Laplace transform :*

(UPTU–2002)

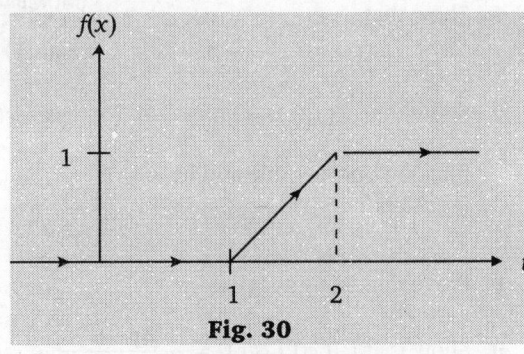

Fig. 30

SOLUTION.   The algebraic form of the function in the figure is

$$F(t) = \begin{cases} 0, & 0 < t < 1 \\ t-1, & 1 < t < 2 \qquad \dots(1) \\ 1, & 2 < t \end{cases}$$

$$= (t-1)[H(t-1) - H(t-2)] + H(t-2)$$
$$= (t-1)H(t-1) - (t-1-1)H(t-2)$$
$$= (t-1)H(t-1) - (t-2)H(t-2)$$

$$\therefore\ L\{F(t)\} = L\{(t-1)H(t-1)\} - L\{(t-2)H(t-2)\}$$

$$= \frac{e^{-p}}{p^2} - L\{(t-2)H(t-2)\} = \frac{e^{-p}}{p^2} - \frac{e^{-2p}}{p^2}.$$

 Exercise-8.48

Express the following functions in terms of unit step function. Also find their Laplace transforms :

1. $F(t) = 2t$ for $0 < t < \pi$, $F(t) = 1$ for $t > \pi$.

2. $F(t) = t^2$ for $0 < t \le 2$, $F(t) = 0$ for $t > 0$.

3. $F(t) = \cos(\omega t + \phi)$ for $0 < t < T$, $F(t) = 0$ for $t > T$.

4. $f(t) = \begin{cases} t, & 0 < t < 2 \\ 0, & t > 2 \end{cases}$.

5. $f(t) = \begin{cases} \sin t, & 0 < t < \pi \\ t, & t > \pi \end{cases}$

6. $f(t) = \begin{cases} 4, & 0 < t < 1 \\ -2, & 1 < t < 3 \\ 5, & t > 3 \end{cases}$

7. Find Laplace transform of $tU_2(t)$.

8. Using unit step function, find the Laplace transform of

(i) $(t-1)^2 u(t-1)$ (ii) $\sin t\,u(t-4)$ (UPTU–2008)

 **Answers**

8. (i) $\dfrac{2e^{-p}}{p^3}$ (ii) $\dfrac{e^{-4p}}{p^3}(4p^2 + 4p + 2)$

Miscellaneous Solved Examples

EXAMPLE 1. *Find $L\left\{\sqrt{t}e^{3t}\right\}$.* (PTU–2009)

SOLUTION. $L\{\sqrt{t}\} = \dfrac{\sqrt{3/2}}{p^{3/2}} = \dfrac{(1/2).\sqrt{t}}{p^{3/2}}$

$\therefore$ By shifting property, we get

$L(e^{3t}\sqrt{t}) = \dfrac{\sqrt{\pi}}{2} = \dfrac{1}{(p-3)^{3/2}}$.

EXAMPLE 2. *Find the Laplace transform of the function*
$$f(t) = |t-1| + |t+1|, t \geq 0.$$ (SVTU–2009)

SOLUTION. Given function is equivalent to

$$f(t) = \begin{cases} 2 & 0 \leq t \leq 1 \\ 2t & t \geq 1 \end{cases}$$

$\therefore \quad Lf(t) = \int_0^1 e^{-pt}(2)dt + \int_1^\infty e^{-pt}(2t)dt$

$$= 2\left[\left|\dfrac{e^{-pt}}{-p}\right|_0^1 + 2\left|\dfrac{t.e^{-pt}}{-p}\right|_1^\infty - \left|\dfrac{e^{-pt}}{(-p)^2}\right|_1^\infty\right]$$

$$= 2\left(\dfrac{e^{-p}}{-p} + \dfrac{1}{p}\right) + 2\left(\dfrac{0-e^{-p}}{-p} - \dfrac{0-e^{-p}}{p^2}\right)$$

$$= \dfrac{2}{p}\left(1 + \dfrac{e^{-p}}{p}\right).$$

EXAMPLE 3. *Find the Laplace trransform of the function $f(t) = [t]$ where [ ] stands for the greatest integer function.* (PTU–2010)

SOLUTION. Given function is equivalent to

$[t] = 0$ in $(0, 1) + 1$ in $(1, 2) + 2$ in $(2, 3) + 3$ in $(3, 4) + ...$

$\therefore \quad L[F(t)] = \int_0^\infty e^{-pt}[f(t)]dt = \int_0^\infty e^{-pt}[t]dt$

$$= \int_0^1 e^{-pt}(0)dt + \int_1^2 e^{-pt}(1)dt$$
$$+ \int_2^3 e^{-pt}(2)dt + \int_3^4 e^{-pt}(3)dt + ... + \infty$$

$$= 0 + \left|\dfrac{e^{-pt}}{-p}\right|_1^2 + 2\left|\dfrac{e^{-pt}}{-p}\right|_2^3 + 3\left|\dfrac{e^{-pt}}{-p}\right|_3^4 + ... + \infty$$

$$= \dfrac{-1}{p}[(e^{-2p} - e^{-p}) + 2(e^{-3p} - e^{-2p})$$
$$+ 3(e^{-4p} - e^{-3p}) + ... + \infty]$$

$$= \dfrac{1}{p}(e^{-p} + e^{-2p} + e^{-3p} ... + \infty)$$

$$= \dfrac{1}{p}\left(\dfrac{e^{-p}}{1-e^{-p}}\right) = \dfrac{1}{p(e^{-p}-1)}.$$

EXAMPLE 4. *Find $L(erf\,2\sqrt{t})$.* (MUMBAI–2006)

SOLUTION. Since we know that $L(erf\sqrt{t}) = \dfrac{1}{p(p+1)}$

$\therefore L(erf\,2\sqrt{t}) = L[erf\sqrt{(4t)}]$

$$= \dfrac{1}{4}.\dfrac{1}{\dfrac{p}{4}\sqrt{\left(\dfrac{p}{4}+1\right)}} = \dfrac{2}{p\sqrt{(p+4)}}.$$

EXAMPLE 5. *Find $L\{t^3e^{-3t}\}$.* (KOTTAYAM–2005)

SOLUTION. Since $L(e^{-3t}) = \dfrac{1}{p+3}$

$\therefore L(t^3e^{-3t}) = (-1)^3\dfrac{d^3}{dp^3}\left(\dfrac{1}{p+3}\right) = -\dfrac{(-1)^3.3!}{(p+3)^{3+1}}$

$$= \dfrac{6}{(p+3)^4}.$$

EXAMPLE 6. *Find $L\left\{(1-e^t)/t\right\}$.* (MADRAS–2000)

SOLUTION. $\because \quad L(1-e^t) = L(1) - L(e^t) = \dfrac{1}{p} - \dfrac{1}{p-1}$

$\therefore L\left(\dfrac{1-e^t}{t}\right) = \int_p^\infty\left(\dfrac{1}{p} - \dfrac{1}{p-1}\right)dp$

$$= \left|\log p - \log(p-1)\right|_p^\infty = \left|\log\left(\dfrac{p}{p-1}\right)\right|_p^\infty$$

$$= -\log\left[\dfrac{1}{1-1/p}\right] = \log\left(\dfrac{p-1}{p}\right).$$

EXAMPLE 7. *Find $L\left\{\dfrac{\cos at - \cos bt}{t} + t\sin at\right\}$.* (VTU–2010)

SOLUTION. Since $L(\cos at - \cos bt) = \dfrac{p}{p^2+a^2} - \dfrac{p}{p^2+b^2}$

and $L(\sin at) = \dfrac{a}{p^2+a^2}$

$\therefore L\left(\dfrac{\cos at - \cos bt}{t}\right) + L(t\sin at)$

$$= \int_p^\infty\left(\dfrac{p}{p^2+a^2} - \dfrac{p}{p^2+b^2}\right)dp - \dfrac{d}{dp}\left(\dfrac{a}{p^2+a^2}\right)$$

$$= \left|\dfrac{1}{2}\log(p^2+a^2) - \dfrac{1}{2}\log(p^2+b^2)\right|_p^\infty - a\dfrac{(-2p)}{(p^2+a^2)^2}.$$

$$= \dfrac{1}{2}\lim_{p\to\infty}\log\dfrac{p^2+a^2}{p^2+b^2} - \dfrac{1}{2}\log\dfrac{p^2+a^2}{p^2+b^2} + \dfrac{2ap}{(p^2+a^2)^2}$$

$$= \dfrac{1}{2}\log\left(\dfrac{1+0}{1+0}\right) - \dfrac{1}{2}\log\left(\dfrac{p^2+a^2}{p^2+b^2}\right) + \dfrac{2ap}{(p^2+a^2)^2}$$

$$= \log\left(\dfrac{p^2+b^2}{p^2+a^2}\right)^{1/2} + \dfrac{2ap}{(p^2+a^2)^2}. \quad [\because \log 1 = 0]$$

EXAMPLE 8. *Find $L\left\{e^{-t}\int_0^t\dfrac{\sin t}{t}dt\right\}$.* (MADRAS–2006)

SOLUTION. $\because \quad L\{\sin t\} = \dfrac{1}{p^2+1}$

$L\left(\dfrac{\sin t}{t}\right) = \int_0^\infty\dfrac{1}{p^2+1}dp$

$$= \dfrac{\pi}{2} - \tan^{-1}p = \cot^{-1}p$$

$\therefore \quad L\left(\int_0^t\dfrac{\sin t}{t}dt\right) = \dfrac{1}{p}\cot^{-1}p$.

Now by shiefting property,

$$L\left\{e^{-t}\left(\int_0^t\dfrac{\sin t}{t}dt\right)\right\} = \dfrac{1}{p+1}\cot^{-1}(p+1).$$

EXAMPLE 9. *Find* $L\left\{t\int_0^t \dfrac{e^{-t}\sin t}{t}\,dt\right\}$.   (PTU–2005)

SOLUTION. $\because$ $L\left(\dfrac{\sin t}{t}\right) = \cot^{-1} p$

$\therefore$ $L\left(e^{-t}\cdot\dfrac{\sin t}{t}\right) = \cot^{-1}(p+1)$

and $L\left\{\int_0^t e^{-t}\cdot\dfrac{\sin t}{t}\,dt\right\} = \dfrac{1}{p}\cot^{-1}(p+1)$

Hence, $L\left\{t\int_0^t e^{-t}\cdot\dfrac{\sin t}{t}\,dt\right\} = -\dfrac{d}{dp}\left\{\dfrac{\cot^{-1}(p+1)}{p}\right\}$

$= -\dfrac{p\left[\dfrac{-1}{1+(p+1)^2}\right] - \cot^{-1}(p+1)}{p^2}$

$= \dfrac{p + (p^2+2p+2)\cot^{-1}(p+1)}{p^2(p^2+2p+2)}$.

EXAMPLE 10. *Find* $L\left\{\int_0^t\int_0^t\int_0^t (t\sin t)\,dt\,dt\,dt\right\}$.   (MUMBAI–2006)

SOLUTION. $\because$ $L(\sin t) = \dfrac{1}{p^2+1}$

$\therefore$ $L(t\sin t) = -\dfrac{d}{dp}\dfrac{1}{(p^2+1)} = \dfrac{2p}{(p^2+1)^2}$

Thus, $L\left\{\int_0^t\int_0^t\int_0^t (t\sin t)\,dt\,dt\,dt\right\} = \dfrac{1}{p^3} L(t\sin t)$

$= \dfrac{1}{p^3}\cdot\dfrac{2p}{(p^2+1)^2} = \dfrac{2}{p^2(p^2+1)^2}$

EXAMPLE 11. *Find* $L\int_0^\infty te^{-3t}\sin t\,dt$.   (VTU–2007)

SOLUTION. $\int_0^\infty te^{-3t}\sin t\,dt = \int_0^\infty e^{-pt}(t\sin t)\,dt$, where $p = 3$

$= L(t\sin t)$, By def.

$= (-1)\dfrac{d}{dp}\left(\dfrac{1}{p^2+1}\right)$

$= \dfrac{2p}{(p^2+1)^2} = \dfrac{2\times 3}{(3^2+1)^2} = \dfrac{3}{50}$.

EXAMPLE 12. *Find* $\int_0^\infty e^{-t}\left(\dfrac{\cos at - \cos bt}{t}\right)dt$.   (MUMBAI–2009)

SOLUTION. $\because$ $L(\cos at) = \dfrac{p}{p^2+a^2}$ and $L(\cos bt) = \dfrac{p}{p^2+b^2}$

$\therefore L\dfrac{\cos at - \cos bt}{t} = \int_p^\infty\left(\dfrac{p}{p^2+a^2} - \dfrac{p}{p^2+b^2}\right)dp$

$= \dfrac{1}{2}\left\{\log\left(\dfrac{p^2+a^2}{p^2+b^2}\right)\right\}_p^\infty = \dfrac{1}{2}\log\left(\dfrac{p^2+b^2}{p^2+a^2}\right)$

$\Rightarrow \int_0^\infty e^{-pt}\left(\dfrac{\cos at - \cos bt}{t}\right)dt = \dfrac{1}{2}\log\left(\dfrac{p^2+b^2}{p^2+a^2}\right)$

Taking $p = 1$, we get

$\int_0^\infty e^{-t}\left(\dfrac{\cos at - \cos bt}{t}\right)dt = \dfrac{1}{2}\log\left(\dfrac{1+b^2}{1+a^2}\right)$.

Miscellaneous Exercise

Find the Laplace transform of the following functions:

1. $e^{2t} + 4t^3 - 2\sin 3t + 3\cos 3t$ (JNTU–2003)
2. $\sin 2t \cos 3t$ (KOTTAYAM–2005)
3. $\sin^5 t$ (MUMBAI–2007)
4. $e^{2t}(3t^5 - \cos 4t)$ (PTU–2007)
5. $e^{-3t}\sin 5t \sin 3t$ (VTU–2006)
6. $e^{2t}\sin^4 t$ (MUMBAI–2007)
7. $\cosh at \sin at$ (DELHI–2002)
8. $\sinh 3t \cos^2 t$ (MADRAS–2000)
9. $t^2 e^{2t}$ (VTU–2008S)
10. $t\sqrt{(1+\sin t)}$ (MUMBAI–2007)
11. $f(t) = \begin{cases} 4, & 0 \le t \le 1 \\ 3, & t > 1 \end{cases}$ (UPTU–2009)
12. $f(x) = \begin{cases} \sin(x - \pi/3), & x > \pi/3 \\ 0, & x < \pi/3 \end{cases}$ (RAJASTHAN–2006)
13. Find the Laplace transform of the saw-toothed wave of period T, given $f(t) = t/T$ for $0 < t < T$.   (VTU–2007)
14. Find the Laplace transform of the square wave function of period a defined as
    $f(t) = k$ when $0 < t < a$
    $= -k$ when $a < t < 2a$.   (VTU–2011)
15. Find the Laplace transform of the triangular wave of period $2a$ given by
    $f(t) = t, 0 < t < a$

    $= 2a - t, a < t < 2a$.
    (NAGARJUNA–2008, VTU–2008S, UPTU–2002)

16. $t\sin^2 t$ (NAGARJUNA–2008)
17. $\sin 2t - 2t\cos 2t$ (ANNA–2003)
18. $te^{2t}\sin 3t$ (MADRAS–2003)
19. $te^{-2t}\sin 4t$ (VTU–2008)
20. $t^2 e^{-3t}\sin 2t$ (MADRAS–2000S)
21. $(e^{-at} - e^{-bt})/t$ (ANNA–2005S)
22. $\dfrac{(\sin t \sin 5t)}{t}$ (MUMBAI–2008)
23. $(1 - \cos 3t)/t$ (VTU–2006)
24. $(1 - \cos t)/t^2$ (HAZARIBAG–2008)
25. $2^t + \dfrac{\cos 2t - \cos 3t}{t} + t\sin t$ (VTU–2004)
26. $\int_0^\infty \dfrac{e^{-\sqrt{2}t}\sin nt \sin t}{t}dt$ (MUMBAI–2005)
27. $\int_0^\infty te^{-2t}\sin 3t\,dt$ (VTU–2008)
28. Prove that $\int_0^\infty \dfrac{e^{-2t}\sin nt}{t}dt = \dfrac{1}{2}\log 3$.   (MUMBAI–2008)
29. Prove that $\int_0^\infty \dfrac{e^{-t}\sin^2 t}{t}dt = \dfrac{1}{4}\log 5$.   (KURUKSHETRA–2006)
30. $L\int_0^t \dfrac{e^t\sin t}{t}dt$ (PTU–2009S, SVTU–2009, BHOPAL–2008)

Answers

1. $\dfrac{1}{p-2}+\dfrac{24}{p^4}+\dfrac{3(p-2)}{p^4+9}$

2. $\dfrac{2(5)}{(p^2+1)(p^2+25)}$

3. $\dfrac{5}{4}\left\{\dfrac{1}{p^2+1}-\dfrac{3/2}{p^2+9}+\dfrac{1/2}{p^2+25}\right\}$

4. $\dfrac{60}{p-2}-\dfrac{p-2}{p^2-4p+20}$

5. $\dfrac{30(p+3)}{(p^2+6p+13)(p^2+6p+73)}$

6. $\dfrac{1}{8}\left\{\dfrac{3}{(p-2)}-\dfrac{4(p-2)}{p^2-4p+8}+\dfrac{p-4}{p^2-8p+32}\right\}$

7. $\dfrac{a(p^2+2a^2)}{p^4+4a^4}$

8. $\dfrac{3}{2}\left[\dfrac{1}{p^2-9}+\dfrac{p^2-13}{p^4-10p^2+169}\right]$

9. $\dfrac{2}{(p+2)^3}$

10. $\dfrac{4(4p^2+4p-1)}{(4p^2+1)^2}$

11. $\dfrac{4}{p}-\dfrac{e^{-p}}{p}$

12. $\dfrac{e^{-\pi p/3}}{p^2+1}$

13. $\left(1/p^2 T\right)-e^{-pT}/p(1-e^{pT})$

14. $(a/p)\tanh(ap/2)$

15. $\left(1/p^2\right)\tanh\dfrac{1}{2}ap$

16. $\dfrac{2(3p^2+4)}{p^2(p^2+4)^2}$

17. $\dfrac{16}{(p^2+4)^2}$

18. $\dfrac{6(p-2)}{(p^2-4p+13)^2}$

19. $\dfrac{8(p+2)}{p^2+4p+20}$

20. $\dfrac{2(p^3+6p^2+9p+2)}{(p^2+4p+5)^3}$

21. $\log\{(p+b)/(p+a)\}$

22. $\dfrac{1}{2}\log\{(p^2+36)/(p^2+16)\}$

23. $\dfrac{1}{2}\log\left(\dfrac{p^2+9}{p^2}\right)$

24. $\cot^{-1}p-\dfrac{1}{2}p\log(1+p^{-2})$

25. $\dfrac{1}{p-\log 2}+\dfrac{2p}{(p^2+1)^2}+\dfrac{1}{2}\log\left(\dfrac{p^2+9}{p^2+4}\right)$

26. $\pi/8$

27. $12/169$

30. $\dfrac{\cot^{-1}(p-1)}{p}$

8.10 THE INVERSE LAPLACE TRANSFORM

If the Laplace transform of a function $F(t)$ is $f(p)$, i.e., if $L\{F(t)\}=f(p)$. Then, $F(t)$ is known as inverse Laplace transform of $f(p)$.

Symbolically, $F(t)=L^{-1}\{f(p)\}$. where, L^{-1} is called the inverse Laplace transformation operator.

For example, If $L\{e^{-2t}\}=\dfrac{1}{p+2}$. Then we can write $L^{-1}\left(\dfrac{1}{p+2}\right)=e^{-2t}$

(1) NULL FUNCTION

A function $N(t)$ of t such that $\int_0^t N(t)\,dt=0,\quad \forall t>0$ is called null function.

(2) UNIQUENESS OF INVERSE LAPLACE TRANSFORMS

Since, we know that the Laplace transform of a null function $N(t)$ is zero. Also, it is clearly that if $L(F(t))=f(p)$, then also

$$L\{F(t)+N(t)\}=f(p)$$

It follows that we can have two different functions with same Laplace transform.

If we allow null functions, we see that the inverse Laplace transform is not unique. It is unique, however, if we disallow null functions.

(3) LEARCH THEOREM

If we restrict ourselves to functions $F(t)$ which are sectionally continuous in every finite interval $0\le t\le N$ and of exponential order for $t>N$, then the inverse Laplace transform of $f(p)$.

i.e., $L^{-1}\{f(p)\}=F(t)$, is unique.

8.10.1 SOME INVERSE LAPLACE TRANSFORMS

S.No.	$f(p)$	$L^{-1}[f(p)]=F(t)$	S.No.	$f(p)$	$L^{-1}[f(p)]=F(t)$
(i)	$\dfrac{1}{p}$	1	(vi)	$\dfrac{p}{p^2+a^2}$	$\cos at$
(ii)	$\dfrac{1}{p^2}$	t	(vii)	$\dfrac{1}{p^2-a^2}$	$\dfrac{\sinh at}{a}$
(iii)	$\dfrac{1}{p^{n+1}}, n=0,1,2\ldots$	$\dfrac{t^n}{n!}$	(viii)	$\dfrac{p}{p^2-a^2}$	$\cosh at$
(iv)	$\dfrac{1}{p-a}$	e^{at}	(ix)	$\dfrac{\Gamma(a+1)}{p^n}$	$t^a,\ a>-1$
(v)	$\dfrac{p}{p^2+a^2}$	$\dfrac{\sin at}{a}$			

8.10.2 Important Properties of Inverse Laplace Transform

(i) Linearity Property.

If C_1 and C_2 are any constants while $f_1(p)$ and $f_2(p)$ are the Laplace transform $F_1(t)$ and $F_2(t)$ respectively, then

$$L^{-1}\{C_1 f_1(p) + C_2 f_2(p)\} = C_1 L^{-1}\{f_1(p)\} + C_2 L^{-1}\{f_2(p)\}$$

Proof. We have

$$L\{C_1 F_1(t) + C_2 F_2(t)\} = C_1 L\{F_1(t)\} + C_2 L\{F_2(t)\} = C_1 f_1(p) + C_2 f_2(p)$$

$$\Rightarrow \qquad L^{-1}\{C_1 f_1(p) + C_2 f_2(p)\} = C_1 F_1(t) + C_2 F_2(t) = C_1 L^{-1}\{f_1(p)\} + C_2 L^{-1}\{f_2(p)\}.$$

(ii) First Translation or Shifting Theorem.

If $\qquad\qquad L^{-1}\{f(p)\} = F(t)$, then

$$L^{-1}\{f(p-a)\} = e^{at} F(t) = e^{at} L^{-1}\{f(p)\}$$

Proof. We have $\qquad f(p) = \int_0^\infty e^{-pt} F(t)\, dt$

$$\Rightarrow \qquad f(p-a) = \int_0^\infty e^{-(p-a)t} F(t)\, dt = \int_0^\infty e^{-pt}\{e^{at} F(t)\}\, dt = L\{e^{at} F(t)\}$$

Hence, $\qquad L^{-1}\{f(p-a)\} = e^{at} F(t) = e^{at} L^{-1}\{f(p)\}.$

(iii) Second Translation or Shifting Theorem.

If $L^{-1}\{f(p)\} = F(t)$, then $L^{-1}\{e^{-ap} f(p)\} = G(t)$, where

$$G(t) = \begin{cases} F(t-a), & t > a \\ 0, & t < a \end{cases}$$

Proof. We know that $\qquad f(p) = \int_0^\infty e^{-pt} F(t)\, dt$

Therefore, $\qquad e^{-ap} f(p) \quad \int_0^\infty e^{-p(t+a)} F(t)\, dt$

$$= \int_0^\infty e^{-px} F(x-a)\, dx, \text{ putting } t + a = x \Rightarrow dt = dx$$

$$= \int_0^a e^{-px} \cdot 0\, dx + \int_a^\infty e^{-px} F(x-a)\, dx = \int_0^a e^{-pt} \cdot 0\, dt + \int_a^\infty e^{-pt} F(t-a)\, dt$$

$$= \int_0^\infty e^{-pt} G(t)\, dt = L\{G(t)\}$$

where, $G(t) = \begin{cases} F(t-a), & t > a \\ 0, & t < a \end{cases}$ shows, $L^{-1}\{e^{ap} f(p)\} = G(t)$

(iv) Change of Scale Property.

If $L^{-1}\{f(p)\} = F(t)$, then $L^{-1}\{f(ap)\} = \dfrac{1}{a} F\left(\dfrac{t}{a}\right)$

Proof. We know that $\qquad f(ap) = \dfrac{1}{a} \int_0^\infty e^{-px} F\left(\dfrac{x}{a}\right) dx \Rightarrow \qquad f(ap) = \int_0^\infty e^{-apt} F(t)\, dt$

Putting $\qquad at = x \qquad \Rightarrow dt = \dfrac{1}{a} dx$, we get

So, $\qquad f(ap) = \dfrac{1}{a} \int_0^\infty e^{-px} F\left(\dfrac{x}{a}\right) dx = \dfrac{1}{a} \int_0^\infty e^{-pt} F\left(\dfrac{t}{a}\right) dt \qquad$ [By the property of definite integral]

$$= \dfrac{1}{a} L\left\{F\left(\dfrac{t}{a}\right)\right\} = L\left\{\dfrac{1}{a} F\left(\dfrac{t}{a}\right)\right\}$$

Hence, $\qquad L^{-1}\{f(ap)\} = \dfrac{1}{a} F\left(\dfrac{t}{a}\right)$

Solved Examples

EXAMPLE 1. *Find the inverse Laplace transforms of the following functions :*

(i) $\dfrac{2p+1}{p(p+1)}$ \qquad (ii) $\dfrac{3p-8}{4p^2+25}$

SOLUTION. (i) We have

$$L^{-1}\left\{\dfrac{2p+1}{p(p+1)}\right\} = L^{-1}\left\{\dfrac{p+(p+1)}{p(p+1)}\right\}$$

$$= L^{-1}\left\{\dfrac{1}{p+1}\right\} + L^{-1}\left\{\dfrac{1}{p}\right\} = e^{-t} + 1$$

(ii) Here, we have

$$L^{-1}\left\{\dfrac{3p-8}{4p^2+25}\right\} = \dfrac{3}{4} L^{-1}\left\{\dfrac{p}{p^2+\left(\dfrac{5}{2}\right)^2}\right\}$$

$$- 2 L^{-1}\left\{\dfrac{1}{p^2+\left(\dfrac{5}{2}\right)^2}\right\}$$

$$= \dfrac{3}{4}\cos\left(\dfrac{5}{2}t\right) - 2 \cdot \dfrac{2}{5}\sin\left(\dfrac{5}{2}t\right)$$

$$= \dfrac{3}{4}\cos\left(\dfrac{5}{2}t\right) - \dfrac{4}{5}\sin\left(\dfrac{5}{2}t\right)$$

EXAMPLE 2. *Show that* $\dfrac{1}{p^{1/2}} = L\left[\dfrac{1}{\sqrt{\pi t}}\right]$ \qquad [UPTU 2005]

SOLUTION. We have to show that $\dfrac{1}{p^{1/2}} = L\left[\dfrac{1}{\sqrt{\pi t}}\right]$

Now, $L^{-1}\left[\dfrac{1}{p^n}\right] = \dfrac{t^{n-1}}{(n-1)!} = \dfrac{t^{n-1}}{\Gamma(n)}$

So, $L^{-1}\left[\dfrac{1}{p^{1/2}}\right] = \dfrac{t^{\frac{1}{2}-1}}{\Gamma(1/2)} = \dfrac{t^{-1/2}}{\Gamma(1/2)} = \dfrac{t^{-1/2}}{\sqrt{\pi}}$

$L^{-1}\left[\dfrac{1}{p^{1/2}}\right] = \dfrac{1}{\sqrt{\pi t}} \Rightarrow \dfrac{1}{p^{1/2}} = L\left[\dfrac{1}{\sqrt{\pi t}}\right].$

EXAMPLE 3. *A function $f(t)$ obey the equation $f(t) + 2\int_0^t f(t)\, dt = \cos h\, 2t$. Find the Laplace transformation of $f(t)$.*   [UPTU 2006]

SOLUTION. We have $f(t) + 2\int_0^t f(t)\, dt = \cos h\, 2t$

Taking Laplace transformation of both the sides, we get

$L\{f(t)\} + 2L\int_0^t f(t)\, dt = L(\cosh 2t)$

$\Rightarrow \qquad F(p) + 2.\dfrac{1}{p}F(p) = \dfrac{p}{p^2-4}$

$\Rightarrow \qquad F(p)\left[1 + \dfrac{2}{p}\right] = \dfrac{p}{p^2-4}$

$\Rightarrow \qquad F(p).\left[\dfrac{p+2}{p}\right] = \dfrac{p}{p^2-4}$

$\Rightarrow \qquad F(p) = \left(\dfrac{p}{p^2-4}\right).\left(\dfrac{p}{p+2}\right)$

$\Rightarrow \qquad F(p) = \dfrac{p^2}{(p^2-4)(p+2)}$

EXAMPLE 4. *Evaluate $L^{-1}\left\{\dfrac{e^{4-3p}}{(p+4)^{5/2}}\right\}$*

SOLUTION. We have

$L^{-1}\left\{\dfrac{1}{(p+4)^{5/2}}\right\} = e^{-4t}L^{-1}\left\{\dfrac{1}{p^{5/2}}\right\}$

$= e^{-4t}\dfrac{t^{(5/2)-1}}{\Gamma\left(\dfrac{5}{2}\right)} = \dfrac{4t^{3/2}e^{-4t}}{3\sqrt{\pi}}$

Therefore,

$L^{-1}\left\{\dfrac{e^{4-3p}}{(p+4)^{5/2}}\right\} = e^4\, L^{-1}\left\{\dfrac{e^{-3p}}{(p+4)^{5/2}}\right\}$

$= \begin{cases} e^4.\dfrac{4}{3\sqrt{\pi}}(t-3)^{3/2}e^{-4(t-3)}, & t>3 \\ 0, & t<3 \end{cases}$

$= \begin{cases} \dfrac{4}{3\sqrt{\pi}}(t-3)^{3/2}\, e^{-4(t-4)}, & t>3 \\ 0, & t<3 \end{cases}$

$= \dfrac{4}{3\sqrt{\pi}}(t-3)^{3/2}e^{-4(t-4)}.H(t-3).$

EXAMPLE 5. *Find the inverse Laplace transform of*

$\dfrac{e^{-cp}}{p^2(p+a)},\ c>0$ [UPTU-2001 SP., 2002]

SOLUTION. We have $L^{-1}\left[\dfrac{e^{-cp}}{p^2(p+a)}\right]$

$= L^{-1}\left[-\dfrac{e^{-cp}}{a^2 p} + \dfrac{e^{-cp}}{ap^2} + \dfrac{e^{-cp}}{a^2(p+a)}\right]$

[By Partial Fractions]

$= L^{-1}\left[-\dfrac{1}{a^2}\dfrac{e^{-cp}}{p} + \left(\dfrac{1}{a}\right)\dfrac{e^{-cp}}{p^2} + \dfrac{1}{a^2}.\dfrac{e^{-c(p+a)}}{e^{-ca}(p+a)}\right]$

$= -\dfrac{1}{a^2}u(t-c) + \dfrac{1}{a}(t-c)u(t-c)$

$\qquad\qquad + \dfrac{1}{a^2 e^{-ca}}.e^{-at}u(t-c)$

$= u(t-c)\left[-\dfrac{1}{a^2} + \dfrac{1}{a}(t-c) + \dfrac{1}{a^2}e^{-a(c+t)}\right]$

Where $u(t-c) = $ unit step function.

EXAMPLE 6. *Evalute $L^{-1}\left\{\dfrac{p+1}{p^2+6p+25}\right\}$*

SOLUTION. We have

$L^{-1}\left\{\dfrac{p+1}{p^2+6p+25}\right\} = L^{-1}\left\{\dfrac{(p+3)-2}{(p+3)^2+16}\right\}$

$= e^{-3t}L^{-1}\left\{\dfrac{p-2}{p^2+16}\right\}$

$= e^{-3}\left[L^{-1}\left\{\dfrac{p}{p^2+4^2}\right\} - L^{-1}\left\{\dfrac{2}{p^2+4^2}\right\}\right]$

$= e^{-3t}\left[\cos 4t - \dfrac{1}{2}\sin 4t\right].$

EXAMPLE 7. *Evalute $L^{-1}\left[\dfrac{e^{-p}-3e^{-3p}}{p^2}\right]$*

SOLUTION. We have

$L^{-1}\left[\dfrac{e^{-p}-3e^{-3p}}{p^2}\right] = L^{-1}\left[\dfrac{e^{-p}}{p^2} - \dfrac{3e^{-3p}}{p^2}\right]$ ... (1)

We know that $Lu(t-a) = \dfrac{e^{-ap}}{p}$

and $L[(t-a)\, u(t-a)] = \dfrac{e^{-ap}}{p^2}$

Using these results in (1), we get

$L^{-1}\left[\dfrac{e^{-p}-3e^{-3p}}{p^2}\right] = (t-1)u(t-1) - 3(t-3)u(t-3).$

EXAMPLE 8. *Show that*

$L^{-1}\left\{\dfrac{p}{p^4+p^2+1}\right\} = \dfrac{2}{\sqrt{3}}\sinh\dfrac{t}{2}.\sin\dfrac{1}{2}\sqrt{3}\, t$

(RAIPUR-2005)

SOLUTION. We have

$L^{-1}\left\{\dfrac{p}{p^4+p^2+1}\right\} = L^{-1}\left\{\dfrac{p}{(p^2+1)^2-p^2}\right\}$

$= L^{-1}\left\{\dfrac{p}{(p^2+p+1)(p^2-p+1)}\right\}$

$= L^{-1}\left\{\dfrac{1}{2}\dfrac{(p^2+p+1)-(p^2-p+1)}{(p^2-p+1)(p^2+p+1)}\right\}$

$= L^{-1}\left\{\dfrac{1}{2(p^2-p+1)} - \dfrac{1}{2(p^2+p+1)}\right\}$

$= \dfrac{1}{2}L^{-1}\left\{\dfrac{1}{\left(p-\dfrac{1}{2}\right)^2+\dfrac{3}{4}}\right\} - \dfrac{1}{2}L^{-1}$

$= \dfrac{1}{2}e^{t/2}L^{-1}\left\{\dfrac{1}{p^2+\left(\dfrac{1}{2}\sqrt{3}\right)^2}\right\} - \dfrac{1}{2}e^{-t/2}L^{-1}$

$\left\{\dfrac{1}{p^2+\left(\dfrac{1}{2}\sqrt{3}\right)^2}\right\}$

$$= \frac{1}{2}e^{t/2}\frac{2}{\sqrt{3}}\sin\left(\sqrt{3}\cdot\frac{t}{2}\right) - \frac{1}{2}e^{-t/2}\frac{2}{\sqrt{3}}\sin\left(\sqrt{3}\cdot\frac{t}{2}\right)$$

$$= \frac{1}{\sqrt{3}}(e^{t/2}-e^{-t/2})\sin\left(\sqrt{3}\cdot\frac{t}{2}\right)$$

$$= \frac{2}{\sqrt{3}}\sinh\frac{t}{2}\sin\left(\sqrt{3}\cdot\frac{t}{2}\right).$$

EXAMPLE 9. *Find* $L^{-1}\left\{\dfrac{5p+3}{(p-1)(p^2+2p+5)}\right\}$

[UPTU-2005, ROHTAK-2009]

SOLUTION. We have

$$\frac{5p+3}{(p-1)(p^2+2p+5)} = \frac{1}{p-1} + \frac{2-p}{p^2+2p+5}$$

[By partial fractions]

$$\therefore L^{-1}\left\{\frac{5p+3}{(p-1)(p^2+2p+5)}\right\}$$

$$= L^{-1}\left\{\frac{1}{p-1} + \frac{2-p}{p^2+2p+5}\right\}$$

$$= L^{-1}\left\{\frac{1}{p-1} - \frac{p+1}{(p+1)^2+4} + \frac{3}{(p+1)^2+4}\right\}$$

$$= L^{-1}\left\{\frac{1}{p-1}\right\} - L^{-1}\left\{\frac{p+1}{(p+1)^2+4}\right\}$$

$$\qquad + 3L^{-1}\left\{\frac{1}{(p+1)^2+4}\right\}$$

$$= e^t - e^{-t}L^{-1}\left\{\frac{p}{p^2+4}\right\} + 3e^{-t}L^{-1}\left\{\frac{1}{p^2+4}\right\}$$

$$= e^t - e^{-t}\cos 2t + \frac{3}{2}e^{-t}\sin 2t$$

$$= e^t - e^{-t}\left(\cos 2t - \frac{3}{2}\sin 2t\right).$$

Exercise-8.49

1. Find the inverse Laplace transform of the following functions :

(a) $\dfrac{1}{p^4}$ (b) $\dfrac{1}{p^2+4}$ (c) $\dfrac{4}{p-2}$ (d) $\dfrac{1}{\sqrt{p}}$

(e) $\dfrac{p}{p^2+2} + \dfrac{6p}{p^2-16} + \dfrac{3}{p-3}$ (f) $\dfrac{2p-5}{p^2-9}$

(g) $\dfrac{6}{2p-3} - \dfrac{3+4p}{9p^2-16} + \dfrac{8-6p}{16p^2+9}$ [UPTU-2001]

(h) $\dfrac{3(p^2-1)^2}{2p^5} + \dfrac{4p-18}{9-p^2} + \dfrac{(p+1)(2-\sqrt{p})}{p^{5/2}}$

(i) $\dfrac{1}{p}\sin\dfrac{1}{p}$

2. Find the inverse Laplace transform of the following functions :

(a) $\dfrac{1}{p^2-6p+10}$ (b) $\dfrac{p+b}{(p+b)^2+a^2}$

(c) $\dfrac{3p+7}{p^2-2p-3}$ (d) $\dfrac{1}{(p+a)^n}$

(e) $\dfrac{p}{(p+1)^5}$ (f) $\dfrac{p^2-2p+3}{(p-1)^2(p+1)}$

(g) $\dfrac{2p^2-1}{(p^2+1)(p^2+4)}$ [UPTU-2004]

(h) $\dfrac{2p^2-6p+5}{p^3-6p^2+11p-6}$ [UPTU-2004; VTU-2007]

3. If $L^{-1}\left\{\dfrac{e^{-1/p}}{p^{1/2}}\right\} = \dfrac{\cos 2\sqrt{t}}{\sqrt{\pi t}}$, evaluate $L^{-1}\left\{\dfrac{e^{-a/p}}{p^{1/2}}\right\}$, $a>0$

4. Find the inverse Laplace transforms of the following functions

(a) $\dfrac{e^{-5p}}{(p-2)^4}$ (b) $\dfrac{e^{-4p}}{(p-3)^4}$

(c) $\dfrac{e^{-3p}}{p^3}$ (d) $\dfrac{p+8}{p^2+8p+5}$

5. Show that

(a) $L^{-1}\left\{\dfrac{pe^{-ap}}{p^2-w^2}\right\} = \cosh w(t-a)\,H(t-a),\ a>0$

(b) $L^{-1}\left\{\dfrac{p+2}{p^2-2p+5}\right\} = e^t\left[\cos 2t + \dfrac{3}{2}\sin 2t\right]$

(c) $L^{-1}\left\{\dfrac{6p^2+22p+18}{p^3+6p^2+11p+6}\right\} = e^{-t} + 2e^{-2t} + 3e^{-3t}$

(d) $L^{-1}\left\{\dfrac{4p+5}{(p-1)^2(p+2)}\right\} = 3te^t + \dfrac{1}{3}e^t - \dfrac{1}{3}e^{-2t}$

(KURUKSHETRA-2005)

(e) $L^{-1}\left\{\dfrac{4p+5}{(p-4)^2(p+3)}\right\} = -\dfrac{1}{7}e^{-3t} + \dfrac{1}{7}e^{4t} + 3te^{4t}$

(f) $L^{-1}\left\{\dfrac{5p^2-15p-11}{(p+1)(p-2)^3}\right\} = -\dfrac{1}{3}e^{-t} + \dfrac{1}{3}e^{2t} + 4te^{2t} + 4te^{2t} - \dfrac{7}{2}t^2e^{2t}$

(g) $L^{-1}\left\{\dfrac{2p+1}{(p+2)^2(p-1)^2}\right\} = \dfrac{1}{3}t(e^t - e^{-2t})$

(h) $L^{-1}\left\{\dfrac{p}{(p^2-2p+2)(p^2+2p+2)}\right\} = \dfrac{1}{2}\sin t\,\sinh t$

6. Show that

(a) $L^{-1}\left\{\dfrac{p^2}{p^4+4a^4}\right\} = \dfrac{1}{2a}(\cosh at\,\sin at + \sinh at\,\cos at)$

(b) $L^{-1}\left\{\dfrac{p^2}{p^4+4a^4}\right\} = \dfrac{1}{2a^2}\sin at\,\sinh at$ (MUMBAI-2008)

(c) $L^{-1}\left\{\dfrac{1}{(p^2+4)(p+1)^2}\right\} = \dfrac{1}{25}\left\{e^{-t}(2+5t) - 2\cos 2t - \dfrac{3}{2}\sin 2t\right\}$

(d) $L^{-1}\left\{\dfrac{3p^3-3p^2-40p+36}{(p^2-4)^2}\right\} = (5t+3)e^{-2t} - 2te^{2t}$

Hints to Selected Problems

1. (g) $L^{-1}\left[\dfrac{6}{2p-3} - \dfrac{3+4p}{9p^2-16} + \dfrac{8-6p}{16p^2+9}\right]$

$$= 3L^{-1}\left(\frac{1}{p-\left(\frac{3}{2}\right)}\right) - \frac{1}{3}L^{-1}\left(\frac{1}{p^2-\left(\frac{4}{3}\right)^2}\right) - \frac{4}{9}L^{-1}$$

$$\left(\frac{p}{p^2 - \left(\frac{4}{3}\right)^2}\right) + \frac{1}{2}L^{-1}\left(\frac{1}{p^2 + \left(\frac{3}{4}\right)^2}\right) - \frac{3}{8}L^{-1}\left(\frac{p}{p^2 + \left(\frac{3}{4}\right)^2}\right)$$

$$= 3e^{3t/2} - \frac{1}{4}\sinh\frac{4t}{3} - \frac{4}{9}\cosh\frac{4t}{3} + \frac{2}{3}\sin\frac{3t}{4} - \frac{3}{8}\cos\frac{3}{4}t$$

2. (a) $L^{-1}\left\{\dfrac{1}{p^2 - 6p + 10}\right\} = L^{-1}\left[\dfrac{1}{(p-3)^2 + 1}\right]$

$$= e^{3t}L^{-1}\left[\frac{1}{p^2 + 1}\right] = e^{3t}.\sin t$$

3. $L^{-1}\left\{\dfrac{e^{-1/p}}{p^{1/2}}\right\} = \dfrac{\cos 2\sqrt{t}}{\sqrt{\pi t}} \Rightarrow L^{-1}\left(\dfrac{e^{-1/pk}}{(pk)^{1/2}}\right) = \dfrac{1}{k}\dfrac{\cos 2\sqrt{t/k}}{\sqrt{\pi t/k}}$.

Now taking $k = 1/a$

5. (a) Since $L^{-1}\left(\dfrac{p}{p^2 - w^2}\right) = \cosh wt$

$\Rightarrow \quad L^{-1}\left[\dfrac{pe^{-ap}}{p^2 - w^2}\right] = \cosh w(t-a)\,H(t-a)$

6. (a) $L^{-1}\left[\dfrac{p^2}{p^4 + 4a^4}\right] = L^{-1}\left[\dfrac{p^2}{(p^2 + 2a^2)^2 - 4a^2p^2}\right]$

$$= L^{-1}\left[\frac{p^2}{(p^2 - 2ap + 2a^2)(p^2 + 2ap + 2a^2)}\right]$$

$$= L^{-1}\left[\frac{1}{4a}\frac{p(p^2 + 2ap + 2a^2) - p(p^2 - 2ap + 2a^2)}{(p^2 + 2ap + 2a^2)(p^2 - 2ap + 2a^2)}\right]$$

$$= L^{-1}\left[\frac{p}{4a(p^2 - 2ap + 2a^2)} - \frac{p}{4a(p^2 + 2ap + 2a^2)}\right]$$

$$= \frac{1}{4a}L^{-1}\left[\frac{(p-a) + a}{(p-a)^2 + a^2}\right] - \frac{1}{4a}L^{-1}\left[\frac{(p+a) - a}{(p+a)^2 + a^2}\right]$$

$$= \frac{e^{at}}{4a}L^{-1}\left[\frac{p+a}{p^2 + a^2}\right] - \frac{1}{4a}e^{-at}L^{-1}\left[\frac{p-a}{p^2 + a^2}\right]$$

Answers

1. (a) $\dfrac{t^3}{6}$; (b) $\dfrac{1}{2}\sin 2t$;   (c) $4e^{2t}$; (d) $\dfrac{1}{\sqrt{\pi t}}$; (e) $\cos\sqrt{2}t + 6\cosh 4t + 3e^{3t}$; (f) $2\cosh 3t - \dfrac{5}{3}\sinh 3t$;

(g) $3e^{3t/2} - \dfrac{1}{4}\sinh\dfrac{4t}{3} - \dfrac{4}{9}\cosh\dfrac{4t}{3} + \dfrac{2}{3}\sin\dfrac{3t}{4} - \dfrac{3}{8}\cos\dfrac{3}{4}t$;   (h) $\dfrac{1}{2} - \dfrac{3}{2}t^2 + \dfrac{1}{16}t^4 - 4\cosh 3t + 6\sinh 3t + 4\sqrt{\dfrac{t}{\pi}} + \dfrac{8}{3}t\sqrt{\dfrac{t}{\pi}} - t$;

(i) $t - \dfrac{t^3}{(3!)^2} + \dfrac{t^5}{(5!)^2} - \dfrac{t^7}{(7!)^2} + \dots$

2. (a) $e^{3t}\sin t$; (b) $e^{-bt}\cos at$;   (c) $4e^{3t} - e^{-t}$;   (d) $e^{-at}\dfrac{t^{n-1}}{(n-1)!}, n \in Z^+$;   (e) $e^{-t}(4t^3 - t^4)/24$;

(f) $\left(t - \dfrac{1}{2}\right)e^t + \dfrac{3}{2}e^{-t}$ (g) $-\sin t + \dfrac{3}{2}\sin 2t$ (h) $\dfrac{1}{2}e^t - e^{2t} + \dfrac{5}{2}e^{3t}$

3. $\cos\dfrac{2\sqrt{at}}{\sqrt{\pi t}}$ **4.** (a) $\dfrac{1}{6}(t-5)^3 e^{2(t-5)}H(t-5)$ (b) $\dfrac{1}{6}(t-4)^3 e^{3(t-4)}.H(t-4)$

(c) $\dfrac{1}{2}(t-3)^2 H(t-3)$ (d) $e^{-4t}\left[\cosh(\sqrt{11}t) + \left(\dfrac{4}{\sqrt{11}}\right)\sinh(\sqrt{11}.t)\right]$

8.10.3 Inverse Laplace Transforms of Derivatives

THEOREM 1.   If $L^{-1}\{f(p)\} = F(t)$, then $L^{-1}\{f^{(n)}(p)\} = (-1)^n.t^n.F(t)$

PROOF.   Since we know that $L\{t^n F(t)\} = (-1)^n f^{(n)}(p)$

Therefore, $t^n F(t) = L^{-1}\{(-1)^n f^{(n)}(p)\} = (-1)^n L^{-1}\{f^{(n)}(p)\}$

Hence, $L^{-1}\{f^{(n)}(p)\} = (-1)^n t^n F(t)$

8.10.4 Division by p

THEOREM 1.   If $L^{-1}\{f(p)\} = F(t)$, then $L^{-1}\left\{\dfrac{f(p)}{p}\right\} = \int_0^t F(u)\,du$

PROOF.   Since we know that $\dfrac{f(p)}{p} = L\left\{\int_0^t F(u)\,du\right\} \Rightarrow L^{-1}\left\{\dfrac{f(p)}{p}\right\} = \int_0^t F(u)\,du$

8.10.5 Multiplication by powers of p

THEOREM 1.   If $L^{-1}\{f(p)\} = F(t)$ and $F(0) = 0$, then $L^{-1}\{pf(p)\} = F'(t)$

PROOF.   We know that

$$L\{F'(t)\} = pL\{F(t)\} - F(0) = pL[F(t)] = p\,f(p) \qquad\qquad [\because F(0) = 0].$$

Hence, $L^{-1}\{p\,f(p)\} = F'(t)$

8.10.6 INVERSE LAPLACE TRANSFORMS OF INTEGRALS

THEOREM 1.   If $L^{-1}\{f(p)\} = F(t)$, then $L^{-1}\left[\int_p^\infty f(x)\,dx\right] = \dfrac{F(t)}{t}$.

PROOF.   We know that

$$L\left\{\frac{1}{t}F(t)\right\} = \int_p^\infty f(x)\,dx \text{ provided } \lim_{t\to 0}\left\{\frac{F(t)}{t}\right\} \text{ exists.}$$

Hence, $L^{-1}\left\{\int_p^\infty f(x)\,dx\right\} = \dfrac{F(t)}{t}$.

 Solved Examples

EXAMPLE 1. Find $L^{-1}\left\{\dfrac{p}{(p^2+a^2)^2}\right\}$.   (SVTU-2009, VTU-2010)

SOLUTION.   We have

$$L^{-1}\left\{\frac{p}{(p^2+a^2)^2}\right\} = L^{-1}\left\{-\frac{1}{2}\frac{d}{dp}\left(\frac{1}{p^2+a^2}\right)\right\}$$

$$= -\frac{1}{2}L^{-1}\left\{\frac{d}{dp}\left(\frac{1}{p^2+a^2}\right)\right\}$$

$$= -\frac{1}{2}t(-1)L^{-1}\left\{\frac{1}{p^2+a^2}\right\} = \frac{t}{2a}\sin at$$

EXAMPLE 2. Find the function whose Laplace transform is

$$\log\left(1+\frac{1}{p}\right).$$ [UPTU-2007]

SOLUTION. $L^{-1}\left[\log\left(1+\frac{1}{p}\right)\right] = -\frac{1}{t}L^{-1}\left[\frac{d}{dp}\log\left(\frac{p+1}{p}\right)\right]$

$$= -\frac{1}{t}L^{-1}\left[\left(\frac{p}{p+1}\right)\left(-\frac{1}{p^2}\right)\right]$$

$$= -\frac{1}{t}L^{-1}\left[-\frac{1}{p(p+1)}\right] = -\frac{1}{t}L^{-1}\left[\frac{1}{p+1}-\frac{1}{p}\right]$$

$$= -\frac{1}{t}[e^{-t}-1] = \frac{1}{t}[1-e^{-t}].$$

EXAMPLE 3. Find the inverse Laplace transform of

$$f(p) = \log\frac{p+a}{p+b}$$ [UPTU-2003, ANNA-2003]

SOLUTION. $L^{-1}\log\left[\frac{p+a}{p+b}\right] = -\frac{1}{t}L^{-1}\left[\frac{d}{dp}\log\frac{p+a}{p+b}\right]$

$$= -\frac{1}{t}L^{-1}\left[\frac{d}{dp}\log(p+a) - \frac{d}{dp}\log(p+b)\right]$$

$$= -\frac{1}{t}L^{-1}\left[\frac{1}{p+a} - \frac{1}{p+b}\right]$$

$$= -\frac{1}{t}\left[e^{-at} - e^{-bt}\right] = \frac{1}{t}\left[e^{-bt} - e^{-at}\right].$$

EXAMPLE 4. Obtain the inverse Laplace transformation

$$\cot^{-1}\left(\frac{p+3}{2}\right).$$ [UPTU-2001 (SP.), 2002]

SOLUTION.   We know that $L^{-1}[f(p)] = -\dfrac{1}{t}L^{-1}\left[\dfrac{d}{dp}f(p)\right]$

$$\therefore\ \left[\cot^{-1}\left(\frac{p+3}{2}\right)\right] = -\frac{1}{t}L^{-1}\left[\frac{d}{dp}\cot^{-1}\left(\frac{p+3}{2}\right)\right]$$

$$= -\frac{1}{t}L^{-1}\left[\frac{-\dfrac{1}{2}}{1+\left(\dfrac{p+3}{2}\right)^2}\right] = \frac{1}{2t}L^{-1}\left[\frac{4}{4+(p+3)^2}\right]$$

$$= \frac{1}{t}L^{-1}\left[\frac{2}{2^2+(p+3)^2}\right] = \frac{1}{t}e^{-3t}L^{-1}\left[\frac{2}{2^2+p^2}\right]$$

$$= \frac{e^{-3t}}{t}\sin 2t.$$

EXAMPLE 5. Evaluate

(i) $L^{-1}\left\{\log\left(1+\dfrac{1}{p^2}\right)\right\}$ [UPTU Q.BANK-2001]

(ii) $L^{-1}\left\{\dfrac{1}{p}\log\left(1+\dfrac{1}{p^2}\right)\right\}$

SOLUTION.   (i) Let $f(p) = \log\left(1+\dfrac{1}{p^2}\right) = -\log\left(\dfrac{p^2}{p^2+1}\right)$

$$= -2\log p + \log(p^2+1)$$

Therefore, $f'(p) = -\dfrac{2}{p} + \dfrac{2p}{p^2+1}$

$\Rightarrow\ L^{-1}\{f'(p)\} = -2 + 2\cos t$

$\Rightarrow\ -tL^{-1}\{f(p)\} = -2(1-\cos t)$

Hence, $L^{-1}\left\{\log\left(1+\dfrac{1}{p^2}\right)\right\} = \dfrac{2(1-\cos t)}{t}$

(ii)   Since

$$L^{-1}\left\{\log\left(1+\frac{1}{p^2}\right)\right\} = \frac{2(1-\cos t)}{t}$$

Therefore,

$$L^{-1}\left\{\frac{1}{p}\log\left(1+\frac{1}{p^2}\right)\right\} = L^{-1}\left\{\frac{1}{p}f(p)\right\}$$

$$= \int_0^t F(x)dx = \int_0^t \frac{2}{x}(1-\cos x)\,dx.$$

 Exercise-8.50

1. Evaluate the following inverse Laplace transforms :

(a) $L^{-1}\left\{\dfrac{p}{(p^2-a^2)^2}\right\}$ (b) $L^{-1}\left\{\dfrac{p}{(p^2-16)^2}\right\}$

(c) $L^{-1}\left\{\dfrac{1}{(p-a)^3}\right\}$ (d) $L^{-1}\left\{\dfrac{p+1}{(p^2+2p+2)^2}\right\}$

(e) $L^{-1}\left\{\dfrac{p^2}{(p^2+4)^2}\right\}$

     [UPTU Q. BANK -2001]

2. Show that :

(a) $L^{-1}\left\{\dfrac{1}{p^3(p+1)}\right\} = 1 - t + \dfrac{t^2}{2} - e^{-t}$

(b) $L^{-1}\left\{\dfrac{1}{p^3(p^2+1)}\right\} = \dfrac{t^2}{2} + \cos t - 1$ (GBTU-2012)

(c) $L^{-1}\left\{\log\dfrac{p+2}{p+1}\right\} = \dfrac{1}{t}(e^{-t} - e^{-2t})$

(d) $L^{-1}\left\{\dfrac{1}{p}\log\dfrac{p+2}{p+1}\right\} = \int_0^t \dfrac{1}{x}(e^{-x} - e^{-2x})dx$

(e) $L^{-1}\left\{\dfrac{1}{p}\log\dfrac{p+3}{p+2}\right\} = \int_0^t \dfrac{1}{x}(e^{-2x} - e^{-3x})\,dx$

(f) $L^{-1}\left(\tan^{-1}\left(\dfrac{2}{p^2}\right)\right) = \dfrac{2}{t}\sin t \sinh t$

[UPTU Q. BANK–2001, MUMBAI–2005; VTU–2011]

3. If $L^{-1}\left\{\dfrac{p}{(p^2+1)^2}\right\} = \dfrac{1}{2}t\sin t$, then show that

$L^{-1}\left\{\dfrac{1}{(p^2+1)^2}\right\} = \dfrac{1}{2}(\sin t - t\cos t)$.

Hints to Selected Problems

1. (a) Since $\dfrac{d}{dp}(p^2 - a^2)^{-1} = -\dfrac{2p}{(p^2-a^2)^2}$,

therefore $\dfrac{p}{(p^2-a^2)^2} = -\dfrac{1}{2}\dfrac{d}{dp}\left(\dfrac{1}{p^2-a^2}\right)$

$\Rightarrow L^{-1}\left(\dfrac{p}{(p^2-a^2)^2}\right) = -\dfrac{1}{2}L^{-1}\left[\dfrac{d}{dp}\left(\dfrac{1}{p^2-a^2}\right)\right]$

$= -\dfrac{1}{2}(-1)'t\,L^{-1}\left(\dfrac{1}{p^2-a^2}\right) = \dfrac{1}{2}t\dfrac{1}{a}\sinh at$

2. (a) $L^{-1}\left[\dfrac{1}{p+1}\right] = e^{-t} = F(t)$ (say)

$\therefore L^{-1}\left[\dfrac{1}{p(p+1)}\right] = \int_0^t F(x)\,dx = \int_0^t e^{-x}dx = 1 - e^{-t}$

$\Rightarrow L^{-1}\left[\dfrac{1}{p^2(p+1)}\right] = \int_0^t (1 - e^{-x})dx = t + e^{-t} - 1$

and $L^{-1}\left[\dfrac{1}{p^3(p+1)}\right] = \int_0^t (x + e^{-x} - 1)dx = 1 - t + \dfrac{t^2}{2} - e^{-t}$

3. We have $L^{-1}\left[\dfrac{p}{(p^2+1)^2}\right] = \dfrac{1}{2}t\sin t = F(t)$ (say)

$\Rightarrow L^{-1}\left[\dfrac{1}{(p^2+1)^2}\right] = L^{-1}\left[\dfrac{1}{p}\cdot\dfrac{p}{(p^2+1)^2}\right] = \int_0^t F(x)dx$

$= \dfrac{1}{2}\int_0^t x.\sin x\,dx = \dfrac{1}{2}(\sin t - t\cos t)$

Answers

1. (a) $\dfrac{t}{2a}\sinh at$;   (b) $\dfrac{t}{8}\sinh 4t$;   (c) $\dfrac{1}{2}t^2 e^{at}$;   (d) $\dfrac{t}{2}e^{-t}\sin t$;   (e) $\dfrac{1}{4}(\sin 2t + 2t\cos 2t)$

8.10.7 Convolution

If $L^{-1}\{f(p)\} = F(t)$ and $L^{-1}\{g(p)\} = G(t)$, where $F(t)$ and $G(t)$ are two functions of class A, then   [UPTU-2002]

$$L^{-1}\{f(p).g(p)\} = \int_0^t F(u)\,G(t-u)\,du = F * G$$

We call $F * G$ the convolution or faulting of F and G.

Proof. $\int_0^t F(x)\,G(t-x)\,dx = H(t)$

Then, $L\{H(t)\} = \int_0^\infty e^{-pt}H(t)\,dt$

$= \int_0^\infty e^{-pt}\left[\int_0^t F(x)G(t-x)dx\right]dt$

$= \int_0^\infty \left[\int_0^t e^{-pt}F(x)G(t-x)dx\right]dt$ ...(1)

The integration being first with respect to x and then t.

The integration (1) is within the region lying below the line OP whose equation is $x = t$ and above OT, t being taken along OT and x along OX, with O is the origin the axes being perpendicular to each other. If the order of integration is changed, the strip will be taken parallel to OT, so that the limits of t are from x to ∞ and of x from 0 to ∞.

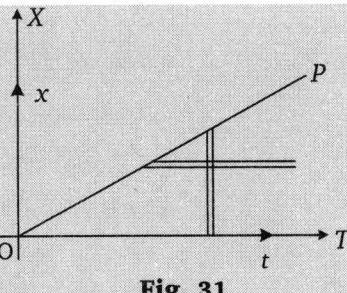

Fig. 31

Therefore, $L\{H(t)\} = \int_0^\infty dx \int_x^\infty e^{-pt}F(x)G(t-x)\,dt$

$= \int_0^\infty e^{-px}F(x)dx \int_x^\infty e^{-p(t-x)}G(t-x)\,dt$

Putting $t - x = \theta \Rightarrow dt = d\theta$

$\Rightarrow$ $L\{H(t)\} = \int_0^\infty e^{-px}F(x)\left\{\int_0^\infty e^{-p\theta}G(\theta)\,d\theta\right\}dx$

$= \int_0^\infty e^{-px}F(x)\,g(p)\,dx = f(p)\,g(p)$

$\Rightarrow$ $L\left\{\int_0^t F(x)\,G(t-x)dx\right\} = f(p)\,g(p)$

$\Rightarrow$ $\int_0^t F(x)\,G(t-x)\,dx = L^{-1}\{f(p)\,g(p)\} = F * G$.

REMARKS

- $F * G$ is commutative, i.e., $F * G = G * F$.
- $F * G$ is associative.
- $F * G$ is distributive over addition.

8.10.8 THE HEAVISIDE EXPANSION FORMULA

THEOREM. If $F(P)$ and $G(P)$ are polynomials in P, the degree of $F(P)$ being less than that of $G(P)$ and if

$$G(p) = (p - \alpha_1)(p - \alpha_2)...(p - \alpha_n)$$

where, $\alpha_1, \alpha_2, ..., \alpha_n$ are distinct constants, real or complex, then

$$L^{-1}\left\{\frac{F(p)}{G(p)}\right\} = \sum_{r=1}^{n} \frac{F(\alpha_r)}{G'(\alpha_r)} e^{\alpha_r . t}$$

PROOF. By the method of partial fractions, let

$$\frac{F(p)}{G(p)} = \frac{A_1}{p - \alpha_1} + \frac{A_2}{p - \alpha_2} + ... + \frac{A_r}{p - \alpha_r} + ... + \frac{A_n}{p - \alpha_n}$$

Multiplying both sides by $(p\ \alpha_r)$ and taking the limit $(p - \alpha_r)$, we get

$$A_r = \lim_{p \to \alpha_r} \frac{F(p)(p - \alpha_r)}{G(p)} = \lim_{p \to \alpha_r} F(p) . \lim_{p \to \alpha_r} \frac{p - \alpha_r}{G(p)}$$

$$= \lim_{p \to \alpha_r} F(p) . \lim_{p \to \alpha_r} \frac{1}{G'(p)} = \frac{F(\alpha_r)}{G'(\alpha_r)} \qquad \text{[By L'Hospital's rule]}$$

Therefore,

$$\frac{F(p)}{G(p)} = \frac{F(\alpha_1)}{G'(\alpha_1)} . \frac{1}{p - \alpha_1} + \frac{F(\alpha_2)}{G'(\alpha_2)} . \frac{1}{p - \alpha_2} ... + \frac{F(\alpha_r)}{G'(\alpha_r)} \frac{1}{p - \alpha_r} + ... + \frac{F(\alpha_n)}{G'(\alpha_n)} . \frac{1}{p - \alpha_n}$$

Hence, $L^{-1}\left[\frac{F(p)}{G(p)}\right] = \frac{F(\alpha_1)}{G'(\alpha_1)} . e^{\alpha_1 t} + \frac{F(\alpha_2)}{G'(\alpha_2)} e^{\alpha_2 t} + ... + \frac{F(\alpha_r)}{G'(\alpha_r)} e^{\alpha_r t} + ... + \frac{F(\alpha_n)}{G'(\alpha_n)} e^{\alpha_n t}$

$$= \sum_{r=1}^{n} \frac{F(\alpha_r)}{G'(\alpha_r)} . e^{\alpha_r . t}$$

Solved Examples

EXAMPLE 1. *Use the convolution theorem to find*

$$L^{-1}\left\{\frac{p^2}{(p^2 + a^2)^2}\right\}$$ (HAZARIBAG-2009)

SOLUTION. We know that $L^{-1}\left\{\frac{p}{(p^2 + a^2)}\right\} = \cos at$

Therefore, by convolution theorem, we have

$$L^{-1}\left\{\frac{p^2}{(p^2 + a^2)^2}\right\} = L^{-1}\left\{\frac{p}{p^2 + a^2} . \frac{p}{p^2 + a^2}\right\}$$

$$= \int_0^t \cos ax \cos a(t - x)\, dx$$

$$= \int_0^t \cos ax (\cos at \cos ax + \sin at \sin ax)\, dx$$

$$= \cos at \int_0^t \cos^2 ax\, dx + \sin at \int_0^t \cos ax \sin ax\, dx$$

$$= \frac{1}{2}\cos at \int_0^t (1 + \cos 2ax)\, dx + \frac{1}{2}\sin at \int_0^t \sin 2ax\, dx$$

$$= \frac{1}{2}\cos at\left[x + \frac{1}{2a}\sin 2ax\right]_0^t + \frac{1}{2}\sin at\left[-\frac{1}{2a}\cos 2ax\right]_0^t$$

$$= \frac{1}{2}\cos at\left[t + \frac{1}{2a}\sin 2a\right] + \frac{1}{4a}\sin at(1 - \cos 2at)$$

$$= \frac{1}{2}t \cos at + \frac{1}{4a}\sin at + \frac{1}{4a}(\sin 2at \cos at - \sin at \cos 2at)$$

$$= \frac{1}{2}t \cos at + \frac{1}{4a}[\sin at + \sin(2at - at)]$$

$$= \frac{1}{2a}[at \cos at + \sin at]$$

EXAMPLE 2. *Evaluate* $L^{-1}\left[\frac{p}{(p^2 + 1)(p^2 + 4)}\right]$ [UPTU-2002]

SOLUTION. We know that

$$L^{-1}\frac{p}{p^2 + 1} = \cos t \text{ and } L^{-1}\frac{2}{p^2 + 2^2} = \sin 2t$$

$$L^{-1}\left[\frac{p}{(p^2 + 1)(p^2 + 4)}\right] = \frac{1}{2}L^{-1}\left[\left(\frac{p}{p^2 + 1}\right) . \left(\frac{2}{p^2 + 4}\right)\right]$$

$$= \frac{1}{2}\int_0^t \sin 2x \cos(t - x)\, dx$$

$$= \int_0^t \sin x \cos x \{\cos t \cos x + \sin t \sin x\}\, dx$$

$$= \int_0^t (\sin x \cos^2 x \cos t + \sin^2 x \cos x \sin t)\, dx$$

$$= \left[-\frac{\cos^3 x}{3}\cos t + \frac{\sin^3 x}{3}\sin x\right]_0^t$$

$$= -\frac{\cos^4 t}{3} + \frac{\sin^4 t}{3} + \frac{\cos t}{3}$$

$$= \frac{1}{3}\left[\sin^4 t - \cos^4 t\right] + \frac{\cos t}{3}$$

$$= \frac{1}{3}(\sin^2 t + \cos^2 t)(\sin^2 t - \cos^2 t) + \frac{\cos t}{3}$$

$$= \frac{1}{3}(\sin^2 t - \cos^2 t) + \frac{\cos t}{3} = -\frac{1}{3}\cos 2t + \frac{\cos t}{3}$$

$$= \frac{1}{3}(\cos t - \cos 2t)$$

EXAMPLE 3. *Using convolution theorem, prove that*

$$L^{-1}\left[\frac{1}{p^3(p^2 + 1)}\right] = \frac{t^2}{2} + \cos t - 1$$

[UPTU-2005, VTU-2007, GBTU-2012]

SOLUTION. We have,

$$L^{-1}\left\{\frac{1}{p^3}\right\} = \frac{t^2}{2!}$$

$$L^{-1}\left\{\frac{1}{p^2+1}\right\} = \sin t$$

Using convolution theorem, we have

$$L^{-1}\left[\frac{1}{p^3(p^2+1)}\right] = \int_0^t \frac{(t-x)^2}{2!}\sin x\, dx$$

$$= \frac{1}{2}\int_0^t (t^2+x^2-2tx)\sin x\, dx$$

$$= \frac{1}{2}\left[(t^2+x^2-2tx)(-\cos x) \right.$$

$$\left. -\int (2x-2t)(-\cos x)dx\right]$$

$$= \frac{1}{2}\left[-\cos x(t^2+x^2-2tx) \right.$$

$$\left. +2\int (x-t)\cos x\, dx\right]_0^t$$

$$= \frac{1}{2}\left[-\cos x(t^2+x^2-2tx) \right.$$

$$\left. +2(x-t)\sin x + 2\cos x\right]_0^t$$

$$= \frac{1}{2}\left[-\cos x(t^2+x^2-2tx) \right.$$

$$\left. +2(x-t)\sin x + 2\cos x\right]_0^t$$

$$= \frac{1}{2}\left[-\cos t(t^2+t^2-2t^2) \right.$$

$$\left. +0+2\cos t+t^2\cos 0-2\cos 0\right]_0^t$$

$$= \cos t + \frac{t^2}{2} - 1 = \frac{t^2}{2} + \cos t - 1.$$

EXAMPLE 4. *Using the convolution theorem, find*

$$L^{-1}\left[\frac{p^2}{(p^2+a^2)(p^2+b^2)}\right],\ a\neq b \qquad \text{[UPTU-2004,}$$

MUMBAI-2007, BHOPAL-2008, UKTU-2011, VTU-2011S]

SOLUTION. We have $\quad L(\cos at) = \dfrac{p}{p^2+a^2}$

$$L(\cos bt) = \frac{p}{p^2+b^2}$$

Hence, by convolution theorem,

$$L\left[\int_0^t \cos ax\, \cos b(t-x)dx\right] = \frac{p^2}{(p^2+a^2)(p^2+b^2)}$$

Therefore $L^{-1}\left[\dfrac{p^2}{(p^2+a^2)(p^2+b^2)}\right]$

$$= \int_0^t \cos ax\, \cos b(t-x)\, dx$$

$$= \frac{1}{2}\int_0^t \{\cos(ax+bt-bx) + \cos(ax-bt+bx)\}dx$$

$$= \frac{1}{2}\int_0^t \cos\{(a-b)x+bt\}\, dx$$

$$+ \frac{1}{2}\int_0^t \cos[(a+b)x-bt]dx$$

$$= \frac{\sin at - \sin bt}{2(a-b)} + \frac{\sin at + \sin bt}{2(a+b)} = \frac{a\sin at - b\sin bt}{a^2-b^2}.$$

EXAMPLE 5. *Evaluate* $L^{-1}\left\{\dfrac{1}{\sqrt{p}}\dfrac{1}{(p-a)}\right\}$ *by the convolution theorem.*

SOLUTION. We know that

$$L^{-1}\left\{\frac{1}{\sqrt{p}}\right\} = L^{-1}\left\{\frac{1}{p^{1/2}}\right\} = \frac{t^{(1/2)-1}}{\Gamma(1/2)}$$

$$= \frac{1}{\sqrt{\pi}.\sqrt{t}} = F_1(t)\ \text{(say)}$$

Also, $L^{-1}\left\{\dfrac{1}{p-a}\right\} = e^{at} = F_2(t)$

Then, by convolution theorem, we have

$$L^{-1}\left\{\frac{1}{\sqrt{p}\,(p-a)}\right\} = F_1(t) * F_2(t)$$

$$= \int_0^t F_1(x).F_2(t-x)\, dx$$

$$= \int_0^t \frac{1}{\sqrt{\pi}}\cdot\frac{1}{\sqrt{x}}e^{a(t-x)}\, dx$$

$$= \frac{e^{at}}{\sqrt{\pi}}\int_0^{\sqrt{at}} \frac{\sqrt{a}}{u}e^{-u^2}\cdot\frac{2u}{a}.du$$

[By Putting] $ax = u^2 \Rightarrow dx = \dfrac{2u\, du}{a}$

$$= \frac{e^{at}}{\sqrt{a}}\cdot\frac{2}{\sqrt{\pi}}\int_0^{\sqrt{(at)}} e^{-u^2}du = \frac{e^{at}}{\sqrt{a}}\, erf\,[\sqrt{(at)}]$$

EXAMPLE 6. *Using convolution theorem, show that*

$$B(m,n) = \int_0^1 x^{m-1}(1-x)^{n-1}dx$$

$$= \frac{\Gamma(m)\,\Gamma(n)}{\Gamma(m+n)},\ \ m>0,\ n>0$$

SOLUTION. Let $\quad F(t) = \int_0^t x^{m-1}(t-x)^{n-1}dx$

$$= \int_0^t F_1(x).F_2(t-x)\, dx$$

where, $F_1(t) = t^{m-1} = F_1 * F_2$ and $F_2(t) = t^{n-1}$

Therefore $L\{F(t)\} = L\{F_1 * F_2\}$

$$= L\{F_1(t)\}\,.\,L\{F_2(t)\}$$

$$= L\{t^{m-1}\}.L\{t^{n-1}\}$$

$$= \frac{\Gamma(m)}{p^m}\cdot\frac{\Gamma(n)}{p^n} = \frac{\Gamma(m)\,\Gamma(n)}{p^{m+n}}$$

$$\Rightarrow \qquad F(t) = L^{-1}\left\{\frac{\Gamma(m).\Gamma(n)}{p^{m+n}}\right\}$$

$$= \Gamma(m).\Gamma(n).L^{-1}\left\{\frac{1}{p^{m+n}}\right\}$$

$$= \frac{\Gamma(m)\,\Gamma(n)}{\Gamma(m+n)}t^{m+n-1}$$

Let $t = 1$, then we have

$$B(m,n) = \int_0^1 x^{m-1}(1-x)^{n-1}dx = \frac{\Gamma(m)\,\Gamma(n)}{\Gamma(m+n)}.$$

EXAMPLE 7. *Using Heaviside's expansion formula, evaluate*

$$L^{-1}\left\{\frac{3p+1}{(p-1)(p^2+1)}\right\}$$

SOLUTION. We have $\quad F(p) = 3p+1$

and $\qquad G(p) = (p-1)(p^2+1)$

$$= (p-1)(p+i)(p-i)$$

Clearly, $G(p)$ has 3 distinct zeroes $\alpha_1 = 1, \alpha_2 = i$ and $\alpha_3 = -i$

Also, $\qquad G'(p) = 3p^2 - 2p + 1$

Using Heaviside's expansion formula, we have

$$L^{-1}\left\{\frac{3p+1}{(p-1)(p^2+1)}\right\} = \frac{F(1)}{G'(1)}e^t + \frac{F(i)}{G'(i)}e^{it} + \frac{F(-i)}{G'(-i)}e^{-it}$$

$$= \frac{4e^t}{2} + \frac{3i+1}{-(2+2i)}e^{it} + \frac{(-3i+1)}{(-2+2i)}e^{-it}$$

$$= 2e^t - \frac{(3i+1)(1-i)}{2(1+i)(1-i)}e^{i.t} + \frac{(3i-1)(1+i)}{2(1-i)(1+i)}e^{-i.t}$$

$$= 2e^t - \frac{1}{2}(i+2)e^{it} + \frac{1}{2}(i-2)e^{-it}$$

$$= 2e^t - \frac{1}{2}i(e^{it} - e^{-it}) - (e^{it} + e^{-it})$$

$$= 2e^t - \frac{1}{2}.i.2i\sin t - 2\cos t = 2e^t + \sin t - 2\cos t.$$

EXAMPLE 8. *Find* $L^{-1}\left[\dfrac{2p^2 - 6p + 5}{p^3 - 6p^2 + 11p - 6}\right]$.   [UPTU-2004]

SOLUTION. Let $f(p) = 2p^2 - 6p + 5$

and $G(p) = p^3 - 6p^2 + 11p - 6$

$$= (p-1)(p-2)(p-3)$$

$$G'(p) = 3p^2 - 12p + 11$$

$G(p) = 0$ has three roots 1, 2, 3.

$\alpha_1 = 1, \alpha_2 = 2, \alpha_3 = 3$

By Heaviside's inverse formula, we have

$$L^{-1}\left[\frac{F(p)}{G(p)}\right] = \sum_{i=1}^{n} \frac{F(\alpha_i)}{G'(\alpha_i)}e^{t\alpha_i}$$

$$L^{-1}\left[\frac{2p^2 - 6p + 5}{p^3 - 6p^2 + 11p - 6}\right] = \frac{F(\alpha_1)}{G'(\alpha_1)}e^{t\alpha_1}$$

$$+ \frac{F(\alpha_2)}{G'(\alpha_2)}e^{t\alpha_2} + \frac{F(\alpha_3)}{G'(\alpha_3)}e^{t\alpha_3}$$

$$= \frac{F(1)}{G'(1)}e^t + \frac{F(2)}{G'(2)}e^{2t} + \frac{F(3)}{G'(3)}e^{3t}$$

$$= \frac{1}{2}e^t + \frac{1}{-1}e^{2t} + \frac{5}{2}e^{3t} = \frac{1}{2}e^t - e^{2t} + \frac{5}{2}e^{3t}.$$

EXAMPLE 9. *Show that* $L^{-1}\left\{\dfrac{1}{\sqrt{p^2 + a^2}}\right\} = J_0(at).$

SOLUTION. Here, we have

$$L^{-1}\left\{\frac{1}{\sqrt{p^2 + a^2}}\right\} = L^{-1}\left\{\frac{1}{p}\left(1 + \frac{a^2}{p^2}\right)^{-1/2}\right\}$$

$$= L^{-1}\left\{\frac{1}{p}\left(1 - \frac{1}{2}\frac{a^2}{p^2} + \frac{1.3}{2.4}\frac{a^4}{p^4} - \frac{1.3.5}{2.4.6}\frac{a^6}{p^6} + \dots\right)\right\}$$

$$= L^{-1}\left\{\frac{1}{p} - \frac{a^2}{2}.\frac{1}{p^3} + \frac{1.3}{2.4}\frac{a^4}{p^5} - \frac{1.3.5}{2.4.6}\frac{a^6}{p^7} + \dots\right\}$$

$$= 1 - \frac{a^2}{2}.\frac{t^2}{2!} + \frac{1.3}{2.4}\frac{a^4 t^4}{4!} - \frac{1.3.5}{2.4.6}\frac{a^6 t^6}{6!} + \dots$$

$$= 1 - \frac{(at)^2}{2^2} + \frac{(at)^4}{2^2.4^2} - \frac{(at)^6}{2^2.4^2.6^2} + \dots = J_0(at).$$

EXAMPLE 10. *Evaluate* $L^{-1}\left\{\dfrac{e^{-\sqrt{p}}}{p}\right\}$ *and hence deduce that*

$$L^{-1}\left\{\frac{e^{x\sqrt{p}}}{p}\right\} = erfc\left(\frac{x}{2\sqrt{t}}\right)$$

SOLUTION. Let $f(p) = e^{-\sqrt{p}}$

Therefore, $F(t) = L^{-1}\{e^{-\sqrt{p}}\}$

$$= L^{-1}\left\{1 - \sqrt{p} + \frac{p}{2!} - \frac{p^{3/2}}{3!} + \frac{p^2}{4!} - \frac{p^{5/2}}{5!} + \dots\right\}$$

$$= L^{-1}\{1\} - L^{-1}\{p^{1/2}\} + \frac{1}{2!}L^{-1}\{p\}$$

$$- \frac{1}{3!}L^{-1}\{p^{3/2}\} + \frac{1}{4!}L^{-1}\{p^2\} - \frac{1}{5!}L^{-1}\{p^{5/2}\} + \dots \quad \dots(1)$$

Now, $L^{-1}\{p^{n+(1/2)}\} = L^{-1}\left\{\dfrac{1}{p^{-n-(1/2)}}\right\}$

$$= \frac{t^{-n-(3/2)}}{\Gamma\left(-n - \dfrac{1}{2}\right)}, \; n \in \mathbf{Z}$$

$$= \frac{(-1)^{n+1}}{\sqrt{n}}\left(\frac{1}{2}\right)\left(\frac{3}{2}\right)\left(\frac{5}{2}\right)\dots\left(\frac{2n+1}{2}\right)t^{-n-(3/2)}$$

Now, from (1), we have

$$F(t) = -\frac{(-1)t^{-3/2}}{\sqrt{\pi}}.\frac{1}{2} - \frac{1}{3!}.\frac{(-1)^2}{\sqrt{\pi}}\left(\frac{1}{2}\right)\left(\frac{3}{2}\right)t^{-5/2}$$

$$- \frac{1}{5!}\frac{(-1)^3}{\sqrt{\pi}}\left(\frac{1}{2}\right)\left(\frac{3}{2}\right)\left(\frac{5}{2}\right)t^{-7/2} + \dots$$

$$= \frac{1}{2\sqrt{\pi}.t^{3/2}}\left[1 - \frac{1}{4t} + \frac{(1/4t)^2}{2!} - \frac{(1/4t)^3}{3!} + \dots\right]$$

$$= \frac{1}{2\sqrt{\pi}.t^{3/2}}e^{-1/4t}.$$

Since $L^{-1}\left\{\dfrac{f(p)}{p}\right\} = \int_0^t F(x)\,dx$

           where $F(t) = L^{-1}\{f(p)\}$

Therefore,

$$L^{-1}\left\{\frac{e^{-\sqrt{p}}}{p}\right\} = \int_0^t \frac{1}{2\sqrt{\pi}\,x^{3/2}}e^{-1/(4x)}.dx$$

$$= -\frac{2}{\sqrt{\pi}}\int_\infty^{1/2\sqrt{t}}e^{-y^2}\,dy$$

       (where $x = \dfrac{1}{4y^2} \Rightarrow dx = -\dfrac{dy}{2y^3}$)

$$= \frac{2}{\sqrt{\pi}}\int_{1/(2\sqrt{t})}^\infty e^{-y^2}\,dy = erf\left(\frac{1}{2\sqrt{t}}\right)$$

Deduction. We have

$$\therefore \quad L^{-1}\left\{\frac{e^{-\sqrt{p}}}{p}\right\} = erf\left(\frac{1}{2\sqrt{t}}\right)$$

$$\therefore \quad L^{-1}\left\{\frac{e^{-\sqrt{(x^2 p)}}}{x^2 p}\right\} = \frac{1}{x^2}erf\left(\frac{1}{2\sqrt{t/x^2}}\right)$$

       (By change of scale property)

or $L^{-1}\left\{\dfrac{e^{-x\sqrt{p}}}{p}\right\} = erf\left(\dfrac{x}{2\sqrt{t}}\right)$

EXAMPLE 11. *Show that* $\int_0^\infty e^{-x^2}dx = \dfrac{\sqrt{\pi}}{2}$

SOLUTION. Let, $F(t) = \int_0^\infty e^{-tx^2}dx$

Then proceed as in example (13), we get

$$L\{F(t)\} = \int_0^\infty L\{e^{-tx^2}\}\,dx = \int_0^\infty \frac{dx}{p + x^2}$$

$$= \left(\frac{}{\sqrt{p}}.\tan \frac{}{\sqrt{p}}\right)_0^\infty = \frac{}{\sqrt{p}}$$

$$\Rightarrow F(t) = \frac{\pi}{2} L^{-1}\left(\frac{1}{\sqrt{p}}\right) = \frac{\pi}{2} \cdot \frac{1}{\sqrt{(\pi t)}} = \frac{1}{2}\sqrt{\frac{\pi}{t}}$$

$$\Rightarrow \int_0^\infty e^{-tx^2} dx = \frac{1}{2}\sqrt{\left(\frac{\pi}{t}\right)}$$

Taking $t = 1$, then we have $\int_0^\infty e^{-x^2} dx = \frac{1}{2}\sqrt{\pi}$

Exercise-8.51

1. Using the convolution theorem, show that

(a) $L^{-1}\left\{\dfrac{1}{(p+1)(p-2)}\right\} = \dfrac{1}{3}[e^{2t} - e^{-t}]$

(b) $L^{-1}\left\{\dfrac{p}{(p^2+a^2)^2}\right\} = \dfrac{1}{2a} t \sin at$ [UPTU-2008]

(c) $L^{-1}\left\{\dfrac{1}{p(p^2+4)^2}\right\} = \dfrac{1}{16}(1 - t\sin 2t - \cos 2t)$

(d) $L^{-1}\left\{\dfrac{1}{(p-2)(p^2+1)}\right\} = \dfrac{1}{5}[e^{2t} - 2\sin t - \cos t]$

(e) $\int_0^t \sin u \cos(t-u)\, du = \dfrac{t}{2}\sin t$

(f) $\int_0^t J_0(u) J_0(t-u)\, du = \sin t$

(g) $L^{-1}\left(\dfrac{p}{(p^2+4)^2}\right) = \dfrac{t}{4}\sin 2t$ [UPTU-2004, GBTU-2010]

(h) $L^{-}\left\{\dfrac{1}{(p^2+1)(p^2+q)}\right\} = \dfrac{1}{8}\left(\sin t - \dfrac{1}{3}\sin 3t\right)$ (MUMBAI-2005S)

(i) $L^{-}\left\{\dfrac{P}{(p^2+1)(p^2+4)(p^2+q)}\right\}$

$$= \frac{1}{12}\cos t - \frac{1}{10}\cos 2t + \frac{1}{60}\cos 3t \quad \text{(MUMBAI-2006)}$$

(j) $L^{-}\left\{\dfrac{1}{(p-2)(p+2)^2}\right\} = \dfrac{1}{16}(e^{2t} - e^{-2t} - 4te^{-2t})$

(MUMBAI-2009)

(k) $L^{-}\left\{\dfrac{P}{(P+2)(P^2+q)}\right\} = \dfrac{1}{13}(3\sin 3t + 2\cos 2t - 2e^{-2t})$

(VTU-2008 S)

(l) $L^{-}\left\{\dfrac{1}{(P^2+4P+13)^2}\right\} = \dfrac{e^{-2t}}{54}(\sin 3t - 3t\cos 3t)$

(MUMBAI-2008)

2. Using the Heaviside formula, show that

(a) $L^{-1}\left\{\dfrac{p^2-6}{p^3+4p^2+3p}\right\} = -2 + \dfrac{5}{2}e^{-t} + \dfrac{1}{2}e^{-3t}$

(b) $L^{-1}\left\{\dfrac{19p+37}{(p+1)(p-2)(p+3)}\right\} = -3e^{-t} + 5e^{2t} - 2e^{-3t}$

(c) $L^{-1}\left(\dfrac{1}{p^3-1}\right) = \dfrac{1}{3}\left[e^t - e^{-\frac{t}{2}}\left\{\cos\left(\dfrac{1}{2}\sqrt{3}t\right) + \sqrt{3}\sin\left(\dfrac{1}{2}\sqrt{3}t\right)\right\}\right]$

(d) $L^{-1}\left(\dfrac{1}{p^3+1}\right) = \dfrac{1}{3}\left[e^{-t} - e^{\frac{t}{2}}\left\{\cos\left(\dfrac{1}{2}\sqrt{3}t\right) - \sqrt{3}\sin\left(\dfrac{1}{2}\sqrt{3}t\right)\right\}\right]$

Hints to Selected Problems

1. (a) $L^{-1}\left(\dfrac{1}{p+1}\right) = e^{-t} = F_1(t)$ (say)

and $L^{-1}\left(\dfrac{1}{p+2}\right) = e^{2t} = F_2(t)$ (say)

Then using convolution theorem.

2. (a) $f(p) = 2p^2 - 6p + 5$

and $G(p) = p^3 - 6p^2 + 11p - 6 = (p-1)(p-2)(p-3)$

$\Rightarrow G'(p) = 3p^2 - 12p + 11$

$\Rightarrow G(p)$ has 3 distinct roots $\alpha_1 = 1$, $\alpha_2 = 2$ and $\alpha_3 = 3$.

Then using Heavyside expansion formula.

2. (c) $f(p) = 19p + 37$

$G(p) = (p+1)(p-2)(p+3)$

$\Rightarrow G(p)$ has three distinct roots $\alpha_1 = -1$, $\alpha_2 = 2$, $\alpha_3 = -3$.

Then by Heavyside's formula, we have

$$L^{-1}\left[\frac{19p+37}{(p+1)(p-2)(p+3)}\right] = \frac{f(-1)}{G'(-1)}e^{-t}$$

$$+ \frac{f(2)}{G'(2)}e^{2t} + \frac{f(-3)}{G'(-3)}e^{-3t} = -3e^{-t} + 5e^{2t} - 2e^{-3t}$$

Miscellaneous Solved Examples

EXAMPLE 1. *Find the inverse transforms of*

$$\frac{p+2}{p^2-4p+13}.$$
     (VTU 2008)

SOLUTION. $L^{-1}\left(\dfrac{p+2}{p^2-4p+13}\right) = L^{-1}\left(\dfrac{p+2}{(p-2)^2+9}\right)$

$$= L^{-1}\left[\frac{p-2+4}{(p-2)^2+3^2}\right]$$

$$= L^{-1}\left[\frac{p-2}{(p-2)^2+3^2}\right] + 4L^{-1}\left(\frac{1}{(p-2)^2+3^2}\right)$$

$$= e^{2t}\cos 3t + \frac{4}{3}e^{2t}\sin 3t .$$

EXAMPLE 2. *Find* $L^{-1}\left\{\dfrac{(p+2)^2}{(p^2+4p+8)^2}\right\}$
     (MUMBAI–2005)

SOLUTION. $L^{-1}\left\{\dfrac{(p+2)^2}{(p^2+4p+8)^2}\right\} = L^{-1}\left\{\dfrac{(p+2)^2}{(p^2+4p+4+4)^2}\right\}$

$$= L^{-1}\left\{\frac{(p+2)^2}{[(p+2)^2+4]^2}\right\}$$

$$= e^{-2t}L^{-1}\left\{\frac{p^2}{(p^2+4)^2}\right\} = e^{-2t}L^{-1}\left\{\frac{p^2+4-4}{(p^2+4)^2}\right\}$$

$$= e^{-2t}L^{-1}\left\{\frac{1}{p^2+4} - \frac{4}{(p^2+4)^2}\right\}$$

$$= \frac{e^{-2t}\sin 2t}{2} - 4e^{-2t}L^{-1}\left\{\frac{1}{(p^2+4)^2}\right\}$$

$$= \frac{e^{-2t}\sin 2t}{2} - 4e^{-2t}\left\{\frac{1}{4}\left(\frac{\sin 2t}{4} - \frac{t\cos 2t}{2}\right)\right\}$$

$$= e^{-2t}\left\{\frac{\sin 2t}{2} - \frac{\sin 2t}{4} + \frac{t\cos 2t}{2}\right\}$$

$$= e^{-2t}\left\{\frac{\sin 2t}{4} + \frac{t\cos 2t}{2}\right\} \quad = e^{-2t}\left\{\frac{\sin 2t}{4} + \frac{t\cos 2t}{2}\right\}$$

EXAMPLE 3. *Find* $L^{-1}\left\{\dfrac{1}{p(p^2+a^2)}\right\}$ (PTU–2003)

SOLUTION. $\therefore \quad L^{-1}\left(\dfrac{1}{p^2+a^2}\right) = \dfrac{1}{a}\sin at$

$$\therefore \quad L^{-1}\left\{\frac{1}{p(p^2+a^2)}\right\} = \int_0^t \frac{1}{a}\sin at\, dt$$

$$= \frac{1}{a^2}[-\cos at]_0^t = \frac{(1-\cos at)}{a^2}.$$

EXAMPLE 4. *Find* $L^{-1}\left\{\dfrac{p+2}{p^2(p+1)(p-2)}\right\}$ (VTU-2003)

SOLUTION. $L^{-1}\left\{\dfrac{p+2}{(p+1)(p-2)}\right\} = \dfrac{4}{3}L^{-1}\left(\dfrac{1}{p-2}\right) - \dfrac{1}{3}L^{-1}\left(\dfrac{1}{p+1}\right)$

$$= \frac{4}{3}e^{2t} - \frac{1}{3}e^{-t}$$

Now, $L^{-1}\left\{\dfrac{p+2}{p(p+1)(p-2)}\right\}$

$$= \int_0^t L^{-1}\left(\frac{p+2}{(p+1)(p-2)}\right) dt$$

$$= \int_0^t \left(\frac{4}{3}e^{2t} - \frac{1}{3}e^{-t}\right) dt = \frac{2}{3}e^{2t} + \frac{1}{3}e^{-t} - 1$$

Now, $L^{-1}\left\{\dfrac{p+2}{p^2(p+1)(p-2)}\right\}$

$$= \int_0^t L^{-1}\left\{\frac{p+2}{p(p+1)(p-2)}\right\} dt$$

$$= \int_0^t \left(\frac{2}{3}e^{2t} + \frac{1}{3}e^{-t} - 1\right) dt = \frac{1}{3}(e^{2t} - e^{-t} - t)$$

EXAMPLE 5. *Find* $\left\{\dfrac{1}{(p^2+4p+5)^2}\right\}$.   (SVTU-2009, PTU-2005)

SOLUTION. $L^{-1}\left\{\dfrac{1}{p^2+4p+5}\right\}$

$$= L^{-1}\left\{\frac{1}{(p+2)^2+1}\right\} = e^{-2t}\sin t$$

Now, $L^{-1}\left\{\dfrac{d}{dp}\left(\dfrac{1}{p^2+4p+5}\right)\right\} = (-1)\,t\cdot e^{-2t}\sin t$

$$L^{-1}\left\{\frac{-(2p+4)}{(p^2+4p+5)^2}\right\} = -t\cdot e^{-2t}\sin t$$

$$L^{-1}\left\{\frac{p+2}{(p^2+4p+5)^2}\right\} = \frac{1}{2}t\cdot e^{-2t}\sin t$$

EXAMPLE 6. *Find* $L^{-1}\left\{\log\dfrac{p+1}{p-1}\right\}$

(BHOPAL–2008, SVTU–2009, UPTU–2009, UKTU–2012)

SOLUTION. It $f(t) = L^{-1}\log\dfrac{p+1}{p-1}$

then, $tf(t) = L^{-1}\left\{\dfrac{-d}{dp}\log\left(\dfrac{p+1}{p-1}\right)\right\}$

$$= -L^{-1}\left\{\frac{d}{dp}\log(p+1)\right\} + L^{-1}\left\{\frac{d}{dp}\log(p-1)\right\}$$

$$= -L^{-1}\left(\frac{1}{p+1}\right) + L^{-1}\left(\frac{1}{p-1}\right)$$

$$= -e^{-t} + e^t = 2\sinh t$$

Thus $f(t) = (2\sinh t)/t$

EXAMPLE 7. *Find* $L^{-1}\left\{\log\dfrac{p^2+1}{p(p+1)}\right\}$ (SVTU-2009, VTU-2008)

SOLUTION. If $f(t) = L^{-1}\log\dfrac{p^2+1}{p(p+1)}$

Then $tf(t) = L^{-1}\left\{-\dfrac{d}{dp}\log\left(\dfrac{p^2+1}{p(p+1)}\right)\right\}$

$$= L^{-1}\left\{\frac{d}{dp}\log(p^2+1)\right\} + L^{-1}\left\{\frac{d}{dp}\log p\right\}$$

$$+ L^{-1}\left\{\frac{d}{dp}\log(p+1)\right\}$$

$$= -L^{-1}\left(\frac{2p}{p^2+1}\right) + L^{-1}\left(\frac{1}{p}\right) + L^{-1}\left(\frac{1}{p+1}\right)$$

$$= -2\cos t + 1 + e^{-t}$$

Thus, $f(t) = \dfrac{1}{t}(1 + e^{-t} - 2\cos t)$

Miscellaneous Exercise

(A) Find the inverse laplace transform of the following functions.

1. $\dfrac{p}{(2p-1)(3p-1)}$ (VTU-2010) **2.** $\dfrac{1}{p^2-5p+6}$ (SVTU-2008)

3. $\dfrac{3p+2}{p^2-p-2}$ (VTU-2010 S) **4.** $\dfrac{1}{p(p^2-1)}$ (NAGARJUNA-2008)

5. $\dfrac{1-7p}{(p-3)(p-1)(p+2)}$ (BPTU-2005 S)

6. $\dfrac{p}{(p^2-1)^2}$ (KURUKSHETRA-2005) **7.** $\dfrac{p^3}{p^4-a^4}$ (KURUKSHETRA-2005)

8. $\dfrac{p^2+2p+3}{(p^2+2p+2)(p^2+2p+5)}$ (MUMBAI-2008)

9. $\dfrac{a(p^2-2a^2)}{p^4+4a^4}$ (MUMBAI-2009) **10.** $\dfrac{1}{p^2(p+5)}$ (MADRAS–2003 S)

11. $\dfrac{p}{a^2p^2+b^2}$ (MADRAS-2000 S) **12.** $\dfrac{p+2}{(p^2+4p+8)^2}$ (MUMBAI-2006)

13. $\dfrac{1}{2}\log\left(\dfrac{p^2+b^2}{p^2+a^2}\right)$ (MUMBAI-2008, VTU-2008)

14. $\log\dfrac{p^2+1}{(p-1)^2}$ (MADRAS-2000 S)

15. $\tan^{-1}\left(\dfrac{2}{p}\right)$ (MUMBAI-2007, PTU-2005)

16. $\cot^{-1}(p)$ (VTU-2005) **17.** $p\log\dfrac{p-1}{p+1}$ (MADRAS-1999)

18. $\dfrac{2p+1}{p^2-4}$ (UPTU (SUM)-2009)

19. $\dfrac{e^{-2\pi p}}{p(p^2+1)}$ (GBTU-2011) **20.** $\dfrac{}{p\,(p\ 7)}$ (GBTU (CO)-2011)

21. $\dfrac{14p+10}{49p^2+28p+13}$ (UPTU (SUM)-2007)

22. $\dfrac{p}{p^2+6p+25}$ (GBTU-2011)

23. $\dfrac{p+1}{p^2-6p+25}$ (GBTU-2010)

24. $\log\left(\dfrac{p^2+4p+5}{p^2+2p+5}\right)$ (MTU (SUM)-2011)

25. $\dfrac{1}{p^4+4}$ (UPTU (SUM)-2007)

26. $\dfrac{1}{p(p+1)(p+2)}$ (UPTU (SUM)-2008)

27. $\log\left(1-\dfrac{a^2}{p^2}\right)$ (UPTU (CO)-2009)

28. $\dfrac{1}{p(p+a)^3}$ (UKTU-2012)

29. $\dfrac{1}{p^2(p^2+1)}$ (UPTU (SUM)-2009)

30. $\dfrac{p^2}{(p^2+w^2)^2}$ (GBTU(CO) 2010)

31. $\dfrac{8p}{(p^2+16)(p^2+1)^2}$ (UPTU (CO)-2009)

(B) Use convolution theorm to find:

1. $L^{-1}\left[\dfrac{1}{(p^2+a^2)^2}\right]$ (UPTU (SUM)-2007, UPTU-2009)

2. $L^{-1}\left[\dfrac{16}{(p-2)(p+2)^2}\right]$ (MTU-2011)

3. $L^{-1}\left[\dfrac{p}{(p^2+a^2)^3}\right]$ (UPTU-2008, MTU-2012)

Answers

(A)

1. $3e^{t/2}+2e^{t/3}$ **2.** $e^{3t}-e^{2t}$ **3.** $\dfrac{1}{3}(8e^{2t}-e^{-t})$ **4.** $\cosh t$ **5.** $e^t+e^{-2t}-2e^{3t}$ **6.** $\dfrac{1}{2}t\sinh t$ **7.** $\dfrac{1}{2}[\cos at+\cosh at]$

8. $\dfrac{1}{3}e^{-t}(\sin t+\sin 2t)$ **9.** $\cos at\sinh at$ **10.** $\dfrac{1}{25}(e^{-5t}+5t-1)$ **11.** $\dfrac{1}{a^2}\cos\left(\dfrac{bt}{a}\right)$ **12.** $\dfrac{1}{2}te^{-2t}\sin 2t$ **13.** $\dfrac{1}{t}(\cos at-\cos bt)$

14. $\dfrac{2}{t}(e^t-\cos t)$ **15.** $\dfrac{\sin 2t}{t}$ **16.** $\dfrac{\sin t}{t}$ **17.** $\dfrac{2(\sinh t-t\cosh t)}{t^2}$ **18.** $2\cosh 2t+\dfrac{1}{2}\sinh 2t$ **19.** $1-\cos tu(t-2\pi)$

20. $-\dfrac{6}{49}+\dfrac{1}{7}t+\dfrac{6}{49}e^{7t}$ **21.** $\dfrac{2}{7}e^{-2t/7}\left(\cos\dfrac{3}{7}t+\sin\dfrac{3}{7}t\right)$ **22.** $e^{-3t}\left(\cos 4t-\dfrac{3}{4}\sin 4t\right)$ **23.** $e^{3t}(\cos 4t+\sin 4t)$

24. $\dfrac{2}{t}(e^{-t}\cos 2t-e^{-2t}\cos t)$ **25.** $\dfrac{1}{4}(\sin t\cosh t-\cos t\sinh t)$ **26.** $\dfrac{1}{2}-e^{-t}+\dfrac{1}{2}e^{-2t}$ **27.** $\dfrac{2}{t}(1-\cosh at)$ **28.** $\dfrac{1}{a^3}-\dfrac{1}{a^3}e^{-at}\left(1+at+\dfrac{a^2t^2}{2}\right)$

29. $t-\sin t$ **30.** $\dfrac{1}{2\omega}\sin\omega t+\dfrac{t}{2}\cos\omega t$ **31.** $\dfrac{60t\sin t-8\cos t+8\cos 4t}{225}$

(B)

1. $\dfrac{1}{2a^3}(\sin at-at\cos at)$ **2.** $e^{2t}-e^{-2t}(1+4t)$ **3.** $\dfrac{t}{8a^3}(\sin at-at\cos at)$

8.11 APPLICATIONS OF LAPLACE TRANSFORM TO SOLUTION OF ORDINARY DIFFERENTIAL EQUATION

8.11.1 SOLUTION OF ORDINARY DIFFERENTIAL EQUATION WITH CONSTANT COEFFICIENTS

Consider a linear differential equation with constant coefficients

$$\frac{d^n y}{dt^n}+A_1\frac{d^{n-1}y}{dt^{n-1}}+...+A_{n-1}\frac{dy}{dt}+A_n y=F(t) \qquad ...(1)$$

where t is the independent variable and $F(t)$ is a function of t.

Let

$$y(0)=C_1,\ y'(0)=C_2,...,y^{n-1}(0)=C_{n-1} \qquad ...(2)$$

be the given initial or boundary conditions, where $C_1, C_2, ..., C_{n-1}$ are constants. Now, taking the Laplace transform of both sides of (1) and using the conditions given by (2), we get an algebraic equation from which $\bar{y}(p)=L\{y(t)\}$ is determined. The required solution is then obtained by finding the inverse Laplace transform of $\bar{y}(p)$.

📑 **REMARKS**

- The algebraic equation, obtained above is known as subsidiary equation.
- The above method is easily extended to higher order differential equation.

WORKING PROCEDURE

Step 1. Taking Laplace transform of both the sides of the given differential equation and use given initial conditions.

Step 2. Solve the equation obtained in step (1) for $L\{y\}$.

Step 3. Taking inverse Laplace transform to find y.

Solved Examples

EXAMPLE 1. *Solve* $\dfrac{d^2y}{dt^2} + y = 0$ *under the condition that*

$$y = 1, \frac{dy}{dt} = 0 \text{ when } t = 0.$$

SOLUTION. Here, the given equation is

$$\frac{d^2y}{dt^2} + y = 0. \qquad \dots (1)$$

Taking the Laplace transform of both sides of the given differential equation, we get

$$L\{y''\} + L\{y\} = 0$$

$$\Rightarrow \quad p^2 L\{y\} - py(0) - y'(0) + L\{y\} = 0$$

$$\Rightarrow \quad (p^2 + 1)\, L\{y\} - p.1 - 0 = 0$$

$$\text{[Using the given conditions]}$$

$$\Rightarrow \quad L\{y\} = \frac{p}{p^2+1}$$

Therefore, $y = L^{-1}\left\{\dfrac{p}{p^2+1}\right\} = \cos t$.

EXAMPLE 2. *Solve* $(D^2 + 1)y = 6\cos 2t$ *if* $y = 3, Dy = 1$ *when* $t = 0$.

SOLUTION. The given equation can be written as

$$y'' + y = 6\cos 2t$$

Taking the Laplace transform of both the sides of the given differential equation, we get

$$L\{y''\} + L\{y\} = 6L\{\cos(2t)\}$$

$$\Rightarrow \quad p^2 L\{y\} - py(0) - y'(0) + L\{y\} = 6\frac{p}{p^2+2^2}$$

$$\Rightarrow \quad (p^2+1)\, L\{y\} - 3p - 1 = \frac{6p}{p^2+4}$$

$$\text{[Using the given conditions]}$$

$$\Rightarrow L\{y\} = \frac{3p}{p^2+1} + \frac{1}{p^2+1} + \frac{6p}{(p^2+1)(p^2+4)}$$

$$= \frac{3p}{p^2+1} + \frac{1}{p^2+1} + \frac{2p[(p^2+4)-(p^2+1)]}{(p^2+1)(p^2+4)}$$

$$= \frac{3p}{p^2+1} + \frac{1}{p^2+1} + 2p\left\{\frac{1}{p^2+1} - \frac{1}{p^2+4}\right\}$$

$$= \frac{5p}{p^2+1} + \frac{1}{p^2+1} - \frac{2p}{p^2+4}$$

Therefore,

$$y = 5L^{-1}\left\{\frac{p}{p^2+1}\right\} + L^{-1}\left\{\frac{1}{p^2+1}\right\} - 2L^{-1}\left\{\frac{p}{p^2+4}\right\}$$

$$\Rightarrow \quad y = 5\cos t + \sin t - 2\cos 2t$$

EXAMPLE 3. *Using Laplace transforms, find the solution of the initial value problem :*

$$y'' + 9y = 6\cos 3t,\ y(0) = 2,\ y'(0) = 0$$

[UPTU–2006]

SOLUTION. The given equation can be written as

$$y'' + 9y = 6\cos 3t,\ y(0) = 2,\ y'(0) = 0 \qquad \dots (1)$$

Taking Laplace transform of (1), we get

$$[p^2 L\{y\} - py(0) - y'(0)] + 9L\{y\} = 6\frac{p}{p^2+9}$$

Putting the value of $y(0)$ and $y'(0)$ in (2), we have

$$p^2 L\{y\} - 2p + 9L\{y\} = \frac{6p}{p^2+9}$$

$$(p^2+9)L\{y\} = 2p + \frac{6p}{p^2+9}$$

$$L\{y\} = \frac{2p}{p^2+9} + \frac{6p}{(p^2+9)^2}$$

$$\Rightarrow \quad y = L^{-1}\left\{\frac{2p}{p^2+9}\right\} + L^{-1}\left\{\frac{6p}{(p^2+9)^2}\right\}$$

$$= 2\cos 3t + 3L^{-1}\frac{d}{dp}\left[-\frac{3}{p^2+9}\right]$$

$$= 2\cos 3t + t\sin 3t$$

EXAMPLE 4. *Solve* $(D^2 + 9)y = \cos 2t$ *if* $y(0) = 1, y\left(\dfrac{\pi}{2}\right) = -1$.

[UPTU–2002, 06, BHOPAL–2008]

SOLUTION. The given equation can be written as

$$y'' + 9y = \cos 2t \qquad \dots (1)$$

Taking the Laplace transform of both the sides of (1), we get

$$L\{y''\} + 9L\{y\} = L\{\cos 2t\}$$

$$\Rightarrow \quad p^2 L\{y\} - py(0) - y'(0) + 9L\{y\} = \frac{p}{p^2+4}$$

$$\Rightarrow \quad (p^2+9)\, L\{y\} - p - C = \frac{p}{p^2+4}, \text{ where }$$

$$C = y'(0)$$

$$\therefore \quad L\{y\} = \frac{p+C}{p^2+9} + \frac{p}{(p^2+9)(p^2+4)}$$

$$= \frac{p}{p^2+9} + \frac{C}{p^2+9} + \frac{p}{5(p^2+4)} - \frac{p}{5(p^2+9)}$$

Therefore,

$$y = L^{-1}\left\{\frac{p}{p^2+9}\right\} + CL^{-1}\left\{\frac{1}{p^2+9}\right\}$$

$$+ \frac{1}{5}L^{-1}\left\{\frac{p}{p^2+4}\right\} - \frac{1}{5}L^{-1}\left\{\frac{p}{p^2+9}\right\}$$

$$= \cos 3t + \frac{1}{3}C\sin 3t + \frac{1}{5}\cos 2t - \frac{1}{5}\cos 3t$$

$$= \frac{4}{5}\cos 3t + \frac{1}{3}C\sin 3t + \frac{1}{5}\cos 2t \qquad \dots (2)$$

Now, since $y\left(\dfrac{\pi}{2}\right) = -1$, therefore, from (2), we have

$$-1 = \frac{4}{5}\cos\frac{3\pi}{2} + \frac{1}{3}C\sin\frac{3\pi}{2} + \frac{1}{5}\cos\pi$$

On solving, we get $C = \dfrac{12}{5}$

Put this value in (2), we get

$$y = \dfrac{4}{5}\cos 3t + \dfrac{4}{5}\sin 3t + \dfrac{1}{5}\cos 2t.$$

EXAMPLE 5. *Solve using Laplace transform method*

$$y''(t) + 4y'(t) + 4y(t) = 6e^{-t}, with$$

$$y(0) = -2,\ y'(0) = 8.$$ [UPTU–2007]

SOLUTION. The given equation can be written as

$$y''(t) + 4y'(t) + 4y(t) = 6e^{-t}$$

Taking Laplace transform on both sides of the given equation, we get

$$[p^2 L\{y\} - py(0) - y'(0)] + 4[pL\{y\}$$

$$-y(0)] + 4L\{y\} = \dfrac{6}{p+1} \qquad \dots (1)$$

Putting $y(0) = -2$ and $y'(0) = 8$ in (1), we get

$$[p^2 L\{y\} - p(-2) - 8] + 4[pL\{y\}$$

$$+2] + 4L\{y\} = \dfrac{6}{p+1}$$

$$\Rightarrow (p^2 + 4p + 4)L\{y\} + 2p = \dfrac{6}{p+1}$$

$$\Rightarrow (p^2 + 4p + 4)L\{y\} = -2p + \dfrac{6}{p+1}$$

$$\Rightarrow (p+2)^2 L\{y\} = \dfrac{-2p^2 - 2p + 6}{(p+1)}$$

$$\Rightarrow L\{y\} = \dfrac{-2p^2 - 2p + 6}{(p+1)(p+2)^2}$$

Let $\dfrac{-2p^2 - 2p + 6}{(p+1)(p+2)^2} = \dfrac{A}{p+1} + \dfrac{B}{p+2} + \dfrac{C}{(p+2)^2}$

$$-2p^2 - 2p + 6 = A(p+2)^2$$

$$+ B(p+1)(p+2) + C(p+1)$$

$$-2 + 2 + 6 = A(-1+2)^2 \Rightarrow A = 6 \text{ [Putting } p=-1]$$

$$-8 + 4 + 6 = C(-2+1) \Rightarrow C = -2 \text{ [Putting } p=-2]$$

Comparing the coefficients of p^2 on both sides, we get

$$-2 = A + B \Rightarrow -2 = 6 + B \Rightarrow B = -8$$

$$L\{y\} = \dfrac{6}{p+1} + \dfrac{-8}{p+2} + \dfrac{2}{(p+2)^2}$$

$$y = L^{-1}\left[\dfrac{6}{p+1} - \dfrac{8}{p+2} - \dfrac{2}{(p+2)^2}\right]$$

Hence, $y = 6e^{-t} - 8e^{-2t} - 2e^{-2t}t$

EXAMPLE 6. *Solve $(D^3 - 2D^2 + 5D)y = 0$ given that*

$$y(0) = 0,\ y'(0) = 1,\ y\left(\dfrac{\pi}{8}\right) = 1.$$

[UPTU Q. BANK–2001]

SOLUTION. The given equation can be written as

$$y''' - 2y'' + 5y' = 0 \qquad \dots (1)$$

Taking the Laplace transforms of both sides of (1), we get

$$L\{y'''\} - 2L\{y''\} + 5L\{y'\} = 0$$

$$\Rightarrow p^3 L\{y\} - p^2 y(0) - py'(0) - y''(0)$$

$$-2[p^2 L\{y\} - py(0) - y'(0)]$$

$$+5[pL\{y\} - y(0)] = 0$$

$$\Rightarrow [p^3 - 2p^2 + 5p]L\{y\} - p$$

$$-C - 2(-1) + 5.0 = 0,$$

where $y''(0) = C$

$$L\{y\} = \dfrac{C - 2 + p}{p^3 - 2p^2 + 5p} = \dfrac{C-2}{p(p^2 - 2p + 5)}$$

$$+ \dfrac{1}{p^2 - 2p + 5}$$

$$= \dfrac{C-2}{5p} - \dfrac{C-2}{5} \cdot \dfrac{p-2}{p^2 - 2p + 5} + \dfrac{1}{p^2 - 2p + 5}$$

$$= \dfrac{C-2}{5p} - \dfrac{C-2}{5} \cdot \dfrac{(p-1)-1}{(p-1)^2 + 4} + \dfrac{1}{(p-1)^2 + 4}$$

$$= \dfrac{C-2}{5p} - \dfrac{C-2}{5} \cdot \dfrac{(p-1)}{(p-1)^2 + 4} + \dfrac{C+3}{10} \cdot \dfrac{2}{(p-1)^2 + 4}$$

Therefore,

$$y = \dfrac{C-2}{5} \cdot L^{-1}\left\{\dfrac{1}{p}\right\} - \dfrac{C-2}{5} L^{-1}\left\{\dfrac{p-1}{(p-1)^2 + 4}\right\}$$

$$+ \dfrac{C+3}{10} L^{-1}\left\{\dfrac{2}{(p-1)^2 + 4}\right\}$$

$$= \dfrac{C-2}{5} - \dfrac{C-2}{5} e^t \cos 2t + \dfrac{C+3}{10} e^t \sin 2t \qquad \dots (2)$$

Now, since $y\left(\dfrac{\pi}{8}\right) = 1$, therefore

$$1 = \dfrac{C-2}{5} - \dfrac{C-2}{5} e^{\pi/8} \dfrac{1}{\sqrt{2}} + \dfrac{C+3}{10} e^{\pi/8} \dfrac{1}{\sqrt{2}}$$

$$\Rightarrow \dfrac{7-C}{5} = \dfrac{e^{\pi/8}}{10\sqrt{2}}(-2C + 4 + C + 3)$$

$$\Rightarrow \left(\dfrac{7-C}{5}\right) \cdot \left(1 - \dfrac{e^{\pi/8}}{2\sqrt{2}}\right) = 0$$

$$\Rightarrow C = 7$$

Put this value of C in (2), we get

$$y = 1 + e^t(\sin 2t - \cos 2t).$$

Exercise-8.52

1. Solve $\dfrac{dy}{dt} + y = 1$ if $y = 2$ when $t = 0$.

2. Show that the general solution of the equation $(D^2 + k^2)y = 0$ is $y = C_1 \cos kt + C_2 \sin kt$.

3. Solve $y''(t) + y(t) = t$ if $y'(0) = 1,\ y(\pi) = 0$.

4. Solve $(D^2 - 1)y = a\cosh nt$ if $y = Dy = 0$, when $t = 0$.

5. Solve $(D^2 + m^2)x = a\cos nt,\ t \neq 0$, where x, Dx equal to and x_1, when $t = 0, n \neq m$.

6. Solve $(D^2 + m^2)y = a\cos nt, t > 0$ if $y = 0 = Dy$, when $t = 0$.

7. Solve $(D^2 + m^2)x = a\sin nt, t > 0$, where x, Dx equal to x_0 and x_1, when $t = 0, n \neq m$.

8. Solve $(D+2)^2 y = 4e^{-2t}$, $y(0) = -1$ and $y'(0) = 4$.

9. Solve $(D^2 + 6D + 9)y = \sin x$, where $y(0) = 1,\ y'(0) = 0$.

10. Solve $(D^2 + 4D + 4)x = \sin \omega t,\ t > 0$, where x_0 and x_1 are the values of x and Dx, when $t = 0$.

11. Solve $(D^2 + 3D + 2)y = 0$, where $y = y_0$ and $Dy = y_1$ at $t = 0$.

12. Solve $(D^2 + 9)y = 18t$, if $y(0) = 0$, $y\left(\dfrac{\pi}{2}\right) = 0$.

13. Solve $(D^2 + 2D + 1)y = 3te^{-t},\ t > 0$ subject to the conditions $y = 4, Dy = 2$, when $t = 0$.

14. Solve $(D^2 + 1)y = \sin t \sin 2t, t > 0$ if $y = 1, Dy = 0$, when $t = 0$. [UPTU–2001(SP.), 02]

15. Solve $(D^2 + n^2)y = a\sin(nt + \alpha)$ if $y = Dy = 0$, when $t = 0$. (GBTU(CO)–2010)

16. Solve $(D^3 + 1)y = 1, t > 0$ if $y = Dy = D^2y = 0$, when $t = 0$.

17. Solve $(D^3 - D)y = 2\cos t$, $y = 3, Dy = 2, D^2y = 1$, when $t = 0$.

18. Solve $(D^3 + D)y = e^{2t}$, $y(0) = y'(0) - y''(0) = 0$.

19. Solve $(D^4 - 1)y = 1$ if $y = Dy = D^2y = D^3y = 0$ at $t = 0$.

20. Solve $(D^4 + 2D^2 + 1)y = 0$ if $y(0) = 0$, $y'(0) = 1$, $y''(0) = 2$ and $y'''(0) = -3$.

21. Solve $(D^2 + D)y = t^2 + 2t$ if $y(0) = 4$, $y'(0) = -2$.

22. Solve $\dfrac{d^3y}{dt^3} - 3\dfrac{d^2y}{dt^2} + 3\dfrac{dy}{dt} - y = t^2e^t$ where $y(0) = 1$,

$y(0) = 1, \left(\dfrac{dy}{dt}\right)_{t=0} = 0, \left(\dfrac{d^2y}{dt^2}\right)_{t=0} = -2$

(UPTU (SUM)–2008, SVTU–2009)

23. Voltage Ee^{-at} is applied at $t = 0$ to a circuit of inductance L and resistance R. Show that the current at time t is $\dfrac{E}{R - aL}(e^{-at} - e^{-Rt/L})$ (UPTU (SUM)–2007, VTU–2000)

24. Solve $y'' + 4y' + 3y = e^{-t}, y(0) = y'(0) = 1$
(VTU–2008 S, KURUKSHETRA–2005)

25. Solve $y'' + y = t, y(0) = 1, y'(0) = 0$ (MUMBAI–2009)

26. Solve $y'' - 3y' + 2y = e^{3t}$ when $y(0) = 1$ and $y'(0) = 0$
(VTU–2010)

27. Solve $(D^2 - 3D + 2)y = 4e^{2t}$ with $y(0) = -3, y(0) = 5$
(MUMBAI–2008)

28. Solve $y'' + 25y = 10\cos 5t$ given that $y(0) = 2, y''(0) = 0$
(SVTU–2008)

29. Solve $\dfrac{d^2y}{dt^2} + 2\dfrac{dy}{dt} - 3y = \sin t, y = \dfrac{dy}{dt} = 0$ when $t = 0$
(KURUKSHETRA–2005, MADRAS–2003)

30. Solve $y'' + 2y' + 5y = 5(t - 2), y(0) = 0$, $y'(0) = 0$
(PTU–2005 S)

31. Solve $\dfrac{d^2x}{dt^2} + 9x = \sin 2t, x(0) = 1, x'(0) = 0$ (GBTU (CO)–2011)

32. Solve $\dfrac{d^2y}{dt^2} + 9x = \sin 3t$, given $y = 0, \dfrac{dy}{dt} = 0$ at $t = 0$
(MTU–2012)

33. Solve $\dfrac{d^2x}{dt^2} + 6\dfrac{dx}{dt} + 8x = e^{-3t} - e^{-5t}$; $x(0) = 0, x'(0) = 0$
(UPTU(CO)–2009)

34. Solve $y'' + 3y' + 2y = te^{-t}$; $y(0) = 1, y'(0) = 0$ (GBTU–2012)

35. Solve $y'' + 2y' + y = te^{-t}$; $y(0) = 1, y'(0) = 2$ (MTU (SUM)–2011)

36. Solve $\dfrac{d^2x}{dt^2} + 3\dfrac{dx}{dt} + 2x = r(t)$ where $r(t) = \begin{cases} e^t & , \ 0 < t < 2 \\ 0 & , \ t > 2 \end{cases}$

and $x(0) = 1, x'(0) = -2$ (GBTU(CO)–2010)

37. Solve $\dfrac{d^2y}{dt^2} + 9y = r(t)$ with intial conditions $y(0) = 0$ and

$y'(0) = 4$ where $r(t) = \begin{cases} 8\sin t & 0 < t < \pi \\ 0 & t > \pi \end{cases}$ (GBTU–2011)

38. A particle moves in a line so that its displacement x from a fixed point O at any time t, is given by

$$\frac{d^2x}{dt^2} + 4\frac{dx}{dt} + 5x = 80\sin 5t$$

Using Laplace transform, find its displacement at any time t if initially particle is at rest at $x = 0$.   (UPTU (CO)–2009)

39. An alternating *e.m.f* $E\sin \omega t$ is applied to circuit with an inductance L and a capacitance C in series. Show that the current in the circuit is

$\dfrac{E\omega}{(n^2 - \omega^2)L}(\cos \omega t - \cos nt)$ where $n^2 = \dfrac{1}{LC}$.   (GBTU–2010)

Hints to Selected Problems

1. Taking the Laplace transform of the given equation, we get

$L(y') + L(y) = L(1)$

$\Rightarrow \qquad pL\{y\} - y\{0\} + L\{y\} = \dfrac{1}{p}$

$\Rightarrow \qquad L\{y\} = \dfrac{2p + 1}{p(p + 1)} = \dfrac{1}{p + 1} + \dfrac{1}{p}$.

Now taking inverse Laplace transform.

3. $L\{y''\} + L\{y\} = L\{t\}$

$\Rightarrow p^2L\{y\} - py(0) - y'(0) + L\{y\} = \dfrac{1}{p^2}$

$\Rightarrow \qquad (p^2 + 1)L\{y\} - pA - 1 = \dfrac{1}{p^2}$. $[\because A = y(0), \ y'(0) = 1]$

$\Rightarrow \qquad L\{y\} = \dfrac{pA}{p^2 + 1} + \dfrac{1}{p^2}$

Now taking inverse Laplace transform.

4. Taking Laplace transform of the given equation and after simplification, we get

$L\{y\} = \dfrac{ap}{(p^2 - 1)(p^2 - n^2)}$

$= \dfrac{ap}{(n^2 - 1)}\left[\dfrac{1}{p^2 - n^2} - \dfrac{1}{p^2 - 1}\right] = \dfrac{1}{(n^2 - 1)}\left[\dfrac{p}{p^2 - n^2} - \dfrac{p}{p^2 - 1}\right]$

Now taking inverse Laplace transform of both the sides.

7. Proceeding as usual, we get

$L\{x\} = \dfrac{px_0}{(p^2 + m^2)} + \dfrac{x_1}{(p^2 + m^2)}$

$+ \dfrac{a_n}{(p^2 + m^2)(p^2 + n^2)} = \dfrac{px_0}{(p^2 + m^2)}$

$+ \dfrac{x_1}{(p^2 + m^2)} + \dfrac{a}{(m^2 - n^2)}\left[\dfrac{n}{p^2 + n^2} - \dfrac{n}{p^2 + m^2}\right]$

Now taking inverse Laplace transform of both the sides.

9. We have $L\{y\} = \dfrac{(p + 6)}{(p + 3)^2} + \dfrac{1}{(p^2 + 1)(p + 3)^2}$

$= \dfrac{1}{(p + 3)} + \dfrac{3}{(p + 3)^2} +$

$\left[\dfrac{3}{50(p + 3)} + \dfrac{1}{10(p + 3)^2} - \dfrac{3p - 4}{50(p^2 + 1)}\right]$

$= \dfrac{1}{50}\left[\dfrac{53}{p + 3} + \dfrac{155}{(p + 3)^2} - \dfrac{3p}{(p^2 + 1)} + \dfrac{4}{(p^2 + 1)}\right]$

Now taking inverse Laplace transform of both the sides.

11. Proceeding as usual, we get

$L\{y\} = \dfrac{p + 3}{(p^2 + 3p + 2)}y_0 + \dfrac{y_1}{(p^2 + 3p + 2)}$

$= \dfrac{(p + 3)}{(p + 1)(p + 2)}y_0 + \dfrac{y_1}{(p + 1)(p + 2)}$

$= \left[\dfrac{2}{p + 1} - \dfrac{1}{p + 2}\right]y_0 + \left[\dfrac{1}{p + 1} - \dfrac{1}{p + 2}\right]y_1$

$= \dfrac{(2y_0 + y_1)}{(p + 1)} - \dfrac{(y_0 + y_1)}{(p + 2)}$

Now taking inverse Laplace transform of both the sides.

14. $(D^2+1)y = \sin t \sin 2t = \dfrac{1}{2}[\cos t - \cos 3t]$

$\Rightarrow \qquad L\{y''\} + L\{y\} = \dfrac{1}{2}\big[L\{\cos t\} - L\{\cos 3t\}\big]$

After simplification, we get

$L\{y\} = \dfrac{p}{p^2+1} + \dfrac{p}{2(p^2+1)^2} - \dfrac{p}{16}\left[\dfrac{(p^2+9)-(p^2+1)}{(p^2+9)(p^2+1)}\right]$

$= \dfrac{p}{p^2+1} - \dfrac{1}{4}\left[\dfrac{d}{dp}\left(\dfrac{1}{p^2+1}\right)\right] - \dfrac{p}{16(p^2+1)} + \dfrac{p}{16(p^2+9)}$

Now taking inverse Laplace transform of both the sides.

15. The given equation can be written as

$y'' + n^2 y = a[\sin nt \cos\alpha + \cos nt \sin\alpha]$

Taking Laplace transform and simplifying, we get

$L\{y\} = a\cos\alpha.\dfrac{n}{(p^2+n^2)^2} + a\sin\alpha\dfrac{p}{(p^2+n^2)^2}$

$\Rightarrow y = a.n\cos\alpha L^{-1}\left\{\dfrac{1}{(p^2+n^2)^2}\right\} + a\sin\alpha L^{-1}\left\{\dfrac{p}{(p^2+n^2)^2}\right\}$

$= a.n\cos\alpha\int_0^t\left(\dfrac{1}{n}\sin nx\right)\dfrac{1}{n}\sin n(t-x)dx$

$\qquad\qquad - \dfrac{a\sin\alpha}{2}L^{-1}\left\{\dfrac{d}{dp}\dfrac{1}{(p^2+n^2)^2}\right\}$

(Using convolution theorem)

$= a\dfrac{\cos\alpha}{2n}\int_0^t[\cos n(t-2x) - \cos nt]dx + \dfrac{a\sin\alpha}{2}t.L^{-1}\left\{\dfrac{1}{p^2+n^2}\right\}$

16. Proceeding as usual, we get

$L\{y\} = \dfrac{1}{p} - \dfrac{1}{3(p+1)} - \dfrac{2\left(p-\dfrac{1}{2}\right)}{3\left[\left(p-\dfrac{1}{2}\right)^2 + \dfrac{3}{4}\right]}$

$\Rightarrow y = 1 - \dfrac{e^{-t}}{3} - \dfrac{2}{3}e^{t/2}L^{-1}\left\{\dfrac{p}{p^2+\left(\sqrt{3}/2\right)^2}\right\}$

Answers

1. $y = e^{-t} + 1$ **3.** $y = \pi\cos t + t$ **4.** $y = \dfrac{a}{n^2-1}(\cosh nt - \cosh t)$ **5.** $x = x_0\cos mt + \dfrac{x_1}{m}\sin mt + \dfrac{a}{m^2-n^2}(\cos nt - \cos mt)$

6. $y = \dfrac{a}{m^2-n^2}(\cos nt - \cos mt)$ **7.** $x = x_0\cos mt + \dfrac{x_1}{m}\sin mt + \dfrac{a}{m^2-n^2}\left(\sin nt - \dfrac{n}{m}\sin mt\right)$ **8.** $y = e^{-2t}(2t^2+2t-1)$

9. $\dfrac{1}{50}[(53+155x)e^{-3x} - (3\cos x - 4\sin x)]$

10. $x = e^{-2t}\left[x_0(1-2t) + (x_1+4x_0) + \dfrac{w}{(4+w^2)}t + \dfrac{4w}{(4+w^2)^2}\right] - \dfrac{4w}{(4+w^2)^2}\cos wt + \dfrac{(4-w^2)}{(4+w^2)^2}\sin wt$

11. $y = (2y_0+y_1)e^{-t} - (y_0+y_1)e^{-2t}$ **12.** $y = \pi\sin 3t + 2t$ **13.** $y = \dfrac{1}{2}e^{-t}.t^3 + 4e^{-t} + 6te^{-t}$

14. $y = \dfrac{15}{16}\cos t + \dfrac{1}{4}t\sin t + \dfrac{1}{16}\cos 3t$ **15.** $y = \dfrac{a}{2n^2}[\cos\alpha\sin nt - nt\cos(\alpha+nt)]$ **16.** $y = 1 - \dfrac{1}{3}e^{-t} - \dfrac{2}{3}e^{t/2}\cos\left(\dfrac{\sqrt{3}t}{2}\right)$

17. $y = 3\sinh t - \sin t + \cosh t + 2$ **18.** $y = \dfrac{1}{3}e^{-t}(\sin 2t + \sin t)$ **19.** $y = -1 + \dfrac{1}{2}\cosh t + \dfrac{1}{2}\cos t$

20. $y = t(\sin t + \cos t)$ **21.** $y = \dfrac{1}{3}t^3 + 2 + 2e^{-t}$ **22.** $y = \left(1 - t - \dfrac{t^2}{2} + \dfrac{t^5}{60}\right)e^t$

24. $y = \dfrac{7}{4}e^{-t} - \dfrac{3}{4}e^{-3t} - \dfrac{1}{2}te^{-t}$ **25.** $y = t - 3\sin t + \cos t$ **26.** $y = 2t + 3 + \dfrac{1}{2}(e^{3t} - e^t) - 2e^{2t}$

27. $y = 4e^{2t}(1+t) - 7e^t$ **28.** $y = 2\cos 5t + t\sin 5t$ **29.** $y = \dfrac{1}{8}e^t - \dfrac{1}{40}e^{-3t} - \dfrac{1}{10}(2\sin t + \cos t)$

30. $y = \dfrac{-12}{5} + \dfrac{12}{5}e^{-t}\cos 2t + \dfrac{7}{10}e^{-t}\sin 2t$ **31.** $x = \cos 3t + \dfrac{1}{5}\sin 2t - \dfrac{2}{15}\sin 3t$ **32.** $y = \dfrac{1}{18}(\sin 3t - 3t\cos 3t)$

33. $x = \dfrac{1}{3}(e^{-2t} - e^{-5t}) - e^{-3t} + e^{-4t}$ **34.** $y = 3e^{-t} - 2e^{-2t} + e^{-t}\left(\dfrac{t^2}{2} - t\right)$

35. $y = e^{-t}\left(1 + 3t + \dfrac{t^3}{6}\right)$ **36.** $x = \dfrac{4}{3}e^{-2t} + \dfrac{1}{6}e^t[1 - u(t-2)] - \dfrac{1}{2}e^{-t} + \dfrac{1}{2}e^{4-t}u(t-2) - \dfrac{1}{3}e^{6-2t}u(t-2)$

37. $y = \sin 3t + \sin t + [\sin(t-\pi) - \dfrac{1}{3}\sin 3(t-\pi)]u(t-\pi)$ **38.** $x = e^{-2t}(2\cos t + 14\sin t) - 2\cos 5t - 2\sin 5t$

8.11.2 Solution of Ordinary Differential Equation with Variable Coefficients by Laplace transforms

The Laplace transform can also be used in solving some ordinary differential equations in which the coefficients are variable. A particular differential equation when the method proves useful is one in which the terms have the form $t^m y^n(t)$ whose Laplace transform is

$$(-1)^m \dfrac{d^m}{dp^m}[L\{y^n(t)\}]$$

Solved Examples

EXAMPLE 1. **Solve** $(tD^2 + D + 4t)y = 0$ **if** $y(0) = 3$, $y'(0) = 0$.

SOLUTION. The given equation can be written as
$$ty'' + y' + 4ty = 0 \qquad \ldots (1)$$
Taking the Laplace transform of both sides of (1), we get

$$L\{ty''\} + L\{y'\} + 4L\{ty\} = 0$$

$$\Rightarrow \quad -\frac{d}{dp}L\{y''\} + L\{y'\} + 4(-1)\frac{d}{dp}L\{y\} = 0$$

$$\Rightarrow \quad -\frac{d}{dp}[p^2 L\{y\} - py(0) - y'(0)] + [pL\{y\}$$
$$-y(0)] - 4\frac{d}{dp}L\{y\} = 0$$

$$\Rightarrow \quad -\frac{d}{dp}[p^2 L\{y\} - 3p] + (pL\{y\} - 3) - \frac{4d[L\{y\}]}{dp} = 0$$

$$\Rightarrow \quad -(p^2 + 4)\frac{d[L\{y\}]}{dp} - pL\{y\} = 0$$

$$\Rightarrow \quad \frac{d[L\{y\}]}{L\{y\}} + \frac{p}{p^2 + 4}dp = 0$$

On integrating, we get
$$\log[L\{y\}] + \frac{1}{2}\log(p^2 + 4) = \log C_1$$

$$\Rightarrow \qquad L\{y\} = \frac{C_1}{\sqrt{p^2 + 4}}$$

Therefore, $y = L^{-1}\left\{\dfrac{C_1}{\sqrt{p^2 + 4}}\right\} = C_1 J_0(2t)$.

EXAMPLE 2. **Solve** $[tD^2 + (t-1)D - 1]y = 0$ **if** $y(0) = 5$, $y(\infty) = 0$.

SOLUTION. The given equation can be written as
$$ty'' + ty' - y' - y = 0 \qquad \ldots (1)$$
Taking the Laplace transforms of both sides of (1), we get

$$L\{ty''\} + L\{ty'\} - L\{y'\} - L\{y\} = 0$$

$$\Rightarrow \quad -\frac{d}{dp}[L\{y''\}] - \frac{d}{dp}[L\{y'\} - [pL\{y\}$$
$$-y(0) - L\{y\}] = 0$$

$$\Rightarrow \quad -\frac{d}{dp}[p^2 L\{y\} - py(0) - y'(0)] - \frac{d}{dp}[pL\{y\}$$
$$-y(0)] - pL\{y\} + 5 - L\{y\} = 0$$

$$\Rightarrow \quad -\frac{d}{dp}[p^2 L\{y\} - 5p - A] - \frac{d}{dp}[pL\{y\} - 5]$$
$$-(p+1)L\{y\} + 5 = 0 \text{, where } A = y'\{0\}$$

$$\Rightarrow \quad \frac{d[L\{y\}]}{dp} + \frac{3p+2}{p^2 + p}L\{y\} = \frac{10}{p^2 + p} \qquad \ldots (2)$$

which is a linear differential equation in $L\{y\}$.
Therefore,

$$\text{I.F.} = e^{\int\left\{\frac{3p+2}{(p^2+p)}\right\}dp} = e^{\int\left(\frac{2}{p}+\frac{1}{p+1}\right)dp}$$
$$= e^{[2\log p + \log(p+1)]} = p^2(p+1).$$

Hence, the solution of equation (2) is given by

$$L\{y\}.p^2(p+1) = C_1 + \int \frac{10}{p^2 + p}.p^2(p+1)dp$$
$$= C_1 + 10\int p\, dp = C_1 + 5p^2$$

$$\Rightarrow \qquad L\{y\} = \frac{C_1}{p^2(p+1)} + \frac{5}{p+1}$$

$$= C_1\left\{\frac{1}{p^2} - \frac{1}{p} + \frac{1}{p+1}\right\} + \frac{5}{p+1}$$

Therefore, $y = C_1 L^{-1}\left\{\dfrac{1}{p^2} - \dfrac{1}{p} + \dfrac{1}{p+1}\right\}$.

$$+ 5L^{-1}\left\{\frac{1}{p+1}\right\} = C_1(t - 1 + e^{-t}) + 5e^{-t}.$$

Now, using the given conditions $y(\infty) = 0$.
We must have $C_1 = 0$. Hence, $y = 5e^{-t}$ is the required solution.

EXAMPLE 3. **Solve** $\dfrac{d^2 y}{dx^2} + 2\dfrac{dy}{dx} + 5y = e^{-x}\sin x$,

where $y(0) = 0$, $y'(0) = 1$.

(UPTU–2004, 08, (SUM)–2009, PTU–2010, MTU–2011)

SOLUTION. $\dfrac{d^2 y}{dx^2} + 2\dfrac{dy}{dx} + 5y = e^{-x}\sin x$

Taking the Laplace transform on both the sides, we get

$$[p^2 L\{y\} - py(0) - y'(0)] + 2[pL\{y\}$$
$$-y(0)] + 5L\{y\} = L\{e^{-x}\sin x\}$$

$$[p^2 L\{y\} - py(0) - y'(0)] + 2[pL\{y\}$$
$$-y(0)] + 5L\{y\} = \frac{1}{(p+1)^2 + 1} \qquad \ldots (1)$$

On substituting the values of $y(0)$ and $y'(0)$ in (1), we get

$$(p^2 L\{y\} - 1) + 2pL\{y\} + 5L\{y\} = \frac{1}{p^2 + 2p + 2}$$

$$(p^2 + 2p + 5)L\{y\} = 1 + \frac{1}{p^2 + 2p + 2} = \frac{p^2 + 2p + 3}{p^2 + 2p + 2}$$

$$L\{y\} = \frac{p^2 + 2p + 3}{(p^2 + 2p + 5)(p^2 + 2p + 2)}$$

On resolving R.H.S. into partial fractions, we get

$$L\{y\} = \frac{2}{3}.\frac{1}{p^2 + 2p + 5} + \frac{1}{3}.\frac{1}{p^2 + 2p + 2}$$

On inversion, we obtain

$$y = \frac{2}{3}L^{-1}\frac{1}{p^2 + 2p + 5} + \frac{1}{3}L^{-1}\frac{1}{p^2 + 2p + 2}$$

$$y = \frac{1}{3}L^{-1}\frac{2}{(p+1)^2 + (2)^2} + \frac{1}{3}L^{-1}\frac{1}{(p+1)^2 + (1)^2}$$

$$\Rightarrow \quad y = \frac{1}{3}e^{-x}\sin 2x + \frac{1}{3}e^{-x}\sin x$$

$$y = \frac{1}{3}.e^{-x}(\sin x + \sin 2x)$$

EXAMPLE 4. **Solve** $(D^2 + 1)y = t\cos 2t$ **subject to the condition** $y = 0$, $\dfrac{dy}{dt} = 0$ **when** $t = 0$.

[UPTU–2005, UKTU–2012, RAIPUR–2005]

SOLUTION. The given equation can be written as
$$y'' + y = t\cos 2t \qquad \ldots (1)$$
Taking the Laplace transform of both sides of (1), we get
$$L\{y''\} + L\{y\} = L\{t\cos 2t\}$$

$\Rightarrow \quad p^2 L\{y\} - py(0) - y'(0) + L\{y\}$
$$= -\frac{d}{dp}[L\{\cos 2t\}]$$

$\Rightarrow (p^2 + 1)L\{y\} = -\frac{d}{dp}\left(\frac{p}{p^2 + 4}\right)$

$$= -\frac{1}{p^2 + 4} + \frac{2p^2}{(p^2 + 4)^2}$$

$\therefore \qquad L\{y\} = \frac{p^2 - 4}{(p^2 + 1)(p^2 + 4)^2}$

$$= -\frac{5}{9(p^2 + 1)} + \frac{5}{9(p^2 + 4)} + \frac{8}{3(p^2 + 4)^2}$$

[Resolving into partial fractions]

$\Rightarrow \quad y = -\frac{5}{9}L^{-1}\left\{\frac{1}{p^2 + 1}\right\} + \frac{5}{9}L^{-1}\left\{\frac{1}{p^2 + 4}\right\}$

$$+ \frac{8}{3}L^{-1}\left\{\frac{1}{(p^2 + 4)^2}\right\}$$

$$= -\frac{5}{9}\sin t + \frac{5}{18}\sin 2t$$

$$+ \frac{8}{3}\int_0^t \frac{1}{2}\sin 2x \cdot \frac{1}{2}\sin 2(t - x)dx$$

[By convolution theorem and using

$$L^{-1}\left\{\frac{1}{p^2 + 4}\right\} = \frac{1}{2}\sin 2t\ ]$$

$$= -\frac{5}{9}\sin t + \frac{5}{18}\sin 2t$$

$$+ \frac{1}{3}\int_0^t \{\cos(2t - 4x) - \cos 2t\}dx$$

$$= -\frac{5}{9}\sin t + \frac{5}{18}\sin 2t$$

$$+ \frac{1}{3}\left[-\frac{1}{4}\sin(2t - 4x) - x\cos 2t\right]_0^t$$

$$= -\frac{5}{9}\sin t + \frac{5}{18}\sin 2t + \frac{1}{12}\sin 2t$$

$$-\frac{1}{3}t\cos 2t + \frac{1}{12}\sin 2t$$

$$= -\frac{5}{9}\sin t + \frac{4}{9}\sin 2t - \frac{1}{3}t\cos 2t$$

EXAMPLE 5. *Solve* $[tD^2 + (1 - 2t)D - 2]y = 0$,

where $y(0) = 1, y'(0) = 2.$ [UPTU–2002, PTU–2002]

SOLUTION. Here, $tD^2 y + (1 - 2t)Dy - 2y = 0$

$\Rightarrow \qquad ty'' + y' - 2ty' - 2y = 0$

Taking Laplace transform of given differential equation, we get

$L\{ty''\} + L\{y'\} - 2L\{ty'\} - 2L\{y\} = 0$

$\Rightarrow \quad -\frac{d}{dp}L\{y''\} + L\{y'\} + 2\frac{d}{dp}L\{y'\} - 2L\{y\} = 0$

$$-\frac{d}{dp}\left[p^2 L\{y\} - py(0) - y'(0)\right] + \left[pL\{y\}\right.$$

$$\left. - y(0)\right] + 2\left[pL\{y\} - y(0)\right] - 2L\{y\} = 0$$

Putting the values of (0) and $y'(0)$, we get

$-\frac{d}{dp}(p^2 L\{y\} - p - 2) + (pL\{y\} - 1)$

$$+ 2\frac{d}{dp}(pL\{y\} - 1) - 2L\{y\} = 0$$

$$[\because y(0) = 1, y'(0) = 2]$$

$\Rightarrow \quad -p^2 \frac{dL\{y\}}{dp} - 2pL\{y\} + 1 + pL\{y\} - 1$

$$+ 2\left(p\frac{dL\{y\}}{dp} + L\{y\}\right) - 2L\{y\} = 0$$

$\Rightarrow \quad -(p^2 - 2p)\frac{dL\{y\}}{dp} - pL\{y\} = 0$

$\Rightarrow \quad -\frac{dL\{y\}}{\bar{y}} - \frac{1}{p - 2}dp = 0$

[Separating the variables]

$\Rightarrow \qquad \int \frac{dL\{y\}}{\bar{y}} + \int \frac{dp}{p - 2} = 0$

$\Rightarrow \qquad \log L\{y\} + \log(p - 2) = \log C$

$\Rightarrow \qquad \log L\{y\}(p - 2) = \log C$

$\Rightarrow \qquad L\{y\}(p - 2) = C$

$\Rightarrow \qquad L\{y\} = \frac{C}{p - 2}$

$\Rightarrow \quad y = CL^{-1}\left\{\frac{1}{p - 2}\right\} \Rightarrow y = Ce^{2t} \qquad ...(1)$

At $\qquad x = 0, y(0) = Ce^0 \qquad ...(2)$

Putting $y(0) = 1$, in (2), we get

$$1 = Ce^0 \Rightarrow C = 1$$

Putting $C = 1$ in (1), we get $y = e^{2t}$. This is the required solution.

EXAMPLE 6. *Using Laplace transform, solve the following differential equation* $y'' + 2ty' - y = t$ *where,* $y(0) = 0$ *and* $y'(0) = 1.$ [UPTU–2003]

SOLUTION. We have $\qquad y'' + 2ty' - y = t \qquad ... (1)$

Taking Laplace transform of (1), we get

$[p^2 L\{y\} - py(0) - y'(0)]$

$$-2\frac{d}{dp}[pL\{y\} - y(0)] - L\{y\} = \frac{1}{p^2} \qquad ... (2)$$

On putting $y(0) = 0$ and $y'(0) = 1$ in (2), we get

$(p^2 L\{y\} - 1) - 2\frac{d}{dp}(pL\{y\} - 0) - L\{y\} = \frac{1}{p^2}$

$\Rightarrow (p^2 L\{y\} - 1) - 2L\{y\} - 2p\frac{dL\{y\}}{dp} - L\{y\} = \frac{1}{p^2}$

$\Rightarrow -2p\frac{dL\{y\}}{dp} + (p^2 - 3)L\{y\} = \frac{1}{p^2} + 1 = \frac{1 + p^2}{p^2}$

$\Rightarrow \frac{dL\{y\}}{dp} - \frac{p^2 - 3}{2p}L\{y\} = \frac{1 + p^2}{-2p^3}$

$\Rightarrow \frac{dL\{y\}}{dp} - \left(\frac{p}{2} - \frac{3}{2p}\right)L\{y\} = -\frac{1}{2p^3} - \frac{1}{2p} \qquad ... (3)$

Thus, (3) is a linear differential equation

I.F. $= e^{\frac{1}{2}\int\left(\frac{3}{p} - p\right)dp} = e^{\frac{1}{2}\left(3\log p - \frac{p^2}{2}\right)} = e^{\frac{p^2}{4}} \cdot p^{3/2}$

Solution of differential equation (3) is

$L\{y\} e^{-p^2/4} \cdot p^{3/2} = \frac{1}{2}\int \left(\frac{1}{p^3} + \frac{1}{p}\right) p^{3/2} \cdot e^{-p^2/4}dp$

$$= -\frac{1}{2}\int \left(\sqrt{p} + \frac{1}{p^{3/2}}\right)e^{-p^2/4}dp$$

Put $p^2 = ut \Rightarrow p = 2\sqrt{t}$ so that $dp = \frac{dt}{\sqrt{t}}$. Then we have

$L\{y\}p^{3/2}e^{-p^2/4} = -\frac{1}{2}\int \left(\sqrt{2}\, t^{1/4} + \frac{1}{2\sqrt{2}}t^{-3/4}\right)e^{-t}\frac{dt}{\sqrt{t}}$

$$= -\frac{1}{\sqrt{2}}\int \left(t^{-1/4} + \frac{1}{4}t^{-5/4}\right)e^{-t}dt$$

$$= -\frac{1}{\sqrt{2}} \int t^{-1/4} e^{-t} dt - \frac{1}{4\sqrt{2}} \int t^{-5/4} e^{-t} dt$$

$$= -\frac{1}{\sqrt{2}} \left[t^{-1/4} \frac{e^{-t}}{-1} + \int \left(-\frac{1}{4} \right) t^{-5/4} e^{-t} dt \right]$$

$$+ \frac{1}{4\sqrt{2}} \int t^{-5/4} e^{-t} dt$$

$$= \frac{1}{\sqrt{2}} e^{-t} . t^{-1/4} = \frac{1}{\sqrt{2}} e^{-p^2/4} \left(\frac{p^2}{4} \right)^{-1/4}$$

$$= \frac{1}{\sqrt{p}} e^{-p^2/4}$$

$$\Rightarrow L\{y\} = \frac{1}{p^2}$$

$$\Rightarrow L\{y\} = \frac{1}{p^2} + C \Rightarrow y = L^{-1} \left\{ \frac{1}{p^2} + C \right\} = t + C.$$

8.11.3 SOLUTION OF SIMULTANEOUS ORDINARY DIFFERENTIAL EQUATIONS BY LAPLACE TRANSFORMS

The Laplace transform can be used to solve two or more simultaneous ordinary differential equations. The procedure is essentially the same as that described in previous sections.

 Solved Examples

EXAMPLE 1. *Solve the simultaneous equation* $\dfrac{dx}{dt} - y = e^t$,

$\dfrac{dy}{dt} + x = \sin t$, *given* $x(0) = 1$, $y(0) = 0$.

[UPTU–2006, Q.BANK–2001, GBTU (SUM)–2010,
UKTU–2011, DELHI–2002]

SOLUTION. Taking Laplace transforms of the given equations, we

get $[p\bar{x} - x(0)] - \bar{y} = \dfrac{1}{p-1}$, where $\bar{x} = L(x)$, $\bar{y} = L(y)$

i.e., $p\bar{x} - 1 - \bar{y} = \dfrac{1}{p-1}$ $[\because x(0) = 1]$

$p\bar{x} - \bar{y} = \dfrac{p}{p-1}$ and $[p\bar{y} - y(0)] + \bar{x} = \dfrac{1}{p^2+1}$

i.e., $\bar{x} + p\bar{y} = \dfrac{1}{p^2+1}$ $[\because y(0) = 0]$ ... (2)

Solving (1) and (2) for $\bar{x}$ and $\bar{y}$, we have

$$\bar{x} = \frac{p^2}{(p-1)(p^2+1)} + \frac{1}{(p^2+1)^2}$$

$$= \frac{1}{2} \left[\frac{1}{p-1} + \frac{p}{p^2+1} + \frac{1}{p^2+1} \right] + \frac{1}{(p^2+1)^2}$$

$$\bar{y} = \frac{p}{(p^2+1)^2} - \frac{p}{(p-1)(p^2+1)}$$

$$= \frac{p}{(p^2+1)^2} - \frac{1}{2} \left[\frac{1}{p-1} - \frac{p}{p^2+1} + \frac{1}{p^2+1} \right]$$

Taking inverse Laplace transform of both sides, we get

$$x = \frac{1}{2} L^{-1} \left\{ \frac{1}{p-1} + \frac{p}{p^2+1} + \frac{1}{p^2+1} \right\} + L^{-1} \left\{ \frac{1}{(p^2+1)^2} \right\}$$

$$= \frac{1}{2} \left[e^t + \cos t + \sin t \right] + \frac{1}{2} (\sin t - t \cos t)$$

$$= \frac{1}{2} \left[e^t + \cos t + 2\sin t - t \cos t \right]$$

$$y = L^{-1} \left\{ \frac{p}{(p^2+1)^2} \right\} - \frac{1}{2} L^{-1} \left\{ \frac{1}{p-1} - \frac{p}{p^2+1} + \frac{1}{p^2+1} \right\}$$

$$= \frac{1}{2} t \sin t - \frac{1}{2} \left[e^t - \cos t + \sin t \right]$$

$$= \frac{1}{2} \left[t \sin t - e^t + \cos t - \sin t \right]$$

Hence, $x = \dfrac{1}{2} (e^t + \cos t + 2\sin t - t \cos t)$

$y = \dfrac{1}{2} (t \sin t - e^t + \cos t - \sin t)$

EXAMPLE 2. *Using Laplace transformation, solve*

$(D-2)x - (D+1)y = 6e^{3t}$

$(2D-3)x + (D-3)y = 6e^{3t}$

Given $x = 3$, $y = 0$ *when* $t = 0$.   [UPTU–2001]

SOLUTION. Taking Laplace transformation of the given equations, we get

$$LDx - 2Lx - LDy - Ly = 6Le^{3t} \Big\}$$
$$2LDx - 3Lx + LDy - 3Ly = 6Le^{3t}$$

$$\Rightarrow \begin{array}{l} p\bar{x} - x(0) - 2\bar{x} - p\bar{y} + y(0) - \bar{y} = 6\dfrac{1}{p-3} \\[2mm] 2p\bar{x} - 2x(0) - 3\bar{x} + p\bar{y} - y(0) - 3\bar{y} = \dfrac{6}{p-3} \end{array} \Bigg\},$$

where $\bar{x} = L(x)$

$$\Rightarrow \begin{array}{l} (p-2)\bar{x} - (p+1)\bar{y} - 3 = \dfrac{6}{p-3} \\[2mm] (2p-3)\bar{x} + (p-3)\bar{y} - 6 = \dfrac{6}{p-3} \end{array} \Bigg\}$$

$$\Rightarrow \begin{array}{l} (p-2)\bar{x} - (p+1)\bar{y} = \dfrac{3p-3}{p-3} \\[2mm] (2p-3)\bar{x} + (p-3)\bar{y} = \dfrac{6p-12}{p-3} \end{array} \Bigg\}$$

$$\Rightarrow \begin{array}{l} (p-3)(p-2)\bar{x} - (p-3)(p+1)\bar{y} \\[1mm] \hspace{3cm} = 3p-3 \\[2mm] (p+1)(2p-3)\bar{x} + (p+1)(p-3)\bar{y} \\[1mm] \hspace{3cm} = \dfrac{(p+1)(6p-12)}{p-3} \end{array} \Bigg\}$$

On adding, we get

$$(3p^2 - 6p + 3)\bar{x} = 3(p-1) + \frac{6(p^2 - p - 2)}{p-3}$$

$$\Rightarrow \bar{x} = \frac{3(p-1)}{3(p-1)^2} + \frac{6(p^2 - p - 2)}{3(p-1)^2(p-3)}$$

$$x = L^{-1} \left\{ \frac{1}{p-1} + \frac{2}{(p-1)^2} + \frac{2}{p-3} \right\}$$

$$= e^t + 2te^t + 2e^{3t}$$

Putting the value of x in (1), we get

$(D-2)(e^t + 2te^t + 2e^{3t}) - (D+1)y = 6e^{3t}$

$\Rightarrow e^t + 2te^t + 2e^t + 6e^{3t} - 2e^t - 4te^t$

$$-4e^{3t} - (D+1)y = 6e^{3t}$$

$\Rightarrow \quad (D+1)y = e^t - 2te^t - 4e^{3t}$ ...(2)

Taking Laplace transform of (2), we get

$p\bar{y} - y(0) + \bar{y} = \dfrac{1}{p-1} - \dfrac{2}{(p-1)^2} - \dfrac{4}{p-3}$

$\Rightarrow \quad (p+1)\bar{y} = \dfrac{1}{p-1} - \dfrac{2}{(p-1)^2} - \dfrac{y}{p-3}$

$\bar{y} = \dfrac{1}{p^2-1} - \dfrac{2}{(p+1)(p-1)^2} - \dfrac{4}{(p+1)(p-3)}$

$\bar{y} = \dfrac{1}{p^2-1} - \dfrac{1/2}{p+1} + \dfrac{1/2}{p-1} - \dfrac{1}{(p-1)^2} + \dfrac{1}{p+1} - \dfrac{1}{p-3}$

$\bar{y} = \dfrac{1}{p^2-1} + \dfrac{1/2}{p+1} + \dfrac{1/2}{p-1} - \dfrac{1}{(p-1)^2} - \dfrac{1}{p-3}$

$\Rightarrow \quad y = L^{-1}\left\{ \dfrac{1}{p^2-1} + \dfrac{1}{2}\dfrac{1}{p+1} + \dfrac{1}{2}\dfrac{1}{p-1} - \dfrac{1}{p-3} - \dfrac{1}{(p-1)^2} \right\}$

$\Rightarrow \quad y = \sinh t + \dfrac{1}{2}e^{-t} + \dfrac{1}{2}e^{t} - e^{3t} - te^{t}$

$\Rightarrow \quad y = \sinh t + \cosh t - e^{-3t} - te^{t}$

Exercise-8.53

Solve the following simultaneous equations :

1. $3\dfrac{dx}{dt} - y = 2t, \dfrac{dx}{dt} + \dfrac{dy}{dt} - y = 0$ with the conditions

$x(0) = y(0) = 0$ [UPTU (SUM)–2008]

2. $\dfrac{dx}{dt} + \dfrac{dy}{dt} + x + y = 1, \dfrac{dy}{dt} = 2x + y; \; x(0) = 0, y(0) = 1$

[MTU (SUM)–2011]

3. $\dfrac{d^2x}{dt^2} - x = y, \dfrac{d^2y}{dt^2} + y = -x$,

given that $t = 0; x = 2, y = -1, \dfrac{dx}{dt} = 0$ and $\dfrac{dy}{dt} = 0$ (PTU–2009S)

4. $3\dfrac{dx}{dt} + \dfrac{dy}{dt} + 2x = 1, \dfrac{dx}{dt} + 4\dfrac{dy}{dt} + 3y = 0;$

given $x = 0, y = 0$ when $t = 0$.   (MADRAS–2003S)

Answers

1. $x = \dfrac{t^2}{2} + \dfrac{t}{2} - \dfrac{3}{4}e^{2t/3} + \dfrac{3}{4}$ **2.** $x = e^{-t} - 1, y = 2 - e^{-t}$ **3.** $x = 2 + t^2/2, y = -1 - t^2/2$

4. $x = \dfrac{1}{10}(5 - 2e^{-t} - 3e^{-6t/11})$, $y = \dfrac{1}{5}(e^{-t} - e^{-6t/11})$

8.11.4 Application of Laplace Transform to Engineering Problems

EXAMPLE 1. *A mass m moves along the x-axis under the influence of a force which is proportional to its instantaneous speed and in a direction opposite to the direction of motion. Assuming that at $t = 0$, the particle is located at $x = 0$ and moving to the right with speed V_0. Find the position where the mass comes to rest.*

SOLUTION. The motion of the particle is described as below :

P————————————•————————————

Fig. 32

By the Newton's second law of motion, we get the equation of motion of the particle, given by

$$m\dfrac{d^2x}{dt^2} = -\mu\dfrac{dx}{dt} \quad ...(1)$$

with initial conditions $x(0) = a$ and $x'(0) = V_0$.

Taking the Laplace transform of both sides of (1), we get

$$mL\left\{\dfrac{d^2x}{dt^2}\right\} = -\mu L\left\{\dfrac{dx}{dt}\right\}$$

$\Rightarrow \quad m[p^2 L\{x\} - px(0) - x'(0)]$

$\qquad = -\mu[pL\{x\} - x(0)]$

$\Rightarrow \quad (mp + \mu p)L\{x\} = m(ap + V) + a\mu$

$\Rightarrow \quad L\{x\} = \dfrac{m(ap + V_0) + a\mu}{p(mp + \mu)}$

$\qquad = \dfrac{mV_0 + \mu a}{\mu p} - \dfrac{mV_0}{\mu\left(p + \dfrac{\mu}{m}\right)}$

Therefore,

$$x = \left(\dfrac{mV_0}{\mu} + a\right) L^{-1}\left\{\dfrac{1}{p}\right\} - \dfrac{mV_0}{\mu} L^{-1}\left\{\dfrac{1}{p + \dfrac{\mu}{m}}\right\}$$

$$= \left(\dfrac{mV_0}{\mu} + a\right) - \dfrac{mV_0}{\mu}.e^{-\mu t/m}. \quad ...(2)$$

If $\dfrac{dx}{dt} = V_0 e^{\mu t/m} = 0$.

Then, from (2), we have $x = \dfrac{mV_0}{\mu} + a$.

Hence, the mass m comes to rest at a distance $\dfrac{mV_0}{\mu} + a$ from the centre.

EXAMPLE 2. *A particle moves along a line so that its displacement X from a fixed point at any time x is given by*

$$X''(t) + 4X'(t) + 5X(t) = 80\sin 5t.$$

Find its displacement at any time $x > 0$, if at $t = 0$, the particle is at rest at $X = 0$.

SOLUTION. Here, the displacement of the particle is given by the differential equation

$$X''(t) + 4X'(t) + 5X(t) = 80\sin 5t \quad ...(1)$$

where $X(0) = 0$ and $X'(0) = 0$.

Taking Laplace transform of both sides of (1), we get

$L\{X''(t)\} + 4L\{X'(t)\} + 5L\{X(t)\} = 80L\{\sin 5t\}$

$\Rightarrow \quad p^2 L\{X(t)\} - pX(0) - X'(0)$

$\qquad + 4[pL\{X(t)\} - X(0)] + 5L\{X(t)\} = 80 \times \dfrac{5}{p^2 + 25}$

$$\Rightarrow \quad (p^2 + 4p + 5)\, L\{X(t)\} = \frac{400}{p^2 + 25}$$

$$\Rightarrow \quad L\{X(t)\} = \frac{400}{(p^2 + 4p + 5)(p^2 + 25)}$$

$$= \frac{-2(p+5)}{p^2 + 25} + \frac{2(p+9)}{p^2 + 4p + 5}.$$

Therefore, $X(t) = -2L^{-1}\left\{\dfrac{p}{p^2+25} + \dfrac{5}{p^2+25}\right\}$

$$+ 2L^{-1}\left\{\frac{(p+2)+7}{(p+2)^2+1}\right\}$$

$$= -2(\cos 5t + \sin 5t) + 2e^{-2t} L^{-1}\left\{\frac{p+7}{p^2+1}\right\}$$

$$\Rightarrow X(t) = -2(\cos 5t + \sin 5t) + 2e^{-2t}(\cos t + 7\sin t).$$

EXAMPLE 3. *A resistance R in series with inductance L is connected with emf E(t). The current i is given by*

$$L\frac{di}{dt} + Ri = E(t)$$

If the switch is connected at $t = 0$ *and disconnected at* $t = a$, *find the current i in terms of t.* [UPTU–2001]

SOLUTION. Conditions under which current i flows are $i = 0$ at $t = 0$.

$$E(t) = \begin{cases} E, & 0 < t < a \\ 0, & t > a \end{cases}$$

Given equation is $L\dfrac{di}{dt} + Ri = E(t)$ ... (1)

Taking Laplace transform of (1), we get

$$L[pL\{i\} - i(0)] + Ri = \int_0^\infty e^{-pt} E(t)\, dt$$

$$LpL\{i\} + Ri = \int_0^\infty e^{-pt} E(t)\, dt \qquad [i(0) = 0]$$

$$(Lp + R)L\{i\} = \int_0^\infty e^{-pt} E(t)\, dt$$

$$= \int_0^a e^{-pt} E\, dt + \int_0^\infty e^{-pt} E\, dt$$

$$= E\left[\frac{e^{-pt}}{-p}\right]_0^a + 0 = \frac{E}{p}\left[1 - e^{-ap}\right] = \frac{E}{p} - \frac{E}{p}e^{-ap}$$

$$\Rightarrow \quad L\{i\} = \frac{E}{p(Lp+R)} - \frac{Ee^{-ap}}{p(Lp+R)}$$

Taking inverse Laplace transform, we obtain

$$i = L^{-1}\left\{\frac{E}{p(Lp+R)}\right\} - L^{-1}\left\{\frac{Ee^{-ap}}{p(Lp+R)}\right\} \qquad ... (2)$$

Now, we have to find the value of $L^{-1}\left\{\dfrac{E}{p(Lp+R)}\right\}$

$$L^{-1}\left\{\frac{E}{p(Lp+R)}\right\} = \frac{E}{L}L^{-1}\left\{\frac{1}{p\left(p+\dfrac{R}{L}\right)}\right\}$$

$$= \frac{E}{L}\cdot\frac{L}{R}L^{-1}\left\{\frac{1}{p} - \frac{1}{p+\dfrac{R}{L}}\right\} = \frac{E}{R}\left[1 - e^{-R/Lt}\right]$$

and $L^{-1}\left\{\dfrac{Ee^{-ap}}{p(Lp+R)}\right\} = \dfrac{E}{R}\left[1 - e^{-\frac{R}{L}(t-a)}\right]u(t-a)$

[By the second shifting theorem]

On substituting the values of the inverse Laplace transforms in (2), we get

$$i = \frac{E}{R}\left[1 - e^{-\frac{R}{L}t}\right] - \frac{E}{R}\left[1 - e^{-\frac{R}{L}(t-a)}\right]u(t-a)$$

Hence, $i = \dfrac{E}{R}\left[1 - e^{-\frac{R}{L}t}\right]$,

for $0 < t < a$, $\quad [u(t-a) = 0]$

$$i = \frac{E}{R}\left[1 - e^{-\frac{R}{L}t}\right] - \frac{E}{R}\left[1 - e^{-\frac{R}{L}(t-a)}\right], \text{ for } t > a,$$

$$[u(t-a) = 1].$$

$$= \frac{E}{R}\left[e^{-\frac{R}{L}(t-a)} - e^{-\frac{R}{L}t}\right] = \frac{E}{R}e^{-\frac{R}{L}t}\left[e^{\frac{Ra}{L}} - 1\right]$$

Exercise-8.54

1. Solve $[tD^2 + (1-2t)D - 2]y = 0$ if $y(0) = 1$, $y'(0) = 2$.

2. Solve $y'' + ty' - y = 0$ if $y(0) = 0$, $y'(0) = 1$.

3. Solve $y''(t) + aty'(t) - 2ay(t) = 1$ if $y(0) = y'(0) = 0$, $a > 0$.

4. Solve $(D-2)x + 3y = 0$; $2x + (D-1)y = 0$, if $x(0) = 8$ and $y(0) = 3$.

5. Solve $(D^2 - 3)x - 4y = 0$, $x + (D^2 + 1)y = 0$, $t > 0$, if $x = y = Dy = 0$, $Dx = 2$, when $t = 0$. [UPTU–2004]

6. Solve $(D-2)x - (D+1)y = 6e^{3t}$, $(2D-3)x + (D-3)y = 6e^{3t}$, if $x = 3$, $y = 0$ when $t = 0$.

7. Solve $(D-2)x - (D-2)y = \sin t$
$(D^2 + 1)x + 2Dy = 0$ if $x = 0 = x'(0) = y(0)$.

8. Using $\dfrac{dy}{dt} + 2x = \sin 2t$, $\dfrac{dx}{dt} - 2y = \cos 2t$ $(t > 0)$ such that $t = 0, x = 1$ and $y = 0$. Show that the particle moves along the curve $4x^2 + 4xy + 5y^2 = 4$. [UPTU–2003, MTU(SUM)–2011]

9. Solve $\dfrac{dx}{dt} + y = \sin t$, $\dfrac{dy}{dt} + x = \cos t$, using $x = 2$, $y = 0$ at $t = 0$. [UPTU–2004, GBTU–2012, KERALA–2005]

10. Solve $\dfrac{dx}{dt} + 4\dfrac{dy}{dt} - y = 0$; $\dfrac{dx}{dt} + 2y = e^{-t}$ with $x(0) = y(0) = 0$. [UPTU 2008]

11. A body falls from rest in a liquid whose density is one fourth that of the body. If the liquid offers resistance proportional to the velocity and the velocity approaches a limiting value of 9 meter/sec, find the distance fallen in 5 seconds. [UPTU Q. BANK–2001]

12. A beam of length l is clamped horizontally at its ends $x = 0$ and is free at the end $x = l$. A point load w is applied at the end $x = l$ in addition to a uniform load w per unit length from $x = 0$ to $x = l/2$. Find the deflection of the beam. [UPTU(SP)–2001]

13. Solve $y^{iv}(t) + y'''(t) = \cos t$, $y(0) = y'(0) = y'''(0) = 0$, $y''(0)$ is arbitrary.

14. Solve $y'''(t) + y''(t) - 4y'(t) - 4y(t) = F(t)$ if $y(0) = y''(0) = 0$ and $y'(0) = 2$.

15. Solve $(D^3 - D^2 + 4D - 4)y = 68e^t \sin 2t$, $y = 1$,

$Dy = -19$, $D^2 y = -37$ at $t = 0$.

16. Solve $y'' - 4y' + 3y = F(t)$ if $y(0) = 1$, $y'(0) = 0$.

17. A particle P of mass 2 grams moves on the x-axis and is

attached towards origin 0 with a force numerically about to $8x$. If it is initially at rest at $x = 10$, find its position at any subsequent time assuming :

(a) no other force acts

(b) a damping force numerically equal to 8 times the instantaneous velocity acts.   [UPTU Q. BANK–2001]

Hints to Selected Problems

1. The given equation can be written as

$ty'' + y' - 2ty' - 2y = 0 \Rightarrow L\{ty''\} + L\{y'\} - 2L\{ty'\} - 2L\{y\} = 0$

$\Rightarrow -\dfrac{d}{dp}L\{y''\} + L\{y'\} + 2\dfrac{d}{dp}L\{y'\} - 2L\{y\} = 0$

$\Rightarrow \dfrac{dL\{y\}}{L\{y\}} + \dfrac{1}{(p-2)}dp = 0$ Now integrate.

2. Proceeding as usual, we get $\dfrac{d}{dp}[L\{y\}] - \left(p - \dfrac{2}{p}\right)L\{y\} = -\dfrac{1}{p}$

which is a linear differential equation of first order with

I.F. $= e^{-\int\left(p - \frac{2}{p}\right)dp} = p^2 e^{-p^2/2}$

The solution of (1) is given by

$L\{y\} = \dfrac{C_1}{p^2}e^{p^2/2} + \dfrac{1}{p^2} = \dfrac{C_1}{p^2}\left[1 + \dfrac{p^2}{2} + \dfrac{1}{2!}\cdot\dfrac{1}{4}p^4 + ...\right] + \dfrac{1}{p^2}$

Now taking inverse Laplace transform.

3. Proceeding same as above, we get

$\dfrac{d}{dp}L\{y\} + \left[-\dfrac{p}{a} + \dfrac{3}{p}\right]L\{y\} = -\dfrac{1}{ap^2}$, which is linear in $L\{y\}$.

7. $L\{x\} = \dfrac{1}{9(p+1)} + \dfrac{1}{3(p+1)^2} + \dfrac{4}{45(p-2)} - \dfrac{(p+2)}{5(p^2+1)}$

and $L\{y\} = \dfrac{1}{p^3 - 3p^2} = \dfrac{1}{9(p+1)} + \dfrac{1}{3(p+1)^2} - \dfrac{1}{9(p-2)}$

Answers

1. $y = e^{2t}$ **2.** $y = t$ **3.** $y = \dfrac{t^2}{2}$ **4.** $y = 5e^{-t} - 2e^{4t}$ **5.** $y = \dfrac{1}{2}(1-t)e^t - \dfrac{1}{2}(1+t)e^{-t}$ **6.** $x = e^t + 2te^t + 2e^{3t}$ and $y = e^t - te^t - e^{3t}$

7. $x = \dfrac{1}{9}e^{-t} + \dfrac{4}{45}e^{2t} - \dfrac{1}{5}\cos t - \dfrac{2}{5}\sin t + \dfrac{1}{3}te^{-t}$, $y = \dfrac{1}{9}e^{-t} - \dfrac{1}{9}e^{2t} + \dfrac{1}{3}e^{-t}.t$ **9.** $x = e^{-t} + e^t$; $y = \sin t + e^{-t} - e^{-t}$

10. $x = -\dfrac{5}{7}e^{-t} + \dfrac{8}{21}e^{3t/4} + \dfrac{1}{4}$; $y = \dfrac{1}{7}e^{-t} - \dfrac{1}{7}e^{\frac{3}{4}t}$ **11.** $x = 34.17$ meters

12. $y = \dfrac{1}{24EI}\left[wx^2\left(x^2 - 2lx + \dfrac{3l^2}{2}\right) - w\left(x - \dfrac{l}{2}\right)^4 u\left(x - \dfrac{l}{2}\right) - 4wx^2(3l - x)\right]$

13. $y = -1 + t + Ct^2 + \dfrac{1}{2}(e^{-t} + \cos t - \sin t)$ **14.** $y = \sinh 2t + \dfrac{1}{12}F(t)(-4e^{-t} + e^{2t} + 3e^{-2t})$

15. $y = \dfrac{1}{5}(e^t + 14\cos 2t - 3\sin 2t) - 2e^t(\cos 2t + 4\sin 2t)$ **16.** $y = \dfrac{3}{2}e^t - \dfrac{1}{2}e^{3t} + \dfrac{1}{2}\int_0^t(e^{3x} - e^x)F(t-x)dx$

17. (a) $X = 10\cos 2t$, (b) $X = 10e^{-2t} + 20te^{-2t}$.

8.12 PARTIAL DIFFERENTIAL EQUATIONS

In this section, we shall discuss the differential equations, with number of independent variables are two or more. In such cases, any dependent variable is likely to be a function of more than one variables, so that it possesses not ordinary derivatives with respect to a single variable but partial derivatives with respect to several variables. The partial differential equation implies necessarily the existence of more than one independent variables. We shall usually take z as dependent variable and x, y as independent variable and throughout the chapter we shall denote the partial derivatives

$\dfrac{\partial z}{\partial x}, \dfrac{\partial z}{\partial y}, \dfrac{\partial^2 z}{\partial x^2}, \dfrac{\partial^2 z}{\partial x \partial y}$ and $\dfrac{\partial^2 z}{\partial y^2}$ by p, q, r, s and t respectively.

Definition. *The equation of the type* $F\left(\dfrac{\partial z}{\partial x}, ..., \dfrac{\partial^2 z}{\partial x^2}, ..., \dfrac{\partial^2 z}{\partial x \partial y}, ...\right) = 0$ *is called a partial differential equation.*

8.12.1 Order and Degree

(1) ORDER OF PDE

The order of the partial differential equation is the order of its highest derivative.

(i) **First order PDE.** A first order partial differential equation for a function $z = f(x, y)$ contains at least one of the partial derivatives

$\dfrac{\partial z}{\partial x}$ or $\dfrac{\partial z}{\partial y}$. But no partial derivative of order higher then one.

For example : $x\dfrac{\partial z}{\partial x} + y\dfrac{\partial z}{\partial y} = 0$.

(ii) Second order PDE. A second order partial differential equation for $z = f(x, y)$ contains at least one of the partial derivatives $\dfrac{\partial^2 z}{\partial x^2}, \dfrac{\partial^2 z}{\partial y^2}, \dfrac{\partial^2 z}{\partial x \partial y}$ but no partial derivatives of order higher than two.

For example :   (i) $\dfrac{\partial^2 \phi}{\partial x^2} + \dfrac{\partial^2 \phi}{\partial y^2} + \dfrac{\partial^2 \phi}{\partial z^2} = 0$ (ii) $\dfrac{\partial z}{\partial t} - C \dfrac{\partial^2 z}{\partial x^2} = 0$.

☞ REMARK

- The second order partial differential equation may also contain first order term, like $\dfrac{\partial z}{\partial x}, \dfrac{\partial z}{\partial y}$ etc.

(2) DEGREE OF PDE

The degree of partial differential equation is the power of the highest order derivative in the equation.

For example :

(i) $\dfrac{\partial^2 \phi}{\partial x^2} + \dfrac{\partial^2 \phi}{\partial y^2} + \dfrac{\partial^2 \phi}{\partial z^2} = 0$ (ii) $\dfrac{\partial z}{\partial t} - C \dfrac{\partial^2 z}{\partial x^2} = 0$ (iii) $x \dfrac{\partial z}{\partial x} + y \dfrac{\partial z}{\partial y} = 0$

(iv) $\dfrac{\partial^2 z}{\partial t^2} = C^2 \dfrac{\partial^2 z}{\partial y^2}$ (v) $\left(\dfrac{\partial z}{\partial x} \right)^3 + \dfrac{\partial z}{\partial x} = 0$.

Equation (i), (ii), (iii) and (iv) are PDEs of degree one, and the equation (v) is a PDE of degree 3.

8.12.2 CLASSIFICATION OF PARTIAL DIFFERENTIAL EQUATIONS

(1) LINEAR AND NON-LINEAR PARTIAL DIFFERENTIAL EQUATION

A partial differential equation is said to be linear if :
  (i) it is of the first degree in the dependent variable and its partial derivatives.
  (ii) it does not contain the product of dependent variables and either of its partial derivatives, and
  (iii) it does not contain any transcendental function.

For example :

(i) $\dfrac{\partial^2 \phi}{\partial x^2} + \dfrac{\partial^2 \phi}{\partial y^2} + \dfrac{\partial^2 \phi}{\partial z^2} = 0$ (ii) $\dfrac{\partial T}{\partial t} - K \dfrac{\partial^2 T}{\partial t^2} = 0$

(iii) $\dfrac{\partial^2 u}{\partial t^2} = C^2 \dfrac{\partial^2 u}{\partial y^2}$ (iv) $\dfrac{\partial^2 u}{\partial x^2} + \dfrac{\partial^2 u}{\partial y^2} = f(x, y)$

The above all equations are linear.

(2) NON-LINEAR PDE

A partial differential equation, which is not linear is called non-linear equation.

For example : $\left(\dfrac{\partial f}{\partial x} \right)^3 + \dfrac{\partial f}{\partial t} = 0$.

(3) QUASI-LINEAR

Consider a non-linear equation $R_1 r + S_1 s + T_1 t = V_1$, where R_1, S_1, T_1 and V_1 are the functions of p and q as well as of x, y and z. Then, we observe that, it has a certain formal resemblance to a linear equation. Due to this resemblance with linear equation, equation (1) is said to be quasi-equation.

☞ REMARKS

- Quasi-linear equation is also called the uniform non-linear equation.
- Quasi-linear equation can be easily solved by Monge's method.

(4) HOMOGENEOUS AND NON-HOMOGENEOUS EQUATIONS

A linear partial differential equation can be classified as follows :
  (i) Homogeneous linear equation.
  (ii) Non-homogeneous linear equation.

  (i) Homogeneous linear equation :
  If each term of a partial differential equation contains either the dependent variable (or unknown function) or one of its partial derivatives, it is said to be homogeneous.

  For example :   (i) $\dfrac{\partial^2 \phi}{\partial x^2} + \dfrac{\partial^2 \phi}{\partial y^2} + \dfrac{\partial^2 \phi}{\partial z^2} = 0$ (ii) $\dfrac{\partial^2 z}{\partial t^2} = c^2 \dfrac{\partial^2 z}{\partial y^2}$.

(ii) Non-homogeneous linear equation :

An equation, which is not homogeneous is called non-homogeneous linear equation.

For example : (i) $\dfrac{\partial^2 z}{\partial x^2} + \dfrac{\partial^2 z}{\partial y^2} = f(x,y)$ (ii) $\dfrac{\partial^3 u}{\partial x^3} + 2\dfrac{\partial^3 u}{\partial x \partial y^2} - 6\left(\dfrac{\partial u}{\partial y}\right)^4 = 0.$

8.12.3 Solution of Partial Differential Equations

A solution of PDE in some region R of the space of independent variables is a function all of whose partial derivatives appearing in the equation exist in some domain containing R and which satisfies the equation everywhere in R.

 REMARKS

- The solution of ODE involves arbitrary constants, the solution of a PDE involves arbitrary functions.
- As we increase the order of partial derivation in a PDE, we must introduce more arbitrary functions.
- If u_1 and u_2 are any linearly independent solution of linear homogeneous PDE, then $u = c_1 u_1 + c_2 u_2$, where c_1 and c_2 are arbitrary constants, is also a solution of that equation. This is known as Principle of superposition and can by extended to the case where n solution of a PDE exists.

 **Exercise-8.55**

Classify the given PDE's by a way order and degree linearity (L)/ non-linearity (NL), homogeneity (H)/ non-homogeneity (NH).

1. $\dfrac{\partial^2 u}{\partial t^2} - c^2 \dfrac{\partial^2 u}{\partial x^2} + 2\beta \dfrac{\partial u}{\partial t} + \alpha\, u = 0.$

2. $\dfrac{\partial^2 u}{\partial t^2} - c^2 \dfrac{\partial^2 u}{\partial x^2} = 0.$

3. $-\dfrac{h^2}{2m}\left(\dfrac{\partial^2 \phi}{\partial x^2} + \dfrac{\partial^2 \phi}{\partial y^2} + \dfrac{\partial^2 \phi}{\partial z^2}\right) = i\,h\dfrac{\partial \phi}{\partial t}.$

4. $\dfrac{\partial \rho}{\partial t} + \rho\left(\dfrac{\partial u}{\partial x} + \dfrac{\partial v}{\partial y} + \dfrac{\partial w}{\partial z}\right) = 0.$

5. $\dfrac{\partial u}{\partial t} - K\left(\dfrac{\partial^2 u}{\partial x^2} + \dfrac{\partial^2 u}{\partial y^2}\right) = 0.$

6. $\dfrac{\partial^2 u}{\partial x^2} + \dfrac{\partial^2 u}{\partial y^2} + \dfrac{\partial^2 u}{\partial z^2} = \dfrac{1}{\varepsilon}\rho(x,y,z).$

7. $x^2 \dfrac{\partial^2 f}{\partial x^2} + y^2 \dfrac{\partial^2 f}{\partial y^2} = 0.$

8. $xy \dfrac{\partial^2 f}{\partial x^2} + x\dfrac{\partial f}{\partial x}\dfrac{\partial f}{\partial y} + y\dfrac{\partial^2 f}{\partial y^2} = x^2 + y^2.$

Answers

1. 2, 1, L, H	**2.** 2, 1, L, H	**3.** 2, 1, L, H	**4.** 1,1, L, H	**5.** 2, 1, L, H
6. 2, 1, L, NH	**7.** 2, 1, L, H	**8.** 2, 1, L, NH		

8.12.4 LINEAR PARTIAL DIFFERENTIAL EQUATION OF FIRST ORDER

A differential equation involving partial derivatives p and q only, no higher derivative is called of order 1. If the degree of p and q is one, then it is called a linear partial differential equation of order one.

(1) COMPLETE INTEGRAL

Let us consider the partial differential equation $f(x,y,z,p,q) = 0$ where x, y are independent variable, and z is dependent, while $p = \dfrac{\partial z}{\partial x}, q = \dfrac{\partial z}{\partial y}$, then a relation of type $F(x,y,z,a,b) = 0$ containing as many arbitrary constants as the number of independent variable in the above partial differential equation is called complete integral.

(2) PARTICULAR INTEGRAL

In the complete integral $F(x,y,z,a,b) = 0$ giving the particular values to the constants a and b, we get the particular integral.

(3) SINGULAR INTEGRAL

The envelope of the surfaces given by the complete integral $F(x,y,z,a,b) = 0$ is called singular integral. Therefore the singular integral is obtained by eliminating a and b from $F(x,y,z,a,b) = 0, \dfrac{\partial F}{\partial a} = 0$ and $\dfrac{\partial F}{\partial b} = 0.$

(4) GENERAL INTEGRAL

Let $u = u(x,y,z)$ and $v = v(x,y,z)$ be two functions of x, y and z then the solution of the differential equation $pP + qQ = R$ of the types $f(u,v) = 0$ is called the general integral. This also can be taken as $u = f(v)$ or $v = f(u)$.

8.12.5 Derivation of Partial Differential Equation by the Elimination of Arbitrary Constants

Consider the equation

$$F(x, y, z, a, b) = 0 \qquad \ldots(1)$$

where a and b are arbitrary constants. Differentiating (1) partially with respect to x, regarding z as a function of two independent variables x and y, we get

$$\frac{\partial F}{\partial x} + p\frac{\partial F}{\partial z} = 0 \qquad \text{and} \qquad \frac{\partial F}{\partial y} + q\frac{\partial F}{\partial z} = 0 \qquad \ldots(2)$$

By the elimination of a and b from (1) and (2), we shall get an equation of the type

$$F(x, y, z, p, q) = 0 \qquad \ldots(3)$$

which is the required partial differential equation of the first order.

☞ **Remark**

- It can be easily shown that if there are more arbitrary constants, then the number of independent variables, the above procedure of elimination will gives rise to partial differential equation of higher order than first.

Solved Examples

EXAMPLE 1. *Construct a partial differential equation, by eliminating a, b and c form*

$$z = a(x + y) + b(x - y) + abt + c. \quad \text{(IAS–1998)}$$

SOLUTION. The given equation is

$$z = a(x + y) + b(x - y) + abt + c \quad \ldots(1)$$

Now, differentiating (1) partially with respect to x, y and z, we get

$$\frac{\partial z}{\partial x} = a + b, \frac{\partial z}{\partial y} = a - b, \frac{\partial z}{\partial t} = ab \quad \ldots(2)$$

Now, using $(a + b)^2 - (a - b)^2 = 4ab$

$$\Rightarrow \qquad \left(\frac{\partial z}{\partial x}\right)^2 - \left(\frac{\partial z}{\partial y}\right)^2 = 4\frac{\partial z}{\partial t}$$

which is the required partial differential equation.

EXAMPLE 2. *Form a partial differential equation by eliminating*

$$a, b, c \text{ from } \frac{x^2}{a^2} + \frac{y^2}{b^2} + \frac{z^2}{c^2} = 1.$$

SOLUTION. The given equation is

$$\frac{x^2}{a^2} + \frac{y^2}{b^2} + \frac{z^2}{c^2} = 1 \qquad \ldots(1)$$

Differentiating (1) partially with respect to x and y, we get

$$\frac{2x}{a^2} + \frac{2z}{c^2} \cdot \frac{\partial z}{\partial x} = 0 \qquad \ldots(2)$$

and $$\frac{2y}{b^2} + \frac{2z}{c^2} \cdot \frac{\partial z}{\partial y} = 0 \qquad \ldots(3)$$

Now, differentiating (2), w.r.t. x and (3) w.r.t. y, we get

$$\frac{2}{a^2} + \frac{2}{c^2}\left\{z\frac{\partial^2 z}{\partial x^2} + \left(\frac{\partial z}{\partial x}\right)^2\right\} = 0$$

$$\Rightarrow \qquad c^2 + a^2\left(\frac{\partial z}{\partial x}\right)^2 + a^2 z\frac{\partial^2 z}{\partial x^2} = 0 \quad \ldots(4)$$

and $$c^2 + b^2\left(\frac{\partial z}{\partial x}\right)^2 + b^2 z\frac{\partial^2 z}{\partial y^2} = 0 \quad \ldots(5)$$

Putting the value of c^2 from (2) in (4), we get

$$\left(-a^2\frac{z}{x}\cdot\frac{\partial z}{\partial x}\right) + a^2\left(\frac{\partial z}{\partial x}\right)^2 + a^2 z\frac{\partial^2 z}{\partial x^2} = 0$$

$$\Rightarrow \qquad -\frac{z}{x}\frac{\partial z}{\partial x} + \left(\frac{\partial z}{\partial x}\right)^2 + z\frac{\partial^2 z}{\partial x^2} = 0$$

$$\Rightarrow \qquad -z\frac{\partial z}{\partial x} + x\left(\frac{\partial z}{\partial x}\right)^2 + xz\frac{\partial^2 z}{\partial x^2} = 0.$$

Similarly, from (3) and (5), we get

$$-z\frac{\partial z}{\partial y} + y\left(\frac{\partial z}{\partial y}\right)^2 + yz\frac{\partial^2 z}{\partial y^2} = 0.$$

8.12.6 Derivation of a Partial Differential Equation by the Elimination of an Arbitrary Function

Let u and v be any two functions of x, y, z connected by the relation

$$\phi(u, v) = 0 \qquad \ldots(1)$$

Now, it is to be shown on the elimination of the arbitrary function ϕ from (1), a partial differential equation will be formed and moreover, this equation will be linear. Differentiating (1) partially with respect to x and y regarding z as independent variables, we have

$$\frac{\partial \phi}{\partial u}\left(\frac{\partial u}{\partial x} + \frac{\partial u}{\partial z}\cdot\frac{\partial z}{\partial x}\right) + \frac{\partial \phi}{\partial v}\left(\frac{\partial v}{\partial x} + \frac{\partial v}{\partial z}\cdot\frac{\partial z}{\partial x}\right) = 0 \qquad \Rightarrow \qquad \frac{\partial \phi}{\partial u}\left(\frac{\partial u}{\partial x} + p\frac{\partial u}{\partial z}\right) + \frac{\partial \phi}{\partial v}\left(\frac{\partial v}{\partial x} + p\frac{\partial v}{\partial z}\right) = 0 \qquad \ldots(2)$$

and $$\frac{\partial \phi}{\partial u}\left(\frac{\partial u}{\partial y} + \frac{\partial u}{\partial z}\cdot\frac{\partial z}{\partial y}\right) + \frac{\partial \phi}{\partial v}\left(\frac{\partial v}{\partial y} + \frac{\partial v}{\partial z}\cdot\frac{\partial z}{\partial y}\right) = 0 \qquad \Rightarrow \qquad \frac{\partial \phi}{\partial u}\left(\frac{\partial u}{\partial y} + q\frac{\partial u}{\partial z}\right) + \frac{\partial \phi}{\partial v}\left(\frac{\partial v}{\partial y} + q\frac{\partial v}{\partial z}\right) = 0 \qquad \ldots(3)$$

Now eliminating $\dfrac{\partial \phi}{\partial u}, \dfrac{\partial \phi}{\partial v}$ between (2) and (3) by the method of determinant, we get

$$\begin{vmatrix} \left(\dfrac{\partial u}{\partial x} + p\dfrac{\partial u}{\partial z}\right) & \left(\dfrac{\partial v}{\partial x} + p\dfrac{\partial v}{\partial z}\right) \\[3mm] \left(\dfrac{\partial u}{\partial y} + q\dfrac{\partial u}{\partial z}\right) & \left(\dfrac{\partial v}{\partial y} + q\dfrac{\partial v}{\partial z}\right) \end{vmatrix} = 0$$

$$\Rightarrow \left(\dfrac{\partial u}{\partial y}\cdot\dfrac{\partial v}{\partial z} - \dfrac{\partial u}{\partial z}\cdot\dfrac{\partial v}{\partial y}\right)p + \left(\dfrac{\partial u}{\partial z}\cdot\dfrac{\partial v}{\partial x} - \dfrac{\partial u}{\partial x}\cdot\dfrac{\partial v}{\partial z}\right)q = \dfrac{\partial u}{\partial x}\cdot\dfrac{\partial v}{\partial y} - \dfrac{\partial u}{\partial y}\cdot\dfrac{\partial v}{\partial x}$$

$$\Rightarrow \qquad \dfrac{\partial(u,v)}{\partial(y,z)}p + \dfrac{\partial(u,v)}{\partial(z,x)}q = \dfrac{\partial(u,v)}{\partial(x,y)}$$

which is the linear PDE of first order and first degree in p and q which can also be written as $Pp + Qq = R$

where, $P = \dfrac{\partial(u,v)}{\partial(y,z)}, Q\dfrac{\partial(u,v)}{\partial(z,x)}$ and $R = \dfrac{\partial(u,v)}{\partial(x,y)}$.

 Solved Examples

EXAMPLE 1. *By means of a partial differential equation, eliminate the arbitrary function from the equation*

$$x + y + z = f(x^2 + y^2 + z^2) \qquad \ldots(1)$$

(SVTU–2007)

SOLUTION. Differentiating (1) partially w.r.t. x and y, we get

$$(1 + p) = f'(x^2 + y^2 + z^2).(2x + 2zp) \qquad \ldots(2)$$

and $(1 + q) = f'(x^2 + y^2 + z^2).(2y + 2zq) \qquad \ldots(3)$

From (2) and (3) we have

$$\dfrac{(1+p)}{(2x + 2zp)} = \dfrac{(1+q)}{(2y + 2zq)}$$

$$\Rightarrow (1+p)(y+zq) = (1+q)(x+zp)$$

$$\Rightarrow (y-z)p + (z-x)q = (x-y), \text{ which is the required PDE.}$$

EXAMPLE 2. *Eliminate the arbitrary functions f and g from $y = f(x - at) + g(x + at)$.* (VTU–2009)

SOLUTION. The given equation is

$$y = f(x - at) + g(x + at) \qquad \ldots(1)$$

$$\Rightarrow \quad \dfrac{\partial y}{\partial x} = f'(x - at) + g'(x + at)$$

and $\dfrac{\partial^2 y}{\partial x^2} = f''(x - at) + g''(x + at) \qquad \ldots(2)$

Now $\dfrac{\partial y}{\partial t} = f'(x - at).(-a) + g'(x + at)(a)$

$$\Rightarrow \dfrac{\partial^2 y}{\partial t^2} = f''(x - at)(-a)^2 + g''(x + at)(a)^2$$

$$= a^2[f''(x - at) + g''(x + at)] = a^2\dfrac{\partial^2 y}{\partial x^2}$$

(Using (2))

$$\Rightarrow \dfrac{\partial^2 y}{\partial t^2} = a^2\dfrac{\partial^2 y}{\partial x^2}, \text{ which is the required PDE.}$$

EXAMPLE 3. *Form the partial differential equation by eliminating the arbitrary function from $z = f(x + it) + g(x - it)$.* (UKTU–2011)

SOLUTION. Let $\quad z = f(x + it) + g(x - it)$

Differentiating z partially w.r.t. z and t we get

$$\dfrac{\partial z}{\partial x} = f'(x + it) + g'(x - it)$$

$$\dfrac{\partial^2 z}{\partial x^2} = f''(x + it) + g''(x - it) \qquad \ldots(1)$$

and $\dfrac{\partial z}{\partial t} = if'(x + it) - ig(x - it)$

$$\dfrac{\partial^2 z}{\partial t^2} = i^2 f''(x + it) + i^2 g''(x - it)$$

$$= -f''(x + it) - g''(x - it) \qquad \ldots(2)$$

On adding (1) and (2) we get

$$\dfrac{\partial^2 z}{\partial x^2} + \dfrac{\partial^2 z}{\partial t^2} = 0, \text{ which is the required partial differential equation.}$$

 Exercise-8.56

A. Form a PDE, by eliminating arbitrary constants for the following equations :

1. $z = (x + a)(y + b).$ 2. $z = ax + by + ab.$
3. $z = ax + a^2 y^2 + b^2$
4. $(x - h)^2 + (y - k)^2 + z^2 = c^2$ (IAS–1996, KOTTAYAM–2005)
5. $z = axe^y + \dfrac{1}{2}a^2 e^{2y} + b.$ 6. $z = (x - a)^2 + (y - b)^2$
7. Find the differential of all spheres of radius λ, having centre in the xy-plane.   (IAS–1996)
8. Find the differential equation by eliminating a and b from $z = (x^2 + a)(y^2 + b)$.   (IAS–1997)
9. Find the differential equation of the set of all right circular cone whose axes coincide with z-axis   (IAS–1998)

B. Form a PDE by eliminating the arbitrary function form the following equations:

1. $z = y^2 + 2f\left(\dfrac{1}{x} + \log y\right)$

(VTU–2010, JNTU–2010, MADRAS–2000)

2. $z = f(y / x);$
3. $f(x + y + z) + (x^2 + y^2 + z^2) = 0.$
4. $lx + my + nz = f(x^2 + y^2 + z^2)$
5. $z = f(x) + xg(y)$ (VTU–2004)
6. $z = f(x^2 - y^2)$ (SVTU–2008)

7. $f(xy + z^2, x + y + z) = 0$ (PTU–2006)
8. $z = x^2 f(y) + y^2 g(x)$ (ANNA–2003)
9. $z = f(x^2 + y^2) + x + y$ (ANNA–2009)
10. $z = f_1(y + 2x) + f_2(y - 3x)$ (KURUKSHETRA–2005)
11. $z = e^{my} \phi(x - y)$ (PTU–2002)

C. Find the differential equation of all surfaces of revolution having z-axis as the axis of rotation   (IAS–1997)

D. Find a partial differential equation by eliminating the arbitrary functions f and g from $z = yf(x) + xg(x)$

E. Find the differential equation of the surface which are the envelope of a one parametric family of planes.   (IAS–1995)

F. Form a partial differential equation by eliminating the arbitrary function ϕ from $\phi(x^2 + y^2 + z^2, z^2 - 2xy) = 0$

G. Form a PDE by eliminating f and g from
$$z = f(x^2 - y) + g(x^2 + y)$$ (IAS–1996)

H. Find the differenmtial equation of all planes which are at a constant distance a from the origin.

(VTU–2009, KURUKSHETRA–2006)

Answers

A. 1. $z = pq$ 2. $z = px + qy + pq$ 3. $q = 2yp^2$ 4. $z^2(p^2 + q^2 + 1) = c^2$ 5. $q = px + p^2$

6. $p^2 + q^2 = z^2$ 8. $4xyz = pq$ 9. $py = xq$

B. 1. $x^2 p + yq = 2y^2$ 2. $px + qy = 0$ 3. $(y + z)p - (z + x)q = x - y$ 4. $(l + np)y + z(lq - mp) = (m + nq)x$

5. $\dfrac{\partial z}{\partial y} = x \dfrac{\partial^2 z}{\partial x \partial y}$ 6. $py + qx = 0$ 7. $xys = px + py - z$ 8. $xyr = 2(px + qy - 2z)$ 9. $qx - py = x + y$

10. $\dfrac{\partial^2 z}{\partial x^2} + \dfrac{\partial^2 z}{\partial x \partial y} - 6 \dfrac{\partial^2 z}{\partial y^2} = 0$ 11. $p + q = mz$

C. $py = xq$ **D.** $xyr = xp + yq - z$ **E.** $(p - q)z = y - x$.   **G.** $x\dfrac{\partial^2 z}{\partial x^2} = p + 4x^3 \dfrac{\partial^2 z}{\partial y^2}$ **H.** $z = px + qy + a\sqrt{1 + p^2 + q^2}$

8.12.7 LAGRANGE'S LINEAR EQUATION

The partial differential equation of the form $Pp + Qq = R$, where P, Q, R are functions of x, y, z in the standard form of the linear partial differential equation of the order one is called Lagrange's linear equation.

8.12.8 SOLUTION OF STANDARD FORMS (NON-LINEAR EQUATION)

In this sub-section, we shall deal with some special types of equations which can be solved easily by some special methods, other than the general method.

(1) STANDARD FORM I : EQUATION INVOLVING ONLY p AND q AND NO x, y, z

The complete integral of equation of the type $f(p, q) = 0$, *i.e.*, in which x, y, z do not occur, is
$$z = ax + by + c \qquad \qquad \dots(1)$$
where a and b are connected by the relation
$$f(a, b) = 0 \qquad \qquad \dots(2)$$

Since, we have $p = \dfrac{\partial z}{\partial x} = a$ and $q = \dfrac{\partial z}{\partial y} = b$, which on substitution in (2) becomes the given equation.

Let us suppose from (2), $b = g(a)$ and replacing c by $\phi(a)$, the general solution is obtained by eliminating 'a' from the following equation
$$z = ax + g(a)y + \phi(a) \qquad \qquad \dots(3)$$
Differentiating (3) with respect to a, we get
$$0 = x + yg'(a) + \phi'(a) \qquad \qquad \dots(4)$$
Now, to find the singular integral, differentiating $z = ax + g(a)y + c$ with respect to a and c, we get
$$0 = x + yg'(a) \qquad \text{and} \qquad 0 = 1 \Rightarrow \text{There is no singular solution.}$$

(2) STANDARD FORM II : EQUATION INVOLVING ONLY p, q AND z

The equation which do not contain x and y, *i.e.*, which are of the form
$$f(z, p, q) = 0 \qquad \qquad \dots(1)$$
Equation (1) can be solved in the following way :

Write $X = x + ay$, where a is an arbitrary constant and assume z to be function of $(x + ay)$, *i.e.*, of X alone
$$\therefore \qquad z = f(X) = f(x + ay)$$
$$\Rightarrow \qquad p = \frac{\partial z}{\partial x} = \frac{dz}{dX} \cdot \frac{\partial X}{\partial x} = \frac{dz}{dX} \cdot 1 \qquad \text{and} \qquad q = \frac{\partial z}{\partial y} = \frac{dz}{dX} \cdot \frac{\partial X}{\partial y} = a\frac{dz}{dX}.$$

Now, the equation (1), becomes $F\left(z, \dfrac{dz}{dX}, a\dfrac{dz}{dX}\right) = 0$, which is an ordinary differential equation of the first order and can be integrated. So the complete integral will be known. If $f = 0$ is the complete integral involving two constants a and b, then replacing b by $f(a)$, the general integral is obtained by eliminating a from $f = 0, \dfrac{\partial f}{\partial a} = 0$. The singular integral is obtained by eliminating a and b, from $f = 0, \partial f / \partial a = 0$ and $\partial f / \partial b = 0$.

Solved Examples

EXAMPLE 1. *Solve $p^2 + q^2 = 1$.*   (OSMANIA–2005)

SOLUTION.   The given equation is of the form $f(p,q) = 0$.

The solution is given by $z = ax + by + c$, where a and b are related by $f(a,b) = 0$

$\Rightarrow a^2 + b^2 = 1 \Rightarrow b = \sqrt{(1 - a^2)}$

Hence, the complete integral is

$$z = ax + \sqrt{(1 - a^2)}\, y + c.$$

For the general integral write $c = \phi(a)$.

Then it is obtained by eliminating a from

$$z = ax + \sqrt{(1 - a^2)}\, y + \phi(a)$$

and $0 = x + \dfrac{-a}{\sqrt{(1 - a^2)}}\, y + \phi'(a).$

EXAMPLE 2. *Solve $x^2 p^2 + y^2 q^2 = z^2$.*

SOLUTION.   The given equation can be written as

$$\left(\frac{x}{z} \cdot \frac{\partial z}{\partial x}\right)^2 + \left(\frac{y}{z} \cdot \frac{\partial z}{\partial y}\right)^2 = 1 \qquad \ldots(1)$$

Putting $\dfrac{1}{z} dz = dZ$, i.e., $z = e^Z$;

$\dfrac{1}{x} dx = dX$,i.e., $x = e^X$

and $\dfrac{1}{y} dy = dY$ i.e., $y = e^Y$

in (1), we get $\left[\dfrac{\partial Z}{\partial X}\right]^2 + \left[\dfrac{\partial Z}{\partial Y}\right]^2 = 1$, which is of the type $f(p,q) = 0$.

Therefore, the complete integral is given by $Z = aX + bY + c_1$, where a and b are related by $a^2 + b^2 = 1$.

$\Rightarrow \quad b = \sqrt{(1 - a^2)} \Rightarrow z = aX + \sqrt{(1 - a^2)}Y + c_1$

$\Rightarrow \quad \log z = a \log x + \sqrt{(1 - a^2)} \log y + c_1.$

To find the general solution put $a = \cos\theta$.

$\Rightarrow \quad \log z = \cos\theta \log x + \sin\theta \log y + \log c$

$\Rightarrow \quad z = c x^{\cos\theta} . y^{\sin\theta}.$

Now, we eliminate θ from

$z = g(\theta) x^{\cos\theta} y^{\sin\theta}$ (take $c = g(\theta)$)

and $0 = g'(\theta) x^{\cos\theta} y^{\sin\theta} + g(\theta) x^{\cos\theta} y^{\sin\theta}$

$(-\sin\theta) \log_e x + g(\theta) x^{\cos\theta} y^{\sin\theta} \cos\theta \log_e y$

which is the required general solution.

To find singular integral, we eliminate θ and c, from

$z = c x^{\cos\theta} . y^{\sin\theta}$

$\Rightarrow \quad \dfrac{\partial z}{\partial \theta} = -c \sin\theta x^{\cos\theta} y^{\sin\theta} \log_e x$

$+ c \cos\theta . x^{\cos\theta} . y^{\sin\theta} \log_e y = 0$

and $\dfrac{\partial z}{\partial c} = x^{\cos\theta} . y^{\sin\theta} = 0$

$\Rightarrow z = 0$ is the singular integral of the given equation.

EXAMPLE 3. *Find the complete integral of*

$$(y - x)(qy - px) = (p - q)^2.$$

SOLUTION.   Substitute $X = x + y$ and $Y = xy$

so that $p = \dfrac{\partial z}{\partial x} = \dfrac{\partial z}{\partial X} \cdot \dfrac{\partial X}{\partial x} + \dfrac{\partial z}{\partial Y} \cdot \dfrac{\partial Y}{\partial x} = \dfrac{\partial z}{\partial X} + y \dfrac{\partial z}{\partial Y}$

and $q = \dfrac{\partial z}{\partial y} = \dfrac{\partial z}{\partial X} \cdot \dfrac{\partial X}{\partial y} + \dfrac{\partial z}{\partial Y} \cdot \dfrac{\partial Y}{\partial y} = \dfrac{\partial z}{\partial X} + x \dfrac{\partial z}{\partial Y}$

in the given equation, we get

$$(y - x)(y - x)\frac{\partial z}{\partial X} = (y - x)^2 \left(\frac{\partial z}{\partial Y}\right)^2$$

$\Rightarrow \dfrac{\partial z}{\partial X} = \left(\dfrac{\partial z}{\partial Y}\right)^2$, which is the standard form (1).

Then, the complete solution is $z = aX + bY + C$, where $a = b^2$.

$\Rightarrow \quad z = b^2(x + y) + bxy + C.$

EXAMPLE 4. *Find the complete integral of $p^3 + q^3 = 27z$.*

SOLUTION.   The given equation is $p^3 + q^3 = 27z$ which is in the standard form $f(p,q,z) = 0$.

Put $X = x + ay \Rightarrow z = f(X) = f(x + ay)$

$\Rightarrow p = \dfrac{\partial z}{\partial x} = \dfrac{dz}{dX}$ and $q = \dfrac{\partial z}{\partial y} = a\dfrac{dz}{dX}$.

We may take $\dfrac{dz}{dX}$ in place of $\dfrac{\partial z}{\partial x}$ because z is a function of x only.

Hence, the given equation reduces to

$$(1 + a^3)\left(\frac{dz}{dX}\right)^3 = 27z$$

$\Rightarrow \quad (1 + a^3)^{1/3} \dfrac{dz}{dX} = 3z^{1/3}$

$\Rightarrow \quad (1 + a^3)^{1/3} . \dfrac{2}{3} z^{-1/3} dz = 2dX.$

On integrating, we get

$z^{2/3}(1 + a^3)^{1/3} = 2X + c = 2(X + b)$

$\Rightarrow \qquad (1 + a^3)z^2 = 8(x + ay + b)^3 \qquad \ldots(1)$

which is the complete integral of the given equation.

To find the singular integral, differentiating (1) partially with respect to a and b, we get

$3a^2 z^2 = 24y(x + ay + b)^2 \qquad \ldots(2)$

and $0 = 24(x + ay + b)^2 \qquad \ldots(3)$

By eliminating a, b from (1), (2) and (3), we get $z = 0$, which is the required singular solution.

EXAMPLE 5. *Find complete solution of $z^2(p^2 z^2 + q^2) = 1$.*

(BHOPAL–2008)

SOLUTION.   The given equation is $z^2(p^2 z^2 + q^2) = 1$.

Putting $z = f(x + ay) = f(X)$,

where $X = x + ay$

So that $p = \dfrac{\partial z}{\partial x} = \dfrac{dz}{dX}$ and $q = \dfrac{\partial z}{\partial y} = a\dfrac{dz}{dX}$

In the given equation, we get

$$z^2\left[\left(\dfrac{dz}{dX}\right)^2 . z^2 + a^2\left(\dfrac{dz}{dX}\right)^2\right] = 1$$

or $\qquad z^2(z^2 + a^2)\left(\dfrac{dz}{dX}\right)^2 = 1.$

On separating the variables, we get

$$z\sqrt{(z^2 + a^2)}\,dz = dX.$$

On integrating, we get $\dfrac{1}{3}(z^2 + a^2)^{3/2} = X + b$.

$\Rightarrow \qquad 9(x + ay + b)^2 = (z^2 + a^2)^3 \qquad \ldots(1)$

Equation (1) gives the required complete solution.

To find the singular integral, differentiating (1) partially with respect to a and b, we get

$$18(x + ay + b)y = 6(z^2 + a^2)^2 . a \qquad \ldots(2)$$

and $\qquad 18(x + ay + b) = 0 \qquad \ldots(3)$

From (2) and (3), we get $a = 0$. Hence, from (1), (3) and (4), we get $z = 0$.

Now since $z = 0 \Rightarrow p = 0, q = 0$ does not satisfy the given equation, therefore, the given equation does not have any singular solution.

Exercise-8.57

1. Solve $q = 3p^2$.

2. Solve $(x + y)(p + q)^2 + (x - y)(p - q)^2 = 1$

   (IAS–1991, BHOPAL–2006, RAJASTHAN–2006, VTU–2003)

3. Find the complete integral of

   $p^m \sec^{2m} x + z^l q^n \csc^{2n} y = z^{lm/m-n}$.

4. Solve $(x^2 + y^2)(p^2 + q^2) = 1$.

5. Solve $p^2 + q^2 = npq$.

6. Solve $\sqrt{p} + \sqrt{q} = 1$.

7. Find the complete integral of $p^2 = zq$.

8. Solve $pz = (1 + q^2)$.

9. Solve $9(p^2 z + q^2) = 4$. (IAS–1988)

10. Solve $p(1 + q^2) = q(z - a)$.

11. Solve $p^2 = z^2(1 - pq)$.

12. Solve $pq = x^m y^n z^l$. (IAS–1989, 94, 2000)

13. Solve $z^2(p^2 + q^2 + 1) = c^2$.

14. Find a complete integral of $p^3 + q^3 - 3pqz = 0$.

Hints to Selected Problems

1. Put $p = a, q = b$ in the given equation.

2. Put $x + y = X^2, x - y = Y^2$.

   Therefore $p = \dfrac{\partial z}{\partial x} = \dfrac{\partial z}{\partial X}\cdot\dfrac{\partial X}{\partial x} + \dfrac{\partial z}{\partial Y}\cdot\dfrac{\partial Y}{\partial x}$

   $\qquad = \dfrac{1}{2X}\dfrac{\partial z}{\partial X} + \dfrac{1}{2Y}\dfrac{\partial z}{\partial Y}.$

   Similarly $q = \dfrac{1}{2X}\dfrac{\partial z}{\partial X} - \dfrac{1}{2Y}\dfrac{\partial z}{\partial Y}$. Finally, putting all these values in the given equation.

3. The given equation can be written as

   $$\left(\dfrac{z^{-l/(m-n)}}{\cos^2 x}\cdot\dfrac{\partial z}{\partial x}\right)^m + \left(\dfrac{z^{-l/(m-n)}}{\sin^2 y}\cdot\dfrac{\partial z}{\partial y}\right)^n = 1.$$

Now putting

$$z^{-l/(m-n)}dz = dZ \quad \Rightarrow \quad Z = \dfrac{m-n}{(m-n-l)}z^{(m-n-l)/(m-n)}$$

$$\cos^2 x\,dx = dX \quad \Rightarrow \quad X = \dfrac{1}{2}\left(x + \dfrac{1}{2}\sin 2x\right)$$

and $\quad \sin^2 y\,dy = dY \quad \Rightarrow \quad Y = \dfrac{1}{2}\left(y - \dfrac{1}{2}\sin 2y\right).$

7. Putting $z = f(x + ay) = f(x)$ in the given equation.

8. Do same as (7).

9. Putting $p = \dfrac{dz}{dX}$ and $q = a\dfrac{dz}{dX}$ in the given equation.

12. Putting $\dfrac{x^{m+1}}{m+1} = X, \dfrac{y^{n+1}}{n+1} = Y$, i.e., $p = x^m\dfrac{\partial z}{\partial X}$ and $q = y^n\dfrac{\partial z}{\partial Y}$ in the given equation.

14. Putting $\dfrac{dz}{dX} = p$ and $a\dfrac{dz}{dX} = q$ in the given equation.

Answers

1. $z = ax + 3a^2 y + c.$

2. $z = a\sqrt{(x + y)} + \sqrt{(1 - a^2)}\sqrt{(x - y)} + c.$

3. $\dfrac{(m-n)}{(m-n-l)}z^{(m-n-l)(m-n)} = \dfrac{a}{2}\left(x + \dfrac{1}{2}\sin 2x\right) + (1 - a^m)^{1/n}.\dfrac{1}{2}\left(y - \dfrac{\sin 2y}{2}\right) + c$

4. $z = \dfrac{a}{2}\log(x^2 + y^2) + \sqrt{(1 - a^2)}\tan^{-1}\dfrac{y}{x} + c$

5. $z = ax + \left(\dfrac{n \pm \sqrt{(n^2 - 4)}}{2}\right).ay + c$

6. $z = ax + (1 - \sqrt{a})^2 y + c$

7. $z = be^{(ax + a^2 y)}$

8. $z^2 \mp [z\sqrt{(z^2 - 4a^2)} - 4a^2\log\{z + \sqrt{(z^2 - 4a^2)}\}] = 4x + 4ay + 2c$

9. $(z + a^2)^3 = (x + ay + b)^2$

10. $4a(z - a) = 4 + (x + ay + b)^2$

11. $\dfrac{1}{\sqrt{a}}\log[z\sqrt{a} + \sqrt{(1 + az^2)}] + \sqrt{(1 + az^2)} = X + C$

12. $\dfrac{z^{-l/2} + 1}{-l/2 + 1} = \dfrac{1}{\sqrt{a}}\left[\dfrac{x^{m+1}}{m+1} + a\dfrac{y^{n+1}}{n+1}\right] + b$

13. $(1 + a^2)(c^2 - z^2) = (x + ay + b)^2$

14. $3a(x + ay) + c = (1 + a^3)\log z.$

(3) STANDARD FORM III : EQUATION OF THE FORM $f_1(x,p) = f_2(y,q)$

If the given equation is of the type $\qquad f_1(x,p) = f_2(y,q)$ ...(1)

then, first write $\qquad f_1(x,p) = f_2(y,q) = c_1$ ...(2)

Now, solving (2) for q and p, we get

$$p = \frac{\partial z}{\partial x} = g_1(x, c_1) \qquad \text{and} \qquad q = \frac{\partial z}{\partial y} = g_2(y, c_1).$$

Now $\qquad dz = pdx + qdy = g_1(x, c_1)dx + g_2(y, c_1)dy$

which gives $\qquad z = \int g_1(x, c_1)dx + \int g_2(y, c_1)dy + b.$

The general solution may be obtained from this complete integral. Also, there is no singular solution.

(4) STANDARD FROM IV : EQUATION OF THE FORM $z = px + qy + f(p,q)$

The equation $\qquad z = px + qy + f(p,q)$ ...(1)

which is analogous to Clairaut's form, has its complete integral

$$z = ax + by + f(a,b)$$ ...(2)

For $\dfrac{\partial z}{\partial x} = p = a$ and $\dfrac{\partial z}{\partial y} = q = b$

In order to obtain the general solution, put $b = g(a)$

Therefore, $\qquad z = ax + y g(a) + f\{a, g(a)\}$ ...(3)

Differentiating (3) with respect to a, we get

$$0 = x + y g'(a) + f'(a)$$ ...(4)

Now, eliminate a from (3) and (4) and get the required general solution.
To obtain the singular solution, differentiating (2) with respect to a and b, which gives

$$0 = x + \frac{\partial f}{\partial a}$$ ...(5)

$$0 = y + \frac{\partial f}{\partial b}$$ ...(6)

Now, eliminate a and b between the equation (2), (5) and (6).

Solved Examples

EXAMPLE 1. **Solve $p^2 + q^2 = x + y$.**   (BHOPAL–2006, MADRAS–2003)

SOLUTION. The given equation can be written as
$$p^2 - x = y - q^2.$$

Let us write $p^2 - x = y - q^2 = a$

$\Rightarrow p = \sqrt{(x + a)}$ and $q = \sqrt{(y - a)}$.

Now, putting the values of p and q in $dz = pdx + qdy$, we get
$$dz = \sqrt{(x + a)}\, dx + \sqrt{(y - a)}\, dy.$$
On integrating, we have
$$z = \frac{2}{3}(x + a)^{3/2} + \frac{2}{3}(y - a)^{3/2} + b.$$

EXAMPLE 2. **Solve $q = xyp^2$.**

SOLUTION. The given equation can be written as $p^2 x = q / y$.

Let us write $p^2 x = \dfrac{q}{y} = a$, which gives $p = \sqrt{(a / x)}$ and $q = ay$.

Now, putting the values of p and q in $dz = pdx + qdy$, we get
$$dz = \sqrt{(a / x)}\, dx + (ay)\, dy.$$

On integrating, we have $z = 2\sqrt{a}\sqrt{x} + \dfrac{a}{2}y^2 + b$

$\Rightarrow (2z - ay^2 - 2b)^2 = 16ax.$

EXAMPLE 3. **Solve $z^2(p^2 + q^2) = x^2 + y^2$.**

SOLUTION. The given equation is
$$z^2(p^2 + q^2) = x^2 + y^2.$$

Replace $zdz = dZ \qquad \Rightarrow \dfrac{z^2}{2} = Z.$

Therefore, the given equation becomes
$$P^2 + Q^2 = x^2 + y^2, \text{ where } P = \frac{dZ}{dx} \text{ and } Q = \frac{dZ}{dy}$$
$$\Rightarrow P^2 - x^2 = y^2 - Q^2.$$

Let us write $P^2 - x^2 = y^2 - Q^2 = a$

$\Rightarrow P = \sqrt{(a + x^2)}; \quad Q = \sqrt{(y^2 - a)}.$

Now, putting the values of P and Q in $dZ = P\,dx + Q\,dy$, we get
$$dZ = \sqrt{(a + x^2)}dx + \sqrt{(y^2 - a)}\, dy.$$
On integrating, we have
$$Z = \frac{x}{2}\sqrt{(a + x^2)} + \frac{a}{2}\log\{x + \sqrt{(a + x^2)}$$
$$+ \frac{y}{2}\sqrt{(y^2 - a)} \quad - \frac{a}{2}\log\{y + \sqrt{y^2 - a}\} + b$$
$$\Rightarrow z^2 = x\sqrt{(a + x^2)} + a\log\{x + \sqrt{(a + x^2)}\}$$
$$+ y\sqrt{(y^2 - a)} - a\log\{y + \sqrt{(y^2 - a)}\} + c$$

EXAMPLE 4. **Solve** $z = px + qy + c\sqrt{(1 + p^2 + q^2)}$.   (ANNA–2009)

SOLUTION. The given equation is of the standard form IV. Therefore, the complete solution is

$$z = ax + by + c\sqrt{(1 + a^2 + b^2)} \qquad \ldots(1)$$

To find the singular solution, differentiating (1) partially with respect to a and b, we have

$$0 = x + \frac{ac}{\sqrt{(1 + a^2 + b^2)}}$$

$$\Rightarrow \quad a = \frac{-x}{\sqrt{(c^2 - x^2 - y^2)}} \qquad \ldots(2)$$

and $\quad 0 = y + \frac{bc}{\sqrt{(1 + a^2 + b^2)}}$

$$\Rightarrow \quad b = \frac{-y}{\sqrt{(c^2 - x^2 - y^2)}} \qquad \ldots(3)$$

which gives $x^2 + y^2 = \dfrac{(a^2 + b^2)\,c^2}{1 + a^2 + b^2}$

$$\Rightarrow \quad (c^2 - x^2 - y^2) = \frac{c^2}{1 + a^2 + b^2}$$

$$\Rightarrow \quad (1 + a^2 + b^2) = \frac{c^2}{(c^2 - x^2 - y^2)} \qquad \ldots(4)$$

Now using (2), (3) and (4), (1) becomes

$$z = \frac{-x^2}{\sqrt{(c^2 - x^2 - y^2)}} - \frac{y^2}{\sqrt{(c^2 - x^2 - y^2)}}$$
$$+ \frac{c^2}{\sqrt{(c^2 - x^2 - y^2)}}$$

$$= \frac{(c^2 - x^2 - y^2)}{\sqrt{(c^2 - x^2 - y^2)}} = \sqrt{(c^2 - x^2 - y^2)}$$

$$\Rightarrow \quad z^2 = c^2 - x^2 - y^2 \Rightarrow x^2 + y^2 + z^2 = c^2.$$

Exercise-8.58

Solve the following equations :

1. $\sqrt{p} + \sqrt{q} = 2x$

2. $pe^y = qe^x$

3. $pq = xy$

4. $py = 2yx + \log q$

5. $z(p^2 - q^2) = (x - y)$

6. Find the complete integral of $z = px + qy + p^2 + q^2$

7. $z = px + qy - 2p - 3q$

8. $z = px + qy - p^2 q$

9. $z = px + qy + pq$ (GBTU(AG)–2012)

10. $x(1 + y)p = y(1 + x)q$

11. $z = px + qy + \sqrt{\alpha p^2 + \beta q^2 + \gamma}$

12. $x^2 y^3 z^{-3} p^2 q = 1$

13. $p - 3x^2 = q^2 - y$

14. $\sin px \cos y - \cos px \sin y = p$

Hints to Selected Problems

1. The given equation can be written as $\sqrt{p} - 2x = -\sqrt{q}$.
   Now, let $\sqrt{p} - 2x = a = -\sqrt{q}$.

2. Putting $pe^{-x} = qe^{-y} = a$.

3. Putting $p = ax, q = y/c$.

4. Putting $p - 2x = \frac{1}{y}\log q = a$.

5. The given equation can be written as
   $$\left(\sqrt{z}\frac{dz}{dx}\right)^2 - \left(\sqrt{z}\frac{\partial z}{\partial x}\right)^2 = x - y\,.$$

   Then putting $\sqrt{z}\,dz = dZ$.

12. The given equation can be written as $x^2 y^3 p^2 q = z^3$.

   Now putting $Z = \log z \Rightarrow \dfrac{\partial Z}{\partial X} = \dfrac{1}{z}\dfrac{\partial z}{\partial X},\ \dfrac{\partial Z}{\partial Y} = \dfrac{1}{z}\dfrac{\partial z}{\partial Y}$

13. Let $p - 3x^2 = q^2 - y = a$

   $\Rightarrow \quad p = a + 3x^3,\ q = \sqrt{a + y}$.

   Putting the values of p and q in $dz = p\,dx + q\,dy$.

Answers

1. $z = \frac{1}{6}(a + 2x)^3 + a^2 y + b$ 2. $z = ae^x + ae^y + b$ 3. $z = \frac{1}{2a}(a^2 x^2 + y^2 + 2ab)$ 4. $z = \frac{1}{a}(ax^2 + a^2 x + e^{ay} + a.b)$

5. $z^{3/2} = (x + a)^{3/2} + (y + a)^{3/2} + c$ 6. $z = ax + by + a^2 + b^2$ 7. $z = ax + by - 2a - 3b$ 8. $z = ax + by - a^2 b$

9. $z = ax + by + ab$ 10. $z + c = a(x + y + \log xy)$ 11. $z = ax + by + \sqrt{(\alpha a^2 + \beta b^2 + \gamma)}$

12. $\log z - \sqrt{a}\log x = \frac{1}{2ay^2} + c$ 13. $z = x^3 + ax \pm \frac{2}{3}(y + a)^{3/2} + b$ 14. $y = x\cos\sin^{-1} c$

8.13   LINEAR AND NON-LINEAR PARTIAL DIFFERENTIAL EQUATIONS

Partial differential equation arise in connection with various type of physical and geometrical problems when the functions are implicit, that is, functions of two or more than two independent variables.

Partial differential equation is an equation having one or more partial derivatives of an unknown function of two or more than two independent variables, we say that a partial differential equation is linear if it is of the first degree in the dependent variable and its partial derivatives. If each term of the differential equation contains either the dependent variable or one of its derivatives, then the equation is called homogeneous, otherwise it is called non-homogeneous.

In this section, we shall discuss some methods : Lagrange's and Charpit's to solve partial differential equations.

8.13.1 Lagrange's Linear Differential Equation

Consider the partial differential equation of the type $Pp + Qq = R$, where P, Q, R are the functions of x, y and z and $p = \partial z / \partial x, q = \partial z / \partial y$. Then this partial differential equation of order one is called Lagrange's Linear Differential Equation.

(1) LAGRANGE'S AUXILIARY EQUATION

Let u and v be two functions of x, y, z which are related by the relation

$$f(u, v) = 0 \qquad \qquad \dots(1)$$

Differentiating (1) partially w.r.t. x and y, we get

$$\frac{\partial f}{\partial u}\left(\frac{\partial u}{\partial x} + \frac{\partial u}{\partial z}\cdot\frac{\partial z}{\partial x}\right) + \frac{\partial f}{\partial v}\left(\frac{\partial v}{\partial x} + \frac{\partial v}{\partial z}\cdot\frac{\partial z}{\partial x}\right) = 0 \quad \text{or} \quad \frac{\partial f}{\partial u}\left(\frac{\partial u}{\partial x} + \frac{\partial u}{\partial z}p\right) + \frac{\partial f}{\partial v}\left(\frac{\partial v}{\partial x} + \frac{\partial v}{\partial z}\cdot p\right) = 0 \qquad \dots(2)$$

and $$\frac{\partial f}{\partial u}\left(\frac{\partial u}{\partial y} + \frac{\partial u}{\partial z}\cdot\frac{\partial z}{\partial y}\right) + \frac{\partial f}{\partial v}\left(\frac{\partial v}{\partial y} + \frac{\partial v}{\partial z}\cdot\frac{\partial z}{\partial y}\right) = 0 \quad \text{or} \quad \frac{\partial f}{\partial u}\left(\frac{\partial u}{\partial y} + \frac{\partial u}{\partial z}q\right) + \frac{\partial f}{\partial v}\left(\frac{\partial v}{\partial y} + \frac{\partial v}{\partial z}\cdot q\right) = 0 \qquad \dots(3)$$

Eliminating $\frac{\partial f}{\partial u}$ and $\frac{\partial f}{\partial v}$ from (2) and (3), we get

From (2), $$\frac{\partial f / \partial u}{\partial f / \partial v} = -\frac{\left(\dfrac{\partial v}{\partial x} + \dfrac{\partial v}{\partial z}p\right)}{\left(\dfrac{\partial u}{\partial x} + \dfrac{\partial u}{\partial z}p\right)} \qquad \qquad \dots(4)$$

From (3), $$\frac{\partial f / \partial u}{\partial f / \partial v} = -\frac{\left(\dfrac{\partial v}{\partial y} + \dfrac{\partial v}{\partial z}q\right)}{\left(\dfrac{\partial u}{\partial y} + \dfrac{\partial u}{\partial z}q\right)} \qquad \qquad \dots(5)$$

From (4) and (5), we get

$$\left(\frac{\partial u}{\partial x} + \frac{\partial u}{\partial z}p\right)\left(\frac{\partial v}{\partial y} + \frac{\partial v}{\partial z}q\right) = \left(\frac{\partial u}{\partial y} + \frac{\partial u}{\partial z}q\right)\left(\frac{\partial v}{\partial x} + \frac{\partial v}{\partial z}p\right)$$

Solving this equation, we get

$$\left(\frac{\partial u}{\partial y}\cdot\frac{\partial v}{\partial z} - \frac{\partial v}{\partial y}\cdot\frac{\partial u}{\partial z}\right)p + \left(\frac{\partial v}{\partial x}\cdot\frac{\partial u}{\partial z} - \frac{\partial u}{\partial x}\cdot\frac{\partial v}{\partial z}\right)q = \left(\frac{\partial u}{\partial x}\cdot\frac{\partial v}{\partial y} - \frac{\partial u}{\partial y}\cdot\frac{\partial v}{\partial x}\right) \quad \text{or} \quad Pp + Qq = R \qquad \dots(6)$$

where $P = \dfrac{\partial u}{\partial y}\cdot\dfrac{\partial v}{\partial z} - \dfrac{\partial v}{\partial y}\cdot\dfrac{\partial u}{\partial z} = \dfrac{\partial(u,v)}{\partial(y,z)}$, $\qquad\qquad$ (Jacobian of u and v w.r.t. y and z)

$Q = \dfrac{\partial v}{\partial x}\cdot\dfrac{\partial u}{\partial z} - \dfrac{\partial u}{\partial x}\cdot\dfrac{\partial v}{\partial z} = \dfrac{\partial(u,v)}{\partial(z,x)}$ $\qquad$ and $\quad R = \dfrac{\partial u}{\partial x}\cdot\dfrac{\partial v}{\partial y} - \dfrac{\partial u}{\partial y}\cdot\dfrac{\partial v}{\partial x} = \dfrac{\partial(u,v)}{\partial(x,y)}$

Thus $f(u, v) = 0$ is the general integral of the differential equation $Pp + Qq = R$. Now we shall determine the values of u and v. For this, let $u = a$ and $v = b$ be two equations, where a and b are arbitrary constants. That is $u(x, y, z) = a$ and $v(x, y, z) = b$.

This implies $du = 0$ and $dv = 0$

But $$du = \frac{\partial u}{\partial x}dx + \frac{\partial u}{\partial y}dy + \frac{\partial u}{\partial z}dz \qquad \text{and} \quad dv = \frac{\partial v}{\partial x}dx + \frac{\partial v}{\partial y}dy + \frac{\partial v}{\partial z}dz$$

Thus, we obtained

$$\frac{\partial u}{\partial x}dx + \frac{\partial u}{\partial y}dy + \frac{\partial u}{\partial z}dz = 0 \qquad \qquad \dots(7)$$

and

$$\frac{\partial v}{\partial x}dx + \frac{\partial v}{\partial y}dy + \frac{\partial v}{\partial z}dz = 0 \qquad \qquad \dots(8)$$

Solving, (7) and (8) by cross multiplication method for dx, dy and dz, we get

$$\frac{dx}{\dfrac{\partial u}{\partial y}\cdot\dfrac{\partial v}{\partial z} - \dfrac{\partial v}{\partial y}\cdot\dfrac{\partial u}{\partial z}} = \frac{dy}{\dfrac{\partial u}{\partial z}\cdot\dfrac{\partial v}{\partial x} - \dfrac{\partial u}{\partial x}\cdot\dfrac{\partial v}{\partial z}} = \frac{dz}{\dfrac{\partial u}{\partial x}\cdot\dfrac{\partial v}{\partial y} - \dfrac{\partial u}{\partial y}\cdot\dfrac{\partial v}{\partial x}} \quad \text{or} \quad \frac{dx}{\dfrac{\partial(u,v)}{\partial(y,z)}} = \frac{dy}{\dfrac{\partial(u,v)}{\partial(z,x)}} = \frac{dz}{\dfrac{\partial(u,v)}{\partial(x,y)}} \quad \text{or} \quad \frac{dx}{P} = \frac{dy}{Q} = \frac{dz}{R} \qquad \dots(9)$$

Thus equation (9) is known as Lagrange's auxiliary equations or Lagrange's subsidiary equations.

(2) GEOMETRICAL INTERPRETATION OF LAGRANGE'S LINEAR DIFFERENTIAL EQUATION

Lagrange's Linear differential equation is $Pp + Qq = R$ ...(1)

where $p = \dfrac{\partial z}{\partial x}$, $q = \dfrac{\partial z}{\partial y}$ and P, Q, R are the functions of x, y and z.

Equation (1) can be written as

$Pp + Qq - R = 0$ or $Pp + Qq + R(-1) = 0$ ...(2)

Lagrange's auxiliary equations are

$$\frac{dx}{P} = \frac{dy}{Q} = \frac{dz}{R}.$$ ...(3)

These equations represent a family of curves and P, Q, R are the direction ratio of the tangent drawn at any point on the curves.

Since $f(u, v) = 0$ represents a surface through these curves, where $u = a$ (constant) and $v = b$ (constant) are the two particular integrals of the equation (3) and are the function of x, y and z. Further since, we know that the direction cosines of the normal to the surface $f(x, y, z) = 0$ at any point on it are proportional to $\dfrac{\partial f}{\partial x} : \dfrac{\partial f}{\partial y} : \dfrac{\partial f}{\partial z}$

Dividing by $\dfrac{\partial f}{\partial z}$, we get $\dfrac{\partial f / \partial x}{\partial f / \partial z} : \dfrac{\partial f / \partial y}{\partial f / \partial z} : 1$ ...(4)

Since $p = \dfrac{\partial z}{\partial x} = -\dfrac{\partial f / \partial x}{\partial f / \partial z}$ and $q = \dfrac{\partial z}{\partial y} = -\dfrac{\partial f / \partial y}{\partial f / \partial z}$, then (4) becomes $-p : -q : 1$ or $p : q : -1$

Thus equation (2) represents that the normal at any point on the surface is perpendicular to the tangent to the curve obtained by equation (3) through which this surface passes. Hence, we say that the equations (1) and (3) give the same equivalent surfaces.

8.13.2 Lagrange's Linear Differential Equation with more than two Independent Variables

Let us consider a Lagrange's linear differential equation with n independent variables :

$$P_1 p_1 + P_2 p_2 + \ldots + P_n p_n = R$$ ...(1)

where $p_i = \dfrac{\partial z}{\partial x_i}$ for $i = 1, 2, 3, \ldots, n$ and $P_1, P_2, \ldots, P_n$ and R are the functions of $x_1, x_2, \ldots, x_n$ and z. Lagrange's auxiliary equations are

$$\frac{dx_1}{P_1} = \frac{dx_2}{P_2} = \ldots = \frac{dx_n}{P_n} = \frac{dz}{R}$$ ...(2)

Let $u_1 = a_1$ (constant), $u_2 = a_2$ (constant), ..., $u_n = a_n$ (constant) be n independent integrals of (2). Then the general integral of (1) is

$$f(u_1, u_2, \ldots, u_n) = 0.$$

WORKING PROCEDURE

To find the solution of the partial differential equation, use the following steps :

Step 1. First change the linear partial differential equation into a standard form $Pp + Qq = R$.

Step 2. Find the Lagrange's auxiliary equations as follows :

$$\frac{dx}{P} = \frac{dy}{Q} = \frac{dz}{R}.$$

Step 3. Now find two independent integrals say $u = a$ (constant) and $v = b$ (constant) from auxiliary equations. Then we obtained the general integral of the given differential equation $f(u, v) = 0$.

REMARKS

- Sometimes we use a set of multipliers to find $u = a$ and $v = b$.
- Set of multipliers may be functions of $x, y, z, \ldots$ etc. or may be constants.

Solved Examples

EXAMPLE 1. *Solve the differential equation $yzp + zxq = xy$.*

SOLUTION. Compare the given partial differential equation with $Pp + Qq = R$

we get $P = yz, Q = zx$ and $R = xy$.

Then the subsidiary equations are

$$\frac{dx}{P} = \frac{dy}{Q} = \frac{dz}{R} \quad \text{or} \quad \frac{dx}{yz} = \frac{dy}{zx} = \frac{dz}{xy}$$ ...(1)

Taking the first two members of (1), we get

$$\frac{dx}{yz} = \frac{dy}{zx} \quad \text{or} \quad x\,dx - y\,dy = 0.$$

Integrating, we get $x^2 - y^2 = c_1$ ...(2)

Now taking second and third members of (1), we get

$$\frac{dy}{zx} = \frac{dz}{xy} \quad \text{or} \quad y\,dy - z\,dz = 0.$$

Integrating, we get $y^2 - z^2 = c_2$ ...(3)

Thus the general solution is

$$f(x^2 - y^2, y^2 - z^2) = 0.$$

EXAMPLE 2. *Solve the partial differential equation*

$$pz - qz = z^2 + (x + y)^2.$$

SOLUTION. Compare the given partial differential equation with the standard partial differential equation $Pp + Qq = R$, we

get
$$P = z, Q = -z, \text{ and } R = z^2 + (x+y)^2$$
The subsidiary equations are given by
$$\frac{dx}{P} = \frac{dy}{Q} = \frac{dz}{R} \Rightarrow \frac{dx}{z} = \frac{dy}{-z} = \frac{dz}{z^2 + (x+y)^2} \quad \ldots(1)$$
Taking first and second ratio of (1), we get
$$\frac{dx}{z} = \frac{dy}{-z} \Rightarrow dx = -dy \Rightarrow dx + dy = 0$$
$$\Rightarrow x + y = c_1 \quad \text{(On integrating)}$$
Now taking first and third ratio of (1), we get
$$\frac{dx}{z} = \frac{dz}{z^2 + (x+y)^2} \quad \text{or} \quad dx = \frac{z\,dz}{z^2 + (x+y)^2}$$
$$\text{or} \quad dx = \frac{z\,dz}{z^2 + c_1^2} \qquad [\because x + y = c_1]$$
On integrating, we get
$$2x = \log(z^2 + c_1^2) + \log c_2 \quad \text{or} \quad e^{2x} = c_2(z^2 + c_1^2)$$
$$e^{2x} = c_2[z^2 + (x+y)^2]$$
$$\Rightarrow \quad c_2 = \frac{e^{2x}}{x^2 + y^2 + 2xy + z^2}$$
Thus the general integral is given by
$$f\left(x+y, \frac{e^{2x}}{x^2 + y^2 + z^2 + 2xy}\right) = 0.$$

EXAMPLE 3. *Solve $xzp + yzq = xy$.*
SOLUTION. Compare the given differential equation with Lagrange's linear differential equation $Pp + Qq = R$, we get
$$P = xz, Q = yz, R = xy.$$
Then, the Lagrange's subsidiary equations are
$$\frac{dx}{P} = \frac{dy}{Q} = \frac{dz}{R} \Rightarrow \frac{dx}{xz} = \frac{dy}{yz} = \frac{dz}{xy} \quad \ldots(1)$$
Taking first and second ratio of (1), we get
$$\frac{dx}{xz} = \frac{dy}{yz} \Rightarrow \frac{dx}{x} = \frac{dy}{y} \Rightarrow \frac{dx}{x} - \frac{dy}{y} = 0.$$
On integrating, we get $\log x - \log y = \log c_1$
or $\dfrac{x}{y} = c_1$.
Now, taking second and third ratio of (1), we get
$$\frac{dy}{yz} = \frac{dz}{xy} \Rightarrow \frac{dy}{z} = \frac{dz}{x}$$
$$\Rightarrow xdy = zdz \Rightarrow c_1 ydy = zdz \qquad [\because x = c_1 y]$$
On integrating, we get $c_1 y^2 - z^2 = c_2$
or $\left(\dfrac{x}{y}\right) y^2 - z^2 = c_2$
or $xy - z^2 = c_2 \qquad \left[\because c_1 = \dfrac{x}{y}\right]$
Thus the general integral is $f\left(\dfrac{x}{y}, xy - z^2\right) = 0.$

EXAMPLE 4. *Solve $p \tan x + q \tan y = \tan z$.*
SOLUTION. Compare the given differential equation with Lagrange's Linear differential equation $Pp + Qq = R$, we get
$$P = \tan x, Q = \tan y, R = \tan z.$$
Then, the Lagrange's subsidiary equations are
$$\frac{dx}{P} = \frac{dy}{Q} = \frac{dz}{R} \Rightarrow \frac{dx}{\tan x} = \frac{dy}{\tan y} = \frac{dz}{\tan z} \quad \ldots(1)$$

Taking first and second ratio of (1), we get
$$\frac{dx}{\tan x} = \frac{dy}{\tan y} \Rightarrow \frac{\cos x}{\sin x} dx = \frac{\cos y}{\sin y} dy.$$
On integrating, we get
$$\log \sin x - \log \sin y = \log c_1$$
or $\log \dfrac{\sin x}{\sin y} = \log c_1$ or $\dfrac{\sin x}{\sin y} = c_1$.
Now taking second and third ratio of (1), we get
$$\frac{dy}{\tan y} = \frac{dz}{\tan z} \Rightarrow \frac{\cos y}{\sin y} dy = \frac{\cos z}{\sin z} dz.$$
On integrating, we get
$$\log \sin y - \log \sin z = \log c_2$$
or $\log \dfrac{\sin y}{\sin z} = \log c_2$ or $\dfrac{\sin y}{\sin z} = c_2$.
Thus, the general integral is $f\left(\dfrac{\sin x}{\sin y}, \dfrac{\sin y}{\sin z}\right) = 0$
or $\dfrac{\sin x}{\sin y} = f\left(\dfrac{\sin y}{\sin z}\right)$.

EXAMPLE 5. *Solve the following differential equation*
$$(z^2 - 2yz - y^2)p + (xy + zx)q = xy - zx.$$
(KERALA–2005)
SOLUTION. Compare the given differential equation with Lagrange's differential equation $Pp + Qq = R$, we get
$$P = z^2 - 2yz - y^2, Q = xy + zx, R = xy - zx.$$
Then, the Lagrange's subsidiary equations are
$$\frac{dx}{P} = \frac{dy}{Q} = \frac{dz}{R}$$
$$\Rightarrow \frac{dx}{z^2 - 2yz - y^2} = \frac{dy}{xy + zx} = \frac{dz}{xy - zx} \quad \ldots(1)$$
Taking second and third ratio of (1), we get
$$\frac{dy}{xy + zx} = \frac{dz}{xy - zx} \quad \text{or} \quad \frac{dy}{y+z} = \frac{dz}{y-z}$$
or $(y-z)dy = (y+z)dz$
or $ydy - zdy = ydz + zdz$
or $ydy - zdy - ydz - zdz = 0$
or $ydy - (zdy + ydz) - zdz = 0$
or $ydy - d(yz) - zdz = 0$.
On integrating, we get $y^2 - 2yz - z^2 = c_1$.
Now taking the multipliers x, y, z, we get
$$\frac{dx}{z^2 - 2yz - y^2} = \frac{dy}{xy + zx} = \frac{dz}{xy - zx}$$
$$= \frac{xdx + ydy + zdz}{x(z^2 - 2yz - y^2) + y(xy + zx) + z(xy - zx)}$$
$$= \frac{xdx + ydy + zdz}{0}.$$
$$\therefore \quad xdx + ydy + zdz = 0.$$
Integrating, we get $x^2 + y^2 + z^2 = c_2$.
Thus the general integral is
$$f(y^2 - 2yz - z^2, x^2 + y^2 + z^2) = 0.$$

EXAMPLE 6. *Solve $(y+z)p + (z+x)q = x+y$.*
SOLUTION. Compare the given differential equation with Lagrange's differential equation $Pp + Qq = R$, we get
$$P = y+z, Q = z+x, R = x+y.$$

Then Lagrange's auxiliary equations are

$$\frac{dx}{P} = \frac{dy}{Q} = \frac{dz}{R} \Rightarrow \frac{dx}{y+z} = \frac{dy}{z+x} = \frac{dz}{x+y} \quad \ldots(1)$$

Equation (1) can also be taken as

$$\frac{dx-dy}{y-x} = \frac{dy-dz}{z-y} = \frac{dx+dy+dz}{2(x+y+z)} \quad \ldots(2)$$

Taking first and second ratio of (2), we get

$$\frac{dx-dy}{y-x} = \frac{dy-dz}{z-y} \quad \text{or} \quad \frac{dx-dy}{x-y} = \frac{dy-dz}{y-z}$$

Let $u = x - y$ and $v = y - z$

$$\therefore \quad du = dx - dy \text{ and } dv = dy - dz$$

Then $\dfrac{dx-dy}{x-y} = \dfrac{dy-dz}{y-z}$ becomes $\dfrac{du}{u} = \dfrac{dv}{v}$

Integrating, we get

$$\log u = \log v + \log c_1$$
$$\log u - \log v = \log c_1$$
$$\log \frac{u}{v} = \log c_1 \Rightarrow \frac{u}{v} = c_1$$
$$\Rightarrow \quad \frac{x-y}{y-z} = c_1 \qquad [\because u = x-y, v = y-z]$$

Now taking second and third ratio of (2), we get

$$\frac{dy-dz}{z-y} = \frac{dx+dy+dz}{2(x+y+z)} \quad \ldots(3)$$

Let $t = y - z$ and $s = x + y + z$

$$\therefore \quad dt = dy - dz \text{ and } ds = dx + dy + dz$$

Hence, equation (3) becomes $\dfrac{dt}{-t} = \dfrac{ds}{2s}$

or $\quad 2\dfrac{dt}{t} + \dfrac{ds}{s} = 0.$

Integrating, we get

$$\log t^2 + \log s = \log c_2$$

or $\quad \log(t^2 s) = \log c_2 \quad$ or $\quad t^2 s = c_2$

$$\Rightarrow (y-z)^2(x+y+z) = c_2 \quad [\because t = y-z, s = x+y+z]$$

Hence, the general integral is

$$f\left(\frac{x-y}{y-z}, (y-z)^2(x+y+z)\right) = 0.$$

EXAMPLE 7. *Find the general solution of the following differential equation*

$$(mz - ny)p + (nx - lz)q = ly - mx.$$

(GBTU(AG)–2012, VTU–2010, SVTU–2009)

SOLUTION. Compare the given differential equation with Lagrange's differential equation $Pp + Qq = R$, we get

$$P = mz - ny, Q = nx - lz, R = ly - mx.$$

Then, Lagrange's auxiliary equations are

$$\frac{dx}{P} = \frac{dy}{Q} = \frac{dz}{R}$$
$$\Rightarrow \quad \frac{dx}{mz - ny} = \frac{dy}{nx - lz} = \frac{dz}{ly - mx} \quad \ldots(1)$$

Taking the multipliers x, y, z then (1) becomes

$$\frac{dx}{mz - ny} = \frac{dy}{nx - lz}$$
$$= \frac{dz}{ly - mx} = \frac{xdx + ydy + zdz}{0}$$

$$\therefore \quad xdx + ydy + zdz = 0.$$

Integrating, we get $x^2 + y^2 + z^2 = c_1$.

Again taking the multipliers l, m, n then (1) becomes

$$\frac{dx}{mz - ny} = \frac{dy}{nx - lz} = \frac{dz}{ly - mx} = \frac{ldx + mdy + ndz}{0}$$

$$\therefore \quad ldx + mdy + ndz = 0.$$

Integrating, we get

$$lx + my + nz = c_2.$$

Hence, the general solution is

$$f(x^2 + y^2 + z^2, lx + my + nz) = 0.$$

EXAMPLE 8. *Find the general solution of the following differential equation*

$$x(y^2 + z)p - y(x^2 + z)q = z(x^2 - y^2).$$

(UPTU–2008)

SOLUTION. Comparing the given differential equation with Lagrange's differential equation $Pp + Qq = R$, we get

$$P = x(y^2 + z), Q = -y(x^2 + z), R = z(x^2 - y^2).$$

Then the Lagrange's auxiliary equations are

$$\frac{dx}{P} = \frac{dy}{Q} = \frac{dz}{R}$$
$$\Rightarrow \frac{dx}{x(y^2 + z)} = \frac{dy}{-y(x^2 + z)} = \frac{dz}{z(x^2 - y^2)} \quad \ldots(1)$$

Taking the multipliers $x, y, -1$, then (1) becomes

$$\frac{dx}{x(y^2 + z)} = \frac{dy}{-y(x^2 + z)} = \frac{dz}{z(x^2 - y^2)}$$
$$= \frac{xdx + ydy - dz}{0}$$

$$\therefore \quad xdx + ydy - dz = 0.$$

Integrating, we have $x^2 + y^2 - 2z = c_1$.

Again taking the multipliers $\dfrac{1}{x}, \dfrac{1}{y}, \dfrac{1}{z}$ then (1) becomes

$$\frac{dx}{x(y^2 + z)} = \frac{dy}{-y(x^2 + z)} = \frac{dz}{z(x^2 - y^2)}$$
$$= \frac{\dfrac{1}{x}dx + \dfrac{1}{y}dy + \dfrac{1}{z}dz}{0}$$

$$\therefore \quad \frac{1}{x}dx + \frac{1}{y}dy + \frac{1}{z}dz = 0.$$

Integrating, we have

$$\log x + \log y + \log z = \log c_2$$

or $\quad xyz = c_2.$

Thus, the general solution is

$$f(x^2 + y^2 - 2z, xyz) = 0.$$

EXAMPLE 9. *Find the general solution of the following differential equation $p + 3q = 5z + \tan(y - 3x).$*

SOLUTION. Compare the given differential equation with Lagrange's differential equation $Pp + Qq = R$, we get

$$P = 1, Q = 3, R = 5z + \tan(y - 3x)$$

Then the Lagrange's auxiliary equations are

$$\frac{dx}{1} = \frac{dy}{3} = \frac{dz}{5z + \tan(y - 3x)} \quad \ldots(1)$$

Taking first and second ratio of (1), we get

$$\frac{dx}{1} = \frac{dy}{3} \qquad \text{or } dy - 3dx = 0$$

Integrating, we get $y - 3x = c_1$.

Now taking first and third ratio of (1), we get

$$\frac{dx}{1} = \frac{dz}{5z + \tan(y - 3x)} \quad \text{or} \quad \frac{dx}{1} = \frac{dz}{5z + \tan c_1}$$

$$[\because y - 3x = c_1]$$

or $\qquad 5dx = \dfrac{5dz}{5z + \tan c_1}$

Integrating, we get

$$5x = \log(5z + \tan c_1) + \log c_2$$

$$5x = \log c_2(5z + \tan c_1) \quad \text{or} \quad e^{5x} = c_2(5z + \tan c_1)$$

or $\qquad e^{5x} = c_2[5z + \tan(y - 3x)] \qquad [\because c = y - 3x]$

Hence, the general solution is

$$e^{5x} = [5z + \tan(y - 3x)]\, f(y - 3x).$$

EXAMPLE 10. *Find the general integral of the following differential equation*

$$x^2(y - z)\,p + (z - x)\,y^2 q = z^2(x - y).$$

(GBTU(AG)–2012, PTU–2009, BHOPAL–2008, SVTU–2007)

SOLUTION. Compare the given differential equation with Lagrange's differential equation $Pp + Qq = R$, we get

$$P = x^2(y - z), \quad Q = (z - x)y^2, \quad R = z^2(x - y)$$

Then the Lagrange's auxiliary equations are

$$\frac{dx}{P} = \frac{dy}{Q} = \frac{dz}{R}$$

$$\frac{dx}{x^2(y - z)} = \frac{dy}{(z - x)y^2} = \frac{dz}{z^2(x - y)} \qquad \text{...(1)}$$

Taking the multipliers $\dfrac{1}{x^2}, \dfrac{1}{y^2}, \dfrac{1}{z^2}$, then (1) becomes

$$\frac{dx}{x^2(y - z)} = \frac{dy}{(z - x)y^2} = \frac{dz}{z^2(x - y)}$$

$$= \frac{\dfrac{1}{x^2}dx + \dfrac{1}{y^2}dy + \dfrac{1}{z^2}dz}{0}$$

$$\therefore \quad \frac{dx}{x^2} + \frac{dy}{y^2} + \frac{dz}{z^2} = 0.$$

Integrating, we get $\dfrac{1}{x} + \dfrac{1}{y} + \dfrac{1}{z} = c_1$.

Now taking the multipliers $\dfrac{1}{x}, \dfrac{1}{y}, \dfrac{1}{z}$ then (1) becomes

$$\frac{dx}{x^2(y - z)} = \frac{dy}{(z - x)y^2} = \frac{dz}{z^2(x - y)}$$

$$= \frac{\dfrac{1}{x}dx + \dfrac{1}{y}dy + \dfrac{1}{z}dz}{0}$$

$$\therefore \quad \frac{1}{x}dx + \frac{1}{y}dy + \frac{1}{z}dz = 0.$$

Integrating, we get $\log x + \log y + \log z = \log c$

or $\log xyz = \log c_2 \quad$ or $\quad xyz = c_2$.

Hence, the general integral is $\dfrac{1}{x} + \dfrac{1}{y} + \dfrac{1}{z} = f(xyz)$.

EXAMPLE 11. *Solve $y^2 p - xyq = x(z - 2y)$.* (SVTU–2008)

SOLUTION. Compare the given differential equation with Lagrange's differential equation $Pp + Qq = R$, we get

$$P = y^2, Q = -xy, R = x(z - 2y).$$

Then the Lagrange's auxiliary equations are

$$\frac{dx}{P} = \frac{dy}{Q} = \frac{dz}{R}$$

$$\Rightarrow \quad \frac{dx}{y^2} = \frac{dy}{-xy} = \frac{dz}{x(z - 2y)} \qquad \text{...(1)}$$

Taking first and second ratio of (1), we get

$$\frac{dx}{y^2} = \frac{dy}{-xy} \quad \text{or} \quad \frac{dx}{y} = \frac{dy}{-x} \quad \text{or} \quad xdx + ydy = 0.$$

Integrating, we get $x^2 + y^2 = c_1$.

Taking second and third ratio of (1), we get

$$\frac{dy}{-xy} = \frac{dz}{x(z - 2y)} \quad \text{or} \quad (z - 2y)dy = -ydz$$

or $zdy + ydz - 2ydy = 0 \quad$ or $\quad d(yz) - 2y\,dy = 0$.

Integrating, we get $yz - y^2 = c_2$.

Hence, the general integral is $f(x^2 + y^2, yz - y^2) = 0$.

EXAMPLE 12. *Solve $(y^2 + z^2)\,p - xyp = -zx$.*

(MTU–2011, PTU–2009, VTU–2009)

SOLUTION. Compare the given differential equation with Lagrange's differential equation $Pp + Qq = R$, we get

$$P = y^2 + z^2, Q = -xy, R = -zx.$$

Then the Lagrange's auxiliary equations are

$$\frac{dx}{P} = \frac{dy}{Q} = \frac{dz}{R}$$

$$\frac{dx}{y^2 + z^2} = \frac{dy}{-xy} = \frac{dz}{-zx} \qquad \text{...(1)}$$

Taking second and third ratio of (1), we get

$$\frac{dy}{-xy} = \frac{dz}{-zx} \quad \text{or} \quad \frac{dy}{y} = \frac{dz}{z}$$

Integrating, we get $\log y = \log z + \log c_1$

or $\qquad \dfrac{y}{z} = c_1$.

Now taking the multipliers x, y, z then (1) becomes

$$\frac{dx}{y^2 + z^2} = \frac{dy}{-xy} = \frac{dz}{-zx} = \frac{xdx + ydy + zdz}{0}$$

$$\therefore \qquad xdx + ydy + zdz = 0.$$

Integrating, we get $x^2 + y^2 + z^2 = c_2$.

Hence, the general integral is $f\left(\dfrac{y}{z}, x^2 + y^2 + z^2\right) = 0$.

EXAMPLE 13. *Solve $(x^2 - y^2 - z^2)p + 2xyp = 2xz$.*

(VTU–2010, ANNA–2009, SVTU – 2008)

SOLUTION. The subsidiary equation of the given equation is

$$\frac{dx}{x^2 - y^2 - z^2} = \frac{dy}{2xy} = \frac{dz}{2xz} \qquad \text{...(1)}$$

From last two fractions, we get $\dfrac{dy}{y} = \dfrac{dz}{z}$

$\Rightarrow \quad \log y = \log z + \log c_1$

$\Rightarrow \quad \dfrac{y}{z} = c_1$

Further using multipliers x, y, z in (1), we get

each fractions $= \dfrac{xdx + ydy + zdz}{x(x^2 + y^2 + z^2)}$

$\Rightarrow \quad \dfrac{2xdx + 2ydy + 2zdz}{x^2 + y^2 + z^2} = \dfrac{dz}{z}$

On integrating we get

$$\log(x^2 + y^2 + z^2) = \log z + \log c_2$$

$$\Rightarrow \qquad \frac{x^2 + y^2 + z^2}{z} = c_2$$

Hence, the general solution is given by

$$f\left(\frac{y}{z}, \frac{x^2 + y^2 + z^2}{z}\right) = 0.$$

EXAMPLE 14. *Solve* $x(x^2 - z^2)p + y(z^2 - x^2)q = z(x^2 - y^2)$

(UPTU–2008)

SOLUTION. Lagrange's subsidiary equation of the given equation is

$$\frac{dx}{x(x^2 - z^2)} = \frac{dy}{y(z^2 - x^2)} = \frac{dz}{z(x^2 - y^2)}$$

Using multiplers, x, y, z, we get
each fraction

$$= \frac{xdx + ydy + zdz}{x^2(y^2 - z^2) + y^2(z^2 - x^2) + z^2(x^2 - y^2)}$$

$$= \frac{xdx + ydy + zdz}{0}$$

$$\Rightarrow \qquad xdx + ydy + zdz = 0$$

$$\text{or} \qquad x^2 + y^2 + z^2 = c_1$$

Again using multipliers $\dfrac{1}{x}, \dfrac{1}{y}, \dfrac{1}{z}$ we get

$$\text{each fraction} = \frac{\dfrac{1}{x}dx + \dfrac{1}{y}dy + \dfrac{1}{z}dz}{y^2 - z^2 + z^2 - x^2 + x^2 - y^2}$$

$$= \frac{\dfrac{1}{x}dx + \dfrac{1}{y}dy + \dfrac{1}{z}dz}{0}$$

$$\Rightarrow \qquad \frac{1}{x}dx + \frac{1}{y}dy + \frac{1}{z}dz = 0$$

On integration, we get

$$\log x + \log y + \log z = \log c_2$$

$$xyz = c_2$$

Hence, the general solution is given by

$$f(x^2 + y^2 + z^2, xyz) = 0$$

EXAMPLE 15. *Solve* $(x^2 - yz)p + (y^2 - zx)q = z^2 - xy$

(GBTU–2010, BHOPAL–2008, VTU–2006, MADRAS–2000)

SOLUTION. The Lagrange's subsidiary equation is given by

$$\frac{dx}{x^2 - yz} = \frac{dy}{y^2 - zx} = \frac{dz}{z^2 - xy}$$

$$\Rightarrow \qquad \frac{dx - dy}{(x - y)(x + y + z)} = \frac{dy - dz}{(y - z)(x + y + z)}$$

$$= \frac{dz - dx}{(z - x)(x + y + z)}$$

Consider

$$\frac{dx - dy}{(x - y)(x + y + z)} = \frac{dy - dz}{(y - z)(x + y + z)}$$

$$\Rightarrow \qquad \frac{dx - dy}{(x - y)} = \frac{dy - dz}{(y - z)}$$

$$\Rightarrow \qquad \log (x - y) = \log (y - z) + \log c_1$$

$$\Rightarrow \qquad \log\left(\frac{x - y}{y - z}\right) = \log c_1$$

$$\Rightarrow \qquad \frac{x - y}{y - z} = c_1$$

Similarly, taking the last two members, we get

$$\frac{y - z}{z - x} = c_2$$

Hence, general solution is given by

$$f\left(\frac{x - y}{y - z}, \frac{y - z}{z - x}\right) = 0.$$

Exercise-8.59

Find the general integrals of the linear partial differential equations :

1. $\left(\dfrac{y - z}{yz}\right)p + \left(\dfrac{z - x}{zx}\right)q = \left(\dfrac{x - y}{xy}\right).$ (BHOPAL–2007)

2. $\dfrac{y^2 z}{x}p + zxq = y^2.$

3. $p + q = \dfrac{z}{a}.$

4. $px(z - 2y^2) = (z - qy)(z - y^2 - 2x^3).$

5. $(y^3 x - 2x^4)p + (2y^4 - x^3 y)q = 9z(x^3 - y^3).$

6. $px(x + y) - qy(x + y) = -(x - y)(2x + 2y + z).$

7. $(y + zx)p - (x + yz)q = x^2 - y^2.$

8. $x(x^2 + 3y^2)p - y(3x^2 + y^2)q = 2z(y^2 - x^2).$

9. $(x + y)(p - q) = z.$

10. $x^2 p + y^2 q = z^2.$

11. $p + q = 1.$

12. $p_2 + p_3 = 1 + p_1$ where $p_i \dfrac{\partial z}{\partial x_i}, i = 1, 2, 3.$

13. $x_2 x_3 z p_1 + x_3 x_1 z p_2 + x_1 x_2 z p_3 = x_1 x_2 x_3.$

14. $x\dfrac{\partial z}{\partial x} + y\dfrac{\partial z}{\partial y} + t\dfrac{\partial z}{\partial t} = az + \dfrac{xy}{t}.$

15. $(y + z + t)\dfrac{\partial t}{\partial x} + (z + x + t)\dfrac{\partial t}{\partial y} + (x + y + t)\dfrac{\partial t}{\partial z} = x + y + z.$

16. $\dfrac{1}{\sqrt{z}}(p_1 + x_1 p_2 + x_1 x_2 p_3) = x_1 x_2 x_3.$

17. $(3x + y - z)p + (x + y - z)q = 2(z - y).$

18. Solve $x\dfrac{\partial u}{\partial x} + y\dfrac{\partial u}{\partial y} + z\dfrac{\partial u}{\partial z} = xyz.$

19. Solve $p_1 + p_2 + p_3 = 4z$, where $p_1 = \dfrac{\partial z}{\partial x_1}, p_2 = \dfrac{\partial z}{\partial x_2}, p_3 = \dfrac{\partial z}{\partial x_3}.$

20. Solve $(2x^2 + y^2 + z^2 - 2yz - 2x - xy)p$

$+ (x^2 + 2y^2 + z^2 - yz - 2zx - xy)q$

$= x^2 + y^2 + 2z^2 - yz - zx - 2xy.$

Hints to Selected Problems

1. The Lagrange's subsidiary equation is given by

$$\frac{dx}{\left(\dfrac{y - z}{yz}\right)} = \frac{dy}{\left(\dfrac{z - x}{zx}\right)} = \frac{dx}{\left(\dfrac{x - y}{xy}\right)}.$$

Now taking the following two sets of multipliers
(i) $1, 1, 1$ (ii) yz, zx, xy

5. Lagrange's auxiliary equations are given by

$$\frac{dx}{y^3x - 2x^4} = \frac{dy}{2y^4 - x^3y} = \frac{dz}{9z(x^3 - y^3)}.$$

Now firstly, taking first and second ratio. For second integral,

taking $\dfrac{1}{x}, \dfrac{1}{y}, \dfrac{1}{3z}$ as a set of multipliers.

6. Lagrange's auxiliary equations are

$$\frac{dx}{x(x+y)} = \frac{dy}{-y(x+y)} = \frac{dz}{-(x-y)(2x+2y+z)}. \qquad ...(1)$$

By taking first and second ratio, we get $xy = C_1$.

Also, from (1)

$$\frac{dx+dy}{(x-y)(x+y)} = \frac{dx+dy+dz}{(x-y)(x+y) - (x-y)(2x+2y+z)}$$

$$= \frac{dx+dy+dz}{(x-y)\{(x+y) - 2x - 2y - z\}}$$

$$\Rightarrow \frac{dx+dy}{x+y} = \frac{dx+dy+dz}{-(x+y+z)}.$$

Now substituting $x + y = u$ and $x + y + z = v$.

7. Here, we have $\dfrac{dx}{y+zx} = \dfrac{dy}{-(x+yz)} = \dfrac{dz}{x^2 - y^2}$. Then taking following two set of multipliers

(i) $y, x, 1$ (ii) $x, y, -z$.

13. The Lagrange's auxiliary equations are

$$\frac{dx_1}{x_2x_3z} = \frac{dx_2}{x_3x_1z} = \frac{dx_3}{x_1x_2z} = \frac{dx_1}{x_1x_2x_3}.$$

Answers

1. $f(x+y+z, xyz) = 0$ **2.** $f(x^3 - y^3, x^2 - z^2) = 0$ **3.** $z = e^{y/a}f(x-y)$ **4.** $z - y^2 + x^3 = xf\left(\dfrac{y}{z}\right)$

5. $f\left(\dfrac{x^3+y^3}{x^2y^2}, xyz^{1/3}\right) = 0$ **6.** $(x+y)(x+y+z) = f(xy)$ **7.** $f(x^2+y^2-z^2, xy+z) = 0$

8. $(x^2+y^2)z = f\left(\dfrac{xy}{z}\right)$ **9.** $(x+y)\log z = x + f(x+y)$ **10.** $\dfrac{1}{x} - \dfrac{1}{y} = f\left(\dfrac{1}{x} - \dfrac{1}{z}\right)$ **11.** $f(x-z, y-z) = 0$

12. $f(x_1+z, x_1+x_2, x_1+x_3) = 0$ **13.** $f(x_2^2 - x_1^2, x_3^2 - x_1^2, z^2 - x_1^2) = 0$ **14.** $f\left(\dfrac{y}{x}, \dfrac{t}{x}, zx^{-a} - \dfrac{y}{t} \cdot \dfrac{x^{1-a}}{1-a}\right) = 0$

15. $f\left[\dfrac{x-y}{y-z}, \dfrac{z-t}{y-z}, (z-t)(x+y+z+t)^{1/3}\right] = 0$ **16.** $f(4\sqrt{z} - x_3^2, 2x_3 - x_2^2, 2x_2 - x_1^2) = 0$.   **17.** $f\left(x - 3y - z, \dfrac{x-y+z}{\sqrt{x+y-z}}\right) = 0$

18. $f\left(\dfrac{x}{y}, \dfrac{y}{z}, xyz - 34\right) = 0$ **19.** $f\left(\dfrac{z}{e^{4x_1}}, \dfrac{z}{e^{4x_2}}, \dfrac{z}{e^{4x_3}}\right) = 0$ **20.** $f\left(\dfrac{x-y}{y-z}, \dfrac{y-z}{z-x}\right) = 0$

8.13.3 CHARPIT'S METHOD

The Charpit's method of solving the partial differential equation

$$f(x, y, z, p, q) = 0 \qquad ...(1)$$

is based on the introduction of a second partial differential equation of the first order. Let this second partial differential equation be

$$F(x, y, z, p, q, a) = 0 \qquad ...(2)$$

which contains an arbitrary constant 'a'. Now solving (1) and (2), to get $p = p(x, y, z, a)$ and $q = q(x, y, z, a)$ and substitute p and q into

$$dz = pdx + qdy \qquad ...(3)$$

The equation (3) is integrable and the integral given by (3) will satisfy the given equation (1), for the values of p and q obtained from it will be same as the values of p and q given in (1). Thus z, p and q may be the functions of x and y and are satisfied by both (1) and (2), Therefore we have

$$\frac{df}{dx} = 0, \quad \frac{df}{dy} = 0 \text{ and } \frac{dF}{dx} = 0, \frac{dF}{dy} = 0.$$

Thus, differentiating (1) and (2) with respect to x and y, we get

$$\frac{\partial f}{\partial x} + \frac{\partial f}{\partial z} \cdot \frac{\partial z}{\partial x} + \frac{\partial f}{\partial p} \cdot \frac{\partial p}{\partial x} + \frac{\partial f}{\partial q} \cdot \frac{\partial q}{\partial x} = 0 \quad \text{or} \quad \frac{\partial f}{\partial x} + \frac{\partial f}{\partial z} \cdot p + \frac{\partial f}{\partial p} \cdot \frac{\partial p}{\partial x} + \frac{\partial f}{\partial q} \cdot \frac{\partial q}{\partial x} = 0 \qquad ...(4)$$

and $\dfrac{\partial f}{\partial y} + \dfrac{\partial f}{\partial z} \cdot \dfrac{\partial z}{\partial y} + \dfrac{\partial f}{\partial p} \cdot \dfrac{\partial p}{\partial y} + \dfrac{\partial f}{\partial q} \cdot \dfrac{\partial q}{\partial y} = 0$ or $\dfrac{\partial f}{\partial y} + \dfrac{\partial f}{\partial z} \cdot q + \dfrac{\partial f}{\partial p} \cdot \dfrac{\partial p}{\partial y} + \dfrac{\partial f}{\partial q} \cdot \dfrac{\partial q}{\partial y} = 0$ ...(5)

and $\dfrac{\partial F}{\partial x} + \dfrac{\partial F}{\partial z} \cdot \dfrac{\partial z}{\partial x} + \dfrac{\partial F}{\partial p} \cdot \dfrac{\partial p}{\partial x} + \dfrac{\partial F}{\partial q} \cdot \dfrac{\partial q}{\partial x} = 0$ or $\dfrac{\partial F}{\partial x} + \dfrac{\partial F}{\partial z} \cdot p + \dfrac{\partial F}{\partial p} \cdot \dfrac{\partial p}{\partial x} + \dfrac{\partial F}{\partial q} \cdot \dfrac{\partial q}{\partial x} = 0$ ...(6)

and $\dfrac{\partial F}{\partial y} + \dfrac{\partial F}{\partial z} \cdot \dfrac{\partial z}{\partial y} + \dfrac{\partial F}{\partial p} \cdot \dfrac{\partial p}{\partial y} + \dfrac{\partial F}{\partial q} \cdot \dfrac{\partial q}{\partial y} = 0$ or $\dfrac{\partial F}{\partial y} + \dfrac{\partial F}{\partial z} \cdot + \dfrac{\partial F}{\partial p} \cdot \dfrac{\partial p}{\partial y} + \dfrac{\partial F}{\partial q} \cdot \dfrac{\partial q}{\partial y} = 0$ ...(7)

Now eliminating $\dfrac{\partial p}{\partial x}$ from (4) and (6). For this multiplying (4) by $\dfrac{\partial F}{\partial p}$ and (6) by $\dfrac{\partial f}{\partial p}$ and subtract, we get

$$\left(\frac{\partial f}{\partial x} + \frac{\partial f}{\partial z}p + \frac{\partial f}{\partial q} \cdot \frac{\partial q}{\partial x}\right)\frac{\partial F}{\partial p} - \left(\frac{\partial F}{\partial x} + \frac{\partial F}{\partial z}p + \frac{\partial F}{\partial q} \cdot \frac{\partial q}{\partial x}\right)\frac{\partial f}{\partial p} = 0$$

and $\dfrac{\partial f}{\partial x}\cdot\dfrac{\partial F}{\partial p}-\dfrac{\partial F}{\partial x}\cdot\dfrac{\partial f}{\partial p}+\left(\dfrac{\partial f}{\partial z}\cdot\dfrac{\partial F}{\partial p}-\dfrac{\partial F}{\partial z}\cdot\dfrac{\partial f}{\partial p}\right)p+\left(\dfrac{\partial f}{\partial q}\cdot\dfrac{\partial F}{\partial p}-\dfrac{\partial F}{\partial q}\cdot\dfrac{\partial f}{\partial p}\right)\dfrac{\partial q}{\partial x}=0$ ...(8)

and eliminating $\partial q\,/\,\partial y$ from (5) and (7), so multiplying (5) by $\partial F\,/\,\partial q$ and multiplying (7) by $\dfrac{\partial f}{\partial q}$ and subtract, we get

$$\left(\dfrac{\partial f}{\partial y}+\dfrac{\partial f}{\partial z}.q+\dfrac{\partial f}{\partial p}\cdot\dfrac{\partial p}{\partial y}\right)\dfrac{\partial F}{\partial q}-\left(\dfrac{\partial F}{\partial y}+\dfrac{\partial F}{\partial z}q+\dfrac{\partial F}{\partial p}\cdot\dfrac{\partial p}{\partial y}\right)\dfrac{\partial f}{\partial q}=0$$

or $\dfrac{\partial f}{\partial y}\cdot\dfrac{\partial F}{\partial q}-\dfrac{\partial F}{\partial y}\cdot\dfrac{\partial f}{\partial q}+\left(\dfrac{\partial f}{\partial z}\cdot\dfrac{\partial F}{\partial q}-\dfrac{\partial F}{\partial z}\cdot\dfrac{\partial f}{\partial q}\right)q+\left(\dfrac{\partial f}{\partial p}\cdot\dfrac{\partial F}{\partial q}-\dfrac{\partial F}{\partial p}\cdot\dfrac{\partial f}{\partial q}\right)\dfrac{\partial p}{\partial y}=0$ ...(9)

Since, we know that $q=\dfrac{\partial z}{\partial y}\Rightarrow\dfrac{\partial q}{\partial x}=\dfrac{\partial^2 z}{\partial x\partial y}$ or $p=\dfrac{\partial z}{\partial x}\Rightarrow\dfrac{\partial p}{\partial y}=\dfrac{\partial^2 z}{\partial y\partial x}$. But $\dfrac{\partial^2 z}{\partial x\partial y}=\dfrac{\partial^2 z}{\partial y\partial x}\Rightarrow\dfrac{\partial q}{\partial x}=\dfrac{\partial p}{\partial y}$

Now adding (8) and (9), we get

$$\dfrac{\partial f}{\partial x}\cdot\dfrac{\partial F}{\partial p}-\dfrac{\partial F}{\partial x}\cdot\dfrac{\partial f}{\partial p}+\left(\dfrac{\partial f}{\partial z}\cdot\dfrac{\partial F}{\partial p}-\dfrac{\partial F}{\partial z}\cdot\dfrac{\partial f}{\partial p}\right)p+\dfrac{\partial f}{\partial y}\cdot\dfrac{\partial F}{\partial q}-\dfrac{\partial F}{\partial y}\cdot\dfrac{\partial f}{\partial q}+\left(\dfrac{\partial f}{\partial z}\cdot\dfrac{\partial F}{\partial q}-\dfrac{\partial F}{\partial z}\cdot\dfrac{\partial f}{\partial q}\right)q=0$$

or $\left(\dfrac{\partial f}{\partial x}+p\dfrac{\partial f}{\partial z}\right)\dfrac{\partial F}{\partial p}+\left(\dfrac{\partial f}{\partial y}+q\dfrac{\partial f}{\partial z}\right)\dfrac{\partial F}{\partial q}+\left(-p\dfrac{\partial f}{\partial p}-q\dfrac{\partial f}{\partial q}\right)\dfrac{\partial F}{\partial z}+\left(-\dfrac{\partial f}{\partial p}\right)\dfrac{\partial F}{\partial x}+\left(-\dfrac{\partial f}{\partial q}\right)\dfrac{\partial F}{\partial y}=0$ ...(10)

Thus, the equation (10) is a linear differential equation for the dependent variable F, so to determine F use the subsidiary equations

$$\dfrac{dx}{-\partial f\,/\,\partial p}=\dfrac{dy}{-\partial f\,/\,\partial q}=\dfrac{dz}{-p\dfrac{\partial f}{\partial p}-q\dfrac{\partial f}{\partial q}}=\dfrac{dp}{\dfrac{\partial f}{\partial x}+p\dfrac{\partial f}{\partial z}}=\dfrac{dq}{\dfrac{\partial f}{\partial y}+q\dfrac{\partial f}{\partial z}}$$ ...(11)

The equations given in (11) are known as Charpit's auxiliary equations. It should be noted that there is no need to use all of Charpit's auxiliary equations, therefore, we find a solution of (11) in which either p or q must occur for $F\equiv0$, then from $f\equiv0$ and $F\equiv0$, find the values of p and q and substitute these values of p and q into

$$dz=pdx+qdy$$

and then integrating, we obtain the solution of (1).

WORKING PROCEDURE

Step 1. Putting all terms of the equation in L.H.S. and denote the whole expression by f.

Step 2. Write Charpit's auxiliary equation.

Step 3. Find the values of different partial derivatives of f and put in Charpit's auxiliary equation.

Step 4. Select two proper fractions such that the resulting integral may come out in the simplest form involving at least one of p and q

Step 5. Solving this simplest relation (obtained in (4)) with the given equation to find p and q.

Step 6. Putting the values of p and q in $dz=pdx+qdy$.

REMARKS

- This method is applicable not only for the solution of the partial differential equations of order one but of any degree.
- This method is applied only when other methods are failed for finding the solution.
- From (11), we obtain $p=a$ (constant) and $q=b$ (constant), then after putting these values in (1), we get the complete integral.

Solved Examples

EXAMPLE 1. *Find the complete integral of the following partial differential equation* $z=px+qy+p^2+q^2$ *by Charpit's method.*

SOLUTION. Let us take $f\equiv z-px-qy-p^2-q^2=0$ ...(1)

and finding partial derivatives of f w.r.t. x,y,z,p and q respectively,

$$\dfrac{\partial f}{\partial x}=-p,\dfrac{\partial f}{\partial y}=-q,\dfrac{\partial f}{\partial z}=1,\dfrac{\partial f}{\partial p}=-x-2p,$$

$$\dfrac{\partial f}{\partial q}=-y-2q.$$

Thus, the Charpit's auxiliary equations are

$$\dfrac{dx}{-\partial f\,/\,\partial p}=\dfrac{dy}{-\partial f\,/\,\partial q}=\dfrac{dz}{-p\dfrac{\partial f}{\partial p}-q\dfrac{\partial f}{\partial q}}$$

$$=\dfrac{dp}{\dfrac{\partial f}{\partial x}+p\dfrac{\partial f}{\partial z}}=\dfrac{dq}{\dfrac{\partial f}{\partial y}+q\dfrac{\partial f}{\partial z}}$$

$$\Rightarrow\dfrac{dx}{x+2p}=\dfrac{dy}{y+2q}=\dfrac{dz}{xp+2p^2+yq+2q^2}$$

$$=\dfrac{dp}{0}=\dfrac{dq}{0}$$ ...(2)

From (1), we get $dp=0$.

$\Rightarrow$ $p=a$ (constant) and $dq=0$

$\Rightarrow$ $q=b$ (constant).

Substituting these values of p and q into (1), we get

$$z=ax+by+a^2+b^2.$$

This is the required complete integral.

EXAMPLE 2. *Find the complete integral of* $2zx-px^2-2qxy+pq=0.$ (RAJASTHAN–2006)

SOLUTION. Assume $f\equiv2zx-px^2-2qxy+pq=0.$ ...(1)

Now finding partial derivatives of f with respect to x,y,z,p and q respectively, we get

$$\dfrac{\partial f}{\partial x}=2z-2px-2qy,\dfrac{\partial f}{\partial y}=-2qx,\dfrac{\partial f}{\partial z}=2x,$$

$$\frac{\partial f}{\partial p} = -x^2 + q, \frac{\partial f}{\partial q} = -2xy + p.$$

Then the Charpit's auxiliary equations are

$$\frac{dx}{-\frac{\partial f}{\partial p}} = \frac{dy}{-\frac{\partial f}{\partial q}} = \frac{dz}{-p\frac{\partial f}{\partial p} - q\frac{\partial f}{\partial q}}$$

$$= \frac{dp}{\frac{\partial f}{\partial x} + p\frac{\partial f}{\partial z}} = \frac{dq}{\frac{\partial f}{\partial y} + q\frac{\partial f}{\partial z}}$$

$$\Rightarrow \frac{dx}{x^2 - q} = \frac{dy}{2xy - p} = \frac{dz}{px^2 - pq + 2xyq - pq}$$

$$= \frac{dp}{2z - 2qy} = \frac{dq}{0} \qquad \ldots(2)$$

From (2), $dq = 0$. Integrating, $q = a$ (constant).
Putting the value of $q = a$ into (1), we get

$$2zx - px^2 - 2axy + pa = 0 \quad \text{or} \quad p = \frac{2x(z - ay)}{x^2 - a}.$$

Now substituting these values of p and q into $dz = pdx + qdy$, we get

$$dz = \frac{2x(z - ay)}{x^2 - a}dx + ady$$

or

$$dz - ady = \frac{2x(z - ay)}{x^2 - a}dx$$

or

$$\frac{dz - ady}{z - ay} = \frac{2x\,dx}{x^2 - a}.$$

Integrating, we get

$$\log(z - ay) = \log(x^2 - a) + \log b$$

or

$$z - ay = b(x^2 - a)$$

or

$$z = ay + b(x^2 - a).$$

EXAMPLE 3. *Solve* $px + qy = pq$. (GBTU(AG)–2012)

SOLUTION. Let us assume $f \equiv px + qy - pq = 0$ $\qquad \ldots(1)$
Differentiating (1) partially w.r.t. x, y, z, p and q, we get

$$\frac{\partial f}{\partial x} = p, \frac{\partial f}{\partial y} = q, \frac{\partial f}{\partial z} = 0, \frac{\partial f}{\partial p} = x - q, \frac{\partial f}{\partial q} = y - p.$$

Then the Charpit's auxiliary equations are

$$\frac{dx}{-\frac{\partial f}{\partial p}} = \frac{dy}{-\frac{\partial f}{\partial q}} = \frac{dz}{-p\frac{\partial f}{\partial p} - q\frac{\partial f}{\partial q}}$$

$$= \frac{dp}{\frac{\partial f}{\partial x} + p\frac{\partial f}{\partial z}} = \frac{dq}{\frac{\partial f}{\partial y} + q\frac{\partial f}{\partial z}}$$

$$\Rightarrow \frac{dx}{q - x} = \frac{dy}{p - y} = \frac{dz}{2pq - xp - yq} = \frac{dp}{p} = \frac{dq}{q} \qquad \ldots(2)$$

Taking last two ratio of (2), we get $\dfrac{dp}{p} = \dfrac{dq}{q}$.
Integrating, $\log p = \log q + \log a$ or $p = aq$.
Putting this value of p into (1), we get

$$aqx + qy - aq^2 = 0 \quad \text{or} \quad ax + y = aq$$

or

$$q = \frac{ax + y}{a} \quad \text{and} \quad p = aq = ax + y.$$

Substitute the values of p and q into the following

equation $dz = pdx + qdy$, we have

$$\therefore \quad dz = (ax + y)dx + \frac{(ax + y)}{a}dy$$

or $adz = (ax + y)(adx + dy)$
Let $u = ax + y$, then $du = adx + dy$.
Then above equation becomes $adz = udu$.
Now integrating, we get

$$2az = u^2 + b \quad \text{or} \quad 2az = (ax + y)^2 + b \quad [\because u = ax + y]$$

This is the required complete integral.

EXAMPLE 4. *Solve* $(p^2 + q^2)y = qz$.

(UPTU–2006, VTU–2007, HISSAR–2005, 11)

SOLUTION. $f(x, y, z, p, q) = 0$ is $(p^2 + q^2)y - qz = 0$. $\qquad \ldots(1)$

$$\frac{\partial f}{\partial x} = 0, \frac{\partial f}{\partial y} = p^2 + q^2, \frac{\partial f}{\partial z} = -q, \frac{\partial f}{\partial p} = 2py,$$

$$\frac{\partial f}{\partial q} = 2qy - z.$$

Now, Charpit's equations are

$$\frac{dx}{-\frac{\partial f}{\partial p}} = \frac{dy}{-\frac{\partial f}{\partial q}} = \frac{dz}{-p\frac{\partial f}{\partial p} - q\frac{\partial f}{\partial q}}$$

$$= \frac{dp}{\frac{\partial f}{\partial x} + p\frac{\partial f}{\partial z}} = \frac{dq}{\frac{\partial f}{\partial y} + q\frac{\partial f}{\partial z}}$$

$$\Rightarrow \frac{dx}{-2py} = \frac{dy}{-2qy + z} = \frac{dz}{-2p^2y - 2q^2y + qz}$$

$$= \frac{dp}{-pq} = \frac{dq}{p^2 + q^2 - q^2}$$

We have to choose the simplest integral involving p and q.

$$\frac{dp}{-pq} = \frac{dq}{p^2} \Rightarrow -\frac{dp}{q} = \frac{dq}{p} \Rightarrow pdp + qdq = 0.$$

Integrating $p^2 + q^2 = a^2$ (say).

Putting for $p^2 + q^2$ in the equation (1), we get

$$a^2y = qz \Rightarrow q = \frac{a^2y}{z},$$

so $\sqrt{a^2 - q^2} = \sqrt{a^2 - \frac{a^4y^2}{z^2}}$

$$p = \frac{a}{z}\sqrt{z^2 - a^2y^2}$$

Now, $dz = p\,dx + q\,dy$.
Putting for p and q in (2), we get

$$dz = \frac{a}{z}\sqrt{z^2 - a^2y^2}dx + \frac{a^2y}{z}dy$$

$$\Rightarrow \quad z\,dz = a\sqrt{z^2 - a^2y^2}dx + a^2y\,dy$$

$$\frac{zdz - a^2ydy}{\sqrt{z^2 - a^2y^2}} = a\,dx$$

Integrating, we get

$$\frac{1}{2} \cdot \frac{2}{1}\sqrt{z^2 - a^2y^2} = ax + b,$$

On squaring, $z^2 - a^2y^2 = (ax + b)^2$.

EXAMPLE 5. *Find the complete integral of the following differential equation $pxy + pq + qy = yz$.*

(JNTU–2006, KURUKSHETRA–2006)

SOLUTION. Assuming $f \equiv pxy + pq + qy - yz = 0$ ...(1)

Now differentiating partially w.r.t. x, y, z, p and q, we get

$$\frac{\partial f}{\partial x} = py, \frac{\partial f}{\partial y} = px + q - z, \frac{\partial f}{\partial z} = -y,$$

$$\frac{\partial f}{\partial p} = xy + q, \frac{\partial f}{\partial q} = p + y.$$

Thus the Charpit's auxiliary equations are

$$\frac{dx}{-\frac{\partial f}{\partial p}} = \frac{dy}{-\frac{\partial f}{\partial q}} = \frac{dz}{-p\frac{\partial f}{\partial p} - q\frac{\partial f}{\partial q}}$$

$$= \frac{dp}{\frac{\partial f}{\partial x} + p\frac{\partial f}{\partial z}} = \frac{dq}{\frac{\partial f}{\partial y} + q\frac{\partial f}{\partial z}}$$

$$\Rightarrow \frac{dx}{-xy - q} = \frac{dy}{-p - y} = \frac{dz}{-xyp - 2pq - yq}$$

$$= \frac{dp}{py - py} = \frac{dq}{px + q - z - qy} \qquad ...(2)$$

Taking fourth ratio with any one ratio, we get

$$dp = 0$$

Integrating we get $p = a$.

Putting the value of $p = a$ in (1), we get

$$axy + aq + qy = yz$$

or $$= \frac{yz - axy}{(a+y)} = \frac{-y(ax-z)}{(a+y)}$$

Now substitute the value of p and q into the following equation $dz = pdx + qdy$

we get $$dz = adx + \frac{-y(ax-z)dy}{a+y}$$

or $$dz - adx = \frac{y(z-ax)dy}{a+y}$$

or $$\frac{dz - adx}{z - ax} + \frac{-y}{a+y}dy = 0.$$

Let $u = z - ax$ then $du = dz - adx$, above equation is then becomes

$$\frac{du}{u} + \frac{-y \, dy}{a+y} = 0 \qquad \text{or} \qquad \frac{du}{u} - dy + \frac{ady}{a+y} = 0.$$

Integrating, we get

$$\log u - y + a\log(a+y) = \log b$$

or $$\log u(a+y)^a - \log b = y$$

or $\log \dfrac{u(a+y)^a}{b} = y$ or $u(a+y)^a = be^y$

or $$(z - ax)(a+y)^a = be^y$$

or $$z - ax = be^y(a+y)^{-a}.$$

This is the required complete integral.

EXAMPLE 6. *Solve $z^2 = pqxy$.*

(ANNA–2009, VTU· 2004)

SOLUTION. Assuming $f \equiv z^2 - pqxy = 0$. ...(1)

Differentiating (1) partially w.r.t. x, y, z, p and q, we get,

$$\frac{\partial f}{\partial x} = -pqy, \frac{\partial f}{\partial y} = -pqx, \frac{\partial f}{\partial z} = 2z,$$

$$\frac{\partial f}{\partial p} = -qxy, \frac{\partial f}{\partial q} = -pxy.$$

Then, the Charpit's auxiliary equations are

$$\frac{dx}{-\frac{\partial f}{\partial p}} = \frac{dy}{-\frac{\partial f}{\partial q}} = \frac{dz}{-p\frac{\partial f}{\partial p} - q\frac{\partial f}{\partial q}}$$

$$= \frac{dp}{\frac{\partial f}{\partial x} + p\frac{\partial f}{\partial z}} = \frac{dq}{\frac{\partial f}{\partial y} + q\frac{\partial f}{\partial z}}$$

$$\Rightarrow \frac{dx}{qxy} = \frac{dy}{pxy} = \frac{dz}{2pqxy}$$

$$= \frac{dp}{-pqy + 2pz} = \frac{dq}{-pqx + 2qz} \qquad ...(2)$$

From (2), we obtain

$$\frac{pdx + xdp}{pqxy - pqxy + 2pxz} = \frac{qdy + ydq}{pqxy - pqxy + 2qzy}$$

or $\dfrac{pdx + xdp}{2pxz} = \dfrac{qdy + ydq}{2qzy}$ or $\dfrac{d(px)}{px} = \dfrac{d(qy)}{qy}$.

Integrating, we get $\log px = \log qy + \log a$.

or $$px = aqy \quad \text{or} \quad p = \frac{ay}{x}q.$$

Putting this value of p in (1) we get

$$z^2 = aq^2y^2$$

or $q^2 = \dfrac{z^2}{ay^2}$ or $q = \dfrac{z}{\sqrt{a}\,y}$

and $p = \dfrac{ay}{x}q = \dfrac{ay}{x}\cdot\dfrac{z}{\sqrt{a}y} \Rightarrow p = \sqrt{a}\dfrac{z}{x}$.

Substituting the values of p and q, into the following equation

$$dz = pdx + qdy$$

we get $dz = \sqrt{a}\dfrac{z}{x}dx + \dfrac{z}{\sqrt{a}}\cdot\dfrac{dy}{y}$

or $$\frac{dz}{z} = \sqrt{a}\frac{dx}{x} + \frac{1}{\sqrt{a}}\cdot\frac{dy}{y}$$

or $\sqrt{a}\dfrac{dz}{z} = a\cdot\dfrac{dx}{x} + \dfrac{dy}{y}$.

Integrating, we get

$$\sqrt{a}\log z = a\log x + \log y + \log b$$

$$z^{\sqrt{a}} = bx^a y.$$

This is the required complete integral.

Exercise-8.60

Using Charpit's method, find the complete integral of the following differential equations:

1. $(p+q)(px+qy) - 1 = 0$.

2. $px^5 - 4q^3x^2 + 6x^2z - 2 = 0$.

3. $yzp^2 = q$.

4. $2(pq + py + qx) + x^2 + y^2 = 0$.

5. $2z + p^2 + 2y^2 + qy = 0$.   (JNTU–2005, KURUKSHETRA–2005)

6. $p^2 - y^2q + x^2 = y^2$.

7. $z = pq$.

8. $p^2 + q^2 - 2px - 2qy + 2xy = 0$.

9. $q - 3y^2 = 0$.

10. $p^2 + q^2 - 2px - 2qy + 1 = 0$.

11. $z(p^2 - q^2) = x - y$.

12. $yp = 2yx + \log q$.

13. $z^2(p^2 + q^2 + 1) = c^2$.

14. $q^2y^2 = z(z - px)$.

15. $p^2 = z^2(1 - pq)$.

16. $p^3 + q^3 = 27z$.

17. $(x^2 + y^2)(p^2 + q^2) = 1$.

18. $x^2p^2 + y^2q^2 = z^2$.

19. $16p^2z^2 + 9q^2z^2 + 4z^2 = 0$.

20. $p(1 + q^2) + (b - z)q = 0$.

21. $p^2x + q^2y = z$.

22. $[z + px + qy] = yp$.

23. $px + qy = z[1 + pq]^{1/2}$.

24. $(x^2 - y^2)pq - xy(p^2 - q^2) - 1 = 0$.

25. $2(y + zq) = q(xp + yq)$.

26. $xp + 3yq = 2(z - x^2q^2)$.

Hints to Selected Problems

2. The Charpit is auxiliary equation becomes

$$\frac{dx}{-x^5} = \frac{dy}{12q^2x^2} = \frac{dz}{-px^5 + 12q^3x^2}$$

$$= \frac{dp}{5px^4 - 8q^3 + 12xz + 6x^2p} = \frac{dq}{6qx^2}$$

Then take first and last ratio.

8. The Charpit's auxiliary equation becomes

$$\frac{dx}{2x - 2p} = \frac{dy}{2y - 2q} = \frac{dz}{2px - 2p^2 + 2qy - 2q^2}$$

$$= \frac{dp}{-2p + 2y} = \frac{dq}{-2q + 2x}$$

$$\Rightarrow \frac{dp + dq}{2(x + y - p - q)} = \frac{dx + dy}{2(x + y - p - q)} \Rightarrow (p - x) + (q - y) = a.$$

Then put this value in the given equation.

11. The Charpit's auxiliary equation becomes

$$\frac{dx}{-2zp} = \frac{dy}{2qz} = \frac{dz}{-2zp^2 + 2q^2z} = \frac{dp}{-1 + p^3 - pq^2} = \frac{dq}{1 + p^2q - q^3}$$

12. The Charpit's auxiliary equation becomes

$$\frac{dx}{-y} = \frac{dy}{1/q} = \frac{dz}{-yp + 1} = \frac{dp}{-2y} = \frac{dq}{p - 2x}$$

Then taking first and fourth ratios.

17. The Charpit's auxiliary equation becomes

$$\frac{dx}{-2p(x^2 + y^2)} = \frac{dy}{-2q(x^2 + y^2)} = \frac{dz}{-(2p^2 + 2q^2)(x^2 + y^2)}$$

$$= \frac{dp}{2x(p^2 + q^2)} = \frac{dq}{2y(p^2 + q^2)}.$$

Then, we have $pdx + qdy = -xdp - ydq \Rightarrow px + qy = 0$.

Answers

1. $z\sqrt{1 + a} = 2\sqrt{ax + y} + b$

2. $z = -\frac{1}{3}a^3e^{9/x^2} + \frac{1}{9} + \frac{1}{3x^2} + (ay + b)e^{3/x^2}$ **3.** $z^2 = \frac{(x + b)^2}{(a - y^2)}$

4. $2z = ax - x^2 + ay - y^2 + \frac{1}{2}(x - y)\sqrt{2(x - y)^2 + a^2} + \frac{a^2}{2\sqrt{2}}\log[\{\sqrt{2}(x - y)\} + \sqrt{2(x - y)^2 + a^2}] + b$

5. $y^2\{(x - a)^2 + y^2 + 2z\} = b$

6. $z = \frac{x}{2}\sqrt{a^2 - x^2} + \frac{a^2}{2}\sin^{-1}\left(\frac{x}{a}\right) - \frac{a^2}{y} - y + b$

7. $2\sqrt{z} = \sqrt{a}.x + \frac{1}{\sqrt{a}}y + b$

8. $2z = x^2 + y^2 + ax + ay + \frac{1}{\sqrt{2}}\left[(x - y)\sqrt{(x - y)^2 - \frac{a^2}{2}} - \frac{a^2}{2}\log\left\{(x - y) + \sqrt{(x - y)^2\frac{a^2}{2}}\right\}\right] + b$; **9.** $z = ax + y^3 + b$.

10. $(a^2 + 1)z = \frac{u^2}{2} + \frac{u}{2}\sqrt{u^2 - (a^2 + 1)} - \frac{a^2 + 1}{2}\log[u + \sqrt{u^2 - (a^2 + 1)}] + b$, where $u = ax + y$

11. $z^{3/2} = (x + a)^{3/2} + (y + a)^{3/2} + b$ **12.** $az = ax^2 + a^2x + e^{ay} + ab$ **13.** $(1 + a^2)(c^2 - z^2) = (x + ay + b)^2$ **14.** $z^{2a^2/(-1 \pm \sqrt{(1 + 4a^2)})} = bxy^a$

15. $\log\left[\frac{\sqrt{1 + az^2} - 1}{z\sqrt{a}}\right] + \sqrt{1 + az^2} = x + b + ay$ **16.** $(1 + a^3)z^2 = 8(x + ay + b)^3$ **17.** $z = \frac{a}{2}\log(x^2 + y^2) + \sqrt{1 - a^2}\tan^{-1}\frac{y}{x} + b$

18. $\sqrt{1 + a^2}.\log z = a\log x + \log y + b$ **19.** $\sqrt{(16a^2 + 9)}\sqrt{1 - z} + 2(ax + y) = b$ **20.** $2\sqrt{[c(z - b) - 1]} = x + cy + c$

21. $\sqrt{(1 + a)}.z = \sqrt{ax} + \sqrt{y} + b$ **22.** $z = \frac{ax}{y^2} - \frac{a^2}{4y^3} + \frac{b}{y}$ **23.** $v = \frac{ax + y}{\sqrt{a}.z}$ **24.** $z = \frac{a}{2}\log(x^2 + y^2) + \frac{1}{a}\tan^{-1}\left(\frac{y}{x} + b\right)$

25. $(2z - ax^2) + \sqrt{(2z - ax^2)^2 - 8y^2} = by^2$ **26.** $z = a^2 + \frac{ay}{x} + bx^2$

8.14 PARTIAL DIFFERENTIAL EQUATIONS OF SECOND ORDER

A partial differential equation is said to be of the second order when it includes at least one of the partial differential coefficients r, s, t but none of higher order, where

$$r = \frac{\partial^2 z}{\partial x^2} = \frac{\partial p}{\partial x}, \ s = \frac{\partial^2 z}{\partial x \partial y} = \frac{\partial p}{\partial y} = \frac{\partial q}{\partial x}, \ t = \frac{\partial^2 z}{\partial y^2} = \frac{\partial q}{\partial y}$$

 Solved Examples

EXAMPLE 1. *Solve* $xr + p = 9x^2 y^3$.

SOLUTION. The given equation can be written as

$$\frac{\partial p}{\partial x} + \frac{1}{x} \cdot p = 9xy^3 \qquad \ldots(1)$$

which is a linear equation in p.

$\therefore$ I.F. $= e^{\int (1/x) dx} = e^{\log x} = x$.

Therefore, solution of (1) is given by

$$px = \int 9x^2 y^3 \, dx + f(y)$$

$$\Rightarrow \quad px = 3x^3 y^3 + f(y)$$

$$\Rightarrow \quad p = \frac{\partial z}{\partial x} = 3x^2 y^3 + \frac{1}{x} f(y).$$

On integrating, we get

$$z = x^3 y^3 + \log x f(y) + F(y).$$

EXAMPLE 2. *Solve* $t - xq = x^2$.

SOLUTION. The given equation can be written as

$$\frac{\partial q}{\partial y} - xq = x^2 \qquad \ldots(1)$$

which is linear in q.

$\therefore \quad$ I.F. $= e^{\int -x \, dy} = e^{-xy}$.

The solution of (1) is given by

$$qe^{-xy} = \int x^2 e^{-xy} \, dy + f(x) = -xe^{-xy} + f(x)$$

$$\Rightarrow \quad q = \frac{\partial z}{\partial y} = -x + f(x) e^{xy}.$$

On integrating, we get

$$z = -xy + f(x) \int e^{xy} \, dy + F(x)$$

$$= -xy + \frac{1}{x} f(x) e^{xy} + F(x).$$

EXAMPLE 3. *Solve* $t = \sin xy$.

SOLUTION. The given equation can be written as

$$\frac{\partial^2 z}{\partial y^2} = \sin xy.$$

On integrating, we have $\frac{\partial z}{\partial y} = -\frac{1}{x} \cos xy + f(x)$.

Again integrating, we get

$$z = -\frac{1}{x^2} \sin xy + y f(x) + F(x).$$

EXAMPLE 4. *Solve* $yt - q = xy$.

SOLUTION. The given equation can be written as

$$\frac{\partial q}{\partial y} - \frac{1}{y} q = x \qquad \ldots(1)$$

which is linear in q.

$$\text{I.F.} = e^{-\int \left(\frac{1}{y}\right) dy} = e^{-\log y} = \frac{1}{y}$$

Therefore, solution of (1) is given by

$$q \cdot \frac{1}{y} = \int x \cdot \frac{1}{y} \, dy + f(x) = x \log y + f(x)$$

$$\Rightarrow \quad q = \frac{\partial z}{\partial y} = xy \log y + y f(x).$$

On integrating, w.r.t. y we get

$$z = x \int y \log y \, dy + f(x) \int y \, dy + F(x)$$

$$= x \left[\frac{y^2}{2} \cdot \log y - \int \frac{1}{y} \cdot \frac{y^2}{2} \, dy \right] + \frac{y^2}{2} f(x) + F(x)$$

$$= \frac{1}{2} xy^2 \log y - \frac{1}{4} xy^2 + \frac{y^2}{2} f(x) + F(x).$$

 Exercise-8.61

Solve the following differential equations:

1. $xys = 1$.

2. $s = 2x + 2y$.

3. $r = 2y^2$.

4. $\log s = x + y$.

5. $s = \frac{x}{y} + a$.

6. $rx = (n-1)p$.

7. $ys + p = \cos(x+y) - y \sin(x+y)$.

8. $p + r + s = 1$.

9. Find a surface satisfying $r + s = 0$ and touching the elliptic paraboloid $z = 4x^2 + y^2$ along its section by the plane $y = 2x + 1$.

Hints to Selected Problems

1. The given equation can be written as $\frac{\partial^2 z}{\partial x \partial y} = \frac{1}{xy}$. Now integrate w.r.t. y and x separately.

2. Do same as (1).

3. The given equation can be written as $\frac{\partial p}{\partial x} = 2y^2$. Then integrating w.r.t. x.

4. Do same as (1).

5. Do same as (1).

8. The given equation can be written as $\frac{\partial z}{\partial x} + \frac{\partial p}{\partial x} + \frac{\partial q}{\partial x} = 1.$ On integrating w.r.t. x we get $p + q = x + f(y) - z.$ Then use Lagrange's subsidiary equation.

9. Here, we have $\frac{\partial p}{\partial x} + \frac{\partial q}{\partial x} = 0.$ On integrating w.r.t. x, we have $p + q = f(y).$ Now, using Lagrange's subsidiary equation.

Answers

1. $z = \log x \log y + \phi(x) + F(y)$

2. $y = x^2 y + xy^2 + \phi(x) + F(y)$

3. $z = x^2 y^2 + xf(y) + F(y)$

4. $z = e^{x+y} + \phi(y) + F(x)$

5. $z = \frac{x^2}{2}\log y + axy + \phi(y) + F(x)$

6. $z = \frac{1}{n}x^n f(y) + F(y)$

7. $yz = y\sin(x+y) + \phi(y) + F(x)$

8. $z = x - y + e^{-y}\phi(y) + e^{-y}F(x-y)$

9. $z + 4x^2 + y^2 - 8xy + 8x - 4y + 2 = 0.$

8.14.1 LINEAR HOMOGENEOUS PARTIAL DIFFERENTIAL EQUATION WITH CONSTANT COEFFICIENTS

A partial differential equation, which is linear with respect to the dependent variable and its derivatives and in which the coefficient are not function of the dependent variable but merely constants is called a linear partial differential equation with constant coefficients, *i.e.*, an equation of the form

$$a_0\frac{\partial^n z}{\partial x^n} + a_1\frac{\partial^n z}{\partial x^{n-1}\partial y} + a_2\frac{\partial^n z}{\partial x^{n-2}\partial y^2} + \dots + a_n\frac{\partial^n z}{\partial y^n} = F(x, y) \qquad \dots(1)$$

where $a_0, a_1, a_2, \dots a_n$ are constants and $F(x, y)$ is a function of nth order with constant coefficients where all the partial derivative are of nth order is known as linear homogeneous partial differential equation with constant coefficients.

Putting $\frac{\partial}{\partial x} = D$ and $\frac{\partial}{\partial y} = D'$, (1) can be written as

$$(a_0 D^n + a_1 D^{n-1}D' + a_2 D^{n-2}D'^2 + \dots + a_n D'^n)z = F(x, y) \qquad \Rightarrow f(D, D')z = F(x, y). \qquad \dots(2)$$

8.14.2 SOLUTION OF LINEAR PARTIAL DIFFERENTIAL EQUATION

The complete solution of (2) consists of two parts :

(a) The complementary function (C.F.) : which is the complete solution of the equation $f(D, D')z = 0.$ It must contain n arbitrary functions, where n is the order of the differential equation.

(b) The particular integral (P.I.) : which is the particular solution of $f(D, D')z = F(x, y).$ Therefore complete solution of (2) is given by $z = $ C.F. + P.I.

8.14.3 RULE FOR FINDING THE COMPLEMENTARY FUNCTION (C.F.)

Consider the equation $a_0\frac{\partial^2 z}{\partial x^2} + a_1\frac{\partial^2 z}{\partial x\,\partial y} + a_2\frac{\partial^2 z}{\partial y^2} = 0$ or $(a_0 D^2 + a_1 DD' + a_2 D'^2)z = 0.$ $\qquad \dots(1)$

WORKING PROCEDURE

Step 1. Putting $D = m$ and $D' = 1$ in (1), we get
$$a_0 m^2 + a_1 m + a_2 = 0. \qquad \dots(2)$$
This is the auxiliary equation.

Step 2. Solve the auxiliary equation (2). Two cases will arise :

Case (i). If the auxiliary equation has the real and different roots, say $m_1, m_2.$
Then, C.F. $= f_1(y + m_1 x) + f_2(y + m_2 x).$

Case (ii). If the auxiliary equation has equal (repeated) roots
Then C.F. $= f_1(y + mx) + x\, f_2(y + mx).$

Solved Examples

EXAMPLE 1. Solve $2r + 5s + 2t = 0.$

SOLUTION. We know that $r = \frac{\partial^2 z}{\partial x^2} = D^2 z, s = \frac{\partial^2 z}{\partial x\partial y} = DD'z$ and

$t = \frac{\partial^2 z}{\partial y^2} = D'^2 z.$

Then the given equation can be written as
$$(2D^2 + 5DD' + 2D'^2)z = 0.$$

The auxiliary equation is $2m^2 + 5m + 2 = 0$

or $(2m+1)(m+2) = 0$ $\quad \therefore \quad m = -\frac{1}{2}, -2.$

Hence, the required solution is

$$z = f\left(y - \frac{1}{2}x\right) + \psi(y - 2x)$$

or $\quad z = \phi(2y - x) + \psi(y - 2x).$

EXAMPLE 2. *Solve* $25r - 40s + 16t = 0$.

SOLUTION. The given equation can be written as
$$(25D^2 - 40DD' + 16D'^2)z = 0.$$
It's A.E. is $25m^2 - 40m + 16 = 0$

or $(5m - 4)^2 = 0$ $\therefore$ $m = \dfrac{4}{5}, \dfrac{4}{5}$.

Hence, the solution is
$$z = f_1\left(y + \frac{4}{5}x\right) + xf_2\left(y + \frac{4}{5}x\right)$$

or $z = \phi(5y + 4x) + x\psi(5y + 4x)$.

EXAMPLE 3. *Solve* $(D^4 - D'^4)z = 0$.   (UPTU(Q. BANK)-2002)

SOLUTION. The given equation is $(D^4 - D'^4)z = 0$.

It's A.E. is $m^4 - 1 = 0$ or $(m^2 - 1)(m^2 + 1) = 0$.

$\therefore$ $m = 1, -1, \pm i$.

Hence, the solution is
$$z = f_1(y + x) + f_2(y - x) + f_3(y + ix) + f_4(y - ix).$$

EXAMPLE 4. *Solve* $(D^3 - 6D^2D' + 11DD'^2 - 6D'^3)z = 0$.

SOLUTION. The given equation is
$$(D^3 - 6D^2D' + 11DD'^2 - 6D'^3)z = 0.$$

It's A. E. is $m^3 - 6m^2 + 11m - 6 = 0$.

or $(m - 1)(m - 2)(m - 3) = 0$.

$\therefore$ $m = 1, 2, 3$.

Hence, the solution is
$$z = f_1(y + x) + f_2(y + 2x) + f_3(y + 3x).$$

EXAMPLE 5. *Solve* $\dfrac{\partial^3 z}{\partial x^3} - 4\dfrac{\partial^3 z}{\partial x^2 \partial y} + 4\dfrac{\partial^3 z}{\partial x \partial y^2} = 0$.

SOLUTION. The equation can be written as
$$(D^3 - 4D^2D' + 4DD'^2)z = 0.$$

It's A.E. is $m^3 - 4m^2 + 4m = 0$

or $m(m - 2)^2 = 0$ $\Rightarrow$ $m = 0, 2, 2$.

Hence, the solution is
$$z = f_1(y) + f_2(y + 2x) + xf_3(y + 2x).$$

EXAMPLE 6. *Solve* $r - 4s + 4t = 0$.   (GBTU (AG)-2010)

SOLUTION. Its A.E is $m^2 - 4m + 4 = 0$ $\Rightarrow$ $m = 2, 2$

C.F. $= f_1(y + 2x) + xf_2(y + 2x)$

P.I. $= 0$

Hence the complete solution is
$$z = \text{C.F.} + \text{P.I} = f_1(y + 2x) + xf_2(y + 2x)$$

Exercise-8.62

Solve the following differential equation :

1. $r = a^2t$.

2. $(4D^2 + 12DD' + 9D'^2)z = 0$.   (PTU-2010)

3. $\dfrac{\partial^3 z}{\partial x^3} - 3\dfrac{\partial^3 z}{\partial x^2 \partial y} + 2\dfrac{\partial^3 z}{\partial x \partial y^2} = 0$.

4. $(D^3 - 6D^2D' + 12DD'^2 - 8D'^3)z = 0$.

5. $\dfrac{\partial^3 z}{\partial x^3} - 7\dfrac{\partial^3 z}{\partial x \partial y^2} + 6\dfrac{\partial^3 z}{\partial y^3} = 0$.

6. $D^4 - 2D^3D' + 2DD'^3 - D'^4 = 0$.   (UPTU(Q. BANK)-2002)

Hints to Selected Problems

1. The given equation can be written as $(D^2 - a^2D'^2)z = 0$

Auxiliary equation is given by $m^2 - a^2 = 0$ $\Rightarrow$ $m = \pm a$

$\Rightarrow$ $z = f_1(y + ax) + f_2(y - ax)$.

2. Auxiliary equation is given by $4m^2 + 12m + 9 = 0$.

3. The given equation can be written as $(D^3 - 3D^2D' + 2DD'^2)z = 0$

Auxiliary equation is $m^3 - 3m^2 + 2m = 0$ $\Rightarrow$ $m = 0, 1, 2$.

4. Auxiliary equation is given by
$$m^3 - 6m^2 + 12m - 8 = 0 \qquad \Rightarrow m = 2, 2, 2.$$

5. The given equation can be written as $(D^3 - 7DD'^2 + 6D'^3)z = 0$

A.E. is given by $m^3 - 7m^2 + 6 = 0$ $\Rightarrow$ $m = 1, 2, -3$.

Answers

1. $z = f_1(y + ax) + f_2(y - ax)$ **2.** $z = f_1(2y - 3x) + xf_2(2y - 3x)$ **3.** $z = f_1(y) + f_2(y + x) + f_3(y + 2x)$

4. $z = f_1(y + 2x) + xf_2(y + 2x) + x^2f_3(y + 2x)$ **5.** $z = f_1(y + x) + f_2(y + 2x) + f_3(y - 3x)$

6. $z = f_1(y - x) + f_2(y + x) + xf_3(y + x) + x^2f_4(y + x)$

8.14.4 Method of Finding Particular Integral of a Linear Homogeneous Partial Differential Equation

The given partial differential equation is $f(D, D')z = F(x, y)$.

Particular integral (P.I.) $= \dfrac{1}{f(D, D')}F(x, y)$.

Case 1. When $F(x, y)$ is the form $x^m y^n$ or a rational integer algebraic function of x and y. Then

$$\text{P.I.} = \dfrac{1}{f(D, D')}x^m y^n = [f(D, D')]^{-1}x^m y^n.$$

Expand $[f(D, D')]^{-1}$ in ascending power of D or D' and operate on $x^m y^n$ term by term.

REMARK

- Here : $\dfrac{1}{D}$ means integration w.r.t. x, $\dfrac{1}{D'}$ means integration w.r.t. y and so on.

Solved Examples

EXAMPLE 1. *Solve* $(D^2 - 2DD' + D'^2) = 12xy$.

SOLUTION. The auxiliary equation is $m^2 - 2m + 1 = 0$.

or $(m-1)^2 = 0$ ∴ $m = 1,1$.

∴ C.F. $= f_1(y+x) + xf_2(y+x)$.

Now P.I. $= \dfrac{1}{D^2 - 2DD' + D'^2} 12xy = \dfrac{1}{(D-D')^2}.12xy$

$= \dfrac{1}{D^2}\left(1 - \dfrac{D'}{D}\right)^{-2}.12xy = \dfrac{1}{D^2}\left(1 + \dfrac{2D'}{D} + ...\right).12xy$

$= \dfrac{1}{D^2}(12xy) + \dfrac{2}{D^3}D'(12xy)$

$= 2x^3y + \dfrac{2}{D^3}12x = 2x^3y + x^4$.

Hence, the required general solution is

$z = $ C.F. + P.I.

or $z = f_1(y+x) + xf_2(y+x) + 2x^3y + x^4$.

EXAMPLE 2. *Solve* $(D^2 + 3DD' + 2D'^2)z = x + y$.   (IAS-1986, 94)

SOLUTION. The A. E. is $m^2 + 3m + 2 = 0$ or $(m+1)(m+2) = 0$

∴ $m = -1, -2$.

∴ C.F. $= f_1(y-x) + f_2(y-2x)$.

Now P.I. $= \dfrac{1}{D^2 + 3DD' + 2D'^2}(x+y)$

$= \dfrac{1}{D^2}\left(1 + \dfrac{3D'}{D} + 2\dfrac{D'^2}{D^2}\right)^{-1}(x+y)$

$= \dfrac{1}{D^2}\left(1 - \dfrac{3D'}{D} + ...\right)(x+y)$

$= \dfrac{1}{D^2}(x+y) - \dfrac{3}{D^3}D'(x+y)$

$= \dfrac{x^3}{6} + \dfrac{x^2}{2}y - \dfrac{3}{D^3}.1 = \dfrac{x^3}{6} + \dfrac{1}{2}x^2y - \dfrac{3x^3}{6}$

$= -\dfrac{1}{3}x^3 + \dfrac{1}{2}x^2y$.

Hence, the complete solution is

$z = f_1(y-x) + f_2(y-2x) - \dfrac{1}{3}x^3 + \dfrac{1}{2}x^2y$.

EXAMPLE 3. *Solve* $r + (a+b)s + abt = xy$.   (GBTU(AG)-2011)

SOLUTION. The given equation can be written as

$\{D^2 + (a+b)DD' + abD'^2\}z = xy$.

It's A. E. is $m^2 + (a+b)m + ab = 0$

or $(m+a)(m+b) = 0 \Rightarrow m = -a, -b$.

∴ C.F. $= f_1(y-ax) + f_2(y-bx)$.

Now P.I. $= \dfrac{1}{D^2 + (a+b)DD' + abD'^2}.xy$

$= \dfrac{1}{D^2}\left\{1 + (a+b)\dfrac{D'}{D} + ab\dfrac{D'^2}{D^2}\right\}^{-1}.xy$

$= \dfrac{1}{D^2}\left\{1 - (a+b)\dfrac{D'}{D} + ...\right\}.xy$

$= \dfrac{1}{D^2}(xy) - (a+b)\dfrac{1}{D^3}D'.xy$

$= \dfrac{1}{6}x^3y - (a+b)\dfrac{1}{D^3}x$

$= \dfrac{1}{6}x^3y - \dfrac{(a+b)x^4}{24}$.

Hence, the complete solution is

$z = f_1(y-ax) + f_2(y-bx) + \dfrac{1}{6}x^3y - \dfrac{1}{24}(a+b)x^4$.

EXAMPLE 4. *Find a real function V of x and y, reducing to zero when y = 0 and satisfying*

$$\dfrac{\partial^2 V}{\partial x^2} + \dfrac{\partial^2 V}{\partial y^2} = -4\pi(x^2 + y^2).$$

SOLUTION. The real function V will be given by the P.I. of the given equation.

Now P.I. $= \dfrac{1}{D^2 + D'^2}\{-4\pi(x^2 + y^2)\}$.

$= \dfrac{1}{D^2}\left[1 + \dfrac{D'^2}{D^2}\right]^{-1}\{-4\pi(x^2 + y^2)\}$

$= \dfrac{1}{D^2}\left[1 - \dfrac{D'^2}{D^2} + ...\right]\{-4\pi(x^2 + y^2)\}$

$= \dfrac{1}{D^2}\{-4\pi(x^2 + y^2)\} - \dfrac{1}{D^4}\{-4\pi.2\}$

$= -4\pi\left\{\dfrac{x^4}{12} + \dfrac{x^2y^2}{2}\right\} - \left\{-8\pi\dfrac{x^4}{24}\right\}$

$= -\dfrac{\pi x^4}{3} - 2\pi x^2 y^2 + \dfrac{\pi x^4}{3} = -2\pi x^2 y^2$.

Hence, the required general solution is $V = -2\pi x^2 y^2$.

 Exercise-8.63

Solve the following differential equations:

1. $r - a^2 t = x^2$.

2. $\dfrac{\partial^2 z}{\partial x^2} + 3\dfrac{\partial^2 z}{\partial x\,\partial y} + 2\dfrac{\partial^2 z}{\partial y^2} = 6(x+y)$.

3. $(2D^2 - 5DD' + 2D'^2)z = 24(y-x)$.

4. $\dfrac{\partial^2 z}{\partial x^2} + 3\dfrac{\partial^2 z}{\partial x\,\partial y} + 2\dfrac{\partial^2 z}{\partial y^2} = 2x + 3y$.

5. $\dfrac{\partial^3 z}{\partial x^3} - \dfrac{\partial^3 z}{\partial y^3} = x^3 y^3$.

6. $\dfrac{\partial^2 z}{\partial x^2} + 2\dfrac{\partial^2 z}{\partial x\partial y} + \dfrac{\partial^2 z}{\partial y^2} = x^2 + xy + y^2$.

7. $\dfrac{\partial^2 z}{\partial x^2} - 2\dfrac{\partial^2 z}{\partial x\,\partial y} + \dfrac{\partial^2 z}{\partial y^2} = x^2 + y$.

Hints to Selected Problems

1. The given equation can be written as $(D^2 - a^2 D'^2)z = x^2$

A.E. $= m^2 - a^2 = 0 \Rightarrow m = \pm a$

C.F. $= f_1(y + ax) + f_2(y - ax)$

P.I. $= \dfrac{1}{(D^2 - a^2 D'^2)} \cdot x^2 = \dfrac{1}{D^2 \left(1 - \dfrac{a^2 D'^2}{D^2}\right)} \cdot x^2$

$= \dfrac{1}{D^2}\left[1 + \dfrac{a^2 D'^2}{D^2} + \dfrac{a^4 D'^4}{D^4} + ...\right] x^2 = \dfrac{1}{D^2}(x^2) = \dfrac{1}{12}x^4.$

2. A.E. $= m^2 + 3m + 2 = 0 \Rightarrow m = -1, -2$

C.F. $= f_1(y - x) + f_2(y - 2x)$

P.I. $= \dfrac{1}{D^2 + 3DD' + 2D'^2} 6(x + y)$

$= \dfrac{1}{D^2}\left[\left(1 + \dfrac{3D'}{D} + \dfrac{2D'^2}{D^2}\right)^{-1}\right] 6(x + y)$. Now expanding.

4. $m = -1, -2 \Rightarrow$ C.F. $= f_1(y - x) + f_2(y - 2x)$

P.I. $= \dfrac{1}{D^2 + 3DD' + 2D'^2}(2x + 3y)$

$= \dfrac{1}{D^2\left(1 + \dfrac{3D'}{D} + \dfrac{2D'^2}{D^2}\right)}(2x + 3y)$

$= \dfrac{1}{D^2}\left[1 - \dfrac{3D'}{D} - \dfrac{2D'^2}{D^2}, ...\right](2x + 3y).$

 Answers

1. $z = f_1(y + ax) + f_2(y - ax) + \dfrac{1}{12}x^4$ **2.** $z = f_1(y - x) + f_2(y - 2x) - 2x^3 + 3x^2 y$

3. $z = f_1(2y + x) + f_2(y + 2x) + 6x^2 y + 3x^3$ **4.** $z = f_1(y - x) + f_2(y - 2x) - \dfrac{7}{6}x^3 + \dfrac{3}{2}x^2 y$

5. $z = f_1(y + x) + f_2(y + \omega x) + f_3(y + \omega^2 x) + \dfrac{x^6 y^3}{120} + \dfrac{x^9}{10080}$ where ω is one of the cube roots of unity.

6. $z = f_1(y - x) + f_2(y - x) + \dfrac{1}{4}(x^4 - 2x^3 y + 2x^2 y^2)$ **7.** $z = f_1(y + x) + xf_2(y + x) + \dfrac{x^4}{12} + \dfrac{x^2 y}{2} + \dfrac{x^3}{3}$

Case 2. When $F(x, y)$ is of the form $e^{ax + by}$ [provided $f(a, b) \neq 0$]. Then, P.I. $= \dfrac{1}{f(D, D')} e^{ax + by} = \dfrac{e^{ax + by}}{f(a, b)}.$ (Put $D = a$ and $D' = b$)

Solved Examples

EXAMPLE 1. *Solve* $\dfrac{\partial^3 z}{\partial x^3} - 3\dfrac{\partial^3 z}{\partial x^2 \partial y} + 4\dfrac{\partial^3 z}{\partial y^3} = e^{x + 2y}.$

(UPTU–2007, GBTU(AG)–2011, BURDWAN–2003)

SOLUTION. The given equation can be written as

$(D^3 - 3D^2 D' + 4D'^3)z = e^{x + 2y}.$

It's A. E. is $m^3 - 3m^2 + 4 = 0$

or $(m + 1)(m - 2)^2 = 0$

or $m = -1, 2, 2.$

∴ C.F. $= f_1(y - x) + f_2(y + 2x) + xf_3(y + 2x).$

Now, P.I. $= \dfrac{1}{D^3 - 3D^2 D' + 4D'^3} e^{x + 2y}$

$= \dfrac{1}{1^3 - 3 \cdot 1^2 \cdot 2 + 4 \cdot 2^3} e^{x + 2y}$

$= \dfrac{1}{1 - 6 + 32} e^{x + 2y} = \dfrac{1}{27} \cdot e^{x + 2y}.$

Hence, the complete solution is

$z = f_1(y - x) + f_2(y + 2x) + xf_3(y + 2x) + \dfrac{1}{27}e^{x + 2y}.$

EXAMPLE 2. *Solve* $(D^2 - 2DD')z = e^{2x} + x^3 y.$

SOLUTION. It's A. E. is $m^2 - 2m = 0$

or $m(m - 2) = 0$ or $m = 0, 2.$

∴ C.F. $= f_1(y) + f_2(y + 2x).$

Now P.I. $= \dfrac{1}{D^2 - 2DD'}(e^{2x} + x^3 y)$

$= \dfrac{1}{D^2 - 2DD'} \cdot e^{2x} + \dfrac{1}{D^2 - 2DD'} \cdot x^3 y$

$= \dfrac{1}{2^2 - 2(2)(0)} \cdot e^{2x} + \dfrac{1}{D^2}\left(1 - \dfrac{2D'}{D}\right)^{-1} \cdot x^3 y$

$= \dfrac{1}{4}e^{2x} + \dfrac{1}{D^2}\left(1 + \dfrac{2D'}{D} + ...\right)x^3 y$

$= \dfrac{1}{4}e^{2x} + \dfrac{1}{D^2}(x^3 y) + \dfrac{2}{D^3}x^3 = \dfrac{1}{4}e^{2x} + \dfrac{x^5 y}{20} + \dfrac{x^6}{60}$

Hence, the complete solution is

$z = f_1(y) + f_2(y + 2x) + \dfrac{1}{4}e^{2x} + \dfrac{x^5 y}{20} + \dfrac{x^6}{60}.$

Case 3.   (Short method). *When $F(x, y)$ is of the form $\phi(ax + by)$.*

If $f(D, D')$ is a homogeneous function of D or D' of degree n, then we have

P.I. $= \dfrac{1}{f(D, D')} \phi(ax + by) = \dfrac{1}{f(a, b)} \int\int ..... \int \phi(v)dv^n$ where $v = ax + by.$

After integrating $\phi(v)$ n times w.r.t. 'v', v must be replaced by $ax + by$.

Case 4. (Exceptional case) *When $F(x,y)$ is of the form $\phi(ax+by)$ and $f(a,b)=0$.*

If $f(D,D')$ is a homogeneous function D or D' of degree n, then we have

$$\text{P.I.} = \frac{1}{(bD-aD')^n}\phi(ax+by) = \frac{x^n}{b^n n!}\phi(ax+by) \qquad\qquad [\text{provided } f(a,b)=0]$$

EXAMPLE 3. *Solve $(D^2+3DD'+2D'^2)z = x+y$.*

(UPTU(Q. BANK)-2002, UPTU(CO)-2008)

SOLUTION. It's A. E. is $m^2+3m+2=0$

or $(m+1)(m+2)=0$ or $m=-1,-2$.

$\therefore$ C.F. $= f_1(y-x)+f_2(y-2x)$.

Now, P.I. $= \dfrac{1}{D^2+3DD'+2D'^2}\cdot(x+y)$

$$[\because D=a=1,\ D'=b=1]$$

$$= \frac{1}{1^2+3.1.1+2.1^2}\iint V\,dV^2$$

where $V = x+y$

$$= \frac{1}{6}\cdot\frac{1}{6}V^3 = \frac{1}{36}(x+y)^3.$$

Hence, the complete solution is

$$z = f_1(y-x)+f_2(y-2x)+\frac{1}{36}(x+y)^3.$$

EXAMPLE 4. *Solve $\dfrac{\partial^2 z}{\partial x^2}+\dfrac{\partial^2 z}{\partial y^2} = \cos mx\,\cos ny$.*

SOLUTION. The given equation can be written as

$$(D^2+D'^2)z = \frac{1}{2}[\cos(mx+ny)+\cos(mx-ny)].$$

It's A. E. is $m^2+1=0$ or $m=\pm i$,

$\therefore$ C.F. $= f_1(y+ix)+f_2(y-ix)$.

Now,

$$\text{P.I.} = \frac{1}{D^2+D'^2}\left[\frac{1}{2}\{\cos(mx+ny)+\cos(mx-ny)\}\right]$$

$$= \frac{1}{2}\frac{1}{D^2+D'^2}\cdot\cos(mx+ny)+\frac{1}{2}\cdot\frac{1}{D^2+D'^2}\cdot\cos(mx-ny)$$

$$= \frac{1}{2}\frac{1}{m^2+n^2}\iint\cos u\,du.du+\frac{1}{2}\frac{1}{m^2+n^2}\iint\cos v.dv.dv$$

where $u = mx+ny$ and $v = mx-ny$

$$= \frac{1}{2}\cdot\frac{1}{m^2+n^2}(-\cos u)+\frac{1}{2}\frac{1}{m^2+n^2}(-\cos v)$$

$$= \frac{1}{2}\frac{1}{m^2+n^2}\{-\cos(mx+ny)\}$$

$$\qquad + \frac{1}{2(m^2+n^2)}\{-\cos(mx-ny)\}$$

$$= -\frac{1}{2(m^2+n^2)}[\cos(mx+ny)+\cos(mx-ny)]$$

$$= -\frac{1}{m^2+n^2}\cos mx\,\cos ny.$$

Hence, the complete solution is

$$z = f_1(y+ix)+f_2(y-ix)-\frac{1}{m^2+n^2}\cos mx.\cos ny.$$

EXAMPLE 5. *Solve the linear partial differential equation.*

$$\frac{\partial^2 z}{\partial x^2}+2\frac{\partial^2 z}{\partial x\partial y}+\frac{\partial^2 z}{\partial y^2} = \sin(2x+3y)$$

(UPTU-2006, Q. BANK–2002)

SOLUTION. We have $\dfrac{\partial^2 z}{\partial x^2}+2\dfrac{\partial^2 z}{\partial x\partial y}+\dfrac{\partial^2 z}{\partial y^2} = \sin(2x+3y)$

$\Rightarrow (D^2+2DD'+D'^2)z = \sin(2x+3y)$,

where $D = \dfrac{\partial}{\partial x}$ and $D' = \dfrac{\partial}{\partial y}$

Put $D = m,\ D' = 1$

The auxiliary equation is $m^2+2m+1=0$

$(m+1)^2=0 \Rightarrow m=-1,-1$.

C.F. $= f_1(y-x)+xf_2(y-x)$.

$$\text{P.I.} = \frac{1}{D^2+2DD'+D'^2}\sin(2x+3y)$$

$$= \frac{1}{-4+2(-6)-9}\sin(2x+3y) = \frac{1}{-25}\sin(2x+3y).$$

Hence, the complete solution is

$z = $ C.F. + P.I.

$$z = f_1(y-x)+xf_2(y-x)-\frac{1}{25}\sin(2x+3y)$$

Case 5. (General Method).

Consider the equation $f(D,D')z = F(x,y)$ ...(1)

when $f(D,D')$ is homogeneous function of D and D'. We use the following result :

$$\frac{1}{(D-mD')}F(x,y) = \int F(x,a-mx)dx, \text{ where } a = y+mx.$$

After performing integration a must be replaced by $y+mx$ respectively.

To find the P.I., factorize $f(D,D')$ into linear factors. Thus from (1)

$$\text{P.I.} = \frac{1}{(D-m_1D')(D-m_2D')...(D-m_nD')}F(x,y). \qquad\qquad ...(2)$$

The value of P.I. is obtained by applying the operations indicated by the factors.

EXAMPLE 6. *Solve* $r + s - 6t = y\cos x.$

(UPTU-2003, ANNA-2005S, BHOPAL-2008, SVTU-2008)

SOLUTION. The equation can be written as

$(D^2 + DD' - 6D'^2)z = y\cos x.$

It's A. E. is $m^2 + m - 6 = 0$

or $(m-2)(m+3) = 0$ or $m = 2, -3.$

$\therefore$ C.F. $= f_1(y + 2x) + f_2(y - 3x).$

Now, P.I. $= \dfrac{1}{D^2 + DD' - 6D'^2}.y\cos x$

$= \dfrac{1}{(D - 2D')(D + 3D')}y\cos x$

$= \dfrac{1}{(D - 2D')}.\int (3x + a)\cos x\, dx$

where $y - 3x = a$

$= \dfrac{1}{(D - 2D')}[(3x + a)\sin x + 3\cos x]$

$= \dfrac{1}{D - 2D'}[y\sin x + 3\cos x]$

$= \int [(b - 2x)\sin x + 3\cos x]dx$

where $y + 2x = b$

$= -b\cos x - 2(-x\cos x + \sin x) + 3\sin x$

$= -(y + 2x)\cos x + 2x\cos x + \sin x$

$= -y\cos x + \sin x.$

Hence, the complete solution is

$z = f_1(y + 2x) + f_2(y - 3x) - y\cos x + \sin x.$

EXAMPLE 7. *Solve* $(D^2 - DD' - 2D'^2)z = (y - 1)e^x.$

(UPTU(Q BANK)-2002, BHOPAL-2006)

SOLUTION. It's A. E. is $m^2 - m - 2 = 0$

or $(m - 2)(m + 1) = 0$ or $m = 2, -1.$

$\therefore$ C.F. $= f_1(y + 2x) + f_2(y - x).$

Now, P.I. $= \dfrac{1}{D^2 - DD' - 2D'^2}.(y - 1)e^x$

$= \dfrac{1}{(D - 2D')}.\dfrac{1}{(D + D')}(y - 1)e^x$

$= \dfrac{1}{(D - 2D')}\int (x + a - 1)e^x dx,$ where $y - x = a$

$= \dfrac{1}{(D - 2D')}[(a - 1)\int e^x\, dx + \int xe^x\, dx]$

$= \dfrac{1}{(D - 2D')}[(a - 1)e^x + (x - 1)e^x]$

$= \dfrac{1}{(D - 2D')}[a + x - 2]e^x$

$= \dfrac{1}{(D - 2D')}(y - 2)\, e^x = \int (b - 2x - 2)e^x dx,$

where $y - 2x = b$

$= (b - 2)\int e^x\, dx - 2\int xe^x\, dx$

$= (b - 2)e^x - 2(x - 1)e^x = (b - 2x)e^x = ye^x.$

Hence, the complete solution is

$z = f_1(y + 2x) + f_2(y - x) + ye^x.$

EXAMPLE 8. *Solve the partial differential equation.*

$$\dfrac{\partial^2 z}{\partial x^2} - 3\dfrac{\partial^2 z}{\partial x\partial y} + 2\dfrac{\partial^2 z}{\partial y^2} = e^{2x-y} + e^{x+y} + \cos(x+2y)$$

(UPTU-2006, GBTU-2012)

SOLUTION. Given equation is

$$\dfrac{\partial^2 z}{\partial x^2} - 3\dfrac{\partial^2 z}{\partial x\partial y} + 2\dfrac{\partial^2 z}{\partial y^2}$$

$$= e^{2x-y} + e^{x+y} + \cos(x+2y)$$

Given equation can be written as

$(D^2 - 3DD' + 2D'^2)z = e^{2x-y} + e^{x+y} + \cos(x+2y)$

The auxiliary equation is

$m^2 - 3m + 2 = 0$

$m^2 - 2m - m + 2 = 0$

$m(m - 2) - 1(m - 2) = 0$

$m = 1, 2.$

Hence, C.F. $= f_1(y + x) + f_2(y + 2x)$

Now, P.I.

$= \dfrac{1}{(D - D')(D - 2D')}\{e^{2x-y} + e^{x+y} + \cos(x+2y)\}$

$= \dfrac{1}{(D - D')(D - 2D')}e^{2x-y} + \dfrac{1}{(D - D')(D - 2D')}e^{x+y}$

$\qquad\qquad + \dfrac{1}{(D - D')(D - 2D')}\cos(x+2y)$

$= I_1 + I_2 + I_3.$

Let, $I_1 = \dfrac{1}{(D - D')(D - 2D')}e^{2x-y}$

(Replacing D by 2 and D' by –1)

$I_1 = \dfrac{1}{(2+1)(2+2)}e^{2x-y} = \dfrac{1}{12}e^{2x-y}$

Now, $I_2 = \dfrac{1}{(D - D')(D - 2D')}e^{x+y}$

$= \dfrac{1}{(D - D')(-1)}e^{x+y} = \dfrac{-1}{(D - D')}e^{x+y}$

$= -x.\dfrac{1}{1}e^{x+y} = -xe^{x+y}.$

Now, $I_3 = \dfrac{1}{(D - D')(D - 2D')}\cos(x+2y)$

$= \dfrac{1}{D^2 - 3DD' + 2D'^2}\cos(x+2y)$

$I_3 = \dfrac{1}{-1 - 3(-2) + 2(-4)}\cos(x+2y)$

(Replacing D^2 by –1; DD' by –2, D'^2 by –4)

$= \dfrac{1}{-1 + 6 - 8}\cos(x+2y) = -\dfrac{1}{3}\cos(x+2y)$

P.I. $= I_1 + I_2 + I_3.$

Thus, required

P.I. $= \dfrac{1}{12}e^{2x-y} - xe^{x+y} - \dfrac{1}{3}\cos(x+2y)$

Hence, the complete solution is

$z = $ C.F. + P.I.

$= f_1(y + x) + f_2(y + 2x) + \dfrac{1}{12}e^{2x-y}$

$\qquad - xe^{x+y} - \dfrac{1}{3}\cos(x+2y).$

EXAMPLE 9. *Solve* $r - t = \tan^3 x\tan y - \tan x\tan^3 y.$

(AGRA (BE)-2001)

SOLUTION. The given equation can be written as

$$(D^2 - D'^2)z = \tan^3 x \tan y - \tan x \tan^3 y.$$

It's A. E. is $m^2 - 1 = 0$

or $(m-1)(m+1) = 0$ or $m = 1, -1$.

$\therefore$ C.F. $= f_1(y+x) + f_2(y-x)$.

Now,

$$\text{P.I.} = \frac{1}{(D^2 - D'^2)}(\tan^3 x \tan y - \tan x \tan^3 y)$$

$$= \frac{1}{(D+D')}\frac{1}{(D-D')}(\tan^3 x \tan y - \tan x \tan^3 y)$$

$$= \frac{1}{(D-D')}\int [\tan^3 x \tan(a-x) - \tan x \tan^3(a-x)]dx,$$

$$\text{where } y+x = a$$

$$= \frac{1}{(D+D')}\int [(\sec^2 x - 1)\tan x \tan(a-x)$$

$$- \tan x \tan(a-x)\{\sec^2(a-x) - 1\}]dx$$

$$= \frac{1}{(D+D')}[\int \tan(a-x)\tan x \sec^2 x\, dx$$

$$- \int \tan x \tan(a-x)\sec^2(a-x)dx]$$

$$= \frac{1}{(D+D')}\Bigg[\tan(a-x)\cdot\frac{\tan^2 x}{2} + \frac{1}{2}\int \sec^2 x(a-x)\tan^2 x\, dx$$

$$+ \tan x \cdot \frac{\tan^2(a-x)}{2} - \frac{1}{2}\int \sec^2 \cdot \tan^2(a-x)dx\Bigg]$$

$$= \frac{1}{2}\frac{1}{(D+D')}[\tan^2 x \tan(a-x) + \tan x \tan^2(a-x)$$

$$- \int [\sec^2 x\{\sec^2(a-x) - 1\} - \sec^2(a-x)(\sec^2 x - 1)dx]$$

$$= \frac{1}{2}\frac{1}{(D+D')}[\tan^2 x \tan(a-x) + \tan x \tan^2(a-x)$$

$$+ \int \{\sec^2 x - \sec^2(a-x)\}dx]$$

$$= \frac{1}{2}\frac{1}{(D+D')}[\tan^2 x \tan y + \tan x \tan^2 y + (\tan x + \tan y)]$$

$$= \frac{1}{2}\frac{1}{(D+D')}[\tan x \sec^2 y + \tan y \sec^2 x]$$

$$= \frac{1}{2}\int [\tan x \sec^2(b+x) + \tan(b+x)\sec^2 x]dx,$$

where $y - x = b$

$$= \frac{1}{2}[\tan x \tan(b+x) - \int \sec^2 x \cdot \tan(b+x)dx$$

$$+ \int \tan(b+x)\sec^2 x\, dx]$$

$$= \frac{1}{2}\tan x \tan y.$$

Hence, the complete solution is given by

$$z = f_1(y+x) + f_2(y-x) + \frac{1}{2}\tan x \tan y$$

 Exercise-8.64

Solve the following differential equations:

1. $(D^2 - 2DD' + D'^2)z = e^{x+2y} + x^3$

2. $\dfrac{\partial^2 z}{\partial x^2} + \dfrac{\partial^2 z}{\partial y^2} = 12(x+y)$ (UPTU-2005)

3. $(D^2 - 5DD' + 4D'^2)z = \sin(4x+y)$ (GBTU(AG)-2011)

4. $(D - 2D')(D - D')^2 = e^{x+y}$

5. $\dfrac{\partial^2 z}{\partial x^2} - 5\dfrac{\partial^2 z}{\partial x\partial y} + 6\dfrac{\partial^2 z}{\partial y^2} = \exp.(3x - 2y)$

6. $(D - DD')z = \cos x \cos 2y$ (BHOPAL-2008S)

7. $\dfrac{\partial^2 z}{\partial x^2} - 3\dfrac{\partial^2 z}{\partial x\partial y} + 2\dfrac{\partial^2 z}{\partial y^2} = e^{2x+3y} + \sin(x - 2y)$

8. $\dfrac{\partial^2 z}{\partial x^2} + \dfrac{\partial^2 z}{\partial x\partial y} - 6\dfrac{\partial^2 z}{\partial y^2} = y\sin x$ (MTU(SUM)-2011, MTU-2012)

9. $(2D^2 - 5DD' + 2D'^2)z = 5\sin(2x+y)$

10. $(D^2 + 2DD' + D'^2)z = 2\cos y - x\sin y$ (PTU-2005)

11. $(D^2 + DD' - 6D'^2)z = x^2\sin(x+y)$

12. $(D^2 + 5DD' + 6D'^2)z = \dfrac{1}{y-2x}$

13. $\dfrac{\partial^3 z}{\partial x^2\partial y} - \dfrac{\partial^3 z}{\partial x\partial y^2} + \dfrac{\partial^3 z}{\partial y^3} = \dfrac{z}{x^2}$

14. $r + s - 2t = (2x+y)^{1/2}$

15. $\dfrac{\partial^2 z}{\partial x^2} - \dfrac{\partial^2 z}{\partial x\partial y} = \sin x \cos 2y$ (UPTU-2003, 08)

16. $(D^2 - 2DD' + D'^2)z = \sin x$ (UPTU-2004, 09)

17. $(D^2 - DD')z = \cos 2y(\sin x + \cos x)$ (UPTU-2003, GBTU-2010)

18. $(D + D' - 1)(D + 2D' - 3)z = 4 + 3x + 6y$ (UPTU(Q. BANK)-2002)

19. $\dfrac{\partial^2 z}{\partial x^2} + \dfrac{\partial^2 z}{\partial y^2} = \cos mx \cos ny + 30(x+y)$ (GBTU(AG)-2011, 12, UPTU(SUM)-2008)

20. $\dfrac{\partial^2 z}{\partial x^2} - \dfrac{\partial^2 z}{\partial x\partial y} = \sin x \cos y$ (GBTU(SUM)-2010)

21. $r + 2s + t = 2(y - x) + \sin(x - y)$ (MTU(SUM)-2011)

22. $(D^2 + 2DD' + D'^2)z = e^{2x+3y}$ (GBTU(AG)-2012)

23. $(D^3 - 4D^2D' + 4DD'^2)z = 6\sin(3x + 2y)$ (UKTU-2012)

24. $(D^2 - 4DD' + 4D'^2)z = e^{2x+y}$ (GBTU(CO)-2010)

25. $(D^2 - 8DD' + 7D'^2)z = \sin(7x + y)$ (UPTU(CO)-2009)

26. $(D^2 - DD')z = \sin(x + 2y)$ (GBTU(AG)-2012)

27. $(D^2 + 4DD' - 5D'^2)z = \sin(2x + 3y)$ (MADRAS-2006)

28. $\dfrac{\partial^3 z}{\partial x^3} - 2\dfrac{\partial^3 z}{\partial x^2\partial y} = 2e^{2x} + 3x^2 y$ (SVTU-2007)

29. $\dfrac{\partial^2 z}{\partial x^2} - \dfrac{\partial^2 z}{\partial x\partial y} - 6\dfrac{\partial^2 z}{\partial y^2} = \cos(2x + y)$ (PTU-2010, SVTU-2009)

30. $(D^2 - 2DD' + D'^2)z = e^{x+y}$ (BHOPAL-2007)

31. $\dfrac{\partial^3 z}{\partial x^3} - 4\dfrac{\partial^3 z}{\partial x^2\partial y} + 5\dfrac{\partial^3 z}{\partial x\partial y^2} - 2\dfrac{\partial^3 z}{\partial y^3} = e^{2x+y}$ (BHOPAL-2008)

32. $\dfrac{\partial^2 z}{\partial x^2} - 2\dfrac{\partial^2 z}{\partial x\partial y} + \dfrac{\partial^2 z}{\partial y^2} = \sin x$ (PTU-2009S)

33. $\dfrac{\partial^3 z}{\partial x^3} - 4\dfrac{\partial^3 z}{\partial x^2\partial y} + 4\dfrac{\partial^3 z}{\partial x\partial y^2} = 2\sin(3x + 2y)$ (SVTU-2007)

34. $(D^3 - 7DD'^2 - 6D'^3)z = \cos(x + 2y) + 4$ (ANNA-2008)

35. $(D^2 - D'^2)z = e^{x-y}\sin(x + 2y)$ (ANNA-2009)

Hints to Selected Problems

1. A.E. is $m^2 - 2m + 1 = 0 \Rightarrow m = 1, 1$

$$P.I. = \frac{1}{(D-D')^2} e^{x+2y} + \frac{1}{(D-D')^2} x^3$$

$$= \frac{e^{x+2y}}{(1-2)^2} + \frac{1}{D^2}\left[1 - \frac{D'}{D}\right]^{-2} x^3$$

2. A.E. is $m^2 + 1 = 0 \Rightarrow m \pm i$

$$P.I. = \frac{1}{(D^2 + D'^2)} 12(x+y) = \frac{1}{1^2 + 1^2} \cdot 12 \frac{(x+y)^3}{6}$$

4. $1, 1, 2$

$$P.I. = \frac{1}{(D - 2D')(D - D')^2} e^{x+y}$$

$$= \frac{1}{(1-2)} \frac{1}{(D - D')^2} e^{x+y} = -\frac{x^2}{2} e^{x+y}$$

6. A.E. is $m^2 - m = 0 \Rightarrow m = 0, 1$

$$P.I. = \frac{1}{(D^2 - DD')} \cdot \cos x \cos 2y$$

$$= \frac{1}{(D^2 - DD')} \cdot \frac{1}{2}\left[\cos(x + 2y) + \cos(x - 2y)\right]$$

8. A.E. $(m+3)(m-2) = 0 \Rightarrow m = 2, -3$

$$P.I. = \frac{1}{(D^2 + DD' - 6D'^2)} y \sin x = I.P. \text{of} \left[\frac{1}{(D^2 + DD' - 6D'^2)} y e^{ix}\right]$$

$$= I.P. \text{ of } \left\{ e^{ix}\left[\frac{1}{(D+i)^2 + (D+i)D' - 6D'^2}\right]\right\} y$$

10. $P.I. = \frac{1}{(D+D')^2}(2\cos y - x \sin y)$

$$= \frac{1}{(D+D')} \int[(2\cos(x+b) - x\sin(x+b)]dx,$$

where $y - x = b$

$$= \frac{1}{D + D'}[2\sin(x+b) + x\cos(x+b) - \sin(x+b)]$$

$$= \frac{1}{D + D'}[\sin(x+b) + x\cos(x+b)]$$

$$= \int \sin(x+b) + x\cos(x+b)dx = x\sin(x+b) = x\sin y$$

12. $P.I. = \frac{1}{(D^2 + 5DD' + 6D'^2)} \frac{1}{(y - 2x)}$

$$= \frac{1}{(D + 3D')(D + 2D')}(y - 2x)^{-1}$$

$$= \frac{1}{(-2+3)(D + 2D')} \int V^{-1} dV, \quad (V = y - 2x)$$

$$= \frac{1}{(D + 2D')} \log(y - 2x) = x\log(y - 2x)$$

Answers

1. $z = f_1(y+x) + xf_2(y+x) + e^{x+2y} + \frac{1}{20}x^5$

2. $z = f_1(y - ix) + f_2(y + ix) + (x + y)^3.$

3. $z = f_1(y+x) + f_2(y + 4x) - \frac{1}{3}x\cos(4x + y)$

4. $z = f_1(y + 2x) + f_2(y + x) + xf_3(y + x) - \frac{x^2}{2}e^{x+y}.$

5. $z = f_1(y + 2x) + f_2(y + 3x) + \frac{1}{63}e^{3x-2y}$

6. $z = f_1(y) + f_2(y + x) + \frac{1}{2}\cos(x + 2y) - \frac{1}{6}\cos(x - 2y).$

7. $z = f_1(y+x) + f_2(y + 2x) + \frac{1}{4}e^{2x+3y} - \frac{1}{15}\sin(x - 2y)$

8. $z = f_1(y + 2x) + f_2(y - 3x) - (y\sin x + \cos x);$

9. $z = f_1(2y + x) + f_2(y + 2x) - \left(\frac{5x}{3}\right)\cos(2x + y)$

10. $z = f_1(y - x) + xf_2(y - x) + x\sin y$

11. $z = f_1(y + 2x) + f_2(y - 3x)$
$\qquad + \frac{1}{4}\left(x^2 - \frac{13}{8}\right)\sin(x + y) - \frac{3}{8}x\cos(x + y)$

12. $z = f_1(y - 2x) + f_2(y - 3x) + x\log(y - 2x)$

13. $z = f_1(x) + f_2(y + x) + xf_3(y + x) - y\log x.$

14. $z = f_1(y + x) + f_2(y - 2x) + \frac{1}{15}(2x + y)^{5/2}.$

15. $z = f_1(x) + f_2(y + x) + \frac{1}{2}\sin(x + 2y) - \frac{1}{6}\sin(x - 2y)$

16. $z = f_1(y + x) + xf_2(y + x) - \sin x$

17. $z = f_1(y) + f_2(y + x) + \frac{1}{2}[\sin(x + 2y) + \cos(x + 2y)] - \frac{1}{6}[\sin(x - 2y) + \cos(x - 2y)]$

18. $z = e^x f_1(y - x) + e^{3x} f_2(y - 2x) + 6 + x + 2y$

19. $z = f_1(y + ix) + f_2(y - ix) - \frac{1}{m^2 + n^2}\cos mx\cos ny + (2x + y)^3$

20. $z = f_1(y) + f_2(y + x) - \frac{x}{2}\cos(x + y) - \frac{1}{4}\sin(x - y)$

21. $z = f_1\left(y + \frac{1}{2}x\right) + xf_2\left(y + \frac{1}{2}x\right) + 2x^2\log(x + 2y)$

22. $z = f_1(y - x) + xf_2(y - x) + \frac{1}{25}e^{2x+3y}$

23. $z = f_1(y) + f_2(y + 2x) + xf_3(y + 2x) + 2\cos(3x + 2y)$

24. $z = f_1(y + 2x) + xf_2(y + 2x) + \frac{x^2}{2}e^{2x+y}$

25. $z = f_1(y + x) + f_2(y + 7x) - \frac{x}{6}\cos(7x + y)$

26. $z = f_1(y) + f_2(y + x) + \sin(x + 2y)$

27. $z = f_1(y + x) + f_2(y - 5x) + \frac{1}{17}\sin(2x + 3y)$

28. $z = f_1(y) + xf_2(y) + f_3(y + 2x) + \frac{1}{60}(15e^{2x} + 3x^5 y + x^6)$

29. $z = f_1(y - 3x) + f_2(y + 2x) + \frac{x}{5}\sin(2x + y) + \frac{1}{25}\cos(2x + y)$

30. $z = f_1(x + y) + xf_2(x + y) + \dfrac{x^2}{2}e^{x+y}$

31. $z = f_1(y + x) + zf_2(y + x) + f_3(y + 2x) - e^{2x+y}$

32. $z = f_1(y + x) + xf_2(y + x) - \sin x$

33. $z = f_1(y) + f_2(y + 2x) + xf_3(y + 2x) + 3x\cos(3x + 2y)$

34. $f_1(yx) + f_2(y - 2x) + f_3(y + 3x) + \dfrac{1}{75}\sin(x + 2y) + \dfrac{2}{3}x^3$

35. $z = f_1(y + x) + f_2(y - x) + \dfrac{3}{28}e^{x-y}[\sin(x + 2y) - 2\cos(x + 2y)]$

8.14.5 Non-Homogeneous Linear Partial Differential Equations with Constant Coefficients

Linear differential equations which are not homogeneous are called non-homogeneous linear equations.

For example, $\dfrac{\partial^2 z}{\partial x^2} + 2\dfrac{\partial^2 z}{\partial x \partial y} + \dfrac{\partial^2 z}{\partial y^2} + 2\dfrac{\partial z}{\partial x} + 2\dfrac{\partial z}{\partial y} + z = 0$ is a non-homogeneous equation.

It can be written in the form $f(D, D') = F(x, y)$. Its complete solution = C.F. + P.I.

8.14.6 Method for Finding the C.F.

Let us consider the equation $(D - mD' - a)z = 0$ or $p - mq = az$.

Applying Lagrange's method, the auxiliary equations are $\dfrac{dx}{1} = \dfrac{dy}{-m} = \dfrac{dz}{az}$.

From the first two members, we have

$$dy + m\,dx = 0 \Rightarrow y + mx = c_1 \text{ (constant)} \qquad \qquad \ldots(1)$$

Again from the first and last members, we get

$$dx = \dfrac{dz}{az} \Rightarrow \log z = ax + \log c_2 \Rightarrow z = c_2 e^{ax}. \qquad \qquad \ldots(2)$$

From (1) and (2), we have $z = e^{ax} f(y + mx)$.

Similarly, the solution of $(D - m_1 D' - a_1)(D - m_2 D' - a_2)\ldots(D - m_n D' - a_n)z = 0$ is

$$z = e^{a_1 x} f_1(y + m_1 x) + e^{a_2 x} f_2(y + m_2 x) + \ldots + e^{a_n x} f_n(y + m_n x).$$

Note. When $f(D, D')$ cannot be factorized into linear factors, then we proceed as follows :

Let the equation be $\qquad (D - D'^2)z = 0 \qquad \qquad \ldots(1)$

The trial solution of (1) be $z = Ae^{hx+ky}$ where A, h and k are constants. $\qquad \qquad \ldots(2)$

From (2), we have

$$Dz = \dfrac{\partial z}{\partial x} = Ahe^{hx+ky}; \qquad\qquad D'z = \dfrac{\partial z}{\partial y} = Ake^{hx+ky}$$

$$D^2 z = \dfrac{\partial^2 z}{\partial x^2} = Ah^2 e^{hx+ky}; \qquad\qquad D'^2 z = \dfrac{\partial^2 z}{\partial y^2} = Ak^2 e^{hx+ky}.$$

With these values, (1) gives

$$Ahe^{hx+ky} - Ak^2 e^{hx+ky} = 0 \qquad \text{or} \qquad A(h - k^2)e^{hx+ky} = 0 \quad \text{or} \quad h = k^2.$$

Putting the value of h in (2), we get $z = Ae^{k^2 x + ky}$.

A more general solution of (1) is given by $z = \sum Ae^{k^2 x + ky}$, where A and k are arbitrary constants in each term and any number of terms may be taken in the above summation.

8.14.7 Method for Finding the P.I. of Non-Homogeneous Equation with Constant Coefficients

The methods of finding particular integrals of non-homogeneous partial differential equation are very similar to those of ordinary differential equation with constant coefficients.

We are considering few cases of finding P.I. of $f(D, D')z = F(x, y)$.

Case I. When $F(x, y) = e^{ax+by}$ and $f(a, b) \neq 0$, then $\quad \text{P.I.} = \dfrac{1}{f(D, D')}e^{ax+by} = \dfrac{1}{f(a, b)}e^{ax+by} \qquad$ (Putting a for D and b for D')

Case II. When $F(x, y) = \sin(ax + by)$ or $\cos(ax + by)$, then P.I. $= \dfrac{1}{f(a, b)}\sin(ax + by)$ is obtained by putting

$D^2 = -a^2, DD' = -ab, D'^2 = -b^2$, provided the denominator is non-zero.

Case III. When $F(x, y) = x^m y^n$, where m and n are positive integers, then we have

$$\text{P.I.} = \dfrac{1}{f(D, D')}x^m y^n = [f(D, D')]^{-1} x^m y^n \text{ with ascending powers of } \dfrac{D'}{D} \text{ or } \dfrac{D}{D'} \text{ or } D \text{ or } D'.$$

Case IV. When $F(x, y) = e^{ax+by} \cdot V$, where V is a function of x and y, then we have

$$\text{P.I.} = \dfrac{1}{f(D, D')}e^{ax+by} \cdot V = e^{ax+by}\dfrac{1}{f(D + a, D' + b)} \cdot V.$$

Solved Examples

EXAMPLE 1. *Solve* $(D^2 - D'^2 + D - D')z = 0.$

SOLUTION. The given equation is $(D^2 - D'^2 + D - D')z = 0.$

or $(D - D').(D + D' + 1)z = 0.$

Hence, the solution is

$$z = f_1(y + x) + e^{-x}f_2(y - x).$$

EXAMPLE 2. *Solve* $(D - 2D' - 1)(D - 2D'^2 - 1)z = 0.$

SOLUTION. The given equation is

$$(D - 2D' - 1)(D - 2D'^2 - 1)z = 0.$$

C.F. corresponding to first factor is

$$z = e^x f_1(y + 2x).$$

Now C.F. corresponding to second factor is $\sum Ae^{hx+ky}$,

where $h - 2k^2 - 1 = 0$ or $h = 2k^2 + 1.$

Hence, the solution is

$$z = e^x f_1(y + 2x) + \sum Ae^{ky + (2k^2+1)x}.$$

EXAMPLE 3. *Solve* $(D^3 - 3DD' + D' + 1)z = e^{2x+3y}.$

SOLUTION. Here, $D^3 - 3DD' + D' + 1$ cannot be resolved into linear factor in D and D'.

Let $z = Ae^{hx+ky}.$

$\therefore \quad (D^3 - 3DD' + D' + 1)z = A(h^3 - 3hk + k + 1)e^{hx+ky}.$

Then $(D^3 - 3DD' + D' + 1)z = 0$

if $\quad h^3 - 3hk + k + 1 = 0.$

$\therefore$ C.F. $= \sum Ae^{hx+ky}$, where $h^3 - 3hk + k + 1 = 0.$

Now, P.I. $= \dfrac{1}{D^3 - 3DD' + D' + 1} . e^{2x+3y}$

$$= \frac{e^{2x+3y}}{2^3 - 3.2.3 + 3 + 1} = -\frac{1}{6}e^{2x+3y}.$$

Hence, the complete solution is

$$z = \sum Ae^{hx+ky} - \frac{1}{6}e^{2x+3y},$$

where $h^3 - 3hk + k + 1 = 0.$

EXAMPLE 4. *Solve* $s + p - q = z + xy.$

SOLUTION. The equation can be written as

$$(DD' + D - D' - 1)z = xy$$

or $\quad (D - 1)(D' + 1) = xy.$

$\therefore$ C.F. $= e^x f_1(y) + e^{-y}f_2(x).$

Now, P.I. $= \dfrac{1}{(D-1)(D'+1)}xy$

$$= -(1-D)^{-1}(1+D')^{-1} . xy$$

$$= -(1 + D + ...)(1 - D' + ...) xy$$

$$= -(1 + D - D' - DD'...)xy$$

$$= -(xy + y - x - 1).$$

Hence, the complete solution is

$$z = e^x f_1(y) + e^{-y}f_2(x) - xy - y + x + 1.$$

EXAMPLE 5. *Solve* $(D^2 + DD' + D' - 1)z = \sin(x + 2y).$

(GBTU-2010, SVTU-2009)

SOLUTION. The given equation can be written as

$$(D + 1)(D + D' - 1)z = \sin(x + 2y).$$

$\therefore$ C.F. $= e^{-x}f_1(y) + e^x f_2(y - x).$

Now, P.I. $= \dfrac{1}{D^2 + DD' + D' - 1} . \sin(x + 2y)$

$$= \frac{1}{-1^2 - 1.2 + D' - 1}.\sin(x + 2y)$$

(On putting -1^2 for D^2 and -1.2 for DD')

$$= \frac{1}{(D'-4)}\sin(x + 2y) = \frac{D'+4}{D'^2 - 16}\sin(x + 2y)$$

$$= (D'+4)\frac{1}{-2^2 - 16} \sin(x + 2y)$$

$$= -\frac{1}{20}(D'+4)\sin(x + 2y) \quad \text{(On putting } -2^2 \text{ for } D'^2)$$

$$= -\frac{1}{20}[2\cos(x + 2y) + 4\sin(x + 2y)]$$

$$= -\frac{1}{10}[\cos(x + 2y) + 2\sin(x + 2y)].$$

Hence, the complete solution is

$$z = e^{-x}f_1(y) + e^x f_2(y - x)$$

$$- \frac{1}{10}[\cos(x + 2y) + 2\sin(x + 2y)].$$

EXAMPLE 6. *Solve* $(D - D'^2 - 3D + 3D')z = xy + e^{x+2y}.$

(UPTU-2002, GBTU(CO)-2011)

SOLUTION. The given equation can be written as

$$(D - D')(D + D' - 3)z = xy + e^{x+2y}.$$

Its C.F. $= f_1(y + x) + e^{3x}f_2(y - x).$

Now, P.I. corresponding to xy

$$= \frac{1}{(D - D')(D + D' - 3)}xy$$

$$= -\frac{1}{3D}\left(1 - \frac{D'}{D}\right)^{-1}\left(1 - \frac{D + D'}{3}\right)^{-1} . xy$$

$$= -\frac{1}{3D}\left(1 + \frac{D'}{D} + ...\right)\left[1 + \frac{D + D'}{3} + \left(\frac{D + D'}{3}\right)^2 + ...\right]xy$$

$$= -\frac{1}{3D}\left(1 + \frac{D'}{D} + ...\right)\left(1 + \frac{D + D'}{3} + \frac{2DD'}{9} + ...\right)xy$$

$$= -\frac{1}{3D}\left(1 + \frac{D}{3} + \frac{D'}{3} + \frac{D'}{D} + \frac{D'}{3} + \frac{2DD'}{9} + ...\right)xy$$

$$= -\frac{1}{3D}\left(xy + \frac{y}{3} + \frac{2x}{3} + \frac{1}{D}x + \frac{2}{9}\right)xy$$

$$= -\frac{1}{3}\left(\frac{x^2y}{2} + \frac{xy}{3} + \frac{x^2}{3} + \frac{x^3}{6} + \frac{2}{9}x\right).$$

Now, P.I. corresponding to e^{x+2y}

$$= \frac{1}{(D + D' - 3)(D - D')}e^{x+2y}$$

$$= \frac{1}{(D + D' - 3)}.\frac{1}{(1 - 2)}e^{x+2y}$$

$$= -\frac{1}{(D + D' - 3)}e^{x+2y}.1$$

$$= -e^{x+2y}\frac{1}{D + 1 + D' + 2 - 3}.1$$

$$= -e^{x+2y}\frac{1}{D + D'}.1 = -e^{x+2y}\frac{1}{D}\left(1 + \frac{D'}{D}\right)^{-1}.1$$

$$= -e^{x+2y}\frac{1}{D}\left(1 - \frac{D'}{D}...\right).1 = -xe^{x+2y}.$$

Hence, the complete solution is

$$z = f_1(y+x) + e^{3x} f_2(y-x)$$

$$-\frac{1}{3}\left(\frac{x^2 y}{2} + \frac{xy}{3} + \frac{x^2}{3} + \frac{x^3}{6} + \frac{2}{9}x\right) - xe^{x+2y}.$$

EXAMPLE 7. *Solve* $(D^2 - DD' + D' - 1)z = \cos(x+2y) + e^y$.

(UPTU(Q. BANK)-2002)

SOLUTION. The given equation can be written as

$$(D-1)(D-D'+1)z = \cos(x+2y) + e^y.$$

$$\therefore \quad \text{C.F.} = e^x f_1(y) + e^{-x} f_2(y+x)$$

Now, P.I. corresponding to $\cos(x+2y)$

$$= \frac{1}{D^2 - DD' + D' - 1}\cos(x+2y)$$

$$= \frac{1}{-1^2 - (-1 \cdot 2) + D' - 1}\cdot\cos(x+2y)$$

$$= \frac{1}{D'}\cos(x+2y) = \frac{D'}{D'^2}\cos(x+2y)$$

$$= \frac{D'}{-2^2}\cos(x+2y) = -\frac{1}{4}\{-2\sin(x+2y)\}$$

$$= \frac{1}{2}\sin(x+2y).$$

P.I. corresponding to e^y

$$= \frac{1}{(D-D'+1)(D-1)}e^y = \frac{1}{(D-D'+1).(0-1)}e^y.1$$

$$= -e^y\frac{1}{D-(D'+1)+1}.1 = -e^y\frac{1}{D-D'}.1$$

$$= -e^y.\frac{1}{D}\left(1-\frac{D'}{D}\right)^{-1}.1 = -e^y.\frac{1}{D}.1 = -xe^y.$$

Hence, the solution is

$$z = e^x f_1(y) + e^{-x} f_2(y+x) + \frac{1}{2}\sin(x+2y) - xe^y.$$

Exercise-8.65

Solve the following equations:

1. $t + s + q = 0.$

2. $DD'(D - 2D' - 3)z = 0.$

3. $D(D - 2D' - 3)z = e^{x+2y}.$

4. $(D + D' - 1)(D + 2D' - 3)z = 4 + 3x + 6y.$

5. $r - s + 2q - z = x^2 y^2.$ (IAS-1993)

6. $[D^2 - DD' + 2D' + 2D - 2D']z = e^{2x+3y} + \sin(2x+y) + xy.$
   (UPTU(B. TECH.)-2003)

7. $(D - D'^2)z = \cos(x - 3y).$

8. $(D - D' - 1)(D - D' - 2)z = \sin(2x+3y).$
   (UPTU(Q. BANK)-2002, UKTU-2011)

9. $(D - D' + 2)(D + D' - 1)z = e^{x-y} - x^2 y.$

10. $(D^2 - DD' - 2D)z = \sin(3x+4y) - e^{2x+y}.$

11. $(D^2 - DD' - 2D'^2 + 2D + 2D')z = \sin(2x+y)$
    (GBTU-2011)

12. $(D + D' - 1)^2 z = xy$ (GBTU(AG)-2011)

13. $(D - 3D' - 2)^3 z = 6\,e^{2x}\sin(3x+y)$ (GBTU(SUM)-2010)

14. $(D^2 + 2DD' + D'^2 - 2D - 2D')z = \sin(x + 2y)$ (UPTU-2004)

15. $\dfrac{\partial^2 z}{\partial x^2} - \dfrac{\partial^2 z}{\partial x \partial y} + \dfrac{\partial z}{\partial y} = x^2 + y^2$ (MADRAS-2000S)

Hints to Selected Problems

1. The given equation can be written as

$$(DD' + D' + D'^2)z = 0 \implies D'(D + D' + 1)z = 0$$

Solution is $z = f_1(x) + e^{-x} f_2(y - x).$

3. C.F. is $z = f_1(y) + e^{3x} f_2(y + 2x)$

$$\text{P.I.} = \frac{1}{D(D-2D'-3)}e^{x+2y} = \frac{e^{x+2y}}{1.(1-2.2-3)} = -\frac{1}{6}e^{x+2y}$$

5. $\text{P.I.} = \dfrac{1}{(D^2 - DD' + 2D' - 1)}x^2 y^2$

$$= -[1 - (D^2 - DD' + 2D')]^{-1} x^2 y^2$$

Now expanding by binomial theorem.

7. Since $D - D'^2$ cannot be resolved into factors, so
   C.F. $= \Sigma A e^{hx+ky} = \Sigma A e^{k^2 x + ky}$, where $h - k^2 = 0$

$$\text{P.I.} = \frac{1}{D - D'^2}.\cos(x-3y) = \frac{1}{D+9}\cos(x+3y)$$

$$= \frac{D-9}{81}\cos(x-3y) = \frac{-\sin(x-3y) - 9\cos(x-3y)}{-1-81}$$

Answers

1. $z = f_1(x) + e^{-x} f_2(y - x).$ 2. $z = f_1(y) + f_2(x) + e^{3x} f_3(y + 2x).$

3. $z = f_1(y) + e^{3x} f_2(y + 2x) - \dfrac{1}{6}e^{x+2y}.$ 4. $z = e^x f_1(y - x) + e^{3x} f_2(y - 2x) + 6 + x + 2y.$

5. $z = \Sigma A e^{hx+ky} - x^2 y^2 - 2y^2 + 4xy - 4x^2 y - 52 - 8x^2 - 16y + 16x$ where $h^2 - hk + 2k - 1 = 0$

6. $z = f_1(y - x) + e^{-2x} f_2(y + 2x) - \dfrac{1}{10}e^{2x+3y} - \dfrac{1}{6}\cos(2x + y) + \dfrac{1}{4}x^2 y - \dfrac{1}{4}xy + \dfrac{3}{8}x^2 - \dfrac{1}{2}x - \dfrac{1}{12}x^3.$

7. $z = \Sigma A e^{k^2+ky} + \dfrac{1}{82}[\sin(x - 3y) + 9\cos(x - 3y)]$ where $h - k^2 = 0$

8. $z = e^x f_1(y + x) + e^{2x}(y + x) + \dfrac{1}{10}[\sin(2x+3y) - 3\cos(2x+3y)].$

9. $z = e^{-2x} f_1(y + x) + e^x f_2(y - x) - \dfrac{e^{x-y}}{4} + \dfrac{1}{2}\left[x^2 y + xy + \dfrac{3y}{2} + 3x + \dfrac{21}{4}\right] + \dfrac{3}{4}x^2.$

10. $z = f_1(y) + e^{2x} f_2(y + x) + \dfrac{1}{15}\sin(3x+4y) + \dfrac{2}{15}\cos(3x+4y) + \dfrac{1}{2}e^{2x+y}.$ 11. $z = f_1(y - x) + e^{-2x} f_2(2x + y) - 1/6 \cos(2x + y)$

12. $z = e^x f_1(y - x) + x e^x f_2(y - x) + xy + 2y + 2x + 6$

13. $z = e^{2x} f_1(y + 3x) + x e^{2x} f_2(y + 3x) + x^2 e^{2x} f_3(y + 3x) + x^3 e^{2x} \sin(3x + y)$

14. $z = \phi_1(y - x) + e^{2x} \phi_2(y - x) + \dfrac{1}{39} [2\cos(x + 2y) - 3\sin(x + 2y)]$ **15.** $z = f_1(y) + e^{-x} f_2(y + x) + \dfrac{1}{3} x^3 - x^2 + xy^2 + 6x$

8.14.8 EQUATIONS REDUCIBLE TO LINEAR PARTIAL DIFFERENTIAL EQUATIONS WITH CONSTANT COEFFICIENTS

A partial differential equation of the form

$$f\left(x\frac{\partial}{\partial x}, y\frac{\partial}{\partial y}\right) = F(x, y)$$

having variable coefficients are reduced to linear form by putting $x = e^X$, $y = e^Y$ so that $X = \log x$ and $Y = \log y$

$\therefore$ $\dfrac{\partial z}{\partial x} = \dfrac{\partial z}{\partial X} \cdot \dfrac{\partial X}{\partial x} = \dfrac{1}{x} \dfrac{\partial z}{\partial X}$ or $x\dfrac{\partial z}{\partial x} = \dfrac{\partial z}{\partial X}$

$\therefore$ $x\dfrac{\partial}{\partial x} \equiv \dfrac{\partial}{\partial X} \equiv D$ (say)

Now, $x\dfrac{\partial}{\partial x}\left(x^{n-1}\dfrac{\partial^{n-1} z}{\partial x^{n-1}}\right) = x^n \dfrac{\partial^n z}{\partial x^n} + (n-1) x^{n-1} \dfrac{\partial^{n-1} z}{\partial x^{n-1}}$ or $x^n \cdot \dfrac{\partial^n z}{\partial x^n} = \left(x\dfrac{\partial}{\partial x} - n + 1\right) x^{n-1} \dfrac{\partial^{n-1} z}{\partial x^{n-1}}.$

Putting $n = 2, 3, \dots$ we get

$$x^2 \frac{\partial^2 z}{\partial x^2} = (D-1)x\frac{\partial z}{\partial x} = D(D-1)z, \ x^3 \frac{\partial^3 z}{\partial x^3} = (D-2)x^2\frac{\partial^2 z}{\partial x^2} = D(D-1)(D-2)z \ \text{etc.}$$

Similarly, $y\dfrac{\partial z}{\partial y} = \dfrac{\partial z}{\partial Y} = D'z; \ y^2\dfrac{\partial^2 z}{\partial y^2} = D'(D'-1)z; \ y^3\dfrac{\partial^3 z}{\partial y^3} = D'(D-1)(D'-2)z$ and $xy\dfrac{\partial^2 z}{\partial x \partial y} = DD'z$

Substituting in the given equation, it reduced to the form $f(DD')z = V$ which is an equation with constant coefficient.

Solved Examples

EXAMPLE 1. **Solve**

$$x^2 \frac{\partial^2 z}{\partial x^2} - 4xy \frac{\partial^2 z}{\partial x \partial y} + 4y^2 \frac{\partial^2 z}{\partial y^2} + 6y \frac{\partial z}{\partial y} = x^3 y^4.$$

(UPTU(Q. BANK)-2002)

SOLUTION. Substituting $x = e^X$, $y = e^Y$

And denoting $\dfrac{\partial}{\partial X}$ and $\dfrac{\partial}{\partial Y}$ by D and D' respectively, then the equation reduces to

$[D(D-1) - 4DD' + 4D'(D'-1) + 6D']z = e^{3X+4Y}$

or $(D - 2D')(D - 2D' - 1)z = e^{3X+4Y}$.

$\therefore$ C.F. $= f_1(Y + 2X) + e^X f_2(Y + 2X)$

$= f_1(\log y + 2\log x) + x f_2(\log y + 2\log x)$

$= f_1(\log yx^2) + x f_2(\log yx^2) = g_1(yx^2) + x g_2(yx^2)$.

Now, P.I. $= \dfrac{1}{(D - 2D')(D - 2D' - 1)} e^{3X+4Y}$

$= \dfrac{e^{3X+4Y}}{(3-8)(3-8-1)} = \dfrac{1}{30} e^{3X+4Y} = \dfrac{1}{30} x^3 y^4$.

Hence, the complete solution is

$$z = g_1(yx^2) + x g_2(yx^2) + \frac{1}{30} x^3 y^4.$$

EXAMPLE 2. **Solve** $x^2 r - y^2 t + xp - yq = \log x$.   (IAS-1993, 97)

SOLUTION. The given equation can be written as

$$x^2 \frac{\partial^2 z}{\partial x^2} - y^2 \frac{\partial^2 z}{\partial y^2} + x\frac{\partial z}{\partial x} - y\frac{\partial z}{\partial y} = \log x.$$

Substituting $x = e^X$, $y = e^Y$ and denoting $\dfrac{\partial}{\partial X}$ and $\dfrac{\partial}{\partial Y}$ by D and D' respectively.

Then equation reduces to

$[D(D-1) - D'(D'-1) + D - D']z = X$

or $(D^2 - D'^2)z = X$ or $(D - D')(D + D')z = X$.

Its   C.F. $= \phi_1(Y + X) + \phi_2(Y - X)$

$= \phi_1(\log y + \log x) + \phi_2(\log y - \log x)$

$= \phi_1(\log xy) + \phi_2(\log y / x)$

$= f_1(xy) + f_2(y / x)$.

Now, P.I. $= \dfrac{1}{(D^2 - D'^2)} \cdot X = \dfrac{1}{D^2}\left(1 - \dfrac{D'^2}{D^2}\right)^{-1} \cdot X$

$= \dfrac{1}{D^2} \cdot X = \dfrac{X^3}{6} = \dfrac{1}{6}(\log x)^3$.

Hence, the complete solution is

$$z = f_1(xy) + f_2(y / x) + \frac{1}{6}(\log x)^3.$$

Exercise-8.66

1. $(x^2 D^2 - y^2 D'^2)z = x^2 y$.

2. $x^2\left(\dfrac{\partial^2 z}{\partial x^2}\right) - y^2\left(\dfrac{\partial^2 z}{\partial y^2}\right) = xy$.   (IAS-1987)

3. $(x^2 D^2 + 2xyDD' + y^2 D'^2)z = x^m y^n$.

4. $(x^2 D^2 + 2xyDD' + y^2 D'^2 - nxD - xyD' + n)z = x^2 y^2$

5. $x^2 r - 3xys + 2y^2 t + px + 2qy = x + 2y$.

6. $x^2 D^2 + 2xyDD' + y^2 D'^2 = (x^2 + y^2)^{3/2}$.

7. $(x^2 D^2 - xyDD' - 2y^2 D'^2 + xD - 2yD')z = \log(y / x) - 1 / 2$

Hints to Selected Problems

1. Putting $x = e^X$, $y = e^Y$ $\Rightarrow$ $X = \log x$, $Y = \log y$

So, $x^2 D^2 = D_1(D_1 - 1)$, $y^2 D'^2 = D'_1(D'_1 - 1)$

where $D_1 = \dfrac{\partial}{\partial X}, D'_1 = \dfrac{\partial}{\partial Y}$

Then, equation (1) reduces to

$[D_1(D_1 - 1) - D'_1(D'_1 - 1)]z = e^{2X+Y}$

$\Rightarrow (D_1 - D'_1)(D_1 + D'_1 - 1) = e^{2X+Y}$

3. Proceed same as (1), we get

$(D_1 + D'_1)(D_1 + D'_1 - 1)z = e^{mX+nY}$

4. The reduced equation is

$\left[(D_1 + D'_1 - 1)(D_1 + D'_1 - n)\right]z = e^{2X} + e^{2Y}$

5. The reduced equation is $\left[(D_1 + D'_1 - 1)(D_1 - 2D'_1)\right]z = e^X + 2e^Y$

C.F. $= \phi_1(Y + X) + \phi_2(Y + 2X)$

$= \phi_1(\log xy) + \phi_2(\log x^2 y) = f_1(xy) + f_2(x^2 y)$

P.I. $= \dfrac{1}{(D_1 - D'_1)(D_1 - 2D'_1)}(e^X + 2e^Y)$

$= \dfrac{e^X}{(1-0)(1-0)} + \dfrac{2e^Y}{(0-1)(0-2)} = e^X + e^Y = x + y$

 Answers

1. $z = xf_1(y/x) + f_2(xy) + \dfrac{1}{2}x^2 y$.   **2.** $z = f_1(xy) + xf_2(y/x) + xy\log x$.   **3.** $z = f_1(y/x) + xf_2(y/x) + \dfrac{x^m y^n}{(m+n)(m+n-1)}$

4. $z = xf_1(y/x) + x^n f_2(y/x) + \dfrac{x^2 + y^2}{2-n}$.   **5.** $z = f_1(xy) + f_2(x^2 y) + x + y$.

6. $z = f_1(y/x) + xf_2(y/x) + \left\{\dfrac{1}{(n^2 - n)}(x^2 + y^2)\right\}^{m/n}$ **7.** $z = f_1(yx^2) + f_2(y/x) + \dfrac{1}{2}(\log x)^2 \log y - \dfrac{1}{4}(\log x)^2$

8.14.9 SOME EXAMPLES UNDER GIVEN GEOMETRICAL CONDITIONS

EXAMPLE 1. *Find the surface passing through two lines $z = x = 0$, $z - 1 = x - y = 0$ satisfying*

$r - 4s + 4t = 0.$ (IAS-1996)

SOLUTION. The given equation can be written as

$(D^2 - 4DD' + 4D'^2)z = 0.$

It's A. E. is $m^2 - 4m + 4 = 0$

or $(m-2)^2 = 0$

or $m = 2, 2.$

$\therefore$ $z = f_1(y + 2x) + xf_2(y + 2x).$ ...(1)

Since the surface passes through the lines $z = x = 0$ and $z - 1 = x - y = 0$.

$\therefore$ $0 = f_1(y).$ ...(2)

and $1 = f_1(y + 2x) + xf_2(y + 2x).$ ...(3)

From (2) and (3), we get

$f_2(y + 2x) = \dfrac{1}{x} = \dfrac{3}{3x} = \dfrac{3}{2x + x} = \dfrac{3}{2x + y}$ $[\because x - y = 0]$

Hence, the surface is $z = x \cdot \dfrac{3}{2x + y}$ or $z(2x + y) = 3x.$

EXAMPLE 2. *Find the surface satisfying the equation $r + t - 2s = 0$ and conditions that $bz = y^2$, when $x = 0$ and $az = x^2$ when $y = 0$.*

SOLUTION. The given equation can be written as

$(D^2 - 2DD' + D'^2)z = 0 \Rightarrow (D - D')^2 z = 0.$

$\therefore$ Its solution is $z = f_1(y + x) + xf_2(y + x).$ ...(1)

Since $z = y^2/b$ when $x = 0$ then from (1),

$\dfrac{y^2}{b} = f_1(y)$

giving $f_1(y + x) = \dfrac{(y + x)^2}{b}.$ ...(2)

Again since $z = \dfrac{x^2}{a}$ when $y = 0$, then from (1)

$\dfrac{x^2}{a} = f_1(x) + xf_2(x).$ ...(3)

But from (2), $f_1(x) = x^2/b$. Then (3) gives

$\dfrac{x^2}{a} = xf_2(x) + \dfrac{x^2}{b}$ or $f_2(x) = \dfrac{b-a}{ab}x$ which gives

$f_2(y + x) = \dfrac{b-a}{ab}(y + x).$ ...(4)

Putting, value of $f_1(y + x)$ and $f_2(y + x)$ from (2) and (4) in (1) the required surface is

$z = \dfrac{b-a}{ab}x(y + x) + \dfrac{(y + x)^2}{b}$

or $z = (x + y)\left(\dfrac{x}{a} + \dfrac{y}{b}\right).$

 Exercise-8.67

1. Find a surface satisfying $r - 2s + t = 6$ and touching the hyperbolic paraboloid $z = xy$ along its section by the plane $y = x$.

2. Find a surface satisfying equation $2x^2 r - 5xys + 2y^2 t + 2(px + qy = 0)$ and touching the hyperbolic paraboloid $z = x^2 - y^2$ along its section by plane $y = 1$.

3. Find a surface satisfying $r + s = 0$ and touching the elliptic paraboloid $z = 4x^2 + y^2$ along its section by the plane $y = 2x + 1$.

1. The given equation can be written as $(D^2 - 2DD' + D'^2)z = 6$,

i.e., $(D - D')^2 z = 6$, C.F. $= \phi_1(y + x) + x\phi_2(y + x)$, P.I. $= 3x^2$.

2. The given equation can be written as

$$2x^2 \frac{\partial^2 z}{\partial x^2} - 5xy \frac{\partial^2 z}{\partial x \partial y} + xy^2 \frac{\partial^2 z}{\partial y^2} + 2\left(x \frac{\partial z}{\partial x} + y \frac{\partial z}{\partial y}\right) = 0.$$

Now putting $x = e^u, y = e^v$ i.e., $u = \log x$ and $v = \log y$.

Answers

1. $z = x^2 - xy + y^2$	**2.** $3z = 4yx^2 - y^4 x^2 - 6\log y - 3$	**3.** $4x^2 - 8xy + y^2 + 8x - 4y + z + 2 = 0$

8.15 APPLICATIONS OF PARTIAL DIFFERENTIAL EQUATIONS TO ENGINEERING PROBLEMS

Many problems in science and engineering, when formulated mathematically, lead to partial differential equations involving one or more unknown functions together with certain prescribed conditions on the functions which arise from the physical situations.

The process of obtaining all solutions of a partial differential equations under given conditions is known as boundary value problem. If time t is regarded as an independent variable and the conditions are stated as $t = 0$, the problem is called initial value problem.

8.15.1 CLASSIFICATION OF PARTIAL DIFFERENTIAL EQUATION

(UPTU-2005, 07)

Any equation $A\dfrac{\partial^2 u}{\partial x^2} + B\dfrac{\partial^2 u}{\partial x \partial y} + C\dfrac{\partial^2 u}{\partial y^2} + F(x, y, u, p, q) = 0$ ... (1)

where A, B, C may be constants or functions of x and y.

This equation (1) will be

(i) elliptic   if $B^2 - 4AC < 0$
(ii) parabolic   if $B^2 - 4AC = 0$
(iii) hyperbolic if $B^2 - 4AC > 0$

1. Parabolic equation : The heat equation $\dfrac{\partial u}{\partial t} = \dfrac{\partial^2 u}{\partial x^2}$

2. Elliptic equation : The Laplace equation $\dfrac{\partial^2 u}{\partial x^2} + \dfrac{\partial^2 u}{\partial y^2} = 0$

3. Hyperbolic equation : The wave equation $\dfrac{\partial^2 u}{\partial t^2} = C^2 \dfrac{\partial^2 u}{\partial x^2}$

8.15.2 PRODUCT METHOD : SOLUTION OF BOUNDARY VALUE PROBLEMS BY THE METHOD OF SEPARATION OF VARIABLES

Let $\qquad A\dfrac{\partial^2 u}{\partial x^2} + B\dfrac{\partial^2 u}{\partial x \partial y} + C\dfrac{\partial^2 u}{\partial y^2} + D\dfrac{\partial u}{\partial x} + E\dfrac{\partial u}{\partial y} + Fu = G \qquad$ ... (1)

be the general linear partial differential equation, where $A, B, \ldots\ldots, G$ are functions of x and y. Here, it must be noted that

(i) If any partial differential equation of second order cannot be put in above form, then it is said to be non-linear.

(ii) If $G = 0$, then equation (1) is called homogeneous. Thus, for non-homogeneous equation $G \neq 0$.

Let $u(x, y) = X(x)\, Y(y)$ ....(2)

be the solution of (1). Here $X(x)$ and $Y(y)$ are respectively the functions of x and y alone.

Using (2), we can find the following values

$\dfrac{\partial^2 u}{\partial x^2}, \dfrac{\partial^2 u}{\partial y^2}, \dfrac{\partial^2 u}{\partial x \partial y}, \dfrac{\partial u}{\partial x}$ and $\dfrac{\partial u}{\partial y}$ in terms of X and Y.

Putting all these values in (1), we get $\dfrac{F(D).X}{X} = \dfrac{g(D').Y}{Y}$, where, $F(D)$ and $g(D')$ are functions of $D = \dfrac{\partial}{\partial x}$ and $D' = \dfrac{\partial}{\partial y}$ respectively.

Thus, we can find a relation in which LHS is a function of x alone while RHS is a function of y alone.

Then putting $\qquad \dfrac{F(D).X}{X} = \dfrac{g(D').Y}{Y} = \lambda$ (say) $\qquad$ ... (3)

Thus, the solution of (1) reduces to be solution of a pair of ODE given by (3).

Solved Examples

EXAMPLE 1. *Solve the boundary value problem* $\dfrac{\partial u}{\partial x} = 4\dfrac{\partial u}{\partial y}$ *with* $u(0, y) = 8e^{-3y}$ *by the method of separation of variables.* (UPTU-2008, JNTU-2006)

SOLUTION. The given equation is

$$\frac{\partial u}{\partial x} = 4\frac{\partial u}{\partial y} \qquad ... (1)$$

with boundary conditions $u(0, y) = 8e^{-3y}$.

Let $u(x, y) = X(x)\,Y(y)$ (where X and Y are respectively the functions of x and y alone) be the solution of (1). Putting the value in (1), we get

$$X'Y = 4XY'$$

$$\Rightarrow \quad \frac{X'}{4X} = \frac{Y'}{Y} \qquad \dots (2)$$

Here, dashes denote derivatives with respect to the relevant variable. Equation (2) is true only when each side is equal to same constant say λ.

Therefore (2) gives $X' - 4\lambda X = 0$ and $Y' - \lambda Y = 0$.

On solving, we get $X(x) = A e^{4\lambda x}$ and $Y(y) = B e^{\lambda y}$.

Thus, $u(x, y) = ABe^{4\lambda x + \lambda y} = C\,e^{\lambda(4x + y)}$,

where $C\,(= AB)$ is any arbitrary constant.

Using given boundary condition

$$u(0, y) = 8e^{-3y} = C\,e^{\lambda y},$$

we get $\qquad C = 8,\ \lambda = -3$.

Hence, the solution of given boundary value problem is

$$u(x, y) = 8e^{-3(4x + y)}$$

EXAMPLE 2. ***Solve by the method of separation of variables***

$$\frac{\partial^2 u}{\partial x^2} - 2\frac{\partial u}{\partial x} + \frac{\partial u}{\partial y} = 0.$$

(UPTU-2005, Q. BANK–2002, PTU-2009 S, BHOPAL-2008)

SOLUTION. The given equation is

$$\frac{\partial^2 u}{\partial x^2} - 2\frac{\partial u}{\partial x} + \frac{\partial u}{\partial y} = 0 \qquad \dots (1)$$

Let $u(x, y) = X(x)\,Y(y)$ be the solution of this equation. Putting the value of u in the given equation, we get

$$X''Y - 2X'Y + XY' = 0 \qquad \dots (2)$$

The dashes denote derivatives with respect to the relevant variable.

Now, relation (2) can be written as

$$\frac{X'' - 2X'}{X} = \frac{Y'}{Y} \qquad \dots (3)$$

This is true if each side is equal to the same constant, say λ. Then, we have

$$X'' - 2X' - \lambda X = 0 \text{ and } Y' - \lambda Y = 0 \qquad \dots (4)$$

$$\Rightarrow (D^2 - 2D - \lambda)X = 0 \text{ and } \frac{dY}{dy} = \lambda Y \qquad \dots (5)$$

The auxiliary equation of (4) is given by

$$m^2 - 2m - \lambda = 0,$$

i.e., $\qquad m = 1 \pm \sqrt{(1 + \lambda)}$.

Thus, $\qquad X(x) = A e^{[1 + \sqrt{(1 + \lambda)}]x} + B e^{[1 - \sqrt{(1 + \lambda)}]x}$

and solution of (5) is given by

$$Y(y) = C e^{\lambda y}$$

Finally, we get

$$u(x, y) = \left\{ ACe^{[1 + \sqrt{(1 + \lambda)}]x} + BCe^{[1 - \sqrt{(1 + \lambda)}]x} \right\} e^{\lambda y}$$

$$= \left\{ A_1 e^{[1 + \sqrt{(1 + \lambda)}]x} + A_2 e^{[1 - \sqrt{(1 + \lambda)}]x} \right\} e^{\lambda y}$$

EXAMPLE 3. ***Use the method of separation of variables to solve the equation*** $\dfrac{\partial^2 u}{\partial x^2} - 2\dfrac{\partial u}{\partial x} + \dfrac{\partial u}{\partial y} = 0$

(UPTU-2009, GBTU (AG)-2012)

SOLUTION. Let $\quad u = XY \qquad \dots (1)$

Where X is a function x only and Y is function of y only.

$$\frac{\partial u}{\partial x} = \frac{\partial}{\partial x}(XY) = Y\frac{dX}{dx}$$

$$\frac{\partial u}{\partial y} = \frac{\partial}{\partial y}(XY) = X\frac{dY}{dy}$$

$$\frac{\partial^2 u}{\partial x^2} = \frac{\partial^2}{\partial x^2}(XY) = Y\frac{d^2X}{dx^2}$$

Putting these values in (1), we get

$$Y\frac{d^2X}{dx^2} - 2Y\frac{dX}{dx} + X\frac{dY}{dy} = 0$$

$$\Rightarrow \quad YX'' - 2YX' + XY' = 0$$

$$\Rightarrow \quad \frac{X'' - 2X'}{X} + \frac{Y'}{Y} = 0$$

$$\Rightarrow \quad \frac{X'' - 2X'}{X} = -\frac{Y'}{Y} = -P^2 \text{ (say)}$$

(i) $\qquad \dfrac{X'' - 2X'}{X} = -P^2$

$$\Rightarrow \quad X'' - 2X' + P^2 X = 0$$

A. E. is $\qquad m^2 - 2m + p^2 = 0$

$$\Rightarrow \quad m = \frac{2 \pm \sqrt{4 - 4P^2}}{2} = 1 \pm \sqrt{1 - P^2}$$

$\therefore$ C. F. $= C_1 e^{(1 + \sqrt{1 - P^2})x} + C_2 e^{(1 - \sqrt{1 - P^2})x}$
P. I. $= 0$

Hence $X = $ C.F. + P. I.

$$= C_1 e^{(1 + \sqrt{1 - P^2})x} + C_2 e^{(1 - \sqrt{1 - P^2})x} \qquad \dots (2)$$

(ii) $\dfrac{-Y'}{Y} = -P^2 \Rightarrow \dfrac{dY}{dy} = p^2 Y \Rightarrow \dfrac{dY}{Y} = p^2 dy$

On integrating, we get $\log Y = p^2 y + \log C_3$

$$Y = C_3 e^{p^2 y}$$

$\therefore u(x, y) = [C_1 e^{(1 + \sqrt{1 - p^2})x} + C_2 e^{(1 - \sqrt{1 - p^2})x}]C_3 e^{p^2 y} \dots (3)$

Exercise-8.72

1. Use the method of separation of variables to solve the equation $\dfrac{\partial^2 V}{\partial x^2} = \dfrac{\partial V}{\partial t}$.
Given that $V = 0$ when $t \to \infty$ as well as $v = 0$ at $x = 0$ and $x = l$. (AMIE-1997)

2. Use the method of separation of variables to solve $\dfrac{\partial u}{\partial x} = 2\dfrac{\partial u}{\partial t} + u$, where $u(x, 0) = 6e^{-3x}$.
(KERALA–2005, UPTU-2006, 17, KURUKSHETRA-2006, VTU-2009, GBTU(SUM)-2010, GBTU(AG)-2011, AMIETE 1997)

3. Use the method of separation of variables to solve the equation $\dfrac{\partial^2 u}{\partial x^2} = \dfrac{\partial u}{\partial y} + 2u$.

4. Solve by the method of separation of variables
$4\dfrac{\partial u}{\partial t} + \dfrac{\partial u}{\partial x} = 3u$, $u = 3e^{-x} - e^{-5x}$ when $t = 0$. (SVTU-2008)

5. Use the method of separation of variables, find the solution of the following differential equation:
$3\dfrac{\partial u}{\partial x} + 2\dfrac{\partial u}{\partial y} = 0$, $u(x, 0) = 4e^{-x}$. (AMIETE-2000, VTU-2008S)

6. Solve by the method of separation of variables.
$py^3 + qx^2 = 0$ (VTU-2011, SVTU-2008)

7. Solve by the method of separation of variables
$x^2\dfrac{\partial u}{\partial x} + y^2\dfrac{\partial u}{\partial y} = 0$ (VTU-2008)

8. Find a solution of the equation $\dfrac{\partial^2 u}{\partial x^2} = \dfrac{\partial u}{\partial y} + 2u$ in the form $u = f(x)g(y)$. Solve the equation subject to the conditions $u = 0$ and $\dfrac{\partial u}{\partial x} = 1 + e^{-3y}$, when $x=0$ for all values of y.

(ANDHRA-2000)

9. Solve by method of separation of variables

$4\dfrac{\partial u}{\partial x} + \dfrac{\partial u}{\partial y} = 3u; u(0,y) = 4e^{-y} - e^{-5y}$ (UKTU-2012)

10. Solve by method of separation of variables

$y^3 \dfrac{\partial u}{\partial x} + x^2 \dfrac{\partial u}{\partial y} = 0$ (GBTU-2011)

11. Solve by method of separation of variables

$\dfrac{\partial^2 u}{\partial x^2} - \dfrac{\partial u}{\partial y} = 0$ (UPTU (SUM)-2007)

12. Solve by method of separation of variables

$\dfrac{\partial u}{\partial x} = \dfrac{\partial u}{\partial y}$ (GBTU-2012)

13. Solve by method of separation of variables

$\dfrac{\partial z}{\partial x} + \dfrac{\partial^2 z}{\partial y^2} = 0; z(x,0) = 0,\ z(x,\pi) = 0, z(0,y) = 4\sin 3y$ (MTU-2012)

14. Solve by method of separation of variables

$\dfrac{\partial^2 u}{\partial x \partial t} = e^{-t}\cos x$

given that $u = 0$ when $t = 0$ and $\dfrac{\partial u}{\partial t} = 0$ when $x = 0$.

[UPTU (SUM)2008, GBTU-2012, GBTU (AG)-2012]

15. Solve by method of separation of variables

$u_{xx} = uy + 2u, u(0,y) = 0, \dfrac{\partial}{\partial x}u(0,y) = 1 + e^{-3y}$

(GBTU-2010, UPTU-2009)

Hints to Selected Problems

1. Let us assume that $V = XT$, where X is a function of x only and T that of t only.

$$\frac{\partial V}{\partial t} = X\frac{dT}{dt} \quad \text{and} \quad \frac{\partial^2 V}{\partial x^2} = T\frac{d^2 X}{dx^2}$$

Substitute in $\dfrac{\partial^2 V}{\partial x^2} = \dfrac{\partial V}{\partial t}$, we get $X\dfrac{dT}{dt} = T\dfrac{d^2 X}{dx^2}$

$\Rightarrow \quad \dfrac{1}{T}\dfrac{dT}{dt} = \dfrac{1}{X}\dfrac{d^2 X}{dx^2}$ ... (1)

Let each side of (1) be equal to constant (p^2)

$$\frac{1}{T}\frac{dT}{dt} = -p^2 \qquad\qquad \frac{1}{X}\frac{d^2 X}{dx^2} = -p^2$$

$\Rightarrow \dfrac{dT}{dt} + p^2 T = 0 \quad \Rightarrow \dfrac{d^2 X}{dx^2} + p^2 X = 0$

$\Rightarrow DT + p^2 T = 0 \quad \Rightarrow D^2 X + p^2 X = 0$

$\Rightarrow (D + p^2)T = 0 \quad \Rightarrow (D^2 + p^2)X = 0$

$\Rightarrow D = -p^2 \quad\quad \Rightarrow D^2 + p^2 = 0$

$\Rightarrow \quad T = C_1 e^{-p^2 t}$ | $D^2 = -p^2 \Rightarrow D = \pm ip$

                $X = C_2 \cos px + C_3 \sin px$

$V = C_1 e^{-p^2 t}(C_2 \cos px + C_3 \sin px)$ ... (2)

On putting $x = 0,\ V = 0$ in (2), we get

$0 = C_1 e^{-p^2 t} C_2 \Rightarrow C_2 = 0$, since $C_1 \neq 0$.

On putting the value of C_2 in (2), we get

$V = C_1 e^{-p^2 t} C_3 \sin px$ ... (3)

Since C_3 cannot be zero

$\sin pl = 0 = \sin n\pi \Rightarrow pl = n\pi \Rightarrow p = \dfrac{n\pi}{l},\ n \in \mathbb{Z}$

On putting the value of p in (3), we get

$V = C_1 C_3 e^{-\frac{n^2 \pi^2 t}{l^2}} \sin\dfrac{n\pi x}{l}$

$V = b_n e^{-\frac{n^2 \pi^2 t}{l^2}} \sin\dfrac{n\pi x}{l}$, where $b_n = C_1 C_2$.

The most general solution is

$V = \displaystyle\sum_{m=1}^{\infty} b_n e^{-\frac{n^2 \pi^2 t}{l^2}} \sin\dfrac{n\pi x}{l}$.

Answers

1. $v = b_n e^{-n^2\pi^2 t/l^2}\sin\dfrac{n\pi x}{l}$, where $b_n = C_1 C_2$

2. $u = 6e^{-(3x+2t)}$

3. $u(x,y) = (C_1 \cos px + C_2 \sin px)C_3\, e^{-(b^2+2)y}$

4. $u(x,t) = 3e^{-x+t} - e^{-5x+2t}$

5. $u = 4e^{-x+3/2y}$

6. $Z = Ce^{4ax^3}.e^{-3ay^4}$

7. $u = Ce^{k(1/y - 1/x)}$

8. $u = \dfrac{1}{\sqrt{2}}\sinh\sqrt{2}x + e^{-3y}\sin x$

9. $u = 4e^{x-y} - e^{2x-5y}$

10. $u = C_1 C_2 e^{p^2\left(\frac{y^2}{4} - \frac{x^3}{3}\right)}$

11. $u(x,y) = C_1 e^{-p^2 y}(C_2\cos px + C_3\sin px)$

12. $u(x,y) = C_1 C_2 e^{k(x+y)}$

13. $z(x,y) = 4e^{ax}\sin 3y$

14. $u(x,t) = \sin x(1 - e^{-t})$

15. $u(x,t) = \dfrac{1}{\sqrt{2}}\sinh\sqrt{2}x + e^{-3y}\sin x$

8.15.3 HYPERBOLIC DIFFERENTIAL EQUATIONS

We know that the most important homogeneous hyperbolic differential equation is the wave equation, which can be written as follows:

$$\frac{\partial^2 u}{\partial t^2} = C^2 \frac{\partial^2 u}{\partial x^2}, \text{ where } C \text{ is the wave speed.}$$

In this section, we shall discuss the derivation and solution of wave equation of one, two and three dimensional.

8.15.4 SOLUTION OF ONE DIMENSIONAL WAVE EQUATION

Consider one dimensional wave equation

$$\frac{\partial^2 u}{\partial x^2} = \frac{1}{C^2} \frac{\partial^2 u}{\partial t^2} \qquad \dots (1)$$

Suppose that solution of (1) is of the form

$$u(x, t) = X(x)\, T(t) \qquad \dots (2)$$

Putting the values from (2) in (1), we get

$$X''T = \frac{1}{C^2} XT''$$

$$\Rightarrow \qquad \frac{X''}{X} = \frac{T''}{C^2 T} = \mu \ (say) \qquad \dots (3)$$

where μ is the separation constant.

From (3), we can deduce that $\quad X'' - \mu X = 0 \qquad \dots (4)$

and $\qquad\qquad\qquad\qquad T'' - C^2 \mu T = 0 \qquad \dots (5)$

Solving (4) and (5), we get

  (i) **When** $\mu = 0$, $X = a_1 x + a_2$, $T = a_3 t + a_4$

  (ii) **When** $\mu = -\lambda^2$, $(\lambda \neq 0)$ then $X = b_1 e^{\lambda x} + b_2 e^{-\lambda x}$ and $T = b_3 e^{C\lambda t} + b_4 e^{-C\lambda t}$

  (iii) **When** $\mu = \lambda^2$, $\lambda \neq 0$ then $X = C_1 \cos \lambda x + C_2 \sin \lambda x$ and $T = C_3 \cos Cpt + C_4 \sin Cpt$

Therefore, the various possible solutions are given by

$$u(x, t) = (a_1 x + a_2)(a_3 t + a_4) \qquad \dots (6)$$

$$u(x, t) = (b_1 e^{\lambda x} + b_2 e^{-\lambda x})(b_3 e^{C\lambda t} + b_4 e^{-C\lambda t}) \qquad \dots (7)$$

and $\qquad u(x, t) = (C_1 \cos \lambda x + C_2 \sin \lambda x)(C_3 \cos Cpt + C_4 \sin Cpt) \qquad \dots (8)$

Finally, since we are dealing with problems on vibration, $u(x, t)$, must be a periodic function of t.

Thus, $u(x, t)$ must involve trigonometric terms. Hence, the solution given by (8) is the only suitable solution.

8.15.6 D'ALEMBERT'S SOLUTION OF WAVE EQUATION

Let $\qquad\qquad\qquad \dfrac{\partial^2 \phi}{\partial x^2} = \dfrac{1}{C^2} \dfrac{\partial^2 \phi}{\partial t^2} \qquad \dots (1)$

be the given wave equation.

Define two new independent variables u and v such that

$$u = x + Ct \text{ and } v = x - Ct \qquad \dots (2)$$

Now, $\qquad \dfrac{\partial \phi}{\partial x} = \dfrac{\partial \phi}{\partial u} \cdot \dfrac{\partial u}{\partial x} + \dfrac{\partial \phi}{\partial v} \cdot \dfrac{\partial v}{\partial x} = \dfrac{\partial \phi}{\partial u} + \dfrac{\partial \phi}{\partial v}$

$$\Rightarrow \qquad \frac{\partial}{\partial x} \equiv \frac{\partial}{\partial u} + \frac{\partial}{\partial v} \qquad \dots (3)$$

Then $\qquad \dfrac{\partial^2 \phi}{\partial x^2} = \dfrac{\partial}{\partial x}\left(\dfrac{\partial \phi}{\partial x}\right) = \left(\dfrac{\partial}{\partial u} + \dfrac{\partial}{\partial v}\right)\left(\dfrac{\partial \phi}{\partial u} + \dfrac{\partial \phi}{\partial v}\right) \qquad \dots (4)$

$$\Rightarrow \qquad \frac{\partial^2 \phi}{\partial x^2} = \frac{\partial^2 \phi}{\partial u^2} + 2 \frac{\partial^2 \phi}{\partial u . \partial v} + \frac{\partial^2 \phi}{\partial v^2}$$

Also, $\qquad \dfrac{\partial \phi}{\partial t} = \dfrac{\partial \phi}{\partial u} \cdot \dfrac{\partial u}{\partial t} + \dfrac{\partial \phi}{\partial v} \cdot \dfrac{\partial v}{\partial t} = C \dfrac{\partial \phi}{\partial u} - C \dfrac{\partial \phi}{\partial v} = C\left(\dfrac{\partial \phi}{\partial u} - \dfrac{\partial \phi}{\partial v}\right)$

$$\Rightarrow \qquad \frac{\partial}{\partial t} \equiv C\left(\frac{\partial}{\partial u} - \frac{\partial}{\partial v}\right) \qquad \dots (5)$$

Thus, $\qquad \dfrac{\partial^2 \phi}{\partial t^2} = \dfrac{\partial}{\partial t}\left(\dfrac{\partial \phi}{\partial t}\right) = C^2\left(\dfrac{\partial}{\partial u} - \dfrac{\partial}{\partial v}\right)\left(\dfrac{\partial \phi}{\partial u} - \dfrac{\partial \phi}{\partial v}\right)$

$$\Rightarrow \qquad \frac{1}{C^2} \frac{\partial^2 \phi}{\partial t^2} = \frac{\partial^2 \phi}{\partial u^2} - 2 \frac{\partial^2 \phi}{\partial u \partial v} + \frac{\partial^2 \phi}{\partial v^2} \qquad \dots (6)$$

Putting the values from (4) and (6) in (1), we get

$$\frac{\partial^2 \phi}{\partial u^2} + 2 \frac{\partial^2 \phi}{\partial u \partial v} + \frac{\partial^2 \phi}{\partial v^2} = \frac{\partial^2 \phi}{\partial u^2} - 2 \frac{\partial^2 \phi}{\partial u \partial v} + \frac{\partial^2 \phi}{\partial v^2}$$

$$\Rightarrow \qquad \frac{\partial^2 \phi}{\partial u . \partial v} = 0$$

On integrating, we get $\qquad \dfrac{\partial \phi}{\partial u} = F(u)$, where $F(u)$ is any arbitrary function of u.

Again integrating w.r.t. u, we get $\qquad \phi = \int F(u)\, du + g(v) = f(u) + g(v)$

$$\Rightarrow \qquad \phi = f(x + Ct) + g(x - Ct)$$

which is the D'Alembert's solution of wave equation.

8.15.6 SOLUTION OF TWO DIMENSIONAL WAVE EQUATION

To find the solution of $\dfrac{\partial^2 u}{\partial t^2} = C^2\left(\dfrac{\partial^2 u}{\partial x^2} + \dfrac{\partial^2 u}{\partial y^2}\right)$ subject to the boundary conditions $u(0,y,t) = u(a,y,t) = u(x,0,t) = u(x,b,t) = 0$ and initial condition $u(x, y, 0) = f(x, y)$ and $\left(\dfrac{\partial u}{\partial t}\right)_{t=0} = g(x, y)$.

The given equation is

$$\frac{\partial^2 u}{\partial x^2} + \frac{\partial^2 u}{\partial y^2} = \frac{1}{C^2}\frac{\partial^2 u}{\partial t^2} \qquad \dots (1)$$

with boundary conditions.

    (a) $u(0, y, t) = 0$, (b) $u(a, y, t) = 0$, (c) $u(x, 0, t) = 0$, and (d) $u(x, b, t) = 0$.

Also, $u(x, y, 0) = f(x, y)$ ... (2)

and $\left(\dfrac{\partial u}{\partial t}\right)_{t=0} = g(x, y)$ ... (3)

Now, suppose that solution of (1) is of the form

$$u(x, y, t) = X(x)\,Y(y)\,T(t) \qquad \dots (4)$$

Putting the values from (4) in (1), we get

$$X''YT + XY''T = \frac{1}{C^2}XYT''$$

$$\Rightarrow \qquad \frac{X''}{X} + \frac{Y''}{Y} = \frac{T''}{C^2 T} \qquad \dots (5)$$

Now, since x, y and t are independent variables, (5) can only be true if each term on each side is equal to a constant, say .

Let $\qquad \dfrac{X''}{X} = \mu_1 \;\Rightarrow\; X'' - \mu_1 X = 0 \qquad \dots (6)$

Now, using (a) and (b), equation (4) gives

$$X(0)\,Y(y)\,T(t) = 0 \quad \text{and} \quad X(a)\,Y(y)\,T(t) = 0 \qquad \dots (7)$$

By assuming $Y(y) \neq 0$, $T(t) \neq 0$, we get

$$X(0) = 0 \quad \text{and} \quad X(a) = 0 \qquad \dots (8)$$

Therefore, we want to solve equation (6) subject to the boundary conditions (8). Now, there are following cases :

Case (I). Let $\mu_1 = 0$. In this case, solution of (6) is given by

$$X(x) = Ax + B \qquad \dots (9)$$

Using (8) in (9), we get $\qquad B = 0$ and $Aa + B = 0$

$\Rightarrow \qquad\qquad\qquad\qquad A = B = 0 \;\Rightarrow\; X(x) = 0$

$\Rightarrow \qquad\qquad\qquad\qquad u = 0$

which does not satisfy (2). Thus, we reject this case.

Case (II). Let $\mu_1 = \lambda_1^2, \lambda_1 \neq 0$. In this case, solution of (6) is given by

$$X(x) = Ae^{x\lambda_1} + Be^{-x\lambda_1} \qquad \dots (10)$$

Using (8) in (10), we get

$$0 = A + B \quad \text{and} \quad 0 = Ae^{a\lambda_1} + Be^{-a\lambda_1} \qquad \dots (11)$$

On solving, we get $\qquad A = B = 0 \Rightarrow u = 0$

$\Rightarrow$ This case is also rejected.

Case III. Let $\mu_1 = -\lambda_1^2, \lambda_1 \neq 0$. In this case, solution of (6) is given by

$$X(x) = A\cos\lambda_1 x + B\sin\lambda_1 x \qquad \dots (12)$$

Using (8) in (12), we get $\qquad 0 = A$ and $0 = A\cos\lambda_1 a + B\sin\lambda_1 a$

$\Rightarrow \qquad\qquad\qquad\qquad A = 0$ and $\sin\lambda_1 a = 0$.

Now, $\sin\lambda_1 a = 0 \;\Rightarrow\; \lambda_1 a = m\pi$

$\Rightarrow \qquad\qquad\qquad\qquad \lambda_1 = \dfrac{m\pi}{a},\quad m = 1, 2, \dots \qquad \dots (13)$

Thus, the non-zero solution $X_m(x)$ are given by

$$X_m(x) = B_n \sin\frac{m\pi x}{a},\quad m = 1, 2, 3, \dots \qquad \dots (14)$$

Further, let $\qquad \dfrac{Y''}{Y} = \mu_2 \;\Rightarrow\; Y'' - \mu_2 Y = 0 \qquad \dots (15)$

Using (c) and (d), (4) gives

$$Y(0) = 0 \quad \text{and} \quad Y(b) = 0 \qquad \dots (16)$$

On solving (15) under boundary condition (16), we get

$$Y_n(y) = D_n \sin\left(\frac{n\pi y}{b}\right),\quad n = 1, 2, 3, \dots \qquad \dots (17)$$

where, $$\mu_2 = -\lambda_2^2 \quad and \quad \lambda_2 = \frac{n\pi}{b}, \ n = 1, 2, 3, \ldots \qquad \ldots (18)$$

Therefore, (5) reduces to

$$\frac{T''}{C^2 T} = \mu_1 + \mu_2 = -\lambda_1^2 - \lambda_2^2 = -\pi^2 \left(\frac{m^2}{a^2} + \frac{n^2}{b^2} \right)$$

$\Rightarrow$ $$T'' + \lambda_{mn}^2 T = 0 \qquad \ldots (19)$$

where, $$\lambda_{mn}^2 = C^2 \pi^2 \left(\frac{m^2}{a^2} + \frac{n^2}{b^2} \right) \qquad \ldots (20)$$

The solution of (19) is given by $$T_{mn}(t) = E_{mn} \cos \lambda_{mn} t + F_{mn} \sin \lambda_{mn} t \qquad \ldots (21)$$

Hence, $$u_{mn}(x, y, t) = X_m(x) Y_n(y) T_{mn}(t) = (A_{mn} \cos \lambda_{mn} t + B_{mn} \sin \lambda_{mn} t) \left(\sin \frac{m\pi x}{a} \right) \sin \frac{n\pi y}{b} \qquad \ldots (22)$$

are solution of (1) satisfying (a) to (d).

Hence, the more general solution is given by $u(x, y, t) = \sum\limits_{m=1}^{\infty} \sum\limits_{n=1}^{\infty} (A_{mn} \cos \lambda_{mn} t + B_{mn} \sin \lambda_{mn} t) \left(\sin \frac{m\pi x}{a} \right) \sin \frac{n\pi y}{b}$

where $$A_{mn} = B_m D_n E_{mn} \quad and \quad B_{mn} = B_m D_n F_{mn}.$$

Solved Examples

EXAMPLE 1. *Find the general solution of one dimensional wave equation* $\dfrac{\partial^2 u}{\partial x^2} = \dfrac{1}{C^2} \dfrac{\partial^2 u}{\partial t^2}$ *and find the particular solution for which* $u = f(x), \dfrac{\partial u}{\partial t} = g(x)$ *at* $t = 0$.

(UPTU (CO)-2008)

SOLUTION. The given equation is $\dfrac{\partial^2 u}{\partial x^2} = \dfrac{1}{C^2} \dfrac{\partial^2 u}{\partial t^2} \qquad \ldots (1)$

Let string be stretched between fixed points $(0, 0)$ and $(a, 0)$. So we are to find $u(x, t)$ subject to the boundary condition

$$u(0, t) = 0, \ u(a, t) = 0 ; \ \forall t \qquad \ldots (2)$$

and initial conditions

$$\left. \begin{array}{l} u(x, 0) = f(x) \\ \left(\dfrac{\partial u}{\partial t} \right)_{t=0} = u_t(x, 0) = g(x) \end{array} \right] \qquad \ldots (3)$$

Suppose that solution of (1) is of the form

$$u(x, t) = X(x) T(t) \qquad \ldots (4)$$

Putting the values from (4) in (3), we get

$$X''T = \frac{1}{C^2} XT''$$

$\Rightarrow$ $$\frac{X''}{X} = \frac{1}{C^2} \frac{T''}{T} = \mu \ (say), \qquad \ldots (5)$$

where μ is a separation constant.

Using (5), we can deduce that

$$X'' - \mu X = 0 \qquad \ldots (6)$$

and $$T'' - \mu C^2 T = 0 \qquad \ldots (7)$$

Using (2) and (4), we get

$$X(0) T(t) = 0 \text{ and } X(a) T(t) = 0 \qquad \ldots (8)$$

$\Rightarrow$ $$X(0) = 0, X(a) = 0 \ [T(t) \neq 0] \qquad \ldots (9)$$

Now, we have to solve (6) under (9). There are following cases :

Case (I). Let $\mu = 0$. In this case, solution of (6) is given by $X(x) = Ax + B \qquad \ldots (10)$

Using (9) in (10), we get

$$B = 0 \text{ and } Aa + B = 0$$

$\Rightarrow A = B = 0 \Rightarrow X(x) = 0$

$\Rightarrow \qquad u = 0$

which does not satisfy (3). Thus, we reject this case.

Case (II). Let $\mu = \lambda^2, \lambda \neq 0$. In this case, solution of (6) is given by

$$X(x) = Ae^{\lambda x} + Be^{-\lambda x} \qquad \ldots (11)$$

Using (9) in (11), we get

$$0 = A + B \text{ and } 0 = Ae^{\lambda a} + Be^{-\lambda a} \qquad \ldots (12)$$

On solving, we get $A = B = 0$

$\Rightarrow X(x) = 0 \qquad \Rightarrow \quad u = 0$

Hence, we reject this case also.

Case (III). Let $\mu = -\lambda^2, \lambda \neq 0$. In this case, solution of (6) is given by

$$X(x) = A \cos \lambda x + B \sin \lambda x \qquad \ldots (13)$$

Using (9) in (13), we get

$$0 = A \text{ and } 0 = A \cos \lambda a + B \sin \lambda a$$

On solving, we get $A = 0$ and $\sin \lambda a = 0 \qquad (B \neq 0)$

Now, $\sin \lambda a = 0 \Rightarrow \lambda a = n\pi$

$\Rightarrow \qquad \lambda = \dfrac{n\pi}{a}, \ n = 1, 2, \ldots \qquad \ldots (14)$

Thus, the non-zero solution of (6) is given by

$$X_n(x) = B_n \sin \left(\frac{n\pi x}{a} \right) \qquad \ldots (15)$$

Now, using (14) in (7), we get

$$T'' + \left(\frac{n^2 \pi^2 C^2}{a^2} \right) T = 0$$

whose general solution is given by

$$T_n(t) = C_n \cos \frac{n\pi Ct}{a} + D_n \sin \frac{n\pi Ct}{a} \qquad \ldots (16)$$

So, $u_n(x, t) = X_n(x) T_n(x)$

$$= \left(E_n \cos \frac{n\pi Ct}{a} + F_n \sin \frac{n\pi Ct}{a} \right) \sin \frac{n\pi x}{a}$$

are solutions of (1) satisfying (2).

Thus, more general solution is given by

$$u(x, t) = \sum_{n=1}^{\infty} u_n(x, t)$$

$$= \sum_{n=1}^{\infty} \left(E_n \cos \frac{n\pi Ct}{a} + F_n \sin \frac{n\pi Ct}{a} \right) \sin \frac{n\pi x}{a} \ \ldots (17)$$

$\Rightarrow \dfrac{\partial u}{\partial t} = \sum\limits_{n=1}^{\infty} \left(-\dfrac{n\pi C E_n}{a} \sin \dfrac{n\pi Ct}{a} \right.$

$$\left. + \frac{n\pi C}{a} F_n \cos \frac{n\pi Ct}{a} \right) \sin \frac{n\pi x}{a} \ \ldots (18)$$

Putting $t = 0$ in (17) and (18) and using (3), we get

$$f(x) = \sum_{n=1}^{\infty} E_n \sin \frac{n\pi x}{a} \qquad \ldots (19)$$

and $\quad g(x) = \sum_{n=1}^{\infty} \frac{n\pi CF_n}{a} \sin\frac{n\pi x}{a}$ ... (20)

which are Fourier sine series expansions for $f(x)$ and $g(x)$ respectively. Thus, we get

$$E_n = \frac{2}{a}\int_0^a f(x)\sin\frac{n\pi x}{a}dx \qquad ...(21)$$

and $\dfrac{n\pi CF_n}{a} = \dfrac{2}{a}\int_0^a g(x)\sin\dfrac{n\pi x}{a}dx$

$\Rightarrow \qquad F_n = \dfrac{2}{n\pi C}\int_0^a g(x)\sin\dfrac{n\pi x}{a}dx \qquad ...(22)$

EXAMPLE 2. *Solve the wave equation* $\dfrac{\partial^2 u}{\partial t^2} = C^2 \dfrac{\partial^2 u}{\partial x^2}$, *where*

$u = p_0 \cos pt$ (p_0 *is a constant*) *when* $x = l$

and $u = 0$ *when* $x = 0$. [AMIE 1997]

SOLUTION. We know that solution of the given equation is

$u(x,t) = (C_1 \cos nx + C_2 \sin nx)$
$\qquad\qquad .(C_3 \cos nCt + C_4 \sin nCt) \quad ...(1)$

Putting $u = 0$ when $x = 0$, we get

$C_1(C_3 \cos nCt + C_4 \sin nCt) = 0$
$\Rightarrow \qquad\qquad C_1 = 0$

Therefore, we have

$u(x,t) = C_2 \sin nx (C_3 \cos nCt + C_4 \sin nCt)$
$\qquad = \sin nx \cos nCt.b_1 + \sin nx \sin nCt.b_2$

where $b_1 = C_2 C_3 \quad and \quad b_2 = C_2 C_4$.

Now, using $u = p_0 \cos pt$, when $x = l$, we get

$p_0 \cos pt = \sin nl \cos nCt.b_1 + \sin nl \sin nCt.b_2$

$\Rightarrow \quad p_0 \cos pt = \sin nl \cos nCt.b_1$ and
$\qquad\qquad 0 = \sin nl \sin nCt.b_2$

$\Rightarrow \qquad\qquad p = nC$. Therefore,
$\qquad p_0 \cos pt = \sin nl \cos pt.b_1$

$\Rightarrow b_1 = \dfrac{p_0}{\sin nl} = \dfrac{p_0}{\sin(pl/C)}$

Hence, $u(x,t) = \sin nx \cos nCt.b_1$

$\qquad\qquad = \dfrac{p_0}{\sin(pl/C)}\cos pt \sin(px/C)$

EXAMPLE 3. *A string is stretched to two fixed points distance l apart. Motion is started by displacing the string in the form* $u = a \sin\left(\dfrac{\pi x}{l}\right)$ *from which it is released at time t = 0. Show that the displacement at any point at a distance x from one end at time t is given by*

$$u(x, t) = a \sin\left(\dfrac{\pi x}{l}\right)\cos\left(\dfrac{\pi Ct}{l}\right)$$

(AMIETE-2003, UPTU-2004, 09, UPTU (SUM)-2009, UKTU-2011, 12, VTU-2010, SVTU-2008, KERALA-2005)

SOLUTION. It is known that the vibrations of the string are governed by the equation $\dfrac{\partial^2 u}{\partial t^2} = C^2 \dfrac{\partial^2 u}{\partial x^2}$

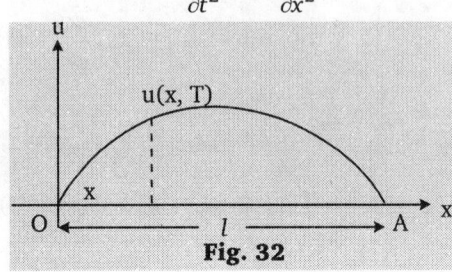

Fig. 32

As per given, the string is fastened at O and A respectively. Therefore,

$\qquad u(0, t) = 0$

and $\quad u(l, t) = 0 ; \ \forall t \qquad\qquad ...(2)$

and $u(x, 0) = a\sin\dfrac{\pi x}{l}$ and $\left(\dfrac{\partial u}{\partial t}\right)_{t=0} = 0 \qquad ...(3)$

The solution of (1) is given by

$\qquad u(x,t) = (C_1 \cos nx + C_2 \sin nx)$
$\qquad\qquad\qquad (C_3 \cos Cnt + C_4 \sin Cnt)$

Now, $u(0, t) = 0$ implies

$\qquad\qquad 0 = C_1(C_3 \cos Cnt + C_4 \sin Cnt)$

$\Rightarrow \qquad\qquad C_1 = 0$

$\Rightarrow \quad u(x, t) = \sin nx\,(b_1 \cos Cnt + b_2 \sin Cnt)$

Thus, $\quad \dfrac{\partial u}{\partial t} = \sin nx \left[-Cnb_1 \sin Cnt + Cn.b_2 \cos Cnt\right]$

$\Rightarrow \left[\dfrac{\partial u}{\partial t}\right]_{t=0} = \sin nx\,[Cn.b_2] = 0 ; \ \forall t$

$\Rightarrow \qquad\qquad b_2 = 0$

Thus, (4) reduces to $u(x, t) = \sin nx \cos Cnt.b_1 \quad ...(5)$

Further, $\quad u(l, t) = 0$ implies

$\sin nl \cos Cnt.b_1 = 0 ; \ \forall t$

$\Rightarrow$ Either $\sin nl = 0 \ $ or $\ b_1 = 0$.

Since $b_1 = 0$ makes the solution trivial, therefore, we take $\sin nl = 0$.

Now, $\quad \sin nl = 0 \Rightarrow nl = m\pi \Rightarrow n = \dfrac{m\pi}{l}, m = 1, 2,...$

Then, from (5), we get

$$u(x, t) = \sin\dfrac{m\pi}{l}x \cos\left(\dfrac{Cm\pi}{l}.t\right).b_1$$

Since $u(x, 0) = a\sin\dfrac{\pi x}{l}$,

therefore $a\sin\dfrac{\pi x}{l} = \sin\dfrac{m\pi x}{l}.b_1$

which is true only when $m = 1$ and $b_1 = a$.

Hence, $\quad u(x, t) = a\sin\left(\dfrac{\pi x}{l}\right)\cos\left(\dfrac{\pi Ct}{l}\right)$

EXAMPLE 4. *A string is stretched and fastened to two points distance l apart. Motion is started displacing the string into the form* $y = k(lx - x^2)$ *from which it is released at time t = 0. Find the displacement of any point on the string at a distance of x from one end at time t.*

[UPTU Q. BANK 2002; IAS 1990,93,97]

Or

If the string is released from rest in the position $y = \dfrac{4t}{l^2}x(l - x)$, *find the displacement y.*

[AMIE-1991,97,2003]

SOLUTION. It is known that the vibrations of the string are governed by the equation

$$\dfrac{\partial^2 y}{\partial t^2} = C^2 \dfrac{\partial^2 y}{\partial x^2} \qquad\qquad ....(1)$$

where $y(0, t) = 0 \ $ and $\ y(l, t) = 0 \qquad\qquad ...(2)$

Also, the initial transverse velocity of any point of the string is zero. Thus, the initial condition is given by

$\left(\dfrac{\partial y}{\partial t}\right)_{t=0} = 0$

$y(x, 0) = k(lx - x^2)$ $\Bigg] \qquad\qquad ...(3)$

We know that solution of (1) is given by
$$y(x,t) = (C_1 \cos px + C_2 \sin px)$$
$$(C_3 \cos Cpt + C_4 \sin Cpt) \qquad ... (4)$$
Now $y(0,t) = 0$
$$\Rightarrow \qquad 0 = C_1(C_3 \cos Cpt + C_4 \sin Cpt) \Rightarrow C_1 = 0$$
$$\therefore \qquad y(x,t) = \sin px \,(C_1' \cos Cpt + C_2' \sin Cpt) \quad ... (5)$$
Therefore,
$$\left(\frac{\partial y}{\partial t}\right)_{t=0} = 0 \Rightarrow 0 = C_2' \sin px \,.(Cp)$$
$$\Rightarrow C_2' = 0 \qquad\qquad [\because C_p \neq 0]$$

Then, (5) gives $y(x,t) = C_1' \sin px \,(\cos Cpt)$

Now, $y(l,t) = 0 \Rightarrow C_1' \sin pl \cos Cpt = 0$
$$\Rightarrow \qquad C_1' = 0 \text{ or } \sin pl = 0$$
Since $C_1' = 0$ makes the solution trivial, thus, we have
$$\sin pl = \sin n\pi, \; n = 1,2,... \Rightarrow \qquad —$$
$$\therefore \qquad y_n(x,t) = C_1' \sin\frac{n\pi x}{l} \cos\frac{n\pi C}{l}.t$$
$$= b_n \sin\frac{n\pi x}{l} \cos\frac{n\pi Ct}{l}$$

Therefore, the complete solution will be of the form
$$f(x,t) = \sum_{n=1}^{\infty} b_n \sin\frac{n\pi x}{l} \cos\frac{n\pi Ct}{l} \qquad ... (6)$$
Further, $f(x,0) = \sum_{n=1}^{\infty} b_n \sin\frac{n\pi x}{l}$
$$\Rightarrow \qquad lx - x^2 = \sum_{n=1}^{\infty} b_n \sin\frac{n\pi x}{l} \qquad ... (7)$$
which is the expansion of $f(x)$ in the form of Fourier sine series. Thus we have
$$b_n = \frac{2}{l}\int_0^l f(x) \sin\frac{n\pi x}{l}dx$$
$$= \frac{2}{l}\int_0^l (lx - x^2) \sin\frac{n\pi x}{l}dx$$
$$= \frac{2}{l}\left[(lx - x^2)\left(-\cos\frac{n\pi x}{l}\right)\frac{l}{n\pi} - (l - 2x)\right.$$
$$\left.\left(-\sin\frac{n\pi x}{l}\right)\frac{l^2}{n^2\pi^2} + (-2)\left(\cos\frac{n\pi x}{l}\right)\frac{l^3}{n^3\pi^3}\right]_0^l$$
$$= \frac{2}{l}\left[(-1)^{n+1}\frac{2l^3}{n^3\pi^3} + \frac{2l^3}{n^3\pi^3}\right] = \frac{8l^2}{n^3\pi^3},$$
$$\text{if } n \text{ is odd.}$$
$$= 0, \text{ if } n \text{ is even.}$$
Putting the value of b_n in (7), we get
$$y(x,t) = \sum \frac{8l^2}{n^3\pi^3}\sin\frac{n\pi x}{l}\cos\frac{n\pi Ct}{l}, \text{ when } n \text{ is odd.}$$

EXAMPLE 5. *A tightly stretched string with fixed end points $x = 0$ and $x = 1$ is initially in a position given by $u = u_0 \sin^3\left(\frac{\pi x}{l}\right)$. If it is released from rest from this position, find the displacement $u(x, t)$.*

[AMIE-1997, GBTU (CO)-2011, RAJASTHAN-2006, VTU-2003; JNTU-2002]

SOLUTION. It is known that the equation of vibrating string is given by
$$\frac{\partial^2 u}{\partial t^2} = C^2\frac{\partial^2 u}{\partial x^2} \qquad ... (1)$$
with boundary conditions $u(0,t) = 0$, $u(l,t) = 0$

$$\left(\frac{\partial u}{\partial t}\right)_{t=0} = 0; \; u(x,0) = u_0\sin^3\left(\frac{\pi x}{l}\right)$$
We know that solution of (1) is given by
$$u(x,t) = (C_1 \cos nx + C_2 \sin nx)$$
$$(C_3 \cos nCt + C_4 \sin nCt) \qquad ... (2)$$
Now, $u(0,t) = 0$
$$\Rightarrow \qquad 0 = C_1(C_3\cos nCt + C_4\sin nCt)$$
$$\Rightarrow \qquad C_1 = 0$$
Then, (2) reduces to
$$u(x,t) = C_2\sin nx \,(C_3\cos nCt + C_4\sin nCt) \quad ... (3)$$
$$\Rightarrow \quad u(x,t) = \sin nx \,(b_1\cos nCt + b_2\sin nCt)$$
where, $b_1 = C_2 C_3$, $b_2 = C_2 C_4$
Further,
$$u(l,t) = 0 \Rightarrow \sin nl \,(b_1\cos nCt + b_2\sin nCt) = 0$$
$$\Rightarrow \quad \sin nl = 0 \Rightarrow \sin nl = \sin n\pi \Rightarrow n = \frac{m\pi}{l}$$
Therefore, $u(x,t) = \sin\frac{m\pi x}{l}$
$$\left(b_1\cos\frac{m\pi Ct}{l} + b_2\sin\frac{m\pi Ct}{l}\right) \quad ... (4)$$
$$\Rightarrow \quad \frac{\partial u}{\partial t} = \sin\left(\frac{m\pi x}{l}\right)$$
$$\left(-\frac{m\pi C}{l}b_1\sin\frac{m\pi C}{l}.t + b_2\frac{m\pi C}{l}\cos\frac{m\pi C}{l}.t\right)$$
Further, $\frac{\partial u}{\partial t} = 0$
$$\Rightarrow \quad 0 = \sin\frac{m\pi x}{l}.b_2\frac{m\pi C}{l} \Rightarrow b_2 = 0$$

Then, (4) gives $u(x,t) = \sin\frac{m\pi x}{l}\cos\frac{m\pi Ct}{l}.b_1$
$$\Rightarrow u_m(x,t) = b_m\sin\frac{m\pi x}{l}\cos\frac{m\pi Ct}{l}$$
Now, $u(x,t) = \sum_{m=1}^{\infty} u_m(x,t)$
$$= \sum_{m=1}^{\infty} b_m\sin\frac{m\pi x}{l}\cos\frac{m\pi Ct}{l} \qquad ... (5)$$
Also, $u(x,0) = u_0\sin^3\left(\frac{\pi x}{l}\right) = \frac{u_0}{4}\left(3\sin\frac{\pi x}{l} - \sin\frac{3\pi x}{l}\right)$
Putting $t = 0$ in (5), we get
$$\sum_{m=1}^{\infty} b_m\sin\frac{m\pi x}{l} = \frac{u_0}{4}\left(3\sin\frac{\pi x}{l} - \sin\frac{3\pi x}{l}\right)$$
where $b_1 = \frac{3u_0}{4}$ and $b_3 = \frac{-u_0}{4}$ remaining b_i's are all zero.
Hence, $u(x,t) = \frac{u_0}{4}$
$$\left(3\sin\frac{\pi x}{l}.\cos\frac{C\pi t}{l} - \sin\frac{3\pi x}{l}\cos\frac{3\pi Ct}{l}\right).$$

EXAMPLE 6. *Find the solution of the wave equation $\frac{\partial^2 y}{\partial t^2} = C^2\frac{\partial^2 y}{\partial x^2}$ such that $y = p_0 \cos pt$, when $x = l$ and $y = 0$ when $x = 0$ (b_0 is a constant).*

(IAS 1992)

SOLUTION. The given equation is
$$\frac{\partial^2 y}{\partial t^2} = C^2\frac{\partial^2 y}{\partial x^2} \qquad ... (1)$$
The solution of (1) is given by
$$y(x,t) = (C_1\cos C\sqrt{k}\,t + C_2\sin C\sqrt{k}\,t) -$$
$$(C_3\cos\sqrt{k}\,x + C_4\sin\sqrt{k}\,x) \qquad ... (2)$$

Now, $y(0, t) = 0$ implies

$0 = (C_1 \cos C\sqrt{k}\, t + C_2 \sin C\sqrt{k}\, t) C_3 \Rightarrow C_3 = 0$

Therefore, (2) becomes

$y(x, t) = (C_1 \cos C\sqrt{k}\, t + C_2 \sin C\sqrt{k}\, t)(C_4 \sin \sqrt{k}\, x)$

$\qquad = C_1' \cos C\sqrt{k}\, t \sin \sqrt{k}\, x + C_2' \sin C\sqrt{k}\, t \sin \sqrt{k}\, x$

$\hspace{10cm} \ldots (3)$

where, $C_1' = C_1 C_4,\ C_2' = C_2 C_4$

Further, $y(x, t) = p_0 \cos pt$, when $x = l$.

Also, $p_0 \cos pt = C_1' \cos C\sqrt{k}\, t$

$\hspace{3cm} \sin \sqrt{k}\, l + C_2' \sin C\sqrt{k}\, t \sin \sqrt{k}\, l$

Equating the coefficients of sin and cos on both the sides, we get

$p_0 = C_1' \sin \sqrt{k}\, l \Rightarrow C_1' = \dfrac{p_0}{\sin \sqrt{k}\, l}$

$0 = C_2' \sin \sqrt{k}\, l \Rightarrow C_2' = 0$

Hence, $y(x, t) = \dfrac{p_0}{\sin \sqrt{k}\, l} \cos C\sqrt{k}\, t . \sin \sqrt{k}\, x$

$\qquad = \dfrac{p_0}{\sin \sqrt{k}\, t} . \cos pt \sin \dfrac{p}{C} x$

$\qquad = \dfrac{p_0}{\sin(pl/C)} \cos pt \sin\left(\dfrac{p}{C} x\right)$

EXAMPLE 7. *A string is stretched between the fixed points (0,0) and (1, 0) and released at rest from the initial deflection given by*

$$f(x) = \dfrac{2k}{l} . x, \quad when \ \ 0 < x < l/2\ ;$$

$$f(x) = \dfrac{2k}{l} . (l - x), \quad when \ \ l/2 < x < l$$

Find the deflection of the string at any time t.

[UPTU Q. BANK-2002]

SOLUTION. We know that the deflection $u(x, t)$ of the string is given by

$$u(x, t) = \sum_{n=1}^{\infty} C_n \cos \dfrac{Cn\pi}{l} t \sin \dfrac{n\pi x}{l}$$

where, $C_n = \dfrac{2}{l} \int_0^l f(x) \sin \dfrac{n\pi x}{l} dx$

$= \dfrac{2}{l}\left[\int_0^{l/2} f(x) \sin \dfrac{n\pi x}{l} dx + \int_{l/2}^l f(x) \sin \dfrac{n\pi x}{l} dx \right]$

$= \dfrac{2}{l}\left[\int_0^{l/2} \dfrac{2k}{l} \sin \dfrac{n\pi x}{l} dx + \int_{l/2}^l \dfrac{2k}{l}(l-x) \sin \dfrac{n\pi x}{l} dx \right]$

$= \dfrac{4k}{l^2}\left[\int_0^{l/2} x \sin \dfrac{n\pi x}{l} dx + \dfrac{4k}{l^2}\int_{l/2}^l (l-x) \sin \dfrac{n\pi x}{l} dx \right]$

$= \dfrac{4k}{l^2}\left[\left(-x \dfrac{l}{n\pi} \cos \dfrac{n\pi x}{l}\right)_0^{l/2} + \dfrac{l}{\pi x}\int_0^{l/2} 1 . \cos \dfrac{n\pi x}{l} dx \right.$

$\left. + \dfrac{4k}{l^2}\left[-(l-x)\dfrac{l}{n\pi}\cos\dfrac{n\pi x}{l} \right]_{l/2}^l - \dfrac{l}{n\pi}\int_{l/2}^l \cos \dfrac{n\pi x}{l} dx \right]$

$= \dfrac{4k}{l^2}\left[-\dfrac{l^2}{2n\pi}\cos \dfrac{n\pi}{2} + \dfrac{l}{n\pi}\left(\dfrac{l}{n\pi}\sin\dfrac{n\pi x}{l}\right)_0^l \right.$

$\left. + \dfrac{4k}{l}\left[-0 + \dfrac{l}{2}\dfrac{l}{n\pi}\cos\dfrac{n\pi}{2} - \dfrac{l}{n\pi}\left(\dfrac{l}{n\pi}\sin\dfrac{n\pi x}{l}\right)_{l/2}^l \right] \right.$

$= \dfrac{4k}{l^2}\left[-\dfrac{l^2}{2n\pi}\cos\dfrac{n\pi}{2} + \dfrac{l^2}{n^2 x^2}\sin\dfrac{n\pi}{2} \right] + \dfrac{4k}{l^2}$

$\left[\dfrac{l^2}{2n\pi}\cos\dfrac{n\pi}{2} - \dfrac{l^2}{n^2\pi^2}\left(\sin n\pi - \sin\dfrac{n\pi}{2}\right) \right]$

$= \dfrac{4k}{l^2}\left[-\dfrac{l^2}{2n\pi}\cos\dfrac{n\pi}{2} + \dfrac{l^2}{n^2 x^2}\sin\dfrac{n\pi}{2} \right.$

$\left. + \dfrac{l^2}{2n\pi}\cos\dfrac{n\pi}{2} - \dfrac{l^2}{n^2\pi^2}\sin n\pi + \dfrac{l^2}{n^2\pi^2}\sin\dfrac{n\pi}{2} \right]$

$= \dfrac{8k}{n^2\pi^2}\sin\dfrac{n\pi}{2}$

Thus, $u(x, t) = \sum_{n=1}^{\infty} \dfrac{8k}{n^2\pi^2}\sin\dfrac{n\pi}{2}\cos\dfrac{Cn\pi}{l}.t.\sin\dfrac{n\pi x}{l}$

$= \dfrac{8k}{\pi^2}\sin\dfrac{\pi}{2}\cos\dfrac{C\pi}{l}.t.\sin\dfrac{\pi x}{l}$

$\qquad + \dfrac{8k}{2^2\pi^2}\sin\pi\cos\dfrac{2C\pi}{l}.t.\sin\dfrac{2\pi x}{l}$

$\qquad + \dfrac{8k}{3^2\pi^2}\sin\dfrac{3\pi}{2}\cos\dfrac{3C\pi}{l}.t.\sin\dfrac{3\pi x}{l} + \ldots$

$= \dfrac{8k}{\pi^2}\left[\sin\dfrac{\pi x}{l}\cos\dfrac{\pi C}{l} - \dfrac{1}{3^2}\sin\dfrac{3\pi x}{l}\cos\dfrac{3\pi C}{l}.t + \ldots \right]$

EXAMPLE 8. *The point of trisection of a string are pulled aside through a distance h on opposite sides of the position of equilibrium, and the string is released from rest. Find an expression for the displacement of the string at any subsequent time and show that the mid point of the string always remains at rest.*

[KERALA-2005, UPTU Q. BANK-2002]

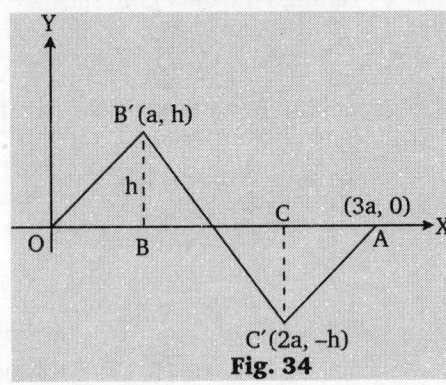

Fig. 34

SOLUTION. Suppose $OBCA$ be the position of the equilibrium of the string of length $l = 3a$. Further, let B and C be the points of trisection of the string be pulled through a distance h on opposite sides and then released. It is known that this is the case in which vibrations are governed by one-dimensional wave equation

$$\dfrac{\partial^2 u}{\partial t^2} = C^2 \dfrac{\partial^2 u}{\partial x^2} \hspace{3cm} \ldots (1)$$

Now, equation of the line joining $O(0, 0)$ to $B'(a, h)$ is given by

$y - 0 = \dfrac{h-0}{a-0} \Rightarrow y = \dfrac{h}{a}x$

Also, the equation of joining

$B'(a, h)$ to $C'(2a, -h)$ is given by

$y - h = \dfrac{-h-h}{2a-a}(x-a) \Rightarrow y = \dfrac{h(3a-2x)}{a}$

Further, equation of the line joining $C'(2a-h)$ to $A(3a-0)$ is

$y - (-h) = \dfrac{0-(-h)}{3a-2a}(x-2a)$

$\Rightarrow \qquad y = \dfrac{h(x-3a)}{a}$

Then, its initial deflection is given by

$$f(x) = \begin{cases} hx/a & ; 0 \le x \le a \\ \dfrac{h(3a-2x)}{a} & ; a \le x \le 2a \\ \dfrac{h(x-3a)}{a} & ; 2a \le x \le 3a \end{cases}$$

and initial velocity is $g(x) = 0$.

So, deflection $u(x, t)$ is given by

$$u(x,t) = \sum_{n=1}^{\infty} C_n \cos \frac{Cn\pi}{3a}.t \sin \frac{n\pi x}{3a}$$

where, $C_n = \dfrac{2}{3a} \int_0^{3a} f(x) \sin \dfrac{n\pi x}{3a} dx$

$$= \frac{2}{3a}\left[\int_0^a \frac{hx}{a} \sin \frac{n\pi x}{3a} dx \right.$$

$$+ \int_a^{2a} \frac{h(3a-2x)}{a} \sin \frac{n\pi x}{3a} dx$$

$$\left. + \int_{2a}^{3a} \frac{h(x-3a)}{a} \sin \frac{n\pi x}{3a}.dx \right]$$

or $\quad C_n = \dfrac{2h}{3a^2}\left[\int_0^a x \sin \dfrac{n\pi x}{3a} dx \right.$

$$+ \int_a^{2a} (3a-2x) \sin \frac{n\pi x}{3a} dx$$

$$\left. + \int_{2a}^{3a} (x-3a) \sin \frac{n\pi x}{3a} dx \right]$$

$$= \frac{2h}{3a^2}\left[\left(-x.\frac{3a}{n\pi} \cos \frac{n\pi x}{3a} \right)_0^a + \frac{3a}{n\pi} \right.$$

$$\int_0^a \cos \frac{n\pi x}{3a} dx + \left[-(3a-2x)\frac{3a}{n\pi} \cos \frac{n\pi x}{3a} \right]_a^{2a}$$

$$-\frac{6a}{n\pi} \int_a^{2a} \cos \frac{n\pi x}{3a} dx + \left[-(x-3a)\frac{3a}{n\pi} \cos \frac{n\pi x}{3a} \right]_{2a}^{3a}$$

$$+ \frac{3a}{n\pi} \int_a^{2a} \cos \frac{n\pi x}{3a} dx$$

$$= \frac{2h}{3a^2}\left[\frac{27a^2}{n^2\pi^2} \sin \frac{n\pi}{3} - \frac{27a^2}{n^2\pi^2} \sin \frac{2n\pi}{3} \right]$$

$$\text{[using } \sin n\pi = 0 \text{]}$$

$$= \frac{18h}{n^2\pi^2}\left[\sin \frac{n\pi}{3} - \sin \left(n\pi - \frac{n\pi}{3} \right) \right]$$

$$= \frac{18h}{n^2\pi^2}\left[1+(-1)^n \right] \sin \frac{n\pi}{3}$$

Thus, $\quad C_n \begin{cases} \dfrac{18h}{n^2\pi^2}.2\sin \dfrac{n\pi}{3} & ; \text{ if } n \text{ is even} \\ 0 & ; \text{ if } n \text{ is odd} \end{cases}$

Further, let $n = 2m$. Then

$$u(x,t) = \frac{36h}{\pi^2} \sum_{m=1}^{\infty} \frac{1}{(2m)^2} \sin \frac{2m\pi}{3}$$

$$\cos \frac{2m\pi Ct}{3} \sin \frac{2m\pi x}{3a}$$

$$= \frac{9h}{\pi^2} \sum_{m=1}^{\infty} \frac{1}{m^2} \sin \frac{2m\pi}{3} \sin \frac{2m\pi}{2a} x \cos \frac{2m\pi Ct}{3a}$$

Finally, to get the displacement of the mid point of the string, we put $x = \dfrac{3a}{2}$ in the above expression, then we get

$$u(x,t) = \frac{9h}{\pi^2} \sum_{m=1}^{\infty} \frac{1}{m^2} \sin \frac{2m\pi}{3} \sin m\pi \cos \frac{2m\pi Ct}{3} = 0$$

Hence, we conclude that the mid point of the string always remains at rest.

Exercise-8.69

1. Find the deflection $u(x, t)$ of the vibrating string whose length is π^2 and $C^2 = 1$ corresponding to zero initial velocity and initial deflection $f(x) = k(\sin x - \sin 2x)$. (VTU-2011)

2. A string of length l has its ends $x = 0$ and $x = l$ fixed. It is released from rest in the position $y = \dfrac{4\lambda x\,[l-x]}{l^2}$. Find an expression for the displacement of the string at any subsequent time.

3. Solve completely the equation $\dfrac{\partial^2 y}{\partial t^2} = C^2 \dfrac{\partial^2 y}{\partial x^2}$, representing the vibration of a string of length l, fixed at both ends, given that $\quad y(0, t) = 0, \ y(l,t) = 0; \quad y(x, 0) = f(x) \quad$ and $\dfrac{\partial}{\partial t} y(x, 0) = 0, \ 0 < x < l$. [UPTU-2005]

4. If a string of length l is initially at rest in equilibrium position and each of its points is given the velocity $\left(\dfrac{\partial y}{\partial t} \right)_{t=0} = b\sin^3 \dfrac{\pi x}{l}$, find the displacement $y(x, t)$. [UPTU-2001,03,06]

5. The vibration of an elastic string is governed by the P.D.E. $\dfrac{\partial^2 u}{\partial t^2} = \dfrac{\partial^2 u}{\partial x^2}$. The length of the string is π and the ends are fixed. The initial velocity is zero and the initial deflection is $u(x, 0) = 2(\sin x + \sin 3x)$. Find the deflection $u(x, t)$ of the vibrating string for $t > 0$.

6. Solve the wave equation $\dfrac{\partial^2 u}{\partial t^2} = a^2 \dfrac{\partial^2 u}{\partial x^2}$ under the conditions $u = 0$, when $x = 0$ and $x = \pi$, $\dfrac{\partial u}{\partial t} = 0$, where $t = 0$ and $u(x, 0) = x$, $0 < x < \pi$.

7. A string is stretched and fastened to two points apart from a distance l. Motion is started by displacing the string into the form $y = k(lx - x^2)$ from which it is released at time $t = 0$. Find the displacement of any point on the string at a distance x from one end at time t. [UPTU-2002, IAS-1993]

8. Find the deflection $u(x, y, t)$ of the square membrane with $a = b = 1$ and $c = 1$, if the initial velocity is zero and the initial deflection is $f(x, y) = A\sin \pi x \sin 2\pi y$. [KURUKSHETRA -2004]

9. Solve the following boundary value problem
$$\frac{\partial^2 u}{\partial t^2} = C^2 \frac{\partial^2 u}{\partial x^2}, \ 0 \le x \le 1, \ t \ge 0$$
subjected to the boundary conditions
$u(0, t) = 0, \ t > 0$, $u_x(1,t) = 0, \ t > 0$
and initial conditions
$$u(x, 0) = \begin{cases} x, & 0 < x < 1/4 \\ \dfrac{1}{2} - x, & 1/4 < x < 1/2 \\ 0, & 1/2 < x < 1 \end{cases}$$
and $u_t(x, 0) = 0, \ 0 < x < 1$.

10. Show how the wave equation $C^2 \dfrac{\partial^2 y}{\partial x^2} = \dfrac{\partial^2 y}{\partial t^2}$ can be solved by the method of separation of variables. If the initial displacement and velocity of a string stretched between $x = 0$ and $x = l$ are given by $y = f(x)$ and $\partial y / \partial t = g(x)$. Determine the constants in the series solution.

[UPTU Q. BANK-2002]

11. A tightly stretched violin string of length l and fixed at both ends is plucked at $x = l/3$ and assumed initially the shape of a triangle of height a. Find the displacement y at any distance x and any time t after the string is released from rest.

[UPTU Q. BANK-2002]

12. Transform the equation $\dfrac{\partial^2 y}{\partial t^2} = C^2 \dfrac{\partial^2 y}{\partial x^2}$ to its normal form using the transformation $u = x + Ct$, $v = x - Ct$ and hence solve it. Show that the solution may be put in the form

$$y = \frac{1}{2}\left[f(x + Ct) + f(x - Ct)\right]$$

Assume initial conditions $y = f(x)$ and $(\partial y / \partial t) = 0$ at $t = 0$.

[UPTU-2003]

13. A tightly stretched string with fixed end points $x = 0$ and $x = l$ is initially at rest in its equilibrium position. If it is set vibrating by giving to each of its points an initial velocity $\lambda x\ (1 - x)$, find the displacement of the string at any distance x from one end at any time t. [MTU(SUM)-2011, (BHOPAL-2008, MADRAS-2006, JNTU-2005, ANNA-2009, UPTU-2002, PTU-2005)]

14. Find the deflection of the vibrating string which is fixed at the ends $x = 0$ and $x = 2$ and the motion is started by displacing the string into the form $\sin^3\left(\dfrac{\pi x}{2}\right)$ and releasing it with zero initial velocity at $t=0$.

(MTU-2012)

15. Find the deflection of the vibrating string of unit length whose end points are fixed if the initial velocity is zero and the initial deflection is given by

$$u(x,0) = \begin{cases} 1 & 0 \le x \le 1/2 \\ -1 & \dfrac{1}{2} < x \le 1 \end{cases}$$

(GBTU-2012)

🪶 Hints to Selected Problems

1. By taking $a = \pi$ and $c^2 = 1$, the required deflection $u(x, t)$ is given by

$$u(x,t) = \sum_{n=1}^{\infty} \cos nt\ \sin nx \qquad \ldots(1)$$

where, $\quad E_n = \dfrac{2}{\pi}\int_0^\pi f(x)\sin nx\, dx$

$$= \dfrac{2k}{\pi}\int_0^\pi (\sin x - \sin 2x)\sin nx\, dx$$

$$= \dfrac{2k}{\pi}\left[\int_0^\pi \sin x\ \sin nx\, dx - \int_0^\pi (\sin x - \sin 2x)\sin nx\, dx\right] \quad \ldots(2)$$

We know that

$$\int_0^\pi \sin px\ \sin qx\, dx = 0 \ \text{ if } \ p \ne q$$
$$= \pi/2 \ \text{ if } \ p = q\,.$$

Using (2) and (3), we get

$$E_1 = k,\ E_2 = -k, \ldots, E_n = 0 \ \ \forall\, n \ge 3$$

Putting all these values in (1), we get

$$u(x,t) = k\cos t\ \sin x - \cos 2t\ \sin 2x$$

2. It is known that the displacement function $y(x, t)$ is the solution of wave equation given by

$$\dfrac{\partial^2 y}{\partial x^2} = \dfrac{1}{C^2}\dfrac{\partial^2 y}{\partial t^2} \qquad \ldots(1)$$

with boundary conditions

$$y(0,t) = y(l,t) = 0 \ \text{ for all } \ t \ge 0 \qquad \ldots(2)$$

and initial conditions $\left(\dfrac{\partial y}{\partial t}\right)_{t=0} = 0 \ \text{ for } \ 0 \le x \le l$

and $\qquad y(x,0) = f(x) = \left\{4\lambda x(l - x)\right\}/l^2 \qquad \ldots(3)$

Proceeding same as above, the solution of (1) satisfying the above boundary and initial conditions is

$$y(x,t) = \sum_{n=1}^{\infty} E_n \sin\dfrac{n\pi x}{l}\cos\dfrac{n\pi Ct}{l} \qquad \ldots(4)$$

where, $\quad E_n = \dfrac{2}{l}\int_0^l f(x)\sin\dfrac{n\pi x}{l}\, dx \qquad \ldots(5)$

Putting the values of $f(x)$ given by (3) in (5), we get

$$E_n = \left(\dfrac{2}{l}\right)\times\left(\dfrac{4\lambda}{l^2}\right)\int_0^l (lx - x^2)\sin\dfrac{n\pi x}{l}\, dx$$

$$= \dfrac{8\lambda}{l^3}\left[(lx - x^2)\left(-\dfrac{l}{n\pi}\cos\dfrac{n\pi x}{l}\right) - (l - 2x)\left(-\dfrac{l^2}{n^2\pi^2}\sin\dfrac{n\pi x}{l}\right)\right.$$

$$\left.\qquad\qquad + (-2)\left(\dfrac{l^3}{n^3\pi^3}\cos\dfrac{n\pi x}{l}\right)\right]_0^l$$

$$= \dfrac{8\lambda}{l^3}\left[-\dfrac{2l^3(-1)^n}{n^3\pi^3} + 2\dfrac{l^3}{n^3\pi^3}\right] = \dfrac{16\lambda}{n^3\pi^3}\left[1 - (-1)^n\right]$$

$$\left[\because \cos n\pi = (-1)^n\right]$$

$$= \begin{cases} 0 & ;\ \text{if } n = 2m,\ m = 1,2,3,\ldots \\[2mm] \dfrac{32\lambda}{(2m-1)^3\pi^3} & ;\ \text{if } n = 2m - 1, m = 1,2,\ldots \end{cases}$$

Finally putting the value of E_n in (4), we get

$$y(x,t) = \dfrac{32\lambda}{\pi^3}\sum_{m=1}^{\infty}\dfrac{1}{(2m-1)^3}\sin\dfrac{(2m-1)\pi x}{l}\cos(2m-1)\dfrac{\pi Ct}{l}$$

3. Here, $\qquad \dfrac{\partial^2 y}{\partial t^2} = C^2\dfrac{\partial^2 y}{\partial x^2} \qquad \ldots(1)$

Let $\qquad y = X(x)\,T(t) \qquad \ldots(2)$

$$\dfrac{\partial^2 y}{\partial t^2} = X\dfrac{d^2 T}{dt^2}$$

$$\dfrac{\partial^2 y}{\partial x^2} = T\dfrac{d^2 X}{dx^2}$$

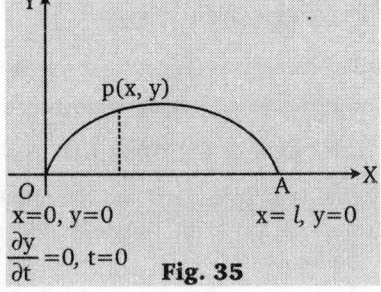

$x=0,\ y=0 \qquad\qquad x= l,\ y=0$

$\dfrac{\partial y}{\partial t} = 0,\ t = 0 \qquad$ **Fig. 35**

Equation (1) becomes

$$X \frac{d^2 T}{dt^2} = C^2 T \frac{d^2 X}{dx^2}$$

Separating the variables

$$\frac{1}{X} \frac{d^2 X}{dx^2} = \frac{1}{C^2 T} \frac{d^2 T}{dt^2} = -p^2 \text{ (let)}$$

$$\frac{1}{X} \frac{d^2 X}{dx^2} = -p^2 \qquad \text{If} \quad \frac{1}{C^2 T} \frac{\partial^2 T}{\partial t^2} = -p^2$$

$$\Rightarrow \qquad \frac{d^2 X}{dx^2} = -Xp^2 \qquad \Rightarrow (D^2 + b^2 C^2) T = 0$$

$$\text{A.E. is } d^2 + p^2 C^2 = 0$$

$$\Rightarrow \quad \frac{d^2 X}{dx^2} + Xp^2 = 0 \qquad \Rightarrow D^2 = -p^2 C^2 \Rightarrow D = \pm pC_i$$

$$\Rightarrow \quad (D^2 + p^2) X = 0 \qquad T = (C_2 \cos pCt + C_4 \sin pCt)$$

$$\Rightarrow \qquad X = (C_1 \cos px + C_2 \sin px)$$

Putting the values of X and T in equation (2), we get

$$y = (C_1 \cos px + C_2 \sin px)(C_3 \cos pCt + C_4 \sin pCt) \qquad \dots (3)$$

Now applying the boundary condition $x = 0$, $y = 0$

Putting these values in (3), we get

$$0 = C_1(C_3 \cos pCt + C_4 \sin pCt) \Rightarrow C_1 = 0$$

Equation (3) becomes $y = C_2 \sin px (C_3 \cos pCt + C_4 \sin pCt)$

$$\dots (4)$$

Putting $x = l$ and $y = 0$ in (4), we get

$$0 = C_2 \sin pl (C_3 \cos pCt + C_4 \sin pCt)$$

$$\Rightarrow \quad \sin pl = 0 = \sin n\pi$$

$$\Rightarrow \qquad pl = n\pi \qquad \Rightarrow \qquad p = \frac{n\pi}{l}$$

On putting $p = \frac{n\pi}{l}$, (4) becomes

$$y = C_2 \sin \frac{n\pi x}{l} \left(C_3 \cos \frac{n\pi Ct}{l} + C_4 \sin \frac{n\pi Ct}{l} \right) \qquad \dots (5)$$

On differentiating (5) w.r.t. t, we get

$$\frac{\partial y}{\partial t} = C_2 \sin \frac{n\pi x}{l} \left(-\frac{n\pi C}{l} C_3 \sin \frac{n\pi Ct}{l} + \frac{n\pi C}{l} C_4 \frac{n\pi C}{l} t \right) \qquad \dots (6)$$

On putting $\frac{\partial y}{\partial t} = 0$ and $t = 0$ in (6), we get

$$0 = C_2 \sin \frac{n\pi x}{l} \cdot \frac{n\pi C}{l} C_4$$

$$\Rightarrow \qquad C_4 = 0$$

On putting $C_4 = 0$, (5) becomes

$$y = C_2 C_3 \sin \frac{n\pi x}{l} \cos \frac{n\pi Ct}{l} \quad \left[\text{let } C_2 C_3 = b_n \right] \qquad \dots (7)$$

Now applying $y = f(x)$ and $t = 0$, (7) becomes

$$f(x) = b_n \sin \frac{n\pi x}{l}$$

$C_2 C_3$ can be calculated using Fourier sine series as

$$b_n = \frac{2}{l} \int_0^1 f(x) \sin \frac{n\pi x}{l} dx$$

The required solution for the given equation is

$$y = b_n \sin \frac{n\pi x}{l} \cos \frac{n\pi Ct}{l}$$

4. The equation for the vibrations of the string is

$$\frac{\partial^2 y}{\partial t^2} = C^2 \frac{\partial^2 y}{\partial x^2} \qquad \dots (1)$$

The solution of equation (1) is

$$y(x, t) = (C_1 \cos Cpt + C_2 \sin Cpt)(C_3 \cos px + C_4 \sin px) \dots (2)$$

Boundary conditions are

$$y(0, t) = 0 \qquad \dots (3)$$
$$y(l, t) = 0 \qquad \dots (4)$$
$$y(x, 0) = 0 \qquad \dots (5)$$
$$\left(\frac{\partial y}{\partial t} \right) = b \sin^3 \frac{\pi x}{l} \text{ at } t = 0 \qquad \dots (6)$$

Putting $x = 0$ and $y = 0$ in (2), we get

$$0 = (C_1 \cos Cpt + C_2 \sin Cpt)C_3 \Rightarrow C_3 = 0$$

Putting the value of C_3 in (2), we get

$$y = (C_1 \cos Cpt + C_2 \sin Cpt)C_4 \sin px \qquad \dots (7)$$

Putting $x = l$ and $y = 0$ in (7), we get

$$0 = (C_1 \cos Cpt + C_2 \sin Cpt) C_4 \sin pl$$

$$\Rightarrow \quad \sin pl = 0 = \sin n\pi \ (n \in I) \Rightarrow p = \frac{n\pi}{l}$$

Putting the value of p in (7), we get

$$y = \left(C_1 \cos \frac{n\pi Ct}{l} + C_2 \sin \frac{n\pi Ct}{l} \right) C_4 \sin \frac{n\pi x}{l} \qquad \dots (8)$$

Putting $t = 0$ and in (8), we get

$$0 = C_1 C_4 \sin \frac{n\pi x}{l} \Rightarrow C_1 = 0$$

Putting the value of C_1 in (8), we get

$$y = C_2 C_4 \sin \frac{n\pi Ct}{l} \sin \frac{n\pi x}{l}$$

$$y = b_n \sin \frac{n\pi Ct}{l} \sin \frac{n\pi x}{l} \quad (\text{where } C_2 C_4 = b_n)$$

The general solution is $y = \sum_1^\infty b_n \sin \frac{n\pi Ct}{l} \sin \frac{n\pi x}{l} \qquad \dots (9)$

On differentiating (9) w.r.t. x^t, we get

$$\frac{\partial y}{\partial t} = \sum_1^\infty b_n \frac{n\pi C}{l} \cos \frac{n\pi Ct}{l} \sin \frac{n\pi x}{l} \qquad \dots (10)$$

On putting the values of $\frac{\partial y}{\partial t} = b \sin^3 \frac{\pi x}{l}$ and $t = 0$ in (10), we get

$$b \sin^3 \frac{\pi x}{l} = \sum_1^\infty b_n \frac{n\pi C}{l} \sin \frac{n\pi x}{l}$$

$$\Rightarrow \quad \frac{b}{4} \left[3 \sin \frac{\pi x}{l} \frac{3\pi x}{l} \right] = b_1 \frac{\pi C}{l} \sin \frac{\pi x}{l} + \frac{2b_2 \pi C}{l} \sin^2 \frac{\pi x}{l}$$

$$+ 3b_3 \frac{\pi C}{l} \sin \frac{3\pi C}{l} + \dots$$

Equating the coefficients of $\sin \frac{\pi x}{l}$, $\sin \frac{2\pi x}{l}$, we get

$$\frac{3b}{4} = b_1 \frac{\pi C}{l} \Rightarrow b_1 = \frac{3bl}{4\pi C}$$

$$b_2 = 0 \text{ and } \frac{3b_3 \pi C}{l} = \frac{b}{4} \Rightarrow b_3 = -\frac{bl}{12\pi C}.$$

Also, $b_4 = 0 = b_5 = \dots\dots$ etc.

Putting the values of b's in equation (9), we get

$$y(x, t) = \frac{3bl}{4\pi C} \sin \frac{\pi Ct}{l} \sin \frac{\pi x}{l} - \frac{bl}{12\pi C} \sin \frac{3\pi Ct}{l} \sin \frac{3\pi x}{l}$$

$$= \frac{bl}{12\pi C} \left[9 \sin \frac{\pi x}{l} \sin \frac{\pi Ct}{l} - \sin \frac{3\pi x}{l} \sin \frac{3\pi Ct}{l} \right].$$

5. The solution of given equation is

$$u(x, t) = (C_1 \cos nx + C_2 \sin nx)(C_3 \cos nt + C_4 \sin nt) \qquad \dots (1)$$

where, $u(0, t) = 0$, $u(\pi, t) = 0$; $\forall t$

$$u(0, t) = 0 \Rightarrow (C_3 \cos nt + C_4 \sin nt) C_1 = 0$$

$$\left(\frac{\partial u}{\partial t}\right)_{t=0} = 0; \quad u(x, 0) = 2(\sin x + \sin 3x) \qquad \ldots (2)$$

which gives $C_1 = 0$.

Thus $\quad u(x, t) = (C_3 \cos nt + C_4 \sin nt) C_2 \sin nx$

$$= (b_1 \cos nt + b_2 \sin nt) \sin nx$$

Again, $\quad u(\pi, t) = 0 \Rightarrow (b_1 \cos nt + b_2 \sin nt) \sin n\pi = 0$

$\Rightarrow \qquad \sin n\pi = 0$

$\Rightarrow n$ must be an integer.

Thus, $\quad u(x, t) = (b_1 \cos nt + b_2 \sin nt) \sin nx$

Then, $\left(\dfrac{\partial u}{\partial t}\right)_{t=0} = b_2 n \sin x \Rightarrow b_2 = 0$

$\Rightarrow \qquad u(x, t) = b_1 \cos nt \sin nx$

Also, $\quad u(x, 0) = 2(\sin x + \sin 3x)$

$\Rightarrow 2(\sin x + \sin 3x) = \sin nx \, b_1$

$\Rightarrow \quad 4 \sin 2x \cos x = \sin x . b_1$

which is true when $b_1 = 4 \cos x$ *and* $n = 2$.

Hence, we have $u(x, t) = 4 \cos x \cos 2t \sin 2x$.

6. The solution is of the form

$$u(x, t) = (C_1 \cos px + C_2 \sin px)(C_3 \cos apt + C_4 \sin apt) \quad \ldots (1)$$

On putting $u = 0$ and $x = 0$ in (1), we get

$$0 = C_1(C_3 \cos apt + C_4 \sin apt) \Rightarrow C_1 = 0.$$

On putting $C_1 = 0$ in (1), we get

$$u(x, t) = C_2 \sin px(C_3 \cos apt + C_4 \sin apt) \qquad \ldots (2)$$

On putting $x = \pi$ and $u = 0$ in (2), we get

$$0 = C_2 \sin p\pi(C_3 \cos apt + C_4 \sin apt)$$

$\Rightarrow \qquad \sin px = 0 = \sin n\pi$

$\Rightarrow \qquad p = n$

On putting $p = n$ in equation (2), we get

$$u(x, t) = C_2 \sin nx(C_3 \cos ant + C_4 \sin ant)$$

and $\quad u(x, t) = \sin nx(b_1 \cos ant + b_2 \sin ant)$

$$[C_2 C_3 = b_1 \text{ and } C_2 C_4 = b_2] \qquad \ldots (3)$$

On differentiating w.r.t. t, we get

$$\frac{\partial u}{\partial t} = \sin nx \left[-ab_1 n \sin ant + ab_2 n \cos ant\right] \qquad \ldots (4)$$

On putting $\dfrac{\partial u}{\partial t} = 0$ and $t = 0$ in (4), we have

$$0 = \sin n\pi(ab_2 n) \Rightarrow b_2 = 0$$

On putting $b_2 = 0$ in (3), we get

$$u(x, t) = \sin nx \, (b_1 \cos ant) \qquad \ldots (5)$$

General solution is

$$u(x, t) = \sum_{n=1}^{\infty} b_n \sin nx \cos ant \qquad \ldots (6)$$

On putting $u = x$ and $t = 0$ in (6), we get

$$x = \sum_{n=1}^{\infty} b_n \sin nx, \text{ where } b_n = \frac{2}{\pi} \int_0^{\pi} x \sin nx \, dx.$$

$$= \frac{2}{\pi} \left[x \left(-\frac{\cos nx}{n}\right)(-1)\left(-\frac{\sin nx}{n^2}\right)\right]_0^{\pi}$$

$$= \frac{2}{\pi} \left[-\frac{\pi}{n} \cos nx\right] = -\frac{2}{n}(-1)^n$$

On putting the value of b_n in (6), we get

$$u(x, t) = -2 \sum_{n=1}^{\infty} \frac{(-1)^n}{n} \sin nx \cos nat$$

7. The vibrations of the string are governed by the equation

$$\frac{\partial^2 u}{\partial t^2} = C^2 \frac{\partial^2 u}{\partial x^2} \qquad \ldots (1)$$

The boundary conditions are

$$u(0, t) = 0 \quad \text{and} \quad u(l, t) = 0 \qquad \ldots (2)$$

Also, the initial transverse velocity of any point of the string is zero, therefore we get

$$\left(\frac{\partial u}{\partial t}\right)_{t=0} = 0 \quad ; \text{ Also, } u(x, 0) = k \, (lx - x^2) \qquad \ldots (3)$$

Proceeding same as usual, the solution of (1) under (2) and (3) is given by

$$u(x, t) = (C_1 \cos px + C_2 \sin px)(C_3 \cos Cpt + C_4 \sin Cpt)$$

Now, $u(0, t) = 0 \Rightarrow (C_1 \cos 0 + C_2 \sin 0)$
$$(C_3 \cos Cpt + C_4 \sin Cpt) = 0$$

$\Rightarrow \quad C_1 (C_3 \cos Cpt + C_4 \sin Cpt) = 0 \Rightarrow C_1 = 0$

Therefore, $u(x, t) = C_2 \sin px(C_3 \cos Cpt + C_4 \sin Cpt)$
$$= \sin px \, (C_3 \cos Cpt + C_4 \sin Cpt)$$

$\Rightarrow \qquad \sin px = 0 = \sin n\pi$, n being integer.

$\Rightarrow \qquad p = \dfrac{n\pi}{l}$.

Therefore,

$$u(x, t) = \sin \frac{n\pi x}{l} \left[C_3' \cos \frac{Cn\pi}{l}.t + C_4' \sin \frac{Cn\pi}{l}.t\right]$$

$$\Rightarrow \quad \frac{\partial u}{\partial t} = \sin \frac{n\pi x}{l} \left[C_3' \frac{Cn\pi}{l} \sin \frac{Cn\pi}{l}t + C_4' \frac{Cn\pi}{l}.\cos \frac{Cn\pi}{l}.t\right]$$

$$\Rightarrow \left(\frac{\partial u}{\partial t}\right)_{t=0} = C_4' \frac{Cn\pi}{l} \sin \frac{n\pi x}{l} = 0 \Rightarrow C_4' = 0$$

$$\Rightarrow \quad u(x, t) = C_3' \sin \frac{n\pi x}{l} \cos \frac{Cn\pi t}{l}$$

Hence, the general solution of (1) is given by

$$u(x, t) = \sum_{n=1}^{\infty} u_n(x, t) = \sum_{n=1}^{\infty} b_n \sin \frac{n\pi x}{l} \cos \frac{Cn\pi t}{l}$$

Also, $u(x, 0) = \sum\limits_{n=1}^{\infty} b_n \sin \dfrac{n\pi x}{l} = k \, (lx - x^2)$, where $C_3' = b_n$

So, $\quad b_n = \dfrac{2}{l} \int_0^l k \, (lx - x^2) \sin \dfrac{n\pi x}{l} \, dx$

8. The deflection of the square membrane is given by the two dimensional wave equation

$$\frac{\partial^2 u}{\partial t^2} = C^2 \left(\frac{\partial^2 u}{\partial x^2} + \frac{\partial^2 u}{\partial y^2}\right)$$

The boundary conditions are

$u(x, 0, t) = 0 = u(x, 1, t)$;

$u(0, y, t) = 0 = u(1, y, t)$

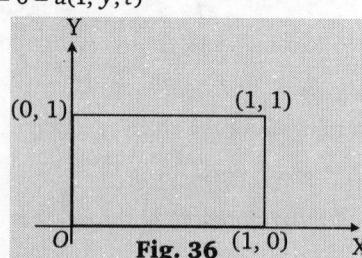

Fig. 36

Also, the initial conditions are

$u(x, y, 0) = f(x, y) = A \sin \pi x \sin 2\pi y$

$\left(\dfrac{\partial u}{\partial t}\right)_{t=0} = 0$

Thus, the deflection is given by

$$u(x, y, t) = \sum_{m=1}^{\infty} \sum_{n=1}^{\infty} A_{mn} \cos k_{mn} t \sin m\pi x \sin n\pi y$$

where, $A_{mn} = 4 \int_0^1 \int_0^1 f(x, y) \sin m\pi x \sin n\pi y \, dx \, dy$ ... (1)

$\qquad = 4 \int_0^1 \int_0^1 \sin \pi x \sin 2\pi y \sin m\pi x \sin n\pi y \, dx dy$

On integrating, we find that $A_{m1} = A_{m3} = A_{m4} = \dots = 0$

But $A_{m2} = 4A \int_0^1 \int_0^1 \sin \pi x \sin m\pi x \sin^2 2\pi y \, dx \, dy$

$\qquad = 2A \int_0^1 \int_0^1 \sin \pi x \sin m\pi x \, (1 - \cos 4\pi y) \, dx \, dy$

$\qquad = 2A \int_0^1 \sin \pi x \sin m\pi x \left(y - \dfrac{1}{4\pi} \sin 4\pi y \right)_0^1 dx$

$\qquad = 2A \int_0^1 \sin \pi x \sin m\pi x \, dx$

Again, on integrating, we get $A_{22} = A_{32} = \dots = 0$.

Further, we find that

$A_{12} = 2A \int_0^1 \sin \pi x \sin \pi x \, dx = A \int_0^1 2 \sin^2 \pi x \, dx$

$\qquad = A \int_0^1 (1 - \cos 2\pi x) \, dx = A \left[x - \dfrac{1}{2\pi} \sin 2\pi x \right]_0^1 = A$

Hence, from (1) we get

$u(x, y, t) = A_{12} \cos k_{12} t \sin \pi x \sin 2\pi y$

$\qquad = A \cos \sqrt{5} \, \pi t \sin \pi x \sin 2\pi y$

$\qquad \left[\because k_{12}^2 = \pi^2(m^2 + n^2) = \pi^2(1^2 + 2^2) = \sqrt{5} \, \pi \right]$

Answers

1. $u(x, t) = k \cos t \sin x - \cos 2t \sin 2x$

2. $y(x, t) = \dfrac{32\lambda}{\pi^3} \sum_{m=1}^{\infty} \dfrac{1}{(2m-1)^3} \sin \dfrac{(2m-1)\pi x}{l} \cos (2m-1) \dfrac{\pi Ct}{l}$

3. $y = b_n \sin \dfrac{n\pi x}{l} \cos \dfrac{n\pi Ct}{l}$, where $b_n = \dfrac{2}{l} \int_0^1 f(x) \sin \dfrac{n\pi x}{l} dx$.

4. $y(x, t) = \dfrac{bl}{12\pi c} \left[9 \sin \dfrac{\pi x}{l} \sin \dfrac{\pi ct}{l} - \sin \dfrac{3\pi x}{l} \sin \dfrac{3\pi ct}{l} \right]$

5. $u(x, t) = 4 \cos x \cot 2t . \sin 2x$

6. $u(x, t) = -2 \sum_{n=1}^{\infty} \dfrac{(-1)^n}{n} \sin nx . \cos nat$

7. $u(x, t) = \sum_{n=1}^{\infty} u_n(x, t) = \sum_{n=1}^{\infty} b_n \dfrac{n\pi x}{l} \cos \dfrac{cn\pi t}{l}$; $u(x, 0) = \sum_{n=1}^{\infty} b_n \sin \dfrac{n\pi x}{l} = k(lx - x^2)$ where, $b_n = \dfrac{2}{l} \int_0^1 k(lx - x^2) \sin \dfrac{n\pi x}{l} dx$.

8. $u(x, y, t) = A_{12} \cos k_{12} t \sin \pi x \sin 2\pi y = A \cos \sqrt{5} \, \pi t \sin \pi x \sin 2\pi y$

9. $u(x, t) = \dfrac{8}{\pi^2} \sum_{n=0}^{\infty} \dfrac{\sin \frac{1}{4}(2n-1)\pi}{(2n-1)^2} \cos \dfrac{(2n-1)\pi Ct}{2} \sin \dfrac{(2n-1)\pi x}{2}$

10. $y(x, t) = \sum_{1}^{\infty} \left(a_n \cos \dfrac{n\pi Ct}{l} + b_n \sin \dfrac{n\pi Ct}{l} \right) \sin \dfrac{n\pi x}{l}$

$a_n = \dfrac{2}{l} \int_0^l f(x) \sin \dfrac{n\pi x}{l} dx$; $b_n = \dfrac{2}{n\pi C} \int_0^l g(x) \sin \dfrac{n\pi x}{l} dx$

11. $y(x, t) = \dfrac{9a}{\pi^2} \sum_{1}^{\infty} \dfrac{1}{n^2} \sin \dfrac{n\pi}{3} \cos \dfrac{n\pi Ct}{l} \sin \dfrac{n\pi x}{l}$

13. $y(x, t) = \dfrac{8\lambda l^3}{c\pi^4} \sum_{m=1}^{\infty} \dfrac{1}{(2m-1)^4} \sin \dfrac{(2m-1)\pi ct}{l} \sin \dfrac{(2m-1)\pi x}{l}$

8.15.7 LAPLACE EQUATION

The heat equation is given by $\qquad \dfrac{\partial U}{\partial t} = C^2 \nabla^2 U$

If temperature are in steady state (*i.e.*, does not depend upon time t), then the heat equation reduces to

$$\nabla^2 u = 0$$

i.e., $\qquad \dfrac{\partial^2 U}{\partial x^2} + \dfrac{\partial^2 U}{\partial y^2} + \dfrac{\partial^2 U}{\partial z^2} = 0$. This is known as Laplace equation.

8.15.8 LAPLACE EQUATION IN TERMS OF POLAR COORDINATES

Let us suppose that the boundary of the region ∂R is a circle. Then, we use the polar coordinates as follows :

Let $\qquad x = r \cos\theta, \; y = r \sin\theta$

Then, $\qquad r^2 = x^2 + y^2$ and $\theta = \tan^{-1} y / x$

Therefore, we can find $\dfrac{\partial r}{\partial x} = \cos\theta, \; \dfrac{\partial r}{\partial y} = \sin\theta, \; \dfrac{\partial \theta}{\partial x} = \dfrac{-\sin\theta}{r}$ and $\dfrac{\partial \theta}{\partial y} = \dfrac{\cos\theta}{r}$ ... (1)

$u = u(r, \theta) \Rightarrow \dfrac{\partial u}{\partial x} = \dfrac{\partial u}{\partial r} \dfrac{\partial r}{\partial x} + \dfrac{\partial u}{\partial \theta} \dfrac{\partial \theta}{\partial x} = \left(\dfrac{\partial u}{\partial r} \cos\theta - \dfrac{\partial u}{\partial \theta} \dfrac{\sin\theta}{r} \right)$

... (2)

Also, $\qquad \dfrac{\partial u}{\partial y} = \dfrac{\partial u}{\partial r} \dfrac{\partial r}{\partial y} + \dfrac{\partial u}{\partial \theta} \dfrac{\partial \theta}{\partial y} = \left(\dfrac{\partial u}{\partial r} \sin\theta + \dfrac{\partial u}{\partial \theta} \dfrac{\cos\theta}{r} \right)$

and $\qquad \dfrac{\partial^2 u}{\partial x^2} = \dfrac{\partial}{\partial x} \left(\dfrac{\partial u}{\partial x} \right) = \dfrac{\partial}{\partial r} \left\{ \dfrac{\partial u}{\partial r} \cos\theta - \dfrac{\partial u}{\partial \theta} \dfrac{\sin\theta}{r} \right\} \cos\theta + \dfrac{\partial}{\partial \theta} \left\{ \dfrac{\partial u}{\partial r} \cos\theta - \dfrac{\partial u}{\partial \theta} \dfrac{\sin\theta}{r} \right\} \left(\dfrac{-\sin\theta}{r} \right)$

$\qquad = \left(\dfrac{\partial^2 u}{\partial r^2} \cos\theta - \dfrac{\partial^2 u}{\partial \theta \partial r} \dfrac{\sin\theta}{r} + \dfrac{\partial u}{\partial \theta} \dfrac{\sin\theta}{r^2} \right) \cos\theta + \left(\dfrac{\partial^2 u}{\partial r \partial \theta} \cos\theta - \dfrac{\partial u}{\partial r} \sin\theta - \dfrac{\partial^2 u}{\partial \theta^2} \dfrac{\sin\theta}{r} - \dfrac{\partial u}{\partial \theta} \dfrac{\cos\theta}{r} \right) \left(\dfrac{-\sin\theta}{r} \right)$ ... (3)

Similarly,
$$\frac{\partial^2 u}{\partial y^2} = \left(\frac{\partial^2 u}{\partial r^2}\sin\theta + \frac{\partial^2 u}{\partial r \partial\theta}\frac{\cos\theta}{r} - \frac{\partial u}{\partial\theta}\frac{\cos\theta}{r^2}\right)\sin\theta + \left(\frac{\partial^2 u}{\partial r\partial\theta}\sin\theta + \frac{\partial u}{\partial r}\cos\theta + \frac{\partial^2 u}{\partial\theta^2}\frac{\cos\theta}{r} - \frac{\partial u}{\partial\theta}\frac{\sin\theta}{r}\right)\left(\frac{\cos\theta}{r}\right) \qquad \dots (4)$$

Adding (3) and (4), we get
$$\frac{\partial^2 u}{\partial x^2} + \frac{\partial^2 u}{\partial y^2} = \frac{\partial^2 u}{\partial r^2} + \frac{1}{r}\frac{\partial u}{\partial r} + \frac{1}{r^2}\frac{\partial^2 u}{\partial\theta^2} = 0$$

which is the required Laplace equation in polar coordinates.

☞ REMARK

* The Laplace equation in Cartesian coordinates has constant coefficients only, whereas in polar coordinates, it has variable coefficients.

8.15.9 LAPLACE EQUATION IN CYLINDRICAL COORDINATES (r, θ, z)

Let (x, y, z) be the cartesian coordinates of the point P whose cylindrical coordinates are given by (r, θ, z) such that
$$x = r\cos\theta, \qquad y = r\sin\theta, \qquad z = z$$
which implies $r^2 = x^2 + y^2$ and $\theta = \tan^{-1} y/x$

Therefore,
$$\frac{\partial r}{\partial x} = \frac{x}{r} = \cos\theta, \quad \frac{\partial r}{\partial y} = \frac{y}{r} = \sin\theta, \quad \frac{\partial\theta}{\partial x} = \frac{1}{1 + y^2/x^2}\left(-\frac{y}{x^2}\right) = \frac{-\sin\theta}{r} \text{ and } \frac{\partial\theta}{\partial y} = \frac{\cos\theta}{r}$$

Now, we have
$$\frac{\partial u}{\partial x} = \frac{\partial u}{\partial r}\frac{\partial r}{\partial x} + \frac{\partial u}{\partial\theta}\cdot\frac{\partial\theta}{\partial x} + \frac{\partial u}{\partial z}\cdot\frac{\partial z}{\partial x} = \frac{\partial u}{\partial r}\cdot\cos\theta + \frac{\partial u}{\partial\theta}\cdot\left(-\frac{\sin\theta}{r}\right)$$

$$\Rightarrow \qquad \frac{\partial}{\partial x} = \cos\theta\frac{\partial}{\partial r} - \frac{\sin\theta}{r}\frac{\partial}{\partial\theta}$$

Therefore,
$$\frac{\partial^2 u}{\partial x^2} = \frac{\partial}{\partial x}\left(\frac{\partial u}{\partial x}\right) = \left(\cos\theta\frac{\partial}{\partial r} - \frac{\sin\theta}{r}\frac{\partial}{\partial\theta}\right)\left(\cos\theta\frac{\partial u}{\partial r} - \sin\theta\frac{\partial u}{\partial\theta}\right)$$

$$= \cos^2\theta\frac{\partial^2 u}{\partial r^2} - 2\frac{\sin\theta\cos\theta}{r}\frac{\partial^2 u}{\partial r\partial\theta} + \frac{2\sin\theta\cos\theta}{r^2}\cdot\frac{\partial u}{\partial\theta} + \frac{\sin^2\theta}{r^2}\frac{\partial u}{\partial r} + \frac{\sin^2\theta}{r^2}\frac{\partial^2 u}{\partial\theta^2} \qquad \dots (1)$$

In a similar manner, we can find
$$\frac{\partial^2 u}{\partial y^2} = \sin^2\theta\frac{\partial^2 u}{\partial r^2} + 2\frac{\sin\theta\cos\theta}{r}\frac{\partial^2 u}{\partial r\partial\theta} - \frac{2\sin\theta\cos\theta}{r^2}\cdot\frac{\partial u}{\partial\theta} + \frac{\cos^2\theta}{r}\frac{\partial u}{\partial r} + \frac{\cos^2\theta}{r^2}\frac{\partial^2 u}{\partial\theta^2} \qquad \dots (2)$$

and
$$\frac{\partial^2 u}{\partial z^2} = \frac{\partial^2 u}{\partial z^2} \qquad \dots (3)$$

On adding (1), (2) and (3), we get
$$\frac{\partial^2 u}{\partial x^2} + \frac{\partial^2 u}{\partial y^2} + \frac{\partial^2 u}{\partial z^2} = \frac{\partial^2 u}{\partial r^2} + \frac{1}{r^2}\frac{\partial^2 u}{\partial\theta^2} + \frac{1}{r}\frac{\partial u}{\partial r} + \frac{\partial^2 u}{\partial z^2} = 0$$

which is the required Laplace equation in cylindrical form.

☞ REMARK

* The above equation can also be written as
$$\frac{1}{r}\frac{\partial}{\partial r}\left(r\frac{\partial u}{\partial r}\right) + \frac{1}{r^2}\frac{\partial^2 u}{\partial\theta^2} + \frac{\partial^2 u}{\partial z^2} = 0$$

8.15.10 LAPLACE EQUATION IN SPHERICAL COORDINATES

Let (x, y, z) be the cartesian coordinates and (r, θ, ϕ) be the spherical coordinates at P such that
$$x = r\sin\theta\cos\phi, \qquad y = r\sin\theta\sin\phi, \qquad z = r\cos\theta$$
then, clearly we have
$$r^2 = x^2 + y^2 + z^2, \qquad \phi = \tan^{-1}(y/x) \text{ and } \theta = \tan^{-1}\left(\frac{\sqrt{x^2 + y^2}}{z}\right)$$

Now,
$$\frac{\partial r}{\partial x} = \frac{x}{r} = \sin\theta\cos\phi, \quad \frac{\partial r}{\partial y} = \frac{y}{r} = \sin\theta\sin\phi, \quad \frac{\partial r}{\partial z} = \frac{z}{r} = \cos\theta,$$

$$\frac{\partial\theta}{\partial x} = \frac{\cos\theta\cos\phi}{r}, \quad \frac{\partial\theta}{\partial y} = \frac{\cos\theta\sin\phi}{r}, \quad \frac{\partial\theta}{\partial z} = \frac{-\sin\theta}{r}, \frac{\partial\phi}{\partial x} = -\frac{\sin\phi}{r\sin\theta}$$

Therefore
$$\frac{\partial u}{\partial x} = \frac{\partial u}{\partial r}\frac{\partial r}{\partial x} + \frac{\partial u}{\partial\theta}\cdot\frac{\partial\theta}{\partial x} + \frac{\partial u}{\partial\phi}\cdot\frac{\partial\phi}{\partial x} = \frac{\partial u}{\partial r}\cdot\sin\theta\cos\phi + \frac{\partial u}{\partial\theta}\cdot\frac{\cos\theta\cos\phi}{r} + \frac{\partial u}{\partial\phi}\left(-\frac{\sin\phi}{r\sin\theta}\right)$$

$$\Rightarrow \qquad \frac{\partial}{\partial x} \equiv \sin\theta\sin\phi\frac{\partial}{\partial r} + \frac{\cos\theta\cos\phi}{r}\frac{\partial}{\partial\theta} - \frac{\sin\phi}{r\sin\theta}\frac{\partial}{\partial\phi}$$

Now, $\dfrac{\partial^2 u}{\partial x^2} = \dfrac{\partial}{\partial x}\left(\dfrac{\partial u}{\partial x}\right) = \left(\sin\theta\cos\phi\dfrac{\partial}{\partial r} + \dfrac{\cos\theta\cos\phi}{r}\dfrac{\partial}{\partial\theta} - \dfrac{\sin\phi}{r\sin\theta}\dfrac{\partial}{\partial\phi}\right)\left(\sin\theta\cos\phi\dfrac{\partial u}{\partial r} + \dfrac{\cos\theta\cos\phi}{r}\dfrac{\partial u}{\partial\theta} - \dfrac{\sin\phi}{r\sin\theta}\dfrac{\partial u}{\partial\phi}\right)$

$= \sin^2\theta\cos^2\phi\dfrac{\partial^2 u}{\partial r^2} + 2\dfrac{\sin\theta\cos\theta\cos^2\phi}{r}\dfrac{\partial^2 u}{\partial r\,\partial\theta} - \dfrac{2\sin\theta\cos\theta\cos^2\phi}{r^2}\cdot\dfrac{\partial u}{\partial\phi} - \dfrac{2\sin\phi\cos\phi}{r}\cdot\dfrac{\partial^2 u}{\partial r\,\partial\phi}$

$\quad + \dfrac{\sin\phi\cos\phi}{r^2}\cdot\dfrac{\partial u}{\partial\phi} + \dfrac{\cos^2\theta\cos^2\phi}{r}\cdot\dfrac{\partial u}{\partial r} + \dfrac{\cos^2\theta\cos^2\phi}{r^2}\cdot\dfrac{\partial^2 u}{\partial\theta^2} - 2\dfrac{\cos\theta\sin\phi\cos\phi}{r^2\sin\theta}\cdot\dfrac{\partial^2 u}{\partial\theta\,\partial\phi} + \dfrac{\cos^2\theta\sin\phi\cos\phi}{r^2\sin^2\theta}\cdot\dfrac{\partial u}{\partial\phi}$

$\quad + \dfrac{\sin^2\phi}{r}\cdot\dfrac{\partial u}{\partial r} + \dfrac{\cos\theta\sin^2\phi}{r^2\sin\theta}\cdot\dfrac{\partial u}{\partial\theta} + \dfrac{\sin^2\phi}{r^2\sin^2\theta}\cdot\dfrac{\partial^2 u}{\partial\phi^2} + \dfrac{\sin\phi\cos\phi}{r^2\sin^2\theta}\dfrac{\partial u}{\partial\phi}$ ... (1)

Also, $\dfrac{\partial u}{\partial y} = \dfrac{\partial u}{\partial r}\dfrac{\partial r}{\partial y} + \dfrac{\partial u}{\partial\theta}\cdot\dfrac{\partial\theta}{\partial y} + \dfrac{\partial u}{\partial\phi}\cdot\dfrac{\partial\phi}{\partial y} = \dfrac{\partial u}{\partial r}\cdot\sin\theta\sin\phi + \dfrac{\partial u}{\partial\theta}\cdot\dfrac{\cos\theta\sin\phi}{r} + \dfrac{\partial u}{\partial\phi}\left(\dfrac{\cos\phi}{r\sin\theta}\right)$

$\Rightarrow \quad \dfrac{\partial}{\partial y} \equiv \sin\theta\sin\phi\dfrac{\partial}{\partial r} + \dfrac{\cos\theta\sin\phi}{r}\dfrac{\partial}{\partial\theta} + \dfrac{\cos\phi}{r\sin\theta}\dfrac{\partial}{\partial\phi}$

which gives

$\dfrac{\partial^2 u}{\partial y^2} = \dfrac{\partial}{\partial y}\left(\dfrac{\partial u}{\partial y}\right) = \left(\sin\theta\sin\phi\dfrac{\partial}{\partial r} + \dfrac{\cos\theta\sin\phi}{r}\dfrac{\partial}{\partial\theta} + \dfrac{\cos\phi}{r\sin\theta}\dfrac{\partial}{\partial\phi}\right)\left(\sin\theta\sin\phi\dfrac{\partial u}{\partial r} + \dfrac{\cos\theta\sin\phi}{r}\dfrac{\partial u}{\partial\theta} + \dfrac{\cos\phi}{r\sin\theta}\dfrac{\partial u}{\partial\phi}\right)$

$= \sin^2\theta\sin^2\phi\dfrac{\partial^2 u}{\partial r^2} + 2\dfrac{\sin\theta\cos\theta\sin^2\phi}{r}\dfrac{\partial^2 u}{\partial r\,\partial\theta} - \dfrac{2\sin\theta\cos\theta\sin^2\phi}{r^2}\cdot\dfrac{\partial u}{\partial\theta} + \dfrac{2\sin\phi\cos\phi}{r}\cdot\dfrac{\partial^2 u}{\partial r\,\partial\phi}$

$\quad - \dfrac{\sin\phi\cos\phi}{r^2}\cdot\dfrac{\partial u}{\partial\phi} + \dfrac{\cos^2\theta\sin^2\phi}{r}\cdot\dfrac{\partial u}{\partial r} + \dfrac{\cos^2\theta\sin^2\phi}{r^2}\cdot\dfrac{\partial^2 u}{\partial\theta^2} + 2\dfrac{\cos\theta\sin\phi\cos\phi}{r^2\sin\theta}\cdot\dfrac{\partial^2 u}{\partial\theta\,\partial\phi} - \dfrac{\cos^2\theta\sin\phi\cos\phi}{r^2\sin^2\theta}\cdot\dfrac{\partial u}{\partial\phi}$

$\quad + \dfrac{\cos^2\phi}{r}\cdot\dfrac{\partial u}{\partial r} + \dfrac{\cos\theta\cos^2\phi}{r^2\sin\theta}\cdot\dfrac{\partial u}{\partial\theta} + \dfrac{\cos^2\phi}{r^2\sin^2\theta}\cdot\dfrac{\partial^2 u}{\partial\phi^2} - \dfrac{\sin\phi\cos\phi}{r^2\sin^2\theta}\dfrac{\partial u}{\partial\phi}$ ... (2)

Now, $\dfrac{\partial u}{\partial z} = \dfrac{\partial u}{\partial r}\dfrac{\partial r}{\partial z} + \dfrac{\partial u}{\partial\theta}\dfrac{\partial\theta}{\partial z} + \dfrac{\partial u}{\partial\phi}\cdot\dfrac{\partial\phi}{\partial z} = \dfrac{\partial u}{\partial r}\cdot\cos\theta + \dfrac{\partial u}{\partial\theta}\left(-\dfrac{\sin\theta}{r}\right)$

$\Rightarrow \quad \dfrac{\partial}{\partial z} \equiv \cos\theta\dfrac{\partial}{\partial r} - \dfrac{\sin\theta}{r}\dfrac{\partial}{\partial\theta}$

Therefore $\dfrac{\partial^2 u}{\partial z^2} = \dfrac{\partial}{\partial z}\left(\dfrac{\partial u}{\partial z}\right) = \left(\cos\theta\dfrac{\partial}{\partial r} - \dfrac{\sin\theta}{r}\dfrac{\partial}{\partial\theta}\right)\left(\cos\theta\dfrac{\partial u}{\partial r} + \left(\dfrac{-\sin\theta}{r}\right)\dfrac{\partial u}{\partial\theta}\right)$

$= \cos^2\theta\dfrac{\partial^2 u}{\partial r^2} - 2\dfrac{\sin\theta\cos\theta}{r}\dfrac{\partial^2 u}{\partial r\,\partial\theta} + \dfrac{2\sin\theta\cos\theta}{r^2}\cdot\dfrac{\partial u}{\partial\theta} + \dfrac{\sin^2\theta}{r^2}\cdot\dfrac{\partial u}{\partial r} + \dfrac{\sin^2\theta}{r^2}\cdot\dfrac{\partial^2 u}{\partial\theta^2}$ ... (3)

On adding (1), (2) and (3), we get

$\dfrac{\partial^2 u}{\partial x^2} + \dfrac{\partial^2 u}{\partial y^2} + \dfrac{\partial^2 u}{\partial z^2} = \dfrac{\partial^2 u}{\partial r^2} + \dfrac{2}{r}\dfrac{\partial u}{\partial r} + \dfrac{1}{r^2}\dfrac{\partial^2 u}{\partial\theta^2} + \dfrac{\cot\theta}{r^2}\dfrac{\partial u}{\partial\theta} + \dfrac{1}{r^2\sin^2\theta}\dfrac{\partial^2 u}{\partial\phi^2} = 0$

which is the required Laplace equation in spherical coordinates.

📧 **REMARK**

• The above equation can also be written as

$$\dfrac{1}{r^2}\dfrac{\partial}{\partial r}\left(r^2\dfrac{\partial u}{\partial r}\right) + \dfrac{1}{r^2\sin\theta}\dfrac{\partial}{\partial\theta}\left(\sin\theta\dfrac{\partial u}{\partial\theta}\right) + \dfrac{1}{r^2\sin^2\theta}\dfrac{\partial^2 u}{\partial\phi^2} = 0$$

8.15.11 SOLUTION OF TWO DIMENSIONAL LAPLACE EQUATION : SEPARATION OF VARIABLES

Consider a two-dimensional Laplace equation in cartesian coordinates

$$\nabla^2 u = \dfrac{\partial^2 u}{\partial x^2} + \dfrac{\partial^2 u}{\partial y^2} = 0$$ ... (1)

Let $\quad u(x, y) = X(x)\ Y(y)$ ... (2)

be the solution of (1).

Using (2) in (1), we get $\ X''Y + Y''X = 0$

$\Rightarrow \qquad \dfrac{X''}{X} = -\dfrac{Y''}{Y} = \lambda$ (say), where λ is a separation constant.

Now, we have the following cases :

Case (i) - Let $\lambda = \mu^2, \mu$ is real. Then $\dfrac{d^2 X}{dx^2} - \mu^2 X = 0$ and $\dfrac{d^2 Y}{dy^2} + \mu^2 Y = 0$

The solution of above equations are given by $X = C_1 e^{\mu x} + C_2 e^{-\mu x}$ and $Y = C_3 \cos \mu y + C_4 \sin \mu y$

Thus, in this case, the required solution is given by $u(x, y) = X(x) \, Y(y) = (C_1 e^{\mu x} + C_2 e^{-\mu x})(C_3 \cos \mu y + C_4 \sin \mu y)$

Case (ii) - If $\lambda = 0$

Then the equation reduces to $\dfrac{d^2 X}{dx^2} = 0$ and $\dfrac{d^2 Y}{dy^2} = 0$

Integrating twice, we get $X = d_1 x + d_2$ and $Y = d_3 y + d_4$

Thus, in this case, the required solution is given by $u(x, y) = (d_1 x + d_2)(d_3 y + d_4)$

Case (iii) - Let $\lambda = -\mu^2$

Proceeding in the same way as in case (i), we get $X = e_1 \cos \mu x + e_2 \sin \mu x$ and $Y = e_3 e^{\mu y} + e_4 e^{-\mu y}$

Hence, in this case, the required solution is given by $u(x, y) = (e_1 \cos \mu x + e_2 \sin \mu x)(e_3 e^{\mu y} + e_4 e^{-\mu y})$.

◄ **REMARK**

- In all the above cases, C_i, d_i and e_i $(i = 1, \ldots, 3)$ are integration constants, which can be calculated by using the given boundary conditions.

8.15.12 SOME PARTICULAR PROBLEMS

(1) Interior Dirichlet's Problem for a Rectangle

The interior Dirichlet's problem for a rectangle is defined as follows :

PDE : $\nabla^2 u = 0$, $\quad 0 \le x \le a$, $0 \le y \le b$ **BCs :** $u(x, b) = u(a, y) = 0, u(0, y) = 0, u(x, 0) = f(x)$

(2) The Neumann Problem for a Rectangle

The Neumann problem for a rectangle is defined as follows :

PDE : $\nabla^2 u = 0, 0 \le x \le a, 0 \le y \le b$ **BCs :** $u_x(0, y) = u_x(a, y) = 0, u_y(x, 0) = 0, u_y(x, b) = f(x)$

(3) Interior Dirichlet's Problem for a Circle

The interior Dirichlet's problem for a circle is defined as follows :

PDE : $\nabla^2 u = 0, 0 \le r \le a, 0 \le \theta \le 2\pi$ **BC :** $u(a, \theta) = f(\theta), 0 \le \theta \le 2\pi$

(4) Exterior Dirichlet's Problem for a Circle

The exterior Dirichlet's problem for a circle is defined as follows :

PDE : $\nabla^2 u = 0$ **BC :** $u(a, \theta) = f(\theta)$

Here, u must be bounded as $r \to \infty$.

8.15.13 SOLUTION OF LAPLACE EQUATION OF THREE DIMENSIONAL

Consider the three dimensional Laplace equation

$$\frac{\partial^2 u}{\partial x^2} + \frac{\partial^2 u}{\partial y^2} + \frac{\partial^2 u}{\partial z^2} = 0 \qquad \ldots (1)$$

Let

$$u(x, y, z) = X(x) \, Y(y) \, Z(z) \qquad \ldots (2)$$

be the solution of (1), where X, Y and Z are functions of x, y and z, respectively.

Putting the value of u [From (2)] in (1), we get $\dfrac{X''}{X} + \dfrac{Y''}{Y} = -\dfrac{Z''}{Z}$ ... (3)

Further, since x, y and z are independent variables, equation (3) can be true if each term on each side is equal to a constant.

Now, we have the following three cases :

Case (i) . If each term in (3) is zero.

In this case, we have

On integrating each twice, we get

$$X'' = 0, Y'' = 0, Z'' = 0$$
$$X = Ax + B, \ Y = Cy + D, \ Z = Ez + F$$

Hence, we get the solution of the form

$$u(x, y, z) = (Ax + B)(Cy + D)(Ez + F) \qquad \ldots (4)$$

Case (ii). Suppose

$$\frac{X''}{X} = \lambda_1^2, \frac{Y''}{Y} = \lambda_2^2 \text{ such that } \lambda_1^2 + \lambda_2^2 = \lambda^2$$

Then, from equation (3), we can find

$$X'' - \lambda_1^2 X = 0, Y'' - \lambda_2^2 Y = 0, Z'' + \lambda^2 Z = 0$$

On solving, we get

$$X = Ae^{x\lambda_1} + Be^{-x\lambda_1} ; Y = Ce^{y\lambda_2} + De^{-y\lambda_2} \text{ and } Z = E \cos \lambda z + F \sin \lambda z$$

Hence, we get the solution of (1) is of the form

$$u(x, y, z) = (Ae^{x\lambda_1} + Be^{-x\lambda_1})(Ce^{y\lambda_2} + De^{-y\lambda_2})(E \cos \lambda z + F \sin \lambda z)$$

Case (iii). In this case, suppose that

$$\frac{X''}{X} = -\lambda_1^2, \frac{Y''}{Y} = -\lambda_2^2, \text{ and } -(\lambda_1^2 + \lambda_2^2) = \lambda^2$$

Then from (3), we can find $\quad X'' + \lambda_1^2 X = 0, \ Y'' + \lambda_2^2 Y = 0, \ Z'' - \lambda^2 Z = 0$

On solving, we get
$$X = A\cos\lambda_1 x + B\sin\lambda_1 x$$
$$Y = C\cos\lambda_2 y + D\sin\lambda_2 y$$
and
$$Z = E\, e^{\lambda z} + F e^{-\lambda z}$$

Hence, the general solution of (1) is given by
$$u(x, y, z) = \sum_{\lambda_1}\sum_{\lambda_2} (A\cos\lambda_1 x + B\sin\lambda_1 x)(C\cos\lambda_2 y + D\sin\lambda_2 y)(E\,e^{\lambda z} + F e^{-\lambda z})$$

 Solved Examples

EXAMPLE 1. *If u be a harmonic function in the interior of a rectangle $0 \le x \le a, \ 0 \le y \le b$ in the XY - plane satisfying Laplace equation*
$$\frac{\partial^2 u}{\partial x^2} + \frac{\partial^2 u}{\partial y^2} = 0 \qquad \ldots (1)$$
with boundary conditions
$$u(0, y) = 0, \ u(a, y) = 0 \qquad \ldots (2)$$
$$u(x, b) = 0, \ u(x, 0) = f(x) \qquad \ldots (3)$$
Obtain the solution of above problem.

[UPTU-2008]

SOLUTION. By the method of separation of variables, we can find a function $u(x, y)$ such that
$$u(x, y) = X(x)\, Y(y) \qquad \ldots (4)$$
Putting this value of u in equation (1), we get
$$X''Y + XY'' = 0$$
$$\Rightarrow \qquad \frac{X''}{X} = -\frac{Y''}{Y} \qquad \ldots (5)$$
For independent x, y, each side of equation (5) must be equal to the same constant say k.
Then (5) reduces to
$$X'' - kX = 0 \qquad \ldots (6)$$
and $\qquad Y'' + kY = 0 \qquad \ldots (7)$
Now, using the given boundary conditions (2) in (4), we get
$$X(0)\, Y(y) = 0 \quad \text{and} \quad X(a)\, Y(y) = 0$$
$$X(0) = 0 \quad \text{and} \quad X(a) = 0 \qquad \ldots (8)$$
$[Y(y) \ne 0$ have been taken, because otherwise $u = 0]$
Now, we have the following cases :

Case (i). Let $k = 0$. Then solution of (6) is given by
$$X(x) = Ax + B$$
Using the boundary conditions given by (8), we get $B = 0$ and $Aa + B = 0$
i.e., $\Rightarrow A = B = 0 \quad \Rightarrow X(x) = 0 \Rightarrow u = 0$
which does not satisfy the given boundary condition $u(x, 0) = f(x)$.
Hence, we reject this case (*i.e.,* $k = 0$).

Case (ii). Let $k = \lambda^2, \lambda \ne 0$. In this case, solution of equation (6) is given by
$$X(x) = Ae^{\lambda x} + Be^{-\lambda x}$$
Using the boundary conditions given by (6), we get
$A + B = 0$ and $Ae^{a\lambda} + Be^{-\lambda a} = 0$
On solving, we get
$A = B = 0 \ \Rightarrow X(x) = 0 \quad \Rightarrow \quad u = 0$
Hence, again, we reject this case.

Case (iii). Let $k = -\lambda^2, \lambda \ne 0$. In this case, the solution of (6) is given by
$$X(x) = A\cos\lambda x + B\sin\lambda x$$
Using the boundary conditions given by (8), we get
$$A = 0 \quad \text{and} \quad A\cos\lambda a + B\sin\lambda a = 0$$
On solving, we get $A = 0$ and $\sin\lambda a = 0$
Here, we have taken $B \ne 0$ because otherwise $X(x) = 0$.
Further, $\sin\lambda a = 0$

$\Rightarrow \lambda a = n\pi, i.e., \ \lambda = \dfrac{n\pi}{a}; \quad n = 1, 2, 3, \ldots\ldots$
Therefore, in this case, non-zero solution $X_n(x)$ of (6) is given by
$$X_n(x) = B_n \sin\left(\frac{n\pi x}{a}\right)$$

Further, using $\quad \mu = -\lambda^2 = -\dfrac{n^2\pi^2}{a^2}$, equation (7) becomes
$$Y'' - \left(\frac{n^2\pi^2}{a^2}\right) Y = 0$$
whose solution is given by
$$Y_n(y) = C_n e^{n\pi y/a} + D_n e^{-n\pi y/a} \qquad \ldots (9)$$
Using the given boundary conditions, we get
$$X(x)\, Y(b) = 0 \quad \Rightarrow \quad Y(b) = 0$$
$$\Rightarrow \qquad Y_n(b) = 0 \qquad \ldots (10)$$
Putting $y = b$ in (9) and using (10), we get
$$0 = C_n e^{n\pi b/a} + D_n e^{-n\pi b/a}$$
$$C_n = -\left.\left\{D_n e^{-n\pi b/a}\right\}\right/e^{n\pi b/a} \qquad \ldots (11)$$
Putting this value in (9), we get
$$Y_n(y) = \frac{D_n\left(e^{-n\pi y/a}\, e^{n\pi b/a} - e^{n\pi y/a}\, e^{-n\pi b/a}\right)}{e^{n\pi b/a}}$$
$$= \frac{D_n\left(e^{n\pi(b-y)/a} - e^{-n\pi(b-y)/a}\right)}{e^{n\pi b/a}}$$
which can also be written as
$$Y_n(y) = 2D_n e^{-n\pi b/a} \sinh\{n\pi(b - y)/a\}$$
$$\left[\because \sinh\theta = \frac{e^\theta - e^{-\theta}}{2}\right]$$
Therefore, $U_n(x, y) = X_n(x)\, Y_n(y)$
$$= F_n \sin(n\pi x/a) \sinh\{n\pi(b - y)/a\}$$
Also, the more general solution is given by
$$U(x, y) = \sum_{n=1}^{\infty} F_n \sin\left(\frac{n\pi x}{a}\right) \sinh\left[\frac{n\pi(b - y)}{a}\right]$$
Putting $y = 0$ and using given boundary conditions $u(x, 0) = f(x)$, we get
$$f(x) = \sum_{n=1}^{\infty} \left\{ F_n \sin\left(\frac{n\pi b}{a}\right)\right\} \sin\left[\frac{n\pi x}{a}\right]$$
which is the half range Fourier sine series of $f(x)$ in $(0, a)$. Hence, we get
$$\Rightarrow \quad F_n \sin\frac{n\pi b}{a} = \frac{2}{a}\int_0^a f(x)\frac{n\pi x}{a} dx$$
$$\Rightarrow \quad F_n = \frac{2}{a\sinh\left(\frac{n\pi b}{a}\right)}\int_0^a f(x)\sin\frac{n\pi x}{a} dx$$

EXAMPLE 2. *Find the steady temperature distribution in a thin plate bounded by the lines x = 0, x = a, y = 0, y = ∞. Assuming that heat can not escape from either surface; the sides x = 0, x = a being kept at temperature zero. The lower edge y = 0 is kept at f(x) and the edge y = ∞ at temperature zero.*

SOLUTION. For steady state, we know that $\dfrac{\partial u}{\partial t} = 0$

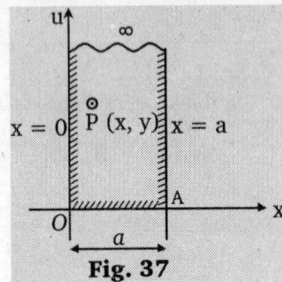

Fig. 37

Therefore the heat equation

$$\frac{\partial^2 u}{\partial x^2} + \frac{\partial^2 u}{\partial y^2} = \frac{1}{k}\frac{\partial u}{\partial t}$$

reduces to

$$\frac{\partial^2 u}{\partial x^2} + \frac{\partial^2 u}{\partial y^2} = 0 \qquad \dots (1)$$

The given boundary conditions can be written as

$$u(0, y) = 0, u(a, y) = 0 \qquad \dots (2)$$

$u(x, y) \to 0$ as $y \to \infty$ and $u(x, 0) = f(x)$ ... (3)

Using the method of separation of variables, let us suppose solution of (1) is of the form

$$u(x, y) = X(x)\, Y(y) \qquad \dots (4)$$

Putting this value in (1), we get

$$X''Y + XY'' = 0$$

$$\Rightarrow \qquad \frac{X''}{X} = -\frac{Y''}{Y} \qquad \dots (5)$$

Now, since x and t are independent, therefore each side of (5) must be equal to the same constant say k. Then (5) reduces to

$$X'' - kX = 0 \qquad \dots (6)$$

and $\qquad Y'' + kY = 0 \qquad \dots (7)$

Using (2) in (4), we get

$$X(0)\,Y(y) = 0 \quad \text{and} \quad X(a)\,Y(y) = 0$$

$$\Rightarrow \qquad X(0) = 0 \quad \text{and} \quad X(a) = 0$$

(By taking $Y(y) \neq 0$) ... (8)

Now, we have to solve equation (6) using the boundary conditions (8) under the following three cases :

Case (i). Let k = 0. In this case, solution of (6) is given by

$$u = Ax + B \qquad \dots (9)$$

Using boundary conditions given by (8) in (9), we get

$$B = 0 \text{ and } Aa + B = 0$$

On solving, we get $A = B = 0 \Rightarrow X(x) = 0, u = 0$

So, we reject this case.

Case (ii). Let $k = \lambda^2, \lambda \neq 0$. In this case, solution of (6) is given by

$$X(x) = Ae^{x\lambda} + Be^{-x\lambda}$$

Using boundary conditions (8), we get

$$0 = A + B \text{ and } 0 = Ae^{a\lambda} + Be^{-a\lambda}$$

On solving, we get $A = B = 0$

$$\Rightarrow \qquad X(x) \equiv 0 \Rightarrow u = 0$$

Hence, we reject $k = \lambda^2$.

Case (iii). Let $k = -\lambda^2, \lambda \neq 0$. In this case, solution of (6) is given by

$$X(x) = A\cos\lambda x + B\sin\lambda x$$

Using boundary conditions (8), we get

$$0 = A \text{ and } 0 = A\cos\lambda a + B\sin\lambda a$$

On solving, we get $A = 0$ and $\sin\lambda a = 0$

(By taking $B \neq 0$)

So, $\quad \sin\lambda a = 0$

$$\Rightarrow \qquad \lambda a = n\pi \quad \Rightarrow \quad \lambda = n\pi/a; \quad n = 1, 2, 3, \dots\dots$$

Therefore, non-zero solutions $X_n(x)$ of (6) are given by

$$X_n(x) = B_n \sin\left(\frac{n\pi x}{a}\right) \qquad \dots (10)$$

Using $k = -\lambda^2 = -\dfrac{n^2\pi^2}{a^2}$, equation (7) becomes

$$Y'' - \left(\frac{n^2\pi^2}{a^2}\right) Y = 0 \qquad \dots (11)$$

On solving (11), we get

$$Y_n(y) = C_n e^{n\pi y/a} + D_n e^{-n\pi y/a} \qquad \dots (12)$$

with $\quad Y_n(y) \to 0 \ as \ y \to \infty$ [Using (3)]

Therefore, we must take $C_n = 0$. Then, from (12), we have

$$Y_n(y) = D_n e^{-n\pi y/a}$$

$$\Rightarrow u_n(x, y) = X_n Y_n = E_n \sin\frac{n\pi x}{a} e^{-n\pi y/a} \quad (E_n = B_n D_n)$$

are solutions of (1) satisfying (2) and (3).

The more general solution is given by

$$u(x, y) = \sum_{n=1}^{\infty} E_n \sin\frac{n\pi x}{a} e^{-n\pi y/a}$$

Now, putting $y = 0$ and using $u(x, 0) = f(x)$, we get

$$f(x) = \sum_{n=1}^{\infty} E_n \sin\frac{n\pi x}{a}$$

which is a Fourier sine series and therefore

$$E_n = \frac{2}{a}\int_0^a f(x) \sin\frac{n\pi x}{a} dx .$$

EXAMPLE 3. *Find the steady state temperature in a rectangular plate bounded by the lines x = 0, x = a, y = 0 and y = b if the edge y = 0 is insulated, the edge x = 0 and x = a are kept at 0°C and the edge y = b is kept at temperature f(x).*

SOLUTION. We know that the temperature $u(x, y)$ in steady state in two-dimensional plate is governed by

$$\frac{\partial^2 u}{\partial x^2} + \frac{\partial^2 u}{\partial y^2} = 0 \qquad \dots (1)$$

As per given, the boundary conditions are

$$u(0, y) = u(a, y) = 0, \text{ for } 0 \leq y \leq b \qquad \dots(2)$$

Also $\quad \left(\dfrac{\partial u}{\partial y}\right)_{y=0} = 0$

(Because the edge $y = 0$ is insulated for $0 \leq x \leq a$) ... (3a)

and $u(x, b) = f(x)$, for $0 \leq x \leq a$ ... (3b)

Now proceeding same as in example (4), we get

$$u(x, y) = \sum_{n=1}^{\infty} u_n(x, y)$$

$$= E_n \sin\left(\frac{n\pi x}{a}\right)\cosh\left(\frac{n\pi y}{a}\right) \qquad \dots (4)$$

Putting $y = b$ in (4) and using (3b), we get

$$f(x) = \sum_{n=1}^{\infty}\left(E_n\cosh\frac{n\pi b}{a}\right)\sin\frac{n\pi x}{a}$$

which is a half range Fourier sine series of $f(x)$ in $(0, a)$. Hence, we get

$$E_n \cosh \frac{n\pi b}{a} = \frac{2}{a} \int_0^a f(x) \sin \frac{n\pi x}{a} dx$$

$$\Rightarrow \qquad E_n = \frac{2}{a \cosh\left(\frac{n\pi b}{a}\right)} \int_0^a f(x) \sin \frac{n\pi x}{a} dx .$$

EXAMPLE 4. *A rectangular metal plate is bounded by the lines $x = 0$, $x = a$, $y = 0$ and $y = b$. The three sides $x = 0$, $x = a$ and $y = b$ are insulated and the side $y = a$ is kept at temperature $u_0 \cos\left(\dfrac{\pi x}{a}\right)$.*

Find the steady state temperature at any point of the plate.

SOLUTION. The governing equation is given by

$$\frac{\partial^2 u}{\partial x^2} + \frac{\partial^2 u}{\partial y^2} = \qquad \ldots (1)$$

Since, $x = 0, x = a$ and $y = b$ are insulated.

Therefore, we have

$$\left(\frac{\partial u}{\partial x}\right)_{x=0} = 0, \quad \left(\frac{\partial u}{\partial x}\right)_{x=a} = 0 \qquad \ldots (2)$$

$$\left(\frac{\partial u}{\partial y}\right)_{y=a} = 0 \qquad \ldots (2)$$

and $\quad u(x, a) \leq \quad \cos\left(-\!\!-\right) \qquad \ldots (3)$

Using the method of separation of variables, suppose (1) has a solution of the form

$$u(x, y) = X(x) \, Y(y) \qquad \ldots (4)$$

Using (4) in (1), we get $X''Y + XY'' = 0$

$$\Rightarrow \qquad \frac{X''}{X} = -\frac{Y''}{Y} = k \text{ (say)} \qquad \ldots (5)$$

$$\Rightarrow \quad X'' - kX = 0 \qquad \ldots (6)$$

and $\quad Y'' + kY = 0 \qquad \ldots (7)$

Also, from (4), $\quad \dfrac{\partial u}{\partial x} = X'(x) \, Y(y) \qquad \ldots (8)$

Using (2) in (8), we get

$$X'(0) = 0 \text{ and } X'(a) = 0 \qquad \ldots (9)$$

Now, we have the following three cases :

Case (i). Let $k = 0$. In this case, solution of equation (6) is given by

$$X(x) = Ax + B \qquad \ldots (10)$$

$$\Rightarrow \qquad X'(x) = A$$

Using (9), we get $A = 0$.

Therefore $X(x) = B$

Case (ii). Let $k = \lambda^2, \lambda \neq 0$. In this case, the solution of (6) is given by

$$X(x) = Ae^{x\lambda} + Be^{-x\lambda}$$

$$\Rightarrow \qquad X'(x) = \lambda(Ae^{\lambda x} - Be^{-\lambda x})$$

Using (9) in the above equation, we get

$$\lambda(A - B) = 0 \text{ and } \lambda (Ae^{a\lambda} - Be^{-a\lambda}) = 0$$

On solving, we get $A = B = 0$

$$\Rightarrow \qquad X(x) \equiv 0$$

$$\Rightarrow \qquad u \equiv 0$$

Case (iii) Let $k = -\lambda^2, \lambda \neq 0$. In this case, solution of (6) is given by

$$X(x) = A\cos\lambda x + B\sin\lambda x \qquad \ldots (11)$$

$$\Rightarrow \qquad X'(x) = -A\lambda \sin\lambda x + B\lambda \cos\lambda x \qquad \ldots (12)$$

Using (9) in (10), we get

$$0 = B\lambda$$

and $\qquad 0 = -A\lambda \sin\lambda a + B\lambda \cos\lambda a \qquad \ldots (13)$

Now, letting $\lambda \neq 0$ and $A \neq 0$, equation (12) gives

$$B = 0 \text{ and } \sin\lambda a = 0$$

$$B = 0 \text{ and } \lambda = \frac{n\pi}{a}, \ n = 1, 2, 3, \ldots \qquad \ldots (14)$$

Therefore, non-zero solutions of (10) is given by

$$X(x) = A\cos\left(\frac{n\pi x}{a}\right); \quad n = 1, 2, 3, \ldots \qquad \ldots (15)$$

$\therefore$ All non-zero solution of (6) are given by

$$X_n(x) = A_n \cos\left(\frac{n\pi x}{a}\right), \ n = 0, 1, 2, 3, \ldots \qquad \ldots (16)$$

Now putting $k = -\lambda^2 = -\dfrac{n^2\pi^2}{a^2}$ in (7), we get

$$Y'' - \left(\frac{n^2\pi^2}{a^2}\right) Y = 0$$

whose general solution is given by

$$Y_n(y) = C_n e^{n\pi x/a} + D_n e^{-n\pi y/a} \qquad \ldots (17)$$

Also, from (4), we have

$$\frac{\partial u}{\partial y} = X(x) \, Y'(y) \qquad \ldots (18)$$

Putting $y = b$ in (17) and using (3), we get

$$0 = X(x) \, Y'(b)$$

$$\Rightarrow \qquad Y'(b) = 0 \qquad \text{[By letting } X(x) \neq 0]$$

$$\Rightarrow \qquad Y_n'(b) = 0 \qquad \ldots (19)$$

From (16), we have

$$Y_n'(y) = C_n \left(\frac{n\pi}{a}\right) e^{n\pi y/a} - D_n \left(\frac{n\pi}{a}\right) e^{-n\pi y/a} \quad \ldots (20)$$

Putting $y = b$ in (19) and using (18), we get

$$0 = \left(\frac{n\pi}{a}\right) (C_n e^{n\pi b/a} - D_n e^{-n\pi b/a})$$

$$\Rightarrow \qquad D_n = \left(C_n e^{n\pi b/a}\right) \Big/ e^{-n\pi b/a}$$

Putting this value in (16), we get

$$Y_n(y) = C_n \left[\frac{(e^{n\pi y/a} e^{-n\pi b/a} + e^{-n\pi y/a} e^{n\pi b/a})}{e^{-n\pi b/a}}\right]$$

$$\Rightarrow Y_n(y) = C_n e^{n\pi b/a} \{e^{n\pi(b-y)/a} + e^{-n\pi(b-y)/a}\}$$

$$= 2C_n e^{n\pi b/a} \cosh\left\{\frac{n\pi(b-y)}{a}\right\}$$

Thus, $u_n(x, y) = X_n(x) \, Y_n(y)$

$$= E_n \cos\frac{n\pi x}{a} \cosh\left\{\frac{n\pi(b-y)}{a}\right\}$$

Further, consider the more general solution given by

$$u(x, y) = \sum_{n=0}^{\infty} u_n(x, y)$$

$$= \sum_{n=0}^{\infty} E_n \cos\left(\frac{n\pi x}{a}\right) \cosh\left\{\frac{n\pi(b-y)}{a}\right\} \quad \ldots (21)$$

Putting $y = b$ and using (3), we get

$$u_0 \cos\left(\frac{\pi x}{a}\right) = \sum_{n=0}^{\infty} E_n \cos\left(\frac{n\pi x}{a}\right) \cosh\left\{\frac{n\pi(b-a)}{a}\right\}$$

$$= E_0 + E_1 \cos\left(\frac{\pi x}{a}\right) \cosh\left\{\frac{\pi(b-a)}{a}\right\}$$

$$+ E_2 \cos\left(\frac{2\pi x}{a}\right) \cosh\left\{\frac{2\pi(b-a)}{a}\right\}$$

$$+ E_3 \cos\left(\frac{3\pi x}{a}\right) \cosh\left\{\frac{3\pi(b-a)}{a}\right\} + \ldots$$

On equating the coefficients of like terms, we get $E_0 = 0$.

$$E_1 \cosh\left\{\frac{\pi(b-a)}{a}\right\} = u_0$$

and $\qquad E_n = 0 \quad$ for $\quad n \geq 1$.

Finally, putting this value in (20), we get

$$u(x, y) = u_0 \sec h\left\{\frac{(b-a)\pi}{a}\right\}$$
$$\cos\left(\frac{\pi x}{a}\right) \cosh\left\{\frac{(b-y)\pi}{a}\right\}.$$

EXAMPLE 5. *An infinitely long uniform plate is bounded by two parallel edges and an end at right angles to them. The breadth is π..The end is maintained at 100°C at all points and the other edges are at 0°C. Find the steady state temperature function u(x, y).*

(UPTU–2008, PTU–2005, JNTU–2005, 08)

SOLUTION. The governing equation for the steady state temperature is given by

$$\frac{\partial^2 u}{\partial x^2} + \frac{\partial^2 u}{\partial y^2} = 0 \qquad \dots (1)$$

with boundary conditions

$$u(0, y) = u(\pi, y) = 0 \text{ for all } y \geq 0 \qquad \dots (2)$$

$$\left.\begin{array}{l} u(x, y) \to 0 \text{ as } y \to \infty, \text{for } 0 \leq x \leq \pi \\ \text{and } u(x, 0) = f(x) = 100, \quad \text{for } 0 \leq x \leq \pi \end{array}\right\} \qquad \dots (3)$$

Now proceeding same as in example (2) by taking $a = \pi$, we get

$$u(x, y) = \sum_{n=1}^{\infty} E_n \sin nx \, e^{-ny} \qquad \dots (4)$$

where $E_n = \frac{2}{\pi} \int_0^{\pi} f(x) \sin nx \, dx \qquad \dots (5)$

$$= \frac{2}{\pi} \int_0^{\pi} 100 \sin nx \, dx = \frac{200}{\pi}\left[-\cos nx\right]_0^{\pi}$$

$$= \frac{200}{\pi}[1 - (-1)^n]$$

$$= \begin{cases} 0 \text{ if } n \text{ is even} \\ \dfrac{400}{(2m-1)\pi}, \text{ if } n = 2m-1, \ m = 1, 2, 3, \dots \end{cases}$$

Finally, putting all these values in (4), we get

$$u(x, y) = \frac{400}{\pi} \sum_{n=1}^{\infty} \frac{\sin(2m-1)x}{2m-1} e^{-(2m-1)y}.$$

EXAMPLE 6. *A thin rectangular homogeneous thermally conducting plate lies in the xy-plane defined by $0 \leq x \leq a$, $0 \leq y \leq b$. The edge y = 0 is held at its temperature Tx(x – a), where T is a constant, while the remaining edges are held at 0°C. The other faces are insulated and no internal sources and sinks are present. Find the steady state temperature inside the plate.*

SOLUTION. As per given, we have no heat sources and sinks are present in the plate. Thus, the steady state temperature function is the solution of

$$\frac{\partial^2 u}{\partial x^2} + \frac{\partial^2 u}{\partial y^2} = 0 \qquad \dots (1)$$

with boundary conditions

$$u(0, y) = 0, u(a, y) = 0, u(x, b) = 0, u(x, 0) = Tx(x - a) \qquad \dots (2)$$

Now proceeding same as example (1), we get

$$u(x, y) = \sum_{n=1}^{\infty} A_m \sin\left(\frac{n\pi x}{a}\right) \sinh\left(\frac{n\pi(y-b)}{a}\right)$$

where, $A_n \sinh\left(\frac{-n\pi b}{a}\right) = \frac{2}{a}\int_0^a f(x) \sin\left(\frac{n\pi x}{a}\right) dx$

Using the boundary conditions

$$U(x, 0) = Tx(x - a) = f(x)$$

we get $A_n \sinh\left(\dfrac{-n\pi b}{a}\right) = \dfrac{2}{a}\int_0^a Tx(x-a) \sin\left(\dfrac{n\pi}{a}x\right) dx$

$$= \frac{2T}{a}\int_0^a x(x-a) \sin\left(\frac{n\pi x}{a}\right) dx$$

$$= \frac{-a}{n\pi} \cdot \frac{2T}{a}\left[\int_0^a x(x-a) . d\left\{\cos\left(\frac{n\pi x}{a}\right)\right\}\right]$$

$$= -\frac{2T}{n\pi}\left[(x-a)\cos\left(\frac{n\pi x}{a}\right)\right]_0^a - \frac{a}{n\pi}\int_0^a (2x-a)d\left\{\sin\left(\frac{n\pi x}{a}\right)\right\}$$

$$= \frac{2aT}{n^2\pi^2}\left[(2x-a)\sin\left(\frac{n\pi x}{a}\right)\right]_0^a - \int_0^a 2\sin\left(\frac{n\pi x}{a}\right) dx$$

$$= \frac{2aT}{n^2\pi^2}\left\{a \sin n\pi + \frac{2a}{n\pi}\left[\cos\left(\frac{n\pi x}{a}\right)\right]_0^a\right\}$$

$$= \frac{2aT}{n^2\pi^2} \cdot \frac{2a}{n\pi}\cos(n\pi - 1) = \frac{4a^2 T}{n^3\pi^3}[(-1)^n - 1]$$

Hence, the required temperature function is given by

$$u(x, y) = \sum_{n=1}^{\infty} \frac{4Ta^2}{n^3\pi^3} \operatorname{cosech}\left(-\frac{n\pi b}{a}\right)$$
$$[(-1)^n - 1]\sin\left(\frac{n\pi x}{a}\right) \sinh\left[\frac{n\pi(y-b)}{a}\right].$$

EXAMPLE 7. *Solve* $\dfrac{\partial^2 u}{\partial x^2} + \dfrac{\partial^2 u}{\partial y^2} = 0$, $\quad 0 \leq x \leq a$, $\quad 0 \leq y \leq b$ *subject to the boundary conditions u(0, y) = 0,* $u(x, 0) = 0$, $u(x, b) = 0$, $\dfrac{\partial u}{\partial x}(a, y) = T \sin^3 \dfrac{\pi y}{a}$.

SOLUTION. Proceeding same as example (3), we get

$$u(x, y) = (C_1 e^{\lambda x} + C_2 e^{-\lambda x})(C_3 \cos \lambda y + C_4 \sin \lambda y)$$

Using $u(x, 0) = 0$, we get

$$u(x, 0) = 0 = C_4 \sin \lambda y (C_1 e^{\lambda x} + C_2 e^{-\lambda x})$$

Again, using $u(x, b) = 0$, we get

$$0 = C_4 \sin \lambda b (C_1 e^{\lambda x} + C_2 e^{-\lambda x})$$

Now, $\quad C_4 \neq 0 \Rightarrow \sin \lambda b = 0$

$$\Rightarrow \quad \lambda b = n\pi \Rightarrow \quad \lambda = \frac{n\pi}{b}; \quad n = 1, 2, 3, \dots\dots$$

$$\Rightarrow \quad u(x, y) = C_4 \sin\left(\frac{n\pi y}{b}\right)(C_1 e^{\lambda x} + C_2 e^{-\lambda x})$$

which can also be written as

$$u(x, y) = \sin\left(\frac{n\pi y}{b}\right) \text{ (By renaming the constants)}$$

$$\left[A \exp\left(\frac{n\pi x}{b}\right) - B \exp\left(\frac{-n\pi x}{b}\right)\right].$$

$n = 1, 2, 3, \dots$

Now, using $(0, y) = 0$, we get

$$0 = \sin\left(\frac{n\pi y}{b}\right)(A + B)$$

$\Rightarrow A + B = 0, \quad i.e., \quad A = -B$

Therefore,

$u(x, y) = A\sin\left(\frac{n\pi y}{b}\right)\left[\exp\left(\frac{n\pi x}{b}\right) - \exp\left(\frac{-n\pi x}{b}\right)\right]$

$\quad = 2A\sin\left(\frac{n\pi y}{b}\right)\sinh\left(\frac{n\pi x}{b}\right), \quad n = 1, 2, 3, \dots \quad \dots (1)$

$\Rightarrow \frac{\partial u}{\partial x} = 2A\frac{n\pi}{b}\sin\left(\frac{n\pi y}{b}\right)\cosh\left(\frac{n\pi x}{b}\right)$

Putting $x = a$ and using last boundary condition, we have

$T\sin^3\frac{\pi y}{a} = 2A\frac{n\pi}{b}\sin\left(\frac{n\pi y}{b}\right)\cosh\left(\frac{n\pi a}{b}\right)$

$\Rightarrow \quad 2A = \dfrac{T\sin^3\dfrac{\pi y}{a}}{\dfrac{n\pi}{b}\sin\left(\dfrac{n\pi y}{b}\right)\cosh\left(\dfrac{n\pi a}{b}\right)}$

Putting this value in (1), the required steady state temperature function is given by

$u(x, y) = \frac{bT}{n\pi}\sin^3\frac{\pi y}{a}\,\mathrm{sec}\,h\frac{n\pi}{b}a\sinh\left(\frac{n\pi x}{b}\right)$

Hence, the general solution is given by

$u(x,y) = \sum_{n=1}^{\infty}\frac{bT}{n\pi}\sin^3\frac{\pi y}{a}\,\mathrm{sech}\left(\frac{n\pi a}{b}\right)\sinh\left(\frac{n\pi x}{b}\right).$

Exercise-8.70

1. Find the steady state temperature distribution in a thin rectangular plate bounded by the lines $x = 0, x = a, y = 0, y = b$. The edges $x = 0, x = a, y = 0$ are kept at temperature zero while the edge $y = b$ is kept at $100°C$.

2. Evaluate the steady temperature in a rectangular plate of length a and width b, the sides of which are kept at temperature zero, the lower end is kept at temperature $f(x)$ and the upper edge is kept insulated.

3. Show that the velocity potential for an irrational flow of an incompressible fluid satisfies the Laplace equation.

4. Find the steady state temperature distribution in a rectangular plate of sides a and b insulated at the lateral surface and satisfying the boundary conditions
$u(0, y) = u(a, y) = 0$ for $0 \le y \le b$
$\qquad u(x, b) = 0$
and $\qquad u(x, 0) = x(a - x)$ for $0 \le y \le a$

5. A rectangular plate with insulated surface 8 cm wide and so long compared to its width that it can be considered infinite in the length without introducing an appreciable error. If the temperature along the short edge $y = 0$ is given by $u(x, 0) = 100\sin\left(\frac{\pi x}{8}\right)$, in $0 < x < 8$, while the two long edges $x = 0$ and $x = 8$ as well as the other short edges are kept at $0°C$. Find the steady state temperature function $u(x, y)$.

6. By separating the variables, show that $\frac{\partial^2 V}{\partial x^2} + \frac{\partial^2 V}{\partial y^2} = 0$ has solutions of the form $A\exp(\pm nx \pm ixy)$, where A and n are constants. Deduce that the function of the form
$V(x, y) = \sum_r A_r \sin\left(\frac{rxy}{a}\right)e^{-(r\pi x)/a}, \quad x \ge 0, \ 0 \le y \le \infty.$
where A_r are constants or plane harmonic functions satisfying the conditions $V(x, 0) = V(x, a) = 0$ and $V(x, y) \to 0$ as $x \to \infty$.

7. The diameter of a semi-circular plate of radius a is kept at $0°C$ and the temperature at the semi-circular boundary is $T°C$. Show that the steady state temperature in the plate is given by
$u(r, \theta) = \frac{4T}{\pi}\sum_{n=1}^{\infty}\frac{1}{2n-1}\left(\frac{r}{a}\right)^{2n-1}\sin(2n-1)\theta$ \qquad (GBTU (CO) 2011)

8. Solve: $\frac{\partial^2 V}{\partial r^2} + \frac{1}{r}\frac{\partial V}{\partial r} + \frac{1}{r^2}\frac{\partial^2 V}{\partial\theta^2} = 0$ with boundary conditions.
(i) V is finite when $r \to 0$
(ii) $V = \sum C_n \cos n\theta$ on $r = a$ \qquad (GBTU-2010)

9. Solve the Laplace equation $\frac{\partial^2 u}{\partial x^2} + \frac{\partial^2 u}{\partial y^2} = 0$ subject to the conditions
$u(0, y) = u(l, y) = u(x, 0) = 0$ and $u(x, a) = \frac{\sin n\pi x}{l}$
(VTU-2011, JNTU-2006, KERALA M. TECH.-2005, UPTU-2004)

Hints to Selected Problems

1. In this problem, we consider the steady state temperature, *i.e.*, it does not depend upon time. Thus the flow is governed by the Laplace equation given by
$\frac{\partial^2 V}{\partial x^2} + \frac{\partial^2 V}{\partial y^2} = 0$ \qquad \dots (1)
As per given, the boundary conditions are
$u(0, y) = 0, \ u(a, y) = 0$ \qquad \dots (2)
$u(x, 0) = 0, \ u(x, b) = 100$ \qquad \dots (3)
Using the method of separation of variables, suppose (1) has a solution of the form
$u(x, y) = X(x)\,Y(y)$ \qquad \dots (4)
Putting this value in equation (1), we get
$X''Y + XY'' = 0$
$\frac{X''}{X} = -\frac{Y''}{Y}$ \qquad \dots (5)
Now, for independent x and t, each side of (5) must be equal to

the constant, say k. Then we have
$\qquad X'' - kX = 0$ \qquad \dots (6)
and $\qquad Y'' + kY = 0$ \qquad \dots (7)
Also, from (2) and (4), we have
$X(0)Y(a) = 0$ and $X(a)\,Y(y) = 0$
$\qquad X(0) = 0$ and $\qquad X(a) = 0$ \qquad \dots (8)
Here, we must take $Y(y) \ne 0$ because otherwise $u = 0$, which does not satisfy the given conditions.
Now, we shall discuss the following three cases :
Case (I). Let $k = 0$. Then solution of equation (6) is given by
$\qquad X(x) = Ax + B$
Using boundary conditions (8), we get
$\qquad B = 0$ and $Aa + B = 0$
$A = B = 0 \ \Rightarrow \ X(x) = 0 \ \Rightarrow \ u = 0$
Which does not satisfy given condition (3). Thus, we reject this case, *i.e.*, $k = 0$.

Case (II). Let $k = \lambda^2, \lambda \neq 0$. In this case, solution of (6) is given by
$$X(x) = Ae^{x\lambda} + Be^{-x\lambda}$$
Using the boundary conditions (8), we get
$$A + B = 0 \text{ and } 0 = e^{a\lambda} + e^{-a\lambda}.$$
On solving, we get
$$A = B = 0 \Rightarrow X(x) = 0 \Rightarrow u = 0$$
Thus, we reject this case also.

Case (III). Let $k = -\lambda^2, \lambda \neq 0$. In this case, the solution of equation (6) is given by
$$X(x) = A\cos\lambda x + B\sin\lambda x$$
Using the boundary conditions (8), we get
$$A = 0 \text{ and } A\cos\lambda a + B\sin\lambda a = 0$$
On solving, we get
$$A = 0 \text{ and } \sin\lambda a = 0 \qquad [\text{By taking } B \neq 0]$$
Further, $\sin\lambda a = 0$
$$\Rightarrow \qquad \lambda = \frac{n\pi}{a}; \quad n = 1, 2, 3, \ldots$$
Thus, the non-zero solution $X_n(x)$ of (6) is given by
$$X_n(x) = B_n \sin\left(\frac{n\pi x}{a}\right)$$
Further, using $k = -\lambda^2 = -\frac{n^2\pi^2}{a^2}$, equation (7) reduces to
$$Y'' - \left(\frac{n^2\pi^2}{a^2}\right)Y = 0$$
The general solution of this equation is given by
$$Y_n(y) = C_n e^{n\pi y/a} + D_n e^{-n\pi y/a} \qquad \ldots (9)$$
Using (3) and (4)
$$0 = X(x)Y(0)$$
$$\Rightarrow \qquad Y(0) = 0 \quad \text{Now,} \quad Y(0) = 0$$
$$\Rightarrow \qquad Y_n(0) = 0 \qquad \ldots (10)$$
Putting $y = 0$ in (9) and using (10), we get
$$0 = C_n + D_n$$
$$D_n = -C_n$$
Then equation (7) becomes
$$Y_n(y) = C_n(e^{n\pi y/a} - e^{-n\pi y/a}) = 2\sinh\left(\frac{n\pi y}{a}\right)$$
Therefore,
$$u(x, y) = X_n(x)Y_n(y) = E_n \sin(n\pi x/a)\sinh(n\pi y/a)$$
where, $E_n = 2B_n C_n$ are arbitrary constants.
The more general solution is given by
$$u(x, y) = \sum_{n=1}^{\infty} E_n \sin\left(\frac{n\pi x}{a}\right)\sinh\left(\frac{n\pi y}{a}\right) \qquad \ldots (11)$$
Putting $y = b$ and using given condition $u(x, b) = 100$, we get
$$100 = \sum_{n=1}^{\infty} E_n \sin\left(\frac{n\pi x}{a}\right)\sinh\left(\frac{n\pi b}{a}\right)$$
which is a Fourier sine series.
Thus, $E_n \sinh\dfrac{n\pi b}{a} = \dfrac{2}{a}\int_0^a 100.\sin\left(\dfrac{n\pi x}{a}\right)dx$
$$= \frac{200}{a}\left[-\frac{\cos(n\pi x/a)}{(n\pi/a)}\right]_0^a$$
So, $E_n = \dfrac{200}{n\pi}\left[1 - (-1)^n\right]\text{cosech}\left(\dfrac{n\pi b}{a}\right)$
$$= \begin{cases} 400\,\text{cosech}\{(2m-1)\pi b/a \text{ if } n = 2m \\ (2m-1)\pi \qquad\qquad\qquad \text{if } n = 2m-1 \end{cases}$$
Putting this value in (11), we get
$$u(x, y) = \sum_{m=1}^{\infty} E_{2m-1} \sin\frac{(2m-1)\pi x}{a}\sinh\frac{(2m-1)\pi y}{a}$$

$$\Rightarrow \qquad u(x, y) = \frac{400}{\pi}\sum_{m=1}^{\infty}\frac{1}{2m-1}\sin\frac{(2m-1)\pi x}{a}$$
$$\sinh\frac{(2m-1)\pi y}{a}\text{cosech}\frac{(2m+1)\pi b}{a}$$

2. We know that temperature $u(x, y)$ in steady state in two-dimensional plate is governed by the Laplace equation.
$$\frac{\partial^2 u}{\partial x^2} + \frac{\partial^2 u}{\partial y^2} = 0 \qquad \ldots (1)$$
According to the given conditions, we have
$$u(0, y) = 0, \ u(a, y) = 0 \qquad \ldots (2)$$
$$\left(\frac{\partial u}{\partial y}\right)_{y=b} = 0 \text{ and } u(x, 0) = f(x) \qquad \ldots (3)$$
Using the method of separation of variables, suppose
$$u(x, y) = X(x)Y(y) \qquad \ldots (4)$$
is solution of (1).
Putting this value of u from (4) in (1), we get
$$X''Y + XY'' = 0$$
$$\Rightarrow \qquad \frac{X''}{X} = -\frac{Y''}{Y} \qquad \ldots (5)$$
Now, since x and t are independent, therefore each side of (5) must be equal to the same constant, say k. Then (5) reduces to
$$X'' - KX = 0 \qquad \ldots (6)$$
and $\qquad Y'' + KY = 0 \qquad \ldots (7)$
with $\qquad X(0) = 0 \text{ and } X(a) = 0. \qquad \ldots (8)$
Now, there are following three cases :

Case (I). Let $k = 0$. In this case, solution of (6) is given by
$$X(x) = Ax + B. \qquad \ldots (9)$$
Using given conditions (8), we get
$$A = B = 0.$$
Hence, we reject this case.

Case (II). Let $k = \lambda^2, \lambda \neq 0$. In this case, solution of (6) is given by
$$X(x) = Ae^{x\lambda} + Be^{-x\lambda} \qquad \ldots (10)$$
Again, using boundary conditions (8) in (10), we get
$$A + B = 0 \text{ and } Ae^{a\lambda} + Be^{-a\lambda} = 0$$
$$\Rightarrow \qquad A = B = 0 \Rightarrow X(x) = 0 \Rightarrow u = 0.$$
Hence, we reject this case also.

Case (III). Let $k = -\lambda^2, \lambda \neq 0$. Then solution of (6) is given by
$$X(x) = A\cos\lambda x + B\sin\lambda x$$
Using boundary conditions (8), we get
$$A = 0 \text{ and } A\cos\lambda a + B\sin\lambda a = 0$$
$$A = 0 \text{ and } \sin\lambda a = 0 \qquad [\text{By taking } B \neq 0]$$
Now, $\sin\lambda a = 0$
$$\Rightarrow \qquad \lambda a = n\pi \Rightarrow \lambda = \frac{n\pi}{a}; \quad n = 1, 2, 3, \ldots$$
Thus, non-zero solution $X_n(x)$ of (6) is given by
$$X_n(x) = B_n \sin\left(\frac{n\pi x}{a}\right)$$
Now, $\qquad k = -\lambda^2 = -\dfrac{n^2\pi^2}{a^2}$,
Then, equation (7) reduces to $Y'' - \left(\dfrac{n^2\pi^2}{a^2}\right)Y = 0$ whose general solution is given by
$$Y_n(y) = C_n e^{n\pi y/a} + D_n e^{-n\pi y/a} \qquad \ldots (11)$$
From (4), we can find
$$\frac{\partial u}{\partial y} = X(x)Y'(y) \Rightarrow \left(\frac{\partial u}{\partial y}\right)_{y=b} = X(x)\,Y'(b)$$
Using (3), we get
$$X(x).Y'(b) = 0$$

$\Rightarrow \qquad Y'(b) = 0$ ... (12)

Differentiating (11) w.r.t. y, we get

$$Y'_n(y) = \left(\frac{n\pi}{a}\right) C_n e^{n\pi y/a} - \left(\frac{n\pi}{a}\right) D_n e^{-n\pi y/a} \qquad ...(13)$$

Putting $y = b$ and using (12), we get

$$0 = \left(\frac{n\pi}{a}\right)\left(C_n e^{n\pi b/a} - D_n e^{-n\pi b/a}\right)$$

$$D_n = C_n e^{n\pi b/a} / e^{-n\pi b/a}$$

Putting this value in (11), we get

$$Y_n(y) = \frac{C_n(e^{n\pi y/a} - e^{-n\pi y/a} + e^{-n\pi y/a} e^{n\pi b/a}}{e^{-n\pi b/a}}$$

$$Y_n(y) = C_n e^{n\pi b/a}\left\{e^{n\pi(b-y)/a} + e^{-n\pi(b-y)/a}\right\}$$

$$= 2C_n\, e^{n\pi b/a} \cosh\left\{\frac{n\pi(b-y)}{a}\right\}$$

Thus, $\quad u_n(x,y) = X_n(x).Y_n(y)$

$$= E_n.\sin\frac{n\pi x}{a}\cosh\left\{\frac{n\pi(b-y)}{a}\right\} \qquad ...(14)$$

Now, the more general solutions are given by

$$u(x,y) = \sum_{n=1}^{\infty} u_n(x,y)$$

$$= \sum_{n=1}^{\infty} E_n \sin\left(\frac{n\pi x}{a}\right)\cosh\left(\frac{n\pi(b-y)}{a}\right) \qquad ...(15)$$

Now, putting $y = 0$ in (15) and using (3), we get

$$f(x) = \sum_{n=1}^{\infty}\left\{E_n \cosh\left(\frac{n\pi b}{a}\right)\right\}\sin\left(\frac{n\pi x}{a}\right)$$

which is the half range Fourier sine series of $f(x)$ in $(0,a)$. Therefore, we get

$$E_n = \frac{2}{a\cosh\left(\dfrac{n\pi b}{a}\right)} \int_0^a f(x).\sin\frac{n\pi x}{a}\, dx$$

3. Let S be a closed surface enclosed a volume V.

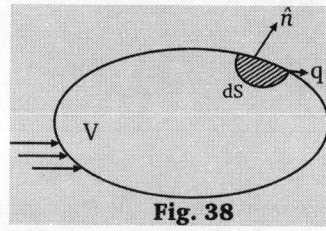

Fig. 38

Let ρ be the density of the fluid. If $\hat{n}$ be a unit normal vector drawn from outside to the surface element ds and $\bar{q}$ be the velocity of the fluid at that point. Then

Inward normal velocity $= -\bar{q}\,\hat{n}$

$\therefore$ Mass of the fluid entering per unit time through the element dS is $(-\bar{q}.\hat{n})\,dS$.

Thus the mass of the fluid entering in the surface S in the unit time is $= -\int\int_S \rho\,(\bar{q}.\hat{n})\,dS$.

Further, the mass of the fluid within S is $= \int\int\int_S \rho dV$

Rate of mass of fluid increasing in S is given by :

$$\frac{\partial}{\partial t}\iiint_V \rho\, dV = \iiint_V \frac{\partial\rho}{\partial t}\, dV.$$

Using the law of conservation of mass, we have

$$\iiint_V \frac{\partial\rho}{\partial t}\, dV = -\iint_S \rho(\bar{q}\,\hat{n})\, dS$$

$$= \iiint_V \nabla(\rho\bar{q})\, dV$$

[By Gauss divergence theorem]

$$\Rightarrow \iiint_V \left[\frac{\partial\rho}{\partial t} + \nabla(\rho\bar{q})\right] dV = 0$$

$$\Rightarrow \frac{\partial\rho}{\partial t} + \nabla(\rho\bar{q}) = 0, \text{ which is known as equation of continuity.}$$

For an incompressible fluid :

$$\rho = \text{constant}$$
$$\nabla.\bar{q} = 0$$

Also, if flow is irrotational

$\bar{q} = -\nabla\phi$, where ϕ is a potential function.

$\nabla.q = \nabla.\nabla\phi = \nabla\phi = 0$, which is a Laplace equation.

4. Proceeding same as example (1) and using $f(x) = x(a-x)$, we get the required temperature.

$$u(x,y) = \sum_{n=1}^{\infty} E_n \sin\frac{n\pi x}{a}\sinh\frac{n\pi(b-y)}{a}$$

where $\quad E_n = \dfrac{2}{a\sinh\left(\dfrac{n\pi b}{a}\right)}\int_0^a f(x)\sin\frac{n\pi x}{a}dx$

$$= \frac{2}{a\sinh\left(\dfrac{n\pi b}{a}\right)}\int_0^a (ax - x^2)\sin\frac{n\pi x}{a}dx$$

$$= \frac{2}{a\sinh\left(\dfrac{n\pi b}{a}\right)}\left[(ax - x^2)\left(-\frac{a}{n\pi}\right)\cos\frac{n\pi x}{a}\right.$$

$$\left.-(a - 2x)\left(-\frac{a^2}{n^2\pi^2}\right)\sin\frac{n\pi x}{a} + (-2)\left(\frac{a^3}{n^3\pi^3}\right)\cos\frac{n\pi x}{a}\right]_0^a$$

$$= \frac{2}{a\sinh\left(\dfrac{n\pi b}{a}\right)}\left[\frac{-2a^3(-1)^n}{n^3\pi^3} + \frac{2a^3}{n^3\pi^3}\right]$$

$$= \frac{4a^2}{n^3\pi^3}[1 - (-1)^n]\text{cosec}\frac{n\pi b}{a}$$

$$= \begin{cases} 0, & \text{if } n = 2m \text{ (even) and } m = 1,2,3,... \\[2mm] \left\{\dfrac{8a^2}{\pi^3(2m-1)^3}\right\}\text{cosech}\left\{\dfrac{(2m-1)\pi b}{a}\right\}, \\[2mm] & \text{if } n = 2m - 1 \text{ and } m = 1,2,3,... \end{cases}$$

Putting this value in (1), we get

$$u(x,y) = \frac{8a^2}{\pi^3}\sum_{n=1}^{\infty}\frac{1}{(2m-1)^3}\sin\frac{(2m-1)\pi x}{a}$$

$$\sinh\frac{(2m-1)(b-y)\pi}{a}\text{cosech}\frac{(2m-1)\pi b}{a}$$

5. Steady state temperature is governed by the following Laplace's equation

$$\frac{\partial^2 u}{\partial x^2} + \frac{\partial^2 u}{\partial y^2} = 0 \qquad ...(1)$$

As per given, the boundary conditions are

$u(0,y) = u(8,y) = 0$ for $0 < y < \infty$

$\left.\begin{array}{l} u(x,y) \to 0 \text{ as } y \to \infty \text{ for } 0 < x < 8 \\[2mm] \text{and} \quad u(x,0) = 100\sin\left(\dfrac{\pi x}{8}\right) \text{ for } 0 < x < 9 \end{array}\right\}$ ...(2)

By taking $a = 8$, we get

$$u(x,y) = \sum_{n=1}^{\infty} E_n \sin\frac{n\pi x}{8} e^{-n\pi y/8} \qquad ...(3)$$

Putting $y = 0$ in (4) and using (3), we get

$$100\sin\left(\frac{\pi x}{8}\right) = \sum_{n=1}^{\infty} E_n \sin\left(\frac{n\pi x}{8}\right)$$

$$= E_1 \sin\left(\frac{nx}{8}\right) + E_2 \sin\left(\frac{2\pi x}{8}\right) + ...$$

Now comparing the coefficients of like terms of both the sides, we get $E_1 = 100$ and $E_n = 0$ for $n \geq 2$.

Hence, the required temperature function is given by

$$u(x, y) = 100 \sin\left(\frac{\pi x}{8}\right) e^{-\pi y/8}$$

6. The given equation is

$$\frac{\partial^2 V}{\partial x^2} + \frac{\partial^2 V}{\partial y^2} = 0 \qquad \ldots (1)$$

Suppose the solution of (1) is of the form

$$V(x, y) = X(x)\, Y(y) \qquad \ldots (2)$$

Putting the value from (2) in (1), we get $X''Y + XY'' = 0$

$$\Rightarrow \qquad \frac{X''}{X} = -\frac{Y''}{Y} = n^2 \quad \text{(say)} \qquad \ldots (3)$$

From (3), we can find

$$X'' - n^2 X = 0 \qquad \ldots (4)$$

and

$$Y'' + n^2 Y = 0 \qquad \ldots (5)$$

whose solutions are given by

$$X = A e^{nx} + B e^{-nx}$$

and

$$Y = C \cos ny + D \sin ny$$

Putting these values in (2), we get

$$V(x, y) = (A e^{nx} + B e^{-nx})(C \cos ny + D \sin ny) \qquad \ldots (6)$$
$$= A.\exp(\pm nx \pm iny) \qquad \ldots (7)$$

Here, the given boundary conditions are

$$V(x, y) \to 0 \quad \text{as} \quad x \to \infty \qquad \ldots (8)$$

and $\quad V(x, 0) = V(x, a) = 0 \qquad \ldots (9)$

Thus, we can write

$$V(x, y) = (E \cos ny + F \sin ny)\, e^{-nx} \qquad \ldots (10)$$

Putting $y = 0$ in (10) and using (9), we get

$$0 = E e^{-nx} \Rightarrow E = 0$$

Further, putting $y = a$ in (10) and using (9), we get

$$0 = F \sin na\, e^{-nx} \text{ for all } .$$

$$\Rightarrow \qquad \sin na = 0 \Rightarrow n = \frac{r\pi}{a}; \ r = 1, 2, 3, \ldots$$

Therefore, non-zero solution of (1) are given by (10) in the form

$$V(x, y) = F \sin\left(\frac{r\pi y}{a}\right) e^{-\left(\frac{r\pi x}{a}\right)}, \ r = 1, 2, 3, \ldots$$

Hence, the more general solution of (1) is of the form

$$V(x, y) = \sum_{r=1}^{\infty} A_r \sin\left(\frac{r\pi y}{a}\right) e^{-\left(\frac{r\pi x}{a}\right)}$$

Answers

1. $\quad u(x, y) = \dfrac{400}{\pi} \sum\limits_{m=1}^{\infty} \dfrac{1}{2m-1} \sin\dfrac{(2m-1)\pi x}{a} \sinh\dfrac{(2m-1)\pi y}{a} \operatorname{cosech}\dfrac{(2m+1)\pi b}{a}$

2. $\quad u(x, y) = \sum\limits_{n=1}^{\infty} u_n(x, y) = \sum\limits_{n=1}^{\infty} E_n \sin\left(\dfrac{n\pi x}{a}\right) \cosh\left\{\dfrac{n\pi(b-y)}{a}\right\}$

$\qquad f(x) = \sum\limits_{n=1}^{\infty} \left\{ E_n \cosh\left(\dfrac{n\pi b}{a}\right)\right\} \sin\dfrac{n\pi x}{a}, \text{ where, } E_n = -\dfrac{2}{a \cosh\left(\dfrac{n\pi b}{a}\right)} \int_0^a f(x) \sin\dfrac{n\pi x}{a}\, dx$

4. $\quad u(x, y) = \dfrac{8a^2}{\pi} \sum\limits_{m=1}^{\infty} \dfrac{1}{(2m-1)^3} \sin\dfrac{(2m-1)\pi x}{a} \sinh\dfrac{(2m-1)(b-y)\pi}{a} \operatorname{cosech}\dfrac{(2m-1)\pi b}{a}$

5. $\quad u(x, y) = 100 \sin\left(\dfrac{\pi x}{8}\right) e^{-\pi y/8}$ **6.** $\quad V(x, y) = \sum\limits_{r=1}^{\infty} A_r \sin\left(\dfrac{r\pi y}{a}\right) . e^{-\left(\frac{r\pi x}{a}\right)}$

7. $\quad u(r, \theta) = \dfrac{4T}{\pi} \sum\limits_{n=1}^{\infty} \dfrac{1}{2n-1}\left(\dfrac{r}{a}\right)^{2n-1} \sin(2n-1)\theta$ **8.** $\quad V = \sum C_n \left(\dfrac{r}{a}\right)^n \cos n\theta$ **9.** $\quad u(x, y) = \dfrac{\sinh(n\pi y/l)}{\sinh(n\pi a/l)} \sin\dfrac{n\pi x}{l}$

8.15.14 PARABOLIC DIFFERENTIAL EQUATIONS

Here, we shall consider a few problems dealing with the simplest of all parabolic equations, namely the one-dimensional heat equation. Under suitable physical assumptions and choice of units, this equation governs the distribution of temperature on a homogeneous thin rod occupying part of all x-axis, the variable t denoting the time.

(1) ONE DIMENSIONAL HEAT EQUATION   (UPTU-2007, UPTU(CO)-2009, A.M.I.E.-1995,97)

Let us consider the flow of heat by conduction in a bar *OA*. Consider an element *PQQ'P'* of the bar. The temperature $u(x, t)$ of the bar at any point *P* is function of x and time t.

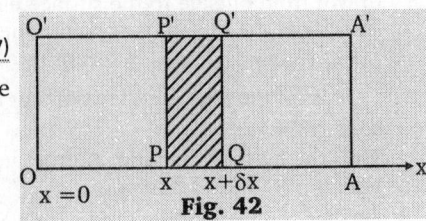

Fig. 42

Now, we have the following assumptions :
  (i) The position of the bar coincides with the *x*-axis.
  (ii) The bar is homogeneous.
  (iii) It is sufficiently thin so that the heat is uniformly distributed over its cross-section at a given time *t*.
  (iv) The surface of the bar is insulated to prevent any loss of heat through the boundary.
  (v) $u(x, t)$ is the temperature at the point *x* at time *t*.
  (vi) The amount of heat crossing any section of the bar is given by $kA\left(\dfrac{\partial u}{\partial x}\right)\delta t$

where, A = Area of the cross-section of the bar

$\qquad \dfrac{\partial u}{\partial x}$ = Temperature gradient at the section

$\qquad \delta t$ = Time of the flow of heat

$\qquad k$ = Thermal conductivity of the material of the bar.

The quantity of heat flowing into the element across the section PP' in time δt.

$$= -kA\left(\frac{\partial u}{\partial x}\right)_x \delta t$$

(The negative sign has been taken because heat flows in the direction of decreasing temperature.)

Also, the quantity of heat flowing out of the element across the section QQ' in time δt.

$$= -kA\left(\frac{\partial u}{\partial x}\right)_{x+\delta x} \delta t$$

Therefore the quantity of heat retained by the element is

$$= -kA\left(\frac{\partial u}{\partial x}\right)_x \delta t + kA\left(\frac{\partial u}{\partial x}\right)_{x+\delta x} \delta t = kA\,\delta t\left\{\left(\frac{\partial u}{\partial x}\right)_{x+\delta x} - \left(\frac{\partial u}{\partial x}\right)_x\right\} \qquad \dots (1)$$

Now, suppose that this heat raises the temperature of the element by a small quantity δu. Therefore, the same quantity of heat is given by

$$= (\rho A\delta x)\sigma\,\delta u \qquad \dots (2)$$

where σ is specific heat of the bar.

Since (1) and (2) are equal, therefore, we have

$$kA\delta t\left\{u(x+\delta x, t) - u(x, t)\right\} = (\rho A\delta x)\,\sigma\,\delta u$$

$$\Rightarrow \qquad k\frac{u(x+\delta x, t) - u(x, t)}{\delta x} = \rho\sigma\frac{\delta u}{\delta t} \qquad \dots (3)$$

As $\delta x \to 0$ and $\delta t \to 0$, we get

$$k\frac{\partial^2 u}{\partial x^2} = \rho\sigma\frac{\partial u}{\partial t}$$

$$\Rightarrow \qquad \frac{\partial u}{\partial t} = k_1\frac{\partial^2 u}{\partial x^2} \qquad \dots (4)$$

where, $k_1 = \dfrac{k}{\rho\sigma}$ is called the diffusivity of the material of the bar. Here, equation (4) is known as one-dimensional heat equation.

REMARK

- Heat equation is also known as Diffusion equation.

(2) SOLUTION OF ONE DIMENSIONAL HEAT EQUATION

(GBTU (CO)-2011, UPTU-2007)

Consider the equation

$$\frac{\partial u}{\partial t} = k\frac{\partial^2 u}{\partial x^2} \qquad \dots (1)$$

For the method of separation of variables, let us assume that solution of (1) is of the form

$$u(x, t) = X(x)\,T(t) \qquad \dots (2)$$

Putting the values from (2) in (1), we get

$$\frac{X''}{X} = \frac{T'}{kT} = \mu \text{ (say), a seperation constant.} \qquad \dots (3)$$

where the dashes denote derivatives with respect to the relevant variable.

From (3), we can find

$$X'' - \mu X = 0 \qquad \dots (4)$$

and

$$T' = \mu k T \qquad \dots (5)$$

Now, we have the following three cases :

Case (I). Let $\mu = 0$ Then solutions of (4) and (5) are given by

$$X = a_1 x + a_2 \text{ and } T = a_3 \qquad \dots (6)$$

Case (II). Let $\mu = \lambda^2, \lambda \neq 0$. Then (4) and (5) reduce to

$$X'' - \lambda^2 X = 0 \text{ and } T' = \lambda^2 k T$$

On solving these equations, we get

$$\left.\begin{array}{l} X = b_1 e^{\lambda x} + b_2 e^{-\lambda x} \\[2mm] T = b_3 e^{\lambda^2 kt} \end{array}\right\} \qquad \dots (7)$$

Case (III). Let $\mu = -\lambda^2, \lambda \neq 0$. Then (4) and (5) reduce to

$$X'' + \lambda^2 X = 0 \text{ and } T' = -\lambda^2 k T$$

On solving, we get

$$\left.\begin{array}{l} X = C_1 \cos\lambda x + C_2 \sin\lambda x \\[2mm] T = C_3 e^{-\lambda^2 kt} \end{array}\right\} \qquad \dots (8)$$

Hence, the various possible solutions are

$$u(x, t) = A_1 x + A_2$$

$$u(x, t) = (B_1 e^{\lambda x} + B_2 e^{-\lambda x})e^{\lambda^2 kt}$$

and

$$u(x, t) = (C_1 \cos\lambda x + C_2 \sin\lambda x)e^{-\lambda^2 kt}$$

(3) SOLUTION OF TWO DIMENSIONAL HEAT EQUATION

The two dimensional heat equation is given by

$$\frac{1}{k}\frac{\partial u}{\partial t} = \left(\frac{\partial^2 u}{\partial x^2} + \frac{\partial^2 u}{\partial y^2}\right) \qquad \ldots (1)$$

Let us suppose (1) has solution of the form $u(x,y,t) = X(x)\,Y(y)\,T(t)$ ... (2)

Putting the values from (2) in (1), we get

$$X''YT + XY''T = \frac{1}{k}XYT'$$

$$\Rightarrow \qquad \frac{X''}{X} + \frac{Y''}{Y} = \frac{1}{k}\frac{T'}{T} \qquad \ldots (3)$$

Now, since x, y and t are independent variables, thus (3) is true if each term on each side is equal to a constant such that

$$\frac{X''}{X} = -n^2, \quad \frac{Y''}{Y} = -m^2 \quad \text{and} \quad \frac{T'}{kT} = -p^2 \qquad \ldots (4)$$

with

$$n^2 + m^2 = p^2$$

The constants may be chosen such that the solution u has the property that $u \to 0 \ as \ t \to \infty$.

Solving (4), we get

$$X_n(x) = A_n \cos nx + B_n \sin nx; \quad Y_m(y) = C_m \cos my + D_m \sin my$$

and

$$T_p(t) = E_p e^{-p^2 kt} = F_{nm} e^{-(n^2+m^2)\,kt}$$

Hence, a suitable solution of (1) is given by $u_{nm}(x,y,t) = F_{nm}(A_n \cos nx + B_n \sin nx)(C_m \cos my + D_m \sin my)\,e^{-(n^2+m^2)kt}$

(4) SOLUTION OF THREE DIMENSIONAL HEAT EQUATION

Three dimensional heat equation is given by

$$\frac{\partial^2 u}{\partial x^2} + \frac{\partial^2 u}{\partial y^2} + \frac{\partial^2 u}{\partial z^2} = \frac{1}{k}\frac{\partial u}{\partial t} \qquad \ldots (1)$$

Let solution of (1) be of the form

$$u(x,y,z,t) = X(x)\,Y(y)Z(z)\,T(t) \qquad \ldots (2)$$

where X, Y, Z, T are respectively the function of x, y, z and t alone.

Putting the values from (2) in (1), we get

$$X''YZT + XY''ZT + XYZ''T = \frac{1}{k}XYZT'$$

$$\Rightarrow \qquad \frac{X''}{X} + \frac{Y''}{Y} + \frac{Z''}{Z} = \frac{1}{k}\frac{T'}{T} \qquad \ldots (3)$$

Now, since x, y, z and t are independent variables, equation (3) is true only when each term on each side is a constant such that

$$\frac{X''}{X} = -n^2, \quad \frac{Y''}{Y} = -m^2, \quad \frac{Z''}{Z} = -l^2 \quad \text{and} \quad \frac{T'}{kT} = -p^2 \qquad \ldots (4)$$

with

$$n^2 + m^2 + l^2 = p^2$$

Further, we have to choose the constants such that solution $u(x, y, z, t)$ has the property that $u \to 0$ as $t \to \infty$.

On solving (4), we get

$$X_n(x) = A_n \cos nx + B_n \sin nx$$

$$Y_m(y) = C_m \cos my + D_m \sin my$$

$$Z_l(z) = E_l \cos lz + F_l \sin lz$$

and

$$T_p(t) = G_p e^{-p^2 kt} = H_{mnl} e^{-(n^2+m^2+l^2)\,kt}$$

which gives $u_{nml}(x,y,z,t) = H_{nml}(A_n \cos nx + B_n \sin nx)(C_m \cos my + D_m \sin my)\,.(E_l \cos lz + F_l \sin lz)\,e^{-(n^2+m^2+l^2)kt}$

which are the required suitable solution of (1).

The general solution of (1) can be obtained by putting $u(x,y,z,t) = \sum\limits_{n=1}^{\infty}\sum\limits_{m=1}^{\infty}\sum\limits_{l=1}^{\infty} u_{nml}(x,y,z,t)$.

(5) TRANSMISSION LINE EQUATIONS

(i) Telegraph Equation : $\dfrac{\partial^2 V}{\partial x^2} = RC\dfrac{\partial V}{\partial t}$ and $\dfrac{\partial^2 i}{\partial x^2} = RC\dfrac{\partial i}{\partial t}$ (UPTU (CO)-2009)

(ii) Radio Equation : $\dfrac{\partial^2 V}{\partial x^2} = LC\dfrac{\partial^2 V}{\partial t^2}$ and $\dfrac{\partial^2 i}{\partial x^2} = LC\dfrac{\partial^2 i}{\partial t^2}$ (UPTU (CO)-2009)

where V = potential, i = current, C = capacitance and L = inductance.

Solved Examples

EXAMPLE 1. *Determine the solution of one dimensional heat equation* $\dfrac{\partial u}{\partial t} = C^2 \dfrac{\partial^2 u}{\partial x^2}$ *subject to the boundary conditions* $u(0, t)=0$, $u(l,t) = 0$ $(t > 0)$ *and the initial condition* $u(x,0) = x$, l *being the length of the bar.* [UPTU-2006]

SOLUTION. We have $\dfrac{\partial u}{\partial t} = C^2 \dfrac{\partial^2 u}{\partial x^2}$ ... (1)

Boundary conditions are

$u(0, t) = 0$

$u(l,t) = 0$ $(t > 0)$

$u(x,0) = x$

On solving (1), we get

$u = C_1 e^{-p^2 C^2 t}(C_2 \cos px + C_3 \sin px)$ ... (2)

Putting $x = 0$ and $u = 0$ in (2), we get

$0 = C_1 e^{-p^2 C^2 t}(C_2) \Rightarrow C_2 = 0$

Putting $C_2 = 0$ in (2), we get

$u = C_1 e^{-p^2 C^2 t} C_3 \sin px$ ... (3)

Again putting $x = l$, $u = 0$ in (3), we get

$0 = C_1 e^{-p^2 C^2 t} C_3 \sin pl$

$\Rightarrow \quad \sin pl = 0 = \sin n\pi$

$\Rightarrow \quad pl = n\pi$

$\Rightarrow \quad p = \dfrac{n\pi}{l}$, n is any integer.

Hence, (3) becomes

$u = C_1 C_3 e^{-\frac{n^2 C^2 \pi^2}{l^2} t} . \sin \dfrac{n\pi x}{l} = b_n e^{-\frac{n^2 C^2 \pi^2}{l^2} t} \sin \dfrac{n\pi}{l} x$ ...(4)

On putting $t = 0$ and $u = x$ in (4), we get

$x = b_n \sin \dfrac{n\pi}{l} x$

General solution is $x = \sum\limits_{n=1}^{\infty} b_n \sin \dfrac{n\pi}{l} x$.

Now, $b_n = \dfrac{2}{l} x \sin \dfrac{n\pi x}{l} dx$

$= \dfrac{2}{l}\left[x . \dfrac{l}{n\pi}\left(-\cos \dfrac{n\pi x}{l}\right) - (1)\left(-\dfrac{l^2}{n^2 \pi^2}\sin \dfrac{n\pi x}{l}\right)\right]_0^l$

$= \dfrac{2}{l}\left[\left(l . \dfrac{l}{n\pi}(-\cos n\pi) + \dfrac{l^2}{n^2 \pi^2}\sin n\pi\right) - 0\right]$

$= \dfrac{2}{l}\left[-\dfrac{l^2}{n\pi}(-1)^n\right] = (-1)^{n+1} . \dfrac{2l}{n\pi}$.

Putting the value of b_n in (4), we get

$u = \dfrac{2l}{\pi}\sum\limits_{n=1}^{\infty} \dfrac{(-1)^{n+1}}{n} \sin \dfrac{n\pi x}{l} e^{-\frac{n^2 C^2 \pi^2}{l^2} t}$.

EXAMPLE 2. *An insulated rod of length l has its ends A and B maintained at 0°C and 100° C respectively until steady state conditions prevail. If B is suddenly reduced to 0°C and maintained at 0°C, find the temperature at a distance x from A at time t.*

[UPTU-2004, 05; GBTU (AG)-2011; UKTU-2011]

SOLUTION. The initial temperature of the rod can be written as

$u(x, t) = 0 + \dfrac{100}{l} x = \dfrac{100}{l} x$

While in steady state, the temperature distribution can be written as

$u(x, t) = 0 + \dfrac{0}{l} x = 0$

To find u in the intermediate period, calculating time from the instant when the end temperature were changed

$u = u_1(x) + u_2(x)$

where $u_2(x)$ is temperature after a sufficient long time and $u_1(x, t)$ is the transient temperature distribution tending to zero as $t \to \infty$. Hence, $u_2(x) = 0$.

Also, $u_1(x, t)$ satisfies one-dimensional heat flow

$C^2 \dfrac{\partial^2 u}{\partial x^2} = \dfrac{\partial u}{\partial t}$

Thus, $u = (C_1 \cos px + C_2 \sin px)e^{-C^2 p^2 t}$ ... (1)

On putting $x = 0, u = 0$ in (1), we get

$0 = C_1 e^{-p^2 C^2 t} \Rightarrow C_1 = 0$

On putting $C_1 = 0$ in (1), we get

$u = C_2 \sin px e^{-C^2 p^2 t}$ ... (2)

On putting $x = l, u = 0$ in (2), we get

$0 = C_2 \sin pl\, e^{-p^2 C^2 t}$

$\Rightarrow \quad \sin pl = 0 = \sin n\pi$

$\Rightarrow \quad pl = n\pi$

$\Rightarrow \quad p = \dfrac{n\pi}{l}$

On putting the value of p in (2), we get

$u = C_2 \sin \dfrac{n\pi x}{l} e^{\frac{n^2 \pi^2 C^2}{l^2} t}$ ... (3)

On putting $t = 0$, $u = \dfrac{100}{l} x$ in (3), we get

$\dfrac{100x}{l} = C_2 \sin \dfrac{n\pi x}{l}$

$C_2 = \dfrac{2}{l}\int_0^l \dfrac{100}{l} . x . \sin \dfrac{n\pi x}{l} dx$

$C_2 = \dfrac{200}{l^2}\int_0^l x \dfrac{\sin n\pi x}{l} dx$

$C_2 = \dfrac{200}{l^2}\left[-\dfrac{xl}{n\pi}\cos \dfrac{n\pi x}{l} - (-1)\dfrac{l^2}{n^2 \pi^2}\sin \dfrac{n\pi x}{l}\right]_0^l$

$C_2 = \dfrac{200}{l^2}\left[-\dfrac{l^2}{n\pi}\cos n\pi\right]$

$\Rightarrow C_2 = -\dfrac{200}{n\pi}(-1)^n$

On putting the value of C_2 in (3), we get

$u = -\dfrac{200}{n\pi}(-1)^n \sin \dfrac{n\pi x}{l} e^{\frac{n^2 \pi^2 C^2}{l^2} t}$

$u = (-1)^n \dfrac{200}{n\pi} . \sin \dfrac{n\pi x}{l} e^{\frac{n^2 \pi^2 C^2}{l^2} t}$.

EXAMPLE 3. *A rod of length l with insulated sides is initially at a uniform temperature μ_0. Its ends are suddenly cooled to 0°C and are kept at that temperature. Show that the temperature* *function u(x, t) is given by*

$$u(x,t) = \sum_{n=1}^{\infty} b_n \sin\frac{n\pi x}{l} e^{-C^2\pi^2 n^2 t/l^2},$$

where b_n is given by $u_0 = \sum_{n=1}^{\infty} b_n \sin\frac{n\pi x}{l}$

(GBTU-2010, 11)

SOLUTION. The heat equation is given by

$$\frac{\partial u}{\partial t} = C^2\frac{\partial^2 u}{\partial x^2} \qquad \dots(1)$$

Suppose that solution of (1) is of the form

$$u(x,t) = X(x)\,T(t) \qquad \dots(2)$$

where X and T are respectively the functions of x and t alone. Putting the value from (2) in (1), we get

$$X\frac{dT}{dt} = C^2 T\frac{d^2 X}{dx^2}$$

$$\Rightarrow \frac{1}{C^2 T}\frac{dT}{dt} = \frac{1}{X}\frac{d^2 X}{dx^2} = -p^2 \text{ (say)} \qquad \dots(3)$$

where p is a separation constant.

From (3), we can find

$$\frac{1}{C^2 T}\frac{dT}{dt} = -p^2$$

$$\Rightarrow \frac{dT}{dt} = -p^2 C^2 T \qquad \dots(4)$$

and $\dfrac{1}{X}\dfrac{d^2 X}{dx^2} = -p^2 \qquad \dots(5)$

Solving (4) and (5), we get

$$T = C_1 e^{-p^2 C^2 t} \text{ and } X = (C_2 \cos px + C_3 \sin px)$$

which gives $u(x,t) = e^{-p^2 C^2 t}(C_1' \cos px + C_2' \sin px)$

Now, $u(0, t) = 0$

$$\Rightarrow \quad 0 = C_1' e^{-p^2 C^2 t} \Rightarrow C_1' = 0$$

$$\Rightarrow u(x,t) = e^{-p^2 C^2 t}\cdot C_2' \sin px$$

Now, $u(l, t) = 0 \Rightarrow 0 = e^{-p^2 C^2 t}(C_1' \cos pl + C_2' \sin pl)$

$$\Rightarrow \quad \sin pl = 0 \quad \Rightarrow \quad p = \frac{n\pi}{l};\ n = 1, 2, \dots\dots$$

Therefore, $u(x,t) = e^{-p^2 C^2 t}\sin\dfrac{n\pi x}{l}C_2'$

$$= e^{-n^2\pi^2 C^2 t/l^2}\sin\frac{n\pi x}{l}C_2'$$

$$= b_n e^{-n^2\pi^2 C^2 t/l^2}\sin\frac{n\pi x}{l}$$

Hence, the general solution is given by

$$u(x,t) = \sum_{n=1}^{\infty} u_n(x,t)$$

$$= \sum_{n=1}^{\infty} b_n e^{-n^2\pi^2 C^2 t/l^2}\sin\frac{n\pi x}{l}$$

Finally, using $u = u_0$ when $t = 0$, we get

$$u_0 = \sum_{n=1}^{\infty} b_n \sin\frac{n\pi x}{l}.$$

EXAMPLE 4. *Find the temperature in a bar of length 2 whose ends are kept at zero and lateral surface insulated if the initial temperature is* $\sin\dfrac{\pi x}{2} + 3\sin\dfrac{5\pi x}{2}$. [MTU-2011, UPTU (CO)-2007, 09]

SOLUTION. Let $u(x, t)$ be the temperature in the bar. The boundary conditions are

$$u(0, t) = 0 = u(2, t) \text{ for any } t \qquad \dots(1)$$

The initial condition is

$$u(x,0) = \sin\frac{\pi x}{2} + 3\sin\frac{5\pi x}{2} \qquad \dots(2)$$

One dimensional heat flow equation is

$$\frac{\partial u}{\partial t} = C^2\frac{\partial^2 u}{\partial x^2} \qquad \dots(3)$$

Its solution is

$$u(x,t) = (C_1 \cos px + C_2 \sin px)C_3 e^{-c^2 p^2 t} \qquad \dots(4)$$

$$u(0,t) = 0 = C_1 C_3 e^{-C^2 p^2 t} \qquad \text{(On using (1))}$$

$$\Rightarrow \quad C_1 = 0$$

$$\therefore \text{ From (4) } u(x,t) = C_2 C_3 \sin px\, e^{-C^2 p^2 t} \qquad \dots(5)$$

$$u_1(2,t) = 0 = C_2 C_3 \sin 2p\, e^{-C^2 p^2 t} \qquad \text{(On using (1))}$$

$$\Rightarrow \quad \sin 2p = 0 = \sin n\pi$$

$$\therefore \qquad p = \frac{n\pi}{2}, n \in I$$

Hence from (5)

$$u(x,t) = b_n \sin\frac{n\pi x}{2} e^{\frac{-n^2\pi^2 C^2 t}{4}} \qquad (\because\ C_2 C_3 = b_n)$$

The most general solution is

$$u(x,t) = \sum_{n=1}^{\infty} b_n \sin\frac{n\pi x}{2} e^{\frac{-n^2\pi^2 C^2 t}{4}} \qquad \dots(6)$$

$$u(x,0) = \sin\left(\frac{\pi x}{2}\right) + 3\sin\left(\frac{5\pi x}{2}\right)$$

$$= \sum_{n=1}^{\infty} b_n \sin\frac{n\pi x}{2}$$

$$= b_1 \sin\left(\frac{\pi x}{2}\right) + b_2 \sin\left(\frac{2\pi x}{2}\right) + \dots + b_5 \sin\left(\frac{5\pi x}{2}\right) + \dots$$

Comparing, we get $b_1 = 1$ and $b_5 = 3$

Hence from (6),

$$u(x,t) = \sin\left(\frac{\pi x}{2}\right)e^{-\pi^2 C^2 t/4} + 3\sin\left(\frac{5\pi x}{2}\right)e^{-25\pi^2 C^2 t/4}$$

EXAMPLE 5. *Solve* $\dfrac{\partial^2 u}{\partial x^2} + \dfrac{\partial^2 u}{\partial y^2} = 0$, *which satisfies the conditions* $u(0, y) = u(l, y) = u(x, 0) = 0$ *and* $u(x, a) = \sin\dfrac{n\pi x}{l}$.

(UPTU-2004, (CO)-2009, GBTU (AG)-2012)

SOLUTION. Consider the heat flow in a metal plate of uniform thickness in the directions parallel to length and breadth of the plate. There is no heat flow along the normal to the plane of the rectangle.

Let $u(x, y)$ be the temperature at any point (x, y) of the plate at time t is given by

$$\frac{\partial u}{\partial t} = C^2\left(\frac{\partial^2 u}{\partial x^2} + \frac{\partial^2 u}{\partial y^2}\right)$$

In the steady state, u does not change with t.

$$\frac{\partial u}{\partial t} = 0$$

Let $\qquad u = X(x)\,.\,Y(y) \qquad \dots(1)$

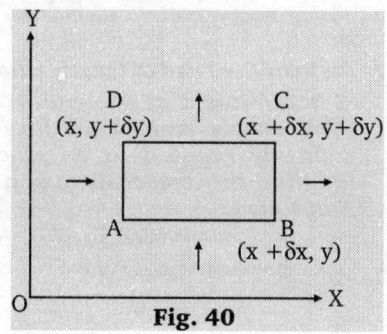

Fig. 40

Putting the values of $\dfrac{\partial^2 u}{\partial x^2}$ and $\dfrac{\partial^2 u}{\partial y^2}$ in (1),

we have $X''Y + XY'' = 0$ ... (2)

$\Rightarrow \quad \dfrac{X''}{X} = -\dfrac{Y''}{Y} = -p^2$ (say)

$D^2 X = -p^2 X$ and $D^2 Y = p^2 Y$

$\Rightarrow \quad D^2 X + p^2 X = 0$ and $D^2 Y - p^2 Y = 0$

$\Rightarrow \quad (D^2 + p^2)X = 0$ and $(D^2 - p^2)Y = 0$

A.E. is $D^2 + p^2 = 0$, A.E. is $D^2 - p^2 = 0$

$\Rightarrow \qquad D^2 = -p^2$ and $D^2 = p^2$

$\Rightarrow \qquad D = \pm ip$ and $D = \pm p$

$X = C_1 \cos px + C_2 \sin px \quad Y = C_3 e^{py} + C_4 e^{-py}$

Putting the values of X and Y in (1), we have

$u = (C_1 \cos px + C_2 \sin px)(C_3 e^{py} + C_4 e^{-py})$ ... (3)

Putting $x = 0, u = 0$ in (3), we have

$0 = C_1(C_3 e^{py} + C_4 e^{-py}) \Rightarrow C_1 = 0$

(3) is reduced to $u = C_2 \sin px(C_3 e^{py} + C_4 e^{-py})$ ... (4)

On putting $x = l, u = 0$ in (4), we have

$0 = C_2 \sin pl(C_3 e^{py} + C_4 e^{-py}) \Rightarrow C_2 \neq 0$

$\therefore \qquad \sin pl = 0 = \sin n\pi \Rightarrow Pl = n\pi$

or $\quad P = \dfrac{n\pi}{l}$

Now, (4) becomes

$u = C_2 \sin \dfrac{n\pi x}{l}\left[C_3 e^{\frac{n\pi y}{l}} + C_4 e^{-\frac{n\pi y}{l}}\right]$ ... (5)

On putting $u = 0$ and $y = 0$ in (5), we have

$0 = C_2 \sin \dfrac{n\pi x}{l}(C_3 + C_4)$

$C_3 + C_4 = 0$ or $C_3 = -C_4$

(5) becomes $u = C_2 C_3 \sin \dfrac{n\pi x}{l}\left(e^{\frac{n\pi y}{l}} - e^{-\frac{n\pi y}{l}}\right)$

On putting $y = a$ and $u = \sin \dfrac{n\pi x}{l}$ in (6), we have

$\sin \dfrac{n\pi x}{l} = C_2 C_3 \sin \dfrac{n\pi x}{l}\left(e^{\frac{n\pi a}{l}} - e^{-\frac{n\pi a}{l}}\right)$

$i.e., \qquad C_2 C_3 = \dfrac{1}{e^{\frac{n\pi a}{l}} - e^{-\frac{n\pi a}{l}}}$

Putting this value in (6), we have

$u = \sin \dfrac{n\pi x}{l} = \dfrac{e^{\frac{n\pi y}{l}} - e^{-\frac{n\pi y}{l}}}{e^{\frac{n\pi a}{l}} - e^{-\frac{n\pi a}{l}}}$

or $\qquad u = \sin \dfrac{n\pi x}{l} \dfrac{\sinh \frac{n\pi y}{l}}{\sin \frac{n\pi a}{l}}.$

EXAMPLE 6. *A thin rectangular plate whose surface is impervious to heat flow, has at $t = 0$ an arbitrary distribution of temperature $f(x, y)$. If four edges $x = 0, x = a, y = 0, y = b$ are kept at zero temperature, find the temperature at a point of the plate as t increases.* [UPTU-2002]

SOLUTION. We know that the two dimensional heat equation is

$\dfrac{\partial u}{\partial t} = C^2\left(\dfrac{\partial^2 u}{\partial x^2} + \dfrac{\partial^2 u}{\partial y^2}\right)$ ... (1)

As per given, the initial temperature of the plate is $f(x, y)$ and the temperature of the four edges of the plate are kept at 0°.

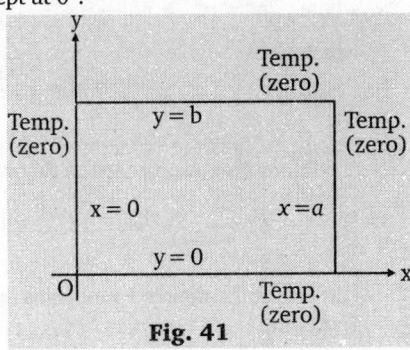

Fig. 41

Therefore, the required boundary conditions are

(i) $u(0, y, t) = 0$, (ii) $u(a, y, t) = 0$,

(iii) $u(x, 0, t) = 0$, (iv) $u(x, b, t) = 0$.

Also, the initial condition is given by

$u(x, y, 0) = f(x, y)$ ... (2)

Suppose that solution of (1) is of the form

$u(x, y, t) = X(x)\, Y(y)\, T(t)$ ... (3)

where X, Y and T are respectively the functions of x, y and t alone. Putting the values of (3) in (1), we get

$\dfrac{1}{C^2 T} = \dfrac{1}{X}\dfrac{d^2 X}{dx^2} + \dfrac{1}{Y}\dfrac{d^2 Y}{dy^2}$ ... (4)

If (3) satisfies (1), we have the following three possibilities

(a) $\dfrac{1}{X}\dfrac{d^2 X}{dx^2} = 0, \quad \dfrac{1}{Y}\dfrac{d^2 Y}{dy^2} = 0, \quad \dfrac{1}{C^2 T}\dfrac{dT}{dt} = 0$

(b) $\dfrac{1}{X}\dfrac{d^2 X}{dx^2} = p_1^2, \dfrac{1}{Y}\dfrac{d^2 Y}{dy^2} = p_2^2, \dfrac{1}{C^2 T}\dfrac{dT}{dt} = p^2$

(c) $\dfrac{1}{X}\dfrac{d^2 X}{dx^2} = -p_1^2, \dfrac{1}{Y}\dfrac{d^2 Y}{dy^2} = -p_2^2, \dfrac{1}{C^2 T}\dfrac{dT}{dt} = -p^2$

where $p^2 = p_1^2 + p_2^2$.

It can be easily verified that differential equation (c) only gives the solution. In this case, the general solution is given by

$X = A_1 \cos p_1 x + B_1 \sin p_1 x$;

$Y = A_2 \cos p_2 y + B_2 \sin p_2 y$; $T = A_3 e^{-C^2 p^2 t}$

Therefore

$u(x, y, t) = (A_1 \cos p_1 x + B_1 \sin p_1 x)$

$(A_2' \cos p_2 y + B_2' \sin p_2 y)\, e^{-C^2 p^2 t}$ ... (5)

Using boundary condition (i), we get

$u(0, y, t) = A_1(A_2' \cos p_2 y + B_2' \sin p_2 y)\, e^{-C^2 p^2 t} = 0$

$\Rightarrow \qquad A_1 = 0$

Now, using boundary condition (ii), we get

$u(a, y, t) = B_1 \sin p_1 a(A_2' \cos p_2 y$

$+ B_2' \sin p_2 y)\, e^{-C^2 p^2 t} = 0$

$\Rightarrow \quad \sin p_1 a = 0 \Rightarrow p_1 a = m\pi$

$p_1 = \dfrac{m\pi}{a}; \ m = 1, 2, 3, \ldots$

Similarly, by using boundary condition (iii) and (iv), we get

$A_2' = 0$ and $p_2 = \dfrac{n\pi}{b}$, $n = 1, 2, 3, \ldots$

Thus, we have

$$u_{mn}(x, y, t) = A_{mn}\, e^{-C^2 p_{mn}^2 t} \cdot \sin \dfrac{m\pi}{a} x \sin \dfrac{n\pi}{b} y$$

where, $p^2 = p_{mn}^2 = \pi^2 \left(\dfrac{m^2}{a^2} + \dfrac{n^2}{b^2} \right)$

which gives

$$u(x, y, t) = \sum_{m=1}^{\infty} \sum_{n=1}^{\infty} A_{mn}\, e^{-C^2 p_{mn}^2 t} \cdot \sin \dfrac{m\pi}{a} x \sin \dfrac{n\pi}{b} y$$

Finally, to find the solution which satisfies the initial conditions also, we proceed as follows :

$$u(x, y, 0) = \sum_{m=1}^{\infty} \sum_{n=1}^{\infty} A_{mn}\, e^{-C^2 p_{mn}^2 t} \cdot \sin \dfrac{m\pi}{a} x \sin \dfrac{n\pi}{b} y$$
$$= f(x, y)$$

The LHS is the double Fourier sine series of $f(x, y)$. Therefore,

$$A_{mn} = \dfrac{2}{a} \cdot \dfrac{2}{b} \int_{x=0}^{a} \int_{y=0}^{b} f(x, y) \sin \dfrac{m\pi}{a} x \sin \dfrac{n\pi}{b} y\, dx\, dy$$

EXAMPLE 7. *Solve the boundary value problem* $\dfrac{\partial^2 u}{\partial x^2} = \dfrac{l}{k} \dfrac{\partial u}{\partial t}$

satisfying the conditions $u(0, t) = u(l, t) = 0$ and $u(x, 0) = x$ when $0 \le x \le l/2$ $u(x, 0) = (l - x)$ when $l/2 \le x \le l$.

SOLUTION. Proceeding same as in Example (4) by taking $a = l$ and we get

$$u(x, t) = \sum_{n=1}^{\infty} E_n \sin \left(\dfrac{n\pi x}{l} \right) e^{-(n^2 \pi^2 k t)/l^2} \quad \ldots (1)$$

It is also given that

$$u(x, 0) = \begin{cases} x & ; \text{ when } 0 \le x \le l/2 \\ l - x & ; \text{ when } l/2 \le x \le l \end{cases} \quad \ldots (2)$$

Putting $t = 0$ in (1), we get

$$u(x, 0) = \sum_{n=1}^{\infty} E_n \sin \dfrac{n\pi x}{l}$$

which is a half range Fourier sine series in $(0, l)$. Thus, E_n is given by

$$E_n = \dfrac{2}{l} \int_0^l u(x, 0) \sin \dfrac{n\pi x}{l} dx$$

$$= \dfrac{2}{l} \left[\int_0^{l/2} u(x, 0) \sin \dfrac{n\pi x}{l} dx \right.$$

$$\left. + \int_{l/2}^l u(x, 0) \sin \dfrac{n\pi x}{l} dx \right]$$

$$= \int_0^{l/2} \dfrac{2x}{l} \sin \dfrac{n\pi x}{l} dx + \int_{l/2}^l \dfrac{2}{l}(l - x) \sin \dfrac{n\pi x}{l} dx$$

$$= \left[\left(\dfrac{2x}{l} \right) \left(-\dfrac{\cos(n\pi x)/l}{(n\pi)/l} \right) - \left(\dfrac{2}{l} \right) \left(-\dfrac{\sin(n\pi x)/l}{(n\pi)^2/l^2} \right) \right]_0^{l/2}$$

$$+ \left[\left(\dfrac{2(l-x)}{l} \right) \left(-\dfrac{\cos(n\pi x)/l}{(n\pi)/l} \right) \right.$$

$$\left. - \left(-\dfrac{2}{l} \right) \left(-\dfrac{\sin(n\pi x)/l}{(n\pi)^2/l^2} \right) \right]_{l/2}^l$$

$$= -\left(\dfrac{l}{n\pi} \right) \cos \left(\dfrac{n\pi}{2} \right) + \left(\dfrac{2l}{n^2 \pi^2} \right) \sin \left(\dfrac{n\pi}{2} \right)$$

$$+ \left(\dfrac{l}{n\pi} \right) \cos \left(\dfrac{n\pi}{2} \right) + \left(\dfrac{2l}{n^2 \pi^2} \right) \sin \dfrac{n\pi}{2}$$

Thus, $E_n = \dfrac{4l}{n^2 \pi^2} \sin \dfrac{n\pi}{2}$

$$= \begin{cases} 0 & ; \text{ if } n = 2m \text{ and } m = 1, 2, 3, \ldots \\ \dfrac{4l}{(2m-1)^2 \pi^2} & ; \text{ if } n = 2m - 1, \ m = 1, 2, \ldots \end{cases}$$

Hence, by (1), we have

$$u(x, t) = \dfrac{4l}{\pi^2} \sum_{m=1}^{\infty} \dfrac{1}{(2m-1)^2}$$

$$\sin \dfrac{(2m-1)\pi x}{l} e^{-\left[(2m-1)^2 \pi^2 k t \right]/l^2}$$

EXAMPLE 8. *Find the solution of one dimensional heat equation satisfying the following boundary conditions*

(i) T is bounded as $t \to \infty$

(ii) $\left. \dfrac{\partial T}{\partial x} \right|_{x=0} = 0 \ ; \ \forall t$

(iii) $\left. \dfrac{\partial T}{\partial x} \right|_{x=a} = 0 \ ; \ \forall t$

(iv) $T(x, 0) = x(a - x); \ 0 < x < a$

SOLUTION. The general acceptable solution of the given partial differential equation is :

$$T(x, t) = \exp(-\alpha \lambda^2 t)(A \cos \lambda x + B \sin \lambda x)$$

$$\Rightarrow \dfrac{\partial T}{\partial x} = \exp(-\alpha \lambda^2 t)(-A\lambda \sin \lambda x + B\lambda \cos \lambda x) \quad \ldots (1)$$

Using boundary condition (ii) in (1), we get $B = 0$.

By using boundary condition (iii), we get

$\sin \lambda a = 0 \ \Rightarrow \ \lambda a = n\pi; \ n = 0, 1, 2, \ldots$

Therefore, the more general solution is given by

$$T(x, t) = \sum A_n \exp(-\alpha \lambda^2 t) \cos \lambda x$$

$$= \sum_{n=0}^{\infty} A_n \exp \left[-\alpha \left(\dfrac{n\pi}{a} \right)^2 t \right] \cos \left(\dfrac{n\pi}{a} \right) x$$

Using boundary condition (iv), we have

$$T(x, 0) = x(a - x)$$

$$= A_0 + \sum_{n=1}^{\infty} A_n \exp \left[-\alpha \left(\dfrac{n\pi}{a} \right)^2 t \right] \cos \left(\dfrac{n\pi}{a} \right) x$$

where, $A_0 = \dfrac{2}{a} \int_0^a (ax - x^2) dx = \dfrac{a^2}{6}$

$$A_n = \dfrac{2}{a} \int_0^a (ax - x^2) \cos \left(\dfrac{n\pi x}{a} \right) dx$$

$$= \dfrac{2a^2}{n^2 \pi^2}(1 + \cos n\pi) = \dfrac{2a^2}{n^2 \pi^2}[1 + (-1)^n]$$

$$\Rightarrow A_n = \begin{cases} -\dfrac{4a^2}{n^2 \pi^2} & ; \text{ if } n \text{ is even} \\ 0 & ; \text{ if } n \text{ is odd} \end{cases}$$

Hence, the required solution is given by

$$T(x, t) = \dfrac{a^2}{6} - \dfrac{4a^2}{\pi^2}$$

$$\sum_{n=2,4,\ldots}^{\infty} \dfrac{1}{n^2} \cos \left(\dfrac{n\pi}{a} \right) x \cdot \exp \left[-\alpha \left(\dfrac{n\pi}{a} \right)^2 t \right]$$

EXAMPLE 9. *Find the current i and voltage V in a transmission line of length l, t seconds after the ends are suddenly grounded given that $i(x, 0) = i_0$, $V(x, 0) = V_0 \sin\left(\dfrac{\pi x}{l}\right)$ and that R and G are negligible.*

[UPTU-2008]

SOLUTION. We have $\dfrac{\partial^2 V}{\partial x^2} = LC \dfrac{\partial^2 V}{\partial t^2}$

Let $V = XT$, where X and T are the functions of x and t respectively.

$$\frac{\partial^2 V}{\partial x^2} = T\frac{\partial^2 X}{\partial x^2} \quad \text{and} \quad \frac{\partial^2 V}{\partial t^2} = X\frac{d^2 T}{dt^2}$$

or $\dfrac{\dfrac{d^2 X}{dx^2}}{X} = LC \dfrac{\dfrac{d^2 T}{dt^2}}{T} = -p^2$ (say)

Since the initial conditions suggest the values of V and i are periodic functions.

$$X = C_1 \cos px + C_2 \sin px$$

and $T = C_3 \cos\dfrac{pt}{\sqrt{LC}} + C_4 \sin\dfrac{pt}{\sqrt{LC}}$

so $V = (C_1 \cos px + C_2 \sin px)$

$$\left(C_3 \cos\frac{pt}{\sqrt{LC}} + C_4 \sin\frac{pt}{\sqrt{LC}}\right) \quad \ldots(1)$$

where, $t = 0$, $V = V_0 \sin\dfrac{\pi x}{l}$

$$V_0 \sin\frac{\pi x}{l} = \left(C_1 \cos px + C_2 \sin px\right) C_3 \quad \ldots (2)$$

On equating the coefficients, we get

$C_1 C_3 = 0 \Rightarrow C_1 = 0$ and $C_2 C_3 = V_0$, $p = \pi / l$

becomes (1)

$$V = \sin\frac{\pi x}{l}\left[V_0 \cos\frac{pt}{\sqrt{LC}} + C_2 C_4 \sin\frac{pt}{\sqrt{LC}}\right] \quad \ldots (3)$$

Now, when $t = 0$, $i = i_0$ (constant).

Hence, $\dfrac{\partial i}{\partial x} = 0$

$$\frac{\partial i}{\partial x} = -C\frac{\partial V}{\partial t}$$

$\therefore \quad \dfrac{\partial V}{\partial t} = 0$ when $t = 0$

Now, $\dfrac{\partial V}{\partial t} = \sin\dfrac{\pi x}{l}\left(\dfrac{p}{\sqrt{LC}}\right)$

$$\left[-V_0 \sin\frac{pt}{\sqrt{LC}} + C_2 C_4 \cos\frac{pt}{\sqrt{LC}}\right] \quad \ldots (4)$$

On putting $\dfrac{\partial V}{\partial t} = 0$ and $t = 0$ in (4), we get

$$C_2 C_4 = 0 \Rightarrow C_4 = 0$$

Now (3) becomes, $V = V_0 \sin\dfrac{\pi x}{l}\cos\dfrac{\pi t}{\sqrt{LC}}$

$$\frac{\partial V}{\partial x} = \frac{\pi}{l}V_0 \cos\frac{\pi x}{l}\cos\frac{\pi t}{\sqrt{LC}} = -L\frac{\partial i}{\partial t} \quad \ldots (5)$$

$$\frac{\partial V}{\partial t} = -\frac{V_0 \pi}{l\sqrt{LC}} \sin\frac{\pi x}{l}\sin\frac{\pi t}{l\sqrt{LC}} = -\frac{1}{C}\frac{\partial t}{\partial x} \quad \ldots (6)$$

Integrating (5) and (6), we get

$$i = -V_0 \sqrt{\frac{C}{L}} \cos\frac{\pi x}{l}\sin\frac{\pi t}{l\sqrt{LC}} + f(x)$$

and $i = -V_0 \sqrt{\dfrac{C}{L}} \cos\dfrac{\pi x}{l}\sin\dfrac{\pi t}{l\sqrt{LC}} + F(t)$

$\therefore$ $f(x)$ and $F(t)$ must be constant only, since $i = i_0$ when $t = 0$

$\therefore$ Constant $= i_0 = f(x)$

Hence, $i = i_0 - V_0 \sqrt{\dfrac{C}{L}} \cos\dfrac{\pi x}{l}\sin\dfrac{\pi t}{l\sqrt{LC}}$.

 Exercise-8.71

1. A rod of length l with insulated sides, is initially at a uniform temperature u_0. Its ends are suddenly cooled at 0°C and are kept at that temperature. Find the temperature distribution $u(x, t)$.

2. Find temperature distribution $y(x, t)$ in a uniform bar of unit length, whose one end is kept at 10°C and the other end is insulated. It is given that $y(x, 0) = 1 - x$, $0 < x < 1$.

[UPTU Q. BANK-2002]

3. The faces $x = 0$ and $x = a$ of an infinite slab are maintained at zero temperature. Given that the temperature $u(x, t) = f(x)$ at $t = 0$. Find the temperature at time t.

4. Find the solution of one dimensional heat equation $\dfrac{\partial \theta}{\partial t} = a^2 \dfrac{\partial^2 \theta}{\partial x^2}$ under the boundary conditions $\theta(0, t) = 0$, $\theta(l, t) = 0$. When $t > 0$ and the initial condition $\theta(x, 0) = x$ when $0 < x < l$, l being the length of bar.

5. The four edges of a thin square plate of area π^2 are kept at temperature zero and the faces are perfectly insulated. The initial temperature is assumed to be $u(x, y, 0) = xy(n - x)(\pi - y)$. Find the temperature $u(x, y, t)$ in the plate.

6. Obtain the temperature $u(x, t)$ in a slab where ends $x = 0$ and $x = l$ are kept at temperature zero and whose initial temperature is given by

$$f(x) = \begin{cases} A & ; \text{ when } 0 < x < l/2 \\ 0 & ; \text{ when } l/2 < x < l \end{cases}$$

[GBTU-2010, GBTU (CO)-2010]

7. Solve $\dfrac{\partial^2 u}{\partial x^2} = h^2\left(\dfrac{\partial u}{\partial t}\right)$ when $u(0, t) = u(l, t) = 0$ and $u(x, 0) = \sin\dfrac{\pi x}{l}$.

8. Solve the onedimensional heat equation $\dfrac{\partial^2 u}{\partial x^2} = \dfrac{1}{k}\dfrac{\partial u}{\partial t}$ in the range $0 \le x \le 2\pi, t \ge 0$, subject to the boundary condition $u(x, 0) = \sin^3 x$, for $0 \le x \le 2\pi$ and $u(0, t) = u(2\pi, t) = 0$ for $t \ge 0$.

9. Solve $\dfrac{\partial^2 z}{\partial x^2} = \dfrac{1}{k}\dfrac{\partial z}{\partial t}$ with the condition $z = 0$ when $x = 0$ and $x = 1$ for all values of t.

10. Find a solution of $\dfrac{\partial V}{\partial t} = k\dfrac{\partial^2 V}{\partial x^2}$ such that $V \ne \infty$ if $t \to \infty$, $V = 100$ if $x = 0$ or π for all values of t; $V = 0$ if $t = 0$ for all values of x between 0 and π.

11. The heat flow equation in a homogeneous rod is $\dfrac{\partial^2 T}{\partial x^2} = \dfrac{1}{\alpha^2} \dfrac{\partial T}{\partial t}$, where T is the temperature and α^2 thermal diffusivity. The rod is of length L with insulated sides. Solve it with boundary conditions $T(x, 0) = f(x), \quad 0 < x < L, T_x(0, t) = T_x(L, t) = 0$. What is the steady temperature of the rod.

12. A conducting bar of uniform cross-section lies among the x-axis with ends at $x = 0$ and $x = L$. It is kept initially at temperature $0°$ and its lateral surface is insulated. There are no heat sources in the bar. The end $x = 0$ is kept at $0°$ and heat is suddenly applied at the end $x = L$ so that there is a constant flux q_0 at $x = L$. Find the temperature distribution in the bar for $t > 0$.

13. Show that the solution of $\dfrac{\partial u}{\partial t} = k \dfrac{\partial^2 u}{\partial x^2}$ subject to the conditions
    (i) u is not infinite for $t \to \infty$.
    (ii) $\dfrac{\partial u}{\partial x} = 0$ for $x = 0$ and $x = l$.
    (iii) $u = lx - x^2$ for $t = 0$ between $x = 0$ and $x = l$ is
    $$u = \frac{1}{6} l^2 - \frac{l^2}{\pi^2} \sum_{n=1}^{\infty} \frac{l}{n^2} \cos \frac{n\pi x}{l} e^{-(4\pi^2 n^2 kt)/l}$$

14. A rectangular plate bounded by the lines $x = 0, y = 0, x = a, y = b$ has an initial distribution given by $V = A \sin\left(\dfrac{\pi x}{a}\right) \sin\left(\dfrac{\pi y}{b}\right)$. The edges are kept at zero temperature and the plane faces are impervious to heat. Find the temperature at any point. [UPTU-2002]

15. A square plate with sides of unit length has its faces insulated and its sides kept at 0°C. If the initial temperature is specified, determine the subsequent temperature at any point of the plate.

16. The temperature distribution in a bar of length. π which is perfectly insulated at ends $x = 0$ and $x = \pi$ is governed by partial differential equation $\dfrac{\partial u}{\partial} \dfrac{\partial u}{\partial}$. Assuming the initial temperature distribution as $u(x, 0) = f(x) = \cos 2x$, find the temperature distribution at any instant of time. [MTU-2011]

17. Solve $\dfrac{\partial u}{\partial t} = a \dfrac{\partial^2 u}{\partial x^2}$; a constant, subject to the boundary conditions $u(0, t) = 0, u(\pi, t) = 0$ and the initial condition $u(x, 0) = \sin 2x$. [MTU-2012]

18. Find the temperature distribution in a rod of length 2m whose end points are fixed at temperature zero and the initial temperature distribution is $f(x) = 100x$. [GBTU-2012]

19. Find the temperature $u(x, t)$ in a homogeneous bar of heat conducting material of length l cm with its ends kept at zero temperature and initial temperature given by $\dfrac{x(L-x)d}{L^2}$. [UPTU (SUM)-2009]

20. Find the temperature in a thin metal rod of length L with both ends insulated (so that there is no passage of heat through the ends) and with initial temperature $\sin \dfrac{\pi x}{L}$ in the rod. [UPTU (SUM)-2009]

21. A homogeneous rod of conducting material of length 'l' has its ends kept at zero temperature. The temperature at the centre is T and falls uniformly to zero at the two ends. Find the temperature distribution. [UKTU-2012]

22. An infinitely long plane uniform plate is bounded by two parallel edges and an end at right angles to them. The breadth is π. This end is maintained at temperature u_0 at all points and the other edges are at zero temperature. Determine the temperature at any point of the plate in the steady state. [GBTU-2012]

23. Solve $\dfrac{\partial^2 u}{\partial x^2} + \dfrac{\partial^2 u}{\partial y^2} = 0, 0 < x < \pi, 0 < y < \pi$ which satisfies the conditions: $u(0, y) = u(\pi, y) = u(x, \pi) = 0$ and $u(x, 0) = \sin^2 x$. [UKTU-2011]

24. Solve the Laplace equation $\dfrac{\partial^2 u}{\partial x^2} + \dfrac{\partial^2 u}{\partial y^2} = 0$ in a rectangle in the xy-plane with $u(x, 0) = 0, u(x, b) = 0, u(0, y) = 0$ and $u(a, y) = f(y)$ plarallel to y-axis. [UPTU (SUM)-2008]

25. Find the steady state temperature distribution in a rectangular thin plate with its two surfaces insulated and with the conditions $u(0, y) = 0, u(x, 0) = 0, u(a, y) = g(y), u(x, b) = f(x)$. [UPTU (SUM)-2007]

26. Solve the following Laplace equation $\dfrac{\partial^2 u}{\partial x^2} + \dfrac{\partial^2 u}{\partial y^2} = 0$ in a rectangle with $u(0, y) = 0, u(a, y) = 0, u(x, b) = 0$ and $u(x, 0) = f(x)$ along x-axis. [UPTU-2008]

27. Solve the boundary value problem.
    $$\dfrac{\partial^2 u}{\partial x^2} + \dfrac{\partial^2 u}{\partial y^2} = 0, \quad 0 \le x \le a, 0 \le y \le b$$
    with the boundary conditions
    $u_x(0, y) = u_x(a, y) = u_y(x, 0) = 0$ and $u_y(x, b) = f(x)$. [MTU (SUM)-2011]

28. Neglecting R and G, find the emf $v(x, t)$ in a line of length l, t seconds after the ends were suddenly grounded, given that $i(x, 0) = i_0$ and $v(x, 0) = e_1 \sin \dfrac{\pi x}{l} + e_5 \sin \dfrac{5\pi x}{l}$ [SVTU-2008, MTU-2012]

29. Solve $\dfrac{\partial^2 V}{\partial x^2} = LC \dfrac{\partial^2 V}{\partial t^2}$ assuming that the initial voltage is $V_0 \sin \dfrac{\pi x}{l}; V_t(x_0) = 0$ and $V = 0$ at the ends $x = 0$ and $x = l$ for all t. [UPTU (SUM)-2007]

30. A steady voltage distribution of 20 volts at the sending end and 12 volts at the receiving end is maintained in a telephone wire of length l. A time $t = 0$, the receiving end is grounded. Find the voltage and current t sec later. Neglect leakance and inductance. [MTU-2011, UPTU (SUM)-2008]

31. A homogeneous rod of conducting malerial of length 100 cm has its ends kept at zero temperature and the temperature initially is
    $$u(x, 0) = x \qquad 0 \le x \le 50$$
    $$= 100 - x \quad 50 \le x \le 100$$
    Find the temperature $u(x, t)$ at any time.
    [BHOPAL-2007, SVTU-2007, KURUKSHETRA-2006]

Hints to Selected Problems

3. The temperature $u(x, t)$ in the given solid is governed by the one dimensional heat equation

$$C\frac{\partial^2 u}{\partial x^2} = \frac{\partial u}{\partial t} \qquad \dots (1)$$

As per given, the boundary conditions are

$$u(0, t) = 0, \ u(a_1 - 1) = 0, \ \forall t \qquad \dots (2)$$

$$u(x, 0) = f(x) \qquad \dots (3)$$

Now, suppose that (1) has the solution of the form

$$u(x, t) = X(x) T(t) \qquad \dots (4)$$

where X and T are respectively the functions of x and t alone. Using the values of (4) in (1), we get

$$\frac{X''}{X} = \frac{T'}{CT} = \mu \ \text{(say)} \qquad \dots (5)$$

where μ is a separation constant.
From equation (5), we can deduce that

$$X'' - \mu X = 0 \qquad \dots (6)$$

and $$T' = \mu CT \qquad \dots (7)$$

Using (2) in (4), we get

$$X(0) = 0 \ \text{and} \ X(a) = 0 \qquad \dots (8)$$

Now, we want to solve (6) subject to the boundary condition (8). Here, we have the following cases :

Case (I). Let $\mu = 0$. Then solution of (6) is given by

$$X(x) = Ax + B \qquad \dots (9)$$

Using (8), we get

$$A = B = 0$$
$$X(x) = 0$$

$u = 0$, which does not satisfy (3).

Case (II). Let $\mu = \lambda^2, \lambda \neq 0$: In this case, solution of (6) is given by

$$X(x) = Ae^{\lambda x} + Be^{-\lambda x}.$$

Using (8), we get $A + B = 0$ and $0 = Ae^{0\lambda} + De^{-a\lambda}$

$\Rightarrow \qquad A = B = 0$
$\Rightarrow \qquad X = 0$
$\Rightarrow \qquad u = 0$

Thus, we reject this case also.

Case (III). Let $\mu = -\lambda^2, \lambda \neq 0$. In this case, solution of (6) is given by $X(x) = A\cos\lambda x + B\sin\lambda x$

Using (8), we get $A = 0$ and $A\cos\lambda a + B\sin\lambda a = 0$
Let $B \neq 0$, then $\sin\lambda a = 0$

$$\lambda a = n\pi, \ n = 1, 2, \dots$$

$$\lambda = \frac{n\pi}{a}, \ n = 1, 2, \dots$$

Therefore, non-zero solution of (6) is given by

$$X_n(x) = B_n \sin\left(\frac{n\pi x}{a}\right) \qquad \dots (10)$$

Putting $\lambda = \frac{n\pi}{a}$ in (7), we get

$$\frac{dT}{T} = -\frac{n^2\pi^2 C}{a^2} dt$$

$$\Rightarrow \qquad \frac{dT}{T} = -C_n^2 dt$$

whose solution is given by

$$T_n(t) = D_n e^{-C_n^2 t} \qquad \dots (11)$$

Thus, we have

$$u_n(x, t) = X_n(x) T_n(t) = E_n \sin\left(\frac{n\pi x}{a}\right) e^{-C_n^2 t} \dots (12)$$

The more general solution of (1) is given by

$$u(x, t) = \sum_{n=1}^{\infty} u_n(x, t) = \sum_{n=1}^{2} E_n \sin\left(\frac{n\pi x}{a}\right) e^{-C_n^2 t} \quad \dots (13)$$

Putting $t = 0$ in (13) and using (3), we get

$$f(x) = \sum_{n=1}^{\infty} E_n \sin\left(\frac{n\pi x}{a}\right)$$

which is a Fourier sine series, thus the constants E_n are given by

$$E_n = \frac{2}{a}\int_0^a f(x) \sin\left(\frac{n\pi x}{a}\right) dx, \ n = 1, 2, 3, \dots$$

4. Proceeding same as example 3, we get

$$u(x, y, t) = \sum_{m=1}^{\infty}\sum_{n=1}^{\infty} F_{mn} \sin mx \sin ny \ e^{-\lambda_{mn}^2 t} \qquad \dots (1)$$

where, $$F_{mn} = \frac{4}{\pi^2}\int_{x=0}^{\pi}\int_{y=0}^{\pi} xy \, (\pi - x)(\pi - y)\sin mx \sin ny \, dx \, dy$$

$$= \frac{4}{\pi^2}\left[\int_0^{\pi}(\pi x - x^2)\sin mx \, dx\right]\left[\int_0^{\pi}(\pi y - y^2)\sin ny \, dy\right] \ \dots (2)$$

Using chain rule of integration by parts, we get

$$\int_0^{\pi}(\pi x - x^2)\sin mx \, dx = \left[(\pi x - x^2)\left(-\frac{\cos mx}{m}\right)\right.$$
$$\left. -(\pi - 2x)\left(-\frac{\sin mx}{m^2}\right) + (-2)\left(\frac{\cos mx}{m^2}\right)\right]_0^{\pi}$$
$$= \frac{2}{m^3}\left[1 - (-1)^m\right]$$

In a similar manner, we get

$$\int_0^{\pi}(\pi y - y^2)\sin ny \, dy = \frac{2}{n^3}\left[1 - (-1)^n\right]$$

Thus, $$F_{mn} = \frac{16}{\pi^2 m^3 n^3}\left[1 - (-1)^m\right]\left[1 - (-1)^n\right]$$

$$= \begin{cases} 0, & \text{when } m = 2p \text{ or } n = 2q \\ \dfrac{64}{\pi^2(2p-1)^3(2q-1)^3}, & \text{when } m = 2p-1, \ n = 2q-1 \end{cases}$$

Hence, the required solution is given by

$$u(x, y, t) = \sum_{p=1}^{\infty}\sum_{q=1}^{\infty} F_{pq} \sin[(2p-1)x]\sin[(2q-1)y]e^{-\lambda_{pq}^2 t}$$

where, $$\lambda_{pq}^2 = C^2\left[(2p-1)^2 + (2q-1)^2\right]$$

and $$F_{pq} = \frac{64}{\pi^2}(2p-1)^3(2q-1)^3.$$

8. Proceeding same as example (3) by taking $a = 2\pi$ and $f(x) = 5m^3 x$, we get

$$u(x, t) = \sum_{n=1}^{\infty} E_n \sin\left(\frac{nx}{2}\right)e^{-(n^2 kt/4)}, \ n = 1, 2, 3, \dots \dots (1)$$

Putting $t = 0$ in (1) and using $u(x, 0) = \sin^3 x$, we get

$$\sum_{n=1}^{\infty} E_n \sin\left(\frac{nx}{2}\right) = \sin^3 x = \frac{3}{4}\sin x - \frac{1}{4}\sin 3x.$$

Because $\sin 3x = 3\sin x - 4\sin^3 x$

$$\Rightarrow \qquad \sin^3 x = \frac{1}{4}(3\sin x - \sin 3x)$$

Putting these values in (1), we get the required solution

$$u(x, t) = \frac{1}{4}(3\sin x e^{-kt} - \sin 3x \, e^{-9kt})$$

12. The given initial boundary value problem is given as follows :

$$\frac{\partial T}{\partial t} = \alpha \frac{\partial^2 T}{\partial x^2}$$

Subject to the boundary conditions

$$T(0,t) = 0,\ t > 0, \quad \frac{\partial T}{\partial x}(L,t) = q_0,\ t > 0$$

with initial conditions

$$T(x,0) = 0, \quad 0 \le x \le L.$$

Let $T(x,t) = T_s(x) + T_1(x,t)$

Where, T_s is a steady part and T is the transient part of the solution.

Thus, we have $\dfrac{\partial^2 T_s}{\partial x^2} = 0 \Rightarrow T_s = Ax + B$.

Using $x = 0$, $T_s = 0$, we get $B = 0$, therefore, $T_s = Ax$.

Further using the boundary condition $\dfrac{\partial T_s}{\partial x} = q_0$, we get $A = q_0$.

Thus the steady solution is given by $T_s = q_0 x$

For the transient part, the boundary and initial condition can be redefined as follows :

(i) $T_1(0,t) = T(0,t) - T_s(0) - 0 - 0 = 0$

(ii) $\dfrac{\partial T_1(x,t)}{\partial x} = \dfrac{\partial T(x,t)}{\partial x} - \dfrac{\partial T_s(L,t)}{\partial x} = q_0 - q_0 = 0$

(iii) $T_1(x,0) = T(x,0) - T_s(x) = -q_0 x, \ 0 < x < L$.

Therefore, for the transient part, we have to solve the given PDE subject to these conditions. We know that the solution of given PDE is $T_1(x,t) = e^{-\alpha \lambda^2 t}(A \cos \lambda x + B \sin \lambda x)$

Using (i), we get $A = 0$

$$T_1(x,t) = B e^{-\alpha \lambda^2 t} \sin \lambda x$$

Again using boundary conditions (ii), we get

$$\frac{\partial T}{\partial x}\bigg|_{x=L} = B\lambda\, e^{-\alpha \lambda^2 t} \cos \lambda L = 0.$$

$\Rightarrow \quad \lambda L = (2n-1)\dfrac{\pi}{2},\ n = 1,2,\ldots$

Thus, the general solution is given by

$$T_1(x,t) = \sum_{n=1}^{\infty} B_n \exp\left[-\alpha \left\{\frac{(2n-1)}{L}\right\}^2 \pi^2 t\right] \sin\left(\frac{2n-1}{2L}\pi x\right)$$

Applying initial condition (iii), we get

$$T_1(x,0) = -q_0 x = \sum_{n=1}^{\infty} B_n \sin\left(\frac{2n-1}{2L}\pi x\right).$$

Now, multiplying both sides by $\sin\left(\dfrac{2n-1}{2L}\pi x\right)$, integrating between 0 to L and using

$$\int_0^L B_n \sin\left(\frac{2n-1}{2L}\right)\pi x \sin\left(\frac{2m-1}{2L}\pi x\right) dx = \begin{cases} 0, & n \neq m \\ \dfrac{B_m L}{2}, & n = m \end{cases}$$

We get $-q_0 \dfrac{4L^2}{(2m-1)^2 \pi^2}\left[\sin\left(\dfrac{2m-1}{2}\pi\right)\right] = B_m \cdot \dfrac{L}{2}$.

$\Rightarrow -q_0 \dfrac{4L^2}{(2m-1)^2 \pi^2}(-1)^{m-1} = B_m \cdot \dfrac{L}{2}$

$\Rightarrow B_m = \dfrac{(-1)^m 8 L q_0}{(2m-1)^2 \pi^2}$

Hence, the required temperature distribution is given by

$$T(x,t) = q_0 x + \frac{8 L q_0}{\pi^2} \sum_{m=1}^{\infty}\left[\frac{(-1)^m}{(2m-1)^2}\right.$$
$$\left.\exp\left[-\alpha\{(2m-1)/2\}\right]^2 \pi^2 t\right] - \sin\left(\frac{2m-1}{2L}\pi x\right)$$

20. $(u_x)_{x=0} = 0 = (u_x)_{x=L} = L; u(x,0) = \sin\dfrac{\pi x}{L}$

21. $u(x,0) = \begin{cases} 2Tx & 0 \le x \le \dfrac{1}{2} \\[2mm] 2T(1-x) & \dfrac{1}{2} \le x \le 1 \end{cases}$

Answers

1. $u(x,t) = x + 2 + \displaystyle\sum_{n=1}^{\infty} E_n \sin n\pi x\, e^{-n^2\pi^2 C^2 t}$;   **2.** $y(x,t) = 10 + \displaystyle\sum_{n=1}^{\infty} E_n \sin\frac{(2n-1)}{2}\pi x\, e^{-C_n^2 t}$

3. $u(x,t) = \displaystyle\sum_{n=1}^{\infty} u_n(x,t) = \sum_{n=1}^{\infty} E_n \sin\left(\frac{n\pi x}{a}\right) e^{-C_n^2 t}$; $f(x) = \displaystyle\sum_{n=1}^{\infty} E_n \sin\frac{n\pi x}{a}$, where $E_n = \dfrac{2}{a}\int_0^a f(x) \sin\left(\dfrac{n\pi x}{a}\right) dx$, $n = 1,2,3,\ldots$

4. $E_n = \dfrac{2}{l}\int_0^l x \sin\dfrac{n\pi x}{l} dx = \begin{cases} \dfrac{2l}{n\pi} & ; n \text{ is odd} \\[3mm] -\dfrac{2l}{n\pi} & ; n \text{ is even} \end{cases}$

5. $u(x,y,t) = \displaystyle\sum_{p=1}^{\infty}\sum_{q=1}^{\infty} F_{pq} \sin[(2p-1)x] \sin[(2q-1)y] e^{-\lambda p^2 q t}$ where $\lambda_{pq}^2 = C^2\left[(2p-1)^2 + (2q-1)^2\right]$ and $F_{pq} = \dfrac{64}{\pi^2}(2p-1)^3 (2q-1)^3$

6. $u(x,l) = \dfrac{4A}{\pi}\displaystyle\sum_{n=1}^{\infty}\dfrac{1}{n}\sin^2\dfrac{n\pi}{4}\sin\dfrac{n\pi x}{l} e^{-(n^2\pi^2 kt)/l^2}$ **7.** $u(x,t) = \sin\left(\dfrac{\pi x}{l}\right) e^{-\pi^2 t/h^2 l^2}$

8. $u(x,t) = \dfrac{1}{4}\left[3\sin x e^{-kt} - \sin 3x e^{-akt}\right]$ **12.** $T(x,t) = q_0 x + \dfrac{8 L q_0}{\pi^2}\displaystyle\sum_{m=1}^{\infty}\left[\dfrac{(-1)^m}{(2m-1)^2}\exp\left[-\alpha\{(2m-1)/2\}^2 \pi^2 t\right]\sin\left(\dfrac{2m-1}{2L}\pi x\right)\right]$

14. $u(x,y,t) = A\sin\left(\dfrac{\pi x}{a}\right)\sin\left(\dfrac{\pi y}{a}\right) e^{-\pi^2 kt\left(\frac{1}{a^2} + \frac{1}{b^2}\right)}$ **16.** $u(x,t) = e^{-4t}\cos 2x$ **17.** $u(x,t) = \sin 2x e^{-4at}$

18. $u(x,t) = \dfrac{-400}{\pi}\displaystyle\sum_{n=1}^{\infty}\dfrac{\cos n\pi}{n}\sin\dfrac{n\pi x}{2} e^{-\left(\frac{c^2 n^2 \pi^2 t}{4}\right)}$ **19.** $u(x,t) = \dfrac{8d}{\pi^3}\displaystyle\sum_{n=1}^{\infty}\dfrac{1}{(2n-1)^3}\dfrac{\sin(2n-1)\pi x}{L} e^{\frac{-(2n-1)^2\pi^2 C^2 t}{L^2}}$

20. $u(x,t) = \dfrac{2}{\pi} - \dfrac{4}{\pi} \sum\limits_{m=1}^{\infty} \dfrac{1}{(4m^2-1)} \cos\left(\dfrac{2m\pi x}{L}\right) e^{\frac{-4m^2\pi^2 C^2 t}{L^2}}$

21. $u(x,t) = \dfrac{8T}{\pi^2} \sum\limits_{m=1}^{\infty} \dfrac{(-1)^{m+1}}{(2m-1)^2} \sin(2m-1)\pi x e^{-[(2m-1)^2\pi^2 c^2 t]}$.

22. $u(x,t) = \dfrac{4u_0}{\pi} \sum\limits_{n=1}^{\infty} \dfrac{1}{(2n-1)} \sin(2n-1)x e^{-(2n-1)y}$

23. $u(x,t) = \dfrac{-8}{\pi} \sum\limits_{m=1,2,3,...}^{\infty} \dfrac{\sin(2m-1)x \sinh(2m-1)(\pi-y)}{(2m-1)\{(2m-1)^2-4\}\sinh(2m-1)\pi}$

24. $u(x,t) = \sum\limits_{n=1}^{\infty} b_n \sin\dfrac{n\pi y}{b} \sinh\dfrac{n\pi x}{b}$ where, $b_n = \dfrac{2}{b\sinh\left(\dfrac{n\pi a}{b}\right)} \int\limits_0^b f(y)\sin\dfrac{n\pi y}{b} dy$

25. $u(x,y) = \sum\limits_{n=1}^{\infty} \left[b_n \sin\left(\dfrac{n\pi x}{a}\right) \sinh\left(\dfrac{n\pi y}{a}\right) + B_n \sin\left(\dfrac{n\pi y}{b}\right) \sinh\left(\dfrac{n\pi x}{b}\right) \right]$ where $b_n = \dfrac{2}{a\sinh\left(\dfrac{n\pi b}{a}\right)} \int\limits_0^a f(x)\sin\dfrac{n\pi x}{a} dx$

and $B_n = \dfrac{2}{b\sinh\left(\dfrac{n\pi a}{b}\right)} \int\limits_0^a g(y)\sin\dfrac{n\pi y}{b} dy$ **26.** $u(x,y) = \sum\limits_{n=1}^{\infty} B_n \sin\left(\dfrac{n\pi x}{a}\right) \sinh\dfrac{n\pi}{a}(b-y)$, where, $B_n = \dfrac{2}{a\sinh\left(\dfrac{n\pi b}{a}\right)} \int\limits_0^a f(x)\sin\dfrac{n\pi x}{a} dx$

27. $u(x,y) = \sum\limits_{n=1}^{\infty} b_n \cos\dfrac{n\pi x}{a} \left(e^{\frac{n\pi y}{a}} - e^{\frac{-n\pi y}{a}} \right)$ where $b_n = \dfrac{1}{n\pi \cosh\dfrac{n\pi b}{a}} \int\limits_0^a f(x)\cos\dfrac{n\pi x}{a} dx$

28. $V = e_1 \sin\dfrac{\pi x}{l} \cos\dfrac{\pi t}{l\sqrt{2C}} + e_5 \sin\dfrac{5\pi x}{l} \cos\dfrac{5\pi t}{l\sqrt{LC}}$ **29.** $V(x,t) = V_0 \sin\dfrac{\pi x}{l} \cos\dfrac{\pi t}{l\sqrt{LC}}$

30. $V(x,t) = \dfrac{20(l-x)}{l} + \dfrac{24}{\pi} \sum\limits_{n=1}^{\infty} \dfrac{(-1)^{n+1}}{n} \sin\dfrac{n\pi x}{l} e^{-n^2\pi^2 t/RCl^2}$ $i(x,t) = \dfrac{20}{lR} + \dfrac{24}{lR} \sum\limits_{n=1}^{\infty} (-1)^n \cos\dfrac{n\pi x}{l} e^{-n^2\pi^2 t/RCl^2}$

31. $u(x,t) = \dfrac{400}{\pi^2} \sum\limits_{n=1}^{\infty} \dfrac{(-1)^n}{(2n+1)^2} - e^{-[(2n+1)c\pi/100]^2 t} \sin\dfrac{(2n+1)\pi x}{100}$

8.16 FOURIER TRANSFORMS

If a function $f(x)$ defined on the interval $]-\infty,\infty[$, and piecewise continuous in each finite partial interval and absolutely integrable in $]-\infty,\infty[$, then

$$F(f(x)) = \tilde{f}(p) = \int_{-\infty}^{\infty} e^{ipx} f(x)dx$$

is defined as Fourier transform of $f(x)$.

The inverse formula for Fourier transform is given by

$$F^{-1}[\tilde{f}(p)] = f(x) = \dfrac{1}{2\pi} \int_{-\infty}^{\infty} \tilde{f}(p)e^{-ipx} dp.$$

☞ REMARK

- We can also define $\tilde{f}(p) = F(f(x)) = \dfrac{1}{\sqrt{2\pi}} \int_{-\infty}^{\infty} e^{-ipx} f(x)dx$ and $F^{-1}[\tilde{f}(p)] = f(x) = \dfrac{1}{\sqrt{2\pi}} \int_{-\infty}^{\infty} e^{-ipx} \tilde{f}(p)dp$.

8.16.1 FOURIER SINE AND COSINE TRANSFORMS

Definition (1): *The infinite Fourier sine transform of the function $f(x)$, $0 < x < \infty$ is defined by $F_s[f(x)]$ or $\tilde{f}_s(p)$ and defined by*

$$\tilde{f}_s(p) = F_s[f(x)] = \sqrt{\dfrac{2}{\pi}} \int_0^{\infty} f(x)\sin px\, dx$$

The inverse formula for infinite Fourier sine transform is given by

$$f(x) = F_s^{-1}[\tilde{f}_s(p)] = \sqrt{\dfrac{2}{\pi}} \int_0^{\infty} \tilde{f}_s(p)\sin px\, dp.$$

Definition (2): *The infinite Fourier cosine transform of $f(x)$, $0 < x < \infty$ is denoted by $F_c[f(x)]$ or $\tilde{f}_c(p)$ and is defined by*

$$\tilde{F}_c[f(x)] = \tilde{f}_c(p) = \sqrt{\dfrac{2}{\pi}} \int_0^{\infty} f(x)\cos px\, dx$$

The inversion formula for Fourier cosine transform is given by

$$f(x) = \sqrt{\dfrac{2}{\pi}} \int_0^{\infty} \tilde{f}_c(p)\cos px\, dp.$$

(1) LINEARITY PROPERTY OF FOURIER TRANSFORMS

Let $\tilde{f}(p)$ and $\tilde{g}(p)$ are Fourier transform of $f(x)$ and $g(x)$ respectively. Then $F\{af(x) + bg(x)\} = a\tilde{f}(p) + b\tilde{g}(p)$ where a and b constants.

(2) CHANGE OF SCALE PROPERTY

THEOREM 1. **(For Complex Fourier Transform).** *If $\tilde{f}(p)$ is the complex Fourier transform of $f(x)$, the complex Fourier transform of $f(ax)$ is given by $\dfrac{1}{a}\tilde{f}\left(\dfrac{p}{a}\right)$.*

PROOF. By definition, we have

$$\int_{-\infty}^{\infty} e^{ipx}f(x)dx = \tilde{f}(p).$$ ...(1)

Consider $\quad \tilde{f}(ap) = \int_{-\infty}^{\infty} e^{ipx}f(x)dx.$

Putting $\quad\quad ax = t \Rightarrow dx = dt$, we get

$$\tilde{f}(ap) = \frac{1}{a}\int_{-\infty}^{\infty} e^{ip(t/a)}f(t)dt = \frac{1}{a}\int_{-\infty}^{\infty} e^{i(p/a)x}f(x)dx = \frac{1}{a}\tilde{f}\left(\frac{p}{a}\right).$$

☛ REMARK

- In a similar way, we can prove that :

  (a) If $\tilde{f}_s(p)$ is the Fourier sine transform of $f(x)$, then Fourier sine transform of $f(ax)$ is given by $\dfrac{1}{a}\tilde{f}_s\left(\dfrac{p}{a}\right)$.

  (b) If $\tilde{f}_c(p)$ is the Fourier cosine transform of $f(x)$, then Fourier cosine transform of $f(ax)$ is given by $\dfrac{1}{a}\tilde{f}_c\left(\dfrac{p}{a}\right)$.

THEOREM 2. **(Shifting Property).** *If $\tilde{f}(p)$ is the complex Fourier transform of $f(x)$, then complex Fourier transform of $f(x-a)$ is $e^{ipa}\tilde{f}(p)$.*

PROOF. By definition, we have

$$\int_{-\infty}^{\infty} e^{ipx}f(x)dx = \tilde{f}(p)$$ ...(1)

Consider $\quad \tilde{f}(x-a) = \int_{-\infty}^{\infty} e^{ipx}f(x-a)dx = \int_{-\infty}^{\infty} e^{ip(t+a)}f(t)dt$ [Putting $x-a=t$]

$$= e^{ipx}\int_{-\infty}^{\infty} e^{ipt}f(t)dt = e^{ipa}\tilde{f}(p)$$

(3) SOME IMPORTANT INTEGRALS (TO BE USED DIRECTLY)

1. $\int e^{ax}\sin bx\,dx = \dfrac{e^{ax}}{a^2+b^2}(a\sin bx - b\cos bx)$ **2.** $\int e^{ax}\cos bx\,dx = \dfrac{e^{ax}}{a^2+b^2}(a\cos bx + b\sin bx)$

3. $\int_0^\infty e^{-ax}\sin bx\,dx = \dfrac{b}{a^2+b^2}$ **4.** $\int_0^\infty e^{-ax}\cos bx\,dx = \dfrac{a}{a^2+b^2}$

5. $\dfrac{d^n}{dx^n}\left(\dfrac{x}{a^2+x^2}\right) = \dfrac{(-1)^n.n!}{(a^2+x^2)^{(n+1)/2}}\cos\left[(n+1)\tan^{-1}\left(\dfrac{a}{x}\right)\right]$

6. $\dfrac{d^n}{dx^n}\left(\dfrac{a}{a^2+x^2}\right) = \dfrac{(-1)^n.n!}{(a^2+x^2)^{(n+1)/2}}\sin\left[(n+1)\tan^{-1}\left(\dfrac{a}{x}\right)\right]$

7. $\int_0^\infty \dfrac{\sin px}{x}dx = \begin{bmatrix} \pi/2 & ; & \text{if } p>0 \\ -\pi/2 & ; & \text{if } p<0 \end{bmatrix}$ **8.** $\int_0^\infty e^{-x^2}dx = \sqrt{\pi}, \int_0^\infty e^{-x^2}dx = \dfrac{\sqrt{\pi}}{2}$

9. $\int_0^\infty \dfrac{e^{ax}-e^{-ax}}{e^{\pi x}-e^{-\pi x}}dx = \dfrac{1}{2}\tan\dfrac{a}{2}, \int_0^\infty \dfrac{e^{ax}+e^{-ax}}{e^{\pi x}-e^{-\pi x}}dx = \dfrac{1}{2}\sec\dfrac{a}{2}$ **10.** $\int_0^\infty \dfrac{\cos px}{1+p^2}dp = \dfrac{\pi e^{-x}}{2}, \int_0^\infty \dfrac{p\sin px}{1+p^2}dp = \dfrac{\pi}{2}e^{-x}$

THEOREM 3. **(Modulation Theorem).** *If $\tilde{f}(p)$ is the complex Fourier transform of $f(x)$ then, the Fourier transform of $f(x)\cos ax$ is $\dfrac{1}{2}[\tilde{f}(p-a)+\tilde{f}(p+a)]$.*

PROOF. By definition, we have

$$\tilde{f}(p) = \int_{-\infty}^{\infty} e^{ipx}f(x)dx$$

Now, $\quad F[f(x)\cos ax] = \int_{-\infty}^{\infty} e^{ipx}f(x)\cos ax\,dx$

$$= \int_{-\infty}^{\infty} e^{ipx}f(x)\frac{e^{iax}+e^{-iax}}{2}dx = \frac{1}{2}\int_{-\infty}^{\infty} e^{i(p+a)x}f(x)dx + \frac{1}{2}\int_{-\infty}^{\infty} e^{i(p-a)x}f(x)dx = \frac{1}{2}\tilde{f}(p+a) + \frac{1}{2}\tilde{f}(p-a)$$

☛ REMARKS

- In a similar way, we can prove that

  If $\tilde{f}_s(p)$ is the complex Fourier transform of $f(x)$, then

  1. Fourier transform of $f(x)\cos ax$ is $\dfrac{1}{2}[\tilde{f}_s(p+a)+\tilde{f}_s(p-a)]$

  2. Fourier transform of $f(x)\sin ax$ is $\dfrac{1}{2}[\tilde{f}_s(p+a)-\tilde{f}_s(p-a)]$

  3. Fourier transform of $f(x)\sin ax$ is $\dfrac{1}{2}[\tilde{f}_c(p-a)-\tilde{f}_c(p+a)]$

(4) SOME MORE IMPORTANT RESULTS

1. Multiple Fourier Transforms

Let $f(x, y)$ be a function of two variables x and y. Temporarily, as a function of x, its Fourier transform is given by

$$\tilde{f}(p, y) = \frac{1}{\sqrt{2\pi}} \int_{-\infty}^{\infty} f(x, y) e^{ipx} dx$$

Also, regarding $\tilde{f}(p, y)$ as a function of y, its Fourier transform is given by

$$\tilde{f}(p, q) = \frac{1}{\sqrt{2\pi}} \int_{-\infty}^{\infty} \tilde{f}(p, y) e^{iqy} dy = \frac{1}{2\pi} \int_{-\infty}^{\infty} \int_{-\infty}^{\infty} f(x, y) e^{i(px + qy)} dx dy$$

2. Inversion Formula : If

$$f(x, y) = \frac{1}{\sqrt{2\pi}} \int_{-\infty}^{\infty} \tilde{f}(p, y) e^{-ipx} dp$$

and

$$\tilde{f}(p, y) = \frac{1}{\sqrt{2\pi}} \int_{-\infty}^{\infty} \tilde{f}(p, q) e^{-iqy} dq$$

Then

$$f(x, y) = \frac{1}{2\pi} \int_{-\infty}^{\infty} \tilde{f}(p, q) e^{-i(px + qy)} dp dq$$

3. Convolution

The function $H(x) = F * G = \frac{1}{\sqrt{2\pi}} \int_{-\infty}^{\infty} F(u) G(x - u) du$ is called the convolution or falting of two integrable functions F and G over the interval $]-\infty, \infty[$.

Sometime it can also be defined as $F * G = \int_{-\infty}^{\infty} F(u) G(x - u) du$.

4. The convolution of Falting Theorem for Fourier Transforms

If $F[f(x)]$ and $F[g(x)]$ are the Fourier transforms of the functions $f(x)$ and $g(x)$ respectively, then the Fourier transform of the convolution of $f(x)$ and $g(x)$ is the product of their Fourier transforms, i.e.,

$$F\{f(x) * g(x)\} = F\{f(x)\}.F\{g(x)\}$$

Proof.　Consider

$$
\begin{aligned}
\text{LHS} &= F\{f(x) * g(x)\} \\
&= F\int_{-\infty}^{\infty} f(u) g(x - u) du \qquad \text{(By definition of convolution)} \\
&= \int_{-\infty}^{\infty} e^{ipx} \left\{ \int_{-\infty}^{\infty} f(u) g(x - u) du \right\} dx = \int_{-\infty}^{\infty} f(u) \left\{ \int_{-\infty}^{\infty} e^{ipx} g(x - u) dx \right\} du \\
&= \int_{-\infty}^{\infty} f(u) \left\{ \int_{-\infty}^{\infty} e^{-ip(u+v)} g(v) dv \right\} du \qquad [\text{Putting } x - u = v \Rightarrow dx = dv] \\
&= \int_{-\infty}^{\infty} e^{ipu} f(u) \left\{ \int_{-\infty}^{\infty} e^{ipv} g(v) dv \right\} du \\
&= \int_{-\infty}^{\infty} e^{ipu} f(u) F\{g(x)\} du = F\{g(x)\} \int_{-\infty}^{\infty} e^{ipu} f(u) du = F\{g(x)\}.F\{f(x)\}.
\end{aligned}
$$

5. Parseval's identity for Fourier Transform: Rayleigh Theorem or Plancharel's Theroem.

If $f(p)$ and $g(p)$ are the complex Fourier transform of $F(x)$ and $G(x)$ respectively, then

(i) $\dfrac{1}{2\pi} \int_{-\infty}^{\infty} f(p) \overline{g(p)} dp = \int_{-\infty}^{\infty} F(x) \overline{G(x)} dx$ 　　　(ii)　$\dfrac{1}{2\pi} \int_{-\infty}^{\infty} |f(p)|^2 dp = \int_{-\infty}^{\infty} |F(x)|^2 dx$

where bar denotes the complex conjugate.

Proof.　(i) By inversion formula, we have

$$G(x) = \frac{1}{2\pi} \int_{-\infty}^{\infty} g(p) e^{-ipx} dp \qquad\qquad \dots(1)$$

$$\Rightarrow \qquad \overline{G(x)} = \frac{1}{2\pi} \int_{-\infty}^{\infty} \overline{g(p)} e^{ipx} dp \qquad\qquad \dots(2)$$

Thus,

$$
\begin{aligned}
\int_{-\infty}^{\infty} F(x) \overline{G(x)} dx &= \int_{-\infty}^{\infty} F(x) \left\{ \frac{1}{2\pi} \int_{-\infty}^{\infty} \overline{g(p)} e^{ipx} dp \right\} dx \\
&= \frac{1}{2\pi} \int_{-\infty}^{\infty} \overline{g(p)} \left\{ \int_{-\infty}^{\infty} F(x) e^{ipx} dx \right\} dp = \frac{1}{2\pi} \int_{-\infty}^{\infty} \overline{g(p)} f(p) dp
\end{aligned}
$$

(ii) Taking $G(x) = F(x)$ in part (i), we get

$$\frac{1}{2\pi} \int_{-\infty}^{\infty} f(p) \overline{f(p)} dp = \int_{-\infty}^{\infty} F(x) \overline{F(x)} dx$$

Hence,

$$\frac{1}{2\pi} \int_{-\infty}^{\infty} |f(p)|^2 dp = \int_{-\infty}^{\infty} |F(x)|^2 dx$$

⌂ REMARKS

- In a similar way, we can prove that

(a)　$\dfrac{2}{\pi} \int_0^{\infty} f_c(p) g_c(p) dp = \int_0^{\infty} F(x) G(x) dx$ 　　　(b)　$\dfrac{2}{\pi} \int_0^{\infty} f_s(p) g_s(p) dp = \int_0^{\infty} F(x) G(x) dx$

(c)　$\dfrac{2}{\pi} \int_0^{\infty} [f_c(p)]^2 dp = \int_0^{\infty} [F(x)]^2 dx$ 　　　(d)　$\dfrac{2}{\pi} \int_0^{\infty} [f_s(p)]^2 dp = \int_0^{\infty} [F(x)]^2 dx$

8.16.2 Relation between Fourier and Laplace Transform

Consider the function $\qquad\qquad f(t) = \begin{cases} e^{-xt} g(t) & ; \quad t < 0 \\ 0 & ; \quad t > 0 \end{cases}$ ...(1)

Then the Fourier transform of $f(t)$ is given by

$$F[f(t)] = \int_{-\infty}^{\infty} e^{ipt} f(t) dt$$

$$= \int_{-\infty}^{0} 0 e^{ipt} dt + \int_{0}^{\infty} e^{-xt} g(t) e^{ipt} dt = \int_{0}^{\infty} e^{(ip-x)t} g(t) dt$$

$$= \int_{0}^{\infty} e^{-xst} g(t) dt = L\{g(t)\} \qquad\qquad\qquad \text{[Putting } x - ip = s]$$

Hence, Fourier transformation of the function $f(t)$ defined by (1) is the Laplace transform of function $g(t)$.

8.16.3 Fourier Transforms of the Derivative of a Function

The Fourier transform of the function $\dfrac{d^n F}{dx^n}$ *is* $(-ip)^n$ *times the Fourier transform of the function* $f(x)$ *provided that the first* $(n-1)$ *derivatives of* $F(x)$ *vanish as* $x \to \pm\infty$, *i.e,* $F\{f^n(x)\} = (-ip)^n F\{f(x)\}$

Proof. By definition, we have

$$F\{f^n(x)\} = \int_{-\infty}^{\infty} f^n(x) e^{ipx} dx = \left[f^{n-1}(x) e^{ipx} \right]_{-\infty}^{\infty} - \int_{-\infty}^{\infty} f^{n-1}(x) ip\, e^{ipx} dx$$

$$= -ip \int_{-\infty}^{\infty} f^{n-1}(x) e^{ipx} dx \qquad\qquad\qquad \left[\because \lim_{x \to \pm\infty} f^{n-1}(x) = 0 \right]$$

Repeating this process of integration by parts $(n-1)$ times more and assuming that

$$\lim_{x \to \pm\infty} f^r(x) = 0 \text{ for } r = 1, 2, 3, \ldots, n-1$$

Thus, finally we have

$$F\{f^n(x)\} = (-ip)^n \int_{-\infty}^{\infty} f(x) e^{ipx} dx$$

Hence, $\qquad\qquad\qquad F[f^n(x)] = (-ip)^n F[f(x)]$.

☑ Remark

- In a similar way, we can prove the following results:

(a) $\tilde{f}_c^{\,2n}(p) = -\sum\limits_{r=0}^{n-1} (-1)^r \alpha_{2n-2r-1} p^{2r} + (-1)^n p^{2n} \tilde{f}_c(p)$ (b) $\tilde{f}_c^{\,2n+1}(p) = -\sum\limits_{r=0}^{n-1} (-1)^r \alpha_{2n-2r} p^{2r} + (-1)^n p^{2n+1} \tilde{f}_s(p)$

(c) $\tilde{f}_c^{\,2n}(p) = -\sum\limits_{r=1}^{n} (-1)^r \alpha_{2n-2r} p^{2r-1} + (-1)^{n+1} p^{2n} \tilde{f}_s(p)$ (d) $\tilde{f}_s^{\,2n+1}(p) = -\sum\limits_{r=1}^{n} (-1)^r \alpha_{2n-2r+1} p^{2r-1} + (-1)^{n+1} p^{2n+1} \tilde{f}_c(p)$

Provided that first $(n-1)$ derivatives of $f(x)$ vanish as $x \to \infty$ and $\dfrac{d^{n+1} f}{dx^{n-1}} \to \alpha_{n-1}$, etc. $x \to 0$.

🗁 Solved Examples

EXAMPLE 1. *Find the complex Fourier transform of* $f(x) = e^{-a|x|}$, *where* $a > 0$ *and* x *belongs to* $]-\infty, \infty[$.

(VTU–2010, SVTU–2008, KOTTAYAM–2005)

SOLUTION. We have $\tilde{f}(p) = \int_{-\infty}^{\infty} f(x) e^{ipx} dx = \int_{-\infty}^{\infty} e^{ipx} e^{-a|x|} dx$

$$= \int_{-\infty}^{0} e^{ipx} e^{ax} dx + \int_{0}^{\infty} e^{ipx} e^{-ax} dx$$

$$= \int_{-\infty}^{0} e^{(a+ip)x} dx + \int_{0}^{\infty} e^{-(a-ip)x} dx$$

$$= \left[\frac{e^{(a+ip)x}}{a+ip} \right]_{-\infty}^{0} + \left[\frac{e^{-(a-ip)x}}{-(a-ip)} \right]_{0}^{\infty}$$

$$= \frac{1}{a+ip} + \frac{1}{a-ip} = \frac{2a}{a^2 + p^2}$$

EXAMPLE 2. *Find the Fourier transform of* $f(x)$ *defined by*

$$f(x) = \begin{cases} 1 & ; \quad |x| < a \\ 0 & ; \quad |x| > a \end{cases} \text{ and hence evaluate}$$

(i) $\int_{-\infty}^{\infty} \dfrac{\sin pa \cos px}{p} dp$ *and* (ii) $\int_{0}^{\infty} \dfrac{\sin p}{p} dp$

(WBTU–2005, MADRAS–2003, PTU–2003, KOTTAYAM–2005)

SOLUTION. We have $|x| < a \Rightarrow -a < x < a$

and $\qquad |x| > a \Rightarrow x < -a \text{ or } x > a$

Thus, $\qquad f(x) = \begin{cases} 1 & ; \quad \text{if } -a < x < a \\ 0 & ; \quad \text{if } x < -a \text{ or } x > a \end{cases}$ ...(1)

So, $\tilde{f}(p) = F[f(x)]$

$$= \int_{-\infty}^{\infty} e^{ipx} f(x) dx$$

$$= \int_{-\infty}^{-a} e^{ipx} f(x) dx + \int_{-a}^{a} e^{ipx} f(x) dx + \int_{a}^{\infty} e^{ipx} f(x) dx$$

$$= 0 + \int_{-a}^{a} e^{ipx} dx + 0 \qquad\qquad \text{[Using (1)]}$$

$$= \left[\frac{e^{ipx}}{ip} \right]_{-a}^{a} = \frac{1}{ip} (e^{ipa} - e^{-ipa}) = \frac{1}{ip} (2i \sin pa)$$

$$\Rightarrow \tilde{f}(p) = [Ff(x)] = \frac{2 \sin pa}{p} \qquad\qquad ...(2)$$

Using corresponding inversion formula, we get

$$\frac{1}{2\pi} \int_{-\infty}^{\infty} \tilde{f}(p) e^{-ipx} dp = f(x)$$

$$\Rightarrow \qquad \int_{-\infty}^{\infty} \frac{2 \sin pa}{p} e^{-ipx} dp = 2\pi f(x)$$

$$\Rightarrow 2\int_{-\infty}^{\infty}\frac{\sin pa}{p}(\cos px - i\sin px)dp = 2\pi f(x)$$

$$\Rightarrow \int_{-\infty}^{\infty}\frac{1}{p}(\sin pa\cos px - i\sin pa\sin px)dp = \pi f(x).$$

Equating real parts of both the sides, we get

$$\int_{-\infty}^{\infty}\frac{\sin pa\cos px}{p}dx = \pi f(x)$$

$$\Rightarrow \int_{-\infty}^{\infty}\frac{\sin pa\cos px}{p}dx = \begin{cases} \pi & ; \ |x| < a \\ 0 & ; \ |x| > a \end{cases}$$

Putting $x = 0$ and $a = 1$, we get

$$\int_{-\infty}^{\infty}\frac{\sin p}{p}dp = \pi \ \text{or} \ 2\int_0^{\infty}\frac{\sin p}{p}dp = \pi$$

Hence, $\int_0^{\infty}\dfrac{\sin p}{p}dp = \dfrac{\pi}{2}$.

EXAMPLE 3. If $f(x) = \begin{cases} 1 & ; \ |x| < a \\ 0 & ; \ |x| > a \end{cases}$ and $\tilde{f}(p) = \dfrac{2\sin pa}{p}$,

where $p \ne 0$ then show that $\int_0^{\infty}\dfrac{\sin^2 ax}{x}dx = \dfrac{\pi a}{2}$.

(VTU–2010, SVTU–2009, UPTU–2008)

SOLUTION. Using Parseval's identity for Fourier transform, we get

$$\int_{-\infty}^{\infty}[f(x)]^2 dx = \frac{1}{2\pi}\int_{-\infty}^{\infty}[\tilde{f}(p)]^2 dp$$

$$\Rightarrow \int_{-\infty}^{-a}[f(x)]^2 dx + \int_{-a}^{a}[f(x)]^2 dx + \int_{a}^{\infty}[f(x)]^2 dx = \frac{1}{2\pi}\int_{-\infty}^{\infty}[\tilde{f}(p)]^2 dp$$

$$\Rightarrow 0 + \int_{-a}^{a}1^2 dx + 0 = \frac{1}{2\pi}\int_{-\infty}^{\infty}\frac{4\sin^2 pa}{p^2}$$

$$\Rightarrow 2a = \frac{2}{\pi}\int_{-\infty}^{\infty}\frac{\sin^2 ax}{x^2}dx \ \text{or} \ \int_{-\infty}^{\infty}\frac{\sin^2 ax}{x^2}dx = \pi a$$

or $2\int_0^{\infty}\dfrac{\sin^2 ax}{x^2}dx = \pi a$ or $\int_0^{\infty}\dfrac{\sin^2 ax}{x^2}dx = \dfrac{\pi a}{2}$.

EXAMPLE 4. Find the Fourier transform of $f(x)$, if

$$f(x) = \begin{cases} x^2 & , \ |x| < a \\ 0 & , \ |x| > a \end{cases} \qquad \text{(SVTU–2008)}$$

SOLUTION. We have

$$\tilde{f}(p) = \frac{1}{\sqrt{2\pi}}\int_{-\infty}^{\infty}f(x)e^{ipx}dx = \frac{1}{\sqrt{2\pi}}\int_{-a}^{a}x^2 e^{ipx}dx$$

$$= \frac{1}{\sqrt{2\pi}}\cdot\left[x^2\cdot\frac{e^{ipx}}{ip} - 2x\cdot\frac{e^{ipx}}{(ip)^2} + 2\cdot\frac{e^{ipx}}{(ip)^3}\right]_{-a}^{a}$$

$$= \frac{1}{\sqrt{2\pi}}\left[\frac{a^2}{ip}(e^{ipa} - e^{-ipa})\right.$$
$$\left. + \frac{2a}{p^2}(e^{ipa} + e^{-ipa}) - \frac{2}{ip^3}(e^{ipa} - e^{-ipa})\right]$$

$$= \frac{1}{p^3\sqrt{2\pi}}\left[\frac{a^2 p^2}{i}\cdot 2i\sin pa\right.$$
$$\left. + 2ap\cdot 2\cos pa - \frac{2}{i}\cdot 2i\sin pa\right]$$

$$= \frac{1}{p^3}\sqrt{\frac{2}{\pi}}\left[(a^2 p^2 - 2)\sin pa + 2ap\cdot\cos pa\right]$$

EXAMPLE 5. Find the Fourier transform of

$$F(x) = \begin{bmatrix} 1 - x^2 & , \ |x| \le 1 \\ 0 & , \ |x| > 1 \end{bmatrix}$$

(VTU–2011, ANNA–2005, 12, MUMBAI –2005)

and hence evaluate $\int_0^{\infty}\left(\dfrac{x\cos x - \sin x}{x^3}\right)\cos\dfrac{x}{2}dx$

SOLUTION. We have

$$\tilde{F}(p) = \frac{1}{\sqrt{2\pi}}\int_{-\infty}^{\infty}e^{ipx}F(x)dx$$

$$= \frac{1}{\sqrt{2\pi}}\int_{-1}^{1}(1 - x^2)e^{ipx}dx$$

$$= \frac{1}{\sqrt{2\pi}}\left(\frac{1-x^2}{ip}e^{ipx}\right)_{-1}^{1} + \frac{2}{\sqrt{2\pi}}\int_{-1}^{1}x\cdot\frac{e^{ipx}}{ip}dx$$

$$= \frac{\sqrt{2}}{ip\sqrt{\pi}}\left[\left(\frac{xe^{ipx}}{ip}\right)_{-1}^{1} - \int_{-1}^{1}1\cdot\frac{e^{ipx}}{ip}dx\right]$$

$$= \frac{\sqrt{2}}{ip\sqrt{\pi}}\left\{\frac{e^{ip} + e^{-ip}}{ip} - \frac{(e^{ipx}) - 1}{(ip)^2}\right\}$$

$$= \frac{\sqrt{2}}{ip\sqrt{\pi}}\left[\frac{2\cos p}{ip} + \frac{e^{ip} - e^{-ip}}{p^2}\right]$$

$$= \frac{\sqrt{2}}{\sqrt{\pi}}\left[-\frac{2\cos p}{p^2} + \frac{2i\sin p}{ip^3}\right]$$

$$= -2\sqrt{\frac{2}{\pi}}\cdot\left[\frac{p\cos p - \sin p}{p^3}\right]$$

If $\tilde{F}(p) = \dfrac{1}{\sqrt{2\pi}}\int_{-\infty}^{\infty}F(x)e^{ipx}dx$

then $F(x) = \dfrac{1}{\sqrt{2\pi}}\int_{-\infty}^{\infty}\tilde{F}(p)e^{-ipx}dp$

$$\therefore -\frac{1}{\sqrt{2\pi}}\int_{-\infty}^{\infty}\frac{2\sqrt{2/\pi}(p\cos p - \sin p)}{p^3}\cdot e^{-ipx}dp$$

$$= \begin{bmatrix} 1 - x^2 & , \ |x| < 1 \\ 0 & , \ |x| > 1 \end{bmatrix}$$

or $-\int_{-\infty}^{\infty}\left(\dfrac{p\cos p - \sin p}{p^3}\right)\cos px\, dp$

or $+i\int_{-\infty}^{\infty}\left(\dfrac{p\cos p - \sin p}{p^3}\right)\sin px\, dp$

$$= \begin{bmatrix} \frac{\pi}{2}(1 - x^2) & , \ |x| < 1 \\ 0 & , \ |x| > 1 \end{bmatrix}$$

or $-\int_{-\infty}^{\infty}\dfrac{p\cos p - \sin p}{p^3}\cos px\, dp$

$$= \begin{bmatrix} \frac{\pi}{2}(1 - x^2) & , \ |x| < 1 \\ 0 & , \ |x| > 1 \end{bmatrix}$$

Now taking $x = \dfrac{1}{2}$, we have

$$-\int_{-\infty}^{\infty}\frac{p\cos p - \sin p}{p^3}\cdot\cos\frac{p}{2}dp = \frac{\pi}{2}\left(1 - \frac{1}{4}\right) = \frac{3\pi}{8}$$

or $2\int_0^{\infty}\dfrac{p\cos p - \sin p}{p^3}\cdot\cos\dfrac{p}{2}dp = -\dfrac{3\pi}{8}$

or $\int_0^{\infty}\left(\dfrac{x\cos x - \sin x}{x^3}\right)\cdot\cos\dfrac{x}{2}dx = -\dfrac{3\pi}{16}$.

EXAMPLE 6. Find the cosine transform of a function of x which is unity for $0 < x < a$ and zero for $x \ge a$. What is the *function whose cosine transform is* $\sqrt{\dfrac{2}{\pi}}\dfrac{\sin ap}{p}$?

(VTU–2010, SVTU–2009 UPTU–2008)

SOLUTION. It is given that $f(x) = \begin{cases} 1 & , \quad 0 < x < a \\ 0 & , \quad x \ge a \end{cases}$

We have $\tilde{f}_c(p) = \sqrt{\dfrac{2}{\pi}}\int_0^\infty f(x)\cos px\,dx$

$= \sqrt{\dfrac{2}{\pi}}\int_0^a \cos px\,dx = \sqrt{\dfrac{2}{\pi}}\dfrac{\sin pa}{p}$

Again, $f(x) = \sqrt{\dfrac{2}{\pi}}\int_0^\infty \tilde{f}_c(p)\cos px\,dp$

$= \sqrt{\dfrac{2}{\pi}}\int_0^\infty \sqrt{\dfrac{2}{\pi}}\dfrac{\sin pa}{p}\cos px\,dp$

$= \dfrac{1}{\pi}\int_0^\infty \dfrac{\sin(a+x)p + \sin(a-x)p}{p}dp$

$= \dfrac{1}{\pi}\int_0^\infty \dfrac{\sin(a+x)p}{p}dp + \dfrac{1}{\pi}\int_0^\infty \dfrac{\sin(a-x)p}{p}dp$

$= \dfrac{1}{\pi}\left(\dfrac{\pi}{2} + \dfrac{\pi}{2}\right) = 1$ if $x < a$

and $= \dfrac{1}{\pi}\left(\dfrac{\pi}{2} - \dfrac{\pi}{2}\right) = 0$ if $x > a$

$$\left[\because \int_0^\infty \dfrac{\sin ax}{x}dx = \dfrac{\pi}{2},\ \text{if } a > 0\right]$$

EXAMPLE 7. *Find the Fourier sine and cosine transform of* $f(x)$, *if*

$$f(x) = \begin{cases} x & , \quad 0 < x < 1 \\ 2-x & , \quad 1 < x < 2 \\ 0 & , \quad x > 2 \end{cases} \quad \text{[JNTU–2006]}$$

SOLUTION. We have

$\tilde{f}_s(p) = \sqrt{\dfrac{2}{\pi}}\int_0^\infty f(x)\sin px\,dx$

$= \sqrt{\dfrac{2}{\pi}}\cdot\left[\int_0^1 x\sin px\,dx + \int_1^2 (2-x)\sin px\,dx\right]$

$= \sqrt{\dfrac{2}{\pi}}\cdot\left[\left(-\dfrac{x}{p}\cos px + \dfrac{1}{p^2}\sin px\right)_0^1\right.$

$\left. + \left\{-\left(\dfrac{2-x}{p}\right)\cos px - \dfrac{1}{p^2}\sin px\right\}_1^2\right]$

$= \sqrt{\dfrac{2}{\pi}}\cdot\left[\dfrac{2}{p^2}\sin p - \dfrac{1}{p^2}\sin 2p\right] = 2\sqrt{\dfrac{2}{\pi}}\cdot\dfrac{\sin p}{p^2}(1-\cos p)$

$\tilde{f}_c(p) = \sqrt{\dfrac{2}{\pi}}\int_0^\infty f(x)\cos px\,dx$

$= \sqrt{\dfrac{2}{\pi}}\left[\int_0^1 x\cos px\,dx + \int_1^2 (2-x)\cos px\,dx\right]$

$= \sqrt{\dfrac{2}{\pi}}\left[\left(\dfrac{x}{p}\sin px + \dfrac{1}{p^2}\cos px\right)_0^1\right.$

$\left. + \left\{\left(\dfrac{2-x}{p}\right)\sin px - \dfrac{\cos px}{p^2}\right\}_1^2\right]$

$= \sqrt{\dfrac{2}{\pi}}\dfrac{1}{p^2}[2\cos p - 1 - \cos 2p] = 2\sqrt{\dfrac{2}{\pi}}\cdot\dfrac{\cos p}{p^2}(1-\cos p)$

EXAMPLE 8. *Find Fourier sine transform of* e^{-ax}/x.

[PTU–2006, VTU–2010, ROHTAK–2005]

SOLUTION. If $f(x) = \dfrac{e^{-ax}}{x}$, then we have

$\tilde{f}_s(p) = \sqrt{\dfrac{2}{\pi}}\int_0^\infty \dfrac{e^{-ax}}{x}\sin px\,dx$

Differentiating both sides w.r.t. p, we have

$\dfrac{d}{dp}\tilde{f}_s(p) = \sqrt{\dfrac{2}{\pi}}\int_0^\infty e^{-ax}\cos px\,dx$

$= \sqrt{\dfrac{2}{\pi}}\left[\dfrac{e^{-ax}}{a^2+p^2}(-a\cos px + p\sin px)\right]_0^\infty = \dfrac{a}{a^2+p^2}\cdot\sqrt{\dfrac{2}{\pi}}$

$\therefore \tilde{f}_s(p) = a\sqrt{\dfrac{2}{\pi}}\dfrac{dp}{a^2+p^2} + c = \sqrt{2/\pi}\tan^{-1}(p/a) + c$

But when $p = 0, \tilde{f}_s(p) = 0$

$\therefore \qquad c = 0$

Hence, $\tilde{f}_s(p) = \sqrt{2/\pi}\cdot\tan^{-1}(p/a)$.

EXAMPLE 9. *Find the Fourier sine transform of*

$$f(x) = \dfrac{1}{x(a^2+x^2)} \qquad \text{(UPTU–2008)}$$

SOLUTION. We have

$\tilde{f}_s(p) = \sqrt{\dfrac{2}{\pi}}\int_0^\infty \dfrac{1}{x(a^2+x^2)}\sin px\,dx \qquad \dots (1)$

Let $I = \int_0^\infty \dfrac{1}{x(a^2+x^2)}\sin px\,dx \qquad \dots (2)$

Then $\dfrac{dI}{dp} = \dfrac{d}{dp}\int_0^\infty \dfrac{\sin px}{x(a^2+x^2)}dx$

$= \int_0^\infty \left[\dfrac{\partial}{\partial p}\left\{\dfrac{\sin px}{x(a^2+x^2)}\right\}\right]dx$

$= \int_0^\infty \dfrac{\cos px}{a^2+x^2}dx \qquad \dots (3)$

$\therefore \quad \dfrac{d^2I}{dp^2} = -\int_0^\infty \dfrac{x\sin px}{a^2+x^2}dx = -\int_0^\infty \dfrac{x^2\sin px}{x(a^2+x^2)}dx$

$= -\int_0^\infty \dfrac{(x^2+a^2)-a^2}{x(a^2+x^2)}\sin px\,dx$

$= -\int_0^\infty \dfrac{\sin px}{x}dx + a^2\int_0^\infty \dfrac{\sin px}{x(a^2+x^2)}dx$

$= -\dfrac{\pi}{2} + a^2 I. \qquad \left[\because \int_0^\infty \dfrac{\sin px}{x}dx = \dfrac{\pi}{2}\right]$

$\therefore \quad \dfrac{d^2I}{dp^2} - a^2 I = -\dfrac{\pi}{2}$

or $(D^2 - a^2)I = -\dfrac{\pi}{2}$, where $D \equiv \dfrac{d}{dp}$

The solution of the above differential equation is

$I = Ae^{-ap} + Be^{ap} + \dfrac{\pi}{2a} \qquad \dots(4)$

$\therefore \quad \dfrac{dI}{dp} = -Aae^{-ap} + Bae^{-ap} \qquad \dots(5)$

Now from (1), when $p = 0$, we have $I = 0$ and from (2), when $p = 0$, we have

$\dfrac{dI}{dp} = \int_0^\infty \dfrac{1}{a^2+x^2}dx = \dfrac{1}{a}\left[\tan^{-1}\dfrac{x}{a}\right]_0^\infty = \dfrac{\pi}{2a}$

So putting $p = 0$ in (4) and (5), we get

$A + B = -\dfrac{\pi}{2a^2} \qquad \dots(6)$

and $a(-A+B) = \dfrac{\pi}{2a^2}$ i.e., $-A+B = \dfrac{\pi}{2a^2}$ ...(7)

Solving (6) and (7), we get $B = 0, A = -\dfrac{\pi}{2a^2}$ ...(8)

Putting the values of A and B in (4), we get

$$I = \int_0^\infty \dfrac{\sin px}{x(a^2+x^2)}dx = -\dfrac{\pi}{2a^2}e^{-ap} + \dfrac{\pi}{2a^2}$$

$$= \dfrac{\pi}{2a^2}(1-e^{-ap}).$$

Now putting the value of I in (1), we get

$$\tilde{f}_s(p) = \sqrt{\dfrac{2}{\pi}} \cdot \dfrac{\pi}{2a^2}(1-e^{-ap}) = \dfrac{1}{a^2}\sqrt{\dfrac{\pi}{2}} \cdot (1-e^{-ap})$$

EXAMPLE 10. *Find the Fourier cosine transform of e^{-x^2}.*

[VTU–2010, RAJASTHAN–2006, ANNA–2009]

SOLUTION. We have

$$\tilde{F}_c\{e^{-x^2}\} = \sqrt{\dfrac{2}{\pi}}\int_0^\infty e^{-x^2}\cos px\,dx = I$$

Differentiating w.r.t. 'p', we have

$$\dfrac{dI}{dp} = -\sqrt{\dfrac{2}{\pi}}\int_0^\infty xe^{-x^2}\sin px\,dx$$

$$= \dfrac{1}{2}\sqrt{\dfrac{2}{\pi}}\int_0^\infty (-2xe^{-x^2}) \cdot \sin px\,dx$$

$$= \dfrac{1}{2}\sqrt{\dfrac{2}{\pi}}\left[(e^{-x^2}\sin px)_0^\infty - p\int_0^\infty e^{-x^2}\cos px\,dx\right]$$

(Integrating by parts taking $\sin px$ as first function.)

$$= -\dfrac{p}{2}I$$

$$\therefore \qquad \dfrac{dI}{I} = -\dfrac{p}{2}dp$$

Integrating, we have

$$\log I = -\dfrac{p^2}{4} + \log A \text{ or } I = Ae^{-p^2/4} \qquad ... (2)$$

But when $p = 0$, from (1),

$$I = \sqrt{\dfrac{2}{\pi}}\int_0^\infty e^{-x^2}dx = \dfrac{1}{\sqrt{2}}$$

$\therefore$ From (2), $A = 1/\sqrt{2}$

Hence $I = F_c\{e^{-x^2}\} = (1/\sqrt{2})e^{-p^2/4}$

Exercise-8.72

1. Find the Fourier complex transform of $f(x)$ if
$$f(x) = \begin{cases} e^{\omega x} & a < x < b \\ 0 & x < a, x > b \end{cases}.$$

2. Find the sine transform of a function of x which is equal to $\sin x$ for $0 < x < a$ and 0 for $x > a$.

3. Find the cosine transform of $f(x)$ if $f(x) = \begin{cases} \cos x & 0 < x < a \\ 0 & x > 0 \end{cases}$.

4. Find the Fourier sine and cosine transform of $f(x)$ if
$$f(x) = \begin{cases} 1 & ; & 0 \le x < 1 \\ 0 & ; & x > 1 \end{cases}$$ [JNTU–2004, KOTTAYAM–2005]

5. Find the Fourier sine transform of $\dfrac{x}{1+x^2}$

6. Find the sine and cosine transform of $\dfrac{e^{ax}+e^{-ax}}{e^{\pi x}-e^{-\pi x}}$

7. Find the Fourier sine transform of $f(x)$ if
$$f(x) = \begin{cases} 0 & ; & 0 < x < a \\ x & ; & a \le x \le b \\ a & ; & x > b \end{cases}$$

8. Find the cosine transform of $\dfrac{1}{x^2+a^2}$. [VTU-2011, ANNA-2009]

9. Find the Fourier transform of $f(x) = e^{-x^2/2}$.

10. Show that $\int_0^\infty \dfrac{\cos xt}{1+t^2}dt = \dfrac{\pi}{2}e^{-x}, x \ge 0$. [VTU–2003]

11. Find $f(x)$ if its cosine transform is $\dfrac{1}{1+p^2}$.

12. Find $f(x)$ if (i) its sine transform is e^{-ap} (ii) its cosine transform is e^{-ap}. (SRM–2009)

13. Find $f(x)$ of its cosine transform is $\begin{cases} \sqrt{\dfrac{1}{2\pi}}\left(a-\dfrac{p}{2}\right) & , \text{ if } p < 2a \\ 0 & , \text{ if } p \ge 2a \end{cases}$

14. Find $f(x)$ if $\tilde{f}_s(p) = \dfrac{e^{-ap}}{p}$. Hence deduce that $F_s^{-1}\left\{\dfrac{1}{p}\right\}$.

15. Find the Fourier transform of $e^{-a^2x^2}$, $a < 0$. Hence deduce that $e^{-x^2/2}$ is self reciprocal in respect to Fourier transform. (KOTTAYAM–2005, MADRAS–2006)

16. Find the Fourier sine and cosine transform of $x^{n-1}, n > 0$. (MADRAS–2006)

17. Solve the integral equation
$$\int_0^\infty f(\theta)\cos\alpha\theta\,d\theta = \begin{bmatrix} 1-\alpha & ; & 0 \le \alpha \le 1 \\ 0 & ; & \alpha > 1 \end{bmatrix}$$
(VTU-2011, KURUKSHETRA-2005)

18. If the Fourier sine transform of $f(x)$ is $\dfrac{1-\cos n\pi}{n^2\pi^2}, 0 \le x \le \pi$, find $f(x)$. (DELHI–2002)

19. Find the Fourier transform of $f(x)$, $f(x) = \begin{cases} a^2 - x^2 & ; & |x| \le a \\ 0 & ; & x > a \end{cases}$ (VTU–2007)

20. Find Fourier sine transform of $f(x) = \begin{cases} 4x & \text{for } 0 < x < 1 \\ 4-x & \text{for } 1 < x < 4 \\ 0 & \text{for } x > 4 \end{cases}$ (VTU–2006)

21. Find the finite Fourier sine and cosine transform of $f(x) = 2x$, $0 < x < 4$. (VTU–2011)

22. Find the Fourier cosine transform of $\left(1-\dfrac{x}{\pi}\right)^2$. (PTU–2006)

23. Solve the integral equation $\int_0^\infty f(x)\sin tx\,dx = \begin{cases} 1 & 0 \le t < 1 \\ 2 & 1 \le t < 2 \\ 0 & t \ge 2 \end{cases}$. (KOTTAYAM–2005)

24. Solve the integral equation $\int_0^\infty f(x)\cos\alpha x\,dx = e^{-a}$.

(SVTU–2009, ROHTAK–2004)

25. Show that the inverse finite Fourier sine transform of

$f(x) = \dfrac{1}{\pi}\left[1 + \cos x\pi - 2\cos\dfrac{x\pi}{2}\right]$ is $f(x) = \begin{cases} 1 & ; \ 0 < x < \pi/2 \\ -1 & ; \ \pi/2 < x < \pi \end{cases}$

(VTU–2008)

Answers

1. $\dfrac{1}{\sqrt{2\pi}}\left[\dfrac{e^{i(p+\omega)a} - e^{i(p+\omega)b}}{p+\omega}\right]$ **2.** $\dfrac{1}{\sqrt{2\pi}}\left[\dfrac{\sin(p-1)a}{p-1} - \dfrac{\sin(p+1)a}{p+1}\right]$ **3.** $\dfrac{1}{\sqrt{2\pi}}\left[\dfrac{\sin(1+p)a}{1+p} + \dfrac{\sin(1-p)a}{1-p}\right]$

4. $\sqrt{\dfrac{2}{\pi}}\dfrac{(1-\cos p)}{p}, \sqrt{\dfrac{2}{\pi}}\left(\dfrac{\sin p}{p}\right)$ **5.** $\sqrt{\dfrac{\pi}{2}}e^{-p}$ **6.** $\dfrac{e^p - e^{-p}}{\sqrt{2\pi}(e^p + e^{-p} + 2\cos a)}, \sqrt{\dfrac{1}{2\pi}}\dfrac{\cos a/2.(e^{p/2} + e^{-p/2})}{2\cos a + e^p + e^{-p}}$

7. $\sqrt{\dfrac{2}{\pi}}\left[-\dfrac{b\cos pb + a\cos pa}{p} + \dfrac{\sin pb - \sin pa}{p^2}\right]$ **8.** $\sqrt{\dfrac{2}{\pi}}.\dfrac{\pi}{2a}e^{-ap}$ **9.** $e^{-p^2/2}$ **11.** $\sqrt{\dfrac{\pi}{2}}.e^{-x}$

12. $\sqrt{\dfrac{2}{\pi}}.\dfrac{a}{a^2+x^2}$ **13.** $\dfrac{1-\cos 2ax}{2\pi x^2}$ **14.** $\sqrt{\dfrac{2}{\pi}}\tan^{-1}\left(\dfrac{x}{a}\right).\sqrt{\dfrac{\pi}{2}}$ **15.** $\dfrac{\sqrt{\pi}}{a}e^{-p^2/4a^2}$ **16.** $\dfrac{\Gamma(n)}{p^n}\sin\dfrac{n\pi}{2}$

17. $f(\theta) = \dfrac{2(1-\cos\theta)}{\pi\theta^2}$ **18.** $\dfrac{2}{\pi^3}\sum_{n=1}^{\infty}\left[\dfrac{1-\cos n\pi}{n^2}\right]n\pi$ **19.** $\dfrac{4}{p^3}\left[\sin p\alpha - p\alpha\cos p\alpha\right]$ **20.** $(2\cos p - \cos 4p - 1)/p^2 - \dfrac{2\sin p}{p}$

21. $-32(-1)^p/p\pi, \dfrac{32(-1)^p - 1}{p^2 x^2}$ **22.** $\dfrac{2}{\pi p^2}$ **23.** $f(x) = (2 + 2\cos x - 4\cos 2x)/\pi x$ **24.** $\dfrac{2}{\pi(1+x^2)}$

8.16.4 FINITE FOURIER TRANSFORMS

(1) FINITE FOURIER SINE TRANSFORMS

Let $f(x)$ be a function, which is sectionally continuous over some finite interval $]0, 1[$ for the variable x. Then finite Fourier sine transforms of $f(x)$ on this interval is given by $\tilde{f}_s(p) = \int_0^l f(x)\sin\dfrac{p\pi x}{l}dx$, where p is any integer.

If the end points of the interval become $x = 0$ and $x = \pi$, then $\tilde{f}_s(p) = \int_0^\pi f(x)\sin px\,dx$.

REMARKS

- $\tilde{f}_s(p)$ is always zero when $p = 0$.
- The function $f(x)$ is called the inverse finite Fourier sine transform of $\tilde{f}_s(p)$, i.e., $f(x) = F_s^{-1}\{\tilde{f}_s(p)\}$.

(2) INVERSION FORMULA FOR SINE TRANSFORM

If $\tilde{f}_s(p)$ is the finite Fourier sine transform of $f(x)$ over the interval $]0, l[$, then the inversion formula for sine transform is given by

$$f(x) = \dfrac{2}{l}\sum_{p=1}^{\infty}\tilde{f}_s(p)\sin\dfrac{p\pi x}{l}$$

or in the interval $]0, \pi[$, we have $f(x) = \dfrac{2}{\pi}\sum_{p=1}^{\infty}\tilde{f}_s(p)\sin px$.

(3) FINITE FOURIER COSINE TRANSFORM

Let $f(x)$ be a function which is sectionally continuous over some finite interval $]0, l[$ of the variable x. Then the finite Fourier cosine transform of $f(x)$ on this interval is defined as $\tilde{f}_c(p) = \int_0^l f(x)\cos\dfrac{p\pi x}{l}dx$, where p is any integer.

(4) INVERSION FORMULA FOR COSINE TRANSFORM

If $\tilde{f}_c(p)$ is the finite Fourier cosine transform of $f(x)$ over the interval $]0, l[$ then the inversion formula for cosine transform is given by

$$f(x) = \dfrac{1}{l}\tilde{f}_c(0) + \dfrac{2}{l}\sum_{p=1}^{\infty}\tilde{f}_c(p)\cos\dfrac{p\pi x}{l} , \text{ where } \tilde{f}_c(0) = \int_0^l f(x)dx .$$

Also, if π is taken as the upper limit for the finite Fourier cosine transform, then inversion formula is given by

$$f(x) = \dfrac{1}{\pi}\tilde{f}_c(0) + \dfrac{2}{\pi}\sum_{p=1}^{\infty}\tilde{f}_c(p)\cos px, \text{ where } \tilde{f}_c(0) = \int_0^\pi f(x)dx .$$

(5) MULTIPLE FINITE FOURIER TRANSFORMS

Let $f(x,y)$ be a function of two variables x and y defined in the square $0 \le x \le \pi$ and $0 \le y \le \pi$. Let us consider $f(x, y)$ temporarily as a function of x, the finite sine transform is given by $\tilde{f}_s(p, y) = \int_0^\pi f(x, y)\sin px\,dx$

and now the finite sine transform of $\tilde{f}_s(p, y)$ which is a function of y is given by

$$\tilde{F}_s(p, q) = \int_0^\pi \tilde{f}_s(p, y)\sin qy\,dy$$

Thus, $\tilde{F}_s(p, q) = \int_0^\pi \int_0^\pi f(x, y)\sin px \sin qy\,dx\,dy$

(6) OPERATIONAL PROPERTIES OF FINITE FOURIER TRANSFORM

(A) Operational Property of Finite Fourier Sine Transform:

The finite Fourier sine transforms resolves the differential form $f''(x)$ into a linear algebraic form in the transform $\tilde{f}_s(p)$ and the boundary value $f(0)$ and $f(\pi)$ such that $F_s[f''(x)] = -p^2\tilde{f}_s(p) + p\{f(0) - (-1)^p f(\pi)\}$ whenever $f(x)$ and $f'(x)$ are continuous and $f''(x)$ is sectionally continuous on the interval $0 \leq x \leq \pi$

(B) Operational Property of Finite Fourier Cosine Transform :

If $f(x)$ and $f'(x)$ are continuous and if $f''(x)$ is sectionally continuous, the finite Fourier transformation resolved the differential form $f''(x)$ into an algebraic form is $\tilde{f}_c(p)$ and the boundary value $f'(0)$ and $f'(\pi)$ such that $F_c[f''(x)] = -p^2\tilde{f}_c(p) - f'(0) + (-1)^p f'(\pi)$

☛ REMARK

- If $\tilde{f}_c(p)$ is the cosine transform of a sectionally continuous function $f(x)$, $0 \leq x \leq \pi$ then

$$F_c^{-1}\left\{\frac{\tilde{f}_c(p)}{p^2}\right\} = \int_0^x \int_t^\pi f(r) dr dt = \frac{\tilde{f}_c(0)}{2\pi}(x - \pi^2) + A,\text{ where } A \text{ is any arbitrary constant.}$$

(7) COMBINED PROPERTIES OF FINITE FOURIER SINE AND COSINE TRANSFORMS

If $f(x)$ is continuous and $f'(x)$ is sectionally continuous, then

(i) $F_s[f'(x)] = -pF_c\{f(x)\}; p = 1, 2, 3...$ and (ii) $F_c[f'(x)] = pF_s\{f(x) - f(0) + (-1)^p f(\pi)\}; p = 0, 1, 2, 3...$

☛ REMARK

- If $H(x)$ is sectionally continuous function, then

$$F_s[H(x)] = -pF_c\left[\int_0^x H(r) dr\right], p = 1, 2...$$

and

$$F_c\left[H(x) - \frac{1}{\pi}\tilde{H}_c(0)\right] = pF_s\left[\int_0^x H(r) dr - \frac{x}{\pi}\tilde{H}_c(0)\right], p = 0, 1, 2...$$

(8) CONVOLUTION

Let $F(x)$ and $G(x)$ be two functions defined on the interval $-2\pi < x < 2\pi$, then the function $F(x) * G(x) = \int_{-\pi}^{\pi} F(x - y)G(y) dy$ is called the convolution of $F(x)$ and $G(x)$ on the interval $-\pi < x < \pi$.

Solved Examples

EXAMPLE 1. Find the finite Fourier sine and cosine transforms of $f(x) = x$.

SOLUTION. We have

$$\tilde{f}_s(p) = \int_0^\pi f(x)\sin px dx = \int_0^\pi x \sin px dx$$

$$= \left(\frac{-x\cos px}{p}\right)_0^\pi + \frac{1}{p}\int_0^\pi \cos px dx$$

$$= \frac{\pi(-1)^{p+1}}{p} + \frac{1}{p}\left[\frac{\sin px}{p}\right]_0^\pi = \frac{\pi(-1)^{p+1}}{p}$$

Similarly,

$$\tilde{f}_c(p) = \int_0^\pi f(x)\cos px dx = \int_0^\pi x \cos px dx$$

$$= \left(\frac{x\sin px}{p}\right)_0^\pi - \frac{1}{p}\int_0^\pi \sin px dx$$

$$= \frac{1}{p}\left[\frac{\cos px}{p}\right]_0^\pi = \frac{(-1)^p - 1}{p^2}, p = 1, 2, 3...$$

Also, if $p = 0$, then $\tilde{f}_c(p) = \int_0^\pi x.1 dx = \frac{\pi^2}{2}$.

EXAMPLE 2. Find the finite sine transform of $f(x)$ if

$$\textbf{(i)} \quad f(x) = \begin{cases} x & ; & 0 \leq x \leq \pi/2 \\ \pi - x & ; & \pi/2 \leq x \leq \pi \end{cases}$$

$$\textbf{(ii)} \quad f(x) = \begin{cases} -x & ; & x < c \\ \pi - c & ; & x > c, \end{cases} \textbf{where } 0 \leq c \leq \pi$$

SOLUTION. (i) We have

$$\tilde{f}_s(p) = \int_0^\pi f(x)\sin px dx$$

$$= \int_0^{\pi/2} x\sin px dx + \int_{\pi/2}^\pi (\pi - x)\sin px dx$$

$$= \left[x\frac{(-\cos px)}{p} + \frac{\sin px}{p^2}\right]_0^{\pi/2}$$

$$+ \left[(\pi - x)\left(\frac{-\cos px}{p}\right) - \frac{\sin px}{p^2}\right]_{\pi/2}^\pi$$

$$= \frac{2}{p^2}\sin\left(\frac{p\pi}{2}\right).$$

(ii) We have $\tilde{f}_s(p) = \int_0^\pi f(x)\sin px dx$

$$= \int_0^C -x\sin px dx + \int_C^\pi (\pi - x)\sin px dx$$

$$= \left[x\frac{\cos px}{p}\right]_0^C - \int_0^C 1.\frac{\cos px}{p} dx + \left[(\pi - x)\right.$$

$$\left.\left(-\frac{\cos px}{p}\right)\right]_C^\pi - \frac{1}{p}\int_C^\pi \cos px dx$$

$$= \frac{C}{p}\cos pC - \frac{1}{p^2}[\sin px]_0^C + \frac{\pi - C}{p}\cos pC - \frac{1}{p^2}[\sin px]_C^\pi$$

$$= \frac{C}{p}\cos pC - \frac{1}{p^2}\sin pC + \frac{\pi - C}{p}\cos pC + \frac{1}{p^2}\sin pC$$

EXAMPLE 3. *Find the finite cosine transform of f(x) if*

$$f(x) = -\frac{\cos k(\pi - x)}{k \sin k\pi}.$$

SOLUTION. We have $\tilde{f}_c(p) = -\int_0^\pi \frac{\cos\{k(\pi - x)\}}{k \sin k\pi} \cos px\, dx$

$$= -\frac{1}{2k \sin k\pi} \int_0^\pi [\cos\{k(\pi - x) + px\}$$
$$+ \cos\{k(\pi - x) - px\}] dx$$

$$= -\frac{1}{2k \sin k\pi} \left[\frac{\sin(k\pi - kx + px)}{p - k} - \frac{\sin(k\pi - kx - px)}{p + k} \right]_0^\pi$$

$$= -\frac{1}{2k \sin k\pi} \left[\frac{\sin p\pi}{p - k} - \frac{\sin(-p\pi)}{p + k} - \frac{\sin k\pi}{p - k} + \frac{\sin k\pi}{p + k} \right]$$

$$= \frac{1}{2k} \left(\frac{1}{p - k} - \frac{1}{p + k} \right) = \frac{1}{p^2 - k^2}, k \neq 0, 1, 2, 3\ldots$$

EXAMPLE 4. *Find f(x) if its finite sine transform is given by*

$$\tilde{f}_s(p) = \frac{1 - \cos p\pi}{p^2 \pi^2}, \text{ where } 0 < x < \pi.$$

SOLUTION. We have

$$f(x) = \frac{2}{\pi} \sum_{p=1}^\infty \tilde{f}_s(p) \sin px = \frac{2}{\pi} \sum_{p=1}^\infty \left(\frac{1 - \cos p\pi}{p^2 \pi^2} \right) \sin px$$

$$= \frac{2}{\pi^3} \sum_{p=1}^\infty \left(\frac{1 - \cos p\pi}{p^2} \right) \sin px$$

EXAMPLE 5. *Find the finite Fourier sine transform of f(x) if*

(i) $f(x) = \dfrac{2}{\pi} \tan^{-1} \dfrac{b \sin x}{1 - b \cos x}$

(ii) $f(x) = \dfrac{2}{\pi} \tan^{-1} \dfrac{2b \sin x}{1 - b^2}$

SOLUTION. (i) If $f(x) = \dfrac{2}{\pi} \tan^{-1} \dfrac{b \sin x}{1 - b \cos x}$, then

$$\tilde{f}_s(p) = \int_0^\pi \frac{2}{\pi} \left[\tan^{-1} \frac{b \sin x}{1 - b \cos x} \right] \sin px\, dx$$

Now let $\tan\theta = \dfrac{b \sin x}{1 - b \cos x}$

Then $\dfrac{i \sin\theta}{\cos\theta} = \dfrac{ib \sin x}{1 - b \cos x}$

Applying componendo and dividend, we have

$$\frac{\cos\theta + i\sin\theta}{\cos\theta - i\sin\theta} = \frac{1 - b\cos x + ib\sin x}{1 - b\cos x - ib\sin x}$$

or $\dfrac{e^{i\theta}}{e^{-i\theta}} = \dfrac{1 - b(\cos x - i\sin x)}{1 - b(\cos x + i\sin x)} = \dfrac{1 - be^{-ix}}{1 - be^{ix}}$

$$\therefore \quad e^{2i\theta} = \frac{1 - be^{-ix}}{1 - be^{ix}}$$

$$\therefore \quad 2i\theta = \log(1 - be^{-ix}) - \log(1 - be^{ix})$$

$$= -\left\{ be^{-ix} + \frac{b^2 e^{-2ix}}{2} + \frac{b^3 e^{-3ix}}{3} + \ldots \right\}$$

$$+ \left\{ be^{ix} + \frac{b^2 e^{2ix}}{2} + \frac{b^3 e^{3ix}}{3} + \ldots \right\}, \text{ if } |b| \leq 1$$

$$= b(e^{ix} - e^{-ix}) + \frac{b^2}{2}(e^{2ix} - e^{-2ix})$$

$$+ \frac{b^3}{3}(e^{3ix} - e^{-3ix}) + \ldots$$

$$= 2ib\sin x + \frac{b^2}{2}.(2i\sin 2x) + \frac{b^3}{3}.(2i\sin 3x) + \ldots$$

$$\therefore \theta = \tan^{-1} \frac{b\sin x}{1 - b\cos x} = b\sin x + \frac{b^2}{2}\sin 2x + \frac{b^3}{3}\sin 3x + \ldots$$

$\therefore$ From (1),

$$\tilde{f}_s(p) = \int_0^\pi \frac{2}{\pi} \left[b\sin x + \frac{b^2}{2}\sin 2x + \frac{b^3}{3}\sin 3x + \ldots \right] \sin px\, dx$$

$$= \int_0^\pi \frac{2}{\pi}.\frac{b^p}{p}\sin^2 px\, dx$$

since all other integrals vanish as
$\int_0^\pi \sin mx \sin nx\, dx = 0$ if m and n are integers and $m \neq n$

$$= \frac{b^p}{\pi p} \int_0^\pi (1 - \cos 2px\, dx) = \frac{b^p}{\pi p} \left[x - \frac{\sin 2px}{2p} \right]_0^\pi$$

$$= \frac{b^p}{\pi p}.\pi = \frac{b^p}{p}, |b| \leq 1$$

(ii) Let $\tan\theta = \dfrac{2b\sin x}{1 - b^2}$ then $\dfrac{i\sin\theta}{\cos\theta} = \dfrac{2ib\sin x}{1 - b^2}$.

Applying componendo and dividend, we have

$$\frac{\cos\theta + i\sin\theta}{\cos\theta - i\sin\theta} = \frac{1 - b^2 + 2ib\sin x}{1 - b^2 - 2ib\sin x} = \frac{1 - b^2 + b(e^{ix} - e^{-ix})}{1 - b^2 - b(e^{ix} - e^{-ix})}$$

$$\therefore \quad \frac{e^{i\theta}}{e^{-i\theta}} = \frac{(1 + be^{ix}) - be^{-ix}(1 + be^{ix})}{(1 - be^{ix}) + be^{-ix}(1 - be^{ix})}$$

or $e^{2i\theta} = \dfrac{(1 + be^{ix})(1 - be^{-ix})}{(1 - be^{ix})(1 + be^{-ix})}$

$$\therefore \quad 2i\theta = \log(1 + be^{ix}) + \log(1 - be^{-ix})$$
$$- \log(1 - be^{ix}) - \log(1 + be^{-ix})$$
$$= \{\log(1 + be^{ix}) - \log(1 - be^{ix})\}$$
$$- \{\log(1 + be^{-ix}) - \log(1 - be^{-ix})\}$$

$$= 2\left\{ be^{ix} + \frac{b^3}{3}e^{3ix} + \frac{b^5}{5}e^{5ix}\ldots \right\}$$

$$- 2\left\{ be^{-ix} + \frac{b^3}{3}e^{-3ix} + \frac{b^5}{5}e^{-5ix} + \ldots \right\} \quad \text{if } |b| \leq 1$$

$$= 2\left[b(e^{ix} - e^{-ix}) + \frac{b^3}{3}(e^{i3x} - e^{i3x}) \right.$$
$$\left. + \frac{b^5}{5}(e^{i5x} - e^{-i5x}) + \ldots \right]$$

$$= 2\left[b(2i\sin x) + \frac{b^3}{3}(2i\sin 3x) + \frac{b^5}{5}(2i\sin 5x) + \ldots \right]$$

$$\therefore \quad \theta = \tan^{-1} \frac{2b\sin x}{1 - b^2}$$

$$= 2\left[b\sin x + \frac{b^3}{3}\sin 3x + \frac{b^5}{5}\sin 5x + \ldots \right] \quad \ldots (1)$$

Now if $f(x) = \dfrac{2}{\pi} \tan^{-1} \dfrac{2b\sin x}{1 - b^2}$, then

$$\tilde{f}_s(p) = \int_0^\pi \frac{2}{\pi} \left[\tan^{-1} \frac{2b\sin x}{1 - b^2} \right] \sin px\, dx$$

$$= \int_0^\pi \frac{2}{\pi}.2 \left[b\sin x + \frac{b^3}{3}\sin 3x + \frac{b^5}{5}\sin 5x\ldots \right] \sin px\, dx$$

[from (1)]

Now if m and n are integers and $m \neq n$ then
$\int_0^\pi \sin mx \sin nx\, dx = 0$

If p is even, then $\tilde{f}_s(p) = 0$

Again if p is odd, then

$$\tilde{f}_s(p) = \frac{2}{\pi}\int_0^\pi \frac{b^p}{p}.2\sin^2 px\,dx = \frac{2}{\pi}\frac{b^p}{p}\int_0^\pi (1-\cos 2px)dx$$

$$= \frac{2}{\pi}\frac{b^p}{p}\left[x - \frac{\sin 2px}{2p}\right]_0^\pi = \frac{2}{\pi}\frac{b^p}{p}.\pi = 2\frac{b^p}{p}.$$

Hence, $\tilde{f}_s(p) = \dfrac{1-(-1)^p}{p}b^p, |b| \le 1$.

Exercise-8.73

1. Find the finite Fourier sine and cosine transform of $f(x) = 1$

2. Show that the finite Fourier sine transform of $\dfrac{x}{\pi}$ is $\dfrac{1}{p}(-1)^{p+1}$.

3. Find the finite Fourier sine transform of $\left(1 - \dfrac{x}{\pi}\right)$ and $\dfrac{x}{4\pi}$.

4. Find the finite sine and cosine transforms of $f(x) = x^2, 0 < x < \pi$.

5. Find the finite cosine transform of $\left(1 - \dfrac{x}{\pi}\right)^2$.

6. Find the finite Fourier sine transforms of $f(x) = \dfrac{\pi}{3} - x + \dfrac{x^2}{2\pi}$.

7. Find the finite sine transforms of $f(x)$ if $f(x) = \cos kx$.

8. Find the finite Fourier cosine transforms of $f(x) = \sin nx$.

9. Find finite Fourier cosine transform of $f(x) = \dfrac{\cosh[c(\pi - x)]}{\sinh(\pi c)}$.

10. Find $f(x)$ if its finite sine transforms is given by

$$\tilde{f}_s(p) = \frac{2\pi(-1)^{p-1}}{p^3}, p = 1, 2, 3 \dots 0 < x < \pi.$$

Answers

1. $\dfrac{1}{p}[1-(-1)^p], 0$ 3. $\dfrac{1}{p}, \dfrac{(-1)^{p+1}}{4p}$ 4. $\pi^2(-1)^{p+1}/p + 2\{(-1)^p - 1\}/p^3, \dfrac{2\pi(-1)^p}{p^2}, p > 0$ 5. $\dfrac{\pi}{3}$ 6. $0, \dfrac{\pi}{2}$

7. $\dfrac{p}{p^2-k^2}[1-(-1)^p\cos k\pi]$ 8. 0 or $\dfrac{2n}{n^2-p^2}$ 9. $\dfrac{c}{c^2+p^2}$ 10. $4\displaystyle\sum_{p=1}^\infty \dfrac{(-1)^{p-1}}{p^3}\sin px$

8.16.5 Applications of Fourier Transform

The finite sine and cosine transforms can be applied when the range of the variable selected for exclusion is 0 to ∞. The choice of sine and cosine transform is decided by the form of the boundary conditions at the lower limit of the variable selected for exclusion.

Hence, we have

$$F_s\left\{\frac{\partial^2 u}{\partial x^2}\right\} = \int_0^\infty \frac{\partial^2 u}{\partial x^2}\sin px\,dx = \left[\frac{\partial u}{\partial x}\sin px\right]_0^\infty - p\int_0^\infty \frac{\partial u}{\partial x}\cos px\,dx = -p\int_0^\infty \frac{\partial u}{\partial x}\cos px\,dx \text{ if } \frac{\partial u}{\partial x} \to 0 \text{ as } x \to \infty$$

$$= -p\left\{[u\cos px]_0^\infty + p\int_0^\infty u\sin px\,dx\right\} = p(u)_{x=0} - p^2\bar{u}_s \qquad \text{[By assuming } u \to 0 \text{ as } x \to \infty]$$

Therefore,

$$F_s\left\{\frac{\partial^2 u}{\partial x^2}\right\} = pu(0,t) - p^2\bar{u}_s(p,t) \qquad \dots(1)$$

Where $u(x, t)$ is a function of two variable x and t and $\bar{u}_s(p,t)$ is the Fourier sine transform of $u(x, t)$ with respect to x.

Further, $F_c\left\{\dfrac{\partial^2 u}{\partial x^2}\right\} = \int_0^\infty \dfrac{\partial^2 u}{\partial x^2}\cos px\,dx = \left[\dfrac{\partial u}{\partial x}\cos px\right]_0^\infty + p\int_0^\infty \dfrac{\partial u}{\partial x}\sin px\,dx = -\left(\dfrac{\partial u}{\partial x}\right)_{x=0} + p\int_0^\infty \dfrac{\partial u}{\partial x}\sin px\,dx$ $\left[\text{Assuming } \dfrac{\partial u}{\partial x} \to 0 \text{ as } x \to \infty\right]$

$$= -\left(\frac{\partial u}{\partial x}\right)_{x=0} + p\left\{[u\sin px]_0^\infty - p\int_0^\infty u\cos px\,dx\right\} = -\left(\frac{\partial u}{\partial x}\right)_{x=0} - p^2\int_0^\infty u(x,t)\cos px\,dx.$$

Then,

$$F_s\left\{\frac{\partial^2 u}{\partial x^2}\right\} = -\left(\frac{\partial u}{\partial x}\right)_{x=0} - p^2\bar{u}_c(p,t) \qquad \dots(2)$$

Where, $\bar{u}_c(p,t)$ is the Fourier cosine transform of $u(x, t)$ with respect to x.

Remarks

- It must be noted that the successful use of a sine transform in removing a term $\dfrac{\partial^2 u}{\partial x^2}$ required $u(0, t)$ i.e ,u at $x = 0$ while the use of a cosine transform for the same purpose requires, $u_x(0,t)$, i.e, $\dfrac{\partial u}{\partial x}$ at $x = 0$.

- The terms $\dfrac{\partial u}{\partial x}$ or any partial derivative of odd order cannot be removed with the help of sine or cosine transforms.

- When one of the variables in a differential equation ranges form $-\infty$ to $+\infty$ then that variable can be excluded with the help of complex Fourier transforms.

Solved Examples

EXAMPLE 1. *Solve* $\dfrac{\partial u}{\partial t} = 2\dfrac{\partial^2 u}{\partial x^2}$ *if* $u(0,t) = 0, u(x,0) = e^{-x}$, $x > 0$, $u(x, t)$ *is bounded where* $x > 0, t > 0$.

[ROHTAK 2006]

SOLUTION. As per given $\dfrac{\partial u}{\partial t} = 2\dfrac{\partial^2 u}{\partial x^2}$ ... (1)

Subject to the boundary conditions
$$u(0,t) = 0, u(x,t) \text{ is bounded} \quad ... (2)$$
and initial condition
$$u(x,0) = e^{-x}, x > 0 \quad ... (3)$$

Since, $u(0, t)$ is given, taking the Fourier sine transform of both sides of (1), we get

$$\int_0^\infty \frac{\partial u}{\partial t} \sin px \, dx = 2 \int_0^\infty \frac{\partial^2 u}{\partial x^2} \sin px \, dx$$

$$\Rightarrow \frac{d}{dt} \int_0^\infty u(x,t) \sin px \, dx$$

$$= 2 \left\{ \left(\frac{\partial u}{\partial x} \sin px \right)_0^\infty - \int_0^\infty \frac{\partial u}{\partial x} p \cos px \, dx \right\}$$

$$\Rightarrow \frac{d\bar{u}_s}{dt} = -2p \int_0^\infty \frac{\partial u}{\partial x} \cos px \, dx \text{ if } \frac{du}{dx} \to 0 \text{ as } x \to \infty$$

$$\left[\text{Assume } \bar{u}_s(p,t) = \int_0^\infty u(x,t) \sin px \, dx \right]$$

$$= -2p \left\{ [u(x,t) \cos px]_0^\infty - \int_0^\infty u(x,t)(-p \sin px) dx \right\}$$

$$= -2p \left\{ 0 - u(0,t) + p \int_0^\infty u(x,t) \sin px \, dx \right\}$$

$$[\because u(x,t) \to 0, \cos x \to \infty]$$

$$= 2pu(0,t) - 2p^2 \bar{u}_s$$

$$\Rightarrow \frac{d\bar{u}_s}{dt} = -2p^2 \bar{u}_s \quad\quad [\because u(0,t) = 0]$$

On separating the variables, we get

$$\frac{d\bar{u}_s}{\bar{u}_s} = -2p^2 dt \Rightarrow \log \bar{u}_s - \log C = -2p^2 t$$

$$\Rightarrow \log \left(\frac{\bar{u}_s}{C} \right) = -2p^2 t \Rightarrow \bar{u}_s(p,t) = Ce^{-2p^2 t} \quad ... (4)$$

Now, taking the Fourier sine transform of both sides of (3), we get

$$\int_0^\infty u(x,0) \sin px \, dx = \int_0^\infty e^{-x} \sin px \, dx$$

$$\Rightarrow \bar{u}_s(p,0) = \left[\frac{e^{-x}}{1+p^2} (-\sin px - p \cos px) \right]_0^\infty$$

$$= \frac{p}{1+p^2} \quad ... (5)$$

Putting $t = 0$ in (4) and (5), we get $\dfrac{p}{1+p^2} = C$

$$\therefore \quad \bar{u}_s(p,t) = \frac{p}{1+p^2} e^{-2p^2 t}.$$

Taking the inverse Fourier sine transform, we get

$$u(x,t) = \frac{2}{\pi} \int_0^\infty \frac{p}{1+p^2} e^{-p^2 t} \sin px \, dx$$

EXAMPLE 2. *Solve* $\dfrac{\partial u}{\partial t} = \dfrac{\partial^2 u}{\partial x^2}, x > 0, t > 0$ *subject to the*

conditions $u(0, t) = 0, u(x,0) = \begin{cases} 1 & ; & 0 < x < 1 \\ 0 & ; & x > 1 \end{cases}$,

$u(x, t)$ *is bounded.*

(UPTU–2003)

SOLUTION. Taking the Fourier sine transform of both side of given PDE, we get

$$\int_0^\infty \frac{\partial u}{\partial t} \sin px \, dx = \int_0^\infty \frac{\partial^2 u}{\partial x^2} \sin px \, dx$$

or $\dfrac{d}{dt} \int_0^\infty u \sin px \, dx = \left[\dfrac{\partial u}{\partial x} \sin px \right]_0^\infty - p \int_0^\infty \dfrac{\partial u}{\partial x} \cos px \, dx$

$$\therefore \frac{d\bar{u}_s}{dt} = -p \int_0^\infty \frac{\partial u}{\partial x} \cos px \, dx \quad \text{if } \frac{\partial u}{\partial x} \to 0 \text{ as } x \to \infty$$

$$= -p \left\{ [u \cos px]_0^\infty + p \int_0^\infty u \sin px \, dx \right\}$$

$$= -pu(0,t) - p^2 \bar{u}_s; \quad \text{if } u \to 0 \text{ as } x \to \infty$$

$$= -p^2 \bar{u}_s.$$

On separating the variables, we get $\dfrac{d\bar{u}_s}{\bar{u}_s} = -p^2 dt$

whose solution is given by

$$\bar{u}_s(p,t) = Ce^{-p^2 t} \quad ...(1)$$

Putting $t = 0$, we get $C = \bar{u}_s(p,0)$ ...(2)

Now, $\bar{u}_s(p,0) = \int_0^\infty u(x,0) \sin px \, dx$

$$= \int_0^1 u(x,0) \sin px \, dx + \int_1^\infty u(x,0) \sin px \, dx$$

$$= \int_0^1 \sin px \, dx$$

Now, from (2)

$$C = \int_0^1 \sin px \, dx = \left[\frac{\cos px}{-p} \right]_0^1 = \frac{1 - \cos p}{p}.$$

Thus, (1) gives $\bar{u}_s(p,t) = \left[\dfrac{(1 - \cos p)}{p} \right] e^{-p^2 t}$.

Finally, taking the inverse Fourier sine transform,

we get $u(x,t) = \dfrac{2}{\pi} \int_0^\infty \dfrac{1 - \cos p}{p} e^{-p^2 t} \sin px \, dp$

which is the required solution.

EXAMPLE 3. *Use the method of Fourier transform to determine the displacement* $y(x,t)$ *of an infinite string, given that the string is initially at rest and that the initial displacement is* $f(x)$, $-\infty < x < \infty$ *show that* $y(x,t) = \dfrac{1}{2}[f(x + Ct) + f(x - Ct)]$.

[ROHTAK–2000]

SOLUTION. It is known that the displacement of a string is governed by one dimensional wave equation

$$\frac{\partial^2 y}{\partial t^2} = C^2 \frac{\partial^2 y}{\partial x^2} \quad ... (1)$$

where $y(x, t)$ is the displacement at any time, $t, -\infty < x < \infty, t > 0$, and $C^2 = \dfrac{T}{\rho}$

Taking the Fourier transform of both sides of (1), we get

$$\frac{1}{\sqrt{2\pi}} \int_{-\infty}^\infty \frac{\partial^2 y}{\partial t^2} e^{-ipx} dx = C^2 \frac{1}{\sqrt{2\pi}} \int_{-\infty}^\infty \frac{\partial^2 y}{\partial x^2} e^{-ipx} dx$$

$$\Rightarrow \frac{d^2}{dt^2} \frac{1}{\sqrt{2\pi}} \int_{-\infty}^\infty y e^{-ipx} dx = C^2 (-ip)^2 \tilde{y}(p,t)$$

$$\Rightarrow \quad \frac{d^2\tilde{y}(p,t)}{dt^2} + C^2p^2\tilde{y}(p,t) = 0$$

whose solution is given by $\tilde{y}(p,t) = A\cos Cpt + B\sin Cpt$

$$...(2)$$

As per given, the string is initially at rest, *i.e.*, $\frac{\partial y}{\partial t} = 0$ at $t = 0$.

Thus, $\frac{1}{\sqrt{2\pi}}\int_{-\infty}^{\infty}\frac{\partial y}{\partial t}e^{-ipx}dx = \frac{d}{dt}\frac{1}{\sqrt{2\pi}}\int_{-\infty}^{\infty}ye^{-ipx}dx$

$$= \frac{d\tilde{y}(p,t)}{dt} = 0 \text{ at } t = 0.$$

Now from (2), we have $0 = BC_p \Rightarrow B = 0$

Also, at $t = 0, y = f(x)$. So, at $t = 0$,

$$\tilde{y}(p,0) = \frac{1}{\sqrt{2\pi}}\int_{-\infty}^{\infty}f(u)e^{-ipu}du = \tilde{f}(p)$$

$\therefore \quad \tilde{y}(p,t) = \tilde{f}(p)\cos Cpt$

Now, taking the inverse Fourier transform, we have

$$y(x,t) = \frac{1}{\sqrt{2\pi}}\int_{-\infty}^{\infty}\tilde{f}(p)\cos Cpt\, e^{-ipx}dp$$

$$= \frac{1}{2\pi}\int_{-\infty}^{\infty}\left[\int_{-\infty}^{\infty}f(u)e^{ipu}du\right]\cos Cpt\, e^{-ipx}dp$$

$$= \frac{1}{4\pi}\int_{-\infty}^{\infty}\left[\int_{-\infty}^{\infty}f(u)e^{ipu}du\right](e^{iCpt} + e^{-iCpt})e^{-ipx}dp$$

$$= \frac{1}{2}\left[\frac{1}{2\pi}\int_{-\infty}^{\infty}\left\{\int_{-\infty}^{\infty}f(u)e^{-i\alpha u}du\right\}\right.$$
$$\left.(e^{-iC\alpha t} + e^{iC\alpha t})e^{i\alpha x}d\alpha\right]$$

Putting $p = -\alpha \Rightarrow dp = -d\alpha$ We get

$$y(x,t) = \frac{1}{2}\left[\frac{1}{2\pi}\int_{-\infty}^{\infty}f(u)e^{-i\alpha u}\right.$$
$$\left\{\int_{-\infty}^{\infty}e^{i\alpha(x+Ct)}d\alpha\right\}du + \frac{1}{2\pi}\int_{-\infty}^{\infty}f(u)e^{-i\alpha u}$$
$$\left.\left\{\int_{-\infty}^{\infty}e^{i\alpha(x-Ct)}d\alpha\right\}du\right].$$

Finally, using Fourier integral formula, we get

$$y(x,t) = \frac{1}{2}[f(x+Ct) + f(x-Ct)].$$

EXAMPLE 4. *A thin membrane of great extent is release from rest in the position $z = f(x, y)$. Show that the displacement at any subsequent time is given by $z(x,y) = \frac{1}{2\pi}\int_{-\infty}^{\infty}\int_{-\infty}^{\infty}F(p,q)$*
$$\cos\{Ct\sqrt{p^2+q^2}\}e^{-i(px+qy)}dpdq$$
where $F(p, t)$ is double Fourier's transform of $f(x, y)$.

(UPTU–2005)

SOLUTION. It is known that the displacement of the membrane is governed by two dimensional wave equation

$$\frac{\partial^2 z}{\partial t^2} = C^2\left(\frac{\partial^2 z}{\partial x^2} + \frac{\partial^2 z}{\partial y^2}\right), \text{ when } C^2 = \frac{T}{\rho}$$

Taking the Fourier transforms of both the sides, we get

$$\frac{1}{2\pi}\int_{-\infty}^{\infty}\int_{-\infty}^{\infty}\frac{\partial^2 z}{\partial t^2}e^{i(px+qy)}dxdy$$

$$= \frac{C^2}{2\pi}\int_{-\infty}^{\infty}\int_{-\infty}^{\infty}\left(\frac{\partial^2 z}{\partial x^2} + \frac{\partial^2 z}{\partial y^2}\right)e^{i(px+qy)}dxdy$$

$$\Rightarrow \quad \frac{d^2\tilde{z}}{dt^2} = C^2(-p^2 - q^2)\tilde{z} = 0, \text{ where}$$

$$\tilde{z} = \frac{1}{2\pi}\int_{-\infty}^{\infty}\int_{-\infty}^{\infty}z.e^{i(px+qy)}.dxdy$$

So, $\frac{d^2\tilde{z}}{dt^2} + C^2(p^2+q^2)\tilde{z} = 0$ whose solution is given by

$$\tilde{z} = A\cos\{C\sqrt{p^2+q^2}.t\} + B\sin[C\sqrt{p^2+q^2}.t]$$

As per given, the initial conditions are $z = f(x,y), \frac{\partial z}{\partial t} = 0$ at $t = 0$.

Taking the Fourier transforms of these conditions, we get

$$\tilde{z} = \frac{1}{2\pi}\int_{-\infty}^{\infty}\int_{-\infty}^{\infty}f(x,y)e^{i(px+qy)}dxdy = F(p,q)$$

and $\frac{d\tilde{z}}{dt} = 0$ at $t = 0$.

$$\therefore \quad 0 = \left(\frac{d\tilde{z}}{dt}\right)_{t=0} = BC\sqrt{p^2+q^2}$$

$$\Rightarrow \quad B = 0$$

Thus, the solution is given by

$$\tilde{z} = F(p,q)\cos\{C\sqrt{p^2+q^2}t\}.$$

Applying the inversion Formula for double Fourier transform, we get

$$z(x,y,t) = \frac{1}{2\pi}\int_{-\infty}^{\infty}\int_{-\infty}^{\infty}F(p,q)$$
$$\cos\{Ct\sqrt{p^2+q^2}\}e^{-i(px+qy)}dpdq.$$

EXAMPLE 5. *Use finite Fourier transform to solve*
$$\frac{\partial u}{\partial t} = \frac{\partial^2 u}{\partial x^2}, \ u(0,t) = 0, u(4,t) = 0, u(x,0) = 2x$$

SOLUTION. Taking the finite Fourier sine transform of both sides of the given partial differential equation, we get

$$\int_0^4\frac{\partial u}{\partial t}\sin\frac{p\pi x}{4}dx = \int_0^4\frac{\partial^2 u}{\partial x^2}\sin\frac{p\pi x}{4}dx$$

$$\Rightarrow \frac{d\tilde{u}_s}{dt} = -\frac{p^2\pi^2}{4}\tilde{u}_s + \frac{p\pi}{4}[u(0,t) - u(4,t)\cos p\pi]$$

where $\tilde{u}_s$ is finite Fourier sine transform of u.

$$\therefore \quad \frac{d\tilde{u}_s}{dt} = \frac{-p^2\pi^2}{16}\tilde{u}_s \Rightarrow \tilde{u}_s = Ae^{-p^2\pi^2 t/16}$$

Now, since $u(x, 0) = 2x$, where $0 < x < 4$, taking finite Fourier sine transform, we have

at $t = 0$, $\tilde{u}_s = \int_0^4 2x.\sin\frac{px\pi}{4}dx$

$$= \left[-2x\frac{4}{p\pi}\cos\frac{p\pi x}{4} + 2.\frac{4}{2\pi}.\frac{4}{p\pi}\sin\frac{p\pi x}{4}\right]_0^4$$

$$= \frac{32(-\cos p\pi)}{p\pi} = \frac{32[-(-1)]^p}{p\pi}$$

$$\therefore \quad \frac{32(-1)^{p+1}}{p\pi} = A,$$

Hence, $\tilde{u}_s = \frac{32(-1)^{p+1}}{p\pi}e^{-tp^2\pi^2/16}$

Finally, taking the inverse finite Fourier sine transform, we get

$$u(x,t) = \frac{2}{4}\sum_{p=1}^{\infty}\frac{32(-1)^{p+1}}{p\pi}e^{-p^2\pi^2 t/16}.\sin\frac{p\pi x}{4}$$

$$= \frac{16}{\pi}\sum_{p=1}^{\infty}\frac{(-1)^{p+1}}{p}e^{-p^2\pi^2 t/16}.\sin\frac{p\pi x}{4}$$

EXAMPLE 6. *Using finite Fourier transform, find the solution of the wave equation* $\dfrac{\partial^2 u}{\partial t^2} = 4\dfrac{\partial^2 u}{\partial x^2}$ *subject to the conditions* $u(0, t) = 0.$

$$u(\pi, t) = 0, u(x,0) = 0.1\sin x + 0.01\sin 4x$$

and $u_t(x, 0) = 0$ *for* $0 < x < \pi, t > 0$

SOLUTION. Taking the finite Fourier transform on both the sides of the given equation, we get

$$\int_0^\pi \frac{\partial^2 u}{\partial t^2}\sin px\,dx = 4\int_0^\pi \frac{\partial^2 u}{\partial x^2}\sin px\,dx$$

$$\Rightarrow \quad \frac{d^2\tilde u_s}{dt^2} = -4p^2\tilde u_s + 4p[u(0,t) - u(\pi,t)\cos p\pi]$$

$$\Rightarrow \quad \frac{d^2\tilde u_s}{dt^2} + 4p^2\tilde u_s = 0$$

$$\Rightarrow \quad \tilde u_s = A\cos 2pt + B\sin 2pt \qquad\qquad \dots(1)$$

Now, at $t = 0, u = (0.1)\sin x + (0.01)\sin 4x$

and $\dfrac{\partial u}{\partial t} = 0$

Taking finite sine transforms, we get

At $t = 0,$

$$\tilde u_s = \int_0^\pi \{(0.1)\sin x + (0.01)\sin 4x\}\sin px\,dx$$

$$= (0.1)\int_0^\pi \sin x\sin px\,dx + (0.01)\int_0^\pi \sin 4x\sin px\,dx$$

and $\dfrac{d\tilde u_s}{dt} = 0$.

Also, from (1), we have

$$(0.1)\int_0^\pi \sin x\sin px\,dx + (0.01)\int_0^\pi \sin 4x\sin px\,dx = A$$

$$\Rightarrow \quad \frac{d\tilde u_s}{dt} = 0 = 2Bp \Rightarrow B = 0.$$

Now,

$$\tilde u_s = \begin{bmatrix} (0.1)\int_0^\pi \sin x\sin px\,dx + (0.01) \\ \int_0^\pi \sin 4x\sin px\,dx \end{bmatrix}\cos 2pt$$

Finally, taking the inverse finite Fourier sine transform, we get

$$u(x,t) = \frac{2}{\pi}\sum_{p=1}^\infty \tilde u_s\sin px$$

$$= \frac{2}{\pi}(0.1)\sum_{p=1}^\infty \left\{\int_0^\pi \sin x\sin px\,dx\right\}\cos 2pt\sin px$$

$$+ \frac{2}{\pi}(0.01)\sum_{p=1}^\infty \left\{\int_0^\pi \sin 4x\sin px\,dx\right\}\cos 2pt\sin px$$

$$= \frac{2}{\pi}(0.1)\left[\int_0^\pi \sin x\sin x\,dx\right]\cos 2t\sin x$$

$$+ \frac{2}{\pi}(0.01)\left[\int_0^\pi \sin 4x.\sin 4x\,dx\right]\cos 8t\sin 4x$$

$$\Rightarrow u(x,t) = (0.1)\cos 2t\sin x + (0.01)\cos 8t\sin 4t.$$

Exercise-8.74

1. Solve the boundary value problem $\dfrac{\partial^2 u}{\partial t^2} = 9\dfrac{\partial^2 u}{\partial x^2}$ subject to all conditions

$$u(0,t) = 0, u(2,t) = 0, u(x,0) = (0.05)\times(2 - x)$$

$$u_t(x,0) = 0, \text{where } 0 < x < 2, t > 0$$

2. Use finite Fourier transform to solve $\dfrac{\partial v}{\partial t} = \dfrac{\partial^2 v}{\partial x^2}, 0 < x < 6, t > 0$

and $v_x(0,t) = 0, v_x(6,t) = 0, v(x,0) = 2x$.

3. A string of density ρ and length π is stretched to a tension ρC^2. At time $t = 0$, one end $(x = 0)$ is given a small oscillation $a\sin\omega t$. If the other end remains fixed, use a finite sine transform to show that the displacement of the point x at time t is

$$a\sin\omega t\sin\frac{\omega(\pi - C)}{C}\csc\frac{\pi\omega}{C} + \left(\frac{2aC\omega}{\pi}\right)$$

$$\sum_{p=0}^\infty (\omega^2 - p^2C^2)^{-1}\sin px\sin pct.$$

4. Solve $\dfrac{\partial u}{\partial t} = \dfrac{\partial^2 u}{\partial x^2}$ if $u_x(0,t) = 0, u(x,0) = \begin{cases} x & ; \ 0\le x\le 1 \\ 0 & ; \ x > 1 \end{cases}$ and $u(x, t)$ is bounded where $x > 0, t > 0$.

5. Solve $\dfrac{\partial u}{\partial t} = 2\dfrac{\partial^2 u}{\partial x^2}$ if $u(0,t) = 0, u(x,0) = e^{-x}, x > 0, u(x,t)$ is bounded where $x > 0, t > 0$.

6. A string is stretched between the two fixed points $(0, 0)$ and $(C, 0)$. If it is displaced into the curve $y = b\sin\left(\dfrac{n\pi x}{C}\right)$ and released from rest in that position at time $t = 0$. Solve the boundary value problem for the displacement $y(x, t)$.

7. A tightly stretched flexible string has its ends fixed at $x = 0$ and $x = l$. At time t = 0, the string is given a shape defined by $F(x) = \mu x(l - x)$ where μ is a constant and then released. Find the displacement of any point x of the string at any point $t = 0$ **(VTU(ME) – 2006)**

8. If the initial temperature of an infinite bar is given by

$$\theta(x) = \begin{cases} \theta_0 & \text{for } |x| < a \\ 0 & \text{for } |x| > a \end{cases}$$

Show that the temperature at any point x at any instant t is given by $\theta(x,t) = \dfrac{\theta_0}{\pi}\left[erf\dfrac{(a + x)}{2c\sqrt t} + erf\dfrac{(a - x)}{2c\sqrt t}\right]$

(SVTU – 2008, ROHTAK –2004)

8.17 Z-TRANSFORMS

Z-transforms plays an important role in discrete analysis as Laplace and Fourier transforms in continuous system. It has many properties similar to those of the Laplace transforms. The z-transforms operate on sequences of the discrete integer valued arguments, not on function of continuous arguments. For every operational rule of Laplace transforms, there is a corresponding operational rule of z-transforms and for every application of the Laplace transforms, there is a corresponding application of z-transforms.

8.17.1 Z-Transforms

If the function $f(n)$ is defined for discrete values ($n = 0, 1, 2, ...$) and $f(n) = 0$, for $n < 0$ then its z-transform is defined by

$$Z(f(n)) = F(z) = \sum_{n=-\infty}^{\infty} f(n) z^{-n} = \sum_{n=-\infty}^{\infty} \frac{f(n)}{z^n}$$

whenever the infinite series converges. Here z is a complex number. Z is an operator and $F(z)$ is the Z-transforms of $\{f(n)\}$.

SOME STANDARD Z-TRANSFORMS

(1) $Z(a^n) = \dfrac{z}{z-a}$ (KOTAYAM–2005)

(2) $Z(n^p) = -z \dfrac{d}{dz} Z(n^{p-1})$, where p is a positive integer.

(3) $Z(1) = \dfrac{z}{z-1}$

(4) $Z(n) = \dfrac{z}{(z-1)^2}$

(5) $Z(n^2) = \dfrac{z^2 + z}{(z-1)^3}$

(6) $Z(n^3) = \dfrac{z^3 + 4z^2 + z}{(z-1)^4}$

(7) $Z(n^4) = \dfrac{z^4 + 11z^3 + 11z^2 + z}{(z-1)^5}$

8.17.2 Properties of Z-transforms

(1) LINEAR PROPERTY

If $<f(n)>$ and $<g(n)>$ are two sequences such that they can be added and a, b are any two scalars, then

$$Z\left[< a f(n) + b g(n) >\right] = aZ\left[<f(n)>\right] + bZ\left[<g(n)>\right]$$

Proof. Consider $\quad Z[<af(n) + bg(n)>] = \sum_{n=-\infty}^{\infty} [a f(n) + b g(n)] z^{-n}$ [By definition of z-transform]

$$= \sum_{n=-\infty}^{\infty} [af(n)z^{-n} + bg(n)z^{-n}] = a \sum_{n=-\infty}^{\infty} f(n)z^{-n} + b \sum_{n=-\infty}^{\infty} g(n)z^{-n}$$

$$= aZ[<f(n)>] + bZ[<g(n)>]$$

(2) CHANGE OF SCALE PROPERTY

If $Z[<f(n)>] = F(z)$, then

(i) $Z[<a^n f(n)>] = F\left(\dfrac{z}{a}\right)$

(ii) $Z[<a^{-n} f(n)>] = F(az)$

Proof. (i) We have $\quad F\left(\dfrac{z}{a}\right) = \sum_{n=-\infty}^{\infty} f(n) \left(\dfrac{z}{a}\right)^{-n}$ [By definition of z-transform]   ...(1)

Also we have

$$Z[<a^n f(n)>] = \sum_{n=-\infty}^{\infty} a^n f(n) z^{-n} = \sum_{n=-\infty}^{\infty} f(n) \left(\dfrac{z}{a}\right)^{-n} = F\left(\dfrac{Z}{a}\right)$$

(ii) We have

$$F(az) = \sum_{n=-\infty}^{\infty} f(n)(az)^{-n}$$

Also, we have

$$Z[\{a^{-n} f(n)\}] = \sum_{n=-\infty}^{\infty} a^{-n} f(n) z^{-n} = \sum_{n=-\infty}^{\infty} f(n)(az)^{-n} = F(az) \text{ (Using equation (2))}$$

☞ REMARK

- Here, the geometric factor a^{-n} when $|a| < 1$ damps the function u_n, hence, above rule is also known as 'damping rule'.

(MADRAS–2006)

(3) APPLICATION OF DAMPING RULE

The application of damping rule gives the following important results :

(i) $Z(na^n) = \dfrac{az}{(z-a)^2}$ (Madras–2000)

(ii) $Z(n^2 a^n) = \dfrac{az^2 + a^2 z}{(z-a)^3}$

(iii) $Z(\cos n\theta) = \dfrac{z(z - \cos\theta)}{z^2 - 2z\cos\theta + 1}$

(iv) $Z(\sin n\theta) = \dfrac{z\sin\theta}{z^2 - 2z\cos\theta + 1}$

(v) $Z(a^n \cos n\theta) = \dfrac{z(z - a\cos\theta)}{z^2 - 2az\cos\theta + a^2}$

(vi) $Z(a^n \sin n\theta) = \dfrac{az\sin\theta}{z^2 - 2az\cos\theta + a^2}$

(4) SHIFTING PROPERTIES

If $Z[\{(f(n)\}] = F(z)$, then

(i) $Z[< f(n \pm k)>] = z^{\pm} F(z)$ *and if $n \geq 0$, then*

(ii) $Z[< f(n+k)>] = z^n F(z) - \sum_{r=0}^{k-1} f(r) z^{k-r}$

(iii) $Z[< f(n-k)>] = z^{-n} F(z) - \sum_{m=1}^{k} f(-m) z^{-k+m}$

Proof. (i) Consider

$$Z[<f(n\pm k)>] = \sum_{n=-\infty}^{\infty} f(n\pm k)z^{-n}$$

$$= \sum_{n=-\infty}^{\infty} f(n\pm k)z^{-(n\pm k)}\cdot z^{\pm n} = z^{\pm k}\sum_{n=-\infty}^{\infty} f(r)z^{-r} \quad \text{(where } n\pm k = r)$$

$$= z^{\pm k}F(z)$$

Hence, $Z[<f(n\pm k)>] = z^{\pm k}F(z)$

(ii) Consider $$Z[<f(n\pm k)] = \sum_{n=0}^{\infty} f(n+k)z^{-n}$$

$$= \sum_{n=0}^{\infty} f(n+k)z^{-(n+k)}\cdot z^n = z^k \sum_{r=k}^{\infty} f(r)z^{-r} \quad \text{(where } n+k=r)$$

$$= z^k\sum_{r=0}^{\infty} f(r)z^{-r} - z^k\sum_{r=0}^{k-1} f(r)z^{-r} = z^kF(z) - \sum_{r=0}^{k-1} f(r)z^{-r} = z^kF(z) - \sum_{r=0}^{k-1} f(r)z^{k-r}$$

Hence, $$Z[<f(n+k)>] = z^kF(z) - \sum_{r=0}^{k-1} f(r)z^{k-r}$$

(iii) Consider

$$Z[<f(n-k)>] = \sum_{n=0}^{\infty} f(n-k)z^{-n} = \sum_{n=0}^{\infty} f(n-k)z^{-(n+k)}\cdot z^{-k}$$

$$= z^{-k}\sum_{r=0}^{\infty} f(r)z^{-r} + z^{-k}\sum_{r=-k}^{-1} f(1)z^{-r} \qquad \text{(when } n+k=r)$$

$$= z^{-k}F(z) + \sum_{r=-k}^{-1} f(r)z^{-k-r} = z^{-k}F(z) + \sum_{m=1}^{k} f(-m)z^{-k+m}$$

Hence, $$Z[<f(n-k)>] = z^{-k}F(z) + \sum_{m=1}^{k} f(-m)z^{-k+m}$$

(5) SOME IMPORTANT DEDUCTIONS

(1) $Z[<f(n+1)>] = z\,F(z) - z\,f(0) = z\,[F(z) - f(0)]$

(2) $$Z[<f(n+2)>] = z^2F(z) - \sum_{r=0}^{1} f(r)z^{2-r}$$

$\Rightarrow$ $Z[<f(n+2)>] = z^2F(z) - z^2f(0) - z\,f(1) = z^2[F(z) - f(0) - z^{-1}f(1)]$

(3) $$Z[<f(n+3)>] = z^3F(z) - \sum_{r=0}^{2} f(r)z^{3-r}$$

$\Rightarrow$ $Z[<f(n+3)>] = z^3F(z) - z^3f(0) - z^2f(2) - z\,f(2)$

$\Rightarrow$ $Z[<f(n+3)>] = z^3[F(z) - f(0) - z^{-1}f(1) - z^{-2}f(2)]$ and so on.

Shifting to the Right

If $Z(f(n)) = F(z)$, then $Z(f(n-k)) = z^{-k}F(z)$ $(k > 0)$

Proof. By definition, we have

$$Z[f(n-k)] = \sum_{n=0}^{\infty} f(n-k)z^{-n} = f_0z^{-k} + f_1z^{-(k+1)} + \dots$$

$$= z^{-k}\sum_{n=0}^{\infty} f_nz^{-n} = z^{-k}F(z)$$

Shifting to the left

If $Z(f(n)) = F(z)$, then $Z(f(n+k) = z^k[F(z) - f_0 - f_1z^{-1} - f_2z^{-2} - \dots - f_{k-1}z^{-(k-1)}])$ (JNTU–2002)

Proof. We have

$$Z[f_{n+k}] = \sum_{n=0}^{\infty} f_{n+k}z^{-n} = z^k\left[\sum_{n=0}^{\infty} f_nz^{-n} - \sum_{n=0}^{k-1} u_nz^{-n}\right]$$

Hence, $Z(f(n+k)) = z^k[F(z) - f_0 - f_1z^{-1} - f_2z^{-2} - \dots - f_{k-1}z^{-(k-1)}]$

8.17.3 Z-TRANSFORM OF nf(n)

If $z[<f(x)>] = F(z)$, then $Z[<nf(n)>] = -z\dfrac{dF(z)}{dz}$

Proof. We have

$$Z[<nf(n)>] = \sum_{n=-\infty}^{\infty} n f(n)z^{-n} = -z\sum_{n=-\infty}^{\infty} [-nf(n)]z^{-n-1}$$

$$= -z \sum_{n=-\infty}^{\infty} f(n)(-nz^{-n-1}) = -z \sum_{n=-\infty}^{\infty} f(n) \frac{d(z^{-n})}{dz}$$

$$= -z \frac{d}{dz} \sum_{n=-\infty}^{\infty} f(n)z^{-n} = -z \frac{d}{dz} F(z)$$

i.e., $$Z[<nf(n)>] = -z \frac{d}{dz} F(z)$$

☞ **Remark**

• The above result can be generalized as follows

$$Z[<n^k f(n)>] = \left(-z \frac{d}{dz}\right)^k F(z)$$

8.17.4 Z-Transform of $f(n)/n$

If $Z[<f(n)>] = F(z)$, then $Z\left[\dfrac{f(n)}{n}\right] = -\int_z \dfrac{1}{z} F(z) \, dz$

Proof. We have

$$Z\left[\frac{f(n)}{n}\right] = \sum_{n=-\infty}^{\infty} \frac{f(n)}{n} z^{-n} = \sum_{n=-\infty}^{\infty} f(n)\left(\frac{1}{n} z^{-n}\right) = -\sum_{n=-\infty}^{\infty} f(n)\int_z z^{-n-1} dz$$

$$= -\int_z \sum_{n=-\infty}^{\infty} f(n) z^{-n-1} dz = -\int_z \frac{1}{z} \sum_{n=-\infty}^{\infty} f(n) z^{-n} dz = -\int_z z^{-1} F(z) \, dz$$

Hence, $$Z\left[\left\{\frac{f(n)}{n}\right\}\right] = -\int_z z^{-1} F(z) \, dz$$

8.17.5 Initial Value Theorem

If $Z[<f(n)>] = F(z), n > 0$, then $f(0) = \lim_{z \to \infty} F(z)$

Proof. Here we have

$$Z[<f(n)>] = \sum_{n=0}^{\infty} f(n) z^{-n} = F(z)$$

or $$f(0) + f(1)z^{-1} + f(2)z^{-2} + \dots = F(z)$$

Taking limits of both the sides, as $z \to \infty$, we get

$$f(0) = \lim_{z \to \infty} F(z)$$

8.17.6 Final Value Theorem

If $Z[<f(n)>] = F(z), n > 0$, then $\lim_{z \to \infty} f(n) = \lim_{z \to 1} (z-1) F(z)$

Proof. We have $$Z[<f(n+1)> - <f(n)>] = \sum_{n=0}^{\infty} [f(n+1) - f(n)] z^{-n}$$

or $$Z[<f(n+1)>] - Z[<f(n)>] = \sum_{n=0}^{\infty} [f(n+1) - f(n)] z^{-n}$$

$$zF(z) - zF(0) - F(z) = \lim_{k \to \infty} \sum_{n=0}^{k} [f(n+1) - f(n)] z^{-n}$$

Taking the limits of both the sides as $z \to 1$, we get

$$\lim_{z \to 1} (z-1) F(z) = f(0) + \lim_{z \to 1} \lim_{k \to \infty} \sum_{n=0}^{k} [f(n+1) - f(n)] z^{-n}$$

$$= f(0) + \lim_{k \to \infty} \sum_{n=0}^{k} \lim_{z \to 1} [f(n+1) - f(n)] z^{-n} \qquad \text{(On changing the order of limits)}$$

$$= f(0) + \lim_{k \to \infty} \sum_{n=0}^{k} [f(n+1) - f(n)]$$

$$= \lim_{k \to \infty} [f(0) - f(0) + f(1) - f(1) + f(2) - f(2) + \dots + f(n+1) - f(n)]$$

$$= \lim_{k \to \infty} (k+1) = \lim_{k \to \infty} f(k) = \lim_{n \to \infty} f(n)$$

Hence, $$\lim_{n \to \infty} f(n) = \lim_{z \to 1} (z-1) F(z)$$

8.17.7 Partial Sum Theorem

If $Z[\{f(n)\}] = F(z)$, then $Z\left[\left\langle \displaystyle\sum_{k=-\infty}^{n} f(k) \right\rangle\right] = \dfrac{F(z)}{1 - z^{-1}}$

Proof. Let $\{u(n)\}$ be a sequence such that $\qquad u(n) = \sum\limits_{k=-\infty}^{n} f(k)$ $\qquad$ ...(1)

We want to find the value of $Z[\{u(n)\}]$.

From (1), we can write $\qquad u(n) - u(n-1) = \sum\limits_{k=-\infty}^{n} f(x) - \sum\limits_{k=-\infty}^{n-1} f(x)$

or $\qquad u(n) - u(n-1) = f(n)$

Taking Z-transforms of both the sides, we get

$$Z[\{u(n)\}] - Z[\{u(n-1)\}] = Z[\{f(n)\}]$$

or $\qquad U(z) - z^{-1} U(z) = F(z)$, where $U(z) = Z[\{u(n)\}]$

$\Rightarrow \qquad (1 - z^{-1}) U(z) = F(z)$

$\Rightarrow \qquad U(z) = \dfrac{F(z)}{1 - z^{-1}}$ $\quad$ *i.e.,* $\quad Z[\{u(n)\} = \dfrac{F(z)}{1 - z^{-1}}$

Hence, $\qquad Z\left[\left\langle \sum\limits_{k=-\infty}^{n} f(k) \right\rangle\right] = \dfrac{F(z)}{1 - z^{-1}}$

8.17.8 Convolution Theorem of Z-transforms

Definition : *Let $< f(n) >$ and $< g(n) >$ be two sequence and let the convolution of $< f(n) >$ and $< g(n) >$ be $< h(n) >$ where $< h(n) > = < f(n) >* < g(n) >$, then $< h(n) >$ is defined as*

$$< h(n) > = < f(n) >* < g(n) >$$

$$= \sum\limits_{n=-\infty}^{\infty} f(n) g(n-k) = \sum\limits_{n=-\infty}^{\infty} g(n) f(n-k) = < g(n) >* < f(n) >$$

Theorem. $\quad$ *If $Z[< f(n) >] = F(z)$ and $Z[< g(n) >] = G(z)$, then $Z[< h(n) >] = Z[< f(n) >* < g(n) >] = F(z) G(z)$*

where the region of convergence of $Z[< h(n) >]$ is the common region of convergence of $F(z)$ and $G(z)$.

PROOF. $\quad$ We have, by defintion $\quad Z[< h(n) >] = \sum\limits_{n=-\infty}^{\infty} \left[\sum\limits_{n=-\infty}^{\infty} f(k) g(n-k) \right] z^{-n} = \sum\limits_{n=-\infty}^{\infty} \sum\limits_{n=-\infty}^{\infty} f(k) g(n-k) z^{-n}$

$$= \sum\limits_{k=-\infty}^{\infty} f(k) z^{-k} \sum\limits_{r=-\infty}^{\infty} g(r) z^{-r} \qquad (n-k=r)$$

$$= \sum\limits_{k=-\infty}^{\infty} f(k) z^{-k} G(z) = F(z).G(z)$$

8.17.9 Inverse Z-transforms

If $F(z)$ is the Z-transform of the sequence $< f(k) >$, then $<f(k)>$ is called the inverse Z-transform of $F(z)$. The operator for inverse Z-transform is denoted by Z^{-1}. Thus, if $Z< f(k) > = F(z)$, then

$$Z^{-1}[F(z)] = < f(k) >$$

Some Useful Z-Transforms

S.No.	Sequence f(n), $n \geq 0$	Z-transform : F(z)	S.No.	Sequence f(n), $n \geq 0$	Z-transform : F(z)
1.	n	$z/(z-1)^2$	12.	$a^n \cos n\theta$	$\dfrac{z(z - a\cos\theta)}{z^2 - 2az\cos\theta + a^2}$
2.	n^2	$(z^2+z)/(z-1)^3$	13.	$\sinh n\theta$	$\dfrac{z\sinh\theta}{z^2 - 2z\cosh\theta + 1}$
3.	n^p	$-zd/dz[Z(n^{p-1})]$, positive integer.	14.	$\cosh n\theta$	$\dfrac{z(z - \cosh\theta)}{z^2 - 2z\cosh\theta + 1}$
4.	$\delta(n) = \begin{cases} 1, & n=0 \\ 0, & n\neq 0 \end{cases}$	1	15.	$a^n \sinh n\theta$	$\dfrac{az\sinh\theta}{z^2 - 2az\cosh\theta + a^2}$
5.	$f(n) = \begin{cases} 1, & n<0 \\ 0, & n\geq 0 \end{cases}$	$z/(z-1)$	16.	$a^n \cosh n\theta$	$\dfrac{z(z - 1\cosh\theta)}{z^2 - 2az\cosh\theta + a^2}$
6.	a^n	$z/(z-a)$	17.	$a^n f(n)$	$(F(z/a))$

7.	na^n	$az/(z-a)^2$	18.	f_{n+1} f_{n+2} f_{n+3}	$z[F(z)-f_0]$ $z^2[F(z)-f_0-f_1z^{-1}]$ $z^3[F(z)-f_0-f_1z^{-1}-f_2z^{-2}]$
8.	n^2a^n	$(az^2+a^2z)/(z-a)^3$	19.	f_{n-k}	$z^{-k}F(z)$
9.	$\sin n\theta$	$\dfrac{z\sin\theta}{z^2-2z\cos\theta+1}$	20.	$n f_n$	$-z\,d/dz\,[F(z)]$
10.	$\cos n\theta$	$\dfrac{z(z-\cos\theta)}{z^2-2z\cos\theta+1}$	21.	$\dfrac{1}{n}f_n$	$-\int_0^z z^{-1}[F(z)]dz$
11.	$a^n\sin n\theta$	$\dfrac{az\sin\theta}{z^2-2az\cos\theta+a^2}$	22.	f_0	$\lim\limits_{z\to\infty}F(z)$
			23.	$\lim\limits_{n\to\infty}f_n$	$\lim\limits_{z\to1}[(z-1)F(z)]$

Some Useful Inverse Z-Transforms

S.No.	F(z)	Inverse Z-transform f(n)	S.No.	F(z)	Inverse Z-transform f(n)
1.	$\dfrac{z}{z-a}$	$a^n u(n)$	4.	$\dfrac{1}{z-a}$	$a^{n-1}u(n-1)$
2.	$\dfrac{z^2}{(z-a)^2}$	$(n+1)a^n u(n)$	5.	$\dfrac{1}{(z-a)^2}$	$(n-1)a^{n-2}u(n-2)$
3.	$\dfrac{z^3}{(z-a)^3}$	$\dfrac{1}{2!}(n+1)(n+2)a^n u(n)$	6.	$\dfrac{1}{(z-a)^3}$	$\dfrac{1}{2}(n-1)(n-2)a^{n-3}u(n-3)$

Solved Examples

EXAMPLE 1. *Find the Z-transforms of the following sequences:*

(i) $<f(n)> = \{15, 10, 7, 4, 1, -1, 0, 3, 6\}$

(ii) $<f(n)> = \{15, 10, 7, 4, 1\}$

(iii) $<f(n)> = 1/3^n$

(iv) $<f(n)> = \dfrac{1}{2^n}$, $-3 \le n \le 2$

(v) $<f(n)> = \{a^n\}$, $n \ge 0$

SOLUTION. (i) We have $Z[<f(n)>]$

$$= F(z) = \sum_{n=-3}^{5} f(n)z^{-n}$$

$$= 15z^3 + 10z^2 + 7z + 4 + \frac{1}{z} - \frac{1}{z^2} + 0 + \frac{3}{z^4} + \frac{6}{z^5}$$

Therefore, $Z[<f(n)>]$

$$= 15z^3 + 10z^2 + 7z + 4 + \frac{1}{z} - \frac{1}{z^2} + 0 + \frac{3}{z^4} + \frac{6}{z^5}$$

(ii) We have $Z[<f(n)>] = F(z)$

$$= \sum_{n=0}^{4} f(n)z^{-n} = 15 + \frac{10}{z} + \frac{7}{z^2} + \frac{4}{z^3} + \frac{1}{z^4}$$

(iii) $Z[<f(n)>] = F(z)$

$$= \sum_{n=-\infty}^{\infty} f(n)z^{-n} = \sum_{n=-\infty}^{\infty} \frac{1}{3^n}z^{-n}$$

$$= \ldots + 27z^3 + 9z^2 + 3z + 1 + \frac{1}{3z} + \frac{1}{9z^2} + \frac{1}{27z^3} + \ldots$$

(iv) $Z[<f(n)>] = F(z) = \sum_{n=-3}^{2} \frac{1}{2^n}z^{-n}$

$$= 8z^3 + 4z^2 + 2z + 1 + \frac{1}{2z} + \frac{1}{4z^2}$$

(v) $Z[\{f(n)\}] = F(z)$

$$= \sum_{n=0}^{\infty} a^n z^{-n} = 1 + \frac{a}{z} + \frac{a^2}{z^2} + \frac{a^3}{z^3} + \ldots$$

$$= \frac{1}{1-\dfrac{a}{z}} = \frac{z}{z-a}$$

EXAMPLE 2. *Find the Z-transforms of the following sequences:*

(i) $<f(n)> = <a^{|n|}>$

(ii) $<f(n)> = \left\{\dfrac{a^n}{n!}\right\}$, $n \ge 0$ (SVTU–2009)

(iii) $<f(n)> = \{C^n\cos(\alpha n)\}$, $n \ge 0$ (UPTU–2004)

(iv) $<f(n)> = \{\sin(3n+5)\}$, $n \ge 0$

                      (VTU–2009, KOTTAYAM–2005)

(v) $<f(n)> = \{C^n\cosh(\alpha n)\}$, $n \ge 0$

SOLUTION. (i) We have $Z[<a^{|n|}>] = \sum_{n=-\infty}^{\infty} a^{|n|}z^{-n}$

$$= \sum_{n=-\infty}^{-1} a^{-n}z^{-n} + \sum_{n=0}^{\infty} a^n z^{-n}$$

$$= [\ldots + a^3z^3 + a^2z^2 + az]$$

$$+ [1 + az^{-1} + a^2z^{-2} + a^3z^{-3} + \ldots]$$

$$= \frac{az}{1-az} + \frac{1}{1-az^{-1}} = \frac{az}{1-az} + \frac{z}{z-a}$$

$$= \frac{az(z-a) + z(1-az)}{(1-az)(z-a)} = \frac{z(1-a^2)}{(1-az)(z-a)}$$

(ii) $Z\left[\left\langle\dfrac{a^n}{n!}\right\rangle\right] = \sum\limits_{n=0}^{\infty}\left(\dfrac{a^n}{n!}\right)z^{-n} = \sum\limits_{n=0}^{\infty}\dfrac{(az^{-1})}{n!}\cdot z^{-n}$

$= \sum\limits_{n=0}^{\infty}\dfrac{(az^{-1})^n}{n!} = 1 + \dfrac{az^{-1}}{1!} + \dfrac{(az^{-1})^2}{2!} + \dfrac{(az^{-1})^3}{3!}$

$= e^{az^{-1}} = e^{a/z}$

(iii) We have $Z\left[\left\langle C^n\cos(\alpha n)\right\rangle\right]$

$= \sum\limits_{n=0}^{\infty}[C^n\cos(\alpha n)]z^{-n}$

$= \sum\limits_{n=0}^{\infty}C^n\left[\dfrac{e^{i\alpha n} + e^{-i\alpha n}}{2}\right]z^{-n}$

$= \dfrac{1}{2}\sum\limits_{n=0}^{\infty}(C^n e^{i\alpha n})z^{-n} + \dfrac{1}{2}\sum\limits_{n=0}^{\infty}(C^n e^{-i\alpha n})z^{-n}$

$= \dfrac{1}{2}\sum\limits_{n=0}^{\infty}(C^n e^{i\alpha}z^{-1})^n + \dfrac{1}{2}\sum\limits_{n=0}^{\infty}(C^n e^{-i\alpha}z^{-1})^n$

$= \dfrac{1}{2}[1 + Ce^{i\alpha}z^{-1} + (Ce^{i\alpha}z^{-1})^2 + (Ce^{i\alpha}z^{-1})^3 + \ldots]$

$+ \dfrac{1}{2}[1 + Ce^{-i\alpha}z^{-1} + (Ce^{-i\alpha}z^{-1})^2 + (Ce^{-i\alpha}z^{-1})^3 + \ldots]$

$= \dfrac{1}{2}\left[\dfrac{1}{1 - Ce^{i\alpha}z^{-1}}\right] + \dfrac{1}{2}\left[\dfrac{1}{1 - Ce^{-i\alpha}z^{-1}}\right] \quad [\because |z| > |C|]$

$= \dfrac{1}{2}\left[\dfrac{1 - Ce^{-i\alpha}z^{-1} + 1 - Ce^{i\alpha}z^{-1}}{(1 - Ce^{i\alpha}z^{-1})(1 - Ce^{-i\alpha}z^{-1})}\right]$

$= \dfrac{1}{2}\left[\dfrac{2 - Cz^{-1}(e^{i\alpha} + e^{-i\alpha})}{1 - Ce^{-i\alpha}z^{-1} - Ce^{i\alpha}z^{-1})C^2 z^{-2}}\right]$

$= \dfrac{1}{2}\left[\dfrac{2 - Cz^{-1}(e^{i\alpha} + e^{-i\alpha})}{1 - Ce^{-i\alpha}z^{-1} - Ce^{i\alpha}z^{-1} + C^2 z^{-2}}\right]$

$= \dfrac{\left[1 - C\left(\dfrac{e^{i\alpha} + e^{-i\alpha}}{2}\right)z^{-1}\right]}{1 - C(e^{i\alpha} + e^{-i\alpha})z^{-1} + C^2 z^{-2}}$

$= \dfrac{1 - (C\cos\alpha)z^{-1}}{1 - (2C\cos\alpha)z^{-1} + C^2 z^{-2}} = \dfrac{z^2 - Cz\cos\alpha}{z^2 - 2Cz\cos\alpha + C^2}$

(iv) We have

$Z[<\sin(3n+5)>] = \sum\limits_{n=0}^{\infty}\sin(3n+5)z^{-n}$

$= \sum\limits_{n=0}^{\infty}\left[\dfrac{e^{i(3n+5)} - e^{-i(3n+5)}}{2!}\right]z^{-n}$

$= \dfrac{1}{2!}\sum\limits_{n=0}^{\infty}e^{i(3n+5)}\cdot z^{-n} - \dfrac{1}{2!}\sum\limits_{n=0}^{\infty}e^{-i(3n+5)}z^{-n}$

$= \dfrac{e^{5!}}{2!}\sum\limits_{n=0}^{\infty}(e^{3i} - z^{-1})^n - \dfrac{e^{-5i}}{2!}\sum\limits_{n=0}^{\infty}(e^{-3i}z^{-1})^n$

$= \dfrac{e^{5i}}{2!}[1 + e^{3i}z^{-1} + (e^{3i}z^{-1})^2 + (e^{3i}z^{-1})^3 + \ldots]$

$- \dfrac{e^{-5i}}{2!}[1 + e^{-3i}z^{-1} + (e^{-3i}z^{-1})^2 + (e^{-3i}z^{-1})^3 + \ldots]$

$= \dfrac{e^{5i}}{2!}\left[\dfrac{1}{1 - e^{3i}z^{-1}}\right] - \dfrac{e^{-5i}}{2i}\left[\dfrac{1}{1 - e^{-3i}z^{-1}}\right] \quad (|Z| > 1)$

$= \dfrac{1}{2!}\left[\dfrac{e^{5i}(1 - e^{-3i}z^{-1}) - e^{-5i}(1 - e^{3i}z^{-1})}{(1 - e^{3i}z^{-1})(1 - e^{-3i}z^{-1})}\right]$

$= \dfrac{1}{2!}\left[\dfrac{(e^{5i} - e^{-5i}) - e^{2i}z^{-1} + e^{-2i}z^{-1}}{1 - e^{3i}z^{-1} - e^{-3i}z^{-1} + z^{-2}}\right]$

$= \dfrac{\left(\dfrac{e^{5i} - e^{-5i}}{2!}\right) - \left(\dfrac{e^{2i} - e^{-2i}}{2!}\right)z^{-1}}{1 - (2\cos 3)z^{-1} + z^{-2}}$

$= \dfrac{\sin 5 - (\sin 2)z^{-1}}{1 - (2\cos 3)z^{-1} + z^{-2}} = \dfrac{z^2\sin 5 - z\sin 2}{z^2 - 2z\cos 3 + 1}$

(v) We have $Z[<C^n\cosh(\alpha n)>]$

$= \sum\limits_{n=0}^{\infty}[C^n\cosh(\alpha n)]z^{-n} = \sum\limits_{n=0}^{\infty}C^n\left[\dfrac{e^{\alpha n} + e^{-\alpha n}}{2}\right]z^{-n}$

$= \dfrac{1}{2}\sum\limits_{n=0}^{\infty}(Ce^{\alpha}z^{-1})^n + \dfrac{1}{2}\sum\limits_{n=0}^{\infty}(Ce^{-\alpha}z^{-1})^n$

$= \dfrac{1}{2}[1 + Ce^{\alpha}z^{-1} + (Ce^{\alpha}z^{-1})^2 + (Ce^{\alpha}z^{-1})^3 + \ldots]$

$+ \dfrac{1}{2}[1 + Ce^{-\alpha}z^{-1} + (Ce^{-\alpha}z^{-1})^2 + (Ce^{-\alpha}z^{-1})^3 + \ldots]$

$= \dfrac{1}{2}\left[\dfrac{1}{1 - Ce^{\alpha}z^{-1}}\right] + \dfrac{1}{2}\left[\dfrac{1}{1 - Ce^{-\alpha}z^{-1}}\right]$

$[\because |Z| > |Ce^{\alpha}| \text{ and } |Z| > |Ce^{-\alpha}|]$

$= \dfrac{1}{2}\left[\dfrac{1 - Ce^{-\alpha}z^{-1} + 1 - Ce^{\alpha}z^{-1}}{1 - Ce^{\alpha}z^{-1} - Ce^{-\alpha}z^{-1} + C^2 z^{-2}}\right]$

$= \dfrac{1}{2}\left[\dfrac{2 - C(e^{\alpha} + e^{-\alpha})z^{-1}}{1 - C(e^{\alpha} + e^{-\alpha})z^{-1} + C^2 z^{-2}}\right]$

$= \dfrac{1 - (C\cosh\alpha)z^{-1}}{1 - (2C\cosh\alpha)z^{-1} + C^2 z^{-2}}$

$= \dfrac{z^2 - Cz\cosh\alpha}{z^2 - 2Cz\cosh\alpha + C^2} = \dfrac{z(z - C\cosh\alpha)}{z^2 - 2Cz\cosh\alpha + C^2}$

EXAMPLE 3. *Show that* $Z\left(\dfrac{1}{n!}\right) = e^{1/z}$.

Hence, evaluate $Z\left[\dfrac{1}{(n+1)!}\right]$ *and* $Z\left[\dfrac{1}{(n+2)!}\right]$.

SOLUTION. We have $Z\left[\dfrac{1}{n!}\right] = \sum\limits_{n=0}^{\infty}\dfrac{1}{n!}z^{-n}$

$= 1 + \dfrac{z^{-1}}{1!} + \dfrac{z^{-2}}{2!} + \dfrac{z^{-3}}{3!} + \ldots = e^{1/z}$

Now, shifting $\left(\dfrac{1}{n!}\right)$ one unit to the left gives

$Z\left[\dfrac{1}{(n+1)!}\right] = z\left[z\left(\dfrac{1}{n!}\right) - 1\right] = z(e^{1/z} - 1)$

Again, shifting $\left(\dfrac{1}{n!}\right)$ two units to the left gives

$Z\left[\dfrac{1}{(n+2)!}\right] = z^2(e^{1/z} - 1 - z^{-1})$

EXAMPLE 4. *Find the Z-transform of* $\cosh\left(\dfrac{n\pi}{2} + \alpha\right)$.

(VTU–2011, UPTU–2008)

SOLUTION. We have

$$F(z) = \sum_{n=0}^{\infty} \cosh\left(\frac{n\pi}{2} + \alpha\right) z^{-n}$$

$$= \sum_{n=0}^{\infty} \left[\frac{e^{(n\pi/2)+\alpha} - e^{-(n\pi/2)+\alpha}}{2}\right] z^{-n}$$

$$= \frac{1}{2}e^{\alpha} \sum_{n=0}^{\infty} (e^{\pi/2}z^{-1})^n + \frac{1}{2}e^{-\alpha} \sum_{n=0}^{\infty} (e^{-\pi/2}z^{-1})^n$$

$$= \frac{1}{2}e^{\alpha}\left[1 + \left(e^{\pi/2}z^{-1}\right) + \left(e^{\pi/2}z^{-1}\right)^2 + ...\right]$$

$$+ \frac{1}{2}e^{-\alpha}\left[1 + \left(e^{-\pi/2}z^{-1}\right) + \left(e^{-\pi/2}z^{-1}\right)^2 + ...\right]$$

[Sum of two G.P's]

$$= \frac{1}{2}e^{\alpha}\left[\frac{1}{e^{-\pi/2}z^{-1}}\right] + \frac{1}{2}e^{-\alpha}\left[\frac{1}{1 - e^{-\pi/2}z^{-1}}\right]$$

$$= \frac{1}{2}\left[\frac{e^{\alpha}(1 - e^{-\pi/2}z^{-1}) + e^{-\alpha}(1 - e^{\pi/2}z^{-1})}{(1 - e^{-\pi/2}z^{-1})(1 - e^{-\pi/2}z^{-1})}\right]$$

$$= \frac{\dfrac{e^{\alpha} + e^{-\alpha}}{2} - \dfrac{e^{\alpha-\frac{\pi}{2}} + e^{-\alpha+\frac{\pi}{2}}}{2}}{1 - e^{\pi/2}z^{-1} - e^{-\pi/2}z^{-1} + z^{-2}}$$

$$= \frac{\cosh\alpha - \cosh(\alpha - \pi/2).z^{-1}}{1 - \left(2\cosh\dfrac{\pi}{2}\right)z^{-1} + z^{-2}}$$

$$= \frac{z^2 \cosh\alpha - z\cosh\left(\dfrac{\pi}{2} - \alpha\right)}{z^2 - 2z\cosh\dfrac{\pi}{2} + 1}$$

EXAMPLE 5. *Find the Z-transform of*

(i) ^nC_k *(ii)* $^{k+n}C_n$ *(iii)* $^{k+n}C_n a^k$

SOLUTION. (i) We have

$$Z\left[\{^nC_k\}\right] = \sum_{k=0}^{n} {}^nC_k z^{-k}$$

$$= 1 + {}^nC_1 z^{-1} + {}^nC_2 z^{-2} + ... + {}^nC_n z^{-k} = (1 + z^{-1})^n$$

(ii) We have

$$Z\left[\{^{k+n}C_n\}\right] = \sum_{k=0}^{\infty} {}^{k+n}C_n z^{-k}$$

$$= \sum_{k=0}^{\infty} {}^{k+n}C_k z^{-k} \quad [\because {}^nC_r = {}^nC_{n-r}]$$

$$= 1 + {}^{n+1}C_1 z^{-1} + {}^{n+2}C_2 z^{-2} + {}^{n+3}C_3 z^{-3} + ...$$

$$= 1 + (n+1)z^{-1} + \frac{(n+2)(n+1)}{2!}z^{-2}$$

(iii) We have

$$Z\left[\{^{k+n}C_n a^k z^{-k}\}\right] = \sum_{k=0}^{\infty} {}^{k+n}C_k a^k z^{-k}$$

$$(\because {}^nC_r = {}^nC_{n-r})$$

$$= \sum_{k=0}^{\infty} {}^{k+n}C_k (az^{-1})^k \qquad |z| > a$$

$$= (1 - az^{-1})^{-(n+1)}$$

EXAMPLE 6. *Find Z-transforms of the following sequences by using the change of scale property*

(i) $<a^n>$, $n \geq 0$ *(ii)* $<c^n \sin(\alpha n)>$, $n \geq 0$

SOLUTION. (i) We have

$$Z[<1>] = \sum_{n=0}^{\infty} 1.z^{-n}$$

$$= 1 + \frac{1}{z} + \frac{1}{z^2} + \frac{1}{z^3} + ... \quad \text{(Infinite G.P.)}$$

$$= \frac{1}{1 - \dfrac{1}{z}} = \frac{z}{z-1}$$

Using change of scale property, *i.e.*, if $Z[<\{f(n)>]$ $= F(z)$, then $Z[<a^n f(n)>] = F(z/a)$, we have

$$Z\left[<a^n>\right] = z\left[<a^n \cdot 1>\right] = \frac{(z/a)}{(z/a) - 1} = \frac{z}{z-a}$$

(ii) We have

$$Z[<\sin\alpha n>] = \sum_{n=0}^{\infty} (\sin\alpha n)z^{-n}$$

$$= \sum_{n=0}^{\infty}\left[\frac{e^{i\alpha n} - e^{-i\alpha n}}{2i}\right]z^{-n}$$

$$= \frac{1}{2i}\sum_{n=0}^{\infty}(e^{i\alpha}z^{-1})^n - \frac{1}{2!}\sum_{n=0}^{\infty}(e^{-i\alpha}z^{-1})^n$$

$$= \frac{1}{2i}\left[1 + e^{i\alpha}z^{-1} + (e^{i\alpha}z^{-1})^2 + ...\right]$$

$$- \frac{1}{2i}\left[1 + e^{-i\alpha}z^{-1} + (e^{-i\alpha}z^{-1})^2 + ...\right]$$

$$= \frac{1}{2i}\left[\frac{1}{1 - e^{i\alpha}z^{-1}}\right] - \frac{1}{2i}\left[\frac{1}{1 - e^{-i\alpha}z^{-1}}\right]$$

$$= \frac{1}{2i}\left[\frac{1 - e^{-i\alpha}z^{-1} - (1 - e^{i\alpha}z^{-1})}{1 - e^{i\alpha}z^{-1} - e^{-i\alpha}z^{-1} + z^{-2}}\right]$$

$$= \frac{1}{2i}\left[\frac{(e^{i\alpha} - e^{-i\alpha})z^{-1}}{1 - (e^{i\alpha} + e^{-i\alpha})z^{-1} + z^{-2}}\right]$$

$$= \frac{(\sin\alpha)z^{-1}}{1 - (2\cos\alpha)z^{-1} + z^{-2}}$$

$$= \frac{z\sin\alpha}{z^2 - 2z\cos\alpha + 1}$$

Now, using the change of scale property, we have

$$Z\left[<c^n \sin\alpha n>\right] = \frac{(z/c)\sin\alpha}{(z/c)^2 - (2z/c)\cos\alpha + 1}$$

$$= \frac{cz\sin\alpha}{z^2 - 2cz\cos\alpha + c^2}$$

EXAMPLE 7. *Find the Z-transforms of the sequence*

$$< f(n) > = \sum_{n=0}^{\infty} 3^n \sum_{n=0}^{\infty} 4^n$$

SOLUTION. We know that

$$Z[<3^n>] = \frac{1}{1 - 3z^{-1}}$$

and

$$Z[<4^n>] = \frac{1}{1 - 4z^{-1}}$$

Therefore

$$Z[<f(n)>] = Z[<3^n>] \cdot Z[<4^n>]$$

(By convolution property)

$$= \left(\frac{1}{1 - 3z^{-1}}\right)\left(\frac{1}{1 - 4z^{-1}}\right)$$

$$= \frac{z^2}{(z-3)(z-4)} = \frac{z^2}{z^2 - 7z + 12}$$

EXAMPLE 8. *If* $F(z) = Z[< f(n) >] = \dfrac{2z^2 + 5z + 14}{(z-1)^4}$, *then find the value of* $f(0)$.

SOLUTION. By initial value theorem, we have

$$f(0) = \lim_{z \to \infty} F(z) = \lim_{z \to \infty}\left[\frac{2z^2 + 5z + 14}{(z-1)^4}\right]$$

$$= \lim_{z \to 0}\frac{1}{z^2}\left[\frac{2 + 5z^{-1} + 14z^{-2}}{(1 - z^{-1})^4}\right] = 0$$

8.17.10 EVALUATION OF INVERSE Z-TRANSFORMS

(1) **Power Series Method :** This is the simplest method of finding the inverse Z-transforms. If $F(z)$ is expressed as the ratio of two polynomials which cannot be factorized, we simply divide the numerator by the denominator and take the inverse Z-transforms of each term in the quotient.

(2) **Partial Fractional Method :** In this method, we decompose $F(z)/z$ into partial fractions and multiply the resulting expansion by z and then inverting the same. We use this method in a similar fashion as that of finding the inverse Laplace transforms, using partial fractions.

(3) **Inversion Integral Method :** The inverse Z-transforms of $F(z)$ is given by the following formula

$$f(n) = \frac{1}{2\pi i}\int_c F(z) z^{n-1} dz = \text{sum of residue of } F(z)\, z^{n-1} \text{ at the poles of } F(z) \text{ which are inside the contour } c \text{ drawn.}$$

 Solved Examples

EXAMPLE 1. *Using convolution theorem, to find*

$$Z^{-1}\left\{\frac{z^2}{(z-a)(z-b)}\right\}$$

SOLUTION. Since, we know that

$$Z^{-1}\left\{\frac{z}{z-a}\right\} = a^n$$

and

$$Z^{-1}\left\{\frac{z}{z-a}\right\} = b^n$$

Therefore

$$Z^{-1}\left\{\frac{z^2}{(z-a)(z-b)}\right\} = Z^{-1}\left\{\frac{z}{z-a}\frac{z}{z-b}\right\}$$

$$= a^n * b^n$$

$$= \sum_{m=0}^{n} a^m . b^{n-m} = b^n \sum_{m=0}^{n}\left(\frac{a}{b}\right)^m, \text{ which in a G.P.}$$

$$= b^n . \frac{(a/b)^{n+1} - 1}{\frac{a}{b} - 1} = \frac{a^{n+1} - b^{n+1}}{a - b}$$

EXAMPLE 2. *Find the inverse Z-transforms of* $\log\left(\dfrac{z}{z+1}\right)$ *by power series method.*

SOLUTION. Putting $z = \dfrac{1}{t}$, $F(z) = \log\left(\dfrac{\frac{1}{t}}{\frac{1}{t}+1}\right) = -\log(1+t)$

$$= -t + \frac{t^2}{2} - \frac{t^3}{3} + \dots = -z^{-1} + \frac{1}{2}z^{-2} - \frac{1}{3}z^{-3} + \dots$$

Hence, $f(n) = \begin{cases} 0, & \text{for } n = 0 \\ (-1)^n / n, & \text{otherwise} \end{cases}$

EXAMPLE 3. *Find the inverse Z-transform of*

(i) $\dfrac{2z^2 + 3z}{(z+2)(z-4)}$ (VTU–2008, SVTU–2007)

(ii) $\dfrac{z^3 - 20z}{(z-2)^3(z-4)}$ (VTU–2011)

SOLUTION. (i) We have

$$F(z) = \frac{2z^2 + 3z}{(z+2)(z-4)}$$

$$\Rightarrow \frac{F(z)}{z} = \frac{2z + 3}{(z+2)(z-4)} = \frac{A}{z+2} + \frac{B}{z-4}$$

Using method of partical fraction, we get $A = 1/6$ and $B = 11/6$.

Therefore $F(z) = \dfrac{1}{6}.\dfrac{z}{z+2} + \dfrac{11}{6}.\dfrac{z}{z-4}$

On inversion, we have

$$f(n) = \frac{1}{6}(-2)^n + \frac{11}{6}(4)^n$$

(ii) We have $F(z) = \dfrac{z^3 - 20z}{(z-2)^3(z-4)}$

$$\Rightarrow \frac{F(z)}{z} = \frac{z^2 - 20}{(z-2)^3(z-4)} = \frac{A + Bz + Cz^2}{(z-2)^3} + \frac{D}{z-4}$$

Then, we get $D = \dfrac{1}{2}$

Mutliplying throughout by $(z-2)^3 (z-4)$, we get

$$z^2 - 20 = (A + Bz + Cz^2)^2(z-4) + D(z-2)^3$$

Putting $z = 0, 1, -1$ successively and solving the resulting simultaneous equations, we get

$$A = 6, B = 0, C = 1/2$$

Hence, $F(z) = \dfrac{1}{2}.\dfrac{12z + z^3}{(z-2)^3} - \dfrac{z}{z-4}$

$$= \frac{1}{2}\frac{z(z-2)^2 + 4z^2 + 8z}{(z-2)^3} - \frac{z}{z-4}$$

$$= \frac{1}{2}\left\{\frac{z}{z-2} + 2\frac{2z^2 + 4z}{(z-2)^3}\right\} - \frac{z}{z-4}$$

On inversion, we get

$$f(n) = \frac{1}{2}(2^n + 2.n^2 2^n) - 4^n$$

$$= 2^{n-1} + n^2 2^n - 4^n$$

EXAMPLE 4. *Find the inverse Z-transform of* $\dfrac{2(z^2 - 5z + 6.5)}{[(z-2)(z-3)^2]}$ *for* $2 < |z| < 3$.

SOLUTION. We have $F(z) = \dfrac{2(z^2 - 5z + 6.5)}{[(z-2)(z-3)^2]}$

$$= \frac{A}{z-2} + \frac{B}{z-3} + \frac{C}{(z-3)^2}$$

Using the method of partial fraction, we get $A=B=C=1$
Therefore

$$F(z) = \frac{1}{z-2} + \frac{1}{z-3} + \frac{1}{(z-3)^2}$$

$$= \frac{1}{2}\left(1-\frac{2}{z}\right)^{-1} - \frac{1}{3}\left(1-\frac{z}{3}\right)^{-1} + \frac{1}{9}\left(1-\frac{z}{3}\right)^{-2}$$

$$\left(\frac{2}{z} < 1, \frac{z}{3} < 1, i.e., 2 < |z| < 3\right)$$

$$= \frac{1}{z}\left[1 + \frac{2}{z} + \frac{4}{z^2} + \frac{8}{z^3} + ...\right] - \frac{1}{3}\left[1 + \frac{z}{3} + \frac{z^2}{9} + \frac{z^3}{27} + ...\right]$$

$$+ \frac{1}{9}\left[1 + \frac{2z}{3} + \frac{3z^2}{9} + \frac{4z^3}{27} + ...\right]$$

$$= \left(\frac{1}{2} + \frac{2}{z^2} + \frac{2^2}{z^3} + \frac{z^3}{z^4} + ...\right) - \left(\frac{1}{3} + \frac{z}{3^2} + \frac{z^2}{3^3} + \frac{z^3}{3^4}\right)$$

$$+ \left(\frac{1}{3^2} + \frac{2z}{3^3} + \frac{3z^2}{3^4} + \frac{4z^3}{3^5} + ...\right)$$

$$= \sum_{n=1}^{\infty} 2^{n-1} z^{-n} - \sum_{n=0}^{\infty} \left(\frac{1}{3}\right)^{n-1} z^n + \sum_{n=0}^{\infty} (n+1)\left(\frac{1}{3}\right)^{n+2} . z^n$$

Finally, on inversion, we get

$$f(u) = 2^{n-1}, \; n \geq 1$$

and $f(n) = -(n+2)3^{n-2}, \; n \leq 0$

EXAMPLE 5. *Find* $Z^{-1}\left\{\dfrac{3z^2 - 18z + 26}{(z-2)(z-3)(z-4)}\right\}$.   (ANNA–2005)

SOLUTION. Clearly, the poles are given by

$$(z-2)(z-3)(z-4) = 0 \Rightarrow z = 2, 3, 4$$

Now, residue at $z = 2$

$$= \left[\frac{(z-2)z^{n-1}(3z^2 - 18z + 26)}{(z-2)(z-3)(z-4)}\right]_{z=2}$$

$$= \left[\frac{3z^{n+1} - 18z^n + 26z^{n-1}}{(z-3)(z-4)}\right]_{z=2}$$

$$= \frac{3.2^{n+1} - 18.2^n + 26.2^{n-1}}{(-1)(-2)}$$

$$= 3.2^n - 9.2^n + 13.2^{n-1} = 2^{n-1}$$

Residue at $z = 3$

$$= \left[\frac{(z-3)z^{n-1}.(3z^2 - 18z + 26)}{(z-2)(z-3)(z-4)}\right]_{z=3}$$

$$= \left[\frac{3z^{n+1} - 18z^n + 26z^{n-1}}{(z-2)(z-4)}\right]_{z=3}$$

$$= \frac{3.3^{n+1} - 18.3^n + 26.3^{n-1}}{1(-1)}$$

and residue at $z = 4$

$$= \left[(z-4)\left[\frac{z^{n-1}(3z^2 - 18z + 26)}{(z-2)(z-3)(z-4)}\right]\right]_{z=4}$$

$$= \left[\frac{3z^{n+1} - 18z^n + 26z^{n-1}}{(z-2)(z-3)}\right]_{z=4}$$

$$= \left[\frac{3.4^{n+1} - 18.4^n + 26.4^{n-1}}{2.1}\right] = 4^{n-1}$$

$$= z^{-1} f(z) = F(n) = \text{sum of residues}$$

$$= 2^{n-1} + 3^{n-1} + 4^{n-1}, n > 0$$

8.17.11 SOLUTION OF DIFFERENCE EQUATION WITH CONSTANT COEFFICIENTS BY Z-TRANSFORM

WORKING PROCEDURE

For solving a linear difference equation with constant coefficient by using Z-transforms, we use the following steps :

Step 1. Take the Z-transforms of both sides of the given difference equation using the formulae of shifting property and given conditions.

Step 2. Transpose all terms without $F(z)$, to the right hand side.

Step 3. Divide the coefficient of $F(z)$, getting $F(z)$ as a function of z.

Step 4. Express this function in terms of the Z-transforms of known functions and take the inverse Z-transforms of both sides, which gives $f(k)$, i.e., $f(k)$ as a function of k which is the required solution.

Solved Examples

EXAMPLE 1. *Solve the difference equation*

$$6f_{n+2} - f_{n+1} + f_n = 0, \text{ which } f(0)=0, f(1)=1, \text{ by}$$

Z-transform.

SOLUTION. The given difference equation is

$$6f_{n+2} - f_{n+1} + f_n = 0 \qquad ...(1)$$

Taking Z-transform of both sides of (1), we get

$$Z[6f_{n+2} - f_{n+1} + f_n] = 0$$

$$\Rightarrow \quad Z(6f_{n+2}) - Z(f_{n+1}) + Z(f_n) = 0$$

$$= 6[z^2 F(z) - z^2 f(0) - z f(1)]$$

$$- [zF(z) - zF(0)] + F(z) = 0 \qquad ...(2)$$

Using the given initial condition in (2), we get

$$6z^2 F(z) - 6z - z F(z) + F(z) = 0$$

$$\Rightarrow \quad (6z^2 - z + 1) F(z) = 6z$$

$$\Rightarrow \quad F(z) = \frac{6z}{6z^2 - z + 1} = \frac{6z}{(3z+1)(2z-1)}$$

$$= \frac{z^{-1}}{\left(1 + \dfrac{z^{-1}}{3}\right)\left(1 - \dfrac{z^{-1}}{2}\right)} = \frac{6/5}{1 - \dfrac{z^{-1}}{2}} - \frac{6/5}{1 + \dfrac{z^{-1}}{3}}$$

$$\Rightarrow \quad f(n) = Z^{-1}\left[\frac{6/5}{1 - z^{-1}/2}\right] - Z^{-1}\left[\frac{6/5}{1 + z^{-1}/3}\right]$$

$$= \frac{6}{5}\left(\frac{1}{2}\right)^n - \frac{6}{5}\left(-\frac{1}{3}\right)^n = \frac{6}{5}\left[\left(\frac{1}{2}\right)^n - \left(-\frac{1}{3}\right)^n\right]$$

EXAMPLE 2. *Solve the difference equation*

$$f_{n+3} - 3f_{n+2} + 3f_{n+1} - f_n = F(n)$$

where $f(0) = f(1) = f(2) = 0$, *by Z-transforms.*

SOLUTION. The given difference equation is

$$f_{n+3} - 3f_{n+2} + 3f_{n+1} - f_n = F(n) \qquad ...(1)$$

Taking Z-transforms of both sides of (1), we get

$$Z(f_{n+3} - 3f_{n+2} + 3f_{n+1} - f_n) = Z[F(n)]$$

$$\Rightarrow Z[f_{n+3}] - 3Z[f_{n+2}] + 3Z[f_{n+1}] - Z[f_n] = Z[F(n)]$$

$$\Rightarrow [z^3 F(z) - z^3 f(0) - z^2 f(1) - z f(2)]$$

$$- 3[z^2 F(z) - z^2 f(0) - z f(1)]$$

$$+ 3[z F(z) - z f(0)] - F(z) = Z[F(n)] \quad ...(2)$$

Using the given initial conditions in (2), we get

$$z^2 F(z) - 3z^2 F(z) + 3z - F(z) = \frac{1}{1 - z^{-1}}$$

$$\Rightarrow \quad (z^2 - 3z^2 + 3z - 1) F(z) = \frac{1}{1 - \frac{1}{z}z^{-1}}$$

$$\Rightarrow \quad (z - 1)^3 F(z) = \frac{1}{1 - z^{-1}}$$

i.e., $\quad F(z) = \frac{1}{(z-1)^3 (1 - z^{-1})} = \frac{1}{z^3 (1 - z^{-1})^3 (1 - z^{-1})}$

$$= z^{-3} (1 - z^{-1})^{-4}$$

$$\Rightarrow \quad f(n) = \text{Ceofficients of } z^{-n} \text{ in } z^{-3} (1 - z^{-1})^{-4}$$

$$= \text{Ceofficients of } z^{-n-3} \text{ in } (1 - z^{-1})^4$$

$$= \frac{(n-2)(n-1)n}{6}, \, n \geq 3$$

EXAMPLE 3. *Using Z-transform, solve*
$$f_{n+2} + 4f_{n+1} + 3f_n = 3^n \text{ with } f(0) = 0, f(1) = 1.$$

                                      (UPTU–2003)

SOLUTION.   If $Z(f_n) = F(z)$, then $Z[f_{n+1}] = z[F(z) - f(0)]$

$$Z[f_{n+2}] = z^2 [F(z) - f(0) - f(1)z^{-1}]$$

Also , $Z(2^n) = \dfrac{z}{z-2}$

Taking Z-transforms of both sides of the given equation, we get

$$z^2 [F(z) - f(0) - f(1)z^{-1}] + 4z[F(z) - f(0)] + 3F(z)$$

$$= \frac{z}{z-3} \quad\quad ...(1)$$

Using the given condition, (1) reduces to

$$F(z)(z^2 + 4z + 3) = z + \frac{z}{z+3}$$

Therefore, $\dfrac{F(z)}{z} = \dfrac{1}{(z+1)(z+3)} + \dfrac{1}{(z-3)(z+1)(z+3)}$

$$= \frac{3}{8}.\frac{1}{z+1} + \frac{1}{24}\frac{1}{z-3} - \frac{5}{12}\frac{1}{z+3}$$

                      (On resolving into partial fractions)

$$\therefore \quad F(z) = \frac{3}{8}\frac{z}{z+1} + \frac{1}{24}\frac{z}{z-3} - \frac{5}{12}\frac{z}{z+3}$$

On inversion, we get

$$f_n = \frac{3}{8} Z^{-1}\left(\frac{z}{z+1}\right) + \frac{1}{24} Z^{-1}\left(\frac{z}{z-3}\right) - \frac{5}{12} Z^{-1}\left(\frac{z}{z+3}\right)$$

$$= \frac{3}{8}(-1)^n + \frac{1}{24}3^n - \frac{5}{12}(-3)^n$$

EXAMPLE 4. *Solve $f_{n+2} + 6f_{n+1} + 9f_n = 2^n$ with $f(0) = f(1) = 0$ using Z-transforms.*

                    (VTU–2011, ANNA–2009, SVTU–2009)

SOLUTION.   If $Z(f_n) = F(z)$, then

$$Z(f_{n+1}) = z[F(z) - f(0)]$$

and $\;Z(f_{n+2}) = Z[F(z) - f(0) - f(1)z^{-1}]$

Also, $\;Z(2^n) = \dfrac{z}{z-2}$

Taking Z-transform of both sides of the given equation, we get

$$z^2 [F(z) - f(0) - f(1)Z^{-1}] + 6z[F(z) - f(0)] + 9F(z) = \frac{z}{z-2} \quad ...(1)$$

Using the given condition in (1), we get

$$F(z)(z^2 + 6z + 9) = \frac{z}{z-2}$$

$$\Rightarrow \frac{F(z)}{z} = \frac{1}{(z-2)(z-3)^2} = \frac{1}{25}\left[\frac{1}{z-2} - \frac{1}{z+3} - \frac{5}{[z+3]^2}\right]$$

Therefore $\;F(z) = \dfrac{1}{25}\left[\dfrac{z}{z-2} - \dfrac{z}{z+3} - 5.\dfrac{z}{(z+3)^2}\right] \quad ...(2)$

On taking inverse Z-transform of both sides of (2), we get

$$f(n) = \frac{1}{25}\left[Z^{-1}\left(\frac{z}{z-2}\right) - Z^{-1}\left(\frac{z}{z+3}\right)\right.$$

$$\left. + \frac{5}{3} Z^{-1}\left(-\frac{3z}{(z+3)^2}\right)\right]$$

$$= \frac{1}{25}\left[2^n - (-3)^n + \frac{5}{3}n(-3)^n\right]$$

$$\left[\because Z^{-1}\left\{\frac{az}{(z-a)^2}\right\} = na^n\right]$$

EXAMPLE 6. *Using the Z-transform, solve*
$$f_{n+2} - 2f_{n+1} + f_n = 3n + 5 \quad\quad \text{(SVTU–2007)}$$

SOLUTION.   The given difference equation is

$$f_{n+2} - 2f_{n+1} + f_n = 3n + 5 \quad\quad ...(1)$$

Taking Z-transform of both sides of (1), we get

$$z^2 [F(z) - f(0) - f(1)z^{-1}] - 2z[F(z) - f(0)] + F(z)$$

$$= 3.\frac{z}{(z+1)^2} + 5.\frac{z}{z-1}$$

or $\;F(z) = \dfrac{5z^2 - 2z}{(z-1)^2} + f(0).\dfrac{z^2 - 2z}{(z-1)^2} + f(1).\dfrac{z}{(z-1)^2}$

Taking inverse Z-transform, we get

$$f(n) = Z^{-1}\left[\frac{5z^2 - 2z}{(z-1)^4}\right] + f(0)Z^{-1}\left[\frac{z^2 - 2z}{(z-1)^2}\right]$$

$$+ f(1)Z^{-1}\left[\frac{z}{(z-1)^2}\right] \quad ...(2)$$

Notting that

$$Z(n) = \frac{z}{(z-1)^2}, \; Z(1) = \frac{z}{z-1}, \; Z(n^2) = \frac{z^2 + z}{(z-1)^3},$$

$$Z(n^3) = \frac{z^3 + 4z^2 + z}{(z-1)^4}$$

We write

$$\frac{5z^2 - 2z}{(z-1)^4} + A.\frac{z^3 + 4z^2 + z}{(z-1)^4} + B.\frac{z^2 + z}{(z-1)^3}$$

$$+ C.\frac{z}{(z-1)^2} + D.\frac{z}{z-1}$$

Equating coefficient of like powers of z, we find

$$A = \frac{1}{2}, \, B = 1, \, C = -\frac{3}{2}, \, D = 0.$$

Therefore, $\;Z^{-1}\left\{\dfrac{5z^2 - 2z}{(z-1)^4}\right\} = \dfrac{1}{2}n^3 + n^2 - \dfrac{3}{2}n$

$$= \frac{1}{2}n(n-1)(n+3)$$

also, $Z^{-1}\left\{\dfrac{z^2-2z}{(z-1)^2}\right\} = Z^{-1}\left\{\dfrac{z}{z-1}\right\} - Z^{-1}\left\{\dfrac{z}{(z-1)^2}\right\}$

and $Z^{-1}\left\{\dfrac{z}{(z-1)^2}\right\} = n$

Putting these values in (2), we get

$$f(n) = \frac{1}{2}n(n-1)(n+3) + f(0)(1-n) + f(1)\cdot n$$

$$= \frac{1}{2}n(n-1)(n+3) + C_0 + C_1 n$$

where $C_0 = f(0)$, $C_1 = f(1) - f(0)$

Exercise-8.75

1. Find the Z-transforms of the following functions for $n \geq 0$:

(i) 2^n

(ii) $\sin \alpha n$

(iii) $\sin(3n+5)$

(iv) $\sinh\dfrac{n\pi}{2}$

(v) $\cosh n\theta$ (VTU–2011)

2. Find the Z-transforms of the following functions for $n \geq 0$:

(i) $\sin\left(\dfrac{n\pi}{2}+\alpha\right)$

(ii) $C^n \sinh \alpha n$

(iii) $\cos\left(\dfrac{n\pi}{2}+\dfrac{\pi}{4}\right)$

(iv) $\dfrac{a^n}{n!}$

3. Find the Z-transforms of the following functions for $n > 0$:

(i) $\sin 5n$

(ii) $e^{\alpha n}$

(iii) $3n \cosh 5n$

4. Find the inverse of Z-transforms of the following functions:

(i) $\dfrac{1}{(z-3)(z-2)}$ for $|z| < 2$

(ii) $\dfrac{1}{(z-5)^3}$, $|z| > 5$

(iii) $\dfrac{3z^2+4z}{z^2-z+1}$, $|z| > 1$

(iv) $\dfrac{ze^{-a}}{(z-e^{-a})^2}$, $|z| > |e^{-a}|$

(v) $\dfrac{z^3-20z}{(z-2)^3(z-4)}$ (VTU–2011)

(vi) $\dfrac{4z^2-2z}{z^3-5z^2+8z-4}$ (VTU–2011)

(vii) $\dfrac{18z^2}{(2z-1)(4z+1)}$ (SVTU–2009)

5. Solve the following difference equations by Z-transform:

(i) $f_{n+1} - 2f_n + f_{n-1} = a_n$, $a \neq 1$

(ii) $f_n - \dfrac{5}{6}f_{n-1} + \dfrac{1}{6}f_{n-2} = F(n)$

(iii) $f_n + \dfrac{1}{a}f_{n-2} = \left(\dfrac{1}{3}\right)^n \cos\dfrac{n\pi}{2}$; $(n \geq 0)$

(iv) $6f_{n+2} + 5f_{n+1} - f_n = 6F(n)$

(v) $6y_{k+2} - y_{k+1} - y_k = 0$, $y(0) = y(1) = 1$ (KOTTAYAM–2005)

6. Show that $Z(\sinh n\theta) = \dfrac{z\sinh\theta}{z^2-2z\cosh\theta+1}$ (VTU–2011)

7. Show that $Z(e^{-an}\sin n\theta) = \dfrac{ze^a \sin\theta}{z^2 e^{2a} - 2ze^a\cos\theta+1}$ (SVTU–2007)

8. Show that $Z(^{n+p}C_p) = \left(1-\dfrac{1}{z}\right)^{-(p+1)}$. Using the damping rule

deduce that $Z(^{n+p}C_p a^n) = \left(1+\dfrac{a}{z}\right)^{-(p+1)}$ (SVTU–2009)

9. Show that $Z\left(\dfrac{1}{n}\right) = z\log\dfrac{z}{z-1}$ (MADRAS–2003)

Answers

1. (i) $\dfrac{z}{z-2}$, $|z| > 2$ (ii) $\dfrac{z\sin\alpha}{z^2-2z\cos\alpha+1}$ (iii) $\dfrac{z^2\sin 5 - z\sin 2}{z^2-2z\cos 3+1}$ (iv) $\dfrac{z\sinh\dfrac{\pi}{2}}{z^2-2z\cosh\dfrac{\pi}{2}+1}$ (v) $\dfrac{z(z-a\cosh\theta)}{z^2-2az\cosh\theta+a^2}$

   (vi) $2+(2)^n+3(n-1)2^n$, $n>1$ (vii) $\dfrac{3}{4}\left(\dfrac{1}{2^{n-1}}+\dfrac{1}{(-4)^n}\right)$

2. (i) $\dfrac{z^2\sin\alpha+z\cos\alpha}{z^2+1}$, $|z|>1$ (ii) $\dfrac{Cz\sinh\alpha}{z^2-2Cz\cosh\alpha+1}$ (iii) $\dfrac{z^2-z}{\sqrt{2}(z^2+1)}$ (iv) $e^{a/z}$

3. (i) $\dfrac{z\sin 5}{z^2-2z\cos 5+1}$ (ii) $\dfrac{1}{1-z^{-1}e^{\alpha n}}$ (iii) $\dfrac{z(z-3\cosh 5)}{z^2-2z\cosh 5+9}$

4. (i) $\left(-\dfrac{1}{3}-\dfrac{z}{3^2}-\dfrac{z^2}{3^3}-\dfrac{z^3}{3^4}-....\right)+\left(\dfrac{1}{2}+\dfrac{z}{2^2}+\dfrac{z^2}{2^3}+\dfrac{z^3}{2^4}+....\right)$ (ii) $\begin{cases} \dfrac{(n-2)(n-1)5^{n-3}}{2}, & n\geq 3 \\ 0 & , n<3 \end{cases}$ (iii) $\left[3\left\{\cos\dfrac{n\pi}{3}\right\}+\dfrac{11}{\sqrt{3}}\sin\dfrac{n\pi}{3}\right]F(n)$

   (iv) $[ne^{-an}]$ (v) $2^{n-1}+n^2 2^n-4^n$

5. (i) $f_n = \dfrac{1}{a}(n+1)F(n) - \dfrac{a}{(a-1)^2}F(n) + \dfrac{a}{(a-1)^2}a^n F(n) + \dfrac{1}{1-a}nF(n-1)$ (ii) $f_n = \left[3-\left(\dfrac{1}{2}\right)^n+\left(\dfrac{1}{3}\right)^n\right]F(n)$

   (iii) $f_n = \left(\dfrac{n+2}{2}\right)\left(\dfrac{1}{3}\right)^n\cos\dfrac{n\pi}{2}F(n)$ (iv) $f_n = \left[\dfrac{6}{7}(n+1)-\dfrac{78}{49}+\dfrac{36}{49}\left(-\dfrac{1}{6}\right)^n\right]F(n)$ (v) $y_k = \dfrac{8}{5}\left(\dfrac{1}{2}\right)^k - \dfrac{3}{5}\left(-\dfrac{1}{3}\right)^k$

ARCHIVE

Solve the following differential equations :

1. $y'' + 4y' + 4y = 3\sin x + 4\cos x, y(0) = 1$ and $y'(0) = 0$ (JNTU–2003)

2. $\dfrac{d^2y}{dx^2} - 4y = x\sin x$ (Madras–2009)

3. $\dfrac{d^2y}{dx^2} - 6\dfrac{dy}{dx} + 9y = 6e^{3x} + 7e^{-2x} - \log 2$ (VTU–2005)

4. $\dfrac{d^2y}{dx^2} + 3\dfrac{dy}{dx} + 2y = 4\cos^2 x$ (Bhopal–2002S)

5. $(D^2 - 4D + 3)y = \sin 3x \cos 2x$ (Madras–2000)

6. $\dfrac{d^3y}{dx^3} + 2\dfrac{d^2y}{dx^2} + \dfrac{dy}{dx} = e^{-x} + \sin 2x$ (VTU–2004)

7. $\dfrac{d^2y}{dx^2} + 2\dfrac{dy}{dx} + y = e^{2x} - \cos^2 x$ (Delhi–2002)

8. $(D^3 - 5D^2 + 7D - 3)y = e^{2x}\cosh x$ (Nagarjuna–2008)

9. $\dfrac{d^2y}{dx^2} - y = e^x + x^2 e^x$ (Nagpur–2009)

10. $(D^3 - D)y = 2x + 1 + 4\cos x + 2e^x$ (Mumbai–2006)

11. $\dfrac{d^2y}{dx^2} - 6\dfrac{dy}{dx} + 25y = e^{2x} + \sin x + x$ (VTU–2006)

12. $(D^2 + 1)^2 y = x^4 + 2\sin x \cos 3x$ (Madras–2006)

13. $\dfrac{d^2y}{dx^2} + 5\dfrac{dy}{dx} + 6y = e^{-2x}\sin 2x$ (Bhopal–2008)

14. $\dfrac{d^2y}{dx^2} + 2\dfrac{dy}{dx} + 3y = e^x\cos x$ (VTU–2010)

15. $(D^2 + 4D + 3)y = e^{-x}\sin x + xe^{3x}$ (Raipur–2005, Anna–2002S)

16. $(D^3 + 2D^2 + D)y = x^2 e^{2x} + \sin^2 x$ (PTU–2003)

17. $\dfrac{d^2y}{dx^2} + 16y = x\sin 3x$ (VTU–2010S)

18. $\dfrac{d^2y}{dx^2} + 3\dfrac{dy}{dx} + 2y = e^{e^x}$ (SVTU–2009)

19. $(D^2 + a^2)y = \tan ax$ (VTU–2005)

20. $x^2\dfrac{d^2y}{dx^2} - 3x\dfrac{dy}{dx} + 4y = (1 + x^2)$ (SVTU-2007)

21. $x\dfrac{d^2y}{dx^2} - \dfrac{2y}{x} = x + \dfrac{1}{x^2}$ (VTU-2005S)

22. $x^2\dfrac{d^2y}{dx^2} + 4x\dfrac{dy}{dx} + 2y = \log x$ (Bhopal 2009)

23. $x^2 y'' + xy' + y = 2\cos^2(\log x)$ (VTU-2011)

24. $x^2\dfrac{d^2y}{dx^2} + 5x\dfrac{dy}{dx} + 4y = x\log x$ (UPTU-2004)

25. $x^2\dfrac{d^2y}{dx^2} + 2x\dfrac{dy}{dx} - 12y = x^3\log x$ (Bhopal-2007)

26. $(2x + 3)^2\dfrac{d^2y}{dx^2} - (2x + 3)\dfrac{dy}{dx} - 12y = 6x$
   (VTU-2007, Kerala-2005, Anita-2025)

27. $(x - 1)^3\dfrac{d^3y}{dx^3} + 2(x - 1)^2\dfrac{d^2y}{dx^2} - 4(x - 1)\dfrac{dy}{dx} + 4y = 4\log(x - 1)$
   (SVTu 2009)

28. $(1 + x)^2\dfrac{d^2y}{dx^2} + (1 + x)\dfrac{dy}{dx} + y = \sin[2\log(1 + x)]$
   (PTU-2006, UTU 2007)

Answers

1. $y = (1 + x)e^{-2x} + \sin x$ 2. $y = C_1 e^{2x} + C_2 e^{-2x} - \dfrac{x}{3}\sinh x - \dfrac{2}{9}\cosh x$ 3. $y = (C_1 + C_2 x)e^{3x} + 3x^2 e^{3x} + \dfrac{7}{25}e^{-2x} - \dfrac{1}{9}\log 2$

4. $y = C_1 e^{-x} + C_2 e^{-2x} + 1 + \dfrac{1}{10}(3\sin 2x - \cos 2x)$ 5. $y = C_1 e^x + C_2 e^{3x} + \dfrac{1}{884}(10\cos 5x - 11\sin 5x) + \dfrac{1}{20}(\sin x + 2\cos x)$

6. $y = C_1 + (C_2 + C_3 x)e^{-x} - \dfrac{x^2}{2}e^{-x} + \dfrac{3}{50}\cos 2x - \dfrac{2}{25}\sin 2x$ 7. $y = (C_1 + C_2 x)e^{-x} + \dfrac{1}{2} + \dfrac{1}{5}(2\sin 2x + \cos 2x)$

8. $y = (C_1 + C_2 x)e^x + C_3 e^{3x} + \dfrac{1}{8}(xe^{3x} - x^2 e^x)$ 9. $y = C_1 e^x + C_2 e^{-x} + \dfrac{e^x}{12}(2x^3 - 3x^2 + 9x)$

10. $y = C_1 + C_2 e^x + C_3 e^{-x} + xe^x - (x + x) - 2\sin x$ 11. $y = e^{3x}(C_1\cos 4x + C_2\sin 4x) + \dfrac{1}{17}e^{2x} + \dfrac{1}{565}(23\sin x + 6\cos x) + \dfrac{x}{25} + \dfrac{6}{625}$

12. $y = (C_1 + C_2 x)\cos x + (C_3 + C_4 x)\sin x + x^4 - 24x^2 + 72 + \dfrac{1}{225}\sin 4x - \dfrac{1}{9}\sin 2x$

13. $y = C_1 e^{-2x} + C_2 e^{-3x} - \dfrac{e^{-2x}}{10}(\cos 2x + 2\sin 2x)$ 14. $y = e^{-x}(C_1\cos\sqrt{2}x + C_2\sin\sqrt{2}x) + \dfrac{e^x}{41}(4\sin x + 5\cos x)$

15. $y = C_1 e^{-x} + C_2 e^{-3x} - \dfrac{e^{-x}}{5}(\sin x + 2\cos x) + \dfrac{e^{3x}}{22}\left(x - \dfrac{5}{11}\right)$

16. $y = C_1 + (C_2 + C_3 x)e^{-x} + \dfrac{e^{2x}}{18}\left(x^2 - \dfrac{7x}{8} + \dfrac{11}{6}\right) + \dfrac{1}{100}(3\sin 2x + 4\cos 2x)$ 17. $y = C_1\cos 4x + C_2\sin 4x + \dfrac{1}{7}\left(x\sin 3x - \dfrac{6}{7}\cos 3x\right)$

18. $y = C_1 e^{-x} + C_2 e^{-2x} + C_3 e^{-3x} + e^{-2x}.e^{e^x}$ 19. $y = C_1\cos ax + C_2\sin ax - \dfrac{1}{a^2}\cos ax\log(\sec ax + \tan ax)$

20. $y = (c_1 + c_2\log x)x^2 + \dfrac{1}{4} + 2x + \dfrac{1}{2}x^2(\log x)^2$ 21. $y = C_1 x^2 + C_2 x^{-1} + \dfrac{1}{3}(x^2 - 1/x)\log x$

22. $C_1 x^{-1} + C_2 x^{-2} + \dfrac{1}{2}\log x - \dfrac{3}{4}$ 23. $y = C_1 x^{-2} + x[C_2\cos(\sqrt{3}\log x) + C_3\sin(\sqrt{3}\log x)] + 8\cos(\log x) - \sin(\log x)$

24. $y = x^{-2}(C_1 + C_2\log x) + \dfrac{x}{9}\left(\log x - \dfrac{2}{3}\right)$ 25. $y = C_1 x^3 + C_2 x^{-4} + \dfrac{x^3}{98}\log x(7\log x - 2)$

26. $y = C_1(2x + 3)^a + C_2(2x + 3)^b - \dfrac{3}{14}(2x + 3) + \dfrac{3}{4}$, where $a, b = \dfrac{3 \pm \sqrt{57}}{4}$

27. $y = C_1(x - 1) + C_2(x - 1)^2 + C_3(x - 1)^{-2} + \log(x + 1) + 1$ 28. $y = C_1\cos\log(1 + x) + C_2\sin\log(1 + x) - \dfrac{1}{3}\sin[2\log(1 + x)]$

❋❋❋❋❋❋

Statistical Methods

9.1 GRAPHICAL REPRESENTATION OF DATA

If some values of a variate are collected in the arbitrary order, in which they occur, we cannot properly grasp the significance of the data.

For example, consider the marks of 50 students in Mathematics, arranged according to their roll numbers, the maximum marks being 100.

9, 70, 75, 15, 0, 33, 69, 66, 37, 99, 81, 12, 31, 22, 60, 79, 46, 73, 46, 79, 75, 65, 85, 22, 8, 12, 41, 87, 82, 72, 50, 22, 87, 50, 89, 28, 29, 50, 40, 36, 40, 30, 28, 87, 81, 90, 22, 15, 30, 35.

The data given in the above form is called ungrouped data. If the data arranged in ascending or descending order of magnitude, it is said to be arranged in an array. If arranged the given data into class intervals 0-10, 10-20, ..., 90-100. Then, this method is known as tally method.

If the identity of the individuals about whom, a particular information is taken is not relevant, nor the order in which the observations arise, then we divide the observed range of variables into a suitable number of class intervals and record the number of observation in each class.

For example in the above case, the data may be expressed as in the following table :

Such a table showing the distribution of the frequencies in the different classes is called a frequency table and the way in which the class frequencies are distributed over the class intervals is called the grouped frequency distribution of the variable.

The following points may be considerable for classifications :

Marks	No. of Students
0–10	3
10–20	4
20–30	7
30–40	7
40–50	5
50–60	3
60–70	4
70–80	7
80–90	8
90–100	2

(i) The class should be clearly defined and should not lead to any ambiguity.

(ii) The number of class should never be less than 6 and not more than 30. Because with less number of classes, the accuracy may be lost, and with more number of classes, the computations become lengthy.

(iii) The observation corresponding to common point of two classes should always be put in the higher class. For example, a number corresponding to the values 20 is to be put up in the class 20-30 and not in 10-20.

(iv) The classes should be of equal width.

MAGNITUDE OF THE CLASS INTERVAL

Having fixed the number of classes, divide the range by it and nearest integer to this value gives the magnitude of the class interval.

CLASS LIMITS

The class limit should be chosen in such a way that the mid value of the class interval and actual average of the observations in that class intervals are as near to each other as possible.

9.1.1 CONTINUOUS FREQUENCY DISTRIBUTION

If we deal with a continuous variable, it is not possible to arrange the data in the class interval.

Let us consider the distribution of age in years. If we take the intervals 15-19, 20-24, then the persons with ages between 19 and 20 years are not taken into consideration. In such a case, we form the class interval as follows.

Age in years   Below 5
                     5 or more but less than 10
                     10 or more but less than 15
                     15 or more but less than 20
                     20 or more but less than 25

where all the persons with any fraction of age are included in one group or the other.

Practically, we re-write it as

0–5; 5–10; 10–15; 15–20; 20–25

This form of frequency distribution is known as continuous frequency distribution.

9.1.2 GRAPHICAL REPRESENTATION

It is often useful to represent a frequency distribution by means of a diagram which makes the unwidely data intelligible and conveys to the eye the general run of the observations. When data of two items is compared with one another, it is always easier to compare through graphs and diagrams. Here, we consider some important types of graphic representations.

(1) HISTOGRAM

In drawing the histograms of a given grouped frequency distribution, first we mark off along a horizontal base line all the class-intervals using a suitable scale, then draw rectangles with the areas proportional to the frequencies of the respective class intervals. For equal class-intervals, the height of the rectangles will be proportional to the frequencies. For unequal class-intervals, the heights of the rectangles will be proportional to the ratios of the frequencies to the width of the corresponding class. Then the diagram of continuous rectangles so obtained is called histogram.

✎ REMARKS

- Histograms are appropriate to cases in which the frequency changes rapidly.
- To draw the histogram for an ungrouped frequency distribution of a variable, assume that the frequency corresponding to the variate value x is distributed over the interval $(x - h/2)$ to $(x + h/2)$, where h is the jump from one value to the next.
- If the grouped frequency distribution is not continuous, convert it into continuous distribution and then draw the histogram.
- The height of each rectangle is proportional to the frequency of the corresponding class, the height of a fraction of the rectangle is not proportional to the frequency of the corresponding fraction of the class, therefore, the histogram cannot be directly used to read frequency over a fraction of a class interval.

Consider the following example.

Marks	No. of Students	Marks	No. of Students
under 10	2	under 60	31
under 20	6	under 70	32
under 30	16	under 80	37
under 40	20	under 90	48
under 50	23	under 100	50

Then the histogram from the above data is :

(2) FREQUENCY POLYGON

The frequency polygon is obtained by plotting points with abscissa as the variate values and the ordinates as the corresponding frequencies and joining the plotted points by a straight line taken in order. In a frequency polygon the variables or individuals of each class are assumed to be concentrated at the mid-points of the class-interval.

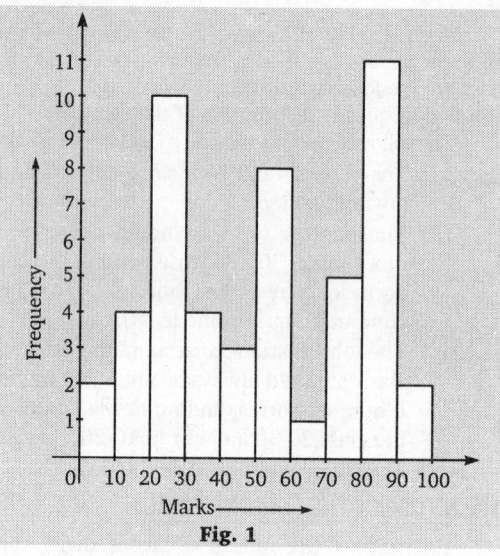

Fig. 1

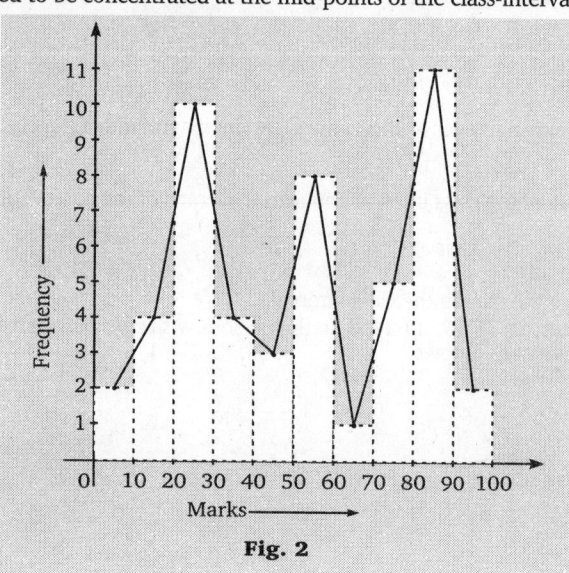

Fig. 2

(3) FREQUENCY CURVE

(UPTU(MCA)–2006)

If through the vertices of a frequency polygon a smooth freehand curve is drawn we get the frequency curve.

(4) COMMULATIVE FREQUENCY CURVE OR OGIVE

If the upper limits of the class taken as x-coordinate and the commulative frequencies as the y-coordinate and the points are plotted, then these points, when joined by a freehand smooth curve given the commulative frequency curve or the ogive.

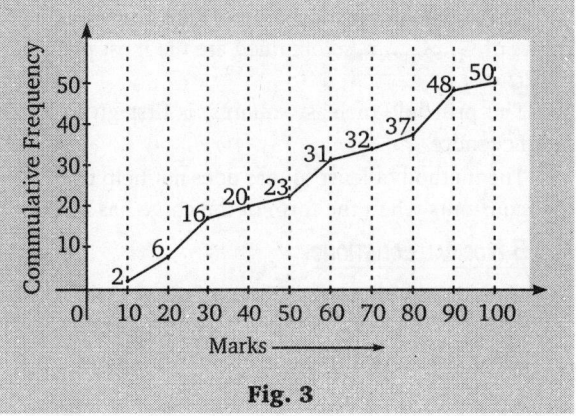

Fig. 3

9.1.3 CURVE FITTING

In applied mathematics, several equations of different types can be obtained to express the given data approximately. But we want to find the equation of the curve of best fit which may be most suitable for predicting the unknown values. The process of finding such an equation of best fit is known as curve fitting.

By curve fitting we means an expression of the relationship between two variables by an equation. If there are n pair of observed values, then it is possible to fit the given data to an equation that contains n arbitrary constants for we can solve n simultaneous equation for n unknowns.

Let us consider m independent linear equations

$$
\left.\begin{array}{l}
a_{11}x_1 + a_{12}x_2 + \ldots + a_{1n}x_n = b_1 \\
a_{21}x_1 + a_{22}x_2 + \ldots + a_{2n}x_n = b_2 \\
\ldots\ldots\ldots \qquad \ldots\ldots\ldots\ldots\ldots \\
\ldots\ldots\ldots \qquad \ldots\ldots\ldots\ldots\ldots \\
a_{m1}x_1 + a_{m2}x_2 + \ldots + a_{mn}x_n = b_m
\end{array}\right\}
\qquad \ldots(1)
$$

where a's and b's are constants and $x_1, x_2, \ldots, x_n$ are n variables. Now, there are two cases :

(i) **If $m = n$**, then there exists a unique set of values satisfying the given system of equations.

(ii) **If $m > n$**, which implies that the number of equations is greater than the number of variables, then there exists no finite solution. In this case we try to find these values of variables $x_1, x_2, \ldots, x_n$ which satisfy as closed as possible to the given equation. These obtained values are called the most plausible values or the best fit values.

REMARKS

- The general problem of finding equation of approximate curves which fit given set of data is called curve fitting.
- 'Curve fitting' is considered very important both from the point of view of theoretical and practical use.
- In theoretical statistics, the line of regression can be regarded as fitting of linear curves.
- The difference between the observed values and expected values is known as residual.

9.1.4 METHOD OF LEAST SQUARES

Consider m independent linear equations in n unknowns $x_1, x_2, \ldots, x_n$ where $m > n$ as

$$
\left.\begin{array}{l}
a_{11}x_1 + a_{12}x_2 + \ldots + a_{1n}x_n = b_1 \\
a_{21}x_1 + a_{22}x_2 + \ldots + a_{2n}x_n = b_2 \\
\ldots\ldots\ldots \qquad \ldots\ldots\ldots\ldots\ldots \\
\ldots\ldots\ldots \qquad \ldots\ldots\ldots\ldots\ldots \\
a_{m1}x_1 + a_{m2}x_2 + \ldots + a_{mn}x_n = b_m
\end{array}\right\}
\qquad \ldots(1)
$$

with constants a's and b's.

Let $x_1, x_2, \ldots, x_n$ be the most plausible values then, we have

$$E_i = (a_{i1}x_1 + a_{i2}x_2 + \ldots + a_{in}x_n - b_i) \quad \text{(known as residual or deviation, or the error } E_i\text{)}$$

$$\Rightarrow \qquad E_i = (a_{i1}x_1 + a_{i2}x_2 + \ldots + a_{in}x_n - b_i) \quad i = 1, 2, \ldots m. \qquad \ldots(2)$$

Let us suppose S is the sum of the squares of E_i.

Then, we get

$$S = \sum_{i=1}^{m}(a_{i1}x_1 + a_{i2}x_2 + \ldots + a_{in}x_n - b_i)^2 = \sum_{i=1}^{m} E_i^2 \qquad \ldots(3)$$

To find the maximum or minimum values of S, we must have

$$\frac{\partial S}{\partial x_1} = 0, \frac{\partial S}{\partial x_2} = 0, \ldots, \frac{\partial S}{\partial x_n} = 0$$

Then (3) $\Rightarrow$ $\sum\limits_{i=1}^{m} a_{i1}E_i = 0$, $\sum\limits_{i=1}^{m} a_{i2}E_i = 0$, ... $\sum\limits_{i=1}^{m} a_{in}E_i = 0$ ...(4)

These n equations given by (4) are known as the normal equations and can be easily solved for the n variables $x_1, x_2, ..., x_n$. The values of $x_1, x_2, ..., x_n$ so obtained are the most plausible or best values.

REMARKS

- The principle of least squares, is first given by Gauss in 1795 but it was named and published for the first time in 1805 by Legendre.
- The method of least square does not help us to choose the degree of the curve to be fitted but helps us in finding the values of the constants when the form of the curve has already been given.

9.1.5 NORMAL EQUATIONS

Consider the curve of nth degree

$$y = a + bx + cx^2 + ... + kx^n \text{ with } (k \neq 0)$$

Then, the normal equation obtained from (4), are

$$\Sigma y = ma + b\Sigma x + ... + k\Sigma x^n$$
$$\Sigma xy = a\Sigma x + b\Sigma x^2 + ... + k\Sigma x^{n+1}$$
$$\Sigma x^2 y = a\Sigma x^2 + b\Sigma x^3 + ... + k\Sigma x^{n+2}$$
$$\Sigma x^n y = a\Sigma x^n + b\Sigma x^{n+1} + ... + k\Sigma x^{2n}$$

When second order partial derivatives are calculated and substituted these values, they gives a positive value of the function, which implies that S is minimum.

In particular, if we take $n = 1$ then we get the equation of straight line

$$y = a + bx$$

and the normal equations are

$$\Sigma y = ma + b\Sigma x + c\Sigma x^2$$
$$\Sigma xy = a\Sigma x + b\Sigma x^2 + c\Sigma x^3$$
$$\Sigma x^2 y = a\Sigma x^2 + b\Sigma x^3 + c\Sigma x^4$$

Solved Examples

EXAMPLE 1. *Find the normal equations and hence find the best fit values of x, y, z in the least square sense from the following equations*

$$x + 2y + z = 1$$
$$2x + y + z = 4$$
$$-x + y + 2z = 4$$
$$4x + 2y - 5z = -7$$

SOLUTION. The given equation can be written as

$$\left.\begin{array}{l} x + 2y + z - 1 = 0 \\ 2x + y + z - 4 = 0 \\ -x + y + 2z - 4 = 0 \\ 4x + 2y - 5z + 7 = 0 \end{array}\right\} \quad ...(1)$$

Now, to obtain the normal equations of x, we multiply these equations by the coefficient of x in that equation and then add.

Then, the normal equations are

$$1(x + 2y + z - 1) + 2(2x + y + z - 4)$$
$$+ (-1)(-x + y + 2z - 4)$$
$$+ 4(4x + 2y - 5z + 7) = 0$$
$$\Rightarrow \qquad 22x + 11y - 19z + 23 = 0 \quad ...(2)$$

Similarly, for the normal equation of y, multiply these equations by the coefficient of y in that equation and then add, we get

$$2(x + 2y + z - 1) + 1(2x + y + z - 4)$$
$$+ 1(-x + y + 2z - 4)$$
$$+ 2(4x + 2y - 5z + 7) = 0$$

$$\Rightarrow \qquad 11x + 10y - 5z + 4 = 0 \quad ...(3)$$

Similarly, the normal equation for z is

$$1(x + 2y + z - 1) + 1(2x + y + z - 4)$$
$$+ 2(-x + y + 2z - 4)$$
$$- 5(4x + 2y - 5z + 7) = 0$$

$$\Rightarrow \qquad 19x + 5y - 31z + 48 = 0 \quad ...(4)$$

Solving (2), (3) and (4) for x, y and z we get

$$x = 0.910, y = -0.378 \text{ and } z = 2.045$$

EXAMPLE 2. *Find the most plausible values of x and y from the following equations* :

$$x + y = 3.00, 2x - y = 0.5, x + 3y = 7.25,$$
$$3x + y = 4.95.$$

SOLUTION. The given equations can be written as

$$\left.\begin{array}{l} x + y - 3.00 = 0 \\ 2x - y - 0.5 = 0 \\ x + 3y - 7.25 = 0 \\ 3x + y - 4.95 = 0 \end{array}\right\} \quad ...(1)$$

Now sum of the square 'S' of the errors is given by

$$S = (x + y - 3.00)^2 + (2x - y - 0.5)^2$$
$$+ (x + 3y - 7.25)^2 + (3x + y - 4.95)^2 = 0$$

For the maxima and minima of S, we must have

$$\frac{\partial S}{\partial x} = 0$$

$\Rightarrow \quad (x + y - 3.00) + 2(2x - y - 0.50)$

$\quad + 1(x + 3y - 7.25) + 3(3x + y - 4.95) = 0$

$\Rightarrow \qquad 15x + 5y - 26.10 = 0 \qquad \dots(2)$

and $\qquad \dfrac{\partial S}{\partial y} = 0$

$\Rightarrow \quad (x + y - 3.00) - (2x - y - 0.50)$

$\quad + 3(x + 3y - 7.25) + 3(3x + y - 4.95) = 0$

$\Rightarrow \qquad 13x + 14y - 39.20 = 0 \qquad \dots(3)$

Solving (2) and (3), we get

$\qquad x = 1.234, \ y = 1.919.$

Exercise-9.1

1. Find the values of x and y which satisfy the following equations most satisfactorily with the help of normal equation of x, y

   $x + 2.5y = 21, 3.2x - y = 28, 4x + 1.2y = 42.04,$

   $1.5x + 6.3y = 40.$

2. Find the most plausible values of x and y from the four equations

   $x - y + 2z = 3, 3x + 2y - 5z = 5, 4x + y + 4z = 21$

   and $-x + 3y + 3z = 14.$

3. Use the method of least squares to find the most plausible values of x and y from the following equations

   $x + y = 3.01, 2x - y = 0.03, x + 3y = 7.03, 3x + y = 4.97$

Answers

1. $x = 9.620, y = 4.064$ **2.** $x = 2.47, y = 3.55, z = 1.92$ **3.** $x = 0.997, y = 2.001$

9.1.6 Method of Curve-Fitting

Let us consider the rth degree curve

$$y = a + bx + cx^2 + \dots + kx^r, \text{ with } k \neq 0 \qquad \dots(i)$$

with the given values $(x_1, y_1), (x_2, y_2), \dots, (x_m, y_m)$.

The curve given by (i) has $(r + 1)$ unknown $a, b, c, \dots, k$ and so if $m = r + 1$, we get $(r + 1)$ equations when the values $(x_1, y_1), (x_2, y_2), \dots, (x_m, y_m)$ are substituted for (x, y) in (i) and thus a unique solution of the values of the unknown $a, b, c, \dots, k$ is possible.

Now let $\qquad y_i' = a + bx_i + cx_i^2 + \dots + kx_i^r \qquad \dots(ii)$

and let y_i be the observed values y for x_i.

Then if u_i be the residual for this point, we have

$$u_i = y_i - y_i' = y_i - (a + bx_i + cx_i^2 + \dots + kx_i^r), \qquad \text{from (ii)}$$

$$u_i = y_i - a - bx_i - cx_i^2 - \dots - kx_i^r \qquad \dots(iii)$$

Now in order to make the sum of the squares of the errors minimum, we define

$$S = \sum_{i=1}^{m} u_i^2 = \sum_{i=1}^{m} (y_i - a - bx_i - cx_i^2 - \dots - kx_i^r)^2 \qquad \dots(iv)$$

$\therefore$ By the principle of maxima and minima, we must have

$$\frac{\partial S}{\partial a} = 0, \frac{\partial S}{\partial b} = 0, \dots, \frac{\partial S}{\partial k} = 0 \qquad \dots(v)$$

which gives the following $(r + 1)$ equations :

$$\frac{\partial S}{\partial a} = -2\sum_{i=1}^{m} (y_i - a - bx_i - cx_i^2 - \dots - kx_i^r) = 0 \quad \Rightarrow \sum_{i=1}^{m} (y_i - a - bx_i - cx_i^2 - \dots - kx_i^r) = 0$$

$$\sum_{i=1}^{m} y_i - \sum_{i=1}^{m} a - \sum_{i=1}^{m} bx_i - \sum_{i=1}^{m} cx_i^2 - \dots - \sum_{i=1}^{m} kx_i^r = 0 \quad \Rightarrow \quad \Sigma y_i = ma + b\Sigma x_i + c\Sigma x_i^2 + \dots + k\Sigma x_i^r \quad \dots(1)$$

Now $\qquad \dfrac{\partial S}{\partial b} = 0 = -2\Sigma(y_i - a - bx_i - cx_i^2 - \dots - kx_i^r)x_i$

or $\qquad \Sigma(y_i - a - bx_i - cx_i^2 - \dots - kx_i^r)x_i = 0$

$\therefore \qquad \Sigma x_i y_i = a\Sigma x_i + b\Sigma x_i^2 + c\Sigma x_i^3 + \dots + k\Sigma x_i^{r+1} \qquad \dots(2)$

Similarly $\qquad \Sigma x_i^2 y_i = a\Sigma x_i^2 + b\Sigma x_i^3 + c\Sigma x_i^4 + \dots + k\Sigma x_i^{r+2} \qquad \dots(3)$

The equations (1), (2) and (3) are known as normal equations and can be solved as simultaneous equations to evaluate $a, b, c, \dots, k$.

9.1.7 FITTING OF SOME SPECIAL CURVES

(1) FITTING OF A STRAIGHT LINE

Let
$$y = a + bx \qquad \text{...(1)}$$
be the equation of the straight line, to be fitted.
Then the normal equations can be obtained as follows.

Let
$$u = (y_i - a - bx_i) \qquad \text{and} \qquad S = u^2 = \Sigma(y_i - a - bx_i)^2$$

for the maxima and minima of S, we must have $\dfrac{\partial S}{\partial a} = 0$

$$\Rightarrow \qquad -2\Sigma(y_i - a - bx_i).1 = 0 \qquad \Rightarrow \qquad \Sigma(y_i - a - bx_i).1 = 0$$

$$\therefore \qquad \Sigma y_i = ma + b\Sigma x_i \qquad \text{...(2)}$$

Now
$$\frac{\partial S}{\partial b} = 0$$

$$-2\Sigma(y_i - a - bx_i)(-x_i) = 0 \qquad \Rightarrow \qquad \Sigma(y_i - a - bx_i)x_i = 0$$

$$\Sigma x_i y_i = a\Sigma x_i + b\Sigma x_i^2 \qquad \text{...(3)}$$

Solving these two equations we can find the values of a and b.

Special Case : If the straight line to be fitted passes through the origin, then the equation of the straight line is given by
$$y = bx$$

$\therefore$ Its normal equation is $\Sigma xy = b\Sigma x^2$

we can find the value of constant b easily.

(2) FITTING OF A PARABOLIC CURVE

Let
$$y = a + bx + cx^2 \qquad \text{...(1)}$$
be the equations of the parabolic curve to be fitted.

Then the normal equations are

$$\begin{aligned} \Sigma y &= ma + b\Sigma x + c\Sigma x^2 \\ \Sigma xy &= a\Sigma x + b\Sigma x^2 + c\Sigma x^3 \\ \Sigma x^2 y &= a\Sigma x^2 + b\Sigma x^3 + c\Sigma x^4 \end{aligned} \right\} \qquad \text{...(2)}$$

Solving above three equations for a, b and c simultaneously we can find the values of a, b and c.

(3) FITTING OF AN EXPONENTIAL CURVE

Let
$$y = ae^{bx} \qquad \text{...(1)}$$
be the equation of the exponential curve to be fitted.

Taking logarithms of both sides of (1) to the base, 10.

We get
$$\log_{10} y = \log_{10} a + bx \log_{10} e \qquad \text{...(2)}$$
This is of the form
$$Y = A + Bx \qquad \text{...(3)}$$

where $Y = \log_{10} y$, $A = \log_{10} a$ and $B = b\log_{10} e$.

Now applying the method of fitting a straight lines we can find the values A and B and hence the values of a and b.

(4) FITTING OF THE CURVE OF THE TYPE $y = ab^x$

Here the method of fitting is the same as given in part 9.1.7.3 above.

(5) FITTING OF THE LOGARITHMIC CURVE OF THE FORM $y = ax^b$

We have
$$y = ax^b$$
Taking log of both the sides, we get
$$\log y = \log a + b \log x$$

Let us put $\log y = Y$, $\log a = A$ and $\log x = X$
Then, we have
$$Y = A + bX \qquad \text{...(1)}$$
Normal equation for (1) are

$$\Sigma Y = An + b\Sigma X$$

$$\Sigma XY = A\Sigma X + b\Sigma X^2$$

(6) FITTING OF THE CURVE $y = a + b/x$

Let put $\dfrac{1}{x} = X$. Then given equations becomes

$$y = a + bX \qquad \text{...(1)}$$

The normal equation for (1) are

$$\Sigma y = an + b\Sigma X \qquad \ldots(2)$$

$$\Sigma Xy = a\Sigma X + b\Sigma X^2 \qquad \ldots(3)$$

Solving these equations we get the required value of a and b.

(7) FITTING OF THE CURVE $y = ax + b/x$

Let the curve $y = ax + \dfrac{b}{x}$ passes through the point $(x_i, y_i), i = 1, 2, \ldots, n$. The error of estimate for i^{th} point (x_i, y_i) is

$$d_i = y_i - ax_i - \frac{b}{x_i}$$

By the principle of least squares, the sum of squares of the error should be minimum.

i.e., $\qquad\qquad\qquad\qquad S = \sum\limits_{i=1}^{n} d_i^2$ should be minimum.

$\Rightarrow \qquad\qquad\qquad\qquad S = \sum\limits_{i=1}^{n} \left(y_i - ax_i - \dfrac{b}{x_i} \right)^2$ should be minimum.

We obtained the normal equations by $\dfrac{\partial S}{\partial a} = 0$ and $\dfrac{\partial S}{\partial b} = 0$.

$\therefore \qquad \dfrac{\partial S}{\partial a} = 0 \Rightarrow -2\sum\limits_{i=1}^{n} x_i \left(y_i - ax_i - \dfrac{b}{x_i} \right) = 0 \Rightarrow \sum\limits_{i=1}^{n} x_i y_i = a\sum\limits_{i=1}^{n} x_i^2 + bn \qquad \ldots(1)$

and $\qquad \dfrac{\partial S}{\partial b} = 0 \Rightarrow -2\sum\limits_{i=1}^{n} \dfrac{1}{x_i} \left(y_i - ax_i - \dfrac{b}{x_i} \right) = 0 \Rightarrow \sum\limits_{i=1}^{n} \dfrac{y_i}{x_i} = an + b\sum\limits_{i=1}^{n} \dfrac{1}{x_i^2} \qquad \ldots(2)$

Solving (1) and (2) we get the required values of a and b.

(8) FITTING OF THE CURVE $y = a + (b/x) + (c/x^2)$

We have $S = \sum\limits_{i=1}^{n} \left(y_i - a - \dfrac{b}{x_i} - \dfrac{b}{x_i^2} \right)$ should be minimum.

Normal equations can be obtained by $\dfrac{\partial S}{\partial a} = 0, \dfrac{\partial S}{\partial b} = 0, \dfrac{\partial S}{\partial c} = 0$

Now $\qquad \dfrac{\partial S}{\partial a} = 0 \Rightarrow -2\sum\limits_{i=1}^{n} \left(y_i - a - \dfrac{b}{x_i} - \dfrac{c}{x_i^2} \right) = 0 \Rightarrow \Sigma y_i = an + b\sum\limits_{i=1}^{n} \dfrac{1}{x_i} + c\sum\limits_{i=1}^{n} \dfrac{1}{x_i^2}$

$\dfrac{\partial S}{\partial b} = 0 \Rightarrow -2\sum\limits_{i=1}^{n} \dfrac{1}{x_i} \left(y_i - a - \dfrac{b}{x_i} - \dfrac{c}{x_i^2} \right) = 0 \Rightarrow \sum\limits_{i=1}^{n} \dfrac{y_i}{x_i} = a\sum\limits_{i=1}^{n} \dfrac{1}{x_i} + b\sum\limits_{i=1}^{n} \dfrac{1}{x_i^2} + c\sum\limits_{i=1}^{n} \dfrac{1}{x_i^3}$

and $\qquad \dfrac{\partial S}{\partial c} = 0 \Rightarrow -2\sum\limits_{i=1}^{n} \dfrac{1}{x_i^2} \left(y_i - a - \dfrac{b}{x_i} - \dfrac{c}{x_i^2} \right) = 0 \Rightarrow \sum\limits_{i=1}^{n} \dfrac{y_i}{x_i^2} = a\sum\limits_{i=1}^{n} \dfrac{1}{x_i^2} + b\sum\limits_{i=1}^{n} \dfrac{1}{x_i^3} + c\sum\limits_{i=1}^{n} \dfrac{1}{x_i^4} \qquad \ldots(3)$

Solving (1), (2) and (3), we get the required values of a and b.

(9) FITTING OF THE CURVE $y = ax^2 + b/x$

We have $\qquad\qquad\qquad S = \sum\limits_{i=1}^{n} \left(y_i - ax_i^2 - \dfrac{b}{x_i} \right)^2$

Now $\qquad \dfrac{\partial S}{\partial a} = 0 \Rightarrow -2\sum\limits_{i=1}^{n} x_i^2 \left(y_i - ax_i^2 - \dfrac{b}{x_i} \right) = 0 \Rightarrow \sum\limits_{i=1}^{n} x_i^2 y_i = a\sum\limits_{i=1}^{n} x_i^4 + b\sum\limits_{i=1}^{n} x_i \qquad \ldots(1)$

$\dfrac{\partial S}{\partial b} = 0 \Rightarrow -2\sum\limits_{i=1}^{n} \dfrac{1}{x_i} \left(y_i - ax_i^2 - \dfrac{b}{x_i} \right) = 0 \Rightarrow \sum\limits_{i=1}^{n} \dfrac{y_i}{x_i} = a\sum\limits_{i=1}^{n} x_i + b\sum\limits_{i=1}^{n} \dfrac{1}{x_i^2} \qquad \ldots(2)$

Solving (1) and (2) we get the values of a and b.

(10) FITTING OF THE CURVE $y = a/x + b\sqrt{x}$

Here we have
$$S = \sum_{i=1}^{n}\left(y_i - a/x_i - b\sqrt{x_i}\right)^2$$

Now
$$\frac{\partial S}{\partial a} = 0 \Rightarrow -2\sum_{i=1}^{n}\frac{1}{x_i}\left(y_i - \frac{a}{x_i} - b\sqrt{x_i}\right) = 0 \Rightarrow \sum_{i=1}^{n}\frac{y_i}{x_i} = a\sum_{i=1}^{n}\frac{1}{x_i^2} + b\sum_{i=1}^{n}\frac{1}{\sqrt{x_i}} \qquad \ldots(1)$$

$$\frac{\partial S}{\partial b} = 0 \Rightarrow -2\sum_{i=1}^{n}\sqrt{x_i}\left(y_i - \frac{a}{x_i} - b\sqrt{x_i}\right) = 0 \Rightarrow \sum_{i=1}^{n}(\sqrt{x_i})y_i = a\sum_{i=1}^{n}\frac{1}{\sqrt{x_i}} + b\sum_{i=1}^{n}x_i \qquad \ldots(2)$$

On solving (1) and (2) we get the required values of a and b.

Solved Examples

EXAMPLE 1. *Fit a straight line to the following data :*

x	1	2	3	4	5
y	5	7	9	10	11

SOLUTION. Here, we have

						Total
x	1	2	3	4	5	$\Sigma x = 15$
y	5	7	9	10	11	$\Sigma y = 42$
xy	5	14	27	40	55	$\Sigma xy = 141$
x^2	1	4	9	16	25	$\Sigma x^2 = 55$

the equation of the line be $y = a + bx$ ...(1)
Normal equation are
$$\Sigma y = ma + b\Sigma x$$
and $\quad \Sigma xy = a\Sigma x + b\Sigma x^2$
Using the given values,
$$42 = 5a + 15b \qquad \ldots(2)$$
$$141 = 15a + 55b \qquad \ldots(3)$$
Solving (2) and (3) we get $b = \dfrac{3}{2}, a = \dfrac{39}{10}$
or $\quad b = 1.5, a = 3.9$
Hence, required straight line is given by
$$y = (3.9) + (1.5)x$$

📋 **REMARK**

• For the sake of convenience, it is sometimes easy to change the origin and scale with the substitution $x = \dfrac{x - A}{h}$ and $y = \dfrac{y - B}{h}$, where A and B are the middle values of x and y series respectively.

EXAMPLE 2. *Fit a straight line to the following data:*

x	0	5	10	15	20	25
y	12	15	17	22	24	30

SOLUTION. We observe that the number of values given is 6, *i.e.*, even and the difference in the values of x is 5. In such a case the calculations can be simplified if we take half the common distance which is 2.5 here as unit of measurement. Also the two mid-values are 10 and 15, so we take their mean, *i.e.*, $\frac{1}{2}(10 + 15)$, *i.e.*, 12.5 as the origin.

Therefore, introduce two new variables

$$u = \frac{x - 12.5}{2.5} \text{ and } v = y - 20 \qquad \ldots(1)$$

Then we have

$$u : \frac{0 - 12.5}{2.5}, \frac{5 - 12.5}{2.5}, \frac{10 - 12.5}{2.5}, \frac{15 - 12.5}{2.5},$$
$$\frac{20 - 12.5}{2.5}, \frac{25 - 12.5}{2.5}$$

							Total
u	−5	−3	−1	1	3	5	0
v	−8	−5	−3	2	4	10	0
uv	40	15	3	2	12	50	122
u^2	25	9	1	1	9	25	70

Now let the equation of line be
$$v = a + bu \qquad \ldots(2)$$
Then its normal equations are
$$\Sigma v = ma + b\Sigma u$$
$$\Sigma vu = a\Sigma u + b\Sigma u^2$$
$\Rightarrow \quad 0 = 6.a + 0.b \qquad \ldots(3)$
$\qquad 122 = 0.a + 70.b \qquad \ldots(4)$
Solving (3) and (4) we get $a = 0$ and $b = 1.743$.
Substituting these values in (2), the equation of the line is
$$v = 0 + (1.743)u$$
or $\quad y - 20 = (1.743)\left[\dfrac{x - 12.5}{2.5}\right]$ [Using (1)]

or $\quad y - 20 = \left(\dfrac{1.743}{2.5}\right)x - \dfrac{1.743 \times 12.5}{2.5}$

$\qquad y - 20 = 0.7x - 8.715$
$\Rightarrow \qquad y = 0.7x + (20 - 8.715)$
$\Rightarrow \qquad y = 0.7x + 11.285$
$\Rightarrow \qquad y = 11.285 + 0.7x$

📋 **REMARK**

• If in such a problem, the number of values given is odd then the calculations are simplified provided we take the common difference as the unit of measurement and the mid-values as the assumed origin.

EXAMPLE 3. *Fit a parabola of the second degree of the following data:*

(UPTU(MCA)-2006, VTU-2009, Bhopal-2008, MDU(BE)-2007)

x	1.0	1.5	2.0	2.5	3.0	3.5	4.0
y	1.1	1.3	1.6	2.6	2.7	3.4	4.1

SOLUTION. Here we observe that the number of values given here is 7 which is odd, so we take the middle value 2.5, *i.e.*, 4th value as our assumed mean and let

$$u = \frac{x - 2.5}{0.5} \quad \text{and} \quad v = y - 2.7$$

Let the second degree curve to be fitted is
$$v = a + bu + cu^2 \quad \text{...(1)}$$
Its normal equations are
$$\Sigma v = na + b\Sigma u + c\Sigma u^2 \quad \text{...(2)}$$
$$\Sigma vu = a\Sigma u + b\Sigma u^2 + c\Sigma u^3 \quad \text{...(3)}$$
$$\Sigma vu^2 = a\Sigma u^2 + b\Sigma u^3 + c\Sigma u^4 \quad \text{...(4)}$$

Now from the given data, we have

Total

u	–3	–2	–1	0	1	2	3	**0**
v	–1.6	–1.4	–1.1	–0.1	0	0.7	1.4	**–14.1**
vu	4.8	2.8	1.1	0	0	1.4	4.2	**14.3**
u^2	9	4	1	0	1	4	9	**28**
vu^2	–14.4	–5.6	–1.1	0	0	2.8	12.6	**–5.7**
u^3	–27	–8	–1	0	1	8	27	**0**
u^4	81	16	1	0	1	16	81	**196**

Substituting the values of $\Sigma u, \Sigma v, \Sigma uv, \Sigma vu^2, \Sigma u^2, \Sigma u^3, \Sigma u^4$ in (2), (3) and (4), we get
$$-2.1 = 7a + 0.b + 28.c \quad \text{...(5)}$$
$$14.3 = 0.a + 28b + 0.c \quad \text{...(6)}$$
$$-5.7 = 28a + 0.b + 196c \quad \text{...(7)}$$
Solving (5), (6) and (7), we get
$$a = -0.04, b = 0.53, c = 0.03.$$
Substituting these values of a, b and c in (1), we get
$$v = -0.04 + 0.53u + 0.03u^2$$
$$\Rightarrow \quad y - 2.7 = -0.04 + 0.53\left(\frac{x - 2.5}{0.5}\right)$$
$$+ 0.03\left(\frac{x - 2.5}{0.5}\right)^2$$
$$\Rightarrow y - 2.7 = -0.04 + \frac{0.53x}{0.5}$$
$$- \frac{2.5 \times 0.53}{0.5} + 0.03\left(\frac{x - 2.5}{0.5}\right)^2$$
$$\Rightarrow \quad y - 2.7 = -0.04 + 1.06x - 2.65$$
$$+ 0.03\left[\frac{x^2 + 6.25 - 5.0x}{0.5}\right]$$
$$\Rightarrow \quad y = -0.04 + 2.7 - 2.65 + 1.06x$$
$$+ 0.06(x^2 - 5x + 6.25)$$
$$\Rightarrow \quad y = 0.01 + 1.06x + 0.06x^2 - 0.30x + 0.375$$
$$\Rightarrow \quad y = 0.385 + 0.76x + 0.66x^2$$

EXAMPLE 4. *If P is a pull required to lift a load W by means of a pulley block, find a linear law of the form P = mW + c connecting P and W, using the following data :*

P	12	15	21	25
W	50	70	100	120

Compute P where W = 150 Kg.

(UPTU–2007, VTU–2002)

SOLUTION. The given equation is $P = mW + c$ ...(1)
The normal equations are
$$\left.\begin{array}{l} \Sigma P = 4c + m\Sigma W \\ \Sigma WP = c\Sigma W + m\Sigma W^2 \end{array}\right\} \quad \text{...(2)}$$
Then, we have

W	P	W^2	WP
50	12	2500	600
70	15	4900	1050
100	21	10000	2100
120	25	14400	3000
Total 340	**73**	**31800**	**6750**

Putting these values in (2), we get
$$73 = 4c + 340m$$
$$6750 = 340c + 31800m$$
which gives $m = 0.1879, c = 2.2785$
Hence, the line of best fit is
$$P = 2.2759 + 0.1879W$$
Now, for $W = 1500$ Kg $\quad P = 30.4635$ Kg

EXAMPLE 5. *The pressure and the volume of a gas are related by the equation $pV^r = K, r$ and K being constants. Fit this equation to the following set of observations.*

P(Kg/cm²)	0.5	1.0	1.5	2.0	2.5	3.0
V(litres)	1.62	1.00	0.75	0.62	0.52	0.46

(VTU–2011)

SOLUTION. We have
$$\log_{10} p + r \log_{10} V = \log_{10} K$$
$$\Rightarrow \quad \log_{10} V = \frac{1}{r}\log_{10} K - \frac{1}{r}\log_{10} P$$
$$\Rightarrow \quad Y = A + BX$$
where $X = \log_{10} p, Y = \log_{10} V, A = \frac{1}{r}\log_{10} K, B = -\frac{1}{r}$
Now

P	V	X	Y	XY	X^2
0.5	1.62	–0.3010	0.2095	–0.0630	0.0906
1.0	1.00	0	0	0	0
1.5	0.75	0.1762	–0.1249	–0.0220	0.0310
2.0	0.62	0.3010	–0.2076	–0.0625	0.0906
2.5	0.52	0.3979	–0.2840	–0.1130	0.1583
3.0	0.46	0.4771	–0.3372	–0.1609	0.2276
Total	1.0511	**–0.7442**	**–0.4214**	**0.5981**	

Now putting all these values in the normal equations and get the required fitted curve.

EXAMPLE 6. *Find a relation of the form* $y = AB^x$ *for the following data by the method of least squares*

x	2	3	4	5	6
y	8.3	15.4	33.1	65.2	126.4

SOLUTION. The curve to be fitted is $y = A(B)^x$.

or, $y = a + bx$ where

$$a = \log_{10} A, b = \log_{10} B \text{ and } Y = \log_{10} y$$

Therefore, the normal equations are $\Sigma Y = 5a + b\Sigma x$ and

$$\Sigma xY = a\Sigma x + b\Sigma x^2.$$

x	y	$Y = \log_{10} y$	x^2	xY
2	8.3	0.1191	4	1.8382
3	15.4	1.1872	9	3.5616
4	33.1	1.5198	16	6.0792
5	65.2	1.8142	25	9.0710
6	127.4	2.1052	36	12.6312
$\Sigma X = 20$		$\Sigma Y = 7.5455$	$\Sigma X^2 = 90$	$\Sigma XY = 33.1812$

Therefore, the normal equations, becomes
$7.5455 = 5a + 20b$ and $33.1812 = 20a + 90b$

$\Rightarrow \quad a = 0.31$ and $b = 0.3$

$\therefore \quad A = $ Antilog $a = 2.04$

and $B = $ Antilog $b = 1.995$

Hence, the required curve is $y = 2.04(1.995)^x$.

EXAMPLE 7. *Determine the constant a and b by the method of least squares such that $y = ae^{bx}$ fits the*

following data :

x	2	4	6	8	10
y	4.077	11.084	30.128	81.897	222.62

SOLUTION. We have $\qquad y = ae^{bx}$

Taking log, we get

$$\log y = \log a + bx$$

i.e., $\qquad Y = A + bx \qquad \qquad$ ...(1)

where $\qquad Y = \log y$

and $\qquad A = \log a$

Normal equations of equation (1) are

$$\Sigma Y = An + b\Sigma x \qquad \qquad ...(2)$$

and $\qquad \Sigma xY = A\Sigma x + b\Sigma x^2 \qquad \qquad$ ...(3)

x	y	$Y = \log y$	xY	x^2
2	4.077	1.4054	2.8107	4
4	11.084	2.4055	9.6220	16
6	30.128	3.4054	20.4327	36
8	81.897	4.4055	35.2437	64
10	222.62	5.4055	54.0547	100
Total 30		**17.0272**	**122.1638**	**220**

Putting all these above values in equations (2) and (3), we get

$$17.0272 = 5A + 30b \qquad \qquad ...(4)$$
$$122.1638 = 30A + 220b \qquad \qquad ...(5)$$

From (4) and (5), we get $b = 0.5, A = 0.4054$, *i.e.,* $A = \log a = 0.4054$

Taking antilog, we get $a = 1.4999 \Rightarrow a = 1.5$ (app.)

9.1.8 Curve Fitting by Sum of Exponentials

Consider the equation

$$y = A_1 e^{\lambda_1 x} + A_2 e^{\lambda_2 x} \qquad \qquad ...(1)$$

It can be seen that the function given by (1) satisfy a differential equation of the type

$$\frac{d^2 y}{dx^2} = a_1 \frac{dy}{dx} + a_2 y \qquad \qquad ...(2)$$

where a_1, a_2 are constants.

Assume that a is the initial value of x we get, by integrating (2) w.r.t. x from a to x.

$$y'(x) - y'(a) = a_1 y(x) - a_1 y(a) + a_2 \int_a^x y(x) dx$$

$\Rightarrow \qquad \qquad y'(x) = \frac{dy}{dx}$

Again integrating (3) w.r.t. x from a to x, we get

$$y(x) - y(a) - (x-a)y'(a) = a_1 \int_a^x y(x)dx - a_1(x-a)y(a) + a_2 \int_a^x \int_a^x y dx dx \qquad \qquad ...(4)$$

Now using the result

$$\int_a^x \cdots \int_a^x f(x)dx = \frac{1}{(n-1)!} \int_a^x (x-t)^{n-1} f(t)dt \qquad \qquad \text{(Convolution theorem of integration)}$$
$$\underbrace{\qquad}_{n \text{ times}}$$

The equation (5) can be written as

$$y(x_1) - y(a) - (x-a)y'(a) = a_1 \int_a^x y dx - a_1(x-a)y(a) + a_2 \int_a^x (x-t)y(t)dt \qquad \qquad ...(6)$$

Now, choosing two points x_1 and x_2 such that $a - x_1 = x_2 - a$, then (6) gives

$$y(x_1) - y(a) - (x_1 - a)y'(a) = a_1 \int_a^{x_1} y(x)dx - a_1(x_1 - a)y(a) + a_2 \int_a^{x_1}(x_1 - t)y(t)dt$$

and $\qquad y(x_2) - y(a) - (x_2 - a)y'(a) = a_1 \int_a^{x_2} y(x)dx - a_1(x_2 - a)y(a) + a_2 \int_a^{x_2}(x_2 - t)y(t)dt$

Adding above two equations and simplifying by using $a - x_1 = x_2 - a$, we get

$$y(x_1) + y(x_2) - 2y(a) = a_1\left[\int_a^{x_1} y(x)dx + \int_a^{x_2} y(x)dx\right] + a_2\left[\int_a^{x_1}(x_1 - t)y(t)dt + \int_a^{x_2}(x_2 - t)y(t)dt\right] \qquad ...(7)$$

Now the equation (7) can be used to setup a linear system of equations for a_1 and a_2 and then obtain λ_1 and λ_2 from the following characteristic equation $\lambda^2 = a_1\lambda + a_2$.

Finally, A_1 and A_2 can be obtained by the method of least square.

Solved Examples

EXAMPLE 1. *Fit a function of the form* $y = A_1 e^{\lambda_1 x} + A_2 e^{\lambda_2 x}$ *to the following data :*

x	1.0	1.1	1.2	1.3	1.4	1.5	1.6	1.7	1.8
y	1.54	1.67	1.81	1.97	2.15	2.35	2.58	2.83	3.11

SOLUTION. To fit the given curve we use the following steps :

Step I : Choose x_1 and x_2 such that
$$a - x_1 = x_2 - a \qquad ...(1)$$
where a is the initial value of x taken from
$$x_i : i = 1, 2, ..., n.$$

Step II : Use
$$y(x_1) + y(x_2) - 2y(a)$$
$$= a_1\left[\int_a^{x_1} y(x)dx + \int_a^{x_2} y(x)dx\right]$$
$$+ a_2\left[\int_a^{x_1}(x_1 - t)y(t)dt + \int_a^{x_2}(x_2 - t)y(t)dt\right] \quad ...(2)$$

Step III : Repeat the step I and II by choosing another set of x_1, x_2 and a such that (2) is true. Therefore, we obtain a linear system of equations for a_1 and a_2. Solve these equation to get a_1 and a_2.

Step IV : Substitute the values of a_1 and a_2 obtained in step III in the characteristic equation $\lambda^2 = a_1\lambda + a_2$

Step V : Finally, use the method of least square to obtain A_1 and A_2 choosing $x_1 = 1.0, a = 1.2$ and $x_2 = 1.4$, we have
Now equation (2) gives

$$1.54 + 2.15 - 3.62$$
$$= a_1\left[\int_{1.2}^{1.0} ydx + \int_{1.2}^{1.4} ydx\right]$$
$$+ a_2\left[\int_{1.2}^{1.0}(1.0 - t)y(t)dt + \int_{1.2}^{1.4}(1.4 - t)ydt\right]$$
$$\Rightarrow 0.07 = a_1\left[-\int_{1.0}^{1.2} ydx + \int_{1.2}^{1.4} ydx\right]$$
$$+ a_2\left[\begin{array}{c}-\int_{1.0}^{1.2}(1.0 - t)ydt \\ +\int_{1.2}^{1.4}(1.4 - t)ydt\end{array}\right]$$

On simplification, we get
$$1.81a_1 + 2.180a_2 = 2.10 \qquad ...(4)$$
Again choosing $x_1 = 1.4, a = 1.6$ and $x_2 = 1.8$, so that we have $a - x_1 = 0.2 = x_2 - a$ and using equation (2), we get
$$2.88a_1 + 3.104a_2 = 3.00 \qquad ...(5)$$
Solving (4) and (5), we get
$$a_1 = 0.03204, a_2 = 0.9364$$
Put these values in (3), we get
$$\lambda^2 - 0.03204\lambda - 0.9364 = 0$$
Now, using the method of least squares, we get
$$A_1 = 0.499, A_2 = 0.491$$
Hence, the required curve for best fit is
$$y = 0.499e^{0.99x} + 0.491e^{-0.96x}.$$

Exercise-9.2

1. Show that the line of fit to the following data :

x	6	7	7	8	8	8	9	9	10
y	5	5	4	5	4	3	4	3	3

is given by $y = -0.5x + 8$.

2. Show that the best-fitting linear function for the points $(x_1, y_1), (x_2, y_2)...(x_n, y_n)$ may be expressed in the form
$$\begin{vmatrix} x & y & 1 \\ \Sigma x & \Sigma y & n \\ \Sigma x^2 & \Sigma y^2 & \Sigma x \end{vmatrix} = 0$$

Show also that the line passes through the mean points $(\bar{x}, \bar{y})$.

3. Find the parabola of the form $a + bx + cx^2$ which fit most closely with the observations : (VTU–2006, JNTU–2000S)

x	–3	–2	–1	0	1	2	3
y	4.63	2.11	0.67	0.09	0.63	2.15	4.58

4. Fit a second degree parabola to the following data

x	1.0	1.5	2.0	2.5	3.0	3.5	4.0
y	1.1	1.1	1.6	2.0	2.7	3.4	4.1

5. The following table gives the results of the measurement of train resistance; V is the velocity in miles per hour, R is the resistance in pounds per ton.

V	20	40	60	80	100	120
R	5.5	9.1	14.9	22.8	33.3	46.0

If R is related to V by the relation $R = a + bV + cV^2$, fit the curve. (UPTU–2002)

6. Fit the relation $R = a + bV^2$ from the following data : (Indore–2008)

V	10	20	30	40	50
R	8	10	15	21	30

7. Fit a least square geometric curve $y = ax^b$ to the following

data :

x	1	2	3	4	5
y	0.5	2	4.5	8	12.5

8. The voltage V across a capacitor at time t seconds is given by the following table :

t	0	2	4	6	8
V	150	63	28	12	5.6

Use the method of least square to fit a curve of the form $V = ae^{kt}$ to the above data.

9. Applying the method of least squares to fit the curves $y = ax^2 + \dfrac{b}{x}$ to the following data :   (Madras–2003)

x	1	2	3	4
y	–1.51	0.99	3.88	7.66

10. Use the method of least square to fit the curve $y = kx^m$ for the following data :

x	1	2	3	4	5
y	7.1	27.8	62.1	110	161

11. Use the method of least square to fit the straight line for the following data :

x	71	68	73	69	67	65	66	67
y	69	72	70	70	68	67	68	64

12. Fit an exponential curve of the form $y = ab^x$ to the following data :   (Tiruchirapalli–2001)

x	1	2	3	4	5	6	7	8
y	1	12	1.8	2.5	3.6	4.7	6.6	9.1

13. Fit the curve $y = ae^{bx}$ to the following data :   (Coimbatore–1997)

x	0	2	4
y	5.1	10	31.1

14. Fit a second degree parabola to the following data :   (UPTU–2009)

x	1989	1990	1991	1992	1993	1994	1995	1996	1997
y	352	356	357	358	360	361	361	360	359

15. Find the best possible curve of the form $y = a + bx$, using method of least squares for the data :   (VTU–2011)

x	1	3	4	6	8	9	11	14
y	1	2	4	4	5	7	8	9

16. Fit a second degree parabola to the following data :   (UPTU–2009)

x	1	2	3	4	5	6	7	8	9	10
y	124	129	140	159	228	289	315	302	263	210

17. The velocity V of a liquid is known to vary with temperature according to a quadratic law $V = a + bT + cT^2$. Find the best values of a, b and c for the following table :   (UPTU(MCA)–2010)

T	1	2	3	4	5	6	7
V	2.31	2.01	3.80	1.66	1.55	1.47	1.41

18. Find the least squares fit of the form $y = a_0 + a_1x^2$ to the following data :   (UPTU–2008)

x	–1	0	1	2
y	2	5	3	0

19. Predict the mean radiation dose at an altitude of 3000 feet by fitting an exponential curve to the given data :

     (SVTU–2007, JNTU–2003)

Amplitude (x)	50	450	780	1200	4400	4800	5300
Dose of radiation (y)	28	30	32	36	51	58	69

20. Fit the curve $y = ax + b/x$ to the following data :   (UPTU–2010)

x	1	2	3	4	5	6	7	8
y	5.4	6.3	8.2	10.3	12.6	14.9	17.3	19.5

21. Predict y at $x = 3.75$, by fitting a power curve $y = ax^b$ to the given data :   (JNTU–2003)

x	1	2	3	4	5	6
y	2.98	4.26	5.21	6.10	6.80	7.50

22. Fit the curve of the form $y = ae^{bx}$ to the following data :   (VTU–2011S, JNTU–2006)

x	77	100	185	239	285
y	2.4	3.4	7.0	11.1	19.6

Answers

3. $y = 1.243 - 0.004x + 0.22x^2$ **4.** $y = 10.4 - 0.198x + 0.244x^2$ **5.** $R = 3.48 - 0.002V + 0.003V^2$

6. $R = 6.32 + 0.0095V^2$ **7.** $a = 0.5012, b = 1.9977$ **8.** $a = 146.3, k = -0.412$ **9.** $a = 0.51, b = -2.04$

10. $k = 7.17, m = 1.95$ **12.** $y = 0.6823(1.384)^x$ **13.** $a = 4.1, b = 0.43$

14. $y = -1000106.41 + 1034.29x - 0.267x^2$ **15.** $a = 0.545, b = 0.636$ **16.** $y = 18.866 + 66.158x - 4.333$

17. $V = 2.593 - 0.326T + 0.023T^2$ **18.** $y = 4.167 - 1.111x^2$ **19.** $y = 44.9$ approx.

20. $a = 3, b = 2$ **21.** $y = 2.978x^{0.5143}; 5.8769$ **22.** $a = 0.1839, b = 0.0221$

9.2 STATISTICAL AVERAGES

In previous section, we have studied about the classification and tabulation of data, but our study about classification and tabulation is not enough to get all the desirable results since when two or more series of same type are under observations, we cannot classify and tabulate them so we need an arithmetic idea or characteristic about the distribution. These characteristic are measure of central tendency, measure of dispersion and skewness and the packedness.

In this section, we study the measure of central tendency or average according to Dr. Bowley statistics may rigidly be called the sequence of averages and averages are statistical constant which enable us to comprehend in a single effort the significance of the whole.

"Average is a point about which all the values of variate cluster."

9.2.1 KINDS OF STATISTICAL AVERAGES

The statistical averages can be divided into five parts.

1. Arithmetic mean (A.M.)   2. Median (Md)   3. Mode (Mo)
4. Geometric mean (G.M.)   5. Harmonic mean (H.M.)

1. Arithmetic mean

"Arithmetic mean is the amount obtained by dividing the sum of values of the items in a series by their number.

The arithmetic mean of some observations is the value which we can obtain by dividing the sum of all the numbers by the total number of items, *i.e.,*

$$\text{A.M.} = \bar{x} = \frac{\text{Sum of all the observations}}{\text{Total number of terms}}$$

Let $x_1, x_2, x_3, \dots x_n$ are the observations. Then arithmetic mean is given by

$$\bar{x} = \frac{x_1 + x_2 + x_3 + \dots\dots + x_n}{n} = \frac{\Sigma x_i}{n}$$

$$\Rightarrow \qquad \bar{x} = \frac{\Sigma x}{n}$$

If the variate $x_1, x_2, \dots x_n$ occurs $f_1, f_2, f_3, \dots f_n$ times, then the arithmetic mean is known as weighted arithmetic mean and given by

$$\bar{x} = \frac{f_1 x_1 + f_2 x_2 + \dots + f_n x_n}{f_1 + f_2 + \dots + f_n}, \quad \bar{x} = \frac{\Sigma f_i x_i}{\Sigma f_i}$$

9.2.2 METHODS OF CALCULATING ARITHMETIC MEAN IN INDIVIDUAL SERIES

We can calculate the arithmetic mean by following three methods :

1. Direct method   2. Short-cut method   3. Step-deviation method

1. Direct Method : In this method, the mean is calculated using the following formula

$$\bar{x} = \frac{x_1 + x_2 + \dots\dots + x_n}{n} = \frac{\sum\limits_{i=1}^{n} x_i}{n}$$

WORKING PROCEDURE

Step 1. Adding all the observations to find Σx_i .
Step 2. Divide this sum Σx_i by total number of observations, *i.e., n*.

2. Short-cut Method : In this method, we assumed a middle number as an assumed mean. Here, we use the following formula

$$\bar{x} = A + \frac{\Sigma d}{n}$$

where, $\bar{x}$ = Arithmetic mean; A= Assumed mean; deviations, $d = (x - A)$; Σd = Sum of deviations

WORKING PROCEDURE

Step 1. Select the assumed mean, *i.e., A* .
Step 2. Calculate the deviation from A, *i.e., $d = x - A$*.
Step 3. Find the sum of deviation as Σd .
Step 4. Using the formula $\bar{x} = A + \dfrac{\Sigma d}{n}$, we get the required mean.

3. Step Deviation Method. Let us assume a number h called scale then $d' = \dfrac{x}{h}$ (x denotes the monthly income)

$$\Sigma d' = \frac{\Sigma x}{h}, \quad \Sigma x = h\Sigma d'$$

$$\frac{\Sigma x}{n} = \frac{h\Sigma d'}{n}, \quad \bar{x} = (h\Sigma d')/n$$

$$\text{A.M.} = \frac{h\,\Sigma d'}{n}$$

 Solved Examples

EXAMPLE 1. *Calculate the arithmetic mean of*

   129, 117, 112, 200, 172, 138, 183

SOLUTION. Arithmetic mean of the above data can be given by :

$$\bar{x}\,(\text{A.M.}) = \frac{129 + 117 + 112 + 200 + 172 + 138 + 183}{7}$$

$$= \frac{1051}{7} = 150.14$$

So, A.M. = 150.14.

EXAMPLE 2. *Find the arithmetic mean of first n natural numbers.*

SOLUTION. Arithmetic mean can be given by :

$$\bar{x}\,(\text{A.M.}) = \frac{\text{Sum of all the observations}}{\text{Total number of terms}}$$

$$= \frac{1 + 2 + 3 + 4 + \ldots\ldots + n}{n} = \frac{\Sigma n}{n}$$

$$= \frac{1}{n}\frac{n(n+1)}{2} \qquad \left(\because \Sigma n = \frac{n(n+1)}{2}\right)$$

$$= \frac{1}{2}(n+1)$$

So the arithmetic mean of first *n* natural numbers is given by $\frac{1}{2}(n+1)$.

EXAMPLE 3. *Show that the arithmetic mean of the series* **1, 2, 2^2, 2^3, 2^4, ... 2^n, *is given by* $(2^{n+1}-1)/n+1$.**

SOLUTION. Arithmetic Mean $= \dfrac{\text{Sum of all the observations}}{\text{Total number of terms}}$

$$= \frac{1 + 2 + 2^2 + 2^3 + \ldots\ldots + 2^n}{n+1} = \frac{\text{Sum of the G.P.}}{n+1}$$

$$= \frac{(2^{n+1}-1)}{(2-1)(n+1)} = \frac{(2^{n+1}-1)}{n+1}$$

Arithmetic mean of the given G.P. $= \dfrac{(2^{n+1}-1)}{n+1}$.

EXAMPLE 4. *A candidate obtain the following marks in an examination in a paper of 100 marks each*

Subject	English	Maths	Physics	Chemistry	Biology
Marks (out of 100)	48	82	70	64	60

It is agreed to give double weight to physics and mathematics as compared to other subjects. What is the arithmetic mean?

(UPTU B.Pharma-2005)

SOLUTION. Since weights are given, we shall calculate weighted arithmetic mean in place of simple arithmetic mean.

Weights are given accordance with the statement given in the example.

Subject	Marks (X)	Weight (W)	WX
English	48	1	48
Maths	82	2	164
Physics	70	2	140
Chemistry	64	1	64
Biology	60	1	60
		ΣW = 7	ΣWX = 476

$$\bar{X}_W = \frac{\Sigma WX}{\Sigma W} = \frac{476}{7} = 68\%.$$

EXAMPLE 5. *A person travels n equal distances with vertices* $V_1, V_2, V_3, \ldots, V_n$. *Show that his average velocity cannot exceed* $(V_1 + V_2 + \ldots + V_n)/n$. *When it will be equal to this value?*

SOLUTION. Total distance = *n* equal distances.
Velocity per distance $V_1, V_2, V_3, \ldots, V_n$

$$\text{Average speed} = \frac{\text{Total distance}}{\text{Total time}}$$

Let equal distance = *x*

$$\text{Time} = \frac{\text{Distance}}{\text{Speed}}$$

$$\text{Per unit distance time} = \frac{\text{Speed}}{\text{Distance}}$$

$$= \frac{V_1 + V_2 + V_3 + \ldots + V_n}{n}$$

This speed is maximum $\dfrac{V_1 + V_2 + V_3 + \ldots + V_n}{n} \le$

total time

Thus maximum average speed can be $\dfrac{V_1 + V_2 + \ldots + V_n}{n}$.

When time taken by each distance is same with same speed then it will be equal.

EXAMPLE 6. *The mean of n numbers of a series is $\bar{x}$ and the sum of first (n−1) numbers is λ. Find the value of the last number.*

SOLUTION. Mean of *n* numbers = $\bar{x}$
Sum of first (*n*-1) number = λ

$$\text{Arithmetic Mean} = \frac{\text{Sum of all the observations}}{\text{Total number of terms}}$$

Sum of all the observations
           = Mean ×Total no. of terms $= \bar{x} \times n$
Sum of all the observations $= n\bar{x}$
Now, sum of first (*n*− 1) numbers = λ
Then the last number = (Sum of *n* terms)
           − (Sum of *n*−1 terms)
           $= n\bar{x} - \lambda$
The value of the last number $= (n\bar{x} - \lambda)$

9.2.3 CALCULATION OF ARITHMETIC MEAN IN A DISCRETE FREQUENCY DISTRIBUTION

1. **Direct Method:** In case of discrete frequency distribution, we multiply the values of a variable (x) by their respective frequencies (f). Then, we use the following formula

$$\overline{x} = \frac{\Sigma f.x}{\Sigma f} = \frac{\Sigma f.x}{n}$$

where, Σfx = The sum of products of observations with their respective frequencies

$\Sigma f = n$ = Total number of frequencies.

WORKING PROCEDURE

Step 1. Multiply the value of variable x by corresponding frequency f to find fx.

Step 2. Calculate sum Σfx of the product obtained in step (1).

Step 3. Putting the values in the formula

$$\overline{x} = \frac{\Sigma f.x}{\Sigma f}$$

2. **Short-cut Method :** Firstly, we shall assume a mean and then take deviation of the variable from this assumed mean. In this method, we use the following formula :

$$\overline{x} = A + \frac{\Sigma fd}{n}$$

where, A = Assumed mean; $d = x - A$; Deviation; f = Frequency

WORKING PROCEDURE

Step 1. Select the assumed mean, i.e., A.

Step 2. Calculate the deviation from A, i.e., $d = x - A$.

Step 3. Calculate f.d.

Step 4. Sum all the deviation to obtain $\Sigma f d$.

Step 5. Using the formula $\overline{x} = A + \frac{\Sigma fd}{n}$.

3. **Step-Deviation Method :** In this method, we divide our deviation by the common factor h. Therefore

$$\overline{x} = A + \frac{\Sigma fd'}{n} \times h,$$

where, h = Common factor of the deviation; $d' = \frac{d}{h} = \frac{x - A}{h}$

9.2.4 CALCULATION OF ARITHMETIC MEAN IN A CONTINUOUS FREQUENCY DISTRIBUTION

In this case, we have to calculate the mid point of the various class intervals and denote it by x. Then, proceed same as above.

Solved Examples

EXAMPLE 1. *Compute the mean of the following data by direct and short-cut method:*

Height (cm)	195	198	201	204	207	210	213	216	219
Children	1	4	5	7	11	10	6	4	2

SOLUTION.

Heigh x (cm)	f	fx	$d = (x - A)$	fd
195	1	195	–12	– 12
198	4	792	–9	– 36
201	5	1005	–6	– 30
204	7	1428	–3	– 21
207	11	2277	0	0
210	10	2100	3	30
213	6	1278	6	36
216	4	864	9	36
219	2	438	12	24
Total	50	10377		27

Here, $A = 207$

By direct method, we can calculate the A.M. by the formula

$$\text{A.M.} = \frac{\Sigma fx}{N} = \frac{10377}{50} = 207.54 \text{ cm}$$

By short-cut method, the formula is given by

$$\text{A.M.} = A + \frac{\Sigma fd}{N} = 207 + \frac{27}{50} = 207 + 0.54 = 207.54.$$

EXAMPLE 2. *Compute the mean of the following distribution by step deviation method*:

Class	0 - 11	11-22	22-33	33-44	44-55	55-66
Frequency	9	17	28	26	15	8

SOLUTION.

Class	Mid Value x	f	$(x - 38.5)$	$d' = \frac{x - 38.5}{11}$	fd'
0-11	5.5	9	–33	–3	–27
11-22	16.5	17	–22	–2	–34

22-33	27.5	28	−11	−1	−28
33-44	38.5	26	0	0	0
44-55	49.5	15	11	1	15
55-66	60.5	8	22	2	16
Total		Σf =103			$\Sigma f d'$ = −58

Here, the assumed mean , $A = 38.5$ and $h = 11$.
A.M. by step deviation method is given by

$$M = A + h \frac{\Sigma fd'}{N} = 38.5 + \frac{11 \times (-58)}{103} = 38.5 - 6.194$$
$$= 32.306$$

Hence, A.M. = 32.306.

EXAMPLE 3. *What is the arithmetic mean of the following data:*

Variate	0	1	2	3	...	n
Frequency	$^nC_0$	$^nC_1$	$^nC_2$	$^nC_3$	...	^nC_n

SOLUTION. The arithmetic mean is given by $= \dfrac{\Sigma fx}{\Sigma f}$

$$\frac{\Sigma fx}{\Sigma f} = \frac{^nC_0.0 + {}^nC_1 \times 1 + {}^nC_2 \times 2 + {}^nC_3 \times 3 + \ldots {}^nC_n \times n}{^nC_0 + {}^nC_1 + {}^nC_2 + {}^nC_3 + \ldots + {}^nC_n}$$

$$= \frac{0 + n.1 + \dfrac{n(n-1)}{2!}.2 + \dfrac{n(n-1)(n-2)}{3!} \times 3 \ldots + 1.n}{(1+1)^n}$$

$$= \frac{n \left[{}^{(n-1)}C_0 + {}^{(n-1)}C_1 + {}^{(n-1)}C_2 + \ldots {}^{(n-1)}C_{n-1} \right]}{2^n}$$

$$= \frac{n \left[{}^{(n-1)}C_0 + {}^{(n-1)}C_1 + {}^{(n-1)}C_2 + \ldots {}^{(n-1)}C_{n-1} \right]}{2^n}$$

$$= \frac{n.(1+1)^{n-1}}{2^n} = \frac{n.2^{n-1}}{2^n} = \frac{n}{2}$$

So the arithmetic mean of the given data $= \dfrac{n}{2}$

9.2.5 Charlier's Accuracy Check

To check the accuracy of the results by short-cut method and step deviation method Charlier suggested the following two formulae

(a) $\Sigma f(x - A) = \Sigma [f\{(x - A) + 1\}] - \Sigma f$ (For short-cut method)

(b) $\Sigma f \dfrac{(x - A)}{h} = \Sigma \left[f \left\{ \dfrac{(x-A)}{h} + 1 \right\} \right] - \Sigma f$ (For step deviation method)

$$\Sigma f(d') = \Sigma [f(d'+1)] - \Sigma f$$

Solved Examples

EXAMPLE 1. *Compute arithmetic mean of the following by both short-cut and step deviation method and apply Charlier's accuracy check for both the result.*

Class	20-30	30-40	40-50	50-60	60-70
Frequency	8	26	30	20	16

SOLUTION.

Class	Mid Value x	f	$(x-A)$	$(x-A)+1$	$f(x-A)$	$f(x-A)+1$
20-30	25	8	−20	−19	−160	−152
30-40	35	26	−10	−9	−260	−234
40-50	45	30	0	1	0	30
50-60	55	20	10	11	200	220
60-70	65	16	20	21	320	336
		100			100	200

$$\text{A.M.} = A + \frac{\Sigma f(x - A)}{\Sigma f} = 45 + \frac{100}{100} = 45 + 1 = 46$$

Applying Charlier's check

$\Sigma f(x - A) = \Sigma [f(x - A) + 1] - \Sigma f = [200] - 100 = 100$

Thus, we get $\Sigma f(x - A) = 100$, which is exactly correct by our table.

Now, applying step deviation method for the given datas

Class	Mid Value x	f	$(x-A)$	$\mu = \dfrac{x-A}{i}$	fd'	$d'+1$	$f(d'+1)$
20-30	25	8	−20	−2	−16	−1	−8
30-40	35	26	−10	−1	−26	0	0
40-50	45	30	0	0	0	1	30
50-60	55	20	10	1	20	2	40
60-70	65	16	20	2	32	3	48
		100			10		110

Applying Charlier's check for step deviation method

$\Sigma fd' = \Sigma [f(d'+1)] - \Sigma f = 110 - 100 = 10$

which is correct since by the table, we see that $\Sigma fd' = 10$

REMARK

- In the given data or frequency of any term is missing and the arithmetic mean of the series is given, then we can calculate the term by using any of three methods (direct method, short-cut method, step deviation method).

EXAMPLE 2. *If the arithmetic mean of the following frequency distribution is 39.25, find the missing term.*

Daily wages	25	30	35	50	60	75
No. of labour	10	?	13	8	5	4

SOLUTION.

Daily wages x	No. of lobours f	fx
25	10	250

30	z	30z
35	13	455
50	8	400
60	5	300
75	4	300
	$\Sigma f = z + 40$	$\Sigma fx = 30z + 1705$

We know that the arithmetic mean is given by

$$\text{A.M.} = \frac{\Sigma fx}{\Sigma f} \Rightarrow 39.25 = \frac{30z + 1705}{z + 40}$$

$$39.25z + 1570 = 30z + 1705$$

$$9.25z = 135, \quad z = \frac{135}{9.25}$$

$$z = 14.59 = 14.6.$$

EXAMPLE 3. *The mean marks of 100 students were found to be 40, later on it was discovered that a score of 53 was misread as 83. Find the corrected mean*

corresponding to the correct score.

SOLUTION. Total number of students, $N = 100$.

The arithmetic mean of the marks of these students is given by $a = 40$.

Then the total marks obtained by the students
$$= N \times a = 100 \times 40$$
$$= 4000$$

Now a score of 53 was misread as 83 so the score calculated will be wrong. Then the correct score can be calculated

Correct score = Total previous marks
$$- \text{wrong score} + \text{correct score}$$
$$= 4000 - 83 + 53 = 3970$$

Now, we have to find out the correct mean corresponding to the correct score.

$$\text{Mean} = \frac{\text{Correct score}}{N} = \frac{3970}{100}$$

$$\text{Mean} = 39.7 \text{ marks}$$

9.2.6 PROPERTIES OF ARITHMETIC MEAN

1. If every variable is increased by a particular value a, then the arithmetic mean is also increased by a.
2. The algebraic sum of the deviations of all the variate values from their arithmetic mean is zero.
3. The sum of the squares of the deviations of all the values taken about their arithmetic mean is minimum.
4. If $M_1, M_2, ..., M_k$ be the arithmetic mean of k distributions with respective frequencies $N = n_1, n_2, ..., n_k$, then the mean M of the whole distribution with frequency $N = (n_1, n_2, ..., n_k)$ is given by $M = \frac{1}{N} \sum_{r=1}^{k} n_r M_r$.
5. Arithmetic mean is not independent of the change of origin and scale.

Solved Examples

EXAMPLE 1. *The weight of 150 students in a certain class is 60 kilograms. The mean weight of boys in the class is 70 kilograms and that of the girls is 55 kilograms. Find the number of boys and girls in the class.*

SOLUTION. $N = 150$

Let mean weight of boys = $\bar{X}_1 = 70$

Let mean weight of girls = $\bar{X}_2 = 55$

Let number of boys = n_1

no. of girls = n_2

$$n_1 + n_2 = N = 150 \qquad ... (1)$$

$$\bar{X}_{12} = 60$$

We know that
$$\bar{X}_{12} = \frac{n_1 \bar{X}_1 + n_2 \bar{X}_2}{n_1 + n_2}$$

$$60 = \frac{70n_1 + 55n_2}{150}$$

$$70n_1 + 55n_2 = 9000$$

$$14n_1 + 11n_2 = 1800 \qquad ...(2)$$

Also, $\qquad 14n_1 + 14n_2 = 2100 \qquad ...(3)$

Subtracting eq. (2) and (3)

$$3n_2 = 300$$

$$n_2 = 100$$

$$n_1 = 150 - 100 = 50$$

Thus, number of boys = 50

Number of girls = 100.

EXAMPLE 2. *The average monthly wages of all the workers in a factory is Rs. 444. If the average wages paid to male and female workers are Rs. 480 and Rs. 360 respectively. Find the percentage of male and female workers employed by factory.*

SOLUTION. Let the total workers be n.

$$\therefore \qquad 444 = \frac{\Sigma x}{n}$$

$$444n = \Sigma x$$

$$480 = \frac{\Sigma x_1}{n_1} \qquad [n_1 = \text{Number of males}]$$

$$\Sigma x_1 = 480 n_1$$

and $\quad \Sigma x_2 = 360 n_2 \qquad [n_2 = \text{Number of females}]$

$$n = n_1 + n_2$$

$$\Sigma x = \Sigma x_1 + \Sigma x_2$$

$$444n = 480n_1 + 360n - 360n_1$$

$$84n = 120n_1$$

$$7n = 10n_1$$

Male percentage $= \frac{n_1}{n} \times 100 = \frac{n_1 \times 7}{10 \times n_1} \times 100 = 70\%$

Female percentage = 30 %.

Exercise-9.3

1. Compute the arithmetic mean of first n natural numbers whose weights are equal to the corresponding number $\frac{1}{3}(2n+1)$.

2. Compute the mean marks of a student from the following table :

Marks	No. of students
Above 0	80
Above 10	77
Above 20	72
Above 30	65
Above 40	55
Above 50	43
Above 60	28
Above 70	16
Above 80	10
Above 90	8
Above 100	0

3. The rainfall of a certain town in centimeters for the first six months of the year are 102, 103, 95, 98, 100, 105. Compute the average rainfall of the town.

4. Compute the arithmetic average in rupees from the data given below :

Salary	100	150	200	250	300	500
No. of labours	30	20	15	10	4	1

5. Find the missing frequency from the following data, it is being given that 19.92 is the average number of the given data :

Tables	4-8	8-12	12-16	16-20	20-24
No. of persons cured	11	13	16	14	?
Tables	24-28	28-32	32-36	36-40	
No. of persons cured	9	17	6	4	

6. Calculate the mean marks of a student from the given data :

Marks	No. of students
Below 10	15
Below 20	35
Below 30	60
Below 40	84
Below 50	96
Below 60	127
Below 70	128
Below 80	250

7. Find the combined average daily wages for the workers of two factories :

No. of workers	250	200
Average	2.00	2.50

8. If the arithmetic average of data given below be 165 rupees, compute the missing term :

Monthly salary	100	150	200	-	300	500
No. of labourers	30	20	15	10	4	1

9. Compute the weighted arithmetic average wage rate of 31 building trade workers from the following table :

Kind of worker	Daily wages (Rs)	Frequency
Masons	15	4
Labourers	8	20
Carpenters	12	5
Painters	10	2

10. Compute the arithmetic average of the marks obtained by 9 students in a test :
75, 43, 52, 65, 48, 35, 40, 70, 40.

11. Compute the missing frequency term from the following data whose arithmetic average is given by 35.64 :

Class	20-25	25-30	30-35	35-40	40-45	45-50
Frequency	18	44	102	—	57	19

12. Compute the arithmetic average of the following data :

0-5	5-8	8-10	10-12	12-15	15-17	17-20	20-25	25-30
2	5	7	5	6	4	4	9	6

13. The arithmetic average of a group of 40 items is 100 and that of another group of 50 items is 70. Find the mean of the combined group of size 90.

14. Compute the arithmetic mean for the following data :

Class	0-5	5-10	10-15	15-20	20-25
Frequency	4	16	2	15	2

15. If the arithmetic average of the following frequency distribution is 7.85. Calculate missing frequency term.

Salary	5	6	7	10	12	15
Labourers	10	—	13	8	5	4

16. The average salary of 500 workers in a factory running in two shifts of 360 and 140 workers respectively is Rs. 70. The average salary of 360 workers working in day shift is Rs. 75. Find the average salary of 140 workers working in the night shift.

17. Find the mean of the following

Height (cm)	65	66	67	68	
Plants	1	4	5	7	
Height (cm)	69	70	71	72	73
Plants	11	10	6	4	2

18. Find the mean of the following distribution :

Class	Frequency
0-7	19
7-14	25
14-21	36
21-28	72
28-35	51
35-42	43
42-49	28

x	f
0	q^n
1	$^nC_1 q^{n-1} p$
2	$^nC_2 q^{n-2} p^2$
3	$^nC_3 q^{n-3} p^3$
...	...
n	p^n

19. If $p + q = 1$, compute the mean of the following :

Answers

2. 51.75	**5.** 250	**6.** 50.4	**7.** 2.22	**8.** 250	
9. 9.68	**11.** 160	**12.** 15.417	**13.** 83.33	**15.** 15.05	**19.** np

9.2.7 COMBINED MEAN

If $\bar{x}_1$ and $\bar{x}_2$ are the mean of two groups of sizes n_1 and n_2, then the mean $\bar{x}$ is the mean of two groups, given by $\bar{x} = \dfrac{n_1 \bar{x}_1 + n_2 \bar{x}_2}{n_1 + n_2}$

Proof. Let $x_1, x_2, ..., x_n$ be the variates of a group of size n_1 and $y_1, y_2, ... y_n$ be the variates of a group of size n_2. Then

$$\bar{x}_1 = \frac{x_1 + x_2 + x_3 + ... + x_n}{n_1} \Rightarrow n_1 \bar{x}_1 = x_1 + x_2 + x_3 + ... + x_n \qquad ...(1)$$

and

$$\bar{x}_2 = \frac{y_1 + y_2 + y_3 + ... + y_n}{n_2} \Rightarrow n_2 \bar{x}_2 = y_1 + y_2 + y_3 + ... + y_n \qquad ...(2)$$

Let $\bar{x}$ be the mean of these two groups, then

$$\bar{x} = \frac{(x_1 + x_2 + x_3 + ... + x_n) + (y_1 + y_2 + y_3 + ... + y_n)}{n_1 + n_2}$$

$$\Rightarrow \qquad \bar{x} = \frac{n_1 \bar{x}_1 + n_2 \bar{x}_2}{n_1 + n_2} \qquad \text{[Using (1) and (2)]}$$

Solved Examples

EXAMPLE 1. *The mean of the marks secured by 25 students of section A of class B.Tech is 47, that of 35 students of section B is 51, and that of 30 students of section C is 53. Find the mean of marks secured by 90 students of class B.Tech.*

SOLUTION. Let n_1, n_2, n_3 be the numbers of students respectively in section A, B and C and $\bar{x}_1, \bar{x}_2$ and $\bar{x}_3$ be the mean of marks secured by them.

$\therefore \quad n_1 = 25, \ n_2 = 35, \ n_3 = 30$
and $\bar{x}_1 = 47, \ \bar{x}_2 = 51$ and $\bar{x}_3 = 53$

$\therefore$ Combined mean $\bar{x} = \dfrac{n_1 \bar{x}_1 + n_2 \bar{x}_2 + n_3 \bar{x}_3}{n_1 + n_2 + n_3}$

$$= \frac{24 \times 47 + 35 \times 51 + 30 \times 53}{25 + 35 + 30} = \frac{4550}{90} = 50.56$$

EXAMPLE 2. *The average score of boys in an examination of a school is 71 and that of girls is 73. The average score of school in that examination is 71.8. Find the ratio of number of boys to the number of girls appeared in the examination.*

SOLUTION. Let there be n_1 boys and n_2 girls in the school.

Here, $\bar{x}_1 = 71, \ \bar{x}_2 = 73$ and $\bar{x} = 71.8$

$\therefore$ Combined mean $\bar{x} = \dfrac{n_1 \bar{x}_1 + n_2 \bar{x}_2}{n_1 + n_2}$

$\Rightarrow \quad 71.8 = \dfrac{n_1 \times 71 + n_2 \times 73}{n_1 + n_2}$

$\Rightarrow \quad 71.8(n_1 + n_2) = 71 n_1 + 73 n_2$

$\Rightarrow \quad 71.8 n_1 + 71.8 n_2 = 71 n_1 + 73 n_2$

$\Rightarrow \quad 0.8 n_1 = 1.2 n_2$

$\Rightarrow \quad 8 n_1 = 12 n_2 \quad \Rightarrow \quad \dfrac{n_1}{n_2} = \dfrac{12}{8} = \dfrac{3}{2}$

Hence, $n_1 : n_2 = 3 : 2$.

Exercise-9.4

1. The mean wage of 150 workers of the first shift in a factory is Rs. 400. The mean wage of 75 workers of the second shift is Rs. 600. Find the combined mean wage of the workers of the factory.

2. There are 50 students in a class out of which 20 are girls. The average weight of 20 girls is 45 Kg and that of 30 boys is 52 Kg. Find the mean weight in Kg of the entire class.

3. The average marks obtained by 30 students of group I is 60 and average marks of 40 students of group II is 55 and that of 30 students of group III is 70. Find the combined average of students of all the three groups.

4. There are 100 students in a class. The mean height of the class is 150 cm. If the mean height of 60 boys is 170 cm. Find

the mean height of the girls.

5. The mean weight of 150 students in a class is 60. The mean weight of boys is 70 Kg and that of girls is 55 Kg. Find the number of boys and girls in the class.

 **Answers**

1. 466.67	2. 49.2 Kg	3. 61	4. 120 cm	5. 50, 100

9.2.8 GEOMETRIC MEAN

Let $x_1, x_2, x_3, ..., x_n$ be the n variates of a variable x, then the geometric mean G of n variables is defined by $G = (x_1. x_2.x_3......x_n)^{1/n}$

If $f_1, f_2, f_3, ..., f_n$ be the frequency of these variables and $N = f_1 + f_2 + f_3 + ... + f_n$

Then.

$$G = (x_1^{f_1}.x_2^{f_2}.x_3^{f_3}......x_n^{f_n})^{1/N}$$

$$\log G = \frac{1}{N}[f_1 \log x_1 + f_2 \log x_2 + f_3 \log x_3 + ... + f_n \log x_n]$$

$$\log G = \frac{1}{N}\left[\sum_{i=1}^{n} f_i \log x_i\right]$$

Thus, we can say that the logarithm of the geometric mean can be calculated by taking weighted mean of the logarithm of the variables x_i.

Solved Examples

EXAMPLE 1. *Calculate the geometric mean of the given data:*
8, 15, 36, 40, 45, 70, 75, 85, 250, 500

SOLUTION.

x	$\log x$
8	0.9031
15	1.1761
36	1.5563
40	1.6021
45	1.6532
70	1.8451
75	1.8751
85	1.9294
250	2.3979
500	2.6990
	17.6373

Now, the geometric mean is given by

$$\log \text{G.M.} = \frac{1}{N} \Sigma \log x_i = \frac{1}{10} \times 17.6373$$

$$\log \text{G.M.} = 1.7637$$

$$\text{G.M.} = \text{Antilog } 1.7637$$

$$\text{G.M.} = 58.03$$

EXAMPLE 2. *Compute the geometric mean from the following data :*

x	11	12	13	14	15
f	3	7	8	5	2

SOLUTION.

x	Frequency (f)	$\log x$	$f \log x$
11	3	log 11 = 1.0414	3.1242
12	7	log 12 = 1.0792	7.5544
13	8	log 13 = 1.1139	8.9112
14	5	log 14 = 1.1461	5.7305
15	2	log 15 = 1.1761	2.3522
	25		27.6725

The geometric mean is given by

$$\log \text{G.M.} = \frac{1}{N} \Sigma f \log x_i = \frac{1}{25} \times 27.6725 = 1.1069$$

Then,   G.M. = Antilog 1.1069= 12.79

Hence,   G.M. = 12.79

EXAMPLE 3. *Find the geometric mean of the given data :*

Marks	1-10	10-20	20-30	30-40	40-50
f	8	12	20	6	4

SOLUTION.

Marks	Mild value x	Frequency (f)	$\log x$	$f \log x$
0-10	5	8	log 5 = 0.6990	5.592
10-20	15	12	log 15 = 1.1761	14.1132
20-30	25	20	log 25 = 1.3979	27.958
30-40	35	6	log 35 = 1.5441	9.2646
40-50	45	4	log 45 = 1.6532	6.6128
		50	17.6373	63.5406

Geometric mean can be calculated by

$$\log \text{G.M.} = \frac{1}{N} \Sigma f \log x_i = \frac{1}{50} \times 63.5406 = 1.2708$$

Then,   G.M. = Antilog 1.2708 = 18.65

So,   G.M = 18.65

EXAMPLE 4. *Calculate the weighted geometric mean of the data given below :*

Articles	Price	Weight
A	125	40
B	150	25
C	100	5
D	122	20
E	75	10

SOLUTION.

Articles	Price x	Weight (w)	$\log x$	$w \log x$
A	125	40	$\log 125 = 2.0969$	83.876
B	150	25	$\log 150 = 2.1761$	54.4025
C	100	5	$\log 100 = 2$	10
D	122	20	$\log 122 = 2.0864$	41.728
E	75	10	$\log 75 = 1.8751$	18.751
		100		208.7575

Weighted geometric mean can be calculated as

$$\log \text{(weighted G.M.)} = \frac{\Sigma w \log x}{\Sigma w} = \frac{1}{100} \times 208.7575$$

$$= 2.0875$$

Weighted G.M. = Antilog 2.0875 = 122.3

So, Weighted G.M. = 122.3

9.2.9 PROPERTIES OF GEOMETRIC MEAN

1. If we put the value of geometric mean in place of the each value of a series, then the product of the value of the series will be unchanged.

2. If G_1 is the geometric mean of the series $x_1, x_2, x_3..., x_n$, G_2 is the geometric mean of the series $y_1, y_2, y_3..., y_n$ and G is the geometric mean of the series obtained by the ratios of corresponding observations. Then G will be equal to G_1 / G_2, *i.e.*, $G = \dfrac{G_1}{G_2}$.

3. Let us consider n series with frequencies $N_1, N_2, N_3, ..., N_n$ respectively and geometric means $G_1, G_2, G_3, ..., G_n$ respectively. Then the combined geometric mean of n series with frequency $N_1 + N_2 + N_3 + ... + N_n$ is given by $G = (G_1^{N_1} G_2^{N_2} G_3^{N_3} ...... G_n^{N_n})^{1/N}$

4. Let us consider n sets of observations whose geometric means are respectively $G_1, G_2, ..., G_n$. Now, if G is the geometric mean of the product of these n sets, then the product of the geometric means of these series will be equal to the value of G.

$$G = G_1 . G_2 . G_3 ... G_n$$

$$\log G = \sum_{i=1}^{n} \log G_i$$

5. Let us consider a series $x_1, x_2, ..., x_p, x_{p+1}...x_n$ whose geometric mean is given by G. In which G is greater than from the each value $x_1, x_2, ..., x_p$ and less than from each of the values $x_{p+1}, x_{p+2}...x_n$, then $G^n = (x_1 x_2...x_p . x_{p+1}...x_n)$

$$\Rightarrow \qquad G^p . G^{n-p} = (x_1 . x_2 ... x_p)(x_{p+1}...x_n)$$

$$\Rightarrow \qquad \frac{G}{x_1} . \frac{G}{x_2} ... \frac{G}{x_p} = \frac{x_{p+1}}{G} . \frac{x_{p+2}}{G} ... \frac{x_n}{G}$$

Exercise-9.5

1. Find the geometric mean of the following data :
   50, 100, 1920, 143740, 204980, 1206740, 154910

2. Find the geometric mean of the series :
   1, 2, $2^2, 2^3, 2^4, ...2^n$

3. The price of certain article rises 5% in first year, 8% in second year and 77% in third year. What is the average change per year?

4. The geometric mean of 10 data are calculated as 16.2. It was later found that one of the data was wrongly read as 12.9, in fact it was 21.9. Calculate the correct geometric mean.

5. Calculate the geometric mean from the following frequency distribution:

Marks obtained	11	12	13	14	15
Frequency	3	7	8	5	2

6. Find the geometric mean of 2, 6, 18, 54, 162.

7. Find the geometric mean of the following series:

Class	0-10	10-20	20-30	30-40	40-50
Frequency	10	15	12	8	5

8. Find the geometric mean from the following table :

Marks obtained	5	7	9	11	13	15
No. of studnets	1	2	3	5	11	9

9. Calculate the average rate of increment in population which is increased by 20% in first year, 25% in second year and 44% in third year.

10. Find out the geometric mean of the following distribution :

Marks obtained	0-10	10-20	20-30	30-40
No. of students	5	8	3	4

Answers

1. 12700	2. $2^{n/2}$	3. 26%	4. 17.08	5. 12.79
6. 18	7. 19.10	8. 11.86	9. 28.02%	10. 14.64

9.2.10 HARMONIC MEAN

The harmonic mean of any series is given by the reciprocal of the arithmetic mean of the reciprocals of the variables.

For different type of series, it can be calculated by different method

(i) For individual series: Let $x_1, x_2, x_3, ..., x_n$ be the n variables, then the harmonic mean of these variables is given by

$$H = \frac{n}{\frac{1}{x_1} + \frac{1}{x_2} + ... + \frac{1}{x_n}}.$$

(ii) For discrete series: Let $x_1, x_2, x_3, ..., x_n$ be the n variables and $f_1, f_2, ..., f_n$ be the frequency of them. Then the harmonic mean is given by

$$\frac{1}{H} = \frac{1}{N} \sum_i^n \left(\frac{f}{x} \right),$$

where

$$N = \Sigma f$$

$$H = N \sum_i^n \left(\frac{x}{f} \right)$$

(iii) For grouped series : When the grouped series are given, we take the mid value of each group and named them as $x_1, x_2, x_3, ..., x_n$ and if the frequencies of these groups are $f_1, f_2, f_3, ..., f_n$, then the harmonic mean can be calculated by

$$H = \frac{N}{\Sigma(f / x)}.$$

Solved Examples

EXAMPLE 1. *Find the harmonic mean of the following data :*
12, 8, 6, 24

SOLUTION. Harmonic mean for individual series is given by

$$H = \frac{n}{\frac{1}{x_1} + \frac{1}{x_2} + ... + \frac{1}{x_n}} = \frac{4}{\frac{1}{12} + \frac{1}{8} + \frac{1}{6} + \frac{1}{24}}$$

$$= \frac{4}{0.0833 + 0.1250 + 0.1666 + 0.0416}$$

$$H = \frac{4}{0.4165} = 9.6038.$$

EXAMPLE 2. *Find the harmonic mean of the following frequency distribution :*

Class	0-10	10-20	20-30	30-40	40-50
Frequency	4	5	11	6	4

SOLUTION.

Class	Mid Value x	Frequency	$1/x$	f/x
0-10	5	4	0.2	0.800
10-20	15	5	0.0666	0.333
20-30	25	11	0.04	0.440
30-40	35	6	0.0285	0.171
40-50	45	4	0.0222	0.088
		30		1.832

Now, harmonic mean for grouped series is given by

$$\frac{N}{\Sigma(f / x)} = \frac{30}{1.832} = 1.63755.$$

EXAMPLE 3. *Find the harmonic mean of the following*

frequency distribution :

Class	11	12	13	14	15
Frequency	3	7	8	5	2

SOLUTION. We have

Marks (x)	Frequency (f)	$1/x$	f/x
11	3	0.0909	0.2727
12	7	0.0833	0.5831
13	8	0.0769	0.6152
14	5	0.0714	0.357
15	2	0.0666	0.1332
	25		1.9612

Now, harmonic mean for grouped series is given by

$$\text{H.M.} = \frac{N}{\Sigma(f / x)} = \frac{25}{1.9612} = 12.7472.$$

EXAMPLE 4. *A man drives a car for three days by covering a distance of 360 km per day. First day he drives for a time of 10 hours and drive with the speed of 36 km/h. On the second day, he drives 15 hours at a speed of 24 km/h and on the third day, he drives for 12 hours at a speed of 30 km/h. Calculate the average speed of the car.*

SOLUTION. It is given that he covers a constant distance of 360 km per day.

His speed on the first day is given by = 36 km/h

His speed on the second day is given by = 24 km/h

His speed on the third day is given by = 30 km/h

Since the distance is given to be constant so the average speed can be calculated by taking harmonic mean of the speeds.

So, average speed $= \dfrac{n}{\dfrac{1}{v_1}+\dfrac{1}{v_2}+\dfrac{1}{v_3}} = \dfrac{3}{\dfrac{1}{36}+\dfrac{1}{24}+\dfrac{1}{30}}$

$= \dfrac{3}{0.0277+0.0416+0.0333}$

$= \dfrac{3}{0.1026} = 29.2397$ km/h

☞ REMARK

- Where the distance in each part of the journey is given to be constant, then average speed will be calculated by harmonic mean. In the case when time being constant, the average is given by arithmetic mean.

9.2.11 PROPERTIES OF HARMONIC MEAN

1. If x_1 and x_2 are any two observations, then A.H. $= G^2$
   where, A = Arithmetic mean; H = Harmonic mean; G = Geometric mean

2. If $x_1, x_2, x_3, ..., x_n$ be the n positive observations then $A \geq G \geq H$.
   The sign of equality will hold if the values of all observations under consideration are same.

3. A variate takes values $a, ar, ar^2, ..., ar^{n-1}$ each with the frequencies one. Then

   $$AM = a(1 - r^n)/n(1 - r)$$
   $$GM = ar^{(n-1)/2}$$
   $$HM = a.n(1 - r)r^{n-1}/(1 - r)^n$$

Exercise-9.6

1. Find the harmonic mean of the following data : 5, 10, 15, 20, 25, 30, 35

2. Find the harmonic mean of the following data:
   0.00002853, 0.0003425, 0.004656, 0.07834, 0.676, 9.45, 78.3, 800

3. Calculate the A.M., G.M. and H.M. of the observations and show that : A.M. > G.M. > H.M.
   37, 32, 36, 35, 43, 39, 41

4. Calculate the geometric mean of the following data:

Marks obtained	5	6	7	8	9
Frequency	3	4	8	7	2

5. Calculate the harmonic mean of the following frequency distribution:

Class	40-50	50-60	60-70	70-80	80-90	90-100
Frequency	19	25	36	72	51	43

6. Calculate the harmonic mean of the following frequency distribution:

Class	0-4	4-8	8-12	12-16	16-20
Frequency	4	12	20	9	5

7. Compute the harmonic mean of the following frequency distribution:

Class	40-50	50-60	60-70	70-80	80-90	90-100
Frequency	12	10	15	17	8	3

8. A car runs at the rate of 15 km/h during the first 30 km, at 20 km/h during the second 30 km and at the rate of 25 km/h during the third 30 km. Find out the average speed of the car.

9. A train starts from rest and travel a distance of 1 km in four parts each of 0.25 km with average speed 12, 16, 24 and 48 km/h. Explain the statement that the average speed over the whole journey of 1 km is 19.2 km/h and not 25 km/h.

10. A variate takes values $1, r, r^2, ..., r^{n-1}$ each with frequency unity. Show that

   $$A = \dfrac{1 - r^n}{n(1 - r)}; \quad G = r^{(n-1)/2}; \quad H = \dfrac{n(1 - r)\, r^{n-1}}{1 - r^n}$$

   From the above observations, also show that $AH = G^2$ and $A > G > H$.

11. Find out the average speed of a car running at the rate of 20 km/h during the first 30 km; at 25 km/h during the second 30 km and at 30 km/h during the third 30 km.

12. Calculate the average speed of a train running at the rate of 20 km/h during the first 100 km, at 25 km/h during the second 100 km and at 30 km/h during the third 100 km.

Answers

1. 13.5030	2. 0.0002095	4. 6.84	5. 82.5669	6. 7.246	7. 32.049
8. 1	9. 15	11. 24.32 km/h		12. 24.39 km/h	

9.2.12 MEDIAN

If we arrange the whole data in ascending or descending order, then the value of the middle variable is known as median.

In case when the number of variables are odd, then the middle value is known as median.

If the number of variables are even, *i.e.*, $(2n)$, the value of the mean of nth, $(n+1)^{th}$ variables will be median.

According to **Connor**, "The median is that value of the variable which divides the group into two equal parts, one part comprising all values greater and the other all values less than the median."

COMPUTATION OF MEDIAN

1. **Formula for individual series :** When the data given are ungrouped, then firstly, we arrange them in ascending or descending order. Then, if number of data are odd number, then the value of the middle variable will be median.

   If number of data are even number ($2n$), then the value of the mean of the nth and $(n+1)^{th}$ variable will give the median.

2. **Formula for discrete series :** Let us assume that $x_1, x_2, ..., x_n$ are the n observations whose frequencies are given by $f_1, f_2, ..., f_n$. To calculate the median of such series first of all we calculate the cumulative frequency and then calculate the sum of the frequency. Now, we calculate the median of series according to the N(sum of the frequency) is odd or even.

3. **Formula for Continuous Series:** In these type of questions all the data are divided into particular classes and their respective frequencies are given. Firstly, we calculate the cumulative frequencies. Then, we calculate the sum of the frequencies (N). According to N is even or odd, we find out the median. The class which contain this median is known as median class.

   Now, the median for this series can be calculated by the formula

$$\text{Median} = l + \frac{\frac{1}{2}N - F}{f} \times i$$

where, l = Lower limit of the median class

             N = Sum of all the frequencies

             F = Sum of all the frequencies preceding the median class

             f = Frequency of median class

and i = Width of the median class

Solved Examples

EXAMPLE 1. *Find the median of the following data: 20, 18, 22, 27, 25, 12, 15.*

SOLUTION. Firstly, we will arrange these data in ascending orders

12, 15, 18, 20, 22, 25, 27

Here, $n = 7$, which is odd.

So, Median = Value of the $\frac{(n+1)}{2}$ th term

= Value of $\frac{(7+1)}{2}$ th term = Value of 4th term.

Hence, Median = 20.

EXAMPLE 2. *Find the median in the following frequency distribution :*

Size	3	5	7	9	11	13	15
Frequency	7	3	12	28	10	9	6

SOLUTION.

(x)	(f)	Cumulative frequency
3	7	7
5	3	10
7	12	22
9	28	50
11	10	60
13	9	69
15	6	75
	$N = 75$	

Here, $N = 75$, which is odd.

So, the median = Value of the $\frac{(75+1)}{2}$ th term

= Value of the 38th term.

In the table, we see that the cumulative frequency 50 contain the 38th term. So the value of x for this column will be the median. Median = 9.

EXAMPLE 3. *Compute the median of the following frequency distribution :*

Wages	0-10	10-20	20-30	30-40	40-50
No. of workers	22	38	46	35	20

SOLUTION.

Wages (x)	Workers(f)	Cumulative frequency
0-10	22	22
10-20	38	60
20-30	46	106
30-40	35	141
40-50	20	161
	$N = 161$	

Here, $N = 161$, which is odd.

So, median is given by the value of $\frac{(N+1)}{2}$ th term.

$\Rightarrow$ $\frac{161+1}{2} = \frac{162}{2} = 81$ th term.

The 81th term is contained in the interval 20-30. So the class 20-30 will be the median class.

Lower limit of median class, $l = 20$

Sum of all the frequencies $N = 161$

Sum of all the frequencies preceding the median class, *i.e.,*

         $F = 60$

Frequency of median class, $f = 46$

width of the median class $i = 10$

Then, median is given by

$$\text{Median} = l + \frac{\frac{1}{2}N - F}{f} \times i = 20 + \frac{\left(\frac{1}{2} \times 161 - 60\right)}{46} \times 10$$

$$= 20 + \frac{(80.5 - 60)}{46} \times 10$$

$$= 20 + \frac{205}{46} = 20 + 4.4565 = 24.4565$$

9.2.13 QUARTILES

If we arrange all the variates into ascending or descending order and this series is divided into four equal parts, then each part of this series is known as a quartile. The first part or first quartile contain first quarter of the series. The second quartile is said to be median and the third quartile is the $\frac{3}{4}$ th term, which is known as upper quartile.

If Q_1, Q_2 and Q_3 are known as first, second and third quartile, then it can be calculated as

$$Q_k = l + \frac{\frac{1}{4}(kN) - F}{f} \times i, \quad k = 1, 2, 3$$

where,
$l =$ Lower limit of the quartile class.
$N =$ Total frequency.
$F =$ Cumulative frequency preceding the quartile class.
$f =$ Frequency of quartile class.
and
$i =$ Class interval of quartile class.

INTERQUARTILE AND SEMI-QUARTILE

If Q_1, Q_2 and Q_3 are the first, second and third quartile, then $Q_3 - Q_1$ is called the interquartile range and $\frac{1}{2}(Q_3 - Q_1)$ is called the semi-interquartile range.

9.2.14 QUANTILES

If we arrange all the variates into ascending or descending order and divide this series into five equal parts, then each part of this series is known as quantiles.

If $Q_{n_1}, Q_{n_2}, Q_{n_3}$ and Q_{n_4} are the first, second, third and fourth quantile, then it can be calculated as

$$Q_{n_k} = l + \frac{\frac{1}{5}(kN) - F}{f} \times i, \quad k = 1, 2, 3, 4$$

where
$l =$ lower limit of the quantile class.
$N =$ Total frequency.
$F =$ Cumulative frequency preceding the quantile class.
and
$i =$ Class interval of quantile class.

9.2.15 DECILE AND PERCENTILE

In an ascending or descending series, if we divide the complete series into ten equal parts, then each part of the series is known as decile and if we divide the whole series into 100 equal parts, then each part of the series is known as percentile or centile. If $D_1, D_2, D_3, ..., D_9$ are the 9 deciles and $P_1, P_2, ..., P_{99}$ are 99 percentiles, then it can be calculated as

$$D_k = l + \frac{\frac{1}{10}kN - F}{f} \times i, \text{ where } k = 1, 2, ..., 9$$

and

$$P_k = l + \frac{\frac{1}{100}kN - F}{f} \times i, \text{ where } = 1, 2, ..., 99$$

where
$l =$ Lower limit of decile or percentile class.
$N =$ Total frequency.
$F =$ Frequency preceding the decile or percentile class.
$f =$ Frequency of decile or percentile class.
and
$i =$ Class width.

Solved Examples

EXAMPLE 1. *Calculate the median, lower quartile and upper quartile of the data given below :*

15, 37, 53, 18, 40, 54, 55, 40, 20, 36, 52, 53, 33, 75, 49, 49, 33, 35, 27, 44, 64, 61, 41, 26, 48, 70, 26, 29, 40, 59

SOLUTION. Firstly, we arrange all the data in increasing order :
15, 18, 20, 26, 27, 29, 33, 33, 35, 36, 37, 40, 40, 40, 41,

44, 48, 49, 49, 52, 53, 53, 54, 55, 59, 61, 64, 70, 75
Here, $n = 30$, which is even.

So, the mean of the values of $\frac{n}{2}$ th term and $\left(\frac{n}{2} + 1\right)$ th term

$$\frac{n}{2} = 15, \quad \frac{n}{2} + 1 = 16$$

$$\text{Median} = \frac{\text{Value of 15th term} + \text{Value of 16th term}}{2}$$

$$= \frac{40+41}{2} = 40.5$$

Now, we have to calculate the lower quartile and upper quartile.

Here, $N = 30$

$$\frac{1}{4}N = \frac{30}{4} = 7.5, \qquad \frac{3}{4}N = \frac{3}{4} \times 30 = 22.5$$

Thus, we see that the lower quartile lies between 7th and 8th term.

$$7^{th} \text{ term} = 29, \quad 8^{th} \text{ term} = 33$$

So, the lower quartile,

$$Q_1 = 29 + \frac{1}{2}(33-29) = 29 + 2$$

$$Q_1 = 31$$

Now, $\frac{3}{4}N = 22.5$. So, the upper quartile will lie between 22nd and 23rd term.

22nd term $= 53, \quad 23^{rd}$ term $= 53$

Hence, the upper quartile will lie between 53 and 53.

So, it will be Q_3 term $= 53$.

EXAMPLE 2. *Calculate the median, lower quartile and upper quartile of the data given below:*

Classes	0-4	4-6	6-8	8-12	12-18	18-20
Frequency	4	6	8	12	7	2

SOLUTION. In this question, the classes are not equally divided. So, firstly, we tabulate them equally.

Classes	Frequency (f)	Cumulative frequency
0-4	4	4
4-8	14	18
8-12	12	30
12-16	5	35
16-20	4	39
	$N = 39$	

Here, $N = 39$, which is odd.

The median number is given by $= \frac{(N+1)}{2} = 20$

Median number 20 lies in the class 8-12. So, 8-12 will be the median class

Now, lower limit of median class, $l = 8$

Frequency of median class, $f = 12$

Frequency preceding the median class, $F = 18$

Width of class interval, $i = 4$.

Then, median $= l + \dfrac{\left(\frac{1}{2}N - F\right) \times i}{f}$

$$= 8 + \frac{\left(\frac{1}{2} \times 39 - 18\right)}{12} \times 4 = 8 + \frac{(19.5 - 18)}{12} \times 4$$

$$\text{Median} = 8.5$$

Now, we will calculate the value of lower quartile and upper quartile.

Firstly, we will calculate the lower quartile number.

Here, $N = 39$.

Lower quartile number $= \frac{1}{4}(39+1) = 10$

The lower quartile number 10 lies in the class 4-8 so it will be lower quartile class.

So, $l = 4, f = 14, F = 4, i = 4$

So,

$$Q_1 = l + \frac{\frac{1}{4}N - F}{f} \times i = 4 + \frac{\left(\frac{1}{4} \times 39 - 4\right) \times 4}{14} = 4 + 1.6428$$

So, $Q_1 = 5.6428$

Now, upper quartile number will be

$$= \frac{3}{4}(N+1) = \frac{3}{4} \times 40 = 30$$

The quartile number 30 lies in the class 8-12. So, it will be the upper quartile class.

For this, $l = 8, f = 12, F = 18, i = 4$

So, $Q_3 = l + \dfrac{\left(\frac{3}{4}N - F\right) \times i}{f} = 8 + \dfrac{\left(\frac{3}{4} \times 39 - 18\right) \times 4}{12}$

$$= 8 + 3.75$$

Hence, $Q_3 = 11.75$

Exercise-9.7

1. Compute the median of the data : 9, 10, 15, 7, 11, 9, 8, 11, 7, 9, 10

2. The marks obtained by the ten students of class 8th is as follows:
   75, 80, 96, 92, 89, 94, 100, 82, 63, 105
   Find the median.

3. In a factory the daily wages of labourers are given by the following frequency distribution. Find the median:

Wages (Rs.)	6	8	10	12	14
No. of labourers	6	3	4	5	2

4. Compute the median for the following frequency distribution:

Age	5-7	8-10	11-13	14-16	17-19
No. of students	7	12	19	10	2

5. Compute the median for the following frequency distribution:

Variable	45-50	50-55	55-60	60-65	65-70
Frequency	2	3	5	7	9

Variable	70-75	75-80	80-85	85-90	90-95
Frequency	11	7	2	3	1

6. Calculate the median for the following frequency distribution:

Variable	0-5	5-10	10-15	15-20	20-25
Frequency	4	16	2	15	2

7. Calculate the median for the following frequency distribution:

Variable	0-10	10-20	20-30	30-40	40-50	50-60	60-70	70-80
Frequency	2	18	30	45	35	20	6	3

8. Compute the median and quartiles of the following frequency distribution :

Age	15-19	20-24	25-29	30-34	35-39	40-44
Frequency	4	20	38	24	10	4

9. Compute the median and quartiles of the following frequency distribution :

Class	0-4	4-6	6-8	8-12	12-18	18-20
Frequency	4	6	8	12	7	2

10. Calculate quartiles, 3rd quantile, 3rd decile and 60th percentile from the following data :

Classes	0-10	10-20	20-30	30-40	40-50	50-60	60-70	70-80
Frequency	5	8	7	12	28	20	10	10

  Answers

1. 9   **2.** 90.5   **3.** Rs. 10   **4.** 11.447   **5.** 7.583   **6.** 36.559
7. 34.45   **8.** Median = 28.49; $Q_1 = 21.64$, $Q_2 = 32.89$ **9.** Median = 8.5; $Q_1 = 5.93$, , $Q_3 = 11.75$
10. $Q_1 = 37.5$, , $Q_2 = 46.43$, $Q_3 = 55.4$, $D_3 = 38.3$, $P_{60} = 50$ Quantile = 50

9.2.16 MODE

The variable whose frequency is maximum, is known as the mode of the distribution. In other words, we can say that the value which occurs most frequently in a distribution is known as the mode of the distribution.

Computation of mode :

1. **For individual series :** For individual series, we can find the mode by inspection only. If number of variables are very large, then we arrange the data into discrete series and then we check the frequency for each variable to know the mode of the series. If frequency for different variables are same in the frequency table, then we use the method of grouping to calculate the mode of the distribution.

2. **For discrete series:** Firstly, we arrange all the data in the frequency table. If the maximum frequency has the unique value, then it will be the mode of the series and if the maximum frequency occurs more than once, then mode can be calculated by grouping of data.

3. **For continuous series :** The class with maximum frequency is known as the modal class and we can obtain the mode of this series by calculating the formula :

$$\text{Mode} = l + \frac{f - f_{-1}}{2f - f_{-1} - f_1} \times i$$

[When $f > f_{-1}$]

     where, l = Lower limit of the modal class.

             f = Frequency of the modal class.

             f_{-1} = Frequency preceding the modal class.

             f_1 = Frequency succeeding the modal class.

             i = Class width.

If f_{-1} and f_1 are both (or one) is greater than f, then we use the following formula

$$M_0 = l + \frac{f_1 \times i}{f_1 + f_{-1}}.$$

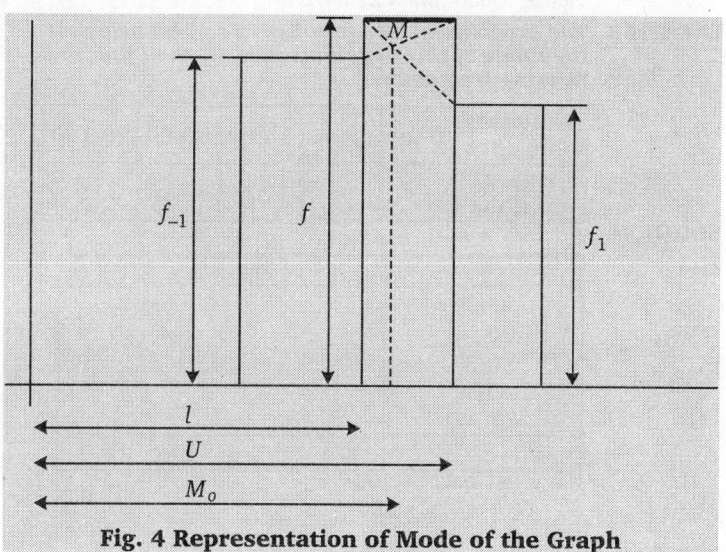

Fig. 4 Representation of Mode of the Graph

EMPIRICAL FORMULA

The empirical relationship between mean, mode and median is given by: Mode = 3 Median – 2 Mean

  Solved Examples

EXAMPLE 1. *Find the mode of the data given below:*
       0, 1, 6, 7, 2, 3, 7, 6, 6, 2, 6, 0, 5, 6, 0

SOLUTION. In the given data, we see that the frequency of 6 is 5, which is maximum and no other frequency is equal to this frequency. So, 6 is the mode of the given data.

EXAMPLE 2. *Find the mode of the following frequency distribution :*

Mid Value	15	20	25	30	35	40	45	50	55
Frequency	2	22	19	14	3	4	6	1	1

SOLUTION. Here, the mid value for each classes is given. So, firstly, we have to convert the given data into grouped frequency. So,

Mid Value	Class	Frequency
15	12.5 - 17.5	2
20	17.5 - 22.5	22
25	22.5 - 27.5	19
30	27.5 - 32.5	14
35	32.5 - 37.5	3
40	37.5 - 42.5	4
45	42.5 - 47.5	6
50	47.5 - 52.5	1
55	52.5 - 57.5	1

Here, by inspection, we see that the maximum frequency is given by = 22. So the class 17.5 – 22.5 will be the modal class.
For the modal class
$f = 22, l = 17.5, f = 22, f_{-1} = 2, f = 19, i = 5$

Then, mode $M_0 = l + \dfrac{f - f_{-1}}{2f - f_{-1} - f_1} \times i$

$= 17.5 + \dfrac{22 - 2}{2 \times 22 - 2 - 19} \times 5$

$= 17.5 + \dfrac{20}{44 - 21} \times 5$

$= 17.5 + \dfrac{20}{23} \times 5 = 17.5 + 4.3478$

Hence, Mode $M_0 = 21.8478$.

EXAMPLE 3. *The expenses of 100 families are given below. If the mode of this distribution is* 48, *then find the missing frequency.*

Expenses (Rs.)	0-20	20-40	40-60	60-80	80-100
No. of families	14	–	27	–	15

SOLUTION.

Expenses	Frequency
0-20	14
20-40	x
40-60	27
60-80	y
80-100	15
	N = 56+x+y

It is given that N = 100
So, 56 + x + y = 100
$x + y = 44$ ... (1)
Now, it is given that mode of this distribution = 48 which falls in the class 40-60. So for this
$l = 40, f = 27, f_{-1} = x, f_1, y, i = 20$
Then mode,

$M_0 = l + \dfrac{f - f_{-1}}{2f - f_{-1} - f_1} \times i$

$48 = 40 + \dfrac{27 - x}{54 - x - y} \times 20$

$8 = \dfrac{27 - x}{54 - x - y} \times 20$

$8(54 - x - y) = 20(27 - x)$

$2(54 - x - y) = 5(27 - x)$

$108 - 2x - 2y = 135 - 5x$

$3x - 2y = 27$ ... (2)

Solving equation (1) and (2)

$5x = 115$

$x = \dfrac{115}{5} = 23$

Put this value in equation (1)
$23 + y = 44, \; y = 21$
Hence, the missing frequency is given by = 23, 21.

EXAMPLE 4. *Calculate the mode of the following data:*

Wages (Rs.)	Below 100	100-200	200-300	300-400	400-500	500 and above
Frequency	8	12	25	15	10	5

SOLUTION. As the frequencies are regular and highest frequency 25 belongs to the class 200-300. Thus 200-300 is the modal class in which $l = 200, f = 25, f_{-1} = 12, f = 15, i = 100$
Using the formula

$\text{Mode} = l + \dfrac{f - f_{-1}}{2f - f_{-1} - f_1} \times i = 200 + \dfrac{25 - 12}{50 - 12 - 15} \times 100$

$= 200 + \dfrac{1300}{23} = 200 + 5.652 = 205.652.$

EXAMPLE 5. *Obtain mean, median and mode of the following classes:*

Class	0-10	10-20	20-30	30-40	40-50
Frequency	13	22	30	20	15

SOLUTION. Calculation of Mean

Class	Mid Value x	f	(x − 25) = d	f.d
0-10	5	13	− 20	−260
10-20	15	22	− 10	−220
20-30	25	30	0	0
30-40	35	20	10	200
40-50	45	15	20	300
		100		20

$a = 25$

Then, Mean $= a + \dfrac{\Sigma fd}{\Sigma f} = 25 + \dfrac{20}{100} = 25.2$

Calculation for Median Value

Class	Frequency f	Cummulative frequency
0-10	13	13
10-20	22	35
20-30	30←Mode class	65←Mode class
30-40	20	85
40-50	15	100

Here, the total frequencies = 100

Size of median = $\dfrac{100}{2} = 50^{th}$

Observation in the data set. This observation lies in the class (20-30).

Thus,

$$\text{Median} = 20 + \dfrac{\left(\dfrac{100}{2}\right) - 35}{30} \times 10$$
$$= 20 + 5 = 25$$

Calculation of Modal Value

In the table, the largest frequency corresponds to the class interval 20-30, therefore it is the mode class. Then, we have

$$l = 20,\ f = 30,\ f_{-1} = 22,\ f_1 = 20,\ h = 10$$

$$\text{Mode} = l + \dfrac{f - f_{-1}}{2f - f_{-1} - f_1} \times h$$

$$= 20 + \dfrac{30 - 22}{60 - 22 - 20} = 20 + \dfrac{80}{18}$$

$$= 20 + 4.44 = 24.44$$

 Exercise-9.8

1. Find the mode of the following data : 4, 5, 8, 6, 9, 8, 8, 6, 5, 11, 9, 8

2. Find the mode of the following data : 0, 1, 6, 7, 2, 3, 7, 6, 6, 2, 6, 0, 5, 6, 0

3. Find the mode of the following data :

Size	2	3	4	5	6	7	8	9	10	11	12	13
Frequency	3	8	10	12	16	14	10	8	17	5	4	1

4. Find the mode of the following data :

Monthly rent	20-40	40-60	60-80	80-100	100-120	120-140	140-160	160-180	180-200
No. of students	6	9	11	14	20	15	10	8	7

5. Find the mode of the following frequency distribution :

Height	120-124	125-129	130-134	135-139	140-144	145-149	150-154
Frequency	2	5	8	15	20	10	5

6. Find the mode of the following frequency distribution :

Marks	10-25	25-40	40-55	55-70	70-85	85-100
Frequency	6	20	44	26	7	1

7. Find the mode for the following frequency distribution :

Age	20-25	25-30	30-35	35-40	40-45	45-50	50-55	55-60
No. of persons	50	70	80	180	150	120	70	50

8. Compute the mode for the following frequency distribution :

Age below (in years)	5	10	15	20	25	30	35	40
No. of persons	2	4	14	27	48	64	72	75

9.2.17 DISPERSION

Sometimes we see that the mean, mode and median of different series give the same value, so it becomes necessary to differentiate these series. For this, we check the deviation of variables from the mean value in both the series. Then the maximum deviation from the mean value in both the series gives the different values. Thus we can differentiate these series and such type of variation is known as dispersion.

For Example :

The marks of six students in I and II tests are given as follows :

I test	43	46	49	50	52	54
II test	38	48	49	51	55	53

The mean in both the cases is given by 49. Now we check that the deviation from mean in first series is given by 49-43 = 6, while in II series, it is given by 49 – 38 = 11. Thus we can differentiate these series with the same value of mean. Such variations in the series is known as dispersions.

According to Griffin, *"A measure of variation or dispersion describes the degree of scatter shown by the observations and is usually measured as an average deviation about some central value or by an order statistics."*

According to Dr. Bowley, *"Dispersion is the measure of the variation of the items."*

MERITS OF A GOOD MEASURE OF DISPERSION

1. It should be easy to calculate.   2. It should be well defined.
3. All the data must be used in calculation.   4. It should be capable of further mathematical and statistical treatment.
5. It should not be affected by the fluctuation of sampling.

MEASURE OF DISPERSION

The dispersion can be calculated by the following two methods :

1. Method of limit.   2. Average deviation method.

In the method of limit, the following measure can be done

1. Range   2. Interquartile range   3. Percentile range

In average deviation method, the following measure can be done :

1. Mean deviation   2. Standard deviation   3. Quartile deviation

9.2.18 RANGE

It is very simple measure of dispersion. It can be defined by the difference with maximum and minimum value of variables, in the distribution.

If $x_1, x_2, ..., x_n$ be any distribution, x_1 is the minimum and x_n is the maximum value of distribution, then range is defined by

$$\text{Range} = x_n - x_1$$

COEFFICIENT OF RANGE

The coefficient of the range is defined by the ratio of the difference between maximum and minimum value with the summation of these values. If $x_1, x_2, ..., x_n$ by any distribution and x_1, x_n be the minimum and maximum values respectively, then coefficient of range is given by

$$\text{Coefficient of Range} = \frac{\text{The difference of max and min value}}{\text{The sum of max and min value}} = \frac{x_n - x_1}{x_n + x_1}$$

MERITS OF RANGE

1. We can easily find the value of range.
2. It is very simple method to know the measure of dispersion.
3. This measure depends only on two values of the given series. So, it is very useful in quick computation.

DEMERITS OF RANGE

1. It suffers much fluctuation due to its dependence on only two (max or min) values of the series.
2. It does not give any idea about the series.
3. It depends only on two extreme values of the series.

Solved Examples

EXAMPLE 1. *Find the range and its coefficient in the given series :*

110, 117, 129, 300, 357, 100, 500, 630, 750

SOLUTION. The largest value of the series, $x_n = 750$
The smallest value of the series, $x_1 = 100$.
Then, Range $x_n - x_1 = 750 - 100$
Range $= 650$
Coefficient of range $= \dfrac{x_n - x_1}{x_n + x_1}$

$= \dfrac{750 - 100}{750 + 100} = \dfrac{650}{850}$

Hence, coefficient of range $= 0.7647$.

EXAMPLE 2. *Find the range and its coefficients in the following frequency distribution :*

x	6	12	18	24	30	36	42
f	4	7	9	18	15	10	5

SOLUTION. The maximum value of the series, $x_n = 42$
The minimum value of the series, $x_1 = 6$
Then,   Range $= x_n - x_1 = 42 - 6$

Range $= 36$

Coefficient of range $= \dfrac{x_n - x_1}{x_n + x_1}$

$= \dfrac{42 - 6}{42 + 6} = \dfrac{36}{48}$

Coefficient of range $= 0.75$.

EXAMPLE 3. *Find the range and its coefficients in the following frequency distribution :*

Marks Group	No. of Students
0-10	5
10-20	7
20-30	10
30-40	16
40-50	11
50-60	7
60-70	3
70-80	2
80-90	2
90-100	0

SOLUTION. The maximum value of the series, $x_n = 100$

The minimum value of the series, $x_1 = 0$

Then,   Range $= x_n - x_1 = 100 - 0 = 100$

                   Range $= 100$

Coefficient of range $= \dfrac{x_n - x_1}{x_n + x_1} = \dfrac{100 - 0}{100 + 0}$

Coefficient of range $= 1$.

9.2.19 Interquartile Range and Semi-Quartile Deviation

If we arrange a series into ascending order and this series is divided into four equal parts, then each part of this series is known as a quartile. If Q_1, Q_2 and Q_3 are the first, second and third quartile, then $Q_3 - Q_1$ is called the interquartile range and $\frac{1}{2}(Q_3 - Q_1)$ is called the semi-interquartile range.

i.e.,   Interquartile range $= Q_3 - Q_1$

                   Semiquartile deviation $= \dfrac{1}{2}(Q_3 - Q_1)$.

☞ Remark

- Semiquartile deviation is also known as quartile deviation.

COEFFICIENT OF QUARTILE DEVIATION

Coefficient of quartile deviation is given by $= \dfrac{\dfrac{1}{2}(Q_3 - Q_1)}{\dfrac{1}{2}(Q_3 + Q_1)}$

Coefficient of quartile deviation $= \dfrac{Q_3 - Q_1}{Q_3 + Q_1}$

Merits
1. It can be easily calculated.
2. It is easy to understand.
3. It does not depend only on the extreme values.

Demerits
1. Fluctuation of sampling affect this deviation very much.
2. We can get only a rough measure from it.
3. All the data are not used in the measure.

Solved Examples

EXAMPLE 1. *Compute the quartile deviation and its coefficient from the following series:*

Item	1	2	3	4	5
Frequency	3	2	5	7	9
Item	6	7	8	9	10
Frequency	5	8	10	2	1

SOLUTION.

Item	1	2	3	4	5	6
Frequency	3	2	5	7	9	5
Cumulative Frequency	3	5	10	17	26	31

Item	7	8	9	10	
Frequency	8	10	2	1	52 Total
Cumulative Frequency	39	49	51	52	

For first quartile, $\dfrac{N}{4} = \dfrac{52}{4} = 13$, which lies in the frequency of 4th item. So Q_1 is given by 4.

$Q_1 = 4$

Now, for third quartile, $\dfrac{3N}{4} = \dfrac{3}{4} \times 52 = 39$ *i.e.,* $Q_3 = 7$

EXAMPLE 2. *Calculate the quartile deviation for the marks of 63 students in mathematics given as follows:*

Marks	0-10	10-20	20-30	30-40	40-50
Frequency	5	7	10	16	11

Marks	50-60	60-70	70-80	80-90	90-100
Frequency	7	3	2	2	0

SOLUTION.

Marks	Frequency	Cumulative frequency
0-10	5	5
10-20	7	12
20-30	10	22
30-40	16	38
40-50	11	49
50-60	7	56
60-70	3	59
70-80	2	61
80-90	2	63
90-100	0	63
	63	

For first quartile, $N = 63$,

which is odd, so $\dfrac{N+1}{4} = \dfrac{63+1}{4} = 16^{th}$ student. which is contained in the group 20-30.

So, 20-30 will be the first quartile class.

$l = 20$, $N = 63$, $F = 12$, $f = 10$, $i = 10$

Then, $Q_1 = l + \dfrac{\frac{1}{4}N - F}{f} \times i$

$= 20 + \dfrac{\frac{1}{4} \times 63 - 12}{10} \times 10 = 20 + 3.75$

$Q_1 = 23.75$ marks

Now, for $Q_3 = \dfrac{3}{4}(N+1) = \dfrac{3}{4}(63+1) = \dfrac{3}{4} \times 64 = 48$

48th student will lie in the class 40-50 so it will be the third or upper quartile class.

For Q_3, $l = 40$, $N = 63$, $F = 38$, $f = 11$, $i = 10$

$$Q_3 = l + \dfrac{\dfrac{3}{4}N - F}{f} \times i = 40 + \dfrac{\dfrac{3}{4} \times 63 - 38}{11} \times 10$$

$$= 40 + 8.4090 = 48.4090$$

Now, we have to calculate the quartile deviation.

Quartile deviation $= \dfrac{1}{2}(Q_3 - Q_1)$

$$= \dfrac{1}{2}(48.4090 - 23.75) = 12.3295 \text{ marks}$$

Coefficients of quartile deviation is given by

Coefficient of quartile deviation $= \dfrac{Q_3 - Q_1}{Q_3 + Q_1}$

$$= \dfrac{48.4090 - 23.75}{48.4090 + 23.75} = \dfrac{24.659}{72.159}$$

Coefficient of quartile deviation = 0.3417 marks.

9.2.20 AVERAGE DEVIATION OR MEAN DEVIATION

In the given series, if we calculate the deviation of each point from a particular value (mean, mode or median), then the arithmetic average of these deviations is known as average deviation or mean deviation.

REMARK
- Each value of the deviation will be positive.

WORKING PROCEDURE

Step 1. For ungrouped data, the mean deviation is given by the formula Mean deviation $= \dfrac{\Sigma |x - M|}{N}$

Step 2. For discrete series, the deviation is given by Mean deviation $= \dfrac{\Sigma f |x - M|}{N}$

Step 3. For continuous series, firstly we calculate the value of x as a mid value of the class and then apply the same formula for discrete series.

Merits of Average Deviation
1. All the terms of the series are used to calculate this deviation.
2. It can be calculated from any of the value, mean, mode and median.
3. It is a simple method to calculate the measure of dispersion.

Demerits of Average Deviation
1. Mathematically, this measure cannot give the absolute value since the signs (+ve and –ve) are ignored.
2. About mode, this measure does not gives any satisfaction.

Solved Examples

EXAMPLE 1. *The weight of five items in Kg is given by 30, 40, 45, 50, 55 Find the deviation from the median.*

SOLUTION. Here $N = 5$ which is odd. So the median is given by $\dfrac{1}{2}(5+1)$ th term.

$$= \dfrac{1}{2} \times 6^{th} \text{ term} = 3^{rd} \text{ term}$$

The 3rd term of the series is = 45

So, the median of the series = 45

x	$x - M_e$	$|x - M_e|$
30	–15	15
40	–5	5
45	0	0
50	5	5
55	10	10
		$\Sigma |x - M_e| = 35$

Then, deviation is given by

$$= \dfrac{\Sigma |x - M_e|}{N} = 35/5$$

So, the mean deviation about mean = 7.

EXAMPLE 2. *Find the mean deviation of the numbers 3, 10,*

9, 9, 4, 7, 14 from the mean.

SOLUTION. Mean $= \dfrac{\text{Sum of all the items}}{\text{Total no. of observations}}$

$$= \dfrac{3 + 10 + 9 + 9 + 4 + 7 + 14}{7} = \dfrac{56}{7}$$

Mean = 8

x	$x - M$	$|x - M|$
3	–5	5
10	2	2
9	1	1
9	1	1
4	–4	4
7	–1	1
14	6	6
		$\Sigma = 20$

Now the mean deviation from mean is given by

$$= \dfrac{\Sigma |x - M|}{N} = \dfrac{20}{7}$$

So, the mean deviation about mean = 2.8571.

EXAMPLE 3. *Calculate the average deviation about median for the following frequency distribution*:

x	6	12	18	24	30	36	42
f	4	7	9	18	15	10	5

SOLUTION. Firstly, we will calculate the median for the given frequency distribution and then find the average deviation.

x	f	Cummulative frequency	$x - M_e$	$f\|x - M_e\|$
6	4	4	– 18	72
12	7	11	– 12	84
18	9	20	– 6	54
24	18	38	0	0
30	15	53	6	90
36	10	63	12	120
42	5	68	18	90
	$N = 68$			$\Sigma = 510$

Here, $N = 68$ which is even so the median is given by the value of $N/2 = 34$ th item.

Value of 34th item = 24

So the median, $M_e = 24$.
Now, the average deviation about mean

$$= \frac{\Sigma f \mid x - M_e \mid}{N} = \frac{510}{68}$$

Average deviation about median = 7.5

EXAMPLE 4. *Calculate the mean deviation about mean for the following frequency distribution*

Class	20-30	30-40	40-50	50-60	60-70
Frequency	120	201	150	75	25

SOLUTION. Firstly, we calculate the arithmetic mean of the given frequency distribution and then we can calculate the mean deviation about mean.

Class	Mid Value x	Frequency f	$d' = \dfrac{x - A}{h}$	fd'	$x - M$ $M = 39$	$f\|x - M\|$
20-30	25	120	– 2	– 240	–14	1680
30-40	35	201	– 1	– 201	– 4	804
40-50	45	150	0	0	6	900
50-60	55	75	1	75	16	1200
60-70	65	25	2	50	26	650
		571		–316		5234

Let the assumed mean = 45 and h (class width) = 10
The arithmetic mean for continuous series is given by the formula

$$M = A + \frac{\Sigma fd'}{N} \times h = 45 + \frac{(-316)}{571} \times 10$$
$$= 45 - 5.5341$$
$$M = 39.4659$$

Let $M = 39$
Now, mean deviation about mean
$$= \frac{\Sigma f \mid x - M \mid}{N} = \frac{5234}{571} = 9.1663$$
Hence, mean deviation about mean = 9.1663.

Exercise-9.9

1. Find the average deviation from mean of the following frequency distribution :

Class	0-6	6-12	12-18	18-24	24-30
Frequency	8	10	12	9	5

2. Find the range and quartile deviation of the data given below :
139, 150, 151, 151, 157, 158, 160, 161, 162, 162, 173, 175

3. Find the range and its coefficients in the following series :

Size	3	4	5	6	7	8	9	10
Frequency	35	30	20	10	6	3	2	1

4. Find out the range and its coefficient in the following series :

Monthly average (Rs.)	100	150	200	250	300	500
Frequency	30	20	15	10	4	1

5. Find the range and its coefficient in the following frequency distribution:

Size	10-60	60-120	120-180	180-240	240-300
Frequency	3	5	6	3	2

6. Find the quartile deviation and its coefficient in the following frequency distribution :

Weight (Kg)	70-80	80-90	90-100	100-110
Frequency	12	18	35	42

Weight (Kg)	110-120	120-130	130-140	140-150
Frequency	50	45	25	8

7. Find the semi-interquartile range and its coefficient from the following data:

Size	6	7	8	9	10	11	12
Frequency	3	6	9	13	8	5	3

8. Find the mean deviation from the mean of the following frequency distribution :

Size	56	63	70	77	84	91	98
Frequency	3	6	14	16	13	6	2

9. Find the mean deviation from median of the following frequency distribution :

Size	4	6	8	10	12	14	16
Frequency	2	4	5	3	2	1	4

10. Find the mean deviation about mode for the following frequency distribution :

Class	14-15	15-16	16-17	17-18	18-19	19-20
Frequency	4	6	10	18	9	3

11. Find the mean deviation from mean of the following frequency distribution :

Age-under	10	20	30	40	50	60	70	80
Frequency	15	30	53	75	100	110	115	125

Marks	0-10	10-20	20-30	30-40	40-50
No. of students	5	8	15	16	6

12. Find the mean deviation from mean of the following frequency distribution :

Answers

1. 6.3	**2.** 36	**3.** 7, 0.53	**4.** 400, 0.666	**5.** 290, 0.93
6. 13.03, 0.117	**7.** 1, 0.11	**8.** 76.53	**9.** 3.24	**10.** 0.986
11. 15.8	**12.** 9.44			

9.2.21 ABSOLUTE AND RELATIVE MEASURE OF DISPERSION

The absolute measure are expressed in terms of the units of the observation such as rupees, kilograms, degree, celcius, etc. These values may be used to compare the variations in the two distributions when the variables are expressed in the same units and are of same average size.

The relative measures of dispersion are useful for comparing two series expressed in different units. They are also useful when we are comparing the variations of two series which have quite different magnitudes even when the units of original measurements are same.

SOME RELATIVE MEASURES OF DISPERSION

(1) Range Coefficient of Dispersion $= \dfrac{\text{Difference between extreme values}}{\text{Sum of extreme values}}$

(2) Quartile Coefficient of Dispersion $= \dfrac{Q_3 - Q_1}{Q_3 + Q_1}$

(3) Coefficient of Mean Dispersion $= \dfrac{\text{Mean deviation about any point } a}{a}$

9.2.22 STANDARD DEVIATION AND ROOT MEAN SQUARE DEVIATION

If in a given series we calculate the deviation of each point from the arithmetic mean then the arithmetic mean of the squares of these deviation is known as variance.

Variance $\qquad \sigma^2 = \dfrac{1}{N} \Sigma(x - M)^2$

Now the square root of this variance is known by the standard deviation and it is denoted by σ

$$\sigma = + \sqrt{\dfrac{1}{N} \Sigma(x - M)^2}$$

Root mean square deviation is the deviation in which we calculate the deviation of each point from an arbitrary point A.

In other words, if in a given series, we calculate the deviation of each point from an arbitrary point, then the arithmetic mean of the squares of these deviation is known as mean square deviation and the root of this mean square deviation is known by root mean square deviation. Hence,

mean square deviation, $\qquad S^2 = \dfrac{1}{N} \Sigma f(x - M)^2$ and root mean square deviation, $S = + \sqrt{\dfrac{1}{N} \Sigma f(x - M)^2}$

REMARK

• The standard deviation is a particular case of mean square deviation when the arithmetic mean is taken in place of arbitrary point.

Merits

1. The standard deviation depends on all the values of observations.
2. In standard deviation, the signs (+ve and –ve) are not ignored.
3. It is very useful to calculate the coefficient of correlation and other statistics calculations.
4. In calculating the standard deviation, we use the mean which is more suitable.

Demerits

1. A person with non-mathematical mind cannot understand it.
2. It wants a long calculation so sometimes it becomes very complicated to solve these problems.
3. It gives more weight to extreme values.

COEFFICIENT OF STANDARD DEVIATION AND VARIATION

1. Coefficient of standard deviation $= \dfrac{\sigma}{M}$

where, $\qquad \sigma$ = Standard deviation

$\qquad M$ = Arithmetic mean

2. Coefficient of variation $= \dfrac{\sigma}{M} \times 100$, where σ = Standard deviation, $\qquad M$ = Arithmetic mean

9.2.23 RELATION BETWEEN STANDARD DEVIATION AND ROOT MEAN SQUARE DEVIATION

We know that the variance is given by

$$\sigma^2 = \frac{1}{N}\Sigma f(x-M)^2, \text{ where } M \text{ is the arithmetic mean.}$$

$$\sigma^2 = \frac{1}{N}\Sigma f[x-A-(M-A)]^2 = \frac{1}{N}\Sigma f[x-A-d]^2 \qquad [\because (M-A)=d]$$

$$= \frac{1}{N}\Sigma f\left[(x-A)^2+d^2-2(x-A)d\right] = \frac{1}{N}\Sigma f(x-A)^2 + \frac{1}{N}\Sigma fd^2 - \frac{2}{N}\Sigma f(x-A).d$$

$$= \frac{1}{N}\Sigma f(x-A)^2 + d^2 - 2.d\frac{1}{N}\Sigma f(x-A) = \frac{1}{N}\Sigma f(x-A)^2 + d^2 - 2d\frac{1}{N}\Sigma fx + 2d\frac{1}{N}\Sigma f.A$$

$$= \frac{1}{N}\Sigma f(x-A)^2 + d^2 - 2d.M + 2d.A = \frac{1}{N}\Sigma f(x-A)^2 + d^2 - 2d(M-A) \qquad (\because \Sigma f = N)$$

We know that $(M-A)=d$.

So,

$$\sigma^2 = \frac{1}{N}\Sigma f(x-A)^2 + d^2 - 2d.d = \frac{1}{N}\Sigma f(x-A)^2 + d^2 - 2d^2 = \frac{1}{N}\Sigma f(x-A)^2 - d^2$$

Hence,

$$\sigma^2 = S^2 - d^2 \qquad \left[\because S^2 = \frac{1}{N}\Sigma f(x-A)^2\right]$$

THEOREM 1. *The sum of the squares of the deviation about arithmetic mean is least.*

PROOF. Consider a discrete distribution

x	x_1	x_2	...	x_n
f	f_1	f_2	...	f_n

Then arithmetic mean, $\bar{x} = \dfrac{\Sigma fx}{\Sigma f} = \dfrac{\Sigma fx}{N} \Rightarrow \Sigma fx = N\bar{x}$, where $\Sigma f = N$

Let A be any arbitrary point and S be the sum of squares of deviation from A, then $S = \Sigma f(x-A)^2$

It is known that S will be minimum if

$$\frac{\partial S}{\partial A} = 0 \quad \text{and} \quad \frac{\partial^2 S}{\partial A^2} > 0$$

Now, $\dfrac{\partial S}{\partial A} = -2\Sigma f(x-A) = 0 \qquad \Rightarrow \qquad -2\Sigma fx + 2A\Sigma f = 0$

$\Rightarrow \quad -\Sigma fx + AN = 0 \Rightarrow \qquad N\bar{x} = AN \qquad \Rightarrow \qquad \bar{x} = A$

Also, $\dfrac{\partial^2 S}{\partial A^2} = 2\Sigma f = 2N > 0$. Hence, S is minimum where $A = \bar{x}$ (Mean).

Solved Examples

EXAMPLE 1. *Calculate the standard deviation of the data given as follows: 3, 4, 9, 11, 13, 6, 8, 10*

SOLUTION. We have

x	x − M	$(x-M)^2$
3	− 5	25
4	− 4	16
9	1	1
11	3	9
13	5	25
6	− 2	4
8	0	0
10	2	4
		84

Firstly, we will calculate the mean of the given data

So, $M = \dfrac{\text{Sum of all observations}}{\text{Total number of items}}$

$= \dfrac{3+4+9+11+13+6+8+10}{8} = 8$

Now, the standard deviation is given by

$$\sigma = \sqrt{\frac{1}{N}\Sigma(x-M)^2} = \sqrt{\frac{84}{8}} = \sqrt{10.5} = 3.25$$

EXAMPLE 2. *Calculate the standard deviation for the following frequency distribution :*

x	0	10	20	30	40	50	60	70	80
f	150	140	100	80	80	70	30	14	0

SOLUTION. Firstly we calculate the mean of the given frequency distribution and then find the standard deviation

x	f	fx	x − M	$(x-M)^2$	$f(x-M)^2$
0	150	0	− 23	529	79350
10	140	1400	−13	169	23660
20	100	2000	−3	9	900
30	80	2400	7	49	3920
40	80	3200	17	289	23120
50	70	3500	27	729	51030
60	30	1800	37	1369	41070
70	14	980	47	2209	30926
80	0	0	57	3249	0
	$\Sigma f =$ 664	$\Sigma fx =$ 15280			$\Sigma f(x-M)^2$ = 253976

Mean of the given distribution will be

$$M = \frac{\Sigma fx}{\Sigma f}$$

$$\Rightarrow \qquad M = \frac{15280}{664} = 23.0120 \approx 23$$

Now the standard deviation is given by

$$\sigma = \sqrt{\frac{\Sigma f(x-M)^2}{N}} = \sqrt{\frac{253976}{664}} = \sqrt{382.4939}$$

$$\sigma = 19.557$$

9.2.24 SHORT CUT METHOD TO CALCULATE THE STANDARD DEVIATION OF DISCRETE SERIES

We know that standard deviation, $\sigma = \sqrt{\dfrac{\Sigma f(x-M)^2}{N}}$

Now, $\sigma_x = \sqrt{\dfrac{1}{N}\Sigma f(x-\bar{x})^2}$ (Let us write $M = \bar{x}$)   ... (1)

Now, let us assume that d is the deviation of each point x from an arbitrary point A.

Then, $d = x - A$, $x = d + A$, $\bar{d} = \bar{x} - A$, $\bar{x} = \bar{d} + A$

Put these values in equation (1)

$$\sigma_x = \sqrt{\frac{1}{N}\Sigma f(d+A-\bar{d}-A)^2} = \sqrt{\frac{1}{N}\Sigma f(d-\bar{d})^2} \Rightarrow \sigma_x = \sqrt{\frac{1}{N}\Sigma fd^2 - \left(\frac{1}{N}\Sigma fd\right)^2}$$

Hence $\qquad \sigma_x = \sigma_d$

Thus, we see that the standard deviation is independent from the change of origin.

STEP DEVIATION METHOD

The standard deviation for a given series is

$$\sigma_x = \sqrt{\frac{1}{N}\Sigma f(x-\bar{x})^2}, \text{ where } \bar{x} = M \qquad \qquad ... (1)$$

Now, let us define a new variable μ such that $\mu = \dfrac{x-a}{h}$, $x = a + h\mu$, $\bar{\mu} = \dfrac{\bar{x}-a}{h}$, $\bar{x} = a + h\bar{\mu}$

Put this value of x in equation (1)

$$\sigma_x = \sqrt{\frac{1}{N}\Sigma f\left[a+h\mu-(a+h\bar{\mu})\right]^2} = \sqrt{\frac{1}{N}\Sigma f(a+h\mu-a-h\bar{\mu})^2} = \sqrt{\frac{1}{N}\Sigma f(\mu-\bar{\mu})^2.h^2}; \; \sigma_x = \sigma_\mu.h.$$

$$\sigma_x = h\,\sigma_\mu \qquad \qquad ... (2)$$

$$\sigma_x = \sqrt{\frac{1}{N}\Sigma f(\mu-\bar{\mu})^2\,h^2} = \sqrt{\frac{1}{N}\Sigma f(\mu^2 - 2\mu\bar{\mu} + \mu^2).h^2}$$

$$= \sqrt{\left(\frac{1}{N}\Sigma f\mu^2 - \frac{2}{N}\Sigma f\mu\bar{\mu} + \frac{1}{N}\Sigma f\bar{\mu}^2\right)h^2} = \sqrt{\left(\frac{1}{N}\Sigma f\mu^2 - 2\mu\bar{\mu} + \bar{\mu}^2\right)h^2} = \sqrt{\left(\frac{1}{N}\Sigma f\mu^2 - \bar{\mu}^2\right)h^2}$$

$$\sigma_n = \sqrt{\frac{1}{N}\Sigma f\mu^2 - \left(\frac{1}{N}\Sigma f\mu\right)^2}$$

Now, by equation (2), we see that the standard deviation is changed by the change of scale.

REMARK

• Using the above two results, we conclude that

(i) Standard deviation is independent from the change of origin.

(ii) Standard deviation is changed by the change of scale.

EXAMPLE 3. *Calculate the standard deviation for the following frequency distribution using short-cut method:*

Class	35-36	36-37	37-38	38-39	39-40	40-41	41-42
Frequency	14	20	42	54	45	18	7

SOLUTION.

Class	Mid value x	f	Cumulative Frequency	$d = x - A$	d^2	fd	fd^2
35-36	35.5	14	14	−3	9	−42	126
36-37	36.5	20	34	−2	4	−40	80
37-38	37.5	42	76	−1	1	−42	42
38-39	38.5	54	130	0	0	0	0
39-40	39.5	45	175	1	1	45	45
40-41	40.5	18	193	2	4	18	72
41-42	41.5	7	200	3	9	7	63
		$\Sigma f = 200$				$\Sigma fx = -54$	$\Sigma fd^2 = 428$

Let us assume mean $A = 38.5$

Then the standard deviation

$$= \left[\frac{1}{N}\Sigma fd^2 - \left(\frac{1}{N}\Sigma fd\right)^2\right]$$

$$= \frac{1}{200} \times 428 - \left[\frac{1}{200}(-54)\right]^2$$

$$= 2.14 - 0.0729$$

$$\sigma^2 = 2.0671$$

$$\sigma = \sqrt{2.0671} = 1.4.$$

EXAMPLE 4. *Calculate the standard deviation for the following frequency distribution using short-cut method:*

Class	5-15	15-25	25-35	35-45	45-55	55-65
Frequency	15	32	51	78	97	109

SOLUTION.

Class	Mid Value x	f	Cumulative Frequency	$d = x - A$	d^2	fd	fd^2
5-15	10	15	15	-25	625	-375	9375
15-25	20	32	47	-15	225	-480	7200
25-35	30	51	98	-5	25	-255	1275
35-45	40	78	176	5	25	390	1950
45-55	50	97	273	15	225	1455	21825
55-65	60	109	382	25	625	2725	68125
		382				3460	109750

Let the assumed mean $A = 35$
Then the standard deviation,

$$\sigma = \sqrt{\frac{1}{N}\Sigma fd^2 - \left(\frac{1}{N}\Sigma fd^2\right)^2}$$

$$\sigma = \sqrt{\frac{1}{382}\times 109750 - \left(\frac{1}{382}\times 3460\right)^2}$$

$$= \sqrt{287.3036 - 82.0383} = \sqrt{205.2653}$$

$$\sigma = 14.3269 .$$

TERMS RELATED TO STANDARD DEVIATION

1. Probable error = $0.6745 \times \sigma$

2. Variance = $\sigma^2 = \dfrac{\Sigma f(x-\bar{x})^2}{N}$

3. Modulus (C) = $\sqrt{\dfrac{2\Sigma f(x-\bar{x})^2}{N}}$ or $\sigma\sqrt{2}$ *i.e.,* $C = \sigma\sqrt{2}$

4. Fluctuation = $C^2 = 2\sigma^2$, where is the modulus or fluctuation = (modulus)^2

5. Precision : $P = \dfrac{1}{C} = \dfrac{1}{\text{modulus}}$ (where σ is standard deviation) *i.e.,* $P = \dfrac{1}{\sigma\sqrt{2}}$

Exercise-9.10

1. Calculate the standard deviation and coefficient of variation for the given frequency distribution

x	1	2	3	4	5	6
f	31	37	33	30	35	34

2. Calculate the standard deviation of the given frequency distribution :

x	0	10	20	30	40	50	60	70	80
f	150	140	100	80	80	70	30	14	0

3. Calculate the standard deviation of the data : 5, 7, 9, 11.

4. Calculate the standard deviation and its coefficient from the given frequency distribution :

Class	20-25	25-30	30-35	35-40	40-45	45-50
Frequency	18	44	102	160	57	19

5. Calculate the standard deviation for the given frequency distribution :

Above x	0	10	20	30	40	50	60	70	80
Frequency f	150	140	100	80	80	70	30	14	0

6. Calculate the standard deviation for the given frequency distribution using step deviation method :

Class	0-10	10-20	20-30	30-40	40-50
Frequency	5	10	15	20	4

7. Calculate the standard deviation for the given frequency distribution using short-cut method :

x	2	3	4	5	9	10	12	13	15
f	25	37	44	59	68	43	31	32	12

8. Calculate the standard deviation and its coefficient for the given frequency distribution using step deviation method :

Class	0-10	10-20	20-30	30-40	40-50	50-60
Frequency	15	17	19	27	19	12

9. Calculate the mean and standard deviation for the following:

Size of item	6	7	8	9	10	11	12
Frequency	3	6	9	13	8	5	4

(VTU–2001)

Answers

1. 1.71, 48.61	2. 19.6	3. 2.23	4. 5.687, 0.159	5. 19.6
6. 10.95	7. 3.76	8. 156, 0.52	9. 9, 1.607	

9.2.25 SUMMARY OF THE MATHEMATICAL PROPERTIES OF STANDARD DEVIATION

1. The sum of the squares of the deviation about mean is least.

2. The standard deviation of a series remain unchanged if each observations of the series is increased or decreased by same constant value.

3. If each observation of a series is multiplied or divided by the same constant value. The standard deviation can also be obtained by dividing or multiplying by the same constant value.

4. Standard deviation is not affected by the change of origin but is not independent of scale.

5. In a symmetrical distribution, the mean, median and mode are coincident. On the basis of mean and S.D., the number of observations lying within specific ranges can be measured in case of normal or symmetrical distribution, which is given as below:

Mean ± 1 S.D. covers 68.27% parts of the distribution.

Mean ± 2 S.D. covers 95.45% parts of the distribution.

Mean ± 3 S.D. covers 99.73% parts of the distribution.

Solved Examples

EXAMPLE 1. *Find the mean, variance and standard deviation for the first n natural numbers.*

SOLUTION. We have to find the mean, variance and standard deviation for the series 1, 2, 3, ..., n.

The mean for this series is given by

$$M = \frac{\text{Sum of all the observations}}{\text{Total number of items}}$$

$$= \frac{1+2+3+...+n}{n} = \frac{\Sigma n}{n} = \frac{n(n+1)}{2.n} = \frac{n+1}{2}$$

So, the mean M is given by $= \frac{n+1}{2}$

Now, we have to find the variance

Variance, $\sigma^2 = \frac{1}{n}\Sigma(x-\bar{x})^2 = \frac{1}{n}\Sigma x^2 - \left(\frac{1}{n}\Sigma x\right)^2$

$$= \frac{1}{n}\cdot\frac{n(n+1)(2n+1)}{6} - \left[\frac{1}{n}\cdot\frac{n.(n+1)}{2}\right]^2$$

$$= \frac{(n+1)(2n+1)}{6} - \frac{(n+1)^2}{4}$$

$$= \frac{(n+1)}{2}\left[\frac{(2n+1)}{3} - \frac{n+1}{2}\right]$$

$$= \frac{(n+1)}{2}\left[\frac{2(2n+1)-3(n+1)}{6}\right]$$

$$= \frac{(n+1)}{2}\left[\frac{4n+2-3n-3}{6}\right] = \frac{(n+1)}{2}\left[\frac{(n-1)}{6}\right]$$

Variance, $\sigma^2 = \frac{(n^2-1)}{12}$

Hence, standard deviation, $\sigma = \left[\frac{(n^2-1)}{12}\right]^{1/2}$

EXAMPLE 2. *In a given distribution, show that the mean deviation is less than the standard deviation about mean.*

SOLUTION. Mean deviation $= \frac{1}{N}\Sigma|x-M|$

and standard deviation $= \sqrt{\frac{1}{N}\Sigma f(x-M)^2}$

To show that the mean deviation is less than the standard deviation, we will show

Standard deviation – Mean deviation ≥ 0

$$\text{S.D.} - \text{M.D.} = \sqrt{\frac{1}{N}\Sigma f(x-M)^2} - \frac{1}{N}\Sigma f|x-M| \geq 0$$

Let $x - M = d$, then

$$\text{S.D.} - \text{M.D.} = \sqrt{\frac{1}{N}\Sigma fd^2} \geq \frac{1}{N}\Sigma f|d|$$

Since we have to prove that S.D $\geq$ M.D

then $\sqrt{\frac{1}{N}\Sigma fd^2} \geq \frac{1}{N}\Sigma f|d|$

Squaring both sides, we get

$$\frac{1}{N}\Sigma fd^2 \geq \left[\frac{1}{N}\Sigma f|d|\right]^2$$

$$N\,\Sigma fd^2 \geq [\Sigma f|d|]^2$$

$$(f_1 + f_2 + f_3 + ... + f_n)(f_1d_1^2 + f_2d_2^2 + f_3d_3^2 + ...)$$

$$\geq [f_1|d_1| + f_2|d_2| + f_3|d_3| + ...]^2$$

$$\Rightarrow (f_1d_1^2 + f_1f_2d_1^2 + f_1f_2d_2^2 + ...)$$

$$+ (f_1f_2d_1^2 + f_2d_2^2 + f_2f_3d_3^2 + ...)$$

$$+ (f_1f_3d_1^2 + f_2f_3d_2^2 + f_3^2d_3^2 + ...)$$

$$\geq (f_1^2d_1^2 + f_2^2d_2^2 + ...) + 2f_1f_2d_1d_2 + 2f_2f_3d_2d_3 + ...$$

$$\Rightarrow f_1f_2(d_1^2 + d_2^2 - 2d_1d_2)$$

$$+ f_1f_3(d_1^2 + d_2^2 - 2d_1d_3) + ... \geq 0$$

$$\sum_{i,j=1}^{n} f_if_j(d_i - d_j)^2 \geq 0$$

Since the square of a number will never be negative, so it proves that the mean deviation is less thatn the standard deviation about mean.

EXAMPLE 3. *Compute the mean deviation from mean and standard deviation of the series a, a +d, a +2d, ..., a + 2nd and verify standard deviation > mean deviation.*

SOLUTION. The arithmetic series is

$a, a + d, a + 2d, ..., a + 2nd$

The summation of this series will be

$$= 2n+1[2a + \{2n+1-1\}.d] = \frac{2n+1}{2}[2a + 2nd]$$

$$= (2n+1)(a+nd)$$

Now, the mean of this series

$$M = \frac{\text{Sum of all the observations}}{\text{Total number of items}} = \frac{(2n+1)(a+nd)}{(2n+1)}$$

$$M = a + nd$$

Now, we will calculate the mean deviation about mean.

Mean deviation about mean

$$= \frac{1}{N}\Sigma|x-M| = \frac{1}{N}\sum_{k=0}^{2n}|x_k - M| \qquad ...(1)$$

The total number of terms, $N = (2n+1)$

We know that x_k takes value from $x = a, ..., a + 2nd$ and mean of this arithmetic series is given by $M = a + nd$.

$$\sum_{k=0}^{2n}|x_k - M| = |a - a - nd|$$

$$+ |a + d - a - nd| + ...|a + nd - a - nd|$$

$$+ |a + (n-1)d - a - nd| + ...|a + 2nd - a - nd| \quad ...(2)$$

$$= nd + (n-1)d + \ldots + d + 0 + d + \ldots nd$$

$$= 2[nd + (n-1)d + \ldots d]$$

$$= 2d\,[n + (n-1) + (n-2) + \ldots 1]$$

$$= 2d\,[1 + 2 + 3 + \ldots (n-2) + (n-1) + n] = 2d.\Sigma n$$

$$\sum_{k=0}^{2n} |x_k - M| = 2d \cdot \frac{n(n+1)}{2} = n(n+1).d$$

Put this value in equation (1), we get

Mean deviation about mean $= \dfrac{1}{2n+1}.n(n+1).d$

$$= \frac{n(n+1)d}{(2n+1)} \qquad \ldots (3)$$

Now, we have to calculate the standard deviation of the given arithmetic series

$$\sigma^2 = \frac{1}{N} \sum_{k=0}^{2n} (x_k - M)^2 \qquad \ldots (4)$$

With the help of equation (2), we can write

$$\sum_{k=0}^{2n} (x_k - M)^2 = (a - a - nd)^2 + (a + d - a - nd)^2$$

$$+ \ldots + (a + nd - a - nd)^2$$

$$+ (a + (n+1)d - a - nd)^2 + \ldots + (a + 2nd - a - nd)^2$$

$$= n^2 d^2 + (n-1)^2 d^2 + \ldots d^2 + 0 + d^2 + \ldots n^2 d^2$$

$$= 2\,[n^2 d^2 + (n-1)^2 d^2 + \ldots + d^2]$$

$$= 2d^2\,[n^2 + (n-1)^2 + (n-2)^2 + \ldots + 3^2 + 2^2 + 1^2]$$

$$= 2d^2\,[1^2 + 2^2 + \ldots + (n-2)^2 + (n-1)^2 + n^2]$$

$$= 2d^2 . \Sigma n^2$$

$$\sum_{k=0}^{2n} (x_k - M)^2 = 2d^2 \cdot \frac{n(n+1)(2n+1)}{6}$$

$$\left[\because \Sigma n^2 = \frac{n(n+1)(2n+1)}{6} \right]$$

But this value in equation (4)

$$\sigma^2 = \frac{1}{(2n+1)} \frac{2d^2.n(n+1)(2n+1)}{6}$$

$$\sigma^2 = \frac{n(n+1).d^2}{3}$$

S.D. $(\sigma) = d\sqrt{\dfrac{n(n+1)}{3}} \qquad \ldots (5)$

Now, we have to show that S.D. > M.D.
Put the value of S.D. and M.D. from equation (3) and (5), we have

$$d\sqrt{\frac{n(n+1)}{3}} > \frac{n(n+1)d}{2n+1}$$

Squaring both sides

$$\frac{n(n+1)}{3} > \frac{n^2(n+1)^2}{(2n+1)^2}$$

$$(2n+1)^2 > 3n(n+1)$$

$$4n^2 + 4n + 1 > 3n^2 + 3n \implies n^2 + n + 1 > 0$$

which is true $\forall\, n > 0$. So the result is proved.

EXAMPLE 4. *Show that in a discrete distribution, the mean deviation is least when measured from the median.*

SOLUTION. Let us suppose that when number of terms are odd, then

$$(x_0 - nh), \ldots, (x_0 - 2h), (x_0 - h),$$
$$(x_0 + h), (x_0 + 2h), \ldots, (x_0 + nh)$$

be the given discrete distribution. By inspection, we observe that x_0 will be the median of this distribution. Let us define S_k such that

$$S_k = \Sigma\,|x_k - x_r|$$

$$S_0 = \Sigma\,|x_0 - x_r|$$

$$= |x_0 - x_0| + |x_0 - x_0 + h| + |x_0 - x_0 + 2h|$$
$$+ \ldots + |x_0 - x_0 + nh| + |x_0 - x_0 - h|$$
$$+ |x_0 - x_0 - 2h| + \ldots + |x_0 - x_0 - nh|$$

$$= h + 2h + \ldots (n-1)h + nh + h + 2h + \ldots (n-1)h + nh$$

$$= 2[h + 2h + \ldots (n-1)h + nh]$$

$$= 2h[1 + 2 + \ldots + n] = 2h\frac{n(n+1)}{2}$$

$$S_0 = n(n+1).h$$

$$S_1 = \Sigma\,|x_1 - x_r|$$

$$= |x_0 + h - x_0| + |x_0 + h - x_0 + h|$$
$$+ |x_0 + h - x_0 + 2h| + \ldots + |x_0 + h - x_0 + nh|$$
$$+ |x_0 + h - x_0 - h| + |x_0 + h - x_0 - 2h|$$
$$+ |x_0 + h - x_0 - 3h| + \ldots + |x_0 + h - x_0 - nh|$$

$$= h + 2h + 3h + \ldots + (n+1)h + h + 2h + \ldots (n-1)h$$

$$= h + 2h + 3h + \ldots + (n-1)h + nh$$
$$+ (n+1)h + h + 2h + \ldots (n-1)h$$

$$= 2[h + 2h + \ldots (n-1)h + nh] + (n+1)h - nh$$

$$S_1 - S_0 = nh(n+1) - nh + (n+1)h - n(n+1)h = h$$

$$S_1 - S_0 = h$$

Thus, by calculating $S_2, S_3, \ldots, S_n$, we get

$$S_2 - S_1 = 3h, \quad S_3 - S_2 = 5h$$

So, $\qquad S_0 < S_1 < S_2$

In the same way, we can find $\quad S_0 < S_{-1} < S_{-2}$

Similarly, we can prove it for even number of terms. Thus, we can say that mean deviation is least when measured from the median.

EXAMPLE 5. *Show that if deviations x are small as compared with the mean M so that $\left(\dfrac{x}{M}\right)^3$ and higher power of $\dfrac{x}{M}$ may be neglected.*

(i) $G = M\left[1 - \dfrac{\sigma^2}{2M^2}\right]$ *(ii)* $M^2 - G^2 = \sigma^2$

(iii) $H = M\left[1 - \dfrac{\sigma^2}{M^2}\right]$ *(iv)* $H + M = 2G$

(v) Mean $\sqrt{X} = \sqrt{M}\left[1 - \dfrac{\sigma^2}{2M^2}\right]$ *(vi)* $MH = G^2$

where the symbols have their usual meanings.

SOLUTION. Let us suppose that $X - M = x \implies X = M + x$

We know that geometric mean

$$G = \left(X_1^{f_1} . X_2^{f_2} . X_3^{f_3} \ldots X_n^{f_n}\right)^{1/N}$$

Taking log of both sides, we get

$$\log G = \log \left(X_1^{f_1} \cdot X_2^{f_2} \cdot X_3^{f_3} \ldots X_n^{f_n} \right)^{1/N}$$

$$= \frac{1}{N} \left[f_1 \log X_1 + f_2 \log X_2 + \ldots + f_n \log X_n \right]$$

$$= \frac{1}{N} \sum_{i=1}^{n} f \log X = \frac{1}{N} \Sigma f \log (M + x)$$

$$= \frac{1}{N} \Sigma f \log \left[M \left(1 + \frac{x}{M} \right) \right]$$

$$= \frac{1}{N} \Sigma f \left[\log M + \log \left(1 + \frac{x}{M} \right) \right]$$

$$= \frac{1}{N} \Sigma f \log M + \frac{1}{N} \Sigma f \log \left(1 + \frac{x}{M} \right)$$

$$= \log M + \frac{1}{N} \Sigma f \left(\frac{x}{M} - \frac{x^2}{2M^2} + \frac{x^3}{3M^3} - \ldots \right)$$

Now, it is given that $\left(\dfrac{x}{M} \right)^3$ and higher power of $\dfrac{x}{M}$ may be neglected then

$$\log G = \log M + \frac{1}{N} \Sigma f \left(\frac{x}{M} - \frac{x^2}{2M^2} \right)$$

$$\because \text{ For } \frac{1}{N} \Sigma f x = 0, \frac{1}{N} \Sigma f x^2 = \sigma^2$$

Then, $\log G = \log M - \dfrac{1}{2M^2} \cdot \sigma^2$

$$\log G - \log M = -\frac{1}{2M^2} \cdot \sigma^2$$

$$\log \frac{G}{M} = -\frac{\sigma^2}{2M^2}$$

$$\Rightarrow \qquad \frac{G}{M} = e^{-\sigma^2/2M^2} \qquad \qquad \ldots (1)$$

$$G = M e^{-\sigma^2/2M^2} = M \left(1 - \frac{\sigma^2}{2M^2} + \ldots \right)$$

Neglecting the higher power of σ/M, we get

$$G = M \left(1 - \frac{\sigma^2}{2M^2} \right)$$

(ii) $M^2 - G^2 = \sigma^2$

From equation (1) $\dfrac{G}{M} = e^{-\sigma^2/2M^2}$

Squaring both sides

$$\frac{G^2}{M^2} = e^{-\sigma^2/M^2} = \left(1 - \frac{\sigma^2}{M^2} - \ldots \right)$$

Neglecting the terms $\dfrac{\sigma^3}{M^2}$ and higher power of $\dfrac{\sigma}{M}$

$$\frac{G^2}{M^2} = \left(1 - \frac{\sigma^2}{M^2} \right)$$

$$G^2 = M^2 - \sigma^2$$

$$M^2 - G^2 = \sigma^2$$

(iii) $\qquad H = M \left[1 - \dfrac{\sigma^2}{M^2} \right]$

We know that harmonic mean H is given by

$$\frac{1}{H} = \frac{1}{N} \Sigma f \left(\frac{1}{X} \right)$$

$$\frac{1}{H} = \frac{1}{N} \Sigma f \left(\frac{1}{M + x} \right) = \frac{1}{N} \Sigma f (M + x)^{-1}$$

$$= \frac{1}{N} \Sigma f \cdot \frac{1}{M} \left(1 + \frac{x}{M} \right)^{-1}$$

$$= \frac{1}{N} \Sigma f \cdot \frac{1}{M} \left(1 - \frac{x}{M} + \frac{x^2}{M^2} - \ldots \right)$$

Neglecting higher power of $\dfrac{x}{M}$, we get

$$= \frac{1}{M} \frac{1}{N} \Sigma f \left(1 - \frac{x}{M} + \frac{x^2}{M^2} \right)$$

$$= \frac{1}{M} \left[\frac{1}{N} \Sigma f - \frac{1}{M} \frac{\Sigma f x}{N} + \frac{1}{M^2} \cdot \frac{\Sigma f x^2}{N} \right]$$

$$= \frac{1}{M} \left[1 - \frac{1}{M} \cdot 0 + \frac{1}{M^2} \cdot \sigma^2 \right]$$

$$\left(\text{For } \Sigma f x = 0, \frac{1}{N} \Sigma f x^2 = \sigma^2 \right)$$

$$\frac{1}{H} = \frac{1}{M} \left[1 + \frac{\sigma^2}{M^2} \right]$$

Then, $\qquad H = \dfrac{M}{\left[1 + \dfrac{\sigma^2}{M^2} \right]} = M \left[1 + \dfrac{\sigma^2}{M^2} \right]^{-1}$

$$= M \cdot \left[1 - \frac{\sigma^2}{M^2} + \frac{\sigma^3}{M^3} - \ldots \right]$$

Neglecting the higher power of $\dfrac{x}{M}$ greater than 2

$$H = M \left[1 - \frac{\sigma^2}{M^2} \right]$$

(iv) $H + M = 2G$

Consider LHS $= H + M$

Put the value of H from part (iii)

$$\text{LHS} = M + M \left[1 - \frac{\sigma^2}{M^2} \right] = M + M - \frac{\sigma^2}{M}$$

$$= 2M - \frac{\sigma^2}{M} = 2M - \frac{2\sigma^2}{2M}$$

$$= 2 \left(M - \frac{\sigma^2}{2M} \right) = 2 \cdot M \left[1 - \frac{\sigma^2}{2M^2} \right] \qquad \ldots (2)$$

By part (i), we observe that

$$M \left[1 - \frac{\sigma^2}{2M^2} \right] = G$$

So, by equation (2)

$$H + M = 2G$$

(v) Mean $\sqrt{X} = \sqrt{M} \left[1 - \dfrac{\sigma^2}{2M^2} \right]$

We know that

$$\text{Mean } \sqrt{X} = \frac{1}{N} \Sigma f \sqrt{X}$$

$$= \frac{1}{N} \Sigma f \sqrt{M + x} \qquad (\because X = M + x)$$

$$= \frac{1}{N}\sqrt{M}\ \Sigma f \sqrt{1+\frac{x}{M}}$$

$$= \sqrt{M}\cdot\frac{1}{N}\Sigma f\left(1+\frac{x}{M}\right)^{1/2}$$

$$= \sqrt{M}\cdot\frac{1}{N}\Sigma f\left(1+\frac{1}{2}\frac{x}{M}-\frac{1}{8}\cdot\frac{x^2}{M^2}+...\right)$$

Neglecting the higher power of $\dfrac{x}{M}$

$$\text{Mean}\ \ \sqrt{X}=\sqrt{M}\cdot\frac{1}{N}\Sigma f\left(1+\frac{1}{2}\cdot\frac{x}{M}-\frac{x^2}{8M^2}\right)$$

$$=\sqrt{M}\left[\frac{1}{N}\Sigma f+\frac{1}{2}\cdot\frac{1}{M}\frac{1}{N}\Sigma fx-\frac{1}{8}\frac{1}{M^2}\frac{1}{N}\Sigma fx^2\right]$$

$$\left[\text{For }\frac{1}{N}\Sigma fx=0,\ \frac{1}{N}\Sigma fx^2=\sigma^2\right]$$

9.2.26 RELATION BETWEEN DIFFERENT MEASURES OF DISPERSION

1. Semi-interquartile range $=\dfrac{2}{3}$ standard deviation

Solved Examples

EXAMPLE 1. *Compute the mean, variance and standard eviation for the series* $1^2, 2^2, 3^2 ... n^2$.

SOLUTION. We know that

$$\text{Mean}=\frac{\Sigma x}{N}=\frac{1^2+2^2+3^2+...+n^2}{n}=\frac{\Sigma n^2}{n}$$

$$=\frac{n(n+1)(2n+1)}{6n}$$

Thus, $\qquad$ Mean $\dfrac{(n+1)(2n+1)}{6}$

Now, $\qquad$ Variance, $\sigma^2=\dfrac{1}{N}\Sigma(x-\overline{x})^2$

$$\sigma^2=\frac{1}{N}\Sigma x^2-\left(\frac{1}{N}\Sigma x\right)^2 \qquad ...(1)$$

We know that

$$\Sigma x=1^2+2^2+3^2+...+n^2=\frac{n(n+1)(2n+1)}{6}$$

$$\Sigma x^2=1^4+2^4+3^4+...+n^4$$

$$=\frac{n(n+1)(2n+1)(3n^2+3n-1)}{30}$$

Putting these values in equation (1)

$$\sigma^2=\frac{1}{n}\,n\,\frac{(n+1)(2n+1)(3n^2+3n-1)}{30}$$

$$-\left(\frac{n(n+1)(2n+1)}{6n}\right)^2$$

$$\sigma^2=\frac{(n+1)(2n+1)(3n^2+3n-1)}{30}-\left[\frac{(n+1)(2n+1)}{6}\right]^2$$

$$=\frac{(n+1)(2n+1)}{6}\left[\frac{3n^2+3n-1}{5}-\frac{(n+1)(2n+1)}{6}\right]$$

$$=\frac{(n+1)(2n+1)}{6}\left[\frac{18n^2+18n-6-5(2n^2+3n+1)}{30}\right]$$

$$\text{Mean}\ \sqrt{X}=\sqrt{M}\left[1+0-\frac{1}{8M^2}\cdot\sigma^2\right] \qquad (\because N=\Sigma f)$$

$$=\sqrt{M}\left[1-\frac{\sigma^2}{2M^2}\right]$$

(vi) $MH=G^2$

From part (iii), put the value of H

$$\text{LHS}=M.M\left[1-\frac{\sigma^2}{M^2}\right]=M^2\left[1-\frac{\sigma^2}{M^2}\right] \quad ...(3)$$

From relation (ii)

$$M^2-G^2=\sigma^2$$

Put in equation (3)

$$MH=M^2\left[1-\frac{M^2-G^2}{M^2}\right]=M^2\frac{\left[M^2-M^2+G^2\right]}{M^2}$$

$$MH=G^2$$

2. Mean deviation $=\dfrac{4}{5}$ standard deviation

$$=\frac{(n+1)(2n+1)}{6}\left[\frac{8n^2+3n-11}{30}\right]$$

$$\sigma^2=\frac{(n+1)(2n+1)(8n^2+3n-11)}{180}$$

$$\sigma=\left[\frac{(n+1)(2n+1)(8n^2+3n-11)}{180}\right]^{1/2}.$$

EXAMPLE 2. *The details of runs gained by two batsman A and B are as follows :*

A	24	79	31	114	14	02	68	01	110	07
B	05	18	42	53	09	47	52	17	81	56

(i) Which of the two batsman is better run scorer?

(ii) Which of the two batsman has more consistency in the number of runs?

SOLUTION. Let us assume the quantity x with respect to the runs of the batsman A and y with respect to the runs of batsman B.

X	$X-M_A$	$(X-M_A)^2$	Y	$Y-M_B$	$(Y-M_B)^2$
24	-21	441	05	-33	1089
79	34	1156	18	-20	400
31	-14	196	42	4	16
114	69	4761	53	15	225
14	-31	961	09	29	841
02	-43	1849	47	9	81
68	23	529	52	14	196
01	-44	1936	17	-21	441
110	65	4225	81	43	1849
07	-38	1444	56	18	324
450		17498	380		5462

For A

The total number of terms $N = 10$

Then, $M_A = \dfrac{\Sigma X}{N} = \dfrac{450}{10} = 45$

For B

Total number of terms $N = 10$

$$M_B = \dfrac{\Sigma Y}{N} = \dfrac{380}{10} = 38$$

Now, by the above calculation, it proves that

$$M_A > M_B$$

So, we can say that A is better scorer.

Now, we will calculate the standard deviation

$$\sigma_A = \sqrt{\dfrac{1}{N}\Sigma(X - M_A)^2}$$

$$= \sqrt{\dfrac{1}{10} \times 17498} = \sqrt{1749.8} = 41.8306 \text{ runs}$$

$$\sigma_B = \sqrt{\dfrac{1}{N}\Sigma(Y - M_B)^2} = \sqrt{\dfrac{1}{10} \times 5462} = \sqrt{546.2}$$

$$\sigma_B = 28.3709 \text{ runs}$$

To check the consistency in number of runs, we have to calculate the coefficient of variation.

$$V_A = \dfrac{\sigma_A}{M_A} \times 100 \qquad V_B = \dfrac{\sigma_B}{M_B} \times 100$$

$$V_A = \dfrac{41.8306}{45} \times 100 \qquad V_B = \dfrac{23.3709}{38} \times 100$$

$$V_A = 92.9568 \qquad V_B = 61.5023$$

We observe that $V_A > V_B$

So, B has the consistency in number of runs.

EXAMPLE 3. *An analysis of the monthly wages paid to workers in two firms A and B, belonging to the same industry gives the following results*

	Firm A	Firm A
Number of wage earners	586	648
Average monthly wages	Rs 525	Rs 475
Variance of the distribution	10,000	12,100

(i) Which firm A or B pays out the larger amount as monthly wages.

(ii) In which firm A or B is there greater

9.2.27 STANDARD DEVIATION OF TWO COMBINED SETS

Let us consider two sets with n_1 and n_2 number of terms and whose standard deviations are given by σ_1 and σ_2, respectively, measured from respective means M_1 and M_2. If we grouped these sets, then standard deviation σ of the combined sets is given by

$$\sigma = \left[\dfrac{n_1\sigma_1^2 + n_2\sigma_2^2}{n_1 + n_2} + \dfrac{n_1 n_2}{(n_1 - n_2)}(M_1 - M_2)^2 \right]^{1/2}$$

PROOF. Let us assume that $x_1, x_2, ..., x_{n_1}$ and $y_1, y_2, ..., y_{n_2}$ be the given two series with total number of terms n_1 and n_2. Then the mean of these series will be

$$M_1 = \dfrac{1}{n_1}\Sigma x_{n_1} \qquad M_2 = \dfrac{1}{n_2}\Sigma x_{n_2}$$

Let S_1^2 and S_2^2 be the mean square deviations of these series and A be the assumed mean.

variability in individual wages.

(iii) **What are measures of average monthly wages and the variability in individual wages of all the workers in the two firms A and B taken together?**

SOLUTION. Given $N_A = 586$, $N_B = 648$

$$\overline{X}_A = \text{Rs. } 525, \quad \overline{X}_B = \text{Rs. } 475$$

Standard deviation $= \sigma_A = \sqrt{10000} = 100$

$$\sigma_B = \sqrt{12100} = 110$$

(i) Total amount paid by firm $A = 525\,(586)$
$$= \text{Rs. } 307{,}650$$

Total amount paid by firm $B = 475\,(648)$
$$= \text{Rs. } 307{,}800$$

$\therefore$ Company B pays larger amount as monthly wages.

(ii) Coefficient of variation for firm A

$$VC_A = \dfrac{\sigma_A}{\overline{X}_A} \times 100 = \dfrac{100}{525} \times 100 = 19.05\%$$

$$VC_B = \dfrac{\sigma_B}{\overline{X}_B} \times 100 = \dfrac{110}{475} \times 100 = 23.16\%$$

Company B has greater variability in individual wages.

(iii) Measure of average taking A and B together

$$\overline{X}_{AB} = \dfrac{N_A \overline{X}_A + N_B \overline{X}_B}{N_A + N_B}$$

$$= \dfrac{586(525) + 648(475)}{586 + 648} = \dfrac{307650 + 307800}{1234}$$

$$= 498.743$$

$$\overline{X}_{AB} = 498.743$$

$$\sigma_{AB} = \sqrt{\dfrac{N_A \sigma_A^2 + N_A(\overline{X}_A - \overline{X}_{AB})^2 + N_B \sigma_B^2 + N_B(\overline{X}_B - \overline{X}_{AB})^2}{N_A + N_B}}$$

$$= \sqrt{\dfrac{586(100)^2 + 586(689.43) + 648(110)^2 + 648(563.73)}{1234}}$$

$$= \sqrt{\dfrac{5860000 + 404005.98 + 7840800 + 365297.04}{1234}}$$

$$= \sqrt{11726.17749} = 108.287$$

Then, $S_1^2 = \dfrac{1}{n_1}\Sigma(x_{n_1} - A)^2$ ... (1)

$$S_2^2 = \dfrac{1}{n_2}\Sigma(x_{n_2} - A)^2 \qquad \text{... (2)}$$

The mean square deviation for the combined series will be

$$S^2 = \dfrac{1}{n_1 + n_2}\left[\Sigma(x_{n_1} - A)^2 + \Sigma(x_{n_2} - A)^2 \right]$$

$$= \dfrac{1}{n_1 + n_2}\left[n_1 S_1^2 + n_2 S_2^2 \right]$$

(with the help of equation (1) and (2))

Since we know that $S^2 = \sigma^2 + d^2$

Then, $S^2 = \dfrac{1}{n_1 + n_2}\left[n_1(\sigma_1^2 + d_1^2) + n_2(\sigma_2^2 + d_2^2) \right]$

$$= \frac{1}{n_1 + n_2}\left[n_1\sigma_1^2 + n_1 d_1^2 + n_2\sigma_2^2 + n_2 d_2^2\right]$$

$$S^2 = \frac{n_1\sigma_1^2 + n_2\sigma_2^2}{n_1 + n_2} + \frac{n_1 d_1^2 + n_2 d_2^2}{n_1 + n_2} \quad \dots (3)$$

For combined series, if $A = M$, then $S^2 = \sigma^2$.

Now, we have to calculate the value of d for grouped series.

The mean for the grouped series $\quad M = \dfrac{n_1 M_1 + n_2 M_2}{n_1 + n_2}$

Also $\quad d_1 = M_1 - A \qquad\qquad d_2 = M_2 - A$

Put these values of d_1 and d_2 in equation (3)

$$S^2 = \frac{n_1\sigma_1^2 + n_2\sigma_2^2}{n_1 + n_2} + \frac{n_1(M_1 - A)^2 + n_2(M_2 - A)^2}{n_1 + n_2} \quad \dots (4)$$

For grouped series $A = M$, then

$$M_1 - A = M_1 - \frac{n_1 M_1 + n_2 M_2}{n_1 + n_2}$$

$$= \frac{n_1 M_1 + n_2 M_1 - n_1 M_1 - n_2 M_2}{n_1 + n_2}$$

$$M_1 - A = \frac{n_2(M_1 - M_2)}{n_1 + n_2}$$

$$M_2 - A = M_2 - \frac{n_1 M_1 + n_2 M_2}{n_1 + n_2}$$

$$M_2 - A = \frac{n_1(M_2 - M_1)}{n_1 + n_2}$$

Put the value of $(M_1 - A)$ and $(M_2 - A)$ in equation (4).

$$\sigma^2 = \frac{n_1\sigma_1^2 + n_2\sigma_2^2}{n_1 + n_2} + \frac{n_1 n_2^2(M_1 - M_2)^2}{(n_1 + n_2)^3} + \frac{n_1^2 n_2(M_2 - M_1)^2}{(n_1 + n_2)^3}$$

$$\sigma^2 = \frac{n_1\sigma_1^2 + n_2\sigma_2^2}{n_1 + n_2} + \frac{n_1 n_2(M_1 - M_2)^2\,[n_2 + n_1]}{(n_1 + n_2)^3}$$

$$\sigma^2 = \frac{n_1\sigma_1^2 + n_2\sigma_2^2}{n_1 + n_2} + \frac{n_1 n_2(M_1 - M_2)^2}{(n_1 + n_2)^2}$$

So the standard deviation for the combined sample is given by

$$\sigma = \left[\frac{n_1\sigma_1^2 + n_2\sigma_2^2}{n_1 + n_2} + \frac{n_1 n_2(M_1 - M_2)^2}{(n_1 + n_2)^2}\right]^{1/2}$$

Solved Examples

EXAMPLE 1. *The mean and standard deviation of two samples are given as follows :*

$n_1 = 50 \qquad M_1 = 54.1 \qquad \sigma_1 = 8$

$n_2 = 100 \qquad M_2 = 50.3 \qquad \sigma_2 = 7$

Find the mean and standard deviation of the combined sample.

SOLUTION. Let M and σ be the mean and standard deviation of the combined sample

Then, $M = \dfrac{M_1 n_1 + M_2 n_2}{n_1 + n_2}$

$$= \frac{54.1 \times 50 + 50.3 \times 100}{50 + 100} \quad \frac{2705 \quad 5030}{150}$$

$M = 51.566$

Now, we have to calculate the standard deviation of combined sample.

We know that S.D. for combined sample

$$\sigma = \left[\frac{n_1\sigma_1^2 + n_2\sigma_2^2}{n_1 + n_2} + \frac{n_1 n_2(M_1 - M_2)^2}{(n_1 + n_2)^2}\right]^{1/2}$$

$$= \left[\frac{50 \times 64 + 100 \times 49}{50 + 100} + \frac{50 \times 100(54.1 - 50.3)^2}{(50 + 100)^2}\right]^{1/2}$$

$$= \left[\frac{3200 + 4900}{150} + \frac{5000 \times 14.44}{(150)^2}\right]^{1/2}$$

$$= \left[\frac{8100}{150} + \frac{19000}{22500}\right]^{1/2} = [54 + 3.2088]^{1/2}$$

$\sigma = \sqrt{57.2088} = 7.555$

EXAMPLE 2. *Out of the two groups the number of observations are given by 100 and 150. The mean and the standard deviations of these groups and of combined sample is as follows.*

First group

$n_1 = 100 \qquad M_1 = ? \qquad \sigma_1 = 7$

Second group

$n_2 = 150 \qquad M = 55 \qquad \sigma = ?$

Combined group

$n = ? \qquad M = 51 \qquad \sigma = \sqrt{130}$

Find the missing values.

SOLUTION. We know that the total number of terms in the combined group $= n_1 + n_2$

$n = 100 + 150 = 250$

Now, the mean of the combined sample

$$M = \frac{M_1 n_1 + M_2 n_2}{n_1 + n_2}$$

$$51 = \frac{M_1 \times 100 + 55 \times 150}{100 + 150}$$

$$51 = \frac{100 M_1 + 8250}{250}$$

$100\,M_1 = 12750 - 8250$

$M_1 = 45$

So the mean of the first group $M_1 = 45$.

Now, we have to calculate the standard deviation of the second group.

S.D. of combined sample,

$$\sigma = \left[\frac{n_1\sigma_1^2 + n_2\sigma_2^2}{n_1 + n_2} + \frac{n_1 n_2(M_1 - M_2)^2}{(n_1 + n_2)^2}\right]^{1/2}$$

$$\sqrt{130} = \left[\frac{100 \times 49 + 150 \times \sigma_2^2}{250} + \frac{100 \times 150(45 - 55)^2}{(250)^2}\right]^{1/2}$$

$$130 = \frac{4900 + 150\sigma_2^2}{250} + \frac{1500000}{(250)^2}$$

$$106 = \frac{4900 + 150\sigma_2^2}{250}$$

$$150\,\sigma_2^2 = 21600 \ i.e. \ \sigma_2^2 = 144$$

$$\Rightarrow \quad \sigma_2 = 12.$$

EXAMPLE 3. *The means of the two sample of size 50 and 100 respectively 54.1 and 50.3 and standard deviations are 8 and 7. Obtain the mean and the standard deviation of combined sample of size 150.*

SOLUTION. $n_1 = 50, \ n_2 = 100$

$\overline{X}_1 = 54.1, \ \overline{X}_2 = 50.3$

$\sigma_1 = 8, \ \sigma_2 = 7$

Combined mean = $\overline{X}_{12} = \dfrac{n_1\overline{X}_1 + n_2\overline{X}_2}{n_1 + n_2}$

$$= \frac{50(54.1) + 100(50.3)}{150}$$

$\overline{X}_{12} = 51.57.$

Combined standard deviation

$\sigma_{12} = \sqrt{\dfrac{n_1\sigma_1^2 + n_1(\overline{X}_1 - \overline{X}_{12})^2 + n_2\sigma_2^2 + n_2(\overline{X}_2 - \overline{X}_{12})^2}{n_1 + n_2}}$

$$= \sqrt{\frac{50(64) + 100(49) + 50(6.4) + 100(1.6129)}{150}}$$

$$= \sqrt{\frac{3200 + 4900 + 320 + 161.20}{150}}$$

$\sigma_{12} = 7.564$

EXAMPLE 4. *For a group of 200 candidates, the mean and standard deviation were found to be 40 and 15, respectively. Find the corrected mean and standard deviation to the correct figures.*

SOLUTION. Incorrect mean = 40.

Incorrect standard deviation = 15.

Correct value = 43

Incorrect value = 53

$$\overline{X} = \frac{\Sigma X}{N}$$

Incorrect : $\Sigma X = \overline{X}N = 400(200) = 8000$

Correct : $\Sigma X =$ Incorrect $\Sigma X -$ Incorrect value + Correct value

$$= 8000 - 53 + 43 = 7990$$

Correct $\overline{X} = \dfrac{7990}{200} = 39.95$

Standard deviation = $\sqrt{\dfrac{\Sigma X^2}{N} - \left(\dfrac{\Sigma X}{N}\right)^2}$

$$(15)^2 = \frac{\Sigma X^2}{200} - (40)^2$$

$$225 + 1600 = \frac{\Sigma X^2}{200}$$

Incorrect $\Sigma X^2 = 200(1825)$

Incorrect $\Sigma X^2 = 365000$

Correct $\Sigma X^2 =$ Incorrect $\Sigma X^2 -$ (Incorrect value)^2 + (Correct value)^2

$$= 365000 - (53)^2 + (43)^2 = 36440$$

Correct S.D. = $\sqrt{\dfrac{364040}{200} - (39.95)^2} = 15.185$

9.2.28 STANDARD VARIABLE

A variable with mean 0, variance unity and defined by $Z = \dfrac{X - M}{\sigma}$ is known as standard variable, in which M and σ are the mean and standard deviation of the variable X.

Solved Examples

EXAMPLE. *The marks of three students in an examination are given as follows :*

Student	Eng.	Math.	Science	Total
A	95	61	70	226
B	69	74	83	226
C	70	82	74	226
Mean	55	50	53	

Find the order of equitable while the standard deviation is given by 16, 11 and 12.

SOLUTION. To calculate the equitable order, we will find the standard variable for the marks obtained by each student in each subject

Student	Eng.	Math.	Science
A	$\dfrac{95 - 55}{16}$	$\dfrac{61 - 50}{11}$	$\dfrac{70 - 53}{12}$
B	$\dfrac{69 - 55}{16}$	$\dfrac{74 - 50}{11}$	$\dfrac{83 - 53}{12}$
C	$\dfrac{70 - 55}{16}$	$\dfrac{82 - 50}{11}$	$\dfrac{74 - 53}{12}$

Student	Eng.	Math.	Science	Total
A	2.5	1	1.416	4.916
B	0.875	2.182	2.5	5.557
C	0.9375	2.909	1.75	5.5965

So, the equitable order will be $A < B < C$.

Exercise-9.11

1. Calculate the standard deviation of the given two series and find which show greater deviation.

   Series A : 192, 288, 236, 229, 184, 260, 348, 291, 330, 243

   Series B : 83, 87, 93, 109, 124, 126, 126, 101, 102, 108

2. In the given series, if d_1 and d_2 represent the deviation from assumed mean 100, then calculate the coefficient of variation

for two series for which

$$n_1 = 150, \ \Sigma d_1 = 100, \ \Sigma d_1^2 = 245320$$

$$n_2 = 200, \ \Sigma d_2 = 250, \ \Sigma d_2^2 = 245320$$

3. The fluctuation in the rate of two items A and B are given as follows. Calculate for which the rate is more consistent.

A	55	54	52	53	56	58	52	50	51	49
B	108	107	105	105	107	104	103	104	106	101

and find the values of standard deviation for each item.

4. The table shows the number of workers in two factories whose weekly earnings are given. Determine the mean value and standard deviations in both the factory.

Range	Factory A	Factory B
4-6	74	71
6-8	376	379
8-10	304	303

10-12	110	112
12-14	18	18
14-16	0	1
16-18	9	3
18-20	9	9
20-22	0	4

5. Show that the variance can be defined in the form of half of mean of the squares of deviations of natural terms

$$\sigma^2 = \frac{1}{2}\left[\frac{1}{n^2}\sum_{i=1}^{n}\sum_{j=1}^{n}(x_i - x_j)^2\right],$$

where $x_1, x_2, ..., x_n$ are n variate value.

6. Discuss the effect of change of origin and scale on standard deviation.

7. Discuss the various measure of dispersion.

8. Establish a relationship between root-mean square deviation and standard deviation.

Answers

1. $\sigma_A = 51.6$, $\sigma_B = 14.96$. Series A shows greater deviation. **2.** 39.9, 14.6

3. $\sigma_A = 2.65$, $V_A = 4.99\%$, $\sigma_B = 2$, $V_B = 1.90\%$, **4.** $M_A = 8.34$, $\sigma_A = 2.34$, $M_B = 8.36$, $\sigma_B = 2.29$

9.2.29 SKEWNESS

Skewness is the measure of asymmetry. A distribution which does not occur equidistant on both sides of the mean value is known as asymmetric or skewed distribution.

Positive and Negative Skewness

A frequency curve is said to be positive skewed if it has a longer tail towards the direction of higher values and the curve which has a longer tail towards the direction of smaller values is known as negative skewed.

For a skewed curve, the value of mean, mode and median will never be equal.

For a positive skewed : Mode < Median < Mean

For a negative skewed : Mean < Median < Mode

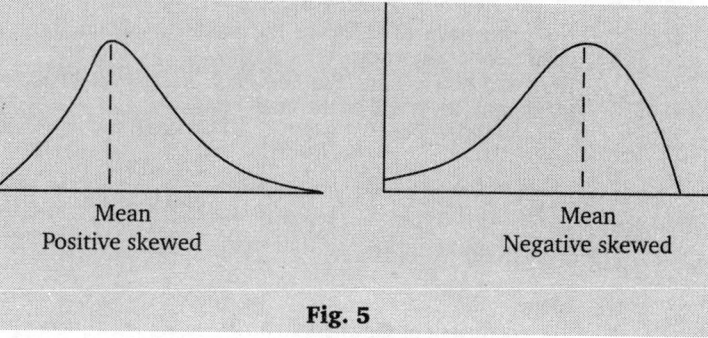

Positive skewed   Negative skewed

Fig. 5

9.2.30 MEASURE OF SKEWNESS

The coefficient of skewness can be calculated by the following two methods

1. According to Karl Pearson, Coeff. of skewness $= \dfrac{\text{Mean} - \text{Mode}}{\text{Standard deviation}} = \dfrac{3(\text{Mean} - \text{Median})}{\text{Standard deviation}}$

REMARK
• The value of this coefficient lies between –3 to +3.

2. The second formula for coefficient of skewness is given by Bowley, which is as follows

$$\text{Coefficient of skewness} = \frac{Q_3 + Q_1 - 2Q_2}{Q_3 - Q_1}$$

where Q_1, Q_2, Q_3 are the first, second and third quartiles. It can also be written as Coefficient of skewness $= \dfrac{(Q_3 - Q_2) - (Q_2 - Q_1)}{(Q_3 - Q_2) + (Q_2 - Q_1)}$

REMARK
• The value of this coefficient lies between –1 to +1.

MEASURE OF SKEWNESS BASED ON MOMENTS

1. If $v_1 = 0$ or if $\mu_3 = 0$, then there will be no skewness.

2. If $v_1 > 0$: positive skewed

3. If $v_1 < 0$: Negative skewed

Solved Examples

EXAMPLE 1. *Find the mean, mode, standard deviation and coefficient of skewness for the following frequency distribution.*

Years	0-10	10-20	20-30	30-40	40-50	50-60
Frequency	15	17	19	27	19	12

SOLUTION.

Class	Mid Value x	Frequency	fx	$(x-M)$	$(x-M)^2$	$f(x-M)^2$
0-10	5	15	75	−25	625	9375
10-20	15	17	255	−15	225	3825
20-30	25	19	475	−5	25	475
30-40	35	27	945	5	25	675
40-50	45	19	855	15	225	4275
50-60	55	12	660	25	625	7500
		109	3265			26125

$$M = \frac{\Sigma fx}{N} = \frac{3265}{109} = 29.95$$

$$M = 29.95 \text{ year}$$

Let $M = 30$

Now,

$$\sigma = \sqrt{\frac{1}{N}\Sigma f(x-M)^2} = \sqrt{\frac{1}{109}(26125)}$$

$$= \sqrt{239.678}; \quad \sigma = 15.48 \text{ year}$$

Now, we have to calculate the mode to obtain the coefficient of skewness.

Here, we observe that the frequency of class 30-40 is maximum. So, it will be the modal class.

$$l = 30, f_{-1} = 19, f = 27, f_1 = 19, i = 10$$

$$\text{Mode} = l + \frac{f-f_{-1}}{2f-f_{-1}-f_1} \times i = 30 + \frac{27-19}{54-19-19} \times 10$$

$$= 30 + \frac{8}{16} \times 10$$

$$\text{Mode} = 35$$

The coefficient of skewness

$$= \frac{\text{Mean} - \text{Mode}}{\text{Standard deviation}} = \frac{29.95 - 35}{15.48} = -0.3262$$

EXAMPLE 2. *Find the Karl Pearson's coefficient of skewness for the following frequency distribution*:

Class	0-4	4-8	8-12	12-16	16-20	20-24
Frequency	5	7	10	15	8	4

SOLUTION.

Class	Mid value x	f	$d = \dfrac{x-A}{h}$	d^2	fd	fd^2
0-4	2	5	−2	4	−10	20
4-8	6	7	−1	1	−7	7
8-12	10	10	0	0	0	0
12-16	14	15	1	1	15	15
16-20	18	8	2	4	16	32
20-24	22	4	3	9	22	36
		49			26	110

Let assumed mean $A = 10$

The mean

$$M = A + \frac{\Sigma fd}{N} \times i = 10 + \frac{26 \times 4}{49} = 12.1224$$

Standard deviation, $\sigma = i\sqrt{\frac{1}{N}\Sigma fd^2 - \left(\frac{1}{N}\Sigma fd\right)^2}$

$$= 4\sqrt{\frac{1}{49} \times 110 - \left(\frac{1}{49} \times 26\right)^2}$$

$$= 4\sqrt{2.2448 - 0.2815}$$

$$\sigma = 4\sqrt{1.9633} = 5.6047$$

Now, we have to calculate the mode of given distribution. We observe that the frequency of class 12-16 is maximum. So it will be the modal class.

$$l = 12, f_{-1} = 10, f = 15, f_1 = 8, i = 4$$

$$\text{Mode} = l + \frac{f - f_{-1}}{2f - f_{-1} - f_1} \times i$$

$$= 12 + \frac{15-10}{30-10-8} \times 4 = 12 + \frac{5}{12} \times 4$$

$$\text{Mode} = 13.66$$

The coefficient of skewness

$$= \frac{\text{Mean} - \text{Mode}}{\text{Standard deviation}} = \frac{12.1224 - 13.66}{5.6047}$$

$$= -0.27$$

EXAMPLE 3. *Calculate the Bowley's coefficient of skewness for the following frequency distribution:*

Class	0-10	10-20	20-30	30-40	40-50
Frequency	2	7	10	5	3

SOLUTION. To calculate the Bowley's coefficient of skewness, firstly we have to calculate the value of Q_1, Q_2 and Q_3.

Class	Frequency f	Cumulative Frequency
0-10	2	2
10-20	7	9
20-30	10	19
30-40	5	24
40-50	3	27
	27	

For Q_1

$$\frac{N}{4} = \frac{27}{4} = 6.75$$

Now, 6.75^{th} item will lie in class 10-20. So, it will be the first quartile class.

$$l = 10, \ f = 7, \ f = 2, \ i = 10$$

$$Q_1 = l + \frac{\frac{1}{4}N - F}{f} \times i = 10 + \frac{\frac{1}{4} \times 27 - 2}{7} \times 10$$

$$= 10 + \frac{4.75 \times 10}{7}$$

$Q_1 = 16.7857$ mark

For Q_2

$$\frac{2N}{4} = \frac{N}{2} = \frac{27}{2} = 13.5$$

13.5^{th} item will lie in class 20-30. So, it will be the second quartile class.

$$l = 20, \; f = 10, \; F = 9, \; i = 10$$

$$Q_2 = l + \frac{\frac{2}{4}N - F}{f} \times i$$

$$= 20 + \frac{13.5 - 9}{10} \times 10 = 20 + 4.5$$

$$Q_2 = 24.5 \text{ mark}$$

For Q_3

$$\frac{3N}{4} = \frac{3 \times 27}{4} = 20.25$$

Now, 20.25^{th} item will lie in class 30-40. So, it will be the third quartile class.

$$l = 30, \; f = 5, \; F = 19, \; i = 10$$

$$Q_3 = l + \frac{\frac{3}{4}N - F}{f} \times i$$

$$= 30 + \frac{20.25 - 19}{5} \times 10 = 30 + 2.5$$

$$Q_3 = 32.5 \text{ mark}$$

9.2.31 MOMENTS

If we calculate the deviation of each point of the given distribution from a particular point A, then the arithmetic mean of the nth power of deviation is known as moment about the point A.

It is defined by
$$\mu'_n = \frac{1}{N} \Sigma f (x - A)^n$$

If we consider the mean M in place of the point A, then the calculated moment will be the moment about mean and defined by

$$\mu_n = \frac{1}{N} \Sigma f (x - M)^n$$

FIRST FOUR MOMENTS ABOUT MEAN

Let $x_1, x_2, \ldots, x_n$ be the n observations with mean $\bar{x}$. Then, r^{th} moment about the actual mean of a variable for both grouped and ungrouped data are given as follows :

(i) For ungrouped data

$$\mu_r = \frac{1}{n} \Sigma (x - \bar{x})^r, \; r = 1, 2, 3, 4$$

$$\Rightarrow \quad \mu_1 = \frac{1}{n} \Sigma (x - \bar{x}) = 0 \; ; \; \mu_2 = \frac{1}{n} \Sigma (x - \bar{x})^2 = \sigma^2 \; ; \; \mu_3 = \frac{1}{n} \Sigma (x - \bar{x})^3 \; ; \; \mu_4 = \frac{1}{n} \Sigma (x - \bar{x})^4$$

(ii) For grouped data

$$\mu_r = \frac{1}{n} \Sigma f (x - \bar{x})^r, \; r = 1, 2, 3, 4$$

$$\Rightarrow \quad \mu_1 = \frac{1}{n} \Sigma f (x - \bar{x}) \; ; \qquad \mu_2 = \frac{1}{n} \Sigma f (x - \bar{x})^2$$

$$\mu_3 = \frac{1}{n} \Sigma f (x - \bar{x})^3 \; ; \qquad \mu_4 = \frac{1}{n} \Sigma (x - \bar{x})^4$$

RELATION BETWEEN THE MOMENTS ABOUT MEAN AND MOMENTS ABOUT A POINT

We know that

$$\mu'_n = \frac{1}{N} \Sigma f (x - A)^n \text{ and } \mu_n = \frac{1}{N} \Sigma f (x - M)^n$$

$$\mu'_0 = \frac{1}{N} \Sigma f \; ; \qquad \mu_0 = \frac{1}{N} \Sigma f$$

$$\mu'_0 = 1 \; ; \qquad \mu_0 = 1$$

$$\mu'_1 = \frac{1}{N} \Sigma f (x - A) \; ; \qquad \mu_1 = \frac{1}{N} \Sigma f (x - M)$$

$$= \frac{1}{N} \Sigma f x - \frac{1}{N} \Sigma f A \qquad = \frac{1}{N} \Sigma f x - \frac{1}{N} \Sigma f . M = M - M$$

$$\mu'_1 = M - A \; ; \qquad \mu_1 = 0$$

$$\mu'_2 = \frac{1}{N} \Sigma f (x - A)^2 \; ; \qquad \mu_2 = \frac{1}{N} \Sigma f (x - M)^2$$

$$\mu'_2 = \frac{1}{N} \Sigma f (x - A)^2 \; ; \qquad \mu_2 = \sigma^2$$

$$\mu'_3 = \frac{1}{N} \Sigma f (x - A)^3 \; ; \qquad \mu_3 = \frac{1}{N} \Sigma f (x - M)^3$$

We know

$$\mu_n = \frac{1}{N}\Sigma f(x-M)^n = \frac{1}{N}\Sigma f(x-A-M+A)^n$$

$$= \frac{1}{N}\Sigma f\left[(x-A)-(M-A)\right]^n = \frac{1}{N}\Sigma f\left[x-A-\mu_1'\right]^n$$

Now expanding by binomial expansion

$$\mu_n = \frac{1}{N}\Sigma f\left[(x-A)^n - {}^nC_1(x-A)^{n-1}.\mu_1' + {}^nC_2(x-A)^{n-2}.\mu_1'^2 + \ldots \ldots + (-1)^n(\mu_1')^n\right]$$

$$\mu_n = \mu_n' - {}^nC_1\mu_{n-1}'\mu_1' + {}^nC_2.\mu_{n-2}'\mu_1'^2 + \ldots + (-1)^n(\mu_1')^n$$

Now in particular cases

$$\mu_1 = \mu_1' - {}^1C_1\mu_0'\mu_1' \quad \Rightarrow \quad \mu_1 = \mu_1' - \mu_1' = 0 \quad \Rightarrow \quad \mu_1 = 0 \qquad\qquad (\because \mu_0' = 1)$$

$$\mu_2 = \mu_2' - 2(\mu_1')^2 + \mu_0'(\mu_1')^2 = \mu_2' - 2\mu_1'^2 + \mu_1'^2 = \mu_2' - \mu_1'^2 \qquad\qquad (\because \mu_0' = 1)$$

$$\mu_3 = \mu_3' - {}^3C_1\mu_2'\mu_1' + {}^3C_2\mu_1'(\mu_1')^2 - {}^3C_3\mu_0'(\mu_1')^3 = \mu_3' - 3\mu_1'\mu_2' + 3\mu_1'^3 - \mu_1'^3 = \mu_3' - 3\mu_1'\mu_2' + 2\mu_1'^3$$

$$\mu_4 = \mu_4' - {}^4C_1\mu_3'\mu_1' + {}^4C_2\mu_2'(\mu_1')^2 - {}^4C_3\mu_1'(\mu_1')^3 + {}^4C_4\mu_0'(\mu_1')^4 = \mu_4' - 4\mu_1'\mu_3' + 6\mu_1'^2\mu_2' - 4(\mu_1')^4 + (\mu_1')^4$$

$$\mu_4 = \mu_4' - 4\mu_1'\mu_3' + 6\mu_1'^2\mu_2' - 3(\mu_1')^4$$

MOMENTS ABOUT ANY POINT A IN TERMS OF MOMENTS ABOUT MEAN

We know that the moment about any point A is

$$\mu_n' = \frac{1}{N}\Sigma f(x-A)^n = \frac{1}{N}\Sigma f(x-M+M-A)^n = \frac{1}{N}\Sigma f(x-M-d)^n$$

Now expanding by binomial expansion

$$\mu_n' = \frac{1}{N}\Sigma f\left[(x-M)^n + {}^nC_1(x-M)^{n-1}d + {}^nC_2(x-M)^{n-2}d^2 + \ldots + d^n\right]$$

$$= \frac{1}{N}\Sigma f(x-M)^n + {}^nC_1\frac{1}{N}\Sigma f(x-M)^{n-1}.d + {}^nC_2\frac{1}{N}\Sigma f(x-M)^{n-2}.d^2 + \ldots \ldots + \frac{1}{N}\Sigma f.d^n$$

$$\mu_n' = \mu_n + {}^nC_1\mu_{n-1}.d + {}^nC_2\mu_{n-2}d^2 + \ldots + d^n \qquad\qquad (\because N = \Sigma f)$$

For particular cases

$$\mu_1' = \mu_1 + {}^1C_1\mu_0.d = \mu_1 + d \qquad\qquad (\because \mu_0 = 1)$$

$$\mu_1' = 0 + (M-A) = M-A$$

$$\mu_2' = \mu_2 + {}^2C_1\mu_1.d + {}^2C_2\mu_0.d^2 = \mu_2 + 2\mu_1.d + d^2$$

$$\mu_2' = \mu_2 + d^2 \qquad\qquad (\because \mu_1 = 0)$$

$$\mu_3' = \mu_3 + {}^3C_1\mu_2.d + {}^3C_2\mu_1d^2 + {}^3C_3\mu_0.d^3 = \mu_3 + 3\mu_2.d + d^3 \qquad\qquad [\because \mu_0 = 1, \ \mu_1 = 0]$$

$$\mu_4' = \mu_4 + {}^4C_1\mu_3.d + {}^4C_2\mu_2.d^2 + {}^4C_3\mu_1d^3 + {}^4C_4\mu_0.d^4 = \mu_4 + 4\mu_3.d + 6\mu_2.d^2 + d^4 \qquad\qquad (\because \mu_0 = 1, \ \mu_1 = 0)$$

$$\mu_4' = \mu_4 + 4\mu_3.d + 6\mu_2.d^2 + d^4$$

EFFECT OF CHANGING THE ORIGIN AND SCALE ON MOMENTS

Let x be the previous variable and a new variable is related to x by

$$\mu = \frac{x-A}{h}, \quad x-A = h\mu, \quad x = A + h\mu$$

We know that

$$\mu_n = \frac{1}{N}\Sigma f(x-\bar{x})^n$$

Put the value of x and $\bar{x}$ in the above equation, we get

$$\mu_n = \frac{1}{N}\Sigma f\,(A+h\mu-A-h\bar{\mu})^n$$

$$\mu_0' = 1 \qquad\qquad \mu_0 = 1$$

$$\mu_1' = d \qquad\qquad \mu_1 = 0$$

$$\mu_2 = \sigma^2$$

$$\mu_n = \frac{1}{N}\Sigma f(h\mu - h\bar{\mu})^n = \frac{1}{N}.h^n\Sigma f(\mu-\bar{\mu})^n$$

$$\mu_n = \frac{h^n}{N}\Sigma f(\mu-\bar{\mu})^n$$

Now, we consider the change on the moment about a point

$$\mu'_n = \frac{1}{N} \Sigma f(x - A)^n$$

Put the value $(x - A) = h\mu$ in the above equation

$$\mu'_n = \frac{1}{N} \Sigma f . h^n \mu^n \Rightarrow \mu'_n = \frac{h^n}{N} \Sigma f . \mu^n$$

Thus, we can say that the central moment remains constant with respect to the change of origin.

PEARSON'S β AND γ COEFFICIENTS

Karl Pearson defined four coefficients β_1, β_2, γ_1 and γ_2 with the help of the central moments. These four coefficients are known by Karl Pearson's coefficient and are defined as follows :

$$\beta_1 = \frac{\mu_3^2}{\mu_2^3}, \beta_2 = \frac{\mu_4}{\mu_2^2}, \ \gamma_1 = \pm\sqrt{\beta_1} \text{ and } \gamma_2 = \beta_2 - 3$$

Indirectly the sign of γ_1 depends on γ_2. These coefficients give some information about the shape of the distribution curve. Here the coefficient β_1 and γ_1 tell us about the skewness of distribution and β_2 and γ_2 are the measure of peakedness or flatness of the top of the curve.

The formula for skewness can be given in the form of the Pearson's coefficient.

$$\text{Skewness} = \frac{\sqrt{\beta_1}\,(\beta_2 + 3)}{2(5\beta_2 - 6\beta_1 - 9)}$$

ABSOLUTE MOMENT

Absolute moment about the mean M is $y_n = \frac{1}{N} \Sigma f \, | x - M |^n$

Absolute moment about the origin is $y'_n = \frac{1}{N} \Sigma f \, | x |^n$

FACTORIAL MOMENT

Factorial moment about the mean M is given by $\mu_{(n)} = \frac{1}{N} \Sigma f(x - M)^{(n)}$

$$\mu_{(n)} = \frac{1}{N} \Sigma f\left[(x - M)(x - M - 1)(x - M - 2)...(x - M - n + 1)\right]$$

Factorial moment about the origin is given by

$$\mu'_{(n)} = \frac{1}{N} \Sigma f(x)^n$$

$$\mu'_{(n)} = \frac{1}{N} \Sigma f\left[x(x - 1)(x - 2)...(x - n + 1)\right]$$

For particular values of n

$$\mu'_{(1)} = \frac{1}{N} \Sigma fx \qquad\qquad \Rightarrow \qquad\qquad \mu'_{(1)} = M$$

$$\mu'_{(2)} = \frac{1}{N} \Sigma f[x(x - 1)] = \frac{1}{N} \Sigma f(x^2 - x) = \frac{1}{N}\Sigma fx^2 - \frac{1}{N}\Sigma fx = \mu'_2 - \mu'_1$$

$$\mu'_{(3)} = \frac{1}{N} \Sigma f[x(x - 1)(x - 2)] = \frac{1}{N} \Sigma f[x(x^2 - 3x + 2)] = \frac{1}{N} \Sigma f[x^3 - 3x^2 + 2x]$$

$$\mu'_{(3)} = \frac{1}{N} \Sigma fx^3 - 3.\frac{1}{N} \Sigma fx^2 + 2.\frac{1}{N} \Sigma f$$

$$\mu'_{(3)} = \mu'_3 - 3\mu'_2 + 2\,\mu'_1$$

$$\mu'_{(4)} = \frac{1}{N} \Sigma f[x(x - 1)(x - 2)(x - 3)] = \frac{1}{N} \Sigma f[(x^2 - x)(x^2 - 5x + 6)]$$

$$= \frac{1}{N} \Sigma f[x^4 - 5x^3 + 6x^2 - x^3 + 5x^2 - 6x] = \frac{1}{N} \Sigma f[x^4 - 6x^3 + 11x^2 - 6x]$$

$$= \frac{1}{N} \Sigma fx^4 - 6.\frac{1}{N} \Sigma fx^3 + 11.\frac{1}{N} \Sigma fx^2 - 6.\frac{1}{N} \Sigma fx$$

$$\mu'_{(4)} = \mu'_4 - 6\mu'_3 + 11\mu'_2 - 6\mu'_1$$

CHARLIER'S CHECK

Charlier suggested the formula known as Charlier check to check the mistakes while calculating mean, variance and other moments. These are as follows

$$\Sigma f(\xi + 1) = \Sigma f\xi + \Sigma f \qquad\qquad\qquad ... (1)$$

$$\Sigma f(\xi+1)^2 = \Sigma f\xi^2 + 2\Sigma f\xi + \Sigma f \qquad \dots (2)$$

$$\Sigma f(\xi+1)^3 = \Sigma f\xi^3 + 3\Sigma f\xi^2 + 3\Sigma f\xi + \Sigma f \qquad \dots (3)$$

$$\Sigma f(\xi+1)^4 = \Sigma f\xi^4 + 4\Sigma f\xi^3 + 6\Sigma f\xi^2 + 4\Sigma f\xi + \Sigma f \qquad \dots (4)$$

If we calculate the values of $\Sigma f(\xi+1)$ and $\Sigma f\xi$, then it will satisfy equation (1). The values of $\Sigma f(\xi+1)^2$, $\Sigma f\xi^2$ and $\Sigma f\xi$ will satisfy equation (2). Thus it is a ready check against the mistakes in calculation.

SHEPPARD'S CORRECTION

While solving the values of mean, mode and variance, etc., in case of class interval the mid point of this interval is assumed to be representative of this class and the whole frequency is assumed to be concentrated on it. Thus the values obtained by these processes are not accurately correct. So W.F. Sheppard suggested the formula to obtain the accurate results. These are as follows :

$$\mu_{2(corrected)} = \mu_{2(calculated)} - \frac{h^2}{10}$$

$$\mu_{3(corrected)} = \mu_{3(calculated)}$$

$$\mu_{4(corrected)} = \mu_{4(calculated)} - \frac{1}{2}h^2\mu_{2(calculated)} + \frac{7}{240}h^4$$

where, h = class interval

Solved Examples

EXAMPLE 1. *The first three moments of a distribution about the point 2 of a variable are 1, 16 and –40.*

Show that the mean is 3, $\sigma^2 = 15$ and $\mu_3 = -86$.

Also find the first three moments about zero.

(VTU-2003 S)

SOLUTION.   It is given that

$A = 2$, $\mu'_1 = 1$, $\mu'_2 = 16$, $\mu'_3 = -40$

The moment about any point A is given by

$$\mu'_n = \frac{1}{N}\Sigma f(x - A)$$

$$\mu'_1 = \frac{1}{N}\Sigma f(x - 2)$$

$$1 = \frac{1}{N}\Sigma fx - \frac{1}{N}\Sigma f.2$$

$$1 = M - 2 ; \qquad M = 3 \qquad (\because N = \Sigma f)$$

We know that

$$\mu_2 = \mu'_2(2) - [\mu'_1(2)]^2 = 16 - 1 = 15$$

$$\mu_3 = \mu'_3 - 3\mu'_1\mu'_2 + 2\mu'^3_1$$

$$= -40 - 3 \times 1 \times 16 + 2 \times 1 = -40 - 48 + 2$$

$$\mu_3 = -86$$

Now, we have to find out the moments about zero.

$$\mu'_1 = \frac{1}{N}\Sigma f(x - 0) = \frac{1}{N}\Sigma fx$$

$$\mu'_1(0) = \text{Mean}$$

$$\mu'_1(0) = 3$$

$$\mu'_2 = \frac{1}{N}\Sigma f(x - 0)^2 \qquad \dots(1)$$

To calculate the value of μ'_2, we will convert the R.H.S. of (1) into the terms of central moments.

$$\mu'_2 = \frac{1}{N}\Sigma f[x - 3 + 3]^2$$

$$= \frac{1}{N}\Sigma f\left[(x-3)^2 + 6(x-3) + 9\right]$$

$$= \frac{1}{N}\Sigma f(x-3)^2 + 6\frac{1}{N}\Sigma f(x-3) + \frac{9 \times 1}{N}\Sigma f$$

$$= \mu_2 + 6\mu_1 + 9$$

$$\mu'_2(0) = 24$$

$$\mu'_3(0) = \frac{1}{N}\Sigma f(x - 0)^3 = \frac{1}{N}\Sigma\left[(x-3)+3\right]^3$$

$$= \frac{1}{N}\Sigma f\left[(x-3)^3 + 27 + 3(x-3) \times 3(x-3+3)\right]$$

$$= \frac{1}{N}\Sigma f\left[(x-3)^3 + 9x(x-3) + 27\right]$$

$$= \frac{1}{N}\Sigma f\left[(x-3)^3 + 9(x-3+3)(x-3) + 27\right]$$

$$= \frac{1}{N}\Sigma f\left[(x-3)^3 + 9(x-3)^2 + 27(x-3) + 27\right]$$

$$= \frac{1}{N}\Sigma f(x-3)^3 + 9.\frac{1}{N}\Sigma f(x-3)^2$$

$$+ 27.\frac{1}{N}\Sigma f(x-3) + 27.\frac{1}{N}\Sigma f$$

$$= \mu_3 + 9.\mu_2 + 27.\mu_1 + 27$$

$$= -86 + 9 \times 15 + 27 \times 0 + 27 = -86 + 135 + 27$$

$$\mu'_3(0) = 76$$

So, the moments about zero is

$$\mu'_1(0) = 3, \quad \mu'_2(0) = 24, \quad \mu'_3(0) = 76$$

EXAMPLE 2. *The first four moments about the value 5 of a variable are given by 2, 20, 40 and 50. Find the moments about mean.*

SOLUTION.   It is given that

$$\mu'_1 = 2, \ \mu'_2 = 20, \ \mu'_3 = 40, \ \mu'_4 = 50$$

We know that

$$\mu'_1 = \frac{1}{N}\Sigma f(x - 5)$$

$$2 = \frac{1}{N}\Sigma fx - \frac{1}{N}\Sigma f.5 = M - 5 \qquad (\Sigma f = N)$$

$$M = 7$$

We know, $\mu_1 = 0$

Second moment about mean

$$\mu_2 = \mu'_2 - \mu'^2_1 = 20 - (2)^2 = 16$$

Third moment about mean

$$\mu_3 = \mu'_3 - 3\mu'_1\mu'_2 + 2\mu'^3_1 = 40 - 3 \times 2 \times 20 + 2 \times (2)^3$$

$$= 40 - 120 + 16 = -64$$

Fourth moment about mean

$$\mu_4 = \mu_4' - 4\mu_3'\mu_1' + 6\mu_2'\mu_1'^2 - 3(\mu_1')^4$$

$$= 50 - 4 \times 40 \times 2 + 6 \times 20 \times 4 - 3 \times 16$$

$$= 50 - 320 + 480 - 48 = 162$$

EXAMPLE 3. *The four moments of a distribution about the value* $-1.5, 17, -30$ *and* 108. *Find the moments about* β_1 *and* β_2.

SOLUTION. Moment about an arbitrary point are

$$\mu_1' = -1.5, \quad \mu_2' = 17; \quad \mu_3' = -30, \quad \mu_4' = 108$$

First moment about mean

$$\mu_1 = \mu_1' - \mu_1' = 0$$

Second moment about mean

$$\mu_2 = \mu_2' - \mu_1'^2 = 17 - (-1.5)^2$$

$$\mu_2 = 14.75$$

Third moment about mean

$$\mu_3 = \mu_3' - 3\mu_1'\mu_2' + 2\mu_1'^3$$

$$= -30 - 3(-1.5)(17) + 2(-1.5)^3$$

$$\mu_2 = 39.75$$

Fourth moment about mean

$$\mu_4 = \mu_4' - 4\mu_1'\mu_3' + 6\mu_1'^2\mu_2' - 3\mu_1'^4 = 142.3125$$

Now, $\beta_1 = \dfrac{\mu_3^2}{\mu_2^3} = \dfrac{(39.75)^2}{(14.75)^3} = \dfrac{1580.0625}{3209.0469} = 0.4923$

$\beta_2 = \dfrac{\mu_4}{\mu_2^2} = \dfrac{142.3125}{(14.75)^2} = \dfrac{142.3125}{217.5625}$

$$= 0.6541224$$

 Exercise-9.12

1. Calculate the coefficient of skewness for the values given below :
25, 40, 23, 23, 15, 27, 25, 20, 25

2. Calculate the coefficient of skewness for the given frequency distribution :

Class	0-5	5-10	10-15	15-20
Frequency	5	20	10	0

Class	20-25	25-30	30-35	35-40
Frequency	5	20	8	7

3. Calculate the coefficient of skewness for the following data :

Marks Above	0	10	20	30	40	50	60	70	80
Frequency	150	140	100	80	80	70	30	14	0

4. Calculate the coefficient of skewness for the following data :

Salary	104.5	105.5	106.5	107.5
Frequency	35	40	48	100

Salary	108.5	109.5	110.5	111.5
Frequency	125	87	43	22

5. Calculate the coefficient of skewness for the given frequency distribution :

Class	0-5	5-10	10-15	15-20
Frequency	2	5	7	13

Class	20-25	25-30	30-35	35-40
Frequency	21	16	8	3

6. Check the skewness and kurtosis of the distribution for which $\mu_1 = 0$, $\mu_2 = 2.5$, $\mu_3 = 0.7$, $\mu_4 = 18.75$.

7. Find the value of mean, median, standard deviation and then the coefficient of skewness for the given frequency distribution :

Class	5-7	8-10	11-13	14-16	17-19
Frequency	7	12	19	10	2

8. For a discrete distribution prove that : (i) $\beta_2 > 1$, (ii) $\beta_2 > \beta_1$

9. Calculate the missing terms and coefficient of skewness for the given frequency distribution for which Mode = 54, A.M. = 53.4 :

Class	0-20	20-40	40-60	60-80	80-100	Total
Frequency	10	–	30	–	14	94

10. Calculate the first four moments about 5 for the given datas :
2, 4, 6, 7, 9

11. Using short-cut method, calculate the mean, variance and third central moment of the following frequency distribution :

Variable	0	1	2	3	4	5	6	7	8
Frequency	1	9	26	59	72	52	29	7	1

12. Calculate the first three central moments :
(VTU-2004, Madras-2003)

Variable	0	1	2	3	4	5	6	7	8
Frequency	1	8	28	56	70	56	28	8	1

13. Calculate the first four moments about mean of the following frequency distribution :

Class	2-4	4-6	6-8	8-10	10-12
Frequency	4	2	8	6	1

14. Calculate the first, second and third central moment about zero and then find these moments about mean :

Variable	0	1	2	3	4	5	6
Frequency	15	38	55	82	60	40	10

15. The first four moments of a distribution about the value 4 is given by –1.5, 17, –30 and 108. Find the central moments.

16. In a frequency distribution, the class width of the interval is given by 3. Then find the corrected value of μ_2, μ_3 and μ_4 where $\mu_2 = 43.353$, $\mu_3 = -9.774$, $\mu_4 = 5508.567$.

17. Calculate the first four central moments and the value of β_1 and β_2 for the given frequency distribution :

Variable	2.0	2.5	3.0	3.5	4.0	4.5	5.0
Frequency	5	38	65	92	70	40	10

18. In a given frequency distribution, the mean of the distribution is given by 10, variance = 16, $\gamma_1 = 1$, $\beta_2 = 4$. Find the first four moments about origin.

19. For any distribution, the value of mean, variance and γ_1 are given by 1, 1 and 1. Find first three moments about zero, when $\mu_3 > 0$.

Answers

1. −0.03   **2.** −0.76   **3.** 0.8   **4.** −0.245   **5.** −0.148   **6.** $\beta_1 = 0.177$, $\beta_2 = 3$

7. Mean = 11.28, Median = 11.947, S.D. = 3.15, Coeff. of skewness = −0.635   **9.** $f_2 = 16, f_4 = 24$, coefficient of skewness = −0.025

10. $\mu_1' = 0.6$, $\mu_2' = 6.2$, $\mu_3' = 9.0$, $\mu_4' = 71$ **11.** 3.973, 1.979, 0.0115   **12.** 0, 2, 0   **13.** 0, 5.3, -4.44, 62.45

14. 2.98, 11.05, 45.64, 0, 2.17, −0.22   **15.** 0, 14.75, 39.75, 142.3125   **16.** $\mu_2 = 42.603$, $\mu_3 = -9.774$, $\mu_4 = 5315.838$

17. 0, 18.13125, 0.0791, 8.033, $\beta_1 = 0.001$, $\beta_2 = 2.44$ **18.** 10, 116, 1544, 23184   **19.** 1, 2, 5

9.3 CORRELATION AND REGRESSION

Correlation is one of the most widely used statistical techniques. It is very useful in the field of biology, economics, agriculture, psychology, etc.

Here, there exists certain relationship between pairs of variables. These relationships enable us to predict certain thing. For example, an increase in rainfall results in increase in agriculture yield or increase in production of nice results in fall in price.

MULTIVARIATE AND BIVARIATE DATA

1. **Multivariate Data:** If for each unit of observation, we record the values of more than one variable, then this observation forms a multivariate data.

2. **Bivariate Data :** If for each unit of observation, we record the values of two variables, then the observation forms a bivariate data.

For example:
1. If we observe the age, height, weight and sex of each student of some college, then the observation so recorded forms a multivariate data.
2. If we measure the height and weight of a certain group of persons, we shall get a bivariate data.

CORRELATION

Whenever two variables are so related that a change in one variable results in direct or inverse change (positive or negative change) in the other and also greater the magnitude of change in the other, then the variables are said to be correlated and the relationship between the variables is known as correlation.

Definition. *When the relationship is of quantitative nature, the appropriate statistical tool for discovering and measuring the relationship and expressing it in brief formula is known as correlation.*

9.3.1 TYPES OF CORRELATION

(I) POSITIVE AND NEGATIVE CORRELATION

The correlation is said to be positive when the values of the two variables increase or decrease together, *i.e.*, an increase in one variable correspond to an increase in the other and a decrease in one variable correspond to a decrease in the other.

On the other hand, it is said to be negative when an increase in one variable corresponds to a decrease in the other and a decrease in one variable correspond to an increase in the other.

☞ REMARK
- The positive correlation is also known as direct correlation.

(2) LINEAR AND NON-LINEAR CORRELATION

If the change in one variable is always in a fixed ratio to the change in the other variable, the correlation is said to be linear. On the other hand, if the change in one variable is in a variable ratio to the change in the other variable, then the correlation is said to be non-linear or curvilinear.

For example:
1. If due to 10% increase in the currency results in a permanent increase of 30% in the general price level, then the correlation is linear.
2. If due to 10% increase in the currency results in increase sometimes of 15%, sometimes of 20% and sometimes of 30% in the general price level, then such a correlation is said to be non-linear.

☞ REMARK
- Graphically, the linear correlation is represented by a straight line.

9.3.2 PERFECT CORRELATION

When two variables change in the same direction and in the same ratio, then there is perfect positive correlation. On the other hand, if two variable changes in the same ratio but in the opposite direction, then there is a perfect negative correlation.

☞ **REMARKS**

- In case of perfect positive correlation, the coefficient of correlation is +1.
- In case of perfect negative correlation, the coefficient of correlation is –1.
- If the change in one variable has no effect on the other variable, then the correlation is completely absent and we say there is no correlation.
- In case of no correlation, the coefficient of correlation is always zero.

9.3.3 METHODS FOR FINDING THE CORRELATION

(1) SCATTER DIAGRAM

It is the simplest way of diagrammatic representation of bivariate data. Thus for the bivariate distribution (x_i, y_i), $i = 1, 2, ..., n$. If the values of the variables X and Y be plotted along the x-axis and y-axis respectively in the XY-plane, then the diagram of dots so obtained is called 'scatter diagram'.

Following are the figures of the scattered data $r > 0, r < 0, r = 0, r = \pm 1$.

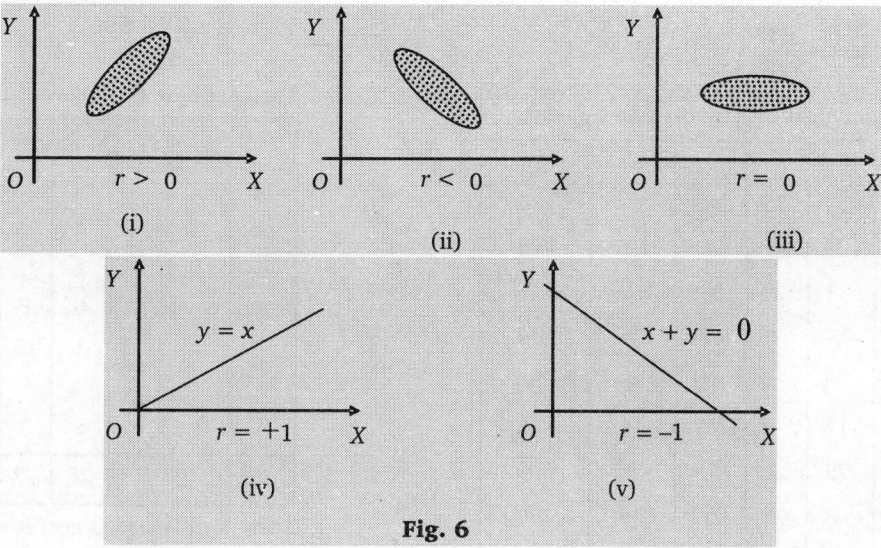

Fig. 6

If $\bar{x}$ and $\bar{y}$ be the means of the values of x and y respectively and if $(\bar{x}, \bar{y})$ be taken as origin of coordinate axis, then the points may be scattered around the origin. Then we observe that for points in first and third quadrants the product $(x - \bar{x})(y - \bar{y})$ is positive and for points in second and fourth quadrants, this is negative.

A measure of correlation between the values of x and y is given by $\sum\limits_{i=1}^{n} (x_i - \bar{x})(y_i - \bar{y})$.

☞ **REMARKS**

- If there is a positive correlation, then there is a cluster of points in first and third quadrants and if there is negative correlation then there is a cluster of points in second and fourth quadrants.
- Large positive or negative values of x correspond to large positive or negative values of y in first and third quadrants and to large negative or positive values of y in second and fourth quadrants.
- This method is not much scientific and does not give an exact amount of correlation found between the two variables.

(2) KARL PEARSON'S COEFFICIENT OF CORRELATION

The coefficient of correlation r between two variables x and y is given by the relation.

$$r = \frac{\Sigma xy}{\sqrt{\Sigma x^2 \cdot \Sigma y^2}} = \frac{Cov(x, y)}{\sqrt{Var(x) \cdot Var(y)}}$$

where x, y are the deviation about their respective means.

☞ **REMARK**

- This is also called the product moment correlation coefficient.

Solved Examples

EXAMPLE 1. *The students got the following percentage of marks in Mathematics and Statistics*:

Roll No.	1	2	3	4	5	6	7	8	9	10
Marks in maths	78	36	98	25	75	82	90	62	65	39
Marks in Statistics	84	51	91	60	68	62	86	58	53	47

Calculate the coefficient of correlation.

SOLUTION. Let marks of two subjects be denoted by X and Y respectively.

Then, Mean for X marks $= \dfrac{650}{10} = 65$ and

Mean for Y marks $= \dfrac{660}{10} = 66$

If x and y are the deviations of X's and Y's from their respective means, then we have the following table:

X	Y	x	y	x^2	y^2	xy
78	84	13	18	169	324	234
36	51	−29	−15	841	225	435
98	91	33	25	1089	625	825
25	60	−40	−6	1600	36	240
75	68	10	2	100	4	20
82	62	17	−4	289	16	−68
90	86	25	20	625	400	500
62	58	−3	−8	9	64	24
65	53	0	−13	0	169	0
39	47	−26	−19	676	361	454
650	660	0	0	5398	2224	2704

Now, $r = \dfrac{\Sigma xy}{\sqrt{\Sigma x^2 \cdot \Sigma y^2}} = \dfrac{2704}{\sqrt{5398 \times 2224}}$

$= \dfrac{2704}{73.4 \times 47.1} = 0.78$.

EXAMPLE 2. *Find the Karl Pearson coefficient of correlation for the following data* :

X	11	10	9	8	7	6	5
Y	20	18	12	8	10	5	4

SOLUTION. Here, we have

$$\bar{X} = \frac{\Sigma X}{n} = \frac{56}{7} = 8 \text{ and} \qquad \bar{Y} = \frac{\Sigma Y}{n} = \frac{77}{7} = 11$$

Then, we have the following table

X	$X - \bar{X} = x$	x^2	Y	$Y - \bar{Y} = y$	y^2	xy
11	3	9	20	9	81	27
10	2	4	18	7	49	14
9	1	1	12	1	1	1
8	0	0	8	−3	9	0
7	−1	1	10	−1	1	1
6	−2	4	5	−6	36	12
5	−3	9	4	−7	49	21
56	0	28	77	0	226	76

Then, Karl Pearson's coefficient of correlation is given by

Now, $r = \dfrac{\Sigma xy}{\sqrt{\Sigma x^2 \cdot \Sigma y^2}} = \dfrac{76}{\sqrt{28 \times 226}}$

$= \dfrac{76}{\sqrt{6328}} = \dfrac{76}{79.55} = 0.96$

(3) STEP-DEVIATION METHOD

If we take A and B as assumed means for the values of x and y respectively, then set $u_i = x_i - A$ and $v_i = y_i - B$, then the coefficient of correlation r is given by $r = \dfrac{\Sigma u_i v_i - \dfrac{1}{n}(\Sigma u_i \, \Sigma v_i)}{\sqrt{\left[\left\{\Sigma u_i^2 - \dfrac{1}{n}(\Sigma u_i)^2\right\}\left\{\Sigma v_i^2 - \dfrac{1}{n}(\Sigma v_i)^2\right\}\right]}}$.

 Solved Examples

EXAMPLE 1. *Find the coefficient of correlation for the following data:*

x	10	14	18	22	26	30
y	18	12	24	6	30	36

SOLUTION. Let us define $v = \dfrac{y_i - 22}{6}$ and $v = \dfrac{y_i - 24}{6}$

where 22 and 24 are the assumed means for x and y respectively.

Now, we have the following table

x	y	u	v	uv	u^2	v^2
10	18	−3	−1	3	9	1
14	12	−2	−2	4	4	4
18	24	−1	0	0	1	0
22	6	0	−3	0	0	9
26	30	1	1	1	1	1
30	36	2	2	4	4	4
Total		−3	−3	12	19	19

Now, $r = \dfrac{\Sigma u_i v_i - \dfrac{1}{n}(\Sigma u_i \, \Sigma v_i)}{\sqrt{\left[\left\{\Sigma u_i^2 - \dfrac{1}{n}(\Sigma u_i)^2\right\}\left\{\Sigma v_i^2 - \dfrac{1}{n}(\Sigma v_i)^2\right\}\right]}}$

$$= \frac{12 - \frac{(-3)(-3)}{6}}{\sqrt{\left[\left\{19 - \frac{(-3)^2}{6}\right\}\left\{19 - \frac{(-3)^2}{6}\right\}\right]}}$$

$$= \frac{12 - 1.5}{19 - 1.5} = \frac{10.5}{17.5} = 0.6.$$

EXAMPLE 2. *Calculate the correlation coefficient from the following data:*

x	9	8	7	6	5	4	3	2	1
y	15	16	14	13	11	12	10	8	9

SOLUTION. Let us define $u = x - 5$ and $v = y - 12$. Then, we have the following table

x	y	u	v	uv	u^2	v^2
9	15	4	3	12	16	9
8	16	3	4	12	9	16
7	14	2	2	4	4	4
6	13	1	1	1	1	1
5	11	0	−1	0	0	1
4	12	−1	0	0	1	0
3	10	−2	−2	4	4	4
2	8	−3	−4	12	9	16
1	9	−4	−3	12	16	9
n = 9		0	0	57	60	60

Putting all these values in

$$r = \frac{\Sigma uv - \frac{1}{n}(\Sigma u\, \Sigma v)}{\sqrt{\left[\left\{\Sigma u^2 - \frac{1}{n}(\Sigma u)^2\right\}\left\{\Sigma v^2 - \frac{1}{n}(\Sigma v)^2\right\}\right]}}$$

$$= \frac{57 - 0}{\sqrt{[\{60 - 0\}\{60 - 0\}]}} = \frac{57}{60} = 0.95.$$

EXAMPLE 3. *It is given that $r = 0.8$, $\Sigma XY = 60$, $\Sigma X^2 = 90$ and $\sigma_y = 2.5$. Find the number of items. Here, X and Y are deviations from the arithmetic mean.*

SOLUTION. It is given that

$$X = x - \bar{x}$$
$$Y = y - \bar{y}$$
$$\Rightarrow \quad \Sigma XY = \Sigma(x - \bar{x})(y - \bar{y}) = 60$$

and $\quad \Sigma X^2 = \Sigma(x - \bar{x})^2 = 90$

$$\sigma_x = \sqrt{\frac{\Sigma X^2}{n}} = \sqrt{\frac{90}{n}}$$

and

$$Cov(X, Y) = \frac{1}{n}\Sigma(x - \bar{x})(y - \bar{y}) = \frac{1}{n}\Sigma XY = \frac{60}{n}$$

Now, using the formula

$$r = \frac{Cov(X, Y)}{\sigma_x \cdot \sigma_y} = \frac{60}{n\sqrt{\frac{90}{n}} \times 2.5}$$

$$\Rightarrow \quad 0.8 = \frac{60}{\sqrt{90n} \times 2.5}$$

$$\Rightarrow \quad 90n \times (2.5)^2 \times (0.8)^2 = (60)^2$$

$$\Rightarrow \quad 360n = 3600$$

$$\Rightarrow \quad n = 10$$

EXAMPLE 4. *A computer while calculating the correlation coefficient between two variables x and y from 25 pairs of observations obtained the following constants*

$$n = 25, \Sigma x = 125, \Sigma x^2 = 650, \Sigma y = 100, \Sigma y^2 = 460,$$
$$\Sigma xy = 508$$

It was, however, later discovered at the time of checking that it had copied down two pairs as

x	6	8
y	14	6

while the correct values are

x	8	6
y	12	8

Obtain the correct value of correlation coefficient. (VTU–2011, SVTU–2009)

SOLUTION. Let us find corrected values by subtracting incorrect values and adding correct values

Correct $\Sigma x = 125 - 6 - 8 + 8 + 6 = 125$

Correct $\Sigma y = 100 - 14 - 6 + 12 + 8 = 100$

Correct $\Sigma x^2 = 650 - 6^2 - 8^2 + 8^2 + 6^2 = 650$

Correct $\Sigma y^2 = 460 - 14^2 - 6^2 + 12^2 + 8^2 = 436$

Correct $\Sigma xy = 508 - 6 \times 14 - 8 \times 6 + 8 \times 12 + 6 \times 8 = 520$

Now, using the formula

$$r = \frac{\Sigma xy - \frac{1}{n}\Sigma x \cdot \Sigma y}{\sqrt{\left\{\Sigma x^2 - \frac{(\Sigma x)^2}{n}\right\}\left\{\Sigma y^2 - \frac{(\Sigma y)^2}{n}\right\}}}$$

$$= \frac{520 - \frac{125 \times 100}{25}}{\sqrt{\left\{650 - \frac{(125)^2}{25}\right\}\left\{436 - \frac{(100)^2}{25}\right\}}}$$

$$= \frac{20}{\sqrt{(650 - 625)(436 - 400)}}$$

$$= \frac{20}{\sqrt{25 \times 36}} = \frac{20}{30} = 0.67.$$

EXAMPLE 5. *Show that the Pearson's coefficient of correlation r lies between –1 and 1.*

SOLUTION. Let $u = \Sigma(y - ax - b)^2$

Using principle of maxima and minima, u will be minimum if

$$\frac{\partial u}{\partial a} = -2\Sigma x(y - ax - b) = 0$$

$$\frac{\partial u}{\partial b} = -2\Sigma(y - ax - b) = 0$$

Since, $\Sigma y = 0 = \Sigma x$, we get $b = 0$

Therefore, $a = \frac{\Sigma xy}{\Sigma x^2}$

$$u = \Sigma(y - ax)^2 = \Sigma y^2 - 2a\Sigma xy + a^2\Sigma x^2$$

$$= \Sigma y^2 - \frac{2\Sigma xy}{\Sigma xy}.\Sigma xy + \left(\frac{\Sigma xy}{\Sigma x^2}\right)^2 \Sigma x^2$$

$$= \Sigma y^2 - \frac{2(\Sigma xy)^2}{\Sigma x^2} + \left(\frac{\Sigma xy}{\Sigma x^2}\right)^2 = \Sigma y^2 - \frac{(\Sigma xy)^2}{\Sigma x^2}$$

$$= \Sigma y^2 \left\{ 1 - \frac{(\Sigma xy)^2}{\Sigma x^2.\Sigma y^2} \right\}$$

$$= \Sigma y^2 (1 - r^2) \qquad \left(\because r = \frac{\Sigma xy}{\sqrt{\Sigma x^2.\Sigma y^2}} \right)$$

Since u is the sum of squares, so will not be negative. Similarly, Σy^2 will not be negative. Thus

$$1 - r^2 \geq 0$$

$$\Rightarrow \quad r^2 \leq 1 \qquad \Rightarrow \quad -1 \leq r \leq 1$$

9.3.4 Coefficient of Correlation for Grouped Distribution

For a bivariate frequency distribution, we define

$$r = \frac{\Sigma fuv - \dfrac{\Sigma fu \, \Sigma fv}{\Sigma f}}{\sqrt{\left[\left\{\Sigma fu^2 - \dfrac{(\Sigma fu)^2}{\Sigma f}\right\}\left\{\Sigma fv^2 - \dfrac{(\Sigma fv)^2}{\Sigma f}\right\}\right]}}$$

where, f = Frequency, u = Deviation of x from any assumed value A.

               v = Deviation of y from any assumed value A.

 Solved Examples

EXAMPLE 1. *Calculate the coefficient of correlation for the following table*

y	0-4	4-8	8-12	12-16
x				
0-5	7			
5-10	6	8		
10-15		5		
15-20		7	3	
20-25			2	9

SOLUTION. Let us choose some convenient origin, say interval 8-12 for x's and 13-15 for y's. Let us find out the deviations u and v (dividing all by common factor). Find out fu, fv, fu^2 and fv^2. Now, we can obtain the second column from the right as

−14 in the row $u = -2$ is obtained $7 \times (-2)$

− 20 in the row $u = -1$ is obtained $6 \times (-2) + 8(-1)$

−5 in the row $u = 0$ is obtained $7(-1) + 2 \times 0$

9 in the row $u = 2$ is obtained 9×1

In a similar manner, we can find all others. Then we get the following table :

x	0-4	4-8	8-12	12-16	f	u	fu	fu²	fv	uΣfv
y										
0-5	④ 7				7	−2	−14	28	−14	28
5-10	6	① 8			14	−1	−14	14	−20	20
10-15		⓪ 5	⓪ 3		8	0	0	0	−5	0
15-20		⊖① 7	⓪ 2		9	1	9	9	−7	−7
20-25				② 9	9	2	18	36	9	18
f	13	20	5	9	47	totals	−1	87	−37	59
v	−2	−1	0	1	total					
fv	−26	−20	0	9	−37					
fv²	52	20	0	9	81					
fu	−20	−1	2	18	−1					
vΣfu	40	1	0	18	59					

Here, $u = \dfrac{y - (10-15)}{5}$; $\quad v = \dfrac{x - (8-12)}{4}$

Now put all these values in

$$r = \frac{\Sigma fuv - \dfrac{\Sigma fu \, \Sigma fv}{n}}{\sqrt{\left[\left\{\Sigma fu^2 - \dfrac{(\Sigma fu)^2}{n}\right\}\left\{\Sigma fv^2 - \dfrac{(\Sigma fv)^2}{n}\right\}\right]}}$$

$$= \frac{59 - \dfrac{(-37)(-1)}{47}}{\sqrt{\left\{87 - \dfrac{1}{47}\right\}\left\{81 - \dfrac{37 \times 37}{47}\right\}}} = \frac{2736}{3156.8} = 0.8.$$

Exercise-9.13

1. Calculate the Karl Pearson's coefficient of correlation from the data given below :

x	4	6	8	10	12
y	2	3	4	6	10

2. Calculate the coefficient of correlation for the following data: (JNTU-2005)

x	78	89	97	69	59	79	68	57
y	125	137	156	112	107	138	123	108

3. Find the correlation between x and y from the following data :

y \ x	10-40	40-70	70-100	Total
0-30	5	20	–	25
30-60	–	28	2	30
60-90	–	32	13	45
Total	5	80	15	100

4. Calculate the coefficient of correlation between x and y from the following data :

$n = 10$, $\Sigma x = 140$, $\Sigma y = 150$, $\Sigma(x-10)^2 = 180$

$\Sigma(y-15)^2 = 215$ and $\Sigma(x-10)(y-15) = 60$

5. Find the coefficient of correlation of the following data :

x	3	5	7	12	20	22	24
y	30	25	24	16	11	9	5

6. Show that the value of Karl Pearson's coefficient of correlation lies between −1 and 1.

7. Find the coefficient of correlation between x and y when $Cov(x, y) = -16.5$, $Var(x) = 2.89$ and $Var(y) = 100$.

8. The following data regarding the heights (y) and weight (x) of 100 college students are given

$\Sigma x = 15000$ $\Sigma x^2 = 2272500$ $\Sigma y = 6800$

$\Sigma y^2 = 463025$ $\Sigma xy = 1022250$

Find the correlation coefficient between height and weight.

Answers

1. 0.95   **2.** 0.96   **3.** 0.4571   **4.** 0.915   **5.** −0.987   **7.** −0.97   **8.** 0.6

9.3.5 Rank Correlation

There are many cases, where the distribution of the variable x and y are unknown but a dependency between these can be observed. Suppose n students are examined in mathematics and the top scorer of marks is assigned a number 1, the second a number 2 and so on. Similarly, the same n students are graded according to the marks obtained by them in another subject. Then we find the correlation coefficient between the graded in the two subjects. This correlation coefficient is known as rank correlation.

Let us assume, no individuals are equal in either classification each of the individual takes the values 1, 2, ..., n. Therefore their arithmetic means are equal.

$$\frac{1+2+...+n}{n} = \frac{n+1}{2}$$

Let $x = X - Y = \left(X - \frac{n+1}{2}\right) - \left(Y - \frac{n+1}{2}\right) = x - y$, where x and y are the deviation from the mean.

$$\Sigma x^2 = \Sigma\left(X - \frac{n+1}{2}\right)^2 = \Sigma X^2 - (n+1)\Sigma X + \Sigma\left(\frac{n+1}{2}\right)^2$$

$$= \frac{n(n+1)(2n+1)}{6} - \frac{(n+1)\,n\,(n+1)}{2} + \frac{n(n+1)^2}{4} = \frac{n(n^2-1)}{12}$$

Clearly, $\Sigma X = \Sigma Y$ and $\Sigma X^2 = \Sigma Y^2$

Now, $\Sigma Y^2 = \frac{n(n^2-1)}{12}$

$\therefore$ $\Sigma d^2 = \Sigma(x-y)^2 = \Sigma x^2 + \Sigma y^2 - 2\Sigma xy$

$$\Sigma xy = \frac{1}{2}(\Sigma x^2 + \Sigma y^2 - \Sigma d^2) = \frac{1}{2}\left(\frac{n(n^2-1)}{6} - \Sigma d^2\right) = \frac{1}{12}n(n^2-1) - \frac{1}{2}\Sigma d^2$$

Putting all these values in $r = \frac{\Sigma xy}{\sqrt{\Sigma x^2 \cdot \Sigma y^2}}$, we get $r = 1 - \frac{6\,\Sigma d^2}{n(n^2-1)}$

☞ REMARKS

- This formula is known as formula for Spearmen rank correlation coefficients.
- It implies substituting for the given quantities their rank. In each series the item with the largest size is ranked 1, next largest 2 and so on.
- If ties occur, then assigned the average of the ranks, they would have received. For example, if two items are tied for 4th rank, each may be ranked $\frac{4+5}{2} = 4.5$.

- If however, some values of x_i are equal, then the coefficient of rank correlation is given by $r = 1 - \dfrac{6\left[\Sigma d^2 + \frac{1}{12}\Sigma(m^3 - m)\right]}{n(n^2 - 1)}$, where m is the number of times a particular x_i is reported.

LIMITS OF COEFFICIENT OF RANK

To show that rank correlation coefficient lies between 1 and –1.

PROOF. Since we know that $r = 1 - \dfrac{6\,\Sigma d^2}{n(n^2 - 1)}$

Now, we observe that r is maximum if Σd^2 is minimum. Since Σd^2 is always positive, so it is minimum if each d is zero.

i.e., $\Sigma d^2 = 0$. Hence, the maximum value of r is +1.

Also, r is minimum if d is maximum. d will be maximum if the ranks of the n individuals are given in the following manner.

x	1	2	3	...	$n-1$	n
y	n	$n-1$	$n-2$	...	2	1

Case 1 : If n is odd

If n is odd, then we can take $n = 2r+1$, then different $d's$ are

$$[(2r+1)-1], [(2r+1-1)-2], ...4, 2, 0, -2, -4, ... -(2r-2), -2r$$

i.e., $2r, 2r-2, 2r-4, ..., 2, 0, -2, -4, ... -(2r-2), -2r$

$$\therefore \quad \Sigma d^2 = 2\left\{(2r)^2 + (2r-2)^2 + ..., 4^2 + 2^2\right\} = 8\left[r^2 + (r-1)^2 + ... + 1^2\right] = \frac{8r(r+1)(2r+1)}{6}$$

$$\therefore \quad \frac{6\,\Sigma d^2}{n(n^2-1)} = \frac{6.8r(r+1)(2r+1)}{6.(2r+1)(4r^2+4r+1-1)} = 2 \quad \Rightarrow \quad r = 1 - \frac{6\,\Sigma d^2}{n(n^2-1)} = 1-2 = -1$$

Case II : If n is even

In this case the values of d are

$$(2r-1), (2r-3), ..., 1, -1, -3, ..., -(2r-3), -(2r-1)$$

$$\therefore \quad \Sigma d^2 = 2\left\{(2r-1)^2 + (2r-3)^2 + ... 1^2\right\} = \left[(2r)^2 + (2r-1)^2 + (2r-2)^2 + ... + 3^2 + 2^2 + 1\right] - \left[(2r)^2 + (2r-2)^2 + ... + 4^2 + 2^2\right]$$

$$= 2\left[1^2 + 2^2 + 3^2 + ... + (2r)^2 + 2^2 r^2 + 2^r(r-1)^2 + ... + 2^2\right] = 2\left[\frac{1}{6}2r(2r+1)(4r+1) - \frac{4}{6}r(r+1)(2r+1)\right] = \frac{2}{3}r(4r^2-1)$$

$$\Rightarrow \quad 1 - \frac{6\,\Sigma d^2}{n(n^2-1)} = 1 - \frac{4r(4r^2-1)}{2r(4r^2-1)} = -1. \text{ Hence, we get } -1 \le r \le 1.$$

🗑 Solved Examples

EXAMPLE 1. *Find the rank correlation coefficient from the following data*

x	10	12	15	14	19
y	40	41	48	60	50

SOLUTION. Here, the table is given as follows :

x	y	Rank in $x = R_1$	Rank in $y = R_2$	$d = R_1 \sim R_2$	d^2
10	40	5	5	0	0
12	41	4	4	0	0
15	48	2	3	–1	1
14	60	3	1	2	4
19	50	1	2	–1	1
				$\Sigma d = 0$	$\Sigma d^2 = 6$

Here, $n = 5$.

Now, $r = 1 - \dfrac{6\,\Sigma d^2}{n(n^2-1)} = 1 - \dfrac{6.6}{5(25-1)} = 1 - \dfrac{36}{120} = 0.7$

EXAMPLE 2. *Two judges ranked 10 beauty contestants as follows*

$$(x_i, y_i) : (6, 4), (4, 1), (3, 6), (1, 7), (2, 5),$$
$$(7, 8), (9, 10), (8, 9), (10, 3), (5, 2)$$

Calculate the coefficient of rank correlation.

SOLUTION.

Rank by 1st judge, x_i	Rank by 2nd judge, x_i	$d = x_i - y_i$	d^2
6	4	−2	4
4	1	−3	9
3	6	3	9
1	7	6	36
2	5	3	9
7	8	1	1
9	10	1	1
8	9	1	1
10	3	−7	49
5	2	−3	9
			$\Sigma d^2 = 128$

Here, $n = 10$.

Now, $r = 1 - \dfrac{6 \Sigma d^2}{n(n^2 - 1)} = 1 - \dfrac{6 \times 128}{10(100 - 1)}$

$= 1 - \dfrac{6 \times 128}{9900} = 1 - 0.077 = 0.923$.

EXAMPLE 3. *Compute rank correlation from the following data*

Marks in Physics	15	20	27	13	45	60	20	75
Marks in Maths	50	30	55	30	25	10	30	70

SOLUTION. Here, we have the following table :

Marks in Physics	Marks in Maths	Rank in Physics x_i	Rank in Maths, y_i	$d = x_i - y_i$	d_i^2
15	50	7	3	4	16
20	30	5.5	5	0.5	0.25
27	55	4	2	2	4
13	30	8	5	3	9
45	25	3	7	−4	16
60	10	2	8	−6	36
20	30	5.5	5	0.5	0.25
75	70	1	1	0	0

Here, $\Sigma d^2 = 81.50$ and $n = 8$.

Also, the marks in Physics $m = 2$ and in Maths $M = 3$

where m is the number of times a particular mark has been repeated. Then

$$r = 1 - \dfrac{6\left[\Sigma d^2 + \frac{1}{12}(m^3 - m) + \frac{1}{12}(M^3 - M)\right]}{n(n^2 - 1)}$$

$$= 1 - \dfrac{6\left[81.5 + \frac{1}{12}(2^3 - 2) + \frac{1}{12}(3^3 - 3)\right]}{8(8^2 - 1)} = 0$$

EXAMPLE 4. *Seven methods of imparting education were ranked by B.Tech students of two universities as follows*

Method of teaching	I	II	III	IV	V	VI	VII
Rank by students of unit A	2	1	5	3	4	7	6
Rank by students of unit B	1	3	2	4	7	5	6

Calculate rank correlation coefficient

SOLUTION. We denote the rank by students of unit A as r_1 and those by unit B by r_2. Then, we have the following table:

Method of teaching	r_1	r_2	$d = r_1 - r_2$	d^2
I	2	1	1	1
II	1	3	−2	4
III	5	2	3	9
IV	3	4	−1	1
V	4	7	−3	9
VI	7	5	2	4
VII	6	6	0	0
$n = 7$				$\Sigma d_i^2 = 28$

Putting all these values in

$$r = 1 - \dfrac{6 \Sigma d_i^2}{n(n^2 - 1)} = 1 - \dfrac{6 \times 28}{7(7^2 - 1)} = 1 - \dfrac{6 \times 28}{48 \times 7}$$

$$= 1 - 0.5 = 0.5.$$

Exercise-9.14

1. Find the coefficient of correlation between the ranks obtained by ten students in mathematics and physics in an examination as given below :

Rank in Maths	1	3	2	5	4	7	6	9	8	10
Rank in Physics	3	5	10	2	1	4	9	7	8	9

2. The coefficient of rank correlation of marks obtained by 10 students in English and Mathematics was found to be 0.5. It was later discovered that the differences in ranks in the two subjects obtained by one of the students was wrongly taken as 3 instead of 7. Find the correct coefficient of rank correlation.

3. Find the rank correlation coefficient of the following data:

x	75	30	60	80	53	35	15	40	38	48
y	85	45	54	91	58	63	35	43	46	44

4. Find rank correlation coefficient of the following data:

x	85	60	73	40	90
y	93	75	65	50	80

5. Ten competitors in a beauty contest are ranked by three judges in the following order. Use the rank correlation coefficient to determine which pair of judges has the nearest approach to common tests in beauty.

Judge X	1	6	5	10	3	2	4	9	7	8
Judge Y	3	5	8	4	7	10	2	1	6	9
Judge Z	6	4	9	8	1	2	3	10	5	7

Answers

1. 0.224   **2.** 0.2676 **3.** 0.685   **4.** 0.8   **5.** X and Z

9.4 REGRESSION

If we measure the heights and weights of a certain number of students, denote the quantity by x and y and plot them on a graph paper referring to two perpendicular axes. For each student, there shall be one point and thus we get scatter diagram.

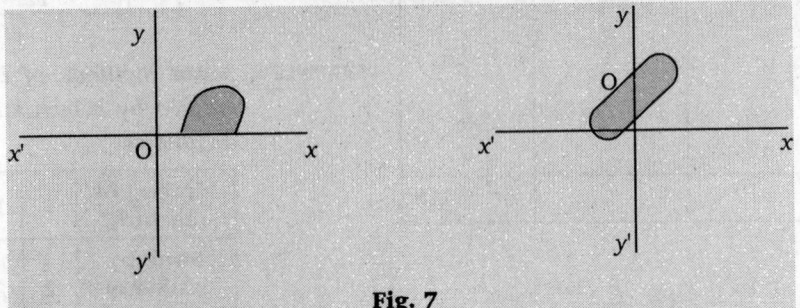

Fig. 7

If the origin of axes is taken as $(\bar{x}, \bar{y})$ where $\bar{x}, \bar{y}$ are the means of the values of x and y respectively, the points may be scattered all around the region. For points lying in I and III quadrants, the product $(x - \bar{x})(y - \bar{y})$ is positive and for those points which are lying in II and IV quadrants, it is negative.

Definition : *Let us suppose that the scatter diagram indicates some relationship between the two variables x and y, the dots of the scatter diagram will be more or less concentrated round a curve. This curve is called the curve regression.*

The straight line about which the various points may be considered as scattered is called the regression line.

✏ REMARK

- It should be noted that one can predict exactly only if the two variables are perfectly related. In that case, there is no scatter in the data and the various points lie exactly on the regression line, but when the correlation is less than perfect, *i.e.*, there is a scatter of points on the scatter diagram, then the regression line is only a representation of the general trend.

EQUATION OF THE LINE OF REGRESSION

Let $y = ax + b$ is the given equation of straight line. The method of least square can be used to fit a straight line to the set of points given on the scatter diagram. Now, transfer the origin to the points $(\bar{x}, \bar{y})$, where $\bar{x}$ and $\bar{y}$ are the means of x-series and y-series, respectively.

Suppose that x, y be the deviation from the respective means $\bar{x}$ and $\bar{y}$.

Therefore, $x = X - \bar{x}$ and $y = Y - \bar{y}$

Let $Y = aX + b$ be the equation of the line of best fit of x. Changing the origin to $(\bar{x}, \bar{y})$, we get the form

$$y = ax + b \text{ ; where } y = Y - \bar{y} \text{ and } x = X - \bar{x}.$$

Let $P(x_r, y_r)$ be any dot, then the difference between P and the line is

$$y_r - ax_r - b$$

Let I denote the sum of the squares of such distances given by

$$I = \Sigma(y - ax - b)^2 \text{ for all values of } r.$$

Now, using the principle of least squares, choose a and b such that I is minimum.

For minima of I, we must have

$$\frac{\partial I}{\partial a} = -2\Sigma x\,(y - ax - b) = 0 \Rightarrow \Sigma xy - a\Sigma x^2 - b\Sigma x = 0$$

and

$$\frac{\partial I}{\partial b} = -2\Sigma(y - ax - b) = 0 \Rightarrow \Sigma y - a\Sigma x - nb = 0$$

Since $\Sigma x = 0$, $\Sigma y = 0$, then we get

$$a = \frac{\Sigma xy}{\Sigma x^2} = \frac{r\sigma_y}{\sigma_x} \quad \text{and } b = 0 \qquad \left(\because \frac{r\sigma_y}{\sigma_x} = \frac{\Sigma xy}{\sqrt{\Sigma x^2 \, \Sigma y^2}} \times \sqrt{\frac{\Sigma y^2}{n} \frac{n}{\Sigma x^2}} \right)$$

Therefore, the line of bit is $y = r\dfrac{\sigma_y}{\sigma_x}.x$

Now, rechanging the origin, we get $Y - \bar{y} = r\dfrac{\sigma_y}{\sigma_x}(X - \bar{x})$. This is known as regression line of Y on X.

If X is taken to be dependent variable, then the regression line is $X - \bar{x} = r\dfrac{\sigma_x}{\sigma_y}(Y - \bar{y})$. This is called the regression line of X on Y.

☛ REMARKS

- If the straight line is chosen such that the sum of squares of deviation parallel to the axis of y is minimum, it is called the line of regression of y on x.
- The coefficients $r\dfrac{\sigma_y}{\sigma_x}$ and $r\dfrac{\sigma_x}{\sigma_y}$ are called the regression coefficients of y on x and of x on y, respectively.
- If $r = \pm 1$, the two regression lines will coincide.

LEAST SQUARE REGRESSION

The approach discussed above, is known as least square regression.

Here if $y = a + bx = f(x)$ is the given equation, then using the principle of least square and principle of maxima and minima, we may find the normal equations, given by

$$\Sigma x_i = na + b\Sigma y_i \quad \text{and} \quad \Sigma x_i y_i = a\Sigma x_i + b\Sigma x_i^2$$

Solving the above equations for a and b, we get

$$a = \frac{\Sigma y_i}{n} - b\frac{\Sigma x_i}{n} = \bar{y} - b\bar{x} \quad \text{and} \quad b = \frac{n\Sigma x_i y_i - \Sigma x_i \Sigma y_i}{\Sigma x_i^2 - (\Sigma x_i)^2}$$

where $\bar{x}, \bar{y}$ are the means of x-series and y-series respectively.

9.4.1 PROPERTIES OF REGRESSION COEFFICIENTS

1. Correlation coefficient is the geometric mean of the regression coefficients.
2. If one of the regression coefficient is greater than unity, the other must be less than unity.
3. Arithmetic mean of regression coefficients is greater than the correlation coefficient.
4. The regression coefficients are independent of the origin but not of scale.
5. The correlation coefficient and the two regression coefficients have same sign.

9.4.2 ANGLE BETWEEN TWO LINES OF REGRESSION

If θ is the acute angle between the two regression lines in the case of two variables x and y, then

$$\tan\theta = \frac{1-r^2}{r} \cdot \frac{\sigma_x \sigma_y}{\sigma_x^2 + \sigma_y^2} \text{ where, } r, \sigma_x, \sigma_y \text{ have their usual meaning.} \qquad \text{[UPTU-2007, VTU-2007]}$$

Proof. Equations of regression lines are given by

$$y - \bar{y} = r\frac{\sigma_y}{\sigma_x}(x - \bar{x}) \qquad \ldots (1)$$

and

$$x - \bar{x} = r\frac{\sigma_x}{\sigma_y}(y - \bar{y}) \qquad \ldots (2)$$

Slopes of (1) and (2) are given by $m_1 = r\dfrac{\sigma_y}{\sigma_x}$ and $m_2 = \dfrac{\sigma_y}{r\sigma_x}$

Now

$$\tan\theta = \pm\frac{m_2 - m_1}{1 + m_2 m_1} = \pm\frac{\dfrac{\sigma_y}{r\sigma_x} - \dfrac{r\sigma_y}{\sigma_x}}{1 + \dfrac{\sigma_y^2}{\sigma_x^2}} = \pm\frac{1-r^2}{r} \cdot \frac{\sigma_y}{\sigma_x} \cdot \frac{\sigma_x^2}{\sigma_x^2 + \sigma_y^2} = \pm\frac{1-r^2}{r} \cdot \frac{\sigma_x \sigma_y}{\sigma_x^2 + \sigma_y^2}$$

Now, since $r^2 \leq 1$ and σ_x, σ_y are positive, therefore, positive sign gives the acute angle between the lines.

Hence,

$$\tan\theta = \frac{1-r^2}{r} \cdot \frac{\sigma_x \sigma_y}{\sigma_x^2 + \sigma_y^2}$$

Also, when $r = 0$, $\theta = \dfrac{\pi}{2}$. Therefore, two lines of regression are perpendicular to each other. Thus, the estimated value of y is the same for all values of x and vice-versa.

When, $r = \pm 1$, $\tan\theta = 0$ or π. Hence, the lines of regression coincide and there is a perfect correlation between two variates x and y.

🎓 Solved Examples

EXAMPLE 1. *Fit a straight line to the following set of data*

x	1	2	3	4	5
y	3	4	5	6	8

SOLUTION. We have

x_i	y_i	x_i^2	$x_i y_i$
1	3	1	3
2	4	4	5
3	5	9	15
4	6	16	24
5	8	25	40
15	26	55	90

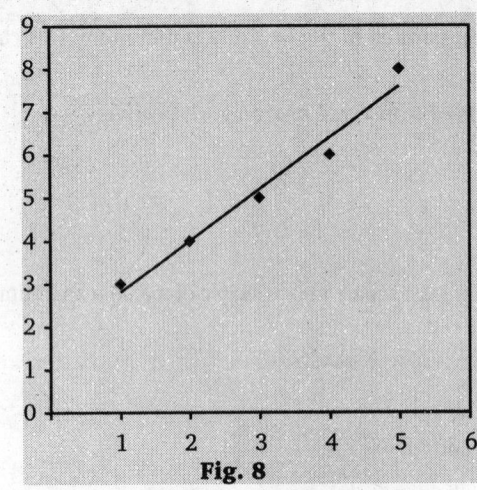

Fig. 8

Now, $b = \dfrac{n\Sigma x_i y_i - \Sigma x_i \Sigma y_i}{n\Sigma x_i^2 - (\Sigma x_i)^2} = \dfrac{5 \times 90 - 15 \times 26}{5 \times 55 - 15 \times 15} = 1.20$

and $a = \dfrac{\Sigma y_i}{n} - b\dfrac{\Sigma x_i}{n} = \dfrac{26}{5} - 1.20 \times \dfrac{5}{5} = 1.60$

Therefore, the linear equation is

$$y = 1.6 + 1.2x$$

which can be shown in the adjoining figure.

EXAMPLE 2. *If the regression coefficients are 0.8 and 0.2, what would be the value of coefficient of correlation.*

SOLUTION. It is known that $r^2 = b_{yx} \cdot b_{xy} = 0.8 \times 0.2 = 0.16$

Since r, b_{xy} and b_{yx} having same sign as both the regression coefficients b_{yx} and b_{xy}.

Hence, $r = \sqrt{0.16} = 0.4$

EXAMPLE 3. *Find linear regression from the following data:*

x	1	2	3	4	5	6	7	8
y	3	7	10	12	14	17	20	24

SOLUTION. It is known that regression coefficients are given by

$$b_{yx} = \dfrac{n\Sigma xy - \Sigma x \Sigma y}{n\Sigma x^2 - (\Sigma x)^2} \qquad \dots (1)$$

$$b_{xy} = \dfrac{n\Sigma xy - \Sigma x \Sigma y}{n\Sigma y^2 - (\Sigma y)^2} \qquad \dots (2)$$

Here, we have the following table

x	y	x^2	y^2	xy
1	3	1	9	3
2	7	4	49	14
3	10	9	100	30
4	12	16	144	48
5	14	25	196	70
6	17	36	289	102
7	20	49	400	140
8	24	64	576	192
36	107	204	1763	599

Also, $n = 8$

Putting these values from above table in (1) and (2), we get

$$b_{yx} = \dfrac{(8 \times 599) - (86 \times 107)}{(8 \times 204) - (36)^2} = 2.7976$$

$$b_{xy} = \dfrac{(8 \times 599) - (36 \times 107)}{(8 \times 1763) - (107)^2} = 0.3540 .$$

EXAMPLE 4. *The regression lines of y on x and x on y are respectively y = ax + b, x = cy + d. Show that*

$$\dfrac{\sigma_y}{\sigma_x} = \sqrt{\dfrac{a}{c}}, \quad \bar{x} = \dfrac{bc+d}{1-ac} \text{ and } \bar{y} = \dfrac{ad+b}{1-ac}.$$

SOLUTION. The regression line of y on x is given by

$$y = ax + b \qquad \dots (1)$$

$\therefore \qquad b_{yx} = a$

Similarly, $b_{xy} = c \qquad \dots (2)$

Now, $\qquad b_{yx} = r\dfrac{\sigma_y}{\sigma_x} \qquad \dots (3)$

and $\qquad b_{xy} = r\dfrac{\sigma_x}{\sigma_y} \qquad \dots (4)$

Using (3) and (4), we get

$$\dfrac{b_{yx}}{b_{xy}} = \dfrac{\sigma_y^2}{\sigma_x^2} \Rightarrow \dfrac{a}{c} = \dfrac{\sigma_y^2}{\sigma_x^2}, \text{ i.e., } \dfrac{\sigma_y}{\sigma_x} = \sqrt{\dfrac{a}{c}}$$

Since, both the regression lines pass through the point $(\bar{x}, \bar{y})$, therefore

$$\bar{y} = a\bar{x} + b, \quad \bar{x} = c\bar{y} + d$$

$\Rightarrow \qquad a\bar{x} - \bar{y} = -b \qquad \dots (5)$

$$\bar{x} - c\bar{y} = d \qquad \dots (6)$$

Multiplying (6) by a and then subtracting from (5), we get

$$(ac-1)\bar{y} = -ad - b \qquad \Rightarrow \qquad \bar{y} = \dfrac{ad+b}{1-ac}$$

Similarly, we get $\bar{x} = \dfrac{bc+d}{1-ac}$.

EXAMPLE 5. *For 10 observations on price x and supply y, the following data were obtained (in appropriate units) :*

$$\Sigma x = 130, \Sigma y = 220, \Sigma x^2 = 2288, \Sigma y^2 = 5506$$

and $\Sigma xy = 3467$

Find the two lines of regression and estimate the supply when the price is 16 units.

(UPTU(MCA)–2010)

SOLUTION. We have $n = 10$, $\bar{x} = \dfrac{\Sigma x}{n} = 13$, $\bar{y} = \dfrac{\Sigma y}{n} = 22$

Regression coefficient of y on x is

$$b_{yx} = \dfrac{n\Sigma xy - \Sigma x \Sigma y}{n\Sigma x^2 - (\Sigma x)^2} = \dfrac{(10 \times 3467) - (130 \times 220)}{(10 \times 2288) - (130)^2}$$

$$= \dfrac{34670 - 28600}{22880 - 16900} = \dfrac{6070}{5980} = 1.015$$

Therefore, regression line of y on x is

$$y - \bar{y} = b_{yx}(x - \bar{x})$$

$\Rightarrow \qquad y - 22 = 1.015(x - 13)$

$\Rightarrow \qquad y = 1.015x + 8.805$

Similarly, regression coefficient of x on y is

$$b_{xy} = \dfrac{n\Sigma xy - \Sigma x \Sigma y}{n\Sigma y^2 - (\Sigma y)^2} = \dfrac{(10 \times 3467) - (130 \times 220)}{(10 \times 5506) - (220)^2}$$

$$= \dfrac{6070}{6600} = 0.9114$$

$\therefore$ Regression line of x on y is

$$x - \bar{x} = b_{xy}(y - \bar{y})$$

$\Rightarrow \quad x - 13 = 0.9114\,(y - 22)$

$\Rightarrow \quad x = 0.9114y - 7.0508$

Now, since we are to estimate supply (y) when price (x) is given.

Therefore, we are to use regression line of y on x.

$\therefore$ When $x = 16$

$$y = 1.015(16) + 8.805 = 24.045$$

EXAMPLE 6. *The two regression equations of the variables x and y are $x = 19.13 - 0.87\,y$ and $y = 11.64 - 0.50\,x$. find (i) mean of x's, (ii) mean of y's and (iii) the correlation coefficient between x and y.*

(V.T.U-2004, Anna-2003, Burdwan-2003)

SOLUTION. Since the mean of x's and the mean of y's line on the two regression lines, we have

$$\bar{x} = 19.13 - 0.87\,\bar{y} \qquad \ldots (1)$$
$$\bar{y} = 11.64 - 0.50\,\bar{x} \qquad \ldots (2)$$

Multiplying (2) by 0.87 and subtracting from (1), we have,

$[1 - (0.87)(0.50)]\bar{x} = 19.13 - (11.64)(0.87)$

or $\qquad\qquad 0.57\bar{x} = 9.00$

or $\qquad\qquad \bar{x} = 15.79$

$\therefore \qquad \bar{y} = 11.64 - (0.50)(15.79) = 3.74$

$\therefore$ regression coefficient of y on x is -0.50 and that of x on y is -0.87.

Now since the coefficient of correlation is the geometric mean between the two regression coefficients.

$\therefore \quad r = \sqrt{[(-0.50)(-0.87)]} = \sqrt{(0.43)} = -0.66$

[–ve sign is taken since both the regression coefficients are negative].

EXAMPLE 7. *In a partially destroyed laboratory record, only the lines of regression of y on x and x on y are available as $4x - 5y + 33 = 0$ and $20x - 9y = 107$ respectively. Calculate $\bar{x}, \bar{y}$ and the coefficient of co-relation between x and y.*

[SVTU-2009, UPTU-2009, VTU-2005, SRM–2014]

SOLUTION. Since the regression lines pass through $(\bar{x}, \bar{y})$.

$\therefore \quad 4\bar{x} - 5\bar{y} + 33 = 0, 20\bar{x} - 9\bar{y} = 107$

Solving these equation, we get $\bar{x} = 13, \bar{y} = 17$

Rewriting the lines of regression of y on x as

$$y = \frac{4}{5}x + \frac{33}{5}, \text{ we get}$$

by $\quad b_{yx} = r\dfrac{\sigma_y}{\sigma_x} = \dfrac{4}{5} \qquad \ldots (1)$

Rewriting the lines of regression of x on y as

$$x = \frac{9}{20}y + \frac{107}{9}, \text{ we get}$$

$$b_{xy} = r\frac{\sigma_x}{\sigma_y} = \frac{9}{20} \qquad \ldots (2)$$

Multiplying (1) and (2), we get

$$r^2 = \frac{4}{5} \times \frac{9}{20} = 0.36$$

$\therefore \quad r = 0.6$

Hence $r = 0.6$, the positive sign being taken as b_{yx} and b_{xy} both are positive.

Exercise-9.15

1. The following table gives the various values of two variables:

x	42	44	58	55	89	98	66
y	56	49	53	58	64	76	58

Determine the regression equations which may be associated with these values.

2. Obtain the line of regression of the following data:

x	1	2	3	4	5	6	7
y	9	8	10	12	11	13	14

3. Find the regression lines using the following data : Mean height = 50.07, Mean age = 9.98, Standard deviation for height = 5.26, Standard deviation of age = 2.59 and $r = 0.898$.

4. Find the mean of the variables x and y and the coefficients of correlation, given the following regression equations $2y - x = 50, 3y - 2x = 10$.

5. Two lines of regression are given by $x + 2y - 5 = 0$ and $2x + 3y - 8 = 0$ and $\sigma_x^2 = 12$. Calculate the mean value of x and y, variance of y and the coefficient of correlation between x and y.

6. Use multiple linear regression to fit:

x_1	1	2	3	4	5
x_2	4	3	2	1	0
y	18	16	16	12	10

Compute coefficients and the error of estimate.

7. Obtain a regression plane to fit the following data:

x_1	5	4	3	2	1
x_2	3	-2	-1	4	0
y	15	-8	-1	26	8

8. The mean of bivariate frequency distribution are at (3, 4) and $r = 0.4$. The line of regression of y on x is parallel to the line $y = x$. Find the two lines of regression and estimate value of x when $y = 1$.

9. Find the multiple linear regression of x_1 on x_2 and x_3 using the following data:

x_1	4	6	7	9	23	15
x_2	15	12	8	6	4	3
x_3	30	24	20	14	10	4

Answers

1. $y = 0.372x + 35.27$, $x = 2.2y - 65.9$

2. $Y = 0.929x + 7.284$, $X = 0.929Y - 6.219$

3. $y = 0.422x - 12.15$, $x = 1.825y + 31.86$

4. Mean of $X = 130$, Mean of $Y = 90$, $r = 0.866$

5. $1, 2$, $\sigma_y^2 = 6$, $r = 0.86$

8. $y = x + 1$, $x = 0.16y + 2.36$, $x = 2.52$

1. The following are scores of two batsmen A and B is a series of innings:

A	12	115	6	73	7	19	119	36	84	29
B	47	12	16	42	4	51	37	48	13	0

Who is the better score getter and who is more consistent?
(VTU-2004)

2. Show that the variance of the first n positive integers is $\frac{1}{12}(n^2 - 1)$. (VTU-2003)

3. The mean of five items of an observation is 4 and the variance is 5.2. If three of the items are 1, 2 and 6, then find the other two (VTU-2002)

4. The following table shows the marks obtained by 100 candidates in an examination. Calculate the mean, median and standard deviation :

Marks obtained :	1-10	11-20	21-30	31-40	41-50	51-60
No. of Candidates :	3	16	26	31	16	8

(Osmania-2003S, VTU-2003S)

5. The following table gives the monthly wages of 72 workers in a factory. Compute the standard deviation, quartile deviation, coefficients of variation and skewness. (VTU-2001)

Monthly wages (in ₹)	No. of Workers	Monthly wages (in ₹)	No. of Workers
12.5 – 17.5	2	37.5-42.5	4
17.5 – 22.5	22	42.5-47.5	6
22.5 – 27.5	19	47.5-52.5	1
27.5 – 32.5	14	52.5-57.5	1
32.5 – 37.5	3		

6. Find Pearson's coefficient of skewness for the following data :
(VTU- 2000S)

Class	10-19	20-29	30-39	40-49
Frequency	5	9	14	20

Class	50-59	60-69	70-79	80-89
Frequency	25	15	8	4

7. Compute the quartile coefficient of skewness for the following distribution:

x	3-7	8-12	13-17	18-22	23-27	28-32	33-37	38-42
f	2	108	580	175	80	32	18	5

(Madras-2002, VTU-2000)

Also compute the measure of skewness based on the third moment.

8. Compute skewness and kurtosis, if the first four moments of a frequency distribution $f(x)$ about the value $x = 4$ are respectively 1, 4, 10 and 45. (Coimbatore, 1999)

9. Psychological tests of intelligence and of engineering ability were applied to 10 students. Here is a record of ungrouped data showing intelligence ratio (I.R.) engineering ratio (E.R.). Calculate the coefficient of correlation.

Student	A	B	C	D	E	F	G	H	I	J
I.R.	105	104	102	101	100	99	98	96	93	92
E.R	101	103	100	98	95	96	104	92	97	94

(Andhra, 2000)

10. The correlation table given below shows that the ages of husband and wife of 53 married couples living together on the census night of 1991. Calculate the coefficient of correlation between the age of the husband and that of the wife.

Age of husband	Age of wife						Total
	15-25	25-35	35-45	45-55	55-65	65-75	
15-25	1	1	–	–	–	–	2
25-35	2	12	1	–	–	–	15
35-45	–	4	10	1	–	–	15
45-55	–	–	3	6	1	–	10
55-65	–	–	–	2	4	2	8
65-75	–	–	–	–	1	2	3
Total	3	17	14	9	6	4	53

11. Establish the formula $r = \dfrac{\sigma_x^2 + \sigma_y^2 - \sigma_{x-y}^2}{2\sigma_x \sigma_y}$

Hence, Calculate r from the following data:

x	21	23	30	54	57	58	72	78	87	90
y	60	71	72	83	110	84	100	92	113	135

(UPTU-2002)

12. Ten participants in a contest are ranked by two judges as follows:

x	1	6	5	10	3	2	4	9	7	8
y	6	4	9	8	1	2	3	10	5	7

Calculate the rank correlation coefficient. (VTU- 2002)

13. Three judges, A, B, C give the following ranks. Find which pair of judges has common approach.

A	1	6	5	10	3	2	4	9	7	8
B	3	5	8	4	7	10	2	1	6	9
C	6	4	9	8	1	2	3	10	5	7

(JNTU-2003)

14. Find the correlation co-efficient and the regression lines of y on x and x on y for the following data:

x	1	2	3	4	5
y	2	5	3	8	7

(VTU-2010)

15. Find two lines of regression and coefficient of correlation for the data given below:

$n = 18, \Sigma x = 12, \ \Sigma y = 18, \Sigma x^2 = 60, \ \Sigma y^2 = 96, \Sigma xy = 48$

(UPTU (MCA) 2009)

16. If the coefficient of correlation between two variables x and y is 0.5 and the acute angle between their lines of regression is $\tan^{-1}(3/8)$, show that $\sigma_x = \frac{1}{2}\sigma y$. (VTU-2004)

17. For two random variables x and y with the same mean, the two regression lines are $y = ax + b$ and $x = ay + b$. Show that $\frac{b}{\beta} = \frac{1-a}{1-\alpha}$. Find also the common mean. (UPTU-2010)

18. Find the rank correlation for the following data:

x	56	42	72	36	63	47	55	49	38	42	68	60
y	147	125	160	118	149	128	150	145	115	140	152	155

(SVTU-2009, JNTU-2003)

Answers

1. *A* is better score getter than *B* and *B* is more consistent than *A*. 3. 4, 7 4. 32, 32.6, 12.4
5. 8.85, 5.25, 0.32, 1.09 6. − 0.2064 7. 0.22, 1.157 8. 0, 2.9 9. 0.59
10. 0.91 (approx.) 11. (a) $\sigma_2^2 = \sigma_x^2 + \sigma_y^2 - 2r\sigma_x\sigma_y$ (b) 0.876 12. 0.6 nearly 13. *A* and *C*
14. $r = 0.81; x = 0.5y + 0.5, y = 1.3x + 1.1$ 15. $r = 0.632; y = 0.467 + 0.8x, x = 0.167 + 0.5y$
17. $m = (\beta - b)/(a - \alpha)$ 18. 0.932

9.5 PROBABILITY AND DISTRIBUTIONS

The word 'probability' is very commonly used in our daily life in different forms. For example, most probably, he will pass with second division in this year or probably there may be very hot today. These sentences convey the sense of probability of happening of an event. Thus, we can define the probability as follows : "*A mathematical measure of uncertainty is known as probability*".

(1) **Sample space :** Consider an experiment whose outcome is not predicted with certainty in advance. However, although the outcome of the experiment will not be known in advance, let us suppose the set of all possible outcomes is known. Then, the set of all possible outcomes of an experiment is known as the sample space.

For example :
  1. If the experiment consists of two dice, then the sample space consists of 36 points, *i.e.*,
  $$S = \{(i, j) : i, j = 1, 2, ..., 6\}$$
  where outcome (i, j) is said to occur if i appears on the left most dice and j on the other dice.
  2. If the outcome of an experiment consist in the determination of the sex of a newborn baby, then the sample space $S = \{g, b\}$, where the outcome g means the child is girl and b is that it is a boy.

(2) **Trial and Events :** Consider an experiment which though repeated under same conditions not give unique results, but may result in many one of the several possible outcomes. Then, the experiment is known as trial and these outcomes are known as events or cases.

For example :
  1. Tossing of a coin is a trial and getting head or tail is an event.
  2. Drawing two cards from a pack of cards is a trial and getting of a specific card (king, joker or queen) are events.

(3) **Exhaustive Events :** A set of events is said to be exhaustive, if it includes all possible events.

Definition. *The total number of possible outcomes in any trial is known as exhaustive events.*

For example :
  1. In tossing a coin, there are two exhaustive cases, either a head or a tail and there is no any other possibility.
  2. In throwing of a dice, there are 6 exhaustive cases, since any one of the 6 faces 1, 2, 3, ..., 6 may come uppermost.

(4) **Mutually Exclusive Events :** The events are said to be mutually exclusive if the happening of any one of them precludes the happening of all others, *i.e.*, if no two or more of them can happen simultaneously in the same trial.

For example :
  1. In tossing of a coin, either head comes up or the tail and both cannot happen together. Thus, these are mutually exclusive events.
  2. In throwing of a coin, all the six faces are mutually exclusive, since if any one of these faces comes, the possibility of other in the same trial is ruled out.

(5) **Equally Likely Events :** When there is no reason to except the happening of one event in preference to the other, then events are known as equally likely events.

For example :

1. In tossing of a coin, the coming of the head or the tail is equally likely.

2. In throwing a unbiased dice, all the six faces are equally likely.

(6) **Independent Events :** Some events are said to be independent if the happening (or non-happening) of an event is not affected by the supplementary knowledge concerning the occurrence of any number of the remaining events. **For example :** In tossing of a coin, the event of getting a head in the first toss is independent of getting a head in the second, third and so on..

(7) **Complimentary Events :** If A denotes an event, then $\bar{A}$ denote an event which includes all the sample points not included in A.

For example : The complimentary event of an odd number falling up in the throw of a dice is the coming up of an even number.

(8) **Compound Event :** When two or more events occur in relation with each other, they are known as compound events.

For example :

1. If a dice is rolled two times, the events of coming up of 1 in the first throw and six in the second throw is a compound event.

2. If four cards are drawn in succession from a pack of cards, the coming of the card of the same suit all the four times is a compound event.

📝 **REMARKS**

- Compound event may be independent or dependent.
- Every compound event can be expressed as the union of simple events corresponding to the same experiments.

(9) **Favourable Events :** The outcomes which make necessary the happening of an event in a trial are called favourable events.

For example : If two dice are thrown, the number of favourable events of getting a sum of 5 is four, *i.e.*, (1, 4), (2, 3), (3, 2) and (4, 1).

9.5.1 CLASSICAL (OR PRIOR) PROBABILITY

If there are n exhaustive mutually exclusive and equally likely events of which m are favourable to an event A, the probability P of the happening of A is $P(A) = m/n$, *i.e.*,

$$P(A) = \frac{\text{Number of favourable cases}}{\text{Number of exhaustive cases}} = \frac{m}{n}$$

📝 **REMARK**

- If m is the number of favourable cases out of n exhaustive mutually exclusive and equally likely cases, then the number of unfavourable cases is $n - m$.

ODDS IN FAVOUR OF AN EVENT A

As there are $(n - m)$ cases in which A will not happen, therefore the chance of A not happening is q or $P(A')$ so that

$$q = \frac{n-m}{n} = 1 - \frac{m}{n} = 1 - p$$

$$\Rightarrow \qquad\qquad P(A') = 1 - P(A)$$

$$\Rightarrow \qquad\qquad P(A) + P(A') = 1$$

Clearly, $P(A)$ and $P(A')$ are non-negative and can not exceed unity.

Hence, $0 \le P(A) \le 1$ and $0 \le P(A') \le 1$.

The probability $\dfrac{m}{n}$ of happening of an event is sometimes expressed as 'the odds' are m to $n - m$ in favour of the event or $(n - m)$ to m against the events.

📝 **REMARKS**

- Odds in favour of an event A is defined as the ratio of number of favourable cases to the number of unfavourable cases.
- If an event is certain to happen, then its probability is unity, while if it is certain not to happen, its probability is zero. In first case, the events are called certain or sure events while the later case, A is called impossible event.

📁 Solved Examples

EXAMPLE 1. *In a single throw with two dice, what is the chance of throwing a sum of* 5?

SOLUTION . Since there are 6 numbers on each dice, therefore, the total number of possible outcome is $6 \times 6 = 36$. The sum of the numbers appearing on the upper faces of two dice can be in four ways, *i.e.*, (1, 4), (2, 3), (3, 2), (4, 1). Thus, the required probability of throwing 5

$$= \frac{\text{No. of favourable cases}}{\text{No. of exhaustive cases}} = \frac{4}{36} = \frac{1}{9}.$$

EXAMPLE 2. *A dice is thrown, find*

    (i) P(*Prime number*) *(ii)* P(*a number* ≥ 3)

   (iii) P(*a number* ≤ 4)

   (iv) P(*a number more than* 6)

   (v) P(*a number less than* 6).

SOLUTION . The number of exhaustive cases = 6.

(i) The no. of favourable cases, $m = 3$

$$(\because 2, 3, 5 \text{ are primes.})$$

$$\therefore P \text{ (prime number)} = \frac{m}{n} = \frac{3}{6} = \frac{1}{2}.$$

(ii) m, the number of favourable cases = 4

$$(\because \text{ possible numbers are } 3, 4, 5, 6.)$$

$$\therefore P \text{ (a number} \geq 3) = \frac{m}{n} = \frac{4}{6} = \frac{2}{3}.$$

(iii) m, the number of favourable cases = 4

$$(\because \text{ possible numbers are } 1, 2, 3, 4.)$$

$$\therefore P \text{ (a number} \leq 4) = \frac{m}{n} = \frac{4}{6} = \frac{2}{3}.$$

(iv) m, the number of favourable cases = 0

$$(\because \text{ There is no number} > 6.)$$

$$\therefore P \text{ (a number} > 6) = \frac{m}{n} = \frac{0}{6} = 0.$$

(v) m, the number of favourable cases = 5

$$(\because \text{ possible numbers are } 1, 2, 3, 4, 5.)$$

$$\therefore P(\text{a number less than } 6) = \frac{m}{n} = \frac{5}{6}.$$

EXAMPLE 3. *A bag contains 5 black and 3 white balls. Two balls are drawn at random. Find the probability of drawing*

    (i) 2 black balls   (ii) Two white balls

SOLUTION . (i) Here, two black balls can be drawn out of 5 in

$$^5C_2 = \frac{5 \times 4}{1 \times 2} = 10 \text{ ways.}$$

Thus, m, the number of favourable cases = 10.

Total number of balls = 5 + 3 = 8.

Two balls can be drawn out of 8 in

$$^8C_2 = \frac{8 \times 7}{1 \times 2} = 28 \text{ ways.}$$

$\therefore n$, the number of exhaustive cases = 28.

Thus, probability of drawing two black balls

$$= \frac{m}{n} = \frac{10}{28} = \frac{5}{14}$$

(ii) Now, two white balls can be drawn out of 3 in $^3C_2$ ways.

i.e., $^3C_2 = \frac{3 \times 2}{1 \times 2} = 2$ ways.

$\therefore$ The number of favourable cases $m = 3$

and the number of exhaustive cases = 28.

                 (same as part (i))

Hence, probability of drawing two white balls

$$= \frac{m}{n} = \frac{3}{28}.$$

EXAMPLE 4. *What is the chance that leap year, selected at random will contain 53 Sundays?* (MADRAS 2003)

SOLUTION . It is known that the total number of days in a leap year is 366. Thus, it contains 52 complete weeks and two extra days. The probability of combination of these two days is given as

(i) Monday and Tuesday

(ii) Tuesday and Wednesday

(iii) Wednesday and Thursday

(iv) Thursday and Friday

(v) Friday and Saturday

(vi) Saturday and Sunday

(vii) Sunday and Monday

Of these 7 equally likely cases, the last two contain Sunday. Therefore, last two are favourable. Hence, the required probability is given by 2/7 .

EXAMPLE 5. *Find the chance of throwing exactly* 10 *in one throw with* 3 *dice.*

SOLUTION . Possible chance of throwing exactly 10 with three dice only once are

$(1, 3, 6), (1, 6, 3), (3, 1, 6), (6, 1, 3), (3, 6, 1), (6, 3, 1), (1, 4, 5),$
$(1, 5, 4), (4, 1, 5), (5, 1, 4), (5, 4, 1), (4, 5, 1), (2, 2, 6), (2, 6, 2),$
$(6, 2, 2), (2, 3, 5), (2, 5, 3), (3, 2, 5), (5, 3, 2), (3, 5, 2),$
$(2, 4, 4), (4, 2, 4), (4, 4, 2)$

Total favourable = 27

Total possible ways = $6 \times 6 \times 6 = 216$

Thus, required chances = $\frac{27}{216} = \frac{9}{72} = \frac{1}{8}$

EXAMPLE 6. *What is the chance that a non leap year selected at random will contain 53 Sundays?*

SOLUTION . We know that a non-leap year consists of 365 days. Therefore in a non-leap year, there are 52 complete weeks and one extra which can be any one of the seven days of a week. Hence, the probability that single day is Sunday is $\frac{1}{7}$.

EXAMPLE 7. *A class consist of 10 boys and 8 girls. Three students are selected at random. Find the probability that the selected group has : (i) all boys, (ii) all girls, (iii) 2 boys and 1 girl.*

SOLUTION . (i) Required probability P (all boys)

$$= \frac{^{10}C_3}{^{18}C_3} = \frac{\frac{10 \times 9 \times 8}{3 \times 2 \times 1}}{\frac{18 \times 17 \times 16}{3 \times 2 \times 1}} = \frac{5}{34}.$$

(ii) Required probability P (all girls)

$$= \frac{^8C_3}{^{18}C_3} = \frac{\frac{8 \times 7 \times 6}{3 \times 2 \times 1}}{\frac{18 \times 17 \times 16}{3 \times 2 \times 1}} = \frac{7}{102}$$

(iii) Required probability (2 boys and 1 girl)

$$= \frac{^{10}C_2 \times {}^8C_1}{^{18}C_3} = \frac{\frac{10 \times 9}{2 \times 1} \times 8}{\frac{18 \times 17 \times 16}{3 \times 2 \times 1}} = \frac{15}{34}.$$

EXAMPLE 8. *A bag contains 50 tickets numbered 1, 2, 3, ..., 50 of which five are drawn at random and arranged in ascending order of magnitude* $(x_1 < x_2 < x_3 < x_4 < x_5)$. *Find the probability that* $x_3 = 30$.

SOLUTION . We know that five tickets out of 50 can be drawn in $^{50}C_5$ ways. Thus, the number of exhaustive cases, $n = {}^{50}C_5$.

Now, since $x_1 < x_2 < x_3 < x_4 < x_5$

and $\qquad\qquad x_3 = 30$

$\Rightarrow \qquad\qquad x_1, x_2 < 30$

Therefore, x_1 and x_2 are to be drawn from 1 to 29, which can happen in $^{29}C_2$ ways.

Further, we have $x_4, x_5 > 30$

Therefore, $x4$ and $x5$ are to be drawn from 31 to 50, which can happen in $^{20}C_2$ ways.

Thus, the number of favourable cases, $m = {}^{29}C_2 \times {}^{20}C_2$

Hence, the required probability

$$= \frac{m}{n} = \frac{{}^{29}C_2 \times {}^{20}C_2}{{}^{50}C_5}$$

$$= \frac{\frac{29 \times 18}{1 \times 2} \times \frac{20 \times 19}{1 \times 2}}{\frac{50 \times 49 \times 48 \times 47 \times 46}{1 \times 2 \times 4 \times 5}} = \frac{551}{15134}$$

EXAMPLE 9. *A bag contain 20 tickets numbered 1 to 20. Two tickets are drawn at random. Find the probability that both the numbers on the tickets are prime.*

SOLUTION . We have

The number of exhaustive cases, $n = {}^{20}C_2$

Also, prime numbers from 1 to 20 are 2, 3, 5, 7, 11, 13, 17, 19, Total = 8

Thus, the number of favourable cases, $m = {}^{8}C_2$

Hence, required probability

$$= \frac{m}{n} = \frac{{}^{8}C_2}{{}^{20}C_2} = \frac{8 \times 7}{20 \times 19} = \frac{14}{95}.$$

EXAMPLE 10. *Four digit numbers are formed by using the digits 1, 2, 3, 4 and 5 without repeating any digit. Find the probability that a number chosen at random is an odd number.*

SOLUTION . We have

total number of exhaustive cases,

$$n = {}^{5}P_4 = 5 \times 4 \times 3 \times 2 \times 1 = 120$$

Let m be the favourable number of cases.

Th	H	T	U
×	×	×	×

Unit's place can be filled by any of 1, 3, 5 and 3 ways. Now, we are left with 4 digits. Therefore, remaining three places can be filled up in $^{4}P_3$ ways, *i.e.*, 24 ways.

Thus, total number of favourable cases $= 3 \times 24 = 72$

Hence, the required probability $= \frac{m}{n} = \frac{72}{120} = \frac{3}{5}.$

Exercise-9.16

1. The letters of the word 'SOCIETY' are placed at random in a row. What is the probability that the three vowels come together.

2. A box contains 5 white and 4 red balls. What is the probability of getting 3 white balls from the box.

3. Two cards are drawn from a well shuffled pack of 52 cards one after the other without replacement. Find the probability that one of these is a queen and other is a king of opposite colour.

4. Two dice are thrown together. What is the probability that the sum of the numbers on the two faces is neither 9 or 11?

5. The letter of the word 'CLIFTON' are placed at random in a row. What is the chance that two vowels come together?

6. Find the probability of getting the sum as a prime number when two dice are thrown together.

7. A box contains 9 balls, two of which are red, three blue and four black. Three balls are drawn from the box at random.

What is the probability that
(i) The three balls are of different colours.
(ii) Two balls are of the same colour and third is different.
(iii) The balls are of the same colour.

8. What is the probability that a number selected from 1, 2, 3, ..., 25 is a prime number if each of the 25 numbers is equally likely to be selected?

9. Three unbiased coins are tossed once. Find the probability of getting

  (i) two heads   (ii) one head or two heads
(iii) at least two heads   (iv) at most two heads

10. Find the probability of event given in the following experiments :
  (i) An even number appears in the toss of a fair dice.
  (ii) One or more heads appear in the toss of three fair coins.
  (iii) A red ball drawn in a random drawing of one ball from a box containing four white, three red and five blue balls.

Answers

1. 1/7   **2.** 5/42   **3.** 2/663 **4.** 5/6   **5.** 2/7   **6.** 5/12   **7.** (i) 2/7   (ii) 55/84   (iii) 5/84
8. 9/25   **9.** (i) 3/8 (ii) 3/4 (iii) 1/2 (iv) 7/8   **10.** (i) 1/2 (ii) 7/8 (iii) 1/4

9.5.2 Theorems on Probability

THEOREM 1. *Let S be a sample space and A be an event in a random experiment. Then*

    (*i*) $P(A) \geq 0$ (*ii*) $P(\phi) = 0$ (*iii*) $P(S) = 1$

SOLUTION . Since S be a sample space and A be an event, therefore $A \subseteq S$

Now, (i) $P(A) = \frac{n(A)}{n(S)} \geq 0$ (ii) $P(\phi) = \frac{n(\phi)}{n(S)} = \frac{0}{n(S)} = 0$ (iii) $P(S) = \frac{n(S)}{n(S)} = 1$

Remarks

• From above theorem, we conclude that
  (i) Probability of occurrence of an event is always non-negative.
  (ii) Probability of an impossible event is 0.
  (iii) Probability of sure event is 1.

THEOREM 2. *If A and B are mutually exclusive events, then $P(A \cap B) = 0$.*

SOLUTION. Let A and B be two mutually exclusive events in a sample space S. Now, since A and B are mutually exclusive. Therefore

$$A \cap B = \phi$$

$$\Rightarrow \quad P(A \cap B) = P(\phi) \Rightarrow P(A \cap B) = 0$$

THEOREM 3. *If A and B are two mutually exclusive events, then $P(A) + P(B) = 1$.*

SOLUTION. Let A and B be two mutually exclusive events of a sample space S. Since A and B are mutually exclusive, therefore

$$S = A \cup B \quad \text{and} \quad A \cap B = \phi$$

Therefore, $\qquad P(A) + P(B) = P(A \cup B) = P(S) = 1$

THEOREM 4. **(Addition Law of Probability).** *If A and B are two events. Then*

$$P(A \cup B) = P(A) + P(B) - P(A \cap B) \quad \text{or} \quad P(A + B) = P(A) + P(B) - P(AB)$$

SOLUTION. Let the number of elements in the sample space S be r and the number of elements in the events A and B be m_1 and m_2 respectively, *i.e.,*

$$n(S) = r, \quad n(A) = m_1 \text{ and } n(B) = m_2 \qquad \text{....(1)}$$

Now, as A and B are two events (not necessarily mutually exclusive) so they will have some common elements and let us suppose there are k common elements between A and B. Therefore, the number of elements in $A \cap B$ is k.

$\Rightarrow$ Number of elements in $A \cup B = (m_1 + m_2) - k$

i.e., $\qquad n(A \cap B) = k \quad \text{and} \quad n(A \cup B) = (m_1 + m_2) - k$

Therefore, $\qquad P(A \cup B) = \dfrac{n(A \cup B)}{n(S)} = \dfrac{m_1 + m_2 - k}{r}$

$$= \frac{m_1}{r} + \frac{m_2}{r} - \frac{k}{r} = \frac{n(A)}{n(S)} + \frac{n(B)}{n(S)} - \frac{n(A \cap B)}{n(S)} = P(A) + P(B) - P(A \cap B)$$

Hence, $\qquad P(A \cup B) = P(A) + P(B) - P(A \cap B)$

REMARKS

- If A and B are mutually exclusive elements, then there is no common element in A and B, *i.e.,* $A \cap B = \phi \Rightarrow P(A \cap B) = 0$. Hence, $P(A + B) = P(A \cup B) = P(A) + P(B)$.

- The above result can be generalized as follows : "If $A_1, A_2,..., A_n$ are n mutually exclusive events, then
$$P(A_1 + A_2 + ... + A_n) = P(A_1 \cup A_2 \cup ... \cup A_n) = P(A_1) + P(A_2) + ... + P(A_n).$$

THEOREM 5. *For any two events A and B, $P(A - B) = P(A) - P(A \cap B)$.*

SOLUTION. Let A and B be two compatible events in a sample space S. By set theory, we have

$$(A - B) \cap (A \cap B) = \phi \text{ and } (A - B) \cup (A \cap B) = A$$

Therefore, $\qquad P(A) = P[(A - B) \cup (A \cap B)] = P(A - B) + P(A \cap B)$

Hence, $\qquad P(A - B) = P(A) - P(A \cap B)$

THEOREM 6. *For each event A, $P(\overline{A}) = 1 - P(A)$, where $\overline{A}$ is the complementary event.*

SOLUTION. Let A be any event in a sample space S. We know that

$$(A - B) \cap (A \cap B) = \phi \qquad \text{and} \qquad (A - B) \cup (A \cap B) = A$$

Then, $\qquad A \subseteq S$

Also, $\qquad A \cap \overline{A} = \phi \text{ and } A \cup \overline{A} = S$

Now, since A and $\overline{A}$ are mutually exclusive, therefore

$$P(A) + P(\overline{A}) = P(A \cup \overline{A}) = P(S) = 1$$

$$\Rightarrow \qquad P(A) = 1 - P(\overline{A})$$

THEOREM 7. *If A and B be two events such that $A \leq B$. Then $P(A) \leq P(B)$.*

SOLUTION. Since $A \subset B$, then

$$B = A \cup (B - A)$$

and $\qquad A \cap (B - A) = \phi$

$$\therefore \qquad P(B) = P[A \cup (B - A)] = P(A) + P(B - A) \qquad \text{...(1)}$$

Now, since $P(B - A) \geq 0$, so from (1) $P(A) \leq P(B)$.

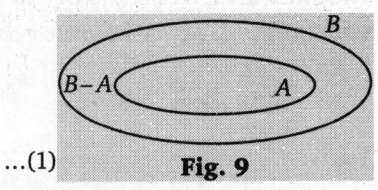

Fig. 9

THEOREM 8. *If A is an event associated with a random experiment, then $0 \leq P(A) \leq 1$.*

SOLUTION. Let A be any event in a sample space S. Then we have $\phi \leq A \qquad \text{and} \qquad A \leq S$

$$\Rightarrow \qquad P(\phi) \leq P(A) \quad \text{and} \quad P(A) \leq P(S) \qquad \text{...(1)}$$

But we know that

$$P(S) = 1 \qquad \text{and} \qquad P(\phi) = 0 \qquad \text{...(2)}$$

From (1) and (2), we conclude that $0 \leq P(A) \leq 1$.

9.5.3 CONDITIONAL PROBABILITY

If there be two events E_1 and E_2, then the happening of E_1 with the condition that E_2 has already happened is denoted by $E_1 | E_2$ and the probability of the event E_1 when the event E_2 has already happened is known as conditional probability. It is denoted by $P(E_1/E_2)$ and read as 'probability of E_1 given that E_2 has occurred.'

In order to calculate $P(E_1/E_2)$, we take elementary events favourable to the occurrence of E_2 as the new sample space and then we find how many of these are favourable to the occurrence of E_1.

Therefore

$$P(E_1 | E_2) = \frac{\text{No. of elementary events favourable to } E_2 \text{ which are also favourable to } E_1}{\text{Number of elementary events favourable to } E_2}$$

9.5.4 DEPENDENT AND INDEPENDENT EVENTS

Let E_1 and E_2 be two events of any sample space S. If there is no effect of the occurrence of any of these events on the probability of the other, then two events are independent of others.

For example : If a bag contains 6 black and 7 white balls and if one ball is drawn at random from the bag and is not put back in the bag and a second ball is drawn, then drawing of the second ball is dependent on that of the first. But if the first ball is put back in the bag before second is drawn then the drawing of the second ball is independent of the drawing of the first.

Definition (1). *Two events are said to be independent if the occurrence of one does not depend upon the occurrence of the other.*

Definition (2). *Two random experiments are said to be independent if for each pair of events E_1 and E_2 associated with the first and second experiment respectively, the probability of simultaneous occurrence of E_1 and E_2 is the product of $P(E_1)$ and $P(E_2)$.*

THEOREM 1. **(Multiplication Law of Probability).** *The probability of combined happening of two events A and B is the product of the probability of the event B and the conditional probability of A on the assumption that B had already happened.*

i.e., $\quad\quad P(A \cap B) = P(B) . P(A | B)$, *where* $B \neq \phi$ *or* $\quad\quad P(AB) = P(B) . P(A | B)$

SOLUTION . Let S be the sample space and A, B be two events. Now, $B \subset S$, $B \neq \phi$ and B has happened.

All the elements of S have not happened and only those elements of S have happened which are included in B, thus, the reduced sample set is B.

Here, we consider the combined happening of the elements A and B so all the elements of A cannot happen and only those element of A can happen which are included in B. The set of such element is $A \cap B$. Therefore, the probability of A when B has already happened, means the probability of $A \cap B$ when the sample space is B.

$$\therefore \quad P(A | B) = \frac{n(A \cap B)}{n(B)} = \frac{\dfrac{n(A \cap B)}{n(S)}}{\dfrac{n(B)}{n(S)}} \quad \Rightarrow \quad P(A | B) = \frac{P(A \cap B)}{P(B)}$$

Hence, $\quad\quad P(A \cap B) = P(A | B) . P(B)$.

☞ REMARKS

- If the events A and B are independent, *i.e.*, if the happening of B does not depend upon A. Then $P(B | A) = P(B)$ and $P(A | B) = P(A)$. Then $P(AB) = P(A) . P(B)$.
- If P_1 and P_2 be the probabilities of happening of two independent events then
  - (i) The probability that the first event happen and the second fails is $p_1(1 - p_2)$.
  - (ii) The probability that both events fail to happen is $(1 - p_1)(1 - p_2)$.
  - (iii) The probability that at least one of the events happen is $1 - (1 - p_1)(1 - p_2)$ (known as cumulative probability).

THEOREM 2. *If A and B are independent, then $P(A \cup B) = 1 - P(\overline{A}) P(\overline{B})$.*

SOLUTION . Since A and B are independent, then

$$P(A \cap B) = P(A) . P(B)$$

Now, $\quad\quad P(A \cup B) = P(A) + P(B) - P(A \cap B)$

$$= P(A) + P(B) - P(A) . P(B) = (1 - P(\overline{A})) + (1 - P(\overline{B})) - (1 - P(\overline{A}))(1 - P(\overline{B}))$$

$$= 1 - P(\overline{A}) + 1 - P(\overline{B}) - 1 + P(\overline{A}) + P(\overline{B}) - P(\overline{A}) P(\overline{B}) = 1 - P(\overline{A}) P(\overline{B})$$

THEOREM 3. *If A and B are two independent events associated with a random experiments, the following pairs of events are also independent*

(i) $\overline{A}, B$ (ii) $A, \overline{B}$ (iii) $\overline{A}, \overline{B}$

SOLUTION . (i) We have $\quad P(B) P(\overline{A}) = P(B)(1 - P(A)) = P(B) - P(B) P(A)$

$$= P(B) - P(B \cap A) \quad\quad\quad\quad\quad\quad (\because A \text{ and } B \text{ are independent events})$$

$$= P(B - (B \cap A)) = P(B - A) = P(B \cap \overline{A})$$

Here, $\overline{A}$ and B are independent.

(ii)   By interchanging A and B in part (i), we get

$$P(A)\,P(\overline{B}) = P(A)\,(1 - P(B)) = P(A) - P(A)\,P(B)$$
$$= P(A) - P(A \cap B)$$
$$= P(A - (A \cap B)) = P(A - B) \; = P(A \cap \overline{B})$$

($\because$ A and B are independent)

Hence, A and $\overline{B}$ are independent events.

(iii)   Since A and B are independent, therefore $\overline{A}$ and B are independent (using part (i)).

Now, on using result (ii) for the events $\overline{A}$ and B, we get $\overline{A}$ and $\overline{B}$ are independent events.

THEOREM 4.   *If the events A and B defined on the sample space S of a random experiment are independent, then*

$$P(A \mid B) = P(A) \quad and \quad P(B \mid A) = P(B)$$

SOLUTION.   We know that if A and B are independent events, then

$$P(A \text{ and } B) = P(A) \cdot P(B)$$

Therefore,

$$P(A \mid B) = \frac{P(A \cap B)}{P(B)} = \frac{P(A) \cdot P(B)}{P(B)} = P(A)$$

Similarly,

$$P(B \mid A) = \frac{P(A \cap B)}{P(A)} = \frac{P(A) \cdot P(B)}{P(A)} = P(B)$$

THEOREM 5.   *Let A and B are events in a sample space S, then*

(i) $P(A) = P(A\overline{B}) + P(AB)$ (ii) $P(B) = P(\overline{A}\,B) + P(AB)$

(iii) $P(A + B) = P(AB) + P(A\overline{B}) + P(\overline{A}B)$ (iv) $P(AB) + P(\overline{A}B) + P(A\overline{B}) + P(\overline{A}\overline{B}) = 1$

SOLUTION.

(i)   If there be two events A and B, then A can happen in two ways, *i.e.*, (i) B happens while A happens and   (ii) B does not happen while A happens.

The probability of happening of B while A happens $= P(A \cap B) = P(AB)$

Also, the probability of not happening of B while A happens $= P(A \cap \overline{B}) = P(A\overline{B})$

Combining these two events, we get $P(A) = P(AB) + P(A\overline{B})$

(ii)   Do same as result (i).

(iii)   It is known that

$$P(A \cup B) = P(A) + P(B) - P(A \cap B)$$
$$\Rightarrow \qquad P(A + B) = P(A) + P(B) - P(AB)$$
$$= \big[(P(AB) + P(A\overline{B})\big] + [P(\overline{A}B) + P(AB)] - P(AB) \qquad \text{[Using (i) and (ii)]}$$
$$= P(AB) + P(A\overline{B}) + P(\overline{A}B)$$

(iv)   Consider

$$P(AB) + P(\overline{A}B) + P(A\overline{B}) + P(\overline{A}\overline{B}) = P(A + B) + P(\overline{A}\overline{B}) \qquad \text{[Using (iii)]}$$
$$= P(A + B) + \text{ Probability of non-occurrence of both events } A \text{ and } B$$
$$= P(A + B) + \{1 - P(A + B)\} = 1$$

THEOREM 6.   *If A and B are two events in a sample space S, then $P(A \mid B) + P(\overline{A} \mid B) = 1$*

SOLUTION.   It is known that $P(AB) = P(A) \cdot P(B \mid A)$

$$\Rightarrow \qquad P(B \mid A) = \frac{P(AB)}{P(A)}$$

Therefore, $P(A \mid B) + P(\overline{A} \mid B) = \dfrac{P(AB)}{P(B)} + \dfrac{P(\overline{A}B)}{P(B)} = \dfrac{P(AB) + P(\overline{A}B)}{P(B)} = \dfrac{P(B)}{P(B)} = 1$

9.5.5 Probability of Happening of at Least One of Independent Events

Let $A_1, A_2, \ldots A_r$ be r independent events. Also, let $p_1, p_1, \ldots, p_r$ be their respective probabilities of happening. Therefore, we have

$$P(A_1) = p_1, \qquad P(A_2) = p_2, \qquad \ldots P(A_r) = p_r$$
$$\Rightarrow \qquad P(\overline{A}_1) = 1 - p_1, \quad P(\overline{A}_2) = 1 - p_2, \; \ldots, \; P(\overline{A}_r) = 1 - p_r$$

Therefore, the probability when no event happens $= (1 - p_1)(1 - p_2)\ldots(1 - p_r)$.

Hence, the probability when at least one event happens $= 1 - (1 - p_1)(1 - p_2)\ldots(1 - p_r) = 1 - P(\overline{A}_1)\,P(\overline{A}_2)\ldots P(\overline{A}_r)$.

9.5.6 Law of Total Probability and Bayes' Theorem

Let S be the sample space B_1, B_2 be two mutually exclusive events and A be the event that occur with B_1 or B_2.

Now, $S = B_1 \cup B_2$ and $S \cap A = (B_1 \cup B_2) \cap A = (B_1 \cap A) \cup (B_2 \cap A)$

$B_1 \cap A$ and $B_2 \cap A$ are mutually exclusive.

Hence, $P(A) = P(B \cap A) = P[(B_1 \cap E) \cup (B_2 \cap E)] = P(B_1 \cap E) + P(B_2 \cap E) = P(B_1) \cdot P(A \mid B_1) + P(B_2) P(A \mid B_2)$

BAYES' THEOREM

If $B_1, B_2, ..., B_n$ *are mutually disjoint events with* $P(B_i) \neq 0$, *then for an arbitrary event A, which is a subset of* $\bigcup\limits_{i=1}^{n} B_i$ *such that* $P(B) > 0$.

We have

$$P(B_i \mid A) = \frac{P(B_i) P(A \mid B_i)}{\sum\limits_{i=1}^{n} P(B_i) P(A \mid B_i)}$$

Proof. As per given, A is a subset of $\bigcup\limits_{i=1}^{n} B_i$, *i.e.*, $A \subset \bigcup\limits_{i=1}^{n} B_i$.

Now, $A = A \cap \left(\bigcup\limits_{i=1}^{n} B_i \right) = \bigcup\limits_{i=1}^{n} (A \cap B_i)$ (By distributivity)

Since $(A \cap B_i) \subset B_i$ $(i = 1, 2, ..., n)$ are mutually exclusive events, then by addition theorem on probability, we have

$$P(A) = P\left(\bigcup\limits_{i=1}^{n} (A \cap B_i) \right) = \sum\limits_{i=1}^{n} P(A \cap B_i) = \sum\limits_{i=1}^{n} P(B_i) P(A \mid B_i)$$ (By multiplication law of probability)

Also, we have $P(A \cap B_i) = P(A) P(B_i \mid A)$

$\Rightarrow$ $P(B_i \mid A) = \dfrac{P(A \cap B_i)}{P(A)} = \dfrac{P(B_i) P(A \mid B_i)}{\sum\limits_{i=1}^{n} P(B_i) P(A \mid B_i)}$

✎ REMARKS

- Bayes' theorem is also known as 'Theorem of inverse probability'.
- Bayes' theorem enables us to connect the conditional probability of B_i given A, with the conditional probability of A, given B_i and the probability of B_i themselves.
- The probabilities $P(B_i)$ are called 'Priori probabilities' because they exist before getting any information from the experiment itself.
- The probabilities $P(A \mid B_i)$, $i = 1, 2, ..., n$ are called 'likelihoods' because they indicate how likely the events A under consideration is to occur given each and every priori probability.
- The probabilities $P(B_i \mid A)$, $i = 1, 2, ..., n$ are called 'posteriori probabilities' because they are determined after the results of the experiment are known.

📁 Solved Examples

EXAMPLE 1. *In a bag there are 6 balls of which 3 are white and 3 are black. They are drawn successively :*
(i) without replacement, (ii) with replacement.
What is the chance that the colours alternate?
It has been supposed that the number of the balls drawn remains the same?

SOLUTION. (i) If balls are chosen without replacement, there will be two situations, first ball is white or first ball is black.
If first ball is white, 2nd black, 3rd white and so on, so the required probability

$$= \frac{3}{6} \times \frac{3}{5} \times \frac{2}{4} \times \frac{2}{3} \times \frac{1}{2} \times \frac{1}{1} = \frac{3}{60} = \frac{1}{20}$$

If first ball is black, 2nd white and so on

$$= \frac{3}{6} \times \frac{3}{5} \times \frac{2}{4} \times \frac{2}{3} \times \frac{1}{2} \times \frac{1}{1} = \frac{1}{20}$$

Both events are mutually exclusive. Only one can occur, so required probability

$$= \frac{1}{20} + \frac{1}{20} = \frac{2}{20} = \frac{1}{10}.$$

(ii) With replacement first white, 2nd black and so on

Probability $= \dfrac{3}{6} \times \dfrac{3}{6} \times \dfrac{3}{6} \times \dfrac{3}{6} \times \dfrac{3}{6} \times \dfrac{3}{6} = \dfrac{1}{64}$

If first is black, 2nd white and so on

Probability $= \dfrac{1}{64}$

Required probability $= \dfrac{1}{64} + \dfrac{1}{64} = \dfrac{1}{32}$.

EXAMPLE 2. *Four persons are chosen at random from a group containing 3 men, 2 women and 4 children. Obtain the chance that at the most 2 of them are children.*

SOLUTION. Four persons can be chosen in the following ways :
(i) 1 child, 2 men, 1 woman

$$P(E) = \frac{{}^4C_1 \times {}^3C_2 \times {}^2C_1}{{}^9C_4}$$

$$= \frac{4 \times 3 \times 2}{9 \times 8 \times 7 \times 6} \times 4 \times 3 \times 2 \times 1 = \frac{8}{42} = \frac{4}{21}$$

(ii) 1 child, 1 man, 2 women

$$P(E) = \frac{{}^4C_1 \times {}^3C_1 \times {}^2C_2}{{}^9C_4}$$

$$= \frac{4 \times 3 \times 1}{9 \times 8 \times 7 \times 6} \times 4 \times 3 \times 2 \times 1 = \frac{4}{42} = \frac{2}{21}$$

(iii) 2 children, 1 man, 1 woman

$$P(E) = \frac{{}^4C_2 \times {}^3C_1 \times {}^2C_1}{{}^9C_4}$$

$$= \frac{4 \times 3}{2 \times 1} \times \frac{3 \times 2}{9 \times 8 \times 7 \times 6} \times 4 \times 3 \times 2 \times 1 = \frac{2}{7}$$

(iv) 0 child, 3 men, 1 woman

$$P(E) = \frac{{}^4C_0 \times {}^3C_3 \times {}^2C_1}{{}^9C_4}$$

$$= \frac{1 \times 1 \times 2}{9 \times 8 \times 7 \times 6} \times 4 \times 3 \times 2 \times 1 = \frac{1}{63}$$

(v) 0 child, 2 men, 2 women

$$P(E) = \frac{{}^4C_0 \times {}^3C_2 \times {}^2C_2}{{}^9C_4} = \frac{1 \times 3 \times 1 \times 4 \times 3 \times 2 \times 1}{9 \times 8 \times 7 \times 6} = \frac{1}{42}$$

EXAMPLE 3. *Two dice are thrown. What is the probability that sum of the numbers appearing on the dice is 7 or 8?* (KURUKSHETRA–2002, SVTU–2004)

SOLUTION. It is known that, the numbers which can appear on each dice are 1, 2, 3, ..., 6 which are 6 in numbers, which lie in equally likely manner.

Total number of outcomes $= 6 \times 6 = 36 = n(S)$, where S is the sample space.

$$S = \{(1, 1), (1, 2), ..., (5, 6), (6, 1), ..., (6, 6)\}$$

where (6, 1) means 6 appearing on the first dice and 1 appearing on second dice.

Let B_1 be the event that sum of the numbers on the dice is 7, then

$$B_1 = \{(1, 6), (2, 5), (3, 4), (4, 3), (5, 2), (6, 1)\}$$

$$\Rightarrow \quad n(B_1) = 6$$

$$\therefore \quad P(B_1) = \frac{n(B_1)}{n(S)} = \frac{6}{36} = \frac{1}{6}$$

Further, if B_2 be the event that the sum of numbers on the dice is 8, then

$$B_2 = \{(2, 6), (3, 5), (4, 4), (5, 3), (6, 2)\}$$

$$\Rightarrow \quad n(B_2) = 5$$

$$\therefore \quad P(B_2) = \frac{n(B_2)}{n(S)} = \frac{5}{36}$$

Also, since the events B_1 and B_2 are mutually exclusive, so the probability that the events B_1 and B_2 may happen

$$= P(B_1 \cup B_2) = P(B_1) + P(B_2) = \frac{6}{36} + \frac{5}{36} = \frac{11}{36}$$

EXAMPLE 4. *A box contains 2 white and 4 black balls. Another box B contains 5 white and 7 black balls. A ball is transferred from the box A to the box B. Then a ball is drawn from the box B. Find the probability that it is white.* (VTU–2004)

SOLUTION. The probability of drawing a white ball from box B will depend on whether the transferred ball is black or white.

If the transferred ball is black, its probability is 4/6. There are now 5 white and 8 black balls in box B. The probability of drawing white ball from box B is 5/13. Therefore, the probability of drawing a white ball from B, if transferred ball is black

$$= \frac{4}{6} \times \frac{5}{13} = \frac{10}{39}.$$

Similarly, the probability of drawing a white ball from B, if transferred ball is white

$$= \frac{2}{6} \times \frac{6}{13} = \frac{2}{13}.$$

Hence, the required probability $= \dfrac{10}{39} + \dfrac{2}{13} = \dfrac{16}{39}$.

EXAMPLE 5. *A problem in Mathematics is given to three students whose chance of solving it are 1/2, 1/3, 1/4. What is the probability in the following cases:*

(i) *that the problem is solved.* (VTU–2004, 10)

(ii) *only one of them solves it correctly.*

(iii) *at least one of them may solve it.*

SOLUTION. Let A, B, C be three events, when a problem in mathematics is solved by three students.

As per given, we have

$$P(A) = \frac{1}{2}, \quad P(B) = \frac{1}{3} \quad \text{and} \quad P(C) = \frac{1}{4}$$

$$\therefore \quad P(\bar{A}) = 1 - \frac{1}{2} = \frac{1}{2}, P(\bar{B}) = \frac{2}{3} \quad \text{and} \quad P(\bar{C}) = \frac{3}{4}$$

(i) The probability that the problem is solved.
   = Probability that the problem is solved by at least one student

$$= 1 - P(\bar{A}) P(\bar{B}) P(\bar{C}) = 1 - \frac{1}{2} \times \frac{2}{3} \times \frac{3}{4} = 1 - \frac{1}{4} = \frac{3}{4}$$

(ii) The probability that only one solves it correctly

$$= P(A)P(\bar{B})P(\bar{C}) + P(\bar{A})P(B)P(\bar{C}) + P(\bar{A})P(\bar{B})P(C)$$

$$= \frac{1}{2} \times \frac{2}{3} \times \frac{3}{4} + \frac{1}{2} \times \frac{1}{3} \times \frac{3}{4} + \frac{1}{2} \times \frac{2}{3} \times \frac{1}{4}$$

$$= \frac{1}{4} + \frac{1}{8} + \frac{1}{12} = \frac{11}{24}$$

(iii) The probability that at least one of them may solve the problem

$$= 1 - P(\bar{A})P(\bar{B})P(\bar{C}) = 1 - \frac{1}{2} \times \frac{2}{3} \times \frac{3}{4} = 1 - \frac{1}{4} = \frac{3}{4}$$

EXAMPLE 6. **(Huyghen's Problem).** *A and B throw alternately with a pair of dice. A wins if he throws 6 before B throws 7 and B wins if he throws 7 before A throws 6. If A begins, find his chance of winning.* (JNTU–2003, MADRAS–2006)

SOLUTION. We observe that the sum 6 can be obtained as follows

(1, 5), (2, 4), (3, 3), (4, 2), (5, 1) = Total 5 ways

Now, the probability of A's throwing 6 with 2 dice is 5/36.

The probability of B's not throwing 6 is $\dfrac{31}{36}$. Similarly, the probability of B's throwing 7 is $\dfrac{6}{36}$, *i.e.*, $\dfrac{1}{6}$.

The probability of B's not throwing 7 is $\dfrac{5}{6}$.

Now, A can win if he throws 6 in the first, third, fifth, seventh, etc. throws.

Hence, the chance of A's winning

$$= \frac{5}{36} + \frac{31}{36} \times \frac{5}{6} \times \frac{5}{36} + \left(\frac{31}{36} \times \frac{5}{6}\right)^2 \times \frac{5}{36} + ...$$

$$= \frac{5}{36}\left[1 + \left(\frac{31}{36} \times \frac{5}{6}\right) + \left(\frac{31}{36} \times \frac{5}{6}\right)^2 + \left(\frac{31}{36} \times \frac{5}{6}\right)^3 + ...\right]$$

$$= \frac{5}{36}\left[\frac{1}{1 - \left(\frac{31}{36} \times \frac{5}{6}\right)}\right] \quad \left[\because \text{Sum of infinite G.P.} = \frac{a}{1-r}\right]$$

$$= \frac{5}{36} \times \frac{36 \times 6}{61} = \frac{30}{61}.$$

Exercise-9.17

1. In a class of 25 students with roll number 1 to 25, a student is picked up at random to answer a question. Find the probability that the roll number of the selected students is either a multiple of 5 or 7 ?

2. In a single throw of two dice, find the probability that neither a doublet nor a total of 10 will appear.

3. Tickets are numbered from 1 to 100. One ticket is picked up at random. Find the probability that the ticket picked up has a number which is divisible by 5 or 8.

4. A box contains 100 bolts and 50 nuts. It is given that 50% bolts and 50% nuts are rusted. Two objects are selected from the box at random. Find the probability that both are bolts or nuts are rusted.

5. Two dice are tossed once. Find the probability of getting an even number on the first dice or a total of 8.

6. In a race, the odds in favour of horses A, B, C, D are $1 : 3, 1 : 4, 1 : 5$ and $1 : 6$ respectively. Find the probability that one of them wins the race.

7. A natural number is chosen at random from amongst the first 500. What is the probability that the number so chosen is divisible by 3 or 5?

8. In a single throw of 2 dice, find the probability of not getting the same number on the two dice.

9. Let A, B, C be three mutually exclusive events associated with a random experiment. Find $P(A)$ given that $P(B) = \frac{3}{2}P(A)$ and $P(C) = \frac{1}{2}P(B)$.

10. One card is drawn from a pack of 52 cards, each of the 52 cards being equally likely to be drawn. Find the probability of (i) the card drawn is red, (ii) the card drawn is a king.

11. From a set of 17 cards numbered 1, 2, 3, 4, ..., 16, 17, one is drawn at random. Show that the chance that its number is divisible by 3 or 7 is 7/17.

12. The probability of student A passing an examination is 2/9 and of student B passing is 5/9. Assuming the two events : 'A passes', 'B passes' as independent, find the probability of (i) only A passing the examination, (ii) only one of them passing the examinations.

13. In a group, there are 3 men and 2 women, 3 persons are selected at random from this group. Find the probability that 1 man and 2 women or 2 men and 1 woman are selected.

14. A bag contains 3 red, 4 black and 2 green balls. Two balls are drawn at random from the bag. Find the probability that both balls are of different colours.

15. Ram is taking up subjects — Mathematics, Physics and Chemistry in the examination. His probability of getting Grade A in these subjects are 0.2, 0.3 and 0.5, respectively. Find the probability that he gets (i) grade A in all subjects, (ii) grade A in no subject, (iii) grade A in two subjects.

16. A bag contains 30 tickets, numbered from 1 to 30. Five tickets are drawn at random and arranged in ascending order. Find the probability that 'the third' number is 20.

17. There are 3 red and 3 black balls in a bag, 3 balls are taken at random from the bag. Find the probability of getting 2 red and 1 black balls or 1 red and 2 black balls.

18. A bag contains 5 red, 6 white and 7 black balls. Two balls are drawn at random. What is the probability that both balls are red or black?

19. A machine manufactured by a firm consists of two parts A and B. Out of 100 A's manufactured, 9 are likely to be defective and of 100 B's 5 are likely to be defective. Find the probability that a machine manufactured by the firm is free of any defect.

20. Three bags contain 7 white, 8 red, 9 white, 6 red and 5 white, 7 red balls respectively. One ball at random is drawn from each bag. Find the probability that all of them are of the same colour.

21. A bag contains 1 red and 4 blue balls. Two balls are drawn at random with replacement. Find the probability of getting one red and one blue ball.

22. A bag contains 4 red and 5 black balls. Another bag contains 3 red and 6 black balls. One ball is drawn from first bag and two balls are drawn from the second bag. Find the probability that out of the three, two are black and one is red.

23. Two cards are drawn from a well shuffled pack of 52 cards, one after another without replacement. Find the probability that one of them is an ace and other is a queen of opposite shade.

24. There are three urns A, B and C. Urn A contains 4 red balls and 3 black balls. Urn B contains 5 red balls and 4 black balls. Urn C contains 4 red balls and 4 black balls. One ball is drawn from each of these urns. What is the probability that the 3 balls drawn consists of 2 red balls and a black ball.

(JNTU–2003)

25. A husband and wife appear in an interview for two vacancies in the same post. The probability of husband's selection is 1/3 and that of wife's selection is 1/5. Show that the probability that only one of them will be selected is 2/5.

26. The probability of A solving a problem is 3/7 and that of B solving it is 1/3. What is the probability that (i) at least one of them will solve the problem, (ii) only one of them will solve the problem.

27. A problem in Mathematics is given to three students whose chances of solving it are (i) 1/2, 1/4 and 1/3, (ii) 1/3, 1/4 and 1/5, (iii) 1/3, 1/5 and 1/6, (iv) 1/2 , 1/3 and 1/5. What is the probability that at least one of them may solve it.

28. A bag contains 1 white and 6 red balls and bag B contains 4 white and 3 red balls. One ball is drawn at random from one of the chosen bags and is found to be white. Find the probability that it was drawn from bag A.

29. In a bolt factory, machine A, B and C are producing 25%, 35% and 40% of total bolts. Of these, 4%, 3% and 2% respectively are defective. A bolt is drawn at random and found to be defective. What is the probability that it is produced by machine A?

30. Let us suppose 5 men out of 100 and 25 women out of 1000 are good orators. An orator is chosen at random. Find the probability that a male person is selected. Assume that there are equal number of men and women.

31. Ram speaks the truth 8 times out of 10 times. A dice is tossed. He reports that it was 5. What is the probability that it was actually 5?

32. A can hit a target 3 times in 5 shots, B 2 times in 5 shots and C 3 times in 4 shots. They fire a volley, what is the probability that
 (i) two shots hits?
 (ii) atleast two shots hits? (AMITE–2003, MADRAS–2003)

Answers

1. 8/25	**2.** 7/9	**3.** 0.3	**4.** 0.58	**5.** 5/9	**6.** 319/420	**7.** 233/500	**8.** 5/6	**9.** 9/13	**10.** 1/2

1. 8/25   **2.** 7/9   **3.** 0.3   **4.** 0.58   **5.** 5/9   **6.** 319/420   **7.** 233/500   **8.** 5/6   **9.** 9/13 **10.** 1/2
11. 1/13   **12.** 8/81, 43/81   **13.** 9/10   **14.** 13/18   **15.** 0.03, 0.28, 0.22   **17.** 9/10   **18.** 31/153
19. 0.86   **20.** 217/900   **21.** 56/121 **22.** 25/54   **23.** 4/663   **24.** 17/42   **26.** 13/21, 10/21
27. 7/10, 3/5, 7/9, 11/15   **28.** 1/5   **29.** 0.26   **30.** 2/3   **31.** 4/7   **32.** (i) –0.45   (ii) 0.63

9.6 RANDOM VARIABLES AND PROBABILITY DISTRIBUTION

If we throw an ordinary dice, we shall get one of the numbers 1, 2, 3, 4, 5, 6 in each throw. Therefore, we get quantitative outcomes. In some cases, the outcome may be qualitative, for example, throw of a coin which may be head or tail. In such type of cases, we may denote the outcome 'head' of the tossing of a coin by 1 and the 'tail' by 0. In this way, each outcome of a random experiment, whether it is qualitative or quantitative, can be expressed by a real number. This numerical value associated with the outcome of a random experiment is known as a random variable.

Definition (1). *The variate which can take certain values depending on chance is called a random variable. It is also known as 'chance variate' or 'stochastic variate'.*

Definition (2). *A real valued function X, defined on a sample space S, there is a real number denoted by X(s), X is called a function defined on S. X(s) is called the value of the function at s.*

Definition (3). *A real valued function X, defined on a sample space S of a random experiment E, is called a random variable which assign to each sample, one and only one real numbers X(s) = x (say), where $s \in S$.*

or

If E is a random experiment and S is a sample space associated with it, a function X(s), $s \in S$ is called a random variable.

Definition (4). *Let X be a random variable which takes the values $x_1, x_2, \dots$, then the probability that $x = x_i$ is denoted by $P(X = x_i)$ or $p(x_i)$ or p_i. The probability of the event X takes values from x_i to x_j is denoted by*

$$P(x_i \leq X < x_j)$$

Also,
$$p(x_i) \geq 0 \quad \text{and} \quad \Sigma p(x_i) = 1 \ .$$

9.6.1 TYPES OF RANDOM VARIABLE

There are two types of random variables

(i) Discrete Random Variables

 If a random variable assumes only a finite number or countably infinite number of values of X, then X is called discrete random variable.

REMARK

- The possible values of X may be taken $x_1, x_2, \dots, x_n$ infinite case and $x_1, x_2, \dots$ in countably infinite case. The countably infinite case will have an infinite sequence of distinct values and that sequence will be countable.

(ii) Continuous Random Variable

 If a random variable assumes any value in some interval, it is called a continuous random variable, i.e., if a variate can take an infinite set of values in the given interval, say $a \leq x \leq b$, it is a continuous random variable and their distribution are accordingly known as continuous distribution.

9.6.2 PROBABILITY DISTRIBUTIONS

The distribution obtained by taking the possible values of a random variable together with their respective probabilities is called probability distribution.

Probability Distribution of a Discrete Random Variable : The probability distribution of a discrete random variable is the set of ordered pair $(x_i, p(x_i))$ provided
 (i) $p(x_i) \geq 0$ (ii) $\Sigma p_i(x_i) = 1$

- $P(x \le x_i) = P(X = x_1) + P(X = x_2) + ... + P(X = x_i) = p_1 + p_2 + ... + p_i$
- $P(x \ge x_i) = P(X = x_i) + P(X = x_{i+1}) + ... + P(X = x_n) = p_{i+1} + p_{i+2} + ... + p_n$

Solved Examples

EXAMPLE 1. *Two cards are drawn one by one without replacement from a well shuffled pack of 52 cards. Find the probability distribution of the number of aces.*

SOLUTION. Let X denote the random variable, *i.e.*, X denote the number of aces. Therefore, X takes values 0, 1, 2.

Then, $P(X = 0) = \dfrac{48}{52} \times \dfrac{47}{51} = \dfrac{188}{221}$

$$P(X = 1) = 2\left(\dfrac{4}{52} \times \dfrac{48}{51}\right)$$
$$= 2\left(\dfrac{1}{13} \times \dfrac{16}{17}\right) = 2\left(\dfrac{16}{221}\right) = \dfrac{32}{221}$$

Also, $P(X = 2) = \dfrac{4}{52} \times \dfrac{3}{51} = \dfrac{1}{13} \times \dfrac{1}{17} = \dfrac{1}{221}$.

Hence, the required probability distribution is given by

X	0	1	2
$P(X)$	$\dfrac{188}{221}$	$\dfrac{32}{221}$	$\dfrac{1}{221}$

EXAMPLE 2. *Find the probability distribution of the number of success in two tosses of a dice, when success is defined of getting a 5 and a 6.*

SOLUTION. We have

Probability of success $= \dfrac{2}{6} = \dfrac{1}{3}$

and Probability of failure $= \dfrac{4}{6} = \dfrac{2}{3}$

Let X be the random variable, therefore

$$P(X = 0) = \dfrac{2}{3} \times \dfrac{2}{3} = \dfrac{4}{9}$$

$$P(X = 1) = \dfrac{1}{3} \times \dfrac{2}{3} + \dfrac{2}{3} \times \dfrac{1}{3} = \dfrac{2}{9} + \dfrac{2}{9} = \dfrac{4}{9}$$

and $P(X = 2) = \dfrac{1}{3} \times \dfrac{1}{3} = \dfrac{1}{9}$

Hence, the required probability distribution is given by

X	0	1	2
$P(X)$	$\dfrac{4}{9}$	$\dfrac{4}{9}$	$\dfrac{1}{9}$

EXAMPLE 3. *A fair dice is tossed twice. If the number appearing on the top is less than 3, it is a success. Find the probability distribution of successes.*

SOLUTION. We have

Probability of success $= \dfrac{2}{6} = \dfrac{1}{3}$

and Probability of failure $= \dfrac{4}{6} = \dfrac{2}{3}$

If X is the random variable, then we have

$$P(0) = \dfrac{2}{3} \times \dfrac{2}{3} = \dfrac{4}{9}$$
$$P(1) = \dfrac{1}{3} \times \dfrac{2}{3} + \dfrac{2}{3} \times \dfrac{1}{3} = \dfrac{2}{9} + \dfrac{2}{9} = \dfrac{4}{9}$$
and $P(2) = \dfrac{1}{3} \times \dfrac{1}{3} = \dfrac{1}{9}$

Hence, the required probability distribution is given by

X	0	1	2
$P(X)$	$\dfrac{4}{9}$	$\dfrac{4}{9}$	$\dfrac{1}{9}$

Exercise-9.18

1. A coin is tossed 5 times, X is the number of heads observed. Find the probability distribution of X.

2. Two cards are drawn successively with replacement from a well shuffled pack of 52 cards. Find the probability distribution of the number of (i) aces, (ii) kings.

3. Find the probability distribution of the number of heads when three coins are tossed.

4. Find the probability distribution of the sum of numbers obtained when two dice are thrown.

5. Four bad eggs are mixed with 10 good ones. If 3 eggs are drawn one by one without replacement, find the probability distribution of the number of bad eggs.

6. A coin is biased so that the head is 3 times likely to occur as a tail. If the coin is tossed twice, find the probability distribution for the number of tails.

7. A random variable X has the following probability distribution :

X	–2	–1	0	1	2	3
$P(X)$	0.1	k	0.2	$2k$	0.3	k^2

Find (i) The value of k, (ii) $P(X \le 1)$, (iii) $P(X \ge 0)$.

8. A bag contains 2 white, 3 red and 4 blue balls. Two balls are drawn at random from the bag. If the random variable X denote the number of white balls among the two balls drawn, find the probability distribution of X.

9. Four rotten mangoes are mixed accidentally with 20 good mangoes. Obtain the probability distribution of the number of rotten mangoes in a random draw of 2 mangoes without replacement.

10. Two cards are drawn successively from a well-shuffled pack of 52 cards. Find the probability distribution of the number of queens one by one.

Answers

1.

X	0	1	2	3	4	5
P(X)	1/32	5/32	10/32	10/32	5/32	1/32

2.

X	0	1	2
P(X)	144/169	24/169	1/169

3.

X	0	1	2	3
P(X)	1/8	3/8	3/8	1/8

4.

X	2	3	4	5	6	7	8	9	10	11	12
P(X)	1/36	2/36	3/36	4/36	5/36	6/36	5/36	4/36	3/36	2/36	1/36

5.

X	0	1	2	3
P(X)	1000/2197	900/2197	270/2197	27/2197

6.

X	0	1	2
P(X)	9/16	3/8	1/16

7. (i) $P(0) = \dfrac{188}{221}$, $P(1) = \dfrac{32}{221}$, $P(2) = \dfrac{1}{221}$ (ii) $P(0) = \dfrac{105}{221}$, $P(1) = \dfrac{96}{221}$, $P(2) = \dfrac{20}{221}$ (iii) $P(0) = \dfrac{19}{34}$, $P(1) = \dfrac{13}{34}$, $P(2) = \dfrac{2}{34}$

8.

X	0	1	2
P(X)	49/81	28/81	41/81

9.

X	0	1	2
P(X)	95/138	40/138	3/138

10.

X	0	1	2
P(X)	144/169	24/169	1/169

9.6.3 BINOMIAL DISTRIBUTION

Let p and q be the probability of the success and failure respectively of an event in one trial, so that $p + q = 1$. If the event can be tried n times and assume that these n trials are independent and the probability p of success is the same in each trial. The number of successor may be $0, 1, 2, ..., n$ in these n trials.

Definition. *The probability of r successes in a series of n trials is given by* $^nC_r\, p^r\, q^{n-r}$, *where r takes any integral value for 0 to n. The probabilities of* $0, 1, 2, ..., r, ..., n$ *successes are given by*

$$q^n, \, ^nC_1 pq^{n-1}, \, ^nC_2 p^2 q^{n-2}, ..., \, ^nC_r p^r q^{n-r}, ..., p^n$$

Then the probability of the number of successes so obtained is called the binomial distribution.

REMARKS

- Binomial distribution contains two independent contents p (or $q = 1 - p$) and n which are known as parameter of Binomial distribution.
- Binomial distribution was discovered by Jacob Bernoulli. Due to this reason, it is also known as Bernoulli distribution.
- If $p = q = 1/2$, then this distribution is said to be symmetric.
- The distribution $f(r) = \, ^nC_r\, p^r\, q^{n-r}$, $r = 0, 1, 2, ..., n$ determines a probability distribution, called the binomial distribution, since the total frequency $p + q = 1$.
- Binomial distribution is a discrete probability distribution.
- Binomial distribution is important not only because of its wide applicability, but because it gives rise to many other probability distribution.

9.6.4 APPLICABILITY OF BINOMIAL DISTRIBUTION

For a binomial distribution, the following conditions must be satisfied :
 (i) There should be a finite number of trials
 (ii) The trials are mutually exclusive and does not depend on each other.
 (iii) Each trial should have only two possibility, either a success or failure.
 (iv) The probability of success or failure is the same for each trial.

9.6.5 MOMENTS OF THE BINOMIAL DISTRIBUTION

MOMENT ABOUT ORIGIN

The probability distribution function for binomial distribution is
$$f(x) = \, ^nC_x\, p^x\, q^{n-x}$$

(1) First Moment about Origin (Mean of Binomial Distribution)

$$\mu_1' = \sum_{x=0}^{n} x \cdot {}^nC_x\, p^x q^{n-x} = 0.q^n + 1.\,{}^nC_1 q^{n-1}.p + 2.\,{}^nC_2 q^{n-2}.p^2 + \ldots + n\,{}^nC_n\, p^n$$

$$= np\,(q^{n-1} + {}^{n-1}C_1 q^{n-2}.p + {}^{n-1}C_2 q^{n-2} p^2 + \ldots + p^{n-1}) = np\,(q+p)^{n-1} \qquad (\because\ p+q=1)$$

Therefore, mean $= \mu_1' = np$

$\Rightarrow$ Mean of the binomial distribution is np .

(2) Second Moment about Origin

$$\mu_2' = \sum_{x=0}^{n} x^2 \cdot {}^nC_x\, p^x q^{n-x} = \sum_{x=0}^{n} [x(x-1)+x]\,{}^nC_x\, p^x q^{n-x} = \sum_{x=0}^{n} x(x-1)\frac{n!}{x!(n-x)!} p^x q^{n-x} + \sum_{x=0}^{n} x \cdot {}^nC_x\, p^x q^{n-x}$$

$$= \sum_{x=1}^{n} x(x-1)\frac{\{n(n-1)(n-2)!\}}{[x(x-1)(x-2)!.\{(n-2)-(x-2)\}!]} \cdot p^2 p^{x-2} q^{((n-2)-(x-2))} + \sum_{x=0}^{n} x \cdot {}^nC_x\, p^x q^{n-x}$$

$$= \sum_{x=1}^{n} n(n-1)\frac{(n-2)!}{(x-2)![(n-2)-(x-2)]!} \cdot p^{x-2} q^{((n-2)-(x-2))} \cdot p^2 + \sum_{x=0}^{n} x \cdot {}^nC_x\, p^x q^{n-x}$$

$$= n(n-1)\, p^2 \sum_{x=1}^{n} {}^{n-2}C_{x-2}\, p^{x-2} q^{((n-2)-(x-2))} + np = n(n-1)\, p^2 (q+p)^{n-2} + np$$

$$= n(n-1)\, p^2 + np = np\,[(n-1)\,p+1] = np\,[np+(1-p)] = np(np+q)$$

$\therefore \qquad \mu_2' = n^2 p^2 + npq$

(3) Third Moment about Origin

$$\mu_3' = \sum_{x=0}^{n} x^3 \cdot {}^nC_x\, p^x q^{n-x} = \sum_{x=0}^{n} \{x(x-1)(x-2)+3x(x-1)+x\}\,{}^nC_x\, p^x q^{n-x} \quad [\because x^3 = x(x-1)(x-2)+3x(x-1)+x]$$

$$= n(n-1)(n-2)p^3(q+p)^{n-3} + 3n(n-1)\, p^2(q+p)^{n-2} + np$$

$$\mu_3' = n(n-1)(n-2)p^3 + 3n(n-1)p^2 + np$$

(4) Fourth Moment about Origin

$$\mu_4' = \sum_{x=0}^{n} x^4 \cdot {}^nC_x\, p^x q^{n-x} = \sum_{x=0}^{n} \{x(x-1)(x-2)(x-3)+6x(x-1)(x-2)+7x(x-1)+x\}\,{}^nC_x\, p^x q^{n-x}$$

$$= n(n-1)(n-2)(n-3)p^4(q+p)^{n-4} + 6n(n-1)(n-2)p^3(q+p)^{n-3} + 7n(n-1)\, p^2(q+p)^{n-2} + np$$

$$\mu_4' = n(n-1)(n-2)(n-3)p^4 + 6n(n-1)(n-2)p^3 + 7n(n-1)p^2 + np$$

MOMENT ABOUT MEAN

(1) First Moment about Mean: $\mu_1 = 0$

(2) Second Moment about Mean (Variance of Binomial Distribution)

$$\mu_2 = \mu_2' - (\mu_1')^2 = (n^2 p^2 + npq) - (np)^2$$

or   Variance $= \mu_2 = npq$

and   S.D. $= \sqrt{\mu_2} = \sqrt{npq}$

(3) Third Moment about Mean

$$\mu_3 = \mu_3' - 3\mu_2'\mu_1' + \mu_1'^3 = np\,[(n-1)(n-2)p^2 + 3(n-1)p+1] - 3(n^2 p^2 + npq)\, np + 2(np)^3$$

$$= np[\{(n-1)(n-2)-3n^2+2n^2\}\, p^2 + \{3(n-1)+3nq\}p+1]$$

$$= np[(-3n+2)p^2 + 3(n-1)p - 3np(1-p)+1] \qquad (\because p=1-q)$$

$$= np\,(2p^2 - 3p+1) = np\,(1-2p)\,(1-p) = np(1-p)\,\{1-2(1-q)\}$$

$$= npq(2q-1) = npq\,[q-(1-q)]$$

$\Rightarrow \qquad \mu_3 = npq\,(q-p)$

(4) Fourth Moment about Mean

We know that $\mu_4 = \mu_4' - 4\mu_3'\mu_1' + 6\mu_2'\mu_1'^2 - 3\mu_1'^4 = [n(n-1)(n-2)(n-3)p^4 + 6n(n-1)(n-2)p^3 + 7n(n-1)\, p^2 + np]$

$$-4[n(n-1)(n-2)p^3 + 3n(n-1)p^2 + np] + 6[n(n-1)p^2 + np]n^2 p^2 - 3(np)^4$$

On simplifying, we get $\mu_4 = 3n^2(p^4 + 2p^3 + p^2) - n(6p^4 - 12p^3 + 6p^2) - np^2 + np$

$$= 3n^2 p^2\,(p^2 - 2p+1) - 6np^2(1-2p+p^2) + np(1-p)$$

$$= 3n^2 p^2(1-p)^2 - 6np^2(1-p)^2 + npq = 3n^2 p^2 q^2 - 6np^2 q^2 + npq = 3n^2 p^2 q^2 + npq(1-6pq)$$

(C) Pearson's Coefficients

(1) $\beta_1 = \dfrac{\mu_3^2}{\mu_2^3} = \dfrac{[nq(q-p)]^2}{(npq)^3} \Rightarrow \beta_1 = \dfrac{(1-2p)^2}{npq}$

(2) $\beta_2 = \dfrac{\mu_4}{\mu_2^2} = \dfrac{3n^2p^2q^2 + npq(1-6pq)}{(npq)^2} = 3 + \dfrac{(1-6pq)}{npq}$

(3) $\gamma_1 = \sqrt{\beta_1} = \dfrac{(q-p)}{\sqrt{npq}}$

(4) $\gamma_2 = \beta_2 - 3 = \dfrac{1-6pq}{npq}$

📝 **REMARK**

- In case of binomial distribution, mean > variance.

9.6.6 MOMENT GENERATING FUNCTION OF BINOMIAL DISTRIBUTION

For binomial distribution, we have $f(x) = {}^nC_x\, p^x\, q^{n-x}$. Therefore, the m.g.f. about origin will be given by

$$M_0(t) = \sum_{x=0}^{n} e^{tx} \cdot {}^nC_x\, p^x\, q^{n-x} = \sum_{x=0}^{n} {}^nC_x\,(pe^t)^x\, q^{n-x} = (q + pe^t)^n$$

$$= \left[q + p\left(1 + t + \frac{t^2}{2!} + \frac{t^3}{3!} + \dots\right)\right]^n = \left[1 + pt + \frac{pt^2}{2!} + \frac{pt^3}{3!} + \dots\right]^n \qquad (\because p + q = 1)$$

∴ The m.g.f. about mean is given by

$$M_m(t) = e^{-mt} M_0(t) \qquad \text{where } m = np$$

$$= e^{-npt}(q + pe^t)^n = [e^{-pt}(q + pe^t)]^n = [qe^{-pt} + pe^{(1-p)t}]^n \qquad \text{where } p + q = 1$$

$\Rightarrow \quad M_m(t) = [qe^{-pt} + pe^{qt}]^n = \left(1 + pq\dfrac{t^2}{2!} + pq(q^2 - p^2)\dfrac{t^3}{3!} + pq(q^3 + p^3)\dfrac{t^4}{4!} + \dots\right)^n$

or $\quad 1 + \mu_1 t + \mu_2\dfrac{t^2}{2!} + \mu_3\dfrac{t^3}{3!} + \mu_4\dfrac{t^4}{4!} + \dots = 1 + npq\dfrac{t^2}{2!} + npq(q-p)\dfrac{t^3}{3!} + npq[1 + 3(n-2)pq]\dfrac{t^4}{4!} + \dots$ ...(1)

📝 **REMARKS**

- We can generate the moments of binomial distribution by equating the coefficients of like powers of t.
- As the number of trials increase indefinitely, $\beta_1 \to 0$ and $\beta_2 \to 3$.
- We can apply the binomial distribution in following types of problems :
  (i) Number of defective items in a sample from production line
  (ii) Estimation of reliability of systems
  (iii) Number of rounds fired from a gun hitting a target
  (iv) Radar detection

9.6.7 CUMULANT GENERATING FUNCTION OF BINOMIAL DISTRIBUTION

(A) About Mean $m = np$

We know that

$$M_m(t) = \sum_{x=0}^{n} e^{t(x-m)} \cdot {}^nC_x p^x q^{n-x} = \sum_{x=0}^{n} e^{tx-npt} \cdot {}^nC_x q^{n-x}$$

$$= e^{-npt} \sum_{x=0}^{n} {}^nC_x (pe^t)^x q^{n-x} = e^{-npt}(q + pe^t)^n = (qe^{-pt} + pe^{(1-p)t}) = (qe^{-pt} + pe^{qt})^n$$

$$= \left[q\left(1 - pt + \frac{p^2t^2}{2!} - \frac{p^3t^3}{3!} + \dots\right) + p\left(1 + qt + \frac{q^2t^2}{2!} + \dots\right)\right]^n$$

$$= \left[(q + p) + \frac{1}{2!}pq(p+q)t^2 + \frac{1}{3!}pq(q^2 - p^2)t^3 + \frac{1}{4!}pq(q^3 + p^3)t^4 + \dots\right]^n$$

$\Rightarrow \quad M_m(t) = \left[1 + \dfrac{1}{2!}pqt^2 + \dfrac{1}{3!}pq(q^2 - p^2)t^3 + \dfrac{1}{4!}pq(q^3 + p^3)t^4 + \dots\right]^n$

Now, $\quad K_m(t) = \log M_m(t) = n\log\left(1 + \dfrac{1}{2!}pqt^2 + \dfrac{1}{3!}pq(q^2 - p^2)t^3 + \dfrac{1}{4!}pq(q^3 + p^3)t^4 + \dots\right)$

which is the cumulative function of the binomial distribution about the mean $m = np$.

Now, from (1), we have

$$K_m(t) = n\left[\left\{\frac{1}{2}pqt^2 + \frac{1}{6}pq(q^2 - p^2)t^3 + \frac{1}{24}pq(q^3 + p^3)t^4 + \dots\right\} - \frac{1}{2}\left\{\frac{1}{4}p^2q^2t^4 + \dots\right\} + \frac{1}{3}\left\{\frac{1}{8}p^3q^3t^6 + \dots\right\} + \dots\right]$$

$$= n\left[pq\frac{t^2}{2!} + pq(q^2 - p^2)\frac{t^3}{3!} + \left\{pq(q^3 - p^3) - 3p^2q^2\right\}\frac{t^4}{4!} + \ldots \right]$$

The coefficients of $t, \dfrac{t^2}{2!}, \dfrac{t^3}{3!}, \dfrac{t^4}{4!}, \ldots$ gives the cumulants as

$$K_1 = 0, \ K_2 = npq, \ K_3 = npq(q - p)$$

and

$$K_4 = npq(q^3 + p^3) - 3np^2q^2 = npq\left[(q + p)(q^2 - qp + p^2) - 3pq\right]$$

$$= npq\left[(q + p)^2 - 4pq\right] = npq\left[(q^2 + p^2)^2 - 6pq\right] = npq\left[(1 - 6pq)\right]$$

Solved Examples

EXAMPLE 1. *For the binomial distribution, prove that*

$$\mu_{r+1} = pq\left(nr\mu_{r-1} + \frac{d\mu_r}{dp}\right).$$

SOLUTION. Since we know that

$$\mu_r = \sum_{x=0}^{n} \{f(x)\}(x - \mu_1')^r = \sum_{x=0}^{n} ({}^nC_x p^x q^{n-x})(x - np)^r$$

$$\therefore \frac{d\mu_r}{dp} = \sum_{x=0}^{n} {}^nC_x \cdot xp^{x-1}(1 - p)^{n-x}(x - np)^r$$

$$+ \sum_{x=0}^{n} {}^nC_x p^x(n - x)(1 - p)^{n-x-1}(-1)(x - np)^r$$

$$+ \sum_{x=0}^{n} {}^nC_x p^x(1 - p)^{n-x}.r(x - np)^{r-1}(-n)$$

$$= \sum_{x=0}^{n} {}^nC_x xp^{x-1}q^{n-x}(x - np)^r$$

$$- \sum_{x=0}^{n} {}^nC_x(n - x)p^x q^{n-x-1}(x - np)^r$$

$$- \sum_{x=0}^{n} {}^nC_x p^x q^{n-x}(x - np)^{r-1}.nr$$

$$= \sum_{x=0}^{n} {}^nC_x p^{x-1}q^{n-x-1}(x - np)^r[xq - (n - x)p]$$

$$- nr \sum_{x=0}^{n} {}^nC_x p^x q^{n-x}(x - np)^{r-1}$$

$$= \sum_{x=0}^{n} {}^nC_x p^{x-1}q^{n-x-1}(x - np)^r[x(q + p) - np]$$

$$- nr\mu_{r-1}$$

$$\Rightarrow \quad pq\frac{d\mu_r}{dp} = \sum_{x=0}^{n} {}^nC_x p^x q^{n-x}(x - np)^r(x - np) - npqr\mu_{r-1}$$

$$= \sum_{x=0}^{n} {}^nC_x p^x q^{n-x}(x - np)^{r+1} - npqr\mu_{r-1}$$

$$= \mu_{r+1} - npqr\,\mu_{r-1}$$

$$\Rightarrow \quad \mu_{r+1} = pq\left(nr\mu_{r-1} + \frac{d\mu_r}{dp}\right).$$

EXAMPLE 2. *Obtain following recurrence relation for the binomial distribution*

$$f(x) = \frac{n - x + 1}{x} \cdot \frac{p}{q} f(x - 1)$$

SOLUTION. For binomial distribution, we have

$$f(x) = {}^nC_x p^x q^{n-x} = \frac{n!}{x!\,(n-x)!} p^x q^{n-x} \qquad \ldots(1)$$

$$\Rightarrow f(x - 1) = \frac{n!}{(x-1)!\,(n-x+1)!}.p^{x-1} q^{n-x+1} \qquad \ldots(2)$$

From (1) and (2),

$$\frac{f(x)}{f(x-1)} = \frac{n - x + 1}{x}\cdot\frac{p}{q} \Rightarrow f(x) = \frac{n - x + 1}{x}\cdot\frac{p}{q} f(x-1).$$

EXAMPLE 3. *The probability that evening college student will graduate is 0.4. Determine the probability that out of 5 students, (a) none, (b) one and (c) at least one will graduate.*

SOLUTION. Given that the probability that evening college student is graduate = 0.4.

$$p = \frac{4}{10} = \frac{2}{5}, \ q = 1 - \frac{2}{5} = \frac{3}{5}, \ n = 5$$

If a random variable X denotes success, then probability of r success in 5 trials = $p(n) = {}^5C_r\, p^r\, q^{n-r}$

(i) The probability of zero success

$$= {}^5C_0\left(\frac{2}{5}\right)^0\left(\frac{3}{5}\right)^{5-0} = 1\times1\times\left(\frac{3}{5}\right)^5 = 0.046$$

(ii) The probability of one success

$$= {}^5C_0\left(\frac{2}{5}\right)^1\left(\frac{3}{5}\right)^{5-1} = 5\times\frac{2}{5}\times\frac{81}{625} = 0.2592$$

(iii) The probability of at least one success

$$= 1 - \text{probability of no success} = 1 - 0.046 = 0.954.$$

EXAMPLE 4. *In sampling a large number of parts are manufactured by a machine, the mean number of defectives in a sample of 20 is 2. Out of 1000 such samples, how many would be expected to contain at least 3 defective parts?* (VTU–2004)

SOLUTION. The mean of the defectives = $2 = np = 20p$

$$\Rightarrow \qquad p = \frac{2}{20} = 0.1$$

i.e., the probability of defective part is $p = 0.1$ and the probability of non-defective part is $q = 0.9$. Therefore, the probability of at least three defectives in a sample of 20

$$= 1 - \text{(prob. that either none or one or two are non-defective parts)}$$

$$= 1 - ({}^{20}C_0(0.9)^{20} + {}^{20}C_1(0.1)(0.9)^{19}$$

$$+ {}^{20}C_2(0.1)^2(0.9)^{18})$$

$$= 1 - (0.9)^{18} \times 4.51 = 0.323$$

Thus the number of sampling having at least three defective parts out of 1000 samples = $1000 \times 0.323 = 323$

EXAMPLE 5. *If 10% of the affecting aircrafts are expected to be shot down before reaching the target, what is the probability that out of 5 aircrafts at least 4 will be shot before they reach the target?*

SOLUTION. Let p and q be the probability of being shot before reaching and safely reaching the target. Then as per given, we have

$$p = \frac{1}{10}, \quad q = 1 - \frac{1}{10} = \frac{9}{10}$$

Thus, required probability of at least 4 aircrafts out of 5 being shot before reaching the target

$$= {}^5C_5\left(\frac{1}{10}\right)^5\left(\frac{9}{10}\right)^{5-5} + {}^5C_4\left(\frac{1}{10}\right)^4\left(\frac{9}{10}\right)^{5-4}$$

$$= \left(\frac{1}{10}\right)^5 + 5\left(\frac{1}{10}\right)^4\left(\frac{9}{10}\right) = \frac{83}{50000}.$$

EXAMPLE 6. *Two persons A and B throw with one dice for a stake of Rs. 11 which is to be won by the player who first throws 6. If A has the first throw, what are their respective expectations?*

SOLUTION. We know that in the first throw, the probability of A's success is $\frac{1}{6}$.

Further, each player must have failed once before A can have a second throw, therefore in second throw, the probability of A's success is

$$\left(\frac{5}{6} \times \frac{5}{6}\right) \times \frac{1}{6} = \left(\frac{5}{6}\right)^2 \times \frac{1}{6}$$

Similarly in his first throw, the probability of his success is $\left(\frac{5}{6}\right)^4 \times \frac{1}{6}$ because each player must failed twice and so on.

Proceeding in the same way, the probability of A's success is the sum of infinite series

$$\frac{1}{6}\left[1 + \left(\frac{5}{6}\right)^2 + \left(\frac{5}{6}\right)^4 + \ldots\right].$$

Similarly, the probability of B's success is the sum of infinite series

$$\frac{5}{6} \times \frac{1}{6}\left[1 + \left(\frac{5}{6}\right)^2 + \left(\frac{5}{6}\right)^4 + \ldots\right]$$

Hence, A's success to B's as 6 is to 5. Thus, their respective probabilities of success are $\frac{6}{11}$ and $\frac{5}{11}$.

Therefore, respective expectations of A and B are Rs. 6 and Rs. 5 respectively.

EXAMPLE 7. *Six dice are thrown 729 times. How many times*

do you expect at least 3 dice to show 5 or 6?

SOLUTION. The probability of throwing 5 or 6

$$= \frac{1}{6} + \frac{1}{6} = \frac{1}{3}.$$

Now, probability of showing 5 or 6 at least 3 dice = Sum of the probabilities of showing 5 or 6 by 3, 4, 5, 6 dice

$$= \sum_{x=3}^{6} {}^6C_x\left(\frac{1}{3}\right)^x\left(1 - \frac{1}{3}\right)^{6-x} = \sum_{x=3}^{6} {}^6C_x\left(\frac{1}{3}\right)^x\left(\frac{2}{3}\right)^{6-x}$$

$$= {}^6C_3\left(\frac{1}{3}\right)^3\left(\frac{2}{3}\right)^3 + {}^6C_4\left(\frac{1}{3}\right)^4\left(\frac{2}{3}\right)^2 + {}^6C_5\left(\frac{1}{3}\right)^5\left(\frac{2}{3}\right)$$

Hence, expected number out of 729 throws

$$= 729 \times \frac{233}{729} = 233.$$

EXAMPLE 8. *A factory A produces 10% defective valves and another factory B produces 20% defectives. A bag contains 4 values of factory A and 5 valves of factory B. If two valves are drawn at random from the bag, find the probability that at least one valve is defective.*

SOLUTION. Let A_1, A_2, A_3 be the events of selecting 2 from the factory A, 1 from factory A, and 1 from B and 2 from factory B respectively.

Then, we have

$$P(A_1) = \frac{{}^4C_2}{{}^9C_2} = \frac{1}{6}, \quad P(A_2) = \frac{{}^4C_1 \cdot {}^5C_1}{{}^9C_2} = \frac{5}{9}$$

$$P(A_3) = \frac{{}^5C_2}{{}^9C_2} = \frac{5}{18}$$

Further, suppose that E be the event that both selections contain at least one defective valve. Then E′ is the event that both selected valves are roots of affections.

Since A_1, A_2, A_3 are mutually exclusive and exhaustive events. Therefore

$$E' = E'A_1 \cup E'A_2 \cup E'A_3$$

$$= P(A_1) P(E' \mid A_1) + P(A_2) P(E' \mid A_2)$$
$$\qquad + P(A_3) P(E' \mid A_3)$$

$$= \frac{1}{6}(0.9)^2 + \frac{5}{9}(0.9)(0.8) + \frac{5}{18}(0.8)^2 = \frac{12.83}{18}$$

Hence, required probability

$$P(E) = 1 - \frac{12.83}{18} = \frac{5.18}{18} = 0.2872$$

Exercise-9.19

1. If on an average 1 vessels in every 10 is wrecked, find the probability that out of 5 vessels expected to arrive, at least 4 will arrive safely. (PTU–2005)

2. Out of 800 families with 5 children each, how many would you expect to have (a) 3 boys, (b) 5 girls, (c) either 2 or 3 boys? Assume equal probability for boys and girls. (VTU–2004)

3. Fit a binomial distribution for the following data and compare the theoretical frequencies with the actual ones.

x	0	1	2	3	4	5
f	2	14	20	34	22	8

(BHOPAL–2006)

4. The probability that a bomb dropped from a plane will strike the target is 1/5. If six bombs are dropped, find the probability that
   (i) exactly two will strike the target
   (ii) at least two will strike the target

5. If 10% of rivets produced by a machine are defective, find the probability that out of 5 rivets chosen at random
   (i) none will be defective
   (ii) one will be defective, and
   (iii) at least two will be defective

6. Find the binomial distribution whose mean is 5 and variance is 10/3.

7. If 10% of bolts produced by a machine are defective, calculate the probability that out of a sample selected at random of 7 bolts, not more than one bolt will be defective.

8. The sum and product of the mean and variance of a binomial distribution are 24 and 128 respectively. Find the distribution.

9. A dice is thrown 6 times. If getting an odd number is a success, find the probability of
   (i) 5 successes
   (ii) at least 5 successes
   (iii) at most 5 successes

10. If the probability that a new born child is a male is 0.6, find the probability that in a family of 5 children there are exactly 3 boys.   (KURUKSHETRA–2005)

11. The probability that a pen manufactured by a company will be defective is 1/10. If 12 such pens are manufactured, find the probability that:
    (a) exactly two will be defective.
    (b) at least two will be defective
    (c) none will be defective.   (VTU–2004, BURDWAN–2003)

12. In 256 sets of 12 tosses of a coin, in how many cases one can except 8 heads and 4 tails.   (JNTU–2003)

13. In a bombing action there is 50% chance that any bomb will strike the target. Two direct hits are needed to destroy the target completely. How many bombs are required to be dropped to give a 99% chance or better of completely destroying the target.   (VTU–2003S)

14. 500 articles were selected at random out of a batch containing 10,000 articles and 30 were found to be defective. How many defectives articles would you reasonably expect to have in the whole batch?   (JNTU–2003)

15. Fit a binomial distribution to the following frequency distribution :

x	0	1	2	3	4	5	6
f	13	25	52	58	32	16	4

(KURUKSHETRA–2009, SVTU–2007)

Answers

1. 45927/50000   2. (a) 250, (b) 25, (c) 500   3. $100(0.432 + 0.568)5$ 4. (i) 0.246, (ii) 0.345   5. (i) 0.59049, (ii) 0.32805,

(iii) 0.08146   6. $P(r)\ ^{15}C_r \left(\dfrac{1}{3}\right)^r \left(\dfrac{2}{3}\right)^{15-r}$ 7. $(1.6)(0.9)^6$ 8. $PP(X = r) = \ ^{32}C_r \left(\dfrac{1}{2}\right)^r \left(\dfrac{1}{2}\right)^{32-r}, r = 0, 1, 2, ..., 32$

9. (i) 3/22 (ii) 7/64 (iii) 63/64   10. 0.3456   11. (a) 0.2301, (b) 0.3412 (c) 0.2833   12. 31 (say),

13. 11   14. 600   15. $200 (0.554 + 0.446)6$

9.6.8 Poisson Distribution

A variable x is said to have the Poisson distribution if it takes the values 0, 1, 2, ... *to* ∞ *with probabilities* $e^{-m}, \dfrac{m}{1!}e^{-m}, \dfrac{m^2}{2!}e^{-m}, \dfrac{m^3}{3!}e^{-m}, ...$ *to* ∞ *respectively. The probability distribution function for Poisson distribution is given by*

$$P(x, m) = P(X = x) = \begin{cases} \dfrac{e^{-m} \cdot m^x}{x!}, & x = 0, 1, 2 \\ 0 & , \text{ otherwise} \end{cases}$$

Remarks

- If Poisson distribution was discovered by French mathematician S. D. Poisson.
- Poisson distribution has only one parameter m.
- Poisson distribution occurs when there are events which do not occur as outcomes of a definite number of trials of an experiment but which occur at random points of time and space where our interest lies only in the number of occurrence of the event, not its non-occurrence.
- It should be noted that $\sum\limits_{x=0}^{\infty} P(X = x) = e^{-m} \sum\limits_{x=0}^{\infty} \dfrac{m^x}{x!} = e^{-m} \cdot e^m = 1$

9.6.9 Limiting form of Binomial Distribution

THEOREM 1. *Poisson distribution is the limiting form of the binomial distribution* $(q + p)^n$ *when p (or q)* $\to 0$ *as* $n \to \infty$ *such that np (or nq) is a finite quantity say m, i.e., np = m.*

PROOF.   We know that, the probability of r success in a binomial distribution is

$$P(r) = \ ^nC_r\, p^r q^{n-r} = \dfrac{n(n-1)(n-2)...(n-r+1)}{r!} p^r q^{n-r}$$

$$= \dfrac{np(np-p)(np-2p)...(np-(r-1)p)}{r!} (1-p)^{n-r}$$

As $n \to \infty$, $p \to 0\ (np = m)$, we get

$$P(r) = \dfrac{m^r}{r!} \lim_{n\to\infty} \dfrac{\left(1 - \dfrac{m}{n}\right)^n}{\left(1 - \dfrac{m}{n}\right)^r} = \dfrac{m^r}{r!} e^{-m}$$

Therefore, the probabilities of 0, 1, 2, ..., r successes in a Poisson distribution are given by

$$e^{-m}, me^{-m}, \frac{m^2 e^{-m}}{2!}, ..., \frac{m^r e^{-m}}{r!}, ...$$

Hence, the limiting form of the binomial distribution $(q+p)^n$ when $p \to 0$ and $n \to \infty$ such that $np = m$ is the Poisson distribution.

9.6.10 MOMENTS OF POISSON DISTRIBUTION

MOMENT ABOUT ORIGIN

(1) First Moment about Origin (Mean of Poisson Distribution)

$$\mu_1' = \sum_{x=0}^{\infty} \frac{e^{-m} m^x}{x!} \cdot x = e^{-m} \cdot 0 + \left(e^{-m} \cdot \frac{m}{1!} \cdot 1\right) + \left(e^{-m} \cdot \frac{m^2}{2!} \cdot 2\right) + ... + \left(e^{-m} \cdot \frac{m^r}{r!} r\right) + ...$$

$$= m \, e^{-m} \left[1 + \frac{m}{1!} + \frac{m^2}{2!} + ... + \frac{m^{r-1}}{(r-1)!} + ...\right] = me^{-m}[e^m] = m$$

$\Rightarrow$ Mean of the Poisson distribution $= \mu_1' = m$.

(2) Second Moment about Origin

$$\mu_2' = \sum_{x=0}^{\infty} \frac{e^{-m} \cdot m^x}{x!} \cdot x^2 = \sum_{x=0}^{\infty} e^{-m} \frac{m^x}{x!} \{x(x-1) + x\}$$

$$= \sum_{x=0}^{\infty} \frac{e^{-m} \cdot m^x}{(x-2)!} + \sum_{x=0}^{\infty} \frac{e^{-m} \cdot m^x}{(x-1)!} = \sum_{x=0}^{\infty} \frac{m^2 e^{-m} \cdot m^{x-2}}{(x-2)!} + \sum_{x=0}^{\infty} \frac{me^{-m} \cdot m^{x-1}}{(x-1)!}$$

$$= m^2 e^{-m} \sum_{x=0}^{\infty} \frac{m^{x-2}}{(x-2)!} + me^{-m} \sum_{x=0}^{\infty} \frac{m^{x-1}}{(x-1)!} = m^2 e^{-m} e^m + me^{-m} e^m = m^2 + m$$

(3) Third Moment about Origin

$$\mu_3' = \sum_{x=0}^{\infty} \frac{e^{-m} \cdot m^x}{x!} \cdot x^3 = \sum_{x=0}^{\infty} e^{-m} \frac{m^x}{x!} \{x(x-1)(x-2) + 3x(x-1) + x\}$$

$\Rightarrow$ $$\mu_3' = m^3 + 3m^2 + m$$

(4) Fourth Moment about Origin

$$\mu_4' = \sum_{x=0}^{\infty} \frac{e^{-m} \cdot m^x}{x!} \cdot x^4 = \sum_{x=0}^{\infty} e^{-m} \frac{m^x}{x!} \{x(x-1)(x-2)(x-3) + 6x(x-1)(x-2) + 7x(x-1) + x\}$$

$\Rightarrow$ $$\mu_4' = m^4 + 6m^3 + 7m^2 + m$$

MOMENT ABOUT MEAN

(1) First Moment about Mean: $\mu_1 = 0$

(2) Second Moment about Mean (Variance of Poisson Distribution)

$$\mu_2 = \mu_2' - (\mu_1')^2 = m^2 + m - m^2 = m$$

$\Rightarrow$ Variance $= \mu_2 = m$

and   S.D. $= \sqrt{\mu_2} = \sqrt{m}$.

(3) Third Moment about Mean

$$\mu_3 = \mu_3' - 3\mu_2'\mu_1' + 2\mu_1'^3 = (m^3 + 3m^2 + m) - 3(m^2 + m)m + 2(m)^3$$

$$= m^3 + 3m^2 + m - 3m^3 - 3m^2 + 2m^3 = m$$

📧 **REMARK**
* For poisson distribution, mean = variance $= m$.

(4) Fourth Moment about Mean

We know that $$\mu_4 = \mu_4' - 4\mu_3'\mu_1' + 6\mu_2'\mu_1'^2 - 3\mu_1'^4$$

$$= (m^4 + 6m^3 + 7m^2 + m) - 4(m^3 + 3m^2 + m)m + 6(m^2 + m)(m)^2 - 3(m)^4$$

$$= m^4 + 6m^3 + 7m^2 + m - 4m^4 - 12m^3 - 4m^3 + 6m^4 + 6m^3 - 3m^4$$

$\Rightarrow$ $$\mu_4 = 3m^2 + m$$

PEARSON'S COEFFICIENTS

(1) $\beta_1 = \dfrac{\mu_3^2}{\mu_2^3} = \dfrac{m^2}{m^3} = \dfrac{1}{m}$

(2) $\beta_2 = \dfrac{\mu_4}{\mu_2^2} = \dfrac{3m^2 + m}{m^2} = 3 + \dfrac{1}{m}$

(3) $\gamma_1 = \sqrt{\beta_1} = \dfrac{1}{\sqrt{m}}$

(4) $\gamma_2 = \beta_2 - 3 = \dfrac{1}{m}$

9.6.11 MOMENT GENERATING FUNCTION (M.G.F.) OF POISSON DISTRIBUTION

The probability distribution function of Poisson distribution is given by $P(x) = \dfrac{e^{-m} m^x}{x!}$

Therefore, m.g.f. about origin will be given by $M_0(t) = \sum\limits_{x=0}^{\infty} e^{tx}\left(\dfrac{e^{-m} m^x}{x!}\right) = e^{-m} \sum\limits_{x=0}^{\infty} \dfrac{(me^t)^x}{x!} = e^{-m} e^{me^t}$ $\qquad \left(\because e^x = \sum\limits_{r=0}^{\infty}\left(\dfrac{x^r}{r!}\right)\right)$

$\Rightarrow \qquad\qquad\qquad\qquad\qquad M_0(t) = e^{m(e^t - 1)}$

Also, we have the mean of Poisson distribution is m and $M_a(t) = e^{-at} M_0(t)$, where $M_0(t)$ is the m.g.f. about the point a.

Therefore, m.g.f. about mean will be given by $M_m(t) = e^{-int} M_0(t) = e^{-mt} e^{m(e^t - 1)} \Rightarrow \qquad M_m(t) = e^{m(e^t - 1 - t)}$

9.6.12 CUMULANT GENERATING FUNCTION OF POISSON DISTRIBUTION

We know that the moment generating function of Poisson distribution is given by $M_m(t) = e^{m(e^t - 1 - t)}$.

Now, cumulant function $K_m(t) = \log M_m(t)$

$$= m(e^t - t - 1) = m\left(1 + t + \dfrac{t^2}{2!} + \dfrac{t^3}{3!} + \dfrac{t^4}{4!} + \dots - t - 1\right) = m\left[\dfrac{t^2}{2!} + \dfrac{t^3}{3!} + \dfrac{t^4}{4!} + \dots\right]$$

Now, the coefficients of $\dfrac{t}{1!}, \dfrac{t^2}{2!}, \dfrac{t^3}{3!}, \dfrac{t^4}{4!}, \dots$ give the cumulant such that

$$K_1 = 0, \ K_2 = m, \ K_3 = m, \ K_4 = m, \dots$$

SOME EXAMPLES OF POISSON VARIATE

(1) Number of suicides reported in a particular city.

(2) Number of air accidents in some unit of time.

(3) Number of printing mistakes at each page of the book.

(4) Number of telephone calls received at a particular telephone exchange in some unit of time.

(5) Number of car passing a crossing per hour during the busy hour of a day.

(6) Number of deaths from a disease.

Solved Examples

EXAMPLE 1. *Find the probability that at most 5 defective fuses will be found in a box of 200 fuses, if experience shows that 2 percent of such fuses are defective.*

SOLUTION. The probability of a fuse being defective is $\dfrac{2}{100}$

Here, $p = \dfrac{2}{100} = \dfrac{1}{50}, \ n = 200$

$\Rightarrow m = np = 200 \times \dfrac{1}{50} = 4$

$\therefore \ e^{-m} = e^{-4} = \dfrac{1}{e^4} = \dfrac{1}{(2.7)^4} = (0.3703)^4 = 0.019$

Now, the required probability of five or less than five defective fuses

$$= \sum\limits_{x=0}^{5} \dfrac{e^{-m} \cdot m^x}{x!} \qquad (\because x \le 5)$$

$$= \sum\limits_{x=0}^{5} \dfrac{e^{-4} \cdot 4^x}{x!} \qquad\qquad (\because m = 4)$$

$$= \dfrac{e^{-4} \cdot 4^0}{0!} + \dfrac{e^{-4} \cdot 4^1}{1!} + \dfrac{e^{-4} \cdot 4^2}{2!}$$

$$+ \dfrac{e^{-4} \cdot 4^3}{3!} + \dfrac{e^{-4} \cdot 4^4}{4!} + \dfrac{e^{-4} \cdot 4^5}{5!}$$

$$= e^{-4}\left[1 + 4 + 8 + \dfrac{32}{3} + \dfrac{32}{3} + \dfrac{128}{15}\right]$$

$$= 0.019\left(\dfrac{643}{15}\right) = 0.814.$$

EXAMPLE 2. *Fit a Poisson distribution to the following and calculate theoretical frequencies.*

Death (x)	0	1	2	3	4
Frequency (f)	122	60	15	2	1

(BHOPAL–2007S, VTU–2004, UPTU–2003)

SOLUTION . Since we know that

$$\frac{\Sigma fx}{\Sigma f} = \frac{\begin{array}{c}(122 \times 0) + (60 \times 1) \\ + (15 \times 2) + (2 \times 3) + (1 \times 4)\end{array}}{122 + 60 + 15 + 2 + 1}$$

$$= \frac{60 + 30 + 6 + 4}{200} = \frac{100}{200} = \frac{1}{2} = 0.5$$

and $e^{-m} = e^{-0.5} = 0.61$.

The probability of x deaths is given by $\dfrac{Ne^{-m}m^x}{x!}$
where $N = 200$ and $m = 0.5$.
Therefore, the required theoretical frequencies of 0, 1, 2, 3 and 4 deaths are respectively given by

$$\frac{200 \times e^{-0.5}(0.5)^0}{0!}, \frac{200 \times e^{-0.5}(0.5)^1}{1!},$$

$$\frac{200 \times e^{-0.5}(0.5)^2}{2!}, \frac{200 \times e^{-0.5}(0.5)^3}{3!},$$

$$\text{and} \quad \frac{200 \times e^{-0.5}(0.5)^4}{4!}$$

i.e.,

$$200 \times 0.61, 200 \times 0.61 \times 0.5, \frac{200 \times 0.61 \times (0.5)^2}{2!}.$$

$$\frac{200 \times 0.61 \times (0.5)^3}{3!} \quad \text{and} \quad \frac{200 \times 0.61 \times (0.5)^4}{4!}$$

$$\Rightarrow \quad 122, 61, 15.25, 2.54 \quad \text{and} \quad 0.31,$$
i.e., $\quad 122, 61, 15, 3, 0$

EXAMPLE 3. *A book of 585 pages contain 43 pages with misprints. If these pages are randomly distributed throughout the book, what is the probability that 10 pages, selected at random will be free from pages with misprint* $(e^{-0.735} = 0.4795)$.

SOLUTION . Let p denote the probability that a page selected at random is a page with misprints.

$$\Rightarrow \quad p = \frac{43}{585} = 0.0735$$

Here, $\quad n = 10$

$$\Rightarrow \quad m = np = 10 \times 0.0735 = 0.735$$

Let X denote the number of pages with misprints in a sample of 10 pages. Then X is a Poisson variate with parameter $m = 0.735$.

$$\therefore \quad P(X = r) = \frac{e^{-0.735} \cdot (0.735)^r}{r!}, \quad r = 1, 2, 3, ..., 10$$

The probability that the sample of 10 pages will be free from pages with misprint $= P(X = 0) = e^{-0.735} = 0.4795$.

EXAMPLE 4. *If m and μ_r denote the mean and r^{th} central moment of a Poisson distribution, then prove that $\mu_{r+1} = rm\mu_{r-1} + \dfrac{md\mu_r}{dm}$. Hence, obtain β_1 and β_2 of Poisson distribution.*

SOLUTION . $\dfrac{d\mu_r}{dm} = \dfrac{d}{dm}\left[\displaystyle\sum_{x=0}^{\infty} (x-m)^r \dfrac{e^{-m}m^x}{x!} \right]$

$$= -r \sum_{x=0}^{\infty} (x-m)^{r-1} \frac{e^{-m}m^x}{x!}$$

$$+ \sum_{x=0}^{\infty} \frac{(x-m)^r}{x!}[xm^{x-1}e^{-m} - m^xe^{-m}]$$

$$= -r\mu_{r-1} + \sum_{r=0}^{\infty} \frac{(x-m)^r e^{-m} m^{x-1}(x-m)}{x!}$$

Multiplying both sides by m

$$m\frac{d\mu_r}{dm} = -mr\mu_{r-1} + \sum_{x=0}^{\infty} \frac{(x-m)^{r+1}e^{-m}m^x}{x!}$$

$$m\frac{d\mu_r}{dm} = -mr\mu_{r-1} + \mu_{r+1}$$

Hence, $\quad \mu_{r+1} = mr\mu_{r-1} + m\dfrac{d\mu_r}{dm}$

Now, $\quad \mu_2 = m\mu_0 + m\dfrac{d}{dm}\mu_1 \qquad [\mu_1 = 0]$

$$= m.1 + 0 = m$$

and $\quad \mu_3 = m.2\mu_1 + m.\dfrac{d}{dm}\mu_2 = 0 + m.\dfrac{dm}{dm} = m$

Now, $\beta_1 = \dfrac{\mu_3^2}{\mu_2^3} = \dfrac{m^2}{m^3} = \dfrac{1}{m}$, $\beta_2 = \dfrac{\mu_4}{\mu_2^2} = \dfrac{3m^2 + m}{m^2} = \left(3 + \dfrac{1}{m}\right)$

EXAMPLE 5. *Derive Poisson distribution as an approximation of binomial distribution.*

SOLUTION . In a binomial distribution

$$P(r) = P(x = r) = {}^nC_r \, p^r \, q^{n-r} = {}^nC_r \, p^r \, (1-p)^{n-r}$$

$$= \frac{n(n-1)(n-2)...(n-r+1)}{r!} p^r \left(1 - \frac{np}{n}\right)^{n-r}$$

$$= \frac{\left(1 - \frac{1}{n}\right)\left(1 - \frac{2}{n}\right)...\left(1 - \frac{r-1}{n}\right)(np)^r}{r!\left(1 - \frac{np}{n}\right)^r} \left(1 - \frac{np}{n}\right)^n$$

$\therefore \; P(r) = $ Probability of r successes in Poisson distribution

$$= \lim_{p \to 0, n \to \infty, np = m} p(x = r)$$

$$= \lim_{n \to \infty} \frac{\left(1 - \frac{1}{n}\right)\left(1 - \frac{2}{n}\right)...\left(1 - \frac{r-1}{n}\right)}{r!} \frac{(np)^r}{\left(1 - \frac{np}{n}\right)^r} \times \left(1 - \frac{np}{n}\right)^n$$

$$= \lim_{n \to \infty} \frac{m^r}{\left(1 - \frac{m}{n}\right)^r r!} \left(1 - \frac{m}{n}\right)^n$$

$$= \frac{m^r}{r!} \lim_{n \to \infty} \left(1 - \frac{m}{n}\right)^n$$

$$P(r) = \frac{m^r e^{-m}}{r!} \qquad \left[\because \lim_{n \to \infty}\left(1 - \frac{m}{n}\right)^n = e^{-m} \right]$$

which is Poisson distribution.

Exercise-9.20

1. If the probability of a bad reaction from a certain injection is 0.001, determine the chance that out of 2000 individuals, more than two will get a bad reaction.

(VTU–2008, KOTTAYAM–2005)

2. In a certain factory turning out razor blades, there is a small chance of 0.002 for any blade to be defective. The blades are supplied in packets of 10. Calculate the approximate number of packets containing no defective, one defective, two defective blades, respectively in a consignment of 10,000 packets. (KURUKSHETRA–2009S, MADRAS–2006, VTU–2004)

3. Fit a Poisson distribution to the following data

x	0	1	2	3	4
f	46	38	22	9	1

(KURUKSHETRA–2009, BHOPAL–2008, VTU–2003S)

4. If a random variable has a Poisson distribution such that $P(1) = P(2)$, then find

(i) mean of distribution, and (ii) $P(4)$. (VTU–2003)

5. A car hire firm has two cars which it hires out day by day. The number of demands for a car on each day is distributed as a Poisson distribution with mean 1.5. Calculate the proportion of days

(i) on which there is no demand

(ii) on which demand is refused (Given, $e^{-1.5} = 0.2231$)

(BHOPAL–2008S; JNTU–2003)

6. Six coins are tossed 6400 times. Using Poisson distribution, obtain the approximate probability of getting 6 heads r times.

7. If X is a Poisson variate, such that $P(X = 1) = 2P(X = 2)$, find

(i) Mean, (ii) Variance, (iii) $P(X = 0)$.

8. The probability that a man aged 50 years will die within a year is 0.01125. What is the probability that out of 12 such men at least 11 will reach their fifty first birthday.

9. If X is Poisson variate such that

$P(X = 2) = 9P(X = 4) + 90P(X = 6)$, find the mean of X.

10. Find the probability that at most 5 defective will be found in a lot of 200 if experience shows that 2% of such components are defective. Also find the probability of more than 5 defective components [given $e^{-4} = 0.018$].

11. A certain screw making machine produces an average of 2 defective screws out of 100 and pack them in boxes of 500. Find the probability that a box contain 15 defective screws.

(KURUKSHETRA–2006)

12. An insurance company found that only 0.01% of the population is involved in a certain type of accident each year. If its 1000 policy holders were randomly selected from the population, what is the probability that not more than two of its clients are involved in such an accident next year. [given $e^{-0.1} = 0.9048$].

13. If the variance of the Poisson distribution is 2, find the probabilities for $r = 1, 2, 3, 4$ from the recurrence relation of the Poisson distribution. Also find $P(x \geq 4)$.

14. For Poisson distribution, Prove that $\mu_2 \gamma_1 \gamma_2 = 1$, where symbols have their usual meanings. (SVTU–2008)

15. Find the expectation of the function $\phi(x) = xe^{-x}$ in a Poisson distribution. (VTU–2003)

16. Fit a Poisson distribution to the following data given the number of yeast cells per square for 400 squares:

No. of cells per squares	0	1	2	3	4	5	6	7	8	9	10
No. of squares	103	143	98	42	8	4	2	0	0	0	0

(SVTU–2007)

 Answers

1. 0.32 **2.** 2 **3.** 44, 43, 27, 7, 1 **4.** (i) 2, (ii) 0.0902 **5.** (i) 0.2231, (ii) 0.1913 **6.** $P(X = r) = \dfrac{e^{-100}(100)^r}{r!}$, $r = 0, 1, 2, \ldots$

7. (i) 1, (ii) 1, (iii) $1/e$ **8.** 0.99166 **9.** Mean = 1 **10.** 0.216 **11.** $\dfrac{(10)^{15} e^{-10}}{15!} = 0.035$ **12.** 0.9998

13. 0.2706, 0.2706, 0.1804, 0.0902, $P(x \geq 4) = 0.1431$. **16.** Theoretical frequencies are 109, 142, 92, 40, 13, 3, 1 , 0, 0, 0, 0.

9.6.13 NORMAL DISTRIBUTION

Definition. *A random variable X is said to follow a normal distribution with parameter* μ *(mean) and* σ^2 *(variance) if its probability distribution f(x) is given by*

$$f(x) = \frac{1}{\sigma\sqrt{2\pi}} e^{-(x-\mu)^2/2\sigma^2}, \quad -\infty < x < \infty$$

- If a random variable X has a normal distribution with mean μ and variance σ^2, then X is said to be normal variate with mean μ and variance σ^2, and can be written as $X \sim N(\mu, \sigma^2)$.
- It was discovered by Karl Pearson and De'Moivre.
- It is a continuous distribution.
- Any quantity whose variation depends on random causes is distributed according to the normal law.
- The graph of $f(x)$ is a bell shaped curve, symmetrical about the mean μ. This curve is known as normal probability curve or normal curve.
- Normal distribution is a limiting case of binomial distribution.
- The equation of the normal curve is $y_x = y_0 e^{-x^2/2\sigma^2}$.

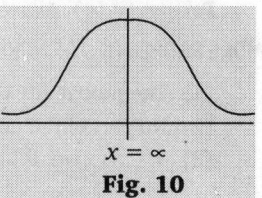

$x = \infty$

Fig. 10

9.6.14 STANDARD FORM OF THE NORMAL CURVE

Consider the equation of the normal curve $y_x = y_0 e^{-x^2/2\sigma^2}$, where origin is taken at mean and the value of y_0 is calculated in such a manner that the total frequency is unity.

Then
$$1 = y_0 \int_{-\infty}^{\infty} e^{-x^2/2\sigma^2}\, dx = 2y_0 \int_0^{\infty} e^{-x^2/2\sigma^2}\, dx$$

$$= 2y_0 \int_0^{\infty} e^{-t^2} \sigma\sqrt{2}\, dt \quad \text{(Take } x = \sigma\sqrt{2}.t \Rightarrow dx = \sigma\sqrt{2}.dt\text{)}$$

$$= 2\sqrt{2}.\sigma y_0 \int_0^{\infty} e^{-t^2} dt = 2\sqrt{2}.\sigma y_0 \left(\frac{1}{2}\sqrt{\pi}\right) \quad \left(\because \int_0^{\infty} e^{-t^2} dt = \frac{\sqrt{\pi}}{2}\right)$$

$$\Rightarrow \qquad y_0 = \frac{1}{\sigma\sqrt{2\pi}}$$

Hence, the standard form of the normal curve is $y = \frac{1}{\sigma\sqrt{2\pi}} e^{-x^2/2\sigma^2}$.

☞ **REMARKS**

- If the total frequency (or the total area under the curve is N, then the corresponding normal curve is given by $y = \frac{N}{\sigma\sqrt{2\pi}} e^{-x^2/2\sigma^2}$.

- If the origin is taken at 0 success, then the equation of normal curve is given by $y = \frac{N}{\sigma\sqrt{2\pi}} e^{-(x-m)^2/2\sigma^2}$, where $m = np$ is the mean.

- The point of inflexion of the normal curve is $x = \pm\sigma$.

9.6.15 PROPERTIES OF THE NORMAL DISTRIBUTION

Property (1). *The mean, median, mode of a normal distribution coincide at the origin.*

Proof. We have, Mean $= \int_{-\infty}^{\infty} y_0 e^{-x^2/2\sigma^2}.x\, dx$ $\qquad\qquad \left(\because y_0 = \frac{1}{\sigma\sqrt{2\pi}}\right)$

$$= y_0 \int_{-\infty}^{\infty} x e^{-x^2/2\sigma^2}.dx = 0 \qquad\qquad (\because \text{it is an odd function of } x.)$$

Since $\qquad y = y_0 e^{-x^2/2\sigma^2}$

$$\Rightarrow \qquad \frac{dy}{dx} = \frac{-y_0}{\sigma^2} x e^{-x^2/2\sigma^2} \qquad \text{and} \qquad \frac{d^2y}{dx^2} = \frac{y_0}{\sigma^2}\left(1 - \frac{x^2}{\sigma^2}\right) e^{-x^2/2\sigma^2}$$

Therefore, $\qquad \frac{dy}{dx} = 0 \Rightarrow x = 0$ also $\left(\frac{d^2y}{dx^2}\right)_{at\ x=0}$ = negative

$\Rightarrow y$ is maximum at $x = 0 \Rightarrow$ The mode is at $x = 0$.

Now, if x_1 be the median, then $\int_{x_1}^{\infty} \frac{1}{\sigma\sqrt{2\pi}} e^{-x^2/2\sigma^2}.dx = \frac{1}{2}$

$$\Rightarrow \qquad \frac{1}{2} = \frac{1}{\sigma\sqrt{2\pi}} \int_{x_1}^{\infty} e^{-x^2/2\sigma^2}.dx = \frac{1}{\sigma\sqrt{2\pi}} \int_{x_1/\sigma\sqrt{2}}^{\infty} e^{-t^2}.\sigma\sqrt{2}\, dt \quad \text{(take } \sigma\sqrt{2} = t\text{)} = \frac{1}{\sqrt{\pi}} \int_{x_1/\sigma\sqrt{2}}^{\infty} e^{-t^2}.dt$$

or $\qquad = \frac{1}{2}\sqrt{\pi} \int_{x_1/\sigma\sqrt{2}}^{\infty} e^{-t^2}.dt = \int_0^{\infty} e^{-t^2}\, dt$

$$\Rightarrow \qquad \frac{x_1}{\sigma\sqrt{2}} = 0 \Rightarrow x_1 = 0$$

Hence, we have that mean, median and mode coincide at $x = 0$.

Property (2). *The curve is symmetrical about y-axis.*

Proof. Since the equation of the normal curve remains unchanged if x is replaced by $-x$, hence the curve is symmetrical about y-axis.

Property (3). *The point of inflexion of the normal curve are given by $x = \pm \sigma$.*

Property (4). *All moments of odd order about the mean vanish.*

MORE PROPERTIES OF NORMAL DISTRIBUTION.

(1) The probability density function is always non-negative.

(2) No part of the normal curve will lie below the x-axis, because the p.d.f. is always greater than or equal to zero.

(3) $\int_{-\infty}^{\infty} f(x)\, dx = 1$, which implies the total area bounded by normal probability curve and x-axis is 1.

(4) Mode and median of normal distribution is equal to m (mean).

(5) The x-axis is the asymptote of the normal curve, because it never tough the x-axis.

(6) The normal curve is concave near the mean value and convex near $x = \mu \pm 3\sigma$.

9.6.16 FITTING OF NORMAL DISTRIBUTION

To fit a normal distribution, first find its mean m and standard deviation σ.

Let us define $t = \dfrac{x - m}{\sigma}$, where x is the mid value of the class interval in case of grouped data. Then the normal curve fitted to the given data is

$$f(x) = \frac{1}{\sigma\sqrt{2\pi}} e^{-(x-\mu)^2/2\sigma^2}, \quad -\infty < x < \infty$$

☞ REMARK

- To find the expected frequencies, we may use any of the following method : (i) Area method, (ii) Ordinate method.

9.6.17 MOMENT GENERATING FUNCTION OF NORMAL DISTRIBUTION

For normal distribution, the p.d.f. is given by $f(x) = \dfrac{1}{\sigma\sqrt{2\pi}} e^{-x^2/2\sigma^2}$

$\therefore$

$$M_0(t) = \frac{1}{\sigma\sqrt{2\pi}} \int_{-\infty}^{\infty} e^{tx}\, e^{-x^2/2\sigma^2}. \, dx = \frac{1}{\sigma\sqrt{2\pi}} \int_{-\infty}^{\infty} \left[\exp\left(tx - \frac{x^2}{2\sigma^2} \right) \right] dx$$

$$= \frac{1}{\sigma\sqrt{2\pi}} \int_{-\infty}^{\infty} \left[\exp\left\{ -\frac{(x - t\sigma^2)^2}{2\sigma^2} + \frac{1}{2}t^2\sigma^2 \right\} \right] dx$$

$$= \frac{1}{\sigma\sqrt{2\pi}} e^{t^2\sigma^2/2} \int_{-\infty}^{\infty} \left[\exp\left\{ -\frac{(x - t\sigma^2)^2}{2\sigma^2} \right\} \right] dx$$

$$= \frac{1}{\sigma\sqrt{2\pi}} e^{t^2\sigma^2/2} \int_{-\infty}^{\infty} e^{-z^2}. \, \sigma\sqrt{2}\, dz \qquad \text{(Take } \frac{x - t\sigma^2}{\sigma\sqrt{2}} = z\text{)}$$

$$= \frac{1}{\sqrt{\pi}} e^{t^2\sigma^2/2} \int_{-\infty}^{\infty} e^{-z^2}\, dz = e^{\frac{t^2\sigma^2}{2}} \qquad \left(\because \int_{-\infty}^{\infty} e^{-z^2} dz = \sqrt{\pi} \right)$$

$\Rightarrow$

$$M_0(t) = e^{t^2\sigma^2/2} = 1 + \left(\frac{t^2\sigma^2}{2} \right) + \frac{1}{2!}\left(\frac{t^2\sigma^2}{2} \right)^2 + \dots + \frac{1}{n!}\left(\frac{t^2\sigma^2}{2} \right)^n + \dots$$

This involves only even powers of t and therefore the coefficient of t^{2n+1} in $M_0(t)$ is zero.

$\Rightarrow \qquad \mu_{2n+1} = 0$

Also, $\qquad \mu_{2n} = $ Coefficient of $\dfrac{t^{2n}}{(2n)!}$ in the expression of $M_0(t)$

$$= \left(\frac{1}{2}\sigma^2 \right)^n \frac{(2n)!}{n!} = \frac{1}{2^n}\sigma^{2n}\frac{2n(2n-1)(2n-2)\dots3.2.1}{n(n-1)(n-2)\dots2.1}$$

$$= \frac{1}{2^n}\sigma^{2n}\frac{[(2n-1)(2n-3)\dots3.1][(2n)(2n-2)\dots4.2]}{n(n-1)(n-2)\dots2.1} = \frac{1}{\sigma}\sigma \frac{[(2n-1)(2n-3)\dots3.1][2 \, .n(n-1)\dots2.1]}{n(n-1)(n-2)\dots2.1}$$

$$= 1.3.5\dots(2n-1)\, \sigma^{2n}$$

In particular, $\quad \mu_2 = 1.\sigma^2 \quad$ and $\quad \mu_4 = 1.3\sigma^4 = 3\sigma^4 \quad$ and $\quad \beta_2 = \dfrac{\mu_4}{\mu_2^2} = \dfrac{3\sigma^4}{(\sigma^2)^2} = 3$

Now, if probability function is given by $f(x) = \dfrac{1}{\sigma\sqrt{2\pi}} e^{-\frac{1}{2}\left(\frac{x-m}{\sigma}\right)^2}$

Then, $M_0(t) = \int_{-\infty}^{\infty} e^{tx} \dfrac{1}{\sigma\sqrt{2\pi}} e^{-\frac{1}{2}\left(\frac{x-m}{\sigma}\right)^2} . dx = \dfrac{1}{\sigma\sqrt{2\pi}} \int_{-\infty}^{\infty} e^{tx} e^{-\frac{1}{2}\left(\frac{x-m}{\sigma}\right)^2} . dx$

$$= \dfrac{1}{\sigma\sqrt{2\pi}} \int_{-\infty}^{\infty} e^{t(m+z\sigma)} e^{-z^2/2}$$

(Put $z = \dfrac{x-m}{\sigma} \Rightarrow x = m + z\sigma$)

$$= \dfrac{e^{mt}}{\sqrt{2\pi}} \int_{-\infty}^{\infty} e^{-\frac{1}{2}z^2 + t\sigma z} . dz = \dfrac{e^{mt}}{\sqrt{2\pi}} \int_{-\infty}^{\infty} e^{-\frac{1}{2}(z^2 - 2t\sigma z + t^2\sigma^2) + \frac{1}{2}t^2\sigma^2} . dz$$

$$= \dfrac{e^{mt + \frac{1}{2}t^2\sigma^2}}{\sqrt{2\pi}} \int_{-\infty}^{\infty} e^{-\frac{1}{2}(z-t\sigma)^2} . dz = \dfrac{e^{mt + \frac{1}{2}t^2\sigma^2}}{\sqrt{2\pi}} \int_{-\infty}^{\infty} e^{-u^2} . \sqrt{2}\, du$$

(Put $z - t\sigma = u\sqrt{2}$)

$$= e^{mt + \frac{1}{2}t^2\sigma^2} \Rightarrow M_0(t) = e^{mt + \frac{t^2\sigma^2}{2}}$$

$\left(\because \int_{-\infty}^{\infty} e^{-u^2} = \sqrt{\pi}\right)$

Also, the m.g.f. about mean m is given by

$$M_m(t) = e^{-mt} M_0(t) = e^{-mt}\left\{e^{mt + \frac{1}{2}t^2\sigma^2}\right\} = e^{\frac{t^2\sigma^2}{2}}$$

9.6.18 Cumulant Generating Function of Normal Distribution

Using above article, we have $M(t)\; e^{mt + \frac{t^2\sigma^2}{}}$.

Now, $K_0(t) = \log M_0(t) = mt + \dfrac{1}{2}t^2\sigma^2$, which is the required cumulative function of normal distribution about origin.

Also, from $K_0(t) = mt + \dfrac{1}{2}t^2\sigma^2$, we get

$$K_1' = m, \; K_2' = \sigma^2, \; K_3' = 0, \; K_4' = 0, \dots \text{ from coefficients of } \dfrac{t}{1!}, \dfrac{t^2}{2!}, \dfrac{t^3}{3!}, \dots$$

The cumulative function about mean m is given in $K_m(t) = \log M_m(t) = \log[e^{t^2\sigma^2/2}] = \dfrac{t^2\sigma^2}{2}$

which gives $K_1 = 0, \; K_2 = \sigma^2, \; K_3 = 0, \; K_4 = 0, \dots$ from the coefficients of $\dfrac{t}{1!}, \dfrac{t^2}{2!}, \dfrac{t^3}{3!}, \dots$

 Solved Examples

EXAMPLE 1. *For the normal distribution* $N(\mu, \sigma^2)$ *show that* $\mu_{2r+2} = \sigma^2 . \mu_{2r} + \sigma^3 \dfrac{d}{d\sigma}\mu_{2r}$ *, where* μ_r *is* r^{th} *moment about mean* μ.

SOLUTION. Since, we know that

$$\mu_{2r} = \dfrac{1}{\sigma\sqrt{2\pi}} \int_{-\infty}^{\infty} (x-m)^{2r} e^{-(x-m)^2/2\sigma^2} . dx \quad \dots(1)$$

$$\Rightarrow \dfrac{d}{d\sigma}(\mu_{2r}) = -\dfrac{1}{\sigma^2\sqrt{2\pi}} \int_{-\infty}^{\infty}(x-m)^{2r} e^{-(x-m)^2/2\sigma^2} . dx$$

$$+ \dfrac{1}{\sigma\sqrt{2\pi}} \int_{-\infty}^{\infty}(x-m)^{2r} e^{-(x-m)^2/2\sigma^2} \dfrac{2(x-m)^2}{2\sigma^3} . dx$$

$$\Rightarrow \sigma^3 \dfrac{d}{d\sigma}(\mu_{2r}) = -\dfrac{\sigma^3}{\sigma^2\sqrt{2\pi}} \int_{-\infty}^{\infty}(x-m)^{2r} e^{-(x-m)^2/2\sigma^2} . dx$$

$$+ \dfrac{1}{\sigma\sqrt{2\pi}} \int_{-\infty}^{\infty}(x-m)^{2r+2} e^{-(x-m)^2/2\sigma^2} . dx$$

$$\Rightarrow \quad \sigma^3 \dfrac{d}{d\sigma}(\mu_{2r}) = -\sigma^2\mu_{2r} + \mu_{2r+2} \quad \text{(Using (1))}$$

$$\Rightarrow \quad \mu_{2r+2} = \sigma^2\mu_{2r} + \sigma^3 \dfrac{d}{d\sigma}\mu_{2r} .$$

EXAMPLE 2. *In a normal distribution, 31% of the items are under 45 and 8% are over 64. Find the mean and standard deviation of the distribution.*

(VTU–2009, SVTU–2008, KURUKSHETRA–2007S)

SOLUTION. Let m and σ denote the mean and standard deviation, respectively. Since 31% of the items are under 45 means area to the left of ordinate $x = 45$.

When $x = 45$, let $t = t_1 \Rightarrow t_1 = \dfrac{45 - m}{\sigma}$

$$\Rightarrow \quad \int_{-\infty}^{t_1} \phi(t)\, dt = 0.31$$

$$\text{or} \quad \int_{-\infty}^{0} \phi(t)\, dt - \int_{t_1}^{0} \phi(t)\, dt = 0.31$$

Therefore $\int_{t_1}^0 \phi(t)dt = \int_{-\infty}^0 \phi(t)dt - 0.31$

$\Rightarrow \quad 0.5 - 0.31 = 0.19$

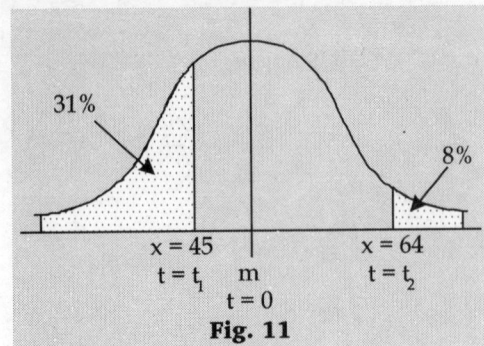

Fig. 11

Now, using table of normal curve
$$t_1 = -0.5 \quad.$$

When $x = 64$, let $t = t_2 \Rightarrow t_2 = \dfrac{(64-m)}{\sigma}$

$\therefore \quad \int_{t_2}^\infty \phi(t)dt = \int_0^\infty \phi(t)dt - 0.08 = 0.5 - 0.08 = 0.42$

Again, using table, we get $t_2 = 1.4$.

Therefore, we conclude that
$$45 - m = -0.5\sigma \qquad \text{...(1)}$$

and $\qquad\qquad 64 - m = 1.4\,\sigma \qquad \text{...(2)}$

Now, solving (1) and (2) for m and σ, we get $m = 50$ and $\sigma = 10$.

EXAMPLE 3. *If the probability of committing an error of magnitude x is given by $y = \dfrac{h}{\sqrt{\pi}} e^{-h^2 x^2}$.*

Compute the probable error from the following data
$$m_1 = 1.305, m_2 = 1.301, m_3 = 1.295,$$
$$m_4 = 1.286, m_5 = 1.318, m_6 = 1.321,$$
$$m_7 = 1.283, \ m_8 = 1.289,$$
$$m_9 = 1.300, \ m_{10} = 1.286$$

(KURUKSHETRA–2005)

SOLUTION. Here, we observe that the data are normally distributed.

Now, we have,

Mean $= \dfrac{1}{10}\Sigma \quad = \dfrac{12.984}{10} = 1.2984$

Also, $\sigma^2 = \dfrac{1}{10}\Sigma(m_i - \text{mean})^2$

$= \dfrac{1}{10}[(0.007)^2 + (0.003)^2 + (0.003)^2$

$\qquad + (0.012)^2 + (0.02)^2 + (0.023)^2$

$\qquad + (0.015)^2 + (0.009)^2 + (0.002)^2 + (0.012)^2]$

$= 0.0001594$

Hence, the probable error $= \dfrac{2}{3}\sigma = 0.0084$.

EXAMPLE 4. *Show that for the normal distribution, the quartile deviation, the mean deviation from mean and standard deviation are*

approximately in the ratio 10 : 12 : 15.

SOLUTION . For a normal distribution, we have
$$f(x) = \frac{1}{\sigma\sqrt{2\pi}} e^{-(x-\mu)^2/2\sigma^2}, \ -\infty < x < \infty$$

Then, Quartile deviation, Q.D. $= \dfrac{2}{3}\sigma$

Mean deviation from mean, M.D. $= \dfrac{4}{5}\sigma$

where, $\sigma = $ standard deviation $=$ S.D.

Hence, Q.D. : M.D. : S.D. $= \dfrac{2}{3}\sigma : \dfrac{4}{5}\sigma : \sigma$

$\qquad\qquad = 10\sigma : 12\sigma : 15\sigma \quad = 10 : 12 : 15$

EXAMPLE 5. *If skills are classified as A, B, C according to the length and breadth index as under : 75 between 75 and 80 and over 80, find approximately (assuming distribution is normal), the mean and standard deviation of a series in which A are 58%, B are 38% and C are 40% being given that if $f(t) = \dfrac{1}{2\sqrt{2\pi}} \int_0^t \exp(-t)^2 dt$.*

SOLUTION. Let m be the mean and σ be the standard deviation of the distribution. Since the total frequency is taken as unity frequency of skill A, whose length and breadth is under 75 is 0.58. The frequency of skill B whose index lies between 75 and 80 is 0.38 and frequency of skill C whose index is under 80 is 0.40.

Therefore, the total area to the left of ordinate RS is 0.58, area between RS and TU is 0.38 and the area to the right of TU is 0.04.

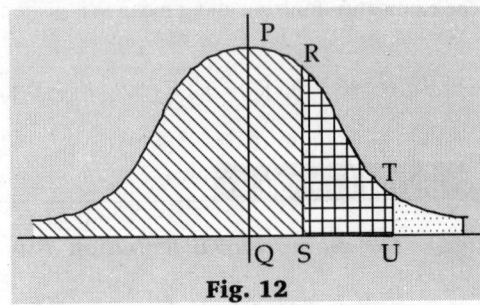

Fig. 12

Area between origin and $x = 75 - m$,

i.e., area $PQRS = 0.58 - 0.5 = 0.08$

and given $f(0.20) = 0.08$

Hence, $\dfrac{75 - m}{\sigma} = 0.20 \qquad \text{...(1)}$

Again, area between origin and $(x = 80 - m)$.

i.e., Area $PQUT = 0.08 + 0.38 = 0.46$

and given $f(1.75) = 0.46 \qquad \text{...(2)}$

Solving equations (1) and (2)
$$m = 74.3, \ \sigma = 3.23.$$

Exercise-9.21

1. In a test on 2000 electric bulbs, it was found that the life of a particular make was normally distributed with an average life of 2040 hrs and S.D. of 60 hrs. Estimate the number of bulbs likely to burn for : (a) more than 2150 hrs, (b) less than 1950 hrs, (c) more than 1920 hrs but less than 2160 hrs.
   (BHOPAL–2008S, UPTU–2008)

2. The mean height of 500 students is 151 cm and the S.D. is 15 cm. Assuming that the heights are normally distributed, find how many students height lie between 120 and 155 cm.
   (BURDWAN–2003)

3. Show that the standard deviation for a normal distribution is approximately 25% more than the mean deviation.

4. In an examination taken by 500 candidates, the average and standard deviation of marks obtained are 40% and 10%. Find (a) How many will pass, if 50% is fixed as a minimum, (b) What should be the minimum if 350 candidates are to pass, (c) How many have scored marks above 60%.

5. Assuming that the diameter of 1000 brass plugs taken consecutively from a machine form a normal distribution with mean 0.7515 cm and standard deviation 0.0020 cm. How many of the plugs are likely to be rejected if the approved diameter is 0.752 ± 0.004 cm.   (BHOPAL–2002)

6. Find the equation of the best fitted normal curve to the following distribution

x	0	1	2	3	4	5
f	13	23	34	15	11	4

7. A random variable x has a standard normal distribution ϕ. Show that $P(|x| < K) = 2\phi(K) - 1$.

8. Assume that the mean height of soldiers to be 68.22 inches with a variance of 10.8 inches square. How many soldiers in a regiment of 10,000 would you expect to be over 6 feet tall?

9. If X is a normal variate with mean 12 and standard deviation 4, find $P(0 \le X \le 12)$.

10. On a final examination in engineering, the mean was 72 and the standard deviation was 15. Determine the standard scores of students receiving grades (i) 60s, (ii) 93s, (iii) 72s.

11. A normal distribution with $\mu = 50$ and $\sigma = 10$ is given. Find the value of X that has (i) 13% of the area to its left, (ii) 14% of the area to its right.

12. The income distribution of workers in a certain factory was found to be normal with mean Rs. 500 and standard deviation Rs. 50. There were 228 persons getting above Rs. 600. How many workers were there in all?

13. The scores of candidates in a certain test are normally distributed with mean 500 and standard deviation 100. What percentage of candidates receives the scores between 350 and 550?

14. Assuming that the diameter of 1000 brass plugs taken consecutively from a machine form a normal distribution with mean 0.7515 cm and standard deviation 0.0020 cm, how many of the plugs are likely to be rejected of the approved diameter is 0.752 ± 0.004 cm.

15. In an intelligence test administered to 1000 students, the average was 42 and S.D. 24. Find (i) the number of students exceeding a score 50, (ii) the number of children lying between 30 and 54.

16. Fit a normal curve to the following distribution.

x	2	4	6	8	10
f	1	4	6	4	1

   (VTU–2001)

17. The income of a group of 10,000 persons was found to be normally distributed with mean Rs. 750 p.m., and standard deviation of Rs.50. Show that of this group, about 95% had income exceeding Rs. 668 and only 5% had income exceeding Rs. 832. Also find the lowest income among the richest 100.
    (UPTU–2004S)

Answers

1. (a) 67, (b) 184, (c) 1904   **2.** 294   **4.** (i) 79, (ii) 35%, (iii) 11   **5.** 52   **6.** $y = \dfrac{100}{\sqrt{3.4\pi}} e^{-\frac{(x-2)^2}{3.4}}$

8. 1251   **9.** 0.4987   **10.** (i) –0.8, (ii) 1.4, (iii) 0   **11.** (i) 38.7, (ii) 60.8   **12.** 1000   **13.** 62.47%

14. 52   **15.** (i) 371,   (ii) 383   **17.** ₹ 866

9.7   SAMPLING AND INFERENCES

The collection of selected number of individuals, objects or results from the parent universe according to the given rule is known as sample. We know that it is not possible to discuss all the members of the universe separately, because it is very costly and time consuming. So, it becomes necessary to find a rule or process under which by examining only a selected part of universe, we get all the information about it. Thus, these selected members are known as sample.

For example: If we buy the sweets from a shop, we examine only a single piece of it. Then this single piece will be treated as a sample.

Size of Sample : Number of objects, results or members in the sample is the size of the sample.

9.7.1 Sampling

A manner in which we can form a sample from the parent universe (population) is known as sampling.

For example :
1. Population of male children born in a particular year.
2. Number of smokers in a particular locality.

9.7.2 TYPES OF SAMPLING

The sample of population can be chosen in four manners or it can be stated as the sampling is of four types :

1. **Random Sampling :** Random sampling is a well known method of sampling. In this method, while choosing the sample, if each member have the same probability of chosen out then this type of sampling is known as random sampling. In short, we can say that when the selection is taken at random, then the sampling is known as random sampling.

   For example : If we throw a dice then each number have the same probability of coming out.

2. **Simple Sampling**

   Simple sampling is a special case of random sampling. In this, each event has the equal probability of chosen out, in which the probability of choosing an event is free from the previous probability of successes or failure of an event.

   For example :

   1. In tossing of a coin, the probability of coming head does not depend upon the previous trial in any manner.
   2. If we want to select a boy from a group of 8 children containing 4 boys and 4 girls, then the probability of choosing a boy is 4/8 and if the child selected is a girl and we does not replace it in the group, then the probability of choosing a boy will be 4/7. Then this sampling is known as simple sampling without replacement. And if we replace the girl to the group and then select the boy again, then this sampling is known as simple sampling with replacement.

3. **Purposive Sampling :** Purposive sampling is a sampling in which samples are taken under a particular consideration. In this sampling, the investigator select that part of universe by which it conclude its desired result. In this sampling personal individuals have a great chance of chosen out.

   For example : If we want to choose the student of 55-60 Kg from a group of students, then the random selection will give the students of all weights. Then, if we choose the students of particular weight, then this is known as purposive sampling.

4. **Stratified Sampling**

   The mixture of random sampling and purposive sampling is known as stratified sampling. Sometimes it is impossible to discuss a population by taking random sampling or purposive sampling as a representative sampling. In such cases, we divide the whole universe into distinct parts and then take the sample randomly according to the size of the part. Then this type of sampling is known as stratified sampling.

   Tippett's Number : A table constructed by L.H.C. Tippett's which solve the problem of making random sampling such that there is no relation among the numbers used is known as Tipptet's number table.

   This table consists of 41600 digit and gives 10400 four figure numbers which are useful in construction of random sampling. The numbers are chosen randomly from this table.

   A small part of this table is shown below :

2952	6641	3992	9792	7979	5911
3170	5624	4167	9525	1545	1396
7203	5356	1300	2693	2370	7483
3408	2762	3563	6107	6913	7691
0560	5246	1112	9025	6008	8126

For example : If we want to select a sample of 8 people from a group of 5000 numbered from 1 to 5000, then we choose the first 8 numbers which not exceed 5000 as

$$2952, 3992, 3170, 4167, 1545, 1396, 1300, 2693$$

This is the random sampling according to Tippett.

9.7.3 CHARACTERISTICS OF A GOOD SAMPLE DESIGN

We can list down the characteristics of a good sample design as under :

1. Sample design must result in a truly representative sample.
2. Sample design must be such which results in a small sampling error.
3. Sample design must be viable in context of funds available for the research study.
4. Sample design must be such so that systematic bias can be controlled in a better way.
5. Sample should be such that the results of the sample study can be applied, in general, for the universe with a reasonable level of confidence.

9.7.4 TECHNIQUES FOR RANDOM SAMPLING

The techniques for random sampling can be of three types :

1. **Random sampling by lottery system :** The lottery system is used by three ways. In first case, we make the pieces of a paper of same size such that we cannot differentiate them. Now, we numbered them according to the individuals and then we mix them. A chit is drawn out and the process will go on until we get the required number of chits equal to the sample size.

   The individuals corresponding to the chits form a sample. This method of random sampling is known as chit method.

   Now, the second method of lottery system is card method. In this, cards are used in place of chits. All the cards are numbered according to the sample and then shuffled. One card is drawn out and then the cards are reshuffled and the process will go on until we get the cards equal to the sample size. Now the individuals corresponding to card form a sample.

   The third method is lottery system. This method is almost same as chit method. In this method, chits are placed in similar containers and these containers are rotated in a rotating drum. Then these containers are picked up one by one until the sample size is obtained. These individuals corresponding to container form a sample. This method is known as lottery system for random sampling.

2. **By arranging the whole numbers according to a rule and then selecting the individuals in a sequence :** Here, at first we arrange the whole universe in a particular manner and then select the individuals in a sequence.

   For example : If we want to select 10 boys from a group of 250, then we arrange the whole group according to a rule (according to height, weight or names). Now, if we select every 25th boy, we get a random sampling of 10 boys.

3. **By random number method :** This method completely depends upon the random number table. Fisher, Kendall, Mahalnobis and Tippett have published such type of tables.

   According to Tippett, if we want to find out a sample of 8 people from a group of 5000, then we number them from 1 to 5000 and then select first 8 numbers which are less than 5000. The people corresponding to these numbers form a random sample.

9.7.5 Stratified Random Sampling

Let us suppose the given population is heterogeneous (non-homogeneous) in nature. Then, entire heterogeneous population is divided into a number of homogeneous group, which is called strata or subpopulation.

Let population of N units be divided into l sub populations of $N_1, N_2, ..., N_l$ units respectively such that each subpopulations (strata) are non-overlapping and together they form the whole population. Thus, we can write

$$\sum_{i=1}^{l} N_i = N$$

When the subpopulation (strata) have been determined a sample is drawn from each stratum. The drawing are made independently, in different strata, the sample sizes within the strata are denoted by $n_1, n_2, ..., n_l$ respectively, *i.e.*,

$$\sum_{i=1}^{l} n_i = n$$

☞ Remarks

- If simple random sampling is taken in each stratum, the whole procedure is known as "Stratified Random Sampling".
- The sample which is the set of all the sampling units drawn from each stratum is known as stratified sampling.

PRINCIPLES OF STRATIFICATION

In stratifying a population, following are the principles :

1. The stratification of population should be done such that strata are homogeneous.
2. The strata should be non-overlapping and together they must form the whole population.
3. Sometimes, when it is difficult to stratify with respect to characteristic under study, administrative convenience may be considered as the basis for stratification.
4. It will be better to treat each subpopulation as a stratum.

9.7.6 Systematic Random Sampling

In this sampling method, we used partly the arbitrariness and partly randomness. Let us suppose N units of populations are numbered 1, 2, ..., N in some order. Let $N = nk$, where n is the sample size and k is an integer (called the sampling interval). Then, a random number less than or equal to k be selected and every k^{th} unit thereafter. The resulting sample is called k^{th} systematic sample and such a procedure is known as linear systematic sampling.

Now, if $N \neq nk$ and every k^{th} unit be included in a circular manner till the whole list is exhausted, it is known as circular systematic sampling.

The systematic random sampling is used if a complete and up-to-date sampling is available. Also, under many situations, systematic sampling provides estimates which are more efficient than those obtained with simple random sampling without replacement.

9.7.7 LIMITATIONS OF SAMPLING

Some limitations of sampling theory are as follows :

1. Proper care should be taken in the planning, otherwise the results obtained might be inaccurate and misleading.
2. If the information is required about each and every unit of the universe, there is no way but to resort to complete enumeration. Also, if time and money are not important, a complete census may be better than any sampling method.
3. In the absence of the services of trained and qualified personnel and sophisticated equipments for its planning, execution and analysis, the result of sample survey are not trustworthy.

 **Exercise-9.22**

1. Write short notes on the following :
   (i) Census method   (ii) Sampling method
   (iii) Systematic random sample
2. Write the advantages and disadvantages of the following :
   (i) Census method   (ii) Sampling method

3. Describe the followings :
   (i) Simple random sampling
   (ii) Stratified random sampling
4. Discuss various methods of selecting a random sample.
5. What are the essentials of sampling.

9.7.8 TEST OF SIGNIFICANCE

Generally the population parameters are unknown and estimate through the sample values. If the sample value are exactly the same as parameter then we accept it and if it is far from our parameter then we reject it. But the problem arises when the sample value neither exactly equal to the parametric value nor too far. In this situation one has to develop some procedure which enables us to decide whether to accept a hypothetical value or not on the basis of sample values. Such a procedure is known as testing of hypothesis.

Population : Any collection of individuals under study is said to be population (or universe). The individuals often called the members or the units of the population.

Sample : A part or small section selected from the population is called a sample and this process of selection is called sampling.

9.7.9 TYPES OF POPULATION

(i) **Hypothetical population :** The population of concrete objects is called an existent population while a hypothetical population may be defined as the collection of all possible ways in which a specified event can happen. The population of 1 2, 3, 4, 5, 6 obtained by rolling a die an infinite numbers of times is a hypothetical one.

(ii) **Real population :** A population of concrete individuals is called a existent or real population.

(iii) **Finite population :** A population containing a finite number of individuals is called a finite population.

(iv) **Infinite population :** A population containing an infinite number of individuals is called an infinite population.

9.7.10 STATISTICAL HYPOTHESIS

A statistical hypothesis is a statement or assertion about a parameter of a population or parameters of two or more populations is denoted by H.

Parametric hypothesis : A statistical hypothesis refers only to the values of unknown parameters of population is usually called a parametric hypothesis.

Simple and Composite Hypothesis : If statistical hypothesis completely specifies a distribution, it is known as simple hypothesis otherwise known as composite hypothesis for example, a normal population $N(\mu, \sigma^2)$ with σ^2 is known. The hypothesis

(i) $H : \mu = \mu_0$ (ii) $H : \mu = 25$ (iii) $H : \mu = 30$

are simple hypothesis and

(i) $H : \mu > \mu_0$ (ii) $H : \mu < \mu_0$ (iii) $H : \mu \neq \mu_0$

are composite hypothesis.

9.7.11 NULL AND ALTERNATIVE HYPOTHESIS

The hypothesis which is natural and non-committal attitude of the statistician or decision makes before the sample observation is known as null hypothesis.

According to Ronald, A. Fisher, "Any hypothesis tested for its possible rejection is called a null hypothesis." The null hypothesis is denoted by H_0 and the hypothesis which provides an alternative of the null hypothesis is known as alternative to the null hypothesis. It is denoted by H_1 or H_A.

For example : Two manufacturing company produces bulbs. Each company claims that the average life of its bulbs is larger to that of the other than there are three hypothesis.

    (i) First manufacturing company produces bulbs of larger average life than second.

    (ii) First manufacturing company produces bulbs of lesser average life than second.

    (iii) Both manufacturing company have same average life of the produce bulbs.

The first two statements appear to be biased since they reflect a preferential attitude to one or other of the two processes. Hence, the best course is to adopt the hypothesis of no difference as stated in (iii).

9.7.12 TESTS OF SIGNIFICANCE

Procedure which enables us to decide, on the basis of sample information whether to accept or reject the hypothesis or to determine whether observed sampling results differ significant from expected results are called tests of significance, rules of decision or test of hypothesis.

9.7.13 LEVEL OF SIGNIFICANCE

The probability level below which will reject the hypothesis is called the level of significance. The levels of significance usually employed in testing of hypothesis are 5% and 1 %.

9.7.14 CRITICAL REGION AND ACCEPTANCE REGION

A region (corresponding to a statistic) is called the sample space. The part of sample space which amounts to rejection of null hypothesis H_0, is called critical region or region of rejection.

If $X = (x_1, x_2, ..., x_n)$ is the random vector observed and W_e is the critical region of the sample space W, then

$$W_a = W - W_e$$

of the space is called the acceptance region.

WORKING PROCEDURE

Step 1. State the null hypothesis H_0 and alternative hypothesis H_1.

Step 2. Make some assumption such as the sample is random, the population is normal, etc.

Step 3. Find the most appropriate test statistic together with its sampling distribution.

Step 4. On the basis of sampling distribution make a decision to either accept or reject the null hypothesis H_0.

Step 5. Take a random sample and compute the test statistic. If the calculated value of the test statistic falls is the acceptance region, then accept the null hypothesis H_0. If it falls in the region of rejection, reject the null hypothesis and accept H_1.

9.7.15 TYPE-I AND TYPE-II ERRORS

There are two types of errors in testing of hypothesis : Type I and Type II errors. When a statistical hypothesis is tested, there are four possible results.

1. The hypothesis is true but our test rejects it.
2. The hypothesis is false but our test accepts it.
3. The hypothesis is true and our test accepts it.
4. The hypothesis is false and our test rejects it.

Obviously, the first two possibilities lead to errors. If we reject a hypothesis when it should be accepted (possibility no.1) we say that a type I error has been made. On the other hand if we accept a hypothesis when it should be rejected (possibility no. 2) we say that a type II error has been made. In either case wrong decision or error in judgement has occurred.

Decision	H_0 : True	H_0 : False
Accept H_0	Correct decision	Type II error
Reject H_0	Type I error	Correct decision

Fig. 1. Two kinds of error in hypothesis testing conditions.

The probability of committing a type I error is designated as α and is called the level of significance. Therefore

$$\alpha = \text{Pr [Type I error]} = \text{Pr [Rejecting } H_0 / H_0 \text{ is true]}$$

must be complement of

$$(1 - \alpha) = \text{Pr [Accepting } H_0 / H_0 \text{ is true]}$$

This probability $(1 - \alpha)$ corresponds to the concept of $100(1 - \alpha)\%$ confidence interval. Our efforts would obviously be to have a small probability of making a type I error, Hence, the objective is to construct the test to minimize α.

Similarly, the probability of committing a type II error is designated by β.

Then $\beta = \text{Pr [Type II error]} = \text{Pr [Accepting } H_0 / H_0 \text{ is false]}$

and $(1 - \beta) = \text{Pr [Rejecting } H_0 / H_0 \text{ is false]}$

This probability $(1 - \beta)$ is known as the power of a statistical test.

The following table gives the possibility associated with each of the four cells shown in the previous table :

The decision is to	The null hypothesis is	
	True	False
Accept H_0	$(1 - \alpha)$ Confidence level	β
Reject H_0	α	$(1 - \beta)$ Confidence level
Sum	1.00	1.00

In order for any tests of hypothesis or rules of decisions to be good, they must be designed as to minimize errors of decision, However, this is not a simple matter, since for a given sample size, an attempt to decrease one type of errors is accompanied in general by an increase in other type of error. The probability of making type I error is fixed in advance by the choice of level of significance employed in the test. We can make the type I error as small as we please, by lowering the level of significance. But by doing so we increase the chance of accepting a false hypothesis i.e., of making a type II error. It follows it is impossible to minimize both errors simultaneously, In the long run errors of type I are perhaps more likely to prove serious research programmes in social science that are errors of type II.

☞ REMARKS
- However, α and β are not independent of each other, nor are they independent of the sample size n. When n is fixed, if α is lowered, then β normally rises and vice versa.
- If n is increased, it is possible for both α and β decrease.

9.7.16 BEST CRITICAL REGION

In testing the hypothesis $H_0 : \theta = \theta_0$ against the alternative $H_1 : \theta = \theta_1$ the critical region is best if the types II error is minimum or the power is maximum when compared to every other possible critical region of size α.

A test defined by this critical region is called most powerful test.

9.7.17 ONE TAIL AND TWO TAIL TESTS

Two tailed test is that where the hypothesis about the population mean is rejected for value of $\overline{X}$ falling into either tail of the sampling distribution. When the hypothesis about population mean is rejected only for value of $\overline{X}$ falling into one of the tails of sampling distribution then it is known as one-tailed test.

If it is the right tail, then it is called right tailed test or one-sided alternative to the right and if it is one of the left tailed to the left and called left tailed test.

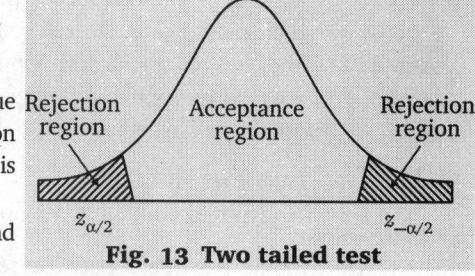

Fig. 13 Two tailed test

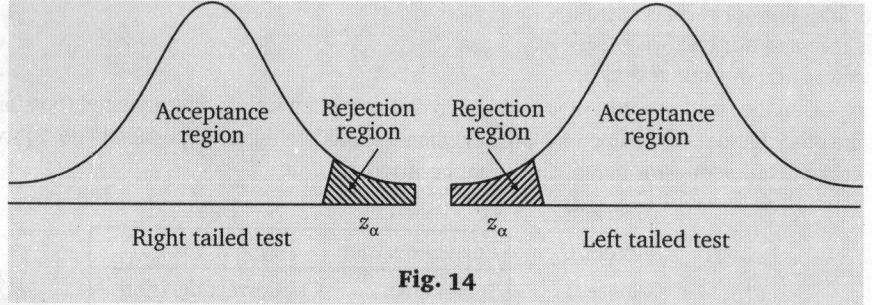

Fig. 14

🛶 Solved Examples

EXAMPLE 1. *It is desired to test a hypothesis $H_0 : P = P_0 = 1/2$ against the alternative hypothesis $H_1 : P = P_1 = 3/4$ on the basis of tossing a coin, where P is the probability of getting ahead in a single trial and agreeing to accept H_0 if a tail appears and to accept H_1 otherwise. Find the values of α and β.*

SOLUTION. We have

α = probability of rejecting H_0 when H_0 is true

$= P(\text{reject } H_0 / H_0) = P(\text{head appears} / P = P_0)$

$= [P]_{P = P_0 = 1/2} = 1/2$

Again, β = probability of accepting H_0 when H_1 is true

$= P(\text{accept } H_0 / H_1) = P(\text{tail appears} / P = P_1)$

$= [P]_{P = P_1 = 3/4} = 1 - \frac{3}{4} = \frac{1}{4}.$

EXAMPLE 2. *Given a binomial distribution*

$$f(x, p) = \begin{cases} {}^nC_x \, p^x \, q^{n-x}, & x = 0, 1, 2, 3, 4 \\ 0 & , & \text{else where} \end{cases}$$

It is desired to test $H_0 : P = P_0 = 1/3$ against

$H_1 : P = P_1 = 1/2$ *by agreeing to accept H_0 if $x \le 2$*
in four trials and to reject otherwise.
What are the probabilities of committing

(a) *type I error;*   (b) *type II error.*

SOLUTION. (a) $P(\text{type I error}) = P(\text{reject } H_0 / H_0 \text{ is true})$

$$= P(x \le 2 \,|\, P = 1/3)$$

$$= \sum_{x=3}^{4} {}^4C_x \left(\frac{1}{3}\right)^x \left(\frac{2}{3}\right)^{4-x}$$

$$= {}^4C_3 \left(\frac{1}{3}\right)^3 \left(\frac{2}{3}\right)^{4-3} + {}^4C_4 \left(\frac{1}{3}\right)^4$$

$$= 4 \times \frac{2}{3^4} + \frac{1}{3^4} = \frac{1}{3^2} = \frac{1}{9}$$

Hence, the hypothesis $H_0 : P = \dfrac{1}{3}$ is being tested at the level of significance.

(b) $\beta = P(\text{type II error})$

$$= P(\text{accept } H_0 \,|\, \text{when } H_1 \text{ is true})$$

$$= P(x \le 2 \,|\, P = 1/2)$$

$$\Rightarrow \quad P = \sum_{x=0}^{2} {}^4C_x \left(\frac{1}{2}\right)^x \left(\frac{1}{2}\right)^{4-x}$$

$$= {}^4C_0 \left(\frac{1}{2}\right)^4 + {}^4C_1 \left(\frac{1}{2}\right)^1 \left(\frac{1}{2}\right)^3$$

$$+ {}^4C_2 \left(\frac{1}{2}\right)^2 \left(\frac{1}{2}\right)^2 = \frac{11}{2^4}.$$

EXAMPLE 3. *An urn contains 4 balls of which 1 or 2 are white and the rest are red. To test the hypothesis H_0 : one ball is white, balls are drawn one after another until a white ball appears. Suggest a good critical region for this test and find the value for α and β.*

SOLUTION. Let X be the number of red balls drawn until a white ball appears. Then the possible test may be to accept H_0 if $X = 3$ and reject otherwise.

$\alpha = P(\text{reject } H_0 \text{ when } H_0 \text{ is true})$

$\quad = P(X \ne 3 / \text{one ball is white})$

$\quad = P(X = 0 \text{ or } 1 \text{ or } 2 / \text{one ball is white})$

$$= \frac{1}{4} + \frac{3}{4} \times \frac{1}{3} + \frac{3}{4} \times \frac{2}{3} \times \frac{1}{2} = \frac{3}{2} .$$

$\beta = P(\text{accept } H_0, \text{ when } H_1 \text{ is true})$

$\quad = P(X = 3 / \text{two balls are white}) = 0.$

The second test may be to accept H_0 if $X = 2$ or 3 and reject otherwise.

$\alpha = P(X = 0 \text{ or } 1 / \text{one ball is white})$

$$= \frac{1}{4} + \frac{3}{4} \times \frac{1}{3} = \frac{1}{2} .$$

$\beta = P(\text{accept } H_0 \text{ when } H_1 \text{ is true})$

$\quad = P[(X = 2 \text{ or } 3) / \text{two balls are white}]$

$$= \frac{3}{4} \times \frac{1}{3} \times \frac{2}{3} + 0 = \frac{1}{6} .$$

9.7.18 TEST OF SIGNIFICANCE OF LARGE SAMPLES

Suppose a large number of samples is classified according to the frequencies of an attribute. It gives rise to a binomial distribution which tends to a normal distribution for large values of n, the number in the sample. It follows therefore a great majority of its members lie within a range $\pm 3\sigma$ on each side of the mean, *i.e.*, of $\pm 3\sqrt{npq}$ on each side of the value np. If the number of successes in a large samples of size n differs from the expected value np by more then $3\sqrt{npq}$, we call the difference highly significant and the truth of the hypothesis is very improbable. Generally we accept the hypothesis as correct and then we calculate np and $\sqrt{npq}$ and apply the above test.

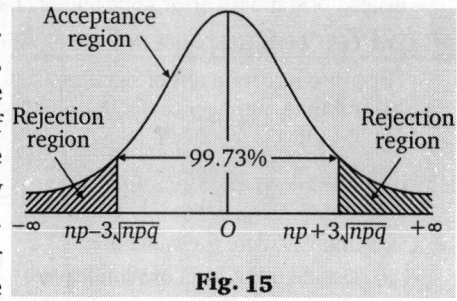

Fig. 15

Thus, by test of significance, we mean a test of hypothesis where we decide to accept or reject a hypothesis on the basis of a random sample. If null hypothesis is rejected we say that the difference $x \sim np$ is significant etc.

9.7.19 STANDARD ERROR

The standard deviation of simple sampling is called standard error. In real sense, the term standard error has wider meaning than merely the standard deviation of simple sampling.

If the difference between the actual and observed frequencies is more than three times the standard error, the difference is said to be significant, and such a difference could not have arisen due to fluctuations of sampling, or the probability of such a difference arising due to change is very low. If the difference is less than three times the standard error, it could have arisen due to fluctuations of sampling. If the difference is less than twice the standard error, the probability of its arising due to chance is fairly high, and it can be ignored. If the difference is more than twice the standard error but less than thrice the standard error, then it could have arisen due to sampling fluctuation, but the low probability of its arising due to chance is very low.

STANDARD ERROR OF A PERCENTAGE

The formula for the estimated standard error of a percentage (S_p) is $S_p = \sqrt{\dfrac{pq}{n}}$

where, S_p = Estimated standard error of a percentage.

$\qquad p$ = Percentage of the universe with the characteristic being studied.

$\qquad q = (100\% - p)$.

$\qquad n$ = Size of sample.

The standard formula for difference between two sample percentage is $S_{\text{difference}} = \sqrt{S_a^2 + S_b^2}$

where, $S_{\text{difference}}$ = Estimated standard error of the difference.

                S_a = Estimated standard error of a sample percentage usage rate.

                S_b = Estimated standard error of b sample percentage usage rate.

STANDARD ERROR OF THE MEAN

When constructing a confidence interval estimate, researchers must also know how the standard error of the mean is related to the standard deviation of the universe. In a simple random sample, the relationships :

$$\sigma_{\bar{x}} = \frac{\sigma}{\sqrt{n}}$$

where, $\sigma_{\bar{x}}$ = Standard error of the mean; σ = Standard deviation of the universe.

and n = Number of observation in the sample.

This formula applies if less than 5% of the universe is included in the sample.

If more than 5% of the universe is included in the sample, then the standard error may be computed as:

$$\sigma_{\bar{x}} = \frac{\sigma}{\sqrt{n}} \sqrt{\frac{N-n}{N}}$$

where, N = Universe size; n = Sample size

Factor $\sqrt{\dfrac{N-n}{N}}$ recognizes that a sample from a finite universe partially exhausts that universe.

☛ REMARK

- This relationship between the standard error of the mean σ_x and the standard deviation of the universe (σ) implies that when large sample size (n) are used the standard error of the mean will be a smaller value, and, therefore, the mean of a given sample is likely to be closer to the universe mean.

9.7.20 PROBABLE ERROR

Instead of standard error same authors have used a quantity called probable error, which is 0.67449 times the standard error.

9.7.21 TEST OF SIGNIFICANCE IN CASE OF ATTRIBUTES (LARGE SAMPLES)

The procedure for a test of significance in case of attributes (large samples) is as follows: .

 (i) Assume null hypothesis H_0 and alternative hypothesis H_1 (though not necessary).

 (ii) Define a test statistic

$$z = \frac{x - np}{\sqrt{npq}} \sim N(0,1) \qquad \text{or} \qquad z = \frac{n}{\sqrt{pq/n}} \sim N(0,1)$$

        and calculate the value of z.

(iii) Decide α the level of significance

   (a) for $\alpha = 5\%$, we reject H_0 if $|z| > 1.96$, two sided

   (b) for $\alpha = 1\%$, we reject H_0 if $|z| > 2.58$, two sided.

   (c) for $\alpha = 0.27\%$, we reject H_0 if $|z| > 3$, two sided.

☛ REMARK

- In testing a given hypothesis or a test of significance, the maximum probability with which we would be willing to risk an error is called the level of significance of the test. In practice a level of significance 0.05 or 0.01 is usually taken.

9.7.22 CONFIDENCE LIMITS OF UNKNOWN p

Let x = no. of success, n = sample size, P = probability of success or proportion is the population, $Q = 1 - P$.

$$p = \frac{x}{n} = \text{sample proportion.}$$

Then, $z = \dfrac{p - P}{\sqrt{PQ/n}} \sim N(0,1)$

Now, $|z| \leq 3$ $\Rightarrow$ $\left| \dfrac{p - P}{\sqrt{PQ/n}} \right| \leq 3$ $\Rightarrow$ $P - 3\sqrt{\dfrac{PQ}{n}} \leq p \leq P + 3\sqrt{PQ/n}$

These are called confidence limits for P at $\alpha = 0.0027$.

However, the limits of np are

$$nP - 3\sqrt{nPQ} \leq nP + 3\sqrt{nPQ} \qquad\qquad \text{or} \qquad\qquad x - 3\sqrt{nPQ} \leq nP \leq x + 3\sqrt{nPQ}$$

Solved Examples

EXAMPLE 1. *A certain cubical die was thrown 9000 times, and a 5 or a 6 was obtained 3240 times. On the assumption of random throwing do the data indicate an unbiased die?* (VTU–2010)

SOLUTION. On the hypothesis of an unbiased die, the chance of throwing a 5 or 6 is $= \dfrac{2}{6} = \dfrac{1}{3} = p$ and hence $q = \dfrac{2}{3}$.

∴ The expected number of successes

$$= np = 9000 \times \frac{1}{3} = 3000$$

The standard error of number of success

$$= \sqrt{npq} = \sqrt{9000 \times \frac{1}{3} \times \frac{2}{3}} = 44.72$$

The difference between observed frequency and expected frequency = 3240 – 3000 = 240 which is nearly 5.4 times the standard error and is therefore most unlikely to appear as a result of simple sampling. We therefore conclude that die is certainly biased one.

Alternative Method:

$$Z = \frac{x - np}{\sqrt{npq}} = \frac{3240 - 3000}{44.72} = \frac{240}{44.72} = 5.37$$

Since $|z| > 3$. Hence, we reject H_0 at 0.27% level of significance. That is the difference $x - np$ is highly significant.

EXAMPLE 2. *A sample of 900 days is taken from meteorological records of a certain district and 100 of them are found to be foggy. What are the probable limits to the percentage of foggy days in the district ?*

SOLUTION. Here $p = \dfrac{100}{900} = \dfrac{1}{9}$ so that $q = \dfrac{8}{9}$.

Standard error

$$= \sqrt{npq} = \sqrt{900 \times \frac{1}{9} \times \frac{8}{9}} = \sqrt{88.2} = 9.4$$
$$3 \text{ S.E.} = 3 \times 9.4 = 28.2$$

∴ Limits are $x \pm 3\sqrt{npq}$, i.e., 100 ± 28.2, i.e., 128.2 and 71.8.

These are the limits out of 900 days. Hence, the limits in percentage are $\dfrac{128.2}{900} \times 100 = 14.2$ and $\dfrac{71.8}{900} \times 100 = 8$ nearly.

EXAMPLE 3. *In some dice throwing experiment, Mohan threw a die 49152 times, and of these 25145 yielded 4, 5 or 6. Is this consistent with the hypothesis that the die was unbiased?*

SOLUTION. The probability of throwing 4, 5 or 6 with one die

$$p = \frac{3}{6} = \frac{1}{2}, q = 1 - \frac{1}{2} = \frac{1}{2}; n = 49152, x = 25145$$

∴ Expected value of number of success

$$E(x) = np = 49152 \times \frac{1}{2} = 24576$$

Standard error $= \sqrt{npq} = \sqrt{49152 \times \dfrac{1}{2} \times \dfrac{1}{2}} = 110.9$

Test statistic, $Z = \dfrac{x - np}{\sqrt{npq}} = \dfrac{25145 - 24576}{110.9}$

$$= \frac{569}{110.9} = 5.1373$$

Since $|z| > 3$, hence hypothesis $H_0 : P = \dfrac{1}{2}$ is rejected.

The difference is not due to sampling fluctuations. Hence, the data is not consistent with the hypothesis that the die was unbiased.

9.7.23 TEST OF SIGNIFICANCE OF MEAN (LARGE SAMPLE OF VARIABLES)

Let X be a random variable with mean μ and standard deviation σ. Let $x_1, x_2, ..., x_n$ be a random sample of size n.

Let $\bar{x}$ = sample mean $= \dfrac{1}{n} \sum_{i=1}^{n} x_i$. Then, $Z = \dfrac{\bar{x} - \mu}{\sigma / \sqrt{n}} \sim N(0,1)$

If Z calculated greater than 3, we reject the null hypothesis $H_0 : \mu = \mu_0$, where $Z = \dfrac{\bar{x} - \mu_0}{\sigma / \sqrt{n}}$.

 Solved Examples

EXAMPLE 1. *A sample of 1600 members found to have a mean 5.7. Could it be reasonably regarded as a simple sample from a large population whose mean is 4.5 and standard deviation is 2.8.*

SOLUTION. Here, $n = 1600, \bar{x} = 5.7, \mu = 4.5, \sigma = 2.8$.

The standard error (S.E.) of the mean for a simple sample of 1600 from such a population is

$$\frac{\sigma}{\sqrt{n}} = \frac{2.8}{\sqrt{1600}} = \frac{2.8}{40} = 0.07 .$$

The deviation of the sample mean from that of the population mean = 5.7 – 4.5 = 1.2, which is 17.1 times the standard error (S.E.)

$$|\bar{x} - \mu| > 3 \text{ S.E.}$$

or $$|Z| = \frac{|\bar{x} - \mu|}{\text{S.E.}} > 3$$

Hence, the deviation is highly significant.

The given sample cannot be regarded as a simple sample from a large population with mean $\mu = 4.5$ and standard deviation 2.8.

EXAMPLE 2. *A sample of 400 male students have a mean height of 168 cm. Can it be reasonably regarded as a sample from a large population with mean height 167.8 cm and standard deviation 3.25 cm.*

SOLUTION. Here $n = 400, \bar{x} = 168$ cm, $\mu = 167.8, \sigma = 3.25$ cm

$$|Z| = \frac{\bar{x} - \mu}{\sigma / \sqrt{n}} = \frac{168 - 167.8}{3.25 / \sqrt{400}} = \frac{0.2 \times 20}{3.25} = 1.23$$

Since $|Z| < 3$, the deviation of sample mean from population mean is not significant. Hence, it can reasonably be regarded as a simple sample from a large population with mean 167.8 cm and standard deviation 3.25 cm.

EXAMPLE 3. *A random sample of 400 months has an average length of 10 cm. Can this be regarded as a sample from a large population with mean of 10.2 cm and a standard deviation of 2.25 cm?*

SOLUTION. Here $N = 400$, $\bar{x} = 10$.

(1) **Null Hypothesis (H_0) :** The sample has been drawn from the normal population with mean (μ) = 10.2 cm and standard deviation (σ) = 2.25 cm *i.e.,* μ = 10.2. Alternative hypothesis (H_1) = $\mu \neq 10.2$ (Two-tailed).

(2) **Competition of test statistic:** Under H_0 (since sample size is large).

$$Z = \frac{\bar{x} - \mu}{\sqrt{\sigma^2/n}} = \frac{10 - 10.2}{\sqrt{(2.25)^2/400}}$$

$$= \frac{-0.2 \times 20}{2.24} = -1.777$$

(3) **Level of significance :** $\alpha = 0.05$

(4) **Critical value :** At 0.05 level of significance, the critical value of $Z = \pm 1.96$ (from the table).

(5) **Decision :** Since the computed value of $(z) = 1.77$ is less than critical value of $Z = 1.96$, it falls in the acceptance region. Hence, the facts are consistent with the null hypothesis which is accepted with 95% confidence and it is concluded that the sample has been drawn from the normal population with mean of 10.2 cm and a standard deviation of 2.25 cm.

EXAMPLE 4. *The mean life time of sample of 100 aquaria fishes is computed to be 1,570 hours with a standard deviation of 120 hours. The shopkeeper claims that the average life of fishes is 1600 hours. Using the level of significance of 0.05, is the claim acceptable?*

SOLUTION. Here, $n = 100$, $\bar{x} = 1570$, and s = sample standard deviation = 120.

1. **Null Hypothesis (H_0) :** $\mu = 1600$, *i.e.,* average life of the fish is 1600 hours. Alternative Hypothesis (H_1) : $\mu_1 \neq 1600$ (two tailed test)

2. **Test statistic:** Under H_0, the test statistic is

$$Z = \frac{\bar{x} - \mu}{\sqrt{\sigma^2/n}} = \frac{\bar{x} - \mu}{\sqrt{s^2/n}}$$

$$Z = \frac{1570 - 1600}{\sqrt{(120)^2/100}} = -\frac{300 \times 10}{120} = -2.5$$

3. **Level of Significance :** $\alpha = 0.05$

4. **Critical Value:** At 0.05 level significance, the critical value of $Z = \pm 1.96$ (As per table).

5. **Inference :** Since the critical value of $Z = 2.5$ is greater than the critical value of $Z = 1.96$, it falls in the rejection region. Hence, the null hypothesis is rejected and it is concluded that the shopkeeper claim of average life of the fishes being 1600 hours is not acceptable.

9.7.24 Testing the Significance of the Difference between the Means of two Large Samples

If two random sample of n_1 and n_2 members respectively have been taken from the sample population of S.D. σ, we would like to know if the difference of their means $(\bar{x}_1 - \bar{x}_2)$, is due to fluctuations of sampling.

On the assumption that the samples are independent and have been taken from the sample population, the S.E. ε of the difference of their means is given by

$$\varepsilon^2 = \varepsilon_1^2 + \varepsilon_2^2$$

where ε_1 and ε_2 are the S.E. of the means of two samples given as

$$\varepsilon_1^2 = \sigma^2/n_1, \quad \varepsilon_2^2 = \sigma^2/n_2$$

Hence,

$$\varepsilon^2 = \sigma^2 \left(\frac{1}{n_1} + \frac{1}{n_2} \right).$$

If the difference $\bar{x}_1 \sim \bar{x}_2$ exceeds 3ε, the difference cannot said to be due to functions of sampling and the assumption $\mu_1 = \mu_2$ is unlikely to be correct.

Alternative method:

$$Z = \frac{(\bar{x}_1 - \bar{x}_2) - (\mu_1 - \mu_2)}{\sqrt{\frac{\sigma^2}{n_1} + \frac{\sigma^2}{n_2}}} \sim N(0, 1)$$

Hence $H_0 : \mu_1 = \mu_2$,

$$Z = \frac{(\bar{x} - \bar{x}_2)}{\sqrt{\frac{\sigma^2}{n_1} + \frac{\sigma^2}{n_2}}}$$

Alternative hypothesis may be $H_1 : \mu_1 > \mu_2$, $\mu_1 < \mu_2$ or $\mu_1 \neq \mu_2$

With the corresponding critical regions

$$Z \geq 2.33, \quad Z \leq -2.33 \qquad |Z| \geq 2.58 \text{ for } \alpha = 0.01$$

$$Z \geq 1.645, \quad Z \leq -1.645 \qquad |Z| \geq 1.96 \text{ for } \alpha = 0.05$$

 Solved Examples

EXAMPLE 1. *Number of fishes from one pond was taken 1000 and from second pond 2000. The mean weight of fishes of two ponds are 67.5 and 68.0 respectively.*

Test of equality of means of the two populations each with, S.D. 2.5 (No credit if the H_0 and H_1 are not stated) Assumptions should be clearly stated.

SOLUTION. As we have been given

$n_1 = 1000, n_2 = 2000, \bar{x}_1 = 67.5, \bar{x}_2 = 68, \sigma_1 = \sigma_2 = \sigma = 2.5$

1. Null Hypothesis : (H_0) $H_0 : \mu_1 = \mu_2$

Alternative hypothesis : $H_0 : \mu_1 \neq \mu_2$

[Two tailed test]

2. Computation of test statistic : Under H_0 the test statistics (for large samples is)

$$Z = \frac{\bar{x}_1 - \bar{x}_2}{\sqrt{\dfrac{\sigma_1^2}{n_1} + \dfrac{\sigma_2^2}{n_2}}} = \frac{\bar{x}_1 - \bar{x}_2}{\sqrt{\dfrac{\sigma^2}{n_1} + \dfrac{\sigma^2}{n_2}}}$$

$$= \frac{67.5 - 68.0}{\sqrt{(2.5)^2 \left\{ \dfrac{1}{1000} + \dfrac{1}{2000} \right\}}}$$

$$= \frac{-0.5}{2.5 \sqrt{\dfrac{3}{2000}}} = \frac{-0.5}{2.5 \times 0.039}$$

$$= \frac{-0.5}{0.0975} = -5.13$$

3. Level of significance : $\alpha = 0.05$

4. Critical value: At 5% level of significance ($\alpha = 0.05$) the critical value of $Z = \pm 1.96$.

5. Interpretation : Since the computed value of $Z = 5.13$ is greater than critical value of $Z = 1.96$, it falls in the rejection region. Hence, the null hypothesis is rejected.

EXAMPLE 2. *Random samples of 500 and 400 have means*

11.5 and 10.9 respectively. Can the samples be regarded as drawn from the population of the standard deviation 5. (Madras–2002)

SOLUTION. With usual notations, where $n_1 = 500, n_2 = 400, \bar{x}_1 = 11.5, \bar{x}_2 = 10.9, \sigma = 5$

$$\varepsilon^2 = \frac{\sigma^2}{n_1} + \frac{\sigma^2}{n_2} = \frac{25}{500} + \frac{25}{400} = \frac{225}{2000} = \frac{9}{80}$$

$$\varepsilon = 0.335$$

Difference of means $= 11.5 - 10.9 = 0.6$.

This difference is 1.79 times the S.E. Hence, it is not significant, *i.e.*, there is no reason to doubt the hypothesis.

EXAMPLE 3. *A potential buyer of light bulbs bought 50 bulbs of each of two brands. Upon testing these bulbs, he found that brand A had a mean life of 1282 hours with a S.D. of 80 hours, where as brand B had a mean life of 1208 hours, with a S.D. of 94 hours. Can the buyer be quite certain that two brands differ in quality ?*

SOLUTION. Here $n_1 = 50 = n_2, \sigma_2^2 = 94^2, \bar{x}_1 = 1282, \bar{x}_2 = 1208.$

Hence $\varepsilon = \sqrt{\dfrac{80^2}{50} + \dfrac{98^2}{50}} = 17.5$ hours.

This difference of means $\bar{x}_1 - \bar{x}_2 = 74 = 4.24\varepsilon$ is greater than 2 times of its S.E. Hence, the difference cannot be said due to the fluctuations of sampling. There is much difference in two brands. Hence, brand A is to be preferred.

9.7.25 TEST OF SIGNIFICANCE OF DIFFERENCE BETWEEN THE SAMPLE PROPORTION

If two samples are drawn from different populations, we may be interested in finding out whether the difference between the proportion of success is significant or not. In this case our hypothesis will be that the difference between proportion of success in one sample ($P1$) and the proportion of success in another sample, is due to fluctuations of random sampling. The standard error of the difference proportions is calculated as follows :

$$\text{S.E.} (p_1 - p_2) = \sqrt{pq \left(\frac{1}{n_1} + \frac{1}{n_2} \right)} \qquad \text{where, } p = \frac{n_1 p_1 + n_2 p_2}{n_1 + n_2} \text{ or } p = \frac{x_1 + x_2}{n_1 + n_2}$$

p is pooled estimate of the actual proportion in the population x_1 and x_2 stand for number of occurrences is the two samples of sizes n_1 and n_2 respectively.

Thus $Z = \dfrac{p_1 - p_2}{S.E.}$ is less than 1.96 S.E.

(5% level of significance), the difference is not significant and the null hypothesis is accepted and the difference is regarded as due to random sampling variation.

 Solved Examples

EXAMPLE. *In a simple random sample of 600 men taken from a big city, 400 are found to be smokers. In another simple random sample of 900 men taken from another city 450 are smokers. Do the data indicate that there is significant difference in the habit of smoking in the two cities?* (JNTU–2003)

SOLUTION. Let us take null hypothesis (H_0) that there is no significant difference in the habit of smoking in the two cities.

$$\text{S.E.} (p_1 - p_2) = \sqrt{pq \left(\frac{1}{n_1} + \frac{1}{n_2} \right)}$$

$$p_1 = \frac{400}{600} = 0.667, \quad p_2 = \frac{450}{900} = 0.5$$

$$p = \frac{x_1 + x_2}{n_1 + n_2}, \quad q = 1 - p$$

where, $n_1 = 600, n_2 = 900, x_1 = 400, x_2 = 450$

$$p = \frac{400 + 450}{600 + 900} = \frac{850}{1500} = \frac{17}{30}, \quad q = 1 - \frac{17}{30} = \frac{13}{30}.$$

$$\text{S.E.} (p_1 - p_2) = \sqrt{\frac{17}{30} \times \frac{13}{30} \left(\frac{1}{600} + \frac{1}{900} \right)}$$

$$= \sqrt{\frac{17 \times 13}{90} \times \frac{1500}{600 \times 900}} = 0.026$$

Thus, $Z = \dfrac{\text{Difference}}{S.E. (p_1 - p_2)} = \dfrac{0.667 - 0.50}{0.026} = 6.42.$

9.7.26 Test of Significance of Difference between the Standard Deviation (Large Samples)

In case of two large random samples each drawn from a normally distributed population, the S.E. of the difference between the standard deviations

$$\text{S.E.} (\sigma_1 - \sigma_2) = \sqrt{\frac{\sigma_1^2}{2n_1} + \frac{\sigma_2^2}{2n_2}}$$

When population standard deviation are not given :

$$\text{S.E.} (s_1 - s_2) = \sqrt{\frac{s^2}{2n_1} + \frac{\sigma_2^2}{2n_2}}$$

The problem is to test if $\sigma_1 = \sigma_2$, *i.e.*, if the sample S.D. differ significantly or not.

Under the null hypothesis (H_0) : that sample standard deviation do not differ significantly, the test statistic is given by

$$Z = \frac{\sigma_1 - \sigma_2}{\text{S.E.} (\sigma_1 - \sigma_2)} = \frac{s_1 - s_2}{\text{S.E.} (s_1 - s_2)}.$$

 Solved Examples

EXAMPLE 1. *One town has poultry farm of State Govt. It is suspected that the efficiency in the poultry is not alike, so a test is carried out by ascertaining the variability of life of the chickens produced by each poultry. The results are as follows :*

	Poultry farm A	Poultry farm B
No. of chickens in samples	100 (n_1)	200 (n_2)
Average life	1100 ($\bar{x}_1$) hrs	900 ($\bar{x}_2$) hrs
Standard deviation	240 (s_1)	220 (s_2)

From the above data, determine whether the difference between the variability of life of chickens from each sample is significant (take 1% level of significance).

SOLUTION. **1. Null hypothesis :** $H_0 : \sigma_1 = \sigma_2$

         Alternative hypothesis: $H_1 : \sigma_1 \neq \sigma_2$

         2. Test statistic : Under H_0, test statistic is

$$Z = \frac{s_1 - s_2}{\sqrt{\frac{\sigma_1^2}{2n_1} + \frac{\sigma_2^2}{2n_2}}} = \frac{s_1 - s_2}{\sqrt{\frac{s_1^2}{2n_1} + \frac{s_2^2}{2n_2}}}$$

$$(\sigma_1^2 = s_1^2 \text{ and } \sigma_2^2 = s_2^2 \text{ for large samples})$$

$$Z = \frac{240 - 220}{\sqrt{\frac{(240)^2}{200} + \frac{(220)^2}{400}}} = \frac{20}{\sqrt{288 + 121}} = 0.96$$

3. Level of significance : $\alpha = 0.01$

4. Critical value : To examine whether σ_1 is greater or lesser than σ_2 we require a two tailed test whether at 0.1 significant level is the extreme 0.005 of the distribution on each side of mean. Therefore critical value of $Z \neq 2.58$ (from table).

5. Inference : Since computed value of $Z(0.9)$ fall in acceptance region, the null hypothesis must be accepted. The result is therefore, not significant, so that it may be concluded that the variability of life of each of the poultry farm is not appreciably different but, on an average, the chickens from one farm have longer life than that of the other farm.

9.7.27 Students 't' Test

Student's t-distribution was given by W.S. Gosset (1908) who wrote under the pen-name of student. The quantity t is defined as

$$t = \frac{\bar{x} - \mu}{S / \sqrt{n}}$$

where n = the number of observations in the sample; $\bar{x} = \frac{1}{n} \sum_{i=1}^{n} x_i$ is the sample mean.

     μ = The mean of the population from which the sample has been drawn.

$$S = \sqrt{\frac{1}{n-1} \sum_{i=1}^{n} (x_i - \bar{x})^2} \text{ is the standard deviation of the sample.}$$

APPLICATIONS OF THE t-DISTRIBUTION

The t-distribution has a wide number of applications in statistics, some are given below :

(i) To test the significance difference between the sample mean $\bar{x}$ and the population mean μ.

(ii) To test the significance difference between two sample mean.

ASSUMPTION FOR STUDENT'S t-TEST

The following assumptions are made in the t-test :

(i) The population from which the sample is drawn in normal.

(ii) All observations in the sample are independent.

(iii) The standard deviation (σ) of the population is unknown.

(iv) The sample values are correctly measured and recorded.

(v) Test for testing the significance difference between same mean μ and the population mean.

Let $x_1, x_2, ..., x_n$ be a random sample of size n $(n < 30)$ taken from normal population with mean μ and variance σ^2 (unknown).

Null hypothesis: There is no significance difference between the sample mean and population mean or the sample has been drawn from the population whose mean is μ_0.

$$H_0 : \mu = \mu_0$$

Alternative hypothesis : H_0 is not true.

Test statistic : Under null hypothesis

$$t_{cal} = \frac{|\bar{x} - \mu_0|}{S / \sqrt{n}} \qquad \text{where} \qquad \bar{x} = \frac{1}{n} \sum_{i=1}^{n} x_i, \, S^2 = \frac{1}{n-1} \sum_{i=1}^{n} (x_i - \bar{x})^2$$

t-Tabulated : Tabulated value are taken from the table at $(n-1)$ degrees of freedom at given level of significance generally 1 %, 5%, 10%.

Conclusion : If $t_{cal} \geq t_{tab}$ then the null hypothesis is accepted.

Degree of freedom : The degree of freedom can be defined as the number of independent variables which make up the statistic (*e.g.*, t, ψ^2). It is denoted by ν.

In general, the number of degrees of freedom is the total number of observations less the number of independent constraints imposed on the observations.

If we have to choose any four numbers whose sum is 70, we can exercise our independent choice for any three numbers only, the fourth being 70 minus the total of the three numbers selected.

Thus, through we were to choose any four numbers, our choice was reduced to three because of one condition imposed. There was only one restriction on our freedom and our degrees of freedom were $4 - 1 = 3$. If two restrictions are imposed, our freedom to choose will be further curtailed and degrees of freedom will be $4 - 2 = 2$ rejected at given level of significance, *i.e.*, there is a significance difference between sample mean and population mean. If $t_{cal} < t_{tab}$, then the null hypothesis may be accepted.

 Solved Examples

EXAMPLE 1. *A manufacturer of dry cells claimed that the life of their cells is 24.0 hours. A sample of 10 cells bad mean life of 22.5 hours with a standard deviation of 3.0 hours. On the basis of available information, test whether the claim of the manufacturer is correct at 5% level of significance.*

SOLUTION. Here we are given

$\mu = 24.0$ hours, $x = 10$, $\bar{} = 22.5$ hours, $S = 3.0$ hours.

Null hypothesis H_0: The claim of manufacturer is correct, *i.e.*, $\mu = 24.0$ hours.

Alternative hypothesis: $H_1 : \mu \neq 24.0$

Test statistic, under null hypothesis

$$t = \frac{|\bar{x} - \mu|}{S / \sqrt{n}} = \frac{|22.5 - 24.0|}{3.0 / \sqrt{10}} = 1.58 .$$

t-tabulated at 5% level at significance with $(10 - 1) = 9$, degree of freedom $= 2.26$.

Conclusion: Since $t_{cal} < t_{tab}$, therefore null hypothesis may be accepted at 5% level of significance, *i.e.*, the claim of manufacturer is correct.

EXAMPLE 2. *A machinist is making engine parts with axle diameter of 0.700 inch. A random sample of 20 parts shows a mean diameter of 0.742 inch with a standard deviation of 0.40 inch. Compute statistic you would use to test the whether the work is meeting the specification at 1% level of significance.* (VTU–2009)

SOLUTION. Here we are given

$\mu = 0.700$ inch, $n = 20$

$\bar{x} = 0.742$ inch, $S = 0.40$

Null Hypothesis : The product is confirming to specification, *i.e.*, $\mu = 0.700$ inch

Alternative hypothesis H_1 : $\mu \neq 0.700$

Test statistic : Under H_0

$$t = \frac{|\bar{x} - \mu|}{S / \sqrt{n}} = \frac{|10.742 - 0.700|}{0.40 / \sqrt{20}} = 0.4696$$

t-tabulated at 1% level of significance with $(20 - 1)$, *i.e.*, 19, d.f. $= 2.86$.

Conclusion: Since $t_{cal} < t_{tab}$, so H_0 may be accepted at 1% level of significance, *i.e.*, the product is confirming to specification.

EXAMPLE 3. *Ten individuals are chosen at random from a population and their heights are found to be in inches 63, 63, 24, 65, 66, 69, 69, 70, 70, 71. Discuss the suggestion that the mean height in the inverse is 65 inches given that for 9 d.f. the value of student's that 5% level of significance is 2.262.*

SOLUTION. **Null hypothesis H_0:** The mean height of the universe is 65

$$H_0 : \mu = 65$$

Alternative hypothesis : $H_1 : \mu \neq 65$

Table for calculation of sample mean and S.D.

x	$d = x - 69$	d^2
63	–6	36
63	–6	36
64	–5	25
65	–4	16
66	–3	9
69	0	0
69	0	0
70	1	1
70	1	1
71	2	4
	$\Sigma d = -20$	$\Sigma d^2 = 128$

$$\bar{x} = A + \frac{\Sigma d}{n} = 69 - \frac{20}{10} = 67$$

$$S^2 = \frac{1}{n-1}\left[\Sigma d^2 - \frac{(\Sigma d)^2}{n}\right]$$

$$= \frac{1}{9}\left[128 - \frac{400}{10}\right] = \frac{1}{9} \times 88$$

$$S = \sqrt{88/9} = 3.13$$

Test statistic : Under H_0

$$t = \frac{|\bar{x} - \mu|}{S/\sqrt{n}} = \frac{|67-65|}{3.13} \times \sqrt{10} = 2.02.$$

t-tabulated at 1 % level of significance with $(10 - 1)$, i.e., 9 d.f. = 3.25.

Conclusion: Since $t_{cal} < t_{tab}$, so H_0 may be accepted at 1% level of significance, i.e., the mean height in the universe is 65 inches.

EXAMPLE 4. *Find the student's t for following variables values in a sample of eight.*

$$-4, -2, -2, 0, 2, 2, 3, 3.$$

9.7.28 F-Test of Equality of Population Variance

Let $x_1, x_2, ..., x_{n_1}$ be a random sample of size n_1 and $y_1, y_2, ..., y_{n_2}$ be another independent random sample of size n_2 taken from two normal populations with mean μ_x and μ_y respectively and having variance σ_x^2 and σ_y^2.

Null Hypothesis (H_0) : The sample have been drawn from normal population with the same variance i.e.,

$$H_0 : \sigma_x^2 = \sigma_y^2 = \sigma^2.$$

Alternative Hypothesis (H_1) : $\sigma_x \neq \sigma_y$ or σ_x (two tailed)

or $\quad\quad\quad \sigma_x > \sigma_y$ (right tailed) $\quad\quad$ or $\quad\quad$ $\sigma_x < \sigma_y$ (left tailed)

Test statistics:

$$F = \frac{S_x^2}{S_y^2} \quad \text{where } S_x^2 = \frac{1}{n_1-1}\sum_{i=1}^{n_1}(x_i - \bar{x})^2 \quad \text{and } S_y^2 = \frac{1}{n_2-1}\sum_{j=1}^{n_2}(y_i - \bar{y})^2$$

where $\bar{x}$ and $\bar{y}$ are mean samples defined as

$$\bar{x} = \frac{1}{n_1}\sum_{i=1}^{n_1} x_i, \quad \bar{y} = \frac{1}{n_2}\sum_{j=1}^{n_2} y_j.$$

F-statistics follows F distribution with $(n_1 - 1, n_2 - 1)$ degrees of freedom.

F-Tabulated: Tabulated value is taken from the table at $(n_1 - 1, n_2 - 1)$ degrees of freedom at given level of significance (generally 1%, 5% and 10%).

Conclusion: If $F_{cal} \geq F_{tab}$ then the null hypothesis is rejected at given level of significance, i.e., there is a significance difference between population variances. Other hand if $F_{cal} \geq F_{tab}$ then the null hypothesis may be accepted.

Assumptions of F-test: The following assumptions are made in the F-test :

(i) The population from which the samples are drawn should be normal.

(ii) The samples should be drawn in random manner.

(iii) The observations should be independent.

(iv) The ratio S_x^2 / S_y^2 should be greater than 1, i.e., $S_x^2 / S_y^2 > 1$

Solved Examples

EXAMPLE 1. *Two samples of size 9 and 8 give the sum of squares of deviations from their respective means equal to 160 inches square and 91 inches square respectively. Can they be regarded as*

---

taking mean of the universe to be zero.

SOLUTION.

Serial No.	x	$x - \bar{x}$	$(x - \bar{x})^2$
1	−4	−4.25	18.0625
2	−2	−2.25	5.0625
3	−2	−2.25	5.0625
4	0	−0.25	0.0625
5	2	1.75	3.0625
6	2	1.75	3.0625
7	3	2.75	7.5625
8	3	2.75	7.5625
	$\Sigma x = 2$		49.5000

$$\bar{x} = \text{mean} = \frac{\Sigma x}{n} = \frac{2}{8} = 0.25$$

$$S = \sqrt{\frac{\Sigma(x-\bar{x})^2}{n-1}} = \sqrt{\frac{49.5}{7}}$$

$$= \sqrt{7.071428} = 2.659$$

H_0 : The mean of the universe $M = 0$, we get

Student's $t = \frac{(\bar{x} - M)}{S}\sqrt{n} = \frac{(0.25 - 0)}{2.659}\sqrt{8} = 0.27$

---

drawn from the normal population with same variance?

(Mumbai–2004, VTU–2002)

SOLUTION. Here we are given

$$n_1 = 9, n_2 = 8, \Sigma(x_i - \bar{x})^2 = 160, \Sigma(y_i - \bar{y})^2 = 91$$

Null Hypothesis (H_0): The samples are drawn from the normal populations with same variance, i.e., $H_0 : \sigma_x^2 = \sigma_y^2$

Alternative Hypothesis (H_1): H_0 is not true, i.e., $H_1 : \sigma_x^2 \neq \sigma_y^2$

Test statistic:

$$F = \frac{S_x^2}{S_y^2}$$

$$S_x^2 = \frac{1}{n_1 - 1} \sum_{i=1}^{n_1} (x_i - \bar{x})^2 = \frac{1}{9-1} \times 160 = 20$$

$$S_y^2 = \frac{1}{n_2 - 1} \sum_{i=1}^{n_2} (y_i - \bar{y})^2 = \frac{91}{8-1} = 13$$

$$F = \frac{S_x^2}{S_y^2} = \frac{20}{13} = 1.54 .$$

F-tabulated at 5% level of significance with $(n_1 - 1, n_2 - 1)$, i.e., (8, 7) degrees of freedom = 3.73.

Conclusion: Since $F_{cal} < F_{tab}$, so null hypothesis may be accepted at 5% level of significance i.e., the samples are drawn from normal populations with same variance.

EXAMPLE 2. *Two random samples are drawn from two normal populations*

| Sample I | 20 | 16 | 26 | 27 | 23 | 22 | 18 | 24 | 25 | 19 | |
| Sample II | 27 | 33 | 42 | 35 | 32 | 34 | 38 | 28 | 41 | 43 | 30 | 37 |

Obtain the estimates of the variances at the population and test whether the two populations have the same variance [given F and 11 and 9 d.f. at 5% level of significance = 311]

SOLUTION. **Null Hypothesis (H_0).** Two populations have the same variance.

$$H_0 : \sigma_x^2 = \sigma_y^2$$

Alternative Hypothesis (H_1) : H_0 is not true, i.e.,

$$\sigma_x^2 \neq \sigma_y^2$$

Test Statistics: $F = S_x^2 / S_y^2$

Sample (x)	$(x - \bar{x})$	$(x - \bar{x})^2$	Sample (y)	$(y - \bar{y})$	$(y - \bar{y})^2$
20	−2	4	27	−8	64
16	−6	36	23	−2	4
26	4	16	42	7	49
27	5	25	35	0	0
23	1	1	32	−3	9
22	0	0	34	−1	1
18	−4	16	38	3	9
24	2	4	28	−7	49
25	3	9	41	6	36
19	−3	9	43	8	64
			30	−5	25
			37	2	4
Total 220		120	420		314

$$\bar{x} = \frac{\Sigma x_i}{n_1} = \frac{220}{10} = 22 ,$$

$$\bar{y} = \frac{1}{n_1} \Sigma y_i = \frac{1}{12} \times 420 = 35 ,$$

$$S_x^2 = \frac{1}{n_1 - 1} \Sigma (x_i - \bar{x})^2 = \frac{1}{9} \times 120 = 13.33 ,$$

$$S_y^2 = \frac{1}{n_2 - 1} \Sigma (y_i - \bar{y})^2 = \frac{1}{11} \times 314 = 28.55$$

So, $$F_{cal} = \frac{S_x^2}{S_y^2} = \frac{13.33}{28.55} < 1$$

So we take $$F_{cal} = \frac{S_x^2}{S_y^2} = \frac{28.55}{13.33} = 2.14$$

F_{tab} at $(n_2 - 1, n_1 - 1) = (11, 9)$ d.f. is 3.11.

Conclusion: Since $F_{cal} < F_{tab}$, so null hypothesis may be accepted at 5% level of significance i.e., the samples are drawn from normal populations with same variance.

Exercise-9.23

1. In one sample of 8 observations the sum of squares of the deviations of the sample values from the sample mean was 84.4 and in another sample of 10 observations, it was 102.6. Test whether the two samples have been drawn from two normal population with same variance 5% (For 7 and 9 d.f. at 5% level of significance = 3.29).

2. Two independents sample of 8 and 7 items respectively had the value of the variable:

Sample I :   9   11   13   11   15   9   12   14

Sample II: 10   12   10   14   9   9   10

Do the two estimate of population variance differ significantly? Given that for (7, 6) d.f. the value of F at 5% level of significance is 4.20 nearly.

Answers

1. Yes

2. No

9.8 ANALYSIS OF VARIANCE (ANOVA)

The analysis of variation is a powerful tool for tests of significance. If we want to test the significance difference between two sample means that we use t-test but if we have to test the significance difference among three or more sample means then we required a repetition of t-test. This is time consuming. This can be done by another technique known as "Analysis of variance".

The term "Analysis of variance" was developed by Prof. R. A. Fisher in 1920's.

Definition. According to Prof. R. A. Fisher, Analysis of Variance (ANOVA) is the *"separation of variation ascribable to one group of causes from the variation ascribable to other group."*

So in this technique we split up the total variance into two parts :

(i) Variance between samples,   (ii) Variance within samples

Analysis of variance based on *F*-statistic which can be defined as the ratio of variance between samples and variance with in sample is symbolically and this ratio always greater than unity.

$$F = \frac{\text{Variance between samples}}{\text{Variance within samples}}$$

If this ratio is less than unity then interchange these variance. So we take numerator value always greater than denominator value.

Assumptions: The following assumptions are made for validity of the *F*-test in ANOVA :

(i) The sample observations are independent.

(ii) The population from which the observation are taken is normal.

(iii) Various effects are additive in nature,

9.8.1 TECHNIQUE OF ANALYSIS OF VARIANCE FOR ONE WAY CLASSIFICATION

When the observations are classified according to any one factor (for example, the application of one or more type of diets may be considered on several man).

Let us suppose that N observations $X_{ij}(i = 1, 2, ..., k; j = 1, 2, ..., n_i)$ are grouped in k classes of sizes $n_1, n_2, ..., n_k$ respectively such that

$$N = \sum_{i=1}^{k} n_i$$

						Total	Means
X_{11}	X_{12}	$\cdots$	X_{1j}	$\cdots$	X_{1n_1}	T_1	$\overline{X}_1$
X_{21}	X_{22}		X_{2j}		X_{2n_2}	T_2	$\overline{X}_2$
X_{i1}	X_{i2}		X_{ij}		X_{in_i}	T_i	$\overline{X}_i$
X_{k1}	X_{k2}		X_{kj}		X_{in_k}	T_k	$\overline{X}_k$

The total variation in the observation can be split in to the following two components :

(i) The variation between the classes or samples.   (ii) The variation within the classes or samples.

These are two methods for analysis of one way classification

(i) Direct method   (ii) Short-cut method

First of all we set up the null hypothesis as

$H_0 : \mu_1 = \mu_2 = ... = \mu_k$

$H_1 : \mu_1 \neq \mu_2 \neq ... \neq \mu_k$

where μ_i $(i = 1, 2, ..., k)$ are mean of the ith class in the population.

(1) DIRECT METHOD

This method have following steps:

(a) Variance among the samples.

(i) Obtain the mean of each sample (column)

$$\overline{X}_1, \overline{X}_2, ..., \overline{X}_k \text{ where, } \overline{X}_1 = \frac{X_{i1} + X_{i2} ... X_{in_1}}{n_i}$$

(ii) Find the grand mean, *i.e.*, mean of samples means as $\overline{\overline{X}} = \dfrac{\overline{X}_1 + \overline{X}_2 + ... + \overline{X}_k}{n_1 + n_2 + ... + n_k} = \dfrac{\sum_{i=1}^{k} \overline{X}_i}{\sum_{i=1}^{k} n_i}$

(iii) Find the deviation of sample means from the grand mean then square these deviations and multiply by the number of items or units in the samples,

i.e., $n_i (\overline{X}_1 - \overline{\overline{X}})^2 + n_2(\overline{X}_2 - \overline{\overline{X}})^2 + ... + n_k (\overline{X}_k - \overline{\overline{X}})^2$

This is known as squariance or deviance.

(iv) This sum of square is divided by its degree of freedom which is $(k-1)$ where k denotes the number of samples. It is called mean sum of squares among samples.

(b) **Variance within samples**

(i) Find the mean of each sample
$$\bar{X}_1, \bar{X}_2, ..., \bar{X}_k.$$

(ii) Find the deviation of various items of the sample from the mean value of that sample these deviation and then add.

(iii) For all the samples repeat this procedure and obtain the total of these deviations.
$$\sum_{j=1}^{n_1}(X_{1j}-\bar{X}_1)^2 + \sum_{j=1}^{n_2}(X_{2j}-\bar{X}_2)^2 + ... + \sum_{j=1}^{n_k}(X_{kj}-\bar{X}_k)^2$$

So we obtain the sum of squares of variation with in sample it is also known sum of squares of error (SSE).

(iv) Now divide the sum of squares of error (SSE) by the degrees of freedom for error which would be
$$(n_1-1)+(n_2-1)...(n_k-1) = n_1+n_2+...+n_k-k$$

(c) **Total sum of squares of variation (SST).** The total sum of squares of variation can be obtained to adding the sum of squares of deviation between the sample (SSC) and sum of squares of deviation within sampler (SSE).

So, SST = SSC + SSE

(d) **Analysis of variance table.** Analysis of variance table have five columns as

Source of variation	Sum of squares (SS)	Degree of Freedom (df)	Mean Sum of Squares (MS)	F-ratio
Between samples	SSC	$k-1$	$MSC = \dfrac{SSC}{k-1}$	$F = \dfrac{MSC}{MSE}$
Within samples (error)	SSE	$N-C$	$MSE = \dfrac{SSE}{N-k}$	
Total	SST	$N-1$		

9.8.2 Variance Ratio of F

The F ratio can be obtained by dividing the mean sum of squares of between samples by mean sum of squares of within samples
$$F = \frac{\text{Variance between samples}}{\text{Variance within samples}} = \frac{MSC}{MSE}$$

Statistic is distributed as F distribution with $(k-1, M-k)$

Definition. *If this ratio is less than unity then interchange the mean sum of squares, i.e., $F = \dfrac{MSE}{MSC}$ and degrees of freedom are adjusted, i.e., F-statistic is distributed as F-distribution with $(N-k, k-1)$ degrees of freedom.*

Tabulated value of F: From the F-table we find the tabulated value of F or $(k-1, N-k)$ degrees of freedom at given level of significance. Generally we take 1%, 5%, 10%.

(e) **Interpretation of F.** If $F_{cal} \geq F_{tab}$ then the null hypothesis is rejected and the difference between the means is significant at given level of significance. Otherwise we may accept the null hypothesis.

Solved Examples

EXAMPLE 1. *The following table gives the yields of 6 fields. In three fields the variety A of seeds and last three fields the variety B is used*

A	20	32	22
B	20	10	16

Prepare the analysis of variance table and discuss there is significance difference between variety of seeds A and B.

SOLUTION. **The null hypothesis:** There is no significance difference between the yields due to variety of A and B i.e.,

$H_0 : \mu_A = \mu_B, \quad H_1 : \mu_A \neq \mu_B,$

$\bar{X}_1 = \dfrac{30+32+22}{3} = \dfrac{84}{3} = 28,$

$\bar{X}_2 = \dfrac{20+18+16}{3} = \dfrac{54}{3} = 18,$

$\bar{\bar{X}} = \dfrac{\bar{X}_1+\bar{X}_2}{k} = \dfrac{28+18}{2} = 23$

Sum of squares between the samples
$$= n_1(\bar{X}_1+\bar{\bar{X}})^2 + n_2(\bar{X}_2+\bar{\bar{X}})^2$$
$$= 3(28-23)^2 + 3(18-23)^2$$
$$= 2\times25+3\times25 = 150$$

Sum of squares within the samples

$$= \Sigma(X_{1f} - \bar{X}_1)^2 + \Sigma(X_{2f} - \bar{X}_2)^2$$
$$= (30-28)^2 + (32-28)^2 + (22-28)^2$$
$$+ (20-18)^2 + (18-18)^2 + (16-18)^2$$
$$= 4 + 16 + 36 + 4 + 0 + 4 = 64$$

Analysis of Variance Table

Source of variation	Sum of squares	Degree of Freedom (df)	Mean Sum of Squares (MS)	F-ratio
Between samples	150	$k-1 =$ $2-1=1$	$\dfrac{150}{1} = 150$	$F = \dfrac{150}{16}$ $= 9.4$
Within samples	64	$N-k$ $= 5-1 = 4$	$\dfrac{64}{4} = 16$	
Total		$N-1$ $= 6-1 = 5$		

F_{tab} at 5% level of significance

$$F_{2.12}(0.05)_{tab} = 3.89$$
$$F_{cal} = 8.51$$

Conclusion: Since $F_{cal} > F_{tab}$ so we reject H_0 at 5% level of significance and conclude the there is a significance difference between the average production of the three varieties A, B and C.

EXAMPLE 2. *To assess the significance of possible variation in body wt. of fishes of different ponds of a town a common survey was conducted for a number of fishes taken at random from the four ponds. The results are as follows :*

Ponds			
A	**B**	**C**	**D**
8	12	18	13
10	11	12	9
12	9	16	12
8	14	6	16
7	4	8	15

SOLUTION.

Pond A		Pond B		Pond C		Pond D	
X_1	X_1^2	X_2	X_2^2	X_3	X_3^2	X_4	X_4^2
8	64	12	144	18	324	13	169
10	100	11	121	12	144	9	81
12	144	9	81	16	256	12	144
8	64	14	196	6	36	16	256
7	49	4	16	8	64	15	225
ΣX_1 $=$ 45	ΣX_1^2 $=$ 421	ΣX_2 $= 50$	ΣX_2^2 $=$ 558	ΣX_3 $=$ 60	ΣX_3^2 $= 824$	ΣX_4 $= 65$	$\Sigma X_4^2 =$ 875

The sum of all items of various samples

$$= \Sigma X_1 + \Sigma X_2 + \Sigma X_3 + \Sigma X_4$$
$$= 45 + 50 + 60 + 65 = 220$$

Correction factor

$$= \frac{T^2}{N} = \frac{(220)^2}{20} = \frac{18400}{20} = 2420$$

Total sum of square

$$= \Sigma X_1^2 + \Sigma X_2^2 + \Sigma X_3^2 + \Sigma X_4^2 - \frac{T^2}{N}$$
$$= 421 + 558 + 824 + 875 - 2420 = 2678 - 2420 = 258$$

Sum of squares between the sample

$$= \frac{(\Sigma X_1)^2}{N} + \frac{(\Sigma X_2)^2}{N} + \frac{(\Sigma X_3)^2}{N} + \frac{(\Sigma X_4)^2}{N} - \frac{T^2}{N}$$
$$= \frac{(45)^2}{5} + \frac{(50)^2}{5} + \frac{(60)^2}{5} + \frac{(65)^2}{5} - 2420$$
$$= \frac{2025}{5} + \frac{2500}{5} + \frac{3600}{5} + \frac{4225}{5} - 2420$$
$$= \frac{12350}{5} - 2420 = 2470 - 2420 = 50$$

Sum of squares within samples :

= Total sum of squares − Sum of squares between sample
$$= 258 - 50 = 208$$

Analysis of Variance (ANOVA) table

Source of variance	Sum of square	Degree of freedom	Mean square
Between sample	50	$(C-1) = (4-1) = 3$	16.7
		$(N-C) = (20-4) = 16$	13.0

$$F = \frac{\text{Variance between samples}}{\text{Variance within samples}}$$
$$= \frac{16.7}{13.0} = 1.235$$

at df (3, 16) the table value of F at 5% level of significance = 3.24.

The calculated value of F is less than the table value hence the difference in mean values of sample is not significant, *i.e.*, the samples could have come from the same universe.

EXAMPLE 3. *The following data give the yields of 12 plots of land in three samples under three varieties of fertilizers*

A	B	C
25	20	24
22	17	26
24	16	30
21	19	20

Is there any significant difference in the average yields of land under the three varieties of fertilizers?

Given that F at df(2, 9) at 5% level = 4.26.

SOLUTION. **Null hypothesis (H_0) :** There is no significant difference in the average yields under the three varieties.

Alternative hypothesis (H_1) : The difference in average yields is significant.

Sample A		Sample B		Sample C	
X_1	X_1^2	X_2	X_2^2	X_3	X_3^2
25	625	20	400	24	576
22	484	17	289	26	676
24	576	16	256	30	900
21	441	19	361	20	400
ΣX_1 = 92	ΣX_1^2 = 2126	ΣX_2 = 72	ΣX_2^2 = 1306	ΣX_3 = 100	ΣX_3^2 = 2552

Now,

$$T = \Sigma X_1 + \Sigma X_2 + \Sigma X_3 = 92 + 72 + 100 = 264$$

Correcting factor $= \dfrac{T^2}{N} = \dfrac{(264)^2}{4+4+4}$

$$= \frac{264 \times 264}{12} = 5808$$

The total sum squares (SST)

$$= \Sigma X_1^2 + \Sigma X_2^2 + \Sigma X_3^2 - \frac{T^2}{N}$$

$$= 2126 + 1306 + 2552 - 2808$$

$$= 5984 - 5808 = 176$$

Sum of squares between the sample (SSB)

$$= \frac{(\Sigma X_1)^2}{N} + \frac{(\Sigma X_2)^2}{N} + \frac{(\Sigma X_3)^2}{N} - \frac{T^2}{N}$$

$$= \frac{(90)^2}{4} + \frac{(72)^2}{4} + \frac{(100)^2}{4} - 5808$$

$$= 2116 + 1296 + 2500 - 5808$$

$$= 5912 - 5808 = 104$$

Degree of freedom (df) $= k - 1 = 3 - 1 = 2$

Mean square between the sample (MSB)

$$= \frac{\text{Sum squares between the samples}}{V_1}$$

$$= \frac{104}{2} = 52$$

Sum square within the samples (SSW)

$$= \text{SST} - \text{SSE} = 176 - 104 = 72$$

Degree of freedom (dfw)

$$= N - k = (n_1 + n_2 + n_3) - k = 12 - 3 = 9$$

Mean square within the samples (MSW) $= \dfrac{\text{SSW}}{V_2}$

$$= \frac{72}{9} = 8$$

Analysis of variance (ANOVA) table

Source of variance	Sum of square (SS)	Degree of freedom (df)	Mean square (MS)	Test Statistic
Between sample	SSB = 104	dft = 2	MSB = $\dfrac{\text{SSB}}{V_1}$ = 52	$F = \dfrac{MSB}{MSW}$ $= \dfrac{52}{8}$
Within samples	SSW = 72	dfw = 9	MSW = $\dfrac{\text{SSW}}{V_2}$ = 8	= 0.5

Total SST = 176, $N - 1 = 11$.

The calculated value of F at 0.05 at degrees of freedom $V_1 = 2$ and $V_2 = 9$ is 6.5. It is much greater than the given value of $F = 4.26$. Hence, we reject the null hypothesis (H_0) at 0.05 level and conclude the difference in average yields under the three varieties is significant.

Exercise-9.24

1. The following table gives the yields on 15 sample plots under three varieties of seed A, B and C:

Varieties of School	Yields				
A	19	19	21	14	18
B	16	18	15	23	13
C	23	26	20	26	30

Test whether the average yield of land under different varieties of seed show significant differences.

2. A tea company appoints four salesman and observers their sales in three seasons — summer, winter and monsoon. The figure (in lakhs) are given in the following table :

Seasons	Salesman				Totals
	A	B	C	D	
Summer	30	30	15	29	104
Winter	22	23	15	26	96
Monsoon	20	22	23	23	88
Total	72	75	63	78	288

Carry out on Analysis of variance.

3. The following figures relate to the production in kg. of three varieties I, II, III of wheat sown in 12 plots :

Variety I	14	16	18	
Variety II	14	13	15	22
Variety III	16	16	19	20

Is there any significance difference in the population of three varieties? Given the tabulated value of F for $v_1 = 2$ and $v_2 = 9$ at 5% level of significance is 4.26.

4. Set up ANOVA table for the following per hectare yield (in kg) for three varieties of heat each grown on four plots

Plot of Land	Variety of Wheat		
	A_1	A_2	A_3
1	6	5	5
2	7	5	4
3	3	3	3
4	8	7	4

Also find *F*-ratio.

5. The following data represent the number of units of production per day turned out by 5 different workmen using different types of machines

Workman	Machines Type			
	A	*B*	*C*	*D*
1	44	38	47	36
2	46	40	52	43
3	34	36	44	32
4	33	38	46	33
5	38	42	49	38

(a) Test whether the mean productivity is the same for the four different machine type.

(b) Test whether 5 men differ with respect to productivity.

6. On an arithmetic reasoning of 31 ten-years old boys and 42 ten year old girls, the following score are made:

	Mean	S.D.	N
Boys	40.39	8.69	31
Girls	35.81	0.33	42

Perform an analysis of variance and show whether mean difference between boys and girls is significant. Given that table value of Fat 5% and at 1 % is 7.01.

9.9 COEFFICIENT OF CONTINGENCY

Various test of significance, such as '*z*' and '*t*' test discussed earlier are parametric test and applied to only quantitative data like length, weight, height, Hb%, consumption, number of seeds per pod etc. These tests were based on the assumption that the samples were drawn from the normally distributed populations. There are many situations in which it is not possible to make any dependable assumption about the distribution from which samples have been drawn. In biological experiments, we also get qualitative data like colour, health, intelligence, curve responses of drug, etc., in which observations are classified in a particular category, class or group. For these qualitative data a non-parametric test called chi-square test is commonly used. In many studies, especially genetic studies, it becomes necessary to test the significance of overall deviation between the observed data and expected frequencies.

Chi-square test was developed by Prof. R. A. Fisher in 1870. Karl Pearson improved Fisher's Chi-square test in its modern form in the year 1900. Chi-square is derived from the Greek Letter (Chi-χ) and pronounced as Ki and Ksy without s).

9.9.1 CHI-SQUARE (χ^2) TEST

A statistical test to determine if the "observed" numbers deviate from those "expected" or "theoretical" number under a particular hypothesis

$$\chi^2 = \sum_{i=1}^{n} \left[\frac{(O-E)^2}{E_i} \right]$$

where, O = observed frequency and E = expected frequency.

9.9.2 APPLICATIONS OF χ^2 DISTRIBUTION

χ^2-distribution is used:

(1) to test the significance of sample variances.

(2) to test the independence of attributes in a contingency table.

(3) to compare a number of frequency distribution.

(4) to test the goodness of fit.

WORKING PROCEDURE

Step 1. Calculate all the expected frequencies *i.e.*, *E* for all value i = 1, 2, 3, ..., *n*.

Step 2. Take the difference between each observed frequency '*O*' and the corresponding expected frequency '*E*' for each value of *i.e.*, (*O*–*E*)

Step 3. Square the difference for each value of *i*, *i.e.*, $(O-E)^2$ for all value of i = 1, 2, 3, ..., *n*.

Step 4. Divide each square difference by the corresponding expected frequency *i.e.*, calculate $\frac{(O-E)^2}{E}$ for all values of i =1, 2, ..., *n*.

Step 5. Add all these quotients obtained in the 'step 4'.

$$\chi^2 = \sum_{i=1}^{n} \left[\frac{(O-E)^2}{E} \right]$$

CHARACTERISTICS OF CHI-SQUARE

1. The value of chi-square is always positive as each pair is squared up.

2. χ^2 (chi-square) will be zero if each pair is zero and it may assume any value extending to infinity, when the difference between the observed frequency and expected frequency in each pair are unequal. Thus chi-square lie between 0 and ∞.

3. Chi-square is a statistic but a parametric.

Solved Examples

EXAMPLE 1. *The standard deviation of a certain dimension of articles produced by a machine is 7.5 over a long period. A random sample of 25 articles gave a standard deviation of 10.0. Is it justifiable to conclude that the variability has increased? Use 5% level of significance.*

SOLUTION. $H_0 : \sigma^2 = 7.5$

$H_1 : \sigma^2 > 7.5$ (Variability has increased)

Test statistic : $\chi^2 = \dfrac{\Sigma(x_i - \overline{x})^2}{\sigma^2} \sim \chi^2 \text{ with } (n-1) \text{ d.f.}$

Computation : $\chi^2 = \dfrac{np^2}{\sigma^2}$, under H_0

$= \dfrac{25 \times 10^2}{(7.5)^2} = 44 \qquad \left[\because s^2 = \dfrac{1}{n}\Sigma(x_i - \overline{x})^2 \right]$

Critical region : Right hand side tail of a χ^2–distribution with x degrees of freedom. Here $x = 24, d = 0.05$.

$$\chi^2_{0.05}(24) = 36.415$$

(from table of χ^2 at different probability level)

Conclusion: Since the value of χ^2 calculated is greater than $\chi^2_{0.05}(24)$, H_0 is rejected. This means there is justification for believing that the variability has increased.

9.9.3 CHI-SQUARE (χ^2) TEST FOR GOODNESS OF FIT

When a coin is tossed 100 times, the theoretical considerations expect 150 heads and 50 tails. But in practice these results are rarely achieved. The magnitude of discrepancy (difference) between theory and experiment (observation) described through a Greek letter χ^2 read as chi-Square. If value of χ^2 is zero then the observed and expected frequencies completely coincide. The greater discrepancy between the observed and expected frequencies, the greater is the value of χ^2. This test is given by Karl Person in 1900.

If $Q_1, Q_2, ..., Q_n$ is a set of observed (experimental) frequencies and $E_1, E_2, ..., E_n$ is the corresponding set of expected (theoretical or hypothetical) frequencies, then null hypothesis (H_0). There is no significance difference between observed and expected (theoretical) frequencies.

Or there is no significance difference between theory and experiment.

Alternative Hypothesis : H_0 is not true.

Test statistic:

$$\chi^2 = \sum_{i=1}^{n} \frac{(O_i - E_i)^2}{E_i}$$

where χ^2 follows χ^2 distribution with $(n-1)$ degree of freedom.

Conclusion: If $\chi^2_{cal} \geq \chi^2_{tab}$ then the null hypothesis (H_0) is rejected at α% level of significance otherwise H_1 may be accepted.

CONDITIONS FOR APPLYING χ^2 TEST

Following are the conditions which should be satisfied before χ^2 test can be applied:

(i) The sample observation should be independent.

(ii) The constraints on the cell frequencies if, any, should be linear, *i.e.*, $\Sigma O_i = \Sigma E_i$.

(iii) N, the total numbers of frequencies should be large. It is difficult to say what constitutes largeness but as an arbitrary figure, we may say that N should be at least 50.

(iv) No expected frequency should be very small.

Here it is difficult to say that constitute smallness, but 5 should be regarded as the very minimum and 10 is better. In case where the expected frequencies fall below these limits, they are to be amalgamated in a single cell and degrees of freedom properly adjusted.

 Solved Examples

EXAMPLE 1. *Calculate χ^2 for the following data*

Class	A	B	C
Observed frequency	37	44	19
Expected frequency	31	38	31

SOLUTION. We have

$$\chi^2 = \sum_{i=1}^{n} \frac{(O_i - E_i)^2}{E_i}$$

$$= \frac{(37-31)^2}{31} + \frac{(44-38)^2}{38} + \frac{(19-31)^2}{31}$$

$$= \frac{36}{31} + \frac{36}{33} + \frac{144}{31} = 6.76$$

EXAMPLE 2. *The following table gives the numbers of aircraft accidents that occurred during the various days of the week. Find whether the accidents are uniformly distributed over the week.*

Day	Sun	Mon	Tue	Wed	Thu	Fri	Sat	Total
No. of accidents	14	16	8	12	11	9	14	84

(χ^2 for 4 d.f. at 5% level of significance = 9.41)

(Hissar–2005)

SOLUTION. **Null Hypothesis.$(H_0)'$:** The accidents are uniformly distributed over the week.

Alternative Hypothesis $(H_1)'$: H_0 is not true.

Expected frequencies of accidents on any day

$$= \frac{84}{7} = 12.$$

Day	Sun	Mon	Tue	Wed	Thu	Fri	Sat	Total
Observed no. of accidents	14	16	8	12	11	9	14	84
Expected no. of accidents	12	12	12	12	12	12	12	84

Since frequencies in some classes are less than 10, we regroup the data

Day	Sun	Mon	Tue and Wed	Thu	Fri and Sat	Total
Observed no. of accidents	14	16	20	11	23	84
Expected no. of accidents	12	12	24	12	24	84

Test statistic :

$$\chi^2 = \sum_{i=1}^{n} \frac{(O_i - E_i)^2}{E_i}$$

$$= \frac{(14-12)^2}{12} + \frac{(16-12)^2}{12} + \frac{(20-24)^2}{24}$$

$$+ \frac{(11-12)^2}{12} + \frac{(23-24)^2}{24}$$

$$= \frac{1}{24}(8 + 32 + 16 + 2 + 1) = \frac{59}{24} = 2.46$$

Degrees of freedom = 5 – 1 = 4.

Tabulated value of χ^2 for 4 degrees of freedom at 5% level of significance = 9.41.

Conclusion: Since calculated value of χ^2 is less tabulated value at 5% level of significance so null hypothesis (H_0) may be accepted, *i.e.*, the accidents are uniformly distributed over the week.

EXAMPLE 3. *In 120 throws of a single die, the following distribution of faces was obtained :*

Faces	1	2	3	4	5	6	Total
f_0	30	25	18	10	22	15	120

Do these results constitute of regulation of the equal probability null hypothesis?

SOLUTION. On the basis of principle of equal probabilities $p = 1/6$, the theoretical frequencies for each face is

$NP = 120 \times \dfrac{1}{6} = 20$. Thus we have

Faces	1	2	3	4	5	6	Total
f_0	30	25	18	10	22	15	120
f_e	20	20	20	20	20	20	120

Since no theoretical frequency is less than 10. Hence,

$$\chi^2 = \sum \frac{(f_0 - f_e)^2}{f_e}$$

$$= \frac{(30-20)^2}{20} + \frac{(25-20)^2}{20} + \frac{(18-20)^2}{20}$$

$$+ \frac{(10-20)^2}{20} + \frac{(22-20)^2}{20} + \frac{(15-20)^2}{20}$$

$$= \frac{1}{20}(100 + 25 + 4 + 100 + 4 + 25)$$

$$= \frac{258}{20} = 12.9$$

Degrees of freedom $v = 6 - 1 = 5$.

$\chi^2_{0.05}(5) = 11.070$ (from table)

Since $\qquad \chi^2_{cal} < \chi^2_{0.05}(5)$

Hypothesis of equal probabilities is rejected.

EXAMPLE 4. *Twelve dice were thrown 4096 times and a throw of 6 was respond as a success, the observed frequencies were as*

No. of Success	0	1	2	3	4	5	6	7 and over	Total
Frequencies	447	1145	1181	796	380	115	24	8	4096

Find the value of χ^2 on the hypothesis that the dice were unbiased and hence show that the data are consistent with the hypothesis so far as the χ^2 test is concerned.

SOLUTION. The probability of getting 6 in a throw of a single die, *i.e.*, probability of success = 1/6 and the probability of failure $= 1 - \dfrac{1}{6} = \dfrac{5}{6}$.

Hence, the theoretical frequencies of 0, 1, 2, 3, ..., 12 successes with 12 dice 4096 times are respectively the successive terms of the binomial expansion

$$4096\left(\frac{5}{6} + \frac{1}{6}\right)^{12} = 4096\left[\left(\frac{5}{6}\right)^{12} + {}^{12}C_1\left(\frac{5}{6}\right)^{11}\right.$$

$$\left(\frac{1}{6}\right) + {}^{12}C_2\left(\frac{5}{6}\right)^{10}\left(\frac{1}{6}\right)^2 + ... + {}^{12}C_{12}\left(\frac{1}{6}\right)^{12}\right]$$

$$= 459 + 1102 + 1212 + 809$$
$$+ 365 + 116 + 27 + 6$$

Thus we have

f_0	447	1145	1181	796	380	115	24	8
f_e	459	1102	1212	809	365	116	27	6

$$\chi^2 = \sum \frac{(f_0 - f_e)^2}{f_e}$$

$$= \frac{(447 - 459)^2}{459} + \frac{(1145 - 1102)^2}{1102}$$

$$+ \frac{(1181 - 1212)^2}{1212} + \frac{(796 - 809)^2}{809}$$

$$+ \frac{(380 - 365)^2}{365}$$

$$+ \frac{(115 - 116)^2}{116} + \frac{(240 - 27)^2}{27} + \frac{(8 - 6)^2}{6}$$

$$= \frac{144}{459} + \frac{1849}{1102} + \frac{961}{1212} + \frac{169}{809} + \frac{225}{365} + \frac{1}{116} + \frac{9}{27} + \frac{4}{5}$$

$$= 0.314 + 1.670 + 0.793 + 0.209 + 0.616$$

$$+ 0.009 + 0.333 + 0.667 = 4.619 \text{ nearly}$$

Degree of freedom $= 8 - 1 = 7$

For 7 degrees of freedom at 5% level of significance

$$\chi^2_{0.05}(7) = 14.027 \qquad \text{(from table)}$$

Since the calculated value (4.619) of χ^2 is much less than $\chi^2_{0.05}(7)$, hence the data are consistent with the hypothesis, the dice were unbiased so far as the χ^2-test is connected.

9.9.4 χ^2-Test for Independence of Attributes

If N individuals or items are classified according two attributes A and B. There are m classification $A_1, A_2, ..., A_i, ..., A_m$ in A and n classifications $B_1, B_2, ..., B_j, ..., B_n$ in B. Then we have a two-way table known as contingency table such as :

A \ B	B_1	B_2	...	B_j	...	B_n	
A_1	O_{11}	O_{12}	...	O_{1j}	...	O_{1n}	R_1
A_2	O_{21}	O_{22}	...	O_{2j}	...	O_{2n}	R_2
⋮	⋮	⋮	⋮	⋮	⋮	⋮	⋮
A_i	O_{i1}	O_{i2}	...	O_{ij}	...	O_{in}	R_i
⋮	⋮	⋮	⋮	⋮	⋮	⋮	⋮
A_m	O_{m1}	O_{m2}	...	O_{mj}	...	O_{mn}	R_m
	C_1	C_2	...	C_j	...	C_n	N

where O_{ij} represent the number of individual or item (observed frequency) belonging to A_i and B_i, $(i = 1, 2, ..., m; \ j = 1, 2, ..., n)$

Here
R_i : Total of ith row $\qquad C_j$: Total of jth column

$$\sum_{i=1}^{m} R_i = N = \sum_{j=1}^{n} C_j$$

Null hypothesis (H_0). The attributes are independent.
There is no association between the attributes under study.
Alternative hypothesis (H_1) : H_0 is not true.

Test statistic :
$$\chi^2_{(m-1) \times (n-1)} = \sum_{i=1}^{m} \sum_{j=1}^{n} \frac{(O_{ij} - E_{ij})^2}{E_{ij}}$$

where E_{ij} represent the expected frequency corresponding to O_{ij} and calculated as

$$E_{ij} = \frac{R_i \times C_j}{N} = \frac{\text{Total of row in which it occurs} \times \text{Total of column in which it occurs}}{\text{Total number of observation}}$$

Conclusion: If calculated value of χ^2 is greater than or equal to tabulated value of $\alpha\%$ level of significance then null hypothesis (H_0), is rejected otherwise null hypothesis (H_0) may be accepted.

Solved Examples

EXAMPLE 1. *Two sample polls of votes for two candidates A and B for a public office are taken, one from among the residents of the rural areas. The results are given in the adjoining table. Examine*

Area	Votes for		Total
	A	**B**	
Rural	620	380	1000
Urban	550	450	1000
Total	1170	830	2000

whether the nature of the area is related to voting performance in this elections.

SOLUTION. H_0 : The nature of the area is independent of the voting

preference in the election.
H_1 : H_0 is not true.
Expected frequencies :

$$E_{(620)} = E_{11} = \frac{R_1 \times C_1}{N} = \frac{1000 \times 1170}{2000} = 585,$$

$$E_{(380)} = E_{12} = \frac{R_1 \times C_2}{N} = \frac{1000 \times 830}{2000} = 415$$

$$E_{(550)} = E_{21} = \frac{R_2 \times C_1}{N} = \frac{1000 \times 1170}{2000} = 585,$$

$$E_{(450)} = E_{22} = \frac{R_2 \times C_2}{N} = \frac{1000 \times 830}{2000} = 415$$

Test statistic :

$$\chi^2 = \sum_{i=1}^{2} \sum_{j=1}^{2} \frac{(O_{ij} - E_{ij})^2}{E_{ij}}$$

$$= \frac{(620-585)^2}{585} + \frac{(380-415)^2}{415}$$

$$+ \frac{(550-585)^2}{585} + \frac{(450-415)^2}{415}$$

$$= 10.089$$

Tabulated value for $(2-1) \times (2-1) = 1$ degrees of freedom is 3.841.

Conclusion: Since calculated χ^2 is greater than at 5% level of significance tabulate value so null hypothesis is at rejected at α% level of significance *i.e.*, nature of area is related to voting preference in the election.

EXAMPLE 2. *A random sample of 300 persons are classified as under*

	Male	Female	Total
Smokers	40	60	100
Non-smokers	110	90	200
Total	150	150	300

It there a relationship between sex and smoking?

SOLUTION. **Null hypothesis (H_0):** There is no relationship between sex and smoking or sex and smoking are independent.

Alternative hypothesis (H_1): H_0 is not true.

Expected frequencies :

$$E_{(H_0)} = E_{11} = \frac{R_1 \times C_1}{N} = \frac{100 \times 150}{300} = 50,$$

$$E_{(60)} = E_{12} = \frac{R_1 \times C_2}{N} = \frac{100 \times 150}{300} = 50$$

$$E_{(110)} = E_{21} = \frac{R_2 \times C_1}{N} = \frac{150 \times 200}{300} = 100,$$

$$E_{(90)} = E_{22} = \frac{R_2 \times C_2}{N} = \frac{200 \times 150}{300} = 100$$

Test statistic :

$$\chi^2 = \sum_{i=1}^{2} \sum_{j=1}^{2} \frac{(O_{ij} - E_{ij})^2}{E_{ij}}$$

$$= \frac{(40-50)^2}{50} + \frac{(60-50)^2}{50} + \frac{(110-100)^2}{100}$$

$$+ \frac{(90-100)^2}{100} = 2 + 2 + 1 + 1 = 6$$

Tabulated value of χ^2 for $(2-1) + (2-1) = 1$ degrees of freedom at 5% level of significance = 4.841.

Conclusion: Since χ^2_{cal} calculate $\chi2$ is greater than tabulated value so null hypothesis is rejected at α% level of significance, *i.e.,* sex and smoking are not independent.

EXAMPLE 3. *In experiments on pea-bruding, Mendel obtained the following frequencies of seeds*

Round and yellow	Wrinkled and yellow	Round and green	Wrinkled and green	Total
315	101	108	32	556

Theory predicts that the frequencies should be in proportions 9 : 3 : 3 : 1. *Examine the correspondence between theory and experiment. Given that the value of* ψ^3 *for 3*

degree of freedom at 5% level and significance is 7.815.

SOLUTION. **Null Hypothesis (H_0):** Theory fits well into the experiment.

Alternative Hypothesis (H_1) : H_0 is not true.

Sum of proportion = 9 + 3 + 3 +1 = 16

So expected frequency of round and yellow (E_1)

$$= \frac{\text{proportion}}{\text{sum of proportion}} \times N$$

$$= \frac{9}{16} \times 556 = 313 \text{ (approx.)}$$

Expected frequency at wrinkled and yellow (E_2)

$$= \frac{3}{16} \times 556 = 104 = 104 \text{ (approx.)}$$

Expected frequency of round and green seeds in (E_3)

$$= \frac{3}{16} \times 556 = 104 \text{ (approx.)}$$

Expected frequency of wrinkled and green seeds is

$$= \frac{1}{16} \times 556 = 35 \text{ (approx.)}$$

					Total
Observed Frequency	315	101	108	32	556
Expected Frequency	313	104	104	35	556

Test Statistic :

$$\chi^2 = \sum_{i=1}^{2} \frac{(O_j - E_1)^2}{E_i}$$

$$= \frac{(315-313)^2}{313} + \frac{(101-104)^2}{104}$$

$$+ \frac{(108-104)^2}{104} + \frac{(32-35)^2}{35}$$

$$= 0.013 + 0.09 + 0.15 + 0.26 = 0.513$$

$\chi^2_{tab} = 7.815$ for 3 d.f. at 5% level of significance.

Conclusion : Since $\chi_{cal} = \chi^2_{tab}$ at 5% level of significance so hypothesis may be accepted. Hence, there is much correspondence between theory and experiment.

EXAMPLE 4. *The following table shows the result of inoculation against cholera*

	Not Attacked	Attacked
Inoculated	431	5
Non-inoculated	291	9

Is there any significant association between inoculation and attack? Given that

$$x = 1 ; \begin{cases} P = 0.047, & \text{for } \chi^2 = 3.2 \\ P = 0.069, & \text{for } \chi^2 = 3.3 \end{cases}$$

SOLUTION. Let us find theoretical frequencies and arrange in the following order :

	Not Attacked	Attacked	Total
Inoculated	431(427.7)	5(8.3)	436
Non-inoculated	291(294.3)	9(5.7)	300
Total	722	14	736

The theoretical frequencies have been shown in brackets.

The first theoretical frequency has been obtained

$$= \frac{436}{736} \times 722 = 427.7$$

$$\chi^2 = (3.3)^2 \left[\frac{1}{427.7} + \frac{1}{8.3} + \frac{1}{294.3} + \frac{1}{5.7} \right] = 3.28$$

Number of degree of freedom = $(2-1)(2-1) = 1$

Since $\chi^2 = 3.2$ corresponds to $P = 0.074$ and $\chi^2 = 3.3$ corresponds to $P = 0.069$ hence by interpolation

$\chi^2 = 3.28$ corresponds to $P = 0.0706$ approximately.

Thus, if hypothesis is true, our data gives results which would be obtained about 7 times in hundred trails. This is infrequent but not very infrequent. We may be unjustified in rejecting the hypothesis but this leads us somewhat to believed that in osculation and attack are associated.

EXAMPLE 5. *A survey of 320 families with 5 children shows the following distribution*

No. of boys and girls	5 Boys and 0 Girl	4 Boys and 1 Girl	3 Boys and 2 Girls	2 Boys and 3 Girls	1 Boy and 4 Girls	0 Boy and 5 Girls	Total
No. of families	18	56	110	88	40	8	320

Given that value of χ^2 for 5 df are 11.0 and 15.1 at 0.05 and 0.01 significance level respectively, test the hypothesis that male and female births are equally probable.

SOLUTION. **Null Hypothesis (H_0):** Male and female births are equally probable.

Alternative Hypothesis (H_1) : H_0 is not true.

The probability of male births $(p) = 1/2$ and female births $(q) = 1/2$.

So the expected frequency of r male births in a family of 5 out of 320 families

$= N \, ^5C_r \, p^r \, q^{5-r}$ (From binomial distribution)

$= 320 \, ^5C_r \, (1/2)^r \, (1/2) \, 5 . r$

$= 320 \, ^5C_r \, (1/2)^5 = 10 \, ^5C_r$

So the expected frequency of 5 boys and 6 girls

$E_1 = 10 \times \, ^5C_5 = 10$

The expected frequency of 4 boys and 1 girl

$E_2 = 10 \times \, ^5C_4 = 50$

The expected frequency of 3 boys and 2 girls

$E_3 = 10 \times \, ^5C_3 = 100$

The expected frequency of 2 boys and 3 girls

$E_4 = 10 \times \, ^5C_2 = 100$

The expected frequency of 1 boy and 4 girls

$E_5 = 10 \times \, ^5C_1 = 50$

The expected frequency of 0 boy and r girls

$E_{16} = 10 \times \, ^5C_0 = 10$

So

							Total
Observed Frequency	18	56	110	88	40	8	320
Expected Frequency	10	50	100	100	50	10	320

Test Statistic : $\chi^2 = \sum\limits_{i=1}^{2} \frac{(O_i - E_i)^2}{E_i}$

$$= \frac{(18-10)^2}{10} + \frac{(56-50)^2}{50}$$

$$+ \frac{(110-100)^2}{100} + \frac{(88-100)^2}{100}$$

$$+ \frac{(40-50)^2}{50} + \frac{(8-10)^2}{10}$$

$$= 6.4 + 0.72 + 1 + 1.44 + 2 + 0.4$$

$$= 11.96.$$

Degrees of freedom = $(n-1) = (6-1) = 5$ given χ^2_{tab} at 5% level of significance = 11.0.

Conclusion: (i) Since $\chi^2_{cal} = \chi^2_{tab}$ at 5% level of significance so null hypothesis is rejected at 5% level of significance *i.e.*, boys and girls births are not equally probable.

Given ψ_{tab} at 1 % of significance = 15.1.

(ii) Since $\psi^2_{cal} < \psi^2_{tab}$ at 1 % level of significance so null hypothesis may be accepted at 1 % level of significance, *i.e.*, boys and girls births are equally probable.

Exercise-9.25

1. Find the value of χ^2 for the following data :

Class	:	A	B
Observed frequency	:	8	29
Expected frequency	:	7	24

2. Find the value of χ^2 for the following data:

Observed frequency	:	10	4	15	18	20	15	5	2	3
Expected frequency	:	10	7	10	15	25	10	5	5	5

3. In 120 throws of a single die the following distribution of faces was obtained:

Faces	:	1	2	3	4	5	6	Total
Frequency	:	30	25	18	10	22	15	120

Find χ^2 on the basis that the die was unbiased.

4. 200 digits were chosen at random from a set of taldes. The frequencies of the digits were:

Digit	:	0	1	2	3	4	5	6	7	8	9
Frequency :		18	19	23	21	16	25	22	20	21	15

Use the chi-square test to assess the correctness of the hypothesis that the digits were distributed in the equal number in the table from which these were chosen (χ^2 for 9 d.f. at 5% level = 16.919).

5. Applying the χ^2 test of goodness of fit to the following data :

Observed frequency	:	1	5	20	28	42	22	15	5	2
Expected frequency	:	1	6	18	22	40	25	18	6	1

(χ^2 for 6 d.f. at 5% level = 12.59)

6. The theory predicts the proportion of beans in the four groups A, B, C and D should be 9 : 3 : 3 : 1. In an experiment among 1600 beans, the numbers in the four groups were 882, 213, 287 and 118. Does the experimental result support the theory (χ^2 for 3 d.f. at 5% level = 7.815).

7. In an experiment on immunization of cattle from tuberculosis, the following results were obtained:

	Affected	Unaffected
Inoculated	12	28
Non-inoculated	13	7

Examine the effect of vaccine in controlling the incidence of the disease.

8. The following data is collected on two characters :

	Cinegores	Non-cinegores
Inoculated	83	57
Non-inoculated	45	68

Based on this, can you conclude that there is no relation between the habit of cinema going and literacy ?

9. In a locality 100 persons are randomly selected and asked about their educational attainments. The results are as under:

	Education			Total
	Middle	High School	College	

Male	10	15	25	50
Female	25	10	15	50
Total	35	25	35	100

Does education depend on sex ($\chi_{0.045}^2(2)5.99, \psi_{0.01}^2(2) = 9.21$) .

10. In a certain sample of 2,000 families : 1,400 families are consumers of tea, out of 1,800 Hindu families, 1,236 families consume tea. Use $\chi2$-test and state whether there is any significant between consumption of tea among Hindu and non-Hindu families.

11. 1072 college students were classified according to their intelligence and economic conditions. Test whether there is any associated between intelligence and economic conditions:

		Excellent	Good	Mediocre	Dull
Econonomic Conditions	Good	48	199	181	82
	Not Good	81	185	190	106

Answers

1. $\chi^2 = 6.76$	**2.** $\chi^2 = 8.80$	**3.** $\chi^2 = 85.81$	**4.** $\chi^2 = 4.3$ accepted
5. $\chi^2 = 12.59$ fit is good	**6.** $\chi^2 = 4.7266$ support theory		**7.** $\chi^2 = 13.89$

ARCHIVE

1. A coin was tossed 400 times and the head turned up 216 times. Test the hypothesis that the coin is unbiased at 5% level of significance. (VTU–2007)

2. In a city A 20% of a random sample of 900 school boys had a certain slight physical defect. In another city B, 18.5% of a random sample of 1600 school boys had the same defect. Is the difference between the proportions significant? (VTU– 2003S)

3. In two large populations there are 30% and 25% respectively of fair haired people. In this difference likely to be hidden in samples of 1200 and 900 respectively from the two populations?

4. A die is tossed 960 times and it falls with 5 upwards 184 times. Is the die biased? (VTU–2006)

5. A machine produces 16 imperfect articles in a sample of 500. After machine is overhauled, it produces 3 imperfect articles in a batch of 100. Has the machine been improved? (Rohtak–2005, Madras–2003)

6. A certain stimulus administered to each of 12 patients resulted in the following increases of blood pressure : 5, 2, 8, –1, 3, 0, –2, 1, 5, 0, 4, 6. Can it be concluded that the stimulus will in general be accompanied by an increase in blood pressure. (VTU–2007)

7. The nine items of a sample have the following values : 45, 47, 50, 52, 48, 47, 49, 53, 51. Does the mean of these differ significantly from the assumed mean of 47.5? (VTU–2010)

8. Eleven students were given a test in statistics. They were given a month's further tuition and a second test of equal difficulty was held at the end of it. Do the marks give evidence that the students have benefitted by extra coaching? (VTU–2011S)

Boys	1	2	3	4	5	6	7	8	9	10	11
Marks I test	23	20	19	21	18	20	18	17	23	16	19
Marks II test	24	19	22	18	20	22	20	20	23	20	17

9. A random sample of 10 boys had the following I.Q. 70, 120, 110, 101, 88, 83, 95, 98, 107, 100. Do these data support the assumption of a population mean I.Q. of 100 (at 5% level of significance)? (VTU–2006, Coimbatore–2001)

10. A random sample of size 25 from a normal population has the mean $\bar{x} = 47.5$ and s.d. $S = 8.4$. Does this information refute the claim that the mean of the population is $\mu = 42.1$. (JNTU–2003)

11. Test runs with 6 models of an experimental engine showed that they operated for 24, 28, 21, 23, 32 and 22 minutes with a gallon of fuel. If the probability of a Type I error is at the most 0.01, is this evidence against a hypothesis that on the average this kind of engine will operate for atleast 29 minutes per gallon of the same fuel. Assume normality. (JNTU–2003)

12. Two horses A and B were tested according to the time (in seconds) to run a particular race with the following results:

Horse A	28	30	32	33	33	29	and 34
Horse B	29	30	30	24	27	and 29	

Test whether you can discriminate between two horses? (Rohtak–2005, Coimbatore–2001)

13. A set of five similar coins is tossed 320 times and the result is

No. of heads	0	1	2	3	4	5
Frequency	6	27	72	112	71	32

Test the hypothesis that the data follow a binomial distribution. (Kottayam–2005, PTU–2005, VTU–2004)

14. Fit a Poisson distribution to the following data and test for its goodness of fit at level of significance 0.05. (VTU–2008)

x	0	1	2	3	4
f(x)	419	352	154	56	19

Answers

2. No	**3.** It is unlike that the real difference will be hidden	**4.** Die is biased
6. No	**7.** No	**5.** No
11. Accept null hypothesis	**8.** No	**9.** $t = 0.62$, yes
	12. Yes with 75% confidence	**10.** Refute the claim
	13. rejected	**14.** Yes, can be fitted

❋❋❋❋❋❋

Numerical Methods

10.1 NUMERICAL SOLUTION OF EQUATIONS

The solution of the equation of the form $f(x) = 0$ is used in science and engineering. In this chapter we shall discuss various useful methods for evaluating the root of any equation having numerical coefficients. To find the roots of an equation $f(x) = 0$ we start with a known approximate solution and apply any method discussed in this chapter.

Definition 1. *An expression of the form*

$$f(x) = a_0 x^n + a_1 x^{n-1} + \dots + a_{n-1}.x + a_n$$

where all a_i's are constants, provided $a_0 \neq 0$ and n is a positive integer called a polynomial in x of degree n.

Definition 2. *An equation of the form $y = f(x)$ is said to be algebraic if it can be expressed in the form*

$$f_n y_n + f_{n-1} y_{n-1} + \dots + f_1 y_1 + f_0 = 0$$

where f_i is an ith order polynomial in x.

Definition 3. *A non-algebraic equation is called a transcendental equation i.e., $f(x)$ is an expression involving some other functions such as trigonometric, logarithmic, exponential etc.*

For example :

(i) $4\sin x - e^x = 0$

(ii) $\log x^3 - 5\tan x = 0$

☞ **REMARK**

- A transcendental equation may have a finite or an infinite number of real roots or may not have real root at all.

Definition 4. *The process of finding the roots of an equation is called the solution of that equation, and the value of x for which $f(x) = 0$ is satisfied is called its root.*

Geometrically, we can say that a root of an equation $f(x) = 0$ is that value of x, where the graph of $y = f(x)$ cuts the x-axis.

10.1.1 PROPERTIES OF THE EQUATIONS AND ITS ROOTS

(1) If $f(x)$ is exactly divisible by $(x - a)$ then a is the root of $f(x) = 0$.

(2) Every algebraic equation of degree n has only n roots real as well as imaginary.

(3) If $f(x)$ is continuous in the interval (a, b) and $f(a), f(b)$ have opposite signs, then the equation $f(x) = 0$ has atleast one root between $x = a$ and $x = b$.

(4) The complex roots of an equation always occur in pairs.

(5) Every equation of odd degree has atleast one real root.

(6) If a polynomial of degree n vanishes for more than n values of x, it must be identically zero.

(7) If an equation of nth degree has at the most p positive roots and at the most of negative roots, then it follows that the equation has atleast $\{n - (p + q)\}$ imaginary roots.

(8) In any algebraic equation $f(x) = 0$, the number of positive roots cannot exceed the number of changes of sign from positive to negative and from negative to positive, the number of negative roots cannot exceed the number of changes of sign in $f(-x)$.
   (Descarte's Rule of signs).

10.1.2 METHODS OF SOLUTION

The methods of finding the solution of an equation includes:

(i) Direct Methods   (ii) Graphical Methods   (iii) Trial and Error Methods   (iv) Iterative Methods.

In some cases, roots can be found by using direct analytical methods. For example, for a quadratic equation $ax^2 + bx + c = 0$, the roots of the equation, obtained by

$$x = \frac{-b \pm \sqrt{b^2 - 4ac}}{2a}$$

Graphical method involves plotting the given function and finding the points where it crosses the *x*-axis. These points represent approximate values of the roots of the function.

The trial and error methods involves a series of guesses for *x*, each time evaluating the function to see whether it is close to zero. The value of *x* that causes the function value closer to zero is one of the approximate roots of an equation.

An iterative technique begins with an approximate value of the root (which is known as initial guess), which is then successively improved by iteration. The zprocess of iteration stops when the desired level of accuracy is obtained.

In this chapter we shall discuss few iterative methods such as Bisection, Regula-Falsi, Newton-Raphoson, Muller's methods, etc. that are commonly used.

10.2 BISECTION METHOD

Bisection method of solving the equation $f(x) = 0$ is based on the following theorem:

Statement. *Let $f(x)$ be a function, which is continuous in the closed interval $[a, b]$ and $f(a)$, $f(b)$ are of opposite signs, then there exist at least one value c, (say) of x where $c \in]a, b[$ satisfy $f(c) = 0$.*

Proof. Let $f(a)$ be negative and $f(b)$ be positive in the interval $]a, b[$ then, at least one root of the equation $f(x) = 0$ lies in the interval $]a, b[$. Let the rounded value of this root be $c = \frac{1}{2}(a + b)$, which is obtained by dividing the distance between the points *a* and b into two equal parts. If $f(c) = 0$, then *c* is the required root of the equation $f(x) = 0$, otherwise root will lie in the interval $]a, c[$ or $]c, b[$ which will depend on whether the value of $f(c)$ is positive or negative.

Let $f(c)$ be positive then as before, we divide the interval $]a, c[$ into two equal parts and let

$d = \frac{1}{2}(a + c)$. The iterative cycle is terminated when the search interval becomes smaller than the prescribed tolerance or the value of the function nearly vanishes at the new *x*-value.

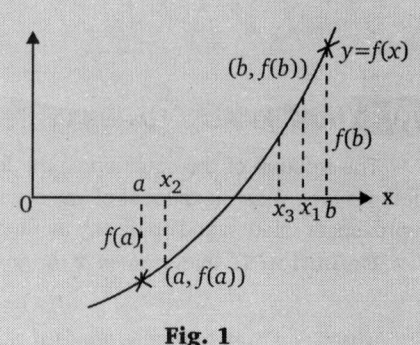

Fig. 1

WORKING PROCEDURE

Step 1. Take initial approximation for x_1 and x_2.

Step 2. Compute $f_1 = f(x_1)$ and $f_2 = f(x_2)$

Step 3. If $f_1 . f_2 > 0$, x_1 and x_2 then there is no root between x_1 and x_2

Step 4. Compute $x_0 = \frac{x_1 + x_2}{2}$; $f_0 = f(x_0)$

Step 5. If $f_1 . f_0 < 0$, then set $x_2 = x_0$ else set $x_1 = x_0$, $f_1 = f_0$

Step 6. Root is given by $(x_1 + x_2)/2$

☞ REMARKS

- If ε is the prescribed tolerance in the required root then the iterative cycle terminates when the absolute error becomes less than or equal to ε, *i.e.*, $|x_1 - x_0| < \varepsilon$.
- For prescribed tolerance ε, the approximate number of iterations required in bisection method, can be determined by the relation.

$$\frac{x_1 - x_0}{2^n} < \varepsilon .$$

10.2.1 PROPERTIES OF BISECTION METHOD

(1) If $f(x)$ is not continuous in the closed interval $[a, b]$ then bisection method is failed.

(2) If $f(x)$ is continuous in a closed interval and does not cut the *x*-axis, then $f(x)$ does not have a real root.

(3) Bisection method gives the real root of $f(x) = 0$.

(4) Bisection method is also named as Bolzano method or interval halving method.

(5) Bisection method is always converges.

10.2.2 ORDER OF CONVERGENCE OF BISECTION METHOD

In Bisection method, at each iteration, the interval in which the root lies is divided into half interval. If we consider the middle points of successive intervals to be the approximation to the root, one half of the current interval (the interval in which the root lies) is the upper bound to the error. Therefore, in bisection method

$$e_{n+1} = \frac{1}{2}e_n \qquad \text{so} \qquad e_{n+1} \propto e_n .$$

Hence, the bisection method is linearly convergent.

REMARK

• Bisection method is always convergent.

Solved Examples

EXAMPLE 1. *Find the root of the equation $x^3 - x - 1 = 0$ lying between 1 and 2 by bisection method.*

SOLUTION. Let $f(x) = x^3 - x - 1 = 0$

since $f(1) = 1^3 - 1 - 1 = -1$, which is negative

and $f(2) = 2^3 - 2 - 1 = 5$, which is positive

Therefore, $f(1)$ is negative and $f(2)$ is positive, so at least one real root will lie between 1 and 2.

First iteration: Now using Bisection method, we can take first approximation

$$x_1 = \frac{1+2}{2} = \frac{3}{2} = 1.5$$

Then, $f(1.5) = (1.5)^3 - 1.5 - 1 = 3.375 - 1.5 - 1 = 0.875$

∴ $f(1.5) > 0$, that is, positive.

Second iteration:

$$x_2 = \frac{1+1.5}{2} = 1.25$$

Then

$$f(1.25) = (1.25)^3 - 1.25 - 1 = 1.953 - 2.25 = -0.297 < 0$$

∴ $f(1.25)$ is negative.

Therefore, $f(1.5)$ is positive and $f(1.25)$ is negative, so that root will lie between 1.25 and 1.5.

Third approximation: The third approxim-ation is given by

$$x_3 = \frac{1.25 + 1.5}{2} = 1.375$$

Now, $f(1.375) = (1.375)^3 - 1.375 - 1 = 0.2246$

∴ $f(1.375)$ is positive

∴ The required root lies between 1.25 and 1.375.

Fourth approximation: The fourth approxi-mation is given by

$$x_4 = \frac{1.25 + 1.375}{2} = 1.313$$

Now, $f(1.313) = (1.313)^3 - 1.313 - 1 = -0.0494$

Therefore, $f(1.313)$ is negative and $f(1.375)$ is positive. Thus, root lies between 1.313 and 1.375.

Fifth approximation: The fifth approximation is given by

$$x_5 = \frac{1.313 + 1.375}{2} = 1.344$$

∴ $f(1.344) = (1.344)^3 - 1.344 - 1 = 0.0837$

∴ $f(1.344) > 0$

∴ $f(1.313)$ is negative and $f(1.344)$ is positive, so root lies between 1.313 and 1.344.

Sixth approximation: The sixth approxim-ation is given by

$$x_6 = \frac{1.313 + 1.344}{2} = 1.329$$

∴ $f(1.329) = (1.329)^3 - 1.329 - 1 = 0.0183$

∴ $f(1.329) > 0$

∴ $f(1.313)$ is negative and $f(1.329)$ is positive, so that the required root lies between 1.313 and 1.329.

Seventh approximation: The seventh appro-ximation is given by

$$x_7 = \frac{1.313 + 1.329}{2} = 1.321$$

∴ $f(1.321) = (1.321)^3 - 1.321 - 1 = -0.0158 < 0$

∴ $f(1.321)$ is negative and $f(1.329)$ is positive, the required root lies between 1.321 and 1.329.

Eighth approximation: The eighth approxi-mation is given by

$$x_8 = \frac{1.321 + 1.329}{2} = 1.325$$

From above iterations, the root of $f(x) = x^3 - x - 1 = 0$ upto two places of decimals is 1.32, which is of desired accuracy.

EXAMPLE 2. *Find a root of the equation $f(x) = x^3 - 4x - 9 = 0$, using the bisection method in four stages.*

(MUMBAI–2003, JNTU–2009)

SOLUTION. Given $f(x) = x^3 - 4x - 9 = 0$ ...(1)

Then $f(2) = (2)^3 - 4(2) - 9 = -9$

and $f(3) = (3)^3 - 4(3) - 9 = 6$

Therefore, $f(2)$ is negative and $f(3)$ is positive, a root lies between 2 and 3.

First approximation: First approximation to the root is given by

$$x_1 = \frac{2+3}{2} = 2.5$$

Thus $f(2.5) = (2.5)^3 - 4(2.5) - 9$ [Using (1)]
$$= 15.625 - 19 = -3.375$$

Therefore, the root lies between 2.5 and 3.

Second approximation: The second approxi-mation to the root is given by

$$x_2 = \frac{2.5 + 3}{2} = 2.75$$

Then $f(2.75) = (2.75)^3 - 4(2.75) - 9$
$$= 20.797 - 20 = 0.797$$

∴ $f(2.75)$ is positive and $f(2.5)$ is negative. Thus root lies between 2.5 and 2.75.

Third approximation: The third approxim-ation to the root is given by

$$x_3 = \frac{2.5 + 2.75}{2} = 2.625$$

Now, $f(2.625) = (2.625)^3 - 4(2.625) - 9$
$$= 18.088 - 19.5 = -1.412$$

∴ $f(2.625)$ is negative while $f(2.75)$ is positive. Then the root lies between 2.625 to 2.75

Fourth approximation: The fourth approxi-mation to the root is given by

$$x_4 = \frac{2.625 + 2.75}{2} = 2.6875$$

Hence, after the four steps the root is 2.6875

approximately.

Exercise-10.1

1. Find a real root of $e^x = 3x$ by bisection method.

2. If $f(x) = 2^x - x - 3$. Find $f(x)$ for $x = -4, -3, -2, -1, 0, 1, 2, 3, 4$ and compute between which integers values roots are lying.

3. Find the root of $\tan x + x = 0$ upto two decimal places which lies between 2 and 2.1 using bisection method

4. Find a root, correct to three decimal places for each of the equations using bisection method.

    (i) $x^3 + x^2 + x + 7 = 0$ (ii) $x^3 - 5x + 3 = 0$

5. Compute the root of $f(x) = \sin 10x + \cos 3x$ by taking initial approximations 4 and 5.

6. Find a positive root of the equation $x^3 + 3x - 1 = 0$ by bisection method.

7. Find the approximate root of the equation $3x - \sqrt{1 - \sin x} = 0$ by bisection method.

8. Compute the real root of the equation $3x + \sin x - e^x = 0$.

9. Find the positive real root of $x - \cos x = 0$ by bisection method upto four places of decimals between 0 and 1.

Answers

1. 1.5121375

2.

x	−4	−3	−2	−1	0	1	2	3	4
f(x)	1.0625	0.125	−0.75	−1.5	−2	−2	−1	2	9

Root lies between (−3, −2) and (2, 3)

3. 2.02875625   4. (i) 0.0552   (ii) 2.279   5. 4.712389 6. 0.322   7. 10.39188   8.   0.3604   9.   0.7391

10.3   SECANT METHOD

This method is an improvement over the method of Regula-falsi method as it does not require the condition that $f(x_0).f(x_1) < 0$. In this method the graph of the function $y = f(x)$ in approximated by a secant line at each iteration. Secant line is nothing but the chord joining the initial limits of the interval. Taking x_0, x_1 as the initial limits of the interval, we write the equation of the chord joining the points $(x_0, f(x_0))$ and $(x_1, f(x_1))$ is given by

$$y - f(x_1) = \frac{f(x_1) - f(x_0)}{x_1 - x_0}(x - x_1)$$

This secant line cuts the x-axis i.e., $y = 0$ then the abscissa of the point is given by

$$x_2 = x_1 - \frac{x_1 - x_0}{f(x_1) - f(x_0)} f(x_1)$$

which is an approximation to the root. In the general way, the successive approximations is given by

$$x_{n+1} = x_n - \frac{x_n - x_{n-1}}{f(x_n) - f(x_{n-1})} f(x_n), n \geq 1$$

REMARKS

- In this method, $f(x_0)$ and $f(x_1)$ is not necessarily of opposite signs.
- In this method, it is not necessary that the interval (x_0, x_1) must contain the root.
- If at any iteration $f(x_n) = f(x_{n-1})$ this method fails and shows that it does not converge necessarily.
- If the secant method converges, its rate of convergence is faster than that of the method of false position.
- The order of convergence of secant method is 1.62.

Solved Examples

EXAMPLE 1. *Determine the root of the equation $f(x) = \cos x - xe^x = 0$ using the secant method upto four decimal places.*

SOLUTION. Since $f(x) = \cos x - xe^x = 0$ ...(i)

Taking the initial approximations $x_0 = 0, x_1 = 1$

So that $f(x_0) = f(0) = \cos 0 - 0(e^0) = 1$

and $f(x_1) = f(1) = \cos 1 - 1(e) = -2.1780$.

Then by secant method, we get

$$x_2 = x_1 - \frac{x_1 - x_0}{f(x_1) - f(x_0)} f(x_1)$$

$$= 1 - \frac{1 - 0}{-2.1780 - 1}(-2.1780)$$
$$= 1 - \frac{1 - 0}{3.1780} - 0.6853$$

$\therefore \quad x_2 = 0.3147$

Now $f(x_2) = f(0.3147)$

$$= \cos(0.3147) - (0.3147)e^{0.3147} \text{ [From (i)]}$$
$$= 0.9509 - 0.4311 = 0.5198$$

Thus the second approximation to the root is

$$x_3 = x_2 - \frac{x_2 - x_1}{f(x_2) - f(x_1)} f(x_2)$$

$$= 0.3147 - \frac{0.3147 - 1}{0.5198 + 2.1780}(0.5198)$$

$$= 0.3147 + \frac{0.3562}{2.6978} = 0.4467$$

Now, $f(x_3) = f(0.4467)$

$$= \cos(0.4467) - (0.4467).e^{0.4467} \text{ [From(i)]}$$

$$= 0.9019 - 0.6983 = 0.2036$$

Then, the third approximation to the root is

$$x_4 = x_3 - \frac{x_3 - x_2}{f(x_3) - f(x_2)} f(x_3)$$

$$= 0.4467 - \frac{0.4467 - 0.3147}{0.2036 - 0.5198}(0.2036)$$

$$= 0.4467 + \frac{0.0269}{0.3162}$$

$$x_4 = 0.5318$$

Now, $f(x_4) = f(0.5318)$

$$= \cos(0.5318) - (0.5318).e^{0.5318} \text{ [From(i)]}$$

$$= 0.8619 - 0.9051 = -0.0432$$

Then, the fourth approximation to the root is

$$x_5 = x_4 - \frac{x_4 - x_3}{f(x_4) - f(x_3)} f(x_4)$$

$$= 0.5318 - \frac{0.5318 - 0.4467}{-0.0432 - 0.2036}(-0.0432)$$

$$= 0.5318 - \frac{0.0037}{0.2468}$$

$$= 0.5318 - 0.0150 = 0.5168$$

Now, $f(x_5) = f(0.5168)$

$$= \cos(0.5168) - (0.5168).e^{0.5168} \text{ [From(i)]}$$

$$= 0.8694 - 0.8665 = 0.0029$$

Then, the fifth approximation to the root is

$$x_6 = x_5 - \frac{x_5 - x_4}{f(x_5) - f(x_4)} f(x_5)$$

$$= 0.5168 - \frac{0.5168 - 0.5318}{0.0029 + 0.0432}(0.0029)$$

$$= 0.5168 + \frac{0.0000435}{0.0461} = 0.5177$$

Now, $f(x_6) = f(0.5177)$

$$= \cos(0.5177) - (0.5177).e^{0.5177} \text{[From (i)]}$$

$$= 0.8690 - 0.8688 = 0.0002$$

Thus, the sixth approximation to the root is

$$x_7 = x_6 - \frac{x_6 - x_5}{f(x_6) - f(x_5)} f(x_6)$$

$$= 0.1577 - \frac{0.5177 - 0.5168}{0.0002 - 0.0029} \times 0.0002$$

$$= 0.5177 + 0.00006 = 0.51776$$

Hence the root of the equation $f(x) = \cos x - xe^x$ is 0.5177 correct to four decimal places.

EXAMPLE 2. *Compute root of the equation $x^2 e^{-x/2} = 1$ in the interval [0, 2] using secant method. The root should be correct to three decimal places.*

(UPTU(MCA)–2005)

SOLUTION. Taking initial approximation $x_0 = 0, x_1 = 2$

We obtain for the secant method

$$f(x) = e^{-x/2}.x^2 - 1$$

$$f(0) = 1, f(2) = 0.4760$$

Apply secant method

$$x_2 = x_1 - \left| \frac{x_1 - x_0}{f_1 - f_0} \right| f_1$$

$$= 2 - \left| \frac{2 - 0}{0.4760 + 1} \right| \times 0.4760 = 1.3550$$

$$\Rightarrow \quad f(x_2) = -0.0665$$

Similarity

$$x_3 = x_2 - \left| \frac{x_2 - x_1}{f_2 - f_1} \right| \times f_2$$

$$= 1.3550 - \left| \frac{1.355 - 2}{-0.0665 - 1} \right| \times (-0.0665)$$

$$= 1.3148$$

10.4 ITERATION METHOD

This method is used when we find the value of the root $x = \alpha$ in terms of a function x, *i.e.,*

$$x = f(x) \text{ from } f(x_0) = 0 \qquad \qquad ...(1)$$

In this method, we first obtain the approximate value of the root, say x_0 of the given equation and then substitute it in $f(x)$ and get second approximation as

$$x_1 = f(x_0) \qquad \qquad ...(2)$$

Now putting $x = x_1$ in $f(x)$ and get the third approximation

$$x_2 = f(x_1) \qquad \qquad ...(3)$$

Continuing this process, we may get

$$x_n = f(x_{n-1}) \qquad \text{or} \qquad x_{n+1} = f(x_n)$$

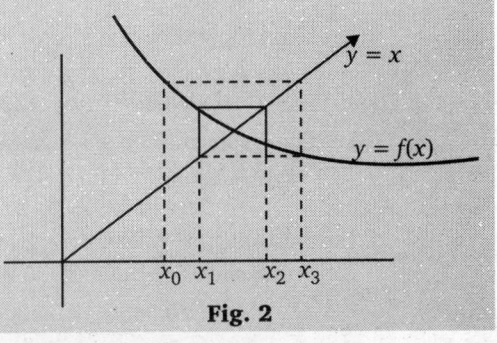

Fig. 2

REMARKS

- If $f(x)$ be a root of $F(x) = 0$ and let I be the interval containing the point $x = a$. Let $f(x)$ and $f'(x)$ be continuous in I when $f(x)$ is given by the equation $x = f(x)$ which is equivalent to $F(x) = 0$. Then if $|f'(x)| < 1 \forall x \in I$, the sequence of approximations $x_0, x_1, ..., x_n$ defined by $x_{n+1} = f(x_n)$ converges to the root a, provided that the initial approximations x_0 is chosen in I.
- The advantages of iterative methods are :
  (i) the use of the same set of instructions will save space in memory unit of the computer.
  (ii) the round off errors are minimised where as these errors are compounded by other method.
- The iterative method $x = g(x)$ is convergent if $|g'(x)| < 1$
  if $|g'(x)| = 1$ and $|g'(x)| < 1$, the number of iterations would be large. But if $g'(x)$ is very small, the number of iteration will be lessor.
- When $|g'(x)| > 1 \Rightarrow g'(x) > 1$ or $g'(x) < -1$, the iterative process is divergent.

10.4.1 RATE OF CONVERGENCE OF ITERATION METHOD

Let $f(x) = 0$ be the equation which is being expressed as $x = g(x)$. The iterative formula for solving the equation is

$$x_{i+1} = g(x_i)$$

If α is the root of the equation $x = g(x)$ lying in the interval $]a, b[, \alpha = g(\alpha)$.

The iterative formula may also be written as $x_{i+1} = g(x + (x_i - \alpha))$

Then, by mean value theorem

$$x_{i+1} = g(\alpha) + (x_i - \alpha)g'(c_i) \qquad \text{where} \quad a < c_i < b$$

But $g(\alpha) = \alpha \Rightarrow \qquad x_{i+1} = \alpha + (x_i - \alpha)g'(c_i)$

$\Rightarrow \qquad\qquad\qquad x_{i+1} - \alpha = (x_i - \alpha)g'(c_i)$ ...(1)

Now, if e_{i+1}, e_i are the error for the approximations x_{i+1} and x_i.

$\therefore \qquad\qquad\qquad e_{i+1} = x_{i+1} - \alpha, e_i = x_i - \alpha$

Using this in (1), we get $e_{i+1} = e_i g'(c_i)$

Here $g(x)$ is a continuous function, therefore, it is bounded

$\therefore \qquad\qquad\qquad |g'(c_i)| \le k, \text{ where } k \in]a, b[\text{ is a constant.}$

$\therefore \qquad\qquad\qquad e_{i+1} \le e_i k$

Since, the index of e_i being 1, the rate of convergence of the iterative method is $x_{i+1} = g(x_i)$ is linear.

 **Solved Examples**

EXAMPLE 1. *Solve $x = 0.21\sin(0.5 + x)$ by iteration method starting with $x_0 = 0.12$.*

SOLUTION. Here, we have

$$x = 0.21\sin(0.5 + x)$$
$$\Rightarrow \quad f(x) = 0.21\sin(0.5 + x)$$

Here, we observe that $|f(x)| < 1$

$\Rightarrow$ Method of iteration can be applied.
Now first approximation of x is given by

$$x_1 = 0.21\sin(0.5 + 0.12) = 0.21\sin(0.62)$$
$$= 0.21(0.58104) = 0.1220$$

The second approximation is given by

$$x_2 = 0.21\sin(0.5 + 0.122) = 0.21\sin(0.622)$$
$$= 0.21(0.58267) = 0.1224$$

The third approximation of x is given by

$$x_3 = 0.21\sin(0.5 + 0.1224) = 0.21\sin(0.622)$$
$$= 0.21(0.58299) = 0.12243$$

The fourth approximation of x is given by

$$x_4 = 0.21\sin(0.5 + 0.12243) = 0.21\sin(0.62243)$$
$$= 0.21(0.58301) = 0.12243$$

Here, we observe that $x_3 = x_4$

Hence, the required root is given by

$$x = 0.12243.$$

EXAMPLE 2. *Find a real root of the equation*

$$\cos x = 3x - 1$$

Correct to three decimal places, using Iteration method.

SOLUTION. Here, we have

$$f(x) = \cos x - 3x + 1 \qquad ...(1)$$

We observe that $f(0) = 2 = +\text{ve}$

and $f(\pi/2) = -3\left(\dfrac{\pi}{2}\right) + 1 = -\text{ve}$

Roots lies between 0 and $\dfrac{\pi}{2}$.

Now, rewriting the given equation as follows

$$x = \frac{1}{3}(\cos x + 1) = g(x) \quad \text{(Say)}$$

Then, we have

$$g'(x) = -\frac{\sin x}{3}$$

$$= |g'(x)| < 1 \text{ in } \left]0, \frac{\pi}{2}\right[$$

= Iteration method can be applied.
Take the first approximation, $x_0 = 0$
Then we can find the successive approximation as follows.

$$x_1 = g(x_0) = \frac{1}{3}[\cos 0 + 1] = 0.667$$

$$x_2 = g(x_1) = \frac{1}{3}[\cos(0.667) + 1] = 0.5953$$

$$x_3 = g(x_2) = \frac{1}{3}[\cos(0.5953) + 1] = 0.6093$$

$$x_4 = g(x_3) = \frac{1}{3}[\cos(0.6093) + 1] = 0.6067$$

$$x_5 = g(x_4) = \frac{1}{3}[\cos(0.6067) + 1] = 0.6072$$

$$x_6 = g(x_5) = \frac{1}{3}[\cos(0.6072) + 1] = 0.6071$$

Now x_5 and x_6 being almost same.

Hence, the required root is given by 0.607.

EXAMPLE 3. *Find the reciprocal of 41 correct to 4 decimal places by iterative formula.*

$$x_{i+1} = x_i \cdot (2 - 41 x_i).$$

SOLUTION. Iterative formula is given by

$$x_{i+1} = x_i \cdot (2 - 41 x_i) \qquad \dots(1)$$

Putting $i = 0$ in equation (1), we get

$$x_1 = x_0(2 - 41 x_0)$$

Let $x_0 = 0.02$

$$x_1 = (0.02)(2 - 0.82) = 0.024$$

Put $i = 1$ in equation (1), we get

$$x_2 = (0.024)[2 - (41 \times 0.024)] = 0.0244$$

Again putting $i = 2$ in equation (1), we get

$$x_3 = 0.02439$$

Hence, Reciprocal of 41 is 0.0244.

EXAMPLE 4. *Suggest a value c so that the iteration formula $f(x) = x + c(x^2 - 3)$ may converge at a good rate. Given that $x = \sqrt{3}$ is a root.*

SOLUTION. Let $f(x) = x + c(x^2 - 3)$

It will converge if

$$|f'(x)| < 1 \quad \Rightarrow \quad -1 < f'(x) < 1$$

i.e., if $\quad -1 < 1 + 2cx < 1$

Also, the convergence will be rapid if $f'(a) = 0$

i.e., if $\qquad 1 + 2ca \approx 0$

i.e., if $\qquad 1 + 2c\sqrt{3} \approx 0$

$$\Rightarrow \qquad c = -\frac{1}{2\sqrt{3}} < -\frac{1}{4}$$

Hence, if we may take $c = -\frac{1}{4}$, then the given iterative

formula may converge at a good rate.

EXAMPLE 5. *If α, β are the root of $x^2 + ax + b = 0$, show that the iteration $x_{n+1} = -\left(\dfrac{ax_n + b}{x_n}\right)$ will converge near $x = \alpha$ if $|\alpha| > |\beta|$ and the iteration $x_{n+1} = -\dfrac{b}{x_{n+a}}$ will converge near $x = \alpha$ if $|\alpha| < |\beta|$.*

(UPTU(MCA)–2005)

SOLUTION. It is given that α, β are the roots of $x^2 + ax + b = 0$.

Then we have $\alpha + \beta = -a$ and $\alpha\beta = b$

Also, the formula $x_{n+1} = -\left(\dfrac{ax_n + b}{x_n}\right)$ which is of the form $x_{n+1} = f(x_n)$ will converge to $x = \alpha$ if

$$\left| \frac{d}{dx}\left\{ \frac{-(ax + b)}{b} \right\} \right|_{x=x_n} < 1$$

$$\Rightarrow \qquad \left| \frac{b}{x_n^2} \right| < 1$$

$$\Rightarrow \quad |x_n^2| > |b| \quad \text{or} \quad x_n^2 > |b|$$

$$\text{or} \quad |\alpha|^2 > |b| \quad \text{as} \quad x_n \to \alpha$$

$$\Rightarrow \quad |\alpha|^2 > |\alpha| \cdot |\beta| \qquad (\because \alpha\beta = b)$$

$$\Rightarrow \quad |\alpha| > |\beta|$$

In a similar way $x_{n+1} = -\dfrac{b}{x_{n+a}}$ will converge to $x = \alpha$ if

$$\left| \frac{d}{dx}\left(\frac{-b}{x + a} \right) \right|_{x=x_n} < 1$$

$$\text{or} \qquad \left| \frac{b}{(x_n + a)^2} \right| < 1$$

$$\Rightarrow \qquad (x_n + a)^2 > |b|$$

$$\Rightarrow \qquad (\alpha + a)^2 > |b| \qquad (\text{as } x_n \to a)$$

$$\Rightarrow \quad \beta^2 > |b| \quad \Rightarrow \quad |\beta|^2 > |\alpha| \cdot |\beta|$$

$$\Rightarrow \quad |\beta| > |\alpha| \quad \text{or} \quad |\alpha| < |\beta|$$

10.4.2 ITERATIVE METHOD FOR THE SYSTEM OF NON-LINEAR EQUATIONS

Consider non-linear equations

$$\left. \begin{array}{l} f(x, y) = 0 \\ g(x, y) = 0 \end{array} \right] \qquad \dots(1)$$

whose real root are to be required within the given degree of accuracy.

Let us take $\qquad x = F(x, y) \quad$ and $\qquad y = G(x, y) \qquad \dots(2)$

provided $\left| \dfrac{\partial F}{\partial x} \right| + \left| \dfrac{\partial F}{\partial y} \right| < 1$ and $\left| \dfrac{\partial G}{\partial x} \right| + \left| \dfrac{\partial G}{\partial y} \right| < 1$ in the neighbourhood of the root.

Let (α, β) be the exact root of (1) and let (x_0, y_0) be the initial approximations.

Then from (2), the successive approximations are given by

$$x_1 = F(x_0, y_0); \qquad\qquad y_1 = G(x_0, y_0)$$

$$x_2 = F(x_1, y_1); \qquad\qquad y_2 = G(x_1, y_1)$$

$$x_3 = F(x_2, y_2); \qquad\qquad y_3 = G(x_2, y_2)$$

and so on, we get following two iterative formulae.

$$x_{n+1} = F(x_n, y_n) \qquad \text{and} \qquad y_{n+1} = G(x_n, y_n)$$

If these formulae converges, then in the limit; $\alpha = F(\alpha, \beta)$ and $\beta = G(\alpha, \beta)$

Hence, (α, β) gives the root of system (1).

Solved Examples

EXAMPLE 1. *Find a real root of the system of the equations by iterative method*

$$x = 0.2x^2 + 0.8; y = 0.3xy^2 + 0.7$$

SOLUTION. Let us assume that

$$\left. \begin{array}{l} F(x, y) = 0.2x^2 + 0.8 \\ \text{and} \quad G(x, y) = 0.3xy^2 + 0.7 \end{array} \right] \qquad \ldots(1)$$

Then $\dfrac{\partial F}{\partial x} = 0.4x, \dfrac{\partial F}{\partial y} = 0, \dfrac{\partial G}{\partial x} = 0.3y^2, \dfrac{\partial G}{\partial y} = 0.6xy$

Let us choose initially $x_0 = \dfrac{1}{2}, y_0 = \dfrac{1}{2}$. Then, we have

$$\left| \frac{\partial F}{\partial x} \right|_{(x_0, y_0)} + \left| \frac{\partial F}{\partial y} \right|_{(x_0, y_0)} = 0.2 < 1$$

and $\left| \dfrac{\partial G}{\partial x} \right|_{(x_0, y_0)} + \left| \dfrac{\partial G}{\partial y} \right|_{(x_0, y_0)} = \dfrac{0.3}{4} + \dfrac{0.6}{4} = \dfrac{0.9}{4} < 1$

Hence, the successive approximations are given by

$$x_1 = F(x_0, y_0) = (0.2)\left(\frac{1}{2}\right)^2 + 0.8 = 0.85$$

$$y_1 = G(x_0, y_0) = (0.3)\left(\frac{1}{2}\right)\left(\frac{1}{2}\right)^2 + 0.7 = 0.74$$

$$x_2 = F(x_1, y_1) = 0.2x_1^2 + 0.8$$
$$= 0.2(0.85)^2 + 0.8 = 0.9445$$

$$y_2 = G(x_1, y_1) = 0.3x_1y_1^2 + 0.7$$
$$= 0.3(0.85)(0.74)^2 + 0.7 = 0.8396$$

Also $x_3 = F(x_2, y_2) = 0.2x_2^2 + 0.8$
$$= 0.2(0.9445)^2 + 0.8 = 0.9784$$

$$y_3 = G(x_2, y_2) = 0.3x_2y_2^2 + 0.7$$
$$= 0.3(0.9445)(0.8396)^2 + 0.7 = 0.8997$$

From these three approximations, we conclude that the root converges to (1, 1). Also from (1), we get
$$1 = F(1, 1) \text{ and } 1 = G(1, 1)$$

Exercise-10.2

1. By iterative method, find $\sqrt{30}$.

2. By iterative method, find the real root of the equation $3x - \log_{10} x = 6$ correct to four significant figures.

3. Find the cube root of 15 correct to four significant figures by iterative method.

4. Find a real root of each of the following equations, using the iterative method.

   (i) $e^x = \cot x$ (ii) $1 + x - x^3 = 0$

(iii) $\sin x = 10(x - 1)$ (iv) $x^2 - 1 = \sin^2 x$

5. If $F(x)$ is sufficiently differentiable and the iteration $x_{n+1} = F(x_n)$ converges, show that the order of convergence is a positive integer.

6. Find a real root of the equation $x^3 + x^2 - 1 = 0$ in (0, 1) with an accuracy of 10^{-4} by iteration method.

Answers

1. 5.477225575   **2.** 2.108   **3.** 2.466

4. (i) 0.5314   (ii) 1.466   (iii) 1.088   (iv) 1.404

10.5 REGULA-FALSI METHOD (OR METHOD OF FALSE POSITION)

This method also known as method of false position is the oldest method of finding the real roots of the equation $f(x) = 0$ and it is somewhat similar to the bisection method.

A graphical description of the false position method is shown in the adjoining fig.3.

Consider the equation $f(x) = 0$. Let x_0 and $x_1(x_0 < x_1)$ be two values of x such that $f(x_0)$ and $f(x_1)$ are of opposite signs. Then the graph of $y = f(x)$ crosses the x-axis at some point between x_0 and x_1.

$\therefore$ The equation of the chord joining two points $A(x_0, f(x_0))$ and $B(x_1, f(x_1))$ is

$$y - f(x_0) = \frac{f(x_1) - f(x_0)}{x_1 - x_0}(x - x_0) \qquad \ldots(1)$$

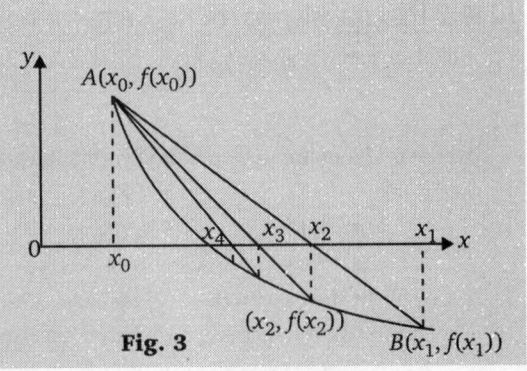

Fig. 3

Since equation (1) intersects the x-axis between the points, where $y = 0$. Thus, we get

$$-f(x_0) = \frac{f(x_1) - f(x_0)}{x_1 - x_0}(x - x_0) \quad \Rightarrow \quad x = \frac{x_0 f(x_1) - x_1 f(x_0)}{f(x_1) - f(x_0)}$$

Therefore, the first approximation of the root is given by

$$x_2 = \frac{x_0 f(x_1) - x_1 f(x_0)}{f(x_1) - f(x_0)} = x_0 - \frac{f(x_0)(x_1 - x_0)}{f(x_1) - f(x_0)} \qquad \qquad ...(2)$$

In general

$$x_{n+1} = x_{n-1} - \frac{f(x_n)(x_n - x_{n-1})}{f(x_n) - f(x_{n-1})} \qquad \qquad ...(3)$$

WORKING PROCEDURE

If $f(x_2)$ and $f(x_0)$ are of opposite sign, then the root lies in between x_0 and x_2, therefore we replace x_1 and x_2 in equation (2) and get the next approximation x_3. On the other hand, if $f(x_2)$ and $f(x_0)$ are of the same sign then $f(x_2)$ and $f(x_1)$ will be of opposite signs and therefore, the root lies between x_2 and x_1. Hence, we replace x_0 by x_2 in equation (2) and get the next approximation x_3. This process is to be repeated, until we get the root to desired accuracy.

REMARKS
- In this method, we choose such points x_0 and x_1 for which $f(x_0)f(x_1) < 0$.
- We choose the points x_0 and x_1 such that they form sufficiently small interval, then in this interval the curve is considered as a straight line. Thus, it is called false position method.
- The rate of convergence of false position method is faster than that of the bisection method.
- It is also called linear interpolation method.

Solved Examples

EXAMPLE 1. *Find a real root of the equation* $f(x) = x^3 - 2x - 5 = 0$ *by the method of false position upto three places of decimals.*

(ROHTAK–2006, MANIPAL–2005, TAMILNADU(MCA)–2005, 08, IAS–2002)

SOLUTION. Given that $f(x) = x^3 - 2x - 5 = 0$ ...(1)

So that $f(2) = (2)^3 - 2(2) - 5 = -1$

and $f(3) = (3)^3 - 2(3) - 5 = 16$

Therefore, a root lies between 2 and 3.

First approximation:

∴ Taking $x_0 = 2, x_1 = 3, f(x_0) = -1, f(x_1) = 16$, then by Regula-Falsi method,

We get $x_2 = x_0 - \dfrac{x_1 - x_0}{f(x_1) - f(x_0)} . f(x_0)$

$= 2 - \dfrac{3-2}{16+1} . (-1) = 2 + \dfrac{1}{17} = 2.0588$

Now, $f(x_2) = f(2.0588)$

$= (2.0588)^3 - 2(2.0588) - 5 = -0.3911$

[Using (1)]

Therefore, root lies between 2.0588 and 3.

Second approximation:

Now, taking $x_0 = 2.0588$

and $x_1 = 3, f(x_0) = -0.3911, f(x_1) = 16$

Then we get

$x_3 = x_0 - \dfrac{x_1 - x_0}{f(x_1) - f(x_0)} . f(x_0)$

$= 2.0588 - \dfrac{3 - 2.0588}{16 + 0.3911} \times (-0.3911)$

$= 2.0588 + 0.0225 = 2.0813$

Now, $f(x_3) = f(2.0813)$

$= (2.0813)^3 - 2(2.0813) - 5$ [Using (1)]

$= 9.0158 - 9.1626 = -0.1468$

∴ Root lies between 2.0813 and 3.

Third approximation : Taking $x_0 = 2.0813$ and

$x_1 = 3, f(x_0) = -0.1468, f(x_1) = 16$. Then by false position method, we get

$x_4 = x_0 - \dfrac{x_1 - x_0}{f(x_1) - f(x_0)} . f(x_0)$

$= 2.0813 - \dfrac{3 - 2.0813}{16 + 0.1468} \times (-0.1468)$

$= 2.0813 + 0.0084 = 2.0897$

Now, $f(x_4) = f(2.0897) = (2.0897)^3 - 2(2.0897) - 5$

[Using (1)]

$= 9.1254 - 9.1794 = -0.054$

Fourth approximation: The root lies between 2.0897 and 3.

Therefore, taking $x_0 = 2.0897, x_1 = 3, f(x_0) = -0.054$ and $f(x_1) = 16$.

Then by false position method, we have

$x_5 = x_0 - \dfrac{x_1 - x_0}{f(x_1) - f(x_0)} . f(x_0)$

$= 2.0897 - \dfrac{3 - 2.0897}{16 + 0.054} \times (-0.054)$

$= 2.0897 + 0.0031 = 2.0928$

Now, $f(x_5) = f(2.0928)$

$= (2.0928)^3 - 2(2.0928) - 5$

$= 9.1661 - 9.1856 = -0.0195$

Fifth approximation: The root lies between 2.0928 and 3 so taking $x_0 = 2.0928, x_1 = 3, f(x_0) = -0.0195$

and $f(x_1) = 16$. Then

$x_6 = x_0 - \dfrac{x_1 - x_0}{f(x_1) - f(x_0)} . f(x_0)$

$= 2.0928 - \dfrac{3 - 2.0928}{16 + 0.0195} \times (-0.0195)$

$= 2.0928 + 0.0011 = 2.0939$

Now, $f(x_6) = f(2.0939) = (2.0939)^3 - 2(2.0939) - 5$

$= 9.1805 - 9.1879 = -0.0074$

Thus, the root lies between 2.0939 and 3.

Sixth approximation:

Taking $x_0 = 2.0939, x_1 = 3, f(x_0) = -0.0074$

and $f(x_1) = 16$. Then, we get

$$x_7 = x_0 - \frac{x_1 - x_0}{f(x_1) - f(x_0)} \cdot f(x_0)$$

$$= 2.0939 - \frac{3 - 2.0939}{16 + 0.0074} \times (-0.0074)$$

$$= 2.0939 + 0.00042 = 2.0943$$

Now $f(x_7) = f(2.0943) = (2.0943)^3 - 2(2.0943) - 5$

$$= 9.1858 - 9.1886 = -0.028$$

Seventh approximation: The root lies between 2.0943 and 3, so taking $x_0 = 2.0943, x_1 = 3, f(x_0) = -0.0028$ and $f(x_1) = 16$. Then

$$x_8 = x_0 - \frac{x_1 - x_0}{f(x_1) - f(x_0)} \cdot f(x_0)$$

$$= 2.0943 - \frac{3 - 3.0943}{16 + 0.0028}(-0.0028)$$

$$= 2.0943 + 0.00016 = 2.0945$$

Hence, the root is 2.094 correct to three decimal places.

EXAMPLE 2. *Find a real root of the equation* $3x + \sin x - e^x = 0$ *by false position method.* (UPTU(MCA)–2007)

SOLUTION. We have $f(x) = 3x + \sin x - e^x = 0$

$$f(0.3) = -0.154 \ (-\text{ve})$$
$$f(0.4) = 0.975 \ (+\text{ve})$$

$\Rightarrow$ Root lies between 0.3 and 0.4.

First Approximation : Using Regula-Falsi method, we have

$$x_2 = x_0 - \frac{x_1 - x_0}{f(x_1) - f(x_0)} f(x_0)$$

$$= 0.3 - \frac{0.4 - 0.3}{(0.0975) - (-0.154)}(-0.154)$$

$$= 0.3 + \left(\frac{0.1 \times 0.154}{0.2515}\right) = 0.3612$$

Also, $f(x_2) = f(0.3612) = 0.0019 \ (+\text{ve})$

$\Rightarrow$ Root lies between 0.3 and 0.3612.

Second Approximation: We have

$$x_3 = x_0 - \frac{x_2 - x_0}{f(x_2) - f(x_0)} f(x_0)$$

$$= 0.3 - \left\{\frac{0.3612 - 0.3}{0.0019 - (-0.154)}\right\}(-0.154)$$

$$= 0.3 + \left(\frac{0.0612}{0.1559}\right)(0.154) = 0.3604$$

Also, $f(x_3) = f(0.3604) = -0.00005 \ (-\text{ve})$

$\Rightarrow$ Root lies between 0.3604 and 0.3612.

Third Approximation: We have

$$x_4 = x_3 - \frac{x_2 - x_3}{f(x_2) - f(x_3)} f(x_3)$$

$$= 0.3604 - \left\{\frac{0.3612 - 0.3604}{0.0019 - (-0.00005)}\right\}(-0.00005)$$

$$= 0.3604 - \left(\frac{0.0008}{0.00195}\right)(0.00005) = 0.36042$$

We observe that x_3 and x_4 are approximately the same. Hence, the required real root is 0.3604 correct to four decimal places.

EXAMPLE 3. *Find the root of the equation* $xe^x = \cos x$ *in the interval (0, 1) using Regula-Falsi method correct to four decimal places.* (BHOPAL–2009)

SOLUTION. We have

$$f(x) = \cos x - xe^x = 0 \qquad \dots(1)$$
$$\therefore \quad f(0) = 1, f(1) = \cos 1 - e = -2.17798$$

$\Rightarrow$ Root lies between 0 and 1.

Then, by Regula-Falsi method.

$$x_2 = x_0 - \frac{(x_1 - x_0)}{f(x_1) - f(x_0)} f(x_0)$$

$$= 0 - \frac{1 - 0}{-3.17798} \cdot 1 = 0.31467$$

Also, $f(x_2) = f(0.31467) = 0.51987$

$\Rightarrow$ Root lies between 0.31487 and 1.

Thus, $x_3 = 0.31487 - \frac{(1 - 0.31487)}{(-2.17798 - 0.51987)}(0.51987)$

$$= 0.44673$$

Also, $f(x_3) = 0.20356$

Repeating this process upto a required number of times, we get the required root is 0.5177 upto 4 decimal places.

EXAMPLE 4. *Using the method of false position, find the root of the equation* $x^6 - x^4 - x^3 - 1 = 0$ *upto four decimal places.* (NAGARJUNA–2001)

SOLUTION. We have $f(x) = x^6 - x^4 - x^3 - 1$

Then $f(1.4) = -0.056$ and $f(1.41) = 0.102$

$\Rightarrow$ Root lies between 1.4 and 1.41

First Approximation : By Regula-Falsi method

$$x_2 = x_0 - \frac{x_1 - x_0}{f(x_1) - f(x_0)} f(x_0)$$

$$= 1.4 - \left(\frac{1.41 - 1.4}{0.102 + 0.056}\right)(-0.056)$$

$$= 1.4 + \left(\frac{0.01}{0.158}\right)(0.056) = 1.4035$$

Now, $f(x_2) = -0.0016$ which is negative.

$\Rightarrow$ Root lies between 1.4035 and 1.41

Second Approximation : Again by Regula-Falsi method

$$x_3 = x_2 - \frac{x_1 - x_2}{f(x_1) - f(x_2)} f(x_2)$$

$$= 1.4035 - \left(\frac{1.41 - 1.4035}{0.102 + 0.0016}\right)(-0.0016)$$

$$= 1.4035 + \left(\frac{0.0065}{0.1036}\right)(0.0016) = 1.4036$$

Now, $f(x_3) = -0.0003$, which is negative.

$\Rightarrow$ Root lies between 1.4036 and 1.41.

Third Approximation : By Regula-Falsi method

$$x_4 = x_3 - \left\{\frac{x_1 - x_3}{f(x_1) - f(x_3)}\right\} f(x_3)$$

$$= 1.4036 - \left(\frac{1.41 - 1.4036}{0.102 + 0.00003}\right)(-0.00003)$$

$$= 1.4036 + \left(\frac{0.0064}{0.10203}\right)(0.00003) = 1.4036$$

Since x_3 and x_4 are approximately equal. Hence, the required root is given by 1.4036.

EXAMPLE 5. *Find a real root of the equation $x \log_{10} x = 1.2$ by Regula-Falsi method correct to four decimal places.* (UPTU(MCA)–2004, AGRA–2006, VTU–2003, 10, JNTU–2008, KOTTAYAM–2005, MUMBAI–2004, BURDWAN–2003)

SOLUTION. We have

$$f(x) = x \log_{10} x - 1.2$$

$$\therefore \quad f(2.74) = -0.0005634$$

and $f(2.741) = -0.0003087$

$\Rightarrow$ **Root Aproximation:** By Regula-Falsi method, We have

First Approximation : By Regula-Falsi method, we have

$$x_2 = x_0 - \left\{ \frac{x_1 - x_0}{f(x_1) - f(x_0)} \right\} \cdot f(x_0)$$

$$= 2.74 - \left\{ \frac{2.741 - 2.74}{0.0003087 - (-0.0005634)} \right\}$$

$$(-0.0005634)$$

$$= 2.74 - \left\{ \frac{0.001}{0.0008721} \right\} (0.0005634)$$

$$= 2.740646027$$

Also, $f(x_2) = -0.00000006016$, *i.e.*, negative

$\Rightarrow$ Root lies between 2.740646027 and 2.741.

Second Approximation : We have

$$x_3 = x_2 - \left\{ \frac{x_1 - x_2}{f(x_1) - f(x_2)} \right\} f(x_2)$$

$$= 2.740646027 - \left(\frac{2.741 - 2.740646027}{0.0003087 + 0.000000060616} \right)$$

$$- (0.000000060616)$$

$$= 2.740646096$$

We observe that x_2 and x_3 agree upto seven decimal places. Hence, the required root correct to four decimal places is 2.7406.

EXAMPLE 6. *Find the smallest positive root of the equation $x - e^{-x} = 0$, using false position method.* (UPTU(MCA)–2003)

SOLUTION. We have $f(x) = x - e^{-x}$

$$\therefore \quad f(0.56) = -0.01121 \text{ and } f(0.58) = 0.201$$

$\Rightarrow$ Root lies between 0.56 and 0.58.
Let $x_0 = 0.56$ and $x_4 = 0.58$
First Approximation: We have

$$x_2 = x_0 - \left\{ \frac{x_1 - x_0}{f(x_1) - f(x_0)} \right\} \cdot f(x_0)$$

$$= 0.56 - \left(\frac{0.58 - 0.56}{0.0201 + 0.1121} \right)(-0.01121)$$

$$= 0.56716$$

Also, $f(x_2) = 0.00002619$, *i.e.*, positive

$\Rightarrow$ Root lies between 0.56 and 0.56716.
Second Approximation: We have

$$x_3 = x_0 - \left\{ \frac{x_2 - x_0}{f(x_2) - f(x_0)} \right\} \cdot f(x_0)$$

$$= 0.56 - \left(\frac{0.56716 - 0.56}{0.00002619 + 0.1121} \right)(-0.01121)$$

$$= 0.567143$$

We observe that x_2 and x_3 agree upto four decimal places. Hence, the required root correct to three decimal places is 0.567.

EXAMPLE 7. *The equation $x^6 - x^4 - x^3 - 1 = 0$ has one real root between 1.4 and 1.5 using false position method find the root (three iterations)* (UPTU(MCA)–2001)

SOLUTION. $f(x) = x^6 - x^4 - x^3 - 1$
$f(1.4) = -2.20736$ and $f(1.5) = 1.953125$
Hence, root lies between 1.4 and 1.5.
First Approximation :
Taking $x_0 = 1.4, x_1 = 1.5, f(x_0) = -2.20736$ in the formula

$$x_2 = x_0 - \frac{(x_1 - x_0)}{f(x_1) - f(x_0)} f(x_0) \qquad \dots(1)$$

$$= 1.4 - \frac{(1.5 - 1.4)(-2.20736)}{1.953125 + 2.20736}$$

$$= 1.453055353$$

$f(x_2) = 0.86419171$
i.e., the root lies between 1.4 and 1.453055353
Second Approximation :
Taking $x_0 = 1.40, x_1 = 1.453055353, f(x_0) = -2.20736$
$f(x1) = 0.8641917$ and putting in (1)

$$x_3 = 1.4 - \frac{(1.453055353 - 1.4)}{0.86419171 + 2.29736}(-2.20736)$$

$$= 1.4 + \frac{(0.053055353)}{3.07155171}(2.20736)$$

$$= 1.438128046$$

$f(x_3) = 0.594926086$
Hence, root lies between 1.4 and 1.438128046.
Third Approximation: Taking

$$x_0 = 1.4, x_1 = 1.438128046, f(x_0) = -2.20736,$$

$f(x_1) = 0.594926086$ becomes (1)

$$x_4 = 1.4 - \frac{0.038128}{2.802286086}(-2.20736)$$

$$= 1.4 + 0.30033415 = 1.430033415$$

$f(x_4) = -0.445778722.$
Hence, root lies between 1.430033415 and 1.438128046.
Hence, approximate root is 1.430.

Exercise-10.3

1. Find a positive root if $xe^x = 2$ by the method of false position. (SVTU–2007)
2. Find real cube root of 18 by Regula-Falsi method.
3. Find the real root of $f(x) = x^3 - 98$ by Regula-Falsi method within $E_8 = 0.01\%$.

4. Find the real root of the following equations:
   (i) $x^4 - x^3 - 2x^2 - 6x - 4 = 0$
   (ii) $x = \tan x$
   (iii) $(5 - x)e^x = 5$ near $x = 5$

(iv) $x^4 + x^3 - 7x^2 - x + 5 = 0$ lying 2 and 3 　　　(vi) $x^2 - 9 = 0$ 　　　(UPTU(MCA)–2008)
(v) $x^3 - 3x + 4 = 0$ between –2 and –3

Answers

1. 0.852605 　　　　　　　　　　　　　**2.** 2.620741394 　　　　　　　　**3.** 4.6104
4. (i) 2.7320506 　　(ii) 4.4934 　　(iii)4.9651142 　　(iv) 2.0608526 　　(v) –2.195823345 　　(vi) 3

10.6 NEWTON-RAPHSON'S METHOD

One of the most widely used methods of solving equations is Newton's method. Like the previous ones, this method is also based on a linear approximation of the function, but does so using a tangent to the curve.

10.6.1 DERIVATION OF NEWTON-RAPHSON METHOD USING TAYLOR'S SERIES EXPANSION

Let x_n be an estimate of a root of the function $f(x)$. Also, let h be a small interval such that

$$h = x_{n+1} - x_n \qquad \qquad ...(1)$$

Now by Taylor's series expansion, we have

$$f(x_{n+1}) = f(x_n + h) = f(x_n) + hf'(x_n) + \frac{h^2}{2!} f''(x_n) + ...$$

Since h is small we can neglect second, third and higher degree terms in h. Therefore, we get

$$f(x_{n+1}) = f(x_n) + hf'(x_n)$$

If x_{n+1} is a root of $f(x)$, then 　　　$f(x_{n+1}) = 0$

which implies 　$f(x_n) + hf'(x_n) = 0$ 　　　　$\Rightarrow$ 　　　$h = -\frac{f(x_n)}{f'(x_n)}$

Now using (1), we get $h = x_{n+1} - x_n = -\frac{f(x_n)}{f'(x_n)}$ 　　$\Rightarrow$ 　　$x_{n+1} = x_n - \frac{f(x_n)}{f'(x_n)}$

WORKING PROCEDURE

Step 1. Take initial approximation say x_0.
Step 2. Find $f(x_0)$ and $f'(x_0)$.
Step 3. Compute next approximation using $x_1 = x_0 - \frac{f(x_0)}{f'(x_0)}$.

Step 4. Check the accuracy of latest approximation.
Step 5. Using above procedure, find other successive approximation x_{i+1}.

REMARKS

- If $f'(x_0)$ is large, *i.e.*, the graph of $y = f(x)$ is nearly vertical to the x-axis, while crossing it, then Newton's method is very useful.
- This method can also be used when the roots are complex.
- The convergence of Newton's method for double root is linear.
- Newton's method is conditionally convergent.
- Newton's method converges if $|f(x)f''(x)| < |f'(x)|^2$.

10.6.2 GEOMETRICAL INTERPRETATION OF NEWTON'S METHOD

Let x_0 be an approximation very close to the root a of the equation $f(x) = 0$. Let $A(x_0, f(x_0))$ be the point on curve $y = f(x)$. Then the equation of the tangent at $A(x_0, f(x_0))$ is given by

$$y - f(x_0) = f'(x_0)(x - x_0)$$

The line cuts the x-axis where $y = 0$ then

$$x_1 = x_0 - \frac{f(x_0)}{f'(x_0)}$$

The point x_1 is a first approximation to the root a. Let $A_1(x_1, f(x_1))$, be a point corresponding to x_1, then the tangent at A_2 cuts the x-axis at x_2 which is second approximation to the root. Continue this process, we get better approximation to the root a.

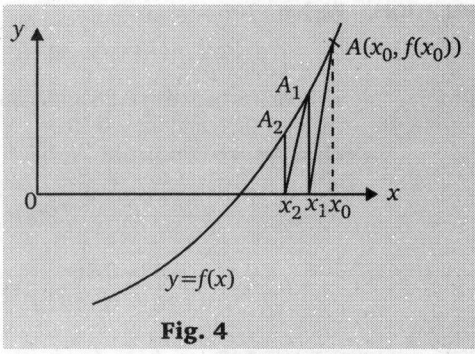

Fig. 4

10.6.3 RATE OF CONVERGENCE OF NEWTON'S METHOD

(UPTU–2002, 07, UPTU(MCA)–02, 03, 06, 08)

Let x_n be different from the root a of $f(x) = 0$ by a small quantity $\in_n$ so that

and 　　　　　$\left.\begin{array}{r} x_n = a + \in_n \\ x_{n+1} = a + \in_{n+1} \end{array}\right\}$ 　　　　　...(1)

Since Newton's Method gives

$$x_{n+1} = x_n - \frac{f(x_n)}{f'(x_n)} \qquad \qquad \text{...(2)}$$

From (1) and (2) we get,

$$a + \epsilon_{n+1} = a + \epsilon_n - \frac{f(a + \epsilon_n)}{f'(a + \epsilon_n)} \quad \Rightarrow \quad \epsilon_{n+1} = \epsilon_n - \frac{f(a + \epsilon_n)}{f'(a + \epsilon_n)}$$

Using Taylor's theorem

$$\Rightarrow \quad \epsilon_{n+1} = \epsilon_n - \frac{f(a) + \epsilon_n f'(a) + \frac{\epsilon_n^2}{2!} f''(a) + ...}{f'(a) + \epsilon_n f''(a) + \frac{\epsilon_n^2}{2!} f'''(a) + ...} = \epsilon_n - \frac{\epsilon_n f'(a) + \frac{\epsilon_n^2}{2!} f''(a) + ...}{f'(a) + \epsilon_n f''(a) + \frac{\epsilon_n^2}{2!} f'''(a) + ...} \qquad (\because f(a) = 0)$$

$$= \frac{\frac{\epsilon_n^2}{2!} f''(a)}{f'(a) - \epsilon_n f''(a)} \qquad \qquad \text{(Neglecting third and higher power of } \epsilon_n)$$

$$= \frac{\epsilon_n^2 f''(a)}{2 f'(a)} \left[1 + \frac{\epsilon_n f''(a)}{f'(a)} \right]^{-1}$$

$$\therefore \quad \epsilon_{n+1} = \frac{\epsilon_n^2 f''(a)}{2 f'(a)} \qquad \text{(Expand by Binomial theorem and neglecting } \epsilon_n^2 \text{ and higher powers of } \epsilon_n)$$

$$\therefore \quad \epsilon_{n+1} \propto \epsilon_n^2$$

This shows that subsequent error is proportional to the square of the previous error. Hence, the convergence is of second order or a quadratic. Hence, rate of convergence of Newton-Raphson method is quadratic.

10.6.4 GENERALISED NEWTON'S METHOD FOR MULTIPLE ROOTS

(UPTU(MCA)–2003)

Let a be the root of the equation $f(x) = 0$ of order m. Then

$$x_{n+1} = x_n - m \frac{f(x_n)}{f'(x_n)}$$

which is called the generalised Newton's Method.

If $m = 1$ this formula gives the Newton-Raphson's method. In this case, the convergence of Newton's Method is $\frac{m-1}{m}$.

10.6.5 LIMITATIONS OF NEWTON-REPHSON'S METHOD

Newton-Raphson method is not applicable in the following situations:
1. Division by zero may occur if $f'(x_i)$ is zero.
2. If the initial guess is too far away from the required root, the process may converge to some other root.
3. A particular value in the iteration sequence may repeat, resulting in an infinite loop.

☞ REMARK

- The Newton-Rephson method approximate the curves of $f(x)$ by tangents. Some complication will arise if the derivatives $f(x_n)$ is zero. In such cases a new initial value for x must be chosen to continue the procedure.

📖 Solved Examples

EXAMPLE 1. *Find a real root of the equation $3x = \cos x + 1$ by Newton's method.* (UPTU(MCA)–2004, ROHTAK–2005)

SOLUTION. Let $f(x) = 3x - \cos x - 1$ ...(1)

So that $f(0) = -2$

and $f(1) = 3 - \cos 1 - 1 = 3 - 0.5403 - 1 = 1.4597$

So a root of the equation $f(x) = 0$ lies between 0 and 1.

Let us take $x_0 = 0.6$

From (1) $f'(x) = 3 + \sin x$ ...(2)

Therefore, the Newton's Method gives

$$x_{n+1} = x_n - \frac{f(x_n)}{f'(x_n)} = x_n - \frac{3x_n - \cos x_n - 1}{3 + \sin x_n}$$

[From (1) and (2)]

$$\therefore \quad x_{n+1} = \frac{x_n \sin x_n + \cos x_n + 1}{3 + \sin x_n} \qquad \text{...(3)}$$

First approximation: Putting $n = 0$, we get first approximation

$$x_1 = \frac{x_0 \sin x_0 + \cos x_0 + 1}{3 + \sin x_0} = \frac{0.6 \sin(0.6) + \cos(0.6) + 1}{3 + \sin(0.6)}$$

$$= \frac{0.6(0.5646) + 0.8253 + 1}{3 + 0.5646} = \frac{2.16406}{3.5646} = 0.6071$$

Second approximation: Putting $n = 1$ into (3), we get second approximation

$$x_2 = \frac{x_1 \sin x_1 + \cos x_1 + 1}{3 + \sin x_1}$$

$$= \frac{0.6071 \sin(0.6071) + \cos(0.6071) + 1}{3 + \sin(0.6071)}$$

$$= \frac{0.6071(0.5705) + 0.8213 + 1}{3 + 0.5705} = \frac{2.1677}{3.5705} = 0.6071$$

Hence $x_1 = x_2$. Then root is 0.6071 correct to four decimal places.

EXAMPLE 2. *Find the real root of the equation* $x \log_{10} x = 1.2$ *by Newton-Raphson's Method.*

(KURUKSHETRA(NIT)–2011, 13, UPTU(AG.)–2004, ROHTAK–2004, 06)

SOLUTION. Let $f(x) = x \log_{10} x - 1.2 = 0$

Then $f(1) = -1.2$

$\quad\quad f(2) = 2\log_{10} 2 - 1.2 = -0.5979$

and $f(3) = 3\log_{10} 3 - 1.2 = 0.2314$

$\therefore$ A root of $f(x) = 0$ lies between 2 and 3.

Let us take $\quad x_0 = 2$

From (1), we have

$$f'(x) = \log_{10} x + \frac{1}{x}.x \log_{10} e = \log_{10} x + 0.4343 \quad ...(2)$$

Now by Newton's Method we have

$$x_{n+1} = x_n - \frac{f(x_n)}{f'(x_n)}$$

$$x_{n+1} = x_n - \frac{x_n \log_{10} x_n - 1.2}{\log_{10} x_n + 0.4343} \quad \text{[From (1) and (2)]}$$

$$x_{n+1} = \frac{0.4343 x_n + 1.2}{\log_{10} x_n + 0.4343}, n = 0,1,2,... \quad ...(3)$$

Putting $n = 0$, we get the first approximation.

First approximation:

$$x_1 = \frac{0.4343 x_0 + 1.2}{\log_{10} x_n + 0.4343}$$

$$= \frac{0.4343(2) + 1.2}{\log_{10}(2) + 0.4343} = \frac{2.0686}{0.7353} = 2.8133$$

Putting $n = 1$ into (3), we get second approxi-mation.

Second approximation:

$$x_2 = \frac{0.4343 x_1 + 1.2}{\log_{10} x_1 + 0.4343}$$

$$= \frac{0.4343(2.8133) + 1.2}{\log_{10} 2.8133 + 0.4343} = \frac{2.4128}{0.8835} = 2.7411$$

Putting $n = 2$ into (3), we get third approximation.

Third approximation:

$$x_3 = \frac{0.4343 x_2 + 1.2}{\log_{10} x_2 + 0.4343}$$

$$= \frac{0.4343(2.7411) + 1.2}{\log_{10} 2.7411 + 0.4343} = \frac{2.3905}{0.8722} = 2.7408$$

Putting $n = 3$ into (3), we get fourth approxim-ation.

Fourth approximation:

$$x_4 = \frac{0.4343 x_3 + 1.2}{\log_{10} x_3 + 0.4343}$$

$$= \frac{0.4343(2.7408) + 1.2}{\log_{10} 2.7408 + 0.4343} = \frac{2.3903}{0.8721} = 2.7408$$

Since, $x_3 = x_4$. Hence, the root of $f(x) = 0$ is 2.7408 correct to four decimal places.

EXAMPLE 4. *Apply Newton's formula to prove that the recurrence formula for finding the nth roots of*

a *is* $x_{i+1} = \dfrac{(n-1)x_i^n + a}{n(x_i^{n-1})}$. *Hence, evaluate*

$(240)^{1/5}$.

SOLUTION. Let $\quad x = a^{1/n} \quad \Rightarrow \quad x^n = a$ or $x^n - a = 0$

Let $\quad f(x) = x^n - a = 0 \quad \Rightarrow \quad f'(x) = nx^{n-1}$

Now, by Newton-Raphson Method, we have

$$x_{i+1} = x_i - \frac{f(x_i)}{f'(x_i)} = x_i - \frac{x_i^n - a}{nx_i^{n-1}}$$

$$\Rightarrow \quad x_{i+1} = \frac{nx_i^n - x_i^n + a}{nx_i^{n-1}} = \frac{(n-1)x_i^n + a}{nx_i^{n-1}} \quad ...(1)$$

Now to find the value of $(240)^{1/5}$.

We know that $(243)^{1/5} = (3^5)^{1/5} = 3$

Take $a = 240$ and $n = 5$ in (1), we get

$$x_{i+1} = \frac{4x_i^5 + 240}{5x_i^4} \quad ...(2)$$

First approximation: Let $i = 0, x_i = x_0 = 2.9$ (say), then from (2), we get

$$x_i = \frac{4x_0^5 + 240}{5x_0^4} = \frac{4(2.9)^5 + 240}{5(2.9)^4}$$

$$= \frac{4(205.111) + 240}{5 \times 70.7281} = \frac{1060.444}{353.6405} = 2.99$$

Second approximation: Let $i = 1, x_i = x_1 = 2.99$ (say), then from (2), we get

$$x_2 = \frac{4x_i^5 + 240}{5x_i^4} = \frac{4(2.99)^5 + 240}{5(2.99)^4}$$

$$= \frac{4(238.977) + 240}{399.627} = 2.9925$$

Hence, the required value of $(240)^{1/5}$ correct to three places of decimal is 2.993.

EXAMPLE 4. *Prove the Chebyshev formula*

$$x_1 = x_0 - \frac{f(x_0)}{f'(x_0)} - \frac{1}{2} . \frac{[f(x_0)]^2 . f''(x_0)}{[f'(x_0)]^3}$$

for the roots of the equation $f(x) = 0$.

SOLUTION. Let $f(x) = 0$ be the given equation, therefore

$$f(x) = f(x + x_0 - x_0) = 0$$

$$= f(x_0) + (x - x_0)f'(x_0)... = 0$$

(On expanding by Taylor's series)

$$\Rightarrow \quad x_1 = x_0 - \frac{f(x_0)}{f'(x_0)}, \text{ which is the first approximation}$$

to the root.

Again by Taylor's series. We have

$$f(x) = f(x_0) + (x - x_0)f'(x_0) + \frac{(x - x_0)^2}{2!}f''(x_0)$$

$$\Rightarrow \quad f(x_1) = f(x_0) + (x - x_0)f'(x_0) + \frac{(x - x_0)^2}{2!}f''(x_0)$$

But $f(x_1) = 0$ as x_1 is an approximation to the root. Therefore,

$$f(x_0) + (x_1 - x_0)f'(x_0) + \frac{1}{2!}(x_1 - x_0)^2 f''(x_0) = 0$$

$$\Rightarrow \quad x_1 = x_0 - \frac{f(x_0)}{f'(x_0)} - \frac{1}{2} . \frac{[f(x_0)]^2 f''(x_0)}{[f'(x_0)]^3}$$

$$\Rightarrow \quad x_1 = x_0 - \frac{f(x_0)}{f'(x_0)} - \frac{1}{2} . \frac{\{f(x_0)\}^2 f''(x_0)}{\{f'(x_0)\}^3}$$

REMARKS
- The above formula is also known as Newton-Raphson's extended formula.
- It can be used interactively.

EXAMPLE 5. *Find a real root of the equation $x = e^{-x}$ by Newton-Raphson method.* (UPTU(MCA)–2008)

SOLUTION. Here, we have
$$f(x) = xe^x - 1 \quad \Rightarrow \quad f'(x) = (1+x)e^x$$
Let $x_0 = 1$

Then $x_1 = 1 - \left(\dfrac{e-1}{2e}\right) = \dfrac{1}{2}\left(1 + \dfrac{1}{e}\right) = 0.6839397$

Now, $f(x_1) = 0.3553424$

and $f'(x_1) = 3.337012$

So that, $x_2 = 0.6839397 - \dfrac{0.3553424}{3.337012} = 0.5774545$

Proceeding in the same way, we get

$x_3 = 0.5672297$ and $x_4 = 0.5671433$

Hence, the required root is 0.5671 correct to four decimal places.

EXAMPLE 6. *Show that the following two sequences, both have convergence of the second order with the same limit $\sqrt{a}$.*

$$x_{n+1} = \frac{1}{2}x_n\left(1 + \frac{a}{x_n^2}\right) \quad \text{and} \quad x_{n+1} = \frac{1}{2}x_n\left(3 - \frac{x_n^2}{a}\right)$$

SOLUTION. We have $x_{n+1} = \dfrac{1}{2}x_n\left(1 + \dfrac{a}{x_n^2}\right)$

$\Rightarrow x_{n+1} - \sqrt{a} = \dfrac{1}{2}x_n\left(1 - \dfrac{a}{x_n^2}\right) - \sqrt{a}$

$\qquad = \dfrac{1}{2}\left(x_n + \dfrac{a}{x_n} - 2\sqrt{a}\right) = \dfrac{1}{2}\left(\sqrt{x_n} - \dfrac{\sqrt{a}}{\sqrt{x_n}}\right)^2$

$\qquad = \dfrac{1}{2x_n}\left(x_n - \sqrt{a}\right)^2$

$\Rightarrow \qquad e_{n+1} = \dfrac{1}{2x_n}e_n^2$

which shows the convergence of second order.

Similarly, $x_{n+1} - \sqrt{a} = \dfrac{1}{2}x_n\left(3 - \dfrac{x_n^2}{a}\right) - \sqrt{a}$

$\qquad = \dfrac{1}{2}x_n\left(1 - \dfrac{x_n^2}{a}\right) + (x_n - \sqrt{a})$

$\qquad = \dfrac{x_n}{2a}(a - x_n^2) + (x_n - \sqrt{a})$

$\qquad = (x_n - \sqrt{a})\left[1 - \dfrac{x_n}{2a}(x_n + \sqrt{a})\right]$

$\Rightarrow \quad e_{n+1} = \dfrac{x_n - \sqrt{a}}{2a}[2a - x_n^2 - x_n\sqrt{a}]$

$\qquad = -\left(\dfrac{x_n - \sqrt{a}}{2a}\right)(x_n - \sqrt{a})(x_n + 2\sqrt{a})$

$\qquad = -\dfrac{(x_n + 2\sqrt{a})}{2a}e_n^2$

which also shows the quadratic convergence.

EXAMPLE 7. *Find the value of p and q so that the rate of convergence of the iterative method is two.*

$$x_{n+1} = px_n + q\cdot\frac{N}{x_n^2} \qquad \text{(UPTU(MCA)–2003)}$$

SOLUTION. Here, we have $x^3 = N$
Let $\quad f(x) = x^3 - N$
If α be the exact root, we have $a^3 = N$
Putting $x_n = \alpha + e_n, x_{n-1} = \alpha + e_{n+1}, N = \alpha^3$

in $x_{n+1} = px_n + q\cdot\dfrac{N}{x_n^2}$

We get $\alpha + e_{n+1} = p(\alpha + e_n) + q\cdot\dfrac{\alpha^3}{(\alpha + e_n)^2}$

$\qquad = p(\alpha + e_n) + q\cdot\dfrac{\alpha^3}{\alpha^2\left(1 + \dfrac{e_n}{\alpha}\right)^2}$

$\qquad = p(\alpha + e_n) + q\alpha\cdot\left(1 + \dfrac{e_n}{\alpha}\right)^{-2}$

$\qquad = p(\alpha + e_n) + q\alpha\cdot\left[1 - 2\dfrac{e_n}{\alpha} + 3\left(\dfrac{e_n}{\alpha}\right)^2 + \dots\right]$

$\qquad = p(\alpha + e_n) + q\alpha - 2qe_n + 3q\dfrac{e_n^2}{\alpha} - \dots$

$\qquad = (p + q - 1)\alpha + (p - 2q)e_n + 0(e_n^2) + \dots$

For the method to become of order as high as possible, *i.e.*, of order 2 we must have
$$p + q = 1 \qquad \text{and} \qquad p - 2q = 0$$

Here $\qquad p = \dfrac{2}{3} \qquad$ and $\qquad q = \dfrac{1}{3}$.

EXAMPLE 8. *Determine p, q and r so that order of the iterative method given by*

$$x_{n+1} = px_n + q\cdot\frac{a}{x_n^2} + r\cdot\frac{a^2}{x_n^5}$$

For computing $a^{1/3}$ becomes as high as possible.

(UPTU(MCA)–2007)

SOLUTION. Let $\qquad x = a^{1/3}$
$$f(x) = x^3 - a = 0$$
$$f'(x) = 3x^2$$
By Newton Raphson method

$$x_{n+1} = x_n - \frac{f(x_n)}{f'(x_0)} = x_n - \frac{x_n^3 - a}{3x_n^2}$$

$$= \frac{3x_n^3 - x_n^3 + a}{3x_n^2} = \frac{2x_n^3 + a}{3x_n^2}$$

$$\therefore \quad x_{n+1} = \frac{2}{3}x_n + \frac{a}{3x_n^2}$$

Comparing the given equation

$$x_{n+1} = px_n + q\cdot\frac{a}{x_n^2} + r\cdot\frac{a^2}{x_n^5}$$

Hence, $p = 2/3, q = 1/3$ and $r = 0$.

Exercise-10.4

1. Using Newton's iterative formula, find the real root of $x \sin x + \cos x = 0$ which is near to $x = \pi$ corect to 3 decimal places.

2. Using Newton's method, find the roots of the following equations.

   (i) $2x - \log_{10} x = 7$ (ii) $x^2 - 25 = 0$

   (iii) $4(x - \sin x) = 1$ (iv) $x^4 + x^2 - 80 = 0$

   (v) $\log x = \cos x$

   (vi) $x^2 + 4 \sin x = 0$ (HAZARIBAGH–2009)

3. Show that $x_{i+1} = \dfrac{1}{3}\left(2x_i + \dfrac{N}{x_i^2}\right)$.

4. Show that the square root can be obtained by the recursion formula $x_{i+1} = x_i\left(1 - \dfrac{x_i^2 - N}{2N}\right)$.

5. Show that the iterative formula for finding the reciprocals of n is $x_{i+1} = x_i(2 - nx_n)$.

6. Show that the modified Newton-Raphson method $x_{n+1} = x_n - \dfrac{2f(x_n)}{f'(x_n)}$ gives a quadratic convergence when $f(x) = 0$ has a pair of double root in nbd of $x = x_n$.

7. Show that the double root of the equation $x^3 - x^2 - x + 1 = 0$ is 1.0001.

8. Show that the equation $f(x) = 1 - xe^{1-x} = 0$ has a double root at $x = 1$. The root is obtained by using the modified Newton-Raphson method with $m = 2$ starting with $x_0 = 0$.

Answers

1. 2.798   2. (i) 3.7892   (ii) 5   (iii) 1.171   (iv) 2.908   (v) 1.303   (vi) –1.934

10.7 COMPLEX ROOTS

If the given equation $f_n(x) = 0$ of degree n has at most M positive roots and almost N negative roots, then the equation $f_n(x) = 0$ has at least $n - (M + N)$ imaginary roots. Here it should be noticed that the imaginary roots always occur in pair.

Solved Examples

EXAMPLE 1. *Solve the equation* $3x^3 - 4x^2 + x + 88 = 0$, *if one root is* $2 + i\sqrt{7}$.

SOLUTION. Here, the given equation is

$$3x^3 - 4x^2 + x + 88 = 0 \qquad \ldots(1)$$

Since, one root is $2 + i\sqrt{7}$ then the other root will be $2 - i\sqrt{7}$. Now we find the equation having these two roots $(x - 2 - i\sqrt{7})(x - 2 + i\sqrt{7}) = 0$

$\Rightarrow \qquad (x - 2)^2 + 7 = 0$

$\Rightarrow \qquad x^2 - 4x + 11 = 0$

Now divide the equation (1) by $x^2 - 4x + 11 = 0$, we get

$$3x + 8 = 0$$

$\therefore \qquad x = -8/3.$

Hence, all the roots are $(2 \pm i\sqrt{7})$ and $-\dfrac{8}{3}$.

10.7.1 NEWTON'S METHOD FOR COMPLEX ROOTS

Let $f(z)$ be a given equation such that $f(z) = 0$.   ...(1)

where $z = x + iy$

Then, (1) can be written as $f(z) = u(x, y) + iv(x, y)$ ...(2)

Where, $u(x, y)$ and $v(x, y)$ are the real and imaginary parts of $f(z)$ respectively.

Thus the problem of finding the roots of the single equation (1) is equivalent to determining the roots of two simultaneous equations

$$\left.\begin{array}{l} u(x, y) = 0 \\ v(x, y) = 0 \end{array}\right] \qquad \ldots(3)$$

Now, assume that (x_i, y_i) is an initial approximation to a solution of (3) and the exact solution is $(x_i + \Delta x, y_i + \Delta y)$

Then, we have

$$\left.\begin{array}{l} u(x_i + \Delta x, y_i + \Delta y) = 0 \\ v(x_i + \Delta x, y_i + \Delta y) = 0 \end{array}\right] \qquad \ldots(4)$$

Expanding (4) by Taylor's series expansion about (x_i, y_i), we get

and

$$\left.\begin{array}{l} u(x_i + \Delta x, y_i + \Delta y) = u(x_i, y_i) + [\Delta x u_x(x_i, y_i)] + [\Delta y u_y(x_i, y_i)] + \ldots = 0 \\ v(x_i + \Delta x, y_i + \Delta y) = v(x_i, y_i) + [\Delta x v_x(x_i, y_i)] + [\Delta y v_y(x_i, y_i)] + \ldots = 0 \end{array}\right] \qquad \ldots(5)$$

where suffixes with respect to x and y represent partial differentiation.

Neglecting the higher powers of Δx and Δy we get

$$\left.\begin{array}{l} u(x_i, y_i) + \Delta x u_x(x_i, y_i) + \Delta y u_y(x_i, y_i) = 0 \\ v(x_i, y_i) + \Delta x v_x(x_i, y_i) + \Delta y v_y(x_i, y_i) = 0 \end{array}\right] \qquad \ldots(6)$$

Now, solving (6) for Δx and Δy we get

$$\Delta x = -\left[\frac{u(x_i,y_i)v_y(x_i,y_i)-v(x_i,y_i)u_y(x_i,y_i)}{J}\right] \quad \text{and} \quad \Delta y = -\left[\frac{u_x(x_i,y_i)v(x_i,y_i)-u(x_i,y_i)v_x(x_i,y_i)}{J}\right]$$

where J is the Jacobian of the functions u and v at (x_i, y_i) and given by

$$J = \begin{vmatrix} u_x & u_y \\ v_x & v_y \end{vmatrix}$$

Thus, an improved approximation (x_{i+1}, j_{i+1}) to the exact solution is given by

$$x_{i+1} = x_i - J^{-1}[u(x_i,y_i)v_y(x_i,y_i)-v(x_i,y_i)u_y(x_i,y_i)]$$

and

$$y_{i+1} = y_i - J^{-1}[u_x(x_i,y_i)v(x_i,y_i)-u(x_i,y_i)v_x(x_i,y_i)] \qquad \text{...(7)}$$

where $i = 0, 1, 2, \ldots$

This is the Newton-Raphson method for two variables and has quadratic convergence.

Let us assume that $f(z)$ is an analytic function of z. Then the function $u(x, y)$ and $v(x, y)$ satisfy the Cauchy-Riemann equation

$$\left.\begin{array}{l} u_x = v_y \\ u_y = -v_x \end{array}\right] \qquad \text{...(8)}$$

Then, the equation (7) becomes

$$x_{i+1} = y_i - \left[\frac{u(x_i,y_i)v_y(x_i,y_i)-v(x_i,y_i)u_y(x_i,y_i)}{u_x^2(x_i,y_i)+v_x^2(x_i,y_i)}\right] \quad \text{and} \quad y_{i+1} = y_i - \left[\frac{v(x_i,y_i)u_x(x_i,y_i)-u(x_i,y_i)v_x(x_i,y_i)}{u_x^2(x_i,y_i)+v_x^2(x_i,y_i)}\right] \qquad \text{...(9)}$$

where $i = 0, 1, 2, \ldots$

Generally, the Newton-Raphson method for (1) may be defined as

$$z_{k+1} = z_k - \frac{f(z_k)}{f'(z_k)}, \quad k = 0, 1, 2, \ldots \qquad \text{... (10)}$$

Then, it is easy to verify that (1) is equivalent to (9) when $f(z)$ is analytic function of z. The initial approximation z_0 in (10) must be a complex number.

Solved Examples

EXAMPLE 1. *Find the complex root of the equation $f(z) = z^3 + 1$ correct to eight decimal places.*

SOLUTION. Here, we have

$$f(z) = z^3 + 1 = 0 \qquad \text{...(1)}$$

Let $z = x + iy$ and $f(z) = u(x, y) + iv(x, y)$
Then (1), reduces to

$$u(x,y) + iv(x,y) = (x + iy)^3 + 1$$

$$\Rightarrow u(x,y) + iv(x,y) = (x^3 - 3xy^2) + i(2x^2y - xy - y^3)$$

Equating real and imaginary part of both the sides, we get

$$u(x,y) = (x^3 - 3xy^2) \quad \text{and} \quad v(x,y) = 2x^2y + xy - y^3$$

Now using the Newton-Raphson method for complex numbers with the initial approximation $z_0 = (0.25, 0.25)$, we get the following successive iteration.

k	z_k	z_{k+1}	$f(z_{k+1})$
0	(0.25, 0.25)	(0.16666667, 2.83333333)	(0.9687, 0.3125(−1))
1	(0.16666667, 2.83333333)	(0.15220505, 1.89374026)	(−0.3009(1), −0.2251(2))
2	(0.15220505, 1.89374026)	(0.19263553, 1.27724322)	(−0.6340, −0.6660(1))
3	(0.19263553, 1.27724322)	(0.31932197, 0.91041889)	(0.6438(−1), −0.1941(1))

4	(0.31932197, 0.91041889)	(0.49252896, 0.83063199)	(0.2385, 0.4761)
5	(0.49252896, 0.83063199)	(0.49983161, 0.86738607)	(0.1000, 0.3140(−1))
6	(0.49983161, 0.86738607)	(0.49999870, 0.86602675)	(−0.3284(−2), −0.2484(−2))
7	(0.49999870, 0.866602675)	(0.50000000, 0.86602540)	(−0.1548(−5), 0.5414(−5))

Which gives that, the second root is

$$(-0.5, -0.86602540).$$

EXAMPLE 2. *Find the roots of $3x^3 - 7x^2 + 10x - 20 = 0$ By taking $x = 3 + i$.*

SOLUTION.
$$f(3+i) = 3(3+i)^3 - 7(3+i)^2 + 10(3+i) - 20$$
$$= 8 + 46i$$
$$f'(x) = 9x^2 - 14x + 10$$

$$\Rightarrow f'(3+i) = 9(3+i)^2 - 14(3+i) + 10$$
$$= 9(8 + 6i) - 42 - 14i + 10 = 40 + 40i$$

Now, $\quad x_1 = x_0 - \dfrac{f(x_0)}{f'(x_0)}$

$$\Rightarrow \quad x_1 = (3+i) - \frac{8+46i}{40+40i} = \frac{72+114i}{40+40i}$$

$$= 2.33 + 0.52i$$
Proceed further in the same manner we can get $x_2, x_3, \ldots$

10.8 LIN-BAIRSTOW'S METHOD

This method is used for finding the complex roots of a polynomial equation $P_n(x) = 0$ with real coefficients, where

$$P_n(x) = a_0 x^n + a_1 x^{n-1} + a_2 x^{n-2} + \ldots + a_{n-1} x + a_n = 0$$

Consider a quadratic polynomial equation

$$P_2(x) = a_0 x^2 + a_1 x + a_2 = 0$$

Clearly, the roots of $P_2(x) = 0$ are given by

$$\frac{-a_1 + \sqrt{a_1^2 - 4a_0 a_2}}{2a_0} \quad \text{and} \quad \frac{-a_1 - \sqrt{a_1^2 - 4a_0 a_2}}{2a_0}$$

Hence, a second degree equation can be solved completely.

10.8.1 TO FIND THE COMPLEX ROOTS OF $P_n(x) = 0, n > 2$

We know that complex roots always occur in pair. If $\alpha + i\beta$ and $\alpha - i\beta$ are two roots of $P_n(x) = 0$, then $(x - \alpha - i\beta)$ and $(x - \alpha + i\beta)$ will be two linear factors of $P_n(x) = 0$.

Therefore, $(x - \alpha - i\beta)(x - \alpha + i\beta)$ or $(x^2 - 2\alpha x + \alpha^2 + \beta^2)$ will be a quadratic factor of $P_n(x)$ which shows that the equation $P_n(x) = 0$ has a pair of complex roots which are obtained from a quadratic factor with real coefficient in polynomial $P_n(x)$.

Since we have to find the complex roots of $P_n(x) = 0$ so if $P_n(x) = 0$ has a real root, then the linear factor corresponding to that real root can be removed by using the method of synthetic division and the remaining polynomial has no real root.

In this method we start with an approximate quadratic factor $x^2 + ax + b$ and find such values of a and b, after applying an iterative process, such that $x^2 + ax + b$ becomes a factor of $P_n(x)$ and then the roots of $x^2 + ax + b$ will be the roots of $P_n(x) = 0$.

If $P_n(x)$ can be written as

$$P_n(x) = (x^2 + ax + b)P_{n-2}(x) + Ax + B \qquad \ldots(1)$$

where $P_{n-2}(x)$ is a polynomial of degree $n - 2$, then $Ax + B$ represents the remainder when $x^2 + ax + b$ is a factor of $P_n(x)$ i.e., when $P_n(x)$ is divided by $x^2 + ax + b$.

From (1) we find that $x^2 + ax + b$ will be a factor of $P_n(x)$ if

$$Ax + B = 0 \qquad \Rightarrow \qquad A = 0 \text{ and } B = 0 \qquad \ldots(2)$$

But both A and B depend on a and b, then the equation (2) can be taken as

$$A(a, b) = 0 \qquad \text{and} \qquad B(a, b) = 0 \qquad \ldots(3)$$

Let $a + \Delta a$ and $b + \Delta b$ be the actual values of a and b that satisfy (3), then

$$\text{and} \qquad \begin{aligned} A(a + \Delta a, b + \Delta b) &= 0 \\ B(a + \Delta a, b + \Delta b) &= 0 \end{aligned} \Biggr\} \qquad \ldots(4)$$

Now expanding (4) by Taylor's series and neglecting the second and higher order term of Δa and Δb, we get

$$\text{and} \qquad \begin{aligned} A(a, b) + \frac{\partial A}{\partial a}\Delta a + \frac{\partial A}{\partial b}\Delta b &= 0 \\ B(a, b) + \frac{\partial B}{\partial a}\Delta a + \frac{\partial B}{\partial b}\Delta b &= 0 \end{aligned} \Biggr\} \qquad \ldots(5)$$

Solving (5) to get the values of Δa and Δb, the values of Δa and Δb thus obtained involving $\frac{\partial A}{\partial a}, \frac{\partial A}{\partial b}, \frac{\partial B}{\partial b}$ and $\frac{\partial B}{\partial b}$.

But A and B are not known explicitly as functions of a and b, so we cannot obtain the values of $\frac{\partial A}{\partial a}, \frac{\partial A}{\partial b}, \frac{\partial B}{\partial a}$ and $\frac{\partial B}{\partial b}$.

Bairstow had developed a method for obtaining the numerical values of $\frac{\partial A}{\partial a}, \frac{\partial A}{\partial b}, \frac{\partial B}{\partial a}$ and $\frac{\partial B}{\partial b}$ which is given as follows :

Since we have

$$P_n(x) = a_0 x^n + a_1 x^{n-1} + a_2 x^{n-2} + \ldots + a_{n-1} x + a_n.$$

If $P_{n-2}(x) = b_0 x^{n-2} + b_1 x^{n-3} + b_2 x^{n-4} + \ldots + b_{n-3} x + b_{n-2}$ then equation (1) becomes

$$a_0 x^n + a_1 x^{n-1} + a_2 x^{n-2} + \ldots + a_{n-1} x + a_n = (x^2 + ax + b)(b_0 x^{n-2} + b_1 x^{n-3} + b_2 x^{n-4} + \ldots + b_{n-3} x + b_{n-2}) + Ax + B$$

Now comparing the coefficients of like powers of x on both sides, we get

$$\text{and} \qquad \left. \begin{aligned} a_0 &= b_0 \\ a_1 &= b_1 + ab_0 \\ a_2 &= b_2 + ab_1 + bb_0 \\ &\cdots\cdots\cdots\cdots\cdots\cdots\cdots \\ a_r &= b_r + ab_{r-1} + ba_{r-2} \\ &\cdots\cdots\cdots\cdots\cdots\cdots\cdots \\ a_{n-2} &= b_{n-2} + ab_{n-3} + bb_{n-4} \\ a_{n-1} &= A + ab_{n-2} + bb_{n-3} \\ a_n &= B + b_{n-2} \end{aligned} \right\} \qquad \ldots(6)$$

Let us setting $b_{-1} = 0 = b_{-2}, b_{n-1} = A$ and $b_n = B - ab_{n-1}$ then equation (5) can be written as

$$a_r = b_r + ab_{r-1} + bb_{r-2}, r = 0, 1, 2, ..., n \quad \text{or} \quad b_r = a_r - ab_{r-1} + bb_{r-2}, r = 0, 1, 2, ..., n \qquad ...(7)$$

Now differentiating (7) partially w.r.t. a and b, we get

$$\frac{\partial b_r}{\partial a} = -b_{r-1} - a\frac{\partial b_{r-1}}{\partial a} - b\frac{\partial b_{r-2}}{\partial a} \qquad ...(8)$$

and

$$\frac{\partial b_r}{\partial a} = -b_{r-2} - a\frac{\partial b_{r-1}}{\partial b} - b\frac{\partial b_{r-2}}{\partial b} \qquad ...(9)$$

Also $b_0 = a_0, b_{-1} = 0, b_{-2} = 0$ so that

$$\frac{\partial b_0}{\partial a} = 0 = \frac{\partial b_0}{\partial b}, \frac{\partial b_{-1}}{\partial a} = 0 = \frac{\partial b_{-1}}{\partial b}, \frac{\partial b_{-2}}{\partial a} = 0 = \frac{\partial b_{-2}}{\partial b} \qquad ...(10)$$

Putting $r = 1$ in (8) and (9), we get

$$\frac{\partial b_1}{\partial a} = -b_0 - a\frac{\partial b_0}{\partial a} - b\frac{\partial b_{-1}}{\partial b} = -b_0 \qquad \text{[Using (10)]} \qquad ...(11)$$

$$\frac{\partial b_1}{\partial a} = -b_{-1} - a\frac{\partial b_0}{\partial b} - b\frac{\partial b_{-1}}{\partial b} = 0 \qquad \text{[Using (10)]} \qquad ...(12)$$

Again put $r = 2$ in (8) and (9), we get

$$\frac{\partial b_2}{\partial a} = -b_1 - a\frac{\partial b_1}{\partial b} - b\frac{\partial b_0}{\partial a} = -b_1 + ab_0 \qquad \text{[Using (10), (11)]} \qquad ...(13)$$

$$\frac{\partial b_2}{\partial b} = -b_0 - a\frac{\partial b_1}{\partial b} - b\frac{\partial b_0}{\partial b} = -b_0 \qquad \text{[Using (10), (12)]} \qquad ...(14)$$

Thus, from (11) and (13), we get

$$\frac{\partial b_2}{\partial b} = \frac{\partial b_1}{\partial a} \qquad ...(15)$$

Again, put $r = 3$ in (9), we get

$$\frac{\partial b_3}{\partial b} = -b_1 - a\frac{\partial b_2}{\partial b} - b\frac{\partial b_1}{\partial b}$$

or

$$\frac{\partial b_3}{\partial b} = -b_1 + ab_0 \qquad \text{[Using (13), (14)]} \qquad ...(16)$$

Thus, from (13) and (16), we get

$$\frac{\partial b_3}{\partial b} = \frac{\partial b_2}{\partial a}$$

Hence, from (15) and (17), we conclude that

$$\frac{\partial b_{r+1}}{\partial b} = \frac{\partial b_r}{\partial a} \text{ for } r = 0, 1, 2, ... n, ...$$

$$\therefore \qquad \frac{\partial b_{r+1}}{\partial b} = \frac{\partial b_r}{\partial a} \text{ for all } r. \qquad ...(18)$$

Now, if we again set $\frac{\partial b_{r+1}}{\partial b} = \frac{\partial b_r}{\partial a} = -c_{r-1}$ for $r = 0, 1, 2, ... n-1, ...$, then from (8) and (9), we get

$$c_{r-1} = b_{r-1} - ac_{r-2} - bc_{r-3} \quad \text{and} \quad c_{r-2} = b_{r-2} - ac_{r-3} - bc_{r-4}.$$

These two relations can be written as a single relation.

$$c_r = b_r - ac_{r-1} - bc_{r-2} \qquad ...(19)$$

where $r = 1, 2, 3, ..., (n-1)$ and $c_{-1} = -\frac{\partial b_0}{\partial a} = 0, c_0 - \frac{\partial b_1}{\partial a} = b_0$.

Next, we shall find the derivatives in terms of c_r's.

Since, we have

$$A = b_{n-1} \quad \text{and} \quad B = b_n + ab_{n-1}$$

Then, we have

$$\frac{\partial A}{\partial a} = \frac{\partial b_{n-1}}{\partial a} = -c_{n-2}, \frac{\partial A}{\partial b} = \frac{\partial b_{n-1}}{\partial b} = -c_{n-3}$$

$$\frac{\partial B}{\partial a} = \frac{\partial b_n}{\partial a} + b_{n-1} + a\frac{\partial b_{n-1}}{\partial b} = -c_{n-1} + b_{n-1} - ac_{n-2} \quad \text{and} \quad \frac{\partial B}{\partial b} = \frac{\partial b_n}{\partial b} + a\frac{\partial b_{n-1}}{\partial b} = -c_{n-2} - ac_{n-3}$$

Now putting the values of $\frac{\partial A}{\partial a}, \frac{\partial A}{\partial b}, \frac{\partial B}{\partial a}$ and $\frac{\partial B}{\partial b}$ in (5), we get.

$$b_{n-1} - c_{n-2}\Delta a - c_{n-3}\Delta b = 0$$

or $$c_{n-2}\Delta a - c_{n-3}\Delta b = b_{n-1} \qquad \text{... (20)}$$

and $$b_n + ab_{n-1} + (-c_{n-1} + b_{n-1} - ac_{n-2})\Delta a + (-c_{n-2} - ac_{n-3})\Delta b = 0$$

or $$b_n - (c_{n-1} - b_{n-1})\Delta a - c_{n-2}\Delta b + a(b_{n-1} - c_{n-2}\Delta a - c_{n-3}\Delta b) = 0$$

or $$b_n - (c_{n-1} - b_{n-1})\Delta a - c_{n-2}\Delta b = 0$$

or $$(c_{n-1} - b_{n-1})\Delta a + c_{n-2}\Delta b = 0 \qquad \text{... (21)}$$

Substituting the values of c_r's and b_r's obtained from (19) and (7) in (20) and (21) and solving them to get the approximate values of Δa and Δb denoted by Δa^* and Δb^* respectively.

Now we shall take the new values $a + \Delta a^*$ and $b + \Delta b^*$ as the initial values and applying the above process again to get the better values of a and b.

In order to find the values of b_r's and c_r's we use the following (synthetic division) scheme :

$a_0(=1)$	a_1	a_2	a_3	...	a_{n-2}	a_{n-1}	a_n	
	$-ab_0$	$-ab_1$	$-ab_2$	...	$-ab_{n-3}$	$-ab_{n-2}$	$-ab_{n-1}$	$-a$
		$-bb_0$	$-bb_1$	...	$-bb_{n-1}$	$-bb_{n-3}$	$-bb_{n-2}$	$-b$
$b_0(=1)$	b_1	b_2	b_3	...	b_{n-2}	b_{n-1}	b_{n-1}	
	$-ac_0$	$-ac_1$	$-ac_2$	...	$-ac_{n-3}$	$-ac_{n-2}$		$-a$
		$-bc_0$	$-bc_1$	...	$-ac_{n-4}$	$-ac_{n-3}$		$-b$
$c_0(=1)$	c_1	c_2	c_3	...	c_{n-2}	c_{n-1}		

Solved Examples

EXAMPLE 1. *Find a quadratic factor of*
$$f(x) = x^4 - 3x^3 + 20x^2 + 44x + 54 = 0$$

SOLUTION. Here, tile quadratic factor will close to $x^2 + 2x + 2$ so we shall start the method with $a = 2, b = 2$.

Now we find b_r's and c_r's by following scheme:

1	-3	20	44	54	
	-2	10	-56	4	-2
		-2	10	-56	-2
1	-5	28	$-2(b_{n-1})$	$2(b_n)$	
	-2	14	-80		-2
		-2	14		-2
1	$-7(c_{n-3})$	$40(c_{n-2})$	$-68(c_{n-1})$		

If Δa and Δb are the corrections in a and b then we have

$$c_{n-2}\Delta a' + c_{n-3}\Delta b = b_{n-1}$$
$$(c_{n-1} - b_{n-1})\Delta a + c_{n-3}\Delta b = b_n$$

Putting the values of $b_n, b_{n-1}, c_{n-1}, c_{n-2}$ and c_{n-3} from table, we get

$$40\Delta a - 7\Delta b = -2 \qquad \text{...(1)}$$
$$-66\Delta a + 40\Delta b = 2 \qquad \text{...(2)}$$

Solving (1) and (2), we get
$$\Delta a = -0.058, \Delta b = -0.0457$$

The new values of a and b are given by
$$a' = a + \Delta a = 2 - 0.058 = 1.942$$

$$b' = b + \Delta b = 2 - 0.0457 = 1.9543$$

Now repeating the same process by taking a' and b'.

1	-3	20	44	54	
	-1.942	9.5974	-53.683	0.0482	-1.942
		-1.9543	9.6582	-54.023	-1.9543

1	-4.942	27.6431	-0.0248	0.0252	
			(b_{n-1})	(b_n)	
	-1.942	13.3687	-75.849		-1.942
		-1.9543	13.4534		-1.9543

1	-6.884	39.0575	-62.4204	
	(c_{n-3})	(c_{n-2})	(c_{n-1})	

If $\Delta a'$ and $\Delta b'$ are the corrections in a' and b', then we have

$$c_{n-2}\Delta a' + c_{n-3}\Delta b = b_{n-1}$$
$$(c_{n-1} - b_{n-1})\Delta a' + c_{n-2}\Delta b' = b_n$$

Putting the values of $b_n, b_{n-1}, c_{n-1}, c_{n-2}, c_{n-3}$ from above table, we get

$$39.0575\Delta a' + 6.884\Delta b' = -0.0248 \qquad \text{... (3)}$$
$$-62.3956\Delta a' + 39.0575\Delta b' = 0.0252 \qquad \text{... (4)}$$

Solving (3) and (4), we get

$$\Delta a' = -0.00073, \Delta b' = -0.00054$$

The new values of a and b are given by
$$a'' = a' + \Delta a' = 1.942 - 0.00073 = 1.94127$$

$$b'' = b' + \Delta b' = 1.9543 - 0.00054 = 1.95376$$

Hence, $x^2 + a''x + b''$,

i.e., $x^2 + 1.94127x + 1.95376$ is the required factor of $f(x)$.

EXAMPLE 2. *Using Lin-Bairstow's method, find the quadratic factor of the polynomial given by*
$$f(x) = x^3 - 2x^2 + x - 2.$$

SOLUTION. Let $x^2 + ax + b$ with $a = 1, b = 1$ be the quadratic factor of $f(x) = x^3 - 2x^2 + x - 2$.

Now we find b_r's and c_r's by the following synthetic scheme:

1	−2	1	−2	
	−1	3	−3	−1
		−1	3	−1
1	−3	$3(=b_{n-1})$	$-2(b_n)$	
	−1	4		−1
		−1		−1
1	−4	6		
$(=c_{n-3})$	$(=c_{n-2})$	$(=c_{n-1})$		

If Δa and Δb are the corrections in a and b, then we have

$$c_{n-2}\Delta a + c_{n-3}\Delta b = b_{n-1}$$

$$(c_{n-1} - b_{n-1})\Delta a + c_{n-2}\Delta b = b_n$$

Putting the values of $b_n, b_{n-1}, c_{n-1}, c_{n-2}$ and c_{n-3} we get

$$-4\Delta a + \Delta b = 3 \qquad \ldots(1)$$

and

$$(6-3)\Delta a - 4\Delta b = -2 \qquad \ldots(2)$$

or

$$3\Delta a - 4\Delta b = -2$$

Solving (1) and (2), we get

$$\Delta a = -0.769, \Delta b = -0.0769$$

Therefore, we get the new values of a and b, given by

$$a' = a + \Delta a = 1 - 0.769 = 0.231$$

$$b' = b + \Delta b = 1 - 0.0769 = 0.9231$$

Now repeating the same process by taking a' and b':

1	−2	1	−2	
	−0.231	0.5154	−0.1368	−0.231
		−0.9231	2.0594	−0.9231
1	−2.231	0.5923	0.0774	
		(b_{n-1})	(b_n)	
	−0.231	0.05687		−0.231
		0.9231		−0.9231
1	−2.462	0.2379		
(c_{n-3})	(c_{n-2})	(c_{n-1})		

If $\Delta a'$ and $\Delta b'$ are the corrections in a' and b', then, we have

$$c_{n-2}\Delta a' + c_{n-3}\Delta b' = b_{n-1}$$

$$(c_{n-1} - b_{n-1})\Delta a' + c_{n-2}\Delta b' = b_n$$

Putting the values of $b_n, b_{n-1}, c_{n-1}, c_{n-2}, c_{n-3}$ we get

$$-2.462\Delta a' - \Delta b' = 0.5923 \qquad \ldots(3)$$

$$-0.3544\Delta a' - 2.462\Delta b' = 0.0774 \qquad \ldots(4)$$

Solving (3) and (4), we get

$$\Delta a = -0.2394, \Delta b' = 0.00289$$

The new values of a' and b' are given by

$$a'' = a' + \Delta a' = 0.231 - 0.2394 = -0.0084 = 0$$

$$b'' = b' + \Delta b' = 0.9231 + .00289 = 0.9256 = 1$$

Hence $x^2 + a''x + b''$, i.e., $x^2 + 1$ is the required quadratic factor of $f(x) = x^3 - 2x^2 + x - 2$.

10.9 GRAEFFE'S ROOT SQUARE METHOD

Sometimes we come across with some algebraic equations which have complex roots. In such cases the root-squaring method of Graeffe is the best to use. This method gives all the roots at once both real and complex.

Graeffe's method is mainly based upon the process in which the given equation is transformed into another whose roots are higher powers of those of the original equation. The roots of the transformed equation are widely separated and because of this fact can easily be found.

10.9.1 THE ROOT-SQUARING PROCESS

In Graeffe's method the given equation is transformed to another by repeated application of a root squaring process. The first step of this process is to separate the even and odd powers of x and then making square of both sides to get second equation whose roots are the squares of the original equation. In the same way the second equation is then transformed into a third equation whose roots are the squares of the roots of second equation. This roots squaring process is continued in this manner until the roots of the last transformed equation are completely separated.

Consider the polynomial equation with real coefficients

$$x^n + a_1 x^{n-1} + a_2 x^{n-2} + a_3 x^{n-3} + \ldots + a_{n-1}x + a_n = 0 \qquad \ldots(1)$$

Separating the even and odd powers of x and squaring, we get

$$(x^n + a_2 x^{n-2} + a_4 x^{n-4} + \ldots)^2 = (a_1 x^{n-1} + a_3 x^{n-3} + \ldots)^2.$$

Putting $x^2 = y$ after simplifying above equation, we get

$$y^n + b_1 y^{n-1} + b_2 y^{n-2} + \ldots + b_{n-1}y + b_n = 0 \qquad \ldots(2)$$

where,

$$\left.\begin{array}{l} b_1 = -a_1^2 + 2a_2 \\ b_2 = a_2^2 - 2a_1 a_3 + 2a_4 \\ \ldots\ldots\ldots\ldots\ldots\ldots\ldots\ldots \\ \ldots\ldots\ldots\ldots\ldots\ldots\ldots\ldots \\ b_n = (-1)^n a_n^2 \end{array}\right\} \qquad \ldots(3)$$

If $\alpha_1, \alpha_2, \alpha_3, \ldots, \alpha_n$ be the roots of equation (1), then the roots of equation (2) are $\alpha_1^2, \alpha_2^2, \alpha_3^2, \ldots, \alpha_n^2$,

Repeating above process m times, we get the new transformed equation as

$$z^n + c_1 z^{n-1} + c_2 z^{n-2} + \ldots c_{n-1}z + c_n = 0. \qquad \ldots(4)$$

The roots of (4) are $\alpha_1^{2m}, \alpha_2^{2m}, \alpha_3^{2m} ..., \alpha_n^{nm}$ if γ_i are the root of (4),then $\gamma_i = \alpha_i^{2m}$ for $i = 1, 2, 3, ..., n$ and $\gamma_i > 0$.

Let us assuming that

$$|\alpha_1| > |\alpha_2| > |\alpha_3| > ... |\alpha_n|$$

then

$$|\gamma_1| >> |\gamma_2| >> |\gamma_3| >> ... >> |\gamma_n|$$

where $>>$ indicates for much greater than.

Therefore, $\dfrac{|\gamma_2|}{|\gamma_1|} = \dfrac{\gamma_2}{\gamma_1}, \dfrac{|\gamma_3|}{|\gamma_2|} = \dfrac{\gamma_3}{\gamma_2}, ..., \dfrac{|\gamma_n|}{|\gamma_{n-1}|} = \dfrac{|\gamma_n|}{|\gamma_{n-1}|} = \dfrac{\gamma_n}{\gamma_{n-1}}$, are negligible as compared to unity.

Now from equation (4), we have

$$\sum_{i=1}^{n} \gamma_i = -c_1 \qquad \text{or} \qquad c_1 = -(\gamma_1 + \gamma_2 + \gamma_3 + ... + \gamma_n) = -\gamma_1\left(1 + \frac{\gamma_2}{\gamma_1} + \frac{\gamma_3}{\gamma_1} + ...\right)$$

$$\sum_{i \neq 1}^{n} \gamma_i\gamma_j = \sum \gamma_1\gamma_2 = c_2 \qquad \text{or} \qquad c_2 = \gamma_1\gamma_2\left(1 + \frac{\gamma_3}{\gamma_1} + ...\right)$$

$$\sum \gamma_1\gamma_2\gamma_3 = -c_3 \qquad \text{or} \qquad c_3 = (-1)^n \gamma_1\gamma_2...\gamma_n$$

···

$$\gamma_1\gamma_2\gamma_3...\gamma_n = (-1)^n C_n \qquad \text{or} \qquad C_n = (-1)^n \gamma_1\gamma_2...\gamma_n.$$

Since $\dfrac{\gamma_2}{\gamma_1}, \dfrac{\gamma_4}{\gamma_3} ...,$ etc. are negligible as compared to unity, then above equations become

$$\gamma_1 \approx c_1$$

$$\left.\begin{array}{l} c_1 \approx -\gamma_1 \\ c_2 \approx \gamma_1\gamma_2 \\ c_3 \approx \gamma_1\gamma_2\gamma_3 \\ ·························· \\ c_n \approx (-1)^n \gamma_1\gamma_2...\gamma_n \end{array}\right\} \Rightarrow \begin{array}{l} \gamma_2 \approx -\dfrac{c_2}{c_1} \\[2mm] \gamma_3 \approx -\dfrac{c_3}{c_2} \\[2mm] ·············· \\[1mm] \gamma_n \approx -\dfrac{c_n}{c_{n-1}} \end{array}$$

But $\gamma_i = \alpha_i^{2m}$ for $i = 1, 2, 3, ..., n$, then

$$\alpha_i = (\gamma_i)^{1/2m}$$

$$|\alpha_i| = (|\gamma_i|)^{1/2m} = \left|\frac{c_i}{c_i - 1}\right|^{1/2m}$$

Hence, we can determine $\alpha_1, \alpha_2, ..., \alpha_n$ the roots of the given equation.

10.9.2 Roots are Real and Numerically Equal

If the magnitude of c_i is half the square of the magnitude of the corresponding coefficient in the preceding equation, then α_i will the double (equal) root of the given equation. We find the double root as follows :

$$\gamma_i = \frac{c_i}{c_i - 1} \text{ and } \qquad\qquad \gamma_{i+1} = -\frac{c_{i+1}}{c_i}$$

$$\therefore \qquad \gamma_i\gamma_{i+1} \approx \gamma_i^2 \approx \left|\frac{c_{i+1}}{c_{i-1}}\right| \qquad \Rightarrow \qquad \alpha_i^{2m} \approx \left|\frac{c_{i+1}}{c_{i-1}}\right| \qquad \Rightarrow \qquad |\alpha_i| \approx \left|\frac{c_{i+1}}{c_{i-1}}\right|^{1/2m} \qquad ...(5)$$

Thus, $|\alpha_i|$ gives the magnitudes of the double root and on substituting it in the given equation we find its sign.

10.9.3 For Complex Roots

When the given equation has only one pair of complex roots and others are real, we first find all real roots.

Let α_r and α_{r+1} be the complex roots, and let $\alpha_r = \xi + i\eta$, and $\alpha_{r+1} = \xi - i\eta$ then we have

$$\alpha_1 + \alpha_2 + ... + \alpha_{r-1} + 2\xi + \alpha_{r+2} + ... + \alpha_n = -a_1. \qquad ...(6)$$

From this equation we find the value of ξ.

If α_r and α_{r+1} form a complex pair $\rho_r e^{\pm i\phi_r}$, then for m (Number of squaring) sufficiently large, we obtain $\rho_r \phi_r$ as follows:

$$\left.\begin{array}{l} \rho_r^{2^{(2m)}} \approx \left|\dfrac{c_{r+1}}{c_{r-1}}\right| \\[4mm] 2\rho_r^m \cos m\phi_r \approx \dfrac{c_{r+1}}{c_{r-1}} \end{array}\right\} \qquad\qquad ...(7)$$

and

Now η is given by $\eta = \sqrt{\rho_r^2 - \xi^2}$.

Solved Examples

EXAMPLE 1. *Solve $x^3 - 8x^2 + 17x - 10 = 0$ by Graeffe's root squaring method.*

SOLUTION. Let $f(x) = x^3 - 8x^2 + 17x - 10 = 2$ ...(1)

Now $f(x) = x^3 - 8x^2 + 17x - 10$
$\qquad\quad + \quad - \quad + \quad -$

Clearly, $f(x)$ has three changes of sign, then by Descarte's rule of sign, $f(x) = 0$ has all the three roots positive and real.

Rewriting (1), we get $x^3 + 17x = 8x^2 + 10$

Squaring both sides, we get

$$(x^3 + 17x)^2 = (8x^2 + 10)^2$$

$$x^6 + 289x^2 + 34x^4 = 64x^4 + 100 + 160x^2$$

or $x^6 - 30x^4 + 129x^2 - 100 = 0$

Putting $x^2 = y$, we get

$$y^3 - 30y^2 + 129y - 100 = 0 \qquad ...(2)$$

Again (2) can be written as $y^3 + 129y = 30y^2 + 100$

Now, squaring both sides, we get

$$y^6 + 16641y^2 + 258y^4 = 900y^4 + 10000 + 60000y^2$$

or $y^6 - 642y^4 + 10641y^2 - 10000 = 0$

Putting $y^2 = z$ we get
$$z^3 - 642z^2 + 10641z - 10000 = 0 \qquad ...(3)$$

rewriting (3), we get
$$z^3 + 10641z = 642z^2 + 10000$$

Squaring both sides, we get
$$(z^3 + 10641z)^2 = (642z^2 + 10000)^2$$

$\Rightarrow z^6 + 113230881z^2 + 21282z^4$
$\qquad = 412164z^4 + 100000000 + 12840000z^2$

or $z^6 - 390882z^4 + 100390881z^2$
$\qquad\qquad\qquad - 100000000 = 0$

Putting $z^2 = u$, we get
$$u^3 - 390882u^2 + 100390881u$$
$$- 100000000 = 0 \quad ... (4)$$

If α_1, α_2 and α_3 are the roots of (1), then
$$\alpha_1^8 = 390882$$

$\Rightarrow \quad |\alpha_1| = (390882)^{1/8} = 5.00041 \approx 5$

$$\alpha_2^8 = \frac{100390881}{390882}$$

$\Rightarrow \quad |\alpha_2| = \left(\frac{100390881}{390882}\right)^{1/8} = 2.00081 \approx 2$

$$\alpha_3^8 = \frac{100000000}{100390881}$$

$\Rightarrow \quad |\alpha_3| = \left(\frac{100000000}{100390881}\right)^{1/8} = 0.999512 \approx 1$

Hence, the required root are 5, 2 and 1.

EXAMPLE 2. *Find all roots of the equation*
$$x^3 - 2x^2 - 5x + 6 = 0$$
by Graeffe's method, squaring thrice.

SOLUTION. Let $f(x) = x^3 - 2x^2 - 5x + 6 = 0$ ...(1)

$\therefore \quad f(x) = x^3 - 2x^2 - 5x + 6 = 0$
$\qquad\qquad + \quad - \quad - \quad +$

Clearly $f(x)$ has two changes of sign, so by Descarte's rule of sign, $f(x) = 0$ will have two positive roots. Also

Clearly, $f(-x)$ has one change of sign, so by Descarte's rule of sign, $f(x) = 0$ will have one negative root. Hence, $f(x) = 0$ has all real roots. Rewriting the equation (1) as
$$x^3 - 5x = 2x^2 - 6$$

Squaring both sides, we get $(x^3 - 5x)^2 = (2x^2 - 6)^2$

or $\quad x^6 + 25x^2 - 10x^2 = 4x^4 + 36 - 26x^2$

or $x^6 - 14x^4 + 49x^2 - 36 = 0$

Putting $x^2 = y$ we get
$$y^3 - 14y^2 + 49y - 36 = 0 \qquad ...(2)$$

Again rewriting (2)
$$y^3 - 49y = 14y^2 + 36$$

Squaring both sides, we get
$$(y^3 + 49y)^2 = (14y^2 + 36)^2$$

or $y^6 + 2401y^2 + 98y^4 = 196y^4 + 1296 + 1008y^2$

or $y^6 - 98y^4 + 1393y^2 - 1296 = 0$ ...(3)

Again putting $y^2 = z$, we get
$$z^3 - 98z^2 + 1393z - 1296 = 0 \quad ...(4)$$

Rewriting (4) as $z^3 + 1393z = 98z^2 + 1296$

Squaring both sides, we get
$$(z^3 + 1393z)^2 = (98z^2 + 1296)^2$$

or $\quad z^6 + 1940449z^2 + 2786z^4$
$$+ 1686433z^2 - 1679616 = 0$$

Putting $z^2 = u$, we get
$$u^3 - 6818u^2 + 1686433u - 1679616 = 0 \quad ...(5)$$

If $\alpha_1, \alpha_2, \alpha_3$ be the roots of the equation (1), then we have

$$\alpha_1^8 = 6818$$

$\Rightarrow \quad |\alpha_1| = (6818)^{1/8} = 3.011443 \approx 3$

$$\alpha_2^8 = \frac{1686433}{6818}$$

$\Rightarrow \quad |\alpha_2| = \left(\frac{1686433}{6818}\right)^{1/8} = 1.991425 \approx 2$

$$\alpha_3^8 = \frac{1679616}{1686433}$$

$\Rightarrow \quad |\alpha_3| = \left(\frac{1679616}{1686433}\right)^{1/8} = 0.999499 \approx 1$

Now, $f(3) = 0, f(-2) = 0, f(1) = 0$

Hence, the required roots are 3, -2, 1.

EXAMPLE 3. *Apply Graeffe's roots square method to find all the roots of the equation $x^4 - 3x + 1 = 0$.*

SOLUTION. Let $f(x) = x^4 - 3x + 1 = 0$ ...(1)
$$f(x) = x^4 - 3x + 1 = 0$$
$\qquad\qquad + \quad - \quad +$

Clearly, $f(x)$ has two changes of sign so by Descarte's rule of sign, $f(x) = 0$ has two positive real roots.

Also, $f(-x) = x^4 + 3x + 1$
$\qquad\qquad\qquad + \quad + \quad +$

Clearly, $f(x)$ has no change of sign so $f(x)$ has no negative real root.

But the degree of $f(x)$ is four so $f(x)$ has a pair of complex root.

Rewriting the equation (1), we get $x^4 + 1 = 3x$

Squaring both sides, we get $(x^4 + 1)^2 = 9x^2$

or $\quad x^8 + 1 + 2x^4 = 9x^2$

or $x^8 + 2x^4 - 9x^2 + 1 = 0$

Putting $x^2 = y$, we get

$$y^4 + 2y^2 - 9y + 1 = 0 \qquad \ldots(2)$$

Again rewriting (2), we get

$$(y^4 + 2y^2 + 1)^2 = (9y)^2 = 81y^2$$

or $y^8 + 4y^4 + 1 + 4y^6 + 2y^4 + 4y^2 = 81y^2$

or $y^8 + 4y^6 + 6y^4 - 77y^2 + 1 = 0$

Putting $y^2 = z$, we get

$$z^4 + 4z^3 + 6z^2 - 77z + 1 = 0 \qquad \ldots(3)$$

Rewriting (3), we get

$$z^4 + 6z^2 + 1 = -4z^3 + 77z$$

Squaring both sides, we get

$$(z^4 + 6z^2 + 1)^2 = (-4z^3 + 77z)^2$$

or $z^8 + 36z^4 + 1 + 12z^6 + 2z^4$
$\qquad\qquad + 12z^2 = 16z^6 + 5929z^2 - 616z^4$

or $z^8 - 4z^6 + 654z^4 - 5917z^2 + 1 = 0$

Putting $z^2 = u$, we get $u^4 - 4u^3 + 65u^2 - 5917u + 1 = 0$

If $\alpha_1, \alpha_2, \alpha_3$ and α_4 be the roots of equation (1), then

$\alpha = \qquad \Rightarrow \quad |\alpha_1| = (4)^{1/8} = 1.1892$

$\alpha_2^8 = \dfrac{654}{4} \qquad \Rightarrow \quad |\alpha_2| = \left(\dfrac{654}{4}\right)^{1/8} = 1.891$

$\alpha_3^8 = \dfrac{5917}{654} \qquad \Rightarrow \quad |\alpha_3| = \left(\dfrac{5917}{654}\right)^{1/8} = 1.3169$

$\alpha_4^8 = \dfrac{1}{5917} \qquad \Rightarrow \quad |\alpha_4| = \left(\dfrac{1}{5917}\right)^{1/8} = 0.3379$

From (3) and (4) we observe that the magnitudes of the coefficients c_1 and c_4 have become constant

i.e., from (4) $\qquad c_1 = -4, c_4 = 1$

and from (3), $\qquad b_1 = -4, b_4 = 1$

which indicates that α_1 and α_4 are the real roots whereas α_2 and α_3 form a pair of complex roots.

Let $\alpha_2 = \xi + i\eta$ and $\alpha_3 = \xi - i\eta$.

Now we find the complex roots of the form

$$\rho_2 e^{\pm i\phi_2} = \xi + i\eta$$

From (4), we have $\rho_2^{2^{(2^3)}} \approx \left|\dfrac{c_2 + 1}{c_2 - 1}\right|$

$\Rightarrow \qquad \rho_2^{16} \approx \left|\dfrac{c_3}{c_1}\right| = \dfrac{5917}{4} = 1479.25$

$\Rightarrow \qquad \rho_2 \approx (1479.25)^{1/16} = 1.5781$

From equation (1), we have

$$\alpha_1 + \alpha_2 + \alpha_3 + \alpha_4 = 0$$

$$\alpha_1 + 2\xi + \alpha_4 = 0$$

$\Rightarrow \quad \xi = -\dfrac{1}{2}(\alpha_1 + \alpha_2) = -\dfrac{1}{2}(1.1892 + 0.3379)$

$\Rightarrow \quad \xi = -0.7636$

and $\eta = \sqrt{\rho_2^2 - \xi^2} = \sqrt{(1.5781)^2 - (-0.7636)^2} = 1.381$

$\therefore$ The complex roots are $-0.7636 \pm 1.381i$.

Hence, all the four roots are

$$1.1892, 0.3379, -0.7636 \pm 1.381i.$$

Exercise-10.5

1. Apply Lin-Bairstow method to find a quadratic factor of the equation $x^4 + 5x^3 + 3x^2 - 5x - 9 = 0$ close to $x^2 + 3x - 5$.

2. Find the roots of the equation $x^4 + 9x^3 + 36x^2 + 51x + 27 = 0$ to three decimal places using Lin-Bairstow method.

3. Find the quadratic factors of the equation $x^4 - 8x^3 + 39x^2 - 62x + 50 = 0$ by using Lin-Bairstow method (upto third iteration)

starting with $a = 0$, $b = 0$.

4. Solve $x^3 - 5x^2 - 17x + 20 = 0$ by Graeffe's method.

5. Apply Graeffe's method to find all the roots of the equation $x^3 - 6x^2 + 11x - 6 = 0$.

6. Find all the roots of the equation $x^3 - 4x^2 + 5x - 2 = 0$ by Graeffe's method, squaring three times.

7. Solve $x^3 - 9x^2 + 18x - 6 = 0$ by Graeffe's method.

Answers

1. $x^2 + 2.9026x - 4.9176$
2. $-0.759, -1.42, -3.411 \pm 2.903i$
3. $(x^2 - 2x + 2)(x^2 - 6x + 25)$
4. $7.018, -2.974, 0.958$
5. $3, 2, 1$
6. $2, 1, 1$
7. $6.3, 2.3, 0.4$

10.9.4 Solution of Simultaneous Linear Algebraic Equations

Simultaneous linear equations have great importance in the field of engineering an Science. In the field of science the analysis of electronic circuits having a number of invariant elements, analysis of a network under sinusoidal steady-state conditions, determination of the output of a chemical plant and finding the cost of reactions are some of the problem which depend on the solution of system of linear algebraic equations. We shall discuss a system of 'm' linear equation with n unknowns. If $m > n$, as a rule the equations cannot be satisfied. If $m < n$, the system of linear equations usually has an infinite number of solution. In this chapter we will discuss the system of linear equation, if $m = n$.

Linear Equations

Let us consider n equation in m variables:

$$\left.\begin{array}{l} a_{11}x_1 + a_{12}x_2 + \ldots + a_{1m}x_m = b_1 \\ a_{21}x_1 + a_{22}x_2 + \ldots + a_{2m}x_m = b_2 \\ a_{31}x_1 + a_{32}x_2 + \ldots + a_{3m}x_m = b_3 \\ \text{...} \\ \text{...} \\ a_{n1}x_1 + a_{n2}x_2 + \ldots + a_{nm}x_m = b_m \end{array}\right\}$$

These equations can be written in a matrix form as follows

$$\begin{bmatrix} a_{11} & a_{12} & \cdots & a_{1m} \\ a_{21} & a_{22} & \cdots & a_{2m} \\ a_{31} & a_{32} & \cdots & a_{3m} \\ \cdots & \cdots & \cdots & \cdots \\ \cdots & \cdots & \cdots & \cdots \\ a_{n1} & a_{n2} & \cdots & a_{nm} \end{bmatrix} \begin{bmatrix} x_1 \\ x_2 \\ x_3 \\ \cdots \\ \cdots \\ x_m \end{bmatrix} = \begin{bmatrix} b_1 \\ b_2 \\ b_3 \\ \cdots \\ \cdots \\ b_m \end{bmatrix}$$

or

$$AX = B$$

Where

$$A = [a_{ij}]_{m \times n}$$

$$X = \begin{bmatrix} x_1 \\ x_2 \\ x_3 \\ \cdots \\ \cdots \\ x_m \end{bmatrix}_{m \times 1} \qquad \text{and} \qquad B = \begin{bmatrix} b_1 \\ b_2 \\ b_3 \\ \cdots \\ \cdots \\ b_m \end{bmatrix}_{m \times 1}$$

If all the $b_i (i = 1, 2, ..., n)$ are zero, then the system of linear equations are called homogeneous. Otherwise it is called non-homogeneous. The solution of such equations can be obtained by direct or iterative methods. Amongst the direct methods, we will describe Gauss elimination, LU decomposition, Jordan's and Crout's methods. About the iterative methods, we will describe the Jocobi, Gauss-Seidel and Relaxation method.

10.10 GAUSS ELIMINATION METHOD

In this method, the variables from the system of linear equations are eliminated successively and the system of equations is therefore reduced to an upper triangular system from which the variable are determined by back substitution. This method is described in chapter 1 in details.

10.11 LU DECOMPOSITION METHOD OR METHOD OF FACTORIZATION

This method is based on the fact that every square matrix A can be expressed as the form LU where L is a unit lower triangular matrix while U is upper triangular matrix and provided all the principal minors of A are non-singular.

That is, If $A = [a_{ij}]_{n \times n}$, then

$$a_{11} \neq 0, \begin{vmatrix} a_{11} & a_{12} \\ a_{21} & a_{22} \end{vmatrix} \neq 0, \begin{vmatrix} a_{11} & a_{12} & a_{13} \\ a_{21} & a_{22} & a_{23} \\ a_{31} & a_{32} & a_{33} \end{vmatrix} \neq 0, \text{ and so on.}$$

For simplicity and understanding this method, let us consider a system of three equations

$$a_{11}x_1 + a_{12}x_2 + a_{13}x_3 = b_1$$
$$a_{21}x_1 + a_{22}x_2 + a_{23}x_3 = b_2$$
$$a_{31}x_1 + a_{32}x_2 + a_{33}x_3 = b_3$$

These equations can be written in matrix form as follows:

$$AX = B$$

Where

$$A = \begin{bmatrix} a_{11} & a_{12} & a_{13} \\ a_{21} & a_{22} & a_{23} \\ a_{31} & a_{32} & a_{33} \end{bmatrix}, X = \begin{bmatrix} x_1 \\ x_2 \\ x_3 \end{bmatrix}, B = \begin{bmatrix} b_1 \\ b_2 \\ b_3 \end{bmatrix} \qquad \qquad ...(1)$$

Let

$$A = LU$$

where

$$A = \begin{bmatrix} 1 & 0 & 0 \\ l_{21} & 1 & 0 \\ l_{31} & l_{32} & 1 \end{bmatrix}, U = \begin{bmatrix} u_{11} & u_{12} & u_{13} \\ 0 & u_{22} & u_{23} \\ 0 & 0 & u_{33} \end{bmatrix} \qquad \qquad ...(2)$$

Then from (2)

$$\begin{bmatrix} a_{11} & a_{12} & a_{13} \\ a_{21} & a_{22} & a_{23} \\ a_{31} & a_{32} & a_{33} \end{bmatrix} = \begin{bmatrix} 1 & 0 & 0 \\ l_{21} & 1 & 0 \\ l_{31} & l_{32} & 1 \end{bmatrix} \begin{bmatrix} u_{11} & u_{12} & u_{13} \\ 0 & u_{22} & u_{23} \\ 0 & 0 & u_{33} \end{bmatrix}$$

or

$$\begin{bmatrix} a_{11} & a_{12} & a_{13} \\ a_{21} & a_{22} & a_{23} \\ a_{31} & a_{32} & a_{33} \end{bmatrix} = \begin{bmatrix} u_{11} & u_{12} & u_{13} \\ l_{21}u_{11} & l_{21}u_{12} + u_{22} & l_{21}u_{13} + u_{23} \\ l_{31}u_{11} & l_{31}u_{12} + l_{32}u_{22} & l_{31}u_{13} + l_{32}u_{23} + u_{33} \end{bmatrix}$$

Comparing the two matrices, we get

(i) $u_{11} = a_{11}, u_{12} = a_{12}, u_{13} = a_{13}$

(ii) $l_{21}u_{11} = a_{21}$ or $l_{21} = \dfrac{a_{21}}{u_{11}} = \dfrac{a_{21}}{a_{11}}$

(iii) $l_{31}u_{11} = a_{31}$ or $l_{31} = \dfrac{a_{31}}{u_{11}} = \dfrac{a_{31}}{a_{11}}$

(iv) $l_{21}u_{12} + u_{22} = a_{22}$ or $l_{21} = \dfrac{a_{22} - u_{22}}{u_{12}}$ or $u_{22} = a_{22} - l_{21}u_{12} = a_{22} - \dfrac{a_{21}}{a_{11}}.a_{12}$

(v) $l_{21}u_{13} + u_{23} = a_{23}$ $\Rightarrow$ $u_{23} = a_{23} - l_{21}u_{13}$ $\Rightarrow$ $u_{23} = a_{23} - \dfrac{a_{21}}{a_{11}}a_{13}.$

(vi) $l_{31}u_{12} + l_{32}u_{22} = a_{32}$ or $l_{32} = \dfrac{1}{u_{22}}\left[a_{32} - \dfrac{a_{31}}{a_{11}}a_{12} \right] = \dfrac{\left[a_{32} - \dfrac{a_{31}a_{12}}{a_{11}} \right]}{\left[a_{22} - \dfrac{a_{21}a_{12}}{a_{11}} \right]}$

(vii) $l_{31}u_{13} + l_{32}u_{23} + u_{33} = a_{33}$

or $u_{33} = a_{33} - l_{31}u_{13} - l_{32}u_{23} = a_{33} - \dfrac{a_{31}}{a_{11}}a_{13} - \dfrac{1}{u_{22}}\left[a_{32} - \dfrac{a_{31}}{a_{11}}a_{12} \right]\left[a_{23} - \dfrac{a_{21}}{a_{11}}a_{13} \right]$

$$= a_{33} - \dfrac{a_{31}a_{13}}{a_{11}} - \dfrac{\left[a_{32} - \dfrac{a_{31}a_{12}}{a_{11}} \right]\left[a_{23} - \dfrac{a_{21}}{a_{11}}a_{13} \right]}{\left[a_{22} - \dfrac{a_{21}a_{12}}{a_{11}} \right]}$$

Thus from above we can find the elements of L and U. Now L and U are therefore obtained. Replacing A by LU in (1), we get

$$LUX = B \qquad ...(3)$$

Now let $$UX = Y \qquad ...(4)$$

where $$Y = \begin{bmatrix} y_1 \\ y_2 \\ y_3 \end{bmatrix}$$

From (3) and (4), we get

$$LY = B$$

or $$\begin{bmatrix} 1 & 0 & 0 \\ l_{21} & 1 & 0 \\ l_{31} & l_{32} & 1 \end{bmatrix}\begin{bmatrix} y_1 \\ y_2 \\ y_3 \end{bmatrix} = \begin{bmatrix} b_1 \\ b_2 \\ b_3 \end{bmatrix}.$$

From this equation, we get

$$y_1 = b_1; y_2 = b_2 - l_{21}y_1; y_3 = b_3 - l_{31}b_1 - l_{32}b_2 \qquad ...(5)$$

Now from (4)

$$UX = Y$$

$$\begin{bmatrix} u_{11} & u_{12} & u_{13} \\ 0 & u_{22} & u_{23} \\ 0 & 0 & u_{33} \end{bmatrix}\begin{bmatrix} x_1 \\ x_2 \\ x_3 \end{bmatrix} = \begin{bmatrix} y_1 \\ y_2 \\ y_3 \end{bmatrix}$$

From this equation, we get

$$x_3 = \dfrac{y_3}{u_{33}}, x_2 = \dfrac{y_2 - u_{23}x_3}{u_{22}}, x_1 = \dfrac{y_1 - u_{12}x_2 - u_{13}x_3}{u_{11}}$$

With the help of L and U, x_1, x_2, x_3 can be calculated.

REMARKS

- LU decomposition method is superior to Gauss elimination method.
- This method is applicable if the coefficient matrix can be expressed as the product of lower and upper triangular matrix.
- This method can also be renamed as Method of factorization.

Solved Examples

EXAMPLE 1. *Solve the following equation by LU decomposition method.*

$$2x_1 + x_2 + x_3 = 2$$
$$x_1 + 3x_2 + 2x_3 = 2$$
$$3x_1 + x_2 + 2x_3 = 2$$

SOLUTION. Given equation are

$$2x_1 + x_2 + x_3 = 2 \qquad ...(1)$$
$$x_1 + 3x_2 + 2x_3 = 2 \qquad ...(2)$$
$$3x_1 + x_2 + 2x_3 = 2 \qquad ...(3)$$

Here, the coefficient matrix A is given by

$$A = \begin{bmatrix} 2 & 1 & 1 \\ 1 & 3 & 2 \\ 3 & 1 & 2 \end{bmatrix}$$

and $\quad X = \begin{bmatrix} x_1 \\ x_2 \\ x_3 \end{bmatrix}, B = \begin{bmatrix} 2 \\ 2 \\ 2 \end{bmatrix}$

$\therefore \qquad AX = B$

Now $\quad A = LU$

$$\begin{bmatrix} 2 & 1 & 1 \\ 1 & 3 & 2 \\ 3 & 1 & 2 \end{bmatrix} = \begin{bmatrix} 1 & 0 & 0 \\ l_{21} & 1 & 0 \\ l_{31} & l_{32} & 1 \end{bmatrix}\begin{bmatrix} u_{11} & u_{12} & u_{13} \\ 0 & u_{22} & u_{23} \\ 0 & 0 & u_{33} \end{bmatrix} \qquad ...(4)$$

$$= \begin{bmatrix} u_{11} & u_{12} & u_{13} \\ l_{21}u_{11} & l_{21}u_{12}+u_{22} & l_{21}u_{13}+u_{23} \\ l_{31}u_{11} & l_{31}u_{12}+l_{32}u_{22} & l_{31}u_{13}+l_{32}u_{23}+u_{33} \end{bmatrix}$$

On comparing two matrices, we get

$$u_{11} = 2, u_{12} = 1, u_{13} = 1$$

$$l_{21}u_{11} = 1 \quad \text{or} \quad l_{21} = \frac{1}{u_{11}} = \frac{1}{2}$$

$$l_{31}u_{11} = 3 \quad \text{or} \quad l_{31} = \frac{3}{u_{11}} = \frac{3}{2}$$

$$l_{21}u_{12} + u_{22} = 3$$

$$\Rightarrow \quad u_{22} = 3 - l_{21}u_{12} = 3 - \frac{1}{2}(1) = 3 - \frac{1}{2} = \frac{5}{2}$$

$$l_{21}u_{13} + u_{23} = 2$$

$$\Rightarrow \quad u_{23} = 2 - l_{21}u_{13} = 2 - \frac{1}{2}(1) = 2 - \frac{1}{2} = \frac{3}{2}$$

$$l_{31}u_{12} + l_{32}u_{22} = 1$$

$$\Rightarrow \quad l_{32} = \frac{1 - l_{31}u_{12}}{u_{22}} = \frac{1 - \frac{3}{2}(1)}{\frac{5}{2}} = -\frac{1}{5}$$

$$l_{31}u_{13} + l_{32}u_{23} + u_{33} = 2$$

$$\Rightarrow \quad u_{33} = 2 - l_{31}u_{13} - l_{32}u_{23} = 2 - \frac{3}{2}(1) - \left(-\frac{1}{5}\right)\left(\frac{3}{2}\right)$$

$$= 2 - \frac{3}{2} + \frac{3}{10} = \frac{8}{10} = \frac{4}{5}$$

Thus $\quad L = \begin{bmatrix} 1 & 0 & 0 \\ l_{21} & 1 & 0 \\ l_{31} & l_{32} & 1 \end{bmatrix} = \begin{bmatrix} 1 & 0 & 0 \\ \dfrac{1}{2} & 1 & 0 \\ \dfrac{3}{2} & -\dfrac{1}{5} & 1 \end{bmatrix}$

and $\quad U = \begin{bmatrix} u_{11} & u_{12} & u_{13} \\ 0 & u_{22} & u_{23} \\ 0 & 0 & u_{33} \end{bmatrix} = \begin{bmatrix} 2 & 1 & 1 \\ 0 & \dfrac{5}{2} & \dfrac{3}{2} \\ 0 & 0 & \dfrac{4}{5} \end{bmatrix}$

Since $\quad LY = B \Rightarrow \begin{bmatrix} 1 & 0 & 0 \\ \dfrac{1}{2} & 1 & 0 \\ \dfrac{3}{2} & -\dfrac{1}{5} & 1 \end{bmatrix}\begin{bmatrix} y_1 \\ y_2 \\ y_3 \end{bmatrix} = \begin{bmatrix} 2 \\ 2 \\ 2 \end{bmatrix}$

From these equation, we get

$$y_1 = 2$$

$$\frac{1}{2}y_1 + y_2 = 2$$

$$\frac{3}{2}y_1 - \frac{1}{5}y_2 + y_3 = 2$$

On solving, we get $y_1 = 2, y_2 = 1, y_3 = -\dfrac{4}{5}$

and since, we have $UX = Y$

$$\begin{bmatrix} 2 & 1 & 1 \\ 0 & \dfrac{5}{2} & \dfrac{3}{2} \\ 0 & 0 & \dfrac{4}{5} \end{bmatrix}\begin{bmatrix} x_1 \\ x_2 \\ x_3 \end{bmatrix} = \begin{bmatrix} y_1 \\ y_2 \\ y_3 \end{bmatrix} = \begin{bmatrix} 2 \\ 1 \\ -\dfrac{4}{5} \end{bmatrix}$$

$\therefore \quad 2x_1 + x_2 + x_3 = 2$

$$\frac{5}{2}x_2 + \frac{3}{2}x_3 = 1$$

$$\frac{4}{5}x_3 = -\frac{4}{5}$$

By back substitution, we get $x_3 = -1, x_2 = 1, x_1 = 1$

Hence, solution is $x_1 = 1, x_2 = 1, x_3 = -1$.

10.12 JORDAN'S METHOD

This method is a modification of Gauss's elimination method. In Jordan's method, the elimination takes places not only below the equations but above also so in this way we get a diagonal matrix. From this values of unknown are found. For understanding this method, let us consider a system of three equations:

$$a_{11}x_1 + a_{12}x_2 + a_{13}x_3 = b_1 \qquad \qquad ...(1)$$

$$a_{21}x_1 + a_{22}x_2 + a_{23}x_3 = b_2 \qquad \qquad ...(2)$$

$$a_{31}x_1 + a_{32}x_2 + a_{33}x_3 = b_3 \qquad \qquad ...(3)$$

Eliminating x_1 from (2) and (3) with the help of (1), we get

$$a_{11}x_1 + a_{12}x_2 + a_{13}x_3 = b_1 \qquad \qquad ...(4)$$

$$b_{22}x_2 + b_{23}x_3 = b_2' \qquad \qquad ...(5)$$

$$c_{32}x_2 + c_{33}x_3 = b_3' \qquad \qquad ...(6)$$

Now eliminating x_2 from (4) and (6) with the help of (5) we get

$$a_{11}'x_1 + a_{13}'x_3 = b_1' \qquad \qquad ...(7)$$

$$b_{22}x_2 + b_{23}x_3 = b_2' \qquad \qquad ...(8)$$

$$c_{33}'x_3 = b_3' \qquad \qquad ...(9)$$

Finally eliminating x_3 from (7) and (8) using (9) we get

$$a_{11}''x_1 = b_1'', \quad b_{22}'x_2 = b_2'', \quad c_{33}'x_3 = b_3''$$

$$\Rightarrow \quad \begin{bmatrix} a_{11}'' & 0 & 0 \\ 0 & b_{22}' & 0 \\ 0 & 0 & c_{33}' \end{bmatrix}\begin{bmatrix} x_1 \\ x_2 \\ x_3 \end{bmatrix} = \begin{bmatrix} b_1'' \\ b_2'' \\ b_3'' \end{bmatrix} \qquad \qquad ...(10)$$

Hence, the coefficient matrix reduced to diagonal matrix. From (10) we can easily find x_1, x_2 and x_3.

Solved Examples

EXAMPLE 1. *Apply Jordan's method to solve*
$$x + 2y + z = 8$$
$$2x + 3y + 4z = 20$$
$$4x + 3y + 2z = 16$$

SOLUTION. First, eliminating x from last two equations using first equation, we get

$$x + 2y + z = 8 \qquad ...(1)$$
$$-y + 2z = 4 \qquad ...(2)$$
$$-5y - 2z = -16 \qquad ...(3)$$

Now eliminating y from (1) and (3) with the help of (2), we get

$$x + 5z = 16 \qquad ...(4)$$
$$-y + 2z = 4 \qquad ...(5)$$
$$-12z = -36 \qquad ...(6)$$

Now eliminating z from (4) and (5) with the help of (6), we get

$$12x = 12 \qquad ...(7)$$
$$-6y = -12 \qquad ...(8)$$
$$-12z = -36 \qquad ...(9)$$

From (7), (8) and (9), we get

$$x = 1, y = 2, z = 3.$$

EXAMPLE 2. *Apply Jordan's method to solve the equations*
$$x + y + z = 9$$
$$2x - 3y + 4z = 13 \qquad \text{(VTU–2009, PTU–2005)}$$
$$3x + 4y + 5z = 40$$

SOLUTION. First of all eliminating x from last two equations with the help of first equation, we get

$$x + y + z = 9 \qquad ...(1)$$
$$-5y + 2z = -5 \qquad ...(2)$$
$$y + 2z = 13 \qquad ...(3)$$

Now eliminating y from (1) and (3) with the help of (2), we get

$$5x + 7z = 40 \qquad ...(4)$$
$$-5y + 2z = -5 \qquad ...(5)$$
$$12z = 60 \qquad ...(6)$$

Further, eliminating z from (4) and (5) using (6), we get

$$60x = 60 \qquad ...(7)$$
$$-30y = -90 \qquad ...(8)$$
$$12z = 60 \qquad ...(9)$$

From (7), (8) and (9), we get

$$x = 1, y = 3, z = 5.$$

10.12.1 Crout's Method

We shall explains this method by considering three equations. Let us consider the equations as follows:

$$\left.\begin{array}{l} a_{11}x_1 + a_{12}x_2 + x_{13}x_3 = b_1 \\ a_{21}x_1 + a_{22}x_2 + x_{23}x_3 = b_2 \\ a_{31}x_1 + a_{32}x_2 + x_{33}x_3 = b_3 \end{array}\right\} \qquad \qquad ...(1)$$

The agumented matrix of (1) is

$$(A \mid B) = \begin{bmatrix} a_{11} & a_{12} & a_{13} & b_1 \\ a_{21} & a_{22} & a_{23} & b_2 \\ a_{31} & a_{32} & a_{33} & b_3 \end{bmatrix} \qquad \qquad ...(2)$$

Now we consider derived matrix as follows:

$$(A' \mid B') = \begin{bmatrix} a_{11}' & a_{12}' & a_{13}' & b_1' \\ a_{21}' & a_{22}' & a_{23}' & b_2' \\ a_{31}' & a_{32}' & a_{33}' & b_3' \end{bmatrix} \qquad \qquad ...(3)$$

WORKING PROCEDURE

Step 1. First column of (3) is same as first column of (2)

   i.e., $\quad a_{11}' = a_{11}, a_{21}' = a_{21}, a_{31}' = a_{31}$ or $\quad a_{i1}' = a_{i1} \ \forall i = 1, 2, 3.$

Step 2. Elements of first row to the right of first column in (3) are given by $a_{12}' = \dfrac{a_{12}}{a_{11}}, a_{13}' = \dfrac{a_{13}}{a_{11}}, b_1' = \dfrac{b_1}{a_{11}}$

   i.e. $\quad, a_{1j}' = \dfrac{a_{1j}}{a_{11}}, j = 2, 3.$

Step 3. Elements of second column except a_{12}' are given by

$$a'_{22} = a_{22} - a'_{12}a'_{21}$$
$$a'_{32} = a_{32} - a'_{12}a'_{31}$$

i.e., $\quad a'_{j2} = a_{j2} - a'_{12}a'_{j1}, j = 2,3$

Step 4. Elements of second row except a'_{21}, a'_{22} are given by $a'_{23} = \dfrac{a_{23} - a'_{13}a'_{21}}{a'_{22}}, b'_2 = \dfrac{b_2 - b'_1a'_{21}}{a'_{22}}$

i.e., $\quad a'_{2j} = \dfrac{a_{1j} - a'_{1j}a'_{21}}{a'_{22}}, j = 3.$

Step 5. Elements of third column except a'_{13}, a'_{23} given by $a'_{33} = a_{33} - a'_{23}a'_{32} - a'_{13}a'_{31}$.

Step 6. Element of third row except $a'_{31}, a'_{32}, a'_{33}$ is $b'_3 = \dfrac{b_3 - b'_2a'_{32} - b'_1a'_{31}}{a'_{33}}$

Thus the solution of the given equation is given by $x_3 = b'_3, x_2 = b'_2 - a'_{23}x_3, x_1 = b'_1 - a'_{13}x_3 - a'_{12}x_2$.

This Crout's method is explained properly like below in which the coefficient a' and constants b' are discussed how they are obtained.

Crout established a method in which Gauss elimination is often performed.

Let us consider a system of three equations.

$$\left.\begin{array}{c} a_{11}x_1 + a_{12}x_2 + x_{13}x_3 = b_1 \\ a_{21}x_1 + a_{22}x_2 + x_{23}x_3 = b_2 \\ a_{31}x_1 + a_{32}x_2 + x_{33}x_3 = b_3 \end{array}\right\} \qquad \ldots(1)$$

Above equations can be written as $\quad AX = B$

where $\quad A = \begin{bmatrix} a_{11} & a_{12} & a_{13} \\ a_{21} & a_{22} & a_{23} \\ a_{31} & a_{32} & a_{33} \end{bmatrix}, X = \begin{bmatrix} x_1 \\ x_2 \\ x_3 \end{bmatrix}, B = \begin{bmatrix} b_1 \\ b_2 \\ b_3 \end{bmatrix} \qquad \ldots(2)$

$\therefore$ Augmented matrix

$$(A \mid B) = \begin{bmatrix} a_{11} & a_{12} & a_{13} & b_1 \\ a_{21} & a_{22} & a_{23} & b_2 \\ a_{31} & a_{32} & a_{33} & b_3 \end{bmatrix} \qquad \ldots(3)$$

Equation (1) becomes after Gauss elimination process as follows:

$$\left.\begin{array}{c} x_1 + a'_{12}x_2 + a'_{13}x_3 = b'_1 \\ x_2 + a'_{23}x_3 = b'_2 \\ x_3 = b'_3 \end{array}\right\} \qquad \ldots(4)$$

Now the first equation of (1) is obtained by multiplication of first equation in (4) by a constant a'_{11} and second equation of (1) is obtained through multiplication of first and second equation in (4) by a'_{21} and a'_{22} respectively, and adding. Similarly, the third equation of (1) is obtained through multiplication of first, second and third equations in (4) by a'_{31}, a'_{32} and a'_{33} and then adding. Thus, we get the following equations

$$\left.\begin{array}{c} a'_{11}b'_1 = b_1 \\ a'_{21}b'_1 + a'_{22}b'_2 = b_2 \\ a'_{31}b'_1 + a'_{32}b'_2 + a'_{33}b'_3 = b_3 \end{array}\right\} \qquad \ldots(5)$$

Let us introduce the matrices P and Q as follows:

$$P = \begin{bmatrix} a'_{11} & 0 & 0 \\ a'_{21} & a'_{22} & 0 \\ a'_{31} & a'_{32} & a'_{33} \end{bmatrix}, Q = \begin{bmatrix} 0 & a'_{12} & a'_{13} \\ 0 & 0 & a'_{23} \\ 0 & 0 & 0 \end{bmatrix}$$

or $\quad P + Q = A'$ (say).

Now equation (1), (4) and (5) take the form

$$\left.\begin{array}{c} AX = B \\ (Q + I)X = B' \\ PB' = B \end{array}\right\} \qquad \ldots(6)$$

From (6), we obtain

$$P(Q + I)X = AX$$

and hence $\quad P(Q + I) = A.$

Augmenting the matrix $(Q + I)$ with new column B' and augmenting the matrix A by column B. Thus we get

$$\begin{bmatrix} a'_{11} & 0 & 0 \\ a'_{21} & a'_{22} & 0 \\ a'_{31} & a'_{32} & a'_{33} \end{bmatrix} \begin{bmatrix} 1 & a'_{12} & a'_{13} & b'_1 \\ 0 & 1 & a'_{23} & b'_2 \\ 0 & 0 & 1 & b'_3 \end{bmatrix} = \begin{bmatrix} a_{11} & a_{12} & a_{13} & b_1 \\ a_{21} & a_{22} & a_{23} & b_2 \\ a_{31} & a_{32} & a_{33} & b_3 \end{bmatrix}$$

From above equation, we get

$$a'_{11} = a_{11}, a'_{21} = a_{21}, a'_{31} = a_{31}, a'_{12} = \frac{a_{12}}{a'_{11}} = \frac{a_{12}}{a_{11}}, a'_{13} = \frac{a_{13}}{a'_{11}} = \frac{a_{13}}{a_{11}}, b'_1 = \frac{b_1}{a_{11}}$$

$$a'_{22} = a_{22} - a'_{21}a'_{12}, a'_{23} = \frac{1}{a'_{22}}(a_{23} - a'_{21}a'_{13})$$

$$a'_{21}b'_1 + a'_{22}b'_2 = b_2 \quad \text{or} \quad b'_2 = \frac{b_2 - a'_{21}b'_1}{a'_{22}}$$

$$a'_{31}a'_{12} + a'_{32} = a_{32} \quad \text{or} \quad a'_{32} = a_{32} - a'_{31}a'_{12}$$

$$a'_{31}a'_{13} + a'_{32}a'_{23} + a'_{33} = a_{33} \quad \text{or} \quad a'_{33} = a_{33} - a'_{31}a'_{13} - a'_{32}a'_{23}$$

$$a'_{31}b'_1 + a'_{32}b'_2 + a'_{33}b'_3 = b_3 \quad \text{or} \quad b'_3 = \frac{b_3 - a'_{31}b'_1 - a'_{32}b'_2}{a'_{33}}$$

Thus after getting all a' and b', with the help of (4), we get the solution given by

$$x_3 = b'_3, x_2 = b'_2 - a'_{23}x_3, x_1 = b'_1 - a'_{12}x_2 - a'_{13}x_3.$$

 Solved Examples

EXAMPLE 1. *Solve the following equations by Crout's method:*

$$x_1 + x_2 + x_3 = 1$$
$$3x_1 + x_2 - 3x_3 = 5$$
$$x_1 - 2x_2 - 5x_3 = 10$$

SOLUTION. Above equations can be written as

$$AX = B \qquad \qquad ...(1)$$

where $A = \begin{bmatrix} 1 & 1 & 1 \\ 3 & 1 & -3 \\ 1 & -2 & -5 \end{bmatrix}, X = \begin{bmatrix} x_1 \\ x_2 \\ x_3 \end{bmatrix}, B = \begin{bmatrix} 1 \\ 5 \\ 10 \end{bmatrix}$

$$\therefore \begin{bmatrix} a_{11} & a_{12} & a_{13} \\ a_{21} & a_{22} & a_{23} \\ a_{31} & a_{32} & a_{33} \end{bmatrix} = \begin{bmatrix} 1 & 1 & 1 \\ 3 & 1 & -3 \\ 1 & -2 & -5 \end{bmatrix}$$

Now derived matrix is given by

$$\begin{bmatrix} a'_{11} & a'_{12} & a'_{13} & b'_1 \\ a'_{21} & a'_{22} & a'_{23} & b'_2 \\ a'_{31} & a'_{32} & a'_{33} & b'_3 \end{bmatrix}$$

where

$$a'_{11} = a_{11} = 1, a'_{21} = a_{21} = 3, a'_{31} = a_{31} = 1$$

and $a'_{12} = \frac{a_{12}}{a_{11}} = \frac{1}{1} = 1, a'_{13} = \frac{a_{13}}{a_{11}} = \frac{1}{1} = 1,$

$$b'_1 = \frac{b_1}{a_{11}} = \frac{1}{1} = 1$$

and $a'_{22} = a_{22} - a'_{12}.a'_{21} = 1 - 1.3 = 1 - 3 = -2,$

$$a'_{32} = a_{32} - a'_{12}.a'_{31} = -2 - 1.1 = -3$$

$$a'_{23} = \frac{a_{23} - a'_{13}.a'_{21}}{a'_{22}} = \frac{-3 - 3}{-2} = 3,$$

$$b'_2 = \frac{b_2 - a'_{21}.b'_1}{a'_{22}} = \frac{5 - 3}{-2} = -1$$

$$a'_{33} = a_{33} - a'_{31}a'_{13} - a'_{32}a'_{23} = -5 - 1 + 9 = 3$$

and $b'_3 = \frac{b_3 - a'_{31}b'_1 - a'_{32}b'_2}{a'_{33}} = \frac{10 - 1 - 3}{3} = \frac{6}{3} = 2$

Thus the solution is

$$x_3 = b'_3 = 2, \; x_2 = b'_2 - a'_{23}x_3 = -1 - 3(2) = -7$$
$$x_1 = b'_1 - a'_{12}x_2 - a'_{13}x_3 = 1 - 1(-7) - 1(2)$$
$$= 1 + 7 - 2 = 6$$

Hence $x_1 = 6, x_2 = -7, x_3 = 2$

EXAMPLE 2. *Solve the following equations by Crout's method:*

$$2x_1 + 3x_2 + x_3 = -1$$
$$5x_1 + x_2 + x_3 = 9$$
$$3x_1 + 2x_2 + 4x_3 = 11$$

SOLUTION. Above equations can be written as

$$AX = B$$

where $A = \begin{bmatrix} 2 & 3 & 1 \\ 5 & 1 & 1 \\ 3 & 2 & 4 \end{bmatrix}, X = \begin{bmatrix} x_1 \\ x_2 \\ x_3 \end{bmatrix}, B = \begin{bmatrix} -1 \\ 9 \\ 11 \end{bmatrix}$

Now derived matrix is given by

And

$$\begin{bmatrix} a'_{11} & a'_{12} & a'_{13} & b'_1 \\ a'_{21} & a'_{22} & a'_{23} & b'_2 \\ a'_{31} & a'_{32} & a'_{33} & b'_3 \end{bmatrix}$$

$$(A \mid B) = \begin{bmatrix} 2 & 3 & 1 & -1 \\ 5 & 1 & 1 & 9 \\ 3 & 2 & 4 & 11 \end{bmatrix} = \begin{bmatrix} a_{11} & a_{12} & a_{13} & b_1 \\ a_{21} & a_{22} & a_{23} & b_2 \\ a_{31} & a_{32} & a_{33} & b_3 \end{bmatrix}$$

Now find a'_{ij} and b_j

$$a'_{11} = a_{11} = 2, a'_{21} = a_{21} = 5, a'_{31} = a_{31} = 3$$

$$a'_{12} = \frac{a_{12}}{a_{11}} = \frac{3}{2}, a'_{13} = \frac{a_{13}}{a_{11}} = \frac{1}{2}, b'_1 = \frac{b_1}{a_{11}} = \frac{-1}{2}$$

$$a'_{22} = a_{22} - a'_{12}a'_{21} = 1 - \frac{3}{2}(5) = 1 - \frac{15}{2} = -\frac{13}{2}$$

$$a'_{32} = a_{32} - a'_{12}a'_{31} = 2 - \frac{3}{2}(3) = 2 - \frac{9}{2} = -\frac{5}{2}$$

$$a'_{23} = \frac{a_{23} - a'_{13}a'_{21}}{a'_{22}} = \frac{1 - \left(\frac{1}{2}\right)5}{-13/2} = \frac{-3/2}{-13/2} = \frac{3}{13}$$

$$b'_2 = \frac{b_2 - a'_{21}b'_1}{a'_{22}} = \frac{9 - 5\left(-\frac{1}{2}\right)}{-13/2} = \frac{23/2}{-13/2} = -\frac{23}{13}$$

$$a'_{33} = a_{33} - a'_{31}a'_{13} - a'_{32}a'_{23}$$

$$= 4 - 3\left(\frac{1}{2}\right) - \left(-\frac{5}{2}\right)\left(\frac{3}{13}\right) = 4 - \frac{3}{2} + \frac{15}{26}$$

$$= \frac{104 - 39 + 15}{26} = \frac{80}{26} = \frac{40}{13}$$

$$b'_3 = \frac{b_3 - a'_{31}b'_1 - a'_{32}b'_2}{a'_{33}}$$

$$= \frac{11 - 3\left(-\dfrac{1}{2}\right) - \left(-\dfrac{5}{2}\right)\left(-\dfrac{23}{13}\right)}{\dfrac{40}{13}}$$

$$= \frac{11 + \dfrac{3}{2} - \dfrac{115}{26}}{\dfrac{40}{13}} = \frac{(286 + 39 - 115)/26}{40/13}$$

$$= \frac{210/26}{40/13} = \frac{210}{80}$$

$$\Rightarrow \quad b'_3 = \frac{21}{8}$$

Thus solution is

$$x_3 = b'_3 = \frac{21}{8}$$

$$x_2 = b'_2 - a'_{23}x_3 = -\frac{23}{13} - \left(\frac{3}{13}\right)\frac{21}{8} = -\frac{19}{8}$$

and $x_1 = b'_1 - a'_{12}x_2 - a'_{13}x_3$

$$= -\frac{1}{2} - \frac{3}{2}\left(-\frac{19}{8}\right) - \frac{1}{2}\left(\frac{21}{8}\right)$$

$$= -\frac{1}{2} + \frac{57}{16} - \frac{21}{16} = \frac{-8 + 57 - 21}{16} = \frac{14}{8} = \frac{7}{4}.$$

Hence, solution is $x_1 = \dfrac{7}{4}, x_2 = -\dfrac{19}{8}, x_3 = \dfrac{21}{8}.$

10.13 ITERATIVE METHODS

The previous methods for finding the solutions of simultaneous linear equations are known as direct methods, because after certain amount of fixed calculation, we obtain the solutions. On the other hand an iterative method is that method in which we assume initial approximation to the solution and obtain better and better solution with the desired degree of accuracy through number of iterations. Thus the amount of calculation depends on the desired degree of accuracy.

In this section, we shall discuss some iterative methods as Jacobi method, Gauss-seidel method and relaxation method.

10.13.1 JACOBI METHOD

Let us consider the system of linear simultaneous equations as

$$\left.\begin{array}{l} a_{11}x_1 + a_{12}x_2 + a_{13}x_3 + \ldots + a_{1n}x_n = b_1 \\ a_{21}x_1 + a_{22}x_2 + a_{23}x_3 + \ldots + a_{2n}x_n = b_2 \\ a_{31}x_1 + a_{32}x_2 + a_{33}x_3 + \ldots + a_{3n}x_n = b_3 \\ \text{\dotfill} \\ \text{\dotfill} \\ a_{n1}x_1 + a_{n2}x_2 + a_{n3}x_3 + \ldots + a_{nn}x_n = b_n \end{array}\right\} \qquad \ldots(1)$$

provided all the diagonal elements are not equal to zero. If so then rearrange the equations in such a way that all $a_{ii} \neq 0$. Now the system of equation (1) can be written as

$$\left.\begin{array}{l} x_1 = \dfrac{1}{a_{11}}(b_1 - a_{12}x_2 - a_{13}x_3 - \ldots - a_{1n}x_n) \\[2mm] x_2 = \dfrac{1}{a_{22}}(b_2 - a_{21}x_1 - a_{23}x_3 - \ldots - a_{2n}x_n) \\[2mm] x_3 = \dfrac{1}{a_{33}}(b_3 - a_{31}x_1 - a_{32}x_2 - \ldots - a_{3n}x_n) \\[2mm] \text{\dotfill} \\ x_n = \dfrac{1}{a_{nn}}(b_n - a_{n1}x_1 - a_{n2}x_2 - \ldots - a_{n(n-1)}x_{n-1}) \end{array}\right\} \qquad \ldots(2)$$

Let us assume the first approximations $x_1^{(1)}, x_2^{(1)}, \ldots, x_n^{(1)}$ for the values of $x_1, x_2, \ldots, x_n$ respectively. Substituting these values on R.H.S. of (2), we get a system of second approximation

$$\left.\begin{array}{l} x_1^{(2)} = \dfrac{1}{a_{11}}(b_1 - a_{12}x_2^{(1)} - a_{13}x_3^{(1)} - \ldots - a_{1n}x_n^{(1)}) \\[2mm] x_2^{(2)} = \dfrac{1}{a_{22}}(b_2 - a_{21}x_1^{(1)} - a_{23}x_3^{(1)} - \ldots - a_{2n}x_n^{(1)}) \\[2mm] x_3^{(2)} = \dfrac{1}{a_{33}}(b_3 - a_{31}x_1^{(1)} - a_{32}x_2^{(1)} - \ldots - a_{3n}x_n^{(1)}) \\[2mm] \text{\dotfill} \\ x_n^{(2)} = \dfrac{1}{a_{nn}}(b_n - a_{n1}x_1^{(1)} - a_{n2}x_2^{(1)} - \ldots - a_{n(n-1)}x_{n-1}^{(1)}) \end{array}\right\} \qquad \ldots(3)$$

Proceeding in the same way, we obtain third, fourth, etc. approximation. Let $x_1^{(n)}, x_2^{(n)}, \ldots, x_n^{(n)}$ be the nth approximation, then we get $(n + 1)$th approximation as

$$x_1^{(n+1)} = \frac{1}{a_{11}}(b_1 - a_{12}x_2^{(n)} - a_{13}x_3^{(n)} - ... - a_{1n}x_n^{(n)})$$

$$x_2^{(n+1)} = \frac{1}{a_{22}}(b_2 - a_{21}x_1^{(n)} - a_{23}x_3^{(n)} - ... - a_{2n}x_n^{(n)})$$

$$x_3^{(n+1)} = \frac{1}{a_{33}}(b_3 - a_{31}x_1^{(n)} - a_{32}x_2^{(n)} - ... - a_{3n}x_n^{(n)})$$

$$x_n^{(n+1)} = \frac{1}{a_{nn}}(b_n - a_{n1}x_1^{(n)} + a_{n2}x_2^{(n)} + ... + a_{n(n-1)}x_{n-1}^{(n)})$$

...(4)

Now equation (2) can be written in matrix form as

$$X^{(2)} = BX + C$$

...(5)

Similarly, equation (4) can be written as

$$X^{(n+1)} = BX^{(n)} + C$$

...(6)

✏ REMARKS
- This method is also known as method of simultaneous-displacement and this method is sufficiently convergent provided $|B| < 1$.
- In the absence of any better approximation, we can take each equal to zero.

🔖 Solved Examples

EXAMPLE 1. *Solve the following equation by Jacobi method.*

$$27x + 6y - z = 85$$
$$6x + 15y + 2z = 72$$
$$x + y + 54z = 110$$

SOLUTION. Above equations can be written as

$$x = \frac{1}{27}(85 - 6y + z)$$
$$y = \frac{1}{15}(72 - 6x - 2z) \qquad ...(1)$$
$$z = \frac{1}{54}(110 - x - y)$$

Let us take the first approximation to be $x^{(1)} = 0$, $y^{(1)} = 0$ and $z^{(1)} = 0$ we obtain second approximation from (1)

$$x^{(2)} = \frac{1}{27}(85 - 6y^{(1)} + z^{(1)}) = \frac{1}{27}(85 - 0 + 0) = \frac{85}{27} = 3.15$$

$$y^{(2)} = \frac{1}{15}(72 - 6x^{(1)} - 2z^{(1)}) = \frac{1}{15}(72 - 0 - 0) = \frac{72}{15} = 4.80$$

and $z^{(2)} = \frac{1}{54}(110 - x^{(1)} - y^{(1)}) = \frac{110}{54} = 2.04$.

Now we obtain third approximation as follows:

$$x^{(3)} = \frac{1}{27}(85 - 6y^{(2)} + z^{(2)})$$
$$= \frac{1}{27}(85 - 6(4.8) + 2.04) = \frac{58.24}{27} = 2.16$$

$$y^{(3)} = \frac{1}{15}(72 - 6x^{(2)} - 2z^{(2)})$$
$$= \frac{1}{15}(72 - 6(3.15) - 2(2.04)) = \frac{49.02}{15} = 3.27$$

$$z^{(3)} = \frac{1}{54}(110 - x^{(2)} - y^{(2)})$$
$$= \frac{1}{54}(110 - 3.15 - 4.80) = \frac{102.05}{54} = 1.89$$

Now fourth approximation is

$$x^{(4)} = \frac{1}{27}(85 - 6y^{(3)} + z^{(3)})$$
$$= \frac{1}{27}(85 - 6(3.27) + 1.89) = \frac{67.27}{27} = 2.49$$

$$y^{(4)} = \frac{1}{15}(72 - 6x^{(3)} - 2z^{(3)})$$
$$= \frac{1}{15}(72 - 6(2.16) - 2(1.89)) = \frac{55.26}{15} = 3.68$$

$$z^{(4)} = \frac{1}{54}(110 - x^{(3)} - y^{(3)})$$
$$= \frac{1}{54}(110 - 2.16 - 3.27) = \frac{104.57}{54} = 1.95$$

The fifth approximation is

$$x^{(5)} = \frac{1}{27}(85 - 6y^{(4)} + z^{(4)})$$
$$= \frac{1}{27}(85 - 6(3.68) + 1.95) = \frac{64.87}{27} = 2.40$$

$$y^{(5)} = \frac{1}{15}(72 - 6x^{(4)} - 2z^{(4)})$$
$$= \frac{1}{15}(72 - 6(2.49) - 2(1.95)) = \frac{53.16}{15} = 3.54$$

$$z^{(5)} = \frac{1}{54}(110 - x^{(4)} - y^{(4)})$$
$$= \frac{1}{54}(110 - 2.49 - 3.68) = \frac{103.83}{54} = 1.92$$

Hence, approximate solution is

$$x = 2.4, y = 3.54, z = 1.92.$$

EXAMPLE 2. *Solve the following equations by Jacobi method:*

$$20x + y - 2z = 17$$
$$3x + 20y - z = -18 \qquad \text{(BHOPAL–2009)}$$
$$2x - 3y + 20z = 25$$

SOLUTION. Above equations can be written as

$$x = \frac{1}{20}(17 - y + 2z)$$

$$y = \frac{1}{20}(-18 - 3x + z) \qquad \qquad ...(1)$$

$$z = \frac{1}{20}(25 - 2x + 3y)$$

Let us take first approximation $x^{(1)} = 0$, $y^{(1)} = 0$ and $z^{(1)} = 0$. Then the second approximation using (1), we get

$$x^{(2)} = \frac{1}{20}(17 - y^{(1)} + 2z^{(1)}) = \frac{17}{20} = 0.85$$

$$y^{(2)} = \frac{1}{20}(-18 - 3x^{(1)} + z^{(1)})$$

$$= \frac{1}{20}(-18 - 0 + 0) = -\frac{18}{20} = -0.9$$

$$z^{(2)} = \frac{1}{20}(25 - 2x^{(1)} + 3y^{(1)})$$

$$= \frac{1}{20}(25 - 0 + 0) = \frac{25}{20} = 1.25 .$$

Substituting these values in R.H.S.of (1), we get third approximation

$$x^{(3)} = \frac{1}{20}(17 - y^{(2)} + 2z^{(2)})$$

$$= \frac{1}{20}(17 + 0.9 + 2(1.25)) = \frac{20.4}{20} = 1.02$$

$$y^{(3)} = \frac{1}{20}(-18 - 3x^{(2)} + z^{(2)})$$

$$= \frac{1}{20}(-18 - 3(0.85) + 1.25) = -\frac{19.3}{20} = -0.965$$

$$z^{(3)} = \frac{1}{20}(25 - 2x^{(2)} + 3y^{(2)})$$

$$= \frac{1}{20}(25 - 2(0.85) + 3(0.9)) = \frac{20.6}{20} = 1.03$$

Now substitute these values into (1), we get fourth approximation

$$x^{(4)} = \frac{1}{20}(17 - y^{(3)} + 2z^{(3)})$$

$$= \frac{1}{20}(17 + 0.965 + 2(1.03)) = \frac{20.025}{20} = 1.0013$$

$$y^{(4)} = \frac{1}{20}(-18 - 3x^{(3)} + z^{(3)})$$

$$= \frac{1}{20}(-18 - 3(1.02) + 1.03) = -\frac{20.03}{20} = -1.0015$$

$$z^{(4)} = \frac{1}{20}(25 - 2x^{(3)} + 3y^{(3)})$$

$$= \frac{1}{20}(25 - 2(1.02) + 3(-0.965)) = \frac{20.065}{20} = 1.0032$$

Substitute these values into (1), we get fifth approximation

$$x^{(5)} = \frac{1}{20}(17 - y^{(4)} + 2z^{(4)})$$

$$= \frac{1}{20}(17 + 1.0015 + 2(1.0032)) = \frac{20.0079}{20} = 1.0004$$

$$y^{(5)} = \frac{1}{20}(-18 - 3x^{(4)} + z^{(4)})$$

$$= \frac{1}{20}(-18 - 3(1.0013) + 1.0032)$$

$$= -\frac{20.0007}{20} = -1.00003$$

$$z^{(5)} = \frac{1}{20}(25 - 2x^{(4)} + 3y^{(4)})$$

$$= \frac{1}{20}(25 - 2(1.0013) + 3(-1.005))$$

$$= \frac{19.9929}{20} = 0.9996$$

Substituting these values into (1), we get sixth approximation

$$x^{(6)} = \frac{1}{20}(17 - y^{(5)} + 2z^{(5)})$$

$$= \frac{1}{20}(17 + 1.00003 + 2(0.9996))$$

$$= \frac{19.99923}{20} = 0.99996$$

$$y^{(6)} = \frac{1}{20}(-18 - 3x^{(5)} + z^{(5)})$$

$$= \frac{1}{20}(-18 - 3(1.0004) + 0.9996)$$

$$= -\frac{20.0016}{20} = -1.00008$$

$$z^{(6)} = \frac{1}{20}(25 - 2x^{(5)} + 3y^{(5)})$$

$$= \frac{1}{20}(25 - 2(1.0004) + 3(-1.00003))$$

$$= \frac{19.99911}{20} = 0.99995$$

The values in the fifth and sixth approximation are nearly same. Hence, the solution is $x = 1, y = -1, z = 1$.

10.13.2 GAUSS-SEIDEL METHOD

This method is a modification of Jacobi Method and sometimes it gives faster convergence. To explain this method, let us consider a system of n equations in which $a_{ii} \neq 0$.

$$a_{11}x_1 + a_{12}x_2 + a_{13}x_3 + ... + a_{1n}x_n = b_1$$

$$a_{21}x_1 + a_{22}x_2 + a_{23}x_3 + ... + a_{2n}x_n = b_2$$

$$a_{31}x_1 + a_{32}x_2 + a_{33}x_3 + ... + a_{3n}x_n = b_3 \qquad ...(1)$$

$$..$$

$$..$$

$$a_{n1}x_1 + a_{n2}x_2 + a_{n3}x_3 + ... + a_{nn}x_n = b_n$$

Above equations can be written as

$$x_1 = \frac{1}{a_{11}}(b_1 - a_{12}x_2 - a_{13}x_3 - \ldots - a_{1n}x_n)$$

$$x_2 = \frac{1}{a_{22}}(b_2 - a_{21}x_1 - a_{23}x_3 - \ldots - a_{2n}x_n)$$

$$x_3 = \frac{1}{a_{33}}(b_3 - a_{31}x_1 - a_{32}x_2 - \ldots - a_{3n}x_n) \qquad \ldots(2)$$

$$\ldots\ldots\ldots\ldots\ldots\ldots\ldots\ldots\ldots\ldots\ldots\ldots\ldots\ldots\ldots\ldots$$

$$x_n = \frac{1}{a_{nn}}(b_n - a_{n1}x_1 - a_{n2}x_2 - \ldots - a_{n(n-1)}x_{n-1})$$

Now let us assume first approximations $x_1^{(1)}, x_2^{(1)}, x_3^{(1)}, \ldots, x_n^{(1)}$. Substitute these values into the first equation of (2), we get

$$x_1^{(2)} = \frac{1}{a_{11}}(b_1 - a_{12}x_2^{(1)} - a_{13}x_3^{(1)} - \ldots - a_{1n}x_n^{(1)}).$$

Now substitute $x_1^{(2)}, x_2^{(1)}, x_3^{(1)}, \ldots, x_n^{(1)}$ into the R.H.S. of second equation of (2), we get $x_2^{(2)} = \frac{1}{a_{22}}(b_2 - a_{21}x_1^{(2)} - a_{23}x_3^{(1)} - \ldots - a_{2n}x_n^{(1)}).$

In the third equation of (2) substitute $x_1^{(2)}, x_2^{(2)}, x_3^{(1)}, \ldots, x_n^{(1)}$ we get $x_3^{(2)} = \frac{1}{a_{33}}(b_3 - a_{31}x_1^{(2)} - a_{32}x_2^{(3)} - \ldots - a_{3n}x_n^{(1)}).$

Proceeding in this way, we get $x_n^{(2)}$ and this completes the first stage of iteration. The whole process is repeated till we get the values of x_1, $x_2, x_3, \ldots, x_n$ to the desired degree of accuracy. Therefore, Gauss-Seidel method is also known as a method of successive displacement.

☞ REMARKS

- Both the methods Jacobi and Gauss-Seidel converge for any type of first approximation if every equation of the system (2) satisfies the condition that the sum of the absolute values of the coefficients $\frac{a_{ij}}{a_{ii}}$ is almost equal to or in atleast one equation less than unity. That is, $\sum\limits_{\substack{j=1 \\ (j \neq i)}}^{n} \left|\frac{a_{ij}}{a_{ii}}\right| \leq 1, i = 1, 2, 3, \ldots n.$

- The sign < holds good in the case of atleast one equation.
- Gauss-Seidel Method converges twice as fast as the Jacobi Method.
- This method is also known as method of successive displacement.

Solved Examples

EXAMPLE 1. *Solve by following equation by Gauss-Seidel method :*

$$27x + 6y - z = 85$$
$$6x + 15y + 2z = 72 \qquad \text{(ANNA–2006)}$$
$$x + y + 54z = 110$$

SOLUTION. Above equations can be written as

$$\left.\begin{array}{l} x = \dfrac{1}{27}(85 - 6y + z) \\[2mm] y = \dfrac{1}{15}(72 - 6x - 2z) \\[2mm] z = \dfrac{1}{54}(110 - x - y) \end{array}\right\} \qquad \ldots(1)$$

Assuming first approximation $x^{(1)} = 0, y^{(1)} = 0, z^{(1)} = 0$. Substituting these values into the first equation of (1), we get

$$x^{(2)} = \frac{1}{27}(85 - 6y^{(1)} + z^{(1)}) = \frac{1}{27}(85) = 3.14$$

Now substitute $x^{(2)} = 3.14, y^{(1)} = 0, z^{(1)} = 0$ into second equation of (1), we get

$$y^{(2)} = \frac{1}{15}(72 - 6x^{(2)} + 2z^{(1)})$$

$$= \frac{1}{15}(72 - 6(3.14) - 0) = \frac{53.16}{15} = 3.54$$

Substitute $x^{(2)} = 3.14, y^{(2)} = 3.54, z^{(1)} = 0$ into third equation of (1), we get

and $z^{(2)} = \frac{1}{54}(110 - x^{(2)} - y^{(2)})$

$$= \frac{1}{54}(110 - 3.14 - 3.54) = \frac{103.32}{54} = 1.91$$

Thus second approximation are $x^{(2)} = 3.14$, $y^{(2)} = 3.54, z^{(2)} = 1.91$. Now, we proceed to obtain third approximations. For this substitute the values $x^{(2)} = 3.14, y^{(2)} = 3.54, z^{(2)} = 1.91$ into the first equation of (1), we get

$$x^{(3)} = \frac{1}{27}(85 - 6y^{(2)} + z^{(2)})$$

$$= \frac{1}{27}(85 - 6(3.54) + 1.91) = \frac{65.67}{27} = 2.43$$

Now, substitute $x^{(3)} = 2.43, y^{(2)} = 3.54, z^{(2)} = 1.91$ into second equation of (1), we get

$$y^{(3)} = \frac{1}{15}(72 - 6x^{(3)} - 2z^{(2)})$$

$$= \frac{1}{15}(72 - 6(2.43) - 2(1.91)) = \frac{53.6}{15} = 3.57$$

Now substitute $x^{(3)} = 2.43, y^{(3)} = 3.57, z^{(2)} = 1.91$ into third equation of (1), we get

$$z^{(3)} = \frac{1}{54}(110 - x^{(3)} - y^{(3)})$$

$$= \frac{1}{54}(110 - 2.43 - 3.57) = \frac{104}{54} = 1.92$$

Thus these values are close to $x^{(2)}, y^{(2)}, z^{(2)}$ respectively. Hence the solution is $x = 2.43$, $y = 3.57$, $z = 1.92$.

EXAMPLE 2. *Find the solution of the system of Gauss-Seidel method:*

$$83x + 11y - 4z = 95$$

$$7x + 52y + 13z = 104 \qquad \text{(HAZARIBAGH–2009)}$$

$$3x + 8y + 29z = 71$$

SOLUTION. Given equations can be written as

$$x = \frac{1}{83}(95 - 11y + 4z) \qquad \dots(1)$$

$$y = \frac{1}{52}(104 - 7x - 13z) \qquad \dots(2)$$

$$z = \frac{1}{29}(71 - 3x - 8y) \qquad \dots(3)$$

Taking first approximation $x^{(1)} = 0, y^{(1)} = 0, z^{(1)} = 0$ and substitute these values in (1), we get

$$x^{(2)} = \frac{1}{83}(95 - 11y^{(1)} + 4z^{(1)})$$

$$= \frac{1}{83}(95 - 0 + 0) = 1.14$$

Put $x^{(2)} = 1.14, y^{(1)} = 0, z^{(1)} = 0$ into (2), we get

$$y^{(2)} = \frac{1}{52}(104 - 7x^{(2)} - 13z^{(1)})$$

$$= \frac{1}{52}(104 - 7(1.14) - 0) = \frac{96.02}{52} = 1.85$$

Put $x^{(2)} = 1.14, y^{(2)} = 1.85, z^{(1)} = 0$ into (3), we get

$$z^{(2)} = \frac{1}{29}(71 - 3x^{(2)} - 8y^{(2)})$$

$$= \frac{1}{29}(71 - 3(1.14) - 8(1.85))$$

$$= \frac{52.78}{29} = 1.82$$

Next, put $x^{(2)} = 1.14, y^{(2)} = 1.85, z^{(2)} = 1.82$ into (1), we get

$$x^{(3)} = \frac{1}{83}(95 - 11y^{(2)} + 4z^{(2)})$$

$$= \frac{1}{83}(95 - 11(1.85) + 4(1.82)) = \frac{81.93}{83} = 0.99$$

Put $x^{(3)} = 0.99, y^{(2)} = 1.85, z^{(2)} = 1.82$ into (2), we get

$$y^{(3)} = \frac{1}{52}(104 - 7x^{(3)} - 13z^{(3)})$$

$$= \frac{1}{52}(104 - 7(0.99) - 13(1.82)) = \frac{73.41}{52} = 1.41$$

Put $x^{(3)} = 0.99, y^{(3)} = 1.41$, into (3), we get

$$z^{(3)} = \frac{1}{29}(71 - 3x^{(3)} - 8y^{(3)})$$

$$= \frac{1}{29}(71 - 3(0.99) - 8(1.41)) = \frac{56.75}{29} = 1.95$$

Again put $x^{(3)} = 0.99, y^{(3)} = 1.41, z^{(3)} = 1.95$ into (1), we get

$$x^{(4)} = \frac{1}{83}(95 - 11y^{(3)} + 4z^{(3)})$$

$$= \frac{1}{83}(95 - 11(1.41) + 4(1.95)) = \frac{87.29}{83} = 1.05$$

Now put $x^{(4)} = 1.05, z^{(3)} = 1.95$ into (2) we get

$$y^{(4)} = \frac{1}{52}(104 - 7x^{(4)} - 13z^{(3)})$$

$$= \frac{1}{52}(104 - 7(1.05) - 13(1.951)) = \frac{71.3}{52} = 1.37$$

Put $x^{(4)} = 1.05, y^{(4)} = 1.37$, into (3), we get

$$z^{(4)} = \frac{1}{29}(71 - 3x^{(4)} - 8y^{(4)})$$

$$= \frac{1}{29}(71 - 3(1.05) - 8(1.37)) = \frac{56.89}{29} = 1.96$$

Thus $x^{(4)}, y^{(4)}, z^{(4)}$ are sufficiently close to $x^{(3)}, y^{(3)}, z^{(3)}$ respectively. Hence, the solution is

$$x = 1.05, y = 1.37, z = 1.96$$

10.14 RELAXATION METHOD

This method was originally established by R.V. Southwll in 1935 for finding the solution related to engineering problems. To explain this method let us consider a system of three equations

$$\left.\begin{array}{l} a_{11}x_1 + a_{12}x_2 + x_{13}x_3 = b_1 \\ a_{21}x_1 + a_{22}x_2 + x_{23}x_3 = b_2 \\ a_{31}x_1 + a_{32}x_2 + x_{33}x_3 = b_3 \end{array}\right\} \qquad \dots(1)$$

First we define the residuals $R_{x_1}, R_{x_2}, R_{x_3}$ by the given relations

$$\left.\begin{array}{l} R_{x_1} = b_1 - a_{11}x_1 - a_{12}x_2 - a_{13}x_3 \\ R_{x_2} = b_2 - a_{21}x_1 - a_{22}x_2 - a_{23}x_3 \\ R_{x_3} = b_3 - a_{31}x_1 - a_{32}x_2 - x_{33}x_3 \end{array}\right\} \qquad \dots(2)$$

Let us take the initial approximation $x_1 = 0, x_2 = 0, x_3 = 0$ and find the initial residuals. These residuals are further reduced step by step, by giving some increments to the value of x_1, x_2 and x_3. At each step the magnitude of the largest residual is reduced to almost zero. In particular to reduce Rx_1, the value of the corresponding variable x_1 is changed, *i.e.*, to reduce R_{x_1} by a, we shall increase x_1 by a/a_{11}. This process is repeated till all the residuals have been reduced to almost zero, then to obtain the solution add all the increments in x_1, x_2, x_3 separately. To find the particular residual by giving increment to the corresponding variable, we use following operating table:

	δR_{x_1}	δR_{x_2}	δR_{x_3}
$\delta x_1 = 1$	$-a_{11}$	$-a_{21}$	$-a_{31}$
$\delta x_2 = 1$	$-a_{12}$	$-a_{22}$	$-a_{32}$
$\delta x_3 = 1$	$-a_{13}$	$-a_{23}$	$-a_{33}$

We observe from the above table that if x_1 is increased by 1 keeping y and z constant, then R_{x_1}, R_{x_2} and R_{x_3} decreases by a_{11}, a_{21} and a_{31} respectively. Similar effect has been shown in above table for the variables y and z respectively.

Remarks

- If $|a_{11}| \geq |a_{12}| + |a_{13}|, |a_{22}| \geq |a_{21}| + |a_{23}|, |a_{33}| \geq |a_{31}| + |a_{32}|$ hold, then relaxation Method can be applied successfully.
- In relaxation method, during the process, if all residuals are not negligible, then there is some mistake and the whole process should be rechecked.

Solved Examples

EXAMPLE 1. *Solve by Relaxation method, the equations :*

$$9x - 2y + z = 50$$
$$x + 5y - 3z = 18$$
$$-2x + 2y + 7z = 19 \qquad \text{(MADRAS–2000S)}$$

SOLUTION. The residuals are given by
$$R_x = 50 - 9x + 2y - z$$
$$R_y = 18 - x - 5y + 3z$$
$$R_z = 19 + 2x - 2y - 7z$$

The operations table is

	δR_x	δR_y	δR_z
$\delta x = 1$	-9	-1	2
$\delta y = 1$	2	-5	-2
$\delta z = 1$	-1	3	-7

The relaxation table is

	R_x	R_y	R_z
$x = y = z = 0$	50	18	19
$\delta x = 5$	5	13	29
$\delta z = 4$	1	25	1
$\delta y = 5$	11	0	-9
$\delta x = 1$	2	-1	-7
$\delta z = -1$	3	-4	0
$\delta y = -0.8$	1.4	0	1.6
$\delta z = 0.23$	1.17	0.69	-0.09
$\delta x = 0.13$	0	0.56	0.17
$\delta y = 0.112$	0.224	0	-0.054

From the above table we observe that intial residuals are 50, 18, 19, out of which 50 is the largest so to reduce it, we give an increment $\delta x = 5$ (or $50/9 \approx 5.5$) and the resulting residuals are 5, 13, 29 out of which 29 is the largest so to reduce it give an increment $\delta z = 4$ and thus 1, 25, 1 are resulting residuals. Continue this process till all the residuals become almost zero. Now find the addition of increments separately. That is,

$$\Sigma \delta x = 5 + 1 + 0.13 = 6.13,$$
$$\Sigma \delta y = 5 - 0.8 + 0.112 = 4.312$$

and $\Sigma \delta z = 4 + 0.23 = 4.23$.

Hence, the solution is $x = 6.13, y = 4.31, z = 4.23$.

EXAMPLE 2. *Solve the following equations by relaxation method:*

$$10x - 2y - 3z = 205$$
$$-2x + 10y - 2z = 154$$
$$-2x - y + 10z = 120 \qquad \text{(VTU–2011S, ROHTAK–2005)}$$

SOLUTION. The residuals are given by
$$R_x = 205 - 10x + 2y + 3z$$
$$R_y = 154 + 2x - 10y + 2z$$
$$R_z = 120 + 2x + y - 10z$$

The operations table is

	δR_x	δR_y	δR_z
$\delta R_x = 1$	-10	2	2
$\delta R_y = 1$	2	-10	1
$\delta R_z = 1$	3	2	-10

Now, the relaxation table is

	R_x	R_y	R_z
$x = y = z = 0$	205	154	120
$\delta x = 20$	5	194	160
$\delta y = 19$	43	4	179
$\delta z = 18$	97	40	-1
$\delta x = 10$	-3	60	19
$\delta y = 6$	9	0	25
$\delta z = 2$	15	4	5
$\delta x = 2$	-5	8	9
$\delta z = 1$	-2	10	-1
$\delta y = 1$	0	0	0

From above table, we observe that the initial residuals are 205, 154 and 120 out of these 205 is the largest so to reduce it give an increment $\delta x = 20$(which is obtained by $205/10 \approx 20$) and the resulting residual are 5, 194 and 160. Out of which 194 is the largest so to reduce it give and increment $\delta y = 19$ (again by $194/10 \approx 19$) and the resulting residual are 43, 4, 179. Proceeding in this way till all the residuals becomes almost zero. Thus

$\Sigma \delta x = 20 + 10 + 2 = 32$

$\Sigma \delta y = 19 + 6 + 1 = 26$

$\Sigma \delta z = 18 + 2 + 1 = 21.$

Hence, the solution is $x = 32, y = 26, z = 21.$

 Exercise-10.6

1. Solve the following equations by Jordan's Method:

(i) $2x - 3y + z = -1$
$x + 4y + 5z = 25$
$3x - 4y + z = 2$ (KERALA–2003)

(ii) $x_1 + 2x_2 + x_3 = 8$
$2x_1 + 3x_2 + 4x_3 = 20$
$4x_1 + 3x_2 + 2x_3 = 16$

(iii) $10x + y + z = 12$
$x + 10y + z = 12$
$x + y + 10z = 12$

2. Solve by LU decomposition Method:

(i) $2x + y + 2z = 2$
$x + 5y + 3z = 4$
$x + y - z = 0$

(ii) $x + y + 3z = 10$
$3x + 2y + 4z = 20$
$3x + 5y - z = 30$

(iii) $2x - 3y + 10z = 3$
$-x + 4y + 2z = 20$
$5x + 2y + z = -12$

3. Solve by Crout's Method:

(i) $2x - 6y + 8z = 24$
$5x + 4y - 3z = 2$
$3x + y + 2z = 16$

(ii) $2x_1 + 4x_2 + x_3 = 5$
$4x_1 + 4x_2 + 3x_3 = 8$
$4x_1 + 8x_2 + x_3 = 9$

(iii) $10x + y + 2z = 13$
$3x + 10y + z = 14$
$2x + 3y + 10z = 15$

4. Solve the equations by Jacobi's method :

(i) $5x - y + z = 10$
$2x + 4y = 12$
$x + y + 5z = -1$

(ii) $5x + 2y + z = 12$
$x + 4y + 2z = 15$
$x + 2y + 5z = 20$

5. Solve the following equations by Gauss-Seidel Method:

(i) $1.2x + 2.1y + 4.2z = 9.9$
$5.3x + 6.1y + 4.7z = 21.6$
$9.2x + 8.3y + z = 15.2$

(ii) $10x_1 - 2x_2 - x_3 - x_4 = 3$
$-2x_1 + 10x_2 - x_3 - x_4 = 15$
$-x_1 - x_2 + 10x_3 - 2x_4 = 27$
$-x_1 - x_2 - 2x_3 + 10x_4 = -9$ (BHOPAL–2009, JNTU–2004)

6. Solve the following equations by relaxation method:

(i) $3x + 9y - 2z = 11$
$4x + 2y + 13z = 24$
$4x - 4y + 3z = -8$ (BHOPAL–2002)

(ii) $-9x + 3y + 4z + 100 = 0$
$x - 7y + 3z + 80 = 0$
$2x + 3y - 5z + 60 = 0$

Answers

1. (i) $x = 8.7, y = 5.7, z = -1.3;$ (ii) $x_1 = 1, x_2 = 2, x_3 = 3;$ (iii) $x = 1, y = 5, z = 1;$

2. (i) $x = 1/5, y = 2/5, z = 3/5$ (ii) $x = 2, y = 5, z = 1$ (iii) $x = -4, y = 3, z = 2$

3. (i) $x = 1, y = 3, z = 5$ (ii) $x_1 = 1/2, x_2 = 3/4, x_3 = 1$ (iii) $x = y = z = 1;$

4. (i) $x = 2.56, y = 1.72, z = -1.06$ (ii) $x = 1.08, y = 1.95, z = 3.16$

5. (i) $x = -13.22, y = 16.76, z = -2.30$ (ii) $x_1 = 0.99, x_2 = 1.99, x_3 = 2.99, x_4 = -0.001$

6. (i) $x = -13.22, y = 16.76, z = -2.30;$ (ii) $x = 52.5, y = 44.5, z = 59.7$

10.15 EIGENVALUE AND EIGENVECTORS

Eigenvalue problems are arises in several engineering situation and applied Mathematics. Let a problem be denoted by a system of equations in matrix form as

$$Y = AX = \lambda X$$

where
$A = $ square matrix of order n

$X = $ a real vector never equal to '0'

$\lambda = $ eigenvalues or latent root of matrix A, which is real and positive,

the vector Y is real, has similar sense and direction when the system has an eigenvalue λ.

Characteristics Equation. Let $AX = \lambda X$ be the given eigenvalue problem then

$$(A - \lambda I)X = 0$$

where

$$A = \begin{bmatrix} a_{11} & a_{12} & \cdots & a_{1n} \\ a_{21} & a_{22} & \cdots & a_{2n} \\ \vdots & \vdots & \vdots & \vdots \\ a_{n1} & a_{n2} & \cdots & a_{nn} \end{bmatrix}$$

$$\lambda I = \begin{bmatrix} \lambda & 0 & \cdots & 0 \\ 0 & \lambda & \cdots & 0 \\ \vdots & \vdots & \vdots & \vdots \\ 0 & 0 & \cdots & \lambda \end{bmatrix}$$

Now, $(A - \lambda I)X = 0 \implies \begin{bmatrix} a_{11} - \lambda & a_{12} & \cdots & a_{1n} \\ a_{21} & a_{22} - \lambda & \cdots & a_{2n} \\ \vdots & \vdots & \vdots & \vdots \\ a_{n1} & a_{n2} & \cdots & a_{nn} - \lambda \end{bmatrix} X = 0$

It will have a non-trivial solution when the coefficient of determinant does not vanishes.

i.e., $\det(A - \lambda I) = \begin{bmatrix} a_{11} - \lambda & a_{12} & \cdots & a_{1n} \\ a_{21} & a_{22} - \lambda & \cdots & a_{2n} \\ \vdots & \vdots & \vdots & \vdots \\ a_{n1} & a_{n2} & \cdots & a_{nn} - \lambda \end{bmatrix} = 0$ $[\because X \neq 0]$

which is known as 'characteristic equation'.

On expanding, it gives the polynomial equation or characteristic equation in λ such that

$P_n(\lambda) = (-1)^n \lambda^n + a_1 \lambda^{n-1} + \dots + a_n = 0$ where $a_1 = -(a_{11} + a_{22} + a_{nn})\dots$ and $a_n = \det A$.

It is of degree n, has n roots say $\lambda_1, \lambda_2, \dots, \lambda_n$ and also $AX_i = \lambda_i X_i$ is known as eigenvector or characteristic vector.

WORKING PROCEDURE

Step 1. First evaluate the characteristic equation by putting
$\det(A - \lambda I) = 0$ ...(1)
Step 2. Find the coefficient or co-factors by expansion of equation (1).
Step 3. Put each value of λ into matrix $(A - \lambda_i)X_i = 0$.
Step 4. Find the different values of x_i for each λ_i.

10.15.1 POWER METHOD

It is an iterative technique which gives largest eigenvalue and its eigenvector conveniently.

Let us suppose A be a square matrix of order n having eigenvalues $\lambda_1, \lambda_2, \lambda_3, \dots, \lambda_n$ such that $|\lambda_1| > |\lambda_2| > \dots > |\lambda_n|$.

Let $X = c_1 x_1 + c_2 x_2 + \dots + c_n x_n$
$\implies$ $AX = c_1 A x_1 + c_2 A x_2 + \dots + c_n A x_n$.

Since $\lambda_1, \lambda_2, \dots, \lambda_n$ be the eigenvalues corresponding to vectors $x_1, x_2, \dots, x_n$, then

$AX = c_1 \lambda_1 x_1 + c_2 \lambda_2 x_2 + \dots + c_n \lambda_n x_n$

$A^2 X = c_1 \lambda_1^2 x_1 + c_2 \lambda_2^2 x_2 + \dots + c_n \lambda_n^2 x_n$

... ...
... ...

$A^m X = c_1 \lambda_1^m x_1 + c_2 \lambda_2^m x_2 + \dots + c_n \lambda_n^m x_n = c_1 \lambda_1^m x_1 + \lambda_1^m \sum_{i=2}^{n} c_i \left(\frac{\lambda_i}{\lambda_1}\right)^m x_i = c_1 \lambda_1^m x_1$

as $m \to \infty \left(\frac{\lambda_i}{\lambda_1}\right) \to 0$.

The power method gives the largest eigenvalue and eigenvector λ_1 and x_1 respectively.

Now avoid this procedure, let us set an iterative procedure as

$Y^{(i)} = AX^{(i-1)} = \lambda^{(i)} X^{(i)}$

Now, we repeat this process till $[X^{(i)} - X^{(i-1)}]$ be equal to zero.

On this condition, $\lambda_1^{(i)}$ will be largest eigenvalue and corresponding eigenvector is $x^{(i)}$.

Also $A^{-1} X = \frac{1}{\lambda} X$ gives the smallest eigenvalue.

Solved Examples

EXAMPLE 1. *Find the largest eigenvalue of the matrix*

$A = \begin{bmatrix} 1 & 2 & 3 \\ 0 & -4 & 2 \\ 0 & 0 & 7 \end{bmatrix}$, *using power method.*

SOLUTION. Let the eigenvector $X = \begin{bmatrix} 1 & 1 & 1 \end{bmatrix}^T = \begin{bmatrix} 1 \\ 1 \\ 1 \end{bmatrix}$.

First Iteration.

$Y^{(1)} = AX = \begin{bmatrix} 1 & 2 & 3 \\ 0 & -4 & 2 \\ 0 & 0 & 7 \end{bmatrix} \begin{bmatrix} 1 \\ 1 \\ 1 \end{bmatrix} = \begin{bmatrix} 6 \\ -2 \\ 7 \end{bmatrix} = 7 \begin{bmatrix} 0.86 \\ -0.28 \\ 1 \end{bmatrix}$

Here $\lambda_1 = 7, X^{(1)} = \begin{bmatrix} 0.86 \\ -0.28 \\ 1 \end{bmatrix}$.

Second Iteration.

$$Y^{(2)} = AX^{(1)} = \begin{bmatrix} 1 & 2 & 3 \\ 0 & -4 & 2 \\ 0 & 0 & 7 \end{bmatrix} \begin{bmatrix} 0.86 \\ -0.28 \\ 1 \end{bmatrix}$$

$$= \begin{bmatrix} 3.3 \\ 3.12 \\ 9 \end{bmatrix} = 7 \begin{bmatrix} 0.47 \\ 0.44 \\ 1 \end{bmatrix}$$

Here $\lambda_2 = 7, X^{(2)} = \begin{bmatrix} 0.47 \\ 0.44 \\ 1 \end{bmatrix}$.

Third Iteration.

$$Y^{(3)} = AX^{(2)} = \begin{bmatrix} 1 & 2 & 3 \\ 0 & -4 & 2 \\ 0 & 0 & 7 \end{bmatrix} \begin{bmatrix} 0.47 \\ 0.44 \\ 1 \end{bmatrix}$$

$$= 7 \begin{bmatrix} 0.62 \\ 0.034 \\ 1 \end{bmatrix} = 7$$

Here $\lambda_3 = 7, X^{(3)} = \begin{bmatrix} 0.62 \\ 0.034 \\ 1 \end{bmatrix}$.

Fourth Iteration.

$$Y^{(4)} = AX^{(3)} = \begin{bmatrix} 1 & 2 & 3 \\ 0 & -4 & 2 \\ 0 & 0 & 7 \end{bmatrix} \begin{bmatrix} 0.62 \\ 0.034 \\ 1 \end{bmatrix}$$

$$= \begin{bmatrix} 3.68 \\ 1.88 \\ 7 \end{bmatrix} = 7 \begin{bmatrix} 0.52 \\ 0.27 \\ 1 \end{bmatrix}$$

Here $\lambda_4 = 7, X^{(4)} = \begin{bmatrix} 0.52 \\ 0.27 \\ 1 \end{bmatrix}$.

Fifth Iteration.

$$Y^{(5)} = AX^{(4)} = \begin{bmatrix} 1 & 2 & 3 \\ 0 & -4 & 2 \\ 0 & 0 & 7 \end{bmatrix} \begin{bmatrix} 0.52 \\ 0.27 \\ 1 \end{bmatrix}$$

$$= \begin{bmatrix} 4.06 \\ 0.92 \\ 7 \end{bmatrix} = 7 \begin{bmatrix} 0.58 \\ 0.13 \\ 1 \end{bmatrix}$$

Here $\lambda_5 = 7, X^{(5)} = \begin{bmatrix} 0.58 \\ 0.13 \\ 1 \end{bmatrix}$.

Sixth Iteration.

$$Y^{(6)} = AX^{(5)} = \begin{bmatrix} 1 & 2 & 3 \\ 0 & -4 & 2 \\ 0 & 0 & 7 \end{bmatrix} \begin{bmatrix} 0.58 \\ 0.13 \\ 1 \end{bmatrix}$$

$$= \begin{bmatrix} 3.84 \\ 1.48 \\ 7 \end{bmatrix} = 7 \begin{bmatrix} 0.55 \\ 0.21 \\ 1 \end{bmatrix}$$

Here $\lambda_6 = 7, X^{(6)} = \begin{bmatrix} 0.55 \\ 0.21 \\ 1 \end{bmatrix}$.

Seventh Iteration.

$$Y^{(7)} = AX^{(6)} = \begin{bmatrix} 1 & 2 & 3 \\ 0 & -4 & 2 \\ 0 & 0 & 7 \end{bmatrix} \begin{bmatrix} 0.55 \\ 0.21 \\ 1 \end{bmatrix}$$

$$= \begin{bmatrix} 3.97 \\ 1.16 \\ 7 \end{bmatrix} = 7 \begin{bmatrix} 0.56 \\ 0.16 \\ 1 \end{bmatrix}$$

Here $\lambda_7 = 7, X^{(7)} = \begin{bmatrix} 0.56 \\ 0.16 \\ 1 \end{bmatrix}$.

Eight Iteration.

$$Y^{(8)} = AX^{(7)} = \begin{bmatrix} 1 & 2 & 3 \\ 0 & -4 & 2 \\ 0 & 0 & 7 \end{bmatrix} \begin{bmatrix} 0.56 \\ 0.16 \\ 1 \end{bmatrix}$$

$$= \begin{bmatrix} 3.88 \\ 1.36 \\ 7 \end{bmatrix} = 7 \begin{bmatrix} 0.55 \\ 0.19 \\ 1 \end{bmatrix}$$

Here $\lambda_8 = 7, X^{(8)} = \begin{bmatrix} 0.55 \\ 0.19 \\ 1 \end{bmatrix}$.

Ninth Iteration.

$$Y^{(9)} = AX^{(8)} = \begin{bmatrix} 1 & 2 & 3 \\ 0 & -4 & 2 \\ 0 & 0 & 7 \end{bmatrix} \begin{bmatrix} 0.55 \\ 0.19 \\ 1 \end{bmatrix}$$

$$= \begin{bmatrix} 3.93 \\ 1.24 \\ 7 \end{bmatrix} = 7 \begin{bmatrix} 0.56 \\ 0.177 \\ 1 \end{bmatrix}$$

Here $\lambda_9 = 7, X^{(9)} = \begin{bmatrix} 0.56 \\ 0.177 \\ 1 \end{bmatrix}$.

Tenth Iteration.

$$Y^{(10)} = AX^{(9)} = \begin{bmatrix} 1 & 2 & 3 \\ 0 & -4 & 2 \\ 0 & 0 & 7 \end{bmatrix} \begin{bmatrix} 0.56 \\ 0.177 \\ 1 \end{bmatrix}$$

$$= \begin{bmatrix} 3.92 \\ 1.28 \\ 7 \end{bmatrix} = 7 \begin{bmatrix} 0.56 \\ 0.18 \\ 1 \end{bmatrix}$$

Here $\lambda_{10} = 7, X^{(10)} = \begin{bmatrix} 0.56 \\ 0.18 \\ 1 \end{bmatrix}$.

Eleventh Iteration.

$$Y^{(11)} = AX^{(10)} = \begin{bmatrix} 1 & 2 & 3 \\ 0 & -4 & 2 \\ 0 & 0 & 7 \end{bmatrix} \begin{bmatrix} 0.56 \\ 0.18 \\ 1 \end{bmatrix} = 7 \begin{bmatrix} 0.56 \\ 0.18 \\ 1 \end{bmatrix}$$

$$\lambda_{11} = 7, X^{(11)} = \begin{bmatrix} 0.56 \\ 0.18 \\ 1 \end{bmatrix}.$$

Here $[X(11) - X(10)] = 0$

Hence, largest eigenvalue is 7 and corresponding

eigenvector is $\begin{bmatrix} 0.56 \\ 0.18 \\ 1 \end{bmatrix}$.

EXAMPLE 2. *Find the largest eigenvalue of the matrix*

$A = \begin{bmatrix} 1 & 2 \\ 3 & 4 \end{bmatrix}$ *using power method.*

SOLUTION . Let eigenvector $X = \begin{bmatrix} 1 & 0 \end{bmatrix}^T = \begin{bmatrix} 1 \\ 0 \end{bmatrix}$.

First Iteration.

$$Y^{(1)} = AX = \begin{bmatrix} 1 & 2 \\ 3 & 4 \end{bmatrix}\begin{bmatrix} 1 \\ 0 \end{bmatrix} = \begin{bmatrix} 1 \\ 3 \end{bmatrix} = 3\begin{bmatrix} 0.3 \\ 1 \end{bmatrix}$$

Here $\lambda_1 = 3, X^{(1)} = \begin{bmatrix} 0.3 \\ 1 \end{bmatrix}$.

Second Iteration.

$$Y^{(2)} = AX^{(1)} = \begin{bmatrix} 1 & 2 \\ 3 & 4 \end{bmatrix}\begin{bmatrix} 0.3 \\ 1 \end{bmatrix} = \begin{bmatrix} 2.3 \\ 4.9 \end{bmatrix} = 4.9\begin{bmatrix} 0.46 \\ 1 \end{bmatrix}$$

Here $\lambda_2 = 4.9, X^{(2)} = \begin{bmatrix} 0.46 \\ 1 \end{bmatrix}$.

Third Iteration.

$$Y^{(3)} = AX^{(2)} = \begin{bmatrix} 1 & 2 \\ 3 & 4 \end{bmatrix}\begin{bmatrix} 0.46 \\ 1 \end{bmatrix} = \begin{bmatrix} 2.46 \\ 5.38 \end{bmatrix} = 5.38\begin{bmatrix} 0.46 \\ 1 \end{bmatrix}$$

Here $\lambda_3 = 5.38, X^{(3)} = \begin{bmatrix} 0.46 \\ 1 \end{bmatrix}$.

Since $[X(3) - X(2)] = 0$.

Hence, largest eigenvalue is 5.38 and corresponding eigenvector $\begin{bmatrix} 0.46 & 1 \end{bmatrix}^T$.

Exercise-10.7

1. Determine the largest eigenvalue and the corresponding eigenvector of the matrices using the power method :

(i) $A = \begin{bmatrix} 2 & -1 & 0 \\ -1 & 2 & -1 \\ 0 & -1 & 2 \end{bmatrix}$ (VTU–2007)

(ii) $A = \begin{bmatrix} 4 & 1 & -1 \\ 2 & 3 & -1 \\ -2 & 1 & 5 \end{bmatrix}$ (VTU–2011)

(iii) $A = \begin{bmatrix} 6 & -2 & 2 \\ -2 & 3 & 1 \\ 2 & -1 & 3 \end{bmatrix}$ (VTU–2011S)

(iv) $A = \begin{bmatrix} 1 & -3 & 2 \\ 4 & 4 & -1 \\ 6 & 3 & 5 \end{bmatrix}$ (ANNA–2005)

(v) $A = \begin{bmatrix} 25 & 1 & 2 \\ 1 & 3 & 0 \\ 2 & 0 & -4 \end{bmatrix}$ (VTU–2008)

(vi) $A = \begin{bmatrix} 1 & 3 & -1 \\ 3 & 2 & 4 \\ -1 & 4 & 10 \end{bmatrix}$ with initial approximation $[1, 1, 0]^T$.

(MADRAS–2006)

Answers

1. (i) 3.41, $[0.74, -1, 0.67]'$ (ii) 6, $[1, 1, -1]'$ (iii) 8, $[1, -0.5, 0.5]'$ (iv) 7, $[2.099/7, 0.467/7, 1]$

(v) 25.182, $[1, 0.045, 0.068]'$ (vi) 11.66, $[0.025, 0.422, 1.000]$

Miscellaneous Exercise

1. By using the bisection method, find an approximate root of the equation $\sin x = 1/x$ that lies between $x = 1$ and $x = 1.5$ (measured in radians). Carry out computations upto the 7th stage. (VTU–2003S)

2. Find the positive root of $x^4 - x = 10$ correct to three decimal places, using Newton-Raphson method.
(JNTU–2008, MADRAS–2006)

3. Find the Newton's method, the real root of the equation $3x = \cos x + 1$. (VTU–2009, SVTU–2007)

4. Find a root of the following equation, using the bisection method correct to three decimal places.
$x^3 - x^2 - 1 = 0$ (JNTU–2009)

5. Using Newton-Raphson method, find a root of the following equations correct to three decimal places.
$x \tan x + 1 = 0$ which is near $x = \pi$.
(JNTU–2006, VTU–2006)

6. Develop a recurrence formula for finding $\sqrt{N}$, , using Newton-Raphson method and hence compute to three decimal places.
(i) $\sqrt{13}$ (UPTU–2008)
(ii) $\sqrt{10}$ (JNTU–2008)

7. Solve $10x - 7y + 3z + 5u = 6, -6x + 8y - z - 4u = 5, 3x + y + 4z + 11u = 2, 5x - 9y - 2z + 4u = 7$ by Gauss elimination method. (SVTU–2007)

8. Solve the equation by Gauss elimination method.
$2x + 2y + z = 12; 3x + 2y + 2z = 8; 5x + 10y - 8z = 10$.
(WBTU–2004)

9. Solve the following equations by Gauss-Jordan method.
(i) $2x + 5y + 7z = 52; 2x + y - z = 0, x + y + z = 9$
(VTU–2010)
(ii) $2x + y + z = 10; 3x + 2y + 3z = 18, x + 4y + 9z = 16$
(VTU–2008)

10. Solve by Jacobi's iteration method, the equations, $\hspace{2cm}$ (VTU–2011, ROHTAK–2005, MADRAS–2003, BHOPAL–2009)

$\quad 10x + y - z = 11.19, x + 10y + z = 28.08, -x + y + 10z =$ $\quad$ **12.** Solve the following equations by Gauss-Seidel method.

$\quad 35.61$, correct to two decimal places. $\hspace{1cm}$ (ANNA–2007) $\qquad 28x + 4y - z = 32, x + 3y + 10z = 24, 2x + 17y + 4z = 35$

11. Apply Gauss-Seidel iteration method to solve the equations. $\hspace{3cm}$ (MUMBAI–2009)

$\quad 20x + y - 2z = 17; 3x + 20y - z = -18, 2x - 3y + 20z = 25$

Answers

1. 1.11328	**2.** 1.856	**3.** 0.6071	**4.** 1.46	**5.** 2.7985

6. $x_{n+1} = 1/2(x_n + N/x_n);$ (i) 3.605 (ii) 3.162 $\quad$ **7.** $u = 1, z = -7, y = 4$ and $x = 5$ $\quad$ **8.** $x = -51/4, y = 115/8, z = 35/4$

9. (i) $x = 1, y = 3, z = 5$ (ii) $x = 7, y = -9, z = 5$ $\qquad$ **10.** $x = 1.23, y = 2.34, z = 3.45$

11. $x = 1, y = -1, z = 1$ $\hspace{4cm}$ **12.** $x = 0.998, y = 1.723, z = 2.024$

10.16 DIFFERENCE SCHEMES

Let $y = f(x)$ be a function (discrete), which takes the values $f(x_0), f(x_0 + h), f(x_0 + 2h)...f(x_0 + nh)$, for the equidistant values $x_0, x_0 + h, x_0 + 2h...x_0 + nh$ of x respectively. Here, the value of the independent variable x is usually called the argument and the corresponding functional value is called entry.

Generally, there are following three types of differences :
(a) Forward differences $\quad$ (b) Backward differences $\quad$ (c) Central differences.

10.16.1 FORWARD DIFFERENCES

The differences $(f(x_0 + h) - f(x_0)), (f(x + 2h) - f(x_0 + h)),...(f(x_0 + nh) - f(x_0 + (n-1)h))$ are called first forward differences and are denoted by $\Delta f(x_0), \Delta f(x_0 + h)...\Delta f(x_0 + (n-1)h)$. Here, Δ is known as forward difference operator.

Therefore, $\hspace{4cm} \Delta f(x_0) = f(x_0 + h) - f(x_0)$

or $\hspace{5cm} \Delta y_0 = y_1 - y_0$

The first forward differences are given by the following formula,

$$\Delta y_k = y_{k+1} - y_k$$

Similarly, the second forward differences are

$$\Delta^2 y_k = \Delta y_{k+1} - y_k = y_{k+2} - 2y_{k+1} - y_k$$

In general $\hspace{4cm} \Delta^r y_k = \Delta^{r-1} y_{k+1} - \Delta^{r-1} y_k$

In function notation, the forward differences are given below:

$$\Delta f(x) = f(x + h) - f(x)$$

$$\Delta^2 f(x) = f(x + 2h) - 2f(x + h) - f(x)$$
$$\Delta^3 f(x) = f(x + 3h) - 3f(x + 2h) + 3f(x + h) - f(x).\therefore \text{ and so on.}$$

The forward differences are usually arranged in form in the following manner.

WORKING PROCEDURE

	(To construct the forward difference table)
Step 1.	Write the value of the independent variable x in column (1).
Step 2.	Write the corresponding value of y for given value of x in column (2).
Step 3.	Subtracting each value of $f(x)$ in column (ii) from the succeding value of $f(x)$ and write this value in column (iii).
Step 4.	The operation of step-3 is applied on the figures of column (iii) to get column (iv) and so on till all the figures in a column become constant.

Forward Difference Table

Argument x	Entry y	First Diff. Δ	Second Diff. Δ^2	Third Diff. Δ^3	Fourth Diff. Δ^4	Fifth Diff. Δ^5
x_0	y_0					
		Δy_0				
$x_1 = x_0 + h$	y_1		$\Delta^2 y_0$			
		Δy_1		$\Delta^3 y_0$		
$x_2 = x_0 + 2h$	y_2		$\Delta^2 y_1$		$\Delta^4 y_0$	
		Δy_2		$\Delta^3 y_1$		$\Delta^5 y_0$
$x_3 = x_0 + 3h$	y_3		$\Delta^2 y_2$		$\Delta^4 y_1$	
		Δy_3		$\Delta^3 y_2$		
$x_4 = x_0 + 4h$	y_4		$\Delta^2 y_3$			
		Δy_4				
$x_5 = x_0 + 5h$	y_5					

In the above table y_0 is called leading term and the difference $\Delta y_0, \Delta^2 y_0, \Delta^3 y_0, \ldots,$ are called the leading differences.

☛ **REMARK**
- The above difference table is known as forward difference table or diagonal difference table.

10.16.2 PROPERTIES OF FORWARD DIFFERENCE OPERATOR Δ

The forward operator Δ satisfies the following properties:

(i) $\Delta[f(x) \pm g(x)] = \Delta f(x) \pm \Delta g(x)$

(ii) $\Delta[a.f(x)] = a.\Delta f(x), a \in \mathbf{R}$

(iii) $\Delta^m \Delta^n f(x) = \Delta^{m+n} f(x) = \Delta^{n+m} f(x) = \Delta^n \Delta^m f(x), m, n \in \mathbf{Z}^+$

(iv) $\Delta[f(x).g(x)] = [\Delta f(x)]g(x) + f(x)[\Delta g(x)]$

10.16.3 BACKWARD DIFFERENCES

The differences $y_1 - y_0, y_2 - y_1, \ldots, y_n - y_{n-1}$ when denoted by $\nabla y_1, \nabla y_2, \ldots, \nabla y_n$ respectively are called the first backward differences, when ∇ is called backward difference operator.

Therefore,

$$\nabla y_1 = y_1 - y_0$$
$$\nabla y_2 = y_2 - y_1$$
$$\nabla y_n = y_n - y_{n-1}$$

Now, the second backward differences are defined as follows :

$$\nabla^2 y_2 = \nabla(\nabla y_2) = \nabla(y_2 - y_1) = \nabla y_2 - \nabla y_1 = (y_2 - y_1) - (y_1 - y_0) = y_2 - 2y_1 + y_0$$
$$\nabla^3 y_3 = \nabla y_3 - \nabla y_2 = y_3 - 2y_2 + y_1$$

and so on.

In general

$$\nabla^n y_k = \nabla^{n-1} y_k - \nabla^{n-1} y_{k-1}$$

In function notation, the backward differences are given below :

$$\nabla f(x) = f(x) - f(x-h)$$
$$\nabla f(x+h) = f(x+h) - f(x)$$
$$\nabla^2 f(x+2h) = f(x+2h) - 2f(x+h) + f(x) \text{ ; where } h \text{ is the interval of differences.}$$

and so on.

The bakcward differences are usually arranged in a tabular from in the following manner.

Backward Differene Table

Argument x	Entry y	First Diff. Δ	Second Diff. ∇^2	Third Diff. ∇^3	Fourth Diff. ∇^4	Fifth Diff. ∇^5
x_0	y_0					
		∇y_1				
$x_1 = x_0 + h$	y_1		$\nabla^2 y_2$			
		∇y_2		$\nabla^3 y_3$		
$x_2 = x_0 + 2h$	y_2		$\nabla^2 y_3$		$\nabla^4 y_4$	
		∇y_3		$\nabla^3 y_4$		$\nabla^5 y_5$
$x_3 = x_0 + 3h$	y_3		$\nabla^2 y_4$		$\nabla^4 y_5$	
		∇y_4		$\nabla^3 y_5$		
$x_4 = x_0 + 4h$	y_4		$\nabla^2 y_5$			
		∇y_5				
$x_5 = x_0 + 5h$	y_5					

10.16.4 CENTRAL DIFFERENCES

If $\qquad y_1 - y_0 = \delta y_{1/2}, y_2 - y_1 = \delta y_{3/2}, y_n - y_{n-1} = \delta y_{(2n-1)/2}$

Then these differences are called central difference and δ is called central difference operator.

Similarly, we can define higher order central differences as

$$\delta y_{3/2} - \delta y_{1/2} = \delta^2 y_1, \delta y_{5/2} - \delta y_{3/2} = \delta^2 y_2$$

and $\qquad \delta^2 y_2 - \delta^2 y_1 = \delta^3 y_{3/2} \qquad$ and so on.

The central difference table is given below.

Central Differene Table

Argument x	Entry y	First Diff.	Second Diff.	Third Diff.	Fourth Diff.	Fifth Diff.
x_0	y_0					
		$\delta y_{1/2}$				
x_1	y_1		$\delta^2 y_1$			
		$\delta y_{3/2}$		$\delta^3 y_{3/2}$		
x_2	y_2		$\delta^2 y_2$		$\delta^4 y_2$	
		$\delta y_{5/2}$		$\delta^3 y_{5/2}$		$\delta^5 y_{5/2}$
x_3	y_3		$\delta^2 y_3$		$\delta^4 y_3$	
		$\delta y_{7/2}$		$\delta^3 y_{7/2}$		
x_4	y_4		$\delta^2 y_4$			
		$\delta y_{9/2}$				
x_5	y_5					

10.16.5 Shift Operator

The operator which increases the argument x by h such that

$$Ef(x) = f(x+h)$$
$$E^2 f(x) = f(x+2h)$$

and
$$E^3 f(x) = f(x+3h) \text{ ... etc.}$$

This operator E is called shift operator. The inverse operator E^{-1} is defined as

$$E^{-1} f(x) = f(x-h), E^{-2} f(x) = f(x-2h) \text{ ... etc.}$$

In general,

$$E^n f(x) = f(x+nh)$$

where n any real number.

Then these differences called central difference and δ is called central difference operator.

10.16.6 Relation between the Operators

We have the following identities :

(i) $\Delta = E - 1$ (ii) $\nabla = 1 - E^{-1}$

(iii) $\delta = E^{1/2} - E^{-1/2}$ (iv) $\Delta = E\nabla = \nabla E$

(v) $E = e^{hD}$

Proof :

(i) Since $\Delta y_x = y_{x+h} - y_x = Ey_x - y_x$ for all x

               $= (E-1)y_x$ for all x

$\therefore$ $\Delta = E - 1$ or $E = 1 + \Delta$

(ii) Since $\nabla y_x = y_x - y_{x-h} = y_x - E^{-1} y_x$ for all x

          $\nabla y_x = (1 - E^{-1})y_x$ $\nabla = 1 - E^{-1}$ for all x

(iii) Since $\delta y_x = y_{x+\frac{n}{2}} - y_{x-\frac{n}{2}} = E^{1/2} y_x - E^{-1/2} y_x$ for all x

             $= (E^{1/2} - E^{-1/2})y_x$ for all x

$\therefore$ $\delta = E^{1/2} - E^{-1/2}$

(iv) Since $E\nabla y_x = E(y_x - y_{x-h}) = Ey_x - Ey_{x-h}$ for all x

             $= y_{x+h} - y_x$ for all x

             $= \Delta y_x$

$\therefore$ $E\nabla = \Delta$

and $\nabla E y_x = \Delta y_{x+h}$ for all x ...(i)

             $= y_{x+h} - y_x$ for all x

$$\therefore \qquad \nabla E = \Delta \qquad \qquad ...(ii)$$

from (i) and (ii), $\qquad \Delta = E\nabla = \nabla E$

(v) Since $\qquad Ef(x) = f(x+h)$

$$= f(x) + h'f(x) + \frac{h^2}{2!}f''(x) + ... \qquad \text{By Taylor's Theorem}$$

$$= f(x) + hD(x) + \frac{h^2}{2!}D^2 f(x) + ... = \left(1 + hD + \frac{h^2}{2!}D^2 + ...\right)f(x)$$

$$Ef(x) = e^{hD}f(x)$$

$$\therefore \qquad E = e^{hD} \qquad\qquad\qquad \text{for all } x.$$

📝 **REMARKS**

- In the following figure (table) $\Delta^k y_0$ lie on a straight line down to right. On the other hand, since $\Delta = E\nabla$, we have $\Delta y_4 = \nabla y5, \Delta^2 y_3 = \nabla^2 y_5, \Delta^3 y_2 = \nabla^3 y_5$ and so on. Further since $\nabla^k y_x$ lie on a straight line sloping downward to the right. Similarly we also have $\Delta = E^{1/2}\delta$ and hence, we have $\Delta^2 y_1 = E\delta^2 y_1 = \delta^2 y_2, \Delta^4 y_0 = \delta^2 y_2$ and so on. In this way, we can observe that $\delta^{2k}y_k$ lie on a horizontal line.

- From all the difference tables, we can see that only the notation changes not the differences. For examples,
$$y_1 - y_0 = \Delta y_0 = \nabla y_1 = \delta y_{1/2}$$

- If we write $y = f(x)$ as $y = f_x$ or $y = y_x$ then the entries corresponding to $x, x = h, x+2h, ...$ are $y_x, y_{x+h}, y_{x+2h}, ...$ respectively and $\Delta y_x = y_{x+h} - y_x, \Delta^2 y_x, = \Delta y_x, -y_x,$ and so on.

- Similarly, $\nabla y_x = y_x - y_{x-h}$ and $\delta y_x = y_{x+\frac{h}{2}} - y_{x-\frac{1}{2}h}$ and so on.

10.17 FACTORIAL NOTATION

A product of the form $x(x-1)(x-2) ... (x-n+1)$ is denoted by $x^{(n)}$ and is called a factorial. In particular,

$$x^{(1)} = x, x^{(2)} = x(x-1), x^{(3)} = x(x-1)(x-2) ... \text{ etc.}$$

If the interval of differencing is h, then,

$$x^{(n)} = x(x-h)(x-2h)...[x-(n-1)h]$$

10.17.1 DIFFERENCE OF $x^{(n)}$

Consider, $\quad \Delta x^{(n)} = (x+h)^n - x^n$

$$= (x+h)x(x-h)...[x-(n-2)h] - x(x-h)...[x-(n-1)h]$$

$$= x(x-h)...[x-(n-1)h]\{x+h-x-(n+1)h\} = x^{(n-1)}.n.h = nhx^{(n-1)}$$

Similarly, $\quad \Delta^2 x^{(n)} = \Delta[\Delta x^{(n)}] = \Delta[nhx^{(n-1)}] = n(n-1)h^2 x^{(n-2)}$

$$\Delta^3 x^{(n)} = n(n-1)(n-2)h^3 x^{(n-3)}$$

In general,

$$\Delta^n x^{(n)} = n(n-1)(n-2)...3.2.1h^n x^{(n-n)} = n!h^n \text{ and } \Delta^{n+1} x(n) = 0$$

📝 **REMARKS**
- The result of difference $x^{(n)}$ is analogous to that of differentiating x^n.
- If $h = 1$, then $\Delta^n x^{(n)} = n!$.

10.17.2 RECIPROCAL FACTORIAL

The reciprocal factorial function $x^{(-n)}$ is defined by

$$x^{(-n)} = \frac{1}{(x+h)(x+2h)...(x+nh)} \qquad \text{where } n \in \mathbf{Z}^+$$

📝 **REMARK**
- Clearly $\Delta x^{(-n)} = (-n)hx^{[-(n+1)]}, \Delta^2 x^{(-n)} = (-1)^2 n(n+1)h^2 x^{[-(n+2)]}$

In general $\Delta^k x^{(-n)} = (-1)^k n(n+1)h^2 x^{[-(n+2)]}$

10.17.3 TO EXPRESS A GIVEN POLYNOMIAL INTO FACTORIAL NOTATION

(Unity being the differencing interval)

Method I.

This can be done by using the finite differences of the function. Let $f(x)$ be the polynomial of degree n which is to be expressed in factorial notation.

Let
$$f(x) = A + Bx^{(1)} + Cx^{(2)} + \ldots + Kx^{(n)} \qquad \ldots(1)$$

where $A, B, C, \ldots, K$ are some unknown constants, $K \neq 0$.

Then
$$\Delta f(x) = \Delta[A + Bx^{(1)} + Cx^{(2)} + \ldots + Kx^{(n)}] = B.1 + C.2x^{(2)} + \ldots + K.nx^{(n-1)}$$

$$\Delta^2 f(x) = \Delta[B + C.2x^{(1)} + \ldots + Kn.x^{(n-1)}] = C.2.1 + D.3.2.x^{(1)} + \ldots + Kn.(n-1)x^{(n-2)}$$

$$\ldots \quad \ldots \quad \ldots \quad \ldots \quad \ldots \quad \ldots \quad \ldots \quad \ldots \quad \ldots$$

$$\Delta^n f(x) = K.n.(n-1)\ldots2.1.x^{(0)} = K.n.!$$

Substitution of $x = 0$ in all above equations yields

$$f(0) = A, \Delta f(0) = \frac{B}{1!}$$

$$\Delta^2 f(0) = C.2! \implies C = \frac{\Delta^2 f(0)}{2!} \; ; \; \Delta^3 f(0) = D.3! \implies D = \frac{\Delta^3 f(0)}{3!}$$

$$\ldots \quad \ldots \quad \ldots \quad \ldots \quad \ldots \quad \ldots \quad \ldots \quad \ldots \quad \ldots \quad \ldots$$

$$\ldots \quad \ldots \quad \ldots \quad \ldots \quad \ldots \quad \ldots \quad \ldots \quad \ldots \quad \ldots \quad \ldots$$

$$\Delta^n f(0) = K.n! \implies K = \frac{\Delta^n f(0)}{n!}$$

Putting the value of $A, B, C, \ldots, K$, in (1) we get

$$f(x) = f(0) + \frac{\Delta f(0)}{1!}x^{(1)} + \frac{\Delta^2 f(0)}{2!}x^{(2)} + \ldots + \frac{\Delta^n f(0)}{n!}x^{(n)}$$

 Solved Examples

EXAMPLE. *Express* $f(x) = 2x^3 - 3x^2 + 3x - 10$ *and its differences in factorial notation, the interval of differencing being unity.*

SOLUTION: **Method I.** The given polynomial is of degree 3. Hence the maximum power of the function in factorial notation can be three only. Thus using upto $x = 3$ only, we get

$$f(x) = f(0) + \Delta f(0)x^{(1)} + \frac{\Delta^2 f(0)}{2!}x^{(2)} + \frac{\Delta^3 f(0)}{3!}x^{(3)}$$

Now to find the values of $f(0), \Delta f(0), \Delta^2 f(0), \Delta^3 f(0)$ we must have four successive values of the function $f(x)$ i.e., at $x = 0$, $x = 1$, $x = 2$ and $x = 3$. From $f(x) = 2x^3 - 3x^2 + 3x - 10$ the values for $x = 0, 1, 2, 3$ respectively are

$$f(0) = -10, f(1) = -8, f(2) = 0 \text{ and } f(3) = 26.$$

Now to find $\Delta f(0)$, $\Delta^2 f(0)$, and $\Delta^3 f(0)$ we can construct the differnce table in the usual manner.

x	$f(x)$	$\Delta f(x)$	$\Delta^2 f(x)$	$\Delta^3 f(x)$
0	-10			
		2		
1	-8		6	
		8		12
2	0		18	
		26		
3	26			

Here $f(0) = -10, \Delta f(0) = 2$,
$\Delta^2 f(0) = 6, \Delta^3 f(0) = 12$.

Hence, we get
$$f(x) = -10 + \frac{2}{1!}x^{(1)} + \frac{6}{2!}x^{(2)} + \frac{12}{3!}x^{(3)}$$
$$= -10 + 2x^{(1)} + 3x^{(2)} + 2x^{(3)}$$

The problem can also be solved by alternative methods given below:

Method II. This can be explained by this example

Let
$$2x^3 - 3x^2 + 3x - 10 = Ax^{(3)} + Bx^{(2)} + Cx^{(1)} + D$$
$$= Ax(x-1)(x-2) + Bx(x-1) + Cx + D \qquad \ldots(i)$$

Now putting $x = 0$ in (i), we get $D = -10$.

Again putting $x = 1$ in (i), we get
$$2 - 3 + 3 - 10 = C + D$$
$$\therefore \quad C = 2 \text{ as } D = -10$$

Similarly putting $x = 2$ in (i), we obtain
$$16 - 12 + 6 - 10 = 2B + 2C + D$$
$$0 = 2B - 6 \text{ or } B = 3.$$

Equating the coefficients of x3 on both the sides of (i) we get A = 2. Putting the values of A, B, C and D in R.H.S. of (i), the required polynomial in factorial notation will be

$$f(x) = 2x^{(3)} + 3x^{(2)} + 2x - 10$$

and by the rule of simple differentiation
$$\Delta f(x) = 6x^{(2)} + 6x + 2$$
$$\Delta^2 f(x) = 12x + 6 \text{ and } \Delta^3 f(x) = 12.$$

WORKING PROCEDURE

Step 1. The given function is expressed term by term in factorial functions with certain unknown coefficients as shown in (i).

Step 2. The values of the unknown coefficients are calculated by putting $x = 0, 1, 2, \ldots$ successively in L.H.S. and R.H.S. of (i) and then these are solved to find the values of $A, B, C,$ etc.

Step 3. The values of $A, B, C,$ etc. are substituted from step 2 in R.H.S. of (i) to get the given polynomial in factorial notation.

The above method though simple is not a convenient method for expressing a polynomial in factorial form. The procedure is lengthly and there is every possibility of committing some error.

Method III. Detached Coefficient Method

Let $2x^3 - 3x^2 + 3x - 10 = Ax^{(3)} + Bx^{(2)} + Cx^{(4)} + D$

$$= Ax(x-1)(x-2) + Bx(x-1) + Cx + D$$

If we divide the given function by x then the remainder will be -10 and the quotient $2x^2 - 3x + 3$. The value of D in (i) is taken as -10.

Again divide the quotient $2x^2 - 3x + 3$ by $(x-1)$, *i.e.*,

$$
\begin{array}{r}
2x - 1 \\
x-1\overline{)2x^2 - 3x + 3} \\
\underline{2x^2 - 2x} \\
-x + 3 \\
\underline{-x + 1} \\
2
\end{array}
$$

∴ The quotient now is $2x - 1$ and the remainder is the value of C, *i.e.*, $C = 2$

Again divide $2x - 1$ by $x - 2$ so that

$$
\begin{array}{r}
2 \\
x-2\overline{)2x - 1} \\
\underline{2x - 4} \\
3
\end{array}
$$

The quotient 2 is the value of A and remainder 3 is B. Thus the required polynomial is

$$2x^{(3)} + 3x^{(2)} + 2x - 10$$

The above method can be simplified by the procedure of detatched coefficients, which is as follows.

Taking the coefficients of the various powers of x in the given polynomial, we have

$$
\begin{array}{cccccc}
1 & 2 & -3 & 3 & & -10 = D & \dots \text{(2)} \\
& 0 & 2 & -1 & & & \\
2 & 2 & -1 & \boxed{2 = C} & & & \dots \text{(3)} \\
& 0 & 4 & & & & \\
3 & 2 & \boxed{3 = B} & & & & \dots \text{(4)} \\
& 0 & & & & & \\
& 2 = A & & & & &
\end{array}
$$

WORKING PROCEDURE

Following are the steps in the method of detached coefficients:

Step 1. Write the coefficients of various power of x in the order starting with the coefficient of the highest power of x.

Step 2. Put 1 in the L.H.S. column and write zero below the coefficient of highest power of x, *i.e.*, write zero below 2 as shown in (2).

Step 3. Multiply 2 by 1 and 0 by 1 and add to get the sum 2 and write it below -3 as shown in. Similarly multiply -3 by 1 and 2 by 1 and add to get -1 and write it below 3. The remainder -10 is the value of D.

Step 4. Add the values of corresponding column of (2) to get 2, -1 and 2 of (3).

Step 5. Write 2 in the left hand column of (3) and repeat the steps (2) and (3) to get $(2 \times 2) + (0 \times 2) = 4$ and write below -1. The remainder 2 of (3) is equal to C.

Step 6. Apply operation of step 4 on (3) to get 2 and 3 of (4).

Step 7. Write 3 in left hand column of (4) and repeat the steps 2, 3 and 4 to get 2 which is equal to A and remainder 3 of (4) is equal to B.

10.18   FUNDAMENTAL THEOREM OF DIFFERENCE CALCULUS

If $y(x)$ is a polynomial of nth degree, then its nth differences are constant and the $(n+1)^{th}$ differences are zero.

(ROHTAK-2007, 09, UPTU-2002, 04; UPTU MCA-2004; VIT-2014)

Proof: Let $\quad y(x) = a_0 x^n + a_1 x^{n-1} + \dots + a_{n-1} x + a_n$

be the polynomial of nth degree, where $a_0, a_1, \dots, a_n$ are constants and $a_0 \neq 0$. Let h be the interval of differencing.

Then $\qquad y(x + h) = a_0(x + h)^n + a_1(x + h)^{n-1} + \dots + a_{n-1}(x + h) + a_n$

Now $\qquad \Delta y(x) = y(x + h) - y(x)$

$\qquad\qquad = a_0[(x + h)^n - x^n] + a_1[(x + h)^{n-1} - x^{n-1}]e + \dots + a_{n-1}[(x + h) - x]$

$\qquad\qquad = a_0 nhx^{n-1} + b'x^{n-2} + c'x^{n-3} + \dots + k'x + l'$ ...(ii)

Where $b', c' ..., k', l'$ are constants.

Clearly (2) is a polynomial of $(n-1)^{th}$ degree.

Now
$$\Delta^2 y(x) = \Delta(\Delta y(x)) = \Delta y(x+h) - \Delta y(x)$$

$$= a_0 nh[(x+h)^{n-1} - x^{n-1}] + b'[(x+h)^{n-2} - x^{n-2}] + ... + k'(x+h-x)$$

$$= a_0 n(n-1)h^2 x^{n-2} + b'' x^{n-3} + c'' x^{n-4} + ... + k''$$

$$= \text{A Polynomial of degree } (n-2) \text{ proceeding in the same way, we get}$$

$$\Delta^n y(x) = a_0 (n-1)(n-2)...(2.1) \quad h^n x^{n-x}$$

$$= a_0 h^n.n! = \text{A constant, (independent of } x)$$

Therefore, the nth difference is constant.

Hence,
$$\Delta^{n+1} y(x) = 0$$
$$\Delta^{n+2} y(x) = 0 \qquad \text{and so on.}$$

☛ REMARK

- The converse of the above theorem is also true, *i.e.*, if the nth differences of a function tabulated at equally spaced intervals are constant, the function is a polynomial of degree n.

Solved Examples

EXAMPLE 1. *Evaluate* **(i)** $\Delta \tan^{-1} x$

(ii) $\Delta^2 \cos 2x$. (PTU-2001)

SOLUTION: (i) $\Delta \tan^{-1} x = \tan^{-1}(x+h) - \tan^{-1} x$

$$= \tan^{-1}\left[\frac{x+h-x}{1+(x+h)x}\right] = \tan^{-1}\left[\frac{h}{1+hx+x^2}\right]$$

(ii) $\Delta^2 \cos 2x = \Delta[\Delta \cos 2x]$

$$= \Delta[\cos(2x+2h) - \cos 2x]$$

$$= \Delta \cos(2x+2h) - \Delta \cos 2x$$

$$= [\cos(2x+4h) - \cos(2x+2h)]$$
$$- [\cos(2x+2h) - \cos 2x]$$

$$= 2\sin(2x+3h).\sin(-h)$$
$$- 2\sin(2x+h).\sin(-h)$$

$$= -2\sinh[\sin(2x+3h) - \sin(2x+h)]$$

$$= -2\sinh[2\cos(2x+2h)\sin h]$$

$$= -4\sin^2 h \cos(2x+2h)$$

$\therefore \Delta^2 \cos 2x = -4\sin^2 h \cos(2x+2h)$

EXAMPLE 2. *Prove that* $e^x = \left(\dfrac{\Delta^2}{E}\right)e^x.\dfrac{Ee^x}{\Delta^2 e^x}$, *the interval of differencing being unity.* (BHOPAL-2009)

SOLUTION: We know that
$$Ef(x) = f(x+1)$$
$$\therefore \quad Ee^x = e^{x+1}$$

and $\Delta^2 e^x = \Delta(\Delta e^x) = \Delta(e^{x+1} - e^x)$

$$\Delta(e-1)e^x = (e-1)\Delta e^x$$
$$= (e-1)(e^{x+1} - e^x) = (e-1)^2 e^x$$

and $\left(\dfrac{\Delta^2}{E}\right)e^x = (\Delta^2 E^{-1})e^x$

$$= (\Delta^2 E^{-1} e^x) = \Delta^2(e^{x-1})$$

$$= \Delta[\Delta(e^{x-1})] = \Delta[e^x - e^{x-1}]$$

$$= \Delta e^x - \Delta e^{x-1} = e^{x+1} - e^x - e^x + e^{x-1}$$

$$= e^x\left(e - 2 + \frac{1}{e}\right) = \frac{(e-1)^2}{e}e^x$$

Now $\left(\dfrac{\Delta^2}{E}\right)e^x.\dfrac{Ee^x}{\Delta^2 e^x} = \dfrac{(e-1)^2}{e}e^x.\dfrac{e^{x+1}}{(e-1)^2 e^x} = e^x$.

EXAMPLE 3. *Evaluate* $\Delta^n \sin(ax + b)$. [UPTU-2003, 04 (CO)]

SOLUTION: Consider $\Delta \sin(ax + b)$

$$= \sin(a(x+h) + b) - \sin(ax+b)$$

$$= 2\sin\frac{ah}{2}\cos\left[a\left(x + \frac{h}{2}\right) + b\right]$$

$$= 2\sin\frac{ah}{2}\sin\left[ax + b + \frac{ah + \pi}{2}\right]$$

$\therefore \quad \Delta^2 \sin(ax+b) = \Delta\left[2\sin\frac{ah}{2}\sin\left(ax + b + \frac{ah+\pi}{2}\right)\right]$

$$= \left(2\sin\frac{ah}{2}\right)\left(2\sin\frac{ah}{2}\right)$$

$$\sin\left[ax + b + \frac{ah+\pi}{2} + \frac{ah+\pi}{2}\right]$$

$$= \left(2\sin\frac{ah}{2}\right)^2 \sin\left[ax + b + 2\left(\frac{ah+\pi}{2}\right)\right]$$

Proceeding in the same way, we get

$$\Delta^3 \sin(ax+b) = \left(2\sin\frac{ah}{2}\right)^3 \sin\left[ax + b + 3\left(\frac{ah+\pi}{2}\right)\right]$$

... ...
... ...

$$\Delta^3 \sin(ax+b) = \left(2\sin\frac{ah}{2}\right)^n \sin\left[ax + b + n\left(\frac{ah+\pi}{2}\right)\right]$$

EXAMPLE 4. *If* $f(x) = e^{ax}$, *evaluate* $\Delta^n f(x)$. (UPTU MCA-2004)

SOLUTION: If the interval of differencing is h, then
$$\Delta f(x) = f(x+h) - f(x)$$

$\Rightarrow \quad \Delta[e^a x] = e^{a(x+h)} - e^{ax} = e^{ax}.e^{ah} - e^{ax}$

$\therefore \quad \Delta e^{ax} = e^{ax}(e^{ah} - 1)$

Again $\Delta^2 f(x) = \Delta[\Delta f(x)]$

$$= \Delta[(e^{ah} - 1)e^{ax}] = (e^{ah} - 1)\Delta e^{ax}$$

$$\Delta^2 e^{ax} = (e^{ah} - 1)^2 e^{ax}$$

Similarly, $\quad \Delta^3 e^{ax} = (e^{ah} - 1)^3 e^{ax}$

$$\cdots \quad \cdots \quad \cdots \quad \cdots \quad \cdots \quad \cdots \quad \cdots \quad \cdots \quad \cdots \quad \cdots \quad \cdots \quad \cdots$$

$$\Delta^n e^{ax} = (e^{ah} - 1)^n e^{ax}$$

EXAMPLE 5. *Find* $\Delta^{10}[(1 - ax)(1 - bx^2)(1 - cx^3)(1 - dx^4)]$.

(UPTU B.TECH (AG.)-2004)

SOLUTION: Let $\quad f(x) = (1 - ax)(1 - bx^2)(1 - cx^3)(1 - dx^4)$

Then $f(x)$ is a polynomial of degree 10 and coefficient of x^{10} is $abcd$. Since, we know that if $f(x)$ is a polynomial of degree n with leading coefficient a_0, then

$$\Delta^n f(x) = a_0 h^n n!$$

Here $n = 10$, $a_0 = abcd$, $h = 1$

Then $\Delta^{10} f(x) = abcd\, 10!$

EXAMPLE 6. *Prove that* $\Delta \log f(x) = \log\left[1 + \dfrac{\Delta f(x)}{f(x)}\right]$.

(RAJASTHAN-2001, 05; BANGLURU-2007)

SOLUTION : We know that

$$\Delta f(x) = f(x + h) - f(x)$$

$\therefore \quad \Delta \log f(x) = \log f(x + h) - \log f(x)$

$$= \log\left[\frac{f(x + h)}{f(x)}\right]$$

$$= \log\left[\frac{Ef(x)}{f(x)}\right] \qquad \because Ef(x) = f(x + h)$$

$$= \log\left[\frac{(1 + \Delta)f(x)}{f(x)}\right] \qquad (\because E = 1 + \Delta)$$

$$= \log\left[1 + \frac{\Delta f(x)}{f(x)}\right]$$

EXAMPLE 7. *Show that* $\Delta^p y_k = \nabla^p y_{k+p}$.

SOLUTION : Since we know that

$$\Delta = \nabla E$$

Then $\Delta^p y_k = (\nabla E)^p y_k = \nabla^p E^p y_k = \nabla^p y_{k+p}$.

EXAMPLE 8. *If* $y = (3x + 1)(3x + 4)\ldots(3x + 22)$

Show that $\Delta^4 y = 136080(3x + 13)(3x + 16)$
$$(3x + 19)(3x + 22).$$

SOLUTION: Clearly, the given equation

$$y = (3x + 1)(3x + 4)\ldots(3x + 22)$$

contains eight factors.

Therefore, $y = 3^8\left(x + \dfrac{1}{3}\right)\left(x + \dfrac{4}{3}\right)\ldots\left(x + \dfrac{22}{3}\right)$

$$= 3^8\left(x + \frac{22}{3}\right)^{(8)}$$

$\Rightarrow \quad \Delta y = 3^8 \cdot 8\left(x + \dfrac{22}{3}\right)^{(7)}$

$$\Delta^2 y = 3^8 \cdot 8 \cdot 7\left(x + \frac{22}{3}\right)^{(6)}$$

$$\Delta^3 y = 3^8 \cdot 8 \cdot 7 \cdot 6\left(x + \frac{22}{3}\right)^{(5)}$$

and $\quad \Delta^4 y = 3^8 \cdot 8 \cdot 7 \cdot 6 \cdot 5\left(x + \dfrac{22}{3}\right)^{(4)}$

Hence, $\Delta^4 y = 11022480\left(x + \dfrac{22}{3}\right)\left(x + \dfrac{22}{3} - 1\right)$

$$\left(x + \frac{22}{3} - 2\right)\left(x + \frac{22}{3} - 3\right)$$

$$= 136080\,(3x + 22)\,(3x + 19)(3x + 16)(3x + 13)$$

EXAMPLE 9. *Show that* $hD = -\log(1 - \nabla) = \sinh^{-1}(\mu\delta)$.

(UPTU-2005, 06)

SOLUTION: $hD = \log E = -\log(E^{-1}) = -\log(1 - \nabla)$

Also, $\mu = \dfrac{1}{2}(E^{1/2} + E^{-1/2})$

$$\delta = E^{1/2} - E^{-1/2}$$

$\therefore \quad \mu\delta = \dfrac{1}{2}(E - E^{-1}) = \dfrac{1}{2}(e^{hD} - e^{-hD}) = \sinh(hD)$

Hence, $\quad hD = \sinh^{-1}(\mu\delta) = -\log(1 - \nabla)$

EXAMPLE 10. *Show that* $(E^{1/2} + E^{-1/2})(1 + \Delta)^{1/2} = 2 + \Delta$.

(BHOPAL-2009, UPTU-2004, 09)

SOLUTION : We have $(E^{1/2} + E^{-1/2})E^{1/2} = E + 1 = 1 + \Delta + 1 = \Delta + 2$.

EXAMPLE 11. *Show that* $\quad \Delta + \nabla = \dfrac{\Delta}{\nabla} - \dfrac{\nabla}{\Delta}$

(UPTU MCA-2003, 07, 09, B.TECH. (AG.)-2004, B. TECH. 2008)

SOLUTION: We have

$$\left(\frac{\Delta}{\nabla} - \frac{\nabla}{\Delta}\right)y_x = \left(\frac{E - 1}{1 - E^{-1}} - \frac{1 - E^{-1}}{E - 1}\right)y_x$$

$$= \left\{\frac{E - 1}{\left(\dfrac{E - 1}{E}\right)} - \frac{\left(\dfrac{E - 1}{E}\right)}{E - 1}\right\}y_x$$

$$= \left(E - \frac{1}{E}\right)y_x = (E - E^{-1})y_x$$

$$= \{(1 + \Delta) - (1 - \nabla)\}y_x = (\Delta + \nabla)y_x$$

Hence, $\quad \dfrac{\Delta}{\nabla} - \dfrac{\nabla}{\Delta} = \Delta + \nabla$.

EXAMPLE 12. *Show that* $\nabla - \Delta = \nabla\Delta$.

(MUMBAI-2005, UPTU MCA-2005)

SOLUTION : We have $\Delta - \nabla = (1 - E^{-1}) - (E - 1)$

$$= \left(\frac{E - 1}{E}\right) - (E - 1)$$

$$= -(E - 1)(E^{-1} - 1)$$

$$= -(E - 1)(1 - E^{-1}) = -\nabla\Delta$$

EXAMPEL 13. *Evaluate* $\quad \log ax$. (PATNA-2007, 09)

SOLUTION : We have

$$\Delta(\log ax) = \log a(x + h) - \log ax \qquad \ldots(1)$$

$\Delta^2[\log(ax)] = \Delta[\Delta \log ax]$

$$= \Delta[\log a(x + h) - \log ax]$$

$$= \Delta \log a(x + h) - \Delta \log ax$$

$$= [\log[a(x + 2h) - \log a(x + h)$$
$$- [\log a(x + h) - \log ax]$$

$$= \log a(x + 2h) - 2\log a(x + h) + \log ax$$

$\ldots(2)$

From (1) and (2), we conclude that

$$\Delta^n \log ax = \log a(x+nh) - n_a \log a(x+(n-1)h) \quad \ldots(3)$$

true for $n = 1, 2$

We shall prove that relation (3) is true for all positive integral value of n.

Now, $\Delta^{n+1} \log ax = \Delta[\Delta^n \log ax]$

$$= \Delta \log a(x+nh) - {}^nC_1 \Delta \log a[x+(n-1)h]$$

$$+\ldots+(-1)^n.{}^nC_r \Delta \log a[x+(n-r)h]$$

$$+\ldots+(-1)^n \Delta \log ax$$

$$= \log[a+(x+(n-1)h] - \log a(x+nh)$$

$$-{}^nC_1[\log a(x+nh) - \log a[x+(n-1)h+\ldots$$

$$+(-1)^r.{}^nC_r[\log a(x+(n-r+1)h]$$

$$-\log a[x+(n-r)h]+\ldots+(-1)^n$$

$$\log a[(x+h) - \log ax]$$

$$= \log a[x+(n+1)h] - (1+{}^nC_1)\log a(x+nh)$$

$$+\ldots+(-1)^r[{}^nC_{r-1}+{}^nC_r[\log a\{x+(n-r+1)h\}]$$

$$+\ldots+ +(-1)^{n+1}\log ax \quad \ldots(4)$$

$(4) \Rightarrow$ Relation (3) is true for $n+1$.

Hence, by principle of mathematical induction, result is true for all positive integral values of n.

EXAMPLE 14. Show that $\sum\limits_{k=\infty}^{n-1} \Delta^2 f_k = \Delta f_n - \Delta f_0$. (UKTU–2014)

SOLUTION : L.H.S. $= \sum\limits_{k=\infty}^{n-1} \Delta^2 f_k = \sum\limits_{k=0}^{n-1} (E-1)^2 f_k$

$$= \sum\limits_{k=0}^{n-1} (E^2 - 2E + 1) f_k$$

$$= \sum\limits_{k=0}^{n-1} (f_{k+2} - 2f_{k+1} + f_k)$$

$$= (f_2 - 2f_1 + f_0) + (f_3 - 2f_2 + f_1)$$

$$+(f_4 - 2f_3 + f_2) + (f_5 - 2f_4 + f_3)$$

$$+(f_{n-1} - 2f_{n-2} + f_{n-3})$$

$$+(f_n - 2f_{n-1} + f_{n-2}) + (f_{n+1} - 2f_n + f_{n-1})$$

$$= f_{n+1} - f_n + f_0 - f_1$$

$$= (f_{n+1} - f_n) - (f_1 - f_0) = \Delta f_n - \Delta f_0$$

EXAMPLE 15. Prove that $u_x = u_{x-1} + \Delta u_{x-2} + \Delta^2 u_{x-3} + \ldots$

$$+ \Delta^{n-1} u_{x-n} + \Delta^n u_{x-n}$$

SOLUTION : Consider

$$u_x - \Delta^n u_{x-n} = u_x - \Delta^n.E^{-n} u_x$$

$$= \left(1 - \frac{\Delta^n}{E^n}\right) u_x = E^{-n} (E^n - \Delta^n) u_x$$

$$= E^{-n}\left[\frac{E^n - \Delta^n}{E - \Delta}\right] u_x \quad (\because E = 1 + \Delta)$$

$$= E^{-n}[E^{n-1} + \Delta E^{n-2} + \Delta^2 E^{n-3} + \ldots + \Delta^{n-1}] u_x$$

$$= [E^{-1} + \Delta E^{-2} + \Delta^2 E^{-3} + \ldots + \Delta^{n-1} E^{-n}] u_x$$

$\therefore u_x - \Delta^n u_{x-n}$

$$= u_{x-1} + \Delta u_{x-2} + \Delta^2 u_{x-3} + \ldots + \Delta^{n-1} u_{x-n}$$

Hence, $u_x = u_{x-1} + \Delta u_{x-2} + \Delta^2 u_{x-3}$

$$+\Delta^{n-1} u_{x-n} + \Delta^n u_{x-n}.$$

EXAMPLE 16. Prove that

(i) $u_0 + \dfrac{u_1 x}{1!} + \dfrac{u_2 x^2}{2!} \ldots$

$$= e^x\left[u_0 + x\Delta u_0 + \frac{x^2}{2!}\Delta^2 u_0 + \ldots\right]$$

(ii) $u_0 + u_1 + u_2 + \ldots + u_n$

$$= {}^{n+1}C_1 u_0 + {}^{n+1}C_2 \Delta u_0$$

$$+{}^{n+1}C_3 \Delta^2 u_0 + \ldots + \Delta^n u_0$$

SOLUTION : (i) $u_0 + \dfrac{u_1 x}{1!} + \dfrac{u_2 x^2}{2!} + \ldots$

$$= u_0 + \frac{xEu_0}{1!} + \frac{x^2 E^2 u_0}{2!} + \frac{x^3 E^3 u_0}{3!} + \ldots$$

$$= \left[1 + \frac{xE}{1!} + \frac{x^2 E^2}{2!} + \frac{x^3 E^3}{3!} + \ldots\right] u_0$$

$$= [e^{xE}] u_0 = [e^{x(1+\Delta)}] u_0 \quad (\because E = 1 + \Delta)$$

$$= (e^x.e^{x\Delta}) u_0$$

$$= e^x\left[1 + \frac{x\Delta}{1!} + \frac{x^2 \Delta^2}{2!} + \frac{x^3 \Delta^3}{3!} + \ldots\right] u_0$$

$$= e^x\left[u_0 + x\Delta u_0 + \frac{x^2}{2!}\Delta^2 u_0 + \frac{x^3}{3!}\Delta^3 u_0 + \ldots\right]$$

(ii) $u_0 + u_1 + u_2 + \ldots + u_n$

$$= u_0 + Eu_0 + E^2 u_0 + \ldots + E^n u_0$$

$$= (1 + E + E^2 + \ldots + E^n) u_0$$

$$= \left[\frac{E^{n+1} - 1}{E - 1}\right] u_0 = \left[\frac{(1+\Delta)^{1+n} - 1}{\Delta}\right] u_0 \quad (\because E = 1 + D)$$

$$= \frac{1}{\Delta}[{}^{n+1}C_0 + {}^{n+1}C_1 \Delta + {}^{n+1}C_2 \Delta^2$$

$$+{}^{n+1}C_3 \Delta^3 + \ldots +{}^{n+1}C_{n+1}\Delta^{n+1} - 1] u_0$$

$$= [{}^{n+1}C_1 + {}^{n+1}C_2 \Delta + {}^{n+1}C_3 \Delta + \ldots + \Delta^n] u_0$$

$$= {}^{n+1}C_1 u_0 + {}^{n+1}C_2 \Delta u_0 + {}^{n+1}C_3 \Delta^2 u_1 + \ldots + \Delta^n u_0.$$

EXAMPLE 17. Prove the following identity:

$$u_{x+n} = u_n + {}^xC_1 \Delta u_{n-1} + {}^{x+1}C_2 \Delta^2 u_{n-2}$$

$$+{}^{x+2}C_3 \Delta^3 u_{n-3} + \ldots$$

(BANGALURU-2002, 03, 07)

SOLUTION : R.H.S. $= u_n + {}^xC_1 \Delta u_{n-1} + {}^{x+1}C_2 \Delta^2 u_{n-2}$

$$+{}^{x+2}C_3 \Delta^3 u_{n-3} + \ldots$$

$$= u_n + {}^xC_1 \Delta E^{-1} u_n + {}^{x+1}C_2 \Delta^2 E^{-2} u_n$$

$$+{}^{x+2}C_3 \Delta^3 E^{-3} u_n + \ldots$$

$$= [1 + {}^xC_1 \Delta E^{-1} + {}^{x+1}C_2 \Delta^2 E^{-2}$$

$$+{}^{x+2}C_3 \Delta^3 E^{-3} + \ldots] u_n$$

$$= [1 - \Delta E^{-1}]^{-x} u_n$$

$$= \left(1 - \frac{\Delta}{E}\right)^{-x} u_n = \left(\frac{E - \Delta}{E}\right)^{-x} u_n$$

$$= \left(\frac{1}{E}\right)^{-x} u_n = E^x u_n = u_{n+1} = \text{L.H.S.}$$

EXAMPLE 18. *Prove the following identity :*

$$u_x - \frac{1}{8}\Delta^2 u_{x-1} + \frac{1.3}{8.16}\Delta^4 u_{x-2}$$

$$- \frac{1.3.5}{8.16.24}\Delta^6 u_{x-3} + \dots$$

$$= u_{x+\frac{1}{2}} - \frac{1}{2}\Delta u_{x+\frac{1}{2}} + \frac{1}{4}\Delta^2 u_{x+\frac{1}{2}} + \dots$$

(BHOPAL-2002, 06)

SOLUTION : LHS

$$= u_x - \frac{1}{8}\Delta^2 u_{x-1} + \frac{1.3}{8.16}\Delta^4 u_{x-2} - \frac{1.3.5}{8.16.24}\Delta^6 u_{x-3} + \dots$$

$$= \left[1 - \frac{1}{8}\Delta^2 E^{-1} + \frac{1.3}{8.16}\Delta^4 E^{-2} - \frac{1.3.5}{8.16.24}\Delta^6 E^{-3} + \dots\right]u_x$$

$$= \left[1 + \left(-\frac{1}{2}\right)\left(\frac{\Delta^2 E^{-1}}{4}\right) + \frac{\left(-\frac{1}{2}\right)\left(-\frac{3}{2}\right)}{2!}\left(\frac{\Delta^2 E^{-1}}{4}\right)^2 \right.$$

$$\left. + \frac{\left(-\frac{1}{2}\right)\left(-\frac{3}{2}\right)\left(-\frac{5}{2}\right)}{3!}\left(\frac{\Delta^2 E^{-1}}{4}\right)^3 + \dots\right]u_x$$

$$= \left[1 + \frac{\Delta^2 E^{-1}}{4}\right]^{-1/2} u_x = E^{1/2}\left[E + \frac{\Delta^2}{4}\right]^{-1/2} u_x$$

$$= E^{1/2}\left[1 + \Delta + \frac{\Delta^2}{4}\right]^{-1/2} u_x \qquad (\because E = 1 + \Delta)$$

$$= E^{1/2}\left[\left(1 + \frac{\Delta}{2}\right)^2\right]^{-1/2} u_x = E^{1/2}\left(1 + \frac{\Delta}{2}\right)^{-1} u_x$$

$$= E^{1/2}\left[1 - \frac{\Delta}{2} + \frac{\Delta^2}{2^2} - \dots\right]u_x$$

$$= u_{x+\frac{1}{2}} - \frac{1}{2}\Delta u_{x+\frac{1}{2}} + \frac{1}{4}\Delta^2 u_{x+\frac{1}{2}} - \dots = \text{R.H.S.}$$

EXAMPLE 19. *Find the function whole first difference is* $9x^2 + 11x + 5$.

SOLUTION : Let $f(x)$ be the required function. As per given

$$\Delta f(x) = 9x^2 + 11x + 5 = 9x(x-1) + Ax + B$$
Putting $x = 0$, we get $B = 5$
Putting $x = 1$, we get $A = 20$
Therefore, $\Delta f(x) = 9x(x-1) + 20x + 5$

$$= 9x^{(2)} + 20x^{(1)} + 5$$

$$\Rightarrow \quad f(x) = \frac{9x^{(3)}}{3} + \frac{20x^{(2)}}{2} + 5x^{(1)} + C$$

$$\Rightarrow \quad f(x) = 3x^{(3)} + 10x^{(2)} + 5x^{(1)} + C.$$

EXAMPLE 20. *Find the relation between* α, β *and* γ *in order that* α + β*x* + γ*x*² *may be expressible in one term of factorial notations.*   (IAS-2004)

SOLUTION : Let $f(x) = \alpha + \beta x + \gamma x^2 = (a + bx)^{(2)}$
where a and b are unknown constants
Now $(a + bx)^{(2)} = (a + bx)(a + b(x-1))$

$$= (a^2 - ab) + (2ab - b^2)x + b^2 x^2$$

Comparing with $\alpha + \beta x + \gamma x^2$ we et

$$\alpha = a^2 - ab, \beta = 2ab - b^2, \gamma = b^2$$
On eliminating a, b from these relations we get
$$\gamma^2 + 4\alpha\gamma = \beta^2 \text{ which is required relation.}$$

EXAMPLE 21. *Express* $f(x) = x^4 - 12x^3 + 24x^2 - 30x + 9$ *and its successive difference in factorial notation, interval of differencing being unity. Hence show that* $\Delta^5 f(x) = 0$.   (HIMACHAL-2006, 10; UPTU (SUM) 2004)

SOLUTION : Let $f(x) = Ax^{(4)} + Bx^{(3)} + Cx^{(2)} + Dx^{(1)} + E$ ...(1)

Using the method of synthetic division, we divide the given $f(x)$ by x, $x-1$, $x-2$ and $x-3$ successively.

1	1	−12	24	−30	9 = E
		1	−11	13	
2	1	−11	13	−17 = D	
		2	−18		
3	1	−9	−5 = C		
		3			
4	1	−6 = B			
	1 = A				

Hence $f(x) = x^{(4)} - 6x^{(3)} - 5x^{(2)} - 17x^{(1)} + 9$

∴ $\Delta f(x) = 4x^{(3)} - 18x^{(2)} - 10x^{(1)} - 17$

[∵ Δ is treated as differential operator on factorial polynomial.]

$$\Delta^2 f(x) = 12x^{(2)} - 36x^{(1)} - 10$$

$$\Delta^3 f(x) = 24x^{(1)} - 36$$

$$\Delta^4 f(x) = 24$$

Hence, $\Delta^5 f(x) = 0$

EXAMPLE 22. *Estimate the missing term in the following table :*

x	0	1	2	3	4
f(x)	1	3	9	?	81

SOLUTION : There are 4 values of $f(x)$, which are given.
Then, we have $\Delta^4 f(x) = 0$ for all x.

Here, the interval of differencing is 1.

Now, $\Delta^4 f(x) = 0$

$\Rightarrow (E - 1)^4 f(x) = 0$ $(\because E = 1 + \Delta)$

$(E^4 - 4E^3 + 6E^2 - 4E + 1)f(x) = 0$

$E^4 f(x) - 4E^3 f(x) + 6E^2 f(x) - 4Ef(x) + f(x) = 0$

$f(x+4) - 4f(x+3) + 6f(x+2) - 4f(x+1) + f(x) = 0$

Putting $x = 0$, we get

$f(4) - 4f(3) + 6f(2) - 4f(1) + f(0) = 0$

$81 - 4f(3) + 6(9) - 4(3) + 1 = 0$

$124 - f(3) = 0 \Rightarrow f(3) = 31.$

EXAMPLE 23. *Estimate the production for 1964 and 1966 from the following*:

Year	1961	1962	1963	1964	1965	1966	1967
Production	200	220	260	--	350	--	430

(UPTU-2003)

SOLUTION : Since five pairs are complete, therefore we assume all the differences of order fifth are zero, *i.e.*, $\Delta^5 y_0 = 0$ and $\Delta^5 y_1 = 0$

$\Rightarrow$ $(E-1)^5 y_0 = 0$ and $(E-1)^5 y_1 = 0$

$\Rightarrow$ $(E^5 - 5E^4 + 10E^3 - 10E^2 + 5E - 1)y_0 = 0$

$y_5 - 5y_4 + 10y_3 - 10y_2 + 5y_1 - y_0 = 0$ ...(1)

and $y_6 - 5y_5 + 10y_4 - 10y_3 + 5y_2 - y_1 = 0$ ...(2)

Putting the values of y_i's from the table in (1) and (2), we get

$y_5 - 1750 + 10y_3 - 2600 + 1100 - 260 = 0$

$\Rightarrow$ $y_5 + 10y_3 = 3450$ ...(3)

$430 - 5y_5 + 3500 - 10y_3 + 1300 - 220 = 0$

$\Rightarrow$ $y_5 + 10y_3 = -5010$ ...(4)

On solving (3) and (4), we get $y_3 = 306$ and $y_5 = 390$.

EXAMPLE 24. *Given that $u_0 + u_8 = 1.9243$, $u_1 + u_7 = 1.9590$ and $u_2 + u_6 = 1.9823$, $u_3 + u_5 = 1.9956$ Find u_4.*

(MUMBAI-2003, DELHI-2005, 09)

SOLUTION : Since there are 8 values of $f(x)$, are given.

Therefore, $\Delta^8 u_0 = 0$

$\Rightarrow$ $(E-1)^8 u_0 = 0$ $(\because E = 1 + \Delta)$

$\Rightarrow$ $({}^8C_0 E^8 - {}^8C_1 E^7 + {}^8C_2 E^6 - {}^8C_3 E^5$
$+ {}^8C_4 E^4 - {}^8C_5 E^3 + {}^8C_6 E^2 - {}^8C_7 E$
$+ {}^8C_8 E^0)u_0 = 0$

$\Rightarrow$ $(E^8 - 8E^7 + 28E^6 - 56E^5 + 70E^4 - 56E^3$
$28E^2 - 8E + 1)u_0 = 0$

$\Rightarrow$ $(u_0 + E^8 u_0) - 8(Eu_0 + E^7 u_0) + 28$

$\Rightarrow$ $(E^2 u_0 + E^6 u_0) - 56(E^3 u_0 + E^5 u_0) + 70E\ u = 0$

$\Rightarrow$ $(u_0 + u_8) - 8(u_1 + u_7) + 28(u_2 + u_6)$
$-56(u_3 + u_5) + 70u_4 = 0$

$\Rightarrow$ $1.9243 - 8(1.9590) + 28(1.9823)$
$-56(1.9956) + 70u_4 = 0$

$\Rightarrow$ $70u_4 + 1.9243 - 15.672 + 55.5044$
$-111.7536 = 0$

$\Rightarrow$ $70u_4 - 69.9969 = 0$

$\Rightarrow$ $u_4 = \dfrac{69.9969}{70}$

$\Rightarrow$ $u_4 = 0.999955$.

10.18.1 DIFFERENCES OF ZERO

Let m and n be positive integers. We know that

$\Delta^n x^m = (E-1)^n x^m$

$= E^n - {}^nC_1 E^{n-1} + {}^nC_2 E^{n-2} + ... + (-1)^{n-1}\, {}^nC_{n-1} E + (-1)^n x^m$

$= (x+n)^m - {}^nC_1 (x+n-1)^m + {}^nC_2 (x+n-2)^m - ... + {}^nC_{n-1}(-1)^{n-1} (x+1)^m + (-1)^n x^m$

Putting $x = 0$, we get

$[\Delta^n x^m]_{x=0} = n^m - {}^nC_1 (n-1)^m\ {}^nC_2 (n-2)^m - ... + {}^nC_{n-1}(-1)^{n-1}$ ...(1)

Here, we write $[\Delta^n x^m]_{x=0}$ as $\Delta^n 0^m$ and $\Delta^n 0^m$ are called the differences of zero.

10.19 DIVIDED DIFFERENCE

Let $f(x_0), f(x_1),..., f(x_n)$ be the values of the function corresponding to the values of $x_0, x_1,..., x_n$ which are not equally spaced. We know that the difference of the function values with respect to the difference of the arguments are called divided differences.

We define the first divided difference of $f(x)$ between x_0 and x_1 as follows

$$f(x_0, x_1) = \underset{x_i}{\Delta}\ f(x_0) = \frac{f(x_1) - f(x_0)}{x_1 - x_0}$$

Similarly, the second divided difference is given by

$$f(x_0, x_1, x_2) = \underset{x_1, x_2}{\Delta^2}\ f(x_0) = \frac{f(x_1, x_2) - f(x_0, x_1)}{x_2 - x_0}$$

and so on.

In general, the nth divided difference is given by

$$f(x_0, x_1,..., x_n) = \underset{x_1, x_2,..., x_n}{\Delta^n}\ f(x_0) = \frac{f(x_1, x_2,... x_n) - f(x_0, x_1,... x_{n-1})}{x_n - x_0}$$

Divided Difference Table

(1) x	(2) $f(x)$	(3) $\Delta f(x)$	(4) $\Delta^2 f(x)$	(5) $\Delta^3 f(x)$
a	$f(a)$			
		$\dfrac{f(b)-f(a)}{b-a} = \underset{b}{\Delta} f(a)$		
			$\dfrac{\underset{c}{\Delta} f(b) - \underset{b}{\Delta} f(a)}{c-a} = \underset{bc}{\Delta^2} f(a)$	
b	$f(b)$			
		$\dfrac{f(c)-f(b)}{c-b} = \underset{c}{\Delta} f(b)$		$\dfrac{\underset{cd}{\Delta^2} f(b) - \underset{bc}{\Delta^2} f(a)}{d-a} = \underset{bcd}{\Delta^2} f(a)$
			$\dfrac{\underset{d}{\Delta} f(c) - \underset{c}{\Delta} f(b)}{d-b} = \underset{cd}{\Delta^2} f(b)$	
c	$f(c)$			
		$\dfrac{f(d)-f(c)}{d-c} = \underset{d}{\Delta} f(c)$		$\dfrac{\underset{de}{\Delta^2} f(c) - \underset{cd}{\Delta^2} f(d)}{c-b} = \underset{cde}{\Delta^2} f(d)$
			$\dfrac{\underset{e}{\Delta} f(d) - \underset{d}{\Delta} f(c)}{e-c} = \underset{de}{\Delta^2} f(c)$	
d	$f(d)$			
		$\dfrac{f(e)-f(d)}{e-d} = \underset{e}{\Delta} f(d)$		
e	$f(e)$			

WORKING PROCEDURE

(Construction of divided difference table)

Step 1. Write the values of x and corresponding values of $f(x)$ in columns (1) and (2) of the table.

Step 2. Calculate the values of column (3) by subtracting the successive values of column (2) and dividing the differences by the corresponding differences of the arguments.

Step 3. Calculate the values of column (4) by taking the difference between two successive values of column (3) and divide each difference by the difference between the lowest and highest values of the arguments corresponding to the two differences, of column (3), *e.g.*, corresponding to the differences $\underset{b}{\Delta} f(a)$ and $\underset{c}{\Delta} f(b)$, the lowest value of the argument is a and the highest value is c, hence in column (4) we get the first observation as $\dfrac{\underset{c}{\Delta} f(b) - \underset{b}{\Delta} f(a)}{c-a}$

Step 4. Repeat the operations of step 3 on the observation of column (4) to get column (5) and so on.

10.19.1 Another Commonly used Notations for Divided Difference

These are $f(a,b), f(a,b,c), f(a,b,c,d)$, etc.

Here $f(a,b) = \dfrac{f(b)-f(a)}{b-a}$ = first order divided difference.

$$f(a,b,c) = \dfrac{f(b,c)-f(a,b)}{c-a} = \text{second order divided difference.}$$

$$f(a,b,c,d) = \dfrac{f(b,c,d)-f(a,b,c)}{d-a} = \text{third order divided difference.}$$

In general,

$$f(a,b,c,...,l,m,n) = \dfrac{f(b,c,...,m,n)-f(a,b,...,l,m)}{n-a} = m^{\text{th}} \text{ order divided difference.}$$

10.19.2 Differences between Divided Difference and Ordinary Difference

The following are the basic differences between divided difference and ordinary differences :

(i) An ordinary difference is not affected by the changes in the values of the argument, whereas in divided difference the changes in the values of the argument are to be taken into consideration.

(ii) An ordinary difference being the difference between two successive values of the entry has numerator only whereas, in the case of divided difference though the numerator will be same as that of ordinary difference but it will also contain a term in denominator which is equal to the difference between the values of the argument corresponding to the entries in the numerator of the difference e.g.,

Given

$x = a$	b	c	$d \dots$
$y = f(a)$	$f(b)$	$f(c)$	$f(d) \dots$

Ordinary difference $\Delta y = f(b) - f(a)$ and Divided difference $\underset{b}{\Delta} y = \dfrac{f(b) - f(a)}{b - a} = \dfrac{\Delta y}{b - a}$

(iii) The operator and the operand of ordinary differences contain no suffixes whereas the suffixes of the operator and the operand in divided difference have special significance *e.g.*, $\underset{b}{\Delta} f(a)$ implies that it is the divided difference of the function $f(x)$ at the point $x = a$ taking into consideration the entry at $x = b$. Similarly $\underset{bc}{\Delta^2} f(a)$ means that it is the divided difference between first order differences at $x = a$ and $x = b$.

Similarly, $\underset{bc}{\Delta} f(a) = \underset{c}{\Delta}\left[\underset{b}{\Delta} f(a)\right] = \underset{c}{\Delta}\left[\dfrac{f(a)}{a-b} + \dfrac{f(b)}{b-a}\right] = \dfrac{\left[\dfrac{f(a)}{a-b} + \dfrac{f(b)}{b-a}\right]}{a-c} + \dfrac{\left[\dfrac{f(c)}{c-b} + \dfrac{f(b)}{b-c}\right]}{c-a}$ ...(2)

In second order divided difference there are three suffixes a, b and c and we consider it as the operation of $\underset{c}{\Delta}$ on $\underset{b}{\Delta} f(a)$, i.e., in the first term of (2) $\underset{b}{\Delta} f(a)$ is written as it is and is divided by $(a - c)$ and in second term put c wherever there is 'a' in $\underset{b}{\Delta} f(a)$ and then divide by $c - a$.

Simplifying (2) we get

$$\underset{bc}{\Delta^2} f(a) = \dfrac{f(a)}{(a-b)(a-c)} + \dfrac{f(b)}{(b-a)(b-c)} + \dfrac{f(c)}{(c-a)(c-b)} = \Sigma \dfrac{f(a)}{(a-b)(a-c)} \qquad ...(3)$$

Now $\underset{bcd}{\Delta^3} f(a) = \underset{d}{\Delta}\left(\underset{bc}{\Delta^2} f(a)\right)$

$$= \dfrac{1}{(a-d)}\left[\dfrac{f(a)}{(a-b)(a-c)} + \dfrac{f(b)}{(b-a)(b-c)} + \dfrac{f(c)}{(c-a)(c-b)}\right] + \dfrac{1}{(d-a)}\left[\dfrac{f(d)}{(d-b)(d-c)} + \dfrac{f(b)}{(b-d)(b-c)} + \dfrac{f(c)}{(c-d)(c-b)}\right]$$

$$= \dfrac{f(a)}{(a-b)(a-c)(a-d)} + \dfrac{f(b)}{(b-a)(b-c)(b-d)} + \dfrac{f(c)}{(c-a)(c-b)(c-d)} + \dfrac{f(d)}{(d-a)(d-b)(d-c)}$$

$$= \Sigma \dfrac{f(a)}{(a-b)(a-c)(a-d)} \qquad ...(4)$$

Similarly we can get the higher order divided differences in terms of the values of the entry as

$$\underset{bc...n}{\Delta^{n-1}} f(a) = \Sigma \dfrac{f(a)}{(a-b)(a-c)...(a-n)} \qquad ...(5)$$

WORKING PROCEDURE

(Construction of divided difference table)

Step 1. Divide $f(a)$ by $(a - b)$.

Step 2. Where there is 'a' in the function $f(x)$ put b and divide the value by $(b - a)$ (*i.e.*, the sign of $(a - b)$ in step (i) is reversed).

Step 3. Add the two values obtained is step (i) and (ii) to get the required value of the divided difference.

10.19.3 SOME IMPORTANT PROPERTIES OF DIVIDED DIFFERENCE

1. Divided differences are symmetric with respect to the arguments.
2. The nth divided differences of a polynomial of degree n are constant.
3. The nth divided difference can be expressed as the quotient of two determinants of order $(n + 1)$.
4. The nth divided difference can be expressed as the product of multiple integrals such that

$$f(x_1, x_2,...,x_n) = \int_0^1 dt_1 \int_0^{t_1} dt_2 \int_0^{t_2} dt_3 ... \int_0^{t_{n-2}} f^{n-1}(u_n) dt_{n-1} \qquad ...(1)$$

where, $u_n = (1 - t_1)x_1 + (t_1 - t_2)x_2 + ... + (t_{n-2} - t_{n-1})x_{n-1} + t_{n-1}x_n$ and $t_1, t_2,...,t_n$ are independent variables, and f^{n-1} means the $(n-1)^{th}$ derivative of f.

Solved Examples

EXAMPLE 1. *Evaluate* $\underset{y}{\Delta} x^2$ *and* $\underset{y, z}{\Delta^2} x^3$.

SOLUTION : Since we know that $\underset{y}{\Delta} f(x) = \dfrac{f(y) - f(x)}{y - x}$

$$\therefore \quad \underset{y}{\Delta} x^2 = \frac{y^2 - x^2}{y - x} = \frac{(y - x)(y + x)}{(y - x)} = y + x$$

and
$$\underset{y,z}{\Delta^2} x^3 = \frac{\underset{z}{\Delta} y^3 - \underset{y}{\Delta} x^3}{z - x} = \frac{\frac{z^3 - y^3}{z - y} - \frac{y^3 - x^3}{y - x}}{z - x}$$

$$= \frac{(z^2 + y^2 + yz) - (y^2 + x^2 + xy)}{z - x}$$

$$= \frac{z^2 - x^2 + yz - xy}{z - y} = \frac{(z - x)(z + x) + y(z - x)}{z - x}$$

$$\therefore \quad \underset{y,z}{\Delta^2} x^3 = x + y + z.$$

EXAMPLE 2. *Show that* $\underset{bcd}{\Delta^3}\left(\dfrac{1}{a}\right) = -\dfrac{1}{abcd}$.

SOLUTION : The divided difference table is as follows :

x	$f(x)$	$\Delta f(x)$
a	$\dfrac{1}{a}$	
		$\dfrac{\frac{1}{b} - \frac{1}{a}}{b - a} = -\dfrac{1}{ab}$
b	$\dfrac{1}{b}$	
		$\dfrac{\frac{1}{c} - \frac{1}{b}}{c - b} = -\dfrac{1}{bc}$
c	$\dfrac{1}{c}$	
		$\dfrac{\frac{1}{d} - \frac{1}{c}}{d - c} = -\dfrac{1}{dc}$
d	$\dfrac{1}{d}$	

$\Delta^2 f(x)$	$\Delta^3 f(x)$
$\dfrac{(-1)\frac{1}{bc} - \frac{1}{ab}}{c - a} = (-1)^2 \dfrac{1}{abc}$	
	$\dfrac{(-1)^2 \frac{1}{bcd} - \frac{1}{abc}}{d - a} = (-1)^3 \dfrac{1}{abcd}$
$\dfrac{(-1)\frac{1}{dc} - \frac{1}{bc}}{d - b} = (-1)^2 \dfrac{1}{bcd}$	

Hence, $\underset{bcd}{\Delta^3}\left(\dfrac{1}{a}\right) = (-1)^3 \dfrac{1}{abcd} = -\dfrac{1}{abcd}$

EXAMPLE 3. *Find the third difference with arguments 2, 4, 9, 10 of the function* $f(x) = x^3 - 2x$.

SOLUTION : The divided difference table is as under :

x	$f(x)$	$\Delta f(x)$
2	4	
		$\dfrac{56 - 4}{4 - 2} = 26$
4	56	
		$\dfrac{711 - 56}{9 - 4} = 131$
9	711	
		$\dfrac{980 - 711}{10 - 9} = 269$
10	980	

$\Delta^2 f(x)$	$\Delta^3 f(x)$
$\dfrac{13 - 26}{9 - 2} = 15$	
	$\dfrac{23 - 15}{10 - 2} = 1$
$\dfrac{269 - 131}{10 - 4} = 23$	

Hence, the third divided difference is 1.

Exercise-10.8

1. Prove the following:

   (i) $(1 + \Delta)(1 - \nabla) = 1$ (ii) $\mu\delta = \dfrac{1}{2}(\Delta + \nabla)$

   (iii) $1 + \dfrac{\delta^2}{2} = \sqrt{1 + \delta^2\mu^2}$ (UPTU MCA-2008)

   (iv) $\nabla = \Delta E^{-1} = E^{-1}\Delta = 1 - E^{-1}$

   (v) $E^{1/2} = \mu + \dfrac{\delta}{2}$

   (vi) $\delta(E^{1/2} + E^{-1/2}) = \Delta E^{-1} + \Delta$

   (vii) $\Delta\nabla = \nabla\Delta = \delta^2$

   (viii) $\delta = \Delta(1 + \Delta)^{-1/2} = \nabla(1 - \nabla)^{-1/2}$

2. Evaluate the following :

   (i) $\Delta \cot 2^x$ (ii) $\dfrac{\Delta^2 x^3}{Ex^3}$ (iii) $\Delta \tan^{-1} a.x$

3. Prove the following :

   (i) For any positive integer m, $\dfrac{(x + 1)^{(m)}}{m!} = \dfrac{x^{(m)}}{m!} + \dfrac{x^{(m-1)}}{(m-1)!}$

   (ii) $\Delta \sin^{-1} x = \sin^{-1}[(x+1)\sqrt{1 - x^2} - x\sqrt{1 - (x+1)^2}]$

   (iii) $\Delta x^{(n)} = nx^{(n-1)}$ for all integers n.

4. Prove the following :

   (i) $y_4 = y_3 + \Delta y_2 + \Delta^2 y_1 + \Delta^3 y_1$

   (ii) $y_4 = y_0 + 4\Delta y_0 + 6\Delta^2 y_{-1} + 10\Delta^3 y_{-1}$

5. Prove that

$$\dfrac{\Delta^2}{E}\sin(x + h) + \dfrac{\Delta^2 \sin(x + h)}{E \sin(x + h)} = 2\cos(h - 1)[\sin(x + h) + 1].$$

6. Find the missing term in the following:

x	100	101	102	103	104
$f(x)$	2.000	2.0043	–	2.0128	2.0170

7. Evaluate the missing term in the following :

x	1	2	3	4	5	6	7
$f(x)$	2	4	8	–	32	64	128

8. Find the missing term in the following table :

x	2.0	2.1	2.2	2.3	2.4	2.5	2.6
$f(x)$	0.135	–	0.111	0.100	–	0.082	0.024

9. Prove the following :

(i) $(2\Delta^2 + \Delta - 1)(x^2 + 2x + 1) = 5h^2 + 2hx + 2h - x^2 - 2x - 1$

(ii) $(\Delta + 1)(2\Delta - 1)(x^2 + 2x + 1) = 5h^2 + 2hx + 2h - x^2 - 2x - 1$

(iii) $(E - 2)(E - 1)(2^{x/h} + x) = -h$

(iv) $(E^2 - 3E + 2)(2^{x/h} + x) = -h$

10. Find the lowest degree polynomial which takes the following values:

x	0	1	2	3	4	5
$f(x)$	0	3	8	15	24	35

11. Find the lowest degree polynomial which satisfying the following numbers:

0, 7, 26, 63, 124, 215, 342, 511

12. Prove the following :

(i) $u_0 + {}^xC_1\Delta u_1 + {}^xC_2\Delta^2 u_2 + {}^xC_3\Delta^3 u_3 + \ldots$
$$= u_x + {}^xC_1\Delta^2 u_{x+1} + {}^xC_2\Delta^4 u_{x+2}$$

(ii) $u_x = u_{x-1} + \Delta u_{x-2} + \Delta^2 u_{x-3} + \ldots + \Delta^{n-1}u_{x-n} + \Delta^n u_{x-n}$

13. Find the eighth term of the series 1, 1.095, 1.179, 1.251, 1.310.

14. Find the seccessive differences of $x^4 - 12x^3 + 42x^2 - 30x + 9$ when the interval of differencing is unity.

15. Prove the following :

(i) $\Delta^n 0^m = n^m - {}^nC_1(n-1)^m + {}^nC_2(n-2)^m + \ldots$

(ii) $\dfrac{\Delta^n 0^m}{n!} = \dfrac{n\Delta^n 0^{m-1}}{n!} + \dfrac{\Delta^{n-1} 0^{m-1}}{(n-1)!}$

(iii) $(n+1)\Delta^n 0^n = 2[\Delta^{n-1}0^n + \Delta^n 0^n]$

(iv) $\Delta^m 0^n + {}^nC_1\Delta^m 0^{n-1} + {}^nC_2\Delta^m 0^{n-2} + \ldots$
$$+ \dfrac{n!}{(n-m)!} = \dfrac{1}{m+1}.\Delta^{m+1}0^{n+1}$$

16. Express the following polynomial into factorial notations :

(i) $x^4 + x + 1$ (ii) $2x^4 + 5x^2 + 4x + 6$

Answers

2. (i) $-\operatorname{cosec} 2^{x+1}$ (ii) $\dfrac{6}{(1+x)^2}$ (iii) $\tan^{-1}\left(\dfrac{1}{1 + a^2x + a^2x^2}\right)$ **6.** 2.0086   **7.** 17   **8.** 0.132, 0.90

10. $x^2 + 2x$ **11.** $x^3 + 3x^2 + 3x$ **13.** 1.399   **14.** $\Delta f(x) = 4x^3 - 30x^2 + 52x + 1$, $\Delta^2 f(x) = 12x^2 - 48x + 26$, $\Delta^3 f(x) = 24x - 36$, $\Delta^4 f(x) = 24$ **16.** (i) $x^{(4)} + 6x^{(3)} + 7x^{(2)} + 2x^{(1)} + 1$ (ii) $2x^{(4)} + 12x^{(3)} + 19x^{(2)} + 11x^{(1)} + 6$

10.20 INTERPOLATION

Any one who has had occasion to consult tables of mathematical functions is familiar with the method of linear interpolation and probably face the situation in which the method of "reading between the lines of the table" has appeared.

For example, if we are to find out the population of India in 1998 when we know that the census in India is done is 1941, 1951, ... 2001, *i.e.*, the figure of population are available. Then the process of finding the figure is known as interpolation.

Definition. *The method of obtaining the value of a function for any intermediate value of the argument from the given set of values of the function for certain values of arguments is know as interpolation. On the other hand, the process of computing the value of the function outside the given range is called extrapolation.*

The study of interpolation is based on the calculus of finite differences. The calculus of finite differences plays in important role in numerical analysis. It deals with the variations in a function when independent variable changes by finite jumps which may be equal or unequal.

ASSUMPTIONS OF INTERPOLATION

(1) The values of the function should be either increasing or decreasing.

(2) The rise or fall in the given values should be uniform.

(3) The given set of observations should be capable of being expressed in a polynomial form.

10.20.1 METHODS OF INTERPOLATION

(1) Method of Graph

Let $y = f(x)$ be a given function. Plot a graph between the values of x and the corresponding values of $y = f(x)$. With the help of the graph we can obtained the unknown values of $f(x)$ for the given values of x.

(2) The method of curve fitting

This method can be used only when the form of the function is known to us. Then by the method of least squares we can fit the curve of known form to the given set of observations and with the help of the fitted curve we can calculate the unknown value.

Merit

The only merit of this method lies in its closer approximation than the graphical method.

Demerits

(1) The form of the function for the given set of observations is assumed to be known.

(2) When some additional observations are included in the data then the calculation for finding the unknown constants are to be done afresh.

(3) The method is not exact.

(4) The method is complicated when the number of observations is sufficiently large.

(3) Finite Difference Method

We can find the value of entry for a given value of argument without actually knowing the form of the function. The use of these methods though approximate have distinct advantages over the methods of graphs and curve fitting. The merits and demerits of this method are as follows.

Merits

(i) The method does not assume the form of the function to be known.

(ii) It is less approximate than the method of graphs.

(iii) The calculations remain simple even if some additional observations are included in the given data.

Demerit

There is no definite way to verify whether the assumptions for the application of finite difference calculus are valid for the given set of observations.

10.20.2 Finite Difference Calculus

We now proceed to study the use of finite difference calculus for the purpose of interpolation. This we shall do in three cases which are as follows :

(1) The value of the argument in given data varies by an equal interval. The technique is known as interpolation with equal intervals.

(2) The values of the argument are not at equal intervals. This is known as interpolation with unequal intervals.

(3) The technique of central differences.

10.21 NEWTON'S FORMULAE FOR INTERPOLATION FOR EQUAL INTERVALS

(1) Newton-Gregory Formula for Forward Interpolation with Equal Intervals.

Newton's-Gregory's formula for forward interpolation with equal interval is

$$f(x_0 + hu) = f(x_0) + u\Delta f(x_0) + \frac{u(u-1)}{2!}\Delta^2 f(x_0) + \frac{u(u-1)(u-2)}{3!}\Delta^3 f(x_0) + \dots + \frac{u(u-1)(u-2)\dots(u-(n-1)h)}{n!}\Delta^n f(x_0)$$

$$where \ \ u = \frac{x - x_0}{h}.$$

Proof. Let $y = f(x)$ be a function which takes the values $f(x_0), f(x_0 + h), f(x_0 + 2h), \dots, f(x_0 + nh)$,

for $x = x_0, (x_0 + h), (x_0 + 2h) \dots (x_0 + nh)$, *i.e*, for $(n+1)$ equidistant values $(x_0), (x_0 + 2h) \dots (x_0 + nh)$ of the independent variable x. Here, we assume that $f(x)$ is a polynomial of nth degree, therefore, $f(x)$ can be written as

$$f(x_0) = A_0 + A_1(x - x_0) + A_2(x - x_0)(x - x_0 - h) + A_3(x - x_0)(x - x_0 - h)(x - x_0 - 2h)\dots$$
$$+ A_n(x - x_0)(x - x_0 - h)\dots(x - x_0 - (n-1)h) \qquad \dots(1)$$

where $A_0, A_1, \dots, A_n$ are constants.

Putting $x = x_0, x_0 + h, x_0 + 2h, \dots$ in succession in (1), we get

$$f(x_0) = A_0 \qquad \dots(2)$$

$$f(x_0 + h) = A_0 + A_1.h$$

$$\Rightarrow \qquad A_1.h = f(x_0 + h) - A_0 = f(x_0 + h) - f(x_0)$$

$$\Rightarrow \qquad A_1 = \frac{f(x_0 + h) - f(x_0)}{h} = \frac{\Delta f(x_0)}{h} \qquad \dots(3)$$

Also, $\qquad f(x_0 + 2h) = A_0 + A_1(2h) + A_2(2h)(h)$

$$\Rightarrow \qquad 2h^2 A_2 = f(x_0 + 2h) - A_0 - 2A_1.h = f(x_0 + 2h) - f(x_0) - 2\Delta f(x_0)$$

$$= f(x_2 + 2h) - f(x_0) - 2f[(x_0 + h) + f(x_0)] = f(x_2 + 2h) - 2f(x_0 + h) + f(x_0) = \Delta^2 f(x_0)$$

$$A_2 = \frac{\Delta^2 f(x_0)}{2!h^2} \qquad \dots(4)$$

Similarly $\qquad A_3 = \frac{\Delta^3 f(x_0)}{3!h^3} \ \dots \dots \ A_n = \frac{\Delta^n f(x_0)}{n!h^n}$ and so on. $\qquad \dots(5)$

Substituting the values of $A_0, A_1, A_2, \dots$ from (2), (3) ... (5) in (1), we get

$$f(x) = f(x_0) + \frac{\Delta f(x_0)}{h} + \frac{\Delta^2 f(x_0)}{2!h^2}(x - x_0)(x - x_0 - h) + \frac{\Delta^3 f(x_0)}{3!.h^3}(x - x_0)(x - x_0 - h)(x - x_0 - 2h)\dots$$

$$+ \frac{\Delta^n f(x_0)}{n!h^n}(x - x_0)(x - x_0 - h)\dots\{x - x_0(n-1)/h\} \qquad \dots(6)$$

Now let $u = \dfrac{x - x_0}{h}$. Then $(u - 1) = \dfrac{x - x_0 - h}{h}$; $(u - 2) = (x - x_0 - 2h)$

$$\vdots \qquad\qquad \vdots$$

$$[u - (n-1)] = [x - x_0 - (n-1)h]$$

Substituting these values in (6), we get

$$f(x + hu) = f(x_0) + u\Delta f(x_0) + \frac{u(u-1)}{2!}\Delta^2 f(x_0) + \frac{u(u-1)(u-2)}{3!}\frac{\Delta^3 f(x_0)}{2} + \dots + \frac{u(u-1)(u-2)\dots(u-(n-1)h)}{n!}\Delta^n f(x_0)$$

which is the required Newton-Gregory's formula for forward interpolation with equal interval.

◄ Remarks

- This formula is particularly useful for interpolating the value of $f(x)$ near the beginning of the set of given values.
- This formula can be expressed in terms of factorials, as follows

$$f(x_0 + hu) = f(x_0) + \frac{u^{(1)}}{1!}\Delta f(x_0) + \frac{u^{(2)}}{2!}\Delta^2 f(x_0) + \frac{u^{(3)}}{3!}\Delta^3 f(x_0) + \dots + \frac{u^{(n)}}{n!}\Delta^n f(x_0)$$

Remainder Term

The remainder term in Newton-Gregory's forward interpolation formula is given by

$$R_n = \frac{u(u-1)(u-2)\dots(u-n)}{(n+1)!}h^{n+1}f^{n+1}(\theta) \text{ where } x_0 \leq \theta \leq x$$

Since R_n contains derivative of $(n+1)^{th}$ order of function $f(x)$ at $x = \theta$ so we may not say something about the error when the form of the function is not known. In this case expressed $f^{n+1}(\theta)$ in terms of the difference of the function

$$\therefore \qquad R_n = \frac{u^{(n+1)}}{(n+1)!}\Delta^{n+1}f(x_0).$$

(2) Newton-Gregory's Backward Interpolation formula with Equal Intervals.

Newton-Gregory's formula for backward interpolation formula with equal interval is

$$f(a + nh + uh) = f(a + nh) + u\Delta f(a + nh) + \frac{u(u+1)}{2!}\nabla^2 f(a + nh) + \dots + \frac{u(u+1)(u+2)\dots(u+n-1)}{n!}\nabla^n f(a + nh).$$ [UPTU MCA-2008 (BP)]

Proof. Let $y = f(x)$ be a function, which assumes the values

$$f(a), f(a+h), f(a+2h), \dots, f(a+nh)$$

corresponding to the values of $x = a, (a+h), (a+2h), \dots, + (a+nh)$ respectively. The values of x are equidistant.

Assume the function $f(x)$ such that

$$f(x) = A_0 + A_1(x - a - nh) + A_2(x - a - nh)[x - a - (n-1)h] + A_3(a - x - nh)(x - a - (n-1)h)[(x - a(n-2)h]$$
$$+ \dots + A_n(x - a - nh)[x - a - (n-1)h]\dots(x - a - h) \qquad \dots(1)$$

where $A_0, A_1, A_2, \dots, A_n$ are constants.

Putting $x = a + nh$ in (1), we get

$$f(a + nh) = A_0 \qquad \dots(2)$$

Now, putting $x = a + (n-1)h$ in (1), we get

$$f[a + (n-1)h] = A_0 + A_1[a + (n+1)h - a - nh] = A_0 - A_1 h$$

$$\Rightarrow \qquad A_1 h = A_0 - f[a + (n+1)h]$$

$$= f(a + nh) - f[a + (n-1)h] = \Delta f(a + nh)$$

i.e., $\qquad A_1 = \dfrac{\nabla f(a + nh)}{h} \qquad \dots(3)$

Similarly, putting $x = a + (n-2)h$ in (1), we get

$$f[a + (n-2)h] = A_0 + A_1[a + (n-2)h - a - nh] + A_2[a + (n-2)h - a - nh][a + (n-2)h - a - (n-1)]$$
$$= A_0 + A_1(-2h) + A_2(-2h)(-h)$$

$$\therefore \qquad A_2 \cdot 2h^2 = f[a + (n-2)h] - A_0 + 2A_1 h$$

$$= f[a + (n-2)h] - f(a + nh) + 2\nabla f(a + nh)$$

$$= f[a + (n-2)h] - f(a + nh) + 2\{f(a + nh) - f[a + (n-1)h]\}$$

$$= \{f[a + (n-2)h - f(a + (n-1)h]\} - \{f[a + (n-1)h] - f(a + nh)\}$$

$$= \nabla f[a + (n-1)h] - \nabla f(a + nh) = \nabla^2 f(a + nh)$$

$$\Rightarrow \qquad A_2 = \dfrac{\nabla^2 f(a + nh)}{h^2 \cdot 2!} \qquad \dots(4)$$

Proceeding in the same way, putting $x = a + (n-3)h, a+(n-4)h \ldots$, etc. in (1), we get

$$A_3 = \frac{\nabla^3 f(a+nh)}{h^3.3!} \ldots A_n = \frac{\nabla^n f(a+nh)}{h^n.n!} \qquad \ldots(5)$$

Substituting all these values of $A_0, A_1, A_2, \ldots$ in (1), we get

$$f(x) = f(a+nh) + (x-a-nh)\frac{\nabla f(a+nh)}{h.1!} + (x-a-nh)[x-a(n-1)h]\frac{\nabla^2 f(a+nh)}{2!} + \ldots$$

$$+ (x-a-nh)[x-a(n-1)h]\ldots(x-a-h)\frac{\nabla^n f(a+nh)}{h^n.n!} \qquad \ldots(6)$$

Let us define $\qquad u = \dfrac{x-(a-nh)}{h}$ so, $x-a-(n-1)h = uh+h = (u+1)h$

$$x-a-h = (a+nh+uh)-a-h = (u+n-1)h$$

Put all these values in equation (6), we get

$$f(a+nh+uh) = f(a+nh) + u\nabla f(a+nh) + \frac{u(u+1)}{2!}\nabla^2 f(a+nh) + \ldots + \frac{u(u+1)\ldots(u+n-1)}{n!}\nabla^2 f(a+nh)$$

which is the required Newton Gregory's backward interpolation formula for equal intervals.

REMARK

- This formula is particularly useful for interpolating the values of $f(x)$ near the end of the set of given values.

WORKING PROCEDURE

Step 1. Put the values of x_i, $f(x_i)$ and value of x for which $f(x)$ is to be calculated.

Step 2. Construct the difference table by using.

$$\nabla^n f(x_i) = \nabla^{n-1} f(x_i+1) - \nabla^{n-1} f(x_i-1)$$

Step 3. Obtain $\quad u = \dfrac{x-(a-nh)}{h}$

Step 4. Put all these values in Newton's backward interpolation formula.

Remainder Term

The remainder term R_n is given by

$$R_n = u(u+1)\ldots(u+n)\frac{h^{n+1}}{(n+1)!}f^{n+1}(\theta) \text{ where } u = \frac{x-x_n}{h}$$

This is the required remainder term of Newton's Backward interpolation formula for finding the error.

REMARK

- If the analytical form of the function $f(x)$ is not known then expressed $f^{n+1}(\theta)$ in term of difference, i.e., $\dfrac{\nabla^{n+1} f(x_n)}{h^{n+1}}$ such that

$$R_n = \frac{\nabla^{n+1} f(x_n)}{(n+1)!}u(u+1)(u+2)\ldots(u+n).$$

Solved Examples

EXAMPLE 1. *From the following table, find the number of students who obtained less than 45 marks:*

Marks	Number of Students
30-40	31
40-50	42
50-60	51
60-70	35
70-80	31

(UKTU, 2011 S; SVTU 2007; MADRAS, 2006)

SOLUTION : The difference table for the given data is as follows:

Marks	No. of student $f(x)$	$\Delta f(x)$	$\Delta^2 f(x)$	$\Delta^3 f(x)$	$\Delta^4 f(x)$
Less 40	31				
		42			
Less 50	73		9		
		51		-25	
Less 60	124		-16		37
		35		12	
Less 70	159		-4		
		31			
Less 80	190				

Here $h = 10$, $a = 40$ and $x = 45$

For Newton-Gregory forward interpolation formula

Let $u = \dfrac{x-a}{h} = \dfrac{45-40}{10} = \dfrac{5}{10} = \dfrac{1}{2}$

Therefore,

$$f(45) = f(40) + \frac{1}{2}\Delta f(40) + \frac{\frac{1}{2}\left(\frac{1}{2}-1\right)}{2!}\Delta^2 f(40) +$$

$$\frac{\frac{1}{2}\left(\frac{1}{2}-1\right)\left(\frac{1}{2}-2\right)}{3!}\Delta^3 f(40) +$$

$$\frac{\frac{1}{2}\left(\frac{1}{2}-1\right)\left(\frac{1}{2}-2\right)\left(\frac{1}{2}-3\right)}{4!}\Delta^4 f(40)$$

$$= 31 + \frac{1}{2} \times 42 - \frac{1}{8} \times 9 - \frac{1}{16} \times 25 - \frac{5}{128} \times 37$$

$$= 31 + 21 - 1.125 - 1.563 \times 1.445$$

$$= 47.867 = 48$$

Hence, the number of the students who obtain less than 45 marks are 48.

EXAMPLE 2. *The following table gives the marks secured by 100 students in the numerical subject:*

Range of Marks	30-40	40-50	50-60	60-70	70-80
No. of Students	25	35	22	11	7

Use Newton's forward difference interpolation formula to find:

(i) the number of students who got more than 55 marks.

(ii) the number of students who secured marks in the range from 36 to 45. (UPTU MCA-2002)

SOLUTION: The difference table for the given data is as follows:

Marks obtained	Number of Students
Less than 40	25
Less than 50	60
Less than 60	82
Less than 70	93
Less than 80	100

(i) Here $a = 40$, $h = 10$, $a + uh = 55$

$\Rightarrow 40 + 10u = 55 \Rightarrow u = 1.5$

Now, we find the number of students who got less than 55 marks. The difference table is given as under:

Marks obtained less than	No. of students, y	Δy	$\Delta^2 y$	$\Delta^3 y$	$\Delta^4 y$
40	25				
		35			
50	60		-13		
		22		2	
60	82		-11		5
		11		7	
70	93		-4		
		7			
80	100				

Then, by Newton's forward interpolation formula, we have

$$y_{55} = y_{40} + u\Delta y_{40} + \frac{u(u-1)}{2!}\Delta^3 y_{40}$$

$$+ \frac{u(u-1)(u-2)(u-3)}{4!}\Delta^4 y_{40}$$

$$= 25 + (-0.4)(35) + \frac{(-4.0)(-1.4)}{2!}(-13)$$

$$+ \frac{(-0.4)(-1.4)(-2.4)}{3!}(2)$$

$$+ \frac{(-0.4)(-1.4)(-2.4)(-3.4)}{4!}(5)$$

$$= 7.864 = 8$$

Similarly

$$y_{45} = y_{40} + u\Delta y_{40} + \frac{u(u-1)}{2!}\Delta^2 y_{40}$$

$$+ \frac{u(u-1)(u-2)}{3!}\Delta^3 y_{40}$$

$$+ \frac{u(u-1)(u-2)(u-3)}{4!}\Delta^4 y_{40}$$

$$= 25 + (0.5)(35) + \frac{(0.5)(-0.5)}{2}$$

$$+ \frac{(0.5)(-0.5)(-1.5)}{6}(2)$$

$$+ \frac{(0.5)(-0.5)(-1.5)(-2.5)}{24}(5)$$

$$= 44.0546 = 44$$

Hence, the number of students who secured marks in the range from 36 to 45 is given by $y_{45} - y_{36} = 44 - 8 = 36$.

EXAMPLE 3. *From the following table of half yearly premium for policies maturing at different ages, estimate the premium for policies maturing at the age of 46.* (UPTU-2010)

Age	45	50	55	60	65
Premium (in rupees)	114.84	96.16	83.32	74.48	68.48

SOLUTION: We have the following difference table.

Age x	Premium in rupees y	Δy	$\Delta^2 y$	$\Delta^3 y$	$\Delta^4 y$
45	114.84				
		-16.68			
50	96.16		5.84		
		-12.84		-1.84	
55	83.32		4		0.68
		-8.84		-1.16	
60	74.48		2.84		
		-6			
65	68.48				

We have $h = 5$, $a = 45$, $a + hu = 46$

$\therefore 45 + 5u = 46 \Rightarrow u = 0.2$

Then, by Newton's forward difference formula, we have

$$y_{46} = y_{45} + u\Delta y_{45} + \frac{u(u-1)}{2!}\Delta^2 y_{45}$$

$$+ \frac{u(u-1)(u-2)}{3!}\Delta^3 y_{45} +$$

$$+ \frac{u(u-1)(u-2)(u-3)}{4!}\Delta^4 y_{45}$$

$$= 114.84 + (0.1)(-18.68)$$
$$+ \frac{(0.2)(0.2-1)}{2!}(5.84)$$
$$+ \frac{(0.2)(0.2-1)(0.2-2)}{3!}(-1.84)$$
$$+ \frac{(0.2)(0.2-1)(0.2-2)(0.2-3)}{4!}(0.68)$$
$$= 110.525632$$

Hence, the premium for policies maturing at the age of 46 is Rs. 110.52.

EXAMPLE 4. *Find the value of the area of the circle of diameter from the following given data.*

(Diameter)	80	85	90	95	100
A (Area)	5026	5674	6362	7088	7854

(VTU 2010)

SOLUTION : The difference table of the above data is given as follows:

d (Dia-meter)	A = f(d) (Area)	$\Delta f(d)$	$\Delta^2 f(d)$	$\Delta^3 f(d)$	$\Delta^4 f(d)$
80	5026				
		648			
85	5674		40		
		688		-2	
90	6362		38		4
		726		2	
95	7088		40		
		766			
100	7854				

Here $a = 80$, $h = 5$ and $x = 82$,

then $u = \dfrac{x-a}{h} = \dfrac{82-80}{5} = \dfrac{2}{5} = 0.4$

From Newton's forward interpolation formula, we get

$$f(82) = f(80) + 0.4\Delta f(80) + \frac{0.4(0.4-1)}{2!}\Delta^2 f(80)$$
$$+ \frac{0.4(0.4-1)(0.4-2)}{3!}\Delta^3 f(80)$$
$$+ \frac{0.4(0.4-1)(0.4-2)(0.4-3)}{4!}\Delta^4 f(80)$$
$$= 5026 + 0.4(648) + \frac{0.4(0.4-1)}{2}(40)$$
$$+ \frac{0.4(0.4-1)(0.4-2)}{6}(-2)$$
$$+ \frac{0.4(0.4-1)(0.4-2)(0.4-3)}{24}(4)$$
$$= 5026 + 259.2 - 4.8 - 0.128 - 0.1664$$
$$= 5280.10$$

Hence, the required area = 5280.10.

EXAMPLE 5. *Given*

x	1	2	3	4	5	6	7	8
f(x)	1	8	27	64	125	216	343	512

Find f(7.5). (MKU (TAMILNADU)-2005)

SOLUTION : The backward difference table is given as follows :

x	f(x)	$\nabla f(x)$	$\nabla^2 f(x)$	$\nabla^3 f(x)$
1	1			
		7		
2	8		12	
		19		6
3	27		18	
		37		6
4	64		24	
		61		6
5	125		30	
		91		6
6	216		36	
		129		6
7	343		42	
		169		
8	512			

Here $a + nh = 8$, $x = 7.5$, $h = 1$

then $u = \dfrac{x-(a+nh)}{h} = \dfrac{7.5-8}{1} = -0.5$

Now using Newton's Backward interpolation formula, we get

$$f(7.5) = f(a+nh) + u\nabla f(a+nh)$$
$$+ \frac{u(u+1)}{2!}\nabla^2 f(a+nh)$$
$$+ \frac{u(u+1)(u+2)}{3!}\nabla^3 f(a+nh)$$
$$f(7.5) = f(8) + (-0.5)\nabla f(8)$$
$$+ \frac{(-0.5)(-0.5+1)}{2}\nabla^2 f(8)$$
$$= 512 - 0.5(169) - \frac{(0.5)(0.5)}{2}(42)$$
$$- \frac{(0.5)(0.5)(1.5)}{6}(6)$$
$$= 512 - 84.5 - 5.25 - 0.375$$

Hence, $f(7.5) = 421.875$.

EXAMPLE 6. *Find the cubic polynomial which takes the following values*:

x	0	1	2	3
f(x)	1	2	1	10

(BHOPAL-2009, ROHTAK-2005, WBTU-2005)

SOLUTION : Here, the difference table is given as under

x	f(x)	$\nabla f(x)$	$\nabla^2 f(x)$	$\nabla^3 f(x)$
0	1			
		1		
1	2		-2	
		-1		12
2	1		10	
		9		
3	10			

Here $x_0 = 0, u = \dfrac{x-0}{h} = x$ $\qquad (\because h = 1)$

By using the Newton's forward interpolation formula, we get

$$f(x) = f(0) + u\Delta f(0) + \frac{u(u-1)}{2!}\Delta^2 f(0)$$
$$+ \frac{u(u-1)(u-2)}{3!}\Delta^3 f(0) + \dots$$

$$= 1 + x.1 + \frac{x(x-1)}{2!}(-2) + \frac{x(x-1)(x-2)}{6}(12)$$

$$= 2x^3 - 7x^2 + 6x + 1$$

which is the required polynomial.

EXAMPLE 7. *Consider the following table*:

x	3	4	5	6	7
$f(x)$	3	6.6	15	22	25

Obtain interpolating polynomial of degree 2 or less using Newton's backward difference interpolation method. Hence compute f(5.5).

[UPTU(MCA)-2007]

SOLUTION : The difference table is

x	$f(x)$	$\Delta f(x)$	$\Delta^2 f(x)$	$\Delta^3 f(x)$	$\Delta^4 f(x)$
3	3				
		3.6			
4	6.6		4.8		
		8.4		−6.2	
5	15		−1.4		3.6
		7		7.4	
6	22		6		
		13			
7	**25**				

Newton's backward difference formula is

$$f(x) = f(x_n) + u\nabla f(x_n) + \frac{u(u-1)}{2!}\nabla^2 f(x_n) + \dots$$

To get the polynomial of degree 2 or less we use only upto second difference term.

Now $\qquad u = \dfrac{x - x_n}{n} = \dfrac{x-7}{1}$

$\therefore \qquad f(x) = 35 + (x-7)1 + \dfrac{(x-7)(x-8)}{2!}6$

$$= 35 + 13x - 91 + 3(x^2 + 56 - 15x)$$

$$= 3x^2 - 32x + 112$$

$$f(5.5) = 3(5.5)^2 - 32 \times 5.5 + 112 = 26.75$$

Exercise-10.9

1. The value of $f(x)$ for $x = 0, 1, 2, \dots 6$ are given below:

x	0	1	2	3	4	5	6
$f(x)$	2	4	10	16	20	24	38

Estimate the value of $f(3.2)$ using only four of given values. Choose the four values that you think will give the best approximations.

2. From the following table find the value of $e^{0.24}$:

x	0.1	0.2	0.3	0.4	0.5
e^x	1.10517	1.22140	13.4986	1.49182	1.64872

3. If p, q, r, s be the successive entries corresponding to equidistant arguments in a table. Show that when third difference are taken into account the entry corresponding to the arguments half way between the arguments of q and r is

$A + \left(\dfrac{B}{24}\right)$, where A is the arithmetic mean of q and r and B is the arithmetic mean of $3q - 2p - s$ and $3r - 2s - p$.

(DELHI-2005, 08)

4. From the following table :

x	10°	20°	30°	40°
$\cos x$	0.9848	0.9397	0.8660	0.7660

x	50°	60°	70°	80°
$\cos x$	0.6428	0.5000	0.3420	0.1737

Calculate cos 25° and cos 73° with Gregory's Newton formula.
(UPTU-2006)

5. Find the number of men getting wages between Rs. 10 and Rs. 15 from the following table :

Wages (in Rs.)	0-10	10-20	20-30	30-40
Frequency	9	30	35	42

6. The table below gives values of tan x for $0.10 \leq x\ 0.30$:

x	0.10	0.15	0.20	0.25	0.30
tan x	0.1003	0.1511	0.2027	0.2523	0.3093

Evaluate tan 0.12 using Newton's forward difference formula.

7. In the following table, values of y are consecutive terms of a series of which 23.6 is the 6th term. Find the first and tenth term of the series. (MDU(B.E.)-2004)

x	3	4	5	6	7	8	9
y	4.8	8.4	14.5	23.6	36.2	52.8	73.9

(ANNA-2007)

8. Find the following table of half yearly premium for policies maturing at different ages, estimate the premium for policy maturing at the age of 63. (UPTU-2010)

Age	45	50	55	60	65
Premium	114.84	96.16	83.32	74.48	68.48

9. Find the value of $e^{-1.9}$ from the following table of values of e^{-x}.

x	1	1.25	1.50	1.75	2.00
e^{-x}	0.3679	0.2865	0.2231	0.1738	0.1353

10. Given

x	1	2	3	4	5	6	7	8
$f(x)$	1	8	27	64	125	216	343	512

find (7.5) using Newton's backward interpolation formula.

 **Answers**

| **1.** 17.28 | **2.** 1.271249088 | **5.** 15 | **6.** 0.1205 | **7.** 3.1, 100 | **8.** 70.585152 | **9.** 0.1496 | **10.** 421.875 |

10.22 NEWTON'S DIVIDED DIFFERENCE FORMULA

(UPTU MCA-2002, 09)

Let $f(x_0), f(x_1), ..., f(x_n)$ be the values of the function $f(x)$ for the values of the arguments $x_0, x_1, ..., x_n$ respectively which are not equally spaced.

The first divided difference of $f(x)$ is given by

$$f(x, x_0) = \frac{f(x_0) - f(x)}{(x_0 - x)} \quad \Rightarrow \quad f(x) = f(x_0) + (x - x_0) f(x, x_0) \quad ...(1)$$

Also, the second divided difference is given by

$$f(x, x_0, x_1) = \frac{f(x_0, x_1) - f(x, x_0)}{(x_1 - x)} \quad \Rightarrow \quad f(x, x_0) = f(x_0, x_1) + (x - x_1) f(x, x_0, x_1)$$

$$\Rightarrow \quad \frac{f(x) - f(x_0)}{(x - x_0)} = f(x_0, x_1) + (x - x_1) f(x, x_0, x_1) \quad \text{[From (1)]}$$

$$\Rightarrow \quad f(x) = f(x_0) + (x - x_0) f(x_0, x_1) + (x - x_0)(x - x_1) f(x, x_0, x_1) \quad ...(2)$$

Similarly, $\quad f(x) = f(x_0) + (x - x_0) f(x_0, x_1) + (x - x_0)(x - x_1) f(x_0, x_1, x_2)$

$$+ (x - x_0)(x - x_1)(x - x_2) f(x, x_0, x_1, x_2) \quad ...(3)$$

Proceeding in the same way, we get

$$f(x) = f(x_0) + (x - x_0) f(x_0, x_1) + (x - x_0)(x - x_1) f(x_0, x_1, x_2) + (x - x_0)(x - x_1)(x - x_2) f(x_0, x_1, x_2, x_3) + ...$$

$$+ (x - x_0)(x - x_1)(x - x_2)...(x - x_{n-1}) f(x_0, x_1, x_2, x_3, ..., x_n) + (x - x_0)(x - x_1)(x - x_2)...(x - x_n) f(x, x_0, x_1, ..., x_{n+1}) \quad ...(4)$$

Since the function $f(x)$ is a polynomial of degree n, therefore $f(x, x_0, x_1, ..., x_{n+1}) = 0$

Hence (4) becomes,

$$f(x) = f(x_0) + (x - x_0) f(x_0, x_1) + (x - x_0)(x - x_1) f(x_0, x_1, x_2) + ... + (x - x_0)(x - x_1)(x - x_2)...(x - x_{n-1}) f(x_0, x_1, x_2, ..., x_n)$$

This is known as Newton's divided difference formula.

REMARK

- Newton's divided difference formula reduces to Newton's Gregory's forward difference formula if the values of the arguments are equally spaced.

WORKING PROCEDURE

Step 1. Put the values of x_i, $f(x_i)$ and values of x for which $f(x)$ is to be calculated.
Step 2. Construct the divided difference table.
Step 3. Substitute the values of divided differences in the Newton's divided difference formula.

Solved Examples

EXAMPLE 1. *By means of Newton's divided Difference formula, find the value of f(8) and f(15) from the following table:* (UPTU MCA-2009, VTU-2008)

x	4	5	7	10	11	13
$f(x)$	48	100	294	900	1210	2028

Solution : Here $x_0 = 4, x_1 = 5, x_2 = 7, x_3 = 10, x_4 = 11, \ x_5 = 13$

The divided difference table is given below :

x	$f(x)$	$\Delta f(x)$	$\Delta^2 f(x)$	$\Delta^3 f(x)$	$\Delta^4 f(x)$
4	48				
		$\dfrac{100-48}{5-4} = 52$			
5	100		$\dfrac{97-57}{7-4} = 15$		
		$\dfrac{294-100}{7-5} = 97$		$\dfrac{21-15}{10-4} = 1$	
7	294		$\dfrac{202-97}{10-5} = 21$		0
		$\dfrac{900-294}{10-7} = 202$		$\dfrac{27-21}{11-5} = 1$	
10	900		$\dfrac{310-202}{11-7} = 27$		0
		$\dfrac{1210-900}{11-10} = 310$		$\dfrac{33-27}{13-7} = 1$	

11	1210		$\dfrac{409-310}{13-10}$ $=33$		
		$\dfrac{2028-1210}{13-11}$ $=409$			
13	2028				

Now using Newton's divided difference formula, we get

$$f(x) = f(x_0) + (x-x_0)f(x_0, x_1)$$
$$+ (x-x_0)(x-x_1)f(x_0, x_1, x_2)$$
$$+ (x-x_0)(x-x_1)(x-x_2)$$
$$f(x_0, x_1, x_2, x_3) \quad ...(1)$$

$$\Rightarrow f(8) = f(4) + (8-4)\times 52 + (8-4)(8-5)15$$
$$+ (8-4)(8-5)(8-7)\times 1$$
$$= 48 + 208 + 180 + 12 = 448$$

and $f(5) = 48 + (15-4)\times 52 (15-4)(15-5)$
$$(15) + (15-4)(15-5)(15-7).1$$
$$= 48 + 572 + 1650 + 880$$
$$\therefore \quad f(5) = 3150.$$

EXAMPLE 2. *Find the polynomial of the lowest possible degree which takes the values 3, 12, 15, –21, when x has the value 3, 2, 1, –1 respectively.*

(ROHTAK-2005, 08)

SOLUTION : The divided difference table is as under :

x	$f(x)$	$\Delta f(x)$	$\Delta^2 f(x)$	$\Delta^3 f(x)$
–1	**–21**			
		$\dfrac{15+21}{1-1}=\mathbf{18}$		
1	15		$-\dfrac{3-18}{2-1}=\mathbf{-7}$	
		$\dfrac{12-15}{2-1}=-3$		$\dfrac{-3+7}{3-1}=\mathbf{1}$
2	12		$\dfrac{-9+3}{3-1}=-3$	
		$\dfrac{3-12}{3-2}=-9$		
3	3			

Here $x_0 = -1, x_1 = 1, x_2 = 2, x_3 = 3$.

Now applying Newton's divided difference formula, we get

$$f(x) = f(x_0) + (x-x_0)f(x_0, x_1)$$
$$+ (x-x_0)(x-x_1)f(x_0, x_1, x_2) + (x-x_0)(x-x_1)$$
$$(x-x_2)f(x_0, x_1, x_2, x_3)$$
$$= f(-1) + (x+1)(18) + (x+1)(x-1)(-7)$$
$$+ (x+1)(x-1)(x-2)(1)$$
$$f(x) = x^3 - 9x^2 + 17x + 6$$

EXAMPLE 3. *Using Newton's divided difference formula, calculate the value of f(6) from the following data:*

(UPTU-2002)

x	1	2	7	8
$f(x)$	1	5	5	4

SOLUTION :

x	$f(x)$	$\Delta f(x)$	$\Delta^2 f(x)$	$\Delta^3 f(x)$
1	1			
		4		
2	5		–2 / 3	
		0		1 / 14
7	5		–1 / 6	
		–1		
8	4			

Put the values in the Newton's divided difference formula

$$f(x) = f(x_0) + (x-x_0)f(x_0, x_1) + (x-x_0)(x-x_1)$$
$$f(x_0, x_1, x_2) + (x-x_0)(x-x_1)(x-x_2)$$
$$f(x_0, x_1, x_2, x_3)$$

$$\Rightarrow f(6) = 1 + (6-1).4 + (6-1)(6-2)(-2/3)$$
$$+ (6-1)(6-2)(6-7)\dfrac{1}{14}$$
$$= 1 + 20 + \dfrac{5\times 4\times -2}{3} + 5\times 4\times -1\times \dfrac{1}{14}$$
$$= 21 - \dfrac{40}{3} - \dfrac{10}{7} = \dfrac{441-280-30}{21} = \dfrac{131}{31} = 6.2.$$

EXAMPLE 4. *Certain corresponding values of x and $\log_{10} x$ are (300, 2.4771), (304. 2.4829), (305, 2.4843) and (307, 2.4871). Find $\log_{10} 301$ using Newton's divided difference formula.*

(UPTU MCA-2001)

SOLUTION : Using the given data, we construct the following divided difference table:

x	$f(x)$	$\Delta f(x)$	$\Delta^2 f(x)$
300	**2.4771**		
		0.00145	
304	2.4829		**– 0.00001**
		0.00140	
305	2.4843		0
		0.00140	
307	2.4871		

Put the values of $f(x)$, $\Delta f(x)$, $\Delta^2 f(x)$ from the table in Newton's divided difference formula, we get

$$f(x) = \log_{10} 301 = f(x_0) + (x-x_0)\Delta f(x_0)$$
$$+ (x-x_0)(x-x_1)\Delta^2 f(x_0)$$
$$= 2.4771 + 0.00145 + (-3)(-0.00001)$$
$$= 2.4786.$$

EXAMPLE 5. *Using the Newton's divided difference formula, find a polynomial function satisfying the following data:*

[UPTU-2004, MDU (B.E.)-07, VTU-2007]

x	–4	–1	0	2	5
$f(x)$	1245	33	5	9	1355

SOLUTION : The divided table is given by

x	$f(x)$	$\Delta\,f(x)$	$\Delta^2 f(x)$	$\Delta^3 f(x)$	$\Delta^4 f(x)$
– 4	**1245**				
		– 404			
– 1	33		**94**		
		– 28		**–14**	
0	5		10		**3**
		2		13	
2	9		88		
		442			
5	1355				

Putting the   values in Newton-divided   difference formula,

$$f(x) = f(x_0)+(x-x_0)\Delta\,f(x_0)+(x-x_0)$$
$$(x-x_1)\Delta^2 f(x_0)+(x-x_0)(x-x_1)$$
$$(x-x_2)\Delta^3 f(x_0)$$
$$+(x-x_0)(x-x_1)(x-x_2)$$
$$(x-x_3)\Delta^4 f(x_0)$$

we get $f(x) = 1245+(x+4)(-404)$
$$+(x+4)(x+1)(94)+(x+4)(x+1)$$
$$x(-14)+(x+4)(x+1)x(x-2)(3)$$
$$= 3x^4-5x^3+6x^2-14x+5.$$

EXAMPLE 6. *Using the following table, find the value of f(x) at x = 4.*

x	1.5	3	6
$f(x)$	– 0.25	2	20

(UPTU MCA (CO)-2003)

SOLUTION : Using the above data, we have the following difference table:

x	$f(x)$	$\Delta f(x)$	$\Delta^2 f(x)$
1.5	**– 0.25**		
		1.5	
3	2		**1**
		6	
6	20		

Putting all these values in Newton's divided difference formula, we get
$$f(x) = -0.25 + (x-1.5)(1.5)+(2-1.5)(x-3)(1)$$
Putting $x = 4$, we get $f(4)=6$

EXAMPLE 7. *Using Newton's divided difference formula, show that*

$$f(x) = f(0) + x\Delta\,f(-1)+\frac{x(x+1)}{2!}\Delta^2\,f(-1)+$$

$$\frac{x(x-1)(x+1)}{3!}\Delta^3 f(-2) + ...$$

SOLUTION : Taking the arguments 0, –1, 1, –2, ... the Newton's divided difference formula is given by
$$f(x) = f(0) + x\underset{-1}{\Delta}\,f(0)+x(x+1)\underset{-1,1}{\Delta^2}\,f(0)+$$

$$x(x+1)(x-1)\underset{-1,1,-2}{\Delta^3}\,f(0)+......(1)$$

$$= f(0)+x\underset{0}{\Delta}\,f(-1)+x(x+1)\underset{0,1}{\Delta^2}f(-1)$$

$$+x(x+1)(x-1)\underset{-1,0,1}{\Delta^3}\,f(-2)+...$$

Now, $\underset{0}{\Delta}\,f(x) = \dfrac{f(0)-f(-1)}{0-(-1)} = \Delta f(-1)$

$$\underset{0,1}{\Delta^2}f(-1) = \frac{1}{1-(-1)}[\underset{1}{\Delta}\,f(0)-\underset{0}{\Delta}\,f(-1)]$$

$$= \frac{1}{2}[\Delta\,f(0)-\Delta\,f(0)] = \frac{1}{2}\Delta^2 f(-1)$$

$$\underset{-1,0,1}{\Delta^3}f(-2) = \frac{1}{1-(-2)}[\underset{0,1}{\Delta^2}f(-1)-\underset{-1,0}{\Delta^2}f(-2)]$$

$$= \frac{1}{3}\left[\frac{\Delta^2 f(-1)}{2}-\frac{\Delta^2 f(-2)}{2}\right]$$

$$= \frac{\Delta^3 f(-2)}{3.2} = \frac{\Delta^3 f(-2)}{3!}$$

.... and so on.

Finally, putting all these values in (1), we get

$$f(x) = f(0)+x\Delta f(-1)+\frac{x(x+1)}{2!}\Delta^2 f(-1)$$

$$+\frac{2(x+1)(x-1)}{3!}\Delta^3 f(-2)+...$$

EXAMPLE 8. *Using Newton's divided difference method, compute f(3) from the following table :*

x	0	1	2	4	5	6
y	1	14	15	5	6	19

[UPTU (MCA)-2005]

SOLUTION : Constructing the divided difference table, we have

x	$f(x)$	$\Delta f(x)$	$\Delta^2 f(x)$	$\Delta^3 f(x)$	$\Delta^4 f(x)$
0	1				
		$\dfrac{14-1}{1}=13$			
1	14		$\dfrac{1-13}{2}=-6$		
		$\dfrac{15-14}{1}=1$		$\dfrac{-2+6}{4}=1$	
2	15		$\dfrac{-5-1}{3}=-2$		**0**
		$\dfrac{5-15}{2}=-5$		$\dfrac{2+2}{4}=1$	
4	5		1 5		**0**
		$\dfrac{6-5}{1}=1$		$\dfrac{6-2}{4}=1$	
5	6		$\dfrac{13-1}{2}=6$		
		$\dfrac{19-6}{1}=13$			
6	19				

From the Newton's divided formula, we get

$$f(x) = f(x_0) + (x - x_0)f(x_0, x_1)$$

$$+ (x - x_0)(x - x_1)f(x_0, x_1, x_2) + (x - x_0)$$

$$(x - x_1)(x - x_2)f(x_0, x_1, x_2, x_3)$$

$$f(3) = 1 + (3 - 0)\,13 + (3-0)(3-1)(-6) + (3-0)$$

$$(3 - 1)\,(3 - 2) \times 1$$

$$= 1 + 39 + 3 \times 2 - 2 \times + 3 \times 2 \times 1$$

$$= 1 + 39(-12) + 6 = 40 + 6 - 12 = 46 - 12 = 34.$$

10.23 LAGRANGE'S INTERPOLATION FORMULA

(TAMILNADU (MKU)-2005; UPTU MCA-2006)

Let $y = f(x)$ be a function, which takes the values $f(x_0), f(x_1), \ldots f(x_n)$, corresponding to the values of $x = x_0, x_1, \ldots, x_n$, not necessarily equally spaced. Then

$$f(x) = \frac{(x - x_1)(x - x_2)\ldots(x - x_n)}{(x_0 - x_1)(x_0 - x_2)\ldots(x_0 - x_n)} f(x_0) + \frac{(x - x_0)(x - x_2)\ldots(x - x_n)}{(x_1 - x_0)(x_1 - x_2)\ldots(x_1 - x_n)} f(x_1) + \ldots + \frac{(x - x_0)(x - x_1)\ldots(x - x_{n-1})}{(x_n - x_0)(x_n - x_1)\ldots(x_n - x_{n-1})} f(x_n)$$

Proof. Let $y = f(x)$ be a function which can assume the values $f(x_0), f(x_1) \ldots f(x_n)$ corresponding to the values of the arguments $x_0, x_1, \ldots, x_n$, respectively.

Now, since there are $(n+1)$ pairs of values of x and y, therefore, we can represent $f(x)$ by a polynomial in x of degree n.

Let this polynomial be of the form

$$f(x) = A_0(x - x_1)(x - x_2)\ldots(x - x_n) + A_1(x - x_0)(x - x_2)\ldots(x - x_n)$$

$$A_2(x - x_0)(x - x_1)(x - x_3)\ldots(x - x_n) + \ldots + A_n(x - x_0)\,(x - x_1)(x - x_2)\ldots(x - x_{n-1}) \qquad \ldots(1)$$

where $A_0, A_1, A_2, \ldots, A_n$ are constants and can be determined by putting

$$f(x) = f(x_0) \text{ at } x = x_0, f(x) = f(x_1) \text{ at } x = x_1 \ldots \text{ etc.}$$

Now put $x = x_0$ and $f(x) = f(x_0)$ in (1), we get

$$f(x_0) = A_0(x_0 - x_1)(x_0 - x_2)\ldots(x_0 - x_n)$$

$$\Rightarrow \qquad A_0 = \frac{f(x_0)}{(x_0 - x_1)(x_0 - x_2)\ldots(x_0 - x_n)}$$

Similarly, putting $f(x) = f(x_1)$ and $x = x_1$ in (1), we get

$$f(x_1) = A_1(x_1 - x_0)(x_1 - x_2)\ldots(x_1 - x_n)$$

$$A_1 = \frac{f(x_1)}{(x_1 - x_0)(x_1 - x_2)\ldots(x_1 - x_n)}$$

Proceeding in the same way, we get $A_2, A_3, \ldots, A_n$,

i.e.,

$$A_i = \frac{f(x_i)}{(x_i - x_0)(x_i - x_1)\ldots(x_i - x_{i-1})(x_i - x_{i+1})\ldots(x_i - x_n)}$$

Put all these values of $A_0, A_1, \ldots, A_n$ in (1), we get

$$f(x) = \frac{(x - x_1)(x - x_2)\ldots(x - x_n)}{(x_0 - x_1)(x_0 - x_2)\ldots(x_0 - x_n)} f(x_0) + \frac{(x - x_0)(x - x_2)\ldots(x - x_n)}{(x_1 - x_0)(x_1 - x_2)\ldots(x_1 - x_n)} f(x_1) + \ldots +$$

$$+ \frac{(x - x_0)(x - x_1)\ldots(x - x_{n-1})}{(x_n - x_0)(x_n - x_1)\ldots(x_n - x_{n-1})} f(x_n)$$

which is the required Lagrange's interpolation formula.

REMARKS
- The Lagrange's formula can be applied whether the values x_i are equally spaced or not.
- If role of x and y are interchanged, then this formula can also be applied.
- This formula can also be used to split the given function into partial fractions.
- Lagrange's interpolation formula is easy to remember but its application is not speedily.
- The main drawback of it is that if another interpolation value is inserted, then the interpolation coefficient are required to be calculated.

WORKING PROCEDURE

Step 1. Put the values of x_i, $f(x_i)$ and values of x for which $f(x)$ is to be calculated.

Step 2. Calculate the values of $\prod_{\substack{r=0 \\ k \neq 0}}^{n} \frac{(x - x_r)}{(x_i - x_r)}$.

Step 3. Find $f(x)$ by using the Lagrange's interpolation formula.

10.23.1 REMAINDER TERM IN LAGRANGE'S INTERPOLATION FORMULA

Let $f(x)$ be a function which is approximated by means of some polynomial $P_n(x)$ in x of nth degree.

Let us suppose $f(x)$ satisfying all the conditions of Rolle's theorem.

Let

$$f(x) = P_n(x) + g(x) \qquad \ldots(1)$$

where $g(x)$ is a polynomial with root $x_0, x_1, ..., x_n$.

$\Rightarrow$ $\qquad f(x) = P_n(x) + k(x)(x - x_0)(x - x_1)...(x - x_n)$ $\qquad$ ...(2)

We want to find the value of $k(x)$.

Let $\qquad \phi(t) = f(t) - P_n(t) - k(x)(t - x_0)(t - x_1)...(t - x_n)$ $\qquad$ ...(3)

We observe that $\phi(t)$ vanishes for $(n+1)$ values of $t = (x_0, x_1,... x_n)$ and $t = x$, which implies $\phi(t)$ vanishes for $(n+2)$ real roots $x, x_0, x_1, ..., x_n$. Then by Rolle's theorem $\phi'(t)$ has at least $(n+1)$ roots lying between the smallest and greatest of the above roots. Similarly, $\phi'(t)$ has at least n roots and $\phi^{n+1}(t)$ has at least one root, say $t = \theta$ in same interval (x_0, x_n).

$$\phi^{n+1}(t) = f^{n+1}(t) - k(x)[(n+1)!] \qquad (\because P_n^{n+1}(t) = 0)$$

$\Rightarrow$ $\qquad \phi^{n+1}(\theta) = f^{n+1}(\theta) - k(x)(n+1)!$

but $\phi^{n+1}(\theta) = 0$

$\Rightarrow$ $\qquad f^{n+1}(\theta) - k(x)(n+1)! = 0$

$\Rightarrow$ $\qquad k(x) = \dfrac{1}{(n+1)!} f^{n+1}(\theta)$ $\qquad\qquad 0 < \theta < x_n$

Now putting the value of $k(x)$ in (2), we get $f(x) = P_n(x) + f^{n+1}(\theta) \dfrac{(x - x_0)(x - x_1)...(x - x_n)}{(n+1)!}$

$\therefore$ The required truncation error in $f(x)$ is $R_n = f(x) - P_n(x)$.

$\Rightarrow$ $\qquad R_n = \dfrac{f^{n+1}(\theta)}{(n+1)!} \prod_{i=0}^{n}(x - x_i)$

10.23.2 MERIT AND DEMERITS OF LAGRANGE'S INTERPOLATION FORMULA

(1) Lagrange's method is simple and easy to remember.
(2) While applying Lagrange's interpolation formula, there is no need to construct the difference table.
(3) The calculation is more complicated than the divided difference formula.
(4) There are more chances of errors in computation.
(5) The calculations provide no check whether the functional values used are taken correctly or not.

Solved Examples

EXAMPLE 1. *The value of x and y are given as below* :

x	5	6	9	11
$f(x)$	12	13	14	16

Find the value of y at x =10.

(UPTU-2003, JNTU-2008, 09)

SOLUTION : Here the values of x are not equally spaced.

We have, $x_0 = 5, x_1 = 6, x_2 = 9, x_3 = 11$

and $y_0 = 12, y_1 = 13, y_2 = 14, y_4 = 16$

$\therefore \quad y_{10} = \dfrac{(10-6)(10-9)(10-11)}{(5-6)(5-9)(5-11)} \times 12$

$\qquad + \dfrac{(10-5)(10-9)(10-11)}{(6-5)(6-9)(6-11)} \times 13$

$\qquad + \dfrac{(10-5)(10-6)(10-11)}{(9-5)(9-6)(9-11)} \times 14 +$

$\qquad + \dfrac{(10-5)(10-6)(10-9)}{(11-5)(11-6)(11-9)} \times 16$

$\qquad = \dfrac{4 \times 1 \times -1}{-1 \times -4 \times -6} \times 12 + \dfrac{5 \times 1 \times -1}{1 \times -3 \times -5} \times 13$

$\qquad + \dfrac{5 \times 4 \times -1}{4 \times 3 \times -2} \times 14 + \dfrac{5 \times 4 \times 1}{6 \times 5 \times 2} \times 16$

$\qquad = 2 - 4.33 + 11.67 + 5.33$

Hence, $y_{10} = 14.67$

EXAMPLE 2. *Find $f(4)$ from the following table* :

x	0	1	2	5
$f(x)$	2	5	7	8

(AVADH 2004, 09)

SOLUTION : Here, $x_0 = 0, x_1 = 1, x_2 = 2, x_3 = 5$

$y_0 = 2, y_1 = 5, y_2 = 7, y_3 = 8$

$\therefore$ By Lagrange's interpolation formula, we get

$f(4) = \dfrac{(4-1)(4-2)(4-5)}{(0-1)(0-2)(0-2)} \times 2$

$\qquad + \dfrac{(4-0)(4-2)(4-5)}{(1-0)(1-2)(1-5)} \times 5 + \dfrac{(4-0)(4-1)(4-5)}{(2-0)(2-1)(2-5)} \times 7$

$\qquad + \dfrac{(4-0)(4-1)(4-2)}{(5-0)(5-1)(5-2)} \times 8$

$\qquad = \dfrac{3.2.(-1)}{(-1)(-2)(-5)} \times 2 + \dfrac{4.2.(-1)}{1.(-1).(-4)} \times 5$

$\qquad + \dfrac{4.3.(-1)}{2.1.(-3)} \times 7 + \dfrac{4.3.2}{5.4.3} \times 8$

$\qquad = 1.2 - 10 + 14 + 3.2$

$\therefore f(4) = 8.4.$

EXAMPLE 3. *Find the value of f(x) for x = 2.5 from the following data:*

x	1	2	3	4
$f(x)$	1	8	27	64

(UPTU MCA-2001)

SOLUTION : Here, $x_0 = 1, x_1 = 2, x_2 = 3, x_3 = 4$

$\qquad y_0 = 1, y_1 = 8, y_2 = 27, y_3 = 64$

Putting all these values in Lagrange's interpolation formula, we get

$f(x) = \dfrac{(x-2)(x-3)(x-4)}{(1-2)(1-3)(1-4)} (1)$

$\qquad + \dfrac{(x-1)(x-3)(x-4)}{(2-1)(2-3)(2-4)} (8) + \dfrac{(x-1)(x-2)(x-4)}{(3-1)(3-2)(3-4)} (27)$

$\qquad + \dfrac{(x-1)(x-2)(x-3)}{(4-1)(4-2)(4-3)} (64)$

$$= -\frac{1}{6}(x-2)(x-3)(x-4)$$

$$+ 4(x-1)\ (x-3)(x-4) - \frac{27}{2}(x-1)(x-2)(x-4)$$

$$+ \frac{32}{3}(x-1)(x-2)(x-3)$$

Putting $x = 2.5$, we get

$$f(2.5) = -\frac{1}{6}(2.5-2)(2.5-3)(2.5-4)$$

$$+ 4(2.5-1)(2.5-3)(2.5-4) - \frac{27}{2}(2.5-1)$$

$$(2.5-2)(2.5-4) + \frac{32}{3}(2.5-1)(2.5-2)(2.5-3)$$

$$= 15.75.$$

EXAMPLE 4. *Find the unique polynomial $P(x)$ of degree 2 such that* $P(1) = 1, P(3) = 27, P(4) = 64$ *using Lagrange's method.* (UPTU MCA-2003)

SOLUTION: We have $x_0 = 1, x_1 = 3, x_2 = 4$

$$f(x_0) = 1, f(x_1) = 27, f(x_2) = 64$$

Putting all these values in Lagrange's interpolation formula, we get

$$P(x) = \frac{(x-3)(x-4)}{(1-3)(1-4)}(1) + \frac{(x-1)(x-4)}{(3-1)(3-4)}(27)$$

$$+ \frac{(x-1)(x-3)}{(4-1)(4-3)}(64)$$

$$= \frac{1}{6}(x^2 - 7x + 12) - \frac{27}{2}(x^2 - 5x + 4)$$

$$+ \frac{64}{3}(x^2 - 4x + 3)$$

$$= 8x^2 - 19x - 12, \text{ which is the required polynomial.}$$

EXAMPLE 5. *Using Lagrange's formula, prove that*

$$y_0 = \frac{1}{2}(y_1 + y_{-1}) - \frac{1}{8}\left[\frac{1}{2}(y_3 - y_1)\right.$$

$$\left. -\frac{1}{2}(y_{-1} - y_{-3})\right].$$

SOLUTION: Here, $x_0 = -3, x_1 = -1, x_2 = 1, x_3 = 3$

Then by Lagrange's formula, we get

$$y_x = \frac{(x+1)(x-1)(x-3)}{(-3+1)(-3-1)(-3-3)} \times y_{-3}$$

$$+ \frac{(x+3)(x-1)(x-3)}{(-1+3)(-1-1)(-1-3)} \times y_{-1}$$

$$+ \frac{(x+3)(x+1)(x-3)}{(1+3)(1+1)(1-3)} \times y_1$$

$$+ \frac{(x+3)(x+1)(x-1)}{(3+3)(3+1)(3-1)} \times y_3$$

Now putting $x = 0$ we have

$$y_0 = -\frac{3}{48}y_{-3} + \frac{9}{16}y_{-1} + \frac{9}{16}y_1 - \frac{3}{48}y_3$$

$$= -\frac{1}{16}y_{-3} + \left(\frac{1}{16} + \frac{8}{16}\right)y_{-1} + \left(\frac{1}{16} + \frac{8}{16}\right)y_1 - \frac{1}{16}y_3$$

$$= \frac{8}{16}(y_1 + y_{-1}) - \frac{1}{16}y_3 + \frac{1}{16}y_1 - \frac{1}{16}y_{-3} + \frac{1}{16}y_{-1}$$

Hence, $y_0 = \frac{1}{2}(y_1 + y_{-1}) - \frac{1}{8}\left[\frac{1}{2}(y_3 - y_1)\right.$

$$\left. -\frac{1}{2}(y_{-1} - y_{-3})\right]$$

EXAMPLE 6. *By Lagrange's formula, prove that*

$$y_1 = y_3 - 0.3(y_5 - y_{-3}) + 0.2(y_{-3} - y_{-5}).$$

SOLUTION: Here, the values of the arguments are given by

$$x_0 = -5, x_1 = -3, x_2 = 3, x_3 = 5$$

and corresponding values of functions are

$$y_{-5}, y_{-3}, y_3, y_5$$

Now using Lagrange's formula, we get

$$y_x = \frac{(x+3)(x-3)(x-5)}{(-5+3)(-5-3)(-5-5)} \times y_{-5}$$

$$+ \frac{(x+5)(x-3)(x-5)}{(-3+5)(-3-3)(-3-5)} \times y_{-3}$$

$$+ \frac{(x+5)(x+3)(x-5)}{(3+5)(3+3)(3-5)} \times y_3$$

$$+ \frac{(x+5)(x+3)(x-3)}{(5+5)(5+3)(5-3)} \times y_5$$

Putting $x = 1$, we get

$$y_1 = -\frac{32}{160}y_{-5} + \frac{48}{96}y_{-3} + \frac{96}{96}y_3 - \frac{48}{160}y_5$$

$$= y_3 - 0.3y_5 + 0.5y_{-3} - 0.2y_{-5}$$

$$= y_3 - 0.3y_5 + 0.3y_{-3} + 0.2y_{-3} - 0.2y_{-5}$$

Hence, $y_1 = y_3 - 0.3(y_5 - y_{-3}) + 0.2(y_{-3} - y_{-5}).$

EXAMPLE 7. *Using Lagrange's method, prove that*

$$y_3 = 0.05(y_0 + y_6) - 0.3(y_1 + y_5)$$

$$+ 0.75(y_2 + y_4).$$

SOLUTION: Here, the arguments are

$$x_0 = 0, x_1 = 1, x_2 = 2, x_3 = 4, x_4 = 5, x_5 = 6$$

and their corresponding values of functions are given by

$$y_0, y_1, y_2, y_4, y_5 \quad \text{and } y_6$$

$\therefore$ Lagrange's formula is given by

$$y_x = \frac{(x-1)(x-2)(x-4)(x-5)(x-6)}{(0-1)(0-2)(0-4)(0-5)(0-6)} y_0$$

$$+ \frac{(x-0)(x-2)(x-4)(x-5)(x-6)}{(1-0)(1-2)(1-4)(1-5)(1-6)} y_1$$

$$+ \frac{(x-0)(x-1)(x-4)(x-5)(x-6)}{(2-0)(2-1)(2-4)(2-5)(2-6)} \cdot y_2$$

$$+ \frac{(x-0)(x-1)(x-2)(x-5)(x-6)}{(4-0)(4-1)(4-1)(4-5)(4-6)} \cdot y_4$$

$$+ \frac{(x-0)(x-1)(x-2)(x-4)(x-6)}{(5-0)(5-1)(5-2)(5-4)(5-6)} \cdot y_5$$

$$+ \frac{(x-0)(x-1)(x-2)(x-4)(x-5)}{(6-0)(6-1)(6-2)(6-4)(6-5)} \cdot y_6$$

Putting $x = 3$, we get

$$y_3 = \frac{12}{240}y_0 - \frac{18}{60}y_1 + \frac{36}{48}y_2 + \frac{36}{48}y_4 + \frac{18}{60}y_5 + \frac{12}{240}y_6$$

$$= \frac{12}{240}(y_0 + y_6) - \frac{18}{60}(y_1 + y_5) + \frac{36}{48}(y_2 + y_4)$$

Hence, $y_3 = 0.05(y_0 + y_6) - 0.3(y_1 + y_5)$

$$+ 0.75(y_2 + y_4).$$

EXAMPLE 8. *The function $y = f(x)$ is given at the point (7, 3), (8, 1), (9,1) and (10, 9) Find $f(9, 5)$ using Lagrange's formula.* (UPTU-2004)

SOLUTION: It is given that

$$x_0 = 7, x_1 = 8, x_2 = 9, x_3 = 10$$

and $y_0 = 3, y_1 = 1, y_2 = 1, y_3 = 9$

Lagrange's formula is given by

$$f(x) = \frac{(x-8)(x-9)(x-10)}{(7-8)(7-9)(7-10)} \times 3$$
$$+ \frac{(x-7)(x-9)(x-10)}{(8-7)(8-9)(8-10)} \times 1$$
$$+ \frac{(x-7)(x-8)(x-10)}{(9-7)(9-8)(9-10)} \times 1$$
$$+ \frac{(x-7)(x-8)(x-9)}{(10-7)(10-8)(10-9)} \times 9$$

Putting $x = 9.5$ we get

$$f(9.5) = \frac{(9.5-8)(9.5-9)(9.5-10)}{(-1)(-2)(-3)} \times 3$$
$$+ \frac{(9.5-7)(9.5-9)(9.5-10)}{(1)(-1)(-2)} \times 1$$
$$+ \frac{(9.5-7)(9.5-8)(9.5-10)}{(2)(1)(-1)} \times 1$$
$$+ \frac{(9.5-7)(9.5-8)(9.5-9)}{(3)(2)(1)} \times 9$$
$$+ \frac{(1.5)(0.5)(-0.5)}{-6} \times 3 + \frac{(2.5)(0.5)(-0.5)}{2} \times 1$$
$$+ \frac{(2.5)(1.5)(-0.5)}{-2} \times 1 + \frac{(2.5)(1.5)(0.5)}{6} \times 9$$
$$= 0.1875 - 0.3125 + 0.9375 + 2.8125$$

Hence, $f(9.5) = 3.625$.

EXAMPLE 9. *Using Lagrange's interpolatioin formula, find the form of the function y(x) from the following table:*   (MDU (B.E.)-2003)

x	0	1	3	4
y	−12	0	12	24

SOLUTION: Here, we have

$$x_0 = 0, \quad y_0 = -12$$
$$x_1 = 1, \quad y_1 = 0$$
$$x_2 = 3, \quad y_2 = 12$$
$$x_3 = 4, \quad y_3 = 24$$

Now applying Lagrange's formula

$$y = f(x) = \frac{(x-x_1)(x-x_2)(x-x_3)}{(x_0-x_1)(x_0-x_2)(x_0-x_3)} y_0$$
$$+ \frac{(x-x_0)(x-x_2)(x-x_3)}{(x_1-x_0)(x_1-x_2)(x_1-x_3)} y_1$$
$$+ \frac{(x-x_0)(x-x_1)(x-x_3)}{(x_2-x_0)(x_2-x_1)(x_2-x_3)} y_2$$
$$+ \frac{(x-x_0)(x-x_1)(x-x_2)}{(x_3-x_0)(x_3-x_1)(x_3-x_2)} y_3$$
$$= \frac{(x-1)(x-3)(x-4)}{(-1)(-3)(-4)} \times (-12)$$
$$+ \frac{(x-0)(x-3)(x-4)}{4 \times 3 \times 1} \times 0$$
$$+ \frac{(x-0)(x-1)(x-4)}{3 \times 2 \times -1} \times 12$$
$$+ \frac{(x-0)(x-1)(x-3)}{4 \times 3 \times 1} \times 24$$
$$= (x-1)(x-3)(x-4) - 2x(x-1)(x-4)$$
$$+ 2x(x-1)(x-3)$$
$$= x^3 - 8x^2 + 19x - 12 - 2x^3 - 8x + 10x^2$$
$$+ 2x^3 - 8x^2 + 6x$$

$$= x^3 - 6x^2 + 17x - 12$$
$$\therefore \quad f(x) = x^3 - 6x^2 + 17x - 12.$$

EXAMPLE 10. *If $y(1) = -3, y(3) = 9, y(4) = 30, y(6) = 132$ find the Lagrange's interpolation polynomial that takes the same values as 'y' at the given points.*
     (MDU(B.E.)-2004; VTU-2006)

SOLUTION: Here, we have

$$x_0 = 1, x_1 = 3, x_2 = 4, x_3 = 6$$
$$y_0 = -3, y_1 = 9, y_2 = 30, y_3 = 132$$

Let the polynomial be of the form

$$y = a_0(x-x_1)(x-x_2)(x-x_3) + a_1(x-x_0)$$
$$(x-x_2)(x-x_3) + a_2(x-x_0)(x-x_1)(x-x_3)$$
$$+ a_3(x-x_0)(x-x_1)(x-x_2)$$
$$= a_0(x-3)(x-4)(x-6) + a_1(x-1)(x-4)$$
$$(x-6) + a_2(x-1)(x-3)(x-6)$$
$$+ a_3(x-1)(x-3)(x-4) \qquad \ldots(1)$$

Now $a_0 = \dfrac{y_0}{(x_0-x_1)(x_0-x_2)(x_0-x_3)}$

$$= \frac{-3}{(1-3)(1-4)(1-6)} = 0.1$$

$$a_1 = \frac{y_1}{(x_1-x_0)(x_1-x_2)(x_1-x_3)}$$

$$= \frac{9}{(3-1)(3-4)(3-6)} = 1.5$$

$$a_2 = \frac{y_2}{(x_2-x_0)(x_2-x_1)(x_2-x_3)}$$

$$= \frac{30}{(4-1)(4-3)(4-6)} = -5$$

$$a_3 = \frac{y_3}{(x_3-x_0)(x_3-x_1)(x_3-x_2)}$$

$$= \frac{132}{(6-1)(6-3)(6-4)} = 4.4$$

Substituting these above values in(1), we get

$$y = 0.1(x^3 - 13x^2 + 54x - 72)$$
$$+ 1.5(x^3 - 11x^2 + 34x - 24)$$
$$- 5(x^3 - 10x^2 + 27x - 18)$$
$$+ 4.4(x^3 - 8x^2 + 9x - 12)$$
$$\therefore \quad y = x^3 - 3x^2 + 5x - 6$$

EXAMPLE 11. *From the given table.*

x	20	25	30	35
y(x)	0.342	0.423	0.500	0.650

Find the value of x from y(x) = 0.390:
     [UPTU (MCA)-2003]

SOLUTION: Here,

$$x_0 = 20 \quad \Rightarrow \quad y_0 = 0.342$$
$$x_1 = 25 \quad \Rightarrow \quad y_1 = 0.423$$
$$x_2 = 30 \quad \Rightarrow \quad y_2 = 0.500$$

and $x_3 = 35 \quad \Rightarrow \quad y_3 = 0.650$

Using Lagrange's formula, we get

$$x = \frac{(y-y_1)(y-y_2)(y-y_3)}{(y_0-y_1)(y_0-y_2)(y_0-y_3)} x_0$$
$$+ \frac{(y-y_0)(y-y_2)(y-y_3)}{(y_1-y_0)(y_1-y_2)(y_1-y_3)} x_1$$
$$+ \frac{(y-y_0)(y-y_1)(y-y_3)}{(y_2-y_0)(y_2-y_1)(y_2-y_3)} x_2$$
$$+ \frac{(y-y_0)(y-y_1)(y-y_2)}{(y_3-y_0)(y_3-y_1)(y_3-y_2)} x_3$$

Now, at $y = 0.390$

$$x = \frac{(0.390 - 0.423)(0.390 - 0.500)(0.390 - 0.650)}{(0.342 - 0.423)(0.342 - 0.500)(0.342 - 0.650)} (20)$$

$$+ \frac{(0.390 - 0.342)(0.390 - 0.500)(0.390 - 0.650)}{(0.432 - 0.342)(0.432 - 0.500)(0.432 - 0.650)} (25)$$

$$+ \frac{(0.390 - 0.342)(0.390 - 0.423)(0.390 - 0.650)}{(0.500 - 0.342)(0.500 - 0.432)(0.500 - 0.650)} (30)$$

$$+ \frac{(0.390 - 0.342)(0.390 - 0.423)(0.390 - 0.500)}{(0.650 - 0.342)(0.650 - 0.423)(0.650 - 0.500)} (35)$$

$$x = \frac{(-0.033)(-0.110)(-0.260)}{(-0.081)(-0.158)(-0.308)} (20)$$

$$+ \frac{(0.048)(-0.110)(-0.260)}{(0.081)(-0.077)(-0.227)} (25)$$

$$+ \frac{(0.048)(-0.033)(-0.260)}{(0.158)(-0.077)(-0.150)} (30)$$

$$+ \frac{(0.048)(-0.033)(-0.110)}{(0.308)(0.227)(0.150)} (35)$$

$$\Rightarrow \quad x = 23.67.$$

EXAMPLE 12. *Given* $\log_{10} 654 = 2.8156$, $\log_{10} 658 = 2.8182$, $\log_{10} 659 = 2.8189, \log_{10} 661 = 2.8202$. *Find* $\log_{10} 656$.

(ROHTAK-2006, 10, BANGLURU-2008, HAZARIBAGH, 2009)

SOLUTION: Putting the given values in Lagrange's interpolation formula, we get

$\log_{10} 656$

$$= \frac{(656 - 658)(656 - 659)(656 - 661)}{(654 - 658)(654 - 659)(654 - 661)} \times (2.8156)$$

$$+ \frac{(656 - 654)(656 - 659)(656 - 661)}{(658 - 654)(658 - 659)(658 - 661)} \times (2.8182)$$

$$+ \frac{(656 - 654)(656 - 658)(656 - 661)}{(659 - 654)(659 - 658)(659 - 661)} \times (2.8189)$$

$$+ \frac{(656 - 654)(656 - 658)(656 - 659)}{(661 - 654)(661 - 658)(661 - 659)} \times (2.8202)$$

$$= \frac{3}{14}(2.8156) + \frac{5}{2}(2.8182) - 2(2.8189) + \frac{2}{7}(2.8202)$$

$$= 2.8170.$$

EXAMPLE 13. *Values of $f(x)$ are given at a, b and c, show that the maximum is obtained by*

$$x = \frac{f(a)(b^2 - c^2) + f(b)(c^2 - a^2) + f(c)(a^2 - b^2)}{[f(a)(b - c) + f(b)(c - a) + f(c)(a - b)]}$$

(DELHI-2009, 10; VANKETSHWARA-2006; NAGPUR-2005, 08, 10)

SOLUTION: For the arguments a, b and c the Lagrange's formula is given by

$$f(x) = \frac{(x - b)(x - c)}{(a - b)(a - c)} f(a)$$

$$+ \frac{(x - a)(x - c)}{(b - a)(b - c)} f(b) + \frac{(x - a)(x - b)}{(c - a)(c - b)} f(c)$$

$$= \frac{x^2 - (b + c)x + bc}{(a - b)(a - c)} f(a)$$

$$+ \frac{x^2 - (a + c)x + ac}{(b - a)(b - c)} f(b)$$

$$+ \frac{x^2 - (a + b)x + ab}{(c - a)(c - b)} f(c)$$

For maxima and minima of $f(x)$ we must have $f'(x) = 0$

i.e., $\quad \dfrac{2x - (b + c)}{(a - b)(a - c)} f(a) + \dfrac{2x - (a + c)}{(b - a)(b - c)} f(b)$

$$\frac{2x - (a + b)}{(c - a)(c - b)} f(c) = 0$$

On solving for x, we get

$$x = \frac{f(a)(b^2 - c^2) + f(b)(c^2 - a^2) + f(c)(a^2 - b^2)}{2[f(a)(b - c) + f(b)(c - a) + f(c)(a - b)]}$$

EXAMPLE 14. *Values of $y(x)$ are given for all integral values of x from 0 to n − 1. Show that y_x is capable of expression in the form*

$$\frac{x!}{(x - n)!(n - 1)!} \left(\frac{y_{n-1}}{x - n + 1} - {}^{n-1}C_1 \frac{y_{n-2}}{x - n + 2} \right.$$

$$\left. + {}^{n-1}C_2 \frac{y_{n-3}}{x - n + 3} + (-1)^{n-1} \cdot {}^{n-1}C_{n-1} \frac{y_0}{x_0} \right)$$

(DELHI-2004, 07)

SOLUTION: We know that

$$y = \frac{(x - 0)(x - 1)(x - 2) - (x - (n - 2))}{[(n - 1 - 0)(n - 1 - 1)(n - 1 - 2) - (n - 1 - (n - 2))]} y_{n-1}$$

$$+ \frac{(x - 0)(x - 1) \dots (x - (n - 3)(x - (n + 1))}{(n - 2 - 0)(n - 2 - 1) \dots (n - 2 - (n - 3)) \cdot (n - 2) - (n - 1)} y_{n-2}$$

$$+ \dots + \frac{(x - 1)(x - 2) \dots (x - (n - 1))}{(0 - 1)(0 - 2) \dots (0 - (n - 1))} y_0$$

$$\Rightarrow \quad y_x \quad \frac{x(x - 1)(x - 2) \dots (x - n + 2)}{(\quad 1)!} y_n$$

$$+ \frac{x(x - 1) \dots (x - n + 3)(x - n + 1)}{(n - 2)!(-1)} y_{n-2}$$

$$+ \frac{x(x - 1) \dots (x - n + 4)(x - n + 2)(x - n + 1)}{(n - 3)!(-1)(-2)} y_{n-3}$$

$$+ \frac{(x - 1)(x - 2) \dots (x - n + 1)}{(-1)(-2) \dots (-(n - 1))} y_0$$

$$y_x = \frac{x(x - 1) \dots (x - n + 2)(x - n + 1)}{(n - 1)!} \cdot \frac{y_{n-1}}{(x - n + 1)}$$

$$+ \frac{x(x - 1) \dots (x - n + 3)(x - n + 2)(x - n + 1)}{(n - 2)!(-1)} \cdot \frac{y_{n-2}}{x - n + 2}$$

$$+ \frac{x(x - 1) \dots (x - n + 1)}{(n - 3)!(-1)(-2)} \cdot \frac{y_{n-3}}{x - n - 3} + \dots$$

$$+ x \frac{(x - 1)(x - 2) \dots (x - n + 1)}{(-1)(-2) \dots (-(n - 1)} \cdot \frac{y_0}{x}$$

$$= \frac{x!}{(x - n)!} \cdot \frac{1}{(n - 1)!} \frac{y_{n-1}}{x - n + 1} + (-1)$$

$$\cdot \frac{x!}{(x - n)!(x - 2)} \cdot \frac{1}{1!} \frac{y_{n-2}}{x - n + 2}$$

$$+ (-1)^2 \frac{x!}{(x - n)!} \frac{1}{0!} \frac{1}{(n - 1)!} \frac{y_0}{x}$$

$$= \frac{x!}{(x - n)!} \cdot \frac{1}{(n - 1)!} \cdot \frac{y_{n-1}}{x - n + 1} + (-1)$$

$$\frac{x!}{(x - n)!} \frac{1}{(n - 1)!} {}^{n-1}C_1 \frac{y_{n-2}}{x - n + 2} + (-1)^2$$

$$\frac{x!}{(x - n)!} \cdot \frac{1}{(n - 1)!} {}^{n-1}C_2 \frac{y_{n-3}}{x - n + 3} + \dots$$

$$+ (-1)^{n-1} \cdot \frac{x!}{(x - n)!} \frac{1}{(n - 1)!} {}^{n-1}C_{n-1} \frac{y_0}{x}$$

$$= \frac{x!}{(x-n)!(n-1)!}\left(\frac{y_{n-1}}{x-n+1}+(-1)^{n-1}C_1\right.$$

$$\left.\frac{y_{n-2}}{x-n+2}\right)+(-1)^2 \cdot {}^{n-1}C_2 \cdot \frac{y_{n-3}}{x-n+3}$$

$$+...+(-1)^{n-1} \cdot {}^{n-1}C_{n-1}\frac{y_0}{x_0}$$

$$= \sum_{r=0}^{n}\frac{Ar}{(x-x_r)} \qquad ...(2)$$

where $A_r = \dfrac{1}{(x_r-x_0)(x_r-x_1)...(x_r-x_{r-1})(x_r-x_{r+1})}$

$$...(x_r-x_n)$$

$$= \frac{1}{\phi'(x_r)} \qquad ...(3)$$

EXAMPLE 15. *Show that the sum of Lagrangian coefficient is unity.*

SOLUTION: Define $\phi(x)$ such that

$$\phi(x) = \prod_{r=0}^{n}(x-x_r) = (x-x_0)(x-x_1)...(x-x_n) \quad ...(1)$$

$$\Rightarrow \frac{1}{\phi(x)} = \frac{1}{(x-x_0)(x-x_1)...(x-x_n)}$$

$$= \frac{A_0}{x-x_0}+\frac{A_1}{x-x_1}+...+\frac{A_n}{x-x_n}$$

Using (3) in (2), we get

$$\frac{1}{\phi(x)} = \sum_{r=0}^{n}\frac{1}{(x-x_r)\phi'(x_r)}$$

$$\Rightarrow \quad 1 = \sum_{r=0}^{n}\frac{\phi(x)}{(x-x_r)\phi(x_r)}.$$

10.24 HERMITE'S INTERPOLATION FORMULA

(UPTU MCA-2008)

The Hermitian interpolation is similar to that of Lagrange's Interpolation. The difference is that in the Lagrange's Interpolation the interpolating polynomial $p(x)$ considers with $f(x)$ at the inerpolation points $x_0, x_1, x_2, ... x_n$, whereas in Hermite interpolation $p(x)$ and $f(x)$ as well as $p'(x)$ and $f'(x)$ coincide. That is,

$$p(x_k) = f(x_k)$$

and $\qquad p'(x_k) = f'(x_k)$ for $k = 1, 2, 3...n$ $\qquad ...(1)$

We have $(2n+2)$ condition given by (1) and we know that a polynomial of degree $2n+1$ has $2n+2$ coefficients to be determined. Thus we have the form

$$p(x) = \sum_{k=0}^{n}u_k(x)f(x_k) + \sum_{k=0}^{n}v_k(x)f'(x_k) \qquad ...(2)$$

Here $u_k(x)$ and $v_k(x)$ are polynomial in x of degree $(2n+1)$. From (1), we obtain,

$$\begin{aligned} u_k(x_i) &= 1, \text{ if } k = i \quad v_k(x_i) = 0 \\ &= 0, \text{ if } k \neq i, \quad v'_k(x_i) = 1, \text{ if } k = i \\ u'_k(x_i) &= 0 \qquad\qquad\quad = 0, \text{ if } k \neq i \end{aligned} \qquad ...(3)$$

Now we may choose

$$u_k(x) = w_k(x)[L_k(x)]^2 \qquad ...(4)$$

and $\qquad v_k(x) = z_k(x)[L_k(x)]^2$

where $\qquad L_k(x) = \dfrac{(x-x_0)(x-x_1)...(x-x_{k-1})(x-x_{k+1})...(x-x_n)}{(x_k-x_0)(x_k-x_1)...(x_k-x_{k-1})(x_k-x_{k+1})...(x_k-x_n)}$

Since $L_k(x)$ is a polynomial of degree n. Then $w_k(x)$ and $z_k(x)$ are both linear. Therefore using condition (3), we get

and $\qquad \begin{aligned} w_k(x_k) &= 1, w'_k(x_k) = -2L'_k(x_k) \\ z_k(x_k) &= 0, z'_k(x_k) = 1 \end{aligned} \qquad ...(5)$

From (5), we find $\qquad w_k(x) = 1 - 2L'_k(x_k)(x-x_k)$ and $z_k(x) = x - x_k$

From (2), (4) and (5), we get

$$p(x) = \sum_{k=0}^{n}\{1-2L'_k(x_k)(x-x_k)\}[L_k(x)]^2 f(x_k) + \sum_{k=0}^{n}(x-x_k)[L_k(x)]^2 f'(x_k)$$

This is Hermite interpolation formula.

☞ **REMARK**
- Hermite interpolation formula is sometimes called osculating interpolation formula.

WORKING PROCEDURE

Step 1. Put the values of $x_i, f(x_i), f'(x_i)$ and value of x for which $f(x)$ is to be determined.

Step 2. Obtained $L_k(x) = \displaystyle\prod_{\substack{r=0 \\ k \neq r}}^{n}\frac{(x-x_k)}{(x_l-x_r)}, l \neq k$,

Step 3. Obtained $L'_k(x)$ at $x = x_i$.

Step 4. Put all these values in Hermite's formula.

Solved Examples

EXAMPLE 1. *Apply Hermite interpolation formula to find the value of sin (1.05) from the following data:*

x	1.00	1.10
sin x	0.84147	0.89121
cos x	0.54030	0.45360

SOLUTION: Here $f(x) = \sin x, f'(x) = \cos x$

and $x_0 = 1.00, x_1 = 1.10$

$$L_0(x) = \frac{x - x_1}{x_0 - x_1} = \frac{x - 1.10}{1.00 - 1.10}$$

$$L_0(x) = -10x + 11$$

$$\therefore \quad L_0'(x) = -10$$

and $L_1(x) = \dfrac{x - x_0}{x_1 - x_0} = \dfrac{x - 1.00}{1.10 - 1.00}$

$$L_0(x) = 10x - 10$$

$$\therefore \quad L_1'(x_1) = 10$$

Since $f(x_0) = 0.84147, f'(x_0) = 0.54030$

$f(x_1) = 0.89121, f'(x_1) = 0.45360$

Now by Hermite interpolation formula, we have

$$P(x) = \sum_{k=0}^{1} \{1 - 2L_k'(x_k)(x - x_k)\}[L_k(x)]^2$$

$$f(x_k) + \sum_{k=0}^{1}(x - x_k)[L_k(x)]^2 f'(x_k)$$

$$= \{1 - 2L_0'(x_0)(x - x_0)\}[L_0(x)]^2 f(x_0)$$

$$+ \{1 - 2L_1'(x_1)(x - x_1)\}[L_1(x)]^2 f(x_1)$$

$$+ (x - x_0)[L_0(x)]^2 f'(x_0) + (x - x_0)[L_1(x)]^2 f'(x_1)$$

$$P(x) = \{1 + 20(x - 1.00)\}(-10x + 11)^2(0.84147)$$

$$+ \{1 - 20(x - 1.10)\}(10x - 10)^2(0.89121)$$

$$+ (x - 1.00)(-10x + 11)^2(0.54030)$$

$$+ (x - 1.10)(10x - 10)^2(0.45360)$$

Putting $x = 1.05$ in this equation, we get

$$P(1.05) = \sin(1.05) = \{(1 + 20(1.05 - 1.00)\}$$
$$(-10(1.05) + 11)^2 (0.84147) + \{1 - 20$$
$$(1.05 - 1.10)\} (10 (1.05) - 10)^2 (0.89121)$$
$$+ (1.05 - 1.00) (-10(1.05) - 11)^2 (54030)$$
$$+ (1.05 - 1.10)\{(1.05) - 10\}^2 (0.45360)$$
$$= (2)(0.25) (0.84147) + (2)(0.25)$$
$$(0.89121) + (0.05) (0.25) (0.54030)$$
$$- (0.5) (0.25) (0.45360)$$
$$= 0.420735 + 0.445605 + 0.00675375$$
$$- 0.00567$$

Hence, $\sin (1.05) = 0.86742$.

EXAMPLE 2. *Apply Hermite's formula to find a polynomial from the following data :*

x	0	1	2
$f(x)$	0	1	0
$f'(x)$	0	0	0

SOLUTION: Here $x_0 = 0, x_1 = 1, x_2 = 2$

and $f(x_0) = 0, f(x_1) = 1, f(x_2) = 0$

and $f'(x_0) = 0, f'(x_1) = 0, f'(x_2) = 0$

$$L_0(x) = \frac{(x - x_1)(x - x_2)}{(x_0 - x_1)(x_0 - x_2)}$$

$$L_0(x) = \frac{(x - 1)(x - 2)}{(0 - 1)(0 - 2)} = \frac{1}{2}(x^2 - 3x + 2)$$

$$\therefore \quad L_0'(x_0) = \frac{1}{2}(2x_0 - 3) = \frac{1}{2}(-3) = -\frac{3}{2}$$

$$L_1(x) = \frac{(x - x_0)(x - x_2)}{(x_1 - x_0)(x - x_2)}$$

$$L_1(x) = \frac{(x)(x - 2)}{(1 - 0)(1 - 2)} = -(x^2 - 2x)$$

$$\therefore \quad L_1'(x_1) = -(2x_1 - 2) = -(2 - 2) = 0$$

$$L_2(x) = \frac{(x - x_0)(x - x_1)}{(x_2 - x_0)(x_2 - x_1)}$$

$$\Rightarrow \quad L_2(x) = \frac{x(x - 1)}{(2 - 0)(x_2 - x_1)} = \frac{1}{2}(x^2 - x)$$

and $L_2'(x) = \dfrac{1}{2}(2x_2 - 1) = \dfrac{1}{2}(4 - 1) = \dfrac{3}{2}$

Now by Hermite interpolation formula, we get

$$P(x) = \sum_{K=0}^{2} \{1 - 2L_k'(x_k)(x - x_k)\}[L_k(x)]^2$$

$$f(x_k) + \sum_{k=0}^{2}(x - x_k)[L_k(x)]^2 f'(x_k)]$$

$$= \{1 - 2L_0'(x_0)(x - x_0)\}[L_0(x)]^2 f(x_0)$$

$$+ \{1 - 2L_1'(x_1)(x - x_1)\}[L_1(x)]^2 f(x_1)$$

$$+ \{1 - 2L_2'(x_2)(x - x_2)\}[L_2(x)]^2 f(x_2)$$

$$+ (x - x_0)[L_0(x)]^2 f'(x_0) + (x - x_1)[L_1(x)]^2$$

$$f'(x_1) + (x - x_2)[L_2(x)]f'(x_2)$$

$$\Rightarrow \quad P(x) = \{1 + 3(x - 0)\}\frac{1}{4}(x^2 - 3x + 2)^2 \times 0$$

$$+ \{1 - 0\}(x^2 - 2x)^2.1 + \{1 - 3(x - 2)\}$$

$$\frac{1}{4}(x^2 - x)^2 \times 0 + (x - 0)\frac{1}{4}(x^2 - 3x + 2)^2$$

$$\times 0 + (x - 1)(x^2 - 2x)^2 \times 0 + (x - 2)$$

$$\left[\frac{1}{4}(x^2 - x)^2\right] \times 0$$

$$P(x) = (x^2 - 2x)^2.$$

This is required polynomial.

 **Exercise-10.10**

1. Find the function u_x in powers of $(x - 1)$

where, $u_0 = 8, u_1 = 11, u_4 = 68, u_5 = 123$.

2. Find u_x in powers of $(x - 4)$

where, $u_0 = 8, u_1 = 11, u_4 = 68, u_5 = 125$.

3. Given the values

x	5	7	11	13	17
$f(x)$	150	392	1452	2366	5202

Find $f_{(9)}$ by using Newton's divided difference formula.

(VTU-2010, PTU-2005, ANNA-2006)

4. Given that

x	1	3	4	6	7
y_x	1	27	81	729	2187

Find f_5. Why it does differ from 3^5?

5. Given that

x	300	304	305	307
$\log_{10}x$	2.4771	2.4829	2.4823	2.4871

Find $\log_{10}310$ by Newton's divided difference formula.

6. Find the value of t when $A = 85$ from the following data, using Lagrange's interpolation formula

t	2	5	8	14
A	94.8	87.9	81.3	68.7

7. The observed value of a function are respectively 168, 120, 72 and 63 at the four positions 3, 7, 9 and 20 of the independent variable. What is the best estimate you can give for values of the function at the position 6 of independent variable.

8. Given that

x	1.2	2.1	2.8	4.1	4.9	6.2
y	4.2	6.8	9.8	13.4	15.5	19.6

Find the value of x corresponding to $y = 12$ using Lagrange's interpolation formula.

9. Given that

x	5	6	9	11
$f(x)$	12	13	14	18

Find $f(x)$ as a polynomial in x using Newton's divided difference formula.

10. Use Lagrange's formula to find $f(0.4)$ given that
$$f(-1) = 1, f(0) = 3, f(1) = 2, f(2) = 5$$

11. Use Lagrange's formula to find u_2 given that
$$u_0 = 6, u_1 = 9, u_3 = 33, u_7 = -15$$

Answers

1. $(x-1)^3 + 2(x-1)^2 + 4(x-1) + 11$

2. $\dfrac{1}{10}[11(x-4)^3 + 117(x-4)^2 + 447(x-4) + 680]$

3. 810

4. 208.82222, 3^x is not a polynomial

5. 2.4786

6. 6.5928

7. 147

8. 3.55

10. 1.88

11. 9

10.25 CENTRAL DIFFERENCES INTERPOLATION FORMULAE

Let $y = f(x)$ be a function, which takes the values ... $y_{-2}, y_{-1}, y_0, y_1, y_2...$ corresponding to the values of $x = x_0 - 2h, x_0 - h, x_0, x_0 + h, x_0 + 2h...$

(i) Newton's Forward Difference Formula.

We want to find the value of the function y_u for $x = x_0 + uh$, where, in general $-1 < u < 1$. Then

$$y_u = E^u y_0 = (1+\Delta)^u y_0 \qquad\qquad (\because E = 1 + \Delta)$$

$$= ({}^uc_0 + {}^uc_1\Delta + {}^uc_2\Delta^2 + {}^uc_3\Delta^3 + ... + {}^uc_4\Delta^4)y_0$$

$$y_u = 1 + u\Delta y_0 + \frac{u(u-1)}{2!}\Delta^2 y_0 + \frac{u(u-1)(u-2)}{3!}\Delta^3 y_0 + ...$$

This is called Newton's forward difference formula.

(ii) Newton's Backard Difference Formula.

This formula is obtained by Newton's forward-difference formula, we have

$$y_u = (E^{-1})^{-u} y_0 = (1-\nabla) \qquad\qquad (\because E^{-1} = 1 - \nabla)$$

$$= [1 + u\nabla + \frac{u(u+1)}{2!}\nabla^2 + \frac{u(u+1)(u+2)}{3!}\nabla^3 y_0 + ...]y_0$$

$$y_u = y_0 + u\nabla y_0 + \frac{u(u+1)}{2!}\nabla^2 y_0 + \frac{u(u+1)(u+2)}{3!}\nabla^3 y_0 + ...$$

This is called Newton's backward-difference formula.

(iii) Gauss-forward Difference Formula.

This Formula is obtained by Newton's forward-difference formula, we have

$$y_u = 1 + u\nabla y_0 + \frac{u(u-1)}{2!}\Delta^2 y_0 + \frac{u(u-1)(u-2)}{3!}\nabla^3 y_0 + ... \qquad ... (1)$$

Since we have

$$\Delta^2 y_0 - \Delta^2 y_{-1} = \Delta^3 y_{-1}$$

or

$$\Delta^2 y_0 = \Delta^2 y_{-1} + \Delta^3 y_{-1} \qquad ... (2)$$

Similarly, $\quad \Delta^3 y_0 = \Delta^3 y_{-1} + \Delta^4 y_{-1}$ ... (3)

$\qquad\qquad \Delta^4 y_0 = \Delta^4 y_{-1} + \Delta^5 y_{-1}$ ... etc ... (4)

Also $\quad \Delta^3 y_{-1} = \Delta^3 y_{-2} + \Delta^4 y_{-2}$

and $\qquad \Delta^4 y_{-1} = \Delta^4 y_{-2} + \Delta^5 y_{-2}$ ... etc. ... (5)

Substituting the values of $\Delta^2 y_0, \Delta^3 y_0, \Delta^4 y_0$, from (2), (3), (4) ... in(1), we get

$$y_u = 1 + u\Delta y_0 + \frac{u(u-1)}{2!}(\Delta^2 y_{-1} + \Delta^3 y_{-1})$$

$$+ \frac{u(u-1)(u-2)}{3!}(\Delta^3 y_{-1} + \Delta^4 y_{-1}) + \frac{u(u-1)(u-2)(u-3)}{4!}(\Delta^4 y_{-1} + \Delta^5 y_{-1}) + \dots$$

$$= 1 + u\Delta y_0 + \frac{u(u+1)}{2!}\Delta^2 y_{-1} + \frac{(u+1)u(u-1)}{3!}\Delta^3 y_{-1} + \frac{(u+1)u(u-1)(u-2)}{4!}\Delta^4 y_{-2}$$

This is called Gauss's-Forward difference formula.

(iv) Gauss's backward Difference Formula.

The Newton's-forward difference formula is given by

$$y_u = y_0 + u\Delta y_0 + \frac{u(u-1)}{2!}\Delta^2 y_0 + \frac{u(u-1)(u-2)}{3!}\Delta^3 y_0 + \dots \qquad\dots(1)$$

Since, we have

$$\Delta y_0 = \Delta y_{-1} + \Delta^2 y_{-1} \qquad\dots (2)$$

$$\Delta^2 y_0 = \Delta^2 y_{-1} + \Delta^3 y_{-1} \qquad\dots (3)$$

$$\Delta^3 y_0 = \Delta^3 y_{-1} + \Delta^4 y_{-1} \dots \text{ etc.} \qquad\dots (4)$$

Also $\quad \Delta^3 y_{-1} = \Delta^3 y_{-2} + \Delta^4 y_{-2}$ ... (5)

$\qquad\qquad \Delta^4 y_{-1} = \Delta^4 y_{-2} + \Delta^5 y_{-2} \dots$ etc. ... (6)

With the help of (1) to (6), we wet

$$y_u = y_0 + u(\Delta y_{-1} + \Delta^2 y_{-1}) + \frac{u(u-1)}{2!}(\Delta^2 y_{-1} + \Delta^3 y_{-1})$$

$$+ \frac{u(u-1)(u-2)}{3!}(\Delta^3 y_{-1} + \Delta^4 y_{-1}) + \frac{u(u-1)(u-2)(u-3)}{4!}(\Delta^4 y_{-1} + \Delta^5 y_{-1}) + \dots$$

or $\quad y_u = y_0 + u\Delta y_{-1} + \frac{(u+1)u}{2!}\Delta^2 y_{-1} + \frac{(u+1)u(u-1)}{3!}\Delta^3 y_{-1} + \frac{(u+1)u(u-1)(u-2)}{4!}\Delta^4 y_{-1} + \frac{u(u-1)(u-2)(u-3)}{4!}\Delta^4 y_{-1}$

$$= y_0 + u\Delta y_{-1} + \frac{(u+1)u}{2!}y_{-1} + \frac{u(u+1)(u-1)}{3!}(\Delta^3 y_{-2} + \Delta^4 y_{-2})$$

$$+ \frac{(u+1)u(u-1)(u-2)}{4!}(\Delta^4 y_{-2} + \Delta^5 y_{-2}) + \dots$$

$$y_u = y_0 + u\Delta y_{-1} + \frac{(u+1)u}{2!}\Delta^2 y_{-1} + \frac{(u+1)u(u-1)}{3!}\Delta^3 y_{-2} + \frac{(u+2)(u+1)u(u-1)}{4!}\Delta^4 y_{-2}\dots$$

This is called Gauss's-backward difference formula.

10.25.1 Stirling's Difference Formula

The Stirling's formula can be obtained by taking the average of Gauss's forward difference formula and Gauss's backward difference formula.

We have, Gauss's-forward difference formula is given by

$$y_u = y_0 + u\Delta y_0 + \frac{u(u-1)}{2!}\Delta^2 y_{-1} + \frac{(u+1)u(u-1)}{3!}\Delta^3 y_{-1} + \frac{(u+1)(u)(u-1)(u-2)}{4!}\Delta^4 y_{-2} + \dots \qquad\dots(1)$$

Also, Gauss's backward formula is given by

$$y_u = y_0 + u\Delta y_{-1} + \frac{(u+1)u}{2!}\Delta^2 y_{-1} + \frac{(u+1)u(u-1)}{3!}\Delta^3 y_{-2} + \frac{(u+2)(u+1)u(u-1)}{4!}\Delta^4 y_{-2}\dots \qquad\dots (2)$$

Taking average of (1) and (2), we get

$$y_u = y_0 + u\left(\frac{\Delta y_0 + \Delta y_{-1}}{2}\right) + \frac{u^2}{2!}\Delta^2 y_{-1} + \frac{u(u^2-1)}{3!}\left(\frac{\Delta^3 y_{-1} + \Delta^3 y_{-2}}{2}\right) + \frac{u^2(u^2-1)}{4!}\Delta^4 y_{-2} + \dots$$

This is called Stirling's difference formula.

10.25.2 Remainder Term in Stirling's Formula

Let the function $f(x)$ be approximated by a polynomial $P_{2n}(x)$.

Assume

$$f(x) = P_{2n}(x) + k(x)(x - x_0)(x - x_1)(x - x_{-1})...(x - x_n)(x - x_{-n}) \qquad ...(1)$$

where $k(x)$ is obtained.

Consider the function,

$$g(t) = f(t) - P_{2n}(t) - k(x)(t - x_0)(t - x_1)(t - x_{-1})...(t - x_n)(t - x_{-n}) \qquad ... (2)$$

Here, $g(t)$ vanishes for $(2n+2)$ values which are given by $t = x, x_0, x_1, x_{-1}, ..., x_n, x_{-n}$

Then, by Rolle's theorem, we $\phi^{2n+1}(\theta)$ has at least one root $t = \theta$ in (x_{-n}, x_n)

But $\qquad \phi^{2n+1}(t) = f^{2n+1}(t) - k(x)\{(2n+1!\}$

$\Rightarrow \qquad \phi^{2n+1}(\theta) = f^{2n+1}(\theta) - k(x)\{(2n+1!\}$

Now $\qquad \phi^{2n+1}(\theta) = 0 \quad \Rightarrow \quad k(x) = \dfrac{f^{2n+1}(\theta)}{(2n+1)!}$ $x_{-n} < \theta < x_n$

Now putting the value of $K(x)$ in (1), we get

$$f(x).P_{2x}(x) + f^{2n+1}(\theta)\left[\frac{(x - x_0)(x - x_1)(x - x_{-1})...(x - x_n)(x - x_{-n})}{(2n+1)!}\right]$$

New error $\qquad R_n = f(x) - P_{2n}(x) = f^{2n+1}(\theta)\dfrac{(x - x_0)(x - x_1)(x - x_{-1})...(x - x_n)(x - x_{-n})}{(2n+1)!}$

Now using, $\qquad u = \dfrac{x - x_0}{h}$ we get $R_n = \dfrac{h^{2n+1}f^{2n+1}(\theta)}{(2n+1)!}u(u^2 - 1^2)(u^2 - 2^2)...(u^2 - n^2)$

Now, the values of R_n, in terms of difference

$$R_n = \frac{\Delta^{2n+1}f(x_{-n-1}) + \Delta^{2n+1}f(x_{-n})}{\{2.(2n+1)!\}}u(u^2 - 1^2)(u^2 - 2^2)...(u^2 - n^2).$$

10.25.3 Choice to Select the Suitable Interpolation Formula

  (i) Newton's forward difference formula. To find a tabulated value near the beginning of the data.

  (ii) Newton's backward difference formula. To find a tabulated value near the end of the data.

  (iii) Stirling formula. If interpolation is required for u, lying between $-\dfrac{1}{4}$ and $\dfrac{1}{4}$.

  (iv) Bessel's and Everett's Formula. If interpolation is required for u lying between $-\dfrac{1}{4}$ and $\dfrac{3}{4}$.

  (v) Lagrange's and Newton's divided difference formula. For unequal intervals.

  (vi) Hermite interpolation formula. To interpolate the value of $f(x)$ and its derivative.

10.25.4 Bessel's Difference Formula

This formula is obtained with the help of Gauss's forward difference formula as follows,

Gauss's forward difference formula is given by

$$y_u = y_0 + u\Delta y_0 + \frac{u(u-1)}{2!}\Delta^2 y_{-1} + \frac{(u+1)u(u-1)}{3!}\Delta^3 y_{-1} + \frac{(u+1)u(u-1)(u-2)}{4!}\Delta^4 y_{-2} + ...$$

Using $\Delta^2 y_{-1} = \Delta^2 y_{-0} - \Delta^3 y_{-1}, \Delta^4 y_{-2} = \Delta^4 y_{-1} - \Delta^5 y_{-2}$ and so on we get,

$$y_u = y_0 + u\Delta y_0 + \frac{u(u-1)}{2!}.\left(\frac{\Delta^2 y_{-1} + \Delta^2 y_0}{2}\right) + \frac{\left(u - \frac{1}{2}\right)u(u-1)}{3!}\Delta^3 y_{-1} + \frac{(u+1)u(u-1)(u-2)}{4!}\left(\frac{\Delta^4 y_{-2} + \Delta^4 y_{-1}}{2}\right) + ...$$

This is called Bessel's difference formula.

✍ Remarks

- Mainly, Bessel's formula is used to find the entry against any arguments between 0 and 2. This formula gives better result if $\dfrac{1}{4} \le u \le \dfrac{3}{4}$.

- Bessel's formula is most convenient for bisection of the interval.

10.25.5 Remainder or Error Term in Bessel's Formula

Since there are $(2n+2)$ terms in Bessel's formula.

Assume $f(x) = P_{2n+1}(x) + k(x)(x-x_0)(x-x_1)(x-x_{-1})...(x-x_n)(x-x_{-n})(x-x_{n+1})$ ...(1)

when $k(x)$ is to be determined.

Now, consider the function

$$g(t) = f(t) - P_{2n+1}(t) - k(x)(t-x_0)(t-x_1)(t-x_{-1})...(t-x_n)(t-x_{-n})(t-x_{n+1})$$ ... (2)

Here, $g(t)$ vanishes at $(2n+3)$ points namely

$$t = x, x_0, x_1, x_{-1}, ..., x_{-n}, x_{n+1}$$

Then, by Rolle's theorem for $\phi^{2n+1}(\theta) = 0$ for $\theta \in]x_{-n}, x_{n+1}[$

Now $\phi^{2n+2}(t) = f^{2n+2}(t) - k(x)[(2n+2)!]$

$$\phi^{2n+2}(\theta) = 0 \Rightarrow k(x) = \frac{f^{2n+2}(\theta)}{(2n+2)!}$$ $x_{-n} < \theta < x_{n-1}$

Putting this value of $k(x)$ in (1), we get

$$f(x) = P_{2n+1}(x) + \frac{f^{2n+2}(\theta)}{(2n+2)!}(x-x_0)(x-x_1)(x-x_{-1})... (x-x_n)(x-x_{-n})(x-x_{n-1})$$

Now error $R_n = f(x) - P_{2n+1}(x)$ $\Rightarrow R_n = \frac{f^{2n+2}(\theta)}{(2n+2)!}(x-x_0)(x-x_1)(x-x_{-1})...(x-x_n)(x-x_{-n})(x-x_{n-1})$

The value of R_n in terms of $u = \left(\frac{x-x_0}{h}\right)$ is given by $R_n = \frac{h^{2n+2}}{(2n+2)!} f^{2n+2}(\theta).u(u-1)(u+1)...(u-1)(u+n)(u-n-1)$.

10.25.6 Everett's Difference Formula

After eliminating the odd differences from Gauss's difference formula we obtain Everett's difference formula. The Gauss's formula is given by

$$y_u = y_0 + u\Delta y_0 + \frac{u(u-1)}{2!}\Delta^2 y_{-1} + \frac{(u+1)u(u-1)}{3!}\Delta^3 y_{-1} + \frac{(u+1)u(u-1)(u-2)}{4!}\Delta^4 y_{-2} + ...$$ ... (1)

Using the differences

$$\Delta y_0 = y_1 - y_0, \Delta^3 y_{-1} = \Delta^2 y_0 - \Delta^2 y_{-1}, \Delta^5 y_{-2} = \Delta^4 y_{-1} - \Delta^4 y_{-2}... \text{ etc.}$$

Then, equation (1) becomes

$$y_u = 1 - uy_0 + uy_1 - \frac{u(u-1)(u-2)}{3!}\Delta^2 y_{-1} + \frac{(u+1)u(u-1)}{3!}\Delta^2 y_0 - \frac{(u+1)u(u-1)(u-2)(u-3)}{5!}\Delta^4 y_{-2} + ...$$

This is known as Everett's difference formula.

Note: All the differences formulae can be explained with help of Fraser or Lozenge diagrams, which is as follows:

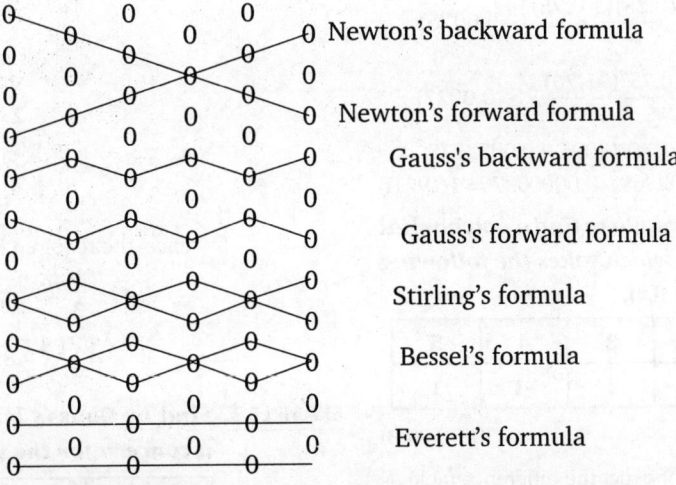

Solved Examples

EXAMPLE 1. *Use Gauss's forward formula to find y_{30} for the following data* $y_{21} = 18.4708, y_{25} = 17.8144,$ $y_{29} = 17.1070, y_{33} = 16.3432, y_{37} = 15.5154.$

(UPTU MCA- 2002, 06)

SOLUTION: From the above data, we have

x	21	25	29
y	18.4708	17.8144	17.1070

x	33	37
y	16.3432	15.5154

Let us take the origin at $x = 29$ and $h = 4$.

To find the value of y for $u = \dfrac{30-29}{4} = 0.25$

Difference Table

x	u	y_u	Δy_u	$\Delta^2 y_u$	$\Delta^3 y_u$	$\Delta^4 y_u$
21	-2	18.4708				
			-0.6564			
25	-1	17.8144		-0.0510		
			-0.7074		-0.0054	
29	0	**17.1070**		**-0.0564**		**-0.0022**
			-0.7638		-0.0076	
33	1	16.3432		-0.0640		
			-0.8278			
37	2	15.5154				

Putting these values in Gauss's Forward difference formulae, we get

$$y_{30} = 17.1070 + (0.25) \times (-0.7638)$$
$$+ \frac{(0.25)(-0.750)}{2}(-0.0564)$$
$$+ \frac{(1.25)(0.25)(-0.75)}{6}(-0.0076)$$
$$+ \frac{(1.25)(0.25)(-0.75)(-1.75)}{24} \times (-0.0022)$$
$$= 17.1070 - 0.19095 + 0.0052875$$
$$+ 0.0002968 - 0.00000375 \approx 16.9216$$

EXAMPLE 2. *Use Gauss forward formula to find a polynomial of degree four or less which takes the following values of the formula f(x).*

x	1	2	3	4	5
$f(x)$	1	–1	1	–1	1

(UPTU-2003)

SOLUTION: Let $u = x - 3$. Then we construct the difference table as follows:

x	u	$f(x)$	$\Delta f(x)$	$\Delta^2 f(x)$	$\Delta^3 f(x)$	$\Delta^4 f(x)$
1	-2	1				
			-2			
2	-1	-1		4		
			2		-8	
3	0	**1**		**4**		**16**
			-2		8	
4	1	-1		4		
			2			
5	2	1				

Now Gauss's forward formula is given by

$$f(x) = f_0 + u\Delta f_0 + \frac{u(u-1)}{2!}\Delta^2 f_{-1}$$
$$+ \frac{(u+1)u(u-1)}{6}\Delta^3 f_{-1}$$
$$+ \frac{u(u-1)(u+1)(u-2)}{24}\Delta^4 f_{-2}$$

$$\Rightarrow f(x) = 1 + u(-2) + \frac{u(u-1)}{2}(-4) + \frac{(u+1)u(u-1)}{6}(8)$$
$$+ \frac{u(u-1)(u+1)(u+2)}{24}(16)$$

$$= 1 - 2u + 2u(u-1) + \frac{4u(u^2-1)}{3}$$
$$+ \frac{2(u^2-1)(u^2-2u)}{3}$$

$$= 1 - 2u - 2u^2 + 2u + \frac{4u^3}{3} - \frac{4u}{3}$$
$$+ \frac{2u^4}{3} - \frac{4u^3}{3} - \frac{2u^2}{3} + \frac{4u}{3}$$

$$= \frac{2}{3}u^4 + \left(\frac{4}{3} - \frac{4}{3}\right)u^3 + \left[-2 + \frac{2}{3}\right]u^2$$
$$+ \left[-2 + 2 - \frac{4}{5} + \frac{4}{3}\right]u + 1$$

$$= \frac{2}{3}u^4 + 0.u^3 + \left(-\frac{8}{3}u^2\right) + 0.u + 1$$

$$= \frac{2}{3}u^4 - \frac{8}{3}u^2 + 1 = \frac{1}{3}[2u^4 - 8u^2 + 3]$$

Hence, the required function is

$$f(x) = \frac{1}{3}(2(x-3)^4 - 8(x-3)^2 + 3)$$
$$= \frac{2}{3}x^4 - 8x^3 + \frac{100}{3}x^2 - 56x + 31$$

EXAMPLE 3. *Find by Gauss's backward formula the sales of a concern for the year 1936, given that*

Year	1901	1911	1921	1931	1941	1951
Sales (in thousand)	12	15	20	27	39	52

(UPTU-2004(SUM)-2005)

SOLUTION: Let us take the origin $= 1931$ and $h = 10$ years then

$$u = \frac{1936 - 1931}{10} = 0.5$$

x	u	y_u	Δy_u	$\Delta^2 y_u$	$\Delta^3 y_u$	$\Delta^4 y_u$	$\Delta^5 y_u$
1901	-3	12					
			3				
1911	-2	15		2			
			5		0		
1921	-1	20		2		3	
			7		3		-10
1931	0	**27**		5		-7	
			12		-4		
1941	1	39		1			
			1				
1951	2	52					

Now using Gauss's backward difference formula

$$y_u = y_0 + {}^4c_1 \Delta y_{-1} + {}^{4+1}c_2 \Delta^2 y_{-1} + {}^{4+1}c_3 \Delta^3 y_{-2} + \ldots$$

We get, $y_u = 27 + 0.5 \times 7 + \dfrac{(1.5) \times (.5)}{2} \times 5$

$$+ \frac{(1.5)(0.5)(-0.5)}{6} \times 3 + \frac{(2.5)(1.5)(0.5)(-0.5)}{24}$$

$$\times (-7) + \frac{(2.5)(1.5)(0.5)(-0.5)}{120} \times (-10)$$

$$= 27 + 3.5 + 1.875 - 0.1875$$

$$+ 0.2734 - 0.11718$$

$$= 32.3437 \text{ thousands.}$$

EXAMPLE 4. *Given that*

$$\sqrt{12500} = 111.803399, \sqrt{12510} = 111.848111$$

$$\sqrt{12520} = 111.892806, \sqrt{12530} = 111.937483$$

Show by Gauss's backward formula that $\sqrt{12516} = 111.8749301$ (RAJASTHAN-2008)

SOLUTION: Let us take the origin at 12520

$$\therefore \quad u = \frac{x - a}{h} = \frac{12516 - 12520}{10} = -\frac{4}{10} = -0.4$$

The difference table is given as below:

u	x	$10^6 f(x)$	$10^6 \Delta f(x)$	$10^6 \Delta^2 f(x)$	$10^6 \Delta^3 f(x)$
-2	12500	11803399			
			44712		
-1	12510	111848111		-17	
			44695		**-1**
0	12520	**11892806**		**-18**	
			44677		
1	12530	**111937483**			

Putting the values in Gauss's backward formula

$$f(u) = f(a) + u \Delta f(-1) + \frac{u(u+1)}{2!} \Delta^2 f(-1)$$

$$+ \frac{u(u+1)(u-1)}{3!} \Delta^3 f(-2) + \ldots$$

we get

$$10^6 f(-4) = 111892806 + (-0.4)(44695)$$

$$+ \frac{(0.6)(-0.4)}{2!}(-18) + \frac{(0.6)(-0.4)(-1.4)}{3!}(-1)$$

$$= 1118749301$$

$$\Rightarrow f(-4) = 111.8749301$$

Hence $\sqrt{12516} = 111.8749301$.

EXAMPLE 5. *Use Stirling's formula to find* y_{35}, *given* $y_{20} = 512, y_{30} = 439, y_{40} = 346$ *and* $y_{50} = 243$.

(UPTU-2004)

SOLUTION: Let us assume origin $= 30$, $h = 10$
Then $a + hu = 35$
$\Rightarrow \quad 30 + 10u = 35$
$\Rightarrow \quad\quad u = 0.5$
The difference table is given as under:

x	u	y	Δy	$\Delta^2 y$	$\Delta^3 y$
20	-1	512			
			-73		
30	0	**439**		-20	
			-93		10
40	1	346		-10	
			-1.3		
50	2	243			

By Stirling's formula, we have

$$f(0.5) = 439 + (0.5)\left(\frac{-93 - 73}{2}\right)$$

$$+ \frac{(0.5)^2}{2!}(-20) + \frac{(1.5)(0.5)(-0.5)}{3!}\left(\frac{10}{2}\right)$$

$$= 394.6875$$

Hence, $y_{35} = 394.6875$

EXAMPLE 6. *Use Stirling's formula to find* y_{28}, *given* $y_{20} = 49225, y_{25} = 48316, y_{30} = 47236$ $y_{35} = 45926, y_{40} = 44306$.   (UPTU (CO)-2004, 2007)

SOLUTION: Let $x = 30$, $h = 5$ then, $u = \dfrac{28 - 30}{5} = -0.4$

x	u	y_u	Δy_u	$\Delta^2 y_u$	$\Delta^3 y_u$	$\Delta^4 y_u$
20	-2	49225				
			-909			
25	-1	48316		-171		
			-1080		-59	
30	0	**47236**		**-230**		**21**
			-1310		-80	
35	1	45926		-310		
			-1620			
40	2	44306				

The Stirling's formula is

$$y_u = y_0 + u\left[\frac{\Delta y_0 + \Delta y_{-1}}{2}\right] + \frac{u^2}{2}[\Delta^2 y_{-1}]$$

$$+ \frac{u(u^2-1)}{6}\left[\frac{\Delta^3 y_{-1}+\Delta^3 y_{-2}}{2}\right]$$

$$+ u^2\frac{(u^2-1)}{24}\Delta^4 y_{-2} \qquad ...(1)$$

On putting the values in (1), we get

$$y_{28} = 47236 \times (-0.4)\left[\frac{-1310-1080}{2}\right]$$

$$+ \frac{(0.16)}{2}(-230)+\frac{(-0.4)(0.16-1)}{6}$$

$$\frac{(-80-59)}{2}+\frac{(0.16)(0.16-1)}{24}\times(-21)$$

$$= 47236 +478-18.4-3.8920+0.1176$$

Hence, $y_{28} = 47692$.

EXAMPLE 7. *Given* $y_{20}=24, y_{24}=32, y_{28}=35, y_{32}=40$. *Find* y_{25} *by Bessel's formula.*

(UPTU-2002, 05, DELHI-2004, 2008)

SOLUTION: Let $x = 24, h = 4$

$\Rightarrow \qquad 4 = \dfrac{25-24}{4}=\dfrac{1}{4}=0.25$

The difference table is given as under

x	u	y_u	Δy_u	$\Delta^2 y_u$	$\Delta^3 y_u$
20	-1	24			
			8		
24	0	**32**		-5	
			3		7
28	1	**34**		2	
			5		
32	2	40			

Now Bessel's formula is

$$y_0 = \frac{1}{2}(y_0+y_1)+(u-\tfrac{1}{2})\Delta y_0$$

$$+ \frac{u(u-1)}{2}\cdot\frac{\Delta^2 y_{-1}+\Delta^2 y_0}{2!}$$

$$+ \frac{\left(u-\tfrac{1}{2}\right)u(u-1)}{3!}\Delta^3 y_{-1}$$

Putting the values in Bessel's formula, we get

$$y_{0.25} = \frac{32+34}{2}+\left(\frac{1}{4}-\frac{1}{2}\right)3$$

$$+ \frac{\frac{1}{4}\left(\frac{1}{4}-1\right)}{2}\left(\frac{-5+2}{2}\right)+\frac{\left(\frac{1}{4}-\frac{1}{2}\right)\left(\frac{1}{4}-1\right)}{6}7$$

$$= 33.5 - 0.75 + 0.1416 + 0.054687= 32.945287$$

EXAMPLE 8. *Find the value of* $f(27.4)$ *from the following table:* (UPTU-2009)

x	25	26	27	28	29	30
$f(x)$	4.000	3.846	3.704	3.571	3.448	3.333

SOLUTION: Let $u = \dfrac{27.4 - 27.0}{1}=0.4$.

(Because origin is at 27.0, $h =1$)

Also $\qquad w = 1-u = 0.6$

Difference table is given as under:

x	u	$10^3 f(u)$	$10^3\Delta f(u)$	$10^3\Delta^2 f(u)$	$10^3\Delta^3 f(u)$	$10^3\Delta^4 f(u)$
25	-2	4000				
			-154			
26	-1	3846		12		
			-142		-3	
27	0	3704		9		4
			-133		1	
28	1	3571		10		-3
			-123		2	
29	2	3448		8		
			-115			
30	3	3333				

Putting these values in Everett's formula, we get

$$10^3 \cdot f(27.4) = (0.4)(3571)+\frac{(1.4)(0.4)(-0.6)}{3!}(10)$$

$$+ \frac{(2.4)(1.4)(0.4)(-0.6)(-1.6)}{5!}(-3)$$

$$+ (0.6)(3704)+\frac{(1.6)(0.6)(-0.4)}{3!}(9)$$

$$+ \frac{(2.6)(1.6)(-0.6)(-0.4)(-1.4)}{5!}(4)$$

$$= 3649.678336$$

Hence, $f(27.4) = 3649.678336$.

EXAMPLE 9. *Given* y_0, y_1, y_2, y_4, y_5 *assuming fifth difference constant and using Bessel's formula, show that*

$$y_{25} = \frac{1}{2}c+\frac{25.(c-b)+3(a-c)}{256}$$

where $a = y_0+y_5, b = y_1+y_4, c = y_2+y_3$.

SOLUTION: Bessel's formula is given by

$$y_u = \frac{1}{2}(y_0+y_1)+\left(u-\frac{1}{2}\right)\Delta y_0$$

$$+ \frac{u(u-1)}{2!}\left(\frac{\Delta^2 y_0+\Delta^2 y_{-1}}{2}\right)+\frac{\left(u-\frac{1}{2}\right)u(u-1)}{3!}\Delta^3 y_{-1}$$

$$+ \frac{(u+1)u(u-1)(u-2)}{4!}\left[\frac{\Delta^4 y_{-1}+\Delta^4 y_{-2}}{2}\right]+...$$

Putting $u = \dfrac{1}{2}$ and taking upto fifth differences, we have

$$y_{1/2} = \frac{1}{2}(y_0+y_1)-\frac{1}{16}(\Delta^2 y_0+\Delta^2 y_{-1})$$

$$+ \frac{3}{256}(\Delta^4 y_{-1}+\Delta^4 y_{-2})$$

Now shifting the origin to 2, we have

$$y_{2\frac{1}{2}} = \frac{1}{2}(y_2+y_3)-\frac{1}{16}(\Delta^2 y_2+\Delta^2 y_1)$$

$$+\frac{3}{256}(\Delta^4 y_1 + \Delta^4 y_0)$$

$$=\frac{1}{2}(y_2 + y_3) - \frac{1}{16}(y_4 - 2y_3 + y_2 + y_3$$

$$-2y_2 + y_1) + \frac{3}{256}(y_5 - 4y_4 + 6y_3$$

$$-4y_2 + y_1 + y_4 - 4y_3 + 6y_2 - 4y_1 + y_0)$$

$$=\frac{1}{2}(y_2 + y_3) - \frac{1}{16}(y_4 - 2y_3 - y_2 + y_1)$$

$$+\frac{3}{256}(y_5 - 3y_4 + 2y_3 + 2y_2 - 3y_1 + y_0)$$

$$=\frac{1}{2}(y_2 + y_3) - \frac{1}{16}[(y_2 + y_4) - (y_2 + y_3)]$$

$$+\frac{3}{256}[(y_0 + y_5) - 3(y_1 + y_4) + 2(y_2 + y_3)]$$

$$=\frac{1}{2}c + \frac{1}{16}(b - c) + \frac{3}{256}(9 - 3b + 2c)$$

$$=\frac{1}{2}c + \frac{1}{256}[16(c - b) + 3(a - 3b + 2c)]$$

$$=\frac{1}{2}c + \frac{1}{256}(16c - 16b + 3a - 9b + 6c)$$

$$=\frac{1}{2}c + \frac{1}{256}(22c - 25b + 3a)$$

$$=\frac{1}{2}c + \frac{1}{256}(25c - 25b - 3c + 3a)$$

$$=\frac{1}{2}c + \frac{1}{256}[25(c - b) + 3(a - c)]$$

 Exercise-10.11

1. By means of Lagrange's formula, Prove that

$$y_0 = \frac{1}{2}(y_1 + y_{-1}) - \frac{1}{8}\left[\frac{1}{2}(y_3 - y_1) - \frac{1}{2}(y_1 - y_3)\right]$$

2. Apply Lagrange's formula to find $f(1.5)$ using the following table:

x	1.0	1.2	1.4	1.6	1.8	2.0
$f(x)$	0.2420	0.1942	0.1497	0.1109	0.079	0.0540

3. Using Lagrange's formula find the cubic polynomial from given data:

x	0	1	4	6
$f(x)$	1	+1	1	−1

4. Obtain the value of t when, $A = 85$ from the following table, using Lagrange's method:

t	2	5	8	14
A	94.8	87.9	81.3	68.7

5. Using Newton's forward formula find the value of $f(1.6)$, if:

x	1	1.4	1.8	2.2
$f(x)$	3.49	4.82	5.96	6.5

6. Determine the values of $f(22)$ and $f(42)$ from the following Table: (JNTU-2007)

x	20	25	30	35	40	45
$f(x)$	354	332	391	260	231	204

7. Apply Newton's backward difference formula to the data given below to obtain a polynomial of degree 4 in x.

x	1	2	3	4	5
$f(x)$	1	−1	1	−1	1

8. Evaluate log 5875 from the following table:

x	40	45	50
$f(x)$	1.60206	1.65321	1.69897

x	55	60	65
$f(x)$	1.74036	1.77815	1.81291

9. Using Hermite formula for points oscillation, drive the following formula:

$$y\left(\frac{1}{2}\right) = \frac{1}{2}(y_0 + y_1) + \frac{1}{3}L(y_0' - y_1') \quad \text{where } L = (x_1 - x_0).$$

10. Given that $f(0) = 9$, $f(1) = 68$, $f(5) = 123$. Find $f(2)$ with the help of Newton's divided difference formula.

11. Apply Newton's divided difference formula to find the value of $f(8)$ if $f(1) = 3, f(3) = 31, f(6) = 223, f(10) = 1011$, $f(11) = 1343$.

12. Find the form of $f(x)$ from the following table:

x	0	1	2	5
$f(x)$	2	3	12	147

13. If $f(x) = \frac{1}{x^2}$ find the divided difference $f(a, b)$ and $f(a, b, c)$.

14. (i) Use Gauss's forward interpolation formula to find $y41$ with the help of following data:

$$y_{30} = 3678.2, y_{35} = 2995.1, y_{40} = 2400.1$$

$$y_{45} = 1876.2, y_{50} = 1416.3$$

(ii) Use Gauss's forward formula to find the annuity value for 27 years from the following data:

Years	15	20	25
Annuity	10.3797	12.4622	14.0939

Years	30	35	40
Annuity	15.3725	16.3742	17.1591

15. From the following table find the value of $f(0.5437)$ by Gauss, Stirling, Bessel and Everett's Formula.

x	0.51	0.52	0.53
$f(x)$	0.529244	0.537895	0.546464

x	0.54	0.55	0.56	0.57
$f(x)$	0.554939	0.663323	0.571616	0.579816

16. Eliminate odd differences from the Gauss's forward formula to derive Everett's formula

$$y_\alpha = (-\alpha)f_0 + \alpha f_1 - \frac{\alpha(\alpha-1)(\alpha-2)}{3!}\delta^2 f_0 +$$

$$\frac{(\alpha+1)(\alpha)(\alpha-1)}{3!}\delta^2 f_1 + ...$$

where $y = f(x_0 + \alpha h), \alpha = \dfrac{x - x_0}{h}$.

17. Apply (i) Stirling's formula, (ii) Bessel's formula to find the value of $f(0.44)$ from the following table which gives the values of $f(t) = \dfrac{1}{\sqrt{2\pi}}\int_0^t \exp\left(-\dfrac{t^2}{2}\right)dt$ at $t = 0.5$ from $t = 0$ to $t = 3$.

x	0	0.5	1.0	1.5
$f(x)$	0	0.19146	0.34134	0.43319

x	2.0	2.5	3.0
$f(x)$	0.47725	0.49379	0.49865

18. Find the missing value of the following data:

x	1	2	3	4	5
$f(x)$	7	–	13	21	37

19. Using Hermite interpolation formula, find polynomial which meets the following requirements:

x	0	1	2
$f(x)$	1	0	9
$f'(x)$	0	0	0

20. Find the polynomial, for following data, using Hermite formula:

x	–1	0	1
$f(x)$	–1	0	1
$f'(x)$	0	0	0

21. Find the cubic polynomial which takes following values $y(0) = 1$, $y(1) = 1$, $y(2) = 1$, and $y(3) = 10$. Hence, or otherwise obtain $y(4)$.

Answers

2. 0.1295   **3.** $\dfrac{1}{30}(-x^3 + 5x^2 - 4x + 30)$ **4.** 6.5928   **5.** 5.54   **6.** 352 ; 219

7. $\dfrac{2}{3}x^4 - 8x^3 + \dfrac{100}{3}x^2 - 56x + 31$ **8.** 3.76905   **9.** 109.5   **11.** 521   **12.** $x^3 + x^2 - x + 2$

13. $\dfrac{-(a+b)}{a^2 b^2}, \dfrac{ab + bc + ca}{a^2 b^2 c^2}$ **14.** (i) $y_{41} = 2290.1$ (ii) 14.643   **15.** 0.558052

17. (i) 0.38891 (ii) 0.38873   **18.** 9.5   **19.** $x^4 - 4x^3 + 4x^2$ **20.** $\dfrac{1}{2}x^3(5 - 3x^2)$

10.26 NUMERICAL DIFFERENTIATION

Let $y = f(x)$ be the given function. The process of evaluating the derivatives of a function $f(x)$ with the help of the given set of values of that functions is known as numerical differentiation. This may be done by first approximating the function by a suitable approximation formula and then differentiating it as many times as desired.

The choice of the formula is the same as discussed for interpolation in chapter 5, *i.e.*, if the derivative at a point near the beginning of a set of values given by a table is required then we use Newton forward formula, and if the same is required at a point near the end of the set of given tabular values, then we use Newton's backward interpolation formula. The central difference formula (Bessel's and Stirling's) used to calculate value for points near the middle of the set of given tabular values. If the values of x are not equally spaced, we use Newton's divided difference interpolation formula or Lagrange's interpolation formula to get the required value of the derivative.

✍ REMARK
- While analytic methods give exact answers, the numerical techniques provide only approximations to derivatives. Numerical differentiation methods are very sensitive to roundoff errors, in addition to the truncation error introduced by the methods themselves.

10.26.1 Derivative Using Newton's Forward Interpolation Formula   (UPTU(MCA)–2005, 07)

Newton's forward interpolation formula is given by

$$y = y_0 + u\Delta y_0 + \frac{u(u-1)}{2!}\Delta^2 y_0 + \frac{u(u-1)(u-2)}{3!}\Delta^3 y_0 + ... \qquad ...(1)$$

where $u = \dfrac{x - x_0}{h}$

Differentiating equation (1) with respect to u, we get

$$\frac{dy}{du} = \Delta y_0 + \frac{2u-1}{2!}\Delta^2 y_0 + \frac{3u^2 - 6u + 2}{3!}\Delta^3 y_0 + ... \qquad ...(2)$$

Now $\dfrac{dy}{dx} = \dfrac{dy}{du} \cdot \dfrac{du}{dx} = \dfrac{1}{h} \cdot \dfrac{dy}{du}$

Therefore

$$\frac{dy}{dx} = \frac{1}{h}\left[\Delta y_0 + \frac{2u-1}{2!}\Delta^2 y_0 + \frac{3u^2-6u+2}{3!}\Delta^3 y_0 + \frac{4u^3-18u^2+22u-6}{4!}\Delta^4 y_0 + \dots\right] \qquad \dots(3)$$

As $x = x_0$, $u = 0$, therefore, putting $u = 0$ in (3), we get

$$\left[\frac{dy}{dx}\right]_{x=x_0} = \frac{1}{h}\left[\Delta y_0 - \frac{1}{2}\Delta^2 y_0 + \frac{1}{3}\Delta^3 y_0 - \frac{1}{4}\Delta^4 y_0 + \dots\right] \qquad \dots(4)$$

Differentiating equation (3) again w.r.t. x, we get

$$\frac{d^2 y}{dx^2} = \frac{d}{du}\left(\frac{dy}{dx}\right)\frac{du}{dx} = \frac{1}{h}\times\frac{d}{du}\left(\frac{dy}{dx}\right) = \frac{1}{h^2}\left[\Delta^2 y_0 + (u-1)\Delta^3 y_0 + \frac{6u^2-18u+11}{12}\Delta^4 y_0 + \dots\right] \qquad \dots(5)$$

Putting $u = 0$ in (5), we get

$$\left[\frac{d^2 y}{dx^2}\right]_{x=x_0} = \frac{1}{h^2}\left[\Delta^2 y_0 - \Delta^3 y_0 + \frac{11}{12}\Delta^2 y_0 - \dots\right] \qquad \dots(6)$$

Similarly

$$\left[\frac{d^3 y}{dx^3}\right]_{x=x_0} = \frac{1}{h^3}\left[\Delta^3 y_0 - \frac{3}{2}\Delta^4 y_0 - \dots\right] \text{ and so on}$$

10.26.2 Derivatives using Newton's Backward Difference Formula

Newton's backward interpolation formula is given by

$$y = y_n + u\nabla y_n + \frac{u(u+1)}{2!}\nabla^2 y_n + \frac{u(u+1)(u+2)}{3!}\nabla^3 y_n + \dots ; \text{ where } \qquad u = \frac{x-x_n}{n} \qquad \dots(1)$$

Differentiating both sides of equation (1) w.r.t. x, we get

$$\frac{dy}{dx} = \frac{1}{h}\left[\nabla y_n + \frac{2u+1}{2!}\nabla^2 y_n + \frac{3u^2+6u+2}{3!}\nabla^3 y_n + \dots\right] \qquad \dots(2)$$

At $x = x_n$, $u = 0$. Therefore putting $u = 0$ in (2), we get

$$\left[\frac{dy}{dx}\right]_{x=x_n} = \frac{1}{h}\left[\nabla y_n + \frac{1}{2}\nabla^2 y_n + \frac{1}{3}\nabla^3 y_n + \dots\right] \qquad \dots(3)$$

Again differentiating both sides of equation (2) w.r.t. x, we get

$$\frac{d^2 y}{dx^2} = \frac{1}{h^2}\left[\nabla^2 y_n + \frac{6u+6}{6}\nabla^3 y_n + \frac{12u^2+36u+22}{24}\nabla^4 y_n + \dots\right]$$

At $x = x_n$, $u = 0$. Therefore putting $u = 0$ in (3), we get

$$\left[\frac{d^2 y}{dx^2}\right]_{x=x_n} = \frac{1}{h^2}\left[\nabla^2 y_n + \nabla^3 y_n + \frac{11}{12}\nabla^4 y_n + \dots\right] \qquad \dots(4)$$

Similarly,

$$\left[\frac{d^3 y}{dx^3}\right]_{x=x_n} = \frac{1}{h^3}\left[\nabla^3 y_n + \frac{3}{2}\nabla^4 y_n + \dots\right] \text{ and so on.} \qquad \dots(5)$$

10.26.3 Derivatives using Stirling's Formula

Here we want to determine the values of the derivatives of the function near the middle of the given set of arguments. We may apply any central difference formula. Therefore using Stirling's formula, we get

$$y_n = y_0 + u\left(\frac{\Delta y_0 + \Delta y_{-1}}{2}\right) + \frac{u^2}{2!}\Delta^2 y_{-1} + \frac{u(u^2-1)}{3!}\left(\frac{\Delta^3 y_{-1} + \Delta^3 y_{-2}}{2}\right) + \frac{u^2(u^2-1)}{4!}\Delta^4 y_{-2} + \dots \qquad \dots(1)$$

where $u = \dfrac{x-x_0}{h}$

Now differentiating w.r.t. x, we get

$$\frac{dy}{dx} = \left[\left(\frac{\Delta y_0 + \Delta y_{-1}}{2}\right) + u\Delta^2 y_{-1} + \frac{(3u^2-1)}{3!}\left(\frac{\Delta^3 y_{-1} + \Delta^3 y_{-2}}{2}\right) + \frac{(4u^3-2u)}{4!}\Delta^4 y_{-2} + \dots\right]\frac{du}{dx}$$

since $u = \dfrac{x-x_0}{h} \qquad \Rightarrow \qquad \dfrac{du}{dx} = \dfrac{1}{h}$

$$\therefore \quad \frac{dy}{dx} = \frac{1}{h}\left[\left(\frac{\Delta y_0 + \Delta y_{-1}}{2}\right) + u\Delta^2 y_{-1} + \frac{3u^2-1}{3!}\left(\frac{\Delta^3 y_{-1} + \Delta^3 y_{-2}}{2}\right) + \frac{(4u^3-2u)}{4!}\Delta^4 y_{-2} + \dots\right] \qquad \dots(2)$$

At $x = x_0$, $u = 0$, therefore, putting $u = 0$ in (2), we get

$$\left[\frac{dy}{dx}\right]_{x=x_0} = \frac{1}{h}\left[\frac{\Delta y_0 + \Delta y_{-1}}{2} - \frac{1}{6}\left(\frac{\Delta^3 y_{-1} + \Delta^3 y_{-2}}{2}\right) + \dots\right]$$

Again differentiating, we get

$$\frac{d^2 y}{dx^2} = \frac{1}{h^2}\left[\Delta^2 y_{-1} + \frac{6u}{6}\left(\frac{\Delta^3 y_{-1} + \Delta^3 y_{-2}}{2}\right) + \frac{12u^2 - 2}{4!}\Delta^4 y_{-2} + \dots\right] \qquad \dots(3)$$

At $x = x_0$, $u = 0$, therefore, putting $u = 0$ in (3), we get

$$\therefore \qquad \left[\frac{d^2 y}{dx^2}\right]_{x=x_0} = \frac{1}{h^2}\left[\Delta^2 y_{-1} - \frac{1}{12}\Delta^4 y_{-2} + \dots\right] \text{ and so on.}$$

10.26.4 Derivative using Newton's Divided Difference Formula

In this case we apply Newton's divided difference formula for finding the successive differentiation at given value of x. Let us consider a function $f(x)$ of degree n, then

$$y = f(x) = f(x_0) + (x - x_0)\Delta f(x_0) + (x + x_0)(x - x_1)\Delta^2 f(x_0)$$

$$\Rightarrow \qquad (x - x_0)(x - x_1)(x - x_2)\Delta^3 f(x_0) + (x - x_0)(x - x_1)\dots(x - x_{n-1})\Delta^n f(x_0)$$

Differentiate this equation w.r.t. 'x' as many times as we require and put $x = x_i$, we get the required derivatives.

10.26.5 Maxima and Minima of Tabulated Function

If we differentiate the Newton's forward interpolation formula with respect to x, then we get

$$\frac{dy}{dx} = \frac{1}{h}\left[\Delta y_0 + \frac{2u-1}{2}\Delta^2 y_0 + \frac{3u^2 - 6u + 2}{3!}\Delta^3 y_0 + \dots\right] \qquad \dots(1)$$

We know that the maximum and minimum values of a function $y = f(x)$ can be found by equating $\frac{dy}{dx}$ to zero and solving it for x.

Put $\frac{dy}{dx} = 0$, then equation (1) gives

$$\Delta y_0 + \frac{2u-1}{2}\Delta^2 y_0 + \frac{3u^2 - 6u + 2}{6}\Delta^2 y_0 + \dots = 0$$

Solving this for u, and substituting $\Delta y_0, \Delta^2 y_0, \Delta^3 y_0$ (using difference table), we get x as $x_0 + ah$ at which y is maximum or minimum.

☞ Remarks

- To determine the values of the derivative of the function near the begining of the given set of arguments for equal interval use the derivative of Newton's forward interpolation formula.
- To find the value of the derivative near the end of the data, use Newton's backward interpolation formula.
- To find the values of the derivatives of the function near the middle of the given set of arguments, use Stirling's formula.
- In case of unequal interval, use the derivatives of Newton's divided difference formula.

Solved Examples

EXAMPLE 1. *Using following table :*

x	1.0	1.1	1.2	1.3	1.4	1.5	1.6
y	7.989	8.403	8.781	9.129	9.451	9.750	10.031

Find $\frac{dy}{dx}$ and $\frac{d^2 y}{dx^2}$ at x = 1.1.

(VTU–2006, MADRAS–2003S, MDU(B.E.)–2007)

SOLUTION. Since the values are at equidistant and we want to find the value of y at $x = 1.1$. Therefore, we apply Newton's forward difference formula.

Difference table

x	y	Δy	$\Delta^2 y$	$\Delta^3 y$	$\Delta^4 y$	$\Delta^5 y$	$\Delta^6 y$
1.0	7.989						
		0.414					
1.1	**8.403**		–0.036				

		0.378		0.006			
1.2	8.781		–0.030		–0.002		
		0.348		0.004		+0.001	
1.3	9.129		–0.026		–0.001		–0.002
		0.322		0.003		0.003	
1.4	9.451		–0.023		–0.002		
		0.299		0.005			
1.5	9.750		–0.081				
		0.281					
1.6	10.031						

We have $\left(\dfrac{dy}{dx}\right)_{x=x_0} = \dfrac{1}{h}\left[\Delta y_0 - \dfrac{1}{2}\Delta^2 y_0 + \dfrac{1}{3}\Delta^3 y_0\right.$

$$\left. -\frac{1}{4}\Delta^4 y_0 + \frac{1}{5}\Delta^5 y_0 - \frac{1}{6}\Delta^6 y_0 + \dots\right]$$

Putting $x_0 = 1.1, \Delta y_0 = 0.378, \Delta^2 y_0 = 0.030$ and so on

and $h = 0.1$, we get

$$\left(\frac{dy}{dx}\right)_{1.1} = \frac{1}{0.1}\left[0.378 - \frac{1}{2}(-0.030) + \frac{1}{3}(0.004)\right.$$

$$\left. -\frac{1}{4}(-0.001) + \frac{1}{5}(0.003)\right]$$

$$= 10[0.378 + 0.015 + 0.0013 + 0.00025$$

$$-0.0002]$$

$$= 10[0.39435] = 3.9435$$

and $\left(\frac{d^2 y}{dx^2}\right)_{x_0} = \frac{1}{h^2}\left[\Delta^2 y_0 - \Delta^3 y_0 + \frac{11}{12}\Delta^4 y_0\right.$

$$\left. -\frac{5}{6}\Delta^5 y_0 + \frac{137}{180}\Delta^6 y_0 + ...\right]$$

$$\Rightarrow \left(\frac{d^2 y}{dx^2}\right)_{1.1} = \frac{1}{(0.1)^2}[-0.030 - 0.004$$

$$-\frac{11}{12} \times 0.001 + \frac{5}{6} \times 0.003\right]$$

$$= 100[-0.030 - 0.004 - 0.0009]$$

$$= 100(-0.0341) = -0.374.$$

EXAMPLE 2. *Find $f'(1.5)$ and $f''(1.5)$ from the following table:*

x	1.5	2.0	2.5	3.0	3.5	4.0
f(x)	3.375	7.000	13.625	24.000	38.875	59.000

(MDU(B.E.)–2005, 06, SVTU–2007)

SOLUTION. **Difference table**

x	f(x)	Δf(x)	Δ²f(x)	Δ³f(x)	Δ⁴f(x)
1.5	**3.375**				
		3.625			
2.0	7.000		**3.000**		
		6.625		**0.75**	
2.5	13.625		3.75		**0**
		10.375		0.75	
3.0	24.000		4.5		0
		14.875		0.75	
3.5	38.875		5.25		
		20.125			
4.0	59.000				

Here $x_0 = 1.5$, $h = 0.5$

$$f'(1.5) = \frac{1}{0.5}\left[\Delta y_0 - \frac{1}{2}\Delta^2 y_0 + \frac{1}{3}\Delta^3 y_0 - \frac{1}{4}\Delta^4 y_0 + ...\right]$$

$$= \frac{1}{0.5}\left[3.625 - \frac{1}{2}(3.000) + \frac{1}{3}(0.75)\right]$$

$$= 2[3.625 - 1.5 + 0.25] = 2[2.375]$$

$$f'(1.5) = 4.75$$

and $f''(1.5) = \frac{1}{(0.5)^2}\left[\Delta^2 y_0 - \Delta^3 y_0 + \frac{11}{12}\Delta^4 y_0\right]$

$$= 4[3.000 - 0.75] = 4[2.25]$$

$$\therefore \quad f''(1.5) = 9.$$

EXAMPLE 3. *Find $f'(1.1)$ and $f''(1.1)$ from the following table :*

x	1.0	1.2	1.4	1.6	1.8	2.0
f(x)	0	0.1280	0.5440	1.2960	2.4320	4.0000

(UPTU–2010, BHOPAL–2009)

SOLUTION. **Difference table**

x	f(x)	Δf(x)	Δ²f(x)	Δ³f(x)	Δ⁴f(x)
1.0	**0**				
		0.1280			
1.2	0.1280		**0.288**		
		0.4160		**0.048**	
1.4	0.5440		0.336		**0**
		0.7520		0.048	
1.6	1.2960		0.384		0
		1.1360		0.048	
1.8	2.4320		0.432		
		1.5680			
2.0	4.0000				

Here, we have to find the derivatives at $x = 1.1$ which lies between given arguments 1.0 and 1.2. So apply Newton's forward formula, we have

$$u = \frac{x - x_0}{h} = \frac{x - 1}{0.2} = 5(x - 1) \qquad ...(1)$$

$$f(x) = f(x_0) + u\Delta f(x_0) + \frac{u(u-1)}{2!}\Delta^2 f(x_0)$$

$$+ \frac{u(u-1)(u-2)}{3!}\Delta^3 f(x_0) + ... \qquad ...(2)$$

Differentiating w.r.t. x, we get

$$f'(x) = \left[\Delta f(x_0) + \frac{(2u-1)}{2!}\Delta^2 f(x_0)\right.$$

$$\left. + \frac{3u^2 - 6u + 2}{3!}\Delta^3 f(x_0)\right]\frac{du}{dx}$$

Also, from (1) $\frac{du}{dx} = 5$

$$\therefore \quad f'(x) = 5\left[\Delta f(x_0) + \frac{(2u-1)}{2}\Delta^2 f(x_0)\right.$$

$$\left. + \frac{3u^2 - 6u + 2}{3!}\Delta^3 f(x_0)\right]$$

At $x = 1.1$, $u = 5(1.1 - 1) = 0.5$

$$f'(1.1) = 5\left[0.128 + \frac{2(0.5) - 1}{2}(0.288)\right.$$

$$\left. + \frac{3(0.5)^2 - 6(5.0) + 2}{6}(0.048)\right]$$

$$= 5[0.128 + 0 - 0.002]$$

$$f'(1.1) = 0.63$$

Differentiating equation (2) again w.r.t. 'x', we get

$$f''(x) = 5\left[\Delta^2 f(x_0) + \frac{6u - 6}{6}\Delta^3 f(x_0)\right]\frac{du}{dx}$$

$$\therefore f''(1.1) = 25[0.288 + (0.5 - 1) \times 0.048]$$

$$= 25[0.288 - 0.024]$$

Hence, $f''(1.1) = 6.6$

EXAMPLE 4. *From the following table, find the values of $\frac{dy}{dx}$ and $\frac{d^2 y}{dx^2}$ at $x = 2.03$.*

x	1.96	1.98	2.00	2.02	2.04
y	0.7825	0.7739	0.7651	0.7563	0.7473

(ANNA–2005)

SOLUTION.

Difference table

x	y	∇y	$\nabla^2 y$	$\nabla^3 y$	$\nabla^4 y$
1.96	0.7825				
		−0.0086			
1.98	0.7739		−0.0002		
		−0.0088		0.0002	
2.00	0.7651		−0		−0.0004
		−0088		−0.0002	
2.02	0.7563		−0.0002		
		−0.0090			
2.04	0.7473				

Here $x_n = 2.04, h = 0.02, u = \dfrac{x - x_n}{h} \Rightarrow \dfrac{du}{dx} = \dfrac{1}{h}$

$\Rightarrow \dfrac{du}{dx} = \dfrac{1}{h}$ at $x = 2.03$

$\therefore$ at $x = 2.03$ $u = \dfrac{2.03 - 2.04}{0.02} = -\dfrac{0.01}{0.02} = -\dfrac{1}{2}$

Then by Newton's backward formula, we have

$$y(x) = y_n + u\nabla y_n + \frac{u(u+1)}{2}\nabla^2 y_n + \frac{u(u+1)(u+2)}{6}\nabla^3 y_n$$

$$+ \frac{u(u+1)(u+2)(u+3)}{24}\nabla^3 y_n + \dots \quad \dots(1)$$

Differentiating w.r.t. x, we get

$$y'(x) = \frac{1}{h}\left[\nabla y_n + \frac{2u+1}{2}\nabla^2 y_n + \frac{3u^2 + 6u + 2}{6}\nabla^3 y_n\right.$$

$$\left. + \frac{4u^3 + 18u^2 + 22u + 6}{24}\nabla^4 y_n + \dots\right] \quad \dots(2)$$

$$y'(2.03) = \frac{1}{0.02}\left[-0.0090 + 0\right.$$

$$+ \frac{3(-1/2)^2 + 6(-1/2) + 2}{6}(-0.0002)$$

$$\left. + \frac{4(-1/2)^3 + 18(-1/2)^2 + 22(-1/2) + 6}{24}(-0.0004)\right]$$

$$= 50[-0.0090 + 0.00008 + 0.000017] = -0.44875$$

Again differentiating equation (2) w.r.t. x,

$$y''(x) = \frac{1}{h^2}\left[\nabla^2 y_n + \frac{6u+6}{6}\nabla^3 y_n\right.$$

$$\left. + \frac{12u^2 + 36u + 22}{24}\nabla^4 y_n + \dots\right]$$

$$y''(2.03) = \frac{1}{(0.02)^2}\left[-0.0002 + \left(-\frac{1}{2} + 1\right)(-0.0002)\right.$$

$$\left. + \frac{12\left(-\frac{1}{2}\right)^2 + 36\left(-\frac{1}{2}\right) + 22}{24}(-0.0004)\right]$$

$$= 2500[-0.0002 - 0.0001 - 0.00012]$$

$$y''(2.03) = -1.05.$$

EXAMPLE 5. *Find f′(5) from the following table :*

x	1	2	4	8	10
f(x)	0	1	5	21	27

(JNTU−2009)

SOLUTION. Here, the arguments are not equally spaced. So we use

Newton's divided difference formula.

Difference table

x	f(x)	$\Delta f(x)$	$\Delta^2 f(x)$	$\Delta^3 f(x)$	$\Delta^4 f(x)$
1	0				
		1			
2	1		1/3		
		2		0	
4	5		1/3		−1/144
		4		−1/16	
8	21		−1/6		
		3			
10	27				

Newton's divided difference formula is given by

$$f(x) = f(x_0) + (x + x_0)\Delta f(x_0)$$

$$+ (x - x_0)(x - x_1)\Delta^2 f(x_0)$$

$$+ (x - x_0)(x - x_1)(x - x_2)\Delta^3 f(x_0)$$

$$+ (x - x_0)(x - x_1)(x - x_2)(x - x_3)\Delta^4 f(x_0) \quad \dots(1)$$

Differentiating (1) w.r.t. x, we get

$$f'(x) = \Delta f(x_0) + (2x - x_0 - x_1)\Delta^2 f(x_0)$$

$$+ [(x - x_1)(x - x_2) + (x - x_0)(x - x_2)$$

$$+ (x - x_0)(x - x_1)]\Delta^3 f(x_0)$$

$$+ [(x - x_1)(x - x_2)(x - x_3)$$

$$+ (x - x_0)(x - x_2)(x - x_3)$$

$$+ (x - x_0)(x - x_1)(x - x_3)$$

$$+ (x - x_0)(x - x_1)(x - x_2)]\Delta^4 f(x_0)$$

At $x = 5$

$$f'(5) = 1 + (10 - 1 - 2)\frac{1}{3} + [(5-2)(5-4)$$

$$+ (5-1)(5-4) + (5-1)(5-2)] \times 0$$

$$+ [(5-2)(5-4)(5-8) + (5-1)(5-4)(5-8)$$

$$+ (5-1)(5-2)(5-8) + (5-1)(5-2)(5-4)]\left(\frac{-1}{144}\right)$$

$$f'(5) = 1 + 7\left(\frac{1}{3}\right) - \frac{1}{144}[-9 - 12 - 36 + 12]$$

$$= 1 + \frac{7}{3} + \frac{45}{144}$$

Hence, $f'(5) = 3.6458$.

EXAMPLE 6. *The table given below reveals the velocity v of a body during the time t. Find its acceleration at t = 1.1.*

(UPTU−2002, JNTU−2009)

x	1.0	1.1	1.2	1.3	1.4
y	43.1	47.7	52.1	56.4	60.8

SOLUTION.

Difference table

t	v	Δ	Δ^2	Δ^3	
1.0	43.1				
		4.6			
1.1	47.7		−0.2		
		4.4		+0.1	
1.2	52.1		−0.1		0.1
		4.3		0.2	
1.3	56.4		−0.1		
		4.4			
1.4	60.8				

We have, $t_0 = 1.1, v_0 = 4.77$ and $h = 0.1$

Then acceleration at $t = 1.1$ is given by

$$\left(\frac{du}{dt}\right) = \frac{1}{0.1}\left[4.4 - \frac{1}{2}(-0.1) + \frac{1}{3}(0.2)\right]$$

(By Newton's forward interpolation formula)

$$= 1 - \left(4.4 + 0.5 + \frac{0.2}{3}\right) = 45.167 \text{(approx.)}$$

EXAMPLE 7. *The distance covered by an athelete for the 50 metre race is given in the following table:*

Time (Sec.)	0	1	2	3	4	5	6
Distance (metre)	0	2.5	8.5	15.5	24.5	36.5	50

Determine the speed of athlete at $t = 5$ sec. correct to two decimals. (UPTU–2009)

SOLUTION. We have to find the derivative at $t = 5$, which is near to the end of the table. Therefore we shall use Newton's backward interpolation formula

Difference table

t	s	∇s	$\nabla^2 s$	$\nabla^3 s$	$\nabla^4 s$	$\nabla^5 s$	$\nabla^6 s$
0	0						
		2.5					
1	2.5		3.5				
		6		-2.5			
2	8.5		1		3.5		
		7		1		-3.5	
3	15.5		2		0		1
		9		1		-2.5	
4	24.5		3		-2.5		
		12		-1.5			
5	36.5		1.5				
		13.5					
6	50						

The speed of athlete at $t = 5$ sec. is given by

$$\left(\frac{ds}{dt}\right)_{t=5} = \frac{1}{h}\left[\nabla s_5 + \frac{1}{2}\nabla^2 s_5 + \frac{1}{3}\nabla^3 s_5 + \frac{1}{4}\nabla^4 s_5 + \frac{1}{5}\nabla^5 s_5\right]$$

$$= \frac{1}{1}\left[12 + \frac{1}{2}(3) + \frac{1}{3}(1) + \frac{1}{4}(0) + \frac{1}{5}(-3.5)\right]$$

$$= 13.1333 = 13.13 \text{ meter/sec.}$$

correct to two decimal places.

10.26.6 ERROR ANALYSIS IN NUMERICAL DIFFERENTIATION

Let $E_r(h)$ is the rounded off error in an approximated derivative, then, we have the total error is

$$E(h) = E_r(h) + E_t(h) \qquad \qquad ...(1)$$

where E_r and E_t denotes the rounding off and truncation error respectively.

By definition of the derivative of a function, we have

$$f'(x) = \frac{f(x+h) - f(x)}{h} = \frac{f_1 - f_0}{h}$$

Let us suppose e_1 and e_0 is the rounding off errors in f_1 and f_0 respectively, then we have

$$f'(x) = \frac{(f_1 + e_1) - (f_0 + e_0)}{h} = \frac{f_1 - f_0}{h} + \frac{e_1 - e_0}{h}$$

Let e be the magnitude of e_1 and e_0, then we have

$$[E\ (h)] \leq -$$

Since, the truncation error for two points formula is

$$|E_t(h)| = -\frac{h}{2}f''(\theta) \quad \Rightarrow \quad |E_t(h)| \leq \frac{Mh}{2} \quad \text{where } M = \max|f''(\theta)|, x \leq \theta \leq x + h$$

Thus, the bound for the total error in the derivative is

$$|E(h)| \leq \frac{Mh}{2} + \frac{2e}{h} \qquad \qquad ...(2)$$

REMARKS

- When the step size h is increased, the truncation error increases and rounding off error decreases.
- From equation (2), we have

$$E'(h) = \frac{M}{2} - \frac{2e}{h^2}$$

From the minima of $E(h)$, we must have

$$\Rightarrow \qquad \frac{M}{2} - \frac{2e}{h^2} = 0$$

which gives the optimum step size i.e., $h_{opt} = 2\sqrt{\frac{e}{M}}$

Put this value of h in eq. (2), we get $E(h_{opt}) = 2\sqrt{eM}$

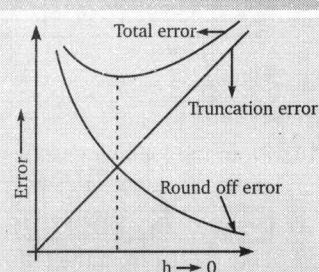

Fig. 6 Error in derivative

Exercise-10.12

1. Use the following data to find $f'(3)$:

x	3	5	11	27	34
$f(x)$	–13	28	899	17315	35606

2. Find $f'(6)$ from the following table :

x	0	1	3	4	5	7	9
$f(x)$	150	108	0	–54	–100	–144	–84

3. Using divided differences, find the value of $f'(8)$, given that $f(6) = 1.556, f(7) = 1.690, f(9) = 1.908, f(12) = 2.158.$

4. Find $f'(5)$ from the following table :

x	2	4	9	10
$f(x)$	4	56	711	980

5. The population of a certain town is shown in tabular form below :

Year	1921	1931	1941	1951	1961
Population (in thousands)	19.96	39.65	58.81	77.21	94.61

Find the rate of growth of population in 1951.

6. Find $y'(0)$ and $y''(0)$ from the following table :

x	0	1	2	3	4	5
y	4	8	15	7	6	2

7. Given that

θ^o	0	10	20	30	40
$\sin \theta^o$	0.000	0.1736	0.3420	0.5000	0.6428

Find $\cos \theta$ when $\theta = 10^o$.

8. A rod is rotating in a plane. The following table given the angle θ in radians through which the rod has turned for various values of the time t second.

t	0	0.2	0.4	0.6	0.8	1.0	1.2
θ	0	0.12	0.49	1.12	2.02	3.20	4.67

Calculate the angular velocity and the angular acceleration of the rod at t = 0.6 second.

(UPTU(MCA)–2002, UPTU–2003, 04, ROHTAK–2004, VTU–2004)

9. The following table of values of x and y is given :

x	0	1	2	3	4	5	6
y	6.9897	7.4036	7.7815	8.1291	8.4510	8.7306	9.0309

Find $\dfrac{d}{dx}$ at

  (i) $x = 1$ (ii) $x = 3$ (iii) $x = 6$

10. From the following values of x and y, find $y'(6)$.

x	4.5	5.0	5.5	6.0	6.5	7.0	7.5
y	9.69	12.90	16.71	21.18	26.37	32.34	39.15

11. Find $f'(10)$ from the following data :

x	3	5	11	27	34
$f(x)$	–13	23	899	17315	35606

12. Find the first derivative for the following table of data at

$x = 0.75, 1.00$ and 1.25. Use h = 0.05 and 0.1

x	0.5	0.7	0.9	1.1	1.3	1.5
y	1.48	1.64	1.78	1.89	1.96	1.00

Compare the result with $h = 0.05$ and $h = 0.1$.

13. Find $\dfrac{dy}{dx}$ and $\dfrac{d^2y}{dx^2}$ at $x = 6$ using the following table :

x	4.5	5.0	5.5	6.0	6.5	7.0	7.5
y	9.69	12.90	16.71	21.18	26.37	32.34	39.15

14. Using the following table, find x for which y to minimum, also find this value of y :

x	0.60	0.65	0.70	0.75
y	0.6221	0.6155	0.6138	0.6170

15. Using the following table, find the value of x, for which y is maximum and find this value of y.

x	1.2	1.3	1.4	1.5	1.6
y	0.9320	0.9636	0.9855	0.9975	0.9996

16. A slider in a machine moves along a fixed straight rod. Its distance x (in cm) along the rod is given at various times t (in secs).

t	0	0.1	0.2	0.3	0.4	0.5	0.6
x	30.28	31.43	32.98	33.54	33.97	33.48	32.13

Evaluate $\dfrac{dx}{dt}$ at t = 0.1 and at t= 05.   (UPTU–2005, VTU–2009)

17. The population of a certain town is given below. Find the rate of growth of the population in 1941 and 1961.

Year	1931	1941	1951	1961	1971
Population in lacs	40.62	60.80	17.95	103.56	132.65

18. Find $f'(7.50)$ from the following table :

x	7.47	7.48	7.49	7.50	7.51	7.52	7.53
$f(x)$	0.193	0.195	0.198	0.201	0.203	0.206	0.208

(JNTU–2006, ROHATK–2005, 10)

19. Find $f'''(15)$ from the following data :

x	2	4	9	13	16	21	29
$f(x)$	57	1345	66340	402052	1118209	4287844	21242820

20. Find the value of $f'(8)$, given that $f(6) = 1.556, f(7) = 1.690, f(9) = 1.908, f(12) = 2.158.$

21. Find $f'(1)$ for $f(x) = \dfrac{1}{(1+x^2)}$ using the following data :

x	1.0	1.1	1.2	1.3	1.4
$f(x)$	0.2500	0.2268	0.2066	0.1890	0.1736

22. Given the following table of values of x and y

x	1.00	1.05	1.10	1.15	1.20	1.25	1.30
$f(x)$	1.000	1.025	1.049	1.072	1.095	1.118	1.140

Find $\dfrac{dy}{dx}$ and $\dfrac{d^2y}{dx^2}$ at   (VTU–2008)

(a) $x = 1.05$ (b) $x = 1.25$

(c) $x = 1.15$

23. Find the values of cos 1.74 from the following table :

x	1.7	1.74	1.78	1.82	1.86
sin x	0.9916	0.9857	0.9781	0.9691	0.9584

24. If $y = f(x)$ and y_n denotes $f(x_0 + nh)$ prove that, if powers of h above h^6 be neglected.

$$\left(\frac{dy}{dx}\right)_{x_0} = \frac{3}{4h}\left[(y_1 - y_{-1}) - \frac{1}{5}(y_2 - y_{-2}) + \frac{1}{45}(y_3 - y_{-3})\right]$$

25. Find the $f'(6)$ from the following data :

x	0	2	3	4	7	8
f(x)	4	26	58	112	466	922

Answers.

1. 1.8828   **2.** –23   **3.** 0.109   **4.** 2097.69   **5.** 1.80 thousands per year **6.** –27.9, 117.67   **7.** 0.9848
8. 3.28 rad/sec., 6.75 rad/sec^2 **9.** (i) 0.3950   (ii) 0.3341   (iii) 0.2719 **10.** 9.66   **11.** 2.33 **13.** 9.64, 2.88
14. 0.692, 0.6137   **15.** 1.58, 1.00   **18.** 0.22666   **19.** 1626 **20.** 0.10859   **21.** –0.5031
22. (a) 0.493, –1.165   (b) 0.4473, –0.1583   (c) 0.4662, –0.2043   **23.** 0.175 **25.** 135

10.27 NUMERICAL QUADRATURE

Given a set of tabulated values of the integrand $f(x)$, to find the value of $\int_{x_0}^{x_n} f(x)dx$ is known as numerical integration. We divide the given interval into a large number of subintervals of equal width h and replace the function tabulated at the points of sub-division by any one of the interpolating polynomial, over each of the subintervals and calculate the integral.

If the integrand has no singular points in the field of the domain then we can calculate the value of the definite integral of that integrand by any numerical method. The process for finding the values of the integral of a function of a single variable, is called quadrature.

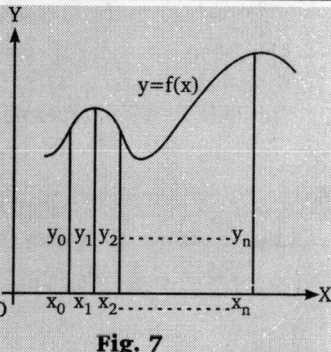

Fig. 7

☞ **REMARKS**

• The problem for finding the value of the integral of $f(x)$ is solved by first approximating the function $f(x)$ by an interpolating polynomial and then integrating it between the desired limits.
• Numerical integration method yield much better results compared to the numerical differentiation. This is due to the fact that the error introduced in separate subintervals tends to cancel each other.

10.27.1 QUADRATURE FORMULA FOR EQUALLY SPACED ARGUMENTS

Let $y = f(x)$ be a function which assume the values $y_0, y_1, ... y_n$ corresponding to the values of the arguments $x = x_0, x_0 + h, x_0 + 2h, ... x_0 + nh$. Divide the interval (a, b) into n subintervals of width h, such that

$$x_0 = a, x_1 = x_0 + h, x_2 = x_0 + 2h, ..., x_n = x_0 + nh = b$$

Consider $I = \int_a^b f(x)dx = \int_{x_0}^{x_0+nh} f(x)dx$ ...(1)

Newton's forward interpolation formula is given by

$$y = f(x) = y_0 + u\Delta y_0 + \frac{u(u-1)}{2!}\Delta^2 y_0 + \frac{u(u-1)(u-2)}{3!}\Delta^3 y_0 + ... \quad \text{where} \quad u = \frac{x-x_0}{h}$$

∴ $\dfrac{du}{dx} = \dfrac{1}{h}dx$ ⇒ $dx = hdu$

∴ Equation (1) becomes $I = h\int\left[y_0 + u\Delta u_0 + \dfrac{u(u-1)}{2!}\Delta^2 y_0 + \dfrac{u(u-1)(u-2)}{3!}\Delta^3 y_0 + ...\right]du$

$$= nh\left[y_0 + \frac{n}{2}\Delta y_0 + \frac{n(2n-3)}{12}\Delta^2 y_0 + \frac{n(n-2)^2}{24}\Delta^3 y_0 + ... + \text{upto } (n+1) \text{ terms}\right]$$

∴ $\int_{x_0}^{x_n} f(x)dx = nh\left[y_0 + \dfrac{n}{2}\Delta y_0 + \dfrac{n(2n-3)}{12}\Delta^2 y_0 + \dfrac{n(n-2)^3}{24}\Delta^3 y_0 + ... \text{ upto } (n+1) \text{ terms}\right]$ ...(2)

This is called general quadrature formula.

10.28 THE TRAPEZOIDAL RULE

Putting $n = 1$ into equation (2) and taking the curve $y = f(x)$ between the points (x_0, y_0) and (x_1, y_1) as a straight line. That is curve is approximated by a polynomial of degree one so that differences of order two and three and so on becomes zero, we get

$$\int_{x_0}^{x_1} f(x)dx = h\left[y_0 + \frac{1}{2}\Delta y_0\right] = h\left[y_0 + \frac{1}{2}(y_1 - y_0)\right] = \frac{h}{2}[y_0 + y_1]$$

Similarly,
$$\int_{x_1}^{x_2} f(x)dx = \frac{h}{2}[y_1 + y_2]$$

$$\int_{x_2}^{x_3} f(x)dx = \frac{h}{2}[y_2 + y_3]$$

....

$$\int_{x_{n-2}}^{x_{n-1}} f(x)dx = \frac{h}{2}[y_{n-2} + y_{n-1}]$$

and
$$\int_{x_{n-1}}^{x_n} f(x)dx = \frac{h}{2}[y_{n-1} + y_n]$$

Now adding these n integrals, we get

$$\int_{x_0}^{x_n} f(x)dx = \frac{h}{2}[y_0 + 2(y_1 + y_2 + ... + y_{n-1}) + y_n] \qquad ...(3)$$

This is known as the trapezoidal rule.

REMARKS

- $\int_a^b f(x)dx$ gives the area of $y = f(x)$ bounded by $x = a$ and $x = b$ and axis of x.
- The shape of each strip between any consecutive points is taken to trapezium. The area of each strip is found separately. That the area bounded by $y = f(x)$ and $x = x_0$ and $x = x_n$ is approximately equal to the sum of the areas of n trapezium. That is how this formula is called trapezoidal rule.
- In trapezoidal rule, $f(x)$ is a linear function of x. It is the simplest rule but least accurate.

WORKING PROCEDURE

Step 1. Obtain the values of $f(x)$, limits of the integral x_0 and x_n and the number of subintervals n.

Step 2. Calculate $\frac{x_n - x_0}{n} = h$.

Step 3. Evaluate the value of x_i as $x_0 + ih$ and corresponding value of $f(x_i)$.

Step 4. Substitute the above values in trapezoidal rule.

REMARK

- The error in trapezoidal rule is $\varepsilon \approx -\frac{b-1}{12}h^2 y''(\xi)$.

10.29 SIMPSON'S 1/3 RULE

Putting $n = 2$ in general quadrature formula and taking the curve through (x_0, y_0), (x_1, y_1) and (x_2, y_2) as a polynomial of degree 2 so that differences of order three and greater than three becomes zero, therefore, we get

$$\int_{x_0}^{x_2} ydx = 2h\left[y_0 + \Delta y_0 + \frac{1}{6}\Delta^2 y_0\right] = 2h\left[y_0 + y_1 - y_0 + \frac{1}{6}(y_2 - 2y_1 + y_0)\right] = \frac{2h}{6}[y_2 - 2y_1 + y_0 + 6y_1]$$

$$= \frac{h}{3}[y_0 + 4y_1 + y_2]$$

Similarly,

$$\int_{x_2}^{x_4} ydx = \frac{h}{3}[y_2 + 4y_3 + y_4]$$

$$\int_{x_4}^{x_6} ydx = \frac{h}{3}[y_4 + 4y_5 + y_6]$$

....

$$\int_{x_{n-4}}^{x_{n-2}} ydx = \frac{h}{3}[y_{n-4} + 4y_{n-3} + y_{n-2}]$$

and
$$\int_{x_{n-2}}^{x_n} ydx = \frac{h}{3}[y_{n-2} + 4y_{n-1} + y_n]$$

Adding all these integrals, we get

$$\int_{x_0}^{x_n} f(x)dx = \int_{x_0}^{x_n} ydx = \frac{h}{3}[y_0 + 2(y_2 + y_4 + y_6 + ... + y_{n-2}) + 4(y_1 + y_3 + y_5 + ... + y_{n-1}) + y_n]$$

This is known as Simpson's 1/3 Rule.

REMARKS

- Simpson's 1/3-Rule requires the division of the interval into an even number of subintervals of width h.
- In this rule, the interpolating polynomial is of degree 2. Therefore this rule is also known as parabolic rule.

Step 1. Obtain the values of $f(x), x_0, x_n$ and n.

Step 2. Calculate $\dfrac{x_n - x_0}{n} = h$.

Step 3. Calculate the value of x_i as $x_0 + ih$ and corresponding value of $f(x)$.

Step 4. Substitute the above obtained values in Simpson's 1/3 Rule.

REMARK

- The error in Simpson's 1/3 rule is $\varepsilon = -\dfrac{b-a}{180} h^4 y^{iv}(\xi)$.

10.30 SIMPSON'S 3/8 RULE

Putting $n = 3$ into equation (2) and taking the curve through the points (x_0, y_0), (x_1, y_1), (x_2, y_2) and (x_3, y_3) as a polynomial of degree 3 so that the differences of order greater than three becomes zero. We get

$$\int_{x_0}^{x_3} f(x)dx = 3h\left[y_0 + \frac{3}{2}\Delta y_0 + \frac{3(3)}{12}\Delta^2 y_0 + \frac{3.1}{24}\Delta^3 y_0\right]$$

$$= 3h\left[y_0 + \frac{3}{2}(y_1 - y_0) + \frac{3}{4}(y_2 - 2y_1 + y_0) + \frac{1}{8}(y_3 - 3y_2 + 3y_1 - y_0)\right] = \frac{3h}{8}[y_0 + 3y_1 + 3y_2 + y_3]$$

Similarly in the intervals $[x_3, x_6], [x_6, x_9]...[x_{n-3}, x_n]$, we get

$$\int_{x_3}^{x_6} f(x)dx = \frac{3h}{8}[y_3 + 3y_4 + 3y_5 + y_6]$$

$$\int_{x_6}^{x_9} f(x)dx = \frac{3h}{8}[y_6 + 3y_7 + 3y_8 + y_9]$$

....

$$\int_{x_{n-6}}^{x_{n-3}} f(x)dx = \frac{3h}{8}[y_{n-6} + 3y_{n-5} + 3y_{n-4} + y_{n-3}]$$

and $\int_{x_{n-3}}^{x_n} f(x)dx = \frac{3h}{8}[y_{n-3} + 3y_{n-2} + 3y_{n-1} + y_n]$

Adding all these integrals, we get

$$\int_{x_0}^{x_n} f(x)dx = \frac{3h}{8}[y_0 + 3(y_1 + y_2 + y_4 + y_5 + ... + y_{n-1}) + 2(y_3 + y_6 + ... + y_{n-3}) + y_n]$$

This is known as Simpson 3/8 rule.

Step 1. Obtain the values of $f(x), x_0, x_n, n$.

Step 2. Calculate $\dfrac{x_n - x_0}{n} = h$.

Step 3. Calculate the value of x_i as $x_0 + ih$ and corresponding value of $f(x_i)$.

Step 4. Substitute the above obtained values in Simpson's 3/8 Rule.

REMARKS
- In this rule, while applying, the number of subintervals should be taken as multiple of 3.
- In Simpson's 3/8 rule, the interpolating polynomial is taken to be of degree 3. That is $y = ax^3 + bx^2 + cx + d = 0$.
- The error in Simpson's 3/8 rule is $\varepsilon = -\dfrac{(b-a)}{80} h^4 y^{iv}(\xi)]$.

10.31 WEDDLE'S RULE

Putting $n = 6$ in general quadrature formula and taking the curve through the points (x_k, y_k); $k = 0, 1, 2, 3, 4, 5, 6$ as a polynomial of degree six so that the differences of order greater than 6 becomes zero. We obtain

$$\int_{x_0}^{x_6} y\,dx = 6h\left[y_0 + 3\Delta y_0 + \frac{9}{2}\Delta^2 y_0 + 4\Delta^3 y_0 + \frac{123}{60}\Delta^4 y_0 + \frac{11}{20}\Delta^5 y_0 + \frac{41}{840}\Delta^6 y_0\right]$$

$$= \frac{3h}{10}[y_0 + 5y_1 + y_2 + 6y_3 + y_4 + 5y_5 + y_6]$$

Similarly, in the intervals $[x_6, x_{12}], [x_{12}, x_{18}], \ldots [x_{n-6}, x_n]$

$$\int_{x_6}^{x_{12}} y\,dx = \frac{3h}{10}[y_6 + 5y_7 + y_8 + 6y_9 + y_{10} + 5y_{11} + y_{12}]$$

....

$$\int_{x_{n-6}}^{x_n} y\,dx = \frac{3h}{10}[y_{n-6} + 5y_{n-5} + y_{n-4} + 6y_{n-3} + y_{n-2} + 5y_{n-1} + y_n]$$

Adding all these integrals, we get

$$\int_{x_0}^{x_n} y\,dx = \frac{3h}{10}[y_0 + 5y_1 + y_2 + 6y_3 + y_4 + 5y_5 + 2y_6 + 5y_7 + y_8 + \ldots]$$

This is known as Weddle's rule.

WORKING PROCEDURE

Step 1. Obtain the values of $f(x), x_0, x_n$ and n.

Step 2. Calculate $\dfrac{x_n - x_0}{n} = h$.

Step 3. Calculate the value of x_i as $x_0 + ih$ and corresponding value of $f(x_i)$.

Step 4. Substitute the above obtained values in Weddle's rule.

☞ REMARKS

- While applying this rule, the number of subintervals should be taken as a multiple of 6.
- In this rule the interpolating polynomial is taken to be of degree 6. That is $y = ax^6 + bx^5 + cx^4 + dx^3 + ex + f = 0$
- The error in Weddle's rule is $\varepsilon = -\dfrac{(b-a)h^6}{840} y^{vi}(\xi)$.

10.32 ROMBERG'S METHOD

We know that, for an interval of width h, the error in trapezoidal rule is

$$= -\frac{(b-a)h^2}{12} f''(\xi), a < \xi < b = Ch^2 \text{ where } C = \frac{-(b-a)}{12} f''(\xi)$$

Suppose, we calculate $I = \int_a^b f(x)dx$ by trapezoidal rule with two different subintervals h_1 and h_2. Let I_1, I_2 be the approximations with errors E_1 and E_2 respectively. Then, clearly

$$I = I_1 + E_1 = I_1 + Ch_1^2 \qquad \text{and} \qquad I = I_2 + E_2 = I_2 + Ch_2$$

Therefore, $I_1 = Ch_1^2 + I_2 + Ch_2^2 \qquad \Rightarrow \qquad C = \dfrac{I_1 - I_2}{h_2^2 - h_1^2}$

$$\therefore \qquad I = I_1 + \left(\frac{I_1 - I_2}{h_2^2 - h_1^2}\right)h_1^2 = \frac{I_1 h_2^2 - I_2 h_1^2}{h_2^2 - h_1^2}$$

which will be a better approximation to I than I_1 or I_2. To calculate systematically we take $h_1 = h$ and $h_2 = \dfrac{h}{2}$, then, we have

$$I = \frac{I_1\left(\dfrac{h}{2}\right)^2 - I_2 h^2}{\left(\dfrac{h}{2}\right)^2 - h^2} = \frac{4I_2 - I_1}{3} = I_2 + \frac{1}{3}(I_2 - I_1)$$

We may find this result by applying Trapezoidal rule two times. By applying the rule several times, we get a sequence of results $A_1, A_2 \ldots$ in which the error is reduced by $\dfrac{1}{4}$ every time.

We apply the formula (1) again to each pair of A's to get improved results $B_1, B_2, \ldots$ etc.

Again applying equation (1) to the pairs $B_1, B_2; B_2, B_3$ etc. we get still better result. Continue this process until two successive values are closed to each other.

☞ REMARK

- The above method is called Richardson's deferred approach to the limit and its systematic improvement is called Romberg integration or method.

10.33 NEWTON-COTE'S FORMULA

The usual strategy is developing formulae of numerical integration is similarly to that for numerical differentiation. We pass a polynomial through points defined by the function, and then integrate this polynomial approximation to the function. This allow us to integrate a function known only as a table of values.

If $f(x)$ has numerical values at certain points, then we can compute the value of the integral of $f(x)$ over (a, b) numerically. As in the problem related to the interpolation, we approximate $f(x)$ by a suitable function $P(x)$, which is usually a polynomial. Suppose the value of $f(x)$ are given at equidistant points of x. Let $f(x)$ be approximated by $P(x)$ given by Lagrange's formula.

$\therefore \qquad f(x) = P(x) = \sum_{k=0}^{n} L_k(x) y_k$

where $\qquad L_k(x) = \dfrac{(x-x_0)(x-x_1)...(x-x_{k-1})(x-x_{k+1})...(x-x_n)}{(x_k-x_0)(x_k-x_1)...(x_k-x_{k-1})(x_k-x_{k+1})...(x_k-x_n)}$

Here $x_k = x_0 + xh$ and $x = x_0 + uh$

$\therefore \qquad L_k = L_k(x_0 + uh) = \dfrac{(x_0+uh-x_0)(x_0+uh-x_1)...(x_0+uh-x_{k-1})(x_0+uh-x_{k+1})...(x_0+uh-x_n)}{(x_k-x_0)(x_k-x_1)...(x_k-x_{k-1})(x_k-x_{k+1})...(x_k-x_n)}$

$\therefore \qquad L_k = \dfrac{u(u-1)...(u-k+1)(u-k-1)...(u-n)}{k(k-1)...(+1)(-1)...(k-n)}$ ...(1)

Now find

$\int_{x_0}^{x_n} f(x)dx = \int_{x_0}^{x_n} \sum_{k=0}^{n} L_k(x)y_k \, dx = \int_{x_0}^{x_n} [L_0(x)y_0 + L_1(x)y_1 + ... + L_n(x)y_n]dx$

$\qquad = y_0 \int_{x_0}^{x_n} L_0(x)dx + y_1 \int_{x_0}^{x_n} L_1(x)dx + ... + y_n \int_{x_0}^{x_n} L_n(x)dx = \sum_{k=0}^{n} y_k \int_{x_0}^{x_n} L_k(x)dx$

Since $x = x_0 + uh$ and $x_n = x_0 + nh$

$\qquad dx = hdu$

$\therefore \qquad \int_{x_0}^{x_n} f(x)dx = \sum_{k=0}^{n} y_k h. \int_0^n L_k du = nh \sum_{k=0}^{n} y_k . \dfrac{1}{n} \int_0^n L_k du$, where L_k is given by eq. (1)

Let $\qquad \dfrac{1}{n} \int_0^n L_k du = C_k^n$, we get $\int_{x_0}^{x_n} f(x)dx = nh \sum_{k=0}^{n} y_k C_k^n$ ...(2)

This is known as Newton-Cote's formula, where C_k^n are called Cote's numbers.

10.33.1 PROPERTIES OF COTE'S NUMBERS

(i) $C_k^n = C_{n-k}^n$ 　　　　(ii) $\sum_{k=0}^{n} C_k^n = 1$ 　　　　$\therefore \qquad C_k^n = C_{n-k}^n .$

PROOF. (i) We known that

$C_k^n = \dfrac{1}{n} \int_0^n L_k du$

$= \dfrac{1}{n} \int_0^n \left[\dfrac{u(u-1)...(u-k+1)(u-k-1)...(u-n)}{k(k-1)...(+1)(-1)...(k-n)} \right] du$

$= \dfrac{1}{n} \int_0^n \dfrac{u(u-1)...(u-k+1)(u-k-1)...(u-n)}{k!(-1)^{n-k}(n-k)!} du$

Let $u - n = -t \Rightarrow u = n - t$ $\therefore$ $du = -dt$, we get

$C_k^n = \dfrac{1}{n} \int_n^0 \dfrac{(n-t)(n-t-1)...(n-t-k+1)(n-t-k-1)...(-t)}{k!(-1)^{n-k}(n-k)!}(-dt)$

$= \dfrac{1}{n} \int_0^n \dfrac{(-1)^n t(t-1)...\{t-(n-k)+1\}\{t-(n-k)-1\}....(t-n)}{(n-k)!k!(-1)^{n-(n-k)}(-1)^n} dt$

$= \dfrac{1}{n} \int_0^n \dfrac{t(t-1)...\{t-(n-k)+1\}\{t-(n-k)-1\}....(t-n)}{(n-k)!k!(-1)^{n-(n-k)}} dt$

$= \dfrac{1}{n} \int_0^n \dfrac{u(u-1)...\{u-(n-k)-1\}\{u-(n-k)+1\}....(u-n)}{(n-k)!k!(-1)^{n-(n-k)}} du$

$\left(\because \int_0^a f(x)dx = \int_0^a f(t)dt \right)$

$= C_{n-k}^n$

(ii) If $n = 1$, then

$\sum_{k=0}^{1} C_k^1 = C_0^1 + C_1^1 = \dfrac{1}{1} \int_0^1 L_0 du + \dfrac{1}{1} \int_0^1 L_1 du$

$\left(\because C_k^n = \dfrac{1}{n} \int_0^n L_k du \right)$

Now $\qquad L_0 = \dfrac{u-1}{0-1} = (1-u)$

$L_1 = \dfrac{u-0}{1-0} = u = \int_0^1 [L_0 + L_1]du$

$\qquad = \int_0^1 [1 - u + u]du = \int_0^1 du = 1$

$\therefore \quad \sum_{k=0}^{1} C_k^1 = 1$

If $n = 2$, then

$\sum_{k=0}^{2} C_k^2 = C_0^2 + C_1^2 + C_2^2$

$= \dfrac{1}{2} \int_0^2 L_0 du + \dfrac{1}{2} \int_0^2 L_1 du + \dfrac{1}{2} \int_0^2 L_2 du = \dfrac{1}{2} \int_0^2 [L_0 + L_1 + L_2]du$

Now $\qquad L_0 = \dfrac{(u-1)(u-2)}{(0-1)(0-2)} = \dfrac{1}{2}(u^2 - 3u + 2) ;$

$L_1 = \dfrac{u(u-2)}{1(1-2)} = -(u^2 - 2u) ; \quad L_2 = \dfrac{u(u-1)}{2(2-1)} = \dfrac{1}{2}(u^2 - u)$

$L_0 + L_1 + L_2 = \dfrac{1}{2}(u^2 - 3u + 2) - (u^2 - 2u) + \dfrac{1}{2}(u^2 - u)$

$= \dfrac{1}{2}[u^2 - 3u + 2 - 2u^2 + 4u + u^2 - u]$

$$= \frac{1}{2}[3u^2 - 3u^2 - 4u + 4u + 2] = 1$$

$$\therefore \sum_{k=0}^{2} C_k^2 = \frac{1}{2} \int_0^2 [L_0 + L_1 + L_2] du$$

$$= \frac{1}{2} \int_1^2 1 \cdot du = \frac{1}{2}(u)_0^2 = \frac{1}{2}(2) = 1$$

$$\therefore \quad \sum_{k=0}^{2} C_k^2 = 1$$

Now taking $n = 3$

$$\sum_{k=0}^{3} C_k^3 = C_0^3 + C_1^3 + C_2^3 + C_3^3 = 2C_0^3 + 2C_1^3$$

$$\left(\because C_{n-k}^n = C_k^n \right)$$

$$= 2[C_0^3 + C_1^3] = 2\left[\frac{1}{3} \int_0^3 L_0 du + \frac{1}{3} \int_0^3 L_1 du \right] = \frac{2}{3} \int_0^3 [L_0 + L_1] du$$

Find

$$L_0 = \frac{(u-1)(u-2)(u-3)}{(0-1)(0-2)(0-3)} = -\frac{1}{6}[u^3 - 6u^2 + 11u - 6];$$

$$L_1 = \frac{u(u-2)(u-3)}{1(1-2)(1-3)} = \frac{1}{2}[u^3 - 5u + 6u]$$

10.33.2 DEDUCTIONS FROM NEWTON-COTE'S FORMULA

We known that $\int_{x_0}^{x_n} f(x)dx = nh \sum_{k=0}^{n} y_k C_k^n$

Putting $n = 1$ into this equation, we get

$$\int f(x)dx = n \sum y_k C_k^1 = h[y_0 C_0^1 + y_1 C_1^1] \qquad \ldots(1)$$

Now find $C_0^1 = \frac{1}{1} \int_0^1 L_0 du = \int_0^1 \frac{u-1}{0-1} du$

$$= \int_0^1 (1-u)du = \left(u - \frac{u^2}{2} \right)_0^1 = 1 - \frac{1}{2} = \frac{1}{2} \quad \left(\because L_0 = \frac{u-1}{0-1} \right)$$

and $C_1^1 = \int_0^1 L_1 du = \int_0^1 \frac{u}{1-0} du = \left(\frac{u^2}{2} \right)_0^2 = \frac{1}{2}$

Substituting the values of C_0^1 and C_1^1 into eq. (1), we get

$$\int_{x_0}^{x_1} f(x)dx = \frac{h}{2}[y_0 + y_1]$$

This is the trapezoidal rule.

Now putting $n = 2$ into Newton-Cote's formula, we get

$$\int_{x_0}^{x_2} f(x)dx = 2h \sum_{k=0}^{2} y_k C_k^2 = 2h[y_0 C_0^2 + y_1 C_1^2 + y_2 C_2^2] \qquad \ldots(2)$$

Find $C_0^2 = \frac{1}{2} \int_0^2 L_0 du$

$$= \frac{1}{2} \int_0^2 \frac{(u-1)(u-2)}{(0-1)(0-2)} du = \frac{1}{2} \cdot \frac{1}{2} \int_0^2 (u^2 - 3u + 2)du$$

$$= \frac{1}{4} \left[\frac{u^3}{3} - \frac{3u^2}{2} + 2u \right]_0^2 = \frac{1}{4} \left[\frac{8}{3} - \frac{12}{2} + 4 \right] = \frac{1}{4} \left[\frac{8}{3} - 2 \right] = \frac{1}{6}$$

and $C_1^2 = \frac{1}{2} \int_0^2 L_1 du$

$$= \frac{1}{2} \int_0^2 \frac{u(u-2)}{1(1-2)} du = -\frac{1}{2} \int_0^2 (u^2 - 2u)du = -\frac{1}{2} \left[\frac{u^3}{3} - u^2 \right]_0^2$$

$$L_0 + L_1 = -\frac{1}{6}(u^3 - 6u^2 + 11u - 6) + \frac{1}{2}(u^3 - 5u^2 + 6u)$$

$$= \frac{1}{6}[-u^3 + 6u^2 - 11u + 6 + 3u^3 - 15u^2 + 18u]$$

$$= \frac{1}{6}[2u^3 - 9u^2 + 7u + 6]$$

$$\therefore \sum_{k=0}^{3} C_k^3 = \frac{2}{3} \int_0^3 \frac{1}{6}(2u^3 - 9u^2 + 7u + 6)du$$

$$= \frac{1}{9} \left[\frac{u^4}{2} - 3u^3 + \frac{7}{2}u^2 + 6u \right]_0^3 = \frac{1}{9} \left[\frac{81}{2} - 81 + \frac{63}{2} + 18 \right]$$

$$= \frac{1}{9}[72 + 18 - 81] = \frac{1}{9}(9) = 1$$

$$\therefore \quad \sum_{k=0}^{3} C_k^3 = 1$$

Hence, in general we can say that $\sum_{k=0}^{n} C_k^n = 1$.

$$= -\frac{1}{2} \left[\frac{8}{3} - 4 \right] = -\frac{1}{2} \left[-\frac{4}{3} \right] = \frac{2}{3}$$

and $\quad C_2^2 = C_0^2 = \frac{1}{6}$ $\qquad (\because C_k^n = C_{n-k}^n)$

Substituting the values of C_0^2, C_1^2, C_2^2 into equation (1), we get

$$\int_{x_0}^{x_2} f(x)dx = 2h \left[\frac{1}{6}y_0 + \frac{2}{3}y_1 + \frac{1}{6}y_2 \right] = \frac{h}{3}[y_0 + 4y_1 + y_2]$$

This gives Simpson's 1/3 Rule.

Similarly, putting $n = 3$ into Newton-Cote's formula, we get

$$\int_{x_0}^{x_3} f(x)dx = 3h \sum_{k=0}^{3} y_k C_k^3 = 3h[y_0 C_0^3 + y_1 C_1^3 + y_2 C_2^3 + y_3 C_3^3]$$

$$\ldots(3)$$

Now find

$$C_0^3 = \frac{1}{3} \int_0^3 L_0 du = \frac{1}{3} \int_0^3 \left[\frac{(u-1)(u-2)(u-3)}{(0-1)(0-2)(0-3)} \right] du$$

$$= \frac{1}{3} \left(-\frac{1}{6} \right) \int_0^3 (u^3 - 6u^2 + 11u - 6)du$$

$$= -\frac{1}{18} \left[\frac{u^4}{4} - 2u^3 + \frac{11u^2}{2} - 6u \right]_0^3$$

$$= -\frac{1}{18} \left[\frac{81}{4} - 54 + \frac{99}{2} - 18 \right] = -\frac{1}{18} \left[\left(\frac{81 + 198}{4} \right) - 72 \right]$$

$$= -\frac{1}{18} \left[\frac{279}{4} - 72 \right] = -\frac{1}{18} \left[\frac{279 - 288}{4} \right]$$

$$= -\frac{1}{18} \left[\frac{-9}{4} \right] = \frac{1}{2} \left(\frac{1}{4} \right) = \frac{1}{8}$$

$$\therefore \qquad C_0^3 = C_3^3 = \frac{1}{8} \qquad (\because C_k^n = C_{n-k}^n)$$

Now

$$C_1^3 = \frac{1}{3} \int_0^3 L_1 du = \frac{1}{3} \int_0^3 \frac{u(u-2)(u-3)}{1(1-2)(1-3)} du = \frac{1}{6} \int_0^3 (u^3 - 5u^2 + 6u)du$$

$$= \frac{1}{6}\left[\frac{u^4}{4} - \frac{5u^3}{3} + 3u^2\right]_0^3$$

$$= \frac{1}{6}\left[\frac{81}{4} - \frac{135}{3} + 27\right] = \frac{1}{6}\left[\frac{81}{4} - 45 + 27\right]$$

$$= \frac{1}{6}\left[\frac{81}{4} - 18\right] = \frac{1}{6}\left[\frac{81-72}{4}\right] = \frac{9}{24} = \frac{3}{8}$$

$$\therefore \quad C_2^3 = C_1^3 = \frac{3}{8}$$

Substituting the values of C_0^3, C_1^3, C_2^3 and C_3^3 in equation (2), we get

$$\int_{x_0}^{x_3} f(x)dx = 3h\left[\frac{y_0}{8} + \frac{3y_1}{8} + \frac{3y_2}{8} + \frac{1}{8}y_3\right]$$

$$= 3h[y_0 + 3y_1 + 3y_2 + y_3]$$

This gives the Simpson's 3/8 Rule.

The Cote's numbers for some values of are given in the table :

n	C_0^n	C_1^n	C_2^n	C_3^n	C_4^n	C_5^n	C_6^n
1	$\frac{1}{2}$	$\frac{1}{2}$					
2	$\frac{1}{6}$	$\frac{4}{6}$	$\frac{1}{6}$				
3	$\frac{1}{8}$	$\frac{3}{8}$	$\frac{3}{8}$	$\frac{1}{8}$			
4	$\frac{7}{90}$	$\frac{32}{90}$	$\frac{12}{90}$	$\frac{32}{90}$	$\frac{7}{90}$		
5	$\frac{19}{288}$	$\frac{75}{288}$	$\frac{50}{288}$	$\frac{50}{288}$	$\frac{75}{288}$	$\frac{19}{288}$	
6	$\frac{41}{840}$	$\frac{216}{840}$	$\frac{27}{840}$	$\frac{272}{840}$	$\frac{27}{840}$	$\frac{216}{840}$	$\frac{41}{840}$

10.34 GAUSS' QUADRATURE FORMULA

Gauss derived a formula for the integration of a function whose values are given for the values of x which are not equally spaced but are symmetrically placed with respect to the middle point of the interval of integration.

Let $I = \int_a^b y\,dx$ where $y = f(x)$ be computed. Now change the variable x to u by the substitution given by

$$x = (b-a)u + \frac{1}{2}(a+b) \qquad \qquad ...(1)$$

Thus the limit of integration becomes $u = -\frac{1}{2}$ where $x = a$ and $u = \frac{1}{2}$ when $x = b$ and also

$$dx = (b-a)du \qquad \text{and} \qquad y = f(x) = f\left[(b-a)u + \frac{1}{2}(a+b)\right]$$

$$y = \phi(u), \text{ (say)}$$

Then the integral becomes

$$I = (b-a)\int_{-1/2}^{1/2}\phi(u)du \qquad \qquad ...(2)$$

Gauss quadrature formula is

$$I = \int_{-1/2}^{1/2}\phi(u)du = R_1\phi(u_1) + R_2\phi(u_2) + ... + R_n\phi(u_n)... \qquad \qquad ...(3)$$

where $u_1, u_2..., u_n$ are the points of subdivision of the interval $\left(-\frac{1}{2}, \frac{1}{2}\right)$. Thus the corresponding values of x are given by

$$x_k = (b-a)u_k + \frac{1}{2}(a+b), \quad k = 1, 2, 3, ... n.$$

Therefore, the value of the integral from equation (2) is obtained as

$$I = \int_a^b f(x)dx = (b-a)[R_1\phi(u_1) + R_2\phi(u_2) + ... + R_n\phi(u_n)] \qquad \qquad ...(4)$$

The derivation of Gauss' formula is out of scope, therefore, we shall not give a detailed derivation but the method for finding the values of $u_1, u_2..., u_n$ and $R_1, R_2, ..., R_n$ be given.

Let us consider a convergent power series $\phi(u)$ in the interval $\left[-\frac{1}{2}, \frac{1}{2}\right]$. We therefore write $\phi(u)$ as follows :

$$\phi(u) = a_0 + a_1u + a_2u^2 + ... + a_mu^m + ... \qquad \qquad ...(5)$$

and also consider that the integral can be expressed as a linear function of the ordinate of the form equation (3).

Now integrating equation (5) from $u = -\frac{1}{2}$ to $\frac{1}{2}$, we get

$$I = \int_{-1/2}^{1/2}\phi(u)du = \int_{-1/2}^{1/2}[a_0 + a_1u + a_2u^2 + ... + a_mu^m + ...]du = \left[a_0u + a_1\frac{u^2}{2} + a_2\frac{u^3}{2} + ... + a_m\frac{u^{m+1}}{m+1} + ...\right]_{-1/2}^{1/2}$$

$$= a_0\left[\frac{1}{2} + \frac{1}{2}\right] + \frac{1}{2}a_1\left[\frac{1}{4} - \frac{1}{4}\right] + \frac{1}{3}a_2\left[\frac{1}{8} + \frac{1}{8}\right] + ... + \frac{a_m}{m+1}\left[\left(\frac{1}{2}\right)^{m+1} - \left(-\frac{1}{2}\right)^{m+1}\right] + ...$$

or $\quad I = a_0 + \frac{1}{12}a_2 + \frac{1}{80}a_4 + \frac{1}{448}a_6 + \frac{1}{2304}a_8 + ... \qquad \qquad ...(6)$

Now from equation (5), we also have

$$\phi(u_1) = a_0 + a_1u_1 + a_2u_1^2 + ... + a_mu_1^m + ...$$

$$\phi(u_2) = a_0 + a_1 u_2 + a_2 u_2^2 + \ldots + a_m u_2^m + \ldots$$

$$\phi(u_3) = a_0 + a_1 u_3 + a_2 u_3^2 + \ldots + a_m u_3^m + \ldots$$

$$\phi(u_n) = a_0 + a_1 u_n + a_2 u_n^2 + \ldots + a_m u_n^m + \ldots$$

Substituting these values into equation (3), we get

$$I = R_1(a_0 + a_1 u_1 + a_2 u_1^2 + \ldots + a_m u_1^m + \ldots) + R_2(a_0 + a_1 u_2 + a_2 u_2^2 + \ldots + a_m u_2^m + \ldots) + \ldots \ldots + R_n(a_0 + a_1 u_n + a_2 u_n^2 + \ldots + a_m u_n^m + \ldots)$$

$$I = a_0(R_1 + R_2 + \ldots + R_n) + a_1(R_1 u_1 + R_2 u_2 + \ldots + R_n u_n) + a_2(R_1 u_1^2 + R_2 u_2^2 + \ldots + R_n u_n^2) + \ldots \ldots + a_m(R_1 u_1^m + R_2 u_2^m + \ldots + R_n u_n^m) \qquad \ldots(7)$$

From eqs. (6) and (7), we get identically the same for all values of $a_0, a_1, \ldots, a_m$ etc.

$$\left. \begin{array}{l} R_1 + R_2 + R_3 + \ldots + R_n = 1 \\[4pt] R_1 u_1 + R_2 u_2 + R_3 u_3 + \ldots + R_n u_n = 0 \\[4pt] R_1 u_1^2 + R_2 u_2^2 + R_3 u_3^2 + \ldots + R_n u_n^2 = \dfrac{1}{12} \\[4pt] R_1 u_1^3 + R_2 u_2^3 + R_3 u_3^3 + \ldots + R_n u_n^3 = 0 \\[4pt] R_1 u_1^4 + R_2 u_2^4 + R_3 u_3^4 + \ldots + R_n u_n^4 = \dfrac{1}{80} \end{array} \right\} \qquad \ldots(8)$$

If we take $2n$ of these equations and solve them simultaneously, we can find them theoretically because of much labour of solving these equations even for small values of n. Therefore we adopt following method. It can be shown that if $\phi(u)$ is a polynomial of degree not higher than $2n - 1$, then $u_1, u_2, \ldots, u_n$ are the roots of Legendre polynomial $P_n(u) = 0$. These roots can be calculated from the equation given by

$$\frac{d^n}{du^n} \left[u^2 - \left(\frac{1}{2} \right)^2 \right]^n = 0$$

All n roots $u_1, u_2, \ldots, u_n$ of this nth degree equation are all real. Substituting these values into (8), we get $R_1, R_2, \ldots, R_n$. Let us take $n = 3$, then eq. (9)

$$\frac{d^3}{du^3} \left[u^2 - \left(\frac{1}{2} \right)^2 \right]^3 = 0 \qquad \Rightarrow \qquad \frac{d^3}{du^3} \left[u^6 - \left(\frac{1}{2} \right)^6 - 3u^4 \left(\frac{1}{2} \right)^2 + 3u^2 \left(\frac{1}{2} \right)^4 \right] = 0$$

or $\quad \dfrac{d^3}{du^3} \left[u^6 - \dfrac{3}{4} u^4 + \dfrac{3}{16} u^2 - \dfrac{1}{64} \right] = 0 \qquad$ or $\qquad \dfrac{d^2}{du^2} \left[6u^5 - 3u^3 + \dfrac{3}{8} u \right] = 0$

or $\quad \dfrac{d}{du} \left[30u^4 - 9u^2 + \dfrac{3}{8} \right] = 0$

or $\quad 120u^3 - 18u = 0 \qquad$ or $\qquad 6u(20u^2 - 3) = 0$

or $\quad u = 0, u = \pm \sqrt{\dfrac{3}{20}} \pm \left(\dfrac{1}{2} \right) \sqrt{\dfrac{3}{5}} \quad$ i.e., $\quad u_1 = -\dfrac{1}{2} \sqrt{\dfrac{3}{5}}, u_2 = 0, u_3 = \dfrac{1}{2} \sqrt{\dfrac{3}{5}}$

Now equation (8) takes the form for $n = 3$,

$$R_1 + R_2 + R_3 = 1$$

$$R_1 u_1 + R_2 u_2 + R_3 u_3 = 0$$

$$R_1 u_1^2 + R_2 u_2^2 + R_3 u_3^2 = \frac{1}{12}$$

Substitute the value of u_1, u_2 and u_3 into above equations, we get

$$R_1 + R_2 + R_3 = 1 ; \qquad\qquad R_3 - R_1 = 0$$

$$(R_1 + R_2) \frac{3}{20} = \frac{1}{12} ; \qquad\qquad R_1 + R_2 = \frac{20}{36}$$

On solving these equations, we get

$$R_1 = \frac{5}{18}, R_2 = \frac{4}{9}, R_3 = \frac{5}{18}$$

From the values of u_1, u_2 and u_3 it is observed that u's are symmetrically placed with respect to the middle point of the interval $[-1/2, 1/2]$. Also we observe that R's are symmetrically placed with respect to each pair of u's. In the above case u_0 is taken to be u_2 and u_1 and u_3 are taken to be u_{-1} and u_{+1} respectively. Thus u_1, u_2, u_3 can be taken as u_{-1}, u_0, u_{+1}. Similarly $u_{\pm 2}$ are taken next pair of symmetric points.

The numerical values of the u's and corresponding R's for $n = 2$ to $n = 10$ are given in the following table, where $u_{\pm k} = N$ means $u_k = N, u_{-k} = -N$.

n	u	R
$n = 2$	$u_1 = \pm 0.2886751346$	$R = 1/2$
$n = 3$	$u_0 = 0$	$R = 4/9$
	$u_1 = \pm 0.3872983346$	$R = 5/18$
$n = 4$	$u_1 = \pm 0.1699905218$	$R = 0.3260725744$
	$u_2 = \pm 0.4305681558$	$R = 0.1739274226$
$n = 5$	$u_0 = 0$	$R = 64/225$
	$u_1 = \pm 0.2692346551$	$R = 0.2393143352$
	$u_2 = \pm 0.4530899230$	$R = 0.1184634425$
$n = 6$	$u_1 = \pm 0.1193095930$	$R = 0.2339569673$
	$u_2 = \pm 0.3306046932$	$R = 0.1803807865$
	$u_3 = \pm 0.4662347571$	$R = 0.0856622419$
$n = 7$	$u_0 = 0$	$R = 0.2089795918$
	$u_1 = \pm 0.2029225757$	$R = 0.1909150253$
	$u_2 = \pm 0.3707655928$	$R = 0.1398526957$
	$u_3 = \pm 0.4745539562$	$R = 0.06474248308$
$n = 8$	$u_1 = \pm 0.0917173212$	$R = 0.1813418917$
	$u_2 = \pm 0.2627662050$	$R = 0.1568533229$
	$u_3 = \pm 0.39883332387$	$R = 0.1111905172$
	$u_4 = \pm 0.4801449282$	$R = 0.05061426815$
$n = 9$	$u_0 = 0$	$R = 0.1651196775$
	$u_1 = \pm 0.1621267117$	$R = 0.1561735385$
	$u_2 = \pm 0.3066857164$	$R = 0.1303053482$
	$u_3 = \pm 0.4180155537$	$R = 0.09032408035$
	$u_4 = \pm 0.4840801198$	$R = 0.4063719418$
$n = 10$	$u_1 = \pm 0.0744371695$	$R = 01477621124$
	$u_2 = \pm 0.2166976971$	$R = 0.1346333597$
	$u_3 = \pm 0.3397047841$	$R = 0.1095431813$
	$u_4 = \pm 0.4325316833$	$R = 0.07472567458$
	$u_5 = \pm 0.4869532643$	$R = 0.3333567215$

REMARKS

- In Simpson's and Weddle's formulae the ordinates are equally spaced while in Gauss formula ordinates are not equally spaced.
- In this formula we shall subdivide the interval (a, b) by means of points which shall not equidistant but shall be symmetrically placed with respect to the middle points of the interval.

Solved Examples

(A) BASED ON TRAPEZOIDAL RULE

EXAMPLE 1. *Evaluate* $\int_0^6 \dfrac{dx}{1+x^2}$. (ROHTAK–2004, 06)

SOLUTION. Divide the integral $[0, 6]$ into six parts each of width $h = 1$. The values of $f(x) = \dfrac{1}{1+x^2}$ are given below :

x	0	1	2	3	4	5	6
$y = f(x)$	1	0.5	0.2	0.1	0.0588	0.0385	0.027

Here $y_0 = 1, y_1 = 0.5, y_2 = 0.2, y_3 = 0.1, y_4 = 0.0588,$ $y_5 = 0.0385$ and $y_6 = 0.027$.

By Trapezoidal Rule

$$\int_0^6 \frac{dx}{1+x^2} = \frac{h}{2}[y_0 + 2(y_1 + y_2 + y_3 + y_4 + y_5) + y_6]$$

We have

$$\int_0^6 \frac{dx}{1+x^2} = \frac{1}{2}[1 + 2(0.5 + 0.2 + 0.1 + 0.0588$$
$$+ 0.0385) + 0.027]$$
$$= \frac{1}{2}[2.8216] = 1.4108$$

EXAMPLE 2. *Calculate the value of the integral* $\int_4^{5.2} \log x\, dx$.

(JNTU–2006, KERALA–20003, VTU–2008)

SOLUTION. Taking $h = 0.2$ and divide $[4, 5.2]$ into six equal parts. Then the values of $\log x$ for each points of subdivision are given below :

x	4	4.2	4.4	4.6	4.8	5.0	5.2
$y = \log x$	1.38629	1.43508	1.48160	1.52605	1.56861	1.60943	1.64865

Here $y_0 = 1.38629, y_1 = 1.43508, y_2 = 1.48160,$
  $y_3 = 1.52605, y_4 = 1.56861, y_5 = 1.60943$
and $y_6 = 1.64865.$
Then by trapezoidal rule

$$\int_{4.0}^{5.2} \log x \, dx = \frac{h}{2}[y_0 + 2(y_1 + y_2 + y_3 + y_4 + y_5) + y_6]$$

$$= \frac{0.2}{2}[1.38629 + 2(1.43508 + 1.48160$$
$$+ 1.52605 + 1.56861 + 1.60943) + 1.64865]$$
$$= 0.1[3.03494 + 2(7.62077)]$$
$$= 0.1[18.27648] = 1.827648$$

EXAMPLE 3. *Evaluate the integral $\int_{0.2}^{1.4}(\sin x - \log_e x + e^x) dx$.*

(KURUKSHATRA–2005, 07, MUMBAI–2005)

SOLUTION. Divide the interval [0.2, 1.41] into 12 equal parts each of width $h = 0.1$. Then $x_0 = 0.2$, $x_1 = 0.3$, $x_2 = 0.4$, $x_3 = 0.5$, $x_4 = 0.6$, $x_5 = 0.7$, $x_6 = 0.8$, $x_7 = 0.9$, $x_8 = 1.0, x_9 = 1.1, x_{10} = 1.2, x_{11} = 1.3, x_{12} = 1.4$. Then the values of $y = f(x) = \sin x - \log_e x + e^x$ are given below :

x	$\sin x$	$+ \log_e x$	e^x	$y = f(x) = \sin x - \log_e x + e^x$
0.2	0.19867	–1.60943	1.22140	$3.02950 = y_0$
0.3	0.29552	–1.20347	1.34986	$2.84935 = y_1$
0.4	0.38942	–0.91629	1.49182	$2.79753 = y_2$
0.5	0.47943	–0.69315	1.64872	$2.82130 = y_3$
0.6	0.56464	–0.51083	1.82212	$2.89759 = y_4$
0.7	0.64422	–0.35667	2.01375	$3.01464 = y_5$
0.8	0.71736	–22314	25.22554	$3.16604 = y_6$
0.9	0.78333	–0.10536	2.45960	$3.34829 = y_7$
1.0	0.84147	0.0000	2.71828	$3.55975 = y_8$
1.1	0.89121	0.09531	3.00417	$3.80007 = y_9$
1.2	0.93204	0.18252	3.32012	$4.06984 = y_{10}$
1.3	0.96356	0.26236	3.66930	$4.37050 = y_{11}$
1.4	0.98545	0.33647	4.05520	$4.70418 = y_{12}$

Then by Trapezoidal Rule

$$\int_{0.2}^{1.4}(\sin x - \log_e x + e^x)dx = \frac{h}{2}[y_0 + 2\{y_1 + y_2$$
$$+ y_3 + y_4 + y_5 + y_6 + y_7 + y_8$$
$$+ y_9 + y_{10} + y_{11}) + y_{12}\}]$$
$$= \frac{0.1}{2}[3.02950 + 2(2.84935 + 2.79753$$
$$+ 2.82130 + 2.89759$$
$$+ 3.01464 + 3.16604 + 3.34829$$
$$+ 3.55975 + 3.80007 + 4.06984$$
$$+ 4.37050) + 4.70418]$$

(B) BASED ON SIMPSON'S RULE

EXAMPLE 6. *Find the value of the integral $\int_0^1 \frac{dx}{1+x^2}$ by using Simpson's $\frac{1}{3}$ and $\frac{3}{8}$ rule. Hence obtain the approximate value of π in each case.*

(ROHTAK–2006, UPTU(MCA)–2006, 07, JNTU–2008,

$$= \frac{0.1}{2}[7.73368 + 2(36.6949)]$$
$$= 0.05(81.12348) = 4.056174$$

EXAMPLE 4. *From the following table, find the area bounded by the curve and the x-axis from x = 7.47 to x = 7.52.* (UPTU–2004)

x	7.47	7.48	7.49	7.50	7.51	7.52
$f(x)$	1.93	1.95	1.98	2.01	2.03	2.06

SOLUTION. The required area is given by the integral
$A = \int_{7.47}^{7.52} f(x) dx$
Here $h = 0.01$, $y_0 = 1.93$, $y_1 = 1.95$, $y_2 = 1.98$, $y_3 = 2.01, y_4 = 2.03$ and $y_5 = 2.06$. Then by trapezoidal rule

$$A = \int_{7.47}^{7.52} f(x)dx = \frac{h}{2}[y_0 + 2(y_1 + y_2 + y_3 + y_4) + y_5]$$

$$= \frac{0.01}{2}[1.93 + 2(1.95 + 1.98 + 2.01 + 2.03) + 2.06]$$

$$= 0.005[3.99 + 2(7.97)] = 0.005[19.93] = 0.09965.$$

EXAMPLE 5. *Evaluate the integral $\int_0^1 \frac{1}{1+x} dx$.*

(UPTU(B. TECH.)–2005, MCA–2008)

SOLUTION. Let $h = 0.125$ and $y = f(x) = \frac{1}{1+x}$, then the values of y are given for the arguments which are obtained by dividing the interval [0, 1] into eight equal parts as given below :

x	0	0.125	0.250	0.375	0.5	0.625	0.750	0.875	1.0
$y = \frac{1}{1+x}$	1.0	0.8889	0.8000	0.7273	0.6667	0.6154	0.5714	0.5333	0.5

Hence $y_0 = 1.0, y_1 = 0.8889, y_2 = 0.8000, y_3 = 0.7273,$ $y_4 = 0.6667, y_5 = 0.6154, y_6 = 0.5714, y_7 = 0.5333,$ $y_8 = 0.5$.
Now by Trapezoidal Rule

$$\int_0^1 \frac{1}{1+x} dx$$

$$= \frac{h}{2}[y_0 + 2(y_1 + y_2 + y_3 + y_4 + y_5 + y_6 + y_7) + y_8]$$

$$= \frac{0.125}{2}[1.0 + 2(0.8889 + 0.8000 + 0.7273 + 0.6667$$
$$+ 0.6154 + 0.5714 + 0.5333) + 0.5]$$

$$= \frac{0.125}{2}[1.5 + 2(4.803)] = \frac{0.125}{2}[11.106] = 0.69413.$$

UPTU–2010, VTU–2007, BHOPAL–2009)

SOLUTION. Divide the interval [0, 1] into six equal parts each of width $h = \frac{1}{6}$. Then the values of $y = f(x) = \frac{1}{1+x^2}$ at each points of subdivisions are given follow :

x	$y = \dfrac{1}{1+x^2}$
0	1.0000
1/6	0.9729
2/6	0.9000
3/6	0.8000
4/6	0.6923
5/6	0.5901
1	0.5000

Here, $y_0 = 1.000, y_1 = 0.9729, y_2 = 0.9000, y_3 = 0.8000$, $y_4 = 0.6923, y_5 = 0.5901$, and $y_6 = 0.5000$.

By Simpson's $\dfrac{1}{3}$ Rule

$\int_0^1 \dfrac{dx}{1+x^2}$

$= \dfrac{h}{3}[y_0 + 4(y_1 + y_3 + y_5) + 2(y_2 + y_4) + y_6)]$

$= \dfrac{1}{18}[1.000 + 4(0.9729 + 0.8000 + 0.5901)$
$\qquad + 2(0.9000 + 0.6923) + 0.5000]$

$= \dfrac{1}{18}[1.5 + 4(2.363) + 2(1.5923)]$

$= \dfrac{1}{18}[1.5 + 9.452 + 3.1846]$

$= \dfrac{1}{18}[14.1366] = 0.785397$

By Simpson's $\dfrac{3}{8}$ Rule

$\int_0^1 \dfrac{dx}{1+x^2} = \dfrac{3h}{8}[y_0 + 3(y_1 + y_2 + y_4 + y_5) + 2y_3 + y_6]$.

$= \dfrac{3 \times \dfrac{1}{6}}{8}[1.0000 + 3(0.9729 + 0.9000 + 0.6923$
$\qquad\qquad + 0.5901) + 2(0.8000) + 0.5000]$

$= \dfrac{1}{16}[1.5 + 3(3.1553) + 1.6]$

$= \dfrac{1}{16}(12.5659) = 0.785395$

But $\int_0^1 \dfrac{dx}{1+x^2} = [\tan^{-1} x]_0^1 = \tan^{-1} 1 - \tan^{-1} 0 = \dfrac{\pi}{4}$

In case of Simpson's $\dfrac{1}{3}$ Rule

$\dfrac{\pi}{4} = 0.785397$

or $\pi = 4(0.785397) = 3.141588$

In case of Simpson's $\dfrac{3}{8}$ Rule

$\dfrac{\pi}{4} = 0.785369$

or $\pi = 4(0.785395) = 3.141476$

EXAMPLE 7. *Evaluate* $\int_0^6 \dfrac{dx}{1+x^2}$ *by Simpson's* $\dfrac{1}{3}$ *Rule.*

(UPTU–2002, MDU(B.E.)–2004, 06, KURUKSHETRA–2004, 07, MUMBAI–2005)

SOLUTION. Divide the interval [0, 6] into 6 equal parts so that width of each subinterval $h = \dfrac{6-0}{6} = 1$. The values of $y = \dfrac{1}{1+x^2}$ at each points of subdivision are given in the table :

x	0	1	2	3	4	5	6
y	1.00000	0.50000	0.20000	0.10000	0.58824	0.03846	0.02702

Here, $y_0 = 1.00000$, $y_1 = 0.50000$, $y_2 = 0.20000$, $y_3 = 0.10000$, $y_4 = 0.58824$, $y_5 = 0.03846$ and $y_6 = 0.02702$.

Then by Simpson's Rule

$\int_0^6 \dfrac{dx}{1+x^2}$

$= \dfrac{h}{3}[y_0 + 4(y_1 + y_3 + y_5) + 2(y_2 + y_6) + y_8]$

$= \dfrac{1}{3}[1.0000 + 4(0.50000 + 0.10000 + 0.03846)$
$\qquad\qquad + 2(0.20000 + 0.58824) + 0.02702]$

$= \dfrac{1}{3}[1.02702 + 4(0.63846) + 2(0.78824)]$

$= \dfrac{1}{3}[1.02702 + 2.55384 + 1.57648]$

$= \dfrac{1}{3}[5.15734] = 1.71911$.

EXAMPLE 8. *Using Simpson's* $\dfrac{3}{8}$ *rule, evaluate* $\int_0^1 \dfrac{1}{1+x} dx$ *with* $h = \dfrac{1}{6}$. (UPTU–2010, VTU–2007)

SOLUTION. Divide the interval [0, 1] into 6 equal parts of width $h = \dfrac{1}{6}$. The values of $y = \dfrac{1}{1+x}$ for each points of sub division are given below :

x	0	1/6	2/6	3/6	4/6	5/6	1
$y = \dfrac{1}{1+x}$	1	0.85714	0.75	0.66667	0.6	0.54545	0.5

Here, $y_0 = 1, y_1 = 0.85714, y_2 = 0.75, y_3 = 0.66667$, $y_4 = 0.6, y_5 = 0.54545$ and $y_6 = 0.5$.

Now by Simpson's $\dfrac{3}{8}$ rule, we get

$\int_0^1 \dfrac{1}{1+x} dx$

$= \dfrac{3h}{8}[y_0 + 3(y_1 + y_2 + y_4 + y_5) + 2y_3 + y_6]$

$= \dfrac{3 \times \dfrac{1}{6}}{8}[1 + 3(0.85714 + 0.75 + 0.6 + 0.54545)$
$\qquad\qquad + 2(0.66667) + 0.5]$

$$= \frac{1}{16}[1.5 + 3(2.75259) + 1.33334]$$

$$= \frac{1}{16}[11.091111] = 0.69319$$

EXAMPLE 9. *Find* $\int_0^6 \frac{e^x}{1+x}$ *approximately using Simpson's* $\frac{3}{8}$ *rule.* (UPTU–2006)

SOLUTION. Divide the given interval of integration into 6 equal subintervals, the arguments and 0, 1, 2, 3, 4, 5, 6 and $h = 1$.

Now, $f(x) = \frac{e^x}{1+x}$, $y_0 = f(0) = 1$,

$$y_1 = f(1) = \frac{e}{2}, y_2 = f(2) = \frac{e^2}{3},$$

$$y_3 = f(3) = \frac{e^3}{4}, y_4 = f(4) = \frac{e^4}{5},$$

$$y_5 = f(5) = \frac{e^5}{6}, y_6 = f(6) = \frac{e^6}{7}$$

Putting all these values in Simpson's $\frac{3}{8}$ rule, we get

$$\int_0^6 \frac{e^x}{1+x} dx$$

$$= \frac{3h}{8}[(y_0 + y_6) + 3(y_1 + y_2 + y_4 + y_5) + 2y_3]$$

$$= \frac{3}{8}\left[\left(1 + \frac{e^6}{7}\right) + 3\left(\frac{e}{2} + \frac{e^2}{3} + \frac{e^4}{5} + \frac{e^5}{6}\right) + \frac{2e^3}{4}\right]$$

$$= \frac{3}{8}[(1 + 57.6327) + 3(1.3591 + 2.463 + 10.9196 + 24.7355 + 2(5.0214)] = 70.1652.$$

EXAMPLE 10. *A train in moving at the speed of 30 m/sec. Suddenly breaks are applied. The speed of the train per second after t second is given by :*

Time (t)	0	5	10	15	20	25	30	35	40	45
Speed (v)	30	24	19	16	13	11	10	8	7	5

Applying Simpson's $\frac{3}{8}$ *rule to determine the distance moved by the train in 45 seconds.*

SOLUTION. Let s be the distance in meter covered in t seconds, then

$$\frac{ds}{dt} = v \quad \Rightarrow \quad [s]_{t=0}^{t=45} = \int_0^{45} v dt$$

Then, using Simpson's $\frac{3}{8}$ rule, we get

$$\int_0^{45} v dt$$

$$= \frac{3h}{8}[(v_0 + v_9) + 3(v_1 + v_2 + v_4 + v_5 + v_7 + v_8) + 2(v_3 + v_6)]$$

$$= \frac{15}{8}[(30 + 5) + 3(24 + 19 + 13 + 11 + 8 + 7) + 2(16 + 10)] = 624.375 \text{ meters.}$$

(C) BASED ON WEDDLE'S RULE

EXAMPLE 11. *Evaluate the integral* $\int_0^6 \frac{dx}{1+x^2}$. (VTU–2008)

SOLUTION. Divide [0, 6] into six equal sub intervals with each of width $h = \frac{6-0}{6} = 1$ and the values of $y = \frac{1}{1+x^2}$ at each points of the sub division are given below :

x	0	1	2	3	4	5	6
y	1.00000	0.50000	0.20000	0.10000	0.05882	0.03846	0.02702

Here $y_0 = 1.00000$, $y_1 = 0.50000$, $y_2 = 0.20000$, $y_3 = 0.10000$, $y_4 = 0.05882$, $y_5 = 0.03846$ and $y_6 = 0.02702$.

By Weddle's rule, we get

$$\int_0^6 \frac{dx}{1+x^2}$$

$$= \frac{3h}{10}[y_0 + 5(y_1 + y_5) + y_2 + y_4 + 6y_3 + y_6]$$

$$= \frac{3}{10}[1.0000 + 5(0.50000 + 0.03846) + 0.20000 + 0.05882 + 6(0.10000) + 0.27002]$$

$$= \frac{3}{10}[1.28584 + 5(0.53846) + 6(0.10000)]$$

$$= \frac{3}{10}[1.28584 + 2.6923 + 0.6] = \frac{3}{10}(4.57814) = 1.37344$$

EXAMPLE 12. *Evaluate the integral* $\int_4^{5.2} \log x dx$, *using Weddle's rule.* (VTU–2008)

SOLUTION. Divide the interval [4, 5.2] into six equal sub-intervals each of width $h = \frac{5.2-4}{6} = 0.2$ and the values of $y = \log x$ are given below at the arguments.

x	4.0	4.2	4.4	4.6	4.8	5.0	5.2
y	1.3862	1.4350	1.4816	1.5261	1.5686	1.6094	1.6486

Here $y_0 = 1.3862$, $y_1 = 1.4350$, $y_2 = 1.4816$, $y_3 = 1.5261$, $y_4 = 1.5686$, $y_5 = 1.6094$ and $y_6 = 1.6486$.

By Weddle's rule, we get

$$\int_4^{5.2} \log x dx$$

$$= \frac{3h}{10}[y_0 + 5(y_1 + y_5) + y_2 + y_4 + 6y_3 + y_6]$$

$$= \frac{3(0.2)}{10}[1.3862 + 5(1.4350 + 1.6094) + 1.4816 + 1.5686 + 6(1.5261) + 1.6486]$$

$$= \frac{0.6}{10}[1.3862 + 5(3.0444) + 1.4816 + 1.5686 + 6(1.5261) + 1.6486]$$

$$= \frac{0.6}{10}[6.085 + 5(3.0444) + 6(1.5261)]$$

$$= \frac{0.6}{10}[6.085 + 15.222 + 9.1566]$$

$$= \frac{0.6}{10}(30.4636) = 1.8278$$

EXAMPLE 13. *A curve passes through the points given by the following table :*

x	1	1.5	2	2.5	3	3.5	4
y	2	2.4	2.7	2.8	3	2.6	2.1

Find the area bounded by the curve, the x-axis and the lines x = 1, x = 4.

SOLUTION. The area $\int_1^4 y dx$. Now divide the interval [1,4] into 6

equal parts of width $h = \frac{4-1}{6} = \frac{3}{6} = 0.5$. Therefore, form the above table, we have
$y_0 = 2, y_1 = 2.4, y_2 = 2.7, y_3 = 2.8, y_4 = 3, y_5 = 2.6$ and $y_6 = 2.1$.

Now by Weddle's rule, we get
$\int_1^4 y dx$

$$= \frac{3h}{10}[y_0 + 5(y_1 + y_5) + y_2 + y_4 + 6y_3 + y_6]$$

$$= \frac{3(0.5)}{10}[2 + 5(2.4 + 2.6) + 2.7 + 3 + 6(2.8) + 2.1]$$

$$= \frac{1.5}{10}[9.8 + 5(5.0) + 6(2.8)] = \frac{1.5}{10}[9.8 + 2.5 + 16.8]$$

(D) BASED ON NEWTON'S-COTE'S FORMULA

EXAMPLE 14. *Find the value of C_3^4, where $C_k^n = \frac{1}{n}\int_0^n L_k du$ and*

$$L_k = \frac{u(u-1)...(u-k+1)(u-k-1)...(u-n)}{k(k-1)...(+1)(-1)...(k-n)}$$

SOLUTION. Since

$$C_k^n = \frac{1}{n}\int_0^n L_k du$$

$\therefore$

$$C_3^4 = \frac{1}{4}\int_0^4 L_3 du$$

and

$$L_3 = \frac{u(u-1)(u-2)(u-4)}{3(3-1)(3-2)(3-4)}$$

$$= \frac{u(u^3 - 7u^2 + 14u - 8)}{3.2.1.(-1)}$$

$$= -\frac{1}{6}(u^4 - 7u^3 + 14u^2 - 8u)$$

$$\therefore C_3^4 = \frac{1}{4}\int_0^4 -\frac{1}{6}(u^4 - 7u^3 + 14u^2 - 8u)du$$

$$= -\frac{1}{24}\left[\frac{u^5}{5} - \frac{7u^4}{4} + \frac{14}{3}u^3 - 4u^2\right]_0^4$$

$$= -\frac{1}{24}\left[\frac{1024}{5} - \frac{1792}{4} + \frac{896}{3} - 64\right]$$

$$= -\frac{1}{24}\left[\frac{12288 - 26880 + 17920 - 3840}{60}\right]$$

$$= -\frac{1}{24}\left[\frac{30208 - 30720}{60}\right]$$

$$= -\frac{1}{24}\left[-\frac{512}{60}\right] = \frac{512}{24 \times 60} = \frac{64}{180} = \frac{32}{90}$$

$\therefore \quad C_3^4 = \frac{32}{90}$

(E) BASED ON GAUSS QUADRATURE FORMULA

EXAMPLE 16. *Evaluate the integral $I = \int_5^{12} \frac{dx}{x}$.* (ROHTAK–2011,12)

SOLUTION. Here taking

$$x = (b-a)u + \frac{1}{2}(a+b)$$

$$= (12-5)u + \frac{1}{2}(12+5)$$

$$x = 7u + 8.5$$

EXAMPLE 15. *Prove that for n = 3; $L_0 + L_1 + L_2 + L_3 = 1$ where*

$$L_k = \frac{u(u-1)...(u-k+1)(u-k-1)...(u-n)}{k(k-1)...(+1)(-1)...(k-n)}$$

SOLUTION. Find L_0, L_1, L_2 and L_3

$\therefore \quad L_0 = \frac{(u-1)(u-2)(u-3)}{(0-1)(0-2)(0-3)}$

$$= -\frac{1}{6}(u^3 - 6u^2 + 11u - 6)$$

$$L_1 = \frac{u(u-2)(u-3)}{1(1-2)(1-3)} = \frac{1}{2}(u^3 - 5u^2 + 6u)$$

$$L_2 = \frac{u(u-1)(u-3)}{2(2-1)(2-3)} = -\frac{1}{2}(u^3 - 4u^2 + 3u)$$

and $L_3 = \frac{u(u-1)(u-2)}{3(3-1)(3-2)} = \frac{1}{6}(u^3 - 3u^2 + 2u)$

Now adding all L_i's we get

$$L_0 + L_1 + L_2 + L_3 = -\frac{1}{6}(u^3 - 6u^2 + 11u - 6)$$

$$+ \frac{1}{2}(u^3 - 5u^2 + 6u) - \frac{1}{2}(u^3 - 4u^2 + 3u)$$

$$+ \frac{1}{6}(u^3 - 3u^2 + 2u)$$

$$= \frac{1}{6}[-u^3 + 6u^2 - 11u + 6 + 3u^3 - 15u^2 + 18u$$

$$-3u^3 + 12u^2 - 9u + u^3 - 3u^2 + 2u]$$

$$= \frac{1}{6}[-4u^3 + 4u^3 + 18u^2 - 18u^2 - 20u + 20u + 6]$$

$$= \frac{1}{6}(6) = 1$$

Hence, $L_0 + L_1 + L_2 + L_3 = 1$.

Since $\quad y = \frac{1}{x} = \frac{1}{7u + 8.5} = \phi(u)$

Now taking $n = 5$, we have

$\therefore \quad y_0 = \phi(u_0) = \frac{1}{8.5} = 0.117647058 \quad (\because u_0 = 0)$

$$y_1 = \phi(u_1) = \frac{1}{7u_1 + 8.5} = \frac{1}{7(0.2692346551) + 8.5}$$

$$(\because u_1 = 0.2692346551)$$

$$= \frac{1}{10.3846426} = 0.0962960439$$

$$y_{-1} = \phi(u_{-1}) = \frac{1}{7u_{-1} + 8.5} = \frac{1}{6.61535741}$$
$$(\because u_{-1} = -0.2692346551)$$
$$= 0.1511634112$$

$$y_2 = \phi(u_2) = \frac{1}{7u_2 + 8.5} = \frac{1}{11.67162946}$$
$$(\because y_2 = 0.4530899230)$$
$$= 0.0856778399$$

$$y_{-2} = \phi(u_{-2}) = \frac{1}{7u_{-2} + 8.5} = \frac{1}{5.32837054}$$
$$(\because y_{-2} = -0.4530899230)$$
$$= 0.187674636$$

For $n = 5$, we have taken the values of $u_0, u_1, u_{-1}, u_2, u_{-2}$ and $R_0, R_1 = R_{-1}$ and $R_2 = R_{-2}$ from table given in previous section.

$$\therefore \quad R_0 = \frac{64}{225}, R_1 = 0.2393143352,$$
$$R_2 = 0.1184634425$$

By Gauss quadrature formula, we get

$$I = \int_5^{12} \frac{dx}{x} = (b-a)[R_0\phi(u_0) + R_1\phi(u_1)$$
$$+ R_{-1}\phi(u_{-1}) + R_2\phi(u_2) + R_{-2}\phi_2(u_{-2})]$$

$$= (12-5)\left[\frac{64}{225}(0.117647058)\right.$$
$$+ 0.2393143352(0.0962960439$$
$$+ 0.151163412) + 0.1184634425$$
$$\left.(0.0856778399 + 0.187674636)\right]$$

$$= 7[0.033464052 + 0.059220595$$
$$+ 0.032382275]$$

$$= 7(0.125066922) = 0.875468454$$

$$\therefore \quad \int_5^{12} \frac{dx}{x} = 0.875468454.$$

EXAMPLE 17. *Find the value of the integral $\int_0^1 x \, dx$.*

SOLUTION. Here taking

$$x = (b-a)u + \frac{1}{2}(a+b)$$

or $\quad x = (1-0)u + \frac{1}{2}(0+1)$

$\Rightarrow \quad x = u + \frac{1}{2}$

Since $\quad y = x = u + \frac{1}{2} = \phi(u)$ ...(1)

Now taking $n = 4$ so that the values of u_1, u_{-1}, u_2, u_{-2} and $R_1 = R_{-1}$ and $R_2 = R_{-2}$ are taken from table given in previous section.

From (1),

$$\therefore \quad \phi(u_{-1}) = u_{-1} + 0.5 = -0.1699905218 + 0.5$$
$$(\because u_{-1} = -0.1699905218)$$
$$= 0.330009479$$

$$\phi(u_1) = u_1 + 0.5 = 0.1699905218 + 0.5$$
$$(\because u_1 = 0.1699905218)$$
$$= 0.669990521$$

$$\phi(u_{-2}) = u_{-2} + 0.5 = -0.4305681558 + 0.5$$
$$(\because u_{-2} = -0.4305681558)$$
$$= 0.069431845$$

$$\phi(u_2) = u_2 + 0.5 = 0.4305681558 + 0.5$$
$$(\because u_2 = 0.4305681558)$$
$$= 0.930568155$$

and values of R's are

$$R_1 = R_{-1} = 0.3260725774$$

and $\quad R_2 = R_{-2} = 0.1739274226$

Now by Gauss quadrature formula, we get

$$\int_0^1 x \, dx$$

$$= (b-a)[R_{-1}\phi(u_{-1}) + R_1\phi(u_1) + R_{-2}\phi(u_{-2}) + R_2\phi(u_2)]$$

$$= (1-0)[0.3260725774(0.330009479 + 0.669990521)$$
$$+ 0.1739274226(0.069431845 + 0.930568155)]$$

$$= 1[0.3260725774(1) + 0.1739274226(1)]$$

$$= 0.499999999.$$

10.35 HIGHER ORDER RULES

10.35.1 BOOLE'S RULE

Putting $n = 4$ in the general quadrature formula and take the differences upto fourth order. Then, we get

$$\int_{x_0}^{x_0+4h} f(x)dx = 4h\left(y_0 + 2\Delta y_0 + \frac{5}{3}\Delta^2 y_0 + \frac{2}{3}\Delta^3 y_0 + \frac{7}{90}\Delta^4 y_0\right) = \frac{2h}{45}(7y_0 + 32y_1 + 12y_2 + 32y_3 + 7y_4)$$

Similarly,

$$\int_{x_0+4h}^{x_0+8h} f(x)dx = \frac{2h}{45}[7y_4 + 32y_5 + 12y_6 + 32y_7 + 7y_8]$$

Adding all such integrals from x_0 to $(x_0 + nh)$, (where n must be a multiple of 4), we get

$$\int_{x_0}^{x_0+nh} f(x)dx = \frac{2h}{45}[7y_0 + 32y_1 + 12y_2 + 32y_3 + 14y_4 + 32y_5 + 12y_6 + 32y_7 + 14y_8 + \ldots$$
$$+ 14y_{n-4} + 32y_{n-3} + 12y_{n-2} + 32y_{n-1} + 7y_n]$$

which is known as Boole's Rule.

10.35.2 ERROR IN BOOLE'S RULE

The error in Boole's Rule is given by $E(h) = -\frac{8h^7}{945} y^{vi}$.

10.35.3 THE EULAR-MACLAURIN'S SUMMATION FORMULA

Let $\Delta F(x) = f(x)$, then we have an operator Δ^{-1} defined by

$$F(x) = \Delta^{-1} f(x)$$

Again we have

$$\Delta F(x) = f(x_0)$$

$$\Rightarrow \qquad F(x_1) - F(x_0) = f(x_0)$$

$$F(x_2) - F(x_1) = f(x_1)$$

....

$$F(x_n) - F(x_{n-1}) = f(x_{n-1})$$

Adding all these, we get

$$F(x_n) - F(x_0) = \sum_{i=0}^{n-1} f(x_i) \qquad \text{...(1)}$$

where $x_0, x_1, ..., x_n$ are the $(n + 1)$ equidistant values of x with interval h.

Now $\quad F(x) = \Delta^{-1} f(x) = (E-1)^{-1} f(x) = (e^{hD} - 1)^{-1} f(x) = \left\{ \left(1 + hD + \dfrac{h^2 D^2}{2!} + \dfrac{h^3 D^3}{3!} + ... \right) - 1 \right\}^{-1} f(x)$

$$= \left(hD + \dfrac{h^2 D^2}{2!} + \dfrac{h^3 D^3}{3!} + ... \right)^{-1} f(x) = (hD)^{-1} \left\{ 1 + \left(\dfrac{hD}{2!} + \dfrac{h^2 D^2}{3!} + ... \right) \right\}^{-1} f(x)$$

$$= \dfrac{1}{h} D^{-1} \left\{ 1 + \left(\dfrac{hD}{2!} + \dfrac{h^2 D^2}{3!} + ... \right) + \dfrac{(-1)(-2)}{2!} \left(\dfrac{hD}{2!} + \dfrac{h^2 D^2}{3!} + ... \right)^2 + ... \right\} f(x)$$

$$= \dfrac{1}{h} D^{-1} \left(1 - \dfrac{hD}{2!} + \dfrac{h^2 D^2}{12} - \dfrac{h^4 D^4}{720} ... \right) f(x) = \dfrac{1}{h} \int f(x) dx - \dfrac{1}{2} f(x) + \dfrac{h}{12} f'(x) - \dfrac{h^3}{720} f'''(x) \qquad \text{...(2)}$$

Between limits $x = x_0$ and $x = x_n$ from eq. (2), we have

$$F(x_n) - F(x_0) = \dfrac{1}{h} \int_{x_0}^{x_n} f(x) dx - \dfrac{1}{2} [f(x_n) - f(x_0)] + \dfrac{h}{12} [f'(x_n) + f'(x_0)] - \dfrac{h^3}{720} [f'''(x_n) - f'''(x_0)] + ... \qquad \text{...(3)}$$

From eqs. (1) and (3), we have

$$\sum_{i=1}^{n-1} f(x_i) = \dfrac{1}{h} \int_0^{x_n} f(x) dx - \dfrac{1}{2} [f(x_n) - f(x_0)] + \dfrac{h}{12} [f'(x_n) - f'(x_0)] - \dfrac{h^3}{720} [f'''(x_n) - f'''(x_0)] + ...$$

But $\sum\limits_{i=0}^{n-1} f(x_i) = \sum\limits_{i=0}^{n} [f(x_i) - f(x_n)]$ and $x_n = x_0 + nh$. Then the above relation reduces to

$$\dfrac{1}{h} \int_{x_0}^{x_0+nh} f(x) dx = \sum_{i=0}^{n} f(x_i) - \dfrac{1}{2} [f(x_0) + f(x_n)] - \dfrac{h}{12} [f'(x_0 + nh) - f'(x)] - \dfrac{h^3}{720} [f'''(x_0 + nh) - f'''(x_0)] - ... \qquad \text{...(4)}$$

$$\Rightarrow \qquad \int_{x_0}^{x_n} f(x) dx = \dfrac{h}{2} [f(x_0) + 2f(x_1) + 2f(x_2) + ... + 2f(x_{n-1}) + f(x_n)]$$

$$- \dfrac{h^2}{12} [f'(x_n) - f'(x_0)] + \dfrac{h^4}{720} [f'''(x_n) - f'''(x_0)] + ...$$

$$\Rightarrow \qquad \int_{x_0}^{x_n} y dx = \dfrac{h}{2} [y_0 + 2y_1 + y_2 + ... + 2y_{n-1} + y_n] - \dfrac{h^2}{12} (y'_n - y'_0) + \dfrac{h^4}{720} (y'''_n - y'''_0) + ...$$

which is known as Euler's Maclaurin's summation formula.

REMARK
- In a simple manner we can deduce the Simpson's rule.

Solved Examples

EXAMPLE 1. *Use Boole's formula to compute* $\int_0^{\pi/2} \sqrt{\sin x} \, dx$.

(UPTU–2008)

SOLUTION. Here $f(x) = \sqrt{\sin x}$

$$\Rightarrow \qquad f_0 = 0$$

$$f_1 = f\left(\dfrac{\pi}{8} \right) = 0.61861$$

$$f_2 = f\left(\dfrac{\pi}{4} \right) = 0.84090$$

$$f_3 = f\left(\dfrac{3\pi}{8} \right) = 0.96119$$

$$f_4 = f\left(\dfrac{\pi}{2} \right) = 1.0$$

$$\therefore \quad \int_{x_0}^{x_0+4h} f(x)dx$$

$$= \frac{\pi}{180}[0 + 32(0.61861 + 0.96119)$$

$$+ 12(0.84090) + 7(1.0)]$$

$$= 1.18062.$$

EXAMPLE 2. *A river is 80 m wide. The depth y of the river at a distance x from bank to given by the following table :*

x	0	10	20	30	40	50	60	70	80
y	0	4	7	9	12	15	14	8	3

Using Boole's rule find the approximate area of cross section of the river.

(ROHTAK–2005, UPTU(MCA)–2002)

SOLUTION. It is known that, the required area of the cross-section of the river

$$= \int_0^{80} ydx$$

Then by Boole's rule, we have

$$\int_0^{80} ydx = \frac{2h}{45}[7y_0 + 32y_1 + 12y_2 + 32y_3 + 7y_4$$

$$+ 7y_4 + 32y_5 + 12y_6 + 32y_7 + 7y_8]$$

$$= \frac{2 \times 10}{45}[7 \times 0 + 32 \times 4 + 12 \times 7 + 32 \times 9$$

$$+ 7 \times 12 + 7 \times 12 + 32 \times 15 + 12 \times 14 + 32 \times 8 + 7 \times 3]$$

$$= 708.$$

EXAMPLE 3. *Use Euler-Maclaurin's formula to prove that* $\sum_1^n x^2 = \dfrac{n(n+1)(2n+1)}{6}.$

SOLUTION. Putting $f(x) = x^2, x_0 = 0, h = 1, x_0 + nh = n, x_i = x_0 + ih = i$ in the Euler-Maclaurin's summation formula, we get

$$\int_0^n x^2 dx = \sum_{i=0}^n i^2 - \frac{1}{2}(n^2 + 0) - \frac{1}{12}(2n - 0)$$

$$\Rightarrow \quad \frac{n^3}{3} = \sum_{i=0}^n i^2 - \frac{1}{2}n^2 - \frac{n}{6}$$

$$\Rightarrow \quad \sum_{i=0}^n i^2 = \frac{n^3}{3} + \frac{1}{2}n^2 + \frac{n}{6} = \frac{2n^3 + 3n^2 + n}{6}$$

$$= \frac{n(2n^2 + 3n + 1)}{6} = \frac{n(n+1)(2n+1)}{6}$$

$$\Rightarrow \quad \sum_{i=0}^n x^2 = \frac{n(n+1)(2n+1)}{6}.$$

EXAMPLE 4. *Evaluate* $\int_0^1 \dfrac{dx}{1+x}$ *to five places of decimal, using Euler-Maclaurin's formula.*

SOLUTION. Let $\quad y = \dfrac{1}{1+x}$

Here, we have $x_0 = 0, n = 10$ and $h = 0.1$

Then we want to evaluate $\int_{x_0}^{x_1} ydx$

where, $y' = -\dfrac{1}{(1+x)^2}$ and $y''' = -\dfrac{6}{(1+x)^4}$

Using Euler-Maclaurin's formula, we get

$$\int_0^1 \frac{dx}{1+x} = \frac{h}{2}(y_0 + 2y_1 + 2y_2 + \dots + y_n)$$

$$- \frac{h^2}{12}[f'(x_{10}) - f'(x_0)] + \frac{h^2}{720}[f''(x_{10}) - f'''(x_0)]$$

$$= \frac{0.1}{2}\left[\frac{1}{1} + \frac{2}{1.01} + \frac{2}{1.02} + \frac{2}{1.03} + \frac{2}{1.04}\right.$$

$$\left. + \frac{2}{1.05} + \frac{2}{1.06} + \frac{2}{1.07} + \frac{2}{1.08} + \frac{2}{1.09} + \frac{1}{2}\right]$$

$$- \frac{(0.1)^2}{1^2}\left[-\frac{1}{2^2} + \frac{1}{1^2}\right] + \frac{(0.1)^4}{720}\left[-\frac{6}{2^4} + \frac{6}{1^4}\right]$$

$$= 0.882661 - 0.000625 + 0.000001$$

$$= 0.882037.$$

EXAMPLE 5. *Find the sum of the series using Euler-Maclaurin's formula*

$$\frac{1}{51^2} + \frac{1}{53^2} + \frac{1}{55^2} + \dots + \frac{1}{99^2}.$$

SOLUTION. Here, we have $y = \dfrac{1}{x^2}, x_0 = 51, n = 24, h = 2$

Then $y' = -\dfrac{2}{x^3}, y''' = -\dfrac{24}{x^5}$ .... and so on.

Using Euler-Maclaurin's formula, we get

$$\int_{51}^{99} \frac{1}{x^2} dx = \frac{h}{2}[y_0 + 2y_1 + 2y_2 + \dots + 2y_{23} + y_{24}]$$

$$- \frac{h^2}{12}[y_{24}' - y_0'] + \frac{h^4}{720}[y_{24}''' - y_0'''] + \dots$$

$$= \left[\frac{1}{51^2} + \frac{2}{53^2} + \frac{2}{55^2} + \dots + \frac{2}{97^2} + \frac{1}{99^2}\right]$$

$$- \frac{4}{12}\left[-\frac{2}{99^3} + \frac{2}{51^3}\right] + \frac{16}{720}\left[-\frac{24}{99^5} + \frac{24}{51^5}\right] + \dots$$

which gives

$$\frac{1}{51^2} + \frac{2}{53^2} + \frac{2}{55^2} + \dots + \frac{2}{97^2} + \frac{1}{99^2}$$

$$= \int_{51}^{99} \frac{1}{x^2} dx + \frac{2}{3}\left[\frac{1}{51^3} - \frac{1}{99^3}\right] - \frac{8}{15}\left[\frac{1}{51^5} - \frac{1}{99^5}\right] + \dots$$

$$\Rightarrow 2\left[\frac{1}{51^2} + \frac{1}{53^2} + \frac{1}{55^2} + \dots + \frac{1}{99^2}\right]$$

$$= \int_{51}^{99} \frac{1}{x^2} dx + \left(\frac{1}{51^2} + \frac{1}{99^2}\right) + \frac{2}{3}$$

$$\left(\frac{1}{51^3} - \frac{1}{99^3}\right) - \frac{8}{15}\left(\frac{1}{51^5} - \frac{1}{99^5}\right) + \dots$$

$$\Rightarrow \frac{1}{51^2} + \frac{1}{53^2} + \frac{1}{55^2} + \dots + \frac{1}{99^2}$$

$$= \frac{1}{2}\left[-\frac{1}{x}\right]_{51}^{99} + \frac{1}{2}\left(\frac{1}{51^2} + \frac{1}{99^2}\right)$$

$$+ \frac{1}{3}\left(\frac{1}{51^3} - \frac{1}{99^3}\right) - \frac{4}{15}\left(\frac{1}{51^5} - \frac{1}{99^5}\right) + \dots$$

$$= 0.00475 + 0.00024 + 0.000002 + \dots = 0.00499.$$

EXAMPLE 6. *Show that* $\sum_{k=1}^n k^7 + \sum_{k=1}^n k^5 = 2\left(\sum_{k=1}^n k^3\right)^2.$

SOLUTION. We know that the Euler-Maclaurin's formula is given by

$$\frac{1}{h}\int_{x_0}^{x_n} f(x)dx = \sum_{x=x_0}^{x_{n-1}} f(x) + \frac{1}{2}(f_n - f_0)$$

$$- \frac{h}{12}[f'(x_n) - f'(x_0)]$$

$$+ \frac{h^3}{720}[f'''(x_n) - f'''(x_n)]$$

$$- \frac{h^5}{30240}[f^v(x_n) - f^v(x_0)] + \dots \quad \dots(1)$$

Put $f(x) = x^7, x_0 = 0$ and $h = 1$ in (1), we get

$$\int_0^n x^7 dx = \sum_{x=0}^{n-1} x^7 + \frac{1}{2}(n^7) - \frac{1}{12}(7n^6)$$

$$+ \frac{1}{720}(210n^4) - \frac{1}{30240}(2520n^2)$$

$$\Rightarrow \sum_{x=0}^{n-1} x^7 = \frac{1}{8}n^8 - \frac{1}{2}n^7 + \frac{7}{12}n^6 - \frac{7}{24}n^4 + \frac{1}{12}n^2$$

$$\Rightarrow \sum_{x=0}^{n} x^7 = \frac{1}{7}n^8 + \frac{1}{2}n^7 + \frac{7}{12}n^6 - \frac{7}{24}n^4 + \frac{1}{12}n^2 \quad \dots(2)$$

[On adding x^7 of both sides]

In a similar way, we may find by substituting $f(x) = x^5, x_0 = 0$ and $h = 1$, we get

$$\sum_{x=1}^{n} x^7 + \sum_{x=1}^{n} x^5 = \sum_{x=0}^{n} x^7 + \sum_{x=0}^{n} x^5$$

$$= \frac{1}{8}n^8 + \frac{1}{2}n^7 + \left(\frac{7}{12} + \frac{1}{6}n^6\right) + \frac{1}{2}n^5 - \left(\frac{7}{24} - \frac{5}{12}\right)n^4$$

$$= \frac{1}{8}n^8 + \frac{1}{2}n^7 + \frac{3}{4}n^6 + \frac{1}{2}n^5 + \frac{1}{8}n^4$$

$$= \frac{1}{8}(n^8 + 4n^7 + 6n^6 + 4n^5 + n^4)$$

$$= \frac{n^4}{8}(n^4 + 4n^3 + 6n^2 + 4n + 1) = \frac{n^4}{8} \times (n+1)^4$$

$$= \frac{2[n^2(n+1)^2]^2}{4} = 2\left(\sum_{x=1}^{n} x^3\right)^2$$

Exercise-10.13

1. Use trapezoidal rule to evaluate $\int_0^1 x^3 dx$ considering five sub intervals.

2. Calculate the approximate value of $\int_{-3}^{3} x^4 dx$ by trapezoidal rule.

3. Use Simpson's rule dividing the range into ten equal parts, to show that $\int_0^1 \frac{\log(1+x^3)}{(1+x^2)} dx = 0.1730$.

4. Evaluate $\int_3^5 \frac{4}{2+x^2} dx$ by dividing the range into eight equal parts.

5. Use Simpson's rule to find the value of $\int_1^5 f(x)dx$ given

x	1	2	3	4	5
f(x)	10	50	70	80	100

6. State and prove Simpson's $\frac{1}{3}$ rule. What is the effect of (i) change of origin and (ii) change of scale on this rule?

7. Calculate $\int_0^{\pi/2} e^{\sin x} dx$ correct to four decimal places by Simpson's $\frac{3}{8}$ rule, dividing the range of integration $(0, \pi/2)$ into 3 equal parts. (UPTU–2007)

8. Find by Weddle's rule the value of the integral $I\int_{0.4}^{1.5} \frac{x}{\sinh x} dx$ by taking 12 subintervals.

9. (i) The velocities of a car running on a straight road at interval of 2 minutes are given below :

Time (in min.)	0	2	4	6	8	10	12
Velocities (in km/hr)	0	22	33	27	18	7	0

Apply Simpson's rule to find the distance covered by car.

(ii) A rocket is launched from the ground. Its acceleration is registered during the first 80 seconds and is given in the table below. Find the velocity of the rocket at $t = 80$ seconds by Simpson's $\frac{1}{3}$ rule.

t (sec)	0	10	20	30	40	50	60	70	80
f (cm/sec²)	30	31.63	33.34	35.47	37.75	40.33	43.25	46.69	50.67

10. Integrate numerically $\int_0^{\pi/2} \sqrt{\cos\theta}\, d\theta$. (ROHTAK–2004)

11. Prove that $\sum_{k=0}^{n} L_k = 1$

where $L_k = \dfrac{u(u-1)\dots(u-k+1)(u-k-1)\dots(u-n)}{k(k-1)\dots(+1)(-1)\dots(k-n)}$

12. For $n = 1$, prove that $\dfrac{1}{L_0} + \dfrac{1}{L_1} = \dfrac{1}{L_0 L_1}$.

13. Prove $C_k^n = C_{n-k}^n$ where $C_k^n = \dfrac{1}{n}\int_0^n L_k\, du$.

14. Prove that $C_1^1 . C_1^2 = \dfrac{C_0^2}{C_0^1}$ where $C_k^n = \dfrac{1}{n}\int_0^n L_k\, du$

15. Compute the integral $\int_0^1 \dfrac{dx}{1+x^2}$ by Gauss quadrature formula, taking $n = 5$.

16. Using Gauss quadrature formula evaluate the integral $I = \int_1^2 \dfrac{dx}{x}$. Taking $n = 5$.

17. Find the value of the integral $\int_1^0 \dfrac{dx}{1+x^2}$ by Chebychev's formula, taking $n = 4$.

18. Evaluate the integral $\int_1^7 \dfrac{dx}{x}$ by Chebychev's formula taking $n = 5$.

19. Evaluate $\int_0^1 \dfrac{dx}{1+x}$ correct to five decimal places by using Euler-Maclaurin's formula.

20. Find the sum of the fifth powers of the first n natural numbers by means of Euler-Maclaurin's formula.

21. Use Boole's Rule, find the value of the integral $I = \int_{0.4}^{1.6} \dfrac{x}{\sinh x} dx$ by taking 12 sub intervals.

22. Using Euler-Maclaurin summation formula, sum the following series :

   (i) $\dfrac{1}{400} + \dfrac{1}{402} + ... + \dfrac{1}{498} + \dfrac{1}{505}$

   (ii) $\dfrac{1}{100} + \dfrac{1}{101} + \dfrac{1}{102} + \dfrac{1}{103} + \dfrac{1}{104}$

23. Show that $\sum_{1}^{n} x^3 = \left(\dfrac{n(n+1)}{2}\right)^2$, applying Euler-Maclaurin's formula.

24. Show that $\sum_{0}^{n} i^4 = \dfrac{n^5}{5} + \dfrac{n^4}{2} + \dfrac{n^3}{3} + \dfrac{n}{30}$, by applying Euler-Maclaurin's summation formula.

25. Evaluate the integral $\int_{0}^{1} \dfrac{x^2}{1+x^3} dx$, using Simpson's rule and hence find the value of $\log_e 3$.

26. The velocity of v of a particle at distance s from a point on its path is given by the following table :

s	0	10	20	30	40	50	60
v	47	58	64	65	61	52	38

Estimate the time taken to travel 60 meters by using Simpson's rule. (UPTU–2007, MADRAS–2003)

27. Evaluate $\int_{0}^{2} \dfrac{dx}{1+x+x^2}$ to three decimal places dividing the range of integration into 8 equal parts.

28. If third differences are constant, prove that

$$\int_{-1}^{1} f(x)dx = \dfrac{2}{3}\left[f(0) + f\left(\dfrac{1}{\sqrt 2}\right) + f\left(\dfrac{-1}{\sqrt 2}\right)\right]$$

29. If $f(x) = a + bx + cx^2$, prove that

$$\int_{1}^{2} f(x)dx = \dfrac{1}{12}[f(0) + 22f(2) + f(4)]$$

30. Obtain the approximate quadrature formula

$$\int_{0}^{1} \dfrac{dx}{1+x^2} = \dfrac{1}{4a}\left(1 + \dfrac{1}{6a}\right) + \sum_{x=1}^{a} \dfrac{a}{a^2+x^2}$$

31. Use the trapezoidal rule to estimate the integral $\int_{0}^{2} e^{x^2} dx$ taking 10 intervals. (UPTU–2008)

32. Use Simpson's 1/3rd rule to find $\int_{0}^{0.6} e^{-x^2} dx$ by taking seven ordinates. (VTU–2011, BHOPAL–2009)

33. The velocity v of a particle at distance s from a point on its linear path is given by the following table :

s (m)	0	2.5	5.0	7.5	10.0	12.5	15.0	17.5	20.0
v (m/sec)	16	19	21	22	20	17	13	11	9

Estimate the time taken by the particle to traverse the distance of 20 meters using Boole's rule. (UPTU–2007)

34. Evaluate $\int_{0}^{6} x \sec x\, dx$ using eight intervals by trapezoidal rule. (UPTU–2009)

35. Evaluate using Simpson's 1/3rd rule $\int_{0}^{2} e^{-x^2} dx$ (take $h = 0.25$). (JNTU–2007)

36. Evaluate using Simpson's 1/3rd rule $\int_{0}^{1} \dfrac{dx}{x^3+x+1}$, choose step length 0.25 (UPTU–2009)

37. Evaluate using Simpson's 1/3rd rule, $\int_{0}^{\pi/2} \sqrt{\cos\theta}\, d\theta$ taking 9 ordinates. (VTU–2009)

38. Evaluate correct to 4 decimal places, by Simpson's 3/8th rule.

$$\int_{0}^{9} \dfrac{dx}{1+x^3}$$ (UPTU(M.TECH)–2010)

39. A curve is drawn to pass through the points given by following table :

x	1	1.5	2	2.5	3	3.5	4
y	2	2.4	2.7	2.8	3	2.6	2.1

Using Weddle's rule, estimate the area bounded by the curve, the x-axis and the lines $x = 1$, $x = 4$. (VTU–2011S)

40. The following table gives the velocity v of a particle at time t :

s (sec)	0	2	4	6	8	10	12
v (m/sec)	4	6	16	34	60	94	136

Find the distance moved by the particle in 12 seconds and also the acceleration at $t = 2$ sec. (SVTU–2007)

Answers

1. 0.26	**2.** 115	**4.** 26.716	**5.** 256.66	**7.** 3.1017
8. 1.01019	**9.** (i) $3\dfrac{5}{9}$ km (ii) 30.87 m/sec	**10.** 1.1873		**15.** 0.785398
16. 0.6931	**17.** 0.785396	**18.** 1.945910	**19.** 0.69315	**20.** $\left(\dfrac{n^6}{6} + \dfrac{n^5}{2} + \dfrac{5n^4}{12} - \dfrac{n^2}{12}\right)$
21. 1.010784	**22.** (i) 0.11382114 (ii) 0.0490291	**25.** 0.23108	**31.** 17.0621	
32. 0.5351	**33.** 1.35 sec	**34.** – 6.436	**35.** 0.635	**36.** 0.6305
37. 1.1873	**38.** 1.1249	**39.** 3.032	**40.** 552m; 3m/sec²	

10.36 DIFFERENCE EQUATIONS

An equation that consists of an independent variable x, a dependent variable y_x and one or more of its difference $\Delta y_x, \Delta^2 y_x, ..., \Delta^n y_x$ is called a difference equation. It is of the form

$$F[x, y_x, \Delta y_x, \Delta^2 y_x, ..., \Delta^n y_x] = 0$$

A difference equation is therefore a relation involving differences.

For example :

(1) $\Delta y_x + 2 y_x = 0$ \qquad\qquad (2) $\Delta^2 y_x + 3 \Delta y_x - 7 y_x = 0$

(3) $\Delta^3 y_x + 3 \Delta^2 y_x - 6 \Delta y_x + y_x = 3x + 2$ \qquad (4) $y_x \Delta^3 y_x = 6$

✎ REMARKS

- It is not necessary that the difference equations are defined over the set of all real numbers. They can be considered as difference equations over some other set.
- One cannot tell by looking at equation (i) to (iv) whether they are difference equations over the set of all real number or over some other set. The information must be given.
- When difference equation defined over some set A, then the relation among the values of y_x, Δy_x, $\Delta^2 y_x$, ... given by the equation and the set of values, denoted by A for which this relation is said to hold.
- The set A consists of either a finite or infinite set of successive integers.
- It is often convenient to have this set start with zero but it is not necessary.

10.36.1 DIFFERENCE EQUATION AS A RELATION AMONG THE VALUE OF y_x

Using $\Delta^k = (E - 1)^k$ and noting that $E^h y_x = y_{x+h}$ under the assumption that the interval of differencing is one. We can express

$$\Delta y_x = (E - 1) y_x = E y_x - y_x = y_{x+1} - y_x$$
$$\Delta^2 y_x = (E - 1)^2 y_x = (E^2 - 2E + 1) y_x = y_{x+2} - 2 y_{x+1} + y_x$$
$$\Delta^3 y_x = (E - 1)^3 y_x = (E^3 - 3E^2 + 3E - 1) y_x = y_{x+3} - 3 y_{x+2} + 3 y_{x+1} - y_x$$

The difference equation (iii) can be written as

$$y_{x+3} - 9 y_{x+1} + 9 y_x = 3x + 2$$

This equation can be written as

$$y(x + 3) - 9 y(x + 1) + 9 y(x) = 3x + 2$$
$$\Rightarrow \qquad (E^3 - 9E + 9) y_x = 3x + 2$$

Similarly we can solve other equation (i), (ii) and (iv), we get

$$(E + 1) y_x = 0 \qquad\qquad \Rightarrow \qquad\qquad (E^2 + E - 9) y_x = 0$$

or $\quad (E^3 - 3E^2 + 3E - 1) y_x^2 = 6$

10.36.2 ORDER OF DIFFERENCE EQUATION

The order of a difference equation is the difference between the highest and the lowest subscripts of the y, it is free from Δ. Thus the order of equation (i) is $(x + 1) - x = 1$ equation (ii) is $(x + 2) - x = 2$ equation (iii) is of order $(x + 3) - x = 3$ and the equation (iv) is of order $(x + 4) - x = 4$.

10.36.3 DEGREE OF DIFFERENCE EQUATION

The degree of difference equation is highest power of y and free from Δ. The degree of equation

$$y_{x+1}^2 y_{x+2}^3 - y_{x+1} y_x - y_x^2 = x$$

is 3 and the order of this equation is 2.

10.36.4 SOLUTION OF DIFFERENCE EQUATION

A function y is a solution of a difference equation over a set A if the value of y make the difference equation a true statement for every point of A.

✎ REMARKS

- A general solution of a difference equation of order n involves n arbitrary constants.
- A particular solution of a difference equation is obtained from the general solution by giving particular values to the constants.

For example: $y_x = A 2^x + B 3^x$ is the general solution to $y_{x+2} - 5 y_{x+1} + 6 y_x = 0$ while $y_x = 2^x$ or $y_x = 3^x$ or $y_x = 5(2^x) + 8(3^x)$ are particular solutions.

Solved Examples

EXAMPLE 1. *Form the difference equation corresponding to the family of curves $y = ax^2 + bx - 3$.*

SOLUTION. The given equation is

$$y_x = ax^2 + bx - 3 \qquad \ldots(1)$$

a and b are arbitrary constants to be determined

$$y_{x+1} = a(x+1)^2 + b(x+1) - 3$$
$$y_{x+2} = a(x+2)^2 + b(x+2) - 3$$

We know that

$$\Delta y_x = y_{x+1} - y_x = 2(x+1)a + b \qquad \ldots(2)$$
$$\Delta^2 y_x = y_{x+2} - 2y_{x+1} + y_x = 2a \qquad \ldots(3)$$
$$\Rightarrow \quad a = \frac{1}{2}\Delta^2 y_x$$

$\therefore$ from equation (2)

$$b = \Delta y_x - \frac{1}{2}(2x+1)\Delta^2 y_x \qquad \ldots(4)$$

Eliminating a, b from equation (1), (3) and (4), we get

$$(x+1)x\Delta^2 y_x - 2x\Delta y_x + 2y_x + 6 = 0$$
$$(x^2 + x)y_{x+2} - 2(x^2 + 2x)y_{x+1}$$
$$+ (x^2 + 3x + 2)y_x + 6 = 0.$$

EXAMPLE 2. *Form the difference equation given that $y_n = A3^n + B5^n$, where A and B are arbitrary constants.*

(Kurukshetra (NIT)–2013)

SOLUTION. Given equation is $y_n = A3^n + B5^n \qquad \ldots(1)$
$$y_{n+1} = A3^{n+1} + B5^{n+1} = 3A3^n + 5B5^n \qquad \ldots(2)$$
$$y_{n+2} = A3^{n+2} + B5^{n+2} = 9A3^n + 25B5^n \qquad \ldots(3)$$
Eliminating A and B from equations (1) to (3), we get

$$\begin{vmatrix} y_n & 1 & 1 \\ y_{n+1} & 3 & 5 \\ y_{n+2} & 9 & 25 \end{vmatrix} = 0$$

or $\quad y_{n+2} - 8y_{n+1} + 15y_n = 0$.

which is the required difference equation.

EXAMPLE 3. *Find the order of the difference equation*
$$y_{x+2} - 7y_x = 5$$

SOLUTION. The given equation is $y_{x+2} - 7y_x = 5$

The difference between the highest and lowest subscripts of y is $x + 2 - x = 2$. Hence the order of the equation is 2.

EXAMPLE 4. *Find the order of the following :*

(i) $y_{x+4} - 5y_{x+2} + 6y_x = 0$

(ii) $\Delta^3 yx + 2\Delta y_x + y_x = x + 3$

SOLUTION. (i) The order of equation $y_{x+4} - 5y_{x+2} + 6y_x = 0$
is $x + 4 - x = 4$.
(ii) The given equation
$$\Delta^3 y_x + 2\Delta y_x + y_x = x + 3$$
$$(y_{x+3} - 3y_{x+2} + 3y_{x+1} - y_x) + 2(y_{x+1} - y_x)$$
$$+ y_x = x + 3$$
$$y_{x+3} - 3y_{x+2} + 5y_{x+1} - 2y_x = x + 3$$

Order of this equation is the difference between the highest and lowest subscript of y is given by $(x + 3) - x = 3$.

EXAMPLE 5. *Show that $y_x = \dfrac{x(x-1)}{2}$ is a solution of the difference equation $y_{x+1} - y_x = x$.*

SOLUTION. We have $\quad y_x = \dfrac{x(x-1)}{2}$

Therefore $\quad y_{x+1} = \dfrac{(x+1)x}{2}$

Substituting these values in

$y_{x+1} - y_x = \dfrac{(x+1)x}{2} - \dfrac{x(x-1)}{2} = x$, we get right hand

side, *i.e.*, $y_x = x(x-1)/2$ satisfy the given difference equation. Hence it is a solution of given difference equation.

EXAMPLE 6. *Show that $y_x = 1 - \dfrac{2}{x}, x = 1, 2, 3, \ldots$ is a solution of the first order difference equation*

$(x + 1)y_{x+1} + xy_x = 2x - 3, x = 1, 2, 3, \ldots$

SOLUTION. We have $y_x = 1 - \dfrac{2}{x}$

$\therefore \qquad y_{x+1} = 1 - \dfrac{2}{x+1}$

Substituting these values in LHS of given equation, we get

$$(x+1)y_{x+1} + xy_x$$
$$= (x+1)\left[1 - \frac{2}{x+1}\right] + x\left[1 - \frac{2}{x}\right]$$
$$= x + 1\left[\frac{x+1-2}{x+1}\right] + x\left[\frac{x-2}{x}\right]$$
$$= x - 1 + x - 2 = 2x - 3 = \text{RHS}$$

$\therefore \quad y_x = 1 - \dfrac{2}{x}$ is the solution of the given first order difference equation.

EXAMPLE 7. *Show that $y_x = C_1 + C_2 2^x - x$ is a solution of the difference equation $y_{x+2} - 3y_{x+1} + 2y_x = 1$.*

SOLUTION. We have $\quad y_x = C_1 + C_2 2^x - x$.
$$y_{x+1} = C_1 + C_2 2^{x+1} - (x+1)$$
and $\quad y_{x+2} = C_1 + C_2 2^{x+2} - (x+2)$
Substituting these values in LHS of given equation, we get
$$y_{x+2} - 3y_{x+1} + 2y_x = (C_1 + C_2 2^{x+2} - x - 2)$$
$$- 3(C_1 + C_2 2^{x+1} - x - 1) + 2(C_1 + C_2 2^x - x)$$
$$= (2^2 - 3.2 + 2)C_2 2^x + 1$$
$$= 1 = \text{RHS}$$

i.e., $y_x = C_1 + C_2 2^x - x$ satisfy the given difference equation. Hence, this is the solution of given difference equation.

 **Exercise-10.14**

1. From the difference equations by eliminating arbitrary constant
   (i) $y = C_1x^2 + C_2x + C_3$ (ii) $y = C_13^x + C_28^x$
   (iii) $y = (C_1 + C_2x)2^x$

2. Find the order of the difference equation.
   (i) $\Delta^3 y_x + \Delta^2 y_x + \Delta y_x + y_x = 0$
   (ii) $y_{x+2} + 3y_x = 2$ (iii) $\Delta^3 y_x + 2\Delta y_x + y_x = x$

3. Show that $C_1 + C_2e^x$ is the solution of the difference equation
   $y_{x+2} - 3y_{x+1} + 2y_x = 0, x = 0, 1, 2, ...$

4. Show that the order of the difference equation $\Delta^2 y_x + 3\Delta y_x - 3y_x = x$ is 2.

5. Prove that $y_x = 3^x(A + Bx)$ satisfy $y_{x+2} - 6y_{x+1} + 9y_x = 0$.

6. Show that the difference equation $y_{x+2} - 4y_{x+1} + 4y_x = 0$, $x = 0, 1, 2, ...$ has the solution $y_x = 2^x(C_1 + xC_2)$, $x = 0, 1, 2, ...$ for any constants. Find the solution satisfying the initial conditions $y_0 = 1$ and $y_1 = 6$.

7. Show that $y_x = \dfrac{C}{1 + Cx}$ is the solution of the difference equation $y_{x+1} = \dfrac{y_x}{1 + y_x}$, $x = 0, 1, 2, ...$

8. Form the difference equation by eliminating the arbitrary constants a and b from the relation
   (i) $y_n = a \cos n\theta + b \sin n\theta$
   (ii) $y_n = an^2 + bn$
        (Kurukshetra–2007, 08, Kurukshetra (NIT)–2009)

Answers

1. (i) $y_{x+3} - 3y_{x+2} + 3y_{x+1} - y_x = 0$ (ii) $y_{x+2} - 11y_{x+1} + 24y_x = 0$ (iii) $y_{x+2} + 4y_{x+1} + 4y_x = 0$
2. (i) 2   (ii) 2   (iii) 3   **8.** (i) $y_{n+2} - 2\cos\theta\, y_{n+1} + y_n = 0$ (ii) $n(n+1)\Delta^2 y_n - 2n\Delta y_n + 2y_n = 0$

10.36.5 Linear Difference Equation

This is a most important type of difference equation and it has the general form

$$a_0 y_{x+n} + a_1 y_{x+n-1} + a_2 y_{x+n-2} + ... + a_n y_x = f(x) \qquad ...(1)$$

where $a_0, a_1, a_2, ... a_n$ and $f(x)$ are each functions of x (but not of y_x)

or
$$L(E)y_x = f(x) \qquad ...(2)$$

where $L(E) = a_0E^n + a_1E^{n-1} + a_2E^{n-2} + ... + a_n$ is a polynomial expression in E and is known an non homogeneous linear equation. If $f(x) = 0$ in equation (2) then

$$L(E)y_x = 0 \qquad ...(3)$$

and is known as homogeneous linear equation.

The reader will notice an obvious analogy now with linear differential equations. In fact the general solution of (1) comprises a particular solution of it combined with the general solution of (3). This follows as an immediate consequence of the following theorems which arise from equation (1) and (3). Their proofs are so similar to those already given for corresponding linear difference equations.

  (i) If $y_x = f_1(x)$ is a solution of equation (3), then $y_x = A_1f_1(x)$ is also a solution of equation (3), where A_1 is any constant.

  (ii) If the homogeneous equation (3) is satisfied by equations $y_x = f_1(x), y_x = f_2(x), ..., y_x = f_n(x)$
     then it is also satisfied by
$$y_x = A_1f_1(x) + A_2f_2(x) + ... + A_rF_r(x)$$
     where $A_1, A_2, ..., A_r$ are constants.

  (iii) If $y_x = f_1(x), y_x = f_2(x), y_x = f_3(x), ... y_x = f_n(x)$
     are n independent solutions of equation (3), then its general solution is
$$y_x = A_1f_1(x) + A_2f_2(x) + ... + A_rf_r(x)$$
     where $A_1, A_2, ..., A_r$ are constants.

  (iv) If $y_x = f_1(x), y_x = f_2(x)$ be solutions of the equation
$$L(E)y_x = g(x), L(E)y_x = h(x)$$
     where $f(x) = g(x) + h(x)$, then $y_x = f_1(x) + f_2(x)$ is a solution of equation (2). This is the superposition principle and is valid only for linear equations.

  (v) If $y_x = u_x$ is a particular solution of (2), then with the conditions of (iv), the general solution of (2) is
$$y_x = u_x + A_1f_1(x) + A_2f_2(x) + ... A_rf_r(x)$$

We first study some homogeneous linear difference equations with constant coefficients, then some difference equations which are reducible to this type and finally some non-homogeneous types.

10.36.6 An Existence and Uniqueness Theorem

Some difference equations have infinitely many solutions whereas others have no solution at all. In the case of linear difference equations we can always find at least one solution. Theorems establishing such results are referred to as existence and uniqueness theorems. Before starting and proving the theorem for the linear difference equations of order n, first look a case of order two. The analysis about the second order difference equation will be helpful to understand the general theorem. The equation of second order difference equation is

$$a_0 y_{x+2} + a_1 y_{x+1} + a_2 y_x = f(x) \quad x = 0, 1, 2\ldots \tag{1}$$

with $a_0 \neq 0$, $a_2 \neq 0$, $x = 0, 1, 2, \ldots$

Now, let y_0 and y_1 be given. Then, with $x = 0$ (1) gives

$$a_0 y_2 + a_1 y_1 + a_2 y_0 = f(0)$$

$$a_0 y_2 = f(0) - a_1 y_1 - a_2 y_0$$

Since $a_0 \neq 0$ for any x, $a_0(0)$ also $\neq 0$ and hence

$$y_2 = \frac{f(0)}{a_0(0)} - \frac{a_1(0)}{a_0(0)} y_1 - \frac{a_2(0)}{a_0(0)} y_0$$

Thus with the help of y_1 and y_0, we can find y_2. Now, we can find y_3. For that put $x = 1$ in (1), we get

$$a_0 y_3 = f(1) - a_1 y_2 - a_2 y_1$$

and since $a_0(1) \neq 0$.

$$y_3 = \frac{f(1)}{a_0(1)} - \frac{a_1(1)}{a_0(1)} y_2 - \frac{a_2(1)}{a_0(1)} y_1$$

Continue this way, generating the unique solution of the second order difference equation.

REMARK

- The linear difference equation of order n

$$a_0 y_{x+n} + a_1 y_{x+n-1} + \ldots + a_{n-1} y_{x+1} + a_n y_x = f(x) \tag{1}$$

over a set A of consecutive integral values of x has one and only one solution y for which values at n consecutive x-values are arbitrary prescribed.

10.36.7 Solution of the Equation $y_{x+1} = Ay_x + B$

The linear first order difference equation is of the form

$$a_0(x) y_{x+1} + a_1(x) y_x = f(x), x = 0, 1, 2, \ldots \tag{1}$$

Over the indicated set of x-values. The functions $a_0(x)$ and $a_1(x)$ according to the definition, are never zero, so if they are constant, they are non-zero.

Dividing (1) by $a_0(x)$, we get

$$y_{x+1} = \frac{-a_1(x)}{a_0(x)} y_x + \frac{f(x)}{a_0(x)}$$

If we now suppose a_0 and a_1 as well as f are constant function, we can write

$$y_{x+1} = Ay_x + B, x = 0, 1, 2, \ldots \tag{2}$$

where A and B are constant and $A \neq 0$.

To find the solution of equation (2), put $x = 0$ in (2), then

$$y_1 = Ay_0 + B$$

At $x = 1$,
$$y_2 = Ay_1 = B$$
$$= A(Ay_0 + B) + B = A^2 y_0 + AB + B = A^2 y_0 + (A+1)B$$

Again at $x = 2$, we get

$$y_3 = Ay_2 + B$$
$$= A[A^2 y_0 + (A+1)B] + B = A^3 y_0 + (1 + A + A^2)B$$

$$y_x = A^x y_0 + (1 + A + A^2 + \ldots + A^{x-1})B$$

We know that $1 + A + A^2 + \ldots A^{x-1}$ is geometric progression, hence

$$1 + A + A^2 + \ldots + A^{x-1} = \frac{1 - A^x}{1 - A}, \text{ if } A \neq 1$$
$$= x \qquad \text{if } A = 1$$

We can write

$$y_x = \begin{cases} A^x y_0 + \dfrac{B(1 - A^x)}{1 - A} & \text{if } A \neq 1 \\ y_0 + Bx & \text{if } A = 1 \end{cases} \tag{3}$$

REMARKS

- The function y given by equation (3) is a solution, and the only solution of the difference equation (2) with y_0 prescribed.
- The linear difference equations

$$y_{x+1} = Ay_x + B, x = i, i+1, i+2 \tag{1}$$

taken over the set of x-values has infinitely many solutions. If y is a solution and C is a constant such that

$$y_x = \begin{cases} CA^{x-1} + B\dfrac{1 - A^{x-1}}{1 - A} & \text{if } A \neq 1 \\ C + B(x - i) & \text{if } A = 1 \end{cases}, x = i, i+1, i+2 \tag{2}$$

if a single value of y is prescribed for one of the h values $i, i+1, i+2, \ldots$, then a unique solution of (1) is determined. In particular, if y_i is prescribed then solution of (1) is given by (2) with $B = y_i$.

Solved Examples

EXAMPLE 1. *Solve the difference equation*

$$y_{x+1} = 2y_x + 3, \ x = 1, 2, 3, \dots \text{ and } y_0 = 0.$$

SOLUTION. Comparing the given equation with $y_{x+1} = Ay_x + B$, where $A = 2$ and $B = 3$.

∴ The solution is

$$y_x = A^x y_0 + B\frac{1-A^x}{1-A} = 2^x.0 + 3\frac{1-2^x}{1-2} \quad (\because y_0 = 0)$$

$$= -3(1-2^x) = 3(2^x - 1), x = 0, 1, 2, \dots$$

Hence, we can write the sequence as 3, 9, 21, …

EXAMPLE 2. *Solve the difference equation over the set of x values 0, 1, 2, …*

$$3y_{x+1} = 2y_x + 3, \ y_0 = 2.$$

SOLUTION. Comparing this equation with $y_{x+1} = Ay_x + B$.

We have $A = \dfrac{2}{3}, B = 1$ therefore we have

$$y_x = A^x y_0 + B\frac{1-A^x}{1-A}$$

$$= \left(\frac{2}{3}\right)^x .2 + 1.\frac{1-\left(\frac{2}{3}\right)^x}{1-\frac{2}{3}}$$

$$= 2\left(\frac{2}{3}\right)^x .2 + 1.\frac{1-\left(\frac{2}{3}\right)^x}{1-\frac{2}{3}} = 2\left(\frac{2}{3}\right)^x + 3\left[1-\left(\frac{2}{3}\right)^x\right]$$

$$y_x = 3 - \left(\frac{2}{3}\right)^x, x = 0, 1, 2, \dots$$

It gives the sequence $3, \dfrac{7}{3}, \dfrac{23}{3}, \dots$

EXAMPLE 3. *Solve $(E-a)(E-b)y_x = 0, \ a \neq b$.*

SOLUTION. If we take $(E-b)y_x = y_x$, then equation reduces to

$$y_x = (E-b)yx = Aa^x$$

$$\Delta^{x+1}\left[\frac{y_{x+1}}{b^{x+1}} - \frac{y_x}{b^x}\right] = Aa^x$$

$$b^{x+1}\Delta\left[\frac{y_x}{b^x}\right] = Aa^x$$

$$\Delta\left(\frac{y_x}{b^x}\right) = \frac{A}{b}\left(\frac{a}{b}\right)^x$$

$$\therefore \quad \frac{y_x}{b^x} = B + \frac{A}{b}\Delta^{-1}\left(\frac{a}{b}\right)^x = B + \frac{A}{b} \cdot \frac{\left(\frac{a}{b}\right)^x}{\frac{a}{b} - 1}$$

$$= B + \frac{A}{a-b}\left(\frac{a}{b}\right)^x$$

$$y_x = \frac{A}{a-b}a^x + Bb^x.$$

EXAMPLE 4. *Solve $y_x y_{x+2} = y_{x+1}^2$*

SOLUTION. We have $y_x y_{x+2} = y_{x+1}^2$

$$\Rightarrow \quad \frac{y_{x+2}}{y_{x+1}} = \frac{y_{x+1}}{y_x}$$

$$\Rightarrow \quad \Delta\left(\frac{y_{x+1}}{y_x}\right) = 0 \quad \Rightarrow \quad \frac{y_{x+1}}{y_x} = A$$

or $(E-A)y_x = 0$

$$y_{x+1} = Ay_x$$

Hence, $\quad y_x = BA^x$

 Exercise-10.15

1. Consider the linear difference equation of order 2. $y_{x+2} - xy_{x+1} - y_x = 0, x = 0, 1, 2, \dots$ over the indicated set of x-values. Show that there is no solution of this difference equation for which $y_0 = 0$ and $y_2 = 1$. Show also that if the prescribed values y_0 and y_2 are equal, there are infinitely many different solutions of the difference equation. Why the existence and uniqueness theorem is not valid by these facts.

2. Prove that $a_0(x) y_{x+2} + a_1(x) y_{x+1} + a_2(x) = f(x), x = 0, 1, 2, \dots, y_0 = \alpha, y_1 = \beta$ has one and only one solution where $a_0, a_1, a_2 \neq 0$. Hence, write down all possible solutions of

$$y_{x+2} + \sinh y_{x+1} + (2x+1)y_x = 0, x = 2, 3, 4, \dots$$

$$y_2 = 0, y_3 = 0.$$

3. Show that $y_x = \dfrac{x(x-1)}{2} + B$ is the solution of the difference equation $y_{x+1} - y_x = x$. Find the particular solution satisfying the initial conditions $y_0 = 1$.

4. Show that $y_x = A_1 + B_1(-1)^x$ is the solution of the difference equation $y_{x+2} - y_x = 0$. Find the particular solution satisfying the initial condition $y_0 = 1, y_1 = 2$.

5. Prove that the function y_x given by

$$y_x = \begin{cases} A^x y_0 + B\left(\dfrac{1-A^x}{1-A}\right) & \text{if } A \neq 1 \\ y_0 + B & \text{if } A = 1 \end{cases}, k = 0,1,2\dots$$

is a solution, and only solution of the following difference equation with y_0 prescribed.

6. Solve the following difference equation over the set of x-values 0, 1, 2, …

   (i) $y_{x+1} = 3y_x - 1, y_0 = 6$ (ii) $y_{x+1} = y_x + 2, y_0 = 2$

   (iii) $y_{x+1} = 2y_x + 1, y_0 = 5$

7. The difference equation $xy_{x+1} - y_x = 0, x = 0, 1, 2, \dots$ is linear but not of first order over the indicated set of x-values. Why? If the initial condition $y_0 = 0$ is given, show that y_1 is not uniquely determined and there are infinitely many solutions of the difference equation with $y_0 = 0$. Show also that if the value of y at any x-values different from zero is given, there is a unique solution of the difference equation.

8. Solve $y_{x+1} - 2\cos\alpha y_x + y_{x-1} = 0$ (Kurukshetra–2013)

9. A series of values of y_n satisfy the relation $y_{n+2} + ay_{n+1} + by_n = 0$, given $y_0 = 0, y_1 = 1, y_2 = y_3 = 2$, show that $y_n = 2^{n/2}\sin\dfrac{n\pi}{4}$. (KURUKSHETRA-2012,KURUKSHETRA (NIT)-2006, 08)

 **Answers**

3. $y_x = \dfrac{x(x-1)}{2} + 1$ **4.** $y_x = \dfrac{3}{2} - \dfrac{1}{2}(-1)^x$ **6.** (i) $y_x = \dfrac{11}{2}3^x + \dfrac{1}{2}$ (ii) $y_x = 2(1+x)$ (iii) $y_x = 6.2^x - 1$

8. $y_n = (1)^x[C_1 \cos(x-1)\alpha + C_2 \sin(x-1)\alpha]$

10.36.8 SOLUTION AS SEQUENCES

The linear first order difference equation

$$y_{x+1} = Ay_x + B, x = 0, 1, 2, \dots \qquad \dots(1)$$

The solution of above equation is

$$y_x = \begin{cases} A^x y_0 + B\dfrac{1-A^x}{1-A} & \text{if} \quad A \neq 1 \\ y_0 + Bx & \text{if} \quad A = 1 \end{cases}, x = 0, 1, 2, \dots \qquad \dots(2)$$

Thus, if the value of y_0, A and B are given, then equation (2) will give a sequence $\langle y_x \rangle$. Two cases are here : Case I, $A = 1$ and Case II, $A \neq 1$.

Case I if $A = 1$. Then

If $\langle y_x \rangle$ is the solution sequence of the difference equation $y_{x+1} = y_x + B$, $x = 0, 1, 2, \dots$ with y_0 prescribed then $\langle y_x \rangle$ is a constant sequence, if $B = 0$ it diverges to $+\infty$ if $B > 0$ and diverges to $-\infty$ if $B < 0$.

PROOF.   If $A = 1$, we have $y_x = y_0 + Bx$ [From equation (2)]

Now, if $B = 0$, then $y_x = y_0$ for $x = 0, 1, 2, \dots$

and $\langle y_x \rangle$ is a constant sequence, as given.

If $B > 0$, we will prove that $\langle y_x \rangle$ diverges to $+\infty$. By definition given, any positive number l, we must find a corresponding integer m such that

$$y_x > l \; \forall \, x > m$$

Suppose integer m is given then

$$l \leq y_0$$
$$y_x = y_0 + Bx > l$$

for all $x > 0$ we will take $m = 1$, therefore $y_x > l$ for all $x \geq m = 1$

Again if $l > y_0$, then we want $y_x = y_0 + Bx > l$

Or $Bx > l - y_0$ for $B > 0$

Therefore, we choose $m > \dfrac{l - y_0}{C}$, we have $y_x > l \; \forall \, x \geq m$

Similarly, we can show when $B < 0$, $\langle y_x \rangle$ diverges to $-\infty$.

Case II if $A \neq 1$

The solution of linear difference equation is

$$y_x = A^x y_0 + \dfrac{B}{1-A} - \dfrac{B}{1-A}A^x = A^x\left(y_0 - \dfrac{B}{1-A}\right) + \dfrac{B}{1-A} = A^x(y_0 - y^*) + y^*$$

$$y_x - y^* = A^x(y_0 - y^*), x = 0, 1, 2, \dots \quad \text{where} \quad y^* = \dfrac{B}{1-A}$$

Behaviour of the solution sequence

$$y_x = A^x(y_0 - y^*) + y^*$$
$$y_{x+1} = Ay_x + B, x = 0, 1, 2, \dots$$

Hypothesis				Conclusion
	A	B	y_0	for $x = 0, 1, 2, \dots$ the sequence $\langle y_x \rangle$
(1)	$A \neq 1$		$y_0 = y^*$	$y_x = y^*$ constant ($= y^*$)
(2)	$A > 1$		$y_0 > y^*$	$y_x > y^*$ monotonic increasing diverges to $+\infty$
(3)	$A > 1$		$y_0 < y^*$	$y_x < y^*$ monotonic decreasing diverges to $-\infty$
(4)	$0 < A < 1$		$y_0 > y^*$	$y_x > y^*$ monotonic decreasing converges to y^*
(5)	$0 < A < 1$		$y_0 < y^*$	$y_x < y^*$ monotonic increasing converges to y^*
(6)	$-1 < A < 1$		$y_0 \neq y^*$	damped oscillatory converges to y^*
(7)	$A = -1$		$y_0 \neq y^*$	divergent, oscillates finitely
(8)	$A < -1$		$y_0 \neq y^*$	$y_x = y_0$ divergent, oscillates infinitely constant ($= y_0$)
(9)	$A = 1$	$B = 0$		
(10)	$A = 1$	$B > 0$		$y_x > 0$ monotonically increasing divergent
(11)	$A = 1$	$B < 0$		$y_x < 0$ monotonically decreasing divergent

Solved Examples

EXAMPLE 1. *Solve* $2y_{x+1} - y_x = 2$, *when* $y_0 = 4$.

SOLUTION. Comparing it with the equation $y_{x+1} = Ay_x + B$, we get $A = 1/2$ and $B = 1$. Then, solution will be

$$y_x = A^x y_0 + B\frac{1-A^x}{1-A} = \left(\frac{1}{2}\right)^x \cdot 4 + 1 \cdot \frac{1-\left(\frac{1}{2}\right)^x}{1-\frac{1}{2}}$$

$$y_x = 4\left(\frac{1}{2}\right)^x + 2\left[1-\left(\frac{1}{2}\right)^x\right] = 2 + 2\left(\frac{1}{2}\right)^x$$

Putting $x = 0, 1, 2, ...$, we have

$$y_0 = 4, \quad y_1 = 2 + 2\left(\frac{1}{2}\right)^1 = 3$$

$$y_2 = 2 + 2\left(\frac{1}{2}\right)^2 = \frac{5}{2} = 2\frac{1}{2}$$

$$y_3 = 2 + 2\left(\frac{1}{2}\right)^3 = 2\frac{1}{4}$$

$$y_4 = 2 + 2\left(\frac{1}{2}\right)^4 = 2\frac{1}{8}$$

... ...
... ...

The graphical representation is

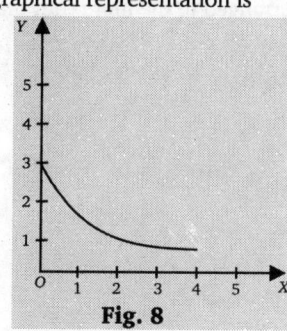

Fig. 8

EXAMPLE 2. *Solve* $y_{x+1} = 3y_x - 1$, $y_0 = 1$.

SOLUTION. On comparing with equation $y_{x+1} = Ay_x + B$, we get $A = 3$ and $B = -1$. The solution is

$$y_x = A^x y_0 + B\frac{1-A^x}{1-A}$$

$$= 3^x \cdot 1 + (-1) \cdot \frac{1-3^x}{1-3}$$

$$y_x = 3^x + \frac{1}{2}\left[1-3^x\right] = 3^x + \frac{1}{2}[1-3^x]$$

$$y_x = \frac{1}{2}\left[3^x + 1\right], x = 0, 1, 2, ...$$

We have $y_0 = \frac{1}{2}\left[3^0 + 1\right] = 1$

$$y_1 = \frac{1}{2}\left[3^1 + 1\right] = 2$$

$$y_2 = \frac{1}{2}\left[3^2 + 1\right] = 5$$

$$y_3 = \frac{1}{2}\left[3^3 + 1\right] = 14$$

... ...
... ...

The graphical representation is

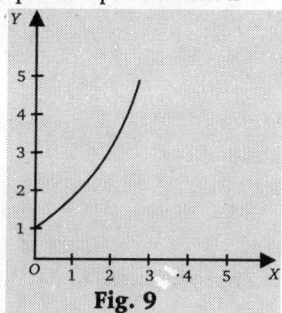

Fig. 9

EXAMPLE 3. *Solve* $y_{x+1} = y_x - 1$, $y_0 = 5$.

SOLUTION. On comparing this equation with $y_{x+1} = Ay_x + B$, we get $A = 1$ and $B = -1$. The solution is

$$y_x = y_0 + Bx$$

$$y_x = 5 + (-1)x = 5 - x, x = 0, 1, 2, ...$$

$$y_0 = 5$$
$$y_1 = 4$$
$$y_2 = 3$$
$$y_3 = 2$$
$$y_4 = 1$$
$$y_5 = 0$$
$$y_6 = -1$$

... ...
... ...

monotonically decreasing sequence diverges to $-\infty$.

 Exercise-10.16

1. Discuss the sequence solution of the difference equation

$$y_{x+1} = \frac{1}{2}y_x + \frac{1}{2}, x = 0, 1, 2, ...$$

2. Define the term sequence and type of sequence.

3. Obtain the difference equation for a probability model of learning.

 **Answers**

1. When $y_0 > 1, <y_x> = \left\{\left(\frac{1}{2}\right)^x (y_0 - 1) - 1\right\}$, When $y_0 < 1, <y_x> = 1 - \left(\frac{1}{2}\right)^x, y_0 = 1, <y_x> = 1$

10.36.9 Linear Homogenous Equation with Constant Coefficients

In order to obtain standard solutions of such equations, we first obtain alternative term for $(E-a)y_x$ and $(E-a)^r y_x$. First consider

$$(E-a)y_x = y_{x+1} - ay_x = a^{x+1}\left(\frac{y_{x+1}}{a^{x+1}} - \frac{y_x}{a^x}\right)$$

Thus, we have,

$$(E - a)y_x = a^{x+1}\Delta \frac{y_x}{a^x} \qquad \qquad \text{...(1)}$$

From (1), we get

$$(E - a)^2 y_x = a^{x+1}\Delta\left\{\left(\frac{1}{a^x}\right)a^{x+1}\Delta\frac{y_x}{a^x}\right\} = a^{x+2}\Delta^2\frac{y_x}{a^x}$$

and so, in general for $r = 1, 2, ...$

$$(E - a)^r y_x = a^{x+r}\Delta^r\frac{y_x}{a^x} \qquad \qquad \text{...(2)}$$

Case I : The equation $\Delta y_x = 0$
Here, $y_{x+1} - y_x = 0$ so that $y_{x+1} = y_x = y_{x-1} ... y_1 = y_0$
          $y_x = $ constant   ...(3)
Case II : The equation $\Delta^r y_x \; 0 \; (r = 1, 2, ...)$
Since, $\Delta^r y_x = 0; \; \Delta^{r-1}y_x = A_1$
                          $\Delta^{r-2}y_x = A_1 x + A_2$
$$\vdots \qquad \qquad \vdots \qquad \qquad \vdots$$
                          $\Delta_x^{-1}(k) = $ constant $+ (x)^{k+1}/k + 1 \; (k = 0, 1, 2, ...)$
So that integration of the last result gives

$$\Delta^{r-3}y_x = A_1\frac{x^2}{2!} + A_2 x + A_3$$

Proceeding, we get

$$\Delta^{r-4}y_x = \frac{A_1 x^3}{3!} + \frac{A_2 x^2}{2!} + A_3 x + A_4$$
$$\vdots \qquad \qquad \vdots \qquad \qquad \vdots \qquad \vdots$$
$$y_x = \frac{A_1 x^{r-1}}{(r-1)!} + \frac{A_2 x^{r-2}}{(r-2)!} + ... + A_{r-1}x + A_r$$

This last equation is a polynomial in x of degree r and contains r constants. It may be rearranged in the more convenient form
$$y_x = B_0 + B_1 x + B_2 x^2 + ... + B_{r-1}x^{r-1} \qquad \qquad \text{...(4)}$$
$B_0, B_1, ..., B_{r-1}$ being constant.
Case III :   The equation $y_{x+1} = \alpha y_x$
can be written as

$$\frac{y_x}{a^x} - \frac{y_x}{a^x} = \qquad\qquad \text{or} \qquad\qquad \Delta\left(\frac{y_x}{a^x}\right) = 0$$

$$\Rightarrow \qquad\qquad \frac{y_x}{a^x} = A \qquad\qquad \Rightarrow \qquad\qquad y_x = Aa^x \qquad \text{...(5)}$$

Case IV : The equation $(E - a)(E - b)y_x = 0 \; (a \neq b)$
Solving for $(E - b)\, y_x = Aa^x$

equation can be written as $b^{x+1}\Delta\left(\frac{y_x}{b^x}\right) = Aa^x$ $\Rightarrow$ $\Delta\frac{y_x}{b^x} = \frac{A/b}{(a/b)^x}$

$$\frac{y_x}{b^x} = B + \frac{A}{b}\sum_{r=0}^{x-1}\left(\frac{a}{b}\right)^r$$

$$\sum_{r=0}^{x-1}\left(\frac{a}{b}\right)^r = \frac{(a/b)^x - 1}{(a/b) - 1} \text{ and so on}$$

$$y_x = B'b^x + \frac{A(a^x - b^x)}{a - b} \qquad \Rightarrow \qquad y_x = B'b^x + A'a^x$$

$$y_x = A'a^x + B'b^x \qquad \qquad \text{...(6)}$$

where $A' = \dfrac{A}{a - b}$ and $B' = B - \dfrac{A}{a - b}$

We can obtain general solution of the difference equations as the sum of the general solution of the homogeneous equation and particular solution of the complete equation. We will first discuss first order difference equation.

📑 **REMARK**
• Let us consider linear first order difference equation with constant coefficients
$$y_{x+1} + a_1 y_x = f(x) \qquad \qquad \text{...(1)}$$
(i) The function Y be given $y_x = \lambda(-a_1)x$ ...(2)
with λ an arbitrary constant is the general solution of the equation $y_{x+1} + a_1 y_x = 0$ ...(3)
(ii) If y' is any particular solution of the complete equation, then $Y + y'$ is the general solution of the complete equation. That is
if y is any solution then there is a value of λ for which
$$y_x = \lambda(a_1)^x + y_x' \qquad \qquad \text{...(4)}$$

10.36.10 LINEARLY INDEPENDENT SOLUTION OR FUNDAMENTAL SET OF SOLUTIONS

Definition. *The rth order homogeneous difference equation*

$$y_{x+r} + a_1 y_{x+r-1} + a_2 y_{x+r-2} + \ldots + a_r y_x = 0 \qquad \ldots(1)$$

$y^{(1)}, y^{(2)}, \ldots, y^{(r)}$ are the solutions of equation (1) are said to form a fundamental set of solutions of (1) if the rth order determinant

$$\begin{vmatrix} y_0^{(1)} & y_0^{(2)} & \cdots & y_0^{(r)} \\ y_1^{1} & y_1^{(2)} & \cdots & y_1^{(r)} \\ \vdots & \vdots & \vdots & \vdots \\ y_{r-1}^{(1)} & y_{r-1}^{(2)} & \cdots & y_{r-1}^{(r)} \end{vmatrix} \text{ is different from zero.}$$

 Solved Examples

EXAMPLE 1. *Solve* $y_{x+1} + 3y_x = 8$.
SOLUTION. The corresponding homogeneous equation is
$$y_{x+1} + 3y_x = 0$$
$$y_{x+1} = -3y_x$$
On combining it with $y_{x+1} = Ay_x + B$, we have $A = -3$
and $B = 0$, the solution is

$$y_x = \lambda A^x + B \frac{1-A^x}{1-A}$$
$$= \lambda(-3)^x = (-3)^x \lambda$$

To find Particular solution of the given equation, we will use hit and trial method. We try to see whether the constant function is a solution or not.
Let $y_x = A$ be a solution of the given equation.
$$y_{x+1} + 3y_x = 8$$
$$A + 3A = 8$$
$$4A = 8$$
$$A = 2$$
Hence, $A = 2$ is a particular solution. Hence the general solution will be $y_x = \lambda(-3)^x + 2$.

 Exercise-10.17

Solve completely the following difference equations:

1. $2y_{x+1} - y_x = 2$

2. $y_{x+1} + y_x = 1$

3. Show that the function $y^{(1)}$ and $y^{(2)}$ given by $y_x^{(1)} = 1$ and $y_x^{(2)} = (-1)^x$ are solutions of the difference equation
$$y_{x+2} - y_x = 0$$

and they form a fundamental set. Find the general solution of the difference equation and a particular solution for which $y_0 = 0$ and $y_1 = 2$.

4. Solve $y_{x+1} - 2y_x = x + 1$.

5. Solve $y_{x+1} + 5y_x = 2x$.

6. Solve $y_{x+1} + 7_x = 2, x = 0, 1, 2$.

7. Solve $2y_{x+1} - 5y_x = 3x + 1$.

 Answers

1. $y_x = \lambda \left(\frac{1}{2}\right)^x + 2$

2. $y_x = \lambda(-3)^x + 2$

3. $C_1 + C_2(-1)^x, y_x = 1 - (-1)^x$

4. $y_x = -x - 2$

5. $y_x = \lambda(-5)^x + \frac{1}{3}x - \frac{1}{18}$

6. $y_x = \lambda(-1)^x + 1, x = 0, 1, 2, \ldots$

7. $y_x = A^x \lambda + B \frac{1-A^x}{1-A}$

10.36.11 GENERAL SOLUTION OF SECOND ORDER HOMOGENEOUS DIFFERENCE EQUATION

The second order homogeneous difference equation is
$$y_{x+2} + a_1 y_{x+1} + a_2 y_x = 0 \qquad \ldots(1)$$
where $a_2 \neq 0$.
Let us suppose $y_x = m^x$ satisfies equation (1), where m is any constant different from zero. Now, from (1), we get
$$m^{x+2} + a_1 m^{x+1} + a_2 m^x = 0$$
$$\Rightarrow \qquad m^2 + a_1 m + a_2 = 0 \qquad \ldots(2)$$

It is auxiliary equation or characteristic equation of (1). If m is a root of (2) then $y_x = m^x$ is a solution of difference equation (1).

The auxiliary equation is quadratic algebraic equation. Therefore, it has two non-zero roots m_1 and m_2. Let $y_x^{(1)} = m_1^x$ and $y_x^{(2)} = m_2^{(x)}$ the solutions. There are three cases that arise :

(i) The roots m_1 and m_2 are real and unequal ($m_1 \neq m_2$).
(ii) The roots m_1 and m_2 are real and equal ($m_1 = m_2$).
(iii) The roots are complex.

Case I : The roots are unequal.
$y_x^{(1)} = m_1^x$ and $y_x^{(2)} = m_2^x$ form a fundamental set, to prove this we will calculate

$$\begin{vmatrix} y_0^{(1)} & y_0^{(2)} \\ y_1^{(1)} & y_1^{(2)} \end{vmatrix} = \begin{vmatrix} 1 & 1 \\ m_1 & m_2 \end{vmatrix} = m_2 - m_1 \neq 0 \qquad [\because m_1 \neq m_2]$$

Hence, m_1 and m_2 are real and unequal, the general solution of difference equation (1) is given by

$$y_x = C_1 m_1^x + C_2 m_2^x$$

Case II : $(m_1 = m_2)$

Now $y_x^{(1)}$ and $y_x^{(2)}$ cannot form a fundamental set because $m_1 = m_2$, the value of determinant will be zero.

We have a function $y^{(1)}$ but find a function $y^{(2)}$ which form a fundamental set with function $y^{(1)}$.

Now, we claim that a function $y_x^{(2)}$ is given as

$$y_x^{(2)} = x \, m_1^x$$

Proof. Let m_1 and m_2 are the roots of second order difference equation, we have

$$m^2 + a_1 m + a_2 = 0$$

and

$$m_1 + m_2 = -a_1$$

$$2m_1 + a_1 = 0 \qquad\qquad (\because m_1 = m_2)$$

Now, we will show that $y_x^2 = x \, m_1^x$ is a solution of the equation is given below

$$y_{x+2}^2 + a_1 y_{x+1}^{(2)} + a_2 y_x^{(2)} = (x+2)m_1^{x+2} + a_1(x+1)m_1^{x+1} + a_2 x m_1^x = x m_1^x [m_1^2 + a_1 m_1 + a_2] + m_1^{x+1}[2m + a_1] = 0$$

Hence, $y_x^2 = x m_1^x$ is solution of the given equation.

Now, we will show $y_x^{(1)} = m_1^x \, y_x^{(2)} = x m_1^x$ form a fundamental set, we have

$$\begin{vmatrix} y_0^{(1)} & y_0^{(2)} \\ y_1^{(1)} & y_1^{(2)} \end{vmatrix} = \begin{vmatrix} 1 & 0 \\ m_1 & m_1 \end{vmatrix} = m_1 \neq 0$$

No root of auxiliary equation is zero. Thus, when $m_1 = m_2$, the general solution will be

$$y_x = C_1 m_1^x + C_2 x m_1^x = m_1^x (C_1 + x C_2)$$

Case III : Complex roots

We know that complex roots in quadratic equation always occur in conjugate pairs then $m_1 \neq m_2$. Let $y_x^{(1)} = m_1^x$ and $y_x^{(2)} = m_2^x$ form a fundamental set. The general solution is

$$y_x = C_1 m_1^x + C_2 m_2^x \qquad\qquad\qquad ...(1)$$

Equation (1) will be complex if m_1 and m_2 are themselves complex. But if C_1 and C_2 are complex then y_x will be real. Let

$$m_1 = \alpha.(\cos\theta + i\sin\theta)$$

$$m_2 = \alpha.(\cos\theta - i\sin\theta)$$

and

$$C_1 = a(\cos B + i\sin B)$$

$$C_2 = a(\cos B - i\sin B)$$

$$m_1^x = \alpha^x(\cos x\theta + i\sin x\theta)$$

$$C_1 m_1^x = a\alpha^x[(\cos(x\theta + B) + i\sin(x\theta + B)]$$

Similarly

$$C_2 m_1^x = a\alpha^x[(\cos(x\theta + B) - i\sin(x\theta + B)]$$

$$y_x = 2a\alpha^2[(\cos(x\theta + B)]$$

$$y_x = A\alpha^x[(\cos(x\theta + B)] \qquad\qquad\qquad [2a = A]$$

$$= \alpha^x[A\cos x\theta + B\sin x\theta]$$

We summarize the above result for second order difference equation

$$y_{x+2} + a_1 y_{x+1} + a_2 y_x = 0$$

Then, A. E is

$$m^2 + a_1 m + a_2 = 0. \text{ Then solution is}$$

$$y_x = C_1 m_1^x + C_2 m_2^x \text{ if } (m_1 \neq m_2)$$

$$y_x = (C_1 + C_2 x)m_1^x \text{ if } m_1 = m_2$$

$$y_x = C_1 \alpha^2 \cos(x\theta + C_2) \; m_1 \text{ and } m_2 \text{ are complex.}$$

$$= \alpha^x(C_1 \cos x\theta + C_2 \sin x\theta)$$

10.36.12 General Solution of the Homogeneous Difference Equation of Order *n*

The homogenous difference equation of nth order is

$$y_{x+n} + a_1 y_{x+n-1} + a_2 y_{x+n-2} + ... + a_n y_x = 0 \,, a_n \neq 0 \qquad\qquad ...(1)$$

The auxiliary equation is given by when $y_x = m^x$

$$m^x + a_1 m^{x-1} + a_2 m^{x-2} + ... + a_n = 0 \qquad\qquad ...(2)$$

Solve equation (1). Let $m_1, m_2, ..., m_n$ are the n roots. The general solution of the homogeneous difference equation depends upon these roots. There are three cases that arise:

Case I : Roots are real and distinct.

In this case general solution will be $y_x = C_1 m_1^x + C_2 m_2^x + ... + C_n m_n^x$

Case II : Roots are Equal.

Some roots are repeated. Let root m_1 is repeated k times then the general solution is $(C_1 + C_2 x + C_3 x^2 + ... + C_k x^{k-1}) m_1^x$

Case III : Roots are Complex.

Let m_1 and m_2 are the complex roots, then solution is $C_1 \rho^x \cos(x\theta + C_2)$ or $\rho^x (C_1 \cos x\theta + C_2 \sin x\theta)$.

Solved Examples

EXAMPLE 1. **Solve** $y_{x+2} - 2y_{x+1} + 2y_x = 0$

SOLUTION. The corresponding auxiliary equation is

$$m^2 - 2m + 2 = 0, \text{ when } y_x = m^x$$

$$\therefore \quad m = \frac{2 \pm \sqrt{4-8}}{2} = 1 \pm i$$

In polar form, $m = 1 \pm i = \rho(\cos\theta \pm i\sin\theta)$

On equating real and imaginary parts of both sides, we get

$$1 = \rho\cos\theta \text{ and } 1 = \rho\sin\theta$$

$$\Rightarrow \quad \rho = \sqrt{2} \text{ and } \theta = \frac{\pi}{4}$$

Then, solution is $y_x = C_1 (\sqrt{2})^x \cos\left(\frac{x\pi}{4} + C_2\right)$

EXAMPLE 2. **Solve the difference equation**

$$y_{x+3} - 3y_{x+2} - 10y_{x+1} + 24y_x = 0.$$

SOLUTION. The auxiliary equation, when $y_x = m^x$ is

$$m^3 - 3m^2 - 10m + 24 = 0$$

$$(m-2)(m+3)(m-4) = 0$$

$$\therefore \quad m = 2, -3, 4$$

Thus, the general solution is

$$y_x = C_1 2^x + C_2(-3)^x + C_3 4^x$$

EXAMPLE 3. **Solve** $y_{x+2} - 7y_{x+1} + 12y_x = 0.$

SOLUTION. The auxiliary equation, when $y_x = m^x$ is

$$m^2 - 7m + 12 = 0$$
$$(m-3)(m-4) = 0$$
$$\therefore \qquad m = 3, 4$$

Thus, solution is $y_x = C_1 3^x + C_2 4^x$

EXAMPLE 4. **Solve the difference equation** $y_{x+2} + y_x = 0$ **with** $y_0 = 0$ **and** $y_1 = 1.$

SOLUTION. When $y_x = m^x$, the auxiliary equation is

$$m^2 + 1 = 0$$

$$m = \pm i = \cos\frac{\pi}{2} + i\sin\frac{\pi}{2}$$

$$\theta = \frac{\pi}{2} \quad \text{and} \quad \rho =$$

Thus, the general solution is

$$y_x = C_1 \cos\left[\frac{x\pi}{2} + C_2\right]$$

When $x = 0$ and $x = 1$, we get

$$y_0 = C_1 \cos C_2 = 0$$

$$y_1 = C_1 \cos\left(\frac{\pi}{2} + C_2\right) = 1$$

Solving these, we get $C_1 = -1$ and $C_2 = \frac{\pi}{2}$

Hence, the solution is

$$y_x = -\cos\frac{\pi}{2}(x+1) = \sin\frac{x\pi}{2}.$$

Exercise-10.18

1. Solve the difference equation $9y_{x+2} - 6y_{x+1} + y_x = 0$.
Also find the particular solution satisfying the initial conditions $y_0 = 0$ and $y_1 = 1$.

2. Solve $y_{x+1} - 2y_x \cos\alpha + y_{x-1} = 0$.

3. Solve the difference equation $3y_{x+2} - 6y_{x+1} + 7y_x = 0$.

4. Solve $y_{x+4} - 4y_{x+3} + 8y_{x+2} - 8y_{x+1} + 4y_x = 0$.

5. Solve the difference equation $y_{x+2} + 6y_{x+1} + 25y_x = 0$.

6. Form the Fibonacci difference equation and solve it.
(Kurukshetra–2009, Kurukshetra(NIT)–2013)

Answers

1. General solution is $y_x = (C_1 + C_2 x)\left(\frac{1}{3}\right)^x$ and particular solution is $y_x = 3x\left(\frac{1}{3}\right)^x$.   **2.** $y_x = C_1 \cos ax + C_2 \sin ax$

3. $y_x = C_1\left(\frac{2}{\sqrt{3}}\right)^x \cos\left(\frac{x\pi}{6} + C_2\right)$ **4.** $y_x = \left\{(C_1 + C_2 x)\cos\frac{\pi x}{4} + (C_3 + C_4 x)\sin\frac{\pi x}{4}\right\}(\sqrt{2})^x$

5. $y_x = C_1(5)^x \cos(\theta x + C_2)$ when $\theta = \tan^{-1}\left(-\frac{4}{3}\right)$ **6.** $y_x = \left[\frac{5-\sqrt{5}}{10}\right]\left[\frac{1+\sqrt{5}}{2}\right]^x + \left[\frac{5+\sqrt{5}}{10}\right]\left[\frac{1-\sqrt{5}}{2}\right]^x$

10.36.13 Particular Solution of the Complete Difference Equation

Let $y_{x+2} + a_1 y_{x+1} + a_2 y_x = f(x)$ is given equation, we can find the general solution of the corresponding homogeneous equation. Now, we will find the particular solution. There are many cases for finding particular solution.

Case I : *When* $f(x) = a^x$, *where* a *is a constant*

$$\phi(E)a^x = (a_0 E^n + a_1 E^{n-1} + ... + a_{n-1}E + a_n)a^x = a_0 a^{x+n} + a_1 a^{x+n-1} + ... + a_{n-1}a^{x+1} + a_n a^x$$

$$= (a_0 a^n + a_1 a^{n-1} + ... + a_{n-1}a + a_n)\,a^x = \phi(a)a^x$$

or $\qquad\qquad\qquad \dfrac{1}{\phi(E)}a^x = \dfrac{1}{\phi(a)}a^x$ provided $\phi(a) \neq 0$.

Case of failure : If $\phi(a) = 0$, then $\phi(E)$ will have any one of the following factor

$$(E-a) \text{ or } (E-a)^2, \text{ etc. or } (E-a)^n$$

Now, let $\qquad\qquad \dfrac{1}{E-a}a^x = y_x \qquad\qquad \Rightarrow \qquad y_{x+1} - ay_x = a^x \qquad \text{or} \qquad a^{-(x+1)}y_{x+1} - a^{-x}y_x = a^{-1}$

$\Rightarrow \qquad\qquad \Delta(a^{-x}y_x) = a^{-1} \qquad\qquad \Rightarrow \qquad a^{-x}y_x = a^{-1}x \qquad\qquad \Rightarrow \qquad\qquad\qquad y_x = xa^{x-1}$

$\Rightarrow \qquad\qquad \dfrac{1}{E-a}a^x = xa^{x-1}$

Similarly, $\qquad\qquad \dfrac{1}{(E-a)^2}a^x = \dfrac{x(x-1)}{2!}a^{x-2}$

In general $\qquad\qquad \dfrac{1}{(E-a)^n}a^x = \dfrac{x(x-1)...(x+n-1)}{n!}a^{x-n}$

Case II : *When* $f(x) = a^x F(x)$, *where* $F(x)$ *is some function of* x.

Noting that $\qquad\qquad E^n a^x F(x) = a^{x+n}F(x+n) = a^x a^n E^n F(x)$

We have $\qquad\qquad \phi(E)\{a^x F(x)\} = a^x(a_0 a^n E^n + a_1 a^{n-1}E^{n-1} + ... + a_{n-1}aE + a_n)F(x) = a^x \phi(aE)F(x)$

The inverse result is $\dfrac{1}{\phi(E)}\{a^x F(x)\} = a^x \dfrac{1}{\phi(aE)}F(x)$

Case III: *When* $r(x)$ *is a polynomial in* x *of degree* r(say)

In this case $\qquad\qquad \dfrac{1}{\phi(E)}f(x) = \dfrac{1}{\phi(1+\Delta)}f(x) = \phi(1+\Delta)^{-1}f(x)$

Here, we expand $[\phi(1+\Delta)^{-1}]$ in ascending powers of Δ and operate on $f(x)$.

Solved Examples

EXAMPLE 1. *Solve the following equation*

$$y_{x+2} - 3y_{x+1} + 2y_x = 1.$$

SOLUTION. On substituting $y_x = m^x$, the auxiliary equation is

$$m^2 - 3m + 2 = 0$$
$$(m-1)(m-2) = 0, m = 1, 2$$

The general solution of the reducible homogeneous equation is

$$y_x = C_1 + C_2 2^x$$

To find the particular solution

$$= \dfrac{1}{(E-1)(E-2)} = -\dfrac{1}{(E-1)^1}1^x \qquad [\because \phi(a) = 0]$$
$$= -\dfrac{x}{1!}1^{x-1} = -x$$

Hence, the general solution of the complete equation is

$$y_x = C_1 + C_2 2^x - x$$

Note : For the particular solution, we can use hit and trial method.

EXAMPLE 2. *Solve* $y_{x+2} - 4y_{x+1} + 4y_x = 3^x + 2^x + 4$.

SOLUTION. Let the auxiliary equation of the given equation is

$$m^2 - 4m + 4 = 0$$
$$(m-2)^2 = 0, m = 2, 2$$

Then general solution is

$$y_x = (C_1 + C_2 x)2^x$$

For the particular solution, we will see

Particular solution of 3^x

$$= \dfrac{1}{E^2 - 4E + 4} = \dfrac{1}{(E-2)^2}3^x = \dfrac{1}{(3-2)^2}3^x = 3^x$$

Particular solution of 2^x

$$= \dfrac{1}{(E-2)^2}2^x = \dfrac{x(x-1)}{2!}2^{x-2}$$

Particular solution of 4

$$= \dfrac{1}{(E-2)^2}4 = 4\dfrac{1}{(E-2)^2}1^x = 4$$

Hence, the complete solution is

$$y_x = (C_1 + C_2 x)2^x + 3^x + x(x-1)2^{x-3} + 4.$$

EXAMPLE 3. *Solve* $y_{x+2} - 3y_{x+1} + 2y_x = a^x$, *where* a *is some constant.*

SOLUTION. The auxiliary equation of the given equation is

$$m^2 - 3m + 2 = 0$$
$$\Rightarrow \quad (m-1)(m-2) = 0$$
$$\Rightarrow \qquad\qquad m = 1, 2$$

The general solution is

$$y_x = C_1 + C_2 2^x$$

Particular solution of $a^x = \dfrac{1}{(E-1)(E-2)}a^x$

$$= \dfrac{1}{(a-1)(a-2)}a^x$$

(provided $a \neq 1$ and $a \neq 2$)

When $a = 1$ or $a = 2$, we will use Case of Failure:

When $a = 1$, particular solution of

$$a^x = \dfrac{1}{(E-1)(E-2)}a^x = \dfrac{1}{(a-2)(E-1)}a^x$$

$$= \frac{1}{(1-2)(E-1)} 1^x = -\frac{1}{1} x = -x$$

When $a = 2$, particular solution of

$$a^x = \frac{1}{(E-1)(E-2)} a^x = \frac{1}{(2-1)(E-2)} 2^x$$

$$= \frac{1}{(E-2)} 2^x = \frac{x.2^x}{2!} = \frac{1}{2} x 2^x$$

Hence, the complete solution is

$$y_x = C_1 + C_2 2^x + \frac{1}{(a-1)(a-2)} a^x$$
$$\text{[When } a \neq 1 \text{ and } a \neq 2]$$

$$y_x = C_1 + C_2 2^x - x \qquad \text{[When } a = 1]$$

$$y_x = C_1 + C_2 2^x + \frac{1}{2} x 2^x$$

Exercise-10.19

1. Solve $y_{x+2} - 4y_{x+1} + 4y_x = 3x + 2^x$
2. Solve $y_{x+2} - 4y_x = 9x^2$.
3. Solve $y_{x+3} - 5y_{x+2} + 8y_{x+1} - 4y_x = x2^x$

4. Solve $y_{x+2} - 6y_{x+1} + 9y_x = 3x$
   (Kurukshetra(NIT)–2008, VTU–2010)
5. Solve $y_{x+2} + y_x = \cos(x/2)$ (Kurukshetra(NIT)–2008)

Answers

1. $y_x = (C_1 + C_2 x)2^x + 6 + 3x + x(x-1)2^{x-3}$
2. $y_x = C_1(-2)^x + C_2 2^x - 3x^2 - 4x - \dfrac{20}{3}$
3. $y_x = C_1 + (C_2 + C_3 x)2^x + \left[\dfrac{x^{(3)}}{6} - x^{(2)}\right]2^{x-2}$
4. $y_x = (C_1 + C_2 x)3^x + \dfrac{1}{2}x(x-1)3^{x-2}$
5. $y_x = A\cos\dfrac{x\pi}{2} + \dfrac{1}{2}\cos\dfrac{x-1}{2}\sec\dfrac{1}{2}$

10.36.14 Solution of Simultaneous Difference Equation

If two or more difference equations are given with the same number of unknown function, we can solve such equations simultaneously.

Solved Examples

EXAMPLE 1. *Solve the simultaneous difference equations*

$$2y_{x+1} - 3y_x + 5z_x = 2$$
$$2y_x + z_{x+1} - 2z_x = 7$$

SOLUTION. We can write these equations in the form

$$(2E - 3)y_x + 5z_x = 2 \qquad ...(1)$$
$$2y_x + (E - 2)z_x = 7 \qquad ...(2)$$

Operate on (1) with $(E - 2)$, on (2) with 5 and subtract to give

$$(2E^2 - 7E - 4)y_x = (E-2)z - 35 = -37$$
$$(2E + 1)(E - 4)y_x = -37$$

The auxiliary equation is

$$(2m + 1)(m - 4) = 0$$
$$m = -\frac{1}{2}, 4$$

General solution of A.E. is $y_x = C_1\left(-\dfrac{1}{2}\right)^x + C_2 4^x$

Particular solution of -37

$$= \frac{1}{(2E+1)(E-4)}(-37) = (-37)\frac{1}{(2E+1)(E-4)}1^x$$

$$= -37\frac{1}{(2+1)(1-4)} = \frac{-37}{3.(-3)} = \frac{37}{9}$$

Hence, the general solution of complete equation is

$$y_x = C_1\left(-\frac{1}{2}\right)^x + C_2 4^x + \frac{37}{9}$$

Now, we have from (1)

$$z_x = \frac{2}{5} - \frac{1}{5}(2E - 3)y_x$$

$$= \frac{2}{5} - \frac{1}{5}(2E - 3)\left[C_1\left(-\frac{1}{2}\right)^x + C_2 4^x + \frac{37}{9}\right]$$

$$= \frac{2}{5} - \frac{2}{5}\left[C_1\left(-\frac{1}{2}\right)^{x+1} + C_2 4^{x+1} + \frac{37}{9}\right]$$

$$+ \frac{3}{5}\left[C_1\left(-\frac{1}{2}\right)^x + C_2 4^x + \frac{37}{9}\right]$$

$$z_x = \frac{11}{9} + \frac{4}{5}\left(-\frac{1}{2}\right)^x C_1 - C_2 4^x.$$

EXAMPLE 2. *Solve the system*

$$y_{x+1} - y_x + 2z_{x+1} = 0$$
$$z_{x+1} - z_x - 2y_x = 2^x$$

SOLUTION. The given system of equation can be written as

$$(E - 1)y_x + 2Ez_x = 0 \qquad ...(1)$$
$$-2y_x + (E - 1)z_x = 2^x \qquad ...(2)$$

Operating on (3) with $E - 1$ and (4) with $2E$, subtracting, we get

$$((E - 1)^2 + 4E)y_x = -2E2^x = -22^{x+1}$$
$$(E + 1)^2 y_x = -4.2^x$$

The auxiliary difference equation is

$$(m + 1)^2 = 0$$
$$m = -1, -1$$

General solution of the homogeneous equation is

$$y_x = (C_1 + C_2 x)(-1)^x$$

Particular solution $= \dfrac{1}{(E+1)^2}(-4.2^x) = -4\dfrac{1}{(E+1)^2}2^x$

$$= -4\frac{2^x}{(2+1)^2} = -\frac{4}{9}2^x$$

Hence, the general solution of the complete equation is

$$y_x = (C_1 + C_2 x)(-1)^x - \frac{4}{9}2^x$$

given $2z_{x+1} = y_x - y_{x+1} = (C_1 + C_2 x)(-1)^x - \frac{4}{9}2^x$

$\qquad - [C_1 + C_2(x+1)(-1)^{x+1} - \frac{4}{9}2^{x+1}]$

$\qquad = (2C_1 + C_2 + 2C_2 x)(-1)^x + \frac{4}{9}2^x$

$z_{x+1} = (C_1 + \frac{1}{2}C_2 + C_2 x)(-1)^x + \frac{2}{9}2^x$

$z_x = \left[C_1 + \frac{C_2}{2} + C_2(x-1) \right](-1)^{x-1} + \frac{2}{9}2^{x-1}$

$z_x = \left[C_1 - \frac{C_2}{2} + C_2 x \right](-1)^{x-1} + \frac{1}{9}2^x.$

 Miscellaneous Exercise

1. Solve the following homogeneous equations :

(i) $y_{x+3} - 2y_{x+2} - y_{x+1} + 2y_x = 0$

(ii) $y_{x+2} - y_x = 0$

(iii) $y_{x+2} + 2y_{x+1} + y_x = 0$

(iv) $y_{x+4} + y_{x+3} - 13y_{x+2} - y_{x+1} + 12y_x = 0$

(v) $y_{x+2} + 8y_{x+1} + 16y_x = 0$

(vi) $y_{x+2} - 5y_{x+1} + 6y_x = 0$

(vii) $y_{x+2} + 2y_{x+1} + 7y_x = 0$

(viii) $y_{x+2} - 3y_{x+1} + 2y_x = 0$

(ix) $y_{x+2} + 3y_{x+1} + y_x = 0$

(x) $y_{x+4} - 9y_{x+3} + 30y_{x+2} - 44y_{x+1} + 24y_x = 0$

(xi) $2y_{x+3} - 7y_{x+2} + 5y_{x+1} + 2y_x = 0$

(xii) $y_{x+4} + y_x = 0$

(xiii) $y_{x+2} - y_{x+1} + y_x = 0$ given $y_0 = 1$ and $y_1 = \frac{1+\sqrt{3}}{2}$

(xiv) $3y_{x+2} - 7y_{x+1} - 6y_x = 0, y_0 = 0, y_1 = 1$

(xv) $9y_{x+2} + 72y_{x+1} + 79y_x = 0, y_0 = 1, y_1 = -7$

(xvi) $y_{x+2} - 5y_{x+1} + 6y_x = 0, y_0 = 1, y_1 = 2$

2. Solve the following difference equations

(i) $y_{x+2} - 6y_{x+1} + 8y_x = 4^x$

(ii) $y_{x+2} - 2y_{x+1} + y_x = 5 + 3x$

(iii) $y_{x+2} - 3y_{x+1} + 2y_x = 1, y_0 = 1, y_1 = -1$

(iv) $y_{x+2} - 3y_{x+1} + 2y_x = 5^x + 2^x$

(v) $8y_{x+2} - 6y_{x+1} + y_x = 2^x$

(vi) $8y_{x+2} - 6y_{x+1} + y_x = 5\sin\frac{x\pi}{2}$

(vii) $y_{x+2} - 4y_{x+1} + 7y_x = 3.2^x + 5y^x$

(viii) $y_{x+3} - 3y_{x+1} + 2y_x = 3^x$

(ix) $y_{x+2} - y_{x+1} + 1/2 \, y_x = 1, y_0 = 2, y_1 = 3$

(x) $y_{x+2} - 4y_{x+1} + 3y_x = 2^x + 3^x + 7$

(xi) $y_{x+2} - 5y_{x+1} + 6y_x = x^2 + x + 1$ (Kurukshetra–2012)

(xii) $(E^2 - 5E + 6)y_x = x^2, y_0 = 1, y(1) = -1$

(xiii) $y_{x+2} - 5y_{x+1} + 6y_x = x^2 - 1$

(xiv) $\Delta^2 y_x + 2\Delta y_x + y_x = 3x + 2$

(xv) $\Delta y_x + \Delta^2 y_x = \cos x$

(xvi) $y_{x+2} + y_{x+1} + 2y_x = 3^x - 10$

(xvii) $y_{x+2} - 4y_x = 9x^2$

(xviii) $y_{x+1} - y_x^2 = y_{x+1}^3$, if $y_1 = 1, y_2 = 2$

3. What do you understand by linear difference equation of the nth order?

4. If $y^{(1)}$ and $y^{(2)}$ constitute a fundamental set for $y_{x+2} + a_1 y_{x+1} + a^2 y_x = 0, x = 0, 1, 2, \ldots$, prove that a necessary and sufficient condition for $\alpha_1 y^{(1)} + \beta_1 y^{(2)}$ and $\alpha_2 y^{(1)} + \beta_2 y^{(2)}$ to be a fundamental set is $\alpha_1 \beta_2 - \alpha_2 \beta_1 \neq 0$.

5. If u_x and v_x are the solutions of the homogeneous difference equation $y(x+2) + p(x)y(x+1) + q(x)y(x) = 0$, then $C_1 u_x + C_2 v_x$ is also a solution of the above equation.

6. Solve the systems of simultaneous equations

(i) $2x_{t+1} - 3x_t + 5y_t = 2$
$2x_t + y_{t+1} - 2y_t = 7$

(ii) $x_{t+1} - y_t = 2(t+1)$
$y_{t+1} - y_t = -2(t+1)$

(iii) $2x_{t+1} + y_{t+1} = x_t + 3y_t$
$x_{t+1} + y_{t+1} = x_t + y_t$

 Answers

1. (i) $y_x = C_1 + C_2(-1)^x + C_3 2^x$ (ii) $y_x = C_1 + C_2(-1)^x$ (iii) $y_x = (C_1 + C_2 x)(-1)^x$

(iv) $y_x = C_1 + C_2(-1)^x + C_3 3^x + C_4(-4)^x$ (v) $y_x = (C_1 + C_2 x)(-4)^x$ (vi) $y_x = C_1 3^x + C_2 2^x$

(vii) $y_x = \left\{ C_1 \cos\frac{2n\pi}{3} + C_2 \sin\frac{2n\pi}{3} \right\} 2^x$ (viii) $y_x = C_1 + C_2 2^x$ (ix) $y_x = C_1 \left(\frac{-3+\sqrt{5}}{2} \right)^x + C_2 \left(\frac{-3-\sqrt{5}}{2} \right)^x$

(x) $y_x = (C_1 + C_2 x + C_3 x^2)2^x + C_4(-3)^x$ (xi) $y_x = C_1 2^x + C_2 \left(\frac{3+\sqrt{17}}{2} \right)^x + C_3 \left(\frac{3-\sqrt{17}}{2} \right)^x$

(xii) $y_x = C_1 \cos\left(\frac{3\pi}{4}x + C_2 \right) + C_3 \cos\left(\frac{\pi x}{4} + C_4 \right)$ (xiii) $y_x = \cos\frac{n\pi}{3} + \sin\frac{n\pi}{3}$ (xiv) $y_x = C_1 \left(\frac{3}{11} \right)^x + C_2 \left(\frac{3}{11} \right)^x$

(xv) $y_x = 3x\left(-\frac{7}{3} \right)^x$ (xvi) $y_x = 3^x - 2^x + 1$

2. (i) $y_x = C_1 2^x + C_2 4^x + \dfrac{x}{8} 4^x$ (ii) $y_x = (C_1 + C_2 x) + x^2 + \dfrac{x^3}{2}$ (iii) $y_x = 1 - 2^x$

(iv) $y_x = C_1 + C_2 2^x + \dfrac{5^x}{12} - x 2^{x-1}$ (v) $y_x = C_1 \left(\dfrac{1}{4}\right)^x + C_2 \left(\dfrac{1}{2}\right)^x + \dfrac{1}{21} 2^x$ (vi) $y_x = C_1 \left(\dfrac{1}{4}\right)^x + C_2 \left(\dfrac{1}{2}\right)^x + \dfrac{5}{3} \sin \dfrac{\pi x}{2}$

(vii) $y_x = (C_1 + C_2 x) 2^x + 3x(x-1) 2^{x-3} + 5.4^{x-1}$ (viii) $y_x = C_1 + C_2 2^x + \dfrac{1}{2} 3^x$ (ix) $y_x = -(2)^{x/2} \cos\left(\dfrac{\pi x}{4} + \dfrac{\pi}{2}\right) + 2$

(x) $y_x = C_1 + C_2 3^x - 2^x + \dfrac{x}{2} 3^{x-1} - \dfrac{7x}{2}$ (xi) $y_x = C_1 2^x + C_2 3^x + \dfrac{1}{4}(2x^2 + 8x + 15)$ (xii) $y_x = 2^{x+2} - 3^{x+1}$

(xiii) $y_x = C_1 2^x + C_2 3^x + 2 + \dfrac{3}{2} x + \dfrac{x^2}{2}$ (xiv) $y_x = C_1 3^x + C_2 (-4)^x - 1 + \dfrac{1}{21} x 3^x$

(xv) $y_x = C_1 2^x + C_2 (-2)^x - (3x^2 + 3x + \dfrac{20}{3})$ (xvi) $y_x = 2^{2x-1} - 1$

6. (i) $x_t = -C_1 4^t + C_2 \left(-\dfrac{1}{2}\right)^t + \dfrac{37}{9}, y_t = -C_1 4^t + \dfrac{4}{5} \left(-\dfrac{1}{2}\right)^t C_2 + \dfrac{11}{9}$ (ii) $x_t = C_1 + C_2 (-1)^t + t, y_t = C_1 - C_2 (-1)^t + (t+1)$

(iii) $x_t = C_1 (-2)^t + C_2, y_t = -C_1 (-2)^t + \dfrac{C_2}{2}$

10.37 NUMERICAL SOLUTION OF ORDINARY DIFFERENTIAL EQUATIONS

Many problems of science and engineering can be formulated in terms of differential equations. A large number of motivation for building early computers came from the need to compute trajectories accurately and quickly. Today computers are used extensively to solve equations of physical situations.

A general equation of first order and first degree is given by

$$y' = f(x, y) \quad \text{with} \quad y(x_0) = y_0 \qquad \dots(1)$$

Many analytical techniques exist for finding the solution of such equations. In most of these methods, we replace the differential equation by a difference equation and then solve it. These methods gives solutions either as a power series in x from which the values of y can be found by direct substitution or a set of values of x and y. The methods of Picard and Taylor series belong to the former class of solution. In these methods y in (1) is approximated by a truncated series, each term of which is a function of x. The information about the curve at one point is utilized and the solution is not interacted. As such, these are referred to as single-step methods. The methods of Euler, Runga-Kutta, Milne, Adams-Bashforth, etc. belong to the latter class of solutions. In these methods, the next point on the curve is evaluated in short steps ahead, by performing interactions till sufficient accuracy is achieved. As such these methods are called step-by-step methods.

Euler and Runga-Kutta methods are used for computing y over a limited range of x-values whereas Milne and Adams methods may be applied for finding y over a wider range for x-values. Therefore Milne and Adams methods required starting values which are found by Picard's, Taylor series or Runga-Kutta methods.

10.37.1 EXISTENCE AND UNIQUENESS OF SOLUTION OF DIFFERENTIAL EQUATION

Consider the differential equation

$$\frac{dy}{dx} = f(x, y) \qquad \dots(1)$$

Let us suppose, the function $f(x, y)$ be analytic and satisfies Lipschitz condition.

Now, by differentiating the given differential equation as many times as needed and using initial conditions $y = y_0$, for $x = x_0$, we get

$$y = y_0 + (x - x_0)y_0' + \frac{1}{2}(x - x_0)^2 y_0'' + \dots$$

It is assumed that $|x - x_0|$ is very small, the series converges to a unique solution.

If $f(x, y)$ is bounded and continuous, the Lipschitz condition

$$|f(x, y) - f(x, z)| < L|y - z| \qquad \dots(2)$$

Where L is a constant is used to find a unique solution. Since, $f(x, y)$ is bounded, therefore $|f(x, y)| < M$.

Integrating (1) between x_0 and x, we get

$$\int dy = \int_{x_0}^x f(x, y) dx \quad \Rightarrow \int_{x_0}^x \frac{dy}{dx} dx = y - y_0 = \int_{x_0}^x f(x, y) dx$$

$$\Rightarrow \qquad y = y_0 + \int_{x_0}^x f(x, y) dx \qquad \dots(3)$$

Thus, the given differential equation has been transformed to an integral equation. Now, if $y_1(x)$ is the first approximation to the solution of (1), then a better solution is obtained by substituting $y_1(x)$ for y in the R.H.S. of (3). Therefore, we obtained an iterative formula.

$$y_{i+1}(x) = y_0 + \int_{x_0}^{x} f[t, y_t(t)]dt \qquad \qquad ...(4)$$

Now, we get

$$\left|y_{i+1}(x) - y_i(x)\right| \le \int_{x_0}^{x}\left|f(t, y_i(t))\right| - \left|f(t, y_{i-1}(t))\right|dt \le L\int\left|y_i(t) - y_{i-1}(t)\right|dt \qquad \text{[Using (2)]}$$

Further, suppose that $x_0 \le x \le X$ and putting $X - x_0 = h$ we get

$$\left|y_2(x) - y_1(x)\right| = \left|y_0 + \int_{x_0}^{x} f[t, y_i(t)]dt - y(x)\right| = \left|y_0 - y_1(x) + \int_{x_0}^{x} f[t, y_i(t)]dt\right|$$

$$= \left|y_1(x) - y_0\right| + \int_{x_0}^{x} Mdt \qquad \qquad [\because f \text{ is bounded by } M.]$$

$$\le 2Mh = N \quad \text{(say)}$$

Similarly, we can obtain

$$\left|y_3(x) - y_2(x)\right| \le L\int_{x_0}^{x} Ndt = NL\int_{x_0}^{x} dt = NL(x - x_0) \le NLh$$

$$\left|y_4(x) - y_3(x)\right| \le L\int_{x_0}^{x} NL(t - x_0)dt = NL^2\frac{(x - x_0)^2}{2!} = \frac{NL^2h^2}{2!}$$

and so on. Finally, we get

$$\left|y_{n+1}(x) - y_n(x)\right| \le L\int_{x_0}^{x}\frac{NL^{n-2}(t - x_0)^{n-2}}{(n-2)!}dt = \frac{NL^{n-1}}{(n-1)!}(x - x_0)^{n-1} \le \frac{NL^{n-1}h^{n-1}}{(n-1)!}$$

But $y_{n+1}(x) = y_1(x) + [y_2(x) - y_1(x)] + ... + [y_{n-1}(x) - y_n(x)]$ ...(5)

Except the first term $y_1(x)$, every term of the series (5) is less than the terms of the series

$$N\left[1 + Lh + \frac{L^2h^2}{2!} + ... + \frac{L^{n-1}h^{n-1}}{(n-1)!}\right]$$

which converges to Ne^{Lh}.

Therefore, the series (5) is absolutely and uniformly convergent to $y(x)$. Also,

$$\left|y(x) - \left(y_0 + \int_{x_0}^{x}[f(t, y(t)dt]\right)\right| = \left|y(x) - y_{n+1}(x) + \int_{x_0}^{x}[f\{t, y(t)\} - f\{t, y_n(t)\}dt]\right| \le \left|y(x) - y_{n+1}(x)\right| + L\int_{x_0}^{x}\left|y(t) - y_n(t)\right|dt$$

which tends to zero because $y_n(x)$ and $y_{n+1}(x)$ converges to $y(x)$ when $n \to \infty$.

Thus, we have

$$y(x) = y_0 + \int_{x_0}^{x} f[t, y(t)]dt \qquad \qquad ...(6)$$

Now, we shall show the uniqueness of this solution. Let if possible $z = z(x)$ be another solution, we get

$$z = y_0 + \int_{x_0}^{x} f[t, z(t)]dt$$

$$y_{n+1} = y_0 + \int_{x_0}^{x} f[t, y_n(t)]dt$$

Thus, $\left|z - y_{n+1}\right| \le \int_{x_0}^{x}\left|f(t, z) - f(t, y_n)\right|dt \le L\int_{x_0}^{x}\left|z - y_n\right|dt$ [Using (2)]

But $\left|z - y_0\right| \le Mh = \dfrac{N}{2}$ and therefore $\left|z - y_1\right| \le \dfrac{N}{2}L(x - x_0) \le \dfrac{N}{2}Lh$

and $\left|z - y_2\right| \le L^2\dfrac{(x - x_0)^2}{2!} \le \dfrac{N}{2} \cdot \dfrac{L^2h^2}{2!}$ and so on.

Finally, we get

$$\left|z - y_{n+1}\right| \le \frac{N}{2}\frac{L^{n+1}h^{n+1}}{(n+1)!} \text{ which tends to zero as } n \to \infty.$$

$\Rightarrow$ $z(x)$ and $y(x)$ coincide.

Hence, the solution given by (6) is unique.

10.38   EULER'S METHOD

The oldest and simplest method was derived by Euler. It gives the basic idea of those numerical methods which seek to determine the change Δy in y corresponding to a small increment in the argument x.

Consider the first order differential equation

$$y' = f(x, y) \text{ with } y = y_0 \text{ where } x = x_0. \qquad \qquad ...(1)$$

We wish to solve (1), for values of y at $x = x_i$ where $x_i = x_0 + ih$, $i = 1, 2, ...$

Integrating (1), we get

$$y_1 = y_0 + \int_{x_0}^{x} f(x, y)dx \qquad \qquad ...(2)$$

Let $\qquad f(x, y) = f(x_0, y_0)$

in the range $x_0 \le x \le x_1$

Now $\qquad y_1 = y_0 + \int_{x_0}^{x_1} f(x_0, y_0)dx = y_0 + (x_1 - x_0)f(x_0, y_0) = y_0 + hf(x_0, y_0)$ (Use $x_1 - x_0 = h$) ...(3)

Similarly for the range $x_1 \le x \le x_2$, we have

$$y_2 = y_1 + hf(x_1, y_1)$$

Proceeding in this manner, finally we obtain

$$y_{n+1} = y_n + hf(x_n, y_n).$$

Thus starting from x_0 when $y = y_0$, we can construct a table of y for given steps.

Working Procedure

A new value of y is estimated using the previous value of y as the initial condition. The term $h.f(x_i, y_i)$ represents the incremental value of y and $f(x_i, y_i)$ is the slope of $y(x)$ at (x_i, y_i) i.e., the new value is obtained by extra olating linearly over the step size h using the slope at its previous value. Hence,

New value = old value × slope × stop size

☞ REMARK

- This process is very slow and to obtain desired accuracy with this method, h should be taken small. If h is not small then the method is too inaccurate.

Solved Examples

EXAMPLE 1. *Given* $y' = \dfrac{y-x}{y+x}$ *with* $y_0 = 1$ *find y for x = 0.1 in 4 steps. (By Euler's method).* (PTU-2001)

SOLUTION. Let $h = \dfrac{0.1}{4} = 0.025$, given $y_0 = 1$, where $x = 0$.

We know that $y_{n+1} = y_n + hf(x_n, y_n)$.

By putting $n = 0, 1, 2, 3$, we obtain

$$y_1 = y_0 + hf(x_0, y_0)$$

$$= 1 + 0.025\frac{(1-0)}{(1+0)} = 1.025$$

$\Rightarrow \qquad y_1 = 1.025$

Again $\qquad y_2 = y_1 + hf(x_1, y_1)$

$$= 1.025 + 0.025\frac{(1.025 - 0.025)}{(1.025 + 0.025)}$$

(where $x_1 = x_0 + h = 0 + 0.025 \Rightarrow x_1 = 0.025$)

$$= 1.0488$$

$\Rightarrow \qquad y_2 = 1.0488$

Now again

$$y_3 = y_2 + hf(x_2, y_2)$$

(where $x_2 = x_0 + 2h = 0 + 2 \times 0.025 \Rightarrow x_2 = 0.05$)

$$= 1.0488 + 0.025\frac{(1.0488 - 0.05)}{(1.0488 + 0.05)} = 1.07152$$

$\Rightarrow y_3 = 1.07152$

$$y_4 = y_3 + hf(x_3, y_3)$$

(where $x_3 = x_0 + 3h = 0 + 3 \times 0.025 \Rightarrow x_3 = 0.075$)

$$= 1.07152 + 0.025\left(\frac{1.07152 - 0.075}{1.07152 + 0.075}\right)$$

$$= 1.09324$$

$\Rightarrow y_4 = 1.09324$

at $x_4 = x_0 + 4h = 0 + 4 \times 0.025 = 0.01$

Hence, $y(0, 1) = 1.0932$

EXAMPLE 2. *Apply Euler's method to initial value problem* $\dfrac{dy}{dx} = x + y, y(0) = 0$, *when* $x = 0$ *to* $x = 1.0$ *taking* $h = 0.2$. (MUMBAI-2005, ROHTAK-2003)

SOLUTION. We have $\quad h = 0.2, x_0 = 0, x_1 = 0.2, x_2 = 0.4,$
$\qquad x_3 = 0.6, x_4 = 0.8, x_5 = 1.0.$

Also, $f(x, y) = x + y$

By Euler's method,

$$y_{n+1} = y_n + hf(x_n, y_n)$$

and, $f(x_0, y_0) = f(0, 0) = 0 + 0 = 0$

$$y_1 = y_0 + hf(x_0, y_0) = 0 + 0.2(0) = 0$$

Now, $f(x_1, y_1) = f(0.2, 0) = 0.2 + 0 = 0.2$

$$y_2 = y_1 + hf(x_1, y_1) = 0 + 0.2(0.2) = 0.04$$

Now, $f(x_2, y_2) = f(0.4, 0.04) = 0.4 + 0.04 = 0.44$

$$y_3 = y_2 + hf(x_2, y_2) = 0.04 + 0.2(0.44) = 0.128$$

Now, $f(x_3, y_3) = f(0.6, 0.128) = 0.6 + 0.128 = 0.728$

$$y_4 = y_3 + hf(x_3, y_3) = 0.128 + 0.2(0.728)$$

$$= 0.2736$$

Now, $f(x_4, y_4) = f(0.8, 0.2736)$

$$= 0.8 + 0.2736 = 1.0736$$

$$y_5 = y_4 + hf(x_4, y_4)$$

$$= 0.2736 + 0.2(1.0736) = 0.48832$$

Now, $f(x_5, y_5) = f(1.0, 0.48832) = 1.48832$

$$y_6 = y_5 + hf(x_5, y_5)$$

$$= 0.48832 + 0.1(1.48832) = 0.63715$$

Hence, $y_{(1)} = y_6 = 0.63715$.

10.39 EULER'S MODIFIED METHOD

Instead of approximating $f(x, y)$ by $f(x_0, y_0)$ in (2) in 10.5.2, the integral in (3) 10.5.2, is approximated by trapezoidal rule to obtain,

$$y_1 = y_0 - \frac{h}{2}[f(x_0, y_0) + f(x_1, y_1)]$$

...(4)

We thus obtain the formula,

$$y_1^{n+1} = y_0 + \frac{h}{2}[f(x_0, y_0) + f(x_1, y_1^{(n)})] \qquad \qquad ...(5)$$

$$(\text{where } n = 0, 1, 2, ...)$$

where $y_1^{(n)}$ is the nth approximation to y_1. The iteration formula (5) uses the initial value $y_1^{(0)}$ from Euler's method

$$y_1^{(0)} = y_0 + hf(x_0, y_0)$$

REMARK

- The modified Euler's method gives a great improvement in accuracy over the original method. In this method we take the average of slopes at (x_0, y_0) and $(x_1, y_1^{(1)})$ instead of slope at (x_0, y_0) as used in Euler's method.

Solved Examples

EXAMPLE 1. *Using the Euler's modified method, solve numerically the equation* $y' = x + \left|\sqrt{y}\right|$ *with* $y(0) = 1$ *for* $0 \leq x \leq 0.6$ *in step of 0.2.*

(MKU-2004, 06, VTU-2007)

SOLUTION. We have $y(0) = 1$ where $x_0 = 0$, and interval $h = 0.2$

By Euler's method $y_1 = y_0 + hf(x_0, y_0)$

$$= 1 + 0.2\left[0 + \left|\sqrt{1}\right|\right] = 1.2 \Rightarrow y_1 = 1.2$$

The value of y_1, thus obtained is improved by modified method,

$$y_1^{n+1} = y_0 + \frac{h}{2}[f(x_0, y_0) + f(x_1, y_1^{(n)})]$$

$$\Rightarrow y_1^{(1)} = 1 + \frac{0.2}{2}[(0 + \left|\sqrt{1}\right|) + (0.2 + \left|\sqrt{1.2}\right|)]$$

$$= 1.2295 \Rightarrow y_1^{(n)} = 1.2295$$

Now $n = 1$ gives

$$y_1^{(2)} = 1 + \frac{0.2}{2}[0 + \left|\sqrt{1}\right| + 0.2 + \left|\sqrt{1.2295}\right|]$$

$$= 1.2309 \Rightarrow y_1^{(2)} = 1.2309$$

Again $n = 2$, gives

$$y_1^{(3)} = 1 + \frac{0.2}{2}[0 + \left|\sqrt{1}\right| + 0.2 + \left|\sqrt{1.2309}\right|]$$

$$= 1.2309 \Rightarrow y_1^{(3)} = 1.2309$$

Hence, we take $y_1 = 1.2309$ at $x = 0.2$ and proceed to compute y at $x = 0.4$

Applying Euler's method, we get

$$y_2 = 1.2309 + 0.2[0.2 + \left|\sqrt{1.2309}\right|] = 1.4979$$

$$\Rightarrow y_2 = 1.4979$$

Now, we apply modified method for more accurate approximations.

$$y_2^{(1)} = 1.2309 + \frac{0.2}{2}[(0.2 + \left|\sqrt{1.2309}\right|)$$

$$+ (0.4 + \left|\sqrt{1.49279}\right|)] = 1.52402$$

$$\Rightarrow y_2^{(1)} = 1.52402$$

$$y_2^{(2)} = 1.2309 + \frac{0.2}{2}[(0.2 + \left|\sqrt{1.2309}\right|)$$

$$+ (0.4 + \left|\sqrt{1.52402}\right|)] = 1.525297$$

$$\Rightarrow y_2^{(2)} = 1.525297$$

$$y_2^{(3)} = 1.2309 + \frac{0.2}{2}[(0.2 + \left|\sqrt{1.2309}\right|)$$

$$+ (0.4 + \left|\sqrt{1.525297}\right|)] = 1.52535$$

$$\Rightarrow y_2^{(3)} = 1.52535$$

$$y_2^{(4)} = 1.2309 + \frac{0.2}{2}[(0.2 + \left|\sqrt{1.2309}\right|)$$

$$+ (0.4 + \left|\sqrt{1.52535}\right|)] = 1.52535$$

$$\Rightarrow y_2^{(4)} = 1.52535$$

Since $y_2^{(3)} = y_2^{(4)} = 1.52535$

To find the value of $y = y_3$ for $x = 0.6$ we apply Euler's method

$$y_3 = 1.52535 + 0.2[0.4 + \left|\sqrt{1.52535}\right|]$$

$$= 1.85236 \Rightarrow y_3 = 1.85236$$

For better approximations, we use Euler's modified method, as follows

$$y_3^{(1)} = 1.52535 + \frac{0.2}{2}[0.4 + \left|\sqrt{1.52535}\right|)$$

$$+ (0.6 + \left|\sqrt{1.855236}\right|)] = 1.88496$$

$$\Rightarrow y_3^{(1)} = 1.88496$$

$$y_3^{(2)} = 1.52535 + \frac{0.2}{2}[0.4 + \left|\sqrt{1.52535}\right|)$$

$$+ (0.6 + \left|\sqrt{1.88496}\right|)] = 1.88615$$

$$\Rightarrow y_3^{(2)} = 1.88615$$

$$y_3^{(3)} = 1.52535 + \frac{0.2}{2}[0.4 + \left|\sqrt{1.52535}\right|)$$

$$+ (0.6 + \left|\sqrt{1.88615}\right|)] = 1.88619$$

$$\Rightarrow y_3^{(3)} = 1.88619$$

$$y_3^{(4)} = 1.52535 + \frac{0.2}{2}[0.4 + \left|\sqrt{1.52535}\right|)$$

$$+ (0.6 + \left|\sqrt{1.88619}\right|)] = 1.886193$$

$$\Rightarrow y_3^{(4)} = 1.886193$$

$$\Rightarrow y_3^{(4)} = 1.88619 \text{ correct to 5 decimal places}$$

$$y = 1.88619 \text{ at } x = 0.6$$

EXAMPLE 2. *Given* $\dfrac{dy}{dx} = x + y$ *with initial conditions* $y(0) = 1$. *Find* $y(0.05)$ *and* $y(0.1)$, *correct to 6 decimal places.*

(MDU(B.E.)-2004, ROHTAK-2005, BHOPAL-2002S, DELHI-2002)

SOLUTION. Using Euler's method, we get

$$y_1^{(0)} = y_1 = y_0 + hf(x_0, y_0) = 1 + 0.5(0 + 1)$$

$$= 1.05$$

Now, we improve this value of by using Euler's modified method

$$y_1^{(1)} = y_0 + \frac{h}{2}[f(x_0, y_0) + f(x_1, y_1^{(0)})]$$

$$= 1 + \frac{0.05}{2}[(0+1)+(0.05+1.05)] = 1.0525$$

$$y_1^{(2)} = 1 + \frac{0.05}{2}[(0+1)+(0.05+1.0525)] = 1.0525625$$

$$y_1^{(3)} = 1 + \frac{0.05}{2}[(0+1)+(0.05+1.0525625)]$$
$$= 1.052564$$

$$y_1^{(4)} = 1 + \frac{0.05}{2}[(0+1)+(0.05+1.052564)]$$
$$= 1.052564$$

We observe that $y_1^{(3)} = y_1^{(4)} = 1.052564$, correct to 6 decimal places. Thus, we take $y_1 = 1.052564$, *i.e.*, we have $y(0.5) = 1.052564$. Again, using Euler's method, we get

$$y_2^{(0)} = y_2 = y_1 + h f(x_1, y_1)$$
$$= 1.052564 + 0.05(1.052564 + 0.5) = 1.1076922$$

Now, we have to improve $y2$ by using Euler's modified method

$$y_2^{(1)} = 1.52564 + \frac{0.5}{2}[(1.052564 + 0.05)$$
$$+ (1.1076922 + 0.1)] = 1.1120511$$

$$y_2^{(2)} = 1.052564 + \frac{0.05}{2}[(1.052564 + 0.05)$$
$$+ (1.1120511 + 0.1)] = 1.1104294$$

$$y_2^{(3)} = 1.052564 + \frac{0.05}{2}[(1.052564 + 0.05)$$
$$+ (1.1104294 + 0.1)] = 1.1103888$$

$$y_2^{(4)} = 1.052564 + \frac{0.05}{2}[(1.052564 + 0.05)$$
$$+ (1.1103888 + 0.1)] = 1.1103878$$

$$y_2^{(5)} = 1.052564 + \frac{0.05}{2}[(1.052564 + 0.05)$$
$$+ (1.1103878 + 0.1)] = 1.1103878$$

Since, $y_2^{(4)} = y_2^{(5)} = 1.1103878$, correct to 7 decimal places. Hence, $y(0.1) = 1.116388$, correct to 6 decimal places.

EXAMPLE 3. *Find $y(0.2)$. Given* $\dfrac{dy}{dx} = f(x,y) = \log_{10}(x+y)$ *with initial condition $y = 1$ for $x = 0$.*
(UPTU–2007)

SOLUTION. As per given, we have

$$\frac{dy}{dx} = f(x,y) = \log_{10}(x+y) \text{ and } h = 0.2$$

By Euler's formula, we have

$$y_1^{(0)} = y_1 = y_0 + h f(x_0, y_0) = 1 + 0.2 \log(0+1) = 1$$

Now, we improve this value by using Euler's modified formula as follows:

$$y_1^{(1)} = y_0 + \frac{h}{2}[f(x_0,y_0) + f(x_1, y_1^{(0)})]$$
$$= 1 + \frac{0.2}{2}[\log(0+1) + \log(0.2+1)] = 1.0079$$

and

$$y_1^{(2)} = 1 + \frac{0.2}{2}[\log(0+1) + \log(0.2+1.0079)] = 1.0082$$

Also,

$$y_1^{(3)} = 1 + \frac{0.2}{2}[\log(0+1) + \log(0.2+1.0082)] = 1.0082$$
$$y_1^{(2)} = y_1^{(3)} = 1.0082$$

Hence, $y(0.2) = 1.0082$.

EXAMPLE 4. *Let* $\dfrac{dy}{dx} = \dfrac{y-x}{x+y}$, *with boundary conditions $y = 1$ when $x = 0$. Find approximately y for $x = 0.1$ by Euler's modified method (5 steps).* (VTU-2007)

SOLUTION. We have $x_0 = 0, y_0 = 1, f(x,y) = \dfrac{y-x}{y+x}$

Taking $h = 0.1$, then $x_1 = 0.1$

$$\therefore f(x_0, y_0) = f(0.1) = \frac{1-0}{1+0} = 1$$

$$\Rightarrow \quad y_1^{(1)} = y_0 + h f(x_0, y_0) = 1 + 0.1(1) = 1.1$$

Thus, $f(x_1, y_1^{(1)}) = f(0.1, 1.1)$

$$= \frac{1.1 - 0.1}{1.1 + 0.1} = \frac{1}{1.2} = 0.8333$$

Now, $y_1^{(2)} = y_0 + \dfrac{h}{2}[f(x_0,y_0) + f(x_1, y_1^{(1)})]$

$$= 1 + \frac{1}{2}(0.1)[1 + 0.8333] = 1.09170$$

$$\Rightarrow f(x_1, y_1^{(2)}) = f(0.1, 1.09170)$$

$$= \frac{1.09170 - 0.1}{10.9170 + 0.1} = \frac{0.9917}{1.1917} = 0.83217$$

So, $y_1^{(3)} = y_0 + \dfrac{h}{2}[f(x_0, y_0) + f(x_1, y_1^{(2)})]$

$$= 1 + \frac{1}{2}(0.1)(1 + 0.83217) = 1.09161$$

$$\Rightarrow f(x_1, y_1^{(3)}) = f(0.1, 1.09161)$$

$$= \frac{1.09161 - 0.1}{1.09161 + 0.1} = 0.83215$$

Now, $y_1^{(4)} = y_0 + \dfrac{h}{2}[f(x_0, y_0) + f(x_1, y_1^{(3)})]$

$$= 1 + \frac{1}{2}(0.1)[1 + 0.83215]$$

So, $f(x_1, y_1^{(4)}) = f(0.1, 1.09167)$

$$= \frac{1.091607 - 0.1}{1.091607 + 0.1} = 0.83215$$

$$\Rightarrow y_1^{(5)} = y_0 + \frac{h}{2}[f(x_0, y_0) + f(x_1, y_1^{(4)})]$$

$$= 1 + \frac{1}{2}(0.1)[1 + 0.83215] = 1.061607$$

We observe that $y_1^{(4)} = y_1^{(5)}$, Hence, the value of y at 0.1 is given by $y(0.1) = 1.091607$.

10.40   SOLUTION BY TAYLOR SERIES

We consider the differential equation with initial condition

$$\left. \begin{array}{c} y' = f(x,y) \\ y(x_0) = y_0 \end{array} \right\}$$

...(1)

where $\qquad\qquad x_0 = 0.$

If $y(x)$ is the exact solution of (1), then the Taylor's series for $y(x)$ about the point $x = x_0$, we get

$$y(x) = y_0 + (x + x_0)y_0' + \frac{(x - x_0)^2}{2!}y_0'' + \ldots \qquad \ldots(2)$$

Putting $x = x_0 + h$ in (2), we obtain

$$y_1 = y_0 + hy_0' + \frac{h^2}{2!}y_0'' + \ldots \qquad \ldots(3)$$

If the values of $y_0', y_0'', \ldots$ are known, then (3) gives a power series for y_1. The coefficient $y_0', y_0'', \ldots$ can be found from (1).

We can write (1) as

$$y' = f(x, y)$$
$$y'' = f' = f_x + f_y y' = f_x + f_y f$$

Similarly, we get,

$$y''' = f'' = f_{xx} + f_{xy}f + f[f_{yx} + f_{yy}f] + f_y[f_x + f_yf] = f_{xx} + 2ff_{xy} + f^2 f_{yy} + f_xf_y + f^2 f_y$$

and other higher derivatives of y. The method is best understood by the following examples.

Solved Examples

EXAMPLE 1. *Using the Taylor series for $y(x)$, find $y(0.1)$ correct to four decimal places if $y(x)$ satisfies $y' = x + (-y^2)$, $y_0 = 1$ where $x_0 = 0$.*

(VTU-2010, MADRAS-2006)

SOLUTION. We know by Taylor series, for $y(x)$ that

$$y_x = 1 + xy_0' + \frac{x^2}{2!}y_0'' + \frac{x^3}{3!}y_0''' + \frac{x^4}{4!}y_0^{iv} + \frac{x^5}{5!}y_0^v + \ldots$$

The derivatives $y_0', y_0'', \ldots$ etc. are obtained thus

$$y'(x) = x - y^2 \qquad \Rightarrow y_0' = -1 \quad (\text{since } x = 0, y_0 = 1)$$
$$y_x'' = 1 - 2xy' \qquad \Rightarrow \quad y_0''' = 3$$

$$y'''_x = -2yy'' - 2y'^2 \quad \Rightarrow \quad y'''_0 = -8$$
$$y_{(x)}^{iv} = -2yy''' - 6y'y'' \quad \Rightarrow \quad y_0^{iv} = 34$$
$$y_{(x)}^v = -2yy^{iv} + 8y'y''' - 6y''^2 \Rightarrow y_0^v = -186$$

Using these values, the Taylor series becomes

$$y_x = 1 - x + \frac{3}{2}x^2 - \frac{4}{3}x^3 + \frac{17}{12}x^4 - \frac{31}{20}x^5 + \ldots$$

At $x = 0.1$, we get

$$y_{(0.1)} = 1 - 0.1 + \frac{3}{2}(0.1)^2 + 0.001\left(\frac{-4}{3}\right)$$
$$+ (0.0001)\left(\frac{17}{12}\right) = 0.9138$$

REMARK

- We can find the range for the value of x for which the above series truncated after term containing x^4 is convert to 4 decimal places.

$$\frac{31}{20}x^2 \le 0.00005 \ i.e., \qquad x \le 0.126.$$

EXAMPLE 2. *Solve the differential equation $\frac{dx}{dy} = x + y$, with $y(0) = 1$, $x \in [0, 1]$ by Taylor series expansion to obtain y for $x = 0.1$.*

(UPTU-2006)

SOLUTION. As per given, we have

$$\begin{array}{lll} y' = x + y & \Rightarrow & y'(0) = 1 \\ y'' = 1 + y' & \Rightarrow & y''(0) = 2 \\ y''' = 0 + y'' & \Rightarrow & y'''(0) = 2 \end{array}$$

Therefore, we get

$$y(x) = 1 + x + \frac{2x^2}{2!} + \frac{2x^3}{3!} + \frac{2x^4}{4!} + \ldots$$

Putting $x = 0.1$, we get

$$y(0.1) = 1 + 0.1 + (0.1)^2 + \frac{2(0.1)^3}{6} + \frac{(0.1)^4}{12}$$
$$= 1 + 0.1 + 0.1 + 0.000333 + 0.0000083$$
$$= 1.11033$$

EXAMPLE 3. *Use Taylor's method, find approximate value of y at $x = 0.2$ for the differential equation $\frac{dy}{dx} = 2y + 3e^x$, $y(0) = 0$. Compare the numerical solution obtained with the exact solution.*

(VTU–2009, PTU–2003)

SOLUTION. We have $x_0 = 0$, $y_0 = 0$ and $y' = \frac{dy}{dx} = 2y + 3e^x$ ...(1)

$$\therefore \quad y'(0) = 2y_0 + 3e^{x_0} = 2 \times 0 + 3e^0 = 3$$
$$y'' = 2y' + 3e^x$$
$$\Rightarrow \quad y''(0) = 2y'(0) + 3 \quad = 9$$
$$y''' = 2y'' + 3e^x$$
$$\Rightarrow \quad y'''(0) = 2y''(0) + 3e^{x_0} = 21$$
$$y'''' = 2y''' + 3e^x$$
$$\Rightarrow \quad y''''(0) = 2y'''(0) = 3e^{x_0} = 45 \ \ldots. \text{etc.}$$

Taylor's series of $y(x)$ about $x = 0$ is given by

$$y(x) = y_0 + xy'(0) + \frac{x^2}{2!}y''(0) + \frac{x^3}{3!}y'''(0)$$
$$+ \frac{x^4}{4!}y''''(0) + \ldots \qquad \ldots(2)$$

$$\Rightarrow \quad y(x) = 3x + \frac{9}{2}x^2 + \frac{7}{2}x^3 + \frac{15}{8}x^4 + \ldots$$

$$\Rightarrow y(0.2) = 3(0.2) + \frac{9}{2}(0.2)^2$$
$$+ \frac{7}{2}(0.2)^3 + \frac{15}{8}(0.2)^4 + \ldots = 0.8110 \qquad \ldots(3)$$

Now exact solution of (1) can be written as

$$\frac{dy}{dx} = 2y + 3e^x \text{ which is a linear differential equation.}$$

$$I.F. = e^{-2x}$$

Solution is $ye^{-2x} = \int e^{-2x}(3e^x)dx + c = 3(-e^{-x}) + c$

$\Rightarrow y(x) = -3e^x + ce^{2x}$

when $x = 0, y = 0$ then $c = 3$

$\therefore y(x) = 3e^{2x} - 3e^x$

$\Rightarrow y(0.2) = 3e^{0.4} - 3e^{0.2} = 0.8112$ ...(4)

From (3) and (4), it is obvious that the approximate value is the same to the exact value upto 3 decimal places.

EXAMPLE 4. *Using Taylor's series method, find the value of y up to five places of decimal when x = 1.02, where $\dfrac{dy}{dx} = xy - 1$, y(1) = 2.* (ROHTAK–2005)

SOLUTION. We have $x_0 = 1, y_0 = 2$

Also $\dfrac{dy}{dx} = xy - = y'$ ...(1)

Now $y'(0) = x_0 y(0) - 1 = 2 - 1 = 1$

$y''(0) = x_0 y'(0) + y(0) = 1 \times 1 + 2 = 3$

$y'''(0) = xy'' + 2y'$

$\Rightarrow y'''(0) = x_0 y''(0) + 2y'(0) = 1 \times 3 + 2 \times 1 = 5$

$y''''= xy''' + 3y''$

$\Rightarrow y''''= x_0 y'''(0) + 3y''(0) = 1 \times 5 + 3 \times 3 = 14$

$y'''''(0) = xy''''+ 4y'''$

$\Rightarrow y'''''(0) = x_0 y''''(0) + 4y'''(0) = 1 \times 14 + 4 \times 5 = 34$

Taylor's series about $x = 1$ is given by

$$y(x) = y_0 + (x-1)y'(0) + \frac{(x-1)^2}{2!}y''(0)$$

$$+ \frac{(x-1)^3}{3!}y'''(0) + \frac{(x-1)^4}{4!}y''''(0)$$

$$+ \frac{(x-1)^5}{5!}y'''''(0) + ... \qquad ...(2)$$

Putting the above values in (2), we get

$$\Rightarrow y(x) = 2 + (x-1) + \frac{3}{2}(x-1)^2 + \frac{5}{6}(x-1)^3$$

$$+ \frac{7}{12}(x-1)^4 + \frac{17}{60}(x-1)^5 + ... \qquad ...(3)$$

$$\Rightarrow y(1.02) = 2 + (0.02) + \frac{3}{2}(0.02)^2 + \frac{5}{6}(0.02)^3$$

$$+ \frac{7}{12}(0.02)^4 + \frac{17}{60}(0.02)^5 + ...$$

$$= 2.02 + 0.0006 + 0.0000067 + 0.0000000093 + ...$$

$$= 2.02061 \text{ approximately.}$$

10.41 PICARD'S METHOD OF SUCCESSIVE APPROXIMATIONS

Consider the differential equation

$$y' = \frac{dy}{dx} = f(x, y) \qquad ...(1)$$

Integrating (1), we get

$$y = y_0 + \int_{x_0}^{x} f(x, y)dx \qquad ...(2)$$

Equation (2) in which the unknown function appears under the integral sign is called as integral equation. Such an equation can be solved by the method of successive approximation in which the first approximation to y is obtained by putting y_0 for y on the right hand side of (2), and we write,

$$y^{(1)} = y_0 + \int_{x_0}^{x} f(x, y_0)dx \qquad ...(3)$$

The integral on the right can now be solved and the resulting $y^{(1)}$ is substituted for y in the integration of (2) to obtain the second approximation $y^{(2)}$.

$$y^{(2)} = y_0 + \int_{x_0}^{x} f(x, y^{(1)})dx$$

Continuing in this manner, we obtain $y^{(3)}, y^{(4)}..., y^{(n-1)}$ and $y^{(n)}$ where

$$y^{(n)} = y_0 + \int_{x_0}^{x} f(x, y^{(n-1)})dx \text{ with } y^{(0)} = y_0.$$

Hence, this method gives a sequence of approximation $y^{(1)}, y^{(2)}...y^{(n)}$ and it can be proved $f(x, y)$ is bounded in some regions containing the point (x_0, y_0) and if $f(x, y)$ satisfies the Lipschitz condition, namely

$$[f(x, y_1) - f(x, y_2)] \le k[y - \bar{y}], \text{ where } k \text{ being a constant,} \qquad ...(4)$$

then the sequence $y^{(1)}, y^{(2)}...$ converges to the solution (2).

Solved Examples

EXAMPLE 1. *Integrate the differential equation $\dfrac{dy}{dx} = x \sin \pi y$ with $y = \dfrac{1}{2}$ at $x = 0$ by Picard's method of successive approximation.*

SOLUTION. We have $y = y_0 + \int_{x_0}^{x} f(x, y)dx$ ...(1)

Putting $y = \dfrac{1}{2}$ in R.H.S. of (1), we get the first approximation y_1 as

$$y_1 = \frac{1}{2} + \int_0^x x \sin\left(\frac{\pi}{2}\right)dx = \frac{1}{2} + \frac{x^2}{2}$$

Similarly,

$$y_2 = \frac{1}{2} + \int_0^x x \sin\left[\pi \frac{(1+x^2)}{2}\right]dx$$

$$= \frac{1}{2} + \int_0^x x \cos\frac{\pi x^2}{2}dx = \frac{1}{2} + \int_0^x x\left(1 - \frac{\pi^2 x^4}{8} + ...\right)dx$$

$$y_3 = \frac{1}{2} + \int_0^x x \sin \pi\left[\frac{1}{2} + \frac{x^2}{2} - \frac{\pi^2 x^6}{48} + ...\right]dx$$

$$= \frac{1}{2} + \int_0^x x \cos \pi\left(\frac{x^2}{2} - \frac{\pi^2 x^6}{48}\right)dx$$

$$= \frac{1}{2} + \int_0^x x \left\{ 1 - \frac{1}{2} \left(\frac{\pi x^2}{2} + \frac{\pi^2 x^6}{48} \right)^2 + ... \right\} dx$$

$$= \frac{1}{2} + \frac{x^2}{2} - \frac{\pi^2 x^6}{48}.$$

Here, we observe that y_2 agree with y_3 upto and including term in x_6.

EXAMPLE 2. *Given* $\dfrac{dy}{dx} = \dfrac{y - x}{y + x}$ *with* $y = 1$, *when* $x = 0$. *Find approximately the value of* y *for* $x = 0.1$ *by Picard's method.*   (PTU–2002)

SOLUTION. We use the following formula
$$y_n = y_0 + \int_{x_0}^x f(x, y_{n-1}) dx$$

where , $f(x, y) = \dfrac{y - x}{y + x}$, $x_0 = 0, y_0 = 1$.

The first approximation is given by y_1 as

$$y_1 = 1 + \int_0^x \frac{1 - x}{1 + x} dx = 1 + 2\log(1 + x) - x$$

$$\Rightarrow \quad y(0.1) = 1 + 2\log(1 + 0.1) - 0.1 = 0.9828$$

Now, the second approximation y_2 is given by

$$y_2 = y_0 + \int_{x_0}^x f(x, y_1) dx$$

$$= 1 + \int_0^x \frac{1 + 2\log(1 + x) - x - x}{1 + 2\log(1 - x) - x + x} dx$$

$$= 1 + \int_0^x \left[1 - \frac{2x}{1 + 2\log(1 + x)} \right] dx$$

$$= 1 + x - 2\int_0^x \frac{x}{1 + 2\log(1 + x)} dx$$

$$= 1 + x - 2\int_0^t \frac{e^{2t}}{1 + 2t} dt + 2\int_0^t \frac{e^t}{1 + 2t} dt,$$

where $t = \log(1 + x)$.

EXAMPLE 3. *Find the series expansion that gives* y *as a function of* x *in the neighbourhood of* $x = 0$, *when* $\dfrac{dy}{dx} = x^2 + y^2$ *with* $y(0) = 0$.   (JNTU–2009)

SOLUTION. As per given, we have
$$f(x, y) = x^2 + y^2, x_0 = 0 \text{ and } y_0 = 0.$$
Also, we have the nth approximation y_n of y as
$$y_n = y_0 + \int_{x_0}^x f(x, y_{n-1}) dx$$
Therefore, the first approximation y_1 is given by
$$y_1 = 0 + \int_0^x (x^2 + 0) dx = \frac{x^3}{3}$$
The second approximation y_2 is given by

$$y = + \int \left[x + \left(\frac{x^3}{3} \right) \right] dx = \frac{x^3}{3} + \frac{x^7}{63}$$

Similarly, the higher order approximations are given by

$$y_3 = 0 + \int_0^x \left[x^2 + \left(\frac{x^3}{3} + \frac{x^7}{63} \right)^2 \right] dx$$

$$= \frac{x^3}{3} + \frac{x^7}{63} + \frac{2x^{11}}{2079} + \frac{x^{15}}{59535} + ...$$

and $y_4 = 0 + \int_0^x x^2 + \left(\frac{x^3}{3} + \frac{x^7}{63} + \frac{2x^{11}}{2079} + \frac{x^{15}}{59535} \right)^2 dx$

$$= \frac{1}{3} x^3 + \frac{1}{63} x^7 + \frac{2}{2079} x^{11} + \frac{13}{218295} x^{15} + ...$$

EXAMPLE 4. *Using Picard's method find a solution of* $\dfrac{dy}{dx} = x^4 y + x$, $y(0) = 3$.

SOLUTION. Let $\dfrac{dy}{dx} = x^4 y + x = f(x, y)$ with $y_0 = 3, x_0 = 0$
The first approximation is

$$y^{(1)} = y_0 + \int_{x_0}^x f(x, y_0) dx = 3 + \int_0^x (x + x^4 y_0) dx$$

$$= 3 + \int_0^3 [x + 3x^4] dx = 3 + \frac{x^2}{2} + \frac{3x^5}{3}$$

The second approximation is

$$y^{(2)} = 3 + \int_0^x f(x, y^{(1)}) dx = 3 + \int_0^x [x + x^4 y^{(1)}] dx$$

$$= 3 + \int_0^x \left[x + x^4 \left(3 + \frac{x^2}{2} + \frac{3x^5}{3} \right) \right] dx$$

$$= 3 + \int_0^x \left[x + 3x^4 + \frac{x^6}{2} + \frac{3x^9}{3} \right] dx$$

$$= 3 + \frac{x^2}{2} + \frac{3x^5}{5} + \frac{x^7}{14} + \frac{3x^{10}}{50}$$

The third approximation is

$$y^{(3)} = 3 + \int_0^x f(x, y^{(2)}) dx = 3 + \int_0^x [x + x^4 y^{(2)}] dx$$

$$= 3 + \int_0^x \left[x + x^4 \left(3 + \frac{x^2}{2} + \frac{3x^5}{5} + \frac{x^7}{14} + \frac{3x^{10}}{52} \right) \right] dx$$

$$= 3 + \int_0^x \left[x + 3x^4 + \frac{x^6}{2} + \frac{3x^9}{3} + \frac{x^{11}}{14} + \frac{3x^{14}}{50} \right] dx$$

$$= 3 + \frac{x^2}{2} + \frac{3x^5}{5} + \frac{x^7}{14} + \frac{3x^{10}}{50} + \frac{x^{12}}{168} + \frac{x^{15}}{250}$$

which is the required solution of the given equation upto third approximation.

Exercise-10.20

1. Use Euler's modified method for :
   (i) Given $y' = x^2 + y$, with $y_0 = 1$ determine $y(0.02), y(0.04)$ and $y(0.06)$.   (ROHTAK–2005)
   (ii) Given that $y' = 2 + \sqrt{xy}$ with $y_{(1)} = 1$.   (ANNA–2004)

   Compute $y_{(2)}$ in steps of (0.2).

2. Using Taylor Series method for :
   (i) The differential equation $y' = x^2 + y^2$ with $y(1) = 0$ obtain $y(1.3)$.
   (ii) The equation $y' = 2xy + 1$ with $y_0 = 0$ and taking $h = 0.2$ at $x = 0.4$.

3. Use Picard's method to solve $y' = 1 + xy$ with $x_0 = 2, y_0 = 0$.

Answers

1. (i) $y(0.2) = 1.0202$, $y(0.04) = 1.0408$, $y(0.06) = 1.0619$

   (ii) $y_2 = 5.0516$, $y_{(0.02)} = 1.6402$, $y_{(0.4)} = 2.3623$, $y_{(0.6)} = 3.1678$, $y_{(0.8)} = 4.0633$

2. (i) $y(1.3) = 0.4158$, (ii) $y(0.2) = 0.21$ app., $y(0.4) = 0.451$

3. $y^{(3)} = \dfrac{x^5}{15} - \dfrac{x^4}{4} + \dfrac{x^3}{3} - \dfrac{x^2}{2} + x - \dfrac{22}{15!}$

10.42 RUNGE-KUTTA METHOD

The Taylor's series method of solving differential equation's numerically is restricted by the labour involved in finding the higher order derivatives. However there is a class of methods known as Runge-Kutta methods which do not require the calculations of higher order derivatives and give greater accuracy. The Runge-Kutta formulae possess the advantage of requiring only the functional values at some selected points. These methods agree with Taylor series solution upto the term h^r where r differs from method to method and is called the order of that method.

10.42.1 RUNGE-KUTTA METHOD OF ORDER ONE

By Euler's method, we have
$$y_1 = y_0 + hf(x_0, y_0) \qquad [\because y' = f(x, y)] \dots (1)$$
Expanding by Taylor's series, we get
$$y_1 = y(x_0 + h) = y_0 + hy_0' + \frac{h^2}{2} \cdot y_0'' + \dots$$
It follows that the Euler's method agrees with the Taylor's series solution upto the term in h.

Hence, the Euler's method is the Runge-Kutta method is of the first order.

10.42.2 RUNGE-KUTTA METHOD OF ORDER TWO

We know that modified Euler's method is given by
$$y_1 = y_0 + \frac{h}{2}[f(x_0, y_0) + f(x_0 + h, y_0)]. \qquad \dots (2)$$
Putting $y_1 = y_0 + hf(x_0, y_0)$ on the right hand side of (2), we obtain
$$y_1 = y_0 + \frac{h}{2}[f_0 + f(x_0 + h, y_0 + hf_0)]. \qquad \dots (3)$$
$$[\text{where } f_0 = f(x_0, y_0)]$$
Expanding Left hand side by Taylor's series, we get
$$y_1 = y(x_0 + h) = y_0 + hy_0' + \frac{h^2}{2!}y_0'' + \frac{h^3}{3!}y_0''' + \dots \qquad \dots (4)$$
Expanding $f(x_0 + h, x_0 + hf_0)$ by Taylor's series for a function of two variables (3) gives
$$y_1 = y_0 + \frac{h}{2}\left[f_0 + \left\{ f(x_0, y_0) + h\left(\frac{\partial f}{\partial x}\right)_0 + hf_0\left(\frac{\partial f}{\partial y}\right)_0 + 0(h^2) \right\}\right]$$
$$= y_0 + \frac{1}{2}\left[hf_0 + hf_0 + h^2\left\{\left(\frac{\partial f}{\partial x}\right)_0 + \left(\frac{\partial f}{\partial y}\right)_0\right\} + 0(h^3)\right]$$
$$= y_0 + hf_0 + \frac{h^2}{2}f_0' + 0(h^3) \qquad \left[\because \frac{df}{dx} = \frac{\partial f}{\partial x} + f\frac{\partial f}{\partial y}\right]$$
$$= y_0 + hy_0' + \frac{h^2}{2!}y_0'' + 0(h^3) \qquad \dots (5)$$

From equation (4) and (5), it follows that the modified Euler's method agrees with the Taylor's series solution upto the term in h^2. Hence, the modifies Euler's method is the Runge-Kutta method of the second order.

Therefore the second order Runge-Kutta formula is
$$y_1 = y_0 + \frac{1}{2}(k_1 + k_2) \text{ where } k_1 = hf(x_0, y_0), \ k_2 = (x_0 + h, y_0 + k_1).$$

10.42.3 RUNGE-KUTTA METHOD OF ORDER THREE

In a similar manner it can be seen that Runge method be agrees with the Taylor's series solution upto the term in h^3.
$$y_1 = y_0 + \frac{h}{6}\left[f(x_0, y_0) + 4f\left(x_0 + \frac{h}{2}, y_0 + \frac{h}{2}f(x_0, y_0)\right) + f(x_0 + h, y_0) + hf(x_0 + h, y_0) + hf(x_0, y_0)\right] \qquad \dots (6)$$
As such, Runge-Kutta method is the Runge-Kutta of the third order.

The third order Runge-Kutta formula is
$$y_1 = y_0 + \frac{1}{6}(k_1 + 4k_2 + k_3)$$

where
$$k_1 = hf(x_0, y_0)$$
$$k_2 = hf\left(x_0 + \frac{h}{2}, y_0 + \frac{k_1}{2}\right)$$
and
$$k_3 = hf(x_0 + h, y_0 + k')$$

10.42.4 Runge-Kutta Method of order four

This method coincides with the Taylor's series solution upto terms of h^4, we know that the Taylor's series solution can be expressed in terms of $f(x, y)$ and its partial derivatives of various orders. But in this method we use a technique which avoids the calculation of derivatives.

We take
$$\begin{aligned} k_1 &= hf(x, y) \\ k_2 &= hf(x + mh, y + mk_1) \\ k_3 &= hf(x + nh, y + nk_2) \\ k_4 &= hf(x + hp, y + pk_3) \end{aligned}\right\} \quad \dots(7)$$

Now
$$y(x + h) = y(x) + ak_1 + bk_2 + ck_3 + dk_4 \quad \dots(8)$$

We know that by Taylor's series
$$y(x + h) = y_x + hy' + \frac{h^2}{2!}y'' + \frac{h^3}{3!}y''' + \frac{h^4}{4!}y^{iv} + 0(h^5) \quad \dots(9)$$

First of all we shall put it in some convenient form
$$F_1 = f_x + ff_y, \quad F_2 = f_{xx} + 2ff_{xy} + f^2 f_{yy}$$
$$F_3 = f_{xxx} + 3ffx_{xxy} + 3f^3 f_{xyy} + f^3 f_{yyy}$$

Now differential equation $y' = \dfrac{dy}{dx} = f(x, y)$, we obtain
$$y'' = f(x) + f_y y' = f_x + ff_y = F_1$$
$$y''' = f_{xx} + 2ff_{xy} + f^2 f_{yy} + f_x f_y + ff_y^2 = (f_{xx} + 2ff_{xy} + f^2 f_{yy}) + f_y(f_x + ff_y) = F_2 + f\,F_1$$

Similarly
$$y^{iv} = F_3 + f_y F_2 + 3F_1(f_{xy} + ff_{yy}) + f_y^2 F_1$$

Putting these values in (7), we get
$$y(x + h) = y_x + hf + \frac{h^2}{2}F_1 + \frac{1}{6}h^3(F_2 + f_y F_1) + \frac{1}{24}h^4[F_3 + f_y F_2 + 3(f_{xy} + ff_{yy})F_1 + f_y^2 F_1] + \dots \quad \dots(10)$$

Using the above notation and the Taylor's theorems, we get,   [where $f = f(x, y)$]
$$k_1 = hf$$
$$k_2 = h\left[f + mhF_1 + \frac{1}{2}m^2 h^2 F_2 + \frac{1}{6}m^3 h^3 F_3 + \dots\right]$$
$$k_3 = h\left[f + mhF_1 + \frac{1}{2}h^2(n^2 F_2 + 2mn + f_y F_1) + \frac{1}{6}h^3\{n^3 F_3 + 3m^2 nf_y F_2 + 6mn^2(f_{xy} + ff_{yy})F_1\} = \dots\right]$$

and
$$k_4 = h\left[f + phF_1 + \frac{1}{2}h^2(p^2 F_2 + 2npf_y F_1) + \frac{1}{6}h^3\{p^3 F_3 + 3n^2 pf_y F_2 + 6np^2(f_{xy} + ff_{yy})F_1 + 6mnpf_y^2 F_1\} + \dots\right]$$

Substituting these values in (8), we get
$$y(x + h) = y_{(x)} + (a + b + c + d)hf + (bm + cn + dp)h^2 F_1 + \frac{1}{2}(bm^2 + cn^2 + dp^2).$$

$$h^2 f_y F_1 + \frac{1}{2}(cm^2 n + dn^2 p)h^4 f_y F_2 + (cmn^2 + dnp^2)h^4(f_{xy} + ff_{yy})F_1 + dmnph^4 f_y^2 F_1 + 0(h^5) \quad \dots(11)$$

Equating this with (10), we get
$$a + b + c + d = 1, \qquad\qquad cmn + dnp = \frac{1}{6}$$
$$bm + cn + dp = \frac{1}{2}, \qquad\qquad cmn^2 + dnp^2 = \frac{1}{8}$$
$$bm^2 + cn^2 + dp^2 = \frac{1}{3}, \qquad\qquad cm^2 n + dn^2 p = \frac{1}{12}$$
$$bm^3 + cn^3 + dp^3 = \frac{1}{4}, \qquad\qquad dmnp \; \overline{\quad 24\quad}.$$

These are eight equations in seven unknowns; A classical solution to these eight equations is
$$m = n = \frac{1}{2}, p = 1, a = d = \frac{1}{6}, b = c = \frac{1}{3}$$

Putting these values in (7) and (8), the Runge-Kutta formulae reduces to

$$k_1 = hf(x, y), k_2 = hf\left(x + \frac{h}{2}, y + \frac{1}{2}k_1\right)$$

$$k_3 = hf\left(x + \frac{h}{2}, y + \frac{1}{2}k_2\right), k_4 = hf(x + h, y + k_3)$$

...(12)

and $\quad y(x + h) = y_{(x)} + \frac{1}{6}(k_1 + 2k_2 + 2k_3 + k_4)$

From this formula, we have

$$y_1 = y(x_0 + h) = y_0 + \frac{1}{6}[k_1 + 2(k_2 + k_3) + k_4]$$

where

$$k_1 = hf(x_0, y_0), \quad k_2 = hf\left(x_0 + \frac{h}{2}, y_0 + \frac{k_1}{2}\right)$$

$$k_3 = hf\left(x_0 + \frac{h}{2}, y_0 + \frac{k_2}{2}\right), \quad k_4 = hf(x_0 + h, y_0 + k_3)$$

...(13)

☞ REMARK

- $o(h^2)$ means 'terms containing second and higher power of h and is read as order of h^2.'

📁 Solved Examples

EXAMPLE 1. *Apply Runge-Kutta method find the solution of the differential equation $y' = 3x + \frac{1}{2}y$ with $y_0 = 1$, at $x = 0.1$.* (VTU–2004)

SOLUTION. We have $h = 0.1$

$k_1 = hf(x_0, y_0)$ (Given $y_0 = 1$, when $x_0 = 0$)

$= 0.1\left(3x_0 + \frac{1}{2}y_0\right)$

$= 0.1\left(3 \times 0 + \frac{1}{2} \times 1\right) = 0.05$

Similarly $k_2 = hf\left(x_0 + \frac{h}{2}, y_0 + \frac{k_1}{2}\right)$

$= 0.1f(0.05; 1.025) = 0.06625$

Now $k_3 = hf\left(x_0 + \frac{h}{2}, y_0 + \frac{k_2}{2}\right)$

$= 0.1f(0.05, 1.033125) = 0.6665625$

and $k_4 = hf(x_0 + h, y_0 + k_3)$

$= 0.1f(0.1, 1.0665625) = 0.0833328125$

$k_4 = 0.0833328125$

Obviously

$k = \frac{1}{6}[k_1 + 2(k_2 + k_3) + k_4] = 0.066652421875$

and $\quad y_{(0.1)} = y_0 + k$

$y_{(0.1)} = 1 + 0.66652421875$

Hence, $y(0.1) = 1.066652421875$.

EXAMPLE 2. *Solve the equation $y' = x + y$ with $y_0 = 1$ by Runge-Kutta method from $x = 0$ to $x = 0.1$ with $h = 0.1$.* (VTU–2009, PTU–2007, SVTU–2007)

SOLUTION. Here $f(x, y) = x + y, h = 0.1$, given $y_0 = 1, x_0 = 0$.

We have

$k_1 = hf(x_0, y_0) = 0.1[0 + 1] = 0.1$

$k_2 = hf\left(x_0 + \frac{h}{2}, y_0 + \frac{k_1}{2}\right) = 0.1[0.05 + 1.05] = 0.11$

$k_3 = hf\left(x_0 + \frac{h}{2}, y_0 + \frac{k_2}{2}\right) = 0.1[0.05 + 1.055] = 0.1105$

and $k_4 = hf(x_0 + h, y_0 + k_3)$

$= 0.1[0.1 + 1.1105] = 0.12105$

Hence, $y_1 = y_{(x=0.1)} = 1 + \frac{1}{6}[0.1 + 0.22 + 0.2210$

$+ 0.12105] = 1.11034$

Similarly for finding $y_2 = y(x = 0.2)$, we get

$k_1 = hf(x_1, y_1) = 0.[(0.1) + 1.11034] = 0.121034$

$k_2 = hf\left(x_1 + \frac{h}{2}, y_1 + \frac{k_1}{2}\right)$

$= 0.1[0.15 + 1.11034 + 0.060517] = 0.13208$

$k_3 = hf\left(x_1 + \frac{h}{2}, y_1 + \frac{k_2}{2}\right)$

$= 0.1[0.15 + 1.11034 + 0.06604] = 0.1323$

$k_4 = hf(x_1 + h, y_1 + k_3)$

$= 0.1[0.20 + 1.11034 + 0.13263] = 0.14429$

Hence

$y_2 = y(x = 0.2) = y_1 + \frac{1}{6}(k_1 + 2k_2 + 2k_3 + k_4)$

$= 1.11034 + \frac{1}{6}[0.121034 + 2(0.13208$

$+ 0.13263) + 0.14429] = 1.2428$

Similarly, for finding $y_3 = y(x = 0.3)$, we have

$k_1 = hf(x_2, y_2) = 0.1[0.2 + 1.2428] = 0.14428$

$k_2 = hf\left(x_2 + \frac{h}{2}, y_2 + \frac{h}{2}\right)$

$= (0.1)(0.25 + 1.2428 + 0.07214) = 0.15649$

$k_3 = hf\left(x_2 + \frac{h}{2}, y_2 + \frac{h}{2}\right)$

$= (0.1)(0.25 + 1.2428 + 0.07824) = 0.15710$

$k_4 = hf(x_2 + h, y_2 + k_3)$

$= (0.1)(0.3 + 1.2428 + 0.15710) = 0.16999$

Hence, $y_3 = 1.2428 + \frac{1}{6}(0.14428 + 2(0.15649$

$+ 0.15710) + 0.16999) = 1.3997$

Similarly for $y_4 = y(x = 0.4)$

$k_1 = (0.1)(0.3 + 1.3997) = 0.16997$

$\Rightarrow \quad k_1 = 0.16997$

$k_2 = (0.1)[0.35 + 1.3997 + 0.8949] = 0.18347$

$\Rightarrow \quad k_2 = 0.18347$

$k_3 = (0.1)[0.3 + 1.3997 + 0.9170] = 0.18414$

$\Rightarrow \quad k_3 = 0.18414$

$k_4 = (0.1)[0.4 + 1.3997 + 0.18414] = 0.19838$

$\Rightarrow \quad k_4 = 0.19838$

$$y_4 = 1.3997 + \frac{1}{6}[0.16997 + 2(0.18347$$
$$+ 0.18414) + 0.19838]$$

$$\Rightarrow \quad y_4 = 1.5836$$

EXAMPLE 3. *Applying Runga-Kutta method to find an approximate value of y for x = 0.2 in step of 0.1 if $\frac{dy}{dx} = x + y^2$. Given that y = 1 when x = 0.*

(VTU–2009, OSMANIA–2007, MADRAS–2000)

SOLUTION. We have $f(x, y) = x + y^2$, $x_0 = 0$, $y(0) = 1$, $h = 0.1$

We have to find $y(0.1)$ and $y(0.2)$.

By Runge-Kutta method, we have

$k_1 = hf(x_0, y_0) = (0.1)f(0, 1) = (0.1)[0 + (1)^2] = 0.1$

$k_2 = hf\left(x_0 + \frac{h}{2}, y_0 + \frac{k_1}{2}\right) = (0.1)f\left(0 + \frac{0.1}{2}, 1 + \frac{0.1}{2}\right)$

$= (0.1)f(0.05, 1.05) = (0.1)[0.05 + (1.05)^2] = 0.11525$

$k_3 = hf\left(x_0 + \frac{h}{2}, y_0 + \frac{k_2}{2}\right) = (0.1)f(0.05, 1.057625)$

$= (0.1)[0.05 + (1.05 + 625)^2] = 0.11686$

and

$k_4 = hf\left(x_0 + h, y_0 + k_3\right) = (0.1)f(0.1, 1.11686)$

$= (0.1)[0.1 + (1.11686)^2] = 0.13474$

Therefore,

$$y(0.1) = y_0 + \frac{1}{6}(k_1 + 2k_2 + 2k_3 + k_4)$$

$$= 1 + \frac{1}{6}(0.1 + 2 \times 0.11525 + 2 \times 0.11686 + 0.13474)$$

$$= 1 + \frac{1}{6}(0.69896) = 1.11649$$

Further, to find $y(0.2)$, take $x_1 = 0.1$, $y_1 = 1.11649$, $h = 0.1$

Then, we have

$k_1 = hf(x_1, y_1) = (0.1)f(0.1, 1.11649)$
$= (0.1)[0.1 + (1.11649)^2] = 0.13465$

$k_2 = hf\left(x_1 + \frac{h}{2}, y_1 + \frac{k_1}{2}\right) = (0.1)f(0.15, 1.118382)$

$= (0.1)[0.15 + (1.18382)^2] = 0.15514$

$k_3 = hf\left(x_1 + \frac{h}{2}, y_1 + \frac{k_2}{2}\right)$

$= (0.1)f(0.15, 1.19406)$

$= (0.1)[0.15 + (1.19406)^2] = 0.15758$

$k_4 = hf(x_1 + h, y_1 + k_3) = (0.1)f(0.2, 1.27407)$

$= (0.1)[0.2 + (1.27407)^2] = 0.18233$

Hence, $y(0.2) = y_1 + \frac{1}{6}[k_1 + 2(k_2 + k_3) + k_4]$

$$= 1.11649 + \frac{1}{6}[0.13465 + 2(0.15514$$
$$+ 0.15758) + 0.18233]$$
$$= 1.11649 + 0.156707 = 1.27356.$$

EXAMPLE 4. *Using Runge-Kutta method of fourth order, solve $\frac{dy}{dx} = \frac{y^2 - x^2}{y^2 + x^2}$ with y(0) = 1 at x = 0.2, 0.4.*

(UPTU(MCA)–2004, UPTU–2010, JNTU–2009, VTU–2008)

SOLUTION. We have

$$f(x) = \frac{y^2 - x^2}{y^2 + x^2}, x_0 = 0, y(0) = 1, h = 0.2$$

Firstly, we shall find $y(0.2)$.

Now, $k_1 = hf(x_0, y_0) = 0.2f(0, 1) = 0.2 \times 1 = 0.2$

$k_2 = hf\left(x_0 + \frac{h}{2}, y_0 + \frac{k_1}{2}\right) = 0.2f(0.1, 1.1)$

$= 0.2 \times \left[\frac{(1.1)^2 - (0.1)^2}{(1.1)^2 + (0.1)^2}\right] = 0.2\left(\frac{1.2}{1.22}\right) = 0.19672$

$k_3 = hf\left(x_0 + \frac{h}{2}, y_0 + \frac{k_2}{2}\right) = 0.2f(0.1, 1.09836)$

$= 0.2\left[\frac{(1.09836)^2 - (0.1)^2}{(1.09836)^2 + (0.1)^2}\right] = 0.2\left(\frac{1.19639}{1.21639}\right) = 0.19671$

$k_4 = hf(x_0 + h, y_0 + k_3) = 0.2f(0.2, 1.19671)$

$= 0.2\left[\frac{(1.19671)^2 - (0.2)^2}{(1.19671)^2 + (0.2)^2}\right] = 0.2\left(\frac{1.39211}{1.47211}\right) = 0.18913$

Therefore,

$$y(0.2) = y_0 + \frac{1}{6}(k_1 + 2k_2 + 2k_3 + k_4)$$

$$= 1 + \frac{1}{6}[0.2 + 2(0.19672 + 0.19671) + 0.18913]$$

$$= 1 + \frac{1}{6}(1.17599) = 1.1960$$

Further, to find $y(0.4)$, take $y_1 = 1.1960$, $x_1 = 0.2$ and $h = 0.2$.
Then
$k_1 = hf(x_0, y_0) = 0.2f(0.2, 1.1960)$

$= (0.2)\left[\frac{(1.960)^2 - (0.2)^2}{(1.960)^2 + (0.2)^2}\right] = (0.2)(0.94560) = 0.18912$

$k_2 = hf\left(x_1 + \frac{h}{2}, y_1 + \frac{k_1}{2}\right) = (0.2)f(0.3, 1.29056)$

$= (0.2)\left[\frac{(1.29056)^2 - (0.3)^2}{(1.29056)^2 - (0.3)^2}\right]$

$= (0.2)\left(\frac{1.57555}{1.75555}\right) = 0.17949$

$k_3 = hf\left(x_1 + \frac{h}{2}, y_1 + \frac{k_2}{2}\right) = (0.2)f(0.3, 1.28575)$

$= (0.2)\left[\frac{(1.28575)^2 - (0.3)^2}{(1.28575)^2 + (0.3)^2}\right]$

$$= (0.2)\left(\frac{1.56315}{1.74315}\right) = 0.17935$$

and $k_4 = hf(x_1 + h, y_1 + k_3) = 0.2f(0.4, 1.37535)$

$$= (0.2)\left[\frac{(1.37535)^2 - (0.4)^2}{(1.37535)^2 + (0.4)^2}\right]$$

$$= (0.2)\left(\frac{1.73159}{2.05159}\right) = 0.16880$$

Hence, $y_2 = y(0.4) = y_1 + \frac{1}{6}[k_1 + 2k_2 + 2k_3 + k_4]$

$$= 1.1960 + \frac{1}{6}[0.1892 + 2(0.17949$$

$$+ 0.17935) + 0.16880]$$

$$= 1.1960 + \frac{1}{6}(1.0756) = 1.37527 \ .$$

EXAMPLE 5. *Using Runge-Kutta method of fourth order, find*

$$y(0.1) \ form \ \frac{dy}{dx} = \frac{x^2 + y^2}{10}, y(0) = 1, \ take \ h = 0.1.$$

(ROHTAK–2004, 05)

SOLUTION. We have

$$f(x, y) = \frac{x^2 + y^2}{10} \ with \ x_0 = 0, y(0) = 1, h = 0.1$$

Then,

$$k_1 = hf(x_0, y_0) = (0.1)f(0,1) = (0.1)\left[\frac{0^2 + 1^2}{10}\right] = 0.01$$

$$k_2 = hf\left(x_0 + \frac{h}{2}, y_0 + \frac{k_1}{2}\right) = (0.1)f(0.05, 1.005)$$

$$= (0.1)\left[\frac{(0.05)^2 + (1.005)^2}{10}\right] = 0.01013$$

$$k_3 = hf\left(x_0 + \frac{h}{2}, y_0 + \frac{k_2}{2}\right) = (0.1)f(0.05, 1.0051)$$

$$= (0.1)\left[\frac{(0.05)^2 + (1.0051)^2}{10}\right] = 0.01013$$

and

$$k_4 = hf(x_0 + h, y_0 + k_3) = (0.1)f(0.1, 1.01013)$$

$$= (0.1)\left[\frac{(0.1)^2 + (1.01013)^2}{10}\right] = 0.01030$$

Hence,

$$y_1 = y(0.1) = y_0 + \frac{1}{6}[k_1 + 2k_2 + 2k_3 + k_4]$$

$$= 1 + \frac{1}{6}[0.01 + 2(0.01013 + 0.01013)$$

$$+ 0.01030] = 1.01014.$$

10.43 SIMULTANEOUS DIFFERENTIAL EQUATIONS

(1) PICARD'S METHOD

The first approximations are

$$y_1 = y_0 + \int f(x, y_0, z_0)dx, z_1 = z_0 + \int g(x, y_0, z_0)dx \qquad \ldots(1)$$

and the second approximations are

$$y_2 = y_0 + \int f(x, y_1, z_1)dx, z_2 = z_0 + \int g(x, y_1, z_1)dx \qquad \ldots(2)$$

In general, the $(n + 1)^{th}$ approximations are

$$y_{n+1} = y_0 + \int f(x, y_n, z_n)dx, z_{n+1} = z_0 + \int g(x, y_n, z_n)dx \text{ and so on.} \qquad \ldots(3)$$

(2) TAYLOR'S SERIES METHOD

If h be the step size, then

$$y_1 = y(x_0 + h), z_1 = z(x_0 + h)$$

Now this method gives

$$y_1 = y_0 + hy_0' + \frac{h^2}{2!}y_0'' + \frac{h^3}{3!}y_0''' + \ldots \qquad \ldots(4)$$

and

$$z_1 = z_0 + hz_0' + \frac{h^2}{2!}z_0'' + \frac{h^3}{3!}z_0''' + \ldots \qquad \ldots(5)$$

Here y_0', y_0'' and y_0''' etc. and z_0', z_0'' and z_0''' etc. are obtained by differentiating (1) and (2) successively.

Similarly, the next approximations are

$$y_2 = y_1 + hy_1' + \frac{h^2}{2!}y_1'' + \frac{h^3}{3!}y_1''' + \ldots \qquad \ldots(6)$$

and

$$z_2 = z_1 + hz_1' + \frac{h^2}{2!}z_1''' + \ldots \qquad \ldots(7)$$

Here y_1', y_1'' and y_1''' etc. and z_1', z_1'' and z_1''', etc. are obtained by the differentiating (1) and (2) successively and putting x_1 for x and y_1 for y and z_1 for z.

Proceeding in this way, we can find other approximate values of x, y and z step by step.

(3) RUNGE-KUTTA METHOD

Let h, k and l be the step sizes for x, y and z respectively. Now starting at (x_0, y_0, z_0) and taking the step sizes h, k and l for x, y and z respectively, Runge-Kutta method gives.

$$k_1 = hf(x_0, y_0, z_0), l_1 = hg(x_0, y_0, z_0)$$

$$k_2 = hf\left(x_0 + \frac{h}{2}, y_0 + \frac{k_1}{2}, z_0 + \frac{l_1}{2}\right), l_2 = hg\left(x_0 + \frac{h}{2}, y_0 + \frac{k_1}{2}, z_0 + \frac{l_1}{2}\right)$$

$$k_3 = hf\left(x_0 + \frac{h}{2}, y_0 + \frac{k_2}{2}, z_0 + \frac{l_2}{2}\right), l_3 = hg\left(x_0 + \frac{h}{2}, y_0 + \frac{k_2}{2}, z_0 + \frac{l_2}{2}\right)$$

$$k_4 = hf(x_0 + h, y_0 + k_3, z_0 + l_3), l_4 = hg(x_0 + h, y_0 + k_3, z_0 + l_3)$$

Hence

$$y_1 = y_0 + \frac{1}{6}[k_1 + 2k_2 + 2k_3 + k_4]$$

and

$$z_1 = z_0 + \frac{1}{6}[l_1 + 2l_2 + 2l_3 + l_4]$$

Again to find y_2 and z_2 we replace x_0, y_0, z_0 by x_1, y_1, z_1 is above formulae.

Using Runge-Kutta method, we can easily solve the simultaneous differential equation.

10.44 SECOND ORDER DIFFERENTIAL EQUATIONS

Let

$$\frac{d^2y}{dx^2} = f\left(x, y, \frac{dy}{dx}\right) \qquad \qquad ...(1)$$

be the second order differential equation with initial condition

$$y(x_0) = y_0 \quad \text{and} \quad y'(x_0) = y_0' \qquad \qquad ...(2)$$

Let

$$y' = \frac{dy}{dx} = z$$

Then (1) reduces to

$$\frac{dy}{dx} \qquad \qquad ...(3)$$

and

$$\frac{dz}{dx} = f(x, y, z) \text{ with initial conditions } y(x_0) = y_0 \text{ and } z(x_0) = z_0. \qquad \qquad ...(4)$$

These equations can be solved easily by any method discussed earlier.

Solved Examples

EXAMPLE 1. *Using Runge-Kutta method, solve*

$$\frac{d^2y}{dx^2} = y'' = xy'^2 - y^2 \text{ for } x = 0.2 \text{ correct to four}$$

decimal places, subject to the initial conditions $y(0) = 1$ *and* $y'(0) = 0.$

SOLUTION. Let $y' = \dfrac{dy}{dx} = z$

Then given equation reduces to

$$\frac{dy}{dx} = z = f(x, y, z) \qquad ...(1)$$

and $\dfrac{dz}{dx} = xz^2 - y^2 = g(x, y, z) \qquad ...(2)$

with initial conditions $y(0) = 1, z(0) = 0.$

Here, we have $x_0 = 0, y_0 = 1, z_0 = 0, h = 0.2$

Using Runge-Kutta method, we have

$$k_1 = hf(x_0, y_0, z_0) = (0.2)f(0, 1, 0) = (2.0) \times 0 = 0$$

$$l_1 = hg(x_0, y_0, z_0) = (0.2)g(0, 1, 0) = (0.2) \times (-1) = -0.2$$

$$k_2 = hf\left(x_0 + \frac{h}{2}, y_0 + \frac{k_1}{2}, z_0 + \frac{l_1}{2}\right)$$

$$= (0.2)f(0.1, 1 - 0.1) = (0.2)(-0.1) = -0.02$$

$$l_2 = hg\left(x_0 + \frac{h}{2}, y_0 + \frac{k_1}{2}, z_0 + \frac{l_1}{2}\right)$$

$$= (0.2)g(0.1, 1 - 0.1) = (0.2)[(0.1)(-0.1)^2 - (1)^2]$$

$$= -0.1998$$

$$k_3 = hf\left(x_0 + \frac{h}{2}, y_0 + \frac{k_2}{2}, z_0 + \frac{l_2}{2}\right)$$

$$= (0.2)f(0.1, 0.99, -0.0999) = (0.2)(-0.0999)$$

$$= -0.01998$$

$$l_3 = hg\left(x_0 + \frac{h}{2}, y_0 + \frac{k_2}{2}, z_0 + \frac{l_2}{2}\right)$$

$$= (0.2)g(0.1, 0.99, -0.0999)$$

$$= (0.2)[(0.1)(-0.0999)^2 - (0.99)^2] = -0.1958$$

$$k_4 = hf(x_0 + h, y_0 + k_3, z_0 + l_3)$$

$$= (0.2)f(0.2, 0.98002, -0.1958)$$

$$= (0.2)(-0.1958) = -0.03915$$

$$l_4 = hg(x_0 + h, y_0 + k_3, z_0 + l_3)$$

$$= (0.2)g(0.2, 0.98002, -0.1958)$$

$$= (0.2)[(0.02)(-0.1958)^2 - (0.98002)^2]$$

$$= -0.1906$$

At $x = 0.2$, we have

$$y(0.2) = y = y(0) + \frac{1}{6}[k_1 + 2k_2 + 2k_3 + k_4]$$

$$= 1 + \frac{1}{6}[0 + 2(-0.02 - 0.01998) + (-0.03916)] = 0.9801$$

and $z = z_0 + \dfrac{1}{6}[l_1 + 2l_2 + 2l_3 + l_4]$

$$= 0 + \frac{1}{6}[-0.2 + 2(-0.1998 - 0.1958) - 0.1906)]$$

$$= -0.1970$$

EXAMPLE 2. *Using Picard's method to find approximate values of y and z at x = 0.1, given that y(0) = 2,*

$$z(0) = 1 \text{ and } \frac{dy}{dx} = x + z, \frac{dz}{dx} = x - y^2.$$

SOLUTION. We have $x_0 = 0, y_0 = 2, z_0 = 1$

and $\dfrac{dy}{dx} = f(x, y, z) = x + z$

$$\frac{dz}{dx} = g(x, y, z) = x - y^2$$

So, $f(x, y_0, z_0) = f(x, 2, 1) = x + 1$

$g(x, y_0, z_0) = g(x, 2, 1) = x - 4$

Using Picard's method the first approximation is given by

$y_1 = y_0 + \int_{x_0}^x f(x, y_0, z_0)dx$

$= 2 + \int_0^x (x+1)dx = 2 + x + \dfrac{x^2}{2}$

$z_1 = z_0 + \int_{x_0}^x g(x, y_0, z_0)dx$

$= 1 + \int_0^x (x-4)dx = 1 + \dfrac{x^2}{2} - 4x$

The second approximations are

$y_2 = y_0 + \int_{x_0}^x f(x, y_1, z_1)dx$

$= 2 + \int_0^x \left(x + 1 + \dfrac{x^2}{2} - 4x\right)dx = 2 + x - \dfrac{3x^2}{2} + \dfrac{x^3}{6}$

and

$z_2 = z_0 + \int_{x_0}^x g(x, y_1, z_1)dx = 1 + \int_0^x \left[x - \left(2 + x + \dfrac{x^2}{2}\right)^2\right]dx$

$= 1 + \int_0^x \left[-4 - 3x - 3x^2 - x^3 - \dfrac{x^4}{4}\right]dx$

$= 1 - 4x - \dfrac{3}{2}x^2 - x^3 - \dfrac{x^4}{4} - \dfrac{x^5}{20}$

Similarly, the third approximations are

$y_3 = y_0 + \int_0^x f(x, y_2, z_2)dx = 2 + \int_0^x (x + z_2)dx$

$= 2 + \int_0^x \left(x + 1 - 4x - \dfrac{3}{2}x^2 - x^3 - \dfrac{x^4}{4} - \dfrac{x^5}{20}\right)dx$

$= 2 + x - \dfrac{3}{2}x^2 - \dfrac{1}{2}x^3 - \dfrac{x^4}{4} - \dfrac{x^5}{20} - \dfrac{x^6}{120}$

and

$z_3 = z_0 + \int_0^x g(x, y_2, z_2)dx = 1 + \int_0^x (x - y_2^2)dx$

$= 1 + \int_0^x \left[x - \left(2 + x + \dfrac{3x^2}{2} + \dfrac{x^3}{6}\right)^2\right]dx$

$= 1 + \int_0^x \left[x - \left(4 + x^2 + \dfrac{9x^4}{4} + \dfrac{x^6}{36} + 4x\right.\right.$

$\left.\left. - 6x^2 + \dfrac{2}{3}x^3 - 3x^2 + \dfrac{1}{3}x^4 - \dfrac{1}{2}x^5\right)\right]dx$

$= 1 + \int_0^x \left[-4 - 3x + 5x^2 + \dfrac{7}{3}x^3\right.$

$\left. - \dfrac{31}{12}x^4 + \dfrac{1}{2}x^5 - \dfrac{1}{36}x^6\right]dx$

$= 1 - 4x - \dfrac{3}{2}x^2 + \dfrac{5}{2}x^3 + \dfrac{7}{12}x^4 - \dfrac{31}{60}x^5 + \dfrac{1}{12}x^6 - \dfrac{1}{252}x^7$

and so on.

Putting $x = 0.1$, we get

$y_1 = 2.105, \qquad z_1 = 0.605,$

$y_2 = 2.08517, \quad z_2 = 0.58397$

$y_3 = 2.08447, \quad z_3 = 0.58672$

Hence, $y(0.1) = 2.0845, z(0.1) = 0.5867$ correct to four decimal places.

Exercise-10.21

1. Given $\dfrac{dy}{dx} = 1 + y^2$ where $y = 0$ when $x = 0$, find $y_{(0.2)}, y_{(0.4)}$ and $y_{(0.6)}$, using Runge-Kutta formula of order four.

2. Use Runge-Kutta formula of fourth order to find the numerical solution at $x = 0.8$ for $\dfrac{dy}{dx} = \sqrt{x+y}$, $y_{(0.4)} = 0.41$. Assume the step length $h = 0.2$.

   (SVTU–2007S)

3. Using Runge-Kutta method to solve $10y' = x^2 + y^2, y(0) = 1$ for the interval $0 < x \le 0.4$ with $h = 0.1$.

4. Use Runge-Kutta fourth order to approximate the solution of the initial value problem, given that $y' = y^2 - x^2$ with $y(0) = 1$, interval $0 \le x \le 0.2$, step length $h = 0.1$.

5. Solve the differential equation $\dfrac{dy}{dx} = \dfrac{2x-1}{x^2}y + 1$ when $x_0 = 1$, $y_0 = 2$, $h = 0.2$. Obtain $y_{(1.2)}$ and $y_{(1.4)}$ using Runge-Kutta method.

6. Using Runge-Kutta method compute the value of y for $x = 0.5, 1, 1.5$ given that $y' = 0.31 + 0.25y + 0.3x^2$ with $y(0) = 0.72$.

7. Use Runge-Kutta quadratic method to calculate three additional points on the solution curve of the problem $y' = 1 - 2xy, y_{(0)} = 0$ with $h = 0.1$.

8. Solve the simultaneous differential equation $y'' = (x^2 + y^2)$ $(1 + y'^2)$ for $x = 0.5$, using Runge-Kutta method, initial value $x = 0, y = 1, y' = 0$. Take $h = 0.5$.

   where $y' = \dfrac{dy}{dx} = z$ so that $y'' = \dfrac{d^2y}{dx^2} = \dfrac{dz}{dx}$.

9. Solve the initial value problem $u' = 2tu^2, u(0) = 1$ with $h = 0.2$ on the interval $(0, 0.4)$. Use Runge-Kutta method.

10. Using fourth order Runge-Kutta method find the value of y when $x = 1$ given that $y = 1$ when $x = 0$ (taking $n = 2$) and $\dfrac{dy}{dx} = \dfrac{y-x}{y+x}$.

    (VTU–2011S)

11. Use Runge-Kutta method to approximate y when $x = 1.1$ given that $y = 1.2$ when $x = 1$ and $\dfrac{dy}{dx} = 3x + y^2$.

12. Using fourth order Runge-Kutta method, compute $y(0.2)$ and $y(0.4)$ when $10\dfrac{dy}{dx} = x^2 + y^2, y(0) = 1$; taking $h = 0.1$.

    (ROHTAK–2003, BHOPAL–2006)

1. $y_{(0.2)} = 0.2027, y_{(0.4)} = 0.4228, y_{(0.6)} = 0.6841$ **2.** $y_{(0.6)} = 0.61035, y_{(0.8)} = 0.84899$

3. $1.0101, 1.0207, 1.0318, 1.0438$ **4.** $y_{(0.1)} = 1.1108, y_{(0.2)} = 1.2470$ **5.** $y_{(1.2)} = 2.658913, y_{(1.4)} = 3.432851$

6. $y_{(0.5)} = 0.8891, y_{(1)} = 0.6674, y_{(1.5)} = 0.5794$ **7.** $y_{(0.1)} = 0.99336, y_{(0.2)} = 1.194751, y_{(0.3)} = 0.282632$

8. $y_{(0.5)} = 1.1423$ **9.** 0.87343769 **10.** 1.5488 **11.** 1.7278 **12.** $1.0207 : 1.038$

10.45 MILNE'S METHODS

Let $y' = f(x, y)$ with $y = y_0, x = x_0$. To find an approximate value of y for $x = x_0 + nh$ by Milne method, we proceed as follows :

The value $y_0 = y(x_0)$, is given, we compute

$y_1 = y(x_0 + h), y_2 = y(x_0 + 2h), y_3 = y(x_0 + 3h)$ by Taylor's series and Picard's method.

Now we solve $f_0 = f(x_0, y_0),$ $f_1 = f(x_0 + h, y_1),$ $f_2 = f(x_0 + 2h, y_2),$ $f_3 = f(x_0 + 3h, y_3)$

Then to find $y_4 = y(x_0 + 4h)$, we putting Newton's forward interpolation formula

$$f(x, y) = f_0 + n\Delta f_0 + \frac{n(n-1)}{2}\Delta^2 f_0 = \frac{n(n-1)(n-2)}{6}\Delta^3 f_{0+\dots}$$

in the relation

$$y_4 = y_0 + \int_{x_0}^{x_0+4h} f(x, y)dx = y_0 + \int_{x_0}^{x_0+4h} \left(f_0 + n\Delta f_0 + \frac{n(n-1)}{2}\Delta^2 f_0 + \dots \right)dx \qquad \text{[Putting } x = x_0 + nh, dx = hdn]$$

$$= y_0 + h\int_0^4 \left(f_0 + n\Delta f_0 + \frac{n(n-1)}{2}\Delta^2 f_0 + \dots \right)dn = y_0 + \left(4f_0 + 8\Delta f_0 + \frac{20}{3}\Delta^2 f_0 + \frac{8}{3}\Delta^3 f_0 + \dots \right)$$

Neglecting fourth and higher order differences and expressing Δf_0, $\Delta^2 f_0$ and $\Delta^3 f_0$ in terms of the function values, we obtain

$$y_4 = y_0 + \frac{4h}{3}(2f_1 - f_2 + 2f_3)$$

is called a predictor formula.

To found y_4 we obtain a first approximation to $f_4 = f(x_0 + 4h, y_4)$

Then a best value of y_4 is found by Simpson's rule given by

$$y_4 = y_2 + \frac{h}{3}(f_2 + 4f_3 + f_4)$$

is called a corrector formula.

☞ REMARK

• Firstly, an improved value of f_4 is computed and again the corrector is applied to find a still better value of y_4. We repeat step until y_4 remains unchanged.

Once y_4 and f_4 are obtained to desired degree of accuracy $y_5 = y(x_0 + 5h)$ is found from the predictor formula as

$$y_5 = y_1 + \frac{4h}{3}(2f_2 - 4f_3 + 2f_4)$$

Now $f_5 = f(x_0 + 5h_1, y_5)$ is solved. Then a better approximation to the value of y_5 is obtained from the corrector as

$$y_5 = y_3 + \frac{h}{3}(f_3 + 4f_4 + f_5)$$

We repeat this step till y_5 becomes same and we then proceed to solved y_6 as before.

Thus in Milne's predictor-corrector method, to ensure greater accuracy, we must first improve the accuracy of the starting values and then sub-divide the intervals.

☞ REMARK

• The general Milne's predictor and corrector formula are

$$y_{n+1} = y_{n-3} + \frac{4h}{3}[2y'_{n-2} - y'_{n-1} + 2y'_n] \quad \text{and} \quad y_{n+1} = y_{n-1} + \frac{h}{3}[y'_{n-1} + 4y'_n + y'_{n+1}]$$

 Solved Examples

EXAMPLE 1. *Using Milne method obtain $y_{(4)}$ from the given table of tabulated values and $y' = y^2 - x^2$.*

x	0	0.1	0.2	0.3
y	1	1.11	1.25	1.42
f	1	1.22	1.52	1.92

SOLUTION. By the predictor formula, we have

$$y_4 = y_0 + \frac{4h}{3}[2f_3 - f_2 - 2f_1]$$

$$= 1.0 + \frac{4 \times 0.1}{3}[2(1.92) - 1.52 + 2(1.22)] = 1.63$$

$\Rightarrow y_4 = 1.63$

Now using this value we compute

$$f_4 = y_4^2 - x_4^2 = (1.63)^2 - (0.4)^2 = 2.50$$

$\Rightarrow f_4 = 2.50$

Now by the corrector formula, we have

$$y_4 = y_2 + \frac{h}{3}[f_2 + 4f_3 + f_4]$$

$$= 1.25 + \frac{0.1}{3}[1.52 + 4(1.92) + 1.50] = 1.64$$

$\Rightarrow y_4 = 1.64$

Hence, $y_{(0.4)} = 1.64$

EXAMPLE 2. *Solve $y' = 2e^x - y$ at $x = 0.4$ and $x = 0.5$ by Milne's method, given their values at the four points.*

x	0	0.1	0.2	0.3
y	2	2.010	2.040	2.090

SOLUTION. We want to find the value of f_1, f_2, f_3.

$$f_1 = 2e^{0.1} - 2.010 = 0.2003$$

$$f_2 = 2e^{0.2} - 2.040 = 0.4028$$

$$f_3 = 2e^{0.3} - 2.090 = 0.6097$$

By Milne's predictor formula, we have

$$y_4 = y_0 + \frac{4h}{3}[2f_1 - f_2 + 2f_3]$$

$$= 2 \times \frac{4 \times 0.1}{3}[2(0.2003)$$

$$- (0.4028) + 2(0.6097)] \approx 2.1623$$

Now

$$f_4 = 2e^{0.4} - 2.1623 = 0.8213494 \approx 0.8213$$

Now by corrector formula, we get

$$y_4 = y_2 + \frac{h}{3}[f_2 + 4f_3 + f_4]$$

$$= 2.04 + \frac{0.1}{3}[0.4028 + 4(0.6097 + 0.8213]$$

$$= 2.162096 \approx 2.1621$$

Again by using predictor formula

$$y_5 = y_1 + \frac{4h}{3}[2f_2 - f_3 + 2f_4]$$

$$= 2.10 + \frac{4 \times 0.1}{3}[2(0.4028) - 0.6097 + 2(0.8215)]$$

$$= 2.2551867 \approx 2.2552$$

then $f_5 = 2e^{0.5} - 2.2552 = 1.0422425 \approx 1.0422$

By corrector formula, we get

$$y_5 = y_3 + \frac{h}{3}[f_3 + 4f_4 + f_5]$$

$$= 2.090 + \frac{0.1}{3}[0.6097 + 4(0.8215) + 1.0422]$$

$$= 2.2545967 \approx 2.255$$

EXAMPLE 3. *Apply Milne's method to find a solution of the differential equation $dy/dx = x - y^2$ in the range $0 \le x \le 1$ with $y(0) = 0$.*

(VTU–2009, ANNA–2005, ROHTAK–2005)

SOLUTION. Here, we use Picard's method to compute y_1, y_2 and y_3. Picard's successive approximations are given by

$$y_n = y_0 + \int_{x_0}^{x} f(x, y_{n-1})dx$$

for $n = 1$, we have

$$y_1 = 0 + \int_0^x x\,dx = \frac{x^2}{2}$$

for $n = 2$,

$$y_2 = 0 + \int_0^x \left[x - \left(\frac{x^2}{2}\right)^2\right]dx = \frac{x^2}{2} - \frac{x^5}{20}$$

Similarly,

$$y_3 = 0 + \int_0^x \left[x - \left(\frac{x^2}{2} - \frac{x^5}{20}\right)^2\right]dx$$

$$= \frac{x^2}{2} - \frac{x^5}{20} + \frac{x^8}{160} - \frac{x^{11}}{4400}$$

Let us take $y = \frac{x^2}{2} - \frac{x^5}{20}$ for finding the various values of y_i's and f_i's

$$y_1 = y(0.2) = 0.019984 = 0.02,$$

$$f_1 = 0.1996$$

$$y_2 = y(0.4) = 0.079488 = 0.0795,$$

$$f_2 = 0.3937$$

$$y_3 = y(0.6) = 0.176112 = 0.176,$$

$$f_3 = 0.5690$$

Now, using predictor formula, we get

$$y_4 = y_0 + \frac{4h}{3}(2f_1 - f_2 + 2f_3)$$

$$= 0 + \frac{4 \times 0.2}{3}(2 \times 0.01996 - 0.3937 + 2 \times 0.5690)$$

$$= 0.3049333 = 0.3049$$

Further, using corrector formula, we get

$$y_4 = y_2 + \frac{h}{3}(f_2 + 4f_3 + f_4)$$

$$= 0.0795 + \frac{0.2}{3}(0.3937 + 4 \times 0.5690 + 0.7070)$$

$[\because f_4 = x_4 + y_4 = 0.8 - (0.3049)^2 = 0.7070359 \approx 0.7070]$

Hence, $y_4 = 0.3046$ at $x = 0.8$ and corrected $f_4 = 0.8 - (0.3046)^2 = 0.7072$

Again, using predictor formula, we get

$$y_5 = y_1 + \frac{4h}{3}(2f_2 - f_3 + 2f_4)$$

$$= 0.2 + \frac{4 \times 0.2}{3}(2 \times 0.3937 - 0.5690 + 2 \times 0.7072)$$

$$= 0.4554133 \approx 0.4554$$

Now, $f_5 = x_5 + y_5 = 1 - (0.4554)^2$

$$= 0.792610 \approx 0.7926$$

Using corrector formula, we get

$$y_5 = y_3 + \frac{h}{3}(f_3 + 4f_4 + f_5)$$

$$= 0.176 + \frac{0.2}{3}(0.5690 + 4 \times 0.7072 + 0.7926)$$

$$= 0.45536 \approx 0.4554$$

Hence, $y(1) = 0.4554$.

Exercise-10.22

1. Solve the differential equation
   $y' = x^3 - y^2 - 2$, by Milne's method for $x = 0.3$ to $x = 0.6$ with initial value are $x(0) = 1$. The values of y for $x = -0.1$, 0.1 and 0.2 are to be computed by series expansion.

2. Using Milne's method, obtain $y_{(0.4)}$ and $y_{(0.5)}$ that satisfy the solution of $y' = 2 + y - 2x$ with $y(0) = 1$.

3. Find the value of y corresponding to $x = 0.08$ and $x = 0.10$ that satisfy the solution of $y' = x + y^2$ with $y_{(0)} = 1 (h = 0.02)$.

4. By Milne's method solve $y' = 2 - xy^2$ with $y(0) = 1$ for $x = 1$ taking $h = 0.2$.

5. By Milne's method solve $y_{(0.3)}$ from $y' = x^2 + y^2$, $y_{(0)} = 1$. Find the initial values $y_{(-0.1)}$, $y_{(0.1)}$ and $y_{(0.2)}$ from the Taylor's series method.   **(ROHTAK–2007, 04)**

6. Find the solution of $y' = x + y$, $y(0) = 0$ for $0.4 \le x \le 1.0$ with $h = 0.1$ by the predictor-corrector formulae.

7. The differential equation $\dfrac{dy}{dx} = x + y, y(0) = 1$, satisfies the

following values :

x	0.1	0.2	0.3
y	1.1103	1.2428	1.3897

8. Solve, by Milne's method, the differential equation,
   $\dfrac{dy}{dx} = y - x^2$, when $y = 1$ at $x = 0$, $y = 1.1219$ at $x = 0.2$, $y = 1.4682$ at $x = 0.4$ and $y = 0.1737$ at $x = 0.6$ and compute at $x = 0.8$.

9. Solve the initial value problem
   $$\frac{dy}{dx} = 1 + xy^2, y(0) = 1$$
   for $x = 0.4$, $x = 0.5$, using Milne's method when it is given that $y(0.1) = 1.105$, $y(0.2) = 1.223$, $y(0.3) = 1.355$.

10. Given $2\dfrac{dy}{dx} = (1 + x^2)y^2$ and $y(0) = 1$, $y(0.1) = 1.06$, $y(0.2) = 1.12$, $y(0.3) = 1.21$. Evaluate $y(0.4)$ by Milne's method.   **(VTU–2011S, MADRAS–2003)**

Answers

1.	x	0.3	0.4	0.5	0.6
	y	0.061493	0.45625	0.29078	0.12566

2. 2.2918, 2.6488   **3.** 1.09035, 1.11649   **4.** 1.6505   **5.** 1.4392
6. 0.0918, 0.1487, 0.2221, 0.3138, 0.4255, 0.5596, 0.7183   **7.** 1.5836   **8.** 2.011   **9.** 1.538   **10.** 1.2797.

10.46   NUMERICAL SOLUTION OF PARTIAL DIFFERENTIAL EQUATIONS

A partial differential equation is a differential equation involving more than one independent variable. These variables determine the behaviour of the dependent variable as described by their partial derivatives contained in the equation. Partial differential equations arise in the study of many branches of applied mathematics, *e.g.*, in the study of displacement of a vibrating string, in heat flow problems, in fluid flow analysis, in analysis of torsion in a bar subjected to twisting and in the study of diffusion of matter and so on. Only a few of these equations can be solved by analytical methods. In most cases, we go in for numerical methods to approximate solution. Of all the numerical methods available, for the solution of partial differential equations, the method of finite differences is most commonly used. In this method, the derivative appearing in the equation and the boundary conditions are replaced by their finite difference approximations. After that, the given equation is changed to a system of linear equations, which are solved by iterative procedures.

10.46.1 DIFFERENCE QUOTIENTS

Definition. *A difference quotient is the quotient obtained by dividing the difference between two values of a function by the difference between the two corresponding values of the independent variable.*

If $u(x)$ is a function, then difference coefficient is given by $\dfrac{u(x+h) - u(x)}{h}$.

REMARK

- The limiting value of above coefficient is the derivative of u w.r.t. x, *i.e.*, du/dx. .
   If $u = u(x, y)$ is a function of two independent variables x and y, then we proeed as follows :
   First, let us consider the difference in x-direction.
   Consider the Taylor's series for $u(x, y_0)$ about the point (x_0, y_0) is

$$u(x, y_0) = u(x_0, y_0) + (x - x_0)u_x(x_0, y_0) + \frac{(x - x_0)^2}{2!}u_{xx}(x_0, y_0) \qquad \qquad ...(1)$$

where $x_0 \le t \le x$.
Put $x = x_0 + h$ in (1), we get

$$u(x_0 + h, y_0) = u(x_0, y_0) + hu_x(x_0, y_0) + \frac{h^2}{2!}u_{xx}(t, y_0) \Rightarrow \frac{u(x_0 + h, y_0) - u(x_0, y_0)}{h} = u_x(x_0, y_0) + \frac{h}{2!}u_{xx}(t, y_0)$$

which gives that

$$u_x(x_0, y_0) = \frac{u(x_0 + h, y_0) - u(x_0, y_0)}{h} - \frac{h}{2!}u_{xx}(t, y_0) \qquad \qquad ...(2)$$

where $x_0 \le t \le x_0 + h$.

If we replace $u_x(x_0, y_0)$ by $\dfrac{u(x_0 + h, y_0) - u(x_0, y_0)}{h}$, then we get the truncation error $= -\dfrac{h}{2!} u_{xx}(t, y_0)$

Therefore, we have the finite difference formula for the first partial derivative at (x_0, y_0) as

$$u_x(x_0, y_0) = \frac{u(x_0 + h, y_0) - u(x_0, y_0)}{h} \qquad \ldots(3)$$

which is called a forward difference approximation to $u_x(x_0, y_0)$.

Similarly by putting $x = x_0 - h$ in (1), we get

$$u_x(x_0, y_0) = \frac{u(x_0, y_0) - u(x_0 - h, y_0)}{h} \qquad \ldots(4)$$

which is known as backward approximation to $u_x(x_0, y_0)$.

Now, to find an approximation for u_{xx}, we use both forward and backward difference. Using forward difference for u_x, we get

$$u_{xx}(x_0, y_0) = \frac{u_x(x_0 + h, y_0) - u_x(x_0, y_0)}{h} \qquad \ldots(5)$$

Since there is a bias in the forward direction, we use backward difference for u_x to avoid this effect, i.e.,

$$u_{xx}(x_0, y_0) = \frac{u(x_0, y_0) - u_x(x_0 - h, y_0)}{h} \qquad \ldots(6)$$

Changing x_0 to $x_0 + h$ in (6), we get

$$u_x(x_0 + h, y_0) = \frac{u(x_0 + h, y_0) - u(x_0, y_0)}{h} \qquad \ldots(7)$$

Now, using (6) and (7) in (5), we get

$$u_{xx}(x_0, y_0) = \frac{u(x_0 + h, y_0) - 2u(x_0, y_0) + u(x_0 - h, y_0)}{h^2}$$

Proceeding in the same way, we have the following formula for the derivatives in y-direction taking the step size as k.

(i) $u_y(x_0, y_0) = \dfrac{u(x_0, y_0 + k) - u(x_0, y_0)}{k}$ (Forward difference formula)

(ii) $u_y(x_0, y_0) = \dfrac{u(x_0, y_0) - 4(x_0, y_0 - k)}{k}$ (Backward difference formula)

(iii) $u_{yy}(x_0, y_0) \quad \dfrac{u(x_0, y_0 + k) - 2u(x_0, y_0) + u(x_0, y_0 - k)}{}$

✎ REMARK

- The truncation error in the above formula will be given by $-\dfrac{k^2}{12} u_{yy}(x_0, \xi)$, where $y_0 - k \le \xi \le y_0 + k$

10.46.2 CLASSIFICATION OF PARTIAL DIFFERENTIAL EQUATIONS

Consider the general linear partial differential equation of the second order in two independent variables as given by

$$A\frac{\partial^2 u}{\partial x^2} + B\frac{\partial^2 u}{\partial x \partial y} + C\frac{\partial^2 u}{\partial y^2} + D\frac{\partial u}{\partial x} + E\frac{\partial u}{\partial y} + Fu = 0 \qquad \ldots(1)$$

where A, B, C, D, E and F are in general functions of x and y.

Then, (1) is said to be

(i) Elliptic if $B^2 - 4AC < 0$ (ii) Parabolic if $B^2 - 4AC = 0$ (iii) Hyperbolic if $B^2 - 4AC > 0$

10.46.3 ELLIPTIC EQUATIONS

Elliptic equations are governed by conditions on the boundary of closed domain. Here, we consider the two most commonly used elliptic equations, namely Laplace equations and Poisson's equation.

The Laplace equation is given by $\nabla^2 u = \dfrac{\partial^2 u}{\partial x^2} + \dfrac{\partial^2 u}{\partial y^2} = 0$

and the Poisson's equation is given by $\dfrac{\partial^2 u}{\partial x^2} + \dfrac{\partial^2 u}{\partial y^2} = f(x, y)$

SOLUTION OF LAPLACE EQUATION

Consider the Laplace equation $\dfrac{\partial^2 u}{\partial x^2} + \dfrac{\partial^2 u}{\partial y^2} = 0$ which can be written as

$$u_{xx} + u_{yy} = 0 \qquad \qquad \dots(1)$$

Relacing the derivative in (1) by their difference approximation, we get

$$\frac{1}{h^2}(u_{i-1,j} - 2u_{i,j} + u_{i+1,j}) + \frac{1}{k^2}(u_{i,j-1} - 2u_{i,j} + u_{i,j+1}) = 0 \qquad \dots(2)$$

Taking a square mesh and putting $h = k$ we get from (2)

$$u_{i,j} = \frac{1}{4}(u_{i+1,j} + u_{i-1,j} + u_{i,j+1} + u_{i,j-1}) \qquad \dots(3)$$

In solving a given boundary value problem we can use the above difference equation (3) to complete the values at interior mesh points in terms of the values on the boundary. The formule given by (3) is known as standard five point formula (SFPF).

✎ REMARKS

- We may also use the formula given below, in place of (3)

$$u_{i,j} = \frac{1}{4}\left[u_{i-1,j-1} + u_{i+1,j-1} + u_{i-1,j+1} + u_{i+1,j+1} \right]$$

which shows that the value of $u_{i,j}$ is the arithmetic mean of its values at the four neighbouring diagonal mesh-points. This is called the diagonal five point formula (DFPF).

- The DFPF is valid, because we know that the Laplace equation remains invariant when the co-ordinate axes are rotated through 45°. But the error in DFPF is four times the error in SFPF. Therefore, we prefer SFPF.

PICTORIAL REPRESENTATION

The SFPF and DFPF are represented in the figures given below :

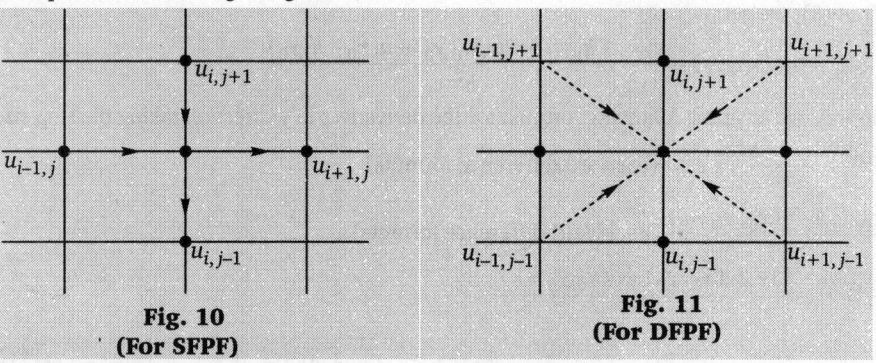

Fig. 10
(For SFPF)

Fig. 11
(For DFPF)

10.46.4 SOLUTION OF LAPLACE EQUATIONS BY LIEBERMANN'S ITERATION PROCESS

Consider the Laplace equation $\dfrac{\partial^2 u}{\partial x^2} + \dfrac{\partial^2 u}{\partial y^2} = 0$ with the given boundary conditions. Let us assume the

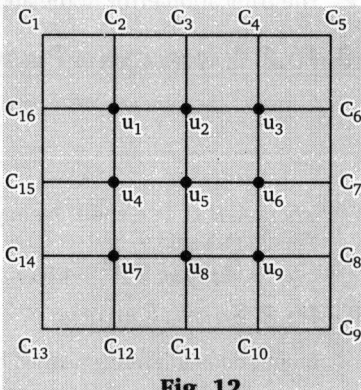

function $u(x, y)$ is required over a rectangular region with boundary C. Divide the rectangle R into a network of small squares of side h. Let the values of $u(x, y)$ on the boundary C be given by C_i and the interior mesh points and boundary points. Now to start the iteration process, initially we find rough values at interior points and then improve it by iterative process.

First of all, we find u_5, at the centre of the square by taking the average of four boundary values (SFPF)

$$\therefore \quad u_5 = \frac{1}{4}\left[C_{15} + C_7 + C_3 + C_{11} \right]$$

Next, find the initial values at the centre of the four large inner squares, using DFPF. Therefore,

$$u_1 = \frac{1}{4}\left[u_5 + C_1 + C_3 + C_{15} \right], u_3 = \frac{1}{4}\left[u_5 + C_5 + C_3 + C_7 \right], u_7 = \frac{1}{4}\left[u_5 + C_{13} + C_{11} + C_{15} \right],$$

$$u_9 = \frac{1}{4}\left[u_5 + C_9 + C_7 + C_{11} \right]$$

Fig. 12

The values of the remaining interior points are obtained by SFPF such that

$$u_2 = \frac{1}{4}\left[u_1 + u_3 + C_3 + u_5 \right], u_4 = \frac{1}{4}\left[C_{15} + u_5 + u_1 + u_7 \right], u_6 = \frac{1}{4}\left[u_5 + C_7 + u_3 + u_9 \right], u_8 = \frac{1}{4}\left[u_7 + u_9 + u_5 + C_{11} \right]$$

✎ REMARK

- To improve the accuracy, we start with u_1 and iterate it using the latest available values of the four adjacent points. Therefore, we iterate all the mesh-points systematically from left to right along successive rows.

(A) ITERATION FORMULAE

(1) Liebermann iteration formula

$$u_{i,j}^{(n+1)} = \frac{1}{4}\left[u_{i-1,j}^{(n+1)} + u_{i+1,j}^{(n)} + u_{i,j-1}^{(n)} + u_{i,j+1}^{(n+1)}\right]$$

(2) Gauss-Seidel method

$$u_{i,j}^{(n+1)} = \frac{1}{4}\left[u_{i-1,j}^{(n+1)} + u_{i+1,j}^{(n)} + u_{i,j+1}^{(n)} + u_{i,j-1}^{(n+1)}\right]$$

(3) Jacobi's method

$$u_{i,j}^{(n+1)} = \frac{1}{4}\left[u_{i-1,j}^{(n)} + u_{i+1,j}^{(n)} + u_{i,j+1}^{(n)} + u_{i,j-1}^{(n)}\right]$$

☞ REMARKS

- Here the subscript denotes the iteration number.
- Initial values may be obtained by either taking diagonal average or cross-average of the adjoining four point.
- The accuracy of calculation depends on the mesh size

Solved Examples

EXAMPLE 1. *Solve the elliptic equation* $\dfrac{\partial^2 u}{\partial x^2} + \dfrac{\partial^2 u}{\partial y^2} = 0$ *for the following square mesh with boundary values given below* : (ROHTAK–2005, VTU–2005)

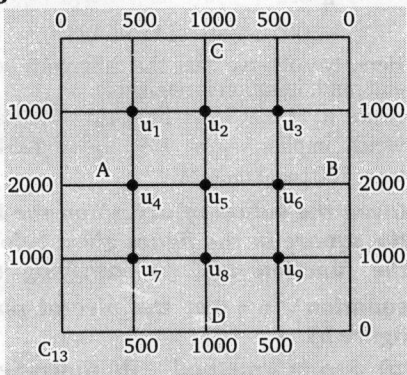

Fig. 13

SOLUTION. Let $u_1, u_2, ..., u_9$ be the values of x at the interior mesh-points. From the above figure, we have the following symmetry :
$u_7 = u_1, u_8 = u_2, u_9 = u_3, u_3 = u_1, u_6 = u_4, u_9 = u_7.$
Therefore, we find the values of u_1, u_2, u_3, u_4 and u_5.

Initial values. The initial values of u_1, u_2, u_3, u_4 and u_5 are given by

$$u_5^{(0)} = \frac{1}{4}[2000 + 2000 + 1000 + 1000] = 15000 \quad \text{(SFPF)}$$

$$u_1^{(0)} = \frac{1}{4}[0 + 1500 + 1000 + 2000] = 1125 \quad \text{(DFPF)}$$

$$u_2^{(0)} = \frac{1}{4}[1125 + 1125 + 1000 + 1500] = 1187.5 \quad \text{(SFPF)}$$

$$u_4^{(0)} = \frac{1}{4}[2000 + 1500 + 1125 + 1125] = 1437.5 \quad \text{(SFPF)}$$

Here, we have the following iteration formulae by SFPF:

$$u_1^{(n+1)} = \frac{1}{4}\left[1000 + u_2^{(n)} + 500 + u_4^{(n)}\right]$$

$$u_2^{(n+1)} = \frac{1}{4}\left[u_1^{(n+1)} + u_1^{(n+1)} + 1000 + u_5^{(n)}\right]$$

$$u_4^{(n+1)} = \frac{1}{4}\left[2000 + u_5^{(n)} + u_1^{(n+1)} + u_1^{(n+1)}\right]$$

$$u_5^{(n+1)} = \frac{1}{4}\left[u_4^{(n+1)} + u_4^{(n+1)} + u_2^{(n+1)} + u_2^{(n+1)}\right]$$

Now, we have the following iterations :

First iteration ($n = 0$)

$$u_1^{(1)} = \frac{1}{4}[1000 + 1187.5 + 500 + 1437.5] = 1031.25$$

$$u_2^{(1)} = \frac{1}{4}[1031.5 + 1031.25 + 1000 + 1500] = 1140.625$$

$$u_4^{(1)} = \frac{1}{4}[2000 + 1500 + 1031.25 + 1031.25] = 1390.625$$

$$u_5^{(1)} = \frac{1}{4}[1390.625 + 1390.625 + 1140.625 + 1140.625]$$
$$= 1265.625$$

Second iteration ($n = 1$)

$$u_1^{(2)} = \frac{1}{4}[1000 + 1149.625 + 500 + 1390.625] = 1007.8125$$

$$u_2^{(2)} = \frac{1}{4}[1007.8125 + 1007.8125 + 100 + 1265.65]$$
$$= 1070.3125$$

$$u_4^{(2)} = \frac{1}{4}[2000 + 1265.625 + 1007.8125 + 1007.8125]$$
$$= 1320.3125$$

$$u_5^{(2)} = \frac{1}{4}[1320.3125 + 1320.3125 + 1070.3125$$
$$+ 1070.3125] = 1195.3125$$

Third iteration ($n = 2$)

$$u_1^{(3)} = \frac{1}{4}[1000 + 1070.3125 + 500 + 1320.3125]$$
$$= 972.6525$$

$$u_2^{(3)} = \frac{1}{4}[972.65625 + 972.65625 + 1000 + 1195.3125]$$
$$= 1035.1563$$

$$u_4^{(3)} = \frac{1}{4}[2000 + 1195.3125 + 972.65625 + 972.65625]$$
$$= 1285.1563$$

$$u_5^{(3)} = \frac{1}{4}[1285.1563 + 1285.1563 + 1035.1563$$
$$+ 1035.1563] = 1160.1563$$

Fourth iteration ($n = 3$)

$$u_1^{(4)} = \frac{1}{4}[1000 + 1035.1563 + 500 + 1285.1563]$$
$$= 955.07815$$

$$u_2^{(4)} = \frac{1}{4}[955.07815 + 955.07815 + 1000 + 1160.1563]$$
$$= 1017.5782$$

$$u_4^{(4)} = \frac{1}{4}[2000 + 1160.1563 + 955.07815 + 955.07815]$$

$$= 1267.5782$$

$$u_5^{(4)} = \frac{1}{4}[1267.5782 + 1267.5782 + 1017.5782$$

$$+ 1017.5782] = 1142.5782$$

Fifth iteration (n = 4)

$$u_1^{(5)} = \frac{1}{4}[1000 + 1017.5782 + 500 + 1267.5782]$$

$$= 946.2891$$

$$u_2^{(5)} = \frac{1}{4}[946.2891 + 946.2891 + 1000 + 1142.5782]$$

$$= 1008.7891$$

$$u_4^{(5)} = \frac{1}{4}[2000 + 1142.5782 + 946.2891 + 946.2891]$$

$$= 1258.7891$$

$$u_5^{(5)} = \frac{1}{4}[1258.7891 + 1258.7891 + 1008.7891$$

$$+ 1008.7891] = 1133.7891$$

Sixth iteration (n = 5)

$$u_1^{(6)} = \frac{1}{4}[1000 + 1008.7891 + 500 + 1258.7891]$$

$$= 941.89455$$

$$u_2^{(6)} = \frac{1}{4}[941.89455 + 941.89455 + 1000 + 1133.7891]$$

$$= 1004.3946$$

$$u_4^{(6)} = \frac{1}{4}[2000 + 1133.7891 + 941.89455 + 941.89455]$$

$$= 1254.3946$$

$$u_5^{(6)} = \frac{1}{4}[1254.3946 + 1254.3946 + 1004.3946$$

$$+ 1004.3946] = 1129.3946$$

Seventh iteration (n = 6)

$$u_1^{(7)} = \frac{1}{4}[1000 + 1004.3946 + 500 + 1254.3946]$$

$$= 939.6973$$

$$u_2^{(7)} = \frac{1}{4}[939.6973 + 939.6973 + 1000 + 1129.3946]$$

$$= 1002.1973$$

$$u_4^{(7)} = \frac{1}{4}[2000 + 1129.3946 + 939.6973 + 939.6973]$$

$$= 1252.1973$$

$$u_5^{(7)} = \frac{1}{4}[1252.1973 + 1252.1973 + 1002.1973$$

$$+ 1002.1973] = 1127.1973$$

Eight iteration (n = 7)

$$u_1^{(8)} = \frac{1}{4}[1000 + 1002.1973 + 500 + 1252.1973]$$

$$= 938.59865$$

$$u_2^{(8)} = \frac{1}{4}[938.59865 + 938.59865 + 1000 + 1127.1973]$$

$$= 1001.0987$$

$$u_4^{(8)} = \frac{1}{4}[2000 + 1127.1973 + 938.59865 + 938.59865]$$

$$= 1251.0987$$

$$u_5^{(8)} = \frac{1}{4}[1251.0987 + 1251.0987 + 1001.0987$$

$$+ 1001.0987] = 1126.0987$$

Ninth iteration (n = 8)

$$u_1^{(9)} = \frac{1}{4}[1000 + 1001.0987 + 500 + 1251.0987]$$

$$= 938.04935$$

$$u_2^{(9)} = \frac{1}{4}[938.04935 + 938.04935 + 1000 + 1126.0987]$$

$$= 1000.5494$$

$$u_4^{(9)} = \frac{1}{4}[2000 + 1126.0987 + 938.04935 + 938.04935]$$

$$= 1250.5494$$

$$u_5^{(9)} = \frac{1}{4}[1259.5494 + 1259.5494 + 1001.5494$$

$$+ 1001.5494] = 1125.5494$$

Here, we observe that the difference between eighth and ninth iteration is negligible.

Hence, $u_1 = 939$, $u_2 = 1001$, $u_4 = 1251$ and $u_5 = 1126$ which implies $u_3 = 939$, $u_6 = 1251$, $u_7 = 939$, $u_8 = 1001$ and $u_9 = 939$.

EXAMPLE 2. *Given the values of u(x, y) on the boundary of the square in the figure given below, calculate the function u(x, y) satisfying the Laplace equation $\nabla^2 u = 0$ at the pivotal points of this figure by*

(a) Jacobi's method   (b) Gauss-Seidel method

(UPTU–2007, BHOPAL–2009, MADRAS–2003)

SOLUTION. (a) To find the initial values of u_1, u_2, u_3, u_4, we have assume that $u_4 = 0$. Then

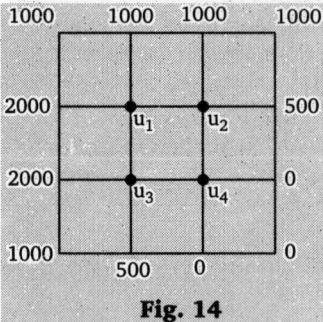

Fig. 14

$$u_1 = \frac{1}{4}[1000 + 0 + 1000 + 2000] = 1000$$

$$u_2 = \frac{1}{4}[1000 + 500 + 1000 + 0] = 625$$

$$u_3 = \frac{1}{4}[2000 + 0 + 1000 + 500] = 875$$

$$u_4 = \frac{1}{4}[875 + 0 + 625 + 0] = 375$$

By using Jacobi's iteration process, we have the following iteration :

$$u_1^{(n+1)} = \frac{1}{4}\left[2000 + u_2^{(n)} + 1000 + u_3^{(n)}\right]$$

$$u_2^{(n+1)} = \frac{1}{4}\left[u_1^{(n)} + 500 + 1000 + u_4^{(n)}\right]$$

$$u_3^{(n+1)} = \frac{1}{4}\left[2000 + u_4^{(n)} + u_1^{(n)} + 500\right]$$

$$u_4^{(n+1)} = \frac{1}{4}\left[u_3^{(n)} + 0 + u_2^{(n)} + 0\right]$$

First iteration ($n = 0$)

$$u_1^{(1)} = \frac{1}{4}\left[2000 + 625 + 1000 + 875\right] = 1125$$

$$u_2^{(1)} = \frac{1}{4}\left[1000 + 500 + 1000 + 375\right] = 719$$

$$u_3^{(1)} = \frac{1}{4}\left[2000 + 375 + 1000 + 500\right] = 969$$

$$u_4^{(1)} = \frac{1}{4}\left[875 + 0 + 625 + 0\right] = 375$$

Second iteration ($n = 1$)

$$u_1^{(2)} = \frac{1}{4}\left[2000 + 719 + 1000 + 969\right] = 1172$$

$$u_2^{(2)} = \frac{1}{4}\left[1125 + 500 + 1000 + 375\right] = 750$$

$$u_3^{(2)} = \frac{1}{4}\left[2000 + 375 + 1125 + 500\right] = 1000$$

$$u_4^{(2)} = \frac{1}{4}\left[969 + 0 + 719 + 0\right] = 422$$

Similarly, we may obtain

$$u_1^{(3)} = 1188, u_2^{(3)} = 774, u_3^{(3)} = 1024, u_4^{(3)} = 438,$$

$$u_1^{(4)} = 1200, u_2^{(4)} = 788, u_3^{(4)} = 1032, u_4^{(4)} = 450,$$

$$u_1^{(5)} = 1204, u_2^{(5)} = 788, u_3^{(5)} = 1038, u_4^{(5)} = 454,$$

$$u_1^{(6)} = 1206.5, u_2^{(6)} = 790, u_3^{(6)} = 1040, u_4^{(6)} = 456.5,$$

$$u_1^{(7)} = 1208, u_2^{(7)} = 791, u_3^{(7)} = 1041, u_4^{(7)} = 458,$$

$$u_1^{(8)} = 1208, u_2^{(8)} = 791.5, u_3^{(8)} = 1041.5, u_4^{(8)} = 458$$

Here, we observe that there is no significant difference between the seventh and eighth iteration.

Hence, $u_1 = 1208, u_2 = 792, u_3 = 1042$ and $u_4 = 458$.

(b) By using Gauss-Seidel method, we have the following iterations :

$$u_1^{(n+1)} = \frac{1}{4}\left[2000 + u_2^{(n)} + 1000 + u_3^{(n)}\right]$$

$$u_2^{(n+1)} = \frac{1}{4}\left[u_1^{(n+1)} + 500 + 1000 + u_4^{(n)}\right]$$

$$u_3^{(n+1)} = \frac{1}{4}\left[2000 + u_4^{(n)} + u_1^{(n+1)} + 500\right]$$

$$u_4^{(n+1)} = \frac{1}{4}\left[u_3^{(n+1)} + 0 + u_2^{(n+1)} + 0\right]$$

First iteration ($n = 0$)

$$u_1^{(1)} = \frac{1}{4}\left[2000 + 625 + 1000 + 875\right] = 1125$$

$$u_2^{(1)} = \frac{1}{4}\left[1125 + 500 + 1000 + 375\right] = 750$$

$$u_3^{(1)} = \frac{1}{4}\left[2000 + 375 + 1125 + 500\right] = 1000$$

$$u_4^{(1)} = \frac{1}{4}\left[1000 + 0 + 750 + 0\right] = 438$$

Second iteration ($n = 1$)

$$u_1^{(2)} = \frac{1}{4}\left[2000 + 750 + 1000 + 1000\right] = 1188$$

$$u_2^{(2)} = \frac{1}{4}\left[1188 + 500 + 1000 + 438\right] = 782$$

$$u_3^{(2)} = \frac{1}{4}\left[2000 + 438 + 1188 + 500\right] = 1032$$

$$u_4^{(2)} = \frac{1}{4}\left[1032 + 0 + 782 + 0\right] = 454$$

Similarly, we may obtain

$$u_1^{(3)} = 1204, u_2^{(3)} = 789, u_3^{(3)} = 1040,$$

$$u_4^{(3)} = 458, u_1^{(4)} = 1207, u_2^{(4)} = 791,$$

$$u_3^{(4)} = 1041, u_4^{(4)} = 458, u_1^{(5)} = 1208,$$

$$u_2^{(5)} = 791.5, u_3^{(5)} = 1041.5, u_4^{(5)} = 458.25$$

Here, we observe that there is no significant difference between the fourth and fifth iteration values.

Hence, $u_1 = 1208, u_2 = 792, u_3 = 1042$ and $u_4 = 458$.

(B) SOLUTION OF POISSON'S EQUATION

Consider the Poisson's partial differential equation

$$\nabla^2 u = f(x, y) \qquad \Rightarrow \qquad \frac{\partial^2 u}{\partial x^2} + \frac{\partial^2 u}{\partial y^2} = f(x, y) \qquad \ldots(1)$$

where $f(x, y)$ is a given function of x and y.

To solve equation (1) numerically, the derivative in (1) are replaced by difference expressions at the points $x = ih, y = jk$ ($h = k$),

Then we get

$$\frac{1}{h^2}\left[u_{i-1,j} - 2u_{i,j} + u_{i+1,j}\right] + \frac{1}{h^2}\left[u_{i,j-1} - 2u_{i,j} + u_{i,j+1}\right] = f(ih, jh)$$

$$\Rightarrow \qquad \left[u_{i-1,j} + u_{i+1,j} + u_{i,j-1} + u_{i,j+1} - 4u_{i,j}\right] = h^2 f(ih, jh) \qquad \ldots(2)$$

Using the formula (2) at each mesh point, we get similar equations in the pivotal values i, j. These equations can be solved by iteration techniques such as Gauss-Seidel method.

REMARK

- The error involved in (1) by difference method is of the order $o(h^2)$.

Solved Examples

EXAMPLE. *Solve the equation* $\nabla^2 u = -10(x^2 + y^2 + 10)$ *over the squares with sides* $x = 0, y = 0, x = 3 = y$ *with* $u = 0$ *on the boundary and mesh length = 1.*
(ANNA–2007, PTU–2007, DELHI–2002)

SOLUTION. The standard five-point formula for the given equation is

$$u_{i-1,j} + u_{i+1,j} + u_{i,j+1} + u_{i,j-1} - 4u_{i,j}$$
$$= -10(i^2 + j^2 + 10) \qquad ...(1)$$

For $u_1(i = i, j = 2)$, equation (1) gives

$$0 + u_2 + 0 + u_3 - 4u_1 = -10(1 + 4 + 10)$$
$$\Rightarrow \quad u_1 = \frac{1}{4}(u_2 + u_3 + 150) \qquad ...(2)$$

For $u_2(i = 2, j = 2)$, equation (1) gives

$$u_2 = \frac{1}{4}(u_1 + u_4 + 180) \qquad ...(3)$$

For $u_3(i = 1, j = 1)$, equation (1) gives

$$u_3 = \frac{1}{4}(u_1 + u_4 + 120) \qquad ...(4)$$

For $u_4(i = 2, j = 1)$, equation (1) gives

$$u_4 = \frac{1}{4}(u_2 + u_3 + 150) \qquad ...(5)$$

Equation (2) and (5) show that $u_4 = u_1$. Therefore, the above equation reduces to

$$u_1 = \frac{1}{4}[u_2 + u_3 + 150], u_2 = \frac{1}{2}[u_1 + 90],$$
$$u_3 = \frac{1}{2}[u_1 + 60]$$

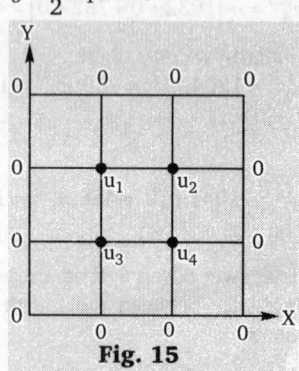
Fig. 15

Now, we solve these equation by Gauss-Seidel method, the iteration formulae are given below :

$$u_1^{(n+1)} = \frac{1}{4}(u_2^{(n)} + u_3^{(n)} + 150), \quad u_2^{(n+1)} = \frac{1}{4}(u_1^{(n+1)} + 90)$$

$$u_3^{(n+1)} = \frac{1}{4}(u_1^{(n+1)} + 60)$$

First iteration. Take initial approximation $u_2 = 0$, $u_3 = 0$, $u_1 = 0$, then we get

$$u_1^{(1)} = 37.5, u_2^{(1)} = \frac{1}{2}[37.5 + 90] = 64,$$

$$u_3^{(1)} = \frac{1}{2}[37.5 + 60] = 49$$

Second iteration.

$$u_1^{(2)} = \frac{1}{4}[64 + 49 + 150] = 66,$$

$$u_2^{(2)} = \frac{1}{2}[66 + 90] = 78,$$

$$u_3^{(2)} = \frac{1}{2}[66 + 60] = 63$$

Third iteration.

$$u_1^{(3)} = \frac{1}{4}[78 + 63 + 150] = 73,$$

$$u_2^{(3)} = \frac{1}{2}[73 + 90] = 82,$$

$$u_3^{(3)} = \frac{1}{2}[73 + 60] = 67$$

Fourth iteration.

$$u_1^{(4)} = \frac{1}{4}[82 + 67 + 150] = 75,$$

$$u_2^{(4)} = \frac{1}{2}[75 + 90] = 82.5,$$

$$u_3^{(4)} = \frac{1}{2}[75 + 60] = 67.5$$

Fifth iteration.

$$u_1^{(5)} = \frac{1}{4}[82.5 + 67.5 + 150] = 75,$$

$$u_2^{(5)} = \frac{1}{2}[75 + 90] = 82.5,$$

$$u_3^{(5)} = \frac{1}{2}[75 + 60] = 67.5$$

Hence, we observe that the values of fifth and fourth iteration are same. Therefore, $u_1 = 75$, $u_2 = 82.5$, $u_3 = 67.5$ and $u_4 = 75$.

10.46.5 PARABOLIC EQUATIONS

A popular case for parabolic type of equation is the study of heat flow in one- dimensional direction in an insulated rod. Such type of problems are governed by both boundary and initial conditions.

Elliptic equations studied previously describe problems that are time independent. Such problems are known as steady-state problems. But we come across problems that are not steady-state. This means that the function is dependent on both space and time. Parabolic equations for which $b^2 - 4ac = 0$ represents the problems that depend on space and time variables.

Definition. *An equation of the type* $\frac{\partial u}{\partial t} = \alpha^2 \frac{\partial^2 u}{\partial x^2}$ *is known as parabolic or heat equation.*

10.46.6 SOLUTION BY FORWARD DIFFERENCE METHOD

The simplest form of finite difference representation is obtained by approximating the time derivative by a forward difference and the space derivative by a central difference.

Now, consider the heat equation $\dfrac{\partial u}{\partial t} = \alpha^2 \dfrac{\partial^2 u}{\partial x^2}$ ...(1)

Substitute the following values in (1)

$$\frac{\partial u}{\partial t}\bigg|_{(i,j)} = \frac{u_{i,j+1} - u_{i,j}}{k} + O(k) \qquad \text{...(2)}$$

$$\frac{\partial^2 u}{\partial x^2}\bigg|_{(i,j)} = \frac{u_{i+1,j} - 2u_{i,j} + u_{i-1,j}}{h^2} + O(h^2) \qquad \text{...(3)}$$

we get $\qquad u_{i,j+1} = r[u_{i+1,j} + u_{i-1,j}] + (1 - 2r)u_{i,j}$ ...(4)

where $r = k/h^2$.

The initial and boundary conditions are given by

$$\left.\begin{aligned} u_{i,0} &= g(ih) = g_1 \\ u_{0,j} &= f_1(jk) = (f_1)_j \\ u_{n,j} &= (f_2)_j \end{aligned}\right\} \qquad \text{...(5)}$$

The numerical computation for the unknown values of u can be directly made by using the equation (4) for u at $(j+1)^{\text{th}}$ time step using the known values of u at j^{th} time step. We will start the computation with $j = 0$ at time $t = 0$.

REMARK

- The forward difference scheme (4) is convergent as well as stable for $r \le 1/2$, i.e., $\dfrac{k}{h^2} \le \dfrac{1}{2}$

10.46.7 SOLUTION BY BENDER-SCHMIDT'S METHOD

Consider one-dimensional heat equation

$$\frac{\partial u}{\partial t} = \alpha^2 \frac{\partial^2 u}{\partial x^2}, \qquad \text{where} \qquad \alpha^2 = \frac{k}{C_\rho} \qquad \text{...(1)}$$

Equation (1) can be written as

$$u_{xx} = au_t, \qquad \text{where} \qquad a = \frac{1}{\alpha^2} \qquad \text{...(2)}$$

Now, we want to solve equation (2) subject to the boundary conditions

$$u_{(0,\,t)} = T_0 \qquad \text{...(3)}$$
$$u_{(l,\,t)} = T_l \qquad \text{...(4)}$$

and the initial conditions

$$u(x,\,0) = f(x) \qquad \text{...(5)}$$

by finite difference method.

Consider the rectangular mesh in x-t plane with spacing h along x-direction and k along t-direction.

Since we know that

$$u_{xx} = \frac{1}{h^2}[u_{i+1,j} - 2u_{i,j} + u_{i-1,j}] \quad \text{and} \quad u_t = \frac{1}{k}[u_{i,j+1} - u_{i,j}]$$

Put all these values in (2), we get

$$\frac{1}{h^2}[u_{i+1,j} - 2u_{i,j} + u_{i-1,j}] = a \cdot \frac{1}{k}[u_{i,j+1} - u_{i,j}]$$

$\Rightarrow \qquad [u_{i,j+1} - u_{i,j}] = \lambda[u_{i+1,j} - 2u_{i,j} + u_{i-1,j}] \qquad\qquad$ where, $\lambda = \dfrac{k}{h^2}a$

$\Rightarrow \qquad u_{i,j+1} = \lambda u_{i+1,j} + (1 - 2\lambda)u_{i,j} + \lambda u_{i-1,j}$ ...(6)

The boundary conditions (3) and (4) can be put in the difference form as follows

$$u_{0,j} = T_0 \qquad \text{...(7)}$$
$$u_{n,j} = T_l \qquad \text{...(8)}$$

where, $j = 1, 2, ..., (nh = l)$

and the initial condition (5) can be written as

$$u_{i,\,0} = f(i,\,h),\; i = 1, 2, ... \qquad \text{...(9)}$$

Equation (6) gives the value of u at $x = ih$ at time t_{j+k} in terms of values of u at $x = (i-1)h$, ih and $(i+1)h$ at a time t_1.

Since $u(x, 0) = f(x)$, u is known at $t = 0$.

Therefore, the recurrence relation (6) allows the evaluation of u at each pivotal point x_i at any t_j.

Hence, (6) is called the Schmidt formula which is valid for $0 \le \alpha \le 1/2$. Pictorial representation of this formula is given follows :

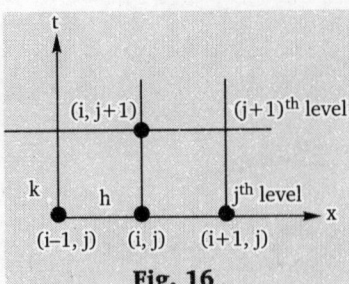

Fig. 16

☞ **REMARK**

- If h and k are chosen such that the coefficient of $u_{i,j}$ vanishes, i.e., $1 - 2\lambda = 0 \Rightarrow \lambda = 1/2$, then (6) gives

$$u_{i,j+1} = \frac{1}{2}[u_{i-1,j} + u_{i+1,j}] \qquad \ldots(10)$$

and

$$k = \frac{a}{2}h^2 \qquad \ldots(11)$$

This formula is known as Bender-Schmidt recurrence equation.

10.46.8 CRANK-NICHOLSON METHOD

Consider the parabolic equation $\qquad\qquad u_{xx} = au_t$ $\qquad\qquad\qquad\ldots(1)$
subject to the conditions $\qquad\qquad\qquad u(0, t) = T_0$ $\qquad\qquad\qquad\ldots(2)$
$\qquad\qquad\qquad\qquad\qquad\qquad\qquad u(l, t) = T_l$ $\qquad\qquad\qquad\ldots(3)$
and $\qquad\qquad\qquad\qquad\qquad\qquad u(x, 0) = f(x)$ $\qquad\qquad\qquad\ldots(4)$

The finite difference approximation for u_{xx}, at point $u_{i,j}$ is given by

$$u_{xx} = \frac{1}{h^2}[u_{i+1,j} - 2u_{i,j} + u_{i-1,j}] \qquad \ldots(5)$$

Also the finite difference approximation for u_{xx} at point $u_{i,j+1}$ is

$$u_{xx} = \frac{1}{h^2}[u_{i+1,j+1} - 2u_{i,j+1} + u_{i-1,j+1}] \qquad \ldots(6)$$

Taking average of (5) and (6), we get

$$u_{xx} = \frac{1}{2h^2}\{u_{i+1,j+1} - 2u_{i,j+1} + u_{i-1,j+1} + u_{i+1,j} - 2u_{i,j} + u_{i-1,j}\} \qquad \ldots(7)$$

For u_t, the forward difference approximation is

$$u_t = \frac{1}{k}[u_{i,j+1} - u_{i,j}] \qquad \ldots(8)$$

Put the value of (7) and (8) in (1), we get

$$\frac{\lambda}{2}u_{i+1,j+1} - (\lambda+1)u_{i,j+1} + \frac{\lambda}{2}u_{i-1,j+1} = -\frac{\lambda}{2}u_{i+1,j} + (\lambda-1)u_{i,j} - \frac{\lambda}{2}u_{i-1,j}$$

(or) $\qquad \lambda\{u_{i+1,j+1} + u_{i-1,j+1}\} - 2(\lambda+1)u_{i,j+1} = 2(\lambda-1)u_{i,j} - \lambda\{u_{i+1,j} + u_{i-1,j}\} \qquad \ldots(9)$

Equation (9) is called Crank-Nicholson difference method.

☞ **REMARKS**

- The Crank-Nicholson formula is convergent for all values of λ.

- If we choose $\lambda - 1$ for λ, then Crank-Nicholson formula becomes $u_{i,j+1} = \frac{1}{4}[u_{i-1,j+1} + u_{i+1,j+1} + u_{i-1,j} + u_{i+1,j}]$

COMPUTATIONAL MODEL

Computational model of Crank-Nicholson difference formula is given below :

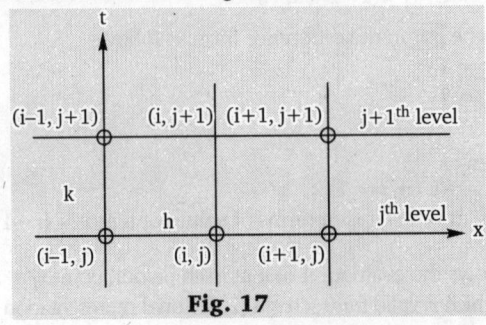

Fig. 17

Crank-Nicholson implicit scheme requires more computations per time step than the explicit scheme. However, a larger value of the time increment k can be used in the scheme, since its time derivative analogue is second order correct and the method is stable for every value of λ. Therefore, much fewer time steps and thus less computations are necessary to compute the values of the dependent variables for a given elapsed time.

10.46.9 DuFort and Frankel's Method

Consider the one-dimensional heat equation

$$\frac{\partial u}{\partial t} = C^2 \frac{\partial^2 u}{\partial x^2} \qquad \qquad ...(1)$$

If we replace the derivatives in (1) by the central difference approximation

$$\frac{\partial u}{\partial t} = \frac{u_{i,j+1} - u_{i,j-1}}{2k} \quad \text{and} \quad \frac{\partial^2 u}{\partial x^2} = \frac{u_{i-1,j} - 2u_{i,j} + u_{i+1,j}}{h^2}$$

we get

$$u_{i,j+1} - u_{i,j-1} = \frac{2kC^2}{h^2}[u_{i-1,j} - 2u_{i,j} + u_{i+1,j}]$$

$$\Rightarrow \qquad u_{i,j+1} = u_{i,j+1} + 2\alpha[u_{i-1,j} - 2u_{i,j} + u_{i+1,j}] \qquad \qquad ...(2)$$

where,

$$\alpha = \frac{kC^2}{h^2}$$

The above 3-level formula given by (2) is called the **Richardson scheme**. Now, if we replace $u_{i,j}$ by the mean of the values $u_{i,j-1}$ and $u_{i,j+1}$, i.e.,

$$u_{i,j} = \frac{1}{2}[u_{i,j-1} + u_{i,j+1}] \text{ in (2) then we get}$$

$$u_{i,j+1} = u_{i,j-1} + 2\alpha[u_{i-1,j} - (u_{i,j-1} + u_{i,j+1}) + u_{i+1,j}]$$

This difference scheme is known as DuFort-Frankel formula.

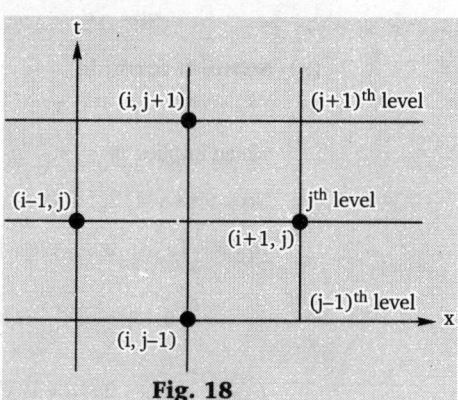

Fig. 18

10.46.10 Iterative Method

By Crank-Nicholson method, we have

$$(1 + \alpha)u_{i,j+1} = \frac{1}{2}\alpha(u_{i-1,j+1} + u_{i+1,j+1}) + u_{i,j} + \frac{1}{2}\alpha(u_{i-1,j} - 2u_{i,j} + u_{i+1,j}) \qquad \qquad ...(1)$$

Let $b_i = u_{i,j} + \frac{\alpha}{2}(u_{i-1,j} - 2u_{i,j} + u_{i+1,j})$ and dropping j's in (1), we get

$$u_1 = \frac{\alpha}{2(1+\alpha)}(u_{i-1} + u_{i+1}) + \frac{b_i}{1+\alpha}$$

which gives

$$u_i^{(n+1)} = \frac{\alpha}{2(1+\alpha)}\left\{u_{i-1}^{(n)} + u_{i+1}^{(n)}\right\} + \frac{b_i}{1+\alpha} \qquad \qquad ...(2)$$

Formula (2) expresses the $(n + 1)^{th}$ iterations in terms of nth iterations only. This is known as Jacobi's iteration formula.

Since the latest value of $u_{i-1}^{(n+1)}$ of u_{i-1} is already available the convergence of the iteration formula (2) can be improved by replacing $u_{i-1}^{(n)}$ by $u_{i-1}^{(n+1)}$. Hence, (2) can be written as

$$u_i^{(n+1)} = \frac{\alpha}{2(1+\alpha)}\left\{u_{i-1}^{(n+1)} + u_{i+1}^{(n)}\right\} + \frac{b_i}{1+\alpha} \qquad \qquad ...(3)$$

which is known as Gauss-Seidel iteration formula.

Solved Examples

EXAMPLE 1. *Solve the equations* $\dfrac{\partial u}{\partial t} = \dfrac{\partial^2 u}{\partial x^2}$ *subject to the conditions*

$$u(x, 0) = \sin\frac{\pi x}{3}, 0 \le x \le 1, u(0, t) = u(1, t) = 0$$

Using (a) Schmidt method, (b) Crank-Nicholson method, (c) DuFort-Frankel method. Carry out computations for two levels taking $h = \dfrac{1}{3}, k = \dfrac{1}{36}$.

(ROHTAK–2003)

SOLUTION. Comparing the given equation to standard equation, we get

$$C^2 = 1, h = \frac{1}{3}, k = \frac{1}{36}$$

$$\Rightarrow \qquad \alpha = \frac{kC^2}{h^2} = \frac{1}{4}$$

Also, $u_{1,0} = \sin\dfrac{\pi}{3} = \dfrac{\sqrt{3}}{2}$,

$$u_{2,0} = \sin\frac{2\pi}{3} = \frac{\sqrt{3}}{2}$$

Then, we have the following figure :

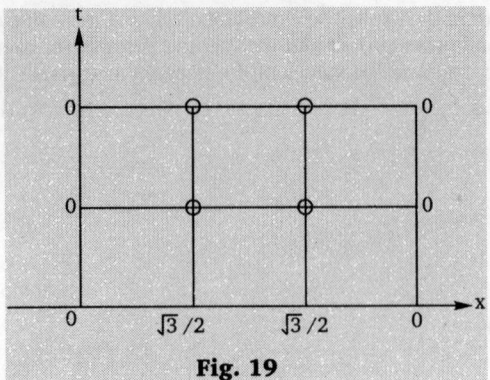

Fig. 19

(a) Schmidt formula

We have $u_{i,j+1} = \alpha u_{i-1,j} + (1-2\alpha)u_{i,j} + \alpha u_{i+1,j}$

which implies $u_{i,j+1} = \dfrac{1}{4}[u_{i-1,j} + 2u_{i,j} + u_{i+1,j}]$

Now, for $i = 1, 2; j = 0$

$u_{1,1} = \dfrac{1}{4}[u_{0,0} + 2u_{1,0} + u_{2,0}]$

$= \dfrac{1}{4}\left[0 + 2\dfrac{\sqrt{3}}{2} + \dfrac{\sqrt{3}}{2}\right] = 0.65$

$u_{2,1} = \dfrac{1}{4}[u_{1,0} + 2u_{2,0} + u_{3,0}]$

$= \dfrac{1}{4}\left[\dfrac{\sqrt{3}}{2} + 2\dfrac{\sqrt{3}}{2}\right] = 0.65$

Similarly for $i = 1, 2, j = 1$, we get

$u_{1,2} = 0.49$ and $u_{2,2} = 0.49$

(b) Crank-Nicholson Method

Here, we have

$-\dfrac{1}{4}u_{i-1,j+1} + \dfrac{5}{2}u_{i,j+1} - \dfrac{1}{4}u_{i+1,j+1}$

$= \dfrac{1}{4}u_{i-1,j} + \dfrac{3}{2}u_{i,j} + \dfrac{1}{4}u_{i+1,j}$

for $i = 1, 2, j = 0$, we get

$-u_{0,1} + 10u_{1,1} - u_{2,1} = u_{0,0} + 6u_{1,0} + u_{2,0}$

$\Rightarrow 10u_{1,1} - u_{2,1} = \dfrac{7\sqrt{3}}{2}$

and $-u_{1,1} + 10u_{2,1} - u_{3,1} = u_{1,0} + 6u_{2,0} + u_{3,0}$

$\Rightarrow -u_{1,1} + 10u_{2,1} = \dfrac{7\sqrt{3}}{2}$

On solving these two equations, we get

$u_{1,1} = u_{2,1} = 0.67$

Similarly, for $i = 2, j = 1$, we get

$u_{1,2} = u_{2,2} = 0.52$.

(c) DuFort Frankel method

Here, we have

$u_{i,j+1} = \dfrac{1}{3}[u_{i,j-1} + u_{i-1,j} + u_{i+1,j}]$

To start the calculation, we need $u_{1,1}$ and $u_{2,1}$, we may take

$u_{1,1} = u_{2,-1} = 0.65$ (By Schmidt formula)

for $i = 1, 2, j = 1$

$u_{1,2} = \dfrac{1}{3}[u_{1,0} + u_{0,1} + u_{2,1}] = \dfrac{1}{3}\left[\dfrac{\sqrt{3}}{2} + 0 + 0.65\right] = 0.5$

$u_{2,2} = \dfrac{1}{3}[u_{2,0} + u_{1,1} + u_{3,1}] = \dfrac{1}{3}\left[\dfrac{\sqrt{3}}{2} + 0.65 + 0\right] = 0.5$

EXAMPLE 2. *Use the Crank-Nicholson implicit method for solving the one- dimensional heat conduction formula* $\dfrac{\partial u}{\partial t} = \dfrac{\partial^2 u}{\partial x^2}$

where $u(x, 0) = \begin{cases} 2x, & 0 \le x \le 1/2 \\ 2(1-x), & 1/2 \le x \le 1 \end{cases}$

$\dfrac{\partial u}{\partial x} = u$ *at* $x = 0$ *and* $\dfrac{\partial u}{\partial x} = -u$ *at* $x = 1 \ \forall \ t.$

Take $h = 0.2$ *and* $r = 1$ *and compute the solution for the first time step.*

SOLUTION. By Crank-Nicholson formula, we have

$-u_{i-1,j+1} + 4u_{i,j+1} - u_{i+1,j+1} = u_{i-1,j} + u_{i+1,j}$ ...(1)

The finite difference representation of the boundary conditions

$\dfrac{\partial u}{\partial x} = u$ at $x = 0$ gives $\dfrac{u_{i+1,j} - u_{i-1,j}}{2h} = u_{ij}$

$\Rightarrow \quad u_{i-1,j} = u_{i+1,j} - 2hu_{ij}, i, j = 0$

$\Rightarrow \quad u_{-1,j} = u_{1,j} - 2hu_{0,j}$ ...(2)

Similarly, $\dfrac{\partial u}{\partial x} = -u$ at $x = 1$ gives

$u_{i+1,j} - u_{i-1,j} - 2hu_{i,j}, i = 5$

$\Rightarrow \quad u_{6,j} = u_{4,j} - 2hu_{5,j}$ ...(3)

Put $j = 0$ and i = 0, 1, 2, 3, 4, 5 in (1), we get

$-u_{-1,1} + 4u_{0,1} - u_{1,1} = u_{-1,0} + u_{1,0}$ ...(4)

$-u_{0,1} + 4u_{1,1} - u_{2,1} = u_{0,0} + u_{2,0}$ ...(5)

$-u_{1,1} + 4u_{2,1} - u_{3,1} = u_{1,0} + u_{3,0}$ ...(6)

$-u_{2,1} + 4u_{3,1} - u_{4,1} = u_{2,0} + u_{4,0}$ ...(7)

$-u_{3,1} + 4u_{4,1} - u_{5,1} = u_{3,0} + u_{5,0}$ ...(8)

$-u_{4,1} + 4u_{5,1} - u_{6,1} = u_{4,0} + u_{6,0}$ ...(9)

Now, substituting $j = 0$ in (2), we get

$u_{-1,0} = u_{1,0}$ ...(10)

Similarly, putting $j = 0$ in (3), we get

$u_{6,0} = u_{4,0}$ ...(11)

Also, for $j = 1$, equation (2) and (3) give

$u_{-1,1} = u_{1,1} - 4u_{0,1}$

$\Rightarrow \qquad u_{-1,1} = u_{1,1} - 0.4u_{0,1}$ ...(12)

$u_{6,1} = u_{4,1} - 0.4u_{5,1}$ ...(13)

Further, the initial condition gives

$u_{0,0} = 0, u_{1,0} = 0.4, u_{2,0} = 0.8, u_{3,0} = 0.8, u_{4,0} = 0.4$ ...(14)

Substituting the relation (10)-(14) in (3)-(9), we get, on simplification

$4.4u_{0,1} - 2u_{1,1} = 0.8$ ...(15)

$-u_{0,1} + 4u_{1,1} - u_{2,1} = 0.8$ ...(16)

$-u_{1,1} + 4u_{2,1} - u_{3,1} = 1.2$ ...(17)

$-u_{2,1} + 4u_{3,1} - u_{4,1} = 1.2$ ...(18)

$-u_{3,1} + 4u_{4,1}u_{5,1} = 0.8$ ...(19)

$$-2u_{4,1} + 3.6u_{5,1} = 0.8 \qquad \dots(20)$$

The system of equation (15) to (20) can be solved easily.

After solving, we get

$$u_{5,1} = 1.5990431, u_{4,1} = 1.6526411,$$

$$u_{3,1} = 2.0259389, u_{2,1} = 1.9738732,$$

$$u_{1,1} = 1.4080155, u_{0,1} = 0.821851$$

EXAMPLE 3. *Using Crank-Nicholson's method, solve* $u_{xx} = 16u_t$, $0 < x < 1$, $t > 0$ *given,* $u(x, 0) = 0$, $u(0, t) = 0$, $u(1, t) = 50t$. *Compute u for two steps in t-direction taking* $h = 1/4$.

SOLUTION. Here, we have $a = 16$, $h = 1/4$

$$k = ah^2 = 16\left(\frac{1}{16}\right) = 1$$

The crank-Nicholson scheme is given by

$$u_{i,j+1} = \frac{1}{4}[u_{i+1,j+1} + u_{i-1,j+1} + u_{i+1,j+1} + u_{i-1,j}] \qquad \dots(1)$$

i $\diagdown$ j	0	0.25	0.5	0.75	1
0	0	0	0	0	0
1	0	u_1	u_2	u_3	50
2	0	u_4	u_5	u_6	100

Applying equation (1) at the mesh-points u_1, u_2, u_3, we

get

$$u_1 = \frac{1}{4}[0+0+0+u_2] = \frac{u_2}{4} \qquad \dots(2)$$

$$u_2 = \frac{1}{4}[0+0+u_1+u_3] = \frac{1}{4}(u_1+u_3) \qquad \dots(3)$$

$$u_3 = \frac{1}{4}[0+0+u_2+50] = \frac{1}{4}(u_2+50) \qquad \dots(4)$$

Substituting (4) and (2) in (3), we get

$$u_2 = \frac{1}{4}\left[\frac{1}{4}u_2 + \frac{1}{4}[u_2+50]\right]$$

$$\Rightarrow \quad 16u_2 = 2u_2 + 50$$

$$\Rightarrow \quad u_2 = 3.5714$$

$$u_1 = 0.89285, u_3 = 13.39285$$

Applying equation (1) again at the mesh-points u_4, u_5 and u_6, we get

$$u_4 = \frac{1}{4}u_5 \qquad \dots(5)$$

$$u_5 = \frac{1}{4}(u_6+u_6) \qquad \dots(6)$$

$$u_6 = \frac{1}{4}(u_5+100) \qquad \dots(7)$$

On solving, we get

$$u_4 = 1.7857, u_5 = 7.1429 \text{ and } u_6 = 26.7857.$$

10.46.11 Solution of Two-Dimensional Heat Equation: ADE Method

Consider a two-dimensional heat equation

$$\frac{\partial u}{\partial t} = C^2\left(\frac{\partial^2 u}{\partial x^2} + \frac{\partial^2 u}{\partial y^2}\right) \qquad \dots(1)$$

Consider a square region $0 \le x \le y \le a$ and assume that u is known at all points within and on the boundary of this square.

Let h be the step size, then a mesh-point $(x, y, t) = (ih, jh, nl)$ may be denoted as simply (i, j, n).

Now, replacing the derivatives in (1) by their finite difference approximation, we get

$$\frac{u_{i,j,n+1} - u_{i,j,n}}{l} = \frac{C^2}{h^2}[(u_{i-1,j,n} - 2u_{i,j,n} + u_{i+1,j,n}) + (u_{i,j-1,mn} - 2u_{i,j,n} + u_{i,j+1,n})]$$

$$\Rightarrow \quad u_{i,j,n+1} = u_{i,j,n} + \alpha(u_{i-1,j,n} + u_{i+1,j,n} + u_{i,j+1,n} + u_{i,j-1,n} - 4u_{i,j,n}) \qquad \dots(2)$$

where,

$$\alpha = \frac{lC^2}{h^2}$$

The above method is known as ADE (Alternating Direction Explicit) method.

Solved Examples

EXAMPLE 1. *Solve the equation* $\frac{\partial u}{\partial t} = \frac{\partial^2 u}{\partial x^2} + \frac{\partial^2 u}{\partial y^2}$ *subject to the initial conditions* $u(x, y, 0) = \sin 2\pi x \sin 2\pi y$, $0 \le x, y \le 1$ *and the conditions* $u(x, y, t) = 0$, $t > 0$, *on the boundaries, using ADE method with* $h = \frac{1}{3}$ *and* $\alpha = \frac{1}{8}$.

SOLUTION. By ADE method, we have

$$u_{i,j,n+1} = u_{i,j,n} + \frac{1}{8}(u_{i-1,j,n} + u_{i+1,j,n} + u_{i,j+1,n} + u_{i,j-1,n} - 4u_{i,j,n})$$

$$\Rightarrow \quad u_{i,j,n+1} = \frac{1}{2}u_{i,j,n} + \frac{1}{8}(u_{i-1,j,n} + u_{i+1,j,n} + u_{i,j+1,n} + u_{i,j-1,n}) \qquad \dots(1)$$

Now, we shall find the mesh-points.

At the Zeroth level ($n = 0$), the initial and boundary conditions are

$$u_{i,j,0} = \frac{1}{2}u_{i,j,0} + \frac{1}{8}(u_{i-1,j,0} + u_{i+1,j,0} + u_{i,j+1,0} + u_{i,j-1,0}) \qquad \dots(2)$$

Step-1. Put $i = j = 1$ in (2), we get

$$u_{1,1,1} = \frac{1}{2}u_{1,1,0} + \frac{1}{8}[u_{0,1,0} + u_{2,1,0} + u_{1,2,0} + u_{1,0,0}]$$

$$= \frac{1}{2}\left(\sin\frac{2\pi}{3}\right)^2 + \frac{1}{8}\left[0 + \sin\frac{4\pi}{3}\sin\frac{2\pi}{3} + \sin\frac{2\pi}{3}\sin\frac{4\pi}{3} + 0\right]$$

$$= \frac{3}{8} + \frac{1}{8}\left(-\frac{\sqrt{3}}{2} \times \frac{\sqrt{3}}{2} - \frac{\sqrt{3}}{2} \times \frac{\sqrt{3}}{2}\right) = \frac{3}{16}$$

Step-2. Put $i = 2, j = 1$ in (2), we get

$$u_{2,1,1} = \frac{1}{2}u_{2,1,0} + \frac{1}{8}[u_{1,1,0} + u_{3,1,0} + u_{2,2,0} + u_{2,0,0}]$$

$$= \frac{1}{2}\sin\frac{4\pi}{3}\sin\frac{2\pi}{3} + \frac{1}{8}\left\{\left(\sin\frac{2\pi}{3}\right)^2 + 0 + 0 + \left(\sin\frac{4\pi}{3}\right)^2 + 0\right\}$$

$$= -\frac{1}{2}\left(\frac{\sqrt{3}}{2}\right)^2 + \frac{1}{8}\left[\left(\frac{\sqrt{3}}{2}\right)^2 + \left(\frac{-\sqrt{3}}{2}\right)^2\right] = -\frac{3}{16}$$

Step-3. Put $i = 1, j = 2$ in (2), we get

$$u_{1,2,0} = \frac{1}{2}u_{1,2,0} + \frac{1}{8}(u_{0,2,0} + u_{2,2,0} + u_{1,3,0} + u_{1,1,0})$$

$$= \frac{1}{2}\sin\frac{2\pi}{3}\sin\frac{4\pi}{3} + \frac{1}{8}\left\{0 + \left(\sin\frac{4\pi}{3}\right)^2 + 0 + \left(\sin\frac{2\pi}{3}\right)^2\right\}$$

$$= -\frac{3}{8} + \frac{1}{8}\left(\frac{3}{4} + \frac{3}{4}\right) = -\frac{3}{16}$$

Step-4. Put $i = 2, j = 2$ in (2), we get

$$u_{2,2,0} = \frac{1}{2}u_{2,2,0} + \frac{1}{8}(u_{1,2,0} + u_{3,2,0} + u_{2,3,0} + u_{2,1,0})$$

$$= \frac{1}{2}\left(\sin\frac{4\pi}{3}\right)^2 + \frac{1}{8}\left(\sin\frac{2\pi}{3}\sin\frac{4\pi}{3} + 0 + 0 + \sin\frac{4\pi}{3}\sin\frac{2\pi}{3}\right)$$

$$= \frac{3}{8} + \frac{1}{8}\left(-\frac{3}{4} - \frac{3}{4}\right) = \frac{3}{16}$$

Similarly we may obtain the mesh values of the second and higher levels.

10.46.12 HYPERBOLIC EQUATION

An equation of the type $a^2 \dfrac{\partial^2 u}{\partial x^2} = \dfrac{\partial^2 u}{\partial t^2}$ or $a^2 u_{xx} - u_{tt} = 0$ is said to the hyperbolic equation.

SOLUTION BY METHOD OF FINITE DIFFERENCE

Consider the equation

$$a^2 \frac{\partial^2 u}{\partial x^2} = \frac{\partial^2 u}{\partial t^2} \qquad \ldots(1)$$

subject to the conditions

$$u(0, t) = 0 \qquad \ldots(2)$$
$$u(l, t) = 0 \qquad \ldots(3)$$

and

$$u(x, 0) = f(x) \qquad \ldots(4)$$
$$u_t(x, 0) = 0 \qquad \ldots(5)$$

Put

$$u_{xx} = \frac{1}{h^2}[u_{i+1,j} - 2u_{i,j} + u_{i-1,j}]$$

$$u_{tt} = \frac{1}{k^2}[u_{i,j+1} - 2u_{i,j} + u_{i,j-1}]$$

in (1), we get

$$\frac{a}{h^2}[u_{i+1,j} - 2u_{i,j} + u_{i-1,j}] - \frac{1}{k^2}[u_{i,j+1} - 2u_{i,j} + u_{i,j-1}] = 0$$

$$\Rightarrow \qquad \lambda^2 a^2[u_{i+1,j} - 2u_{i,j} + u_{i-1,j}] - [u_{i,j+1} - 2u_{i,j} + u_{i,j-1}] = 0$$

where, $\lambda = \dfrac{k}{h}$.

$$\Rightarrow \qquad u_{i,j+1} = 2(1 - \lambda^2 a^2)u_{i,j} + \lambda^2 a^2(u_{i+1,j} + u_{i-1,j}) - u_{i,j-1} \qquad \ldots(6)$$

The boundary condition (2) and (3) can be put in the difference form as

$$u_{0,j} = 0 = u_{n,j}; j = 1, 2, 3$$

The initial condition (4) as

$$u_{i,0} = f(ih), i = 1, 2, \ldots \qquad \ldots(7)$$

and (5) as

$$\frac{1}{k}(u_{i,j+1} - u_{i,0}) = 0 \quad \text{when } t = 0, \text{ i.e., } j = 0$$

$$\therefore \qquad u_{i,1} - u_{i,0} = 0 \qquad \ldots(8)$$

$$\Rightarrow \qquad u_{i,1} = u_{i,0} = f(ih) \qquad \ldots(9)$$

Now, equation (7) and (8) give the values of u on the first two rows $j = 0$ and $j = 1$. Putting $j = 1$ in (6), we get

$$u_{i,2} = 2(1 - \lambda^2 a^2)u_{i,1} + \lambda^2 a^2(u_{i+1,1} + u_{i-1,1}) - u_{i,0} \qquad \ldots(10)$$

Now, RHS of (10) involves the values of u on the first two rows $j = 0$ and $j = 1$. These are known from the initial condition (7) and (8). Hence, we can find $u_{i,2}$.

REMARK
- If $k < h$ solution(6) is convergent.

Solved Examples

EXAMPLE 1. *Use the finite difference method to find the solution of the problem*

$$\frac{\partial^2 u}{\partial t^2} - \frac{\partial^2 u}{\partial x^2} = 0$$

with $u(x, 0) = x(\pi - x), \frac{\partial u}{\partial t}(x, 0) = 0$;

$$u(0, t) = u(\pi, t) = 0$$

SOLUTION. Here, we consider a square grid with spacing $h = k = \frac{\pi}{13}$.

The given problem is symmetric over the interval $[0, \pi]$. Therefore, we find the solution in the interval $[0, \pi/2]$. The value of u_{10} are given by the initial condition $u_{10} = x_i(\pi - x)$.

$u_{00} = 0, u_{10} = 0.518, u_{20} = 0.975, u_{30} = 1.371,$
$u_{40} = 1.706, u_{50} = 1.980, u_{60} = 2.193, u_{70} = 2.346,$
$u_{80} = 2.437, u_{90} = 2.467$

To find the solution for the next time step we use the Taylor series. We write

$$u_{i1} = u_{i0} + k\frac{\partial u_{i0}}{\partial t} + \frac{k^2}{2}\frac{\partial^2 u_{i0}}{\partial t^2}$$

$$\frac{\partial u_{i0}}{\partial t} = 0, \frac{\partial^2 u_{i0}}{\partial t^2} = \frac{\partial^2 u_{i0}}{\partial x^2} = -2.$$

$$\therefore \quad u_{i1} = u_{i0} - k^2 = u_{i0} - 0.03048$$

Therefore, the values of u_{i1} are

$u_{01} = 0, u_{11} = 0.487, u_{21} = 0.944,$
$u_{31} = 1.340, u_{41} = 1.675, u_{51} = 1.950,$
$u_{61} = 2.163, u_{71} = 2.315, u_{81} = 2.406,$
$u_{91} = 2.437$

We now compute the values of $u_{i, j+1}$ for $j = 1, 2, \dots$ using the finite difference representation

$$u_{i, j+1} - 2u_{ij} + u_{i, j-1} = u_{i+1, j} - 2u_{i, j} + u_{i-1, j}$$

$$\Rightarrow u_{i, j+1} = u_{i+1, j} + u_{i-1, j} - u_{i, j-1}$$

Therefore,

$u_{22} = u_{31} + u_{11} - u_{20} = 1.340 + 0.487 - 0.975 = 0.853$

$u_{32} = u_{41} + u_{21} - u_{30} = 1.675 + 0.944 - 1.371 = 1.249$

$u_{42} = u_{51} + u_{31} - u_{40} = 1.950 + 1.340 - 1.706 = 1.584$

$u_{52} = u_{61} + u_{41} - u_{50} = 2.163 + 1.675 - 1.980 = 1.858$

$u_{62} = u_{71} + u_{51} - u_{60} = 2.315 + 1.950 - 2.346 = 2.224$

$u_{82} = u_{91} + u_{71} - u_{80} = 2.437 + 2.315 - 2.437 = 2.315$

$u_{92} = u_{101} + u_{81} - u_{90} = 2.406 + 2.406 - 2.467 = 2.346$

EXAMPLE 2. *Solve* $u_{tt} = 4u_{xx}$ *with the boundary conditions*

$u(0, t) = 0 = u(4, t), t(x, 0)$ *and* $u(x, 0) = x(4 - x)$.

SOLUTION. Here, the given equation is

$$\frac{\partial^2 u}{\partial t^2} = 4\frac{\partial^2 u}{\partial x^2} \quad \dots(1)$$

Comparing with the standard equation, we get

$a^2 = 4 \quad \Rightarrow \quad a = 2.$

Taking $h = 1 \quad \Rightarrow \quad k = \frac{h}{a} = \frac{1}{2} = 0.5$.

From the initial conditions

$u(0, t) = 0 \quad \Rightarrow \quad u = 0$ along line $x = 0$.
$u(4, t) = 0 \quad \Rightarrow \quad u = 0$ along line $x = 4$.

which can be written in difference form as follows

$u_{0, j} = 0 \quad$ and $\quad u_{4, j} = 0, \forall j$

Now, $u(x, 0) = x(4 - x)$

$\Rightarrow u(0, 0) = 0, u(1, 0) = 3, u(2, 0) = 4,$
$u(3, 0) = 3, u(4, 0) = 0$

In difference notation

$u_{i, 0} = u(i, 0) = i(4 - i)$, for different i.

Putting $i = 0, 1, 2, 3, 4,$ we get

$u_{0, 0} = 0, u_{1, 0} = 3, u_{2, 0} = 4,$
$u_{3, 0} = 3, u_{4, 0} = 0$

The condition $u_t(x, 0)$ gives

$$\frac{1}{k}(u_{i, j+1} - u_{i, j}) = 0$$

when $j = 0, u_{i, 1} = u_{i, 0} \forall i$

$\Rightarrow u$ on the first two rows are equal.

Consider the recurrence relation

$$u_{i, j+1} = u_{i+1, j} + u_{i-1, j} - u_{i, j-1}$$

If we put $j = 1$, we get

$$u_{i, 2} = u_{i+1, 1} + u_{i-1, 1} - u_{i, 0}$$

Putting $i = 1, 2, 3, \dots$ successively, we get

$u_{1, 2} = u_{2, 1} + u_{0, 1} - u_{1, 0} = 4 + 0 - 3 = 1$
$u_{2, 2} = u_{3, 1} + u_{1, 1} - u_{2, 0} = 3 + 3 - 4 = 2$
$u_{3, 2} = u_{4, 1} + u_{2, 1} - u_{3, 0} = 0 + 4 - 3 = 1.$

In the similar way, we can fill in the remaining rows as shown in following table :

i / j	0	1	2	3	4
0	0	3	4	3	0
1	0	3	4	3	0
2	0	1	2	1	0
3	0	–1	–2	–1	0
4	0	–3	–4	–3	0

10.46.13 SOLUTION BY THE METHOD OF CHARACTERISTICS

In order to understand the method of characteristics, consider a quasi-linear PDE of the form

$$a\frac{\partial^2 u}{\partial x^2} + b\frac{\partial^2 u}{\partial x \partial y} + c\frac{\partial^2 u}{\partial y^2} + h = 0 \quad \dots(1)$$

where, the coefficients a, b, c and h may be functions of $u, \frac{\partial u}{\partial x}$ and $\frac{\partial u}{\partial y}$ only.

Here, we use the following notation

$$\frac{\partial u}{\partial x} = p, \frac{\partial u}{\partial y} = q, \frac{\partial^2 u}{\partial x^2} = r, \frac{\partial^2 u}{\partial x \partial y} = s, \frac{\partial^2 u}{\partial y^2} = t \quad \dots(2)$$

Then, (1) can be written as

$$ar + bs + ct + h = 0 \qquad \qquad \text{...(3)}$$

Let C be the curve in the x-y plane such that the function representing C satisfies the above equation. The total differentials of p and q in the direction tangential to C are given by

$$dp = \frac{\partial p}{\partial x}dx + \frac{\partial p}{\partial y}dy \quad \Rightarrow \quad dp = rdx + sdy \qquad \qquad \text{...(4)}$$

$$dq = \frac{\partial q}{\partial x}dx + \frac{\partial q}{\partial y}dy \quad \Rightarrow \quad dq = sdx + tdy \qquad \qquad \text{...(5)}$$

Eliminate r and t from (3) using (4) and (5), we get

$$\frac{a}{dx}(dp - sdy) + bs + \frac{c}{dy}(dq - sdx) + h = 0$$

$$\Rightarrow \quad s\left(-a\frac{dy}{dx} + b - c\frac{dx}{dy}\right) + a\frac{dp}{dx} + c\frac{dq}{dy} + h = 0$$

$$\Rightarrow \quad s(am^2 - bm + c) - \left(am\frac{dp}{dx} + c\frac{dq}{dx} + hm\right) = 0 \qquad \qquad \text{...(6)}$$

where, $m = \dfrac{dy}{dx}$.

Equation (6) is independent of r and t. In order to make (6) independent of s, also, we choose the curve C such that the slope of the tangent at every point on it satisfies the quadratic equation

$$am^2 - bm + c = 0 \qquad \qquad \text{...(7)}$$

The two directions given by (7) are real and distinct when $b^2 - 4ac > 0$, *i.e.*, the second order partial differential equation (1) is hyperbolic. The two curves characterised by the above two directions are called characteristic curves. The differential equation (3) along a characteristic curve, then, reduces to the simple form

$$am + \frac{dp}{dx} + c\frac{dq}{dx} + hm = 0 \qquad \qquad \text{...(8)}$$

The two families of characteristic are termed as f-characteristic and g-characteristic, where

$$f = \frac{b + \sqrt{b^2 - 4ac}}{2a} = \text{constt.} \qquad \qquad \text{...(9)}$$

$$g = \frac{b - \sqrt{b^2 - 4ac}}{2a} = \text{constt.} \qquad \qquad \text{...(10)}$$

Let the value of u, p and q be prescribed on the initial curve C_1, which does not belong to any characteristic family. Let E, F, G be any three neighbouring points on C_1.

Let the f-characteristic through E intersect the g-characteristic through F at P. Similarly, the f-characteristic through F intersects the g-characteristic at Q and so on. The points P, Q would lie on a neighbouring curve, say C_2. Let the coordinate of P be (x_p, y_p) which are to be determined. Since E and F are close points. The curve EP and FP are approximated as straight line. If (x_E, y_E), (x_F, y_F) are the coordinates of E and F respectively, then we have

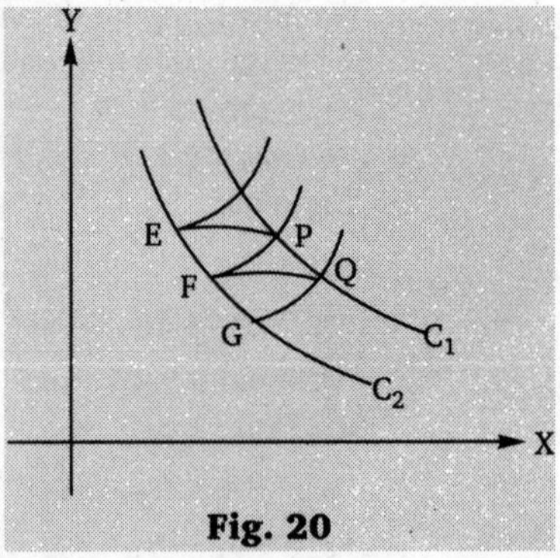

$$\frac{y_p - y_E}{x_p - x_E} = f_E \qquad \qquad \text{...(11)}$$

$$\frac{y_p - y_F}{x_p - x_F} = g_F \qquad \qquad \text{...(12)}$$

Fig. 20

Since m ($= f_E$ or g_E) satisfies (8), we have

$$a_E f_E(p_P - p_E) + C_E(q_P - q_E) + hf_E = 0 \qquad \qquad \text{...(13)}$$

$$a_F g_F(p_P - p_F) + C_F(q_P - q_F) + hg_F = 0 \qquad \qquad \text{...(14)}$$

Using (11), (12) in (13) and (14) respectively, we have two equations for the two unknowns p_P and q_P. We can next determine u_P, the value p using

$$du = pdx + qdy \qquad \qquad \text{...(15)}$$

This can be approximated as

$$u_P - u_E = \frac{p_P - p_E}{2}(x_p - x_E) + \frac{q_P - q_E}{2}(y_p - y_E) \quad \Rightarrow \quad u_P = u_E + \frac{p_P - p_E}{2}(x_p - x_E) + \frac{q_P - q_E}{2}(y_p - y_E) \qquad \text{...(16)}$$

The value of u at the point Q can be determined in a similar manner. Thus, we can obtain the solution along a neighbouring non-characteristic curve C_2 using the values u, p, q along C_2, we can proceed to determine the values along the next non-characteristic curve C_3 and so on.

REMARK

- The method of characteristic cannot be applied to parabolic and elliptic partial differential equations, since the characteristic directions are coincident for parabolic and are imaginary for elliptic differential equations.

Solved Examples

EXAMPLE. *Use the method of characteristic to solve the boundary value problem*

$$\frac{\partial^2 u}{\partial x^2} + \frac{\partial^2 u}{\partial x \partial y} - 2\frac{\partial^2 u}{\partial x^2} + 1 = 0$$

$$u = x, \frac{\partial u}{\partial y} = x \ for \ y = 0, \ 0 \le x \le 1.$$

SOLUTION. Comparing the given equation, with the standard equation, we get $a = 1, b = 1, c = -2, h = 1$.

The characteristic directions are given by

$$m^2 - m - 2 = 0$$

$$\Rightarrow \quad m = \frac{1 \pm 3}{2} \text{ which gives } f = 2, g = -1.$$

From the given condition, the initial curve C_1 is the straight line $y = 0$ and

$$u = x, \frac{\partial u}{\partial y} = x \text{ along } y = 0.$$

$$\therefore \quad p = \frac{\partial u}{\partial x} = 1, q = \frac{\partial u}{\partial y} = x \text{ along } y = 0.$$

Let $x_E = 0.4$ and $x_F = 0.5$

$\therefore \quad q_E = 0.4 \qquad q_F = 0.5$

$u_E = x_E = 0.4, \qquad u_F = x_F = 0.5.$

Now, $y_P = y_E + f_E(x_P - x_E) = 0 + 2(x_P - 0.4)$
$\qquad = 2x_P - 0.8,$

Also, $y_P = y_F + g_F(x_P - x_F)$
$\qquad = 0 + (-1)(x_P - 0.5) = -x_P + 0.5$

Equating these two equations, we get

$2x_P - 0.8 = -x_P + 0.5 \quad \Rightarrow \quad x_P = 0.433$

Hence, $y_P = 0.5 - 0.433 = 0.067$

Now, $a_E f_E(p_P - p_E) + c_E(q_P - q_E) + h(y_P - y_E) = 0$

Therefore, $2(p_P - 1) - 2(q_P - 0.4) + 0.067 = 0$

$\Rightarrow \quad p_P - q_P = 0.4665$

and $a_F q_F(p_P - p_F) + c_F(q_P - q_F) + h_F(y_P - y_F) + (P_P - 1) = 0$

$\Rightarrow \quad -p_P - 2(q_P - 0.5) = -0.067$

$\Rightarrow \quad p_P + 2q_P = 2.067$

Solving $q_P = 0.5335$ and $p_P = 1$

$$\therefore u_P = u_E + \frac{p_P + p_E}{2}(x_P - x_E) + \frac{q_P + q_E}{2}(y_P - y_E)$$

$$\Rightarrow u_P = 0.4 + \frac{1+1}{2}(0.433 - 0.4) + \frac{0.5335 + 0.4}{2}(0.067)$$

$$= 0.46425$$

Hence, the value of u at P is 0.46427 to a first approximation. The coordinates of p are $x_p = 0.433$, $y_p = 0.067$.

Exercise-10.23

1. Classify the following equations :
   (i) $3u_{xx} + u_{xy} - 4u_{yy} + 3u_y = 0$
   (ii) $u_{xx} - 6u_{xy} + 9u_{yy} - 17u_y = 0$

2. Solve the elliptic equation $u_{xx} + u_{yy} = 0$ for the following square mesh with boundary values as shown in the figure. Iterate until the maximum difference between two successive values at any point is less than 0.001.

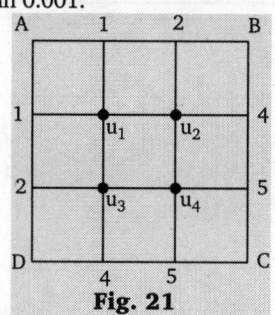

Fig. 21

3. Solve $u_{xx} - u_{yy} = 0$ over the square mesh of side four units satisfying the following boundary conditions :
   (i) $u(0, y) = 0$ for $0 \le y \le 4$
   (ii) $u(4, y) = 12 + y$ for $0 \le y \le 4$
   (iii) $u(x, 0) = 3x$ for $0 \le x \le 4$
   (iv) $u(x, 4) = x^2$ for $0 \le x \le 4$

4. Solve $\dfrac{\partial^2 u}{\partial x^2} + \dfrac{\partial^2 u}{\partial y^2} = 8x^2 y^2$ for square mesh, given $u = 0$ on the four boundaries dividing the square into 16-subsquares of length one unit. (JNTU–2004S)

5. Find the values of $u(x, y)$ satisfying the Laplace equation

$\nabla^2 u = 0$ at the pivotal points of a square region with boundary values given below :

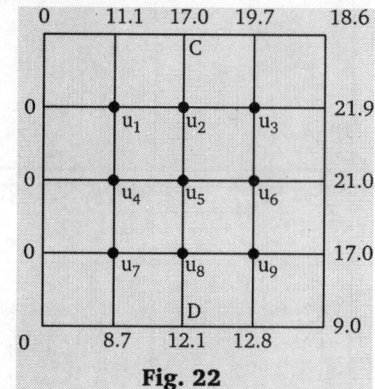

Fig. 22

7. Compute u for one time step by Crank-Nicholson method if $u_t = u_{xx}; 0 < x < 5, t > 0; u(x, 0) = 20, u(0, t) = 0$ and $u(5, t) = 100$. (ANNA–2006)

8. Find the numerical solution to solve $u_t = u_{xx}, 0 \le x \le 1, t \ge 0$ under the conditions that $u(0, t) = u(1, t)$ and

$$u(x, 0) = \begin{cases} 2x, & \text{for } 0 \le x \le 1/2 \\ 2(1 - x), & \text{for } 1/2 \le x \le 1 \end{cases}$$

9. Solve the hyperbolic partial differential equation for one half period of oscillation taking $h = 1$.

$u_{tt} = 25u_{xy}, u(0, t) = u(5, t) = 0, u_t(x, 0) = 0$

$$u(x, 0) = \begin{cases} 2x, & \text{if } 0 \le x \le 2.5 \\ 10 - x, & \text{if } 2.5 \le x \le 5 \end{cases}$$

10. Evaluate the pivotal values for the following equation taking $h = 1$ and upto one half of the period of vibration $16u_{xx} = u_{tt}$, given that $u(0, t) = u(5, t) = 0, u(x, 0) = x^2(x - 5)$ and $u_t(x, 0) = 0$. (MADRAS–2006)

Answers

1. (i) Hyperbolic; (ii) Parabolic

2. $u_1 = 1.999$, $u_2 = 2.999$, $u_4 = 3.999$

3. 2.37, 5.59, 9.87, 2.88, 6.13, 9.88, 3.01, 6.16, 9.51

4. –3, –2, –3, –2, –2, –2, –3, –2, –3

5. 7.9, 13.7, 17.9, 6.6, 11.9, 16.3, 6.6, 11.2, 14.3

6.

j \ i	0	0.25	0.5	0.75	1
0	0	0	0	0	0
1/16	0	0.00116	0.004464	0.01674	1/16
1/8	0	0.005899	0.019132	0.052771	1/8

7.

j \ i	0	1	2	3	4	5
0	0	20	20	20	20	100
1	0	9.80	20.19	30.72	59.92	100

8.

j \ i	0	0.2	0.4	0.6	0.8	1.0	0.8	0.6	0.4	0.2	0
0	0	0.1936	0.3689	0.5400	0.6461	0.6291	0.6461	0.5400	0.3689	0.1936	0
1	2	0.1989	0.3956	0.5834	0.7381	0.7691	0.7381	0.5834	0.3956	0.1989	0

9.

j \ i	0	1	2	3	4	5
0	0	2	4	4	2	0
0.2	0	2	4	4	2	0
0.4	0	2	2	2	2	0
0.6	0	0	0	0	0	0
0.8	0	–2	–2	–2	–2	0
1.0	0	–2	–4	–4	–2	0

10.

j \ i	0	1	2	3	4	5
0	0	4	12	18	16	0
1	0	4	12	18	16	0
2	0	8	10	10	2	0
3	0	6	6	–6	–6	0
4	0	–2	–10	–10	–8	0
5	0	–16	–18	–12	–4	0

ARCHIVE

1. Express $y = 2x^3 - 3x^2 + 3x - 10$ in a factorial notation and hence show that $\Delta^3 y = 12$.   (BHOPAL–2007, PTU–2005)

2. Find the missing values in the following table:

x	45	50	55	60	65
y	3.0	–	2.0	–	2.4

(BHOPAL–2007, VTU–2001)

3. Find the missing term in the table:

x	2	3	4	5	6
y	45.0	49.2	54.1	–	67.4

(UPTU–2008)

4. Prove that

(i) $\mu^2 = 1 + \dfrac{\delta^2}{4}$ (UPTU–2009)

(ii) $\Delta f_k^2 = (f_k + f_{k+1})\Delta f_k$ (JNTU MCA–2006)

5. Find the missing term in the following table:

x	0	1	2	3	4
f(x)	1	3	9	–	81

(SVTU-2007)

6. Find the missing term in the following data:

x	1	1.5	2	2.5	3	3.5	4
f(x)	6	?	10	20	?	1.5	5

(UPTU-2010)

7. Find the missing term in the following table:

x	0	1	2	3	4	5	6
f(x)	5	11	22	40	–	140	–

(VTU-2006)

8. Estimate the value of $f(22)$ and $f(42)$ from the following available data:

x	20	25	30	35	40	45
f(x)	354	332	291	260	231	204

(JNTU-2007)

9. Find the polynomial interpolating the data:

x	0	1	2
f(x)	0	5	2

(UPTU-2008)

10. Construct the difference table for the following data:

x	0.1	0.3	0.5	0.7	0.9	1.1	1.36
f(x)	0.003	0.067	0.148	0.248	0.370	0.518	0.697

Evaluate $f(0.6)$. (JNTU-2007)

11. Estimate from following table f(3.8) to three significant figures using Gregory Newton backward interpolation formula:

x	0	1	2	3	4
f(x)	1	1.5	2.2	3.1	4.6

(UPTU-2009)

12. The following table gives the population of a town during the last six censuses. Estimate the increase in the population during the period from 1976 to 1978.

year	1941	1951	1961	1971	1981	1991
Population (in thousands)	12	15	20	27	39	52

(UPTU-2009)

13. Interpolate by means of Gauss's backward formula, the population of a town for the year 1974, given that:

year	1939	1949	1959	1969	1979	1989
Population (in thousands)	12	15	20	27	39	52

(KOTTAYAM–2005, MADRAS–2003)

14. Apply Stirling's formula to compute $y_{12.2}$ from the following table:

$x°$	10	11	12	13	14
$10^5 y_x$	23.967	28.060	31.788	35.209	38.368

(VTU-2004)

15. Using Gauss's backward difference formula, find $y(8)$ from the following table:

x	0	5	10	15	20	25
y	7	11	14	18	24	32

(JNTU-2007)

16. From the following table:

x	1.00	1.05	1.10	1.15
e^x	2.7183	2.8577	3.0042	3.1582

x	1.20	1.25	1.30
e^x	3.3201	3.4903	3.6693

(UPTU-2006)

17. Calculate the value of $f(1.5)$ using Bessel's interpolation formula, from the following table:

x	0	1	2	3
f(x)	3	6	12	15

(UPTU-2008)

18. Apply Everett's formula to obtain u_{25}, given $u_{20} = 854$, $u_{24} = 3162$, $u_{28} = 3544$, $u_{32} = 3992$. (SVTU-2007)

19. Curve passes through the point (0,18), (1,10), (3, –18) and (6, 90). Find the slope of the curve at $x = 2$. (JNTU-2009)

20. The following are the measurements T made on a curve recorded by oscilograph representing a change of current I due to change in the conditions of an electric current.

T	1.5	2.0	2.5	3.0
I	1.36	0.58	0.34	0.20

(JNTU-2009)

Using Lagrange's formula find I at $T = 16$

21. Find the third divided difference with arguments 2, 4, 9, 10 of the function $f(x) = x^3 - 2x$. (UPTU-2005)

22. Use Newton's divided difference method to compute $f(5.5)$ from the following data:

x	0	1	4	5	6
f (x)	1	14	15	6	3

(UPTU-2010)

23. Obtain the Newton's divided difference interpolation polynomial and hence find $f(6)$:

x	3	7	9	10
f (x)	168	120	72	63

(UPTU-2007)

24. Using the following table, find $f(x)$ as a polynomial in

x	–1	0	3	6	7
f (x)	3	–6	39	822	1611

(UPTU-2009)

25. Find by Taylor's series method the value of y at $x = 0.1$ and $x = 0.2$ to five places of decimals from $\dfrac{dy}{dx} = x^2y - 1$, $y(0) = 1$.   (VTU–2009, ROHTAK–2005)

26. Solve $y' = x + y$ given $y(1) = 0$. Find $y(1.1)$ and $y(1.2)$ by Taylor's method. Compare the result with its exact value.   (JNTU–2008–ANNA–2005)

27. Using modified Euler's method, Find $y(0.2)$ and $y(0.4)$ given $y' = y + e^x$, $y(0) = 0$.   (JNTU–2009)

28. Given $y' = x + \sin y$, $y(0) = 1$. Compute $y(0.2)$ and $y(0.4)$ with $h = 0.2$ using Euler's modified method.   (JNTU–2007)

29. Given that $\dfrac{dy}{dx} = x^2 + y$ and $y(0) = 1$. Find an approximate value of $y(0.1)$ taking $h = 0.05$ by modified Euler's method.   (VTU–2010)

30. Find $y(0.1)$ and $y(0.2)$ using Runge–Kutta 4th order formula, given that $y' = x^2 - y$ and $y(0) = 1$.   (JNTU–2006)

31. Use fourth order Runge–Kutta method to find y at $x = 0.1$, given that $\dfrac{dy}{dx} = 3e^x + 2y$, $y(0) = 0$ and $h = 0.1$.   (VTU–2006)

32. Using Runge–Kutta method of order 4, find y for $x = 0.1$, 0.2, 0.3 given that $\dfrac{dy}{dx} = xy + y^2$, $y(0) = 1$. Continue the solution at $x = 0.4$ using Milne's method.

                      (VTU–2008,SVTU–2007, MADRAS–2006)

33. From the data given below, find y at $x = 1.4$, using Milne's predictor–corrector formula :

$$\frac{dy}{dx} = x^2 + \frac{y}{2}$$

x	1	1.1	1.2	1.3
y	2	2.2156	2.4549	2.7514

34. Find the value of $\Delta(e^x \log 2x)$.   (NIT(Kurukshetra)–2012)

35. Prove that $\nabla^2 = h^2D^2 - h^3D^3 + \dfrac{7}{12}h^4D^4$.

                      (MTU–2011, NIT(Kurukshetra)–2009)

36. Prove that $\nabla y_{n+1} = h\left(1 + \dfrac{\nabla}{2} + \dfrac{5}{12}\nabla^2 + ...\right) y_n'$

                      (NIT(Kurukshetra)–2008, 09)

37. Prove that

$$1.2.3 + 2.3.4 + 3.4.5 + ... \text{ to } n \text{ terms} = \frac{n(n+1)(n+2)(n+3)}{4}.$$

                      (NIT(Kurukshetra)–2008, 11, 12)

38. Prove that Newton's Cote formula.   (NIT(Kurukshetra)–2013)

39. Form the difference eqution generating by the following relations :

  (i) $y_x = (A + Bx)3x$ (Kurukshetra–2008, 10, 12, 13)
  (ii) $y = ax + bx^2$ (NIT(Kurukshetra)–2005, 2013)

40. Solve $y_{n+1} - 2\cos\alpha . y_n + y_{n-1}$.   (Kurukshetra–2013)

41. State and prove Newton's Raphson method.

                      (UKTU–2011, Kurukshetra–2007, 08, 13)

42. Find the order of Convergence of Newton's Raphson method.

                      (UKTU–2013, UPTU–2013, 14, MTU–2012,

                      Kurukshetra–2006, 07, 09)

43. Find the root of the equation $xe^{x'} = \cot x$ using the Regula-Falsi method correct to four decimal places.

                      (NIT(Kurukshetra)–2008, 09, 12)

44. Using Muller's method, find the root of the equation $x^3 - x - 1 = 0$.   (UKTU–2011, 13)

45. Find the root of the equation $\tan x + \tanh x = 0$ correct to three significant figures using an iterative formula.   (MTU–2012)

46. Compute the real root of the equation $xe^x - 2 = 0$ using Regula-Falsi method.   (SVTU–2007, NIT(Kurukshetra)–2007)

47. Form the Fibonacci difference equation and solve it.

                      (NIT(Kurukshetra)–2013)

48. State and prove Newton's divided difference formula.

                      (NIT(Kurukshetra)–2013)

Answers

1. $y = 2[x]^3 + 3[x]^2 + 2[x] - 10$ **2.** 2.925, 0.225   **3.** 60.05   **5.** 31

6. $f(1.5) = 0.222, f(5) = 22.022$ **7.** $y(4) = 74, y(6) = 261$ **8.** 352; 219   **9.** $f(x) = 9x - 4x^2$

10. 0.1955   **11.** 4.219   **12.** 2530   **13.** 32.345 thousands approx.

14. 0.32497   **15.** 12.826   **16.** 3.2219   **17.** 9   **18.** 3250.875

19. –16   **20.** 0.89   **21.** 1   **22.** 3.09   **23.** 133.19

24. $f(x) = x^4 - 3x^3 + 5x^2 - 6$ **25.** $y(0.1) = 0.90033, y(0.2) = 0.80227$

26. $y(1.1) = 0.1103, y = (1.2) = 0.2428$. Exact $y(1.1) = 0.1103$ and $y(1.2) = 0.2428$

27. $y(0.2) = 0.2468$ and $y(0.4) = 0.6031$ **28.** $y(0.2) = 1.2046, y(0.4) = 1.4644$ **29.** 2.2352,

30. $y(0.1) = 0.9052, y(0.2) = 0.8213$ **31.** 0.3487 **32.** $y(0.4) = 1.8392$ **33.** $y(1.4) = 3.0794$

❋❋❋❋❋❋

CHAPTER
Complex Numbers
11

Let x and y be two real numbers, then the set of ordered pairs (x, y) are called the system of complex numbers if it satisfies the following definitions of equality, addition, subtraction, multiplication and division.

(i) **Equality:** The two complex numbers (x, y) and (α, β) are equal if and only if $x = \alpha, y = \beta$.

(ii) **Addition:** The sum of two complex numbers is defined by the equation : $(x, y) + (\alpha, \beta) = (x + \alpha, y + \beta)$

(iii) **Subtraction :** The subtraction is defined by the equation : $(x, y) - (\alpha, \beta) = (x - \alpha, y - \beta)$

(iv) **Multiplication :** The multiplication is defined by the equation : $(x, y) \times (\alpha, \beta) = (x\alpha - y\beta, x\beta + y\alpha)$

(v) **Division :** The division $(x, y) \div (\alpha, \beta)$ is defined as the complex number (c, d) given by $(\alpha, \beta) \times (c, d) = (x, y)$, provided that such a number exists.

(vi) We also define $(x, 0)$ to be the real number x and $(0, 0)$ to be the real number 0. It is easy to verify that definitions (i) to (v) given above reduce to the ordinary rules of algebra when the numbers concerned are of the form $(x, 0)$.

11.1.1 NOTATION

It is convenient to denote the complex number (x, y) by the compound symbol $(x + iy)$, where $i^2 = -1$, for then we can obtain the sum, difference, product, etc. of complex numbers by usual rules of algebra (such as are applicable to real numbers). In this notation

$$x + iy + \alpha + i\beta = (x + \alpha) + i(y + \beta)$$

$$x + iy - (\alpha + i\beta) = (x - \alpha) + i(y - \beta)$$

$$(x + iy)(\alpha + i\beta) = (x\alpha - y\beta) + i(x\beta + y\alpha) \qquad (\because i^2 = -1)$$

and these results are in agreement with the definitions given above for sum, difference and product of two complex numbers.

REMARKS
- The complex number $x + iy$ is also denoted by z, x is called the real part of z, i.e., $R(z)$ and y the imaginary part of z, i.e., $I(z)$.
- In the case of real numbers $a \times a \times a \ldots$ to n factors is denoted by a^n, similarly $z \times z \times z \times \ldots$ to n factors is denoted by z^n, where n is any positive integer.

11.1.2 REDUCTION OF COMPLEX NUMBERS INTO THE STANDARD FORM $r(\cos \theta + i \sin \theta)$

Let $z = x + iy$ be the complex number.

If (x, y) be taken as the cartesian co-ordinates of a point P and (r, θ) the polar co-ordinate of the same point, we have

$$x = r\cos\theta, y = r\sin\theta \qquad \ldots(A)$$

$$\therefore \qquad z = x + iy = r(\cos\theta + i\sin\theta) \qquad \ldots(1)$$

where

$$r = +\sqrt{x^2 + y^2}, \tan\theta = \frac{y}{x}$$

or

$$\theta = \tan^{-1}\frac{y}{x} \qquad \ldots(2)$$

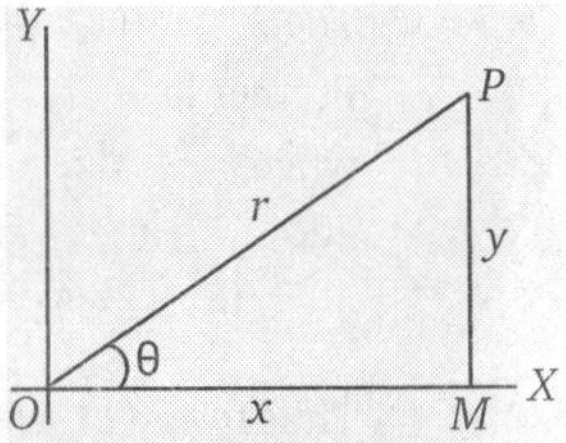

We notice that all complex numbers can be put in the form $r(\cos \theta + i \sin \theta)$. Here r be the positive square root of $x^2 + y^2$ called the absolute value or modulus of the complex number $x + iy$ and is designated by $|z|$ or $|x + iy|$. The circular measure of θ which satisfies conditions (A), viz.

$$\cos\theta = \frac{x}{\sqrt{(x^2 + y^2)}}, \sin\theta = \frac{y}{\sqrt{x^2 + y^2}}$$

Fig. 1

is called the argument or amplitude of $x + iy$. Its value must be taken between $-\pi$ and $+\pi$. The factor $\cos\theta + i\sin\theta$ sometimes abbreviated as cis θ is called the direction factor of the complex number. Its modulus is unity. The form $r(\cos\theta + i\sin\theta)$ is called modulus-amplitude form or polar form or trigonometrical form.

11.1.3 THE GEOMETRICAL REPRESENTATION OF COMPLEX NUMBERS

The complex number $z = x + iy$ is represented by a point P whose cartesian co-ordinates are (x, y) referred to rectangular axes OX and OY, usually called real and imaginary axes respectively. If r is the modulus and θ the argument of the complex number z, then clearly the polar co-ordinates of the point P are (r, θ). The plane whose points are represented by complex numbers is called argand's plane or argand's diagram after Argand. (This is also called Complex plane or Gaussian plane).

The complex number z is called the affix of the point (x, y) which represent it.

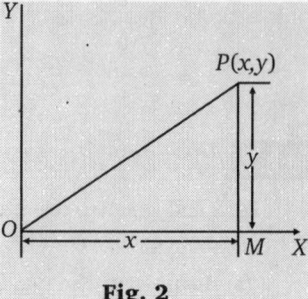

Fig. 2

(I) VECTOR REPRESENTATION OF COMPLEX NUMBERS

Let P be a point (x, y) on the argand plane corresponding to the complex number $z = x + iy$ referred to OX and OY as co-ordinate axes. The modulus and argument of z are represented by the length (or modulus) and direction of the vector $\overrightarrow{OP}$ respectively and vice versa.

(II) THE SUM AND DIFFERENCE OF TWO COMPLEX NUMBERS

(A) Sum: Let the two complex numbers $z_1 = x_1 + iy_1$ and $z_2 = x_2 + iy_2$ be represented by the points $P(x_1, y_1)$ and $Q(x_2, y_2)$ respectively on the argand plane.

Complete the parallelogram $OPRQ$. The middle points of the diagonals PQ and OR are same. But the mid point of PQ is

$$\left(\frac{1}{2}(x_1 + x_2), \frac{1}{2}(y_1 + y_2)\right)$$

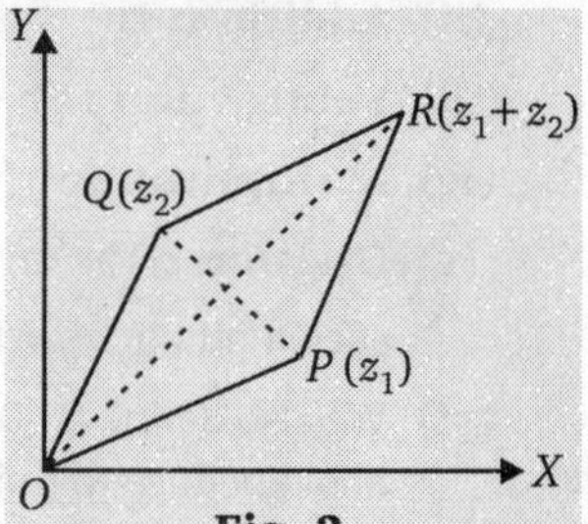

Since the point O is $(0, 0)$ and therefore the co-ordinates of R are $(x_1 + x_2, y_1 + y_2)$. Hence the sum of the two complex numbers z_1 and z_2 is represented by the point R.
In vectorial notation

$$z_1 + z_2 = \overrightarrow{OP} + \overrightarrow{OQ} = \overrightarrow{OP} + \overrightarrow{PR} = \overrightarrow{OR}$$

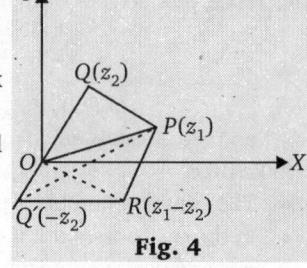

Fig. 3

(B) Difference: We represent $-z_2$ by Q' so that O is the middle point of QQ'. Complete the parallelogram $OPRQ'$. The middle points of PQ' and OR are the same. But the middle point of PQ' is

$$\left(\frac{1}{2}(x_1 - x_2), \frac{1}{2}(y_1 - y_2)\right).$$

Therefore the co-ordinates of R are $(x_1 - x_2, y_1 - y_2)$. Hence the difference $z_1 - z_2$ of the two complex numbers z_1 and z_2 is represented by the point R.
Since OQ is equal and parallel to RP, so $ORPQ$ is parallelogram and hence $\overrightarrow{OR} = \overrightarrow{QP}$. Thus in vectorial notation, we have

$$z_1 - z_2 = z_1 + (-z_2) = \overrightarrow{OP} + \overrightarrow{OQ'} = \overrightarrow{OP} + \overrightarrow{QO} = \overrightarrow{OP} + \overrightarrow{PR} = \overrightarrow{OR} = \overrightarrow{QP}$$

Therefore, it follows that the complex number $z_1 - z_2$ is represented by the vector $\overrightarrow{QP}$.

Fig. 4

11.1.4 PRODUCT AND QUOTIENT OF TWO COMPLEX NUMBERS

(A) *If z_1 and z_2 be two complex numbers then*

(i) $|z_1 z_2| = |z_1| \cdot |z_2|$ *and*   (ii) $\arg(z_1 z_2) = \arg z_1 + \arg z_2$

Let $z_1 = x_1 + iy_1$ and $z_2 = x_2 + iy_2$

Then $z_1 z_2 = (x_1 + iy_1)(x_2 + iy_2) = (x_1 x_2 - y_1 y_2) + i(x_1 y_2 + x_2 y_1)$

Further let $x_1 = r_1\cos\theta_1, y_1 = r_1\sin\theta_1$ and $x_2 = r_2\cos\theta_2, y_2 = r_2\sin\theta_2$

where $|z_1| = r_1, |z_2| = r_2$

and $\arg z_1 = \theta_1, \arg z_2 = \theta_2$

$z_1 z_2 = (r_1\cos\theta_1 + ir_1\sin\theta_1) \cdot (r_2\cos\theta_2 + ir_2\sin\theta_2) = r_1 r_2[(\cos\theta_1\cos\theta_2 - \sin\theta_1\sin\theta_2) + i(\sin\theta_1\cos\theta_2 + \cos\theta_1\sin\theta_2)]$

$= r_1 r_2[\cos(\theta_1 + \theta_2) + i\sin(\theta_1 + \theta_2)]$ ...(1)

Thus from (1), we clearly have

$|z_1 z_2| = r_1 r_2 = |z_1| \cdot |z_2|$ and $\arg(z_1 z_2) = \theta_1 + \theta_2 = \arg z_1 + \arg z_2$

📝 **REMARK**

• In a similar manner as above, we have in general

$|z_1 z_2 \ldots z_n| = |z_1||z_2|\ldots|z_n|$ and $\arg(z_1 z_2 \ldots z_n) = \arg z_1 + \arg z_2 + \ldots + \arg z_n$

Thus we can state that the modulus of the product of complex number is equal to the product of the moduli of these complex numbers and the arg of the product of the complex numbers is equal to the sum of their arguments.

11.1.5 Conjugate of a Complex Number

If $z = x + iy$ is any complex number, then the complex number $x - iy$ is called the conjugate to z and is written as $\bar{z}$. Clearly conjugate to $z_1 + z_2$ is $\bar{z}_1 + \bar{z}_2$ and conjugate of $z_1 z_2$ is $\bar{z}_1 . \bar{z}_2$. We have

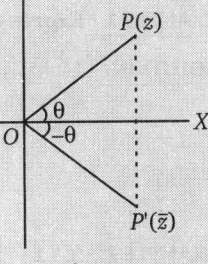

$$|z|^2 = z\bar{z} = |\bar{z}|^2, \qquad \therefore \quad |z| = |\bar{z}|$$
$$2R(z) = z + \bar{z} \qquad \text{and} \qquad 2iI(z) = z - \bar{z}$$

Also,
$$\overline{(\bar{z})} = z$$

Geometrically, the reflection (or image) of z in the real axis is the conjugate of z. thus if (r, θ) are the polar co-ordinates of P, then the polar co-ordinates of its reflection P' are $(r, -\theta)$.

Clearly
$$|z| = |\bar{z}| = r = \sqrt{(x^2 + y^2)}$$
$$\text{amp } z = \theta, \text{ amp } (\bar{z}) = -\theta \quad \text{and} \quad \text{amp } (z) = -\text{ amp } (\bar{z})$$

Fig. 5

11.1.6 Properties of Moduli

THEOREM 1. *The modulus of the sum of the two complex numbers can never exceed the sum of their moduli.*

PROOF. It is required to prove that $|z_1 + z_2| \le |z_1| + |z_2|$

We know that $|z|^2 = z\bar{z}$

$$\therefore \qquad |z_1 + z_2|^2 = (z_1 + z_2)\overline{(z_1 + z_2)} = (z_1 + z_2)(\bar{z}_1 + \bar{z}_2) = z_1\bar{z}_1 + z_1\bar{z}_2 + z_2\bar{z}_1 + z_2\bar{z}_2 \qquad \ldots(1)$$

But $z_1\bar{z}_1 = |z_1|^2, z_2\bar{z}_2 = |z_2|^2$

$$z_1\bar{z}_2 + z_2\bar{z}_1 = (x_1 + iy_1)(x_2 - iy_2) + (x_2 + iy_2)(x_1 - iy_1) = 2(x_1 x_2 + y_1 y_2) = 2R(z_1\bar{z}_2)$$

Substituting the values in (1), we have

$$|z_1 + z_2|^2 = |z_1|^2 + |z_2|^2 + 2R(z_1\bar{z}_2) \le |z_1|^2 + |z_2|^2 + 2|z_1||z_2| \qquad [\because |z_1\bar{z}_2| \ge \mathbf{R}(z_1\bar{z}_2)]$$

$$(\because \text{ real part of a complex number can never exceed its modulus})$$

or $|z_1 + z_2|^2 \le |z_1|^2 + |z_2|^2 + 2|z_1||z_2| \qquad [\because |z_1\bar{z}_2| = |z_1||\bar{z}_2| = |z_1||z_2|]$

or $|z_1 + z_2|^2 \le (|z_1| + |z_2|)^2 \qquad\qquad \text{or} \qquad\qquad |z_1 + z_2| \le |z_1| + |z_2|$

REMARKS

- By the use of above theorem, we have $|z_1 + z_2 + z_3| \le |z_1 + z_2| + |z_3| \le |z_1| + |z_2| + |z_3|$.

- In general, by induction, we have $\left| \sum\limits_{m=1}^{n} z_m \right| \le \sum\limits_{m=1}^{n} |z_m|$.

THEOREM 2. *The modulus of the difference of two complex numbers can never be less than the difference of their moduli.*

i.e., $|z_1 - z_2| \ge |z_1| - |z_2|$

PROOF.
$$|z_1 - z_2|^2 = (z_1 - z_2)\overline{(z_1 - z_2)} = (z_1 - z_2)(\bar{z}_1 - \bar{z}_2) = |z_1|^2 + |z_2|^2 - 2R(z_1\bar{z}_2)$$

$$\ge |z_1|^2 + |z_2|^2 - 2|z_1\bar{z}_2| = |z_1|^2 + |z_2|^2 - 2|z_1||\bar{z}_2| = (|z_1| - |z_2|)^2$$

$$\therefore \quad |z_1 - z_2| \ge |z_1| - |z_2|$$

THEOREM 3. *The modulus of the sum of two complex numbers is always greater than or equal to the difference of their moduli.*

i.e., $$|z_1 + z_2| \ge |z_1| - |z_2|$$

PROOF. From theorem 1 above, we have
$$|z_1 + z_2|^2 = |z_1|^2 + |z_2|^2 + 2\mathbf{R}(z_1\bar{z}_2) \qquad \ldots(1)$$

But $$|z_1 z_2| \ge \mathbf{R}(z_1\bar{z}_2)$$

$$\therefore \qquad -|z_1 z_2| \le \mathbf{R}(z_1\bar{z}_2) \quad \text{or} \quad -2|z_1 z_2| \le 2\mathbf{R}(z_1\bar{z}_2) \qquad \ldots(2)$$

Adding $|z_1|^2 + |z_2|^2$ to both sides of (2), we get

$$|z_1|^2 + |z_2|^2 - 2|z_1 z_2| \le |z_1|^2 |z_2|^2 + 2\mathbf{R}(z_1 z_2)$$

or $$(|z_1| - |z_2|)^2 \le |z_1 + z_2|^2$$

or $$|z_1 + z_2| \ge |z_1| - |z_2|. \qquad\qquad\qquad \text{[Using (1)]}$$

Solved Examples

EXAMPLE 1. *Express $\dfrac{5+7i}{2-3i}$ in the form of $x + iy$.*

SOLUTION. Multiplying the denominator and numerator by $2 + 3i$, the complex conjugate of denominator, we have

$$\frac{5+7i}{2-3i} = \frac{5+7i}{2-3i} \times \frac{2+3i}{2+3i}$$

$$= \frac{10-21+i(14+15)}{4+9} = -\frac{11}{13} + i\frac{29}{13}$$

EXAMPLE 2. *If $(1+i)(1+2i)(1+3i)...(1+ni) = A + iB$. show that $2 . 5 . 10 ...(1+n^2) = A^2 + B^2$.*

SOLUTION. We have

$$(1+i)(1+2i)(1+3i) ... (1+ni) = A + iB \qquad ...(1)$$

Taking conjugate of (1), we have

$$(1-i)(1-2i)(1-3i)...(1-ni) = A - iB \qquad ...(2)$$

Multiplying (1) and (2) pair-wise, we have

$$(1-i^2)(1-2^2 i^2)(1-3^2 i^2)...(1-n^2 i^2) = A^2 - i^2 B^2$$

or $(1+1)(1+4)(1+9)...(1+n^2) = A^2 + B^2$

or $2.5.10...(1+n^2) = A^2 + B^2$.

EXAMPLE 3. *Using argand-diagram find the product of $(2 + i)$ and $(2 – i)$.*

SOLUTION. Let $z_1 = 2 + i = r_1[\cos\theta_1 + i\sin\theta_1]$

$$z_2 = 2 - i = r_2[\cos\theta_2 + i\sin\theta_2]$$

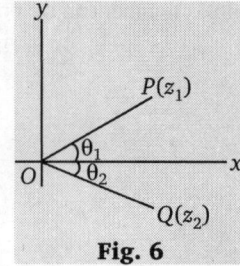

Fig. 6

Then $r_1 = \sqrt{5}, \theta_1 = \tan^{-1}\left(\dfrac{1}{2}\right)$.

$$r_2 = \sqrt{5}, \theta_2 = \tan^{-1}\left(-\dfrac{1}{2}\right).$$

Now

$$z_1 . z_2 = r_1 r_2 [\cos(\theta_1 + \theta_2) + i\sin(\theta_1 + \theta_2)]$$

$\therefore$ $|z| = |z_1| . |z_2| = \sqrt{5} \times \sqrt{5} = 5$

and $\arg z = \theta_1 + \theta_2 = 0$

EXAMPLE 4. *Prove that $|z_1 + z_2|^2 + |z_1 - z_2|^2 = 2|z_1|^2 + 2|z_2|^2$. Also, deduce that*

$$|\alpha + \sqrt{(\alpha^2 - \beta^2)}| + |\alpha - \sqrt{(\alpha^2 - \beta^2)}| = |\alpha + \beta| + |\alpha - \beta|$$

all numbers involved being complex.

SOLUTION. We have

$$|z_1 + z_2|^2 = (z_1 + z_2)\overline{(z_1 + z_2)} = (z_1 + z_2)(\bar{z}_1 + \bar{z}_2)$$

$$= z_1 \bar{z}_1 + z_2 \bar{z}_2 + z_1 \bar{z}_2 + z_2 \bar{z}_1 \qquad ...(1)$$

and $|z_1 - z_2|^2 = (z_1 - z_2)(\bar{z}_1 - \bar{z}_2)$

$$= z_1 \bar{z}_1 + z_2 \bar{z}_2 - z_1 \bar{z}_2 - z_2 \bar{z}_1 \qquad ...(2)$$

Adding (1) and (2) we have

$$|z_1 + z_2|^2 + |z_1 - z_2|^2 = 2z_1 \bar{z}_1 + 2z_2 \bar{z}_2 = 2|z_1|^2 + 2|z_2|^2$$

$$...(3)$$

Further let $z_1 = \alpha + \sqrt{(\alpha^2 - \beta^2)}, z_2 = \alpha - \sqrt{(\alpha^2 - \beta^2)}$

Now from equation (3), we have

$$|z_1|^2 + |z_2|^2 = \frac{1}{2}|z_1 + z_2|^2 + \frac{1}{2}|z_1 - z_2|^2$$

$$= \frac{1}{2}|2\alpha|^2 + \frac{1}{2}|2\sqrt{(\alpha^2 - \beta^2)}|^2$$

$$= 2|\alpha|^2 + 2|\alpha^2 - \beta^2| \qquad ...(4)$$

Again,

$$\{|z_1| + |z_2|\}^2 = |z_1|^2 + |z_2|^2 + 2|z_1||z_2|$$

$$= 2|\alpha|^2 + 2|\alpha^2 - \beta^2| + 2|\alpha^2 - (\alpha^2 - \beta^2)|$$

$$\text{[Using 4]}$$

$$= 2|\alpha|^2 + 2|\alpha^2 - \beta^2| + 2|\beta|^2$$

$$= 2|\alpha|^2 + 2|\beta|^2 + 2|\alpha + \beta||\alpha - \beta|$$

$$= |\alpha + \beta|^2 + |\alpha - \beta|^2 + 2|\alpha + \beta||\alpha - \beta|$$

$$\text{[Using 3]}$$

$$= \{|\alpha + \beta| + |\alpha - \beta|\}^2$$

$\therefore$ $|z_1| + |z_2| = |\alpha + \beta| + |\alpha - \beta|$

or $|\alpha + \sqrt{(\alpha^2 - \beta^2)}| + |\alpha - \sqrt{(\alpha^2 - \beta^2)}|$

$$= |\alpha + \beta| + |\alpha - \beta|.$$

EXAMPLE 5. *Prove that the area of the triangle whose vertices are the points represented by the complex numbers z_1, z_2, z_3 on the argand plane is $\sum\left[\dfrac{|z_1|^2(z_2 - z_3)}{4iz_1}\right]$*

SOLUTION. Let $z_1 = x_1 + iy_1, z_2 = x_2 + iy_2, z_3 = x_3 + iy_3$

Hence, the required area of the triangle

$$= \frac{1}{2}\begin{vmatrix} x_1 & y_1 & 1 \\ x_2 & y_2 & 1 \\ x_3 & y_3 & 1 \end{vmatrix} = \frac{1}{2i}\begin{vmatrix} x_1 & iy_1 & 1 \\ x_2 & iy_2 & 1 \\ x_3 & iy_3 & 1 \end{vmatrix}$$

Now applying $C_2 \to C_2 + C_1$

$$= \frac{1}{2i}\begin{vmatrix} x_1 & x_1 + iy_1 & 1 \\ x_2 & x_2 + iy_2 & 1 \\ x_3 & x_3 + iy_3 & 1 \end{vmatrix} = \frac{1}{2i}\begin{vmatrix} x_1 & z_1 & 1 \\ x_2 & z_2 & 1 \\ x_3 & z_3 & 1 \end{vmatrix}$$

Expanding by the column

$$= \frac{1}{2i}\Sigma[x_1(z_2 - z_3)]$$

$$= \frac{1}{2i}\Sigma\left[\frac{1}{2}(z_1 + \bar{z}_1)(z_2 - z_3)\right] \quad \left[\because x_1 = \frac{1}{2}(z_1 + \bar{z}_1)\right]$$

$$= \frac{1}{4i}[\Sigma\{z_1(z_2 - z_3) + \Sigma\{\bar{z}_1(z_2 - z_3)\}]$$

$$= \frac{1}{4i}\left[\Sigma(z_1 z_2) - \Sigma(z_1 z_3) + \Sigma\left\{\frac{z_1 \bar{z}_1(z_2 - z_3)}{z_1}\right\}\right]$$

$$= \frac{1}{4i}\left[\Sigma(z_1 z_2) - \Sigma(z_1 z_2) + \Sigma\left\{\frac{|z_1|^2(z_2 - z_3)}{z_1}\right\}\right]$$

$$[\because \Sigma z_1 z_2 = \Sigma z_1 z_3]$$

$$= \Sigma\left[\frac{|z_1|^2(z_2 - z_3)}{4iz_1}\right].$$

EXAMPLE 6. *Show that the triangle whose vertices are the points represented by the complex numbers z_1, z_2, z_3 on the argand plane is equilateral if and only if*

$$\frac{1}{z_2 - z_3} + \frac{1}{z_3 - z_1} + \frac{1}{z_1 - z_2} = 0$$

that is iff $z_1^2 + z_2^2 + z_3^2 - z_1 z_2 - z_2 z_3 - z_3 z_1 = 0$

SOLUTION. Let the vertices A, B, C of a $\triangle ABC$ be represented by the complex numbers z_1, z_2, z_3 respectively, then

$BC = |z_2 - z_3|, CA = |z_3 - z_1|, AB = |z_1 - z_2|$

Now suppose

$$z_2 - z_3 = \alpha , z_3 - z_1 = \beta, z_1 - z_2 = \gamma$$

$\therefore \quad \alpha + \beta + \gamma = 0 \qquad \qquad \ldots(1)$

$$\overline{\alpha + \beta + \gamma} = 0$$

or $\quad \overline{\alpha} + \overline{\beta} + \overline{\gamma} = 0 \qquad \qquad \ldots(2)$

The condition is necessary. Let the $\triangle ABC$ be equilateral.

Then $BC = CA = AB$

or $\quad |z_2 - z_3| = |z_3 - z_1| = |z_1 - z_2|$

or $\quad |\alpha| = |\beta| = |\gamma|$ or $|\alpha|^2 = |\beta|^2 = |\gamma|^2$

or $\quad \alpha\overline{\alpha} = \beta\overline{\beta} = \gamma\overline{\gamma} = k$ (say) $\qquad \ldots(3)$

Using (3) in (2), we have

$$\frac{k}{\alpha} + \frac{k}{\beta} + \frac{k}{\gamma} = 0 \quad \text{or} \quad \frac{1}{\alpha} + \frac{1}{\beta} + \frac{1}{\gamma} = 0$$

or $\quad \dfrac{1}{z_2 - z_3} + \dfrac{1}{z_3 - z_1} + \dfrac{1}{z_1 - z_2} = 0 \qquad \ldots(4)$

or $\quad z_1^2 + z_2^2 + z_3^2 - z_1 z_2 - z_2 z_3 - z_3 z_1 = 0$

This proves the necessary condition.

The condition is sufficient. Let the condition (4) hold, then to prove that the $\triangle ABC$ is equilateral.

From (4), we have

$$\frac{1}{\alpha} + \frac{1}{\beta} + \frac{1}{\gamma} = 0 \qquad [\because z_2 - z_3 = \alpha \text{ etc.}]$$

or $\quad \beta\gamma + \alpha\gamma + \alpha\beta = 0$

or $\quad \beta\gamma + (\gamma + \beta)\alpha = 0$

or $\quad \beta\gamma + (-\alpha)\alpha = 0 \qquad \qquad$ [Using (1)]

$\quad \alpha^2 = \beta\gamma \qquad \qquad \ldots(5)$

Conjugate of (5) is,

$$\overline{(\alpha^2)} = \overline{(\beta\gamma)} \quad \text{or} \quad \overline{\alpha}^2 = \overline{\beta}\,\overline{\gamma} \qquad \ldots(6)$$

Multiplying (5) and (6), we get

$$\alpha^2 \overline{\alpha}^2 = \beta\gamma\overline{\beta}\,\overline{\gamma}$$

or $\quad (\alpha\overline{\alpha})^2 = \beta\overline{\beta}\gamma\overline{\gamma}$ or $(\alpha\overline{\alpha})^3 = \alpha\overline{\alpha}\beta\overline{\beta}\gamma\overline{\gamma} \qquad \ldots(7)$

Similarly, $(\beta\overline{\beta})^3 = \alpha\overline{\alpha}\beta\overline{\beta}\gamma\overline{\gamma} = (\gamma\overline{\gamma})^3 \qquad \ldots(8)$

From (7) and (8), we get

$$(\alpha\overline{\alpha})^3 = (\beta\overline{\beta})^3 = (\gamma\overline{\gamma})^3$$

or $\quad \alpha\overline{\alpha} = \beta\overline{\beta} = \gamma\overline{\gamma}$

$$|\alpha|^2 = |\beta|^2 = |\gamma|^2$$

or $\quad |z_2 - z_3|^2 = |z_3 - z_1|^2 = |z_1 - z_2|^2$

or $\quad BC^2 = CA^2 = AB^2$

or $\quad BC = CA = AB$

Therefore the $\triangle ABC$ is equilateral.

EXAMPLE 7. *If $|z_1| = |z_2| = \ldots = |z_n| = 1$, prove that*

$$|z_1 + z_2 + \ldots + z_n| = \left| \frac{1}{z_1} + \frac{1}{z_2} + \ldots + \frac{1}{z_n} \right|$$

SOLUTION. We have $|z_i| = 1 \Rightarrow |z_i|^2 = 1$ for $i = 1, 2, 3, \ldots n$

$$\Rightarrow \qquad z_i \overline{z_i} = 1 \qquad \qquad \ldots(1)$$

We know that $\quad |z| = |\overline{z}|$

Hence $|z_1 + z_2 + \ldots + z_n|$

$$= \overline{|z_1 + z_2 + \ldots + z_n|} = |\overline{z_1} + \overline{z_2} + \ldots + \overline{z_n}|$$

$$= \left| \frac{1}{z_1} + \frac{1}{z_2} + \ldots + \frac{1}{z_n} \right|. \qquad \text{[Using (1)]}$$

Exercise-11.1

1. Express the following complex numbers in the form $r(\cos\theta + i\sin\theta)$:

   (i) $\sqrt{3} - \sqrt{-1}$ (ii) $2 + \sqrt{3} + i$

   (iii) $-5 - 12i$ (iv) $1 - i$

   (v) $-\sqrt{3} + i$ (vi) $\dfrac{(1+i)(2-i)}{3+i}$

2. Put the following in $A + iB$ form:

   (i) $(2 + 3i)^2$ (ii) $(5 - 6i)^2$

   (iii) $\dfrac{2 + 5i}{7 - 7i}$

3. Find the numbers A and B if

   (i) $A + iB = \dfrac{1}{(1 - 2i)(2 + 3i)}$ (ii) $A + iB = \dfrac{3 - 2i}{7 + 4i}$

4. If $\left| \dfrac{z-1}{z+1} \right| = 2$ prove that the locus if z on the argand plane is a circle whose centre has affix $\left(-\dfrac{5}{3}, 0 \right)$ and whose radius is $\dfrac{4}{3}$.

5. If $\left| \dfrac{z-i}{z+i} \right| = 5$ prove that the locus of z on the argand diagram is a circle, whose centre has affix $\left(0, -\dfrac{13}{12} \right)$ and radius is $\dfrac{5}{12}$.

6. If $\left| \dfrac{2z+5}{z-3} \right| = 1$, show that the locus of z on the argand plane is $3x^2 + 3y^2 + 26x + 16 = 0$.

7. A variable complex number $z = x + iy$ is such that the amplitude of the fraction $\dfrac{z-1}{z+1}$ is always equal to $\pi/4$, show that $x^2 + y^2 - 2y = 1$.

8. If $\text{amp}\left(\dfrac{z-i}{z+i} \right) = \dfrac{\pi}{4}$, prove that locus of z on the argand-diagram is a circle whose centre has affix $(-1, 0)$ and radius is $\sqrt{2}$.

9. Show that $\dfrac{|z|}{\overline{z}} + \dfrac{|\overline{z}|}{z} = 2$.

10. Show that the triangle whose vertices are the points -1, 1 and $i\sqrt{3}$ in the argand plane is equilateral.

11. If z_1, z_2 and z_3 are the vertices of an isosceles triangle, right

angled at the vertices z_2 prove that

$$z_1^2 + 2z_2^2 + z_3^2 = 2z_2(z_1 + z_3).$$

12. Find the locus of the complex number z, if $\arg\left(\dfrac{z - 3i}{z - 3}\right) = \dfrac{\pi}{3}$.

13. Prove that if x, y are real number such that $x^2 + y^2 \neq 0$. Then

$$\left|\frac{x - iy}{x + iy}\right| = 1.$$

14. If z_1, z_2 and z_3 are the vertices of an equilateral triangle, prove that $z_1^2 + z_2^2 + z_3^2 = z_1 z_2 + z_2 z_3 + z_3 z_1$.

15. Find the complex number z if $\arg(z + 1) = \pi/6$ and $\arg(z - 1)$ = $2\pi/3$.   (MUMBAI–2009)

16. Find the locus of $P(z)$ when
(i) $|z - a| = k$ (ii) $\text{amp}(z - a) = \alpha$
where k and α are constant.   (GORAKHPUR–1999)

17. If $|z_1 + z_2| = |z_1 - z_2|$, prove that the difference of amplitudes of z_1 and z_2 is $\pi/2$.

18. Express $1/(z + 1)^2 - 1/(z - 1)^2$ in the modulus amplitude form.   (VTU–2011S)

19. If $\alpha - i\beta = 1/(a - ib)$, prove that $(\alpha^2 + \beta^2)(a^2 + b^2) = 1$.   (MUMBAI–2008S)

20. If $|z^2 - 1| = |z|^2 + 1$, prove that z lies on the imaginary axis.   (MUMBAI–2007)

Hints to Selected Problems

1. (i) We can write $\sqrt{3} - \sqrt{(-1)} = \sqrt{3} - i = r(\cos\theta + i\sin\theta)$

$\Rightarrow r\cos\theta = \sqrt{3}$ and $r\sin\theta = -1$ $\Rightarrow r = 2, \theta = \dfrac{11}{6}\pi$

4. $\left|\dfrac{z - 1}{z + 1}\right| = 2$ $\Rightarrow$ $\left|\dfrac{x + iy - 1}{x + iy + 1}\right| = 2$.

On solving, we get $x^2 + y^2 + \dfrac{10}{3}x + 1 = 0$

Comparing this equation with $x^2 + y^2 + 2gx + 2fy + c = 0$, we get $g = 5/3, f = 0, c = 1$.

Hence, the centre of the circle = $(-g, -f)$ = $(-5/3, 0)$ and radius is $4/3$.

7. On putting $z = x + iy$ in the given equation and after some simplification, we get

$$r\cos\theta = \frac{x^2 + y^2 - 1}{\{(x + 1)^2 + y^2\}}, r\sin\theta = \frac{2y}{\{(x + 1)^2 - y^2\}}$$

$\Rightarrow \tan\theta = \dfrac{2y}{x^2 + y^2 - 1}$ According to given question, we have

$\tan\theta = \tan\dfrac{\pi}{4} = 1$ $\Rightarrow$ $\dfrac{2y}{x^2 + y^2 - 1} = 1$, i.e., $x^2 + y^2 - 2y = 1$.

9. L.H.S $= \dfrac{|z|}{|\bar{z}|} + \dfrac{|\bar{z}|}{|z|} = \dfrac{|z||z| + |\bar{z}||\bar{z}|}{|\bar{z}|.|z|}$

Now, using $z = x + iy$ and $z = x - iy$

$\Rightarrow$ $|z| = \sqrt{x^2 + y^2} = |\bar{z}|$

11. Since $BA = BC$

$\Rightarrow |z_1 - z_2| = |z_3 - z_2|$

$\Rightarrow |z_1 - z_2|^2 = |z_3 - z_2|^2$

$\Rightarrow (z_1 - z_2)(\bar{z}_1 - \bar{z}_2) = (z_3 - z_2)(\bar{z}_3 - \bar{z}_2)$ ...(1)

Also $\angle ABC = --$. Thus $\arg\left(\dfrac{z_1 - z_2}{z_3 - z_2}\right) = \dfrac{\pi}{2}$.

So that $\dfrac{z_1 - z_2}{z_3 - z_2}$ is purely imaginary

$\dfrac{z_1 - z_2}{z_3 - z_2} + \dfrac{\bar{z}_1 - \bar{z}_2}{\bar{z}_3 - \bar{z}_2} = 0$

$\Rightarrow \dfrac{z_1 - z_2}{z_3 - z_2} + \dfrac{z_3 - z_2}{z_1 - z_2} = 0$

On simplifying, we have
$$z_1^2 + 2z_2^2 + z_3^2 = 2z_2(z_1 + z_3).$$

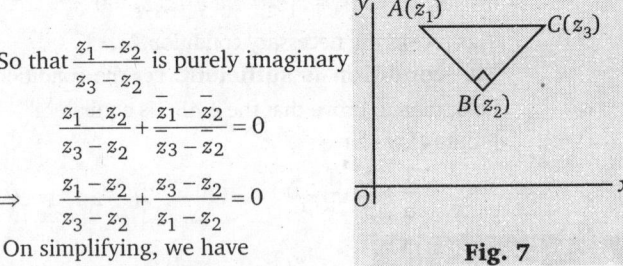

Fig. 7

Answers

1. (i) $2\left(\cos\dfrac{11\pi}{6} + i\sin\dfrac{11\pi}{6}\right)$ (ii) $r = 2\sqrt{2 + \sqrt{3}}, \theta = \tan^{-1}(2 - \sqrt{3})$ (iii) $r = 13, \theta = \tan^{-1}\left(\dfrac{12}{5}\right)$ (iv) $\sqrt{2}\left(\cos\dfrac{7\pi}{4} + i\sin\dfrac{7\pi}{4}\right)$

(v) $2\left(\cos\dfrac{5\pi}{6} + i\sin\dfrac{5\pi}{6}\right)$ (vi) 1 **2.** (i) $-5 + 12i$ (ii) $-11 - 60i$ (iii) $-\dfrac{21}{98} + \dfrac{1}{2}i$

3. (i) $\dfrac{8}{65}, \dfrac{1}{65}$ (ii) $\dfrac{13}{65}, \dfrac{-26}{65}$ **12.** $x^2 + y^2 - (3 - \sqrt{3})x - (3 - \sqrt{3})y - 3\sqrt{3} = 0$ **15.** $z = \dfrac{1}{2} + \dfrac{\sqrt{3}}{2}i$

16. (i) Circle with centre $A(a)$ and radius k (ii) Straight line through $A(a)$ making an $\angle\alpha$ with OX. **18.** $-8i/25$

11.2 LIMITS AND CONTINUITY

We are very well familiar with the definition of limits and continuity for functions of a real variable. Here we shall deal with corresponding definitions of functions of a complex variable and these definitions will be just analogous to the definitions we are already familiar with. We have already defined complex numbers and their geometrical representation on the argand plane.

Definition 1. *Any collection of points in the complex plane is called a (two dimensional) point set and each point is called a member or element of the set.*

Definition 2. *Let z_0 be a point in the argand diagram then the neighbourhood of this point z_0 is defined as the set of all those point z such that $|z - z_0| < \varepsilon$ where ε is an arbitrary small positive number. This ε is called the radius of the neighbourhood of z_0.*

Definition 3. *If from the neighbourhood of a point z_0 defined by $|z - z_0| < \varepsilon$ we exclude the point z_0 then such a neighbourhood is called the deleted neighbourhood of the point z_0, i.e.,*

$$0 < |z - x_0| < \varepsilon$$

Definition 4. *The set of all points z such that $|z| > k$ where k is any positive real number is called a neighbourhood of the point at infinity.*

Definition 5. *Let S be a set of points in the argand plane then a point z_0 is called a limit point of the set S if every ε neighbourhood of z_0 contains one points of S then it contains an infinite number of points of S.*

☑ **REMARK**

- The limit point may or may not belong to the set, *e.g.,* consider the set of points $S = \left\{ \dfrac{3 + 2n}{2 + n} : n = 1, 2, 3 \ldots \right\}$

  The limit point of the above set is clearly 2i which does not belong to the set S. Consider another example and let us choose the set of points defined by $|z| < r$. Evidently all points on the circle $|z| = r$ are the limit points of this set and they do not belong to the set. Also all points within the circle $|z| = r$ are also limit points of the set defined by $|z| < r$ and they belong to the set. Form above example we conclude that the limit point of a set may or may not belong to the set.

Definition 6. *A point z_0 is said to be an interior point of a set S if there exists an ε-neighbourhood of z_0 all of whose points belong to S.*

On the other hand if every ε-neighbourhood of z_0 contains points belonging to S and also points not belonging to S then z_0 is called a boundary point. If a point z_0 is neither an interior point nor a boundary point of the set S then it is called exterior point of the set S. With reference to the example given above we can say that all points inside the circle $|z| = r$ are inner points of the set defined by $|z| < r$ whereas all points on the circumference of the circle $|z| = r$ are the boundary points of the above set $|z| < r$.

Definition 7. *A set S is said to be open if it contains only the interior points.* As a very simple example we can say that the set of points given by $|z| < r$ is an open set because only those limit points which are inner points, belong to this set.

A set S is said of be closed if every limit point (both interior and boundary points) of S belong to it.

In accordance with the definition we can say that the set of points defined by $|z| < r$ is a closed set.

☑ **REMARK**

- There may be sets which are both open and closed sets just as the empty set and the whole plane are such but there can be sets which are neither open nor closed. As an example we can say that the set of points defined by $|z| < r$ and the point $z = r$ is neither open nor closed in accordance with our definitions given above.

Definition 8. *A set of points S is said to be bounded if a constant M can be found such that $|z| \leq M$ for every point z of the set.*

In case the set is not bounded then it is said to be unbounded. A set S is said to be compact if it is both bounded and closed.

Definition 9. *If any two points of a given set can be joined by a continuous polygonal arc (i.e., a path consisting of straight line segment) such that all points of the arc are the points of the set then such a set is said to be a connected set.*

Definition 10. *An open connected set is called an open domain. If however the boundary points of S are also included then it is called a closed domain.*

Definition 11. *If all the limit points of a given set S are added to it then the new S thus formed is defined as the closure of the set S.* Evidently the closure S' of a given set S is a closed set.

Definition 12. *Let x(t) and y(t) be continuous real valued functions of a real variable t defined in the range $\alpha \leq t \leq \beta$, then the set of points z in the argands plane given by the equation $z = x(t) + iy(t)$ is called a continuous arc. A point z_1 is said to be a multiple point to the arc if the equation $z_1 = x(t) + iy(t)$ is satisfied by more than one value of t in the given range. In case the above equation is satisfied by two values of t in the given range then the point z_1 is called a double point. A Jordan arc is defined as a continuous arc without multiple points.*

Definition 13. *If a continuous arc has only one multiple point which is a double point corresponding to the two terminal values of t, i.e., $t = \alpha$, $t = \beta$, then the arc is called a simple Jordan closed curve. In this case the end points $t = \alpha$ and $t = \beta$ of the continuous arc are coincident.*

☑ **REMARK**

- Every simple closed Jordan curve divides the argand plane into two disjoint open sets such that the curve is the boundary of these sets. One set the interior of the curve is bounded whereas the other exterior of the curve is unbounded. **(Jordan Curve Theorem)**

For example we can say that the circle $|z| = r$ divides the complex plane into two open domains $|z| < 1$ which is bounded and is the interior of the circle $|z| = 1$ and other $|z| > 1$ is unbounded and is the exterior of the circle $|z| = 1$.

Definition 14. *If a symbol z takes any one of the values of a set of complex numbers then z is called a complex variable.*

11.3 FUNCTIONS OF A COMPLEX VARIABLE

If corresponding to each value of a complex variable z, there are one or more then one values of another complex variable w, then w is said to be a function of z and is expressed as $w = f(z)$. Here, z is called independent variable and w the dependent variable.

Again we know that $z = x + iy$ and if $w = f(z) = u + iv$ then u and v are functions of two real variables x and y and we may write $w = f(z) = u + iv = u(x, y) + iv(x, y)$ where x, y are real.

11.3.1 SINGLE VALUED AND MANY VALUED FUNCTIONS

w is said to be single valued function of z if it takes only one value for each value of z in a given range D. If however w takes two or more than two values for some or all the values of z in the given domain D then w is said to be many valued function of z e.g. $w = f(z) = z$ is a single valued function but $w = f(z) = z^{1/2}$ is a many valued function.

11.3.2 Limit of a Function

Let $w = f(z)$ be any single valued function defined in the deleted neighbourhood of $z = a$. We say that $f(z)$ tends to limit l as z tends to a along any path in a defined region, if to each positive arbitrary number ε, however small there corresponds a positive number δ depending upon ε such that $|f(z) - l| < \varepsilon$ for all points of the region for which

$$0 < |z - a| < \delta.$$

In other words, it means that there exists a deleted neighbourhood of the point $z = a$ in which $|f(z) - l|$ can be made as small as we please. Symbolically we write $\lim_{z \to a} f(z) = l$.

LIMIT IN TERMS OF REAL AND IMAGINARY PARTS

Let $z = x + iy$, $w = f(z) = u + iv$ where $u = u(x, y)$ and $v = v(x, y)$

Now, it can be prove that if $a = a_1 + ia_2$ and $l = l_1 + il_2$, then $\lim_{z \to a} f(z) = l$

$$\Leftrightarrow \qquad \lim_{(x,y) \to (a_1, a_2)} u(x, y) = l_1 \qquad \text{and} \qquad \lim_{(x,y) \to (a_1, a_2)} v(x, y) = l_2 \qquad \qquad \ldots(1)$$

We have $\qquad\qquad |f(z) - l| = |u(x, y) + iv(x, y) - (l_1 + il_2)| = (u - l_1) + i(v - I_2)$

Also $\qquad\qquad\qquad |u - l_1| \le |f(z) - l|, |v - l_2| \le |f(z) - l| \qquad\qquad\qquad\qquad \ldots(2)$

and $\qquad\qquad\qquad |f(z) - l| \le |u - l_1| + |v - l_2| \qquad\qquad\qquad\qquad\qquad\qquad \ldots(3)$

Hence, result (1) may be proved with the help of (2) and (3).

✎ REMARK

- From above, it is clear that $\lim R [f(z)] = R(l)$ and $\lim I (f(z)) = I(l)$, $\lim f(z) = l \qquad \Rightarrow \qquad \lim \overline{f(z)} = l$

ALGEBRA OF LIMITS

Similar as in case of real numbers, we have the following results on limits :

if $\qquad\qquad \lim_{z \to a} f(z) = l \qquad$ and $\qquad \lim_{z \to a} g(z) = m$

Then

(1) $\lim_{z \to a} [f(z) \pm g(z)] = l \pm m$ $\qquad\qquad\qquad$ (2) $\lim_{z \to a} [f(z) . g(z)] = l.m$

(3) $\lim_{z \to a} \left| \dfrac{f(z)}{g(z)} \right| = \dfrac{l}{m} (m \ne 0)$ $\qquad\qquad\qquad$ (4) $\lim_{z \to a} |k.f(z)| = k.l, k \in R$

11.3.3 Continuity of a Function

A functions $w = f(z)$ of a complex variable z defined for a certain region D is said to be continuous at the point $z = a$ of D, if given a positive number ε, we can find a number $\delta(\delta > 0)$ depending on ε such that $|f(z) - f(a)| < \varepsilon$ for all points z of D satisfying the condition $0 < |z - a| < \delta$.

From above definition it is clear that $f(z)$ will be continuous at $z = a$ if $\lim_{z \to a} f(z) = f(a)$.

Again a function $f(z)$ is said to be continuous in a certain domain D if it is continuous at every point of D.

Also if $f(z) = u(x, y) + iv(x, y)$ then $f(z)$ is continuous if and only if $u(x, y)$ and $v(x, y)$ are separately continuous functions of x and y.

11.3.4 Uniformly Continuous Function

If $|f(z) - f(z_0)| < \varepsilon$ for every pair of points z, z_0 of the domain D such that $|z - z_0| < \delta$. (δ depending upon ε) then the function $f(z)$ is said to be uniformly continuous in D.

11.3.5 Differentiabililty of a Function

A function $w = f(z)$ defined in a certain domain D is said to be differentiable at $z = a$ if the increment ratio

$$\frac{\Delta w}{\Delta z} = \frac{f(a + \Delta z) - f(a)}{\Delta z}$$

tends to a unique finite limit as z tends to a (i.e., as $\Delta z \to 0$) along any path of the domain D. This unique limit is then called the derivative or differential coefficient of $f(z)$ at $z = a$ and is denoted by $f'(a)$.

✎ REMARK

- If this limit is different for different paths of D i.e., if it depends on amp Δz then we would say that $f(z)$ is not differentiable at $z = a$. we will generally choose different paths for Δz to be along real and imaginary axes. In other words Δz would be taken as purely real or purely imaginary.

Hence, for differentiability we must have unique and finite limit independent of the path.

We may also say that

$$f'(z) = \lim_{\Delta z \to 0} \frac{f(z + \Delta z) - f(z)}{\Delta z}, f'(z_0) = \lim_{z \to z_0} \frac{f(z) - f(z_0)}{z - z_0}$$

or $\qquad f'(z) = \lim_{h \to 0} \frac{f(z + h) - f(z)}{h}.$

✑ NOTE

- We may also choose z in place of Δz for finding $f'(0)$ i.e. $f'(z) = \lim_{z \to 0} \frac{f(z) - f(0)}{z}$

THEOREM 1. *Continuity is the necessary but not a sufficient condition for the existence of a finite derivative.*

PROOF. **Case I. Differentiability implies continuity**

i.e, A function which is differentiable is necessarily continuous. Let $f(z)$ be differentiable at $z = z_0$ so that $f'(z_0)$ exists and we will show that it is continuous also

Now $\qquad f(z_0 + h) - f(z_0) = \frac{f(z_0 + h) - f(z_0)}{h}.h \quad$ where $h \neq 0$.

$\therefore \qquad \lim_{h \to 0}[f(z_0 + h) - f(z_0)] = \lim_{h \to 0}\frac{f(z_0 + h) - f(z_0)}{h}. \lim_{h \to 0} h = F'(z).0 = 0$

$\therefore \qquad \lim_{h \to 0} f(z_0 + h) = f(z_0)$

Above relation shows that $f(z)$ is continuous.

Case II. A function which is continuous is not necessarily differentiable.

We shall prove the above result by taking a suitable example.

For example (1) Consider the function $f(z) = \bar{z} = x - iy$ at $z = 0$

Now $\qquad | f(z) - f(0) | = | \bar{z} - 0 | = | z | < \varepsilon$ when $| z - 0 | < \varepsilon$

Above relation shows that the given function is continuous at $z = 0$

Again $\dfrac{f(0 + h) - f(0)}{h} = \dfrac{f(h) - f(0)}{h}$ where $h = \Delta z = p + iq$ say

$$= \frac{\bar{h} - 0}{h} = \frac{p - iq}{p + iq}$$

$$\lim_{h \to 0} \frac{f(0 + h) - f(0)}{h} = \lim_{h \to 0} \frac{p - iq}{p + iq}$$

Now choosing different paths for Δz to be along real and imaginary axes.

i.e., $h = p$ along real axis and $h \to 0$ when $p \to 0$

or $h = iq$ along imaginary axis and $h \to 0$ when $q \to 0$

$\therefore \lim_{h \to 0} \dfrac{f(0 + h) - f(0)}{h} = -1$ when $p \to 0$ when $q \to 0$

Since the limit is not unique therefore we conclude that the given function is not differentiable even though it is continuous.

Exactly as above we can shows that $f(z) = \bar{z}$ is continuous for every z but is not differentiable for any z.

 Solved Examples

EXAMPLE 1. *Prove that the function $f(z) = |z|^2$ is continuous every where but no where differentiable except at the origin.*

SOLUTION. We have $f(z) = |z|^2 = x^2 + y^2$

The continuity of the above function is evident because of the continuity of $x^2 + y^2$. Let us consider its differentiability

$$f'(z_0) = \lim_{\Delta z \to 0} \frac{f(z_0 + \Delta z) - f(z_0)}{\Delta z}$$

$$= \lim_{\Delta z \to 0} \frac{|z_0 + \Delta z|^2 - |z_0|^2}{\Delta z} = \lim_{\Delta z \to 0} \left[\frac{z_0 \overline{\Delta z} + \bar{z}_0 \Delta z + \Delta z \overline{\Delta z}}{\Delta z} \right]$$

$$= \lim_{\Delta z \to 0} \frac{(z_0 + \Delta z)(\bar{z}_0 + \overline{\Delta z}) - z_0 \bar{z}_0}{\Delta z}$$

$$= \lim_{\Delta z \to 0} \bar{z}_0 + \overline{\Delta z} + z_0 \frac{\overline{\Delta z}}{\Delta z}$$

$$= \lim_{\Delta z \to 0} \overline{z_0} + z_0 \frac{\overline{\Delta z}}{\Delta z} \quad (\because \overline{\Delta z} \to 0 \text{ as } \Delta z \to 0)$$

Now at $z_0 = 0$ the above limit is clearly zero so that $f'(0) = 0$

Let us now choose that $z_0 \neq 0$ and let

$$\Delta z = r(\cos \theta + i \sin \theta) = re^{i\theta}$$

$\therefore \qquad \overline{\Delta z} = r(\cos \theta - i \sin \theta) = re^{-i\theta}$

$\therefore \qquad \dfrac{\overline{\Delta z}}{\Delta z} = e^{-2i\theta} = \cos 2\theta - i \sin 2\theta$

Clearly, it does not tend to a unique limit as this limit depends on θ, i.e., amp Δz. Therefore the given function is not differentiable at any other non-zero value of z.

EXAMPLE 2. *If $f(z) = \dfrac{x^3 y(y - ix)}{x^6 + y^2}, z \neq 0, f(0) = 0$*

Prove that $\dfrac{f(z) - f(0)}{z} \to 0$ **as** $z \to 0$ **along any**

radius vector but not as $z \to 0$ **in any manner.**

SOLUTION. We have $y - ix = -i^2 y - ix = -i(x + iy) = -iz$

$\therefore \qquad f(z) = -\dfrac{x^3 yiz}{x^6 + y^2}, f(0) = 0$ given.

$$\lim_{z \to 0} \dfrac{f(z) - f(0)}{z} = \lim_{x \to 0} -\dfrac{x^3 yi}{x^6 + y^2}$$

Now if $z \to 0$ along any radius vector say $y = mx$ then

$$\lim_{z \to 0} \dfrac{f(z) - f(0)}{z} = \lim_{x \to 0} -\dfrac{x^4 mi}{x^6 + m^2 x^2}$$

$$= \lim_{x \to 0} \dfrac{-imx^2}{x^4 + m^2} = 0$$

Now let us suppose that $z = 0$ along the curve $y = x^3$ then

$$\lim_{z \to 0} \dfrac{f(z) - f(0)}{z} = \lim_{x \to 0} -\dfrac{x^6 i}{x^6 + x^6} = -\dfrac{i}{2}.$$

11.4 BRANCH LINE AND BRANCH POINT

A point is called a branch point of a function f(z) if some of the branches interchanges as the independent variable z describes a closed path about it.

Let $w = f(z) = z^{1/2}$ be a multivalued complex function. Since $z = re^{i\theta}$, therefore

$f(z) = r^{1/2} e^{i\theta/2}$. Now make a complete circuit in anticlockwise direction around the origin O starting from any point $P(z_1)$ whose amplitude is θ_1 say.

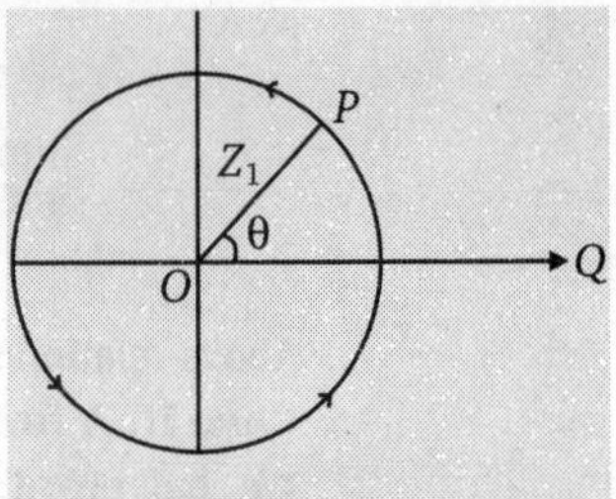

$\therefore \qquad\qquad w = r_1^{1/2} e^{i\frac{1}{2}\theta_1}$ at P.

After a complete circuit we again arrive at P so that $\theta = \theta_1 + 2\pi$

or $\qquad\qquad w = r_1^{1/2} e^{i(\theta_1 + 2\pi)/2} = r_1^{1/2} e^{i\left(\frac{1}{2}\theta_1 + \pi\right)} = r_1^{1/2} e^{-i\frac{1}{2}\theta_1}.$

Above shows that we have not obtained the same value of w with which we started.

Again after a second round when we arrive at P again then $\theta = \theta_1 + 4\pi$

or $\qquad\qquad w = r_1^{1/2} e^{i(\theta_1 + 4\pi)/2} = r_1^{1/2} e^{i\left(\frac{1}{2}\theta_1 + 2\pi\right)} = r_1^{1/2} e^{i\frac{1}{2}\theta_1}$

Above shows that we have not obtained the same value of w with which we started.

Hence we are on one branch of the function $w = z^{1/2}$ when $0 \le \theta \le 2\pi$ and on the second branch when $2\pi \le \theta \le 4\pi$.

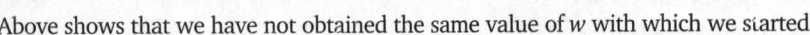

Fig. 8

Evidently each branch of the function is single valued. For keeping the function single valued an artificial barrier OQ where Q is supposed at infinity is made. The artificial barrier is called branch line or branch cut or cross cut and the point O is called the branch point. We may however have any other line from O as an artificial barrier in place of OQ.

In a similar manner we can show that $f(z) = \log z$ has a branch point at $z = 0$, $z = re^{i\theta}$ $\therefore$ $w = f(z) = \log z = \log r + i\theta$. $\omega = \log r_1 + i\theta_1$ at P or $\omega = \log r_1 + i(\theta_1 + 2\pi)$ after one complete circuit $= \log r_1 + i(\theta_1 + 4\pi)$ after two complete circuits.

Above shows that after each complete circuit we are at a different branch of the function and as such the point $z = 0$ is a branch point.

 Exercise-11.2

1. Define the following
   (i) Point set   (ii) Closure, interior and exterior
   (iii) Limit of a complex function
   (iv) Continuity at a point of a complex function
   (v) Uniform continuity

2. Show that $f(z) = \bar{z}$ is continuous for every z.

3. If $\lim\limits_{z + a} f(z) = l \ne \infty$ and $\lim\limits_{z \to a} g(z) = \infty$. Then show that
   $$\lim_{z \to a}[f(z) + g(z)] = \infty.$$

4. If $\lim\limits_{z \to a} f(z) = l \ne 0$ and $\lim\limits_{z \to a} g(z) = \infty$ the show that
   $$\lim_{z \to a}[f(z).g(z)] = \infty$$

11.5 DE-MOIVRE'S THEOREM

We have already studied that if z_1 and z_2 are any two complex numbers, then the modulus of the product of z_1 and z_2 is the product of their moduli and the argument of the product of z_1 and z_2 is the sum of their arguments. In case $z_1 = z_2 = z$ (say) then $|z_1| = |z_2| = 1$, it follows that the modulus of z^2 is unity and argument of z^2 is twice that of the argument of z. we can express this fact as

$$(\cos \theta + i\sin \theta)^2 = \cos 2\theta + i\sin 2\theta.$$

✎ **REMARK**
- The above equality was first of all observed by Abraham De Moivre (1667-1734) — a French mathematician and he proved the following theorem, which goes by his name.

THEOREM 1. *If n is any integer, positive or negative. Then*

$$(\cos \theta + i \sin \theta)^n = \cos n\theta + i \sin n\theta$$

and if n is a fraction, positive or negative, then cos nθ + i sin nθ is one of the values of (cos θ + i sin θ)ⁿ.

or if n is any rational number, cos nθ + i sin nθ is a value of (cos θ + i sin θ)ⁿ.

PROOF. **Case I.** If n be a positive integer. Then

$$(\cos \alpha + i \sin \alpha)(\cos \beta + i \sin \beta) = \cos \alpha \cos \beta - \sin \alpha \sin \beta + i(\sin \alpha \cos \beta + \cos \alpha \sin \beta) = \cos(\alpha + \beta) + i \sin(\alpha + \beta)$$

Similarly, multiplying by $(\cos \gamma + i \sin \gamma)$ on both sides, we get

$$(\cos \alpha + i \sin \alpha)(\cos \beta + i \sin \beta)(\cos \gamma + i \sin \gamma) = \cos(\alpha + \beta + \gamma) + i \sin(\alpha + \beta + \gamma)$$

Proceeding in this way and continuing upto n factors

$$(\cos \alpha + i \sin \alpha)(\cos \beta + i \sin \beta)(\cos \gamma + i \sin \gamma) \ldots \text{to } n \text{ factors}$$
$$= \cos(\alpha + \beta + \gamma + \ldots \text{to } n \text{ terms}) + i \sin(\alpha + \beta + \gamma + \ldots n \text{ terms})$$

Put $\alpha = \beta = \gamma \ldots = \theta$

Then $(\cos \theta + i \sin \theta)^n = \cos n\theta + i \sin n\theta$

Case II. If n is a negative integer.

Let n be equal to $-m$, where m is a +ve integer.

$$\therefore \quad (\cos \theta + i \sin \theta)^n = (\cos \theta + i \sin \theta)^{-m} = \frac{1}{(\cos \theta + i \sin \theta)^m} = \frac{1}{\cos m\theta + i \sin m\theta}$$

$$= \frac{\cos m\theta - i \sin m\theta}{(\cos m\theta + i \sin m\theta)(\cos m\theta - i \sin m\theta)} = \frac{\cos m\theta - i \sin m\theta}{\cos^2 m\theta - i^2 \sin^2 m\theta} = \cos m\theta - i \sin m\theta$$

$$= \cos(-m\theta) + i \sin(-m\theta) = \cos n\theta + i \sin n\theta$$

Case III. Suppose n is a fraction and equal to p/q where p is an integer (positive or negative) and q is a positive integer.

Now $\left(\cos \dfrac{\theta}{q} + i \sin \dfrac{\theta}{q}\right)^q = \cos q.\dfrac{\theta}{q} + i \sin q.\dfrac{\theta}{q} = (\cos \theta + i \sin \theta) \Rightarrow \cos \dfrac{\theta}{q} + i \sin \dfrac{\theta}{q}$ is one of the qth roots of $(\cos \theta + i \sin \theta)$

or $\cos \dfrac{\theta}{q} + i \sin \dfrac{\theta}{q}$ is one of the values of $(\cos \theta + i \sin \theta)^{1/q}$.

Now raising each of these quantities of pth power.

$\left(\cos \dfrac{\theta}{q} + i \sin \dfrac{\theta}{q}\right)^p$ is one of the values of $(\cos \theta + i \sin \theta)^{p/q}$.

or $\left(\cos \dfrac{p\theta}{q} + i \sin \dfrac{p\theta}{q}\right)$ is one of the values of $(\cos \theta + i \sin \theta)^{p/q}$.

or $(\cos n\theta + i \sin n\theta)$ is one of the values of $(\cos \theta + i \sin \theta)^n$ as $p/q = n$.

REMARKS

- $(\sin \theta + i \cos \theta)^n \neq \sin n\theta + i \cos n\theta$ and $(\cos \theta + i \sin \phi)^n \neq \cos n\theta + i \sin n\phi$
- If we write $-\theta$ for θ, we see at once that $\cos n\theta - i \sin n\theta$ is a values of $(\cos \theta - i \sin \theta)^n$ where n is any rational number.

 Solved Examples

EXAMPLE 1. *Prove that*

(i) $\left(\dfrac{1 + \cos \phi + i \sin \phi}{1 + \cos \phi - i \sin \phi}\right)^n = \cos n\phi + i \sin n\phi$

(ii) $\left(\dfrac{1 + \sin \theta + i \cos \theta}{1 + \sin \theta - i \cos \theta}\right)^n$
$= \cos\left(\dfrac{n\pi}{2} - n\pi\right) + i \sin\left(\dfrac{\pi n}{2} - n\theta\right)$ (SVTU–2006)

SOLUTION. (i) L.H.S. $= \left\{\dfrac{2\cos^2 \dfrac{\phi}{2} + 2i \sin \dfrac{\phi}{2} \cos \dfrac{\phi}{2}}{2\cos^2 \dfrac{\phi}{2} - 2i \sin \dfrac{\phi}{2} \cos \dfrac{\phi}{2}}\right\}^n$

$= \left\{\dfrac{\cos \dfrac{\phi}{2} + i \sin \dfrac{\phi}{2}}{\cos \dfrac{\phi}{2} - i \sin \dfrac{\phi}{2}}\right\}^n = \left\{\left(\cos \dfrac{\phi}{2} + i \sin \dfrac{\phi}{2}\right)^2\right\}^n$

$= \cos n\phi + i \sin n\phi$

(ii) Put $\dfrac{\pi}{2} - \theta$ for ϕ in (i).

EXAMPLE 2. *If* $2\cos \theta = x + \dfrac{1}{x}$, $2\cos \phi = y + \dfrac{1}{y}$ *prove that one of the values of*

(i) $x^m y^n \div \dfrac{1}{y^m y^n}$ *is* $2\cos(m\theta + n\phi)$ (SVTU-2007)

and (ii) $\dfrac{x^m}{y^n} + \dfrac{y^n}{x^m}$ *is* $2\cos(m\theta - n\phi)$ (NAGPUR–2009)

SOLUTION. Now $x^2 - 2x \cos \theta = -1$
$x^2 - 2x \cos \theta + \cos^2\theta = -1 + \cos^2\theta = -\sin^2\theta$
$(x^2 - \cos \theta)^2 = -\sin^2\theta$
$x - \cos \theta = i \sin \theta$
$x = \cos \theta + i \sin \theta,$
$\dfrac{1}{x} = \cos \theta - i \sin \theta$

Similarly, $y = \cos \phi + i \sin \phi$
$\dfrac{1}{y} = \cos \phi - i \sin \phi$

(i)
$$x^m y^n = (\cos\theta + i\sin\theta)^m (\cos\phi + i\sin\phi)^n$$
$$= (\cos m\theta + i\sin m\theta)(\cos n\phi + i\sin n\phi)$$
$$= \cos(m\theta + n\phi) + i\sin(m\theta + n\phi)$$
$$\frac{1}{x^m} \times \frac{1}{y^n} = x^{-m} \times y^{-n} = \cos(m\theta + n\phi) - i\sin(m\theta + n\phi)$$

$$\therefore \quad x^m y^n + \frac{1}{x^m \cdot y^n} = 2\cos(m\theta + n\phi)$$

and

(ii)
$$\frac{x^m}{y^n} = \frac{(\cos\theta + i\sin\theta)^n}{(\cos\phi + i\sin\phi)^n} = \frac{\cos m\theta + i\sin m\theta}{\cos n\phi + i\sin n\phi}$$
$$= \cos(m\theta - n\phi) + i\sin(m\theta - n\phi)$$

Similarly, $\dfrac{y^n}{x^m} = \cos(m\theta - n\phi) - i\sin(m\theta - n\phi)$

Adding, we get $\dfrac{x^m}{y^n} + \dfrac{y^n}{x^m} = 2\cos(m\theta - n\phi)$.

EXAMPLE 3. *If* $(a_1 + b_1 i)(a_2 + b_2 i) \dots (a_n + b_n i) = A + Bi$

Prove that

$$(a_1^2 + b_1^2)(a_2^2 + b_2^2)\dots(a_n^2 + b_n^2) = A^2 + B^2$$

and $\tan^{-1}\dfrac{b_1}{a_1} + \tan^{-1}\dfrac{b_2}{a_2} + \dots + \tan^{-1}\dfrac{b_n}{a_n} = \tan^{-1}\dfrac{B}{A}$

SOLUTION. Let $\left.\begin{array}{l} a_k = r_k\cos\theta_k \\ b_k = r_k\sin\theta_k \end{array}\right\}$, $k = 1, 2, 3, \dots n$

Then from $(a_1 + b_1 i)(a_2 + b_2 i)\dots(a_n + b_n i) = A + iB$
We have $r_1 r_2 \dots r_n (\cos\theta_1 + i\sin\theta_1)$
$(\cos\theta_2 + i\sin\theta_2)\dots(\cos\theta_n + i\sin\theta_n)$
$$= A + iB$$
or $r_1 r_2 \dots r_n [\cos(\theta_1 + \theta_2 + \dots + \theta_n)$
$+ i\sin(\theta_1 + \theta_2 + \dots + \theta_n)] = A + iB$
Equating real and imaginary parts, we get
$$A = r_1 r_2 \dots r_n \cos(\theta_1 + \theta_2 + \dots + \theta_n)$$
$$B = r_1 r_2 \dots r_n \sin(\theta_1 + \theta_2 + \dots + \theta_n)$$
$$\therefore \quad \frac{B}{A} = \tan(\theta_1 + \theta_2 + \dots + \theta_n)$$

or $\tan^{-1}\dfrac{B}{A} = \theta_1 + \theta_2 + \dots + \theta_n$

$$= \tan^{-1}\frac{b_1}{a_1} + \tan^{-1}\frac{b_2}{a_2} + \dots + \tan^{-1}\frac{b_n}{a_n}$$

and $A^2 + B^2 = r_1^2 . r_2^2 \dots r_n^2$

$$= (a_1^2 + b_1^2)(a_2^2 + b_2^2)\dots(a_n^2 + b_n^2)$$

EXAMPLE 4. *If* $\cos\alpha + \cos\beta + \cos\gamma = \sin\alpha + \sin\beta + \sin\gamma = 0$
Prove that $\cos 3\alpha + \cos 3\beta + \cos 3\gamma$
$$= 3\cos(\alpha + \beta + \gamma)$$
and $\sin 3\alpha + \sin 3\beta + \sin 3\gamma = 3\sin(\alpha + \beta + \gamma)$

SOLUTION. We know that if $x + y + z = 0$
$$x^3 + y^3 + z^3 = 3xyz$$
Let $x = \operatorname{cis}\alpha$, $y = \operatorname{cis}\beta$, $z = \operatorname{cis}\gamma$
$\qquad\qquad$ (cis α denotes $\cos\alpha + i\sin\alpha$)
hence $x + y + z$
$$= (\cos\alpha + \cos\beta + \cos\gamma) + i(\sin\alpha + \sin\beta + \sin\gamma)$$
$$= 0, \text{ by given conditions.}$$
$x^3 + y^3 + z^3 = 3xyz$ gives
$$(\operatorname{cis}\alpha)^3 + (\operatorname{cis}\beta)^3 + (\operatorname{cis}\gamma)^3 = 3(\operatorname{cis}\alpha)(\operatorname{cis}\beta)(\operatorname{cis}\gamma)$$
or $\operatorname{cis}(3\alpha) + \operatorname{cis}(3\beta) + \operatorname{cis}(3\gamma) = 3\operatorname{cis}(\alpha + \beta + \gamma)$

$$(\cos 3\alpha + \cos 3\beta + \cos 3\gamma)$$
$$+ i(\sin 3\alpha + \sin 3\beta + \sin 3\gamma)]$$
$$= 3[\cos(\alpha + \beta + \gamma) + i\sin(\alpha + \beta + \gamma)]$$
Equating real and imaginary parts, we get the result.

EXAMPLE 5. *If* $\cos\alpha + \cos\beta + \cos\gamma = \sin\alpha + \sin\beta + \sin\gamma = 0$
Prove that $\cos^2\alpha + \cos^2\beta + \cos^2\gamma$
$$= \sin^2\alpha + \sin^2\beta + \sin^2\gamma$$
and $\cos 2\alpha + \cos 2\beta + \cos 2\gamma$
$$= \sin 2\alpha + \sin 2\beta + \sin 2\gamma = 0$$

(MUMBAI–2009)

SOLUTION. We know that if $x + y + z = 0$ ...(1)

and $\qquad \dfrac{1}{x} + \dfrac{1}{y} + \dfrac{1}{z} = 0$ ...(2)

then $\qquad x^2 + y^2 + z^2 = 0$ ...(3)
Let $\quad x = \operatorname{cis}\alpha, y = \operatorname{cis}\beta, z = \operatorname{cis}\gamma$
then we know that (1) and (2) are satisfied due to given conditions. Hence, the relation (3) is also satisfied. Now we have
$$x^2 + y^2 + z^2 = 0$$
or $(\operatorname{cis}\alpha)^2 + (\operatorname{cis}\beta)^2 + (\operatorname{cis}\gamma)^2 = 0$
or $\operatorname{cis}(2\alpha) + \operatorname{cis}(2\beta) + \operatorname{cis}(2\gamma) = 0$
or $(\cos 2\alpha + \cos 2\beta + \cos 2\gamma)$
$$+ i(\sin 2\alpha + \sin 2\beta + \sin 2\gamma) = 0$$
$$\therefore \quad \cos 2\alpha + \cos 2\beta + \cos 2\gamma = 0,$$
$$\sin 2\alpha + \sin 2\beta + \sin 2\gamma = 0$$
Hence, we get
$$\cos^2\alpha - \sin^2\alpha + \cos^2\beta - \sin^2\beta\cos^2\gamma - \sin^2\gamma = 0$$
$$\cos^2\alpha + \cos^2\beta + \cos^2\gamma = \sin^2\alpha + \sin^2\beta + \sin^2\gamma.$$

EXAMPLE 6. *If* $x_r = \operatorname{cis}\dfrac{\pi}{2^r}$ *prove that*
$$x_1 x_2 x_3 \dots \text{ ad inf.} = \cos\pi = -1$$

(SVTU–2009, MUMBAI–2007)

SOLUTION. We have $x_1 x_2 x_3 \dots$ ad inf.

$$= \left[\operatorname{cis}\frac{\pi}{2}\right]\left[\operatorname{cis}\frac{\pi}{2^2}\right]\left[\operatorname{cis}\frac{\pi}{2^3}\right]\dots \text{ upto infinite terms}$$

$$= \operatorname{cis}\left[\frac{\pi}{2} + \frac{\pi}{2^2} + \frac{\pi}{2^3} + \dots\infty\right] = \operatorname{cis}\left[\frac{\pi/2}{1 - \frac{1}{2}}\right] = \operatorname{cis}\pi$$

$$= \cos\pi = -1. \qquad\qquad [\text{Since } \sin\pi = 0]$$

EXAMPLE 7. *If* α, β *are roots of the equation* $x^2 - 2x + 4 = 0$

prove that $\alpha^n + \beta^n = 2^{n+1}\cos\dfrac{n\pi}{3}$ (DELHI–2002)

SOLUTION. $x^2 - 2x + 4 = 0$ or $x = 1 \pm i\sqrt{3}$
Let $\alpha = 1 + i\sqrt{3}$ and $\beta = 1 - i\sqrt{3}$
then we have to simply to prove that
$$\alpha^n + \beta^n = (1 + i\sqrt{3})^n + (1 - i\sqrt{3})^n$$
$$= 2^{n+1}\cos\frac{n\pi}{3} \qquad\qquad ...(1)$$

let $1 = r\cos\theta, \sqrt{3} = r\sin\theta$

$$\therefore \quad r = 2, \theta = \frac{\pi}{3}$$

$\therefore$ L.H.S. of (1)
$$= r^n(\cos\theta + i\sin\theta)^n + r^n(\cos\theta - i\sin\theta)^n$$
$$= 2r^n\cos n\theta = 2^{n+1}\cos\frac{n\pi}{3}.$$

Exercise-11.3

1. Simplify

   (a) $\dfrac{(\cos 2\theta - i\sin 2\theta)^7.(\cos 3\theta + i\sin 3\theta)^{-5}}{(\cos 4\theta + i\sin 4\theta)^{12}.(\cos 5\theta + i\sin 5\theta)^6}$

   (b) $[(\cos\theta + \cos\phi) + i(\sin\theta + \sin\phi)]^n$
   $+ [(\cos\theta + \cos\phi) - i(\sin\theta + \sin\phi)]^n$

   (c) $[1 + \cos\theta + i\sin\theta]^n + [1 + \cos\theta - i\sin\theta]^n$

   (d) $[(\cos\theta - \cos\phi) + i(\sin\theta - \sin\phi)]^n$
   $+ [(\cos\theta - \cos\phi) - i(\sin\theta - \sin\phi)]^n$

   (e) $\dfrac{(\cos 3\theta + i\sin 3\theta)^5.(\cos\theta - i\sin\theta)^3}{(\cos 5\theta + i\sin 5\theta)^7.(\cos 2\theta - i\sin 2\theta)^5}$

   (f) $\dfrac{(\cos\theta + i\sin\theta)^8.(\cos 3\theta - i\sin 3\theta)^2}{(\cos 2\theta + i\sin 2\theta)^5.(\cos 4\theta - i\sin 4\theta)^7}$

2. If x, y, z, u stand for $(\cos\alpha + i\sin\alpha)$, $(\cos\beta + i\sin\beta)$, $(\cos\gamma + i\sin\gamma)$ and $(\cos\delta + i\sin\delta)$ respectively. Find the value of

   (a) $(x + y)(z + u)$ \qquad (b) $\dfrac{1}{(x - y)(z - u)}$

   (c) $xy + zu$

3. Prove that $(a + ib)^{m/n} + (a - ib)^{m/n}$
   $= 2(a^2 + b^2)^{m/2n}.\cos\left(\dfrac{m}{n}\tan^{-1}\dfrac{b}{a}\right)$

4. If $p = \cos\theta + i\sin\theta$ and $q = \cos\phi + i\sin\phi$, show that
   $$\dfrac{p - q}{p + q} = i\tan\left(\dfrac{\theta - \phi}{2}\right).$$ \hfill (KURUKSHETRA–2005)

5. If a denotes $\cos 2\alpha + i\sin 2\alpha$ with similar expression for b, c, d prove that

   (a) $\sqrt{(abcd)} + \dfrac{1}{\sqrt{(abcd)}} = 2\cos(\alpha + \beta + \gamma + \delta)$

   (b) $\sqrt{\left(\dfrac{ab}{cd}\right)} + \sqrt{\left(\dfrac{cd}{ab}\right)} = 2\cos(\alpha + \beta - \gamma - \delta)$

   (c) $\sqrt{(a^p b^q c^r d^s)} + \dfrac{1}{\sqrt{(a^p b^q c^r d^s)}} = 2\cos(p\alpha + q\beta + r\gamma + s\delta)$

   (d) $\dfrac{(a + b)(b + c)(c + a)}{abc}$ is real and equal to $8\cos(\alpha - \beta)\cos(\beta - \gamma)$
   $\cos(\gamma - \alpha)$.

6. If $2\cos\theta = x + \dfrac{1}{x}$, $2\cos\phi = y + \dfrac{1}{y}$ prove that

   (i) $2\cos(\theta + \phi + ...) = xyz... + \dfrac{1}{xyz}...$

   (ii) $2\cos(p\theta + q\phi + ...) = x^p y^q z^r... + \dfrac{1}{x^p y^q z^r}$

7. If $x + \dfrac{1}{x} = 2\cos\theta$, show that :

   (i) $x^7 + \dfrac{1}{x^7} = 2\cos 7\theta$ \qquad (ii) $x^n + \dfrac{1}{x^n} = 2\cos n\theta$

   (iii) $x^n - \dfrac{1}{x^n} = 2i\sin n\theta$

8. If α and β are the roots of the equation $x^2 - 2x\cos\theta + 1 = 0$ form an equation whose roots are α^n and β^n.

9. If $\cos\alpha + \cos\beta + \cos\gamma = \sin\alpha + \sin\beta + \sin\gamma = 0$, prove that $\cos^2\alpha + \cos^2\beta + \cos^2\gamma = \sin^2\alpha + \sin^2\beta + \sin^2\gamma = \dfrac{3}{2}$.

10. If $\cos\alpha + 2\cos\beta + 3\cos\gamma = 0 = \sin\alpha + 2\sin\beta + 3\sin\gamma$, prove that
    $\cos 3\alpha + 8\cos 3\beta + 27\cos 3\gamma = 18\cos(\alpha + \beta + \gamma)$
    and $\sin 3\alpha + 8\sin 3\beta + 27\sin 3\gamma = 18\sin(\alpha + \beta + \gamma)$

11. If $(1 + x)^n = p_0 + p_1 x + p_2 x_2 + ...$, show that
    $$p_0 - p_2 + p_4... = 2^{n/2}\cos\dfrac{n\pi}{4}$$
    and $$p_1 - p_3 + p_5... = 2^{n/2}.\sin\dfrac{n\pi}{4}$$

12. If $z_n = \cos\dfrac{\pi}{2^n} + i\sin\dfrac{\pi}{2^n}$, prove that
    $$\lim_{n\to\infty}(z_1, z_2, z_3, ..., z_n) = -1$$

13. If $2\cos\theta = x + \dfrac{1}{x}$, prove that

    (i) $2\cos r\theta = x^r + \dfrac{1}{x^r}$ \qquad (ii) $\dfrac{x^{2n} + 1}{x^{2n-1} + x} = \dfrac{\cos n\theta}{\cos(n-1)\theta}$
    \hfill (MADRAS–2000S)

14. If $\sin\alpha + \sin\beta + \sin\gamma = \cos\alpha + \cos\beta + \cos\gamma = 0$, prove that
    (i) $\sin 2\alpha + \sin 2\beta + \sin 2\gamma = 0$
    (ii) $\sin 3\alpha + \sin 3\beta + \sin 3\gamma = 3\sin(\alpha + \beta + \gamma)$
    (iii) $\sin 4\alpha + \sin 4\beta + \sin 4\gamma = 2\,\Sigma\sin 2(\alpha + \beta)$
    (iv) $\sin(\alpha + \beta) + \sin(\beta + \gamma) + \sin(\gamma + \alpha) = 0$
    \hfill (MUMBAI–2009)

15. If $\cos\alpha + \cos\beta + \cos\gamma = 0$ and $\sin\alpha + \sin\beta + \sin\gamma = 0$, prove that $\cos(\alpha + \beta) + \cos(\beta + \gamma) + \cos(\gamma + \alpha) = 0$
    \hfill (MUMBAI–2009, SVTU–2008)

Hints to Selected Problems

3. Let $a + ib = r(\cos\theta + i\sin\theta)$
   $\Rightarrow a - ib = r(\cos\theta - i\sin\theta)$
   $\Rightarrow r\cos\theta = a$, $r\sin\theta = b$. Also $r = (a^2 + b^2)^{1/2}$
   and $\theta = \tan^{-1}\dfrac{b}{a}$
   Therefore $(a + ib)^{m/n} + (a - ib)^{m/n}$
   $= r^{m/n}(\cos\theta + i\sin\theta)^{m/n} + r^{m/n}(\cos\theta - i\sin\theta)^{m/n}$.
   Now simplifying this equation.

4. Consider $\dfrac{p - q}{p + q} = \dfrac{(\cos\theta + i\sin\theta) - (\cos\phi + i\sin\phi)}{(\cos\theta + i\sin\theta) + (\cos\phi + i\sin\phi)}$

   $= \dfrac{[(\cos\theta - \cos\phi) + i(\sin\theta - \sin\phi)]}{[(\cos\theta + \cos\phi) + i(\sin\theta + \sin\phi)]}$

   $= \dfrac{2\sin\dfrac{(\theta+\phi)}{2}.\sin\dfrac{(\phi-\theta)}{2} + 2i\cos\dfrac{\theta+\phi}{2}\sin\dfrac{(\theta-\phi)}{2}}{2\cos\dfrac{(\theta+\phi)}{2}.\cos\dfrac{(\theta-\phi)}{2} + i\sin\dfrac{(\theta+\phi)}{2}.\cos\dfrac{(\theta-\phi)}{2}}$

Now simplifying this equation to get the required result.

7. (i) As per given

$$x + \frac{1}{x} = 2\cos\theta \Rightarrow x^2 - 2x\cos\theta + 1 = 0$$

$$\Rightarrow \quad x = \frac{2\cos\theta \pm \sqrt{4\cos^2\theta - 4}}{2}$$

$$= \cos\theta \pm \sqrt{(-\sin^2\theta)} = \cos\theta \pm i\sin\theta$$

Taking positive sign only, we get

$$x = \cos\theta + i\sin\theta \Rightarrow \quad \frac{1}{x} = \cos\theta - i\sin\theta$$

$$\Rightarrow \quad x^7 + \frac{1}{x^7} = (\cos\theta + i\sin\theta)^7 + (\cos\theta - i\sin\theta)^7$$

$$= \cos 7\theta + i\sin 7\theta + \cos 7\theta - i\sin 7\theta = 2\cos 7\theta.$$

8. Proceed same as (7)(i), we get

$$x = \cos\theta \pm i\sin\theta$$

Let $\alpha = \cos\theta + i\sin\theta$, $\beta = \cos\theta - i\sin\theta$

Then required equation is

$$x^2 - (\text{sum of the roots})\, x + \text{product of the roots} = 0$$

i.e. $\quad x^2 - (\alpha^n + \beta^n)x + \alpha^n\beta^n = 0$

Answers

1. (a) $\cos 107\,\theta - i\sin 107\,\theta$ 　　(b) $2^{n+1}\cos^n\dfrac{1}{2}(\theta - \phi)\cos\dfrac{n}{2}(\theta + \phi)$ 　　(c) $2^{n+1}\cos^n\left(\dfrac{\theta}{2}\right)\cos\left(\dfrac{n\theta}{2}\right)$

(d) $2^{n+1}\sin\left(\dfrac{\theta - \phi}{2}\right)\left[\cos n\left(\dfrac{\pi + \theta + \phi}{2}\right)\right]$ 　　(e) $\cos 13\theta - i\sin 13\theta$ 　　(f) $\cos 20\,\theta + i\sin 20\,\theta$

2. (a) $4\cos\left(\dfrac{\alpha - \beta}{2}\right)\cos\left(\dfrac{\gamma - \delta}{2}\right)\left[\cos\left(\dfrac{\alpha + \beta + \gamma + \delta}{2}\right) + i\sin\left(\dfrac{\alpha + \beta + \gamma + \delta}{2}\right)\right]$

(b) $\dfrac{1}{4}\operatorname{cosec}\left(\dfrac{\alpha - \beta}{2}\right)\operatorname{cosec}\left(\dfrac{\gamma - \delta}{2}\right)\left[-\cos\left(\dfrac{\alpha + \beta + \gamma + \delta}{2}\right) + i\sin\left(\dfrac{\alpha + \beta + \gamma + \delta}{2}\right)\right]$

(c) $2\cos\left(\dfrac{\alpha + \beta - \gamma - \delta}{2}\right)\left[\cos\left(\dfrac{\alpha + \beta + \gamma + \delta}{2}\right)\quad \sin\left(\dfrac{\alpha + \beta + \gamma + \delta}{2}\right)\right]$

11.6 EXPANSIONS OF COS nθ AND SIN nθ IN POWERS OF COS θ AND SIN θ (n BEING POSITIVE INTEGER)

From De Moivre's theorem, we have

$$(\cos\theta + i\sin\theta)^n = \cos n\theta + i\sin n\theta \qquad \qquad \dots(1)$$

Also by binomial theorem, we have

$$(\cos\theta + i\sin\theta)^n = \cos^n\theta + {}^nC_1\cos^{n-1}\theta(i\sin\theta) + {}^nC_2\cos^{n-2}\theta(i\sin\theta)^2 + {}^nC_3\cos^{n-3}\theta(i\sin\theta)^3$$

$$+ \dots + {}^nC_{n-1}\cos\theta(i\sin\theta)^{n-1} + (i\sin\theta)^n \qquad \qquad \dots(2a)$$

$$= [\cos^n\theta - {}^nC_2\cos^{n-2}\theta\sin^2\theta + \dots] + i[{}^nC_1\cos^{n-1}\theta\sin\theta - {}^nC_3\cos^{n-3}\theta\sin^3\theta + \dots] \qquad \qquad \dots(2b)$$

From (1) and (2b), we have

$$\cos n\theta + i\sin n\theta = [\cos^n\theta - {}^nC_2\cos^{n-2}\theta\sin^2\theta + \dots] + i\,[{}^nC_1\cos^{n-1}\theta - {}^nC_3\cos^{n-3}\theta\sin^3\theta + \dots]$$

Equating real and imaginary parts

$$\cos n\theta = \cos^n\theta - {}^nC_2\cos^{n-2}\theta\sin^2\theta + \dots \qquad \qquad \dots(A)$$

$$\sin n\theta = {}^nC_1\cos^{n-1}\theta\sin\theta - {}^nC_3\cos^{n-3}\theta\sin^3\theta + \dots \qquad \qquad \dots(B)$$

The terms in either series are alternately positive and negative and the expansions are valid whether n is even or odd, but the last terms are different in the two cases. The last two terms in (2a) are ${}^nC_{n-1}\cos\theta(i\sin\theta)^{n-1}$ and $(i\sin\theta)^n$, which are respectively real and imaginary when n is odd and imaginary and real when n is even.

Case I. n is odd

then $\qquad {}^nC_{n-1}\cos\theta(i\sin\theta)^{n-1} = i^{n-1}.n\cos\theta\sin^{n-1}\theta = (-1)^{(n-1)/2}.n\cos\theta\sin^{n-1}\theta$

and $\qquad (i\sin\theta)^n = i.i^{n-1}\sin^n\theta = (-1)^{(n-1)/2}.i\sin^n\theta$

Case II. n is even

then $\qquad {}^nC_{n-1}\cos\theta(i\sin\theta)^{n-1} = ni(-1)^{(n-2)/2}\cos\theta\sin^{n-1}\theta$,

and $\qquad (i\sin\theta)^n = (-1)^{n/2}\sin^n\theta$

Thus last term in the series for $\cos n\theta$ is

$$(-1)^{(n-1)/2}n\cos\theta\sin^{n-1}\theta \qquad\qquad \text{or} \qquad\qquad (-1)^{n/2}\sin^n\theta$$

according as n is odd or even, and last term in the series for $\sin n\theta$ is

$$(-1)^{(n-1)/2}\sin^n\theta \qquad\qquad \text{or} \qquad\qquad (-1)^{(n-2)/2}n\cos\theta\sin^{n-1}\theta$$

11.6.1 Expansision of tan $n\theta$

$$\tan n\theta = \frac{\sin n\theta}{\cos n\theta} = \frac{{}^nC_1 \cos^{n-1}\theta \sin\theta - {}^nC_3 \cos^{n-3}\theta \sin^3\theta + \dots}{\cos^n\theta - {}^nC_2 \cos^{n-2}\theta \sin^2\theta + \dots}$$

Now dividing the numerator and denominator of the right hand side by $\cos^n\theta$. We have

$$\tan n\theta = \frac{{}^nC_1 \tan\theta - {}^nC_3 \tan^3\theta + \dots}{1 - {}^nC_2 \tan^2\theta + \dots}$$

☞ Remark

* It is easy to note that if n is odd, the last term of the numerator is $(-1)^{(n-1)/2}\tan^n\theta$ and the last term of the denominator is $n(-1)^{(n-1)/2}\tan^{n-1}\theta$. If n is even, the last term of the numerator is $n(-1)^{(n-2)/2}\tan^{n-1}\theta$ and that of denominator $(-1)^{n/2}\tan^n\theta$.

11.6.2 Expansion of tan $(\theta_1 + \theta_2 + \dots \theta_n)$

Expressions for sine, cosine and tangent of the sum of any number of unequal angles in terms of these angles can be obtained from De Moivre's theorem.

We have seen that

$$(\cos\theta_1 + i\sin\theta_1)(\cos\theta_2 + i\sin\theta_2)\dots(\cos\theta_n + i\sin\theta_n) = \cos(\theta_1 + \theta_2 + \dots + \theta_n) + i\sin(\theta_1 + \theta_2 + \dots + \theta_n)$$

$$\cos(\theta_1 + \theta_2 + \dots + \theta_n) + i\sin(\theta_1 + \theta_2 + \dots + \theta_n)$$

$$= \cos\theta_1(1 + i\tan\theta_1)\cos\theta_2(1 + i\tan\theta_2)\dots\cos\theta_n(1 + i\tan\theta_n)$$

$$= \cos\theta_1\cos\theta_2\dots\cos\theta_n\{(1 + i\tan\theta_1)(1 + i\tan\theta_2)\dots(1 + i\tan\theta_n)\}$$

$$= \cos\theta_1\cos\theta_2\dots\cos\theta_n\{1 + i(\tan\theta_1 + \dots + \tan\theta_n) + i^2(\tan\theta_1\tan\theta_2 + \tan\theta_2\tan\theta_3 + \dots)$$

$$+ i^3(\tan\theta_1\tan\theta_2\tan\theta_3\dots)\}$$

$$= \cos\theta_1\cos\theta_2\cos\theta_3\dots\cos\theta_n[1 + is_1 - s_2 - is_3 + s_4\dots]$$

where $s_1 = \Sigma\tan\theta_1, s_2 = \Sigma\tan\theta_1\tan\theta_2,\dots, s_n = \Sigma\tan\theta_1\tan\theta_2\dots\tan\theta_n$

Equating real and imaginary parts on both sides,

$$\cos(\theta_1 + \theta_2 + \dots + \theta_n) = \cos\theta_1\cos\theta_2\dots\cos\theta_n[1 - s_2 + s_4\dots]$$

$$\sin(\theta_1 + \theta_2 + \dots + \theta_n) = \cos\theta_1\cos\theta_2\dots\cos\theta_n[1 - s_3 + s_5\dots]$$

Hence by division, $\tan(\theta_1 + \theta_2 + \dots + \theta_n) = \dfrac{s_1 - s_3 + s_5 - s_7 + \dots}{1 - s_2 + s_4 - s_6 + \dots}$

☞ Remark

* It is easy to note that if n is even, the last terms of numerator and denominator are respectively $(-1)^{(n-2)/2}S_{n-1}$ and $(-1)^{n/2}S_n$ in case n is odd, they will be $(-1)^{(n-1)/2}S_n$ and $(-1)^{(n-1)/2}S_{n-1}$ respectively.

Solved Examples

EXAMPLE 1. *Express cos 3θ and sin 3θ in terms of sin θ and cos θ.*
(MADRAS–2002)

SOLUTION. By De-Moivre's theorem, we have
$$(\cos\theta + i\sin\theta)^3 = \cos 3\theta + i\sin 3\theta \quad \dots(1)$$
Since n is a positive integer, on expanding $(\cos\theta + i\sin\theta)^3$ by binomial theorem, we have
$$(\cos\theta + i\sin\theta)^3$$
$$= \cos^3\theta + {}^3C_1\cos^2\theta(i\sin\theta) + {}^3C_2\cos\theta(i\sin\theta)^2$$
$$+ {}^3C_3(i\sin\theta)^3$$
$$= \cos^3\theta + 3i\cos^2\theta\sin\theta - 3\cos\theta\sin^2\theta - i\sin^3\theta$$
$$= (\cos^3\theta - 3\cos\theta\sin^2\theta) + i(3\cos^2\theta\sin\theta - \sin^3\theta) \quad \dots(2)$$
From (1) and (2) by equating real and imaginary parts, we have
$$\left.\begin{array}{l} \text{and } \cos 3\theta = \cos^3\theta - 3\cos\theta\sin^2\theta \\ \sin 3\theta = 3\cos^2\theta\sin\theta - \sin^3\theta \end{array}\right\}$$

EXAMPLE 2. *If $\alpha, \beta, \gamma, \delta$ are the roots of the equation $x^4 - x^3\sin 2\theta + x^2\cos 2\theta - x\cos\theta - \sin\theta = 0$, prove that*
$$\tan^{-1}\alpha + \tan^{-1}\beta + \tan^{-1}\gamma + \tan^{-1}\delta = n\pi + \frac{\pi}{2} - \theta$$

SOLUTION. We have $x^4 - x^3\sin 2\theta + x^2\cos 2\theta - x\cos\theta - \sin\theta = 0$, $\alpha, \beta, \gamma, \delta$ are the roots.
$$s_1 = \Sigma\alpha = \sin 2\theta \qquad s_3 = \Sigma\alpha\beta\gamma = \cos\theta,$$
$$s_2 = \Sigma\alpha\beta = \cos 2\theta, \quad s_4 = \alpha\beta\gamma\delta = -\sin\theta$$
Now $\tan(\tan^{-1}\alpha + \tan^{-1}\beta + \tan^{-1}\gamma + \tan^{-1}\delta)$
$$= \frac{s_1 - s_3}{1 - s_2 + s_4} = \frac{\sin 2\theta - \cos\theta}{1 - \cos 2\theta - \sin\theta}$$
$$= \frac{\cos\theta(2\sin\theta - 1)}{\sin\theta(2\sin\theta - 1)} = \cot\theta = \tan\left(\frac{\pi}{2} - \theta\right).$$
$$\therefore \quad \tan^{-1}a + \tan^{-1}\beta + \tan^{-1}\gamma + \tan^{-1}\delta = n\pi + \frac{\pi}{2} - \theta.$$

Exercise-11.4

Prove that : (Qus 1 – 4)

1. (a) $\sin 4\theta = 4(\cos^3\theta \sin\theta - \cos\theta \sin^3\theta)$
    (b) $\cos 4\theta = \cos^4\theta - 6\cos^2\theta \sin^2\theta + \sin^4\theta$

2. (a) $\sin 5\theta = 5\cos^4\theta \sin\theta - 10\cos^2\theta \sin^3\theta + \sin^5\theta$
    (b) $\cos 5\theta = \cos^5\theta - 10\cos^3\theta \sin^2\theta + 5\cos\theta \sin^4\theta$
    $$= 16\cos^5\theta - 20\cos^3\theta + 5\cos\theta$$

3. (a) $\cos 8\theta = \cos^8\theta - 28\cos^6\theta \sin^2\theta + 70\cos^4\theta \sin^4\theta$
    $$- 28\cos^2\theta \sin^6\theta + \sin^8\theta$$
    (b) $\sin 6\theta = 6\cos^5\theta \sin\theta - 20\cos^3\theta \sin^3\theta + 6\cos\theta \sin^5\theta$

4. (a) $\tan 5\theta = \dfrac{5\tan\theta - 10\tan^3\theta + \tan^5\theta}{1 - 10\tan^2\theta + 5\tan^4\theta}$

    (b) $\tan 4\theta = \dfrac{4\tan\theta - 4\tan^3\theta}{1 - \tan^2\theta + \tan^4\theta}$

5. Prove that the equation $ah \sec\theta - bk \operatorname{cosec}\theta = a^2 - b^2$ has four roots, and that the sum of the four values of θ, which satisfy it, is equal to an odd multiple of π radians.

6. (a) Show that $\cos\dfrac{r\pi}{3}$, $r = 1, 3, 5, 7, 9, 11$ are the roots of the equation
    $$64 x^6 - 32 x^5 - 80 x^4 + 32 x^3 + 24 x^2 - 6x - 1 = 0$$
    (b) Find the equation whose roots are $s_1^2, s_3^2, s_5^2, s_7^2, s_9^2, s_{11}^2$ where $s_r = \sin\dfrac{r\pi}{26}$.

7. Show that $\cos\dfrac{2\pi}{9}, \cos\dfrac{4\pi}{9}, \cos\dfrac{6\pi}{9}, \cos\dfrac{8\pi}{9}$ are the roots of
    $$16x^4 + 8x^3 - 12x^2 - 4x + 1 = 0.$$

8. Prove that the equation $a^2 \cos^2\theta + b^2 \sin^2\theta + 2ga \cos\theta + 2fb \sin\theta + c = 0$ has four roots and that the sum of the values of θ which satisfy it is an even multiple of π radians.

Or

If the circle $x^2 + y^2 + 2gx + 2fy + c = 0$ meets the ellipse $x = a\cos\theta$, $y = b\sin\theta$ in points whose eccentric angels are $\theta_1, \theta_2, \theta_3$ and θ_4. Show that $\theta_1 + \theta_2 + \theta_3 + \theta_4 = 2n\pi$ where n is a positive integer.

9. Form an equation whose roots are
    $$\tan^2\frac{\pi}{9}, \tan^2\frac{2\pi}{9}, \tan^2\frac{3\pi}{9}, \tan^2\frac{4\pi}{9}$$
    and deduce the value of $\cot^2\dfrac{\pi}{9} + \cot^2\dfrac{2\pi}{9} + \cot^2\dfrac{3\pi}{9} + \cot^2\dfrac{4\pi}{9}$.

10. If $\alpha, \beta, \gamma, \delta$ are the values of x which satisfies $\cos 2x + a\cos x + b\sin x + c = 0$, no two of which differ by a multiple of π, prove that $\alpha + \beta + \gamma + \delta$ is a multiple of 2π.

11. Prove that $\pm\tan\dfrac{\pi}{5}, \pm\tan\dfrac{2\pi}{5}$ are the roots of the equations $x^4 - 10x^2 + 5 = 0$

12. Prove that $\cos\dfrac{\pi}{15} + \cos\dfrac{7\pi}{15} + \cos\dfrac{11\pi}{15} + \cos\dfrac{13\pi}{15} = -\dfrac{1}{2}$.

13. Prove that the roots of the equations $16x^4 + 8x^3 - 12x^2 - 4x + 1 = 0$ are $\cos\dfrac{2\pi}{9}, \cos\dfrac{4\pi}{9}, \cos\dfrac{6\pi}{9}, \cos\dfrac{8\pi}{9}$.

14. Prove that the roots of the equation $x^3 - 3\sqrt{3}x^2 - 3x + \sqrt{3} = 0$ are $\tan\left(\dfrac{\pi}{9}\right), \tan\left(\dfrac{4\pi}{9}\right), \tan\left(\dfrac{7\pi}{9}\right)$.

15. Use De Moivre's theorem to solve the equation
    $$x^4 - x^3 + x^2 - x + 1 = 0.$$ (PTU–2005)

16. Find the 7th roots of unity and prove that the sum of their nth powers always vanishes unless n be a multiple number of 7, n being an integer, and then the sum is 7.
                              (MUMBAI–2008, KURUKSHETRA–2005)

 Hints to Selected Problems

1. We have $(\cos\theta + i\sin\theta)^4 = \cos 4\theta + i\sin 4\theta$
    Now by Binomial theorem, we have
    $$(\cos\theta + i\sin\theta)^4 = \cos^4\theta + {}^4C_1\cos^3\theta(i\sin\theta) + {}^4C_2\cos^2\theta$$
    $$(i\sin\theta)^2 + {}^4C_3\cos\theta(i\sin\theta)^2 + {}^4C_4(i\sin\theta)^4$$
    $$= \cos^4\theta + 4i\cos^3\theta\sin\theta - 6\cos^2\theta\sin^2\theta$$
    $$- 4i\cos\theta\sin^3\theta + \sin^4\theta$$
    $\Rightarrow \cos 4\theta + i\sin 4\theta = \cos^4\theta - 6\cos^2\theta\sin^2\theta$
    $$\sin^4\theta + i(4\cos^3\theta\sin\theta - 4\cos\theta\sin^3\theta)$$

Now equating real and imaginary parts of both sides, we get the required result.

6. (b) If $13\theta = (2n + 1)\pi$, $\cos 7\theta = -\cos 6\theta$

13. Let $\quad 9\theta = 2n\pi$ ...(1)
    $$\theta = \frac{2n\pi}{9} = \frac{\text{an even multiple of }\pi}{9}$$ ...(2)

    By giving to $n = 0, 1, 2, 3, 4$ the values of $\cos\theta$ obtained from equation (1) and (2) are
    $$\cos 0, \cos\frac{2\pi}{9}, \cos\frac{4\pi}{9}, \cos\frac{6\pi}{9}, \cos\frac{8\pi}{9}$$

Answers

8. (b) $4096 x^6 - 11264 x^5 + 11520 x^4 - 5376 x^3 + 1120 x^2 - 84x + 1 = 0$ 9. $y^4 - 36 y^3 - 126 y^2 - 84 y + 9 = 0$, $28/3$

15. $\cos\dfrac{\pi}{5} \pm i\sin\dfrac{\pi}{5}, \cos\dfrac{3\pi}{5} \pm i\sin\dfrac{3\pi}{5}$

11.6.3 $\cos^n\theta$ AND $\sin^n\theta$ IN TERMS OF z

If $z = \cos\theta + i\sin\theta$, $\dfrac{1}{z} = \dfrac{1}{\cos\theta + i\sin\theta} = (\cos\theta + i\sin\theta)^{-1} = \cos\theta - i\sin\theta$

$$\therefore \qquad z + \frac{1}{z} = 2\cos\theta \qquad \text{and} \qquad z - \frac{1}{z} = 2i\sin\theta$$

Again by De Moivre's theorem, we have

$$z^n = (\cos\theta + i\sin\theta)^n = \cos n\theta + i\sin n\theta$$

$$\frac{1}{z^n} = (\cos\theta + i\sin\theta)^{-n} = \cos n\theta - i\sin n\theta$$

$$\therefore \qquad z^n + \frac{1}{z^n} = 2\cos n\theta \quad \text{and} \qquad z^n - \frac{1}{z^n} = 2i\sin n\theta$$

☑ REMARK

- Throughout the chapter, n will be taken as a positive integer unless otherwise stated.

11.6.4 EXPANSION OF $\cos^n\theta$ IN A SERIES OF COSINES OF MULTIPLES OF θ (n BEING A POSITIVE INTEGER)

If $z = \cos\theta + i\sin\theta$, we have,

$$(2\cos\theta)^n = \left(z + \frac{1}{z}\right)^n = z^n + {}^nC_1 z^{n-1}.\frac{1}{z} + {}^nC_2 z^{n-2}.\frac{1}{z^2} + \ldots + {}^nC_{n-2} z^2.\frac{1}{z^{n-2}} + {}^nC_{n-1} z.\frac{1}{z^{n-1}} + \frac{1}{z^n}\ldots$$

(Expanding by binomial theorem)

Since $\qquad {}^nC_1 = {}^nC_{n-1}, {}^nC_r = {}^nC_{n-r}$ etc.

$$(2\cos\theta)^n = \left(z^n + \frac{1}{z^n}\right) + {}^nC_1\left(z^{n-2} + \frac{1}{z^{n-2}}\right) + {}^nC_2\left(z^{n-4} + \frac{1}{z^{n-4}}\right) + \ldots$$

$$= 2\cos n\theta + {}^nC_1.2\cos(n-2)\theta + {}^nC_2.2\cos(n-4)\theta + \ldots$$

For writing the last term of the series, two cases arise.

Case I. When n is even, number of terms in (1) will be $(n+1)$ an odd number and $\left(\frac{n}{2} + 1\right)$ th term will be the middle term which will be independent of z, i.e., it shall be ${}^nC_{n/2}$

So when n is even,

$$2^n\cos^n\theta = 2\cos n\theta + n.2\cos(n-2)\theta + \frac{n(n-1)}{1.2}2\cos(n-4)\theta + \ldots + {}^nC_{n/2}$$

or $\qquad 2^{n-1}\cos^n\theta = \cos n\theta + n\cos(n-2)\theta + \frac{n(n-1)}{1.2}\cos(n-4)\theta + \ldots + \frac{1}{2}.{}^nC_{n/2}.$

Case II. When n is odd, there will be even number of terms in (1) and the two middle terms will be $\frac{n+1}{2}$ th and $\frac{n+3}{2}$ th. They will respectively be ${}^nC_{(n-1)/2}.z$ and ${}^nC_{(n+1)/2}.\frac{1}{z}$ which when combined will be ${}^nC_{(n-1)/2}\left(z + \frac{1}{z}\right)$, i.e., $\frac{n!}{[(n-1)/2]![(n+1)/2]!}.2\cos\theta$.

Here if n is odd,

$$2^n\cos^n\theta = 2\cos n\theta + n\cdot 2\cos(n-2)\theta + \frac{n(n-1)}{1\cdot 2}2\cdot\cos(n-4)\theta + \ldots + \frac{n!}{[(n-1)/2]![(n+1)/2]!}.2\cos\theta$$

$$2^{n-1}\cos^n\theta = \cos n\theta + n\cos(n-2)\theta + \frac{n(n-1)}{1.2}\cos(n-4)\theta + \ldots + \frac{n!}{(n-1)/2!(n+1)/2!}\cos\theta$$

☑ REMARK

- It will be convenient for the student to remember the binomial coefficient in the expansion of the form $(a + x)^n$, where n is a positive integer.

It is interesting to note that the number of sets of binomial coefficients may be quickly reproduced according to the following scheme,

1	1							$(n = 1)$
1	2	1						$(n = 2)$
1	3	3	1					$(n = 3)$
1	4	6	4	1				$(n = 4)$
1	5	10	10	5	1			$(n = 5)$
1	6	15	20	15	6	1		$(n = 6)$
1	7	21	35	35	21	7	1 etc,	$(n = 7)$

each number being formed as the sum of one immediately above it and the preceding one. The first and the last coefficients in each case being '1'.

Thus in forming the fifth row ($n = 4$) we have

1	+	3	+	3	+	1	($n = 3$)
		↓		↓		↓	
1		4		6		4 1	($n = 4$)
1		5		10		10 5 1	($n = 5$)

i.e., $1, 1 + 3 = 4, 3 + 3 = 6, 3 + 1 = 4,$

and $1, 1 + 4 = 5, 4 + 6 = 10, 6 + 4 = 10, 4 + 1 = 5, 1$ etc.

Solved Examples

EXAMPLE 1. *Prove that* $32 \cos^6\theta = \cos 6\theta + 6\cos 4\theta + 15\cos 2\theta + 10.$

SOLUTION. If $z = \cos\theta + i\sin\theta$

$$(2\cos\theta)^6 = \left(z + \frac{1}{z}\right)^6 = z^6 + 6z^4 + 15z^2 + 20 + 15$$

$$\cdot \frac{1}{z^2} + 6 \cdot \frac{1}{z^4} + \frac{1}{z^6}$$

$$= \left(z^6 + \frac{1}{z^6}\right) + 6\left(z^4 + \frac{1}{z^4}\right) + 15\left(z^2 + \frac{1}{z^2}\right) + 20$$

$$= 2\cos 6\theta + 6.2\cos 4\theta + 15.2\cos 2\theta + 20$$

$$\therefore 2^5\cos^6\theta = \cos 6\theta + 6\cos 4\theta + 15\cos 2\theta + 10.$$

EXAMPLE 2. *Expand* $\cos^9\theta$ *in a series of multiple of* θ.

SOLUTION. If $z = \cos\theta + i\sin\theta$

$$2^9\cos^9\theta = \left(z + \frac{1}{z}\right)^9 = z^9 + 9.z^7 + 36z^5 + 84z^3$$

$$+ 126z + 126 \cdot \frac{1}{z} + \frac{84}{z^3} + \frac{36}{z^5} + \frac{9}{z^7} + \frac{1}{z^9}$$

$$= \left(z^9 + \frac{1}{z^9}\right) + 9\left(z^7 + \frac{1}{z^7}\right)$$

$$+ 36\left(z^5 + \frac{1}{z^5}\right) + 84\left(z^3 + \frac{1}{z^3}\right) + 126\left(z + \frac{1}{z}\right)$$

$$= 2\cos 9\theta + 9.2\cos 7\theta + 36.2\cos 5\theta$$

$$+ 84.2\cos 3\theta + 84\cos 3\theta + 126.2.\cos\theta$$

$$\therefore 2^8\cos^9\theta = \cos 9\theta + 9\cos 7\theta + 36\cos 5\theta$$

$$+ 84\cos 3\theta + 126\cos\theta$$

Hence $\cos^9\theta = \dfrac{1}{2^8}(\cos 9\theta + 9\cos 7\theta$

$$+ 36\cos 5\theta + 84\cos 3\theta + 126\cos\theta).$$

11.6.5 EXPANSION OF $\sin^n\theta$ IN A SERIES OF COSINES OF MULTIPLES OF θ (IF n IS EVEN) AND IN TERMS OF SINES OF MULTIPLES OF θ (IF n IS ODD)

If $z = \cos\theta + i\sin\theta$, then

$$(2i\sin\theta)^n = \left(z - \frac{1}{z}\right)^n$$

$$2^n i^n \sin^n\theta = z^n - {}^nC_1 z^{n-1}.\frac{1}{z} + C_2 z^{n-2}.\frac{1}{z^2} - \dots + (-1)^{(n-1)}.\,{}^nC_{n-1}z.\frac{1}{z^{n-1}} + (-1)^n.\,{}^nC_n.\frac{1}{z^n} \qquad \dots(1)$$

Case I. If n is even

$$2^n(-1)^{n/2}\sin^n\theta = z^n - n.z^{n-2} + \frac{n(n-1)}{1.2}z^{n-4} - \dots - n.\frac{1}{z^{n-2}} + \frac{1}{z^n} = \left(z^n + \frac{1}{z^n}\right) - n\left(z^{n-2} + \frac{1}{z^{n-2}}\right) + \dots$$

The middle term in (1) will be $(-1)^{n/2}.\,{}^nC_{n/2}$ (independent of z)

Therefore if n is even

$$2^n(-1)^{n/2}\sin^n\theta = 2\cos n\theta - n.2\cos(n-2)\theta + \frac{n(n-1)}{1.2}.2\cos(n-4) + (-1)^{n/2}.2^nC_{n/2}$$

Hence, $2^{n-1}(-1)^{n/2}\sin^n\theta = \cos n\theta - n\cos(n-2)\theta + \dfrac{n(n-1)}{1.2}\cos(n-4)\theta - \dots + (-1)^{n/2}.\dfrac{1}{2}\,{}^nC_{n/2}$

Case II. If n is odd

Then since $i^n = i.(i^2)^{(n-1)/2} = i(-1)^{(n-1)/2}$, we have

$$2^n i(-1)^{(n-1)/2}\sin^n\theta = z^n - {}^nC_1 z^{n-2} + \dots + {}^nC_{n-1}\frac{1}{z^{n-2}} - \frac{1}{n}$$

and the two middle terms will be

$$(-1)^{(n-1)/2}.\,{}^nC_{(n-1)/2}z \quad\text{and}\quad (-1)^{(n+1)/2}.\,{}^nC_{(n+1)/2}.\frac{1}{z}$$

which when combined will be

$$(-1)^{(n-1)/2}.\,{}^nC_{(n-1)/2}.\left(z - \frac{1}{z}\right).$$

Therefore if n is odd

$$2^n i(-1)^{(n-1)/2}\sin^n = 2i\sin n\theta - n2i\sin(n-2)\theta + \dots + \frac{(-1)^{(n-1)/2}.n!2i\sin\theta}{[(n-1/2]![(n+1)/2]!}$$

Hence, $2^{n-1}(-1)^{(n-1)/2}\sin^n\theta = \sin n\theta - n.\sin(n-2)\theta + \dfrac{n(n-1)}{1.2}\sin(n-4)\theta - ... + \dfrac{(-1)^{(n-1)/2}.n!\sin\theta}{[(n-1)/2]![(n+1)/2]!}$

Solved Examples

EXAMPLE 1. *Expand $\sin^6\theta$ in a series of cosine of multiple of θ.*

SOLUTION. We have $(2i\sin\theta)^6 = \left(z - \dfrac{1}{z}\right)^6$

or $\quad 2^6 i^6 \sin^6\theta = z^6 - 6z^4 + 15z^2 - 20$
$\qquad\qquad\qquad + \dfrac{15}{z^2} - \dfrac{6}{z^4} + \dfrac{1}{z^6}$

$\qquad = \left(z^6 + \dfrac{1}{z^6}\right) - 6\left(z^4 + \dfrac{1}{z^4}\right) + 15\left(z^2 + \dfrac{1}{z^2}\right) - 20$

$\qquad = 2\cos 6\theta - 6\times 2\cos 4\theta + 15\times 2\cos 2\theta - 20$

or $\sin^6\theta = -\dfrac{1}{2^5}[\cos 6\theta - 6\cos 4\theta + 15\cos 2\theta - 10]$

EXAMPLE 2. *Prove that $2^6 \cos^7\theta = \cos 7\theta + 7\cos 5\theta + 21 \cos 3\theta + 35\cos\theta$.* (MADRAS–2003S)

SOLUTION. We have

$(2\cos\theta)^7 = \left(z + \dfrac{1}{z}\right)^7 = z^7 + 7z^5 + 21z^3 + 35z$
$\qquad\qquad\qquad + \dfrac{35}{z} + \dfrac{21}{z^3} + \dfrac{7}{z^5} + \dfrac{1}{z^7}$

$\qquad = \left(z^7 + \dfrac{1}{z^7}\right) + 7\left(z^5 + \dfrac{1}{z^5}\right) + 21\left(z^3 + \dfrac{1}{z^3}\right)$
$\qquad\qquad\qquad + 35\left(z + \dfrac{1}{z}\right)$

$\qquad = 2\cos 7\theta + 7\times 2\cos 5\theta + 21\times 2\cos 3\theta$
$\qquad\qquad\qquad + 35\times 2\cos\theta$

or $\quad 2^6 \cos^7\theta = \cos 7\theta + 7\cos 5\theta + 21\cos 3\theta$
$\qquad\qquad\qquad + 35\cos\theta$.

EXAMPLE 3. *Prove that $64(\cos^8\theta + \sin^8\theta) = \cos 8\theta + 28\cos 4\theta$*

$+\ 35$.

SOLUTION. If $z = \cos\theta + i\sin\theta$

$2^8 \cos^8\theta = \left\{z + \dfrac{1}{z}\right\}^8 = z^8 + 8z^6 + 28z^4 + 56z^2$
$\qquad\qquad + 70 + \dfrac{56}{z^2} + \dfrac{28}{z^4} + \dfrac{8}{z^6} + \dfrac{1}{z^8}$

$\qquad = \left(z^8 + \dfrac{1}{z^8}\right) + 8\left(z^6 + \dfrac{1}{z^6}\right) + 28\left(z^4 + \dfrac{1}{z^4}\right)$
$\qquad\qquad + 56\left(z^2 + \dfrac{1}{z^2}\right) + 70$

$\qquad = 2\cos 8\theta + 8.2\cos 6\theta + 28.2\cos 4\theta$
$\qquad\qquad + 56.2\cos 2\theta + 70$

$\therefore\ 2^7 \cos^8\theta = \cos 8\theta + 8.\cos 6\theta + 28\cos 4\theta$
$\qquad\qquad + 56.\cos 2\theta + 35 \qquad ...(1)$

Again $2^8 i^8 \sin^8\theta$

$\qquad = \left(z - -\right) = z^8 - 8z^6 + 28z^4 + 70$

$\qquad\qquad - \dfrac{56}{z^2} + \dfrac{28}{z^4} - \dfrac{8}{z^6} + \dfrac{1}{z^8} - 56z^2$

$\qquad = \left(z^8 + \dfrac{1}{z^8}\right) - 8\left(z^6 + \dfrac{1}{z^6}\right) + 28$

$\qquad\quad \left(z^4 + \dfrac{1}{z^4}\right) - 56\left(z^2 + \dfrac{1}{z^2}\right) + 70$

$\qquad = 2\cos 8\theta - 8.2\cos 6\theta + 28.2\cos 4\theta$
$\qquad\qquad - 56.2\cos 2\theta + 70$

$2^7.\sin^8\theta = \cos 8\theta - 8.\cos 6\theta + 28.\cos 5\theta$
$\qquad\qquad - 56.\cos 2\theta + 35 \qquad ...(2)$

Adding (1) and (2)

$2^7(\cos^8\theta + \sin^8\theta) = 2\cos 8\theta + 2.28\cos 4\theta + 70$

$\therefore\ 64(\cos^8\theta + \sin^8\theta) = \cos 8\theta + 28\cos 4\theta + 35$.

11.6.6 EXPANSION OF EXPRESSIONS OF THE FORM $\sin^m\theta\ \cos^n\theta$

The method of expressing the expression of the form $\sin^m\theta\ \cos^n\theta$ is illustrated by following solved examples.

Solved Examples

EXAMPLE 1. *Expand $\cos^5\theta.\sin^3\theta$ in terms of sines of multiples of θ.*

SOLUTION. If $\quad z = \cos\theta + i\sin\theta$, we have

$2^5\cos^5\theta\ 2^3.i^3\sin^3\theta = \left(z + \dfrac{1}{z}\right)^5\left(z - \dfrac{1}{z}\right)^3 \quad ...(1)$

To obtain the coefficients of the various powers of z in (1), we first write the binomial coefficients in expansion of $\left(z + \dfrac{1}{z}\right)^5$ and then multiplying by $\left(z - \dfrac{1}{z}\right)$ three times in succession, as in the following scheme :

Binomial coefficients $\left(z + \dfrac{1}{z}\right)^5$ in are

$\qquad 1 \quad 5 \quad 10 \quad 10 \quad 5 \quad 1$

Binomial coefficients in $\left(z + \dfrac{1}{z}\right)^5\left(z - \dfrac{1}{z}\right)$ are obtained

by subtracting from each the preceding, taking zeroes, to be the extremes.

$(1-0),\ (5-1),\ (10-5),\ (10-10),\ (5-10),\ (1-5),\ (0-1)$, *i.e.,* $1,\ 4,\ 5,\ 0,\ -5,\ -4,\ -1$

Coefficients in $\left(z + \dfrac{1}{z}\right)^3\left(z - \dfrac{1}{z}\right)^2$ are

$(1-0),\ (4-1),\ (5-4),\ (0-5),\ (-5-0),\ [-4-(-5)]$,

$[-1-(-4)],\ [0-(-1)]$, *i.e.,* $1,\ 3,\ 1,\ -5,\ -5,\ 1,\ 3,\ 1$

Similarly coefficients in $\left(z + \dfrac{1}{z}\right)^5\left(z - \dfrac{1}{z}\right)^3$ are

$\qquad 1,\ 2,\ -2,\ -6,\ 0,\ 6,\ 2,\ -1$

$\therefore\quad 2^8 i^3 \cos^5\theta\ \sin^3\theta = \left(z + \dfrac{1}{z}\right)^5\left(z - \dfrac{1}{z}\right)^3$

$$= z^8 + 2z^6 - 2z^4 - 6z^4 + 0 + 6 \cdot \frac{1}{z^2} + \frac{2}{z^4} - \frac{2}{z^6} - \frac{1}{z^8}$$

$$= \left(z^8 - \frac{1}{z^8}\right) + 2\left(z^6 - \frac{1}{z^6}\right) - 2\left(z^4 - \frac{1}{z^4}\right) - 6\left(z^2 - \frac{1}{z^2}\right)$$

$$\therefore \quad 2^8(-1)\, i\cos^5\theta\, \sin^3\theta = 2i\sin 8\theta + 4i\sin 6\theta$$
$$- 4i\sin 4\theta - 12i\sin 2\theta$$

$$\therefore \quad \cos^5\theta\, \sin^3\theta = \frac{1}{2^7}\{\sin 8\theta + 2\sin 6\theta$$
$$- 2\sin 4\theta - 6\sin 2\theta\}$$

🖝 REMARK

- The above scheme can be written in short as coefficients of:

$$\left(z + \frac{1}{z}\right)^5 \text{ are } 1, 5, 10, 10, 5, 1;\ \left(z + \frac{1}{z}\right)^5\left(z - \frac{1}{z}\right) \text{ are } 1, 4, 5, 0, -5, -4, -1$$

$$\left(z + \frac{1}{z}\right)^5\left(z - \frac{1}{z}\right)^2 \text{ are } 1, 3, 1, -5, -5, -1, +3, 1;\ \left(z + \frac{1}{z}\right)^5\left(z - \frac{1}{z}\right)^3 \text{ are } 1, 2, -2, -6, 0, 6, 2, -2, -1$$

EXAMPLE 2. *Express* $\sin^7\theta\, \cos^2\theta$ *in a series of sines of multiples of* θ.

SOLUTION. If $z = \cos\theta + i\sin\theta$

$$2^7 \cdot i^7 \sin^7\theta \cdot 2^2 \cos^2\theta = \left(z - \frac{1}{z}\right)^7\left(z + \frac{1}{z}\right)^2$$

Now binomial coefficients in the expansion of

$$\left(z - \frac{1}{z}\right)^7 \text{ are } 1, -7, 21, -35, 35, -21, 7, -1$$

$$\left(z - \frac{1}{z}\right)^7\left(z + \frac{1}{z}\right) \text{ are } 1, -6, 14, -14, 0, 14, -14, 6, -1$$

$$\left(z - \frac{1}{z}\right)^7\left(z + \frac{1}{z}\right)^2 \text{ are } 1, -5, 8, 0, -14, 14, 0, -8, 5, -1.$$

$$\therefore \quad \left(z - \frac{1}{z}\right)^7\left(z + \frac{1}{z}\right)^2$$

$$= z^9 - 5z^7 + 8z^5 + 0 - 14z + \frac{14}{z} + 0 - \frac{8}{z^5} + \frac{5}{z^7} - \frac{1}{z^9}$$

$$\therefore \quad 2^9 \cdot (-i) \sin^7\theta\, \cos^2\theta$$

$$= \left(z^9 - \frac{1}{z^9}\right) - 5\left(z^7 - \frac{1}{z^7}\right) + 8\left(z^5 - \frac{1}{z^5}\right)$$

$$- 14\left(z - \frac{1}{z}\right)$$

$$= 2i\sin 9\theta - 10\, i\sin 7\theta + 16\, i\sin 5\theta - 28i\sin\theta$$

$$\therefore \quad 2^8 \sin^7\theta\, \cos^2\theta = -[\sin 9\theta - 5\sin 7\theta$$
$$+ 8\sin 5\theta - 14\sin\theta]$$

🗒 Exercise-11.5

1. Expand $\sin^7\theta$ in a series of sines of multiples of θ.

2. Expand $\sin^8\theta$.

3. Prove that $2^9 \cos^{10}\theta = \cos 10\theta + 10\cos 8\theta + 45\cos 6\theta + 120\cos 4\theta + 210\cos 2\theta + 126$.

4. Express $\cos^4\theta\, \sin^3\theta$ in a series of sines of multiples of θ.

5. Express $\sin^5\theta\, \cos^2\theta$ in a series of sines of multiples of θ.
   (MADRAS–2003)

6. Expand $\cos^5\theta\, \sin^7\theta$ in a series of sines of multiples of θ.
   (MADRAS–2002)

7. Prove that $2^5 \sin^4\theta\, \cos^2\theta \doteq \cos 6\theta - 2\cos 4\theta - \cos 2 + 2$.

8. Show that $128\cos^8\theta = \cos 8\theta + 8\cos 6\theta + 28\cos 4\theta + 56\cos 2\theta + 35$.

9. Prove that $\cos^5\theta = 1/16\ [\cos 5\theta + 5\cos 3\theta + 10\cos\theta]$.

10. Express $\sin^8\theta\, \cos^2\theta$ in terms of cosines of multiples of θ.

11. If $\tan^{-1}x + \tan^{-1}y + \tan^{-1}z = \pi/2$ show that $xy + yz + zx = 1$.
    (PTU–2003)

12. Prove that $\cos^6\theta - \sin^6\theta = \frac{1}{16}(\cos 6\theta + 15\cos 2\theta)$
    (MUMBAI–2007)

🗒 Hints to Selected Problems

4. Suppose $x = \cos\theta + i\sin\theta$ and $1/x = \cos\theta - i\sin\theta$

$$\Rightarrow \quad x + \frac{1}{x} = 2\cos\theta,\ x - \frac{1}{x} = 2i\sin\theta$$

which gives $(2\cos\theta)^4 (2i\sin\theta)^3 = \left(x + \frac{1}{x}\right)^4 \cdot \left(x - \frac{1}{x}\right)^3$.

Now proceed as usual.

🗒 Answers

1. $-\dfrac{1}{64}(\sin 7\theta - 7\sin 5\theta + 21\sin 3\theta - 35\sin\theta)$

2. $\dfrac{1}{2^7}[\cos 8\theta - 8\cos 6\theta + 28\cos 4\theta - 56\cos 2\theta + 35]$

4. $-\dfrac{1}{2^6}[\sin 7\theta + \sin 5\theta - 3\sin 3\theta - 3\sin\theta]$

5. $\dfrac{1}{2^6}[\sin 7\theta - 3\sin 5\theta + \sin 3\theta + 5\sin\theta]$

6. $-\dfrac{1}{2^{11}}[\sin 12\theta - 2\sin 10\theta - 4\sin 8\theta + 10\sin 6\theta + 5\sin 4\theta + 20\sin 2\theta]$

10. $\dfrac{1}{2^7}[\cos 8\theta - 4\cos 6\theta + 4\cos 4\theta + 4\cos 2\theta - 5]$

11.6.7 EXPONENTIAL SERIES OF COMPLEX NUMBERS

The exponential series for all real values of x is given by

$$e^x = 1 + x + \frac{x^2}{2!} + \frac{x^3}{3!} + \dots \text{ ad. inf.} \qquad \dots (1)$$

But where x is complex, the expression e^x has no meaning at present. The series (1) is absolutely convergent for all finite values of x. Now consider the series

$$E(z) = 1 + z + \frac{z^2}{2!} + \frac{z^3}{3!} + \dots \text{ ad. inf.} \qquad \dots (2)$$

where $z = x + iy = r(\cos\theta + i\sin\theta)$ and therefore $|z| = r > 0$
Let the series of the moduli be

$$1 + |z| + \frac{|z^2|}{2!} + \frac{|z^3|}{3!} + \dots + \frac{|z^n|}{n!} + \dots = 1 + r + \frac{r^2}{2!} + \frac{r^3}{3!} + \dots + \frac{r^n}{n!} + \dots$$

This is a series of positive numbers and convergent and hence the series (2) is absolutely convergent for all finite values of z.
In particular if $z = x + i0$ which corresponding to the real number x, $E(z)$ assumes the values

$$1 + \frac{x}{1!} + \frac{x^2}{2!} + \dots + \frac{x^n}{n!} + \dots$$

which corresponds to exp. (x) or e^x, which e stands for,

$$\left(1 + \frac{1}{1!} + \frac{1}{2!} + \frac{1}{3!} + \dots + \frac{1}{n!} + \dots\right) \text{ and } e^x \text{ means } \left(1 + \frac{1}{1!} + \frac{1}{2!} + \dots + \frac{1}{n!} + \dots\right)^x \text{ for all real values of } x.$$

Hence, the series (2) is usually written as exp. (z) or e^z in close analogy to the exponential series for real number, and also because the fact that

$$E(z) = 1 + z + \frac{z^2}{2!} + \dots + \frac{z^n}{n!} + \dots$$

where $z \equiv x + iy$ when $z \equiv x + i0$ corresponds to exp. (x) or e^x.
It should be clearly understood that the series

$$1 + z + \frac{z^2}{2!} + \dots + \frac{z^n}{n!} + \dots$$

is written exp (z) or e^z by definition only, and that it does not mean, unless it is so proved, that the exp (z) or e^z stands for

$$\left(1 + \frac{1}{1!} + \frac{1}{2!} + \dots + \frac{1}{n!} + \dots\right)^z$$

If z is complex. Then by definition if $z = x + iy$

$$\exp(z) = 1 + z + \frac{z^2}{2!} + \dots + \frac{z^n}{n!} + \dots$$

THEOREM 1. *If z_1 and z_2 are any two complex numbers, then $e^{z_1} \cdot e^{z_2} = e^{z_1 + z_2}$ or $\exp(z_1) \times \exp(z_2) = \exp(z_1 + z_2)$.*

PROOF. By definition we have

$$\exp(z_1) \times \exp(z_2) = \left[1 + z_1 + \frac{z_1^2}{2!} + \dots + \frac{z_1^n}{n!} + \dots\right] \times \left[1 + z_2 + \frac{z_2^2}{2!} + \dots + \frac{z_2^n}{n!} + \dots\right]$$

$$= \left[1 + (z_1 + z_2) + \frac{1}{2!}(z_1^2 + 2z_1 z_2 + z_2^2) + \dots + \frac{1}{n!}\left(z_1^n + n z_1^{n-1} z_2 + \frac{n(n-1)}{2!} z_1^{n-2} z_2^2 + \dots + z_2^n\right) + \dots\right]$$

$$= \left[1 + (z_1 + z_2) + \frac{(z_1 + z_2)^2}{2!} + \frac{(z_1 + z_2)^3}{3!} + \dots + \frac{(z_1 + z_2)^n}{n!} + \dots\right] = \exp(z_1 + z_2).$$

REMARK

- The series on the R.H.S. is absolutely convergent if exp (z_1) exp (z_2) are absolutely convergent.

THEOREM 2. *If z is a complex number, then $(e^z)^m = e^{mz}$.*

PROOF. If m is positive integer, we have by repeated application of theorem 1,

$$\exp(z_1) \exp(z_2) \dots \exp(z_m) = \exp(z_1 + z_2 + \dots + z_m)$$

If $z_1 = z_2 = \dots = z_m = z$ we have $(\exp z)^m = \exp(mz)$.

THEOREM 3. *$E(z) \neq 0$ for any value of z.*

PROOF. By the addition theorem, we have

$$E(z) \cdot E(-z) = E\{z + (-z)\} = E(0) = 1$$

Since $E(z)$ is well defined for all values of z, therefore it follows that $E(z) \neq 0$, for any value of z.

☞ **REMARK**
- $\{E(z)\}^{-1} = E(-z)$

11.6.8 EULER'S EXPONENTIAL VALUE

We have $\qquad\qquad e^z = 1 + z + \dfrac{z^2}{2!} + \dfrac{z^3}{3!} + \dots$

Put $z = i\theta$ where θ is real

$$e^{i\theta} = 1 + i\theta + \dfrac{i^2\theta^2}{2!} + \dfrac{i^3\theta^3}{3!} + \dots = 1 - \dfrac{\theta^2}{2!} + \dfrac{\theta^4}{4!} - \dfrac{\theta^6}{6!} + \dots + i\left[\theta - \dfrac{\theta^3}{3!} + \dfrac{\theta^5}{5!} - \dots\right] \qquad (\because i^2 = -1)$$

$$= \cos\theta + i\sin\theta \qquad \qquad \dots(1)$$

$\therefore \qquad\qquad e^{i\theta} = \cos\theta + i\sin\theta \qquad\qquad \dots(2)$

and $\qquad\qquad e^{-i\theta} = \cos\theta - i\sin\theta$

By adding of (1) and (2) we get

(a) $\quad \cos\theta = \dfrac{e^{i\theta} + e^{-i\theta}}{2}$

By subtracting (2) from (1) we get

(b) $\quad \sin\theta = \dfrac{e^{i\theta} - e^{-i\theta}}{2i}$

These results are known as Euler's exponential values.

(c) $\quad \tan\theta = \dfrac{\sin\theta}{\cos\theta} = \dfrac{(e^{i\theta} - e^{-i\theta})}{i(e^{i\theta} + e^{-i\theta})}$;   (d) $\qquad \cot\theta = \dfrac{\cos\theta}{\sin\theta} = \dfrac{i(e^{i\theta} + e^{-i\theta})}{(e^{i\theta} - e^{-i\theta})}$

☞ **REMARK**
- Since $e^{x+iy} = e^x.[\cos y + i\sin y]$, this method helps in breaking an exponential function into real and imaginary parts.

11.7 CIRCULAR FUNCTION OF COMPLEX QUANTITIES

For real values of x

$$\sin x = x - \dfrac{x^3}{3!} + \dfrac{x^5}{5!} + \dots + \dfrac{(-1)^n x^{2n+1}}{(2n+1)!} + \dots$$

and $\qquad\qquad \cos x = 1 - \dfrac{x^2}{2!} + \dfrac{x^4}{4!} - \dots + \dfrac{(-1)^n x^{2n}}{(2n)!} + \dots$

These definitions are extended for the complex quantity $z = x + iy$ where x and y are real.

$$\sin z = z - \dfrac{z^3}{3!} + \dfrac{z^5}{5!} - \dots + \dfrac{(-1)^n z^{2n+1}}{(2n+1)!} + \dots$$

and $\qquad\qquad \cos z = 1 - \dfrac{z^2}{2!} + \dfrac{z^4}{4!} - \dots + \dfrac{(-1)^n z^{2n}}{(2n)!} + \dots$

Similarly, $\qquad\qquad \tan z = \dfrac{\sin z}{\cos z}, \cot z = \dfrac{\cos z}{\sin z}, \sec z = \dfrac{1}{\cos z}$ and $\operatorname{cosec} z = \dfrac{1}{\sin z}$.

From these definitions, we can deduce the fundamental properties of two functions :

(a) $\cos z + i\sin z = e^{iz}$; $\cos z - i\sin z = e^{-iz}$

$\therefore \quad \cos^2 z + \sin^2 z = e^{iz} e^{-iz} = e^{-0} = 1$

(b) $\cos z = \dfrac{1}{2}[e^{iz} + e^{-iz}]; \sin z = \dfrac{1}{2i}[e^{iz} - e^{-iz}]$

(c) To prove that $\sin(z_1 + z_2) = \sin z_1 \cos z_2 + \cos z_1 \sin z_2$

$$\text{R.H.S.} = \dfrac{e^{iz_1} - e^{-iz_1}}{2i} \times \dfrac{e^{iz_2} + e^{iz_2}}{2} + \dfrac{e^{iz_1} + e^{-iz_1}}{2} \cdot \dfrac{e^{iz_2} - e^{-iz_2}}{2i}$$

$$= \dfrac{1}{4i}[2e^{i(z_1+z_2)} - e^{i(z_2-z_1)} - 2e^{-(z_1+z_2)} + e^{i(z_1-z_2)} - e^{i(z_1-z_2)} + e^{i(z_2-z_1)}] = \dfrac{1}{2i}[e^{i(z_1+z_2)} - e^{-i(z_1+z_2)}] = \sin(z_1 + z_2).$$

(d) To prove that $\sin 3z = 3\sin z - 4\sin^3 z$.

$$\text{R.H.S} = 3\left(\dfrac{e^{iz} - e^{-iz}}{2i}\right) - 4\left(\dfrac{e^{iz} - e^{-iz}}{2i}\right)^3 = \dfrac{1}{2i}(3e^{iz} - 3e^{-iz}) - 4.\left(\dfrac{e^{3iz} - 3e^{2iz}.e^{-iz} + 3e^{iz}e^{-2iz} - e^{-3iz}}{8i^3}\right)$$

$$= \frac{1}{2i}(3e^{iz} - 3e^{-iz} + e^{3iz} - 3e^{iz} + 3e^{-iz} - e^{-3iz}) = \frac{1}{2i}(e^{3iz} - e^{-3iz}) = \sin 3z.$$

Similarly we can derive other results. These results show the generality of trigonometrically formulae.

11.7.1 PERIOD OF COMPLEX CIRCULAR FUNCTIONS

$$\cos(z + 2n\pi) = \cos z \cos 2n\pi - \sin z \sin 2n\pi = \cos z \qquad \text{[If } n \text{ is an integer]}$$
$$\sin(z + 2n\pi) = \sin z \cos 2n\pi + \cos z \sin 2n\pi = \sin z \qquad \text{[}n \text{ being an integer]}$$

and $$\tan(z + n\pi) = \frac{\sin(z + n\pi)}{\cos(z + n\pi)} = \frac{\pm \sin z}{\pm \cos z} = \tan z \qquad \text{[According as } n \text{ is even or odd integer.]}$$

Hence the periods of $\cos z$, $\sin z$ and $\tan z$ are real and are the same (*i.e.*, 2π in case of $\cos z$ and $\sin z$ and π in case of $\tan z$) as the periods of the circular functions of a real numbers.

11.7.2 PERIOD OF EXP(z)

If $z = x + iy$, then

$$e^{z + 2n\pi i} = e^x \cdot e^{i(y + 2\pi)} = e^x[\cos(y + 2\pi) + i\sin(y + 2\pi)] = e^x[\cos y + i\sin y] = e^x \cdot e^{iy} = e^{x+iy} = e^z$$

Hence $e^{z + 2n\pi i} = e^z$.

Thus the period of $\exp(z)$ is $2\pi i$.

Solved Examples

EXAMPLE 1. Show that $\exp\left(\pm i\dfrac{\pi}{2}\right) = \pm i$.

SOLUTION. Since $\exp(\pm i\theta) = \cos\theta \pm i\sin\theta$ we have

$$\exp\left(\pm i\frac{\pi}{2}\right) = \cos\frac{\pi}{2} \pm i\sin\frac{\pi}{2} = \pm i.$$

EXAMPLE 2. Prove $\sin(\alpha + n\theta) - e^{i\alpha}\sin n\theta = e^{-in\theta}\sin\alpha$.

SOLUTION. L.H.S $= \sin(\alpha + n\theta) - (\cos\alpha + i\sin\alpha)\sin n\theta$
$$= \sin\alpha\cos n\theta + \cos\alpha\sin n\theta$$
$$\qquad - \cos\alpha\sin n\theta - i\sin\alpha\sin n\theta$$
$$= \sin\alpha(\cos n\theta - i\sin n\theta) = \sin\alpha e^{-in\theta} = \text{R.H.S}$$

EXAMPLE 3. Prove that $\{\sin(\alpha - \theta) + e^{-\alpha i}\sin\theta\}^n$
$$= \sin^{n-1}\alpha\{\sin(\alpha - n\theta) + e^{-\alpha i}\sin n\theta\}$$

SOLUTION. L.H.S $\{\sin(\alpha - \theta) + e^{-\alpha i}\sin\theta\}^n$
$$= \{\sin\alpha\cos\theta - \cos\alpha\sin\theta + (\cos\alpha - i\sin\alpha)\sin\theta\}^n$$
$$= \{\sin\alpha\cos\theta - i\sin\alpha\sin\theta\}^n = \sin^n\alpha\{\cos\theta - i\sin\theta\}^n$$
$$= \sin^n\alpha\{\cos n\theta - i\sin n\theta\} \quad \text{[by De Moivre's theorem]}$$
Again, R.H.S
$$= \sin^{n-1}\alpha\{\sin(\alpha - n\theta) + e^{-i\alpha}\sin n\theta\}$$
$$= \sin^{n-1}\alpha\{\sin\alpha\cos n\theta - \cos\alpha\sin n\theta$$
$$\qquad\qquad + (\cos\alpha - i\sin\alpha)\sin n\theta\}$$
$$= \sin^{n-1}\alpha\{\sin\alpha\cos n\theta - i\sin\alpha\sin n\theta\}$$
$$= \sin^n\alpha\{\cos n\theta - i\sin n\theta\} = \text{L.H.S}$$

$\therefore$ R.H.S = L.H.S

EXAMPLE 4. If $\cos\theta + i\sin\theta = x$, $\sqrt{(1 - c^2)} = nc - 1$, prove that $1 + c\cos\theta = \dfrac{c}{2n}(1 + nx)\left(1 + \dfrac{n}{x}\right)$.

SOLUTION. We have $x = \cos\theta + i\sin\theta$
$$\therefore \ x^{-1} = \cos\theta - i\sin\theta$$
$$\therefore \ \text{R.H.S.} = \frac{c}{2n}(1 + nx)\left(1 + \frac{n}{x}\right)$$
$$= \frac{c}{2n}[\{1 + n(\cos\theta + i\sin\theta)\}\{1 + n(\cos\theta - i\sin\theta)\}]$$
$$= \frac{c}{2n}[\{1 + ne^{i\theta}\}\{1 + ne^{-i\theta}\}] = \frac{c}{2n}[1 + ne^{-i\theta} + ne^{i\theta} + n^2]$$
$$= \frac{c}{2n}[1 + n(e^{i\theta} + e^{-i\theta}) + n^2] = \frac{c}{2n}[1 + 2n\cos\theta + n^2]$$
$$= \frac{c}{2n}(1 + n^2) + c\cos\theta \qquad\qquad ...(1)$$

But $\sqrt{1 - c^2} = nc - 1$ or $1 - c^2 = (nc - 1)^2$
Hence form (1), we have
$$\Rightarrow c^2(1 + n^2) = 2nc$$
$$\frac{c}{2n} \times 2n + c\cos\theta = 1 + c\cos\theta.$$

Exercise-11.6

Prove the following :

1. (a) $\cos(-z) = \cos z$
   (b) $\sin(-z) = -\sin z$
2. (a) $\cos 2z = \cos^2 z - \sin^2 z = 1 - 2\sin^2 z$
   (b) $\sin 2z = 2\sin z\cos z$
   (c) $\cos 3z = 4\cos^3 z - 3\cos z$
   (d) $\sin 3z = 3\sin z - 4\sin^3 z$
3. (a) $\sin x + \sin y = 2\sin\left(\dfrac{x+y}{2}\right)\cos\left(\dfrac{x-y}{2}\right)$
   (b) $\cos x - \cos y = 2\sin\dfrac{(x+y)}{2}\sin\dfrac{1}{2}(y - x)$

4. (a) $\sin(x + y) = \sin x\cos y + \cos x\sin y$
   (b) $\sin(x - y) = \sin x\cos y - \cos x\sin y$
   (c) $\cos(x + y) = \cos x\cos y - \sin x\sin y$
   (d) $\cos(x - y) = \cos x\cos y + \sin x\sin y$
5. $[\sin(\alpha + \theta) - e^{\alpha i}\sin\theta]^n = \sin^n a e^{-n\theta i}$
6. $\sin(a + n\theta) - e^{i\alpha}\sin n\theta = e^{-in\theta}\sin\alpha$
7. If $\tan^{-1}(e^{ix}) - \tan^{-1}(e^{-ix}) = \tan^{-1} i$, find x.

Hints to Selected Problems

5. L.H.S. $= \sin\alpha\cos\theta + \cos\alpha\sin\theta - (\cos\alpha + i\sin\alpha)\sin\theta$
$$= \sin^n\alpha(\cos\theta - i\sin\theta)^n = \sin^n a e^{-in\theta}$$

7. $\tan^{-1}\dfrac{e^{ix}-e^{-ix}}{1+e^{ix}.e^{-ix}}=\tan^{-1}i$ $\Rightarrow$ $x=n\pi+(-1)^n\dfrac{\pi}{2}.$

or $i\sin x=i$ $\Rightarrow$ $\sin x=1=\sin\dfrac{\pi}{2}$

11.8 HYPERBOLIC FUNCTIONS

We have proved that for all values of the argument y (real or complex).

$$\cos y=1-\frac{y^2}{2!}+\frac{y^4}{4!}-\frac{y^6}{6!}+\dots \qquad\qquad\dots\text{(A)}$$

$$\sin y=y-\frac{y^3}{3!}+\frac{y^5}{5!}-\frac{y^7}{7!}+\dots \qquad\qquad\dots\text{(B)}$$

We notice that in each of these series, the terms are alternatively positive and negative. If we place the positive sign before all the terms, we get two functions of y defined by indefinite series, which are related to the circular functions $\cos y$ and $\sin y$ by interesting properties. These functions are known as hyperbolic cosine and hyperbolic sine of y and are indicated for shortness by $\cosh y$ and $\sinh y$. Thus

$$\cosh y=1+\frac{y^2}{2!}+\frac{y^4}{4!}+\dots \qquad\qquad\dots\text{(1)}$$

$$\sinh y=y+\frac{y^3}{3!}+\frac{y^5}{5!}+\frac{y^7}{7!}+\dots \qquad\qquad\dots\text{(2)}$$

$\therefore$ $\cosh y+\sinh y=1+y+\dfrac{y^2}{2!}+\dfrac{y^3}{3!}+\dots=e^y$ $\qquad\qquad\dots\text{(3)}$

and $\cosh y-\sinh y=e^{-y}$ $\qquad\qquad\dots\text{(4)}$

$\therefore$ $\cosh y=\dfrac{1}{2}[e^y+e^{-y}]$ and $\sinh y=\dfrac{1}{2}[e^y-e^{-y}].$

Definition. *The quantity* $\dfrac{e^y-e^{-y}}{2}$, *whether* y *be real or complex is called the hyperbolic sine of* y *and is written as* $\sinh y$.

Similarly $\dfrac{e^y+e^{-y}}{2}$ *is known as hyperbolic cosine of* y *and is written as* $\cosh y$.

☛ **REMARK**
- The hyperbolic tangent, secant, cosecant and cotangent can be obtained with the help of hyperbolic sine and cosine.

$$\tanh y=\frac{\sinh y}{\cosh y}=\frac{e^y-e^{-y}}{e^y+e^{-y}};\quad \operatorname{cosech} y=\frac{1}{\sinh y}=\frac{2}{e^y-e^{-y}};\quad \operatorname{sech} y=\frac{1}{\cosh y}=\frac{2}{e^y+e^{-y}};\quad \coth y=\frac{\cosh y}{\sinh y}=\frac{e^y+e^{-y}}{e^y-e^{-y}}$$

11.8.1 RELATION BETWEEN HYPERBOLIC AND CIRCULAR FUNCTIONS

Hyperbolic functions can be expressed in terms of corresponding circular functions.

We know that $\sin x=\dfrac{e^{ix}-e^{-ix}}{2i}$ put $x=iy$ $\Rightarrow$ $\sin iy=\dfrac{e^{i^2y}-e^{-i^2y}}{2i}=\dfrac{i[e^{-y}-e^y]}{2i^2}$

$$\sin iy=\frac{i[e^y-e^{-y}]}{2}=i\sinh y$$

Similarly, $\cos iy=\dfrac{e^{i^2y}+e^{-i^2y}}{2}=\dfrac{e^y+e^{-y}}{2}=\cosh y$

and $\tan iy=\dfrac{\sin iy}{\cos iy}=\dfrac{i\sinh y}{\cosh y}=i\tanh y$

From (3) and (4), we have

$$(\cosh y+\sinh y)^n=e^{ny}=\cosh ny+\sinh ny \qquad\qquad\dots\text{(5)}$$

and $(\cosh y-\sinh y)^n=e^{-ny}=\cosh ny-\sinh ny$ $\qquad\qquad\dots\text{(6)}$

☛ **REMARK**
- These results are analogous to De Moivre's theorem.

11.8.2 PERIOD OF HYPERBOLIC FUNCTIONS

We know that $\cos i\theta=\cosh\theta$

Therefore $\cosh(x+iy)=\cos[i(x+iy)]=\cos(xi-y)=\cos[-2\pi+ix-y]=\cos[(2\pi i+x+iy)i]=\cosh[2\pi i+x+iy]$

Similarly, $\cosh(x+iy)=\cosh[4\pi i+x+iy].$

Hence, the hyperbolic cosine is periodic, its period being imaginary and equal to $2\pi i$.

Similarly it can be shown for $\sinh(x + iy)$ that its period is $2\pi i$ and of $\tanh(x + iy)$ is πi.

Aliter. These results can be obtained in a simpler way also.

Since
$$e^{2n\pi i} = \cos 2n\pi + i \sin 2n\pi = 1.$$

$\therefore$
$$e^{z + 2n\pi i} = e^z \text{ and } e^{-z - 2n\pi i} = e^{-z}.$$

$\therefore$
$$e^{z + 2n\pi i} + e^{-z - 2n\pi i} = e^z + e^{-z} \text{ and } e^{z + 2n\pi i} - e^{-z - n\pi i} = e^z - e^{-z}.$$

$\therefore$
$$\cosh(z + 2n\pi i) = \cosh(z) \text{ and } \sinh(z + 2n\pi i) = \sinh z$$

Next
$$e^{n\pi i} = \cos n\pi + i \sin n\pi = (-1)^n,$$
$$e^{-n\pi i} = \cos n\pi - i \sin n\pi = (-1)^n.$$

$\therefore$
$$\tanh(z + n\pi i) = \frac{e^{z + n\pi i} - e^{-z - n\pi i}}{e^{z + n\pi i} + e^{-z - n\pi i}} = \frac{e^z - e^{-z}}{e^z + e^{-z}} = \tanh z.$$

Hence, $\cosh z$ and $\sinh z$ have an imaginary period of $2\pi i$ and $\tanh z$ an imaginary period of πi.

✒ REMARK

- The hyperbolic functions differ from the circular functions in having imaginary periods.

11.8.3 GEOMETRICAL ANALOGY BETWEEN CIRCULAR AND HYPERBOLIC FUNCTIONS

If $x = a \cosh u$, and $y = a \sinh u$, then point P whose co-ordinates are (x, y) satisfy the rectangular hyperbola
$$x^2 - y^2 = a^2$$

Area APN is given by

$$\int_0^x y\,dx = \int_0^x a \sinh u \, d(a \cosh u) = \int_0^u a^2 \sinh^2 u \, du = \frac{1}{2} a^2 \int_0^u (\cosh 2u - 1)\,du = \frac{1}{2} a^2 \left[\frac{1}{2} \sinh 2u - u \right]$$

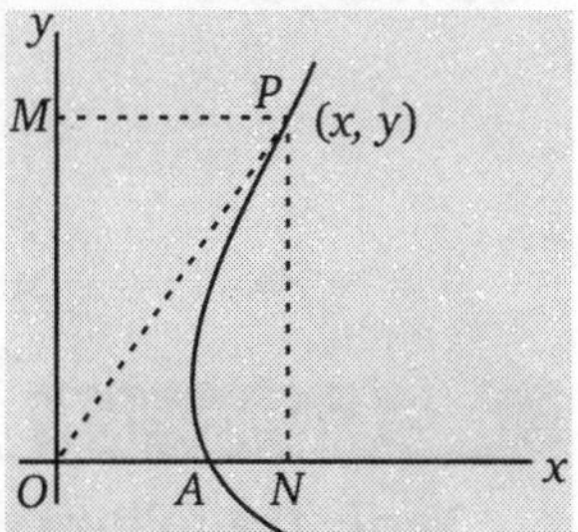

Area of triangle
$$OPN = \frac{1}{2} ON \times NP = \frac{1}{2} a \cosh u \times a \sinh u = \frac{1}{4} a^2 \sinh 2u$$

Hence, area of the sector
$$OAP = \triangle OPN - \text{area } APN = \frac{1}{2} a^2 u$$

Denoting the area of sector OAP by S, we have

$$\frac{1}{2} a^2 u = S \qquad \text{or} \qquad u = \frac{2S}{a^2}$$

Hence co-ordinates of any point on hyperbola $x^2 - y^2 = a^2$ may be represented by

$$x = a \cosh \frac{2S}{a^2}, y = a \sinh \frac{2S}{a^2}$$

Fig. 9

We notice that this is similar to the expressions for the co-ordinates of a point on the circle $x^2 + y^2 = a^2$,

where
$$x = a \cos \frac{2S}{a^2}, y = a \sin \frac{2S}{a^2} \quad S \text{ being the area of sector } OAP = \frac{1}{2} a^2 \theta$$

so that $\dfrac{2S}{a^2} = \theta = $ circular measure of the angle AOP.

It is thus apparent that the hyperbolic functions are connected in the same way with the rectangular hyperbola as the circular functions are with the circle.

Solved Examples

EXAMPLE 1. *If* $\tan y = \tan \alpha \tanh \beta$, $\tan z = \cot \alpha \tanh \beta$, *prove that* $\tan(y + z) = \sinh 2\beta \operatorname{cosec} 2\alpha$.

SOLUTION. $\tan(y + z) = \dfrac{\tan y + \tan z}{1 - \tan y \tan z}$

$$= \frac{\tan \alpha \tanh \beta + \cot \alpha \tanh \beta}{1 - \tan \alpha \tanh \beta \times \cot \alpha \tanh \beta}$$

$$= \frac{\tanh \beta \left[\dfrac{\sin \alpha}{\cos \alpha} + \dfrac{\cos \alpha}{\sin \alpha} \right]}{1 - \tanh^2 \beta}$$

$$= \frac{\dfrac{\sinh \beta}{\cosh \beta}}{1 - \dfrac{\sinh^2 \beta}{\cosh^2 \beta}} \times \frac{1}{\sin \alpha \cos \alpha}$$

$$= \frac{\sinh \beta \cosh \beta}{\sin \alpha \cos \alpha} \times \frac{2}{2} = \frac{\sinh 2\beta}{\sin 2\alpha}$$

$$= \sinh 2\beta \operatorname{cosec} 2\alpha.$$

EXAMPLE 2. *If* $\cosh x = x \sec \theta$, *prove that* $\tanh^2 \dfrac{x}{2} = \tan^2 \dfrac{\theta}{2}$.

SOLUTION. We know

$$\cosh x = \frac{1 + \tanh^2 x/2}{1 - \tanh^2 x/2} = \sec \theta = \frac{1}{\cos \theta}$$

Apply componendo and dividendo, we have

$$\left[i.e, \text{if } \frac{a}{b} = \frac{c}{d} \text{ then } \frac{a-b}{a+b} = \frac{c-d}{c+d} \right]$$

$$\frac{2 \tanh^2 x/2}{2} = \frac{1 - \cos \theta}{1 + \cos \theta} = \tan^2 \frac{\theta}{2}$$

$$\therefore \qquad \tanh^2 \frac{x}{2} = \tan^2 \frac{\theta}{2}.$$

EXAMPLE 3. *If* θ *is acute and* $x = \log \tan\left(\dfrac{\pi}{4} + \dfrac{\theta}{2}\right)$*, show that* $\cos\theta\cosh x = 1.$

SOLUTION. $\qquad x = \log \tan\left(\dfrac{\pi}{4} + \dfrac{\theta}{2}\right)$

$\therefore \qquad e^x = \tan\left(\dfrac{\pi}{4} + \dfrac{\theta}{2}\right) = \dfrac{1 + \tan\theta/2}{1 - \tan\theta/2}.$

Now $\cosh x = \dfrac{1}{2}[e^x + e^{-x}]$

$= \dfrac{1}{2}\left[\dfrac{1 + \tan\theta/2}{1 - \tan\theta/2} - \dfrac{1 - \tan\theta/2}{1 + \tan\theta/2}\right]$

$= \dfrac{1}{2}\left[\dfrac{(1 + \tan\theta/2)^2 + (1 - \tan\theta/2)^2}{1 - \tan^2\theta/2}\right]$

$= \dfrac{1 + \tan^2\theta/2}{1 - \tan^2\theta/2} = \sec\theta.$

$\therefore \quad \cosh x \cos\theta = 1.$

EXAMPLE 4. *If* $u = \log \tan\left(\dfrac{\pi}{4} + \dfrac{\theta}{2}\right)$*, prove that* $\tanh\dfrac{u}{2} = \tan\dfrac{\theta}{2}.$

SOLUTION. Given $\qquad u = \log \tan\left(\dfrac{\pi}{2} + \dfrac{\theta}{2}\right)$

or $\qquad e^u = \tan\left(\dfrac{\pi}{2} + \dfrac{\theta}{2}\right)$

or $\qquad \dfrac{e^{u/2}}{e^{-u/2}} = \dfrac{1 + \tan\theta/2}{1 - \tan\theta/2}$

By componendo and dividendo, we have

$\dfrac{e^{u/2} - e^{-u/2}}{e^{u/2} + e^{-u/2}} = \dfrac{2\tan\theta/2}{2} = \tan\theta/2$

or $\qquad \tanh\dfrac{1}{2}u = \tan\dfrac{\theta}{2}.$

EXAMPLE 5. *If* $u = \log \tan\left(\dfrac{\pi}{4} + \dfrac{\theta}{2}\right)$ *prove that*

$$\theta = -i\log \tan\left(\dfrac{\pi}{4} + \dfrac{iu}{2}\right) \qquad \text{(KURUKSHETRA–2006}$$

MUMBAI–2008, PTU–2006, MADRAS–2003)

SOLUTION. We have $e^u = \tan\left(\dfrac{\pi}{4} + \dfrac{\theta}{2}\right)$

or $\qquad \dfrac{e^{u/2}}{e^{-u/2}} = \dfrac{1 + \tan\theta/2}{1 - \tan\theta/2}$

By componendo and dividendo, we get

$\dfrac{e^{u/2} - e^{-u/2}}{e^{u/2} + e^{-u/2}} = \tan\theta/2$

i.e., $\quad \tanh\dfrac{u}{2} = \tan\theta/2$

or $\quad \dfrac{1}{i}\tan\dfrac{iu}{2} = \dfrac{1}{i}\tanh\dfrac{i\theta}{2}$

or $\quad \dfrac{i\theta}{2} = \tanh^{-1}\left(\tan\dfrac{iu}{2}\right) = \dfrac{1}{2}\log\dfrac{1 + \tan iu/2}{1 - \tan iu/2}$

or $\quad \theta = \dfrac{1}{i}\log \tan\left(\dfrac{\pi}{4} + \dfrac{iu}{2}\right) = -i\log \tan\left(\dfrac{\pi}{4} + \dfrac{iu}{2}\right)$

Exercise-11.7

Verify the following :

1. (a) $\sinh(x - y) = \sinh x \cosh y - \cosh x \sinh y$
   (b) $\cosh(x - y) = \cosh x \cosh y - \sinh x \sinh y$
   (c) $1 - \coth^2 x = -\operatorname{cosech}^2 x$

2. (a) $\sinh 2y = 2\sinh y \cosh y$
   (b) $\sinh 3y = 3\sinh y + 4\sinh^3 y$
   (c) $\cosh 3y = 4\cosh^3 y - 3\cosh y$

3. $\tanh(x + y) = \dfrac{\tanh x + \tanh y}{1 + \tanh x \tanh y}$

4. (a) $\sinh x - \sinh y = 2\cosh\dfrac{x + y}{2}\sinh\dfrac{x - y}{2}$

   (b) $\cosh x + \cosh y = 2\cosh\dfrac{x + y}{2}\cosh\dfrac{x - y}{2}$

Prove that :

5. $\sinh(x + y)\cosh(x - y) = \dfrac{1}{2}[\sinh 2x + \sinh 2y]$

6. $\cosh 2x + \cosh 5x + \cosh 8x + \cosh 11x$

   $= 4\cosh\dfrac{13x}{2}\cosh 3x \cosh\dfrac{3x}{2}$

7. $\sinh x + n\sinh 2x + \dfrac{n(n-1)}{1.2}\sinh 3x + \dots + \text{to } (n+1) \text{ terms}$

   $= 2^n \cosh^n\dfrac{x}{2}\sinh\left(\dfrac{n}{2} + 1\right)x$

8. $\sinh\beta\sin\alpha + i\cosh\beta\cos\alpha = i\cos(\alpha + i\beta)$

9. $\sin 2\alpha + i\sinh 2\beta = 2\sin(\alpha + i\beta)\cos(\alpha - i\beta)$

10. $\cos(\alpha + i\beta) + i\sin(\alpha + i\beta) = e^{-\beta}(\cos\alpha + i\sin\alpha)$

11. $\dfrac{1 + \tanh x}{1 - \tanh x} = \cosh 2x + \sinh 2x.$

12. $\cos(\alpha - i\beta) + i\sin(\alpha - i\beta) = e^{-\beta}(\cos\alpha - i\sin\alpha)$

13. If $\cosh\alpha = \sec\theta$, show that $\alpha = \log_e \tan(\pi/4 + \theta/2)$

14. Prove that

    (a) $\dfrac{1}{2}[\sinh x + \sin x] = x + \dfrac{x^5}{5!} + \dfrac{x^9}{9!} + \dots \text{ ad. inf.}$

    (b) $\dfrac{1}{2}[\cosh x + \cos x] = 1 + \dfrac{x^4}{4!} + \dfrac{x^8}{8!} + \dots$

15. If $\tan\theta = \tanh x \cot y$ and $\tan\phi = \tanh x \tan y$, prove that

    $\dfrac{\sin 2\theta}{\sin 2\phi} = \dfrac{\cosh 2x + \cos 2y}{\cosh 2x - \cos 2y}$

16. If n be a positive integer, show that

    $2^{n-1}\cosh^n\theta = \cosh\theta + {}^nC_1\cosh(n-2)\theta + {}^nC_2\cosh(n-4)\theta + \dots$

17. If $\cosh u = \sec\theta$, prove that $\sinh u = \tan\theta$.

18. Find $\tanh x$, if $5\sinh x - \cosh x = 5$ (MUMBAI–2004)

11. L.H.S. $= \dfrac{1+\tanh x}{1-\tanh x} = \dfrac{1+\sinh x/\cosh x}{1-\sinh x/\cosh x} = \dfrac{\cosh x+\sinh x}{\cosh x-\sinh x}$

$= \dfrac{e^x}{e^{-x}} = e^{2x} = \cosh 2x + \sinh 2x$

15. $\dfrac{\sin 2\theta}{\sin 2\phi} = \dfrac{2\tan\theta}{(1+\tan^2\theta)} \cdot \dfrac{(1+\tan^2\theta)}{2\tan\phi}$

16. $(2\cosh\theta)^n = (e^\theta + e^{-\theta})^n = e^{n\theta} + {}^nC_1 e^{(n-1)\theta} e^{-\theta}$

$+ {}^nC_2 e^{(n-2)\theta} e^{-2\theta} + \dots + {}^nC_{n-1} e^\theta \cdot e^{-(n-2)\theta} + e^{-2\theta}]$

$= (e^{n\theta} + e^{-n\theta}) + {}^nC_1(e^{(-2)\theta} + e^{-(-2)\theta} + \dots$

$= \cosh n\theta + {}^nC_1 \cosh(n-2)\theta + \dots$

18. $\dfrac{4}{5}$ or $-\dfrac{3}{5}$

11.9 SEPARATION INTO REAL AND IMAGINARY PARTS

(i) $\cos(\alpha \pm \beta) = \cos\alpha\cos i\beta \pm \sin\alpha\sin\beta i = \cos\alpha\cosh\beta \pm i\sin\alpha\sinh\beta$

(ii) $\cot(\alpha - \beta i) = \dfrac{\cos(\alpha - \beta i)}{\sin(\alpha - \beta i)} = \dfrac{\cos(\alpha - \beta i)}{\sin(\alpha - \beta i)} \times \dfrac{2\sin(\alpha + \beta i)}{2\sin(\alpha + \beta i)} = \dfrac{\sin 2\alpha + \sin 2\beta i}{\cos 2\beta i - \cos 2\alpha} = \dfrac{\sin 2\alpha + i\sinh 2\beta}{\cosh 2\beta - \cos 2\alpha} = \dfrac{\sin 2\alpha}{\cos 2\beta - \cos 2\alpha} + i\dfrac{\sinh 2\beta}{\cosh 2\beta - \cos 2\alpha}$

(iii) $\sinh(\alpha + \beta i) = \dfrac{\sin i(\alpha + \beta i)}{i} = i\left[\dfrac{\sin i\alpha\cos(-\beta) + \cos i\alpha\sin(-\beta)}{i^2}\right] = \sinh\alpha\cos\beta + i\cosh\alpha\sin\beta$

(iv) $\cosh(\alpha + i\beta) = \cos i(\alpha + i\beta) = \cos(i\alpha - \beta) = \cos i\alpha\cos\beta + \sinh i\alpha\sin\beta = \cosh\alpha\cos\beta + i\sinh\alpha\sin\beta$

(v) $\exp[\cos(x + iy)] = \exp[\cos x\cosh y - i\sin x\sinh y] = \exp[\cos x\cosh y].\exp[-i\sin x\sinh y]$

$= \exp[\cos x\cosh y].[\cos(\sin x\sinh y) - i\sin(\sin x\sinh y)] = A + iB$

where $A = \exp(\cos x\cosh y).[\cos(\sin x\sinh y)],\ B = -\exp(\cos x\cosh y).[\sin(\sin x\sinh y)]$

EXAMPLE 1. *Separate $e^{i\beta}/(1 + e^{i\alpha})$ into real and imaginary parts.*

SOLUTION. $\dfrac{e^{i\beta}}{1 + e^{i\alpha}} = \dfrac{e^{i\beta}(1 + e^{-i\alpha})}{(1 + e^{i\alpha})(1 + e^{-i\alpha})} = \dfrac{e^{i\beta} + e^{i(\beta - \alpha)}}{2 + (e^{i\alpha} + e^{-i\alpha})}$

$= \dfrac{\cos\beta + i\sin\beta + \cos(\beta - \alpha) + i\sin(\beta - \alpha)}{2 + 2\cos\alpha}$

$\therefore$ Real part $= \dfrac{\cos\beta + \cos(\beta - \alpha)}{2 + 2\cos\alpha}$

and Imaginary part $= \dfrac{\sin\beta + \sin(\beta - \alpha)}{2 + 2\cos\alpha}$

EXAMPLE 2. *Resolve $e^{\sin(x + iy)}$ into real and imaginary parts.*

SOLUTION. We have

$e^{\sin(x + iy)} = e^{\sin x\cos iy + \cos x\sin iy}$

$= e^{\sin x\cosh y + i\cos x\sinh y} = e^{\sin x\cosh y}.e^{i\cos x\sinh y}$

$= e^{\sin x\cosh y}[\cos(\cos x\sinh y) + i\sin(\cos x\sinh y)]$

Here real part $= e^{\sin x\cosh y}\cos(\cos x\sinh y)$

and imaginary part $= e^{\sin x\cosh y}\sin(\cos x\sinh y)$

EXAMPLE 3. *Resolve $\sin^2(x + iy)$ into real and imaginary parts.*

SOLUTION. We have

$\sin^2(x + iy) = \dfrac{1}{2}[1 - \cos 2(x + iy)]$

$= \dfrac{1}{2}[1 - \{\cos(2x)\cos(2iy) - \sin(2x)\sin(2iy)\}]$

$= \dfrac{1}{2}[1 - \{\cos 2x\cosh 2y - i\sin 2x\sinh 2y\}]$

$= \dfrac{1}{2}(1 - \cos 2x\cosh 2y) + \dfrac{1}{2}i(\sinh 2x\sinh 2y)$

$\therefore$ Real part $= \dfrac{1}{2}(1 - \cos 2x\cosh 2y)$

and Imaginary part $= \dfrac{1}{2}(\sinh 2x \cdot \sinh 2y)$

Exercise-11.8

Separate the following into real and imaginary parts.

1. $\sin(\alpha \pm i\beta)$

2. $\tan(\alpha \pm i\beta)$

3. $\sec(\alpha \pm i\beta)$

4. $\text{cosec}(\alpha \pm i\beta)$

5. $\cot(\alpha + i\beta)$

6. $\tanh(\alpha \pm i\beta)$

7. $\text{sech}(\alpha \pm i\beta)$

8. $\cos^2(x + iy)$

9. $e^{\sinh(x + iy)}$

10. $e^{\cosh(x \pm iy)}$

11. If $\left(\dfrac{x - a + iy}{x + a + iy}\right) = P + iQ$ find P and Q.

1. $\sin\alpha\cosh\beta \pm i\cos\alpha\sinh\beta$

2. $\dfrac{\sin 2\alpha}{\cos 2\alpha + \cosh 2\beta} \pm \dfrac{i\sinh 2\beta}{\cos 2\alpha + \cosh 2\beta}$

3. $\dfrac{2\cos\alpha\cosh\beta}{\cos 2\alpha + \cosh 2\beta} \pm \dfrac{2i\sin\alpha\sinh\beta}{\cos 2\alpha + \cosh 2\beta}$

4. $\dfrac{2\sin\alpha\cosh\beta \mp 2i\cos\alpha\sinh\beta}{\cosh 2\beta - \cos 2\alpha}$ **5.** $\dfrac{\sin 2\alpha \pm i\sinh 2\beta}{\cosh 2\beta - \cos 2\alpha}$ **6.** $\dfrac{\sin 2\alpha \pm i\sin 2\beta}{\cosh 2\alpha + \cos 2\beta}$ **7.** $\left[\dfrac{2\cosh\alpha\cos\beta \mp i\sinh\alpha\sin\beta}{\cosh 2\alpha + 2\cos 2\beta}\right]$

8. $\dfrac{1}{2}[(1 + \cos 2x\cosh 2y) - i\sin 2x\sinh 2y]$ **9.** $e^{\sinh x\cos y}\times[\cos(\cosh x\sin y) + i\sin(\cosh x\sin y)]$

10. $e^{\cosh x\cos y}\times[\cos(\sinh x\sin y) \pm i\sin(\sinh x\sin y)]$ **11.** $P = \dfrac{x^2 + y^2 - a^2}{(x+a)^2 + y^2}$ and $Q = \dfrac{2ay}{(x+a)^2 + y^2}$

More Solved Examples

EXAMPLE 1. If $\tan(\theta + \phi i) = \sin(x + iy)$, then $\coth y\sinh 2\phi = \cot x\sin 2\theta$.   (SVTU–2006)

SOLUTION. $\tan(\theta + \phi i) = \dfrac{\sin(\theta + \phi i)}{\cos(\theta + \phi i)}$

$= \dfrac{\sin(\theta + \phi i)}{\cos(\theta + \phi i)} \times \dfrac{2\cos(\theta - i\phi)}{2\cos(\theta - i\phi)}$

$= \dfrac{\sin 2\theta - \sin 2i\phi}{\cos 2\theta + \cosh 2\phi i}$

or $\tan(\theta + i\phi) = \dfrac{\sin 2\theta}{\cos 2\theta + \cosh 2\phi} + i\dfrac{\sinh 2\phi}{\cos 2\theta + \cosh 2\phi}$

and $\sin(x + iy) = \sin x\cos iy + \cos x\sin iy$
$\qquad\qquad\qquad = \sin x\cosh y + i\cos x\sinh y$

Equating real and imaginary parts,

$\sin x\cosh y = \dfrac{\sin 2\theta}{\cos 2\theta + \cosh 2\phi}$

$\cos x\sinh y = \dfrac{\sinh 2\phi}{\cos 2\theta + \cosh 2\phi}$

Dividing, $\tan x\coth y = \dfrac{\sin 2\theta}{\sinh 2\phi}$

or $\sinh 2\phi\coth y = \cot x\sin 2\theta$.

EXAMPLE 2. If $\sin(A + iB) = x + iy$, prove that

(i) $\dfrac{x^2}{\cosh^2 B} + \dfrac{y^2}{\sinh^2 B} = 1$

(ii) $\dfrac{x^2}{\sin^2 A} - \dfrac{y^2}{\cos^2 A} = 1$ (PTU–2010)

SOLUTION. $\sin(A + iB) = x + iy$

$\Rightarrow \quad \sin A\cos iB + \cos A\sin iB = x + iy$
$\Rightarrow \quad \sin A\cosh B + i\cos A\sinh B = x + iy$

Equating real and imaginary parts,

$\sin A\cosh B = x$ ...(1)
$\cos A\sinh B = y$ ...(2)

From (1) and (2),

$\sin A = \dfrac{x}{\cosh B}, \cos A = \dfrac{y}{\sinh B}$

Squaring and adding,

$\dfrac{x^2}{\cosh^2 B} + \dfrac{y^2}{\sinh^2 B} = \sin^2 A + \cos^2 A = 1$

To get second result, eliminate B between (1) and (2),

$\cosh B = \dfrac{x}{\sin A}, \sinh B = \dfrac{y}{\cos A}$

Squaring and subtracting,

$\dfrac{x^2}{\sin^2 A} - \dfrac{y^2}{\cos^2 A} = \cosh^2 B - \sinh^2 B = 1$

EXAMPLE 3. If $\tan(\theta + i\phi) = \cos\alpha + i\sin\alpha$ or $\tan^{-1}(e^{i\alpha}) = \theta + i\phi$ and $\alpha \neq \pi/2$ prove that

(i) $\theta = \dfrac{n\pi}{2} + \dfrac{\pi}{4}$ (SVTU–2007, ROHTAK–2005)

(ii) $2\phi = \log\tan\left(\dfrac{\pi}{4} + \dfrac{\alpha}{2}\right)$

(AGRA–1998, KANPUR–1994, KUMAON–1991, MPPSC–1979, 82, UPPCS–1984)

SOLUTION. (i) Given that

$\tan(\theta + i\phi) = \cos\alpha + i\sin\alpha$

$\therefore \tan(\theta - i\phi) = \cos\alpha - i\sin\alpha$

$\tan 2\theta = \tan\{(\theta + i\phi) + (\theta - i\phi)\}$

$= \dfrac{\tan(\theta + i\phi) + \tan(\theta - i\phi)}{1 - \tan(\theta + i\phi)\tan(\theta - i\phi)}$

$= \dfrac{\cos\alpha + i\sin\alpha + \cos\alpha - i\sin\alpha}{1 - (\cos^2\alpha + \sin^2\alpha)}$

$= \dfrac{(2\cos\alpha)}{0} = \infty = \tan\dfrac{\pi}{2}$ $\left[\because \alpha \neq \dfrac{\pi}{2}\right]$

$\therefore \quad 2\theta = n\pi + \dfrac{\pi}{2}$ (general value)

or $\qquad \theta = \dfrac{n\pi}{2} + \dfrac{\pi}{4}$.

(ii) $\tan(2i\phi) = \tan[(\theta + i\phi) - (\theta - i\phi)]$

$= \dfrac{\tan(\theta + i\phi) - \tan(\theta - i\phi)}{1 + \tan(\theta + i\phi)\tan(\theta - i\phi)}$

$= \dfrac{2i\sin\alpha}{1 + 1} = i\sin\alpha$

$\therefore \quad \tanh 2\phi = \sin\alpha$

or $\dfrac{e^{2\phi} - e^{-2\phi}}{e^{2\phi} + e^{-2\phi}} = \dfrac{\sin\alpha}{1}$

or $\dfrac{2e^{2\phi}}{-2e^{-2\phi}} = \dfrac{\sin\alpha + 1}{\sin\alpha - 1}$ (By componendo an dividendo)

or $e^{4\phi} = \dfrac{1 + \sin\alpha}{1 - \sin\alpha} = \dfrac{1 - \cos\{(\pi/2 + \alpha\}}{1 + \cos\{(\pi/2 + \alpha\}}$

$= \dfrac{2\sin^2\left(\dfrac{\pi}{4} + \dfrac{\alpha}{2}\right)}{2\cos^2\left(\dfrac{\pi}{4} + \dfrac{\alpha}{2}\right)} = \tan^2\left(\dfrac{\pi}{4} + \dfrac{\alpha}{2}\right)$

$\therefore \quad e^{2\phi} = \tan\left(\dfrac{\pi}{4} + \dfrac{\alpha}{2}\right)$

$\therefore \quad 2\phi = \log\tan\left(\dfrac{\pi}{4} + \dfrac{\alpha}{2}\right)$

EXAMPLE 4. If $\sin(x + iy)\sin(\theta + i\phi) = 1$, then prove that

$\tanh^2 y\cosh^2\phi = \cos^2\theta$

SOLUTION. From given relation, we have

$$\sin(x+iy) = \frac{1}{\sin(\theta+i\phi)} = \frac{\sin(\theta-i\phi)}{\sin(\theta+i\phi)\sin(\theta-i\phi)}$$

or $\sin x \cosh y + i \cosh x \sinh y$

$$= \frac{2[\sin\theta\cosh\phi - i\cos\theta\sinh\theta]}{\cos 2\theta - \cosh 2\phi} \qquad \ldots(1)$$

Squaring imaginary part of (1), we have

$$\cos x \sinh y = -\frac{2\cos\theta\sinh\phi}{\cos 2\theta - \cosh 2\phi} \qquad \ldots(2)$$

Also $\cos(x+iy)$

$$= \sqrt{1-\sin^2(x+iy)} = \sqrt{\left(1 - \frac{1}{\sin^2(\theta+i\phi)}\right)}$$

$$= \pm i \frac{\cos(\theta+i\phi).2\sin(\theta-i\phi)}{\sin(\theta+i\phi).2\sin(\theta-i\phi)}$$

or $\cos x \cosh y - i \sin x \sinh y = \pm i \frac{(\sin 2\theta - i\sinh 2\phi)}{\cos 2\theta - \cosh 2\phi}$

$$\qquad \ldots(3)$$

Separating real part of (3), we have

$$\cos x \cosh y = \pm \frac{\sinh 2\phi}{\cos 2\theta - \cosh 2\phi} \qquad \ldots(4)$$

Dividing (2) by (4), we have

$$\tanh y = \pm \frac{2\cos\theta\sinh\phi}{\sinh 2\phi} = \pm\frac{2\cos\theta\sinh\phi}{2\sinh\phi\cosh\phi} = \pm\frac{\cos\theta}{\cosh\phi}$$

or $\tanh^2 y \cosh^2\phi = \cos^2\theta$

Given relation can also be written as

$$\sin(\theta+i\phi) = \frac{1}{\sin(x+iy)} \qquad \ldots(5)$$

EXAMPLE 5. **If $\cos(\theta+i\phi) = r(\cos\alpha + i\sin\alpha)$, prove that**

$$\phi = \frac{1}{2}\log\frac{\sin(\theta-\alpha)}{\sin(\theta+\alpha)} \qquad \text{(VTU–2006)}$$

SOLUTION. We have, $\cos(\theta+i\phi) = r(\cos\alpha + i\sin\alpha)$

or $\cos\theta\cosh\phi - i\sin\theta\sinh\phi = r(\cos\alpha + i\sin\alpha)$

Equating real and imaginary parts, we have

$$\cos\theta\cosh\phi = r\cos\alpha \qquad \ldots(1)$$
$$\sin\theta\sinh\phi = -r\sin\alpha \qquad \ldots(2)$$

We have from (1) and (2)

$$\frac{\cos\theta\cosh\phi}{\sin\theta\sinh\phi} = \frac{\cos\alpha}{-\sin\alpha}$$

$$\frac{\cosh\phi}{\sinh\phi} = \frac{\sin\theta\cos\alpha}{-\cos\theta\sin\alpha}$$

Applying componendo and dividendo, we have

$$\frac{\cosh\phi + \sinh\phi}{\cosh\phi - \sinh\phi} = \frac{\sin\theta\cos\alpha - \cos\theta\sin\alpha}{\sin\theta\cos\alpha + \cos\theta\sin\alpha}$$

or $\dfrac{\frac{1}{2}(e^\phi + e^{-\phi}) + \frac{1}{2}(e^\phi - e^{-\phi})}{\frac{1}{2}(e^\phi + e^{-\phi}) - \frac{1}{2}(e^\phi - e^{-\phi})} = \dfrac{\sin(\theta-\alpha)}{\sin(\theta+\alpha)}$

or $e^{2\phi} = \dfrac{\sin(\theta-\alpha)}{\sin(\theta+\alpha)} \Rightarrow 2\phi = \log\left[\dfrac{\sin(\theta-\alpha)}{\sin(\theta+\alpha)}\right]$

or $\phi = \dfrac{1}{2}\log\left[\dfrac{\sin(\theta-\alpha)}{\sin(\theta+\alpha)}\right]$.

Exercise-11.9

1. If $\sin(\theta+i\phi) = \cos\alpha + i\sin\alpha$, prove that
  (i) $\cos^2\theta = \pm\sin\alpha$ (ii) $\sinh^2\phi = \pm\sin\alpha$

2. If $\cosh u = \sec\theta$, show that $u = \log\tan\left(\frac{\pi}{4}+\frac{\theta}{2}\right)$.

3. If $\cos(\theta+\phi i) = r[\cos\alpha + i\sin\alpha]$, prove that
  (a) $\tan\theta\tanh\phi + \tan\alpha = 0$ (b) $\frac{1}{2}[\cos 2\theta + \cosh 2\phi] = R^2$

4. If $\tanh(u+iv) = \sin(x+iy)$, then prove that $\sinh 2u$ cosec $2v = \tan x \coth y$.

5. If $\tan(\theta+\phi i) = \tan\alpha + i\sec\alpha$, prove that $e^{2\phi} = \pm\cot\frac{\alpha}{2}$ and

  that $2\theta = n\pi + \frac{\pi}{2} + \alpha$ (NAGPUR–2009, SVTU–2008)

6. If α, β be the imaginary cube roots of unity, prove that
  $$\alpha e^{\alpha x} + \beta e^{\beta x} = -e^{-x/2}\left[\sqrt{3}\sin\frac{\sqrt{3}}{2}x + \cos\frac{\sqrt{3}}{2}x\right].$$

7. If $\cos(x+iy) = \cos\alpha + i\sin\alpha$, prove that
  $\cosh 2y + \cos 2x = 2$.

8. If $\cos(\theta+i\phi)\cos(\alpha+i\beta) = 1$, prove that
  $\tanh^2\phi \cosh^2\beta = \sin^2\alpha$,
  $\tanh^2\beta \cosh^2\phi = \sin^2\theta$.

9. If $A+iB = C\tan(x+iy)$, then show that
  $\tan 2x = \dfrac{2CA}{C^2 - A^2 - B^2}$.

10. If $x = 2\cos\alpha\cosh\beta$ and $y = 2\sin\alpha.\sinh\beta$, then prove that
  (i) $\sec(\alpha+i\beta) - \sec(\alpha-i\beta) = \dfrac{4iy}{x^2+y^2}$

  (ii) $\sec(\alpha+i\beta) + \sec(\alpha-i\beta) = \dfrac{4x}{x^2+y^2}$.

11. If $\cos(x+iy) = \cos\alpha + i\sin\alpha$ prove that
  (a) $\cosh 2y + \cos 2x = 2$ (b) $\sin^4 x = \sin^2\alpha$
  (c) $\sinh^4 y = \sin^2\alpha$

12. If $\cosh u = \sec\theta$, prove that $u = \log\tan\left(\frac{\pi}{4}+\frac{\theta}{2}\right)$
  and $\tanh^2\frac{u}{2} = \tan^2\frac{\theta}{2}$.

13. If $x = \cos\alpha\cosh\beta$ and $y = \sin\alpha\sinh\beta$, prove that
  $(x^2+y^2)[\cos(\alpha+i\beta) + \cos(\alpha-i\beta)]$
    $= \cos\alpha\cosh\beta(\cos 2\alpha + \cosh 2\beta)$

14. If $\tan(x+iy) = \cosh(\theta+i\phi)$, prove that
  $\tanh\theta\tan\phi = $ cosec $2x\sinh 2y$.

15. If $\cosh(u+iv) = x+iy$, prove that
  $$\frac{x^2}{\cosh^2 u} + \frac{y^2}{\sinh^2 u} = 1 \qquad \text{(PTU–2009S)}$$
  $$\frac{x^2}{\cos^2 v} - \frac{y^2}{\sin^2 v} = 1 \qquad \text{(MADRAS–2000)}$$

16. If cosec $\left(\frac{\pi}{4}+ix\right) = u+iv$, prove that
  $(u^2+v^2) = 2(u^2-v^2)$ (MUMBAI–2009)

17. If $\cos(\alpha+i\beta) = r(\cos\theta + i\sin\theta)$, prove that
  $$e^{2\beta} = \frac{\sin(\alpha-\theta)}{\sin(\alpha+\theta)}. \qquad \text{(KURUKSHETRA–2005, MADRAS–2003)}$$

Hints to Selected Problems

1. Using the result
$$\sin(\theta + i\phi) = \cos\alpha + i\sin\alpha$$
$$\Rightarrow \quad \cos\theta\cos i\phi + \cos\theta\sin i\phi = \cos\alpha + i\sin\alpha$$
$$\Rightarrow \quad \cos\theta\cosh\theta + i\cos\theta\sinh\theta = \cos\alpha + i\sin\alpha$$
Equating real and imaginary parts of both the sides, we get
$$\sin\theta\cosh\phi = \cos\alpha$$
$$\cos\theta\sinh\phi = \sin\alpha$$

2. $\cosh u = \dfrac{1}{\cos\theta} \quad \Rightarrow \quad \dfrac{1+\tan^2 u/2}{1-\tan^2 u/2} = \dfrac{1+\tan^2\theta/2}{1-\tan^2\theta/2}$

Now using componendo and dividendo.

5. $\tan(\theta + i\phi) = \tan\alpha + i\sec\alpha$
$$\Rightarrow \quad \tan(\theta - i\phi) = \tan\alpha - i\sec\alpha$$
Therefore $\tan 2\theta = \tan[(\theta + i\phi) + (q - i\phi)]$. Now using the formula of $\tan(A + B)$.

8. As per given
$$\cos(\theta + i\phi)\cos(\alpha + i\beta) = 1$$

$\therefore \quad \cos(\theta + i\phi) = \dfrac{1}{\cos(\alpha + i\beta)} = \dfrac{2\cos(\alpha - i\beta)}{2\cos(\alpha + i\beta)\cos(\alpha - i\beta)}$

or $\quad \cos\theta\cosh\phi - i\sin\theta\sinh\phi$
$$= \dfrac{2\cos\alpha\cosh\beta + 2i\sin\alpha\sinh\beta}{\cos 2\alpha + \cosh 2\beta}$$

9. As per given $A + iB = C\tan(x + iy)$

or $\quad \tan(x + iy)\dfrac{A+iB}{C} \Rightarrow \tan(x - iy) = \dfrac{A-iB}{C}$

or $\quad \tan 2x = \tan(x + iy + x - iy)$.

13. $x^2 + y^2 = \cos^2\alpha\cosh^2\beta + \sin^2\alpha\sinh^2\beta$
$$\Rightarrow \quad 4(x^2 + y^2) = (1 + \cos 2\alpha)(1 + \cosh 2\beta)$$
$$+ (1 - \cos 2\alpha)(\cosh 2\beta - 1)$$
$$\Rightarrow \quad (x^2 + y^2) = \dfrac{1}{2}(\cos 2\alpha + \cosh 2\beta)$$
also $\cos(\alpha - i\beta) = 2\cos\alpha\cosh\beta$.

11.10 INVERSE CIRCULAR FUNCTION OF COMPLEX QUANTITIES

(i) If $\cos z = \cos(x + iy) = u + iv$, then $x + iy$ is called the inverse cosine of $u + iv$ and written as $\cos^{-1}(u + iv) = x + iy$ since $\cos(x + iy) = \cos[2n\pi \pm(x + iy)]$ where n is any integer, $2n\pi \pm (x + iy)$ is also inverse cosine of $u + iv$ and this shows the many valued nature of the inverse cosine. These values are called general values of $\cos^{-1}(u + iv)$ and are denoted as

$$\cos^{-1}(u + iv) = 2n\pi \pm (x + iy) = 2n\pi \pm \cos^{-1}(u + iv)$$

i.e., to denote general values first letter is written in capital.

The principal value of $\cos^{-1}(u + iy)$ that value of which is its real part lies between 0 and π.

(ii) Similarly if $\sin(x + iy) = u + iv$, then general value of inverse sine of $u + iv$ is denoted as

$$\sin^{-1}(u + iv) = n\pi + (-1)^n(x + iy) = (n\pi) + (-1)^n\sin^{-1}(u + iv)$$

And the principal value is that for which its real part lies between $-\dfrac{1}{2}\pi$ and $\dfrac{1}{2}\pi$.

(iii) Again if $\tan(x + iy) = u + iv$, then general values of inverse tangent of $u + iv$ is denoted as

$$\tan^{-1}(u + iv) = n\pi + (x + iy) = n\pi + \tan^{-1}(u + iv)$$

and the principal value is that for which its real part lies between $-\dfrac{1}{2}\pi$ and $\dfrac{1}{2}\pi$.

☞ **REMARK**
- The inverse cosine, inverse sine, etc. with first letter capital always denote general value and with small letters principal value.

EXAMPLE 1. *Express* $\tan^{-1}(\alpha + i\beta)$ *in the form* $(A + iB)$

SOLUTION. Suppose $\tan^{-1}(\alpha + i\beta) = A + iB$ ...(1)
$$\tan^{-1}(\alpha - i\beta) = A - iB \quad ...(2)$$

Adding $\tan^{-1}(\alpha + \beta i) + \tan^{-1}(\alpha - \beta i) = 2A$

or $\quad \tan^{-1}\left[\dfrac{2\alpha}{1 - (\alpha^2 + \beta^2)}\right] = 2A$

$\therefore \quad A = \dfrac{1}{2}\tan^{-1}\left[\dfrac{2\alpha}{1 - (\alpha^2 + \beta^2)}\right]$

Subtracting (2) from (1),
$$\tan^{-1}(\alpha + \beta i) - \tan^{-1}(\alpha - \beta i) = 2iB$$

$\therefore \quad \tan^{-1}\dfrac{2\beta i}{1 + \alpha^2 + \beta^2} = 2iB$

or $\quad \dfrac{2\beta i}{1 + \alpha^2 + \beta^2} = i\tanh 2B$

$\therefore \quad B = \dfrac{1}{2}\tanh^{-1}\left(\dfrac{2\beta}{1 + \alpha^2 + \beta^2}\right)$

$\therefore \quad \tan^{-1}(\alpha + i\beta) = \dfrac{1}{2}\tan^{-1}\dfrac{2\alpha}{1 - \alpha^2 - \beta^2}$
$$+ i.\dfrac{1}{2}\tanh^{-1}\left(\dfrac{2\beta}{1 + \alpha^2 + \beta^2}\right)$$

General value of $\tan^{-1}(\alpha + i\beta)$ is denoted by $\tan^{-1}(\alpha + i\beta)$ and is equal to

$$n\pi + \dfrac{1}{2}\tan^{-1}\dfrac{2\alpha}{1 - \alpha^2 - \beta^2} + \dfrac{1}{2}i\tanh^{-1}\dfrac{2\beta}{1 + \alpha^2 + \beta^2}.$$

11.10.1 INVERSE HYPERBOLIC FUNCTIONS

Let $w = u + iv$ and $z = x + iy$.

1. w is called inverse hyperbolic sine of z and is denoted as $\sinh^{-1} z$ if z and w are connected by the relation

$$\sinh w = z \qquad \qquad \ldots(a)$$

and $\qquad \cosh w = \sqrt{z^2 + 1} \qquad \qquad \ldots(b)$

Adding (a) and (b),

$$\sinh w + \cosh w = z + \sqrt{z^2 + 1}$$

or $\quad \dfrac{e^w - e^{-w}}{2} + \dfrac{e^w + e^{-w}}{2} = z + \sqrt{z^2 + 1} \qquad$ or $\qquad e^w = z + \sqrt{z^2 + 1}$

or $\qquad w = \log[z + \sqrt{z^2 + 1}] \qquad$ or $\qquad \sinh^{-1} z = \log[z + \sqrt{z^2 + 1}]$

2. w is called inverse hyperbolic cosine of z and is denoted as $\cosh^{-1} z$ if z and w are connected by the relation

$$\cosh w = z \qquad \qquad \text{and} \qquad \sinh w = \sqrt{z^2 - 1}$$

By addition,

$$\cosh w + \sinh w = z + (z^2 - 1)^{1/2}$$

or $\quad \dfrac{e^w + e^{-w}}{2} + \dfrac{e^w - e^{-w}}{2} = z + \sqrt{z^2 - 1} \qquad$ or $\qquad e^w = z + \sqrt{z^2 - 1}$

or $\qquad w = \log[z + \sqrt{z^2 - 1}] \qquad$ or $\qquad \cosh^{-1} z = \log[z + \sqrt{z^2 - 1}]$

3. $\qquad \tanh w = z \qquad$ or $\qquad \dfrac{\sinh w}{\cosh w} = z \qquad$ or $\qquad \dfrac{e^w - e^{-w}}{e^w + e^{-w}} = z$

Applying componendo and dividendo, we get

$$\dfrac{2e^w}{2e^{-w}} = \dfrac{1 + z}{1 - z} \qquad \text{or} \qquad e^{2w} = \dfrac{1 + z}{1 - z} \qquad \text{or} \qquad w = \tanh^{-1} z = \dfrac{1}{2} \log \dfrac{1 + z}{1 - z}$$

🖉 REMARK

- Since $\sinh[n\pi i + (-1)^n w] = \sinh w = z$, , $\cosh[2n\pi i \pm w] = \cosh w = z$ and $\tanh[n\pi i + w] = \tanh w = z$.

It follows that general values of inverse hyperbolic sine, inverse hyperbolic cosine, inverse hyperbolic tangent are given by

$$n\pi i + (-1)^n \log[z + \sqrt{1 + z^2}] ; \quad 2n\pi i \pm \log[z + \sqrt{z^2 - 1}] ; \text{ and } n\pi i + \dfrac{1}{2} \log \dfrac{1 + z}{1 - z} \text{ respectively.}$$

The principal values of these functions are those values for which their imaginary parts lie between $-\dfrac{i\pi}{2}$ and $\dfrac{i\pi}{2}$ in case of $\sinh^{-1} z$, 0 and $i\pi$ in case of $\cosh^{-1} z$, $i\dfrac{\pi}{2}, -i\dfrac{\pi}{2}$ in case of $\tanh^{-1} z$.

EXAMPLE 1. *Prove that* $\tanh^{-1} x = \sinh^{-1} \dfrac{x}{\sqrt{(1 - x^2)}}$

SOLUTION. Let $\quad \tanh^{-1} x = y$

$\therefore \qquad x = \tanh y$

Now $\sinh^{-1} \dfrac{x}{\sqrt{(1 - x^2)}} = \sinh^{-1} \left[\dfrac{\tanh y}{\sqrt{(1 - \tanh^2 y)}} \right]$

By putting the values of

$$x = \sinh^{-1} \left[\dfrac{\sinh y}{\cosh y \operatorname{sech} y} \right]$$

$$= \sinh^{-1}[\sinh y] = y = \tanh^{-1} x .$$

11.10.2 RELATION BETWEEN INVERSE CIRCULAR AND HYPERBOLIC FUNCTIONS

1. If $\qquad x = \sinh y$, then $ix = i \sinh y = \sin(iy)$

$\therefore \qquad iy = \sin^{-1}(ix) \qquad$ or $\qquad y = \dfrac{1}{i} \sin^{-1}(ix) = -i \sin^{-1}(ix) \qquad$ or $\qquad y = \sinh^{-1} x = -i \sin^{-1}(ix) .$

2. If $\qquad x = \cosh y$, then $x = \cos (iy)$

or $\qquad y = \dfrac{1}{i} \cos^{-1}(x) = -i \cos^{-1}(x) \qquad$ or $\qquad y = \cosh^{-1}(x) = -i \cos^{-1}(x) .$

3. If $\qquad x = \tanh y$, then

$ix = i \tanh y = \tan (iy) \qquad$ or $\qquad iy = \tan^{-1}(ix)$

$y = -i \sin^{-1}(ix) \qquad$ or $\qquad y = \tanh^{-1}(x) = -i \tan^{-1}(ix) .$

Solved Examples

EXAMPLE 1. *Show that* $\sinh^{-1} x = \tanh^{-1} \dfrac{x}{\sqrt{(1+x^2)}}$.

SOLUTION. Let $\sinh^{-1} x = y$; then $x = \sinh y$

$\therefore$ R.H.S. $= \tanh^{-1} \dfrac{\sinh y}{\sqrt{(1+\sinh^2 y)}}$

$= \tanh^{-1} \dfrac{\sinh y}{\cosh y} = \tanh^{-1}(\tanh y) = y = \sinh^{-1} x$

EXAMPLE 2. *Show that* $\sin^{-1}(ix) = n\pi + i(-1)^n \log[x + \sqrt{1+x^2}]$

SOLUTION. Let $\sin^{-1}(ix) = n\pi + (-1)^n \sin^{-1}(ix)$

$\sin^{-1}(ix) = z$

$ix = \sin z$

$\therefore \quad \cos z = \sqrt{1 - \sin^2 z} = \sqrt{1 + x^2}$

$\therefore \quad \cos z + i \sin z = \sqrt{1 + x^2} - x$

or $\quad e^{iz} = \sqrt{1 + x^2} - x$

$z = -i \log[\sqrt{1+x^2} - x]$

$= i \log[\sqrt{1+x^2} - x]^{-1} = i \log[\sqrt{1+x^2} + x]$

$\therefore \sin^{-1}(ix) = n\pi + (-1)^n i \log[\sqrt{1+x^2} + x]$

EXAMPLE 3. *Prove that* $\tan^{-1}(\cos\theta + i\sin\theta)$

$= \dfrac{n\pi}{2} + \dfrac{\pi}{4} + \dfrac{i}{2} \log\tan\left(\dfrac{\pi}{4} + \dfrac{\theta}{2}\right)$

$= \dfrac{n\pi}{2} + \dfrac{\pi}{4} - \dfrac{i}{2} \log\tan\left(\dfrac{\pi}{4} - \dfrac{\theta}{2}\right).$ (MUMBAI–2009)

SOLUTION. Let $\tan^{-1}(\cos\theta + i\sin\theta) = x + iy$

then $\cos\theta + i\sin\theta = \tan(x + iy)$ ...(1)

and $\cos\theta - i\sin\theta = \tan(x - iy)$ ...(2)

Now, $\tan 2x = \tan[(x + iy) + (x - iy)]$

$= \dfrac{\cos\theta + i\sin\theta + \cos\theta - i\sin\theta}{1 - [(\cos\theta + i\sin\theta)(\cos\theta - i\sin\theta)]}$

$= \dfrac{2\cos\theta}{0} = \infty$

$\therefore \quad 2x = n\pi + \dfrac{\pi}{2}, x = \dfrac{n\pi}{2} + \dfrac{\pi}{4}.$

Again, $\tan 2iy = \tan[x + iy - (x - iy)]$

$= \dfrac{\cos\theta + i\sin\theta - \cos\theta + i\sin\theta}{1 + [(\cos\theta + i\sin\theta)(\cos\theta - i\sin\theta)]}$

$i\tanh 2y = \dfrac{2i\sin\theta}{2}$

$i \dfrac{e^{2y} - e^{-2y}}{e^{2y} + e^{-2y}} = i\sin\theta$

or $\dfrac{e^{2y} + e^{-2y}}{e^{2y} - e^{-2y}} = \dfrac{1}{\sin\theta}$.(By componendo and divideno)

$\dfrac{e^{2y}}{e^{-2y}} = \dfrac{1 + \sin\theta}{1 - \sin\theta}$

or $e^{4y} = \dfrac{1 - \cos\left(\dfrac{\pi}{2} + \theta\right)}{1 + \cos\left(\dfrac{\pi}{2} + \theta\right)} = \tan^2\left(\dfrac{\pi}{4} + \dfrac{\theta}{2}\right).$

$\therefore \quad e^{2y} = +\tan\left(\dfrac{\pi}{4} + \dfrac{\theta}{2}\right)$ (Taking +ive sign)

$\therefore \quad y = \dfrac{1}{2} \log\tan\left(\dfrac{\pi}{4} + \dfrac{\theta}{2}\right)$

$\therefore \tan^{-1} e^{(i\theta)} = \dfrac{n\pi}{2} + \dfrac{\pi}{4} + \dfrac{i}{2} \log\tan\left(\dfrac{\pi}{4} + \dfrac{\theta}{2}\right)$ ...(1)

Since $\log\tan\left(\dfrac{\pi}{4} + \dfrac{\theta}{2}\right) = \log\left[\tan\left(\dfrac{\pi}{4} + \dfrac{\theta}{2}\right)\right]^{-1}$

$= -\log \dfrac{1}{\tan\left(\dfrac{\pi}{4} + \dfrac{\theta}{2}\right)}$

$= -\log \dfrac{\tan\left(\dfrac{\pi}{4} - \dfrac{\theta}{2}\right)}{\tan\left(\dfrac{\pi}{4} + \dfrac{\theta}{2}\right)\tan\left(\dfrac{\pi}{4} - \dfrac{\theta}{2}\right)}$

$= -\log\tan\left(\dfrac{\pi}{4} - \theta\right)$

Hence $\tan^{-1}(e^{i\theta}) = \dfrac{n\pi}{2} + \dfrac{\pi}{4} - \dfrac{i}{2} \log\tan\left(\dfrac{\pi}{4} - \dfrac{\theta}{2}\right).$

EXAMPLE 4. *Show that* $\tan^{-1}\left(i\dfrac{x-a}{x+a}\right) = -\dfrac{1}{2} i \log\dfrac{a}{x}$

SOLUTION. Let $\tan^{-1}\left(i\dfrac{x-a}{x+a}\right) = z$; then $i\dfrac{x-a}{x+a} = \tan z$

or $\quad -\dfrac{x-a}{x+a} = i\tan z$

or $\quad \dfrac{a-x}{a+x} = \dfrac{e^{iz} - e^{-iz}}{e^{iz} + e^{-iz}}$

By componendo and dividendo, we have

$\dfrac{2a}{-2x} = \dfrac{2e^{iz}}{-2e^{-iz}}$ or $\dfrac{a}{x} = e^{2iz}$

$\therefore \quad 2iz = \log\dfrac{a}{x}$ or $z = \dfrac{1}{2i}\log\dfrac{a}{x} = -\dfrac{1}{2} i \log\dfrac{a}{x}$.

EXAMPLE 5. *Prove that* $\sin^{-1}(\cos\theta + i\sin\theta) = \cos^{-1}\sqrt{\sin\theta}$

$+ i\log\{\sqrt{\sin\theta} + \sqrt{1 + \sin\theta}\}.$

SOLUTION. Let $\sin^{-1}(\cos\theta + i\sin\theta) = x + iy$

$\cos\theta + i\sin\theta = \sin(x + iy)$

$= \sin x \cosh y + i\cos x \sinh y$

Equating the real and imaginary parts, we get

$\cos\theta = \sin x \cosh y$ ...(1)

$\sin\theta = \cos x \sinh y$ ...(2)

Squaring and adding, we get

$1 = \sin^2 x \cosh^2 y + \cos^2 x \sinh^2 y$

or $\quad 1 = \sin^2 x (1 + \sinh^2 y) + \cos^2 x \sinh^2 y$

or $\quad \cos^2 x = \sinh^2 y$

$\therefore \quad \sinh y = \cos x$

Hence from (2), we have $\cos^2 x = \sin\theta$

[Since θ is acute and positive.]

or $\quad \cos x = \pm\sqrt{\sin\theta}$

Since x being real part of $\sin^{-1}(\cos\theta + i\sin\theta)$ lies between $-\pi/2$ and $\pi/2$.

$\therefore \quad x = \cos^{-1}\sqrt{\sin\theta}$ ...(3)

From (2) $\sin\theta = \sqrt{\sin\theta}\, \sinh y$ or $e^y - e^{-y} = 2\sqrt{\sin\theta}$.

$\therefore e^{2y} - 2\sqrt{\sin\theta}\, e^y - 1 = 0.$

Solving $e^y = \sqrt{\sin\theta} \pm \sqrt{1+\sin\theta}$

$\therefore \qquad y = \log[\sqrt{\sin\theta} + \sqrt{1+\sin\theta}]$

The negative sign is neglected since e^y cannot be negative, y being real. Hence

$$\sin^{-1}(\cos\theta + i\sin\theta) = \cos^{-1}\sqrt{\sin\theta} + i\log[\sqrt{\sin\theta} + \sqrt{1+\sin\theta}].$$

EXAMPLE 6. *Prove that* $\tan^{-1}(\sinh\theta) = -i\log\tan\left(\dfrac{\pi}{4} + i\dfrac{\theta}{2}\right)$.

SOLUTION. Let $\tan^{-1}(\sinh\theta) = z$

$\therefore \qquad \sinh\theta = \tan z \quad$ or $\quad -i\sin i\theta = \dfrac{e^{iz} - e^{-iz}}{i(e^{iz} + e^{-iz})}$

or $\qquad \dfrac{e^{iz} - e^{-iz}}{e^{iz} + e^{-iz}} = \sin i\theta$

By componendo and dividendo,

$\dfrac{e^{iz}}{e^{-iz}} = \dfrac{1+\sin i\theta}{1-\sin i\theta}$

or $\qquad e^{2iz} = \dfrac{1 - \cos\left(\dfrac{\pi}{2} + i\theta\right)}{1 + \cos\left(\dfrac{\pi}{2} + i\theta\right)} = \tan^2\left(\dfrac{\pi}{4} + \dfrac{i\theta}{2}\right)$

$\therefore \qquad e^{iz} = \tan\left(\dfrac{\pi}{4} + \dfrac{i\theta}{2}\right)$

or $\qquad z = \dfrac{1}{i}\log\tan\left(\dfrac{\pi}{4} + \dfrac{i\theta}{2}\right) = -i\log\tan\left(\dfrac{\pi}{4} + \dfrac{i\theta}{2}\right)$.

EXAMPLE 7. *If A, B, C are the angles of triangle, prove that*

$$\begin{vmatrix} e^{2iA} & e^{-iC} & e^{-iB} \\ e^{-iC} & e^{2iB} & e^{-iA} \\ e^{-iB} & e^{-iA} & e^{2iC} \end{vmatrix}$$

is purely real, and compute its simplest value.

SOLUTION. Given $A + B + C = \pi$, now

$$\begin{vmatrix} e^{2iA} & e^{-iC} & e^{-iB} \\ e^{-iC} & e^{2iB} & e^{-iA} \\ e^{-iB} & e^{-iA} & e^{2iC} \end{vmatrix}$$

$= e^{2iA}[e^{2i(B+C)} - e^{-2iA}] + e^{-iC}$
$\qquad [e^{-i(A+B)} - e^{iC}] + e^{-iB}[e^{-i(A+C)} - e^{iB}]$

$= e^{2iA}[e^{2i\pi - 2iA} - e^{-2iA}] + e^{-iC}[e^{-i\pi + iC} - e^{iC}]$
$\qquad\qquad\qquad + e^{-iB}[e^{-i\pi + iB} - e^{iB}]$

$= e^{2\pi i} - 1 + e^{-\pi i} - 1 + e^{-\pi i} - 1$

$= (\cos 2\pi + i\sin 2\pi) + 2[\cos\pi - i\sin\pi] - 3$

$= 1 - 2 - 3 = -4.$

Exercise-11.10

1. Separate into their real and imaginary parts $\tanh^{-1}(x + iy)$
(SVTU–2009)

2. Express $\cos^{-1}(x + iy)$ and $\sin^{-1}(x + iy)$ in the form of $A + iB$.

3. (a) Put $\sin^{-1}(\operatorname{cosec}\theta)$ in the form $A + iB$.
(b) Show that $\sin^{-1}(2) = n\pi + (-1)^n(\pi/2) + i(-1)^n\log(2 + \sqrt{3})$.

4. Prove that $\log\tan\left[\dfrac{\pi}{4} + \dfrac{ix}{2}\right] = i\tan^{-1}(\sinh x)$.

5. Prove that $\tan^{-1}(\cot\theta\tanh\phi) = \dfrac{1}{2i}\log\dfrac{\sin(\theta + \phi i)}{\sin(\theta - \phi i)}$.

6. Prove that $\tan^{-1}\dfrac{\tan 2\theta + \tanh 2\phi}{\tan 2\theta - \tanh 2\phi} + \tan^{-1}\dfrac{\tan\theta - \tanh\phi}{\tan\theta + \tanh\phi}$
$\qquad = \tan^{-1}(\cot\theta\coth\phi)$.

7. If $\sin^{-1}(u + iv) = \alpha + i\beta$, then show that $\sin^2\alpha$ and $\cosh^2\beta$ are roots of $x^2 - x(1 + u^2 + v^2) + u^2 = 0$.

8. Show that $\sin^{-1}(i) = 2n\pi - i\log(\sqrt{2} - 1)$.

9. If $\cos^{-1}(x + iy) = \theta + i\phi$ prove that

(i) $x^2\sec^2\theta - y^2\operatorname{cosec}^2\theta = 1$

(ii) $x^2\operatorname{sech}^2\phi + y^2\operatorname{cosech}^2\phi = 1$

10. If $\sin^{-1}(x + iy) = \tan^{-1}(u + iv)$ show that
$$[(x-1)^2 + y^2][(x+1)^2 + y^2] = \left(\dfrac{x^2 + y^2}{u^2 + v^2}\right)^2.$$

11. If $x > y$, then show that $\tan^{-1}\left(\dfrac{x + iy}{x - iy}\right) = \dfrac{\pi}{4} + \dfrac{i}{2}\log\left(\dfrac{x+y}{x-y}\right)$.

12. Show that $\tan^{-1}\dfrac{3 - 2i}{3 + 2i} = \dfrac{\pi}{4} - \dfrac{i}{2}\log 5$.

13. If $\tan(\alpha + i\beta) = \tan^{-1}(x - iy)$ where $x^2 + y^2 \ne 1$ prove that,
$$\tan[\log(\alpha^2 + \beta^2)] = \dfrac{2x}{1 - x^2 - y^2}.$$

14. Prove that $\coth^{-1}x = \dfrac{1}{2}\log\left(\dfrac{x+1}{x-1}\right)$.

Hints to Selected Problems

1. Let $\tanh^{-1}(x + iy) = \alpha + i\beta$

$\qquad x + iy = \tanh(\alpha + i\beta) = \dfrac{1}{i}\tan\{i(\alpha + i\beta)\}$

or $\qquad i(x + iy) = \tan(i\alpha - \beta)$

$\Rightarrow \qquad ix - y = -\tan(\beta - i\alpha)$

$\Rightarrow \qquad y - ix = \tan(\beta - i\alpha)$ or $\beta - i\alpha = \tan^{-1}(y - ix)$

$\Rightarrow \qquad \beta + i\alpha = \tan^{-1}(y + ix)$

Now adding these two equations.

4. R.H.S. $= i\tan^{-1}(-i\sin i\theta)$
$= -i.i\tanh^{-1}(\sin i\theta) = \tanh^{-1}(\sin i\theta)$
$= \dfrac{1}{2}\log\dfrac{1 + \sin i\theta}{1 - \sin i\theta}$.

9. $\cos^{-1}(x + iy) = \theta + i\phi$
$\Rightarrow \cos(\theta + i\phi) = x + iy$.
Now, eliminating ϕ and θ, we get the required result.

10. We have $\sin^{-1}(x+iy) = \tan^{-1}\left[\dfrac{x+iy}{\sqrt{1-(x+iy)^2}}\right] = \tan^{-1}(u+iv)$

Also, $\sin^{-1}(x-iy) = \tan^{-1}\left[\dfrac{x-iy}{\sqrt{1-(x-iy)^2}}\right] = \tan^{-1}(u-iv)$

13. $\log(\alpha + i\beta) = \tan^{-1}(x-iy)$

$\therefore \quad \log(\alpha - i\beta) = \tan^{-1}(x+iy)$

On adding, we get $\log(\alpha+i\beta) + \log(\alpha-i\beta)$

$$= \tan^{-1}(x-iy) + \tan^{-1}(x+iy)$$

$\Rightarrow \log(\alpha+i\beta)(\alpha-i\beta) = \tan^{-1}\left[\dfrac{x-iy+x+iy}{1-(x-iy)(x+iy)}\right]$

$$= \tan^{-1}\left[\dfrac{2x}{1-x^2-y^2}\right]$$

$\Rightarrow \quad \log(\alpha^2+\beta^2) = \tan^{-1}\left[\dfrac{2x}{1-x^2-y^2}\right].$

14. Let $\cosh^{-1}x = z$　$x = \coth z$

$\Rightarrow \qquad \dfrac{x}{1} = \cosh z = \dfrac{e^z + e^{-z}}{e^z - e^{-z}}$

Now, using componendo and dividendo.

Answers

1. $\dfrac{1}{2}\tanh^{-1}\left(\dfrac{2x}{1+x^2+y^2}\right) + i\dfrac{1}{2}\tan^{-1}\left(\dfrac{2y}{1-x^2-y^2}\right)$

2. $A = \dfrac{1}{2}\cos^{-1}[(x^2+y^2) - \sqrt{(1-x^2+y^2)^2 + 4x^2y^2}]$, $B = \dfrac{1}{2}\cosh^{-1}[(x^2+y^2) + \sqrt{(1-x^2+y^2)^2 + 4x^2y^2}]$

and $A = \dfrac{\pi}{2} - \dfrac{1}{2}\cos^{-1}[(x^2+y^2) - \sqrt{(1-x^2+y^2) - 4x^2y^2}]$, $B = -\dfrac{1}{2}\cosh^{-1}[(x^2+y^2) + \sqrt{(1-x^2+y^2) + 4x^2y^2}]$

3. (a) $n\pi + (-1)^n\left[\dfrac{\pi}{2} + i\log\cot\dfrac{\theta}{2}\right]$.

11.11　LOGARITHM OF COMPLEX QUANTITIES

We know that if x and y are real quantities and $e^x = y$, then x is said to be the logarithm of y to the base e and is written as

$$x = \log_e y$$

Similarly if $e^{x+iy} = u + iv$, then $x + iy$ is called the logarithm (Napierian) of $u + iv$ to the base e and is written as

$$\log_e(u+iv) = x + iy \qquad \qquad ...(1)$$

Since $e^{2n\pi i} = 1$ (where n is an integer or zero), we have

$$e^{x+iy+2n\pi i} = e^{x+iy}$$

giving that $\log_e(u+iv) = 2n\pi i + x + iy = x + i(2n\pi + y)$ ...(2)

This shows that the logarithm of a complex quantity has an infinite number of values and hence is many valued function. These values are called general values of $\log_e(u+iv)$.

This is known as the general value of logarithm, the principal value of the logarithm is obtained by putting $n = 0$ in (2).

In order to distinguish between the general value of the logarithm as given by (2) and the principal value as given by (1), we get by putting $n = 0$ in (2), general value is written as 'Log' and the principal values of 'log'.

Since n can take any integral values, there are an infinite number of logarithms of $x + iy$ and they differ from each other by $2\pi i$.

☞ REMARK

* The base of a logarithm will be e, unless stated otherwise.

11.11.1 LOGARITHM OF A REAL NUMBER

If x and y are real quantities and $e^x = y$, then $x = \log_e y$

Also $e^x \cdot e^{2n\pi i} = y$

Since $e^{2n\pi i} = 1$, hence, $\mathrm{Log}_e y = x + 2n\pi i$ ...(3)

It follows that a real quantity has an infinite number of imaginary logarithms and only one real logarithms which we get by putting n equal to zero in (3).

11.11.2 LOGARITHM OF x + iy IN THE FORM OF α + iβ

Let $x + iy = r(\cos\theta + i\sin\theta) = re^{i\theta}$ then $x = r\cos\theta, y = r\sin\theta$

so that, $r = \sqrt{x^2 + y^2}$ and $\theta = \tan^{-1}y/x$

Now $\log_e(x+iy) = \log_e re^{i\theta} = \log_e r + \log_e e^{i\theta} = \log_e r + i\theta = \log_e\sqrt{x^2+y^2} + i\tan^{-1}y/x$

$\therefore \qquad \log_e(x+iy) = \dfrac{1}{2}\log_e(x^2+y^2) + i\tan^{-1}\dfrac{y}{x}$ ...(4)

The general value

$$\log_e(x+iy) = \log_e(x+iy) + 2n\pi i = \dfrac{1}{2}\log_e(x^2+y^2) + i\left(\tan^{-1}\dfrac{y}{x} + 2n\pi\right) \qquad ...(5)$$

where $-\pi < \tan^{-1}y/x < \pi$

REMARKS

- In principal values of a logarithm, the coefficient of i should lie between $-\pi$ and π i.e. $-\pi < \tan^{-1}\dfrac{y}{x} < \pi$
- Results (4) and (5) should be remembered as they give directly log $(x + iy)$ into the form $A + iB$ when principal and general value of logarithm are considered.

11.11.3 LOGARITHM OF A NEGATIVE QUANTITY

If x is a real quantity, then

$$\log_e(-x) = \log_e(-1)(x) = \log_e(-1) + \log_e x = \log_e(e^{i\pi}) + \text{Log}_e x = \pi i + \log_e x + 2n\pi i = \log_e x + i\pi(2n+1)$$

or $\log_e(-x) = \log_e x + i\pi$

Hence, the principal value of the logarithm of a negative quantity is logarithm of the positive quantity added with πi.

11.11.4 SOME IMPORTANT RESULTS

(i) $\log(z_1 z_2) = \log z_1 + \log z_2$

(ii) $\log\dfrac{z_1}{z_2} = \log z_1 - \log z_2$

Let $z_1 = r_1 e^{i\theta_1}, z_2 = r_2 e^{i\theta_2}$

Now $\log z_1 + \log z_2 = [\log r_1 + i(2m_1\pi + \theta_1)] + [\log r_2 + i(2m_2\pi + \theta_2)] = (\log r_1 + \log r_2) + i(\theta_1 + \theta_2 + 2n\pi)$ ...(1)

Where m_1 and m_2 are integers, and $n = m_1 + m_2$

Also $\log z_1 z_2 = \log r_1 r_2 \, e^{i(\theta_1 + \theta_2)} = \log r_1 r_2 + i(\theta_1 + \theta_2 + 2m\pi)$ ...(2)

Since n and m can take up any integral values, it it clear that every value of Log $(z_1 z_2)$ is equal to some value of $\log z_1 + \log z_2$ and that every value of latter is equal to some value of the former.

$\therefore \qquad \log z_1 z_2 = \log z_1 + \log z_2$ ...(A)

Similarly it can be proved that

$$\log\dfrac{z_1}{z_2} = \log z_1 - \log z_2 \qquad \text{...(B)}$$

REMARK

- It is important to note that, $\log z_1 z_2 = \log z_1 + \log z_2$ and $\log\dfrac{z_1}{z_2} = \log z_1 - \log z_2$ the principal values of the two sides of these equations need not necessarily be equal, for the simple reason that amp. $(z_1) \pm$ amp. (z_2) need not necessarily lie between $-\pi$ and $+\pi$, whereas amp. $(z_1 z_2)$ and amp. $\left(\dfrac{z_1}{z_2}\right)$ must lie between $-\pi$ and $+\pi$.

Solved Examples

EXAMPLE 1. *Find the values of* $\log(-1+\sqrt{3}i)+\log(-\sqrt{3}+i)$ *and* $\log[(-1+\sqrt{3}i)(-\sqrt{3}+i)]$ *and compare their principal values.*

SOLUTION. Let $-1+\sqrt{3}i = r_1(\cos\theta_1 + i\sin\theta_1)$

$\therefore \qquad r_1 = 2, \theta_1 = \dfrac{2}{3}\pi$

and $-\sqrt{3}+i = r_2(\cos\theta_2 + i\sin\theta_2)$

$\therefore \qquad r_2 = 2, \theta_2 = \dfrac{5\pi}{6}$

$\therefore \log(-1+\sqrt{3}i)+\log(-\sqrt{3}+i)$

$\qquad = \log 2 + \log 2 + i\left(\dfrac{2\pi}{3} + \dfrac{5\pi}{6} + 2n\pi\right)$

$\qquad = \log 4 + i\left(\dfrac{3\pi}{2} + 2n\pi\right)$

$\therefore$ The principal value is $\log 4 + \dfrac{3\pi i}{2}$ ...(1)

Also, $\log(-1+\sqrt{3}i)(-\sqrt{3}+i) = \text{Log}(-4i)$

$\qquad = \log 4\left[\cos\left(-\dfrac{\pi}{2}\right) + i\sin\left(-\dfrac{\pi}{2}\right)\right]$

$\qquad = \log 4 + \left(2m\pi - \dfrac{\pi}{2}\right)i$

$\therefore$ The principal value is, in this case,

$$\log 4 - \dfrac{\pi}{2}i \qquad \text{...(2)}$$

which is different from (1)

Thus, although

$\log(-1+\sqrt{3}i)+\log(-\sqrt{3}+i) = \log(-1+\sqrt{3}i)(-\sqrt{3}+i)$

yet the principal values of the two are not equal.

EXAMPLE 2. *Find the general value of* $\log_e(1 + \cos\theta + i\sin\theta)$.

SOLUTION. $\log_e(1 + \cos\theta + i\sin\theta)$

$\qquad = \log_e\left(2\cos^2\dfrac{\theta}{2} + 2i\sin\dfrac{\theta}{2}\cos\dfrac{\theta}{2}\right)$

$\qquad = \log_e 2\cos\dfrac{\theta}{2}\left(\cos\dfrac{\theta}{2} + i\sin\dfrac{\theta}{2}\right)$

$\qquad = \log_e 2\cos\dfrac{\theta}{2} + \log_e e^{\theta/2}$

$\qquad = \log_e 2\cos\dfrac{\theta}{2} + i\left(\dfrac{\theta}{2} + 2n\pi\right).$

where n is any integer (positive or negative) or zero.

EXAMPLE 3. *Prove that* $\log_e \tan\left(\dfrac{\pi}{4} + \dfrac{x}{2}i\right) = i\tan^{-1}\sinh x$.

SOLUTION. L.H.S. $= \log_e \tan\left(\dfrac{\pi}{4} + \dfrac{x}{2}i\right) = \log_e \dfrac{\sin\left(\dfrac{\pi}{4} + \dfrac{x}{2}i\right)}{\cos\left(\dfrac{\pi}{4} + \dfrac{x}{2}i\right)}$

$$= \log_e \left\{ \dfrac{2\sin\left(\dfrac{\pi}{4} + \dfrac{x}{2}i\right)\cos\left(\dfrac{\pi}{4} - \dfrac{x}{2}i\right)}{2\cos\left(\dfrac{\pi}{4} + \dfrac{x}{2}i\right)\cos\left(\dfrac{\pi}{4} - \dfrac{x}{2}i\right)} \right\}$$

$$= \log_e \dfrac{\sin\dfrac{\pi}{2} + \sin xi}{\cos\dfrac{\pi}{2} + \cos xi} = \log_e \dfrac{1 + i\sinh x}{\cosh x}$$

$$\text{(As } \sin ix = i \sinh x\text{)}$$

$$= \dfrac{1}{2}\log_e \dfrac{1 + \sinh^2 x}{\cosh^2 x} + i\tan^{-1}(\sinh x)$$

$$= \dfrac{1}{2}\log_e \dfrac{\cosh^2 x}{\cosh^2 x} + i\tan^{-1}(\sinh x)$$

$$= i\tan^{-1}(\sinh x) .$$

EXAMPLE 4. *If* $u = \log \tan\left(\dfrac{\pi}{4} + \dfrac{\theta}{2}\right)$ *prove that*

$$\theta = -i\log\tan\left(\dfrac{\pi}{4} + \dfrac{iu}{2}\right)$$

SOLUTION. Given $\log\tan\left(\dfrac{\pi}{4} + \dfrac{\theta}{2}\right) = u$ or $\tan\left(\dfrac{\pi}{4} + \dfrac{\theta}{2}\right) = e^u$

$$\therefore \quad \dfrac{1 + \tan\dfrac{\theta}{2}}{1 - \tan\dfrac{\theta}{2}} = \dfrac{e^{u/2}}{e^{-u/2}}$$

Applying dividendo and componendo, we have

$$\tan\dfrac{\theta}{2} = \dfrac{e^{u/2} - e^{-u/2}}{e^{u/2} + e^{-u/2}} = \tanh\dfrac{u}{2}$$

or $\quad \tan\dfrac{\theta}{2} = -i\tan\dfrac{ui}{2}$

$$\therefore \quad \dfrac{e^{i\theta/2} - e^{-i\theta/2}}{i(e^{i\theta/2} + e^{-i\theta/2})} = -i\tan\dfrac{ui}{2}$$

or $\quad \dfrac{e^{i\theta/2} - e^{-i\theta/2}}{(e^{i\theta/2} + e^{-i\theta/2})} = \dfrac{\tan ui/2}{1}$

Applying again componendo and dividendo, we get

$$\dfrac{e^{i\theta/2}}{e^{-i\theta/2}} = \dfrac{1 + \tan(ui/2)}{1 - \tan(ui/2)}$$

or $\quad e^{i\theta} = \tan\left(\dfrac{\pi}{4} + \dfrac{ui}{2}\right)$

$$\therefore \quad \theta = -i\log\tan\left(\dfrac{\pi}{4} + \dfrac{ui}{2}\right)$$

EXAMPLE 5. *Show that* $\tan\left(i\log\dfrac{a - ib}{a + ib}\right) = \dfrac{2ab}{a^2 - b^2}$

SOLUTION. Let $a = r\cos\theta$, $b = r\sin\theta$

$$\therefore \quad \dfrac{a - ib}{a + ib} = \dfrac{r(\cos\theta - i\sin\theta)}{r(\cos\theta + i\sin\theta)} = \dfrac{e^{-i\theta}}{e^{i\theta}} = e^{-2i\theta}$$

$$\therefore \quad \tan\left[i\log\dfrac{a - ib}{a + ib}\right] = \tan[i(-2i\theta)] = \tan 2\theta = \dfrac{2\tan\theta}{1 - \tan^2\theta}$$

$$= \dfrac{2\dfrac{b}{a}}{1 - \dfrac{b^2}{a^2}} = \dfrac{2ab}{a^2 - b^2} .$$

Exercise-11.11

Prove that : (Ques 1 – 3)

1. (a) $\log_e(-3) = \log 3 + i(\pi + 2n\pi)$

   (b) $\log i = \dfrac{1}{2}(4n + 1)\pi i$

   (c) $\log 3i = \log 3 + \left(2n\pi + \dfrac{1}{2}\pi\right)i$

   (d) $\log\left(\dfrac{a + ib}{a - ib}\right) = 2i\tan^{-1}\left(\dfrac{b}{a}\right)$ **(PTU–2006)**

2. $\log(1 + i) = \dfrac{1}{2}\log 2 + i\left(2n\pi + \dfrac{\pi}{4}\right)$

3. $\log(1 + i\tan\theta) = \log_e \sec\theta + i\theta$

4. Show that $\log_e \dfrac{1}{1 - e^{i\alpha}} = \log_e\left[\dfrac{1}{2}\text{cosec}\dfrac{\alpha}{2}\right] + i\left(\dfrac{\pi}{2} - \dfrac{\alpha}{2}\right)$

5. Show that $\log\log(x + iy) = \dfrac{1}{2}\log(\alpha^2 + \beta^2) + \tan^{-1}\dfrac{\beta}{\alpha}$

   where $2\alpha = \log_e(x^2 + y^2)$ and $\beta = \tan^{-1}\dfrac{y}{x}$

6. If $(a_1 + ib_1)(a_2 + ib_2)\ldots(a_n + ib_n) = A + iB$, prove that

$$\tan^{-1}\dfrac{b_1}{a_1} + \tan^{-1}\dfrac{b_2}{a_2} + \ldots + \tan^{-1}\dfrac{b_n}{a_n} = \tan^{-1}\dfrac{B}{A}$$

   and $\quad (a_1^2 + b_1^2)(a_2^2 + b_2^2)\ldots(a_n^2 + b_n^2) = A^2 + B^2 .$

7. Prove that $\log\dfrac{\cos(x - iy)}{\cos(x + iy)} = 2i\tan^{-1}(\tan x \tanh y) .$

8. Prove that the value of $\log\log\sin(x + iy)$ is

$$\dfrac{1}{2}\log(u^2 + v^2) + i\tan^{-1}\dfrac{v}{u}$$

   where $\quad u = \dfrac{1}{2}\log\dfrac{\cosh 2y - \cos 2x}{2}$

   and $\quad v = \tan^{-1}(\cot x \tanh y) .$

9. If $\log_e \log_e \log_e(\alpha + i\beta) = p + iq$, prove that

   (i) exp. $(e^p . \cos q)\cos(e^p \sin q) = \dfrac{1}{2}\log(\alpha^2 + \beta^2)$

   (ii) exp. $(e^p \cos q)\sin(e^p \sin q) = \tan^{-1}\dfrac{\beta}{\alpha} .$

10. Show that $\log\cos(x + iy)$

$$= \dfrac{1}{2}\log\left\{\dfrac{1}{2}(\cosh 2y + \cos 2x)\right\} - i\tan^{-1}(\tan x \tanh y) .$$

11. Prove that $\log\left[\dfrac{\sin(x + iy)}{\sin(x - iy)}\right] = 2i\tan^{-1}(\cot x \tanh y) .$

                        (MUMBAI–2007)

12. Find the general value of

   (i) $\log(6 + 8i)$ **(ROHTAK–2006)**

   (ii) $\log(-1)$ **(JNTU–2003)**

Hints to Selected Problems

3. L.H.S. $= \log(1 + i\tan\theta)$

$$= \log\left(1 + i\frac{\sin\theta}{\cos\theta}\right) = \log\left(\frac{\cos\theta + i\sin\theta}{\cos\theta}\right)$$

$$= \log\left(\frac{e^{i\theta}}{\cos\theta}\right) = \log\left(\sec\theta.e^{i\theta}\right)$$

$$= \log\sec\theta + \log e^{i\theta} = \log\sec\theta + i\theta$$

4. L.H.S. $= \log\dfrac{1}{1 - \cos\alpha - i\sin\alpha} = \log\dfrac{1}{2\sin^2\dfrac{\alpha}{2} - 2i\sin\dfrac{a}{2}\cos\dfrac{\alpha}{2}}$

$$= \log\left[\frac{1}{2}\operatorname{cosec}\frac{\alpha}{2}\right] - \log\left[\sin\frac{\alpha}{2} - i\cos\frac{\alpha}{2}\right]$$

$$= \log\left(\frac{1}{2}\operatorname{cosec}\frac{\alpha}{2}\right) - \log\left[\cos\left(\frac{\pi}{2} - \frac{\alpha}{2}\right) - i\sin\left(\frac{\pi}{2} - \frac{\alpha}{2}\right)\right]$$

5. Using log $\log(x + iy) = \log\left[\dfrac{1}{2}\log(x^2 + y^2) + i\left(\tan^{-1}\dfrac{y}{x}\right)\right]$.

7. $\log\dfrac{\cos x - iy}{\cos x + iy} = \log[\cos x\cosh y + i\sin x\sinh y]$

$$- \log[\cos x\cosh y - i\sin x\sinh y]$$

10. $\log\cos(x + iy) = \log(\cos x\cosh y - i\sin x\sinh y)$

$$= \frac{1}{2}\log(\cos^2 x\cosh^2 y + \sin^2 x\sinh^2 y)$$

$$- i\tan^{-1}\left(\frac{\sin x\sinh y}{\cos x\cosh y}\right)$$

$$= \frac{1}{2}\log\left[\frac{1}{2}(1 + \cos 2x)\cosh^2 y + \frac{1}{2}(1 - \cos 2x)\right.$$

$$\left. \sinh^2 y - i\tan^{-1}(\tan x\tanh y)\right]$$

 Answers

12. (i) $\log_e 10 + i\left[i\tan^{-1}\left(\dfrac{4}{3}\right) \pm 2n\pi\right]$ (ii) $\log_e 1 + i(\pi + 2n\pi)$

11.11.5 THE GENERAL EXPONENTIAL FUNCTION

If z is any real or complex quantity, we have

$$a^z = e^{z\log_e a} = 1 + \frac{z\log_e a}{1!} + \frac{(z\log_e a)^2}{2!} + \frac{(z\log_e a)^3}{3!} + \dots$$

If principal values are considered. The general value of a^z is given by

$$a^z = e^{z\log_e a} = 1 + \frac{z\log_e a}{1!} + \frac{(z\log_e a)^2}{2!} + \frac{(z\log_e a)^3}{3!} + \dots$$

It may be noted that the general exponential function is a multi-valued function.

$$a^z = \exp(z\log_e a) = \exp[z(\log_e a + 2n\pi i)]$$

If $a = r(\cos\theta + i\sin\theta)$ and $z = x + iy$, we have

$$a^z = \exp[(x + iy)\log re^{i\theta}] = \exp[(x + iy)\{\log r + i(\theta + 2n\pi)\}]$$

$$= \exp[x\log r - y(\theta + 2n\pi) + i\{y\log r + x(\theta + 2n\pi)\}]$$

Since n can take all integral value or zero, general exponential function has infinite number of values, the principal value being when $n = 0$.

11.11.6 REAL AND IMAGINARY PART OF $(\alpha + i\beta)^{p + iq}$

Let $(\alpha + i\beta)^{p + iq} = A + iB$

Now, $(\alpha + i\beta)^{p + iq} = \exp\{(p + iq)\log_e(\alpha + i\beta)\} = \exp\left[(p + iq)\left\{\frac{1}{2}\log(\alpha^2 + \beta^2) + i\left(\tan^{-1}\frac{\beta}{\alpha} + 2n\pi\right)\right\}\right]$

$$= \exp\left[\frac{1}{2}p\log(\alpha^2 + \beta^2) - q\left(\tan^{-1}\frac{\beta}{\alpha} + 2n\pi\right) + i\left\{\frac{1}{2}q\log_e(\alpha^2 + \beta^2) + p\left(\tan^{-1}\frac{\beta}{\alpha} + 2n\pi\right)\right\}\right]$$

Hence, $(\alpha + i\beta)^{p+iq} = e^x(\cos y + i\sin y)$

where $x = \dfrac{1}{2}p\log_e(\alpha^2 + \beta^2) - q\left(\tan^{-1}\dfrac{\beta}{\alpha} + 2n\pi\right)$ and $y = \dfrac{1}{2}q\log_e(\alpha^2 + \beta^2) + p\left(\tan^{-1}\dfrac{\beta}{\alpha} + 2n\pi\right)$

The real part $A = e^x\cos y$ and the imaginary part $B = e^x\sin y$.

☞ **REMARK**

- If the principal value only is required, we put $n = 0$ in the above result.

11.11.7 LOGARITHMS OF ANY BASE

If z, w and σ be any three complex numbers, and if

$$\sigma^w = z \qquad \qquad \dots(1)$$

then we define that w is a logarithm of z to the base σ, and we write

$$\log_\sigma z = w \qquad \qquad \dots(2)$$

but we have already defined σ^w as $e^w \log_e \sigma$

$\therefore$ $\qquad e^w \log_e \sigma = z$ $\qquad$ or $\quad w \log_e \sigma = \log_e z$ $\qquad$ or $\qquad w = \dfrac{\log_e z}{\log_e \sigma}$ $\qquad$ ...(3)

From (2) and (3) we have

$$\log_\sigma z = \frac{\log_e z}{\log_e \sigma} \qquad\qquad\qquad ...(A)$$

With the help of formula (A), we can write logarithm of any base to base 'e'.

The principal value of $\text{Log}_\sigma z$ is defined by

$$\log_\sigma z = \frac{\log_e z}{\log_e \sigma} \qquad\qquad\qquad ...(B)$$

 Solved Examples

EXAMPLE 1. *Prove that* $\log_i i = \dfrac{4n+1}{4m+1}$ *where m and n are integers.*

SOLUTION. We know that, $\log_i i = \dfrac{\log_e i}{\log_e i}$

$$= \frac{\log_e\left[\cos\dfrac{\pi}{2} + i\sin\dfrac{\pi}{2}\right]}{\log_e\left[\cos\dfrac{\pi}{2} + i\sin\dfrac{\pi}{2}\right]} = \frac{\left(2n\pi + \dfrac{\pi}{2}\right)i}{\left(2m\pi + \dfrac{\pi}{2}\right)i}$$

Taking general values of numerator and denominator

$$= \frac{4n+1}{4m+1}.$$

EXAMPLE 2. *Find the general and principal values of* $(i)^i$. *Hence show that values of* i^i *can be arranged in a geometric progression.*

SOLUTION. We have $(i)^i = \exp\{i\log_e i\}$

$$= \exp\{i\log_e(\cos\pi/2 + i\sin\pi/2)\}$$
$$= \exp\{i(2n\pi i + \pi i/2)\}$$
$$= \exp\{-(2n+\tfrac{1}{2})\pi\}$$

Putting $n = 0, 1, 2 \ldots$, the various values of $(i)^i$ are

$$e^{-\pi/2}, e^{-5\pi/2}, e^{-9\pi/2}, e^{-13\pi/2}, \ldots$$

which is a G.P with common ratio $e^{-2\pi}$.

EXAMPLE 3. *If* $a^{\alpha+i\beta} = (x+iy)^{p+iq}$, *principal values only being considered, prove that*

(a) $\alpha = \dfrac{1}{2}p\log_a(x^2+y^2) - q\tan^{-1}\left(\dfrac{y}{x}\right)\log_a e$

(b) $\dfrac{2(\alpha p + \beta q)}{p^2 + q^2} = \log_a(x^2+y^2)$

SOLUTION. (a) We have

$$a^{\alpha+i\beta} = (x+iy)^{p+iq}$$

or $(\alpha + i\beta)\log_e a = (p+iq)\log_e(x+iy)$

$$= (p+iq)\left\{\frac{1}{2}\log(x^2+y^2) + i\tan^{-1}\frac{}{}\right\} \quad ...(1)$$

Equating real parts on both sides, we have

$$\alpha\log a = \frac{1}{2}p\log_e(x^2+y^2) - q\tan^{-1}\frac{y}{x}$$

or $\alpha = \dfrac{1}{2}\cdot\dfrac{p\log_e(x^2+y^2)}{\log_e a} - \dfrac{q\tan^{-1}\dfrac{y}{x}}{\log_e a}$

$$= \frac{1}{2}p\log_e(x^2+y^2) - q\tan^{-1}\frac{y}{x}\log_a e$$

since $\log_e b = \log_e a \times \log_a b$

and $\log_e a \times \log_a e = 1$

(b) From equation (1), we have

$(\alpha + i\beta)\log_e a$

$$= (p+iq)\left\{\frac{1}{2}\log_e(x^2+y^2) + i\tan^{-1}\frac{y}{x}\right\}$$

or $\dfrac{(\alpha+i\beta)}{p+iq} = \dfrac{\dfrac{1}{2}\log_e(x^2+y^2)}{\log_e a} + i\dfrac{\tan^{-1}y/x}{\log_e a}$

or $\dfrac{(\alpha+i\beta)(p-iq)}{p^2+q^2} = \dfrac{1}{2}\log_a(x^2+y^2) + i$

$$\tan^{-1}\frac{y}{x}\log_a e$$

or $\dfrac{\alpha p + \beta q + i(\beta p - \alpha q)}{p^2+q^2} = \dfrac{1}{2}\log_a(x^2+y^2)$

$$+ i\tan^{-1}\frac{y}{x}\log_a e$$

Equating real parts on both sides, we get

$$\frac{\alpha p + \beta q}{p^2+q^2} = \frac{1}{2}\log_a(x^2+y^2).$$

EXAMPLE 4. *If* $i^{i^{i\ldots ad\,inf.}} = A + iB$ *principal values only being considered, prove that* $\tan\dfrac{1}{2}\pi A = \dfrac{B}{A}$ *and* $A^2 + B^2 = e^{-\pi B}$. $\qquad$ (SVTU–2006S)

SOLUTION. We have $\qquad i^{i^{i\ldots ad\,inf}} = A + iB$

or $\qquad \exp.\{(A+iB)\log_e i\} = A + iB$

or $\qquad \exp\left\{(A+iB)\dfrac{i\pi}{2}\right\} = A + iB$

or $\qquad \exp\left\{Ai\dfrac{\pi}{2} - \dfrac{B\pi}{2}\right\} = A + iB$

or $e^{-B\pi/2}\left\{\cos\dfrac{A\pi}{2} + i\sin\dfrac{A\pi}{2}\right\} = A + iB$

Equating real and imaginary parts on both sides, we get

$$e^{-B\pi/2}\cos\frac{A\pi}{2} = A \qquad\qquad ...(1)$$

and $\qquad e^{-B\pi/2}\sin\dfrac{A\pi}{2} = B \qquad\qquad ...(2)$

Squaring and adding (1) and (2), we get

$$A^2 + B^2 = e^{-B\pi}$$

and dividing (2) by (1), $\tan\dfrac{A\pi}{2} = \dfrac{B}{A}$.

Exercise-11.12

1. If $(i^i)^i = \cos\theta - i\sin\theta$ show that $\theta = \frac{1}{2}\pi(4n+1)$.

2. Show that the real part of $(i)^{\log(1+i)}$ is $e^{-\pi^2/8}\cos\left(\frac{1}{4}\pi\log 2\right)$.

3. If $\tan\log(x+iy) = a+ib$ and $a^2+b^2 \ne 1$, show that
$$\log(x^2+y^2) = \tan^{-1}\frac{2a}{1-a^2-b^2}.$$

4. Show that the principal value of $\dfrac{(a+ib)^{p+iq}}{(a-ib)^{p-iq}}$
is $\cos 2(p\alpha + q\log r) + i\sin 2(p\alpha + q\log r)$
where $r = (a^2+b^2)^{1/2}$ and $a = \tan^{-1} b/a$.

5. Show that the ratio of the principal values of $(1+i)^{1-i}$ and $(1-i)^{1+i}$ is $\sin(\log 2) + i\cos(\log 2)$.

6. If $\left(\dfrac{a+x+iy}{a-x-iy}\right)^{\lambda+\mu i} = X+Yi$ prove that one of the values of
$$\tan^{-1}\frac{Y}{X} \text{ is } \lambda\tan^{-1}\left(\frac{2ay}{a^2-x^2-y^2}\right) + \frac{\mu}{2}\log\frac{(a+x)^2+y^2}{(a-x)^2+y^2}$$

7. Criticize the fallacy : $e^\theta = (e^{-\theta i})^i = [e^{(2\pi-\theta)i}]^i = e^{\theta-2\pi}$.

8. Show that the principal value of $(x+iy)^{a+ib}$ is wholly real or imaginary according as
$$\frac{1}{2}b\log(x^2+y^2) + a\tan^{-1}\left(\frac{y}{x}\right)$$
is an even or odd multiple of $\frac{1}{2}\pi$.

9. Prove that the general value of $(1+i\tan\alpha)^{-i}$ is
$$e^{\alpha+2m\pi}[\cos(\log\cos\alpha) + i\sin(\log\cos\alpha)]$$
where m is any integer.

10. If $\log\log(x+iy) = p+ir$, prove that
$$y = x\tan\left\{\frac{1}{2}\tan r\log(x^2+y^2)\right\}.$$

11. If $x+iy$ be a non zero complex number, prove that
$$\sin\left\{i\log\frac{x-iy}{x+iy}\right\} = \frac{2xy}{x^2+y^2}.$$

12. If $(1+i)(1+2i)(1+3i)\ldots(1+ni) = x+iy$, prove that
$$2.5.10\ldots(1+n^2) = x^2+y^2.$$
$$\tan^{-1}1 + \tan^{-1}2 + \tan^{-1}3 + \ldots + \tan^{-1}x = \tan^{-1}\frac{y}{x}$$

 Hints to Selected Problems

1. Taking log of both sides.

3. $\tan\log(x+iy) = a+ib$
$\Rightarrow \tan^{-1}\log(x-iy) = (a-ib)$
Now L.H.S $= \tan\{\log(x^2+y^2)\} = \tan[\log(x+iy)(x-iy)]$
$= \tan[\log(x+iy) + \log(x-iy)]$

6. De Moivre's theorem is applicable when n is real *i.e.*,
$$(\cos\theta + i\sin\theta)^n = \cos n\theta + i\sin n\theta = (e^{i\theta})^n = e^{ni\theta}$$
∴ If $n = i$ (imaginary), then
$$(e^{i\theta})^i \ne e^{-\theta} \Rightarrow (e^{-i\theta})^i \ne e^\theta$$

9. $(1+i\tan\alpha)^i = e^{-i\log(1+i\tan\alpha)}$

$= e^{-i[2m\pi i + \log(1+i\tan\alpha)]}$
$= e^{-i[2m\pi i + \log\{1 + i(\sin\alpha/\cos\alpha)\}]}$
$= e^{-i[2m\pi i + \log\{(\cos\alpha + i\sin\alpha)/\cos\alpha\}]}$
$= e^{-i[2m\pi i + \log e^{i\alpha} - \log\cos\alpha]} = e^{-i(2m\pi i + i\alpha - \log\cos\alpha)}$
$= e^{(2m\pi + \alpha) + i\log\cos\alpha}$
$= e^{2m\pi + \alpha} \cdot e^{i\log\cos\alpha}$
$= e^{2m\pi + \alpha}[\cos(\log\cos\alpha) + i\sin(\log\cos\alpha)]$

11. Put $x = r\cos\theta$, $y = r\sin\theta$, then
L.H.S. $= \sin\{i\log e^{-2i\theta}\} = \sin 2\theta = \dfrac{2\tan\theta}{1+\tan^2\theta}$ etc.

11.12 GREGORY'S SERIES

If θ lies within the closed interval $(-\pi/4, \pi/4)$, i.e., if $-\pi/4 \le \theta \le \pi/4$, then
$$\theta = \tan\theta - \frac{1}{3}\tan^3\theta + \frac{1}{5}\tan^5\theta - \frac{1}{7}\tan^7\theta + \ldots \text{ ad. inf.}$$

Proof. We have $(1 + i\tan\theta) = \left(1 + i\dfrac{\sin\theta}{\cos\theta}\right) = \dfrac{1}{\cos\theta}(\cos\theta + i\sin\theta) = \sec\theta.e^{i\theta}$

Taking logarithm of both sides (considering only principal values)
$$\log(1 + i\tan\theta) = \log\sec\theta + \log e^{i\theta}$$
∴ $\log\sec\theta + i\theta = \log(1 + i\tan\theta)$ ...(1)

Now since $-\pi/4 \le \theta \le \pi/4$, $\tan\theta$ is numerically not greater than unity.
Expanding R.H.S of (1), we have
$$\log\sec\theta + i\theta = i\tan\theta - \frac{i^2\tan^2\theta}{2} + \frac{i^3\tan^3\theta}{3} - \frac{i^4\tan^4\theta}{4} + \ldots \qquad ...(2)$$
$$= i\tan\theta + \frac{\tan^2\theta}{2} - \frac{i\tan^3\theta}{3} - \frac{\tan^4\theta}{4} + \frac{i\tan^5\theta}{5}\ldots$$

Equating imaginary parts on both sides, we get
$$\theta = \tan\theta - \frac{\tan^3\theta}{3} + \frac{\tan^5\theta}{5}\ldots \text{ ad. inf.} \qquad ...(3)$$

This is called Gregory's series after the name of James Gregory.

☞ **REMARK**

- If we put $\tan \theta = x$, so that $\theta = \tan^{-1} x$, we have another from of the Gregory's series.

$$\tan^{-1} x = x - \frac{x^3}{3} + \frac{x^5}{5} - \frac{x^7}{7} + \dots \text{ ad. inf.} \qquad \dots (4)$$

when x lies between -1 and 1 and $-\pi/4 \le \tan^{-1} x \le \pi/4$

Equating real parts on both sides of (2), we have $\log \sec \theta = \frac{1}{2} \tan^2 \theta - \frac{1}{4} \tan^4 \theta + \frac{1}{6} \tan^6 \theta - \dots$

THEOREM 1.   *If θ lies between $n\pi - \pi/4$ and $n\pi + \pi/4$, both limits being inclusive*

i.e., $$n\pi - \frac{\pi}{4} \le \theta \le n\pi + \frac{\pi}{4}$$

then $$\theta - n\pi = \tan \theta - \frac{\tan^3 \theta}{3} + \frac{\tan^5 \theta}{5} - \frac{\tan^7 \theta}{7} + \dots$$

PROOF.   Let $\theta = n\pi + \phi$ where ϕ lies between $-\pi/4$ and $\pi/4$, then

$$1 + i \tan \theta = 1 + i \tan (n\pi + \phi) = 1 + i \tan \phi = 1 + \frac{i \sin \phi}{\cos \phi} = \frac{1}{\cos \phi}(\cos \phi + i \sin \phi) = \sec \phi . e^{i\phi}$$

Taking logarithm of both sides, we get

$$\log (\sec \phi . e^{i\phi}) = \log (1 + i \tan \theta)$$

Under the given conditions $\tan \theta$ is not numerically greater than unity, hence

$$\log \sec \phi + i\phi = i \tan \theta - \frac{i^2 \tan^2 \theta}{2} + \frac{i^3 \tan^3 \theta}{3} - \dots = i \tan \theta + \frac{\tan^2 \theta}{2} - \frac{i \tan^3 \theta}{3} + \dots$$

$$\therefore \qquad \phi = \theta - n\pi = \tan \theta - \frac{\tan^3 \theta}{3} + \frac{\tan^5 \theta}{5} - \dots \text{ ad. inf.}$$

11.12.1 VALUE OF π

Gregory's series has been used for evaluating the value of π. We have seen that

$$\tan^{-1} x = x - \frac{x^3}{3} + \frac{x^5}{5} - \frac{x^7}{7} + \dots$$

Putting $x = 1$, we get $$\frac{\pi}{4} = 1 - \frac{1}{3} + \frac{1}{5} - \frac{1}{7} + \dots$$

From this series the values of π can be calculated; but as the successive terms do not rapidly become small a very large number of terms would have to be taken to obtain the value of π correct to a certain decimal place. On account of this other series have been found out.

11.12.2 EULER'S SERIES

It is easy to show that

$$\tan^{-1} \frac{1}{2} + \tan^{-1} \frac{1}{3} = \tan^{-1} 1 = \pi/4 = \left(\frac{1}{2} - \frac{1}{3} . \frac{1}{2^2} + \frac{1}{5} . \frac{1}{2^5} - \dots \right) + \left(\frac{1}{3} - \frac{1}{3} . \frac{1}{3^3} + \frac{1}{5} . \frac{1}{3^5} - \dots \right)$$

$$= \left(\frac{1}{2} + \frac{1}{3} \right) - \frac{1}{3} \left(\frac{1}{2^3} + \frac{1}{3^3} \right) + \frac{1}{5} \left(\frac{1}{2^5} + \frac{1}{3^5} \right) - \dots$$

This series is more rapidly convergent than the preceding, but more than eleven terms of the series of $\tan^{-1} 1/2$ would have to be taken to get the value of π correct to 7 places of decimal.

11.12.3 MACHIN'S SERIES

We can prove that

$$\frac{\pi}{4} = 4 \tan^{-1} \frac{1}{5} - \tan^{-1} \frac{1}{239} = 4 \left\{ \frac{1}{5} - \frac{1}{3} . \frac{1}{5^3} + \frac{1}{5} . \frac{1}{5^5} - \frac{1}{7} . \frac{1}{5^7} + \dots \right\} - \left\{ \frac{1}{239} - \frac{1}{3} . \frac{1}{(239)^3} + \frac{1}{5} . \frac{1}{(239)^5} - \dots \right\}$$

$$\therefore \qquad \pi = 16 \left\{ \frac{2}{10} - \frac{1}{3} . \frac{2^3}{10^3} + \dots \right\} - 4 \left\{ \frac{1}{239} - \frac{1}{3} . \frac{1}{(239)^3} + \dots \right\}$$

This series is clearly more rapidly convergent than Euler's series.

11.12.4 RUTHERFORD'S SERIES

Since $$\tan^{-1} \frac{1}{70} - \tan^{-1} \frac{1}{99} = \tan^{-1} \frac{1}{239} \text{, we have form the identity}$$

$$\frac{\pi}{4} = 4 \tan^{-1} \frac{1}{5} - \tan^{-1} \frac{1}{70} + \tan^{-1} \frac{1}{99}$$

$$\therefore \qquad \frac{\pi}{4} = 4 \left(\frac{1}{5} - \frac{1}{3} . \frac{1}{5^3} + \frac{1}{5} . \frac{1}{5^5} - \dots \right) - \left(\frac{1}{70} - \frac{1}{3} . \frac{1}{70^3} - \dots \right) + \left(\frac{1}{99} - \frac{1}{3} . \frac{1}{99^3} + \dots \right)$$

This is still more rapidly convergent than Machin's series.

Solved Examples

EXAMPLE 1. *Assuming that*

$$\theta - n\pi = \tan\theta - \frac{\tan^3\theta}{3} + \frac{\tan^5\theta}{5} - \dots$$

when θ *lies between* $n\pi - \frac{\pi}{4}, n\pi + \frac{\pi}{4}$ *write down the value of* n *when* θ *lies between* $\frac{19\pi}{4}$ *and* $\frac{21\pi}{4}$.

SOLUTION. θ lies between $\frac{19\pi}{4}$ and $\frac{21\pi}{4}$, *i.e.*, between $\left(5\pi - \frac{\pi}{4}\right)$ and $\left(5\pi + \frac{\pi}{4}\right)$

Therefore $n = 5$.

EXAMPLE 2. *If* $x < \sqrt{2} - 1$, *prove that* $2\left(x - \frac{x^3}{3} + \frac{x^5}{5} - \dots \text{ad inf.}\right)$

$$= \frac{2x}{1-x^2} - \frac{1}{3}\left(\frac{2x}{1-x^2}\right)^3 + \frac{1}{5}\left(\frac{2x}{1-x^2}\right)^5 - \dots ad.inf.$$

SOLUTION. Here $x < \sqrt{2} - 1$, *i.e.*, $x < 1$

Therefore, $2\left(x - \frac{x^3}{3} + \frac{x^5}{5} - \dots\right) = 2\tan^{-1}x = \tan^{-1}\frac{2x}{1-x^2}$

Again, $x < \sqrt{2} - 1$ or $x + 1 < \sqrt{2}$

or $x^2 + 2x + 1 < 2$ or $2x < 1 - x^2$

or $\frac{2x}{1-x^2} < 1$.

Therefore, $\frac{2x}{1-x^2} - \frac{1}{3}\left(\frac{2x}{1-x^2}\right)^3 + \dots = \tan^{-1}\frac{2x}{1-x^2}$

Hence, L.H.S = R.H.S

EXAMPLE 3. *Prove that* $\pi = 2\sqrt{3}\left[1 - \frac{1}{3^2} + \frac{1}{5\sqrt{3^4}} - \frac{1}{7\sqrt{3^6}} + \dots\right].$

SOLUTION. R.H.S. $= 2\sqrt{3}\left[1 - \frac{1}{3(\sqrt{3})^2} + \frac{1}{5(\sqrt{3})^4} - \frac{1}{7(\sqrt{3})^6} + \dots\right]$

$$= 6\left[\frac{1}{\sqrt{3}} - \frac{1}{3(\sqrt{3})^3} + \frac{1}{5(\sqrt{3})^5} - \frac{1}{7(\sqrt{3})^7} + \dots\right]$$

$$= 6\tan^{-1}\left(\frac{1}{\sqrt{3}}\right) = 6.\frac{\pi}{6} = \pi = \text{L.H.S.}$$

EXAMPLE 4. *Prove that*

(i) $\frac{\pi}{4} = \frac{17}{21} - \frac{713}{81 \times 342} + $

$$\dots + \frac{(-1)^{n+1}}{2n-1}\left(\frac{2}{3}.9^{1-n} + 7^{1-2n}\right) + \dots$$

(ii) $\frac{\pi}{4} = \left(\frac{2}{3} + \frac{1}{7}\right) - \frac{1}{3}\left(\frac{1}{3^3} + \frac{1}{7^3}\right) + \left(\frac{2}{3^5} + \frac{1}{7^5}\right) - \dots$

SOLUTION. (i) T_n of right hand side is obviously given by

$$T_n = \frac{(-1)^{n+1}}{2n-1}\left[2.\frac{1}{3^{2n-1}} + \frac{1}{7^{2n-1}}\right]$$

$$T_1 = \left[\frac{2}{3} + \frac{1}{7}\right]$$

$$T_2 = -\left[\frac{1}{3}.\frac{2}{3^3} + \frac{1}{3.7^3}\right]$$

$$T_3 = \frac{1}{5}\left[\frac{2}{3^5} + \frac{1}{7^5}\right]$$

$$T_4 = -\frac{1}{7}\left[\frac{2}{3^7} + \frac{1}{7^7}\right]$$

$$\therefore S = 2\left[\frac{1}{3} - \frac{1}{3.3^3} + \frac{1}{5.3^5} - \frac{1}{7.3^7} + \dots\right]$$

$$+ \left[\frac{1}{7} - \frac{1}{3.7^3} + \frac{1}{5.7^5} - \frac{1}{7.7^7} - \dots\right]$$

$$= 2\tan^{-1}\frac{1}{3} + \tan^{-1}\frac{1}{7} = \tan^{-1}\frac{2.\frac{1}{3}}{1 - \frac{1}{3^2}} + \tan^{-1}\frac{1}{7}$$

$$= \tan^{-1}\frac{3}{4} + \tan^{-1}\frac{1}{7} = \tan^{-1}\frac{\frac{3}{4} + \frac{1}{7}}{1 - \frac{3}{4}.\frac{1}{7}} = \tan^{-1}1 = \frac{\pi}{4}.$$

(ii) Same as part (i).

EXAMPLE 5. *Find the sum of the following series :*

$$\frac{7}{1.3.5} + \frac{19}{5.7.9} + \frac{31}{9.11.13} + \dots ad \text{ inf.}$$

SOLUTION. Let $S = 7 + 19 + 31 + \dots + T_n$

$\underline{\quad S = \quad 7 + 19 + \dots + T_{n-1} + T_n}$

$0 = 7 + 12 + 12 + \dots$ to $n-1$ terms $- T_n$

$\therefore T_n = 7 + 12 + 12 + \dots$ to $n - 1$ terms $= 12n - 5$

Hence n^{th} term of given series is

$$T_n = \frac{12n-5}{(4n-3)(4n-1)(4n+1)}$$

$$= \frac{1}{2}\left[\frac{1}{4n-3} + \frac{1}{4n-1} - \frac{2}{4n+1}\right]$$

$$T_1 = \frac{1}{2}\left[1 + \frac{1}{3} - \frac{2}{5}\right]$$

$$T_2 = \frac{1}{2}\left[\frac{1}{5} + \frac{1}{7} - \frac{2}{9}\right]$$

$$T_3 = \frac{1}{2}\left[\frac{1}{9} + \frac{1}{11} - \frac{2}{13}\right]$$

$$T_{n-1} = \frac{1}{2}\left[\frac{1}{4n-7} + \frac{1}{4n-5} - \frac{2}{4n-3}\right]$$

$$T_n = \frac{1}{2}\left[\frac{1}{4n-3} + \frac{1}{4n-1} - \frac{2}{4n+1}\right]$$

Adding, we get

$$S = T_1 + T_2 + T_3 \dots + T_n + \dots$$

$$= \frac{1}{2}\left[1 + \frac{1}{3} - \frac{1}{5} + \frac{1}{7} - \frac{1}{9} + \frac{1}{11} - \frac{1}{13} + \dots\right]$$

$$= \frac{1}{2} - \frac{1}{2}\left[-\frac{1}{3} + \frac{1}{5} - \frac{1}{7} + \frac{1}{9} - \frac{1}{11} + \dots\right]$$

$$= \frac{1}{2} - \frac{1}{2}\left[\frac{\pi}{4} - 1\right] = 1 - \frac{\pi}{8}$$

Exercise-11.13

1. If $\theta - n\pi = \tan\theta - \dfrac{1}{3}\tan^3\theta + \dfrac{1}{5}\tan^5\theta - \ldots$
   write down the value of n when θ lies between
   (i) $\dfrac{11\pi}{4}$ and $\dfrac{13\pi}{4}$ 　　　(ii) $-\dfrac{15\pi}{4}$ and $-\dfrac{17\pi}{4}$

   Prove that

2. $\dfrac{\pi}{8} = \dfrac{1}{1.3} + \dfrac{1}{5.7} + \dfrac{1}{9.11} + \ldots$

3. $\dfrac{\pi}{4} = \dfrac{1}{2} - \dfrac{1}{3.2^3} + \dfrac{1}{5.2^5} - \dfrac{1}{7.2^7} + \ldots + \dfrac{1}{3} - \dfrac{1}{3}.\dfrac{1}{3^3} + \dfrac{1}{5}.\dfrac{1}{3^5} - \ldots$

4. $\dfrac{\pi}{12} = \left(1 - \dfrac{1}{3^{1/2}}\right) - \dfrac{1}{3}\left(1 - \dfrac{1}{3^{3/2}}\right) + \dfrac{1}{5}\left(1 - \dfrac{1}{3^{5/2}}\right) - \ldots$

5. If $x > 0$, prove that
   $$\tan^{-1} x = \dfrac{\pi}{4} + \dfrac{x-1}{x+1} - \dfrac{1}{3}\left(\dfrac{x-1}{x+1}\right)^3 + \dfrac{1}{5}\left(\dfrac{x-1}{x+1}\right)^5 - \ldots$$

6. If θ lies between 0 and $\pi/2$, prove that
   $$\tan^{-1}\dfrac{1-\cos\theta}{1+\cos\theta} = \tan^2\dfrac{\theta}{2} - \dfrac{1}{3}\tan^6\dfrac{\theta}{2} + \dfrac{1}{5}\tan^{10}\dfrac{\theta}{2} - \ldots$$

7. If $\theta < \pi/4$ prove that
   $$\log\sec\theta = \dfrac{1}{2}\tan^2\theta - \dfrac{1}{4}\tan^4\theta + \dfrac{1}{6}\tan^6\theta - \ldots$$

8. Prove that if θ and $\tan^{-1}(\sec\theta)$ both lie between 0 and $\dfrac{1}{2}\pi$
   $$\tan^{-1}(\sec\theta) = \dfrac{\pi}{4} + \tan^2\dfrac{\theta}{2} - \dfrac{1}{3}\tan^6\dfrac{\theta}{2} + \ldots$$

9. Expand $\tan^{-1}\left(\dfrac{\cos\theta + \sin\theta}{\cos\theta - \sin\theta}\right)$ as a power series in $\tan\theta$.

10. Prove that $1 - 2\left[\dfrac{1}{3.5} + \dfrac{1}{7.9} + \ldots\infty\right] = \dfrac{\pi}{4}$

Hints to Selected Problems

2. $\text{R.H.S} = \dfrac{1}{1.3} + \dfrac{1}{5.7} + \dfrac{1}{9.11} + \ldots$

   $= \dfrac{1}{2}\left(1 - \dfrac{1}{3}\right) + \dfrac{1}{2}\left(\dfrac{1}{5} - \dfrac{1}{7}\right) + \dfrac{1}{2}\left(\dfrac{1}{9} - \dfrac{1}{11}\right) + \ldots$

   $= \dfrac{1}{2}\left[\left(1 - \dfrac{1}{3} + \dfrac{1}{5} - \dfrac{1}{7} + \dfrac{1}{9} - \dfrac{1}{11} + \ldots\right)\right]$

   $= \dfrac{1}{2}\tan^{-1}1 = \dfrac{1}{2} \times \dfrac{\pi}{4} = \dfrac{\pi}{8}$

8. $\text{R.H.S.} = \dfrac{1}{4}\pi + \tan^{-1}\left(\tan^2\dfrac{\theta}{2}\right) = \tan^{-1}\dfrac{1+\tan^2\dfrac{\theta}{2}}{1-\tan^2\dfrac{\theta}{2}} = \tan^{-1}(\sec\theta)$

   since $0 < \theta < \dfrac{\pi}{2}, 0 < \tan^2\dfrac{\theta}{2} < 1$

9. $\tan^{-1}\left(\dfrac{\cos\theta + \sin\theta}{\cos\theta - \sin\theta}\right) = \tan^{-1}\left[\dfrac{\cos\theta\sin\dfrac{\pi}{4} + \sin\theta\cos\dfrac{\pi}{4}}{\cos\theta\cos\dfrac{\pi}{4} - \sin\theta\sin\dfrac{\pi}{4}}\right]$

   $= \tan^{-1}\left[\dfrac{\sin\left(\dfrac{\pi}{4} + \theta\right)}{\cos\left(\dfrac{\pi}{4} + \theta\right)}\right]$

10. $\text{L.H.S.} = 1 - 2\left[\dfrac{1}{3.5} + \dfrac{1}{7.9} + \ldots\infty\right]$

   $= 1 - 2\left[\dfrac{1}{2}\left\{\left(\dfrac{1}{3} - \dfrac{1}{5}\right)\right\} + \left(\dfrac{1}{7} - \dfrac{1}{9}\right)\right] + \ldots\infty$

   $= 1 - \dfrac{1}{3} + \dfrac{1}{5} - \dfrac{1}{7} + \ldots\infty = \tan^{-1}1 = \dfrac{\pi}{4}$

Answers

1. (i) $n = 3$ 　　　(ii) $n = -4$ 　　　9. $r\pi + \dfrac{\pi}{4} + \tan\theta - \dfrac{1}{3}\tan^3\theta + \dfrac{1}{5}\tan^5\theta - \ldots$

11.13 SUMMATION OF TRIGONOMETRICAL SERIES

In this section, we shall deal with the summation of some trigonometrical series with a finite or infinite number of terms. It may be noted that all the series with infinite number of terms are not summable and hence due care should be taken regarding the convergence of such series.

11.13.1 GENERAL C + iS METHOD

A general method for the summation of series of the type
$$c_1 \cos\theta + c_2 \cos(\theta + \alpha) + c_3 \cos(\theta + 2\alpha) + \ldots \qquad \ldots(1)$$
$$c_1 \sin\theta + c_2 \sin(\theta + \alpha) + c_3 \sin(\theta + 2\alpha) + \ldots \qquad \ldots(2)$$
is to denote the series (1) by C, since the coefficients of C's in this series are cosines of the angles increasing in arithmetic progression and the series (2) by S, being a sine series. Multiplying (2) by i and adding,
$$C + iS = c_1(\cos\theta + i\sin\theta) + c_2\{\cos(\theta + \alpha) + i\sin(\theta + \alpha) + c_3\{\cos(\theta + 2\alpha) + i\sin(\theta + 2\alpha)\} + \ldots$$
$$= c_1 e^{i\theta} + c_2 e^{i(\theta + \alpha)} + c_3 e^{i(\theta + 2\alpha)} + \ldots = e^{i\theta}\{c_1 + c_2 e^{i\alpha} + c_3 e^{2i\alpha} + \ldots\} \qquad \ldots(3)$$
Now the series with in the brackets of (3) may be either
   (i) series in geometrical progression,
   (ii) binomial series or one which can be reduced to it,
   (iii) exponential series or the allied series *e.g.* sine or cosine series, and
   (iv) logarithmic series or the allied series *e.g.* Gregory's series.

The real part of (3) gives C and the imaginary part gives S. If either of the series C or S is given, the other series, known as auxiliary series, can be formed and the sum of $C + iS$ is found. The real part of the sum so found is equal to C and the imaginary part equal to S.

11.13.2 Use of Geometric Series

Trigonometric series is given by

$$\alpha + \alpha z + \alpha z^2 + \alpha z^3 + \dots \text{ inf.}$$

Sums of n terms and infinity of the above series are given by

$$S_n = \alpha \frac{(z^n - 1)}{z - 1} \text{ if } |z| > 1 \qquad \dots(1)$$

$$= \alpha \frac{(1 - z^n)}{z - 1} \text{ if } |z| < 1 \qquad \dots(2)$$

$$S_\infty = \frac{\alpha}{1 - z} \text{ provided } |z| < 1 \qquad \dots(3)$$

REMARK

- The series becomes divergent if $|z| > 1$; hence sum to infinity in this case is not possible.

Solved Examples

EXAMPLE 1. *Sum the following series to n terms and to infinity* $1 + x \cos\theta + x^2 \cos 2\theta + \dots ad\ inf$ *where x is less than unity.*

SOLUTION. Let $C = 1 + x\cos\theta + x^2\cos 2\theta + \dots$

$S = x\sin\theta + x^2\sin 2\theta + \dots$

$C + iS = 1 + x(\cos\theta + i\sin\theta)$
$$+ x^2(\cos 2\theta + i\sin 2\theta) + \dots$$
$$= 1 + xe^{i\theta} + x^2 e^{2i\theta} + \dots$$

This is a geometric series with common ratio $xe^{i\theta}$, modulus of which is less than 1, hence

$$C_n + iS_n = \frac{1 - (xe^{i\theta})^n}{1 - xe^{i\theta}} = \frac{1 - x^n e^{in\theta}}{1 - xe^{i\theta}}$$

$$= \frac{(1 - x^n e^{in\theta})(1 - xe^{-i\theta})}{(1 - xe^{i\theta})(1 - xe^{-i\theta})}$$

multiplying the numerator and denominator by complex conjugate of denominator

$$= \frac{1 - xe^{-i\theta} - x^n e^{in\theta} + x^{n+1} e^{i(n-1)\theta}}{1 + x^2 - x(e^{i\theta} + e^{-i\theta})}$$

$$= \frac{1 - x(\cos\theta - i\sin\theta) - x^n(\cos n\theta + i\sin n\theta)}{1 - 2x\cos\theta + x^2} {+ x^{n+1}\{\cos(n-1)\theta + i\sin(n-1)\theta\}}$$

Equating real and imaginary parts, we get

$$C_n = \frac{1 - x\cos\theta - x^n\cos n\theta + x^{n+1}\cos(n-1)\theta}{1 - 2x\cos\theta + x^2}$$

$$S_n = \frac{x\sin\theta - x^n\sin n\theta + x^{n+1}\sin(n-1)\theta}{1 - 2x\cos\theta + x^2}$$

Since $x < 1$, $x^n, x^{n+1} \to 0$ as $n \to \infty$

$$C_\infty = \frac{1 - x\cos\theta}{1 - 2x\cos\theta + x^2}$$

and $$S_\infty = \frac{x\sin\theta}{1 - 2x\cos\theta + x^2}$$

REMARK

- If sum to n terms is not required and only sum to infinity is required, we can proceed as under.

$$C_\infty + iS_\infty = \frac{1}{xe^{i\theta}} = \frac{1 - xe^{-i\theta}}{(1 - xe^{i\theta})(1 - xe^{-i\theta})} = \frac{1 - x\cos\theta + ix\sin\theta}{1 - 2x\cos\theta + x^2}$$

$$\therefore \qquad C_\infty = \frac{1 - \cos\theta}{1 - 2x\cos\theta + x^2} \text{ and } S_\infty = \frac{x\sin\theta}{1 - 2x\cos\theta + x^2}$$

EXAMPLE 2. *Sum the series*

(i) $\cos\alpha + c\cos(\alpha + \beta) + c^2\cos(\alpha + 2\beta) + \dots$ *to n terms and to infinity*

(ii) $\sin\alpha + c\sin(\alpha + \beta) + c^2\sin(\alpha + 2\beta) + \dots$ *to n terms and to infinity.*

SOLUTION. Let $C_n = \cos\alpha + c\cos(\alpha + \beta) + c^2\cos(\alpha + 2\beta)$
$$+ \dots \text{ to } n \text{ terms}$$
$$S_n = \sin\alpha + c\sin(\alpha + \beta) + c^2\sin(\alpha + 2\beta)$$
$$+ \dots \text{to } n \text{ terms}$$
$$\therefore C_n + iS_n = e^{i\alpha} + ce^{i(\alpha + \beta)} + c^2 e^{i(\alpha + 2\beta)}$$
$$+ \dots \text{to } n \text{ terms}$$

This is geometric series with first term $e^{i\alpha}$ and common ratio $ce^{i\beta}$, hence

$$C_n + iS_n = \frac{e^{i\alpha}(1 - c^n e^{i\beta n})}{(1 - ce^{i\beta})}$$

$$= \frac{e^{i\alpha}(1 - c^n e^{i\beta n})(1 - ce^{-i\beta})}{(1 - ce^{i\beta})(1 - ce^{-i\beta})}$$

$$= \frac{e^{i\alpha} - ce^{i(\alpha - \beta)} - c^n e^{i(\alpha + n\beta)}}{1 - c(e^{i\beta} + e^{-i\beta}) + c^2} {+ c^{n+1} e^{i(\alpha + (n-1)\beta)}}$$

Equating real and imaginary parts, we get

$$C_n = \frac{\cos\alpha - c\cos(\alpha - \beta) - c^n\cos(\alpha + n\beta)}{1 - 2c\cos\beta + c^2} {+ c^{n+1}\cos[\alpha + (n-1)\beta]}$$

$$S_n = \frac{\sin\alpha - c\sin(\alpha - \beta) - c^n\sin(\alpha + n\beta)}{1 - 2c\cos\beta + c^2} {+ c^{n+1}\sin[\alpha + (n-1)\beta]}$$

Now for sum to infinity, we have since $|ce^{i\beta}| = c < 1$, hence c^n and $c^{n+1} \to 0$ as $n \to \infty$

$$\therefore \quad C_\infty = \frac{\cos\alpha - c\cos(\alpha - \beta)}{1 - 2c\cos\beta + c^2}$$

and $$S_\infty = \frac{\sin\alpha - c\sin(\alpha - \beta)}{1 - 2c\cos\beta + c^2}$$

EXAMPLE 3. *Sum the series*

$$1 + \frac{\cos\theta}{\cos\theta} + \frac{\cos 2\theta}{\cos^2\theta} + \frac{\cos 3\theta}{\cos^3\theta} + \dots n \text{ terms}$$

SOUTION. Let $C_n = 1 + \dfrac{\cos\theta}{\cos\theta} + \dfrac{\cos 2\theta}{\cos^2\theta} + \dfrac{\cos 3\theta}{\cos^3\theta} + \dots$ to n terms

and $S_n = \dfrac{\sin\theta}{\cos\theta} + \dfrac{\sin 2\theta}{\cos^2\theta} + \dfrac{\sin 3\theta}{\cos^3\theta} + \dots$ to n terms

$\therefore C_n + iS_n = 1 + \dfrac{\cos\theta + i\sin\theta}{\cos\theta} + \dfrac{\cos 2\theta + i\sin 2\theta}{\cos^2\theta}$

$\qquad\qquad\qquad + \dfrac{\cos 3\theta + i\sin 3\theta}{\cos^3\theta}$

$\qquad\qquad = 1 + \sec\theta\, e^{i\theta} + \sec^2\theta\, e^{2i\theta} + \dots$ to n terms

$$= \frac{(\sec\theta e^{i\theta})^n - 1}{\sec\theta e^{i\theta} - 1}$$

Since the series on R.H.S. in G.P. with common ratio $\sec\theta\, e^{i\theta} \dots$, the modulus of which is greater than unity, hence $C_n + iS_n$

$$= \frac{\sec^n\theta e^{in\theta} - 1}{\sec\theta(\cos\theta + i\sin\theta) - 1} = \frac{e^{in\theta} - \cos^n\theta}{i\sin\theta \cdot \cos^{n-1}\theta}$$

$$= \frac{\cos n\theta + i\sin\theta - \cos^n\theta}{i\sin\theta \cdot \cos^{n-1}\theta}$$

$$= \frac{\sin n\theta + i(\cos^n\theta - \cos n\theta)}{\sin\theta \cdot \cos^{n-1}\theta}$$

$$\therefore\quad C_n = \frac{\sin n\theta}{\sin\theta \cdot \cos^{n-1}\theta}.$$

📝 **REMARK**

- Since $|\sec\theta\, e^{i\theta}| = \sec\theta > 1$, the series is divergent and sum to infinity is not possible.

11.13.3 A PARTICULAR METHOD (SUM OF FINITE SERIES WHOSE ANGLES ARE IN A.P.)

Now we give a very useful and easy method for obtaining the sum of cosine and sine series whose angles are in the arithmetic progression.

(i) Cosine series.

Let the series be $\cos\alpha + \cos(\alpha + \beta) + \cos(\alpha + 2\beta) + \dots$ to n terms and let S_n be the sum to n terms of the series.

Multiplying both sides by $2\sin\frac{1}{2}\beta$, we have

$$2\sin\frac{\beta}{2}S_n = 2\sin\frac{\beta}{2}\cos\alpha + 2\sin\frac{\beta}{2}\cos(\alpha+\beta) + 2\sin\frac{\beta}{2}\cos(\alpha+2\beta) + \dots + 2\sin\frac{\beta}{2}\cos\{\alpha + (n-1)\beta\}$$

$$= \sin\left(\alpha + \frac{1}{2}\beta\right) - \sin\left(\alpha - \frac{1}{2}\beta\right) + \sin\left(\alpha + \frac{3}{2}\beta\right) - \sin\left(\alpha + \frac{1}{2}\beta\right) + \dots + \sin\left\{\alpha + \left(n - \frac{3}{2}\right)\beta\right\} - \sin\left\{\alpha + \left(n - \frac{5}{2}\right)\beta\right\}$$

$$+ \sin\left\{\alpha + \left(n - \frac{1}{2}\right)\beta\right\} - \sin\left\{\alpha + \left(n + \frac{3}{2}\right)\beta\right\}$$

$$= \sin\left\{\alpha + \left(n - \frac{1}{2}\right)\beta\right\} - \sin\left(\alpha - \frac{1}{2}\beta\right) = 2\cos\left\{\alpha + \frac{1}{2}(n-1)\beta\right\}\sin\frac{1}{2}n\beta$$

$$\therefore\qquad S_n = \frac{\cos\left\{\alpha + \frac{1}{2}(n-1)\beta\right\}\sin\frac{1}{2}n\beta}{\sin\frac{1}{2}\beta}$$

Cor. 1. When $\beta = \alpha$, we have

$$\cos\alpha + \cos 2\alpha + \cos 3\alpha + \dots + \cos n\alpha = \frac{\cos\frac{n+1}{2}\alpha\sin\frac{1}{2}n\alpha}{\sin\frac{1}{2}\alpha}$$

Cor. 2. Also if $\beta = \dfrac{2\pi}{n}$

$$\cos\alpha + \cos\left(\alpha + \frac{2\pi}{n}\right) + \cos\left(\alpha + \frac{4\pi}{n}\right) + \dots + \cos\left\{\alpha + 2(n-1)\frac{\pi}{n}\right\} = \frac{\cos\left\{\alpha + (n-1)\frac{\pi}{n}\right\}\sin\pi}{\sin\frac{\pi}{n}} = 0.$$

(ii) Sine series. Consider the series

$$S_n = \sin\alpha + \sin(\alpha + \beta) + \sin(\alpha + 2\beta) + \dots + \sin\{\alpha + (n-1)\beta\}$$

$$2\sin\frac{\beta}{2}S_n = 2\sin\frac{\beta}{2}\sin\alpha + 2\sin\frac{\beta}{2}\sin(\alpha+\beta) + 2\sin\frac{\beta}{2}\sin(\alpha+2\beta) + \dots + 2\sin\frac{\beta}{2}\sin\{\alpha + (n-1)\beta\}$$

$$= \cos\left(\alpha - \frac{\beta}{2}\right) - \cos\left(\alpha + \frac{\beta}{2}\right) + \cos\left(\alpha + \frac{\beta}{2}\right) - \cos\left(\alpha + \frac{3\beta}{2}\right)$$

$$+ \dots + \cos\left\{\alpha + \left(n - \frac{5}{2}\right)\beta\right\} - \cos\left\{\alpha + \left(n + \frac{3}{2}\right)\beta\right\} + \cos\left\{\alpha\left(n - \frac{3}{2}\right)\beta\right\} - \cos\left\{\alpha + \left(n - \frac{1}{2}\right)\beta\right\}$$

$$= \cos\left(\alpha - \frac{1}{2}\beta\right) - \cos\left\{\alpha + \left(n - \frac{1}{2}\right)\beta\right\} = 2\sin\left\{\alpha + \frac{1}{2}(n-1)\alpha\right\}\sin\frac{1}{2}n\beta$$

$$\therefore \quad S_n = \frac{\sin\left\{\alpha + \frac{1}{2}(n-1)\beta\right\}\sin\frac{1}{2}n\beta}{\sin\frac{1}{2}\beta}$$

Cor. 1. In case $\alpha = \beta$, we have

$$\sin\alpha + \sin 2\alpha + \sin 3\alpha + \ldots + \sin n\alpha = \frac{\sin\frac{1}{2}(n-1)\alpha \sin\frac{1}{2}n\alpha}{\sin\frac{1}{2}\alpha}$$

Cor. 2. Also if $\beta = \dfrac{2\pi}{n}$, we have $\sin\alpha + \sin\left(\alpha + \dfrac{2\pi}{n}\right) + \ldots + \sin\left\{\alpha + 2(n-1)\dfrac{\pi}{n}\right\} = 0$

 REMARK

- The sum of the cosine and sine series obtained above may be committed to memory in the following form :

$$\cos\alpha + \cos(\alpha + \beta) + \cos(\alpha + 2\beta) + \ldots + \cos\{\alpha + (n-1)\beta\} = \cos\left[\frac{\text{first angle} + \text{last angle}}{2}\right]\frac{\sin\left(n \times \dfrac{\text{diff.}}{2}\right)}{\sin\left(\dfrac{\text{diff}}{2}\right)}$$

$$\sin\alpha + \sin(\alpha + \beta) + \sin(\alpha + 2\beta) + \ldots + \sin\{\alpha + (n-1)\beta\} = \sin\left[\frac{\text{first angle} + \text{last angle}}{2}\right]\frac{\sin\left(n \times \dfrac{\text{diff.}}{2}\right)}{\sin\left(\dfrac{\text{diff}}{2}\right)}$$

Solved Examples

EXAMPLE 1. *Sum to n terms the series,*
$$\sin\alpha - \sin(\alpha + \beta) + \sin(\alpha + 2\beta) - \sin(\alpha + 3\beta) + \ldots$$

SOLUTION. The given series can be written as
$$\sin\alpha + \sin(\alpha + \beta + \pi) + \sin(\alpha + 2\beta + 2\pi)$$
$$+ \sin(\alpha + 3\beta + 3\pi) + \ldots + n \text{ terms}$$
which is the sum of the sines of the angles in A.P. with common difference $\pi + \beta$. Hence by the above article,
$$S_n = \frac{\sin\left\{\alpha + \frac{1}{2}(n-1)(\pi + \beta)\right\}\sin\frac{1}{2}\pi(\beta + \pi)}{\cos\frac{1}{2}\beta}$$

EXAMPLE 2. *Sum the series*
$$\cos^2\alpha + \cos^2\left(\alpha + \frac{\pi}{2}\right) + \cos^2\left(\alpha + \frac{2\pi}{2}\right) + \ldots$$
$$+ \text{ to n terms}$$

SOLUTION. Since $2\cos^2 x = \cos 2x + 1$, we have if S_n is the sum of the given series
$$2S_n = 2\cos^2\alpha + 2\cos^2\left(\alpha + \frac{1}{2}\pi\right) + 2\cos^2\left(\alpha + \frac{1}{2}2\pi\right)$$
$$+ \ldots + 2\cos^2\left\{\alpha + \frac{1}{2}(n-1)\pi\right\}$$
$$= [\cos 2\alpha + \cos(2\alpha + \pi) + \cos(2\alpha + 2\pi)$$
$$+ \ldots + \cos\{2\alpha + (n-1)\pi\}] + n$$
$$= n + \cos\left[\frac{2.2\alpha + (n-1)\pi}{2}\right]\frac{\sin\frac{1}{2}n\pi}{\sin\frac{1}{2}n\pi} \quad \text{[By art. 55.57]}$$

[For complete proof prove this]
$$= n + \cos\left[2\alpha + \frac{1}{2}(n-1)\pi\right]\sin\frac{1}{2}n\pi$$
$$\therefore \quad S_n = \frac{1}{2}n + \frac{1}{2}n\cos\left[2\alpha + \frac{1}{2}(n-1)\pi\right]\sin\frac{1}{2}n\pi$$

EXAMPLE 3. *Sum the series $\sin A + \sin 2A + \ldots + \sin nA$ and deduce the result*
$$1 + 2 + 3 + \ldots + n = \frac{n(n+1)}{2}.$$

SOLUTION. We have already seen in the article that (prove this for complete proof)
$$\sin A + \sin 2A + \ldots + \sin nA$$
$$= \frac{\sin\{(n+1)A/2\}\sin\{n(A/2)\}}{\sin A/2}$$
Now expanding all the terms in the series on both sides, we have
$$\left\{A - \frac{A^3}{3!} + \frac{A^5}{5!} - \ldots\right\} + \left\{2A - \frac{(2A)^3}{3!} + \ldots\right\}$$
$$+ \ldots + \left\{nA - \frac{(nA)^3}{3!} + \ldots\right\}$$
$$= \left\{(n+1)\frac{A}{2} - \frac{(n+1)^3 A^3}{3!8} + \ldots\right\}\left\{\frac{nA}{2} - \frac{(nA)^2}{3!8} - \ldots\right\}$$
$$\frac{(n+1)A/2 \cdot nA/2}{A/2}\left\{1 - \frac{(n+1)^2 A^2/4}{3!}\right\}$$
$$\times \left\{\frac{A}{2} - \frac{(A/2)^3}{3!} + \ldots\right\}^{-1} \times \left\{1 - \frac{(nA)^2}{3!} + \ldots\right\}$$
$$\left\{1 + \frac{(A/2)^2}{3!} + \ldots\right\}$$

Equating the coefficients of A on both sides, we have
$$1 + 2 + 3 + \ldots + n = \frac{n(n+1)}{2}$$

11.13.4 SUMMATION DEPENDING ON ARITHMETICO-GEOMETRIC SERIES

Now we give examples which will illustrate the method of solving with arithmetico-geometric series.

Solved Examples

EXAMPLE 1. *Sum the series* $3 \sin \alpha + 5 \sin 2\alpha + 7 \sin 3\alpha + \ldots$ *to n terms.*

SOLUTION. Let

$S = 3 \sin \alpha + 5 \sin 2\alpha + 7 \sin 3\alpha + \ldots$ to n terms

$C = 3 \cos \alpha + 5 \cos 2\alpha + 7 \cos 3\alpha + \ldots$ to n terms

$\therefore C + iS = 3 e^{i\alpha} + 5 e^{2i\alpha} + 7 e^{3i\alpha} + \ldots + (2n + 1) e^{in\alpha}$

and $e^{i\alpha}[C + iS] = 3 e^{2i\alpha} + 5 e^{3i\alpha} + \ldots$

$\qquad\qquad\qquad\qquad + (2n-1)e^{in\alpha} + (2n+1)e^{i(n+1)\alpha}$

$\therefore \quad (1 - e^{i\alpha})[C + iS] = 3 e^{i\alpha} + 2[e^{2i\alpha} + e^{3i\alpha}$

$\qquad\qquad\qquad + \ldots - (n-1)\text{terms}] - (2n+1)e^{i(n+1)\alpha}$

$\qquad = [e^{i\alpha} - (2n+1)e^{i(n+1)\alpha}] + \dfrac{2e^{i\alpha[1-e^{in\alpha}]}}{1 - e^{i\alpha}}$

$\qquad = [e^{i\alpha}(2n+1)e^{i(n+1)\alpha}] + \dfrac{2[1-e^{in\alpha}]}{e^{-i\alpha} - 1}$

$\therefore \quad C + iS = \dfrac{[e^{i\alpha} - (2n+1)e^{i(n+1)\alpha}]}{1 - e^{i\alpha}} - \dfrac{2[1-e^{in\alpha}]}{(1-e^{-i\alpha})(1-e^{i\alpha})}$

$\qquad = \dfrac{[e^{i\alpha} - (2n+1)e^{i(n+1)\alpha}][1-e^{-i\alpha}]}{(1-e^{i\alpha})(1-e^{-i\alpha})} - \dfrac{2(1-e^{in\alpha})}{(1-e^{i\alpha})(1-e^{-i\alpha})}$

$\qquad = \dfrac{e^{i\alpha}(2n+1)e^{i(n+1)\alpha} + (2n+3)e^{in\alpha} - 3}{2 - 2\cos\alpha}$

Hence, equating imaginary parts,

$S = \dfrac{\sin\alpha - (2n+1)\sin(n+1) + (2n+3)\sin n\alpha}{2(1-\cos\alpha)}$

EXAMPLE 2. *If* S_n *denotes the sum of n terms of the series*

$\qquad \sin x + \sin 2x + \ldots n$ *terms*

show that $\lim\limits_{n\to\infty} \dfrac{(S_1 + S_2 + \ldots S_n)}{n} = \dfrac{1}{2}\cot\dfrac{x}{2}$.

SOLUTION. We have $S_n = \sin x + \sin 2x + \ldots \sin nx$

$\therefore C_n + iS_n = e^{ix} + e^{2ix} + \ldots e^{inx} = \dfrac{e^{ix}[1-e^{inx}]}{1-e^{ix}}$

$\qquad = \dfrac{e^{ix}[1-e^{inx}][1-e^{-ix}]}{[1-e^{ix}][1-e^{-ix}]} = \dfrac{e^{ix} - 1 - e^{i(n+1)x} + e^{inx}}{2[1-\cos x]}$

So, $\quad S_n = \dfrac{\sin x - \sin(n+1)x + \sin nx}{2[1-\cos x]}$

$\qquad = \dfrac{\sin x}{2[1-\cos x]} - \dfrac{\cos\left(n+\dfrac{1}{2}\right)x \sin\dfrac{x}{2}}{[1-\cos x]}$

Now $\dfrac{S_1 + S_2 + \ldots S_n}{n} = \dfrac{\sum\limits_{k=1}^{n} S_k}{n} = \dfrac{\sin x}{2[1-\cos x]}$

$\qquad - \dfrac{\sin x/2}{(1-\cos x)} \sum\limits_{k=1}^{n} \dfrac{\cos\left(k+\dfrac{1}{2}\right)x}{n}$

Also, $\sum\limits_{k=1}^{n} \cos\left(k+\dfrac{1}{2}\right)x = \cos\dfrac{3}{2}x + \cos\dfrac{5}{2}x + \ldots\cos\left(n+\dfrac{1}{2}\right)x$

Let $C = \cos\dfrac{3}{2}x + \cos\dfrac{5}{2}x + \ldots\cos\left(n+\dfrac{1}{2}\right)x$

Then $C + iS = e^{\frac{3}{2}ix} + e^{\frac{5}{2}ix} + \ldots e^{\left(n+\frac{1}{2}\right)ix}$

$\qquad = e^{\frac{3}{2}ix}\dfrac{[1-e^{inx}]}{1-e^{ix}} = \dfrac{e^{\frac{3}{2}ix}[1-e^{inx}][1-e^{-ix}]}{[1-e^{ix}][1-e^{-ix}]}$

$\qquad = \dfrac{e^{\frac{3}{2}ix} - e^{i\left(n+\frac{3}{2}\right)x} - e^{i\frac{x}{2}} + e^{i\left(n+\frac{1}{2}\right)x}}{2[1-\cos x]}$

$\therefore \quad C = \dfrac{\cos\dfrac{3}{2}x - \cos\left(n+\dfrac{3}{2}\right)x - \cos\dfrac{x}{2} + \cos\left(n+\dfrac{1}{2}\right)x}{2[1-\cos x]}$

Now

$\qquad = \lim\limits_{n\to\infty} \dfrac{S_1 + S_2 + \ldots S_n}{n} = \dfrac{2\sin\dfrac{x}{2}\cos\dfrac{x}{2}}{2 \times 2\sin^2\dfrac{x}{2}}$

$\qquad = \lim\limits_{n\to\infty} \dfrac{1}{n}\left[\dfrac{\sin\dfrac{x}{2}}{2[1-\cos x]^2}\left\{ \cos\dfrac{3}{2}x - \cos\left(n+\dfrac{3}{2}\right)x \right.\right.$

$\qquad\qquad\qquad\qquad \left.\left. - \cos\dfrac{x}{2} + \cos\left(n+\dfrac{1}{2}\right)x \right\}\right]$

$\qquad = \dfrac{1}{2}\cot\dfrac{x}{2} - \lim\limits_{n\to\infty} \dfrac{1}{n}\text{ [finite quantity] } = \dfrac{1}{2}\cot\dfrac{x}{2}$

Exercise-11.14

Sum the following series to n terms : (Ques 1–2)

1. $\sin A + \sin 3A + \sin 5A + \ldots$ and deduce the sum of $1 + 3 + 5 \ldots$ to n terms.

2. $\cos^2\theta + \cos^2 2\theta + \cos^2 3\theta + \ldots$

3. Prove that the two series $\sin\dfrac{\pi}{14} + \sin\dfrac{2\pi}{14} + \sin\dfrac{3\pi}{14} + \ldots$ to 28 terms, and $\cos\dfrac{\pi}{14} + \cos\dfrac{2\pi}{14} + \cos\dfrac{3\pi}{14} + \ldots$ to 28 terms have the same sum. What is the magnitude of the sum?

4. Sum the series $\cos\theta \, \cos 2\theta + \cos 2\theta \cos 3\theta + \cos 3\theta \cos 4\theta + \ldots$ to 20 terms

5. Find the value of
$$\dfrac{\sin\alpha - \sin(\alpha+\beta) + \sin(\alpha+2\beta) - \sin(\alpha+3\beta) + \ldots n \text{ terms}}{\cos\alpha - \cos(\alpha+\beta) + \cos(\alpha+2\beta) - \cos(\alpha+3\beta) + \ldots n \text{ terms}}$$

6. Sum the following series :
   (i) $\cos x - \sin 2x - \cos 3x + \sin 4x + \ldots$ to n terms
   (ii) $\sin^3\alpha + \sin^3 3\alpha + \sin^3 5\alpha + \ldots$ to n terms
   (iii) $\sin^3\alpha + \sin^3(\alpha + \beta) + \sin^3(\alpha + 2\beta) + \ldots$ to n terms
   (iv) $\cos^3\theta + \cos^3\left(\theta - \dfrac{2\pi}{n}\right) + \cos^3\left(\theta - \dfrac{4\pi}{n}\right) \ldots$ to n terms

7. Sum the following series to n terms :

(i) $\sin^2\alpha + \sin^2(\alpha + \beta) + \sin^2(\alpha + 2\beta) + \ldots$

(ii) $\sin^4\alpha + \sin^4 2\alpha + \sin^4 4\alpha + \ldots$

(iii) $\cos^4\alpha + \cos^4\left(\alpha + \dfrac{2\pi}{n}\right) + \cos^4\left(\alpha + \dfrac{4\pi}{n}\right) + \ldots$

8. Show that $\cos^2\alpha + \cos^2\left(\alpha + \dfrac{\pi}{n}\right) + \cos^2\left(\alpha + \dfrac{2\pi}{n}\right) + \ldots$ to n terms

$= n/2$.

9. Sum to infinity the following series :

(i) $\sin\alpha + \dfrac{1}{2}\sin 2\alpha + \dfrac{1}{2^2}\sin 3\alpha + \ldots$ ad. Inf.

(ii) $\dfrac{\sin\alpha}{\tan\beta} - \dfrac{\sin 2\alpha}{\tan^2\beta} + \dfrac{\sin 3\alpha}{\tan^3\beta} - \ldots$ad inf

where $\tan\beta > 1$ numerically.

10. Find the sum of the following series to n terms and hence deduce the sum to infinity given $\alpha \neq \pi/2$

(i) $\cos\alpha\ \sin\alpha + \cos^2\alpha\ \sin 2\alpha + \cos^3\alpha\ \sin 3\alpha + \ldots$

(ii) $\cos\alpha\ \cos\alpha + \cos^2\alpha\ \cos 2\alpha + \cos^3\alpha\ \cos 3\alpha + \ldots$

(iii) $\cos\alpha\ \sin\alpha + \cos 2\alpha\ \sin^2\alpha + \cos 3\alpha\ \sin^3\alpha + \ldots$

(iv) $\sin\alpha\ \sin\alpha + \sin 2\alpha\ \sin^2\alpha + \sin 3\alpha\ \sin^3\alpha + \ldots$

(v) $1 + c\cos\alpha + c^2\cos 2\alpha + c^3\cos 3\alpha + \ldots, |c| < 1$.

11. Sum the following series :

(i) $\cos\alpha + 2\cos 2\alpha + 3\cos 3\alpha + \ldots n\cos n\alpha$

(ii) $1 - 2\cos\alpha + 3\cos 2\alpha - 4\cos 3\alpha + \ldots$to n terms

(iii) $1 + \dfrac{2}{2}\cos\alpha + \dfrac{3}{2^2}\cos 2\alpha + \dfrac{4}{2^3}\cos 3\alpha + \ldots$ ad inf

12. Prove that $2[\sin\theta + \sin 2\theta + \sin 3\theta + \ldots \infty] = \sin\theta/(1 - \cos\theta)$.

13. Prove that

$\log(\cos\theta) + \log\left(\cos\dfrac{\theta}{2}\right) + \log\left(\cos\dfrac{\theta}{2^2}\right) + \ldots = \log\left(\dfrac{\sin 2\theta}{2\theta}\right)$.

14. Prove that $\displaystyle\sum_{n=1}^{m}\cos^n\theta\cos n\theta = \dfrac{\cos^{m+1}\theta\sin\theta}{\sin\theta}$.

Hints to Selected Problems

5. $C = \cos\alpha - \cos(\alpha + \beta) + \cos(\alpha + 2\beta) - \ldots$

$S = \sin\alpha - \sin(\alpha + \beta) + \sin(\alpha + 2\beta) - \ldots$

$\therefore\ C + iS = e^{i\alpha} - e^{i(\alpha + \beta)} + e^{i(\alpha + 2\beta)} - \ldots$

$= e^{i\alpha}(1 - e^{i\beta} + e^{2i\beta} - \ldots \text{ upto } n \text{ terms})$

which is G.P.

7. (ii) $\sin^4\alpha = \left(\dfrac{1}{2}\right)^3(\cos 4\alpha - 4\cos 2\alpha + 3)$

11. (i) $C_n = \cos\alpha + 2\cos 2\alpha + 3\sin 3\alpha + \ldots$ to n terms

and $S_n = \sin\alpha + 2\sin 2\alpha + 3\sin 3\alpha + \ldots$ to n terms

$\Rightarrow C_n + iS_n = e^{i\alpha} + 2e^{2i\alpha} + 3e^{3i\alpha} + \ldots + ne^{ni\alpha}$ ...(1)

Also $e^{i\alpha}(C_n + iS_n) = e^{2i\alpha} + 2e^{3i\alpha} + 3e^{4i\alpha} + \ldots + ne^{ni\alpha + k}$ ...(2)

Now subtracting (2) from (1)

12. Suppose $S_\infty = \sin\theta + \sin 2\theta + \sin 3\theta + \ldots \infty$

and $C_\infty = \cos\theta + \cos 2\theta + \cos 3\theta + \ldots \infty$

$\therefore\ C_\infty + iS_\infty = e^{i\theta} + e^{2i\theta} + e^{3i\theta} + \ldots\infty = \dfrac{e^{i\theta}}{1 - e^{i\theta}}$

After some simplification and taking only imaginary part, we get required result.

Answers

1. $(\sin^2 nA)/\sin A;\ n^2$ **2.** $\dfrac{1}{2}\{n + \cos(n + 1)\theta\sin n\theta\ \mathrm{cosec}\ \theta\}$ **5.** $\tan\left[\alpha + \dfrac{1}{2}(n - 1)(\pi + \beta)\right]$

6. (i) $\dfrac{\sin\left[\dfrac{n+1}{2}\left(\dfrac{\pi}{2} + x\right)\right]\sin\left(\dfrac{n\pi}{4} + \dfrac{nx}{2}\right)}{\sin\left(\dfrac{\pi}{4} + \dfrac{x}{2}\right)}$ (ii) $\dfrac{3}{2}\sin^2 n\alpha\ \mathrm{cosec}\ \alpha - \dfrac{1}{4}\sin^2 3n\alpha\ \mathrm{cosec}\ 3\alpha$

(iii) $\dfrac{3}{4}\sin\left[\alpha + \dfrac{n-1}{2}\beta\right]\sin\dfrac{n}{2}\beta\ \mathrm{cosec}\ \dfrac{\beta}{2} - \dfrac{1}{4}\sin\left[3\alpha + \dfrac{3(n-1)}{2}\beta\right]\sin\dfrac{3n}{2}\beta\ \mathrm{cosec}\ \dfrac{3\beta}{2}$ (iv) 0

7. (i) $\dfrac{1}{2}[n - \cos\{2\alpha + (n - 1)\beta\}\sin n\beta \times \mathrm{cosec}\ \beta]$ (ii) $\dfrac{3}{8}n - \dfrac{1}{2}\cos(n + 1)\alpha n\alpha\ \mathrm{cosec}\ \alpha + \dfrac{1}{8}\cos 2(n + 1)\alpha\sin 2n\alpha\ \mathrm{cosec}\ 2\alpha$ (iii) $\dfrac{3}{8}n$

9. (i) $\dfrac{4\sin}{5 - 4\cos\alpha}$ (ii) $\dfrac{\sin\alpha\cot\beta}{1 + 2\cos\alpha\cot\beta + \cot^2\beta}$

10. (i) $\cot\alpha\ (1 - \cos^n\alpha\cos n\alpha).\cot\alpha$ (ii) $\cot\alpha.\cos^n\alpha\sin n\alpha,\ 0$

(iii) $C_n = \dfrac{\sin\alpha\cos\alpha + \sin^2\alpha - \sin^n\alpha\cos n\alpha + \sin^{n+1}\alpha\cos(n - 1)\alpha}{1 + \sin^2\alpha - 2\sin\alpha\cos\alpha},\ \infty$ $\dfrac{\sin\alpha\cos\alpha + \sin\ \alpha}{1 + \sin\ \alpha - 2\sin\alpha\cos\alpha}$

(iv) $S_n = \dfrac{\sin^2\alpha - \sin^2\alpha\sin n\alpha + \sin^{n+1}\alpha\sin^2(n - 1)\alpha}{1 + \sin^2\alpha - 2\sin\alpha\cos\alpha},\ S_\infty = \dfrac{\sin^2\alpha}{1 + \sin^2\alpha - 2\sin\alpha\cos\alpha}$

(v) $C_n = \dfrac{1 - c\cos\alpha - c^n\cos n\alpha + c^{n+1}\cos(n - 1)\alpha}{1 - 2c\cos\alpha + c^2},\ C_\infty = \dfrac{1 - c\cos\alpha}{1 - 2c\cos\alpha + c^2}$

11.13.5 USE OF BINOMIAL SERIES

Following are the binomial expansions when z is complex.

(i) If n is positive integer,
$$(1+z)^n = 1 + {}^nC_1 z + {}^nC_2 z^2 + \ldots + z^n$$

(ii) If n is any quantity (say negative integer) or a positive or negative fraction.
$$(1+z)^n = 1 + nz + \frac{n(n-1)}{2!}z^2 + \frac{n(n-1)(n-2)}{3!}z^3 + \ldots \text{ad inf. provided } |z| < 1,$$

when $|z| = 1$, this result is still true if (i) $n > 0$ or if (ii) $-1 < n < 0$ and $z \neq -1$

With the help of results (i) and (ii) we can recognize all binomial expansions without remembering any more formula. The method is illustrated in the following solved examples.

Solved Examples

EXAMPLE 1. *Sum the series*

$$1 + \frac{1}{2}\cos 2\alpha - \frac{1}{2.4}\cos 4\alpha + \frac{1.3}{2.4.6}\cos 6\alpha - \ldots \text{ ad inf.}$$

where α lies between $-\pi/2$ and $\pi/2$.

SOLUTION. Let

$$C = 1 + \frac{1}{2}\cos 2\alpha - \frac{1}{2.4}\cos 4\alpha + \frac{1.3}{2.4.6}\cos 6\alpha - \ldots$$

$$S = \frac{1}{2}\sin 2\alpha - \frac{1}{2.4}\sin 4\alpha + \frac{1.2}{2.4.6}\sin 6\alpha - \ldots$$

$$\therefore C + iS = 1 + \frac{1}{2}e^{2i\alpha} - \frac{1}{2.4}e^{4i\alpha} + \frac{1.3}{2.4.5}e^{6i\alpha} - \ldots$$

Comparing this with

$$(1+z)^n = 1 + nz + \frac{n(n-1)}{2!}z^2 + \ldots$$

We have $nz = \frac{1}{2}e^{2i\alpha}$ ...(1) $\quad \frac{n(n-1)}{2}z^2 = -\frac{1}{8}e^{4i\alpha}$ ...(2)

$$\therefore \quad \frac{n(n-1)}{2n^2} = -\frac{1}{8} \times \frac{4}{1} \quad \text{[Dividing (2) by square (1)]}$$

$$\therefore \quad \frac{n-1}{n} = -1 \text{ or } n = \frac{1}{2}$$

$$\therefore \quad z = e^{2i\alpha}$$

Hence $C + iS = (1 + e^{2i\alpha})^{1/2}$

$$\text{[here } n > 0 \text{ and } |e^{2i\alpha}| = 1]$$

$$= (1 + \cos 2\alpha + i\sin 2\alpha)^{1/2}$$

$$= \sqrt{2\cos\alpha}[\cos\alpha + i\sin\alpha]^{1/2}$$

$$= \sqrt{2\cos\alpha}\left[\cos\frac{\alpha}{2} + i\sin\frac{\alpha}{2}\right]$$

Equating real and imaginary parts, we have

$$C = \sqrt{2\cos\alpha}.\cos\frac{\alpha}{2} = \sqrt{\cos\alpha(1 + \cos\alpha)}$$

$$S = \sqrt{2\cos\alpha}.\sin\frac{\alpha}{2} = \sqrt{\cos\alpha(1 - \cos\alpha)}$$

EXAMPLE 2. *Sum the series*

$$\sin\alpha + \frac{1}{2}\sin 3\alpha + \frac{1.3}{2.4}\sin 5\alpha + \ldots \text{ ad inf.}$$

SOLUTION. Let $\quad S = \sin\alpha + \frac{1}{2}\sin 3\alpha + \frac{1.3}{2.4}\sin 5\alpha + \ldots$

$$C = \cos\alpha + \frac{1}{2}\cos 3\alpha + \frac{1.3}{2.4}\cos 5\alpha + \ldots$$

$$C + iS = e^{i\alpha} + \frac{1}{2}e^{3i\alpha} + \frac{1.3}{2.4}e^{5i\alpha} + \ldots$$

$$= e^{i\alpha}\left[1 + \frac{1}{2}e^{i\alpha} + \frac{1.3}{2.4}e^{4i\alpha} + \ldots\right]$$

Comparing the series within square brackets with

$$(1+z)^n = 1 + nz + \frac{n(n-1)}{2!}z^2 + \ldots$$

We have,

$$nz = \frac{1}{2}e^{2i\alpha}, \frac{n(n-1)}{2!}z^2 = \frac{3}{8}z^{4i\alpha}$$

Solving, we get $n = -\frac{1}{2}, z = -e^{2i\alpha}$, hence

$$C + iS = e^{i\alpha}(1 - e^{2i\alpha}) - 1/2$$

Provided $\alpha \neq m\pi$, m being an integer

$$= [\cos\alpha + i\sin\alpha][1 - \cos 2\alpha - i\sin 2\alpha]^{-1/2}$$

$$= (\sqrt{2\sin\alpha})^{-1/2}[\cos\alpha + i\sin\alpha][\sin\alpha - i\cos\alpha]^{-1/2}$$

$$= (\sqrt{2\sin\alpha})^{-1/2}[\cos\alpha + i\sin\alpha]$$

$$\left[\cos\left(\frac{\pi}{2} - \alpha\right) - i\sin\left(\frac{\pi}{2} - \alpha\right)\right]^{-1/2}$$

$$= (\sqrt{2\sin a})^{-1/2}[\cos\alpha + i\sin\alpha]$$

$$\left[\cos\left(\frac{\pi}{4} - \frac{\alpha}{2}\right) - i\sin\left(\frac{\pi}{2} - \frac{\alpha}{2}\right)\right]$$

Equating imaginary parts, we get

$$S = (\sqrt{2\sin\alpha})^{-1/2}\left[\cos\alpha\sin\left(\frac{\pi}{4} - \frac{\alpha}{2}\right)\right.$$

$$\left. + \sin\alpha\cos\left(\frac{\pi}{4} - \frac{\alpha}{2}\right)\right]$$

$$= (\sqrt{2\sin\alpha})^{-1/2}\left[\sin\left(\alpha + \frac{\pi}{4} - \frac{\alpha}{2}\right)\right]$$

$$= (\sqrt{2\sin\alpha})^{-1/2}\sin\left(\frac{\pi}{4} + \frac{\alpha}{2}\right)$$

EXAMPLE 3. *Sum the following series* :

$$1 - \frac{1}{2}\cos\theta + \frac{1.3}{2.4}\cos 2\theta - \frac{1.3.5}{2.4.6}\cos 3\theta + \ldots \text{ ad inf.}$$

$(-\pi < \theta < \pi)$　　　　　　　　　　(SVTU–2009)

SOLUTION. Since in the above series, the cosines of the angles are increasing in A.P. we denote it by C. The auxiliary series is given by

$$S = -\frac{1}{2}\sin\theta + \frac{1.3}{2.4}\sin 2\theta - \frac{1.3.5}{2.4.6}\sin 3\theta + \ldots \text{ ad inf.}$$

So that $C + iS$

$$= 1 - \frac{1}{2}(\cos\theta + i\sin\theta) + \frac{1.3}{2.4}(\cos 2\theta + i\sin 2\theta)$$

$$- \frac{1.3.5}{2.4.6}(\cos 3\theta + i\sin 3\theta) + \ldots$$

$$= 1 - \frac{1}{2}e^{i\theta} + \frac{1.3}{2.4}e^{2i\theta} - \frac{1.3.5}{2.4.6}e^{3i\theta} + \dots$$

Comparing with the series

$$(1+z)^n = 1 + nz + \frac{n(n-1)}{2!}z^2 + \dots$$

We have $nz = -\frac{1}{2}e^{i\theta}, \frac{n(n-1)}{2!}z^2 = \frac{3}{8}e^{4i\theta}$

Solving $n = -\frac{1}{2}, z = e^{i\theta}$

$\therefore C + iS = [1 + e^{i\theta}]^{-1/2} = [1 + \cos\theta + i\sin\theta]^{-1/2}$

$$= \left[2\cos^2\frac{\theta}{2} + 2i\sin\frac{\theta}{2}\cos\frac{\theta}{2}\right]^{-1/2}$$

or $C + iS = \left(2\cos\frac{\theta}{2}\right)^{-1/2}\left\{\cos\frac{\theta}{2} + i\sin\frac{\theta}{2}\right\}^{-1/2}$

$$= \left(2\cos\frac{\theta}{2}\right)^{-1/2}\left\{\cos\frac{\theta}{4} - i\sin\frac{\theta}{4}\right\}$$

Equating real parts on both sides, we get

$$C = \cos\frac{\theta}{4}\left(2\cos\frac{\theta}{2}\right)^{-1/2}.$$

11.13.6 USE OF EXPONENTIAL SERIES

The following are important results for complex z :

(i) $e^z = 1 + z + \frac{z^2}{2!} + \frac{z^3}{3!} \dots$ ad inf

(ii) $e^{-z} = 1 - z + \frac{z^2}{2!} - \frac{z^3}{3!} + \dots$ ad inf

(iii) $\sin z = z - \frac{z^3}{3!} + \frac{z^5}{5!} - \dots$ ad inf

(iv) $\cos z = 1 - \frac{z^2}{2!} + \frac{z^4}{4!} - \dots$ ad inf

(v) $\cosh z = 1 + \frac{z^2}{2!} + \frac{z^4}{4!} + \dots$ ad inf

(vi) $\sinh z = z + \frac{z^3}{3!} + \frac{z^5}{5!} + \dots$ ad inf

Solved Examples

EXAMPLE 1. Sum the series

$$\cos^n\alpha - n\cos^{n-1}\alpha\cos\alpha$$
$$+ \frac{n(n-1)}{1.2}\cos^{n-2}\alpha\cos 2\alpha + \dots \text{ to } (n+1) \text{ terms.}$$

SOLUTION. Let $C = \cos^n\alpha - n\cos^{n-1}\alpha\cos\alpha$
$$+ \frac{n(n-1)}{1.2}\cos^{n-2}\alpha\cos 2\alpha + \dots$$
to $(n+1)$ terms

Then $C + iS$

$$= \cos^n\alpha - n\cos^{n-1}\alpha e^{i\alpha} + \frac{n(n-1)}{1.2}\cos^{n-2}\alpha e^{2i\alpha} + \dots$$
to $(n+1)$ terms
$$= [\cos\alpha - e^{i\alpha}]^n = [\cos\alpha - \cos\alpha - i\sin\alpha]^n$$
$$= (-i)^n\sin^n\alpha = (-1)^n i^n\sin^n\alpha$$

$\therefore C = (-1)^{n/2}\sin^n\alpha$, when n is even.
$= 0$, when n is odd.

EXAMPLE 2. Sum the series

$$1 - \cos\alpha\cos\beta + \frac{\cos^2\alpha\cos 2\beta}{2!} - \frac{\cos^3\alpha\cos 3\beta}{3!} + \dots$$

ad inf.

SOLUTION. Let $C = 1 - \cos\alpha\cos\beta$
$$+ \frac{\cos^2\alpha\cos 2\beta}{2!} - \frac{\cos^3\alpha\cos 3\beta}{3!} + \dots$$
$$S = -\cos\alpha\sin\beta$$
$$+ \frac{\cos^2\alpha\sin 2\beta}{2!} - \frac{\cos^3\alpha\sin 3\beta}{3!} + \dots$$

$\therefore C + iS = 1 - \cos\alpha\, e^{i\beta}$
$$+ \frac{\cos^2\alpha e^{2i\beta}}{2!} - \frac{\cos^3\alpha e^{3i\beta}}{3!} + \dots$$
$$= \exp\{-\cos\alpha\, e^{i\theta}\}$$
$$= \exp\{-\cos\alpha\,(\cos\beta + i\sin\beta)\}$$
$$= \exp\{-\cos\alpha\cos\beta\}[\cos(\cos\alpha\sin\beta)$$
$$- i\sin(\cos\alpha\sin\beta)]$$
$\therefore\quad C = \exp\{-\cos\alpha\cos\beta\}\cos(\cos\alpha\sin\beta)$

EXAMPLE 3. Sum the following series :

$$\sin\theta\cos\theta + \frac{\sin 2\theta\cos^2\theta}{2!} + \frac{\sin 3\theta\cos^3\theta}{3!} + \dots ad\ inf.$$

SOLUTION. Let

$$S = \sin\theta\cos\theta + \frac{\sin 2\theta\cos^2\theta}{2!} + \frac{\sin 3\theta\cos^3\theta}{3!} + \dots$$
ad inf

The auxiliary series is given by
$$C = \cos\theta.\cos\theta$$
$$+ \frac{\cos 2\theta\cos^2\theta}{2!} + \frac{\cos 3\theta\cos^3\theta}{3!} + \dots$$

so that $C + iS = \cos\theta\,(\cos\theta + i\sin\theta)$
$$+ \frac{\cos^2\theta}{2!}(\cos 2\theta + i\sin 2\theta)$$
$$+ \frac{\cos^3\theta}{3!}(\cos 3\theta + i\sin 3\theta) + \dots$$
$$= e^{i\theta}\cos\theta + \frac{e^{2i\theta}\cos^2\theta}{2!} + \frac{e^{3i\theta}\cos^3\theta}{3!} + \dots$$
$$= \exp\{e^{i\theta}\cos\theta\} - 1 = \exp\{(\cos\theta + i\sin\theta)\cos\theta\} - 1$$
$$= \exp\{\cos^2\theta + i\sin\theta\cos\theta\} - 1$$
$$= e^{\cos^2\theta}\{\cos(\sin\theta\cos\theta) + i\sin(\sin\theta\cos\theta)\} - 1$$
$$S = e^{\cos^2\theta}\sin(\sin\theta\cos\theta).$$

EXAMPLE 4. Sum the following series :

$$1 - \frac{c^2}{2!}\cos 2\theta + \frac{c^4\cos 4\theta}{4!} - \frac{c^6\cos 6\theta}{6!} + \dots ad\ inf.$$

SOLUTION. Let $C = 1 - \frac{c^2}{2!}\cos 2\theta + \frac{c^4\cos 4\theta}{4!} - \frac{c^6\cos 6\theta}{6!} + \dots$ad inf

The auxiliary series is given by

$$S = -\frac{c^2}{2!}\sin 2\theta + c^4\frac{\sin 4\theta}{4!} - c^6\frac{\cos 6\theta}{6!} + \dots$$

so that

$$C + iS = 1 - \frac{c^2}{2!}(\cos 2\theta + i \sin 2\theta)$$

$$+ \frac{c^4}{4!}(\cos 4\theta + i \sin 4\theta)$$

$$- \frac{c^6}{6!}(\cos 6\theta + i \sin 6\theta) + \dots$$

$$= 1 - \frac{c^2}{2!}e^{2i\theta} + \frac{c^4}{4!}e^{4i\theta} - \frac{c^6}{6!}e^{6i\theta} + \dots$$

The above series is a cosine series and hence

$$C + iS = \cos(ce^{i\theta}) = \cos(c\cos\theta + ic\sin\theta)$$
$$= \cos(c\cos\theta)\cos(ci\sin\theta) - \sin(c\cos\theta)\sin(ci\sin\theta)$$
$$= \cos(c\cos\theta)\cosh(c\sin\theta) - i\sin(c\cos\theta)\sinh(c\sin\theta)$$

$$\therefore \quad C = \cos(c\cos\theta)\cosh(c\sin\theta).$$

11.13.7 USE OF LOGARITHMIC AND GREGORY'S SERIES

If $|z| \leq 1$, but $z \neq -1$, then $\log(1 + z) = z - \frac{z^2}{2} + \frac{z^3}{3} - \dots$ ad. inf   ...(1)

If $|z| \leq 1$, but $z \neq 1$, then $\log(1 - z) = -\left(z + \frac{z^2}{2} + \frac{z^3}{3} \dots \text{ad.inf.} \right)$ ...(2)

From (1) and (2), we have

$$\log(1 + z) + \log(1 - z) = -2\left[\frac{z^2}{2} + \frac{z^4}{4} + \frac{z^6}{6} + \dots \right] \text{ and } \quad \log(1 + z) - \log(1 - z) = 2\left[z + \frac{z^3}{3} + \frac{z^5}{5} + \dots \right].$$

Gregory's series is, if $-1 \leq |z| \leq 1$, then $\tan^{-1} z = z - \frac{z^3}{3} + \frac{z^5}{5} \dots$ ad inf

Solved Examples

EXAMPLE 1. *Sum the following series.*

  (i) $\cos\alpha - \frac{1}{2}\cos 2\alpha + \frac{1}{3}\cos 3\alpha - \dots \text{ad.inf}.$

                                  (SVTU–2006)

  (ii) $\sin\alpha - \frac{1}{2}\sin 2\alpha + \frac{1}{3}\sin 3\alpha - \dots \text{ad.inf}.$

SOLUTION. Let series (i) and (ii) be denoted by C and S respectively, then

$$C + iS = e^{i\alpha} - \frac{1}{2}e^{2i\alpha} + \frac{1}{3}e^{3i\alpha} - \dots$$
$$= \log(1 + e^{i\alpha}) \text{ provided } e^{i\alpha} \neq 1 \text{ or } \alpha \neq (2n + 1)\pi$$
$$= \log(1 + \cos\alpha + i\sin\alpha)$$
$$= \log\left[2\cos\frac{\alpha}{2}\left(\cos\frac{\alpha}{2} + i\sin\frac{\alpha}{2} \right) \right]$$
$$= \log\left[2\cos\frac{\alpha}{2}e^{i(\alpha/2)} \right] = \log\left(2\cos\frac{\alpha}{2} \right) + i\frac{\alpha}{2}$$

$$\therefore \quad C = \log\left(2\cos\frac{\alpha}{2} \right), S = \frac{\alpha}{2}$$

Condition is that $\alpha \neq (2n + 1)\pi$.

EXAMPLE 2. *Find the sum of the series*

  (i) $a\cos\alpha - \frac{1}{2}a^2\cos 2\alpha + \frac{1}{3}a^3\cos 3\alpha - \dots \text{ad. inf.}$

  (ii) $a\sin\alpha - \frac{1}{2}a^2\sin 2\alpha + \frac{1}{3}a^3\sin 3\alpha - \dots \text{ad inf.}$

                            (KURUKSHETRA–2005)

SOLUTION. Let series (i) and (ii) be denoted by C and S respectively; then

$$C + iS = ae^{i\alpha} - \frac{1}{2}a^2e^{2i\alpha} + \frac{1}{3}a^3e^{3i\alpha} - \dots$$
$$= \log(1 + ae^{i\alpha}) \text{ provided } |a| \leq 1,$$
$$\text{and } ae^{i\alpha} \neq -1$$
$$= \log(1 + a\cos\alpha + ai\sin\alpha)$$
$$= \frac{1}{2}[1 + a^2 + 2a\cos\alpha] + i\tan^{-1}\frac{a\sin\alpha}{1 + a\cos\alpha}$$

$$\left[\text{Since } \log(x + iy) = \frac{1}{2}\log(x^2 + y^2) + i\tan^{-1}\frac{y}{x} \right]$$

Equating real and imaginary parts, we get

$$C = \frac{1}{2}\log[1 + a^2 + 2a\cos\alpha],$$

$$S = \tan^{-1}\frac{a\sin\alpha}{1 + a\cos\alpha}.$$

Conditions of validity are

(i) $|ae^{i\alpha}| \leq 1$, i.e, $|a| \leq 1$ numerically

and (ii) α is not odd multiple of π when $a = 1$ and even multiple of π when $a = -\alpha$, i.e, $ae^{i\alpha} \neq -1$.

EXAMPLE 3. *Sum the following series to infinity :*

$$\cos\theta - \frac{1}{3}\cos 3\theta + \frac{1}{5}\cos 5\theta - \dots$$

SOLUTION. Let $\quad C = \cos\theta - \frac{1}{3}\cos 3\theta + \frac{1}{5}\cos 5\theta - \dots$

$$S = \sin\theta - \frac{1}{3}\sin 3\theta + \frac{1}{5}\sin 5\theta - \dots$$

so that $C + iS$

$$= (\cos\theta + i\sin\theta) - \frac{1}{3}(\cos 3\theta + i\sin 3\theta)$$

$$+ \frac{1}{5}(\cos 5\theta + i\sin 5\theta) - \dots$$

$$= e^{i\theta} - \frac{1}{3}e^{3i\theta} + \frac{1}{5}e^{5i\theta} - \dots$$

This is Gregory's series. Hence

$$C + iS = \tan^{-1}e^{i\theta} = \tan^{-1}(\cos\theta + i\sin\theta)$$
$$\therefore \quad \tan(C + iS) = \cos\theta + i\sin\theta$$
$$\tan(C - iS) = \cos\theta - i\sin\theta$$

$$\tan\{(C + iS) + (C - iS)\} = 1 - \frac{\tan(C + iS) + \tan(C - iS)}{\tan(C + iS)\tan(C - iS)}$$

or $\tan 2C = \dfrac{2\cos\theta}{1 - (\cos\theta + i\sin\theta)(\cos\theta - i\sin\theta)} = \dfrac{2\cos\theta}{1 - 1}$

$= +\infty$ if $\cos\theta \neq 0$ and $+$ ive

$= -\infty$ if $\cos\theta \neq 0$ and is a $-$ive quantity.

Hence, $C = \dfrac{1}{2}\tan^{-1}(+\infty) = \dfrac{\pi}{4}$ if $\cos\theta$ is positive.

and $C = \dfrac{1}{2}\tan^{-1}(-\infty) = -\dfrac{\pi}{4}$ if $\cos\theta$ is negative.

If $\cos\theta$ is zero, the cosines of all the odd multiples of θ are zero and hence the sum of the given series is zero. Hence, the given series is $\pi/4$ or $-\pi/4$ as $\cos\theta$ is positive or negative and is equal to zero if $\cos\theta$ is zero.

Exercise-11.15

Sum the series :

1. $\cos\alpha + \dfrac{\sin\alpha\cos 2\alpha}{1!} + \dfrac{\sin^2\alpha\cos 3\alpha}{2!} + \ldots$ ad. inf.

2. $\cos\alpha + \dfrac{\cos\alpha\cos 2\alpha}{1!} + \dfrac{\cos^2\alpha\cos 3\alpha}{2!} + \ldots$ ad. inf.

3. $\sin\alpha\sin\alpha - \dfrac{1}{2}\sin^2\alpha\sin 2\alpha + \dfrac{1}{3}\sin^3\alpha\sin 3\alpha + \ldots$ ad. inf

4. $\sin\alpha\sin\beta + \dfrac{1}{2}\sin 2\alpha\sin 2\beta + \dfrac{1}{3}\sin 3\alpha\sin 3\beta + \ldots$ ad. inf

5. $\cos\dfrac{\pi}{3} + \dfrac{1}{2}\cos\dfrac{2\pi}{3} + \dfrac{1}{3}\cos\dfrac{3\pi}{3} + \ldots$ ad. inf.

6. $\dfrac{1}{2}\sin\theta + \dfrac{1.3}{2.4}\sin 2\theta + \dfrac{1.3.5}{2.4.6}\sin 3\theta + \ldots$ ad. inf.

7. $1 + \dfrac{1}{2}\cos\theta + \dfrac{1.3}{2.4}\cos 2\theta + \dfrac{1.3.5}{2.4.6}\cos 3\theta + \ldots$ ad. inf.

8. (i) $c\cos\alpha + \dfrac{1}{2}c^2\cos 2\alpha + \dfrac{1}{3}c^3\cos 3\alpha + \ldots$ ad. inf.

   (ii) $c\cos\alpha - \dfrac{1}{2}c^2\cos 2\alpha + \ldots$ ad. inf.

9. $c\sin\alpha + \dfrac{c^2}{2}\sin 2\alpha + \dfrac{c^3}{3}\sin 3\alpha + \ldots$ ad. inf.

10. $\cos\alpha - \dfrac{\cos(\alpha+2\beta)}{3!} + \dfrac{\cos(\alpha+4\beta)}{5!} - \ldots$ ad. inf.

11. $\cos\alpha + c\cos(\alpha+\beta) + \dfrac{c^2}{2!}\cos(\alpha+2\beta) + \ldots$ ad. inf.

12. $\sin\alpha + c\sin(\alpha+\beta) + \dfrac{c^2}{2!}\sin(\alpha+2\beta) + \ldots$ ad. inf.

13. $\dfrac{\sin\theta}{1!} + \dfrac{\sin 2\theta}{2!} + \dfrac{\sin 3\theta}{3!} + \ldots$ ad. inf.

14. $c\sin\alpha + \dfrac{1}{2}c^2\sin(\alpha+\beta) + \dfrac{c^3}{3}\sin(\alpha+2\beta) + \ldots$ ad. inf.

15. $\sin\alpha + n\sin(\alpha+\beta) + \dfrac{n(n-1)}{2!}\sin(\alpha+2\beta) + \ldots$ to $n + 1$ terms

16. $c\sin\alpha + \dfrac{1}{3}c^3\sin 3\alpha + \dfrac{1}{5}c^5\sin 5\alpha + \ldots$ ad inf. $C < 1$

17. $c\cos\alpha + \dfrac{1}{3}c^3\cos 3\alpha + \dfrac{1}{5}c^5\cos 5\alpha + \ldots$ ad. inf.

18. $c\sin^2\alpha - \dfrac{1}{2}c^2\sin^2 2\alpha + \dfrac{1}{3}c^3\sin^2 3\alpha - \ldots$ ad. inf.

19. $2x\cos\theta + \dfrac{3}{2}x^2\cos^2\theta + \dfrac{4}{3}x^3\cos^3\theta + \dfrac{5}{4}x^4\cos^4\theta \ldots$ ad. inf.

   if $|x| < 1$.

20. $\sin\theta - \dfrac{1}{3}\sin 3\theta + \dfrac{1}{5}\sin 5\theta - \ldots = v$, show that $\tanh 2v = \sin\theta$.

21. Sum the series

   (a) $a\sin a - \dfrac{1}{3}a^3\sin(\alpha+2\beta) + \dfrac{1}{5}a^5\sin(\alpha+4\beta) - \ldots$ ad inf.

   (b) $a\cos a - \dfrac{1}{3}a^3\cos(\alpha+2\beta) + \dfrac{1}{5}a^5\cos(\alpha+4\beta) - \ldots$ ad inf.

22. If $C = 1 + z\cos\theta + \dfrac{z^2\cos 2\theta}{2!} + \dfrac{z^3\cos 3\theta}{3!} + \ldots$ ad. inf.

   and $S = z\sin\theta + \dfrac{z^2\sin 2\theta}{2!} + \dfrac{z^3\sin 3\theta}{3!} + \ldots$ ad. inf. Show that

   $z\sin\theta = \tan^{-1}\dfrac{S}{C}$ and $z\cos\theta = \dfrac{1}{2}\log(C^2 + S^2)$.

23. $1 + \dfrac{\cos\alpha}{\cos\alpha} + \dfrac{\cos 2\alpha}{2!\cos^2\alpha} + \dfrac{\cos 3\alpha}{3!\cos^3\alpha} + \ldots\infty$

24. (i) $1 + \dfrac{1}{3}c\cos\alpha + \dfrac{1.4}{3.6}c^2\cos 2\alpha + \dfrac{1.4.7}{3.6.9}c^3\cos 3\alpha \ldots$ ad inf.

   (ii) $1 + n\cos\alpha + \dfrac{n(n+1)}{1.2}\cos 2\alpha$

     $+ \dfrac{n(n+1)(n+2)}{1.2.3}\cos 3\alpha + \ldots + \infty$

   (iii) $n\sin\alpha + \dfrac{n(n+1)}{1.2}\sin 2\alpha + \ldots$

25. Sum the series : $1 + \dfrac{c^2\cos 2\theta}{2!} + \dfrac{c^4\cos 4\theta}{4!} + \ldots$

26. Sum the series : $\dfrac{a\sin\theta}{1!} + \dfrac{a^3\sin 3\theta}{3!} + \ldots$

27. Find the sum : $\sin\theta\cos\theta + \dfrac{\sin 2\theta\cos^2\theta}{2!} + \dfrac{\sin 3\theta\cos^3\theta}{3!} + \ldots$

28. Find the sum of the series : $\sin\theta - \dfrac{\sin 2\theta}{2!} + \dfrac{\sin 3\theta}{3!} - \ldots\infty$

29. Find the sum of the series: $\sin\theta + \dfrac{\sin(\theta+2\phi)}{3!} + \dfrac{\sin(\theta+4\phi)}{5!} + \ldots\infty$

30. Sum the series : $\sin\alpha - \dfrac{\sin 3\alpha}{2!} + \dfrac{\sin 5\alpha}{4!} - \ldots\infty$

31. Sum the series : $\cos^2\theta - \dfrac{1}{3}\cos^3\theta\cos 3\theta + \dfrac{1}{5}\cos^5\theta\cos 5\theta - \ldots\infty$

32. Sum the following series :

   (a) $\sin^2\theta - \dfrac{1}{2}\sin 2\theta\sin^2\theta + \dfrac{1}{3}\sin 3\theta\sin^3\theta$

     $- \dfrac{1}{4}\sin 4\theta\sin^4\theta + \ldots\infty$ (PTU–2010, VTU–2006S)

   (b) $\cos\theta + \sin\theta\cos 2\theta + \dfrac{\sin^2\theta}{1.2}\cos 3\theta + \ldots\infty$ (PTU–2005)

   (c) $1 - \dfrac{1}{2}\cos\theta + \dfrac{1.3}{2.4}\cos 4\theta - \dfrac{1.3.5}{2.4.6}\cos 6\theta + \ldots\infty$

     (KURUKSHETRA–2006)

   (d) $\sin\alpha + \sin(\alpha+\beta) + \sin(\alpha+2\beta) + \ldots\sin(\alpha+(n-1)\beta)$

     (PTU–2009S)

   (e) $\cos\alpha + \cos(\alpha+\beta) + \cos(\alpha+2\beta) + \ldots$ to n terms.

     (KURUKSHETRA–2006)

Answers

1. $e^{\sin\alpha\cos\alpha}.\cos(\alpha+\sin^2\alpha)$

2. $e^{\cos^2\alpha}.\cos(\alpha+\sin\alpha\cos\alpha)$

3. $\tan^{-1}\left[\dfrac{\sin^2\alpha}{1+\sin\alpha\cos\alpha}\right]$

4. $\dfrac{1}{2}\left[\log\operatorname{cosec}\dfrac{\alpha-\beta}{2}-\log\operatorname{cosec}\dfrac{\alpha+\beta}{2}\right]$

5. 0

6. $\left(2\sin\dfrac{\theta}{2}\right)^{-1/2}\cos\dfrac{1}{4}(\pi-\theta)$

7. $\left(2\sin\dfrac{\theta}{2}\right)^{-1/2}\cos\dfrac{1}{4}(\pi-\theta)$

8. (i) $-\dfrac{1}{2}\log(1-2c\cos\alpha+c^2)$ (ii) $\dfrac{1}{2}\log[1+c^2+2c\cos\alpha]$

9. $\tan^{-1}\dfrac{c\sin\alpha}{1-c\cos\alpha}$

10. $[\cos(\alpha-\beta)\sin(\cos\beta)\cosh(\sin\beta)-\sin(\alpha-\beta)\cos(\cos\beta)\sinh(\sin\beta)]$

11. $e^{c\cos\beta}.\cos(\alpha+c\sin\beta)$

12. $e^{c\cos\beta}.\sin(\alpha+c\sin\beta)$

13. $e^{\cos\theta}.\sin(\sin\theta)$

14. $-\dfrac{1}{2}\sin(\alpha-\beta)\log(1-2c\cos\beta+c^2)+\cos(\alpha-\beta)\tan^{-1}\left\{\dfrac{c\sin\beta}{1-c\cos\beta}\right\}$

15. $2^n\cos^n\dfrac{\beta}{2}\sin\left(\alpha+\dfrac{1}{2}n\beta\right)$

16. $\dfrac{1}{2}\tan^{-1}\dfrac{c\sin\alpha}{1+c\cos\alpha}+\dfrac{1}{2}\tan^{-1}\dfrac{c\sin\alpha}{1-c\cos\alpha}$

17. $\dfrac{1}{4}\log\left(\dfrac{1+c^2+2c\cos\alpha}{1+c^2-2c\cos\alpha}\right)$

18. $\dfrac{1}{2}\log[(1+c)/\sqrt{(1+c^2+2c\cos2\alpha)}]$

19. $\dfrac{x\cos\theta}{1-x\cos\theta}-\log(1-n\cos\theta)$

21. (a) $\dfrac{1}{2}\sin(\alpha-\beta)\tan^{-1}\dfrac{2a\cos\beta}{1-a^2}+\dfrac{1}{2}\cos(\alpha-\beta)\tanh^{-1}\dfrac{2a\sin\beta}{1+a^2}$ (b) $\dfrac{1}{2}\cos(\alpha-\beta)\tan^{-1}\dfrac{2a\cos\beta}{1-a^2}-\dfrac{1}{2}\sin(\alpha-\beta)\tanh^{-1}\dfrac{2a\sin\beta}{1+a^2}$

23. $e.\cos(\tan\alpha)$

24. (i) $C=r^{-1/3}\cos\theta/3$ where $r=(1-2c\cos\alpha+c^2)$ and $\theta=\tan^{-1}$ and $\theta\{(c\sin\alpha)/1-c\cos\alpha\}$

(ii) $2^{-n}\sin^{-n}\dfrac{\alpha}{2}.\sin\dfrac{n}{2}(\pi-\alpha)$ (iii) $2^{-4}\sin^{-4}\left(\dfrac{\alpha}{2}\right).\sin\left\{n\left(\dfrac{\pi}{2}-\dfrac{\alpha}{2}\right)\right\}$

25. $\cosh(c\cos\theta)\cos(c\sin\theta)$

26. $\dfrac{1}{2}\tan^{-1}\dfrac{a\sin\theta}{1+a\cos\theta}+\dfrac{1}{2}\tan^{-1}\dfrac{a\sin\theta}{1-a\cos\alpha}$

27. $e^{\cos^2\theta}\sin\left(\dfrac{1}{2}\sin2\theta\right)$

28. $e^{-\cos\theta}\sin(\sin\theta)$

29. $\cos(\theta-\phi)\cos(\cos\phi)\sin(\sin\phi)+\sin(\theta-\phi)\sin(\cos\phi)\cos(\sin\phi)$

30. $\sin\alpha\cos(\cos\alpha)\cosh(\sin\alpha)-\cos\alpha\sin(\cos\alpha)\sinh(\sin\alpha)$

31. $\dfrac{1}{2}\tan^{-1}(\cot^2\theta)$

32. (a) $\tan^{-1}\left(\dfrac{\sin^2\theta}{1+\cos\theta\sin\theta}\right)$

(b) $e^{\sin\theta\cos\theta}\cos(\theta+\sin^2\theta)$

(c) $(2\cos\theta)^{-1/2}\cos\theta/2$

(d) $\sin\left(\alpha+\dfrac{n-1}{2}\beta\right)\sin\dfrac{n\beta}{2}\operatorname{cosec}\dfrac{\beta}{2}$

(e) $\dfrac{\cos\left\{\alpha+\dfrac{1}{2}(n-1)\beta\right\}\sin\dfrac{n\beta}{2}}{\sin\dfrac{1}{2}\beta}$

11.13.8 Hyperbolic Series

A series of hyperbolic sines and cosines can be summed up either by writing them in terms of either exponential values or directly.

Solved Examples

EXAMPLE 1. *Find the sum of the following series* :

$$\frac{\sinh\alpha}{1!}+\frac{\sinh2\alpha}{2!}+\frac{\sinh3\alpha}{3!}+...\,ad\,inf.$$

SOLUTION. Writing the above series in terms of exponential values of hyperbolic sines, we have the sum of the given series.

$$S=\frac{1}{1!}\left(\frac{e^\alpha-e^{-\alpha}}{2}\right)+\frac{1}{2!}\left(\frac{e^{2\alpha}-e^{-2\alpha}}{2}\right)+...$$

$$=\frac{1}{2}\left[\left(\frac{e^\alpha}{1!}+\frac{e^{2\alpha}}{2!}+\frac{e^{3\alpha}}{3!}+...\right)\right.$$

$$\left.\left(\frac{e^{-\alpha}}{1!}+\frac{e^{-2\alpha}}{2!}+\frac{e^{-3\alpha}}{3!}+...\right)\right]$$

$$=\frac{1}{2}[(e^{e^\alpha}-1)-e^{e^{-\alpha}}+1]=\frac{1}{2}(e^{e^\alpha}-e^{e^{-\alpha}})$$

$$=\frac{1}{2}[e^{(\cosh\alpha+\sinh\alpha)}-e^{(\cosh\alpha-\sinh\alpha)}]$$

$$=\frac{1}{2}e^{\cosh\alpha}(e^{\sinh\alpha}-e^{-\sinh\alpha})=\frac{1}{2}e^{\cos\alpha}\sinh(\sinh\alpha).$$

EXAMPLE 2. *Sum the following series.*

$$\sinh\alpha-\frac{1}{2}\sinh2\alpha+\frac{1}{3}\sinh3\alpha-...\,ad.\,inf.$$

SOLUTION. We know that $\sinh\alpha=\dfrac{e^\alpha-e^{-\alpha}}{2}$

Hence the sum of the given series

$$S=\frac{1}{2}\left[(e^\alpha-e^{-\alpha})-\frac{1}{2}(e^{2\alpha}-e^{-2\alpha})+\frac{1}{3}(e^{3\alpha}-e^{-3\alpha})-...\right]$$

$$=\frac{1}{2}\left[\left(e^\alpha-\frac{e^{2\alpha}}{2}+\frac{e^{3\alpha}}{3}\right)-\left(e^{-\alpha}-\frac{e^{-2\alpha}}{2}+\frac{e^{-3\alpha}}{3}-...\right)\right]$$

$$= \frac{1}{2}[\log_e(1+e^\alpha) - \log_e(1+e^{-\alpha})] = \frac{1}{2}\left(\log_e \frac{1+e^\alpha}{1+e^{-\alpha}}\right)$$

$$= \frac{1}{2}\left[\log_e \frac{e^\alpha(1+e^{-\alpha})}{1+e^{-\alpha}}\right] = \frac{\alpha}{2}.$$

EXAMPLE 3. *Sum the series :* $a\cosh\alpha + a^2\cosh 2\alpha + a^3\cosh 3\alpha$
$+ ...ad.\ inf.$

SOLUTION. Let C be the sum of this series, then

$$C = a\left[\frac{e^\alpha + e^{-\alpha}}{2}\right] + a^2\left[\frac{e^{2\alpha}+e^{-2\alpha}}{2}\right]$$

$$= \frac{1}{2}\left[ae^\alpha + a^2 e^{2\alpha}\right] + \frac{1}{2}\left[e^{-\alpha} + a^2 e^{-2\alpha} + ...\right]$$

$$= \frac{1}{2}\frac{ae^\alpha}{1-ae^\alpha} + \frac{1}{2}\frac{ae^{-\alpha}}{1-ae^{-\alpha}}$$

$$\text{provided } |ae^\alpha| < 1 \text{ and } |ae^{-\alpha}| < 1$$

$$= \frac{a}{2}\left[\frac{e^\alpha + e^{-\alpha} - 2a}{(1-ae^\alpha)(1-ae^{-\alpha})}\right] = a\left[\frac{\cosh\alpha - a}{1-2a\cosh\alpha + a^2}\right].$$

EXAMPLE 4. *Find the sum of the series*
$\cosh^2\alpha + \cosh^2(\alpha + \beta) + \cosh^2(\alpha + 2\beta) + ...$ *to n terms.*

SOLUTION. Let $C = \cosh^2\alpha + \cosh^2(\alpha + \beta) + \cosh^2(\alpha + 2\beta)$
$+ ...$ to n terms

$$= \frac{1}{2}[1+\cosh 2\alpha] + \frac{1}{2}[1+\cosh(2\alpha + 2\beta)]$$
$$+ \frac{1}{2}[1+\cosh 2(\alpha + 2\beta)] + ...\ n \text{ terms}$$

$$= \frac{n}{2} + \frac{1}{2}[\cosh 2\alpha + \cosh 2(\alpha+\beta)$$
$$+ \cosh 2(\alpha + 4\beta) + ... \text{ to } n \text{ terms}]$$

$$= \frac{n}{2} + \frac{1}{4}[e^{2\alpha} + e^{-2(\alpha+\beta)} + e^{2(\alpha+2\beta)} + ...$$
$$\text{to } n \text{ terms}]$$

$$+ \frac{1}{4}[e^{-2\alpha} + e^{-2(\alpha+\beta)} + e^{-2(\alpha+2\beta)} + ... \text{ to } n \text{ terms}]$$

$$= \frac{n}{2} + \frac{1}{4}\left\{\frac{e^{2\alpha}[1-2^{2n\beta}]}{1-e^{2\beta}} + \frac{e^{-2\alpha}[1-e^{-2n\beta}]}{1-e^{-2\beta}}\right\}$$

$$\text{or } C = \frac{n}{2} + \frac{1}{4}\left[\frac{\begin{array}{l}(e^{2\alpha}+a^{-2\alpha})-(e^{2(\alpha-\beta)})+e^{-2(\alpha+\beta)}\\ -(e^{2(\alpha+n\beta)})+e^{-2(\alpha+n\beta)}+e^{2(\alpha+\overline{n-1}\beta)}\\ +e^{-2(\alpha+\overline{n-1}\beta)}\end{array}}{2-(e^{2\beta}+e^{-2\beta})}\right]$$

$$= \frac{n}{2} + \frac{1}{4}\left[\frac{\cos 2\alpha - \cosh 2(\alpha-\beta) - \cosh 2(\alpha-n\beta)}{1-\cosh 2\beta}\right]$$
$$+ \cosh 2(\alpha + \overline{n-1}\beta)$$

Exercise-11.16

Sum the following series : (Qus 1 – 5)

1. $\cosh x + \cosh(x + y) + \cosh(x + 2y) + ...\ n$ terms.

2. $\cosh\theta - \frac{1}{2}\cosh 2\theta + \frac{1}{3}\cosh 3\theta - ...ad.\ inf.$

3. $1 + \frac{\cosh\alpha}{1!} + \frac{\cosh 2\alpha}{2!} + \frac{\cosh 3\alpha}{3!} + ...ad.\ inf.$

4. $\sinh\theta + {}^nC_1\sinh(\theta+\phi) + {}^nC_2\sinh(\theta + 2\phi) ...$ to $n + 1$ terms

5. $x\cosh\theta + \frac{x^2}{2!}\cosh 2\theta + \frac{x^3}{3!}\cosh 3\theta + ...ad.\ inf.$

6. Prove that

$$\tanh\theta + \frac{1}{3}\tanh^3\theta + \frac{1}{5}\tanh^5\theta + ...$$

$$= \tan\theta - \frac{1}{3}\tan^3\theta + \frac{1}{5}\tan^5\theta - ... \text{ where } -\frac{\pi}{4} < \theta < \frac{\pi}{4}.$$

7. Show that

$$C\sinh\theta + C^2\sinh 2\theta + C^3\sinh 3\theta + \infty = \frac{C\sinh\theta}{1-2C\cosh\theta + C^2}.$$

Answers

1. $\cosh\left[x + \frac{n-1}{2}y\right]\sinh\frac{ny}{2}\operatorname{cosech}\frac{y}{2}$

2. $\log\left(2\cosh\frac{\theta}{2}\right)$

3. $\frac{1}{2}[e^{e^\alpha} + e^{e^{-\alpha}}]$

4. $\sinh\left(\theta + \frac{n\phi}{2}\right)\left(2\cosh\frac{\phi}{2}\right)^n$

5. $\frac{1}{2}[e^{xe^\theta} + e^{xe^{-\theta}}] - 1$

11.13.9 THE DIFFERENCE METHOD

Sometimes it is easier to sum the series by expressing each term as the difference of two terms, so that the expressions into successive terms cancel out, leaving only one or two terms. No particular method can be given for splitting the terms and it generally depends upon the practice and chance in many cases.

Let T_n be the nth term of a series and let it expressed in the form

$$T_n = C[f(n + 1) - f(n)]$$

Then

$$S_n = T_1 + T_2 + ...+ T_n = C[f(2) - f(1) + f(3) - f(2) + ... + f(n + 1) - f(n)] = C[f(n + 1) - f(1)]$$

since the intermediate terms all cancel out.

If series is convergent and sum to infinity is required, we deduce the sum as below :

$$\lim_{n\to\infty} S_n = \lim_{n\to\infty}[f(n+1) - f(1)]$$

REMARK

- If a series is such that its nth term is of the form $\tan^{-1}\left(\dfrac{a}{b}\right)$ then we put $T_n = \tan^{-1}\left(\dfrac{a}{b}\right) = \tan^{-1}x - \tan^{-1}y = \tan^{-1}\dfrac{x-y}{1+xy}$

i.e., $\left.\begin{array}{c} x - y = \alpha \\ xy = b - 1 \end{array}\right\}$ $\therefore\ (x + y) = \sqrt{a^2 + 4(b-1)}$

Solving, we get x and y. Then putting $x = 1, 2, 3,\ldots$ and adding, we get the required sum.

Solved Examples

EXAMPLE 1. *Sum the series to n terms :*

$$\tan^{-1}\frac{4}{1+3.4} + \tan^{-1}\frac{6}{1+8.9} + \tan^{-1}\frac{8}{1+15.16}\ldots$$

SOLUTION. Here $T_n = \tan^{-1}\dfrac{2(n+1)}{1+(n+2)n(n+1)^2}$

$= \tan^{-1}x - \tan^{-1}y = \tan^{-1}\dfrac{x-y}{1+xy}$

$\therefore\ x - y = 2(n+1),\ xy = n(n+2)(n+1)^2$

$\therefore\ x + y = 2(n+1)^2$

$\therefore\ \ \ x = (n+2)(n+1),\ y = n(n+1)$

$\therefore\ \ \ \ T_n = \tan^{-1}[(n+2)(n+1)] - \tan^{-1}[(n+1)n]$

Now giving values of n as $1, 2, 3, \ldots n$, we have

$T_1 = \tan^{-1}3.2 - \tan^{-1}2.1$

$T_2 = \tan^{-1}4.3 - \tan^{-1}3.2$

$T_3 = \tan^{-1}5.4 - \tan^{-1}4.3$

$\ldots\ \ \ldots\ \ \ldots\ \ \ldots\ \ \ldots\ \ \ldots$

$T_n = \tan^{-1}[(n+2)(n+1)] - \tan^{-1}[(n+1)n]$

Adding, we get

$S_n = \tan^{-1}[(n+2)(n+1)] - \tan^{-1}2.1$

$= \tan^{-1}(n^2+3n+2) - \tan^{-1}2$

$= \tan^{-1}\dfrac{n^2+3n}{1+2(n^2+3n+2)}$

$= \tan^{-1}\dfrac{n^2+3n}{2n^2+6n+5}.$

EXAMPLE 2. *Sum the series :*

$$\tan^{-1}\frac{1}{3+3.1+1^2} + \tan^{-1}\frac{1}{3+3.2+2^2}$$
$$+ \tan^{-1}\frac{1}{3+3.3+3^2} + \ldots + \tan^{-1}\frac{1}{3+3n+n^2}.$$

SOLUTION. Let

$T_n = \tan^{-1}\dfrac{1}{3+3n+n^2} = \tan^{-1}\dfrac{1}{1+2+3n+n^2}$

$= \tan^{-1}x - \tan^{-1}y = \tan^{-1}\dfrac{x-y}{1+xy}$

$\therefore\ x - y = 1$

$\ \ \ \ \ xy = n^2 + 3n + 2$

$\ \ \ \ x + y = \sqrt{1 + 4(n^2+3n+2)} = 2n + 3$

$\therefore\ \ \ x = n + 2,\ y = n + 1$

$\ \ \ \ T_n = \tan^{-1}(n+2) - \tan^{-1}(n+1)$

Putting $n = 1, 2, 3,\ldots n$ and adding, we get

$S_n = \tan^{-1}(n+2) - \tan^{-1}2.$

EXAMPLE 3. *Sum the series :*

$$\tan^{-1}\frac{1}{3} + \tan^{-1}\frac{1}{7} + \tan^{-1}\frac{1}{13} + \ldots n \text{ terms.}$$

SOLUTION. **Method I.** The given series can be put in the form

$S_n = \tan^{-1}\dfrac{2-1}{1+2.1} + \tan^{-1}\dfrac{3-2}{1+3.2}$

$\ \ \ \ \ \ \ \ \ \ + \tan\dfrac{4-3}{1+4.3} + \ldots + \dfrac{(n+1)-n}{1+(n+1)n}$

$= (\tan^{-1}2 - \tan^{-1}1) + (\tan^{-1}3 - \tan^{-1}2)$
$\ \ \ \ \ + (\tan^{-1}4 - \tan^{-1}3) + \ldots + (\tan^{-1}(n+1)$

$= \tan^{-1}(n) - \tan^{-1}(n+1) - \tan^{-1}1 = \tan^{-1}\dfrac{n}{n+2}.$

Method II. To find n^{th} term of given series, let

$\ \ \ \ \ S = 3 + 7 + 13 + \ldots + T_n$

Again $\ \ \ \underline{S = \ \ \ \ \ \ \ 3 + 7 + \ldots + T_{n-1} + T_n}$

Subracting, $0 = 3 + 4 + 6 + \ldots$ to n terms $- T_n$

$\therefore\ T_n = 1 + (2 + 4 + 6 + \ldots \text{ to } n \text{ terms}) = 1 + (n+1)n$

$\therefore\ T_n$ of given series

$= \tan^{-1}\dfrac{1}{1+(n+1)n} = \tan^{-1}(n+1) - \tan^{-1}n$

Now put $n = 1, 2, 3, \ldots n$ and add to get the required sum.

Exercise-11.17

Sum the following series :

1. $\text{cosec } \theta + \text{cosec } 2\theta + \text{cosec } 2^2\theta + \ldots n$ terms.

2. $\tan\alpha\tan(\alpha + \beta) + \tan(\alpha + \beta)\tan(\alpha + 2\beta) + \tan(\alpha + 2\beta)\tan(\alpha + 3\beta) + \ldots$ to n terms

3. $\tan\theta\tan^2\dfrac{\theta}{2} + 2\tan\dfrac{\theta}{2}\tan^2\dfrac{\theta}{2^2} + 2^2\tan\dfrac{\theta}{2^2}\tan^2\dfrac{\theta}{2^3} + \ldots$ to n terms.

4. $\tan^2\theta\tan 2\theta + \dfrac{1}{2}\tan^2 2\theta\tan 4\theta + \dfrac{1}{2^2}\tan^2 4\theta\tan 8\theta + \ldots$ to n terms.

5. $\tan^{-1}\dfrac{2}{4} + \tan^{-1}\dfrac{2}{9} + \tan^{-1}\dfrac{2}{16} + \ldots$ to n terms.

6. $\tan^{-1}x + \tan^{-1}\dfrac{x}{1+1.2.x^2} + \tan^{-1}\dfrac{x}{1+2.3.x^2} + \ldots$ to n terms.

7. $\cot^{-1}3 + \cot^{-1}7 + \cot^{-1}13 + \ldots \cot^{-1}(1+n+n^2)$.

8. Sum the series $\displaystyle\sum_{1}^{n}\tan^{-1}\dfrac{4}{4n^2+3}$.

9. $\tan^{-1}\dfrac{1}{1+1+1^2} + \tan^{-1}\dfrac{1}{1+2+2^2} + \tan^{-1}\dfrac{1}{1+3+3^2} \ldots$ to n terms.

10. (a) $\cot^{-1}(2.1)^2 + \cot^{-1}(2.2^2) + \cot^{-1}(2.3^2) + \ldots$ ad. inf.

Deduce its sum to infinity

(b) $\tan^{-1}\dfrac{1}{2.1^2}+\tan^{-1}\dfrac{1}{2.2^2}+\tan^{-1}\dfrac{1}{2.3^2}+\ldots\tan^{-1}\dfrac{1}{2n^2}$

Deduce its sum to infinity.

form and hence or otherwise sum the infinite series

$$\cot^{-1}\left(2^2+\dfrac{1}{2}\right)+\cot^{-1}\left(2^3+\dfrac{1}{2^2}\right)+\ldots+\cot^{-1}\left(2^{n+1}+\dfrac{1}{2^n}\right)+$$

11. $\cot^{-1}\left(1^2+\dfrac{3}{4}\right)+\cot^{-1}\left(2^2+\dfrac{3}{4}\right)+\cot^{-1}\left(3^2+\dfrac{3}{4}\right)+\ldots$

15. Find the sum $\sum\limits_{m=1}^{n}\sqrt{1+\sin mx}$.

12. $\dfrac{1}{\cos\theta+\cos3\theta}+\dfrac{1}{\cos\theta+\cos5\theta}+\dfrac{1}{\cos\theta+\cos7\theta}\ldots$ to n terms.

16. Sum the series : $c\sin\alpha-\dfrac{c^2}{2}\sin2\alpha+\dfrac{c^3}{3}\sin3\alpha-\ldots$ad. inf.

13. $\dfrac{1}{\sin\theta\cos2\theta}-\dfrac{1}{\cos2\theta\sin3\theta}+\dfrac{1}{\sin3\theta\cos4\theta}-\ldots$ to n terms.

17. Sum the following series :

$$\dfrac{1}{\sin\theta\sin2\theta}+\dfrac{1}{\sin2\theta\sin3\theta}+\dfrac{1}{\sin3\theta\sin4\theta}+\ldots n \text{ terms.}$$

14. Reduce the expression $\tan^{-1}(2^{n+1})-\tan^{-1}(2^n)$ to its simplest

18. $\tan\theta\sec2\theta+\tan2\theta\sec4\theta+\tan4\theta\sec8\theta+\ldots n$ terms.

Hints to Selected Problems

4. $\tan^2\theta\tan2\theta=\tan2\theta-2\tan\theta$

13. Putting $\theta=\phi-\pi/2$, the series becomes

$\sec\alpha\sec2\alpha+\sec2\alpha\sec3\alpha+\sec3\alpha\sec4\alpha+\ldots$

15. $\sqrt{1+\sin mx}=\sin\dfrac{mx}{2}+\cos\dfrac{mx}{2}$

Answers

1. $\cot\dfrac{\theta}{2}-\cot2^{n-1}\theta$ **2.** $\cot\beta\,[\tan(\alpha+n\beta)-\tan\alpha]-n$ **3.** $\tan\theta-2^n\tan\dfrac{\theta}{2^n}$ **4.** $\dfrac{1}{2^{n-1}}\tan2^n\theta-2\tan\theta$

5. $\tan^{-1}(n+2)+\tan^{-1}(n+1)-\tan^{-1}2-\tan^{-1}1$ **6.** $\tan^{-1}nx$ **7.** $\tan^{-1}\dfrac{n}{n+2}$ **8.** $\tan^{-1}\dfrac{4n}{2n+5}$

9. $\tan^{-1}(n+1)-\tan^{-1}1$ **10.** $\dfrac{\pi}{4}$ **11.** $\cot^{-1}\dfrac{1}{2}$ **12.** $\dfrac{1}{2\sin\theta}\{\tan(n+1)\theta-\tan\theta\}$

13. $\dfrac{1}{\sin\phi}\{\tan(n+1)\phi-\tan\phi\}$ **14.** $\cot^{-1}(2^{n+1}+2^{-n});\cot^{-1}(2)$ **15.** $\sqrt{2}\sin\dfrac{1}{4}\{\pi+(n+1)x\}\sin\dfrac{nx}{4}\operatorname{cosec}\dfrac{x}{4}$

16. $\tan^{-1}\dfrac{c\sin\alpha}{1+c\cos\alpha}$ **17.** $\operatorname{cosec}\theta[\tan(n+1)\theta-\tan\theta]$ **18.** $\tan2^n\theta-\tan\theta$

11.14 CALCULUS OF COMPLEX FUNCTIONS

We know that a single valued function $f(z)$ in a domain D is said to be analytic at a point $z=a$ if there exists a neighbourhood $|z-a|<\delta$ at all points of which the function is differentiable, i.e., $f'(z)$ exists.

Also, if the above function $f(z)$ is differentiable at every point of a domain D except possilbly at a finite number of exceptional points then the function is said to be analytic in the domain D. These exceptional points at which $f'(z)$ does not exist are called singular points or singularities of the funciton. We also know that if a function $f(z)$ be such that $f'(z)$ exists at every point of the domain D then $f(z)$ is said to be regular in D.

REMARK

• The terms regular and holomorphic are also sometimes used as synonyms for analytic.

11.14.1 CAUCHY-RIEMANN PARTIAL DIFFERENTIAL EQUATIONS

THEOREM 1. **(Necessary condition for f(z) to be analytic)** *The necessary condition for $w=f(z)=u(x,y)+iv(x,y)$ to be analytic (i.e., differentiable) at any point $z=x+iy$ of its domain D is that the four paritial derivatives u_x, u_y, v_x and v_y should exist and satisfy the Cauchy-Riemann partial differential equations*

$$u_x=v_y \quad and \quad u_y=-v_x$$

i.e.,

$$\dfrac{\partial u}{\partial x}=\dfrac{\partial v}{\partial y} \quad and \quad \dfrac{\partial u}{\partial y}=-\dfrac{\partial v}{\partial x}$$

PROOF. Let $f(z)=u(x,y)+iv(x,y)$ be analytic at any point z of its domain therefore $f'(z)=\lim\limits_{\delta z\to0}\dfrac{f(z+\delta z)-f(z)}{\delta z}$ exists and is unique, i.e., it is independent of the path along which $\delta z\to0$

Also $z=x+iy$ $\therefore$ $\delta z+\delta x+i\delta y$ and as $\delta z\to0$, δx and δy also $\to0$.

$\therefore$

$$f'(z)=\lim_{\substack{\delta x\to0\\\delta y\to0}}\dfrac{[u(x+\delta x,y+\delta y)+iv(x+\delta x,y+\delta y)]-[u(x,y)+iv(x,y)]}{\delta x+i\delta y}$$

$$=\lim_{\delta x\to0,\,\delta y\to0}\left[\dfrac{u(x+\delta x,y+\delta y)-u(x,y)}{\delta x+i\delta y}+i\dfrac{v(x+\delta x,y+\delta y)-v(x,y)}{\delta x+i\delta y}\right] \qquad \ldots(1)$$

Now let us consider two possible approaches in which $dz \to 0$.

In the first case take δz to be purely real so that $\delta z = \delta x$, $\delta y = 0$ and $\delta x \to 0$.

Hence from 1, we get

$$f'(z) = \lim_{\delta x \to 0} \left[\frac{u(x+\delta x, y) - u(x, y)}{\delta x} + i \frac{v(x+\delta x, y) - v(x, y)}{\delta x} \right]$$

or $$f'(z) = \frac{\partial u}{\partial x} + i \frac{\partial v}{\partial x} = u_x + iv_x \qquad \qquad \dots (2)$$

Since $f'(z)$ exists therefore the above limit exists which in other words means that u_x and v_x exists.

In the second case let $\delta z \to 0$ along the imaginary axis so that δz is purely imaginary and hence $\delta z = i\delta y$, $\delta x = 0$ and $\delta y \to 0$.
Hence from 1, we get

$$f'(z) = \lim_{\delta \to} \frac{u(x, y+\delta y) - u(x, y)}{i\delta y} + i \frac{v(x, y+\delta y) - v(x, y)}{i\delta y}$$

$$f'(z) = \frac{1}{i} \frac{\partial u}{\partial y} + \frac{\partial v}{\partial y} = -i \frac{\partial u}{\partial y} + \frac{\partial v}{\partial y} = -iu_y + v_y \qquad \qquad \dots (3)$$

Since $f'(z)$ exists therefore the above limit exists which in other words means that u_y and v_y exists.

Also by definition we know that the limit should be unique and hence the two limits obtained in (2) and (3) should be identical.

$$\therefore \qquad u_x + iv_x = -iu_y + v_y$$

Equating real and imaginary parts, we get

$$u_x = v_y \text{ and } u_y = -v_x$$

Above equations are known as Cauchy Riemann partial differential equations.

THEOREM 2. **(Sufficient condition for $f(z)$ to be analytic.)** *The function $w = f(z) = u(x, y) + iv(x, y)$ is analytic in a domain D if*

(1) *u, v are differentiable in D and $u_x = v_y$, $u_y = -v_x$.*

(2) *The partial derivatives u_x, u_y, v_x and v_y are all continuous in D.*

PROOF. Let $w = u + iv$ $\therefore$ $\qquad \delta w = \delta u + i\delta v$ $\qquad \qquad \dots (1)$

$$\delta u = u(x+\delta x, y+\delta y) - u(x, y) = [u(x+\delta x, y+\delta y) - u(x, y+\delta y)] + [u(x, y+\delta y) - u(x, y)] \qquad \dots (2)$$

In the above step we have subtracted and added $u(x, y+\delta y)$. Again by mean value theorem we know that if $f(x)$ is continuous in $a \le x \le b$ and differentiable in $a < x < b$ then

$$f(a+h) - f(a) = h \, f'(a+\theta h) \text{ where } 0 < \theta < 1.$$

Applying the result of above theorem in both the breackets in (2), we get

$$\delta u = \delta x . u_x(x+\theta\delta_x, y+\delta y) + \delta y . u_y(x, y+\theta'\delta y) \,; \; 0 < \theta < 1 \text{ and } 0 < \theta' < 1 \qquad \dots (3)$$

Again u_x and u_y are given to be continuous

$$\therefore \qquad |u_x(x+\theta\delta x, y+\delta y) - u_x(x, y)| < \varepsilon \text{ and } |u_y(x, y+\theta\delta y) - u_x(x, y)| < \eta$$

Now choosing $\varepsilon_1 < \varepsilon$ and $\eta_1 < \eta$ we have from above

$$u_x(x+\theta\delta x, y+\delta y) - u_x(x, y)| = \varepsilon_1 \text{ and } u_y(x, y+\theta'\delta y) - u_y(x, y)| = \eta_1$$

Hence, from (3) by the help of above relation, we get

$$\delta u = (u_x(x, y) + \varepsilon_1)\delta x - u_y(x, y) + \eta_1)\delta y \qquad \qquad \dots (4)$$

Proceeding exactly as above, we get $\delta v = (v_x(x, y) + \varepsilon_2)\delta x + (v_y(x, y) + \eta_2)\delta y \qquad \dots (5)$

Putting the values of δu and δv from (4) and (5) in (1) and writing $u_x(x, y)$ simply as u_x and similarly for others we get

$$\delta w = [(u_x + \varepsilon_1)\delta x + (u_y + \eta_1)\delta y] + i[(y_x + \varepsilon_2)\delta x = (v_y + \eta_2)\delta y]$$

or $$\delta w = (u_x + iv_x)\delta x + (u_y + iv_y)\delta y + (\varepsilon_1 + i\varepsilon_2)\delta x + (\eta_1 + i\eta_2)\delta_y \qquad \dots (6)$$

Now by Cauchy Riemann equations $u_x = v_y$ and $u_y = -v_x = i^2 v_x$ and choosing $\varepsilon_3 = \varepsilon_1 + i\varepsilon_2$ and $\eta_3 = \eta_1 + i\eta_2$.
Hence (6) can be written as

$$\delta w = (u_x + iv_x)\delta x + i(iv_x + u_x)\delta y + \varepsilon_3\delta x + \eta_3\delta y$$

$$= (u_x + iv_x)(\delta x + i\delta y) + \varepsilon_3\delta x + \eta_3\delta y \qquad \qquad \dots (7)$$

Dividing throughout by $\delta z = \delta x + i\delta y$, we get

$$\frac{\delta w}{\delta z} = u_x + iv_x + \frac{\varepsilon_3\delta x + \eta_3\delta y}{\delta x + i\delta y}$$

Taking limit when $\delta z \to 0$ so that $\delta x \to 0$, $\delta y \to 0$, $\varepsilon_3 \to 0$, $\eta_3 \to 0$, we get

$$f'(z) = \frac{dw}{dz} = \lim_{\delta z \to 0} \frac{\delta w}{\delta z} = u_x + iv_x$$

Since u_x, v_x exist and are unique therefore from above we conclude that $f'(z)$ exists. Therefore $f(z)$ is analytic at an arbitrary point z of D and hence it is analytic in the domain D.

NOTE

- From (1) we have

$$\frac{dw}{dz} = u_x + iv_x = \frac{\partial u}{\partial x} + i\frac{\partial v}{\partial x} = \frac{\partial}{\partial x}(u + iv) = \frac{\partial w}{\partial x} \qquad \dots(8)$$

Again by using Cauchy Riemman equation

$$\frac{dw}{dz} = u_x + iv_x = v_y - iu_y = \frac{1}{i}[iv_y - i^2 u_y] = \frac{1}{i}[u_y + iv_y] = \frac{1}{i}\left[\frac{\partial u}{\partial y} + i\frac{\partial v}{\partial y}\right] = \frac{1}{i}\frac{\partial}{\partial y}(u + iv) = \frac{1}{i}\frac{\partial w}{\partial y} \qquad \dots(9)$$

11.14.2 CONJUGATE FUNCTIONS

Let

$$w = f(z) = u(x, y) + iv(x, y)$$

If $f(z)$ be analytic then the two functions u and v of two real variables are called conjugate functions.

Since

$$z = x + iy \text{ therefore } \bar{z} = x - iy$$

$$\therefore \qquad x = \frac{1}{2}(z + \bar{z}), y = \frac{1}{2i}(z - \bar{z}) \qquad \dots (1)$$

Hence u and v may now be regarded as function of two independent variables z and $\bar{z}$ so that $w = w(z, \bar{z})$. Now if u and v have first order continuous partial derivatives, then the condition for w to be independent of $\bar{z}$ is

$$\frac{\partial w}{\partial \bar{z}} = 0 \text{ or } \frac{\partial}{\partial \bar{z}}(u + iv) = 0 \qquad \text{or} \qquad \frac{\partial u}{\partial \bar{z}} + i\frac{\partial v}{\partial \bar{z}} = 0$$

or

$$\left(\frac{\partial u}{\partial x}\cdot\frac{\partial x}{\partial \bar{z}} + \frac{\partial u}{\partial y}\cdot\frac{\partial y}{\partial \bar{z}}\right) + i\left(\frac{\partial v}{\partial x}\cdot\frac{\partial x}{\partial \bar{z}} + \frac{\partial v}{\partial y}\cdot\frac{\partial y}{\partial \bar{z}}\right) = 0$$

or

$$\left(\frac{1}{2}\frac{\partial u}{\partial x} - \frac{1}{2i}\frac{\partial u}{\partial y}\right) + i\left(\frac{1}{2}\frac{\partial v}{\partial x} - \frac{1}{2i}\frac{\partial v}{\partial y}\right) = 0 \qquad \text{or} \qquad \frac{\partial u}{\partial x} + i\frac{\partial u}{\partial y} + i\frac{\partial v}{\partial x} - \frac{\partial v}{\partial y} = 0,$$

or

$$\left(\frac{\partial u}{\partial x} - \frac{\partial u}{\partial y}\right) + i\left(\frac{\partial u}{\partial y} + \frac{\partial v}{\partial x}\right) = 0,$$

Equating real and imaginary parts, we get

$$\frac{\partial u}{\partial x} - \frac{\partial v}{\partial y} = 0, \frac{\partial u}{\partial y} + \frac{\partial v}{\partial x} = 0 \qquad \text{or} \qquad u_x = v_y \; ; \; u_y = -v_x$$

Above are Cauchy Riemman equations.

REMARK

- If $f(z)$ is an analytic function of z, then x and y can occur in $f(z)$ only in the combination of $x + iy$.

THEOREM 3. **(Laplace's Differential Equations)** *Real and imaginary parts u and v of analytic function $f(z)$ satisfy Laplace's equation, i.e.,*

$$\frac{\partial^2 u}{\partial x^2} + \frac{\partial^2 u}{\partial y^2} = 0 \quad and \quad \frac{\partial^2 v}{\partial x^2} + \frac{\partial^2 v}{\partial y^2} = 0 \quad or \quad \nabla^2 u = 0 \quad and \quad \nabla^2 v = 0$$

PROOF. Since $f(z) = u + iv$ is analytic then we have the following Cauchy-Riemann equations

$$\frac{\partial u}{\partial x} = \frac{\partial v}{\partial y} \text{ and } \frac{\partial u}{\partial y} = -\frac{\partial v}{\partial x}$$

Let us suppose that the second order partial derivatives of u and v exist and are continuous functions of x and y. Defferentiationg the above equtions w.r.t. x and y respectively and adding

$$\frac{\partial^2 u}{\partial x^2} + \frac{\partial^2 u}{\partial y^2} = \frac{\partial^2 v}{\partial x \partial y} - \frac{\partial^2 v}{\partial y \partial x} = 0 \quad \text{or} \qquad \nabla^2 u = 0$$

Again differentiating the equaitons in (1) w.r.t. y and x and subtracting.

$$\frac{\partial^2 u}{\partial y \partial x} - \frac{\partial^2 u}{\partial x \partial y} = \frac{\partial^2 v}{\partial y^2} - \left(-\frac{\partial^2 v}{\partial x^2}\right)$$

or

$$\frac{\partial^2 v}{\partial x^2} + \frac{\partial^2 v}{\partial y^2} = 0 \qquad \text{or } \nabla^2 v = 0$$

11.14.3 HARMONIC FUNCTIONS

Any function of x and y having continuous partial derivatives of first and second order and also satisfying Laplace's equation is called a harmonic function.

Hence from above we conclude that both u and v are harmonic functions if $f(z) = u + iv$ be analytic.

ORTHOGONAL SYSTEM

Two families of curves $u(x, y) = c_1$ and $v(x, y) = c_2$ are said to form an orthogonal system if they intersect at right angle at each point of their intersection.

THEOREM 4. *If $w = f(z) = u + iv$ be analytic then families of curve given by $u = c_1$ and $v = c_2$ form an orthogonal system.*

PROOF. Differentiating $u = c_1$ w.r.t x, we get

$$\frac{\partial u}{\partial x} + \frac{\partial u}{\partial y} \cdot \frac{dy}{dx} = 0 \qquad \therefore \qquad m_1 = \frac{dy}{dx} = -\frac{u_x}{u_y}$$

Differentiating $v = c_2$ w.r.t x, we get

$$\frac{\partial v}{\partial x} + \frac{\partial v}{\partial y} \cdot \frac{dy}{dx} = 0 \qquad \therefore \qquad m_2 = \frac{dy}{dx} = -\frac{v_x}{v_y}.$$

The two families of curves will intersect orthogonally if $m_1 m_2 = -1$. But $m_1 m_2 = \left(\frac{-u_x}{u_y}\right)\left(-\frac{v_x}{v_y}\right) = -1$ because by Cauchy

Riemman equations, we have

$$u_x = v_y \text{ and } u_y = -v_x$$

Therefore the families of curves $u = c_1, v = c_2$ where c_1 and c_2 are parameters form an orthogonal system.

Verification in the case of $f(z) = \sin z$

$$u + iv = f(z) = \sin z \text{ where } z = x + iy.$$
$$\therefore \qquad u + iv = \sin(x + iy) = \sin x \cos iy + \cos x \sin iy$$
or $\qquad u + iv = \sin x \cos hy + i \cos x \sin hy \qquad \qquad \qquad \qquad \qquad \dots (1)$
$$u = \sin x \cos hy = c_1 \text{ (say)}$$
$$v = \cos x \sin hy = c_2 \text{ (say)} \qquad \qquad \qquad \qquad \qquad \dots (2)$$

where c_1 and c_2 are parameters for the two families of curves given by $u = c_1$, $v = c_2$.

Differentiating (1) and (2) w.r.t. x we get

$$\cos x \cosh y + \sin x \sinh y \left(\frac{dy}{dx}\right)_{(1)} = 0$$

$$-\sin x \sinh y + \cos x \cosh y \left(\frac{dy}{dx}\right)_{(2)} = 0$$

$$\therefore \qquad \left(\frac{dy}{dx}\right)_{(1)} \times \left(\frac{dy}{dx}\right)_{(2)} = -\frac{\cos x \cosh y}{\sin x \sinh y} \cdot \frac{\sin x \sinh y}{\cos x \cosh y} = -1$$

Hence the two families of curves $u = c_1, v = c_2$ form an orthogonal system.

11.14.4 POLAR FORM OF CAUCHY RIEMANN EQUATIONS

THEOREM 5. *If $w = f(z) = u + iv$ is an analytic function, then in polar form the Cauchy Riemann equations are $\dfrac{\partial u}{\partial r} = \dfrac{1}{r}\dfrac{\partial v}{\partial \theta}$ and $\dfrac{\partial v}{\partial r} = -\dfrac{1}{r}\dfrac{\partial u}{\partial \theta}$*

(UPTU–2008, VTU–2006)

PROOF. Cauchy Riemman equations are

$$\frac{\partial u}{\partial x} = \frac{\partial v}{\partial y} \qquad \text{and} \qquad \frac{\partial u}{\partial y} = -\frac{\partial v}{\partial x} \qquad \qquad \qquad \qquad \dots (1)$$

Now $x = r \cos \theta, y = r \sin \theta$.

$$\therefore \qquad r = \sqrt{(x^2 + y^2)}, \theta = \tan^{-1} y / x \qquad \Rightarrow \qquad \frac{\partial r}{\partial x} = \frac{1}{2\sqrt{(x^2 + y^2)}} \cdot 2x = \frac{x}{r}, \frac{\partial r}{\partial y} = \frac{y}{r}$$

$$\left.\begin{array}{l} \dfrac{\partial \theta}{\partial x} = \dfrac{1}{1 + y^2/x^2}(-y/x^2) = -\dfrac{y}{x^2 + y^2} = -\dfrac{r \sin \theta}{r^2} = \dfrac{\sin \theta}{r} \\[3mm] \dfrac{\partial \theta}{\partial y} = \dfrac{1}{1 + y^2/x^2}(1/x) = \dfrac{x}{x^2 + y^2} = \dfrac{r \cos \theta}{r^2} = \dfrac{\cos \theta}{r} \end{array}\right\} \qquad \dots (A)$$

Now $\qquad \dfrac{\partial u}{\partial x} = \dfrac{\partial u}{\partial r} \cdot \dfrac{\partial r}{\partial x} + \dfrac{\partial u}{\partial \theta} \cdot \dfrac{\partial \theta}{\partial x} = \cos \theta \dfrac{\partial u}{\partial r} - \dfrac{\sin \theta}{r} \dfrac{\partial u}{\partial \theta}$ by (A).

$\dfrac{\partial v}{\partial y} = \dfrac{\partial v}{\partial r} \cdot \dfrac{\partial r}{\partial y} + \dfrac{\partial v}{\partial \theta} \cdot \dfrac{\partial \theta}{\partial y} = \sin \theta \dfrac{\partial v}{\partial r} + \dfrac{\cos \theta}{r} \dfrac{\partial v}{\partial \theta}$ by (A).

But from (1) $\dfrac{\partial u}{\partial x} = \dfrac{\partial v}{\partial y}$ and hence from above, we obtain

$$\cos\theta\,\dfrac{\partial u}{\partial r} - \dfrac{\sin\theta}{r}\dfrac{\partial u}{\partial \theta} = \sin\theta\,\dfrac{\partial v}{\partial r} + \dfrac{\cos\theta}{r}\dfrac{\partial v}{\partial \theta} \qquad \text{... (2)}$$

Again $\quad \dfrac{\partial u}{\partial y} = \dfrac{\partial u}{\partial r}\cdot\dfrac{\partial r}{\partial y} + \dfrac{\partial u}{\partial \theta}\dfrac{\partial \theta}{\delta y} = \sin\theta\,\dfrac{\partial u}{\partial r} + \dfrac{\cos\theta}{r}\dfrac{\partial u}{\partial \theta},\ \dfrac{\partial v}{\partial x} = \dfrac{\partial v}{\partial r}\cdot\dfrac{\partial r}{\partial x} + \dfrac{\partial v}{\partial \theta}\dfrac{\partial \theta}{\delta x} = \cos\theta\,\dfrac{\partial v}{\partial r} + \dfrac{\sin\theta}{r}\dfrac{\partial v}{\partial \theta}$

But from (1) $\dfrac{\partial u}{\partial y} = -\dfrac{\partial v}{\partial x}$ and hence from above we obtain

$$\sin\theta\,\dfrac{\partial u}{\partial r} + \dfrac{\cos\theta}{r}\dfrac{\partial u}{\partial \theta} = -\cos\theta\,\dfrac{\partial v}{\partial r} + \dfrac{\sin\theta}{r}\dfrac{\partial v}{\partial \theta} \qquad \text{... (3)}$$

Multiplying (2) by $\cos\theta$ and (3) by $\sin\theta$ and adding we get

$$\dfrac{\partial u}{\partial r} = \dfrac{1}{r}\dfrac{\partial v}{\partial \theta}$$

Again multiplying (2) by $\sin\theta$ and (3) by $\cos\theta$ and subtracting we get

$$-\dfrac{1}{r}\dfrac{\partial u}{\partial \theta} = \dfrac{\partial v}{\partial r} \quad \text{or} \quad \dfrac{\partial v}{\partial r} = -\dfrac{1}{r}\dfrac{\partial u}{\partial \theta} \qquad \text{... (5)}$$

Equations (4) and (5) give the corresponding polar form of Cauchy Riemann equations.

11.14.5 Derivative of w in Polar Form

THEOREM 6. $\quad \dfrac{dw}{dz} = (\cos\theta - i\sin\theta)\dfrac{\partial w}{\partial r} = \dfrac{-i}{r}(\cos\theta - i\sin\theta)\dfrac{\partial w}{\partial \theta}.$

PROOF. We have proved that in cartesian form

$$\dfrac{dw}{dz} = \dfrac{\partial w}{\partial x} = \dfrac{1}{r}\dfrac{\partial w}{\partial y}$$

$$\therefore \qquad \dfrac{dw}{dz} = \dfrac{\partial w}{\partial x} = \dfrac{\partial w}{\partial r}\cdot\dfrac{\partial r}{\partial x} + \dfrac{\partial w}{\partial \theta}\cdot\dfrac{\partial \theta}{\partial x} = \dfrac{\partial w}{\partial r}\cos\theta + \left(\dfrac{\partial u}{\partial \theta} + i\dfrac{\partial v}{\partial \theta}\right)\left(\dfrac{-\sin\theta}{r}\right) \text{ by } A \text{ and } w = u+iv.$$

We want the result in terms of $\dfrac{\partial w}{\partial r}$ and hence we shall put for $\dfrac{\partial u}{\partial \theta}$ and $\dfrac{\partial v}{\partial \theta}$ from Cauchy-Riemann equations (4) and (5) polar form.

$$\therefore \qquad \dfrac{dw}{dz} = \dfrac{\partial w}{\partial r}\cos\theta + \left[-r\dfrac{\partial v}{\partial r} - ir\dfrac{\partial u}{\partial r}\right]\left[\dfrac{-\sin\theta}{r}\right] = \dfrac{\partial w}{\partial r}\cos\theta - \left[i^2\dfrac{\partial v}{\partial r} + i\dfrac{\partial u}{\partial r}\right]\sin\theta$$

$$= \dfrac{\partial w}{\partial r}\cos\theta - i\sin\theta\dfrac{\partial}{\partial r}(u+iv) = \dfrac{\partial w}{\partial r}\cos\theta - i\sin\theta\dfrac{\partial w}{\partial r} = (\cos\theta - i\sin\theta)\dfrac{\partial w}{\partial r}$$

Again from (1) we have $\dfrac{dw}{dz} = \left(\dfrac{\partial u}{\partial r} + i\dfrac{\partial v}{\partial r}\right)\cos\theta + \dfrac{\partial w}{\partial \theta}\left(\dfrac{-\sin\theta}{r}\right)$ by A and $\overline{w} = +iv$

We want the result in terms of $\dfrac{\partial}{\partial \theta}$ and hence we shall put for $\dfrac{\partial u}{\partial r}$ and $\dfrac{\partial v}{\partial r}$ from Cauchy-Riemman equation (4) and (5) polar form.

$$\therefore \qquad \dfrac{\partial w}{\partial z} = \left(\dfrac{1}{r}\dfrac{\partial v}{\partial \theta} - \dfrac{i}{r}\dfrac{\partial u}{\partial \theta}\right)\cos\theta - \dfrac{\partial w}{\partial \theta}\dfrac{\sin\theta}{r} = \dfrac{1}{r}\left(-i^2\dfrac{\partial v}{\partial \theta} - i\dfrac{\partial u}{\partial \theta}\right)\cos\theta - \dfrac{\partial w}{\partial \theta}\dfrac{\sin\theta}{r}$$

$$= \dfrac{-i}{r}\dfrac{\partial}{\partial \theta}(u+iv)\cos\theta + i^2\dfrac{\partial w}{\partial \theta}\dfrac{\sin\theta}{r} = \dfrac{-i}{r}(\cos\theta - i\sin\theta)\dfrac{\partial w}{\partial \theta}$$

Results (2) and (3) give the derivative of w in polar form and the corresponding derivatives in cartesian form are given in (1).

Solved Examples

EXAMPLE 1. *If n is real, show that $w = r^n(\cos n\theta + i\sin n\theta)$ is analytic except possibly when $r = 0$. Also, show that its derivative is*

$$nr^{n-1}[\cos(n-1)\theta + i\sin(n-1)\theta]$$

SOLUTION. We have $w = r^n(\cos n\theta + i\sin n\theta) = u+iv$

$\therefore \quad u(r,\theta) = r^n\cos n\theta$

$\qquad v(r,\theta) = r^n\sin n\theta$

$\therefore \quad \dfrac{\partial u}{\partial r} = nr^{n-1}\cos n\theta, \dfrac{\partial v}{\partial r} = nr^{n-1}\sin n\theta$

$$\dfrac{\partial u}{\partial \theta} = -nr^n\sin n\theta, \dfrac{\partial v}{\partial \theta} = nr^n\cos n\theta$$

Clearly $\dfrac{\partial u}{\partial r} = \dfrac{1}{r}nr^n\cos n\theta = \dfrac{1}{r}\dfrac{\partial v}{\partial \theta}$

$$\dfrac{\partial v}{\partial r} = \dfrac{1}{r}nr^n\sin n\theta = -\dfrac{1}{r}\dfrac{\partial u}{\partial \theta}$$

Thus Cauchy Riemman equations (polar form) are satisfied. Also from result (2) we know that

$$\dfrac{dw}{dz} = (\cos\theta - i\sin\theta)\dfrac{\partial w}{\partial r}$$

$$= e^{-i\theta} n r^{n-1} (\cos n\theta + i \sin n\theta)$$
$$= n r^{n-1} e^{i(n-1)\theta}$$
$$= n r^{n-1} [\cos(n-1)\theta + i \sin(n-1)\theta].$$

EXAMPLE 2. *If* $f(z) = \dfrac{x^3 y (y - ix)}{x^6 + y^2}, z \neq 0$ *and* $f(0) = 0$ *show*

that $\dfrac{f(z) - f(0)}{z} \to 0$ *as* $z \to 0$ *along any radius*

vector but not as $z \to 0$ *in any manner.*

SOLUTION. We can write $\dfrac{f(z) - f(0)}{z - 0} = \dfrac{f(z) - 0}{z} = \dfrac{f(z)}{z}$

$$= \dfrac{x^3 y (y - ix)}{(x^6 + y^2)(x + iy)}$$

$$= \dfrac{-i x^3 y (x + iy)}{(x^6 + y^2)(x + iy)} = -i \dfrac{x^3 y}{x^6 + y^2}$$

Let $z \to 0$ along the radius vector $y = mx$. Then, we have

$$\lim_{z \to 0} \dfrac{f(z) - f(0)}{z - 0} = \lim_{x \to 0} \dfrac{-i x^3 . mx}{x^6 + (mx)^2}$$

$$= \lim_{x \to 0} \dfrac{-i m x^2}{x^4 + m^2} = 0$$

Similarly, let $z \to 0$ along $y = x^3$ then

$$\lim_{z \to 0} \dfrac{f(z) - f(0)}{z - 0} = \lim_{x \to 0} \dfrac{-i x^3 . x^3}{x^6 + (x^3)^2} = \dfrac{-i}{2} \pm 0.$$

11.14.6 CONSTRUCTION OF ANALYTIC FUNCTIONS

Let $f(z) = u + iv$ be an analytic function where both u and v are conjugate functions. If one of these say $u(x, y)$ be given then to determine the other.

Method I. If v is a function of x, y.

$\therefore$
$$dv = \dfrac{\partial v}{\partial x} dx + \dfrac{\partial v}{\partial y} dy$$

$$dv = -\dfrac{\partial u}{\partial y} dx + \dfrac{\partial u}{\partial x} dy \text{ by Cauchy Riemman Equations.} \qquad \dots (1)$$

R.H.S. is of the form $M dx + N dy$.

Where
$$= -\dfrac{\partial}{\partial} \text{ and } N = \dfrac{\partial u}{\partial x}$$

$\therefore$
$$\dfrac{\partial M}{\partial y} = -\dfrac{\partial^2 u}{\partial y^2} \text{ and } \dfrac{\partial N}{\partial x} = \dfrac{\partial^2 u}{\partial x^2} \qquad \dots (2)$$

But as u satisfies Laplace's equation so that $\dfrac{\partial^2 u}{\partial x^2} + \dfrac{\partial^2 u}{\partial y^2} = 0$ or $\dfrac{\partial^2 u}{\partial x^2} = -\dfrac{\partial^2 u}{\partial y^2}$

Hence, from (2) we get $\dfrac{\partial M}{\partial y} = \dfrac{\partial N}{\partial x}$ which is the condition of exactness.

Therefore the equation (1) can be integrated and v can be found. Now both u and v are known and as such $f(z) = u + iv$ can be determined.

2nd Method: Milne-Thomsom Method

$$z = x + iy \qquad \therefore \qquad \bar{z} = x - iy$$

$\therefore$
$$x = \dfrac{1}{2}(z + \bar{z}), y = \dfrac{1}{2i}(z - \bar{z})$$

Now
$$w = f(z) = u(x, y) + iv(x, y) = u\left(\dfrac{z + \bar{z}}{2}, \dfrac{z - \bar{z}}{2i}\right) + iv\left(\dfrac{z + \bar{z}}{2}, \dfrac{z - \bar{z}}{2i}\right)$$

The above relation can be regarded as formal identity in two independent variables z and $\bar{z}$. By setting $\bar{z} = z$ we obtain.
$$y = 0, x = z. \text{ So that } f(z) = u(z, 0) + iv(z, 0)$$

Now
$$f'(z) = \dfrac{dw}{dz} = \dfrac{\partial w}{\partial x} = \dfrac{\partial u}{\partial x} + i \dfrac{\partial v}{\partial x} \qquad \dots (1)$$

Since u is known but v is not known so that we replace $\dfrac{\partial v}{\partial x}$ by $-\dfrac{\partial u}{\partial y}$ in (1) from Cauchy-Riemann equations.

$\therefore$
$$f'(z) = \dfrac{\partial u}{\partial x} - i \dfrac{\partial u}{\partial y}$$

Now choose
$$\dfrac{\partial u}{\partial x} = \phi_1(x, y) = \phi_1(z, 0) ; \dfrac{\partial u}{\partial y} = \phi_2(x, y) = \phi_2(z, 0)$$

$\therefore$
$$f'(z) = \phi_1(z, 0) - i\phi_2(z, 0)$$

Integrating, we get
$$f(z) = \int [\phi_1(z, 0) - i\phi_2(z, 0)] dz + c$$

Above gives the construction of function $f(z)$ when u is given.

Similarly if v is given and u is unknown then we replace $\dfrac{\partial u}{\partial x}$ by $\dfrac{\partial v}{\partial y}$ in (1) from Cauchy-Riemann equations.

$\therefore$ $$f'(z) = \frac{\partial v}{\partial y} + i\frac{\partial v}{\partial x}$$

Now choose $$\frac{\partial v}{\partial y} = \psi_1(x,y) = \psi_1(z,0) \quad \text{and} \quad \frac{\partial v}{\partial x} = \psi_2(x,y) = \psi_2(z,0)$$

$\therefore$ $$f'(z) = \psi_1(z,0) + i\psi_2(z,0)$$

Integrating we get $$f(z) = \int[\psi_1(z,0) + i\psi_2(z,0)]dz + c$$

 Solved Examples

EXAMPLE 1. *Prove that the function u given as under is harmonic and find its harmonic conjugate and the corresponding analytic function f(z) in terms of z.*

(a) $u = x^3 - 3xy^2$ *(b)* $u = y^3 - 3x^2y$

(c) $u = \dfrac{1}{2}\log(x^2 + y^2)$ (UPTU–2010)

SOLUTION. (a) $u = x^3 - 3xy^2$

$$\frac{\partial u}{\partial x} = 3x^2 - 3y^2, \frac{\partial u}{\partial y} = -6xy$$

$$\frac{\partial^2 u}{\partial x^2} = 6x, \frac{\partial^2 u}{\partial y^2} = -6x$$

Clearly $\dfrac{\partial^2 u}{\partial x^2} + \dfrac{\partial^2 u}{\partial y^2} = 0$ and hence, u is harmonic.

Let the conjugate of u be v. Then

$$dv = \frac{\partial v}{\partial x}dx + \frac{\partial v}{\partial y}dy = \frac{\partial u}{\partial y}dx + \frac{\partial u}{\partial x}dy \quad \text{(C.R.Eq.)}$$

or $dv = 6xy\,dx + (3x^2 - 3y^2)dy$

Above is exact equation and hence on integrating.

$$v = \int 6xy\,dx + \int -3y^2\,dy + c$$

y constant only those which do not contain x

or $v = 3x^2y - y^3 + c$

$\therefore f(z) = u + iv = x^3 - 3xy^2 + i(3x^2y - y^3) + ic$

$$= x^3 + 3x^2(iy) + 3x(iy)^2 + (iy)^3 + ic$$

or $f(z) = (x + iy)^3 + c' = z^3 + c'$

you may try to obtain the result by Milne Thomson method also.

(b) $u = y^3 - 3x^2y$

Proceed as above, we get

$v = -3xy^2 + x^2 + c, f(z) = iz^3 + ic$

(c) $u = \dfrac{1}{2}\log(x^2 + y^2)$

$$\frac{\partial u}{\partial x} = \frac{x}{x^2 + y^2}, \frac{\partial u}{\partial y} = \frac{y}{x^2 + y^2}$$

$$\frac{\partial^2 u}{\partial x^2} = \frac{y^2 - x^2}{(x^2 + y^2)^2}, \frac{\partial^2 u}{\partial y^2} = \frac{x^2 - y^2}{(x^2 + y^2)^2}$$

Clearly u satisfies Laplace's equation $\dfrac{\partial^2 u}{\partial x^2} + \dfrac{\partial^2 u}{\partial y^2} = 0$,

and hence u is harmonic.

Let the conjugate of u be v then

$$dv = \frac{\partial v}{\partial x}dx + \frac{\partial v}{\partial y}dy = -\frac{\partial u}{\partial y}dx + \frac{\partial u}{\partial x}dy \quad \text{(C.R.Eq.)}$$

$$dv = -\frac{y}{x^2 + y^2}dx + \frac{x}{x^2 + y^2}dy$$

Above equation is exact and hence on integrating

$$v = \int_{y\,const.} -\frac{y}{x^2 + y^2}dx + \int 0\,dy + c$$

or $v = -y\dfrac{1}{y}\tan^{-1}\dfrac{x}{y} + c = \tan^{-1}\left(-\dfrac{x}{y}\right) + c$

$$= \frac{\pi}{2} - \cot^{-1}\left(-\frac{x}{y}\right) + c$$

$$= \cot^{-1}\left(\frac{x}{y}\right) + k = \tan^{-1}\frac{y}{x} + c'.$$

We may also write

$$dv = \frac{xdy - ydx}{x^2 + y^2} = \frac{xdy - ydx}{x^2}\cdot\frac{1}{1 + (y/x)^2}$$

$$= \frac{1}{1 + (y/x)^2}.d(y/x)$$

$\therefore$ $v = \tan^{-1} y/x + c'$

$\therefore$ $f(z) = u + iv$

$$= \frac{1}{2}\log(x^2 + y^2) + i\tan^{-1} y/x + ic'$$

$$= \log(x + iy) + ic' = \log z + ic'$$

EXAMPLE 2. *If $u = x^2 - y^2, v = \dfrac{-y}{x^2 + y^2}$ prove that both u and v satisfy Laplace's equaiton but u+iv is not an analytic function of z.*

SOLUTION. We have $\dfrac{\partial u}{\partial x} = 2x, \dfrac{\partial u}{\partial y} = -2y$

$$\frac{\partial v}{\partial x} = \frac{2xy}{(x^2 + y^2)^2}, \frac{\partial v}{\partial y} = \frac{y^2 - x^2}{(x^2 + y^2)^2}$$

$$\frac{\partial^2 u}{\partial x^2} = 2, \frac{\partial^2 u}{\partial y^2} = -2$$

$$\frac{\partial^2 v}{\partial x^2} = \frac{2y(y^2 - 3x^2)}{(x^2 + y^2)^3}, \frac{\partial^2 v}{\partial y^2} = \frac{2y(y^2 - 3x^2)}{(x^2 + y^2)^3}$$

From above it is clear that

$$\frac{\partial^2 u}{\partial x^2} + \frac{\partial^2 u}{\partial y^2} = 0 \quad \text{and} \quad \frac{\partial^2 v}{\partial x^2} + \frac{\partial^2 v}{\partial y^2} = 0$$

Above shows that both u and v satisfy Laplace's equation. But from above we conclude that

$$\frac{\partial u}{\partial x} \neq \frac{\partial v}{\partial y} \quad \text{and} \quad \frac{\partial u}{\partial y} \neq -\frac{\partial v}{\partial x}$$

which shows that Cauchy Riemman equations are not satisfied and hence the function $f(z) = u+iv$ is not analytic.

EXAMPLE 3. *If $u - v = (x - y)(x^2 + 4xy + y^2)$ and $f(z) = u + iv$ is an analytic function of $z = x + iy$ find $f(z)$ in terms of z.* (MUMBAI–2008, VTU–2007, WBTU–2005)

SOLUTION. We have $f(z) = u + iv$ $\therefore$ $if(z) = iu - v$

$\therefore$ $(1+i)f(z) = (u-v) + i(u+v) = U + iV$ say where

U and V are functions of u and v and hence of x and y.

$F(z) = (1+i)f(z) = U + iV$ where $U = u - v$ is known.

$$\frac{\partial U}{\partial x} = 1(x^2 + 4xy + y^2) + (x - y)(2x + 4y)$$

$$= 3x^2 + 6xy - 3y^2 = \phi_1(x, y)$$

$\therefore$ $\phi_1(z, 0) = 3z^2$

$$\frac{\partial U}{\partial y} = -1(x^2 + 4xy + y^2) + (x - y)(4x + 2y)$$

$$= 3x^2 - 6xy - 3y^2 = \phi_2(x, y)$$

$\therefore$ $\phi_2(z, 0) = 3z^2$

$$F'(Z) = \frac{dW}{dz} = \frac{\partial W}{\partial x} = \frac{\partial U}{\partial x} + i\frac{\partial V}{\partial x} = \frac{\partial U}{\partial x} - i\frac{\partial U}{\partial y}$$

$$= \phi_1(z, 0) - i\phi_2(z, 0) = (1-i)3z^2.$$

On integrating, we get

$$F(z) = \int (1-i)3z^2 dz + c = (1-i)z^3 + c$$

$\therefore$ $(1+i)f(z) = (1-i)z^3 + c$

or $f(z) = \dfrac{1-i}{1+i}z^3 + \dfrac{c}{1+i} = -iz^3 + c'$ because

$$\frac{(1-i)}{(1+i)} = \frac{1-i}{1+i}\cdot\frac{1-i}{1-i} = \frac{1+i^2-2i}{1-i^2} = \frac{-2i}{2} = -i$$

EXAMPLE 4. *If $u - v = \dfrac{\cos x + \sin x - e^{-y}}{2\cos x - e^{y} - e^{-y}}$ and $f(z) = u+iv$ is an analytic function of $z = x+iy$, find $f(z)$ subject to the condition $f(\pi/2) = 0$.*

(AMIETE–2005, OSMANIA–2003, MUMBAI–2007)

SOLUTION. We have $f(z) = u + iv$ $\therefore$ $if(z) = iu - v$

$\therefore$ $(1+i)f(z) = (u-v) + i(u+v) = U + iV$ say

Where U and V are function of u and v and hence of x and y.

$F(z) = (1+i)f(z) = U + iV$ say where $U = u - v$ is known.

$$U = \frac{1}{2}\frac{\cos x + \sin x - (\cosh y - \sinh y)}{\cos x - \cosh y}$$

$$= \frac{1}{2}\left[\frac{(\cos x - \cosh y) + (\sin x + \sinh y)}{\cos x - \cosh y}\right]$$

$$U = \frac{1}{2}\left[1 + \frac{\sin x + \sinh y}{\cos x - \cosh y}\right]$$

or $\dfrac{\partial U}{\partial x} = \dfrac{1}{2}\left[\dfrac{\begin{array}{c}(\cos x - \cosh y)\cos x\\ -(\sin x + \sinh y)(-\sin x)\end{array}}{(\cos x - \cosh y)^2}\right]$

$$= \frac{1}{2}\left[\frac{1 + \sin x \sinh y - \cos x \cosh y}{(\cos x - \cosh y)^2}\right] = \phi_1(x, y)$$

$\therefore$ $\phi_1(z, 0) = \dfrac{1}{2}\left[\dfrac{1 - \cos z}{(\cos z - 1)^2}\right]$

$$= \frac{1}{2}\frac{1}{(1-\cos z)} = \frac{1}{4}\mathrm{cosec}^2 z/2$$

$$\frac{\partial U}{\partial y} = \frac{1}{2}\left[\frac{-1 + \sin x \sinh y + \cos x \cosh y}{(\cos x - \cosh y)^2}\right] = \phi_2(x, y)$$

$\therefore$ $\phi_2(z, 0) = \dfrac{1}{2}\left[\dfrac{\cos z - 1}{(\cos z - 1)^2}\right]$

$$= -\frac{1}{2}\cdot\frac{1}{1-\cos z} = -\frac{1}{4}\mathrm{cosec}^2\frac{z}{2}$$

$\therefore$ $F'(z) = \dfrac{dW}{dz} = \dfrac{\partial W}{\partial x} = \dfrac{\partial U}{\partial x} + i\dfrac{\partial V}{\partial x} = \dfrac{\partial U}{\partial x} - i\dfrac{\partial u}{\partial y}$

$$= \phi_1(z, 0) - i\phi_2(z, 0) = \frac{1}{4}\mathrm{cosec}^2\frac{z}{2}(1+i)$$

Integrating we get

$$F(z) = \frac{1}{4}(1+i)(-2\cot z/2) + c$$

or $(1+i)f(z) = -\dfrac{1}{2}(1+i)\cot\dfrac{z}{2} + c$

$\therefore$ $f(z) = -\dfrac{1}{2}\cot\dfrac{z}{2} + \dfrac{c}{1+i} = -\dfrac{1}{2}\cot\dfrac{z}{2} + c'$

when $z = \pi/2, \cot z/2 = 1$ and $f(\pi/2) = 0$

$\therefore$ $c' = \dfrac{1}{2}$

$\therefore$ $f(z) = \dfrac{1}{2}(1 - \cot z/2).$

EXAMPLE 5. *If $u + v = \dfrac{2\sin 2x}{e^{2y} + e^{-2y} - 2\cos 2x}$ and $f(z) = u + iv$ is an analytic function of $z = x + iy$, find $f(z)$ in terms of z.* (PTU–2002)

SOLUTION. We have $f(z) = u + iv$ $\therefore$ $if(z) = iu - v$

$\therefore$ $F(\bar{z}) = (1+i)f(z) = (u-v) + i(u+v) = U + iV$

where U and V are functions of u and v and hence of x and y and also V is known.

Now $V = \dfrac{2\sin 2x}{2\cosh 2y - 2\cos 2x} = \dfrac{\sin 2x}{\cosh 2y - \cos 2x}$

$$F'(z) = \frac{dW}{dz} = \frac{\partial W}{\partial x} = \frac{\partial U}{\partial x} + i\frac{\partial V}{\partial x} = \frac{\partial V}{\partial y} + i\frac{\partial V}{\partial x}$$

$$= \psi_1(x, y) + i\psi_2(x, y)$$

$$\frac{\partial V}{\partial y} = \frac{2\sin 2x \sinh 2y}{(\cosh 2y - \cos 2x)^2} = \psi_1(x, y)$$

$\therefore$ $\psi_1(z, 0) = 0$

$$\frac{\partial V}{\partial x} = \frac{2\cos 2x(\cosh 2y - \cos 2x) - \sin 2x(2\sin 2x)}{(\cosh 2y - \cos 2x)^2}$$

$$= \frac{2\cos 2x \cosh 2y - 2}{(\cos 2y - \cos 2x)^2} = \psi_2(x, y)$$

$\psi_2(z, 0) = \dfrac{2\cos 2z - 2}{(1 - \cos 2z)^2} = \dfrac{-2}{(1-\cos 2z)} = -\mathrm{cosec}^2 z$

$$F(z) = \int[\psi_1(z, 0) + i\psi_2(z, 0)]dz + c$$

or $(1+i)f(z) = \int 0 - i\,\mathrm{cosec}^2 z\,dz + c = i\cot z + c$

$\therefore$ $f(z) = \dfrac{i}{1+i}\cot z + \dfrac{c}{1+i} = \dfrac{1}{2}(1+i)\cot z + c'$

$\therefore$ $\dfrac{i}{1+i} = \dfrac{i(1+i)}{1+i^2+2i} = \dfrac{1}{2}(1+i)$

EXAMPLE 6. *If $f(z) = u+iv$ is an analytic function of $z = x+iy$ and f is any function of x and y with differential coefficients of first and second order then*

(a) $\left(\dfrac{\partial \phi}{\partial x}\right)^2 + \left(\dfrac{\partial \phi}{\partial y}\right)^2 = \left\{\left(\dfrac{\partial \phi}{\partial u}\right)^2 + \left(\dfrac{\partial \phi}{\partial v}\right)^2\right\} |f'(z)|^2$

(b) $\dfrac{\partial^2 \phi}{\partial x^2} + \dfrac{\partial^2 \phi}{\partial y^2} = \left(\dfrac{\partial^2 \phi}{\partial u^2} + \dfrac{\partial^2 \phi}{\partial v^2}\right) |f'(z)|^2$

(c) $\left(\dfrac{\partial^2}{\partial x^2} + \dfrac{\partial^2}{\partial y^2}\right) |f(z)|^2 = 4 |f'(z)|^2$

(JNTU–2006, KOTTAYAM–2005, MADRAS–2006)

SOLUTION. (a) $\dfrac{\partial \phi}{\partial x} = \dfrac{\partial \phi}{\partial u}\dfrac{\partial u}{\partial x} + \dfrac{\partial \phi}{\partial v}\dfrac{\partial v}{\partial x}$

$\dfrac{\partial \phi}{\partial y} = \dfrac{\partial \phi}{\partial u}\dfrac{\partial u}{\partial y} + \dfrac{\partial \phi}{\partial v}\dfrac{\partial v}{\partial y}$ ... (1)

$\qquad = \dfrac{\partial \phi}{\partial u}\left(-\dfrac{\partial v}{\partial x}\right) + \dfrac{\partial \phi}{\partial v}\left(\dfrac{\partial u}{\partial x}\right)$ by C.R. Equations ...(2)

Squaring and adding (1) and (2), we get

$\left(\dfrac{\partial \phi}{\partial x}\right)^2 + \left(\dfrac{\partial \phi}{\partial y}\right)^2 = \left[\left(\dfrac{\partial \phi}{\partial u}\right)^2 + \left(\dfrac{\partial \phi}{\partial v}\right)^2\right]\left[\left(\dfrac{\partial u}{\partial x}\right)^2 + \left(\dfrac{\partial v}{\partial x}\right)^2\right]$

$\qquad = \left[\left(\dfrac{\partial \phi}{\partial u}\right)^2 + \left(\dfrac{\partial \phi}{\partial v}\right)^2\right] |f'(z)|^2$

$\because \quad f'(z) = \dfrac{dw}{dz} = \dfrac{\partial w}{\partial x} = \dfrac{\partial u}{\partial x} + i\dfrac{\partial v}{\partial x}$

$\therefore \quad |f'(z)|^2 = \left(\dfrac{\partial u}{\partial x}\right)^2 + \left(\dfrac{\partial v}{\partial x}\right)^2$

(b) We know $w = f(z) = u + iv$

$\therefore \qquad \bar{w} = f(\bar{z}) = u - iv$

$\therefore \qquad u = \dfrac{1}{2}(w + \bar{w}), v = \dfrac{1}{2i}(w - \bar{w})$

$\therefore \qquad \dfrac{\partial}{\partial w} = \dfrac{\partial}{\partial u}\dfrac{\partial u}{\partial w} + \dfrac{\partial}{\partial v}\dfrac{\partial v}{\partial w} = \left(\dfrac{1}{2}\dfrac{\partial}{\partial u} + \dfrac{1}{2i}\dfrac{\partial}{\partial v}\right)$

or $\qquad \dfrac{\partial}{\partial w} = \dfrac{1}{2}\left(\dfrac{\partial}{\partial u} - i\dfrac{\partial}{\partial v}\right) \quad \because \dfrac{1}{i} = -i$

Similarly $\dfrac{\partial}{\partial \bar{w}} = \dfrac{1}{2}\left(\dfrac{\partial}{\partial u} + i\dfrac{\partial}{\partial v}\right)$

$\therefore \quad 4\dfrac{\partial^2}{\partial w \partial \bar{w}} = \left(\dfrac{\partial}{\partial u} - i\dfrac{\partial}{\partial y}\right)\left(\dfrac{\partial}{\partial u} + i\dfrac{\partial}{\partial v}\right)$

or $4\dfrac{\partial^2}{\partial w \partial \bar{w}} = \dfrac{\partial^2}{\partial u^2} + \dfrac{\partial^2}{\partial v^2}$

or $4\dfrac{\partial^2 \phi}{\partial w \partial \bar{w}} = \dfrac{\partial^2 \phi}{\partial u^2} + \dfrac{\partial^2 \phi}{\partial v^2}$ ...(4)

Again as $z = x + iy, \bar{z} = x - iy$

i.e., $\qquad x = \dfrac{1}{2}(z + \bar{z}), y = \dfrac{1}{2i}(z - \bar{z})$

Hence, proceeding as above we have

$4\dfrac{\partial^2 \phi}{\partial z \partial \bar{z}} = \dfrac{\partial^2 \phi}{\partial x^2} + \dfrac{\partial^2 \phi}{\partial y^2}$ ... (5)

Now $\quad 4\dfrac{\partial^2}{\partial w \partial \bar{w}} = 4\left(\dfrac{\partial}{\partial z}\dfrac{dz}{dw}\right)\left(\dfrac{\partial}{\partial \bar{z}} \cdot \dfrac{d\bar{z}}{dw}\right)$

$4\left(\dfrac{z}{f'(z)}\dfrac{\partial}{\partial z}\right)\left(\dfrac{1}{f'(z)}\right) \quad \because f'(z) = \dfrac{dw}{dz}$

$\qquad = \dfrac{1}{f'(z)\overline{f'(z)}} \cdot 4\dfrac{\partial^2}{\partial z \partial \bar{z}}$

$\therefore \quad |f'(z)|^2 \cdot 4\dfrac{\partial^2 \phi}{\partial w \partial \bar{w}} = 4\dfrac{\partial^2 \phi}{\partial w \partial \bar{w}}$

$\because \qquad f'(\bar{z}) = \overline{f'(z)}$

Hence putting the values of $4\dfrac{\partial^2 \phi}{\partial w \partial \bar{w}}$ and $4\dfrac{\partial^2 \phi}{\partial z \partial \bar{z}}$ from

(4) and (5) in the above we get result (b).

(c) L.H.S. $= \left(\dfrac{\partial^2}{\partial x^2} + \dfrac{\partial^2}{\partial y^2}\right) |f(z)|^2 = 4\dfrac{\partial^2}{\partial z \partial \bar{z}}[f(z)f(\bar{z})]$

$\qquad\qquad\qquad\qquad\qquad\qquad\qquad\qquad$ by (5)

$\qquad = 4\dfrac{\partial}{\partial z}[f(z)f'(\bar{z})]$

$\qquad = 4f'(z)f'(\bar{z}) = 4f'(z)\overline{f'(z)} = 4|f'(z)|^2$

EXAMPLE 7. *Prove that an analytic function with constant modulus is constant.*

(MADRAS–2003, BHOPAL–2002S, UPTU–2008, MUMBAI–2005S)

SOLUTION. We have $\quad f(z) = u + iv$

$\therefore \qquad |f(z)|^2 = u^2 + v^2 = c^2$ say.

$\therefore \qquad u\dfrac{\partial u}{\partial x} + v\dfrac{\partial v}{\partial x} = 0$ ...(1)

$\qquad\qquad u\dfrac{\partial u}{\partial y} + v\dfrac{\partial v}{\partial y} = 0$

or $\qquad -u\dfrac{\partial v}{\partial x} + v\dfrac{\partial u}{\partial x} = 0$ ... (2)

$\qquad\qquad\qquad\qquad\qquad$ by C.R. Equ.

Squaring and adding 1 and 2 we get

$(u^2 + v^2)\left[\left(\dfrac{\partial u}{\partial x}\right)^2 + \left(\dfrac{\partial v}{\partial x}\right)^2\right] = 0$

or $c^2[\,|f'(z)|^2 = 0]$

$\because \quad f'(z) = \dfrac{\partial w}{\partial x} = \dfrac{\partial u}{\partial x} + i\dfrac{\partial v}{\partial x}$, i.e., $f'(z) = 0$ as $c \neq 0$

Hence $f(z)$ is constant.

EXAMPLE 8. *If $f(z) = u + iv$ is a regular function of z in any domain prove that*

(i) $\left(\dfrac{\partial^2}{\partial x^2} + \dfrac{\partial^2}{\partial y^2}\right) |f(z)|^r = p^2 |f(z)|^{p-2} |f'(z)|^2$

(KERALA–2005)

(ii) $\left(\dfrac{\partial^2}{\partial x^2} + \dfrac{\partial^2}{\partial y^2}\right) |u|^p = p(p-1) |u|^{p-2} |f'(z)|^2$

SOLUTION. We know that

$\dfrac{\partial^2}{\partial x^2} + \dfrac{\partial^2}{\partial y^2} = 4\dfrac{\partial^2}{\partial z \partial \bar{z}}$

Also $|f(z)|^2 = u^2 + v^2 = f(z)f(\bar{z})$

Also $f(z) + f(\bar{z}) = (u + iv) + (u - iv) = 2u$

$\therefore \quad |u|^2 = \left|\dfrac{f(z) + f(\bar{z})}{2}\right|^2$

Therefore L.H.S of (1) by the help of above results is

$$4\frac{\partial^2}{\partial z \partial \bar{z}}\{f(z)f(\bar{z})\}^{p/2} = 4\frac{\partial^2}{\partial z \partial \bar{z}}[f(z)^{p/2} \cdot f(\bar{z})^{p/2}]$$

Differentiating first w.r.t $\bar{z}$ taking $f(z)$ as constant and then w.r.t z taking $f(\bar{z})$ as constant we have

L.H.S. $= 4\frac{\partial}{\partial z}[f(z)^{p/2} \cdot p/2f(\bar{z})^{p/2-1} \cdot f'(\bar{z})]$

$= 4.[p/2f(z)^{p/2-1} \cdot f'(z)p/2f(\bar{z})^{p/2-1} \cdot f'(\bar{z})]$

$= 4.p^2/4[|f(z)f(\bar{z})^{(p-2)/2}f'(z)f'(\bar{z})|]$

$= p^2[|f(z)|^2]^{(p-2)/2}|f'(z)|^2$

$= p^2|f(z)|^{p-2}|f'(z)|^2 =$ R.H.S. of 1.

(ii) L.H.S. of (ii) by the help of result written above is

$$4\frac{\partial^2}{\partial z \partial \bar{z}}\left\{\frac{1}{2}|(f(z)+f(\bar{z}))\right\}|^p$$

$$= \frac{4}{2^p}\frac{\partial^2}{\partial z \partial \bar{z}}|\{f(z)+f(\bar{z})\}^2|^{p/2}$$

$$= \frac{4}{2^p}\frac{\partial^2}{\partial z \partial \bar{z}}[\{f(z)+f(\bar{z})\}\{f(\bar{z})+f(z)\}]^{p/2}$$

$$= \frac{4}{2^p}\frac{\partial^2}{\partial z \partial \bar{z}}[\{f(z)+f(\bar{z})\}^2]^{p/2}$$

$$= \frac{4}{2^p}\frac{\partial^2}{\partial z \partial \bar{z}}[f(z)+f(\bar{z})]^p$$

Differentiating (as explained above) we get

$$= \frac{4}{2^p}\frac{\partial}{\partial z}p[f(z)+f(\bar{z})]^{p-1}f'(\bar{z})$$

$$= \frac{4}{2^p}p(p-1)[f(z)+f(\bar{z})]^{p-2}f'(z)f'(\bar{z})$$

$$= \frac{4}{2^p}p(p-1)\left[\frac{f(z)+f(\bar{z})}{2}\right]^{p-2}|f'(z)|^2$$

$$= p(p-1)\left[\left\{\frac{f(z)+f(\bar{z})}{2}\right\}^2\right]^{(p-2)/2} \cdot |f'(z)|^2$$

$$= p(p-1)\left[\left|\frac{f(z)+f(\bar{z})}{2}\right|^2\right]^{(p-2)/2}|f'(z)|^2$$

$$= p(p-1)[|u|^2]^{(p-2)/2}|f'(z)|^2$$

$$= p(p-1)|u|^{p-2}|f'(z)|^2.$$

EXAMPLE 9. *Prove that function* $f(z) = u + iv$ *where*

$$f(z) = \frac{x^3(1+i) - y^3(1-i)}{x^2+y^2} \; z \neq 0, \; f(0) = 0$$

is continuous and that Cauchy Riemman equations are satisfied at the origin. Yet $f'(z)$ *does not exist there.* (SVTV–2009, VTU–2001)

SOLUTION. We have $f(z) = \frac{x^3-y^3}{x^2+y^2} + i\frac{(x^3+y^3)}{x^2+y^2}$

$= u(x,y) + iv(x,y)$

For $z \neq 0$, we observe that both u and v are rational functions of x and y with non zero denominators and hence they are continuous at all those points where $z \neq 0$. Therefore $f(z)$ is continuous when $z \neq 0$.

For $z = 0$. On changing to polar, we get

$u = r(\cos^3\theta - \sin^3\theta)$ and $v = r(\cos^3\theta + \sin^3\theta)$

When $r \to 0$ both u and v tend to zero whatever values

may be associated to θ. Also it is given that $f(0) = 0$ which means that actual values of u and v at the origin are zero. Thus the actual values of u and v and their limiting values at the origin are the same and hence u and v are continuous at the origin. Hence $f(z)$ is continuous for all values of z.

Cauchy Riemman equations are satisfied at the origin

i.e., $\frac{\partial u}{\partial x} = \frac{\partial V}{\partial y}$ and $\frac{\partial u}{\partial y} = -\frac{\partial v}{\partial x}$ at origin

$\frac{\partial u}{\partial x} = \lim_{x \to 0}\frac{u(x,0) - u(0,0)}{x} = \lim_{x \to 0}\frac{x-0}{x} = 1$

$\frac{\partial v}{\partial x} = \lim_{y \to 0}\frac{v(x,0) - v(0,0)}{x} = \lim_{x \to 0}\frac{x-0}{x} = 1$

$\frac{\partial v}{\partial y} = \lim_{y \to 0}\frac{v(0,v) - v(0,0)}{y} = \lim_{y \to 0}\frac{y-0}{y} = 1$

From the above four values we conclude that the Cauchy Riemman equations are satisfied

Now $f'(0) = \lim_{z \to 0}\frac{f(z) - f(0)}{z}$ (choosing z in place of Δz)

$= \lim_{\substack{x \to 0 \\ y \to 0}}\frac{(x^3-y^3) + i(x^3+y^3)}{x^2+y^2} \cdot \frac{1}{x+iy}$

Now let $z \to 0$ along $y = x$ then

$f'(0) = \lim_{x \to 0}\frac{0 + i.2x^3}{2x^2.x(1+i)}$

$= \frac{i}{1+i} = \frac{i(1+i)}{1+i^2+2i} = \frac{1}{2}(1+i)$ ... (1)

Again choosing that $z \to 0$ along x axis.

$\therefore \qquad f'(0) = \lim_{x \to 0}\frac{x^3 + ix^3}{x^3} = 1+i$ ... (2)

From (1) and (2) we conclude that the limit is not unique and hence $f'(0)$ does not exist, i.e., $f(z)$ is not differentiable at $z = 0$.

EXAMPLE 10. *Prove that the function* $f(z) = \sqrt{[|xy|]}$ *is not analytic at the origin even though Cauchy Riemman equations are satisfied at that point.*

(AMIETE–2005 S, OSMANIA–2003)

SOLUTION. Here $u(x,y) = \sqrt{[|xy|]}, v(x,y) = 0$

It is easy to observe as in part (a) that all the four derivatives $\frac{\partial u}{\partial x}, \frac{\partial u}{\partial y}, \frac{\partial v}{\partial x}, \frac{\partial v}{\partial y}$ are zero at the origin and hence Cauchy Riemman quations are satisfied at the origin.

$f'(0) = \lim_{z \to 0}\frac{f(z) - f(0)}{z}$ (Choosing z in place of Δz)

$= \lim_{\substack{x \to 0 \\ y \to 0}}\frac{\sqrt{[(|xy|)]} - 0}{x+iy}$

Let us assume that $z \to 0$ along a line $y = mx$.

$\therefore f'(0) = \lim_{x \to 0}\frac{\sqrt{[(|mx^2|)]}}{x(1+im)} = \frac{\sqrt{|m|}}{1+im}$

The above limit is not unique as it depends upon m, therefore $f'(0)$ does not exist, i.e., the function $f(z)$ is not differentiable at $z = 0$.

EXAMPLE 11. *Exammine the nature of the function*

$$f(z) = \frac{x^2 y^5 (x+iy)}{x^4 + y^{10}} \text{ for } z \neq 0$$

$f(z) = 0$ for $z= 0$ in a region including the origin. (ROHTAK–2004)

SOLUTION. $u(x,y) = \dfrac{x^3 y^6}{x^4 + y^{10}}, v(x,y) = \dfrac{x^2 y^6}{x + y^{10}}$

It is easy to observe that all the four derivatives $\dfrac{\partial u}{\partial x}, \dfrac{\partial u}{\partial y}, \dfrac{\partial v}{\partial x}, \dfrac{\partial v}{\partial y}$ are zero at the origin and hence Cauchy Riemman equations are satisfied at the origin.

$$f'(0) = \lim_{z \to 0} \frac{f(z) - f(0)}{z}$$

choosing z in place of Δz

$$= \lim_{z \to 0} \frac{x^2 y^5 (x+iy)}{(x^4 + y^{10})(x+iy)} = \lim_{z \to 0} \frac{x^2 y^5}{x^4 + y^{10}}$$

Let us assume that $z \to 0$ along the line $y = x$ so that

$$f'(0) = \lim_{x \to 0} \frac{x^7}{x^4 + x^{10}} = \lim_{z \to 0} \frac{x^3}{1 + x^6} = 0$$

Again suppose that $z \to 0$ along the curve $y^5 = x^2$.

$$\therefore \ f'(0) = \lim_{x \to 0} \frac{x^2 x^2}{x^4 + x^4} = \frac{1}{2}$$

The above limits calculated along different paths are different therefore $f'(0)$ does not exist, *i.e.*, the function is not differentiable at $z = 0$.

EXAMPLE 12. *Show that the function f(z) = xy + iy is every where continuous but is not analytic.*

(OSMANIA–2003S)

SOLUTION. Here $u = xy$, $v = y$. continuity of $f(z)$ follows from continuity of u and v. Also

$$\frac{\partial u}{\partial x} = y, \frac{\partial u}{\partial y} = x, \frac{\partial v}{\partial x} = 0, \frac{\partial v}{\partial y} = 1$$

$$\therefore \quad \frac{\partial u}{\partial x} \neq \frac{\partial v}{\partial y} \text{ and } \frac{\partial u}{\partial y} \neq \frac{\partial v}{\partial x}$$

Above shows that Cauchy Riemman equations are not satisfied and hence $f(z)$ is not an analytic function.

EXAMPLE 13. *Show that a harmonic function satisfies the differential equation $\dfrac{\partial^2 z}{\partial z \partial \bar{z}} = 0$.*

SOLUTION. Let $z = x + iy \ \Rightarrow \ \bar{z} = x + iy$

Therefore $x = \dfrac{1}{2}(z + \bar{z}) \quad y = \dfrac{1}{2i}(z - \bar{z})$

If u is a harmonic function then it satisfies the Laplace equation

$$\therefore \quad \frac{\partial^2 u}{\partial x^2} + \frac{\partial^2 u}{\partial y^2} = 0$$

Now $\dfrac{\partial u}{\partial \bar{z}} = \dfrac{\partial u}{\partial x} \cdot \dfrac{\partial x}{\partial \bar{z}} + \dfrac{\partial u}{\partial y} \cdot \dfrac{\partial y}{\partial \bar{z}} = \dfrac{1}{2}\left[\dfrac{\partial u}{\partial x} - \dfrac{1}{2i}\dfrac{\partial u}{\partial y}\right]$

$$\frac{\partial^2 u}{\partial z \partial \bar{z}} = \frac{\partial}{\partial z}\left(\frac{\partial u}{\partial \bar{z}}\right)$$

$$= \frac{1}{2}\frac{\partial^2 u}{\partial x^2}\frac{\partial x}{\partial z} + \frac{1}{2}\frac{\partial^2 u}{\partial y \partial x}\frac{\partial y}{\partial z} - \frac{1}{2i}\frac{\partial^2 u}{\partial x \partial y}\frac{\partial x}{\partial z} - \frac{1}{2i}\frac{\partial^2 u}{\partial y^2}\cdot\frac{\partial y}{\partial z}$$

$$= \frac{1}{4}\frac{\partial^2 u}{\partial x^2} + \frac{1}{4i}\frac{\partial^2 u}{\partial y \partial x} - \frac{1}{4i}\frac{\partial^2 u}{\partial x \partial y} + \frac{1}{4}\frac{\partial^2 u}{\partial y^2}$$

$$= \frac{1}{4}\left(\frac{\partial^2 u}{\partial x^2} + \frac{\partial^2 u}{\partial y^2}\right) = \frac{1}{4} \times 0 = 0.$$

EXAMPLE 14. *If $w = u + iv$ represents the complex potential field and $v = x^2 - y^2 + \dfrac{x}{x^2 + y^2}$. Find the function u.* (VTU–2011, MUMBAI–2008, BHOPAL–2002S)

SOLUTION. We have $v = x^2 - y^2 + \dfrac{x}{x^2 + y^2}$

$$\Rightarrow \quad \frac{\partial v}{\partial y} = -2y - \frac{2xy}{(x^2 + y^2)^2}$$

and $\dfrac{\partial v}{\partial x} = 2x - \dfrac{2x^2}{(x^2 + y^2)^2}$

By C-R equations, we have

$$\frac{\partial u}{\partial x} = \frac{\partial v}{\partial y} \text{ and } \frac{\partial u}{\partial y} = -\frac{\partial v}{\partial x}$$

$$\therefore \quad \frac{\partial u}{\partial x} = -2y - \frac{2xy}{(x^2 + y^2)^2}$$

On integrating w.r.t x, we get

$$u = -2xy + \frac{y}{x^2 + y^2} + \phi(y)$$

where $\phi(y)$ is the constant of integration.

Further $\dfrac{\partial u}{\partial y} = -\dfrac{\partial v}{\partial x}$

$$\Rightarrow \quad -2x + \frac{x^2 - y^2}{(x^2 + y^2)^2} + \phi'(y) = -2x + \frac{x^2 - y^2}{(x^2 + y^2)^2}$$

$$\Rightarrow \quad \phi'(y) = 0 \text{ i.e., } \phi(y) = c$$

Hence $u = -2xy + \dfrac{y}{x^2 + y^2} + c$

EXAMPLE 15. *If f(z) = u + iv is an analytic function in a domain D. Prove that f(z) is constant in D if any one of the following conditions hold.*

(i) *f'(z) vanishes identically in D*
(ii) *R(f(z)) = u = constant*
(iii) *I(f(z)) = v = constant*
(iv) *|f(z)| = constant*
(v) *arg(f(z)) = constant.*

SOLUTION. It is given that $f(z)$ is analytic in D, therefore

$$\frac{\partial u}{\partial x} = \frac{\partial v}{\partial y}, \frac{\partial u}{\partial y} = -\frac{\partial v}{\partial x}$$

(i) $f'(z) = \dfrac{\partial u}{\partial x} + i\dfrac{\partial v}{\partial x} = \dfrac{\partial v}{\partial y} - i\dfrac{\partial u}{\partial y}$

If $f'(z) = 0$. Then we have $\dfrac{\partial u}{\partial x} + i\dfrac{\partial v}{\partial x} = 0$ and

$$\frac{\partial v}{\partial y} - i\frac{\partial u}{\partial y} = 0$$

$$\Rightarrow \quad \frac{\partial u}{\partial x} = 0, \frac{\partial v}{\partial x} = 0, \frac{\partial v}{\partial y} = 0 \text{ and } \frac{\partial u}{\partial y} = 0$$

$\Rightarrow$ u and v are constant.

$\Rightarrow f(z)$ is constant in D.

(ii) $R(f(z)) = u = $ constant

$$\Rightarrow \frac{\partial u}{\partial x} = 0 = \frac{\partial u}{\partial y}$$

$$\therefore \; f'(z) = \frac{\partial u}{\partial x} + i\frac{\partial v}{\partial x} = \frac{\partial u}{\partial x} - i\frac{\partial u}{\partial y} = 0$$

$\Rightarrow f(z)$ is constant in D.

(iii) $I(f(z)) = v = $ constant $\Rightarrow \dfrac{\partial v}{\partial x} = \dfrac{\partial v}{\partial y} = 0$

$$\Rightarrow \; f'(z) = \frac{\partial u}{\partial x} + i\frac{\partial v}{\partial x} = \frac{\partial v}{\partial y} + i\frac{\partial v}{\partial x} = 0$$

$\Rightarrow f(z)$ is constant in D

(iv) $|f(z)| = $ constant $\Rightarrow u^2 + v^2 = $ constant

$$\Rightarrow u\frac{\partial u}{\partial x} + v\frac{\partial v}{\partial x} = 0 \text{ and } u\frac{\partial u}{\partial y} + v\frac{\partial v}{\partial y} = 0$$

$$\Rightarrow u\frac{\partial u}{\partial x} + v\frac{\partial v}{\partial x} = 0 \text{ and } -u\frac{\partial v}{\partial x} + v\frac{\partial u}{\partial x} = 0$$

(By C-R equation)

$$\Rightarrow \frac{\partial u}{\partial x} = 0 \text{ and } \frac{\partial v}{\partial x} = 0 \quad \text{provided } u^2 + v^2 \neq 0$$

$\Rightarrow u = $ constant and $v = $ constant

$\Rightarrow f(z)$ is constant in D.

(v) We have $\arg.(f(z)) = \tan^{-1}\dfrac{v}{u}$

$$\arg f(z) = c \text{ (constant)}$$

$$\Rightarrow \tan^{-1}\frac{v}{u} = c \quad \Rightarrow \quad - \tan$$

$$\Rightarrow u = v\cot c \quad \Rightarrow \quad u = kv \quad \text{where } k = \cot c$$

$\Rightarrow u - kv = 0$

Clearly $u - kv$ is the real part of $(1 + ik)f$

$\Rightarrow (1 + ik)f = $ constant (By part (iii))

Hence f is constant is D.

EXAMPLE 16. *If f(z) is an analytic function of z, prove that*

$$\left(\frac{\partial^2}{\partial x^2} + \frac{\partial^2}{\partial y^2}\right)|Rf(z)|^2 = 2|f'(z)|^2.$$

SOLUTION. Let $f(z) = u + iv$ when $z = x + iy$

Clearly $Rf(z) = u$

Now $\dfrac{\partial}{\partial x}(u^2) = 2u\dfrac{\partial u}{\partial x}$

and $\dfrac{\partial^2}{\partial x^2}(u^2) = 2\left(\dfrac{\partial u}{\partial x}\right)^2 + 2u\dfrac{\partial^2 u}{\partial x^2}$

Therefore, $\dfrac{\partial^2}{\partial x^2}(u^2) + \dfrac{\partial^2}{\partial y^2}(u^2)$

$$= 2\left[\left\{\left(\frac{\partial u}{\partial x}\right)^2 + \left(\frac{\partial u}{\partial y}\right)^2\right\} + u\left(\frac{\partial^2 u}{\partial x^2} + \frac{\partial^2 u}{\partial y^2}\right)\right]$$

or $\left(\dfrac{\partial^2}{\partial x^2} + \dfrac{\partial^2}{\partial y^2}\right)u^2 = 2\left\{\left(\dfrac{\partial u}{\partial x}\right)^2 + \left(\dfrac{\partial u}{\partial y}\right)^2\right\} + 0$

$(\because u$ is a hormonic function$)$

$$\Rightarrow \left(\frac{\partial^2}{\partial x^2} + \frac{\partial^2}{\partial y^2}\right)u^2 = 2\left[\left(\frac{\partial u}{\partial x}\right)^2 + \left(\frac{\partial v}{\partial x}\right)^2\right]$$

(By C-R equation)

$$\Rightarrow \left(\frac{\partial^2}{\partial x^2} + \frac{\partial^2}{\partial y^2}\right)|u^2| = 2|f'(z)|^2$$

since $f'(z) = \dfrac{\partial u}{\partial x} + i\dfrac{\partial v}{\partial x}$

Hence $\left(\dfrac{\partial^2}{\partial x^2} + \dfrac{\partial^2}{\partial y^2}\right)|Rf(z)|^2 = 2|f'(z)|^2$

EXAMPLE 17. *Show that the curves* $r^n = \alpha \sec n\theta$ *and* $r^n = \beta \operatorname{cosec} n\theta$ *orthogonally.*

(MUMBAI–2005, JNTU–2003)

SOLUTION. Writing $u(r,\theta) = r^n \cos n\theta = \alpha$ and $v(r,\theta) = r^n \sin n\theta = \beta$

Now, $u(r,\theta) + iv(r,\theta) = \alpha + i\beta$

$$= r^n(\cos n\theta + i\sin n\theta) = r^n.e^{in\theta} = (re^{i\theta})^n = z^n$$

This is an analytic function.

Thus $f(z) = u + iv$, gives the curve $u = a$ and $v = b$ which cut orthogonally.

Exercise-11.18

1. Check the analyticity of the following functions.
   (i) $f(z) = \bar{z}$ (JNTU–2003)
   (ii) $f(z) = e^z$

2. Show that the real and imaginary parts of $w = \sin z$ satisfy Cauchy-Riemman equations.

3. Show that the function $f(z) = z^n$, where n is a positive integer is an analytic function.   (VTU–2010)

4. Show that the function $e^x(\cos y + i\sin y)$ is holomorphic.

5. Show that an analytic function with constant argument is constant.

6. If $u = e^x(x\cos y - y\sin y)$. Find the analytic function $u + iv$.

7. Find the analytic function whose real part is given by
   (i) $e^x \sin y$ (ii) $\sin x \cosh y$

8. Find the analytic function whose imaginary parts is given by
   (i) $\dfrac{x - y}{x^2 + y^2}$ (ii) $\tan^{-1}\dfrac{y}{x}$

9. Prove that following functions are harmonic
   (i) $2x - x^3 + 3xy^2$ (ii) $e^{-x}(x\cos y + y\sin y)$

10. If $f(z) = u + iv$ is an analytic function, regular in a domain D where $f(z) \neq 0$. Prove that the curves $u = $ constant, $v = $ constant form two orthogonal families.

11. $f(z) = z + 2\bar{z}$ is not analytic anywhere in the complex plane.   (JNTU–2003)

12. Determine p such that the function
    $f(z) = \dfrac{1}{2}\log_e(x^2 + y^2) + i\tan^{-1}(px/y)$ be an analytic function.
    (MUMBAI–2007, JNTU–2003)

13. Show that $f(z) = \begin{cases} xy^2(x + iy) \div (x^2 + y^4), & z \neq 0 \\ 0, & z = 0 \end{cases}$ is not analytic at $z = 0$, although $C - R$ equations are satisfied at the origin.   (JNTU–2003)

14. For the function $f(z)$ defined by $f(z)^2 = (\bar{z})^{2/z}, z \neq 0; f(0) = 0$, show that the $C–R$ equations are satisfied at $(0, 0)$, but $f(z)$ is not differentiable at $(0, 0)$.   (PTU–2010)

15. Determine the analytic funaction whose real part is
    (i) $y + e^x \cos y$ (SVTU–2008, VTU–2006)
    (ii) $e^{2x}(x\cos 2y - y\sin 2y)$
    (VTU–2008; MUMBAI, 2005; KOTTAYAM, 2005)

(iii) $x \sin x \cosh y - y \cos x \sinh y$ (VTU, 2006)

(iv) $e^x [(x^2 - y^2) \cos y - 2xy \sin y]$

     (VTU–2010 S; ROHTAK–2005)

16. Prove that $u = x^2 - y^2$ and $v = \dfrac{y}{x^2 + y^2}$ are harmonic functions of (x, y) but are not harmonic conjugates.

     (UPTU–2004S)

17. For $w = exp\ (z^2)$, find u and v, and prove that the curves $u(x, y) = C_1$ and $v(x, y) = C_2$ where C_1 and C_2 are constants,

cut orthogonally.   (JNTU–2003)

18. If $f(z)$ is an analytic funation of z, show that

$$\left\{ \frac{\partial}{\partial x} |f(z)| \right\}^2 + \left\{ \frac{\partial}{\partial y} |f(z)| \right\}^2 = |f'(z)|^2$$

     (UPTU–2009; VTU–2008S; PTU–2005)

19. Verify if $f(z) = \dfrac{xy^2(x+iy)}{x^2 + y^4}, z \neq 0\ ; f(0) = 0$ is analytic or not.

     (UPTU–2008)

Hints to Selected Problems

1. (i) $f(z) = u + iv$, $\bar{z} = x - iy$

$\Rightarrow\quad u = x, v = -y$

$\therefore\quad \dfrac{\partial u}{\partial x} = 1, \dfrac{\partial v}{\partial x} = 0, \dfrac{\partial u}{\partial y} = 0, \dfrac{\partial v}{\partial y} = -1$

2. $\dfrac{\partial u}{\partial x} = \cos x \cosh y, \dfrac{\partial u}{\partial y} = \sin x \sinh y, \dfrac{\partial v}{\partial x} = -\sin x \sinh y$

$\dfrac{\partial v}{\partial y} = \cos x \cosh y$

3. Prove that $f'(z)$ exists for all finite value of z.

4. $\dfrac{\partial u}{\partial x} = -e^x \cos y, \dfrac{\partial u}{\partial y} = -e^x \sin y, \dfrac{\partial v}{\partial x} = e^x \sin y, \dfrac{\partial v}{\partial y} = e^x \cos y$

6. Use $dv = \dfrac{\partial v}{\partial x} dx + \dfrac{\partial v}{\partial y} dy = -\dfrac{\partial u}{\partial y} dx + \dfrac{\partial u}{\partial x} dy$

7. (i) $\dfrac{\partial u}{\partial x} = e^x \sin y, \dfrac{\partial u}{\partial y} = e^x \cos y$

8. (i) $v = \dfrac{x - y}{x^2 + y^2} \Rightarrow \dfrac{\partial u}{\partial x} = \dfrac{y^2 - x^2 + 2xy}{(x^2 + y^2)^2}, \dfrac{\partial v}{\partial y} = \dfrac{y^2 - x^2 - 2xy}{(x^2 + y^2)^2}$

10. For orthogonality of two curves, prove the result

$$\left(\frac{-\partial u / \partial x}{\partial u / \partial y} \right) \cdot \left(\frac{-\partial v / \partial x}{\partial v / \partial y} \right) = -1$$

Answers

1. (i) not analytic   (ii) analytic

6. $f(z) = ze^z + ci$

7. (i) $f(z) = e^x \sin y + i(c - e^x \cos y)$

(ii) $f(z) = \sin x \cosh y + i(\cos x \sinh y + c)$

8. $f(z) = \left(\dfrac{x + y}{x^2 + y^2} + c \right) + i\left(\dfrac{x - y}{x^2 + y^2} \right)$

(ii) $f(z) = \log z + c$

15. (i) i/z (ii) $ze^{2z} + ic$ (iii) $z \sin z$ (iv) $x^2 e^z + ic$ **19.** $f(z)$ is not analytic at origin although $C - R$ equations are satisfied there.

11.15 CONFORMAL REPRESENTATIONS

We are familiar with the term graph of real valued function $f(x)$, i.e., the relation $y = f(x)$ gives rise to correspondence between the points on axis of x and on axis of y. If we plot these points (x, y) and join them by a smooth curve then we obtain the graph of the function.

We shall now consider the complex valued function $w = f(z) = u + iv$ (say) where $z = x + iy$,

and $u = u(x, y), v = v(x, y)$ ... (1)

As in the case of real valued function we shall not have a graphical representation of the above complex valued function as in this case planes known as z-plane and w-plane are equired to represent each of the complex variable z and w respectively. We will however establish a correspondence between the points on z plane and on w plane by the help of equations (1). From (1) corresponding to each point (x, y) in the z plane we shall have a point (u, v) in the w plane and these corresponding points are called images of each other. Because of the above correspondence a curve or a region of z-plane is said to be transformed into or mapped upon or represented by the corresponding curve or regions in the w-plane.

One-one correspondence: If to each point (x, y) of the z-plane there corresponds one end point (u, v) of the w-plane and conversely then we say that the correspondence is one-one.

11.15.1 JACOBIAN OF A TRANSFORMATION

In general the transformation

          $w = f(z)$,   i.e., $u = u(x, y), v = v(x, y)$ ...(1)

where $z = x + iy$ and $w = u + iv$

maps a closed region R of the z plane into a closed region R' of the w plane. If Δ_z and Δ_w be the areas of these regions respectively and if u and v are continuously differentiable then it can be easily shown that

$$\lim_{\Delta_z \to 0} \frac{\Delta_w}{\Delta_z} = \left| \frac{\partial(u, v)}{\partial(x, y)} \right|$$

The determinant $J = \dfrac{\partial(u,v)}{\partial(x,y)} = \begin{vmatrix} \dfrac{\partial u}{\partial x} & \dfrac{\partial u}{\partial y} \\ \dfrac{\partial v}{\partial x} & \dfrac{\partial v}{\partial y} \end{vmatrix} = u_x v_y - u_y v_x$ is called the Jacobian of the transformation.   ...(2)

In case $w = f(z) = u + iv$ be analytic function of z then Cauchy Riemman equations will hold good

i.e., $u_x = v_y$ and $u_y = -v_x$ ...(3)

Hence from (2) by the help of (3)

$$J = \frac{\partial(u,v)}{\partial(x,y)} = u_x(u_x) + v_x(v_x) = u_x^2 + v_x^2 = |u_x + iv_x|^2 = \left|\frac{\partial u}{\partial x} + i\frac{\partial v}{\partial x}\right|$$

$$= \left|\frac{\partial}{\partial x}(u + iv)\right|^2 = \left|\frac{\partial w}{\partial x}\right|^2 = \frac{dw^2}{dx^2}$$

or $J = |f'(z)|^2$ provided $f(z)$ is analytic.

Conformal Transformation

Let the transformation $u = u(x, y)$, $v = v(x, y)$ map a point $z_0 = (x_0, y_0)$ of the z plane to a point $w_0 = (u_0, v_0)$ of the w plane. Further let the two curves C_1 and C_2 intersecting at z_0 be mapped on two curves Γ_1 and Γ_2 intersecting at w_0. Then

(i) Isogonal transformation

If the angle between the intersecting curves C_1 and C_2 at z_0 is equal to the angle between the intersecting curves Γ_1 and Γ_2 at w_0 then the transformation is called isogonal.

(ii) Conformal transformation

In case the sense of rotation as well as the magnitude of the angle is preserved, then the transformation is said to be conformal. With reference to the figures given below the first figure represents isogonal transformation as only angle is preserved and not the sense. In the second figure the transformation is conformal as both the magnitude as well as the sense of description of the angle is preserved.

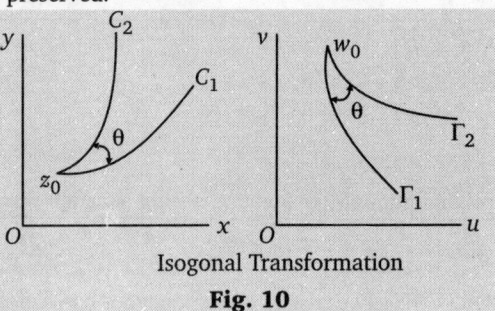

Isogonal Transformation
Fig. 10

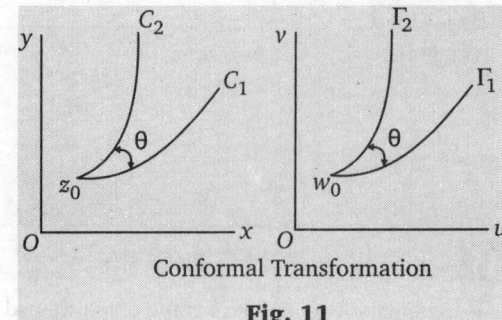

Conformal Transformation
Fig. 11

11.15.2 Sufficient Condition for $w = f(z)$ to represent a Conformal Mapping

THEOREM 1. *At each point z of a domain where $f(z)$ is analytic and $f'(z) \neq 0$, the mapping $w = f(z)$ is conformal.*

PROOF. Let $w = f(z)$ be an analytic be an analytic function of z in a domain D of the z plane and z_0 be an interior point of D. Further suppose that C_1 and C_2 be two continuous curves intersecting at z_0, the tangents at which to the two curves make angles α_1 and α_2 respectively with real axis. Let z_1 and z_2 be the points on the curves C_1 and C_2 respectively at the same distance r from z_0 where r is small. Hence we can write

$$z_1 - z_0 = re^{i\theta_1}, z_2 - z_0 = re^{i\theta_2}$$ ... (1)

Clearly as $r \to 0$ the line $z_1 z_0$ and $z_2 z_0$ will tend to tangents to the curves C_1 and C_2 at z_0 and consequently we can say that

$$\theta_1 \to \alpha_1, \theta_2 \to \alpha_2, \text{ when } r \to 0$$

Let w_0, w_1, w_2, be the images in w-plane of the points z_0, z_1, and z_2 respectively in the z-plane. As the point z_0 moves to z_1 along C_1 the image point moves along the curve Γ_1 from w_0 to w_1 and similarly as z_0 moves to z_2 along C_2 the image point moves, along the curve Γ_2 from w_0 to w_2.

As below suppose that tangent at w_0 to the curves Γ_1 and Γ_2 make angles β_1 and β_2 with the real axis and let.

$$w_1 - w_0 = \rho_1 e^{i\phi_1} \text{ and } w_2 - w_0 = \rho_2 e^{i\phi_2}$$

∴ $\phi_1 \to \beta_1$ as $\rho_1 \to 0$ and $\phi_2 \to \beta_2$ as $\rho_2 \to 0$

Now $f'(z_0) = \lim\limits_{z_1 \to z_0} \dfrac{f(z_1) - f(z_0)}{z_1 - z_0} = \lim\limits_{z_1 \to z_0} \dfrac{w_1 - w_0}{z_1 - z_0}$

$$= \lim\limits_{z_1 \to z_0} \frac{\rho_1 e^{i\phi_1}}{r e^{i\theta_1}} = \lim\limits_{z_1 \to z_0} \frac{\rho_1}{r} e^{i(\phi_1 - \theta_1)}$$

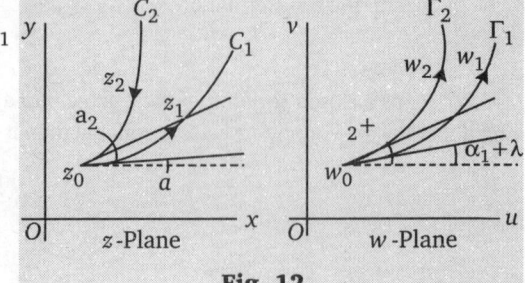

Fig. 12

Now $f(z)$ being analytic it follows that $f'(z_0) \neq 0$ and as such we write $f'(z_0) = Re^{i\lambda}$ so that $R = |f'(z_0)|$ and $\lambda = \text{amp}[f'(z_0)]$

$$\therefore \quad Re^{i\lambda} = \lim_{z_1 \to z_0} = \frac{\rho_1}{r} e^{i(\phi_1 - \theta_1)}$$

Equating modulus and amplitude on both sides, we get

$$R = \lim_{z_1 \to z_0} \frac{\rho_1}{r} \text{ and } \lim(\phi_1 - \theta_1) = \lambda \text{ or } \beta_1 - \alpha_1 = \lambda$$

$$\beta_1 = \lambda + \alpha_1$$

Proceeding exactly as above we can show that $\beta_2 = \lambda + \alpha_2$

Clearly $\qquad \beta_1 - \beta_2 = (\lambda + \alpha_1) - (\lambda + \alpha_2) = \alpha_1 - \alpha_2$.

Above relation show that angle between the curves Γ_1 and Γ_2 at w_0 is the same as the angle between the curves C_1 and C_2 at z_0. Also from the figure it is clear that these angles have the same sense of description. Hence the mapping $w = f(z)$ preserves both the magnitude and sence of the angle between the tangents to the curves and as such it is a conformal mapping.

Geometrical Interpretation of $R = |f'(z_0)|$ and $\beta_1 = \alpha_1 + \lambda$ where $f'(z_0) = Re^{i\lambda}$

We have seen above that $Re^{i\lambda} = \lim \dfrac{\rho_1}{r} e^{i(\phi_1 - \theta_1)}$

$$\therefore \qquad \lim \frac{\rho_1}{r} = R = |f'(z_0)|$$

and $\qquad \lambda = \lim(\phi_1 - \theta_1) = \lim \phi_1 - \lim \theta_1 = \beta_1 - \alpha_1 \text{ or } \beta_1 = \alpha_1 + \lambda$

where $\qquad \lambda = \text{amp}[f'(z)]$

Hence to obtain a figure in w plane corresponding to a figure in z plane it should be rotated through an angle $\lambda = \text{amp}[f'(z_0)]$

and subjected to magnification (*i.e.*, the distance of a point z_1 on c_1 from z_0 is magnified) by an amount $\lim \dfrac{\rho_1}{r} = R = |f'(z_0)|$

This magnification is independent of the direction, but varies from point to point. From above we conclude that the shape of the image of a small figure near z_0 is the same as that of the origonal figure under a conformal transformation $w = f(z)$.

11.15.3 Necessary Condition For $w = f(z)$ to represent a Conformal Mapping

THEOREM 1. *If the mapping $w = f(z)$ is conformal then the function $f(z)$ is an analytic function of (z).*

PROOF. Let $w = f(z) = u + iv$ where $u = u(x, y), v = v(x, y)$

are both differentiable equations defining conformal transformation from z plane to w plane.

Let ds and $d\sigma$ denote the elements in z plane and w plane respectively.

$$\therefore \qquad ds^2 = dx^2 - dy^2, d\sigma^2 = du^2 + dv^2$$

But $\qquad du = \dfrac{\partial u}{\partial x} dx + \dfrac{\partial u}{\partial y} dy, dv = \dfrac{\partial v}{\partial x} dx + \dfrac{\partial v}{\partial y} dy$

Squaring and adding we get

$$d\sigma^2 = du^2$$

$$dv^2 = \left[\left(\frac{\partial u}{\partial x}\right)^2 + \left(\frac{\partial v}{\partial x}\right)^2\right] dx^2 + \left[\left(\frac{\partial u}{\partial y}\right)^2 + \left(\frac{\partial v}{\partial y}\right)^2\right] dy^2$$

$$2\left[\frac{\partial u}{\partial x}\frac{\partial u}{\partial y} + \frac{\partial v}{\partial x}\frac{\partial v}{\partial y}\right] dx\, dy \qquad \qquad \dots (1)$$

Also $\qquad ds^2 = dx^2 + dy^2 \qquad \qquad \dots (2)$

Again as the mapping is conformal therefore the ratio $d\sigma^2 : ds^2$ is independent of direction. Hence comparing (1) and (2) we have

$$\frac{\left(\dfrac{\partial u}{\partial x}\right)^2 + \left(\dfrac{\partial v}{\partial x}\right)^2}{1} = \frac{\left(\dfrac{\partial u}{\partial y}\right)^2 + \left(\dfrac{\partial v}{\partial y}\right)^2}{1} = \frac{\dfrac{\partial u}{\partial x}\dfrac{\partial u}{\partial y} + \dfrac{\partial v}{\partial x}\dfrac{\partial v}{\partial y}}{0}$$

$$\therefore \qquad \left(\frac{\partial u}{\partial x}\right)^2 + \left(\frac{\partial v}{\partial x}\right)^2 = \left(\frac{\partial u}{\partial y}\right)^2 + \left(\frac{\partial v}{\partial y}\right)^2 \quad \text{or} \quad u_x^2 + v_x^2 = u_y^2 + v_y^2 \qquad \dots (3)$$

$$\frac{\partial u}{\partial x} \cdot \frac{\partial u}{\partial y} + \frac{\partial v}{\partial x} \cdot \frac{\partial v}{\partial y} = 0 \qquad \qquad \text{or} \qquad u_x u_y + v_x v_y = 0 \qquad \dots (4)$$

The equation (3) is satisfies if

$$u_x = v_y \qquad \qquad \text{and} \qquad \qquad v_x = -u_y \qquad \dots (5)$$

Above are Cauchy Riemma equations and hence the function $w = f(z) = u + iv$ where $z = x + iy$ is an analytic function of z.

But the equation (4) is satisfied if

$$u_x = -v_y \qquad \qquad \text{and} \qquad \qquad v_x = u_y \qquad \dots (6)$$

Now equation (6) reduces to equation (5) if we write $-v$ for v.

In other words we have to take as image figure found by reflection in the real axis of the w-plane. (*i.e.*, in this case magnitude of angle will be preserved but directio will be charged). Hence equations (6) correspond to an isogonal but not conformal transformation. Thererore it is established that if the transformation. $w = f(z)$ is conformal then $f(z)$ must be an analytic function of z.

☛ REMARK

- A small arc in the z-plane through the point z_0 is magnified in the ratio $|f'(z_0)| : 1$ in the w-plane under the transformation $w = f(z)$ where $f(z)$ is an analytic funciton and $f'(z) \neq 0$.

11.15.4 CERTAIN TYPE OF TRANSFORMATIONS

(a) Let $w = f(z)$ and $\xi = \phi(w)$ where both $f(z)$ and $\phi(w)$ are analytic functions then ξ is an analytic function of z. Therefore if a region D of z plane is conformally repesented on a region D' of the w-plane and the D' is represented on a region D'' of the of ξ-plane then the transformation from the z plane to ξ plane will also be conformal transformation.

(b) $w = f(\bar{z})$ transformation is isogonal.

We know that if $z = x + iy$ then $\bar{z} = x - iy$.

The above transformation maps every point into its reflection in the real axis and hence only the anlgle are preserved but not their signs. In general we can say that every transformation of the type $w = f(\bar{z})$ is isogonal and not conformal where $f(z)$ is analytic. The above transformation can be taken as the resultant of the following two transformations.

(i) $\xi = \bar{z}$ (ii) $w = f(\xi)$

The first transformation preserves the angles while their signs are changed whereas the second preserves both the angles and transformation and hence in the resultant transformation $w = f(\bar{z})$ only angles are preserved not their signs. Therefore the above transformation is isogonal.

(c) Translation. $w = f(z) = z + \alpha$.

The above transformation will dislace or translate every point in the z plane along the direction of α through a distance equal to $|\alpha|$.

From the figure it is clear that if $\overrightarrow{OP} = z$ and $\overrightarrow{OA} = \alpha$ then completing the $||m$ fourth vertex Q represents the point $z + \alpha$, *i.e.*, $\overrightarrow{OQ} = z + \alpha$

Since $\overrightarrow{PQ} = \overrightarrow{OA} = \alpha$ therefore Q is obtained by moving the point P through a distance $|\alpha|$ in a direction parallel to that of α. The above result is true for every point P in any region of z-plane. Therefore the image of the region in w-plane will be just a translation of this region and hence the two regions in the z-plane and w-plane will have the same shape, size and orientation as will be clear from the examples to follow.

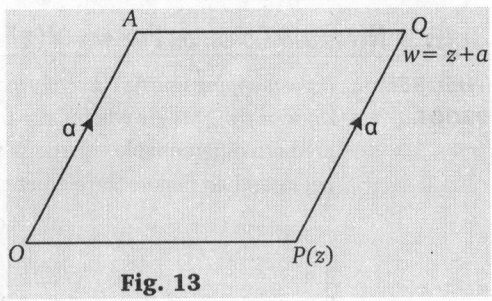

Fig. 13

(d) Magnification and rotation. $w = f(z) + \beta z$.

Let P and B represent the variable and fixed point z and β respectively, then as discussed in the points $w = \beta z$ is obtained by rotating the line OP through an angle equal to amp of β, then talking point Q on it such that $OQ = OB. OP = |\beta| \ OP$, *i.e.*, by magnifying OP in the ratio $|\beta|$.

Hence the above transformation consists of both rotation and magnification. Rotation is positive (anticlockwise) if amp $\beta > 0$ (and negative clockwise) if amp $\beta < 0$. Also magnification becomes contraction if $|\beta| < 1$ and expansion if $|\beta| > 1$.

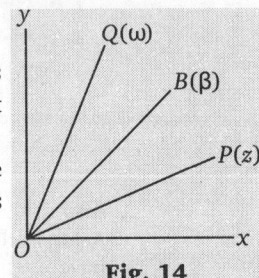

Fig. 14

(i) Condition for only rotation.

$w = \beta z$ transformation will simply be that of rotation if $|\beta| = 1$, *i.e.*, if β is unimodular, *i.e.*, β is of the form $e^{i\theta}$.

(ii) Condition for only magnification.

$w = \beta z$ transformation will simply be that of magnification if amp $\beta = 0$, *i.e.*, if β lies along the real axis and is positiive.

(e) Inversion $w = \dfrac{1}{z}$ (Inversion in the real axis and the unit circle.)

Let $z = re^{i\theta}$ $\therefore$ $w = \dfrac{1}{z} = \dfrac{1}{r}e^{-i\theta}, \ \bar{z} = re^{-r\theta}$ $\therefore$ $\dfrac{1}{z} = \dfrac{1}{r}e^{i\theta}$

$\therefore$ $|w| = \dfrac{1}{r}$ and amp $w = -\theta = $ amp $\bar{z}$

Hence the transformation $w = \dfrac{1}{z}$ may be considered as the resultant of the two transformations.

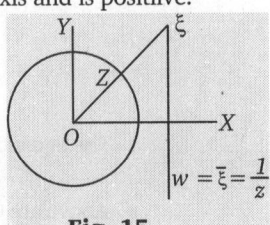

Fig. 15

(i) $\xi = \dfrac{1}{\bar{z}} = \dfrac{1}{r}e^{i\theta}$ and (ii) $w = \bar{\xi} = \left[\dfrac{1}{r}e^{-i\theta} = \dfrac{1}{z}\right]$

Since $|\xi||\bar{z}| = \frac{1}{r} \cdot r = 1$ the first of the above transformation is an inversion with regard to unit circle $|z| = 1$. This means that the

point ξ lies on the radius drawn through the point z and its distance from the centre of the circle is such that
$\xi = |z| = 1$. This inversion is followed by reflection.
$w = \xi$ in the real axis.

11.15.5 THE BILINEAR TRANSFORMATION OR MOBIUS TRANSFORMATION

The transformation T defined by

$$w = T(z) = \frac{az+b}{cz+d} \qquad \ldots (1)$$

where a, b, c, d are complex constants and w and z are complex variables is called a bilinear or linear fractional transformation. This transformation associates a unique point of the w plane to each point of the z plane and conversely.

The above transformation may be written as $cwz + wd - az - b = 0$ which is linear both in w and z and hence the name bilinear. It is also known as linear fractional transformation or Mobius transformation after the name of A.F. Mobius (1790 -1808) who first studied these transformations.

Descriminant.

Let $$w = \frac{az + b/a}{cz + d/c}$$

then if $b/a = d/c$ or if $ad - bc = 0$ then for every value of z we have same value of w and we say that w is merely constant. The expression $ad - bc$ is called descriminant of bilinear transformation.

Condition for one-one correspondence.
$$ad - bc \neq 0$$

Let $$w_1 = \frac{az_1 + b}{cz_1 + d}, w_2 = \frac{az_2 + b}{cz_2 + d}$$

$\therefore$ $$w_1 - w_2 = \frac{(ad - bc)(z_3 - z_1)}{(cz_1 + d)(cz_2 + d)} = 0 \text{ if } ad - bc = 0 \text{ provided } z_1 \neq d/c \text{ and } z \neq d/c.$$

i.e., $w_1 = w_2$. In other words it means that w is constant and if $ad - bc = 0$ and either $z_1 = -d/c$ or $z_2 = -d/c$ then w becomes meaningless. Hence $ad - bc \neq 0$ is the necessary condition for the billinear transformation T to set up a one-one correspondence between the points of the closed z plane and closed w-plane.

Critical points. If $$w = \frac{az+b}{cz+d} \text{ then } z = -\frac{dw+b}{cw+a} \qquad \ldots (2)$$

$\therefore$ $$\frac{dw}{dz} = \frac{ad - bc}{(cz + d)^2}$$

Now if $z = -d/c$ then $\frac{dw}{dz} = \infty$ and if $z = \infty$ then $\frac{dw}{dz} = 0$.

These points $z = -d/c$ and $z = \infty$ are the critical points where the conformal property does not hold good. Also from (2) it is clear that for each $z \neq -d/c$ we have a value of w and for each $w \neq a/c$ there corresponds a value of z and the correspondence between w and z is one-one. The exceptional points $z = -d/c$ and $w = a/c$ are mapped into the points $w = \infty$ and $z = \infty$ respectively.

Extended complex plane. These points will not remain exceptions if we adjoin a new point called point at infinity denoted by ∞ to the complex plane and the complex plane in this case is called extended complex plane. Thus the critical point $z = \infty$ corresponds to the point $w = a/c$.

Hence we can now say that the bilinear transformation sets up a one-one correspondence between the points of entire extended complex z-plane upon the entire extended complex w-plane.

PRODUCT OR RESULTANT OF TWO BILINEAR TRANSFORMATIONS

Let $$\xi = T_1(z) = \frac{a_1 z + b_1}{c_1 z + d_1} \text{ s.t. } a_1 d_1 - b_1 c_1 \neq 0 \qquad \ldots (1)$$

and $$w = T_2(\xi) = \frac{a_2 \xi + b_2}{c_1 \xi + d_2} \text{ s.t. } a_2 d_2 - b_2 c_2 \neq 0 \qquad \ldots (2)$$

(1) sets up a one-one correspondence between the points of z plane and the points of ξ plane whereas (2) sets up a one-one correspondence between points of ξ plane and points of w-plane. Now we shall establish a transformation which sets up a one-one correspondence between the points of z-plane and w-plane.

$$w = T_2(\xi) = T_2(T_1 z) = T_2\left(\frac{a_1 z + b_1}{c_1 z + d_1}\right) \text{ by (1)}$$

$$= \frac{a_2\left(\dfrac{a_1 z + b_1}{c_1 z + d_1}\right) + b_2}{c_2\left(\dfrac{a_1 z + b_1}{c_1 z + d_1}\right) + d_2} \text{ by (2)} = \frac{(a_2 a_1 + b_2 c_1)z + (a_2 b_1 + b_2 d_1)}{(c_2 a_1 + d_2 c_1)z + (c_2 b_1 + d_2 d_1)}$$

or $\qquad w = T_2 T_1(z) = \dfrac{Az + B}{Cz + D}$ $\qquad\qquad\qquad$ ... (3)

The determinant of (3) is $\begin{vmatrix} A & B \\ C & D \end{vmatrix} = AD - BC$

$$= (a_2 a_1 + b_2 c_1)(c_2 b_1 + d_2 d_1) - (a_2 b_1 + b_2 d_1)(c_2 a_1 + d_2 c_1)$$
$$= a_2 a_1 d_2 d_1 + b_2 b_1 c_2 c_1 - a_2 b_1 d_2 c_1 - b_2 d_1 c_2 a_1$$
$$= (a_1 d_1 - b_1 c_1)(a_2 d_2 - b_2 c_2) \neq 0 \qquad\qquad \text{by (1) and (2).}$$

Therefore (3) also represents a bilinear transformation $T_2 T_1$ which is the resultant of the two bilinear transformations T_1 and T_2.

THEOREM 1.   *Every bilinear transformation is the resultant of bilinear transformation with simple geometric imports.*

PROOF.   Let $w = \dfrac{az + b}{cz + d}$ where $ad - bc \neq 0, a \neq 0$

By actual division $w = \dfrac{a}{c} + \dfrac{bc - ad}{c} \cdot \dfrac{1}{cz + d}$ $\qquad$ or $\qquad$ $w = \dfrac{a}{c} + \dfrac{bc - ad}{c^2} \dfrac{1}{z + a/c}$

Consider the following transformations $z_1 = z + a/c$, *i.e.*, $z + \alpha$. Translation $z_2 = \dfrac{1}{z_1}$ Inversion $z_3 = \dfrac{bc - ad}{c^z} z_2$, *i.e*, $z_3 = \beta z_2$

rotation and magnification.

Hence $w = \dfrac{a}{c} + z_3$ which is again a translation.

The three transformations involved above are of the type $w = z + \alpha, w = \beta z, w = \dfrac{1}{z}$ which have been discussed earlier.

THEOREM 2.   *The set of all bilinear transformation forms a non-abelian group under the product of transformations.*

PROOF.

   (i)   Closure property: Since the product of two bilinear transformations is again a bilinear transformation. Therefore closure property is satisfied.

   (ii)   Associativity : Clearly $(T_1 T_2)T_3 = T_1 (T_2 T_3)$

   (iii)   Existence of Identity: The identity mepping I defined by $w = I(z) = z$ is a bilinear transformation and hence I serves as the identity element.

   (iv)   Existence of inverse: Let T is a bilinear transformation defined by $w = T(z) = \dfrac{az + b}{cz + d}$

$$ad - bc \neq 0. \text{ Then } T^{-1}(w) = z = \dfrac{dw - b}{-cw + a}$$

we have $\qquad T^{-1}T(z) = T^{-1}\left(\dfrac{az + b}{cz + d}\right) = \dfrac{d\left(\dfrac{az + b}{cz + d}\right) - b}{-c\left(\dfrac{az + b}{cz + d}\right) + a} = \dfrac{adz + db - bcz - bd}{-caz - cb + acz + ad} = \dfrac{(ad - bc)z}{ad - bc} = z$

Similarly, we may show that $TT^{-1}(w) = z$

$\Rightarrow \qquad\qquad\qquad TT^{-1} = T^{-1}T = I$

Hence, the set of all bilinear transformation forms a group.

☞ REMARK

- In general, the above group is not abelian.

11.15.6 FIXED POINTS AND CROSS RATIO

Fixed points: The points which coincide with their transformatons are called fixed or invariant points of the transformation.

In other words the fixed points of the transformation $w = f(z)$ are obtained from the equation $z = F(z)$. As an example the fixed points of $w = z^2$ or $z = z^3$ are the solutions of the equations $z = z^2$ and $z = z^3$ respectively.

Cross ratio: If z_1, z_2, z_3, z_4 are distinct points taken in the order in which they are written then the cross ratio of these point is defined as

$$\dfrac{z_1 - z_2}{z_2 - z_3} \left| \dfrac{z_4 - z_1}{z_3 - z_4} \right. \qquad \text{or} \qquad \dfrac{\overset{I}{(z_1 - z_2)}\overset{III}{(z_3 - z_4)}}{\underset{II}{(z_2 - z_3)}\underset{IV}{(z_4 - z_1)}}$$

▌WORKING PROCEDURE

Rule to write the cross ratio (z_1, z_2, z_3, z_4)

Write down the four successive cyclic differences

$\qquad\qquad\qquad$ I $\qquad$ II $\qquad$ III $\qquad$ IV

$\qquad\qquad z_1 - z_2, \; z_2 - z_3, \; z_3 - z_4, \; z_4 - z_1$

Write I and III in the numerator and II and IV in the denominator.

- The cross ratio will change if the order of the factors is changed, *i.e.*, for writing (z_3, z_1, z_2, z_4) we write down the four successive cyclic differences

$$\begin{array}{cccc} I & II & III & IV \end{array}$$
$$z_3 - z_1,\ z_1 - z_2,\ z_2 - z_4,\ z_4 - z_3,$$

Now write down I and III in the numerator and II, IV in the denominator

$$\therefore \qquad (z_3, z_1, z_2, z_4) = \frac{(z_3 - z_1)(z_2 - z_4)}{(z_1 - z_2)(z_4 - z_3)} \neq (z_1, z_2, z_3, z_4)$$

NUMBER OF DISTINCT CROSS RATIOS

Since the four letters z_1, z_2, z_3, z_4 can be arranged in $4! = 24$ ways, there will be 24 cross ratios but as a matter of fact there will be only six distinct cross ratios. This is so because if we interchange any two letters and then interchange remaining two, the cross ratios of the letters in this new order will be the same e.g. interchange z_1 and z_3 and then z_2 and z_4 then

$$(z_1, z_2, z_3, z_4) \text{ becomes } (z_3, z_4, z_1, z_2)$$

$$\begin{array}{cccc} I & II & III & IV \end{array}$$
$$\text{write down } \quad z_3 - z_4,\ z_4 - z_1,\ z_1 - z_2,\ z_2 - z_3,$$

$$\therefore \qquad (z_3, z_4, z_1, z_2) = \frac{(z_3 - z_4)(z_1 - z_2)}{(z_4 - z_1)(z_2 - z_3)}$$

Above is same as (z_1, z_2, z_3, z_4). In all there will be 4 cross ratios each equal to (z_1, z_2, z_3, z_4).

The six distinct cross ratios (each having four equivalents by the above rule) are written down below. In each of them z_1 is fixed and the remaining three are arranged in six ways.

$$(z_1, z_2, z_3, z_4), (z_1, z_2, z_4, z_3), (z_1, z_3, z_2, z_4)$$
$$(z_1, z_3, z_4, z_2), (z_1, z_4, z_2, z_3), (z_1, z_4, z_3, z_2)$$

CONDITION OF CROSS RATIO TO BE REAL

$$(z_1, z_2, z_3, z_4) = \frac{(z_1 - z_2)(z_3 - z_4)}{(z_2 - z_3)(z_4 - z_1)} = \frac{\dfrac{(z_1 - z_2)}{(z_2 - z_3)}}{\dfrac{(z_4 - z_1)}{(z_3 - z_4)}}$$

$$\text{Amp } (z_1, z_2, z_3, z_4) = \text{Amp}\frac{z_1 - z_2}{z_2 - z_3} - \text{Amp}\frac{z_4 - z_1}{z_3 - z_4} \qquad \qquad \ldots(1)$$

Clearly α, $\beta = 0$ or $\pm\pi$ depending on position of the four points if and only if these four points are concyclic. In case Amp $(z_1, z_2, z_3, z_4) = 0$ or $\pm \pi$ then (z_1, z_2, z_3, z_4) is purely real. Hence cross ratio is real if and only if the four points are concyclic.

THEOREM 1. **(Preservance of cross ratio).** *Cross ratio remains invariant under a bilinear transformation.*

PROOF. Let

$$w = \frac{az + b}{cz + d}, ad - bc \neq 0 \qquad \qquad \ldots(1)$$

and w_1, w_2, w_3, w_4 be the images of the four points z_1, z_2, z_3, z_4 under the bilinear transformation defined by (1). We have to prove that

$$(w_1, w_2, w_3, w_4) = (z_1, z_2, z_3, z_4)$$

$$\text{L.H.S.} = \frac{(w_1 - w_2)(w_3 - w_4)}{(w_2 - w_3)(w_4 - w_1)}$$

$$w_1 - w_2 = \frac{az_1 + b}{cz_1 + d} - \frac{az_2 + b}{cz_2 + d} = \frac{(ad - bc)(z_1 - z_2)}{(cz_1 + d)(cz_2 + d)}$$

and

$$w_3 - w_4 = \frac{(ad - bc)(z_3 - z_4)}{(az_3 + d)(cz_4 + d)} \text{ etc.}$$

$$\therefore \qquad \text{L.H.S} = \frac{(ad - bc)^2(z_1 - z_2)(z_3 - z_4)}{(ad - bc)^2(z_2 - z_3)(z_4 - z_1)} \text{ as denominator cancel}$$

$$= \frac{(z_1 - z_2)(z_3 - z_4)}{(z_2 - z_3)(z_4 - z_1)} = (z_1, z_2, z_3, z_4)$$

 Solved Examples

EXAMPLE 1. **Find a bilinear transformation which maps the points $z_1 = 2$, $z_2 = i$, $z_3 = -2$ into the point $w_1 = 1$, $w_2 = i$ and $w_3 = -1$.**

SOLUTION. The required transformation is given by

$$(w, w_1, w_2, w_3) = (z, z_1, z_2, z_3) \text{ or } (w, 1, i, -1) = (z, 2, i, -2)$$

or $\dfrac{(w-1)(i+1)}{(1-i)(-1-w)} = \dfrac{(z-2)(i+2)}{(2-i)(-2-z)}$

or $\dfrac{(w-1)(1+i)^2}{(1-i^2)(w+1)} = \dfrac{(z-2)(2+i)^2}{(4-i^2)(z+2)}$

or $\dfrac{(w-1)2i}{(w+1)} = \dfrac{(z-2)(3+4i)}{(z+2).5}$

or $\dfrac{w-1}{w+1} = \dfrac{z-2}{z+2}\dfrac{(3+4i)}{5i} = \dfrac{(z-2)(3i-4)}{(z+2)(-5)}$

$\because \qquad i^2 = -1$

or $\dfrac{w-1}{w+1} = \dfrac{(3i-4)z+(8-6i)}{-5z-10}$

Apply componendo and dividendo we get

$$\dfrac{2w}{2} = \dfrac{(3i-9)z-2(1+3i)}{(-1-3i)z-6(3-i)}$$

or $w = \dfrac{3z(i-3)+2i(i-3)}{i(i-3)z+6(i-3)}$ we have put $-1 = i^2$

or $w = \dfrac{3z+2i}{iz+6}$ is the required transformation.

EXAMPLE 2. *Find transformation which maps outside* $|z| = 1$ *on the half plane* $R(w) \geq 0$ *so that the points* $z = 1, -i, -1$ *correspond to* $w = i, 0, -i$ *respectively.*

SOLUTION. Let $z_1 = 1, z_2 = -i, z_3 = -1, w_1 = i, w_2 = 0, \quad w_3 = -i$
We know that cross ratio is preserved under a bilinear transformation and so

$$\dfrac{(w-w_1)(w_2-w_3)}{(w_1-w_2)(w_3-w)} = \dfrac{(z-z_1)(z_2-z_3)}{(z_1-z_2)(z_3-z)} \qquad \ldots (1)$$

Putting the given values in (1), we get

$\dfrac{(w-i)(0+i)}{(i-0)(-i-w)} = \dfrac{(z-1)(-i+1)}{(1+i)(-1-z)}$

$\Rightarrow \quad \dfrac{w-i}{w+i} = -i\left(\dfrac{z-1}{z+1}\right) = \dfrac{-iz+i}{z+1}$

$\Rightarrow \quad \dfrac{(w-i)+(w+i)}{(w-i)-(w+i)} = \dfrac{-iz+i+z+1}{-iz+i-(z+1)}$

$\Rightarrow \quad \dfrac{2w}{-2i} = \dfrac{z(1-i)+(1+i)}{-z(1+i)-(1-i)}$

$\Rightarrow \quad w = i\left(\dfrac{1-i}{1+i}\right)\left[\dfrac{z+\left(\dfrac{1+i}{1-i}\right)}{z+\left(\dfrac{1-i}{1+i}\right)}\right]$

$= i(-i)\left[\dfrac{z+i}{z-i}\right] = \dfrac{z+i}{z-i}$

which is the required bilinear transformation

Now by $w = \dfrac{z+i}{z-i}$, we get

$w(z-i) = z+i$

or $\quad z = i\left(\dfrac{w+1}{w-1}\right)$

$|z| \geq 1$ is transformed into $\left|\dfrac{w+1}{w-1}\right||i| \geq 1$

or $\quad |w+1|^2 \geq |w-1|^2$

or $\quad (u+1)^2 + v^2 \geq (u-1)^2 + v^2$

$\Rightarrow \quad R(w) = 4 \geq 0$

Hence the exterior of the circle $|z| = 1$ is transformed into half plane $R(w) \geq 0$.

THEOREM 2. **(Theorem on fixed (invariant) points).** *In general there are two values of z for which $w = z$ (invariant or fixed points) but there is only one if $(a-d)^2 + 4bc = 0$.*

Also if there are distinct invariant points p and q the transformation may be put in the form $\dfrac{w-p}{w-q} = k\left(\dfrac{z-p}{z-q}\right)$

and that if there is only one invariant point p, the transformation may be put in the form $\dfrac{1}{w-p} = \dfrac{1}{z-p} + k$.

The above form of bilinear transformation is called normal form.

PROOF. Let the bilinear transformation $w = \dfrac{az+b}{cz+d}$ ... (1)

For invariant points $w = z$, $\therefore \quad \dfrac{az+b}{cz+d} = z$,

or $\qquad cz^2 - (a-d)z - b = 0$ ... (2)

Above in genral gives two values of z and hence two fixed points. In case the discriminant of (2), *i.e.*, $(a-d)^2 + 4bc = 0$ then there will be only one fixed point.

2nd part: If p and q be two fixed points then they are the roots of (2)

$\therefore \quad cp^2 - (a-d)p - b = 0 \qquad\qquad cq^2 - (a-d)q - b = 0$

$\therefore \quad cp^2 - ap = b - pd \qquad\qquad\qquad cq^2 - dp = b - qd$ ... (3)

$\therefore \qquad w - p = \dfrac{az+b}{cz+d} - p = \dfrac{(a-pc)z+b-dp}{cz+d} = \dfrac{(a-pc)z+p(cp-a)}{cz+d}$ by (3)

or $\qquad\qquad (w-p) = \dfrac{(a-pc)(z-p)}{cz+d}$

Similarly $\qquad\qquad (w-q) = \dfrac{(a-qc)(z-q)}{cz+a}$

$\dfrac{w-p}{w-q} = \dfrac{a-pc}{a-qc}\dfrac{z-p}{z-q} = k\left(\dfrac{z-p}{z-q}\right)$

3rd part : In case there is only one fixed point then

$$= q = \frac{p+q}{2} = \frac{a-d}{2c} \text{ from (1)} \qquad \dots(5)$$

$$\therefore \quad \frac{1}{w-p} = \frac{cz+d}{(a-pc)(z-p)} \qquad \text{by (4)}$$

$$= \frac{cz+a-2cp}{(a-pc)(z-p)} \qquad \text{by (5)}$$

$$= \frac{c(z-p)+(a-pc)}{(a-pc)(z-p)} = \frac{c}{a-pc} + \frac{1}{z-p} = k + \frac{1}{z-p}$$

This proves the third part.

PARABOLIC, ELLIPTIC AND HYPERBOLIC TRANSFORMATIONS

(a) A bilinear transformation $w = \frac{az+b}{cz+d}$ having only one fixed point is called a parabolic transformation and as shown above the transformation is of the form $\frac{1}{w-p} = \frac{1}{z-p} + k$, where p is the fixed point.

(b) A bilinear transformation having two fixed points as p,q can be put in the form $\frac{w-p}{w-q} = \frac{z-p}{z-q}k$.

If $|k| = 1$ then it is called elliptic and if k is real then it is called hyperbolic.

(c) A bilinear transformation which is neither hyperbolic nor elliptic or parabolic is called loxodromic.

 Solved Examples

EXAMPLE 1. *Consider the transformation* $w = T_1(z) = \frac{z+2}{z+3}$, $w = T_2(z) = \frac{z}{z+1}$. *Then find* $T_1^{-1}(w), T_2^{-1}(w),$ $T_2 T_1(z), T_1 T_2(z)$ *and* $T_2^{-1}T_1(z)$.

SOLUTION. Given $w = \frac{z+2}{z+3}$

On solving we get $z = \frac{2-3w}{w-1}$

$\Rightarrow \quad T_1^{-1}(w) = \frac{2-3w}{w-1}$

Similarly $\quad T_2^{-1}(w) = -\frac{w}{w-1}$

and $\quad T_2 T_1(z) = T_2\left(\frac{z+2}{z+3}\right) = \dfrac{\frac{z+2}{z+3}}{\frac{z+2}{z+3}+1} = \frac{z+2}{2z+5}$

and $\quad T_1 T_2(z) = T_1\left(\frac{z}{z+1}\right) = \dfrac{\frac{z}{z+1}+2}{\frac{z}{z+1}+3} = \frac{3z+2}{4z+3}$

Also, $\quad T_2^{-1}T_1(z) = T_2^{-1}\left(\frac{z+2}{z+3}\right) = -\dfrac{\frac{z+2}{z+3}}{\frac{z+2}{z+3}-1} = z+2.$

EXAMPLE 2. *Find the fixed point and the normal form of the following bilinear transformations and classify their nature.*

(a) $w = \frac{3z-4}{z-1}$ (b) $w = \frac{z-1}{z+1}$ (MADRAS–2003)

(c) $w = \frac{(2+i)z-2}{z+i}$

SOLUTION. (a) Putting $w = z$ for fixed points we get $(z-2)^2 = 0$.

Thus $z = 2$ is the only fixed point so that transformation is parabolic.

Normal form:

$$w - 2 = \frac{3z-4}{z-1} - 2 = \frac{z-2}{z-1}$$

$$\therefore \quad \frac{1}{w-2} = \frac{z-1}{z-2} = \frac{z-2+1}{z-2} = \frac{1}{z-2} + 1$$

Above is the required normal form.

(b) $w = \frac{z-1}{z+1}$

For fixed points $z = \frac{z-1}{z+1}$

or $\quad z^2 = -1 \quad \therefore z = \pm i$

$$w - i = \frac{z-1}{z+1} - i = \frac{z(1-i)-i(1-i)}{z+1} = \frac{(z-i)(1-i)}{z+1}$$

$$w + i = \frac{z-1}{z+1} + i = \frac{z(1+i)+i(1+i)}{z+1} = \frac{(z+i)(1+i)}{z+1}$$

$$\therefore \quad \frac{w-i}{w+i} = \frac{z-i}{z+i}\frac{1-i}{1+i} = \frac{z-i}{z+i}k$$

Above is the normal form where $|k| = \left|\frac{1-i}{1+i}\right| = 1$ and hence the transformation is elliptic.

(c) $w = \frac{(2+i)z-2}{z+i}$

For fixed points $w = z$ and hence $z^2 - 2z + 2 = 0$

$\therefore \quad z = 1+i, 1-i$

$$w - (1+i) = \frac{(2+i)z-2}{z+i} - (1+i) - \frac{z-(1+i)}{z+i}$$

$$w - (1-i) = \frac{(2+i)z-2}{z+i} - (1-i)$$

$$= \frac{2z+iz-2-z-i+iz-1}{z+i}$$

$$= \frac{z-(1-i)+2i(z-(1-i))}{z+i}$$

$$= \frac{[z-(1-i)](1+2i)]}{z+i}$$

$$\therefore \frac{w-(1+i)}{w-(1-i)} = \frac{z-(1+i)}{z-(1-i)} \cdot \frac{1}{1+2i} = \frac{z-(1+i)}{z-(1-i)} \cdot k$$

Above is normal form where

$$k = \frac{1}{1+2i} = \frac{1-2i}{1+4} = \frac{1}{5}(1-2i)$$

$\therefore |k| = 1/5\sqrt{(1+4)} = \sqrt{5}$, *i.e.*, $\neq 1$ and also k is not real. Hence the transformation is neither elliptic nor hyperbolic.
It is therefore loxodromic.

11.15.7 CIRCLE AND STRAIGHT LINES UNDER BILINEAR TRANSFORMATION

We have already proved the following results in relation to a circle:

(1) Equation of circle with centre a and radius r is $|z-a|^2 = r^2$.

(2) General Equation of a circle $bz\bar{z} + b\bar{z} + z + c = 0$ where c is real and b complex. Centre is $-b$ and radius $\sqrt{(b\bar{b}-c)}$
In case the coeffficient of $z\bar{z}$ be some constant, we should first divide by it.

(3) Inverse points.
Two points $P(z_1)$ and $Q(z_2)$ are said to be inverse points with respect to a circle with centre C if
(a) P and Q are collinear with C and are on the same side of C.
(b) $CP \cdot CQ = (\text{radius})^z$
In the circle be $z\bar{z} + b\bar{z} + \bar{b}z + c = 0$ then z_1 and z_2 are inverse points if

(a) $|z_1+b||z_2+b| = b\bar{b} - c = r^2$ (b) $\text{amp}(z_1+b) = \text{amp}(z_2+b)$

Both these can be combined into one condition as $z_1\bar{z}_2 + b\bar{z}_2 + \bar{b}z_1 + c = 0$ *i.e.*, Replace z by z_1 and $\bar{z}$ by $\bar{z}_2$ or replace z by z_2 and $\bar{z}$ by $\bar{z}_1$.

Straight line. Line through two points z_1 and z_2

$$\arg \cdot \frac{z-z_1}{z_1-z_2} = 0 \text{ or } \pi.$$

or $z\bar{b} + \bar{z}b + c = 0$ where $b = i(z_1 - z_2)$ and c is real.

☛ REMARKS

• If z_1 and z_2 are two given points and $k > 0$ is real constant then the equation

$$\left|\frac{z-z_1}{z-z_2}\right| = \lambda (\lambda > 0)$$

represents a family of circles of which z_1 and z_2 are inverse points and conversely the equation of a circle with respect to which z_1 and z_2 are inverse points can be put in the form (1).

• The circle $\text{amp}\left(\frac{z-z_1}{z-z_2}\right) = \lambda$, λ real and $\left|\frac{z-z_1}{z-z_2}\right| = \lambda(\lambda \neq 1)$ are called Steniner circles of first and second kinds.

• Every bilinerar transformation transforms circles into circles and inverse points into inverse points. In case straight line it is transformed to circles whose inverse points correspond to symmetrical points about the line.

• The equation $\left|\frac{z-z_1}{z-z_2}\right| = \lambda(\lambda > 0)$ represents a family of circles of which z_1 and z_2 are inverse points and if $\lambda = 1$ then it represents a family of straight lines of which z_1 and z_2 are symmetrical points.

• For the circle $|w| = c$ the inverse points as a limiting case may be taken $w = 0$ and $w = \infty$ and the corresponding points z may be regarded as inverse points of the circle or symmetrical points of the straight line in the z-plane.

• If $z = x + iy$ be a point then its smmetrical points about the real axis $y = 0$ is $\bar{z} = x - iy$ and about imaginary axis.

$$x = 0 \text{ is } -x + iy = -(x-iy) = -\bar{z}$$

If $|z| = 1$ then z may be written as $z = e^{i\lambda}$ where λ is real.

(a) If z_1, z_2 are inverse points w.r.t. unit circle $|z| = 1$ then $z_1 \cdot \bar{z}_2 = 1$ or $z_2\bar{z}_1 = 1$ *i.e.*, $z_1 = \frac{1}{\bar{z}_2}$ or $z_2 = \frac{1}{\bar{z}_1}$

Hence the inverse points z_1, z_2 becomes $z_1, \frac{1}{\bar{z}_1}$ or $z_2, \frac{1}{\bar{z}_2}$.

(b) Inverse of a point 'a' w.r.t. circle $|z-c| = r$ is the point $c + \frac{r^2}{\bar{a}-\bar{c}}$.

Solved Examples

EXAMPLE 1. *Find the condition that the transformation*
$$w = \frac{az+b}{cz+d}$$
transforms the unit circle in the w-plane into stright line in the z-plane.

SOLUTION. We can write

$$w = \frac{a}{c}\left(\frac{z+b/a}{z+d/c}\right)$$

$$\therefore |w| = 1 = \frac{|a|}{|c|}\frac{|z+b/a|}{|z+d/c|}$$

for unit circle in w plane.

It will represent a straight line in z-plane if $\left|\dfrac{a}{c}\right| = 1$, i.e.,

$|a| = |c|$. Because we know that $\left|\dfrac{z - z_1}{z - z_2}\right| = 1$ represents

a straight line with respect to which the points z_1 and z_2 the symmetrical points.

EXAMPLE 2. *Show that the transformation* $w = \dfrac{2z + 3}{z - 4}$ *transforms the circle* $x^2 + y^2 - 4x = 0$ *into the straight line* $4u + 3 = 0$ *and explain why the curve obtained is not a circle.*

(MUMBAI–2007, JNU–2003, BHOPAL–2002)

SOLUTION. **First Method**

$z = x + iy$ ∴ $z\bar{z} = x^2 + y^2, z + \bar{z} = 2x$
$w = u + iv$ ∴ $w + \bar{w} = 2u$

From the given transformaton $w = \dfrac{2z + 3}{z - 4}$, we have

$$z = \frac{4w + 3}{w - 2} \qquad \dots (1)$$

Circle $x^2 + y^2 - 4x = 0$

or $z\bar{z} - 2(z + \bar{z}) = 0 \qquad \dots (2)$

Putting for z from 1 in (2) we get the transformed equation in w-plane as

$$\frac{4w + 3}{w - 2} \cdot \frac{4\bar{w} + 3}{\bar{w} - 2} - 2\left[\frac{4w + 3}{w - 2} \cdot \frac{4\bar{w} + 3}{\bar{w} - 2}\right] = 0$$

$$[16w\bar{w} + 12\bar{w} + 12w + 9] - 2[4w\bar{w}$$

$$+ 3\bar{w} - 8w - 6] + 4w\bar{w} - 8\bar{w} + 3w - 6] = 0$$

$$2(w + \bar{w}) + 3 = 0 \quad \text{or} \quad 2(2u) + 3 = 0$$

or $4u + 3 = 0$

Above equation represents a straight line.

Explanation. We know that $\left|\dfrac{z - p}{z - q}\right| = \lambda$ represents a

circle ($\lambda > 0$) and in case $\lambda = 1$ then it is a straight line. In other words it means that straight line is a particular case of circle. We may also say that the equation $az\bar{z} + b\bar{z} + \bar{b}z + c =$ represents in general a circle and in particular when $a = 0$, it reduces to a straight line.

Second Method

From the given transformation

$$w = \frac{2z + 3}{z - 4} \quad \text{or} \quad w = \frac{(2x + 3) + i2y}{(z - 4) + iy}$$

In order to make the denominator real multiply above and below by conjugate of denominator.

$$w = \frac{(2x + 3) + i2y}{(x - 4) + iy} \cdot \frac{(x - 4) - iy}{(x - 4) - iy}$$

$$= \frac{(2x + 3)(x - 4) + 2y^2}{+ i\{2y(x - 4) - y(2x + 3)\}}{(x - 4)^2 + y^2}$$

or $w = \dfrac{2(x^2 + y^2) - 5x - 12 - i11y}{x^2 + y^2 - 4x - 4x + 16}$

$= \dfrac{2.4x - 5x - 12 - i11y}{4(x - 4)}$

∵ $x^2 + y^2 - 4x = 0$

or $u + iv = \dfrac{3(x - 4) - i11y}{-4(x - 4)}$

Equating real part, we get

$$u = \frac{3}{-4} \quad \text{or} \quad 4u + 3 = 0$$

About represents straight line in the w-plane.

EXAMPLE 3. *Show that both the transformations* $w = \dfrac{1 + z}{1 - z}$ *and* $w = \dfrac{z + 1}{z - 1}$ *transform* $|w| \le 1$ *into half plane* $Re(z) \le 0$.

SOLUTION. We have $w\bar{w} - 1 = \dfrac{1 + z}{1 - z} \cdot \dfrac{1 + \bar{z}}{1 - \bar{z}} - 1$

$$= \frac{2(z + \bar{z})}{|1 - z|^2} = \frac{4Re(z)}{|1.z|^2} \qquad \dots (1)$$

The same result will be obtained if we choose the second transformation

Now $|w|^2 - 1 = \dfrac{4x}{|1' - z|^2}$ ∵ $Re(z) = x$

Now $|w| = 1$ corresponds to $x = 0$, i.e., imaginary axis of z-plane and $|w| < 1$ corresponds to $x < 0$, i.e, left half of z plane.

Hence $|w| < 1$ is transformed into half plane $Re(z) \le 0$.

EXAMPLE 4. *In the transormation* $w = i\dfrac{1 - z}{1 + z}$ *show that the interior of the circle* $|z| = 1$ *is represented in the w-plane by the plane above the real axis, the upper semi-circle into positive half of real axis and lower semi-circle into negative half of the real axis.* (OSMAINIA–2003S, VTU–2001)

SOLUTION. We have

$$w = u + iv = i\frac{1 - (x + iy)}{1 + (x + iy)} = \frac{i(1 - x) - i^2 y}{(1 + x) + iy}$$

$$w = \frac{y + i(1 - x)}{(1 + x) + iy} \times \frac{(1 + x) - iy}{(1 + x) - iy}$$

$$u + iv = \frac{2y - i(x^2 + y^2 - 1)}{(1 + x)^2 + y^2}$$

∴ $u = \dfrac{2y}{(1 + x)^2 + y^2}, v = -\dfrac{x^2 + y^2 - 1}{(1 + x)^2 + y^2}$

Now $v = 0$ corresponds to $x^2 + y^2 - 1 = 0$, i.e., $|z| = 1$. Above shows that $|z| = 1$ corresponds to real axis $v = 0$ on w-plane.

Also for upper semi-circle in z plane y is positive and hence u is positive, i.e., positive half of real axis in w-plane. Again for lower semi-circle in z-plane. y is positive and hence u is negative, i.e., negative half of real axis in w-plane.

Now for interior of the circle $|z| = 1$ we have $x^2 + y^2$ i.e., $x^2 + y^2 - 1$ is negative and hence from $v =$ positive, i.e., half of the w-plane above the real axis. The same may be exhibited as in the figure follow.

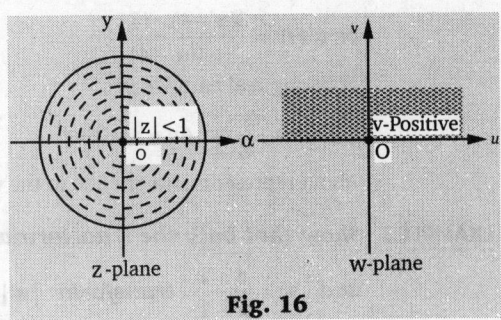

Fig. 16

EXAMPLE 5. *Find the bilinear transformation which maps*
z=1, i, –1 respectively onto w = i, 0 –i.
For this correspondence find the images of
(i) $|z| \leq 1$ (MUMBAI–2006; DELHI–2002)
(ii) *concentric circles* $|z| = r \ (r > 1)$

SOLUTION. $w = \dfrac{i-z}{i+z}$ or $z = i\dfrac{1-w}{1+w}$

(i) $|z| \leq 1$ or $\left| i\dfrac{1-w}{1+w} \right| \leq 1$

or $|1-w| \leq |1+w|$

or $(1-w)(1-\overline{w}) \leq (1+w)(1+\overline{w})$

or $2(w+\overline{w}) \geq 0$ or $4u \geq 0$

or $u \geq 0$ or $Re(w) \geq 0$

i.e., right half of the *w*-plane. Hence interior of circle $|z| = 1$ is transformed to right half of *w*-plane.

(ii) $|z| = r$ or $\left| i\dfrac{1-w}{1+w} \right| = r$

or $|1-w|^2 = r^2 |1+w|^2$

or $(1-w)(1-\overline{w}) = r^2(1+w)(1+\overline{w})$

or $w\overline{w} - \dfrac{1+r^2}{1-r^2}(w+\overline{w}) + 1 = 0$

or $u^2 + v^2 - \dfrac{1+r^2}{1-r^2} 2u + 1 = 0$

Above represents a circle in *w*-plane with centre $\left(\dfrac{1+r^2}{1-r^2}, 0 \right)$ and radius

$$\sqrt{\left\{ \left(\dfrac{(1+r^2)^2}{(1-r^2)^2} \right) - 1 \right\}} = \sqrt{\left(\dfrac{4r^2}{(1-r^2)^2} \right)} = \dfrac{2r}{1-r^2}.$$

Exercise-11.19

1. Find a Bilinear transformation which transforms the unit circle $|z| = 1$ into the real axis in such a way that point $z_1 = 1$ is mapped into $w_1 = 0$, the point $z_2 = i$ is mapped into $w_2 = 1$ and the point $z_3 = -1$ is mapped into $w_3 = \infty$.

2. Find the bilinear transformation which maps the points $0, -i, -1$ into the point $i, 1, 0$.

3. Find the bilinear transformation which maps $z = 1, 0, -1$ of z-plane into $w = i, 0, -i$ of w-plane.

4. Find the bilinear transformation which maps the points $z_1 = 1 + i$, $z_2 = -i$ and $z_3 = \alpha - i$ into the points $w_1 = 0$, $w_2 = 1$, $w_3 = i$.

5. Find the bilinear transformation which maps the points $z = 0, -1, \infty$ into the point $w = -1, -z, -i, i$ respectively.

6. Find the bilinear transformation which maps $z_1 = 1$, $z_2 = 1$ and $z_3 = \infty$ into the points $w_1 = -i$, $w_2 = -1$, and $w_3 = i$.

7. Find the bilinear transformation which maps
(i) $1, -i, z$ onto $0, z, -i$ respectvely.
(ii) $1, -i, \infty$ onto $1+i, 1-i, 1$ respectvely.

8. Find the fixed points and the normal form of the following billnear transformation :
(i) $w = \dfrac{z-1}{z+1}$ (ii) $w = \dfrac{(2+i)z-2}{z+i}$

9. Show that if $(a-d)^2 + 4bc \neq 0$ then for the transformation.
$w = \dfrac{az+b}{cz+d}$ there are two unequal numbers, α, β such that $\dfrac{w-\alpha}{w-\beta} = \lambda \dfrac{z-\alpha}{z-\beta}$ such that $\dfrac{w-\alpha}{w-\beta} = \lambda \dfrac{z-\alpha}{z-\beta}$ where λ is constant.

10. Find the image of circle $|z-2| = 2$ under the mobius transformation $w = \dfrac{z}{z+1}$.

11. Show that $w = \dfrac{1+iz}{i+z}$ maps the part of the real axis between $z = 1$ and $z = -1$ on a semi-circle in the *w*-plane.

12. Find the transformation which maps the points $-1, i, 1$ of the z-plane onto $1, i, -1$ of the w-plane respectively. Also find the invariant points.   (VTU–2011)

13. Find the bilinear transformation which maps $1, i, -1$ to $2, i, -2$ respectively. Find the fixed and critical points of the transformation.   (SVTU–2008; MUMBAI–2007; VTU–2006)

14. Determine the bilinear transformation that maps the points $1-2i, 2+i, 2+3i$ respectively into $2+2i, 1+3i, 4$.
     (JNTU–2003, COIMBATORE–1999)

15. Find the bilinear transformation which maps.
(i) The points $z=1, i, -1$ into the points $w=0, 1, \infty$.
     (VTU–2008; MUMBAI–2007)
(ii) The points $z = 0, 1, i$ into the points $w=1+i, -i, 2-i$.
     (VTU–2010S)

16. Show that under the transformation $w=(z-i)/(z+i)$, real axis in the z-plane is mapped into the circle $|w|=1$, which portion of the z-plane corresponds to the interior of the circle?   (JNTU–2003)

17. Determine the region of the *w*-plane into which the following regions are mapped by the transformation $w=z^2$.
(i) first quadrant of z-plane.   (JNTU–2000)
(ii) region bounded by $x = 1, y = 1, x + y = 1$.
     (KOTTAYAM–2005; VTU–2005)
(iii) the region $1 \leq x \leq 2$ and $1 \leq y \leq 2$
     (OSMANIA–2003, VTU–2000)

18. Under the transformation $w = 1/2$, find the image of
(a) the circle $|z-2i| = 2$.
     (BHOPAL–2009; KERALA(M.TECH)–2005)
(b) the straight line $y - x + 1 = 0$ (PTU–2007)
(c) the hyperbola $x^2 - y^2 = 1$.   (MUMBAI–2005; JNTU–2005)

19. Show that the transformation $w = z + a^2/z$ transforms circles with origin at the centre in the z-plane into co-axial concentric, confocal ellipses in the w-plane.

(KURUKSHETRA 2005, JNTU 2005)

20. Prove that the transformation $w = \sin z$, maps the families of lines $x =$ constant and $y =$ constant into two families of comfocal central conics.

(JNTU–2003)

 Hints to Selected Problems

1. Taken $z_1 = 1, z_2 = i, z_3 = -1, w_1 = 0, w_2 = 1, w_3 = \infty$ and use the definition of cross ratio.

2. Putting $z_1 = 0, z_2 = -i, z_3 = -1$ and $w_1 = i, w_2 = 1, w_3 = 0$ in

$$\frac{(w-w_1)(w_2-w_3)}{(w_1-w_2)(w_3-w)} = \frac{(z-z_1)(z_2-z_3)}{(z_1-z_2)(z_3-z)}$$

3-6. Do same as 2.

8. Solve the given transformation for z by putting $w = z = T(z)$

10. $w = \dfrac{z}{z+1} \Rightarrow wz + w = z \Rightarrow z = \dfrac{w}{1-w}$ ...(1)

$|z-2| = 2 \Rightarrow (z-2)(\bar{z}-2) = 2^2$

Also $\bar{z} - 2 = \dfrac{3\bar{w}-2}{1-\bar{w}}$ ...(2)

Putting the value from (2) in (1).

 **Answers**

1. $z = -\dfrac{w-i}{w+i}$ **2.** $w = i\left(\dfrac{1+z}{1-z}\right)$ **3.** $w = iz$ **4.** $w = \dfrac{2i(z-1-i)}{5-3i-z(1+i)}$ **5.** $w = \dfrac{iz-2}{2+z}$

6. $w = \dfrac{iz-2+i}{z+(1-2i)}$ **7.** (i) $w = \dfrac{2(z-1)}{(1+i)(z-2)}$ (ii) $w = \dfrac{z+i}{2}$

8. (i) $z = \pm i, \dfrac{w-i}{w+i} = -i\dfrac{z-i}{z+i}$ (ii) $z = 1 \pm i, \dfrac{w-(1+i)}{w-(1-i)} = \dfrac{1-2i}{5}\left(\dfrac{z-(1+i)}{z-(1-i)}\right)$

10. Image is a circle with centre $\left(\dfrac{2}{5}, 0\right)$ and radius $\dfrac{2}{5}$ in w-plane   **12.** $w = -1/2$

13. $w = (2i - 6z)/(iz - 3)$; Fixed points $z = i, 2i$; no critical points   **14.** $w = \dfrac{(20+18i)-(32+12i)z}{(29+17i)-(11-3i)z}$

15. (i) $w = i(1-z)/(1+z)$ (ii) $w = \dfrac{(4i-2)-(5-3i)z}{2i+(1+i)z}$ **16.** Upper half

17. (i) $I(w) > 0$;   (ii) Region bounded by the parabolas $v^2 = 4(1 \pm u)$ and $u^2 = 1 - 2v$

(iii) Region bounded by the parabolas $v^2 = 1/4 \pm u, v^2 = 4(1 \pm u)$ **18.** (a) Line $4v + 1 = 0$

11.16 COMPLEX INTEGRATION

We are fully familiar with the process of integration of a function of a real variable both indefinite and definite integration. The indefinite integration was regarded as an operation inverse to that of differentiation whereas definite integral was regarded as the limit of a sum.

Here in this chapter we shall deal with integration of a function of complex variable. The concept of indefinite integral for the function of a complex variable is the same as for the function of a real variable, *i.e.*, inverse of differentiation provided the function is analytic. Thus if $f(z)$ be a given analytic function of z and if $\int f(z)dz = f(z)$ then differentiation of $F(z)$ is equal to $f(z)$.

But in the case of definite integral of a complex variable we can just call it as an extension of the concept of definite integral of a function of real variable. Take for example the definite integrals $\int_a^b f(x)dx$ or $\int_a^b f(y)dy$ for the case of real variables the path of integration is always along the x-axis from $x = a$ to $x = b$ or y-axis from $y = a$ to $y = b$.

However in the case of definite integration of a function of complex variable $\int_a^b f(z)dz$ the path of definite integral may be along any curve from $z = a$ to $z = b$, *i.e.*, value of definite integral in this case depends upon the path (curve) of integration. But it will be seen that this variation in the values of definite integral can be made to disappear if the different curves (paths) from a to b are regular curves. In other words we shall prove that if $f(z)$ is analytic in a simply connected region D, then $\int_a^b f(z)dz$ is independent of the path in D joining a and b where a and b are points in D.

11.16.1 SOME BASIC DEFINITIONS

Definition (1)

Let z be a point on an arc L such that

$$z = z(t) = \phi(t) + i\psi(t) \qquad \qquad ...(1)$$

when
$$x = \phi(t) \qquad \qquad ...(a)$$
$$y = \psi(t) \qquad \qquad ...(b)$$

Now if both $\phi(t)$ and $\psi(t)$ are real continuous functions of the real variable, 't' defined in the range $\alpha \le t \le \beta$, then the arc L is said to be a continuous arc.

Definition (2). *If the above equations (a) and (b) are satisfied by more than one value of t in the assigned region then we say that the point z or (x, y) is a multiple point of the arc L.*

Definition (3). *A continuous arc without multiple points is called a Jordan arc.*

Definition (4). *An arc of a Jordan curve is said to be regular arc if $\phi'(t)$ and $\psi'(t)$ are also continuous in the assigned range.*

Definition (5). *An arc L is said to be differentiable if $z'(t)$ exist and is continuous. We also say that L is continuously differentiable. Again if in addition to existence of $z'(t)$ we have that $z'(t) \neq 0$ then L is said to be regular or smooth.*

Definition (6). *A chain of finite number of continuous arcs is called Continuous Jordan Curve.*

Definition (7). *By contour we mean a Jordan curve consisting of continuous chain of a finite number of regular arcs. If P be the starting point of first arc and Q the end point of the last arc then we write the integral along such a curve as*

$$\int_{PQ} f(z)dz$$

Definition (8). *If the starting point P of the first arc and the end point Q of the last arc coincide then the contour is said to be closed. The integral along the closed contour C is written as $\int_C f(z)dz$. It does not indicate the direction along the curve but by convention we take anti-clockwise direction to be positive unless stated otherwise.*

Definition (9). *A set of points S in the argand plane is said to be connected set if any two of its points can be joined by a continuous curve all of whose points belong to S. Also an open connected set is called an open domain but if we include the boundary points of S in the open domain then it is termed as closed domain.*

Definition (10). *A domain in which every closed curve can be shrunk to a point without passing out of the region is called simply connected domain. If the domain is not simply connected then it is called multiply connected domain.*

Definition (11). *Let [a, b] be a closed interval which is divided into n sub-intervals.*

$$[a = t_0], [t_1, t_2], [t_2, t_3]...[t_{n-1}, t_n = b] \qquad ...(1)$$

by inserting n – 1 intermediate points $t_1, t_2, ..., t_{n-1}$ such that

$$a = t_0 < t_1 < t_2 < t_3 ... < t_{n-1} < t_n = b.$$

Then the set $P = [t_0, t_1, t_2... t_n]$ is called a partition of the interval [a, b] and the greatest of the numbers

$$t_1 - t_0, t_2 - t_1... t_n - t_{n-1} \qquad ...(2)$$

is called the norm of the partition P and is denoted by $|P|$ which represents, the maximum length of the sub-intervals given in (1).

Definition (12). *Let an arc L be defined by equations (1) and (a), (b). The equation $x = \phi(t)$*

and $y = \psi(t)$ where $\alpha \leq t \leq \beta$ represent a plane cuve. Now consider any partition

$$P = [\alpha = t_0, t_1, t_2, ..., t_n = \beta]$$

corresponding to $z_0, z_1, z_2, ..., z_n$ of the given arc, then the points on the curve corresponding to these values of t be denoted by $P_0, P_1, P_2, ..., P_n$.

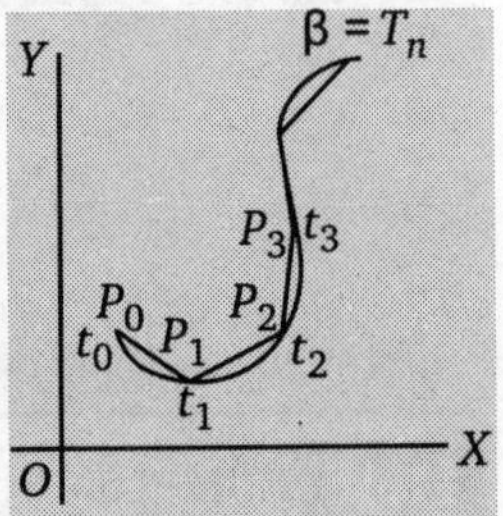

The length of the polygonal arc $P_0, ..., P_n$ which is obtained by drawing straight lines P_0 to P_1, P_1 to P_2 and so on is measured by

$$s = \sum_{r=1}^{n} |z_r - z_{r-1}| = \sum_{r=1}^{n} |(x_r + iy_r) - (x_{r-1} + iy_{r-1})| = \sum_{r=1}^{n} |(x_r - x_{r-1}) + (iy_r - iy_{r-1})| = \sum_{r=1}^{n} \{(x_r - x_{r-1})^2 + (iy_r - iy_{r-1})^2\}^{1/2}$$

We will say that arc L is rectifiable if the least upper bound, i.e., supremum of the sums (l) taken over all partitions P is finite say l then l is called the length of the curve.

Fig. 17

Solved Examples

EXAMPLE 1. *Evalate the following integrals.*

(a) $\int_C z\,dz$ (b) $\int_C dz$

(c) $\int_C |dz|$

SOLUTION. (a) We know that

$$\int_C f(z)dz = \lim_{n=\infty} \left[\sum_{r=1}^{n} (z_r - z_{r-1}) f(\xi_r) \right] \qquad ...(1)$$

where $z_{r-1} \leq \xi_r \leq z_r$

(a) $\int_C z\,dz$ here $f(z) = z$

Since ξ_r is arbitrary and hence putting $\xi_r = z_r$ and z_{r-1} successively in (1), we get

$$\int_C z\,dz = \lim_{n \to \infty} \sum_{r=1}^{n} (z_r - z_{r-1})z_r \qquad ...(2)$$

$$\int_C z\,dz = \lim_{n \to \infty} \sum_{r=1}^{n} (z_r - z_{r-1})z_{r-1} \qquad ...(3)$$

Adding (2) and (3) we get

$$2\int_C z\,dz = \lim_{n \to \infty} \sum_{r=1}^{n} (z_r - z_{r-1})(z_r + z_{r-1})$$

$$= \lim_{n \to \infty} \sum_{r=1}^{n} (z_r^2 - z_{r-1}^2)$$

$$= \lim_{n \to \infty} (z_1^2 - z_0^2) + (z_2^2 - z_1^2) + ... + (z_n^2 - z_{n-1}^2)$$

$$= \lim_{n \to \infty} [z_n^2 - z_0^2] = \beta^2 - \alpha^2$$

$$\therefore \quad \int_C z\,dz = \frac{1}{2}(\beta^2 - \alpha^2)$$

Because $z_n = \beta$, $z_0 = \alpha$ where α and β are the end point of A and B of the curve C.

(b) Here $f(z) = 1$ and hence from (1) we have

$$\int_C 1.dz = \lim_{n \to \infty} \left[\sum_{r=1}^{n} (z_r - z_{r-1}).1 \right]$$

$$= \lim_{n \to \infty} [(z_1 - z_0) + (z_2 - z_1) + ...(z_n - z_{n-1})]$$

$$= \lim_{n \to \infty} [z_n - z_0] = \beta - \alpha \qquad \text{as } z_n = \beta \text{ and } z_0 = \alpha$$

$$= \text{Chord } AB \qquad \because \ \overline{AB} = P.V \text{ of } B - P.V \text{ of } A$$

☞ **REMARK**

- In case the curve C is a closed curve then the end points α and β coincide and as such $\int_C f(z)dz = 0$ where C is a closed curve.

(c) Here $f(z) = 1$ and in place of dz we have $|dz|$.
Hence from (1) we have

$$\int_C |dz| = \lim_{n \to \infty}\left[\sum_{r=1}^n |z_r - z_{r-1}|1\right]$$
$$= \lim_{n \to \infty}[|z_1 - z_0| + |z_2 - z_1| + ... + |z_n - z_{n-1}|]$$
$$= \lim_{n \to \infty}[\text{chord } z_1 z_0 + \text{chord } z_2 z_1 + ...]$$
$$= \text{arc } z_1 z_0 + \text{arc } z_2 z_1 + ... \text{arc } z_n z_{n-1}$$
$$= \text{arc length of } C \text{ between the points } z_0 \text{ and } z_n.$$

EXAMPLE 2. *Find the value of the integral $\int_0^{1+i}(x - y + ix^2)dz$*

(a) *Along the straight line from $z = 0$ to $z = 1+i$, i.e., OM.*

(b) *Along the real axis from $z = 0$ to $z = 1$ i.e. OL and then along a line parallel to the imaginary axis from $z = 1$ to $z = 1 + i$, i.e., LM.* (UPTU–2003)

SOLUTION. (a) Along $OM, y = x, z = x + iy = (1 + i)x$
$\therefore \quad dz = (1 + i)dx$
Also x varies from 0 to 1 along OM.

$$\therefore \quad \int_{OM}(x - y + ix^2)dz = \int_0^1 (ix)^2(1+i)dx$$

$$= (i + i^2)\left[\frac{1}{3}x^3\right]_0^1 = -\frac{1+i}{3}.$$

(b) Along $OL, y = 0, z = x + iy = x, dz = dx, x$ varies from 0 to 1.

Along $LM, x = 1, z = x + iy = 1 + iy, \quad dz = idy.$
Also y varies from 0 to 1.

$$\therefore \quad \int_{OLM}(x - y + ix^2)dz$$

$$= \int_{OL}(x - y + ix^2)dz + \int_{LM}(x - y + ix^2)dz$$

$$= \int_0^1 (x + ix^2)dx + \int_0^1 [(1+i) - y]idy$$

$$= \left[\frac{x^2}{2} + i\frac{x^3}{3}\right]_0^1 + (i + i^2)[y]_0^1 - \left[\frac{y^2}{2}\right]_0^1$$

$$= \frac{1}{2} + \frac{i}{3} + (i - 1) - i.\frac{1}{2} = -\frac{1}{2} + \frac{5i}{6}.$$

EXAMPLE 3. *Show that the integral of $\bar{z}$ along a semi-circular arc $|z| = 1$ from -1 to $+1$ has the value $-\pi i$ or πi according as the arc lies above or below the real axis.* (ROHTAK–2005)

SOLUTION. We have $|z| = 1$ so $z = 1.e^{i\theta} \Rightarrow dz = ie^{i\theta}d\theta$
Also, $\bar{z} = e^{-i\theta}$

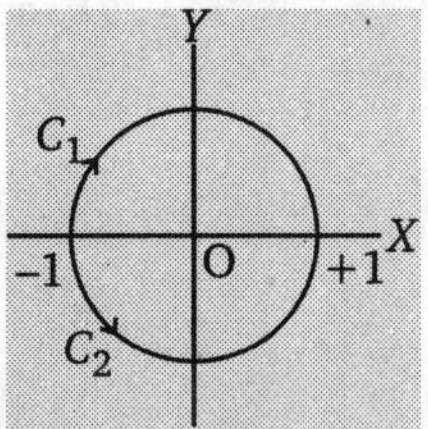

$$\int_{C_1}\bar{z}dz = \int_\pi^0 e^{-i\theta}.ie^{i\theta}.d\theta$$

$$= i[\theta]_\pi^0 = -\pi i$$

$$\int_{C_2}\bar{z}dz = \int_\pi^{2\pi} e^{-i\theta}.ie^{i\theta}.d\theta$$

$$= i[\theta]_\pi^{2\pi} = \pi i.$$

From above we conclude that the integral round the complete circle in a counter clockwise direction.

$$= \pi i - (-\pi i) = 2\pi i$$

Fig. 18

Also for the unit circle $z\bar{z} = 1$ $\therefore \bar{z} = \frac{1}{z}$

$$\therefore \int_C \frac{1}{z}dz = 2\pi i \text{ where } C \text{ is complete circle.}$$

11.16.2 ELEMENTARY PROPERTIES OF COMPLEX INTEGRALS

Prop. 1. $\int_C [f(z) + \phi(z)]dz = \int_C f(z)dz + \int_C \phi(z)dz$

The above property can be generalized for the finite number of functions.

Prop. 2. $\int_C f(z)dz = \int_{C_1} f(z)dz + \int_{C_2} f(z)dz$ where C_1 and C_2 are two parts of C.

Prop. 3. $\int_C f(z)dz = -\int_{-C} f(z)dz$ where $-C$ indicates the curve C traversed in the opposite direction or if we denote by C_1 the contour C when described in opposite direction then above can be written as $\int_{C_1} f(z)dz = -\int_C f(z)dz$

Prop. 4. $\int_{C_1} kf(z)dz = k\int_{-C} f(z)dz$ where k is a constant.

11.16.3 AN INEQUALITY FOR COMPLEX INTEGRALS

THEOREM 1. *If a function $f(z)$ is continuous on a contour C of length l, and M be the upper bound of $|f(z)|$ on C then $\left|\int_C f(z)dz\right| \le Ml$*

PROOF. We know that $\int_C f(z)dz = \lim_{n \to \infty}\sum_{r=1}^n (z_r - z_{r-1})f(\xi_r)$ where $z_{r-1} \le \xi_r \le z_r$. ...(1)

Let the equation of the curve C be $x = \phi(t), y = \psi(t)$.

$$\therefore \quad l = \int ds = \int\left[\left(\frac{dx}{dt}\right)^2 + \left(\frac{dy}{dt}\right)^2\right]^{1/2} dt \quad \text{...(2)}$$

Also $z = x + iy$ $\therefore$ $dz = dx + idy$

$\therefore \quad |dz| = |dx + idy| = [(dx)^2 + (dy)^2]^{1/2}$

$$\therefore \quad \int_C |dz| = \int[(dx)^2 + (dy)^2]^{1/2} = \int\left[\left(\frac{dx}{dt}\right)^2 + \left(\frac{dy}{dt}\right)^2\right]^{1/2} dt = l \text{ by (2)}$$

$$\therefore \qquad \int_C |dz| = l$$

Again we know that modulus of the sum of n complex quantitites is always less than or equal to sum of their moduli.

$$\therefore \qquad \left| \sum_{r=1}^{n} (z_r - z_{r-1}) f(\xi_r) \right| \le \sum_{r=1}^{n} \left| (z_r - z_{r-1}) f(\xi_r) \right| = \sum_{r=1}^{n} \left| (z_r - z_{r-1}) \right| \left| f(\xi_r) \right| . \qquad \dots(3)$$

Now by tending n to infinity the above inequality may be written as

$$\left| \int_C f(z)dz \right| \le \int_C |f(z)| |dz| \le M \int_C |dz|$$

$$\because \quad M \text{ is upper bound for } |f(z)| \le Ml \text{ as by (3) } \int_C |dz| = l .$$

$$\therefore \qquad \left| \int_C f(z)dz \right| \le Ml.$$

STATEMENT. **(Cauchy Theorem).** *If $f(z)$ is analytic function of z if $f'(z)$ is continuous at each point within and on a closed contour C, then*

$$\int_C f(z)dz = 0.$$

PROOF. In proving the above result we shall make use of the following two results, *i.e.*, Green's theorem for a plane and Cauchy Riemman equations.

Green's Theorem. If $P(x, y)$, $Q(x, y)$, $\dfrac{\partial P}{\partial y}$, $\dfrac{\partial Q}{\partial x}$ are all continuous functions within a domain D and if C is any closed contour in D, then Green's theorem states that

$$\int_C (Pdx + Qdy) = \iint_D \left(\frac{\partial Q}{\partial x} - \frac{\partial P}{\partial y} \right) dxdy \qquad \dots(1)$$

Cauchy Riemman Equations.

If $f(z) = u(x, y) + iv(x, y)$ be an analytic function then

$$\frac{\partial u}{\partial x} = \frac{\partial v}{\partial y} \qquad \text{and} \qquad \frac{\partial u}{\partial y} = -\frac{\partial v}{\partial x} \qquad \dots(2)$$

or $\qquad u_x = v_y \qquad \text{and} \qquad u_y = -v_x.$

Proof of the Theorem.

Since $f(z) = u + iv$ is analytic and has a continuous derivative

$$f'(z) = \frac{\partial u}{\partial x} + i\frac{\partial v}{\partial x} = \frac{\partial v}{\partial y} - i\frac{\partial u}{\partial y} \text{ by (2)}$$

From above it follows that u, v and the partial derivatives

$$\frac{\partial u}{\partial x} = \frac{\partial v}{\partial y} \quad \text{and} \qquad \frac{\partial u}{\partial y} = -\frac{\partial v}{\partial x}$$

are all continuous inside and on C. Thus the Green's theorem can be applied.

$$\therefore \qquad \int_C f(z)dz = \int (u + iv)(dx + idy) = \int (udx - vdy) + i(vdx + udy)$$

$$= \iint_D \left(-\frac{\partial v}{\partial x} - \frac{\partial u}{\partial y} \right) dxdy + i\iint_D \left(\frac{\partial u}{\partial x} - \frac{\partial v}{\partial y} \right) dxdy \qquad \text{by (1)}$$

$$= -\iint_D \left(\frac{\partial v}{\partial x} + \frac{\partial u}{\partial y} \right) dxdy + i\iint_D \left(\frac{\partial u}{\partial x} - \frac{\partial v}{\partial y} \right) dxdy = 0 + i.0 \qquad \text{by (2)}$$

Hence, $\qquad \int_C f(x)dz = 0 .$

☛ **REMARK**

• In the above from of Cauchy theorem we had the assumption that the derived function $f'(z)$ continuous. It was the famous Mathematician Goursat who first established that the above condition of continuity of $f'(z)$ is unnecessary and can be removed from the hypothesis. Hence Cauchy theorem holds only if $f(z)$ is analytic within and on C.

THEOREM 2. (Cauchy-Goursat Theorem).

STATEMENT. *If a function $f(z)$ is analytic and one valued inside and on a simple closed contour C then*

$$\int_C f(z)dz = 0.$$

PROOF. In order to prove the theorem we shall first prove a Lemma known as Goursat's Lemma.

STATEMENT OF LEMMA. *Given $\in > 0$, it is possible to divide the region inside C into finite number of meshes, either complete squares, or part of squares, such that within each mesh there exists a point z_0 such that*

$$\left\{ \frac{f(z) - f(z_0)}{z - z_0} - f'(z_0) \right\} < \in$$

for all values of z in the mesh.

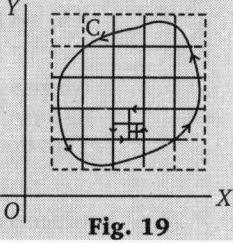

Fig. 19

Proof of the Lemma. Let us suppose that the Lemma is false. This means that it fails at least in one mesh. Subdivide this mesh by lines joining the middle point of the opposite sides. In case there still remains say part which do not satisfy the condition (1) we shall again subdivide them in the same way. The above process may end either after a finite number of steps when the condition (1) is satisfied for every subdivision or else the process may go on indefinitely. In the second case we obtain a sequence of squares (each contained in the proceeding one) which has z_0 as its limit point at which the condition (1) is not satisfied. As the condition (1) is not satisfied at the point z_0 therefore $\left\{\dfrac{f(z)-f(z_0)}{z-z_0}-f'(z_0)\right\} \nleqslant \in$

where $|z-z_0|$ is small.

Above relation shows that $f(z)$ is not differentiable at z_0 which in other words means that $f(z)$ is not analytic at z_0. But this contradicts the hypothesis that $f(z)$ is analytic at all the points within and on the contour C. Therefore the Lemma is true,

i.e., $\left\{\dfrac{f(z)-f(z_0)}{z-z_0}-f'(z_0)\right\} < \in$

or $\dfrac{f(z)-f(z_0)}{z-z_0}-f'(z_0) = \eta(z)$ where $|\eta| < \in$

and $\eta \to 0$ as $z \to z_0$

$\therefore \qquad f(z) = f(z_0) + (z-z_0)f'(z_0) + (z-z_0)\eta(z).$ ...(2)

Proof of the Main Theorem.

Let $\in > 0$ be given then by the Lemma proved above the given closed contour C can be divided into squares and partial squares whose boundaries are C_r ($r = 1, 2 \ldots n$). Hence a point z_0 exists for which result (1) of our Lemma holds.

Again let z be any point on the boundary C_r. The value of $f(z)$ at any point z_0 of C_r can be given by result (2) of our Lemma as under

$f(z) = f(z_0) + (z_0-z)f'(z_0) + (z-z_0)\eta(z)$ ...(3)

where $|\eta| < \in$.

Fig. 20

Suppose the integral has been taken in the counter clockwise sense around each C_r, then the sum of these integrals will be the integral around the closed curve C in the anticlock wise sense, i.e.,

$\int_C f(z)dz = \sum\limits_{r=1}^{n} \int_{C_r} f(z)dz$ ...(4)

Now the line integral along the boundary which is the common boundary line of every adjacent sub-region cancel each other, the sense of integral in a region is opposite to that of the other as it evident from the figure. Hence the only parts of the sum that remain are integrals along the arcs which are the parts of the curve. Now putting for $f(z)$ from (3) in (4), we get

$\int_C f(z)dz = \sum\limits_{r=1}^{n} \int \{f(z_0)-(z_0-z)f'(z_0)+(z-z_0)\eta(z)\}dz$

$= \sum\limits_{r=1}^{n} [\{f(z_0)-z_0 f'(z_0)\}]\int_{C_r} dz + f'(z_0)\int_{C_r} zdz + \int_{C_r}(z-z_0)\eta dz$ ...(5)

We know that $\int_C zdz = 0$ and $\int_C dz = 0$ where C is a closed curve.

Hence from (5) we get

$\int_C f(z)dz = \sum\limits_{r=1}^{n} \int_{C_r}(z-z_0)\eta dz$ ...(6)

$\therefore \qquad \left|\int_C f(z)dz\right| \le \sum\limits_{r=1}^{n}\left|\int_{C_r}(z-z_0)\eta dz\right| \le \sum\limits_{r=1}^{n}\int_{C_r}|z-z_0||\eta||dz| \le \in \sum\limits_{r=1}^{n}\int_{C_r}|z-z_0||dz|$ ...(7)

Because $|\eta| < \in$

Clearly each boundary C_r coincides either with a complete square or part of it and let l_r be length of the side of a square. Therefore the diagonal of the square is $\sqrt{2}l_r$. Now the point z is on C_r and z_0 may be either on the boundary or within the square and as such $|z-z_0|$ cannot be greater than the length of the diagonal.

$\therefore \qquad \int_C |z-z_0||dz| \le \sqrt{2}l_r \int_{C_r}|dz|$ by (7) ...(8)

Again we know that $\int_{C_r}|dz|$ is the length of the region C_r. If it is a complete square it is equal to $4l_r$ and it cannot exceed $(4l_r + L_r)$. If C_r is a partial square where L_r is the length of the arc of the contour C which constitutes the part of C_r a square and the area of this square is A_r, then

$\int_C |z-z_0||dz| \le 4\sqrt{2}l^2 = 4\sqrt{2}A_r$ ...(9)

If C_r is a partial square, then

$$\int_{C_r} |z - z_0||dz| \le \sqrt{2} l_r (4 l_r + L_r) < 4\sqrt{2} A_r + 2 L_r S \qquad \ldots(10)$$

where S is the length of the side of the square enclosing the whole curve C as well as the squares which cover C. Hence sum of all the A_r^S cannot exceed S^2.

Thus from (6), (7), (8) and (9) we observe that

$$\left| \int_{C_r} f(z) dz \right| < \epsilon (4\sqrt{2} S^2 + \sqrt{2} SL) = \epsilon \beta$$

where β is any constant. Since ϵ is arbitrary and small.

$$\therefore \qquad \int_C f(z) dz = 0.$$

This completes the proof of the theorem.

☞ REMARKS

- **Extension of Cauchy Theorem** Let $f(z)$ be analytic in a simply connected domain D, then the integral along any rectifiable in D joining any two given points of D is the same. In other words it means that it does not depend upon the curve joining two points.

- If C is a closed curve and $C_1, C_2, C_3 \ldots$ the other closed curves which lie inside C, and if a function $f(z)$ is analytic in the region between these curves and continuous on C, then

$$\int_C f(z) dz = \int_{C_1} f(z) dz + \int_{C_2} f(z) dz + \int_{C_3} f(z) dz + \ldots$$

where integral along each curve is taken in the anti-clockwise direction.

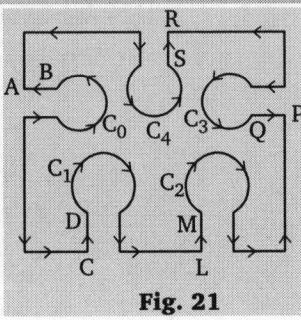

Fig. 21

11.16.4 DERIVATIVE OF f(z)

To show that $F(z)$ (the indefinite integral of $f(z)$) is an analytic function of upper limit z.

PROOF.   Using previous theorem, we have

$$F(z) = \int_{z_0}^z f(z) dz$$

$$F(z + \Delta z) = \int_{z_0}^{z + \Delta z} f(z) dz$$

$$\therefore \qquad F(z + \Delta z) - F(z) = \int_{z_0}^{z + \Delta z} f(z) dz - \int_{z_0}^z f(z) dz = \int_{z_0}^{z + \Delta z} f(z) dz + \int_z^{z_0} f(z) dz = \int_z^{z + \Delta z} f(Z) dZ$$

The R.H.S. is the indefinite integral of $f(z)$ where z is a fixed point and Z is any point on the path joining z to $z + \Delta z$.

Let the path to integration from z to $z + \Delta z$ be a straight line then

$$F(z + \Delta z) - F(z) = \int_z^{z + \Delta z} \{f(Z) - f(z)\} dZ + f(z) \int_z^{z + \Delta z} dZ = \int_z^{z + \Delta z} \{f(Z) - f(z)\} dZ + f(z) \Delta z$$

$$\therefore \qquad \frac{F(z + \Delta z) - F(z)}{\Delta z} - f(Z) = \frac{1}{\Delta z} \int_z^{z + \Delta z} \{f(Z) - f(z)\} dZ$$

As $f(z)$ is continuous $|f(Z) - f(z)| < \epsilon$ when $|Z - z| < \delta$

In particular; When $|Z - z| < \delta$ we have

$$\left[\frac{F(z + \Delta z) - F(z)}{\Delta z} - f(z) \right] \le \left| \frac{1}{\Delta z} \right| \int_z^{z + \Delta z} |f(Z) - f(z)| |dZ| \le \frac{1}{|\Delta z|} \cdot \epsilon \int_z^{z + \Delta z} |dZ| = \frac{1}{|\Delta z|} \cdot \epsilon |\Delta z| = \epsilon$$

Above relation shows that the derivative of $F(z)$ exists and that $F'(z) = f(z)$ at every point of the region D. Thus we prove that $F(z)$ is analytic at every point z of the region.

Thus we have established that $F(z)$ the indefinite integral of $f(z)$ is an analytic function of its upper limit.

11.16.5 CAUCHY'S INTEGRAL FORMULA

STATEMENT.   *If $f(z)$ is analytic within and on a closed contour C and if a is any point within C, then*

$$f(a) = \frac{1}{2\pi i} \int_C \frac{f(z) dz}{z - a}.$$

PROOF.   Consider a circle γ about the point $z = a$ and lying entirely within C.

The function $\dfrac{f(z)}{z - a}$ is analytic in the region between C and γ. Therefore by Cauchy Theorem for multi-connected region we have

$$\int_C \frac{f(z)}{z - a} dz = \int_\gamma \frac{f(z)}{z - a} dz$$

where both C and γ are traversed in the anti-clockwise direction.

or $$\int_C \frac{f(z)}{z - a} dz - \int_\gamma \frac{f(a)}{z - a} dz = \int_\gamma \frac{f(z) - f(a)}{z - a} dz$$

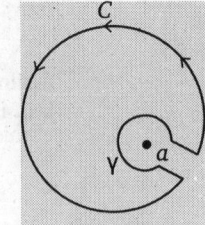

Fig. 22

or
$$\int_C \frac{f(z)}{z-a}dz - f(a)\int_\gamma \frac{1}{z-a}dz = \int_\gamma \frac{f(z)-f(a)}{z-a}dz$$

or
$$\int_C \frac{f(z)}{z-a}dz - f(a)2\pi i = \int_\gamma \frac{f(z)-f(a)}{z-a}dz$$

or
$$\left|\int_C \frac{f(z)}{z-a}dz - 2\pi i f(a)\right| = \left|\int_\gamma \frac{f(z)-f(a)}{z-a}dz\right| \le \int_\gamma \frac{|f(z)-f(a)|}{|z-a|}|dz| \le \in \int_\gamma \frac{|dz|}{|z-a|}$$

$$(\because |f(z)-f(a)| < \in, \text{ as } f(z) \text{ is continuous at } z = a)$$

$$= \frac{\in}{r}\int|dz| \text{ as } |z-a| = r \text{ for } z \text{ on } r$$

$$= \frac{\in}{r}2\pi r \qquad \qquad \therefore \int_\gamma |dz| = \text{Arc of } \gamma$$

$$= 2\pi \in \to 0 \text{ as } \in \to 0$$

Hence,
$$\int_C \frac{f(z)}{z-a}dz - 2\pi i f(a) = 0 \qquad \text{or} \qquad f(a) = \frac{1}{2\pi i}\int_C \frac{f(z)}{z-a}dz$$

Above is known as Cauchy's Integral formula.

REMARK
- The above Cauchy's integral formula expresses the value of an analytic function at any point within the contour C in terms of the value of the function at the boundary only.

11.16.6 GAUSS'S MEAN VALUE THEOREM

STATEMENT. *If $f(z)$ is an analytic function on a domain D and if the circular region $|z-a| \le r$ is contained in D then*
$$f(a) = \frac{1}{2\pi i}\int_0^{2\pi} f(a+re^{i\theta})d\theta.$$

PROOF.
In other words it means that the value of the function at the point a is equal to the average of its values on the boundary of the circle.
Let γ denote the circle $|z-a| = r$ and this equation can be written as $z = a + re^{i\theta}$ $(0 \le \theta \le 2\pi)$
$dz = rie^{i\theta}.d\theta$. Hence by Cauchy's integral formula
$$f(a) = \frac{1}{2\pi i}\int_\gamma \frac{f(z)}{z-a} = \frac{1}{2\pi i}\int_0^{2\pi} \frac{f(a+re^{i\theta})}{re^{i\theta}}re^{i\theta}d\theta = \frac{1}{2\pi}\int_0^{2\pi} f(a+re^{i\theta})d\theta.$$

11.16.7 EXTENSION OF CAUCHY'S INTEGRAL FORMULA TO MULTIPLE CONNECTED REGION

STATEMENT. *If $f(z)$ is analytic in the region bounded by two closed curves C and C' and a is a point in the region, then*
$$f(a) = \frac{1}{2\pi i}\int_C \frac{f(z)dz}{z-a} - \frac{1}{2\pi i}\int_{C'} \frac{f(z)dz}{z-a}$$
where C is the outer contour.

PROOF.
Draw a small circle γ of radius λ centered about the point $z = a$ and lying entirely within C.

The function $\dfrac{f(z)}{z-a}$ is analytic in the region bounded by the three contours C, C' and γ as $z - a$ is not zero for any value of z in this annulus and it is also analytic on these curves.

Thererfore by Cauchy Theorem for multi-connected region, we have
$$\int_C \frac{f(z)dz}{z-a} = \int_{C'} \frac{(z)dz}{z-a} + \int_\gamma \frac{(z)dz}{z-a}$$

The integration round each curve is so taken that the angular region always lies to the left.

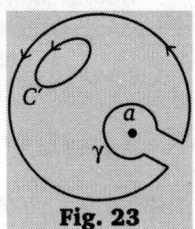

Fig. 23

$$\therefore \qquad \int_C \frac{f(z)dz}{z-a} = \int_{C'} \frac{f(z)dz}{z-a} + 2\pi i f(a) \text{ by Cauchy's formula}$$

or
$$f(a) = \frac{1}{2\pi i}\int_C \frac{f(z)dz}{z-a} - \frac{1}{2\pi i}\int_{C'} \frac{f(z)dz}{z-a}$$

The above formula can be extended if there be more curves $C'', C''', ...$ then as above we may say that
$$f(a) = \frac{1}{2\pi i}\int_C \frac{f(z)dz}{z-a} - \frac{1}{2\pi i}\int_{C'} \frac{f(z)dz}{z-a} - \frac{1}{2\pi i}\int_{C''} \frac{f(z)dz}{z-a} - ...$$

11.16.8 CAUCHY'S INTEGRAL FORMULA FOR THE DERIVATIVE OF AN ANALYTIC FUNCTION

THEOREM 1. *If a function $f(z)$ is analytic in a region D then its derivative at any point $z = a$ of D is also analytic in D and is given by*
$$f'(a) = \frac{1}{2\pi i}\int_C \frac{f(z)dz}{(z-a)^2}$$
where C is any closed contour in D surrounding the point $z = a$.

PROOF.   Let $a + h$ be a point in the neighbourhood of the point a, then by Cauchy's intgral formula we have

$$f(a) = \frac{1}{2\pi i}\int_C \frac{f(z)dz}{z-a} \quad \text{and} \quad f(a+h) = \frac{1}{2\pi i}\int_C \frac{f(z)dz}{z-a-h}$$

$$\therefore \quad f(a+h) - f(a) = \frac{1}{2\pi i}\int_C \left[\frac{1}{z-a-h} - \frac{1}{z-a}\right]f(z)dz = \frac{1}{\pi i}\int_C \frac{h}{(z-a-h)(z-a)}f(z)dz$$

or

$$\frac{f(a+h) - f(a)}{h} = \frac{1}{2\pi i}\int_C \frac{f(z)dz}{(z-a-h)(z-a)}$$

or

$$\lim_{h \to 0}\frac{f(a+h) - f(a)}{h} = \lim_{h \to 0}\frac{1}{2\pi i}\int_C \frac{f(z)dz}{(z-a-h)(z-a)}$$

or

$$f'(a) = \frac{1}{2\pi i}\int_C \frac{f(z)dz}{(z-a)^2}.$$

Now a is any point of the region D and hence from above we conclude that $f'(a)$ is analytic in D. Thus we conclude that the derivative of an analytic function is an analytic function of z.

11.16.9 HIGHER ORDER DERIVATIVES

If a function $f(z)$ is analytic in a region D, then $f(z)$ has at any point $z = a$ of D, derivative of all order, all of which are again analytic functions in D. Their values are given by

$$f^n(a) = \frac{n!}{2\pi i}\int_C \frac{f(z)dz}{(z-a)^{n+1}}$$

where C is any closed contour in D surrounding the point $z = a$.

Proof. We know that $$f'(a) = \frac{1}{2\pi i}\int_C \frac{f(z)dz}{(z-a)^2}$$

Above result shows that the theorem is true for $n = 1$. Let us assume that result is true for $n = m$ so that

$$f^m(a) = \frac{m!}{2\pi i}\int_C \frac{f(z)dz}{(z-a)^{m+1}}$$

$$\therefore \quad \frac{f^m(a+h) - f^m(a)}{h} = \frac{1}{h}\frac{m!}{2\pi i}\left[\int_C \frac{f(z)dz}{(z-a-h)^{m+1}} - \int_C \frac{f(z)dz}{(z-a)^{m+1}}\right] = \frac{1}{h}\frac{m!}{2\pi i}\int_C \frac{1}{(z-a)^{m+1}}\left\{\left(1 - \frac{h}{(z-a)}\right)^{-(m+1)} - 1\right\}f(z)dz$$

$$= \frac{1}{h}\frac{m!}{2\pi i}\int_C \frac{1}{(z-a)^{m+1}}\left\{1 + (m+1)\frac{h}{z-a} + \frac{(m+1)(m+2)}{2!}\frac{h^2}{(z-a)^2} + \dots - 1\right\}f(z)dz$$

Taking limit when $h \to 0$, we get

$$\lim_{h \to 0}\frac{f^m(a+h) - f^m(a)}{h} = \frac{m!(m+1)}{2\pi i}\int_C \frac{f(z)dz}{(z-a)^{m+2}}$$

or

$$f^{m+1}(a) = \frac{(m+1)!}{2\pi i}\int_C \frac{f(z)dz}{(z-a)^{(m+1)+1}}.$$

Above result shows that the required result holds goods for $n = m - 1$.

Also, as shown earlier the result is true for $n = 1$ and it hence must be true for $n = 2$ and so on. Thus we conclude that it holds goods for any positive integral value of n.

$$\therefore \qquad f^n(a) = \frac{n!}{2\pi i}\int_C \frac{f(z)dz}{(z-a)^{n+1}} \qquad \qquad \dots(1)$$

Again $f'(a)f^2(a)f^3(a)\dots$ all exist, therefore the derivatives of $f(z)$ of all order are analytic if $f(z)$ is analytic.

THEOREM 1.   *If C be a closed contour containing the origin inside it, then $\dfrac{a^n}{n!} = \dfrac{1}{2\pi i}\int_C \dfrac{e^{az}dz}{z^{n+1}}$.*

PROOF.   We know that $$f^n(a) = \frac{n!}{2\pi i}\int_C \frac{f(z)dz}{(z-a)^{n+1}}$$

$$\therefore \qquad f^n(0) = \frac{n!}{2\pi i}\int_C \frac{f(z)dz}{z^{n+1}}$$

Now choose $f(z) = e^{az} \therefore f^n(z) = a^n e^{az}$ or $f^n(0) = a^n$

$$\therefore \qquad a^n = \frac{n!}{2\pi i}\int_C \frac{e^{az}dz}{z^{n+2}} \qquad \text{or} \qquad \frac{a^n}{n!} = \frac{1}{2\pi i}\int_C \frac{e^{az}dz}{z^{n+1}}$$

11.16.10 Morera's Theorem

If $f(z)$ is continuous in a domain D and if for every closed contour C in the domain D, $\int_C f(z)dz = 0$ then f is analytic within D.
(*The above theorem is more or less converse of Cauchy theorem*)

Proof. Let z_0 be a fixed point and z a variable point inside the domain D then the value of the integral $\int_{z_0}^{z} f(z)dz$ is independent of the curve joining z_0 to z and is a function of the upper limit z. Therefore we may write

$$F(z) = \int_{z_0}^{z} f(t)dt \text{ Taking the variable of intetgration as } t \qquad \ldots(1)$$

$$F(z+h) = \int_{z_0}^{z+h} f(t)dt$$

$$F(z+h) - F(z) = \int_{z_0}^{z+h} f(t)dt - \int_{z_0}^{z} f(t)dt = \left[\int_{z}^{z_0} f(t)dt + \int_{z_0}^{z+h} f(t)dt \right] \text{Prop.}$$

$$= \int_{z}^{z+k} f(t)dt \qquad \ldots(2)$$

Now the integral in (2) is independent of the path and may be taken along a straight line segment joining z to $z + h$. Hence

$$\frac{F(z+h)-F(z)}{h} - f(z) = \frac{1}{h}\int_{z}^{z+h} f(t)dt - \frac{f(z)}{h}h = \frac{1}{h}\int_{z}^{z+h} f(t)dt - f(z)\int_{z}^{z+h} dt = \frac{1}{h}\int_{z}^{z+h}[f(t)-f(z)]dt \qquad \ldots(3)$$

As $f(t)$ is continuous at z therefore given $\epsilon > 0$ there exist a $\delta > 0$ such that $|f(t) - f(z)| < \epsilon$ for every t satisfying $|t - z| < \delta$.
Let us now choose h such that $|h| < \delta$ so that the relation
$$|f(t) - f(z)| < \epsilon$$
is satisfied for every point t on the line segment joining z to $z + h$.
Hence from (3), we observe that

$$\left| \frac{F(z+h)-F(z)}{h} - f(z) \right| \leq \frac{1}{|h|}\int_{z}^{z+h}|f(t)-f(z)||dt| < \frac{1}{|h|}\epsilon\int_{z}^{z+h}|dt| = \frac{1}{|h|}\epsilon|h| = \epsilon$$

Since ϵ is arbitrary we get from above

$$\lim_{h \to 0} \frac{F(z+h)-F(z)}{h} = f(z)$$

Above relation shows that $F(z)$ exists for all values of z in D and $F'(z) = f(z)$. Hence, $F(z)$ is analytic in D. Also we know that derivative of an analytic function is analytic therefore $F'(z)$, *i.e.*, $f(z)$ is also analytic in D.

11.16.11 Cauchys' Inequality

If $f(z)$ is analytic within and on a circle C given by $|z - a| = R$ and if $|f(z)| \leq M$ for every z on C then $\left| f^n(a) \right| \leq n!\dfrac{M}{R^n}$.

Proof. We know that $$f^n(a) = \frac{n!}{2\pi i}\int_C \frac{f(z)dz}{(z-a)^{n+1}}$$

$$\therefore \qquad \left| f^n(a) \right| = \frac{n!}{2\pi i}\int_C \frac{f(z)dz}{(z-a)^{n+1}} \leq \frac{n!}{|2\pi i|}\int_C \frac{|f(z)||dz|}{(z-a)^{n+1}}$$

The equation of the circle $|z - a| = R$ may be written as $z - a = Re^{i\theta}$ so that $dz = Re^{i\theta} i\, d\theta$.

$$\therefore \qquad \left| f^n(a) \right| \leq \frac{n!}{2\pi i}\int_C \frac{M}{R^{n+1}}\left|Re^{i\theta}\right|id\theta = \frac{n!}{2\pi}\frac{M}{R^{n+1}}\int_0^{2\pi}Rd\theta = n!\frac{M}{R^n}$$

$$\therefore \qquad \left| f^n(a) \right| \leq n!\frac{M}{R^n}.$$

11.16.12 Liouville's Theorem

Definition. *A function which is analytic in every finite region of the z plane is called an Integral function or Entire function.*

THEOREM 1. *If a function $f(z)$ is analytic for all finite values of z and is bounded then $f(z)$ is constant.*

PROOF. Let z_1, z_2 be any two points of the z-plane. Choose the contour C to be a large circle of radius R centered at origin and containing the point z_1 and z_2. Therefore, $|z_1| < R$ and $|z_2| < R$.
Also as $f(z)$ is bounded therefore there exists a positive constant M such that $|f(z)| \leq M \; \forall \; z$.
By Cauchy's Integral formula, we have

$$f(z_1) = \frac{1}{2\pi i}\int_C \frac{f(z)dz}{z-z_1} \quad \text{and} \quad f(z_2) = \frac{1}{2\pi i}\int_C \frac{f(z)dz}{z-z_2}$$

$$\therefore \qquad f(z_1) - f(z_2) = \frac{1}{2\pi i}\int_C \frac{f(z)dz}{z-z_1} - \frac{1}{2\pi i}\int_C \frac{f(z)dz}{z-z_2} = \frac{1}{2\pi i}\int_C \frac{z_1-z_2}{(z-z_1)(z-z_2)}f(z)dz$$

$$\therefore \qquad |f(z_1) - f(z_2)| = \left| \frac{1}{2\pi i}\int_C \frac{(z_1-z_2)f(z)dz}{(z-z_1)(z-z_2)} \right| \leq \left| \frac{1}{2\pi i} \right|\int_C \frac{|z_1-z_2||f(z)||dz|}{|z-z_1||z-z_2|}$$

$$\leq \frac{1}{2\pi}|z_1-z_2|M\int_C \frac{|dz|}{(|z|-|z_1|)(|z|-|z_2|)} \quad \text{as } |f(z)| < M$$

$$= \frac{1}{2\pi} \frac{|z_1 - z_2| M}{(R - |z_1|)(R - |z_2|)} \int_C |dz| \qquad\qquad \because |z| = R.$$

Again $|z| = R \Rightarrow z = Re^{i\theta}$ or $dz = Re^{i\theta} i \, d\theta$

$$\therefore \quad |dz| = Rd\theta = \frac{1}{2\pi} \frac{|z_1 - z_2| M.R}{R^2 \left(1 - \frac{|z_1|}{R}\right)\left(1 - \frac{|z_2|}{R}\right)} \int_0^{2\pi} d\theta = \frac{1}{2\pi} \frac{|z_1 - z_2| M.2\pi}{R \left(1 - \frac{|z_1|}{R}\right)\left(1 - \frac{|z_2|}{R}\right)} \to 0 \text{ as } R \to \infty.$$

$$\therefore \qquad f(z_1) - f(z_2) = 0 \qquad \text{or} \qquad f(z_1) = f(z_2)$$

As z_1 and z_2 are arbitrary it follows that $f(z)$ is constant.

11.16.13 Poisson's Integral Formula

If $f(z)$ be analytic in a region including the circle $|z| \le R$ for $0 < r < R$. Then

$$f(re^{i\theta}) = \frac{1}{2\pi} \int_0^{2\pi} \frac{R^2 - r^2}{R^2 - 2Rr\cos(\theta - \phi) + r^2} f(Re^{i\phi}) d\phi$$

where $a = re^{i\theta}$ is any point of the domain $|z| < R$.

Proof. We are given that $f(z)$ is analytic within and on the circle C for which $|z| = R$, therefore if a is a point within the circle then by Cauchy's integral formula.

$$f(a) = \frac{1}{2\pi i} \int_C \frac{f(z) dz}{z - a} \qquad\qquad\qquad ...(1)$$

Again we know that the inverse of the point a with respect to the circle $|z| = R$ is $\dfrac{R^2}{\bar{a}}$ and lies outside the circle. Hence the function $\dfrac{f(z)}{z - \dfrac{R^2}{\bar{a}}}$

is analytic within and on C. Hence by Cauchy's theorem $\dfrac{1}{2\pi i} \int_C \dfrac{f(z) dz}{z - \dfrac{R^2}{\bar{a}}} = 0$ ...(2)

From (1) and (2) by subtracting we get

$$f(a) - 0 = \frac{1}{2\pi i} \int_C \left[\frac{f(z)}{z - a} - \frac{f(z)}{z - \dfrac{R^2}{\bar{a}}} \right] dz = \frac{1}{2\pi i} \int_C \frac{\left(a - \dfrac{R^2}{\bar{a}} \right)}{(z - a)\left(z - \dfrac{R^2}{\bar{a}} \right)} f(z) dz$$

$$= \frac{1}{2\pi i} \int_C \frac{(a\bar{a} - R^2)}{(z - a)(z\bar{a} - R^2)} f(z) dz = \frac{1}{2\pi i} \int_C \frac{(R^2 - a\bar{a})}{(z - a)(R^2 - \bar{a}z)} f(z) dz \qquad ...(3)$$

Now in the result (3) put $z = Re^{\phi}$ as $|z| = R$ and $a = re^{i\theta}$ so that $\bar{a} = re^{-i\theta}$

$$\therefore \qquad R^2 - a\bar{a} = R^2 - re^{i\theta}re^{-i\theta} = R^2 - r^2 \qquad\qquad ...(a)$$

$$(z - a)(R^2 - z\bar{a}) = (Re^{i\phi} + re^{i\theta})(R^2 - re^{i\phi}re^{-i\theta}) = R(Re^{i\phi} - re^{i\theta})(R - re^{i(\phi - \theta)})$$

$$= R[R^2 e^{i\phi} - rR(e^{i(2i-\theta)}) + r^2 e^{i\theta}] = Re^{i\phi}[R^2 - rR(e^{-i(\phi-\theta)}e^{i(\phi-\theta)} + r^2]$$

$$= Re^{i\phi}[R^2 - rR(2\cos(\phi - \theta) + r^2] = Re^{i\phi}[R^2 - 2rR\cos(\theta - \phi) + r^2] \qquad ...(b)$$

$$z = Re^{i\phi} \qquad \therefore \quad dz = Re^{i\phi} i d\phi \qquad\qquad ...(c)$$

Hence from (3) by the help of results (a), (b) and (c), we get

$$f(r^2 e^{i\theta}) = \frac{1}{2\pi i} \int_0^{2\pi} \frac{(R^2 - r^2) f(Re^{i\phi}).Re^{i\phi}.id\phi}{Re^{i\phi}(R^2 - 2Rr\cos(\theta - \phi) + r^2)} = \frac{1}{2\pi} \int_0^{2\pi} \frac{(R^2 - r^2) f(Re^{i\phi})}{R^2 - 2Rr\cos(\theta - \phi) - r^2} d\phi \qquad ...(4)$$

☞ **REMARK**
- Differentiating result (4) w.r.t. θ, we get

$$f'(re^{i\theta}) = \frac{1}{2\pi} \int_0^{2\pi} \frac{R(R^2 - r^2)(f(Re^{i\phi}\sin(\phi - \theta))}{[R^2 - 2Rr\cos(\theta - \phi) + r^2]^2} d\theta \;.$$

Solved Examples

EXAMPLE 1. *Verify Cauchy's theorem for the function $z^3 - iz^2 - 5z + 2i$ if c is the circle $|z - 1| = 2$.*

SOLUTION. Given $f(z) = z^3 - iz^2 - 5z + 2i$, which is being a polynomial is analytic within C.
On C choose $z - 1 = 2e^{i\theta}$, $0 \le \theta \le 2\pi$
$\Rightarrow z = 1 + 2e^{i\theta} \Rightarrow dz = 2ie^{i\theta} d\theta$

$\therefore \int_C f(z) dz$

$= \int_0^{2\pi} [(1 + 2e^{i\theta})^3 - i(1 + 2e^{i\theta})^2 - 5(1 + 2e^{i\theta}) + 2i] 2ie^{i\theta} d\theta$

$= 2i \int_0^{2\pi} [8e^{4i\theta} + 4(3 - i)e^{3i\theta} - 4(1 + i)e^{2i\theta} + (-4 + i)e^{i\theta}] d\theta$

$= 0 \qquad\qquad (\because \int_0^{2\pi} e^{im\theta} = \text{if } m \ne 0)$

Hence Cauchy's theorem is verified for the given function $f(z)$ and contour C.

EXAMPLE 2. *Evaluate* $\int_C \dfrac{1}{z(z-1)}dz$ *where C is the circle* $|z| = 3$.

SOLUTION. As per given, $|z| = 3$ is a circle with centre at origin and radius 3.

Now $z(z-1) = 0 \Rightarrow z = 0, 1$

Let $O = O(0, 0)$

$\quad A = (1, 0)$

Since, both points are within C, therefore,

we can write $f(z) = 1$ which is analytic everywhere

and $\dfrac{1}{z(z-1)} = \dfrac{A}{z} + \dfrac{B}{z-1}$.

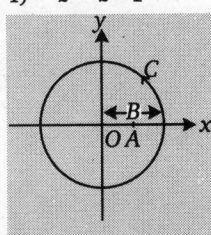

Fig. 24

Using the method of partial fractions, we can write

$$\frac{1}{z(z-1)} = \frac{1}{z-1} - \frac{1}{z}$$

$$\therefore \int_C \frac{1}{z(z-1)}dz = \int_C \frac{f(z)}{z(z-1)}dz$$

$$= \int_C \frac{f(z)}{z-1}dz - \int_C \frac{f(z)}{z}dz \qquad \ldots(1)$$

Now, by Cauchy's integral formula, we have

$$\int \frac{f(z)}{z-1}dz = 2\pi i f(1)$$

and $\int_C \dfrac{f(z)}{z}dz = \int_C \dfrac{f(z)}{z-0}dz = 2\pi i f(0) = 2\pi i$

$$(\because f(0) = 1)$$

Hence, from (1), we have

$$\int_C \frac{1}{z(z-1)}dz = 2\pi i - 2\pi i = 0.$$

EXAMPLE 3. *Evaluate* $\int_C \dfrac{e^z}{z-2}dz$ *where C is the circle*

$\quad$ **(i)** $|z| = 3$ $\qquad\qquad$ **(ii)** $|z| = 1$

SOLUTION. (i) Let $f(z) = e^z$

Clearly, e^z is analytic in the z-plane and point $a = 2$ lies inside the circle $|z| = 3$. Then by Cauchy's integral formula

We get $\int_C \dfrac{e^z}{z-2}dz = 2\pi i f(2) = e^2$

(ii) Let $f(z) = \dfrac{e^z}{(z-2)}$

Since $\dfrac{e^z}{(z-2)}$ is analytic within and on the circle $|z| = 1$ as $|z| = 2$ lies outside $|z| = 1$. Then by Cauchy's integral formula, we have

$$\int_C \frac{e^z}{z-2}dz = 0.$$

EXAMPLE 4. *Using Cauchy's integral formula, evaluate*

$$\int_C \frac{dz}{z(z+\pi i)} \text{ where } C \text{ is } |z+3i| = 1$$

SOLUTION. We have $\dfrac{1}{z(z+\pi i)} = \dfrac{1}{\pi i}\left(\dfrac{1}{z} - \dfrac{1}{z+\pi i}\right)$

Now, clearly the points $z = 0$, $z = -\pi i$ lies inside C.

Let $f(z) = 1 \Rightarrow f(0) = 1, f(-\pi i) = 1$

$$\therefore \int_C \frac{dz}{z(z+\pi i)} = \frac{1}{\pi i}\int_C\left(\frac{1}{z} - \frac{1}{z+\pi i}\right)dz = \frac{1}{\pi i}\int_C\frac{dz}{z} - \frac{1}{\pi i}\int_C\frac{dz}{z+\pi i}$$

$$= \frac{1}{\pi i}\int_C\frac{f(z)}{z}dz - \frac{1}{\pi i}\int_C\frac{f(z)}{z+\pi i}dz$$

$$= 2[f(0) - f(-\pi i)] = 2(1-1) = 0$$

EXAMPLE 5. *Evaluate* $\int_C \dfrac{e^{2z}}{(z+1)^4}dz$ *where the path of integration C is* $|z| = 3$. $\qquad$ (UPTU–2008)

SOLUTION. Let $f(z) = e^{2z}$ which is analytic within and on C by Cauchy's Integral formula, we have

$$f^n(a) = \frac{n!}{2\pi i}\int_C\frac{f(z)}{(z-a)^{n+1}}dz$$

$$\Rightarrow \quad f^3(-1) = \frac{3!}{2\pi i}\int_C\frac{e^{2z}}{(z+1)^{3+1}}dz$$

$$\Rightarrow \quad \int_C\frac{e^{2z}}{(z+1)^4}dz = \frac{2\pi i}{6}f^3(-1) = \frac{\pi i}{3}[f^3(z)]_{z=-1}$$

$$= \frac{\pi i}{3}[2^3.e^{2z}]_{z=-1} = \frac{\pi i}{3}8e^{-2} = \frac{8\pi i}{3e^2}.$$

11.16.14 Taylor's Theorem

STATEMENT. *If a function $f(z)$ is analytic at all the points inside a circle C, with its centre at the point a and radius R, then for each point z inside C, we have*

$$f(z) = \sum_{n=0}^{\infty} a_n(z-a)^n \text{ where } a_n = \frac{f^n(a)}{n!}, \text{ i.e., } \quad f(z) = \sum_{n=0}^{\infty}\frac{f^n(a)(z-a)^n}{n!}.$$

PROOF. Let $f(z)$ be analytic within and on the circle C of radius R centered at a so that its equation is

$$|z-a| = R.$$

Draw another circle C_1 concentric with C and of radius r enclosing z as shown in figure 25.

$\therefore \qquad\qquad |z-a| = r$ where $r < R$

Hence by Cauchy's integral formula, we have

$$f(z) = \frac{1}{2\pi i}\int_{C_1}\frac{f(t)dt}{(t-z)} \qquad\qquad \ldots(1)$$

Now $\qquad t - z = (t-a) - (z-a) = (t-a)\left[1 - \dfrac{z-a}{t-a}\right]$

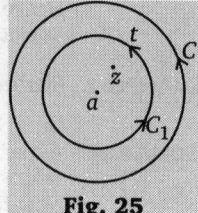

Fig. 25

$$\therefore \quad \frac{1}{t-z} = \frac{1}{t-a}\left[1 - \frac{z-a}{t-a}\right]^{-1}$$

Now by actual division we know that

$$\frac{1}{1-x} = 1 + x + x^2 + \dots + x^{n-1} + \frac{x^n}{1-x}$$

$$\therefore \quad \frac{1}{t-z} = \frac{1}{(t-a)}\left[1 + \frac{z-a}{t-a} + \left(\frac{z-a}{t-a}\right)^2 + \dots + \left(\frac{z-a}{t-a}\right)^{n-1} + \frac{(z-a)^n}{(t-a)^n} \cdot \frac{1}{1-(z-a)/(t-a)}\right]$$

or $$\frac{1}{t-z} = \frac{1}{t-a} + \frac{(z-a)}{(t-a)^2} + \frac{(z-a)^2}{(t-a)^3} + \dots + \frac{(z-a)^{n-1}}{(t-a)^n} + \frac{(z-a)^n}{(t-a)^n} \cdot \frac{1}{t-z}$$

Multiplying both sides by $f(t)$ and writting in sigma notation.

$$\frac{f(t)}{t-z} = \sum_{r=0}^{n-1} \frac{(z-a)^r f(t)}{(t-a)^{r+1}} + \left(\frac{z-a}{t-a}\right)^n \frac{f(t)}{t-z} \qquad \dots(2)$$

Now the general term of the series on the R.H.S. of (2) is

$$\frac{(z-a)^{n-1}}{(t-a)^n} f(t)$$

$$\therefore \quad \left|\frac{(z-a)^{n-1}}{(t-a)^n} f(t)\right| = \frac{|z-a|^{n-1}}{|(t-a)|^n}|f(t)| \leq M \frac{r_1^{n+1}}{r^n} = \frac{M}{r}\left(\frac{r_1}{r}\right)^{n-1} \qquad \dots(3)$$

where M is the maximum value of $f(t)$ on C_1 and $|z-a| = r_1$ say $< r$ as clear from the figure given above. Also $\frac{r_1}{r} < 1$.

Now the series of positive constants $\sum \frac{M}{r}\left(\frac{r_1}{r}\right)^{n-1}$ is convergent as $\frac{r_1}{r} < 1$.

Therefore by Weierstrass M-test the series on R.H.S. of (2) can be integrated term by term. Also we know that

$$f^n(a) = \frac{n!}{2\pi i}\int_{C_1}\frac{f(t)dt}{(t-a)^{n+1}} \qquad \dots(4)$$

$$\therefore \quad f^n(a) = \frac{1}{2\pi i}\int_{C_1}\frac{f(t)dt}{(t-a)^{n+1}} \text{ when } n = 0, 1, 2, 3, \dots$$

Now integrating (2) term by term and making use of the above results, we get

$$\int_{C_1}\frac{f(t)dt}{t-z} = \sum_{r=1}^{n-1}\int_{C_1}\frac{(z-a)^r f(t)}{(t-r)^{r+1}}dt + (z-a)\int_{C_1}\frac{f(t)dt}{(t-a)^n(t-z)}$$

Making use of results (1) and (4) in the above we get

$$2\pi i f(z) = \sum_{r=1}^{n-1}(z-a)2\pi i\frac{f^r(a)}{r!} + (z-a)^n\int_{C_1}\frac{f(t)dt}{(t-a)^n(t-z)}$$

$$\therefore \quad f(z) = \sum_{r=0}^{n-1}(z-a)^r\frac{f^r(a)}{r!} + \frac{(z-a)^n}{2\pi i}\int_{C_1}\frac{f(t)dt}{(t-a)^n(t-z)}$$

or $$f(z) = \sum_{r=0}^{n-1}\frac{(z-a)^r f^r(a)}{r!} + R_n \qquad \dots(5)$$

where $$R_n = \frac{(z-a)^n}{2\pi i}\int_{C_1}\frac{f(t)dt}{(t-a)^n(t-z)}$$

$$|R_n| = \frac{|z-a|^n}{|2\pi i|}\int_{C_1}\frac{f(t)dt}{(t-a)^n(t-z)} \leq \frac{r_1^n}{2\pi}\int_{C_1}\frac{|f(t)||dt|}{|t-a|^n|t-z|} \leq \frac{r_1^n}{2\pi}\frac{M}{r^n}\int_{C_1}\frac{|dt|}{|t-z|}$$

$$= \frac{M}{2\pi}\left(\frac{r_1}{r}\right)^n\int_{C_1}\frac{|dt|}{|(t-a)-(z-a)|} \leq \frac{M}{2\pi}\left(\frac{r_1}{r}\right)^n\int_{C_1}\frac{|dt|}{|t-a|-|z-a|}$$

$$= \frac{M}{2\pi}\left(\frac{r_1}{r}\right)^n\int_{C_1}\frac{|dt|}{(r-|z-a|)} = \frac{M}{2\pi}\left(\frac{r_1}{r}\right)^n\frac{1}{(r-|z-a|)}2\pi r$$

Now as $\frac{r_1}{r} < 1$ $\quad \therefore \lim_{n\to\infty}\left(\frac{r_1}{r}\right)^n = 0$

$$\therefore \quad \lim_{n\to\infty}|R_n| = 0 \qquad \Rightarrow \qquad R_n \to 0$$

Hence from (5) we get $\quad f(z) = \sum\limits_{r=0}^{\infty} \dfrac{(z-a)^r}{r!} f^r(a) = \sum\limits_{n=0}^{\infty} \dfrac{f^n(a)}{n!}(z-a)^n$ ...(6)

$\therefore \qquad\qquad f(z) = f(a) + \dfrac{(z-a)}{1!}f'(a) + \dfrac{(z-a)^2}{2!}f''(a) + \dfrac{(z-a)^3}{3!}f'''(a) + ...$

✎ REMARKS

- If the centre of the circle C is chosen at the origin then writing $a = 0$, we get

$$f(z) = \dfrac{f^n(0)}{n!}z^n \qquad \text{or} \qquad f(z) = f(0) + \dfrac{z}{1!}f'(0) + \dfrac{z^2}{2!}f''(0) + \dfrac{z^3}{3!}f'''(0) + ...$$

- Above is known as Maclaurin's Series.
- Again z is a point within C_1 for which $|t - a| = r$ so that we can choose $z = a + h$ or $z - a = h$ and hence from (5) we get

$$f(a+h) = f(a) + \dfrac{h}{1!}f'(a) + \dfrac{h^2}{2!}f''(a) + \dfrac{h^3}{3!}f'''(a) + ...$$

This is another form of Taylor's Series.

11.16.15 LAURENT'S THEOREM

If function $f(z)$ is analytic in the annulus (ring shaped region) between two concentric circles C and C' with centre at the point $z = a$ and radii R and R' then at any point z within the annulus.

$$f(z) = \sum_{n=0}^{\infty} a_n(z-a)^n + \sum_{n=1}^{\infty} b_n(z-a)^{-n}$$

where $\qquad\qquad a_n = \dfrac{1}{2\pi i}\int_C \dfrac{f(t)dt}{(t-a)^{n+1}}, b_n = \dfrac{1}{2\pi i}\int_{C'}\dfrac{f(t)dt}{(t-a)^{-n+1}}.$

Proof. Choosing z to be any point within the annulus region we have by Cauchy integral formula for multiply connected region, such that

$$f(z) = \dfrac{1}{2\pi i}\int_C \dfrac{f(t)dt}{t-z} - \dfrac{1}{2\pi i}\int_{C'}\dfrac{f(t)dt}{t-z} \qquad ...(1)$$

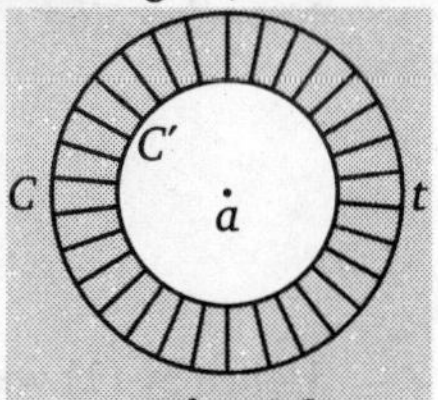

Equation of the circle C is $|t - a| = R$

and that of C' is $|t - a| = R'$ and let $|z - a| = r$.

Fig. 26

$$\therefore \qquad \left.\begin{array}{l}\left|\dfrac{z-a}{t-a}\right| = \dfrac{r}{R} < 1 \text{ if } t \text{ lies on } C \\[3mm] \left|\dfrac{t-a}{z-a}\right| = \dfrac{R'}{r} < 1 \text{ if } t \text{ lies on } C'\end{array}\right\} \qquad ...(2)$$

and

i.e., we have intercahanged t and z.

Also let M and M' be the absolute values of $|f(t)|$ on C and C' respectively.

Now for all t on C we have a in Taylor's series such that

$$t - z = (t - a) - (z - a) = (t - a)\left[1 - \dfrac{z-a}{t-a}\right]$$

$$\therefore \qquad \dfrac{1}{t-z} = \dfrac{1}{t-a}\left[1 - \dfrac{z-a}{t-a}\right]^{-1}$$

Now by actual division, we know that

$$\dfrac{1}{1-x} = 1 + x + x^2 + ... + x^{n-1} + \dfrac{x^n}{1-x}$$

$$\therefore \quad \dfrac{1}{t-z} = \dfrac{1}{(t-a)}\left[1 + \dfrac{z-a}{t-a} + \left(\dfrac{z-a}{t-a}\right)^2 + ... + \left(\dfrac{z-a}{t-a}\right)^{n-1} + \dfrac{(z-a)^n}{(t-a)^n}\cdot\dfrac{1}{1-(z-a)/(t-a)}\right]$$

or $\quad \dfrac{1}{t-z} = \dfrac{1}{t-a} + \dfrac{(z-a)}{(t-a)^2} + \dfrac{(z-a)^2}{(t-a)^3} + ... + \dfrac{(z-a)^{n-1}}{(t-a)^n} + \dfrac{(z-a)^n}{(t-a)^n}\cdot\dfrac{1}{t-z}$ ...(3)

Upto this stage the procedure is exactly the same as of the last article of Taylor's Theorem

Above result holds good for all t on C. In order t get the corresponding result for all t on C' we have to interchange t and z in (3).

$$\therefore \quad \dfrac{1}{z-t} = \dfrac{1}{(z-a)} + \dfrac{(t-a)}{(z-a)^2} + \dfrac{(t-a)^2}{(z-a)^3} + ... + \dfrac{(t-a)^{n-1}}{(z-a)^n} + \dfrac{(t-a)^n}{(z-a)^n}\cdot\dfrac{1}{z-t} \qquad ...(4)$$

Multiplying both sides of (3) and (4) by $f(t)$ and writting $z - t$ as $-(t - z)$ in the L.H.S. of (4) we have in sigma notation.

$$\dfrac{f(t)}{t-z} = \sum_{r=0}^{n-1}\dfrac{(z-a)^r f(t)}{(t-a)^{r+1}} + \left(\dfrac{z-a}{t-a}\right)^n \dfrac{f(t)}{t-z} \quad \forall t \text{ on } C \qquad ...(5)$$

$$\frac{f(t)}{t-z} = \sum_{r=0}^{n-1} \frac{(t-a)^r f(t)}{(z-a)^{r+1}} + \left(\frac{t-a}{z-a}\right)^n \frac{f(t)}{z-t} \quad \forall\, t \text{ on } C' \tag{6}$$

Now the general term of the series on the R.H.S. of (5) is $\dfrac{(z-a)^{n-1}}{(t-a)^n} f(t)$

$$\therefore \qquad \left|\frac{(z-a)^{n-1}}{(t-a)^n} f(t)\right| = \frac{|z-a|^{n-1}}{|(t-a)|^n} |f(t)| \le M \frac{r^{n+1}}{r^n} = \frac{M}{r}\left(\frac{r}{R}\right)^{n-1} \text{ by (2)}$$

Similarly the general term of the series on the R.H.S. of (6) is $\dfrac{(t-a)^{n-1}}{(z-a)^n} f(t)$.

$$\therefore \qquad \left|\frac{(t-a)^{n-1}}{(z-a)^n} f(t)\right| = \frac{|t-a|^{n-1}|f(t)|}{|z-a|^n} \le M' \frac{R'^{n-1}}{r^n} = \frac{M'}{r}\left(\frac{R'}{r}\right)^{n-1} \text{ where both } \frac{r}{R} \text{ and } \frac{R'}{r} \text{ are less than unity.}$$

Thus the series of positive constants

$$\sum \frac{M}{r}\left(\frac{r}{R}\right)^{n-1} \qquad \text{and} \qquad \sum \frac{M'}{r}\left(\frac{R'}{r}\right)^{n-1}$$

are both convergent. Hence by Weierstrass M-test the series in the R.H.S. of both (5) and (6) can be integrated term by term.

$$\therefore \qquad \frac{1}{2\pi i}\int_C \frac{f(t)dt}{t-z} \quad \frac{1}{2\pi i}\int_{C'} \frac{f(t)dt}{t-z} = \sum_{r=0}^{n-1} \frac{(z-a)^r}{2\pi i}\int_C \frac{f(t)dt}{(t-a)^{r+1}} + \frac{(z-a)^n}{2\pi i}\int_C \frac{f(t)dt}{(t-a)^n(t-z)}$$

$$+ \sum_{r=0}^{n-1} \frac{1}{2\pi i(z-a)^{r+1}}\int_{C'}(t-a)^r f(t)dt + \frac{1}{2\pi i(z-a)^n}\int_{C'}\frac{(t-a)^n f(t)dt}{z-t}$$

Now choosing

$$a_r = \frac{1}{2\pi i}\int_C \frac{f(t)dt}{(t-a)^{r+1}}, b_r = \frac{1}{2\pi i}\int_{C'}(t-a)^{r-1}f(t)dt \ , b_{r+1} = \frac{1}{2\pi i}\int_{C'}(t-a)^r f(t)dt$$

$$U_n = \frac{1}{2\pi i}\int_C \frac{(z-a)^n}{(t-a)^n}\frac{f(t)dt}{t-z}, V_n = \frac{1}{2\pi i}\int_C \left(\frac{t-a}{z-a}\right)^n \frac{f(t)dt}{z-t}$$

Hence by (1) we get from above

$$f(z) = \sum_{r=0}^{n-1} a_r(z-a)^r + \sum_{r=1}^{n} b_r(z-n)^{-r} + U_n + V_n \tag{7}$$

$$|U_n| = \left|\frac{1}{2\pi i}\int_C \frac{(z-a)^n}{(t-a)^n}\frac{f(t)dt}{t-z}\right| \le \frac{1}{2\pi i}\int_C \left(\frac{r}{R}\right)^n M \frac{|dt|}{R-r} = \left(\frac{r}{R}\right)^n \cdot \frac{M}{R-r} \cdot \frac{2\pi R}{2\pi} \text{ by (2)}$$

$\because \int_C |dt| = \text{Arc of } C - 2\pi R$

and

$$|t-z| = |(t-a)-(z-a)| \ge |t-a| - |z-a| = R-r$$

$$\therefore \qquad |U_n| = \frac{M}{1-\dfrac{r}{R}}\left(\frac{r}{R}\right)^n \to 0 \text{ when } n \to \infty \text{ as } \frac{r}{R'} < 1$$

Similarly $|V_n| \to 0$ when $n \to \infty$ as $\dfrac{R'}{r} < 1$

$$\therefore \qquad \lim_{n\to\infty} U_n = 0, \lim_{n\to\infty} V_n = 0$$

Hence tending n to infinity in (7), we have

$$f(z) = \sum_{r=0}^{\infty} a_r(z-a)^r + \sum_{r=1}^{\infty} b_r(z-a)^{-r}$$

or

$$f(z) = \sum_{n=0}^{\infty} a_n(z-a)^n + \sum_{n=1}^{\infty} b_n(z-a)^{-n} \tag{8}$$

where a_n and b_n have values as given.
Since the integral is analytic in the annulus.

$$R' < |z-a| < R,$$

therefore, if C_0 is any circle of radius R_0 with centre at a such that $R' < R_0 < R$ then we can write

$$a_n = \frac{1}{2\pi i}\int_{C_0}\frac{f(t)dt}{(t-a)^{n+1}} \qquad \text{and} \qquad b_n = a_{-n} = \frac{1}{2\pi i}\int_{C_0}\frac{f(t)dt}{(t-a)^{-n+1}}$$

The series (8) can now be expressed as

$$f(z) = \sum_{n=-\infty}^{\infty} a_n (z-a)^n \text{ where } a_n = \frac{1}{2\pi i} \int_{C_1} \frac{f(t)dt}{(t-a)^{n+1}}$$

REMARK

- Suppose that we have obtained in any manner or as the definition of the formula.

$$f(z) = \sum_{n=-\infty}^{\infty} A_n (z-a)^n, R' < |z-a| < R$$

Then the series necessarily identical with the Laurent's series for $f(z)$.

11.16.16 CERTAIN IMPORTANT RESULTS

The following results may be used directly in expansion of certain functions by Taylor's or Laurent's series.

(a) We have already shown both in Taylor's as well as Laurent's series that

$$\frac{1}{1-x} = 1 + x + x^2 + \dots + x^n + \frac{x^{n+1}}{1-x}$$

$$\therefore \quad \frac{1}{z+a} = \frac{1}{a}\left[1 - \frac{z}{a} + \frac{z^2}{a^2} + \dots + \frac{(-1)^n z^n}{a^n} + \frac{(-1)^{n+1} z^{n+1}}{a^{n+1}} \cdot \frac{1}{z+a}\right]$$

$$\frac{1}{z-a} = \frac{1}{a}\left[1 + \frac{z}{a} + \frac{z^2}{a^2} + \dots + \frac{z^n}{a^n} + \frac{z^{n+1}}{a^{n+1}} \cdot \frac{1}{z-a}\right]$$

(b) If C be the circle of radius r centred at C, i.e., $|z-a| = r$ then $\int_C \frac{dz}{z-a} = 2\pi i$

and $\int_C \frac{dz}{(z-a)^k} = 0$ if $k \neq 1$ as shown below.

Since $|z-a| = r$ $\quad \therefore z-a = re^{i\theta}$ $\quad \therefore dz = re^{i\theta} id\theta$

$$\therefore \quad I = \int_C \frac{re^{i\theta} \cdot id\theta}{r^k e^{ik\theta}} = \frac{1}{r^{k-1}} \int_0^{2\pi} e^{i(1-k)\theta} id\theta = \frac{1}{r^{k-1}}\left[\frac{e^{i(1-k)\theta}}{1-k}\right]_0^{2\pi} = 0 \text{ as } k \neq 1$$

and $\left[e^{im\theta}\right]_0^{2\pi} = e^{im2\pi} - e^0 = (\cos 2m\pi + i\sin 2m\pi) - 1 = 1 - 1 = 0$

In particular for the circle $|z| = r$ we have

$$\int_C \frac{dz}{z} = 2i \text{ and } \int_C \frac{dz}{z^k} = 0 \text{ if } k \neq 1$$

(c) $\int_C \frac{1}{z^{n+1}} \cdot \frac{1}{z+a} dz = \frac{(-1)^n}{a^{n+1}}\left[2\pi i + \int_C \frac{dz}{z+a}\right]$ for $n \neq 0$.

and in particular when $n = 0$ then

$$\int_C \frac{1}{z} \cdot \frac{1}{z+a} dz = \frac{2\pi i}{a}$$

$$I = \int_C \frac{1}{z^{n+1}} \cdot \frac{1}{z+a} dz = \int_C \frac{1}{z^{n+1}} \cdot \frac{1}{a}\left[1 - \frac{z}{a} + \frac{z^2}{a^2} + \dots + \frac{(-1)^n z^n}{a^n} + \frac{(-1)^{n+1} z^{n+1}}{a^n} \cdot \frac{1}{z-a}\right] dz \text{ by (a)}$$

Now by (b), we know that

$\int_C \frac{dz}{z^k} = 0$ if $k \neq 1$, $= 2\pi i$ when $k = 1$

$$\therefore \quad I = \frac{(-1)^n}{a^{n+1}} \int_C \frac{dz}{z} + \frac{(-1)^{n+1}}{a^{n+1}} \int_C \frac{dz}{z+a} = \frac{(-1)^n}{a^{n+1}}\left[2\pi i - \int_C \frac{dz}{z+a}\right] \qquad \because k = 1 \text{ for the first integral}$$

Particular case: When $n = 0$

$$I = \int_C \frac{1}{z} + \frac{1}{z+a} dz = \int_C \frac{1}{z} \cdot \frac{1}{a}\left[1 - \frac{z}{a} + \frac{z^2}{a^2} - \frac{z^3}{a^3} + \dots\right] dz$$

$$= \frac{1}{a} \int_C \frac{1}{z} dz + \text{ all other integrals vanish by (b) as } k \neq 1 = \frac{1}{a} 2\pi i \text{ by (b)}$$

(d) $\int_C \frac{1}{z^{n+1}} \cdot \frac{1}{z-a} dz = -\frac{1}{a^{n+1}}\left[2\pi i + \int_C \frac{dz}{z-a}\right]$ for $n \neq 0$

and in particular when $n = 0$, then

$$\int_C \frac{1}{z} \cdot \frac{1}{z-a} dz = -\frac{2\pi i}{a}$$

$$I = \int_C \frac{1}{z^{n+1}} \cdot \frac{1}{z-a} dz = \int_C -\frac{1}{z^{n+1}} \frac{1}{a} \left[1 + \frac{z}{a} + \frac{z^2}{a^2} + \ldots + \frac{z^n}{a^n} + \frac{z^{n+1}}{a^n} \cdot \frac{1}{z-a} \right] dz \text{ by (a)}$$

Now by (b) we know that $\int_C \frac{dz}{z^k} = 0$ if $k \neq 1$; $= 2\pi i$ when $k = 1$

$$\therefore \qquad I = -\frac{1}{a^{n+1}} \int_C \frac{dz}{z} - \frac{1}{a^{n+1}} \int_C \frac{dz}{z-a} = -\frac{1}{a^{n+1}} \left[2\pi i + \int_C \frac{dz}{z-a} \right]$$

Particular case : When $n = 0$

$$I = \int_C \frac{1}{z} \cdot \frac{1}{z-a} dz = -\frac{1}{a} \int_C \frac{1}{z} \left(1 + \frac{z}{a} + \frac{z^2}{a^2} + \ldots \right) dz = -\frac{1}{a} \int_C \frac{dz}{z} + \text{all other integrals vanish by (b) as } k \neq 1$$

$$= -\frac{1}{a} 2\pi i \text{ by (b)}$$

(e) $\qquad \int_C z^{n-1} \frac{1}{z+a} dz = 0, \int_C z^{n-1} \frac{1}{z-a} dz = 0$

except when $n = 0$

(i) $\quad I = \int_C z^{n-1} \frac{1}{a} \left[1 - \frac{z}{a} + \frac{z^2}{a^2} + \ldots \right] dz = 0$ except when $n = 0$

(ii) $\quad I = \frac{-1}{a} \int_C z^{n-1} \left[1 + \frac{z}{a} + \frac{z^2}{a^2} + \ldots \right] dz = 0$ except when $n = 0$

these reduce to particular cases of (d).

Solved Examples

EXAMPLE 1. *If C is a closed contour around the origin, prove that*

$$\left(\frac{a^n}{n!} \right)^2 = \frac{1}{2\pi i} \int_C \frac{a^n e^{az} dz}{n! \cdot z^{n+1}}$$

Hence deduce that $\displaystyle\sum_{n=0}^{\infty} \left(\frac{a^n}{n!} \right)^2 = \frac{1}{2\pi} \int_0^{2\pi} e^{2a\cos\theta} d\theta$

SOLUTION. Let $f(z) = e^{az}$ $\therefore f^n(z) = a^n e^{az}$
Putting $z = 0$ in the above, we get

$$a^n = f^n(0) = \frac{1}{2\pi i} n! \int_C \frac{f(z) dz}{z^{n+1}}, \text{ for } a = 0$$

Multiplying both sides by $\dfrac{a^n}{(n!)^2}$ we get

$$\left(\frac{a^n}{n!} \right)^2 = \frac{1}{2\pi i} \int_C \frac{a^n e^{az} dz}{n! z^{n+1}} \qquad \therefore f(z) = e^{az}$$

Above proves first part of the question

$$\therefore \sum_{n=0}^{\infty} \left(\frac{a^n}{n!} \right)^2 = \sum_{n=0}^{\infty} \frac{1}{2\pi i} \cdot \frac{1}{n!} \int_C \frac{a^n e^{az}}{z^{n+1}} dz$$

$$= \frac{1}{2\pi i} \int_C \sum_{n=0}^{\infty} \frac{a^n e^{az}}{n! \cdot z^{n+1}} dz = \frac{1}{2\pi i} \int_C e^{az} \left[\sum_{0}^{\infty} \left(\frac{a}{z} \right)^2 \cdot \frac{1}{n!} \right] \frac{dz}{z}$$

$$= \frac{1}{2\pi i} \int_C e^{az} \cdot e^{a/z} \cdot \frac{dz}{z} \qquad \text{by exponential}$$

$$= \frac{1}{2\pi i} \int_C e^{a(z+1/z)} \frac{dz}{z}$$

Now choose C to be circle of unit radius centered at origin, *i.e.*, $|z| = 1$.
So that $z = ie^{i\theta}$ and $dz = ie^{i\theta}$.
Also $z + 1/z = e^{i\theta} + e^{-i\theta} = 2 \cos \theta$

$$\therefore \sum_{n=0}^{\infty} \left(\frac{a^n}{n!} \right)^2 = \frac{1}{2\pi} \int_0^{2\pi} e^{2a\cos\theta} \cdot \frac{i.e^{i\theta} d\theta}{e^{i\theta}} = \frac{1}{2\pi} \int_0^{2\pi} e^{2a\cos\theta} d\theta .$$

EXAMPLE 2. *Prove that* $\cosh(z + 1/z) = a_0 + \displaystyle\sum_{n=1}^{\infty} a_n(z^n + 1/z^n)$

where $a_n = \dfrac{1}{2\pi} \int_0^{2\pi} \cos n\theta \cosh(2\cos\theta) d\theta$

SOLUTION. The given function $f(z) = \cosh(z + z^{-1})$ is analytic every where in the finite part of z-plane except at $z = 0$. Therefore $f(z)$ is analytic in the annulus $r \leq |z| \leq R$ where r is small and R is large.
Hence as before $f(z)$ can be expanded by Laurent's series.

$$\therefore f(z) = \cosh(z + z^{-1}) = \sum_{n=0}^{\infty} a_n z^n + \sum_{n=0}^{\infty} b_n z^{-n} \quad \ldots(1)$$

where $a_n = \dfrac{1}{2\pi i} \int_C \cosh(z + z^{-1}) \dfrac{dz}{z^{n+1}}$

$$b_n = \frac{1}{2\pi i} \int_C \cosh(z + z^{-1}) z^{n-1} dz .$$

Now C being any circle centred at origin, we choose it to be of radius unity so that $|z| = 1$ or $z = e^{i\theta}$, $z^{-1} = e^{-i\theta}$, $z + z^{-1} = 2 \cos \theta$.

$$\therefore a_n = \frac{1}{2\pi i} \int_0^{2\pi} \cosh(2\cos\theta) \frac{e^{i\theta} . id\theta}{e^{i(n+1)\theta}}$$

$$= \frac{1}{2\pi i} \int_0^{2\pi} \cosh(2\cos\theta) e^{-in\theta} d\theta$$

$$= \frac{1}{2\pi i} \int_0^{2\pi} \cosh(2\cos\theta)(\cos n\theta - i \sin n\theta) d\theta$$

or $a_n = \dfrac{1}{2\pi} \int_0^{2\pi} \cosh(2\cos\theta) \cos n\theta d\theta + 0 \qquad \ldots(2)$

$\because \int_0^{2\pi} F(\theta) d\theta = 0$ if $F(2\pi - \theta) = -F(\theta)$

which is true for $F(\theta) = \cosh(2 \cos \theta) \sin n\theta$.

Hence a_n has the value as given.

From the value of a_n and b_n it is easy to observe that if we replace n by $-n$ the value of a_n, we get the value of b_n.

$$\therefore \; b_n = a_{-n} = \frac{1}{2\pi}\int_0^{2\pi}\cosh(2\cos\theta)\cos(-n\theta)d\theta \quad \text{[by (2)]}$$

$$= \frac{1}{2\pi}\int_0^{2\pi}\cosh(2\cos\theta)\cos n\theta\, d\theta = a_n$$

$$\therefore \; \cosh(z+1/z) = \sum_{n=0}^{\infty} a_n z^n + \sum_{n=1}^{\infty} a_n z^{-n} \qquad \because b_n = a_n$$

$$= a_0 + \sum_{n=1}^{\infty} a_n(z^n + z^{-n})$$

where a_n has the value as given in (2).

EXAMPLE 3. *Expand* $f(z) = \dfrac{1}{(z+1)(z+3)}$ *in a Laurent's series valid for the reasons.*

  (i) $|z| < 1$ **(ii)** $1 < |z| < 3$

  (iii) $|z| > 3$ **(iv)** $0 < |z+1| < 2$

SOLUTION. Given that $f(z) = \dfrac{1}{(z+1)(z+3)} = \dfrac{1}{2(z+1)} - \dfrac{1}{2(z+3)}$

              (By resolving into partial fractions)

(i) If $|z| < 1$. Then we have

$$f(z) = \frac{1}{2}(1+z)^{-1} - \frac{1}{6}\left(1+\frac{z}{3}\right)^{-1}$$

$$= \frac{1}{2}[1 - z + z^2 - z^3 + \ldots] - \frac{1}{6}\left[1 - \frac{z}{3} + \left(\frac{z}{3}\right)^2 - \left(\frac{z}{3}\right)^3 + \ldots\right]$$

$$= \left(\frac{1}{2} - \frac{1}{6}\right) - \left(\frac{1}{2} - \frac{1}{18}\right)z + \left(\frac{1}{2} - \frac{1}{54}\right)z^2 - \ldots$$

$$= \frac{1}{3} - \frac{4}{9}z + \frac{13}{27}z^2 - \ldots$$

(ii) If $1 < |z| < 3$ then $\dfrac{1}{|z|} < 1$ and $\dfrac{|z|}{3} < 1$

$$\therefore \; f(z) = \frac{1}{2z\left(1+\dfrac{1}{z}\right)} - \frac{1}{6\left(1+\dfrac{z}{3}\right)}$$

$$= \frac{1}{2z}\left(1+\frac{1}{z}\right)^{-1} - \frac{1}{6}\left(1+\frac{z}{3}\right)^{-1}$$

$$= \frac{1}{2z}\left(1 - \frac{1}{z} + \frac{1}{z^2} - \frac{1}{z^3} + \ldots\right) - \frac{1}{6}\left(1 - \frac{z}{3} + \frac{z^2}{9} - \frac{z^3}{27} + \ldots\right)$$

$$= \frac{1}{6} - \frac{1}{18}z + \frac{1}{54}z^2 - \frac{1}{162}z^3 + \ldots$$

Hence, the required Laurent's series is given by

$$f(z) = \ldots + \frac{1}{2z^3} - \frac{1}{2z^2} + \frac{1}{2z} + \frac{1}{6} - \frac{1}{18}z$$

$$+ \frac{1}{54}z^2 - \frac{1}{162}z^3 + \ldots$$

(iii) If $|z| > 3 \Rightarrow \left(\dfrac{3}{|z|}\right) < 1$

$$\therefore \; f(z) = \frac{1}{2z}\left(1+\frac{1}{z}\right)^{-1} - \frac{1}{2z}\left(1+\frac{3}{z}\right)^{-1}$$

$$= \frac{1}{2z}\left(1 - \frac{1}{z} + \frac{1}{z^2} - \frac{1}{z^3} + \ldots\right)$$

$$- \frac{1}{2z}\left(1 - \frac{3}{z} + \left(\frac{3}{z}\right)^2 - \left(\frac{3}{z}\right)^3 + \ldots\right)$$

$$= \frac{1}{z^2} - \frac{4}{z^3} + \frac{13}{z^4} - \frac{40}{z^5} + \ldots$$

(iv) If $0 < |z+1| < 2$

Let us write $z + 1 = u$

$$\Rightarrow \quad 0 < |u| < 2$$

$$\therefore \; f(z) = \frac{1}{(z+1)(z+3)} = \frac{1}{u(u+2)} = \frac{1}{2u}\left(1+\frac{u}{2}\right)^{-1}$$

$$= \frac{1}{2u}\left(1 - \frac{u}{2} + \left(\frac{u}{2}\right)^2 - \left(\frac{u}{2}\right)^3 + \ldots\right)$$

$$= \frac{1}{2u} - \frac{1}{4} + \frac{u}{8} - \frac{u^2}{16} + \ldots$$

$$= \frac{1}{2(z+1)} - \frac{1}{4} + \frac{z+1}{8} - \frac{(z+1)^2}{16} + \ldots$$

EXAMPLE 4. *Represent the function* $f(z) = \dfrac{4z+4}{z(z-3)(z+2)}$ *in Laurent's series.*

  (a) *within* $|z| = 1$.

  (b) *in the annulus region within* $|z| = 2$ *and* $|z| = 3$.

  (c) *Exterior to* $|z| = 3$.

SOLUTION. Splitting into partial fractions, we get

$$f(z) = \frac{1}{2z} - \frac{1}{2(z+2)} + \frac{1}{z-3} \qquad \ldots(1)$$

Clearly $f(z)$ is analytic everywhere except at $z = 0, -2$ and 3.

(a) $0 < |z| < 1$.

$$f(z) = -\frac{1}{2z} - \frac{1}{4}\left(1+\frac{z}{2}\right)^{-1} - \frac{1}{3}\left(1-\frac{z}{3}\right)^{-1}$$

$$= -\frac{1}{2z} - \frac{1}{4}\sum_0^{\infty}(-1)^n\left(\frac{z}{2}\right)^n - \frac{1}{3}\sum_0^{\infty}\left(\frac{z}{3}\right)^n$$

$$= -\frac{1}{2z} + \sum_0^{\infty}\left[(-1)^{n+1}\frac{1}{2^{n+1}} - \frac{1}{3^{n+1}}\right]z^n$$

Above being a series of positive and negative powers of z in Laurent's expansion in the annulus $0 < |z| < 1$.

(b) $2 < |z| < 3$

Here we write $f(z)$ given by (1) in a manner so that the binomial expansion is valid for

$$2 < |z| < 3, \text{ i.e., } \frac{2}{|z|} < 1 \text{ and } \frac{|z|}{3} < 1$$

$$\therefore \; f(z) = -\frac{1}{2z} - \frac{1}{2z}\left(1+\frac{2}{z}\right)^{-1} - \frac{1}{3}\left(1-\frac{3}{z}\right)^{-1}$$

$$= -\frac{1}{2z} - \frac{1}{2z}\sum_0^{\infty}(-1)^n\left(\frac{2}{z}\right)^n - \frac{1}{3}\sum_0^{\infty}\left(\frac{z}{3}\right)^n$$

Above being a series of positive and negative powers of z is Laurent's expansion of $f(z)$ in the annulus $2 < |z| < 3$.

(c) When $|z| > 3$

Here we write $f(z)$ given by (1) in a manner so that the binomial expansion is valid for

$$|z| > 3 \text{ or } \frac{1}{|z|} < \frac{1}{3} < 1 \text{ or } \frac{3}{|z|} < 1 \text{ and } \frac{1}{|z|} < 1$$

$$\therefore \quad f(z) = -\frac{1}{2z} - \frac{1}{2z}\left(1 + \frac{2}{z}\right)^{-1} - \frac{1}{3}\left(1 - \frac{3}{z}\right)^{-1}$$

$$= -\frac{1}{2z} - \frac{1}{2z}\sum_{0}^{\infty}(-1)^n\left(\frac{2}{z}\right)^n - \frac{1}{3}\sum_{0}^{\infty}\left(\frac{z}{3}\right)^n$$

Above being a series of positive and negative powers of z is Laurent's expansion in the annulus $3 < |z| < R$.

EXAMPLE 5. *Show that*

$$\log z = (z - 1) - \frac{(z-1)^2}{2} + \frac{(z-1)^3}{3} - \dots$$

When $|z - 1| < 1$

SOLUTION. Let $f(z) = \log z$

By Taylor's theorem, we have

$$f(z) = f(z_0) + (z - z_0)f'(z_0) + \frac{(z-z_0)^2}{2!}f''(z_0) + \dots \qquad \dots(1)$$

Let us take $z_0 = 1$ then we have

$$f(z) = \log(z) \quad \Rightarrow \quad f(1) = \log 1 = 0$$

$$f'(z) = \frac{1}{z} \quad \Rightarrow \quad f'(1) = 1$$

$$f''(z) = -\frac{1}{z^2} \quad \Rightarrow \quad f''(1) = -1$$

$$f'''(z) = \frac{2}{z^3} \quad \Rightarrow \quad f'''(1) = 2$$

... and so on.

Substituting all these values in (1), we get

$$\log z = f(1) + f'(1)(z - 1) + f''(1)\frac{(z-1)^2}{2!} + \dots$$

$$= (z - 1) - \frac{1}{2}(z - 1)^2 + \frac{1}{3}(z - 1)^3 + \dots$$

EXAMPLE 6. *Find the Taylor's or Laurent's series which represent the function* $\dfrac{1}{(1 + z^2)(z + 2)}$ *when*

(a) $|z| < 1$ (b) $1 < |z| < 2$

(c) $|z| > 2$

SOLUTION. $\quad f(z) = \dfrac{1}{(1 + z^2)(z + 2)} = \dfrac{1}{5}\left[\dfrac{1}{z+2} - \dfrac{z-2}{1+z^2}\right] \quad \dots(1)$

(a) $|z| < 1$

$$f(z) = \frac{1}{10}\left(1 + \frac{z}{2}\right)^{-1} + \frac{(2-z)}{5}(1 + z^2)^{-1}$$

Binomial expansion is valid for $|z| < 1$.

$$\therefore \quad f(z) = \frac{1}{10}\sum_{0}^{\infty}(-1)^n\left(\frac{z}{2}\right)^n + \frac{(2-z)}{5}\sum_{0}^{\infty}(-1)^n z^{2n}$$

Above being a series of positive terms represents the Taylor's expansion of $f(z)$ for the given region.

(b) $1 < |z| < 2$

Here we write $f(z)$ given by (1) in a manner so that the binomial expansion is valid for

$$1 < |z| < 2, \text{ i.e., } \frac{1}{|z|} < 1, \frac{|z|}{2} < 2$$

$$\therefore \quad f(z) = \frac{1}{10}\left(1 + \frac{z}{2}\right)^{-1} - \frac{z-2}{5} \cdot \frac{1}{z^2}\left(1 + \frac{1}{z^2}\right)^{-1}$$

$$= \frac{1}{10}\sum_{n=0}^{\infty}(-1)^n\left(\frac{z}{2}\right)^n + \frac{2-z}{5z^2}\sum_{n=0}^{\infty}(-1)^n\left(\frac{1}{z^2}\right)^n$$

Above being a series positive and negative powers of z in Laurent's expansion of $f(z)$ in the annulus $1 < |z| < 2$.

(c) $|z| > 2$ or $\dfrac{2}{|z|} < 1$ or $\dfrac{1}{|z|} < \dfrac{1}{2} < 1$.

$$\therefore \quad f(z) = \frac{1}{5} \cdot \frac{1}{z}\left(1 + \frac{2}{z}\right)^{-1} - \frac{1}{5}(z - 2) \cdot \frac{1}{z^2}\left(1 - \frac{1}{z^2}\right)^{-1}$$

$$= \frac{1}{5z}\sum_{0}^{\infty}(-1)^n\left(\frac{2}{z}\right)^n - \frac{(z-2)}{5z^2}\sum_{0}^{\infty}(-1)^n\left(\frac{1}{z^2}\right)^n$$

Above is also Laurent's expansion within the annulus $2 < |z| < R$ where R is large.

EXAMPLE 7. *For the function* $f(z) \quad \dfrac{2}{z} \quad \dfrac{1}{z}$ *find*

(a) *A Taylor's series valid in the neighbourhood of the point i.*   (PTU–2003)

(b) *A Laurent's series valid within the annulus of which centre is origin.*

SOLUTION. (a) $f(z) = \dfrac{2z^3 + 1}{z^2 + z} = 2(z - 1) + \dfrac{1}{z+1} + \dfrac{1}{z}$

or $\quad f(z) = f_1(z) + f_2(z)$ say $\qquad \dots(1)$

We have to determine Taylor's expansion in the neighbourhood of $z = i$

$$\therefore \quad f_1(z) = 2(z - 1) = \sum_{0}^{\infty}a_n(z - 1)^n$$

where $a_n = \dfrac{f_1^n(i)}{n!}$.

$$f_1(z) = 2(z - 1) \quad \therefore \quad f_1'(z) = 2, f_1''(z) = 0$$

In general $f_1^n(z) = 0$ for $n \geq 2$

$$\therefore \quad a_n = 0 \text{ for } n \geq 2 \text{ and } a_1 = \frac{f_1^1(i)}{1!} = 2$$

Also $a_0 = \dfrac{f_1(i)}{0!} = 2(i - 1)$

$$\therefore \quad f_1(z) = a_0 + a_1(z - i) = 2(i - 1) + 2(z - i) \qquad \dots(2)$$

$$f_2(z) = \frac{1}{z+1} = \sum_{0}^{\infty}a_n(z - i)^n \quad \text{where } a_n = \frac{f_2^n(i)}{n!}$$

$$f_2^n(z) = \frac{(-1)^n n!}{(z+1)^{n+1}}$$

$$\therefore \quad \frac{f_2^n(i)}{n!} = \frac{1}{n!} \cdot \frac{(-1)^n n!}{(i+1)^{n+1}} = \frac{(-1)^n}{(i+1)^{n+1}}$$

$$\therefore \quad f_2(z) = \sum_{0}^{\infty}\frac{(-1)^n}{(i+1)^{n+1}}(z - i)^n \qquad \dots(3)$$

$$f_2(z) = \frac{1}{z} = \sum_{0}^{\infty}a_n(z - i)^n \quad \text{where } a_n = \frac{f_3^n(i)}{n!}$$

$$f_3^n(z) = \frac{(-1)^n n!}{z^{n+1}}$$

$$\therefore \quad \frac{f_3^n(i)}{n!} = \frac{1}{n!} \cdot \frac{(-1)^n n!}{i^{n+1}} = \frac{(-1)^n}{i^{n+1}}$$

$$\therefore \quad f_3(z) = \sum_0^\infty \frac{(-1)^n}{i^{n+1}} (z-i)^n \qquad \dots(4)$$

Hence from (1) by the help of (2), (3) and (4) we have

$$f(z) = 2(i-1) + 2(z-i) + \sum_0^\infty (-1)^n (z-i)^n \cdot \left[\frac{1}{(1+i)^{n+1}} + \frac{1}{i^{n+1}} \right]$$

Above is the required Taylor's series for $f(z)$ in the neighbourhood of point i.

(b) From $f(z)$ is clear that its singularities are at

$z = 0$ and $z = 1$ and as such $f(z)$ is analytic in the region $0 < |z| < 1$.

Therefore it can be expanded in Laurent's series.

$$f(z) = 2(z-1) + \frac{1}{z} + (1+z)^{-1}$$

$$= 2(z-1) + \frac{1}{z} + \sum_{n=2}^\infty (-1)^n z^n .$$

The terms corresponding to $n = 0$ and $n = 1$ are 1 and $-z$ respectively.

$$\therefore \quad f(z) = 2(z-1) + \frac{1}{z} + 1 - z + \sum_{n=2}^\infty (-1)^n z^n$$

$$= -1 + z + \frac{1}{z} + \sum_{n=2}^\infty (-1)^n z^n .$$

Above is the required Laurent's expansion.

Exercise-11.20

1. Show that $\int_C e^{-2z} dz$ is independent of the path C joining the points $1 - \pi i$ and $2 + 3\pi i$ and determine its value.

2. If $f(z) = \frac{z^2 + 5z + 6}{z - 2}$, does Cauchy's theorem apply
   (i) when the path of integration C is a circle of radius 3 with origin as centre.
   (ii) when C is a circle of radius 1 with origin as centre.

3. Without using integral formula, evaluate the following integrals :
   (i) $\int_C \frac{1}{z - z_0} dz$ (ii) $\int_C \frac{1}{(z - z_0)^2} dz$

4. Verify Cauchy's theorem for the function $5 \sin 2z$ if c is the square with vertices at $1 \pm i, -1 \pm i$.

5. Evaluate $\int_C \frac{z - 3}{z^2 + 2z + 5} dz$ where C is a circle.
   (i) $|z| = 1$ (ii) $|z + 1 - i| = 2$

6. Evaluate the following integrals by using Cauchy's intergral formula
   (i) $\int_C \frac{\sin \pi z^2 + \cos \pi z^2}{(z-1)(z-2)} dz$

   (VTU–2010, ANNA–2003S, UPTU–2002, 10)

   (ii) $\frac{1}{2\pi i} \int_C \frac{e^{zt}}{z^2 + 1} dz, t > 0$ (UPTU–2009)
   where C represents the circle $|z| = 3$.

7. Evalutate $\int_C \frac{e^{az}}{z^2 + 1} dz$ where C is the circle $|z| = 2$.

8. Show that $\int_C \frac{\sin^6 z}{\left(z - \frac{\pi}{6}\right)^3} dz = \frac{21}{16} \pi i$, where C is the circle $|z| = 1$.

   (ROHTAK–2005)

9. Using Cauchy's integral formula, prove the following :
   (i) $\int_C \frac{z.dz}{(9 - z^2)(z + i)} = \frac{\pi}{5}$, where C is circle $|z| = 2$ described in positive sense.

   (ii) $\int_C \frac{\cosh(\pi z)}{z(z^2 + 1)} dz = 4\pi i$, where C is the circle $|z| = 2$.

   (iii) $\int_C \frac{e^{9z}.dz}{(z - \pi i)} = 0$, where C is the ellipse $|z - 2| + |z + 2| = 6$.

   (iv) $\int_C \frac{dz}{z - 2} = 1$ where C is $|z| = 3$.

10. Prove that $\frac{1 + 2z}{z^2 + z^3} = \frac{1}{z^2} + \frac{1}{z} - 1 + z - z^2 + z^3 \dots$ where $0 < |z| < 1$.

11. Find two Laurent's series expansions in powers of z of the function $f(z) = \frac{1}{z(1 + z^2)}$.

12. Find the Laurent's series of the function $f(z) = \frac{1}{z^2(1 - z)}$ about $z = 0$.

13. Show that
    (i) $\int_C \frac{\tan\left(\frac{\pi}{2}\right)}{(z - z_0)^2} dz = \pi i \sec^2\left(\frac{z_0}{2}\right)$ where C is the boundary of the square whose sides lie along the lines $x = \pm 2, y = \pm 2$ and it is described in positive sense, where $|z_0| < 2$.

    (ii) $\int_C \frac{dz}{z^2 + 2z + 2} = -\pi$ where C is the square having vertices at $(0, 0), (-2, 0), (-2, -2), (0, -2)$ obtained in anticlockwise direction.

    (iii) $\int_C \frac{\sin z}{(z - \pi/4)^3} dz = -\frac{\pi i}{\sqrt{2}}$ where C is $\left| \frac{z - \pi}{4} \right| = \frac{1}{2}$.

14. For the function $f(z) = \frac{2z^3 + 1}{z^2 + z}$, find
    (i) Taylor's series valid in the neighbourhood of the point $z = i$
    (ii) a Laurent series valid within the annulus of which centre is the origin.

15. Find the Laurent's series of the function $f(z) = \frac{1}{(z^2 - 4)(z + 1)}$ valid in the region $1 < |z| < 2$.

16. Expand $\sin z$ in a Taylor's series about $z = \frac{\pi}{4}$.

17. Expand $\frac{1}{z}$ as a Taylor's series about $z = 1$.

18. Prove that $e^{\frac{u}{2}(z + vz)} = \sum_{n=0}^\infty a_n z^n + \sum_{n=1}^\infty b_n z^{-n}$

    where $a_n = \frac{1}{2\pi} \int_0^{2\pi} \exp\{(u + v)\cos\theta\} \cos\{(v - u)\sin\theta - n\theta\} d\theta$

and $b_n = \frac{1}{2\pi}\int_0^{2\pi} \exp\{(u+v)\cos\theta\}\cos\{(u-v)\sin\theta - n\theta\}d\theta$

19. Find the Laurent expansion of $\dfrac{z}{(z+1)(z+2)}$ about $z = -2$.

20. Find the Taylor's series expansion of the function $f(z) = \dfrac{z}{z^4 + 9}$

around $z = 0$. Also find the radius of convergence.

21. Find the value of the integral $\int \dfrac{1}{z-a}dz$ round the circle whose

equation is $|z - a| = \rho$.

22. Expand $f(z) = \dfrac{z+3}{z(z^2 - z - 2)}$ in powers of z when

(a) $|z| < 1$ (b) $1 < |z| < 2$

(c) $|z| > 2$

23. Expand $\dfrac{1}{z(z^2 - 3z + 2)}$ for the regions

(a) $0 < |z| < 1$ (b) $1 < |z| < 2$

(c) $|z| > 2$

24. Find the Taylor's and Laurent's series which represents the

function $\dfrac{z^2 - 1}{(z+2)(z+3)}$.

(a) $|z| < 2$ (b) $2 < |z| < 3$

(c) $|z| > 3$

25. Prove that

(i) $\int_C \dfrac{dz}{z-a} = 2\pi i$

(ii) $\int_C (z-a)^n dz = 0$ [n, any intger $\neq -1$]

where C is the circle $|z - a| = r$.   (UPTU–2003)

26. Evelute $\int_0^{2+i}(\bar{z})^2 dz$, along

(i) the line $y = x/2$.   (BHOPAL–2007, UPTU–2002)

(ii) the real axis to 2 and then vertically to $2 + i$.   (SVTU–2009, PTU–2008S, MUMBAI–2006)

27. Evaluate $\int_C (z^2 + 3z + 2)dz$ where C is the arc of the cycloid

$x = a(\theta + \sin\theta)$, $y = a(1 - \cos\theta)$ between the points

$(0, 0)$ and $(\pi a, 2a)$.   (ROHTAK–2004)

28. Evaluate :

(i) $\int_0^{1+i}(x^2 - iy)dz$ along the paths

     (a) $y = x$

and   (b) $y = x^2$ (UPTU–2010)

(ii) $\int_{1-i}^{2+i}(2x + iy + 1)dz$, along the two paths :   (UPTU–2010)

     (a) $x = t + 1, y = 2t^2 - 1$

     (b) the straight line joining $1 - i$ and $2 + i$

              (UPTU–2006)

(iii) $\int_{1-i}^{2+3i}(z^2 + z)dz$ along the line joining the points $(1, -1)$ and

     $(2, 3)$.   (VTU–2004)

29. Evaluate $\int_C \dfrac{z^2 - z + 1}{z - 1}dz$, where C is the circle

(i) $|z| = 1$

(ii) $||\ \ -$ (SVTU–2007)

30. Evaluate $\int_C \dfrac{e^z}{(z^2 + \pi^2)^2}dz$, where C is $|z| = 4$.

     (UPTU–2008, JNTU–2000)

31. Verify Cauchy's theorem by integrating e^{iz} along the boundary of the triangle with the vertices at the points $1 + i, -1 + i$, and $-1 - i$.   (UPTU–2006)

32. If $F(\zeta) = \int_C \dfrac{4z^2 + z + 5}{z - \zeta}dz$ where C is the ellipse $\left(\dfrac{x}{2}\right)^2 + \left(\dfrac{y}{3}\right)^2 = 1$

find the value of

(a) $F(3.5)$;   (b) $F(i)$; $F''(-1)$ and $F''(-i)$

     (BHOPAL–2009, MARATHWADA–2008, MUMBAI–2006)

33. Evaluate

(i) $\int_C \dfrac{e^z}{z^2 + 1}dz$, where C is the circle $|z| = 1/2$.   (PTU–2010)

(ii) $\oint_C \dfrac{e^{3iz}}{(z+\pi)^3}dz$, where C is the circle $|z - \pi| = 3$.

             (UPTU–2007)

34. Use Cauchy's integral formula to calculate $\oint_C \dfrac{\sin\pi z + \cos\pi z}{(z-1)(z-2)}dz$

where C is $|z| = 4$.   (UPTU–2008)

35. Evaluate, using Cauchy's intgral formula :

(i) $\int_C \dfrac{e^z}{z^2 + 1}dz$ where C is $|z - 2| = 1/2$.

     (UPTU–2009, HISSAR–2007, MADRAS–2000)

(ii) $\int_C \dfrac{\log z}{(z-1)^3}dz$, where C is $|z - 1| = 1/2$.   (JNTU–2003)

36. Verify Cauchy's theorem for the integral of z^3 taken over the boundary of the

(i) rectangle with vertices $-1, 1, 1 + i, -1 + i$

(ii) triangle with vertices $(1, 2), (1, 4), (3, 2)$ (VTU–2003, 10)

37. Expand $\sin z$ in a Taylor's series about $z = 0$ and determine the region of convergence.   (PTU–2009S)

38. Find Taylor's expansion of $f(z) = \dfrac{1}{(z+1)^2}$ about the point $z = -i$.

              (VTU–2009S)

39. Expand $f(z) = 1/[(z - 1)(z - 2)]$ in the region :

(a) $|z| < 1$ (b) $1 < |z| < 2$

(c) $|z| > 2$ (d) $0 < |z - 1| < 1$

     (UPTU–2010, VTU–2010, BHOPAL–2009, 16)

40. Find the Laurent's expansion of $f(z) = \dfrac{7z - 2}{(z+1)z(z-2)}$ in the

region $1 < z + 1 < 3$.   (SVTU-2009, ANNA-2003, 14, VTU-2003)

41. Find the first three terms of the Taylor's series expansion of $f(z) = 1/(z^2 + 4)$ about $z = -i$. Also find the region of convergence.   (UPTU–2006)

42. Find the Laurent's series expansion of $\dfrac{z^2 - 1}{z^2 + 5z + 6}$ about $z = 0$

in the region $2 < |z| < 3$.   (VTU–2011S, OSMANIA–2003)

43. Find the nature and location of the singularities of the following functions :

(i) $\sin\left(\dfrac{1}{z}\right)$ (UPTU–2009)

(ii) $\tan\left(\dfrac{1}{z}\right)$ (PTU–2006)

(iii) $\dfrac{\cot\pi z}{(z-a)^2}$ (UPTU–2008)

Hints to Selected Problems

2. Given function is not analytic at $z = 2$.

When the path of integration is circle $|z| = 3$, the point $z = 2$ lies inside C so $f(z)$ is not analytic within C.

3. (i) $\int \frac{1}{z - z_0} dz = \int_0^{2\pi} \frac{1}{re^{i\theta}} ire^{i\theta} d\theta \quad z = z_0 + re^{i\theta}$

$$= i\int_0^{2\pi} id\theta = i[\theta]_0^{2\pi} = 2\pi i$$

5. We have $\dfrac{z - 3}{z^2 + 2z + 5} = \dfrac{z - 3}{(z + 1 + 2i)(z + 1 - 2i)}$

$$= \frac{\frac{1}{2} - i}{z + 1 + 2i} + \frac{\frac{1}{2} + i}{z + 1 - 2i}$$

6. (i) If $f(z) = \sin \pi z^2 + \cos \pi z^2$

$\therefore \int_C \dfrac{\sin \pi z^2 + \cos \pi z^2}{(z - 1)(z - 2)} dz = \int \dfrac{f(z)}{z - 2} dz - \int \dfrac{f(z)}{z - 1} dz$

7. $|z| = 2$ is a circle with centre at origin and origin and radius 2.

$z^2 + 1 = 0 \qquad \Rightarrow \qquad z = \pm i$

Point $(0, -1)$ and $(0, 1)$

If $f(z) = e^{az} \quad \therefore \quad \int_C \dfrac{e^{az}}{(z^2 + 1)} dz = \dfrac{1}{2i} \int_C \dfrac{f(z)}{z - i} dz - \int_C \dfrac{f(z)}{z + i} dz$

8. $f(z) = \sin^6 z$

Taken $n = z, a = \dfrac{\pi}{6}$ and using $f^n(a) = \dfrac{n!}{2\pi i} \int \dfrac{f(z)}{(z - a)^{n+1}} dz$

10. $f(z) = \dfrac{1 + 2z}{z^2 + z^3} = \dfrac{1 + 2z}{z^2(1 + z)} = \dfrac{1}{z^2} + \dfrac{1}{z} - \dfrac{1}{1 + z}$

11. $f(z) = \dfrac{1}{z(1 + z^2)} = \dfrac{1}{z} - \dfrac{z}{1 + z^2}$

12. $f(z) = \dfrac{1}{z^2}(1 - z)^{-1}$

14. $f(z) = \dfrac{2z^3 + 1}{z^2 + z} = 2(z - 1) + \dfrac{1}{z} + \dfrac{1}{z + 1}$

15. $f(z) = \dfrac{1}{(z^2 - 4)(z + 1)} = \dfrac{1}{12(z - 2)} + \dfrac{1}{4(z + 2)} - \dfrac{1}{3(z + 1)}$

16. $f(z) = \sum\limits_{n=0}^{\infty} a_n \left(z - \dfrac{\pi}{4} \right)^n$ whose $a_n = \dfrac{f^n\left(\dfrac{\pi}{4} \right)}{n!}$

17. Let $f(z) = \dfrac{1}{z}$

By Taylor's expansion, we have

$$f(z) = \sum\limits_{n=0}^{\infty} a_n(z - 1)^n \text{ whose } a_n = \dfrac{f^n(1)}{n!}.$$

19. $f(z) = \dfrac{z}{(z + 1)(z + 2)} = \dfrac{2}{z + 2} - \dfrac{1}{z + 1}$

20. $f(z) = \dfrac{z}{z^4 + 9} = \dfrac{z}{9} \left(1 + \dfrac{z^4}{9} \right)^{-1} = \dfrac{z}{9} \sum\limits_{n=0}^{\infty} (-1)^n \left(\dfrac{z^4}{9} \right)^n$

Answers

1. 1   **2.** (i) Not applicable   (ii) Applicable   **3.** (i) $2\pi i$ (ii) 0   **4.** Applicable

5. (i) 0   (ii) $\pi(-2 + i)$ **6.** (i) $4\pi i$ (ii) sint   **7.** $2\pi i \sin a$ **12.** $f(z) = \dfrac{1}{z^2} + \dfrac{1}{z} + 1 + \sum\limits_{n=1}^{\infty} z^n$

14. (i) $f(z) = 2(i - 1) + 2(z - i) + \sum\limits_{n=0}^{\infty} (-1)^n \dfrac{(z - i)^n}{i^{n+1}} + \sum\limits_{n=0}^{\infty} (-1)^n \dfrac{(z - i)^n}{(1 + i)^{n+1}}$ (ii) $f(z) = 2(z - 1) + \dfrac{1}{z}(1 - z + z^2 - z^3 + \ldots)$

15. $f(z) = -\dfrac{1}{24} \sum\limits_{n=0}^{\infty} \left(\dfrac{z}{2} \right)^n + \dfrac{1}{8} \sum\limits_{n=0}^{\infty} (-1)^n \left(\dfrac{z}{2} \right)^n - \dfrac{1}{3z} \sum\limits_{n=0}^{\infty} \dfrac{(-1)^n}{z^n}$ **16.** $f(z) = \sum\limits_{n=0}^{\infty} \sin \left(\dfrac{\pi}{4} + \dfrac{n\pi}{2} \right) \dfrac{\left(z - \dfrac{\pi}{4} \right)^n}{n!}$

19. $f(z) = \dfrac{2}{2 + z} + \sum\limits_{n=0}^{\infty} (z + 2)^n$ **20.** $f(z) = \sum\limits_{n=0}^{\infty} (-1)^n \dfrac{z^{4n+1}}{3^{2n+2}}$, Radius of Convergence $= \sqrt{3}$ **21.** $2\pi i$

22. (a) $-\dfrac{3}{2z} + \dfrac{2}{3} \sum\limits_0^{\infty} (-1)^n z^n - \dfrac{5}{12} \sum\limits_0^{\infty} \left(\dfrac{z}{2} \right)^n$ (b) $-\dfrac{3}{2z} + \dfrac{2}{3z} \sum\limits_0^{\infty} (-1)^n \dfrac{1}{z^n} - \dfrac{5}{12} \sum\limits_0^{\infty} \left(\dfrac{z}{2} \right)^n$ (c) $-\dfrac{3}{2z} + \dfrac{2}{3z} \sum\limits_0^{\infty} (-1)^n \dfrac{1}{z^n} - \dfrac{5}{6z} \sum\limits_0^{\infty} \left(\dfrac{2}{z} \right)^n$

23. (a) $\dfrac{1}{2z} + \sum\limits_0^{\infty} z^n - \dfrac{1}{4} \sum\limits_0^{\infty} \left(\dfrac{z}{2} \right)^n$ (b) $\dfrac{1}{2z} - \dfrac{1}{z} \sum\limits_0^{\infty} z^{-n} - \dfrac{1}{4} \sum\limits_0^{\infty} \left(\dfrac{z}{2} \right)^n$ (c) $\dfrac{1}{2z} - \dfrac{1}{z} \sum\limits_0^{\infty} z^{-n} - \dfrac{1}{2z} \sum\limits_0^{\infty} \left(\dfrac{2}{z} \right)^n$

24. (a) $1 + \dfrac{3}{z} \sum\limits_0^{\infty} (-1)^n \left(\dfrac{z}{2} \right)^n - \dfrac{8}{3} \sum\limits_0^{\infty} (-1)^n \left(\dfrac{z}{3} \right)^n$ (b) $1 + \dfrac{3}{z} \sum\limits_0^{\infty} (-1)^n \left(\dfrac{2}{z} \right)^n - \dfrac{8}{3} \sum\limits_0^{\infty} (-1)^n \left(\dfrac{z}{3} \right)^n$ (c) $1 + \dfrac{3}{z} \sum\limits_0^{\infty} (-1)^n \left(\dfrac{2}{z} \right)^n - \dfrac{8}{3} \sum\limits_0^{\infty} (-1)^n \left(\dfrac{3}{z} \right)^n$

26. (i) $\dfrac{5}{3}(2 - i)$ (ii) $\dfrac{1}{3}(14 + 11i)$ **27.** $\dfrac{\pi a}{6}(2\pi^2 a^2 + 9\pi a + 12) + \dfrac{a^3}{3}(\pi + 2i)^3 + \dfrac{3a^2}{2}(\pi + 2i)^2 + 4ia$

28. (i)(a) $(5 - i)/6$ (b) $(5 + i)/6$ (ii)(a) $4 + \left(\dfrac{25}{3} \right) i$ (b) $4 + 8i$ (iii) $\dfrac{1}{6}(64i - 103)$

29. (i) $2\pi i$ (ii) 0   **30.** $\dfrac{i}{\pi}$ **32.** (a) 0   (b) $2\pi(i - 1), -14\pi i$ and $16\pi i$ **33.** (i) 0   (ii) 0   **34.** $4\pi i$

35. (i) $4\pi i$ (ii) $-\pi i$ **37.** $\sin z = \sum\limits_{n=1}^{\infty} a_n (z-0)^{2n-1}$ where $a_n = \dfrac{(-1)^{n-1}}{(2n-1)!}$ and the region of convergence is all reals

38. $\dfrac{i}{2}\left[1 + \sum\limits_{n=1}^{\infty} (-1)^n \dfrac{(n+1)(z+i)^n}{(1-i)^n}\right]$ **39.** (a) $f(z) = \dfrac{1}{2} + \dfrac{3}{4}z + \dfrac{7}{8}z^2 + \dfrac{15}{16}z^3 + ...$

39. (b) $f(z) = ... - z^{-4} - z^{-3} - z^{-2} - z^{-1} - \dfrac{1}{2} - \dfrac{1}{4}z - \dfrac{1}{8}z^2 - \dfrac{1}{16}z^3 - ...$ (c) $f(z) = ... + 7z^{-4} + 3z^{-3} + z^{-2} + ...$

(d) $f(z) = ... - (z-1)^{-1} - [1 + (z-1) + (z-1)^2 + (z-1)^3 + ...]$

40. $f(z) = \dfrac{-2}{z+1} + \dfrac{1}{(z+1)^2} + \dfrac{1}{(z+1)^3} + ...\infty - \dfrac{2}{3}\left[1 + \dfrac{z+1}{3} + \dfrac{(z+1)^2}{3^2} + \dfrac{(z+1)^3}{3^3} + ...\infty\right]$

41. $f(z) = \dfrac{1}{3} + \dfrac{u}{9}(z+i) + -\dfrac{7}{27}(z+1)^2 + ...$ Region of convergence is $|z+i| < 1$

42. $1 + \dfrac{3}{z}\left(1 - \dfrac{2}{z} + \dfrac{2^2}{z^2} - \dfrac{2^3}{z^3} + ...\right) - \dfrac{8}{3}\left(1 - \dfrac{z}{3} + \dfrac{z^2}{3^2} - \dfrac{z^3}{3^3} + ...\right)$ **43.** (i) $z = 0$ is an isolated essential singularity.

43. (ii) $z = 0$, is a non-isolated essential singularity.   (iii) $z = a$ is a double pole and $z = 0, \pm 1, \pm 2, ...$ are simple poles.

11.17   THE ZEROS OF AN ANALYTIC FUNCTION

Definition 1. *A zero of an analytic function $f(z)$ is a value of z for which $f(z) = 0$.*

Definition 2. *If $f(z)$ is analytic in a domain D and a is any point in D then by Taylor's theorem $f(z)$ can be expanded about $z = a$, i.e.,*

$$f(z) = \sum_{n=0}^{\infty} a_n (z-a)^n, \text{where } a_n = \frac{f^n(a)}{n!}$$

If $a_0 = a_1 = a_2 ... = a_{m-1} = 0$ but $a_m \neq 0$, i.e., then we say that $f(z)$ has a zero of order m at $z = a$

In this case $f(z) = a_m(z-a)^m + a_{m+1}(z-a)^{m+1} + ... = (z-a)^m[a_m + a_{m+1}(z-a) + ...]$

or $f(z) = (z-a)^m \phi(z)$ say

where $\phi(z)$ is analytic and non-zero at and in the neighbourhood of $z = a$.

It is also clear that $\phi(a) = a_n$, i.e., $\phi(z) = a_m$ at $z = a$.

Definition 3. Simple zero. *In case $f(z)$ has a zero of order one at $z = a$ then $f(z)$ is said to have a simple zero at $z = a$.*

11.17.1 SINGULARITIES OF AN ANALYTIC FUNCTION

A singularity of an analytic function $f(z)$ is the point at which the function ceases to be analytic.

For example The function $f(z) = \dfrac{1}{z-a}$ is analytic except at the point $z = a$. Thus the point $z = a$ is singularity of $f(z)$.

11.17.2 TYPE OF SINGULARITIES

(i)   Isolated, non-Isolated singularity.

Let $z = a$ be a singularity of $f(z)$ and if there is no other singularity within a small circle surrounding the poinit $z = a$, then this point is said to be an isolated singularity and otherwise it is termed as non-isolated singularity.

For example. Let $f(z) = \dfrac{z+1}{(z-1)(z-2)}$

The above function is analytic everywhere except at $z = 1$, $z = 2$ which therefore are its singularities. Also there are no other singularities of $f(z)$ in the neighbourhood of these points and as such these are isolated singularities.

Similarly the function $f(z) = \dfrac{z+3}{z^2(z^2+2)}$ possesses three isolated singularities at $z = 0$, $z = \sqrt{2i}$ and $z = -\sqrt{2i}$

Again consider the function $f(z) = \dfrac{1}{\sin\left(\dfrac{\pi}{z}\right)}$. The above function is not analytic at the points where $z = \pm\dfrac{1}{n}$ ($n = 1, 2, 3$).

Thus it has an infinite number of isolated singularities all lying on the segment of real axis from $z = -1$ to $z = 1$.

Also at $z = 0$ there is a singularity but it is not an isolated because every neighbourhood of $z = 0$ contains other singularities of the function.

Principal part of $f(z)$ at the isolated singularity $z = a$.

Let $f(z)$ be an analytic function within a domain D except at the point $z = a$ which is an isolated singularity. Now draw a circle C' centred at $z = a$ and of radius as small as we please. Draw another concentric circle of any radius say R lying wholly within the domain D. The function $f(z)$ is analytic within the annulus between these two circles. Hence by Laurent's theorem, we have

$$f(z) = \sum_{n=0}^{\infty} a_n(z-a)^n + \sum_{n=1}^{\infty} b_n(z-a)^{-n}, 0 < |z-a| < R$$

The second term in the R.H.S., i.e., $\sum\limits_{n=1}^{\infty} b_n(z-a)^{-n}$ is called the principal part of $f(z)$ at the isolated singularity $z = a$.

(ii) Removable Singularity.

If the principal part of $f(z)$ as stated above contains no term, i.e., $b_n = 0 \ \forall \ n$ then the singularity $z = a$ is called removable singularity of $f(z)$.

Another definition.

A singularity $z = a$ is called removable singularity of $f(z)$ if $\lim\limits_{z \to a} f(z)$ exists

For example : Let $\ f(z) = \dfrac{\sin(z-a)}{(z-a)} = \dfrac{1}{(z-a)}\left[(z-a) - \dfrac{1}{3!}(z-a)^3 + \dfrac{1}{5!}(z-a)^5 ...\right] = 1 - \dfrac{(z-a)^2}{3!} + \dfrac{(z-a)^4}{5!} ...$

It has no term containing negative powers of $(z-a)$, i.e., the principal part of $f(z)$ has no term and as such by definition it is a case of removable singularity.

Alternatively $\lim\limits_{z \to a} \dfrac{\sin(z-a)}{(z-a)} = 1$. Therefore $z = a$ is a removable singularity.

Thus the type of singularity which can be made to disappear by defining the function suitably is termed as removable singularity. Thus in the above case the singularity can be removed and the function made analytic by defining the function

$$f(z) = \dfrac{\sin(z-a)}{(z-a)} = 1 \text{ at } z = a.$$

(iii) Pole.

If in the principal part of $f(z)$ as stated above from and after certain fixed term all the terms are zero i.e., $b_n = 0 \qquad \forall \ n > m$, i.e., the principal part contains a finite number of terms say m then the singularity $z = a$ is called a pole of order m of $f(z)$. However a pole of order one is called simple pole.

$$\therefore \qquad\qquad f(z) = \sum\limits_{n=0}^{\infty} a_n(z-a)^n + \sum\limits_{n=1}^{\infty} b_n(z-a)^{-n} \text{ where } z = a \text{ is a pole of order } m.$$

For example. Let

$$f(z) = \dfrac{\sin(z-a)}{(z-a)^4} = \dfrac{1}{(z-a)^4}\left[(z-a) - \dfrac{1}{3!}(z-a)^3 + \dfrac{1}{5!}(z-a)^5 ...\right]$$

$$= \dfrac{1}{(z-a)^3} - \dfrac{1}{3!}\dfrac{1}{(z-a)} + \dfrac{1}{5!}(z-a) - \dfrac{1}{7!}(z-a)^3 ...$$

The principal part of $f(z)$ in which negative powers of $(z-a)$ occur contains only two terms and hence $z = a$ is a pole of order 2.

☞ Remark

- If there exists a positive integer m such that $\lim\limits_{z \to a}(z-a)^m f(z) = A \neq 0$ then $z = a$ is called a pole of order m.

(iv) Essential singularity.

If the principal part of $f(z)$ as stated above contains an infinite number of terms, then the singularity $z = a$ is called an essential singularity.

In other words it means that if there exists no finite value of n such that $\lim\limits_{z \to a}(z-a)^n f(z) = A = $ finite non-zero constant, then $z = a$ is called an essential singularity.

For example. Consider $\ f(z) = e^{1/z} = 1 + \dfrac{1}{z} + \dfrac{1}{2!}.\dfrac{1}{z^2} + \dfrac{1}{3!}.\dfrac{1}{z^3} + ...$

Above is an infinite series of negative powers of $z - 0$ and as such $z = 0$ is an essential singularity.

Similarly $\qquad\qquad f(z) = \sin\left(\dfrac{1}{z-a}\right) = \dfrac{1}{z-a} - \dfrac{1}{3!}.\dfrac{1}{(z-a)^3} + \dfrac{1}{5!}.\dfrac{1}{(z-a)^5} ...$

has infinite number of terms in negative powers of $z - a$ and hence $z = a$ is an essential singularity.

(v) Meromorphic function.

A function $f(z)$ which has poles as its only singularities in the finite part of the plane is called a meromorphic function.

11.17.3 Certain Theorems

THEOREM 1. *If $f(z)$ has a pole at $z = a$ then $|f(z)| \to \infty$ as $z \to a$.*

PROOF. Let $z = a$ be a pole of order m then by Laurent's theorem

$$f(z) = \sum\limits_{n=0}^{\infty} a_n(z-a)^n + \sum\limits_{n=1}^{m} b_n(z-a)^{-n} = \sum\limits_{n=0}^{\infty} a_n(z-a)^n + \dfrac{b_1}{z-a} + \dfrac{b_2}{(z-a)^2} + ... + \dfrac{b_m}{(z-a)^m}$$

$$= \sum\limits_{n=0}^{\infty} a_n(z-a)^n + \dfrac{1}{(z-a)^m}[b_m + b_{m-1}(z-a) + ... + b_2(z-a)^{m-2} + b_1(z-a)^{m-1}] \qquad ...(A)$$

The expression within brackets on the R.H.S. tends to b_m as $z \to a$ and consequently R.H.S. tend to infinity.

$\therefore$ $|f(z)| \to \infty$ as $z \to a$.

THEOREM 2. *If an analytic function $f(z)$ has a pole of order m at $z = a$, then $\dfrac{1}{f(z)}$ has a zero of order m at $z = a$ and conversely.*

PROOF.   We have

$$f(z)(z-a)^m = \sum_{n=0}^{\infty} a_n(z-a)^{m+n} + b_m + b_{m-1}(z-a) + \ldots + b_2(z-a)^{m-2} + b_1(z-a)^{m-1}$$

or $f(z)(z-a)^m = \phi(z)$ (say)   ...(1)

where $\phi(z)$ is analytic and $\phi(a) \neq 0$.

From (1) we have $\dfrac{1}{f(z)} = \dfrac{(z-a)^m}{\phi(z)}$ ...(2)

Let $z \to a$ in (2) and we know that $\phi(a) \neq 0$, we get $\dfrac{1}{f(z)} = 0$ as $z \to a$.

Above relation shows that $\dfrac{1}{f(z)}$ has a zero of order m at $z = a$.

Conversely $\dfrac{1}{f(z)}$ has a zero of order m and as such it can be written as

$$\frac{1}{f(z)} = (z-a)^m \psi(z) \text{ where } \psi(z) \text{ is analytic and } \psi(a) \neq 0.$$

$\therefore$ $(z-a)^m f(z) = \dfrac{1}{\psi(z)} = \phi(z)$ say.   ...(4)

Since $\psi(z)$ is analytic therefore $\phi(z)$ is also analytic. Also as $\psi(a) \neq 0$ therefore $\phi(a)$ is also not zero.

The form (4) shows that $f(z)$ has a pole of order m at $z = a$.

Deductions.

(i) **Zeros are isolated.**

Let $z = a$ be a zero of order m for the function $f(z)$ which is analytic in the neighbourhood of $z = a$

$\therefore$ $f(z) = (z-a)^m \phi(z)$ where $\phi(a) \neq 0$.

Now $\phi(z)$ is analytic and non zero at $z = a$ and its neighbourhood.

Also $(z-a)^m \neq 0$ for values of $z \neq a$.

From above we conclude that there exists no other point in the neighbourhood of $z = a$ at which $f(z)$ vanishes so that zero $z = a$ is isolated. The same is true for every zero of $f(z)$ and as such we conclude that zeros of $f(z)$ are isolated.

(ii) **Poles are isolated.**

Let $z = a$ be a pole of order m for the analytic function $f(z)$ and hence by last theorem $\dfrac{1}{f(z)}$ is analytic and has a zero of order m as $z \to a$. But as proved above zeros are isolated and hence poles are also isolated.

(iii) **The point at infinity.**

In order to study the behaviour of the point $z = \infty$ we make the substitution $z = 1/t$ in $f(z)$. Then the behaviour of $f(z)$ at $z = \infty$ is determined by the behaviour of $f(1/t)$ at $t = 0$.

Thus $f(z)$ has a zero or pole at $z = \infty$ according as $f(1/t)$ has the corresponding property at $t = 0$.

THEOREM 3. *The order of a zero of a polynomial equals the order of its first non-vanishing derivatives.*

PROOF.   Let $P(z)$ be a given polynomial and $z = a$ is a zero of order m of $P(z)$.

Then, we can write $P(z) = (z-a)^m Q(z), Q(a) \neq 0$

$\Rightarrow$ $P'(z) = m(z-a)^{m-1}Q(z) + (z-a)^m Q'(z)$

          $P''(z) = m(m-1)(z-a)^{m-2}Q(z) + 2m(z-a)^{m-1}Q'(z) + (z-a)^m Q''(z)$

          $P^m(z) = m!Q(z) + {}^mC_1 m!(z-a)Q'(z) + \ldots + (z-a)^m Q^m(z)$

Substituting $z = a$ in the above relations, we get

          $P(a) = P'(a) = \ldots P^{m-1}(a) = 0$

and $P^m(a) = m!Q(a) \neq 0$

Hence, the order of zero of a polynomial equals the order of its non-vanishing derivative.

📑 **REMARKS**

- If all the zeroes of a polynomial lie in a half plane then all the zeroes of its derivatives also lie in the same half plane (This is called Luca's theorem).

- A function $f(z)$ is a polynomial of degree n if and only if $f(z)$ has no singularities in the finite part of the plane and has a pole of order n at infinity.

11.17.4 LIMIT POINT OF ZEROS

THEOREM 1. *Let $f(z)$ be an analytic function in a simply connected region D. Let a_1, a_2 ... a_n be a sequence of zeros having a as its interior point of D. Then either $f(z)$ vanishes identically or else has an isolated essential singularity.*

PROOF. Since $f(z)$ is analytic in a simply connected domain D, therefore it is continuous in D. Let a_1, a_2, a_3 ... be an infinite set of zeros of $f(z)$ which must have at least one limit point say a which may or may not be a point of the set.

If a is a point of the set then it is a zero of $f(z)$ and since it is also a limit point of the set of zeros, it should have in its neighbourhood a cluster of zeros. But we know that zeros are isolated, *i.e.*, zeros of a function must not have any other zeros around it. Therefore a cannot be a zero of $f(z)$ unless the function is identically zero in the domain D.

In case the function does not vanish identically in D, then a is not a zero of $f(z)$ while being surrounded by many zeros. Thus a is a singularity which is not a pole as $f(z)$ does not tend to infinity in the neighbourhood of a (it tends to zero as a matter of fact). Hence a is an essential singularity. But the singularity is isolated because in the neighbourhood of a the function $f(z)$ is analytic tending to zero every where in the neighbourhood. Thus a is an isolated essential singularity.

✎ REMARK

- If we are to establish that a certain point a is an isolated essential singularity of $f(z)$ it will be sufficient if we prove that a is a limiting point of zeros of $f(z)$.

THEOREM 2. **(Identity Theorem).** *If $f(z)$ and $g(z)$ are analytic in a domain D and if $f(z) = g(z)$ on a subset E which has a limit point in D, then $f(z)$ is identically equal to $g(z)$.*

PROOF. Let $F(z) = f(z) - g(z)$. We are given that $f(z) = g(z)$ on F, therefore $F(z)$ vanishes on E. Again if α be a limit point of E, then $F(z)$ must vanish at an infinite number of points in every arbitrarily small neighbourhood of α. Also from continuity of $F(z)$ at α, we conclude that $F(\alpha) = 0$. But as zeros are isolated, α cannot be a zero of $F(z)$ unless $F(z)$ vanishes identically in D which in turn means that $f(z) = g(z)$ in the whole of D.

THEOREM 3. **(The Limit Point of Poles).** *The limit point of a sequence of poles of a function $f(z)$ is a non-isolated essential singularity.*

PROOF. Let a_1, a_2, a_3... be an infinite set of the poles of $f(z)$ which necessarily must have at least one limit point say a which may or may not be a point of the set. Now if a is a point of the set, it must itself be a pole of $f(z)$ and since it is also a limit point, it should have a cluster of poles around it. But this is not possible as poles are isolated. Therefore a cannot be a pole.

Again a cannot be a zero of $f(z)$ because the function is not analytic (has poles) in the neighbourhood of a. Therefore a is an essential singularity. But this singularity is not isolated as there are poles around a. We shall call such a singularity as a non-isolated essential singularity or simply an essential singularity of $f(z)$.

THEOREM 4. **[Removable singularity (Riemann)]** *If $z = a$ is an isolated singularity of $f(z)$ and if $|f(z)|$ is bounded on some deleted neighbourhood of a, then a is a removable singularity.*

PROOF. Let $|f(z)|$ be bounded on some deleted neighbourhood $N(a)$ of a, then $|f(z)| \leq M$ where M is the maximum value of $f(z)$ on a circle C defined by $|z - a| = r$.

The radius r is chosen so small that C lies entirely within $N(a)$. By Laurent's expansion,

$$f(z) = \sum_{n=0}^{\infty} a_n(z-a)^n + \sum_{n=1}^{\infty} b_n(z-a)^{-n} \qquad ...(1)$$

where
$$b_n = \frac{1}{2\pi i}\int_C (z-a)^{n-1} f(z)dz$$

$\therefore$
$$\left|b_n\right| \leq \frac{M}{2\pi i}\int_C |z-a|^{n-1}|dz| = \frac{Mr^{n-1}}{2\pi}2\pi r = Mr^n$$

or
$$\left|b_n\right| \leq Mr^n \to 0 \text{ as } r \to 0 \qquad \therefore b_n = 0 \ \forall \ n.$$

Thus we conclude that principal part of Laurent's expansion for $f(z)$ contains no term. Hence by definition we conclude that $z = a$ is a removable singularity.

11.17.5 THE BEHAVIOUR OF A FUNCTION NEAR AN ESSENTIAL SINGULARITY

THEOREM. **(Weirstrass's Theorem).** *If $z = a$ is an essential singularity of $f(z)$. Then for given any positive numbers r, $\in$ and any arbitrary constont c, there is a point in the circle $|z - a| < r$ at which $|f(z) - c| < \in$.*

Another form :

In every arbitrary neighbourhood of an essential singularity there exists a point (and therefore an infinite number of points), at which the function differs as little as we please from any pre-assigned number.

PROOF. Let us assume that the theorem is false.

Then for given $\in$, $r > 0$ and a number c there exists a point in the circle $|z - a| < r$ at which $|f(z) - c| > \in$

or
$$\frac{1}{|f(z)-c|} < \frac{1}{\in}$$

Therefore by Riemanns Theorem we conclude that the function $\dfrac{1}{f(z)-c}$ has a removable singularity at a. Hence by the definition

of removable singularity we conclude that the principal part of Laurent's expansion for $\dfrac{1}{f(z)-c}$ contains no negative powers of

$z-a$.

$$\therefore \qquad \frac{1}{f(z)-c} = \sum_{n=0}^{\infty} a_n(z-a)^n \qquad \qquad \text{...(I)}$$

If $a_0 \neq 0$ we define $\dfrac{1}{f(a)-c} = a_0$ or $f(a) = c + \dfrac{1}{a_0}$

Thus $\dfrac{1}{f(z)-c}$ becomes analytic and non zero for $z=a$ and as a result $f(z)$ is itself analytic at $z=a$. But this is contrary to the

assumption that $z=a$ is an essential singularity of $f(z)$. Next we suppose that $a_n = 0$ for $n = 0, 1, 2 \ldots m-1$ then by I

$$\frac{1}{f(z)-c} = \sum_{n=m}^{\infty} a_n(z-a)^n = a_m(z-a)^m + a_{m+1}(z-a)^{m+1} \ldots$$

$$= (z-a)^m(a_m + a_{m+1}(z-a)^1 + a_{m+2}(z-a)^2 + \ldots) = (z-a)^m \sum_{n=0}^{\infty} a_{m+n}(z-a)^n$$

Above form shows that the point $z=a$ is a zero of order m of the function $\dfrac{1}{f(z)-c}$ which in turn means that $f(z)-c$ has a

pole of order m at $z=a$. Since c is merely a constant, therefore $f(z)$ also has a pole of order m at $z=a$ which is again a
contradiction. Hence our assumption that the theorem is false is wrong and as such the theorem as stated is true.

RULES TO DETECTING THE TYPES OF SINGULARITIES

1. If $\lim\limits_{z \to a} f(z)$ exists finitely, then $z=a$ is a removable singularity

2. If $\lim\limits_{z \to a} f(z) = \infty$ then $z=a$ is a pole of $f(z)$

3. If there are finite number of terms say m in the principal part of f(z), i.e., there are only m terms in the negative powers of $z-a$ then $z=a$ is a pole of order m.

4. If $\lim\limits_{z \to a} f(z)$ does not exist then $z=a$ is an essential singularity.

5. If the principal part of $f(z)$ contains an infinite number of terms then $z=a$ is an isolated essential singularity.
   (i) Isolated essential singularity. Limit point of zeros is an isolated essential singularity.
   (ii) Non-isolated essential singularity. Limit point of poles is a non-isolated essential singularity.

Solved Examples

EXAMPLE 1. *Prove that the function e^z has an isolated essential singularity at $z = \infty$.*

SOLUTION. The behaviour of the function $f(z)=e^z$ at $z=\infty$ is the same as the behaviour of the function $f\left(\dfrac{1}{t}\right) = e^{1/t}$ at $t = 0$.

Now $\lim\limits_{t \to 0} f(1/t) = \lim\limits_{t \to 0} e^{1/t} = \lim\limits_{t \to 0}\left[1 + \dfrac{1}{t} + \dfrac{1}{2!.t^2} + \ldots\right]$
This limit does not exist. We may also say that the principal part of $f\left(\dfrac{1}{t}\right)$ contains an infinite number of terms in the negative powers of t. Hence $z = 0$ is an isolated essential singularity of $e^{1/t}$ or $z = \infty$ is an isolated essential singularity of e^z.

EXAMPLE 2. *Prove that the function e^{-1/z^2} has no singularities.*

SOLUTION. $f(z) = e^{-1/z^2} = \dfrac{1}{e^{1/z^2}}$. Poles of $f(z)$ are given by equating to zero the denominator of $f(z)$, i.e., by $e^{1/z^2} = 0$. This is not possible for any value of z real or complex and hence no poles.

Again zero of $f(z)$ are given by $e^{-1/z^2} = 0 = e^{-\infty}$

$\therefore \quad 1/z^2 = \infty$ or $z^2 = 0$, i.e., $z = 0$ (twice).
Thus $z = 1$ is a zero of order two and as such there is no limit of zeros and hence non-singularity. Therefore $f(z)$ is free from any singularity.

EXAMPLE 3. *The only singularities of a single valued function $f(z)$ are poles of order 1 and 2 at $z = -1$ and $z = 2$ with residues at these poles 1 and 2 respectively. If $f(0) = \dfrac{7}{4}, f(1) = \dfrac{5}{2}$, determine the function and expand it in a Laurent's series valid in $1 < |z| < 2$.*

SOLUTION. Under given condition of poles with given residues we can say that the principal part of $f(z)$ has the following terms

$$\frac{1}{z+1} + \frac{2}{z-2} + \frac{b}{(z-2)^2}$$

Hence $f(z) = a_0 + \sum\limits_{n=1}^{\infty} a_n z^n + \dfrac{1}{z+1} + \dfrac{2}{z-2} + \dfrac{b}{(z-2)^2}$ ...(1)

Again $z = \infty$ is not a pole of $f(z)$ so that $t = 0$ is not a pole of $f\left(\dfrac{1}{t}\right)$.

Therefore the principal part of $f\left(\dfrac{1}{t}\right)$ should not contain any term in negative powers of t. From (1) the principal part of $f\left(\dfrac{1}{t}\right)$ is $\sum\limits_{n=1}^{\infty} a_n t^{-n}$.

Since it is not to contain any term therefore $a_n = 0$ for $n=1, 2, 3 \ldots$

$$\therefore \quad f(z) = a_0 + \frac{1}{z+1} + \frac{2}{z-2} + \frac{b}{(z-2)^2} \qquad \ldots (2)$$

We are given that $f(0) = \dfrac{7}{4}$ and $f(1) = \dfrac{5}{2}$.

$$\therefore \quad \frac{7}{4} = a_0 + 1 - 1 + \frac{b}{4} \qquad \therefore \quad 4a_0 + b = 7$$

$$\frac{5}{2} = a_0 + \frac{1}{2} - 2 + b \qquad \therefore \quad a_0 + b = 4$$

Solving we get $a_0 = 1$, $b = 3$.

$$\therefore \quad f(z) = 1 + \frac{1}{z+1} + \frac{2}{z-2} + \frac{3}{(z-2)^2} \qquad \ldots (3)$$

Laurent's Expansion of $f(z)$

$$1 < |z| < 2 \quad \therefore \quad \frac{1}{|z|} < 1 \text{ and } \frac{|z|}{2} < 1$$

$$\therefore \quad f(z) = 1 + \frac{1}{z}\left(1 + \frac{1}{z}\right)^{-1} - \left(1 - \frac{z}{2}\right)^{-1} + \frac{3}{4}\left(1 - \frac{z}{2}\right)^{-2}$$

$$= 1 + \frac{1}{z}\sum_{n=0}^{\infty} (-1)^n \frac{1}{z^n}$$

$$\quad - \sum_{n=0}^{\infty}\left(\frac{z}{2}\right)^n + \frac{3}{4}\sum_{n=0}^{\infty}(n+1)\left(\frac{z}{2}\right)^n$$

$$= 1 + \sum_{n=0}^{\infty}(-1)^n \frac{1}{z^{n+1}} + \sum_{n=0}^{\infty}\left(\frac{z}{2}\right)^n\left[\frac{3}{4}(n+1) - 1\right]$$

$$= 1 + \sum_{n=0}^{\infty}(-1)^n \frac{1}{z^{n+1}} + \sum_{n=0}^{\infty}\frac{3n-1}{4}\left(\frac{z}{2}\right)^n$$

Above is the required Laurent's expansion of $f(z)$.

11.17.6 Number of Zeros and Poles of a Meromorphic Function : Argument Principle

THEOREM 1. *If $f(z)$ is analytic within and on a closed contour C except at a finite number of poles and is not zero on C, then* $\dfrac{1}{2\pi i}\int_C \dfrac{f'(z)}{f(z)}dz = N - P$

where N is the number of zeros and P the number of poles inside C (A pole or zero of order m must be counted m times).

PROOF. Let $z = a$ be a zero of order n then in the neighbourhood of this point $f(z) = (z-a)^n \phi(z)$

where $\phi(z)$ is analytic and $\phi(z)$ is not zero at $z = a$ and its neighbourhood.

$$\frac{1}{f(z)}f'(z) = \frac{n}{(z-a)} + \frac{\phi'(z)}{\phi(z)} \qquad \ldots (1)$$

Now $\phi(z)$ is analytic and as such $\phi'(z)$ is also analytic.

$$\therefore \quad \frac{\phi'(z)}{\phi(z)} \text{ is analytic at } z = a.$$

Hence from (1) we conclude that $\dfrac{f'(z)}{f(z)}$ has a simple pole at $z = a$ with residue n.

Similarly if $z = b$ be a pole of order p then in the neighbourhood of this point $(z-b)^p f(z) = \psi(z)$ where $\psi(z)$ is analytic and $\psi(a) \neq 0$.

or $$f(z) = \frac{\psi(z)}{(z-b)^p}$$

Taking log and differentiating, we get

$$\frac{f'(z)}{f(z)} = \frac{\psi'(z)}{\psi(z)} - \frac{p}{z-b} = \frac{\psi'(z)}{\psi(z)} + \frac{-p}{z-b}$$

Arguing as above we conclude that $\dfrac{f'(z)}{f(z)}$ has a simple pole at $z = b$ with residue $-p$.

Therefore by Cauchy's residue theorem for the function $\dfrac{f'(z)}{f(z)} = \dfrac{1}{2\pi}\int \dfrac{f'(z)}{f(z)}dz = $ Sum of all the residues within C.

$$= \Sigma n - \Sigma p = N - P$$

where N is the number of zeros and P the number of poles.

THEOREM 2. **(Argument Principle).** *Let a function $f(z)$ be analytic within and on a closed contour C, having N zeros inside C but no zero on C then.*

$$N = \frac{1}{2\pi}\Delta c \arg f(z)$$

where $\Delta c \arg f(z)$ denotes the variation in the value of $\arg f(z)$ as z moves round the closed contour C(a zero of order n must be counted n times).

PROOF. By Theorem (1) we know that

$$N - P = \frac{1}{2\pi i}\int_C \frac{f'(z)}{f(z)}dz$$

Put $f(z) = Re^{i\theta}$ $\qquad \therefore \quad R = |f(z)|$ and $\theta = \arg f(z)$

$$\therefore \qquad f'(z)dz = df(z) = d(Re^{i\theta}) = e^{i\theta}(dR + iRd\theta) = \frac{f(z)}{R}(dR + iR\,d\theta)$$

$$\therefore \qquad N - P = \frac{1}{2\pi i}\int_C \frac{f'(z)}{f(z)}dz = \frac{1}{2\pi i}\int\left(\frac{dR}{R} + id\theta\right)$$

$$N - P = \frac{1}{2\pi i}\int_C \frac{dR}{R} + \frac{1}{2\pi}\int_C d\theta \qquad\qquad \dots (1)$$

Now $$\int_C \frac{dR}{R} = [\log R]_C = 0$$

Because $\log R$ returns to its original value as z moves once round C.

$$\int_C d\theta = [\theta]_C = \Delta C \arg f(z)$$

Because we know that $\arg f(z)$ does not return to its original value as z moves round C and hence $\Delta c \arg f(z)$ is not necessarily zero.

$$N - P = 0 + \frac{1}{2\pi}\Delta c \arg f(z).$$

In other words the excess of the number of zero over the number of poles of a meromorphic function is equal to $\frac{1}{2\pi}$ times the increase in $\arg f(z)$.

In particular if the function $f(z)$ has no poles then putting $P = 0$ we have $N = \frac{1}{2\pi}\Delta c \arg f(z)$.

THEOREM 3. (Rouche's Theorem). *If $f(z)$ and $g(z)$ are analytic within and on a closed curve C and $|g(z)| < f(z)$ on C, then $f(z)$ and $f(z) + g(z)$ have the same number of zeros inside C.*

PROOF. First of all we establish that neither $f(z)$ nor $f(z) + g(z)$ have a zero on C. Let if possible for some point a on C we have $f(a) = 0$ then as $|g(a)| < |f(a)|$ we conclude that $g(a) = 0$. Therefore $f(a) = g(a)$ which is contradiction as we are given that $|g(z)| < |f(z)|$ on C. In a similar manner if $f(a) + g(a) = 0$ then also $|f(a)| = |g(a)|$ which is again a contradiction.

Thus we establish that neither $f(z)$ nor $f(z) + g(z)$ have a zero on C.

Now let $F(z) = \dfrac{g(z)}{f(z)}$

Since $\dfrac{|g(z)|}{|f(z)|} < 1$ on C $\qquad \therefore |F(z)| < 1$ on C.

Hence it follows that $g(z)$ and $f(z)$ are not zero on C.
Now $g(z) = F(z)f(z)$ or $g = Ff$
$$\therefore \quad g' = F'f + f'F \qquad\qquad \dots(1)$$
Let N_1 and N_2 be the number of zeros of $f + g$ and f respectively inside C and by hypothesis both these functions do not have any poles inside C therefore by Theorem (1).

$$N - P = \frac{1}{2\pi i}\int_C \frac{f'(z)}{f(z)}dz$$

$$N_1 - 0 = \frac{1}{2\pi i}\int_C \frac{f' + g'}{f + g}dz$$

$$N_2 - 0 = \frac{1}{2\pi i}\int_C \frac{f'}{f}dz$$

$$\therefore \quad N_1 - N_2 = \frac{1}{2\pi i}\int_C \frac{f' + F'f + f'F}{f + fF}dz - \frac{1}{2\pi i}\int_C \frac{f'}{f}dz \qquad\qquad \text{by (1)}$$

$$= \frac{1}{2\pi i}\int_C\left[\frac{f'(1 + F)}{f(1 + F)} + \frac{F'f}{f(1 + F)}\right]dz - \frac{1}{2\pi i}\int_C \frac{f'}{f}dz$$

$$= \frac{1}{2\pi i}\int_C \frac{f'}{f}dz + \frac{1}{2\pi i}\int_C \frac{F'}{1 + F}dz - \frac{1}{2\pi i}\int_C \frac{f'}{f}dz$$

or $$N_1 - N_2 = \frac{1}{2\pi i}\int_C \frac{F'}{1 + F}dz \qquad\qquad \dots(2)$$

Now $(1 + F) \neq 0$ in C because $|F(z)| < 1$ by hypothesis.
Also we know that the derivative of an analytic function is analytic and F is analytic on C so is F'. As a result therefore we conclude that $\dfrac{F'}{1 + F}$ is analytic on C.

Hence by Cauchy's Theorem $\int_C f(z)dz = 0$

$$\therefore \qquad N_1 - N_2 = \frac{1}{2\pi i}\int_C \frac{F'}{1 + F}dz = 0 \quad \therefore N_1 = N_2$$

11.17.7 Schwarz's Lemma

THEOREM 1. *If $f(z)$ is analytic in a domain D defined by $|z| < R$ and satisfies the conditions $|f(z)| \leq M \, \forall \, z \in D$ and $f(0) = 0$ then $f(z) \leq \dfrac{M}{R}|z|$. If equality occurs for any z, then $f(z) = \dfrac{M}{R}ze^{i\alpha}$ where α is a real constant.*

(VTU–2013)

PROOF.   Let us assume that C denote the circle $|z| = r, r < R$.

By Taylor's theorem, we can write

$$f(z) = \sum_{n=0}^{\infty} a_n z^n = a_0 + a_1 z + a_2 z^2 + ... \qquad ...(1)$$

If is given that $f(0) = 0$

$\Rightarrow \qquad f(0) = a_0 + a_1.0 + a_2.0 + ... \qquad \Rightarrow \qquad f(0) = a_0 = 0$

Then from (1)

$$f(z) = a_1 z + a_2 z^2 + ...$$

Consider a function

$$F(z) = \frac{f(z)}{z} = a_1 + a_2 z + a_3 z^2 + ...$$

Clearly the function $F(z)$ has a singularity at z=0 which can be removed, if we define $F(z) = a_1$ for $z = 0$.

$$F(0) = a_1.$$

Using, maximum modulus principle, $|f(z)|$ attains its maximum value at some point $z = a$ on C, not within C.

$$\therefore \qquad |F(a)| = \left|\frac{f(a)}{a}\right| = \text{ maximum value of } |F(z)|. \qquad ... (2)$$

$$= \text{Maximum value of } \left|\frac{f(z)}{z}\right| \leq \frac{M}{r}$$

If z lies inside C, we have

$$|F(z)| < \text{Maximum vlaue of } |F(z)| \leq \frac{M}{r}$$

$$\Rightarrow \qquad \left|\frac{f(z)}{z}\right| < \frac{M}{r} \qquad \Rightarrow \qquad |f(z)| < \frac{M}{r}.|z|$$

Clearly inequality holds for all $r < R$.

Then from (3) when $r \to R$

$$|f(z)| < \frac{M}{R}|z| \, \forall z \text{ such that } |z| < R$$

Again from (2) $\qquad |f(a)| = \dfrac{M}{R}|a|$

When $r \to R$, we have

$$|f(a)| = \frac{M}{R}|a| \qquad \Rightarrow \qquad |f(a)| = \frac{M}{R}ae^{i\lambda} \qquad \Rightarrow \qquad f(z) = \frac{M}{R}ze^{i\alpha}$$

☞ Remark

- If $f(z)$ is analytic for $|z| < 1$ and satisfies the conditions $|f(z)| \leq 1, f(0) = 0$ then $|f(z)| \leq |z|$ and $|f'(0)| \leq 1$. Also, equality sign holds only if $f(z)$ is a linear transformation $w = e^{i\alpha}.z$ where α is a real constant. **(Alternative form of Schwarz's Lemma)**

THEOREM 2. **(The fundamental Theorem of Algebra).** *Every polynomial of degree n has exactly n zeros.*

PROOF.   We shall make use of Rouche's Theorem to prove the above result. Let the polynomial be

$$P(z) = a_0 + a_1 z + a_2 z^2 + ... + a_n z^n$$

Choose $\qquad f(z) = a_n z^n$

$$g(z) = a_0 + a_1 z + a_2 z^2 + ... + a_{n-1} z^{n-1}$$

Let C be a circle centred at the origin and radius R>1 then on C we have

$$\frac{|g(z)|}{|f(z)|} = \frac{|a_0 + a_1 z + a_2 z^2 + ... + a_{n-1} z^{n-1}|}{|a_n z^n|} \leq \frac{|a_0| + |a_1|R + |a_2|R^2 + ... + |a_{n-1}|R^{n-1}}{|a_n|R^n}$$

$$\leq \frac{|a_0|R^{n-1} + |a_1|R^{n-1} + |a_2|R^{n-1} + ... + |a_{n-1}|R^{n-1}}{|a_n|R^n} \leq \frac{|a_0| + |a_1| + |a_2| + ... + |a_{n-1}|}{|a_n|R}$$

Now R being arbitrary, we therefore choose R such that

$$R > \frac{|a_0| + |a_1| + |a_2| + ... + |a_{n-1}|}{|a_n|} \quad \text{ or } \quad \frac{|a_0| + |a_1| + |a_2| + ... + |a_{n-1}|}{|a_n|R} < 1$$

$$\therefore \qquad \frac{|g(z)|}{|f(z)|} < 1 \text{ on } C.$$

Hence by Rouche's Theorem $f(z)$ and $f(z)+g(z)$ have the same number of zeros. But $f(z) = a_n z^n$ has clearly zeros located at $z=0$ and hence $f(z)+g(z) = P(z)$ also has n zeros.

THEOREM 3.   **(Maximum Modulus Principle).** *If $f(z)$ be analytic within and on a simple closed contour C, then the point giving the maximum of $|(f(z))|$ can lie on the boundary C and within it.*

        Another form.

        If M is maximum value of $|f(z)|$ on the within C then unless f is a constant $|f(z)| < M$ for every point z within C.

PROOF.   We are given that $f(z)$ is analytic and hence continuous within and on C. Therefore it attains its maximum vlaue M within or on C.

        Let if possible $|f(z)|$ attain its maximum value M at a point a lying inside C and not on C then $|f(a)| = M$ with a lying within C. Now draw a small circle γ centred at a lying entirely within C.

        As $f(z)$ in continuous at the point a therefore for $\epsilon > 0 \, |f(z)| < M - \epsilon$ for all z on the circle γ.

        Again $f(z)$ is analytic within and on γ and a is a point inside γ and hence by Cauchy integral formula, we have

$$f(a) = \frac{1}{2\pi i} \int_\gamma \frac{f(z)}{z-a} dz$$

$$\therefore \quad |f(a)| = \left| \frac{1}{2\pi i} \int_\gamma \frac{f(z)}{z-a} dz \right| \le \frac{1}{2\pi} \int_\gamma \frac{|f(z)|}{|z-a|} |dz|$$

$$\le \frac{1}{2\pi} \int_\gamma \frac{M-\epsilon}{r} |dz| \qquad \text{where } |z-a| = r \, .$$

$$= \frac{M-\epsilon}{2\pi r} \int_r |dz| = \frac{M-\epsilon}{2\pi r} . 2\pi r = M - \epsilon$$

$$\therefore \quad M \le M - \epsilon \text{ where } \epsilon > 0.$$

        Above relation shows that a number M is less than a lesser number which is absurd and hence our supposition is wrong.

        Therefore the maximum value of $f(z)$ is not reached within C but on C.

THEOREM 4.   **(Minimum Modulus Principle).** *If $f(z)$ is analytic within and on a closed contour C and $f(z) \ne 0$ inside C. Suppose further that $f(z)$ is not constant then $|f(z)|$ attains its minimum value at a point on the boundary of C. In other words if m is the minimum of $|f(z)|$ inside and on C then $|f(z)| > m \; \forall \; z$ inside C.*

PROOF. $f(z)$ is analytic within and on C and $f(z) \ne 0$ inside C. Therefore $\dfrac{1}{f(z)}$ is analytic within C. Hence by maximum modulus principle $\dfrac{1}{|f(z)|}$ attains its maximum value on the boundary of C. So that $|f(z)|$ attains its minimum value on the boundary of C.

☑ **REMARK**

- In some of the examples to follow we will have to show that some roots of a given equation lie in a particular quadrant of the z-plane. We shall treat the whole of the z-plane on a circular disc with centre at the origin and radius R very large. The four quadrants of the z plane are respectively identical with the four quadrants of this disc when its radius R tends to infinity. The complete boundary of each quadrant of the disc may be treated as a closed curve C consisting of the arc of a large circle $|z| = R$ and the two radii joining its ends to the origin.

Solved Examples

EXAMPLE 1.   **Prove that all the roots of $z^7 - 5z^3 + 12 = 0$ lie between the circle $|z|=1$ and $|z|=2$.**

SOLUTION.   Let us denote by C_1 and C_2 the circles $|z| = 1$ and $|z| = 2$.

We are to prove that all the zeros lie between these two circles and as such we must first establish that no zeros lie within the samller circle $|z| = 1$, *i.e.*, C_1.

Choose $f(z) = 12$ and $g(z) = z^7 - 5z^3$.

Clearly both $f(z)$ and $g(z)$ are analytic within and on C_1.

On C_1 we have

$$\frac{|g(z)|}{|f(z)|} = \frac{|z^7 - 5z^3|}{12} \le \frac{|z|^7 + 5|z|^2}{12}$$

$$= \frac{1+5}{12} \text{ as } |z| = 1 \text{ on } C_1$$

$$= \frac{1}{2}$$

$$\therefore \quad \frac{|g(z)|}{|f(z)|} < 1 \text{ on } C_1$$

Therefore by Rouches theorem, both $f(z)+g(z) = z^7 - 5z^3 + 12$ and $f(z) = 12$ have the same number of zeros inside C_1. But $f(z) = 12$ has no zero and hence $f(z)+g(z) = z^7 - 5z^4 + 12 = 0$ has no zero inside C_1.

Next we consider the circle $|z| = 2$ and we shall prove that all the zeros lie within C_2.

Choose $f(z) = z^7$ and $g(z) = -5z^3 + 12$.

Both $f(z)$ and $g(z)$ are analytic within and on C_2.

On $C_2, \dfrac{|g(z)|}{|f(z)|} = \dfrac{|12 - 5z^3|}{|z^7|} < \dfrac{12 + 5|z|^3}{|z|^7}$

$$= \frac{12 + 5.8}{2^7} \text{ on } C_2, \text{ i.e., } |z| = 2$$

$$= \frac{52}{128} < 1$$

$$\therefore \quad \frac{|g(z)|}{|f(z)|} < 1 \text{ on } C_2$$

Therefore by Rouche's theorem $f(z) + g(z) = z^7 - 5z^3 + 12$ and $f(z) = z^7$ have the same number of zeros inside C_2. But $f(z) = z^7$ has seven zeros inside C_2 and hence all the seven zeros of $f(z) + g(z) = z^7 - 5z^3 + 12$ lie inside $|z| = 2$.

Thus we conclude that all the seven zeros of the equation $z^7 - 5z^3 + 12$ lie inside C_2 but outside C_1 so that they lie within the circle C_1 and C_2 i.e., $|z| = 1$ and $|z| = 2$.

EXAMPLE 2. *Show that the equation $z^3 + iz + 1 = 0$ has a root in each of the first, second and fourth quadrants*

SOLUTION. Let $f(z) = z^3 + iz + 1 = 0$

Let C be the curve $OABO$

for $OA \qquad z = x$.

$$\therefore \quad f(z) = x^3 + 1 + ix$$

$$\text{Arg } f(z) = \tan^{-1} \frac{x}{x^3 + 1}$$

for arc AB, $z = Re^{i\theta}$

$$f(z) = R^3 e^{3i\theta} \left[1 + \frac{i}{R^2 e^{2i\theta}} + \frac{1}{R^3 e^{3i\theta}} \right]$$

As $R \to \infty$, above tends to $R^3 e^{3i\theta}$.

$\therefore$ Arg $f(z) = 3\theta$ and θ varies from 0 to $\pi/2$.

for $BO \quad z = iy$

$$f(z) = i^3 y^3 + i^2 y + 1 = 1 - y - iy^3$$

$$\therefore \quad \arg f(z) = \tan^{-1} \frac{-y^3}{1 - y} = \tan^{-1} \frac{y^3}{y - 1}$$

Also y varies from ∞ to 0.

Hence by argument principle.

$$N = \frac{1}{2\pi} \Delta C \arg f(z)$$

$$= \frac{1}{2\pi} [\Delta OA + \Delta AB + \Delta BO] \arg f(z)$$

$$= \frac{1}{2\pi} \left[\left\{ \tan^{-1} \frac{x}{x^3 + 1} \right\}_0^\infty + \{3\theta\}_0^{\pi/2} \left\{ \tan^{-1} \frac{y^3}{y - 1} \right\}_\infty^0 \right]$$

$$= \frac{1}{2\pi} \left[0 + 3 \cdot \frac{\pi}{2} + \frac{\pi}{2} \right] = 1.$$

Explanation for $\left[\tan^{-1} \dfrac{y^3}{y - 1} \right]_\infty^0 = \pi / 2$.

The numerator is zero when $y = 0$ and denominator is zero when $y = 1$.

$f(z)$ along $BO = (1 - y) - iy^3 = u + iv$ say.

Thus when $1 < y < \infty$ both u and v are negative (3rd quadrant).

At $y = 1$, $y = 0$, $v = -1$ and when $0 < y < 1$, u is positive and v negative (4th quadrant).

At $y = 0$, $u = 1$ and $v = 0$.

From above we observe that as y varies from ∞ to 0 the point (u, v) starting from ∞ in the third quadrant moves to $(0, -1)$ and then after remaining in the fourth quadrant comes to the point $(1, 0)$.

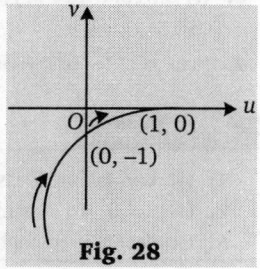
Fig. 28

Therefore as z moves from ∞ to 0 along positive y axis variation in arg $f(z) = \pi/2$.

2nd quadrant : $OBA'O$.

$$N = 1/2\pi \, [\Delta OBA'O] \arg f(z)$$

$$= \frac{1}{2\pi} [\Delta OB + \Delta BA' + \Delta A'O] \arg f(3)$$

$$= \frac{1}{2\pi} \left[\left\{ \tan^{-1} \frac{y^3}{y - 1} \right\}_{y=0}^\infty + \{3\theta\}_{\pi/2}^\pi \left\{ \tan^{-1} \frac{x}{x^3 + 1} \right\}_{-\infty}^0 \right]$$

$$= \frac{1}{2\pi} \left[\frac{\pi}{2} + 3 \left(\pi + \frac{\pi}{2} \right) + 0 \right] = \frac{1}{2\pi} 2\pi = 1$$

Hence one zero in 2nd quadrant.

4th quadrant : $OB'AO$.

$$N = 1/2\pi \, [\Delta OB' + \Delta B'A + \Delta AO] \arg f(z)$$

$$= \frac{1}{2\pi} \left[\left\{ \tan^{-1} \frac{y^3}{y - 1} \right\}_0^{-\infty} + \{3\theta\}_{\theta=3\pi/2}^{2\pi} \left\{ \tan^{-1} \frac{x}{x^3 + 1} \right\}_\infty^0 \right]$$

$$= \frac{1}{2\pi} \left[\frac{\pi}{2} + 3 \left(\frac{2\pi}{3\pi / 2} \right) + 0 \right] = \frac{1}{2\pi} 2\pi = 1$$

Hence one zero in 4th quadrant.

Thus the given equation has one root each in first, second and 4th quadrants.

Exercise-11.21

1. What kind of singularity have the following functions
   (i) $\cot z$ at $z = \infty$ (ii) $\tan z$

2. Discuss the nature of the singularities of the following functions :
   (i) $\dfrac{1}{z(1 - z^2)}$ (ii) $\dfrac{\sin z}{(z - \pi)^2}$

3. Discuss the function represented by the series
   $$\sum_{n=0}^\infty \frac{1}{n!} \cdot \frac{1}{1 + (a^n z)^2}, a > 0 .$$

4. Using Rouche's theorem, find the number of zeroes of the polynomial
   $$P(z) = z^{10} - 6z^7 + 3z^3 + 1 \text{ in } |z| < 1.$$

5. Using Rouche's theorem, find the number of zeroes of the equation $z^8 - 4z^5 + z^2 - 1 = 0$ that lie inside the circle $|z| = 1$.

6. In which quadrants, do the roots of the equation $z^4 + 4z^3 + 8z^2 + 8z + 4 = 0$ lie?

7. What type of singularity have the following functions

(i) $\dfrac{e^{2z}}{(z-1)^4}$　　　　(ii) $\dfrac{e^{1/2}}{z^2}$　　　　(UPTU–2009)

Hints to Selected Problems

1. (i) $f(z) = \cot z = \dfrac{\cos z}{\sin z}$, for poles put $\sin z = 0$

2. (i) Singularities of $f(z)$ are given by $z(1 - z^2) = 0$
$\Rightarrow z = 0, -1, 1$

4. $P(z) = z^{10} - 6z^7 + 3z^3 + 1$.

Let us take $f(z) = -6z^7$, $g(z) = z^{10} + 3z^3 + 1$

Then $P(z) = f(z) + g(z)$

5. Take $f(z) = -4z^5$, $g(z) = z^8 + z^2 - 1$,

then, $\left|\dfrac{g(z)}{f(z)}\right| < 1$ for $|z| = 1 \Rightarrow |g| < |f|$

Answers

1. (i) Non-isolated essential singularity　　　(ii) $z = (2n + 1)\pi/2 \; n \in Z$. Simple pole.

2. (i) $z = 0, -1, 1$ simple pole　　　(ii) $z = \pi$ is a pole of order 2　　　**4.** 7　　　**5.** 5

6. One root lie in each quadrant　　　**7.** (i) $z = 1$ is a pole of 4th order　　　(ii) Essential singularity at $z = 0$

11.18　THE CALCULUS OF RESIDUES

We have already defined that a function $f(z)$ is said to be meromorphic in a region D if it is analytic in the region D except at a finite number of poles.

A function $f(z)$ which has poles as its only singularities the finite port of the plane is called a meromorphic function.

11.18.1 Residue at a Pole

Let a single valued function $f(z)$ has a pole of order m at $z = a$ and γ be any circle of radius r centred at $z = a$ which excludes all other singularities. Therefore $f(z)$ is analytic within the annulus $0 < |z - a| < r$ and as such it can be expanded in Laurent's series such that its principal part contains only m terms.

$\therefore$
$$f(z) = \sum_{n=0}^{\infty} a_n(z-a)^n + \sum_{n=1}^{\infty} b_n(z-a)^{-n} \qquad \ldots(1)$$

where
$$b_n = \frac{1}{2\pi i}\int_\gamma (z-a)^{n-1} f(z)dz . \text{ In particular } b_1 = \frac{1}{2\pi i}\int_\gamma f(z)dz .$$

The coefficient b_1 is called the residue of $f(z)$ at the pole $z = a$ whatever be the order of the pole and is expressed as Res $(z = a)$.

$\therefore$
$$\text{Res}(z = a) = b = \frac{}{}\int f(z)dz .$$

Again the circle γ is arbitrary and as such it can be replaced by any closed contour C (small or large) which. contains no other singularities except $z = a$. Thus we can say

$$\text{Res}(z = a) = b_1 = \frac{1}{2\pi i}\int_C f(z)dz . \qquad \ldots(2)$$

The integration along C being taken in anti-clockwise direction. The value of b_1 given by (2) does not depend upon the order of the pole and therefore it represents a general definition of the residue at a pole.

The case when $z = a$ is a simple pole.

(i) In this case Laurent's expansion (1) becomes $f(z) = \sum_{n=0}^{\infty} a_n(z-a)^n + \dfrac{b_1}{z-a}$

$\therefore$
$$\lim_{z \to a}(z-a)f(z) = b_1$$

$\therefore$
$$\text{Res}(z = a) = \lim_{z \to a}(z-a)f(z) = b_1 = \frac{1}{2\pi i}\int_C f(z)dz .$$

(ii) If $f(z)$ is of the form

$$f(z) = \frac{\phi(z)}{\psi(z)} \text{ where } \psi(z) = 0 \text{ but } \phi(a) \neq 0$$

then
$$\lim_{z \to a}(z-a)f(z) = \lim_{z \to a}\frac{(z-a)\phi(z)}{\psi(z)} = \lim_{z \to a}(z-a)\frac{[\phi(a) + (z-a)\phi'(a) + \ldots]}{\psi(a) + (z-a)\psi'(a) + \ldots} \text{ by Taylor's theorem.}$$

$$= \lim_{z \to a}\frac{(z-a)\phi(a) + (z-a)^2\phi'(a) + \ldots}{(z-a)\psi'(a) + \ldots} \qquad \because \psi(a) = 0$$

Hence,　　　Residue $= \dfrac{\phi(a)}{\psi'(a)}$.

THEOREM 1.　(Residue at a pole of order m).

　　　If $f(z)$ has a pole of order m at $z = a$ then the residue at $z = a$ is the limit

$$\frac{1}{(m-1)!}\frac{d^{m-1}}{dz^{m-1}}[(z-a)^m f(z)] \text{ as } z \to a.$$

PROOF. Since $f(z)$ has a pole of order m at $z = a$ therefore $f(z)$ is expressible as $(z-a)^m f(z) = \phi(z)$. Where $\phi(z)$ is analytic and $\phi(a) \neq 0$.

$$\therefore \qquad f(z) = \frac{\phi(z)}{(z-a)^m}$$

Residue of $f(z)$ at $z = a$ is b_1 where

$$b_1 = \frac{1}{2\pi i}\int_C f(z)dz = \frac{1}{2\pi i}\int_C \frac{\phi(z)}{(z-a)^m}dz = \frac{1}{(m-1)!} \cdot \frac{(m-1)!}{2\pi i}\int_C \frac{\phi(z)dz}{(z-a)^{m-1+1}} = \frac{1}{(m-1)!}\phi^{m-1}(a).$$

By Cauchy integral formula for $\phi(z)$, we have

$$f^n(a) = \frac{n!}{2\pi i}\int_C \frac{f(z)dz}{(z-a)^{n+1}}$$

$$\therefore \qquad b_1 = \text{Res}(z = a) = \lim_{z \to a} \frac{1}{(m-1)!}\frac{d^{m-1}}{dz^{m-1}}[[z-a]^m f(z)] \qquad \text{by (1)}$$

11.18.2 RESIDUE AT A POLE $z = a$ OF ANY ORDER, i.e., SIMPLE OR OF ORDER m

We have stated before that the expansion of $f(z)$ about $z = a$ in a Laurent's series is

$$f(z) = \sum_0^m a_n(z-a)^n + \frac{b_1}{(z-a)} + \frac{b_2}{(z-a)^2} + ... + \frac{b_m}{(z-a)^m}$$

where b_1 by definitions is the residue of $f(z)$ at the pole $z = a$ what ever be the order of the pole, i.e,. simple or of order m. Now if we put $z - a = t$, i.e,. $z = a + t$ where t is small then

$$f(a+t) = \sum_0^m a_n t^n + \frac{b_1}{t} + \frac{b_2}{t^2} + ... + \frac{b_m}{t^m}.$$

From above we observe that b_1 is the coefficient of $\dfrac{1}{t}$ in the above expansion.
Hence we have the following rule.

WORKING PROCEDURE

Put $z = a + t$ in $f(z)$ and expand it. In powers of t where t is small, then coefficient of $\dfrac{1}{t}$ is the residue of $f(z)$ at $z = a$.

11.18.3 RESIDUE AT INFINITY

If $f(z)$ has an isolated singularity $z = \infty$ or is analytic there, then residue at $z = \infty$ is defined as

$$\text{Res }(z = \infty) = -\frac{1}{2\pi i}\int_C f(z)dz$$

where C is any closed contour which encloses all the finite singularities of $f(z)$. The integral being taken in positive direction, i.e., anticlockwise.

REMARK

- The function may be regular at $z = \infty$ and yet has a residue there. Consider the function $f(z) = \dfrac{b}{z-a}$

$$\text{Res }(z = \infty) = \frac{1}{2\pi i}\int_C f(z)dz = -\frac{1}{2\pi i}\int_C \frac{b}{z-a}dz = -\frac{1}{2\pi i}b.2\pi i = -b.$$

Also $z = a$ is a simple pole of $f(z)$ and its residue there is $\dfrac{1}{2\pi i}\int_C f(z)dz = \dfrac{1}{2\pi i}\int_C \dfrac{b}{z-a}dz = \dfrac{1}{2\pi i}b.2\pi i = b$

or else, $\text{Res}(z = a) = \lim_{z \to a}(z-a)f(z) = b$

$\therefore \qquad \text{Res}(z = a) = b = -\text{Res}(z = \infty)$

THEOREM 1. *Residue of $f(z)$ at $z = \infty$ is the negative of the coefficient of $\dfrac{1}{z}$ in the expansion of $f(z)$ for values of z in the neighbourhood of $z = \infty$.*

PROOF. Let us suppose that $f(z)$ has a pole of order m at $z = \infty$ then $f\left(\dfrac{1}{z'}\right)$ has a pole of order m at $z' = 0$. Now by Laurent's expansion

of $f\left(\dfrac{1}{z'}\right)$ about $z' = 0$,

we have
$$f\left(\frac{1}{z'}\right) = \sum_{n=0}^\infty a_n z'^n + \sum_{n=1}^\infty b_n z'^{-n} \qquad ...(1)$$

Now putting $\dfrac{1}{z'} = z$ we get

$$f(z) = \sum_{n=0}^\infty a_n z^{-n} + \sum_{n=1}^\infty b_n z^n \qquad ...(2)$$

$$\therefore \qquad \int_C f(z)dz = \int_C \left(\sum_{n=0}^\infty a_n z^{-n}\right)dz + \int_C \left(\sum_{n=1}^\infty b_n z^n\right)dz \qquad ...(3)$$

$$= \sum_{n=0}^{\infty} a_n \int_C z^{-n} dz + \sum_{n=0}^{m} b_n \int_C z^n dz = a_1 \int_C z^{-1} dz = a_1 \int_C \frac{1}{z} dz \text{ (All other integrals vanish)} = a_1 2\pi i$$

$$\therefore \qquad \frac{1}{2\pi i} \int_C f(z) dz = a_1 \qquad \qquad \therefore \qquad -\frac{1}{2\pi i} \int_C f(z) dz = -a_1$$

or $\qquad$ Res $(z = \infty) = \dfrac{1}{2\pi i} \int_C f(z) dz = -a_1$

But $-a_1$ is the negative of the coefficient of $\dfrac{1}{z}$ in the expansion of $f(z)$ in the neighbourhood of $z = \infty$.

THEOREM 2. *Residue of $f(z)$ at $z = \infty$ is* $\lim\limits_{z \to \infty} \{-z f(z)\}$ *provided $f(z)$ is analytic at $z = \infty$.*

PROOF. $\qquad$ If $f(z)$ is analytic at $z = \infty$ then from (2)

$\qquad \qquad b_n = 0 \ \forall \ n$ such that $1 \le n \le m$.

$$\therefore \qquad\qquad f(z) = \sum_{n=0}^{\infty} a_n z^{-n} = a_0 + \frac{a_1}{z} + \frac{a_2}{z^2} + \dots \ \Rightarrow \ z f(z) = a_1 + \frac{a_2}{z} + \frac{a_3}{z^2} + \dots$$

If $a_0 = 0$

$\qquad \therefore \qquad\qquad \lim[z f(z)] = a_1$

$\qquad \therefore \qquad\qquad \lim[-z f(z)] = -a_1 = \text{Res}(z = \infty)$ provided $a_0 = 0$ and the limit has a definite value.

Solved Examples

EXAMPLE 1. **Find the residue of** $\dfrac{z^2}{z^2 + a^2}$ **at** $z = ia$.

$\qquad\qquad\qquad\qquad\qquad\qquad\qquad$ (PTU–2009S)

SOLUTION. Poles of $f(z)$ are given by $z^2 + a^2 = 0$, *i.e.* $z = \pm ia$.

$\qquad \therefore \qquad z = ia$ is a simple pole.

$\text{Res}(z = ia) = \lim\limits_{z \to ia} (z - ia) f(z)$

$$= \lim_{z \to ia} (z - ia) \frac{z^2}{(z - ia)(z + ia)} = \lim_{z \to ia} \frac{z^2}{z + ia} = \frac{i^2 a^2}{2ia} = \frac{1}{2} ia.$$

2nd Method

$f(z) = \dfrac{z^2}{z^2 + a^2} = \dfrac{\phi(z)}{\psi(z)}$ where $\psi(ia) = 0$ but $\phi(ia) \ne 0$.

$$\therefore \quad \text{Res}(z = ia) = \frac{\phi(ia)}{\psi'(ia)} = \left[\frac{z^2}{2z}\right] \text{ at } z = ia$$

$$= \frac{1}{2} ia.$$

EXAMPLE 2. **Find the residue of** $f(z) = \dfrac{z^3}{(z-1)^4 (z-2)(z-3)}$

at $z = 1$.

SOLUTION. Let us choose $\phi(z) = \dfrac{z^3}{(z-2)(z-3)}$

$$\therefore \qquad f(z) = \frac{\phi(z)}{(z-1)^4}$$

$\therefore \quad z = 1$ is a pole of order 4.

$\text{Res}(z = 1) = \dfrac{\phi^2(1)}{3!}$, $z = 1$ being a pole of order 4.

$$\phi(z) = \frac{z^3}{(z-2)(z-3)} = z + 5 - \frac{8}{z-2} + \frac{27}{z-3}$$

$\qquad\qquad\qquad\qquad\qquad\qquad$ (Partial fractions)

$$\phi^1(z) = 1 + \frac{8}{(z-2)^2} - \frac{27}{(z-3)^2}$$

$$\phi^2(z) = \frac{-16}{(z-2)^3} + \frac{54}{(z-3)^3}$$

$$\phi^3(z) = \frac{48}{(z-2)^4} - \frac{162}{(z-3)^4}$$

$$\therefore \quad \phi^3(1) = 41 - \frac{162}{16} = \frac{303}{8}$$

$$\therefore \quad \text{Res}(z = 1) = \frac{\phi^3(1)}{3!} = \frac{1}{6} \cdot \frac{303}{8} = \frac{303}{48} = \frac{101}{16}$$

2nd Method

Putting $(z - 1) = t$, *i.e.*, $z = 1 + t$ in $f(z)$, residue at $z = 1$ is the coefficient of $\dfrac{1}{t}$ in the expansion of $f(z)$ where t is small.

$$f(z) = \frac{1}{t^4} \left[\frac{(1+t)^3}{(t-1)(t-2)} \right]$$

$$= \frac{1}{2t^4} \left[(1+t)^3 \cdot (1-t) \left(1 - \frac{t}{2}\right)^{-1} \right]$$

$$= \frac{1}{2t^4} \Big[(1 + 3t + 3t^2 + t^3)(1 + t + t^2$$

$$+ t^3 + \dots) \left(1 + \frac{t}{2} + \frac{t}{4} t^2 + \frac{t}{8} t^3 \dots \right) \Big]$$

$$= \frac{1}{2t^4} \Big[(1 + 3t + 3t^2 + t^3)$$

$$\cdot \left(1 + \frac{3}{2} t + \frac{7}{4} t^2 + \frac{15}{8} t^3 + \dots \right) \Big]$$

$$\therefore \text{Coefficient of } 1/t = \frac{1}{2} \left(1 + \frac{9}{2} + \frac{21}{4} + \frac{15}{8}\right) = \frac{101}{16}.$$

EXAMPLE 3. **Find the residue of** $f(z) = \dfrac{z^3}{z^2 - 1}$ **at** $z = \infty$

SOLUTION. We have $f(z) = \dfrac{z^3}{z^2} \left(1 - \dfrac{1}{z^2}\right)^{-1}$

$$= z \left(1 + \frac{1}{z^2} + \frac{1}{z^4} + \frac{1}{z^6} + \dots \right)$$

$$= z + \frac{1}{z} + \frac{1}{z^3} + \frac{1}{z^5} + \dots$$

$\therefore \quad$ Res $(z = \infty) = -$ coefficient of $\dfrac{1}{z} = -(1) = -1$.

EXAMPLE 4. *Find the residue of* $\dfrac{z^2}{(z-a)(z-b)(z-c)}$ *at*

infinity.

SOLUTION. The given function is analytic at $z = \infty$ as $f\left(\dfrac{1}{z}\right)$ is

analytic at $z = 0$.

$$\text{Res}(z = \infty) = \lim_{z \to \infty}\left[-z f(z)\right]$$

$$= \lim_{z \to \infty}\left[-z \frac{z^2}{(z-a)(z-b)(z-c)}\right] = -1$$

From above example we conclude that a function though being analytic at infinity yet has a residue there.

THEOREM 1. **(Cauchy's Residue Theorem.)** *Let $f(z)$ be single valued and analytic within and on a closed contour C, except at a finite number of poles z_1, z_2,... z_n within C and if R_1, R_2,... R_n, be the residues of $f(z)$ at these poles then $\int_C f(z)dz = 2\pi i(R_1 + R_2 + ... + R_n)$ or $\int_C f(z)dz = 2\pi i \Sigma R$ where $\Sigma R = $ (Sum of the residues at the poles within C).*

PROOF. Let γ_1, γ_2, γ_3 ... γ_n be the circles with centres at z_1, z_2,... z_n respectively and radii so small that they lie entirely within the closed curve C and do not overlap. Now $f(z)$ is analytic within the region enclosed by the curve C and these circles. Therefore by Cauchy's Theorem for multi-connected region.

$$\int_C f(z)dz = \int_{\gamma_1} f(z)dz + \int_{\gamma_2} f(z)dz + ... + \int_{\gamma_n} f(z)dz$$

But by definition, we know that $\dfrac{1}{2\pi i}\int_{\gamma_1} f(z)dz = \text{Res}(z = z_1) = R_1$ etc.

$\therefore \quad \int_C f(z)dz = 2\pi i(R_1 + R_2 + ... + R_n)$

THEOREM 2. *If a function $f(z)$ is analytic except at finite number of singularities (including that at infinity) then the sum of the residues at these singularities is zero.*

PROOF. Let C be a closed contour which encloses all the singularities of $f(z)$ except that at infinity then by Theorem 1

$$\int_C f(z)dz = 2\pi i \Sigma R \quad \text{or} \quad \frac{1}{2\pi i}\int_C f(z)dz = \Sigma R$$

Also by definition we know that

$$-\frac{1}{2\pi i}\int_C f(z)dz = \text{Res}(z = \infty)$$

Adding the above two we get

$$\Sigma R + \text{Res}(z = \infty) = 0$$

i.e., Sum of the residues at these singularities is zero.

11.18.4 EVALUATION OF REAL DEFINITE INTEGRALS BY CONTOUR INTEGRATION

In this section we shall deal with the evaluation of real definite integrals. These integrals can however be evaluated of real definite integrals. These integrals can however be evaluated by our usual known methods but we shall here evaluate them by simpler method by using the Cauchy's theorem of residues. We shall choose a closed curve C and find the pole of $f(z)$ and calculate residues at those poles which lie within C. Then by Cauchy's theorem of residues, we have

$$\int_C f(z)dz = 2\pi i \text{ (Sum of the residues of } f(z) \text{ at the poles within } C)$$

The chosen closed curve is usually called a contour which may be a circle, semicircle or a quadrant of a circle. The process of integrating along a contour is termed as contour integration.

11.18.5 INTEGRATION ROUND THE UNIT CIRCLE

Here we shall consider integrals of the type

$$\int_0^{2\pi} \phi(\cos\theta.\sin\theta)d\theta$$

where ϕ is a rational function of $\cos\theta$ and $\sin\theta$.

we put $\quad z = e^{i\theta}$

$\therefore \qquad\qquad dz = ie^{i\theta}d\theta \qquad \text{or} \qquad d\theta = \frac{1}{i}\frac{dz}{z}$

Also $\qquad\qquad \cos\theta = \frac{1}{2}(e^{i\theta} - e^{-i\theta}) = \frac{1}{2}\left(z + \frac{1}{z}\right)$

$$\sin\theta = \frac{1}{2i}(e^{i\theta} - e^{-i\theta}) = \left(z - \frac{1}{z}\right)$$

Also $|z| = 1$, *i.e.,* C is the unit circle $|z| = 1$.

$\therefore \quad \int\limits_0^{2\pi} \phi(\cos\theta.\sin\theta)d\theta = \int \phi\left\{\frac{1}{2}(z + z^{-1})\frac{1}{2i}(z - z^{-1})\right\}\frac{dz}{iz} = \int_C f(z)dz$

where $f(z)$ is a rational function of z and the contour C is the unit cicle $|z| = 1$.

$\int_C f(z)dz$ shall be evaluated by using Cauchy's residues theorem.

THEOREM 1. If $\lim\limits_{z \to 0}(z-a)f(z) = A$ and if C is the arc $\theta_1 \leq \theta \leq \theta_2$ of the circle $|z-a| = r$ then

$$\lim\limits_{z \to 0}\int_C f(z)dz = iA(\theta_2 - \theta_1)$$

PROOF. Since $\lim\limits_{z \to 0}(z-a)f(z) = A$, therefore for a given ε we can find δ depending upon ε such that

$$|(z-a)f(z) - A| < \varepsilon \text{ for } |z-a| < \delta$$

But $|(z-a)| = r$ and hence if we choosed $r < \delta$ then

$$|(z-a)f(z) - A| < \varepsilon \text{ on the arc } C$$

$$\therefore \quad |z-a| f(z) = A + \eta \text{ where } |\eta| < \varepsilon$$

or $$f(z) = \frac{A+\eta}{z-a}$$

$$\therefore \quad \int_C f(z)dz = \int_C \frac{A+\eta}{re^{i\theta}}dz = \int_{\theta_1}^{\theta_2}\frac{A+\eta}{re^{i\theta}}re^{i\theta}.id\theta \qquad (\because \ z-a = re^{i\theta})$$

$$(A+\eta)i\int_{\theta_1}^{\theta_2}d\theta = (\theta_2-\theta_1)iA + (\theta_2-\theta_1)i\eta$$

$$\therefore \quad |\int_C f(z)dz - iA(\theta_2-\theta_1)| = |(\theta_2-\theta_1)i\eta| = (\theta_2-\theta_1)|\eta| < (\theta_2-\theta_1)\varepsilon$$

Taking limit when $\varepsilon \to 0$ and consequently $z = 0$, we have

$$\lim\limits_{z \to 0}\int_C f(z)dz = iA(\theta_2-\theta_1)$$

▶ REMARK

- $\lim\limits_{z \to 0}(z-a)f(z) = \text{Res}(z=a)$ for a simple pole.

THEOREM 2. If C is an arc $\theta_1 \leq \theta \leq \theta_2$ of the circle $|z| = R$ and if $\lim\limits_{R \to \infty} zf(z) = A$ then $\lim\limits_{R \to \infty}\int_C f(z)dz = iA(\theta_2-\theta_1)$.

PROOF. Since $\lim\limits_{R \to \infty}zf(z) = A$, we can choose R so large that

$$|zf(z)| - A < \varepsilon \text{ on the arc } C \qquad \text{or } zf(z) = A+\eta \text{ where } |\eta| < \varepsilon$$

$$\therefore \quad \int_C f(z)dz = \int_C \frac{A+\eta}{z}dz.$$

Put $z = Re^{i\theta}$

$$\therefore \quad \int_C f(z)dz = \int_{\theta_1}^{\theta_2}\frac{(A+\eta)}{Re^{i\theta}}Re^{i\theta}.id\theta = (A+\eta)i\int_{\theta_1}^{\theta_2}d\theta = (A+\eta)i(\theta_2-\theta_1)$$

$$\therefore \quad |\int_C f(z)dz - iA(\theta_2-\theta_1)| = |(\theta_2-\theta_1)i\eta| = (\theta_2-\theta_1)|\eta| < (\theta_2-\theta_1)\varepsilon$$

Taking limit when $\varepsilon \to 0$ and consequently $R \to \infty$, we get $\int_C f(z)dz = iA(\theta_2-\theta_1)$

In particular if $\lim\limits_{R \to \infty}zf(z) = 0$, then from above we have $\lim\limits_{R \to \infty}\int_C f(z)dz = 0$.

REMARK

$$\lim\limits_{R \to \infty}[-zf(z)] = \text{Res}(z = \infty)$$

Solved Examples

EXAMPLE 1. **Show that** $\int_0^{2\pi}\dfrac{d\theta}{2+\cos\theta} = \dfrac{2\pi}{\sqrt{3}}$ (PTU 2010)

SOLUTION. We shall use contour integration and apply Cauchy's theorem of residues to evaluate.

$$I = \int_0^{2\pi}\frac{d\theta}{2+\cos\theta}$$

We put $z = e^{i\theta}$ $\therefore dz = ie^{i\theta}d\theta$ or $\dfrac{1}{i}\cdot\dfrac{dz}{z} = d\theta$

Also $\cos\theta = \dfrac{1}{2}(e^{i\theta} + e^{-i\theta}) = \dfrac{1}{2}\left(z + \dfrac{1}{z}\right)$

$$\therefore \quad I = \frac{1}{i}\int_C \frac{dz}{z\left[2 + \dfrac{1}{2}\left(z + \dfrac{1}{z}\right)\right]} \text{ where } C \text{ is the unit circle}$$

$|z| = 1$

$$I = \frac{1}{i}\int_C \frac{2dz}{z^2 + 4z + 1}$$

$$= \frac{2}{i}\int_C \frac{1}{z^2+4z+1}dz = \frac{2}{i}\int_C f(z)dz \qquad \ldots (1)$$

Poles of $f(z)$ are given by $z^2 + 4z + 1 = 0$

$$\therefore \quad z = -2 \pm \sqrt{3}$$

Now $z = -2 + \sqrt{3}$ is the only pole which lies inside C.

$$\text{Res}(z = -2 + \sqrt{3}) = \frac{\phi(z)}{\psi'(z)}$$

at $z = -2 + \sqrt{3}$

$$= \frac{1}{2z + 4} \text{ at } z = -2 + \sqrt{3} = \frac{1}{2\sqrt{3}}$$

$$\therefore \quad \int_C f(z)dz = 2\pi i$$

(Sum of residues of poles which lie within C)

$$= 2\pi i . \frac{1}{2\sqrt{3}} = \frac{\pi i}{\sqrt{3}}$$

$$I = \frac{2}{i} . \frac{\pi i}{\sqrt{3}} = \frac{2\pi}{\sqrt{3}} \text{ by (1)}$$

EXAMPLE 2. **Show that** $\int_0^{2\pi} \dfrac{d\theta}{a + b\cos\theta} = \int_0^{2\pi} \dfrac{d\theta}{a + b\sin\theta}$

$$= \frac{2\pi}{\sqrt{[(a^2 - b^2)]}}, a > b > 0$$

SOLUTION. Put $z = e^{i\theta} \Rightarrow dz = e^{i\theta}i d\theta \therefore d\theta = \dfrac{dz}{iz}$

$$\therefore \quad I = \frac{1}{i}\int_C \frac{dz}{z\left[a + \frac{1}{2}b\left(z + \frac{1}{z}\right)\right]}, \text{ where } C \text{ is unit circle}$$

$|z| = 1$

or $\quad I = \dfrac{2}{i}\int_C \dfrac{dz}{bz^2 + 2az + b} = \dfrac{2}{i}\int_C f(z)dz \qquad ...(1)$

The poles of $f(z)$ are given by $bz^2 + 2az + b = 0$

$$z = \frac{-a \pm \sqrt{(a^2 - b^2)}}{b}$$

or $\quad z = \alpha = \dfrac{-a + \sqrt{(a^2 - b^2)}}{b}$;

$$z = \beta = \frac{-a - \sqrt{(a^2 - b^2)}}{b}$$

Now $a > b > 0 \therefore |\beta| > 1$.

Also $\alpha\beta = 1$ or $|\alpha| \; |\beta| = 1$

Since $|\beta| > 1$ therefore $|\alpha| < 1$. Hence $z = \alpha$ is the only pole which lies within C, i.e., $|z| = 1$.

$$\text{Res}(z = \alpha) = \lim_{z \to \alpha}(z - \alpha)f(z)$$

$$= \lim_{z \to \alpha}(z - \alpha).\frac{1}{(bz^2 + 2az + b)}$$

$$= \lim_{z \to \alpha}(z - \alpha).\frac{1}{b(z - \alpha)(z - \beta)} = \frac{1}{b(\alpha - \beta)}$$

$$= \frac{1}{b\left(2\dfrac{\sqrt{(a^2 - b^2)}}{b}\right)} = \frac{1}{2\sqrt{(a^2 - b^2)}}$$

$\therefore \int_C f(z)dz = 2\pi i$ (sum of the residues of the poles within C)

$$= 2\pi i.\frac{1}{2\sqrt{(a^2 - b^2)}} = \frac{\pi i}{\sqrt{(a^2 - b^2)}}$$

Hence from (1)

$$I = \frac{2}{i}\int_C f(z)dz = \frac{2}{i}\frac{\pi i}{\sqrt{(a^2 - b^2)}} = \frac{2\pi}{\sqrt{(a^2 - b^2)}}$$

Exactly as above we can evaluate the other integral.

SOLUTION. Proceeding as above.

$$I = \frac{4}{i}\int_C \frac{z \, dz}{(bz^2 + 2az + b)^2} = \frac{4}{i}\int_C f(z)dz$$

where $f(z) = \dfrac{z}{(bz^2 + 2az + b)^2} = \dfrac{z}{b^2(z - \alpha)^2} \dfrac{1}{(z - \beta)^2}$

where a and b have values written in part a. Clearly $z = \alpha$ is a pole of order two which lies within C, i.e., $|z| = 1$

Now Residue at pole of order m is

$$\lim_{z \to 0}\frac{1}{(m-1)!}\frac{d^{m-1}}{dz^{m-1}}[(z - a)^m f(z)]. \text{ Here } m = 2$$

$$\therefore \quad \text{Res}(z = \alpha) \lim_{z \to \alpha}\frac{1}{1!}$$

$$\frac{d}{dz}\left[(z - \alpha)^2.\frac{z}{b^2(z - \alpha)^2(z - \beta)^2}\right]$$

$$= \frac{1}{b^2}\frac{d}{dz}\frac{z}{(z - \beta)^2} \text{ at } z = \alpha.$$

$$= \frac{1}{b^2}\frac{(z - \beta)^2.1 - z.2(z - \beta)}{(z - \beta)^4} \text{ at } z = \alpha.$$

$$= \frac{1}{b^2}\frac{z - \beta - 2z}{(z - \beta)^3}$$

at $(z = \alpha) = -\dfrac{1}{b^2}\dfrac{(\alpha + \beta)}{b^2(\alpha - \beta)^3}$

$$= -\frac{1}{b^2} \times \frac{-2a}{b}\left(\frac{2\sqrt{(a^2 - b^2)}}{b}\right)^3$$

$$= \frac{a}{4(a^2 - b^2)^{\frac{3}{2}}}$$

$\therefore \int_C f(z)dz = 2\pi i$ (Sum of the residues of the poles which lie within C)

$$= 2\pi i\left(\frac{a}{4}\frac{1}{(a^2 - b^2)^{\frac{3}{2}}}\right)$$

$$\therefore \quad I = \frac{4}{i}\int_C f(z)$$

$$= \frac{4}{i}.2\pi i\left[\frac{a}{4}\frac{1}{(a^2 - b^2)^{\frac{3}{2}}}\right] = \frac{2\pi a}{(a^2 - b^2)^{\frac{3}{2}}}.$$

EXAMPLE 3. **Prove that** $\int_0^{2\pi} \dfrac{\cos 2\theta \, d\theta}{5 + 4\cos\theta} = \dfrac{\pi}{6}$. (UPTU–2010)

SOLUTION. Let $I = \int_0^{2\pi} \dfrac{\cos 2\theta \, d\theta}{5 + 4\cos\theta}$ = R.P. of $\int_0^{2\pi} \dfrac{e^{2i\theta}}{5 + 2(e^{i\theta} + e^{-i\theta})}d\theta$

Put $z = e^{i\theta} \therefore \dfrac{dz}{iz} = d\theta$

$$\therefore \quad I = \text{R.P. of } \int_C \frac{z^2}{5 + 2\left(z + \dfrac{1}{z}\right)}.\frac{dz}{iz}$$

$$= \text{R.P. of } \frac{1}{2} \int_C \frac{z^2}{(2z^2 + 5z + 2)} dz$$

$$= \text{R.P. of } \frac{1}{i} \int_C f(z) dz \qquad \dots (1)$$

where $f(z) = \dfrac{z^2}{2z^2 + 5z + 2}$ and C is the unit circle $|z| = 1$.

Poles of $f(z)$ are given by

$2z^2 + 5z + 2 = 0$ or $(z+2)(2z+1) = 0$.

$\therefore \quad z = -2, -\dfrac{1}{2}$

Clearly $z = -\dfrac{1}{2}$ is the pole lying within C

$$\text{Res}\left(z = -\frac{1}{2}\right) = \lim_{z \to -\frac{1}{2}}\left(z + \frac{1}{2}\right) f(z)$$

$$= \lim_{z \to -\frac{1}{2}}\left(z + \frac{1}{2}\right) \frac{z^2}{2(z+2)\left(z + \frac{1}{2}\right)}$$

$$= \lim_{z \to -\frac{1}{2}} \frac{z^2}{2(z+2)} = \frac{\frac{1}{4}}{2\left(\frac{3}{2}\right)} = \frac{1}{12}$$

Hence by Cauchy's residue theoerm.

$$\therefore \int_C f(z) dz = 2\pi i \text{ (Sum of residues of poles within C)}$$

$$= 2\pi i. \frac{1}{12} = \frac{\pi i}{6}$$

Also, from (1), we get $\text{I} = \text{R.P. of } \dfrac{1}{i}\left(\dfrac{\pi i}{6}\right) = \dfrac{\pi}{6}$

EXAMPLE 4. *Prove that* $\int_0^\pi \dfrac{a\, d\theta}{a^2 + \sin^2\theta} = \dfrac{\pi}{\sqrt{(1+a^2)}} (a > 0)$

(SVTU–2009)

SOLUTION. Let $I = \int_0^\pi \dfrac{2a\, d\theta}{2a^2 + 2\sin^2\theta} = \int_0^\pi \dfrac{2a\, d\theta}{2a^2 + 1 - \cos 2\theta}$

Put $2\theta = t$ $\quad \therefore \quad 2d\theta = dt$ and adjust the limits.

$$\therefore \quad I = \int_0^{2\pi} \frac{a\, dt}{2a^2 + 1 - \cos t}$$

Put $z = e^{i\theta}$ $\quad \therefore \quad \dfrac{dz}{iz} = dt$ and $\dfrac{1}{2}\left(z + \dfrac{1}{z}\right) = \cos t$

$$\therefore \quad I = \int_C \frac{a}{2a^2 + 1 - \frac{1}{2}\left(z + \frac{1}{z}\right)} \frac{dz}{iz} \text{ where } C \text{ is}$$

the unit circle $|z| = 1$.

$$= \frac{2a}{i} \int_C \frac{1}{(4a^2 + 2)z - (z^2 + 1)} dz$$

$$= \frac{2ai}{-i^2} \int_C \frac{1}{z^2 - 2z(2a^2 + 1) + 1} dz$$

$$= 2ai \int_C f(z) dz$$

where $\quad f(z) = \dfrac{1}{z^2 - 2z(2a^2 + 1) + 1}$

Poles of $f(z)$ are given by $z^2 - 2z(2a^2 + 1) + 1 = 0$.

$$\therefore \quad z = \frac{2(2a^2 + 1) \pm \sqrt{[4(2a^2 + 1)^2 - 4]}}{2}$$

$$z = (2a^2 + 1) + 2a\sqrt{(a^2 + 1)},$$

$$(2a^2 + 1) - 2a\sqrt{(a^2 + 1)} = \alpha, \beta \text{ say}$$

Clearly $|\alpha| > 1$ and since $\alpha\beta = 1$ or $|\alpha| - |\beta| = 1$

$\therefore \qquad |\beta| < 1$

Hence $z = \beta$ is the only pole lying inside C.

$$\text{Res}(z = \beta) = \lim_{z \to \beta}(z - \beta)f(z)$$

$$= \lim_{z \to \beta}(z - \beta) \frac{1}{(z - \alpha)(z - \beta)}$$

$$= \frac{1}{\beta - \alpha} = \frac{1}{-4a\sqrt{(a^2 + 1)}}$$

Hence by Cauchy's theorem

$$\int_C f(z) dz = 2\pi i \text{ (Sum of residues of poles within C)}$$

$$= 2\pi i\left[-\frac{1}{4a\sqrt{(a^2 + 1)}}\right] = \frac{-\pi i}{2a\sqrt{(a^2 + 1)}}$$

$$\therefore \quad I = 2ai\int_C f(z) dz = 2ai \frac{-\pi i}{2a\sqrt{(a^2 + 1)}} = \frac{\pi}{\sqrt{(a^2 + 1)}}.$$

EXAMPLE 5. *Use the method of contour integration to prove that* $\int_0^{2\pi} \dfrac{d\theta}{1 + a^2 - 2a\cos\theta} = \dfrac{2\pi}{1 - a^2}, 0 < a < 1$

(HISSAR–2007; MUMBAI–2006; KERALA–2005; JNTU–2006; MADRAS–2006, ANNA–2003)

SOLUTION. Let $I = \int_0^{2\pi} \dfrac{d\theta}{1 + a^2 - a(e^{i\theta} + e^{-i\theta})}$

Put $z = e^{i\theta}$ $\quad \therefore \quad \dfrac{dz}{iz} = d\theta$

$$\therefore I = \int_C \frac{1}{1 + a^2 - a\left(z + \frac{1}{z}\right)} \frac{dz}{iz}$$

$$= -\frac{1}{ia} \int_C \frac{dz}{z^2 - z\left(a + \frac{1}{a}\right) + 1}$$

$$= -\frac{1}{ia} \int_C \frac{dz}{(z - a)(z - 1/a)} = -\frac{1}{ia}\int_C f(z) dz \quad \dots (1)$$

Poles of $f(z)$ are $z = a, \dfrac{1}{a}$. Clearly $z = a$ is the pole which lies within C, *i.e.*, $|z| < 1$ as $a < 1$.

$$\therefore \quad \text{Res}(z = a) = \lim_{z \to 0}(z - a)f(z)$$

$$= \lim_{z \to 0}(z - a)\frac{1}{(z - a)\left(z - \frac{1}{a}\right)} = \frac{1}{a - \frac{1}{a}} = \frac{a}{a^2 - 1}$$

Hence by Cauchy's residue theorem

$$\int_C f(z) = 2\pi i \text{ (Sum of residues of poles within C)}$$

$$= 2\pi i\frac{a}{a^2 - 1}$$

Hence from (1), we get

$$I = -\frac{1}{ai} 2\pi i.\frac{a}{a^2 - 1} = \frac{-2\pi}{a^2 - 1} = \frac{2\pi}{1 - a^2}.$$

EXAMPLE 6. *Evaluate* $\int_0^\infty \dfrac{\cos ax}{x^2+1}\,dx.$ (UPTU–2006; DELHI–2002)

SOLUTION. Consider $\int_C \dfrac{e^{iaz}}{z^2+1}\,dz = \int_C f(z)\,dz$

where C is the contour consisting of the semi-circle C_R of radius R together with the part of real axis from $-R$ to R. The integral has simple poles at $z=i$ and $z=-i$, of which $z=i$ only lies inside C.

∴ By residue theorem,

$\int_C f(z)\,dz = 2\pi i\,\mathrm{Res}\,f(i) = 2\pi i \lim\limits_{z\to i}[(z-i)f(z)]$

$\quad = 2\pi i \lim\limits_{z\to i}\dfrac{(z-i)e^{iaz}}{z^2+1} = 2\pi i \lim\limits_{z\to i}\dfrac{e^{iaz}}{z+i} = \pi e^{-a}$ ... (1)

Also $\int_C f(z)\,dz = \int_{-R}^R f(x)\,dx + \int_{C_R} f(z)\,dz$ ...(2)

Now $|z|=R$ on C_R and $|z^2+1| \ge R^2-1$

Also $|e^{iaz}| = |e^{ia(x+iy)}| = |e^{iax}\cdot e^{-ay}| = e^{-ay} < |$ [∵ $y>0$]

∴ $\left|\dfrac{e^{iaz}}{z^2+1}\right| = |e^{iaz}|\cdot\dfrac{1}{|z^2+1|} < 1\cdot\dfrac{1}{R^2-1}$

Thus $\int_{C_R} f(z)\,dz = \left|\int_{C_R}\dfrac{e^{iaz}}{z^2+1}\,dz\right|$

$\qquad < \int_{C_R}\dfrac{1}{R^2-1}\,|dz| < \dfrac{\pi R}{R^2-1}$

which $\to$ to 0 as $R\to\infty$ ...(3)

Hence from (1), (2) and (3), we get

$\pi e^{-a} = \int_{-\infty}^\infty f(x)\,dx + 0$

or $\int_{-\infty}^\infty \dfrac{e^{iax}}{x^2+1}\,dx = \pi e^{-a}$

Equating real parts from both sides, we get

$$\int_{-\infty}^\infty \frac{\cos ax}{x^2+1}\,dx = \pi e^{-a}$$

∵ $\cos ax/(x^2+1)$ is an even function of x, we have

$2\int_0^\infty \dfrac{\cos ax}{x^2+1}\,dx = \pi e^{-a}$ or $\int_0^\infty \dfrac{\cos ax}{x^2+1}\,dx = \dfrac{\pi}{2}e^{-a}$

EXAMPLE 7. *By integrating around a unit circle, evaluate*

$\int_0^{2\pi} \dfrac{\cos 3\theta}{5-4\cos\theta}\,d\theta.$

(SVTU–2009; UPTU–2009; MADRAS–2003)

SOLUTION. Putting $z=e^{i\theta}$,

i.e., $d\theta = \dfrac{dz}{iz}$, $\cos\theta = \dfrac{1}{2}\left(z+\dfrac{1}{z}\right)$

and $\cos 3\theta = \dfrac{1}{2}(e^{3i\theta}+e^{-3i\theta}) = \dfrac{1}{2}\left(z^3+\dfrac{1}{z^3}\right)$

∴ The given integral $I = \int_C \dfrac{\dfrac{1}{2}\left(z^3+\dfrac{1}{z^3}\right)}{5-2\left(z+\dfrac{1}{z}\right)}\cdot\dfrac{dz}{iz}$

$= \dfrac{-1}{2i}\int_C \dfrac{z^6+1}{z^3(2z^2-5z+2)}\,dz = \dfrac{-1}{2i}\int_C \dfrac{(z^6+1)\,dz}{z^3(2z-1)(z-2)}$

$= \dfrac{-1}{2i}\int f(z)\,dz,$

where C is the unit circle $|z|=1$.

Now $f(z)$ has a pole of order 3 at $z=0$ and simple poles at $z=\dfrac{1}{2}$ and $z=2$. Of these only $z=0$ and $z=\dfrac{1}{2}$ lie within the unit circle.

∴ $\mathrm{Res}\,f\left(\dfrac{1}{2}\right) = \lim\limits_{z\to 1/2}\dfrac{\left(z-\dfrac{1}{2}\right)(z^6+1)}{(2z-1)(z-2)}$

$\qquad = \lim\limits_{z\to 1/2}\left\{\dfrac{z^6+1}{2z^3(z-2)}\right\} = \dfrac{-65}{24}$

$\mathrm{Res}\,f(0) = \dfrac{1}{(n-1)!}\left\{\dfrac{d^{n-1}}{dz^{n-1}}[(z-0)^n f(z)]\right\}_{z=0}$

$\hfill$ where $n=3$

$= \dfrac{1}{2}\left\{\dfrac{d^2}{dz^2}\left(\dfrac{z^6+1}{2z^2-5z+2}\right)\right\}_{z=0}$

$= \dfrac{d}{dz}\left[\dfrac{(2z^2-5z+2)6z^5-(z^6+1)(4z-5)}{2(2z^2-5z+2)^2}\right]$ at $z=0$.

$= \left\{\dfrac{d}{dz}\left(\dfrac{8z^7-25z^6+12z^5-4z+5}{2(2z^2-5z+2)^2}\right)\right\}_{z=0}$

$= \left[\dfrac{(2z^2-5z+2)^2(56z^6-150z^5+60z^4-4)}{2(2z^2-5z+2)^4}\right.$
$\qquad \dfrac{-(8z^7-25z^6+12z^5-4z+5)}{}$
$\qquad \left.\dfrac{2(2z^2-5z+2)(4z-5)}{2(2z^2-5z+2)^4}\right]_{z=0}$

$= \dfrac{21}{8}$

Hence $I = \dfrac{-1}{2i}\{2\pi i\,[\mathrm{Res}\,f(1/2)+\mathrm{Res}\,f(0)]\}$

$= -\pi\left[\dfrac{-65}{24}+\dfrac{21}{8}\right] = \dfrac{\pi}{12}.$

EXAMPLE 8. *Determine the poles of the function* $f(z) = z^2/(z-1)^2(z+2)$ *and the residue at each pole, Hence evaluate* $\oint_C f(z)\,dz$ *where C is the circle* $|z| = 2.5.$ (SVTU–2008; JNTU–2005)

SOLUTION. $\lim\limits_{z\to-2}\{(z+2)f(z)\} = \lim\limits_{z\to-2}\dfrac{z^2}{(z-1)^2} = \dfrac{4}{9}$

which is finite and non-zero, the function has a simple pole at $z=-2$ and $\mathrm{Res}\,f(-2) = 4/9$.

Also $\lim\limits_{z\to 1}[(z-1)^2 f(z)]$ is finite and non-zero, $f(z)$ has a pole of order two at $z=1$.

∴ $\mathrm{Res}\,f(1) = \dfrac{1}{1!}\left[\dfrac{d}{dz}\{(z-1)^2 f(z)\}\right]_{z=1}$

$= \left[\dfrac{d}{dz}\left(\dfrac{z^2}{z+2}\right)\right]_{z=1} = \left[\dfrac{z^2+4z}{(z+2)^2}\right]_{z=1} = \dfrac{5}{9}$

[Otherwise writing $z=1+t$,]

$f(z) = \dfrac{(1+t)^2}{t^2(3+t)} = \dfrac{1}{3t^2}(1+t)^2(1+t/3)^{-1}$

$$= \frac{1}{3t^2}(1+t)^2\left[1-\frac{t}{3}+\frac{t^2}{9}\cdots\right]$$

$$= \frac{1}{3t^2}\left[1+\frac{5t}{3}+\frac{4t^2}{9}+\cdots\right] = \frac{1}{3t^2}+\frac{5}{9t}+\frac{4}{27}+\cdots \quad \ldots(1)$$

$$\therefore \ \text{Res} f(1) = \text{coefficient of } \frac{1}{t} \text{ in } (1) = \frac{5}{9}$$

Clearly $f(z)$ is analytic on $|z| = 2.5$ and at all points inside except the poles $z = -2$ and $z = 1$. Hence by residue theorem.

$$\oint_C f(z)dz = 2\pi i[\text{Res} f(-2) + \text{Res} f(1)]$$

$$= 2\pi i\left[\frac{4}{9}+\frac{5}{9}\right] = 2\pi i.$$

11.18.6 EVALUATION OF THE INTEGRALS OF THE FORM $\int_{-\infty}^{\infty} f(x)dx$

THEOREM 1. *If $f(z)$ is a function which is analytic in the upper half of the z-plane except at a finite number of poles in it, having again no poles on the real axis and if further $zf(z)$ tends to zero as $|z|$ tends to infinity then by contour integration*

$$\int_{-\infty}^{\infty} f(x)dx = 2\pi i\Sigma R^+$$

where ΣR^+ represents the sum of the residues of the poles in the upper half plane.

PROOF. Above type of integrals are evaluated by integrating $f(z)$ round a contour C consisting of a semi-circle Γ of radius R large enough to include all the poles of $f(z)$ and the part of real axis from $-R$ to $+R$. By given conditions of the question the only singularities of the function $f(z)$ in the upper half of the z-plane are poles.

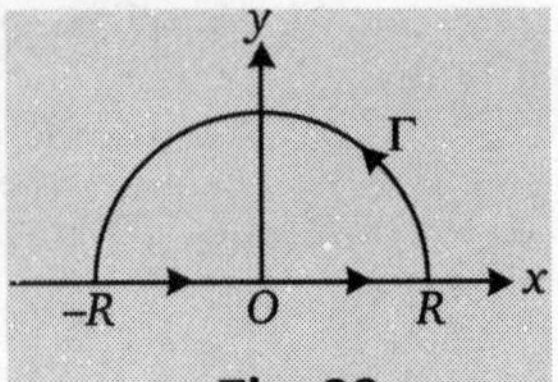

Therefore by Cauchy's residues theorem.

$$\int_C f(z)dz = 2\pi i \text{ (sum of the residues of the poles within } C)$$

Now C consists of semicircle Γ and real axis for which $z = x$.

$$\therefore \quad \int_{-R}^{R}f(z)dx + \int_\Gamma f(z)dz = 2\pi i\Sigma R^+ \quad \ldots(1)$$

where ΣR^+ denotes the sum of residues of the poles of $f(z)$ in the upper half plane and there being no poles on the real axis. Here we shall make use of the following result that

If C is arc $\theta_1 \le \theta \le \theta_2$ of the circle $|z| = R$ and if $\lim\limits_{z\to\infty} zf(z) = A$ then $\lim\limits_{R\to\infty}\int_C f(z)dz = iA(\theta_2 - \theta_1)$

and in particular if $\lim\limits_{z\to\infty} zf(z) = 0$ then $\lim\limits_{R\to\infty}\int_C f(z)dz = 0$ on putting $A=0$

Here we are given that $\lim zf(z) = 0$ as $|z|\to\infty$ therefore $\int_\Gamma f(z)dz = 0$ by the theorem stated above.

Also $\lim\limits_{R\to\infty}\int_{-R}^{R}f(x)dx = P\int_{-\infty}^{\infty}f(x)dx$; where P denotes the principal value of the integral.

By given conditions the integral $\int_{-\infty}^{\infty}f(x)dx$ is convergent and as such $P\int_{-\infty}^{\infty}f(x)dx = \int_{-\infty}^{\infty}f(x)dx$

Therefore proceeding to limits when $R \to \infty$ we have from (1).

$$\int_{-\infty}^{\infty}f(x)dx = 2\pi i\Sigma R^+.$$

Fig. 29

Solved Examples

EXAMPLE 1. *Prove that* $\int_0^\infty \dfrac{dx}{1+x^2} = \dfrac{\pi}{2}$.

SOLUTION. Let us consider $\int_C f(z)dz$ where $f(z) = \dfrac{1}{1+z^2}$ where C is the contour consisting of a large semi-circle Γ of radius R along with the part of real axis from $x = -R$ to $x = +R$.

$$\therefore \quad \int_C f(z)dz = \int_{-R}^{R}\frac{f(x)}{1+x^2}dx + \int_\Gamma \frac{dz}{1+z^2} = 2\pi i\Sigma R^+ \quad \ldots(1)$$

Now, $\lim\limits_{z\to\infty} zf(z) = \lim\limits_{z\to\infty}\dfrac{z}{1+z^2} = 0$

Also $\lim\limits_{R\to\infty}\int_{-R}^{R}\dfrac{dx}{1+x^2} = \int_{-\infty}^{\infty}\dfrac{dx}{1+x^2}$

Therefore from (1) by Theorem (1) we have

$$\int_{-\infty}^{\infty}\frac{dx}{1+x^2} = 2\pi i\Sigma\mathbf{R}^+ \quad \ldots(2)$$

Now poles of $f(z) = \dfrac{1}{1+z^2}$ are $z = \pm i$ out of which

only one pole $z=i$ lies inside C.

$$\text{Res}(z = i) = \lim\limits_{z\to i}(z-i)f(z)$$

$$= \lim\limits_{z\to i}(z-i)\frac{1}{(z-i)(z+i)} = \frac{1}{2i}$$

$$\therefore \quad \int_{-\infty}^{\infty}\frac{dx}{1+x^2} = 2\pi i\frac{1}{2i} = \pi \quad \text{[by (2)]}$$

or $2\int_0^\infty \dfrac{dx}{1+x^2} = \pi$ or $\int_0^\infty \dfrac{dx}{1+x^2} = \dfrac{\pi}{2}$

EXAMPLE 2. *Prove that* $\int_0^\infty \dfrac{dx}{(1+x^2)^2} = \dfrac{\pi}{4}$.

SOLUTION. Proceeding exactly as above we arrive at result

i.e., $\int_{-\infty}^{\infty}\dfrac{dx}{(1+x^2)^2} = 2\pi i\Sigma R^+$

Now poles of $f(z) = \dfrac{1}{(1+z^2)^2}$ are $z = \pm i$ (twice) out of which only pole $z=i$ (of order two) lies inside C.

Now $f(z) = \dfrac{1}{(z-i)^2(z+i)^2} = \dfrac{\phi(z)}{(z-i)^2}$

where $\phi(z) = \dfrac{1}{(z+i)^2}$

$\therefore \quad \text{Res}(z = i) = \dfrac{\phi'(z)}{1!} = -\dfrac{2}{(z+i)^3} = -\dfrac{2}{(2i)^3} = \dfrac{1}{4i}$

$\because \quad \text{Res } (z = a) \text{ (order } n) \text{ of } \dfrac{\phi(z)}{(z-a)^n} \text{ is } \dfrac{\phi^{n-1}(a)}{(n-1)!}$

$\therefore \quad \int_{-\infty}^{\infty} \dfrac{dx}{(1+x^2)^2} = 2\pi i \dfrac{1}{4i} = \dfrac{\pi}{2}$

or $2\int_0^{\infty} \dfrac{dx}{(1+x^2)^2} = \dfrac{\pi}{2}$ or $\int_0^{\infty} \dfrac{dx}{(1+x^2)^2} = \dfrac{\pi}{4}$

EXAMPLE 3. *Evaluate* $\int_0^{\infty} \dfrac{x^2 dx}{(x^6+1)}$. (ROHTAK–2006)

SOLUTION. Here $f(z) = \dfrac{z^2}{(z^6+1)}$ and $\lim\limits_{z \to \infty} zf(z) = \lim\limits_{z \to \infty} \dfrac{z^3}{z^6+1} = 0$

Therefore proceeding as in Ex.1 we arrive at result (2).

$$\int_{-\infty}^{\infty} \dfrac{x^2 dx}{x^6+1} = 2\pi i \Sigma R^+$$

Now poles of $f(z) = \dfrac{z^3}{z^6+1}$ are given by $z^6 + 1 = 0$

$\therefore \quad z = (-1)^{1/6} = \dfrac{\cos(2n\pi + \pi)}{6}$

$\qquad\qquad + i\dfrac{\sin(2n\pi + \pi)}{6}, n = 0, 1, 2, 3, 4, 5$

or $z = e^{\pi i/6}, e^{3\pi i/6}, e^{5\pi i/6}, e^{7\pi i/6}\ e^{9\pi i/6}, e^{11\pi i/6}$

Out of these six poles the amplitude of first three is less than π and as such they lie within C.

If α denotes any of the above poles then $\alpha^6 = -1$.

$\text{Res}(z = \alpha) = \lim\limits_{z \to \alpha}(z - \alpha)f(z)$

$\qquad = \lim\limits_{z \to \alpha}(z - \alpha)\dfrac{z^2}{z^6+1}$

$\qquad = \lim\limits_{z \to \alpha}\dfrac{z^3 - \alpha z^2}{z^6 - \alpha^6} \left(\text{form } \dfrac{0}{0}\right)$

$\qquad = \lim\limits_{z \to \alpha}\dfrac{3z^2 - 2\alpha z}{6z^5} = \dfrac{\alpha^2}{6\alpha^5}$

$\qquad = \dfrac{\alpha^3}{6\alpha^6} = -\dfrac{\alpha^3}{6}$.

Now putting $\alpha = e^{\pi i/6}, e^{3\pi i/6}, e^{5\pi i/6}$ we get the residues at the three poles which lie within C as

$-\dfrac{1}{6}e^{\pi i/2} = -\dfrac{1}{6}.i, -\dfrac{1}{6}e^{3\pi i/2}$

$\qquad = -\dfrac{1}{6}(-i), -\dfrac{1}{6}e^{5\pi i/2} = -\dfrac{1}{6}e^{\pi i/2} = -\dfrac{1}{6}(i)$

$\therefore \quad$ Sum of the residues $= -\dfrac{1}{6}(i - i + i) = \dfrac{-1}{6}i$

$\therefore \quad \int_{-\infty}^{\infty} \dfrac{x^2 dx}{x^4+1} = 2\pi i \Sigma R^+ = 2\pi i\left(-\dfrac{1}{6}i\right) = \dfrac{\pi}{3}$

or $2\int_0^{\infty} \dfrac{x^2 dx}{x^6+1} = \dfrac{\pi}{3}$ or $\int_0^{\infty} \dfrac{x^2 dx}{x^6+1} = \dfrac{\pi}{6}$.

EXAMPLE 4. *Use the method of contour integration, prove*

that $\int_0^{\infty} \dfrac{x^6 dx}{(a^4+x^4)^2} = \dfrac{3\pi\sqrt{2}}{16a}, a > 0.$

SOLUTION. Here $f(z) = \dfrac{z^6}{(a^4+x^4)^2}$ and

$$\lim\limits_{z \to \infty} zf(z) = \lim\limits_{z \to \infty}\dfrac{z^7}{(a^4+z^4)^2} = 0$$

Therefore proceeding an in Ex. 1, we arrive at result (2)

$$\int_{-\infty}^{\infty}\dfrac{x^6 dx}{(a^4+x^4)^2} = 2\pi i \Sigma R^+$$

Now poles of $f(z)$ are given by $(a^4 + z^4)^2 = 0$

$\therefore \quad z = ae^{\pi i/4}$ and $ae^{3\pi i/4}, ae^{5\pi i/4}, ae^{7\pi i/4}$ each repeated twice.

Out of these the amplitude of first two is less than π and as such they lie within C.

$\therefore \quad \alpha = ae^{\pi i/4}$ and $\beta = ae^{3\pi i/4}$ are the two poles each of order two whose residues we are to calculate. For this we shall put $z = \alpha + t$ in $f(z)$ and collect the coefficient of $1/t$.

$f(\alpha + t) = \dfrac{(\alpha + t)^6}{[a^4 + (\alpha + t)^4]^2}$

$\qquad = \dfrac{(\alpha + t)^6}{[a^4 + \alpha^4 + 4\alpha^3 t + 6\alpha^2 t^2 + ...]^2}$

$\qquad = \dfrac{(\alpha + t)^6}{[4\alpha^3 t + 6\alpha^2 t^2 + ...]^2}$ as $a^4 + \alpha^4 = 0$

$\qquad = \dfrac{(\alpha + t)^6}{16\alpha^6 t^2}\left[1 + \dfrac{3}{2\alpha}t\right]^{-2}$

$\qquad = \dfrac{(\alpha^5 + 6\alpha^5 t + ...)\left(1 - \dfrac{3}{\alpha}.t - ...\right)}{16\alpha^6 t^2}$

$\therefore$ Coefficient of $\dfrac{1}{t}$ is $\dfrac{1}{16\alpha^6}[6\alpha^5 + 3\alpha^5] = \dfrac{3}{16\alpha}$.

Similarly $\text{Res}(z = \beta) = \dfrac{3}{16\beta}$

$\therefore$ Sum of the residues of poles within

$C = \dfrac{3}{16}\left(\dfrac{1}{\alpha} + \dfrac{1}{\beta}\right) = \dfrac{3}{16a}\left[e^{\frac{-\pi i}{4}} + e^{\frac{-3\pi i}{4}}\right]$

$\qquad = \dfrac{3}{16a}\left[e^{\frac{-\pi i}{4}} - e^{\frac{\pi i}{4}}\right]$ as $e^{-\pi i} = -1$

$\qquad = \dfrac{3}{16a}\left[-2i\sin\dfrac{\pi}{4}\right] = -\dfrac{3i\sqrt{2}}{16a}$

$\therefore \quad \int_{-\infty}^{\infty}\dfrac{x^6 dx}{(a^4+x^4)^2} = 2\pi i \Sigma R^+$

$\qquad = 2\pi i\left(-\dfrac{3i\sqrt{2}}{16a}\right) = \dfrac{3\pi\sqrt{2}}{8a}$

or $2\int_0^{\infty}\dfrac{x^6 dx}{(a^4+x^4)^2} = \dfrac{3\pi\sqrt{2}}{8a}$

or $\int_0^{\infty}\dfrac{x^6 dx}{(a^4+x^4)^2} = \dfrac{3\pi\sqrt{2}}{16a}$.

EXAMPLE 5. ***Prove that*** $\int_{-\infty}^{\infty} \dfrac{x^2 - x + 2}{x^4 + 10x^2 + 9} dx = \dfrac{5\pi}{12}$.

(AMIETE–2003; DELHI–2002)

SOLUTION. Here $f(z) = \dfrac{z^2 - z + 2}{z^4 + 10z^2 + 9}$ and $\lim\limits_{z \to \infty} zf(z) = 0$

Therefore proceeding as usual we arrive at result

$$\int_{-\infty}^{\infty} \frac{x^2 - x + 2}{x^4 + 10x^2 + 9} dx = 2\pi i \Sigma R^+ \qquad \ldots(1)$$

Poles of $f(z)$ are given by $z^4 + 10z^2 + 9 = 0$

or $(z^2 + 9)(z^2 + 1) = 0$

$\therefore \quad z = \pm i, \pm 3i$ are the four poles out of which only two poles given by $z = i$ and $z = 3i$ lie within C.

11.18.7 JORDAN'S INEQUALITY

If $0 \le \theta \le \dfrac{\pi}{2}$ then the inequality $\dfrac{2}{\pi} \le \dfrac{\sin \theta}{\theta} \le 1$ or $\dfrac{2\theta}{\pi} \le \sin \theta \le \theta$ is known as Jardon's inequality.

THEOREM 2. **(Jordan's Lemma).** *If (z) is analytic except at finite number of singularities and if $f(z) \to 0$ uniformly as $|z| \to \infty$ then*

$$\lim_{R \to \infty} \int_{\Gamma} e^{imz} f(z) dz = 0 \quad m > 0$$

where Γ denotes the semicircle $|z| = R, I(z) > 0$.

PROOF. R is chosen so large that all the singularities of $f(z)$ lie within the semicircle Γ and none on its boundary.

Since $f(z) \to 0$ uniformly as $|z| \to \infty$. Therefore, $\exists \in > 0$ such that $|f(z)| < \in \forall z$ on Γ.

Now $\left| \int_{\Gamma} e^{imz} f(z) dz \right| \le \int_{\Gamma} |e^{imz}| \, |f(z)| \, |dz|$.

Put $z = Re^{i\theta}$

$$< \int_0^\pi |e^{imR(\cos\theta + i\sin\theta)}| \, |Re^{i\theta} id\theta| = \in \int_0^\pi 1 \cdot e^{-mR\sin\theta} \cdot Rd\theta = 2\in R \int_0^{\pi/2} e^{-mR\sin\theta} \cdot d\theta$$

$$\le 2 \in R \int_0^{\pi/2} e^{-\left(\frac{mR2\theta}{\pi}\right)} d\theta \text{ as } \frac{2\theta}{\pi} \le \sin\theta \le \theta$$

$$= \frac{\pi \in}{m} [1 - e^{-2mR}] \to 0 \text{ as } R \to \infty \text{ and } \in \to 0$$

$\therefore \quad \lim\limits_{R \to \infty} \int_{\Gamma} e^{imz} f(z) dz = 0$

AN IMPORTANT INTEGRAL

Evaluation of the integral of the form

$$\int_{-\infty}^{\infty} \frac{P(x)}{Q(x)} \sin mx \, dx, \int_{-\infty}^{\infty} \frac{P(x)}{Q(x)} \cos mx \, dx, m > 0$$

where
(i) $P(x)$ and $Q(x)$ are polynomials
(ii) degree of $Q(x)$ exceeds that of $P(x)$
(i) $Q(x) = 0$ has no real roots.

$\text{Res}(z = i) = \lim\limits_{z \to i} (z - i) f(z)$

$$= \lim_{z \to i} (z - i) \frac{z^2 - z + 2}{(z - i)(z + i)(z^2 + 9)}$$

$$= \frac{i^2 - i + 2}{2i(i^2 + 9)} = \frac{1 - i}{16i}.$$

Similarly $\text{Res}(z = 3i) = \dfrac{7 + 3i}{48i}$

$\therefore$ Sum of the residues $= \dfrac{1 - i}{16i} + \dfrac{7 + 3i}{48i} = \dfrac{10}{48i} = \dfrac{5}{24i}$

Hence, from (1), we have

$$\int_{-\infty}^{\infty} \frac{x^2 - x + 2}{x^4 + 10x^2 + 9} dx = 2\pi i \cdot \frac{5}{24i} = \frac{5}{12}\pi.$$

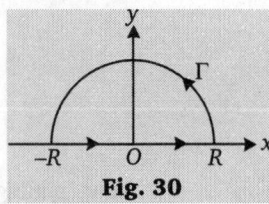

Fig. 30

Above types of integrals are evaluated by integrating $\int_C e^{imz} f(z) dz$ where $f(z) = \dfrac{P(z)}{Q(z)}$ round a contour C consisting of a semi-circle Γ of radius R large enough to include all the poles of integrand in the upper half plane and also part of the real axis from $x = -R$ to $x = +R$. Therefore by Cauchy's residues theorem

$\int_C f(z) dz = 2\pi i$ (Sum of residues of the poles of the integrand within C)

Now C consists of semi circle Γ and real axis for which $z = x$ from $x = -R$ to $x = +R$.

$\therefore \qquad \int_{-R}^{R} e^{imx} f(x) dx + \int_{\Gamma} e^{imz} f(z) dz = 2\pi i \Sigma R^+ \qquad \ldots (1)$

where ΣR^+ denotes the sum of the residues of poles.

Now by Jordan's Lemma theorem (2), we know that

$$\lim_{R \to \infty} \int_{\Gamma} e^{imx} f(z) dz = 0 (m > 0)$$

As $f(z) = \dfrac{P(z)}{Q(z)}$ and by given condition $f(z) \to 0$ as $z \to \infty$.

Hence when $R \to \infty$ we have from (1) the principal value of
$$\int_{-\infty}^{\infty} e^{imx} f(x) dx = 2\pi i \Sigma R^+$$

or
$$\int_{-\infty}^{\infty} f(x)(\cos mx + i \sin mx) dx = 2\pi i \Sigma R^+$$

or
$$\int_{-\infty}^{\infty} f(x) \cos mx dx + i \int_{-\infty}^{\infty} f(x) \sin mx dx = 2\pi i \Sigma R^+$$

Equating real and imaginary parts we get the values of the given integrals.

 Solved Examples

EXAMPLE 1. *By the method of contour integration, prove that* $\int_{0}^{\infty} \dfrac{\cos mx}{a^2 + x^2} dx = \dfrac{\pi}{2a} e^{-ma} (m \geq 0)$

and deduce the following

(a) $\int_{0}^{\infty} \dfrac{\cos x}{1 + x^2} dx = \dfrac{\pi}{2e}$

(b) $\int_{0}^{\infty} \dfrac{x \sin mx}{a^2 + x^2} dx = \dfrac{\pi}{2} e^{-ma}$

(c) $\int_{0}^{\infty} \dfrac{x \sin x}{x^2 + a^2} dx = \dfrac{\pi}{2} e^{-a}$

SOLUTION. Consider the integral

$$\int_C \frac{e^{miz}}{a^2 + z^2} dz = \int_C e^{miz} f(z) dz = \int_C \psi(z) dz$$

where C is the contour consisting of a semicircle Γ of radius R large enough to include all the poles of integrand in the upper half plane and also part of the real axis from $x = -R$ to $x = +R$.

Hence by Cauchy's residues theorem

$$\int_C e^{miz} f(z) dz = \int_{-R}^{R} e^{mix} f(x) dx + \int_\Gamma e^{miz} f(z) dz$$

$$= 2\pi i \Sigma R^+ \qquad \ldots(1)$$

where ΣR^+ denotes the sum of the residues of the poles of $\psi(z)$ which lie within C.

Now $\lim\limits_{z \to \infty} f(z) = \lim\limits_{z \to \infty} \dfrac{1}{a^2 + z^2} = 0$ and hence by Jordan's

Lemma $\lim\limits_{z \to \infty} \int_\Gamma e^{miz} f(z) dz = 0 \qquad \ldots(2)$

Hence as $R \to \infty$ we have from (1) and (2)

$$\int_{-\infty}^{\infty} e^{mix} f(x) dx = 2\pi i \Sigma R^+ \qquad \ldots(3)$$

Now poles of the integrand $\psi(z) = e^{miz} f(z) = e^{miz} \dfrac{1}{a^2 + z^2}$ are given by $a^2 + z^2 = 0$, i.e, $z = \pm ai$ out of which only pole $z = ai$ lies inside C, i.e., upper half plane.

$$\text{Res } (z = ai) = \lim_{z \to ai} (z - ai) \psi(z)$$

$$= \lim_{z \to ai} (z - ai) \frac{e^{miz}}{(z - ai)(z + ai)} = \frac{1}{2ai} e^{-ma} .$$

Hence from (3) we have

$$\int_{-\infty}^{\infty} e^{miz} \frac{1}{a^2 + x^2} dx = 2\pi i . \frac{1}{2ai} e^{ima} = \frac{\pi}{a} e^{-ma}$$

$$\int_{-\infty}^{\infty} (\cos mx + i \sin mx) \frac{1}{a^2 + x^2} dx = \frac{\pi}{a} e^{-ma}$$

Equating real and imaginary parts

$$\int_{-\infty}^{\infty} \frac{\cos mx}{a^2 + x^2} dx = \frac{\pi}{a} e^{-ma} \qquad \ldots(4)$$

and $\int_{-\infty}^{\infty} \dfrac{\sin mx}{a^2 + x^2} dx = 0$

Again from (4) we have

$$2 \int_{0}^{\infty} \frac{\cos mx}{a^2 + x^2} dx = \frac{\pi}{a} e^{-ma}$$

or $\int_{0}^{\infty} \dfrac{\cos mx}{a^2 + x^2} dx = \dfrac{\pi}{2a} e^{-ma} \qquad \ldots(5)$

Deductions :

(a) If $m = 1$ and $a = 1$ then from (5) we have

$$\int_{0}^{\infty} \frac{\cos x}{1 + x^2} dx = \frac{\pi}{2} e^{-1} = \frac{\pi}{2e}$$

(b) Differentiate (5) w.r.t. m we get

$$\int_{0}^{\infty} - \frac{x \sin mx}{a^2 + x^2} dx = \frac{\pi}{2a} e^{-ma} (-a)$$

or $\int_{0}^{\infty} \dfrac{x \sin mx}{a^2 + x^2} dx = \dfrac{\pi}{2} e^{-ma} \qquad \ldots(6)$

(c) Putting $m = 1$ in (6) we have

$$\int_{0}^{\infty} \frac{x \sin x}{a^2 + x^2} dx = \frac{\pi}{2} e^{-a}$$

EXAMPLE 2. *Prove that*

$$\int_{-\infty}^{\infty} \frac{\cos x dx}{(x^2 + a^2)(x^2 + b^2)}$$

$$= \frac{\pi}{a^2 - b^2} \left(\frac{e^{-b}}{b} - \frac{e^{-a}}{a} \right) (a > b > 0).$$

SOLUTION. Consider the integral

$$\int_C \psi(z) dz = \int_C e^{iz} f(z) dz = \int_C e^{iz} \frac{1}{(z^2 + a^2)(z^2 + b^2)} dz$$

where C is the contour of Ex. 1.

Hence by Cauchy's residues theorem

$$\int_C e^{iz} f(z) dz = \int_{-R}^{R} e^{ix} f(x) dx + \int_\Gamma e^{iz} f(z) dz$$

$$= 2\pi i \Sigma R^+ \qquad \ldots(1)$$

Now $\lim\limits_{z \to \infty} f(z) \lim\limits_{z \to \infty} = \dfrac{1}{(z^2 + a^2)(z^2 + b^2)} = 0$

and hence by Jordan's Lemma

$$\lim_{R \to \infty} \int_\Gamma e^{imz} f(z) dz = 0 \qquad \ldots(2)$$

Hence as $R \to \infty$ we have from (1) and (2)

$$\int_{-\infty}^{\infty} e^{ix} . \frac{1}{(x^2 + a^2)(x^2 + b^2)} dx = 2\pi i \Sigma R^+ \qquad \ldots(3)$$

Now poles of the integrand $\psi(z)$ are given by

$$(z^2 + a^2)(z^2 + b^2) = 0$$

or $z = \pm ai, \pm bi$. Out of those four poles only two $z = ai$ and $z = bi$ lie within C.

$$\text{Res}(z = ai) = \lim_{z \to ai} (z - ai)\psi(z)$$

$$= \lim_{z \to ai} (z - ai) \cdot \frac{e^{iz}}{(z^2 + a^2)(z^2 + b^2)}$$

$$= \frac{e^{iz}}{(z - ai)(z + ai)(z^2 + b^2)} \qquad = \frac{e^{-a}}{2ai(b^2 - a^2)}$$

Replacing a by b we get

$$\text{Res}(z = bi) = \frac{e^{-b}}{2bi(a^2 - b^2)} .$$

$\therefore$ Sum of the residues

$$= \frac{1}{2i(a^2 - b^2)}\left[\frac{e^{-b}}{b} - \frac{e^{-a}}{a}\right] = \Sigma R^+$$

Hence from (3) by equating real part on both sides we get by the help of above

$$\int_{-\infty}^{\infty} \frac{\cos x}{(x^2 + a^2)(x^2 + b^2)} dx$$

$$= 2\pi i \frac{1}{2i(a^2 - b^2)}\left[\frac{e^{-b}}{b} - \frac{e^{-a}}{a}\right]$$

$$= \frac{\pi}{a^2 - b^2}\left[\frac{e^{-b}}{b} - \frac{e^{-a}}{a}\right].$$

11.18.8 THE CASE WHEN POLES LIE ON THE REAL AXIS AS WELL

In the last two exercises we had done integrals of certain functions $f(z)$ and we had seen that in no questions the poles of $f(z)$ were lying on the real axis. Here we are going to discuss the integral of certain functions whose poles lie on real axis as well as poles within the contour C. In these types of questions we shall exclude the poles on the real line by enclosing it with a semi-circle centred at the pole and of small radius drawn in the upper half on the plane. The above method is known as indenting the semi-circular contour. After we have indented the semi-circular contour of radius R by smaller circles centred at poles on the real axis then $f(z)$ is regular along the modified contour C. Hence by Cauchy's residue theorem

$$\int_C f(z)dz = 2\pi i \Sigma R^+ .$$

IMPORTANT FACTS

The following points should be kept in mind.

- $\lim_{z \to \infty} zf(z) = 0$ then $\int_\Gamma f(z)dz = 0$.

- $\lim_{z \to \infty} f(z) = 0$ then $\lim_{R \to \infty} \int_\Gamma e^{imz} f(z)dz = 0$ (*Jordan's Lemma*).

- If C is the arc of the circle $|z - a| = r$ such that $\theta_1 \le \theta \le \theta_2$ and $\lim_{z \to \infty} (z - a)f(z) = A$ then $\lim_{z \to 0} \int_C f(z)dz = iA(\theta_2 - \theta_1)$.

- If C is the arc of the circle $|z| = R$ such that $\theta_1 \le \theta \le \theta_2$ and $\lim_{z \to \infty} f(z) = A$ then $\lim_{R \to \infty} \int_C f(z)dz = iA(\theta_2 - \theta_1)$.

Solved Examples

EXAMPLE 1. *Apply the calculus of residues to prove that*

$$\int_0^\infty \frac{\sin mx}{x} dx = \frac{\pi}{2}. \qquad \text{(KERALA–2005, UPTU–2007)}$$

SOLUTION. Consider the integral $\int_C f(z)dz$ where $f(z) = \dfrac{e^{imz}}{z}$.

The given function has a singularity at $z = 0$ on the real axis. We choose the contour C to be a large semi circle $|z| = R$ indented at $z = 0$ and r be the radius of this small semi circle of indentation. There is no singularity within C.

Hence by Cauchy's residue theorem

$$\int_C f(z)dz = 2\pi i \Sigma R^+ = 0$$

or $\int_{-R}^{-r} f(x)dx + \int_\gamma f(z)dz + \int_r^R f(x)dx + \int_\Gamma f(z)dz = 0$

or $\qquad I_1 + I_2 + I_3 + I_4 = 0$

$$I_4 = \int_\Gamma f(z)dz = \int_\Gamma e^{imz} \frac{1}{z} dz.$$

Since $\lim_{z \to \infty} \dfrac{1}{z} = 0 \qquad \therefore \qquad \lim_{R \to \infty} \int_\Gamma e^{imz} \dfrac{1}{z} dz = 0$

$I_2 = \int_\gamma f(z)dz \quad \gamma$ is described in clockwise direction

and γ is the circle $|z| = r$

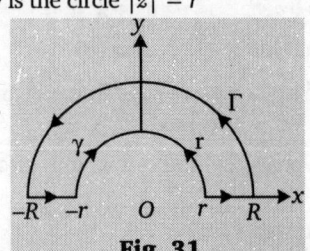

Fig. 31

$$\lim_{z \to 0} zf(z) = \lim_{z \to 0} z \cdot \frac{e^{imz}}{z} = 1$$

$$\Rightarrow \quad \lim_{r \to 0} \int_\gamma f(z) = -i.1(\pi - 0) = -\pi i.$$

We have chosen the negative sign because of the clockwise direction.

Hence making $R \to \infty$ and $r \to 0$ we have from (1)

$$\int_{-\infty}^0 f(x)dx - \pi i + \int_0^\infty f(x)dx + 0 = 0$$

or $\qquad \int_{-\infty}^\infty \dfrac{e^{imx}}{x} dx = \pi i.$

Equating imaginary part on both sides we get

$$\int_{-\infty}^{\infty} \frac{\sin mx}{x}dx = \pi \quad \text{or} \quad \int_0^{\infty} \frac{\sin mx}{x}dx = \frac{\pi}{2}$$

In particular if $m = 1$ then $\int_0^{\infty} \frac{\sin x}{x}dx = \frac{\pi}{2}$

EXAMPLE 2. *Show that if $a \geq b \geq 0$ then*

$$\int_0^{\infty} \frac{\cos 2ax - \cos 2bx}{x^2}dx = \pi(b - a)$$

SOLUTION. Consider the integral $\int_C f(z)dz$ where

$$f(z) = \frac{e^{i2az} - e^{i2bz}}{z^2}$$

The given function has a pole at $z = 0$ on the real axis. We choose the contour C to be a large semi circle $|z| = R$ indented at $z = 0$ and r be the radius of this small circle of indentation. There is no pole within C. The figure is same as in Ex. 1.

Hence by Cauchy's residue theorem we have

$$\int_C f(z)dz = 2\pi i \Sigma R^+ = 0$$

or $\int_{-R}^{-r} f(x)dx + \int_\gamma f(z)dz + \int_r^R f(x) + \int_\Gamma f(z)dz = 0$ ...(1)

or $I_1 + I_2 + I_3 + I_4 = 0$

$$I_4 = \int_\Gamma f(z)dz = \int_\Gamma (e^{i2az} - e^{i2bz})\frac{1}{z^2}dz.$$

Since $\lim_{z \to \infty} \frac{1}{z^2} = 0$ $\therefore$ $\lim_{R \to \infty} \int_\Gamma (e^{i2az} - e^{i2bz})\frac{1}{z^2}dz = 0$

$I_2 = \int_\gamma f(z)dz$, γ is described in clockwise direction and γ is the circle $|z| = r$

$$\lim_{z \to 0} zf(z) = \lim_{z \to 0} z\frac{e^{i2az} - e^{i2bz}}{z^2}$$

$$= \lim_{z \to 0} \frac{\left(1 + 2aiz + \frac{4a^2i^2z^2}{2}\right) - \left(1 + 2biz + \frac{4b^2i^2z^2}{2}\right)}{z}$$

$$= \lim_{z \to 0} 2(a - b)i - 2(a^2 - b^2)z + ... = 2(a - b)i$$

$\therefore$ $\lim_{r \to 0} \int_\gamma f(z)dz = -i(\pi - 0)2(a - b)i = 2\pi(a - b).$

We have chosen negative sign because γ is described in clockwise direction.
Hence making $R \to \infty$ and $r \to 0$, we have from (1)

$$\int_{-\infty}^0 f(x)dx + 2\pi(a - b) + \int_0^{\infty} f(x)dx + 0 = 0$$

or $\int_{-\infty}^{\infty} \frac{e^{i2ax} - e^{i2bx}}{x^2}dx = -2\pi(b - a)$

Equating real parts in both sides, we get

$$\int_{-\infty}^{\infty} \frac{\cos 2ax - \cos 2bx}{x^2}dx = 2\pi(b - a)$$

Hence,

$$\int_0^{\infty} \frac{\cos 2ax - \cos 2bx}{x^2}dx = \pi(b - a) = -\pi(a - b) = \pi(b - a)$$

EXAMPLE 3. *Prove that*

$$\int_0^{\infty} \frac{\sin mx\,dx}{x(x^2 + a^2)^2} = \frac{\pi}{2a^4} - \frac{\pi}{4a^3}e^{-ma}\left(m + \frac{2}{a}\right)$$

where $m > 0$, $a > 0$.

SOLUTION. Proceed exactly as in previous Ex.2 with the same contour C indented at the pole $z = 0$ on the real axis.

$$f(z) = \frac{e^{imz}}{z(z^2 + a^2)^2}$$

The poles are given by $z(z^2 + a^2)^2 = 0$
i.e., $z = 0$, $z = ai$, $z = -ai$ (The last two being of order two).
Out of these the pole $z = ai$ (of order two lies within C).
In order to get the residue of the pole $z = ai$ (order two) put $z = t + ai$ in $f(z)$ and the coefficient of $1/t$ shall be the required residue.

$$f(t + ai) = \frac{e^{im(t+ai)}}{(t + ai)[(t + ai)^2 + a^2]^2}$$

$$= \frac{e^{-ma}e^{tmi}}{ai\left(1 + \frac{t}{ai}\right)(t^2 + 2ait)^2}$$

$$= \frac{e^{-ma}.e^{imt}}{-ai.4a^2t^2\left(1 - \frac{it}{a}\right)\left(1 - \frac{it}{2a}\right)^2}$$

$$= -\frac{e^{-ma}}{i.4a^3t^2}e^{tmi}.\left(1 - \frac{it}{a}\right)^{-1}\left(1 - \frac{it}{2a}\right)^{-2}$$

$$= -\frac{e^{-ma}}{i.4a^3t^2}(1 + imt + ...)\left(1 + \frac{it}{a}...\right)\left(1 + \frac{it}{a}...\right)$$

$$= -\frac{e^{-ma}}{i.4a^3t^2}(1 + imt)\left(1 + 2i\frac{t}{a} + ...\right)$$

Coefficient of $\frac{1}{t}$ in the above is

$$-\frac{e^{-ma}}{i.4a^3}\left(im + \frac{2i}{a}\right) = -\frac{e^{-ma}}{4a^3}\left(m + \frac{2}{a}\right)$$

By Cauchy's residue theorem we have

$$\int_C f(z)dz = 2\pi i \Sigma R^+$$

or $\int_{-R}^{-r} f(x)dx + \int_\gamma f(z)dz + \int_r^R f(x)dx + \int_\Gamma f(z)dz$

$$= -2\pi i.\frac{e^{-ma}}{4a^3}\left(m + \frac{2}{a}\right)$$

$$I_1 + I_2 + I_3 + I_4 = -2\pi i.\frac{e^{-ma}}{4a^3}\left(m + \frac{2}{a}\right) \quad ...(1)$$

$$I_4 = \int_\Gamma f(z)dz = \int_\Gamma e^{imz}.\frac{1}{z(z^2 + a^2)^2}dz.$$

Since $\lim_{z \to \infty} \frac{1}{z(z^2 + a^2)^2} = 0$

$\therefore$ $\lim_{R \to \infty} \int_\Gamma \frac{e^{imz}}{z(z^2 + a^2)}dz = 0$

$I_2 = \int_\gamma f(z)dz$, γ is described in clockwise direction and

$$\lim_{z \to 0} zf(z) = \lim_{z \to 0} z.\frac{e^{imz}}{z(z^2 + a^2)^2} = \frac{1}{a^4}$$

$\Rightarrow$ $\int_\gamma f(z)dz = -i.(\pi - 0).\frac{1}{a^4} = -\frac{\pi i}{a^4}.$

We have chosen the negative sign as γ is described in clockwise direction. Hence making $R \to \infty$ and $r \to 0$, we have from (1)

$$\int_{-\infty}^{0} f(x)dx - \frac{\pi i}{a^4} + \int_{0}^{\infty} f(x)dx + 0 = -2\pi i \frac{e^{-ma}}{4a^2}\left(m + \frac{2}{a}\right)$$

or $\int_{-\infty}^{\infty} f(x)dx = \frac{\pi i}{a^4} - \frac{\pi i}{2a^3}e^{-ma}\left(m + \frac{2}{a}\right)$

or $\int_{-\infty}^{\infty} \frac{e^{imx}}{x(x^2+a^2)^2}dx = \frac{\pi i}{a^4} - \frac{\pi i}{2a^3}e^{-ma}\left(m + \frac{2}{a}\right)$

Equating imaginary parts on both sides, we get

$$\int_{-\infty}^{\infty} \frac{\sin mx \, dx}{x(x^2+a^2)^2} = \frac{\pi}{a^4} - \frac{\pi}{2a^3}e^{-ma}\left(m + \frac{2}{a}\right)$$

or $\int_{0}^{\infty} \frac{\sin mx \, dx}{x(x^2+a^2)^2} = \frac{\pi}{2a^4} - \frac{\pi}{4a^3}e^{-ma}\left(m + \frac{2}{a}\right)$

EXAMPLE 4. *Prove by the help of contour integration that*

$$\int_{0}^{\infty} \frac{x^a}{1+x^2}dx = \frac{\pi}{2}\sec\frac{\pi a}{2}, (-1 < a < 1).$$

SOLUTION. We consider the integral $\int_{C} f(z)dz$

where $f(z) = \dfrac{z^a}{1+z^2}(-1 < a < 1)$.

$z = 0$ is a branch point of $f(z)$ and its poles are given by $z = \pm i$, or the contour C is the same as of last two examples consisting of Γ the upper half of the large circle $|z| = R$ and the real axis from $-R$ to R indented at $z = 0$ by a small circle γ of radius r.

The only pole lying within C is given by $z = i$

$$\text{Res}(z = i) = \lim_{z \to i}(z-i)\frac{z^a}{(z+i)(z-i)} = \frac{i^a}{2i}$$

But $i = \cos\dfrac{\pi}{2} + i\sin\dfrac{\pi}{2} = e^{i\frac{\pi}{2}}$

$$\therefore \quad \text{Res}(z = i) = \frac{e^{i\pi a/2}}{2i}$$

Hence by Cauchy's residue theorem, we have

$$\int_{C} f(z)dz = 2\pi i \Sigma R^+ = 2\pi i \frac{e^{\frac{\pi i a}{2}}}{2i} = \pi e^{\frac{i\pi a}{2}}$$

or $\int_{-R}^{-r} f(x)dx + \int_{\gamma} f(z)dz$

$$+ \int_{r}^{R} f(x)dx + \int_{\Gamma} f(z)dz = \pi e^{\frac{i\pi a}{2}}$$

or $\quad I_1 + I_2 + I_3 + I_4 = \pi e^{\frac{i\pi a}{2}}$ ...(1)

$$\lim_{z \to 0} z f(z) = \lim_{z \to 0} z \cdot \frac{z^a}{1+z^2} = \lim_{z \to 0} \frac{z^{a+1}}{1+z^2} = 0.$$

as $a + 1 > 0$.

$\therefore \lim_{z \to 0} \int_{\gamma} f(z)dz = -i(\pi - 0).0 = 0 = I_2$.

Again $\left| \int_{\Gamma} f(z)dz \right| \le \int_{\Gamma} |f(z)||dz|$

where $z = Re^{i\theta}|z| = R|dz| = Rd\theta$

$$\le \int_{\Gamma} \frac{|z|^a}{|z|^2-1}|dz| = \frac{R^a}{R^2-1}\int_{0}^{\pi} Rd\theta = \frac{\pi R^{a+1}}{R^2-1} \to 0$$

when $R \to \infty$ as $a < 1$
Hence $I_4 = 0$.
Thus making $R \to \infty$ and $r \to 0$ we have from (1)

$$\int_{-\infty}^{0} \frac{x^a}{1+x^2}dx + 0 + \int_{0}^{\infty} \frac{x^a}{1+x^2}dx + 0 = \pi e^{\frac{i\pi a}{2}}$$

Putting $-y$ for x in 1st and adjusting the limits, we have

$$\int_{\infty}^{0} \frac{y^a(-1)^a}{1+y^2}(-dy) + \int_{0}^{\infty} \frac{x^a}{1+x^2}dx = \pi e^{\frac{i\pi a}{2}}$$

or $\int_{0}^{\infty} \frac{y^a}{1+y^2}e^{i\pi a}dy + \int_{0}^{\infty} \frac{x^a}{1+x^2}.dx = \pi e^{\frac{i\pi a}{2}}$

or $\quad (1 + e^{i\pi a})\int_{0}^{\infty} \frac{x^a}{1+x^2}dx = \pi e^{\frac{i\pi a}{2}}$.

Equating real parts on both sides, we have

$$(1 + \cos\pi a)\int_{0}^{\infty} \frac{x^a}{1+x^2}dx = \pi\cos\frac{\pi a}{2}$$

$$\therefore \quad \int_{0}^{\infty} \frac{x^a}{1+x^2}dx = \frac{\pi\cos\dfrac{\pi a}{2}}{2\cos^2\dfrac{\pi a}{2}} = \frac{1}{2}\pi\sec\frac{\pi a}{2}.$$

11.18.9 Expansion of a Meromorphic Function

We have already define that a function $f(z)$ is said to meromorphic in a region D if it is analytic in D except at a finite number of poles.

THEOREM 1. (Mittag Lefler's expansion Theorem).

STATEMENT. *Let $f(z)$ be a function whose only singularities in the finite part of the plane are simple poles at $a_1, a_2, a_3, ... a_n$ arranged in order of increasing absolute value, i.e., $0 < |a_1| < |a_2|... < |a_n|$*

with residues $b_1, b_2, ... b_n$ respectively at these poles. Further let there be a sequence of closed contours either circles or square $c_1, c_2, c_3 ... c_n$ such that

(i) *The contour C_n encloses $a_1, a_2, a_3, ..., a_n$ but no other poles.*

(ii) *The contour C_n must be such that minimum distance R_n of C_n from origin tends to infinity as n tends to infinity.*

(iii) *Length L_n of the contour C_n is $O(R_n)$,*

i.e., $\dfrac{L_n}{R_n} = \lambda(\lambda \neq 0)$ *as $n \to \infty$*

(iv) *On $C_n, f(z) = 0(R_n)$, i.e., $\dfrac{f(z)}{R_n} = 0$ when $n \to \infty$*

(The above condition will be satisfied if $f(z)$ is bounded on the sequence of contours).

Then for all values of z except at poles

$$f(z) = f(0) + \sum_{n=1}^{\infty} b_n\left(\frac{1}{z - a_n} + \frac{1}{a_n}\right)$$

PROOF. Let $\qquad I = \int_C \dfrac{f(t)}{t(t-z)} dt$ where z is a point inside C_n. $\qquad$ ...(1)

The poles of the integrand are at the values of t for which $f(t) = \infty$ and $t(t-z) = 0$.

But by hypothesis $f(t) = \infty$ at $t = a_1, a_2, ... a_n$.

Hence the poles of the integrand inside C_n are given by

$$t = a_m \ (m = 1, 2, 3, ... n) \text{ and at } t = z \text{ and at } t = 0.$$

$$\text{Res. } t = a_m \text{ is } \lim_{t \to a_m} (t - a_m) \frac{f(t)}{t(t-z)} = \frac{b_m}{a_m(a_m - z)}$$

because a_m being the residue of $f(t)$ at $t = a_m$ is given by

$$b_m = \lim_{t \to a_m} (t - a_m) f(t)$$

Again Res at $t = z$ is $\lim\limits_{t \to z} (t-z) \dfrac{f(t)}{t(t-z)} = \dfrac{f(t)}{z}$

Res at $t = 0$ is $\lim\limits_{t \to 0} (t-z) \dfrac{f(z)}{t(t-z)} = \dfrac{f(0)}{-z}$

Therefore by Cauchy's residue theorem, we have

$$I = 2\pi i \left[\sum_{m=1}^{n} \frac{b_m}{a_m(a_m - z)} + \frac{f(z)}{z} - \frac{f(0)}{z} \right] \qquad ...(2)$$

Now if we establish that $I \to 0$ when $n \to \infty$ we shall prove the required result.

$$|I| = \left| \int_{C_n} \frac{f(t)}{t(t-z)} dt \right| \le \int_{C_n} \frac{|f(t)||dt|}{|t|(|t| - |z|)} \le \frac{M_n}{R_n(R_n - |z|)} \int_{C_n} |dt|$$

where M_n is value of $|f(t)|$ on C_n.

$$= \frac{M_n}{R_n(R_n - |z|)} L_n = \frac{L_n}{R_n(R_n - |z|)} . M_n \to 0 \text{ as } n \to \infty$$

Because as $n \to \infty$ we have $M_n = $ max. value of $|f(t)|$ on $C_n = 0$ by given condition (iv).

Hence from (2)

$$0 = \lim_{n \to \infty} \left\{ \sum_{m=1}^{n} \frac{b_m}{a_m(a_m - z)} + \frac{f(z)}{z} - \frac{f(0)}{z} \right\}$$

or $\qquad f(z) = f(0) + \sum\limits_{n=1}^{\infty} b_n \left(\dfrac{1}{z - a_n} + \dfrac{1}{a_n} \right)$.

Solved Examples

EXAMPLE 1. *Prove that* $\cot z = \dfrac{1}{z} + 2z \sum\limits_{n=1}^{\infty} \dfrac{1}{z^2 - n^2\pi^2}$.

SOLUTION. Consider the function $f(z) = \cot z - \dfrac{1}{z}$

or $\quad f(z) = \dfrac{z \cos z - \sin z}{z \sin z} \qquad$ ...(1)

The poles of $f(z)$ are given by $\sin z = 0$.

$\therefore \quad z = n\pi$ where $n = \pm 1, \pm 2, \pm 3$.

$z = 0$ is not a pole of $f(z)$ because

$$\lim_{z \to 0} f(z) = \lim_{z \to 0} \frac{z \cos z - \sin z}{z \sin z}$$

$$= \lim_{z \to 0} \frac{z\left(1 - \dfrac{z^2}{2!} + ...\right) - \left(z - \dfrac{z^3}{3!}...\right)}{z\left(z - \dfrac{1}{3!}z^3 ...\right)}$$

$$= \lim_{z \to 0} \frac{\left(-\dfrac{1}{2} + \dfrac{1}{6}\right)z + ...}{1 - \dfrac{1}{3!}z^2 + ...} = 0$$

Therefore $f(z)$ has removable singularity at $z = 0$ and hence we can define $f(0) = 0$.

$$\text{Res}(z = n\pi) = \left[\frac{z \cos z - \sin z}{d/dz(z \sin z)} \right]_{\text{at } z = n\pi}$$

$$= \left[\frac{z \cos z - \sin z}{z \cos z + \sin z} \right]_{\text{at } z = n\pi}$$

$$\frac{n\pi \cos n\pi - \sin n\pi}{n\pi \cos n\pi + \sin n\pi} = 1$$

In accordance with the terminology of the theorem

$\text{Res}(z = n\pi) = bn$ and $z = n\pi = an$

Thus $\text{Res}(z = b_n)$ $= a_n = 1$

Let C_n be the square $ABCD$ with centre at the origin with corners at

$$z = \left(n + \frac{1}{2}\right)$$

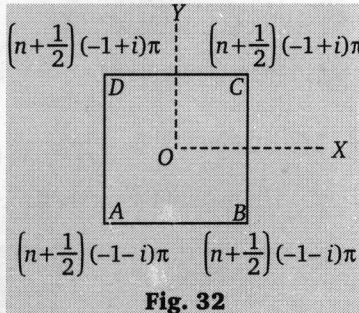

$\left(n+\dfrac{1}{2}\right)(-1+i)\pi \qquad \left(n+\dfrac{1}{2}\right)(-1+i)\pi$

$\left(n+\dfrac{1}{2}\right)(-1-i)\pi \qquad \left(n+\dfrac{1}{2}\right)(-1-i)\pi$

Fig. 32

($\pm 1 \pm i$) so that each side is of length $(2n + 1)\pi$ as shown in the figure.

Now, we observe that the following four conditions of the theorem are satisfied.

(i) C_n encloses the poles $z = n\pi (n = \pm 1, \pm 2, ...)$ and no other poles.

(ii) The minimum distance R_n of C_n from the origin is $\left(n + \dfrac{1}{2}\right)\pi$ which tends to infinity when $n \to \infty$.

(iii) The length L_n of C_n, i.e., perimeter of the sequence is $8n\pi + 4\pi$ and $R_n = \left(n + \dfrac{1}{2}\right)\pi$. Also

$$\frac{L_n}{R_n} = \frac{8n\pi + 4\pi}{\left(n + \dfrac{1}{2}\right)\pi} = 8 \text{ when } n \to \infty, \text{ i.e., } L_n = O(R)$$

(iv) We have to prove that $f(z)$ is bounded along the perimeter of the square

i.e., $\dfrac{f(z)}{R_n} = 0$ When $n \to \infty$ i.e., $f(z) = 0(R_n)$.

Now $\cot z = i \dfrac{e^{iz} + e^{-iz}}{e^{iz} - e^{-iz}} = i \dfrac{e^{2iz} + 1}{e^{2iz} - 1}$ or $i \dfrac{1 + e^{-2iz}}{1 - e^{-2iz}}$

$\therefore |\cot z| = \left| i \dfrac{e^{2i(x+iy)} + 1}{e^{2i(x+iy)} - 1} \right|$ or $\left| i \dfrac{1 + e^{-2i(x+iy)}}{1 - e^{-2i(x+iy)}} \right|$

$= \left| \dfrac{1 + e^{-2y}}{1 - e^{-2y}} \cdot \dfrac{e^{2ix}}{e^{2ix}} \right|$ or $\left| \dfrac{1 + e^{2y}}{1 - e^{2y}} \cdot \dfrac{e^{2ix}}{e^{-2ix}} \right|$

$< \dfrac{1 + e^{-2y}}{1 - e^{-2y}}$ or $\dfrac{1 + e^{2y}}{1 - e^{2y}}$

The above two forms are to be chosen according as y is positive or negative

Now along CD, $y = \left(n + \dfrac{1}{2}\right)\pi$ and along AB

$y = -\left(n + \dfrac{1}{2}\right)\pi$.

Hence both along AB and CD we have (y is positive along CD and negative along AB)

$$|\cot z| < \frac{1 + e^{-2(n+1/2)\pi}}{1 - e^{-2(n+1/2)\pi}}$$

R.H.S. is finite when n is finite and clearly tends to unity when $n \to \infty$. Thus we conclude z is bounded along

both CD and AB.

Now along BC and AD, y is both positive and negative.

$\therefore |\cot z| < \dfrac{1 + e^{-2y}}{1 - e^{-2y}}$ or $\dfrac{1 + e^{2y}}{1 - e^{2y}}$

on BC and AD which is finite when y is finite chosen according as z is positive or negative and tends to unity when $y \to \infty$ or $-\infty$. In fact $y \to \infty$ as $n \to \infty$. Thus we conclude that $\cot z$ is bounded on BC and AD.

Also $|1/z| \to 0$ on C_n as $n \to \infty$.

From above we conclude that $|f(z)| \leq 1$ on C_n as $n \to \infty$. Thus $f(z)$ is bounded. Hence by Mittag Leffer's theorem, we have

$$f(z) = f(0) + \sum_1^\infty b_n \left(\frac{1}{z - a_n} + \frac{1}{a_n} \right)$$

where $b_n = 1$ and $a_n = \pm n\pi$.

$$= 0 + \sum_{n=1}^\infty 1\left(\frac{1}{z - n\pi} + \frac{1}{n\pi} \right) + \sum_{n=-1}^{-\infty} \left(\frac{1}{z - n\pi} + \frac{1}{n\pi} \right)$$

$$= \sum_{n=1}^\infty \left[\left(\frac{1}{z - n\pi} + \frac{1}{n\pi} \right) + \left(\frac{1}{z + n\pi} - \frac{1}{n\pi} \right) \right]$$

$$= \sum_{n=1}^\infty \left[\frac{1}{z - n\pi} + \frac{1}{z + n\pi} \right] = \sum_{n=1}^\infty \frac{2z}{z^2 - n^2\pi^2}$$

or $\cot z - \dfrac{1}{z} = \sum_{n=1}^\infty \dfrac{2z}{z^2 - n^2\pi^2}$

or $\cot z = \dfrac{1}{z} + \sum_{n=1}^\infty \dfrac{2z}{z^2 - n^2\pi^2}$.

Deduction 1. Replacing z by πz in the above result, we get

$$\cot \pi z = \frac{1}{\pi z} + \sum_{n=1}^\infty \frac{2\pi z}{\pi^2 z^2 - n^2\pi^2}$$

or $\pi \cot \pi z = \dfrac{1}{z} + \sum_{n=1}^\infty \dfrac{2z}{z^2 - n^2}$ ...(2)

Deduction 2.

$\pi \cot \pi z = \dfrac{1}{\pi} + \dfrac{z}{\pi} \Sigma \left(\dfrac{1}{z - n} + \dfrac{1}{z + n} \right)$ where accent in

indicates that $n = 0$ is omitted.

Multiply both sides of (2) by $\dfrac{z}{\pi}$ and writing $\dfrac{2z}{z^2 - n^2}$ as

$\dfrac{1}{z - n} + \dfrac{1}{z + n}$.

We get $z \cot \pi z = \dfrac{1}{\pi} + \dfrac{z}{\pi} \sum_{n=1}^\infty \left(\dfrac{1}{z - n} + \dfrac{1}{z + n} \right)$.

Exercise-11.22

1. Find the residue of $\dfrac{1}{(z^2 + 1)^3}$ at $z = i$.

2. Find the residue of $\dfrac{1}{(z^2 + a^2)^2}$ at $z = ai$.

3. Find the residues of the function $\dfrac{\cot \pi z}{(z - a)^2}$.

4. Find the residues of $e^z \csc^2 z$ at all its poles in the finite plane.

5. Evaluate $\int_C \dfrac{z - 3}{z^2 + 2z + 5} dz$, where C is the circle.

  (i) $|z| = 1$ (ii) $|z + 1 - i| = 2$

(iii) $|z + 1 + i| = 2$

6. Find the residue of $\dfrac{z^2}{(z - a)(z - b)(z - c)}$ at infinity.

7. Evaluate the residues of $\dfrac{z^3}{(z - 1)(z - 2)(z - 3)}$ at $z = 1, 2, 3$ and infinity and show that their sum is zero.

8. Show that $\int_0^\pi \dfrac{a}{a^2 + \cos^2\theta} d\theta, a > 0 = \dfrac{\pi}{\sqrt{a^2 + 1}}$

9. Show that $\int_0^{2\pi}\dfrac{\cos^2 3\theta}{5-4\cos^2\theta}d\theta=\dfrac{3\pi}{8}$

10. Show that $\int_0^\infty e^{-\cos\theta}.\cos(n\theta+\sin\theta)d\theta=\dfrac{2\pi}{n!}(-1)^n$ where $n\in Z^+$.

11. Show that $\int_0^\infty\dfrac{dx}{(x^2+b^5)^{n+1}}=\dfrac{\pi}{2^{2n+1}.b^{2n+1}}.\dfrac{2n!}{(n!)^2}$

12. Show that $\int_0^\infty\dfrac{\cos^2 x}{(1+x^2)^2}dx=\dfrac{\pi}{4}(1+3e^{-2})$

13. By integrating $\dfrac{e^{iz}}{z+ai}$ round a suitable contour prove that
$$\int_{-\infty}^\infty\dfrac{-a\cos x+x\sin x}{x^2+a^2}dx=0$$

14. Prove that $\int_{-\infty}^\infty\dfrac{\cos mx}{x^4+a^4}dx=\dfrac{\pi}{a^3}e^{-ma/\sqrt2}.\sin\left(\dfrac{ma}{\sqrt{}}+\dfrac{\pi}{}\right)$

15. By contour integration, prove that $\int_0^\infty\dfrac{\cosh ax}{\cosh x}dx=\dfrac{\pi}{2}\cos\left(\dfrac{\pi a}{2}\right)$ if $|a|<1$.

16. Prove that $\tan z=\sum_{n=1}^\infty\dfrac{2z}{\left(n+\dfrac{1}{2}\right)^2\pi^2-z^2}$

17. Find the sum of the residues of $f(z)=\dfrac{\sin z}{z\cos z}$ at its poles inside the circle $|z|=2$.   (ROHTAK–2004)

18. Evaluate $\oint_C\dfrac{z-3}{z^2+2z+5}dz$, where C is the circle

(i) $|z|=1$ (ii) $|z+1-i|=2$
(iii) $|z+1+i|=2$ (JNTU–2003)

19. Evaluate $\oint_C\dfrac{e^z}{\cos\pi z}dz$, where C is the unit circle $|z|=1$.

(ROHTAK–2006)

20. Evaluate $\oint_C\tan z\,dz$, where C is the circle $|z|=2$.

(VTU–2010S)

21. Show that $\int_0^{2\pi}\dfrac{\cos 2\theta d\theta}{1-2a\cos\theta+a^2}=\dfrac{2\pi a^2}{1-a^2},(a^2<1)$

(BHOPAL–2009, ROHTAK–2003)

22. Evaluate $\int_{-\infty}^\infty\dfrac{x^2 dx}{(x^2+1)(x^2+4)}$.   (UPTU–2008)

23. Apply the calculas of residues to prove that

(i) $\int_0^{2\pi}\dfrac{\sin^2\theta d\theta}{a+b\cos\theta}=\dfrac{2\pi}{b^2}\left[a-\sqrt{a^2-b^2}\right],(0<b<a)$

(JNTU–2003)

(ii) $\int_{-\infty}^\infty\dfrac{x^2 dx}{(x^2+a^2)(x^2+b^2)}=\dfrac{\pi}{a+b}(a,b>0)$

(PTU–2007, MUMBAI–2006, ANNA–2003)

(iii) $\int_{-\infty}^\infty\dfrac{dx}{x^4+1}=\dfrac{\pi}{\sqrt2}$ (JNTU–2006)

(iv) $\int_0^\infty\dfrac{1-\cos x}{x^2}dx=\dfrac{\pi}{2}$ (PTU–2005)

Hints to Selected Problems

1. $f(z)=\dfrac{1}{(z^2+1)^3}=\dfrac{1}{(z+i)^3(z-i)^3}=\dfrac{\phi(z)}{(z-i)^3}$

where $\phi(z)=\dfrac{1}{(z+i)^3}$.

$\Rightarrow$ $z=i$ is a pole of order 3 of $f(z)$. So residue $=\dfrac{1}{2!}\phi''(i)$.

2. $f(z)=\dfrac{1}{(z^2+a^2)^2}=\dfrac{1}{(z+ia)^2(z-ia)^2}=\dfrac{\phi(z)}{(z-ia)^2}$

where $\phi(z)=\dfrac{1}{(z+i)^2}$.

3. $f(z)=\dfrac{\cot\pi z}{(z-a)^2}=\dfrac{\cos\pi z}{\sin\pi z(z-a)^2}$

Poles of $f(z)$ are given by $(z-a)^2\sin\pi z=0$.

4. $f(z)=e^z\,\mathrm{cosec}^2 z=\dfrac{e^z}{\sin^2 z}$ poles of $f(z)$ are given by $\sin^2 z=0$.

5. $f(z)=\dfrac{z-3}{z^2+2z+15}$

Poles of $f(z)$ are given by $z^2+2z+5=0$

$\Rightarrow$ $z=-1\pm2i$.

7. $f(z)=\dfrac{z^3}{(z-1)(z-2)(z-3)}$

$\Rightarrow$ $z=1,2,3$ are the simple poles of $f(z)$.

The residue at $z=1$ is $\lim_{z\to1}(z-1)f(z)=\lim_{z\to1}\dfrac{z^3}{(z-2)(z-3)}=\dfrac{1}{2}$

Similarly, find the residue at $z=2,3$.

12. $\cos^2 x=\dfrac{1+\cos 2x}{2}=$ Real part of $\dfrac{1+e^{2ix}}{2}$

choose $\psi(z)=\dfrac{1+e^{2iz}}{2}.\dfrac{1}{(1+z^2)^2}$.

Answers

1. $-\dfrac{3i}{16}$ 2. $\dfrac{-i}{4a^3}$ 3. $\dfrac{1}{\pi(n-a)^2}$ 4. $e^{m\pi}$ 5. (i) 0   (ii) $\pi(-2+i)$ (iii) $\pi(2+i)$

6. -1 7. 0   17. 0   18. (i) 0   (ii) $\pi(i-2)$ (iii) $\pi(2+i)$ 19. $-4i\sinh\dfrac{1}{2}$ 20. $-4\pi i$ 22. $\pi/3$

ARCHIVE

1. Using C-R equations, show that $f(z) = |z|^2$ is not analytic at any point. (MTU–2012)

2. Show that the function $f(z) = u + iv$, where

$$f(z) = \begin{cases} \dfrac{x^3(1+i) - y^3(1+i)}{x^2 + y^2} & , \ z \neq 0 \\ 0 & , \ z = 0 \end{cases}$$

satisfies C-R equations at $z = 0$. (ROHTAK–2009)

3. State and prove the C-R equation in Polar form. (ROHTAK–2010, BHOPAL(RGPV)–2009)

4. Find the imaginary part of the analytic function whose real part is $x^3 - 3xy^2 + 3x^2 - 3y^2$. (BHOPAL(RGPV)–2005, 08, 09)

5. Show that $e^x(x \cos y - y \sin y)$ is a harmonic function. (UPTU–2009, BHOPAL–2004, 10)

6. Prove that an analytic function $f(z)$ ceases to be conformal at the points where $f'(z) = 0$. (UPTU–2006)

7. Verify Cauchy's theorem for the function $f(z) = e^{iz}$ along the boundary of the triangle with vertices at the point $1 + i$, $-1 + i$ and $-1 - i$. (GBTU–2012)

8. Expand the function $f(z) = \tan^{-1}z$ in powers of z. (UPTU–2010)

9. State and prove Laurent's theorem. (UPTU–2009)

10. Show that the function e^z has an isolated essential singularity at $z = \infty$. (BHOPAL–2003)

11. State and prove Baye's theorem.

12. Find the moment generating function of Binomial distribution about the origin. (GBTU–2012)

13. Find the real root of the equation $x^4 - x - 9 = 0$ by Newton-Raphson method.

14. Find the missing terms in the following table :

x	1	2	3	4	5	6	7	8
$f(x)$	2	4	8	–	32	–	128	256

(GBTU–2012)

15. Calculate the nth divided difference of $f(x) = \dfrac{1}{x}$.

16. Evaluate $\int_0^1 \dfrac{dx}{1+x^2}$ by using Simpson's 1/3 rule.

17. Using Picard's method to approximate y when $x = 0.2$ given that $y = 1$ when $x = 0$ and $\dfrac{dy}{dx} = x - y$.

18. Apply Runga-Kutta method to solve

$$5.\dfrac{dy}{dx} = x^2 + y^2, y(0) = 1$$

and find y in the interval $0 \leq x \leq 0.2$ taking $h = 0.1$. (GBTU–2012)

19. If the sum of the mean and variance of a Binomial distribution of 5 trials is $9/5$, find $P(x \geq 1)$. (GBTU–2013)

20. Using Gauss-Seidal method, solve

$20x + y - 2z = 17, 3x + 20y - z = -18, 2x - 3y + 20z = 25$

❋❋❋❋❋❋

APPENDIX

Table 1. Poisson Distribution

x/λ	0.1	0.2	0.3	0.4	0.5	0.6	0.7	0.8	0.9	1.0
0	.9048	.8187	.7408	.6703	.6065	.5488	.4966	.4493	.4066	.3679
1	.0905	.1637	.2222	.2681	.3033	.3293	.3476	.3595	.3659	.3679
2	.0045	.0164	.0333	.0536	.0758	.0988	.1217	.1438	.1647	.1839
3	.0002	.0011	.0033	.0072	.0126	.0198	.0284	.0383	.0494	.0613
4	.0000	.0001	.0002	.0007	.0016	.0030	.0050	0.0077	.0111	.0153
5	.0000	.0000	.0000	.0001	.0002	.0004	.0007	.0012	.0020	.0031
6	.0000	.0000	.0000	.0000	.0000	.0000	.0001	.0002	.0003	.0005
7	.0000	.0000	.0000	.0000	.0000	.0000	.0000	.0000	.0000	.0001

x/λ	1.1	1.2	1.3	1.4	1.5	1.6	1.7	1.8	1.9	2.0
0	.3329	.3012	.2725	.2466	.2231	.2019	.1827	.1653	.1496	.1353
1	.3662	.3014	.3543	.3452	.3347	.3230	.3106	.2957	.2842	.2707
2	.2014	.2169	.2303	.2417	.2510	.2584	.2640	.2678	.2700	.2707
3	.0738	.0867	.0998	.1128	.1255	.1378	.1496	.1607	.1710	.1804
4	.0203	.0260	.0324	.0395	.0471	.0551	.0636	.0723	.0812	.0902
5	.0045	.0062	.0084	.0111	.0141	.0176	.0216	.0260	.0309	.0361
6	.0008	.0012	.0018	.0026	.0035	.0047	.0061	.0078	.0098	.0120
7	.0001	.0002	.0003	.0005	.0008	.0011	.0015	.0020	:0027	.0034
8	.0000	.0000	.0000	.0001	.0001	.0002	.0003	.0005	.0006	.0009
9	.0000	.0000	.0000	.0000	.0000	.0000	.0001	.1001	.0001	.0002

x/λ	2.1	2.2	2.3	2.4	2.5	2.6	2.7	2.8	2.9	3.0
0	.1225	.1108	.1003	.0907	.0821	.0743	.0672	.0608	.0550	.0498
1	.2572	.2438	.2306	.2177	.2052	.1931	.1815	.1703	.1596	.1494
2	.2700	.2681	.2652	.2613	.2505	.2510	.2450	.2384	.2314	.2240
3	.1890	.1966	.2033	.2090	.2138	.2176	.2205	.2225	.2237	.2240
4	.0992	.1082	.1169	.1254	.1336	.1414	.1488	.1557	.1622	.1680
5	.0417	.0476	.0538	.0602	.668	.0735	.0804	.0872	.0940	.1008
6	.0146	.0174	.0206	.0241	.0278	.0319	.0362	.0407	.0455	.0504
7	.0044	.0055	.0068	.0083	.0099	.0118	.0139	.0163	.0188	0.216
8	.0011	.0015	.0019	.0025	.0031	.0038	.0047	.0057	.0068	.0081
9	.0003	.0004	.0005	.0007	.0009	.0011	.0014	.0068	.0022	.0027
10	.0001	.0001	.0001	.0002	.0002	.0003	.0004	.0005	.0006	.0008
11	.0000	.0000	.0000	.0000	.0000	.0001	.0001	.0001	.0002	.0002
12	.0000	.0000	.0000	.0000	.0000	.0000	.0000	.0000	.0000	.0001

x/λ	3.1	3.2	3.3	3.4	3.5	3.6	3.7	3.8	3.9	4.0
0	.0450	.0408	.0369	.0334	.0302	.0273	.0247	.0224	.0224	.0183
1	.1397	.1304	.1217	.1135	.1057	.0984	.0915	.0850	.0789	.0733
2	.2165	.2087	.2008	.1929	.1850	.1771	.1692	.1615	.1539	.1465
3	.2237	.2226	.2209	.2186	.2158	.2125	.2087	.2046	.2001	.1954
4	.0734	.1781	.1823	.1858	.1888	.1912	.1931	.1944	.1951	.1954
5	.1075	.1140	.1203	.1264	.1322	.1377	.1429	.1477	.1522	.1563
6	.0555	.0608	.0662	.0716	.0771	.0826	.0881	.0936	.0989	.1042
7	.0246	.0278	.0312	.0348	.0385	.0425	.0466	.008	.0551	.1595
8	.0095	.0111	.0129	.0148	.0159	.0191	.0215	.0241	.0269	.0298
9	.0033	.0040	.0047	.0056	.0066	.0076	.0089	.0102	.0116	.0132
10	.0010	.0013	.0016	.0019	.0023	.0028	.0033	.0039	.0045	.0053
11	.0003	.0004	.0005	.0006	.0007	.0009	.0011	.0013	.0016	.0019
13	.0000	.0000	.0000	.0000	.0001	.0001	.0001	.0001	.0002	.0002
14	.0000	.0000	.0000	.0000	.0000	.0000	.0000	.0000	.0000	.0001

Table I. Poisson Distribution (Contd.)

x/λ	4.1	4.2	4.3	4.4	4.5	4.6	4.7	4.8	4.9	5.0
0	.0166	.0150	.0136	.0123	.0111	.0101	.0091	.0082	.0074	.0067
1	.0679	.0630	.0583	.0540	.0500	.0462	.0427	.0395	.0365	.0337
2	.1393	.1323	.1254	.1188	.1125	.1063	.1005	.0948	.0894	.0842
3	.1904	.1852'	.1798	.1743	.1687	.1631	.1574	.1517	.1460	.1404
4	.1951	.1944	.1933	.1917	.1898	.1875	.1849	.1820	.1789	.1775
5	.1600	.1633	.1662	.1687	.1708	.1725	.1738	.1747	.1753	.1755
6	.1093	.1143	.1191	.1237	.1281	.1323	.1362	.1398	.1432	1462
7	.0640	.0686	.0732	.0778	.0824	.0869	.0914	.0959	.1002	.1044
8	.0328	.0360	.0393	.0428	.0463	.0500	.0537	.0537	.0614	.0653
9	.0150	.0168	.0188	.0209	.0232	.0255	.0280	.0307	.0334	.0363
10	.0061	.0071	.0081	.0092	.0104	.0118	.0132	.0147	.0164	.0181
11	.0023	.0027	.0032	.0037	.0043	.0049	.0056	.0064	.0073	.0082
12	.0008	.0009	.0011	.0014	.0016	.0019	.0022	.0026	.0030	.0034
13	.0002	.0003	.0004	.0005	.0006	.0007	.0008	.0009	.0011	.0013
14	.0001	.0001	.0001	.0001	.0002	.0002	.0003	.0003	.0004	.0005
15	.0000	.0000	.0000	.0000	.0001	.0001	.0001	.0001	.0001	.0002

x/λ	5.1	5.2	5.3	5.4	5.5	5.6	5.7	5.8	5.9	6.0
0	.0061	.0055	.0050	.0045	.0041	.0037	.0033	.0030	.0027	.0025
1	.0311	.0287	.0265	.0244	.0225	.0207	.0191	.0176	.0162	.0149
2	.0793	.0746	.0701	.0659	.0618	.0580	.0544	.0509	.0477	.0446
3	.1348	.1293	.1239	.1185	.1133	.1082	.1033	.0985	.0938	.0892
4	.1719	.1681	.1641	.1600	.1558	.1515	.1472	.1428	.1383	.1339
5	.1753	.1748	.1740	.1728	.1714	.1697	.1678	.1656	.1632	.1606
6	.1490	.1515	.1537	.1555	.1571	.1584	.1594	.1601	.1605	.1606
7	.1086	.1125	.1163	.1200	.1234	.1267	.1298	.1326	.1353	.1377
8	.0692	.0731	.0711	.0810	.0849	.0887	.0925	.0962	.0998	.1033
9	.0362	.0423	.0454	.0486	.0519	.0552	.0586	.0620	.0654	.0688
10	.0200	.0220	.0241	.0262	.0285	.0309	.0334	.0359	.0386	.0413
11	.0093	.0104	.0116	.0129	.0143	.0157	.0173	.0190	.0207	.0225
12	.0039	.0045	.0051	.0058	.0065	.0073	.0082	.0092	.0102	.0113
13	.0015	.0104	.0021	.0024	.0028	.0032	.0036	.0041	.0046	.0052
14	.0006	.0007	.0008	.0009	.0011	.0013	.0015	.0017	.0019	.0022
15	.0002	.0002	.0003	.0003	.0004	.0005	.0006	.0007	.0008	.0009
16	.0001	.0001	.0001	.0001	.0001	.0002	.0002	.0002	.0003	.0003
17	.0000	.0000	.0000	.0000	.0000	.0000	.0001	.0001	.0001	.0001

x/λ	6.1	6.2	6.3	6.4	6.5	6.6	6.7	6.8	6.9	7.0
0	.0022	.0020	.0018	.0017	.0015	.0014	.0012	.0011	.0010	.0000
1	.0137	.0126	.0116	.0106	.0098	.0090	.0082	.0076	.0070	.0064
2	.0417	.0390	.0364	.0340	.0318	.0296	.0276	.0258	.0240	.0223
3	.0848	.0806	.0765	.0726	.0688	.0652	.0617	.0584	.0552	.0521
4	.1294	.1249	.1203	.1162	.1118	.1076	.1034	.0992	.0952	.0912
5	.1579	.1549	.1519	.1487	.1454	.1420	.1385	.1349	.1314	.1277
6	.1605	.1601	.1595	.1586	.1575	.1562	.1546	.1529	.1511	.1490
7	.1399	.1418	.1435	.1450	.1462	.1472	.1480	.1486	.1489	.1490
8	.1066	.1099	.1130	.1160	.1188	.1215	.1240	.1263	.1284	.1304
9	.0723	.0757	.0791	.0825	.0858	.0891	.0923	.0954	.0985	.1014
10	.0441	.0469	.0498	.5285	.0558	.0588	.0618	.0679	.0679	.0710
11	.0245	.0265	.0285	.0307	.0330	.0353	.0377	.0401	.0426	.0452
12	.0124	.0137	.0150	.0164	.0179	.0194	.0210	.0227	:0245	.0264
13	.0058	.0065	.0073	.0081	.0089	.0098	.0108	.0119	.0130	.0142
14	.0025	.0029	.0033	.0037	.0041	.0046	.0052	.0058	,0064	.0071
15	.0010	.0012	.0014	.0016	.0018	.0020	.0023	.0026	.0029	.0033
16	.0004	.0005	.0005	.0006	.0007	.0008	.0010	.0011	.0013	.0014
17	.0001	.0002	.0002	.0002	.0003	.0003	.0004	.0004	.0005	.0006
18	.0000	.0001	.0001	.0001	.0001	.0001	.0001	.0002	.0002	.0002
19	.0000	.0000	.0000	.0000	.0000	.0000	.0000	.0001	.0001	.0001

Table 2. Area Under the Standard Normal Distribution for Negative Values of Z

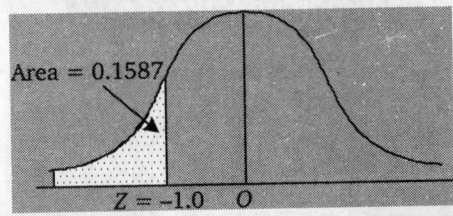

Area = 0.1587

Z = −1.0   O

Z to First Decimal	Second Decimal									
	.00	.01	.02	.03	.04	.05	.06	.07	.08	.09
−3.0	.0014	.0013	.0013	.0012	.0012	.0011	.0011	.0011	.0010	.0010
−2.9	.0019	.0018	.0018	.0017	.0016	.0016	.0015	.0015	.0014	.0014
−2.8	.0026	.0025	.0024	.0023	.0023	.0022	.0021	.0021	.0020	.0019
−2.7	.0035	.0034	.0033	.0032	.0031	.0030	.0029	.0028	.0027	.0026
−2.6	.0047	.0045	.0044	.0043	.0041	.0040	.0039	.0038	.0037	.0036
−2.5	.0062	.0060	.0059	.0057	.0055	.0054	.0052	.0051	.0049	.0048
−2.4	.0082	.0080	.0078	.0075	.0073	.0071	.0069	.0068	.0066	.0064
−2.3	.0107	.0104	.0102	.0099	.0096	.0094	.0091	.0089	.0087	.0084
−2.2	.0139	.0136	.0132	.0129	.0126	.0122	.0119	.0116	.0113	.0110
−2.1	.0179	.0174	.0170	.0166	.0162	.0158	.0154	.0150	.0146	.0143
−2.0	.0228	.0222	.0217	.0212	.0207	.0202	.0197	.0192	.0188	.0183
−1.9	.0287	.0281	.0274	.0268	.0262	.0256	.0250	.0244	.0238	.0233
−1.8	.0359	.0352	.0344	.0336	.0329	.0322	.0314	.0307	.0300	.01.94
−1.7	.0446	.0436	.0427	.0418	.0409	.0401	.0392	.0384	.0375	.0367
−1.6	.0548	.0537	.0526	.0516	.0505	.0495	.0485	.0475	.0465	.0455
−1.5	.0668	.0655	.0643	.0630	.0618	.0606	.0594	.0582	.0570	.0559
−1.4	.0808	.7938	.0778	.0764	.0749	.0735	.0722	.0708	.0694	.0681
−1.3	.0968	.0951	.0934	.0918	.0901	.0855	.0869	.0853	.0838	.0823
−1.2	.1151	.1131	.1112	.1093	.1075	.1056	.1038	.1020	.1003	.0985
−1.1	.1357	.1335	.1314	.1292	.1271	.1251	.1230	.1210	.1190	.1170
−1.0	.1587	.1562	.1529	.1515	.1492	.1469	.1446	.1423	.1401	.1379
−0.9	.1841	.1814	.1785	.1762	.1736	.1711	.1685	.1660	.1635	.1611
−0.8	.2119	.2090	.2061	.2033	.2005	.1977	.1949	.1922	.1894	.1867
−0.7	.2420	.2389	.2358	.2327	.2297	.2266	.2236	.2206	.2177	.2143
−0.6	.2743	.2709	.2676	.2643	.2611	.2578	.2546	.2514	.2483	.2451
−0.5	.3085	.3050	.3015	.2981	.2946	.2912	.2877	.2843	.2810	.2776
−0.4	.3446	.3409	.3372	.3336	.3300	.3264	.3228	.3192	.3156	.3121
−0.3	.3821	.3783	.3745	.3707	.3669	.3632	.3594	.3557	.3520	.3483
−0.2	.4207	.4168	.4129	.4090	.4052	4013	.3974	.3936	.3897	.3859
−0.1	.4602	.4562	.4522	.4483	.4443	.4404	.4364	.4325	.4286	.4247
−0.0	.5000	.4960	.4920	.4880	.4840	.4801	.4761	.4721	.4681	.4641

Table 3. Area Under the Standard Normal Distribution from Extreme Left to Positive Values of Z

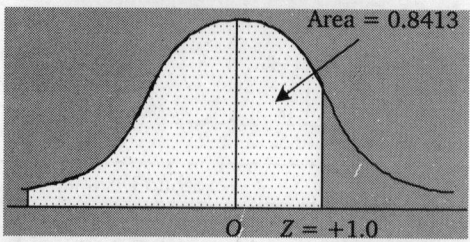

Area = 0.8413

$O \quad Z = +1.0$

Example. To find the area under the curve from extreme left $Z = -\infty$ to a point $Z = 1.0$ to the right of the mean, look up the value opposite 1.0 in the table; 0.8413 of the area under the curve lies between the extreme left ($Z = -\infty$) to a z value.

Z to First Decimal	Second Decimal									
	.00	.01	.02	.03	.04	.05	.06	.07	.08	.09
0.0	.5000	.5040	.5080	.5120	.5160	.5199	.5239	.5279	.5319	.5359
0.1	.5398	.5438	.5478	.5517	.5557	.5596	.5636	.5675	.5714	.5753
0.2	.5793	.5832	.5871	.5910	.5948	.5987	.6026	.6064	.6103	.6141
0.3	.6179	.6217	.6255	.6293	.6331	.6368	.6406	.6443	.6480	.6517
0.4	.6554	.6591	.6628	.6664	.6700	.6736	.6772	.6808	.6841	.6879
0.5	.6915	.6950	.6985	.9019	.7054	.7088	.7123	.7157	.7190	.7224
0.6	.7257	.7291	.7324	.7357	.7389	.7422	.7454	.7406	.7517	.7549
0.7	.7580	.7611	.7642	.7673	.7703	.7734	.7764	.7794	.7823	.7852
0.8	.7881	.7910	.7939	.7967	.7995	.8023	.8051	.8078	.8106	.8133
0.9	.8159	.8186	.8212	.8238	.8264	.8289	.8315	.8340	.8365	.8389
1.0	.8413	.8438	.8461	.8485	.8508	.8531	.8554	.8577	.8599	.8621
1.1	.8643	.8665	.8686	.8708	.8729	.8749	.8770	.8790	.8910	.8830
1.2	.8849	.8869	.8888	.8907	.8925	.8944	.8962	.8980	.8997	.9015
1.3	.9032	.9049	.9066	.9082	.9099	.9115	.9131	.9147	.9162	.9177
1.4	.9192	.9207	.9222	.9236	.9251	.9265	.9278	.9292	.9306	.9319
1.5	.9332	.9345	.9357	.9370	.9382	.9394	.9406	.9418	.9430	.9441
1.6	.9452	.9463	.9474	.9485	.9495	.9505	.9515	.9525	.9535	.9545
1.7	.9554	.9564	.9573	.9582	.9591	.9599	.9608	.9616	.9625	.9633
1.8	.9641	.9649	.9656	.9664	.9671	.9678	.9686	.9693	.9700	.9706
1.9	.9713	.9719	.9726	.9732	.9783	.9744	.9750	.9756	.9762	.9767
2.0	.9772	.9778	.9783	.9788	.9793	.9798	.9803	.9808	.9812	.9817
2.1	.9821	.9826	.9830	.9834	.9838	.9842	.9846	.9850	.9854	.9857
2.2	.9861	.9865	.9868	.9871	.9874	.9878	.9881	.9884	.9887	.9890
2.3	.9893	.9896	.9898	.9901	.9904	.9906	.9909	.9911	.9913	.9916
2.4	.9918	.9920	.9922	.9924	.9926	.9928	.9930	.9932	.9934	.9936
2.5	.9938	.9940	.9941	.9943	.9944	.9946	.9932	.9949	.9951	.9952
2.6	.9953	.9955	.9956	.9957	.9958	.9960	.9961	.9962	.9963	.9964
2.7	.9965	.9966	.9967	.9968	.9969	.9970	.9971	.9972	.9973	.9974
2.8	.9974	.9975	.9976	.9977	.9977	.9979	.9979	.9979	.9980	.9981
2.9	.9981	.9982	.9982	.9983	.9984	.9984	.9985	.9985	.9986	.9986
3.0	.9986	.9987	.9987	.9988	.9988	.9988	.9989	.9989	.9990	.9990

Appendix

Table 4. Propositions of Total Area Under the Normal Curve from ∞ to t, where t = (x − mμ)σ

t	ψ(t)	t	ψ(t)	t	ψ(t)	t	ψ(t)
0.00	0.5000	0.65	0.7422	1.30	0.9032	1.95	0.9744
0.01	0.5040	0.66	0.7454	1.31	0.9049	1.96	0.9750
0.02	0.5080	0.67	0.7486	1.32	0.9066	1.97	0.9756
0.03	0.5120	0.68	0.7517	1.33	0.9082	1.98	0.9761
0.04	0.5160	0.69	0.7549	1.34	0.9099	1.99	0.9767
0.05	0.5199	0.70	0.7580	1.35	0.9115	2.00	0.9772
0.06	0.5239	0.71	0.7611	1.36	0.9131	2.02	0.9783
0.07	0.5279	0.72	0.7642	1.37	0.9147	2.04	0.9793
0.08	0.5319	0.73	0.7673	1.38	0.9162	2.06	0.9803
0.09	0.5359	0.74	0.7703	1.39	0.9177	2.08	0.9812
0.10	0.5398	0.75	0.7734	1.40	0.9192	2.10	0.9821
0.11	0.5438	0.76	0.7764	1.41	0.9207	2.12	0.9830
0.12	0.5478	0.77	0.7794	1.42	0.9222	2.14	0.9838
0.13	0.5517	0.78	0.7823	1.43	0.9236	2.16	0.9846
0.14	0.5557	0.79	0.7852	1.44	0.9251	2.18	0.9854
0.15	0.5596	0.80	0.7881	1.45	0.9265	2.20	0.9861
0.16	0.5636	0.81	0.7910	1.46	0.9279	2.22	0.9868
.0.17	0.5675	0.82	0.7939	1.47	0.9292	2.24	0.9875
0.18	0.5714	0.83	0.7967	1.48	0.9306	2.26	0.9881
0.19	0.5753	0.84	0.7995	1.49	0.9319	2.28	0.9887
0.20	0.5793	0.85	0.8023	1.50	0.9332	2.30	0.9893
0.21	0.5832	0.86	0.8051	1.51	0.9345	2.32	0.9898
0.22	0.5871	0.87	0.8078	1.52	0.9357	2.34	0.9904
0.23	0.5910	0.88	0.8106	1.53	0.9370	2.36	0.9909
0.24	0.5948	0.89	0.8133	1.54	0.9382	2.38	0.9913
0.25	0.5987	0.90	0.8159	1.55	0.9394	2.40	0.9918
0.26	0.6026	0.91	0.8186	1.56	0.9406	2.42	0.9922
0.27	0.6064	0.92	0.8212	1.57	0.9418	2.44	0.9927
0.28	0.6103	0.93	0.8238	1.58	0.9429	2.46	0.9931
0.29	0.6141	0.94	0.8264	1.59	0.9441	2.48	0.9934
0.30	0.6179	0.95	0.8289	1.60	0.9252	2.50	0.9938
0.31	0.6217	0.96	0.8315	1.61	0.9463	2.52	0.9941
0.32	0.6255	0.97	0.8340	1.62	0.9474	2.54	0.9945
0.33	0.6293	0.98	0.8365	1.63	0.9484	2.56	0.9948
0.34	0.6331	0.99	0.8389	1.64	0.9495	2.58	0.9951
0.35	0.6368	1.00	0.8413	1.65	0.9505	2.60	0.9953
0.36	0.6406	0.01	0.8438	1.66	0.9515	2.62	0.9956
0.37	0.6443	1.02	0.8461	1.67	0.9525	2.64	0.9959
0.38	0.6480	1.03	0.8485	1.68	0.9535	2.66	0.9961
0.39	0.6517	1.04	0.8508	1.69	0.9545	2.68	0.9963
0.40	0.6554	1.05	0.8531	1.70	0.9554	2.70	0.9965
0.41	0.6591	1.06	0.8554	1.71	0.9564	2.72	0.9967
0.42	0.6628	1.07	0.8577	1.72	0.9573	2.74	0.9969
0.43	0.6664	1.08	0.8599	1.73	0.9582	2.76	0.9971
0.44	0.6700	0.09	0.8621	1.74	0.9591	2.78	0.9973
0.45	0.6736	1.10	0.8643	1.75	0.5999	2.80	0.9974
0.46	0.6772	1.11	0.8665	1.76	0.9608	2.82	0.9976
0.47	0.6808	1.12	0.8686	1.77	0.9616	2.84	0.9977
0.48	0.6844	1.13	0.8708	1.78	0.9625	2.86	0.9979
0.49	0.6879	1.14	0.8729	1.79	0.9633	2.88	0.9980
0.50	0.6915	1.15	0.8749	1.80	0.9641	2.90	0.9981
0.51	0.6950	1.16	0.8770	1.81	0.9649	2.92	0.9982
0.52	0.6985	1.17	0.8190	1.82	0.9656	2.94	0.9984
0.53	0.7019	1.18	0.8810	1.83	0.9664	2.96	0.9985
0.54	0.7054	1.19	0.8830	1.84	0.9671	2.98	0.99116
0.55	0.7088	1.20	0.8849	1.85	0.9678	3.00	0.99865
0.56	0.7123	1.21	0.8869	1.86	0.9686	3.20	0.99931
0.57	0.7190	1.22	0.8888	1.87	0.9693	3.40	0.99966
0.58	0.7190	1.23	0.8907	1.88	0.9699	3.60	0.999841
0.59	0.7224	1.24	0.8925	1.89	0.9706	3.80	0.999928
0.60	0.7257	1.25	0.8944	1.90	0.9713	4.00	0.999968
0.61	0.7291	1.26	0.8962	1.91	0.9719	4.50	0.999997
0.62	0.7324	1.27	0.8980	1.92	0.9726	5.00	0.999997
0.63	0.7357	1.28	0.8997	1.93	0.9732		
0.64	0.7389	1.29	0.9015	1.94	0.9738		

Table 5. Values of χ^2 with various values of p and v

v \ P	0.99	0.95	0.50	0.10	0.05	0.01
1	0.0002	0.0039	0.455	2.706	3.841	6.635
2	0.0201	0.103	1.386	4.605	5.991	9.210
3	0.115	0.352	2.366	6.251	7.815	11.34
4	0.297	0.711	3.357	7.779	9.488	13.28
5	0.554	1.145	4.351	9.236	11.07	15.09
6	0.872	1.635	5.348	10.64	12.59	16.81
7	1.239	2.167	6.346	12.02	14.07	18.48
8	1.646	2.733	7.344	13.36	15.51	20.09
9	2.088	3.325	8.343	14.68	16.92	21.67
10	2.558	3.940	9.342	15.99	18.31	23.21
11	3.053	4.575	10.34	17.28	19.68	24.72
12	3.571	5.226	11.34	18.55	21.03	26.22
13	4.107	5.892	12.34	19.81	22.36	27.69
14	4.660	6.571	13.34	21.06	23.68	29.14
15	5.229	7.261	14.34	22.31	25.00	30.58
16	5.812	7.962	15.34	23.54	26.30	32.00
17	6.408	8.672	16.34	24.77	27.59	33.41
18	7.015	9.390	17.34	25.99	28.87	34.80
19	7.633	10.12	18.34	27.20	30.14	36.19
20	8.260	10.85	19.34	28.41	31.41	37.57
21	8.897	11.59	20.34	29.62	32.67	38.93
22	9.542	12.34	21.34	30.81	33.92	40.29
23	10.20	13.09	22.34	32.01	35.17	41.64
24	10.86	13.85	23.34	33.20	36.42	42.98
25	11.52	14.61	24.34	34.38	37.65	44.31
26	12.20	15.38	25.34	35.56	38.88	45.64
27	12.88	16.15	26.34	36.74	40.11	46.96
28	13.57	16.93	27.34	37.92	41.34	48.28
29	14.26	17.71	28.34	39.09	42.56	49.59
30	14.95	18.49	29.34	40.26	43.77	50.89

Table 6. Values of F (F - distribution) for 5% and 1% level, where v_1 is the number of degree of freedom for greater estimate of variance and v_2 for the smaller

v	P	1	2	3	4	5	6	8	12	24	∞
1	5%	161.4	199.5	215.7	224.6	230.2	234.0	238.9	243.9	249.0	254.0
	1%	4052	4999	5403	5625	5764	5849	5981	6016	6234	6366
2	5%	18.51	19.00	19.16	19.25	19.30	19.32	19.37	19.41	19.45	19.50
	1%	98.49	99.00	99.17	99.25	99.30	99.33	99.36	99.42	99.46	99.50
3	5%	10.13	9.55	9.28	9.12	9.01	8.94	8.84	8.74	8.64	8.53
	1%	34.12	30.82	29.46	28.71	28.24	27.91	27.49	27.05	26.60	26.12
4	5%	7.71	6.94	6.59	6.39	6.26	6.16	6.04	5.91	5.77	5.63
	1%	21.20	18.00	16.69	15.98	15.52	15.21	14.80	14.37	13.93	13.46
5	5%	6.61	5.79	5.41	5.19	5.05	4.95	4.82	4.68	4.53	4.36
	1%	16.26	13.27	12.06	11.39	10.97	10.67	10.27	9.89	9.47	9.02
6	5%	5.99	5.14	4.76	4.53	4.39	4.28	4.15	4.00	3.84	3.67
	1%	13.74	10.92	9.78	9.15	8.75	8.47	8.10	7.72	7.31	6.88
7	5%	5.59	4.74	4.35	4.12	3.97	3.87	3.73	3.57	3.41	3.23
	1%	12.25	9.55	8.45	7.85	7.46	7.19	6.84	6.47	6.07	5.65
8	5%	5.32	4.46	4.07	3.84	3.89	3.58	3.44	3.28	3.12	2.93
	1%	11.26	8.65	7.59	7.01	6.63	6.37	6.03	5.67	5.28	4.86
9	5%	5.12	4.26	3.86	3.63	3.48	3.37	3.23	3.07	2.90	2.71
	1%	10.56	8.02	6.99	6.42	6.06	5.80	5.47	5.11	4.73	4.31
10	5%	4.96	4.10	3.71	3.48	3.33	3.22	3.07	2.91	2.74	2.54
	1%	10.04	7.56	6.55	5.99	5.64	5.39	5.06	4.71	4.33	3.91
12	5%	4.75	3.88	3.49	3.26	3.11	3.00	2.85	2.69	2.50	2.30
	1%	9.33	6.93	5.95	5.41	5.06	4.82	4.50	4.16	3.78	3.36
14	5%	4.60	3.74	3.34	3.11	2.96	2.85	2.70	2.53	2.35	2.13
	1%	8.86	6.51	5.56	5.03	4.69	4.46	4.14	3.80	3.43	3.00
16	5%	4.49	3.63	3.24	3.01	2.85	2.74	2.59	2.42	2.24	2.01
	1%	8.53	6.23	5.29	4.77	4.44	4.20	3.89	3.55	3.18	2.75
18	5%	4.41	3.55	3.16	2.93	2.77	2.66	2.51	2.34	2.15	1.92
	1%	8.28	6.01	5.09	4.58	4.25	4.01	3.71	3.37	3.01	2.57
20	5%	4.35	3.49	3.10	2.87	2.71	2.60	2.45	2.28	2.08	1.84
	1%	8.30	5.85	4.94	4.43	4.10	3.87	3.56	3.23	2.86	2.42

INDEX